RSGB YEARBOOK 2024

Editor
Mike Browne, G3DIH

Advertising
Chris Danby, G0DWV, Danby Advertising

Front Cover
Kevin Williams, M6CYB

Production
Mark Allgar, M1MPA

Published by the
Radio Society of Great Britain,
3 Abbey Court, Fraser Road,
Priory Business Park, Bedford MK44 3WH
Website: *www.rsgb.org*
Tel: 01234 832700
Fax: 01234 831496

ISBN: 9781 9101 9375 4
ISSN: 1460-454X
Printed in Great Britain

Contents

Foreword

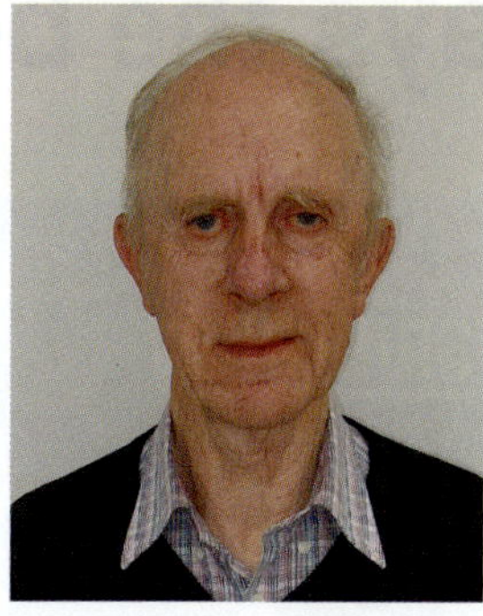

Welcome to the 2024 edition of the RSGB Yearbook. The coming year promises to be a very interesting one for Amateur Radio in the UK. In June, Ofcom issued a consultation document entitled "Updating the Amateur Radio Licensing Framework". This document heralds a generational shift in the amateur licence with proposals for changes to the callsign structure, increases in permitted power levels, the removal of many restrictions and much more. By the time you read this the consultation should have closed and we will be awaiting the outcome of the process. What an exciting time, I know some will be unhappy with the changes, some will welcome them but whatever happens you can be assured that the RSGB will continue to work with the Regulator to get the best outcome for our members and Amateur Radio in general.

The world situation continues to be in flux, over the past few years we suffered from the COVID pandemic and then the Ukraine situation came upon us with the resulting humanitarian crisis. The war has caused huge price increases resulting in inflation well beyond anything seen in recent times. Consequently, the RSGB has been under significant financial pressure which has led to a necessary increase in fees, we do appreciate that everyone is also under these pressures, and we will continue to monitor the situation closely to keep costs under control.

It is not only costs that are rising but also the threat on our spectrum, especially in the 23cm band. Last year G3YSX reported on the issues with Galileo, the global positioning system, which has channels overlapping much of our allocation in 23cms. Unfortunately, Galileo receivers are not very good in the presence of other signals in the band and therefore restrictions are being proposed on Amateur use of the spectrum which could impact greatly on our use of the band. Our HF frequencies are also under scrutiny as global threats to communications continue to rise and HF is now being returned to as a back-up medium when other means fail. Higher order microwave bands continue to be looked on enviously by mobile operators and again we will have to be on our guard.

As a major part of our efforts in defence of our spectrum, we will have an RSGB delegate present to represent our interests at the forthcoming ITU World Radiocommunications Conference (WRC-23). These conferences are held every three to four years to review, and, if necessary, revise the Radio Regulations, the international treaty governing the use of the radio-frequency spectrum and the geostationary-satellite and non-geostationary-satellite orbits. Revisions are made based on an agenda determined by the ITU Council, which considers recommendations made by previous world radiocommunication conferences. The outcome of WRC-23 will shape the future use of spectrum for years to come and having a delegate there is both an honour and an onerous responsibility, we wish them well in their work.

As previously mentioned, financial pressures continue to rise, and the way that you can help to enhance and protect amateur radio is to remain a member of the RSGB, despite the increasingly tough calls on your finances. It would also be most helpful for you to encourage others to join the RSGB and keep it strong in these challenging times Your membership helps the RSGB to fund services which enable you to develop your amateur radio skills. Membership fees also support the vital work as described above with the international community, the IARU and the ITU.

Volunteers are the lifeblood of the Society so please help us by answering our calls for help. We are also always looking for assistance across a range of our activities and giving up some of your time to protect the future of amateur radio in the United Kingdom and world-wide can at times be frustrating, but it is also most rewarding.

When I joined the Board about six months ago, I thought I knew a lot about the Society having been chair of a committee for twelve years. Being a Board member and now President has opened a much wider picture to me and demonstrated the huge range of activities that we are involved with, and which need leadership to take forward. Please do consider putting your name forward for office when the time comes, as someone else once said your country, or in this you're your Society, needs you!

I look forward to meeting with many of you at some of the events in the coming year and perhaps we will have a QSO along the way too.

I do hope that you enjoy this edition of the RSGB Yearbook and join me in thanking Mike Browne, G3DIH, and the team of RSGB volunteers and HQ staff that compiled this edition.

John McCullagh MBE, GI4BWM
RSGB President

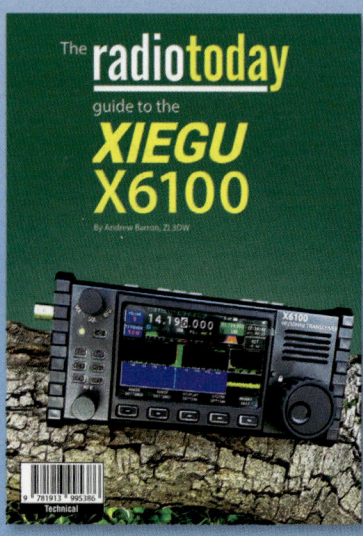

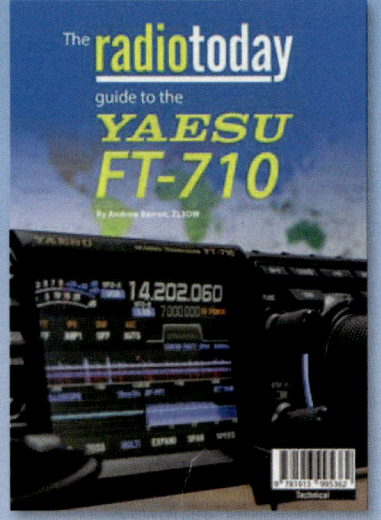

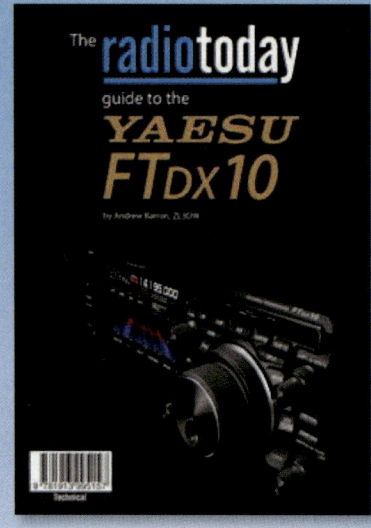

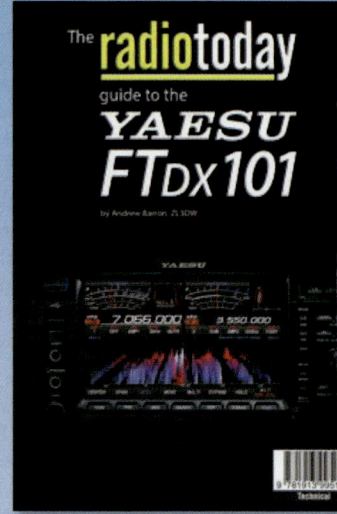

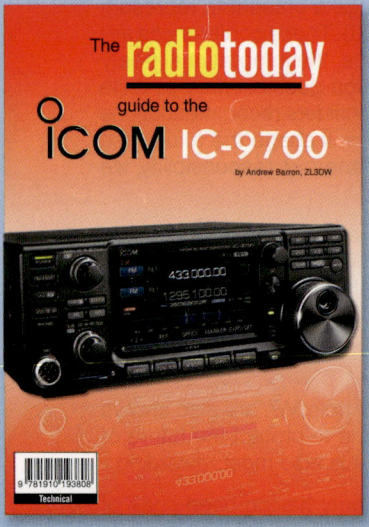

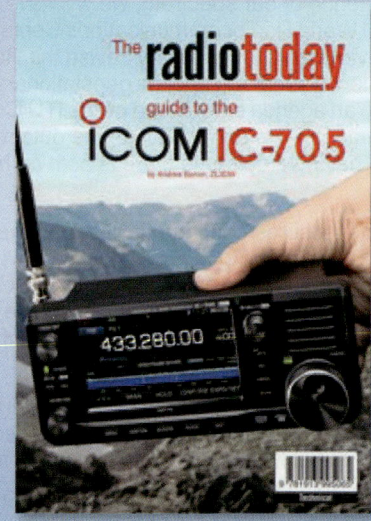

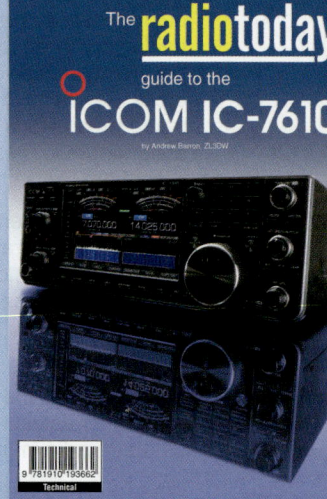

Introduction

Welcome to the 2023/24 edition of the RSGB Yearbook and UK Callbook, the information contained in this edition is accurate as far as possible but with a publication lasting until December 2023, changes will occur during the lifetime of this edition – so please check the pages of RadCom or the RSGB and Club websites for information which became available after we closed for press.

This particularly applies to items such as GB2RS newsreaders, local QSL sub-managers, and national clubs and societies, also the repeater listings and beacon listings etc.

Many events are back on the calendar after years of disrubtion due to the pandemic but please check the various club and contest websites for further information.

Club Information pages

For administrative purposes the Society divides itself into regions, each comprising of four to six districts. The RSGB identifies these regions as, Regional Representatives (RR) and District Representatives (DR).

All the local information on all the clubs and societies can be found on pages 46-60 of this Yearbook.

Beacons

This year we have managed to include listings of the VHF/UHF and Microwave Beacons as well as the HF Beacons, also included is a UK only Beacons in Frequency and by Callsign sections, as always you can keep up to date with any changes to the listing at:

https://rsgb.org/main/operating/beacons-and-repeaters/vhf-uhf-and-microwave-beacons/

Validating your Licence

Revalidate your licence to avoid revocation

Ofcom has advised the Society that plans will be drawn up to revoke licences that have not been revalidated as required by the licence conditions at intervals of not more than 5 years.

Therefore it is important, so as to retain your licence, and to remain legal to operate on the amateur bands, to re-validate your licence every 5 years.

To validate your licence or amend your licence, log in to: *https://ofcom.force.com/licensingcomlogin* or email: *amateur.validations@ofcom.org.uk* If you need assistance in the process, Ofcom staff are available to help, but please be patient during times of heavy workload.

If you have not yet registered to use the Ofcom Online Licensing Service you will need to do so in order to access your licence online. When registering for the first time you will need to have details of your lifetime licence number, which can be found on page 1 of your Licence document.
There are several other publishers of the data, so it is vital that you contact the source of data so that it is either blocked or released in *all* callbooks. The Society cannot accept direct input regarding an individual's entry.

UK Callsign entries

The Callsigns listed in this *Yearbook* reflect the official records held by Ofcom. The Callsign data was output mid July 2021 and therefore any callsign changes made after this date will not appear in the Callsign listing pages. They show correspondence addresses, **not** licence addresses, so do not be misled into thinking that if a foreign address is quoted the station is operating from there. The amateur concerned will have a different station address when in the UK.

Please Note that some callbooks, CDs and websites rely entirely on updates being notified to them by individual licensees, so the information can be many years out of date. To make this book readable and compact, we make a few standard abbreviations, but no changes are made to the substance of any entries.

Please see the article on Callsign Formatting on page 180 of the Callsign section of this Yearbook for more information.

Details withheld

The words "Details withheld" mean exactly what they say, because entries are not within the control of the Society. In some cases there is a perfectly good reason for a licensee not to want his or her address publishing, but in others it is possible that the 'callsign book consent' box was unchecked in error.
Should you wish your details to be withheld from publication, or released if they are not, or there is an error in the substance of your entry, you can update the entry online or write to Spectrum Licensing (*and not to the RSGB*) and request them to take the necessary action. Their address is:

Spectrum Licensing,
Riverside House,
2a Southwark Bridge Road,
London SE1 9HA
Tel: 020 7981 3131

Email: *spectrum.licensingenquiries@ofcom.org.uk*
Web: *www.ofcom.org.uk/licensing/olc/*

International listings

Amateurs sometimes ask about their entry in the *International Listings* of the Radio Amateur Call Book, sometimes known as the *DX Listings*. This is an entirely separate work published annually by an independent company. If you want to be listed in their publication, please write direct to them, and not to the RSGB. Corrections are free, but they do make a small charge for special entries. Their address is:

Radio Amateur Callbook,
mfcit GmbH
P.O.Box 1170
34216 Baunatal
Germany
E-mail: *contact20@callbook.biz*
Web: *www.callbook.biz*

Acknowledgments

Finally my thanks go to all the contributors to the Yearbook, this includes a number of RSGB Staff, Committee Members, and Club Secretaries.

Mike Browne, G3DIH
Editor

At Your Service!

The Society provides a broad range of services for its members through its professional staff and management at Headquarters and a country-wide force of skilled, dedicated and knowledgeable volunteers. To get the best out of the RSGB it is important that you approach the correct part of the Society. On these pages you will find a practical guide to finding the right person for your enquiry.

YOUR RSGB

The Society's affairs are directed by the Board, supported by the Volunteer Leadership Team comprising the Regional Representatives, Committee Chairs and Honorary Officers. The Board Members and the Regional Representatives are elected by the membership in a postal and electronic ballot.

The day to day running of the Society is the responsibility of the General Manager, supported by the Board Members who liaise with each of the committees.

Each [Director/Board Member] has responsibilities for one or more Committees. The Board Liaison Member for each Committee ensures that communication between the Committees and the Board is maintained. In addition, each Director has a specific strategic focus on one of the 2022 goals.

Full details of Board and Committees' terms of reference can be found on our website.

For HQ staff, both email addresses and telephone details are provided, including the option to select when dialling through the RSGB switchboard (01234 832 700).

Committee Chairs & Honorary Officers

These are all volunteers and give their time freely to support the Society. Members should respect the fact that many also have full time day jobs, and so email is the appropriate method of communication.

THE RSGB BOARD

President

John McCullagh, GI4BWM
email: *gi4bwm@rsgb.org.uk*

Chair

Stewart Bryant, G3YSX
email: *g3ysx@rsgb.org.uk*

Directors

Len Paget, GM0ONX,
email: *gm0onx@rsgb.org.uk*

Dave Wilson, M0OBW
email: *m0obw@rsgb.org.uk*

Paul Nicholls, M0PVN
email: *m0pvn@rsgb.org.uk*

Tony Miles, MM0TMZ
email: *mm0tmz@rsgb.org.uk*

Ian Shepherd, G4EVK
email: *g4evk@rsgb.org.uk*

Note: The General Manager, Company Secretary and Acting Honorary Treasurer are not Directors, but are in attendance at Board Meetings.

General Manager

Steve Thomas, M1ACB
email: *steve.thomas@rsgb.org.uk*

Honorary Treasurer

Chris Wood FCCA, GD6TWF
email: *hon.treasurer@rsgb.org.uk*

Company Secretary

Stephen Purser, GW4SHF
email: *company.secretary@rsgb.org.uk*

WEBSITE

Main website: *www.rsgb.org*
Members Pages: Log in using your callsign as the user name and your Membership number, without the leading zeros (see your *RadCom* address label) as the password.
If you need to update your Membership details, please log into Membership Services at: *www.rsgb.org/members*

REGIONAL REPRESENTATIVES

Information on the Regional Representatives (RR) and District Representatives (DR), may be found elsewhere in the RSGB Yearbook and on the RSGB website, *www.rsgb.org*

Region 1 – Tony Miles, MM0TMZ , rr1@rsgb.org.uk

Region 2 – Jim Campbell, MM7BIW, rr2@rsgb.org.uk

Region 3 – Martyn Bell, M0TEB, rr3@rsgb.org.uk

Region 4 – Ian Bowman, G7ESY, rr4@rsgb.org.uk

Region 5 – Neil Yorke, M0NKE, rr5@rsgb.org.uk

Region 6 – Liz Cabban, GW0ETU, rr6@rsgb.org.uk

Region 7 – Brian Jones, GW6ZYI, rr7@rsgb.org.uk

Region 8 – Micheal Na bPiob, MI0HOZ, rr8@rsgb.org.uk

Region 9 – Ron White, G6LTT (Temp), rr9@rsgb.org.uk

Region 10 –Keith Bird, G4JED, rr10@rsgb.org.uk

Region 11 – Andrew Jenner, G7KNA, rr11@rsgb.org.uk

Region 12 – David De La Haye, M0MBD, rr12@rsgb.org.uk

Region 13 – Bob Hambly, M0HAF, rr13@rsgb.org.uk

SPECIALIST AREAS

The many different activities of the Society are run by its committees, honorary officers and full-time staff. If you wish to take advantage of one of these services or have an administrative enquiry about any one of them, contact details are listed below.

Abuse and Poor Operating

Operating Advisory Service (OAS),
Ian Suart, GM4AUP, Coordinator
email: *oas@rsgb.org.uk, www.rsgb.org/oas/*

Amateur Radio Direction Finding

Bob Titterington, G3ORY,
Chair, ARDF Committee,
email: *ardf.chairman@rsgb.org.uk, www.rsgb.org/ardf/*

Awards

Lindsay Pennell, G8PMA, Awards Manager
email: *awards@rsgb.org.uk, www.rsgb.org/awards/*

Beyond Exams

Mark Burrows, 2E0SBM, BE Coordinator
email: *be.coordinator@rsgb.org.uk*

Contests

Ian Pawson, G0FCT, Chair,
Contest Support, email: *csc.chair@rsgb.org.uk, www.rsgb.org/radiosport/*

Nick Totterdell, G4FAL,
HF Contest Committee
email: *hfcc.chair@rsgb.org.uk*

Andy Cook, G4PIQ,
VHF Contest Committee
email: *vhfcc.chair@rsgb.org.uk*

EMC

John Rogers, M0JAV, EMC Committee,
email: *emc.chairman@rsgb.org.uk, www.rsgb.org/emc/*

Examination Standards Committee

Tony Kent, G8PBH, Chair
email: *esc.chair@rsgb.org.uk*

EQAM Exam Quality Assurance Manager

Dave Wilson, M0OBW EQA Manager
email: *eqam@rsgb.org.uk*

General Technical Matters

Andy Talbot, G4JNT, Chair,
Technical Forum,
email: *tech.chair@rsgb.org.uk, www.rsgb.org/technicalmatters/*

General Spectrum & Regulatory Matters

Murray Niman, G6JYB, Chair,
Spectrum Forum,
email: *spectrum.chairman@rsgb.org.uk www.rsgb.org/spectrumforum/*

GB2RS News Service Management

Steve Richards, G4HPE, GB2RS Manager
email: *gb2rs.manager@rsgb.org.uk*
(GB2RS news items should be sent to: *radcom@rsgb.org.uk)*

HF Matters
Ian Greenshields, G4FSU, HF Manager, email: *hf.manager@rsgb.org.uk*

Intruders to the Amateur Bands
Richard Lamont, G4DYA, IW Coordinator email: *iw@rsgb.org.uk, www.rsgb.org/intruders/*

Legacy Committee
Ian Shepherd, G4EVK, Committee Chair email: *legacy.chair@rsgb.org.uk www.rsgb.org/legacy/*

Microwave Matters
Barry Lewis, G4SJH, Microwave Manager, email: *microwave.manager@rsgb.org.uk*

Planning Advice
John Mattocks, G4TEQ, Chair, email: *pac.chairman@rsgb.org.uk, www.rsgb.org/planning/*

Propagation Studies
Steve Nichols, G0KYA, Chair, Propagation Studies Committee, email: *psc.chairman@rsgb.org.uk, www.rsgb.org/psc/*

Repeater and Data Communications
Andrew Barrett, G8DOR, Chair, ETCC, email: *etcc.chairman@rsgb.org.uk, www.ukrepeater.net*

Regional Forum
David De La Haye, M0MBD, Chair email: *rr12@rsgb.org.uk*

Special Interest Group
Philip Hosey, MI0MSO, Manager email: *sig.manager@rsgb.org.uk*

Developing Amateur Radio Group
DARG Lead, Vacant email: *darg.lead@rsgb.org.uk*

Trophies
Jacqui Goodey, G6XSY, Trophy Manager email: *trophy.manager@rsgb.org.uk*

VHF Matters
John Regnault, G4SWX, VHF Manager email: *vhf.manager@rsgb.org.uk*

Youth Champion
Liam Robbins, G5LDR, Youth Champion email: *youth.champion@rsgb.org.uk www.rsgb.org/youth/*

Details of the Society's volunteer officers can be found in this Yearbook and on the RSGB website, *www.rsgb.org*

HEADQUARTERS STAFF

For HQ staff below, both email addresses and telephone details are provided, including the option to select when dialling through the RSGB switchboard (01234 832 700).

Subscription renewals
Telephone: 01234 832 700, Option 1

Sales department
(Membership, books and other products)
email: *sales@rsgb.org.uk*
Telephone: 01234 832 700, Option 2

Amateur Radio Examinations
email: *exams@rsgb.org.uk*
Telephone: 01234 832 700, Option 3

Technical Amateur Radio Enquiries
email: *AR.dept@rsgb.org.uk*
Telephone: 01234 832 700, Option 4

Amateur Radio Licensing Enquiries
email: *AR.dept@rsgb.org.uk*
Telephone: 01234 832 700, Option 4

GB2RS and Club News
email: *radcom@rsgb.org.uk*

Telephone: 01234 832 700, Option 5

RadCom
(news items, feature submissions, etc)
Elaine Richards, G4LFM email: *radcom@rsgb.org.uk*
Telephone: 01234 832 700, Option 5

General Manager
email: *GM.dept@rsgb.org.uk*

HEADQUARTERS AND REGISTERED OFFICE

3 Abbey Court, Fraser Road,
Priory Business Park, Bedford MK44 3WH
Telephone: 01234 832 700
Fax: 01234 831 496

Main website: *www.rsgb.org*

Log in using your callsign as the user name and your membership number, without the leading zeros (see your *RadCom* address label) as the password.

QSL BUREAU ADDRESS

PO Box 5, Halifax HX1 9JR, England
Telephone: 01422 359 362
email: *qsl@rsgb.org.uk, www.rsgb.org/qsl*

PLAY YOUR PART IN YOUR RSGB

Have Your Say
Let us know how we're doing! Through 'Have Your Say' you can let us know your views and you will receive a reply from the General Manager or a Board Member.
email: *haveyoursay@rsgb.org.uk*
www.rsgb.org/haveyoursay

Consultations
From time to time you will find we are consulting the Membership on aspects of Society policy. You can find current consultations at *www.rsgb.org/consultations/*

National Radio Centre
Don't forget to tell your friends about the National Radio Centre at Bletchley Park. Full details at *www.rsgb.org/nrc/*

RSGB Members can enter Bletchley Park for free by downloading the personalised voucher available from the *www.rsgb.org* home page

Licensing & Special Event Stations
Licensing and Notices of Variation (NoVs) for special event stations are handled by Ofcom, Tel: 0300 123 1000, *www.ofcom.org.uk,*
email: *Spectrum.Licensing@ofcom.org.uk*

FAQs
The RSGB has compiled the questions most frequently asked by Members at: *www.rsgb.org/faq/*

Band Plan
The latest version of the band plan is always available on the website at: *www.rsgb.org/band-plans/*

Good Operating Practice
The RSGB fully supports the code of conduct and encourages all amateurs to read the advice at *www.rsgb.org/op-guidelines*

RSGB Shop
All RSGB goods - books, filters, clothing etc - can be purchased online at: *www.rsgbshop.org/*

Club Finder
Use the website to find your nearest radio club and check out the facilities they have to offer. *www.rsgb.org/clubsandtraining/*

Yearbook
If you have moved home, if you would like your name and address to be withheld from future editions of the Yearbook (or released, for use in it), or if your callsign is not listed, you can **only** make the necessary changes to the database via Ofcom, direct by phone or via Ofcom's website below:
Tel: 0300 123 1000
www.ofcom.org.uk

RSGB HQ and Registered Office:
3 Abbey Court, Fraser Road,
Bedford MK44 3WH
Tel: 01234 832700
Fax: 01234 831496
Web site: *www.rsgb.org*

Radio Society of Great Britain

Benefits that keep you on the air

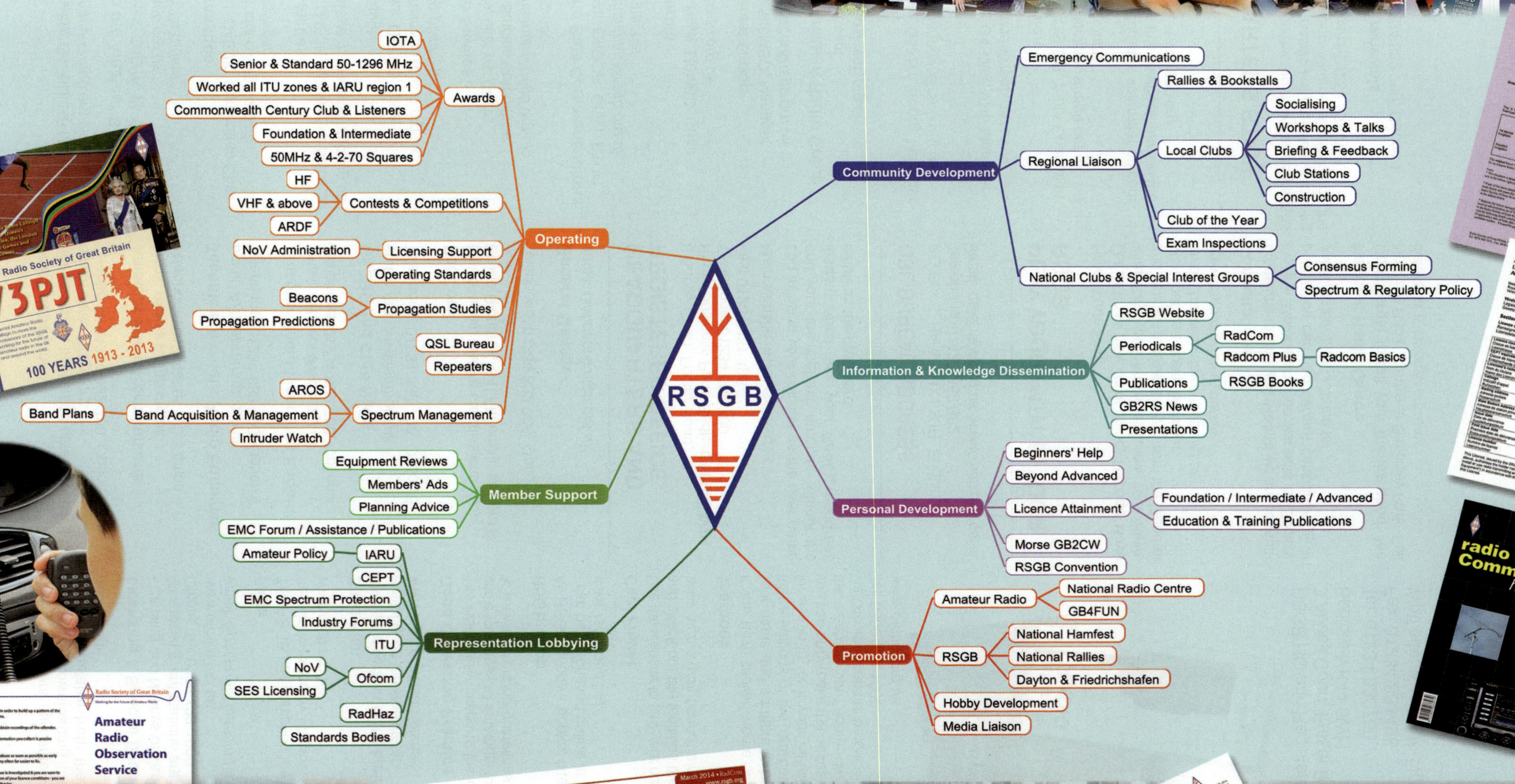

National Radio Centre

The RSGB National Radio Centre (NRC) is a public showcase for radio communications technology. The Centre provides the opportunity for members of the public to get 'up close and personal' with the history and technology of radio communications.

This world-class radio communications education centre is situated at Bletchley Park in Buckinghamshire. From the first inventors in the late 19th century through to future radio developments, visitors will find films, interactive displays and hands-on experiments. Visitors learn about the basic principles of radio and discover the history of radio communication. They see how different parts of the radio spectrum have differing uses, can explore how radio works and experiment with the building blocks of a radio system. The NRC also allows visitors to find out about the role of radio amateurs - who push technology to the limits and have fun at the same time.

You can watch some interesting videos about the NRC in the RSGB National Radio Centre playlist on the RSGB YouTube channel: youtube.com/theRSGB

Visitor numbers

The NRC welcomed over 76,000 visitors in 2022 and we expect a slightly larger number during 2023. Whilst some of the visitors are radio amateurs specifically coming to the NRC, many are not, and it is a pleasure to introduce Bletchley Park museum visitors to amateur radio, many for the first time.

The NRC experience

Starting in the Reception area, visitors learn of the importance of wireless as well as the role of the Voluntary Interceptors and Military WI (Y) Stations in intercepting the WWII Morse code messages brought to Bletchley Park for decrypting. In the main Theatre Zone, a short film describes the importance and many uses of wireless technology in society today – something easily overlooked in this world of internet and instant mobile phone communication.

Interactive displays

The RSGB National Radio Centre boasts a collection of interactive displays - both hardware and software - and experiments that show visitors the workings of a radio communications system. Interactive touchscreen presentations take visitors through key areas of radio technology, while hardware displays allow them to explore and discover the technologies that come together to make radio work.

New displays

The NRC has two new displays this year – one focuses on how to get started in amateur radio, and the other on receiving amateur TV signals via the QO-100 geostationary satellite. Both have attracted a lot of visitor attention. We are also delighted to have received the generous donation of the radio equipment used by His late Majesty, King Hussein of Jordan, JY1 whilst in the UK.

GB3RS

Showcasing amateur radio to visitors, the NRC has a state-of-the-art demonstration radio station, GB3RS. Operational from 80m through to 70cm, GB3RS can communicate on the LF, HF, VHF & UHF bands using CW, SSB, FT8, JT65, FM and D-STAR. With supporting software, the NRC can demonstrate many modern communication aids such as WSJT-X, DX Maps, KST Chat and PSK Reporter. Additionally, the station can track and communicate via a number of polar-orbiting amateur satellites, and the geostationary Oscar100 satellite, as well as monitoring the International Space Station (ISS). If you're a UK radio amateur, you are welcome to bring your licence (along with photo ID) and operate GB3RS.

International visitors

Over the years, the NRC has welcomed visiting radio amateurs from around the world. Many of the visitors come from Australia, Canada, Germany, Netherlands, New Zealand and the USA. If they bring a copy of their Full, Extra or Advanced Licence they can enjoy operating the GB3RS demonstration station.

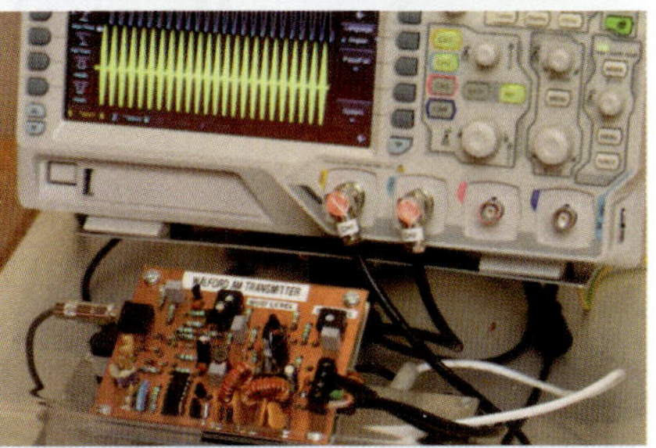

Special events

The NRC is proud to host a range of special events throughout the year including the international Enigma Reloaded event, the Arkwright Trust Foundation Licence training days, and several 'Build a Radio' workshops with support from the Radio Communications Foundation (RCF). We also host the Special Operations Executive demo station, GB1SOE each year and when the ISS astronauts link-up with UK schools using amateur radio, this always proves popular with visitors.

Support and visitor entry

The RSGB is grateful to both individuals and manufacturers who have donated amateur radio equipment to the NRC, making it a state-of-the-art centre. The RSGB National Radio Centre is located at Bletchley Park, Milton Keynes MK3 6DS. Entry to Bletchley Park museum and the NRC is free to RSGB Members on production of a printed downloaded voucher from the RSGB website (www.rsgb.org/bpvoucher). For opening times and other details, see www.nationalradiocentre.com

For non-members, entry price and travel details are available on the Bletchley Park website, see www.bletchleypark.org.uk

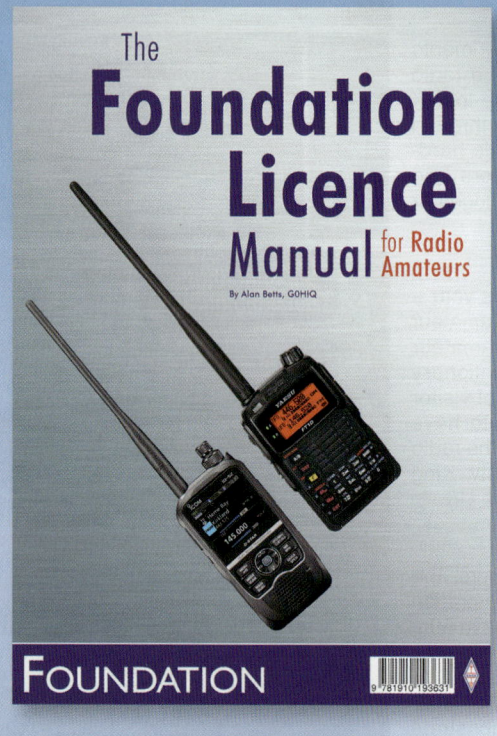

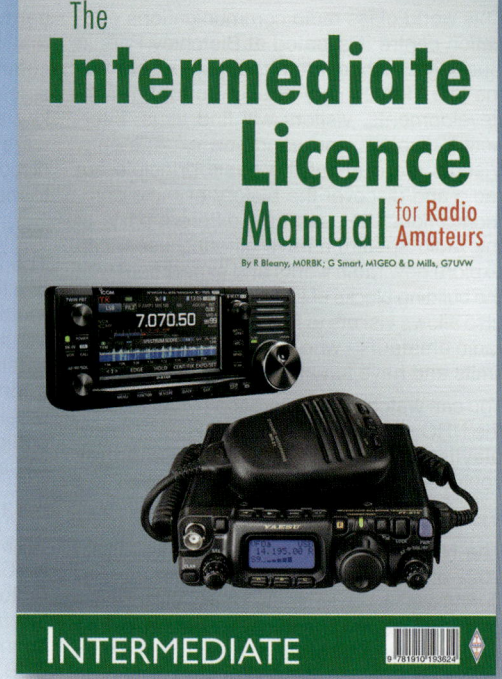

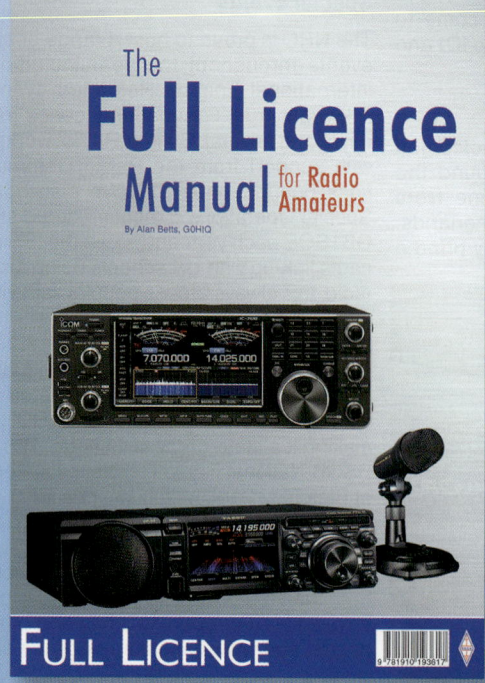

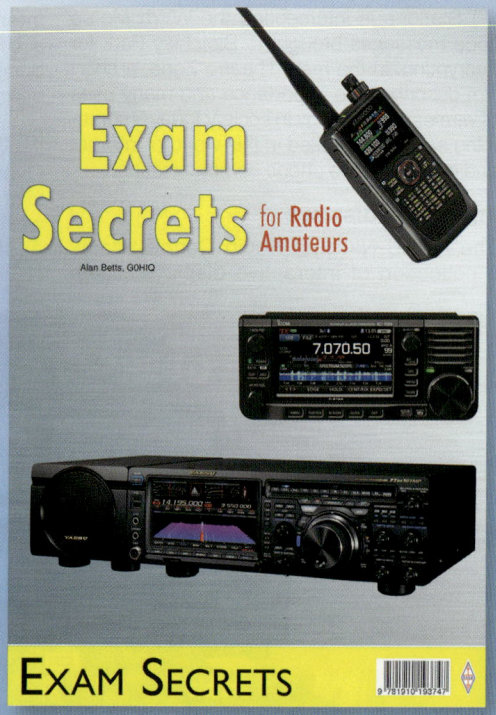

Intruder Watch

RSGB Intruder Watch (IW) collects reports from licensed amateurs in the UK and Crown Dependencies about HF intruders. An intruder is a non-amateur transmission in an amateur band that is not entitled to be there, such as a military data link or over-the-horizon radar.

Fortunately most observed intruders disappear within minutes or hours of their arrival. However when an intruder in a primary amateur allocation is regular and persistent and has been observed by UK amateurs on three or more occasions (e.g. one amateur on three occasions or three amateurs once each), IW may report it to Ofcom's Spectrum Management Centre at Baldock, which has engineering staff on duty 24/365.

Following a report from IW, usually Ofcom will monitor the signal and attempt to get a 'fix' on the transmitter location with direction finders. In some cases an intrusion will last long enough for a formal complaint to be made by Ofcom to the administration concerned. From previous experience some of these transmissions will subsequently cease but others will persist.

Over-the-horizon radar continues to be a problem. Such transmissions are nearly always too short-lived for IW to ask Ofcom to investigate, but they are aware of the problem nevertheless. These radar intrusions come mainly from Russia, China, Iran and a UK base on Cyprus.

RSGB Intruder Watch forms part of Region 1 of the IARU Monitoring System. It has an online database which enables national coordinators and a limited number of collaborators to store observations of intruders in a searchable, shared log.

Information about intruders is also exchanged via a mailing list and a monthly newsletter. There is a comprehensive web site with links to these resources at: *https://www.iaru-r1.org/about-us/committees-and-working-groups/iarums/*

While input from other countries provides a useful stream of tip-offs about intruder activity, IW can only make reports to Ofcom on the basis of intruders heard by licensed amateurs in the UK and Crown Dependencies. Amateurs in other countries should report intruders via their own national monitoring systems.

There is no formal process for observers to sign up with IW. Anyone can submit reports to IW on a one-off, occasional or regular basis as they wish.

The best way to submit reports of suspected intruders is by email to the RSGB Intruder Watch Coordinator at: *iw@rsgb.org.uk*

Abbey Court

The Radio Society of Great Britain continues to be one of a few radio societies in the world to maintain a full time staff. The Society is now administered from a modern, two-storey, open-plan office situated on the prestigious Priory Business Park in Bedford. No. 3 Abbey Court houses the General Manager's Department, Sales and Accounts, Examinations Department, Website Management, IT, *RadCom* and *RadCom Plus*.

Office hours are Monday to Friday, 8.30am to 4.30pm

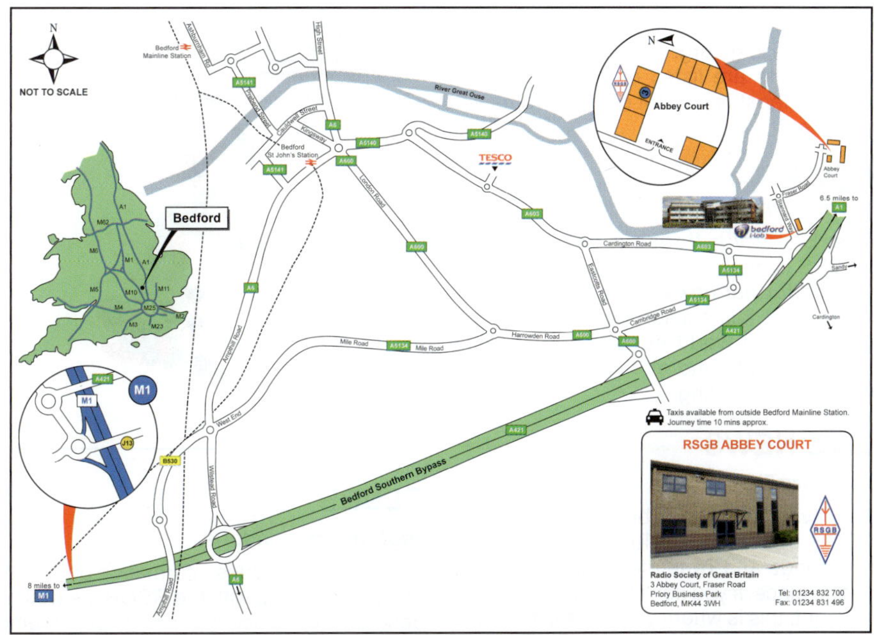

RSGB
3 Abbey Court, Fraser Road, Priory Business Park, Bedford MK44 3WH
Tel: 01234 832 700 **Fax**: 01234 831 496 **Web**: *www.rsgb.org*

RSGB online

In 2023, approximately 150,000 web pages were served to visitors to the RSGB website every month and we welcomed thousands of new users. There are hundreds of pages of information and links to resources from around the world, plus the very latest news from the world of amateur radio.

PORTAL

rsgb.org

Go straight to key areas of the website from our tablet and mobile-friendly front page. Access the latest news headlines, the main site index, guidance for newcomers and information on training and operating.

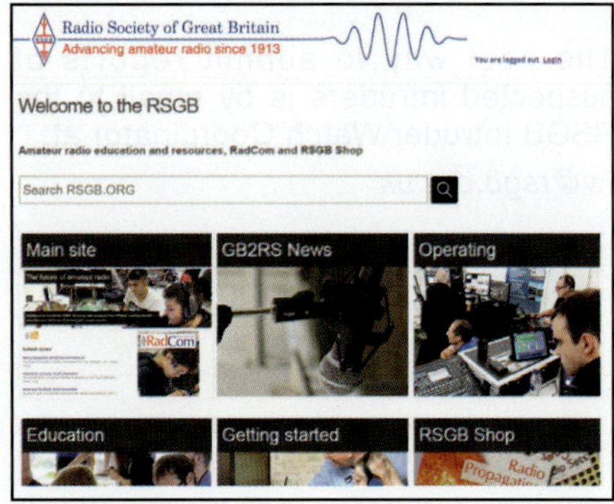

MAIN SITE

rsgb.org/main

The main site provides access to all the content in rsgb.org. The latest updates to the website are listed on this page along with our most important current events and activities.

NEWS

rsgb.org/news

The Society's flagship news bulletin GB2RS is published here every Friday and we add further news items throughout the week. Visit this section for local, regional, national and world news, plus upcoming special events and all the latest from the world of contesting.

LIVE NEWS

rsgb.org/live

The live news page brings together the RSGB newsfeed, YouTube channel, Facebook page and Twitter feed all in one easy-to-find place. If you don't have a social media account, this is where you can see the latest news, videos and discussions online and keep up to date.

ABOUT US

rsgb.org/main/about-us

A summary of who we are and what we do. You will find information about how we are organised, including details of our teams and regions.

RSGB BOARD ***rsgb.org/board*** You'll find a list of the current Board Directors and their areas of responsibility. If you have a question or concern about any of those areas, please contact the Board Director responsible via their email address shown on the page.

COMMITTEES The different committees can be found on the website by typing in rsgb.org/ and then adding the abbreviation of the committee name e.g. *rsgb.org/pac* for the Planning Advisory Committee. These committees offer a wealth of knowledge and support on a wide variety of amateur radio topics.

HONORARY OFFICERS ***rsgb.org/main/about-us/honorary-officers/*** RSGB Honorary Officers, Managers and Advisors help to organise and run particular areas of interest within the hobby. You'll find a list on this page with information about their areas of expertise and how to contact them.

REGIONAL TEAM ***rsgb.org/regional-team*** The Society is organised on a Regional basis; in each Region there are small teams of volunteers made up of District Representatives led by an elected Regional Representative. On these pages you can see who the representatives are in your area and how to contact them.

Regions and Regional Team

The Society is organised on a Regional basis; in each Region there are small teams of volunteers made up of District Representatives led by an elected Regional Representative

Tap or click on a Region to meet your Regional Team

1 2 3 4 5 6 7 8 9 10 11 12 13

CLUBS

rsgb.org/main/clubs

This section includes information on how to affiliate your club with the RSGB, as well as details on the Society's insurance scheme for affiliated clubs. There is information on planning events, as well as details of special interest groups. You can use Club Finder to locate UK amateur radio clubs. Enter your location or postcode and select a travel distance to display clubs near you. Click the markers to display meeting and contact information.

TRAINING

rsgb.org/training

These pages bring you essential educational resources for all three levels of examinations (Foundation, Intermediate and Full). There is also information on the Direct to Full licence examinations which were introduced in January 2023. You'll find support material and advice for trainers as well as information about courses, exams and exam fees. We offer remote invigilation and paper exams and you can find all the information you need as well as the online booking process in this section.

GET STARTED

rsgb.org/getting-started

Everything you need to know in one place if you are new to amateur radio, from getting licensed to setting up your first radio shack.

BRICKWORKS

rsgb.org/brickworks

Passing the Foundation Exam is just a first step to discovering the great diversity of amateur radio. As you progress to the Intermediate and Full licences, there is a range of activities that you can enjoy along the way whether you are new to amateur radio or want to try a different aspect of the hobby. Our Brickworks scheme is great for doing with your local club, or you could take a look at the Individual/Discovery Scheme if you prefer to try activities and new skills on your own.

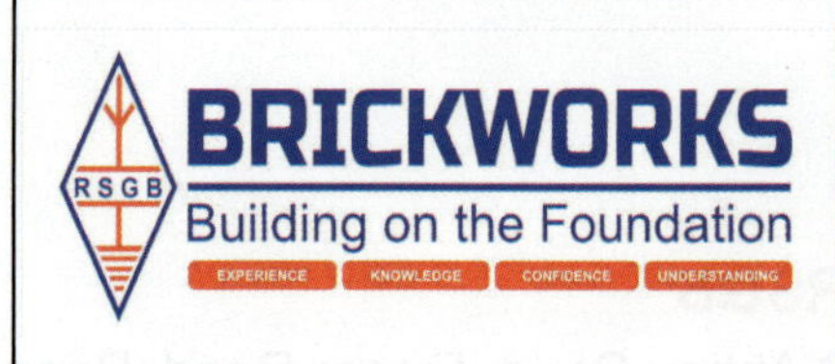

OPERATING

rsgb.org/operating

Information on band plans, awards, beacons and repeaters, emergency comms, the QSL bureau, planning, CW and NoVs, including online applications for selected NoVs.

TECHNICAL

rsgb.org/technical

Home to the EMC and propagation pages and a wide range of other specialist information, including space and satellites and microwave operation. There are also links to technical forums and useful apps.

YOUTH

rsgb.org/youth

We have dedicated web pages covering all things amateur radio for young people. Find out about the radio clubs based at UK schools and universities, as well as Youngsters on the Air (YOTA) events and activity resources. There are also details about who to contact if you would like to get involved in youth activities.

SOCIAL MEDIA

facebook.com/theRSGB

twitter.com/theRSGB

youtube.com/theRSGB

Our social media channels have a combined following of 29,000 people. Join us on Facebook and Twitter for breaking news, extra material and discussions. With multiple playlists on our YouTube channel you can enjoy a wide range of video content whatever your licence level or particular interests within amateur radio. You can also browse our *Tonight@8* live webinars, along with promotional videos for the RSGB National Radio Centre and a range of special events.

VIDEOS

www.rsgb.org/video

Our video library contains a wealth of videos from amateur radio promotional films to celebrations of special events. Our Foundation Practicals and Useful Practical Skills suites of videos, especially for new licensees, have been watched by thousands who have welcomed the clear and inspiring content. We have guidance and support videos to help with the new EMF regulations, archive footage, video content aimed at young amateurs, as well as over 80 RSGB Convention lecture videos which clubs are welcome to use for their meetings.

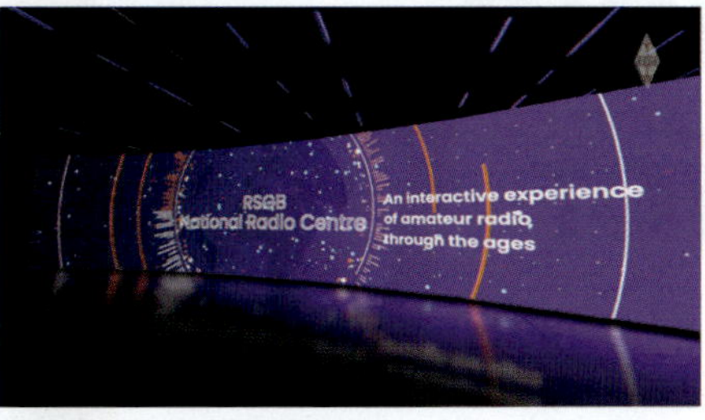

PUBLICATIONS

rsgb.org/main/publications-archives

Members can read the digital edition of RadCom and search the RadCom archive. Members can also read RadCom Basics and RadCom Plus, our two supplements that offer additional content for newer or more technical radio amateurs. You can also browse and purchase the full range of RSGB books.

CONSULTATIONS

rsgb.org/main/rsgb-consultations

A list of current active consultations on matters of importance to the amateur radio community. You are invited to participate.

ARCHIVES

rsgb.org/main/archive

The Photo Archive contains a century of fascinating amateur radio photography, whilst the Events Archive includes preserved material from major RSGB events. Go to the Publications Archive for back issues of our publications, including RadCom Basics and RadCom Plus. You'll find material relating to closed consultations in the Consultations Archive.

FAQs

rsgb.org/faq

If you have an amateur radio-related question, chances are it has been asked and answered before. In this section we answer your most common questions on amateur radio, DBS checking, exams and how to become a radio amateur.

YOUR MEMBERSHIP PREFERENCES

rsgb.org/members

Keep your details up to date via the Membership Services portal. Here you can subscribe to email updates, such as online events, RadCom alerts, GB2RS news and Membership service updates. This is also where you can update your RSGB account details, renew Membership, reset your login password, update your roles and preferences, read our digital publications and download a free admission voucher to Bletchley Park.

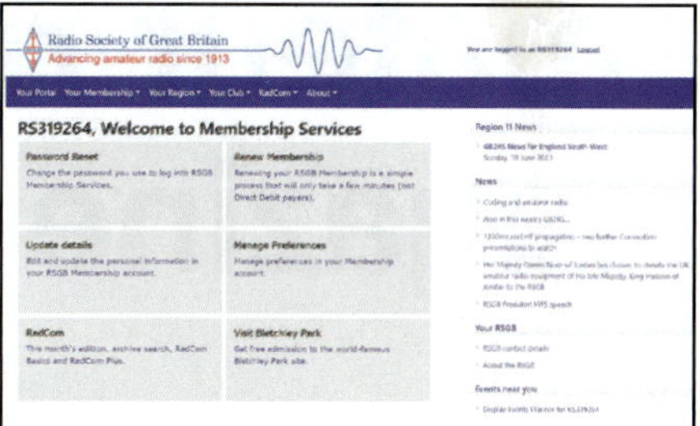

CONTACT

rsgb.org/main/contact

The RSGB's address, phone numbers, and departmental email addresses.

JOIN

rsgb.org/join

Sign up for RSGB Membership and become part of our 21,000-plus community working for the future of amateur radio.

SHOP

rsgb.org/shop

Here you can join the Society, renew your Membership and buy RSGB and other publications with a members' discount. The shop also supplies EMC-related components, RSGB-themed polo shirts and baseball caps. All major cards accepted.

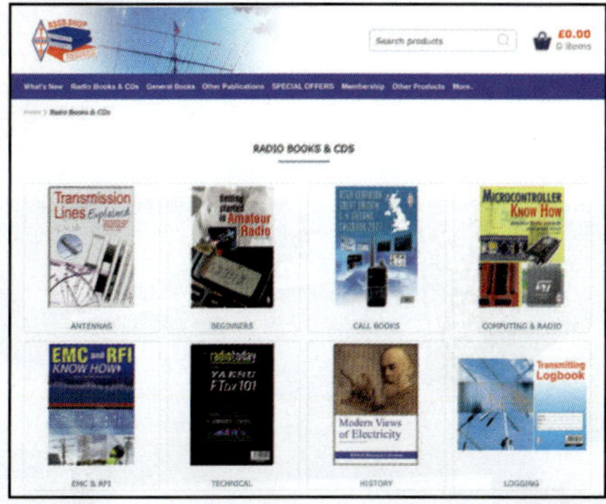

RSGB website: *www.rsgb.org* • RSGB Shop: *www.rsgbshop.org*

Radio Society of Great Britain

Advancing amateur radio since 1913

JOIN US TODAY and get the best amateur radio magazine for FREE

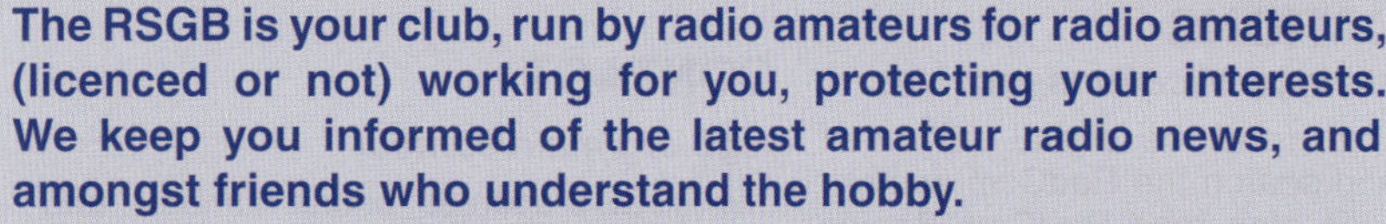

The RSGB is your club, run by radio amateurs for radio amateurs, (licenced or not) working for you, protecting your interests. We keep you informed of the latest amateur radio news, and amongst friends who understand the hobby.

Being part of the RSGB means that we post direct to your door each month the biggest and best amateur radio magazine, *RadCom*. Only available to RSGB members for less than the price of some other high street radio magazines, there is no better way to stay in touch with the world of radio.

Being a member of the RSGB is much more than a subscription to our magazine. You become part of the club that provides all the great benefits shown overleaf.

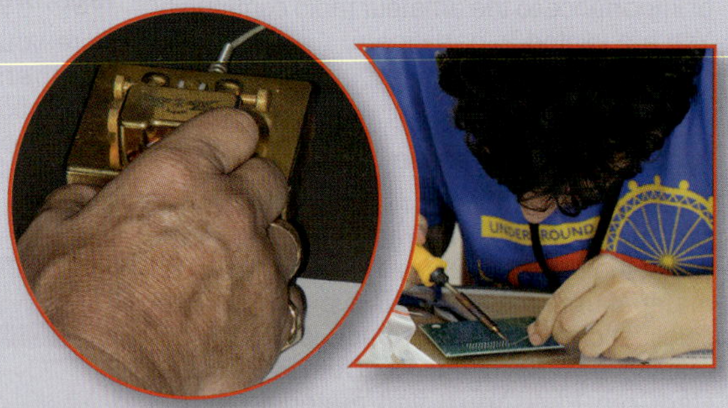

If you want to get the most out of amateur radio, there is simply no better way to do that than by joining the Radio Society of Great Britain (RSGB).

RSGB QSL Bureau

Whilst sending cards for a much-prized contact will always be quicker by direct mail, QSLing via the RSGB Bureau remains an extremely cost effective option, indeed the RSGB QSL Bureau enables members to exchange cards worldwide in the cheapest practical way.

How it works

QSL cards arriving at the central bureau are initially separated into UK and Foreign destinations. Overseas cards are sent in bulk to other member societies of the International Amateur Radio Union (IARU). Cards for stations within the UK are sorted into separate callsign groups and sent to the appropriate volunteer collection managers, on a quarterly schedule. They place cards in stamped addressed envelopes (SAEs) provided to them by the call holders.

Who can use The Bureau?

Unlike the RSGB, many other national societies make extra charges for using their QSL service. The RSGB QSL Bureau is an inclusive membership benefit as follows:

- **UK RSGB members,** including Channel Islands and Isle of Man only, can send and receive their personal cards without additional charges, subject to the conditions shown here.
- **UK non-RSGB members** can collect their personal cards only by using the '*Pay-to-Receive*' service but cannot send cards via the bureau. See RSGB website for details
- **Overseas RSGB members** can send their outgoing cards to the RSGB QSL bureau for distribution. UK call holders should collect in the normal way directly from their UK call sign. Non-UK call holders should arrange collection via a UK-based QSL manager, who must also be a member.
- **Overseas non-RSGB members** may send cards addressed to UK-based stations only.
- **Affiliated Societies and independent QSL Managers** can send their own cards and those for club members or stations for whom they act, but must include current membership or affiliated club confirmation for every station or group whose cards they wish to send. Cards included from overseas stations and intended for delivery outside the UK will not processed without proof of membership and will not be returned.

Available Destinations

A full list of IARU partner QSL bureaus can be found at: *www.iaru.org/iaruqsl.html* Keeping an up-to-date copy to hand is vital when deciding which route to send your card. For example, there are currently no bureaus in, Egypt, Kazakhstan, Morocco, and Mauritius, Sudan and several other African and Caribbean countries, plus many more smaller destinations.

Activity also relates to the frequency with which cards can be dispatched to a particular destination. This may range from monthly to annually, according to demand and is something to consider before sending your card via the bureau.

Responsible QSLing

The Bureau handles approximately 1.5 million cards per year and is one of the busiest in the world. The Society has a policy of discouraging the sending of cards when they are not wanted or cannot be received.

Active Amateurs, GB and Special events, all Clubs and DXpeditions are strongly advised that 100% QSL outgoing is no longer desirable or cost effective.

Transporting large volumes of cards between bureaus, only to have them ultimately destroyed, returned or uncollected, is disappointing and not eco-friendly.

Tip: Ask yourself… Do I need to send a card for every contact before QSLing? Always ask the other station if they can receive a bureau card, before sending.

Log Book of the World (LOTW)

Receiving a nice card for a memorable contact is always a thrill, never matched by an electronic confirmation via the Internet. However, do consider the alternatives, uploading your logs to Logbook of The World can automatically confirm some contacts, such as for contests and award purposes etc.

Confirmations via LOTW are easy and work well for everyone, if a few simple steps are followed. See: *www.arrl.org/logbook-of-the-world*

OQRS systems - the future of QSLing?

Many stations and most DXpeditions and rare calls are now using the worldwide OQRS network and only responding to requests for QSL cards. This online system means there is now no need to automatically send a card, to receive one via the bureau, or direct.

Using OQRS also speeds up the system so that it can now be only half the time it presently takes to send and receive a card, with the added benefit of not needing to send yours. Simply put cards are only sent in response to OQRS requests for a card. So if you are sending QSL cards you or your QSL manager will receive an email to generate a genuinely wanted card. This saves time and waste for both the user and the QSL system in general and is therefore recommended as good practice.

In the UK we are fortunate to have the free to use ClubLog, courtesy of Michael Wells, G7VJR and his team. Simply go to, *www.clublog.org* for more information or to register your call, club, GB station or event and start uploading your logs

Sending cards via the Bureau

Cards from RSGB members for both UK and worldwide should be sent, suitably packed, to the main UK bureau address: **RSGB QSL Bureau PO Box 5, Halifax HX1 9JR, England.** Members, clubs or DX groups wishing to send large or heavy packages to the Bureau via carriers other than Royal Mail should contact the Bureau for an alternative delivery address.

Fair Usage Policy

As part of their subscription, each Member can send up to 15kgs of cards through the Bureau each year (about 5000 standard cards). Each Affiliated Club can send up to 20kgs through the Bureau each year. Additional cards will be charged at £6 per kilo or part thereof.

In the interests of fairness to others, members can only send to the Bureau a maximum of 1kg of cards (approx, 300) for any single country or DXCC entity per month.

More than 1kg of cards must be sent directly to the bureau in the relevant country (see IARU list on line) or direct to the call sign manager.

Heavy users such as, DXpeditions, affiliated clubs and groups should send large volumes

Responsible QSLer...

Help us to speed up processing and cut waste for everyone,

- 10 simple things you can do.....
- Ask new contacts, "If I send you a QSL card, do you collect and how? – every time
- If you don't QSL, be polite but honest "Thanks but no thanks" is all it takes
- Please don't say you do when you don't, or ignore the other guy's kind offer
- QSLing 100% outgoing is costly for everyone. Half your cards could be wasted, please check before you send
- Create your own OQRS system or use ClubLog *www.clublog.org* to reply to incoming requests with a real card, save time and money for you or your club
- Make your QRZ.com QSL details clear, honest and visible
- Amend your online details if you change your QSL status - don't leave it
- QSL info– Direct or via…' is confusing. Please clarify what you really mean
- Clubs Calls – Collect only from the calls sub group. For, 'QSL-Direct,' show an address not a callsign as this can change, it's often confusing
- Always collect your cards. - even if you never send one, they will arrive

Be responsible, it only costs a stamp!

The bureau operates a 24-hour message line for members' QSL enquiries. Tel: 01422 359362. When calling, please leave contact details, a brief message and if possible an email address. email: qsl@rsgb.org.uk (please put your callsign series in the subject box)

direct to overseas bureaux. please see – 'Responsible QSLing' advice before QSLing 100% outgoing.

Every package of cards should contain ...

- Proof of current membership; that is an original RadCom address label, taken from the magazine wrapper or printed insert, showing: address, callsign and membership number, not more than 3 months older than the membership expiry date on the label.

  As Clubs receive the RSGB Yearbook each year in lieu of RadCom, they should include sufficient information for a check to be made against the Affiliated Societies' register, ideally in the form of a club letterhead, showing the membership number and renewal date. To speed status checking, clubs and groups are asked to ensure that they register club call and contact details at *My Account* directly in the group's name and not as secondary to a personal callsign, or qsl manager.
- Special event stations (GB) and single letter Abbreviated/Contest callsigns should include the membership number and call of the NoV license holder or affiliated club, for contact purposes.

Checklist for sending cards

We need your help to sort more than a million cards each year and reduce delays.

A ***First, place your cards into three piles...***

1. UK destinations

Pre-sort G, M and 2 as per the Sub Managers list.

2. USA destinations

Sort by number only, 0-9, regardless of prefix. Separate cards for Alaska, Hawaii and Puerto Rica.

3. Rest of the World

In DXCC callsign prefix order.

4. Calls with numbers first

Sort in digit order, i.e. 3A, 4X, 8P, 9H etc.

5. Calls with one letter then one number

These come before two letter prefixes, i.e. S5 before SM, etc.

B ***Check ALL cards for possible ' Via' destinations***

Re-sort if necessary, and never rely on your computer print log, for example: F5/G3UGF isn't a French destination.

Africa, Caribbean and DX destinations are mostly QSL Direct only, or via a QSL Manager. Check *www.qrz.com*

C ***Pack your cards securely and don't forget***

A recent RadCom address wrapper as proof of membership.

Your callsign and return address on the package.

If you put more than ten cards in a C5 envelope, check the dimensions and weight at the Post Office, before sending - don't just post.

Whatever the quantity, never send unsorted cards!

Other important points

- Clubs and QSL Managers sending a bulk dispatch to the bureau should ensure that all callsign holders for whom they send cards are current members of RSGB and should enclose current membership details for every callsign with every batch of cards. All Clubs and Groups must include their Affiliated Society membership number.
- Members who operate from another station, typically a foreign club call or that of an individual overseas amateur, may send cards for contacts made from that station, provided they clearly identify themselves as the operator and state their UK callsign and membership number on each card.
- Listener report QSLs need sufficient information to be of genuine value to the transmitting amateurs. Reception reports relating to broadcasting stations cannot be accepted. *see*' Card Issues..' for details of acceptable card details.

  The bureau system accepts standard cards only no letters, SAEs or money orders.
- All cards, whatever the quantity, should be pre-sorted into alphabetical and numerical country DXCC order (see the Prefix List pages and/or the *RSGB Prefix Guide* which also contains a complete cross reference and awards section).

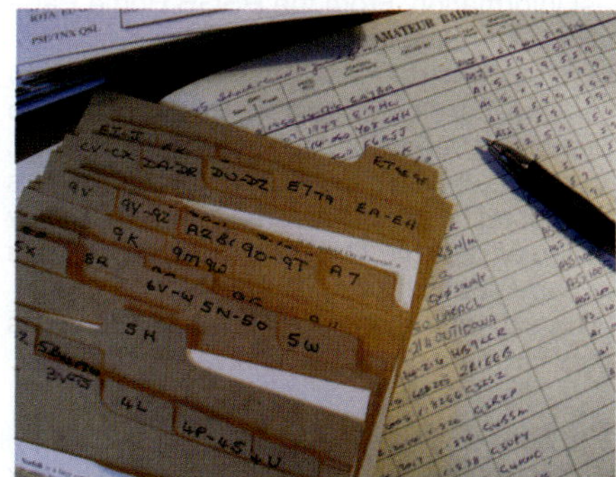

Hand written cards can be easily pre-sorted, using a simple card index box.

- Countries with more than one prefix should be placed together. For example, JA and 7J cards (destined for Japan), F and TK-TM cards (destined for France) and SP, HF and 3Z (destined for Poland) may be grouped together.
- Cards for the USA need be sorted separately into call areas (numbers 0-9), regardless of the prefix letters.

NB: Exceptionally, cards in the number 4 series with either one or two letters before the number are handled by different bureaus and need to be separated, as are cards for Alaska (KL), Hawaii (KH6-7) and Puerto Rico (KP3-4).

- Cards for Russian Federation and former Soviet countries were traditionally grouped together. They now need to be separated into five individual groups; RA-RZ, UA-UI, UJ-UM, UN-UQ and UR-UZ, as they are no longer sent to a single destination in Moscow. See: *www.iaru.org/iaruqsl.html*
- Cards for UK delivery need to be provided separately to foreign destinations. Our UK sorters currently have to split cards into 40 alpha-numeric categories. For this reason cards should be supplied to us pre-sorted as per the Sub Manager list on the following pages.
- Envelopes, paper or card dividers to separate countries or call groups are not required, as removing these can sometimes slow down the distribution process.
- Cards sent in date/time/logbook or random order are not acceptable, as they typically take up to five times longer to process. Similarly, those with small print or hand written callsigns can be very difficult to process, resulting in delays for other users. The bureau reserves the right, at its discretion, to reject unsorted cards or those with callsign or routing information in small or difficult to read print. The minimum print size requirement is 12 point.

Tip: If you are unsure about your handwriting, why not ask someone else to check the cards to see if they can easily read the callsigns?

Before you send, please check that we can deliver!

Check all Vias before you send, as your card may come back to you, or it may never arrive.

There are many world destinations, but only 190 IARU member and associate Bureaus worldwide.

The following IARU Bureaus are currently closed.

3B Mauritius	D4 Cape Verde
3DA Swaziland	HH 4V Haiti
4J-K Azerbaijan	HV Vatican City
7P Lesthoto	PZ Suriname
9L Sierra Leone	ST Sudan
A3 Tonga	SU, 6A-B Egypt
C2 Nauru	V3 Belize
C5 Nauru	V4 St Kitts and Nevis
C6 Bahamas	XY-XZ Myanmar
CN, 5C-G Morocco	Z2 Zimbabwe

Download your own Bureau list from: *www.iaru.org/iaruqsl.html*

Post outgoing cards to: RSGB QSL Bureau, PO Box 5, Halifax HX1 9JR.

Card issues and some good advice

For economic shipment and to avoid possible transit damage, all cards need be single page, IARU recommended size and weight. That is, standard postcard size (140m x 90 mm) and no larger. Card weight and thickness is important and needs to be in the range, 230-350 grams.

Note: Per 10 cards, a single card should have an average weight not more than 3 grams.

The bureau reserves the right not to process or return, repeat batches of cards falling outside these guidelines and where a sender has been previously advised.

Large or unusual shaped cards are extremely difficult to process and most easily damaged when packed, or folded with others.

Thin, small, and paper cards are slow and extremely difficult to handle. They often stick to cards for other destinations, as do homemade cards using photo print or heat laminated paper. This type of card does not travel well, is difficult to write on and is very easily damaged when subjected to humidity or damp – not recommended!

Multi-page and cards not meeting the above IARU recommendations must not be sent via the bureau system as they increase workload and overheads. In fairness to others they should be sent direct only.

QSL Routing & QSL Direct

It sounds obvious, but the Bureau can only process outgoing cards if there is a destination to which they can be sent. Before sending cards, therefore, (particularly to rare stations or DXpeditions) please check the recipient's QSL policy. This is usually available on QRZ.com or via a websearch.

Many DXpeditions and rare callsigns only QSL direct, or respond to an OQRS request often via a QSL Manager who may be in another country. These stations are most often located where there is no bureau service and are operated by visiting non-resident Amateurs from another country. Some stations do not QSL at all, so it is vital to check before sending, whatever route you choose.

Please note that outgoing bureau cards where no destination bureau is available, or no clear 'Via' route is indicated, will be recycled.

Tip:

- Ask for the other station's QSL details at the time of the QSO, or by an Internet search before posting.
- Consider posting your most wanted cards direct or to an overseas bureau, if it's active. This helps to speed replies as most bureaus world-wide have backlogs. The IARU world-bureau list can be downloaded at: *www.iaru.org/qsl-bureas.html*. It's good practice to check the listings for changes at least twice a year.
- Always search the web and check *www.qrz.com* first before posting.
- Make sure that any "Via" information on your cards appears directly below, or next to, the station callsign, to avoid being missed. Using a different coloured ink for this purpose is a great help.

For guidance on what information to include on your card (and where), *see the example card on page 20*.

Using printed labels

Avoid cramming too much information on small printed labels. For health and safety reasons all callsigns should be a minimum of 12 point print size and in common, easy to read fonts such as Arial, Times New Roman, or clear, block capital hand written letters.

The bureau system is for the exchange of QSL cards only. Envelopes containing letters, photographs, money, stamped addressed envelopes (SAEs), awards, certificates and other items will not be processed and should be sent by other means.

Part of the overseas side of the bureau

Heavy users

Those sending more than a few thousand cards per year should send their largest volumes of cards directly to their top ten destination countries. The remaining balance can be sent via the normal bureau system.

The aim of this is to share some of the burden of cost, without penalising others who may only occasionally send a few more cards than normal. The bureau weighs and notes regular large consignments. Members or clubs may be contacted if their usage becomes excessive, with a request to follow the guidelines above. The IARU bureau list can be found at: *www.iaru.org/iaruqsl.html*

Packing and posting your cards

The bureau receives many damaged envelopes and packages from both UK and foreign amateurs. It also receives a significant number of requests each month from Royal Mail for payment of additional postage, which are always rejected.

Having first separated pre-sorted, UK destination cards from the rest of the world, please read on…

- Never post loose cards in lightweight or thin envelopes, as they will often cut through the edge of the envelope in transit.
- Always print return address details and callsign on your package, in case it arrives damaged.
- Secure batches of cards with a rubber band or - better still - a banknote style band of thin paper strip, folded around the cards.
- Never place two or more packs of cards side by side in a C5, A4 or larger envelope, as it will fold in transit and split down the middle, allowing the cards to spill out.
- Using lightweight 'Mail-Tuff' style plastic or 'Mail-Lite' style padded bags or Post-Pack envelopes usually avoids this problem.
- Always check the size and weight of your envelopes and packages, before posting as the Post Office now charge by volume as well as by weight.
- The current weight limit for a First Class stamp is 100g, but the package size is limited to 240mm x 165mm and the package should fit through a postal slot only 5mm in height.
- It is possible to send a large envelope A4, or a smaller envelope over 5mm in thickness. This type of envelope is considered to be a 'Large Letter.' Large Letter stamps costs more, but allows the letter thickness to be up to 25mm. 'Second Class Large' offers better value
- The Post Office can supply a paper/card copy of their pricing slot guide for a small charge. Frequent users are advised to obtain a plastic Helix HP5 'Pricing in Proportion' Ruler. It has postal slots built in, to check your packets.

Sending small numbers of cards in separate envelopes is not cost effective for the sender and means much more time spent opening, sorting and checking in the bureau. Sending not more than one pack per month, with your *RadCom* label, resolves many issues and can save you money .

Recorded delivery is not cost effective. We receive many packages and we are not always asked to sign for individual items, secure packing and a return address offers better value.

Receiving cards from the Bureau

RSGB is extremely fortunate to have around 30 dedicated volunteer Sub Managers who give freely of their time to support the work of the Bureau and in the service of their fellow radio amateurs. Members' cards are sent to the Sub Managers for onward distribution.

Sub Managers details are subject to change, so it a good idea to check the QSL section of the RSGB website from time to time for the latest information. From the RSGB Home Page click,'Operating' and follow the links.

Our system relies upon those wishing to receive cards depositing SAEs with Sub Managers, ready for each quarterly despatch. Members should use SAEs, as Sub Managers are not authorised or insured to accept money in lieu of postage stamps. RSGB is not liable in case of any loss or dispute.

The scheme is open to all RSGB members plus UK-based, pay-to-receive subscribers.

Collection Envelopes

- Envelopes need to be C5 size (160mm x 230mm) and of strong material (*see earlier*).

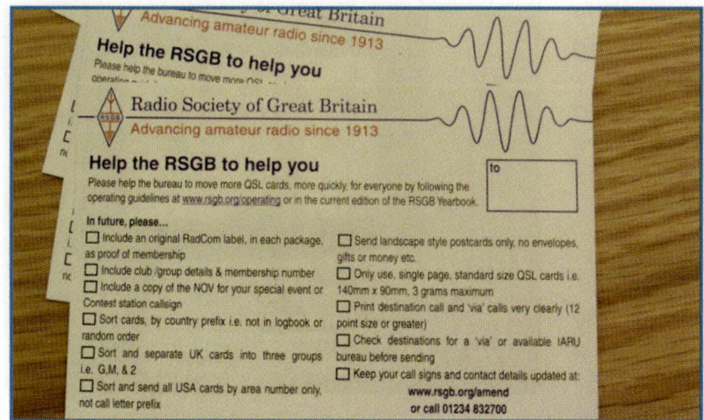

RSGB's new feedback card, is designed to help speed QSL throughput at one of the world's busiest bureaus. - If you receive one... "Please help us to help you."

- Callsign or Listener number should be printed in the top left hand corner, followed by the a current membership number, immediately below.
- Print the name, delivery address and postcode clearly, as normal.
- Number each envelope sent to the Manager (eg '1 of 6', '2 of 6', '3 of 6', etc) always mark one of them 'Last', so that you will know when a fresh batch should be sent.
- Envelopes are normally despatched every quarter, subject to card availability.
- Always use stamps worded Second or First Class, rather than a numerical amount, as these will be honoured if the postal rate changes.

No delivery in that quarter means 'no cards waiting'.

Cards for amateurs who have not lodged envelopes are not returned to sender and will, at the Sub Manager's discretion, be recycled after a period of three months.

Checklist for receiving cards

1. Register <u>all</u> your callsigns, past/present and always collect from the callsign as used, don't divert

Do this via the RSGB website at, 'Your membership ' page, or phone 01234 832700.

2. Send C5 stamped envelopes to each Sub Manager

In addition to your name and address...
Write your callsign and RSGB membership number at the top left.
Number each envelope at the bottom left.
Mark 'Last envelope' at the bottom left of the last envelope.
N.B. Sub Managers are not authorised to accept cash in lieu of SAEs.

3. Holidays and portable activations

We don't automatically divert cards to your home call.
For temporary prefix operation (e.g. GM, MW, 2I etc), lodge separate envelopes with the relevant Sub Manager to collect your cards.
N.B. The Channel Islands and Isle of Man use different stamps.

4. Special event (GB) and abbreviated contest calls (G1A, etc)

No diversions apply, so please see the Sub Manager list.

5. Special Prefix NoV callsigns

For GR, MQ, 2O etc, no diversions apply. See the Sub Manager details.

Multiple callsigns can be listed on the same envelope.

Always keep your envelopes up to date. Many volunteer Sub Managers now operate their own websites, with links from the RSGB website, giving cards waiting, envelope status and next anticipated delivery details. RSGB requires these lists to be confidential. Members permission to display their callsign and details to others, is a condition of inclusion on any such listing, operated by a volunteer.

It is a good idea to note in your diary to check your Manager's list every quarter.

UK amateurs who do not wish to collect cards or those who use a separate QSL Manager are asked to notify the appropriate Sub Manager as a matter of courtesy and also make this clear at: *www.qrz.com*

More than one callsign?

Always collect from the callsign used.
Stations changing their callsign as a result of a licence upgrade, or other reason should inform RSGB of their change of status. Contact membership services direct or via the web site. Log in using current call and membership number and enter personal details at the, 'Update Details' page.

Amend the primary callsign and list all previous calls in the additional category, as soon as possible. Cards must not be diverted from one call to another. Envelopes should be maintained with both new and old QSL sub-Managers. Typically, envelopes for the old callsigns and membership number need to be available for up to five years after the old call is no longer the primary call.

Club stations should enter their callsign details in the club's name and not as a secondary call of a member or QSL manager, as this gives rise to confusion. Please avoid registering calls using optional club identifiers, such as X,S,C,N.

Stations operating from a different prefix, for example G9ABC as GW9ABC/P or GU9ABC/P, need to lodge envelopes with the appropriate Sub Managers for every area of operation, as cards may not be forwarded to the home call.

UK mainland stamps are not valid when sent from the Isle of Man or Channel Islands. Local stamps should be obtained during the period of operation, for use later.

When operating outside the UK under CEPT rules, e.g. F/G9ABC, or more importantly with another callsign, it is vital to tell the QSO partner to 'QSL via G9ABC' and not simply state 'via home call'.

Registering any foreign calls separately, together with the QSL route and contact email address at: *www.qrz.com* is extremely helpful to others in these cases.

Requesting a 'Via' call route

In recent years there has been an explosion in the use of 'Via' requests from members, significantly those using multiple call signs, clubs and others using a QSL manager.

With so many cards passing through the bureau at any one time this massive growth in via activity contributes card delays ie. slowing down the process of sorting and delivering cards.

Members can help to speed up the QSL process as follows by:

- Always collecting from the call sign used at the time of the contact.
- Not giving divert information during a contact, - see exceptions below.
- Not listing divert information at QRZ.com.

The following are not subject to any diverts:

- Personal QSL cards, including to a members other call signs.
- Club and Affiliated group calls,
- Contest calls
- Special Prefix calls (R.Q.O.V etc)

NB. Separate C5 collection envelopes, need to be lodged with the appropriate sub-manager for every call sign held. -see list

The following cards may be diverted:

- Calls used in another UK pre-fix location, back to the home call.
- Members personal overseas call signs sent to the UK home call.

Remember:
Even if you never send a QSL care, someone, somewhere, sometime will send a card to you. It would be a shame not to receive it so, please send an SAE to your RSGB volunteer sub manager for each callsign you hold.

Card design

Whether you are designing and making your own QSL or having it made professionally, ***size, quality and design are the most important factors*** if you are hoping for a reply. Gone are the days when cards were printed in a single colour (black), with only a callsign and basic information and which took several weeks to produce. The advent of high resolution digital photography and computers has changed everything. High quality commercial QSL cards are now more interesting, colourful, easier, quicker and much cheaper to produce or change than ever before. What's more, professionally printed cards can and often do work out cheaper than making your own. All the more reason to consider having a distinctive card that gives not just your station details, but perhaps reflects your radio and other interests, family, pets, location or some other part of your life. Cards can be simple, beautiful, artistic, funny, technical or even something completely unexpected. They make a statement about you - so what does your card say?

The range of choice has never been greater, so just use your imagination. Above all, make your QSL card something of quality that stands out; something that the other station will want to keep and display. If you are sending or receiving a 'gift', make it memorable. It's now possible to collect special interest cards showing planes, trains, ships, cars,

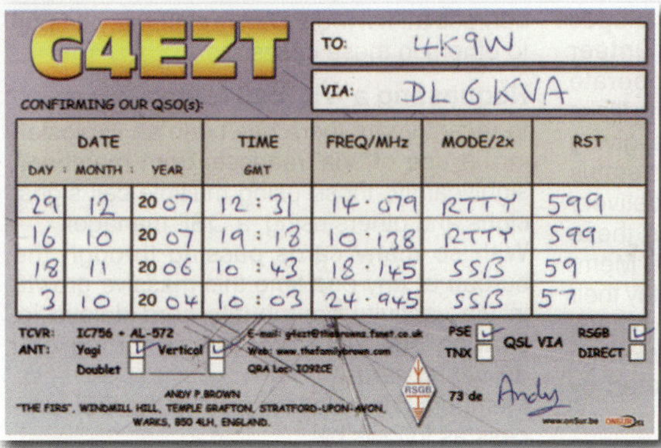

Example of a QSL card that's well laid out, easy to read and easy to sort.

families, pets, castles, churches, windmills, lighthouses, motorcycles and many other things, in addition to antenna farms, radios, vintage gear and shack interiors.

RSGB Bureau reserves the right not to accept, process, or return, cards from any source, containing images or content not directly relevant to Amateur Radio and which in its opinion may be likely to, or does cause offence to those handling or receiving them.

Tip: Remember to tidy up before you take a photo of your station!

The business side of the card is also very important. Here, simple clarity is the key to a good card and to receiving a reply. Use a clear type face that is easy to read. Don't put too much information or too many logos on the card, unless it's a special event when background information is always nice to see. Remember that English is not always a first language, so keep it simple, keep it relevant. Allow enough space to write or print the contact information clearly on the card, ensuring that the destination call is at the ***top right, with any via routing details immediately below.***

Many cards now have space to log more than one contact. This is a great eco-friendly idea. *See example card from G4EZT.*

Where to buy cards

The RSGB doesn't endorse any particular producer. Take a look at the cards you receive, as they will often include maker's details. Apart from your local printer and checking with friends, there are now a whole range of specialist online makers offering superb, correctly sized card. We regularly see cards from UK stations being sent to us that have been designed online, some produced in other countries, and many are simply stunning. It is possible to download card making software from the Internet, but so much depends on the actual equipment used to make the card that the results are often disappointing or uneconomic, unless you have access to specialist print and cutting machinery. However, where practical, they do make possible one-off special, individual and personalised cards, for QSLing direct.

Remember: If you have invested time and energy on your station, isn't it right to do the same with your QSL card? Send something you would be pleased to receive.

Using www.QRZ.com

There's no doubt that QRZ.com is the go-to place for station information. Completing your outgoing cards either as a computer run or handwritten, it's wise to check the QSL choices for the intended station, to avoid waste and disappointment.

- The information is only as good as the person who entered it so remember to keep your own up to date and easy to understand. n.b. every page also shows the last amended date.
- For, 'QSL bureau cards.' Avoid confusion, always state , 'QSL via RSGB Bureau.'
- For 'QSL Direct' – Always send to the station address, never to the page manager or to the bureau with a, via. It may be wrong or out of date
- If you wish to receive direct cards write "QSL direct to address above." To avoid confusion.
- Those giving no QSL details it must be assumed do not want or accept incoming cards.
- Always read the whole page as some put their QSL info at the end of the page, not at the top.

The modern way to display your cards, using a digital photo frame. Scan the best of your collection, Create albums of interest, Antennas, Ships, Stations, Landscapes, people, places, DX, IOTA and more. Just use your imagination

RSGB QSL Bureau Sub Managers

All details correct at time of press, but may be subject to change. For the latest information visit the QSL pages at the RSGB website

Abbreviated & Contest Calls

Mr. M. G. Coomber. G0NBI
3, Dolly Grove,
Blackdown Heights, Crimchard.
Somerset. TA20 1PF
grahamG0NBI@gmail.com

G0 Series

Mr. R. Rogerson G0OUC
8, Dearne Street,
Darton,
Barnsley
South Yorkshire S75 5 HL
Email: g0qsl@yahoo.com

G1 & G2 Series

Mr L. Pennell, G8PMA
182, Northampton Road,
Wellingborough,
Northamptonshire NN8 3PJ
g8pma@pennell.eu

G3A-F

Mr P J Pasquet, G4RRA
Honey Blossom Cottage Spreyton
Devon EX17 5AL
gee4rra@gmail.com

G3G-L

Mr L Pennell, G8PMA
182 Northampton Road,
Wellingborough
Northamptonshire NN8 3PJ
g8pma@pennell.eu

G3M-S

Mr G Coomber, G0NBI
3, Dolly Grove,
Blackdown Heights, Crimchard.
Somerset. TA20 1PF
grahamg0nbi@gmail.com

G3T -V

Mr. N. S. Cawthorne. G3TXF.
Dormers, Hinton Charterhouse,
Bath, Somerset BA2 7TJ
nigel@g3txf.com

G3W-Z

Mr J Peden, G3ZQQ
51A, Bewdley Road, Kidderminster.
Worcestershire DY11 6RL
g3zqq@yahoo.co.uk

G4A-F

Mr J J Pascoe, G4ELZ
3 Aller Brake Road, Newton Abbot,
Devon TQ12 4NJ
g4elz@blueyonder.co.uk

G4G-L

Mr I N Fugler, G4IIY
Lees Hill Farm, Lees Hill Brampton,
Cumbria CA8 2BB
ian.g4iiy@zen.co.uk

G4M-S

Mr C G Rowe, G4MAR
29 Lucknow Road, Willenhall,
West Midlands WV12 4QF
cliff1.g4mar@gmail.com

RSGB QSL Bureau Sub Managers

All details correct at time of press, but may be subject to change. For the latest information visit the QSL pages at the RSGB website

G4TAA-ZZZ Series
Mr P. Rivers, G4XEX
34 Coales Gardens,
Market Harborough,
Leicestershire.
LE16 7NY g4taa.g4zzz@gmail.com

G5 Series
Mr P J Pasquet, G4RRA
Honey Blossom Cottage, Spreyton, Devon
EX17 5AL gee4rra@gmail.com

G6 Series
Mr S Wellon, G6DMG
71 Toftdale Green, Lyppard Bourne,
Worcester WR4 0PE
g6dmg@hotmail.co.uk

G7 Series
Mr C. Flanagan, G7NRO
2 Wynyard House, Durham Road, Wolviston.
Billingham, TS22 5LP
g77nro@gmail.com

G8 Series
Mr D Helliwell, G6FSP
1 Beechfield Avenue, Torquay TQ2 8HU
dave@g6fsp.com

GBxAAA-ZZZ
Mrs D Williams, M0LXT
20 Neale Close Wollaston Northamptonshire
NN29 7UT
qsltrek@hotmail.co.uk
www.gb-special-event-qsl-status.webs.com

GD, MD & 2D Series
Mr A Ames GD4SVD
20, Sunnybank Avenue, Onchan. Isle of Man
IM3 3BW gd.md.2dcards@outlook.com

GI, MI & 2I Series
Dr E H Squance, GI4JTF
11 Ballymenoch Road, Holywood,
Co Down, Northern Ireland BT18 0HH
gi4jtf@gmx.com

GJ, MJ & 2J Series
Mr M Roche, MJ0ASP
Flat 1 Stratscombe House, Le Quai Bisson,
St Brelade, Jersey JE3 8JT
mathieu.roche@hotmail.com

GM0 - GM3 Series
Mr F A Roe, GM0ALS
74 Willow Grove, Livingston,
West Lothian EH54 5NA
fred.roe190@googlemail.com

GM4-8 Series
Mr P Rose. GM3ZZA
4 Heatherfield Glade
Adambrae,
Livingston EH54 9JE
gm48.qsl@btinternet.com

GU, MU & 2U Series
Mr P F H Cooper, GU0SUP
1 Clos au Pre, Hougue du Pommier,
Castel, Guernsey GY5 7FQ
pcooper@guernsey.net

GW- Series
Mr. J. L. Lewis. GW0RAD
189,Heol y Gors, Cwmgors, Ammanford,
Carmarthenshire Wales SA18 1RF
gwmanager@sky.com

2E Series
Mr R Maltby, 2E1DFI
1 Briar Close Southfields, London Road,
Sleaford, Lincolnshire NG34 7NT
ray2e1dfi@aol.co.uk

2M Series
Mr. R. Roberts. MM0CPZ
16, Swanston Avenue,
Edinburgh
EH10 7BX
Email ; mmand2m@yahoo.com

2W Series
Mr G Coomber, G0NBI
3, Dolly Grove,
Blackdown Heights, Crimchard.
Somerset. TA20 1PF
grahamg0nbi@gmail.com

M0A-L
Mr D E Mappin, G4EDR
13 Willow Close, Filey,
North Yorks YO14 9NY
g4edr@yahoo.com

M0M-Z
Mrs V Bates, G6MML
The Anvil.4 Eastgate,
North Newbald,
York YO43 4SD
g6mml@btinternet.com

M1 Series
Mr R Taylor, M0RRV
2 Chadwick Road, Moorends,
Thorne, DN8 4NG South Yorkshire
roytaylor187@btinternet.com
groups.yahoo.com/group/M6_QSL/

M3 Series
Mr R Taylor, M0RRV
2 Chadwick Road, Moorends,
Thorne, DN8 4NG South Yorkshire
roytaylor187@btinternet.com
groups.yahoo.com/group/M6_QSL/

M5 series
Mr R Taylor, M0RRV
2 Chadwick Road,
Moorends, Thorne.
South Yorkshire DN8 4NG
Email: roytaylor187@btinternet.com
Website: groups.yahoo.com/group/M6_QSL/

M6 - M7 Series
Mr R Taylor, M0RRV
2 Chadwick Road, Moorends,
Thorne, DN8 4NG South Yorkshire
roytaylor187@btinternet.com
groups.yahoo.com/group/M6_QSL/

MM Series
Mr. R. Roberts. MM0CPZ
16, Swanston Avenue,
Edinburgh
EH10 7BX
Email ; mmand2m@yahoo.com

MW Series
Mr G Coomber, G0NBI
3, Dolly Grove, Blackdown Heights,
Crimchard. Somerset. TA20 1PF
grahamg0nbi@gmail.com

RS Receiving stations
Mr R Small, RS8841
13 Rydall Close, Stowmarket,
Suffolk IP14 1QX
rob@g3ali.co.uk

Special UK Prefixes - NoV Call Holders
R. Royal Weddings, **Q**. Queen's Jubilee.
O. Olympic Games. **V**. RSGB Centenary.
Mr J. Peden, G3ZQQ
51A Bewdley Road, Kidderminster.
Worcestershire DY11 6RL
g3zqq@yahoo.co.uk

Note 1. The sub group for 2 letter suffix G Callsigns, has closed. All 2 letter calls are now sorted and distributed with 3 letter calls.

Amateur radio in the media

It is great to see amateur radio mentioned in the mainstream media and to see stories about radio amateurs. Here we bring together just some of the publicity we have gained or shared during the last 18 months.

The Secret Genius of Modern Life

In March 2022, the RSGB was approached by the BBC to find experts to contribute to its "The Secret Genius of Modern Life" series. Episode one, which aired in November, looked at the bank card and Neil Smith, G4DBN re-created the Great Seal Bug – a wooden seal, gifted to the US embassy in Moscow in 1945, which contained a covert listening device.

The technology developed for the bug was an early example of RFID, which is what allows contactless card payments to work. You can catch the full programme again on BBC iPlayer if you missed that episode. Neil put in a huge amount of work for this opportunity and you can see some videos of the process on his YouTube channel: *https://youtube.com/MachiningandMicrowaves*

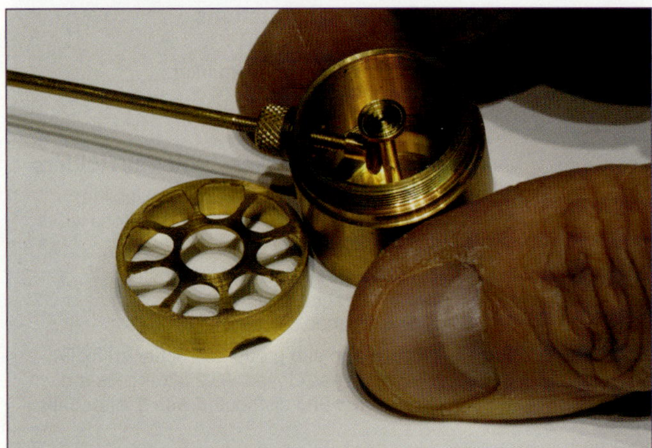

Practical Boat Owner

Also in March, Practical Boat Owner magazine contacted us for some general information about amateur radio and how to get involved. It was used in the May issue as part of an interesting article by Tony Preedy, G3LNP about the use of amateur radio instead of satellite phones whilst at sea.

PBO PRACTICAL

The budget rival to satellite phone connectivity

Radio amateurs can maintain contact worldwide for the cost of a radio and an antenna, explains Tony Preedy

These days it's relatively easy to maintain phone contact almost anywhere in the world – but to do so from a boat utilising a satellite phone is expensive. Radio amateurs, meanwhile, can maintain contact between themselves anywhere on the planet for only the cost of their radio and an antenna.

My daughter Liz and I recently completed a one year cruise of the islands of the Atlantic and Caribbean. By using amateur radio, for which I've been licensed for 66 years, we were able to keep in touch with radio friends throughout the world during our journey.

Casual radio contacts invariably wanted to know about the boat's radio equipment. From what I have seen of other cruising boats of similar size to ours, we shared similar facilities for living on board and for navigation with maximum independence. But our radio set-up is less conventional.

Wild Bird, our 23-year-old glassfibre Vancouver 38p, had been taken by previous owners around the world and across the Atlantic several times so we have great confidence in her. She has a single aluminium mast, a mainsail and two foresails which make her cutter rigged. All the working lines are extended to the cockpit from where she may be sailed single-handed.

While on the lookout for other vessels and icebergs Liz, who is a competent sailor, and I operated a four-hour watch system. This leaves plenty of time for radio.

ABOVE You don't need the latest, expensive satellite phone for worldwide connectivity if you join the amateur radio community

Our equipment

Wild Bird is equipped with a Navtex receiver for shipping and weather reports, an AIS receiver that provides information about nearby vessels, a general coverage SSB/MF/HF/ VHF radio receiver, a marine VHF transceiver, a VHF emergency rescue transmitter, A UHF satellite location beacon, RADAR, a domestic broadcast receiver/CD player and an ICOM 706 Mk11G ham radio transceiver.

The latter I had converted to cover both amateur and marine frequencies. With an SCS 7400 Pactor4 modem, we were also equipped for high frequency (HF) internet and for receiving weatherfax charts in conjunction with a laptop computer.

***Wild Bird* is set up to maintain radio contact. Note the Ampair, towed prop generator, Hydrovane wind-powered steering device and the backstay high frequency (HF) antenna**

72 Practical Boat Owner • www.pbo.co.uk

STAYING IN TOUCH PBO

Wild Bird, a 23-year-old glassfibre Vancouver 38p, ashore at Lanzarote

North Atlantic Ocean

NORTH AMERICA

SOUTH AMERICA

EUROPE

AFRICA

ABOVE LEFT The tubular housings for the watermaker membranes behind the generator's acoustic enclosure. **ABOVE RIGHT** Our twin boomed headsails which we set for the 18-day easterly trade wind crossing from Sao Vicente in Cape Verde to St Lucia

Weather charts for the eastern half of the North Atlantic were received from DDH in Hamburg and those for the western side and Caribbean from NMG in New Orleans. The radio shack is also my sleeping cabin.

Small boat considerations

Problems that arise when you try to install and use HF radio equipment on a small boat are:

- Battery capacity for transmitting.
- Fitting an efficient antenna without compromising sailing ability.
- Electromagnetic compatibility (EMC) or interference between the radio and the boat's electronic or electrical equipment. There are, for example, a surprising total of 22 electric motors that can be connected to the 12V system on *Wild Bird*.

Supplying the power

To feed these and the electronic systems, we have five 110Ah AGM batteries. One is dedicated to starting engines. They can be charged either via splitting diodes from an 80A main engine alternator, a 4kVA diesel generator, a 100A shore powered battery charger, a wind turbine or a 100W towed generator that's driven by a propeller on the end of a trailing rope. We were not, therefore, short of primary power for the radio.

Many cruising boats rely on solar panels to keep their batteries topped up but, on a boat, they can only produce a few per cent of their potential because they have to be mounted horizontally, or frequently adjusted towards the sun, can sometimes be shaded by sails and can only be irradiated for about half a day.

I considered adding a linear amplifier to boost my radio signal because this could have been powered by the auxiliary generator rather than batteries but it transpired that 100W output power was enough to get me through to most of the world. Besides, the generator may then have needed electrical suppression.

The Ampair towed generator could just about make up the average 6A load imposed on the batteries by the refrigeration compressor, providing we were making more than 6 knots. With the boat steered by an electric autopilot, it

ABOUT THE AUTHOR

Tony Preedy's autism shaped his young life, leading him to enter Technical College. He studied to be a Chartered Engineer while a REME apprentice, at Bristol Polytechnic, then the BBC. He was elected a Fellow of the Institution of Engineering and Technology. He has worked as an engineer in broadcasting, spending time in Arabia- inspiration for his book *Mustafa's Last Well* – and in law in the USA. Widower Tony married Jean while living on Ascension Island and had two daughters. They moved to Shropshire after retirement. He enjoys sailing, restoring classic boats and vehicles, walking, and amateur radio.

Practical Boat Owner • www.pbo.co.uk 73

Countryfile

In April 2022, we began discussions with the BBC who wanted our input for a planned episode of "Countryfile" in which they wanted to feature amateur radio. I introduced them to the SOTA organisation and to Tom Read, M1EYP who is the SOTA Publicity Officer. He brought a great SOTA team together and on Flat Holm Island Ben Lloyd, GW4BML set up a portable station to contact SOTA activators on hill and mountain summits around the UK. The programme aired in June and the enthusiasm shown by presenters Ellie Harrison and Matt Baker was unmistakable when the contacts were made around the country.

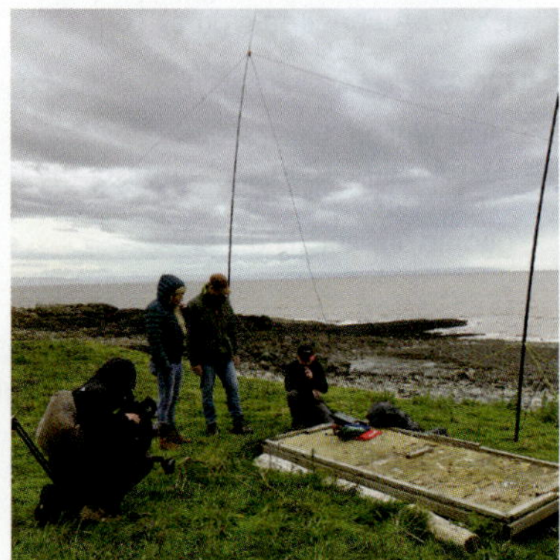

BBC Radio Wales

In September, BBC Radio Wales invited callers to join in an amateur radio feature. The wide-ranging discussion included conversations with 8-year-old Isabella and her dad Matt, M0LMK talking about their recent ISS contact, as well as with Jeffrey from Swansea ARC and Glyn, GW0ANA from Barry ARS.

Powerboat and RIB Magazine

Peter Talbot, G5AIB was featured in the November/December issue of "Powerboat and RIB Magazine" in an article about small boat adventuring. Peter mentions how useful amateur radio is for his adventures and how he combines his interest in SOTA with boating in places like Windemere. You can read the full feature on pages 126-131 if you sign up for the free digital edition: *https://powerboatandrib.com/interactive-digital-edition/*

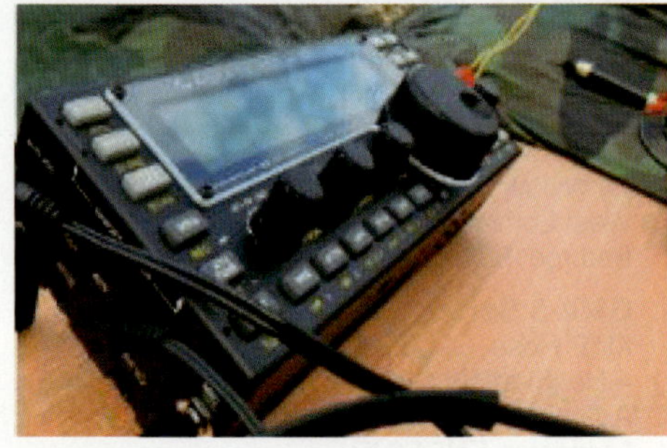

Australia Overnight

In December, John Quarmby, G3XDY was interviewed by Tony Moclair on Australia Overnight. John is the Secretary of the UK Microwave Group and gave a friendly and clear explanation about microwaves and amateur radio in general. Search for Australia Overnight on the internet or via Apple Podcasts if you'd like to listen

The Times

We were delighted to have a feature in The Times in January 2023. It was another great opportunity to highlight amateur radio and the RSGB in the mainstream media. I spent a week liaising with Rhys Blakely, the Science Correspondent, and coordinating the interviewees. I put forward a whole range of ideas and possible interviewees and would like to thank everyone who supported this opportunity, even though some contributions didn't make it into the final piece. The feature was called "Ham radio tunes into a new generation" and an external media monitoring agency has confirmed that the potential online audience for this feature was over 42 million. You can still read it online if you search for the feature title on *www.thetimes.co.uk*

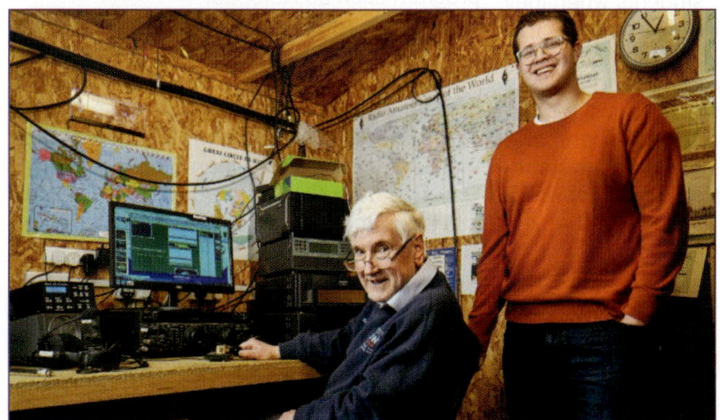

Amateur radio in media

The RSGB is constantly looking for opportunities to highlight amateur radio in the media, and I respond to many requests for help from the BBC and other organisations as you will see from some of the examples above. Sometimes our input is acknowledged, as it was in the credits on BBC's Countryfile, but unfortunately that isn't always the case and we have to accept that is the nature of working with the media. The important thing each time is that there has been an opportunity to present amateur radio to a different, mainstream audience.

Building relationships with the media

The examples above are just from the last 18 months, but we have been building relationships with the media for a number of years now. The massive media campaign we ran for 'Get on the air to care' (GOTA2C) during the pandemic and in the year after, led to us being in contact with a wide variety of local and national media.

The GOTA2C campaign was monitored by an external agency and confirmed as having a potential reach of over 38 million people. As a result of that campaign, many people heard about amateur radio for the first time and many lapsed radio amateurs were inspired to return to it.

You can still see some of the media coverage on our website: *rsgb.org/gota2c-media*

Have you seen the mention of amateur radio in...?

This is a question I get asked quite frequently and the answer is often not just "Yes", but "Yes, I helped to make it happen"!

The RSGB is constantly looking for opportunities to highlight amateur radio in the media, and I respond to many requests for help from the BBC and other organisations as you will see from some of the examples above. Sometimes our input is acknowledged, as it was in the credits on Countryfile, but unfortunately that isn't always the case and we have to accept that is the nature of working with the media. The important thing each time is that there has been an opportunity to present amateur radio to a different, mainstream audience. However, I can assure you that we are always working hard to raise the profile of amateur radio and do our best to be heard amongst the competing headlines

Local stories can grab media attention

Personally, I love to see radio amateurs contacting their local media about an event they're putting on, an activity they've tried or when they take the opportunity to talk about amateur radio in a 'hobby of the week' feature, for example. The RSGB certainly has an important part to play, but a local person contacting their local media about something they are doing in their community is very powerful and will often capture the media's imagination in a different way.

Working together to increase media awareness

What I would love to happen more often, is that when a radio amateur or a club gets some media coverage, they share that with us. If we know about it, we can share it even more widely and inspire others. If anyone wants to tell us about media coverage they've received, just email *comms@rsgb.org.uk* and we'll be happy to help spread the news of what you're doing!

Equally, I would encourage you to use the national coverage that the RSGB achieves to raise awareness via your own club or group websites or your social media platforms – you are welcome to put links on your websites to the media stories we have on the RSGB site, and if you use social media, please retweet/share our media news to your own followers. Individual radio amateurs are more likely to have non-radio amateurs in their circle of followers than we do, so you will help to raise awareness amongst new audiences.

If we work together, as individual radio amateurs and the national society, we can multiply the media mentions and reach new audiences in ways that demonstrate the relevance of amateur radio to people with a wide range of interests.

Newsroom

In the coming year we will be launching a new section of the website called 'Newsroom' where we will draw media coverage together in one place, along with news of what radio amateurs are doing. Look out for more information on our comms channels.

Heather Parsons
RSGB Communications Manager
comms@rsgb.org.uk

Morse Tuition, Practice and Assessment GB2CW

Whilst a Morse qualification is not needed by the present day licence, amateurs are realising that they are missing out on a lot of fun and DX by not using Morse and limiting themselves to part of the total amateur bandwidth available.

Introduction

Since the RSGB introduced the Certificate of Competency, there have been a number awarded. Some have been publicised in the *RadCom*, so the scheme is enjoying a considerable degree of success. Making the Certificate an 'award' rather than something mandatory in order to gain a licence has encouraged a lot of newly licensed amateurs to take up the challenge and achieve a skill that will enable them to work more DX, plus pass on their achieved skills to others. It can be compared to chasing DX certificates where effort is rewarded for the time spent in achieving the necessary qualifications.

The Morse certificate is no different - some will wish to obtain it and some won't. In the same vein, some clubs will embrace the idea and others will not. However, the skill will stay with you for the rest of your life.

International recognition is not an aspiration of the initial scheme, but if that develops later in a non-contentious manner it could be a clear additional benefit for some (provided that it is not abused to gain a higher class of licence overseas than is held here in the UK).

GB2CW volunteer, Malcolm Prestwood, G3PDH

A Certificate of Proficiency for students when they pass a Morse test, this is not a legal licensing requirement, but does provide stimulus for further improvement and gives the student something to be proud of which can be displayed in the shack.

The initial assessment will require the candidate to receive and send text, including some punctuation, for 3 minutes with no more than three uncorrected errors and will also include some figure groups (receiving and sending). The lowest speed for which the Certificate will be issued is 5WPM, although most people don't bother to start that low. Success in this will merit issuing the Certificate of Proficiency, after which endorsements (or a new certificate) may be obtained at 12, 15, 20, 25, 30WPM, etc. The test will be conducted with an Assessor only; an Adjudicator is no longer required. All assessments may be taken using equipment chosen by the candidate and appropriate for the speed being examined, including straight keys, paddles, bugs and semi automatic keys. Reception can be either on paper using a pen/pencil or using a computer with a suitable program such as Notepad.

To obtain an endorsement or a new certificate at higher speed, further assessments will also include receiving and sending proficiency in a basic rubber-stamp type QSO. This applies also to those wishing to take their initial assess ment at a speed higher than 5WPM.

Havingpassed the test, the Certificate will be mailed to the student as a PDF and he can print it out himself.

GB2CW Schedule

HF Transmissions

Day	Time	Frequency	Call	Location
Monday	20.15	3.555	G4BSW	Kent (Margate)
Tuesday	20.00	3.555	GW0KZW	Prestatyn
Thursday	09.00	3.605	G3UKV	Telford
Friday	20.00	3.563	GW0KZW	Prestatyn
Sunday	19.00	3.555	GM5AUG	Glasgow

VHF Transmissions

Day	Time	Frequency	Call	Location
Monday	10.00	145.250	Headcopy Class, G3LDI	
	18.30	145.250	M0APY	Leeds
	20.00	144.600	GM5AUG	Glasgow
Tuesday	09.00	145.250	G4OOC	Pontefract
	10.00	145.250/GB3NB	G4CCX	Norwich area (Intermediate)
	18.00	145.250	M0HAZ	Skegness***
	18.30	145.250	M0APY	Leeds
	19.00	145.250	G5FM	Glastonbury
	19.00	145.250	MI0WWB	Newtownards
	20.00	GB3NB	G3YLA	Norwich
Wednesday	09.00	145.250	G4OOC	Pontefract
	19:00	145.250	G0VCW	Lowestoft
	19.30	145.250 (USB)	G3XVL	Ipswich
Thursday	09.00	145.250	G4OOC	Pontefract
	10.00	GB3NB	G4CCX	Norwich (Intermediate)
	11.30	145.250	MX0NCA	East Runton
	18.00	145.250	M0HAZ	Skegness
	18.30	145.250	M0APY	Leeds
	19.00	145.250	GM0EDJ	Johnstone
	19.00	145.250	G3XNE	Bude
	19.15	145.250	GM0UOU/ GM0EDJ	Elderslie, Renfrewshire *****
	19.30	145.250	G0TDJ	Crayford (Kent)
	20.00	144.600	GM5AUG	Glasgow
	20.00	GB3NB	G3PDH	Norwich area (Advanced))
Friday	18.30	145.250	M0APY	Leeds
Saturday	08.00	145.250	G4OOC	Pontefract
Sunday	19:00	145.250	G0VCW	Lowestoft
	20.00	145.250 Voice / 145.250 CW	G4PVB	St Albans

Modes: A1A/J3E 144.2508 and all HF transmissions
F2A/F3E All VHF transmissions

*** Excluding first Tuesday in each month

**** (The EARS club call sign MM0PYR will be used for the broadcast)

Regardless of speed, a requirement of every assessment taken will be that all sending will be pre-recorded, to guarantee the speed and ensure integrity of the assessment process. All assessments may be taken using equipment chosen by the candidate and appropriate for the speed being examined, including straight keys, paddles, bugs and semi automatic keys.

Tests over Skype

This will be via an online sound and video program such as Skype or similar

Training

The Society is not prescriptive about the method of training used to achieve the Certificate. There are numerous methods of learning Morse code and it is a personal choice as to the method used. Instructors and students will have preferences and individual teaching and learning styles. No written rules are made, but once the code has been learned it is absolutely necessary for candidates to practice regularly. This may be done in a group, such as at a club, by listening to Morse on the bands or by using one of the numerous computer programs available.

The student can supplement individual or group training at clubs; and (ideally) be further supported by the use of an active and well promoted GB2CW broadcast schedule. Regular attendance to a weekly tutorial on the air using GB2CW in an interactive way is extremely beneficial. 2m FM is preferred to achieve this activity and it is normally a lot of fun, especially with mutual competition with other students in the same class.

There is more comprehensive information which adds to and builds on this in the RSGB book *Morse Code for Radio Amateurs*, by Roger Cooke, G3LDI.

Assessments

When a candidate is ready to be tested they should complete the online application form on the RSGB Website. This can be found in the Operating Section by selecting "Morse"

The application form includes the option for a candidate to express a preference for a face-to-face or online test. It also enables them to choose their preferred Assessor and location.

On receipt of the application, the Morse Test Coordinator will contact the applicant's chosen Assessor and e-mail their details to the Assessor, The Assessor will then contact the applicant to make mutually agreeable arrangements for the Test to take place.

Following a successful test, the Assessor will contact the Morse Test Coordinator who will e-mail a Certificate of Proficiency to the applicant. Full details of the testing scheme can be found on the RSGB Website. The address for any enquiries is Phillip Brooks G4NZQ by email to: *morse.tests@rsgb.org.uk.*

Consistency and integrity

Consistency, integrity and development of the scheme is monitored by a joint committee of the Regional Team and Amateur Radio Development Committee. The terms of reference for the ARDC encompass training and testing, and ensuring that the scheme is conducted in a thoroughly professional and competent manner. This is intended to guarantee that the Certificate is both desirable and that its reputation is respected both in the UK and internationally.

It may also be desirable that the ARDC will in due time extend its focus to further encourage the use of Morse.

Learning Morse

Unlike the Foundation licence, where a course of a few hours learning will probably produce a pass for the candidate, learning Morse Code and becoming a proficient operator is akin to learning a musical instrument. Attending a class once a week will not produce results. Occasional listening on the air is also a waste of time. The student has to be motivated and disciplined. Learning a musical instrument requires constant practice, and that does not mean just ten minutes a day. If you aspire to become a top-notch CW operator, consider at least one or two hours per day practice, EVERY day, not just once a week. This must carry on for a few months to reach acceptable speeds. If you cannot meet those requirements, then Morse is not for you. This cannot be stressed enough. The results you obtain will be well worth that effort.

There is a new way of encouraging learning. In East Anglia, the Norfolk ARC started running two Bootcamps a year, and have around 15 students attend and it is normally a whole day event, running it in three classes and some on-air activity as well. Both sending and receiving are catered for with a range of keys, from straight to the latest in technology, the 9A5N paddle, and Begali paddles too.

Essex CW Club copied this and was extremely successful such that they had to hire a village hall! Lots of fun can be had and XYLs usually provide a range of cakes and so on, with lots of tea and coffee. Try it in your Club if you can and it might catch on and provide incentive for more learning and practice, The latest one in Norfolk can be seen in the picture.

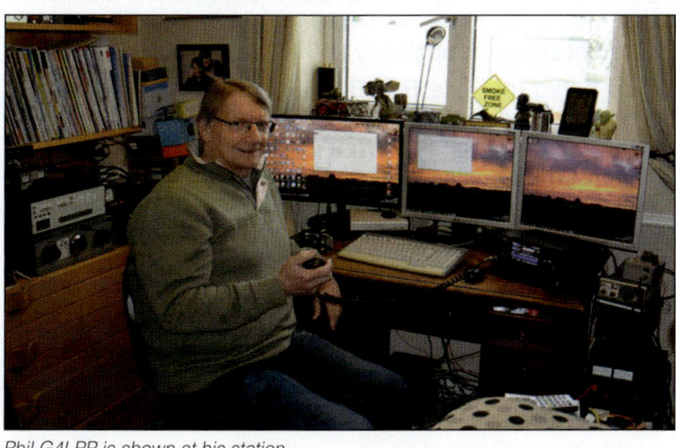

Phil G4LPP is shown at his station.

Volunteers for the GB2CW scheme

Norwich now has six classes running each week. Roger G3LDI runs a headcopy class on Monday mornings at 1000. This is a rapid fire decoding of a mixture of sending, including QSO format, cut numbers, EISH and 5 groups. The other classes cater for varying speeds, from raw beginners to those up to 30 wpm. One of my students has done so well that he is now a tutor running a beginner's class. Phil G4LPP is shown at his station.

Two of the classes are on GB3NB and the others on 145.250MHz. Other tutors locally are Malcolm G3PDH, Chris G4CCX and Jim G3YLA.

We were lucky to have a few new volunteers join in the past year including Roger MI0WWB, our first one from Northern Ireland.

If you feel like becoming a volunteer yourself you would be most welcome. There can never be enough volunteers and using the GB2CW scheme it is so easy to do and also a lot of fun.

It really isn't that much to ask and the rewards come to both the student and tutor when progress is made. Additional volunteers are always needed to run GB2CW broadcasts, especially in some of the more remote parts of the UK. Broadcasts can take place on several bands, ranging from 3.5MHz to 50MHz. It may only take an hour of your time per week to ensure that amateur radio continues to have a flourishing pool of CW operators to ensure the future of the mode, so please consider helping.

As interest grows, more tutors are needed. It would be very nice to see volunteer instructors in every Club in the UK, and that is what I would like to see as Coordinator. There is a long way to go to achieve anything like that but in order to maintain this quintessential mode used in amateur radio we need a lot more tutors. Remember, some Elmer taught you, so now it's your turn to be an Elmer!

Also, there are gaps in the coverage of Assessors. Assessors are needed in Regions 8 (Northern Ireland), and 11 (South West England and the Channel Islands). If you live in one of these areas, please consider joining this most worthwhile scheme. We have been fortunate in filling a few areas in the last year or so. However, more are always needed, not only for the vacant areas, but all areas, to act as backup. Full details, including an application form, can be found on the RSGB website at: *https://rsgb.org/main/operating/morse/certificate-of-competency/the-morse-test/*

Volunteers are scarce and are perceived by some to be those with super human skills and speed in excess of 30WPM. This is far from the truth and if you have a good average skill level of around 15 to 20WPM, you could take

G4PVB teaching with one of the local Norfolk ARC Bootcamp, now a bi-annual event, spring and autumn. This is highly popular and normally has about 16 students attend an all-day session, all ranges from beginners to Full.

on the role of instructor to that level anyway.

Computer programs are used for instruction so therefore the Morse sent is perfectly formed so it is completely straightforward to implement on the air.

Two New Bootcamps In 2019

Bootcamps are catching on at last. Rich G4FAD held his first one in Hereford and it was a great success. This is what he had to say about it.

We held our first Hereford Morse Bootcamp with the kind permission of Geoff G8BPN at his QTH which used to be Geoff's electronic factory and it was an ideal building for our purposes and easily had room for the instructors and the 23 students. In the past every town or village had a person who used Morse code in their job and could help an aspiring CW operator. Today this

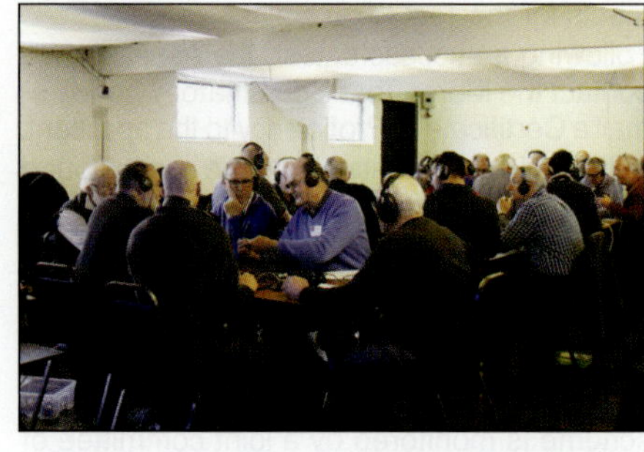

is no longer true and a Bootcamp helps to fill that void.

Judging by the number of people attending, CW is not a dying mode at all. This is one of the most enjoyable ways of learning Morse whilst having a good sociable day out. Try one in your club, you might be surprised.

There was also another new Bootcamp, this time in Scotland run by Gavin, GM0GAV. This too was successful with 12 students attending.

As the Covid situation carries on this year again, there are no Bootcamps running. However, it is hoped that in 2022 they will resume again and also increase. They provide an excellent training feature and also some stimulus to those attending to do even more to become proficient operators. This year the Al Slater G3FXB Memorial Award was presented to three Bootcamps for their contribution to CW. A silver salver was sent to me on behalf of the Norwich Club as we started Bootcamps nearly ten years ago now.

To offer your assistance, please contact the scheme's co-ordinator (details below).

The Morse Test

Taking the test is a simple three-stage process

Step 1 – When a candidate is ready to be tested, they should click the button below to complete the application form.

The application form includes the option for a candidate to express a preference for either a face-to-face or an online test.

Step 2 – The Morse Test Coordinator will contact an Approved Assessor and email contact details to the candidate who can then get in touch with the assessor and arrange the test.

Step 3 – Following a successful test, the assessor will notify the Morse Test Coordinator who will email a Certificate of Competency to the candidate.

Test structure

The lowest speed for which the Certificate is issued is 5wpm.

This has been chosen as the threshold to encourage learning the Code, to provide early confidence and to encourage moving forward to higher speeds.

It is specifically designed for those who are most comfortable with this rate of progress.

There is, however, no barrier to those who wish to enter the scheme at a higher speed.

Tests are available for 10, 12, 15, 20, 25 and 30 wpm.

The Certificate of Competency Morse test will:

- Require the receiving and sending of plain language*, for three minutes with no more than four uncorrected errors and;
- Require the receiving and sending of numbers, in five figure groups, for one minute with no more than three uncorrected errors.

*The plain language section of the test will include some numbers as well as the more commonly used punctuation marks such as oblique stroke and question mark.

Success in this test will merit issuing the Certificate of Competency, after which further certificates can be obtained for a successful test at a higher speed.

A requirement of every test taken—regardless of speed—will be that all sending by the assessor must be prerecorded, or generated by suitable computer software, application or keyer, at guaranteed speeds to ensure integrity of the testing process and thus prevent appeals in this regard.

Sending tests may be taken using equipment chosen by the candidate and appropriate for the speed being examined including straight keys, paddles, bugs and semi-automatic keys, etc.

Keyboard-generated Morse is not acceptable for sending tests.

The process of taking the test online is as follows:

- This will be via an online sound and video program such as Skype or similar
- The candidate and assessor must ensure that the video cameras are sited so that the assessor and candidate have a clear view of each other—in particular the assessor must be able to watch the candidate keying during the sending part of the test
- In the event of a failure of video during the test, it may continue using sound only, at the discretion of the assessor
- On completion of a successful test, the assessor will email confirmation to the Morse Test Coordinator who will email a Certificate of Competency to the candidate

It should be noted that The RSGB will not pay any expenses incurred by the candidate in connection with undertaking a Morse test.

If any of this causes any difficulties, please contact the Morse Test Coordinator via: *morse.tests@rsgb.org.uk*

Co-ordinator email: *morse.tests@rsgb.org.uk*

Latest GB2CW broadcast schedule:

www.rsgb.org/main/operating/morse/certificate-of-competency/gb2cw-broadcast-schedule/

Featured Clubs

Cambridge & District ARC ZOOM Meeting

Harwell Amateur Radio Society VHF Field Day

Throughout the UK there are over 500 local clubs and societies affiliated to the RSGB. They vary from small local clubs, through repeater groups , online clubs to the large contest groups. Below are the details of some of these clubs, telling you a little about what they do for amateur radio and their members, and what they plan for 2023/24.

We are sorry that we cannot include every club that submitted information or every photo from those that did.

Aberdeen ARS

Regular Schedules set up during the Covid pandemic, continue to be popular, especially on 80 meters. In fact, these have attracted several new members to the club and attracted participants from all over Scotland, and beyond.

A gift of several FM1000 radios from the Shetland Amateur Radio Club has led to an increased interest in the 4-meter band. One of our members developed software for these radios during the winter and now many members of the club are able to join in the clubs regular 4-meter nets. Another winter project was to construct simple Delta loop antenna for 4 meters and this has increased activity on this band locally.

Activity on a wholly renewable energy station for /P activity continues, and the club has taken part in several RSGB events, such as the NFD in June and QRP Field day in September, using this very satisfying approach. (Photo).The challenge this summer is to put a 100-watt station on the air /P solely powered by the sun.

The club has also increasingly taken part in providing information and talks to local schools. It is hoped that in the not to distant future one of these schools will set up its own Amateur Radio Station. Interestingly, one of the aspects of Amateur Radio attracting attention during school visits is CW, which is now regularly included in talks. The club also has an active CW training programme, with several members now operating in that mode.

Thursday evening meetings continue to be popular with a mixture of Junk Sales, talks, demonstrations and our monthly equipment workshops Full details at; *www.aars.org.uk*

Cambridge and District ARC

For over one hundred years Cambridge and CDARC have provided a focus for radio activities in the city. We meet both in-person and online every two weeks for talks and presentations, equipment demonstrations and social activities. Many club members also participate in weekly 2m, 80m and DMR on-air nets. Visitors are welcome at all our events.

We are very active in training and hold well-attended Foundation and Intermediate courses a couple of times each year. Mentoring is available for those studying for the advanced exam as is a buddy scheme for the newly licenced.

The club also provides internet connected remote transceiver/receivers at an electri-

cally quiet location which can be used by our members who are troubled with noise issues or when away from their own shacks.

CDARC is active in contesting and field days, and we participate in various events such as Museums and Mills on the Air.

Our regular meetings are on the 2nd and 4th Friday evenings of each month at 7:30pm at Coleridge Community College, Radegund Road, Cambridge CB1 3RJ and online via Zoom. Anyone with an interest in amateur radio is welcome at our meetings, so please come along or contact us for further information.

Find us on Facebook or at our website *www.cdarc.org.uk.*

Email contacts are: *publicity@cdarc.org.uk* & *training@cdarc.org.uk*

Carmarthen ARS

Carmarthen ARS (CARS) is a medium sized club in Zone 7 (South Wales) which is run in accordance with its constitution, whose members go out of their way to provide a friendly and helpful atmosphere at their two club nights each month, who provide training at all three UK licence levels and who partake in all sorts of events ranging from field days to contests to rallies to BBQ's.

The club meet at 18:30 on the 1st and 3rd Tuesday of every month at the Cwmduad Community Centre, Cwmduad, Carmarthenshire, SA33 6XN (which is on the A484), and is located just north of Carmarthen at 51° 57' 20.96" N, 4° 21' 56.13" W for those with satnavs.

The 1st Tuesday is usually a social and 'on air' radio night, whilst the 3rd Tuesday tends to be an activity night, either a talk, or another form of group activity.

Although CARS do run training for all three UK licence levels, we try NOT to run training courses on club nights, preferring instead to run these on other days to avoid distractions for students. We are also an exam centre.

The club call (GW4YCT) is in use on club nights allowing members access to the HF, VHF and UHF bands using our club radio equipment and antennas.

Visitors and potential new members are made most welcome. Mask wearing remains individually optional. Light refreshments are usually available.

We continue to hold a number of regular nets during each week to allow members to maintain good contact with each other, experiment with different modes, and non-members are welcome to join in.

Anybody wishing to visit or join should check initially with Andy, GW0JLX on 07768 282880 or by email to *carsmembershipsec@gmail.com*

Colchester Radio Amateurs

CRA celebrated 60 years of amateur radio in 2023 with a return to a full training program, presentations and activities including an Island Activation, Field Days, contesting and club activities. 2024 will be no different with similar activities planned including another Island Activation and training.

2024 will also see an extension of our outside events including, but not limited to, GB6WLB to support Walton Lifeboat and GB6NT the Naze Tower activation.

Anyone wishing to join should have a good sense of humour and a wish to advance their knowledge.

We look forward to welcoming you, anyone wishing to visit or join can contact us by email at: *secretary@g3co.uk*

Cornish Radio Amateur Club

Our club has been running successfully for over 80 years. Being a club so rich in history we are well established in the county of Cornwall and known throughout the UK.

The Cornish Radio Amateur Club meets on the 1st Thursday of the month and membership is open to anyone with an interest in

radio communications, computing or indeed electronics. We have members young and old in the club all enjoying various aspects of the hobby. These range from traditional methods such as CW and voice to the more modern day digital modes there's something for everyone.

We try to put on as many special event stations as we can to promote our hobby and we like to make our club meetings a truly sociable evening for everyone. We have taken part in SOS week, Railways On The Air, Museums On The Air and JOTA to name a few. When possible we will always try to attend local events too. Another enjoyable event has been for the Cornwall Air Ambulance Trust and use this opportunity to promote the fact that it was Cornwall who had the very 1st Air Ambulance.

Our biggest and proudest event we run is International Marconi Day. In 1988 Norman Pascoe G4USB came up with the idea of running an event to Celebrate the work and birthday of Guglielmo Marconi. He ran the idea past a friend, got a small team together and this was the beginning of International Marconi Day as we know it today. This was to be held on the weekend closest as possible to Marconi's birthday, 25th April. Norman organised this worldwide event every year until he sadly became Silent Key in 2016. Since then Steve G7VOH (G4CRC Chairman) has taken on the role and coordinates the event each year. In 2023 we had 63 stations registered with us from all over the world, which as always made this a truly major event in the radio amateur's calendar. As the organisers of the event we also run our own station GB4IMD.

We are also very proud to run Cornish Radio Amateur Club's Rally once a year at Penair School in Truro.

We also offer training for Foundation, Intermediate and Full. Most of our training is done via Zoom but we can also offer 1 to 1 training in person if required. We have a large selection of training videos for all levels on our YouTube channel, which are proving to be extremely useful to newcomers, and also to those already licensed who wish to advance. We have had fantastic feedback from people up and down the country who have successfully passed their exams using them. These are available for anyone to use free of charge and will be updated as the syllabus changes.

To find out more about our club please visit our website *www.gx4crc.com*

Crawley ARC

Crawley ARC has about 70 members and benefits from having sole use of its own premises offering a meeting room, tea bar, modern lecture facilities and well-equipped contest standard radio shack with tower, beams and wire antennas.

Club members, some of whom are professional engineers, bring their combined technical expertise in RF engineering, electronics design & construction, antenna design, microwave engineering and electronic servicing to the membership. This experience is invaluable to members starting in the hobby.

The Ashdown Forest Repeater Group is an associated organisation well supported by our members who provide a wealth of ex-

perience with design updates and ongoing maintenance for both the D-Star/DMR/Fusion and FM repeaters. Call signs GB7MH and GB3MH respectively are co-located in the nearby village of Turners Hill and each have internet linking capability.

CARC hosts one of the four annual Microwave Round Tables bringing together microwave enthusiasts from a wide area with lectures and sale of components and general items of surplus equipment.

The Club's biggest outside event of every year is participation in the VHF National Field Day Contest in association with another local club, which involves the establishment of multiple stations to cover all bands and modes.

We also support Museums on the Air, Mills on the Air and Jamboree on the air.

In support of these activities, twice weekly meetings in our own premises cater for operating the club station, socialising, contesting, and monthly interesting talks by guest speakers.

We have a lot to offer, and we are a friendly helpful bunch! Please contact us at: *secretary@carc.org.uk* or come and visit on a Wednesday at 20:00hrs or on a Sunday at 11:00hrs local time.

Essex CW ARC

Essex CW ARC has as its mission to encourage the use and preservation of the CW mode on the amateur bands. Based and formed in Essex, but open to all licensed radio amateurs worldwide who are interested in Morse code

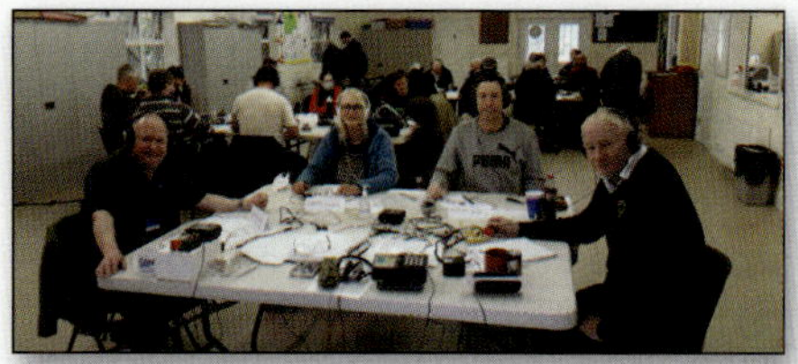

(CW). The club further aims to support those who strive to learn and improve their CW skills, as well as operators who are already competent at any level. Membership is free and on-air activities are publicised via the clubs web site and via a regular electronic newsletter. There are weekly on-the-air training nets on both 80m and 2m, as well as a weekly GB2CW QRS transmission on 80m. Each October we hold a one-day "boot camp" where CW enthusiasts of all abilities meet to practise and share their passion for CW. In November is the annual activity week, which is a great opportunity for CW practice and to qualify for our operating award. For further information please visit the club's web site at: *https://essexcw.uk/*

Greenock and District Scouts ARC

Every year we participate in Jamboree on the Air, Thinking Day on the Air and other Scout and Guide events where we provide a radio station using a range of special event callsigns.

We offer communications based activities relating to the Cub, Scout, Brownie and Guide Activity Badges and parts of the Beaver, Cub and Scout Challenge Awards. We also offer fun activities for Beavers and Rainbows. During the year we deliver this programme to our local Groups on a demand basis and we continue to develop Workbooks and support material to help with the delivery.

All this material is available on our website, additionally, we prepare radio related activities for Explorer Scouts and Senior Section Guides.

We have provided members for the Amateur Radio Teams at International and National Scout and Guide events.

Our regular meeting nights feature a range of amateur radio related presentations and workshops.

We meet at Greenock and District Scout HQ, 159A Finnart Street, Greenock, Inverclyde, PA16 8HZ.

Contact us via Email at: *mm0tsg@gmail.com* or our Website: *https://mm0tsg.org/*

You can also follow us on Twitter: *@mm0tsgs*

Guernsey ARS

Guernsey Amateur Radio Society was first formed sometime in the 1940's or 50's, and is currently located in an old WWII bunker at Beau Sejour Leisure Centre in St Peter Port, Guernsey, GY1 2DL.

The club meets every Friday at 7.30pm, and we have a lounge and a small, separate shack with an IC7300 and several VHF/UHF radios.

There are weekly nets on Tuesdays on 2m FM on 145.525MHz, and on UHF DMR on TG23515 every Thursday, both at 8pm.

There is a UHF FM repeater (GB3GU) on 435.525MHz.

We only have a small membership, but we try to be as active as possible. Every 5 years, we celebrate the Liberation of the Channel Islands on May 9th 1945, and put on a special call.

The club attends the Vintage Agricultural Show in the south of the island each August, where we put on a public demonstration of what amateur radio is all about to try and interest new members.

Our members have a wide range of skills, including computing, electronics and surface mount technology, antenna building, etc, and we also have a diverse range of interests, from QRP, CW, SSB, digital modes and satellite working, including QO-100.

Being a separate DXCC entity, means we often enjoy large pile-ups, and some visitors have been amazed to find that they can work DXCC easily in a weekend.

Visitors are most welcome to come along on a Friday and join in our discussions and have a go on the radio.

All contacts are logged using Cloudlog, and then uploaded to LoTW.

For more information, please contact Phil GU0SUP on: 07781 151747

You can also find us online here

gu3hfn@suremail.gg FaceBook - GsyARS

Twitter: @Guernsey_ARS

Website: gars.org.gg

Instagram: @guernseyradioclub

YouTube: search Guernsey Amateur Radio Society Website: gars.org.gg

Guildford & District RS

Founded in 1918 as the Guildford Wireless Alliance, club meetings are on the second, fourth and fifth Fridays of the month and are held at the Guildford Model Engineer's building, Burchett's Gate entrance, Stoke Park, London Road, Guildford GU1 1TU.

Guildford town is twinned with Freiburg. At Thursday lunchtimes GDRS members can be heard talking to friends in Freiburg. Other nets also operate and details can be found on our website *www.gdrs.net* together with other information, such as our programme of events.

GDRS can also be heard operating during various contests as "The Guildford Contest Group" using the call-sign G5RS.

We have the use of a recently modernized, extended clubhouse with access for the disabled and inside toilets, together with a patio area just outside and also a large area of grass, for BBQs etc. We have an Altron CM35 mast and a wire doublet for HF, a 2m beam and a 6m LFA Innov antenna. We have a club rig (Icom IC-7300) complete with antenna tuner, Daiwa keyer and Kent paddle key.

Members gather from about 7:30 pm (but not much before) and meetings start at 8:00 pm. We are a very friendly club and visitors are very welcome. We have plenty of free parking at the club house.

Coffee, tea and a selection of soft drinks and confectionary are available at most meetings.

Following a rule change GDRS can now offer Associate Membership for those who live a significant distance from the club house and attend meetings infrequently. For full details see our updated Constitution and Rules document, which can be found on our web site: *www.gdrs.net*

For enquiries please contact the secretary, Timothy Dabbs, G7JYQ Tel: 020 8241 9396 or via email: *secretary@gdrs.net*

Hilderstone RS

The Hilderstone Radio Society is made up of licensed amateur radio enthusiasts, students studying for their licence, short wave listeners, electronics & computer hobbyists.

We have a wide variety of interests including construction of electronic projects, computers, coding and contests. We are keen to advance our knowledge of radio communications as well as passing on that knowledge to local schools and organisations. Learning for members doesn't stop after passing licence exams nor is it limited by them alone.

We have been a RSGB regional & national club award winner.

In 2016 we helped Wellesley House school in Broadstairs with their amateur radio contact with Tim Peake while he was on the International Space Station. We hope to assist another school to again contact the ISS in 2023! As well as schools we have an established relationship with the local Scouting & Guiding groups supporting both JOTA/JOTI and Thinking Day on the Air. Members also support a local group of young girls who look at STEM subjects independent of school, the STEaMettes have regularly made presentations to the club on their projects.

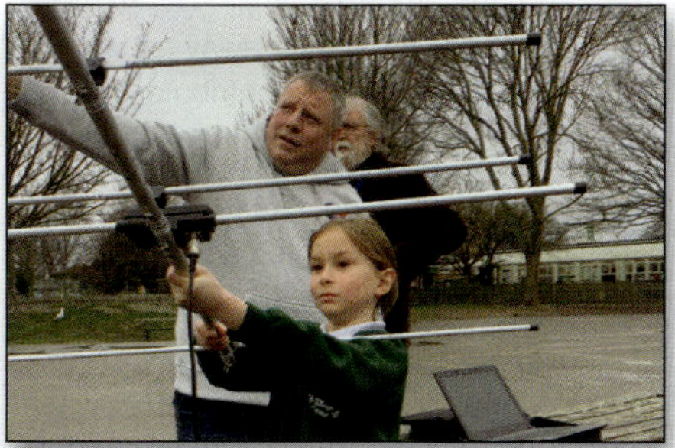

Now fully established in our 'new' meeting place in a local Scout Hall we have invested in a club shack, equipped for HF, VHF & UHF, both for members to use & as a resource for training.

As training is an important thread within the club, we have recently acquired a second club call, G7HTI (Hilderstone Trainees & Instructors), to support that activity alongside the established club call of G0HRS. Face to face licence training as well as BRICKWORKS activity are promoted. A regular feature of club nights is to have a speaker both from within our own membership & guests on a wide variety of topics, not always amateur radio related but usually with a tangential link.

The club puts on several Special Event Stations throughout the year both to support local organisations & charities but also to demonstrate the hobby to a wider audience. The club is in the planning stage to get equipment to demonstrate the amateur satellite & amateur television service at these events, modes which particularly hold the attention of young people, almost more than operating a Morse key!

The club membership has been constantly growing over the past decade, amazingly including during the covid years.

We encourage new members to learn more about amateur radio, electronics or computing so come along and meet us. You will be sure of a warm welcome.

Contact: *secretary@g0hrs.org*

Isle of Man ARS

The Society was affiliated to the RSGB on the 13 April 1948 which made 2023 the club's 75th anniversary.

The club has seen great changes in operating, from CW to AM to SSB and FM through to digital modes prevalent today and rigs from home-built valve equipment to the modern all singing and dancing radios from many manufacturers.

With currently about 30 active members, the society has had it's meetings at various venues over the years, and is currently at the Scout hut in Mill Road, Ballasalla. This venue has a large hall for meetings and demonstrations, accessible facilities, small mezzanine for classes and kitchen facilities and suits the Club very well. Meetings are held weekly on Wednesday evenings from 7pm.

We are registered for training and examinations.

The Club has another venue, more central on the Island, at the Sea Cadet hall in Tromode in Douglas and meets once a month on the second Tuesday from 7 pm. This smaller venue is ideal for talks and presentations.

To keep in touch with members on the Island and further afield, Nets are on the GB3IM repeater network after the 11:00 hours GB2RS news bulletin and talk-in Wednesday night at 18:40, Sunday on 3.715 Mhz and 7.115 at 09:30 and 22:00 hours. Visitors are always assured of a warm welcome at both venues.

Club news and events are on the website: *www.iomars.im*

Norfolk Tank Museum Radio Group

Discover the world of military radio technology with the Norfolk Tank Museum's Radio Group. We are based at the Norfolk Tank Museum website: http://norfolktankmuseum.co.uk/ near Long Stratton in South Norfolk.

The Radio Group has a strong focus on military radio history and technology, and we're committed to preserving and showcasing this important aspect of military history. Our volunteers share a passion for military radio which they are keen to share with the public and fellow radio enthusiasts. Our membership is an eclectic mix of radio enthusiasts of diverse backgrounds including military and professional communications, academia and the electronic industry.

We aim to give visitors an insight into the importance of radio communication, from both the historical and modern perspective. The opportunity to try Morse Code operating is also available to younger (and older!) visitors.

The radio exhibits are housed in the front section of a Quonset Hut within the Norfolk Tank Museum's site and include examples of World War 2 interception receivers, famous radios such as the RAF R1155/T1154 and the Wireless set 18 and 19, together with NATO, Warsaw Pact, Indian and South East Asian equipment dating from WW2 to the 1980s.

We're always looking for new volunteers to join us. Whether you're a beginner seeking to progress or a licensed radio operator looking to share your knowledge you'll be warmly

welcomed. The Radio Group is hands on with the radio collection's exhibits, transmitting by voice and CW modes using the group's permanent special event callsign GB2NTM whenever possible when the Museum is open to the public every Tuesday – Thursday and Sunday between Easter and the end of October, including additional external displays at the Museum's popular Armourfest weekend in August. We are usually on-site Wednesdays and Sundays if you are planning to visit specifically for the radio collection.

We are privileged to be an integral part of the Museum, enabling us to offer unique operating opportunities, including those from radio-equipped vehicles such as the Sultan Armoured Command Vehicle, using its original configuration radios.

In addition to a variety of aerial mountings on the Quonset Hut, the acres of open space around the museum provide tremendous potential to erect a range of both military and amateur radio configured aerials, always being aware of the somewhat unusual field day planning need to avoid snaring a passing tracked and armoured vehicle of course!

You can contact us through our QRZ.com page at *https://www.qrz.com/db/GB2NTM* or the museum to learn more about our military radio collection. Club/group visits are welcome, and it may be possible to arrange these outside normal opening hours; please do let us know in advance, we may even be able to host your club "away day" operating event! The Museum and Radio Group Volunteers look forward to meeting you; both as a visitor and as a new volunteer.

Nuneaton & District ARC

We are a busy and sociable club and have recently moved to an excellently located and appointed sports and social club on the Hartshill Ridge, at 140m ASL, overlooking North Warwickshire and South-West Leicestershire. Antennas have been added to the site and we now have a 2m/70cm collinear, 4m and 6m halos and a multiband

40m EFHW antenna set up with permanent connections into the club room. Our offering is as varied as our members, and include operating from our QTH, practical workshops from antenna building to instrumentation, talks on all aspects of radio, an annual project competition, exams, bring and buys, an active VHF contest group as well as group visits. We meet at least twice a month on first Mondays and third Fridays from 7pm, with drinks (and Jaffa cakes!) available.

New members and visitors at any level are welcome.

Contact us via our email: *secretary@ndarc.co.uk* join our Facebook page, or drop in and meet us on a club night at the Windmill Sports and Social Club, Mancetter Road, Nuneaton, Warwickshire CV10 0HW.

Pontefract and District ARS

Pontefract and District Amateur Radio Society welcomes members from all over Yorkshire including Leeds and as far away as Hull.

For a small club, PDARS contributes much to amateur radio. Around 20 members make up the club some newer to the hobby than others and more would be welcomed.

The club meets in it's rented accommodation at the Grange Community Centre, Carleton, a short distance from Pontefract, on Tuesday and Thursday evenings from 6.30 onwards. As the rooms are rented we can have 24/7 access,and all the equipment is ready to use. This equipment includes an Icom 7300, a Yaesu 7800, a 160m delta loop antenna as well as the portable gear for our many events.

Every year we hold Special Event Stations, mainly in the community, for events such as Ackworth Vintage Steam Rally, Battle of Britain commemoration and Yorkshire Day. The club takes part in WAB contests but treats them as an enjoyable time on the air (including having rag chews) rather than serious contesting. Entering is fun, if we win it's a bonus. Thursday Club evenings consist of rag chews on many different subjects and aspects of Amateur Radio, while Tuesdays are for more formal education or construction projects. A few members have other hobbies apart from radio, including photography, motorbikes and computer science.

We also have a long history of construction. However mainly due to a lack of members we don't do much at the moment. The same applies to training for the radio licence exams. Although we don't run courses, we will help with topics on which the candidates are struggling, and the exams are held at the club.

We have recently re-arranged the main room creating a reconfigured space to give us more room so that we can accommodate more members.

Our aim is to encourage more members, including family members and more ladies. We have a new website promoting our cause and hope to continue introducing amateur radio to the local community. All are welcome

For more information visit our website: *www.pdars.co.uk*

South Dorset RS

The club has enjoyed a busier year following the ending of Covid lockdown restrictions and has made good use of its new Clubhouse QTH situated in the Weymouth-West Scout Hut for both monthly meetings and special-event activations.

Special events have included a Dxpedition to The Isle of Wight (GB0IOW), RAFARS Airfields On The Air (RAF Chickerell) (GB0CAF) and Marconi Day (GB0MPB) with future plans to activate the Battle of Britain Airfield RAF Warmwell in August 2023, Railways On The Air in September and JOTA in October.

Every Tuesday (almost) without fail we host a

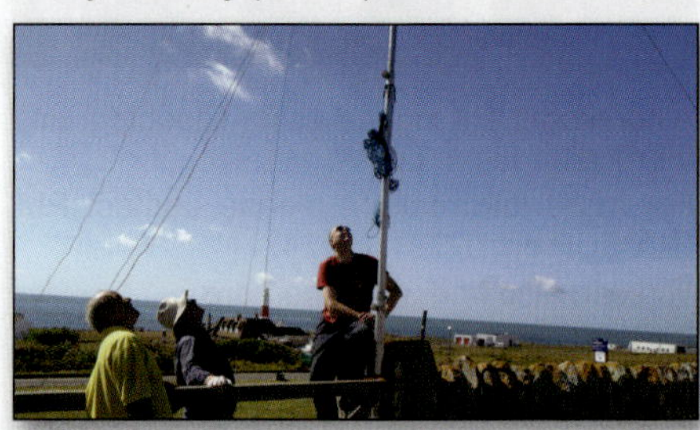

club net (G3XDS) from 11.00 to 12.00 local time on or near 3.784MHz LSB, preceded by a 15 minute AM net on 3.615MHz at 10.45 for club members who like to try out their vintage military sets. As conditions improve, we may migrate back to the 40m band where we used to use a frequency of 7.150MHz before the solar minimum made things difficult.

During our monthly meetings held on the second Friday of each month we have maintained a stimulating program of presentations and demonstrations

All of our activities, presentations, and news items are faithfully documented on our well managed website: *www.SDRGS.org.uk* which also gives information on the South Dorset repeaters, managed by the South Dorset Repeater Group.

South Normanton, Alfreton & District ARC

Whilst possibly being known for having the longest club's name in England it is best known for promoting the Junction 28 Radio Rally in June every year. We meet just one mile from Junction 28 of the M1 motorway in Derbyshire about halfway between Derby and Sheffield. The Post Mill Centre is a modern local Community Centre with comfortable surroundings including bar and coffee lounge.

With over 60 members we are always busy, meeting on Mondays about 8pm except Bank Holidays. We book a room twice a month and meet socially other weeks. Most months our program is: First Monday - technical night, either hands on or a talk/video. We have our own radios and mast with various aerials for members to use. Third, most months- Junk Sale. Other Mondays are a social natter night in the lounge area. Groups tend to gather based on various interests as Amateur Radio has so many facets.

We run local Nets every Tuesday about 7pm on 2M which are open to all. Most Saturdays during the better weather will find us out and about at Field Days from local beauty spots, such as Crich Stand Memorial and Highoredish with a take-off of over 300m ASL. These enable members to try out their own equipment construction and practices as well as allowing newer members to use club equipment, or their own, in a supervised environment at

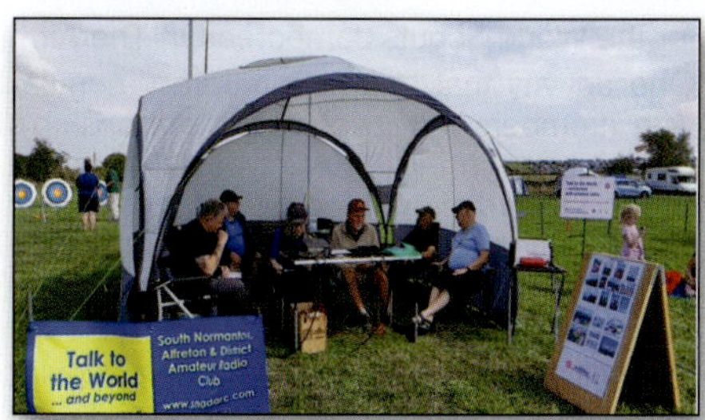

higher power than the Foundation licence allows. We also enjoy several Special Event Stations and JOTA events.

We are a licenced Examination Centre but since the move to online training and exams most of our training is mentoring rather than formal courses. We welcome anybody with an interest in radio as well as those wanting to qualify or those considering returning to the hobby.

For more information see *www.snadarc.com* or look for 'snadarc' on Facebook.

Stockport Radio Society

Here in Stockport our aim is to provide a centre whereby enthusiasts can meet and share their experiences as well as being able to learn from each other. The sense of community means alot to us as we enhance our reputation both locally and nationally as a forward looking society.

Not only do we meet regularly at our well established town centre HQ in the centre we also continue to utilize online facilities allowing those interested the opportunity to join in our activities from further afield. Our meetings feature practical sessions as well as interesting and topical presentations plus the opportunity to meet with fellow enthusiasts from around the area. The kettle is a very popular and well used piece of equipment.

Keen to support local events and take the hobby to the general public, our marquee and

stand can be found regularly at such diverse venues as Avro Woodford, The Runway Visitor Park Manchester Airport and Bramhall Park for the annual Duck Race. Also to be found out in the open are our contestors who continue to compete in many of the major RSGB and worldwide events from a well positioned hill top site.

Back at base, in recent years we have invested heavily in renewing the equipment available to use. Fully equipped HF, VHF and digital stations allow members to sample the delights of what's on offer and the recent addition of a QO-100 Ground station is proving to a very popular investment.

Information about us and our activities can be found by visiting our website or our social media platforms, particularly YouTube. We welcome visitors at our HQ, online meetings, or club nets.

Website: *www.g8srs.co.uk*

Email: *info@g8srs.co.uk*

Phone: 07598 892795

Facebook: Stockportradiosociety1920

YouTube: @g8srs Twitter: @g8srs

Thames Amateur RG

Thames Amateur Radio Group is a South Essex based club, in regular discussion with other Essex clubs and RSGB Region 12 on making the most of all of our expertise and contacts. We have a lot of expertise and experience across our membership and a wish to help and develop others interested in the hobby. You are just as likely to find people who will delight in chatting to you about the hobby or life in general as you are to be steeped in radio activities alone, as we are a sociable group.

We live-stream all the face to face meetings we can, and have a good handful-plus of joiners through that route if they can't get out, or live at too great a distance. This development has been well received.

We meet in person on each month's 1st and 3rd Fridays, the 1st being presentation and Q&A based, the 3rd an activities and hands-on session where members bring in their gear, their questions on their radio projects, challenges or proud solutions, or their keen interest in our chosen themes. On Fridays in each month when we don't meet in person, we hold a net on GB3DA. Output 145.725 / Input 145.125 / CTCSS 110.90. Our back up if GB3DA isn't possible is GB3ER: 434.675MHz Input, 433.075MHz Output, CTCSS 110.9Hz.

Our Friday night nets are welcoming gatherings – anybody with an amateur radio licence who can reach us is welcome to join. A rota of members hosts them, so there is variety in styles, and nobody has to do the task more than once every couple of months. We get double figures of participants on most nets. We run them with the welcome use of the Essex Repeater Group's repeaters.

In the months since our AGM last November, to give you only a flavour of what we do, we have had a lively and very supportive CW evening, and talks on SSB, the use of and problems with radio in the emergency services, the joys of satellite tracking, microwave frequencies' set up, reach and versatility, tracking air traffic communications through

your radio, the many uses of APRS, and what "going mobile" takes and why it's such fun – among many more! We will be part of the Military Machines event in mid-August 2023, and will take part in Gateways on the Air (GOTA.) We are found running radio activities at events around the Southend Airport based Vulcan bomber. We regularly support radio rallies, the Scouts' Jamborees on the Air (JOTA,) RAYNET deployments, and much more.

Contact: *Chair@thamesARG.org.uk* or *Secretary@thamesARG.org.uk*

Website: *www.thamesarg.org.uk*

Trowbridge & District ARC

We are a friendly Club that's been established for 40 years. Whether you already have an interest in Amateur Radio or a new member wishing to find out more please come and join us at one of our meetings or contact us if you prefer, everyone is welcome at our Club.

Our monthly talks and activities cover a wide range of topics, our members interests span from HF to microwaves with a range of operating modes. Our hobby is truly fun, exciting and diverse with an interest for everyone.

As a Club we aim to be proactive and diverse in various activities throughout the year. We take part in a number of activities, quite a number of members take place in the RSGB UKAC's on all bands from 6m to 3cm. In June we have an annual 2m direction finding competition. In November we hold our constructors cup competition for kit built and "built from scratch" projects and we award a trophy to the winner of each category. We must not forget another jewel in our Crown which is RSGB VHF NFD in July, this is a full weekend which can be enjoyed in many ways, this ranges from setting up and operating in the Contest, enjoying the beautiful views of the Mendip hills or simply attending and enjoying the company and banter amongst each other (there is even a nice hearty breakfast if nothing else tempts you!) We also provide CW training at every level of competence & experience

Naturally we have a daily morning net and Sunday evening net on 2m FM.

For any enquiries please don't hesitate to contact us, we look forward to hearing from you.

G0BKU Secretary, Trowbridge & District Amateur Radio Club

email: *secretary.tdarc@gmail.com*

Warrington ARC

Warrington Amateur Radio Club recently celebrated its 75th anniversary and a lot has changed.

We find ourselves now the other side of Covid but it has changed us. For the better. We now hold our main meetings as mixed format – face to face and zoom at the same time – it allows members across the country and beyond to participate.

It has had other effects too. In 2023 we expanded the HF and satellite stations and as a result have found ourselves lacking space in the club's shack so have moved the test bench out into the main room. Zoom has been good in this regard as it has freed up space in the main room – we are after all enjoying a communications-based hobby.

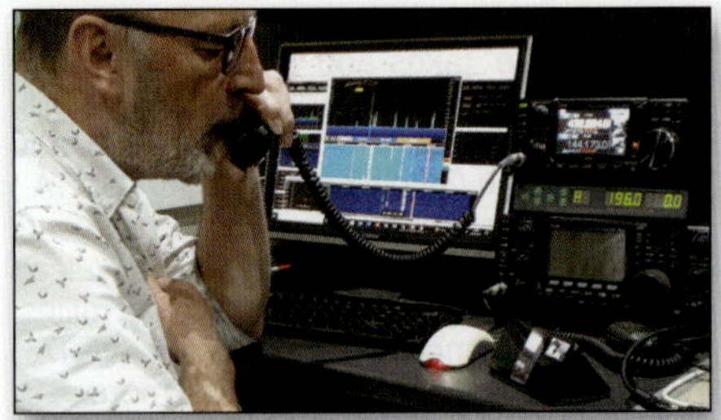

In 2024 we plan to continue regular talks interspersed with DF hunts, BBQs and trips – and are hoping to get back into contesting. Having repurposed our QO-100 SSB ground station from the NARSA rally as a club facility, we can now look forward to more operations via the satellite as well as on HF

If you want to join us, call in on a Tuesday night for the main meeting, or Thursday morning for a coffee and a chat or Sunday afternoon for a technical session. We've also re-instigated club nets but with a twist – these are via YSF & P25 – more details on the club website: *warc.org.uk*

West Kent ARS

A thriving club, with over 40 members, mainly from the Tonbridge, Tunbridge Wells & Crowborough areas, who enjoy a wide range of interests ranging from QRP HF CW with homebuilt gear to EME with a 3m dish! We hold monthly meetings involving amateur radio or social events at Bidborough Village Hall, where we now have HF & VHF antennas for live demonstrations. Members also meet on the first and third Friday evenings for practical sessions, building hardware & its firmware. Our nets are very popular and a sizeable portion of the Society support us from home in RSGB contests and at the HF &

VHF Field Days & Trophy contests, for which the Society recently upgraded its HF contest gear with a FTdx10. By invitation, we support Nutley Windmill annually, running HF, VHF, QRP CW & FT8 stations for the Mills on the Air weekend as GB2NW. We are also seeking opportunities to extend our special event activities to Museums & Railways.

Formed on 7 April 1948, in 2023 the Society celebrated its 75th Anniversary with a gala dinner at Salomons Estate with its museum of the renowned inventor 2nd Baronet Sir David Salomons, famous for the introduction of domestic electrical lighting and, as a founder of the RAC, responsible for removing many of the legal restrictions on the use of the motor car in England. The Society call-signs GX1WKS & GX3WKS, using a /75 suffix, enabled individual members to celebrate & promote the Society's 75th in 2023, with our Foundation & Intermediate licensees hosted by a Full Licence holder,with great success. A summer BBQ in conjunction with our VHF Field Day operation completed the celebrations.

Visitors are always welcome at our meetings, details of all the Society activities being on our website: *www.wkars.org.uk.*

West Manchester ARC

We are based at Astley and Tyldesley Miners Welfare Club which has a bar, meeting rooms and snooker (if you play). We are a friendly club and enjoy a beer and a chat when we are not operating in our shack. We have excellent facilities and operate on most bands from our tower (with multiple antennas). We have also built and operate a local UHF Repeater (Yaesu System Fusion) call sign GB7WM. The repeater is connected to the North West Fusion Group and is also accessible via the Wires-X fusion network.

Visitors are welcome to come down to the club 3 times as a guest before joining the club. We are one of the cheapest clubs in the UK with annual membership of just £20 (or £10 if you are over 60). This cost also includes Membership to the Miners Welfare Club with a subsidised bar and other activities.

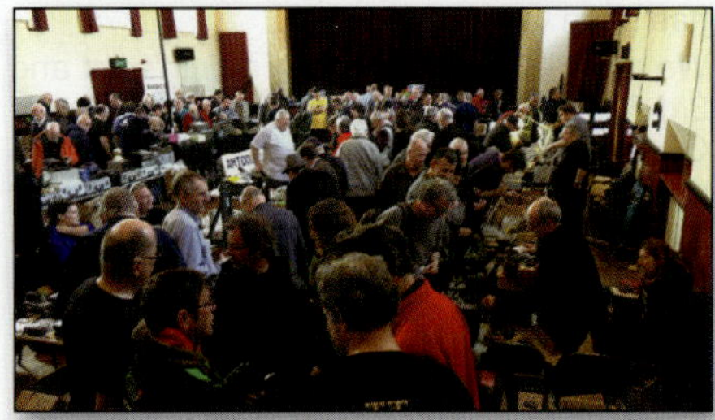

We have a good mix of abilities from beginners studying for the foundation license through to intermediate and full license holders. During the year we run a number of events including; The Red Rose Radio Rally, field trips, Museums on the Air (from Astley Mining Museum), annual bowl & BBQ and the mandatory Christmas Party!

Our web site (wmrc.co.uk) includes lots of information about the club, past and future events plus photos. We are always looking for new members so why not pop down for a drink and meet us one Thursday evening. We also have a Facebook page for members plus a public group called 'West Manchester Radio Group' if you want to connect.

Club venue: Astley and Tyldesley Miners Welfare Club, Gin Pit Village, Astley M29 7DW *secretary@wmrc.co.uk* or *wmrc.co.uk*

Wirral and District ARC

Wirral and District Amateur Radio Club meets twice monthly at Irby Cricket Club. Last year was a busy year as we fully restarted activities, post Covid. 2024 is set to continue this.

Twice we have been RSGB Regional Club of The Year.

Our youngest Foundation licence holder so far was 8 years old when first licenced, but we also have members more than ten times this age. Everyone is welcome.

We regularly run special event stations such as GB4LL for International Lighthouses on The Air weekend, Youngsters On The Air and many others. We also always put on a station for the World Scouts Jamboree On The Air.

One activity that has become very popular with members in recent years is the Club's participation in the UKAC weekly contests. Around a third of our members regularly take part in these challenges on different bands. In each of the last few years we have finished in the top 3 places nationally.

Throughout the year, members hold regular Direction Finding competitions across Wirral and North Wales as well as an Annual Construction Challenge, often featuring some of the projects that have been the subject of talks at our meeting evenings. In addition to the twice monthly meetings we also have social gatherings on the weeks when we don't have meetings.

Our members' interests range right across the full Amateur Radio spectrum from the lowest frequencies right up to microwave and include all the data modes, digital and television. For the more traditional, one of our members also runs Morse training sessions. Whatever you want to get involved with, there will be a club member who can help and give advice.

We are affiliated to the Northern Amateur Radio Societies Association (NARSA) and our members are involved annually with the NARSA (Blackpool) Rally.

WADARC is a very active club and becoming more and more so as time goes on. We are always pleased to see new-comers. Most of our meetings are radio orientated evenings but there are also occasional, general interest talks with twice yearly equipment sales and joint events with our neighbours in the Chester Radio Society.

Yet more details and useful links can be found on our excellent club website and we can also be contacted via there.

www.wadarc.com or via 0151 601 3269

Worthing and District ARC

The Worthing and District Amateur Radio Club meets every Tuesday at TS Vanguard (the headquarters of the Worthing Sea Cadets) 9a Broadwater Road, Worthing BN14 8AD at 19.00 hours. We are a friendly club and prospective new members, amateurs visiting the area or those just interested in radio are always welcome to attend – tea and biscuits are usually provided.

The club was founded in 1948 and we are still going strong. For example, we have recently installed a complete radio shack with antennas and new, mainly Icom, radios for all amateur bands that members can use while at the clubhouse.

We also have just finished installing a new FlexRadio system, complete with a Maestro Control Console.

A major benefit of this system is that remote access to it is available to all members who wish to use it to get on the air using just their own PC, laptop, Mac, tablet or smartphone. They can use it to operate from their home, from work, in fact, from wherever they want so long as they have an internet connection and the appropriate software.

If you would like to find out more about our club and its activities, please visit our website at wadarc.org.uk

Cambridgeshire Repeater Group

The Cambridgeshire Repeater Group (CRG) was formed out of the old Pye Telecoms Radio Club in Cambridge. The club designed and operated the UK's first 2m FM Repeater – GB3PI, and the first 70cm FM Repeater – GB3PY.

The group now boasts a fleet of 14 analogue and digital repeaters covering 6m to 23cm.

Our annual rally, which for the past few years has been held at Foxton, near Cambridge, is a regular feature in the calendar and attracts many visitors & traders from across the UK.

In 2006, "The Camb-Hams" was launched as the social side of the CRG, with the aim of providing opportunities for a wider range of radio activities and experiences. The first Camb-Hams event was a casual entry in VHF NFD. Since then, the group has gone on to greater things – winning the AFS Super League on several occasions, as well as taking part in UKACs and NFDs throughout the year.

2007 saw the first group trip to Ham Radio Friedrichshafen – a tradition that is still going strong to this day; as are our regular DXpeditions to the Scottish Isles, along with a brief sortie to the Isle of Man. With our multi-station approach, GS3PYE has put several thousand IOTA contacts into logbooks across the world.

Probably our most famous "Member" is our mobile radio shack, "Flossie". She accompanies us on our DXpeditions and field days, as well as enabling us to support other events, such as the RSGB Convention and the National Hamfest. Her 60ft mast makes her an unmistakable sight wherever she goes!

Alongside all these activities, the Camb-Hams still holds traditional monthly "Pye & Pint" meetings at The White Horse in Milton, near Cambridge. We also have a very active email reflector – just search for "Camb-Hams" on groups.io

You can find out more about our repeaters at *www.cambridgerepeaters.net*; or visit *www.camb-hams.com* to learn more about our social activities.

Central Radio Amateur Circle

In 2023 our group organised an Expedition to Malta which went well, we had several operators from around the world meet in Malta and run a station for a week. Due to this success, we are looking at doing it again during 2024 with more operators, date to be arranged please keep an eye on Radcom for article.

As a group we run several plug and play days during 2023 and will be doing more during 2024. All are welcome and because we play on Barr Beacon it has easy access as you can drive to the top.

At the time of writing this we are in the process of sorting out a new website, so please check the club finder to see the new website details.

Our group has had a bit of a change over the past few years, and we now meet every other Saturday in Bloxwich, Walsall. All are welcome and we have the added advantage of being a free club, which we have found does help with membership retention.

Our group run a net on 2m every Tuesday night and alternate Saturdays, we have found that some members prefer on air as we have many seniors who like most are still concerned about the Covid situation.

We look forward to seeing you all during 2024.

Contact: Martin Hallard, G1TYV, *radio-circle@live.co.uk*

Chelmsford ARS

The Society holds monthly meetings at Danbury Village Hall on the first Tuesday of the month, usually taking the form of a technical talk by a Club member, or a visiting speaker. CARS is a Brickworks Accredited Club, and our programme also includes regular meetings with a focus on basic skills, often with a lively discussion. The meetings also offer a "table top" sale area for equipment and the obligatory tea and biscuits. Full details of the meetings are published monthly on the CARS website and in the newsletter.

Open Club nets are regularly held on other Tuesday evenings using VHF, UHF and HF - in turn.

With some 80 paid up members CARS is probably one of the largest and oldest clubs in Essex if not the Country. Formed in 1936 by engineers from the company then known as the Marconi's Wireless Telegraph Co., this is reflected in our well-known Club callsign G0MWT - all part of keeping the Marconi legacy alive in Chelmsford, the birthplace of radio. Louis Varney G5RV was one of our founder members.

CARS typically runs a number of special event stations during the year. Recent activity has included operating GR0MWT to celebrate the Coronation of King Charles III, and operating GX0MWT from the former Marconi Research Centre in Great Baddow to commemorate International Marconi Day.

CARS have been actively involved with training since at least 1960. Courses for Foundation, Intermediate and Full licence exams including theory and practical assessments have been continued using training material designed and delivered by club members based on the current RSGB syllabus. The CARS website has a section dedicated to training and training materials are available to download. The materials are updated in line with RSGB updates to the syllabus. The Society's reputation for quality training is well deserved and our very experienced training team has achieved numerous successes at Foundation, Intermediate and Full level. Our registered exam centre in Danbury provides invigilators to support the RSGB written and on-line exam sessions. Our Exam Secretary and training coordinator John M0JOC is an experienced volunteer, as he is also a Remote Invigilator for the RSGB online exams. CARS offers face to face training sessions at Danbury for all three levels based on our very successful "Fast Track" evening format for revision and pre exam preparation.

Interest in Morse code continues to thrive and is as popular as ever, classes are held weekly by Skype for the Beginner or for a social CW chat.

Our monthly newsletter gives excellent coverage of all our events and much more.

Our website *www.g0mwt.org.uk* has full information regarding future activities, training, venues, dates etc.

On social media you can follow us on Twitter @ChelmsfordARS, @TrainWithCARS, and Facebook; *www.facebook.com/ChelmsfordARS.*

Cockenzie & Port Seton ARC

The Cockenzie and Port Seton ARC is structured in a non-traditional way that seems to work. Since 1984 the club has operated without official positions, formal membership or politics that sometimes invade formal structures. Its members like it that way, and the results give evidence to that. Financial affairs are transparent and healthy. An extensive inventory of rigs, antennae, towers, and other equipment has been accumulated over years to enable the club to mount a wide variety of activity.

In 2022 the club held regular monthly meetings, successfully participated in contests, , Mini-Rally, training and more. CW Training is being run weekly via ZOOM. Training (Foundation, Intermediate & Full) are now back to Face to Face. Members come from a wider geographic spread than might be expected for an essentially local club

A central activity of the club is training, with courses at all levels being conducted,

CPSARC believes that qualification is only the beginning, and actively encourages new licensees to be active.

New licensees are strongly encouraged in contesting, and some of younger members have become very competent contesters. The fact that having less experienced operators in contests might reduce winning chances is seen as a price worth paying.

Members are active from LF through to UHF. HF IOTA and CQWW are our most attended contests.

We are a supporter of the British Heart Foundation, and its cumulative donations continue to climb, reaching £18,699.40 in 2022.

Our website at *www.cpsarc.com* where you will find news, discussion forums, etc, archives of our award-winning monthly newsletter 'Elements' which is free to download . Website receives >50,000 visits a year

Facebook page "CPSARC" where up to the minute changes on events, etc can be found

Activity Nights are run as a regular monthly event and have become extremely popular encouraging all levels of Licences to participate and take part.

DF nights are lots of fun with 2 a year.

Construction nights are also held

The Club meets each month on the first Friday of every month in the lounge of the Thorntree Inn (Lounge) from 7pm till late.

Cwmbran and district ARS

Cwmbran and district amateur radio society would like to extend a warm welcome to all amateurs and members past and present.

We meet every Wednesday evening at the community hall, Henllys village road Cwmbran, opposite the dorallt public house. We meet at 19.00 – 21.00 hrs (Entry is through the rear doors)

At present we cover the whole of the Eastern and Western valleys including Newport Cwmbran Pontypool abergavenny and Eastern Cardiff

We run and maintain the GB3YA repeater and also the echolink system connected to it.

As an affiliated RSGB club we are the only club in the area that are able to do tuition at all levels and arrange the appropriate exam at the end of it in a classroom environment.

We can also boast a fully equipped shack with equipment capable of all bands and modes.

Also we have field day through the year including the Practicle wireless QRP contest, Mills on the air,

Canals and summits on the air plus others including railways on the air. We offer advice and help to install and set up your shack.

Tea coffee and drinks are available and snacks in the clubhouse are available. A small club shop is also available for all your peripherals like plugs connectors coax and also an equipment loan facility for things like analysers and swr meters.

Contact us on *www.mc0yad.club* or call Ken on 07411 595 103 for information and anything you need to know.

Echelford ARS

The Society was formed in Ashford, Middlesex, UK in 1964, using the old English town name 'Echelford' to avoid confusion with other Ashfords in the UK.

The Zoom meetings that we started during the COVID lockdowns have continued in popularity and attract a number of members who spend part or all of their time abroad, including Northern Ireland, the USA and Australia – it's good to be able to keep up with their news. Meanwhile, we continue to hold regular meetings twice a month which now offer free chocolate biscuits in addition to the free tea and coffee. The Sunday nets on Top Band and 2m have become noticeably more popular, and smaller nets are held on most days of the week, including a "Slow Morse" practice net. Despite a sadly high SK attrition rate in recent years, we have managed to maintain our membership which at the time of writing numbers 52. Participation in "club-oriented" contests has also increased, eg. the RSGB 80m CC or AFS series and with the guidance of our Contest Co-ordinator we have improved our standing in the league tables considerably in the last year. We also make an event of both the RSGB CW and SSB Field Day weekends and have just retired our battered and elderly steel framed tent for a more modern "pump-up" unit. We were able to erect a full-size half-wave dipole for 160m for NFD in 2023 and, with the acquisition of a new and enthusiastic expert CW operator and contester, improved our QSO count beyond our wildest dreams. Non-members are always welcomed to our nets and events and we do not make a charge for the latter – instead we hope that our hospitality will persuade them to join, as is regularly the case.

We have a number of keen DF'ers who enter and administer qualifying events for the national championships. Our members have held morse events and demonstrated on-air operating at Scout, Guide and ChOTA events just prior to COVID. Members have given talks to a number of local clubs. There has also been increasing interest in construction projects within the club, which will add variety to our annual construction contest and new names to the list of trophy winners.

We have a monthly newsletter ("All EARS"), a Twitter account, a YouTube channel and a "groups.io" discussion forum, the latter open to members only. More information is available at our website: *https://qsl.net/g3ues/*

We are looking forward to celebrating our 60th anniversary year in 2024. If you are looking for a friendly, active society in the North Surrey area, do not hesitate to contact us.

Horsham ARC

The Horsham Amateur Radio Club meet at the Guide Hall, Denne Road, Horsham, West Sussex, on the first Thursday of each month at 20:00. We also have a social meeting on a Thursday evening later on in each month at a pub in the area.

Monthly meetings usually consist of a talk on a radio-related subject or a surplus equipment sale and the year ends with the AGM in December. One popular meeting is 'Bring, Show & Tell' in which members can bring along some items of interest (not necessarily radio related) and give a brief talk. Junk sales are held in March and October and entry is free.

We hold regular direction-finding exercises on 2m on either

Sunday mornings or Thursday evenings.

Local club nets may be found on 2m FM on 144.725MHz on Tuesday and Saturday evenings at 21:30. On Sunday mornings there is the Billingshurst and District net on 80m on 3.722MHz at 10:00 local.

For further information, please take a look at our website *www.harc.org.uk*

Lincoln Short Wave Club

Lincoln Short Wave Club (*www.g5fz.co.uk*) was founded on the 10th February 1921 and we are now over 100 years old. The members meet every Wednesday evening and Saturday morning at our shack in the village of Aisthorpe, near RAF Scampton. We also have a 2m or 70cms net on the Lincoln repeaters at 8pm on Thursday evenings. Wednesday meetings are general meetings with visiting speakers normally arranged once a month. Other Wednesday evening meetings include feedback to the members from Committee meetings and general discussion and social evenings.

The Club holds a "surplus equipment" sale twice each year which is held in the adjacent Aisthorpe Village Hall. Saturday mornings are given up to mentoring new licensees and discussion with other members on any matters of interest from repairing equipment, to demonstrations of the Club's extensive range of radio equipment. CW, SSB, FM, data modes, D*Star, DMR, Yaesu Fusion (C4FM) and ATV are available for demonstration and use by arrangement. Also, with the recent growing interest in working Amateur Satellites the club is setting up a full satellite station using a G5500 rotator, Yaesu FT-847 and Wimo X-Quad antennas for 2m & 70cm, with more bands being added in the future. The club currently has a portable setup for working narrowband modes through the QO100 satellite, which can be deployed upon request, and for use at shows and special events.

Training for Foundation and Intermediate examinations regularly takes place on Saturday afternoons and examinations are arranged as needed. Additionally, support is provided for new licensees to enable them to fully enjoy the various aspects of Amateur Radio with equipment advice and training on the Club's own equipment.

The Club provides support to the local repeater group which maintains GB3LM (2M), GB3LS (70cms), GB3VL (23cms TV) and GB3LX 10GHz) which are all located high above the City on Lincoln Cathedral.

We also maintain a radio shack at the Lincolnshire Heritage Aviation Museum (*www.lincsaviation.co.uk*) at East Kirkby (GB2CWP). The shack includes working equipment from WW11 including a complete R1155/T1154 station. The shack is normally open on the third Thursday of the month throughout the summer and also when the museum holds special event days with "Just Jane" – their own taxiing Lancaster aircraft. The shack can be opened by special arrangement by contacting the Club Secretary (via *www.g5fz.co.uk*)

The club regularly arranges special events each year including in 2023 the very special Dambusters Memorial GB80DAM celebrating 80 years since the Dambusters raid which was run in collaboration with the International Bomber Command Centre. The club has also been involved with various scouts and guides events including The Lincolnshire Poacher which is held every four years.

We also enjoy operating stations for VHF NFD and SSB NFD. The Club's extensive array of portable equipment, including two trailer masts and enthusiastic members participation, come in very useful at these events.

LSWC is pleased to help with the National Hamfest each year, sponsored by the RSGB, which takes place at the end of September at the Newark Showground (*www.nationalhamfest.org.uk*). The event organisation includes volunteers from many of the local clubs without whose help the event would struggle.

Otley ARS

Otley amateur radio society (OARS), callsign G3XNO, is located just north of the town of Otley, near Leeds in Yorkshire. We meet on a Tuesday night at Clifton with Newall village hall (LS21 2ES) where members have use of a dedicated shack. Our club nights are varied, members and guests offer talks on various subjects, such as amateur satellite communications, digital modes, and getting started with microcontrollers as examples. We also host practical nights, allowing members to share their skills and teach others about mak-

ing antennas, learn soldering skills, and how to get to grips with their new radios.

Back in 2020 the club acquired a trailer mast, ordinarily used as a lighting rig by the fire and rescue service. It has been transformed into a mobile antenna set up, with a rotating Cushcraft A3S tri-band beam. This trailer mast has been a big help, and allows us to run more special event stations, with a quicker set up and take down, leaving more time to play radio!

The club is active in participating in special events. We operate the jamboree on the air, JOTA, with a local Scout group, enabling them to talk to other Scouts across the world. We contest in the 2m Practical Wireless contest from 'the Shack on the hill, Yorkshire Dales,' 1600ft asl. We operate Churches on the air from a local church at Farnley, and Windmills on the air from the east coast at Speeton. You can find us at the Otley Vintage Transport Extravaganza, and at the Otley Science Fair, sharing the joys of amateur radio with the public.

We support the Yorkshire Air Ambulance and operate the special event call sign, GB0YAA, during international air ambulance week. We have raised over £3500 for the air ambulance through shack sales and raffles at Christmas.

New members are always welcome, licensed, or not, and we support prospective amateurs through any training leading up to the radio exams.

The club runs a net on our repeater, GB3WF on a Friday evening, and a 2m local net on a Sunday evening. Anyone is welcome to call in.

Contact: M. Ross 2E0SNZ, *michael.ross@ashfieldprimary.co.uk*, tel: 07768996370

St Tybie ARS

St Tybie Amateur Radio Society was founded in 1996 in the village of Llandybie, Carmarthenshire. Llandybie is named after the Church of St Tybie, an early Christian Martyr and the daughter of a Local Chieftain, as we met in the church hall, STARS was born.

In 2018 we decided to move to a more radio friendly location, this being Saron Hall, off Saron Road, Ammanford. There are several acres of land and being out of the village centre, has an extremely low RF noise floor. The Club now enjoys this hilltop position, a purpose built shack, a good selection of antennas and room for a mast in the future.

A number of members are experts in their own field, and we operate from Topband to 24GHz and can do all modes. We regularly compete in the RSGB 80m CC, UKAC 6m-SVHF and 50MHz contests amongst others. The club has held the Special Contest Call GW7W since 2020, this has really helped with our contest endeavours. Gareth, GW4JPC activates this call most regularly as he lives close to the HQ station and had a great night in the 80m CC SSB leg on 19 April 2023 when he made 123 QSO's and 624 points for the club, fantastic. In our HF contesting there is a strong contingent of members using 100W and wire antennas to good effect.

Our members look forward to the RSGB 80m/40m AFS SSB and RSGB SSB Field Day when we enjoy the social side, and focus on the BBQ nearly as much as the technical challenges of these events.

We have an extensive library of books and magazines, and have many donated radios and accessories in the shack with such delights as the Ten Tec Century 22 and Trio TS520 legacy radios. The main contesting radio is a Kenwood TS590SG.

Our Club is growing again and we are seeing pre-covid membership numbers, which is very encouraging. We had many interesting presentations in the past on club nights and I'm sure the time is right to bring them back. Ninety percent of our members have full licences, so I think a morse code refresher course is always on the cards, and I think we will be back to full throttle with our contest activities in 2024.

Contact: Gareth Woods, GW4JPC, *gw4jpc@yahoo.co.nz*

Warrington ARC

Warrington Amateur Radio Club recently celebrated its 75th anniversary and a lot has changed.

We find ourselves now the other side of Covid but it has changed us. For the better. We now hold our main meetings as mixed format – face to face and zoom at the same time – it allows members across the country and beyond to participate.

It has had other effects too. In 2023 we expanded the HF and satellite stations and as a result have found ourselves lacking space in the club's shack so have moved the test bench out into the main room. Zoom has been good in this regard as it has freed up space in the main room – we are after all enjoying a communications-based hobby.

Where in 2024? we plan to continue regular talks interspersed with DF hunts, BBQs and trips – and are hoping to get back into contesting.

Having repurposed our QO-100 SSB ground station from the NARSA rally as a club facility, we can now look forward to more operations via the satellite as well as on HF (We recently replaced our aged Stepp-IR HF beam with a new Innov antenna).

If you want to join us, call in on a Tuesday night for the main meeting, or Thursday morning for a coffee and a chat or Sunday afternoon for a technical session. We've also re-instigated club nets but with a twist – these are via YSF & P25 – more details on the club website: *warc.org.uk*

Wythall Radio Club

We are a lively club with a wide range of activities. We aim to "have fun with RF", so our radio activities also have a strong social element bringing together many of our 40+ members. The Club meets every Tuesday at Wythall House, a large community centre set in a park. Club members can use our HF and VHF rigs, and our antennas - beams for 6, 4, 2 and 70 on a 60' Versatower, a collinear for 2/70 and a 240' doublet for HF. We also host the GB3WL 70cms repeater.

The Club's Christmas and Easter contests are eagerly anticipated by members. Run over several days, members contact each-other, with bonuses available for using different bands, modes, and portable and mobile operation. During the summer, our 'Plug and Play' events in Wythall Park are well supported. The Club provides a generator to power members' rigs, and catapulting cord over the park's trees provides the opportunity to install a variety of wire antennas. We enjoy a bar-b-q or fish and chips to round-off a fun day.

An annual DXCC table gives a focus for members' radio activities. This is supplemented with ad hoc 'challenges' – in 2023, the challenge is to work all US states. Other events are suggested by members, including a 'How low can you go?' week in which only QRP was used and a 'legacy' (SSB/CW) vs. digital mode challenge on 10m. Several members operate portable, including SOTA activations and VHF backpacker contests. During 2023 there has been considerable activity using FT817s with portable antennas, including comparing what members can work with the 'Miracle whip'.

The Club provides mentoring to new licensees and those progressing to intermediate and full licences. Although we no longer provide training courses, the Club's Training Coordinator (Chris G0EYO) is a tutor on the Bath Based Distance Learning programme.

We organise or participate in special event stations during the year, including GB0BUS for Museums on the Air at the nearby Transport Museum Wythall, GB22GE Birmingham Commonwealth Games and the GB70E Queen's Jubilee. The Club supports local events, including providing marshalling communications for the Wythall Fun Run. We have improved our promotion of amateur radio at these events by using 4m high feather flags and small information cards. We also have an excellent club website (see below) and use social media (FB and Groups.io) and Zoom for virtual meetings extensively.

A further part of our outreach to the community is the communication badge sessions for Beavers, Cubs, Guides or Scouts. These are highly participative, giving young people the opportunity to pass messages in Morse code or by licence-free handhelds using phonetics, as well as send greetings messages over the air. This year, the Beavers were able to talk to radio club members in Seattle, USA – one youngster was heard telling his parents it was "really cool" to talk to people in the States!

We welcome visitors and prospective members. Our activities and contact details are provided on our website: wythallradioclub.co.uk

Worthing Radio Events Grp

This year we'll add helping with JOTA for our local scouts, putting on special event stations for the public to see amateur radio and getting back to field days.

Our group has almost as many interests as we have members, from vintage radio restoration to kite antennas, morse code to digital modes. As a result, every coffee morning brings something new to see and learn.

We are a small group of individualists; however, we are all alike in our desire to help out others in the best way that we can.

Contact: *secretary@m0reg.co.uk*

Online Clubs

Essex Ham (Region 12)

Essex Ham is more than a club, we're an online amateur radio community, and from our humble beginnings as a blog over ten years ago, the group now has over 3,400 members around the world.

It's fair to say that things have changed for clubs as a result of the pandemic, with amateurs and clubs now doing things a little differently. We've seen an increase in interest for video content, especially for newcomers, and at field events, we've met several new licence-holders who've studied online but now need some practical help and support.

We've recently formed a solid link with the East Essex Hackspace, which encourages people in Essex to gets hands-on with "making stuff" – the space has rooms for 3D printing, laser cutting, electronics, woodwork and metalwork, and as there's a natural overlap into our hobby, we're regularly on-air from the Hackspace to show off what we do, and interact with other "makers".

Our free online Foundation training courses continue to be a large part of what we do. We've now trained over 10,500 students. There's little doubt that the ability to train and sit exams online has made a massive difference to take-up of the hobby, due to the improved access to courses and exams, and we'd also like to acknowledge the hard-working volunteer RSGB exam invigilators and the exam team at RSGB HQ for helping to make it easy for students to take their exams.

We've taken part in some fun and interesting initiatives. Broadcasting started in Essex back in 1922, and we were on-air to commemorate the moment that the first broadcast took place, with live video and audio streams. We were also proud to support one of our members, Andrew M0ONH with a campaign to raise awareness for Prostate Cancer, a cause that affects 1 in 8 men. We ran several events for GB8PCA, and were joined by the GB1NHS team. We supported one of the largest JOTA events in the region, adding a "blindfold radio maze" challenge to the fun, using radio headsets from dBD Communications, providers of comms to the rail industry. Field events included our regular St George's Day get-together, a Jubilee event, and participation in the county-wide "Essex 2m Activity Days". Our weekly "Monday Night Nets" continue to be a popular on-air meeting place.

As well as being there for our members, we also actively support clubs around the UK. Our training slides, videos and handouts are available free to all tutors, and we have almost 200 tutors in our independent trainer's group who receive our regular Tutor's Newsletter. We also work hard to promote and champion the hobby.

Regardless of where you live, we'd love you to join our growing online community. Membership is free and open to all. Join us at *www.essexham.co.uk*. You'll also find us on Twitter, Facebook, YouTube and Discord.

Wales Digital Radio Group. (Region 7)

The WDRG is an online group with members in Wales, England, Scotland and Cyprus. We meet online via Google Meet every Thursday at 20:00. Where we have a general chat all things radio, as well as some technical help and advice.

The aim of the group is to encourage outdoor operating, either portable or from the back garden. In a group or on your own. The WDRG organise some field operations to take part in events like SOS Week, promoting the RNLI, and Railways On The Air celebrating the local rail heritage.

Contact : Daryll . MW0TTF
Darylldmellow@gmail.com

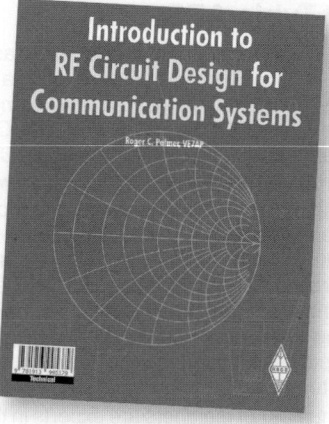

National Affiliated Clubs & Societies

AMSAT-UK *(G0AUK)*

This is the UK national society specialising in amateur radio satellite matters. It has approximately 700 members, produces a regular publication, Oscar News, for its members four times a year. It maintains a web site http://amsat-uk.org and holds weekly nets on 3,780 KHz +/- QRM on Sundays at 10.00 local UK time and also on 10,489,780 KHz on the QO-100 satellite, also on Sundays, normally at around 10.30 UK time. You do not have to be a member of AMSAT-UK to join either net - anyone with an interest in satellites is welcome.

In addition, AMSAT-UK organises an annual Colloquium. In recent years these have been held at the same dates and venue as the RSGB Convention. In 2020 and 2021, due to COVID restrictions, it held a very successful virtual webinar events. Both types of event provide an excellent chance to rub shoulders with experienced satellite operators and satellite designers. It is hoped in 2022 to hold a face to face event in October in Milton Keynes.

AMSAT-UK has been responsible for the development of a number of operational CubeSat spacecraft that follow the FUNcube format. These combine linear amateur transponders with STEM outreach telemetry Membership of AMSAT-UK is by donation, for which there is a suggested minimum. Any funds remaining after running expenses, are used to fund satellite design, building, and launching. Two forms of membership are available, either postal or electronic. Those with postal membership pay a little extra (to cover postage costs) and have a printed copy of Oscar News sent to them by Royal Mail.

Members with electronic membership down load Oscar News as a pdf file.

AMSAT-UK runs a small online shop, http://shop.amsat-uk.org, at which members, or new members, can make their donation. The shop also has a few items of interest to satellite operators. Enquiries and applications forms for membership can be sent with an sase to the Hon Sec, Jim Heck, G3WGM, Pickles Orchard, Great Hampden, Great Missenden, Bucks, HP16 9RE. He can also be contacted via email at: *g3wgm@amsat.org*

If you use amateur radio satellites be prepared to pay something to the organisation that builds and designs them – AMSAT!!

British Amateur Radio Teledata Group *(BARTG – G4ATG, GB2ATG)*

We are a web based datacoms group. Our aim is to encourage and promote the use of all types of amateur radio datacoms (such as RTTY, PSK, WSPR, Olivia and JT modes).

We run four Contests each year. In January we run our 24 hour RTTY Sprint Contest, in March we run our our 48 hour HF RTTY Contest, in April we run our 4 hour 75 Baud RTTY Sprint Contest and in September we run our 4 hour PSK63 Sprint Contest.

All Single Operator All Band (SOAB) results count towards the BARTG Championship. The leading station overall in the Championship will become the BARTG Diddler of the Year and will receive the prestigious BARTG FlexRadio Trophy kindly sponsored by FlexRadio.

Our Awards Scheme is open to both licensed amateurs and listeners. Its entry level will appeal to the occasional operator and its upgrades will appeal to the serious DX chaser. We have awards available for all data modes ranging from RTTY to the latest JT modes.

Our Quarter Century award is for making contact with 25 different ARRL DXCC countries. We have a series of Continent awards (African, Asian, European, North American, South American and Oceania) for contacting countries in those continents. Logs submitted to our contests can be used to apply for any of our awards. Award certificates are available for a modest fee through our website.

We have a small fund (dependent on income from fees for our award certificates and on donations to us) that we use to sponsor a few carefully chosen DXpeditions each year.

We are happy, via our For Sale/Wanted web page, to help people give away, sell or obtain datacoms equipment or seek technical advice.

Our website is central to everything we do so please visit it for more information about us, including our contests and awards.

www.bartg.org.uk

Secretary: Ian Brothwell G4EAN, 56 Arnot Hill Road, Arnold, Nottingham, NG5 6LQ.

bartg@bartg.org.uk

British Amateur Television Club *(BATC – RS38114)*

The BATC is the world's largest television technology club, based in the UK, with more than 1,400 members around the world. The club is open to all who are interested in television, from programme making and production, to building equipment for transmission and reception. Amateur TV can be transmitted on most bands above 50 MHz, and standard definition TV pictures are routinely exchanged over distances of up to 50 km on 437 MHz, but DX of hundreds of km can be achieved under exceptional propagation conditions. TV pictures can also be transmitted and received through the QO-100 satellite to most of Europe, Africa and Asia using just a 1.2 metre dish.

The club sponsors a number of hardware and software projects (such as the Portsdown Transceiver and the Ryde receiver) to enable the construction of Amateur TV equipment by newcomers to this aspect of amateur radio.

In addition to producing a quarterly colour magazine (available to members as a hard copy or a download), the club runs a number of online services including a forum https://forum.batc.org.uk/, a Wiki https://wiki.batc.org.uk/ and a video streaming site https://batc.org.uk/live/ where ATV repeater outputs and members' streams can be viewed online. The club has a range of "hard to get" components and a selection of printed circuit boards for some of the club's projects available in our online shop: https://batc.org.uk/shop/

The BATC also sponsor "Conventions for Amateur Television" (CATs) at various venues around the UK each year which are also streamed online for overseas members and those unable to travel. The Conventions include technical lectures, specialist products for sale, bring and buy sales and a chance to meet other like-minded individuals.

The Club offers free membership (with access to online copies of CQ-TV) to students in full time education. For more information, or to join our club, see our website: https://batc.org.uk/ or email the secretary: *memsec@batc.t*

British Railways ARS (G4LMR)

We are a small and friendly society founded in 1966 by railway workers who were interested in amateur radio. Membership is open to anyone, licensed or listener, interested in railways of any gauge (including model railways and light rail) and amateur radio.

Our Newsletter is published several times a year depending on content and is sent by post to our members. Articles and photos from our members are always welcomed by our editor. We operate weekly HF nets according to band conditions.

Our members run several special event stations, usually with a railway theme, during the year, often at or near railways (usually heritage railways). We aim to organise a Spring Meeting weekend each year somewhere in the U.K. (usually near a railway line) as a weekend social event for our members and spouses. Radio stations and railway stations are usually involved in these Spring Meetings.

BRARS is affiliated to FIRAC (Federation Internationale des Radio Amateurs Cheminots - International Federation of Railway Radio Amateurs) which organises various events including an Annual Congress during the year.

To find out more about us please go to our website or contact our Membership secretary:

Richard Waterman G4KRW, 170 Station Road, Mickleover, DERBY, DE3 9FJ

membership@BRARS.info

www.BRARS.info

British Young Ladies Amateur Radio Association *(BYLARA – M0BYL)*

BYLARA was formed on 29th April 1979 to further YL operation in Great Britain and to promote friendships with YLs and OMs worldwide. The association aims to encourage good operating, as well as persuade all YLs across the Great Britain and the World on to the air and to explore more of the hobby, renewing old friendships and making new ones, across the airwaves.

We would also like to encourage good operating techniques and courtesy to other operators at all times and also aim to encourage and introduce up and coming Guides, Brownies, Scouts as well as other young people who would like to go into the Hobby. We now have Members worldwide and are growing all the time.

We have had our Callsign (M0BYL) since 1998 and more information can be found on our Web Page. Our Members take part in various Special Events, Dxpeditions, Contests and have our own Net Night. We are always open to new members, OMs and family groups.

For more information, contact:
The Chairmam – Carol, 2E1RBH on
01305 820400 or

email: *carolhodges1@btinternet.com*
www.bylara.org.uk

CDXC – The UK DX Foundation *(M0CDX)*

CDXC is the UK's premier DX foundation and the biggest in Europe. We are dedicated to encouraging excellence in DXing and contest operating and have around 750 members – a quarter of whom are overseas. Formed in the 1980s, CDXC has grown substantially with our members sharing a common interest in HF/6m DX and contesting and has become one of the largest and most respected DX groups in the world.

There are FIVE reasons (at least!) why you will benefit from being a member of CDXC:

1. The CDXC community. Become part of the UK's active DX and HF contest community. You will belong to a Club of like-minded people who are also passionate about your favourite hobby!

2. Help sponsor DXpeditions. CDXC sponsors DXpeditions to rare and hard to get to DXCC and IOTA entities and part of your subscription goes towards making significant CDXC contributions to Club Log, WRTC, and YOTA. You will see us on the sponsors' pages of the entire major DXpeditions taking place in 2022/23.

3. The CDXC Digest magazine and Website. Receive the best DX magazine every three months - the CDXC Digest is available in printed and on-line formats - crammed with articles on things DX and Contest oriented keeping you up-to-date and informed. You can see the front cover and find a list of contents for the latest edition on our website (*www.cdxc.org.uk*). You can also access the CDXC DIGEST ARCHIVE to read previous digests at the following link (note no login required): *http://www.cdxc.org.uk/CDXC_Digest_Archive.* On the Club's Web Site there is lots of club and members info PLUS members For Sale / Wanted ads.

4. Learn from other CDXC members. Improve your DXing and contesting performance by learning from members' experiences. Via CDXCs active reflector get rapid advice and guidance 24/7 and debate the hot topics of the moment.

5. CDXC social events. Great social events and get-togethers: CDXC annual convention and dinner, stands at the RSGB convention and Friedrichshafen Hamfest - attended by well-known DXers and Contesters.

But most of all, CDXC is about being a member of a DX Community, of joining with like-minded people who want to see high standards of operating and want to work HF and 6m DX and contests on all modes and with anything from wires in trees (most of us!) to beams at 60ft.

You can sign up on line. Just go to our website *www.cdxc.org.uk* > Join CDXC. You can simply pay your membership fee by PayPal. Follow us on Twitter: @cdxcuk

The Duxford Radio Trust

The Duxford Radio Trust (DRT) is based in the East Anglia region of the UK and comprises an active group of radio specialists, plus supporters, many of whom are from the radio communications industry or Armed Forces. The group researches, collects, conserves, restores, displays, and demonstrates historic radio communica- tions, radar, and navigation equipment used in military conflict and civil emergencies, for the education of the public.

DRT also operates the permanent exhibition radio station G0PZJ/GB2IWM, using both modern and historic military equipment.

DRT was formerly the Duxford Radio Society (DRS), and previously provided the Radio Section at the Imperial War Museums, Duxford for 32 years. The group was founded in 1986 by Richard Pope G4HXH and Major John Brown G3EUR, renowned wartime designer of S.O.E. and clandestine radio equipment.

DRT is currently in the process of establishing new public exhibition and engineering facilities- at the Signals Museum RAF Henlow following its departure from IWM, prompted by a change of policy and direction at IWM Duxford and HMS Belfast, London.

DRT is entirely self-funded by its local and distant supporters and in addition to its public exhibition activities, normally publishes three issues of a historical/technical journal each year and a monthly newsletter.

The Trust welcomes new active members who share an interest in the history, technology, conservation, and restoration of military communication equipment, and who wish to explain and demonstrate this subject to the public and operate the radio transmitting station, using both vintage and modern equipment. DRT also welcomes distant supporters who are not able to act as active explainers, operators or restorers in the Cambridgeshire or Bedfordshire area.

Details can be found at *www.duxfordradio.org* and DRT can be contacted at: *duxfordradio@gmail.com*

Essex CW ARC

(ECWARC – G1FCW, G4C)

Based and formed in Essex, but open to all licensed radio amateurs worldwide who are interested in Morse code (CW), Essex CW ARC has as its mission to encourage the use and preservation of the CW mode on the amateur bands. It further aims to support those who strive to learn and improve their CW skills (associate members) as well as operators who are already competent at any level (full members). Membership is free and most on-air activities are publicised via the club's website and via a regular email bulletin to members. Activities include club CW nets on the HF bands, ad-hoc communication using recommended ECWARC frequencies on each HF band and team efforts in various CW contests throughout the year. Learning opportunities for those starting out includes a weekly GB2CW QRS transmission on 80m as well as advice for those who are running classes or practice sessions for others in their own areas.

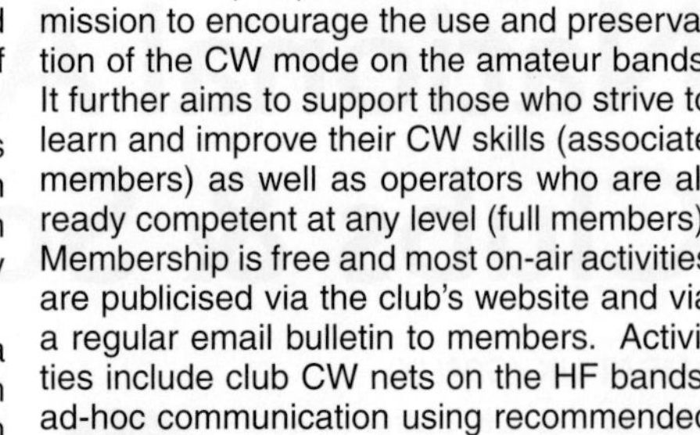

For further information please visit the club's website at: *http://www.essexcw.org.uk*

First Class CW Operators' Club

(FOC – G4FOC)

The First Class CW Operators' Club (FOC) was founded in 1938.

FOC is limited to approximately 500 members, who will typically meet both on air and in person. The Club holds a wide range of social gatherings throughout the year, in many different countries, with members also often meeting at major events such as Dayton, Friedrichshafen and the RSGB Convention. On-air activities include the FOC Marathon in February, and two FOC QSO Parties, which are normally held in March and September.

In 2023, FOC celebrated its 85th anniversary. The aim of the Club is to promote good CW (Morse code) operating, activity, friendship and socialising. The Club is UK-based, with many international members. The character of the club is best expressed in its motto, "A man should keep his friendship in constant repair" (Samuel Johnson, 1755).

Membership of FOC is achieved through a process of nomination by existing members, based on on-air activity and other criteria which are explained on the Club's website at *www.g4foc.org.*

The Secretary email: *secretary@g4foc.org*

FISTS CW Club

(GX0IPX, GX3ZQS & MX5IPX)

FISTS is probably the largest open-membership CW club in the world. Founded as a UK group in 1987 by Geo Longden, G3ZQS (SK) over 37 years it has grown into an international club. As well as the UK-based FISTS Europe, there is a FISTS chapter in the Americas, thereís FISTS Down Under in New Zealand and Australia, and FISTS East Asia.

The club actively promotes the use of Morse code on the air. FISTS runs a variety of activities, including friendly, competitive events and challenges every year. There is a great deal of activity on 80m and 40m ñ 3.558 and 7.028MHz ñ especially on Ladder Sundays every month with UK and European members taking part. FISTS actively supports EuCW and its on-air activities as well. Our comprehensive club website is *www.fists.co.uk.*

FISTS members attend many rallies every year. New members are always welcome to FISTS, wherever you may be and whatever your CW speed or proficiency. Our members are beginners, improvers and experts; newcomers, old timers and returners to the hobby. There is just one criterion for membership of FISTS: a love of Morse.

FISTS publishes our quarterly journal, Key Note, available on paper, delivered by post, or digitally. The club also encourages personal improvement and development of Morse skills, and operating CW on the amateur bands.

To find out more about FISTS visit *www.fists.co.uk* or contact *members@fists.co.uk*

GMDX Group - Scotland's DX Association *(GM5A, GS8VL)*

The GMDX Group, based in Scotland, caters for all radio amateurs who have an interest in DXing, contesting and award chasing. The group has its own award scheme and produces a quarterly magazine, The GMDX Digest, which is sent out to members electronically. The Group has supported most of the major international DXpeditions and criteria for sponsorship is outlined on the group website.

There are usually two meetings per year, including a DX Convention held in April in the Stirling area. Membership is open to all radio amateurs or SWLs with an interest in Contesting and DXing.

Full details from Robert W Ferguson, GM3YTS, 19 Leighton Avenue, Dunblane, Perth FK15 0EB. Tel: 01786 824199. email: *gm3yts@btinternet.com*

website: *www.gmdx.org.uk*

G-QRP Club

The club has recently passed its 45th year in existence. The club specialises in low power operation (hence the QRP), primarily on the HF bands but can be found on LF/VHF/UHF and SHF as well. The club produces a quarterly magazine for its members called Sprat that focuses on construction and operation but that also includes articles and comments from members about their activities. Membership is growing and is in excess of 4,000 and contains many active members from DX locations not only the UK. Many members take part in all sorts of radio related activities, especially DXpeditions, buildathons, contests and field day activities to name but a few. There are many, many members who are active SOTA, IOTA and DXCC activators and chasers.

Activities cover the whole radio spectrum and many different modes. It is not uncommon to find members active on bands from LF to SHF. There are many members involved in helping others enjoy the hobby through training and construction, several members hold and have held NoV for novel operations.

Further details can be obtained from the chairman by emailing *g0fuw@tiscali.co.uk* or from their website: *www.gqrp.com*

Ham Radio Network *(G5HAM)*

Ham Radio Network (HRN) is a virtual radio club formed during the very first Covid-19 lockdown in 2020. Its overarching purpose was to alleviate potential loneliness and a feeling of isolation for many like-minded radio enthusiasts who like to share, collaborate and experiment.

It was, and remains, intended to be a club that would offer a chance to meet via regular Zoom meetings, thereby attracting talks and demonstrations from all around the world, and to offer support and interesting topics via HRN's main interface on Facebook: *https://www.facebook.com/groups/517224945921950*

HRN continues to attract good discussions, talks, demonstrations, help and support, especially (but not restricted to) new licensees and returners to the hobby. still growing. HRN welcomes all amateurs from across the globe.

For more info: *www.hrn.world*

International Short Wave League *(ISWL – G4BJC, M1SWL)*

Known As the ISWL, the League was founded in October 1946 and caters for all interested in both the amateur and broadcast bands. Membership is open to both short wave listeners and licensed amateurs.

All members receive the monthly journal Monitor which includes columns of interest to both Short Wave Listeners and Licensed amateurs. The pages include Members Mailbag, reception reports for HF/VHF/UHF; SW BC Bands; Transmitting Topics; "In Days Gone By", a review of receivers from the past; articles of technical interest submitted by members; additional members contributions include "Meet the Member" and articles of interest; Stations to be looked for during the month together with QSL Managers and Addresses; information of general interest world-wide. A monthly report of the ISWL Nets activities during each month plus reservations of licensed members each month.

A full Contest programme and a comprehensive awards programme is open to all members, the latter free of charge to members, also available to non-members at a charge.

Weekly Nets are held on 80m and 40m SSB Sundays, Tuesdays and Saturdays; 160m SSB on Sundays; 80m or 40m CW Mondays.

Full details and a membership application can be found on the web: *www.iswl.org.uk* or available from the Membership Secretary Cliff Jobling, G-13557/G4YHP, "Joycliff", 20a Poplar Road, Healing, Grimsby, DN41 7RD, email *ch.jobling@ntlworld.com*, or, from The Hon Secretary, Peter Whiffing, G-13102/M0PGW, 38 Green Close, Stannington, Morpeth, Northumberland, NE61 6PE, email:

peter.whiffing@yahoo.co.uk

Martello Tower Group

The Martello Tower Group is an amateur radio group which was originally formed in the late 1980s under the name ClackPak.

In recent times the group has set up an SSTV repeater (MB7TV), an FSTV beacon (GB3CZ) and the first licensed amateur radio to internet gateway in the UK (MB7UIV).

ClackPak was reborn in May 2009 as the Martello Tower Group with ambitious plans for other ground-breaking radio projects in the Tendring area.

In 2015 we installed a 70cms DMR repeater, GB7CL as well as a D-STAR repeater, GB7TE on 2m at the Martello Tower and in October 2017 it was moved to our Thorpe-le-Soken site.

Although affiliated to the RSGB, we are not a 'club' in the traditional sense. The Martello Tower where we meet is very limited in space hence we have to keep our membership numbers down.

Membership is strictly by invitation only.

www.martellotowergroup.com

Northern ARS's Association

NARSA (Northern Amateur Radio Societies Association)

NARSA is an association of radio clubs who have been organising an annual rally in the North West of England since 1962.

The first rallies were held at Belle Vue in Manchester but since 1988 we have been at the Norbreck Hotel Conference centre at Backpool.

NARSA is club focussed and any radio club can affiliate for a small fee which entitles them to put on a stand at the rally. The rally attracts dealers, manufacturers and suppliers of anything related to amateur radio. There is an RSGB membership stand and an RSGB book stall at the rally and representatives of many of the RSGB specialist committees attend.

We encourage the sharing and distribution of technical knowledge and information between member clubs.

We aim to showcase the benefits of belonging to a club and being part of the wider radio amateur community.

NARSA is a non profit making organisation run entirely by volunteers and the next rally will be on 21st April 2024 in Blackpool.

For further information visit our web site:

http://narsa.org.uk/

National Hamfest (Lincoln) Ltd

The show attracts manufacturers, dealers, and suppliers of everything related to Ham Radio from both the UK and internationally. The event is organised in association with, and supported by the Radio Society of Great Britain (RSGB) who have the RSGB book stall and representatives of many of the RSGB specialist committees in attendance. Many of the special interest groups and clubs from around the UK also support the event.

www.nationalhamfest.org.uk

North West Fusion Group

North West Fusion Group operate a network of repeaters and gateways all connected to the NWFG room. This means we can offer superb coverage across the North West as well as enabling radio amateurs the ability to access our infrastructure from anywhere with internet access via Wires-X room 41755 or a hotspot connected to our reflector FCS00428

We are primarily based in the North West of England but we do have members all over the world.

If you are interested in adding a repeater or

gateway to the network, or require any more information please email us at:

admin@nwfg.online and we will be happy to help.

Online Amateur Radio Community

We are a UK online based community, here to have fun and experiment with amateur radio – An amateur radio club providing Ham Support, Advice, and Community, without Committee.

For more information please checkout our website: *www.oarc.uk*

Radio Fraternity Lodge No. 8040

Radio Fraternity Lodge No. 8040 was Consacrated on Tuesday 23rd. November 1965. Many famous Amateur Radio Station Callsigns formed the bedrock of the Lodge and we, of the present generaton, have come to regard their legacy with pride and affection.

The beginnings were closely linked with the Radio Society of Great Britain (RSGB) and the London Members Luncheon' Club to which many of the famous Callsigns were active officers and members. For more information: *secretary@radiofraternitylodge.co.uk*

www.radiofraternity.co.uk

Radio Amateur Old Timers' Association *(RAOTA – G2OT)*

Our motto is "HONOUR the PAST, ENJOY the PRESENT, ENSURE the FUTURE".

We aim to keep alive the history, pioneer spirit and traditions of amateur radio, not only for the benefit of to-day's radio amateur but also for the benefit of future generations of radio amateur. We do this via personal and radio contact and via the pages of our magazine and books. We are also mindful of members with special needs.

Our Associate Membership is open to anyone who has an active interest in amateur radio Irrespective of whether or not they hold an amateur radio licence.

Our FULL membership is open to anyone who has been actively involved in amateur radio for 25 years or more Irrespective of whether or not they hold an amateur radio licence. Full members have the right to vote in the running of RAOTA.

Associate members are welcome to transfer to Full membership when they achieve the 25 year qualifying period. There is no difference in subs.

All members are entitled to join in our activities and social events. All members receive our magazine, entitled OTNews, which is edited and printed to a high standard of which we are justly proud. OTN is also available as an audio magazine.

We operate several regular nets under the callsign G2OT using a variety of modes and frequencies.

We have an awards scheme which is open to both members and non-members.

More information about all our activities can be found on our web site.

Applications for membership and requests for a sample copy of OTN should be addressed to: RAOTA, 65 Montgomery Street, Hove, East Sussex BN3 5BE.

email: *memsec@raota.org*

website: *www.raota.org*

Radio Officers Association

(ROARS - M0ROA)

The Association was formed in 1995 when the commercial use of Morse for maritime communications was being phased out. Some members of Portishead Radio and other ex-Radio Officers conceived the idea of forming a society to preserve the historical importance of this form of communication which had been in operation for nearly 100 years and had saved many lives.

It was discovered that a number of members were Radio Amateurs and the Radio Officers' Amateur Radio Section was formed to promote the use of Morse on the bands and communicate with members using the form of transmission that we all knew so well. Any ex-R/O, whether from maritime, aeronautical or coast stations reading this piece who is interested in joining us please get in touch. The Association has two get-togethers each year; the meeting in April incorporates the AGM and the September one is purely social. They are usually held in a hotel over three days in an appropriate city with a maritime connection.

We have two CW nets each week, the main one being on Thursday evening at 19.00 local on 3565kHz run by net controller Mike M6MPC and consists of weather reports, plain language and code groups, followed by a series of questions related to our time at sea. The Tuesday net is at 10.00 local on 3538kHz and is just a general natter net.

RSGB and other contests. Maritime Radio Day in April is a popular event that allows contact with ex-R/Os and Coast station Officers from around the world.

The Association also has a website: *www.radioofficers.com* with a wealth of information and pictures showing the ships and radio rooms of our time at sea as well as details of the nets in the ROARS section.

Membership Secretary - John Chalmers *johnnchalmers994@outlook.com*

Chairman ROARS - Peter Gavin M0URL *petertgavin@gmail.com*

Radio Security Services Memorial ARS

We are a virtual RSGB affiliated amateur radio club, callsign GI5TKA. Membership is open to any licensed radio amateur, or to anyone with an interest in amateur radio. As a virtual club we do not plan to introduce membership fees, but prefer to operate on a volunteer basis. Our activities will include three or four lecture talks per annum, which can be accessed live via Zoom. Occasionally on air activities will be arranged on the amateur radio frequencies and announced via this web site. We also organise an annual field day event to mark the role of the Radio Security Service. George Busby, author of 'Spies at Gilnahirk' is Patron of the Society.

For more information:

mi0wwb@btinternet.com or https://rssmars.com

Royal Air Force Amateur Radio Society

(RAFARS – G8FC, G8RAF, G3RAF, G4RAF, G6RAF, G7RAF)

Formed in 1938, RAFARS, now with over 600 members worldwide, is an international society that aims to promote amateur radio activities within the Royal Air Force.

Through amateur radio it maintains and fosters the existing close bonds between radio amateurs still serving and those who have retired from, or have close associations with, the Royal Air Force, Commonwealth, NATO or Allied Air Forces. Additionally, all radio amateurs and short wave listeners are welcome to join as Associate members if they have an interest in military aviation and the RAF, and are willing to assist the Society in achieving its published aims.

The society runs its own QSL bureau and publishes its own call book, QRZ. An in-house Journal 'QRV', is published twice a year.

A monthly Newsletter will be found on our website. The Newsletter can also be emailed to subscribers.

The RAFARS website (*www.rafars.org*) contains information about the Society, including an abridged membership list, an application form to join, net schedules, monthly newsletters, details of awards and competitions Prospective members are welcome to join one of our daily or weekly nets.

Further information is available from:

The Secretary, G4DQP, HQ RAFARS,
RAF Cosford, Wolverhampton WV7 3EX.

Tel: 01902 372722 (answerphone).

email: *rafars.secretary@gmail.com*

Royal Naval Amateur Radio Society

(RNARS – GB3RN, GB0SUB, G3CRS, G1BZU, G3BZU, G7DOL)

The RNARS was formed in 1960 to promote amateur radio as an aid to technical education within the Royal Naval Service. In 1964 the society was considered to be sufficiently established to invite the Captain of HMS Mercury to become its first president. When HMS Mercury closed down the Signal School was transferred to HMS Collingwood.

The last Commanding Officer of HMS Mercury, Cdre Paul Sutermeister DL RN is our current President. The society HQ is at HMS Collingwood, Fareham and the HQ Shack callsign is GB3RN. Membership is open to all Radio Amateurs with an interest in maritime affairs but particularly serving or veterans from:

Royal Navy, Royal Marines, Women's Royal Naval Service, Royal Naval Reserve, Royal Naval Auxiliary Service, Merchant Service, Sea Cadets, Sea Scouts, Nautical Training Corps, MOD in a civilian capacity, Commonwealth & other Navies.

A Free membership tier is available for those 25 and under.

We publish a newsletter four times a year, and radio nets are run on 2m, HF and Digital Modes. A number of awards are also avail-

able. Please see our Website for news and information:

www.rnars.org.uk

Contact the Secretary: Leading Seaman Martin Longbottom, M0EHL. *secretary@rnars.uk*

Membership Secretary: Joe Kirk, G3ZDF. *g3zdf@btinternet.com*

RNARS, Building 512, HMS COLLINGWOOD, Newgate Lane, Fareham, Hants, PO14 1AS. 01329 717627 (24hr answerphone)

RSGB Contest Club

The Contest Club is for any RSGB members who are interested in taking part in amateur radio contests but is particularly aimed at those who are not already members of a radio club or group that encourages contesting. Membership is open to all RSGB Members.

Contest Club members who are not members of another participating Affiliated Society are welcome to enter any of the relevant contest series as representatives of the Contest Club – which is in the General Category of clubs. To join the Contest Club, you do not need any previous experience of contesting and you will not be obliged to enter any of the Affiliated Society events on behalf of the club

The Contest Club is the custodian of five historic RSGB call signs: G6XX, G6ZZ, G5WS, G5AT and G3DR. All five are held as club callsigns by the RSGB Contest Club. Members of the Contest Club who are full licensees may use these historic callsigns, or their regional variants, on behalf of RSGB, by prior arrangement; see "How to Join" below.

QSL details for all call signs and their regional variants are available on QRZ.com. The logs are regularly uploaded to LoTW and Clublog.

The Contest Club takes an active role in organising Radio Marathons on behalf of RSGB. These are International Events that encourage QSO activity, particularly on the HF bands.

If you would like to be involved in the RSGB HQ station in future years please get in touch using the contact details below. This is a great opportunity to develop operating expertise in a team context.

The club is affiliated to the RSGB as well as being organised from within the RSGB contest committees. The Club committee consists of the Chair of the RSGB HF Contest Committee, the Chair of the RSGB VHF Contest Committee and the Chair of the RSGB Contest Support Committee.

Membership is open to any RSGB Member and there are no joining fees or subscriptions.

If you are an RSGB member and would like to join the RSGB Contest Club please send an email with your RSGB membership number to: *contestclub@rsgbcc.org*

Royal Signals Amateur Radio Society *(RSARS – G4RS)*

Formed in 1961 under the chairmanship of the late Major General Eric Cole CB CBE, G2EC, membership is open to: serving and past members of the British Regular and Territorial Army, Cadet Forces, civilian staff who have supported army telecommunications, Commonwealth Army signallers and licensed amateurs from other countries who have proven military connections, subject to status.

Members receive a high quality magazine, *Mercury*, three times a year, and the society runs its own contests and awards scheme. More information from: The Membership Secretary, (G3VBE), RSARS, 65 Montgomery Street, Hove, Sussex BN3 5BE.

Tel: **01273 703680**.

email: *RSARSMemSec@virginmedia.com*
website: *www.rsars.org.uk*

RAYNET-UK

RAYNET-UK is the UK's only national voluntary communications service provided for the community by licensed radio amateurs and other supporting volunteers.

RAYNET was formed in 1953 following the severe East coast flooding. It provides a way of organising the valuable resource that Amateur Radio is able to provide to the community.

Since then, it has grown into a very active organisation with around 2000 members, providing communication assistance to many hundreds of events each year.

RAYNET-UK is a company limited by guarantee in England (2771954). Charity registered in England and Wales (1047725) and in Scotland (SC046184).

Further details can be found on the Internet at: *www.raynet-uk.net*

Radio Security Service Memorial ARS

The patron of the Society is George Busby, author of 'Spies at Gilnahirk', the specific aims of the Society shall be to preserve and memory of Voluntary Interceptors (VIs) and staff employed in the RSS, to educate others in the amateur radio community of the role they played in defence of King and Country during the Second World War and to pass on their legacy to the next generation.

Given our very specific remit we are keen to research the background of radio amateurs who served as VIs and to endure that their contribution is preserved. Membership is open to all who share this specific interest.

The RSSM ARS is RSGB Affiliated and it operates as a virtual nation-wide club, with quarterly Zoom meetings at which there will be a special speaker. Annually each May the club activates a special event station GB0GLS at the site of a former Y Station at Gilnahirk. Inquiries about the Society can be made to: *mi0wwb@btinternet.com.*

https://rssmars.com

Scarlett Point ARC, GT0SP

The Scarlett Point Amateur Radio Society seeks and promotes members from across the UK, currently having members in, GM, GD, GI, G and GW to not only operate radio in a portable settings throughout the UK but to keep the watch tower located at Scarlett Point, Isle of Man on air as part of the clubs operations. The club has a station there for visiting amateurs to operate (within CEPT regulations etc) including members ability to operate. The station is funded and maintained solely by the club and visitors and new members are welcome.

Scarlett Point Tower, Castletown, Isle of Man

Please contact the secretary Billy GM6DX at *gm6dx@outlook.com* or email the club at *gt0sp@outlook.com* for a member to reply.

Suffolk RED

Suffolk RED is a collaboration event for amateur radio and electronics enthusiasts in Region 12.

RED stands for Radio and Electronics Development, and the events are aimed at anyone with an interest in any aspect of electronics and radio engineering and operation.

Whether you are just starting out in the hobby, taking your foundation licence or long established and looking for something new to try.

Experience not required just a passion to learn! Suffolk RED is every other month commencing January of each year.

On average each event has 50 people in attendance, up to six live demo tables and at least one construction table where people can take away something they have built during the evening.

Demos range from basic advice for absolute beginners, handhelds or rigs which is best for getting started on air, know your HF and UHF, direction finding demos, SSTV, FT8 and Pi, through to setting up your own kit for expeditions and field days and so much more.

Constructions are kept to the basic antennas, aimed primarily at portable operation or stations set up with very little space – slim-jims, flower pots, halos and the like.

RED is not a traditional club, nor is it a virtual club or special interest/contest group. It is an event. Formality is kept to a minimum – all organisers are volunteers, and no club or individual has a majority control in RED.

To clarify for those new and existing in the hobby: Suffolk RED does have a membership consisting of just four members. This is so legal formalities such as insurance and bank accounts can be held, these members are the elected officials: Chairman, Deputy, Secretary and Treasurer.

These four members are the RED organisers, who work collaboratively with other clubs in order to populate the event, not organise the event.

Those who participate at a RED event are formally classed as attendees not members.

There is no entry charge, and refreshments plus a 'chat' area are always available. To cover the costs of hall rental, however, we do appreciate you making a donation for your tea and biscuits!

Find out more via: *suffolkred.co.uk*

email: *redadmin@suffolkred.co.uk*

or follow RED on social media: Facebook.com/SuffolkRED and Twitter.com #SuffolkRED

The UK Microwave Group *(UKuG) GX3EEZ*

The UK Microwave Group acts as the focal point for radio amateurs and SWLs interested in frequencies above 1GHz. This includes optical communication - "nanowaves".

The primary aim of the Group is to develop and promote all forms of amateur interest in the microwave spectrum. This is not limited to traditional amateur radio activities, members

are involved in a wide range of interests ranging from casual hilltopping, amateur television, data networking, propagation research, tropo and EME dxing, to monitoring deep space probes and radio astronomy.

All levels of UK licence allow operation on some, or all, of our microwave allocations. Microwaving, in all its forms, can be a fulfilling amateur radio activity for newcomers and experienced radio amateurs alike.

As a member of the RSGB Spectrum Forum, the Group works closely with the Society on band planning and licensing issues.

To focus activity, the Group organises Microwave Contests, and runs an Award Scheme which recognises all levels of achievement. It encourages microwave meetings (Roundtables) a mixture of flea market, social event, equipment test facility, and lecture stream.

A regionally based Technical Support Team is available, in which experienced members spread across the UK, with excellent test gear, provide facilities to those either beginning their exploration of the microwave spectrum, or wishing to improve existing equipment.

The Group's e-newsletter, 'Scatterpoint' has a worldwide reputation for providing excellent practical articles on microwave techniques, construction and operation, as well as news of recent activities. 'Scatterpoint' is accessible to all: contributors represent a cross section of the amateur microwave community, and the magazine reflects that.

The *www.beaconspot.uk* website is a unique facility supported and funded by UKuG. This aggregates 'spots' on the DXCluster, and inputs from other sources to provide a remarkably complete and up-to-date resource of information on beacons on the bands above 30MHz.

Beacons are an important feature of microwave operation, and the Group offers financial support on a 'payment by results' basis to individuals and groups building beacons.

The has also published designs for 2.4GHz and 10GHz equipment for using the QO-100 satellite.

Building equipment is aided by the Group's 'Chipbank' service which provides a means of obtaining some specialised components at zero cost. Membership of the Group is open to all those interested in amateur microwaves or a remarkably small subscription – currently £6.00 annually. For members under 21 years, there is no subscription.

For further information, please visit the Group's website: *www.microwavers.org* or contact our secretary – John Quarmby G3XDY 12 Chestnut Close Rushmere St Andrew Ipswich IP5 1ED Tel 01473 717830 or email: *secretary@microwavers.org*

You can also follow us on Twitter @UKGHz and on the YouTube channel UKMicrowave- Group

UK Six Metre Group *(G5KW)*

The UKSMG was formed in 1982 with the primary aim of promoting 50MHz activity by amateurs worldwide. Today the group is the largest organisation in the world dedicated to 50MHz.

Through its quarterly journal Six News, the group provides comprehensive information on all aspects of the band, including DX and beacon news, propagation, awards, contests, technical articles, equipment reviews, QSL information and DXpedition reports. For further information on "Six News" and to submit items for publication, contact John Rivers, G0GCQ (email: *editor@uksmg.org*).

A major objective of the group is to actively support 50MHz operation from new countries and from DX locations, as well as promoting the establishment of beacons in various parts of the world. Further information on sponsorship can be obtained from Trevor Day, EA5ISZ (email:*sponsorship@uksmg.org*).

The UKSMG website at *www.uksmg.org* carries a wealth of information about the 50MHz band and is guaranteed to provide interesting browsing for newcomers and old hands alike.

Further information about the group and membership details can be obtained from the website or from David Bondy, G4NRT. 19 Harriet Drive, Rochester, ME1 1DY (email: *secretary@uksmg.org*)

Vintage & Military Amateur Radio Society *(VMARS - M0VMW)*

The VMARS is an international society based in the UK with nearly 400 members. Its main aims are to restore and preserve historic communication and electronic equipment, from military, commercial and amateur sources, encourage and enable the use of historic equipment on the amateur bands, encourage the use of historic modes such as AM on the amateur bands and encourage research into radio history.

These aims are realised through the very wide range of VMARS members' interests. Whereas many are engaged in the operation of "vintage" military and amateur equipment on the air (19-sets, T1154/R1155, T1509, 52-sets, TCS TX/RX, LG300, Heathkit, KW equipment, RA17, AR88 etc) with many a rare item in evidence, and much of which is lovingly restored, there is a very strong interest in military man-packs, military wireless vehicles, spy-sets, radar, history and museums and not least in amateur radio construction of valved equipment. The Society runs a weekly AM net on 80m at 08.30 local time on Saturdays on 3615kHz ± QRM. All suitably licensed amateurs (whether VMARS members or not) are welcome. During the week, members may be found around 3615kHz at almost any time. There is also SSB activity, on Wednesdays, around 3615kHz and starting at 20.00 local time, primarily for those with ex-military USB-only equipment and on Fridays, a general SSB net, again around 3615kHz and starting at 19.30 local time.

The Society's website includes a virtual library of manuals for vintage equipment, many of which can be downloaded at no charge. This library is supplemented by many paper documents, for which copies are available to members. Further details may be obtained from the website or the Hon. Secretary, Peter Chadwick, G3RZP, Three Oaks, Braydon, Swindon, SN5 0AD. tel: +44 (0) 1666 860423 email: g8on@btinternet.com or honsec@vmars.org.uk Membership information may be obtained from the Membership Secretary, Ron Swinburne M0WSN, 32 Hollywell Road, Sheldon, Birmingham B26 3BX +44(0)1217 421808 *memsec@vmars.org.uk*

Hon Secretary is Peter Chadwick, G3RZP, Three Oaks, Braydon, Swindon, SN5 0AD.

email: *honsec@vmars.org.uk*

website: *www.vmars.org.uk.*

Worked All Britain Awards Group *(WAB – G3ABG, G4WAB, G7WAB)*

The group was founded in 1969 by the late John Morris, G3ABG, to encourage greater amateur radio interest in Britain. The group promotes an award programme and on air activity including contests and nets and makes regular donations to organisations such as the RAIBC who help the less fortunate members of the amateur radio community and other national charities.

The award scheme, which is open to licensed amateurs and SWLs, is based on the OS / OSNI grid squares of the UK and UK Crown Dependencies. QSL cards are not required, only log entries, and special tracker spreadsheets are available to assist in the claiming of awards. Full details and checklists of all the squares and WAB awards are contained on the WAB website: *www.worked*-all-britain.org.uk and in the WAB Book/USB Key. For further details please email the membership secretary: Dave Dolling at: *g0fvh@worked*-all-britain.org.uk

World Association of Christian Radio Amateurs & Listeners *(WACRAL – M1CRA)*

Dedicated to 'Friendship and Christian Fellowship among Radio Amateurs Worldwide', membership is open to committed Christians of all denominations and all nationalities.

Founded by the late Rev. Arthur Shepherd, G3NGF, members are encouraged to follow his example in spreading the 'Good News', fostering international relations, supporting disadvantaged amateurs and SWLs and providing a good role model for the new generations.

WACRAL publish a newsletter four times a year, featuring the news and activities of members around the globe. A members' QSL Bureau is provided and a variety of awards are available.

An annual activity period is open to all on HF and VHF, including CHOTA (Churches and Chapels On The Air) and conferences take place each year.

The popular WACRAL 'Good News' Nets meet at 07:30 UK local time every Sunday and Wednesday morning on 3.747kHz SSB and a schedule of national and international Christian nets is promoted on various bands on Saturday afternoons – all subject to propagation.

You are invited to go to their website at: *www.wacral.org* for full details of the Nets, and the latest news and details of membership, or contact the Membership Secretary:

Richard Paul G7KMZ
07519 508147

RSGB Club Affiliation

Many amateur radio societies and clubs choose to affiliate with the Radio Society of Great Britain because they see it as an effective way of demonstrating their support for the aims and aspirations of the National Society. The Society welcomes this support because it can only strengthen its claim to speak on behalf of amateurs and amateur radio.

The Society recognises that much of the vitality of amateur radio lies within clubs and wishes to encourage all clubs, societies and groups to join us to advance amateur radio. There are tangible benefits for the affiliated society. These include:

- Publicity for club activities through 'club news' in RadCom, via broadcasts on the Society's news service GB2RS, and on the RSGB website..
- Full facilities of the RSGB QSL Bureau for cards bearing the club station call.
- Purchase of publications at a discount with the RSGB.
- Receipt of the *RSGB Yearbook*.
- Freedom to participate in RSGB Affiliated Societies' contests.
- Third party liability insurance up to £10m whilst taking part in club events such as field days.
- Freedom to borrow RSGB DVDs, tapes and display materials. (This facility is also available to certain non-affiliated groups such as schools).

How to affiliate your club, group or society to the RSGB

Any UK club, group, society or emergency communication group may affiliate to the RSGB, provided it fulfills just a few requirements. The affiliation fee is currently £51.00 (with a £4 discount for those who pay by Direct Debit) and includes receipt of monthly email newsletters.

Procedure

(i) Please contact RSGB HQ or visit the website for an Affiliated Club membership application form. If your organisation has a callsign, please let us know on the application form. If it does not, we will issue a receiving station number for reference purposes.

Note that for UK clubs, groups and societies the RSGB region and RSGB district will be determined by the address given on the application form. Clubs, groups and societies near to, or spanning county boundaries should decide carefully with which county they wish to be associated and insert the appropriate choice in the address of the club they are entering on the application form.

Please also note that once your club address is on our files, we will regard it as information that can be freely given out to those seeking to contact clubs.

(ii) Please send to the appropriate Regional Manager the following:

- Your completed application form signed by the chairman or secretary
- A copy of the club's constitution or rules. There is no prescribed form of constitution but an example is provided for guidance if required.
- A list of current officers of the club
- A statement of the number of members, and the proportion who are RSGB members.

A list of Regional Managers is published on the RSGB website and in the RSGB Yearbook. The Regional Manager will vet your constitution/rules and if suitable will countersign your application form. He or she will then return the form and your constitution or rules to you.

Note that only the Regional Manager may countersign an application for a club, group or society. Overseas organisations should send their application form and constitution/rules direct to RSGB HQ addressed to 'The RSGB Secretary'.

(iii) Finally, please send your countersigned application form, constitution and remittance to:

Membership Secretary, 3 Abbey Court, Fraser Road, Priory Business Park, Bedford, MK44 3WH.

Notes on Example Constitution (overleaf)

(1) It is recommended that the words 'amateur radio' appear in its title.

(2) It is useful to specify which groups have voting rights and whether reduced subscriptions apply, particularly for students in full time education.

(3) Alternatively, the subscription may be recommended by the Committee for ratification at the AGM.

(4) This period perhaps should not exceed one to three years to avoid placing an undue burden on future Committees

(5) There are great advantages in running the Club's finances on a strict basis, although a less formal arrangement may still be effective.

(6) There are two methods for electing the Committee: the more common is for the meeting to elect the Committee members and for the latter in turn to elect the officers from within the Committee; alternatively, the members may elect individuals to specific offices. The method adopted will need to be specified.

(7) The number of Ordinary Committee members should be related to the size of the Club. Remember that being a committee member is an essential part of the training of the future officers of the Club.

(8) These can replace elected Committee members who have left the Committee

(9) These can be people who need to be familiar within the work of the Committee such as the editor of the Club magazine or the press officer.

(10) This can be expressed either as a fixed number or, for example, as at least half or two-thirds of the full membership of the committee.

(11) This can be set either as a fixed number or a fixed percentage of the membership (state which members are to be included), or both "whichever is the smaller/greater". It is probably safer to make the numbers on the small side so as to ensure that the meeting can take place.

(12) Such as, among its members, to a charity, or to a club of similar interest.

Downloadable application form: https://rsgb.org/main/clubs/club-affiliation/

Example Constitution for Affiliated Clubs

Guidance intended for those writing a constitution for their local club or society which will be acceptable for RSGB affiliation.

1. Name

The Club (1) shall be known as the ....

2. Aims

The aims of the Club shall be to further the interests of its members in aspects of amateur radio and directly associated activities.

3. Membership

Membership shall be open, subject to the discretion of the Committee, to all persons interested in the aims of the Club

(a) **Full members** Full members must be 16 years of age or over.

(b) **Honorary members** Honorary Life Membership may be granted to any person, who, in the opinion of the Committee, has rendered outstanding service to the Club, either directly or indirectly. Such membership shall carry the rights of full membership but shall be free from subscriptions.

(c) **Guests** Members may invite guests to meetings. No visitor may attend more than three meetings in each year.

All members shall abide by the constitution of the Club. The Committee shall have power to expel any member whose conduct, in the opinion of at least three-quarters of the full Committee, renders that person unfit to be a member of the Club. No Member shall be expelled without first having been given an opportunity to appear before the Committee.

4. Subscriptions

(a) The annual subscriptions for membership shall be set by the Committee (3).

(b) All subscriptions shall be due and payable at the beginning of the financial year. Members in arrears have no voting rights.

(c) The financial year shall be determined by the Committee

(d) A member shall be deemed to have resigned from the Club, if, by the end of the financial year, the subscription has not been paid.

(e) The Committee shall have the power to waive or reduce subscriptions in special circumstances for a period not exceeding...years at a time (4).

5. Finance

All money received by the club shall be promptly deposited in the Club's bank account. Withdrawals require the signature of the Club's Treasurer and one other nominated officer of the Club (5).

6. Membership of the Club's Committee

The Club's affairs shall be administered by a Committee elected at the Annual General Meeting (6). The Committee, in whom the Club's property shall be vested, shall consist of:

(a) A Chair who will preside at all meetings at which he is present.

(b) A Vice-Chair who will act as Chair in the absence of the Chair.

(c) A Club Secretary who will be responsible for:

- (i) keeping the minutes of all meetings of the Club.
- (ii) ensuring that all correspondence is correctly handled.
- (iii) maintaining the definitive register of members and honorary members.
- (iv) maintaining a register of Club equipment.

(d) A Treasurer, who will be responsible for:

- (i) keeping the Club's accounts.
- (ii) advising the Committee on all financial matters.
- (iii) preparing the accounts for audit and presenting them at the AGM.

(e) ......Ordinary Committee Members (8).

(f) Not more than......co-opted members who have full voting powers (8), and not more than......who are not permitted to vote (9).

7. Committee Standing Orders

(a) The quorum for the Committee shall be..... (10).

In the absence of a quorum, business may be dealt with but any decisions taken only become valid after ratification at the next meeting at which a quorum exists.

(b) Committee meetings may be called by the Chair, the Secretary or any vote.

8. Annual General Meeting

(a) The Annual General Meeting shall normally be held at the beginning of each financial year. At least 21 days notice shall be given to each member in writing.

(b) The quorum for the meeting shall be...... (11).

(c) The agenda for the meeting shall be:

- (i) Apologies for absence
- (ii) Minutes of the previous AGM
- (iii) Chair's report
- (iv) Club Secretary's report
- (v) Treasurer's report
- (vi) Election of the new Committee
- (vii) Election of auditors
- (viii) Other business

(d) Items (i) to (v) shall be chaired by the outgoing Chair, item (vi) by an acting Chair who is not standing for election to office, and the remaining business by the newly elected Chair.

(e) Nominations for Committee members will only be valid if confirmed by the nominee at the meeting or previously in writing.

(f) Items to be raised by members under other business must be notified to the Club Secretary not less than 21 days before the AGM.

9. Extraordinary General Meeting

(a) Extraordinary General Meetings may be called by the Committee or not less than......members of the Club, the date of the meeting being the earliest convenient as decided by the Committee. At least 28 days notice in writing must be given to the Secretary, who in turn shall give members at least 14 days notice in writing of the agenda. No other business may be transacted at the EGM.

(b) The quorum for the EGM shall be......(11).

10. Amendments to the Constitution

The constitution may be amended only at an EGM called for that purpose.

11. Winding-up of the Society

(a) The decision to wind up the Club may be taken only at an EGM.

(b) The funds of the Club shall, after the sale of all assets and the payment of all outstanding debts, be disposed of as directed by members at the final EGM (12).

* Club is used to denote any club or similar organisation wishing to apply for affiliation

Downloadable Model Constitution: www.rsgb.org/main/clubs/club-affiliation

Structure and Representation

For administrative purposes the Society divides itself into regions, each comprising four to six districts. Each region has an elected Regional Representatives (RR) and each district has an appointed District Representatives (DR). The map below identifies the RSGB Regions. On the following pages, listed by Region, you will find details of the relevant Regional Representatives (RR) and the District Representatives (DR) along with contact information on local clubs and societies, examination centres and Emergency Comms Groups.

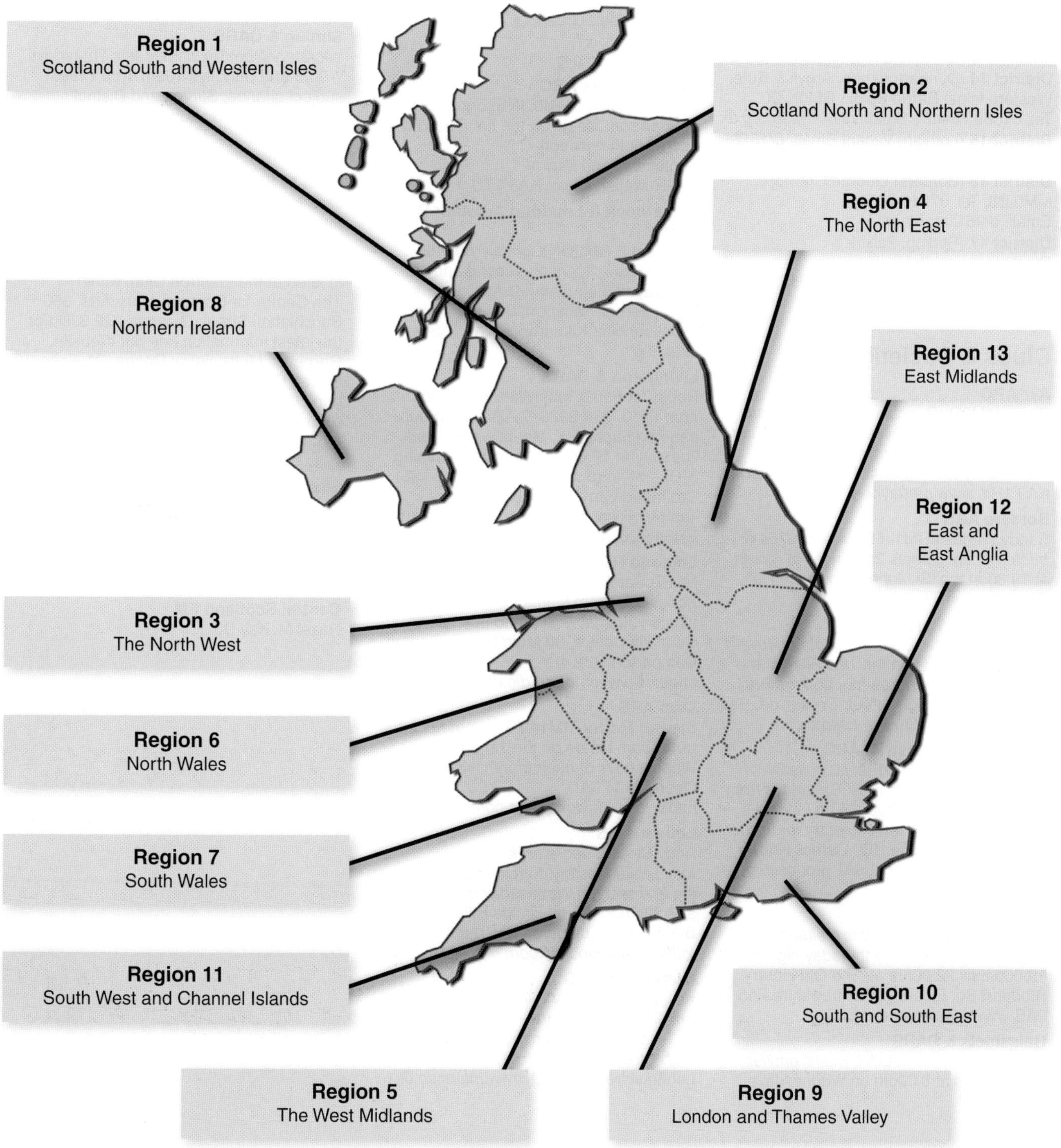

Region 1
Scotland South and Western Isles

Region 2
Scotland North and Northern Isles

Region 4
The North East

Region 8
Northern Ireland

Region 13
East Midlands

Region 12
East and
East Anglia

Region 3
The North West

Region 6
North Wales

Region 7
South Wales

Region 11
South West and Channel Islands

Region 10
South and South East

Region 5
The West Midlands

Region 9
London and Thames Valley

Region 1 - Scotland South & Western Isles

(http://rsgb.org/region1)

RR

Tony Miles, MM0TMZ
Tel: 07702 134188
Email: *rr1@rsgb.org.uk*

DRs

District 11 (Renfrewshire and Inverclyde) William McCue, MM0ELF Tel: 01389 755758 Email: *dr11@rsgb.org.uk*
District 12 (Glasgow, Central and Lanarkshire) Harry Mcdonald, GM1VFR Tel: 07742 059257 Email: *dr12@rsgb.org.uk*
District 13 (Ayrshire, Dumfries & Galloway) Vacant – Email: *dr13@rsgb.org.uk*
District 14 (Dunbartonshire, Argyll & Bute, Western Isles) Barrie Spink, GM0KZX Tel: 01389 764401 Email: *dr14@rsgb.org.uk*
District 15 (Lothian) Vacant Email: *dr15@rsgb.org.uk*
District 16 (Borders) George Crawford, MM0JNL Tel: 07946 753985 Email: *dr16@rsgb.org.uk*
District 17 (Stirling, Falkirk & Clackmannanshire) Nathan Nuttall, 2M0OCC Tel: 07952 819626 Email: *dr17@rsgb.org.uk*

Clubs & Societies

Ayr ARG
Derek MM0OVD, *derek.secaarg@gmail.com* 07482 994614 Meets 7.15pm-9:30pm on alternate Wednesdays Prestwick Community Hall, Briarhill Road, Prestwick, KA9 1HY *www.gm0ayr.org*

Borders ARS
Sandy Weddell GM1JFF *a.weddell764@btinternet.com* Meets 2nd Friday in month at 19.30 at the, St. John Ambulance Hall, Berwick-upon-Tweed, TD15 1NG

Cockenzie & Port Set
Mr Bob Glasgow, GM4UYZ. *bob.gm4uyz@talktalk.net* Meets on the 1st Friday of the month at the, Thorntree Inn, Lounge Bar, Old Cockenzie High Street, Cockenzie, East Lothian EH32 0DQ *www.cpsarc.com*

Edinburgh and District ARC
Secretary, 07740 946192; *jacobite@btinternet.com* Meets most Saturdays from 7pm. Corstorphine Club net on Mondays from 8pm. Contact the secretary for further details, Corstorphine, 160 Carrick Knowe Drive, Edinburgh, EH12 8AT groups.io/g/gs4ham

Elderslie ARS
John French, 2MOJSF *info@elderslie-*ars.org.uk Meets every thursday night at 19:00pm till 22:00pm at, The Old Library, Stoddart Sq, Elserslie, Renfrewshire PA5 9AS *www.elderslie-*ars.org.uk

Galashiels & DARS
Mr Jim Keddie, GM7LUN, *mail@gm7lun.co.uk* Meets at 8.00pm on Wednesdays at the, Focus Centre, Livingston Place, Galashiels, Scottish Borders TD1 1DQ *www.galaradioclub.co.uk*

Glasgow & Clyde Raynet
Paul Lucas, MM3DDQ Tel: 01389 499972, *mm3ddq@yahoo.co.uk*, Meets 1st Tuesday of each month at 7.30pm, Braehead Shopping Centre, Glasgow, G51 4BN

Greenock and District Scouts ARC
mm0tsg@gmail.com Meets every Friday during school term time 7-10pm, Greenock and District Scout Headquarters, 159A Finnart Street, Greenock, Inverclyde PA16 8HZ
https://mm0tsg.org

Jaggy Thistles ARC
Charles Stewart MM0GNS, *cgnstewart@hotmail.com* Meets on the first Tuesday of each month between
September and May at
, 46 Rowallan Drive, KA3 1TU

Kilmarnock & Loudoun Amateur Radio Club
Len Paget, GM0ONX *gm0onx@gmail.com* Meets 7.30pm on second and fourth Tuesday of the month, KLARC Clubhouse, EAC Internal Transport, 34a Main Street, Crookedholm, Kilmarnock KA3 6JS *www.klarc.org*

Livingston & DARS
Robin Tannahill: robin*_tannahill@hotmail.com* Tel: 07880 992377 Also see group page on Facebook, Meets weekly on a Tuesday at 19.00, for Meetings, Training & Operating (disabled access available)., Crofthead Farm Community Centre, Templar Rise, Livingston, West Lothian EH54 6DG https://*www.ladars.org.uk*

Lomond RC
Mr BP Spink,GM0KZX,*gm0kzx@googlemail.com* Meets for weekly club nights Thursday 7.30pm to 9pm, John Connolly Centre, 30 Main Street, Renton, Dumbarton G82 4LY
https://*www.lomondradio.club*

Lorn ARS
Stewart McIver MM1AVR, *lornradioclub@gmail.com* Meets on the 1st and 3rd Wednesdays of each month, Kelvin Hotel, Oban, PA34 5AD
https://gm0lra.wordpress.com/

Lothians RS
Andy Sinclair *secretary@lothianradiosociety.com* meets normally on the 2nd and 4th Wednesday of the month at 7.30pm for 8pm., Braids Hills Hotel, 134 Braid Road, Edinburgh EH10 6JD *www.lothiansradiosociety.com*

Mid Lanark ARS
Kevin Mair, 2M0KVM Chair *mlarsclub@gmail.com* Meets every Friday 18.30 - 21.30, Newarthill Community Ed. Cent., High Street, Newarthill, Motherwell, Lanarkshire ML1 5JU *www.mlars.co.uk*

Na Fir Chlis ARC
Angus MacLeod, *mm0nfc@hotmail.com* Meets The club meets on an irregular basis., Heathbank Hotel, Northbay, Isle of Barra, Outer Hebrides HS9 5YQ
https://*www.ms0nfc.yolasite.com*

Paisley ARC
Stuart McKinnon, MM0PAZ *mm0paz@gmail.com* Meets every Thursday at 19:30, Trinity Room of the Paisley Methodist Church Halls, 2 Gauze Street, Paisley, PA1 1EP
http://gm0pym.wix.com/paisleyarc

Stirling & DARS
secretary@gm6nx.com Meets Thursdays 7-9pm and Sundays 10am-1pm Times subject change, Stirling and District Amateur Radio Society Unit 6, Bandeath Industrial Estate, Throsk, FK7 7XY
https://*www.gm6nx.com*

West of Scotland [Glasgow] ARS
Jack Hood, GM4COX, *info@wosars.club* Meets Fridays 8-10pm. Also technical support and licence guidance & support on Wednesday nights as SOLDER GROUP, located at the Electron Club, within The Centre for Contemporary Arts, 350 Sauchiehall Street, Glasgow, G2 3JD For the latest information see our Website, Garnethill Multicultural Centre, 21 Rose Street, Glasgow, Scotland G3 6RE
https://*info@wosars.club*

Wigtownshire ARC
Bob Bower, GM4DLG: *gm4dlg@gmail.com* Meets 7.30pm on Thursdays at, Aird Unit, Stranraer Academy, Stranraer, (entrance from Cairnport Road), DG9 8BY
www.gm4riv.org

Repeater Groups

Central Scotland FM
Hazel McKay D3

Region 2 - Scotland North & Northern Isles

(http://rsgb.org/region2)

RR

Jim Campbell, MM7BIW
Tel: 07749 520320
Email: *rr2@rsgb.org.uk*

DRs

District 21 (Highlands) Vacant
Email: *dr21@rsgb.org.uk*
District 22 (Aberdeenshire, Moray)
Peter Thomson, GM1XEA, 01224 740091
Email: *dr22@rsgb.org.uk*
District 23 (Angus, Fife. Perth & Kinross)
Martin Krawczyk, 2M0KAU Tel: 07763 708933 Email: *dr23@rsgb.org.uk*
District 24 (Orkney) David Wishart, MM5DWW Tel: 01856 721422
Email: *dr24@rsgb.org.uk*
District 25 (Shetland) Peter Bruce, GM0CXQ Tel: 01595 880241
Email: *dr25@rsgb.org.uk*

Clubs & Societies

Aberdeen ARS
Fred Gordon GM3ALZ fred*_gordon@btinternet.com* Meets 19.30 - 21.30 on Thursdays at, 25th Scout Group, Oakhill Crescent Lane, Aberdeen, AB15 5HY *www.aars.org.uk*

Caithness ARS
Alistair Ross (2M0WTN). *ms0fnr@outlook.com* Meets 1st & 3rd Thursday of each month 19:30 to 21:30, Craigwell Farm, Skirza, Caithness, KW1 4XX *www.radioclubs.net/c.a.r.s./*

Dundee ARC
Jim Wilson *secretary@dundee*-amateur-radio.co.uk Meets every Tuesday throughout the year 19.00 to 21.00 during term time, Dundee and Angus College, Old Glamis Road, Dundee, DD3 8LE *www.dundee*-amateur-radio.co.uk

Glenrothes & DARC
Ron Murray 2M0DSU, *murrayphone@aol.com* Meets in New Football Pavilion Station Road, Thornton, Fife, KY1 4AX *www.gdarc.org.uk*

Inverness DARC
Adrian Hart MM0DHY *dhyclimber@yahoo.co.uk* Meets 2nd and 4th Wednesday at 7.30pm-9.30pm, DHM Blacksmiths, 7 Carsegate Road, Inverness, IV3 8EX

Montrose Air Station Heritage Centre
Ewan Cameron, MM0BIX, *rafmontrose@aol.com* Meets Sundays at 12.00pm at, Montrose Air Museum, Waldron Road, Broomfield, Montrose, Angus DD10 9BB *www.rafmontrose.org.uk*

Moray Firth ARS
Paul Furness *MFARS.secretary@gmail.com* Meets on first and second Tuesday in every month at 7.30pm, Reserve Forces Centre, Edgar Rd, Elgin, IV30 6YQ *www.mfars.co.uk*

Museum of Communication ARS
Ken Horne GM3YBQ Meets Wednesday and Saturday at 11am - 4am at, 131 High Street, Burntisland, Fife, KY3 9AA

Orkney ARC
Mr David Wishart MM5DWW, *oarc@live.co.uk* Meets on the first Wednesday of each month at the, Sea Cadets HQ, TS Vanguard, Scapa Beach, Kirkwall, Orkney Islands KW15 1SD *http://eu009.webplus.net/*

Sutherland & District A.R.C
Frank Dinger GM0CSZ. *sadarc@sutherland*-arc.org.uk Meets evey Friday at 7:00pm at, Dunrobin Farm, Golspie, Highland, KW10 6RH

Repeater Groups

Grampian Repeater Group
gb3gn@btinternet.com The Grampian Repeater Group maintains the 2m Amateur Radio Repeaters GB3GN (Banchory) and GB3NG (Fraserburgh) located in NE of Scotland. We have no fixed meeting place or regular meetings. Our AGM is usually before the end of March each year in Aberdeen. Meeting Information: April, Once per yearInverurie, AB51 4FL https://*www.grampianrepeatergroup.co.uk*

Region 3 - North West

(http://rsgb.org/region3)

RR

Vacant
Email: *rr3@rsgb.org.uk*

DRs

District 31 (Cumbria) Martyn Bell, M0TEB
Tel: 01229 833571
Email: *dr31@rsgb.org.uk*
District 32 (Lancashire) Vacant
Email: *dr32@rsgb.org.uk*
District 33 (Greater Manchester) Vacant
Email: *dr33@rsgb.org.uk*
District 34 (Cheshire) Vacant
Email: *dr34@rsgb.org.uk*
District 35 (Isle of Man) Stuart Hill, GD0OUD Tel: 01624 613226
Email: *dr35@rsgb.org.uk*
District 36 (Merseyside) Vacant
Email: *dr36@rsgb.org.uk*

Clubs & Societies

Ashton In Makerfield ARC
Peter Williams M0RGN, *mx0htr@gmail.com* Meets fortnightly from 20:00 to 22:00, The Diamond Community Centre, Grey Road, Ashton-in-Makerfield, Wigan WN4 9QW https://*www.aimarc.co.uk*

Bolton Wireless Club
Mark Bryant M0UFC, *boltonwireless@gmail.com* Meets at 7:00pm every 2nd and 4th, Monday (Bank Holidays excepted)

, Ladybridge Community Centre, Beaumont Drive, Bolton, BL3 4RZ https://*www.boltonwireless.org.uk*

Bury RS
Mike G4GSY *mail@buryradiosociety.org.uk* Meets every Tuesday at 7.30pm, Hollins Social Club, brook lane, Bury, Gtr Manchester BL9 8BA https://*www.buryradiosociety.org.uk*

Central Lancs ARC
Nonie Sinclair, G3TNN *secretary.clarc@gmail.com* Meets at the clubhouse at Ribble Steam railway, visits by arrangement only. Find us on Facebook. 'Central Lancs Amateur Radio Club And Friends'
, Ribble Steam Railway Musuem, Chain Caul Road, Ashton On Ribble, Preston PR2 2PD https://*www.facebook.com/groups/443941206406534*

Chester & DRS
Philip Hughes G8IPT, *secretary@chesterdars.org.uk* Meets 1st, 3rd, 4th (&5th) Tues of each month except August, The Burley Memorial Hall, Common Lane, Waverton, Chester, or The Waverton Institute, Waverton, Chester CH3 7QN *www.chesterdars.org.uk*

East Lancs Pendle Radio Club
Neil Mooney M0NFI, *info@elrc.uk* Meets Monday evenings 19:00-21:00, Closed Bank

Holidays, Higham Village Hall, Higham Hall Road, Higham, Burnley, Lancashire BB12 9EU *www.eastlancsradioclub.co.uk*

Furness ARS
Martyn Bell M0TEB. *info@fars.org.uk* Meets on the 2nd and 4th Wednesdays 7.30pm for practical evenings, talks and presentations. Social meetings every Monday at the Farmers arms, Newton-in-Furness LA13 0NB. FM Net around 145.325 Fridays at 8pm and some Wednesdays we chat on DMR TG23525. Regular breakfast meetings Wednesdays at the Furness Railway from 9.30am. During the summer we usually have outdoor activities., Hawcoat Park Sports Club, Hawcoat Lane, Barrow-in-Furness LA14 4HF *www.fars.org.uk*

Gilmore Radio Club
Heather Stanley: *info@m0juw.co.uk* Meets 1st Thursday of each month, St Marys church Hall, St Marys Drive, stockport, SK9 3PA https://*www.m0juw.co.uk*

Isle of Man ARS
info@iomars.im Meets every Wednesday from 7pm onwards at the Scout HQ, Mill Road, Ballasalla, IM92EG and on 2nd Tuesday at Sea Scout HQ Tromode, Scout HQ, Mill Road, Ballasalla, Isle of Man IM9 2EG https://iomars.im/

Leyland And District Amateur Radio
Phil 2E0DGP: *ladar@mail.com* Meets 2nd and 4th Wednesdays of the month 7-9pm , Chorley Sea Cadets, Heapy Road, Chorley, PR6 9BQ https://*www.ladar.club*

Macclesfield & DRS
Greg Acton M0TXX, *info@gx4mws.uk* Meets 19.30pm on Mondays, The Pack Horse Bowling Club, Abbey Road, Macclesfield, SK10 3AU *www.gx4mws.com*

Manchester Wireless Society
secretary@g5ms.com Meets on the first Monday at 8pm Cleveland Public House, Wilton Road, Crumpsall, M8 4WQ *www.g5ms.com*

Marine Radio Museum Society (Wallasey
W.H Cross 0151 207 1959 *g0elz@yahoo.com* Meets Tuesday and Friday each week 10am-4pm, Tug France Hayhurst Albert, Dock Harbourmasters Office, Liverpool L3 4AE

Merseyside Amateur Radio Society
Alan Birch G4NXG *g4nxg@btinternet.com* Meets 7.30pm on the 1st and 3rd Monday of the month, Mel Inn, 513 Hawthorne Road, Bootle, Merseyside L20 6JJ https://*www.marsradio.uk*

Mid Cheshire ARS
Peter Paul Fox, G8HAV,*midcars@woollysheep.org* Meets every Wednesday evening 19:30-22:30, Cotebrook Village Hall, Stable Lane, Cotebrook, nr Tarporley, Cheshire (NGR: SJ 571 655) CW6 0JJ *www.midcars.org*

Morecambe Bay ARS
Jon Allen M0IXK, *joninglemere@gmail.com* Meets 8:00pm on every Tuesday, Trimpell Sports & Social Club, Outmoss Lane, Morecambe, Lancs LA4 4UP https://*www.mbars.uk*

Oldham ARC
president@oarc.org.uk Meets Thursdays between 19:30-22:00 hours. Training available on demand., 1855Squadron (Royton)ATC, Park Lane, Royton, Oldham OL2 6RE https://*www.oarc.org.uk*

Quantum Amateur Radio & Technology Society
Alison Hughes M6COV *info@quantumtech.club* Meets 1st & 3rd Thursday of each month at 19.30 for 20.00 to 22.00, Cottage Lane Mission, Ormskirk, Lancashire, L39 3NE *www.quantumtech.club*

Radio Millenium Lodge
Mr N Stackhouse, email: *g1scl@ntlworld.com* Meets the first Friday of Feb, Apr, Jun, Oct & Dec at 6:00pm, 1881 Building, 15 Westbourne Road, Urmston, Manchester, M41 0XQ https://*www.rml9709.org.uk*

Rochdale & DARS
Dave Carden *dave@cardens.me.uk* Meets 2nd and 4th Thursdays of the month from 19:30 until 21:30, Rochdale and District Amateur Radio Society, Crimble Croft Community Centre, Aspinall Street, Heywood OL10 4HW https://*www.g0roc.co.uk*

Sands Amateur Radio Communications Group
Brian Watson, G0RDH, *g0rdh@m0scg.org.uk* Meets every other Monday at 8pm at, The Owls Nest, Bare Lane, Morcambe, LA4 6DD *www.m0scg.org.uk*

South Cheshire Amateur Radio Society

Ronald Carter G7RC *info@g6tw.org.uk* Meets every 2nd & 4th Wednesday of the month at 7.30pm, Wilson House, Ford Lane, Crewe, CW1 3EH *www.g6tw.co.uk*

South Manchester Radio Club
David Crowe G4MVU *contact@smrcc.org.uk* Meets every Thursday. 20.00 to 22.00., The Woodheys Club., 299 Washway Road, Sale, Cheshire M33 4EE *www.smrcc.org.uk*

Southport & DARC
Matt Kerwin 2E0TXO Meets 8.00pm on 3rd Monday in the month at, St Marks Church Hall, Scarisbrick, Lancs, L40 9RE *www.sadarc.org.uk*

Stockport Radio Society
Tony Smithies M0SAV *info@g8srs.co.uk* Meets: 1st & 4th Tuesdays of the month(7pm) @ Walthew House, 3rd Tuesday of the month online (7.15pm), 2nd Sunday of the month (10am) @ Walthew House, Walthew House, 112 Shaw Heath, Stockport, Greater Manchester SK2 6QS https://*www.g8srs.co.uk*

Thornton Cleveleys Amateur Radio Society
Dave Ward G8KBH, *tcars.radio@gmail.com* Meets 8pm every Monday evening Except Bank Holidays, Cleveleys Community Centre and Church, Kensington Road, Cleveleys, Lancashire FY5 1ER https://*www.tcars.club*

Warrington & ARC
Dave Roberts G8KBB *dave.roberts@btinternet.com* Meets Tuesday 8 pm Thursday 10am and most Sundays 1:30pm , Grappenhall Community Centre, Bellhouse Lane, Grappenhall, Warrington, Cheshire WA4 2SG *www.warc.org.uk*

West Manchester Radio Cub
secretary@wmrc.co.uk Meets 8.00pm on Thursdays at the, WMRC (Astley and Tyldesley Miners Welfare Club), Gin Pit Village, Astley, Manchester M29 7DW *www.wmrc.org.uk*

Widnes & Runcorn ARC
Mark Keilty G8AA Meets every other Tuesday, Lostock Sports & Social Club, Works Lane, Northwich, Cheshire CW9 7NW *www.wararc.co.uk*

Wigan-Douglas Valley
D Snape, G4GWG, *daveg4gwg@gmail.com* 01942 211397 Meets every Wednesday at 8.00pm, 13 Walthew Green, Roby Mill, Upholland, WN8 0QT *www.radioclubs.net/wigandvars*

Wirral & DARC
Simon Richards G6XHF: *secretary@wadarc.com* Meets 8.00pm 2nd & 4th Wednesdays of month at, Irby Cricket Club, Mill Hill Road, Wirral, CH61 4XQ *www.wadarc.com*

Wirral ARS
William Davies *secretary.g3nwr@virginmedia.com* Meets on Tuesday & Wednesday evenings 19:00 - 21:00. Open seven days 10:00-14:00 for pre-arranged visitors, 24/7 for members, c/o William Davies, Building 27D Hooton Park, Airfield Way, Hooton, Cheshire CH65 1BQ *www.g3nwr.org.uk*

Wirral Peninsula ARC
wirralpeninsulararc@outlook.com Meets at 19.00 hrs every Wednesday, Unit 18 Peninsula Business Park, Reeds Lane, Moreton, CH46 1DW https://*www.wparc.co.uk*

Workington and District ARC
Peter Webster G8RZ, *pwebster220@gmail.com* Meets 1st & 3rd Wed of each month at 7pm Moorclose Community Centre, Needham Drive, Moorclose, Workington, Cumbria, CA14 3SE *www.mx0wrc.org*

Repeater Groups

UKFM Group Western
Julian Wolvern M0JPW
www.ukfmgw.org.uk

Contest Groups

350 DX Club
Alan Birch G4NXG *g4nxg@btinternet.com*, Litherland Town Hall, Hontin Hill Road, L21

Duddon Contest Team
g3dct@outlook.com VHF, UHF, Microwave Contest club on the Duddon Valley area, LA16 7JG

Tall Trees Contest Group
Mr Bran Gale G3UJE *Bgale111s@googlemail.com*,

Travelling Wave Contest Group
Keith Haywood G8HXE, Skype

Region 4 - North East

(http://rsgb.org/region4)

RR

Ian Bowman, G7ESY Tel: 07710 992961
Email: *rr4@rsgb.org.uk*

DRs

District 41 (Northumberland, Tyne & Wear, Cleveland, Co Durham, Northallerton) Ian Bowman, G7ESY Tel: 07710 992961
Email: *dr41@rsgb.org.uk*
District 42 (East Yorkshire) John Baines, M0JBA Tel: 01482 842430
Email: *dr42@rsgb.org.uk*
District 43 (West Yorkshire) Vince Greatwood, M0VHG Tel: 07740 346777
Email: *dr43@rsgb.org.uk*
District 44 (South Yorkshire, NE Lincolnshire) Chris Drury, G1WSA Tel: 07821 761514
Email: *dr44@rsgb.org.uk*
District 45 (North Yorkshire) Vacant
Email: *dr45@rsgb.org.uk*

Clubs & Societies

93 Contest Group
John Spurgeon, G4LKD Tel: 01405 704136 Meets last Saturday of each 14.00, Whitgift House, Whitgift, Goole, DN14 8HL

Angel of the North ARC
anarc.club@gmail.com Meets Mondays, fortnightly, 7-9pm, Whitehall Road Church, Bensham, Gateshead, NE8 4LH
https://*www.anarc.net*

Barnsley & DARC
David Gillot, G4TMZ, *g6aj@outlook.com* Meets every Monday 19.00 to 21.30, Higham Cricket club, Pog Well Lane, Barnsley, S Yorks S75 1PH

Bishop Auckland Radio Amateurs' Club
S.Bowman G7ESX *ian.bowman70@yahoo.co.uk* Meets 8.00pm Thursday evenings at the 1900 - 2200, Stanley Crook Village Hall, Rear High Road, Stanley, Crook, Co Durham DL15 9SN *www.barac.org.uk*

Brimham Contest Group
Neil Clark G8MC Meets 1st Tuesday of every Month at 7.30pm, Brimham Lodge, Brimham Rocks Road, Ripley, Harrogate, HG3 3HE *www.stevebb.co/brimham.htm*

Colburn & Richmondshire DARS
Chris Kirby, G4FZN *crdars@mailbox01.co.uk* Meets first and third Wednesday at 19.30, Hudswell Village Halll, Hudswell, Nr Richmond, N. Yorkshire DL11 6BL *www.crdars.org*

Denby Dale & DARS
Darran, G0BWB *g6ldinfo@gmail.com* Meets 1st & 3rd Wednesdays at 20.00, Pie Hall, Denby Dale, West Yorkshire, HD8 8RX *www.g4cdd.net*

Durham & District Amateur Radio Society
Michael Wright, G7TWX *dadars@gmx.com* Meets 18.30 - 21.00 every Wednesday Bowburn Community Centre, Durham Road, Bowburn, Co. Durham, DH6 5AT
https://g4euz.com/

East Ardsley Radio Society
paul@m0xzt.radio Meets Wednesdays 6:30-8pm and Fridays 7-9pm, East Ardsley Cricket Club, Jeffrey Field, Royston Hill, East Ardsley WF3 2HB
http://www.ears.radio

Finningley ARS
M Hotchin M0HOM, *martin.m0hom@gmail.com* Meets Tuesdays 6-9pm, Sat 12-6pm at The Hurst Communications Centre, Belton Road, Sandtoft, North Lincolnshire, DN8 5SX https://*www.g0ghk.com*

Goole R & ES
Ken G6YYN or Richard G0GLZ Meets every Wednesday. First Wednesday of the month at the Black Swan Inn, Main Street, Asselby DN17 7HE, White Horse Inn, Gilberdyke, HU15 2UP *www.gooleradioclub.btck.co.uk*

Grimsby ARS
Darren Hughes 2E0GUH *2e0guh@gmail.com* Meets Thursdays 7.30pm
, Cromwell Club, Cromwell Road, Grimsby, North East Lincolnshire DN31 2BA https://*www.gars.uk*

Guisborough District ARC
Mark Dutton (G0UOK) *rm.dutton@virginmedia.com* Meets Monday @ 7.30pm, Guisborough Cons Club, Chapel St, TS14 6QE

Halifax District Amateur Radio Society
Martin, M0GQB (Chairman) and Darren, M0WIT (Sec.) Meets every Tuesday between 19.00-21.00, at the Church of the Good Shepherd, New Road, Mytholmroyd, Halifax, West Yorkshire HX7 5EA
www.hadars.org.uk

Hambleton ARS
John Earland M6BHP *jearland@me.com* Meets at 7.30pm on alternative Wednesdays at, The Mencap Centre, Northallerton, N Yorks, DL6 1EG
hambletonars.wordpress.com

Hartlepool Amateur Radio Club
Tom Dyer / Trevor Sherwood 07469 710637 *hartlepoolclub@gmail.com*, Meets every Friday 7pm - 9pm, HQ Scout Centre, 236 Stockton road, Hartlepool TS25 1JW

Hornsea ARS
John Wresdell G3XYF *jhwresdell@gmail.com* Meets on Wednesday nights 7.30 for 8pm, Outdoor Bowling Club, Atwick Road, Hornsea, East Yorkshire HU18 1EL
www.hornseaarc.co.uk

Houghton-le-Spring
Mr George Thompson, M5GHT *m5ght@hotmail.co.uk* Meets Weekly on Tuesdays from 18.30 at the, Dubmire Royal British Legion, Dubmire, Fencehouses, Tyne & Wear DH4 6LJ

Hull & DARS
Richard Pike M0RKK *m0rrk@yahoo.com* Meets alternate Wednesdays at 19:30, Nellies, 173 -175 New Bridge Road, Hull HU9 2LR *www.hadars.co.uk*

Humber Fortress DX ARC
John G6LNV *club@hfdxarc.com* Meets Fridays of each week at 7pm, Mill House, Haven Road, Patrington, HU12 0PS
http://hfdxarc.co.uk/

Keighley ARS
Meets 8.00pm Thursdays at the, The Old Silent Inn, Hobb Lane, Stanbury, Keighley BD22 0HW *www.keighleyradio.co.uk*

Mexborough & DARS
Mrs S Saiger M0BOH *M0boh@aol.com* Meets Fridays 19.00-22.00 at, The Place, Castle Street, Conisborough, Doncaster DN12 3HH *www.madars.net*

Northumbria ARC
Mike Smith *mandmsmith5@talktalk.net*, 01670 861751 Meets Thursday evenings 7pm, Old Telephone Exchange, Cresswell Road, Ellington, Morpeth, Northumberland NE61 5HR
https://gx4aax.wordpress.com/

Otley ARS
M. Ross 2E0SNZ *michael.ross@ashfieldprimary.co.uk* Tel: 07768 996370, Meets 7pm-10pm Tuesdays at, The Clifton with Newall Village Hall, Newall Carr Road, Newall with Clifton, Otley, West Yorkshire LS21 2ES
www.otleyradio.org

Pontefract & District ARS
colin.g0nqe@btinternet.com or via website Meets Thursdays at the, Carleton Community Centre, Pontefract, West Yorkshire, WF8 3RJ *www.pdars.com*

Ripon & DARS
Mr David Cutter G3UNA. *d.cutter@ntlworld.com* Meets 7.30pm Thursdays at, The Bunker, rear of Ripon Town Hall, 21 Water Skellgate, Ripon, North Yorkshire HG4 1BH
www.ripon.org.uk

Scarborough Amateur Radio Society
Janet Porte, 2E0SCN Tel 01723 354502 Meets 7:30pm alternate Mondays, The Westover Club, 3 Westover Road, Scarborough, North Yorks, YO12 5AA
www.g4bp.org

Scunthorpe Steel ARC
Peter Jackson G3KNU, *peter.jackson17@ntlworld.com* Meets every other Tuesday at 8.00pm, App Frod Athletic Club, Off Ashby Road, Scunthorpe, North East Lincolnshire DN16 1AA *www.g4fuh.co.uk*

Sheffield And District Wireless Society
Pat Davies M7PAT, secretary Meets every Wednesday of each month, Heeley Green Community Centre, 304 Gleadless Road, Heeley, Sheffield S2 3AJ
https://sheffieldwireless.org

Sheffield ARC
Steve Webster M1ERS, *enquiries@sheffieldarc.org* Meets 7.00pm on Mondays at, Local repeater, Transport Sports Club, Greenhill Main Rd, Lowedges, Sheffield S8 7RH
https://www.sheffieldamateurradioclub.co.uk

South Tyneside ARS
Brian Burdis G0WZB, starsradio *club@yahoo.co.uk* Meets every Monday (Excl.

Bank Holidays). 19.30 - 21.30., St. Peter's Church Centre, Jarrow, Tyne-and-Wear NE32 5LP
www.starsradioclub.co.uk

Spen Valley ARS
Mr J R Wilde, G0FOI, *russell@wildegardens.co.uk* Meets first and third Thursday in the month at, Old Bank Club, Old Bank Road, MIRFIELD, West Yorkshire WF14 0HY *www.svars.org.uk*

SW Durham Raynet Grp
Mr Ian Bowman, G7ESY, *Ian.bowman@raynet*-uk.net *ian.bowman70@yahoo.co.uk* Meets 8.00pm 2nd Monday in the month at, Stanley Crook Village Hall, rear of High Road, Stanley Crook, Co. Durham DL15 9SN *http://g4ttf.uhp.me.uk/raynet.php*

Tynemouth ARC
Club Secretary. Email: *grahamerrington@sky.com* Meets 7pm to 9pm Fridays at, Tynemouth Scouts HQ, 31-35 Norfolk St, North Shields, Tyne & Wear NE30 1NQ https://*www.facebook.com/g0nwm/*

Wearside Electronics & Amateur Radio Society
Barry Young G0SCI *g0bnk@outlook.com* Meets on Mondays 6-10pm at, Herrington 2nd Scout Headquarters, Crow Lane, Herrington, Sunderland SR3 3TE https://facebook.com/groups/512400014309112/

York Amateur Radio Society
John Nowell G4FUO *radtronix@gmail.com* Meets 8pm Fourth Tuesday of each month , Guppy's Enterprise Club, 17 Nunnery Lane, York, YO23 1AB

York Radio Club
Mr Alan Hemenway G1VIZ *alan.hemenway@gmail.com* Meets 7.30pm Thursdays, Heworth and Bootham Conservative Club, 16-18 East Parade, York, YO31 7YK

Yorkshire Radio Friends
Steve Hall, M0HTH, *s.hall21@btinternet.com* Meets Tuesday 6.30 pm, Hatfield Woodhouse, Doncaster, DN7 6BP

Contest Groups

807 ARO
J Robert Brown M0JRB M0OTO,

Northern Fells Contest Group
Clive Davies, G4FVP, clive_*davies@ntlworld.com*,

Yorkshire Operator Brigade
Darran Chappell G0BWB *g0bwb@gobwb.comOnline* Virtual Club, *http://yobsradio.co.uk*

Region 5 - West Midlands

(http://rsgb.org/region5)

RR

Neil Yorke, M0NKE Tel: 07812 212898
Email: *rr5@rsgb.org.uk*

DRs

District 51 (Staffordshire, Warwickshire)
Craig Langman, M7LAN Tel: 07882 996604
Email: *dr51@rsgb.org.uk*

District 52 (Central & East Birmingham)
Martin Hallard, G1TYV Tel: 07496 847642
Email: *dr52@rsgb.org.uk*

District 53 (Shropshire, North Worcestershire & West Birmingham)
Martyn Vincent, G3UKV, Tel: 07421 001166
Email: *dr53@rsgb.org.uk*

District 54 (Gloucestershire, Hereford & South Worcestershire)
Giles Herbert, G0NXA Tel: 07769 658041
Email: *dr54@rsgb.org.uk*

District 55 (Gloucestershire, Hereford & South Worcestershire)
Leigh Preece, M5GWH Tel: 07534 640123
Email: *dr55@rsgb.org.uk*

Clubs & Societies

Birmingham, South Radio Society
gemmagordon.m6gkg@gmail.com Meets 8.00pm every Mon, Wed and Fri nights at c/o West Heath comm Association, West Heath, Birmingham, B31 3QY https://*www.radioclubs.net/southbirmgham*

Bromsgrove & DARC
Alan G4LVK. *alkel@btinternet.com* Meets Every Friday evening from 8pm local time, Avoncroft Arts Centre, Redditch Road, Stoke Heath, Worcestershire. B60 4JR
www.radioclubs.net/bdarc/

Burton ARC
Bob McCracken *info@burton*-arc.co.uk Roger Smith Meets on each Wednesday of the month at 7.30pm at, Stapenhill Institute Club, 23 Main Street, Stapenhill, Burton Upon Trent DE15 9AP
www.burtonarc.co.uk

Central Radio Amateur Circle
Martin Hallard, G1TYV, radio-*circle@live.co.uk* Meets Alt Saturdays 12:00 - 15:00, Bloxwich Memorial Club, 5 Harrison Street, Bloxwich, Walsall West Midlands WS3 3HP https://*www.theradioclub.co.uk*

Cheltenham ARA
Derek Thom, G3NKS *secretary@g5bk.uk* 01242 241099 Meets 7.30 for 8pm on the third Thursday of the month at, Robins Nest, Cheltenham Football Club, Whaddon Road, Cheltenham Gl52 5NA *www.caranet.org*

Coventry ARS
Mr John Beech, G8SEQ. *john@g8seq.com* Meets 1st, 2nd & 4th Friday each month at 20.00 3rd Fri are outdoor events or 2m net, St Bartholomews Church Hall, Brinklow Road, Coventry, CV3 2DT *www.coventryradio.org.uk*

Dudley and District ARS
Kevin Cartwright M0KCC *secretary@dadars.com* Meets 7.30pm every Tuesday at Ruiton Windmill, Vale Street, Dudley, West Midlands, DY3 3XF *www.dadars.co.uk*

Gloucester A R & E S
Dave Tunnicliffe *garesg4aym@aol.com* Meets twice a month on Monday evenings 7.30-9.30, Down Hatherley Village Hall, Down Hatherley Lane, Down Hatherley, Gloucester GL2 9QB
https://*www.g4aym.org.uk*

Hereford Amateur Radio Society
enquiries@herefordradioclub.uk Meets 1st Friday of the month, 19:30 to 21:00, Hill House, Newton, nr Leominster, HR6 0PF https://*www.herefordradioclub.uk*

Malvern Hills RAC
Mike Allenson G3TGD *mike.g3tgd@gmail.com* Meets 8.00pm on 2nd Tuesday in the month at the, The Town Club, 30 Worcester Road, Great Malvern, Worcestershire WR14 4QW *www.mhrac.org*

Mid Warwickshire ARS
Don Darkes, G4CYG. *midwarwicks@gmail.com* Meets on 2nd and 4th Tuesday of the month 19.30 in Spring/Summer and 14.00 in Autumn/Winter, Warwick Ambulance Association HQ, 61 Emscote Road, Warwick, CV34 5QR

Midlands ARS
Ron Swinburne *M0WSN@aol.com* Meets every Wednesday evening from 7 - 9pm, Selly Park Baptist Church, 1041 Pershore Rd, Stirchley, Birmingham B29 7PS
www.radioclubs.net/mars

Moorlands & DARS
Ian King 2E0IDK *m6idk@yahoo.com* Meets 8.30pm on Thursdays at the, Foxfield Railway, Caverswall Road Station, Caverswall Road, Blythe Bridge, Stoke-on-Trent, Staffs ST11 9BG

Nuneaton & District ARC
info@NDARC.co.uk Meets 1st Monday of the month as social and operating night, 3rd Friday of the month is club night, Windmill Sports and Social Club, 145 Mancetter Rd, Nuneaton, Warwickshire CV10 0HP *www.ndarc.co.uk*

Rugby ATS
Mr Stephen Tompsett G8LYB *stephen@tompsett.net* Meets every Tuesday from 20.00 to 22.30 and every Saturday from 14.00 to 18.00, 12th Rugby Scout Headquarters, Broughton-Leigh Community Junior School, Wetherell Way, Brownsover, Rugby CV21 1LT *www.rugbyats.co.uk*

Salop ARS
Les Griffiths M5LMG, *salopamateurradio@gmail.com* Meets Thursdays 8:00pm. Please check club website, as we occasionally have activities away our usual venue., The Telepost Club, Railway Lane, Abbey Foregate, Shrewsbury SY2 6BT *www.salopradiosociety.org*

Solihull ARS
Stuart Hammonds G4KUR *solihullradioclub@gmail.com* Meets every 3rd Thursday in the at 19.45 to 21.45, 1st Solihull Scouts HQ, Lode Lane, Solihull B91 2HZ
www.solihullradioclub.co.uk

Stafford & DARS
Nick Barnes G4KQK, *cnb@doctors.org.uk* Meets 7.30pm Wednesday at the, Wildwood Community Centre, Wildwood Gate, Stafford ST17 4RA *www.g3sbl.org.uk*

Staffordshire Portable ARC
Neville Briggs, M0VSP, 01922 449668 Meets Tuesdays from 7pm., Bolehall Manor Club, Amington Road, Tamworth, Lichfield, Staffs B77 3LH https://m0spa.co.uk

Stourbridge & DRS
Mr John Clarke Meets 8.00pm 1st & 3rd Mondays in month at the, Old Swinford Hospital/School, Stourbridge, West Midlands, DY8 1QX *www.g6oi.org.uk*

Stratford-upon-Avon
Clive Ousbey *sdrsinfo@talktalk.net* Meets second and fourth Mondays of most months 7.30pm for 8pm, Foundation House, Masons Road, Warwickshire CV37 9NF *www.stratfordradiosociety.freeserve.co.uk*

Sutton Coldfield ARS
Mr Robert Bird 2E0ZAP, *spirit.guide@hotmail.co.uk* Meets from 7 30pm till 10 30pm 2nd & 4th Mondays in the month (except bank hoildays), Sutton Coldfield Rugby Club, 160 Walmley Road, Nr Sutton Coldfield, Birmingham B76 2QA *www.g3rsc.co.uk*

Tamworth ARS
TG 'Robbie' Robertson *robbiebarb204@btinternet.com* Meets Drayton Village Club, Drayton Lane, Drayton Bassett, Tamworth, Staffordshire B78 3TX *www.tamworth*-ars.org.uk

Telford & DARS
John Humphreys, *M0JZH@yahoo.co.uk*, 01952 457234 Meets Wednesdays at 7.00pm. Events are at 8:00pm, Village Hall, Malthouse Bank, Little Wenlock, Telford TF6 5BG *www.tdars.org.uk*

Vulture Squadron C.G
Iain Kelly M0PCB *iain@m0pcb.co.uk* Meets 2nd Monday of each month at 7:30pm, Meeting venue varies, Old Chapel House, Malvern Road, Staunton, Glos GL19 3NZ

Wolverhampton ARS
A. Atkinson 2E0YEZ *secretary@wolverhmaptonars.com* Meets 7.30pm every Wednesday, The Royal British Legion Club, Vicarage Road, Wednesfield, Wolverhampton WV11 1SF *www.wolverhamptonars.co.uk*

Wythall RC
Chris Skelcher, G3YHF *chris.bham@gmail.com* Meets weekly Tuesday and Friday 19.30 to 22.30, C/O Wythall Community Association, Wythall House, Silver Street, Wythall, Birmingham B47 6LZ *www.wythallradioclub.co.uk*

Repeater Groups

Kidderminster Repeater Group
Ms P Dowie, G8PZT, *g8pzt@gb3kd.org.uk* 3rd Tuesday, monthly at 8pm at Queens Head Wolverley Village Nr Kidderminster Worcs DY11 5XB

Contest Groups

Bad Weather DX Group
Nicholas Pearce, G4WLC, chairmanMeets Tuesday 6pm, 4 Dunster Grove, Cheltenham, Glous.GL51 0PE *www.badwxdx.club*

Radio GaGa Contest Group
John Warburton G4IRN
g4irn@dxdx.co.uk,

Redditch AR
Roy Jones G0RMG Meeting Information: On Air Only 144.700 most nights, B97 5NW *http://reddicams.co.uk/radio/about.htm*

Region 6 - North Wales

(http://rsgb.org/region6)

RR

Liz Cabban, GW0ETU
Tel: 01690 710257
Email: *rr6@rsgb.org.uk*

DRs

District 61 (Flintshire, Wrexham & Powys)
Mark Harper, MW1MDH, Tel: 07967 517892
Email: *dr61@rsgb.org.uk*

District 62 (Conwy, Denbigh)
Liz Cabban, GW0ETU Tel: 01690 710257
Email: *dr62@rsgb.org.uk*

District 63 (Gwynedd, Anglesey (Ynys Mon))
John Martin, MW0VTK, Tel: 07772 720099
Email: *dr63@rsgb.org.uk*

District 64 (Ceredigion & North Powys)
Ray Ricketts, GW7AGG, Tel: 01970 611853
Email: *dr64@rsgb.org.uk*

Clubs & Societies

Aberystwyth & DARS
Abby Pugh MW0ZXY *mw0zxy@abbypugh.uk* Meets 2nd Thursday of the month (except August) at 8.00pm-10pm we sometimes have visits or events elsewhere. Waunfawr Community Hall, Brynceinion, Waunfawr, Aberystwyth SY23 3PN *http://adars.org.uk*

Dragon ARC
Simon Taylor MW0NWM, *darc.secretary@gmx.co.uk* Tel: 07904 874652, Meets 7.30 (for 8) 1st and 3rd Monday each month, at the Canolfan Esceifiog, Gaerwen, Llanfairpwll, Isle Of Anglesey LL6 0DE *http://www.radioclubs.net/dragonarc*

Halkyn Radio Group
halkynradiogroup@yahoo.com Meets Tuesday evenings between 20.00 - 2200hrs, Britannia Inn X, Pentre Halkyn, Halkyn, Flintshire CH8 8BS
https://*www.gw0hrg.co.uk*

Holy Island Amateur Radio Club
Cath Thorley *mw0kbn@aol.com* Meets at 19.00 on 2nd and 4th Tuesdays each month, Boathouse Hotel, Newry Beach, Beach Road, Holyhead, Anglesey LL65 1YF

Meirion ARS
Mr R Smith GW0AYQ *monbob.bayq34@btinternet.com* Meets 1st Wednesday of each month (except August), Bryn Eiddion, Rhydymain, Dolgellau, LL40 2AS *www.meirion*-ars.info

Powys ARC
Michael.Lawton, *michael.lawton1@outlook.com* Meets 8.00pm 1st Thursday in the month, Berriew Community Centre, Welshpool, SY21 8AZ *www.parc.care4free.net*

Rhyl And District Amateur Radio Club
secretary@radarc.uk
Meets for club net & activity nights - Sundays at 7pm 145.375 MHz FM, Rhyl Rugby Sports & Social Club, Tynewydd road, Rhyl, LL18 4AQ https://radarc.uk/

Wrexham and Marches ARS
Mr William Taylor 01244571425 Meets second and fourth Thursdays of each month at 19.30hrs at Black Park Community Centre, Halton, Chirk. LL14 5BB.Facebook page. *wrexhammarchesars@gmail.com*, Black Park Community Centre, Lon Graig, Halton, Chirk LL14 5BB *www.wrexham-ars.com/*

Contest Groups

Mike-Whiskey DX Group
Online Team meetings 20.00 Monthly,

North Wales Amateur Radio Group
secretary@nwarg.org Every Tuesday 7-9pm at Mochdre Community Centre, Old Conway Rd, Colwyn Bay LL28 5HU *www.nwarg.org*

Region 7 - South Wales

(http://rsgb.org/region7)

RR

Brian Jones, GW6ZYI Tel: 07904 327633
Email: *rr7@rsgb.org.uk*

DRs

District 71 (South Powys, Pembrokeshire, Camarthenshire, Swansea & West Glamorgan) Andy Digby, GW0JLX
Tel: 07768 282880
Email: *dr71@rsgb.org.uk*

District 72 (Mid Glamorgan, East Glamorgan, Cardiff)
Brian Jones, GW6ZYI Tel: 07904 327633
Tel: 07422 661551
Email: *dr72@rsgb.org.uk*

District 73 (Monmouthshire, Newport)
Nigel Paull, GW1CUQ Tel: 02920 892580
Email: *dr73@rsgb.org.uk*

Clubs & Societies

Aberdare and DARS
Andy Hall *gw7rkc@gmail.com* Meets every other Friday 18.30 to 21.00, Hirwaun YMCA, Manchester Place, Hirwaun, CF44 9RB

Aberkenfig And District ARC
Brian price GW0DVB Meets every Wednesday 19:00 to 22:00
, Tondu Railway Canteen, Station Approach, Tondu, Bridgend CF32 9DY

Barry ARS
Glyn Jones. *glyndxis@talktalk.net* Meets 7.30pm on Tuesdays at, Sully Sports & Leisure Club, South Road, Sully, S Glamorgan CF64 5SP

Blackwood & DARS
L.W.Wright GW8UAM *wynnwright7@aol.com* Meets on Fridays 1900 to 2100 School term time only 1900 to 2100, Islwyn High School, Waterloo, Oakdale, Blackwood Gwent NP12 0NU *www.gw6gw.co.uk*

Blaenau Gwent Radio Group ARS
Chris Taylor 2W0DOE *bgrg@gmx.co.uk* Meets 7pm every Tuesday, Bryn Farm Community House, 13 Heol Onen, Brynmawr, Gwent NP23 4TS

Brecon And Radnor ARS
Adam Tofarides Meets first Thursday of the month 7pm Llanddew village Hall
, Llanddew village Hall, Llanddew, Brecon, Powys LD3 9ST

Carmarthen ARS
Mr Andy Digby, GW0JLX
carsmembershipsec@gmail.com
Meets 1830 (6:30 pm) onwards 1st and 3rd Tuesdays each month, The Cwmduad Community Centre, Cwmduad, Carmarthenshire, which is on the A484. 51° 57' 20.96 N, 4° 21' 56 SA33 6XN

Chepstow & DARS
Dan Taylor GW0EGH *gw0egh@hotmail.com* Meets first and third Tuesday of the month at 19.30 - 22.00 pm, Chepstow Athletic Club, Chepstow, NP16 5JT *www.gw4lwz.co.uk*

Cleddau ARS
Ian Rogers MW0RRW perfect-*pcs@btconnect.com* Meets on Zoom, SA72 6DG *https://www.cleddau-ars.org.uk*

Cwmbran DARS
Ken Smith - *mc0yad.cadars@gmail.com* Meets 7-9pm every Wednesday Henllys Village Hall, Henllys Village Road, Cwmbran, NP44 6HX *www.mc0yad.co.uk*

Highfields ARC
Darren Jenks - MW0MPD Meets 7.00pm Tuesdays, Meets 7.00pm Tuesdays, Rhiwbina Sports and Social Club, Lon-Y-Dail, Rhiwbina, Cardiff CF14 6EA *www.highfields*-arc.co.uk

Newport ARS
Mr Paul Nicholls. Email: *nars@gw4ezw.org.uk* Meets every Thursday evening 19.00-21.00 at the chapel, Tabernacle Congregational Chapel, Rhiwderin, Nr Bassaleg, Newport NP10 8RH *www.gw4ezw.org.uk*

No1 Welsh Wing ATC Amateur Radio Society
Mr Chris Stubbs, MW0LZZ, *onewingshotoff@hotmail.co* Meets 1st Wednesday of each month at 7.00pm, HQ 1344 Sqn ATC, Ty Walter Cleall GC, Maindy Barracks, Cardiff CF14 3YE

Pencoed Amateur Radio Club
Ieuan Jones Meets Tuesdays at 7pm at Pencoed Rugby Club, Felindre Road, Pencoed, NR. BRIDGEND, CF35 5PB

Rhondda ARS
Mr John Howells, GW4BUZ *secretary@gw2fof.co.uk* Meets every other Tuesdays at 7.30 pm, Ystrad, Rhondda Cynon Taff CF41 7SY

Sparks Amateur Radio Club
admin@mw0sml.co.uk Meets Wednesdays at 1400 & 1700, Old Station Yard, Station Road, Letterston, Pembrokeshire SA41 3UF

St. Tybie ARS
Gareth Woods, GW4JPC, *gw4jpc@yahoo.co.nz* Meets fortnightly at Saron Hall at 7.30pm 01269 597524, Saron Hall, Saron Road, Saron, Ammanford, Carmarthenshire SA18 3LN *http://gc0vpr.clubbz.com*

Swansea & District Amateur Radio Club
Jeff Downer, GW6TYJ, *jeffdowner2017@gmail.com* Meets Tuesdays 7.00pm to 8.30pm, Llansamlet Community Centre, Church Rd, Llansamlet, Swansea SA7 9RH

Taff Vale ARC
Ashley Burns 01685 389434 *aburns02@btinternet.com* Meets every Monday 19.00-21.00, St Johns Ambulance Hall, Gwaun Farren, Merthyr Tydfil, CF47 8LX

Uskside Amateur Radio Club
Mr John Davies GW6RTV Meets Every Wednesday from 1800 - 2000 hrs, Newport Indoor Bowls Centre, Glebelands Stadium, Bank Street, Newport NP19 7HF
https://Coming Soon

Repeater Groups

Tenby Radio Repeater Group
John Rees, GW0JRF Meets 1st Saturday of the month at 1pmMan Shed, Pembroke Port Gate 4Fort Road, Pembroke Dock, SA72 6TH

Contest Groups

The Gower/Gwyr Contest Club
Paul Valerio, GW4KTT *paulvalerio@gmail.comEvery* VHF/UHF/SHF Contest evenings HF every major event, East Pilton Farm, Gower/GWYR, near Rhossili Swansea SA3 1PQ
www.gw4cc.wales

Region 8 - Northern Ireland

(http://rsgb.org/region8)

RR

Micheal Na bPiob, MI0HOZ Tel: 07955 208203 Email: *rr8@rsgb.org.uk*

DRs

District 81 (Co Antrim)
John Campbell, MI0WJC
Tel: 07712 115791 Email: *dr81@rsgb.org.uk*

District 82 (Co Down)
Roger Bradley, MI0WWB
Tel: 07788 207215 Email: *dr82@rsgb.org.uk*

District 83 (Co Londonderry, Co Tyrone)
Trevor Campbell, MI5TCC Tel: 07710 468835 Email: *dr83@rsgb.org.uk*

District 84 (Belfast)
Vacant Email: *dr84@rsgb.org.uk*

District 85 (Co Fermanagh, Co Armagh)
Vacant Email: *dr85@rsgb.org.uk*

Clubs & Societies

Antrim & District Amateur Radio Society
Darren Brown MI0YPT *drbniradio@gmail.com* Meets on the 2nd Friday of each month at 7.30pm at, Greystone Community Centre, 30 Ballycraigy Road, Antrim, BT41 1PW *www.adars.co.uk*

Ballymena ARC
Hugh Kernohan (Secretary) 07715539481 *hkernohan@aol.com*, Meets Thursdays night at 70 Nursery Road, Gracehill, Ballymena, Co. Antrim, BT42 2QA *http://gi3fff.synthasite.com/*

Bangor & DARS
Harry Squance GI4JTF, *gi4jtf@gmx.com* Meets 7.30pm to 9.30pm 1st Thursday of the month at the, Castle Park, 5 Abbey Street, Bangor, BT20 4JE *www.bdars.com*

Belfast RSGB Group
David Gillespie MI0FBI, *gwocni@hotmail.com* Meets 8.00pm on 3rd Wednesdays of each month September to June at the, Maple Leaf Club, Park Avenue, Standtown, Belfast BT4 1PU

Bushvalley Amateur Radio Club
Victor Mitchell Gi4ONL, *bushvalleyarc@gmail.com* Meets every 8pm 3rd Thursday of each month at the United Services Club, 8, Roe Mill Road, Limavady, Co. Londonderry, BT49 9DF *www.bushvalley*-arc.co.uk

Carrickfergus ARG
John Roberts GI0USX Meets every Tuesday at the, Downshire Community School, Downshire Road, Carrickfergus, BT38 7DA *www.radioclubs.net/carg/*

City of Belfast Radio Amateur Society
Frank Hunter GI4NKB *fthunter@virginmedia.com* Meets first Monday of each month at 8pm at, Shorts Social Club, Holywood Road, Belfast, BT4 1SL

Lagan Valley ARS
Mr Andrew Mulholland, *MI0BPB.gi4gty@hotmail.com* Meets every Wednesday 8pm at The Society Shack, Ballynahinch Road, Lisburn, Co Antrim, BT27 5LX *https://www.lvars.uk*

Lough Erne ARC
Alan Gault GI6PYP, *alan.r.gault@btinternet.com* Meets at the SHARE Centre., Smith's Strand, Lisnaskea, Co.Fermanagh, BT92 0EQ *https://www.lougherneradioclub.co.uk*

Mid Ulster ARC
Hazel McMullen *muarc.secretary@yahoo.co.uk* Meets Second Sunday of each month at 3pm, Ulster Aviation Society, Gate 3 – Maze Long Kesh, 94-b Halftown Road, Lisburn, BT27 5RF *www.muarc.com*

North West Amateur R.C
Mícheal, MI0HOZ *info@nwgarc.net* Meets every 6 weeks at 19.30pm, Lisneal College, 70 Crescent Link, Londonderry, BT47 5FQ

Strabane ARS
Terry White GI7THH *gi7thh@hotmail.com* Meets the second Thursday of each month at 20.00hrs at, 3a Park Road, Strabane, Co Tyrone, Northern Ireland BT82 8EL

Tri-County Amateur Radio Club
Stephen McFarland GI4RNP Meets Sunday weekly at 2pm, Mark Evans Plant Limited, 7 Sheepwalk Road, BT28 3RD

West Tyrone ARC
Philip Hosey MI0MSO *info@wtarc.org.uk* Meets 2nd Wednesday of each month at 20:00-21:30, Order of Malta Hall, Brook Street, Omagh BT78 5HE *www.wtarc.org.uk*

Contest Groups

Orchard County DX Club
Mrs Edith Simpson, MI0PRM *edithmioprm.es.45@gmail* All training and meetings are listed on *http://ocdclub.webs.com*, We are a web based club, BT62 2DD https://*www.mn0ocg.co.uk*

Sperrin Amateur Radio Club
Mark Robinson MI0LNL *secretary@sarc.club* Meets at 19.30 last Wednesday of the month, The Hub, 14 Burn Road, Cookstown *www.sarc.club*

Region 9 - London & Thames Valley

(http://rsgb.org/region9)

RR

Ron White, G6LTT Tel: 07800 950175
Email: *rr9@rsgb.org.uk*

DRs

District 91 (North, East & West London)
Ron White, G6LTT
Tel: 07800 950175 Email: *dr91@rsgb.org.uk*

District 92 (Berkshire)
Alison Johnson, G8ROG Tel: 0118 9545368
Email: *dr92@rsgb.org.uk*

District 93 (Oxfordshire)
Malcolm Andrew, G8NRP
Tel: 01235 524844 Email: *dr93@rsgb.org.uk*

District 94 (Bedfordshire & Stevenage)
Terry Baldwin, G4UEM Tel: 07789 555514
Email: *dr94@rsgb.org.uk*

District 95 (South West London)
Garo Molozian, G0PZA
Tel: 07765 657542 Email: *dr95@rsgb.org.uk*

District 96 (Buckinghamshire)
Tom Cotton, M7BRW Tel: 07552 670486
Email: *dr96@rsgb.org.uk*

Clubs & Societies

Aylesbury Vale RG
Cathy Clark G1GQJ *cmc5146@gmail.com* Meets on Fridays from 8pm at, Hardwick Village Hall, Lower Road, Hardwick, Nr Aylesbury, HP22 4EA

Aylesbury Vale RS
Mr V Gerhardi, G6GDI, *avrs@rakewell.com* Meets every 2nd Wednesday of the month at 20.00 - 22.00, The Dog House Inn, Broughton Crossing, Broughton, Aylesbury HP22 5AR *www.avrs.org.uk*

B.A.R.S Banbury ARS
Stephen McGuigan *Stephen@mcguigan.com* Meets 1st & 3rd Wednesday of each month at 7pm - 9pm at, 169 Bloxham Rd, Banbury, Oxfordshire, OX16 9JU *www.banburyares.co.uk*

Bedford & DARC
Mr Glen Loake G0GBI *g0gbi.glen@gmail.com* Meets every Tuesday at 7.30 pm at, The Shack opposite the Plantation, Ravensden, Bedfordshire MK44 2RJ *www.badarc.net*

Bracknell ARC
Colin Ashley 2E0XDA *secretary@g4bra.org.uk* Meets on the 2nd Wednesday of each month at 8pm. We also hold weekly Radio Nets at 8pm each Wednesday on 145.375 FM and each Sunday at 5pm via the GB3BN Repeater, Shepherds Lane,

Bracknell, Berkshire RG42 2BU
https://g4bra.org.uk

Burnham Beeches RC
Dave Chislett G4XDU, *bbradioclub@gmail.com*, Meets 8.00pm on 1st and 3rd Mondays in the month at the, Farnham Common Village Hall, Victoria Road, Farnham Common, Bucks SL2 3NL
www.bbrc.info

Chesham & District ARS
Malcolm Appleby G3ZNU Meets second and fourth Wednesdays in the month, 8:00pm, Ashley Green Memorial hall, Two Dells Lane, Ashley Green, Bucks HP5 3PN
http://www.g3mdg.org.uk/

Drowned Rats Radio Group
Carl Ratcliffe *info@drownedrats.uk* Meets second Wednesday of the month 20.00pm - 22.30pm, The Coy Carp, Coppermill Lane, Harefield, Middlesex UB9 6HZ
www.g3rat.com

Dunstable Downs Radio Club
Mike Scarlett G4CAK *mikescarl@btinternet.com* Meets every friday at 8pm, St Fremunds Church Hall, 185 Westfield Road, Dunstable, LU6 1DR
www.dunstabledownsradioclub.org

Edgware & DRS
Steve SlaterG0PQB *g0pqb25@gmail.com* Meets 8.00pm on 2nd and 4th Thursdays in the month at the, Watling Community Centre, 145 Orange Hill Road, Burnt Oak, Edgware, Middlesex HA8 0TR
www.g3asr.co.UK

Farringdon Radio Group
secretary: *g0cag@hotmail.com* Meets 18.00 every Tuesday, c/o Hanwha Phasor, 4th Floor, Record Hall, 16 Baldwins Gardens, London EC1N 7RJ

Harwell Amateur Radio Society
Ann Stevens G8NVI *secretary@g3pia.net* Meets at 19.45 on the 2nd Thursday of each month 19.45 for 20.00hrs, Chilton Village Hall, Church Hill, Chilton, OX11 0SH
www.g3pia.net

LHS Radio Club
info@m0hsl.uk Meets London Hackspace is currently between premises so there are no regular radio club events but check back or join the Discord channel or follow the Google Group https://groups.google.com/g/lhs-radio, London Hackspace, 388 High Road, Wembley, W13 8AG
http://london.hackspace.org.uk

Maidenhead & DARC
Peter Hicks, G4KCX *G4KCX@talktalk.net* Meets 7.45 on 1st Thursday and 3rd Tuesday in the month at the, The Friends Meeting House, 14. West Street, Maidenhead, Berkshire SL6 1RL
https://www.madarc.org

Milton Keynes ARS (MKARS)
Francis Hennigan: *information@mkars.org.uk* Meets Mondays 19.30. Zoom Coffee Chats 10.30 on Fridays, Block H, Bletchley Park, Bletchley, Milton Keynes MK3 6EB
https://*www.mkars.org.uk*

Newbury & DARS
Phill G6EES 07771 504738 *secretary@nadars.org.uk* Meets 4th Wednesday of every month (apart from December) at 19.30hrs, Acland Hall, Hermitage Rd, Cold Ash, Berkshire RG18 9JF
https://*www.nadars.org.uk*

Oxford And District ARS
Graham Diacon G8EWT, *G8EWT@diacon.co.uk* Meets 7.30 for 8.00pm on the 1st and 3rd Tuesday of every month, at The Gladiator Club, 263 Iffley Road, Oxford, OX4 1SJ *www.odars.org.uk*

Reading and District ARC
https://radarc.org/contact-us/ Meets 2nd and 4th Tuesday of each month for social and practical sessions. Some events are held at Reading RFC, Sonning, 2nd Woodley Scout Group Hut Vauxhall Park, Vauxhall Drive, Woodley, Reading RG5 4EA
https://*www.radarc.org*

RNARS London Group
D Goodison, G0LUH 0208 8470260 *info@gb2rn.org.uk*, Meets last Thursday of every alternate month, at 15.00hrs on HMS Belfast, Battleship Lane, Tooley Street, London SE1 2JH
www.gb2rn.org.uk

RS of Harrow
Linda, G7RJL *info@G3EFX.org.uk* Meets 1st and 3rd Fridays of the month at 20.00 20.00 to 22.00, Blackwell Hall, Uxbridge Road, Stanmore, HA3 6DQ
www.g3efx.org.uk

Shefford & District ARS
D Lloyd,G8UOD *davidg8uod@virginmedia.com* Meets at 8pm Thursdays (with summer break) at the, Community Hall, Ampthill Road, Shefford, Beds SG17 5AX
www.sadars.co.uk

Silverthorn RC
Robin Bernard, M0HVC *m0hvc@protonmail.com* Meets every Friday evening at 7.30pm to 10.30pm on Fridays at, Friday Hall, 56 Friday Hill East, Chingford, London E4 6JT *www.silverthornradioclub.org.uk*

Southgate Amateur Radio Club
Keith Mendum G8RPA Tel: 02083603614 Meets 1st Monday of month at 19.30 hrs, 14th Southgate Scout Group Hut, 81 Green Road, Oakwood, London N14 4AP
http://southgatearc.uk

Stevenage & DARS
Mr Rob McTait G2BKZ, rob*_g2bkz@talktalk.net* Meets 7.30pm on Tuesdays at the, Stevenage Resource Centre, Chells Way, Stevenage, Herts SG2 0LT
www.sadars.org

Triple B Amateur Radio Contest Group
M Goodey, G0GJV - Tel: 07770938478 Meets Fridays 10pm, Jack O'Mewbury, Terrace Road South, Binfield, RG42 5PH

Verulam ARC
Greg Beacher: *secretary@verulam*-arc.org.uk Meets on the 3rd Monday of the month, The Boot (Inn / public house), Conservatory, High Street, 2 miles East of Harpenden, Hertfordshire SG4 8PT
http://www.verulam-arc.org.uk/

Contest Groups

AMC Radio Club
David Millward M0PTP *David.m@pebbleltd.co.uk* Meets once every quarter and every day online, 69 Primrose Hill, Kings Langley, Hertfordshire WD4 8HX

Beds, Mid Contest A
Fred Handscombe
g4mbc@homeshack.freeserve.co.uk, IP28 8LQ

Region 10 - South & South East

(http://rsgb.org/region10)

RR

Keith Bird, G4JED Tel: 01732 446331
Email: *rr10@rsgb.org.uk*

DRs

District 101 (Surrey, London South of the Thames) Alun Cross, G4WGE Tel: 07779 079503, Email: *dr101@rsgb.org.uk*
District 102 (Wiltshire)
Simon Harris, G4WQG Tel: 07498 213585
Email: *dr102@rsgb.org.uk*
District 103 (West Sussex)
Sean Pryer, 2E0XBT Tel: 07505 546810
Email: *dr103@rsgb.org.uk*
District 104 (Hampshire)
Alan Ball, G3UQW Tel: 07770 536975
Email: *dr104@rsgb.org.uk*
District 105 (Isle of Wight)
Fred Dawson, G1HCM Tel: 07885 634518
Email: *dr105@rsgb.org.uk*
District 106 (Kent)
Dave Lee, G8ZZK Tel: 07739 549822
Email: *dr106@rsgb.org.uk*
District 107 (East Sussex) Vacant
Email: *dr107@rsgb.org.uk*

Clubs & Societies

Andover RAC
Paul Philips, G4KZY, *arac@arac.org.uk* Meets first and third Tuesday of each month at 19:30 to 21:30, Tangley Parish Village Hall, Wildhern, Andover, Hants SP11 0JE
https://*www.arac.org.uk*

Barbarian Drifters Radio Society
Philip Bourke M0IMA *barbariandrifters@gmail.com* Meets Saturday of each week at 12 noon, Meadow Bank, Rye Lane, Otford, Kent TN14 5JF

Basingstoke ARC
telephone 01256 883838 Meets 3rd

Monday of the month,7:30 meet in the bar for an 8pm start, May's Bounty Cricket Club, Fairfields Road, Basingstoke, Hampshire RG21 3DR
https://basingstokearc.wordpress.com/

Brede Steam ARS
secretary@bsars.co.uk Meets 1st and 3rd Saturday of each month. Please check with secretary before visiting, as we might be participating in an event., Brede Steam Amateur Radio Society, The Scout Hut, Stubbs Lane, Brede, Nr Rye TN31 6EH
www.bsars.co.uk

Bredhurst Rx & Tx Soc
secretary@brats-qth.org Meets Brats Shack, Brats Shack, Hurstwood Rd, Bredhurst, Gillingham ME7 3JZ
www.brats-qth.org

Bromley & DARS
Andy Brooker *enquiries@bdars.co.uk* Meets 3rd Tuesday in the month 19:45 for 20:00, Victory Social Club, Kechill Gardens, Hayes, Bromley BR2 7NG
www.bdars.wordpress.com

Caterham RG
Mr P N Lewis, G4APL. *catrad@theskywaves.net* Meets on alternate Fridays evenings at members' houses, CR3 5EL
www.theskywaves.net

Chippenham and District ARC
Brian Tanner G6HUI *chairman@chippenhamradio.club* Meets every Tuesday from 8pm at the Tuesday 19.30 - 21.30, Kington Langley Village Hall, Church Road, Kington Langley, Chippenham SN15 5NJ
http://chippenhamradio.club

Coulsdon Amateur Transmitting Society
Glenn Rankin G4FVL *secretary@catsradio.org* Meets 8.00pm on 2nd Mon in the month St Swithuns Church Hall, Grovelands Road, Purley, Surrey CR8 4LA
https://www.catsradio.org

Crawley Amateur Radio Club
Phil Moore M0TZZ *secretary@carc.org.uk* Meets 8.00pm Wednesdays and, 11.00 am Sundays at the, Tilgate Forest Rec. Centre, Hut 18, Tilgate Forest, Crawley, West Sussex RH11 9BQ *www.carc.org.uk*

Cray Valley RS
Dave Lee, G8ZZK. *secretary@cvrs.org* Meets 1st & 3rd thursday of the month 19.30 - 21.30 at the, 1st Royal Eltham Scouts HQ, Rear of 61 - 71 Southend Crescent, Eltham, London SE9 2SD
www.cvrs.org

Crystal Palace Radio And Electronics Club
Damien Nolan, 2E0EUI *cprec.g2lw@gmail.com* Meets 7.30pm on 1st Friday each month at the, All Saints Church, Beulah Hill, London SE19 3LG https://*www.cprec.org*

Darenth Valley RS
Phil Bourke M0IMA *info@darenthvalleyrs.org.uk* Meets 2nd & 4th Wednesday of the month 20.00 - 22.30, Crockenhill Village Hall, Stones Cross Road, Crockenhill, Swanley BR8 8LT https://*www.darenthvalleyrs.org.uk/joining*-or-listening-to-the-darenth-valley-club-net/

Dorking & District Radio Society
Sue Ellinor, *ddrs.secretary@gmail.com* Meets 7.45pm on 4th Tuesday of each Month at 19.45 pm, at The Friends Meeting House, Butter Hill, South St, Dorking, Surrey, RH4 2LE *www.ddrs.org.uk*

Dover Radio Club
Nathan Friend 2E0NFX, *secretary@darc.online* Meets on every Thursday 7-9pm. Talks and demos planned every 4 weeks or so, please visit the club calendar for more information, St Peter & St Pauls Church Hall, Minnis Lane, River, Kent CT17 0RG https://*www.darc.online*

DSTL Radio Club
Peter Allcock, *dstlradioclub@dstl.gov.uk* Meets 2nd Wednesday each month 12 noon, DSTL Porton Down, Crossley House, Porton Down, Wiltshire SP4 0JQ

East Kent Radio Soc.
Alan Perkins G7RBB *perkins.alan@gmail.com* Meets 8.00pm on the 2nd Wednesday (summer months only), The Herne Mill, Mill Lane, Herne Bay, Kent CT6 7DR
www.g0ekr.co.uk

Echelford ARS
Philip Miller Tate, M1GWZ *m1gwz@icloud.com* Meets 7:30pm on 2nd & 4th Wednesdays of month, St Hildas Church Hall, Stanwell Road, Ashford, Surrey TW15 3QL *https://qsl.net/g3ues/*

Fareham & District Amateur Radio Club
C. Jenkins-Powell G7MFR, *chris@jenkins*-powell.com Meets 7.30pm on Wednesdays at the, Fareham Motorboat & Sailing Club, Lower Quay, Fareham, Hants PO16 0RA
www.fareham-darc.co.uk

Farnborough & DRS
C. Andrews M7WJC, *treasurer@farnboroughradio.org.uk* Meets every 2nd and 4th Wed of the month at 7.30pm, The Aldershot Military Museum, Off Queens Avenue, Aldershot, Hants GU11 2LG *www.farnboroughradio.org.uk*

Fort Purbrook ARC
Chris Bryant G3WIE *g3wie@fparc.org.uk* Meets at 19.00 - 21.00 on the last Friday each month, Peter Ashley Lane, Portsdown Hill Road, Nr Cosham, Portsmouth, Hampshire PO6 1BJ *www.fparc.org.uk*

Guildford & District Radio Society
Timothy Dabbs, G7JYQ *secretary@gdrs.net* Meets on 2nd, 4th and 5th Fridays of each month. 1930 for 2000hrs, Guildford Model Engineers HQ, Stoke Park, London Road, Guildford, Surrey GU1 1TU *www.gdrs.net*

Hastings Electronics & Radio Club
John G0OZY, *herc.hastings@gmail.com* Meets every Wednesday in the local area for /P radio-related activities. The location and time changes from time to time, visit page 2 of our website on Wednesday mornings to find details for that day. TN38 0LQ
www.hastings-electronics-radio-club.com

Hilderstone Radio Society
Ian Lowe, G0PDZ *secretary@g0hrs.org* Meets every Thursday (except August) 7pm for 7:30pm start. Visitors welcomed. Onsite parking & level access for the disabled, 1st Margate (St John's) Scout Group Hall, Durban Road, Margate, Kent CT9 2TE
https://g0hrs.org

Hog's Back Amateur Radio Club
Simon Lambert M0XIE *hogsbackarc@gmail.com* Meets 2nd and 4th Mondays of the month (excluding Bank holidays), Scout Centre, Pankridge Street, Crondall, Hampshire, Surrey GU10 5RQ
https://www.hogsback-arc.org.uk

Horndean & DARC
Mr Stuart Swain, *G0FYX@msn.com* Meets 1st & 3rd Friday in the month from 1830 to 2130, Deverell Hall, 84 London Road, Purbrook, Waterlooville, Hants PO7 5JU
www.hdarc.co.uk

Horsham ARC
Mr Alister Watt, G3ZBU. *info@harc.org.uk* Meets 8pm 1st Thur in month at the, Guide Hall, Denne Road, Horsham, West Sussex RH12 1JF
www.harc.org.uk

Isle of Wight RS
Steve Cownley, 2E0RQD. *iowradsoc@gmail.com* Meets 7.00pm Fridays at, Haylands Farm, Salters Road, Haylands, Ryde, Isle of Wight PO33 3HU
www.iowrs.org

Itchen Valley ARC
Chris Ash, G7LWV. *chris.g7lwv@gmail.com* Meets 8.00pm on the 2nd and 4th Fridays of the month, Otterbourne Village Hall, Cranbourne Road, Otterbourne, Hampshire SO21 2ET *www.ivarc.org.uk*

Kent Active Radio Amateurs
Debbie Richardson G7EOZ *avckev@btinternet.com* Meeting will be regional and some regular meets the PS Medway Queen, Gillingham Pier, Pier Approach Rd, Gillingham, ME7 1RX

Kent Weald Radio Club
Mr P J Blunt, G0UXG *palybl@btinternet.com* Meets on the last Saturday of each month at 19.00 hrs at, Headcorn Airfield, Briefing Room, Headcorn, Ashford, Kent TN27 9HX *https://www.qrz.com/m0kwa*

Medway ARTS
secretary@marts.org.uk Meets weekly on Fridays from 19:30, Location Tunbury Hall, Catkin Close Chatham. ME59HP. Please see website for more detail and meeting schedule. Daily net at 09:30 on 144.650 FM, Tunbury Hall, Catkin Close, Walderslade, Chatham, Kent ME5 9HP *https://marts.org.uk*

Newhaven Fort Amateur Radio Group
Roger Parish, M1RPY, *newhavenarg@gmail.com* Meets daily at 10 am at, Newhaven Fort, Fort Road, Newhaven, East Sussex BN9 9DS

North Kent RS
Mr Stephen Osborn, G8JZT, *secretary@nkrs.org.uk* Meets 7.30pm on 1st and 3rd Tuesday of each month 19.30 to 22.00 hrs, Hurst Community Centre, Hurst Road, Bexley, Kent DA5 3LH *www.nkrs.org.uk*

Southampton Amateur Radio Club
Nigel Phillips, G7POC *sotonarc@gmail.com* Meets at 7.30pm Third Wednesday of each month in 19.30-21.30hrs, Lounge Bar, The Peartree Inn, Peartree Road, Southampton SO19 7GZ
https://*www.southampton*-arc.org.uk

Southampton University Wireless Soc
Andrew Barrett-Sprot *contact@suws.org.uk* Meets every Thurs at 18.00-20.00, University of Southampton, Buliding 40,

University Road, Highfield, Southampton SO17 1BJ *www.suws.org.uk*

Southdown ARS
Tom Pitcher, *secretary@sars.club* Meets Every first Friday of the month, Old Town Community Centre, 1a Central Ave, Eastbourne, East Sussex BN20 8PL *http://sars.club*

Surbiton Heritage AR & Elec Society
Tony M0SHA email: *RSGB@m0sha.com* Meets 7.45 pm on 1st Wednesday of the month at the, The Coffee Bar, 1st Floor, Surbiton Hill Methodist Church, 39 Ewell Road, Surbiton, KT6 6AF *https://www.m0sha.com*

Surrey EARS
Caspar B Pierce, *radio@surreyears.co.uk* Meets Wednesdays 17:00 to 21:00. Open only to University students and staff, University of Surrey, 46-AB-05 EARS SHACK, 388 Stag Hill, Guildford, Surrey GU2 7XH https://activity.ussu.co.uk/ears

Surrey Electronics Makers & Radio Club
Revd Graham Smith G4NMD Meets 3rd Wednesday of the month at 1930-2130, Grafham Room, Horsham Road, Grafham, GU5 0LJ *https://www.facebook.com/hamradiobuilders*

Surrey Radio Contact Club
Quin Collier G3WRR *secretary@srcc.uk* Meets 7.45pm - 9.30pm on 1st and 3rd Mondays, Jubilee Room, St. Paul's United Reformed Church, Croham Park Avenue, South Croydon CR2 7HF *www.srcc.uk*

Sussex 4x4 Response
David Green G4OTV *dave.w.green@btinternet.com* Meets last Friday of each 2nd month 8pm, Hassocks Hotel, Station Approach East, Hassocks, West Sussex BN6 8HN

Sussex Mid ARS
contact.msars1@gmail.com Meets 2nd & 4th Friday of the month, 19.30 - 21.30 GMT and 19.45 - 21.45, Cyprus Hall, Millfield Suite, Cyprus Road, Burgess Hill, West Sussex RH15 8DX *https://www.midsussexars.org.uk*

Sutton & Cheam RS
Chris Howard M0TCH. *secretary@scrs.org.uk* Meets Monday nights at 8pm, 7th Banstead Scout Hall, Woodgavil, Banstead, SM7 1DA *www.scrs.org.uk*

Swindon & DARC
Den, M0ACM *secretary@sdarc.net* Meets 7pm - 10pm every Thursday except July and August, Pinetrees Centre, Pinehurst Circle, Swindon, SN2 1RF *www.sdarc.net*

Trowbridge & DARC
Shaun Coles, G0BKU. *secretary.tdarc@gmail.com* Meets 1st & 3rd Wednesdays monthly at 8PM, Southwick Village Hall, Frome Road, Southwick, Wiltshire BA14 9QN *sites.google.com/prod/view/tdarc/home*

Waterside New Forest
Mr Tim Williams, G4YVY, 02380894278 Meets 8pm on 1st & 3rd Tues of month (not Aug) at, Applemore Scout HQ, near Hythe, Southampton, SO45 4RQ *www.watersideears.org.uk*

West Kent ARS
Phil Parkman G3MGQ, *secretary@wkars.org.uk* Meets 8pm on the 2nd Monday of the month, at Bidborough Village Hall, Bidborough, Kent, TN3 0XD *www.wkars.org.uk*

Wings Museum
B.Bloomfield, G4OKB, *bloomfieldbarrie@hotmail.com* Meets first Meets every Wed of each month at 11am, Wings Museum, Brantridge Lane, Balcombe, West Sussex RH17 6JT

Worthing & District ARC
secretary@wadarc.org.uk Meets every Tuesday doors open 7pm, TS Vanguard, 9a Broadwater Road, Worthing, W. Sussex BN14 8AD *www.wadarc.org.uk*

Worthing Radio Events Group
secretary@m0reg.co.uk Meets 1st Monday of the month at 8pm. Please contact the secretary for confirmation before visiting, Goring Conservative Club, 49 Mulberry Lane, Worthing, BN12 4RA *http://www.m0reg.co.uk*

Repeater Groups

Guildford UHF Repeater Group
Mr Alex Morris, G6ZPR *morris.alex@btconnect.comHolds* Yearly AGM Sanford ArmsEpsom Road, Guildford, Surrey *http://gb3gf.co.uk*

Kent Repeater Grp
chairman@krg.org.uk

Ridgeway Repeater Group
Robert Loss G4XUT *g4xut@rrg.org.uk* *https://www.rrg.org.uk*

Contest Groups

A1 Contest Group
Mr N Wilson, G4VVZ. *g4zap@aol.com*,

Addiscombe ARC
Mike Franklin G3VYI,

Invicta Contest Group
Ian Hope, *ian@dr.com* or Ian Lowe G0PDZ *g0pdz.ian@gmail.com* A Kent Based mainly VHF and above contest group. We normally enter contests such as the March 144/432, The May 144, 50mhz Trophy, VHF NFD, The Aug 144/432 Low power, the 144 trophy in Sep and the December 144 AFS. We currently do not enter HF contests. No Experience Required, Any Licence Level welcome. DA11 7EB *https://www.invictacg.co.uk*

Three A's Contest
Ian Pritchard, G3WVG, *g3wvg@btinternet.com*,

Vecta Contest Group ARS
Fred Dawson *fred.wp.dawson@googlemail.com* Church Crookham, Fleet Hants GU52 8LD

Region 11 - South West & Channel Islands

(http://rsgb.org/region11)

RR

Andrew Jenner, G7KNA
Email: *rr11@rsgb.org.uk*

DRs

District 111 (Cornwall and Scillies) Callum Macleod, G5XDX Tel: 07933 034528 Email: *dr111@rsgb.org.uk*

District 112 (North Cornwall and North Devon) John Lovell, G3JKL, Tel: 01237 478410 Email: *dr112@rsgb.org.uk*

District 113 (South Gloucestershire, Bristol and South and East Somerset) Andrew Jenner, G7KNA Tel: 07838 695471 Email: *dr113@rsgb.org.uk*

District 114 (Dorset) Vacant Email: *dr114@rsgb.org.uk*

District 115 (Jersey) Peter Bertram, GJ8PVL Tel: 07829 722722 Email: *dr115@rsgb.org.uk*

District 116 (Guernsey) Jerry Bligh, MU0ZVV Tel: 01481 243322 Email: *dr116@rsgb.org.uk*

District 117 (East Devon and West Somerset) Vacant Email: *dr117@rsgb.org.uk*

District 118 (South Devon) Nigel Bennetts, M7NGL Tel: 07966 133537 Email: *dr118@rsgb.org.uk*

Clubs & Societies

Appledore & DARC
John Lovell, G3JKL (*john@g3jkl.co.uk*) Meets 7.30pm on 3rd Monday in the month at the, Appledore Football Club, EX39 1PA *www.adarc.co.uk*

Bath and District Amateur Radio Club
P D Carter, G4PDC *badarc@protonmail.com* Meets at 7.30 on last Wednesday of each month, Farmborough Memorial Hall, Bath BA2 0AH https://badarc.webs.com

Blackmore Vale ARS
Keith Chadwick M0TMO; *keith.m0tmo@btinternet.com* Meets 7.30pm every Tuesday Apr-Oct 2nd and 4th Tues Nov-Mar 19.30 to 21.30, New Remembrance Hall, Remembrance field, Charlton, Shaftsbury, SP7 0PL *http://www.bvars.org.uk/*

Bristol RSGB Group
Paul Roberts G0OER (*secretary@g6yb.org*) Meets on the last Tuesday of the month, 589 Southmead Road, Bristol, BS34 7RG *https://www.g6yb.org*

Burnham ARC - Mike Lang
Mike Lang G4DVK *mikelangphotography@gmail.com* Meets 1st & 3rd Wed of the month 7pm to 9pm, The Bay Centre Cassis Close, Highbridge, Burnham on Sea, Somerset TA8 1NN *http://burnhamradioclub.co.uk/*

Callington ARS
secretary@callingtonradiosociety.org.uk Meets on the first Wednesday of each month, Callington Town Hall, PL17 7BD. Every Monday 8pm: Weekly net - GB3JL repeater. Every Wednesday: Lunch, The Engine House Café PL17 8EA from 1:30pm. Everyone welcome to all events, Council Chamber, Callington Town Hall, Callington, Cornwall PL17 7BD *https://www.callingtonradiosociety.org.uk/*

Christchurch ARS
Martin Clack M0KZC, *martinclacksp6@gmail.com* Meets 8.00pm on Thursdays at, The Clubhouse, adjacent to East Christchurch Sports, and Social Club, Grange Road, Christchurch, Dorset BH23 4JE *http://christchurchars.org.uk*

Cornish Radio Amateur Club
Steven Holland G7VOH, *g7voh@btinternet.com* Meets on the 1st & 3rd Thur of the month from 7.30pm at Gweal an Top School, School Lane, Redruth, TR15 2ER *http://gx4crc.com/*

Dartmoor Radio Club
Please get in touch via https://dartmoorradioclub Meets at the Yelverton War Memorial Hall on the last Thursday of the month. On all other Thursdays we hold virtual meetings via Zoom, Yelverton War Memorial Village, Meavy Lane, Yelverton, Devon PL20 6AL *https://dartmoorradioclub.uk*

Exeter ARS
Mr John Rooke, G4AP *rooke906@btinternet.com* Meets 1st and 3rd Tuesdays at 19:00, America Hall, De La Rue Way, Exeter, EX4 8PX *www.exeterars.co.uk*

Exmoor Radio Club
Simon Fouracres G8MBE *sfour@me.com* Meets last Tuesday of the month 19.30 - 21.00, Black Cock Inn, South Molton, EX36 3NW

Flight Refuelling
Sue Macdonald M0PSZ. *info@frars.co.uk* Meets twice a week Wednesdays and Sundays 19:00 to 22:00, Cobham Sports and Social Club, Merley Park Road, Merley, Wimborne, Dorset BH21 3DA *https://www.frars.co.uk*

Gordano ARG
Malcolm Pitt *mal@g4kpm.co.uk* Meets fourth Wednesday of the month at 20:00, The Ship Inn, 310 Down Road, Portishead, Bristol BS20 8JT

Guernsey ARS
Phil Cooper GU0SUP 07781 151747 Meets 7.30pm the every Friday at The Bunker, Beau Sejour, Amherst, St Peter Port., GY1 2DL *www.gars.org.gg*

Holsworthy ARC
Ken Sharman G7VJA, *ken@g7vja.co.uk* Meets the 1st Wednesday of each month at 7.30pm at the, Milton Damerel Parish Hall, EX22 6PS *http://m0omc.co.uk/*

Jersey Amateur Radio Society
Peter Bertram, GJ8PVL (*gj8pvl@hotmail.com*) Meets 8.00pm every Friday, The German Signal Station, Le Chemin des Siganux, Rue Baal, La Moye, St. Brelade, Jersey JE3 8LQ *https://jerseyars.org.je/*

Mid Somerset ARC
Chris lavis G6HIQ *christopherlavis3@gmail.com* Meets Shepton Mallet Baptist Church: 2nd Monday of the month at 19:30 also meets Quarter Jack: last Monday of the month at 14:00, The Prayer Room, Shepton Mallet Baptist Church, Commercial Road, Shepton Mallet, BA4 5BU, Quarter Jack, 18 Priory Road, Wells BA5 1SY *www.midsarc.org.uk*

Newquay & DARS
Terry 2E0XTM *newquayradioclub@gmail.com* Meets 2nd Wednesday of the month, Tel: 01841 540142, The Treviglas Community College, Bradley Road, Newquay TR7 3JA *https://newquayradioclub.co.uk*

North Bristol Amateur Radio Club
Dave Bendrey *g7byn@blueyonder.co.uk* Meets every Friday at 19.00, SHE7 Building, Braemar Crescent, Northville, Filton, Bristol BS7 0TD *www.nbarc.org.uk*

Plymouth Radio Club
Martin Mills M0MLZ *martinmills8448@gmail.com* Meets on the 2nd Tuesday of each month 19.00 hrs, Weston Mill Oak Villa Social Club, Ferndale Road, Weston Mill, Plymouth PL2 2EL *www.radioclubs.net/g3prc*

Poldhu ARC
secretary@gb2gm.org Tel: 01326 241 656 Meets on the 2nd Tuesday of the month, Marconi Centre, Poldhu Cove, Mullion, Cornwall TR12 7JB *http://gb2gm.org/*

Poole Radio Society
Charles Riley, G4JQX (*charles@fernpatch.com*) Meets 8pm every Thursday evening (except at Christmas and Easter) at, St Aldhelms Centre, Poole Road, Branksome, Poole BH13 6BT *www.g4prs.org.uk*

Radio Operators Cornwall
info@g8roc.org.uk Callum G5XDX 07933034528 We hold a variety of events, on air, on line and in person each month. Please check our website and social media for the next event. ALL are welcome, TR26 3LY *https://www.g8roc.org.uk*

Riviera ARC
Ian Nelson M0IDP *rivieraARC@gmail.com* Meets first and third Thursdays of every month Precinct Centre 19.00 - 21.00, Church Road, St Marychurch, Torquay, Devon TQ1 4QY *www.rivieraarc.org.uk*

Saltash District Arc
Mark Chanter M0WMB, *m0wmbsadarc@gmail.com* Meets second Tuesday of month at 7.30pm (Not Aug or Dec), Burraton Community Centre, Grenfell Avenue, Saltash, Cornwall PL12 4JB *www.sadarc.co.uk*

Shirehampton ARC
Mr Chidgey *secretary@shirehampton-*arc.org.uk Meets 7.30pm on Fridays at, T.S Enterprise, Station Road, Shirehampton, Bristol, BS11 9XA *https://www.shirehampton-arc.org.uk*

Sidmouth Amateur Radio Society
Dave Lee G6XUV Meets on Wednesdays at 19.00 throughout the year, our meetings vary during the year depending on seasonal variation., SARS, Sidford Community Hub, (rear of tennis courts), Byes Lane, Sidford Sidmouth EX10 9TB *https://www.facebook.com/groups/1207964666698144*

South Bristol ARC
secretariat@sbarc.co.uk - Meets 1st and 3rd Thursdays of the month from 19:30. Check our website (*www.sbarc.co.uk/calendar*) for event details, Novers Park Community Association, Rear of 124 Novers Park Road, Bristol, Avon BS4 1RN *https://www.sbarc.co.uk*

South Dorset RS
Ray Coles M0XDL, *raycoles960@btinternet.com* Meets 2nd Friday of each month at 19:30, unless otherwise specified on the website, Weymouth West Air Scout Group Hall, Granby Close, Weymouth, Dorset DT4 0SW *www.sdrg.co.uk.*

T.S.W.A.R.C
Tim Hugill G4FJK *tim@whitnole.com* Meets 8pm every Tuesday at, ROC Site, 7 Crosses, Tiverton, EX16 8JR *https://www.facebook.com/G4TSW/*

Taunton & District ARC
Peter Robinson G0EYR, *g0eyr@hotmail.com* meets the first Tuesday of each month, Tangier Scout and Guide Centre, Tangier, Castle Street, Taunton TA1 4AS *https://tauntonarc.wixsite.com/taunton-radio-club*

Thornbury & Sth Glos
T J Humphreys, M0TJX, *secretary@tsgarc.uk* Meets Every Wed Evening 1930 - 2130, The Chantry, 52 Castle Street, Thornbury, BS35 1HB *https://www.tsgarc.uk/*

Torbay ARS
John *membsec@tars.org.uk* Meets on Fridays from 19.30 at, Teignbridge District Scout Headquarters, Wolborough Street, Newton Abbot, Devon TQ12 1LJ *www.tars.org.uk*

Watcombe Radio Club
Mr Pete Worlledge M0BHJ, *peteworlledge@hotmail.com* Meets 7.30pm every Monday at Torbay Scout Camp Site, Easterfield Lane, Torquay, TQ1 4SW

Weston Super Mare Radio Society
The Secretary (*westonradiosociety@gmail.com*) Meets Mondays at 7:30pm, Devonshire Road Social Club., Weston-super-Mare BS23 4LG *www.g4wsm.club*

Yeovil ARC
Darren Mallinson 2E0EVU, *secretary@yeovil-*arc.com Meets 19.30 on Thursdays, Abbey Community Centre, The Forum, Abbey Manor Park, Yeovil, BA20 2BE *http://yeovil-arc.com*

Contest Groups

Castel Contest Club
ichard Allisette GU4CHY
www.castelcontestclub.com

Isle of Avalon Amateur Radio Club
Matt Morse - 2E0FNT - *2e0fnt@gmail.com* Meets Fridays from 6pm. We monitor 145.475 FM, Ivor's Paddock and Yard, (contact Matt 2E0FNT for how to get there, this is a field not a building) Glastonbury BA6 9AF https://*www.avalonarc.org.uk*

Repeater Groups

JJersey AR Repeater Group
Peter Bertram, GJ8PVL (*gj8pvl@hotmail.com*)Meets 8.00pm every Friday at Old German Signal Station, Rue Baal, Moye, Jersey JE3 8LQ *www.radioclubs.net/gb3gi*

Kidderminster Repeater Group
Ms P Dowie, G8PZT, *g8pzt@gb3kd.org.uk* Meets 3rd Tuesday of month at 8pm. Queens HeadWolverley Village, Nr Kidderminster, Worcs DY11 5XB
www.gb3kd.org.uk

Mid Cornwall Repeater Group
Paul Andrews G6MNJ

Region 12 - East & East Anglia

(http://rsgb.org/region12)

RR

David De La Haye, M0MBD
Tel: 07980 165172
Email: *rr12@rsgb.org.uk*

DRs

District 121 (Cambridgeshire)
Mervyn Foster, G4KLE, Tel: 01480 878111
Email: *dr121@rsgb.org.uk*

District 122 (Norfolk)
Vacant, Email: *dr122@rsgb.org.uk*

District 123 (Essex - North) Dave Cutts, M0TAZ (temp) Tel: 07506 035599
Email: *dr123@rsgb.org.uk*

District 124 (Essex - South)
Dave Cutts, M0TAZ, Tel: 07506 035599
Email: *dr124@rsgb.org.uk*

District 125 (Suffolk)
Iain Moffatt, G0OZS, Tel: 01449 766089
Email: *dr125@rsgb.org.uk*

Clubs & Societies

Bishops Stortford
Tony Judge, G0PQF, *g0pqf@hotmail.com* Meets 8.00pm on third Monday of the month (check on website for occassional changes), Farnham Village Hall, Rectory Lane, Farnham, Essex CM23 1HU
www.bsars.org

Bittern DX Group
Linda Leavold G0AJJ *bdxg.secretary@gmail.com* Meets last Thursday in every month 19.30 to 21.30, Erpingham Arms, Eagle Road, Erpingham, Norfolk NR11 7QA
www.bittern-dxers.org.uk

Braintree & DARS
Geoff Nurse G1GNQ *secretary@badars.co.uk* Meets 8.00pm on 2nd & 4th Tuesdays in the month at the, St Peters Church Hall, St Peters Road, Braintree, Essex CM7 9AR
www.badars.co.uk

Bury St Edmunds ARS
Mr Melvin Green, *m0iid@bsears.co.uk*
Meets on the third Wednesday each month at the, Rougham Tower Museum, Rougham, IP32 7QB *www.bsears.co.uk*

Cambridge & DARC
Bryan Davies, M0IPO *publicity@cdarc.org.uk* Meets on 2nd and 4th Fridays of the month 19.30 - 21.30, Coleridge College, Radegund Road, Cambridge, CB1 3RJ
www.cdarc.org.uk

Cambridge Univ WS
Dan McGraw M0WUT, *chairman@g6uw.org* Meets Thursdays 8pm at the, Maypole Public House, Park Street, Cambridge, CB5 8AS *www.g6uw.org*

Chelmsford Amateur Radio Society
Paul Tittensor *secretary@g0mwt.org.uk*
Meets first Tuesday of each month at 19.45 to 22.30, Danbury Village Hall, Main Road, Danbury, nr Chelmsford, Essex CM3 4NQ
https://*www.g0mwt.org.uk*

Colchester Radio Amateurs
Graeme Chalklin G3CO *treasurer@g3co.uk* Meets 7.30pm on the third Thursday of each month at 19.30 - 21.30 facebook.com/groups/1928805160709742, Wilson Marriage Centre, Barrack Street, Colchester, Essex, CO1 2LR https://g3co.uk

Dengie Hundred ARS
Roger Jones G7RGR *rjonesa@aol.com*
Meet Thursday's 09:30 - 13:00, Oak Tree Bungalow, The Endway, Althorne, Essex CM3 6DU *www.dhars.org.uk*

Felixstowe & DARS
Mark Riley M5BOP *m5bop@hisimage.co.uk* Meets 7:30pm on first, third (and if present, fifth) Mondays in month, Suffolk Aviation Heritage Museum, RAF Foxhall, Foxhall Road, IP16 4JU *www.fdars.org.uk*

Harlow & DARS
The Secretary, *secretary@g6ut.com* Meets 8.00pm every Friday until late, Mark Hall Barn, First (Mandela) Avenue, Harlow, Essex CM20 2LE https://dartmoorradioclub.uk

Harwich ARIG
Kevin Francis M0JVC *g0rgh@amsat.org*
Meets at 8pm on the 2nd Wednesday of each month at, Park Pavillion, Barrack Lane, Dovercourt, Harwich, Essex CO12 3NS *www.harig.org.uk*

Havering & DARC
Alan Paul, *HaveringRadioClub@gmail.com* Meets every Wednesday 20.00 to 22.00, Fairkytes Arts Centre, 51 Billet Lane, Hornchurch, Essex RM11 1AX
www.haveringradioclub.co.uk

Huntingdonshire ARS
Shaun Hagan, 2E0FFK: *hars.secretary@gmail.com* Meets on 2nd and 4th Thursdays in the month, 7.30pm - 9.30pm at, Buckden Village Hall, Burberry Road, Buckden, St Neots PE19 5UY
http://hunts-hams.co.uk

Ipswich Radio Club
John Gee, G4BAV, *g4irc@icloud.com*
Meets every Wednesday at 19.30 except 1st Wed of month at our contest site, Shrubbery Farm, Otley, IP6 9PD
www.qrz.com/db/G4IRC/

Kings Lynn ARC
Mr Eric Allison G4JNQ 01485600587 Meets every Thursday 19:30 to 22:00, The Scout Hall, Chequers Lane, North Runcton, Kings Lynn, Norfolk PE33 0QN *www.klarc.org.uk*

Leiston ARC
Debbie Lucock 2E0ICX, *secretary@larc.org.uk* Meets 2nd Tuesday of the month at, Leiston Community Centre, Sizewell Road, Leiston, Suffolk IP16 4JU *www.larc.org.uk*

Loughton & Epping Forest (LEFARS)
David Priest, M0VID, *info@lefars.org.uk*
Meets fortnightly on Fridays at 19.45 . Zoom Video Conference on alternate Fridays, All Saints House, Romford Road, Chigwell Row, Essex IG7 4QD *www.lefars.org.uk*

Lowestoft District & Pye ARC
Tim Ward,2E0TJW, *secretary@ldparc.co.uk*
Meets Thursdays at 20.00 at, Clubhouse Victoria Road, Lowestoft, Suffolk, NR33 9LY
http://ldparc.co.uk/index.htm

Martlesham RS
John Quarmby, G3XDY *g3xdy@btinternet.com* Meets last Saturday of every month at 9.30AM, IP3 9RZ, Also quarterly lecture programme at, BT Adastral Park, IP5 3RE
www.mrsap.org

Norfolk ARC
David Palmer G7URP *radio@dcpmicro.com* Meets every Wednesday evening 19.00-21.30 at CNS School or via our online show NARC Live which starts at 19.30. Our website has details of our programme and meeting places. Watch live on BATC: https://batc.org.uk/live/NARC or via Facebook Live: https://*www.facebook.com/norfolkamateurradioclub*, City of Norwich School, Sixth form common room, Eaton Road, Norwich, Norfolk NR4 6PP
www.norfolkamateurradio.org

Norfolk Coast ARS
Richard Leeds, G4RZN Tel: 07767 492456
Meets most Thursdays 10.00 to 14.00, Mill Lane, East Runton, Norfolk NR27 9PH
www.norfolkcoastamateurs.co.uk

Norfolk Tank Museum
Tony Francis M0XTF *m0xtf@outlook.*

com Meets most Wednesday mornings at 11:30am between April and October, Norfolk Tank Museum, Station Road, Forncett St Peter, Norwich, Norfolk NR16 1HZ *http://norfolktankmuseum.co.uk/radiocollection*

North Norfolk ARG
01263 824275, *g4nre.uk@gmail.com* Meets Thursdays at 10am, Mr Ward, Cromer, 88 Central Road, Norfolk NR27 9BW https://*www.facebook.com/groups/134778936544632/*

Peterborough And District ARC
Alan Ralph *secretary@padarc.co.uk* Meets second & fourth Wednesday of the month Doors open at 19.00 to 21.30, Mace Road Church Hall, Mace Road, Stanground, Peterborough, Cambs PE2 8RQ *www.padarc.co.uk*

South Essex ARS
sears.enquiries@gmail.com Meets 2nd Tuesday of the month at 7.30pm, St Michaels Church, St Michaels Road, Daws Heath, Benfleet Essex SS7 2UW *https://g4rse.co.uk/*

Sudbury & DRA
Tony Harman G8LTY, *contact@sudburyradioamateurs.co* Meets 2nd Wednesday in the month at 8PM, Room 1, The Stevenson Centre, Stevenson Approach, Broom Street, Great Cornard, Sudbury CO10 0WD *https://sudburyradioamateurs.co.uk*

Thames Amateur Radio Group
Andy Atkinson M0IXY 07832 978681 Meets 1st Friday of the month at 20:00, Radioactive Night Meeting 3rd Friday of the month at 20:00. Club nets on GB3DA (or 145.300 if GB3DA not available) on other Fridays of the month at 20:00, Jubilee Hall, Waterside Farm Sports Centre, Somnes Avenue, Canvey Island SS8 9RA *www.thamesarg.org.uk*

Vange ARS
vars@live.co.uk Meets 1st Thursday of the month at 8:00pm, St Gabriels Youth And Community Centre, Rectory Road, Basildon, Essex SS13 2AA *https://www.vangeradio.org.uk*

Wisbech AR & Elec. C
Alan Bridgeland, M0DUQ *m0duq@talktalk.net* Meets Mondays 7.30 pm Elme Hall Hotel, 69 Elm High Rd, Wisbech, PE14 0DQ *www.warec.org.uk*

Repeater Groups

Cambridgeshire Repeater Gp
Phil Nice, G8IER Email: *secretary@cambridgerepeaters.net* Meets 7.30pm the first Wednesday of the month at The Carpenters Arms, Great Wilbraham, CB21 5JD. *www.cambridgerepeaters.net*

Essex Repeater Group
Mr Murray Niman, G6JYB *secretary@essexrepeatergroup.org.uk* Meets Danbury - Church Green c/o Danbury Village Hall, Chelmsford, CM3 4NQ *https://www.essexrepeatergroup.org.uk*

Contest Groups

Blackwater AR Contest Group
Clive Bennet M0BRT *cbenn10453@aol.com, www.barcg.org.uk*

Magnetic Fields Contest Group
Paul Marchant,

Secret Nuclear Bunker Contest Group
George Smart, *george@george*-smart.co.ukThe group holds informal meetings at the J. J. Moons Wetherspoons Public House in Hornchurch, Essex., Kelvedon Hatch Secret Nuclear Bunker, Crown Buildings, Kelvedon Hall Lane,Brentwood Essex CM14 5TL *http://www.gb0snb.com/wordpress/snbcg*

Region 13 - East Midlands

(http://rsgb.org/region13)

RR

Bob Hambly, M0HAF Tel: 07447 945601
Email: *rr13@rsgb.org.uk*

DRs

District 131 (Leicestershire & Rutland) David Jacobs, 2E0DFJ, Tel: 07407 716445 Email: *dr131@rsgb.org.uk*

District 132 (South Derbyshire & South Nottinghamshire) Simon Strange, M0SYS Tel: 07856 809130 Email: *dr132@rsgb.org.uk*

District 133 (North Nottinghamshire & North Derbyshire) Dr John Rogers, M0JAV Tel: 07836 731544 Email: *dr133@rsgb.org.uk*

District 134 (Northamptonshire) Bob Hambly, M0HAF Tel: 07447 945601 Email: *dr134@rsgb.org.uk*

District 135 (Lincolnshire South) Graham Boor, G8NWC Tel: 07754 619701 Email: *dr135@rsgb.org.uk*

District 136
Vacant Email: *dr136@rsgb.org.uk*

District 137 (North of South Lincolnshire) Andrew Gilfillan, G0FVI Tel: 07909 680047 Email: *dr137@rsgb.org.uk*

Clubs & Societies

Bolsover ARS
Mr Alvey Street, G4KSY *streeta@suevin.com* Meets 1st & 3rd Wednesday of the month from 8pm, Bainbridge hall, Chapel street, Bolsover, Derbyshire S44 6PX https://*www.g4srb.co.uk*

Buxton RA
Mrs C Royle M0NHG *royle495@btinternet.com* Meets 8.00pm on 2nd and 4th Tuesday in the month at the, Leewood Hotel, Buxton, SK17 6TQ *https://buxtonradioamateurs.wixsite.com/buxton-radio-club*

Hinckley ARES
Mark Burrows 2E0SBM *hinckleyradiosociety@gmail.com* Meets every 2nd and 4th Wednesday from 20.00, Britannia Scout Hut, Britannia Road, Burbage, Hinckley LE10 2HE *http://hinckleyares.co.uk*

Hucknall Rolls Royce ARC
Mark Attenborough, *Secretary@hrrarc.com* Meets at 8.30pm evey Friday but doors usually open for 8pm, Hucknall Rolls Royce Leisure Association, Gate 1, Watnall Road, Hucknall, Nottingham NG15 6BU *www.hrrarc.com*

Kettering & DARS
Les Moyles 2E0MNZ *les@moylehousehold.co.uk* Meets at 7:00pm on Tuesday evenings and from 10:00-12.00 Sundays, Harrington Aviation Museum, Sunnyvale Farm Nursery, Off Lamport Road, Harrington, Northants NN6 9PF *https://www.g5kn.org*

Leicester RS
Sandra Morley *nbcymar4@hotmail.co.uk* Meets every Monday at 19.00 - 22.00 pm, Gilroes Cottage, Groby Road, Leicester, LE3 9QJ *http://g3lrs.org.uk*

Lincoln Short Wave
Mrs Pam Rose, G4STO Meets 8.00pm Wednesdays (and from 9.00m to 12 noon on Saturdays in shack), LSWC C/o BSA Social Club, Village Hall Lane, Aisthorpe, Lincolnshire LN1 2SG *www.g5fz.co.uk*

Loughborough & DARC
Mr Chris Walker, G1ETZ, *g1etz@aol.com* Meets Tuesday evening from 19.30, Glenmore Community Centre, Thorpe Road, Shepshed, LE12 9LU *www.radioclub.org.uk*

Melton Mowbray ARS
Graham Mason G4PTK - 01664410733 - 07511416613 Meets 19.30 on 3rd Friday of the month, except July & August, The Edge Community Centre, Dalby Road (corner of Queesway), Melton Mowbray, Leics LE13 0BQ *https://www.facebook.com/groups/383420802558219*

Northampton Radio Club
Richard Kellow *M0RKJsecretary@northamptonradioclub* Meets on a Thursday from 20:00. If visiting please email *secretary@northamptonradioclub.co.uk* beforehand. Non-members are welcome to visit up to 3 times., The Grangewood Club, 50 Barn Owl Close, Northampton, NN4 0SL *www.northamptonradioclub.co.uk*

Northampton Scout
NSARG Team *nsarg@nsarg.co.uk* Meets 3rd Saturday every month, Overstone Scout Activity Cntr., Northampton, NN6 0AF *http://nsarg.servehttp.com/*

Nunsfield House ARG
Mr Adrian Price, G1OXH. *sec@nharg.org.uk* Meets 7.45pm on Fridays at the, Nunsfield House, 33 Boulton Lane, Alvaston, Derby DE24 0FD *www.nharg.org.uk*

RAF Waddington ARC
Mr Bob Pickles, G3VCA Meets 7.30pm Thursdays at, The Pyewipe, Fossebank, Saxilby Road, Lincoln, LN1 2BG *www.g0raf.co.uk*

Sherwood Amateur Radio Club
c/o Edward Rippon *sherwoodarc@gmail.com* Meets at, 319 Beechdale Road, Nottingham, NG8 3FF *http://mx0gzd.org*

South Derby & Ashby Woulds ARG
sdawarg.chairman@gmail.com Meets 7.00pm on Wednesdays 19.00 to 21.00, Moira Replan Centre, 17 Ashby Road, Moira, Swadlincote, Derbyshire DE12 6DJ *www.sdawarg.org*

South Kesteven ARS
Stave Marsh Meets 1st and 3rd Saturdays of each month at 10am, Rookery Lane, Sudbrook, Grantham, Lincolnshire NG32 3RU *www.skars.co.uk*

South Normanton, Alfreton And District ARC
Alan Jones, M0OLT *secretary@snadarc.com* Meets Mondays 7:30pm, except Bank Holidays, Post Mill Centre, Market Street, South Normanton, Derbyshire DE55 2EJ *https://www.snadarc.com*

South Notts ARC
secretary@snarc.org.uk Contact Page on the website Meets 7.00pm Mondays see the website, Mapperley Plains Recreation and Social Club, Plains Rd, Mapperley, Nottingham NG3 5RH *www.snarc.org.uk*

Spalding & DARS
Graham Boor, G8NWC. *secretary@sdars.org.uk* Meets Every Friday at 19.30, West Pinchbeck Village Hall, Six House Bank, West Pinchbeck, Nr Spalding PE11 3GQ *www.sdars.org.uk*

The Gliding Centre Radio Amateur Society
email *gx1idr@outlook.com* Meet every 1st Saturday of the month 1400 onwards, The Gliding Centre, Husbands Bosworth Airfield Sibbertoft Road, Lutterworth, Leicestershire LE17 6JJ

Welland Valley ARS
Peter Rivers G4XEX, *info@wvars.com* Meets 7.30pm on third monday in the month at the, Great Bowden village hall, Great Bowden, Market Harborough, Leics LE16 7EU *www.wvars.com*

Worksop ARS
G.Lebond-Carroll M7WSP *gayle@lebond.co.uk* Meets on Tuesdays 6.30pm till 11.00pm, 59/61 West Street, Worksop, Notts, England S80 1JP *https://www.facebook.com/g3rcw*

Repeater Groups

Leicestershire Repeater Group
Phil Taylor M0VSE *phil@m0vse.uk* *www.leicestershirerepeatergroup.org.uk*

Contest Groups

Blacksheep Contest and DX ARS
Stephen Purser G4SHF *stephen@blacksheep.org*,

Five Bells Group
Mr B K Tatnall, G4ODA, Active in VHF/UHF contests

Leicestershire Foxes Contest Group
Adam Moss/Dabid Carter: *adyg6ad@gmail.comMeets* Thursdays at 20.00, Gilroes Cottages, The Chantry, Leicester LE3 9QT

Parallel Lines
g4lip@plcg.orgMeeting Information: Ad-hoc meeting times about once a month, The Hollybush, Main Street, Ashby ParvaLeicestershire LE17 5HS *https://www.plcg.org*

Weekend Contest Group (WCG)
Steve Hambleton G0EAK *g0eaksteve@gmail.com*
As this is a contest group we don't meet very often. The group was formed a few years ago mainly for VHF and up RSGB weekend contests. The members are mostly active UKAC contesters who submit their UKAC logs for their main club but that club doesn't do weekend contests. We mainly operate in the RSGB AFS Superleague and VHF Championship. Further information available from G0EAK or G1PPA, DN22 7DX

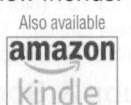

Training and Exams

Who was to know back in March 2020 just how great an impact Covid-19 would have on all of our lives and how it changed the way that those with an interest in amateur radio were able to join what is a fascinating hobby. All amateur radio exams were suspended with little idea of when they might resume. However, after a couple of weeks the Society was able to introduce a scheme centred on remotely invigilated online exams, initially at Foundation level with Intermediate and Full being introduced a little later, that has proven to be hugely successful and has resulted in twice as many candidates taking the amateur radio examinations.

Exams, Post Covid

Before the 2020 lockdown anyone wishing to obtain a Foundation amateur radio transmitting licence was required to undertake some mandatory practical assessments and having completed those assessments take a multiple-choice exam. This would usually happen under the watchful eye of club assessors and invigilators in a club environment. Anyone wishing to progress to an Intermediate licence were similarly required to complete a number of practical activities essentially based around construction, before taking an Intermediate exam, again under the watchful eye of club assessors and invigilators. To take that final step up to a Full licence it was simply a case of taking a 58-question theory-only exam, still under the watchful eye of club invigilators.

Lockdown and social distancing rules meant however that clubs were unable to meet, either for their usual club meetings but also for any sort of training or exam sessions

Since, as agreed by Ofcom, it's the RSGB which has the responsibility for making amateur radio exams available in the UK and in order to be able to resume exams, initially at Foundation level, the RSGB had to make a few quick decisions. Firstly, what to do about practical assessments which under the Covid restrictions couldn't take place and secondly how might it be possible to start running exams again using remote invigilators? After some discussions between various individuals and groups, including Ofcom, and after a period of testing a solution was found and put in place. In answer to the first challenge it was agreed that all practical assessments would be suspended. As time went on it was confirmed that the practical assessments at both Foundation and Intermediate would be scrapped altogether. As regards how to run exams using remote invigilation this required only a slight adjustment to the TestReach online exam system that the Society first introduced way back in March 2017.

Clearly because of lockdown and social distancing it wasn't, for the foreseeable future, going to be possible to run exams in the traditional exam room environment but it was agreed by the relevant parties that it would be possible for candidates to take an online exam in the safety and comfort of their own homes, generally on a day and time to suite them, using remote invigilation. Prior to lockdown clubs ran online exams in exactly the same way that they ran paper exams ie with candidates taking their exam sitting in an open room whilst being monitored by invigilators. Remote invigilators carry out those exact same functions as a face to face invigilator would do, the only difference being that they aren't in the same room as the candidate but are still able, using video conferencing facilities (Zoom, Webex etc), to keep a watchful eye on candidates taking an exam. The success of such a scheme depended on being able to recruit a team of trusted volunteers to take on the actual remote invigilation and in this respect some 30 volunteers offered their services.

The TestReach exam software allows students to get used to the system ahead of their exam through an inbuilt tutorial. The software employs various techniques that prevent students from simply using their computer to look up answers online. The exams themselves are conducted under strict conditions, with the remote invigilator carrying out a virtual 'tour' of the student's exam area using the student's own webcam before the exam is allowed to start, and then continues to watch and listen to the student throughout their exam, again using the candidate's own webcam. The whole exam session is recorded for audit purposes, the recordings being destroyed 5 days after the exam.

On instruction from the invigilator, the exam starts, students are monitored throughout the process, and as soon as the student finishes their exam, they are presented with the outcome of their exam (pass or fail) their score, and as observed by many of the remote invigilators this often results in huge smiles, thumbs up and sighs of relief.

Confirmation of the exam pass is sent out from RSGB HQ to candidates via Royal Mail including a pass certificate, and at the same time the RSGB sends the candidate's data to Ofcom. Within a matter of just a few days, successful candidates can log into their Ofcom account, choose their new callsign, download a PDF of their licence, and be on the air.

The Intermediate and Full exams require the use of a second webcam, for example, using the video feature on a smartphone, to give an additional and different view of the candidate while they are taking their exam.

The changing landscape

The fact that twice as many newcomers joined the hobby in 2020 and 2021 compared to the previous year indicates the sheer popularity and convenience of being able to sit an exam at home.

A recent survey of over 1,000 newcomers conducted by the online group Essex Ham shows that 49% of respondents rated the ability to take exams at home as a key reason for deciding to join the hobby, with "spare time due to the lockdown" coming in at 34%. Over 95% of newcomers said that it was easy to arrange their online exam, and over 99% reported a positive experience with the online invigilator.

Whilst the numbers joining the hobby for the first time has hit new hight's there are other benefits to be considered, not least progression. In a similar way to the increase in newcomers to the hobby it's been observed that progression through the 3 levels has increased

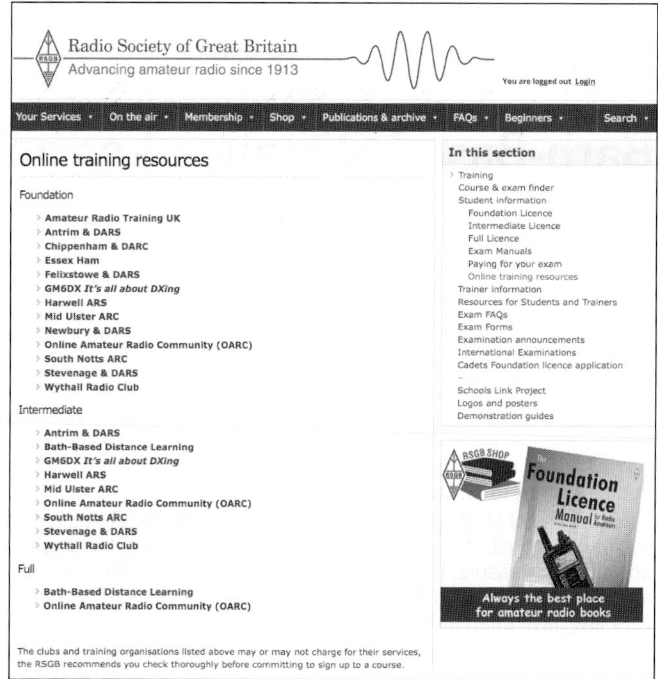

Online training resources

significantly. The 2020 Intermediate numbers, bearing in mind that exams only resumed in August 2020, are over double what they were in 2019 and in a similar way the 2020 Full numbers equate to double what they were in 2019.

Distance Learning

Traditionally, the RSGB's exam syllabus has been taught to prospective new amateurs in village halls, scout huts and community centres across the UK by a network of volunteers at amateur radio clubs. A Foundation course could typically be taught over a weekend or spread across a few weekday evenings. A structured club course would normally include the hands-on mandatory practical activities, mixed with "chalk-and-talk" presentations by club members.

The Covid-19 lockdown put paid to all 'in-person' training, and with exams being possible to be taken at home, there was a sizeable demand for online training. Distance learning courses were nothing new, and the Bath-based Distance Learning Course, pioneered by Steve Hartley G0FUW, had been successfully teaching the Full course nationally for some time and was responsible for over 28% of all Full passes between 2011 and 2019. The online group Essex Ham had launched a remote Foundation course in 2015 to supplement the work of local clubs. The Online Amateur Radio Community (OARC) and GM6DX's "It's all about Dxing", as well as offering online Foundation courses also offer online Intermediate courses.

You can find a list of known online course providers on the RSGB site:

https://rsgb.org/main/clubs-training/for-students/online-training-resources-for-students/

Groups providing online training saw a massive increase in the number of candidates, and many clubs that were unable to teach in the classroom quickly adapted to the online method, with training being done via video conferencing or in online virtual classrooms. Many of the online courses are free, as previous overheads such as venue hire costs don't apply online. With around a dozen clubs offering online training, many believe that getting started in amateur radio has never been easier, cheaper or more convenient.

The advantages of online study include the ability to study at a time of the students' choosing, the option to review course material again, and the opportunity to study "on the go" from a smartphone or tablet. Course videos have proven especially useful for those with dyslexia, who need extra help or find it hard to study from the written word alone.

During 2020 the RSGB launched a series of videos to help those that had taken their remotely invigilated online exam without being able to do the practical assessments – see link below. In 2021 a series of "Useful practical skills videos" was introduced, whilst many of the *Tonight@8* videos cover topics that newcomers should find useful – see links below

Online study isn't for everyone, and for some, face-to-face time with a non-virtual tutor will be the preferred option, but there's little doubt that during the lockdown, thousands have studied and passed online in record time, without the need to leave home.

Of course, there's still very much a place for clubs in the new online world, as many students still prefer the personal touch, and potentially need extra help and support. Those who've studied online will no doubt still be keen to meet up with others, ask questions, get hands-on with radio equipment, and enjoy the social interaction that only a friendly local amateur radio club can offer.

Getting Results

Since the introduction of remotely invigilated online exams in April 2020, over 6,400 people have sat a remotely-invigilated exam, with almost 4,300 newcomers passing their entry-level Foundation exam online. In July 2020, under the same scheme, the Intermediate exam was made available and since then almost 1,400 have sat the Intermediate exam. Finally in October 2020 the Full level exam was made available and to date over 750 candidates have sat the Full exam online again using remote invigilation.

What about candidates with special needs such as visual impairment, mobility issues etc? As has always been the case candidates with special requirements are encouraged to make contact with the exam office (01234 832 717 or *exams@rsgb.org.uk*) directly to discuss options that are open to them. Each case is looked at individually.

There's little doubt that online training and exams have been a significant success. 2020 saw 2,774 newcomers get their first licence – over double the number of Foundation passes in 2019. The move to remotely invigilated online exams has brought many people into the hobby who've previously not been able to get involved, including those who don't have a club near to them, are restricted in terms of travel or mobility, have busy lives with little free time, or who appreciate the convenience of being able to study and pass at home at a time of their choosing.

For further information regarding training and examinations check out the Education pages on the RSGB web pages:

www.rsgb.org/main/clubs-training

Bath-Based Distance Learning

Background

The Bath-Based Distance Learning (BBDL) courses have been developed over a number of years by the Bath Radio Classes Team for use in parallel with our classroom training. The Distance Learning Team includes around fifteen volunteer tutors spread throughout the UK, led by Steve Hartley, G0FUW.

The process is tried and tested and has been used by many Distance Learning students; it has helped nearly 1000 students to achieve exam passes at the Full Licence level since the first course started in Jan 2011. Between 2011 and 2019 some 28% of all UK exam passes at that level were BBDL students.

Our successful students have come from a diverse cross-section of society and have included GCSE, A-level and university students, university professors, a street cleaner, a retired cat breeder, a welder, a member of a famous Brit-pop band, and a national TV presenter; very few have been electronics wizards or RF engineers!

Our track record shows that the vast majority of students who see the course through to the end pass the exam; we have consistently had pass rates higher than the national average.

How It Works

The team took a year out to rework their training material and align it with the 2019 syllabus and associated RSGB textbooks. There are now 16 weeks of study allocated to each of our Intermediate and Full level courses; this includes some revision weeks and the examination. We have no plans to offer any specific training for the Direct to Full examination.

Weekly work packages are posted using a Virtual Learning Environment called Moodle. We provide 'how to' guidance on how to use the Moodle system; no prior knowledge of the system is required.

We try to bring the theory to life with practical exercises where possible. These can involve making up some simple circuits and carrying out measurements, or making use of on-line resources. These are not mandatory, as they are not required for the exam, but we include them because students have told us that they are extremely useful in understanding the theory set out in the textbooks.

BBDL does not charge any fees for the courses, we do it for the love of amateur radio, and because we all had folk that helpsed us when we were learning.

Pre-Course Work

Before each course, we set up We have a pre-course classroom set up in Moodle for anyone wanting to join the courses. This allows us to check that potential students are able to access and use our systems and it allows potential students to decide if the format suits them.

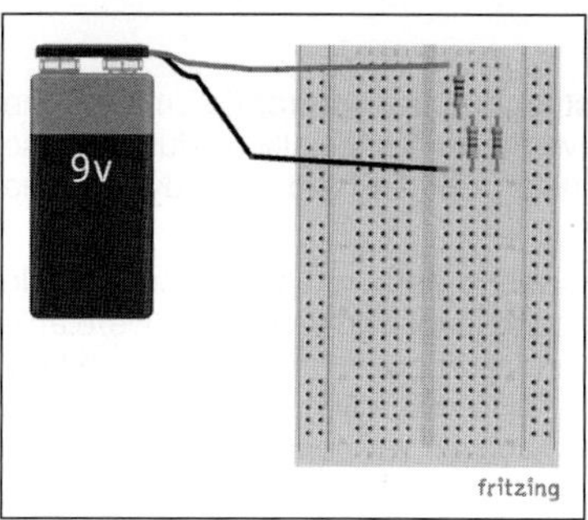

Volts and Amps in Series and Parallel

The pre-course classroom has some tutorial videos to watch, some reading, and a quiz; if potential students have watched the video and read the words, they should be able to answer the quiz questions. This work covers topics from the previous level that have been added since 2019, so anyone passing before then is able to get up to speed.

We have a limited number of places available and our courses have always been very popular. If we are over-subscribed, we now allocate places based on results in the pre-course classroom.

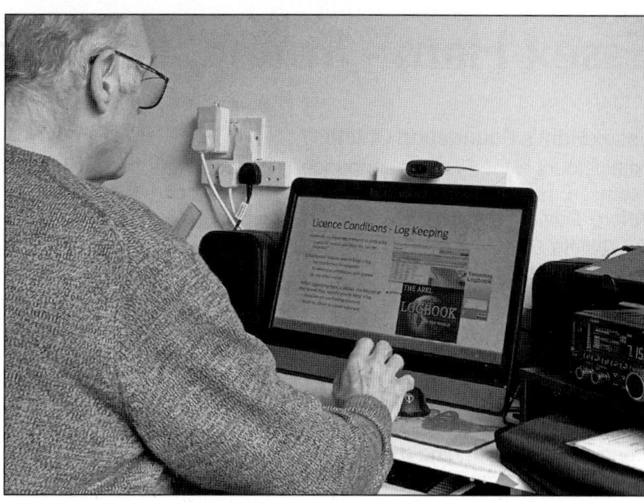

Phil Miles, M7XCQ (now M1XCQ), accessing a BBDL tutorial. Phil was one of many who have completed both the Intermediate and Full Licence courses with BBDL.

Student Commitments

Students must have internet access and are required to have their own copy of the RSGB textbook, a scientific calculator and a copy of the Exam Reference Data Booklet.

Students need to complete the weekly work packages within the allocated timescales. This normally involves:

- Attending the virtual tutorials, or watching the recordings later.
- Reading some pages of the RSGB textbook
- Carrying out any additional exercises and/or watching recommended videos.
- Answering around 10-20 revision questions based on the previous week's work.

Every third fourth week there are additional revision questions looking back to the beginning of the course and, at the end of the course, we provide a number of Mock Exams.

Potential students should not underestimate the time that is required to complete the weekly work packages; most students spend somewhere between three and four hours studying every week, some have said they spend up to eight hours a week studying and revising; doing a distance learning course is not a walk in the park and being able to dedicate time to studying is absolutely essential.

Tutor Commitments

Tutors commit to provide feedback and the 'worked answers' to quizzes within a few days of student submissions. Tutors answer specific questions about course material within a reasonable timescale, subject to the time they have available to do this voluntary work. The Tutors also monitor progress of all students and identify any weak areas for revision time.

Examinations

Students are responsible for making appropriate registration arrangements and paying for their exam. We advise against booking an exam until after Week 9; by then Tutors have a good idea if students are going to be ready, or not.

Timetable

The plan going forward is to run an Intermediate course January to April and a Full Licence course August to December, but that may change.

Up to date information can be obtained by sending an e-mail to *g0fuw@tiscali.co.uk* **Steve Hartley**, G0FUW

Scotland-Based Distance Learning - GM6DX

Based in Central Scotland GM6DX is run by Billy McFarland who's callsign this is. Training changed dramatically in 2020. Previously Billy had been used to conducting basic classes at Stirling and District Amateur Radio Society and spending some hours taking trainees through the practical assessment aspect of the training. With the lack of club attendance due to a national pandemic another platform was needed to provide my notes for people to access. GM6DX chose the thinkifc platform to access courses, mainly because its main purpose is online training and that it offers a lot of features at no cost.

After 6 weeks of continually editing and creating notes based on the Foundation examination syllabus this took the form of PDF documents and videos which helps explain the examined subjects. At the end of each section a small quiz allows the students to identify potential training gaps. As the student comes to the end of their course, they will find 10 or so mock examinations to give them a feel for what the actual exam may be like. What GM6DX didn't expect when creating this online course was the level of take up nationally.

From creation well over eight thousand people have signed up for the Foundation and Intermediate Courses with a very high success rate at both levels. Support to students who prefer a virtual classroom was created as a Facebook group, known as UK Intermediate Amateur Radio Licence. This group now numbers around 1000 members from various backgrounds. People post syllabus related questions which are explained in depth a true virtual classroom for all.

GM6DX now concentrates training mainly on Foundation and Intermediate although Billy is starting to offer Full licence for small group numbers when time permits. Overall, 2022 was a great success on the previous year and he hopes that many people make use at this free offering of amateur radio courses.

To enrol for GM6DX free licence courses then visit:
https://gm6dx.thinkific.com

Essex Ham - Foundation Online Courses

Essex Ham's Foundation Online

Within hours of the RSGB's announcement about remotely-invigilated exams, Essex Ham saw a massive spike in course enrolments. Before Covid, around 100 enquiries per month was common, but this quickly rose to 750 enquiries in the first month. Within that initial year, over 3,800 enrolled on a course and started their journey into amateur radio.

The course originally launched in 2015, aimed at students with no prior experience of radio, electronics and RF theory. The course structure follows the same modular format as the RSGB's Foundation Licence Manual, for ease of cross-referencing. Three modules are released per week with a quick test at the end of each module, and mock tests along the way. With nine modules in total, the course runs for three weeks, also a 'Fast Track' version is available too.

The RSGB's popular 'Train the Trainers' sessions highlight that students study in different ways, which is why the course material is presented in different ways – in written format, in the form of bullet-point slides that can be printed, and a series of videos covering the entire course content, as well as several topics outside of the exam syllabus that every new radio amateur needs to know about. The course uses a web-based virtual classroom that allows for group discussion of topics, plus private messaging if extra help is required.

The course is supplemented by frequent webinars where the training team answer common questions about the syllabus and exam, as well as demonstrating amateur radio with live contacts (simplex, and via a repeater), radio operation, data modes and operating 'nets.

The advantages of online study include the ability to study at a time of the students' choosing, the option to review course material again, and the opportunity to study "on the go" from a smartphone or tablet. The course videos have proven especially useful for those with dyslexia, who need extra help or find it hard to study from the written word alone.

Online study isn't for everyone, and in-person tuition is once again possible at local clubs, but during the lockdown thousands were able to study for, and pass, their Foundation exam online in record time without the need to leave home. With restrictions now lifted, Essex Ham continues to see high demand for its popular online course into 2022.

Also you can book a place for a free Foundation Online course at:

www.hamtrain.co.uk

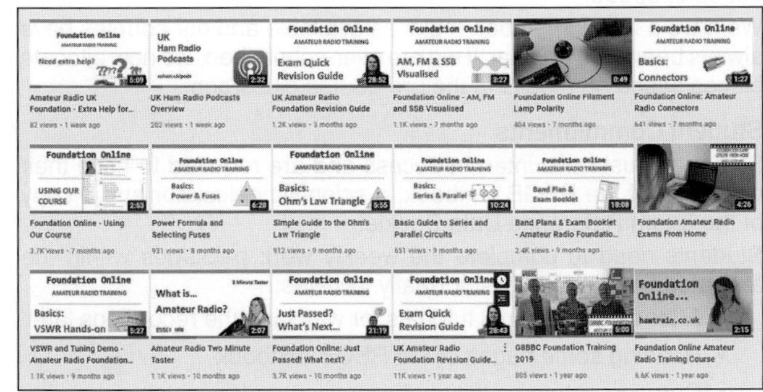

Essex Ham have a range of YouTube Training Videos for the Foundation Course

Online Amateur Radio Community (OARC)

Our courses are primarily delivered through weekly Zoom sessions, lasting up to two hours, supplemented by assigned readings from the RSGB licence manuals, interactive quizzes on Google Classroom, and vibrant discussions on dedicated Discord channels. Following the successful implementation of the Intermediate training, there arose a need for Foundation and Full courses. Despite having limited prior experience, our dedicated team of volunteer trainers diligently crafted these courses, incorporating engaging PowerPoint presentations and comprehensive mock exam questions. Presently, we provide continuous training for all levels of licences, with an impressive pass rate estimated to exceed 80%.

To date, we have successfully conducted 7 Foundation, 10 Intermediate, and 7 Full courses, with support for embrace the Direct to Full exam students who may elect to complete all or some of our course which run consecutively and offering additional pre-exam support on request. The current schedule is to run 2 courses at each level per year is planned. To find out more info please go to OARC.UK

The online amateur radio community thrives on various platforms, including Discord servers, where members come together to engage in a wide range of activities. These platforms serve as a hub for hams to connect, collaborate, and learn from one another. Within these Discord servers, hams actively participate in projects that push the boundaries of amateur radio. One notable project is the National Packet Radio Network, which utilize TNCs (Terminal Node Controllers) and other software modems for data communication. This project aims to establish a robust and reliable network for exchanging data among hams across the country. Additionally, hams contribute to an ADBS (Automatic Dependent Surveillance–Broadcast) flight tracking network, which involves monitoring and tracking aircraft positions using amateur radio technology.

To further enhance the social aspect of the community, monthly meetups on platforms like Zoom are arranged. These virtual gatherings bring hams together, allowing them to interact face-to-face, share stories, and forge deeper connections. It provides an opportunity to celebrate achievements, discuss ongoing projects, and plan for future endeavours.

Morse net, a weekly event, holds a special place in the online amateur radio community. It serves as a platform for hams to practice and improve their skills in Morse code communication.

To help spread the message about amateur radio and also to record other information, the community also has a youtube channel which has many talks as well as community nights where we discuss whats happening within the community and through them you can get a flavour of what the community is all about: *https://www.youtube.com/@OnlineAmateurRadioCommunity*

Many members joining from Ireland meant that we have also set up an Ireland section of the community which has also got an Irish club callsign EI2OARC, this is as well as are UK club call of M0OUK.

Recently, the online amateur radio community, achieved a significant milestone by surpassing 1700 members.

As a prominent example of the online amateur radio community, offers a welcoming environment for hams of all skill levels and backgrounds, whether one is a seasoned expert or a beginner just starting their journey in amateur radio. OARC provides a wealth of resources and guidance to help individuals get started and actively participate in the community, this includes special channels within discord but also to the community WIKI where there is a number of articles on may topics: *https://wiki.oarc.uk/doku.php*

By visiting *oarc.uk* newcomers can access a plethora of information, including guides, tutorials, and resources tailored to different areas of interest.

So, if you are interested in amateur radio, don't hesitate to visit: *oarc.uk* and start your journey today.

For more information on Online training and resources and also FAQ please follow the links below:

https://rsgb.org/main/clubs-training/for-students/

https://rsgb.org/main/clubs-training/for-students/online-training-resources-for-students/

https://rsgb.org/main/clubs-training/exam-faq/

https://rsgb.org/main/clubs-training/course-exam-finder/

Operating Abroad

Assuming that the country allows Amateur operation, there are generally two routes that may apply to operating in it. For temporary operating a number of countries allow this under a European Conference of Postal and Telecommunication Administrations (CEPT) agreement. The second, that allows permanent operating, is to obtain a reciprocal licence.

CEPT agreement

CEPT is a group of countries from Europe near to Europe. Note that Russia and Belarus are currently suspended from CEPT.

Amongst their many actions, many but not all have agreed a common standard of amateur radio licence and rules (T/R 61-01) so as to facilitate temporary operation when visiting a fellow CEPT country, as well as a number of countries outside CEPT who have agreed and met the conditions to use T/R 61-01. Once each CEPT member's country has confirmed that their amateur radio licence conforms to the T/R 61-01 minimum standard, then its amateurs may operate temporarily in other countries which have also confirmed the recommendation.

CEPT T/R 61-01 operation does not replace reciprocal licensing - rather it supplements it. Only temporary operation (up to three months) is permitted under T/R 61-01, eg from holiday accommodation or mobile. Therefore, if you seek either longer-term (over three months) residence or additional facilities, you will still need to apply for a licence. Operating under CEPT regulations means that you are restricted by the regulations of the foreign country; thus the RSGB recommends that you get a copy of the licensing regulations for the countries in which you plan to operate.

Providing you hold a Full UK licence you can operate in countries which have implemented CEPT Recommendation T/R 61-01 in accordance with the terms of that recommendation.

To underline the point licences excluded from this arrangement are Full (Temporary Reciprocal) and Full (Club) along with the Intermediate and the Foundation licence holders.

If you satisfy the above then you must also ensure that at all times you comply with the requirements applicable to the use of amateur radio equipment at the location in the country concerned. Some vary from ours, so it is vital to check before your journey. Also, remember that if requested you must present your licence documentation plus any other documentation required to the relevant authorities in that country. Unless told otherwise you must use the call sign specified in section 1 of your licence after the appropriate host country callsign prefix.

To get the up-to-date list of countries (both CEPT and non-CEPT countries) that have implemented CEPT T/R 61-01, and to be sure you have the latest information check out the ERO website which has at the top of the page a downloadable version of CEPT Recommendation T/R 61-01. Another useful source of up-to-date information on licensing is OH2MCN's website.

Reciprocal licensing

A reciprocal licence is a licence issued by a foreign country to you because that country recognises the standards of the UK Full exam. A licence may be issued under the provisions of CEPT Recommendation T/R 61-02 or it may be issued under a bilateral reciprocal agreement between the UK and the other country.

Under T/R 61-02, the UK may issue a HAREC (Harmonised Amateur Radio Exam Certificate). This certificate confirms that you have passed the UK Full exam. It is recognised in the other countries that participate in T/R 61-02. The other country may issue its equivalent of the UK Full Licence against a UK HAREC. A UK HAREC is now appended to each Full Exam certificate that we issue. If you do not have a UK HAREC but have passed the UK Full exam, you may request a HAREC from Ofcom.

Countries that don't participate in T/R 61-02 but with which the UK has a bilateral agreement are listed in the UK amateur radio licence application form. Check with the licensing authorities in the overseas country what their licensing process is.

The callsign issued with a reciprocal licence may be your own callsign with the foreign country's suffix or prefix or may be allocated in the country's normal series of callsigns.

Overseas radio amateurs in the UK get a normal Full call sign if their licence is issued against a HAREC. Radio amateurs from other countries get a Temporary Reciprocal Licence.

Due to overseas post and administration delays, it is generally best to allow at least two to three months for your application to be processed - longer if it is a Third World country where amateur radio is not so sympathetically regarded. The use of air mail certainly helps in this regard.

Like operating under the CEPT T/R 61-01 agreement, HARECs are not available for Temporary Reciprocal, Foundation or Intermediate licence holders.

Other information

Most countries have a national society which looks after the wellbeing of that country's amateurs. A list may be found on the IARU website. Few are of the size of the RSGB - indeed many are staffed entirely by volunteers. Nevertheless, they will all give you as much assistance as they can.

There is usually little problem with customs. It certainly helps to be able to show that the equipment was purchased abroad and is not being exported. Unfortunately, neither a reciprocal licence nor operation under CEPT regulations is deemed an exemption from customs formalities. If in doubt, you should seek additional advice about importing/exporting equipment.

Ofcom's advice on operating abroad can also be found at:
https://www.ofcom.org.uk/__data/assets/pdf_file/0023/51296/fkm.pdf

Useful resources on operating in foreign countries can be found on the ARRL website at:

www.arrl.org/reciprocal-permit

and OH2MCN's website at:

www.qsl.net/oh2mcn/license.htm and also at: European Radiocommunications Office website:

https://www.cept.org/ecc/topics/radio-amateurs

Ofcom has published advice at:

https://www.ofcom.org.uk/__data/assets/pdf_file/0026/109547/guidance-become-radio-amateur.pdf

Section 2.20 et seq of the online Guidance:
https://www.ofcom.org.uk/__data/assets/pdf_file/0026/82637/amateur_radio_licence_guidance_for_licensees.pdf

see also the RSGB website

Operating Abroad:
https://rsgb.org/main/operating/licensing-novs-visitors/operating-abroad/

Operating for Visitors to the UK:
https://rsgb.org/main/operating/licensing-novs-visitors/operating-for-visitors/

Details of IARU Societies (on the IARU website): www.iaru.org/iaru-soc.html
European Radiocommunications Office website: https://www.cept.org/ecc/topics/radio-amateurs

Ofcom

Regulatory Principles

Ofcom was established as a public corporation by the Office of Communications Act 2002. Ofcom is the regulator for the UK communications industries, with responsibilities across television, radio, telecommunications and wireless communications services.

Ofcom's Statutory Duties Under the Communications Act 2003

"3(1) It shall be the principal duty of Ofcom, in carrying out their functions;

(a) to further the interests of citizens in relation to communications matters; and

(b) to further the interests of consumers in relevant markets, where appropriate by promoting competition"

Ofcom's specific duties fall into six areas:

1. Ensuring the optimal use of the electro magnetic spectrum
2. Ensuring that a wide range of electronic communications services - including high speed data services - is available through-out the UK
3. Ensuring a wide range of TV and radio services of high quality and wide appeal
4. Maintaining plurality in the provision of broadcasting
5. Applying adequate protection for audiences against offensive or harmful material
6. Applying adequate protection for audiences against unfairness or the infringement of privacy

Ofcom's Regulatory Principles

- Ofcom will regulate with a clearly articulated and publicly reviewed annual plan, with stated policy objectives.
- Ofcom will intervene where there is a specific statutory duty to work towards a public policy goal which markets alone cannot achieve.
- Ofcom will operate with a bias against intervention, but with a willingness to intervene firmly, promptly and effectively where required.
- Ofcom will strive to ensure its interventions will be evidence-based, proportionate, consistent, accountable and transparent in both deliberation and outcome.
- Ofcom will always seek the least intrusive regulatory mechanisms to achieve its policy objectives.
- Ofcom will research markets constantly and will aim to remain at the forefront of technological understanding.
- Ofcom will consult widely with all relevant stakeholders and assess the impact of regulatory action before imposing regulation upon a market.

Licences are issued on a lifetime basis. This is subject to the licence being validated with Ofcom at least once every five years. Amateur licensees are therefore reminded that their licence (now 'lifetime') must be validated at least once every 5 years in accordance with the terms, conditions and limitations of their licence. If your licence has been amended (e.g. by notifying Ofcom of a change of address) this will count as a validation. It only takes a few minutes to register your details and validate or make any necessary changes to your licence on-line (*https://ofcom.force.com/licensingcomlogin*) If you are experiencing difficulties or need assistance in processing your licence on-line, you can call: 0300 123 1000 or 020 7981 3131 or Textphone: 020 7981 3043 or 0300 123 2024 (please note that Textphone numbers only work with special equipment used by people who are deaf or hard of hearing).

Applying for an Amateur radio licence

You can apply for, vary, re-validate or surrender an amateur radio licence by either using the online system (subject to conditions), or by completing a paper based application form (new applications are subject to a small administrative fee, unless you are 75 years of age or over).

The free online system has eased administration and reduced costs for amateur users. This approach makes it easier for users to comply with legal obligations. More information on how to apply for an amateur radio licence is available on the Ofcom website (see below).

If you have any questions about your licence please contact:

Spectrum Licensing
PO Box 1285
Warrington WA1 9GL
Tel: 020 7981 3131
Email: *spectrum.licensing@ofcom.org.uk*
Website: https://*www.ofcom.org.uk/manage-your-licence*

Varying the terms and conditions of a licence

Any licence changes affecting all amateurs are subject to Ofcom publishing a Notice of Variation (NoV) or a consultation (depending on the changes proposed). These are announced by Ofcom in the manner specified in the licence and will also appear on the Ofcom website.

Ofcom and the RSGB

Ofcom works closely with the RSGB and meets on a regular basis to ensure that where feasible, the amateur licence meets the needs of today's amateur. Ofcom continues to work closely with the Society to facilitate regulatory requirements on a number of ongoing developments in amateur radio.Spectrum allocations are listed in the Licence, which is published on the Ofcom website: *https://www.ofcom.org.uk/__data/assets/pdf_file/0015/214116/emf-amateur-licence-terms-and-conditions.pdf* Special Contests Call signs are available for individuals and clubs who regularly participate in Special Contests. These are administered for Ofcom by the RSGB and further information is available at: *http://rsgb.org/main/operating/licensing-novs-visitors/online-nov-application/application-for-a-special-contest-call-sign/* Ofcom also supports the Society in the international scene (see below).

International

There are three main areas of interest; Spectrum, Commonality of licensing conditions and Reciprocal agreements as detailed below.

Spectrum

Ofcom is responsible for administering the International Telecommunication Union (ITU) Radio Regulations in the UK, and ensures the effective management of spectrum taking into account international obligations.

Commonality of licensing conditions

Ofcom represents the UK within the European Conference of Postal and Telecommunications Administrations (CEPT) and continually works towards harmonisation of licence conditions and mutual recognition including countries outside Europe. The UK has not adopted any recognition agreements for the Foundation or Intermediate level licences.

Reciprocal agreements

A number of reciprocal agreements have already been negotiated which allow UK amateurs who have passed the 'Full' examination (or past equivalents) to operate abroad, although there are still a number of countries with which no agreement currently exists. In general, Ofcom prefers reciprocal operation under the CEPT arrangements. Non-participating countries (including those not in CEPT) may apply to CEPT to be included in the CEPT Reciprocal arrangement, governed by Recommendations T/R 61-01 and 61-02. This route is considered as being more effective and efficient than the process of numerous individual bi-lateral arrangements between two administrations. However, Ofcom will continue to review needs for new agreements as appropriate.

Due to the many differing arrangements in countries at levels below the full licence level, Ofcom does not recognise reciprocal agreements at Foundation and Intermediate level.

Spectrum Licensing website: https://www.ofcom.org.uk/manage-your-licence/radiocommunication-licences/online-licensing-service and also the online service portal at https://ofcom.force.com/licensingcomlogin

Repeater network

The processing of applications and issuing of the Notices of Variation (NoV) is carried out by Ofcom, with initial assessments undertaken by the Society. Clearance involves consultation with Government departments. The NoV holder remains responsible for compliance with the amateur licence.

Packet Data and Internet (voice) Linking

Ofcom is responsible for the issuing of Notice of Variations (NoVs) for Packet Data Nodes and Internet Gateways, including those linked to Voice Repeaters. Initial assessments of applications are undertaken by the Society. The NoV may or may not permit unattended operation, but the NoV holder remains responsible for compliance with the amateur licence.

Special Event Stations

Ofcom is responsible for the issuing of NoVs for Special Event Stations. Applications for a standard SES can be made and issued via the Ofcom online licensing system. Please see elsewhere in the Yearbook for full details. For further details regarding NoVs, please contact Spectrum Licensing at Ofcom.

Enforcement

The number of people who misuse amateur radio is fortunately small. However, there are some individuals who cause considerable problems to other amateurs or to other licensed radio users by transmitting in unauthorised frequency bands, using obscene language or generally using radio in an antisocial way. The majority of abuse is directed at the repeater network.

Repeater keeper responsibilities include taking steps to prevent and stop abuse. Details of repeater abuse should therefore be sent to the ETCC Chairman, c/o RSGB HQ. Other cases of abuse should also be taken up with Ofcom through the Operating Advisory Service (OAS). Subject to priorities, Ofcom may be able to take action, where there is a breach of licence terms and conditions. Having obtained evidence against an offender, Ofcom may issue formal warnings or Conformity Notices, initiate prosecution pro- ceedings and/or revoke a licence, depending on the seriousness of the offence.

UK amateur radio technical information

The UK Interface Requirement (IR) for Amateur Radio is IR 2028 and includes the technical details of radio equipment for Foundation licensees. The IR is available at: *https://www.ofcom.org.uk/__data/assets/pdf_file/0028/53965/ir2028.pdf* Ofcom plans to update this IR in the next year.

Provision of Information

Ofcom's policy is to provide its information in electronic form via its website, *www.ofcom.org.uk* Some of these information sheets are available free in paper form. The following relate to the amateur service.

Amateur application forms

Of346a Amateur Licence amendment, validation, surrender form

OfW 287 Application for a Notice of Variation for a Special Event Callsign

OfW 306 Application for an Amateur Radio Special Research Permit

In its 2023/24 Plan of Work, Ofcom committed to review amateur radio licences. You can keep up-to-date with this review and what it means for the Amateur Radio Licence by subscribing to e-mail updates from Ofcom

Licences

There are three levels of qualification for gaining an amateur radio licence; the three-tier structure consisting of Foundation, Intermediate and Full Amateur radio licence.

The licensing structure in the United Kingdom is now progressive, which means all new entrants have to enter the hobby at the Foundation level, progressing through the Intermediate licence and finally to the Full Amateur radio licence.

Foundation Amateur Radio Licence

Introduced in January 2002, the Foundation licence is designed as an entry point into amateur radio and forms the first part of the three-tier amateur licensing structure.

A Foundation licence can be obtained after a short course of study lasting some 12-15 hours. At the end of the course the candidate sits a short, multiple-choice exam, which lasts 60 minutes.

Available online, with no in-course practical assessment.

There are 26 straightforward questions to answer. The pass mark is 19 correctly answered questions. A 'pass' will allow operation of an existing Full licensee's station as a fully qualified Foundation Amateur immediately. You will be able to operate your own station when you receive your licence and M7 callsign.

Foundation Syllabus

1) Amateur Radio Licensing Conditions
2) Technical Basics
3) Transmitters and Receivers
4) Feeder and Antenna
5) Propagation
6) EMC
7) Operating Practices and Procedures
8) Safety

Intermediate Amateur Radio Licence

The second level in the three-tier structure. Candidates must have completed the requirements of the Foundation licence syllabus and passed the associated examination as a prerequisite to sitting the Intermediate Licence Examination.

To obtain the Intermediate licence it is advisable to take a training course. This is a little longer than the Foundation course, lasting some 20-25 hours. The aim is to teach many of the fundamentals of radio in a stimulating way, building on the experience gained as a Foundation licence holder.

After completing the practical assessments, a candidate will be ready to sit the Intermediate amateur radio licence examination. Once again this examination is multiple-choice, this time with 46 questions 90 mins and pass mark 28 correctly answered questions.

Intermediate Syllabus

1) Amateur Radio Licensing Conditions
2) Technical Basics
3) Transmitters and Receivers
4) Feeder and Antenna
5) Propagation
6) EMC
7) Operating Practices and Procedures
8) Safety

Licence types and basic privileges

Type	Foundation	Intermediate	Full
Prefix(es)	M3, M6, M7	2*0, 2*1	G, M0, M1, M5
Amateur bands	136kHz-430MHz, 10GHz	(not 472kHz & 5MHz)	All
Maximum output power	10W [1]	50W [1]	400W [1]
Permitted TX equipment	Commercial, kit	Commercial, kit, home-made	Commercial, kit, home-made
TX modes	All permitted	All permitted	All permitted
Act as mailbox/BBS	No	No	Yes
Act as repeater?	No	No	Yes
Low power device/beacon?	Yes	Yes	Yes
CEPT reciprocal operation?	No	No	Yes (but check)
Other licensees allowed to operate?	Yes [2]	Yes [2]	Yes
Disaster comms permitted?	Yes	Yes	Yes
Control station remotely?	Yes [3]	Yes [3]	Yes

* Regional identifier letter. [1] Some bands have lower power limits.

[2] Other licensee must hold UK licence. [3] Restrictions apply.

Full Exam (formerly Advanced, or RAE C&G)

This is the highest level of licence that you can obtain. To gain a Full licence it is necessary to pass the Full radio communications examination, which contains 58 questions, pass mark 35 and lasts two hours. Once again the examination covers radio theory and licence conditions, but because holding an Full licence enables you to use 400 watts power output from your transmitter such subjects as Electro Magnetic Compatability (EMC), antenna design and safety issues are covered in some depth. The licence allows access to all the amateur allocations with full power.

When studying for the Full radio communications examination there is no requirement to take a formal training course because there is no practical element. It is possible to study at home on your own if you so wish, but you should recognise that a good understanding of the material is required.

Many local amateur radio clubs and societies and technical colleges run courses specifically for the Full radio communications examination. Alternatively, there are some correspondence and Internet courses available

Full Syllabus

1) Amateur Radio Licensing Conditions
2) Technical Aspects
3) Transmitters and Receivers
4) Feeder and Antenna
5) Propagation
6) EMC
7) Operating Practices and Procedures
8) Safety
9) Measurements

How and where to find a training course

The quickest ways to find a training course is to check on the RSGB website Clubs and Training web page or look at the Local Information section in this Yearbook.

Tutors may run courses using their own personal station. Local amateur radio clubs, Scouts, Guides or other organisations such as the Air Training Corps may also run courses. Youth organisations are likely to run courses solely for their own members, but it is always worth asking. If you are at school and your school does not run a course, suggest to someone that they do so. The RSGB will be happy to advise on how this might be done and point to sources of assistance and training.

In addition RSGB has also been developing a 'Direct-to-Full' examination syllabus. This will satisfy the requirement that an applicant for a Full Licence must have passed all three levels of examinations. It is currently expected this maybe introduced sometime during 2022 for technically proficient candidates, who may for example have a professional background in electronics or telecomms.

Costs

There is likely to be a small charge for training course administration. It is best to ask before committing to attend, but the costs should not be high and certainly should not be a reason to miss out. Examinations and assessments cost £27.50 for the Foundation licence, £32.50 for the Intermediate licence and £37.50 for the Full licence.

Study material

The following titles are available from the RSGB Shop.

The Foundation Licence Manual for Radio Amateurs

The Intermediate Licence Manual for Radio Amateurs

The Full Licence Manual for Radio Amateurs

Exam Secrets for Radio Amateurs

Audio & Braille resources

UK Amateur radio has a charity dedicated to supporting the needs of people with disabilities – RAIBC :

RAIBC website: *www.raibc.org.uk*

Contact: Sandra Brown M6SDS, RAIBC Secretary, email: gensec@raibc.org.uk, Phone: 0800 028 8660

RAIBC Spoken word recordings

Exam Manuals: The RAIBC maintains an Audio Library which contains the Foundation, Intermediate and Full/Advanced manuals. Updated versions are available for Syllabus 2019. Please note that these are not available to download but are available on audio disk

https://rsgb.org/main/clubs-training/training-resources/audio-braille-resources/

Special needs candidates

Arrangements can be made for candidates with special needs who would otherwise have difficulties taking any of the amateur radio licence examinations. The key requirement is proper advice on what provisions should be made for the candidates concerned. This must be from an appropriate health, educational or other professional. The need is not for a statement of the candidate's circumstances, but what should be done in order to provide equality of access to the examinations.

Where a disability prevents a candidate from carrying out a practical task then advanced permission should be sought, again supported by proper advice, to waive that specific requirement. Not all requirements can be waived and this will be dealt with on a case by case basis.

It is important to start this process well in advance of requesting and exam, as it may take some weeks to obtain appropriate professional advice. Advice should be sought from the RSGB Examination Department

RSGB Clubs and Training web page: ***www.rsgb.org/clubsandtraining***

* Daisy players are available from the RNIB's Talking Book service and allow a visually impaired or blind person to listen to many hours of pre-recorded material. The player allows skipping between tracks, or in this case, pages, and bookmarks can be set to allow an easy return to a study point.

** Symphony players are available from the British Wireless for the Blind Fund (BWBF) and allow the listener to pause and return to the same point, provided the CD has not been changed.

Special Contest Callsigns

The holder of any UK amateur radio Full or Full (Club) licence may apply for a Special Contest Callsign, for use in a number of contests.

Ofcom Policy and Procedure

The holder of a UK Amateur Radio Full (Club) Licence or an Amateur Full Licence may apply for a special contests such as those shown in **Table 1**. The callsign will consist of G or M, a chosen digit and a chosen suffix letter, eg G8Z or GW8Z etc. RSLs may be inserted into an SCC. 520 callsigns are available:

G*(number) A-Z, and M*(number) A-Z.

The RSGB administers SCCs for Ofcom. Applicants should apply on the form at *http://rsgb.org/main/operating/licensing-novs-visitors/online-nov-application/application-for-a-special-contest-call-sign*, giving three choices of callsign in order of preference. Current availability may be checked by contacting the RSGB

Allow up to four weeks for processing.

Individual licensees or club licensees (holder of the club licence) will be expected to provide evidence of having entered at least five of the contests listed in Table 1 within the last five years and having achieved at least one third of the number of contacts of the leader in the appropriate table. However, in a contest where the licensee achieves more than one half of the number of contacts of the leader, (rather than one third), this contest will count as two contests towards the requirement to have entered five contests within the last five years. Individual licensees wishing to apply for a special contest callsign should do so with achievements gained under their own individual callsign.

Details of the application will be passed to the RSGB's Contest Support Committee (CSC) for a recommendation on issuing a NoV. Ofcom will treat the RSGB-CSC response only as a recommendation. The NoV to the licence is issued by Ofcom to the holder of the relevant licence.

Licence variations currently being granted are valid until 31st December 2024. If the licensee wishes to renew the callsign that was issued, a fresh application must be made within the period of two months immediately prior to the expiry date of the NoV. Unless a fresh application has been submitted in this period, the callsign will be withdrawn for a period of two years prior to it being made available for general re-issue. Please note that 'Licensees' who hold a special contests callsigns NoV can use the following RSGB's Contest Support Committee email: *csc.chair@rsgb.org.uk*. Whilst the RSGB has offered this facility, please note that the onus is on the licensee in case of a reminder not being received.

The RSGB will send NoVs directly to successful applicant.

Extract of Terms relating to the NoV for use of Special Contests Callsigns:

1. Terms and expressions defined in the Licence shall have the same meaning herein except where otherwise stated or the context requires otherwise.
2. The Contest Call Sign may only be used to identify the station when the station is participating in an Amateur Radio Contest.
3. Where the licence relates to an Amateur Full (Club) Licence, only members of the club may use the special contests callsign, subject to the conditions in this Licence.
4. Where a log is kept under Clause 12 of Section 2 of the Licence, the licensee must maintain a separate log in respect of the Contest Call Sign.
5. A Regional Secondary Locator (as provided for in Clause 2(2) and in Note (c) of the Licence) may also be used in conjunction with the Contest Call Sign.
6. The Contest Call Sign may be used only within the United Kingdom, Guernsey, Isle of Man or Jersey.
7. This Variation shall be read as an integral part of the Licence and the following additional terms shall apply in respect of the Station.

ARRL DX CW	ARRL DX SSB	ARRL 1.8MHz CW only
ARRL 28MHz	ARRL RTTY Roundup	BARTG HF RTTY (March)
BARTG SPRINT (January)	BARTG SPRINT75 (April)	BARTG SPRINT75 (September)
CQ WPX CW	CQ WPX RTTY	CQ WPX SSB
CW WW CW	CW WW RTTY	CW WW SSB
CQ WW 160m CW	CQ WW 160m SSB	IARU HF Championship
Russian DX	WAE DX CW	WAE DX RTTY
WAE DX SSB	RSGB IOTA (July)	RSGB Commonwealth Contest
RSGB 1st 1.8MHz	RSGB 2nd 1.8MHz	RSGB DX
RSGB 50MHz Trophy (June)	RSGB 144MHz Trophy (Sept)	144MHz Marconi Memorial
RSGB March 144MHz/432MHz (counts as one contest)	RSGB May 432MHz – 248GHz (counts as one contest)	RSGB 432MHz – 248GHz (Oct) (counts

Table 1: *List of qualification contests*

For a list of current Special Contest Callsigns, see the Callsign Listings section of this Yearbook

RSGB Contest Support Committee (RSGB CSC)
https://rsgb.org/main/about-us/committees/contest-support-committee/

The application form may be found and printed from the Internet at:
https://rsgb.services/public/nov/sccs/docs/application-for-a-special-contest-call-sign-v2019.1.docx

Special Event Callsigns

Do you really need a GB callsign?

All club stations are able to pass greetings messages sent by a non-licensed third party. This means that applying for a GB callsign no longer holds any advantages for a club!

Provided the club uses the prefix letters (see below), the club station is allowed to pass greetings messages and to operate simultaneously on more than one band. The club prefixes are very distinctive and create interest through their rarity. If a club regularly operates a special event station using the club callsign, this will help increase the club's identity. It will also benefit by being able to print its QSL cards in larger, more economic quantities.

Another advantage is that any suitably licensed and authorised club member may operate the club station. This gives greater flexibility over a GB callsign which is just a variation to an individual's licence.

Best of all, you don't have to fill in any forms or give 28 days notice of operation!

Old club prefix	New club prefix
G/M	GX/MX
GD/MD	GT/MT
GI/MI	GN/MN
GJ/MJ	GH/MH
GM/MM	GS/MS
GU/MU	GP/MP
GW/MW	GC/MC

GB callsigns

Ofcom issues and despatches Notices of Variation authorising special event (GB) callsigns. Consequently, all enquiries and correspondence should be addressed to Ofcom and not to the Society. Ofcom has stated that GB callsigns are issued for special event stations and that as such, they should normally be open to viewing by members of the public.

Applying for a GB callsign

You may apply for an NoV for a Special Event Stations online, via the licensing portal *https://ofcom.force.com/licensingcomlogin* These NoVs are available immediately. No charge is currently made by Ofcom for a special event callsign.

If you make a manual application, the form must be sent in at least 28 days prior to the start of the event.

Paper applications are normally processed shortly after receipt. If nothing has been received 14 days prior to the event, please contact Ofcom immediately. Please note that no authority exists until a Notice of Variation has been received.

A Notice of Variation will only be issued to licensees who hold a current Full, including a Club licence (ie not Foundation, Intermediate or Temporary Reciprocal). It will be valid for a maximum of 28 days. The station may only be established and operated at the location. This must be the address stated on the application form which must be detailed enough for anyone to find easily. Operation of a special event station from a licensee's home address is not normally permitted. This is because an SES is a showcase for the hobby.

Only the person responsible for the station need sign the form, as the authorisation is by Notice of Variation to that individual's licence. This person is required to be present to supervise the correct operation of the station. Additional operators need only sign and write their callsigns in the logbook.

If you have not used the callsign before, you can avoid last-minute disappointment by first contacting Ofcom, who can check that it is available and reserve it for you. A GB callsign may be reserved for up to six months in advance. When a GB callsign has been used it will not normally be re-issued to another amateur for use at a different event for a period of two years.

The holder of a Full Licence must apply for a GB callsign.

Subject to availability, special event callsigns are available in the following formats:

GB0 + 2 or 3 letters	GB1 + 2 or letters
GB2 + 2 or 3 letters	GB4 + 2 or 3 letters
GB5 + 2 or 3 letters	GB8 + 2 or 3 letters
GB6 + 2 or 3 letters	

Greetings messages

Ofcom licence conditions 3(4) and 3(5) enable a Full or Full (Club) licence to facilitate a non-licensed person to pass a message on air. Such 'greetings messages' are on the basis that:

> a non-licensed person may speak into the microphone but the licensed radio amateur must identify the station and operate the transmitter controls at all times.

Charitable events

It is recognised that some special event stations will be established at certain charitable events where a major concern will be the raising of funds.

Ofcom has agreed that the charity (if one is involved) or the reason for establishing the special event station may be mentioned 'on-air' provided that under no circumstances may a donation be requested during the contact, and sending of QSL cards must not be conditional upon the pledge of a donation. It is in the interests of everyone who holds a special event station licence that operators keep within the spirit of this by not asking for any money over the air.

The station may be sponsored per contact, ie the licensee may in advance of the event seek from his/her friends and relatives sponsorship assurances under the usual arrangements for sponsorship. You must not seek sponsors 'on-air' at any time.

QSL information

Special event stations generate many QSL cards, so it is important that you use the QSL Bureau correctly.

For instructions, see the 'QSL Bureau' pages in this edition of the *RSGB Yearbook*.

JOTA and Radio Scouting

Jamboree on the Air (JOTA) is an annual event designed to allow Scouts to send greetings messages to each other. Started in 1957, it now involves approximately 600,000 Scouts and Guides, with the help of over 23,000 radio amateurs in over 100 countries.

JOTA takes place on the third full weekend of October each year, officially between 00.00 Saturday and 24.00 Sunday, although most stations run for a period within these hours to suit their own requirements. The event is organised by the Scout movement, supported by radio amateurs or clubs. Their aim is to bring Scouts around the world closer together and to introduce them to the capabilities of amateur radio.

All amateur bands are used. Most stations use a special event or a club call, allowing the Scouts to pass greetings messages over the air. JOTA information packs are sent to all participating GB stations at the beginning of October. Any clubs taking part in JOTA wishing to receive this information pack should contact the SES Administrator at RSGB HQ.

The interest fostered by JOTA and World Jamboree has spread and many Scout camps and campsites boast amateur radio facilities. A number of proficiency badges in the radio, electronics and computer fields are available for Scouts. Several countries have permanent Scout Headquarters stations – for example the World Scout Bureau in Geneva has the callsign HB9S and Gilwell Park in the UK operates under the callsign GB2GP.

Many countries run periodic Scout nets. There are regular weekly UK and European nets aimed at Scouters who are also radio amateurs.

Thinking Day On The Air

Thinking Day On The Air (TDOTA) is organised by The Guide Association on the third full weekend in February, to celebrate the birthdays of the founder of the movement,

Usual Scout Net Frequencies

Band	SSB (Phone)	CW
80m	3.740 and 3.940*	3.590
40m	7.090	7.030
20m	14.290	14.070
17m	18.140	18.080
15m	21.360	21.140
12m	24.960	24.910
10m	28.990	28.190

The UK Scout net is on Saturdays, 3.740MHz at 0900 local time. The European Scout net is on Saturdays, 14.290MHz at 09:30GMT.

* USA only

Lord Baden-Powell, and of his wife, Lady Baden-Powell, the World Chief Guide, on 22 February.
The aim of TDOTA is to encourage the girls to make Guiding friendships with members of other units and to introduce them to amateur radio. Station organisers are asked to keep these objectives in mind.
Guide amateur radio stations rely on the goodwill of radio amateurs in setting up stations, though the association has an increasing number of members of all ages holding callsigns.
Guiders interested in organising a TDOTA station can apply for a comprehensive information pack with suggestions for activities, logos for certificates and posters to report forms. Further information is published from time to time in the association's magazines.
Stations are requested to complete a brief report which is sent to Girlguiding UK HQ. All the information is collated into a National Report, which is sent to those who took part and contributed to the report. Copies of the current report are available on receipt of an A4 SASE or from the website (see below).
Amateur radio has a place in the programme for all age groups, encouraging girls to embrace the technical aspects and international perspectives of a world-wide movement. Girl- guiding UK supports the revised amateur radio licence structure and particularly welcomes the Foundation Licence.
While the main focus remains TDOTA, Guides can be heard on the air at other times of the year from camps, activity days and leader training courses.

Short-form call signs

Ofcom will assign a callsign with only two trailing letters or which starts with 'G2' only if the applicant previously held it.

For a list of permanent Special Event Callsigns, see the Callsign Listings section of this Yearbook

---

Prefixes and suffixes

Below is a list of Prefix characters for UK Callsigns for individual countries within the UK also Suffixes to be used. Approximate year of issue of UK callsigns is also listed below.

Prefixes

1st Character: Foundation: M3, M6, M7
Intermediate: 2
Full: G or M (except M3, M6 or M7)

2nd Character:

	Intermediate	Foundation/ Full	Club (Full)
England	E	(none)	X
Isle of Man	D	D	T
Northern Ireland	I	I	N
Jersey	J	J	H
Scotland	M	M	S
Guernsey	U	U	P
Wales	W	W	C

GB3 + 2 letters:	Repeaters
GB3 + 3 letters:	Beacons
GB7 + 2 letters:	Data repeaters
GB7 + 3 letters:	Data mailboxes
	Special event stations (class of call sign normally corresponds to the appropriate M format)
M/foreign call:	Reciprocal CEPT

Suffixes

-/A	Alternative address
-/M	Mobile (includes inland waterways and pedestrian)
-/P	Portable (temporary location)
-/MM	Maritime Mobile

Approximate Issue Dates for UK Callsigns

Full Licence

Two letters

Callsign	Date
G2AA	1920-39
G3AA	1937-38
G4AA	1938-39
G5AA	1921-39
G6AA	1921-39
G8AA	1936-37

Three letters

Callsign	Date
G2AAA	Pre-war
G3AAA	1946
G3CAA	1947
G3EAA	1948
G3GAA	1949-50
G3HAA	1950-51
G3IAA	1951-52
G3JAA	1952-54
G3KAA	1954-56
G3LAA	1956-57
G3MAA	1957-58
G3NAA	1958-60
G3OAA	1960-61
G3PAA	1961-62
G3QAA	not issued
G3RAA	1962-63
G3SAA	1963-64
G3TAA	1964-65
G3UAA	1965-66
G3VAA	1966-67
G3WAA	1967
G3XAA	1967-68
G3YAA	1968-69
G3ZAA	1969-71
G4AAA	1971-72
G4BAA	1972-73
G4DAA	1974-75
G4EAA	1975-76
G4GAA	1977
G4IAA	1979
G4MAA	1981
G4QAA	not issued
G4RAA	1982
G4SAA	1983
G4WAA	1984
G0AAA	1985
G0EAA	1986
G0HAA	1987
G0JAA	1988
G0LAA	1989
G0MAA	1990
G0NAA	1991
G0SAA	1992
G0TAA	1993
G0VAA	1994
G0WAA	1995
M0AAA	1996
M0BAA	1997-98
M0CAA	1998-00
M5AAA	1999-00
G8AAA	1964-67
G8BAA	1967-68
G8CAA	1968-69
G8DAA	1969-70
G8EAA	1970-71
G8FAA	1971-72
G8HAA	1973
G8IAA	1973-74
G8JAA	1974-75
G8KAA	1975
G8MAA	1976-77
G8NAA	1977
G8OAA	1977-78
G8PAA	1978
G8QAA	not issued
G8TAA	1979
G8ZAA	1981
G6CAA	1981
G6QAA	not issued
G6RAA	1982
G1AAA	1983
G1DAA	1984
G1LAA	1985
G1QAA	not issued
G1SAA	1986
G1XAA	1987
G7AAA	1988
G7EAA	1989
G7FAA	1990
G7HAA	1991
G7MAA	1992
G7OAA	1993
G7SAA	1994
G7TAA	1995
G7WAA	1996
M1AAA	1996
M1CAA	1997
M1DAA	1998-99
M1EAA	1999-00

Intermediate

Callsign	Date
2E0AAA	1991
2E1AAA	1991
2E1BAA	1992
2E1CAA	1993
2E1DAA	1994
2E1EAA	1995-97
2E1GAA	1997-99
2E1HAA	1999-00

Foundation

Callsign	Date
M3AAA	2002
M6AAA	2008
M7AAA	2018

Note: *From April 2000, out-of-sequence callsigns could be requested, so calls in later series may be heard.*

GB2RS News

GB2RS is the weekly news service of the RSGB, delivering information to members and the wider radio community. It all began at 10am on Sunday 25th September 1955, when the Post Office authorised the first ever broadcast on 3600kHz from the home of Frank Hicks-Arnold, G6MB, in Walton-on-Thames, using the special callsign GB2RS. Today, the news is broadcast every Sunday by a team of over 100 volunteer readers, using a wide range of amateur bands and modes.

The latest schedule of GB2RS broadcasts can be downloaded from *https://rsgb.org/gb2rsschedule* National transmissions, on 160m, 60m and 40m, can be heard throughout the UK and in parts of western Europe. Many localised broadcasts take place on VHF and UHF, either simplex or via analogue and digital repeaters, and there is an international edition via the QO-100 satellite. Items of local news are read out as appropriate to the coverage area of the transmission. As well as traditional spoken readings, GB2RS is available via amateur television transmissions, and on the internet as both video and audio. There is also a digital voice broadcast on 80m.

About the broadcast

GB2RS carries items of interest to radio amateurs and listeners, together with current contest, special event and propagation information, provided by the RSGB's specialist teams or by direct submission. The news script is prepared each week by the RadCom editorial team at RSGB HQ and posted to the RSGB website on Friday afternoons. The majority of the GB2RS broadcasts take place on Sunday mornings, although a number extend throughout the rest of the day to suit listeners' free time and to make best use of propagation. The on-air broadcast frequencies are allocated to appointed newsreaders by the RSGB on behalf of Ofcom.

The frequencies are shown in the Schedule and all amateurs (including contest operators) are respectfully asked to avoid using these on Sundays so that the news transmissions can be heard as widely as possible. The readers are provided with a special GB2RS Notice of Variation which permits them to broadcast the news bulletin and, while using this callsign, two-way communication is not permitted. Therefore, readers will often convene nets before and after the news broadcast, using their own callsign. All stations are very welcome to call in. Listeners can also catch up with the news at their leisure. The GB2RS podcast from Jeremy G4NJH is available at *https://gb2rs.podbean.com* A catch-up visual version from Alison G8ROG can be seen at *http://gb2rs.apj1.co.uk/* and TX Factor provides a listen-again reading at *http://txfactor.co.uk/gb2rs-news.html*

Newsreaders

The organisation of the GB2RS News Service and the network of newsreaders is the responsibility of the GB2RS News Manager. Contact *gb2rs.manager@rsgb.org.uk* New readers are always needed. Interested amateurs need to be RSGB members and must hold a Full or Intermediate licence. The GB2RS News Manager will provide an application form and then discuss your news-reading proposal with you. Newsreaders must be able to deliver the script fluently and their station capability should match the proposed coverage area. The GB2RS News Manager will guide on time and frequencies of operation and arrange the required Notice of Variation.

Submitting news

News items for GB2RS, RadCom magazine and the RSGB website should be sent to the editorial team as far in advance as possible by email to: *radcom@rsgb.org.uk* This is also the destination for club and local news. The deadline is 10am on Thursdays.

GB2RS News Broadcast Schedule

The following broadcasts are made every Sunday

Time	Freq	Mode	Location	Reader(s)
National Transmissions				
11000 UK	7.127	LSB	HOY	GM8OFQ
			EXMOOR	G6ASK
1500 UTC	5.3985	USB	ROYSTON	G4HPE
			ST HELENS	G4MWO
			HOY	GM8OFQ
			MEPPERSHALL	G4JBD
		Note: Read in rotation		
2130 UK	1.99	LSB	ROYSTON	G4HPE
			WEST BUTTERWICK	G8JET
			OSWESTRY	G4IOQ
			HOY	GM8OFQ
			MEPPERSHALL	G4JBD
		Note: Read in rotation		
Northern England				
0900 UK	50.8	FM	VIA GB3WY WAKEFIELD	G0TKF
0900 UK	145.525	FM	TYNE, WEAR, TEES	G4OLK G7MFN
0930 UK	145.525	FM	BRADFORD	M0JPA
			CLECKHEATON	M0WIT
			HALIFAX	G6NTI
			HALIFAX	G6YGV
1000 UK	145.525	FM	HULL	G3GJA
1000 UK	1308	DATV	VIA GB3EY HULL	G3GJA
		Note: 144.7750 ATV talkback frequency is also monitored		
		Also streamed live via BATC website		
			https://batc.org.uk/live/gb3ey	
1030 UK	70.425	FM	BURY	G4GSY
			DUKINFIELD	G0NAJ
			WIGAN	M0HDE
1030 UK	145.525	FM	BURY	G4GSY
			DUKINFIELD	G0NAJ
			WIGAN	M0HDE
1200 UK	145.525	FM	CHIRNSIDE	MM0JNL
			DUNS	2M0YTN
		Note: for Northumberland		
2100 UK	70.425	FM	BURY	G4GSY
			DUKINFIELD	G0NAJ
			WIGAN	M0HDE
2100 UK	145.525	FM	BURY	G4GSY
			DUKINFIELD	G0NAJ
			WIGAN	M0HDE
Midlands				
0800 UK	3.6500	AM	MEPPERSHALL	G4JBD
			STOWMARKET	G0OZS
0930 UK	3.65	LSB	PERSHORE	G8BGT
			WORCESTER	G4IDF
1100 UK	145.525	FM	MANSFIELD	G3XDS
1200 UK	430.875	DMR	GB7DC	G4SEB
	439.65		GB7IN	
	439.6125		GB7RR	
		Note: Talk Group TG23590 - Time Slot 1		
1800 UK	145.525	FM	STOKE ON TRENT	G0VVT
			NEWCASTLE-UNDER-LYME	M5GWH
1800 UK	70.425	FM	STOKE ON TRENT	G0VVT
1830 UK	50.79	FM	VIA GB3SX STOKE ON TRENT	G0VVT
1830 UK	439.525	DMR	VIA GB7ST STOKE ON TRENT	G0VVT
		Note: Talk Group TG8 (local only) - Time Slot 2		
1830 UK	433.525	FM	STOKE ON TRENT	G0VVT
1900 UK	145.6375	FM	VIA GB3IN HUTHWAITE	G4TSN
				G0LCG
2030 UK	145.525	FM	NUNEATON	G8VHI
			BURTON ON TRENT	G8EKG
			NUNEATON	G4AEH
2030 UK	144.25	USB	NUNEATON	G8VHI
		Note: 10 ele, beaming north		
2030 UK	433.525	FM	NUNEATON	G8VHI
			BURTON ON TRENT	G8EKG
South East / East Anglia				
0800 UK	3.65	AM	MEPPERSHALL	G4JBD
			STOWMARKET	G0OZS
0900 UK	3.65	LSB	BRISTOL	G4TRN G6UWK
			READING	G8ROG G4RDC
0900 UK	3.643	DV	BURES	G6WPJ
		Note: FreeDV @ 700 b/s - set receiver to LSB		
0900 UK	51.53	FM	READING	G8ROG G4RDC
0900 UK	70.425	FM	READING	G8ROG G4RDC
0900 UK	433	FM	VIA GB3BN BRACKNELL	G8ROG
			READING	G4RDC
0900 UK	3408	DATV	VIA GB3HV FARNHAM	G8ROG
			READING	G4RDC
		Note: Also streamed live via BATC website		
			https://batc.org.uk/live/gb2rs	
0900 UK	10355	WFM	READING	G8ROG G4RDC
		Note: Four paths, each antenna at 60 deg beamwidth		
0900 UK	145.525	FM	HAINAULT	M0MBD
			CHIGWELL	M0XTA

**For GB2RS News Online, there are links from the RSGB main page.*

Time	Freq	Mode	Location	Reader(s)
0930 UK	145.525	FM	EAST PRESTON	M0KEL
			LANCING	G0TLU
			LITTLEHAMPTON	M0RDV
0930 UK	145.525	FM	STOWMARKET	G0OZS
			WOODBRIDGE	G7CIY
0930 UK	145.525	FM	FELIXSTOWE	G4YQC
			IPSWICH	G0DVJ
		Note: Also covers parts of Norfolk, Essex and N Kent		
1000 UK	70.425	FM	DUNTON	G4OXY
			LANGFORD	G1GSN
1000 UK	145.525	FM	DUNTON	G4OXY
			LANGFORD	G1GSN
			HITCHIN	G4OXD
1015 UK	433.225	FM	VIA GB3IW ISLE OF WIGHT	M0KEL M0RDV G0TLU
1100 UK	433.525	FM	CLACTON ON SEA	G0NAD
1800 UK	433.525	FM	FOLKSTONE	G4IMP
1900 UK	145.625	FM	VIA GB3NB NORWICH	G4DYC G7URP G3LDI G3YLA
1900 UK	70.425	FM	CHIGWELL	M0XTA
			ABBEYWOOD	2E0HNH
			CLACTON ON SEA	G0NAD
1930 UK	145.525	FM	EASTBOURNE	G0NQZ G1FBH M0LRE
South West				
0900 UK	3.65	LSB	BRISTOL	G4TRN G6UWK
			READING	G8ROG G4RDC
0900 UK	3.643	DV	BURES	G6WPJ
		Note: FreeDV @ 700 b/s - set receiver to LSB		
0900 UK	51.53	FM	READING	G8ROG G4RDC
0900 UK	70.425	FM	READING	G8ROG G4RDC
0900 UK	145.525	FM	EXMOOR	G6ASK
0900 UK	433	FM	VIA GB3BN BRACKNELL	G8ROG
			READING	G4RDC
0900 UK	3408	DATV	VIA GB3HV FARNHAM	G8ROG
			READING	G4RDC
		Note: Also streamed live via BATC website	**https://batc.org.uk/live/gb2rs**	
0900 UK	10355	WFM	READING	G8ROG G4RDC
		Note: Four paths, each antenna at 60 deg beamwidth		
0930 UK	145.725	FM	VIA GB3NC ST. AUSTELL	G4BHD G4OCO G0PNM G5XDX
0930 UK	430.825	FM	VIA GB3ZB BRISTOL	G4TRN
	430.925		and GB3FI CHEDDAR	G7NJX
1030 UK	145.525	FM	CENTRAL BRISTOL	G4TRN
			SANDFORD	G7NJX
1100 UK	145.3375	FM	VIA MB7IPN WATERLOOVILLE	G7TEM
Channel Islands				
0900 UK	3.65	LSB	BRISTOL	G4TRN G6UWK
			READING	G8ROG
			READING	G4RDC
0900 UK	3.643	DV	BURES	G6WPJ
		Note: FreeDV @ 700 b/s - set receiver to LSB		
0900 UK	145.525	FM	JERSEY	GJ0PDJ GJ8PVL
0930 UK	145.525	FM	GUERNSEY	GU0SUP GU8ITE GU6EFB
1830 UK	145.5250	FM	Caernarfon	GW0AQR, GW4KAZ
Wales				
0900 UK	3.65	LSB	BRISTOL	G4TRN G6UWK
			READING	G8ROG G4RDC
0900 UK	3.643	DV	BURES	G6WPJ
		Note: FreeDV @ 700 b/s - set receiver to LSB		
Northern Ireland				
0930 UK	145.525	FM	CARRICKFERGUS	MI0AWL
1000 UK	3.64	LSB	DUNGIVEN	GI0AZA GI0AZB
			STRABANE	MI0HOZ
1200 UK	439.5875	DMR	VIA GB7HZ STRABANE	MI0HOZ
	439.6	DMR	VIA GB7WT OMAGH	
	430.8625	DMR	VIA GB7CX COLERAINE	
	439.7375	DMR	VIA GB7HI LISBURN	
	439.4875	DMR	VIA GB7MW CARRICKFERGUS	
		Note: Talk Group TG2354 - Time Slot 2		
1930 UK	430.95	DMR	VIA GB3OM OMAGH	MI0RWY
	439.6625	DMR	VIA GB7LY DERRY	2I0SJV
	439.525	DMR	VIA GB7UL CARRICKFERGUS	MI0HOZ
	439.625	DMR	VIA GB7HB TANDRAGEE	MI0RYL
	439.5875	DMR	VIA GB7HZ STRABANE	M1AIB
	439.675	DMR	VIA GB7MF MAGHERAFELT	2E0SHZ
	439.575	DMR	VIA GB7KP COMBER	2I0FVX
		Note: Talk Group TG880 - Time Slot 2		
Isle Of Man				
1100 UK	430.825	FM	VIA GB3IM-C DOUGLAS	GD6ICR GD6AFB 2D0PEY MD0MAN
1100 UK	430.875	FM	VIA GB3IM-P PEEL	GD6ICR GD6AFB 2D0PEY MD0MAN
1100 UK	430.825	FM	VIA GB3IM-R RAMSEY	GD6ICR GD6AFB 2D0PEY MD0MAN
1100 UK	430.125	FM	VIA GB3IM-S DOUGLAS	GD6ICR GD6AFB 2D0PEY MD0MAN
Scotland				
0900 UK	145.525	FM	PERTH	GM6MEN
0930 UK	70.425	FM	EDINBURGH	GM4DTH
0930 UK	145.525	FM	EDINBURGH	GM4DTH
			EAST WEMYSS	MM0TGB
0930 UK	433.525	FM	EDINBURGH	GM4DTH
0930 UK	145.65	FM	VIA GB3OC KIRKWALL	GM1BAN MM3YHA
		Note: Repeater covers Caithness and Orkney		
1000 UK	145.525	FM	ELGIN	GM4ILS
1000 UK	70.425	FM	LENZIE	MM0TMZ
1000 UK	145.525	FM	BEARSDEN	GM3VTB
			CARLUKE	GM4COX
			LENZIE	MM0TMZ
			WHITECRAIGS	MM0TSS
1000 UK	433.525	FM	LENZIE	MM0TMZ
			CLYDEBANK	2M0PIJ
1000 UK	TG23559	DMR	CLYDEBANK	2M0PIJ
		Note: Talk Group TG23559 during the reading - Time Slot 2		
1200 UK	145.525	FM	CHIRNSIDE	MM0JNL
			DUNS	2M0YTN
1830 UK	145.525	FM	SOUTH WEST GLASGOW	GM5AUG
Satellites				
0800 UTC	10489.855	USB	Es'hail-2 / QO-100	GU6EFB G4BIP G0NAD
		Note: Can also be monitored via WebSDR	https://eshail.batc.org.uk/nb/	
Podcasts And Web				
Weekly			https://gb2rs.podbean.com	G4NJH DD5LP
0900 UK			https://batc.org.uk/live/gb2rs	G8ROG
		Note: Live TV news reading		
Weekly			http://gb2rs.apj1.co.uk/	G8ROG
		Note: Catch-up TV reading		
Weekly			http://txfactor.co.uk/gb2rs-news.html	G1IAR G0FGX 2E0FGQ

When submitting news items...

Do:

- Submit items by email, to: *radcom@rsgb.org.uk* This is the ONLY email address for all submissions.
- Always give a contact name, callsign (if any) and phone number.
- Say if a phone number is daytime only or evening only.
- Send GB2RS any last-minute details of your rally.
- Give the proper name of your club – there is a Wirral and District Amateur Radio Club and a Wirral Amateur Radio Society, so saying “the Wirral club” could lead to confusion.
- Always give the callsign of a speaker if he/she has one, not just ‘Talk by John on aerials’.
- Provide full dates, not ‘Last Friday in month’.
- Listen to GB2RS to hear for yourself the format used.

Don't:

- Forget to include the venue and opening time of a rally, details of talk-in (if any) and a contact name, callsign, and phone number.
- Send in ‘to be confirmed’ items. If they are not confirmed we assume that they are not taking place and they will therefore not be broadcast. Only send your item in when it has been confirmed.
- Give more than one contact person or telephone number. There is only time to broadcast one, so you decide which one to use.
- Mix regular club meetings with main news items, such as rallies and special event stations.
- Use GB2RS and *RadCom* as the only means of publicising club events to your own members – the main purpose of GB2RS should be to inform casual listeners and members of other clubs of the exciting things your club is doing.
- Use cryptic titles for talks, or ‘in-jokes’. If we don't know what you mean, it is unlikely that anyone else will.

DO NOT SEND TO MORE THAN ONE EMAIL ADDRESS because, paradoxically, this increases the chances of your item getting lost. ONLY use *radcom@rsgb.org.uk*

Got a news item?
email: *radcom@rsgb.org.uk*

Got a network enquiry?
email: *gb2rs.manager@rsgb.org.uk*

Planning Advice

Many, if not most, radio amateurs never see the need to apply for planning permission for their aerials. After all the aerials work just as well without it and there is a school of thought that if you don't ask for planning permission the Planning Department can't be tempted to say no. This might seem an attractive argument if you use small visually unobtrusive wire aerials, but if you have aspirations of anything more substantial you are likely to fall foul of the local Planning Department.

Urban Myths

Unfortunately, holding an amateur radio licence in the United Kingdom does not convey any special 'rights' under planning legislation to have an aerial and there are a number of urban myths circulating regarding the need for planning permission.

Amateur radio aerials and masts are generally treated as householder development, exactly the same as a garage or conservatory, and will require planning permission unless they come under one of the following categories

Temporary

Unlike non-residential land which has a limit of 28 days, there are no specific time limits on how long a mast or aerial can be present and still be classed as temporary. It is the degree of permanence that is the deciding factor. The fact that the mast or aerial is installed in a ground socket and can be easily removed is not enough for it to be classed as temporary if it is in regular use.

De Minimis

This latin term means that something is of minor significance or is 'trifling' and should not need permission. A thin long wire would normally be regarded as such, but it is subject to the interpretation of the Planning Department. Also, in planning law, something which has 'no material effect on the external appearance of the building' is not classed as development and does not require planning permission. A wire dipole stretched between a house and a tree usually comes within that category which may also apply to roof or gable mounted verticals. If you have permission for a mast it is a 'building' and changing the aerials may not constitute a 'material effect' but it is better to seek advice before doing so.

Permitted Development

Minor development within the curtilage of a dwelling house, alterations and/or extensions to the dwelling are classed as 'permitted development' and does not require planning permission as long as certain conditions are met. The limits are set in what is called 'The General Permitted Development Order'. Different Orders apply in each of the four countries of the United Kingdom. Although no references are made in the Orders to amateur radio aerials and masts, some radio amateurs have successfully argued that an aerial, mast or pole to the rear of and attached to a dwelling house is an 'enlargement, improvement or alteration of a dwellinghouse' provided the aerial or mast does not protrude above the ridge of the roof. In England, it must also be of similar appearance to the existing dwelling. Any free-standing 'structure or erection' in the rear garden of a house is permitted up to 3m. high (2.5m. if within 2m of a boundary).

There is a class in the Order which refers to 'microwave antennas' which is intended to apply to satellite dishes and includes size restrictions. Should this be referred to by the Planning Department you may need to explain that there is no provision for HF or other aerials

Mobile installation

The legal position regarding mobile, trailer mounted, masts is uncertain. It very much depends upon the circumstances of the case. Any installation must be truly mobile and not fixed to the ground or stayed in any way. It should be possible to demonstrate that it can be folded down and removed from site easily. If it cannot, and there is any degree of permanence it is likely that planning permission will be required. In such circumstances, especially if an enforcement notice is served, you should seek immediate advice from the Planning Advisory Committee.

4 year rule

If your house is not a listed building and you have had your aerials and masts present and unchanged for 4 years or more, no enforcement action can be taken against you. You may be required to prove that the installation has been there for 4 years or more, but this need only be a letter of confirmation from your immediate neighbours or a receipt if it was commercially installed. It also makes sense to take some dated digital photographs of the new system and to note the log. Remember, if you change any part of the installation, e.g. the aerial, the clock starts again for the part you have changed.

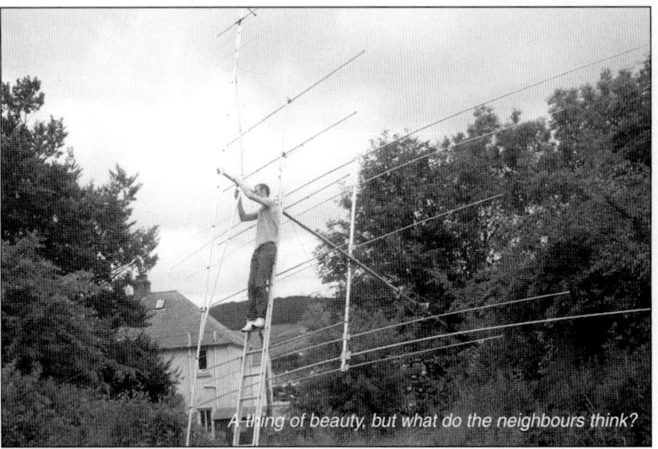
A thing of beauty, but what do the neighbours think?

A Certificate of Lawfulness for your aerials and mast can be obtained from the Planning Department after four years if you want one, but there is no legal requirement to do this.

Applying for planning permission

Most planning applications are now made electronically using the Planning Portal at *www.planningportal.co.uk* but each local authority will also have their own planning permission application forms, but they generally follow a similar style. They will typically require you to complete a Householders Planning Application form, a site location plan(s) and a development plan(s) showing the dimensions of the proposed aerial and/or mast and the distances to your property and the boundary with neighbouring properties. The number of copies and scale for these plans will be specified by the Planning Department in their planning pack. The drawings need not be professionally prepared, as long as it is clear what your proposals are and they are to the scale specified by the Planning Department. If you forget to show the aerial on your planning drawings you may receive planning permission for the mast only, without permission to attach any aerials.

You will also need to complete a neighbourhood notification form, detailing your 'notifiable neighbours'. A notifiable neighbour is someone who shares a boundary with your property or directly face any part your property from across the road. It is worth discussing your proposals with them before making your submission, so that when the official notice comes through their door it will not be a surprise. If you have TVI issues get these resolved first, as although TVI is not part of the planning process experience has shown neighbours will just object on other grounds, usually visual amenity.

Before formally submitting your planning application, ask if you can discuss the submission with your Case Officer. Minor changes at this stage may alleviate any concerns he/she may have, giving your application a better chance of success. You can also contact RSGB HQ to ask to be put in contact with a member of the Planning Advisory Committee, to discuss your proposals prior to submission. A letter of support from the RSGB for your proposed aerial or mast is also available on request.

Refusal to grant planning permission

Sadly, not all planning applications are successful and there is sometimes no apparent reason why one Planning Department will grant

For help and advice, email: pac.chairman@rsgb.org.uk
Planning Permission; Advice to Members: www.rsgb.org/main/operating/planning-matters/

planning permission for an aerial and mast in one area and another in a neighbouring area will refuse planning permission for a near identical installation.

You will be told why your application was refused. Usually it's on the grounds of visual amenity. Consider if the Planning Department has a valid point. To a radio amateur a large beam is a thing of beauty and a joy to own, but what do your neighbours think? Does it overly dominate the area? The Planning Department has to weigh-up the rights of all involved, not simply take sides. You will usually be able to resubmit a revised application free of charge if it is less than 12 months from the original application. If appropriate, reconsider a less ambitious proposal.

If however you believe the Planning Department has treated your application unfairly you have the right of appeal to the Planning Inspectorate (England & Wales), the Planning Appeals Commission (Northern Ireland) or The Directorate for Planning and Environmental Appeals (DPEA), (Scotland).

The appeal must be made within 12 weeks from the planning decision and is usually made in the form of 'Written Submission'. No charge is made for the appeal. Most appeals are submitted electronically using the Planning Portal although it is still possible to fill in the appropriate form and to submit your evidence in writing.

To be successful you must state why you believe the original decision was unsound. Simply saying you disagree or that it will curtail your operations as a licensed radio amateur is not enough. You must establish that the Planning Department has failed to comply with planning law, policy or guidelines, or has sought to impose a different standard on your application than it has done for others.

The RSGB's Planning Advisory Committee can provide guidance to members in the preparation of a planning appeal if required. If you require assistance, contact RSGB HQ who will put you in contact with your nearest Committee member

If your appeal is not upheld and you have not used up your free resubmission, you can submit a revised proposal free of change if it is still less than 12 months from the original application.

Enforcement notices

If you have put up aerials without permission you might, at any time, receive a rather threatening letter from your local planning department along the lines of 'either apply for planning permission or take the aerials down within 28 days, or else'. Whatever you do, don't touch them, especially if they have been there unaltered for more than 4 years. Contact the Planning Advisory Committee for advice.

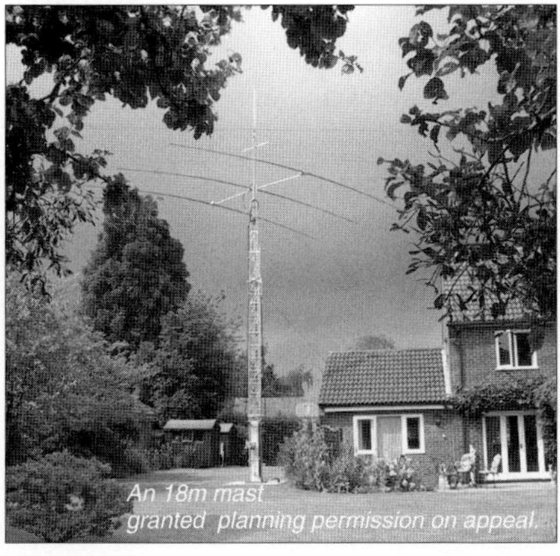
An 18m mast granted planning permission on appeal.

The Planning Department are likely to take enforcement action against you in two circumstances:

1. Where you have erected an aerial or mast which, in the Planning Department's opinion, requires permission and you have not obtained it.
2. Where the Planning Department alleges that you have breached a condition attached to the planning permission they have issued (for example, to keep a mast wound down when not in use).

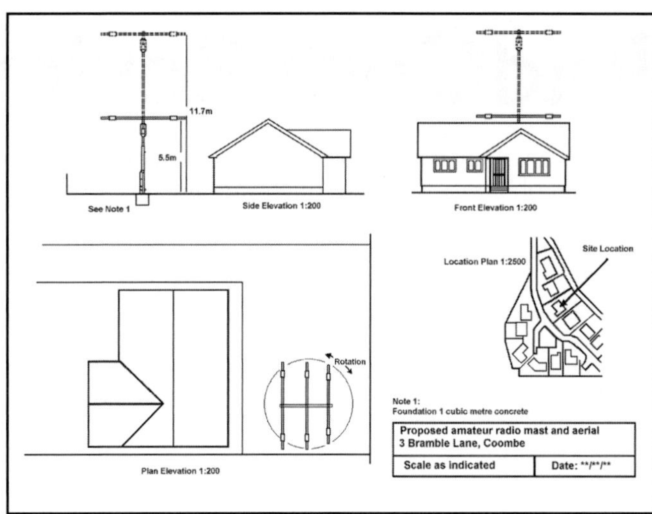

The kind of drawing that a council will want you to submit with your application.

The first is the most common. If you have not already submitted an application and had it refused the Planning Department will normally write to invite you to submit an application. It is usually worth doing so unless you want to argue that you have permitted development rights for the aerial or they are de minimalist. The Planning Department may serve on you a Planning Contravention Notice. This requires you to give certain information as to ownership or to attend the Planning Department's Offices at a specific date and time to give details of your installation and why you believe it does not need planning permission (for example, because it's permitted development or de minimis). You must comply with the Notice, because if you fail to do so you may be prosecuted.

If the Planning Department is not satisfied with your explanation they may elect to issue you an Enforcement Notice. Planning Departments can only do this if they can give reasons why they would not consider granting planning permission and may have to justify their decision to the Planning Inspectorate.

If an Enforcement Notice is issued it will set out what the Planning Department want you to do. Usually this will require you to remove the aerial and/or mast. Should you be served an Enforcement Notice you have two choices:

1. Comply by removing the offending aerial, mast, etc.
2. Appeal.

You must appeal within 28 days of receiving the Notice. Details on how to appeal are available from the Planning Inspectorate, Scottish Government and the Northern Ireland Planning Appeals Commission websites listed below. If the notice relates to a breach of conditions, the Planning Department may serve on you an ordinary enforcement notice, (against which you can appeal as above), or alternatively a Breach of Condition Notice, against which there is no appeal.

Failure to comply with an Enforcement Notice quickly can lead to legal action being taken against you, so don't ignore them. If the Planning Department considers that the aerial/mast has a severe environmental concern which requires immediate action they can apply to the Court for an injunction. If such an injunction is granted, you must comply or you will be prosecuted.

Planning Advisory Committee

The Planning Advisory Committee exists to assist RSGB members with planning applications, enforcement notices and planning appeals. Committee members will not actually prepare your planning application or submit an appeal on your behalf, but can check your application or provide you with a suggested appeal strategy.

The Committee also provides a guide to the planning process. This is available free of charge from RSGB HQ, or as a download from the member's only website.

Tenancy matters

For tenants, both planning permission and landlord consent are likely to be needed. Obtaining planning permission does not mean the landlord has to agree, so you could find yourself having incurred the expense of obtaining planning permission only to find that the landlord does not agree and so you cannot implement the permission.

Especially for private sector tenants, failure to obtain consent may give the landlord grounds to terminate the tenancy.

The Society cannot become involved in legal disputes between landlords and tenants, but will try to provide advice or to signpost members to other bodies who can help. You may therefore want to contact the PAC Chair before starting - contact details below.

	Online planning information	Appeals and enforcement notices
England/Wales	www.planningportal.gov.uk/	www.planning-inspectorate.gov.uk
Scotland	www.eplanning.scotland.gov.uk	www.dpea.scotland.gov.uk
Northern Ireland	www.planningni.gov.uk	www.pacni.gov.uk

RadCom Reviews

Items of amateur radio equipment are frequently reviewed in RadCom. The types of equipment range from antennas, through budget 'handies', software and ancilliary equipment, to top-of-the-line transceivers. Listed below are the reviews that have taken place in RadCom since January 1995.
Please refer to the relevant edition, if you need more information on any of the items.

A

AADE DFD (digital frequency display kit)	Apr 1998
AADE IIB (digital L/C meter kit)	Apr 2005
AAlog (logging software)	May 2006
ACE-HF (propagation prediction software)	Oct 2006
ACE-HF PRO propagation prediction software	May 2016
Acom 1000 (160m-6m linear)	Mar 2001
Acom 1010 (HF linear amplifier)	Aug 2005
Acom 1500 (160m-6m linear)	Dec 2012
Acom 2000A (HF auto-tuning linear)	Mar 2001
ADI AR-446 (70cm FM mobile)	Mar 1997
ADI AT-600D (2m+70cm handheld)	Apr 1997
Adonis AM-308 (desk microphone)	Jan 2006
Adonis AM-508 (desk microphone)	Jan 2006
Adonis AM-708 (desk microphone)	Jan 2006
AirNav RadarBox (virtual radar)	Mar 2009
Airspy SDR Dongle	Sep 2015
Aerial-51 ALT-512 QRP transceiver	Mar 2020
AKD HF3E (communications receiver)	Jun 1998
AKD Target HF3 (VLF-HF receiver)	Nov 1996
Ailunce HS-2 1.8MHz to 432MHz SDR transceiver	Dec 2021
Albrecht AE485S (10m multimode mobile)	Jan 2001
Alinco DM330MW PSU	Jun 2021
Alinco DJ-596 (2m+70cm FM handheld)	Dec 2004
Alinco DJ-C1 (2m FM micro handheld)	Feb 1998
Alinco DJ-C4 (70cm FM micro handheld)	Feb 1998
Alinco DJ-C5 (2m+70cm FM mini handheld)	Jul 1998
Alinco DJ-C7 (2m+70cm FM handheld)	Jan 2005
Alinco DJ-G5 (2m+70cm FM handheld)	Apr 1997
Alinco DJ-G7 (2m+70cm+23cm FM handheld)	Aug 2009
Alinco DJ-MD5	Feb 2019
Alinco DJ-V5E (2m+70cm FM handheld)	Sep 2000
Alinco DJ-V17 (2m FM handheld)	Jul 2008
Alinco DR-610E (2m+70cm FM mobile)	Feb 1999
Alinco DX-70TH (HF+6m multimode mobile)	Aug 1999
Alinco DX-SR8E(100W HF transceiver)	Jan 2014
Alinco DJ-VX50 dual band handheld	Apr 2020
Alinco switch mode power supplies	Jan 2016
Alpha 4510 (HF power/SWR meter)	Nov 2006
Alpha DX-Jr (lighweight portable 40m - 6m ant)	Feb 2013
Alpha SPID Rak (azimuth rotator)	Mar 2007
Alpin 100 (HF+6m linear amplifier)	Apr 2011
Ameritron ALS-500M (HF linear)	Dec 2005
Amlog 3 (logging software)	Sep 1996
Anan 8000DLE SDR 2000W HF Transceiver	Nov 2017
Anytone AT-779 70MHz mobile transceiver	May 2022
Anytone AT-D878UV Plus handheld	Sep 2019
Anytone AT-D578UVPRO dual band mobile	Mar 2020
Anytone AT-779UV 2m/70cm transceiver	May 2021
AOR AR-DV1 Digital Voice Receiver	May 2016
AOR AR7030 (VLF-HF receiver)	Jul 1996
AOR ARD9000 Digital Voice (digital voice adapter)	Oct 2005
AOR ARD9800 Fast Data Modem (digital voice adapter)	Jul 2004
Arno Elettronica E-H Antennas (small HF antennas)	Sep 2003
Arno Elettronica Venus-80 (small HF antenna)	Aug 2005
Arno Elettronica Venus-160 (small HF antenna)	Aug 2005
Array Solutions Bandmaster (universal band decoder)	Jan 2011
Array Solutions PowerMaster (wattmeter)	Feb 2006
Ascel AE20401 frequency counter and power meter kit	May 2020
Astatic SWR Power Meter	Sept 2021
ATM Motion Picture (video grabber)	Nov 1996
Avair AV-20 (HF power/SWR meter)	Oct 2006
Avair AV-40 (2m-70cm SWR/power meter)	Jun 2006

B

Badger Boards Receiver (kit for Novice course)	Apr 1998
Basicomm CW Touch Paddle	Jan 2016
Begali HST (single lever paddle key)	Jul 2009
Begali Sculpture (iambic paddle key)	Feb 2008
Begali Simplex Mono (single lever paddle key)	Feb 2008
bhi Compact In-Line Noise Eliminating Module	Dec 2015
bhi 5W DSP noise cancelling inline module	Jan 2022
bhi DSPKR (noise reduction speaker)	Feb 2010
bhi groundbreaker	June 2019
bhi DSPKR 10W (noise reduction speaker)	Jan 2011
bhi ParaPro EQ20 DSP	Sep 2017
bhi NEDSP1061 (add-on DSP module for FT-817)	Dec 2003
bhi NEDSP1062 (add-on DSP module)	Jul 2005
bhi NES 10-2 (noise eliminating speaker)	Dec 2002
bhi NES 10-2 Mk2 (noise eliminating speaker)	Nov 2005
bhi NEIM1031 (noise eliminating inline module)	Mar 2004
bhi 1042 (switch box)	Mar 2004
bhi NES10-2 Mk 4 noise reduction speaker	May 2020
Bilal Isotron (small antennas for 80m and 40m)	Aug 2010
Buddipole (portable HF-VHF antenna)	Mar 2005
Butternut HF2V (HF multiband vertical antenna)	Mar 2005

C

Ciro-Mazzoni 'Stealth' antenna	Jult 2019
Ciro-Mazzoni 'Baby' Loop	Dec 2019
Cloud IQ SDR	June 2016
Comet CHA-250B (wideband vertical antenna)	Dec 2006
Comet CSW-201G (antenna switch)	Jul 2007
Comet CAT-300 ATU	Feb 2023
CommSlab μ-Modem (multimode radio modem)	Sep 1995
Crazy Daisy (Mag loop Antenna)	May 2014
CRKITS HT-1A Dual Band QRP Transceiver Kit	Aug 2019
Cross Country Wireless SDR (single band receivers)	Aug 2010
CRT SS6900 (10m multimode mobile)	Aug 2011
CT (contest logging software)	Nov 1999
Cushcraft 13B2 (2m beam)	Mar 2005
Cushcraft MA5B (compact HF beam)	Nov 1999
Cushcraft MA8040V (80m/40m vertical antenna)	Sep 2005
Cushcraft R8 (40m-6m vertical antenna)	Jun 2000
Cushcraft R7000 (HF vertical antenna)	Jan 1997
Cushcraft X7 ('Big Thunder' HF beam)	Nov 1998
CWMORSE 3D printers keys and paddles	May 2021

D

Daiwa CS-201A (antenna switch)	Jul 2007
DB6NT 13cm transverter (kit)	Jan 1998
Derek Stillwell Morse Key (handmade straight key)	Jun 1995
Diamond CA-35RS (lightning arrestor)	May 2006
Diamond CX-210A (antenna switch)	Jul 2007
Diamond SD300 (3-30MHz screwdriver antenna)	Nov 2011
Diamond SX20C (HF power/SWR meter)	Oct 2006
Diamond SX40C (2m-70cm SWR/power meter)	Jun 2006
Discovery TX-500 160-6m QRP transceiver	Nov 2021
DG8SAQ VNWA3 (vector network analyser)	Dec 2011
Digirig Mobile	Sept 2022
DK2DB 13cm PA (kit)	Jan 1998
DPRE4-6VL (cavity filters for 2m repeater)	Jun 2009
DSP-10 (DSP 2m multimode transceiver kit)	Feb 2000
DX4Win (station logkeeping software)	Aug 2000
DV Dongle (D-Star adapter)	Dec 2008
DV Access Point Dongle (2m 10mW D-Star node)	Mar 2011
DXAID 5.0 (propagation prediction software)	May 2004
DX Commander Rapide 7m antenna kit	Oct 2021
DX Engineering HEXX-5TAP-2 (5-band 2-ele hex beam)	Mar 2011

E

Elad FDM77 (software defined radio)	Nov 2005
Elecraft K1 (HF QRP CW transceiver kit)	Sep 2001
Elecraft K2 (HF QRP CW transceiver kit)	Mar 2003
Elecraft K3S	April 2016
Elecraft KX3 (HF/6m allmode transceiver)	Apr 2013
Elecraft KX2 (80-10m 10W allmode transceiver)	Jan 2017
Elecraft K4D HF and 50MHz transceiver	Jan 2022

Elecraft KPA500 (Solid State Amplifier)	Jan 2013
Elecraft KRC2 (band decoder kit)	Jan 2005
Elecraft N-gen (wideband noise generator)	Nov 2006
Elecraft P3 Panadapter & K3 Transceiver revisited	Oct 2012
Elecraft T1 (HF-6m QRP auto ATU)	Mar 2007
Elecraft XG2 (receiver test osc. / S-meter calibrator)	Nov 2006
ETO/Alpha 91B (HF linear)	Feb 1997
ETO/Alpha 87A (HF linear)	Feb 1997
EZMaster (SO2R interface)	Mar 2006
EZNEC 2 (antenna modelling software)	Sep 1998
F	
FA-VA5 Antenna Analyser	Sep 2018
Flexradio SDR-1000 (software defined transceiver)	Jun 2006
Flexradio Flex-1500 (HF-6m software defined xcvr)	Apr 2011
Flexradio Flex-3000 (HF-6m software defined xcvr)	Aug 2009
FlexRadio Systems Maestro console	Aug 2016
Flexradio Flex-5000A (HF-6m software defined xcvr)	Jan 2008
Flexradio Flex-5000A upgrades (auto ATU and 2nd rx)	Mar 2009
Force 12 XR6/XR6C Yagi (11 ele six band ant)	May 2015
FoxRex 3500 ARDF Receiver	Sep 2017
FunCube Dongle Pro +	Feb 2013
G	
G1MFG ATV modules	May 2003
G3LIV Isoterm (data interfaces)	Sep 2009
G3WDG (23cm transverter kit)	Jun 2000
G4HUP L-C Meter Kit (test instrument)	Jul 2009
G4TPH Magnetic Loop Antennas (QRP loops)	Oct 2008
G4ZPY 3-in-1 Combo Keyer (electronic keyer)	Jun 1998
Garth 70cm and 23cm bandpass filters	Sep 2007
Gemini 23 (1296MHz linear amplifier)	Feb2016
GH Engineering PA1.6-16 (1.3GHz PA kit)	Jul 2006
Global CX201 (antenna switch)	Jul 2007
Goodwinch TDS (electric tower winches)	Jun 2011
Green Heron RT-21 (digital rotator controller)	Oct 2009
H	
HamGadgets MasterKeyer MK-1 (Morse keyer)	Oct 2011
HamGadgets PicoKeyer-Plus (Morse keyer kit)	Oct 2011
Ham Radio Deluxe (station control & logging software)	Apr 2005
Hamware AT-502 (remote auto ATU)	Mar 2009
Hamware AT-515 (remote auto ATU)	Mar 2009
Hands RTX/AMP (broadband HF linear amp kit)	Jan 1995
Hatley Crossed-Field Loop (antenna)	May 2002
Heil Proset Headphones (headphones with boom mics)	Aug 1996
Heil Proset 5 (headphones with boom mic)	Mar 2006
Heil Pro 7 communications headset	June 2015
Hexbeam Folding Antennas (20-10m multiband ant)	April 2015
HFx (propagation prediction software)	Oct 1999
HFx (propagation prediction software)	Jun 2000
High Sierra 1800/Pro (HF mobile antenna)	Aug 2004
Hilberling PT-8000A (HF-2m transceiver)	Nov 2013
Howes ASL5 (audio filter kit)	Jun 1995
Howes AT160, VF160 & MA4 (kits)	May 1997
Howes DC2000 (receiver kit)	Jul 1997
Howes DXR20 (receiver kit)	May 1995
Howes Tx2000 (transmitter kit)	Mar 1998
hupRF DG8 preamp and DCI-V bias injector	Oct 2016
I	
ICEPAK (propagation prediction software)	Jun 2000
Icom IC-207H (2m+70cm FM mobile)	Feb 1999
Icom IC-703 (HF-6m multimode portable)	Oct 2003
Icom IC-706 (HF-2m multimode mobile)	Nov 1995
Icom IC-706 Mk2 (differences from Mk1 version)	Jun 1997
Icom IC-705	Apr 2021
Icom IC-7100 (HF-70cm multimode)	Feb 2014
Icom IC-736 (HF+6m multimode base station)	May 1995
Icom IC-738 (HF multimode base station)	May 1995
Icom IC-746 (HF-2m multimode base station)	Mar 1998
Icom IC-756 (HF+6m multimode base station)	May 1997
Icom IC-756 Pro (HF-6m multimode base station)	Mar 2000
Icom IC-756 Pro II (HF-6m multimode base station)	Jun 2002
Icom IC-756 Pro III (HF-6m multimode base station)	Feb 2005
Icom IC-775DSP (HF multimode base station)	Jan 1996
Icom IC-821H (2m+70cm multimode base station)	Jan 1998
Icom IC-910 (2m-23cm multimode base station)	Jul 2001
Icom IC-7000 (HF-70cm multimode mobile)	Apr 2006
Icom IC-7200 (HF-6m multimode rugged /P/base)	Jan 2009
Icom IC-7300 (HF-6m multimode base station)	Aug 2016
Icom IC-7400 (HF-2m multimode base station)	Oct 2002
Icom IC-7410 (HF-6m multimode base station)	Jan 2012
Icom IC-7600 (HF-6m multimode base station)	Jun 2009
Icom IC-7700 (HF-6m multimode base station)	Jun 2008
Icom IC-7800 (HF-6m multimode base station)	May & Aug 2004
Icom IC-7851 HF-6m transceiver	Nov 2015
Icom IC-9100 (HF-23cm base station)	Apr 2011
Icom IC-9100 (HF-23cm base station)	May 2012
Icom IC-E7 (2m+70cm handheld)	Jul 2006
Icom IC-E80 (2m+70cm FM + D-Star handheld)	Feb 2011
Icom IC-E90 (6m+2m+70cm FM handheld)	Dec 2004
Icom IC-E91 (2m+70cm FM handheld)	Jan 2007
Icom IC-E92 (2m+70cm FM + D-Star handheld)	Nov 2008
Icom IC-E2820 (2m+70cm FM + D-Star mobile)	Mar 2008
Icom IC-R3 (HF-microwave handheld receiver/TV)	Nov 2001
Icom IC-R20 (HF-microwave handheld receiver)	Sep 2004
Icom IC-T3H (2m handheld)	Jul 2002
Icom IC-T7E (2m+70cm handheld)	Apr 1997
Icom IC-T70E (2m+70cm handheld)	Nov 2010
Icom IC-T81E (6m-23cm FM handheld)	Sep 2000
Icom IC-V80E (2m handheld)	Nov 2010
Icom IC-V82 (2m FM/digital handheld)	Jan 2006
Icom IC-T10	Apr 2023
Icom ID-52E VHF/UHF D-Star transceiver,	Aug 2022
Icom ID-E880 (2m+70cm D-Star mobile)	Feb 2011
Icom PCR-1000 (remotable receiver)	Dec 1997
Idiom Press Rotor-EZ (rotator controller)	May 2001
IK Telecom DPRE4-6VL (cavity filters)	Jun 2009
Index Labs QRP+ (HF SSB/CW QRP transceiver)	Nov 1995
InnovAntennas 9-ele 2m LFA Yagi (antenna)	Mar 2012
InnovAnetnnas 5-element 15m OP-DES Yagi (antenna)	Aug 2012
Innovantennas 20/15/10 DESpole (antenna)	Dec 2013
INRAD W1 headset	Jan 2020
International Radio Roofing Filter (for FT-1000MP)	Jan 2005
I-PRO Home (multiband vertical dipole)	Jul 2011
I-PRO Traveller (portable vertical HF dipole)	Jun 2010
J	
JPS ANC-4 (antenna noise canceller)	Aug 1996
JRC JST-245 (HF+6m multimode base station)	Oct 1997
JRC NRD-630 (professional MF/HF receiver)	Feb 2008
K	
Kanga Finningley (80m SDR kit)	Aug 2011
Kanga Foxx3 (QRP transceiver kit)	Sep 2010
Kanga QRP Pocket Transmatch 30	Jul 2023
Kenwood LF-30A (HF low pass filter)	May 2007
Kenwood TH-79E (2m+70cm handheld)	Apr 1997
Kenwood TH-D7E (2m+70cm FM handheld)	Sep 2000
Kenwood TH-F7E (2m+70cm FM handheld)	Dec 2004
Kenwood TH-D74E	Mar 2017
Kenwood TM-D710E (2m+70cm FM mobile)	Nov 2007
Kenwood TM-D710E + AvMap Geosat 5 Blu	Feb 2009
Kenwood TM-G707E (2m+70cm FM mobile)	Feb 199
Kenwood TS-480HX (HF+6m multimode mobile/base stn)	Mar 2004
Kenwood TS-570D (HF multimode base station)	Dec 1996
Kenwood TS-590S (HF+6m multimode base station)	Jan 2011
Kenwood TS-590SG (HF and 50MHz)	Mar 2015
Kenwood TS-870S (HF multimode base station)	Apr 1996
Kenwood TS-890S	Apr 2019
Kenwood TS-990S (HF+6m Flagship base station)	Jun 2013
Kenwood TS-2000 (HF-23cm multimode base station)	Apr 2001
Kinetics SBS-1eR (virtual radar)	Nov 2009
Kinetic SBS-3 (virtual radar)	Aug 2012
KK7P DSPx & KDSP-10 (DSP kit)	Jul 2005
Kuhne 23cm transverter	Dec 2007
Kuhne MKU23 G4 13cm band transverter	Sep 2016
Kuhne MKU LNA 131A HEMT (23cm preamp)	Jan 2008
Kuhne MKU10 G3 (10GHz transverter)	May 2008
Kuhne MKU 432 G2 (70cm transverter)	Apr 2011
Kuhne 2400MHz upconverter & 10GHz downconverter	Feb 2020
L	
Lake DTR7-5 (HF QRP transceiver)	Oct 1995
Lake Novice Receiver (kit)	Feb 2001
LAMCO DU1500L (HF ATU)	Feb 2012
LAMCO DU1500T (HF ATU)	Feb 2012
LDG AT-100 (automatic ATU)	Feb 2006
LDG Z11 (automatic ATU kit)	Aug 2002
LDG Z11 Pro (automatic ATU)	Jun 2009
Linear Amp UK Challenger (HF linear)	Feb 1997
Linear Amp UK Discovery 64 (6m/4m linear)	Apr 2010

Linear Amp UK Explorer 1200 (HF linear)	Feb 1997
Linear Amp UK Ranger 811 kit (HF linear)	Sep 2004
Little Tarheel II (80m-6m mobile antenna)	Aug 2008
LimeRFE front end and PA for LimeSDR	May 2022
Low cost digital oscilloscope kit	May 2019
Logger (station logkeeping software)	Aug 2000
Log-EQF (station logkeeping software)	Aug 2000
M	
M2 LEO pack 2m/70xm satellite antenna system	Nov 2022
Maha MH-C777 Plus-II (intelligent battery charger)	Jun 2007
Maldol HF, VHF & UHF mobiles (antennas)	Jan 2003
Maldol HVU-8 (80m-70cm vertical antenna)	Jan 2004
Maldol MFB-300 (broadband vertical antenna)	Apr 2007
Mastrant rope and antenna equipment	Sept 2020
MFJ-226 (graphical impedance analyser)	Nov 2015
MFJ-260C (1-650MHz 300W dummy load)	Feb 2008
MFJ-269 (HF-UHF SWR analyser)	May 2000
MFJ-270 (lightning arrestor)	May 2006
MFJ-393 (headphones with boom mic)	Mar 2006
MFJ-402 Nano keyer	Mar 2019
MFJ-461 (pocket Morse code reader)	May 2002
MFJ-704 (HF low pass filter)	May 2007
MFJ-781 (multimode DSP data filter)	Sep 1997
MFJ-784B (tunable DSP filter)	Feb 1996
MFJ-805 (RF current meter)	Jul 2006
MFJ-817C (2m-70cm SWR/power meter)	Jun 2006
MFJ-854 (RF current meter)	Jul 2006
MFJ-860 (HF power/SWR meter)	Oct 2006
MFJ-890 (DX beacon monitor)	Jul 2003
MFJ-935 (small loop tuner)	Apr 2005
MFJ-939	Mar 2016
MFJ-935B (small loop tuner)	Apr 2006
MFJ-936 (small loop tuner)	Apr 2005
MFJ-941 Tuner	Oct 2020
MFJ-989D (HF QRO manual ATU)	Jan 2007
MFJ-991B (automatic ATU)	Feb 2006
MFJ-993B (automatic ATU)	Jun 2009
MFJ-1234 RigPi Radio Station Server	Jan 2021
MFJ-1702C (antenna switch)	Jul 2007
MFJ-1786X Super Hi-Q Loop (HF loop antenna)	May 2007
MFJ-1897 (HF vertical antenna)	Sep 1995
MFJ 9020, 9420, 9140 & 9040 (HF QRP transceivers)	Oct 1995
MFJ 'Cub' (HF QRP transceiver kit)	Feb 2001
Microham Micro Keyer II (multimode digital interface)	Jan 2008
Microham Station Master & Six Switch (antenna switch)	Apr 2009
MicroKeyer (PIC-based electronic keyer)	Mar 1996
Microset CF-300 (DC-1GHz 300W dummy load)	Feb 2008
Microset PTS-124 (13.8V DC power supply)	Dec 2005
Microtelecom Perseus (software defined receiver)	Mar 2008
Microtelecom Perseus (software defined receiver)	May 2010
Miniprop Plus (propagation prediction software)	Jun 2000
miniVNA Tiny (network analyser)	Feb 2015
Miracle Ducker (antenna for FT-817 etc)	Nov 2004
Miracle Whip (antenna for FT-817 etc)	Feb 2002
Mizuho MX-14S (HF QRP transceiver)	Oct 1995
Moonraker HT-90E (2m FM handheld)	Sep 2011
Moonraker MT-270M (2m/70cm FM Mobile)	Dec 2015
Moonraker SPX-200 (HF-6m mobile antenna)	Jan 2007
Moonraker HT-500D (2m/70cm FM Mobile)	Oct 2017
Moonraker SHK-1 Straight hand key	Dec 2022
Morphy Richards 27024 (domestic DAB+DRM receiver)	May 2008
MW0JZE seven-band wide-spaced Hexbeam	July 2015
MyDEL AnyTone AT-5189 (4m FM mobile)	Apr 2011
MyDEL CG3000 (HF auto ATU)	Nov 2006
MyDEL HB-1A (compact 3-band QRP CW transceiver)	Oct 2010
MyDEL Multi-Trap Dipole (HF antenna)	Dec 1995
MyDEL SB-2000 (radio interface)	Sep 2010
MyDel SWR-001 & SWR-006 (SWR & Power Meters)	Aug 2017
MyDel JPC-7 and JPC-12 portable antenna kit	Oct 2022
MYDEL Windcamp Gipsy HF antenna	Feb 2023
M0CVO HW-20HP (off-centre fed dipole)	Jan 2012
N	
NA (contest logging software)	Nov 1999
nanoVNA-H series Vector Network Analyser	Feb 2021
O	
Optibeam OB9-5 (9-ele 5-band HF beam)	Aug 2003
Optibeam OB10-3W (10-element 20/17/15m beam)	Mar 2005
Optibeam OB10-5 (10-element 5-band HF beam)	Mar 2007

Outbacker Joey (QRP HF portable antenna)	May 2008
P	
Palstar AT1KM (HF ATU)	May 2005
Palstar AT1500CV (HF QRO manual ATU)	May 2005
Palstar AT1500CV (HF QRO manual ATU)	Jan 2007
Palstar AT-AUTO (HF auto ATU)	Sep 2006
Palstar KH-6 (6m FM handheld)	Oct 1997
Palstar ZM30 (digital antenna impedance bridge)	Sep 2005
Peak Electronics 'Atlas' Analysers (L, C & R meters)	Sep 2005
Peak Atlas DCA PRO (semiconductor analyser)	Mar 2013
Peak Atlas ZEN50 intelligent Zener diode tester	July 2015
Peet Bros Ultimeter 2100 (weather station)	Jun 2008
Piccolo 6m Transceiver Kit (synthesised FM)	Jun 1996
Pico Balun and Pico Tuner kits	July 2016
Pixie QRP transceiver kit	Jan 2016
PJ-80 (DF receiver)	Feb 2006
PJ-80 (DF receiver)	Apr 2007
PK-4 (Auto CW Pocket Keyer)	Nov 2015
PocketDigi (datamode software for PDA/smartphone)	Dec 2006
PolyPhaser IS-50UX-C0 (lightning arrestor)	May 2006
PolyPhaser IS-B50HN-C2 (lightning arrestor)	May 2006
PolyPhaser VHF50HN (lightning arrestor)	May 2006
Powerex MH-C9000 (charger for AA and AAA cells)	May 2010
Prepp Comm DMX-40 CW transceiver	Apr 2022
Procom DPF 2/33 & DPF 70/6 (2m & 70cm duplexers)	Sep 2009
Primetec Primesat Controller (satellite stn controller)	Dec 2005
Pro-Am HF Mobile Antennas (mobile whips)	Aug 1995
Pro-Am MM-3401 (mobile antenna mag-mount)	Aug 1996
Pro Antennas DMV Pro (portable antenna system)	May 2009
Pro Antennas Dual Beam Pro (5-band non-resonant beam)	May 2011
Pro Antennas I-pro Traveller (portable antenna system)	Jun 2010
Q	
Qpak Precision Tuner (mini ATU)	Oct 2004
QRP Labs QCX-mini QRP transceiver kit	Oct 2022
QDX Digital Transceiver	Dec 2022
R	
Red Pitaya STEMlab oscilloscope	Oct 2018
Ranger RCI-2950DX (10m/12m multimode mobile)	Feb 2012
RFSpace SDR-IQ (software defined receiver)	Mar 2008
RF Explorer spectrum analyser WSUB1	Oct 2021
RFinder B1 Dual Band DMR and analogue transceiver	Jan 2021
RigExpert AA-1000 antenna analyser (Antennas)	Aug 2012
RIGblaster Advantage (PC-radio interface)	Mar 2012
RigExpert TI-5000 Interface	Jun 2022
RIGrunner 4005 (12V distribution panel)	Mar 2012
Rigol DS2000 series (oscilloscope)	Aug 2013
Rigol DSA815-TG (spectrum analyser)	May 2013
Rigol DSA815 and RSA3030 spectrum analysers	Mar 2019
Rig Expert AA200 (antenna analyser)	May 2008
Rock-Mite QRP (40m Xtal controlled Kit)	May 2014
Rohde & Schwarz FSH3 (spectrum analyser)	Sep 2004
S	
Samlex SEC-1223 (13.8V DC power supply)	Dec 2005
SD (contest logging software)	Nov 1999
SDR play	Mar 2016
SDRplay RSP2	Apr 2017
SDRplay RSPduo	July 2018
SDRPlay RSPdx	Jan 2020
SGC ADSP2 (noise eliminating speaker)	Nov 2003
SGC ADSP2 Mk2 (noise eliminating speaker)	Nov 2005
SGC SG-211 (internally powered HF-6m auto ATU)	Jun 2005
SGC SG-231 (HF-6m auto ATU)	Feb 2000
SGC SG-239 (budget HF auto ATU)	Jun 2005
SGC SG-500 (HF linear)	Dec 2005
SGC SG-2020 (HF 20W SSB/CW transceiver)	Mar 1999
SGC Stealth Kit (antenna)	May 2003
SG-Lab 2.3GHz transverter	Jan 2017
Shacklog (logging software)	Jan 1995
Shacklog (station logkeeping software)	Aug 2000
Sharman multiCOM AV-6075NF PSU	Jul 2022
Sharman AV-508 desktop microphone	Sep 2022
Sharman HLP-270 Halo antenna	Nov 2022
Shure 522 (desk microphone)	Jun 2007
Shure 550L (desk microphone)	Jun 2007
Shure 572B (fist microphone)	Jun 2007
Siglent SDG1062X waveform generator	Jan 2020
Siglent SDS 1202X-E oscilloscope	Nov 2019
Siglent SVA-1032X spectrum & network analyser	Jul 2020

Signal Hound SA44B (spectrum analyser)	Sep 2011
Signal Hound TG44 (tracking generator)	Sep 2011
SkySweeper (datamodes software)	Apr 2005
SkySweeper professional (datamodes software)	Feb 2008
Softrock V6.2 (HF SDR receiver)	Mar 2007
Sony ICF-SW100E (mini communications receiver)	Jul 1996
SOTA Beams 2m Portable Yagi (antenna)	Jul 2004
SOTAbeams LASERBEAM-DUAL CW filter	Nov 2016
SOTAbeams aerials, log book & battery monitor	Oct 2013
SOTAbeams WOLFWAVE	Oct 2019
SOTAbeams WSPRlite Antenna Tester	Jun 2017
SOTABEAMS 2m bandpass filter	Dec 2022
SPE Expert 1K-FA (linear amplifier)	Jun 2007
SPE Expert 1.3K-FA linear amplifier	July 2015
Spectran HF-6085 (hand-held spectrum analyser)	Jan 2010
Spiderbeam (5-band antenna kit)	Mar 2010
Spiderbeam 160-18-4WTH (160m vertical antenna)	Sep 2013
Spiderbeam Aerial-51 Model 404-UL (port ant 40m-6m)	May 2015
SRW CobbWebb (HF antenna)	Jun 1993
Standard C108 (2m FM handheld)	Jan 1997
Standard C408 (70cm FM handheld)	Jan 1997
Standard C568 (2m+70cm handheld)	Apr 1997
Standard C5900D (6m+2m+70cm FM mobile)	Jul 1997
StationMaster (station logkeeping software)	Aug 2000
SteppIR 3-element Yagi (14-52MHz beam)	Feb 2004
SteppIR 4-element Yagi (14-52MHz beam)	Oct 2007
SV4401A handheld vector network analyser,	Jan 2023
SunSDR2 PRO (HF to VHF transceiver)	Dec 2015
Super Antenna MP-1 (portable HF-VHF dipole)	Jun 2008
Super Keyer 3 (electronic keyer)	Jan 1997
Super Antenna MP1B Super-Stick	Jan 2013
T	
Talksafe (Bluetooth adapter)	Sep 2007
Telecom 23CM150 (23cm linear amplifier)	Oct 2009
Tennadyne T10 (13-30MHz log periodic HF beam)	Jan 2002
Teensy SWR/Power meter	Nov 2020
TenTec Argo (HF QRP transceiver)	Oct 1995
TenTec Eagle 599 (HF-6m compact multimode base stn)	Jul 2011
TenTec Jupiter 538 (HF multimode base station)	Jan 2004
TenTec Omvi VII (HF-6m multimode base station)	Sep 2007
TenTec Orion 565 (HF multimode base station)	Jun 2004
TenTec Orion 2 566 (HF multimode base station)	Aug 2006
TenTec 506 Rebel (CW QRP transceiver)	Dec 2015
TenTec RX340 (professional HF DSP receiver)	Mar 2002
TenTec 1320 (20m CW transceiver kit)	Aug 2000
The tinySA Spectrum analyser	Mar 2022
Thamway TX-2200A (136kHz transmitter)	Feb 2010
Timewave DSP-59+ (tunable DSP filter)	Feb 1996
Timewave TZ-900 (antenna analyser)	Nov 2009
Tokyo Hy-Power HL-1KFX (HF linear amplifier)	Oct 2005
Tokyo Hy-Power HL-2KFX (HF linear amplifier)	Oct 2005
Tokyo Hy-Power HL-50B (linear amplifier)	Sep 2002
Tokyo Hy-Power HL-100BDX (HF linear amplifier)	Oct 2005
Toyocom MS-5 (mobile hands-free kit)	Mar 2009
Trident 6M5L (6m long yagi antenna)	Jun 2003
TROPIC T-R (sequencer)	Aug 2012
TRlog (contest logging software)	Nov 1999
Turbolog III (software)	Nov 1997
Turbolog (station logkeeping software)	Aug 2000
TYT TH-UVF1 (2m+70cm handheld)	Dec 2010
Tytera MD380 DMR handheld	July 2016
TYT TH-UFV9 (dual band handheld)	Dec 2012
U	
uBITX V6 QRP transceiver	Dec 2020
V	
Vargarda 11EL2 (2m beam)	Mar 2005
Vectronics 1010K (10m FM receiver kit)	Feb 2001
Vectronics DL-300M (DC-150MHz 300W dummy load)	Feb 2008
Vectronics LP-30 (HF low pass filter)	May 2007
Vero VR-N7500 Dual band radio	Aug 2022
Videologic DRX-601E/ES (digital radio tuner)	Oct 2001
Vine Antennas LFA Yagis (loop fed VHF antennas)	Nov 2009
Vortex Whirlwind 6M4(6m delto loop)	Dec 2011
W	
Walford Berrow (QRP transceiver kit)	Apr 2014
Walford Brent (single band CW transceiver kit)	Feb 2005
Walford Compton (direct conversion receiver kit)	Jun 2002
Walford Langport (80m+20m CW+SSB transceiver kit)	Apr 2000
Walford Radio Today Chedzoy (receiver kit)	Feb 2001
Watson AT-715 (five-in-one power station)	Nov 2009
Watson CS-600 (antenna switch)	Jul 2007
Watson PBX-100 (portable HF antenna)	Nov 2002
Watson Power-Mite-NF (power supply)	Jun 2008 & Sept 2021
Watson SP-350V (lightning arrestor)	May 2006
Watson VAA-1 (antenna analyser)	Oct 2015
Watson W-25SM (13.8V DC power supply)	Dec 2005
Watson W-184 (headphones with boom mic)	Mar 2006
Watson W-8682 (radio controlled weather centre)	Sep 2008
Watson WM-S (mobile microphone system)	Nov 2005
Wavecom W-Code (datamodes decoding software)	Jul 2012
Wedmore 80m QRP Transceiver (kit)	Aug 1997
Wellbrook ALA1530 (active receiving loop)	Jan 2012
Whistler TRX-1 and TRX-2 scanners	Aug 2020
Wiber Mini 1300 digital antenna analyser	Feb 2022
WinCAP Wizard II (propagation prediction software)	Jun 2000
WinCAP Wizard III (propagation prediction software)	Dec 2002
WinRadio WR-G1DDC Excalibur (SDR receiver)	Oct 2010
WinRadio WR-G313i (PC-controlled receiver)	Mar 2005
Wimo Big-Wheel antenna	Jan 2016
Wonder Wand (antenna for FT-817 etc)	Jun 2004
Wouxun KG-699E (4m handheld)	Sep 2010
Wouxun KG-UV6D Pro Pack (2m/70cm handheld)	Jun 2012
Wouxun KG-UVD1P (2m/70cm handheld)	Sep 2010
Wouxan KG-UV8H 70/144 and 144/430MHz handhelds	Dec 2020
WriteLog (contest logging software)	Nov 1999
WSPR Desk Transmitter	Feb 2022
W2IYH (range of audio products)	Jan 2009
X	
Xiegu X5105 HF-6M QRP transceiver	Nov 2018
Xiegu G90 HF multi-mode transceiver	Feb 2020
Xiegu X6100 HF and 50MHz SDR transceiver	Jul 2022
Xiegu GNR1 digital noise reduction and filter	Oct 2022
Xiegu G106 HF transceiver	Nov 2022
Y	
Yaesu ATAS-100 (mobile antenna)	May 1999
Yaesu FT-5DE 144/430MHz C3FM/analogue handheld	Feb 2022
Yaesu FT-50R (2m+70cm handheld)	Apr 1997
Yaesu FT-60R (2m+70cm handheld)	Dec 2004
Yaesu FT-100 (HF-70cm multimode mobile)	Jun 1999
Yaesu FT-450 (HF-6m compact multimode base stn)	Oct 2007
Yaesu FT-450D (HF-6m compact multimode base stn)	Nov 2011
Yaesu FT-710 review	June 2023
Yaesu FT-817 (160m-70cm multimode portable)	Jun 2001
Yaesu FT-847 (HF-70cm multimode base station)	Aug 1998
Yaesu FT-857 (HF-70cm mobile)	Jun 2003
Yaesu FT-891 (HF+6m multimode)	Mar 2017
Yaesu FT-897 (HF-70cm multimode /P/base station)	Apr 2003
Yaesu FT-920 (HF+6m multimode base station)	Aug 1997
Yaesu FT-950 (HF+6m multimode base station)	Dec 2007
Yaesu FT-991 (HF/VHF/UHF transceiver)	Feb 2016
Yaesu FT-1000MP (HF multimode base station)	Jan 1996
Yaesu FT-1000MP Mark-V (HF multimode base stn)	Oct 2000
Yaesu FT-1000MP Mark-V Field (HF multimode base)	Oct 2002
Yaesu FT-2000 (firmware upgrade)	May 2009
Yaesu FT-2000D (HF-6m multimode base station)	Mar 2008
Yaesu FT-8100R (2m+70cm FM mobile)	Feb 1999
Yaesu FTDX101D	Oct 2019
Yaesu FT-DX1200 (HF-6m multimode base station)	Mar 2014
Yaesu FT-DX3000 (HF-6m multimode base station)	Jan 2014
Yaesu FT-DX5000 (HF-6m multimode base station)	Jun 2010
Yaesu FT-DX9000D (HF-6m multimode base station)	Oct 2005
Yaesu FT-DX9000D (HF-6m multimode base station)	Dec 2006
Yaesu FTM-10R (2m+70cm FM mobile)	Feb 2008
Yaesu FTM-400DE dual band, dual mode mobile	Jan 2015
Yaesu FTM300 dual band mobile	Feb 2021
Yaesu VR-5000 (HF-microwave multimode receiver)	Aug 2001
Yaesu VX-5R (6m/2m/70cm FM handheld)	Sep 2000
Yaesu VX-7R (6m/2m/70cm FM handheld)	Oct 2003
YouKits FG-01 Antenna Analyser	Sep 2012
YP-3 (6-band 3-ele portable yagi)	May 2009
Z	
Zeus ZS-1 (SDR HF transceiver)	Dec 2013
ZUMspot RPi Digital Voice Hotspot	Mar 2021
ZUMspot-RPi Elite 3.5" LCD	May 2023

Radio Communications Foundation

We were established in 2002 by the RSGB and formally incorporated as a Registered Charity in 2003. Although the RCF was established by the RSGB we are independent and we work to create a fund which can support efforts to heighten awareness of the importance of radio communications.

Mission

Our Trust Deed sets out our objective as "to advance the education of the public in the science and practice of radio communications and to promote the wider benefits to the public resulting from such education and training". In practical terms this means that we seek to increase the engagement of people, especially young people, in radio communications technology. We therefore encourage and assist students to pursue relevant higher education courses, leading to them being employed in the radio communications sector, and also work to raise public interest and involvement in radio communications and its associated science and technologies, including amateur radio.

The interested young person of today is the radio amateur of tomorrow and the engineer of the future.

Near and medium term objectives:

The objectives are to advance the RCF mission by engaging with four key groups:

- Those at school (and also those in uniformed groups such as the Scouts, Guides, Army Cadet Force, Royal Air Force Air Cadets, Air Training Corps and Ambulance Brigades) in order to develop an interest in radio communications
- Those planning or undertaking a university or higher education course to encourage the study of radio communications options
- Those planning or considering employment in radio communications
- The public more generally, including those with an interest in amateur radio

Radio communication is so widely practised that it is almost taken for granted, but a greater public understanding of it is vital for the UK economy. There is a serious shortage of radio communications scientists, engineers and technicians, all needed to exploit a myriad of commercial opportunities.

Radio communication is one of the vital technologies for the 21st century. It provides the backbone technology for the information economy. Every member of the public uses it. Many innovations were developed by scientists and engineers who had their interest aroused by a hands-on demonstration, perhaps at school, perhaps at an exhibition, perhaps through a demonstration by a radio amateur. Amateur radio is an underlining supporting strategy for us, and we see encouraging young people into amateur radio and helping them develop as not only supporting the longer term health of amateur radio itself, but directly underpinning achievement of our objectives.

Fund Raising

To the end of 2022, we have raised over £400k That money came from:

- Members of the RSGB. Many members already make donations, some with their membership renewals, and donations can be increased when Gift Aid is applied to them
- Through bequests. Members can make bequests to the Foundation in their Wills and the Foundation rigorously respects any instructions made in a legacy
- Fund raising or club events
- Approaches to industry and public sector sources of grant aided money. We seek to work with industry by raising public awareness of the opportunities for jobs and careers in radio

Projects

The following are the sort of projects and activities the Foundation has supported in the past:

Grants to individual clubs or educational institutions

Examples of such funding include funding radio kit building workshops for young scientists, providing kits to school children in support of Tim Peake's Principia mission, supporting the International Youngsters On The Air (YOTA) event at Gilwell Park and helping to re-establish an amateur radio club at Swansea University.

As a direct result of our collaboration with the Smallpiece Trust over 60 young people have gained Foundation licences. Some have progressed and are enjoying amateur radio as part of their school/university life.

In 2019 we initiated a joint project working with the RSGB to promote radio within schools as part of British Science Week. Nearly 200 radio kits were built and many schools added other activities, like operating Special Event Stations and using Morse Code. This was seen to be a great success. We repeated the exercise again in 2020, with some good coverage in both Radcom and Practical Wireless.

In August 2019 we approved a request for the funding of a radio antenna from a school radio club in Somerset. The delivery was delayed but the antenna is now in place and ready for the children to return after lockdown. Similarly, in June 2019 we approved the funding of a radio demonstration trailer from the Royal Signal Museum Radio Club. This was intended to be taken out to county shows and similar public events.

Two transceivers have been donated, one to a school and one to the ??? Scout group, who have been very active in encouraging youngsters to try amateur radio for the first time.

Bursaries and scholarships

In partnership with the Smallpeice Trust, we continued to sponsor Arkwright Engineering Scholars. These students are approaching higher education and thinking about what universities they would like to attend. They have undergone a rigorous interview and selection process to become an Arkwright scholar and already boast impressive CVs. One of our Scholars told us about an induction powered projectile launcher project he was undertaking and several are involved in national robotics championships. Two of our past Scholars have secured engineering apprenticeships with James Dyson's company and a number have

RCF Scholars

2022 Arkwright Scholars received their RCF sponsorships at a ceremony at the IET

The RCF donated radio in action with the Rotherham Scouts

been selected to represent the UK at international Youngsters on the Air (YOTA) summer camps.

https://commsfoundation.org/arkwrightscholarships/rcf-arkwright-scholars/

We partner with the UK Engineering Skills

Foundation to promote a competition at universities to recognise the best final year projects in the field of radio communications.

https://commsfoundation.org/ projects-2/ukesf-rcf-partnership/

The first prize winner in 2018 was a radio amateur who is now working in satellite communications. This illustrates the benefits of learning through amateur radio as a route to a rewarding career.

Furthermore, we can help fund students' university projects with grants towards test equipment or software. More details can be found on our website:

www.commsfoundation.org

How to donate

There are several ways:

- Through RSGB membership renewals
- Through the RSGB shop - *http://www.rsgbshop.org/acatalog/RCF_Donations.html*
- By the payroll giving scheme for Charities, which some employers offer as a route for regular donations
- By a bequest in a Will, so that an interest in amateur radio lives on for the benefit of others

Gift Aid

For every £1 you give to us, we get an extra 25p from the Inland Revenue. You must pay an amount of income tax and/or capital gains tax at least equal to the tax that the charity reclaims on your donations in the tax year for the RCF to receive this.

Registered charity number 1100694

http://commsfoundation.org/

Radio Communications Foundation,
3 Abbey Court, Fraser Road,
Bedford MK44 3WH

Website: *www.commsfoundation.org*

For more information,
email: *secretary@commsfoundation.org*

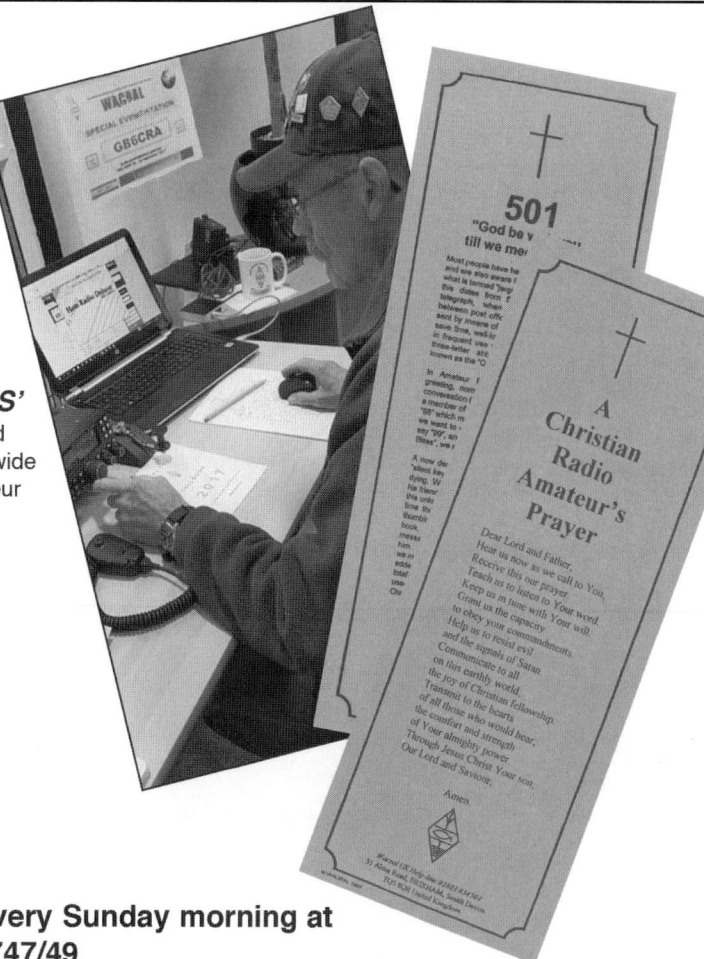

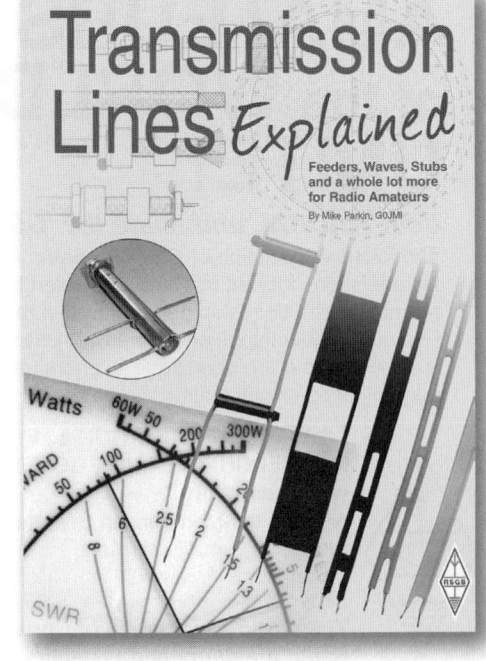

Operating Advisory Service

The Operating Advisory Service (OAS) is a service of the RSGB which is intended to assist radio amateurs or others who may be affected by problems which occur within the amateur bands or which develop on other frequencies as a result of amateur transmissions.

OAS works with a team of named volunteers, trained to offer support to the Regional Teams and to RSGB members and provide better education of amateurs who might be affected at some point, rather than focussing on those causing the problems. The aim is to get everyone better prepared and aware, thereby minimising the effect that these people have on the enjoyment of the amateur service by the overwhelming majority.

OAS is led by the OAS Coordinator, assisted by a team of OAS Advisers, located around the country and trained to help. The whole team communicates via an RSGB email list and so they can support each other in determining how best to resolve various problems. In this way, previous successful outcomes will assist others with similar problems to solve.

The team has developed good-practice guidelines for download by members and non-members alike. OAS additionally offers direct advice to RSGB members. They make themselves available to clubs in their region, to give presentations about what to do if problems occur. Requests from clubs for such a presentation are welcome, even if there is no immediate problem: prevention is always better than cure. Being prepared also means a swifter response by those directly affected, containing the problem early. Planning what to do if a fire breaks out is better than having a fire and then trying to find the buckets!

Getting support from OAS

The Regional Team is the first contact that most members will make to request any form of help, including a request for a club talk. This has the advantage that the Regional Team may be able to help directly but they can also help in a co-ordinated way. For example, they might want to bring EMC committee members to work alongside OAS members to help address your problem. They will also know who the ROA nearest to you is.

You can also contact the OAS Coordinator, at oas@rsgb.org.uk. They will find a nearby Regional Operating Advisor and can additionally raise your problem with all the ROAs, to obtain the benefit of their collective experience. An acknowledgement of your request will be given swiftly, though it may take a little longer to consider how best to address the problem.

Information

A report to OAS should contain details of the issue and should if possible include dates, times, modes and details of what was heard or what happened to cause disruption.

It is important not to send sound or video files, nor to include a name, call sign or address of any alleged perpetrator of malpractice. These would be counter to RSGB's GDPR practice and will not be accepted.

Support offered

General guidance has been developed for downloading from the RSGB OAS web pages. Anyone is free to do this, whether or not they are RSGB members.

We strongly encourage those having problems to use OAS advice to formulate a plan and encourage local amateurs to cooperate in carrying out the plan. In most cases this will be key to obtaining a satisfactory result.

Given that most problems centre on repeater misuse, active OAS support will require the repeater keeper to be willing to work with OAS. The co-operation of regular users of that repeater will also greatly improve the outcome. OAS cannot solve problems for you but it can guide a committed local team to improve the chance of success.

Regional Operating Advisers are recruited and trained to help. They are expected to work courteously with you and will expect to be treated courteously in return. Like you, they are licensed users of the amateur service but they have no magic powers to solve any given issue. Please help them to help you.

OAS: 3 Abbey Court, Fraser Road, Bedford MK44 3WH. Email: *oas@rsgb.org.uk*

Your Time

We rely on the active support of a myriad of volunteers to enable the RSGB to provide the range of member services that we do. Indeed, we can only operate with the active support of dedicated and committed people who believe in the future of amateur radio and are prepared to give their time to making that future actually happen.

People like you

Putting something back into the hobby or passing on your experience, knowledge and specialised skill is a rewarding and fulfilling experience.

The RSGB has many areas in which your skills might be used to help others. Some of them require a level of specialised knowledge: EMC, Planning, Repeater Management and Data Communications. In other areas, enthusiasm and commitment is all that is required: GB2RS newsreading, Emergency Services or Deputy RSGB Regional Manager, to name but a few.

When it comes to training, why not consider becoming an RSGB Registered Instructor? There is a constant need for Instructors for the Foundation, Intermediate and Full amateur radio licence courses.

You decide when and where you can help (please note that all applications for Registered Instructors are, for security reasons, subject to a vetting procedure).

If you can't find time to volunteer to work for the Society, remember the other area where you can make a real difference - mentoring.

Too often we hear of newly licensed amateurs who feel 'on their own' after completing their studies. You can pass on your experience and skill by 'taking under your wing' a new or prospective amateur and ensuring that he or she learns appropriate operating practice and behaviour. In that way we can ensure that our bands are properly used, to the greater enjoyment of everyone, and newcomers can make the most of amateur radio.

For further information on becoming a volunteer for the Society, contact the

RSGB General Manager,
Tel: **01234 832700** or
email: *GM.dept@rsgb.org.uk*

website: *www.rsgb.org/volunteer*

Public Service and Emergency Comms

RAYNET-UK

The definition of a disaster varies from person to person but whatever the definition Amateur Radio has a long history of supporting the community in times of need. The International Telecommunications Union recognises and encourages Radio Amateurs to prepare to assist their communities. To provide the best services to the community it helps to be trained, recognised and organised and this is where RAYNET-UK comes in as an affiliated organisation to the RSGB.

Emergencies

While the UK is not frequently struck by natural disasters, both natural and man-made events can have significant consequences on the communities affected feeling like a disaster to those concerned. No matter the cause, the quickest way for any crisis to turn into a disaster is to lose communications. As recognised over the Winter of 2022/23, the normal networks used by the public and the emergency services are vulnerable in varying degrees to loss of power, fallen pylons etc and it is here that the Radio Amateur with their flexibility and independence from other networks and utilities can provide valuable support to our services and communities with the many frequencies and tools available to us.

Amateur Radio emergency communications in the UK are usually provided by organised groups under the generic but trademarked name of 'RAYNET'. RAYNET-UK is a Registered Charity, affiliated to RSGB and recognised as the UK's principal organisation comprising radio amateurs who provide voluntary radio communications in support of the activities of the User Services and to local communities in times of disaster and emergencies as well as providing communications support to local community events. It is also the main conduit of amateur radio representation to the User Services specified in the Amateur Radio Licence.

RAYNET-UK groups, like others in the voluntary sector are an attractive option to those responsible for emergency planning - because they are quick to respond, flexible, have skills and capabilities that augment the 'business as usual' requirements, and, perhaps above all at a time of diminishing resource they are FREE!

Amateur Radio brings more capabilities to emergency response because of the wide range of spectrum available to us and the relatively unrestricted ability to deploy voice or data communications locally, nationally and internationally. To understand better the needs of local users RAYNET groups work with their Local Resilience Forums throughout the country. These forums have the responsibility of planning 'resilience' or 'the ability to bounce back quickly' in the event of any type of disaster, co-ordinating the professional and volunteer response. They value dealing with organisations like RAYNET-UK who provide a single point of contact that can be tasked with dealing with a particular issue and whose members are a 'known quantity' providing a particular service.

An example of this is the increasing collaboration between RAYNET-UK Groups and '4x4 Responders'. Two skilled organisations working together to provide services in severe weather conditions helping our communities. This has added benefits for the hobby with some 4x4 group members taking out amateur licenses to increase their skill level and the service they can offer.

Community Events

While the number of disasters and emergencies in the UK is relatively low, this brings challenges in maintaining skills and training. Recognising this the Amateur Radio was relaxed many decades ago to allow Radio Amateurs to support our User Services by providing communications support to a range of local events such as fun runs, overnight hikes etc. These events are used by RAYNET-UK groups as a means of practising procedures, testing equipment and honing skills, etc. The events also provide a 'shop window' for Amateur Radio to demonstrate our abilities to Emergency Planning Officers, Police Officers etc. and also for RAYNET to learn from those services about how they approach situations. While any Radio Amateur or group can undertake these activities, being part of a larger organisation allows the most benefit to be obtained from the (hopefully) close relationships built over years with the Emergency Services as well as the Event Organisers.

The amateur radio licence normally restricts the use of a station to self-training in radio communications and as a leisure activity. The latter is interpreted to allow the use of amateur radio in support of community events. The licence also allows any amateur to use his/her station to send traffic on behalf of specified bodies, referred to as 'User Services' in paragraph 17(1)(qq) of the licence.

However, there is a limit to what an individual amateur can achieve, so most will join local clubs or groups and pool their equipment and skills. All amateur radio operators are responsible for the safe operation of their station when in a public area.

RAYNET-UK groups enjoy the protection of public liability and personal accident insurance, as well as cover for the services we provide, but such cover is not to be taken for granted and may be invalidated if carelessness is proven.

No limits?

Emergencies know no boundaries and technically we should not limit ourselves in our response. While voice communications will always be the backbone of communication, our user services increasingly look for the ability to send email, pictures etc. just as they would do on a 'normal' day. RAYNET operators are leveraging 'normal' amateur radio modes like QO-100 geostationary satellite capabilities, independent mesh data networks and HF weak signal data modes to provide services to the community. While our communications must always be reliable, a central DMR server has also been built to bring together operators interested in digital voice modes. If you have an interest in data, video or microwaves, please get in touch with a local group and bring your specialist skills to emergency communications.

Since we may be asked to provide communications ranging from the very local to international, IARU and UK band plans allocate the following frequencies to emergency communications. VHF/UHF frequencies tend to be 'spot' frequencies while at HF they are defined as 'Centres of Activity' since no one, even emergency communications has an absolute right to a frequency. However, given the potential urgency of the communications, and that our User Services are likely to be listening, the co-operation of all Radio Amateurs is requested to allow a clear channel for emergency communications. Even on what may sound like the quietest of public service events, if something goes wrong an emergency message will need to get through quickly and without error: *(see opposite)*

Most RAYNET activity takes place on those bands for which equipment is readily available, namely 2m and 70cm, although use is also made of 4m, 6m and 23cm. More use of HF is also being made to supplement VHF & UHF, particularly for distance and hostile terrain. Increasingly, both 10m and 6m are being used for cross-band talk-through, in addition to the more usual 2m and 70cm. bands. These temporary repeaters are identified by a unique number, issued via permit by RAYNET-UK to RAYNET groups.

RAYNET is not restricted to UK operation and the lessons learned from emergency groups are shared internationally through IARU Emergency Communications meetings.

6m: 51.65-51.75, 51.97 & 51.99 MHz
4m: 70.350, 70.375, 70.400 MHz
2m: 144.625-144.675MHz,144.775, 145.200, 145.225, 144.800 (APRS), 144.260 (SSB)
70cm: 433.700-433.775MHz
In-Band 7.6MHz split temp talkthrough
430.800 (mobile)
438.400 (base)
In-Band 1.6MHz split temp talkthrough
434.375 (mobile)
432.775 (base)

International frequencies:

- 21.360 MHz – Global
- 18.160 MHz – Global
- 14.300 MHz – Global
- 7.110 MHz – IARU Region 1
- 3.760 MHz – IARU Region 1
- 3.663 MHz - UK Only.

The organisation

RAYNET-UK is affiliated to the RSGB and there is regular liaison between the two, promoting a coordinated approach to emergency communications. As the RSGB recognises RAYNET-UK's specialist nature, in return the RSGB is recognised as the national body representing all UK radio amateurs and the main route of amateur radio representation to the UK licensing authorities and bodies such as CEPT and the ITU. The RSGB is also the representative society of radio amateurs throughout the United Kingdom to the International Amateur Radio Union (IARU) who also co-ordinate emergency communications activities between countries.

The RSGB co-ordinate matters relating to changes in amateur radio licensing or band plans where such changes have an effect on the functioning of communication in community events, emergencies and disaster situations.

There are over 100 groups affiliated to RAYNET-UK around the country and at the time of writing there are approximately 1800 members who are committed to using their hobby to help others. The organisation is administered by a Committee of Management, comprising Zonal Co-ordinators who represent the nations and regions of the UK. Each group or county has its own organisational structure meeting their local needs.

To cover administration costs a small annual charge is levied on each member, and in return offers combined liability and personal accident insurance, discounted supplies, as well as technical and training backup.

RAYNET-UK provides a wide range of services to support Members, their Groups and liaisons with our User Services. This support is delivered at Member, Group and National level and is organised by the Committee of Management.

Becoming Involved

If you feel you would like to become involved on a regular basis, then you should consider joining a RAYNET-UK Group. First of all, find a Group near to you and contact them. Groups usually suggest that you come along to a few events and observe, and then perhaps apply to join. As lockdown restrictions come to an end it will be good to get away from endless videoconference training sessions and get back into the field! You do not necessarily need to have an amateur radio licence, as there are often plenty of jobs that non-licensed persons can do and you may get radio experience by using a PMR446 licence free radio or even CB. ID cards are issued to members to identify them to User Services as a member of RAYNET-UK. Some groups adopt a readily identified dress code, which is often appreciated by User Services, giving them a quick means of identifying members on site at an incident. RAYNET-UK operates a Supplies service for all their members.

RAYNET-UK Groups in the UK

Contact information for RAYNET Groups can be found at: *www.raynet-uk.net*

EMC (Electromagnetic Compatibility)

The RSGB can offer help to members on EMC matters through its EMC Committee, which consists of volunteers who have professional as well as amateur radio experience in the field of EMC.

Introduction

Operating an amateur radio station in the 21st century in an urban or suburban environment presents particular challenges. Not only may there be limited space for antennas but the presence nearby of other electronic devices can result in emissions raising the noise level on the amateur bands, as well as breakthrough from amateur transmissions into other devices. EMC, or 'Electromagnetic Compatibility' is the term used to describe the ability of devices to co-exist without excessive interaction.

Fortunately, cases of breakthrough from amateur transmissions are becoming less frequent, and by following the "Good Radio Housekeeping" guidance can generally be managed. See Avoiding Interference below.

On the other hand emissions from the increasing number of electronic devices in every home have led to a marked increase of cases of interference to reception.

Particular threats

Almost any electronic device has the potential to cause emissions of some sort. Most are benign and conform to relevant Standards, but some have significant potential to cause problems:

- PLT/Powerline devices
- xDSL wired internet
- Plasma TVs
- Switch-mode power supplies (SMPSU)
- PV Solar Panels
- Wind Farms

– Plus a plethora of other electronic devices, such as:

- Remote controlled lamps
- LED low voltage lighting modules
- RF-excited lighting modules

Many of the sources of interference are familiar to members. The RSGB, through members of its EMC Committee, is represented on international standards bodies working to achieve standards which should allow coexistence of electronic devices with radio communications systems.

The Social Side

Many complaints of EMC problems can only be solved with the active cooperation of both parties. This requires diplomacy and tact. Complaints from neighbours of interference may be related to environmental impact of antennas, so see the Planning Advice pages in this Yearbook and discuss planning issues with the Planning Advisory Committee. Interference problems are not often understood by complainants or the owners of the offending apparatus, so help them to understand - be a good radio neighbour and be sensitive to their point of view.

Data Transmission Systems Using Telephone Lines & Electricity Cables

Technologies which use the telephone lines and the electricity cables to carry high speed data signals have been a source of concern to radio amateurs for many years. These notes give a brief outline of the radio interference (RFI) threat that may be expected from the various technologies.

Dial-up modems

These use audio signals on the phone line and have now been almost completely superseded by DSLs and fibre optic links. There are a whole family of DSLs, but the only ones of interest to us are ADSL and VDSL.

ADSL (Asymmetric Digital Subscriber Line)

Techniques and Frequencies used

Generally up to 1.1MHz, but could be up to 2.2MHz. ADSL is usually fed into the phone line at the local exchange, which could be up to 5km from the customer's premises.

Deployment

It was widely deployed in the UK with many millions of customers. It has now been largely superseded by VDSL.

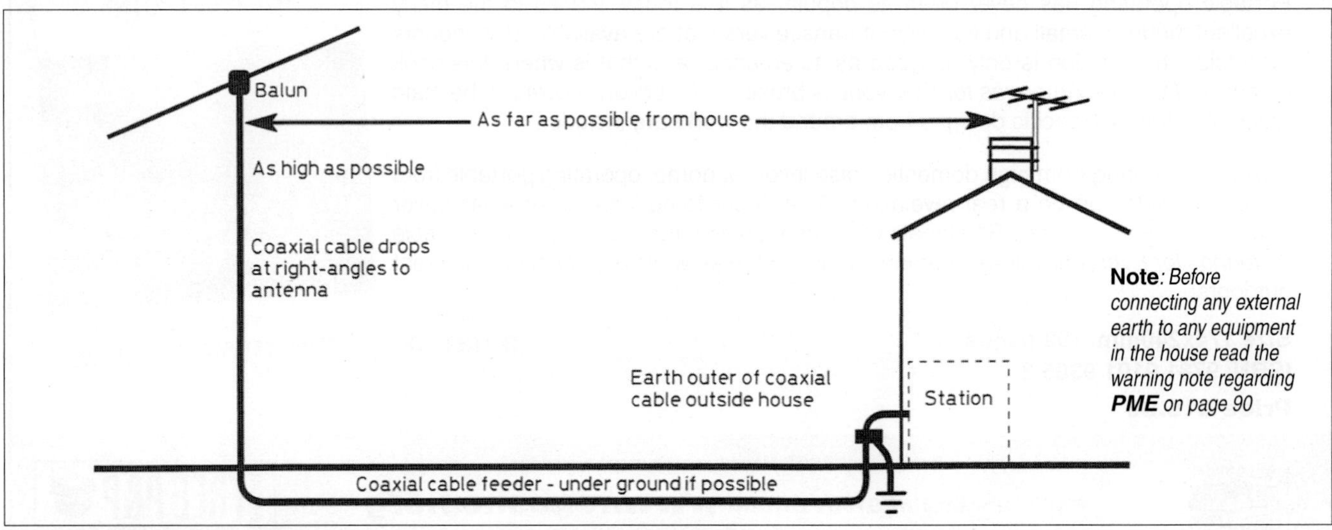

Good radio housekeeping – site your antenna and feeder system well away from the house.

Interference Potential

In general, interference from ADSL is not a problem to amateur radio but there have been a number of reports of breakthrough to ADSL by amateur transmissions. More information can be found in the EMC Columns of RadCom.

VDSL (Very High Speed Digital Subscriber Line)

Techniques and Frequencies used

VDSL operates at up to about 17MHz and is launched into the telephone lines at the street cabinet (it is sometimes called FTTC Fibre-To-The-Cabinet). Since only a relatively short length of telephone cable from the street cabinet to the customer is involved (1km maximum), data speeds of up to 40Mb/s are possible.

Deployment

Deployment is well advanced in the UK and the service is now available in most urban and many rural locations.

Interference potential

Interference from VDSL is fairly common and usually takes the form of noise which sounds very much like white noise. This masks weak signals giving the effect of poor receiver sensitivity but with a higher than normal background noise reading on the "S" meter. Investigation and mitigation of VDSL interference has become a major issue for the EMC Committee. There is a large amount of information in the EMC column in RadCom and on the EMC web pages. Installations with underground connections seldom exhibit problems.

Identifying VDSL Interference

A key step is to identify whether the interference you may be experiencing, is from VDSL as opposed to other sources. The presence of VDSL interference can be proven using a software tool called 'Lelantos' developed by Dr Martin Sach, G8KDF—a member of the RSGB EMCC.

Lelantos is a stand-alone Microsoft Windows application. If you have an SDR receiver and a PC running Microsoft Windows, you can download it from the RSGB website. Further details of how Lelantos works can be found in the instruction manual and in a November 2018 RadCom article which is online along with the software at:

https://rsgb.org/main/technical/emc/vdsl-interference-reporting/

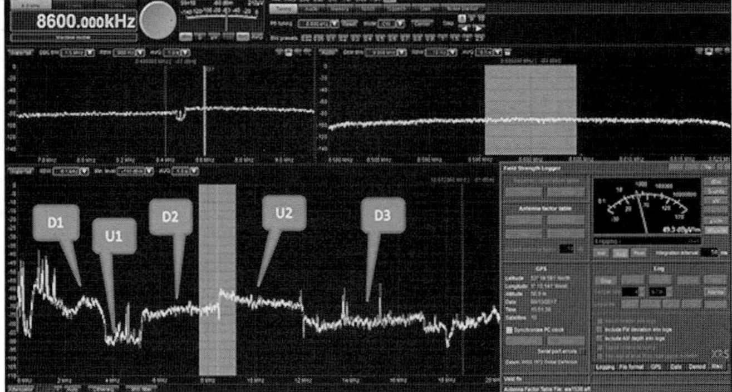

VDSL Interference Downlink and Uplink bands

Systems using electricity cables

This is known as Power Line Telecommunications (PLT or PLC). In the USA it is usually known as BPL, Broadband over Power Lines. Low frequency signalling on the electricity mains has a long history, but so far as radio amateurs are concerned PLT refers to Internet access and computer networking.

In-House PLT

This makes use of modems which plug into a mains socket and communicate with one another via the electricity wiring in the house. The modems are called Power Line Adapters (PLAs).

Frequencies used

Systems vary, traditionally 4 to 28MHz. But newer devices go up to about 70MHz.

Deployment

Apart from computer networking, Power Line Adapters are widely used for video distribution in Internet TV systems (IPTV).

Interference potential

All PLAs reduce their launch power in the international amateur bands. This is known as 'notching'. This seems to be reasonably effective, though some filling due to intermodulation has been observed. Without the notching the interference on the amateur bands would be intolerable. Discussions on an EMC Standard specific to PLT have resulted in two new Standards EN50561-1 below 30MHz and EN50561-3 above 30MHz. Both seem to give a reasonable degree of protection to the amateur bands. This is too big a subject for these short notes and further information can be found on the RSGB website.

Other Potential Sources of Interference

The emphasis on preservation of the Environment has resulted in many schemes aimed at reducing the use of energy and harvesting of renewable sources. These have inevitably resulted in consequential environmental impact.

Solar panels

The Government incentives offered to house-holders and industrial users have encouraged many electrical power users to install Photo Voltaic (PV) panels on their roofs. These installations are a potential source of RFI. From the outset, it must be said that there are good RFI-free installations, and of course the converse is true.

An installation consists of the solar panels on the roof, and much more importantly an inverter, usually placed somewhere below the roof, which are connected by cabling. The inverter is the source of RFI and the cables are potentially the antenna that radiates the energy. The current UK Government and Ofcom view is that solar PV installations are comprised of separate items of apparatus rather than being an integrated fixed installation. The RSGB's view is that even so, an installer is responsible for ensuring the apparatus meets the EMC compliance requirements when the apparatus is first taken into service (see the section on the EMC Directive). In any case any member contemplating a Solar Energy Harvesting system should check that the installer understands the requirements of the EMC regulations. The industry has given some recognition to the potential RFI problems and lightweight inverters which can be installed within the roof space have been introduced. The interconnection between these is usually quite short and results in very low antenna efficiency, and low radiation. At the same time, much greater care has been taken to ensure that the leakage of RFI from the units is minimised. However, it is also true to say that the move towards so called 'transformer less' inverters has presented new challenges. These inverters, using solid state commutation to create the 50Hz AC signal, produce high frequency spikes which leak more readily from the unit housing.

Wind Farms

It is not necessary to travel very far in the UK to see a hilltop wind farm installation, and members have expressed concerns regarding how these will affect the amateur bands. An installation consists of the wind turbine itself and a complex control system at the base of the mast. There are a number of arrangements available for feeding 50Hz energy into the National Grid. Almost all of these involve complex electrical conversion of the voltages and current, with the inevitable switched mode power convertor playing an important part.

The most probable cause of RFI from a wind farm is from the electrical control systems at the base of the tower, with once

again the cables connected to the top acting as an antenna. Although these are usually screened within the metal structure, at ground level there may be feeds to the control systems that radiate.

The EMC Committee is gathering information from members and David Lauder will be making measurements on actual installations. These will be reported in his regular column.

Utility Services

A regular source of complaints to the EMC Committee comes from members who live in a rural, normally quiet location. An unexpected high noise level appears on the lower HF bands. In almost all cases the mains power feed is overhead.

As well as the possibility of arcing on the power line itself, frequently the cause has been found to be thyristor controlled motors installed in pumping stations operated by the water or sewage utility. Overhead power lines accentuate the radiation, acting as long wire antennas. Fortunately, the RFI is evident on MW broadcast stations, and is easily demonstrated with a portable radio as coming from an enclosure housing a pump. The advice from the EMCC has been to contact the Utility company who will usually be sympathetic to the problem.

Putting the RFI in context

Background noise on the HF bands

How much noise would one expect in a typical residential location?

Situations where there are continuous, high level broadband sources of interference are unusual in residential areas, though they are common in industrial/commercial premises. In residential locations broadband noise is usually relatively low, with occasional periods of high level noise. In addition there may be high levels of narrowband interference on specific frequencies. Where there is continuous broadband noise in a residential location it is likely to be something specific like an alarm system or some device such as a switch mode power supply.

There is an EMC Leaflet: EMC 16 Background Noise on the HF Bands. This can be found along with the other EMC Leaflets on the EMCC website at:

https://rsgb.org/main/technical/emc/emc-publications-and-leaflets/

Complaints Procedure

TV and Radio interference

The BBC has responsibility for investigating complaints of interference to domestic radio and television. All complaints should be made to the BBC. You can find the BBC's diagnostic guidance at the following address:

www.radioandtvhelp.co.uk/interference/rtis_tv/radcom_tools

This page also carries useful commentary for any of your neighbours who may be affected by your transmissions. There is also a facility to contact the Radio & Television Investigation Service (RTIS) where the basic diagnostic guidelines have not helped. If, following the investigation by the BBC, there is evidence of interference caused by something which is unlawful, the BBC may refer your case back to Ofcom for possible enforcement action.

Interference to amateur radio

Amateurs often mistakenly believe that the 'non-protected' status of the Amateur Radio Service means they are not entitled to any action in the case of interference caused to them. In fact, 'non-protected' is only in respect of interference from other authorised services operating in the same bands. Amateurs are as entitled to protec- tion from external interference as any other radio user, although it must be accepted that Ofcom will have to give priority to safety of life and business radio users.

Avoiding Interference

Avoiding interference from the transmitter

Spurious Emissions

At one time, complaints of interference to TV from harmonics of amateur transmitters were a major concern in amateur radio. Nowadays complaints of this type are rare, mainly because TV is now transmitted Digitally via satellite or terrestrial signals, but also because transceivers, whether home brew or commercial, are designed with reduction of harmonics and other spurious emissions in mind.

Spurious emissions do still occasionally cause problems, for instance when harmonics of a 2m transmitter fall onto a UHF TV frequency or a harmonic of an HF or 50MHz transmitter might fall on a VHF radio frequency. Such cases are easy to identify by considering the frequency of the station being interfered with and the operating frequency of the amateur station. The solution is to check that the transmitter is working correctly and if necessary fit a low pass or band pass filter. Further information on spurious emissions can be found in the Radio Communication Handbook.

It is worth noting that interference to digital TV will not cause the typical picture and audio degradation which was associated with analogue TV, but will cause the picture to 'freeze', appear as blocks, or possibly disappear altogether until the receiver re-synchronises. These effects can also be caused by a number of signal degradation situations not related to amateur radio.

Breakthrough

When the fundamental signal from the transmitter gets into radio and electronic devices and causes interference it is usually called "breakthrough" to emphasise the fact that it not caused by a fault at the transmitter but a lack of immunity of the victim equipment. Breakthrough can be to either radio or non-radio equipment such as telephones or audio units. Most cases of HF interference to radio and TV are actually breakthrough, with the fundamental of the amateur signal getting in via the braid of the antenna coax or the mains lead, and causing overloading and inter-modulation effects.

There are two ways of tackling breakthrough problems.

By taking care to operate the amateur station so as to minimise RF energy getting into nearby radio and electronic equipment. This has been called good radio housekeeping and is covered in more detail in EMC leaflet **10**.

By increasing the immunity of the affected equipment.

Increasing Immunity

The simplest and, in most cases, the only way of increasing the immunity of radio or electronic equipment is by the use of ferrite chokes on the leads to the affected device. A choke is made by winding the lead onto a suitable ferrite ring. Where a lead comprises a pair of wires such as an audio lead the ferrite choke attenuates the common-mode currents picked up from the nearby transmitter while the wanted differential currents are not affected. Where the lead is a coaxial cable the same effect takes place and the wanted signals pass down the coax unaffected while the current on the braid is attenuated. This type of choke is often used on TV aerial leads. In this case they are called braid-breakers. The low-loss coax used for TV downleads is not suitable for winding on a ferrite ring so it is usual to use a short length of thinner 75 Ohm coax with connectors at each end. If possible 12 to 14 turns should be wound onto the core, though it is not necessary for the cable to be tight on the core. It is only necessary for the cable to pass through the ring to make a "turn". Ferrite rings available from the RSGB are about 12.7mm thick and one ring is sufficient. At one time thinner rings were popular and two of these were stacked together to make a thicker core. More information of ferrite chokes can be found on the EMCC website.

In cases of breakthrough to neighbours' equipment it is particularly important to be diplomatic. Quite often complaints about breakthrough are exacerbated by other grievances such as unsightly antennas (from the neighbours point of view) or by unrelated causes of friction. Leaflets EMC 01, EMC 02, EMC 05 and EMC 08 are written with minimal technical jargon so that they can be given to neighbours if appropriate

Avoiding interference to amateur radio reception

There are three ways of dealing with interference to reception

1. Tackling the interference at source

This is the best option and should always be considered first. The object is to track down the source of interference and then persuade the owner to take action to suppress it or modify the use of the offending device so as to minimise the effect on your amateur operation. This will probably not be too much of a problem if the device is in your own home, but may be much more difficult if it is in a neighbouring property. Possible actions depend on whether the device is compliant with EMC regulations or not, but the golden rule is that any approach to neighbours should be diplomatic. It is not possible in these notes to do justice to this difficult subject. Further information can be found in EMC Leaflets **04** and EMC **09**. Contact the EMC helpline if you need specific help.

2. Reducing the coupling

The term good radio housekeeping was coined to cover breakthrough situations and especially to publicise the need to operate an amateur station with 'due care and attention' and to bear in mind the reasonable expectations of neighbours *see* EMC Leaflet **10**. For the purposes of these notes, good radio housekeeping has been expanded to include a discussion of the application of these principles to minimising received interference.

When the station is located in a residential area, siting the antenna in relation to surrounding properties is of major importance. Antennas should be as far from your own and neighbouring houses as possible, and as high as practical. This applies to both transmitting and receiving, since situations which cause breakthrough will also couple noise from the same wiring back to the antenna.

Some HF antennas can function near ground level, but this is not a good policy from the EMC point of view.

On HF there is, however, one big difference between transmission and reception. This is that, regardless of any local noise, there is an ambient noise level on the HF band, which greatly outweighs the thermal noise generated in the receiver front end. So, unless the antenna is very inefficient, the ambient noise dictates the received noise level. This means that it might be better for reception to mount a small, relatively inefficient, antenna in a place where local interference is least; high up and far from buildings. In special circumstances it might be worth considering an active antenna. Apart from this, good housekeeping rules for HF receiving and transmitting antennas are the same. They should be:

a – Horizontally Polarised. House wiring tends to look like an earthed vertical antenna and is more susceptible to vertical radiation. Likewise - but for rather more complex reasons - vertical receiving antennas tend to be noisier than horizontal ones.

b – Balanced. Out-of-balance currents on feeders generate vertically polarised radiation and likewise tend to pick-up vertically polarised noise.

c – Compact. So that one end is not much closer to the house than the other.

For most of us it is not possible to fulfil both conditions (b) and (c) at the lower HF frequencies, unless we have a very large or oddly shaped garden. However, they illustrate what to consider when making a compromise.

With VHF antennas there is a trade-off between antenna siting and feeder loss.

Where a high gain antenna is used, careful consideration must be given to the effective radiated power (ERP) and the proximity of nearby houses.

3. Actions to reduce the effects of interference at the receiver

It is usually better to tackle the interference at source, but if this is not possible the only option is to attempt to minimise the effect of the interference at the receiving end. First look at your radio housekeeping and at the same time check the whole antenna/earth and feeder installation for corroded joints. These can cause passive intermodulation products (PIPs), which, though not really interference, have the effect of increasing background noise. Interference can enter a receiver from the mains by unexpected common impedances and tests with a battery-operated receiver may give clues to what is happening. Don't forget that, in the absence of a signal, the receiver AGC will pull up the interference to a more or less constant level. This often leads to false conclusions.

If all else fails there are anti-interference measures which can be used at the receiver itself. Most amateurs are familiar with the function of the noise blanker and its much less effective grandfather the noise limiter. Modern transceivers include digital signal processing (DSP) which can be very effective with some types of interference.

In difficult cases it might be worth considering interference cancelling.This can be tricky to set up and operate but when functioning correctly is remarkably effective.

EMC leaflets: *www.rsgb.org/main/technical/emc/emc-publications-and-leaflets/*

Links to RSGB and Ofcom EMC problem reporting web pages:
www.rsgb.org/main/technical/emc/

EMC Helpdesk

If you are having problems with interference, the EMC help desk is there to help, and can be contacted by e-mail on:

helpdesk.emc@rsgb.org

The helpdesk, coordinated by Ken Underwood G3SDW, operates via e-mail, calling on many years of experience of dealing with interference problems. Please note that we are not able to carry out home visits except in very exceptional circumstances.

Note: Our helpdesk service is operated by volunteers in their own time via e-mail. Requests to correspond via the telephone, especially to mobile numbers are to be avoided.

The RSGB does not have any powers of enforcement, so if this is required, the regulatory body, Ofcom will have to be involved.

Additional Advice

Locating the source of interference, particularly if it is outside your own premises, will require a certain amount of "leg work". If you are unable to do this, you may be able to enlist the help of another amateur or local club.

The RSGB EMC Committee has produced a number of leaflets designed to help you, these can be found at *https://rsgb.org/main/technical/emc/emc-publications-and-leaflets*

Often, the problem is closer to home than you might think, so it is important to ensure that the source of the interference is not within your own premises. Resolving interference, particularly in urban environments, may at first seem a virtually impossible task. Nevertheless, by taking a logical approach, in most situations it should be possible to identify the source (or sources) of the interference. EMC Leaflet-04 explains how to do this.

You are likely to need the co-operation of your neighbours, so it is important to be on reasonably good terms with them.

Earthing and the Radio Amateur

EMC07 Basic Leaflet

1 **Disclaimer**

The leaflet EMC07 is intended for members of the RSGB who have passed the Radio Amateur's Examination or who are studying for it. It assumes a knowledge of electrical principles and safety practice.

RSGB Leaflets are made available on the understanding that any information is given in good faith and the Society cannot be held responsible for any misuse or misunderstanding. Where any doubt exists a suitably qualified electrical contractor shall be consulted.

UK Domestic Installations

With very few exceptions, electricity Installations in the UK will be one of three types. These are:

TN-S, TN-C-S and TT.

The letters stand for:

T = Terre, meaning earth in French coming from the Latin for earth terra, as in "terra firma".

N = Neutral

C = Combined, and

S= Separate

The three configurations are shown below

Prior to the late 1970s many properties in the UK were wired to the TN-S system where the earth and neutral are electrically separated all the way back to sub-station (Fig 1).

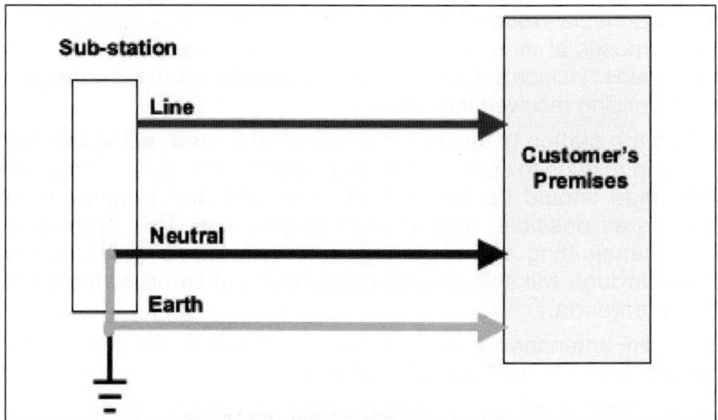

Fig 1: The TN-S Configuration

WARNING: Protective Multiple Earthing (PME)

Many houses in the UK are wired on what is known as the PME system. In this system the earth conductor of the consumer's installation is bonded to the neutral, close to where the supply enters the premises, and there is no separate earth conductor going back to the sub-station. Under certain rare supply fault conditions, a shock or fire risk could occur where external conductors such as antennas or earths are connected. For this reason, the supply regulations require additional bonding and similar precautions in PME systems.

Many houses in the UK were wired on the old TN-S system, where a separate earth goes back to the sub-station. In such systems there were no problems with connecting an external radio antenna or earth. It has recently become evident that changes to maintenance and installation practice mean that the inherent safety of the old TN-S systems cannot always be guaranteed. The situation is being reviewed. Until further information is available all installations should be treated as if they were PME.

Read EMC Leaflet 07 Earthing and the Radio Amateur before connecting any earth or antenna system to equipment inside the house.

If in doubt consult a qualified electrician

EMC Leaflet 07 is available on request from RSGB, or from the RSGB EMC Committee website.

The RSGB website has extensive guidance on EMC Matters: *www.rsgb.org/emc*

EMF – (Electromagnetic Field Exposure)

In May 2021, Ofcom issued new Amateur Radio Licence conditions which now require an assessment of **EMF compliance** for all station **equipment configurations** that you are currently using. Implementation has been rolled out over a period of 18 months, moving downward in frequency, so all bands down to 1.8MHz will be covered from November 2022 onwards.

What is EMF compliance?

Our radio transmitting stations communicate by generating electromagnetic fields (**EMF** for short) that propagate to other stations, often over considerable distances. This 'far field' becomes weaker with increasing distance, but closer to the antenna in the 'near field' there are potential hazards due to interactions of strong EM fields with the human body. Ofcom now requires us to comply with the recommendations of **ICNIRP**, the International Commission on Non-Ionizing Radiation Protection, and requires us to demonstrate compliance by making an EMF assessment for each **equipment configuration** that we currently use. You need to have these results available in case of any inspection.

What is an "equipment configuration"?

An **equipment configuration** is a list of the main factors that affect the EMFs around your station. These include:

- Frequency (typically, the middle of the amateur band in use)
- RF power delivered to the antenna (PEP output of transmitter, minus losses in feedlines, ATU etc)
- Mode of transmission and transmit/receive times (to calculate the RF power averaged over any rolling 6-minute period)
- Type and properties of the antenna (which radiates the EM field)
- How and where the antenna is installed, relative to nearby dwellings and other places where the general public may have access.

A change in any of the above will create a new equipment configuration and may well require a new assessment. However, if you can demonstrate compliance at the highest averaged RF power level that you currently use, that equipment configuration will also be compliant at lower power levels.

You do not need to update the EMF assessment or the equipment configuration if, for example, you replace your 100W transceiver with a different make and/or model. For these purposes, "100W is 100W" regardless of how that RF power is generated. However, you would need to review the assessment if you make a significant improvement to the feedline, because that increases the power delivered to the antenna.

What is an EMF assessment?

An EMF assessment is the combination of two steps:

1. A calculation of EM field strengths to define an **EMF Exclusion Zone** where no member of the general public should remain while you are transmitting.

This is combined with:

2. A practical way to ensure that no member of the general public can enter the Exclusion Zone while you are transmitting.

Most real-life situations are very straightforward. For example:

- "My antenna is on the chimney, inaccessible without a ladder."
- "Height of mast or tower makes the Exclusion Zone completely inaccessible."
- "My vertical antenna is ground-mounted, but the Exclusion Zone is entirely in my back garden and under my control. I can easily verify that no-one is present while I am transmitting."
- "**I have determined the extent of the Exclusion Zone around my vehicle [step 1 above]. I can thus make informed decisions on whether or not to transmit in any given circumstance.**"

> EMF assessments introduce many new concepts that can only be briefly outlined here. For full details, downloadable calculators and practical advice, follow the RSGB EMF web pages at:
> *www.rsgb.org/emf*

RSGB's minimum separation guideline is 2.4m

For effective communication and to minimise interference, RSGB always recommends that antennas are installed as clear of obstructions as possible (see the EMC section above). To help meet the new requirements for EMF compliance – which begins with avoiding the risks of anyone touching the antenna while you are transmitting – RSGB now recommends that you should **aim for a minimum separation of 2.4m (8 feet) between any person and any part of the antenna**.

For example, if all parts of the antenna are more than 2.4m above ground, there is very little risk of anyone touching the antenna by accident. This minimum separation will also be a good start in ensuring compliance about the radiated EM fields. It obviously cannot guarantee compliance in every case, but a wide range of detailed assessments have shown 2.4m to be a useful guideline for time-averaged transmitter powers up to 100W.

Low-power compliance

The 2.4m separation guideline obviously cannot apply to hand-held or body-worn radios. For these and other low power situations, Ofcom allows compliance to be demonstrated a different way, by showing that the time-averaged EIRP is less than 10W (providing peak power is also less than 100W). If so, no further assessment is required – but you do have to calculate the EIRP and record that fact.

This same compliance route can often be used for low power operation on any band with simple low-gain antennas – but even at low power, operation on the VHF/UHF or microwave bands using high-gain beam antennas will very often exceed the 10W EIRP level, so more detailed assessment will be needed. This is where EMF calculators come in.

EMF Calculators

Ofcom and RSGB have both produced online EMF calculators to help you complete your assessments. The RSGB EMF calculator has been developed to offer extra help in numbers of areas:

- Help with entering the basic data that are pertinent to your equipment configuration;
- Help with navigating the ICNIRP guidelines, which vary considerably across the amateur bands from 1.8MHz to the high microwaves;
- A quick route to claim the low power exemption for typical VHF/UHF hand-held radios and some other kinds of low power operation;
- Formatted, downloadable copies of the completed calculations.

If the low power exemption does not apply, the RSGB calculator will then help you to choose the most appropriate calculator:

Further Information can be found at: **Ofcom**: *www.ofcom.org.uk/emf* **RSGB**: *www.rsgb.org.uk/emf*

- A simplified calculator that gives compatible results to the Ofcom calculator;
- Guidance and links to more advanced methods, eg for higher-power stations, users of beam antennas, and stations in heavily built-up areas.

The "results" of any EMF compliance calculator are expressed as the size of the **EMF Exclusion Zone** (**EZ**) as noted earlier. Unfortunately, the use of any simplified EMF calculator (including either the Ofcom calculator or the compatible RSGB calculator) comes at a cost. For some equipment configurations, these calculators can over-estimate the size of the Exclusion Zone, giving a pessimistic impression about the possibilities for compliance.

If an overestimate of the size of the Exclusion Zone does not cause you any practical difficulty, then that's fine – save the results and you're done. But many UK amateurs operate in heavily built-up areas where an overestimate of even a few metres could cause practical difficulties… and that is why RSGB volunteers are developing advanced methods that can give more accurate results when needed.

Pre-Assessed Equipment Configurations (PAECs)

Advanced methods include the use of **Pre-Assessed Equipment Configurations** (**PAECs**), which are station configurations that have been assessed in great detail to identify the true size of the Exclusion Zone, using methods that are acceptable to Ofcom. When embodied in a calculator these methods are the simplest way to get an accurate prediction of your own local EZ.

PAECs are continually under development and the latest version of the RSGB EMF calculator will signpost you to the options that are currently available. Download the latest version from the RSGB EMF web pages at *www.rsgb.org/emf* and then follow the instructions.

RSGB EMF calculator

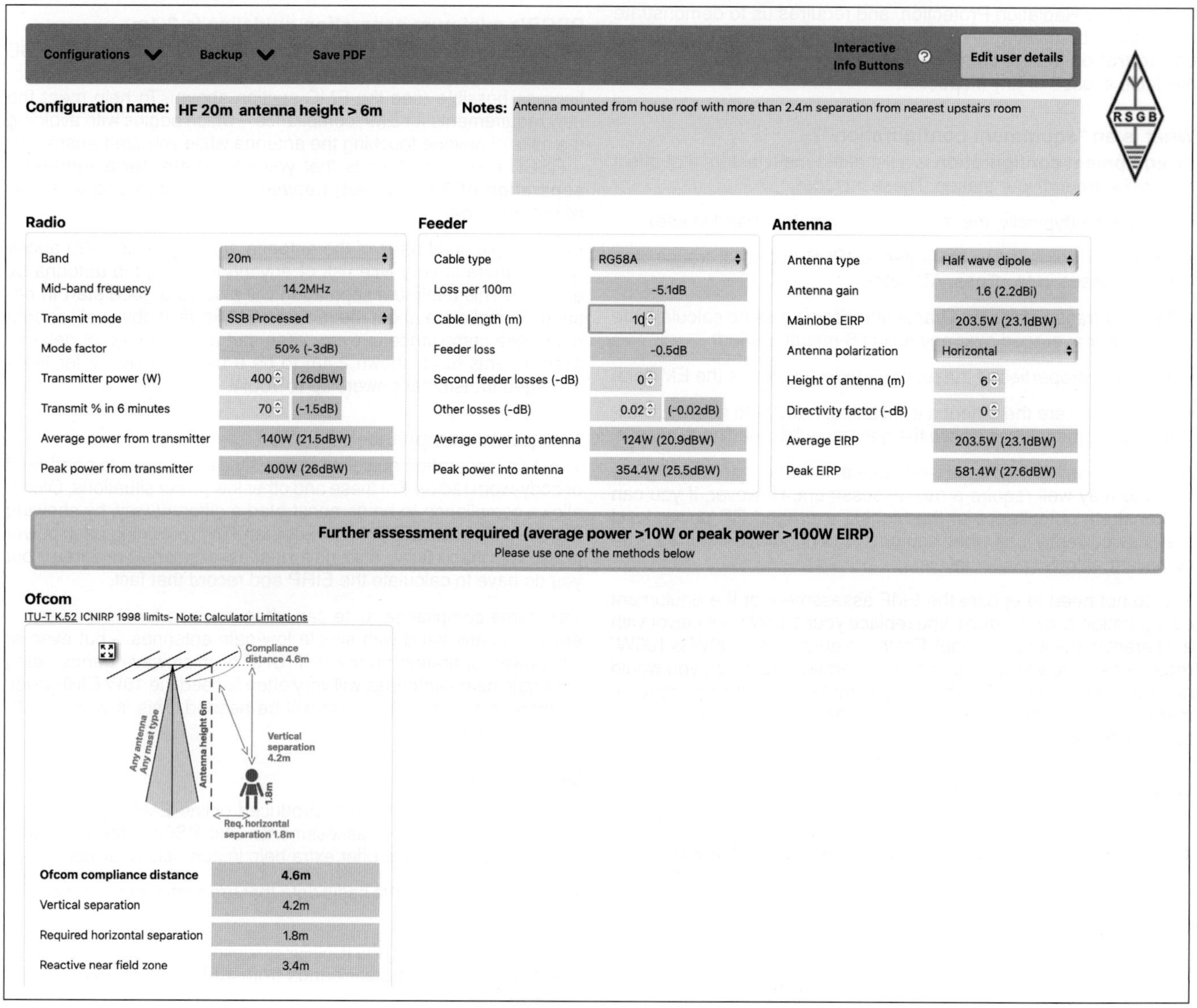

Configurations | Backup | Save PDF | Interactive Info Buttons | Edit user details

Configuration name: HF 20m antenna height > 6m

Notes: Antenna mounted from house roof with more than 2.4m separation from nearest upstairs room

Radio

Band	20m
Mid-band frequency	14.2MHz
Transmit mode	SSB Processed
Mode factor	50% (-3dB)
Transmitter power (W)	400 (26dBW)
Transmit % in 6 minutes	70 (-1.5dB)
Average power from transmitter	140W (21.5dBW)
Peak power from transmitter	400W (26dBW)

Feeder

Cable type	RG58A
Loss per 100m	-5.1dB
Cable length (m)	10
Feeder loss	-0.5dB
Second feeder losses (-dB)	0
Other losses (-dB)	0.02 (-0.02dB)
Average power into antenna	124W (20.9dBW)
Peak power into antenna	354.4W (25.5dBW)

Antenna

Antenna type	Half wave dipole
Antenna gain	1.6 (2.2dBi)
Mainlobe EIRP	203.5W (23.1dBW)
Antenna polarization	Horizontal
Height of antenna (m)	6
Directivity factor (-dB)	0
Average EIRP	203.5W (23.1dBW)
Peak EIRP	581.4W (27.6dBW)

Further assessment required (average power >10W or peak power >100W EIRP)
Please use one of the methods below

Ofcom
ITU-T K.52 ICNIRP 1998 limits- Note: Calculator Limitations

Compliance distance 4.6m
Any antenna Any mast type
Antenna height 6m
Vertical separation 4.2m
1.8m
Req. horizontal separation 1.8m

Ofcom compliance distance	4.6m
Vertical separation	4.2m
Required horizontal separation	1.8m
Reactive near field zone	3.4m

Input page of the RSGB EMF calculator including results (lower left) compatible with the Ofcom calculator.

Always use the latest version for the RSGB EMF calculator, from *www.rsgb.org/emf*
Upgraded versions are likely to appear during the lifetime of this Yearbook, and may be different from the example shown here. (Ofcom will accept calculations that have already been made using earlier versions of the Ofcom or RSGB calculators.)

Locator System

The IARU Locator System, usually just called 'Locator', provides a means of pinpointing stations throughout the world. It is most often used by operators above 30MHz, as a means of calculating the distance between two stations. It is also used on the 136kHz band for the same reason. For use by operators on the upper microwave bands, it can have eight digits, though only the first six are dealt with here. The system is based upon latitude and longitude.

As the map and diagrams show, there are three sizes of 'rectangle'. The largest, known as a 'field', is 20° of longitude (east-west) by 10° latitude (north-south), and is designated by two letters. Most of Britain is in IO field. The next rectangle, known as a 'square' (though it is actually neither truly square nor rectangular!) is 2° of longitude by 1° of latitude. One hundred squares make up one field and, as the map shows, these are given numbers 00 in the south-west corner to 99 in the north-east. Dublin is in IO63. Finally, each square is divided into 576 'sub-squares', 5 minutes of longitude by 2.5 minutes of latitude, and given letters from AA to XX.

To find out your locator, first use a map of your area to determine your exact latitude and longitude, then use the map on this page and the squares diagram opposite to pinpoint your locator. Computer programs and online calculators are available to do this more easily, especially for those who operate from various locations.

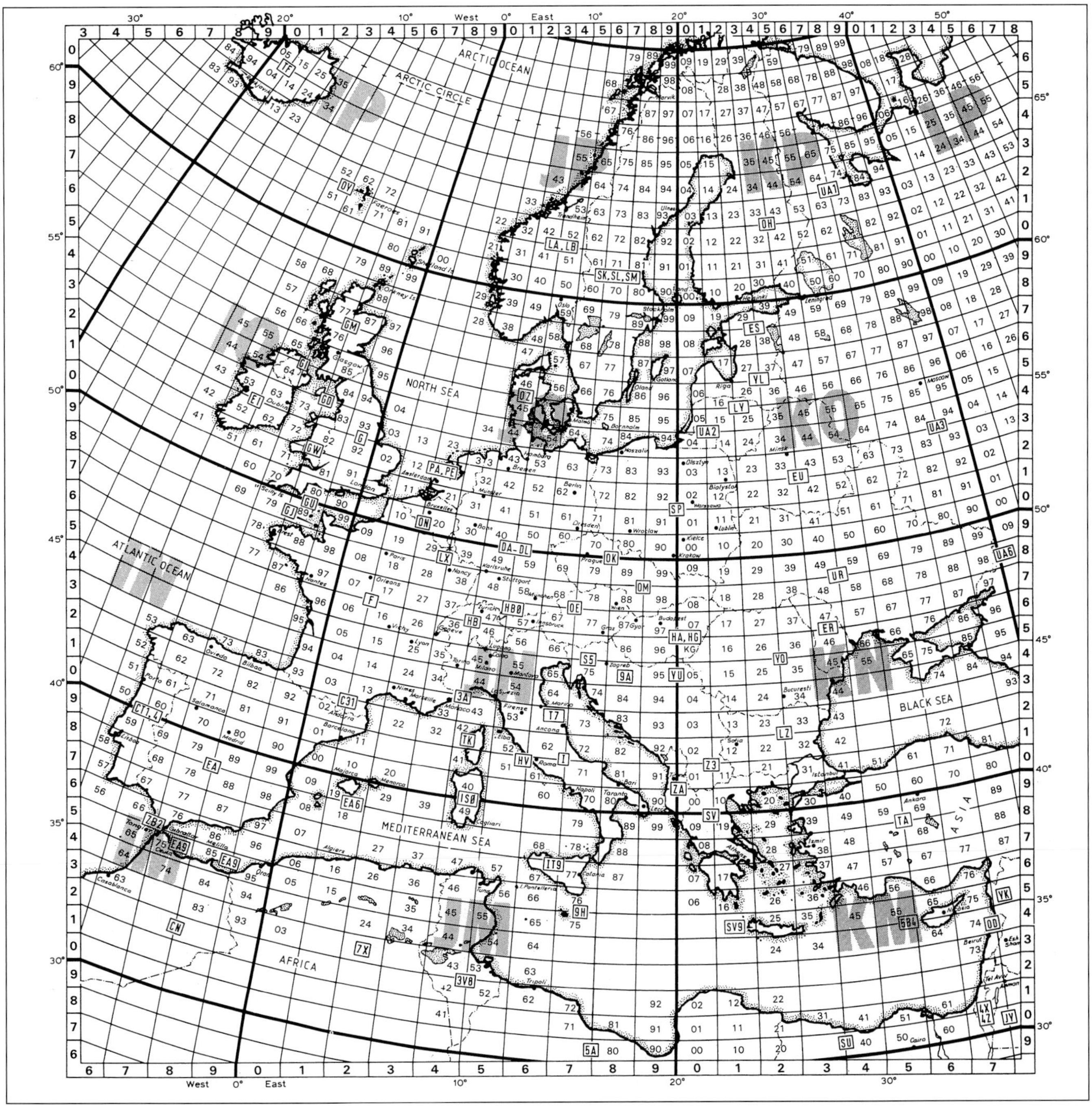

On-line Lat+Long to/from Locator calculators: www.arrl.org/locate/grid.html and www.amsat.org/cgi-bin/gridconv

On-line NGR to Locator calculator: www.ntay.com/contest/NGR2Loc.html

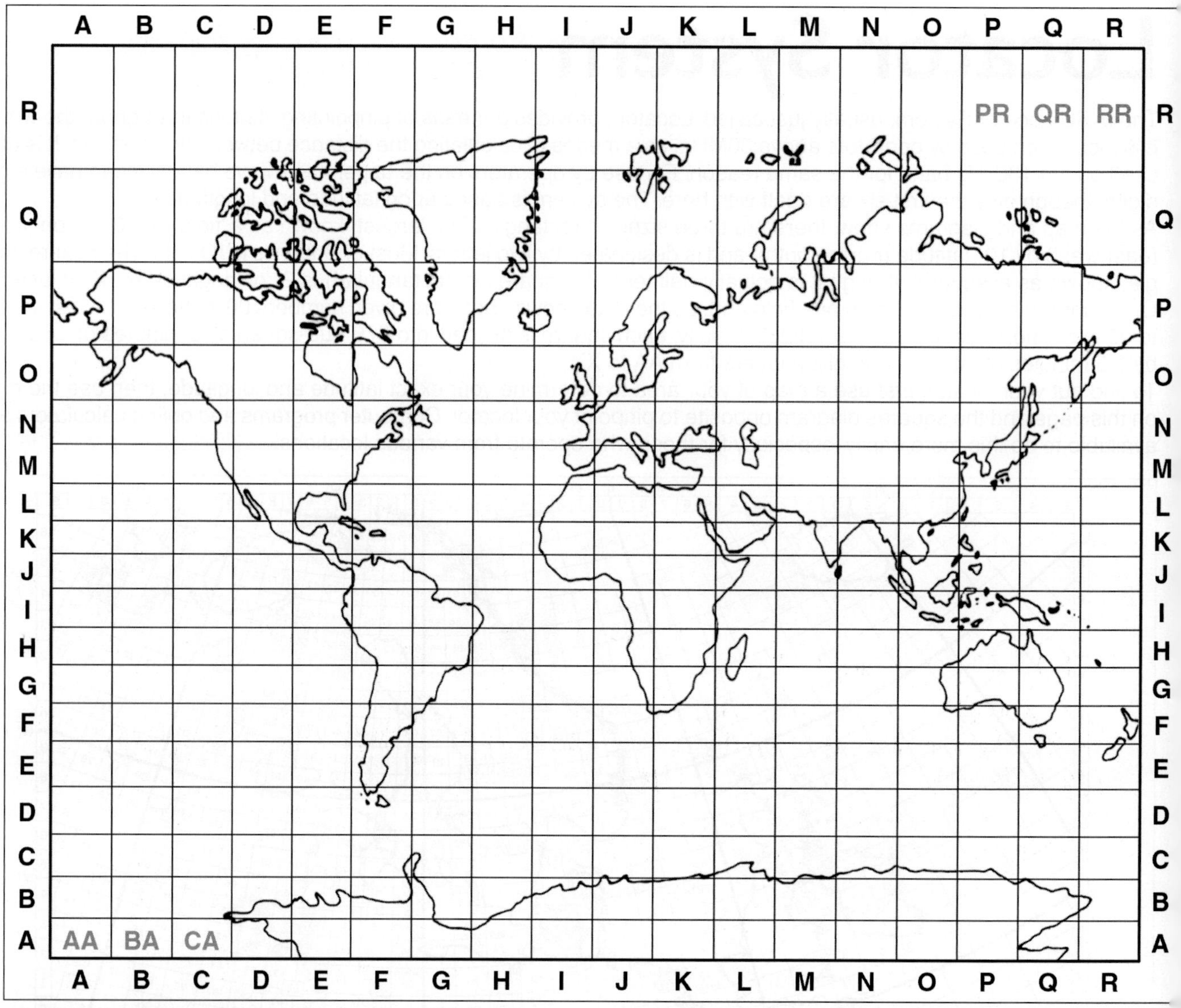

The IARU Locator system may be used throughout the world without repeats. The map above shows the fields that make up the first two letters of the Locator. Examples are shown at two of the corners. The map left shows numbering of squares within the fields.

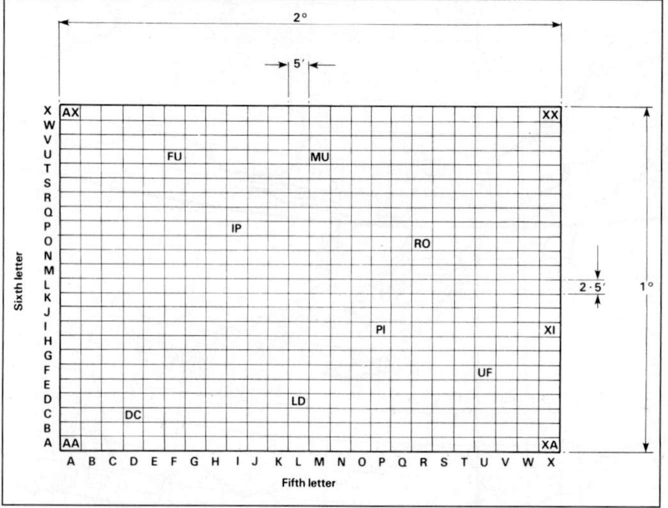

A square (the numbered part of the Locator) is divided into 576 sub-squares, designated AA to XX. Each sub-square is 5' W-E and 2.5' N-S.

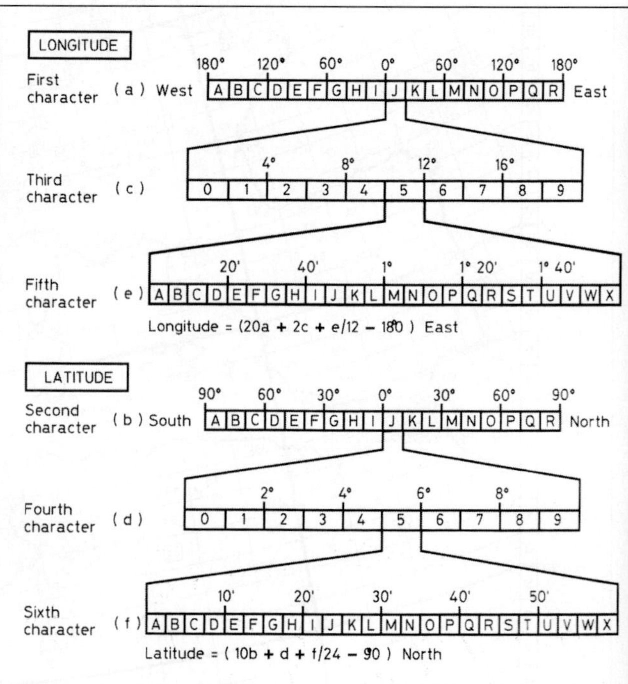

The final two letters may be calculated thus.

A large Locator map of Europe is available from the RSGB
See www.rsgbshop.org or phone 01234 832700

Datacommunications

The Emerging Technology Co-ordination Committee (ETCC) exists to deal with all matters concerning amateur radio repeaters and data communications on behalf of the RSGB. It Liaise with Ofcom in the processing of applications for Notices of Variation (NoV's) for repeaters, Internet Gateways and mailboxes. It is also responsible for the coordination of all requests for site and frequency clearances prior to submission to Ofcom.

The ETCC was formed in 2007 by the amalgamation of the Repeater Management Committee and the Data Communications Committee. Because of the continuing and accelerating convergence of radio and computing and the emergence of amateur digital voice protocols it was considered that interests would be best served by having just one committee responsible for both analogue and digital proposals.
From 1999 to 2007 one of the ETCC's predecessors, the Data Communications Committee, acted as the body responsible for facilitating—by means of frequency coordination—simplex internet voice gateways.
At the time of writing there are operational gateways on 51MHz, 70MHz, 145MHz, 430MHz, 431MHz, 435MHz and 1297MHz.

What is Packet Radio?

Packet radio is digital communications via radio. It began in 1978 in Canada and was introduced into the UK in the 1980s. Packet radio mailboxes (BBS) were first licensed in the UK in about 1988. The numbers grew until the late 1990's, when numbers then started to decline due to the widespread availability of broadband Internet to most of the UK population. However, some mailboxes do still operate in UK, so the following information should help any newcomers to the mode.

What can I do on Packet?

Mailboxes

Mailboxes allow amateurs to connect with their local mailbox and send and receive text messages. These messages can be sent as personal messages to another amateur anywhere in the world (or in space!). Alternatively, messages can be sent as a bulletin for any amateur to read.
Within the UK your messages will normally be relayed via other mailboxes using Internet gateways. These are used to forward mail to more distant continental mailboxes, so it is possible to exchange mail with amateurs on the other side of the world by simply logging into your local mailbox using a low power VHF transceiver and a simple aerial system, as well as a Terminal Node Controller (TNC) and computer.

File Transfer

Packet radio also allows you to be able to transfer files between amateur packet stations in both text and binary format.

DXCluster

There are a few DXCluster stations around the UK and, alongside, the 'automatic position reporting' component of the network. The provision of near real-time on a truly global network exchange of DX, band condition and stations heard/worked information is highly valued by the world's DX community.

APRS

Automatic Packet Reporting System (APRS) was developed by Bob Bruninga, WB4APR, to track mobile GPS stations with two-way radio. APRS can be used in a number of applications for data, communications and telemetry.
In the UK, the late Roger Barker, G4IDE, authored a protocol-compatible variant called 'UI-View'. This is extremely popular and is well supported with regular updates and third-party extensions that add increased functionality to the base product. Examples of add-ons include 'DXCluster spy', satellite telemetry decoding, rig control, rotator control the list is long! For full details see the website at www.ui-view.org
APRS remains one of the most active packet radio modes. ETCC issue NoV's for the operation of APRS digipeaters and internet gateways.
NoV's are now required for any unattended packet radio operation. Please see the ETCC website for details, as well as the on-line forms.

What will I need?

As well as a VHF or UHF FM radio transceiver, you will also need a TNC and a terminal or computer with some form of terminal software program or a specialist packet radio software program. And finally... a great deal of patience and willingness to learn.

Other data modes

RTTY or Radio Teletype

A frequency-shift-keying mode that has been in use longer than any other digital mode (except Morse). It uses a simple five-bit code to represent the alphabet, numbers, a few punctuation marks and a few control codes. As there is no error-correction, QRM and QRN can have seriously detrimental effects on copy. Despite all that, it is still the most popular digital mode. The bandwidth of a RTTY signal is 230Hz, and at 45.45 Baud it gives about 60 WPM throughput.

PSK31

This mode has the advantage of having a very narrow bandwidth and was designed for 'real-time' keyboard-to-keyboard QSOs. It can be used to send almost all of the characters shown on your keyboards and has even been used to send small pictures. If only lower case letters are used, you get about 53 WPM, but with all capitals, this is reduced to about 39 WPM. The bandwidth is as the name suggests 31Hz.

MFSK16

Uses 16 tones and has forward error correction, where it sends all data twice with an interleaving technique to reduce errors from such things as static crashes. It has a comparatively wide bandwidth of 316Hz, which allows faster baud rates, and has greater immunity to multipath phase shifts. This wider bandwidth gives you around 42 WPM.

Hellschrieber

Uses facsimile techniques to transmit and receive characters. It has been in use since the 1920s but has come on in leaps and bounds thanks to modern day DSP soundcard processing. Characters are painted on the screen in ticker-tape like fashion and are read directly from the screen, as opposed to being decoded and printed. Although this mode has a relatively small bandwidth of about 75Hz, it can handle about 35 WPM.

MT63

An excellent mode for sending text over propagation paths that suffer from fading and interference from other signals. It works by encoding text with a matrix of 64 tones over time and frequency. Although this is rather complicated, it does provide error correction at the receiving end, and gives about 100WPM. MT63 has a wide bandwidth of 1kHz.

Throb

Uses either 5 or 9 tones, depending on the version in use. The latest is the 9-tone version, and has speeds of 1, 2 and 4 Baud, enabling data rates of 10, 20 and 40WPM respectively. It appears to be quite good under poor propagation conditions and although it isn't commonly heard, it does seem to be gaining popularity.

23Hz RTTY

A relatively new thing, and hasn't proved as popular as hoped. It sounds similar to PSK31 but is a bit narrower in bandwidth. Although several programs now include this mode, it hasn't taken off well.

FT8 Digital Mode

FT8 is a weak signal mode for HF DXing. FT8 is simple to set up and works well even if the noise level is high.

FT4 Protocol for Digital Contesting

FT4 is an experimental digital mode designed specifically for radio contesting. Like FT8, it uses fixed-length transmissions, structured messages with formats optimized for minimal QSOs, and strong forward error correction. T/R sequences are 6 seconds long, so FT4 is 2.5 × faster than FT8 and about the same speed as RTTY for radio contesting. FT4 can work with signals 10 dB weaker than needed for RTTY, while using much less bandwidth.

Where to find the other data communication signals

In recent years there has been an explosion in the number of digital modes, with many different people writing software to decode them. Some of the modes may be familiar to readers, others perhaps not. Many of you will know the sound of RTTY, and maybe that of

PSK31, but do you know what Throb or MFSK sound like or where to find their signals in the amateur bands?

Within the amateur bands the data communications signals are to be found around the frequencies shown in the bandplans.

Throb, Hellschrieber and MFSK16 tend to congregate just below the RTTY segment.

These frequencies are approximate, but will give you an idea of where to start looking. A full set of datacommunications bandplans is available on the DCC website.

Getting started

The advent of soundcard software for the PC heralded a new era in the digital modes, especially since most of the software is free!

The advice given here is intended for people who are newcomers to the data modes and is therefore presented at a beginner's level. It does not explore the theory behind the data modes in any depth. Neither does it cover all aspects. Once you have understood the basic principles, you can start to learn the more advanced features at your own pace.

If you are not confident in wiring-up cables then you should either buy one of the commercial interfaces or get someone to do it for you.

These explanations will refer mainly to RTTY and PSK31, as they are the most popular of the digital modes in use at present. However, details are given of other modes present on the bands.

Background

Computer soundcards use Digital Signal Processing (DSP) to handle sound, and these techniques lend themselves very well to the processing and decoding of the audio data signals from the output of your radio.

Although RTTY, AmTOR and PacTOR have been around for quite some time, it is the general feeling that a renewed interest in the digital modes really came about when Peter Martinez, G3PLX, created the PSK31 program for the Windows operating system.

At first, it proved to be quite difficult to tune in a PSK signal, so much so that many gave up before they had even watched a QSO in progress.

For those of us who persevered, however, it proved to be something of an enlightenment. At that point, tuning was aided only by turning the tuning knob, with the aid of the phase scope, and it took a great deal of patience and careful finger work to tune in one of these new sounds. You only needed to be off by as little as 5Hz to get garbage on the screen. After several months, Peter came up with a version that included a waterfall display, and that really made a big difference. It was probably a key turning point. Now you could see which way to turn the dial, and you had a fair chance of getting it right quite quickly. Later versions allowed you to point and click on a signal on the waterfall and it was tuned in instantly.

On the bands one was quite likely to meet up with G3PLX, and although he wouldn't hesitate to tell you that you were over-driving the soundcard, if indeed you were, he would offer suggestions, and between you, it was possible to adjust the levels during the QSO to an optimum state. Since then the number of operators has increased considerably. Unfortunately, many have not taken the advice offered by those with greater experience, especially with respect to the adjustments of sound levels and transmitter power. One common mistake is to leave the speech processor switched on, which will definitely cause you problems with the transmitted tones.

In the early days it was the norm to use between 5 and 10 watts of transmitted power and many used much less than that. This level of signal is perfectly adequate for world-wide communications, although these days many stations seem to use several hundred watts into large beams. Apart from the fact that it is not necessary, it also tends to reduce the bandwidth available to others.

Some new modes, such as MFSK16, have been developed in the past few years with the idea of replacing RTTY. Although they do have a following, RTTY has remained at the forefront of the digital modes. PSK31 has gained respect mainly because it is so good at low power levels, making it ideal for QRP work.

Operating the Data Modes

When you are ready to begin using the digital modes, if you are new to data communications it is suggested that initially you spend some time just listening and watching QSOs in progress. This will give you an idea of how they are conducted, what sort of phrases are commonly used and general etiquette.

When you are set up and as you begin operating you may find that RTTY and PSK give you greater scope for learning, mainly because they are more common, but also because they are easier to operate.

The world is waiting for you. It is quite feasible to achieve DXCC on digital modes and the good news is that there are many rare DX countries out there that regularly appear on one of the digital modes. A lot of the DXpeditions these days include RTTY and/or PSK31.

Remember that RTTY is 100% duty-cycle, so please watch your output power! PSK31 is about 80 per cent duty cycle, but as it is used at much lower power this is of less importance. Curiously, Hellschrieber has a duty cycle of only about 21 per cent, so is a lot more 'equipment friendly'.

If you change the output power when you change mode, you may need to adjust the mic gain, as the ALC may well have altered. Just hit the transmit key and adjust the gain so that the ALC is showing slightly, then back it off a touch. It shouldn't make any difference to the indicated output power, but your signal will be cleaner.

If you use the DX Cluster while you are using RTTY, you may well find that the spots are not where they are listed. This is because operators in Europe tend to use different standards from the rest of the world. In Europe, the norm is to use 'low tones' (1275 and 1445Hz) on USB whilst everyone else seems to use 'high tones' (2125 and 2295Hz) on LSB. Although the tone values won't make any difference to your receiving, the sideband will.

Many of the RTTY programs cater for the American market, which means that 'Normal' equates to LSB. If you intend using USB, you may need to find the 'Reverse' button to invert the tones. If you have a RTTY signal with the audio tuned in nicely but only get garbage on the screen, try hitting the 'Reverse' button and you will probably get clear text after that. It can be quite common to see someone operating 'Inverted', and although it will work equally well providing the other station is also inverted, it can reduce the chance of getting a reply to a CQ call.

If you want to work DX and increase your country count, a good way to do this is to enter one of the many RTTY contests. There are about 14 or 15 such contests per year, and many have sections for single operators with low power. No matter how many contacts you make, always try and submit your log, as this helps the contest organiser get an idea of popularity (it also allows them to verify other logs). If you are unsure about your log, then simply send it as a checklog.

One thing you will regularly see in PSK31 is the use of all capitals in the transmitted text. PSK uses an alphabet called 'Varicode', which has shorter codes for the more common letters (similar to Morse) and which also includes both upper and lower case letters. The lower case letters, being more common, have the shorter codes, so any given sentence takes longer to send in capitals. There really isn't any need to use capital letters at all in PSK, even for callsigns. RTTY uses ITC2 (5-element) code that doesn't include lower case, so all capitals is the only option.

Many operators use abbreviations much like in a CW QSO, and it really depends on the type of QSO you are having as to how you will operate. Don't worry about typing mistakes, as these are normal, and anyway who can tell if it wasn't a burst of static that caused his screen to mis-represent what you just typed?

If you are trying to contact a rare DX station, listen for a while and see if the operator is working in any kind of pattern. Listen to the callers and see how they are working. It may be that timing is the key to making the contact. Unless you have big antennas way up high, plus a big linear, and can drown everyone else out, don't try to be first in the pile; wait and let your call be the last one to be heard. That can often get you a response!

Don't forget that with most of the digital modes, if the transmissions of two or more stations overlap, all you get on screen is garbage. It is no good transmitting over the top of a QSO that is in progress, even though it may be nothing to do with the DX station, you will not be read and you will disrupt the QSO in progress.

When you do get through, the other station really doesn't need to know your working conditions or your life history, so keep your QSO short and leave space for others to have a go. It's a good idea to create a macro just for responding to a DX station, perhaps something like this: HISCALL DE MYCALL - TNX - UR ALSO 599 599, VY 73 ES GD DX DE MYCALL. That is all that is needed and anything more is only likely to get lost in the pile-up that will follow.

PSK31 is an excellent mode if you have a limited set-up and can operate only low power. 5 to 10 watts is quite practical with this mode and you can easily obtain DXCC at QRP levels.

To Sum Up:

1. Check and re-check your volume settings.
2. Listen BEFORE you transmit.
3. Watch your power levels.
4. Don't worry too much about spelling mistakes.

Amateur Radio Direction Finding

Amateur direction finding in the UK goes back to the time between the wars when competitions were run using the 1.8MHz band. The arrival of the commercial VHF hand-held radio in the 1970s spawned a different kind of direction finding competition, generally using the 144MHz band. These events were organised by local clubs, many of which were affiliated to the RSGB. Usually a club member parked up in the countryside and made a series of transmissions using a mobile 144MHz radio. Other club members attempted to locate him and the evening often concluded comparing notes in a local hostelry.

Using cars in ARDF competitons

A lot has changed in the past fifteen years or so, both in terms of the Regulations attached to the Road Traffic Act and also the exclusions attached to the majority of motor insurance policies. As a responsible national society it is a role of the RSGB to draw the attention of its members and Affiliated Societies to factors that affect car based direction finding competitions, so that they can make their own decisions about the format of any event they wish to organise.

Section 12 of the Road Traffic Act says that both the competitor and the organiser of a race or trial of speed on the public highway is guilty of an offence.

The Regulations attached to the Road Traffic Act have changed in the last two decades and the Department of Transport has authorised the Royal Automobile Club Motor Sports Association Ltd to operate the Motor Vehicles (Competitions and Trials) Regulations 1969 on behalf of the Department. These Regulations cover many varieties of competitive activities on the public roads. A relatively new kind of event has been authorised and this is called a 'Navigational Scatter Event'.

The good news is that these are automatically authorised (so no permission from anyone is required) and participants are free to roam wherever they like on the public highways - exactly as one would need for a DF Competition.

Unfortunately there are three restrictions:

a. Timing is banned for any event which uses the public highway in part or in whole,

b. minimum distance covered is also not allowed in determining the winner,

c. participants must not all visit an identical set of control points - in motor sport typically 75% of the control points, as selected by individual participants, have to be visited.

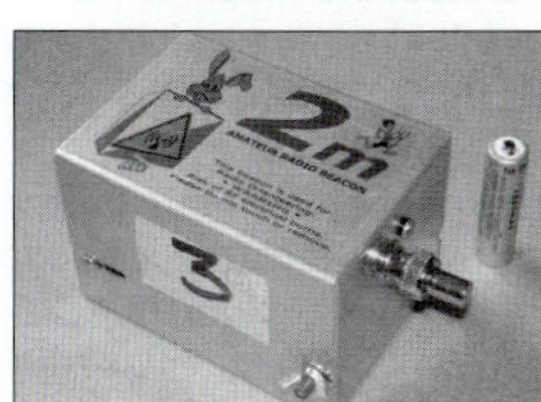

Fig 1: A typical 2m ARDF transmitter with an AA cell for size comparison.

Now it is possible to have a car based DF event and keep within these restrictions. For example, with one transmitter, five 'control points' could be specified. Participants would be required to visit three of these (of their choice) and take a bearing of the transmitter from each one. Competitors then have to give the grid reference of the transmitter location based on their bearings. The closest to the actual location can then be declared the winner.

There has been another development in the last fifteen years concerning third party motor insurance cover. The motor insurance market has become highly competitive. One way the insurers can offer cheaper cover is to reduce their risk. A way of doing this is to make participation in 'competitions' an exclusion on most policies (that is to say the policy is invalid when participating in any kind of competition). Of course it is possible to obtain third party cover without having this exclusion but that will come at the cost of an increased premium.

This is a requirement affecting the individual driver but only a small minority (if any) of the members of the average radio club are likely to have this cover. As a result it makes it very very difficult to run a fully legal car based DF competition. Readers may have noticed that there are very few traditional 'treasure hunts' taking place these days, for exactly the same reasons of third party insurance cover.

TX	Minute 1	Minute 2	Minute 3	Minute 4	Minute 5	Minute 6
No.1	MOE (one dot)	Silent	Silent	Silent	Silent	MOE (one dot)
No.2	Silent	MOI (two dots)	Silent	Silent	Silent	Silent
No.3	Silent	Silent	MOS (three dots)	Silent	Silent	Silent
No.4	Silent	Silent	Silent	MOH (four dots)	Silent	Silent
No.5	Silent	Silent	Silent	Silent	MO5 (five dots)	Silent
	1 x 5-minute cycle (all five transmitters operate on the same freq)					

Table 1: The timing sequence of the five hidden transmitters.

Club Competitions

It is still perfectly possible to have ARDF on the summer programme of your radio club. The RSGB has a set of 3.5MHz direction finding equipment available on loan. This comprises five transmitters suitable for the 'classic' format with each one transmitting for 1 minute in a 5 minute cycle and all transmitters on the same frequency. In addition ten 3.5MHz DF receivers are provided. The RSGB pays to have the kit shipped to you and your Club has to pay for the return. If you manage to break or lose anything your Club is expected to pay to have it repaired or replaced. The range of the transmitters is about 400m, so the kit is very suitable for use in an event on foot in a small country park and the limited distances make it attractive for a majority of Club members.

Contact *ardf.chairman@rsgb.org.uk* to arrange to borrow the kit.

International development of ARDF

After the end of WW2 things developed along a different track on the Continent. Back then, only in Great Britain, Eire, Gibralter and Czechoslovakia were amateurs allowed to operate on 1.8MHz, so countries wishing to introduce surface wave direction finding simply used the lowest frequency band available to them, which was 3.5MHz. Today, this choice is embodied in the set of rules supported by the IARU and also in thousands of DF receivers for this band across the world. Region 1 of the IARU has an ARDF Working Group responsible for the formulation of rules, since by far the greatest interest is in Europe. It is the custom for Regions 2 and 3 to adopt the rules originating in Region 1. The situation today is that ARDF is extremely vibrant in Europe. There have been nineteen bi-annual World Championships and twenty two Region 1 Championships, also a bi-annual event but in odd numbered years. Competitions are organised using the 3.5MHz band, where propagation is predictable and good bearings are generally obtained, and the 144MHz band, which exhibits significant multi-path propagation, sometimes leading to misleading bearings being obtained. The competitions take place entirely on foot and no motor vehicles

are involved. In Great Britain the RSGB made rather a late start to this international style of ARDF, with the first UK event being held in 2002. Many European countries have over 50 years of experience, especially in the old Soviet Bloc countries where there used to be significant state support for activities like ARDF that were also useful militarily. State support has ebbed away since 1989 but the Eastern European countries have inherited a long tradition of direction finding that makes them dominant at World level.

Introduction to the IARU Rules

Transmitters and timing

Five low power transmitters (3W output if 3.5MHz is being used and 800mW if 144MHz is chosen) are deployed in the area to be used. A typical transmitter is shown in Fig 1. All the transmitters operate on the same frequency, but not all at the same time. They transmit in sequence and send an identifier in Morse code for one minute each. Before the reader freaks out at the mention of Morse code, it should be pointed out that the identifier is simply a matter of dot counting.

The first transmitter sends the letters MOE in Morse. The first two letters are long ones in Morse and serve to keep the transmitter on the air for a while, to allow the competitor to swing the aerial carried and assess the direction of the transmitter. The last letter is a single dot, so one dot denotes transmitter 1. This transmitter sends for one minute before shutting down.

The second transmitter then radiates the morse sequence MOI. The last letter (i) is two dots, to denote transmitter number 2. Transmitters 3, 4 and 5 transmit MOS, MOH and MO5 respectively. The sequence is shown pictorially in Table 1.

In the UK it is a licence condition that the callsign of the supervising licensed amateur for the unattended transmitters, is radiated at the end of each transmission, so the one minute transmission terminates with a burst of higher speed Morse, which is this callsign.

Men	Women
M19	W19
M21	W21
M40	W35
M50	W45
M60	W55
M70	W65

Table 2: Age categories for competitors.

In addition to the five hidden transmitters there is a beacon transmitter operating on a different frequency, which radiates the letters MO repeatedly in Morse and is interrupted at intervals with the callsign of the supervising licensed amateur. This transmission is continuous and enables competitors who get hopelessly lost to simply DF the beacon to find their way to the finish.

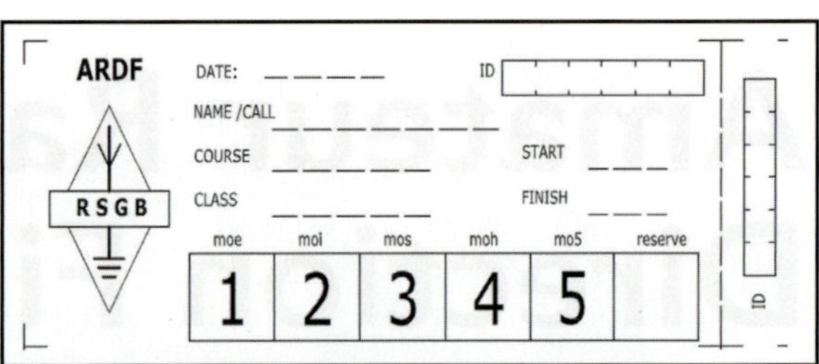

Fig 2: Control Card for a five transmitter DF hunt.

All the transmitters use some form of omnidirectional antenna. For 3.5MHz an 8m vertical wire with an 8m counterpoise is frequently deployed, while on 144MHz a pair of crossed horizontal dipoles (aka a turnstile antenna) at a height of about 3m is commonplace.

Proof of finding the transmitters

Clearly it is necessary for the competitor to demonstrate that each assigned transmitter has been visited. This can be done in one of two ways:

1. The competitor carries a control card (see Fig 2) with a space for each of the five hidden transmitters plus one for the beacon if the latter is to be registered. At each transmitter there is a needle punch (see Fig 3) which is used to mark a unique pattern of needle holes in the card. In international competition the beacon will also be 'punched' but practice varies in domestic races.

2. Electronic timing equipment may be used. Each competitor carries a microchip (see Fig 4), which is inserted into a unit at each transmitter. The timing unit writes the identity of the transmitter plus the time of the visit to the microchip. On completion of the course, the competitor punches at the finish and then downloads all the data to a computer, which is able to print the time taken, the transmitters visited and all the split times.

Age categories

ARDF is organised into a series of age categories and the adult age categories in force are shown in Table 2. To explain how the system works, consider the age group M21. The M denotes a male age group. A man enters the M21 class on 1 January of the year in which he becomes 21 and leaves it on 1 January of the year in which he becomes 40 (M40 being the next age group).

There are a total of twelve adult age groups, with the older age groups hunting fewer transmitters over shorter distances than the younger age groups.

The result of this is that competition is against one's peers and this considerably broadens the appeal of this radio sport.

Fig 3: Pin punch found at the transmitter and used to mark the appropriate box on the control card. The punches at each transmitter carry a unique pattern of pins.

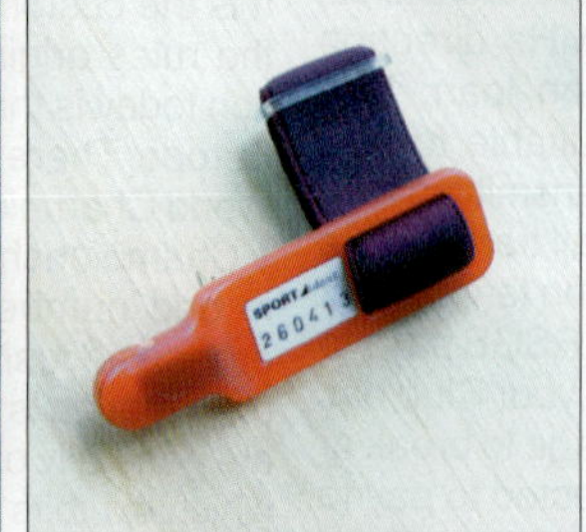

Fig 4: The electronic punching 'dibber' is carried on a finger of the competitor by an elastic strap. This is inserted into the unit at each transmitter to register that the competitor has been there.

Transmitter placement and the time limit

There are three further rules, which can be of great significance, depending on the shape and size of the area used for the competition. No transmitters can be placed within 750m of the start. In domestic competition this distance is frequently reduced to 400m, to avoid 'sterilising' a large part of a small wood as far as transmitter placement is concerned.

The second restriction is that there can be no transmitter within 400m of the finish. Finally, transmitters must be placed at least 400m apart.

A time limit is rigorously enforced for competitions, with the rule that any competitor over time is placed below a competitor who has found at least one transmitter and is within the time limit. It can be rather galling to find all five transmitters and finish one minute outside the time, to be beaten by someone who took nearly two hours to find just one transmitter. The time limit is decided by the course planner, but two hours is a frequent choice, with 90 minutes for easier areas. The object of this rule is to constrain those competitors who are determined to find all the transmitters at any cost, even if this sees them still hunting as darkness falls.

Equipment

It is obvious that competitors will require a receiver for the frequency band being used and a directional antenna for that band. These two items are normally combined into one unit and at UK events there is usually equipment available on loan, although it is sensible to confirm this with the organiser beforehand.

The receiver should be an AM receiver, since the direction to the hidden transmitters will be determined by swinging the antenna from side to side and noting changes in signal amplitude. The amplitude limiter in an FM receiver makes it less suitable for this task, although it can still be made to work in this application.

There is a need to plot bearings on the map provided at the start. Beginners usually plot more bearings than experienced competitors. To do this, the map should be taped to a lightweight board of some kind and either a spirit pen or a chinagraph (wax) pencil can then be used to mark the map. Lines can be drawn by both of these markers on any clear plastic covering used to protect the map. Neither of them will run if they later become wet, but only the chinagraph will make a satisfactory mark if the plastic is already wet.

A compass will be required to measure the bearings. The type with a rectangular base plate also doubles as a protractor. The compass is best looped round the wrist with the cord normally provided.

Proving the visit made to each transmitter involves carrying either a Control Card or an

SI dibber. This 'dibber' is a plastic encased microchip used for electronic 'punching'. It is on a small elastic strap which allows it to be attached to the index finger. The control card is best pinned with safety pins to the front of the clothing. Many control cards are printed on tough, waterproof Tyvek paper and require no protection or strengthening. Control cards that are printed on thin card should be covered with Sellotape to both waterproof and strengthen them.

Moving to the desirable rather than the essential, a whistle should be carried. In some competitions is mandatory with a 'no whistle – no run' policy. The emergency signal is six blasts of the whistle at one minute intervals. Also in the desirable category is a circle stencil to mark the circles around the start and the finish, in which no controls can be placed. Obviously a different stencil is required for a map at 1:10,000 scale to one at 1:15,000 scale.

A check list is shown in Table 3.

Competition hints

Pre-start

There is normally five minutes after being given the map and before getting the signal to start, in which the competitor is able to:

a. waterproof and protect the map as deemed appropriate for the weather conditions,

b. on the map, draw a 750m circle around the start and a 400m circle around the finish (in domestic competition the 750m start circle is often reduced to 400m),

c. study the map to identify height features in particular.

Start + 5 minutes

After being given the start signal, the rules oblige the competitor to keep moving to the end of the start funnel. Once at the end the aim should be to listen to each transmitter in turn, assess the strength of the signal and plot the bearing – all within the 60 seconds that it is on the air. Prior practice at this procedure will pay big dividends for the beginner. If the 144MHz band is being used, bearings taken from high spots are more accurate than those taken from valleys. It may pay to sacrifice a complete transmitter cycle and climb to the top of a nearby hill or spur from which more accurate bearings can be obtained.

Decision time

Based on the information gained by listening just once to each of the transmitters, the most important decision of the day must be made. This is the choice of the first transmitter to be visited. In the case of a co-located start/finish, the penalties for getting it wrong are not as severe compared to a split start/finish. With a co-located start and finish, a poor choice can often be rectified on the route back from the furthest transmitters to the finish. When the start and finish are at separate locations, a bad decision may mean a lot of 'back tracking' and hence wasted time.

Bearing quality

The surface wave propagation on 3.5MHz during daytime leads to bearings which are generally pretty accurate. While it is wise to avoid wire fences and overhead power lines when taking bearings, the accuracy of the plotted bearing is determined by the equipment and skill of the competitor. This results in fast runners being able to get to the transmitters first, assuming that they also have reasonable direction finding skills.

Item carried	Note(s)
Receiver	Usually fixed to the antenna.
Antenna	Usually fixed to the receiver.
Map	Normally issued at the start line, 5 or 10 minutes before starting.
Lightweight rigid board for the map	To deal with wet weather conditions, the competitor will need waterproofing (a plastic folder or sticky backed plastic film) to cover the map and possibly tape to fix the map to the board.
Compass	The type with a rectangular backplate doubles as a protractor.
Spirit pens and/or wax pencils	Will not run if it rains, but note that only wax pencils will write satisfactorily on plastic film that is already wet.
Circle stencil	750 and 400m circles at the map scale in use.
Control Card or SI 'dibber'	To register that the competitor has visited each assigned transmitter.
Whistle	Emergency signal is 6 blasts at 1 minute intervals.

Table 3: A checklist of the items you require for a competition.

On 144MHz there is a lot of multi-path propagation, with the signal being reflected or scattered from steep hillsides, rock outcrops and even the edges of wooded areas. The bearings obtained vary greatly in 'quality'. A sharp, clear peak in the signal as the antenna is swung from side to side is indicative of a single path signal and this is often the direct path from the transmitter. Multi-path propagation most often reveals itself as a rather diffuse bearing as the antenna is swung. Sometimes there may be more than one distinct peak to the signal and this is where an antenna with very low side and back responses comes into its own, to differentiate between the direct and the multi-path signals.

All this interpretation of bearings coupled with the need to view the bearings against the background of any high ground in the vicinity; leads to the winner needing to process all this information quickly and accurately. Hence, being able to run fast is no longer such a key quality to gain victory.

Your first event

Event information

Finding out about competitions is clearly the first step and there are currently around 15-20 events held in the UK each year. There is usually a break around Christmas and January, with the 'season' generally commencing in February and the last event in November or December. The RSGB website is the gateway to information about ARDF events. The URL is given at the end of this article.

The events organised by the RSGB fall into two categories:

'National Events' in which there are usually two competitions in the day, with a Classic 144MHz event before lunch and then a less physically taxing event on 3.5MHz after lunch. There are about six of these in the year

In addition there are some introductory Regional events using the classic format on 3.5 MHz. In these events, newcomers are required to take bearings on all five transmitters before starting. One of the mentors present will check these bearings and give a 'good to go' message if the bearings plotted look OK. After this, the newcomers can cross the start line to commence the timed competition, secure in the knowledge that he or she has already taken and plotted five reasonable accurate bearings.

These introductory events are not available in all RSGB Regions. The ARDF Committee would welcome interest from any member willing to organise events in the Regions which currently have these events. There is a lot of support available.

Clothing and equipment

The issue of equipment has already been covered above. As far as clothing is concerned, for a first outing, stout shoes and outdoor attire is sufficient. If you become more committed, then studded orienteering shoes, gaiters to protect the lower leg against brambles and nettles and an orienteering suit are more appropriate. On occasions when the weather is particularly inclement (thankfully few) then a cagoule or other waterproof garment will be needed.

Registration

Events that are run in conjunction with an orienteering event will benefit from the direction signs to that event. ARDF events that are freestanding are not likely to be extensively signed, so the competitor should ensure that a copy of the Grid Reference and of the

Andrew, G4KWQ 'flies the flag' at the World ARDF Championships.

RSGB ARDF page: https://rsgb.org/main/about-us/committees/ardf-committee/

map extract that is frequently given with the event details, are carried on the journey to the venue.

Once there, it is necessary to register and pay the event fee (usually of the order of £5- £6). This provides for entry for all the competitions taking place on the day. There is usually a full scale event in the morning, followed by a more relaxed competition after lunch. All of this provides an excellent day of radio sport and makes it well worthwhile to travel a fair distance for a full day of radio in the open air.

At registration you may have to make a choice regarding the number of transmitters you wish to hunt. Details of the frequencies of the transmitters (hidden transmitters on one frequency and the homing beacon on a second frequency), the radius of the zone around the start in which transmitters may not be placed and your individual start time will be given to you. Finally, the time limit for the event should be noted.

If electronic timing is being used, it will probably be necessary for you to hire an electronic 'chip' (colloquially known as a dibber).

If pin punching is in use, then you will be given a control card. Fill this out with your details. If it is made of plain paper or thin card, protect and strengthen it with Sellotape and finally pin it to the front of your clothing with safety pins.

The map

An orienteering style map will be in use and the most common scale in domestic competition is 1:10,000. These maps show much more ground detail than an Ordnance Survey map. The first thing to note is that the white bits are trees – quite the opposite to an Ordnance Survey map. The white parts of the map denote runnable forest and various shades of green show less runnable areas, with dark green being really impenetrable and well worth avoiding. Fortunately you are very unlikely to have transmitters located in these latter areas.

Open and semi-open areas are shown in a yellow ochre colour.

Only the start (a triangle) and the finish (a double circle with a smaller circle inside a larger one) will be marked on the map.

After the start

In simple terms, listen to all of the transmitters to get a bearing and an idea of the signal strength of each one, decide which transmitter you wish to visit first and then head for the one selected.

Finding your very first hidden transmitter is a great moment and one to be remembered for a very long time. Keep an eye on the clock, so that you get back to the finish inside the time, and see if more transmitters can be located.

Newcomers are usually a bit erratic in their first events, as is to be expected. For some a brilliant performance at the first outing can be followed by poor and disappointing results at subsequent events. Experience tells us that it takes about six outings for the majority of competitors to settle in and be able to locate all the assigned transmitters inside the time on a reliable basis. In other words don't get discouraged by a few poor results; it will all come together for you with a bit of experience.

After finishing there is the opportunity to compare notes with other competitors and to get some tips on how to avoid any mistakes at future events.

Robert Vickers G3ORI shows that ARDF can be enjoyed by amateurs of any age.

Other Formats

There are two variants of the basic format. The first is Foxoring, which is a hybrid of Orienteering and Radio Direction Finding. A large number of very low power 3.5MHz transmitters are deployed and a circle is marked on the map within which each one will be audible. The competitor uses orienteering techniques to navigate to the area of the circle. Once the signals from the transmitter are picked up, direction finding enables the transmitter to be located.

The second variant is the sprint format. This provides two clusters of five transmitters operating on two different frequencies in the 3.5MHz band and keyed at different speeds. Each transmission lasts just 12 seconds and competitors have to return to a spectator beacon after finding all the transmitters in the first group and before they set out to find transmitters in the second group. The transmitters are keyed with the usual MOE, MOI etc. The format is very fast and furious and winning times of less than 15 minutes are not unknown.

UK coverage

There are many RSGB Regions where there is little or no ARDF taking place beyond the occasional Club based event. ARDF really lends itself to being a Regional activity and, for example in Region 13 (East Midlands), there are a series of five weekend events on 80m, one each month from May to September. There are never enough people in the average Radio Club who are interested in having more than an occasional evening event but across a whole Region this is no longer the case.

To promote this activity the RSGB Legacy Fund has financed the provision of two sets of transmitters and receivers available on long term loan (a year or so) to any Region wishing to run a similar series of events. Contact the ARDF Committee Chairman (ardf.chairman@rsgb.org.uk) if you have an interest in running such a series of events in your Region.

Further Information

A description of competitive direction finding cannot be exhaustive in the space available here. The RSGB book 'Radio Orienteering – The ARDF Handbook', goes into much fuller detail and is essential reading for the beginner. Event information is available on the RSGB website: www.rsgb.org>main>on the air>activity ardf. The ARDF pages also include the results of competitions, details of the big international events and information about sources of suitable equipment.

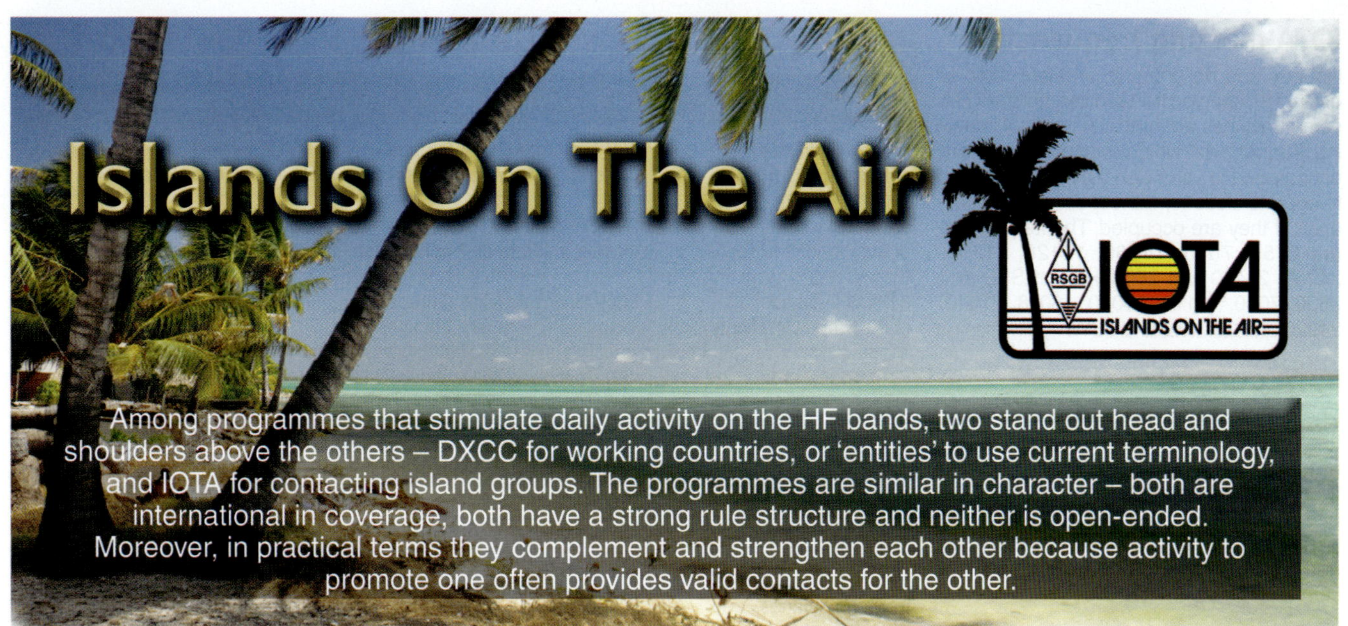

Islands On The Air

Among programmes that stimulate daily activity on the HF bands, two stand out head and shoulders above the others – DXCC for working countries, or 'entities' to use current terminology, and IOTA for contacting island groups. The programmes are similar in character – both are international in coverage, both have a strong rule structure and neither is open-ended. Moreover, in practical terms they complement and strengthen each other because activity to promote one often provides valid contacts for the other.

IOTA, or the Islands On The Air Programme to give it its full title, was created in 1964 by the late Geoff Watts, a leading British short wave listener and the only SWL in the DX Hall of Fame. When the programme was taken over, at Geoff's request, by the RSGB in 1985 it was already a favourite for many DXers. Its popularity has since grown each year, not only among ever-increasing numbers of island chasers but also among a rapidly expanding band of amateurs attracted by the possibilities for operating portable from islands. For both it is a fun pastime adding much enjoyment to on-the-air activity.

The basic building block for IOTA is the IOTA Group. The oceans' islands have been corralled into some 1200 IOTA Groups with, for reasons of geography, varying numbers of 'counters', i.e. qualifying islands, in each. Only in very few cases do the rules of IOTA allow single islands to count separately, DXCC island entities such as Barbados being one. The number of groups is now capped and further changes are expected to be minimal.

Each group activated has been issued with an IOTA reference number, for example EU-005 for Great Britain. Part of the fun of IOTA is that it is an evolving programme with new groups being activated for the first time. Currently some 1136 of the 1200 groups have confirmed numbers.

The objective, for the island chaser, is to make radio contact with at least one counter in as many of these groups as possible and, for the DXpeditioner, to provide island contacts. A wide range of separate certificates, graded in difficulty, is currently available for island chasers as well as two prestigious awards for high achievement (see the table overleaf). Applicants may be any licensed radio amateur (or SWL on a 'heard' basis) who has had confirmed contacts with the required number of IOTA Groups listed.

IOTA Directory

The latest IOTA Directory, 18th Edition, published in May 2018, gives a full listing of IOTA Groups together with the names of 15,000 qualifying islands. You can order a copy on-line on the IOTA website at: *www.iota-world.org/iota-shop.html.*

Applying for an Award

IOTA has since 2017 run software which allows award credits to be given by electronic confirmation of contact by QSO matching with logs on Club Log and since May 2020 with logs on the ARRL's Logbook of the World (LoTW). The system however continues to accept confirmation by QSL cards. Award applicants should prepare and submit their applications electronically on the Internet. Full details of the application procedure and a list of checkpoints can be found on the IOTA website at *www.iota-world.org*. After signing off your application on-line for processing by your checkpoint, you should immediately send him any cards required for confirmation with return postage.

Island-chasing

1000 or more IOTA Groups may seem an enormous target. If you are a long time DXer who has worked it all and are looking for something new, you will already have amassed a very respectable IOTA score from among your DXCC contacts. If, however, you are new to the bands or one of the many amateurs who adopt a more relaxed approach to their operating, you can take full advantage of a very high level of IOTA activity, comprising easy and semi-rare groups, to launch you on your way. Well over 600 IOTA groups are usually activated over a three year period with, during a typical summer weekend, some 20/25 IOTA Groups being heard around the IOTA meeting frequencies. An enthusiast should be able to gain the IOTA Plaque of Excellence for working 750 groups in about six years, operating mainly at weekends. This must be a reasonable target to go for – after all, how long does it take to get to the top of the DXCC Honour Roll?

IOTA is one of the few award programmes that has an annual Honour Roll and other performance listings. These create a great deal of interest when they are published each spring. Many IOTA enthusiasts are more interested in participating in these listings than in collecting the certificates. All you need to enable you to participate is a registered score of at least 100 Island groups.

Operating from an island

Many amateurs are fortunate enough to live on an island and to be able to give out an IOTA every time they make a contact. Others are not so lucky. For both there is the lure of operating portable from a rare or semi-rare group – the fun of being at the other end of a pile-up for a few days. Many islands lie within a few hours' reach and, subject to the availability of suitable equipment, could be put on the air relatively easily.

Those amateurs lucky enough to be able to activate a rare or semi-rare IOTA Group can expect to generate huge pile-ups with thousands of contacts during even a short 2/3 day period. Rare groups are not all remote and difficult to access. Even in Europe and North America there are many that are needed by the chasers. For those interested, a list of most wanted IOTA Groups in each continent, ranked by rarity, can be viewed on the IOTA website.

Categories of application

IOTA began as an award for single operators working on the HF bands (1.8 to 30MHz). However, in response to demand, the IOTA Committee subsequently introduced categories specifically for club stations and for working on VHF/UHF (50MHz and above).

Address: Islands on the Air (IOTA) Ltd, La Quinta, Station Road, Chobham, Woking, Surrey GU24 8AR
Email: info@iota-world.org

IOTA meeting frequencies

Nobody and no group in amateur radio is entitled to reserved frequencies, but the IOTA community has adopted a number of 'meeting frequencies' which island stations are encouraged to use when they are free – and to operate close to, without causing interference, if they are occupied. The frequencies are 3755, 7055, 14260, 18128, 21260, 24950, 28460 and 28560kHz on SSB and 3530, 7030, 10115, 14040, 18098, 21040, 24920 and 28040kHz on CW.

IOTA contest

The IOTA Contest, first held in July 1993, has become enormously popular and now regularly attracts more than 2000 entries. It provides an opportunity annually, at the end of July, to work large numbers of rare and semi-rare IOTA Groups. Contest rules and results are available from the RSGB HF Contest Committee website.

IOTA Annual Listings

The following pages show the IOTA Annual Listings as of February 2023. The lists are divided as follows:

The Honour Roll is a list of the callsigns of stations with a checked score equalling or exceeding 50% of the total of numbered IOTA groups, excluding those with provisional numbers, at the time of preparation.

The Annual Listing is a list of the callsigns of stations with a checked score of 100 or more IOTA groups but less than the qualifying threshold for entry into the Honour Roll.

The Club Listing is a list of the callsigns of club or multi-operator stations with a checked score of 100 or more IOTA groups.

The VHF/UHF Listing is a list of the callsigns of stations with a checked score of 100 or more IOTA groups on the VHF/UHF bands.

The SWL Listing is a list of SWLs with a checked score of 100 or more IOTA groups.

Listing in the 2023 tables was restricted to those participants who had updated their scores since February 2018. IOTA rules limit inclusion in the listings to those participants who have updated their scores at least once in the preceding five years and have opted to have their scores published. All participants should be reminded that the final decision on acceptance of credits is made at IOTA HQ and that this can mean downward adjustments to scores at any time to reflect corrections of one sort or another. Data-cleansing work is on-going and covers every participant's complete record, not just the latest credits added. Although efforts are made to alert participants to score changes, this cannot be guaranteed to happen in each case. Remember, the line of communication is via your checkpoint, so please do not route queries direct to IOTA HQ or the IOTA Manager. Always check first to see if the answer is in the IOTA Directory.

Annual Update Deadline

IOTA enthusiasts are reminded that the last date for submitting applications or updates

IOTA Awards

Award	All Band Categories	VHF /UHF Categories
IOTA 100 Islands of the World Certificate	100 Confirmed IOTA Groups including 1 from all 7 continents	100 Confirmed IOTA Groups including 5 continents
IOTA 200 Islands of the World Certificate	200 Confirmed IOTA Groups including 1 from all 7 continents	200 Confirmed IOTA Groups including 5 continents
IOTA 300 Islands of the World Certificate	300 Confirmed IOTA Groups including 1 from all 7 continents	
IOTA 400 Islands of the World Certificate	400 Confirmed IOTA Groups including 1 from all 7 continents	
IOTA 500 Islands of the World Certificate	500 Confirmed IOTA Groups including 1 from all 7 continents	
IOTA 600 Islands of the World Certificate	600 Confirmed IOTA Groups including 1 from all 7 continents	
IOTA 700 Islands of the World Certificate	700 Confirmed IOTA Groups including 1 from all 7 continents	
IOTA 800 Islands of the World Certificate	800 Confirmed IOTA Groups including 1 from all 7 continents	
IOTA 900 Islands of the World Certificate	900 Confirmed IOTA Groups including 1 from all 7 continents	
IOTA 1000 Islands of the World Certificate	1000 Confirmed IOTA Groups including 1 from all 7 continents	
IOTA 1100 Islands of the World Certificate	1100 Confirmed IOTA Groups including 1 from all 7 continents	
IOTA 750 Islands Plaque (with shields for each additional 25 IOTA Groups)	750 Confirmed IOTA Groups including 1 from all 7 continents	300 Confirmed IOTA Groups including 5 continents
IOTA 1000 Islands Trophy (with shields for each additional 25 IOTA Groups)	1000 Confirmed IOTA Groups including 1 from all 7 continents	
IOTA Africa Certificate	75 African IOTA Groups	50 African IOTA Groups
IOTA Antarctica Certificate	75% of Antarctic IOTA Groups	50% of Antarctic IOTA Groups
IOTA Asia Certificate	75 Asian IOTA Groups	50 Asian IOTA Groups
IOTA Europe Certificate	75 European IOTA Groups	50 European IOTA Groups
IOTA North America Cert.	75 North American IOTA Groups	50 North American IOTA Groups
IOTA Oceania Certificate	75 Oceanian IOTA Groups	50 Oceanian IOTA Groups
IOTA South America Certificate	75 South American IOTA Groups	50 South American IOTA Groups
IOTA World Diploma	50% of the IOTA Groups in all 7 continents or 50 IOTA Groups for the continents where there are more than 100 IOTA Groups	
IOTA Arctic Islands Certificate	75 Arctic Island Groups	50 Arctic Island Groups
IOTA British Isles Certificate	75% of the British Isles Groups	50% of the British Isles Groups
IOTA West Indies Certificate	75% of the West Indies Groups	50% of the West Indies Groups

Further information and updates: *www.iota-world.org*

2023 Honour Roll

Pos.	Callsign	Total	Pos.	Callsign	Total	Pos.	Callsign	Total	Pos.	Callsign	Total	Pos.	Callsign	Total	Pos.	Callsign	Total
1	9A2AA	1132	88	JF1SEK	1083	179	UA4CC	1033	265	IZ4CZE	978	357	I4KMN	917	445	DL9UBF	840
2	I2YDX	1131	88	N6AWD	1083	179	VE7SMP	1033	265	JH4GJR	978	357	JJ0NCC	917	445	RA6ATZ	840
3	I1JQJ	1130	88	RZ1OA	1083	179	W5ZPA	1033	265	R7KM	978	357	R7KW	917	448	JF1RDH	839
4	G3KMA	1129	88	UA4HBW	1083	182	HA1RW	1032	271	OE2VEL	976	360	SM3DMP	916	449	WD8PKF	837
5	K9PPY	1128	94	VE7DP	1082	182	IK2QPR	1032	272	JA3FYC	975	360	SM3GSK	916	450	SM7DXQ	836
5	W1NG	1128	95	DL8FL	1081	184	HK3JJH	1031	272	JA3UCO	975	360	UX2IQ	916	451	IZ2AMW	835
7	I1SNW	1127	95	ON5NT	1081	185	JA9IFF	1030	272	OZ1HPS	975	363	EA3NT	915	451	IZ8EJB	835
7	I4LCK	1127	97	F6GCP	1080	186	I5HOR	1029	275	K0DEQ	974	363	S57L	915	453	IF9ZWA	833
7	W5BOS	1127	98	UA1OIZ	1079	187	DL6KVA	1028	276	7N1GMK	972	365	UT1UY	914	453	JA6FIO	833
10	HB9AFI	1126	99	G3OCA	1078	187	W4ABW	1028	276	IK4HPU	972	366	JM1XCW	913	453	JH4DYP	833
10	ON6HE	1126	99	JE1DXC	1078	189	AG9S	1027	278	RA3RGQ	969	366	UA3DPM	913	456	CU3EJ	832
12	VE3XN	1125	101	G0DQS	1077	189	K9MUF	1027	279	DL3APO	968	368	I2ZBX	912	457	MD0CCE	831
12	W9DC	1125	101	JA8RJE	1077	191	9A2NO	1026	279	W5PF	968	368	OM5FM	912	458	K6FW	830
14	HA0DU	1124	101	OZ1BUR	1077	192	AH6HY	1025	281	DL2RU	967	370	I2PQW	908	458	UX1AA	830
14	OM3JW	1124	101	R6AF	1077	193	9A2EU	1024	281	RM0F	967	371	AB5EB	906	460	RU3SD	829
14	YT7DX	1124	105	DL6MST	1076	194	DL2VPF	1022	281	SM5BFJ	967	371	HA3HP	906	461	DL2OE	828
17	EA8AKN	1123	106	HA5KG	1075	194	R7DX	1022	284	R7KC	966	371	IK2OVC	906	461	I2AOX	828
17	I2YBC	1123	106	JA1EY	1075	194	RU3EQ	1022	284	VK3UY	966	374	EA7TV	905	461	JH1MXV	828
17	N8JV	1123	106	PA3EXX	1075	194	VK3QI	1022	286	I5OYY	965	375	ON4CAS	904	464	RW5C	827
17	OE3WWB	1123	109	DL8MLD	1074	198	HA8IB	1021	286	K8CW	965	375	UT7UW	904	465	OE3RPB	825
21	I8XTX	1122	109	KD6WW	1074	199	DL2RNS	1020	286	SP6CIK	965	377	JH2IEE	902	466	F6HQP	824
21	IK8FIQ	1122	111	EA4MY	1073	200	UA9LP	1019	289	JA8BNP	963	377	ON7TK	902	467	ON4VT	823
21	K9AJ	1122	111	HA5AGS	1073	201	OH2BU	1018	289	RU3FM	963	379	JG3LGD	901	468	JA2IHL	822
24	AA5AT	1121	113	OZ4RT	1072	202	IZ8DBJ	1017	291	EA3GP	962	380	IK8BQE	897	469	DL5AN	821
24	W1DIG	1121	113	SM3NXS	1072	202	JR2KDN	1017	291	OZ1ACB	962	380	W7BEM	897	470	DL3JPN	820
26	F6DLM	1120	113	W4PKU	1072	202	SP7GAQ	1017	293	I0SYQ	960	382	UA3EDQ	896	471	N6AR	819
27	F6BFH	1119	113	WC6DX	1072	205	G3RTE	1015	294	YO7LCB	959	383	JA1NLX	895	471	RZ3DX	819
28	F2BS	1118	117	UR3IFD	1071	205	JA1BPA	1015	295	S55SL	958	383	R3AW	895	473	CT1DKS	817
28	N5JR	1118	118	KD1CT	1069	205	UA9YJO	1015	296	G0WRE	957	383	UA0FO	895	473	JH1OCC	817
28	W4DKS	1118	119	DL7CM	1068	208	DL8DSL	1014	296	OH2BCK	957	383	W5FKX	895	473	R0AZ	817
31	G3ZAY	1117	119	VE3JV	1068	208	K9RR	1014	296	UY5XE	957	387	OM7CA	894	476	GM7TUD	815
31	K8NA	1117	121	IK1AIG	1067	208	UA3AGW	1014	299	I5YDO	956	388	DL1CL	893	476	VA7ZT	815
31	VE6VK	1117	122	F5PAC	1066	211	UT5UGR	1012	300	DK2BR	954	389	DL4FDM	892	476	W2FB	815
34	DL8NU	1116	122	UT7QF	1066	212	DH5VK	1011	300	G3XPO	954	390	EA1EAU	888	479	N8DX	814
34	IK1ADH	1116	124	JA8MS	1065	212	VE3EXY	1011	300	JN3SAC	954	390	SV1GYG	888	480	IK8VRH	813
34	N5UR	1116	125	IK4HLU	1064	214	DL6ATM	1010	300	RN3QN	954	390	W3TN	888	480	OE3KKA	813
34	OE3SGA	1116	125	IT9EJW	1064	214	HB9CEX	1010	304	US4EX	953	393	DL3JON	887	482	CT1BOY	811
34	ON4AAC	1116	125	VE7YL	1064	214	JA7DOT	1010	305	OZ8BZ	951	394	JE3GUG	885	482	K8AJK	811
34	SM3EVR	1116	128	9A5CY	1062	214	SP8HXN	1010	305	RA1OD	951	395	DL5BUT	883	482	RA9CMO	811
34	SM6CVX	1116	128	SM5FWW	1062	218	HA9PP	1009	305	UY0ZG	951	396	DK1BX	881	485	N9GKE	809
34	W1JR	1116	129	JA5IU	1061	218	JA7BWT	1009	305	W2YC	951	397	AC0A	880	486	VE3ZZ	807
42	ON4IZ	1114	129	JF4VZT	1061	220	IK6DLK	1008	309	5B4MF	950	397	JL7BRH	880	487	IZ5JMZ	806
43	DL8USA	1113	132	DK6NJ	1060	220	JR0DLU	1008	309	DL2DXA	950	399	I0MOM	878	487	JE3AGN	806
44	IK4WMA	1112	132	HA5WA	1060	220	PA0ZH	1008	311	HB9ICC	949	400	KJ3L	876	489	SM4AZQ	804
45	UA9YE	1111	132	OE6IMD	1060	223	CT4NH	1007	312	DL1EJA	948	400	LX1NO	876	490	DL6MHG	803
46	F6FHO	1110	132	ON4ON	1060	223	SM3NRY	1007	312	R7NB	948	402	EA3WL	874	490	IZ1ANU	803
46	IK1JJB	1110	132	R3OK	1060	225	HA5DA	1006	314	JQ1ALQ	947	402	JH1IED	874	492	DK2LO	802
46	N6VR	1110	132	S51RU	1060	225	IT9YRE	1006	315	EA3BT	945	404	IK2VUC	873	492	F4FEP	802
46	PY7ZZ	1110	138	RA9YN	1058	227	I1FY	1005	315	SM5AQD	945	405	EA7TG	870	492	JG1OWV	802
46	VE3LYC	1110	139	DL5CT	1056	227	LZ1BJ	1005	317	N4MM	944	405	RW3XZ	870	495	LA8DW	800
51	G3NDC	1109	140	DL1BKI	1055	227	W5RQ	1005	318	CT1CJJ	943	407	HA0HW	869	496	OM3DX	799
51	I4GAS	1109	141	DJ3XG	1054	230	DJ9HX	1004	319	JH1QVW	939	407	JA2CEJ	869	497	F8AMV	798
51	IK8DDN	1109	142	G3SJX	1053	230	VE6WQ	1004	320	HB9BIN	938	407	RW3RN	869	497	RZ3DJ	798
54	CT1EEB	1108	142	UR3HC	1053	232	SM6DHU	1003	321	9A1DX	936	410	HB9AMO	868	499	JA8ZO	797
55	SM0AJU	1106	144	UA3AKO	1052	233	HA0IH	1002	322	N6FX	935	410	UA4PT	868	500	JA9TWN	796
56	4Z4DX	1105	144	VE7KDU	1052	234	UA0CW	1001	323	I5ZGQ	934	410	VK8NSB	868	501	K6PJ	794
56	F6AJA	1105	144	W1CU	1052	234	UA3TCJ	1001	323	R9OK	934	413	HB9BGV	867	501	R8DX	794
56	K5MT	1105	147	JA4UQY	1051	234	UR5ZEL	1001	323	SM6BZV	934	413	JF6WTY	867	503	I5KG	791
59	IK5IWU	1104	147	N7GR	1051	237	DL5MX	1000	326	VA3DXA	933	415	5B4AHJ	866	504	OZ4O	789
59	K8SIX	1104	149	K1HTV	1050	237	RA3DX	1000	327	JA1WPX	932	415	SM6CUK	866	504	UA9JLL	789
61	SM0CXS	1102	150	JH4IFF	1049	237	SM6CMU	1000	328	JM1PXG	932	417	DK3GG	865	506	IK2RPE	787
62	VE3LDT	1101	150	N4WW	1049	240	DK1FW	999	329	IK4DRR	931	418	G4NXG/M	864	506	UA4PF	787
63	GM3ITN	1100	152	F6DZU	1048	240	JA6LCJ	999	330	K0AP	930	418	M0OXO	864	508	PE1NCP	786
63	K1OA	1100	152	K5MK	1048	242	JA1GHH	996	331	JE8TGI	929	418	N6PF	864	508	UX1UA	786
65	IT9DAA	1099	152	PT7WA	1048	243	DL2CHN	995	331	RA1QY	929	421	IK2ILH	862	510	JA1GRM	785
65	N7RO	1099	152	UT5URW	1048	243	VE3VHB	995	331	W2KKZ	929	422	JH7VHZ	860	511	RV1CC	783
65	VE7QCR	1099	156	IK5ACO	1045	245	ON4CD	994	334	AD6W	928	423	CT1EGW	858	512	DL3KZA	782
68	SP9FKQ	1097	156	JA2KVB	1045	246	DK5WL	992	334	DF6QP	928	424	K8GI	857	513	SM5ELV	781
69	AD5A	1095	156	K2VV	1045	247	IN3ASW	991	334	IK2ZJN	928	424	UY5BC	857	513	WX4G	781
69	DL5ME	1095	159	AB6QM	1043	247	RA6AR	991	334	OK1DH	928	426	KC6AWX	855	515	I5JRR	780
69	JA1QXY	1095	159	DL6ZXG	1043	249	OM3XX	990	338	DJ4GJ	927	426	OE3JHC	855	515	JH7DFZ	780
72	N5ET	1094	159	EU7A	1043	250	F5HNQ	988	338	ON7DR	927	428	AI9Y	854	517	HB9DPZ	779
73	EA3JL	1093	162	JA1SKE	1042	251	CT1BXX	986	338	UY5ZZ	927	429	JI3MJK	853	518	DL5KUD	777
73	G3OAG	1093	162	LZ1HA	1042	251	RA3CQ	986	341	KD3CQ	926	429	WB5JID	853	518	EA3NW	777
75	HB9BZA	1091	162	RZ3EC	1042	253	I4GAD	985	341	OE3EVA	926	431	DL1AMQ	852	518	F8GB	777
75	I4MKN	1091	162	SM4CTT	1042	253	I8DVJ	985	341	UA0SFN	926	432	DF7GK	851	518	G0JHC	777
77	DF9ZN	1089	166	DL5DSM	1041	253	IK5PWQ	985	344	IK8JVG	925	433	RN1ON	850	518	SM1TDE	777
77	OK1ADM	1089	167	7K3EOP	1040	253	IZ8EFB	985	345	JA9BEK	923	433	SM7NGH	850	523	ER1OO	776
77	WB2YQH	1089	168	DJ5AV	1037	253	JH8JYV	985	345	RW0LT	923	433	SV1DPI	850	524	JA5AQC	774
80	G0ANH	1087	168	G4BWP	1037	253	RJ3AA	985	347	IK8CVZ	922	433	UA9FAR	850	524	JA8COE	774
80	S52KM	1087	170	IT9FXY	1036	259	K6VVA	984	348	DL6KR	921	437	F5PAL	849	524	JK1TCV	774
80	SP6BOW	1087	170	N6JV	1036	260	OH2BF	983	348	JA1AML	921	437	IK7NXM	849	527	ON5JV	772
83	DK1RV	1085	170	R4GM	1036	260	RZ3FW	983	348	JA9GPG	921	439	RA0FF	847	528	SP5APW	770
83	DL1BKK	1085	170	RY7G	1036	260	SP5TZC	983	348	JE7JIS	921	440	RL6M	846	529	F4WBN	769
83	IK8PGC	1085	170	WI8A	1036	263	DL6JGN	981	352	JE1LFX	920	441	BA4DW	845	530	JH3CUL	768
83	OH2BLD	1085	175	IK8TWV	1035	263	G3UAS	981	352	RK6AM	920	441	UX2KA	845	531	OZ3SK	767
83	UA0ZC	1085	175	JA3FGJ	1035	265	CT1EKY	978	352	SV1FJA	920	443	NA5AR	841	532	IZ8FFA	766
88	DL4MCF	1083	175	N9BX	1035	265	EA1DFP	978	355	JR2UJT	919	443	WT2O	841	533	UR5LCZ	765
88	G3HTA	1083	178	AB5EU	1034	265	EI7CC	978	356	RW4K	918	445	DL7VSN	840	534	JH2RMU	762

2023 Honour Roll

Pos.	Callsign	Total	Pos.	Callsign	Total	Pos.	Callsign	Total	Pos.	Callsign	Total	Pos.	Callsign	Total	Pos.	Callsign	Total
535	F4EUG	761	581	DL9MKA	726	627	G3LUW	688	673	DJ1OJ	645	719	UA3AB	599			
535	G3TXF	761	581	EC1AIJ	726	628	HL4CBX	687	674	JA6CBG	643	720	CT1DIZ	598			
535	K9RHY	761	583	HB9TKS	724	628	OE1WEU	687	675	DL5ZL	641	721	DL3SUG	597			
538	IK8TMI	760	583	JA1FGB	724	630	IT9RZR	686	675	I8LWL	641	722	IK2PZC	595			
538	KD7H	760	583	K1ZN	724	631	JA4GXS	683	677	DB3LO	640	723	SM5CBM	594			
540	IK1WGX	759	586	DL6MIG	723	631	K7ACZ	683	677	WA1ZIC	640	724	KJ6P	593			
540	JA2FGL	759	587	DL6CNG	722	631	UT5ZY	683	679	HA5VZ	639	725	IZ8XQC	592			
542	DL6ZFG	758	587	JN6RZM	722	634	RJ9I	681	680	DL2VFR	636	725	JA3VPA	592			
542	R2DO	758	589	DS5ACV	721	635	DJ8VC	678	680	UA6EX	636	727	IZ3ETU	591			
544	HB9DKZ	757	589	GW0IWD	721	635	DK3DUA	678	682	DL7VKD	635	728	RU6AI	590			
544	OZ8SW	757	591	JF2OZH	720	637	EV1R	677	682	JA6WJL	635	728	VK5MAV	590			
546	E73Y	756	592	W3UR	719	637	LY3BG	677	682	N3RW	635	730	KE4DH	589			
546	I2BUH	756	593	DF5JE	717	637	SP3CJS	677	682	W4HG	635	731	AA0FT	588			
546	RW9LL	756	593	DL5CW	717	640	IZ2DVI	676	686	G3KYF	633	731	F4HDR	588			
549	IV3RAV	755	595	HA1DAE	715	641	DL1NAX	672	687	G4KFT	632	733	I4YCE	587			
549	IZ1BII	755	596	W5VFO	714	641	JH2XQY	672	687	IK4DRY	632	733	RA6MQ	587			
549	N3NT	755	597	JA3KZV	713	641	UN7ECA	672	687	JA2JRG	632	733	SV9AHZ	587			
549	OE2LCM	755	597	W9ILY	713	644	UR8IDX	671	687	UW5IM	632	736	DL4FCS	586			
553	JF2UPM	754	599	DK7MD	711	645	HB9IIO	670	687	VE2BR	632	737	UT4EK	585			
554	JJ1CZR	753	600	DL4FAY	710	646	DJ9IN	669	692	HA5JP	630	738	ER5DX	584			
555	IN3NJB	751	600	JH1OAI	710	646	RT0F	669	692	IK0CNA	630	738	K4HB	584			
555	JA4LKB	751	600	UR7EZ	710	648	DL8ZBA	667	694	AA0MZ	629	740	US0LW	582			
555	JI3DST	751	603	I7PXV	709	648	DS4DRE	667	695	7K3QPL	628	741	DL7UXG	581			
558	R7LV	750	603	IN3QCI	709	650	A65BR	665	695	EA3IM	628	742	RM2A	580			
558	RV9CX	750	603	JH3GFA	709	650	W5GAI	665	697	DL1ROJ	624	743	SP6MLX	578			
560	AA4V	749	606	DL8UAT	708	652	E72U	664	697	UW5ZM	624	743	US0YA	578			
561	UR5EDX	745	606	SV2DGH	708	652	IZ8DFO	664	699	DL4SZB	623	743	YO3APJ	578			
562	IK8YTA	741	608	DH2PC	707	654	DK3DG	662	699	IK2GPQ	623	746	IK2DOT	577			
563	CT1JOP	740	608	IW9HII	707	655	IK1RAE	661	701	HB9AAA	621	747	DL5XL	573			
563	JA7BEW	740	610	JA5CEX	706	655	JA3DAY	661	702	G3KZR	619	747	RA6FG	573			
563	PB1TT	740	611	UA3ECJ	705	657	JO1CRA	658	703	SM4DDS	617	749	G4JFS	572			
563	SM0MPV	740	611	YB5QZ	705	657	SM7DAY	658	704	IW3SSA	615	749	IZ4MJP	572			
567	DL1FU	737	613	G4XRX	704	659	4Z5AV	657	705	DL3TC	614	749	JH9AUB	572			
567	W4UM	737	613	N6UK	704	659	HB9DDZ	657	705	JG1UKW	614	752	DK3BT	570			
569	PA7RA	735	613	WW8W	704	659	JF6XQJ	657	707	JH7CFX	612	752	K5WAF	570			
569	UA9LBQ	735	616	JE3GRQ	703	659	RU5A	657	707	W8JRK	612	754	DL4BBH	569			
571	JO3AXC	733	617	DL2BQV	701	663	OH2BN	655	709	DL1TRK	609	755	JE2VFX	568			
571	W1OW	733	617	IW2FND	701	663	SP7BCA	655	709	DL2YY	609						
573	JP3AYQ	732	619	F4BKV	700	665	G4VMX	654	711	DL9MRF	607						
573	ON7LX	732	619	IK8CNT	700	665	IT9RTA	654	712	UT3IW	605						
575	G4IUF	731	619	OE1ZL	700	667	JJ1KZZ	652	713	EA6VQ	603						
576	F5BOY	730	622	F8NAN	699	668	HB9BXE	650	714	DL6JZ	602						
576	IK3OYU	730	623	F6ACV	698	669	I5HLK	649	714	K2AJY	602						
576	PA3C	730	623	HS0ZIV	698	669	PT7ZT	649	716	JH3GCN	601						
579	IK4MSV	729	625	RK9UN	697	671	SM5LNE	647	717	DL5DF	600						
580	DL3BRE	727	626	OH2FT	690	672	DL4MN	646	717	UA4PCM	600						

to checkpoints (and mailing cards and fees) for inclusion in the 2024 Honour Roll and other performance tables is 31 January 2024. If submitted/postmarked after that date, they will be processed in the normal way but the scores will be held over to the following year's listing. It is important that members who have not updated since the 2019 annual listings and wish to remain listed should make a submission on or before 31 January 2024.

Official sources of information

IOTA's website (*www.iota-world.org*) provides the following:

- A List of Frequently Asked Questions (FAQ) on IOTA
- IOTA Programme Rules
- A list of authorised IOTA checkpoints
- A detailed listing of IOTA groups by continent, region and country
- A listing of IOTA groups by short title only
- A listing of the most wanted IOTA groups by continent
- Information on how to obtain the IOTA directory

The IOTA website allows you to search the island listings by IOTA group number or island name – and returns the relevant listing together with details of rarity and past and future operations.

Radio Communication (RadCom), the monthly journal of the RSGB (see its website: *www.rsgb.org*), covers IOTA in the monthly HF column.

Sources of information on IOTA activity

DX-World.net

A very attractive website run by a dedicated team of DXers and IOTA enthusiasts, which is updated daily by Col McGowan, MM0NDX. It features a web-form for readers to submit their own DX or IOTA information.

DXnews.com

This is another great website with an IOTA section run by a team headed by Al Teimurazov 4L5A. Well worth bookmarking. As it says, "More than just DX News".

425 DX News

A weekly email DX bulletin issued by IOTA Adviser Mauro Pregliasco, *i1jqj.mauro@gmail.com*. This carries IOTA news – official news releases and listings, DXpedition activity reports, a calendar of forthcoming events – as well as a host of other useful material. For more information, check the 425 DX News website: *www.425dxn.org*.

The Daily DX

The first subscription-based email DX bulletin, published by Bernie McClenny, W3UR (email: *bernie@dailydx.com*). This provides a comprehensive daily commentary on the DX scene with a prominent IOTA section listing island activity. You may prefer to subscribe to The Weekly DX, which is also available. For full details, check Bernie's home page (*www.dailydx.com*).

Internet reflectors and forums

IOTA-chasers@groups.io

This provides a meeting-place for IOTA enthusiasts to exchange information and share views about different aspects of the IOTA Programme including island operations, QSL queries, etc.

Other information sources

The DX Summit web cluster *www.dxsummit.fi* is an invaluable way of keeping an eye on the bands when you are away from your QTH. It offers a listing of the last 50, 100, 500 and 1000 spots logged from clusters worldwide, as well as propagation reports. The search facility is a particularly valuable aid to IOTA DXing as you can check for spots for callsigns and IOTA numbers.

An increasing number of IOTA operators and expeditioners maintain home pages with IOTA information and features. Try the pages of W9DC (*www.w9dc.com*).

2023 Annual Listing

Pos.	Callsign	Total	Pos.	Callsign	Total	Pos.	Callsign	Total	Pos.	Callsign	Total	Pos.	Callsign	Total	Pos.	Callsign	Total
756	G4MFX	567	837	UR5WCW	494	916	W1RM	411	999	JA4CZM	342	1080	DL5NO	302	1158	JJ1HHJ	245
756	GW4BKG	567	838	G4GIR	490	919	CT1AVR	410	1000	EA7BUU	341	1080	JA9APS	302	1162	JH8RZJ	244
758	JR1WCT	566	839	CT1JOH	488	920	N8OC	409	1000	JH1IFS	341	1080	RA3AOS	302	1162	RA4PQ	244
758	RA3TAR	566	840	CU3AC	487	921	G3LDI	408	1000	M5KJM	341	1080	RA4DAR	302	1164	DL2RZG	243
760	IK1NLZ	565	840	DK6WA	487	922	UT5EL	407	1003	JA2ACI	340	1084	AB1J	301	1164	G4WGE	243
760	JH6JMM	565	840	W6OUL	487	923	IZ8QPA	406	1004	RM3DA	339	1084	DL8JDX	301	1166	SM3OMO	242
760	JM1GHT	565	843	DL2DQL	486	923	VK3GA	406	1005	HB9DHG	338	1084	DL9NEI	301	1167	DJ3XA	241
760	KN7D	565	844	JH3VWN	485	925	DL7UGO	405	1006	G0FUV	336	1084	F5VKT	301	1167	OE1HHB	241
764	M0AID	564	844	K0JGH	485	925	JA3QOS	405	1006	N1EN	336	1084	HL2KV	301	1167	W6AER	241
764	UT7UU	564	844	W3LL	485	927	9A1AA	404	1006	N9EAJ	336	1084	UA6CEY	301	1170	IK5SRF	240
766	DF8HS	563	847	AA1QD	484	927	DJ7YM	404	1006	SV1VS	336	1090	DK8IZ	300	1171	IZ0FVD	239
767	DF7FC	562	847	K4KKL	484	927	JA6EXO	404	1010	AA6RE	335	1090	HB9EHJ	300	1171	JA0RQV	239
767	VE1AI	562	849	DL3ZZ	482	927	N3RC	404	1010	JA3PNN	335	1090	HK3W	300	1171	W5VY	239
769	JA1XZF	561	850	IU7QBB	480	927	OH1TM	404	1012	JH4JNG	334	1090	IK5BSC	300	1171	W7FN	239
770	JP1EWY	559	850	UR7LY	480	927	SM1NJC	404	1012	JH7XRG	334	1090	M0BUI	300	1175	IW0HQE	238
771	AD1C	558	852	JK1KSB	479	933	DS5DNO	403	1012	JI1HNC	334	1090	SV1CNS	300	1175	K0HB	238
772	VK5CE	557	853	LZ2PEP	477	933	G3YJQ	403	1015	W1GWN	333	1090	W7MAE	300	1175	LA9VFA	238
773	RU6B	556	853	MM0EAX	477	935	WA5KBH	403	1016	DL4CF	332	1097	I2YPY	297	1175	OZ0J	238
773	UA1OMS	556	855	DF1ZN	476	936	DL6FBR	402	1017	G3TTC	331	1097	N7QT	297	1175	SQ8LUV	238
775	DK7YY	555	855	G3USR	476	936	G3ZSS	402	1018	RX3DTN	330	1097	UR0IG	297	1180	G8AJM	237
776	JH0JQS	554	855	HB9JOE	476	938	DL3JXN	401	1019	JA8NSF	329	1100	DK2FW	296	1180	JJ0NSL	237
776	R2RZ	554	858	G1VDP	475	938	PY6HD	401	1020	JA9CHJ	328	1101	JA2MNB	295	1182	CT1APN	236
776	UN5J	554	859	I4KDJ	474	938	UR5ZTH	401	1020	JL3MCM	328	1102	HB9AIJ	294	1182	G4NAQ	236
779	SM6BZE	552	860	JR6CSY	468	941	DL2ASB	400	1020	W6KGP	328	1102	M0BJL	294	1182	K4KGG	236
779	VE2ACP	552	861	JI1FXS	467	941	DL2GAC	400	1023	DL1HBT	327	1104	F1BFD	293	1185	IZ2USP	235
781	RD0L	551	862	EI7BA	466	941	PT7YV	400	1023	DS5TOS	327	1104	XE1EE	293	1186	IK3DVY	234
782	G3SJH	549	862	JO7KMB	466	941	RX3AEX	400	1023	IW1ARB	327	1106	DH0JAE	291	1186	IK6XEJ	234
782	JG3SKK	549	862	KM6HB	466	941	SV5DKL	400	1023	JA0LFV	327	1106	N7DED	291	1188	DL2RUM	233
784	OE1PMU	545	865	R7AL	464	946	HL2IFR	398	1027	I2KBD	326	1106	XE1KK	291	1188	N2DPF	233
785	DL6MKA	543	865	UT5IP	464	947	EA3CCN	395	1028	DL9WO	325	1109	F1RAF	287	1188	RX3X	233
786	N7BT	542	867	WQ5C	463	947	JL3LSF	395	1029	JI1BJB	324	1110	G4FCI	285	1191	DL1HTW	232
787	JE3SSL	539	868	IK2RLS	462	949	DL8BFV	394	1029	JR9LKE	324	1110	I4FYV	285	1191	JH1CTV	232
788	IZ8FQI	538	869	DM3PKK	459	949	M0KCM	394	1031	DL2YBG	323	1112	4L6QL	284	1191	JK1UNZ	232
789	DM3ZF	537	869	JJ6DGP	459	951	JN1RFY	392	1031	JA0HWF	323	1112	DL4FAP	284	1191	N2ADE	232
789	JF7RJM	537	869	W1WBB	459	952	F4GYM	389	1031	K9OT	323	1112	JN1FRL	284	1195	DL1BSN	229
791	EA1N	534	872	DJ6OI	458	952	JA1WTI	389	1034	IT9BLB	321	1112	K2DSW	284	1195	DL7JRD	229
791	PA0MBD	534	872	DL2DWC	458	952	JR3CNQ	389	1035	IZ1LBG	320	1116	CT1EEQ	283	1195	HA2MM	229
791	XE1RBV	534	874	H44MS	456	955	JH3KAI	388	1035	IZ8FDH	320	1116	PG7M	283	1195	I8KRC	229
794	N1RR	532	875	G3SBP	454	956	IZ5FSA	387	1035	JA3AOP	320	1118	DJ2II	282	1199	VA7DXX	228
795	JA1FVS	530	875	JG3WCZ	454	957	G4ZCS	386	1035	JG8IBY	320	1118	JA6CMQ	282	1200	F5AAR	227
796	CT1BLE	528	875	VA2IG	454	957	W4KVS	386	1039	F5VHQ	319	1118	N1LID	282	1200	JG0CQK	227
797	HB9BQB	527	878	HL2WA	453	959	IZ2IPF	385	1040	RD4A	318	1121	9A5CB	278	1200	LA6OP	227
798	RN3CT	526	878	I5PLS	453	959	R5DB	385	1041	IW7DOL	317	1122	DL8ZAJ	277	1203	WA3WZR	226
799	I1YDT	525	880	IZ8CCW	452	961	JG8FWH	384	1041	JE2UFF	317	1122	F6HDH	277	1204	IZ1MHY	225
799	JM2LEI	525	881	DL7UVO	450	961	RV6ANI	384	1041	JH1FVE	317	1124	LB5WB	276	1204	JA1TBX	225
801	M0URX	524	882	JF1MTV	446	963	JE6HCL	383	1041	JK1FNN	317	1125	JH1QKG	274	1204	K0TRL	225
802	JP7EIP	522	883	EA2WD	444	964	IK7EOT	382	1045	JF0EBM	316	1126	JL1QDO	273	1204	VK3KTT	225
803	DL8JS	520	884	IV3VBM	443	965	IZ2KPE	380	1046	JA1MZL	315	1127	IU8HEP	270	1204	W6ZL	225
804	UA9LAO	518	884	OZ4ZT	443	966	JR3ADB	379	1046	JA2AYP	315	1127	JA1NQU	270	1209	DL9ZWG	224
805	DL9NC	517	886	RN3RQ	440	967	JH1LPZ	377	1048	CT1IUA	314	1129	JE1UMG	268	1209	FK8CE	224
805	JL1ELQ	517	886	RN3RY	440	968	G0THF	376	1048	G4FFN	314	1130	I4JED	267	1211	DL1JPF	223
807	W9RN	516	888	JG4OOU	438	969	F5AKL	374	1050	KO8SCA	313	1130	JA7ZP	267	1211	IZ2ESV	223
808	JR7FRW	515	889	NW7M	435	969	IT9VDQ	374	1051	HL4CJG	311	1132	VE5MX	266	1211	KN7Y	223
809	7K1CPT	514	890	F5RAB	434	971	G3XLF	373	1051	JH4HMG	311	1133	G4FKA	264	1214	JE6HID	222
809	DK1EI	514	891	JH1BSJ	433	972	DK7TX	370	1051	WA4MIT	311	1134	WA3FRP	263	1215	HB9RUZ	221
809	G3UHU	514	892	IK4YCQ	431	972	IK0HFO	370	1054	HB9DDO	310	1135	JE6TSP	262	1215	HB9TRR	221
812	JA5CPJ	511	893	DL1AY	430	972	IK8GYS	370	1054	JE2LUN	310	1136	DS1JFY	261	1217	7K4VPV	220
813	DL8WEM	510	893	MW0RLJ	430	975	G4PVM	368	1054	KB9LIE	310	1137	G4DBW	260	1217	IZ2CSX	220
814	DG5LAC	509	893	W1AL	430	976	DL3LBM	367	1054	OM8FR	310	1137	RD3AAD	260	1217	RZ0AF	220
814	JH8GEU	509	896	JE1BJT	429	976	UA0CID	367	1058	G0DWV	309	1139	G4AYU	259	1220	HL1VAU	219
814	K6UM	509	897	EA8AXT	426	978	DL9FCY	365	1058	G3SVD	309	1140	DG8HJ	257	1220	JR3XUH	219
817	UA1OIW	508	897	W6AFA	426	978	IK4AUY	365	1058	R6AW	309	1140	G4DDL	257	1222	DD5MA	218
817	VK2DX	508	899	VE6BMX	424	978	JA3HZT	365	1058	VK5GR	309	1142	WX2CX	256	1222	UT7KF	218
819	CU3BL	507	899	YO9FLD	424	981	DJ3CS	362	1062	DL3FCG	308	1143	CT1DRB	255	1224	JA0CJK	216
819	UT3IB	507	901	IK6HRB	423	982	HB9MEJ	360	1062	HB9ARF	308	1144	IC8TEM	253	1224	N6VH	216
821	N0ODK	506	902	UA9CGL	422	982	IK0PEA	360	1062	JA3UNA	308	1144	JA6WIF	253	1224	SM3CZS	216
822	EA5HEU	505	903	JA5ALE	421	982	JA3AVO	360	1062	PY1NP	308	1144	OE3CHC	253	1224	WB1ASL	216
822	M0YTT	505	903	NL7V	421	985	DL6UAA	359	1066	VE7JH	307	1144	VE1VOX	253	1228	IK8DNJ	215
824	UX3IA	504	905	NI0C	420	986	JA1VSL	358	1067	IK6ZDF	306	1148	DH2PG	251	1228	JF1WLK	215
825	KM4HI	503	906	RA9HM	418	986	NQ7R	358	1067	JA1VRY	306	1148	KM2O	251	1230	DL3MB	213
825	US3LR	503	907	IK2CMN	417	988	UT5ZC	357	1067	KB1DMX	306	1148	ND4V	251	1230	IW2CAM	213
825	W7YAQ	503	908	SP6DVP	416	989	LB2TB	356	1070	DO4DXA	304	1151	9A2GA	250	1230	PA2TMS	213
828	VK4CAG	502	909	IK2YGZ	415	990	RL8C	355	1070	JE2RBK	304	1151	LA3MHA	250	1233	RN1CW	212
829	HB9DOT	501	910	K3VAR	414	991	I3VJW	353	1070	N4UOZ	304	1151	NF1G	250	1233	US5CB	212
829	IV3ARJ	501	910	M5BFL	414	992	G0PHY	352	1073	G0AHC	303	1154	IK2GAJ	249	1233	VK3OHM	212
829	K1NU	501	912	BA4TB	413	993	DL7HKL	351	1073	G4POF	303	1155	WA1NXC	248	1236	IW2ENA	211
829	YO9FNP	501	912	GW4TSG	413	994	DL1BSH	350	1073	IZ8GCP	303	1156	DF9VJ	246	1237	IZ4IRO	210
833	DL5XAT	500	912	IZ0RVI	413	994	KS1J	350	1073	JE1QYI	303	1156	EI4BZ	246	1237	JE8VZK	210
833	NN7A	500	912	YO5BRZ	413	996	WA2BCK	349	1073	JI1IXW	303	1158	EA5UJ	245	1237	N1AM	210
833	UA9CVQ	500	916	9A6W	411	997	DS4AOW	347	1073	K5KUA	303	1158	EI3CTB	245	1237	SP7XK	210
836	IK5HHA	498	916	JF2AXT	411	998	AA5JF	345	1073	YL2TD	303	1158	IZ1QLT	245	1237	SQ8GBG	210

2023 Annual Listing

Pos.	Callsign	Total
1242	DL2DQN	209
1242	F5MZE	209
1244	DD0VU	207
1244	EI8IU	207
1244	IK7FPV	207
1244	NH6T/W4	207
1248	IZ0DIB	206
1248	JA4GZK	206
1248	JL2KEQ	206
1248	W7GSV	206
1252	F1VEV	205
1252	G8GHD	205
1252	IU4CHE	205
1252	KF6HI	205
1252	PA3BFH	205
1257	IZ1XEE	204
1257	JA1RRA	204
1259	N2AE	203
1259	N2CJ	203
1259	WW7Q	203
1262	HB9FKK	202
1262	IK5AEQ	202
1262	IU8FRF	202
1262	JA3IWB	202
1262	R0CAF	202
1267	HL4CCM	201
1267	JA2NSH	201
1267	JS1ERB	201
1267	W1FNB	201
1267	YO3IPR	201
1267	YO8CRU	201
1273	DK2AJ	200
1273	HL2DBP	200
1273	JR7COH	200
1273	K6EGF	200
1273	K6KZM	200
1273	K8WHA	200
1273	N7AME	200
1273	WL7CG	200
1273	WU4B	200
1282	IZ2MHT	199
1283	JL1DLQ	197
1283	SV2AEL	197
1285	K1ZE	196
1285	KB2TGU	196
1287	DL8MF	195
1287	K5FUV	195
1287	SM0LPO	195
1290	VE5EL	193
1291	DF2GH	190
1291	LA5SJA	190
1293	G4VWI	189
1294	DL8UD	188
1295	K9BO	187
1296	DL2RTU	185
1297	JP1KOA	184
1297	K6VXI	184
1297	W4PID	184
1300	AF2F	182
1300	IW3GJF	182
1302	IK1TTD	181
1303	5B4AIX	180
1303	IU8IYE	180
1305	N2SO	178
1306	EA3FZT	177
1306	K4OY	177
1308	N2IGW	176
1309	JG3KMT	174
1310	IK2LDA	172
1310	JF2VAX	172
1310	W8AKS	172
1313	JE1WBA	171
1313	KY6J	171
1313	UA3NFI	171
1316	K3FRK	170
1316	W7TLV	170
1318	HB9DWR	169
1318	JA5CDL	169
1320	DK3CG	168
1321	K4PWS	164
1322	G4SJX	163
1322	IW3QRM	163
1322	IZ3AYS	163
1322	JH6SCA	163
1322	LA9RY	163
1322	OK2CSU	163
1328	EI2II	161
1328	G0BPK	161
1328	JA3AER	161
1328	JG5RVQ	161
1332	DH7RG	160
1332	SV2CLJ	160
1334	JI6BEN	159
1335	G4NBS	155
1335	JA1CCJ	155
1335	JK1VXE	155
1338	IC8WIC	154
1338	VK3AWG	154
1338	W3MRL	154
1341	IZ3QFG	153
1342	G3WKL	152
1342	KC7CS	152
1342	OE6CLD	152
1345	AH0U	151
1346	JH1IHO	150
1347	KF7RO	149
1348	RD3TBQ	147
1349	CT2FEY	145
1350	DM4EZ	143
1350	EA5KY	143
1350	G4BLI	143
1350	IZ1JIZ	143
1354	G5CL	142
1354	I3MDU	142
1356	EA4HKF	141
1356	G0VAX	141
1356	JA9EJG	141
1359	JJ0AEB	140
1360	DL2DCX	138
1360	IK5WOB	138
1360	IZ0INX	138
1360	M0NKR	138
1360	NI4Y	138
1365	DM2FX	137
1365	K9DWS	137
1367	DD0VE	136
1367	EI3HA	136
1367	IK3MLF	136
1370	N4FN	135
1370	SM6TKG	135
1372	DM4TJ	134
1373	AC0CU	133
1374	KB3LAN	132
1375	JH1OZV	130
1376	6K5BXQ	129
1376	DK5LQ	129
1376	IW7DMH	129
1376	JJ3DKQ	129
1380	JH1GBO	127
1380	YO8AAZ	127
1382	JA8XOD	126
1382	PV8AAS	126
1384	G6GLP	124
1384	VK3YR	124
1384	W0RMS	124
1387	AK0MR	123
1387	GI3SG	123
1389	R6FFB	122
1390	KJ8O	121
1391	JL1JVT	120
1392	F5PLR	119
1393	KM4VI	118
1393	PY7BEL	118
1395	G0YCE	116
1395	W3DVY	116
1397	JE6JZP	115
1398	7K1LUE	114
1399	DF4DT	113
1399	DM1HR	113
1399	EA4EMC	113
1399	HL2DCM	113
1399	JL1CNY	113
1399	JR7ASO	113
1399	OH4UI	113
1399	PV8RR	113
1399	SV3QUP	113
1399	W7APM	113
1409	K0UD	112
1410	DL3DUE	111
1411	DH0GME	110
1411	DK3ME	110
1411	EC1DD	110
1411	F4FDA	110
1411	G8HXE	110
1411	HA3OU	110
1411	HB9EFJ	110
1411	I1RJP	110
1411	IK2WSO	110
1411	IT9YOZ	110
1411	IU8DKG	110
1411	IW5BMS	110
1411	JA1UAV	110
1411	JA1XEC	110
1411	JA2KAK	110
1411	JA4NIJ	110
1411	JA5OXV	110
1411	JA6MWW	110
1411	JA7DNO	110
1411	JE1NVD	110
1411	JE6AVT	110
1411	JF1CCH	110
1411	JG1RYQ	110
1411	JH3FEN	110
1411	JI3CJP	110
1411	JI4WHS	110
1411	JK7LXU	110
1411	JM8FEI	110
1411	JO7BTV	110
1411	JS1IFK	110
1411	K4DS	110
1411	KF7ZN	110
1411	NB3R	110
1411	NS4C	110
1411	R0JF	110
1411	RA3AV	110
1411	RV3BV	110
1411	SP6FXY	110
1411	SQ2TOM	110
1411	UA3SAQ	110
1411	UA7R	110
1411	UR4QFP	110
1411	US5CAO	110
1411	US8UA	110
1411	VE3TG	110
1411	W4ANT	110
1411	W9OO	110
1411	XE3TT	110
1459	JA7EPO	109
1459	K4WSB	109
1459	VA3VF	109
1462	2I0WAI	108
1462	G8GNI	108
1462	JA2QPD	108
1462	JJ1PFC	108
1462	N4LKB	108
1467	IK1HJR	107
1467	JI1NNE	107
1469	DL9YCS	106
1469	JA2FYO	106
1469	JL1KBS	106
1469	PV8ABC	106
1469	WV7S	106
1474	AC9GK	105
1474	KN4DXT	105
1474	N7WEJ	105
1477	IU4DTT	104
1477	JS6TWW	104
1477	PY2AB	104
1480	IC8AJU	103
1480	RN0F	103
1482	7N4GIB	102
1482	CE4WT	102
1482	M0TWB	102
1482	VK4COZ	102
1482	WC6Y	102
1487	AA5H	101
1487	EA3HRE	101
1487	HK5NLJ	101
1487	JA7NGE	101
1487	JE3WQU	101
1487	PG4I	101
1487	UN7AW	101
1487	W5BR	101
1495	3D2KM	100
1495	BI8CKU	100
1495	DC5IMM	100
1495	IU0LFQ	100
1495	JA3MAT	100
1495	JI5USJ	100
1495	JQ1CIV	100
1495	JS6SRY	100
1495	K5UHF	100
1495	KB6IGK	100
1495	KB7JJG	100
1495	OK1NYD	100
1495	R1TB	100
1495	RX1AL	100
1495	UR5ECW	100
1495	UW7LL	100
1495	XE2SI	100

2023 Club Listing

Pos.	Callsign	Total
1	UT7WZA	1099
2	DK0EE	1035
3	9A1CCY	1033
4	SL0ZG	906
5	DL0IOA	840
6	DK0PM	795
7	9A1HBC	701
8	IQ2VA	686
9	RO2E	617
10	RN3D	524
11	RY9C	505
12	HB9G	419
13	K6LY	355
14	HB9VC	333
15	UR4CWQ	309
16	DL0AVH	143
17	M0RSC	135
18	SK6LK	120
19	IQ2PB	117

2023 VHF-UHF Listing

Pos.	Callsign	Total
1	IW9HII	227
2	IK4WMA	174
3	SM6CMU	173
4	JH6BPG	164
5	G0JHC	163
5	IT9RZR	163
7	JA1UAV	161
8	JH4IFF	156
9	EA6VQ	154
10	EA3GP	143
10	HB9RUZ	143
12	OZ1BUR	141
13	JE3GRQ	137
14	IK1UWL	135
15	G3KMA	133
16	JA6WJL	132
17	JA4LKB	127
18	JI1IXW	125
19	OK1RD	116
20	G4BWP	115
21	JA6LCJ	113
21	W1JR	113
23	SM0AJU	111
24	JA2FGL	110
24	JF4VZT	110
26	DL5ME	108
27	JP7EIP	105
27	UT5URW	105
29	JA1SKE	104
29	W4UM	104
31	JA4GXS	103
31	N4MM	103
33	W1NG	102
34	MD0CCE	101
35	DL1EJA	100
35	I5KG	100

2023 SWL Listing

Pos.	SWL Number	Total
1	I1-21171	1132
2	BRS8841	1097
3	W1-7897	927
4	F-59706	867
5	R1A-644/MM	686
6	SM4-3434	645
7	JA4-4665	574
8	SM2-7734	196
9	JA1-22456	129

Trophy Winners

The Society is fortunate to have a large number of trophies

Many of these are awarded to winners of various contests, whilst others give public recognition to some particular aspect of Society work.

They are presented at a number of events - typically the RSGB Convention and the AGM.

Below you will find details of some of the many trophy winners, presented during 2022/2023

The Board

Calcutta Key
For work associated with international friendship through amateur radio
Nick Totterdell G4FAL

Founders Trophy
For outstanding service to the society
David Palmer G7URP & Tammy Palmer M0TC

Special RSGB Award
For exceptional service to the RSGB and its members
Ken Underwood G3SDW

Training & Education Committee

Kenwood Trophy
For making a significant contribution to training and development in amateur radio within the UK
Casper Pierce 2E0KRH

ARDF Committee

3.5MHz Trophy
(Not Awarded)

144MHz Plate
(Not Awarded)

Sprint
(Not Awarded)

HF Contest Committee

1930 Committee Cup
International LP Contest Leading 10W Fixed Unassisted
Dave Cree G3TBK

AFS Club Data Challenge Trophy
AFS Data Contest winning team
Three As CG "A"

AFS Data Challenge Trophy
AFS Data Contest winner single-op
Andy Cook G4PIQ

Ariel Trophy
Club Calls Contest winning single-op
Andy Cook G4PIQ

Braaten Trophy
ARRL DX CW Contest Leading SO G
Nick Totterdell G6XX (G4FAL)

Bristol Trophy
HF NFD Winner LP Assisted Portable
De Montfort University ARS

CDXC Geoff Watts Trophy
Leading Non-UK Fixed M1 or M2
Monteverde CG

Col. Thomas Rose Bowl
Comonwealth Contest winer O, SO, Unassisted
Don Beattie G3BJ

Commonwealth Medal
Commonwealth Contest most contribution
Kevin Smith VK6T (VK6LW)

Cyril Leyden G4RYY - Memorial Trophy
Leading UK&CD DXpedition QRP
Oliver Droese MM/DH8BQA

David Hill G4IQM Memorial Trophy
Club Calls Contest winning team
Camb-Hams "A"

David King G3PFS Trophy
Leading UK&CD Non-DXpedition Mixed
Don Beattie G5W (G3BJ)

Edgware Trophy
AFS CW Contest winning team
Three As CG "A"

Flight Refuelling ARS Trophy
AFS SSB Contest winning team
Three As CG "A"

G2QT Cup
HF Championship winner
Dave Cree G3TBK

G3MZV Memorial Cup
HF NFD winner LP Unassisted Portable SO
Peter Gavin M0URL

G3PSH Memorial Trophy
SSB FD winner QRP
(Not Awarded)

G3XTJ Memorial Trophy
RoLo CW Contest highest placed error-free entrant
Dave Cree G3TBK

G5MY Trophy
RoLo Contests highest combined score
Graham Bubloz G4FNL

G5RV Memorial Shield
Club Championship winning AFS local club
Norfolk ARC

G8KW Trophy
CQ WW DX CW Contest leading SO Unassisted
Dave Lawley G4BUO

GM5VG Trophy
Leading UK&CD DXpedition M1 HP
Gordon Gray & Mark Jones MM1E (MM0GOR & M0UTD)

GMDX (high-power) Trophy
Leading UK&CD DXpedition HP
Ian Pritchard MN5A (G3WVG)

GMDX (low-power) Trophy
Leading UK&CD DXpedition LP
Stephen Kopetsch MJ0X/P (M0RTI)

GW4BLE Memorial Cup
ARRL DX SSB Contest leading SO
Andy Goldsmith G1A (M0NKR)

Henry Lewis G3GIQ Memorial Cup
Leading UK&CD Non-DXpedition SSB
Richard Brokenshaw MD7C (M5RIC)

Horace Freeman Trophy
Club Championship winning AFS General Club
Three As CG "A"

Houston Fergus Trophy
International LP Contest leading 10W Portable Unassisted
Aberdeen ARS

International 10w Trophy
International LP Contest leading 10W Non-UK Unassisted
Holger Wilhelm DL9EE

International QRP Trophy
International LP Contest leading 5W Portable Unassisted
Klaus Brokmeier DB0DH (DL3CQ)

IOTA Contest Manager Trophy
Leading Non-UK DXpedition M1/M2
Radio Klub Dubrovnik 9A5D/P

John Dunnington G3LZQ Trophy
Commonwealth Contest leading R, SO, Unassisted
John Cockrill G4CZB

Junior Rose Bowl
Commonwealth Contest winner R,SO,Unassisted
Mike Franklin 9H6YI (G3VYI)

Lichfield Trophy
AFS SSBContest leading SO
Martin Platt MD4K (GD4XUM)

Marconi Cup
AFS CW Contest leading SO
Martin Platt MD4K (GD4XUM)

Milne Trophy
ARRL DX CW Contest leading SO Non-G
Allan Duncan GM4Z (GM4ZUK)

Newbury Trophy
Autumn Series winning AFS local club
Norfolk ARC

NFD Shield
HF NFD winning LP Unassisted, Portable
North of Scotland CG

Northumbria Trophy
SSB NFD winner LP Unassisted Portable
Sussex Downs

Powditch Transmitting Trophy
DX Contest leading SO Mixed
Nick Totterdell M5DX (G4FAL)

Reading QRP Shield
HF NFD winner QRP Portable
Loxlot Club

Ross Cary G3DYY Memorial Trophy
Leading UK&CD Non-DXpedition CW
John Warburton G4IRN

Ross Cary Rose Bowl
Commonwealth Contest leading 12Hr, O, SO, Unassisted
Dave Lawley G4BUO

RSGB 80m Challenge Trophy
Club Championship highest scoring individual
Dave Cree G3TBK

RSGB CDXC Cup
CQ WW DX SSB Contest leading SO Unassisted
Eddy Howells MW0YVK

RSGB HF Contest Committee Trophy
International LP Contest leading 5W Fixed Unassisted
Paul Tittensor G4PVM

RSGB-IOTA Dxpedition Multi-2 Trophy
Leading UK&CD DXpedition M2 HP
North of Scotland CG

RSGB-IOTA Multi-Op Fixed Trophy
Leading UK&CD Fixed M1/M2
Wisbech ARC

RSGB-IOTA Multi-Op LP Dxpedition Trophy
Leading UK&CD DXpedition LP M1/M2
Addiscombe ARC

RSGB-IOTA Non-UK Dxpedition HP Trophy
Leading Non-UK DXpedition HP
Petar Milicic 9A6A

RSGB-IOTA Non-UK Dxpedition M1 HP Trophy
Leading Non-UK DXpedition HP M1
Milan Pivk 9A/S58MU

RSGB-IOTA Non-UK Dxpedition M2 HP Trophy
Leading Non-UK DXpedition HP M2
Dietmar Kasper OJ0DX

RSGB-IOTA Non-UK Dxpedition QRP Trophy
Leading Non-UK DXpedition QRP
Fred Jan-Cooremans EJ/ON6QR

RSGB-IOTA Non-UK Non-Dxpedition CW Trophy
Leading Non-UK Non-DXpedition CW
Pertti Simovaara OH0R (OH2PM)

RSGB-IOTA Non-UK Non-Dxpedition Mixed Mode Trophy
Leading Non-UK Non-DXpedition Mixed
Ken Widelitz VY2TT (K6LA)

RSGB-IOTA Non-UK Non-Dxpedition SSB Trophy
Leading Non-UK NonDXpedition SSB
Norman Banks 5B4AIF (5b4AIE)

RSGB-IOTA Single-Op HP Trophy
Leading Non-UK SO HP
Dragan Radanovic E77EA

RSGB-IOTA Single-Op LP Trophy
Leading Non-UK SO LP
Kaspar Uztics YL1ZF

RSGB-IOTA Single-Op QRP Trophy
Leading Non-UK SO QRP
Harry Ashickyan LZ6BE

Senior Rose Bowl
Commonwealth Contest winner O,SO,Unassisted
Colin Smithers ZF2CA (G4CWH)

Somerset Trophy
1st 1.8MHz Contest winner
Clive Penna GM3X (GM3POI)

Southgate Trophy
Low Power Contest leading 5W Portable Unassisted
John Cockrill G4CZB

Summer Isles Trophy
Leading Non-UK DXpedition LP
Guiseppe Piparo IF9A (IT9PPG)

T E Wilson G6VQ Trophy
DX Contest leading SO CW
Dave Cree M7W (G3TBK)

The Lilliput Cup
Commonwealth Contest winner LP, SO, Unassisted
Alan Horne M7R (G0TPH)

The Rosebery Shield
Commonwealth Contest leading SO Assisted
John Sluymer VE3EJ

Verulam Silver Jubilee Trophy
RoLo SS Contest highest placed error-free log
Andy Cook G4PIQ

Victor Desmond Trophy
2nd 1.8MHz Contest winner
Clive Penna GM3X (GM3POI)

VP8GQ Trophy
Commonwealth Contest leading Non-UK, 12Hr, O, SO Unassisted
Barry Simpson VK2BJ

Whitworth Trophy
DX Contest leading SO SSB
Dave Keston M0K (G8FMC)

VHF Contest Committee

10GHz Trophy
10GHz Trophy Contest winner
Colchester RA

144MHz BackPackers Trophy
144MHz Backpackers Contest winner 5B Section
Steve Clements G1YBB

1951 Council Cup
432MHz Trophy Contest winner
John Quarmby G3XDY

Arthur Watts Trophy
VHF NFD winner LP
Andover RAC

Bolton Wireless Club Trophy
Local Club in UKAC leading AFS
Hereford ARS

Bryn Llewellyn G4DEZ Trophy
UKAC leading SO
Pete Lindsay G4CLA

Cockenzie Quaich
VHF NFD leading restricted resident GM
Lothians RS

Denis Jones G3UVR Trophy
SHF UKAC leading SO
Pete Lindsay G4CLA

Foundation Shield
144MHz UKAC leading foundation
Donna Buck M7DON

Four Metre Cup
70MHz Trophy Contest leading SO
Keith Tattnall G4ODA

G0ODQ Trophy
432MHz AFS Contest leading AFS
Camb-Hams "A"

G3JYP Memorial Award
70MHz CW Contest winner
Phil Guttridge G3TCU/P

G3MEH Trophy
144MHz AFS Contest leading AFS
Camb-Hams "A"

G5BY Trophy
VHF NFD winner M&M
Northern Fells CG

G6NB Trophy
144MHz UKAC leading local club
Hereford ARS

G6ZR Memorial Microwave Trophy
2.3GHz Trophy Contest winner
John Quarmby G3XDY

Hadley Wood Contest Group Trophy
AFS Super League leading AFS local club
Camb-Hams "A"

Intermediate Shield
144MHz in UKAC leading intermediate
Pete Millard 2E0NEY

John Pilags Memorial Trophy
VHF Championship leading SO Fixed
Keith Tattnall G4ODA

Low Power VHF Championship Trophy
VHF Championship leading SO Fixed LP
Andy Bloomer M5TKA

Martlesham Trophy
VHF NFD winner restricted
Warrington CG

Mitchell-Milling Trophy
144MHz Trophy Contest leading O
Blacksheep CG

Racal Radio Cup
VHF Championship leading O
Colchester RA

Scottish Trophy
VHF NFD leading LP GM
Not Awarded

SMC Six Metre Cup
50MHz Trophy Contest leading SO
Allan Duncan GM4ZUK

Surrey Trophy
VHF NFD winner O
Reigate ATS & Crawley ARC

Tartan Trophy
VHF NFD leading resident GM O
Aberdeen VHF Group

Telford Trophy
50MHz Trophy Contest winning O
Blacksheep CG

Thorogood Trophy
144MHz Trophy Contest leading SO
Allan Duncan GM4ZUK

UKAC 144MHz Club Trophy
144MHz UKAC leading AFS general club
807 ARO

UKAC Club Challenge Trophy
UKAC leading AFS general club
Northern Fells CG

VHF Championship AFS Trophy
VHF Championship leading AFS club
Hereford ARS

VHF Contests Committee Cup
1.2GHz Trophy Contest winner
John Quarmby G3XDY

VHF Manager's Trophy
70MHz Trophy Contest winner O
Colchester RA

Spectrum Forum

(HF Manager)

G5RP
For outstanding and consistent DX work
Ian Justice M0RNH

ROTAB
For greatest progress in the field of HF DX
Keith Evans G3VKW

(VHF Manager)

1962 Committee Cup
For outstanding amateur development at VHF/UHF
Not Awarded

Don Cameron G4STT Award
For outstanding contribution to low power radio communications
Dennis Anderson G6YBC

Fraser Shepherd Award
For research into microwave applications for radio communications
FRARS Group G4RFR

Harold Rose Trophy
For outstanding contribution to 50MHz
Not Awarded

Louis Varney Cup
For advances in space communication
Phil Crump M0DNY

Technical Forum

Bennett Prize G8PF
Significant contribution furthering the art of radio communications
Gwyn Griffiths G3ZIL & Glenn Elmore N6GN

Courtenay Price Trophy
For outstanding technical contribution to amateur radio
Peter Deneef AE7PD

Norman Keith Adams Prize
For the most original article published in RadCom
Barry Chambers G8AGN

Ostermeyer Trophy
For the description of a home constructed equipment in RadCom
Bryan Mussell G4CXJ

Wortley-Talbot Trophy
For most outstanding experimental work in amateur radio
Brian Coleman G4NNS

Propagation Committee

Les Barclay Memorial Trophy
For contribution to propagation research
Pascal Grandjean F5LEN

UK Microwave Group

24GHz Cumulative Trophy
24GHz Trophy Contest leader
Roger Ray G8CUB

47GHz Cumulative Trophy
47GHz Trophy Contest leader
Roger Ray G8CUB

Dain Evans, G3RPE Memorial Cup
10GHz Cumulatives Open winner
John Lemay G4ZTR

Dave Cox, G0RRJ Memorial Trophy
24GHz Cumulatives winner
Martyn Vincent G3UKV

G4EAT Trophy
UKuG Low Band Championships leading 1.3GHz
Combe Gibberlets M0HNA

Jack Brooker, G3JMB Trophy
10GHz Cumulatives Open winner
Barry Lewis G4SJH

Tim Leighfield, G3KEU Trophy
5.7GHz Cumulatives leader
Telford & DARS G6ZME

G3EEZ Memorial Trophy
For Contributions to Microwave Communications
Neil Smith G4DBN

G3JVL Award
The newcomer who has made the greatest contribution to microwave communication
Dave Newman G4GLT

G3VVB Trophy
For the best home-constructed microwave equipment exhibited at a microwave roundtable or convention
Mark Hughes GM4ISM

Les Sharrock, G3BNL Trophy
For innovation or technical development of microwave equipment or techniques
Dave Austen G1EHF

Other trophies awarded at the AGM

Jock Kyle Memorial Award
Scottish club, society or member done most for scottish VHF amateur radio
Scottish Microwave Round Table

Jack Wyllie Trophy
Scottish club, society or member done most for scottish amateur radio
Dr Barry Beggs GM3YEH

Club of the Year

Club of the Year - Large Club
(not awarded)

Club of the Year - Small Club
(not awarded)

Amateur Television

Radio Amateurs have been transmitting and receiving TV pictures for over 60 years. In most cases, simplified versions of the broadcast standards of the day have been used, perhaps tailored to a reduced bandwidth to fit within the amateur bands. All amateur bands above 50MHz are suitable for amateur TV, and the recently developed advances in reduced-bandwidth DVB-T transmission may allow that mode to be used for impressive DX contacts on the lower VHF bands.

Licensing

No extra licence is required to transmit amateur TV, you just have to adhere to the existing rules of your licence. This does mean that Foundation licensees can transmit ATV in the 432MHz and 10GHz bands if their equipment meets the required standards. The regulations on transmitted content and station identification are the same as for voice, so most transmissions are pictures of the operators themselves, their equipment, their latest project or their surroundings. The station callsign needs to be transmitted every 15 minutes, either on-screen or in the digital station identification.

Transmission Modes

Although amplitude modulation was used for many years, it is now rarely employed because there is insufficient bandwidth to accommodate it within the 432MHz amateur band, and it is difficult to generate efficiently on the higher bands. As AM TV is no longer used commercially in the UK, receiving equipment is also harder to find.

Frequency modulation, as used for early commercial satellite TV transmissions, is still popular in the 1296MHz band and above. Generally amateur transmissions use half the deviation level of satellite transmissions (to reduce bandwidth occupancy and increase signal to noise ratio) with an FM audio subcarrier at 6MHz. The exception to this is at 5.6GHz, where re-purposed hobby drone TV transmitters can be used without modification on 5665MHz.

Digital television modes, as used for current commercial satellite TV transmissions, have proved to be very robust and efficient for amateur television transmissions. The DVB-S and DVB-S2 modes are both used, and can be generated from software defined radios driven by user-friendly Linux or Windows computers (including the Raspberry Pi). These transmissions can be received on a domestic satellite TV receiver (with an up-converter or down-converter for bands other than 1296MHz), or a specialised receiver such as the MiniTiouner.

The published standard for Satellite TV does not include transmissions of less than 1.2MHz wide (generated using a digital symbol rate of less than 1 MSymbol/sec), however the amateur-developed MiniTiouner receiver can decode signals of far lower bandwidths, and compatible transmissions are easily generated using computers and software defined radios. The advantage of this is that the reduced bandwidth transmissions can be decoded from weaker signals than the wider bandwidth transmissions. These signals are referred to as Reduced-Bandwidth TV (RB-TV) and typical symbol rates used are 500 kS, 333 kS, 250 kS, 125 kS and 66 kS.

Commercial terrestrial transmissions in the UK use the DVB-T transmission standard; this usually has a fixed 6 or 8MHz bandwidth, making it difficult to fit into some amateur bands and requiring higher power than the narrower DVB-S and DVB-S2 modes. However, recent developments in commercial tuner technology have enabled amateurs to develop reduced-bandwidth receivers for DVB-T and software defined radios have been developed to enable compatible reduced-bandwidth transmissions. The use of reduced-bandwidth DVB-T should provide advantages in situations where multipath is a problem, such as DX contacts on 50MHz or mobile operation on 437MHz.

Receiving Digital Amateur TV

By far the most popular receive system is the MiniTiouner. This uses a Satellite TV Receiver module with a USB interface to a computer running a receiver program. F6DZP has made the comprehensive "MiniTioune" software available for free use on Windows 10, and a program for Linux computers, known as LongMynd, has been developed by M0HMO and is available either stand-alone (from GitHub) or as part of the BATC Portsdown DATV System. The MiniTiouner and Longmynd software is also used in the "Ryde" set-top box DATV receiver developed by BATC members. The MiniTiouner hardware can be self-built or purchased as a pre-assembled unit. The hard-to-purchase components are available from the British Amateur Television Club shop for self-builders and the total parts cost is about £75.

The MiniTiouner input frequency range is 144MHz to 2650MHz so, with the addition of preamplifiers, it will receive on 4 amateur bands directly and it can be used with down-converters for the other bands. Being a satellite tuner, it is designed to operate with 50 dB of gain between it and the aerial, so it always benefits from a preamplifier for amateur use.

Transmitting Digital Amateur TV

The simplest option for transmitting Digital ATV is to use BATC Portsdown software on a Raspberry Pi 4 with a touchscreen and LimeSDR Mini or Pluto software defined radio (SDR). An alternative for Windows is the DATV Express software with either of these SDRs. All of these solutions are broadband allowing the generation of signals anywhere in the frequency range 70MHz to 3450MHz. The signal from the SDR at about 0 dBm (1 mW) can either be amplified directly in a (very linear) power amplifier, or up-converted to the frequency required. For many microwave

A typical ATV Picture from M0DTS

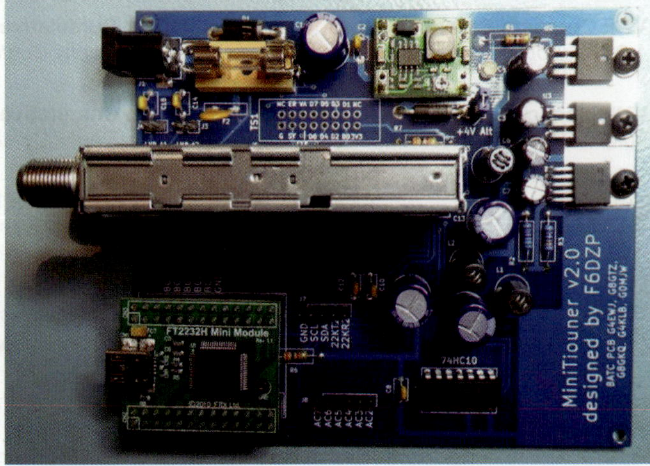

The MiniTiouner Hardware

operators, existing transverters can be used with minimal modification. The MiniTiouner can be tuned to the normal receive IF, and the transmit signal generated at the same IF.

Operating Standards

Commonly used operating frequencies and modes are listed below. Talkback is generally on 144.75MHz FM or the DX Spot website https://dxspot.batc.org.uk/ can be used for chat at longer ranges.

Amateur TV By Satellite

The launch of Es'hail-2 with its QO-100 wideband transponder has enabled amateur TV the possibility of contacts between stations in the UK and Europe, Africa and parts of Asia and South America. Reception of these transmissions is relatively easy with an 80 cm dish, a commercial LNB and a MiniTiouner.

The wideband transponder downlinks between 10490.5MHz and 10499.5MHz using horizontal polarisation, and a normal "Universal" LNB converts this down to 740.5 – 749.5MHz when 18v is supplied to the LNB. This signal can then be directly demodulated by a MiniTiouner. A good first signal to look for is the beacon on 10491.5MHz. The MiniTiouner parameters for this will be a frequency of 741.5MHz, symbol rate 1500, DVB-S2. Once the beacon has been received, other signals found on the BATC/AMSAT-UK online spectrum monitor https://eshail.batc.org.uk/wb can be tuned in.

The uplink for the wideband transponder is from 2401.0MHz to 2410.0MHz Right Hand Circular Polarised. Uplink transmissions should be DVB-S or DVB-S2 at less than 2 MS. Typically 30 W into a 1.2 meter dish is required to uplink a 333 kS digital ATV signal.

Further Information

The British Amateur Television Club (BATC) (https://batc.org.uk/) publish a quarterly magazine with all the latest ATV news and construction projects. They also maintain a Wiki (https://wiki.batc.org.uk/) with lots of useful information, and run a forum about amateur television (https://forum.batc.org.uk/). Membership of the BATC can be purchased online and is free to students in full-time education.

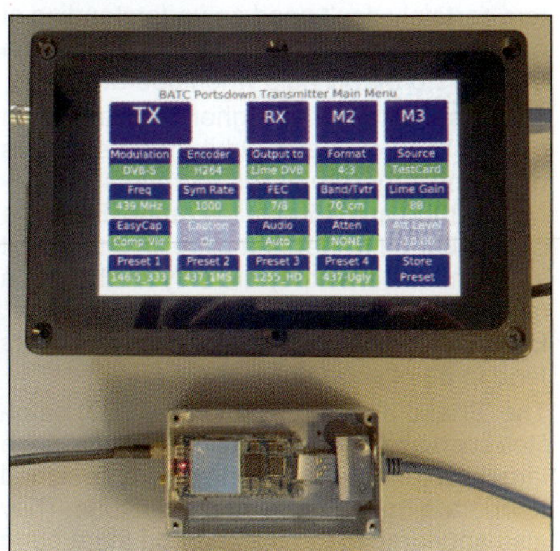

The BATC Portsdown Transmitter and LimeSDR

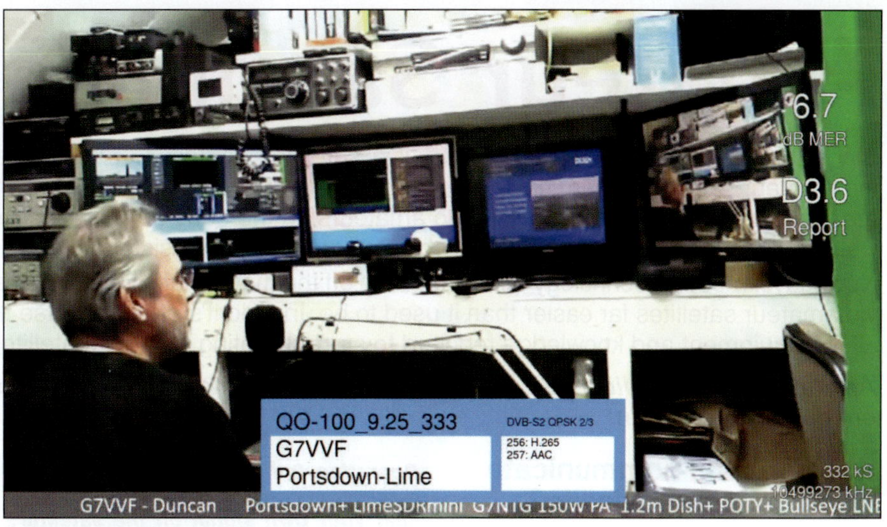

The Ryde Received Picture with on-screen-display

Frequency	Mode	Parameters	Notes
29.25 MHz	DVB-T	333 kHz QPSK	Max 500 kHz bandwidth
51.7 MHz	DVB-T	333 kHz QPSK	Max 500 kHz bandwidth
71.0 MHz	DVB-T	333 kHz QPSK	NoV required. 70.5 - 71.5 MHz
	DVB-S2	333 kS QPSK	
146.5 MHz	DVB-S2	333 kS QPSK	NoV required. 146.0 – 147.0 MHz
437.0 MHz	DVB-S2	333 kS QPSK	
		1 MS QPSK	Band Plan 436.0 – 438.0 MHz
1255.0 MHz	DVB-S2	Various SRs	FM ATV being replaced by DATV
	FM ATV	-	Caution not to interfere with Primary User
2395.0 MHz	DVB-S2	Various SRs	
3405.0 MHz	DVB-S2	333 kS	
5665.0 MHz	FM ATV	Wideband FM	Using FPV Drone equipment
5762.5 MHz	DVB-S2	333 kS	Using NB Transverters (from 146.5)
10370.5 MHz	DVB-S2	333 kS	Using NB Transverters (from 146.5)
24047.5 MHz	DVB-S2	333 kS	Using NB Transverters (from 143.5)
47090.5 MHz	DVB-S2	333 kS	Using NB Transverters (from 146.5)
75978.5 MHz	DVB-S2	333 kS	Using NB Transverters (from 146.5)

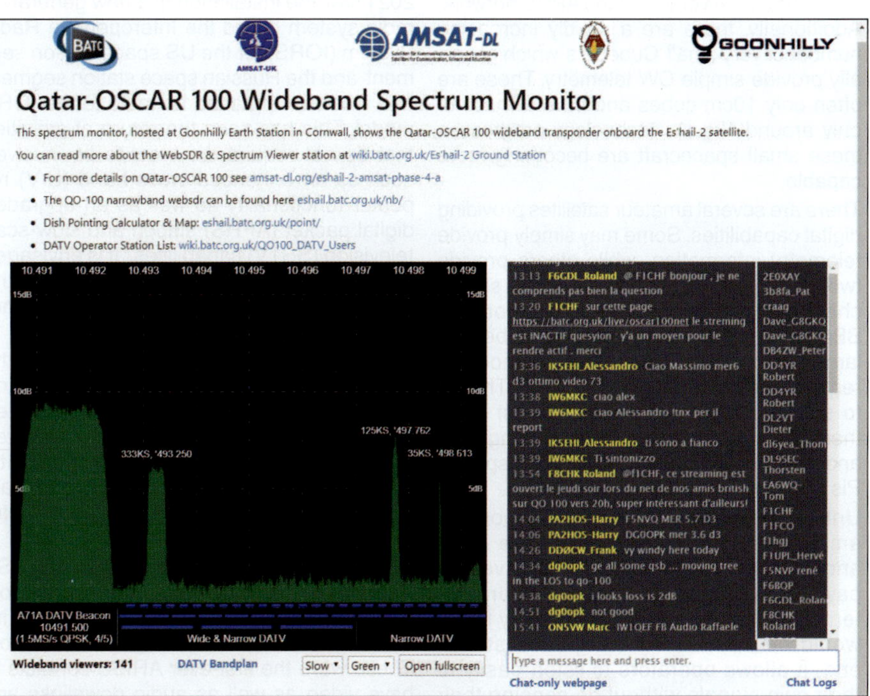

The Online Spectrum Monitor for the QO-100 Wideband Transponder

Amateur Satellites

Since soon after the launch of the first artificial satellite, "Sputnik-1", radio amateurs have been constructing and operating amateur satellites. There have been more than 100 amateur satellites launched since then, of which more than 20 are currently operating and available to amateurs. In the early days, only a few well-equipped stations with operators well versed in orbital mechanics were able to make QSOs consistently.

Nowadays, modern technology takes the strain off the mathematics and physics, making listening to and communicating via amateur satellites far easier than it used to be. Indeed it's often the case that amateurs find that they already have all the equipment and knowledge required to start operating amateur satellites.

All radio amateurs licensed in the UK can communicate via satellites, Foundation licensees are actively encouraged to join in!

How far can I communicate via an amateur satellite?

How far a station can expect to communicate using an amateur satellite depends on a number of factors, but it is possible to operate almost globally using amateur satellites. One benefit of using amateur satellites as a communication medium is that it is possible to predict very accurately and consistently when a QSO can be made. Because satellite communication almost always demands line of sight between the ground station and the satellite, the vagaries of propagation present in many other DX modes can generally be ignored.

What's onboard an amateur satellite?

Different amateur satellites have differing capabilities in their payloads. Analogue satellites provide voice and CW transponders. Some analogue transponders simply receive a single channel FM signal on one band and retransmit the channel on another band. Others carry linear transponders designed for CW and SSB QSOs. Some satellites support an international version of the APRS network. Additionally, there are a rapidly increasing number of very small 'CubeSats' which generally provide simple CW telemetry. These are often only 10cm cubes and have a mass of only around 1kg. As technology progresses these small spacecraft are becoming more capable.

There are several amateur satellites providing digital capabilities. Some may simply provide telemetry information, while others provide two-way communications from simple single-channel digipeating to store-and-forward BBSs. Over the years, there have been a large number of different hardware modems required for digital satellite operation. Thanks to software DSP techniques, almost all of these modems are now emulated using a PC and a soundcard and, increasingly, Raspberry Pis and similar hardware.

Unlike a terrestrial repeater, almost all amateur satellites receive on one band and retransmit on another. This saves on payload weight on the satellite where potentially expensive, large and heavy filters would be needed. At the ground station end, it allows operators to listen easily to their own signals without de-sensing their receivers.

Operating full duplex is recommended practice on amateur satellites. If you can't hear your own signal on the satellite's downlink, it's difficult to know whether you are on frequency or even making it to the satellite at all.

ARISS

As well as autonomous spacecraft, amateur radio also has a permanent presence in space onboard the crewed International Space Station. ARISS (Amateur Radio on the International Space Station) is an international organisation comprised of national amateur radio societies, national AMSAT societies and five space agencies (NASA, ESA, CSA, JAXA and Roscosmos). It is designed to let students worldwide experience the excitement of talking directly with crew members of the International Space Station, inspiring them to pursue interests in careers in science, technology, engineering and mathematics (STEM), and engaging them with radio science technology through amateur radio. It is also possible for individual amateurs to hold random QSOs with astronauts - surely a highlight for any amateur's career!

2021 saw the installation of a new generation radio system called the Interoperable Radio System (IORS) for the US space station segment and the Russian space station segment has recently upgraded its radio to the IORS model. This has seen the return of activities that do not require the crew to be involved such as an enhanced cross band (U/V) repeater functionality as well as an upgraded digital packet (APRS) station and slow-scan television (SSTV) capabilities. It is envisaged that the cross band repeater mode and the packet/APRS mode will be active simultaneously in the near future.

Future upgrades and enhancements to the IORS are in various stages of design and development. These include an upgraded Ham Video system L-band (uplink) repeater, ground command operations capability, LimeSDR signal reception, a microwave 'Ham Communicator,' and a Lunar Gateway prototype experiment.

During 2016, Tim Peake, the first British ESA Astronaut, made contact with ten schools throughout the UK. Most of these also included live video from the HamTV installation. These were the first-ever ARISS contacts to have video as well as audio downlinks and this facility contributed greatly to the impact of the contacts. Great support was given by the ARISS UK Operations Team, the UK Space Agency and the STEM outreach teams of the RSGB itself.

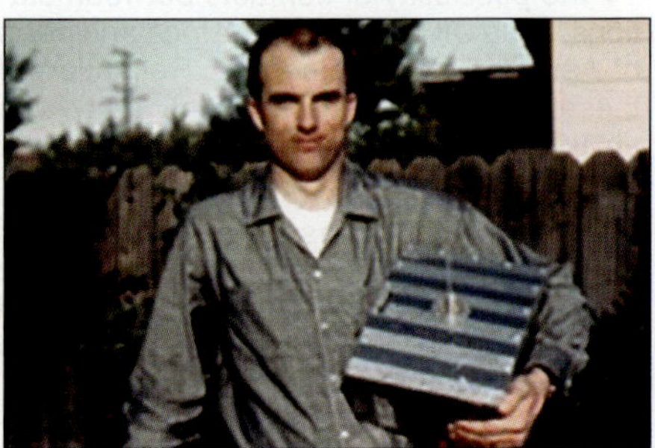

Lance Ginner, K6GSJ, poses with the flight model of Amateur Radio's first satellite, OSCAR I. The blue, stick-on label on the top of the spacecraft reads: "OSCAR I – AMATEUR RADIO BEACON SATELLITE"

ARISS school contacts continue to occur in the UK at a rate of on average once every year. Competition for inclusion is high, and schools that apply are expected to pursue a course of activities and learning opportunities for students in STEM/STEAM related subjects, including amateur radio. Applications to take part in the worldwide ARISS program occur in "windows" from January to March and September to November each year. Successful applicants can normally expect their contact to be carried out within approximately 12 months Application forms and more information can be found at *http://www.ariss-eu.org*

Amateur satellite orbits

Although, these days, amateur satellite operators certainly don't need an understanding of orbital mechanics, knowing some basics of a satellite's orbit will help when operating.

Some satellites orbit higher in space than others. The Low Earth Orbit (LEO) satellites have coverage areas (or 'footprints') of between 3000 and 4500 miles range. Because of their low orbit, they complete an entire orbit in 90+ minutes, so they are only "visible" for a maximum of about fifteen minutes each orbit. Because of the short 'pass', QSOs and overs tend to be of short duration.

The 'Phase 3' High Earth Orbit (HEO) satellites such as AO-40 had the benefit of a much larger footprint because they were designed to operate from a much higher altitude. They also appeared to the observer on the ground to be hovering around for several hours at a

Blue Peter presenter Lindsey Russell, with competition winner Troy and UK ESA astronaut, Tim Peake with the patch.

time, perfect for a ragchew. These spacecraft were in a highly elliptical orbit but, at the time of writing, there are no operational Phase 3 orbiting satellites.

The recent launch of a "Phase 4" geostationary satellite, called Es'hail-2 or Oscar 100, represents a step-change for amateur satellite operation. It is in a geosynchronous orbit, located at 25.9° East and covers more than 1/3rd of the earth. It has been developed as a result of co-operation between the Es'HailSat Company, the Qatar Amateur Radio Society and AMSAT-DL. Its amateur payload comprises of two linear transponders.

These transponders operate with 2400MHz uplinks and 10450MHz downlinks. They comprise of a 500kHz bandwidth linear transponder intended for conventional analogue operations and an 8MHz bandwidth transponder for experimental digital modulation schemes and amateur television using DVB-S and DVB-S2 modulation at a variety of symbol rates.

This spacecraft appears stationary in the sky and is therefore available for use 24/7. It also does not require tracking antennas so the ground station implementation can be quite simple. More information about this unique amateur spacecraft, and how to use it, can be found at: *https://eshail.batc.org.uk* You can also listen to the downlink using the web SDR at the Goonhilly Earth Station *https://eshail.batc.org.uk/nb/*

How do I find a satellite?

To figure out when a LEO satellite is passing over, and where it will be in the sky at a given moment in time, prediction software is used. If the station has directional antennas, most prediction software can also steer the antennas in real-time. It's important to ensure that prediction software has up-to-date orbital parameters, called Keplerian Elements, or often simply as 'Keps' or TLEs (Two Line Elements). These parameters are downloadable from the Internet and most software can be configured to obtain these updates automatically.

Because satellites are moving relatively rapidly compared to the ground station, there will be some degree of Doppler shift in the receiving and transmitting frequencies. Doppler shift is experienced in daily life when, for example, the tone of an emergency vehicle's siren drops as it passes. When operating amateur satellites, Doppler shift may be adjusted by manual tuning, or alternatively most prediction software can update the radio's frequency in real-time.

Ground Stations for operating amateur satellites

It is not necessary to have a large or expensive station to operate satellites. In its simplest form, a dual-band handheld FM transceiver can be used to make satellite contacts. When used with a handheld Yagi antenna, this simple configuration can make an effective station for portable contacts.

Satellite Orbits

Apogee
GTO
Geosynchronous
OSCAR-100
GEO
Polar LEO
LEO
MEO
Perigee
Most Other Amateur Satellites

Typical orbits used by the Amateur Satellite Service

At the other end of the scale, a station may have a steerable antenna array under automatic computer control and a radio with satellite-specific functionality such as full-duplex operation, SSB capability and tracking uplink and downlink VFOs. The benefit of the larger station is that it allows the operator to use more satellites more consistently.

A well-equipped satellite station has antennas of similar size to a standard domestic TV and FM Band II antenna configuration, with perhaps eight elements on 70cm and four on 2m. This makes it surprisingly easy to make a neighbour-friendly and capable amateur satellite station. These antennas are usually crossed dipoles and can either be steerable in azimuth only or, for more consistent QSOs, in elevation too.

Rather than purchasing expensive Az/El rotators, antenna pointing is very often done manually with the antennas at ground level, especially in temporary or portable configurations. Because the satellite is above the horizon in the sky, locations often considered ineffective for terrestrial radio communication can be effective for amateur satellites.

As technology progresses, more satellites will be operating on the higher bands. For microwave use, a small dish is employed and because of their narrower beamwidths, it's essential to be able to point the antennas accurately both in azimuth and in elevation.

Perhaps the most important rule of thumb in satellite operation is to concentrate on the station's receiving equipment before investing time, money and effort in the transmitting side. The nature of satellite communications means that the old adage 'if you can't hear them, you can't work them' is especially true. It is often tempting to improve your signal by increasing your station's ERP. It is likely that the transponder may already be limiting (eg, in linear transponders, the transponder's AGC has started to attenuate the passband so that its output can be maintained in the linear region), so more ERP is not going to be beneficial. Masthead preamps are always beneficial.

How are amateur satellites launched?

Historically, amateur satellites have been launched by generous space agencies with space available on their rockets. Often this space would otherwise have been dead weight ballast. Recently most of these free rides have dried up, but the launch costs for CubeSats can often be managed. The collaboration between radio amateurs and universities is also leading to having shared missions as described below.

Who makes amateur satellites?

Amateur satellites continue to be made by many organisations throughout the world, by individual national AMSAT societies, or a collection of national AMSAT societies. Many educational establishments have also launched amateur satellites.

AMSAT-UK is at the forefront of satellite build-

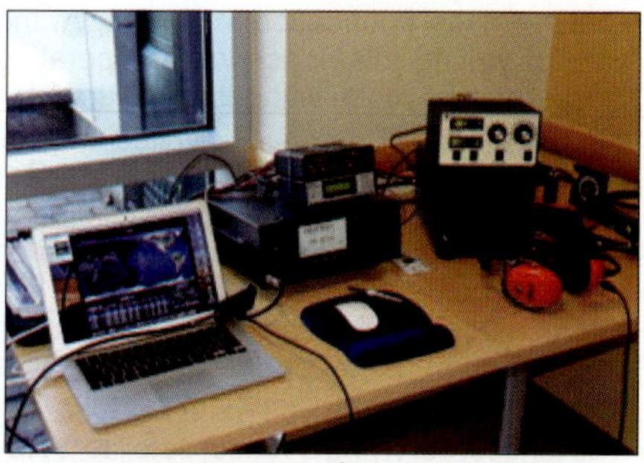
Picture from N1FD showing a typical, well equipped amateur satellite ground station.

Picture from Makezine.com showing Diana Eng with her portable ground station and homemade yagi antenna.

ing and AMSAT-UK has already created the FUNcube-1 CubeSat in collaboration with AMSAT-NL which was successfully launched in late 2013. This also carries the name Oscar AO73 and acts as a linear 70cm to 2m transponder at night and during weekends and holidays. At other times it provides telemetry data for educational outreach for schools and colleges. The sharing of this resource is intended to encourage the uptake of STEM (Science Technology Engineering & Mathematics) subjects as well as increasing knowledge about and interest in amateur radio.

This spacecraft continues to operate nominally and has provided more than 1.6GB of data which is stored on a central Data Warehouse and which is available for research purposes.

AMSAT-UK also provided a similar FUNcube-2 payload for the UKube-1 spacecraft which was developed for the UK Space Agency. This spacecraft is also in orbit and although it has completed its science mission the transponder continued to function until mid-2018. Additionally, early 2017 saw the successful launch of Nayif-1 which provides a further addition to the FUNcube constellation.

Their latest launch, in late 2018, was JY1Sat. This is based on the FUNcube mission of having a transponder for radio amateurs and educational out-reach, with images in SSDV format, for schools and colleges. It was developed under the auspices of the Crown Prince Foundation in Jordan. At the time of writing this spacecraft is operating in transponder mode without any telemetry.

All artificial satellites have a limited lifetime. Solar panels gradually become less efficient as they are bombarded by the radiation in space, from which we are protected here on Earth. On other occasions, it might be that the rechargeable batteries fail first. Because of their inherent limited lifetime, there is a continual need for replacement satellites.

The longest surviving amateur satellite is AO-7, which was launched in 1974. AO-7 went silent for over two decades until Pat Gowen, G3IOR, heard it again in 2002. It is believed that AO-7 originally stopped functioning due to the battery going short circuit. After many years of the cumulative electrical and chemical stresses on the battery from the solar cells attempting to charge them, one of the cells in the battery went open circuit and now AO-7 is available again - although only when it is in sunlight. It is understood to be THE oldest earth orbiting spacecraft still in active use!

Where to find more information on amateur satellites

The AMSAT-UK and AMSAT-NA websites are very good sources of the most up-to-date information about satellites, and there is also a very active AMSAT reflector available on the internet. To join, send an email with the following text on the first line of your message: subscribe AMSAT- your request to *majordomo@amsat.org*

The RSGB has also published "HAMSATS & AMSATS", a book by Andre Barron, ZL3DW, which provides lots of useful information.

Who pays for amateur satellites?

Funding for amateur satellites comes from a number of sources, including the national radio societies and national AMSAT societies.

In the United Kingdom, AMSAT-UK represents the interests of amateur satellite operators, and, as discussed above, AMSAT-UK members have been instrumental in the development, manufacture and funding of several amateur satellites. With the generous donations and subscriptions of its members, AMSAT-UK helps to keep amateur satellites in space. So if you use amateur satellites, remember to contribute by joining AMSAT-UK! See *https://www.amsat-uk.org*

Which amateur satellites are currently operational?

The list can change very quickly with new launches taking place and some failing or burning up and de-orbiting... It is therefore not possible to publish a sensibly up to date list here. The *http://www.amsat.org/status/* page shows a large amount of detail for each spacecraft and this is updated by the hour. More details of each spacecraft can be found here *http://www.dk3wn.info/p/?page_id=29535* As will be seen, most satellites presently use frequencies in 2m or 70cm bands for which equipment is very readily available. A simple on-line satellite prediction service is provided here *http://www.amsat.org/track/* Be sure to enter your location first and then select the "amateur satellites" tab

Try amateur satellites for yourself!

Getting started on amateur radio satellites is straight forward, it is recommended that you read the AMSAT-UK an AMSAT 'beginners' web pages and the Oscar status page at: *http://www.amsat.org/status/*

Both operating schedules and modes of satellite operation regularly change. Over time, batteries, solar panels and other hardware can and do fail, and there's little opportunity to repair satellites once they're in space.

For first timers it is likely you will start on the FM satellites, the most common being AO-91 and AO-92. Both of these are equipped with single channel FM bent pipe transponders with a 70cm uplink and 2m downlink, however you will need a CTCSS tone of 67Hz to access, it's also worth noting that AO-92 also contains a 23cm uplink which is operated on a regular schedule.

To use either of these two satellites it's recommended that you operate in duplex mode, that is you can hear the satellite downlink at the same time as transmitting, this reduces interference to other users and increases your chances of a contact, Duplex can either be achieved with a handheld radio which supports it or two separate radios.

For an antenna, it's very beneficial to have a dual band antenna with some directional gain such as the Arrow or Elk antenna, if you prefer, you'll find plenty of online designs - a popular choice is the dual band satellite yagi design which is on M1GEO's website: *https://www.george-smart.co.uk/antennas/dual_band_satellite_yagi/*

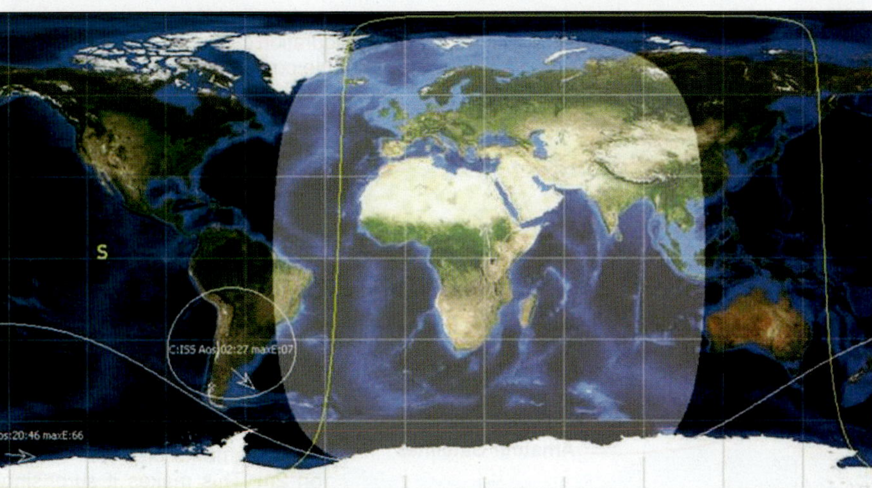

Earth Coverage of the Es'HailSat-2 also known as Oscar 100

The picture shows the well-engineered antenna system of Matthias DD1US on a sunny but, cold day.

Once you have those items, you will need a method to find when the satellite will pass over you, a simple method is the online prediction tool by AMSAT (*https://amsat.org/track/*) or you can use one of the many mobile applications for example Heavens Above or ISS Detector.

Before you transmit it's a good idea to follow some best practices that help all users, especially when a transponder is a single channel.

1. Share the Pass - FM satellites are just repeaters, only one person can work it at a time, so don't monopolise a pass, remember to let others have a go.
2. Let other QSOs Finish - Always let a QSO finish, transmitting over an on-going contact causes frustration and means there's less time for other QSOs.
3. Minimise Repeat QSOs - As FM satellites only have a single channel, it's best that on busy passes, once you have worked someone once that you do not call them on every pass. This gives everyone else a chance to work them, of course if the pass is quiet, a repeat contact is fine.
4. Don't call CQ - Please don't call "CQ Satellite" on an FM satellite. It's the same as calling CQ on a repeater; you just don't do it. Usually with FM satellites they are busy during the day time so you can just directly call someone, if not just announce your callsign.
5. Use Phonetics - Always use phonetics, ideally NATO phonetics that everyone recognises, just saying your callsign can cause delay or mis-understanding and incorrect logging.

The footprint of Saudisat 1C (SO-50) showing co-visibility in both north east USA and the UK

6. Rare/Portable Stations take Priority - It is common for satellite operators to take their equipment with them to portable locations, to transmit from rare grid squares or other DX countries. Courtesy should be extended to these stations; they are providing a rare location to all satellite operators and will be at that location for a limited time. If you hear a station on from a rare grid or DXCC entity, use good judgement before calling stations in more common grids
7. Use only the Minimum Power Required - Generally 2.5-5W from a handheld with a directional yagi is more than enough to access a satellite, while it's tempting to run more it can damage the satellite longer term.
8. Work the New Stations - Remember to always listen and work the new stations, everyone remembers the excitement of their first satellite contact, remember to help others achieve theirs.

AO-91 Frequency Chart

AOS 2	435.240 MHz	145.960 MHz
AOS 1	435.245 MHz	145.960 MHz
TCA	435.250 MHz	145.960 MHz
LOS 1	435.255 MHz	145.960 MHz
LOS 2	435.260 MHz	145.960 MHz

AO-92 Frequency Chart

AOS 2	435.340 MHz	145.880 MHz
AOS 1	435.345 MHz	145.880 MHz
TCA	435.350 MHz	145.880 MHz
LOS 1	435.355 MHz	145.880 MHz
LOS 2	435.360 MHz	145.880 MHz

Practice receiving the satellite for a few passes first. Remember to open the squelch of your receiver. Although they are not very weak, AO-91 & AO-92 will appear to suffer fading, and you will find that leaving the squelch open will aid reception. The AO-91 and AO-92 downlinks on 2m will mean you don't need to adjust for doppler although the uplink will need careful adjustment on 70cm and will need to be tuned in 5kHz steps for the correct Doppler. At the start of the pass it will be 10kHz lower and by the end of the pass 10kHz higher.

The biggest knack to learn is how to manoeuvre the antenna for the best signal. There is no hard or fast rule here, but keep in mind that you should know approximately where the satellite is in the sky, and that you are aiming to orient your antenna at the satellite. Also consider that because satellites spin like a gyroscope, their antenna polarisation changes. You'll find turning the antenna to try to match polarisations will prevent deep fades. It takes a couple of passes to get the hang of it. Keeping the pass listing, compass and your watch readily available is very useful here.

You may notice that both satellites have a background noise on their audio. This is their sub audible telemetry, which allows telemetry and voice at the same time. Also note that AO-91 & AO-92 require 67Hz CTCSS tone to be transmitted by the user to activate the transponder, if the satellite doesn't hear any CTCSS for 60 seconds the transponder will time out and you will hear just a voice announcement.

When it comes to conducting a QSO, 5W will easily be sufficient as long as the 'alligators' aren't about. This term is often used in the amateur satellite community to describe an operator who's all mouth and no ears. It is obvious from monitoring the downlink that these operators usually cannot hear the satellite at all! Both these satellites can be operated with less than 1W if no one is using high power.

Because you'll be operating full-duplex, take along some headphones. This will prevent feedback via the satellite when you're transmitting. If you have one, take a digital recording device with you for logging purposes. There's a lot to do, and recording the QSOs will allow you to keep your hands available for things other than logging. After the pass, you can write the QSOs into your logbook.

You will find that with satellite passes the more you practice the easier it becomes, a good way to look at a pass is like a contest, quick exchanges to get as many QSOs as possible in a 15 minute window.

Once you have honed your skills you will find that it's possible to work across Europe with ease and during periods of apogee (when the satellites furthest away from earth) you will be able to make QSOs into North America, Canada, Greenland and further into Europe.

Howard Long, G6LVB, demonstrates amateur satellite operation with nothing more than a dual-band handheld transceiver and an 'Arrow' antenna, so-called because its elements are made from arrow shafts.

AMSAT-UK represents the amateur satellite community in the UK whose members not only operate amateur satellites, but also help to design, build and fund them

AMSAT-UK address: J D Heck, G3WGM, "Pickles Orchard", 30 Memorial Road, Great Hampden, Buckinghamshire HP16 9RE
email: *g3wgm@amsat-uk.org* (Jim Heck) website: *www.amsat-uk.org*

AMSAT-NA website: *www.amsat.org*

HF Awards

RSGB Awards

One of the most exciting and broad ranging facets of amateur radio operating is awards chasing. Awards are an area that everyone, whether individuals or clubs, participates in by simply being on air. Every contact you make has the potential to check off another requirement on many of the awards that are available through the RSGB. It is a major motivating force of so many QSOs that occur on the bands day after day. You should also realise that one QSO may qualify for more than that one RSGB award you are working towards. There are many awards worldwide that can enhance your enjoyment of your on-air interests.

Aside from the fun of operating itself, awards chasing is also a good way to get maximum performance from your station, become familiar with propagation, and learn about the geography, history or culture of places near and far.

Whilst the traditional method of proof of contact by QSL cards is still acceptable, LoTW confirmation is also available using the unique Record ID number. We now offer a QSL Card Checking service on request. If you require a card checker, the RSGB Awards Manager will contact you to arrange this.

When your claim has been approved by the RSGB Awards Manager you will be given a code number to be entered on the Order page to obtain your certificate.

Price and ordering information are available at the RSGB Shop at: *http://www.rsgbshop.org/acatalog/Awards.html*

NOTE 1: Information on how to apply for awards is on the RSGB Website *https://www.rsgb.org/awards/*

NOTE 2: Spreadsheets are very useful to track progress towards an award, and these have been made available for all of the RSGB's HF awards (as Check Sheets). Note that these have been updated in 2021 so it is wise to download the latest version from the awards website when you start working towards an award, and check for updates.

RSGB HF Awards

The HF Awards Programme aims to make the attainment of awards enjoyable, progressive, challenging and achievable in a reasonable time frame.

RSGB Youth Award

This RSGB Award is designed to encourage activity and exploration of the breadth of amateur radio, broadening the interest and experience of amateurs who are aged up to 26.

There are four classes of award:

Bronze: Use two modes with a minimum of 50 contacts

Silver: Use three modes with a minimum of 100 contacts

Gold: Use four modes with a minimum of 150 contacts

Platinum: Use five modes with a minimum of 300 contacts

Any Band or mode may be used

https://rsgb.org/main/operating/amateur-radio-awards/youth-award/

Foundation Award

This award is available to holders of UK Foundation Licences and is designed to encourage activity across four designated bands during their first year of being licensed. Contacts may be made using CW, SSB or FM modes, as appropriate on 40m, 20m, 17m and 2m with a minimum of 10 contacts and maximum of 25 contacts per band. Multiple contacts with the same amateur do not count.There are three levels of award :

Bronze: 40 contacts
Silver: 70 contacts
Gold: 100 contacts

Intermediate 100 Award

This award is available to holders of UK Intermediate Licences and is designed to encourage activity by making 100 contacts across four designated HF bands during their first year of being licensed.

Contacts may be made using CW, SSB or Data modes, as appropriate, on 80m, 40m, 30m, 17m with a maximum of 25 contacts per band. Multiple contacts with the same amateur do not count.

IARU Region 1 Award

The purpose of this award is to generate activity on the HF bands whilst recognising the work of the International Amateur Radio Union Region 1. The award may be claimed by any licensed radio amateur who can produce evidence of having contacted amateur radio stations in the required number of countries whose national societies are members of the Region 1 Division of the International Amateur Radio Union (IARU).

There are 3 classes for contacts as follows:

Class 1 102 member societies
Class 2 60 member countries
Class 3 40 member countries

Confirmation of two-way contacts is required either by QSL cards or LoTW for contacts on 160, 80, 40, 30, 20, 17, 15, 12 and 10m.

IARU Region 1 28MHz Award

This special single band award is available under the same rules and 3 class structure as the IARU Region 1 Award. All contacts must be within the 10m band and have been made since 1st July 1983.

IARU Region 2 and IARU Region 3 also administer a regional award. Information can be found at their respective websites.

The Commonwealth Century Award (CCA)

This award may be claimed by any licensed radio amateur producing evidence of having contacted amateur radio stations in 100 (Century), 70 (Silver) and 40 (Bronze) Commonwealth call areas on the list current at the time of application.

The award can be claimed for contacts on the traditional 5 bands (80m, 40m, 20m, 15m and 10m) or over all 9 HF bands, (160m to 10m). *http://rsgb.org/main/files/2020/01/Commonwealth-Century-Award-2020.pdf*

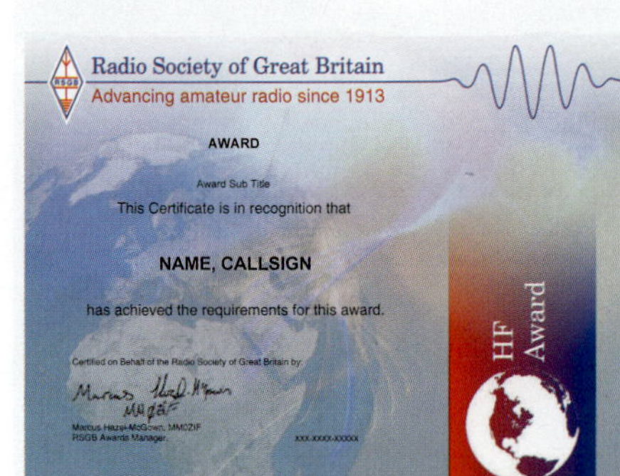

New style Award for all Class of Certificate

Commonwealth Century Club Awards in 2022

VA3OKG	Century Bronze Award (5-band)
VA7QI	Century Award (9-band)
VA7QI	Century Silver Award (5-band)
DL9MWG	Century Silver Award
G6BFP	Century Bronze Award (5-band)
G6BFP	Century Bronze Award (9-band)
M0IFT	Century Bronze Award (5-band)
M0IFT	Century Award (9-band)
G8GNI	Century Silver Award (5-band)

Worked ITU Zones Award

The purpose of this award is to generate activity on the HF bands whilst recognising the work of the International Telecommunications Union.

The award may be claimed by any licensed radio amateur who can produce evidence of having contacted amateur radio stations in the required number of ITU zones.

There is a choice of 4 categories:

- Classic Worked ITU Zones
- 5 Band Worked ITU Zones
- WARC Band Award
- Top Band Award

Confirmation of two-way contacts is required either by QSL cards or LoTW for contacts on 160, 80, 40, 30, 20, 17, 15, 12 and 10m.

Worked All Continents Awards

This award, issued by IARU headquarters, may be claimed by any licensed radio amateur in the UK, Channel Isles or Isle of Man who can produce evidence of having contacted amateur radio stations in each of the 6 continents : North America, South America, Europe, Africa, Asia and Oceania. All contacts must be made from the same country or separate territory within the same continent. Various endorsements including "all 1.8MHz" are available. Applicants should send QSL cards or scanned images to the RSGB HF Awards Manager who will certify the claim to the IARU headquarters society (ARRL) for issue of the award.

Email: awards@rsgb.org.uk Web site: *http://www.rsgb.org/awards*

5-Band CCC Awards (not WARC bands)

Award Number	500 Areas Supreme	450 Areas Class 1	400 (min 50/band) Class 2	300 (min 40) Class 3	200 (min 30) Class 4
1	G3TJW	G3UML	K2SHZ	G3UML	G8JR
2	G3GIQ	UP1BZZ	G4GIR	K2SHZ	K2SHZ
3	OK3EY	G4GIR	-	G3SWH	G3SJX
4	VK9NS	I8SAT	G3TBK	G3KWK	YU2OB
5	A92BE	G3TBK	CT3FT	G0EHO	G3IFB
6	G3UML	G3SWH	G3SWH	VK2NU	G3EZZ
7	G4BWP	G3IAF	G3KWK	G0EHO	G3SWH
8	LY2ZZ	G3VKW	G3AKU	G3LPS	G3KWK
9	G3IFB	-	G3VKW	G3VKW	G0EHO
10	G3TBK	-	-	G3AKU	G3LPS
11	G3SJX	-	-	-	G3WW
12	G3KWK	-	-	-	-
13	G3LZQ	-	-	-	-
14	G3TXF	-	-	-	-
15	G3SWH	-	-	-	-
16	G3IAF	-	-	-	-
17	G3VKW	-	-	-	-

WARC CCC Endorsements

Award Number	300 Call Areas Supreme	275 Class 1	250 (min 50) Class 2	200 (min 40) Class 3	150 (min 30) Class 4
1	G3TBK	G3TBK	G3SWH	G3SWH	G3EZZ
2	G4BWP	G3SWH	G3TBK	G3TBK	G3SWH
3	G3SWH	G4BWP	G4BWP	G3SJX	G3TBK
4	G3TXF	G3IAF	G3AKU	G4BWP	G4BWP
5	G3SJX	-	G3LZQ	G3LZQ	G0EHO
6	G3IAF	-	-	G3VKW	G3KWK
7	-	-	-	-	VK2NU
8	-	-	-	-	G3VKW

160m CCC Endorsements

Award Number	70 Areas Supreme	60 Areas Class 1	50 Areas Class 2	40 Areas Class 3	30 Areas Class 4
1	G4BWP	G4BWP	G4BWP	G4BWP	SM5HV/HK7
2	G3LZQ	G3LZQ	G3LZQ	G3KWK	G3SJX
3	-	G3SJX	G3SJX	G3LZQ	G3TBK
4	-	-	-	G3SJX	G4BWP
5	-	-	-	-	G3KWK
6	-	-	-	-	G3TXF

IARU Region 1 Class 1 Award 2021

HB9JOE	IARU Region 1 Class 1 Award

IARU Region 1 Award (Classes 2/3) 2021-22

Callsign	Award
MQ0KUH	IARU Region 1 Class 3
M0DMC	IARU Region 1 Class 3
JJ1VCJ	IARU Region 1 Class 3
G6BFP	IARU Region 1 Class 3
YC4RWH	IARU Region 1 Class 3
G0WJK	IARU Region 1 Class 3
WM3PEN	IARU Region 1 Class 3
JA0PZJ	IARU Region 1 Class 2
G8GFF	IARU Region 1 Class 2

IARU Region 1 28MHz Award 2020-21

Call	Class
9Z4Y	3
G6MXL	3
G6NHU	2
G8PMA	3

Worked All ITU Zones Award 2021-22

UA6CEY	5 Band Worked ITU Zones Class 2
UA6CEY	Classic Worked ITU Zones
RC2A	5 Band Worked ITU Zones Class 2
RC2A	Classic Worked ITU Zones

Foundation, Intermediate & Youth Awards

2E0HVK	Intermediate 100 Award
MM7DMW	Foundation Award Gold
G4SEB	Youth Award

Members of IARU Region 1 (from February 2020)

Prefix	Country	Nat. Soc.
3A	Monaco	ARM
3B8	Mauritius	MARS
3DA	Swaziland / Eswatini	RSE
3V	Tunisia	CAST
3X	Republic of Guinea	ARGUI
4J/K	Azerbaijan	FRS
4L	Georgia	NARG
4O	Montenegro	MARP
4X	Israel	IARC
5B	Cyprus inc UK Sov Bases	CARS
5H	Tanzania	TARC
5N	Nigeria	NARS
5X	Uganda	UARS
5Z	Kenya	ARSK
6W	Senegal	ARAS
7P	Lesotho	LARS
7X	Algeria	ARA
9A	Croatia	HRS
9G	Ghana	GARS
9H	Malta	MARL
9J	Zambia	RSZ
9K	Kuwait	KARS
9L	Sierra Leone	SLARS
9Q5	Congo DR	ARAC
9U	Burundi	ABART
9X	Rwanda	RARU
A2	Botswana	BARS
A4	Oman	ROARS
A6	United Arab Emirates	EARS
A7	Qatar	QARS
A9	Bahrain	
C3	Andorra	URA
C5	Gambia	RSTG
C9	Mozambique	LREM
CN	Morocco	ARRAM
CT	Portugal inc.CU CT3	REP
DL	Germany	DARC
E7	Bosnia & Herzegovina	ARABIH
EA	Spain	URE
EI	Ireland	IRTS
EK	Armenia	FRRA
EL	Liberia	LRAA
ER	Moldova	ARDM
ES	Estonia	EARU
ET	Ethiopia	EARS
EU	Belarus	BFRR
EX	Kyrgyz Republic	
EZ	Turkmenistan (ex UH)	LRT
EZ	Turkmenistan (ex UH)	LRT
F	France inc. TK	REF
G etc	United Kingdom	RSGB
HA	Hungary	MRASZ
HB	Switzerland	USKA
HB0	Liechtenstein	AFVL
HZ	Saudi Arabia	SARS
I	Italy	ARI
J2	Djibouti	ARAD
JT	Mongolia	MRSF
JY	Jordan	RJARS
LA	Norway	NRRL
LX	Luxembourg	RL
LY	Lithuania	LRMD
LZ	Bulgaria	BFRA
OD	Lebanon	RAL
OE	Austria	OEVSV
OH	Finland	SARL
OK	Czech Republic	CRC
OM	Slovakia	SARA
ON	Belgium	UBA
OY	Faeroe Islands	FRA
OZ	Denmark	EDR
PA	Netherlands	VERON
R	Russian Federation	SRR
S5	Slovenia	ZRS
S7	Seychelles	SARA
SM	Sweden	SSA
SP	Poland	PK
SU	Egypt	EARA
SV	Greece	RAAG
T7	San Marino	ARRSM
TA	Turkey	TRAC
TF	Iceland	IRA
TJ	Cameroon	ARTJ
TN	Congo	URAC
TR	Gabon	AGRA
TU	Ivory Coast	ARAI
TZ	Mali	CRAM
UN	Kazakstan	KFRR
UR	Ukraine	UARL
V5	Namibia	NARL
XT	Burkina Faso	ARBF
YI	Iraq	IARS
YK	Syria	SSTARS
YL	Latvia	LRAL
YO	Romania	FRR
YU	Serbia	SRS
Z2	Zimbabwe	ZARS
Z3	North Macedonia	RSM
Z6	Kosovo	SHRAK
ZA	Albania	AARA
ZB	Gibraltar	GARS
ZS	South Africa	SARL

VHF/UHF Awards

RSGB VHF/UHF Awards Programme

The RSGB VHF/UHF Awards Programme recognises successful operating achievements that depend on the special propagation modes which can be experienced at these frequencies.

http://rsgb.org/main/operating/amateur-radio-awards/vhfuhf-awards/

50MHz Squares, Continents and Countries Awards

The purpose of these awards is to encourage activity on the "magic" band and to recognise personal operating achievements. Stations are eligible for awards as: Fixed stations, Portable stations, and Mobile stations, but these categories cannot be mixed.

Squares awards start at 25 squares confirmed, and increase with incremental levels of 50, 75, 100, 200, 300, 400, 500, 600.

At the discretion of the Awards Manager, interim and higher levels can be added (a handful of stations in the past have achieved upto 800 squares and received commensurate awards).

Countries awards start at 10 countries confirmed and increment in 10s to 160. A new Continents & Countries Award, launched in 2021, additionally proceeds in increments of combined continents and countries from a start of 2/30, followed by incremental levels of 2/40, 2/50, 3/45, 3/50, 4/55, 4/65, 5/60, 5/65, 5/75, which can be extended at the discretion of the Awards Manager.

4-2-70 Squares and Countries

These are the traditional '4-2-70 Squares Awards' covering 4m, 2m and 70cms and are intended to mark successful vhf/uhf achievement. Initially, one certificate will be issued.

Further certificates will be issued as additional squares and countries are claimed. The title of each award gives the number of locator squares and countries needed to qualify for the award.

For example, to obtain the 144MHz 40/10 award you must have confirmed contact with 40 locator squares including 10 countries on 144MHz. Eligible countries are those shown in the countries list printed in the 'Year Book'. Stations are eligible for awards as:

Fixed stations, Portable stations, any location or Mobile stations, any location. Categories cannot be mixed.

Band	Award
4m band	
70MHz	40/10
70MHz	50/15
70MHz	60/20
70MHz	70/25
70MHz	80/30
70MHz	90/35
2m band	
144MHz	40/10
144MHz	100/20
144MHz	200/30
144MHz	300/40
144MHz	400/50
144MHz	475/50
70cm band	
432MHz	40/10
432MHz	60/15
432MHz	100/15
432MHz	140/20
432MHz	160/20
432MHz	180/20

Radio Surfer Award

In 2021 a new award was introduced by the Youth team, the Radio Surfer Award, which is issued to enthusiasts, regardless of whether they are licensed, who complete a selection of tasks from a list published on the Awards website pages.

Each task is worth a certain number of points, and the award is achieved when the applicant achieves a number of task-points equal to or greater than their age in years.

The tasks are to be supervised by a suitably qualified person (safety-vetted if the applicant is a child), and the achievement certified by them in the application.

VHF/UHF Activity Award

In 2022 a new award was created, the VHF/UHF Activity Award, which recognises operating activity and achievement for radio to radio (simplex) contacts on the vhf and uhf bands, particularly targeted at newcomers to the hobby but open to all classes of licensee:

Bronze: 30 Simplex contacts on 2m plus 20 Simplex contacts on 70cms using FM or SSB or a combination of these modes

Silver: 50 Simplex contacts on 2m plus 40 Simplex contacts on 70cms using any combination of FM, SSB, CW

Gold: 75 Simplex contacts on 2m, 60 Simplex contacts on 70cms, and at least 50 Simplex contacts on either of 6m or 4m, using any combination of FM, SSB, CW

Platinum: 100 Simplex contacts on three of the following bands: 6m, 4m, 2m, 70cm using at least 2 of the following modes on each band: FM, SSB, CW

Diamond: 100 Simplex contacts on all of the following bands: 6m, 4m, 2m, 70cms, plus 25 simplex contacts on 23cms using at least 2 of the following modes: FM, SSB, CW

Since introducing the award, Andy Callaghan, M0XTY and Ian Evans, GI0AZB, have received Bronze and Silver levels of award; more participation is strongly encouraged, to bolster activity on these bands and because they are a good fiirst step into VHF award chasing.

A more detailed overview of all awards, including application forms and rules, are available on the

RSGB website at:

http://www.rsgb.org/awards

VHF/UHF Award Achievements Table

50MHz Squares / Countries in 2021/22:

Callsign	Award
M0XTY	V/UHF Activity Award (silver)
G4RHR	50MHz Continents & Countries (3/60)
2E0NEY	144MHz Squares and Countries (150/20)
GM0SCA	50MHz Squares (200)
M0DMC	50MHz Countries (20)
M0DMC	50MHz Squares (50)
G6MXL	50MHz Squares (150)
G8GNI	50MHz Countries (70)
G8GNI	50MHz Squares (300)
MM7SWM	V/UHF Activity Award (Bronze)
G3SAO	50MHz Countries (50)
G3SAO	50MHz Continents & Countries (2/50)
DF1ZN	50MHz Continents & Countries
G4NMD	V/UHF Activity Award (Bronze)
G0UUU/P	50MHz Countries (20)
G8VVY	50MHz Squares (225)

4-2-70 awards:

70MHz 45/10: G8CQR
70MHz 55/10: MM0INH
144MHz 40/10: G4PDF

Historic Achievements:

70MHz 100/35: G8PNN
144MHz 125/20: G8PNN

Microwave Awards

As of 2010, RSGB microwave awards were transferred to the UK Microwave Group (UKuG).

The following awards are intended to mark achievement on the microwave bands. Successful applicants will initially receive a certificate and one sticker. Further stickers will be issued as later claims are received.

Microwave Squares Awards

Existing RSGB records have been transferred so that claims for extra stickers to add to existing RSGB Squares Awards can be accepted. The existing sticker design has been retained.
Awards are available in 5 square increments on the following bands:

1.3GHz: 5 to 150 Locator squares
2.3GHz: 5 to 75 Locator squares
3.4GHz: 5 to 75 Locator squares
5.7GHz: 5 to 75 Locator squares
10GHz: 5 to 75 Locator squares
24GHz: 5 to 25 Locator squares

- Initial claims will require submission of QSL cards for all contacts claimed.
- Subsequent increments will need the additional cards and the countersigned check list from the previous claim.
- QSL cards must include the IARU QTH locator of the station worked.
- eQSL confirmations can be accepted from stations that have completed eQSL verification
- Contacts must have been made after 31 December 1978.
- All contacts must be two-way on the band in question, cross-band contacts are not eligible.
- Awards are available for fixed or portable/mobile stations but these categories cannot be mixed.
- Claims for portable operation must be for contacts made from one site, defined as anywhere within a 5km radius of the point operated from.
- Cards should be listed and sorted in IARU QTH Locator alphanumeric order.
- A self-addressed envelope must be included for return of the QSL cards, with sufficient postage value in UK stamps or IRCs.
- Certificates are free for members of the UK Microwave Group.
- Claims from non-members must include payment of £3 or $5 when an initial certificate is requested.
- Subsequent stickers will be enclosed with returned QSL cards and checklists.

Claims should be made using the application form and squares checklist available on the UK Microwave Group website: *www.microwavers.org* and sent to the address given below.

UKuG/SOTA Microwave Distance Awards

These new awards are jointly issued by UKuG and SOTA, and recognise achievement in working distance. They replace the previous RSGB/UKuG distance awards, and are available in 50km endorsements for all bands 1.3GHz and above.
For further information, please see: *http://www.sota-shop.co.uk/microwave.html*

Recording and recognising 'Firsts'

The UK Microwave Group has an award that recognises the achievements of British stations in making first contacts with other countries, and maintains a list of firsts as a historic record of the development of microwave operating techniques in Great Britain.
Should you not wish to claim a certificate we would still be very grateful for details of your First to ensure the records are accurate.
Certificates will be awarded to stations that can demonstrate that they completed the first contact between one of the countries in Great Britain and Northern Ireland (Prefixes M, MM, MI, MW, MU, MJ, MD - and the G and 2E equivalents) and any other country on a particular band above 1GHz. Contacts within Great Britain are valid for this award (eg Scotland to Guernsey).
Only one 'First' will be recognised per band, irrespective of the propagation mode (eg tropo, rainscatter or EME).
Claims should be made using the application form available on the UK Microwave Group website.
A QSL will be required for award of a certificate (but are not required to provisionally enter a First in the database). QSL cards should not be sent until requested. eQSL confirmation will be accepted from stations that have completed eQSL verification.
All claims should be sent to John Quarmby, G3XDY. Details below.

Please note that data provided for the above awards will be held by the UK Microwave Group for the purposes of providing a published list on the UKuG website and in the group's newsletter *Scatterpoint*, of achievements by stations in Great Britain and Northern Ireland. Data will not be used for any other purposes.

You do not have to be a member of the UK Microwave Group to lodge a claim.
The decision of the UK Microwave Group committee is final on the validity of claims.
Prior to making a claim, please check the Firsts database on the UK Microwave Group website at: *www.microwavers.org.uk* for existing contacts.

Trophies & Awards

In addition to the certificates and distance/squares awards, and the trophies awarded for microwave contests by the RSGB Contest Committee, the UK Microwave Group also awards a number of cups and trophies, both for contest operating successes and for contribution to the facet of the hobby in both technical and supportive aspects. Most of these awards are presented annually at the UK Microwave Group Round Table event held at Martlesham, Suffolk, each April.
Photos of all the trophies can be found online at: *www.microwavers.org/trophies.htm*

Operating Trophies

5.7GHz - G3KEU

In memory of Tim Leighfield, (SK 2002), this cup is awarded annually to the winner of the UKuG 5.7GHz cumulative contest. See: *www.g3pho.free-online.co.uk/microwaves/keu.html*

10GHz - G3JMB

Awarded in memoriam to Jack Brooker MBE, (SK 2004) is awarded annually to the winner of the Restricted section of the 10GHz Cumulative contest. *www.r-type.org/g3jmb/*
This trophy is an attractive glass plaque, but due to the inadvisability of engraving it, each winner is awarded a small plaque which they retain in perpetuity.

10GHz - G3RPE

Awarded to the winner overall of the 10GHz Cumulative contest, this cup commemorates Dain S Evans, BSc, PhD, FIM and RSGB President in 1978. Dain was one of the prime movers of the expansion in 10GHz construction and activity in the UK and was the first, with G3ZGO, to break the 150km distance barrier, setting the bar for those who followed.

24GHz - G0RRJ

This cup, in memory of Dave Cox (SK 2009), is awarded to the winner of the 24GHz Cumulative contest.

24GHz and 47GHz - Trophy

These two cups are awarded to the winners of the 24 and 47GHz Trophy events (as distinct from the 24GHz Cumulative sessions).

Contribution Awards

There are four awards in this area:

G3BNL

In memory of Les Sharrock, this ornate decanter trophy is awarded by the UK Microwave Group committee for innovation or technical development of microwave equipment or techniques.

Send microwave award applications to the awards manager: awards@rsgb.org.uk

Fraser-Shepherd Award

This award, for research into microwave applications for radio communication, is presented by the RSGB on the nomination of the UK Microwave Group committee. Fraser Shepherd, GM3EGW was an early pioneer of UHF and microwave operation and techniques, and was a participant in the first GM 432MHz moonbounce activity.

G3EEZ

Awarded for contributions to Microwave Communications, this cup is in memory of Alan Wakeman.

G3VVB

Presented for the best piece of home constructed microwave equipment exhibited at a Round Table, this magnificent trophy is in memory of Cyril James, a keen home constructor who produced microwave artefacts of very high quality workmanship. The entries and judging of this event are a regular feature of Microwave Round Tables, with the trophy awarded at the RSGB Convention in October.

G3JVL

This award, introduced in 2020, is presented to the best newcomer each year who the committee considers has made the greatest contribution; be that in developing hardware/ software or operating on the microwave bands.

Send microwave award applications to: John Quarmby, G3XDY, 12 Chestnut Close, Rushmere St Andrew, Ipswich IP5 1ED email: g3xdy@btinternet.com

Microwaves

This an aspect of amateur radio where experimentation and construction are still alive and well. And it's never been easier to get started.

What are microwaves?

In amateur radio terms, frequencies above 1000MHz (1GHz).

Are microwaves only for line-of-sight communication?

Absolutely not! Try telling those operators who routinely work hundreds of kilometres with relatively modest antennas and power that they can't do it. They can, and so could you!

Microwaves is just another aspect of amateur radio which can be learned. You can take it in so many directions ...

All amateur radio equipment costs money, but a simple system of antenna and transverter could cost less than a new headset! Less if you are prepared to build your own from a kit.

So what can we do with microwaves?

Lots! That question is probably better answered by asking what we can't do! The microwave bands support a wide variety of propagation modes, some of which will be completely unfamiliar to people who haven't ventured above VHF.

On most bands troposcatter – scattering from the upper part of the lowest layer of the atmosphere, the troposphere, is important. This will reliably support contacts over several hundred kilometres with good modern equipment. With modest kit, perhaps a couple of watts to a discreet yagi antenna a couple of metres long on 1.3 or 2.3GHz, from a site with open horizons – troposcatter path losses are very dependent on the vertical angle of your horizon - will produce regular contacts up to around 200km without much effort. A similar power level to a repurposed satellite TV dish will provide contacts on 10GHz over the same range, or further.

The troposphere is involved in another way. At times the masses of air from different sources, some from the cold dry Arctic, some, perhaps, from the warm, wet Atlantic, move over each other. In regions sometimes thousands of kilometres across, these boundaries are able to 'duct' microwave signals over long distances. Ducting produces openings on the microwave bands just like those on VHF/UHF, and signals are often propagated over paths in excess of 1000km.

Other ways of propagating microwave signals include scatter from rainfall (and other forms of precipitation) and scattering from aircraft. Although rainscatter can occasionally be heard on 1.3 and 2.3GHz, it really starts to be useful at 3.4GHz and above. 'Aircraft scatter' can allow all year propagation up to about 700km.

On the higher frequency bands, at 24GHz and above absorption by water in the atmosphere becomes a challenge but it has not deterred UK microwavers from having a go. The UK DX record on the 24GHz band is almost 400km. Contacts have been made by UK radioamateurs on all of our allocated bands up to, and including 241GHz.

Microwave moonbounce (EME) currently produces worldwide DX from the UK on all of our bands up to 24GHz. Some operators apply for HF-style awards, such as WAC. Some UK stations are advancing towards DXCC on 1.3GHz. On the microwave bands, EME is practical with quite modest antennas. Significant numbers of contacts can be made on 23cms with single yagis or 2m diameter dishes. On 13cms and up even smaller antennas can be used. 10GHz has become a favourite for EME. There 50W and a 2.4m dish will allow a station to hear its own Moon echoes on SSB. 24GHz is starting to make headway as a moonbounce band. Bands up to 76GHz have been used experimentally.

Amateur microwavers can be found throughout the world, especially in the Americas, Western Europe, Australasia and Japan. Regular EME contacts are made between all of these areas, and there even DXpeditions bringing activity to other areas.

Amateur television has found its way into earth orbit with DATV transmissions being made from the ISS at 2.4GHz, while the geostationary Es'hailsat2 satellite is due to launch in 2017 with digital and linear transponders using 2.4GHz uplinks, and 10GHz downlinks.

A few UK radioamateurs have heard deep space satellites such as Voyager at the very edge of our Solar System. Others have discovered the joy of amateur Radio Astronomy.

Microwaving can be challenging, it can be fun, it can be educational, it can be social. Finally it's what you make it. It's a rapidly developing area of our hobby.

How did amateur microwaves come about?

As early as 1894 to 1896, Jagadish Chandra Bose, an Indian physicist, experimented on 60GHz over a one mile distance using primitive semiconductors! He was a truly remarkable man who is now seen as the 'Father of Microwaves'. It may seem strange, but microwaves pre-date HF radio!

In 1946 the first amateur microwave contacts were recorded in the USA, when W1LZV/2 worked W2JN over two miles on 10GHz and W1NVL/2 worked W9SAD/2 on 21.9GHz (800 feet!). The World 10GHz record was then set at 7.6 miles by W4HPJ/4 and W6IFE/3.

The first UK amateur microwave contact was made in 1949 by G3BAK and G3LZ over 27 miles, at the time a new world 10GHz record.

For the next twenty years operation on the bands above 3.4GHz centred mainly on the use of tubes, such as klystrons as local oscillators and transmitters with waveguide mounted diode mixers in wideband FM systems. Gunn diodes replaced the klystrons during the '70s, and the '80s saw a sharp spike in activity on 10GHz. Activity on the lower frequency microwave bands also grew, with many people using NBFM, or even AM, generated by solid-state varactor multipliers from 70cm.

A modern 10 and 24GHz home system

At the beginning of the '80s, more advanced microwavers started to move to an approach more like that on the lower frequency bands, with CW and SSB becoming more common. The problem with low-power, wideband systems is that it is very difficult, particularly above 3.4GHz, to cover many non-line-of-sight paths.

More recently, the field has continued to develop, with narrowband standards in use on all bands. All DX operation is now on narrowband, often using weak-signal modulation schemes, such as K1JT's WSJT. Transmitters capable of several watts output are commonplace on 10GHz.

In addition commercial kits and ready-made modules have become available.

In the last decade, much focus has been on increasing frequency stability by locking oscillators to GPS stabilised standards. SDR technology operating directly at microwave frequencies is becoming more common.

The current State of Play: UK Microwave Terrestrial Records

Band (GHz)	Distance (km)
1.3	2617
2.3	1389
3.4	1137
5.7	1244
10	1429
24	408
47	203
76	129
122	35.9
134	35.6
145	1.29
241GHz	9.3
Light (red)	129.1

RSGB Spectrum Forum website: www.rsgb.org/committees/spectrumforum/

Although much has been made above about the DX potential of the microwave bands, and many enthusiasts concentrate on this aspect of the hobby, there are other important user groups.

Conventional voice repeaters/beacons operate in the 23cm band. These have recently been joined by a number of digital voice repeaters using the 'Dstar' modulation scheme.

Amateur Television is very well established above 1GHz, with a network of repeaters. Many now employ Digital Video technology, although many analogue FM TV repeaters remain in service.

Digital Networking is an interest of an increasing number amateur radio operators, and several long haul networks exist using IEEE802 standards at both 2.4 and 5.6GHz. In many countries on the European mainland extensive amateur owned and operated high-speed data networks, linked to the Internet, operate on 1.3 and 2.3GHz.

More about the Microwave Bands

Many bands are shared with other users and are increasingly under threat from commercial interests. For example parts of the 2.3 and 3.4GHz bands have been subject to Public Sector Spectrum Release (PSSR) which has seen loss of access to some frequencies and other changes to the main UK amateur licence schedule. Users of parts of 2.3GHz (shown by ***) are now also specifically required to register with Ofcom. There are both limits to times of activity and to location, which make continued sharing with the main User possible. NoVs for new additional bands such as 2300-2302MHz and >275GHz are also available on request. (N)

Many of the bands shown in the following list contain amateur beacons and repeaters which are useful for frequency calibration and as indicators of propagation conditions. A more detailed spectrum allocation can be found in the Band plans Section of this Yearbook and on the RSGB website.

23cm	1240.0-1325.0MHz
13cm	2300.0-2302.0MHz(N)
13cm	2310.0-2350.0MHz***
13cm	2390.0-2450.0MHz
9cm	3400.0-3410.0MHz
6cm	5650.0-5680.0MHz
6cm	5755.0-5765.0MHz
6cm	5820.0-5850.0MHz
3cm	10.000-10.125GHz
3cm	10.225-10.500GHz
12mm	24.000-24.250GHz
6mm	47.00-47.2GHz
4mm	75.50-81.0GHz
2.5mm	122.25-123,
2.2mm	134-141
1.25mm	241-250GHz
also	>250GHz (N)

Ways into microwaves

The 1.3GHz (23cm) and the 10GHz (3cm) bands are the easiest ones on which to make a start. There is a lot of ready-made equipment, including antennas for these bands. The on-line auction and flea-market sites can be a great source of microwave surplus.

Simple 10GHz wideband gear

The days of simple Gunn oscillator transceivers are now past, if only because the intruder alarm modules which many people used in the past are no longer easily and cheaply available. The Gunn oscillator diodes used are obsolescent as low cost items. Modern intruder alarm modules are much less easy to modify. However, if old wideband gear is available, perhaps lurking in someone's attic, it can still be used, if you can find someone to cooperate with. In this case, it would be sensible to check that the gear is operating in a part of the band to which we still have access.

Satellite TV LNBs make very acceptable receive converters when combined with a cheap DATV "Dongle" or the more expensive Funcube Dongle. Many are good enough for narrowband use, but there is no obvious easily available companion transmitter. If you have a beacon or ATV repeater locally, a LNB, particularly if mounted in a dish, would make a great tool for initially exploring microwave propagation at 10GHz. You could also take part in UkuG contests, as 'one-way' contacts are allowed for 50% of the points of a two-way!

Another way of getting going, assuming that your licence allows it, is to consider using 5.6GHz and to exploit (cheap!) video-sender equipment. These often consist of a wideband, synthesised FM transmitter and a companion receiver, which can be tuned to a channel within the amateur band. Combine these with a suitable antenna, also available cheaply, and you have a simple transceiver which is capable of surprisingly good performance over line-of-sight paths.

There are groups around the UK using video sender technology, but if you can't locate one (try asking on 'ukmicrowaves') or they are too far away and you know someone close-by who is interested in experimenting, go ahead and try!

As with all worthwhile activities, there is a learning curve. You might not work very far initially, so it's worth finding someone with similar interests to experiment with. It's still an excellent, fun and cheap way to 'wet your feet' in microwaves. Don't forget you will also need something for talkback liaison, be it a mobile phone, computer with internet access, or even 2m/70cm!

Other bands for a beginner

1296MHz is the other band which can be recommended for people taking their first steps into the microwaving world. The technology is not too dissimilar to that used on VHF/UHF, and equipment needed to make a start is easily available. Recently, the Company owned by Bulgarian microwaver, Hristyiyan, LZ5HP, has been producing a simple 2W transverter – not a kit - of good performance, at a very attractive price *www.sg-lab.com/TR1300/tr1300.html*. This is small enough, and light enough to be mounted in a waterproof box close to the feed point of a 1296MHz antenna. From many locations this will allow regular QSOs up to around 250km – much further in a tropo opening.

M0EYT portable.

Microwave Antennas

On 1296MHz and 2.3GHz, yagi or sometimes dish antennas are the norm. High gain dishes and horns are the most commonly used antennas at 3.4GHz and above.

Microwave antennas give much higher gains than HF ones. 100 microwatts of 10GHz CW into a 60cm dish will be heard over any line-of-sight path in the UK by a receiver using a similar antenna.

On 10GHz, a 60cm ex-Satellite TV dish with a suitable feed horn will have a gain of around 35dB. This will give a potential range greater than 144MHz, for similar transmit power.

The extra gain comes from concentrating the transmitted energy into a narrow beam; an antenna with 35dB gain will have a -3dB beamwidth of about ±1.5°, so you have to point the dish more accurately than you would, say, a 2m yagi. The ability to go to a site, and to work out where to accurately point a dish is another skill which has to be learned as a microwaver!

A more modern satellite TV LNB.

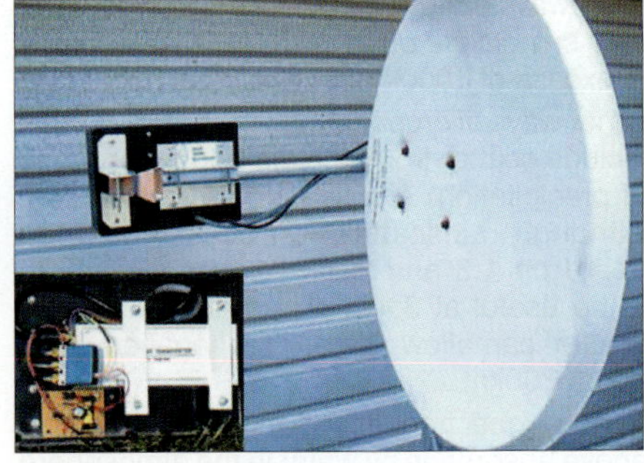
The 'Quickstart' microwave system

A small dish microwave EME system for 13cms

Equipment

Ready built transverters are available from a number of sources, such as Kuhne Electronik (DB6NT) in Germany, DEMI in the USA, and LZ5BP in Bulgaria. Kuhne and DEMI, along with Mini-Kits in Australia also market kits.

There is a market for second-user microwave equipment. A request on one of the online groups, such as:

https://groups.io/g/UKMicrowaves could well produce results.

Making QSOs, Is it like the lower frequency bands?

No! Perhaps on 1.3GHz, it can be, but at higher frequencies the beamwidth of antennas is so small that the chances of hearing another station randomly are not large. So, the use of talkback has developed. UK stations use 144.390 MHz for on-air talkback but also use online tools. SSB. In most of the rest of Europe, 432MHz SSB was used. More recently, the ON4KST chat server *www.on4kst.com* has become very popular along with the use of Zello SHF Chat. While there are those who regret the way in which the Internet has replaced a 'pure' amateur radio talkback channel, it has its advantages.

The first is that as 2m propagation is often not the same as that at higher frequencies: the microwave bands are often 'open' when 2m is not, for well understood reasons. More prosaically, the use of a public talkback channel, rather than a semi-private VHF link, discourages people from trying to complete QSOs or even make QSOs over talkback.

On the microwave (and VHF/UHF) bands the completion of a QSO is defined rather differently to the usual practice on the lower frequencies. On the band in which you are operating you must exchange callsigns, some piece of unknown information, like a report or a QTH locator. Both stations must acknowledge the receipt of of the callsigns and unknown information from the other station. If you don't keep to that procedure, you haven't made a contact which would be seen as valid.

There are three sets of parameters you need to get right, if you are to work another station; the beam heading, the frequency, and who is to transmit and when.

The beam heading is easily obtained from a number of sources. Most conveniently this can be found by a couple of clicks in the ON4KST interface. But there are a number of apps available for all of the major operating systems which will convert an input in the shape of two QTH locators into a bearing and distance. The majority of antenna rotators – particularly those aimed at HF operators - do not have particularly accurate azimuth indications and some modification will be needed. Fortunately, some European manufacturers of rotators understand the need for more accurate calibration. For portable operation, a simple compass rose which can be calibrated by reference to a local landmark can produce an elegant solution.

Frequency is very important. Older transverters used either a free-running crystal oscillator or a crystal oscillator partially stabilised with a simple temperature control oven. That resulted in the frequent need to tune over perhaps 20kHz to find another station, even if both stations were fairly confident of their frequency. Many weak and transient signals were missed because of this. Modern practice is to phase-lock the transverter local oscillator to a GPS stabilised frequency reference, and even at 10GHz frequency accuracies of a very few Hz can be obtained.

If this all sounds very different to what you're used to, you can get help?'

The UK Microwave Group (UKuG) has a policy of supporting newcomers. Its runs Technical Support and free chip component services, covering a large part of the UK. This is a voluntary service to members – not a right of membership – and depends on the goodwill of the volunteers freely to provide (within sensible limits) their expertise, and often very extensive test equipment to support other members. Many of those providing this service are involved professionally with microwaves and have excellent laboratory facilities. In the commercial world the time they give would cost lots of money!

To find out more about UK Microwave Group you can find out more information by looking on the National Affiliated Clubs & Societies pages in this Yearbook or by searching their website:

www.microwavers.org

'Tell me more about UKuG'

The UK Microwave Group (GX3EEZ) was founded as a representative body for Amateur Radio Microwave activity in the late 1990s. It is a UK-wide national body affiliated to the RSGB, and works with that Society on issues related to the amateur radio allocations above 1GHz. It is a membership organisation, run by a Committee elected annually at the Group's AGM.

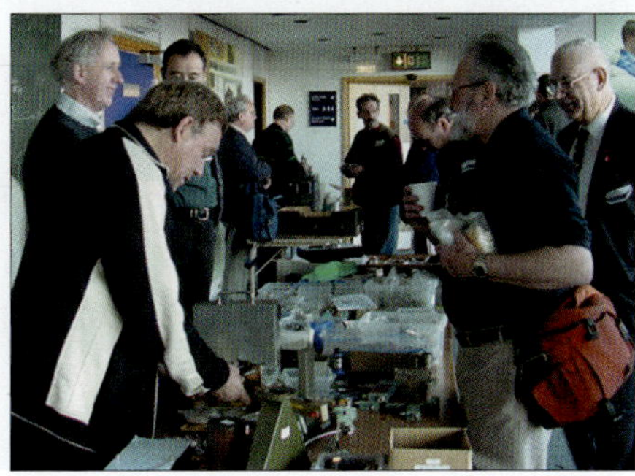

The Martlesham Round Table Fleamarket.

New-technology 10GHz PCB modules from GW4DGU

Apart from the important work with RSGB, and through that with Ofcom, the Group offers a number of services.

'Scatterpoint' is a monthly newsletter, usually of about 30pages, distributed electronically to UkuG members. It contains notices, operating and technical articles.

'Microwave Roundtables' are run at several locations around the UK each year. Here members can meet and socialise, exchange ideas, listen to lectures on various aspects of microwaving, and exchange and trade equipment and components.

'The Chip Bank' is a free-to-members benefit providing small quantities of components from an increasingly comprehensive stock donated by well-wishers in the industry. Even postage is free!

'UKuG Project Support'. Do you need some help with technical issues, or even finance relating to the development of a bona-fide microwave project, such as a beacon? UkuG has a considerable collective pool of knowledge and experience available. Financial support is considered in advance on a project-by-project basis on a 'cash by results' basis. However, the Group finances are not large enough to be able to provide ongoing support eg. for the running costs of a beacon.

'UkuG Technical Support' is a volunteer service provided by experienced microwavers who give their time and facilities to assist others.

It's a bargain at just £6.00 per year in the UK but membership is free for those under 21. If you want to join, please go to: *www.microwavers.org* and click on 'Membership'.

Have fun on microwaves!

Propagation

An explanation of the solar and geophysical information published by the RSGB.

Each week the GB2RS news bulletin includes a brief solar and propagation report and forecast prepared by members of the Society's Propagation Studies Committee. As carried on most GB2RS broadcasts and as posted on the Internet, these reports look back on the week up to the Thursday before transmission, and ahead to the week from the Sunday of transmission.

The usual format for the bulletins is a review of solar activity, geomagnetic activity and ionospheric data during the week. This is followed by the forecast.

Spacecraft data is used to compile the weekly propagation report, including imagery from the STEREO (Solar TErrestrial RElations Observatory) twin orbiters, which were launched in 2006. Unfortunately one of them has since developed a fault and imagery is currently only available from the STEREO Ahead spacecraft. The Solar Dynamics Observatory (SDO), which was launched in February 2010, also helps us understand the Sun's influence on the Earth and near-Earth space.

Although a lot can happen in a few days (new regions appear, old ones decay), from now on more accurate forecasting will be possible.

Solar Activity

The general level of solar activity for a 24-hour period from midnight to midnight is described as:

Very low
Either no solar flares or only A- or B-class flares.

Low
C-class flares, which usually have little or no impact on propagation.

Moderate
Between one and four isolated M-class flares.

High
Five or more M-class flares or isolated (1-4) M5 or greater flares, including X-class solar flares.

Very high
Several flares of M5 or greater magnitude, including X-class flares.

A flare is a sudden eruption of energy on the solar disc emitting radiation and particles, which can last anything from a few minutes to some hours. Flares are classified as 'A', 'B', 'C', 'M' or 'X' according to their X-ray energy level. This is measured by satellites in terms of megaelectron volts (MeV).

There are four energy thresholds: 2, 10, 50 and 100MeV, and flares are classified by numbers, ie M3, X4 etc. Generally the broadcasts mention only M- or X-class events, as they are the most likely to have an impact on propagation.

However, flares are also classified by their optical importance, which gives a measure of their size and brilliance. Size is indicated by a number between 1 and 4, while brilliance is either 'F' = faint, 'N' = normal, 'B' = bright or 'S' meaning sub (so 'SF' indicates sub-faint). The energy and optical indicators combine to give the complete flare data, eg 'M3/2B'.

Major Flares

Flares above M9/3B and all X-class flares can be very disruptive to the ionosphere. They can lead to severely-degraded HF propagation and any associated Earth-facing coronal mass ejections (CMEs) can cause auroral events, usually after a lapse of between 30 and 50 hours, particularly if they are on the Sun's limbs or have just passed the central meridian. Flare effects may be reported as:

Sudden Ionospheric Disturbance (SID) or Short-Wave Fadeout (SWF)
HF propagation blacked out or degraded for between a few minutes and many hours, with the lower bands affected first and most severely, the higher bands affected less and recovering first, but LF (<500kHz) and VLF DX signals may be enhanced.

(Sudden) Storm Commencement
Increase or decrease in the northward component of the geomagnetic field, marking the beginning of a geomagnetic storm. The onset may be very sudden (SSC) or more gradual (SC).

Noise Bursts or Noise Storms
Enhanced emissions from the Sun at radio wavelengths, associated with major flare events or complex solar active regions. They may last only a few minutes or for many hours.

Proton Events
These may be mentioned if they have an energy level exceeding 10MeV. Proton events cause high absorption in the D region of the ionosphere, particularly affecting transpolar propagation due to polar cap absorption, which can be degraded for days or even weeks following such an event.

Coronal Holes

Coronal holes are holes in the Sun's outer corona through which material is ejected by various means. There are always holes at the Sun's polar regions but tongues sometimes extend to the equatorial regions, or small holes can form. The passage of these can cause a magnetic disturbance. This is particularly so if the interplanetary magnetic field is southerly, as this couples to the Earth's northward field. What have become known as 'Scottish' type auroras can generally be attributed to the passage of a coronal hole. If known about, coronal holes are always referred to in the text due to their importance. Coronal holes become more geoeffective when at low solar latitudes and are more numerous around the time of solar maximum and during the first few declining years after solar maximum.

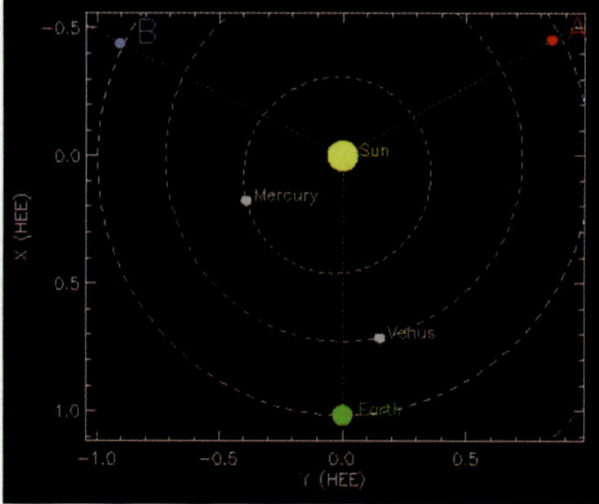

Fig 1: *Present locations of the STEREO spacecraft.*

Other Solar Events

Now and again reference is made to solar filaments. They appear as prominences on the solar limbs and as dark snaking strings of material against the limb as viewed in the light of hydrogen alpha. Occasionally the magnetic fields that hold filaments together break apart and fling the filamentary material into space. Filaments can last for several solar rotations before fading or erupting. These events can be sudden and are mostly unpredictable and can cause widespread auroras, ionospheric blackouts, and worldwide disruption to radio communication. Sometimes eruptive prominences are reported, but because they are located on the solar limbs they are not so geoeffective and therefore not so disruptive.

Satellite Data

Increasingly, new forms of data from satellites are supplementing or replacing traditional forms of ground-gathered data in explaining and predicting the 'propagation weather'. Some or all of the following may feature in a bulletin:

X-ray background flux
A more sensitive indicator than solar flux. It is reported on a rising scale of A1-9, B1-9 and C1-9. GB2RS reports the weekly average and any unusual levels.

>2MeV electron fluence
Referred to as 'high', 'normal' or 'low.' High levels adversely affect the HF bands.

Solar wind speed
The ACE satellite measures the speed of the flux of solar particles and magnetic fields moving outwards from the sun. Normal velocities are around 350-400km/s, though speeds exceeding 1000km/s have been recorded.

Particle densities
Under 10 per cm^3 are 'low'; 10-25 'moderate' and above 25 'high'.

Bz
The orientation of the interplanetary magnetic field measured in nanoteslas. A southerly (-ve) orientation, coupling with Earth's northern orientation, results in HF disturbances and auroras. A northerly (+ve) orientation has little or no effect.

K	0	1	2	3	4	5	6	7	8	9
A	0	3	7	15	27	48	80	140	240	400

Table 1: *Geomagnetic activity look-up table for a typical middle-latitude magnetic observatory.*

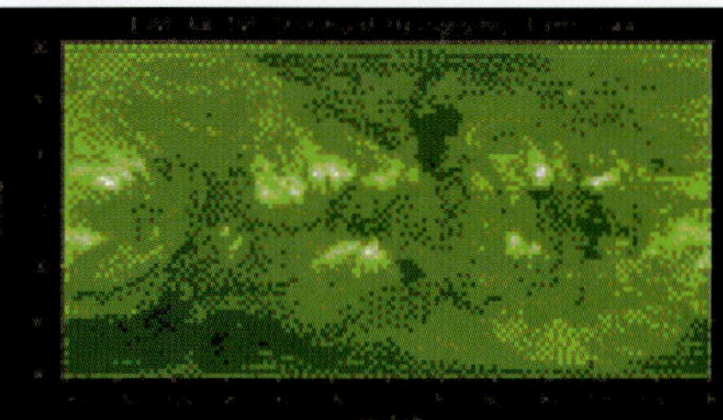

Fig 2: *A whole Sun image, created by combining images from Earth, STEREO-A and STEREO-B.*

Solar Flux

This is the 2800MHz radio noise output from the Sun at midday. This frequency is chosen because the radio Sun looks the same size as the visible Sun. The figure given is that obtained at Penticton (BC, Canada), which is the world standard. The level varies from (at the cycle's minimum) about 64 units to a maximum of around 300 units. The higher the level, the more intense is the Sun's ionising radiation and the higher the frequency that can be reflected from the ionosphere. Good HF band conditions require a high solar flux but the level of magnetic activity will also be a crucial factor. A 90-day average of flux levels is given, as this has been found to be best for home computer prediction programs.

Geomagnetic Activity

References to geomagnetic activity are made in terms of the worldwide or 'planetary' A index, expressed as 'Ap' units. During magnetic storms, the A-index may reach levels as high as 100. During severe storms, the A-index may exceed 200. Great 'rogue' storms may produce index values in excess of 300, although storms associated with indices this high are very rare indeed. Generally, an Ap index of 0-10 is considered 'quiet', 11-20 is 'unsettled', 21-50 as 'sub-storm', 51-80 'storm', 81 and above 'severe storm' or 'major storm'. High levels of geomagnetic activity - say roughly over 30 - are associated with poor HF conditions, especially on the higher bands. The greater the index over about 50, the greater the likelihood of aurora. And the higher the index, the further south auroral working may be possible. While auroral events, including auroral-E, are most likely to be found at 50, 70 and 144MHz, they can sometimes be found at 28MHz and, during major disturbances, contacts may be possible at 432MHz.

The Ap index is linear, unlike the alternative K index, used by WWV among others, which is quasi-logarithmic and open ended. Each observatory uses a look-up table created for that specific location, to convert an amplitude range into an associated K-index value. The table above shows the look-up table for a typical middle-latitude magnetic observatory.

Ionospheric Data

The critical frequency is the highest frequency reflected back from the ionosphere from a signal sent vertically upwards by an ionosonde. The maximum frequency that can be used for normal communication at equal latitudes is very roughly about three times the critical frequency (so a critical frequency of 7.0MHz indicates that it should be possible to work due east at single hop distance - 3000km - at 21.0MHz). For southerly working the multiplication factor would be higher but over northerly paths it would be lower. However, the actual level attainable also depends on the level of geomagnetic activity, the season and the time of day.

During the hours of darkness, the normally relatively smooth ionospheric layers can break up during magnetically disturbed periods. This is referred to as **Spread F**. The break-up can be vertical or horizontal, or both at once. The resulting holes give rise to deep fading. Northern circuits are more prone to this effect, which is more likely to occur during the early morning. References in the bulletins usually refer to the number of hours when Spread F has been present, or if it was very bad, on any particular day.

Blanketing E means that the E layer is so intensely ionised that the ionosonde cannot see through it. This effect is often associated with summertime sporadic-E or, for northern stations, with auroral conditions.

Absorption

Sometimes, for northern stations, complete absorption of the ionosonde signal occurs. This suggests that the D region is so heavily ionised that it is absorbing all but the strongest radio signals. These events can be associated with proton events, or high energy 'M' or 'X' flare events, or by electron precipitation from the Earth's radiation belts.

Seasonal Changes

The daily highs tend to be higher in winter and lower in summer. The darkness hour lows vary in the opposite way – high in summer and low in winter. The weekly average variations are balanced against these seasonal changes, and reference is made to any discrepancy when this applies. For the HF bands, the higher the daily 'highs', the better is the chance of DX on the higher HF bands.

The average times of the highest and lowest frequency recorded are given. The high times vary with the season, being around midday in the winter and early evening (about 2000UTC) in the summer. The low times do not vary much, being usually about 0400UTC.

Forecasts

Each week the bulletins include a forecast of expected events for the seven days following the Sunday of broadcast. This includes expected levels of solar flux, geomagnetic activity and the passage of any expected coronal holes. Maximum Usable Frequencies (MUFs) during daylight hours at equal latitudes are estimated for southern England. Scotland and Northern England will generally be down on these levels by around 3MHz at equal latitudes due to factors such as geomagnetic activity, which may affect northern areas more. In general, north-south paths will tend to be more readily workable than east-west, especially around the equinoxes. Bulletins usually include a forecast for one or more path.

The MUF given in these forecasts indicates the frequency up to which the path should be workable on 50% of days. On the better days, therefore, this value should be exceeded - at times by a considerable margin.

It also indicates a lower and more consistently reliable Optimum Working Frequency. This is the frequency on which it should be possible to operate on 90% of days. During months when sporadic-E propagation is prevalent - roughly May to August and around Christmas and the New Year - reminders are given of the likelihood of openings on 2, 4, 6 and 10m, and brief reports of major openings may be included.

Solar activity may be mentioned in the form "the quiet side of the Sun" or "the active side of the Sun"; which refers to the best chance of flare activity. For instance, the forecasts will attempt to predict the best chance of solar activity reaching moderate or high levels, therefore that being the active side of the Sun. This works in reverse for the quiet side.

The table published each month in *RadCom* shows path predictions from the UK to 31 locations around the world, 27 being short path and four being long path. All are F-layer predictions. The numbers indicate the expected reliability of the circuit; with a '1' representing between 1 and 19% of days, '2' between 20 and 29% of days, etc. The colours represent relative signal strength; a dot being no signal, black being a weak signal, blue being a fair signal, and red being a strong signal.

While every effort is made to ensure the table is as accurate as possible, it should always be remembered that propagation prediction is not an exact science. There will be times, for example during magnetically disturbed periods, when the table may be unduly optimistic. Alternatively, and especially during the peak of the solar cycle, conditions may for a time considerably exceed predicted values.

It should also be noted that these HF predictions are for F-layer propagation. They do not take into account either sporadic-E which, particularly during the summer, enlivens 28MHz, or auroral-E which occurs occasionally at HF (and VHF) during geomagnetic disturbances. They also do not include 'greyline' long-distance openings, which are a feature of the lower bands during the periods around dusk and dawn, as these are of too short a duration to appear in the two-hourly steps used in the tables. Neither can they take into the reckoning other short-lived phenomena like back-scatter and side-scatter. All these make the task of the forecaster more complex - but also make day-to-day working on the bands more varied and challenging. Fuller explanations of these and other solar and propagation phenomena, as well as near-real-time data, can be found on the Internet, notably at the Propagation Studies Committee (PSC) page.

RSGB PSC: www.rsgb.org/psc/

GB2RS propagation enquiries: Steve Nichols, G0KYA. Email: psc.chairman@rsgb.org.uk (or QTHR)

HF Propagation in 2024

2023 will be remembered for its unsettled geomagnetic conditions, a healthy return of sunspots and better upper HF band propagation.

The year was characterised by a large increase in the number of sunspots, but also an increase in solar flares and coronal mass ejections. These often pushed the geomagnetic Kp index high with a corresponding lowering of maximum useable frequencies at times. We even had a Kp index of eight earlier in the year which sparked visible aurora as far south as Cornwall and the Channel Islands.

So the goal was really to look for DX when the SFI was high and the Kp index was low - a message that I'll repeat for 2024!

We also had a prediction that solar maximum might arrive earlier than first predicted. Dr Scott McIntosh, a solar physicist, says the Sun may reach the peak of its current activity cycle in 2024, one year ahead of official predictions.

The original prediction McIntosh and his colleagues made was based on the expectation that the terminator event ending Cycle 24 would arrive in mid-2020, which would suggest a very strong Cycle 25.

Cycle 24, however, ended up lingering for a year and a half longer, with its magnetic field eventually completely disappearing in December 2021.

Only time will tell if that prediction is correct, but the message really is: "Work the DX while you can!"

Sometimes it looks like FT8 has taken over from SSB and CW with a general reduction in activity in the older modes and where the bands seem otherwise quiet at times.

Autumn tends to be the best time overall for HF propagation, with paths to North America and beyond opening up in late September.

The winter months, with lower levels of daylight, tend to be associated with good conditions on the lower bands, such as 160, 80, and 40 metres.

What we can also say with some confidence is that the equinox months are best for North-South paths.

But you can prepare your HF activity with a range of tools available on the web. It is worth reminding you what they are.

Check out www.rsgb.org/voacap and see www.rsgb.org/proppy, which offer two alternative methods of predicting propagation to various locations around the world. These now allow you to prepare your own RadCom-type predictions, based on your preferred mode, power and antenna choice.

If you want access to the full-blown VOACAP prediction tool then go to www.voacap.com. Or if you want an alternative, which uses the newer ITURHFPROP engine to produce results, go to Proppy at https://soundbytes.asia/proppy/.

Now let's look at the whole year, season-by-season, band-by-band. From a propagation perspective, conditions are dependent upon the angle the Sun makes with the ionosphere, so the periods around both equinoxes are likely to be similar. There is a gradual change from one season type to another, so the periods listed below should not be taken too literally.

Winter period

(Jan-Feb / Nov-Dec)

These periods are when the low bands (160m, 80m and 40m) come into their own. While solar maximum is not the best time for low band propagation there will still be plenty of DX to be worked. Generally, winter is a good time for East-West paths on HF too.

160m (1.8MHz or Top Band)

Solar absorption will prevent skip during daylight hours. You should be able to work other UK stations out to about 50-80 miles via ground wave. The band will start to come alive around sunset and openings up to around 1,300 miles should be possible, with frequent openings up 2,300 miles. DX openings to the east from the UK should be possible around midnight and to the west before sunrise for well-equipped stations.

80m (3.5MHz)

Expect a similar pattern to Top Band, with DX openings at night with peaks at midnight and around sunrise (greyline openings). Openings around the UK and out to around 500 miles should be possible during the day and between 750-2,300 miles at night. A low, horizontal antenna will be useful for relatively local, NVIS (Near Vertical Incidence Skywave) signals, but lower angle radiation, such as obtained with a vertical, will be required for DX.

40m (7MHz)

Another great DX band at this time of year. 40m should open for DX in an easterly direction during the late afternoon and towards the south at sunset. Paths during the afternoon may also include W6 (west coast USA) in mid-winter. Openings to the west, including long path to VK/ZL, should be possible after midnight and should peak just before sunrise. Relatively local contacts should be possible during the day as the daytime critical frequency is higher with the rising number of sunspots.

20m (14MHz)

This is likely to provide great DX openings during the hours of daylight. Peak conditions will be a couple of hours after sunrise for paths to the east and a couple of hours before sunset for paths to the west. Contacts up to 2,300 miles should be possible during daylight hours, and the band may remain open after sunset if the solar flux remains high. Occasional DX openings towards South America may be possible after nightfall. Watch out for higher levels of D-layer absorption around local noon, which may mean that you are better off heading for the higher bands at this time.

17m/15m (18MHz/21MHz)

Should provide good to excellent DX openings during daylight hours. The period from noon to late afternoon may be best, but both bands are likely to close soon after sunset and remain closed until some time after sunrise the following day, unless the solar flux is very high.

12m/10m (24MHz/28MHz)

If the solar flux index remains higher than 120 or so, which it should, good DX could be possible during daylight hours. A brief spell of Sporadic-E can sometimes occur in the New Year, resulting in very strong, but short-lived propagation on 10m out to around 1,300 miles.

Equinox periods

(Feb-May / Aug-Nov)

The equinox periods provide longer daytime periods than winter, but logically, shorter night-time periods too. These tend to be the best months for working North-South paths, such as UK to South Africa.

160m (1.8MHz or Top Band)

Look for short-skip and DX openings at night. Again, no daylight skip is possible due to absorption, but openings out to 1,300 miles and occasionally further afield can be expected at night, with conditions peaking around midnight and again at sunrise (greyline).

80m (3.5MHz)

Will generally follow the characteristics of Top Band at night, but will also provide good openings out to around 250 miles during the day. These will lengthen to around 500-2,300 miles at night, with fairly good DX opportunities at times.

40m (7MHz)

Should open to DX in an easterly direction at sunset. Openings to the west should be possible after midnight and should peak just before sunrise. Contacts should be possible during the day, and the higher critical frequencies caused by the rising solar flux index will mean that it is possible to work other UK stations via NVIS – Near Vertical Incidence Skywave. Also, look for good paths near local midnight and at sunrise (greyline).

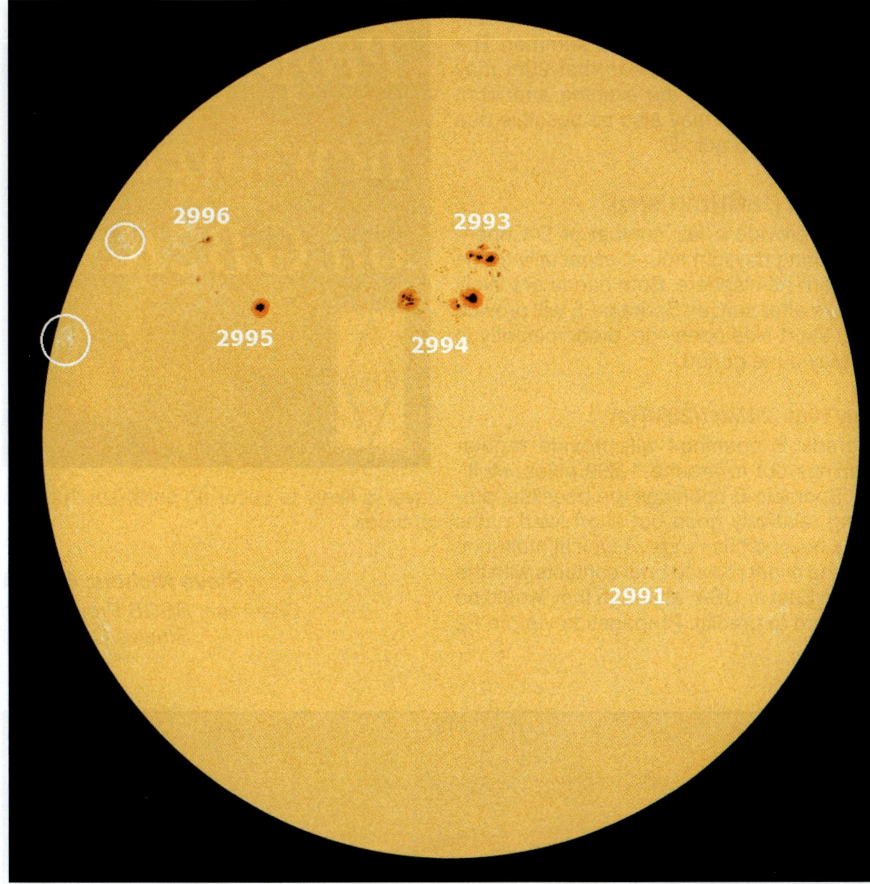

This image from the solar dynamics observatory (SDO) shows the sun in visible light and with a large number of sunspots

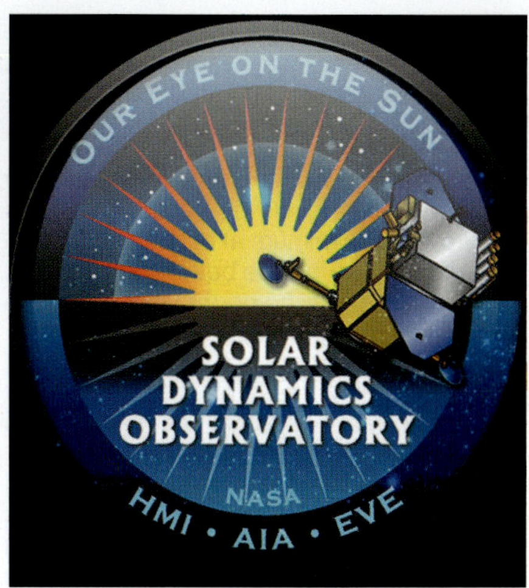

The Solar Dynamics Observatory mission logo.

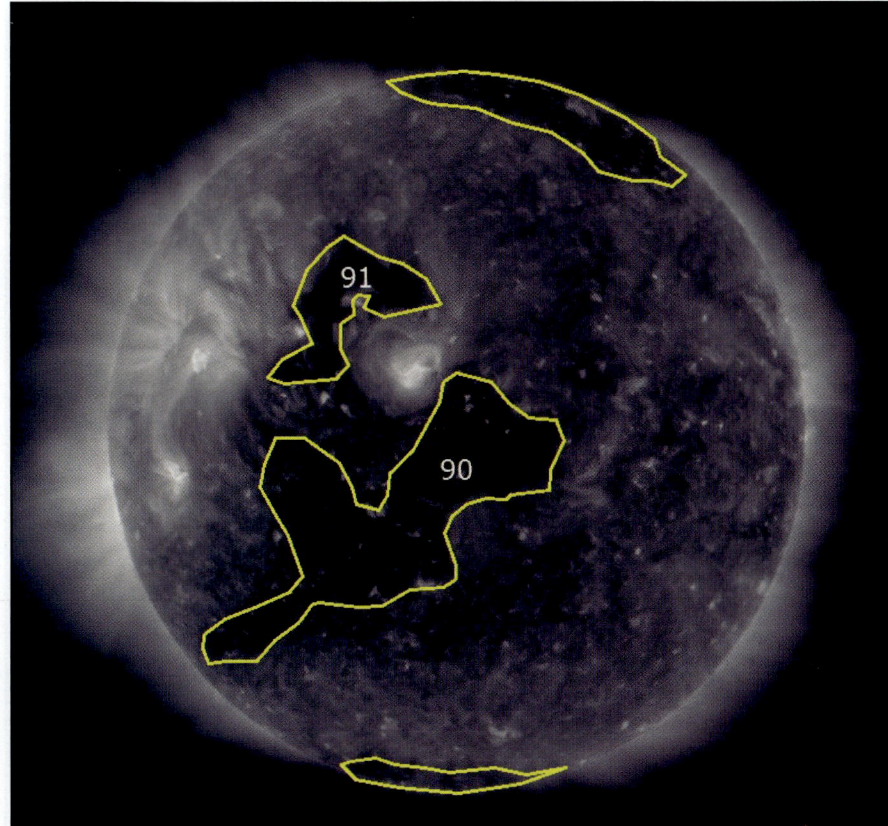

But in extreme ultra-violet this SDO image shows the extensive coronal holes, responsible for poor HF conditions.

20m (14MHz)

Likely to be the best DX band between sunrise and sunset. The band should remain open after dark if the SFI is high, perhaps giving openings to the southern hemisphere. Good openings will be possible during daylight hours out to around 2,300 miles.

17m/15m (18MHz/21MHz)

Should provide fairly good DX openings during daylight hours, especially to Africa and South America, with 17m being open more often than 15m. Once again, 15m could provide good openings if the SFI stays above about 100-120. Both bands are likely to close after sunset.

12m/10m (24MHz/28MHz)

If the solar flux heads above 120-130+ then openings will occur on both bands, although 24MHz will open first. If it keeps breaking the 130 mark then expect to see some excellent DX openings on 10m, especially in early spring/late autumn. Autumn should offer fine openings to the USA/Caribbean, even in the FM portion of the band around 29.6MHz.

Summer Solstice period

(May-Aug)

Daytime MUFs are likely to be lower than those of winter. The so-called 'Seasonal Anomaly' is thought to be due to a large summer electron loss rate caused by an increase in the molecular/atomic composition of the ionosphere and the reaction rates being temperature sensitive.

It is not all bad news though. Night-time MUFs may be higher in summer than those in winter. Note that DX on the low bands, if possible, is unlikely to occur until around midnight or the early hours, due to the late sunset.

160m (1.8MHz or Top Band)

High levels of static and solar absorption mean that the band will not really support sky-wave contacts during the day. During darkness, short-skip openings may occur, but DX may be a rarity. Occasional open-

ings can occur during the hours of darkness, especially around local midnight/early hours. Not the best season for Top Band.

80m (3.5MHz)

Will generally follow the characteristics of Top Band with high levels of static. Absorption will grow to a maximum at midday for inter-G contacts, so you may be better going to 40m. DX capabilities will be poor to fair during the hours of darkness, compared with the winter.

40m (7MHz)

Will suffer from high static, caused by high numbers of thunderstorms. Nevertheless, night-time openings should be reliable from sunset to sunrise. Local daytime openings will be possible on the whole with 40m being the band of choice for contacts around the UK. Night-time skip distances are likely to be between 300 and 2,300 miles.

20m (14MHz)

Still likely to be a good DX band around the clock, although the band will be noisier than the winter period and perhaps not as reliable for long-haul contacts in the summer. The higher MUFs at night mean that 20m may remain open during the evening and night to DX. Short skip may also be possible due to summer Sporadic-E.

17m/15m (18MHz/21MHz)

Should provide a fair number of DX openings during daylight hours, especially to the southern hemisphere. Both bands are likely to close after sunset. Sporadic-E will provide good short-skip openings, predominantly in the May-June period.

12m/10m (24MHz/28MHz)

Sporadic-E openings will provide regular openings out to around 1,300 miles. Multi-hop Sporadic-E openings are possible, providing relatively good but short-lived paths to DX beyond this range. A typical multi-hop opening might provide brief contacts with the Middle East or USA, although they would be very hard to predict. Propagation via the F2 layer is likely to occur on north-south paths at times.

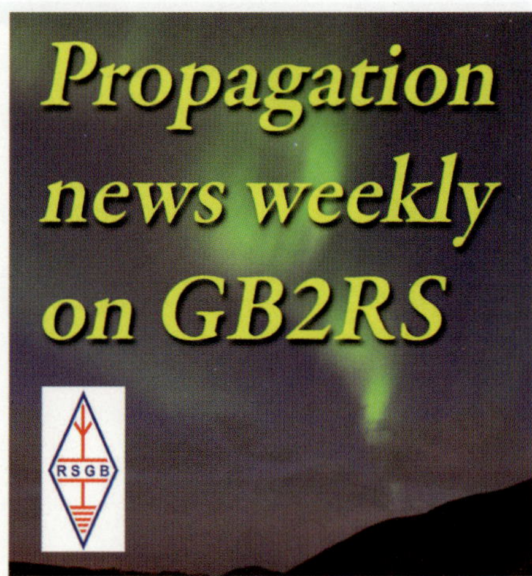

Steve Nichols, G0KYA
Chairman, RSGB Propagation Studies Committee

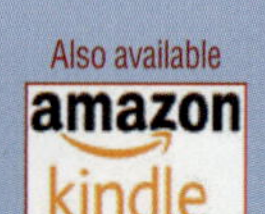

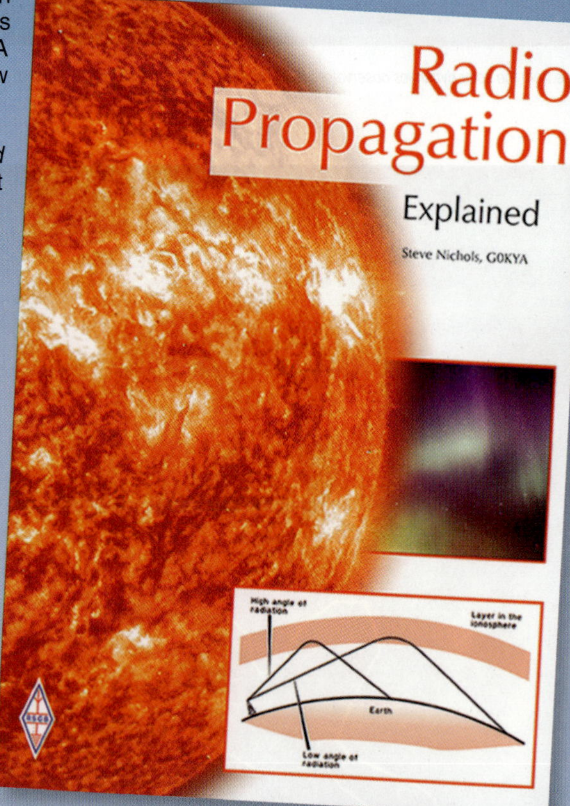

Contest Calendars

Contests are sporting events between amateur stations on specific bands and modes, conducted according to published rules. The activity appeals mainly to those with a competitive instinct, but construction and station optimisation are also important. For more information on the RSGB Contest Club *see* page 41.

2024 RSGB HF Contest Calendar

Date (2024)	Time (UTC)	Contest Name	Sections
Sat 6 Jan	1300-1700	**RSGB AFS 80m-40m Contests CW**	400W, 100W, 10W, Non-UK
Sun 14 Jan	1300-1700	**RSGB AFS 80m-40m Contests Datamodes**	400W, 100W, 10W, Non-UK
Sat 20 Jan	1300-1700	**RSGB AFS 80m-40m Contests Phone**	400W, 100W, 10W, Non-UK
Mon 5 Feb	2000-2130	80m CC SSB	100W Assisted, 10W Assisted
Sat 10 Feb	1900-2300	**1st 1.8MHz Contest**	UK Assisted, UK Unassisted, Non-UK Assisted, Non-UK Unassisted
Wed 14 Feb	2000-2130	80m CC DATA	100W Assisted, 10W Assisted, 100W Unassisted, 10W Unassisted
Thu 22 Feb	2000-2130	80m CC CW	100W Assisted, 10W Assisted, 100W Unassisted, 10W Unassisted
Mon 26 Feb	2000-2130	FT4 Series #1	100W UK, 100W Open UK, 10W UK, 100W Non-UK, 100W Open Non-UK, 10W Non-UK
Mon 4 Mar	2000-2130	80m CC DATA	100W Assisted, 10W Assisted, 100W Unassisted, 10W Unassisted
Sat 9-Sun 10 Mar	1000-1000	**Commonwealth Contest**	Open SO (12/24hr), Open SOA (12/24hr), Open Remote (12/24hr), Restr SO (12/24hr), Restr SOA (12/24hr), QRP SO, QRP SOA, Multi-Op, HQ
Wed 13 Mar	2000-2130	80m CC CW	100W Assisted, 10W Assisted, 100W Unassisted, 10W Unassisted
Mon 18 Mar	2000-2130	FT4 Series #2	100W UK, 100W Open UK, 10W UK, 100W Non-UK, 100W Open Non-UK, 10W Non-UK
Thu 28 Mar	2000-2130	80m CC SSB	100W Assisted, 10W Assisted
Sat 6 April	0800-2000	**FT4 International Activity Day**	100W UK, 100W Open UK, 10W UK, 100W Non-UK, 100W Open Non-UK, 10W Non-UK
Mon 8 Apr	1900-2030	80m CC CW	100W Assisted, 10W Assisted, 100W Unassisted, 10W Unassisted
Wed 18 Apr	1900-2030	80m CC SSB	100W Assisted, 10W Assisted
Thu 25 Apr	1900-2030	80m CC DATA	100W Assisted, 10W Assisted, 100W Unassisted, 10W Unassisted
Sat 27-Sun 28 Apr	1200-1200	UKEI DX CW	SO, SOA, 12hr, 24hr, High, Low, QRP
Mon 29 Apr	1900-2030	FT4 Series #3	100W UK, 100W Open UK, 10W UK, 100W Non-UK, 100W Open Non-UK, 10W Non-UK
Mon 13 May	1900-2030	80m CC SSB	100W Assisted, 10W Assisted
Mon 20 May	1900-2030	FT4 Series #4	100W UK, 100W Open UK, 10W UK, 100W Non-UK, 100W Open Non-UK, 10W Non-UK
Wed 22 May	1900-2030	80m CC DATA	100W Assisted, 10W Assisted, 100W Unassisted, 10W Unassisted
Thu 30 May	1900-2030	80m CC CW	100W Assisted, 10W Assisted, 100W Unassisted, 10W Unassisted
Sat 1-Sun 2 Jun	1500-1500	NFD	QRP Unassisted Portable, QRP Renewable Energy, Low Power Unassisted Portable, Low Power Assisted Portable, Fixed
Mon 3 Jun	1900-2030	80m CC DATA	100W Assisted, 10W Assisted, 100W Unassisted, 10W Unassisted
Wed 12 Jun	1900-2030	80m CC CW	100W Assisted, 10W Assisted, 100W Unassisted, 10W Unassisted
Mon 24 Jun	1900-2030	FT4 Series #5	100W UK, 100W Open UK, 10W UK, 100W Non-UK, 100W Open Non-UK, 10W Non-UK
Thu 27 Jun	1900-2030	80m CC SSB	100W Assisted, 10W Assisted
Mon 1 Jul	1900-2030	80m CC CW	100W Assisted, 10W Assisted, 100W Unassisted, 10W Unassisted
Wed 10 Jul	1900-2030	80m CC SSB	100W Assisted, 10W Assisted
Mon 15 Jul	1900-2030	FT4 Series #6	100W UK, 100W Open UK, 10W UK, 100W Non-UK, 100W Open Non-UK, 10W Non-UK
Sun 21 Jul	0900-1600	**International Low Power Contest**	UK Fixed 10W, UK Portable 10W, UK Fixed 5W, UK Portable 5W, Non-UK 10W, Non-UK 5W
Thu 25 Jul	1900-2030	80m CC DATA	100W Assisted, 10W Assisted, 100W Unassisted, 10W Unassisted
Sat 27-Sun 28 Jul	1200-1200	IOTA Contest	SO, SOA, Multi-1, Multi-2, SSB only, CW only, Mixed (SSB & CW), 12hr, 24hr, Island Fixed, Island DXpedition, World
Sat 31 Aug - Sun 1 Sep	1200-1200	UKEI DX SSB	SO, SOA, 12hr, 24hr, High, Low, QRP
Mon 2 Sep	1900-2030	Autumn Series SSB	100W Assisted, 10W Assisted
Sat 7-Sun 8 Sep	1300-1300	SSB Field Day	QRP Unassisted Portable, QRP Renewable Energy, Low Power Unassisted Portable, High Power Assisted Portable, Fixed
Wed 11 Sep	1900-2030	Autumn Series CW	100W Assisted, 10W Assisted, 100W Unassisted, 10W Unassisted, 100W QRS, 10W QRS
Mon 16 Sep	1900-2030	FT4 Series #7	100W UK, 100W Open UK, 10W UK, 100W Non-UK, 100W Open Non-UK, 10W Non-UK
Thu 26 Sep	1900-2030	Autumn Series DATA	100W Assisted, 10W Assisted, 100W Unassisted, 10W Unassisted
Mon 7 Oct	1900-2030	Autumn Series CW	100W Assisted, 10W Assisted, 100W Unassisted, 10W Unassisted, 100W QRS, 10W QRS
Wed 16 Oct	1900-2030	Autumn Series DATA	100W Assisted, 10W Assisted, 100W Unassisted, 10W Unassisted
Thu 24 Oct	1900-2030	Autumn Series SSB	100W Assisted, 10W Assisted
Mon 28 Oct	2000-2130	FT4 Series #8	100W UK, 100W Open UK, 10W UK, 100W Non-UK, 100W Open Non-UK, 10W Non-UK
Mon 4 Nov	2000-2130	Autumn Series DATA	100W Assisted, 10W Assisted, 100W Unassisted, 10W Unassisted
Sat 9 Nov	2000-2300	**Club Calls (1.8MHz AFS)**	32W SSB only, 32W Mixed (SSB & CW)
Wed 13 Nov	2000-2130	Autumn Series SSB	100W Assisted, 10W Assisted
Sat 16 Nov	1900-2300	**2nd 1.8MHz Contest**	UK Assisted, UK Unassisted, Non-UK Assisted, Non-UK Unassisted
Mon 18 Nov	2000-2130	FT4 Series #9	100W UK, 100W Open UK, 10W UK, 100W Non-UK, 100W Open Non-UK, 10W Non-UK
Thu 28 Nov	2000-2130	Autumn Series CW	100W Assisted, 10W Assisted, 100W Unassisted, 10W Unassisted, 100W QRS, 10W QRS 10W Non-UK

HF Key to Special Rules

HF Championship	Special Rules for HF Championship
S1	Affiliated Societies contest
S2	Commonwealth Contest
AFS Super League 2020-21	Special Rules for AFS Super League 2020-21
AFS Super League 2022-23	Special Rules for AFS Super League 2022-23

HF Key to Multipliers

M1	DXCCs worked on each band
M2	UK Districts per band and mode
M3	DXCC & UK District Bonus
M4	IOTA Points

2024 RSGB VHF Contest Calendar

Date (2024)	Time (UTC)	Contest Name	Sections
Every 1st Tuesday	1900-1955 (L)	144MHz FMAC	10W 50W
Every 1st Tuesday	2000-2230 (L)	144MHz UKAC	AO AR AL
Every 2nd Tuesday	1900-1955 (L)	432MHz FMAC	10W 50W
Every 2nd Tuesday	2000-2230 (L)	432MHz UKAC	AO AR AL
Every 3rd Tuesday	2000-2230 (L)	1.3GHz UKAC	AO AR AL
Every 4th Tuesday (Jan-Nov)	1930-2230 (L)	SHF UKAC	SAO SAR
Every 1st Wednesday	1700–2100 or 1900-2100	144MHz FT8AC	AO AR AL, (2 or 4 hour)
Every 2nd Wednesday	1700–2100 or 1900-2100	432MHz FT8AC	AO AR AL, (2 or 4 hour)
Every 2nd Thursday	2000-2230 (L)	50MHz UKAC	AO AR AL
Every 3rd Thursday	2000-2230 (L)	70MHz UKAC	AO AR AL
Sun 4 Feb	0900-1300	432MHz AFS	SF O
Sat 2 - Sun 3 Mar	1400-1400	**March 144 432MHz**	O 6O SF SO 6S
Sat 20 - Sun 21 Apr	1400-1400	MGM Contest	UKO EUO UKL EUL
Sat 4 May	1400-2200	432MHz Trophy Contest	O SF
Sun 5 May	0800-1400	10GHz Trophy Contest	O SF
Sat 4 - Sun 5 May	1400-1400	**May 432MHz-245GHz Contest**	O SF
Sun 12 May	0900-1200	70MHz Contest CW	AO AR AL
Sat 18 - Sun 19 May	1400-1400	144MHz May Contest	O SF SO 6S 6O
Sun 19 May	1100-1500	1st 144MHz Backpackers	5B 25H
Sun 9 Jun	0900-1300	2nd 144MHz Backpackers	5B 25H
Sat 15 - Sun 16 Jun	1400-1400	**50MHz Trophy Contest**	O SF SO 6O 6S Overseas
Sun 23 Jun	0900-1200	50MHz Contest CW	AO AR AL
Sat 6 - Sun 7 Jul	1400-1400	VHF NFD	O R L MM MS FSO FSR
Sun 7 Jul	1100-1500	3rd 144MHz Backpackers	5B 25H
Sat 20 Jul	1400-2000	**70MHz Trophy Contest**	O SO SF
Sat 3 Aug	1400-1800	4th 144MHz Backpackers	5B 25H
Sat 3 Aug	1400-1800	**144MHz Low Power Contest**	O SF SO
Sun 4 Aug	0800-1200	**432MHz Low Power Contest**	O SF SO
Sat 7 - Sun 8 Sep	1400-1400	144MHz Trophy Contest	O SF SO 6O 6S
Sun 8 Sep	1100-1500	5th 144MHz Backpackers	5B 25H
Sun 22 Sep	0900-1200	70MHz AFS Contest	SF O
Sat 5 Oct	1400-2200	**2.3GHz Trophy**	O SF
Sat 5 Oct	1400-2200	**1.2GHz Trophy**	O SF
Sat 5 - Sun 6 Oct	1400-1400	October 432MHz-245GHz Contest	O SF 6O(432) 6S(432)
Sun 20 Oct	0900-1300	50MHz AFS Contest	SF O
Sat 2 - Sun 3 Nov	1400-1400	144MHz CW Marconi	SF O 6S 6O
Sun 1 Dec	1000-1400	144MHz AFS	SF O
Tue 26 Dec	1400-1600	50MHz Christmas Contest	AR AO AL
Wed 27 Dec	1400-1600	70MHz Christmas Contest	AR AO AL
Thu 28 Dec	1400-1600	144MHz Christmas Contest	AR AO AL
Fri 29 Dec	1400-1600	432MHz Christmas Contest	AR AO AL

Events in **bold** qualify for the VHF Championship
L = Local

HF Key to Sections

A	10W Fixed
ALL	ALL
B	10W Portable
C	3W Fixed
D	3W Portable
HQ	HQ
LOW	100W maximum output power
LowPower	Low Power
Multi Operator	
Multi-Op	Multi Operator
Non-UK Open	Non-UK Open
Non-UK QRP	Non-UK QRP
Non-UK Restricted	Non-UK Restricted
Non-UK-Assisted	(b) Non-UK
Non-UK-Unassisted	(b) Non-UK
OPEN-SOA	Open - Single Operator Assisted
OPEN-SOU	Open - Single Operator Unassisted
Open	Open
Open (SSBFD)	Open
QRP	10W maximum output power
RESTRICTED-SOA	Restricted - Single Operator Assisted
RESTRICTED-SOU	Restricted - Single Operator Unassisted
Restricted	Restricted
Restricted (SSBFD)	Restricted
Single Operator Assisted	
Single Operator Unassisted	
UK Open	UK Open
UK QRP	UK QRP
UK Restricted	UK Restricted
UK-Assisted	(a) UK stations
UK-Unassisted	(a) UK stations

VHF/UHF Key to Sections

10H	10W Hill Toppers
3B	2m 3W Backpacker
6O	6 hours others
6S	6 hours Single Op Fixed
A	All
AL	UKAC Low Power
ALL	All
AO	UKAC Open
AR	UKAC Restricted
AX	UKAC DXers
FSO	Fixed Station Sweepers Open
FSR	Fixed Station Sweepers Restricted
L	Low Power Section of VHF NFD
LP	Single Operator, Fixed, 25W - Single Antenna
MM	Mix and Match Section of VHF NFD
MS	Single Transmitter Section of VHF NFD
O	Open
Open	Open Section of VHF NFD
Overseas	Overseas
R	Restricted Section of VHF NFD
SAO	SHF UKAC Open
SAR	SHF UKAC Restricted
SF	Single Op Fixed
SO	Single Op Others
Sweeper	VHF NFD Overall Sweeper Results

RSGB Contest Committee
website: *www.rsgbcc.org*

Beacons

Beacons provide an essential service as propagation indicators, but also for development and testing – the later aspect being of particular importance in the microwave bands. There is an extensive network of coordinated propagation beacons in both the HF and VHF/Microwave bands. They feature in a wide range of band plans where segments are designated for beacons and where transmissions by any other users which may interfere with them are discouraged.

In the HF bands, systems such as the International Beacon Project (IBP) transmit CW at regular intervals across a range of bands. For HF there is also a strong IARU recommendation that with few exceptions, there should be no beacons below 14MHz

In the VHF/Microwave bands, beacon technology has advanced considerably and often incorporates GPS-locked frequency sources and various digital Machine Generated Modes (MGM) to facilitate automated monitoring.

The trend towards modern frequency sources at VHF and upwards has also resulted in beacons increasingly using Frequency Shift Keying, rather than Amplitude keying, for the CW ident along with a period of plain carrier. The latter can be useful in SDR waterfall displays where spreading or Doppler from aircraft scatter or other effects can be more readily identified. Alternate time periods may also feature transmission modes for the beacon callsign using MGM such JT65, JT4 or PI-4. These can be automatically detected at low signal levels by monitoring stations and fed into the DX cluster where sites such as Beaconspot.uk can automatically aggregate them and indicate openings.

Reception reports for VHF/Microwave illustrate that these bands are far from just simple line-of-sight. The GB3VHF 144MHz beacon can often be heard across the UK.

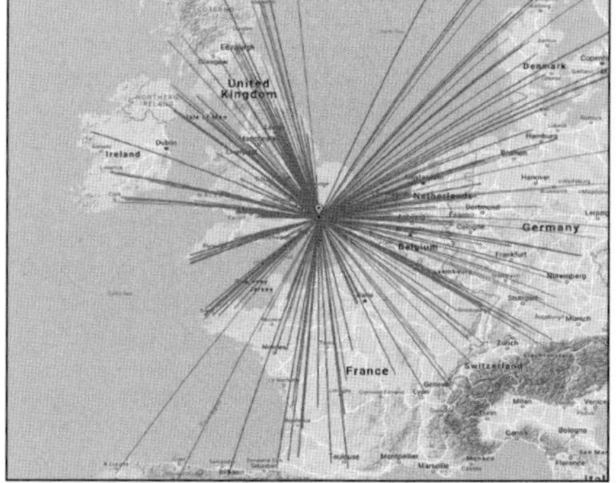

Figure 1 - GB3VHF on 144MHz Reception Reports aggregated and mapped by www.Beaconspot.uk

Timed / Synchronised beacons

HF- IBP

On HF there is a well-established network known as the International Beacon Programme (IBP). This was constructed by the NCDXF in cooperation with IARU

The network of 18 beacons (**Fig 2**) transmit on a repeating 3 minute cycle at staggered times on 14.100, 18.110, 21.150, 24.930, and 28.200MHz.

50MHz - SBP

A new network, the Synchronised Beacon Project (SBP) is gradually being deployed in the lower part of the 50MHz band, coordinated by the IARU Region-1 VHF Committee. This has been designed to accommodate a high number of sophisticated beacons that can accommodate the highly variable nature of propagation in the 50MHz band from flat tropo conditions to aurora and Sporadic-E.

The SBP network is configured as a time shared multiplex, with ten frequencies available each of which has five 60s timeslots. Some beacons such as GB3MCB and GB3NGI when not transmitting on their assigned SBP time slot will additionally transmit further up the band in the 50.4-50.5MHz range. The mode most commonly used for SBP automated reception is PI4 - PharusIgnis4 - a digital modulation specifically developed for beacon purposes.

Setting up a beacon

The UK licence has a specific conditions and Frequency Schedule-2 associated with beacons. It permits private low power ones mainly in microwave bands, as well as a provision for Direction Finding completions. However even this provision has a number of exclusion zones that are 50km radius around a number of national facilities.

Establishing a permanent beacon (which are assigned GB3xxx callsigns by Ofcom) can be a complex and lengthy undertaking. Such an exercise is more suited to a group rather than an individual as they require a team of closedown operators (similar to repeaters). Another aspect when planning a beacon is that its site should be relatively remote from any other radio system or user as their 24/7 transmissions can be quite strong and de-sense local receivers. Other factors to be considered are that the 100% transmission duty cycle can be a significant consumer of electricity and IARU frequency coordination is typically required.

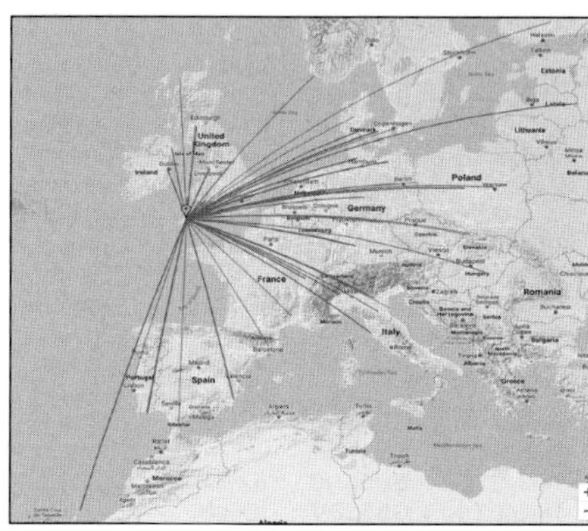

Figure 3 - GB3MCB on 50MHz is now one of a number of newer SBP systems and has reception distances enhanced by Sporadic-E and the use of PI-4 modulation

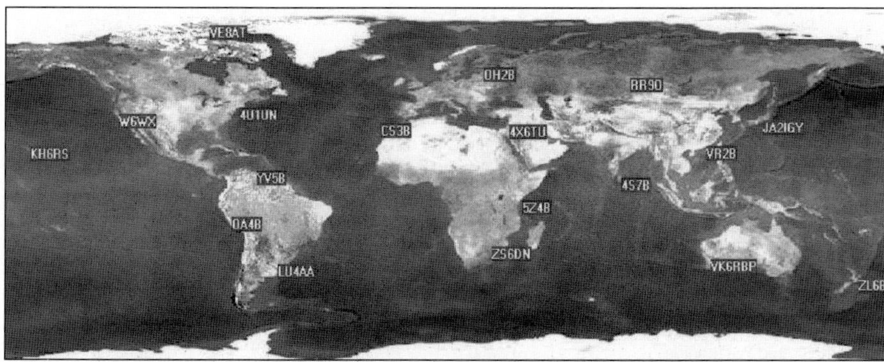

Figure 2 - The 18 sites used for the HF International Beacon Project (IBP)

Further Information:-

There is a UK technical discussion group for supporting Beacon Keepers and developers at: *https://groups.io/g/UKBeacons*

NoV applications and renewals for UK beacons are handled via the RSGB-ETCC website, *https://www.ukrepeater.net/*. However liaison with RSGB Spectrum Managers is recommended in the first instance.

HF

- IARU List: *https://iaruhfbeacons.wordpress.com/*
- NCDXF/IBP: *http://www.ncdxf.org/pages/beacons.html*

VHF/UHF/Microwave

- BeaconSpot: *https://www.beaconspot.uk/* for UK/European reception reports/maps
- IARU-R1 list: *https://iaru-r1-c5-beacons.org/*

List of UK Beacons (by Frequency)

Freq, MHz	Call	Locator	Nearest Town	Antenna	Height agl	Bearing	Polarisation	ERPw	Mode	Keeper
28.215	**GB3MCB**	IO70OJ	Roche	Vertical	10	OMNI	V	14	F1	G7KFQ
28.247	**GB3NWB**	JO02PP	Norwich	Dipole	3		H	3	F1	M0ZAH
28.260	**GB3MAT**	IO82VO	Bilbrook	1/2 Wave Vertical	3	OMNI	H	7	F1	M0IAW
28.268	**GB3AAX**	IO95FF	Ashington	Vertical	5	OMNI	H	1	F1	G4NAB
28.287	**GB3XMB**	IO83SV	Waddington	Dipole	10	OMNI	V	10	F1	G0SXC
50.000	**GB3BUX**	IO93BF	Buxton	Turnstile	7	OMNI	H	14	A1	G7EKY
50.002	**GB3RMK**	IO87AX46	Golspie	Dipole	7	N/S	H	14	F1	GM3WOJ
50.004	**GB3NGI**	IO65VB	Loghguile	Vertical	3	OMNI	V	15	F1	GI6ATZ
50.005	**GB3MCB**	IO70OJ	St Austell	Dipole	30	90/270	H	16	F1	G7KFQ
50.006	**GB3BAA**	IO91PS	Tring Herts	Dipole	35	OMNI	V	14	F1	M0ZPU
50.408	**GB3MBA**	IO93JC	Mansfield	Rh Circular Pol Dips / Similar	5	0	H	26	A1	G3PHO
50.416	**GB3BAA**	IO91PS	Tring Herts	Dipole	35	OMNI	V	14	F1	M0ZPU
50.443	**GB3MCB**	IO70OJ	St Austell	Dipole	30	90/270	H	16	F1	G7KFQ
50.460	**GB3RMK**	IO87AX46	Golspie	Dipole	7	N/S	H	14	F1	GM3WOJ
50.462	**GB3NGI**	IO65VB	Loghguile	Vertical	3	OMNI	V	15	F1	GI6ATZ
50.464	**GB3LER**	IP90JD	Lerwick	Dipole	8	0/180	H	16	F1	GM7AFE
70.000	**GB3BUX**	IO93BF	Buxton	Turnstile	7	OMNI	H	14	A1	G7EKY
70.007	**GB3WSX**	IO81VC	Mere Wiltshire	5 Element Yagi	10	70 DEGREES	H	17	F1	G3ZXX
70.016	**GB3BAA**	IO91PS	Tring Herts	Dipole	40	OMNI	H	14	F1	M0ZPU
70.020	**GB3ANG**	IO86MN	Dundee	3 El Yagi	15	170	H	20	F1	GM4ZUK
70.025	**GB3MCB**	IO70OJ	St Austell	2 El Yagi	27	45	H	16	F1	G7KFQ
70.027	**GB3CFG**	IO74BS	Carrickfergus	2 X 3 El Yagi	15	45/135	H	14	F1	GI0GDP
144.430	**GB3VHF**	JO01EH	Gravesend	Two 3-El Yagis	40	288 & 348	H	22	F1	G0FDZ
144.453	**GB3ANG**	IO86MN	Dundee	4 El Yagi	15	170	H	20	F1	GM4ZUK
144.462	**GB3SEV**	IO82UI	Stourport	Big Wheel	6		H	20	F1	M0VXX
144.469	**GB3MCB**	IO70OJ	St Austell	Two 3 El Yagis	25	45 AND 180	H	16	F1	G7KFQ
144.482	**GB3NGI**	IO65VB	Loghguile	Twin Yagis	10	45 & 135	H	21	A1	GI6ATZ
144.487	**GB3WGI**	IO64BL	Derrygonnelly Ni	2 X 6 Ele Yagi	35	280	H	30	F1	GI6ATZ
432.430	**GB3UHF**	JO01EH	Gravesend	Two 3-El Yagis	40	288 & 348	H	22	F1	G0FDZ
432.445	**GB3FNY**	IO93NN	Doncaster	Two Yagis	10	90 & 180	H	17	F1	G3AAF
432.453	**GB3ANG**	IO86MN	Dundee	9 El Yagi	15	170	H	17	F1	GM4ZUK
432.470	**GB3MCB**	IO70OJ	St Austell	4 El Yagi	20	45	H	13	F1	G7KFQ
432.482	**GB3NGI**	IO65VB	Loghguile	Yagi	10	60 & 135	H	24	F1	GI6ATZ
432.490	**GB3LEU**	IO92IQ	Markfield	Twin Yagis	10	135 & 300	H	17	F1	G3TQF
1296.830	**GB3MHZ**	JO02PB	Martlesham Heath	2 X 16 Slots	65	OMNI	H	27	F1	G7OCD
1296.850	**GB3FRS**	IO91LC	Four Marks	Yagi Uhf-12-1200	30	NNW	H	14	F1	G3TCU
1296.860	**GB3MCB**	IO70OJ	St Austell	15+15 Yagi	20	28	H	17	F1	G7KFQ
1296.870	**GB3USK**	IO81VC	Mere Wiltshire	Flat Plate Wimo Pa-23R-16	15	90	H	20	F1	G3TCT
1296.890	**GB3DUN**	IO91SV	Dunstable	Alford Slot	6	OMNI	H	16	F1	G3ZFP
1296.905	**GB3CFG**	IO74BS	Carrickfergus	Alford Slot	15	OMNI	H	15	F1	GI0GDP
1296.905	**GB3NGI**	IO65VB	Loghguile	Alford Slot	10	OMNI	H	15	F1	GI6ATZ
1296.965	**GB3ANG**	IO86MN	Dundee	15/15 Yagi	15	170	H	16	F1	GM4ZUK
1296.985	**GB3CSB**	IO75XX	Kilsyth	Wimo Pa23R	7	150	H	14	F1	GM6BIG
1296.990	**GB3EDN**	IO85JV	Edinburgh	Cloverleaf	10	OMNI	H	14	F1	GM8BJF
2320.830	**GB3MHZ**	JO02PB	Martlesham Heath	Slotted Waveguide	65	OMNI	H	14	F1	G7OCD
2320.870	**GB3ZME**	IO82RP	Telford	Slotted Waveguide	10	OMNI	H	19	F1	G3UKV
2320.890	**GB3ANT**	JO02PP	Norwich	Alford Slot	52	OMNI	H	7	F1	G8VLL
2320.905	**GB3SCS**	IO80UU	Blandford	Alford Slot	12	OMNI	H	8	F1	G4JNT
2320.920	**GB3FNM**	IO91OF	Farnham	Slotted Waveguide	20	NNW/SSE	H	16	F1	G4EPX
2320.925	**GB3BSS**	IO81SR	Bristol	Slotted Waveguide	65	OMNI	H	16	F1	G4CJZ
2320.955	**GB3LES**	IO92IQ	Markfield	Alford Slot	12	OMNI	H	15	F1	G3TQF
2320.985	**GB3CSB**	IO75XX	Kilsyth	Wimo Pa13R	7	150	H	14	F1	GM6BIG
3400.830	**GB3MHZ**	JO02PB	Martlesham Heath	Slotted Waveguide	65	OMNI	H	14	F1	G7OCD
3400.900	**GB3OHM**	IO92AJ	Northfield	Slotted Waveguide	12	OMNI	H	20	F1	G8IFT
3400.905	**GB3SCF**	IO80UU	Blandford	Alford Slot	12	OMNI	H	11	F1	G4JNT
3400.910	**GB3ZME**	IO82RP	Telford	Slotted Waveguide	10	OMNI	H	19	F1	G3UKV
3400.935	**GB3LPC**	IO91FR	Bampton	Slotted Waveguide	30	OMNI	H	10	F1	G3NNG
3400.955	**GB3LEF**	IO92IQ	Markfield	Alford Slot	12	OMNI	H	8	F1	G3TQF
3400.985	**GB3CSB**	IO75XX	Kilsyth	Slotted Waveguide	7	150	H	14	F1	GM6BIG
5760.830	**GB3MHZ**	JO02PB	Martlesham Heath	Slotted Waveguide	65	OMNI	H	14	F1	G7OCD
5760.900	**GB3OHM**	IO92AJ	Northfield	Alford Slot	12	OMNI	H	17	F1	G8IFT
5760.905	**GB3SCC**	IO80UU	Blandford	Slotted Waveguide	12	OMNI	H	8	F1	G4JNT
5760.910	**GB3ZME**	IO82RP	Telford	Slotted Waveguide	10	OMNI	H	19	F1	G3UKV
5760.920	**GB3FNM**	IO91OF	Farnham	Slotted Waveguide	20	NNW/SSE	H	14	F1	G4EPX
5760.925	**GB3KEU**	IO93NN	Doncaster	Slotted Waveguide	10	OMNI	H	14	F1	G3AAF
5760.985	**GB3CSB**	IO75XX	Kilsyth	Slotted Waveguide	7	150	H	14	F1	GM6BIG
10368.750	**GB3CAM**	IO92WI	Cambridge	Slotted Waveguide	30	OMNI	H	7	F1	G4HJW
10368.752	**GB3FNY**	IO93NN	Doncaster	Slotted Waveguide	10	OMNI	H	10	F1	G3AAF
10368.780	**GB3OSW**	IO82KV	Oswestry	Wg16 Slot	8	OMNI	H	10	F1	G3SMT
10368.790	**GB3UVR**	IO83KC	Mold	180 Deg Slot Antenna	11	NORTH EAST	H	13	F1	GW0MDQ
10368.830	**GB3MHZ**	JO02PB	Martlesham Heath	Slotted Waveguide	65	OMNI	H	8	F1	G7OCD
10368.850	**GB3SEE**	IO91VG	Reigate	Slotted Waveguide	20	OMNI	H	5	F1	G0OLX
10368.870	**GB3KBQ**	IO80LV97	Trull	Slotted Waveguide	185	OMNI	H	10	F1	G4UVZ
10368.895	**GB3NGI**	IO65VB	Loghguile	Slotted Waveguide	10	OMNI	H	13	F1	GI6ATZ
10368.905	**GB3SCX**	IO80UU	Blandford	Slotted Waveguide	12	OMNI	H	7	F1	G4JNT
10368.910	**GB3RPE**	IO71UU	Carmarthen	Slotted Waveguide	39	OMNI	H	8	F1	GW8KCY
10368.935	**GB3GCT**	IO91IJ	Newbury	Slotted W/G	10	OMNI	H	10	F1	G1EHF
10368.945	**GB3PKT**	JO01MT	St Osyth	Slotted Waveguide	12	OMNI	H	3	F1	G0MBA
10368.955	**GB3LEX**	IO92IQ	Markfield	Slotted Waveguide	12	OMNI	H	10	F1	G3TQF
10368.960	**GB3CMS**	JO01GQ	Chelmsford	Slotted Waveguide	15	OMNI	H	7	F1	G6JYB
10368.980	**GB3MCB**	IO70OJ	St Austell	Sectoral Horn	20	67	H	10	F1	G7KFQ

Freq, MHz	Call	Locator	Nearest Town	Antenna	Height agl	Bearing	Polarisation	ERP	Mode	Keeper
10368.985	**GB3CSB**	IO75XX	Kilsyth	Slotted Waveguide	7	150	H	14	F1	GM6BIG
24048.830	**GB3MHZ**	JO02PB	Martlesham Heath	Slotted Wg + Sect Horn	65	OMNI	H	14	F1	G7OCD
24048.850	**GB3FNY**	IO93NN	Doncaster	Slotted Waveguide	10	OMNI	H	13	F1	G3AAF
24048.870	**GB3CAM**	IO92WI	Cambridge	Slotted Waveguide	30	OMNI	H	14	F1	G4HJW
24048.890	**GB3DUN**	IO91SV	Dunstable	2 X Horn	60	135 & 315	H	0	F1	G3ZFP
24048.905	**GB3SCK**	IO80UU	Blandford	Slotted Waveguide	12	OMNI	H	11	F1	G4JNT
24048.910	**GB3ZME**	IO82RP	Telford	Slotted Waveguide	10	OMNI	H	19	F1	G3UKV
24048.920	**GB3FNM**	IO91OF	Farnham	Slotted Waveguide	20	NNW/SSE	H	13	F1	G4EPX
24048.940	**GB3AMU**	IO81JN	Cardiff	Sectoral Horn	10	135	H	-3	F1	GW3TKH
24048.945	**GB3PKT**	JO01MT	St Osyth	Slotted Waveguide	10	OMNI	H	14	F1	G0MBA
24048.960	**GB3SEE**	IO91VG	Reigate	Slotted Waveguide	20	OMNI	H	14	F1	G0OLX
24048.985	**GB3CSB**	IO75XX	Kilsyth	Horn	7	150	H	14	F1	GM6BIG
47088.905	**GB3SCQ**	IO80UU	Blandford	Horn	12	75DEG TRUE	H	6	F1	G8BKE
47088.985	**GB3CSB**	IO75XX	Kilsyth	Horn	7	150	H	14	F1	GM6BIG

List of UK Beacons (by Callsign)

Call	Freq	Locator	Nearest Town	Antenna	Height agl	Bearing	Polarisation	ERP	Mode	Keeper
GB3AAX	28.268	IO95FF	Ashington	Vertical	5	OMNI	H	1	F1	G4NAB
GB3AMU	24048.940	IO81JN	Cardiff	Sectoral Horn	10	135	H	-3	F1	GW3TKH
GB3ANG	70.020	IO86MN	Dundee	3 El Yagi	15	170	H	20	F1	GM4ZUK
GB3ANG	144.453	IO86MN	Dundee	4 El Yagi	15	170	H	20	F1	GM4ZUK
GB3ANG	432.453	IO86MN	Dundee	9 El Yagi	15	170	H	17	F1	GM4ZUK
GB3ANG	1296.965	IO86MN	Dundee	15/15 Yagi	15	170	H	16	F1	GM4ZUK
GB3ANT	2320.890	JO02PP	Norwich	Alford Slot	52	OMNI	H	7	F1	G8VLL
GB3BAA	50.006	IO91PS	Tring Herts	Dipole	35	OMNI	V	14	F1	M0ZPU
GB3BAA	50.416	IO91PS	Tring Herts	Dipole	35	OMNI	V	14	F1	M0ZPU
GB3BAA	70.016	IO91PS	Tring Herts	Dipole	40	OMNI	H	14	F1	M0ZPU
GB3BSS	2320.925	IO81SR	Bristol	Slotted Waveguide	65	OMNI	H	16	F1	G4CJZ
GB3BUX	50.000	IO93BF	Buxton	Turnstile	7	OMNI	H	14	A1	G7EKY
GB3BUX	70.000	IO93BF	Buxton	Turnstile	7	OMNI	H	14	A1	G7EKY
GB3CAM	10368.750	IO92WI	Cambridge	Slotted Waveguide	30	OMNI	H	7	F1	G4HJW
GB3CAM	24048.870	IO92WI	Cambridge	Slotted Waveguide	30	OMNI	H	14	F1	G4HJW
GB3CFG	70.027	IO74BS	Carrickfergus	2 X 3 El Yagi	15	45/135	H	14	F1	GI0GDP
GB3CFG	1296.905	IO74BS	Carrickfergus	Alford Slot	15	OMNI	H	15	F1	GI0GDP
GB3CMS	10368.960	JO01GQ	Chelmsford	Slotted Waveguide	15	OMNI	H	7	F1	G6JYB
GB3CSB	1296.985	IO75XX	Kilsyth	Wimo Pa23R	7	150	H	14	F1	GM6BIG
GB3CSB	2320.985	IO75XX	Kilsyth	Wimo Pa13R	7	150	H	14	F1	GM6BIG
GB3CSB	3400.985	IO75XX	Kilsyth	Slotted Waveguide	7	150	H	14	F1	GM6BIG
GB3CSB	5760.985	IO75XX	Kilsyth	Slotted Waveguide	7	150	H	14	F1	GM6BIG
GB3CSB	10368.985	IO75XX	Kilsyth	Slotted Waveguide	7	150	H	14	F1	GM6BIG
GB3CSB	24048.985	IO75XX	Kilsyth	Horn	7	150	H	14	F1	GM6BIG
GB3CSB	47088.985	IO75XX	Kilsyth	Horn	7	150	H	14	F1	GM6BIG
GB3DUN	1296.890	IO91SV	Dunstable	Alford Slot	6	OMNI	H	16	F1	G3ZFP
GB3DUN	24048.890	IO91SV	Dunstable	2 X Horn	60	135 & 315	H	0	F1	G3ZFP
GB3EDN	1296.990	IO85JV	Edinburgh	Cloverleaf	10	OMNI	H	14	F1	GM8BJF
GB3FNM	2320.920	IO91OF	Farnham	Slotted Waveguide	20	NNW/SSE	H	16	F1	G4EPX
GB3FNM	5760.920	IO91OF	Farnham	Slotted Waveguide	20	NNW/SSE	H	14	F1	G4EPX
GB3FNM	24048.920	IO91OF	Farnham	Slotted Waveguide	20	NNW/SSE	H	13	F1	G4EPX
GB3FNY	432.445	IO93NN	Doncaster	Two Yagis	10	90 & 180	H	17	F1	G3AAF
GB3FNY	10368.752	IO93NN	Doncaster	Slotted Waveguide	10	OMNI	H	10	F1	G3AAF
GB3FNY	24048.850	IO93NN	Doncaster	Slotted Waveguide	10	OMNI	H	13	F1	G3AAF
GB3FRS	1296.850	IO91LC	Four Marks	Yagi Uhf-12-1200	30	NNW	H	14	F1	G3TCU
GB3GCT	10368.935	IO91IJ	Newbury	Slotted W/G	10	OMNI	H	10	F1	G1EHF
GB3KBQ	10368.870	IO80LV97	Trull	Slotted Waveguide	185	OMNI	H	10	F1	G4UVZ
GB3KEU	5760.925	IO93NN	Doncaster	Slotted Waveguide	10	OMNI	H	14	F1	G3AAF
GB3LEF	3400.955	IO92IQ	Markfield	Alford Slot	12	OMNI	H	8	F1	G3TQF
GB3LER	50.464	IP90JD	Lerwick	Dipole	8	0/180	H	16	F1	GM7AFE
GB3LES	2320.955	IO92IQ	Markfield	Alford Slot	12	OMNI	H	15	F1	G3TQF
GB3LEU	432.490	IO92IQ	Markfield	Twin Yagis	10	135 & 300	H	17	F1	G3TQF
GB3LEX	10368.955	IO92IQ	Markfield	Slotted Waveguide	12	OMNI	H	10	F1	G3TQF
GB3LPC	3400.935	IO91FR	Bampton	Slotted Waveguide	30	OMNI	H	10	F1	G3NNG
GB3MAT	28.260	IO82VO	Bilbrook	1/2 Wave Vertical	3	OMNI	H	7	F1	M0IAW
GB3MBA	50.408	IO93JC	Mansfield	Rh Circular Polar Dip / Similar	5	0	H	26	A1	G3PHO
GB3MCB	28.215	IO70OJ	Roche	Vertical	10	OMNI	V	14	F1	G7KFQ
GB3MCB	50.005	IO70OJ	St Austell	Dipole	30	90/270	H	16	F1	G7KFQ
GB3MCB	50.443	IO70OJ	St Austell	Dipole	30	90/270	H	16	F1	G7KFQ
GB3MCB	70.025	IO70OJ	St Austell	2 El Yagi	27	45	H	16	F1	G7KFQ
GB3MCB	144.469	IO70OJ	St Austell	Two 3 El Yagis	25	45 AND 180	H	16	F1	G7KFQ
GB3MCB	432.470	IO70OJ	St Austell	4 El Yagi	20	45	H	13	F1	G7KFQ
GB3MCB	1296.860	IO70OJ	St Austell	15+15 Yagi	20	28	H	17	F1	G7KFQ
GB3MCB	10368.980	IO70OJ	St Austell	Sectoral Horn	20	67	H	10	F1	G7KFQ
GB3MHZ	1296.830	JO02PB	Martlesham Heath	2 X 16 Slots	65	OMNI	H	27	F1	G7OCD

Call	Freq	Locator	Nearest Town	Antenna	Height agl	Bearing	Polarisation	ERP	Mode	Keeper
GB3MHZ	2320.830	JO02PB	Martlesham Heath	Slotted Waveguide	65	OMNI	H	14	F1	G7OCD
GB3MHZ	3400.830	JO02PB	Martlesham Heath	Slotted Waveguide	65	OMNI	H	14	F1	G7OCD
GB3MHZ	5760.830	JO02PB	Martlesham Heath	Slotted Waveguide	65	OMNI	H	14	F1	G7OCD
GB3MHZ	10368.830	JO02PB	Martlesham Heath	Slotted Waveguide	65	OMNI	H	8	F1	G7OCD
GB3MHZ	24048.830	JO02PB	Martlesham Heath	Slotted Wg + Sect Horn	65	OMNI	H	14	F1	G7OCD
GB3NGI	50.004	IO65VB	Loghguile	Vertical	3	OMNI	V	15	F1	GI6ATZ
GB3NGI	50.462	IO65VB	Loghguile	Vertical	3	OMNI	V	15	F1	GI6ATZ
GB3NGI	144.482	IO65VB	Loghguile	Twin Yagis	10	45 & 135	H	21	A1	GI6ATZ
GB3NGI	432.482	IO65VB	Loghguile	Yagi	10	60 & 135	H	24	F1	GI6ATZ
GB3NGI	1296.905	IO65VB	Loghguile	Alford Slot	10	OMNI	H	15	F1	GI6ATZ
GB3NGI	10368.895	IO65VB	Loghguile	Slotted Waveguide	10	OMNI	H	13	F1	GI6ATZ
GB3NWB	28.247	JO02PP	Norwich	Dipole	3		H	3	F1	M0ZAH
GB3OHM	3400.900	IO92AJ	Northfield	Slotted Waveguide	12	OMNI	H	20	F1	G8IFT
GB3OHM	5760.900	IO92AJ	Northfield	Alford Slot	12	OMNI	H	17	F1	G8IFT
GB3OSW	10368.780	IO82KV	Oswestry	Wg16 Slot	8	OMNI	H	10	F1	G3SMT
GB3PKT	10368.945	JO01MT	St Osyth	Slotted Waveguide	12	OMNI	H	3	F1	G0MBA
GB3PKT	24048.945	JO01MT	St Osyth	Slotted Waveguide	10	OMNI	H	14	F1	G0MBA
GB3RMK	50.002	IO87AX46	Golspie	Dipole	7	N/S	H	14	F1	GM3WOJ
GB3RMK	50.460	IO87AX46	Golspie	Dipole	7	N/S	H	14	F1	GM3WOJ
GB3RPE	10368.910	IO71UU	Carmarthen	Slotted Waveguide	39	OMNI	H	8	F1	GW8KCY
GB3SCC	5760.905	IO80UU	Blandford	Slotted Waveguide	12	OMNI	H	8	F1	G4JNT
GB3SCF	3400.905	IO80UU	Blandford	Alford Slot	12	OMNI	H	11	F1	G4JNT
GB3SCK	24048.905	IO80UU	Blandford	Slotted Waveguide	12	OMNI	H	11	F1	G4JNT
GB3SCQ	47088.905	IO80UU	Blandford	Horn	12	75DEG TRUE	H	6	F1	G8BKE
GB3SCS	2320.905	IO80UU	Blandford	Alford Slot	12	OMNI	H	8	F1	G4JNT
GB3SCX	10368.905	IO80UU	Blandford	Slotted Waveguide	12	OMNI	H	7	F1	G4JNT
GB3SEE	10368.850	IO91VG	Reigate	Slotted Waveguide	20	OMNI	H	5	F1	G0OLX
GB3SEE	24048.960	IO91VG	Reigate	Slotted Waveguide	20	OMNI	H	14	F1	G0OLX
GB3SEV	144.462	IO82UI	Stourport	Big Wheel	6		H	20	F1	M0VXX
GB3UHF	432.430	JO01EH	Gravesend	Two 3-El Yagis	40	288 & 348	H	22	F1	G0FDZ
GB3USK	1296.870	IO81VC	Mere Wiltshire	Flat Plate Wimo Pa-23R-16	15	90	H	20	F1	G3TCT
GB3UVR	10368.790	IO83KC	Mold	180 Deg Slot Antenna	11	NORTH EAST	H	13	F1	GW0MDQ
GB3VHF	144.430	JO01EH	Gravesend	Two 3-El Yagis	40	288 & 348	H	22	F1	G0FDZ
GB3WGI	144.487	IO64BL	Derrygonnelly Ni	2 X 6 Ele Yagi	35	280	H	30	F1	GI6ATZ
GB3WSX	70.007	IO81VC	Mere Wiltshire	5 Element Yagi	10	70 DEGREES	H	17	F1	G3ZXX
GB3XMB	28.287	IO83SV	Waddington	Dipole	10	OMNI	V	10	F1	G0SXC
GB3ZME	2320.870	IO82RP	Telford	Slotted Waveguide	10	OMNI	H	19	F1	G3UKV
GB3ZME	3400.910	IO82RP	Telford	Slotted Waveguide	10	OMNI	H	19	F1	G3UKV
GB3ZME	5760.910	IO82RP	Telford	Slotted Waveguide	10	OMNI	H	19	F1	G3UKV
GB3ZME	24048.910	IO82RP	Telford	Slotted Waveguide	10	OMNI	H	19	F1	G3UKV

OK0EU

OK0EU is a cluster of five high-stability transmitters with 5Hz separation, running for multipoint measurement. The callsign is sent 40Hz up to identify ionospheric echoes. Output power 1W to a magnetic loop radiating N-S.

Transmitters are on:

3.594.492 Vackov (JO60EF),
3.594.496 Diouha Louka (JO60TP),
3.594.500 Panksa Ves (JO70GM),
3.594.504 Pruhonice (JN79GX),
3.594.508 Kasperske Hory (JN69SD).

DK0WCY/DRA5

DK0WCY 3579 operates 0720-0900UTC and 1600-1900UTC LOCAL time (ie UTC+1 winter and UTC+2 summer) DRA5 and DK0WCY on 10144 are 24/7. Transmissions on 3579 are cw only, with beacon id interrupted at ten minute intervals by a datagram giving the Kiel K, current fof2 and MUF at Juliusruh and indications of current solar-geophysical events. DRA5 and DK0WCY on 10MHz carry the CW datagrams but, at 10 minutes after the hour the datagram transmission is on RTTY and contains additional information on X-ray background and hf2, using RTTY. The H+50 transmission is PSK31(BPSK). During auroral events the carrier at the end of the ID changes to a series of dots.
See also: *www.dk0wcy.de*

Schedule of IBP/NCDXF Beacon Transmissions

International Beacon Project beacons transmit for ten seconds on each frequency in turn in the sequence shown below. The whole cycle takes three minutes and then repeats. They send callsign at 22WPM and 100 watts, then four 1-second dashes at 100W, 10W, 1W and 0.1W. Equipment is a Kenwood TS-50, Cushcraft R-5 multiband vertical and a Trimble Navigation GPS receiver to ensure sychronization, with a control unit built by NCDXF.

LA2BCN

A2BCN operates 10-min cycle. Min 0-1t

Schedule of IBP/NCDXF Beacon Transmissions

		Frequency (kHz)				
Location	**Call**	**14100**	**18110**	**21150**	**24930**	**28200**
United Nations NY	4U1UN	00.00	00.10	00.20	00.30	00.40
Northern Canada	VE8AT	00.10	00.20	00.30	00.40	00.50
USA (CA)	W6WX	00:20	00.30	00:40	00.50	01:00
Hawaii	KH6WO	00.30	00.40	00.50	01.00	01.10
New Zealand	ZL6B	00.40	00.50	01.00	01.10	01.20
West Australia	VK6RBP	00.50	01.00	01.10	01.20	01.30
Japan	JA2IGY	01.00	01.10	01.20	01.30	01.40
Siberia	RR9O	01.10	01.20	01.30	01.40	01.50
China	VR2HK	01.20	01.30	01.40	01.50	02.00
Sri Lanka	4S7B	01.30	01.40	01.50	02.00	02.10
South Africa	ZS6DN	01:40	01.50	02:00	02:10	02:20
Kenya	5Z4B	01.50	02.00	02.10	02.20	02.30
Israel	4X6TU	02:00	02:10	02:20	02.30	02:40
Finland	OH2B	02:10	02:20	02:30	02:40	02:50
Madeira	CS3B	02.20	02.30	02.40	02.50	00.00
Argentina	LU4AA	02:30	02:40	02:50	00.00	00:10
Peru	OA4B	02.40	02.50	00.00	00.10	00.20
Venezuela	YV5B	02:50	00.00	00:10	00:20	00:30

DK0WCY: *www.dk0wcy.de* NCDXF: *www.ncdxf.org/beacons.html*

Worldwide List of HF Beacons

(Updated June 2023)

The list is maintained by Dennis Green, ZS4BS, Region 1 HF Beacon Coordinator
Please notify errors/changes to *zs4bs@iaru-r1.org*

? = Activity uncertain; **INT** = Intermittent; **PT**=Part-time; **IRREG** = Irregular; **UC** = Under Construction; (T) **Non Op** = (Temporarily) NotOperational; **OP?** = Operational?; **V** = Varies; **EXP** = Experimental; **WkEnd** = Weekends during daylight; **LT** = Local Time;
SYNCH = frequency sharing synchronised beacons: SYNCHn current sequence N4ESS (0,00 seconds), N4ES (0,20), WB4WOR (0,30), K7EK (0,40), N4ES (0,50). SYNCHx: K4UKB ID at 00 seconds and K4FUM at 30 seconds. ## K6FRC transmits on 28 245, 28 250, 28 300
+++ 10 minutes constant carrier followed by ID
zzz GB3WES at H+1, H+16 etc and GB3ORK at H+2, H+17 etc. Transmissions have a stepped power sequence.

IARU Region 1 discourages beacon operation on 1,8MHz
IARU Region 1 discourages beacon operation on 3,5MHz (DKOWCY, OK0EU OK!
IARU Region 1 discourages beacon operation on 7MHz (OK0EU, OK0EP OK!
IARU Region 1 discourages beacon operation on 10MHz (DKOWCY excepted)

Freq	Call	Nearest Town	Locator	ERPw	Antenna	Direction	Mode	Status
160 m								
1810.50	YR2TOP	Zlatita	KN94RU	100	Inv Vee		A1	24
1853.00	OK0EV	Near Prague	JN79EV	0.1	25m Vert	Omni	A1	PT
1875.00	DL3KR		JO63LV	5	Dipole		A1	PT
80 m								
3570.00	YO8RIX	Near Dorohoi	KN37FW	300mw	Lazi Loop	Omni	A1	24
3576.80	IZ3DVW	Monselice	JN55VF	0.5	Inv Vee		A1	IRREG
3579.00	DKOWCY	Scheggerott	JO44VQ	30	Dipole		A1	PT/zz??
3579.80	SM2IUF	Kalix	KP15NU	QRPP				15 - 07 UTC
3594.50	OK0EU	Panska Ves	JO70GM	100 mw	Mag Loop	N - S	A1	24
3600.00	OK0EN	Near Kladno	JO70AC	150mw	LW 41 m	SE - NW	A1	24
60 m								
5195.00	DRA5	Scheggerott	JO44VQ	30	Dipole		A1	06 - 24 LT
5195.00	DRA5	Scheggerott	JO44VQ	30	Dipole		PSK,RTTY	24
5205.25	LX0HF	Junglinster	JN39DR	5	Dipole		A1	24
5289.50	OV1BCN	10 km S Soroe	JO55SI	30	32 m F, Dip @ 1 m		USB/MT63	H + 4, 19, 34, 49
5290.00	ZS6SRL	Randburg	KG33WV	?			WSPR	
5290.50	OV1BCN	?						
5290.00	GB3ORK	Orkney	IO89JA	10<158 uw	Inv Vee		JT9A	24
5291.00	HB9AW	Sursee	JN43BA	10/5/1/,1/,001	1/2 Dip		A1 + PSK	5 m cycle
40 m								
7003.00	ES1VHF		KO29IK	?				
7034.80	PT9BCN		GG29RN	12			A1	OP?
7038.00	ZS1AGI	Kanaberg, Mossel Bay	KF06XB	500 mW	Inverted V		A1	
7038.50	OK0EU	Panska Ves	JO70GM	1	Mag Loop	N - S	A1	OP?
7039.10	SK7CQ	TEST						
7039.20	IK1HGI	Cerano	JN61MM	0.1	QRSS			
7039.40	OK0EPB	Prague	JO70EC	10	Dipole@22m		A1	24
7039.80	IZ3DVW	Monselice Padua	JN55VF	0.5	Inv Vee		A1	24
7047.50	YD0MWK		OI33MQ	5	Inv Vee @ 15 m		A1	Planned
30 m								
10130.00	OK1IF	Liberec	JO70HG	0.5			A1	?
10133.00	SA6RR	Oxaback	JO67KI	0.5	GP	Omni	A1	INT
10133.50	HB4FV			0.5	H-pole		A1	?
10137.20	IK3NWX	Near Monselice	JN55VB	4.2	Rot Dipole	E - W	A1	24
10138.70	WSPR beacons around here							
10140.00	WSPR beacons around here							
10142.51	IK1HGI	Tricati	JN45IK	0.1	Dipole		QRSS3	OP?
10144.00	DKOWCY	Scheggerott	JO44VQ	30	Dipole		A1, PSK, RTTY	24 zz
10145.00	OK0EF			0.5				OP?
20 m								
14097.00	WSPR beacons around here							
14100.00	4U1UN	UN NY	FN30AS	100 - 0,1	Vertical	Omni	A1	IBP Cycle
14100.00	VE8AT	Eureka, Nunavut	EQ79AX	100 - 0,1	Vertical	Omni	A1	IBP Cycle
14100.00	W6WX	Mt Umunhum, CA	CM97BD	100 - 0,1	Vertical	Omni	A1	IBP Cycle
14100.00	KH6RS	Laie, Oahu	BL10TS	100 - 0,1	Vertical	Omni	A1	IBP Cycle
14100.00	ZL6B	near Masterton	RE78TW	100 - 0,1	Vertical	Omni	A1	IBP Cycle
14100.00	VK6RBP	Rolystone	OF87AV	100 - 0,1	Vertical	Omni	A1	IBP Cycle
14100.00	JA2IGY	Mt Asama	PM84JK	100 - 0,1	Vertical	Omni	A1	IBP Cycle
14100.00	RR9O	Novosibirsk	NO14KX	100 - 0,1	Vertical	Omni	A1	IBP Cycle
14100.00	VR2B	Hong Kong	OL72CQ	100 - 0,1	Vertical	Omni	A1	IBP Cycle
14100.00	4S7B	Colombo	NJ06CR	100 - 0,1	Vertical	Omni	A1	IBP Cycle
14100.00	ZS6DN	Vereniging	KG33XI	100 - 0,1	Vertical	Omni	A1	IBP Cycle
14100.00	5Z4B	Kiambu	KI88MX	100 - 0,1	Vertical	Omni	A1	IBP Cycle
14100.00	4X6TU	Tel Aviv	KM72JB	100 - 0,1	Vertical	Omni	A1	IBP Cycle
14100.00	OH2B	Lohja	KP20BM	100 - 0,1	Vertical	Omni	A1	IBP Cycle
14100.00	CS3B	Santo da Serra	IM12OR	100 - 0,1	Vertical	Omni	A1	T Nonop
14100.00	LU4AA	Buenos Aires	GF05TJ	100 - 0,1	Vertical	Omni	A1	IBP Cycle
14100.00	OA4B	Lima	FH17MW	100 - 0,1	Vertical	Omni	A1	T Nonop
14100.00	YV5B	Caracas	FK06NK	100 - 0,1	Vertical	Omni	A1	T Nonop
17 m								
18095.00	YO8RIX	near Dorohoi	KN37FW	150mw	Lazi Loop	Omni	A1	24
18095.50	HP1AVS	Cerro Jefe	FJ09HD	2	Inv Vee		A1	24

Freq	Call	Nearest Town	Locator	ERPw	Antenna	Direction	Mode	Status
18100.00	IK6BAK	Montefelcino	JN63KR	0.1	Inv Vee	Omni	A1	?
18102.00	I1M	Bordighera	JN33UT	10	Vert	Omni	A1	24
18104.60	WSPR beacons around here							
18110.00	4U1UN	UN, NY	FN30AS	100 - 0,1	Vertical	Omni	A1	IBP Cycle
18110.00	VE8AT	Eureka, Nunavut	EQ79AX	100 - 0,1	Vertical	Omni	A1	IBP Cycle
18110.00	W6WX	Mt Umunhum, CA	CM97BD	100 - 0,1	Vertical	Omni	A1	IBP Cycle
18110.00	KH6RS	Laie, Oahu	BL10TS	100 - 0,1	Vertical	Omni	A1	IBP Cycle
18110.00	ZL6B	near Masterton	RE78TW	100 - 0,1	Vertical	Omni	A1	IBP Cycle
18110.00	VK6RBP	Rolystone	OF87AV	100 - 0,1	Vertical	Omni	A1	IBP Cycle
18110.00	JA2IGY	Mt Asama	PM84JK	100 - 0,1	Vertical	Omni	A1	IBP Cycle
18110.00	RR9O	Novosibirsk	NO14KX	100 - 0,1	Vertical	Omni	A1	IBP Cycle
18110.00	VR2B	Hong Kong	OL72CQ	100 - 0,1	Vertical	Omni	A1	IBP Cycle
18110.00	4S7B	Colombo	NJ06CC	100 - 0,1	Vertical	Omni	A1	IBP Cycle
18110.00	ZS6DN	Vereniging	KG33XI	100 - 0,1	Vertical	Omni	A1	IBP Cycle
18110.00	5Z4B	Kiambu	KI88MX	100 - 0,1	Vertical	Omni	A1	IBP Cycle
18110.00	4X6TU	Tel Aviv	KM72JB	100 - 0,1	Vertical	Omni	A1	IBP Cycle
18110.00	OH2B	Lohja	KP20BM	100 - 0,1	Vertical	Omni	A1	IBP Cycle
18110.00	CS3B	Santo da Serra	IM12OR	100 - 0,1	Vertical	Omni	A1	T Nonop
18110.00	LU4AA	Buenos Aires	GF05TJ	100 - 0,1	Vertical	Omni	A1	IBP Cycle
18110.00	OA4B	Lima	FH17MW	100 - 0,1	Vertical	Omni	A1	T Nonop
18110.00	YV5B	Caracas	FK60NK	100 - 0,1	Vertical	Omni	A1	T Nonop
15 m								
21068.00	SK7OB				V,Dipole			IRREG
21094.60	WSPR beacons around here							
21145.70	IZ3DVW	near Monselice	JN55VF	2.6	Inv Vee, Dipole		A1	24
21149.00	F5ZHL		JO10SI					IRREG
21149.00	IT9ATQ/B	Catania	JM77MM	2	Bent Dipole		A1	24
21150.00	4U1UN	UN, NY	FN30AS	100 - 0,1	Vertical	Omni	A1	IBP Cycle
21150.00	VE8AT	Eureka, Nunavut	EQ79AX	100 - 0,1	Vertical	Omni	A1	IBP Cycle
21150.00	W6WX	Mt Umunhum, CA	CM97BD	100 - 0,1	Vertical	Omni	A1	IBP Cycle
21150.00	KH6RS	Laie, Oahu	BL10TS	100 - 0,1	Vertical	Omni	A1	IBP Cycle
21150.00	ZL6B	near Masterton	RE78TW	100 - 0,1	Vertical	Omni	A1	IBP Cycle
21150.00	VK6RBP	Rolystone	OF87AV	100 - 0,1	Vertical	Omni	A1	IBP Cycle
21150.00	JA2IGY	Mt Asama	PM84JK	100 - 0,1	Vertical	Omni	A1	IBP Cycle
21150.00	RR9O	Novosibirsk	NO14KX	100 - 0,1	Vertical	Omni	A1	IBP Cycle
21150.00	VR2B	Hong Kong	OL72CQ	100 - 0,1	Vertical	Omni	A1	IBP Cycle
21150.00	4S7B	Colombo	NJ06CC	100 - 0,1	Vertical	Omni	A1	IBP Cycle
21150.00	ZS6DN	Vereniging	KG33XI	100 - 0,1	Vertical	Omni	A1	IBP Cycle
21150.00	5Z4B	Kiambu	KI88MX	100 - 0,1	Vertical	Omni	A1	IBP Cycle
21150.00	4X6TU	Tel Aviv	KM72JB	100 - 0,1	Vertical	Omni	A1	IBP Cycle
21150.00	OH2B	Lohja	KP20BM	100 - 0,1	Vertical	Omni	A1	IBP Cycle
21150.00	CS3B	Santo da Serra	IM12OR	100 - 0,1	Vertical	Omni	A1	T Nonop
21150.00	LU4AA	Buenos Aires	GF05TJ	100 - 0,1	Vertical	Omni	A1	IBP Cycle
21150.00	OA4B	Lima	FH17MW	100 - 0,1	Vertical	Omni	A1	T Nonop
21150.00	YV5B	Caracas	FK60NK	100 - 0,1	Vertical	Omni	A1	T Nonop
21151.00	I1M	Bordighera	JN33UT	10	2 5/8 Vert	Omni	A1	24
12 m								
24920.00	IY4M	Bologna	JN54QK				A1	H+30>H+60m
24930.00	4U1UN	UN, NY	FN30AS	100 - 0,1	Vertical	Omni	A1	IBP Cycle
24930.00	VE8AT	Eureka, Nunavut	EQ79AX	100 - 0,1	Vertical	Omni	A1	IBP Cycle
24930.00	W6WX	Mt Umunhum, CA	CM97BD	100 - 0,1	Vertical	Omni	A1	IBP Cycle
24930.00	KH6RS	Laie, Oahu	BL10TS	100 - 0,1	Vertical	Omni	A1	IBP Cycle
24930.00	ZL6B	near Masterton	RE78TW	100 - 0,1	Vertical	Omni	A1	IBP Cycle
24930.00	VK6RBP	Rolystone	OF87AV	100 - 0,1	Vertical	Omni	A1	IBP Cycle
24930.00	JA2IGY	Mt Asama	PM84JK	100 - 0,1	Vertical	Omni	A1	IBP Cycle
24930.00	RR9O	Novosibirsk	NO14KX	100 - 0,1	Vertical	Omni	A1	IBP Cycle
24930.00	VR2B	Hong Kong	OL72CQ	100 - 0,1	Vertical	Omni	A1	IBP Cycle
24930.00	4S7B	Colombo	NJ06CC	100 - 0,1	Vertical	Omni	A1	IBP Cycle
24930.00	ZS6DN	Vereniging	KG33XI	100 - 0,1	Vertical	Omni	A1	IBP Cycle
24930.00	5Z4B	Kiambu	KI88MX	100 - 0,1	Vertical	Omni	A1	IBP Cycle
24930.00	4X6TU	Tel Aviv	KM72JB	100 - 0,1	Vertical	Omni	A1	IBP Cycle
24930.00	OH2B	Lohja	KP20BM	100 - 0,1	Vertical	Omni	A1	IBP Cycle
24930.00	CS3B	Santo da Serra	IM12OR	100 - 0,1	Vertical	Omni	A1	T Nonop
24930.00	LU4AA	Buenos Aires	GF05TJ	100 - 0,1	Vertical	Omni	A1	IBP Cycle
24930.00	OA4B	Lima	FH17MW	100 - 0,1	Vertical	Omni	A1	T Nonop
24930.00	YV5B	Caracas	FK60NK	100 - 0,1	Vertical	Omni	A1	T Nonop
24931.00	7Z1CQ	Jeddah	KL91ON	5	Vert, Dip	Omni	A1	IRREG
10 m								
28163.00	VK3XPT	Lang Lang	QF21SR		1/4 Vert			
28169.00	ZB2TEN	Gibraltar	IM76HD	4	1/4 Vert	Omni	A1	24?
28171.00	XE1FAS	Publa, PU	EK09UB	12	Dipole		A1	24
28173.00	IZ1EPM	27 km NE Turin	JN35WD	20	1/2 Vert	Omni	A1	24
28174.00	VE1VDM		FN85AA				A1	?
28175.00	VE3TEN	Ottawa, ON	FN25	10	GP	Omni	A1	24
28176.90	HP1RCP	Cerro Jefe	FJ09HD	5	Slope Dipole		A1	24
28177.00	IW1AVR	Cravanzana		5	Vertical		A1	?
28178.00	IQ0GV		JN61TR				A1	?
28180.30	I1M	Bordighera	JN33UT	5 W / 20 W	2 × 5/8 Vert	Omni	A1	30/60min
28182.40	SV3AQR	Amalias	KM07QS	4	GP	Omni	A1	24
28183.20	XE1RCS	Cerro Gordo	EK09OS	8	AR10	Omni	A1	24
28184.00	VE2REA	Quebec, QC	FN46IT				A1	24
28185.00	VA3SRC	Burlington, ON	FN03BH	5	Dipole		A1	PT
28187.60	VE7KC	Penticton, BC	DN09EL				A1	24
28188.00	OE3XAC	Kaiserkogel	JN78SB	20	7/8 GP @ 750 m	Omni	A1	24
28188.10	JE7YNQ	Fukushima	QM07					24

For the latest HF beacon listing, see: *RSGB Website for link*

Freq	Call	Nearest Town	Locator	ERPw	Antenna	Direction	Mode	Status
28188.90	SV5TEN	Raad	KM46CK	5	Vertical	Omni		24
28189.50	LU2DT	Mar del Plate	GF12	5	Vert, Dipole	Omni	A1/PSK	
28189.80	LU8XW	Ushuaia	FD55UE					?
28190.00	LU3HFA	Cordoba, CD	FF78UP	5	Vertical	Omni	A1	24?
28190.00	LU4VI	Villa Regina	FF60VI				A1	24
28191.50	A62ER	Sharjah, UAE	LL75QI				A1	OP?
28192.00	EP4HR	Shiraz	LL69GP	20/2/0,2	Dipole		A1	24
28193.00	EI7GR		IO53MG					?
28192.90	VE4ARM	Austin, MB	EM09HW	5	GP	Omni	A1	24
28193.50	A47RB	Oman	LL93FO	10	Vertical	Omni	A1	OP?
28193.50	LU2XPK		FF66DE				A1	?
28193.90	IW4EIR		JN54AS	1.5			A1	?
28194.00	IW4ERI		JN54AS					
28195.10	IY4M	Bologna	JN54QK	20	5/8 GP	Omni	A1	H>H+30m
28196.00	VA3ITA	Bramton, ON	FN03CW				A1	IRREG
28196.10	LU4JJ	Concordia, ER	GF08XO				A1	24
28196.70	LU5FB	Rosario, SF	FF97PB				A1	24
28197.00	VE7MTY	Vancouver, BC	CN89	5	Vertical	Omni	A1	24
28197.70	IK3OTW						A1	?
28193.00	LU2ERC	Ensenada	GF15AD	10	Vertical	Omni	A1	24
28198.00	LU9FE		FF98GR					?
28199.00	LW6DJD	La Plata	GF15AC	5			A1	?
28199.30	LU1FHH	El Trebol, SF					A1	24
28200.00	4U1UN	UN, NY	FN30AS	100 - 0,1	Vertical	Omni	A1	IBP Cycle
28200.00	VE8AT	Eureka, Nunavut	EQ79AX	100 - 0,1	Vertical	Omni	A1	IBP Cycle
28200.00	W6WX	Mt Umunhum, CA	CM97BD	100 - 0,1	Vertical	Omni	A1	IBP Cycle
28200.00	KH6RS	Laie, Oahu	BL10TS	100 - 0,1	Vertical	Omni	A1	IBP Cycle
28200.00	ZL6B	near Masterton	RE78TW	100 - 0,1	Vertical	Omni	A1	IBP Cycle
28200.00	VK6RBP	Rolystone	OF87AV	100 - 0,1	Vertical	Omni	A1	IBP Cycle
28200.00	JA2IGY	Mt Asama	PM84JK	100 - 0,1	Vertical	Omni	A1	IBP Cycle
28200.00	RR9O	Novosibirsk	NO14KX	100 - 0,1	Vertical	Omni	A1	IBP Cycle
28200.00	VR2B	Hong Kong	OL72CQ	100 - 0,1	Vertical	Omni	A1	IBP Cycle
28200.00	4S7B	Colombo	NJ06CC	100 - 0,1	Vertical	Omni	A1	IBP Cycle
28200.00	ZS6DN	Vereniging	KG33XI	100 - 0,1	Vertical	Omni	A1	IBP Cycle
28200.00	5Z4B	Kiambu	KI88MX	100 - 0,1	Vertical	Omni	A1	IBP Cycle
28200.00	4X6TU	Tel Aviv	KM72JB	100 - 0,1	Vertical	Omni	A1	IBP Cycle
28200.00	OH2B	Lohja	KP20BM	100 - 0,1	Vertical	Omni	A1	IBP Cycle
28200.00	CS3B	Santo da Serra	IM12OR	100 - 0,1	Vertical	Omni	A1	T Nonop
28200.00	LU4AA	Buenos Aires	GF05TJ	100 - 0,1	Vertical	Omni	A1	IBP Cycle
28200.00	OA4B	Lima	FH17MW	100 - 0,1	Vertical	Omni	A1	T Nonop
28200.00	YV5B	Caracas	FK60NK	100 - 0,1	Vertical	Omni	A1	T Nonop
28200.50	VA3GMT	Toronto					A1	?
28200.80	AC7AV	Oak Harbour, WA					A1	?
28201.00	N7JUB						A1	?
28201.00	WB9OTX		EN55				A1	?
28201.30	PU2SUT	Sao Paulo	GG66TB	20	Dipole		A1	?
28202.00	WN2WNC	New Berlin, NY	FN22IO				A1	
28202.50	KA3BWP	Stafford, VA	FM18GK				A1	24
28203.00	K4MTP	Tannersville, PA					A1	
28203.00	PY2WFG	Ipisanga, SP	GG77FF				A1	24
28203.00	KB1QZY	Springfield, MA	FN32QC	2	Imax2000 V	Omni	A1	24
28203.00	KG8CO	Clinton, MI	EN82AB	5	Vertical	Omni	A1	24
28203.00	N6DXX	Sacramento, CA	CM98FM				A1	?
28203.00	K4MTP		FN21TA				A1	
28203.50	K6LLL	Mission Viejo, CA	DM13EO				A1	24
28202.00	SV2HNE		KN10LL	5	GP			?
28204.00	WA2NTK	Big Flats, NY	FN12NE			E-W	A1	24
28204.00	WL7N	Ward Cove, AK	CO45KK				A1	?
28204.00	KE4TWI	Watertown, TN	EM66VO	5			A1	24
28204.00	W0WF	St Charles, MO	EN02QB				A1	?
28204.00	W6CF	San Francisco	CM87UU				A1	24
28204.00	KA1KNW	Windsor, CT	FN31RU	10			A1	?
28205.00	DL0IGI	Hohenpeissenberg	JN57MT	Varies	1/4 Vert	Omni	A1	24
28205.20	VN3NIA	Ridgway, PA	FN01PK	4	Dipole		A1	24
28205.90	HS0BBD	Bangkok	OK13					?
28206.10	VA3GRR	Brampton, ON	FN03	1.75	1/2 Vert	Omni	A1	24
28206.30	K9EJ	Toledo	EM59UG	2	Vert @ 4 m	Omni	A1	24
28206.50	HP1RIS	Panama City	FJ09GA	3	Vert Dipole	Omni	A1	24
28207.00	ON0RY	Binche	JO20CK	5	Vertical	Omni	A1	24
28207.30	KW7HR	Pasco, WA	DN06KG		A1	24?		
28207.80	W4CND	Jemison, AL	EM63QA	2	Vertical	Omni	A1	24
28208.00	KE6TE	Elk Grove, CA	CM98HK				A1	OP?
28208.00	AK2F	Randolph, NJ	FN20QT				A1	share
28208.00	WN2A	Budd Lake, NJ	FN20OU				A1	AK2F
28208.10	JR0YAN	near Toyama	PM86JW	25	Hor,Loop		A1	24+++
28208.20	IZ3LCJ	St Lucia di Piave	JN65DU	5	1/4GP@15m	Omni	A1	QRT?
28208.50	NB7A	Reno, NV	DM09BM				A1	24
28208.70	N8PVL	Livonia. MI	EN82GJ				A1	24?
28209.00	KH6AP	Kikei Maui, HI	BL10SS	20	3/8 Vert	Omni	A1	24
28209.50	K9CW	Thomasboro. IL	EN50WF	2	AR99@15	Omni	A1	24
28209.80	KV6Q	San Diego	DM12JS	3			A1	?
28210.00	PT2SSB	Brasilia	GH64CI				A1	?
28210.00	KB9UGA	Egg Harbour, WI					A1	24
28210.00	NJ4R						A1	?
28210.20	SB7W							?
28210.40	NT4F	Wilmington, NC	FM14AE	5			A1	24

Freq	Call	Nearest Town	Locator	ERPw	Antenna	Direction	Mode	Status
28210.50	VE4TEN	Kelowna, BC		2	1/2 Vert	Omni	A1	24
28211.00	DB0FKS	near Frankfurt/M	JN40IT	0.2	DV27 Vert	Omni	A1	OP?
28211.00	K5ARC	Galvez, LA	EM40	20	Vertical	Omni	A1	24?
28211.00	CE1TUW	Antofagasta	FF46RQ				A1	?
28211.10	LA4TEN	near Hellvik	JO28WL	250	Vertical	Omni	A1	24
28211.50	VK4ADC	New Beith, QLD		10/5/2/1	1/4 Vert	Omni	A1	24
28211.50	VK4WSS	Mt Cotton, QLD						24
28211.80	AC7GZ	Chandler, AZ	DM43				A1	?
28212.00	W4AMA		EM63WA				A1	?
28212.00	7Z1AL	Dammam	LL56BK	10	Vertical		A1	OP?
28212.50	K0KP	Fredenberg, MN	EN36VW	0.5	GP	Omni	A1	OP?
28212.50	KJ4QYB	Rainbow City, AL	EM63WO				A1	?
28212.60	LU7DQP	Lanos Oest, BA	GF05TH				A1	24
28213.00	WA5SAT		EL09				A1	?
28213.30	KD8RKJ	Cleveland, OH	EN91CK	2	Vertical	Omni	A1	24
28213.50	KE4KAA	Big Stone Gap, VA	EM86OV	5			A1	24
28213.50	W3IK	Gray, TN	EM86				A1	24
28213.80	KF5KBZ	Austin, TX	EM10FB				A1	?
28213.80	WA5SAT		EL09					?
28214.00	N4PAL	Longwood, FL	EL98HQ	5	Vert@4,5m	Omni	A1	24?
28214.00	LA9TEN	Snertingdal	JP50EV	10	5/8 GP	Omni	A1	24
28214.50	FR1GZ	Reunion Island	LG79RC					IRREG
28215.00	LU5EGY	Buenos Aires	GF05QI				A1	24
28215.00	N8MIE		EM44				A1	?
28215.00	SR5TDM		KO01KX				A1	?
28215.00	KA9SZX	Paxton, IL	EN50VD	1	Antron99	Omni	A1	24
28215.00	W4JPL	Liberty. NC	FM05FV				A1	24
28215.00	GB3MCB	Cornwall	IO70OJ	25	Vertical	Omni	F1	24
28215.30	XE3D	Merida, YUC	EL50EG				A1	24
28215.50	KD5CKP	Olive Branch, MS	EM54BW	3	Vertical	Omni	A1	24
28215.80	K6WKX	Santa Cruz, CA	CM86XX	10	Horiz Dipole	Omni	A1	24
28216.00	K3FX	Neptune City, NJ	FN20XE	7	1/2 Vert	Omni	A1	24
28216.00	N7MA	Cataldo, ID	DN17SN	5	Vertical	Omni	A1	24
28217.00	LA2BCN	Telemark	JO48GX	8	5/8 Vert	Omni	A1	24$$$
28126.10	LA2BCN	Telemark	JO48GX	8	5/8 Vert	Omni	WSPR	24
28217.00	WB0FTL	Alden, MN	EN33FP	5	AR10@25	Omni	A1	24
28217.00	W6GY	Star, ID	DN13RP				A1	24
28217.50	WA1LAD	West Warick, RI	FN41FQ	4.5	J-Pole	Omni	A1	24
28217.50	W8MI	Mackinaw MI	EN75	0.5	Vertical	Omni	A1	24
28218.00	IQ5MS	Marina da Massa	JN54AA		Vertical	Omni	A1	24
28218.30	AC0KC	Fort Lupton, CO	DN70OA	qrpp			A1	?
28218.30	KD8RKJ	Cleveland, OH		2	Vertical	Omni	A1	?
28218.30	KJ4LAA	Decatur, AL	EM64LN	3	Vertical	Omni	A1	?
28218.50	W5RDW	Murphy, TX	EM12QX	5	AR10@25	Omni	A1	24
28218.60	KA4RRU	Catlett, VA	FM18EN				A1	24
28218.80	KN8DMK	Amanda, OH	EM89OO	3	Slope Dipole	NW/SE	A1	?
28219.60	PY2UEP		GG58GA				A1	?
28219.90	KB9DJA	Mooresville, IN	EM69RO	35	GP	Omni	A1	24
28220.00	5B4CY	Mandria	KM64KU	15	GP	Omni	F1	24
28220.00	W8VO	Sterling Hts, MI	EN82NU	5	Vert@50	Omni	A1	OP?
28220.00	K4AQXI		FM03IL				A1	?
28220.40	WK4DS	Trenton, GA	EM74FU	2	Dipole	NE-SW	A1	24
28220.50	YM7KK	Giresun	KN90IV	4			A1	24
28220.50	N5FUN	Caroltton, TX	EM12				A1	?
28220.80	SV2MCG	Thessaloniki	KN10FC				A1	?
28221.00	WE4S	Rock Springs, CA					A1	24
28221.50	KC0TKS	Sedalia MO	EM38IQ	5	J-poleVert	Omni	A1	24
28221.50	GW7HDS		IO81IP	3			A1	IRREG
28221.90	W1DLO	Calais, ME	FN65JE	10	A99 Vert	Omni	A!	24
28222.00	TP2CE	Strasbourg	JN38VO	450mw	GP-7000	Omni	A1	24
28222.00	K6JCA	Carmel Valley CA	CM96DN	A1				QRT?
28222.00	IZ0KBA	Castel Madama	JN61KX	4	GP 350 m ASL	Omni	A1	24
28222.00	ZS1TEN	Cape Town	JF96FB	5	Vertical	Omni	A1	
28222.20	W4KLP						A1	?
28222.80	N4QDK	Sauratown Mt, NC	EM95	3	dipole		A1	24
28223.00	WY5B	Biloxi, MS	EM50NK				A1	24
28223.00	9H1LO	Malta	JM75	8	GP		A1	Irreg
28223.00	KP3FT	Ponce, PR	FK68	4	Vertical	Omni	A1	24
28223.30	XE3ACB	Hecelchakan	EL40				A1	24?
28223.50	KK6RE	Chico, CA	CM96VG				A1	?
28223.60	PB5A		JO21DE				A1	?
28224.00	WD0AKX	Albert Lea, MN	EN33	5	Vertical	Omni	A1	24
28224.00	WA3RNC	Lewistown, PA	FN10FO				A1	24
28224.20	YB9BWN	Denpasar, Bali	OJ13FK	2	Dipole	EU/VK	A1	24
28224.70	HA5BHA	near Budapest	JN97KO	5		Omni		24
28224.80	KW7Y	Marysville, WA	CN88SD	4	Vertical	Omni	A1	24
28224.80	IT9EJW	Sicily	JM77NN	3			A1	
28224.80	NT6T	Goleta, CA					A1	?
28225.00	YM7TEN		KN91RB	1	Vertical	Omni	A1	24
28225.00	K6FRC	Angels Common, CA	CM97GP		Vertical	Omni	A1	24??
28225.00	K5GJR	Corpus Christi TX	EL17HR				A1	24
28225.40	KC0JCA	St Louis, MO					A1	?
28225.50	W2DLL	near Buffalo, NY	FN02PP	8	1/2 Vert	Omni	A1	24
28225.60	WB0LYV	Beatrice, NE	EN10		Delta Loop		A1	?
28226.00	ED1YCA		IN73AL	5	H Loop	Omni	A1	24
28226.00	PU4CBX	Baraco de Cocais	GH80DB				A1	Irreg

Freq	Call	Nearest Town	Locator	ERPw	Antenna	Direction	Mode	Status
28226.20	WA6HXW	Westn Covina, CA	DM14AB				A1	irreg
28226.50	N7MSH	North Powder, OR	DN15				A1	24
28226.60	KC6WGN	Las Vegas, NV	DM26LD	10	Omni	A1	24	
28226.60	PY2RFF	Sco Pedro, SP	GG67AL	1.5	1/4 Vert	Omni	A1	24
28226.70	KU4A	Lexngton, KY	EM78SB				A1	?
28227.00	VE9AT	White Head Island		0.1	Dipole		A1	OP?
28227.00	KJ4HYV	Zellwood, FL					A1	24
28227.50	KC5MO	Austin, TX	EM10BF	2	Dipole		A1	24
28227.70	IW3FZQ	Monselice, PD	JN55VF	5	J-Pole@20m	Omni	A1	24
28228.00	ZL3TEN	Rolliston	RE66	10	1/2 Vert	Omni	A1	TNonOp
28228.00	OH5TEN	Kouvola	KP30HV	4	Horiz	Dip	A1	24
28228.30	TG9TEN		EK44				A1	OP?
28228.80	N3PV	Spring Valley CA	DM12LP				A1	24
28229.00	ZL2MHF	near Wellington	RE78NU	10	1/2 Vert	Omni	F1	24
28229.00	NG9Y	Vevay, IN	EM78LR				A1	?
28229.30	KA2LIM	Pine Valley, NY	FN12NP	NY			A1	?
28229.70	IQ8CZ	Catanzo	JM88HV	10	GP	Omni	A1	24
28230.00	WA4ZKO	Dry Ridge, KY	EM78PP	4	A99	Omni	A1	24
28230.50	AA0RQ	Pine, CO					A1	?
28230.00	KI4AED	Ocoee, FL	EL98FN	5	Antron 99	Omni	A1	24
28230.00	KQ4TG	Leland, NC	FM40XF				A1	24
28230.50	HP6RCP	Santiago	EJ98MB	3	AR99	Omni	A1	IRREG
28231.00	F5ZEH		IN88VA	,5/5/50	1/2 V ,3-el Yagi		A1	24
28231.80	WA4FC	Prince George, VA	FM17HD	5	Ringo@200	Omni	A1	24
28232.00	N1FSX	Simla, CO	DM89AC	5	Vertical	Omni	A1	24
28232.60	N9BPE	Tuscaloosa, IL	EM59UT	2	1/2 Dip @20	Omni	A1	24
28232.70	N2MH	West Orange, NJ	FN20UT				A1	24
28233.00	N2UHC	St Paul, KS	EM27JM	4	Vert Dipole	Omni	A1	24
28231.60	SV2AHT	Hortiatis	KN10NO				A1	24
28232.30	W7SWL	Tucson, AZ		5	Vertical	Omni	A1	?
28233.00	I0KNQ	Genzano di Roma	JN61FU	2	Turnstile	Omni	A1	24
28233.00	KB9GSY	Hammond, IN	EN61FP				A1	24
28234.00	K4DP	Covington, VA	FM07AR				A1	IRREG
28234.30	K4DXY	Birmingham, AL	EM63PP	2			A1	?
28234.80	VE1CBZ	Keswick Ridge, NB	FN65	2	A99 Vert	Omni	A1	24
28235.00	KI4AED	Ocoee, FL	EL98FN	5	Antron 99	Omni	A1	24
28235.00	KK6RE	Chico, CA					A1	?
28235.00	NP4LW	San Sebastian	FK68MI	15	Vertical		A1	?
28235.00	KQ4FM	Southlake, TX	EM12JW	5	GP		A1	24
28235.00	OY6BEC	Faroe Islands	IP62	20	Yagi		A1	24
28235.10	KI4HOZ	Pickens, SC	EM83XM				A1	?
28235.60	VE3GOP	Mississauga, ON	FN03GD	0.2	A99 Vert	Omni	A1	24
28236.00	W8YT	Martinsburg, WV	FM19AJ	5	Vertical	Omni	A1	24
28236.50	W0KIZ	near Denver, CO	DM79	5	Vertical	Omni	A1	24
28237.00	K7TIA	Houston, TX					A1	?
28237.50	WA2NEW	Beach Haven, NJ					A1	?
28237.60	LA5TEN	near Oslo	JO59JP	15	1/2 Vert	Omni	A1	24
28237.80	K7ZSA	Alger, WA	CN88UO	5	Vertical	Omni	A1	24?
28238.00	KB2SEO	Eton, GA	EM74OT	5	1/4 GP	Omni	A1	24
28239.00	VA7PL	Crystal Mountain	DM09	5	GP	Omni	A1	24
28239.00	PP6AJM	Nosso Senhora da Socorro	HH19LD					?
28239.20	AL7FS	Anchorage, AK	BP51BD	3	1/2 Vert	Omni	A1	PT
28239.50	WA3HGT	Montoursville PA	FN11MG					
28239.80	N4LEM	Cocoa, FL	EL98	50	Vertical	Omni	A1	?
28239.80	IZ8RVA	Agropoli, SA	JN70LI				A1	24?
28240.00	I0KNQ	Rome	JN61FU				A1	24
28240.00	XE3OAX	Ocotlan, OAX	EK17PA	0,5?	+ sev,spurii		A1	24
28240.11	WE6Z	Granite Bay, CA	CM98JS	4	Vertical	Omni	A1	24
28240.50	N2DWS	Port Republic NJ	FM29SM				A1	24
28240.50	W4RKC	Winchester, VA	FM09VD	5	1/2 Vert	Omni	A1	24
28240.60	YO2X	Timisoara	KN05PS	2	GP	Omni	A1	09 - 15 UTC
28240.70	AJ8T	Sturgis, MI	EN71HS				A1	24
28241.50	F5ZUU	Malataverne	JN24IL	5	1/2 Vert	Omni	A1	24
28241.50	K5DZE	Erlanger, KY	EM78QS	5			A1	24
28242.00	IZ8DXB	Naples	JN70LN	6			A1	24?
28242.50	WD9CVP	Elgin, IL	EN52UA				A1	24
28242.70	F5ZWE	Foix	JN02TW	15	Vert Dipole	Omni	A1	24
28243.00	AA1SU	VT	FN34KL				A1	24
28244.00	WA6APQ	Long Beach, CA	DM13	30	Vertical	Omni	A1	24
28244.00	GB3TEN	Fleetwood	IO83LV	0.4	Dipole	Omni	F1A	24
28244.50	DV2FQN		KN10FC	5	GP			24?
28245.00	EB4YAK		IN80FK					24?
28245.00	DB0TEN	Bomlitz	JO42TW	2	1/2 GP	Omni	A1	24?
28245.60	SV2AHT	Hortiatis	KN1ONO	24				
28246.00	VE9BEA	Crabbe Mountain, NB	FN66	6	AR10@43	Omni	A1	QRT
28246.00	KG2GL	Nutley, NJ	FN20UT	5	R5 Vert @4'	Omni	A1	INT
28246.20	KI4LEV	Clarksville, TN	EM66	5			A1	?
28247.00	K6EMI		DM13AU				A1	?
28247.90	N1ME	Bangor, ME	FN54PS	5			A1	24
28248.00	K5DZE	Newman, GA					A1	?
28248.50	K5DDJ	San Antonio, TX	EL09	0.5	GP	Omni	A1	24
28249.00	N7LT	Bozeman. MT	DN45LQ	4/,4/,04	1/2 GP	Omni	A1	24
28249.00	KA3JOE	Bensalem, PA	FN29MB				A1	?
28249.10	ER1TEN	Chisinau	KN47IB	4	Vertical	Omni	A1	24
28249.50	PY3PSI	Porto Alegre	GF49KX	2.8	GP	Omni	A1	IRREG
26249.80	W4CJB	Santa Rosa Beach. FL	EM60WR				A1	?

Freq	Call	Nearest Town	Locator	ERPw	Antenna	Direction	Mode	Status
28249.90	W3ATV	Trevose, PA	FN20	1	Dipole		A1	24
28250.00	K1GND	Johnston, RI	FN41FT				A1	
28250.00	UB6LGR		JN54				A1	IRREG
28250.00	K8NDB	Somerton, AZ	DM22QQ	4	1/4 Vert	Omni	A1	24
28250.00	N4ES	Tampa, FL	EL88TA	20/2/,2/,02	Horiz		A1	SYNCH)n
28250.00	N4ESS	Zephyrhills, FL	EL88VG	A1			A1	SYNCH)n
28250.00	WB4WOR	Greensboro, NC	FM06BT	20/2/,2/,02	Horiz	Hor	A1	SYNCH)n
28250.00	K7EK	Graham, WA	CN87TB	25	1/2 Vert	Omni	A1	SYNCH)n
28250.00	N4ES	Clearwater, FL	EL88PA	20/2/,2/,02	Horiz	Horiz	A1	SYNCH)n
28250.00	K6FRC	Sutter Buttes, CA	CM97GP	10	Vertical	Omni	A1	24??
28250.00	K0HTF	Des Moines, IA	EN31DO	3	Inv,V@10		A1	OP?
28251.00	AC0MO	Hutchinson, KS	EM18				A1	?
28251.15	WA4GEH	Clayton NC	FM05SN				A1	OP?
28251.10	ED4YAK		IN80FK	5	Vertical	Omni	A1	24
28251.50	KE5JXC	Pecan Island, LA	EL39SP	5	Vertical	Omni	A1	24
28252.00	WA2DVU	Cape May, NJ	FM29NC	NJ			A1	24
28252.50	K7OC	Fort Worth, TX	EL29	3.3	Vertical		A1	24
28252.00	WW9EE	Tremont, IL	EN50GK		A1	?		
28252.50	W6PC/4	Ocala, FL	EL89VD	10	Dipole	Omni	A1	24
28253.00	N3BSQ	Bethel Park, PA					A1	24?
28253.00	ED5YAU		IM98WN	5	Vertical	Omni	A1	24
28253.00	KG4YUV	Crandall, GA	EM74OW	7	A99 Vert	Omni	A1	24
28253.00	K8HWW	Warren, MI	EN82MN	3			A1	24?
28253.80	XE1USG	Puebla	EK09VB				A1	?
28254.00	W4CJB	Point Washington, Fl	EM60WR				A1	24?
28254.00	PI7BXM	Baron de la Marckstraat 7, 6095 AW BAexem		JO21WF	1,5 Watt	1/2 wave vertical		Omni A1
28254.30	N1FCU	Windham, ME	FN43ST				A1	24?
28254.50	K4JEE	Louisville, KY	EM78				A1	24?
28254.50	K5AHH	Broken Bow, OK		OK			A1	?
28255.00	N0AR	St Paul, MN	EM73SW	0.5	l/2 Vert	Omni	A1	24
28255.50	K8HWW	Stirling Heights, MI	EN82MN	3	Vertical	Omni	A1	24
28256.00	C30P	Andorra	JN02SM	10	R5 Vert@2m	Omni	A1	24
28256.00	WI5V	Oklahoma City	EM15	0.5			A1	24
28256.50	VK3RMH	25 km NE Melbourne	QF22OH	20-Feb	1/2 Vert	Omni	A1	24
28256.50	K9JHQ	O'Fallon, IL	EM58AM	10	Vertical	Omni	A1	24?
28257.00	KB4UPI	Bessemer, AL	EM63MG				A1	24
28257.30	WA2DVU	Cape May, NJ		10	Mosley 57	45	A1	24
28257.50	N5WYN	Seven Points, TX	EM12VI				A1	
28257.50	DK0TEN	Sipplingerberg, near Überlingen			10			F1A
28257.80	WY5I	Port St Kucie, FL	EL97TF	5	7 dB Coll	Omni	A1	07 - 22:00 LT
28258.00	EA7JNC	La Linea de la Conception	IM76IE	8				24
28258.00	NM5TW	Albuquerque, NM	DM65RD	5	Vert Dipole	Omni	A1	24
28258.30	N1YPM	Corea, ME	FN64				A1	24
28259.00	K5TLL	Hattiesburg, MS	EM51GG				A1	?
28259.00	AB8CL	Arcanum, OH	EN79RA				A1	?
28259.00	AA4AN	Brentwood, TN	EM65NW	4	Vertical	Omni	A1	24
28259.00	F5ZVM	Valenciennes	JO10PA	5	3dbi Vert	Omni	A1	24
28259.30	VK5W1	Adelaide	PF95HG	10	GP	Omni	A1	24
28259.30	AF6PI	Indio, CA					A1	
28260.00	AD5KO	Mena, AR	EM24VS	20	Vert@20	Omni	A1	24
28260.10	W7LFD/0	Shell Knob, MO		5	Vertical	Omni	A1	?
28260.80	NJ3T	Somerset, PA	FM09LX				A1	24?
28260.80	W5TXR	Schertz, TX	EL09VP				A1	?
28261.00	N7LF	Corbett, OR	CN85VI				A1	24
28261.50	N4VBV	Sumter, SC	EM93TW	5	Attic Dipole		A1	24
28261.60	RK3XWA	Kaluga	KO84DM	24				
28261.80	VK2RSY	Sydney	QF56MH	25	1/2 Vert	Omni	A1	24
28262.00	N4HFA	Ocala, FL	EL89VP	3	5/8 GP @ 25'	Omni	A1	24
28262.30	K8TK	Clarklake, MI	EN72TC	2	GP@15	Omni	A1	24
28262.50	WF4HAM	Altamonte Springs, FL	EL95HP	6	A99@40	Omni	A1	INT
28263.00	VK3RRU	Mildura	QF15AT	20			A1	?
28263.00	ED4YBA	Cuneca	IN80WC	5	GP	Omni	A1	24
28263.50	N5YEY	Kilgore, TX	EM22OJ				A1	24
28263.50	W4JPL	Liberty, NC	FM05EW	4			A1	24
28264.00	AB8Z	Parma, OH	EN91DJ	5	5/8 Vert	Omni	A1	24
28264.00	VK6RWA	Carine, WA	OF78WB	20	5/8 Vert	Omni	A1	24
28264.50	K7NWS	Kent, WA	CN87TK	1	GP	Omni	A1	24
28264.50	W5ZA	Shreveport, LA	EM32DJ	3	Vert Dipole	Omni	A1	24
28265.00	DF0ANN	Moritzberg Hill	JN59PL	5	Dipole	E-W	A1	24
28265.00	VK4RRC	Woody Point QD	QG62NS	10	Vertical	Omni	A1	24?
28265.00	KJ3P	Schwenksville PA	FN20GG	5			A1	?
28265.00	PT9BCN	Campo Grande, MS	GG29RN	12	1/2 Vert		A1	24
28265.40	KR4HO	Lake City, FL	EM80QG	1	Vertical	Omni	A1	?
28265.00	NC4SW	Zebulon, NC	FM05				A1	24
28265.00	N7SCQ	Dixon, CA	CM98CK				A1	24
28266.20	KB3ZI	Bloomsberg, PA					A1	24
28266.50	KA1EKS	Millinocket, ME	FN55OO	4	A99 GP	Omni	A1	24
28266.50	W5DJT	Pocola, OK	EM25SH				A1	24
28266.60	WN5KNY	Radium Springs NM	DM62LP	24			A!	24
28267.00	VK7RAE	TAS	QE38DT	10	Vertical	Omni	A1	24
28267.50	W5EFR	Houston, TX	EL29EW	2.75			A1	TQRT
28267.60	OH9TEN	Pirttikoski	KP36OI	20	1/2 GP	Omni	A1	24
28268.00	KB0QZ	Centralia, MO	EM39WE	5	Vertical	Omni	A1	24
28268.00	NM0R	St Genevieve, MO	EM47UV		A1		A1	
28268.30	VK8VF	Darwin, NT	PH57KP	25	1/4 Vert	Omni	A1	24
28268.50	KG4GXS	Coral Springs FL	EL96UG	3	Dip@23	E-W	A1	24

Freq	Call	Nearest Town	Locator	ERPw	Antenna	Direction	Mode	Status
28268.60	K7ZS	Hillsboro, OR	CN85MM				A1	24
28268.80	KD5ITM	Spring, TX		4	G5RV@50		A1	
28269.00	WA2SFT	Cookville, TN	FN02OU				A1	
28268.90	AA1TT	Claremont, NH	FN33	5			A1	24
28268.90	SV6DBG	Ioannina	KM09KQ	2	5/8 Vertical	Omni	A1/ RTTY	24
28269.50	W3HH	near Ocala, FL	EL89VB	6	Hamstick	Omni	A1	24
28270.00	VK4RTL	Townsville	QH30JS	5	Vertical	Omni	F1	24
28270.50	PY4MAB	Pocos de Caldas	GG68RE	10	Vertical		A1	24
28271.00	OZ7IGY	Jystrup	JO55WM	10	Halo@90m	Omni	P14/CW	24
28271.70	W4TIY	Dallas, GA	EM73OW	4	¼ over 5/8	Omni	A1	24
28271.80	SV2HQL	Katakali-Grevens	KM09UV	5	5/8 GP	Omni	A1	24
28272.30	N1KON	Centerville, IN	EM79LT	5	Vertical	Omni	A1	24
28271.90	AC0RR	Springfield, MO	EM37IE				A1	?
28272.00	PY1RJ	San Goncalo, RJ	GG87ME	4			A1	24
28272.50	K5BTV	Cumming, GA	EM74	0.25	HF6V Vert	Omni	A1	24
28273.00	AC4DJ	Eustis, FL	EL98EU	20	Ringo	Omni	A1	24
28273.00	WF4HAM	Altamont Spr, FL		10	Ringo		A1	?
28273.00	DB0BER	Berlin	JO62QL	5	Dipole	Omni	A1	24
28274.00	LW1DZ	Escobar, BA	GF05OQ	10	Loop		A1	24
28274.70	N0UD	Halliday, ND	DN87SH				A1	24
28275.00	PY2EMG	Jacarel, SP	GG76AQ				A1	?
28275.00	NP2SH	St John, VI	FK78OI				A1	?
28275.00	KG4GVV	Summerville, SC	EM93		Vertical	Omni	A1	24
28276.00	K4UKB	Danville, KY	EM77NP	10	5/8 Vert	Omni	A1	SYNCHx
28276.00	K4FUM	Stone Mountain, GA	EM73WU				A1	SYNCHx
28276.50	XE2YBG	Victoria Tamaulipas	EL03				A1	?
28277.00	WB7RBN	Pasco, WA	DN06IG				A1	24
28277.00	WI4L	Dalton, GA	EM74MS				A1	24
28277.00	WD8AQS	Fremont, MI	EN73AL				A1	24
28277.30	KD4MZM	Sarasota, FL	EL87RG	3	Ringo@15	Omni	A1	24
28277.60	DM0AAB	near Kiel	JO54GH	12	GP	Omni	F1	24
28278.00	WA4OTD	Carmel, IN	EM69	5	Indoor	Dip	A1	24
28278.20	KE4IFI	Lexington, SC					A1	24
28278.50	WA6MHZ	Crest, CA			Ringo@20	Omni	A1	24
28279.00	DB0UM	Schwedt	JO73CE	2	SlopeDipV	Omni	A1	24
28280.00	KA3NXN	Charlottesville	FM08SA				A1	24
28280.00	K5AB	Goldthwaite, TX	EM01	20	5/8 GP @ 45'	Omni	A1	24
28280.00	PU5AAD	Nova Brasilia	GG51PS	10			A1	24
28280.00	N6SPP	Concord, CA	CM97	10	Vert Dipole	Omni	A1	24
28280.50	WB6FYR	Quartz Hill, CA	DM04VP	10	5/8 Vert	Omni	A1	24
28281.00	W8EH	Middletown, OH	EM79TL	7.5	Vert@40	Omni	A1	
28281.20	IK6ZEW	Pescara	JN72OK				A1	?
28281.50	W4HEW	Millegeville GA	EM94LX				A1	24
28282.00	LA6TEN	Kirkenes	KP49XQ	10/1/,1	Omni		A1	OP?
28282.00	HP1ATM	Santiago	EJ98				A1	24
28282.00	XE2ES	Mexicali	DM22RP				A1	24
28282.00	N2IFC	Allamuchy					A1	?
28282.60	OK0EG	Hradec Kralove	JO70WE	10	GP	Omni	F1	24
28282.80	W0ERE	Fordlan, MO	EM36	5	Vertical	Omni	A1	24
28283.00	K7YSP	Gainsville, GA	EM84AH				A1	24
28283.60	KC9GNK	Madison, WI	EN53	4	Inv Vee		A1	24
28283.80	W5OM		DN28	98	3-el		A1	IRREG
28284.00	K2XG	Monticello, KY	EM76NU	5	V,Dip@25	Omni	A1	24
28284.80	WD8AQS	Fremont, MI	EN73AL	5			A1	24
28284.80	WA3IIA	Bloomsberg, PA	FN11TA				A1	24
28284.50	KB9NK	Hudsonville, MI	EN72BU				A1	24?
28285.00	VP8ADE	Adelaide Is	FC52WK	10	1/4 Vert	Omni	A1	24
28285.00	W7IEW	Olympia, WA	CN87MC				A1	24?
28285.00	KD0GZJ	Loveland, CO	DN70KJ	1	Vertical		A1	24
28285.00	PT9BCN	Campo Grande, MS	GG29RN	1	Vertical	Omni	A1	24
28285.30	K5DRG	Lago Vista, TX	EM10CM			A1	?	
28285.80	WA4ROX	Largo, FL	EL87	0.75		A1	24	
28285.80	W0ILO	Fargo, ND	EN16					A1
28286.00	WI6J	Bakersfield, CA	DM05JJ	5	Vertical	Omni	A1	24
28286.00	N5AQM	Chandler, AZ	DM43AH	2	Vertical	Omni	A1	24
28286.00	N2PD	Middletown, NY	FN21	5			A1	?
28286.70	K3XR	Sinking Spring PA	FN10XH				A1	?
28287.00	GB3XMB	near Waddington	IO83SV	10	V,Dip@800	Omni	F1A	24
28287.00	W6WTG	Bakersfield, CA	DM05MJ				A1	24
28287.30	W2SDX	Buffalo, NY	FM18LV				A1	
28287.50	N8PUM	Ishpeming, MI		0.5	Loop		A1	24
28288.00	WA7LNW	Harmony Mesa, UT	DM37	5	F Wave Loop	EU/PAC	A1	24
28288.00	K4LJP	W Palm Beach FL	EL96	5	AR99@30	Omni	A1	24
28288.00	RA3ATX		KO85NX					OP?
28288.50	ND3E	Newcastle, DE					A1	24
28289.00	KB9WA	Egg Harbour, WI					A1	
28289.00	WB5BXZ	Hattiesburg, MS					A1	?
28289.00	WJ5O	near Columbus, AL	EM72NE	2	Yagi	NE	A1	24
28289.50	N1KXR	Medway, MA	FN32	5			A1	24
28289.80	W0ERE	Highlandsville MO	EM36HX	Varies			A1	24
28289.00	PS8RF	Teresina, PI	GI84OW				A1	24
28290.00	N6UN	San Diego, CA	DM12	5			A1	24
28290.30	WB4WOR	Randleman, NC	FM06BT	3	Vertical	Omni	A1	24
28291.00	K5TLJ	Trumann, AR	EM45RQ	20			A1	?
28291.00	W6NIF	Fresno, CA			GP @ 25 Ft	Omni	A1	?

Freq	Call	Nearest Town	Locator	ERPw	Antenna	Direction	Mode	Status
28291.80	N5MAV	Midland, TX					A1	
28292.00	VA3VA	Windsor, ON	EN82	5	Horiz Dipole		A1	24
28292.30	NH6HI	Kaleheo, HI		2			A1	24
28292.50	KM4GS	Kentucky Lake KY	EM56	0.5	Vertical	Omni	A1	24
28292.50	SK0CT	Sollentuna	JO89XK	10	GP	Omni	A1	24
28292.80	K7GFH	Damascus, OR	CN85SJ	3	Attic Dipole		A1	24
28293.40	ND4Z	Gilbert, SC	EM94JA	5	5/8@40	Omni	A1	24
28293.70	W4DJD	Woodbridge, VA	FM18HP				A1	24
28294.00	K7RON	Peoria, AZ	DM33				A1	24
28294.00	KE4IAP	Woodbridge, VA	FM18HP				A1	24
28295.00	KD1ZX	Central Falls RI	FN41HV	4	Vert@10	Omni	A1	24
28295.00	PU5ATX	Santa Catarina	GG51PR		Dipole		A1	06 to 23:50
28295.10	SK2TEN	Kristineberg	KP08FC	5	Vertical	Omni	A1	OP?
28295.40	K1SPD	La Vergne, TN	EM65RX			A1	24	
28295.50	K4IT	Flatwoods, KY	EM88PM	3.5		Dipole	A1	24
28295.50	IZ0CWW	Cervaro, FR	JN61VL	3				24
28295.70	W9MUP	WI	EN52				A1	?
28295.80	W3APL	Laurel, MD	FM19NE	10	Hor,Dipole	NE/SW	A1	24
28296.00	KA7BGR	Central Point OR	CN82				A1	24
28296.20	W5JDG	Washington, TX					A1	24
28297.00	NS9RC	Northfield, IL	EN62CC	5	1/2V@30	Omni	A1	24
28297.80	WA3BM	Valencia, PA	FN00SQ				A1	24
28298.00	V73TEN	Roi Namur I	RJ39RJ		Horiz OA50	Omni		24?
28298.00	K5TLL	Hattiesburg, MS	EM51GG	5	Vertical	Omni	A1	24
28298.15	WZ8D	Blachester OH	EM89OO				A1	24
28298.10	SK7GH	Bor	JO77BF	5			A1	24
28298.10	K7PO	Tenopah AZ	DM32	0.5	Vertical	Omni	A1	24
28298.50	K7FL	Battle Ground, WA	CN85SS	4	Horiz	Loop	A1	24
28298.60	K4JDR	Raleigh, NC	FM05	10	Vertical	Omni	A1	24
28299.00	N1SCA	Palm Bay, FL	EL97QX				A1	24
28300.00	K6FRC	Sutters Mountain, CA	CACM99				A1	24##
28300.00	HL1ZLL	Seoul	PM37MM	85	1/4 Vert	Omni	A1	24
28300.00	IK6ZEW	Pescara	JN72CK	0.4	Vertical		A1	?
28300.00	KF4MS	Tallahassee, FL	EM70VM	10	A99@25	Omni	A1	24
28300.30	PU5ZAA	Novegentes	GG53QE	3			A1	24
28301.00	PI7ETE	Amersfoort	JO22QD	0.5	Vertical	Omni	F1A	24
28320.97	IZ8HUJ	Pignola PZ	JN70VN	0.4	Windom			?
28321.00	I1YRB	Torre Bert (TO)	JN35UB	0.2	Vertical	Omni	QRSS3	24
28321.20	IS0GOV	Cagliari	JM49NF	0.3	Vertical	Omni	QRSS3	24
28321.20	IK3ERY	Vittorio, Veneto	JN65DX	0.1	Vertical	OMNI	QRSS3	24
28321.47	IZ1GJH	Casarza Lig, GE	JN44RG	0.1	Vertical	Omni	QRSS3	24
28321.65	IN3KLQ	near Trento	JN56RG	0.3	Vertical	Omni	QRSS3	24
28321.70	IW4EMG	Ferrara	JN54RU	0.1			QRSS3	24
28321.80	I3GNQ	Tencarola, PD	JN55VJ	0.4	GP	Omni	QRSS3	24
28321.85	IK0IXI	Aprilia LT	JN62VB	0.1	Dipole@30m	NE/SW	A1/F1	24
28321.86	IZ8JFA	Cosenza	JM89CH	0.3		N/NE	QRSS3	24
28321.94	IW9BAJ	Sicily	JM77NO	1			QRSS3	24
28321.95	IW0HK	Civitavecchia	JN52WD	1/,1	1/2 Vert	Omni	QRSS3	24
28321.70	IZ1TAA	near la Spezia	JN54AC	0.1	Dipole	NW	QRSS3	24
28322.00	IQ3QR		JN55XR	0.5			QRSS3	24
28322.00	IZ7AVU	Brindisi	JN80XP					?
28322.20	IZ5ILK		JN63BN	0.05	5/8 Vert	Omni	?	?
28322.00	IZ0ANE	Cassino		0.1				?
28322.01	IW1QIF	Davagna-Genova	JN44LL	0.1	Long Wire		QRSS3	24
28322.04	IT9YAF	Canicatti	JM67WI	0.1	Vertical	Omni	QRSS3	24
28322.05	IK1HGI	Trecate	JN45IK	0.1	Dipole	NNW/SSE	QRSS3	24
28322.08	IS0GSR	near Cagliari	JM49IN	0.1	Dipole	N-S	QRSS3	24
28322.11	IK8YTN	Salerno	JN70LG	0.1	Int Vee	E-W	QRSS3	24
28322.18	IZ0HCC	Roma	JN61FT	0.1	Vertical	Omni	QRSS3	?
28322.20	IK8SUT	Salerno	JN70JQ	0.1	Inv Vee	E - W	QRSS3	24
28322.20	IW7DEC	near Bari	JN81GF	0.1	1/2 Vert	Omni	QRSS3	24
28322.32	IW9FRA	Trapani, Sicily	JM68HA	0.1	Dipole	N-S	QRSS3	24
28322.36	IW3SGT	Trieste	JN65VP				QRSS3	24
28322.45	IW4EMG	near Ferrara	JN54RW		Vertical	Omni	QRSS3	?
28322.62	IK1BPL	Novara	JN45HK	0.1	GP	NNW/SSE	QRSS3	24
28322.63	IK0VVE	near Latina	JN61KN	0.1	Dipole	NNW/SSE	QRSS3	24
28322.70	G3ZJG	near Leicester					QRSS3	PT
29010.00	DU1EV	Metro, Manila		100	Vert GP	Omni		

Notes:

? = Activity uncertain; **INT** = Intermittent; **PT**=Part-time; **IRREG** = Irregular; **UC** = Under Construction; (T) **Non Op** = (Temporarily) NotOperational; **OP?** = Operational?; **V** = Varies; **EXP** = Experimental; **WkEnd** = Weekends during daylight; **LT** = Local Time; **SYNCH** = frequency sharing synchronised beacons: SYNCHn current sequence N4ESS (0,00 seconds), N4ES (0,20), WB4WOR (0,30), K7EK (0,40), N4ES (0,50). SYNCHx: K4UKB ID at 00 seconds and K4FUM at 30 seconds. ## K6FRC transmits on 28 245, 28 250, 28 300
+++ 10 minutes constant carrier followed by ID
zzz GB3WES at H+1, H+16 etc and GB3ORK at H+2, H+17 etc. Transmissions have a stepped power sequence.

IARU Region 1 VHF/UHF and Microwave Beacons

This list of 50MHz Beacons is compiled for IARU Region 1 by G3USF. Thanks are also due to the VHF/UHF/Microwave Managers of radio societies across Region 1, beacon keepers, beacon co-ordinators and VHF/UHF DXers too numerous to mention.

Note that this list includes information regarding beacons notified to the R1 coordinator which operate (or are Planned to operate) outside of the IARU R1 band Planned beacon segments due to local licensing requirements.

Some member Societies may be unable to comply for local regulatory reasons. As a general principle, all beacons should move to band Planned frequencies as soon as possible.

Freq	Call	Nearest Town	Locator	ERPw	Antenna	Direction	H/V	Mode	Keeper
6M									
50000.000	YM1SIX	Tekirdag	KN30QT	16	Halo	omni	H	A1A	TA2NC
50000.000	YM2SIX	Ankara	KM69IV	16	Halo	omni	H	A1A	TA2NC
50000.000	YM3SIX	Izmir	KM38OJ	16	Halo	omni	H	A1A	TA2NC
50000.000	YM5SIX	Adana	KM76QX	16	Halo	omni	H	A1A	TA2NC
50000.000	YM6SIX	Merzifon	KN70QV	16	Halo	omni	H	A1A	TA2NC
50000.000	YM8SIX	Gaziantep	KM87QB	16	Halo	omni	H	A1A	TA2NC
50004.000	EI0SIX	Kilmolin, Enniskerry, Co. Wicklow	IO63VE	50	Ho Loop	omni	H	F1	EI7BIB
50005.000	PI7SIX	Rotterdam	JO21VF68	17	Crossed Dipole	Omni	H	F1A	PA2M
50006.000	IW9GDC	Messina	JM78SD	10	Big Weel	Omni	H	A1A	IW9GDC
50006.000	IW9GDC/B	Messina	JM78SD	10	Big Weel	Omni	H	A1A	IW9GDC
50012.000	OX3SIX	Tasiilaq city	HP15EO	25	Dipole		H	F1A	OZ1DJJ
50016.000	SV5SIX	Profitis Ilias Mt. / Rhodes	KM36XG	3	3el yagi	dir 330	H	A1A	SZ5RDS
50017.000	OH0SIX	Stålsby	JP90XI	3	Dipole	Omni	V	A1A	OH0JFB
50023.000	LX0SIX	Burscheid	JN39AV	10	Loop	omni	H	A1A	RL
50025.000	OH2SIX	Lohja	KP20DH	50	1/2 Vertical	Omni	V	A1A	OH2LNM
50039.000	FY7THF	Guyane	GJ35IG	10	Verticale	omni	V	A1A	FY1FL
50045.000	OX3VHF	Prince Christiansund	GP80KB	25	Dipole		H	F1A	OZ1DJJ
50046.000	SV9GPV	Rethimno / Crete	KM25EH	5	2el yagi	dir 330	H	F1A	SV9GPV
50047.000	OX6M	North East Greenland	HQ90AL	25	Dipole	N/S	H	F1A	OZ1DJJ
50052.000	SK2CP/B	Kiruna / Esrange Space Center	KP07MU	30		Omni	H	A1A	
50058.000	OE3XLB	Hohe Wand, Kleine Kanzel	JN78SB	10	Straight Dipole	N-S	H	A1A	OE3GWC
50060.000	SK6QW/B	Mariestad	JO68WR	8		Omni	H	A1A	
50066.000	OE3XAC	Kaiserkogel	JN78SB	10	M2 HoLoop	OMNI	H	A1A	OE3KLU
50067.000	OH9SIX	Rovaniemi Pirttikoski	KP36OI	35	2x X-Dipole	Omni	H	A1A	OH9LA
50069.000	FM1ZAC	Martinique	FK94NL	15	Turnstile	omni		A1A	FM1HM
50080.000	FK8SIX	Noumea	RG37GT	10	Verticale	omni	V	F1A	FK8HA
50400.500	IW3FZQ/B	Monselice PD	JN55VF	8	5/8 Vert.	Omni	V	A1A	IW3FZQ
50401.000	ON0VRN	DIKSMUIDE	JO11JA	5		omni	?	F1A	ON8SD
50401.000	IQ8KK/B	Cosenza	JM89DH	2	Halo	Omni	H	A1A	U
50402.000	OY6BEC	Sornfelli	IP62MB	15	Quad Loop	omni	H	F1A	OY9JD
50403.000	YM4SIX	Tepe, Alanya	KM66AO02	40	2 stacked triangle dipoles	omni	H	F1A	PE2M
50405.000	OE7XBI	Ranggerköpfl/Tirol	JN57OF	10	Groundplane	Omni	V	F1A	OE7NCI
50406.000	SV2JAO	Vermion Mt. / Imathia	KN10CL	10	3el yagi	dir 310	H	F1A	SV2JAO
50406.000	F5ZHQ	Le Pradet	JN33BC	10	Verticale	omni	V		F6FCE
50408.000	IK5ZUL/B	Poggio Spada GR	JN52JW	3	Dipole	90/270°	H	A1A	IK5ZUL
50412.000	OH1SIX	Ikaalinen	KP11QU	50	2x X-Dipole	Omni	H	A1A	
50414.000	OK0SIX	Ostrava Poruba	JN99CT	5	Groundplane	Omni	V	A1A	OK2WA
50415.000	SR5TDM	Tarczyn	KO01KX01	3.0 W PEP	cross dipole	OMNI	H	A1A	SP5XMU
50418.000	F1ZFE	Erching	JN39OC	8	Boucle	omni			?
50419.000	IZ1EPM/B	Saronsella TO	JN35WD	10	GH50 5/8	Omni	V	A1A	IZ1EPM
50420.000	IQ4FE/B	Fidenza PR	JN54AS	6	AR-6	Omni		A1A	IK4CIE
50422.000	S55ZRS	Kum	JN76MC	5	GP	omni	V	A1A	ZRS
50422.000	F5ZMT		IN88GN	10	Halo	omni	H	A1A	F5NLG
50425.000	PI7SIX	Rotterdam	JO21VF68	17	Crossed Dipole	Omni	H	F1A	PA2M
50427.000	IW0DTK/B	Latina	JN61TG	3	5/8 Vert.	Omni	V	A1A	IW0DTK
50430.000	IZ3OHR/B	Verona	JN55LJ	5	Dipole h.	Omni	H	A1A	IZ3OHR
50431.000	9A0BLH	Moslavacka gora	JN85JO	3.2	X dipole	omni	H	A1A	9A5R
50432.000	F5ZKY	Mt Alouettes	IN96LV	10	Verticale	omni	V	A1A	F6DBA
50433.000	OH7SIX		KP52			Omni	H	A1A	OH7GGX
50434.000	F8BHU	Nevers	JN17NA	1	GP	omni	V		F8BHU
50436.000	D4C	Monteverde, Mindelo	HK76MU79VB	30	5/8 Vertikal	OMNI	V	A1A	D4C (HB9DUR)
50437.000	ES0SIX	Vilsandi/Isl.Saaremaa	KO08WJ32MG	15.84	folded dipole	240/30	H	A1A	ES2PW
50438.000	OE5XHE	Sternstein/Mkr.	JN78DN	10	OA 50	OMNI	H	A1A	OE5ANL
50441.000	ON0SIX	VIEUX GENAPPE	JO20EP	50		omni	H	F1A	ON4KJV
50445.000	JW5SIX	Hopen	KQ26MM	10	Dipole	Omni	V	A1A	LA5RIA
50447.000	JW7SIX	Isfjord Radio	JQ68TB	30	3el Yagi	180	H	A1A	LA0BY
50448.000	FX4SIX	Neuville	JN06CQ	30	Turnstile	omni		A1A	?
50449.000	IQ0AM/B	M.te Serpeddi CA	JM49PI	5	GP	Omni	V	A1A	IS0BSR
50450.000	SV3BSF	Patras	KM08VA	5	2el yagi	dir 320	H	F1A	SV3BSF
50451.000	LA7SIX	Målselv	JP99EC	100	4el Yagi	190	H	A1A	LA5TFA

Freq	Call	Nearest Town	Locator	ERPw	Antenna	Direction	H/V	Mode	Keeper
50455.000	LA8SIX	Hvasser	JO59FB	50	Dipole	Dipole 0° / 180°	H	A1A	LA6LCA
50455.000	IQ0HV/B	Cecchina Roma	JN61HQ	8	Horned	NW-SE	H	A1A	IK0BDO
50456.000	YM0SIX	Bozcaada	KM39AT28TD	5	Halo	Omni	H	A1A	TA0G
50457.000	TF1VHF	Myrar, 80Km NNW Reykjavik	HP84WL	25	1/2 Dipole	Dipole NW / SO	H	F1A	TF1A
50458.000	IQ4AD/B	Parma	JN54DT	8	5/8 Vert.	Omni	V	A1A	I4YMB
50459.000	LA9SIX	Gjøvik	JP50EV	25	5/8 GP	omni	V	A1A	LB3RE
50468.000	SK3SIX	Östersund	JP73HC17RO	15	Crossed Dipole	Omni	H	F1A	SM3PXO
50471.000	OZ7IGY	Jystrup	JO55WM54UA	25	Big Wheel	omni	H	F1A	OZ7IS
50474.000	IQ5MS/B	Massa MS	JN54AA	3	Vert.	Omni	V	A1A	IW5ECP
50475.000	JW9SIX	Bjørnøya	JQ94LM	10	dipole	Omni	V	A1A	LA5RIA
50477.000	IZ3GWJ/B	Lendinara RO	JN55TC	5	J-pole ¼ wl	Omni	V	A1A	IZ3GWJ
50479.000	JX7SIX	Jan Mayen	IQ50RX	10	Dipole	150/330	H	A1A	LA7DFA
50481.000	F1ZFB	Ste Lheurine	IN95TM	10	dipole horiz.	Dipole 0° / 180°	H	A1A	F1MMR
50483.000	DB0DUB	nr Gangelt	JO31AA	1	Halo	omni	H	A1A	DL1KBL
50483.000	DB0ANN	Moritzberg	JN59PL	2	Ho-Loop	OMNI	H	A1A	DL8ZX
50485.000	SV1SIX	Athens	KM17UX	30	5/8 vertical	omni	V	F1A	SZ1SV
50485.000	OH5SHF	Kouvola	KP30HV	100	2 dBd	180	H	F1	OH5IY
50485.000	ED7YAN	Grenada	IM87GC	5	Loop	Omni	H	F1	EA7UU
50488.000	LA2SIX		JP53EG	25			V	A1A	
50488.000	YM7SIX	Ordu	KN90AW	16	Halo	omni	H	A1A	TA2NC
50490.000	F6IKY	Louhans	JN26OP	3	commut H/V	omni	H/V		F6IKY
50493.000	YM9SIX	Kars	LN10PM	16	Halo	omni	H	A1A	TA2NC
50495.000	SV9SIX	Vasiliko Mt. / Crete	KM25NH	25	Halo	omni	H	F1A	SZ1SV
50495.000	V53SIX	Swakopmund, Namibia	JG77II	H 20.0\V 10.0"	Ringo	omni		F1A	V51DM
60013.000	EI1KNH	Kilmolin, Enniskerry, Co. Wicklow	IO63VE	25	vertikal folded dipole	omni	V	F1A	EI4GNB
4M									
70008.000	OH1FOUR								
70012.000	OX4MB	Kangerlussuaq	GP47TA	25	Dipole		H	F1A	OZ1DJJ
70014.000	S55ZRS	Kum	JN76MC	5	ho-loop	omni	H	A1A	ZRS
70016.000	SV5FOUR	Profitis Ilias Mt. / Rhodes	KM36XG	5	3el yagi	dir 330	H	F1A	SZ5RDS
70018.000	OH2FOUR	Lohja	KP20DH	30	Dipole	Omni	V	A1A	OH2LNM
70021.000	OZ7IGY	Jystrup	JO55WM54UA	25	Big Wheel	omni	H	F1A	OZ7IS
70029.000	S55ZMB	Starše	JN76VK	5	4	omni	H	A1A	S51DI
70031.000	9A0BFH	Moslavacka gora	JN85JO	4	X dipole	omni	H	A1A	9A5R
70033.000	OH5RBG	Kouvola	KP30HW	15	7 dBi	232	H	F1	OH5IY
70035.000	OY6BEC	Sornfelli	IP62MB	20	Quad Loop	omni	H	F1A	OY9JD
70037.000	ES1VHF	Tallinn	KO39IK50RK	10	dipole	240°/30°	H	F1A	ES1NI
70040.000	SV1FOUR	Athens	KM17UX	15	5/8 vertical	omni	V	F1A	SZ1SV
70047.000	OX4M	North East Greenland	HQ90AL	25	Dipole	N/S	H	F1A	OZ1DJJ
70055.000	SV9FOUR	Rethimno / Crete	KM25EH	5	3el yagi	dir 330	H	F1A	SV9GPV
70057.000	TF1VHF	Myrar, 80Km NNW Reykjavik	HP84WL	25	1/2 Dipole	Dipole NW / SO	V	F1A	TF1A
70063.000	LA2VHF	Vassfjellet	JP53EG	25	GP	omni	V	A1A	LA1K
70065.000	LA5VHF	Kristiansand	JO48AD	15	GP	omni	V	A1A	LA4YGA
70067.000	LA9VHF	Gjøvik	JP50EV	25	5/8 GP	omni	V	A1A	LB3RE
70070.000	PI7RAZ	Zoetermeer	JO22FB	10			H		
70070.000	PI7RTD	Rotterdam	JO21FV68	10					PE1GHG
70075.000	LA4VHF	Egersund	JO28WL	50	Dipole	N/S	H	A1A	LA3EQ
70081.000	LA7VHF	Målselv	JP99EC	40	4-ele Yagi	190	H	A1A	LA5TFA
70087.000	SV2JAO	Vermion Mt. / Imathia	KN10CL	10	4el yagi	dir 310	H	F1A	SV2JAO
70114.000	ON0HVL	Heuvelland	JO10KS	10	Big Wheel 2 dBi	omni	H	F1A	ON7FLY
70115.000	SR5TDM	Tarczyn	KO01KX01	3.0 W PEP	cross dipole	OMNI	H	A1A	SP5XMU
70130.000	EI4RF	Kilmolin, Enniskerry, Co. Wicklow	IO63VE	30	2 Yagi System	Yagi E & S	H	F1	EI4GNB
70161.000	LX0FOUR	Burscheid	JN39AV	10	Loop	omni	H	A1A	RL
70242.000	ON0RCL	Leuven	JO20IV	5		omni	H	F1A	ON5CMB
70244.000	ON0BR	Montigny-Le-Tilleul	JO20EI				H	F1A	ON4UQ
2M									
144150.000	UR0UBA	Kyiv	KO50FK	5	Crossed Dipole	OmniH	H	F1A	UT4UWJ
144401.000	SV2HQL	Katakali / Grevena	KM09UV	10	Yagi	120° 315°	H	F1A	SV2HQL
144402.000	OY6BEC	Sornfelli	IP62MB	100	2el. HB9CV	SE	H	F1A	OY9JD
144403.000	YM4VHF	Tepe, Alanya	KM66AO02	400	2 x 4x3ele Yagi	30° 184° / 344°	H	F1A	PE2M
144404.000	DB0THE	Ohmgebirge - Birkenberg	JO51EL		4 x 2 Element	OMNI	H	"A1A	DF7AP
144404.000	IW1AU/B	Giaveno TO	JN35PA	5	Big wheel	Omni	H	A1A	IW1AU
144405.000	F5ZRB	Quistinic	IN87KW	40	Yagi 7 elts	Yagi 7 elts 135°	H	F1A	F6ETI
144406.000	SK6VHF	Tjörn Island	JO57TX	10	M2 loop	Omni	H	A1A	SM6CEN
144407.000	GB3SSS	Helston	IO70IA	200	8 OVER 8 SLOT YAGI	Yagi 300°	H		G7THT
144409.000	F5ZSF	Lannion	IN88GS	5	9 elts	9 elts 60°	H	A1A	F6DBI
144410.000	DB0MFI	Hainsfarth	JN58HW	2	Big Wheel	- OMNI	H	A1A	DK1MFI
144410.000	IK1FJI/B	Sant'Olcese GE	JN44LL	6	eggbeater	Omni	H	A1A	IK1FJI
144412.000	SK4MPI	Borlänge	JP70NJ	200	2 X 5 EL YAGIS	NV+NO	H	A1A	SM4HFI
144412.500	SZ1SV	Parnis Mt. / Athens	KM18UE	0.5	1/4 vertical-indoor	omni	V	F1A	SZ1SV
144415.000	DB0JW	Hellenthal	JO30EK	10	2 x Big Wheel	OMNI	H	A1A	DL3KAT
144415.000	SR5TDM	Tarczyn	KO01KX01	5.0 W PEP	cross dipole	OMNI	H	A1A	SP5MX
144415.000	IQ2MI/B	M.te Rosa VC	JN35WW	1	Dipole Kathrein	110°	H	A1A	IZ2JGB
144416.000	PI7CIS	Den Haag	JO22DC	17			H	FSK	PA0C

Freq	Call	Nearest Town	Locator	ERPw	Antenna	Direction	H/V	Mode	Keeper
144417.000	OH9VHF	Rovaniemi Pirttikoski	KP36OI	160	7 dBd	200	H	A1A	OH9LA
144417.000	F5ZXT		JN33	5					F6FCE
144418.000	ONOVHF	Louvain-La-Neuve	JO20HP	25		omni	H	F1A	ON4IV
144419.000	IQ2CY/B	Cremona	JN55AD	10	Big wheel	Omni	H	A1A	IK2THZ
144420.000	RB1CA	Priozersk/Russia	KP51BA	5	3 Quad Style Yagi 3 ele	135°	H	A1A	UA1CCB
144421.000	SK3VHF	Östersund	JP73HC17RO	500	2 X 3el Quad	60° 180°	H	F1A	SA3AZK
144422.000	OH0	Åland							
144425.000	SR9VHK	Siemianowice	JO90MH01	3.0 W PEP	Dipole	120° / 300°	H	F1A	SP9BGS
144425.000	F5ZAM	Blaringhem	JO10EQ	10	Trefle	omni	H	A1A	F6BPB
144425.000	I5WBE/B	Mt. Fiore PT	JN53LT28SV	3	2 x Halo	Omni	H	A1A	I5WBE
144428.000	DB0JT	Wildberg	JN67JT	30	4 Dipole	OMNI	H		DJ8QP
144429.000	IQ3MF/B	Cormons GO	JN65RW	3	2 Turnstile	Omni	H	A1A	IV3HWT
144430.000	GB3VHF	Fairseat,Kent	JO01EH08	30 / 100	2 X 3 Ele. Yagi	Yagi 288° / 348°	H	F1A	G0FDZ
144431.000	PI7BRG	Zevenbergen	JO21HP	16			H	FSK	PA3FSY
144431.000	9A0BVH	Moslavacka gora	JN85JO	3.8	V dipole	omni	H	F1A	9A5R
144432.000	GB3SEV	Stourport-on-Severn	IO82UI	10	Big Wheel	Omni	H	F1A	M0VXX
144433.000	OH7VHF	Tuupovaara / Joensuu	KP52HL	50	2x BigWheel	Omni	H	A1A /F1B	OH7DI
144433.000	IZ3DVW/B	Monselice PD	JN55VF	10	3 el. Yagi	30°	H	A1A	IZ3DVW
144434.000	DB0LBV	Leibzig	JO61EH	0.4	2 Dipole	OMNI	H		DC6WM
144435.000	IQ5MS/B	Massa	JN54AB	3	Big wheel	Omni	H	A1A	IW5ECP
144436.000	D4C	Monteverde, Mindelo	HK76MU79VB	50	3 X stacked dipoles	OMNI	H	A1A	D4C (HB9DUR)
144437.000	F1ZXK	Montigny	JN18AS	30/10/3/1.5	Trefle	omni	H	A1A	F4BUC
144438.000	SR8ZBW	Jasło	KN09RR67	5.0 W PEP	Crossed Dipole	OMNI	H	A1A	SP8WJW
144440.000	DB0OHZ	Osterholz/Scharmbeck	JO43JF	7.9	M² HOLOOP	OMNI	H	A1A	DL1BEB
144441.000	LA4VHF	Egersund	JO28WL	3	2 x 3el Yagi	40/220	H	A1A	LA1YCA
144442.000	IK4PNJ/B	Pianoro BO	JN54QK	10	2x Dipole	Omni	H	A1A	IK4PNJ
144443.000	OH2VHF	Inkoo	KP20BB	300/300	9/9dBi	230/20	H	A1A	OH2LNM
144444.000	DB0FGB	Schneeberg	JO50WB	15	4 x 3el. Yagi	OMNI	H	A1A	DG4NBI
144444.000	IQ5LU/B	Lucca	JN53GW	6	Big wheel	Omni	H	A1A	IK5AMB
144447.000	SK1VHF	Klintehamn	JO97CJ	10	BIG WHEEL	Omni	H	A1A	
144450.000	DM0HVL		JO62KI	7.8	Winkeldipol	OMNI	H		DL7AIG
144450.000	F5ZVJ	Le Pin	JN24GB	10	2 Trefles	omni	H	F1A	F5IHN
144451.000	LA7VHF	Tromsø	JP99KQ	1000	15el Yagi	185	H	A1A	LA5TFA
144451.000	LZ0STZ	YASTREBOVO	KN22UG	2	4 X HB9CH	omni	H	F1A	LZ1RT
144452.000	IQ7FG/B	San Severo FG	JN71TQ	10	Dipolo	Omni	H	A1A	IK7UXW
144453.000	GB3ANG	Dundee Scotland	IO86MN	20	4 ele Yagi	Yagi 160°	H	A1A	GM4ZUK
144455.000	DB0MMO	Breitsol	JN49RV	7.5	M² HoLoop	OMNI	H	F1A	DH4FAJ
144455.000	OE3XAC	Kaiserkogel	JN78SB	10	folded dipole	Dipole 270°	H	A1A	OE3KLU
144455.000	OH5ADB	Hamina	KP30NN	0,1	Dipole	NW/SE	H	A1A	
144455.000	F5ZXV	Nancy	JN38CO	2,5	Halo	omni	H	A1A	F5OOM
144456.000	OM0MVC	Skalka	JN98LR	5	MOXON	270	H	A1	OM7AC
144457.000	SK2VHF	Vindeln / Buberget	JP94TF		10 el Que-Deedir SSW	N+SV	H	A1A	
144457.000	IQ0HV/B	Cecchina Roma	JN61HQ	3	Big weel	Omni	H	A1A	IK0BDO
144458.000	F1ZAT	Brive	JN05VE	3	Trefle	omni	H	A1A	F1HSU
144461.000	SK7VHF	Sjöbo	JO65UQ	10	HALO	Omni	H	A1A	SM7DTT
144461.000	IW0FFK/B	Ostia RM	JN61DS	3	2 x Hentenna	Omni	H	A1A	IW0FFK
144463.000	LA2VHF	Vassfjellet	JP53EG	500	10el Yagi	15	H	A1A	LA1K
144464.000	F1ZDU	Pierre St Martin	IN92OX	1/0.25	Dipole	Dipole 0° / 180°	H		F6ENL
144465.000	DB0ANN	Moritzberg	JN59PL	0.3	Ho-Loop	OMNI	H	A1A	DL8ZX
144468.000	LA6VHF	Kirkenes	KP59AL	300	9el Yagi	210	H	A1A	LA4OO
144468.000	F1ZAW	Vercel	JN37EE	10	Trefle	omni	H	A1A	F4CQG
144469.000	GB3MCB	St. Austell, Cornwall	IO70OJ	8	3 ele Yagi	Yagi 45°	H	F1A	G7KFQ
144469.000	IQ9UI/B	M.te Lauro RG	JM77IA	10	2*4 el.dk7zb	355°	H	A1A	IT9CJC
144470.000	OH2VHH	Vantaa	KP20MH	1	Dipole	0/180	H	A1A	OH2FTJ
144470.000	ED7YAN	Grenada	IM87GC	5	Loop	Omni	H	A1A	EA7UU
144471.000	OZ7IGY	Jystrup	JO55WM54UA	50	2xBig Wheel	omni	H	F1A	OZ7IS
144471.000	IZ8EDE/B	Monte Pierfaone PZ	JN70VM	2	Halo	Omni	H	A1A	IZ8EDE
144473.000	YM0VHF	Bozcaada	KM39AT28TD	5	Halo	Omni	H	A1A	TA0G
144475.000	DB0HRF	Großer Feldberg Ts.	JO40FF	15	Turnstyle	OMNI	H	F1A	DK7FU
144475.000	OM0MVA	Bratislava	JN88NE	0,4	Dipole	225/45	H	A1	OM3ID
144475.000	IZ8GMP/B	Gambarie RC	JM78WE	10	2el.	310°	H	A1A	IZ8GMP
144476.000	F5ZAL	Pic Neulos	JN12LL	10	Halo	omni	H	A1A	F6HTJ
144477.000	IQ9GD/B	Monte Bonifato TP	JM67LX				H	A1A	IT9CIT
144478.000	LA3VHF	Mandal	JO38SA	120	9el Yagi	180	H	A1A	LA3PRA
144478.000	S55ZRS	Kum	JN76MC	1	X Dipol	omni	H	A1A	ZRS
144478.000	9A0BVS	Goli	JN75BA	4	Dipole	dipole N / S	H	A1A	9A4QV
144479.000	OE3XTR	Hohe Wand	JN87AT	10	2el. Quad	120° 90°	H	A1A	OE3KLU
144480.000	LA8VHF	Stavern	JO48XX	100	3 x 2el Yagi	150	H	A1A	LA6LCA
144480.000	IZ8IBC/B	Nocera Inferiore SA	JN70HR	2	Halo	Omni	H	A1A	IZ8IBC
144481.000	SR3VHX	Pozna	JO82KL80	6.0 W PEP	4 X 2el Yagi	OMNI	H	A1A	SO3Z
144482.000	GB3NGI	Slieve Anorra	IO65VB27	20 / 125	2 x 4 ele. Yagi	Yagi 45° / 135°	H	A1A	GI6ATZ
144485.000	TK5ZMK	Coti Chiavari	JN41JS	5	Trefle	omni	H	A1A	TK5EP
144486.000	DM0PR		JO44JH	5	2 X 6el Yagi	N-S	H		DJ8ES
144486.000	SV3BSF	Patras	KM08VA	5	3el yagi	dir 320	H	F1A	SV3BSF

Freq	Call	Nearest Town	Locator	ERPw	Antenna	Direction	H/V	Mode	Keeper
144487.000	GB3WGI	Derrygonnelly NI	IO64BL	85 / 1000	2 x 6 ele. Yagi	Yagi 280°	H	F1A	GI6ATZ
144489.000	IT9GRR/B	Caltagirone CT	JM77GF	2	Halo	Omni	H	A1A	IT9GRR
144490.500	F6ABJ	P. de Beaurepaire	JN25NJ	50	2 X Trefle	omni	H	F1A	F6ABJ
144491.000	DB0XIT		JN39MI	3	2 x Winkeldipol	OMNI	H		DK4XI
144492.000	ON0VRN	DIKSMUIDE	JO11JA	5		omni	H	WSPR	ON8SD
144492.000	DB0LY	Retzow	JO63PF		2 times stacked BIG Wheel	Omni	H	WSPR	DL9SAU
144492.000	DM0ADA	Wöbbelin	JO53SJ04					WSPR	DM1AD
144847.000	SV5VHF	Profitis Ilias Mt. / Rhodes	KM36XG	1	5el yagi	dir 330	H	A1A	SZ5RDS
340888.000	OM0MSA	Bratislava	JN88NE	0,25	Slot	225/45	H	A1	OM3ID
70cm									
432400.000	OE3XAC	Kaiserkogel	JN78SB	10	7-El Yagi	53° 360°	H	A1A	OE3KLU
432401.000	SK2UHF	Buberget, Vindeln	JP94WG		2 x 5 el Que-Dee	S	H	A1A	
432401.000	F5ZBU		JN18KF	5	4 x 9 elts	omni	H		F2AI
432402.000	OY6BEC	Sornfelli	IP62MB	250	4xdipole	SE	H	F1A	OY9JD
432404.000	DB0THE	Ohmgebirge - Birkenberg	JO51EL	15	4 x 3 Element	OMNI	H	"A1A	DF7AP
432404.000	F5ZZI	Hyeres	JN33BD	5	Trefle	omni	H		F6FCE
432405.000	SK1UHF	Klintehamn	JO97CJ	30	ALFORD SLOT	Omni	H	A1A	
432405.000	ED3YBF	Sant Fost De Campsentelles	JN11CL	2.5	Big Wheel	omni	H	A1	EA3URC
432408.000	F5ZPH	Quistinic	IN87KW	15	4 elts	Yagi 135°	H	A1A	F6ETI
432410.000	DB0ZW		JN69AS	1	Schlitz	OMNI	H		DC9RK
432410.000	DB0JW	Hellenthal	JO30EK	10	4 x Doppelquad	OMNI	H	A1A	DL3KAT
432410.000	IW1AU/B	Giaveno TO	JN35PA	3	Big wheel	Omni	H	A1A	IW1AU
432412.000	DB0JG		JO31HS	1	Clover Leaf	OMNI	H		DL3QP
432412.000	SK6UHF	Varberg / Veddige	JO67EH	10	5 el yagi	210 deg	H	A1A	SM6ESG
432413.000	F5ZTX	Lacapelle	JN14EB	10/5/2.5	2 x 3 elts	Yagi 45° / 90°	H		F5AXP
432415.000	SR5TDM	Tarczyn	KO01KX01	5.0 W PEP	cross dipole	OMNI	H	A1A	SP5MX
432416.000	PI7CIS	Den Haag / Scheveningen	JO22DC	17.6			H	A1A	PA0C
432417.000	OH9UHF	Rovaniemi Pirttikoski	KP36OI	70	9 dBd	200	H	A1A	OH9LA
432418.000	F1ZQT	Moragne	IN95OX	1	Trefle	omni	H		F1MMR
432420.000	F5ZAS	Cerdagne	JN12BL	10	Trefle	omni	H	A1A	F6HTJ
432425.000	DB0MMO	Breitsol	JN49RV	10	M² HoLoop	OMNI	H	F1A	DH4FAJ
432430.000	DB0MFI	Hainsfarth	JN58HW	1	Big Wheel	- OMNI	H	A1A	DK1MFI
432430.000	DB0OHZ	Osterholz/Scharmbeck	JO43JF	4.5	Big Wheel	OMNI	H	A1A	DL1BEB
432432.000	D4C	Monteverde, Mindelo	HK76MU79VB	50	9 ele. Yagi	40° 40°	H	A1A	D4C (HB9DUR)
432432.000	OH6UHF	Uusikaarlepyy / Nykarleby	KP13GM	7	3x BigWheel	Omni	H	A1A	OH6NVQ
432432.000	PI7HVN	Heerenveen	JO22WW	0.5	Omni		H		
432435.000	IQ5MS/B	Massa	JN54AB	3	Big wheel	Omni	H	A1A	IW5ECP
432436.000	F5ZAA	Nerignac	JN06IH	20	Trefle	omni	H	F1A	F5EAN
432440.000	F1ZTV	Cloutons	JN24WX	2	Boucle	omni	H	A1A	F4AII
432440.000	IQ3ZB/B	M.te Cesen TV	JN65AW			Omni	H	A1A	IK3HHG
432441.000	LA5UHF	Flekkefjord	JO38HH	200	6el Yagi	180	H	A1A	LA3EQ
432443.000	OH2UHF	Inkoo	KP20BB	250/250	10/10 dBi	230/20	H	A1A	OH2LNM
432444.000	DB0FGB	Schneeberg	JO50WB	15	4 x Doppelquad	OMNI	H	A1A	DG4NBI
432447.000	DB0IH		JN39HJ	15	HoLoop	OMNI	H	CW	DF1VW
432447.000	S55ZRS	Kum	JN76MC	1	Omni	omni	H	A1A	ZRS
432449.000	OZ1UHF		JO57GH					F1A	
432449.000	I5WBE/B	Mt. Serra PI	JN53GS60KF	3	2x Ho-loop	Omni	H	A1A	I5WBE
432450.000	DM0HVL		JO62KI	6.5	2 x MalteserKreuz	OMNI	H		DL7AIG
432451.000	LZ0STZ	Yastrebovo	KN22UG	4	Big Wheel	omni	H	F1A	LZ1RT
432454.000	F5ZZY	Nancy	JN38CO	3.5	Halo	omni	H		F5OOM
432455.000	OM0MUC	Skalka	JN98LR	7	MOXON	270	H	A1	OM7AC
432455.000	SK3UHF	Nordingrå / Rävsön	JP92FW	50		Omni	H	A1A	
432458.000	IQ0HV/B	Cecchina Roma	JN61HQ	3	2x Ho-loop	Omni	H	A1A	IK0BDO
432459.000	F5ZHG	.	JO10UH	25	Halo	omni	H		F5HMS
432460.000	DB0LB		JN48OV	1	V- Dipol	OMNI	H		DK3PS
432460.000	SK4BX/B	Garphyttan / Storstenshöjden	JO79LH	50	4 X LOG PER	NESW	H	A1A	
432463.000	LA2UHF	Vassfjellet	JP53EG	150	10el Yagi	180	H	A1A	LA1K
432465.000	DB0ANN	Moritzberg	JN59PL	1	Ho-Loop	OMNI	H	A1A	DL8ZX
432467.000	IZ3DVW/B	Monselice PD	JN55VF	3,5	Dipole	45/225°	H	A1A	IZ3DVW
432468.000	LA6UHF	Kirkenes	KP59AL	40	15el Yagi	210	H	A1A	LA4OO
432470.000	ED7YAN	Grenada	IM87GC	5	Loop	Omni	H	A1A	EA7UU
432471.000	OZ7IGY	Jystrup	JO55WM54UA	75	3xBig Wheel	omni	H	F1A	OZ7IS
432473.000	ON0HVL	Heuvelland	JO10KS	10	Dipole	Dipole 180°	H	F1A	ON7FLY
432475.000	ES0VHF	Panga/Isl.Saaremaa	KO18ND46MQ	39.8	2xDouble Diamond	omni	H	A1A	ES0OK
432475.000	DB0XY	Boxberg/Harz	JO51EU25	15		omni	H		DF4OL
432475.000	IZ8GMP/B	Gambarie RC	JM78WE	10	3 EL. YAGI	320°	H	A1A	IZ8GMP
432478.000	LA3UHF	Mandal	JO38SA	50	13el Yagi	180	H	A1A	LA3PRA
432480.000	LA8UHF	Hvasser	JO59FB	50	8el Yagi	90/180	H	A1A	LA6LCA
432485.000	LA4UHF	Haugesund	JO29PJ	15	3el Yagi	250	H	A1A	LA9RY/NQ
432488.000	DB0AD	Salzburg/Westerwald	JO40AQ	1	2 X Yagi + LogPer	omni	H	A1A	DM8MM
432812.500	SZ1SV	Parnis Mt. / Athens	KM18UE	0.5	1/4 vertical-indoor	omni	V	F1A	SZ1SV
432838.000	9A0BVU	Moslavacka gora	JN85JO	1.0	V dipole	omni	H	A1A	9A5R
432875.000	OH7UHF	Kuopio	KP32TV	15	6 dBd	225	H	A1A	OH7VM
432886.000	OK0EP	Praded / Bruntal	JO80OB	2 X 5.0	2 X 4 Ele. Yagi	40° / 40° 280° / 150°	H	F1A	OK1VPZ

Freq	Call	Nearest Town	Locator	ERPw	Antenna	Direction	H/V	Mode	Keeper
432888.000	OM0MUA	Bratislava	JN88NE	0,08	Dipole	225/45	H	A1	OM3ID
502945.000	ON0ODR	Oudenaarde	JO10TT	25		omni	?	F1A	ON4PN
23cm									
1269875.000	ON0VHF	Louvain-La-Neuve	JO20HP	10		omni	H	F1A	ON4IV
1269985.000	DB0JW	Hellenthal	JO30EK	10	Wimo PA-23R	Beam 35°	H	A1A	DL3KAT
1296000.000	ON0EME	Lille	JO21JG	200			LHCP	F1A	ON7UN
1296050.000	S55ZSE	Kokoš	JN65WP	0.3	Slot	omni	H	A1A	S53MV
1296090.000	S55ZMS	Dolina	JN86CR	0.3	Slot	omni	H	A1A	S53M
1296380.000	S55ZRS	Kum	JN76MC	0.2	Slot	omni	H	A1A	ZRS
1296456.000	IZ1ERR/B	Bagnolo CN	JN34OS60UJ	0,2	Quad	45°	H	A1A	IK1YWB
1296800.000	DB0HEG		JN59JA	0.5	4*Schlitz	OMNI	H		DL2QQ
1296800.000	OE3XAC	Kaiserkogel	JN78SB	10	WG Slot	OMNI	H	A1A	OE3KLU
1296800.000	SZ1SV	Hymettus Mt. / Athens	KM17VW	10	2X 5/8 vertical	omni	V	F1A	SZ1SV
1296800.000	SK6MHI	Hönö	JO57TQ	30	Alford Slot	Omni	H	A1A	SM6EAN
1296805.000	DB0RIG		JN48WQ	50	4fach Kasten	OMNI	H		DG9SQ
1296805.000	SK6UHI	Tjörn Island	JO57TX	30	Alford Slot	Omni	H	A1A	SM6CEN
1296810.000	DB0ZW		JN69AS	1	Schlitz	OMNI	H		DC9RK
1296812.000	F1ZBI	Petit Ballon	JN37NX	0.8	Quad	180°	H	F1A	?
1296815.000	DB0VI		JN39NK	0.2		OMNI	H		DL3CM
1296815.000	SR5TDM	Tarczyn	"KO01KX01	4.0 W PEP	Alfrod Slot	OMNI	H	A1A	SP5MX
1296816.000	F1ZTF	Segonzac	IN95VO	10	Big Wheel	omni	H	F1A	F1MMR
1296820.000	LA5SHF	Karmøy	JO29OE	10	3-ele-yagi	180	H	A1A	LA9RY
1296820.000	IQ0RM/B	Roma	JN61FW	1	Alford	Omni	H	A1A	IW0CZC
1296825.000	F5ZRS	Chamrousse	JN25WD	0.1	Corner		H	A1A	F5LGJ
1296835.000	SKØEN/B	Väddö	JO99JX	4	Alford Slot	Omni	H	A1A	
1296840.000	DB0FGB	Schneeberg	JO50WB	1	3xDoppelquad	OMNI	H	A1A	DG4NBI
1296840.000	OH6SHF	Uusikaarlepyy / Nykarleby	KP13GM	30	4x EIA	180	H	A1A	OH6NVQ
1296840.000	IK5CON/B	Camaiore LU	JN53CV	10	Alford	Omni	H	A1A	IK5CON
1296845.000	DB0LBV	Leipzig	JO61EH	2	4*Schlitz	OMNI	H		DC6WM
1296846.000	I5MDE/B	Montignoso FI	JN53KM94HB	5	2x 11 el. Yagi	30°- 290	H	A1A	I5BLH
1296847.000	F5ZBM	Nangis	JN18MN	10	Slot	omni	H	F1A	F6ACA
1296850.000	DM0UB		JO62KK	4.5	Schlitzstrahler	OMNI	H		DL7AIG
1296850.000	OH3SHF	Tampere	KP11TM	50	6 dBi	Omni	H	A1A	OH3LWP
1296850.000	IQ3VO/B	Verona	JN55LL	2	Big weel	Omni	H	A1A	I3LDP
1296854.000	DB0JO		JO31SL	350	4*15 Yagi	W	H		DF1VB
1296854.000	F1ZBK	Nancy	JN38CS	4	Slot	omni	H		F6CXA
1296855.000	SK3UHG	Nordingrå / Rävsön	JP92FW	30	Slotted Waveguide	Omni	H	A1A	
1296857.000	ON0NR	NAMUR (Wépion)	JO20KJ	10		omni	H	F1A	ON6YH
1296859.000	F1ZAK	Istres	JN23MM	20	Slot	omni	H	F1A	F1AAM
1296860.000	DB0LB		JN48OV	1	Big Wheel	OMNI	H		DK3PS
1296860.000	LA8SHF	Hvasser	JO59FB	60	13dB Horn	180	H	A1A	LA6LCA
1296863.000	I7IWN/B	Martano LE	JN90DE	5	2x 26 el. Yagi	320°	H	A1A	I7IWN
1296865.000	DB0JK		JO30LW	40	4*8el Yagi	OMNI	H		DK2KA
1296869.000	IQ0AH/B	Monte Plebi OT	JN40RX	2	23 elem	30°	H	A1A	IS0YPW
1296870.000	LB2SHF	Mandal	JO38RB	90	Horn 8dBi	150	H	A1A	LB2S
1296872.000	F1ZMT	Le Mans	JN07CX	10	Panneau Trefle	180°	H		F1BJD
1296876.000	IZ8EDE/B	M.te Pierfaone PZ	JN70VM	10	Alford Slot	Omni	H	A1A	IZ8EDE
1296877.000	LA3SHF	Tromøy	JO48JK	100	2 x 15el Yagi	150	H	A1A	LA7LW
1296880.000	DB0MOT	Kleiner Feldberg im Taunus	JO40FF33	15	Big Wheel	Omni	H	F1A	DL1ZB
1296880.000	ON0SHF	Ellignies Sainte Anne	JO10UN	5		omni	H	F1A	ON5PX
1296883.000	DB0INN		JN68GI	1	Schlitz	OMNI	H		DL3MBG
1296885.000	DB0TUD		JO61UA		Quad	OMNI	H		DL4DTU
1296885.000	OY6BEC	Sornfelli	IP62MB	50	Hybrid quad	SE	H	F1A	OY9JD
1296886.000	F1ZBC	Adriers	JN06JG	15	A. Slot	omni	H		F1AFJ
1296888.000	OM0MLA	Bratislava	JN88NE	0,18	Dipole	225/45	H	A1	OM3ID
1296890.000	LA4SHF	Egersund	JO28XJ	45	collinear	180	H	A1A	LA3EQ
1296895.000	F5ZAN	Pic Neulos	JN12LL	10	Slot	omni	H	A1A	F6HTJ
1296900.000	OZ5SHF	Yding	JO45VX	10	Double Quad	omni	H	F1A	OZ9ZZ
1296900.000	OH1SHF	Salo	KP10NJ	2	6x 5/8	Omni	V	A1A	OH1BOI
1296900.000	IQ3ZB/B	Monte Cesen TV	JN65AW	10	Slot Alford 16	Omni	H	A1A	IK3HHG
1296903.000	OE3XTR	Hohe Wand	JN87AT	2	WG Slot	OMNI	H	A1A	OE3KLU
1296904.000	DB0THE	Ohmgebirge - Birkenberg	JO51EL		Ho-Loop	OMNI	H	A1A	DF7AP
1296905.000	DB0AD		JO40AQ						DM8MM
1296910.000	DB0UX	Kerlsruhe / Bergwald	JN48FX	1	4 fach Kastenantenne	OMNI	H	F1A	DK2DB
1296910.000	DB0XY	Boxberg/Harz	JO51EU25	15	4xDoppelquad	omni	H		DF4OL
1296910.000	LA1SHF	Drammen	JO59BR	3	Big Wheel	omni	H	A1A	LA3PNA
1296910.000	ES0VHF	Panga/Isl.Saaremaa	KO18ND46MQ	39.8	Alford Slot	omni	H	A1A	ES0OK
1296915.000	TK5ZMV	Coti Chiavari	JN41JS	10	Yagi	315°	H	A1A	TK5EP
1296918.000	PI7ALK	Alkmaar	JO22IP61HS	5		omni	H	F1A	PH0V
1296920.000	DB0VC	Schönwalde / Bungsberg	JO54IF60	15	2*Big Wheel	OMNI	H	A1A	DC6UW
1296920.000	IT9CIT/B	M.te Bonifato TP	JM67LX	14	2 x biquad	0/330°	H	A1A	IT9CIT
1296925.000	DB0AAT		JN67HU	1	Vertikal	OMNI	H		DL8MCG
1296928.000	OH2SHF	Inkoo	KP20BB	50	10/10 dBi	230/20	H	A1A	OH2LNM
1296930.000	OZ7IGY	Jystrup	JO55WM54UA	90	Slot WG	omni	H	F1A	OZ7IS

Freq	Call	Nearest Town	Locator	ERPw	Antenna	Direction	H/V	Mode	Keeper
1296933.000	F5ZBT	Pessac	IN94QT	10			H		F6CBC
1296935.000	DB0YI		JO42XB	3	Big Wheel	OMNI	H		DL4AS
1296935.000	LX0AO	Differdange	JN29WM	1.6	Yagi	36° NE	H	A1A	ADRAD
1296940.000	DB0MFI	Hainsfarth	JN58HW	5	4xBig Wheel	- OMNI	H	A1A	DK1MFI
1296945.000	DB0AJA		JN59AS	10		OMNI	H		DF6NA
1296945.000	LA7SHF	Bergen	JP20QJ	15		180		A1A	LA3QMA
1296945.000	OH9SHF	Rovaniemi Pirttikoski	KP36OI	30	10 dBd	200	H	A1A	OH9LA
1296950.000	SK1UHG	Klintehamn	JO97CJ	30	Alford Slot	Omni	H	A1A	
1296950.000	ON0TB	Weismes/Waimes (Robertville)	JO30BM	10		omni	H	?	ON6GPS
1296950.000	DB0GW	Duisburg / Universität	JO31JK	10	Slot 8	Omni	H	A1A	DL3YDP
1296955.000	OZ1UHF		JO57GH						
1296960.000	SK4BX/B	Garphyttan / Ånnaboda	JO79LI		Alford Slot			A1A	
1296963.000	LA2SHF	Vassfjellet	JP53EG	30	Dipole	bidir. 15/195	H	A1A	LA1K
1296965.000	DB0ANN	Moritzberg	JN59PL	0.5	4xDoppelquad	OMNI	H	A1A	DL8ZX
1296970.000	SR3LHY	Pozna	JO82LJ83	10.0 PEP	Alfrod Slot	OMNI	H	F1A	SO3Z
1296975.000	ON0AZ	Antwerpen	JO21FE	10		omni	H	F1A	ON7BPS
1296980.000	LA9SHF	"Kolsås	JO59GW	10	Doppelquad	160°	H	A1A	LA8GKA
1296983.000	F5ZWX	Grand Cap	JN23XE	1	Slot	omni	H	F1A	F6FCE
1296985.000	SK2SHF	Vännäs / Granlundsberget	JP93VU	10	Alford Slot	Omni	H	A1A	
1296995.000	DB0WOS		JN68ST	5	4xDoppelquad	OMNI	H		DF8RU
1444905.000	ON0ODR	Oudenaarde	JO10TT	25		omni	H	WSPR	ON4PN
13cm									
2320050.000	S55ZSE	Kokoš	JN65WP	0.1	Slot	omni	H	A1A	S53MV
2320090.000	S55ZMS	Dolina nr. Murska Sobota	JN86CR	0.1	Slot	omni	H	A1A	S53M & S51ZO
2320800.000	SK6MHI	Göteborg	JO57XQ	150	Slotted Waveguide	Omni	H	A1A	SM6EAN
2320805.000	DB0ZEH		JO62PW	10		360?	H		DL2RUD
2320805.000	IW1GLM/B	Bagnolo CN	JN34OS60UJ	5	8 Slot	60°	H	A1A	IW1GLM
2320810.000	DB0ZW		JN69AS	1	6fach Schlitz	OMNI	H		DC9RK
2320810.000	DB0NCO		JN59JD	2	Dipol	45Grad	H		DL6NCO
2320815.000	DB0VI		JN39NK	0.2		OMNI	H		DL3CM
2320815.000	SR5TDM	Tarczyn	KO01KX01	3.0 W PEP	Waveguide 2X16	OMNI	H	A1A	SP5MX
2320816.000	F1ZQU	Segonzac	IN95VO	25	Slot	omni	H		F1MMR
2320820.000	IQ0RM/B	Roma	JN61FW	1	Alford	Omni	H	A1A	IW0CZC
2320825.000	DB0FKS		JN49IT	0.8	Doppelhelix 6dB	OMNI	H	A1A	DL2OCB
2320832.000	OH6SHF	Uusikaarlepyy / Nykarleby	KP13GM	40	7 dBd	180	H	A1A	OH6NVQ
2320833.000	DB0FGB	Schneeberg	JO50WB	15	8-fach Schlitz	OMNI	H	A1A	DG4NBI
2320835.000	F5ZAC	Cerdagne	JN12BL	5	Slot	omni	H	F1A	F6HTJ
2320840.000	DB0ODW		JN49LN	10		OMNI	H		DK2NO
2320840.000	F1ZYY	Mauroux	JN03KV	1	Panneaux	315° / 135°	H	F1A	F1MOZ
2320842.000	OH3SHF	Tampere	KP11VK	125	6 dBi	Omni	H	A1A	OH3LWP
2320845.000	DB0LBV	Leipzig	JO61EH	1.5	Doppelquad	SW-S	H		DC6WM
2320850.000	DM0UB		JO62KK	0.6	4fach Kastenant	OMNI	H		DL7AIG
2320850.000	DB0GW	Duisburg / Universität	JO31JK	10	Slot 8	Omni	H	A1A	DL3YDP
2320850.000	LA4SHF	Jæren	JO28UO	10	Log periodic	225	H	A1A	LA3EQ
2320850.000	I3EME/B	M.te Tomba TV	JN55WV	1	6 slot	180°	H	A1A	I3EME
2320852.000	DB0APD	Apolda	JO51SA25VI	0,1	WG Slot 20	Omni	H	A1A	DF4AE
2320855.000	DB0SHF		JN48WP	6	6er Feld	260?	H		DL1SBE
2320855.000	F1ZUM	Orleans	JN07WV	2		omni	H	A1A	F1JGP
2320857.000	PI7RTD	Rotterdam	JO21FV68	15.0	Slot	Omni	H		PE1GHG
2320860.000	LA8SHF	Hvasser	JO59FB	50	13dB Horn	180	H	A1A	LA6LCA
2320864.000	F5ZVY	.	IN93GJ	3.6	Slot	omni	H	F1A	F2CT
2320875.000	ON0VHF	Louvain-La-Neuve	JO20HP	10		omni	H	F1A	ON4IV
2320876.000	IZ8EDE/B	Monte Pierfaono PZ	JN70VM	3	4 Dipole H.	Omni	H	A1A	IZ8EDE
2320880.000	DB0YI		JO42XB	3	Big Wheel	OMNI	H		DL4AS
2320880.000	LA3SHF	Tromøy	JO48JK	1	6dB Horn	150	H	A1A	LA7LW
2320880.000	DB0MOT	Kleiner Feldberg im Taunus	JO40FF33	15	Big Wheel	Omni	H	F1A	DL1ZB
2320883.000	DB0INN		JN68GI	1	Schlitz	OMNI	H		DL3MBG
2320885.000	DM0TUD		JO61UC	10	Schlitz	OMNI	H		DG0DI
2320886.000	F5ZMF	Adriers	JN06JG	8	Slot	omni	H	F1A	F5BJL
2320895.000	PI7RMD	Melick	JO31AD	5	Doublequad	80° 270°	H	A1A	PE1KXH
2320900.000	DB0UX	Karlsruhe / Bergwald	JN48FX	1	4 fach Kastenantenne	OMNI	H	F1A 800Hz	DK2DB
2320900.000	DB0MJ		JO31UB	1	6-fach Schlitz	OMNI	H		DH1MJ
2320900.000	SK3UHH	Nordingrå / Rävsön	JP92FW			220 deg.	H	A1A	
2320900.000	F6DWG		JN19BQ	10	Slot	omni	H	F1A	F6DWG
2320900.000	F1ZCC	Maurepas	JN08XS	1	Slot	omni	H	F1A	F1PDX
2320903.000	IQ5FI/B	M.te Secchieta FI	JN53SR	5	Alford Slot	Omni	H	A1A	I5KRD
2320910.000	DB0UM		JO73CE	1	Schlitz	OMNI	H		DG1BHA
2320910.000	DB0XY	Boxberg/Harz	JO51EU25	15	4x Doppelquad		H		DF4OL
2320910.000	DB0WML	Reken	JO31MU						DK8QU
2320910.000	IW0REF/B	M.te San Pancrazio TR	JN62HK14PE	11	13dB	Omni	H	A1A	IW0REF
2320919.000	IQ9GD/B	Mte Bonifato TP	JM67LX	0,9	6 slot		H	A1A	IT9CIT
2320920.000	DB0VC	Schönwalde / Bungsberg	JO54IF60	15	Big Wheel	OMNI	H	A1A	DC6UW
2320920.000	PI7ALK	Alkmaar	JO22IP61HS	5		omni	H	F1A	PH0V

Freq	Call	Nearest Town	Locator	ERPw	Antenna	Direction	H/V	Mode	Keeper
2320928.000	OH2SHF	Inkoo	KP20BB	150/150	10/10 dBi	230/20	H	A1A	OH2LNM
2320933.000	F5ZEN	Pessac	IN94QT	5	cornets	?	H	F1A	F6CBC
2320940.000	DB0MFI	Hainsfarth	JN58HW	5	4xBig Wheel	- Omni	H	A1A	DK1MFI
2320960.000	DB0AJA		JN59AS	10		OMNI	H		DF6NA
2320960.000	SK4BX/B	Garphyttan / Ånnaboda	JO79LI		SLOTTED WAVEGUIDE		H	A1A	
2320965.000	DB0ANN		JN59PL	5	4xDoppelquad	OMNI	H		DL8ZX
2320975.000	DB0JL		JO31MC	2	Schlitz	OMNI	H		DB5KN
2320983.000	F5ZHX	Grand Cap	JN23XE	10	Slot	omni	H	F1A	F6FCE
2320985.000	SK2SHF	Vännäs / Granlundsberget	JP93VU	10	Parabol	190	H	A1A	
2400900.000	OZ5SHF	Yding	JO45VX	10	Double Quad	omni	H	F1A	OZ9ZZ
2400930.000	OZ7IGY	Jystrup	JO55WM54UA	30	Slot WG	omni	H	F1A	OZ7IS
9cm									
3400007.000	DM0HVL		JO62KI	7.5	12fach Schlitz	OMNI	H		DL7AIG
3400050.000	DB0JL		JO31MC	1	Helical	OMNI	H		DB5KN
3400050.000	S55ZSE	Kokoš	JN65WP	0.5	Slot	omni	H	A1A	S53MV
3400090.000	S55ZMS	Dolina nr. Murska Sobota	JN86CR	0.5	Slot	omni	H	A1A	S53M & S51ZO
3400800.000	DB0MOT	Kleiner Feldberg im Taunus	JO40FF33	15	Slot 12	Omni	H	F1A	DL1ZB
3400800.000	LA8SHF	Hvasser	JO59FB					A1A	LA6LCA
3400800.000	OH3SHF	Tampere	KP11VK	120	6 dBi	Omni	H	A1A	OH3LWP
3400810.000	DB0NCO		JN59JD	2	SBF	45?	H		DL6NCO
3400840.000	DB0ODW		JN49LN	10		OMNI	H		DK2NO
3400850.000	DB0GW	Duisburg / Universität	JO31JK	10	Slot 8	Omni	H	A1A	DL3YDP
3400850.000	LA4SHF	Jæren	JO28UO	2	Log periodic	225	H	A1A	LA3EQ
3400850.000	DB0KK	Berlin Lichtenberg	JO62RM76OP	13	Slot	Omni	H	A1A	DC7YS
3400883.000	DB0INN		JN68GI	1	Schlitz	OMNI	H		DL3MBG
3400885.000	DM0TUD		JO61UC	10	Schlitz	OMNI	H		DG0DI
3400900.000	OZ5SHF	Yding	JO45VX	2	Double Quad	omni	H	F1A	OZ9ZZ
3400910.000	DB0WML	Reken	JO31MU						DK8QU
3400912.000	DB0XY	Boxberg/Harz	JO51EU25	15		omni	H	A1AN	DF4OL
3400920.000	PI7RTD	Rotterdam	JO21FV68	14.5	Slot	Omni	H		PE1GHG
3400925.000	PI7ALK	Alkmaar	JO22IP61HS		Slotted Waveguide 6dBd	omni	H	F1A	PH0V
3400930.000	OZ7IGY	Jystrup	JO55WM54UA	50	Slot WG	omni	H	F1A	OZ7IS
3400940.000	DB0MFI	Hainsfarth	JN58HW	4	2X10 Slot	OMNI	H		DK1MFI
3400945.000	DB0AJA		JN59AS	10		OMNI	H		DF6NA
3400955.000	OZ1UHF		JO57GH						
3400975.000	DB0HRF	Großer Feldberg Ts.	JO40FF			OMNI	H		DF7KU
3456855.000	DB0SHF		JN48WP	0.5	Horn	260?	H		DL1SBE
3456965.000	DB0ANN		JN59PL				H		DL8ZX
4324875.000	F1ZBY	Roc Blanc	JN13TV	5	Trefle	omni	H	A1A	F4DVR
6cm									
5760010.000	S55ZSE	Kokoš	JN65WP	0.5	Slot	omni	H	A1A	S53MV
5760045.000	S55ZRS	Kum	JN76MC	0.2	Slot	omni	H	A1A	ZRS
5760060.000	F1ZAO	Plougonver	IN88HL	1	Fentes	omni	H	F1A	F1LHC
5760070.000	DB0JL		JO31MC	0.8	Schlitz	OMNI	H		DB5KN
5760090.000	S55ZMS	Dolina nr. Murska Sobota	JN86CR	0.5	Slot	omni	H	A1A	S53M & S51ZO
5760800.000	DB0MOT	Kleiner Feldberg im Taunus	JO40FF33	15	Slot 14	Omni	H	F1A	DI1ZB
5760800.000	SK4BX/B	Garphyttan / Ånnaboda	JO79LI				H	A1A	
5760800.000	SK6MHI	Göteborg	JO57XQ	10	SLOTTED WAVEGUIDE	Omni	H	A1A	SM6EAN
5760800.000	OH3SHF	Tampere	KP11VK	150	15 dBi	Omni	H	A1A	OH3LWP
5760800.000	IQ4AD/B	Monte Cassio PR	JN54AO	0,5	6 slot	35°	H	A1A	I4YMB
5760804.000	F1ZMF	Lure	JN24VC	10	Fentes	omni	H	F1A	F1OW
5760805.000	DB0RIG		JN48WQ	15		OMNI	H		DG9SQ
5760810.000	DB0NCO		JN59JD	2	Schlitz	OMNI	H		DL6NCO
5760816.000	IK1YWB/B	Bagnolo CN	JN34OS60UJ	0,6	Slot 16	55°	H	A1A	IK1YWB
5760820.000	F5ZBE		JN18HN	12	Fentes	omni	H	F1A	F5HRY
5760833.000	DB0FGB	Schneeberg	JO50WB	15	8-fach Schlitz	OMNI	H	A1A	DG4NBI
5760840.000	DB0ODW		JN49LN	10		OMNI	H		DK2NO
5760841.000	SK7MW/B	Gladsax	JO75DN	40		Omni	H	FSK	SM7DTE
5760845.000	F1ZBD	Orleans	JN07WV	2	Fentes	omni	H	F1A	F1JGP
5760850.000	DM0UB		JO62KK	9	12fach Schlitz	OMNI	H		DL7AIG
5760850.000	DB0GW	Duisburg / Universität	JO31JK	10	Slot 8	Omni	H	A1A	DL3YDP
5760850.000	LA4SHF	Jæren	JO28UO	2	Log periodic	225	H	A1A	LA3EQ
5760855.000	DB0SHF		JN48WP	0.5	Array	260?	H		DL1SBE
5760855.000	OE8XGQ	Gerlitzen, Carinthia	JN66XQ	3.2	Slot 10	Omni	H	A1A	OE8WOZ
5760860.000	DB0RDH		JN68KW	2.5		SW bis SO	H		DL2RDH
5760860.000	LA8SHF	Hvasser	JO59FB	25	13dB Horn	180	H	A1A	LA6LCA
5760860.000	F5ZUO	Pic Neulos	JN12LL	1	Fentes	omni	H	F1A	F6HTJ
5760880.000	LA3SHF	Tromøy	JO48JK	1		150	H	A1A	LA7LW
5760880.000	ON0VHF	Louvain-La-Neuve	JO20HP	10		omni	H	F1A	ON4IV
5760883.000	DB0INN		JN68GI	1	Schlitz	OMNI	H		DL3MBG
5760883.000	F5ZWY	Grand Cap	JN23XE	1	Fentes	omni	H	F1A	F6FCE
5760885.000	DM0TUD		JO61UC		Schlitz	OMNI	H		DL4DTU
5760888.000	I3EME/B	Monte Tomba TV	JN55WV	0,22	Slot 8	170°	H	A1A	I3EME

Freq	Call	Nearest Town	Locator	ERPw	Antenna	Direction	H/V	Mode	Keeper
5760889.000	F5ZIE		IN93GJ	0.1	Fentes	omni	H	F1A	F2CT
5760893.000	OE1XGA	Wien/Kahlenberg	JN88EG	7	Slot 10	Omni	H	A1A	OE4WOG
5760900.000	OZ5SHF	Yding	JO45VX	2	WG slot	omni	H	F1A	OZ9ZZ
5760900.000	ON0TB	Weismes/Waimes (Robertville)	JO30BM	0.2		omni	H		ON6GPS
5760900.000	IQ5FI/B	M.te Secchieta AR	JN53SR84XO	1	Slot,8 + 8		H	A1A	I5KRD
5760903.000	SKØCT/B	Väddö	JO99JX	80		Omni	H	A1A	
5760903.000	F6DWG		JN19BQ	10	Fentes	omni	H	F1A	F6DWG
5760910.000	DB0UM		JO73CE	1	Schlitz	OMNI	H		DG1BHA
5760910.000	DB0WML	Reken	JO31MU						DK8QU
5760910.000	IW0REF/B	M.te San Pancrazio TR	JN62HK14PE	1	10dB	Omni	H	A1A	IW0REF
5760915.000	PI7RTD	Rotterdam	JO21FV68	17	Slot	Omni	H		PE1GHG
5760917.000	F1ZMH	Berneuil	IN95QP	2.5	Fentes	omni	H	F1A	F1MMR
5760925.000	F5ZOI	Ste Fortunade	JN05VE	4	Fentes	omni	H	F1A	F6ETI
5760928.000	OH2SHF	Helsinki	KP20LE	65	10 dBi	Omni	H	A1A	OH2LNM
5760930.000	OZ7IGY	Jystrup	JO55WM54UA	50	Slot WG	omni	H	F1A	OZ7IS
5760930.000	F1ZWJ	Lacapelle	JN14EB	2	Fentes	omni	H	F1A	F1BOH
5760933.000	F5ZPR	Pessac	IN94QT	8	Cornet	130°	H	F1A	F6CBC
5760935.000	OH5TEN	Kouvola	KP30HV	32	12.9 dBd	240	H	A1A	OH5IY
5760935.000	PI7ALK	Alkmaar	JO22IP61HS	1	Slotted Waveguide 10dBd	omni	H	F1A	PH0V
5760940.000	DB0MFI	Hainsfarth	JN58HW	2	12fach Schlitz	OMNI	H		DK1MFI
5760945.000	DB0AJA		JN59AS	10		OMNI	H		DF6NA
5760965.000	DB0ANN		JN59PL				H		DL8ZX
5760975.000	ON0GHZ	Tielt-Winge	JO20KV	5		omni	H	F1A	ON4IY
5769955.000	OZ1UHF		JO57GH						
10GHz									
10368033.000	I3CLZ/B	M Falcone VI	JN55OQ	0,6	Slot	Omni	H	A1A	I3CLZ
10368050.000	S55ZRS	Kum	JN76MC	0.8	Slot	omni	H	A1A	ZRS
10368057.000	I4BER/B	Piane di Mocogno MO	JN54IG	0,07	Slot 16		H	A1A	I4BER
10368070.000	S55ZMS	Dolina	JN86CR	0.8	Slot	omni	H	A1A	S53M
10368120.000	DB0JL		JO31MC	0.15	Schlitz	OMNI	H		DB5KN
10368282.000	I1TEX/B	Rossana CN	JN34QM58AU	1	Slot 8	65°	H	A1A	I1TEX
10368760.000	ON0NIV	Tielt-Winge	JO20DO	1		omni	H	F1A	
10368796.000	IQ0RM/B	Roma	JN61FW	0,2	Slot	0/180°	H	A1A	IW0CZC
10368800.000	SK6MHI	Göteborg	JO57XQ	1	SLOTTED WAVEGUIDE	Omni	H	A1A	SM6EAN
10368800.000	OH3SHF	Tampere	KP11VK	80	15 dBi	Omni	H	A1A	OH3LWP
10368800.000	IW3FZQ/B	Colli Euganei PD	JN55TG	0,5	Slot 14+14	60°-240°	H	A1A	IW3FZQ
10368805.000	DB0UL		JN48XK	4		OMNI	H		DL1SBM
10368805.000	ON0TNR	NAMUR (Temploux)	JO20KJ	0.2		omni	H	F1A	ON4TLW
10368805.000	DB0MVP	Schwerin	JO53QP		Slot 10dB	omni	H		DG3TP
10368809.000	IQ1GE/B	Monte Fasce GE	JN44MJ	1	Slot	340° - 160°	H	A1A	IZ1EVF
10368810.000	DB0GHZ	Helgoland	JO43UP	0.15	Slot	OMNI	H	A1A	DK7LJ
10368810.000	DB0ANU		JN59GG	10	Schlitz	OMNI	H		DL3NDX
10368815.000	DB0VI		JN39NK	0.3		OMNI	H		DL3CM
10368815.000	SR5TDM	Tarczyn	"KO01KX01	4.0 W PEP	Waveguide 2X8	OMNI	H	A1A	SP5MX
10368819.000	IZ1ERR/B	Bagnolo CN	JN34OS	1	Slot 20	55°	H	A1A	IZ1ERR
10368820.000	DB0MOT	Kleiner Feldberg im Taunus	JO40FF33	15	Slot 12	Omni	H	F1A	DL1ZB
10368820.000	LA5SHF	Stavanger	JO28UX	12	10dB Slot	240	H	A1A	LA3EQ
10368825.000	DB0HRO		JO64AD	0.2	Schlitz	OMNI	H		DL6KWN
10368826.000	IQ5FI/B	M.te Secchieta AR	JN53SR84XO	1,5	Slot 16	Omni	H	A1A	I5KRD
10368833.000	DB0FGB	Schneeberg	JO50WB	8	8-fach Schlitz	OMNI	H	A1A	DG4NBI
10368835.000	ON0VHF	LOUVAIN-LA-NEUVE	JO20HP	10		omni	H	F1A	ON4IV
10368840.000	DB0MMO	Breitsol	JN49RV	2	8fach-Schlitz	OMNI	H	F1A	DH4FAJ
10368841.000	SK7MW/B	Gladsax	JO75DN	40		Omni	H	FSK	SM7DTE
10368845.000	DO0CWC		JO60MT	10	8fach-Schlitz	OMNI	H	A1A	DO1CTL
10368847.000	SKØEN/B	Väddö	JO99JX	10000	33 dB dish	360 deg	H	A1A	
10368850.000	DB0GG		JN48NR	1	Schlitz	OMNI	H		DL5AAP
10368850.000	DM0UB		JO62KK	8.7	12fach-Schlitz	OMNI	H		DL7AIG
10368850.000	DB0GW	Duisburg / Universität	JO31JK	10	Slot 8	Omni	H	A1A	DL3YDP
10368850.000	LA4SHF	Jæren	JO28UO	25	2 x 13dB Horn	180/0	H	A1A	LA3EQ
10368850.000	SK1SHH	Klintehamn	JO97CJ	10000	SLOT WG	360 deg	H	A1A	
10368850.000	IQ6AN/B	Ancona	JN63QN	4	Horn	355°	H	A1A	IW6ATU
10368852.000	DB0APD	Apolda	JO51SA25VI	0,01	WG Slot 20	Omni	H	A1A	DF4AE
10368853.000	DL0WY		JN67AQ	1.5	12 Fach-Schlitz	O-N-	H		DJ8VY
10368855.000	DB0SHF		JN48WP	0.3	Horn	260?	H		DL1SBE
10368855.000	OE8XGQ	Gerlitzen, Carinthia	JN66XQ	0.6	Slot 8	Omni	H	A1A	OE8WOZ
10368855.000	IK7UXW/B	Martina Franca TA	JN80OQ	4	Slot 2x16	Omni	H	A1A	IK7UXW
10368860.000	DB0RDH		JN68KW	1.3		SW bis SO	H		DL2RDH
10368860.000	LA8SHF	Hvasser	JO59FB	10	13dB Horn	180	H	A1A	LA6LCA
10368865.000	DB0JK		JO30LW	200	Schlitz	OMNI	H		DK2KA
10368865.000	I3EME/B	Monte Tomba TV	JN55WV	2,2	Slot 12	170°	H	A1A	I3EME
10368868.000	IZ0CZW/B	Monte Calvo LT	JN61QI	0,25	Slot	0°- 180°	H	A1A	IZ0CZW
10368870.000	LB2SHF	Odderøya	JO48AD	25	Sector horn	unidir. 180	H	A1A	LB2S
10368870.000	IQ0AH/B	Monte Plebi OT	JN40RX	1,5	Slot 16	35°	H	A1A	IW0UAM

Freq	Call	Nearest Town	Locator	ERPw	Antenna	Direction	H/V	Mode	Keeper
10368875.000	DB0HW		JO51GT	0.04	Schlitz	WNO	H		DL3AAS
10368875.000	ON0EME	LILLE	JO21JG	10		omni	H	F1A	ON7UN
10368876.000	IZ8EDE/B	M.te Pierfaone PZ	JN70VM	1	Slot	Omni	H	A1A	IZ8EDE
10368878.000	IQ2CF/B	Bovezzo BS	JN55DO	0,25	Horn	180°	H	A1A	IK2BVP
10368880.000	DB0ODW		JN49LN	10	Schlitz	OMNI	H		DK2NO
10368880.000	OE1XGA	Wien/Kahlenberg	JN88EG	7	Slot 12	Omni	H	A1A	OE4WOG
10368880.000	IW9ARO/B	Caltagirone CT	JM77GF	6	Slot	Omni	H	A1A	IW9ARO
10368880.000	9A0BXP	Psunj	JN85QJ	0.2	Slot	omni	H	A1A	9A5R
10368883.000	DB0INN		JN68GI	1	Schlitz	OMNI	H		DL3MBG
10368885.000	DM0TUD		JO61UC	5	Schlitz	OMNI	H		DG0DI
10368885.000	OY6BEC		IP61PJ	750	Horn	SE	H	F1A	OY9JD
10368890.000	LA9SHF	Kolsås	JO59GW	10	10dB Horn	180	H	A1A	LA8GKA
10368890.000	OM0MXA	Bratislava	JN88NE	0,12	Slot	225/45	H	A1	OM3ID
10368890.000	9A0BXV	Velebitska pljesivica	JN74LT	0.2	Slot	omni	H	A1A	9A4QV
10368892.000	S55ZKP	Slavnik	JN65XM	0.4	Slot	omni	H	A1A	S51JN
10368895.000	LX0DB	Sandweiler	JN39CO	4	Slot	omni	H	F1A	RL
10368900.000	DB0ECA		JN58XD	1.5 / 8.0	8-Fach Schlitz	OMNI	H		DC8EC
10368900.000	DB0ELS		JO43FF	0.2	Schlitz 11dB	OMNI	H		DL9BL
10368900.000	DB0UX	Karlsruhe / Bergwald	JN48FX	1	Schlitz	OMNI	H	F1A 800Hz	DK2DB
10368900.000	OE3XTR	Hohe Wand	JN87AT	5	WG Slot	OMNI	H	A1A	OE3KLU
10368900.000	OZ5SHF	Yding	JO45VX	25	Slot WG	omni	H	F1A	OZ9ZZ
10368900.000	OH1SHF	Salo	KP10NJ	1	Corner reflector	100	H	A1A	OH1BOI
10368900.000	DB0UY	Karlsruhe							DL5IN
10368900.000	IW8PNY/B	Contessa Soprana CS	JM89CK	6	Slot 20+20	199°	H	A1A	IW8PNY
10368904.000	PI7RTD	Rotterdam	JO21FV68	10	Slot	Omni	H		PE1GHG
10368905.000	DB0SCS	Nürnberg							DG7NDV
10368910.000	DB0UM		JO73CE	1	Schlitz	OMNI	H		DG1BHA
10368910.000	DB0XY	Boxberg/Harz	JO51EU25	15	Schlitz		H		DF4OL
10368910.000	DB0WML	Reken	JO31MU	10	Slot 10dB	omni	H	A1A	DK8QU
10368910.000	IW0REF/B	M.te San Pancrazio TR	JN62HK14PE	0,7	Slot 2	Omni	H	A1A	IW0REF
10368915.000	ON0GBN	FLOBECQ/VLOESBERG	JO10US	1		omni	H	F1A	ON7FLY
10368920.000	DB0VC	Schönwalde / Bungsberg	JO54IF60	15	WG Slot 10	OMNI	H	A1A	DC6UW
10368920.000	PI7ALK	Alkmaar	JO22IP61HS	1		omni	H	F1A	PH0V
10368925.000	ON0GHZ	TIELT-WINGE	JO20KV	2.5		omni	H	F1A	ON4IY
10368928.000	OH2SHF	Helsinki	KP20LE	30	13.8 dBi	Omni	H	A1A	OH2LNM
10368930.000	OE3XAC	Kaiserkogel	JN78SB	10	WG Slot	OMNI	H	A1A	OE3KLU
10368930.000	OZ7IGY	Jystrup	JO55WM54UA	50	Slot WG	omni	H	F1A	OZ7IS
10368930.000	IK3TCH/B	San Mauro di Saline VR	JN55NO	1	Slot 16	Omni	H	A1A	IK3TCH
10368935.000	OH5TEN	Kouvola	KP30HV	23	15.9 dBd	240	H	A1A	OH5IY
10368935.000	DM0MAX	Holsen	JO42IG	15		omni	H	A1A	DL6XB
10368940.000	DB0DON		JN58JR	1	Schlitz	OMNI	H		DK1MFI
10368940.000	SR3XHY	Pozna	JO82LJ83	6.0 PEP	Waveguide 2X8	OMNI	H	F1A	SO3Z
10368945.000	LA7SHF	Bergen	JP20QJ	10	17 dB Horn	180	H	A1A	LA3QMA
10368946.000	DB0AJA		JN59AS	10	Schlitz	OMNI	H		DF6NA
10368950.000	DB0OFG		JN48CO47	0.4	Schlitz	OMNI	H		DC5GF
10368952.000	IT9CIT/B	M.te Bonifato TP	JM67LX	1,16	Slot 12	0°	H	A1A	IT9CIT
10368955.000	OZ1UHF		JO57GH						
10368960.000	DB0MFI	Hainsfarth	JN58HW	1	Schlitzstrahler	- OMNI	H	A1A	DK1MFI
10368960.000	SK4BX/B	Garphyttan / Ånnaboda	JO79LI		Slotted Waveguide		H	A1A	
10368965.000	DB0ANN		JN59PL	0.2	12fach-Schlitz	OMNI	H		DL8ZX
10368965.000	ON0BRU	BRUSSEL	JO20EU			omni	H	F1A	ON4BHM
10368972.000	DB0MU	Nottuln	JO31QX	7	Slot 8	Omni	H	A1A	DJ2QZ
10368985.000	ON0HVL	Heuvelland	JO10KS	2	Dipole	Slot Antenna 10dBi	H	F1A	ON7FLY
24GHz									
23209354.000	LX0AOK	Differdange	JN29WM	1	Yagi	36° NE	H	A1A	ADRAD
24048020.000	IQ5FI/B	M.te Secchieta FI	JN53SR84XO	0,8	Slot 8+8	Omni	H	A1A	I5KRD
24048033.000	I3CLZ/B	M Falcone VI	JN55OQ	0,6	Slot	135°	H	A1A	I3CLZ
24048044.000	S55ZMS	Dolina	JN86CR	0.17	Slot	omni	H	A1A	S53M & S51ZO
24048050.000	ON0GHZ	TIELT-WINGE	JO20KV	2.5		omni	H	F1A	ON4IY
24048065.000	IZ1OTT/B	Frassinetto TO	JN35TK24SI	0,6	20 dB Horn	105°	H	A1A	IZ1OTT
24048076.000	DB0QQ		JO31RI	0.13		OMNI	H		DJ5VW
24048078.000	I3EME/B	Monte Tomba TV	JN55WV	0,18	Slot 12	170°	H	A1A	I3EME
24048180.000	LX0CDF	Sandweiler	JN39CO	1	Slot	omni	H	F1A	RL
24048210.000	S55ZKP	Slavnik	JN65XM	0.4	Slot 10	omni	H	A1A	S56RGA
24048618.000	IK1YWB/B	Bagnolo CN	JN34OS60UJ	0,8	Horn	50°	H	A1A	IK1YWB
24048800.000	LA1SHF	Drammen	JO59BR					A1A	LA3PNA
24048800.000	SK6MHI	Göteborg	JO57XQ	10	Slotted Waveguide	Omni	H	A1A	SM6EAN
24048800.000	OH3SHF	Tampere	KP11VK	30	18 dBi	Omni	H	A1A	OH3LWP
24048805.000	IW3RMR/B	M.te Bernadia	JN66OF	0,1	Slot 10	180°	H	A1A	IW3RMR
24048810.000	DB0NCO		JN59JD	0.8	Schlitz	OMNI	H		DL6NCO
24048816.000	DB0VI		JN39NK	0.3		OMNI	H		DL3CM
24048820.000	DB0JL		JO31MC	0.01	Schlitz	OMNI	H		DB5KN
24048820.000	LA5SHF	Stavanger	JO28UX	0.2	10 dB Horn	180	H	A1A	LA3EQ

Freq	Call	Nearest Town	Locator	ERPw	Antenna	Direction	H/V	Mode	Keeper
24048820.000	PI7RTD	Rotterdam	JO21FV68	10	Slot	Omni	H		PE1GHG
24048825.000	DB0MMO	Breitsol	JN49RV	1	8-fach Schlitz	OMNI	H	F1A	DH4FAJ
24048832.000	DB0FGB	Schneeberg	JO50WB	5	8-fach Schlitz	OMNI	H	A1A	DG4NBI
24048833.000	IW3FZQ/B	Colli Euganei PD	JN55TG	0,16	Sector	210°	H	A1A	IW3FZQ
24048835.000	ON0VHF	Louvain-La-Neuve	JO20HP	1		omni	H	F1A	ON4IV
24048840.000	DB0ODW		JN49LN	10		OMNI	H		DK2NO
24048840.000	DB0WOL	Wollenberg / Barnim	JO62XR	12	Sector Horn	90° 240°	H	A1A	DL7VTX
24048843.000	SK7MW/B	Gladsax	JO75DN	70		Omni	H	FSK	SM7DTE
24048850.000	DB0GW	Duisburg / Universität	JO31JK	10	Slot 8	omni	H	A1A	DL3YDP
24048850.000	DB0KK	Berlin Lichtenberg	JO62RM76OP	6	Horn	270°	H	A1A	DC7YS
24048852.000	DB0MOT	Kleiner Feldberg im Taunus	JO40FF33	15	Slot 12	Omni	H	F1A	DL1ZB
24048852.000	DL0WY		JN67AQ	12	Sektorhorn	O-N-	H		DJ8VY
24048855.000	OE8XGQ	Gerlitzen, Carinthia	JN66XQ	0.9	Slot 15	Omni	H	A1A	OE8WOZ
24048860.000	LA8SHF	Hvasser	JO59FB	5	10dB Horn	120° +-90°	H	A1A	LA6LCA
24048864.000	DB0JK		JO30LW	1	2xH-Horn	OMNI	H		DK2KA
24048875.000	ON0EME	LILLE	JO21JG	10		omni	H	F1A	ON7UN
24048883.000	SK6SHG	Tjörn Island	JO57TX	2x1W	2xhorn	N / S	H	A1A	SM6CEN
24048884.000	DM0TUD		JO61UC	2	Schlitz	OMNI	H		DG0DI
24048900.000	DB0ECA		JN58XD	0.5 / 5.0	E-Horn 50	E	H		DC8EC
24048900.000	OZ5SHF	Yding	JO45VX	7	Slot WG	omni	H	F1A	OZ9ZZ
24048902.000	9A5AA	Sljeme	JN75XV35	0.3	Slot	OMNI	H	F1A	9A5AA
24048910.000	DB0WML	Reken	JO31MU						DK8QU
24048910.000	IW0REF/B	M.te San Pancrazio TR	JN62HK14PE	0,15	Slot 2	Omni	H	A1A	IW0REF
24048912.000	DB0HHB		JN69LF	1	10 Slots	Omni	H		DL6NCO
24048920.000	DB0VC	Schönwalde / Bungsberg	JO54IF60	15	WG Slot 12	Omni	H	A1A	DC6UW
24048925.000	PI7ALK	Alkmaar	JO22IP61HS	1	slotted waveguide 12dBd	Omni	H	F1A	PH0V
24048928.000	OH2SHF	Helsinki	KP20LE	20	24-slot WG	Omni	H	A1A	OH2LNM
24048930.000	OZ7IGY	Jystrup	JO55WM54UA	20	Slot WG	Omni	H	F1A	OZ7IS
24048960.000	OE1XGA	Wien/Kahlenberg	JN88EG	10	Slot 12	Omni	H	A1A	OE4WOG
24048964.000	DB0ANN	Moritzberg	JN59PL			Omni	H	A1A	DL8ZX
24048985.000	ON0HVL	Heuvelland	JO10KS	1	Dipole/Slot Antenna 15dBi		H	F1A	ON7FLY
47GHz									
47088080.000	DB0AJA		JN59AS	10	Schlitz	OMNI	H		DF6NA
47088100.000	DB0QQ		JO31RI	10		3° 13°	H		DJ5VW
47088800.000	DB0FGB	Schneeberg	JO50WB	4	Sektorhorn	Beam 0°	H	A1A	DG4NBI
47088800.000	LA1SHF	Drammen	JO59BR					A1A	LA3PNA
47088800.000	OE1XGA	Wien/Kahlenberg	JN88EG	2.5	Sector Horn	“?° ÖW 100°”	H	A1A	OE4WOG
47088820.000	LA5SHF	Stavanger	JO28UX	0.2	10 dB Horn	180	H	A1A	LA3EQ
47088840.000	DB0WOL	Wollenberg / Barnim	JO62XR	12	Sector Horn	90° 240°	H	A1A	DL7VTX
47088850.000	DB0KK	Berlin Lichtenberg	JO62RM76OP	0.5	Slot	omni	H	A1A	DC7YS
47088850.000	DB0MOT	Kleiner Feldberg im Taunus	JO40FF33	10	Dish	2.6° 360° remote	H	F1A	DL1ZB
47088850.000	DB0GW	Duisburg / Universität	JO31JK	5	2 x Sectorhorn	0° / 180°	H	A1A	DL3YDP
47088853.000	DL0WY		JN67AQ	0.5	Sektorhorn	O-N	H		DJ8VY
47088865.000	DB0JK		JO30LW	0.1	2xH-Horn	OMNI	H		DK2KA
47088884.000	DM0TUD		JO61UC	1	Horn	SW	H		DG0DI
47088910.000	DB0WML	Reken	JO31MU						DK8QU
47088910.000	DB0OHL								
47088920.000	DB0VC	Schönwalde / Bungsberg	JO54IF60	10	WG Slot 15	Omni	H	A1A	DC6UW
47088925.000	PI7ALK	Alkmaar	JO22IP61HS	0.01	sector horn 17dBd	Horn 135°	H	F1A	PH0V
48088900.000	LX0CDE	Sandweiler	JN39CO	1	Slot	Omni	H	F1A	RL
76GHz									
76032050.000	DB0AJA		JN59AS	0,2	Slot 8	Omni	H	A1A	DF6NA
76032100.000	DB0QQ		JO31RI	10		3° 13°	H		DJ5VW
76032800.000	DB0FGB	Schneeberg	JO50WB	3	Sektorhorn	Beam 0°	H	A1A	DG4NBI
76032825.000	DB0MOT	Kleiner Feldberg im Taunus	JO40FF33	10	Dish	1.8° 360°: remote	H	F1A	DL1ZB
76032840.000	DB0WOL	Wollenberg / Barnim	JO62XR	12	Sector Horn	90° 240°	H	A1A	DL7VTX
76032850.000	DB0GW	Duisburg / Universität	JO31JK	2	2 x Sectorhorn	0° / 180°	H	A1A	DL3YDP
76032865.000	OE1XGA	Wien/Kahlenberg	JN88EG	“0.1	Sector Horn	?° ÖW 100°	H	A1A	OE4WOG
76032900.000	LX0CBC	SANDWEILER	JN39CO	0.1	Slot	Omni	H	F1A	RL
76032910.000	DB0WML	Reken	JO31MU						DK8QU
76032910.000	DB0OHL								

Abbreviations & Codes

Numerous abbreviations and codes are used by radio amateurs, especially when using Morse (CW) or Datamodes. The abbreviations listed below are internationally recognised and cross language barriers, enabling people without a common language to communicate. Listed are many of the more common ones in use.

For a list of acronyms, abbreviations and conventions as used in the Yearbook, RadCom and other RSGB publications please go to the RSGB website: *http://rsgb.org/main/publications-archives/radcom/supplementary-information/abbreviations-and-acronyms/*

Abbreviations

73 Best regards
88 Love and kisses

AA All after (used after question mark to request a repetition)
AB All before (similarly)
ARRL American Radio Relay League
ABT About
ADS Address
AGN Again
ANT Antenna

BN All between
BK Break (to pause transmission of a message, say)
BUG Semiautomatic mechanical key

C Yes
CBA Callbook address
CFM Confirm
CLG Calling
CQ Calling any station
CS Callsign
CUL See you later
CUZ Because
CW Continuous wave
CX Conditions

DE From
DX Distance (sometimes refers to long distance contact)

ES And

FB Fine business (Analogous to "OK")
FCC Federal Communications Commission
FER For
FM From
FREQ Frequency

GA Good afternoon or Go ahead (depending on context)
GB Good Bye
GE Good evening
GM Good morning
GND Ground (ground potential)
GUD Good

HI Laughter
HR Here
HV Have
HW How

LID Poor operator

MILS Milliamperes

N No
NIL Nothing
NR Number
NW Now

OB Old boy
OC Old chap
OM Old man (any male amateur radio operator is an OM)
OO Official observer
OP Operator
OT Old timer
OTC Old timers club
OOTC Old old timers club

PSE Please
PWR Power

QCWA Quarter Century Wireless Association

R I acknowledge or decimal point (depending on context. The origin of "Roger")
RCD Received
RCVR Receiver (radio)
REF Refer to, Reference, Referring to.
RFI Radio Frequency Interference
RIG Radio apparatus
RPT Repeat or report (depending on context)
RPRT Report
RST Signal report format (Readability-Signal Strength-Tone)
RTTY Radioteletype
RX Receive

SAE Self-addressed envelope
SASE Self-addressed, stamped envelope
SED Said
SEZ Says
SIG Signal or signature
SIGS Signals
SKED Schedule
SN Soon
SMS Short message service
SRI Sorry
STN Station

TEMP Temperature
TFC Traffic
TMW Tomorrow
TNX Thanks
TT That
TU Thank you
TVI Television Interference
TX Transmit, transmitter
TXT Text

U You
UR Your or You're (depending on context)
URS Yours

VY Very

WA Word after
WB Word before
WD Word
WDS Words
W Watts
WDS Words
WKD Worked
WL Will
WUD Would
WX Weather

XMTR Transmitter
XYL Wife

YL Young lady (used for any female)

Q Codes

There are a huge number of three-letter Q codes. Radio amateurs use only a small percentage of them, as many are of relevance only to shipping, aircraft, the police etc. They fall into the following pattern.

QAA-QNZ Aeronautical, QOA-QQZ Maritime, QRA-QUZ All services, QZA-QZZ Other

All Q codes follow the form of a question and answer. The list below gives details of those in common use by radio amateurs.

Q-code	Question	Answer	Colloquial use (if different)/explanation
QRB	How far are you from my station?	The distance between our stations is ...	
QRG	What is my exact frequency?	Your frequency is ...	Frequency of operation
QRH	Does my frequency vary?	Your frequency varies	
QRI	How is the tone of my transmission?	The tone of your transmission is ...	
QRL	Are you (or is the frequency) busy?	I am (or the frequency is) busy	
QRM	Are you suffering interference?	I am suffereing interference	Man-made interference
QRN	Are you troubled by static?	I am troubled by static	Natural interference (atmospherics)
QRO	Shall I increase power?	Increase power	High power
QRP	Shall I decrease power?	Decrease power	Low power
QRQ	Shall I send faster?	Send faster	High speed
QRS	Shall I send more slowly?	Send slower	Low speed
QRT	Shall I stop transmitting?	Stop transmitting	To close down
QRU	Have you anything for me?	I have nothing for you	
QRV	Are you ready to transmit?	I am ready to transmit	
QRX	When will you call again?	I will call you again at ...	To stand-by
QRZ	Who is calling me?	You are being called by ...	
QSA	What is the strength of my signal?	The strength of your signal is ...	
QSB	Does the strength of my signals vary?	The strength of your signals varies	Fading
QSD	Is my keying defective?	Your keying is defective	
QSK	Can you hear me between your signals (and if so can I break-in)?	I can hear you between my signals (and it is OK to break-in on my transmission)	Break-in Morse operation
QSL	Can you acknowledge receipt?	I am acknowledging receipt	A card to confirm contact
QSO	Can you communicate with ... ?	I can communicate with ...	A contact
QSP	Will you relay to ...?	I will relay to ...	
QSX	Will you listen for ... (callsign) on ...?	I am listening for ... on ...	Split frequency operation
QSY	Shall I change frequency?	Change frequency to ...	
QTF	Will you give me the position of my station according to bearings taken?	The position of your station according to bearings taken is...	Beam heading (in degrees)
QTH	What is your position (location)?	My position (location) is ...	
QTR	What is the exact time?	The exact time is ...	

Phonetic Alphabet

The Phonetic Alphabet used by radio amateurs today was developed by NATO in the 1950s to be intelligible (and pronounceable) to all NATO allies. It replaced several other phonetic alphabets and is now widely used in business and telecommunications across Europe and North America.

A	Alpha	N	November
B	Bravo	O	Oscar
C	Charlie	P	Papa
D	Delta	Q	Quebec
E	Echo	R	Romeo
F	Foxtrot	S	Sierra
G	Golf	T	Tango
H	Hotel	U	Uniform
I	India	V	Victor
J	Juliet	W	Whiskey
K	Kilo	X	X-ray
L	Lima	Y	Yankee
M	Mike	Z	Zulu

ASCII Code

The American Standard Code for Information Interchange is the code used in computing and for packet radio. Standard ASCII consists of seven data bits, which gives 128 possible combinations. The first 32 characters are used for control and the remaining 96 are representable characters.

Dec	Hex	Chr	Ctrl	Dec	Hex	Chr	Dec	Hex	Chr	Dec	Hex	Chr
0	0	NUL	^@	32	20	SP	64	40	@	96	60	‘
1	1	SOH	^A	33	21	!	65	41	A	97	61	a
2	2	STX	^B	34	22	“	66	42	B	98	62	b
3	3	ETX	^C	35	23	#	67	43	C	99	63	c
4	4	EOT	^D	36	24	$	68	44	D	100	64	d
5	5	ENQ	^E	37	25	%	69	45	E	101	65	e
6	6	ACK	^F	38	26	&	70	46	F	102	66	f
7	7	BEL	^G	39	27	‘	71	47	G	103	67	g
8	8	BS	^H	40	28	(	72	48	H	104	68	h
9	9	HT	^I	41	29	)	73	49	I	105	69	i
10	0A	LF	^J	42	2A	*	74	4A	J	106	6A	j
11	0B	VT	^K	43	2B	+	75	4B	K	107	6B	k
12	0C	FF	^L	44	2C	,	76	4C	L	108	6C	l
13	0D	CR	^M	45	2D	-	77	4D	M	109	6D	m
14	0E	SO	^N	46	2E	.	78	4E	N	100	6E	n
15	0F	SI	^O	47	2F	/	79	4F	O	111	6F	o
16	10	DLE	^P	48	30	0	80	50	P	112	70	p
17	11	DC1	^Q	49	31	1	81	51	Q	113	71	q
18	12	DC2	^R	50	32	2	82	52	R	114	72	r
19	13	DC3	^S	51	33	3	83	53	S	115	73	s
20	14	DC4	^T	52	34	4	84	54	T	116	74	t
21	15	NAK	^U	53	35	5	85	55	U	117	75	u
22	16	SYN	^V	54	36	6	86	56	V	118	76	v
23	17	ETB	^W	55	37	7	87	57	W	119	77	w
24	18	CAN	^X	56	38	8	88	58	X	120	78	x
25	19	EM	^Y	57	39	9	89	59	Y	121	79	y
26	1A	SUB	^Z	58	3A	:	90	5A	Z	122	7A	z
27	1B	ESC		59	3B	;	91	5B	[	123	7B	{
28	1C	FS		60	3C	<	92	5C	\	124	7C	|
29	1D	GS		61	3D	=	93	5D	]	125	7D	}
30	1E	RS		62	3E	>	94	5E	^	126	7E	~
31	1F	US		63	3F	?	95	5F	_	127	7F	DEL

Morse Code

The standard alphabet and numbers, as required for the Foundation Licence Morse Assessment.

A	• ▬	N	▬ •
B	▬ • • •	O	▬ ▬ ▬
C	▬ • ▬ •	P	• ▬ ▬ •
D	▬ • •	Q	▬ ▬ • ▬
E	•	R	• ▬ •
F	• • ▬ •	S	• • •
G	▬ ▬ •	T	▬
H	• • • •	U	• • ▬
I	• •	V	• • • ▬
J	• ▬ ▬ ▬	W	• ▬ ▬
K	▬ • ▬	X	▬ • • ▬
L	• ▬ • •	Y	▬ • ▬ ▬
M	▬ ▬	Z	▬ ▬ • •

1	• ▬ ▬ ▬ ▬	6	▬ • • • •
2	• • ▬ ▬ ▬	7	▬ ▬ • • •
3	• • • ▬ ▬	8	▬ ▬ ▬ • •
4	• • • • ▬	9	▬ ▬ ▬ ▬ •
5	• • • • •	0	▬ ▬ ▬ ▬ ▬

Special Morse Characters

There are numerous procedural and punctuation characters. The following are those most commonly used by Morse operators.

$\overline{AR}$ (+)	[di-dah-di-dah-dit]	End of message (used before the final calls and is written as 'AR' or '+')
$\overline{CT}$	[dah-di-dah-di-dah]	Preliminary call
$\overline{BT}$ (=)	[dah-di-di-di-dah]	Separation signal (used in text and is written as 'BT' or '=')
$\overline{KN}$	[dah-di-dah-dah-dit]	Transmit only the station called (used after the final calls and is written as 'KN')
$\overline{VA}$	[di-di-di-dah-di-dah]	Transmission ends (written as 'VA' or 'SK')
?	[di-di-dah-dah-di-dit]	Question mark (written as 'IMI' or '?')
/	[dah-di-di-dah-dit]	Oblique stroke (can be used as part of a callsign and is written as '/')
.	[di-dah-di-dah-di-dah]	Full stop
Error	[di-di-di-di-di-di-di-dit]	Erases the word in which a mistake has been made
@	[di-dah-dah-di-dah-dit]	As used in Email addresses

Five-unit Code

The so-called Murray Code is used by RTTY operators and telex machines. It consists of one 'start' bit, five 'data' bits, then 1.5 'stop' bits. As there are only 32 possible combinations of code, a limited character set is available (e.g. no lower case). Traditionally, amateur RTTY is sent at 45.45 Bauds, which results in an element length of 22ms. It is known officially as the International Telegraphic Alphabet No.2.

Binary	Dec	Hex	Octal	Letter	Figure
00000	0	00	00	[Blank]	
00001	1	01	01	T	5
00010	2	02	02	[Carr. Return]	
00011	3	03	03	O	9
00100	4	04	04	[Space]	
00101	5	05	05	H	# (note)
00110	6	06	06	N	,
00111	7	07	07	M	.
01000	8	08	10	[Line Feed]	
01001	9	09	11	L	)
01010	10	0A	12	R	4
01011	11	0B	13	G	& (note)
01100	12	0C	14	I	8
01101	13	0D	15	P	0
01110	14	0E	16	C	:
01111	15	0F	17	V	;
10000	16	10	20	E	3
10001	17	11	21	Z	"
10010	18	12	22	D	$
10011	19	13	23	B	?
10100	20	14	24	S	Bell
10101	21	15	25	Y	6
10110	22	16	26	F	! (note)
10111	23	17	27	X	/
11000	24	18	30	A	-
11001	25	19	31	W	2
11010	26	1A	32	J	'
11011	27	1B	33	[Figure Shift]	
11100	28	1C	34	U	7
11101	29	1D	35	Q	1
11110	30	1E	36	K	(
11111	31	1F	37	[Letter Shift]	

Note:
The letters F, G and H in the Figures mode are not allocated internationally. Each country is free to use them as they see fit. The American usage is shown in the table above.

Repeater Listings

There are over 900 repeaters licensed in the United Kingdom which range in frequency from 28MHz. Many are traditional FM units, but there are also several for Amateur Television and increasingly Digital modes such as D-STAR, FUSION, DMR and other protocols. Many voice repeaters also offer internet access via computers or linking to other repeaters, using services such as Echolink and IRLP (Internet Radio Linking Project). These allow stations to communicate widely using radio into a local repeater or gateway, or by computer through a VoIP gateway. There is a network of APRS stations (Automatic Position Reporting System) which uses AX25 packet data to report the GPS location of users on a common frequency of 144.8000MHz. More information is available at: *https://ukrepeater.net*

Voice Repeaters (by output frequency)

Band	CH	Out (MHz)	In (MHz)	Callsign	CTCSS	Location	Keeper
10M	10M	29.2100	50.5200	GB3WX	77	Mere Wiltshire	G3ZXX
10M	10M	29.6200	29.5200	GB3CQ	88.5	Worthing	G4WTV
10M	10M	29.6400	29.5400	GB3CJ	77	Northampton	G1IRG
10M	10M	29.6500	29.5500	GB3TX	94.8	Norwich	M0ZAH
10M	10M	29.6900	29.5900	GB3HF	103.5	Llangollen	GW0NIS
6M	R50-01	50.7200	51.2200	GB3EF	110.9	Stowmarket	G0OZS
6M	R50-02	50.7300	51.2300	GB3XD	71.9	Louth	G7AJP
6M	R50-02	50.7300	51.2300	GB3GC	77	Gunnislake	M0WMB
6M	R50-02	50.7300	51.2300	GB3SL	103.5	Kilsyth	GM4COX
6M	R50-03	50.7400	51.2400	GB3UM	77	Markfield	M1NAS
6M	R50-04	50.7500	51.2500	GB3LP	77	Liverpool	M1SWB
6M	R50-05	50.7600	51.2600	GB3HX	103.5	Hastings	M0HOW
6M	R50-06	50.7700	51.2700	GB3FH	77	Somerset	G4RKY
6M	R50-06	50.7700	51.2700	GB3DB	110.9	Danbury	G6JYB
6M	R50-07	50.7800	51.2800	GB3PX	77	Royston	M0ZPU
6M	R50-07	50.7800	51.2800	GB3TY	110.9	Carrickfergus	GI6DKQ
6M	R50-08	50.7900	51.2900	GB3SX	103.5	Stoke On Trent	G4SCY
6M	R50-09	50.8000	51.3000	GB3WY	82.5	Wakefield	G1XCC
6M	R50-09	50.8000	51.3000	GB3ZY	77	Bristol	G4RKY
6M	R50-10	50.8100	51.3100	GB3ZW	103.5	Newtown Powys	GW4IQP
6M	R50-10	50.8100	51.3100	GB3FX	82.5	Farnham	G4EPX
6M	R50-11	50.8200	51.3200	GB3HM	71.9	Belper	G8IQP
6M	R50-13	50.8400	51.3400	GB3AM	77	Amersham	G0WTZ
6M	R50-14	50.8500	51.3500	GB3XS	77	Northampton	G7SYT
6M	R50-15	50.8600	51.3600	GB3VI	67	Dudley	G8PYT
2M	RV48	145.6000	145.0000	GB7RW	88.5	Whitby	G4EQS
2M	RV48	145.6000	145.0000	GB3RW	88.5	Worthing	G4WTV
2M	RV48	145.6000	145.0000	GB3CF	77	Markfield Leics	M0VSE
2M	RV48	145.6000	145.0000	GB3NZ	94.8	Catton	M0ZAH
2M	RV48	145.6000	145.0000	GB3EL	82.5	London	G4RZZ
2M	RV48	145.6000	145.0000	GB3LY	110.9	Limavady	MI1VOX
2M	RV48	145.6000	145.0000	GB3FF	103.5	Upper Largo	GM7LUN
2M	RV48	145.6000	145.0000	GB3SS	67	Elgin Moray	GM4ILS
2M	RV48	145.6000	145.0000	GB3WR	94.8	Cheddar	G4RKY
2M	RV48	145.6000	145.0000	GB3AS	77	Langholm	GM6LJE
2M	RV49	145.6125	145.0125	GB3OA	82.5	Southport	G4EID
2M	RV49	145.6125	145.0125	GB3NA	71.9	Barnsley	G4LUE
2M	RV49	145.6125	145.0125	GB3TE	67	Frinton On Sea	G0MBA
2M	RV49	145.6125	145.0125	GB3SV	94.8	Stafford	G7PFT
2M	RV49	145.6125	145.0125	GB3EW	77	Exeter	G6ATJ
2M	RV49	145.6125	145.0125	GB7LV	CC1	Livingston	GM7HHB
2M	RV49	145.6125	145.0125	GB3EI	88.5	Lochmaddy	GM8SAU
2M	RV49	145.6125	145.0125	GB3VM	103.5	Tenbury Wells	G8XYJ
2M	RV50	145.6250	145.0250	GB3KS	103.5	Dover	M1CMN
2M	RV50	145.6250	145.0250	GB3PB	71.9	Poole	G7ICH
2M	RV50	145.6250	145.0250	GB3GD	110.9	Snaefell Iom	GD4HOZ
2M	RV50	145.6250	145.0250	GB3HG	88.5	Thirsk	G8IMZ
2M	RV50	145.6250	145.0250	GB3MH	88.5	East Grinstead	G3NZP
2M	RV50	145.6250	145.0250	GB3NB	94.8	Norwich	G8VLL
2M	RV50	145.6250	145.0250	GB3NG	67	Fraserburgh	GM4ZUK
2M	RV50	145.6250	145.0250	GB3PA	103.5	Paisley	GM7OAW
2M	RV50	145.6250	145.0250	GB3SI		Helston	M1ERD
2M	RV50	145.6250	145.0250	GB3NW	67	Worcester	G4IDF
2M	RV50	145.6250	145.0250	GB3NF	77	Nottingham	M0VUB
2M	RV51	145.6375	145.0375	GB3DN	77	Stibb Cross	G1BHM
2M	RV51	145.6375	145.0375	GB7RN		Fareham	G3ZDF
2M	RV51	145.6375	145.0375	GB3VO	110.9	Wrexham	GW0WZZ
2M	RV51	145.6375	145.0375	GB3GJ	71.9	St Helier	GJ8PVL
2M	RV51	145.6375	145.0375	GB3ZN	118.8	South Shields	M0OMT
2M	RV51	145.6375	145.0375	GB3IN	71.9	Alfreton	G4TSN
2M	RV51	145.6375	145.0375	GB7DE	CC1	Edinburgh	GM7RYR
2M	RV51	145.6375	145.0375	GB7SW	CC2	Chippenham	G0RMA
2M	RV51	145.6375	145.0375	GB7CT	CC3	Tring	G0WTZ
2M	RV52	145.6500	145.0500	GB3AY	103.5	Dalry	MM0YET
2M	RV52	145.6500	145.0500	GB3OC	77	Kirkwall	GM0HQG
2M	RV52	145.6500	145.0500	GB3PO	110.9	Ipswich	G7CIY
2M	RV52	145.6500	145.0500	GB3SB	118.8	Selkirk	GM0FTJ
2M	RV52	145.6500	145.0500	GB3TR	94.8	Torquay	G8XST
2M	RV52	145.6500	145.0500	GB3WH	118.8	Swindon	G4LDL
2M	RV52	145.6500	145.0500	GB3MN	82.5	Disley	G8LZO
2M	RV52	145.6500	145.0500	GB7KB	CC3	Heathfield	G1ERJ
2M	RV52	145.6500	145.0500	GB7IP	CC1	Leicester	M1FJB
2M	RV52	145.6500	145.0500	GB3HS	88.5	Hull	G3GJA
2M	RV53	145.6625	145.0625	GB3KI	103.5	Herne Bay	G4TKR
2M	RV53	145.6625	145.0625	GB7IC-C		Herne Bay	G4TKR
2M	RV53	145.6625	145.0625	GB3PK	110.9	Ballycastle	MI0CRR
2M	RV53	145.6625	145.0625	GB3DW	110.9	Harlech	MW0VTK
2M	RV53	145.6625	145.0625	GB3CO	77	Corby	G1DIW
2M	RV53	145.6625	145.0625	GB3FE	103.5	Stirling	GM0MZB
2M	RV53	145.6625	145.0625	GB3AA	94.8	Bristol	G4CJZ
2M	RV53	145.6625	145.0625	GB7AM		Higher Poynton	G1DVA
2M	RV53	145.6625	145.0625	GB7PB		Durham	G7VIL
2M	RV53	145.6625	145.0625	GB3SH	71.9	Southampton	G4MYS
2M	RV53	145.6625	145.0625	GB3LZ	77	Lapford	G4MQQ
2M	RV53	145.6625	145.0625	GB7IL	CC5	Inverness	GM6GRE
2M	RV54	145.6750	145.0750	GB3ES	103.5	Hastings	G6ZZX
2M	RV54	145.6750	145.0750	GB7RB-C		Cowbridge	GW6CUR
2M	RV54	145.6750	145.0750	GB7NA-C		Barnsley	G4LUE
2M	RV54	145.6750	145.0750	GB3LD	110.9	Lancaster	G3VVT
2M	RV54	145.6750	145.0750	GB3RD	118.8	Reading	G8DOR
2M	RV54	145.6750	145.0750	GB3AE	118.8	Allendale	M0UKB
2M	RV54	145.6750	145.0750	GB3PR	94.8	Perth	GM8KPH
2M	RV54	145.6750	145.0750	GB7IE	77	Plymouth	G1LOE
2M	RV54	145.6750	145.0750	GB3LU	77	Lerwick	GM0GFL
2M	RV54	145.6750	145.0750	GB3PE	94.8	Peterborough	M0ZPU
2M	RV54	145.6750	145.0750	GB3BX	146.2	Cleobury North	G4VZO
2M	RV55	145.6875	145.0875	GB3SJ	103.5	Northwich	M0WTX
2M	RV55	145.6875	145.0875	GB3CN	94.8	Newcastle Emlyn	GW6JSO
2M	RV55	145.6875	145.0875	GB3DC	71.9	Derby	G7NPW
2M	RV55	145.6875	145.0875	GB3EA	110.9	Wickhambrook	G1YFF
2M	RV55	145.6875	145.0875	GB3KE	103.5	Glasgow	GM7SVK
2M	RV55	145.6875	145.0875	GB3WE	94.8	Weston-S-Mare	G4SZM
2M	RV55	145.6875	145.0875	GB7GD	CC1	Aberdeen	MM0RDM
2M	RV55	145.6875	145.0875	GB3CD	118.8	Crook	G0OCB
2M	RV55	145.6875	145.0875	GB3XP	82.5	Morden	M0SGL
2M	RV55	145.6875	145.0875	GB7DK	103.5	Stranraer	GM0HPK
2M	RV55	145.6875	145.0875	GB7WA	CC2	Grimsby	G7EOG
2M	RV56	145.7000	145.1000	GB3AR	110.9	Caernarfon	GW4KAZ
2M	RV56	145.7000	145.1000	GB3BT	118.8	Berwick On Tweed	GM1JFF
2M	RV56	145.7000	145.1000	GB3EV	77	Dufton	G7ITT
2M	RV56	145.7000	145.1000	GB3HH	71.9	Buxton	G7EKY
2M	RV56	145.7000	145.1000	GB3HI	103.5	Isle Of Mull	MM0JRM
2M	RV56	145.7000	145.1000	GB3WD	77	Ivybridge	M0ZCP
2M	RV56	145.7000	145.1000	GB3VA	118.8	Brill	G8GQJ
2M	RV56	145.7000	145.1000	GB3KN	103.5	Maidstone	G3VFC
2M	RV56	145.7000	145.1000	GB7CM	71.9	Blandford	M0MRP
2M	RV56	145.7000	145.1000	GB3BB	94.8	Brecon	MW0XDD
2M	RV57	145.7125	145.1125	GB3SW	77	Sidmouth	G6XUV
2M	RV57	145.7125	145.1125	GB7CG	77	Woburn Sands	G0WTZ
2M	RV57	145.7125	145.1125	GB3FG	94.8	Carmarthen	GW8KCY
2M	RV57	145.7125	145.1125	GB3YA	94.8	Cwmbran	MW0YAC
2M	RV57	145.7125	145.1125	GB3KY	94.8	Kings Lynn	G1SCQ
2M	RV57	145.7125	145.1125	GB3LA	103.5	Sanquhar	GM3SAN
2M	RV57	145.7125	145.1125	GB3MI	82.5	Manchester	G0TOG
2M	RV57	145.7125	145.1125	GB7OK	CC3	Beckenham	G1HIG
2M	RV57	145.7125	145.1125	GB3BM	67	Dudley	G8PYT
2M	RV58	145.7250	145.1250	GB3TP	82.5	Shipley	G8ZMG
2M	RV58	145.7250	145.1250	GB3DA	110.9	Danbury	G6JYB
2M	RV58	145.7250	145.1250	GB3VT	103.5	Stoke On Trent	G8NSS
2M	RV58	145.7250	145.1250	GB3AG	94.8	Forfar	MM0XET
2M	RV58	145.7250	145.1250	GB3CG	118.8	Gloucester	G3LVP
2M	RV58	145.7250	145.1250	GB3NI	110.9	Belfast N.I.	MI1VOX
2M	RV58	145.7250	145.1250	GB3SN	71.9	Alton Hants	G4EPX
2M	RV58	145.7250	145.1250	GB3NC	77	Roche	G4WVD
2M	RV58	145.7250	145.1250	GB3TW	118.8	Gateshead	G7UUR
2M	RV58	145.7250	145.1250	GB3LM	71.9	Lincoln	M0TEF
2M	RV58	145.7250	145.1250	GB3BI	67	Muir Of Ord	MM0GKB
2M	RV59	145.7375	145.1375	GB3CP	110.9	Fermanagh	GI6JPO
2M	RV59	145.7375	145.1375	GB7SF-C		Sheffield	M1ERS
2M	RV59	145.7375	145.1375	GB3ZA	118.8	Hereford	G0JWJ
2M	RV59	145.7375	145.1375	GB7NM		Attlebourgh	G0LGJ
2M	RV59	145.7375	145.1375	GB7PD	94.8	Maenclochog	GW4OZU
2M	RV59	145.7375	145.1375	GB3TA	67	Tamworth	M0TSD
2M	RV59	145.7375	145.1375	GB3AL	77	Amersham	G0WTZ
2M	RV59	145.7375	145.1375	GB3DR		Weymouth	G3VPF
2M	RV59	145.7375	145.1375	GB3AI	103.5	Hawick	GM8SJP
2M	RV60	145.7500	145.1500	GB3HA	118.8	Corbridge	G7UUR
2M	RV60	145.7500	145.1500	GB3BC	94.8	Pontypridd	GW0UXJ
2M	RV60	145.7500	145.1500	GB3FU	71.9	Nottingham	M5ADU
2M	RV60	145.7500	145.1500	GB3PI	77	Royston	M0ZPU
2M	RV60	145.7500	145.1500	GB3FK		Folkestone	M1CMN
2M	RV60	145.7500	145.1500	GB3WS	88.5	Crawley Sussex	G4EFO
2M	RV60	145.7500	145.1500	GB3CS	103.5	Motherwell	GM8HBY
2M	RV60	145.7500	145.1500	GB3MP	110.9	Prestatyn	M0OBW
2M	RV60	145.7500	145.1500	GB7NI	CC1	Carrickfergus	GI6DKQ
2M	RV60	145.7500	145.1500	GB3YC	88.5	Filey	M0KXQ
2M	RV61	145.7625	145.1625	GB3IK	103.5	Rochester	G6CKK
2M	RV61	145.7625	145.1625	GB3IR	88.5	Richmond Yorks	G4FZN
2M	RV61	145.7625	145.1625	GB3NE	118.8	Newbury	G6IBI
2M	RV61	145.7625	145.1625	GB3PL	77	E.Cornwall	M0WMB
2M	RV61	145.7625	145.1625	GB3EM	77	Cold Ashby	M1FJB

Voice Repeaters (by output frequency)

Band	CH	Out (MHz)	In (MHz)	Callsign	CTCSS	Location	Keeper
2M	RV61	145.7625	145.1625	GB7SG	CC1	Liverpool	M1SWB
2M	RV61	145.7625	145.1625	GB3DU	118.8	Duns	GM7LUN
2M	RV62	145.7750	145.1750	GB7TE		Clacton On Sea	G0MBA
2M	RV62	145.7750	145.1750	GB7DA	CC1	Airdrie	GM4AUP
2M	RV62	145.7750	145.1750	GB7IV		Southampton	G4MYS
2M	RV62	145.7750	145.1750	GB3FR	71.9	Spilsby Lincs.	M1FJB
2M	RV62	145.7750	145.1750	GB3GN	67	Banchory	GM4ZUK
2M	RV62	145.7750	145.1750	GB3IG	88.5	Stornoway	GM0LZE
2M	RV62	145.7750	145.1750	GB3NL		North London	G4DFB
2M	RV62	145.7750	145.1750	GB3PW	103.5	Newtown Powys	GW4IQP
2M	RV62	145.7750	145.1750	GB3WT	110.9	Omagh	GI3NVW
2M	RV62	145.7750	145.1750	GB7ER	77	Exeter	M0ZZT
2M	RV62	145.7750	145.1750	GB3DG	103.5	Newton Stewart	GM8CJG
2M	RV62	145.7750	145.1750	GB3WK	67	Leamington Spa	G6FEO
2M	RV62	145.7750	145.1750	GB3RF	82.5	Accrington	M0NFI
2M	RV62	145.7750	145.1750	GB3WW	94.8	Port Talbot	MW0HAC
2M	RV62	145.7750	145.1750	GB7MS	10	Newcastle	M0DMP
2M	RV63	145.7875	145.1875	GB3GO	110.9	Llandudno	M0BUQ
2M	RV63	145.7875	145.1875	GB3JB	103.5	Mere Wiltshire	G3ZXX
2M	RV63	145.7875	145.1875	GB3JL	77	Liskeard	G4RKY
2M	RV63	145.7875	145.1875	GB3LB	118.8	Lauder	GM7LUN
2M	RV63	145.7875	145.1875	GB3YW	82.5	Wakefield	G1XCC
2M	RV63	145.7875	145.1875	GB7YL	1	Lowestoft	M0JGX
2M	RV63	145.7875	145.1875	GB3EB	88.5	Newhaven	G0TJH
70CM	RV63	145.7875	145.1875	GB3BF	77	Bedford	G8MGP
70CM	DVU-R20	430.2500	439.2500	GB7KS	CC2	Stroud	G1MAW
70CM	DVU-R21	430.2625	439.2625	GB7RC		Cheltenham	M0URF
70CM	DVU-R21	430.2625	439.2625	GB7RO		Haslingden	M0LMN
70CM	DVU-R22	430.2750	439.2750	GB7CC	CC2	Cheltenham	G8WVW
70CM	DVU-R22	430.2750	439.2750	GB7FO	CC1	Blackpool	G0WDA
70CM	DVU-R23	430.2875	439.2875	GB7TH	CC5	Margate	M0LMK
70CM	DVU-R23	430.2875	439.2875	GB7CV	CC5	Coventry	G4USP
70CM	DVU-R24	430.3000	439.3000	GB7SA	CC5	Port Talbot	MW0IBF
70CM	DVU-R24	430.3000	439.3000	GB7KR	CC8	Kidderminster	G0EWH
70CM		430.8000	438.4000	GB3MF	118.8	Ledbury	M0TRY
70CM	RU65	430.8125	438.4125	GB3BW	82.5	Wakefield	M0YDG
70CM	RU65	430.8125	438.4125	GB3KR	118.8	Stourport On Severn	G7SAI
70CM	RU65	430.8125	438.4125	GB7LK	CC5	Liskeard	G4RKY
70CM	RU65	430.8125	438.4125	GB3CV	67	Coventry	G4USP
70CM	RU65	430.8125	438.4125	GB3FV	88.5	Stirling	GM0WUR
70CM	RU65	430.8125	438.4125	GB3VW	82.5	Wymondham	M0ZAH
70CM	RU66	430.8250	438.4250	GB3WU	118.8	Worcester	G8TIC
70CM	RU66	430.8250	438.4250	GB7NV	CC1	Hucknall	M0NGT
70CM	RU66	430.8250	438.4250	GB3IM	110.9	Douglas Iom	GD4HOZ
70CM	RU66	430.8250	438.4250	GB3IM	71.9	Ramsey Iom	GD4HOZ
70CM	RU66	430.8250	438.4250	GB7CP	77	Slough	M0WAQ
70CM	RU66	430.8250	438.4250	GB3ZB	77	Bristol	G4RKY
70CM	RU66	430.8250	438.4250	GB3ZX	88.5	Eastbourne	G6ZZX
70CM	RU66	430.8250	438.4250	GB3KW	103.5	Glasgow	GM7SVK
70CM	RU66	430.8250	438.4250	GB3CI	77	Corby	G7HPE
70CM	RU66	430.8250	438.4250	GB3RO	82.5	Haslingden	M0LMN
70CM	RU66	430.8250	438.4250	GB3NY	88.5	Harpham	M0DPH
70CM	RU67	430.8375	438.4375	GB7EU	94.8	Edinburgh	GM7RYR
70CM	RU67	430.8375	438.4375	GB3OR	103.5	Llandudno	MW0JWP
70CM	RU68	430.8500	438.4500	GB3NX	88.5	Crawley Sussex	G4EFO
70CM	RU68	430.8500	438.4500	GB3BS	118.8	Bristol	G4SDR
70CM	RU68	430.8500	438.4500	GB3FC	82.5	Blackpool	M0DKR
70CM	RU68	430.8500	438.4500	GB3IE	77	Plymouth	G7DQC
70CM	RU68	430.8500	438.4500	GB3RR	67	Dudley	G8PYT
70CM	RU68	430.8500	438.4500	GB7JI	CC2	St Helier	GJ8PVL
70CM	RU68	430.8500	438.4500	GB3BZ	110.9	Braintree	G0DEC
70CM	RU68	430.8500	438.4500	GB7EI	88.5	Lochmaddy	GM8SAU
70CM	RU68	430.8500	438.4500	GB3GR	71.9	Grantham	G0OJF
70CM	RU69	430.8625	438.4625	GB7CX	12	Coleraine	MI0PKO
70CM	RU69	430.8625	438.4625	GB3CM	94.8	Carmarthen	GW8KCY
70CM	RU69	430.8625	438.4625	GB3LR	88.5	Newhaven	G0TJH
70CM	RU69	430.8625	438.4625	GB3BE	118.8	Duns	GM7LUN
70CM	RU69	430.8625	438.4625	GB3RC	103.5	Cheltenham	M0URF
70CM	RU69	430.8625	438.4625	GB3CY	88.5	York	G4FUO
70CM	RU69	430.8625	438.4625	GB3PT	94.8	Prescot	M0NFI
70CM	RU69	430.8625	438.4625	GB3DP	88.5	Baston	G6YCA
70CM	RU69	430.8625	438.4625	GB3SC	67	Burton-Upon-Trent	G8EKG
70CM	RU70	430.8750	438.4750	GB7KL	110.9	Kirkby Lonsdale	M1JCB
70CM	RU70	430.8750	438.4750	GB3IM	110.9	Peel	GD4HOZ
70CM	RU70	430.8750	438.4750	GB7DC	71.9	Derby	G7NPW
70CM	RU70	430.8750	438.4750	GB3JM	118.8	Leominister	M0WYP
70CM	RU70	430.8750	438.4750	GB3BL	77	Bedford	G8MGP
70CM	RU70	430.8750	438.4750	GB3WA	103.5	Dalry	GM7GDE
70CM	RU70	430.8750	438.4750	GB7HN	82.5	Leigh	M0HOY
70CM	RU70	430.8750	438.4750	GB3CH	77	Liskeard	G4RKY
70CM	RU70	430.8750	438.4750	GB7XG	103.5	Glenrothes	MM0DXE
70CM	RU70	430.8750	438.4750	GB3WG	94.8	Port Talbot	MW0HAC
70CM	RU70	430.8750	438.4750	GB7NQ	10	Eastnor	M0VNA
70CM	RU71	430.8875	438.4875	GB7TQ	94.8	Torquay	G8XST
70CM	RU71	430.8875	438.4875	GB3EK	103.5	Margate	M0LMK
70CM	RU71	430.8875	438.4875	GB3WQ	77	Whitehaven	G0NIS
70CM	RU71	430.8875	438.4875	GB3NP	77	Towcester	G1IRG
70CM	RU71	430.8875	438.4875	GB7FI	CC3	Axbridge	G4RKY
70CM	RU71	430.8875	438.4875	GB3XL	82.5	Shipley	M0IRK
70CM	RU71	430.8875	438.4875	GB3HO	88.5	Horsham	G4EFO
70CM	RU71	430.8875	438.4875	GB3KA	118.8	Kidderminster	G7SAI
70CM	RU71	430.8875	438.4875	GB7FU		Nottingham	G0LCG
70CM	RU71	430.8875	438.4875	GB3IS	94.8	Isle Of South Uist	GM8SAU
70CM	RU72	430.9000	438.5000	GB3MG		Pontypridd	GW0UXJ
70CM	RU72	430.9000	438.5000	GB3HY	88.5	Haywards Heath	G6DGK
70CM	RU72	430.9000	438.5000	GB3GL	103.5	Glasgow	GM3SAN
70CM	RU72	430.9000	438.5000	GB3BK	103.5	Bromley	G0WYG
70CM	RU72	430.9000	438.5000	GB7GJ	88.5	St Helier	GJ8PVL
70CM	RU72	430.9000	438.5000	GB3PM	71.9	Sheffield	M0HOY
70CM	RU72	430.9000	438.5000	GB3MT	110.9	Magherafelt	MI0GRN
70CM	RU72	430.9000	438.5000	GB3PY	77	Cambridge	M0ZPU
70CM	RU72	430.9000	438.5000	GB3PZ	82.5	Dukinfield	G4ZPZ
70CM	RU72	430.9000	438.5000	GB3KC	67	Stourbridge	G0EWH
70CM	RU73	430.9125	438.5125	GB3UP	94.8	Fishguard	GW6TKK
70CM	RU73	430.9125	438.5125	GB3XX	77	Daventry	G8KHF
70CM	RU73	430.9125	438.5125	GB3SR	71.9	Skegness	G7UVD
70CM	RU73	430.9125	438.5125	GB7LZ	77	Lapford	G4MQQ
70CM	RU73	430.9125	438.5125	GB3SG	118.8	Stroud	G1MAW
70CM	RU73	430.9125	438.5125	GB3EG	82.5	Wigan	M1EGH
70CM	RU73	430.9125	438.5125	GB3ML	103.5	Staines Upon Thames	M0TIG
70CM	RU74	430.9250	438.5250	GB7CA	CC2	Douglas	GD4HOZ
70CM	RU74	430.9250	438.5250	GB3DM	103.5	Dumbarton	GM7GDE
70CM	RU74	430.9250	438.5250	GB7PY	CC3	Cambridge	M0ZPU
70CM	RU74	430.9250	438.5250	GB3HE	103.5	Hastings	G3TCG
70CM	RU74	430.9250	438.5250	GB7EC	67	Aberdeen	GM7RYR
70CM	RU74	430.9250	438.5250	GB3XN	71.9	Worksop	G3XXN
70CM	RU74	430.9250	438.5250	GB3VN	103.5	Ludlow	G4OYX
70CM	RU74	430.9250	438.5250	GB7EE	CC1	Edinburgh	GM7RYR
70CM	RU74	430.9250	438.5250	GB7BR	CC3	Ramsey Iom	GD4HOZ
70CM	RU74	430.9250	438.5250	GB3FI	77	Cheddar	G4RKY
70CM	RU74	430.9250	438.5250	GB3LN	118.8	Leicester	M0HEL
70CM	RU74	430.9250	438.5250	GB7KK	CC1	Ballycastle	MI0CRR
70CM	RU74	430.9250	438.5250	GB7NT	118.8	Newbury	G3WYW
70CM	RU75	430.9375	438.5375	GB7SZ	71.9	Poole	M1EZF
70CM	RU75	430.9375	438.5375	GB3LC	71.9	Louth	G7AJP
70CM	RU75	430.9375	438.5375	GB7EH	CC8	Banbury	G8NDT
70CM	RU75	430.9375	438.5375	GB3WP	82.5	Glossop	G6YRK
70CM	RU76	430.9500	438.5500	GB3OY	82.5	Buckhurst Hill	G7UZN
70CM	RU76	430.9500	438.5500	GB7FG	94.8	Carmarthen	GW8KCY
70CM	RU76	430.9500	438.5500	GB3EX	77	Silverton	G7NBU
70CM	RU76	430.9500	438.5500	GB7RY	103.5	Rye	M0HOW
70CM	RU76	430.9500	438.5500	GB3FJ	71.9	Asgarby	G3ZPU
70CM	RU76	430.9500	438.5500	GB3NH	77	Northampton	G1IRG
70CM	RU76	430.9500	438.5500	GB3OM	110.9	Omagh	GI4SXV
70CM	RU76	430.9500	438.5500	GB7BD	CC3	Bristol	G4RKY
70CM	RU76	430.9500	438.5500	GB3WL	67	Wythall	G3YXM
70CM	RU76	430.9500	438.5500	GB3OH	94.8	Linlithgow	GM0MZB
70CM	RU76	430.9500	438.5500	GB3UO	110.9	Wrexham	GW0WZZ
70CM	RU76	430.9500	438.5500	GB3PN	82.5	Preston	M0VXT
70CM	RU77	430.9625	438.5625	GB7JL	82.5	Wigan	G1EFU
70CM	RU77	430.9625	438.5625	GB7SH	71.9	Sheffield	M1ERS
70CM	RU77	430.9625	438.5625	GB3EH	67	Banbury	G4OHB
70CM	RU77	430.9625	438.5625	GB3CC	88.5	Chichester	G3UEQ
70CM	RU77	430.9625	438.5625	GB3MD	103.5	Llanfair Caereinion	GW4BML
70CM	RU77	430.9625	438.5625	GB3SA	82.5	Stamford	G8IOA
70CM	RU78	430.9750	438.5750	GB3JR	71.9	Mansfield	M5ADU
70CM	RU78	430.9750	438.5750	GB3BY	82.5	Blackpool	G0WDA
70CM	RU78	430.9750	438.5750	GB3DQ	77	Polperro	G1YDQ
70CM	RU78	430.9750	438.5750	GB3EZ	110.9	Wickhambrook	G1YFF
70CM	RU78	430.9750	438.5750	GB3KV	103.5	Kilsyth	GM3SAN
70CM	RU78	430.9750	438.5750	GB3KK	110.9	Ballycastle	MI0CRQ
70CM	RU78	430.9750	438.5750	GB3IP	82.5	South Nutfield	M0ZAH
70CM	RU78	430.9750	438.5750	GB3WO	88.5	Worthing	M0IAD
70CM	RU78	430.9750	438.5750	GB3AC	94.8	Bristol	G4CJZ
70CM	RU78	430.9750	438.5750	GB3ZI	103.5	Stafford	G4YFF
70CM	RU79	430.9875	438.5875	GB3DZ	71.9	Derby	G7NPW
70CM	RU79	430.9875	438.5875	GB3HL	88.5	Hull	M0ORH
70CM	RB0	433.0000	434.6000	GB3DT	71.9	Blandford	G0ZEP
70CM	RB0	433.0000	434.6000	GB3BN	118.8	Bracknell	G8DOR
70CM	RB0	433.0000	434.6000	GB3CK	103.5	Charing Kent	M0ZAA
70CM	RB0	433.0000	434.6000	GB3NR	94.8	Norwich	G8VLL
70CM	RB0	433.0000	434.6000	GB3PU	94.8	Perth	GM8KPH
70CM	RB0	433.0000	434.6000	GB3MK	77	Milton Keynes	G6GEI
70CM	RB0	433.0000	434.6000	GB3NT	118.8	Newcastle	G7UUR
70CM	RB0	433.0000	434.6000	GB3US	103.5	Sheffield	M0RFT
70CM	RB0	433.0000	434.6000	GB3SO	71.9	Withernsea	M0MFP
70CM	RB0	433.0000	434.6000	GB3LL	94.8	Llangollen	G7RPG
70CM	RB01	433.0250	434.6250	GB3HJ	118.8	Harrogate	G4MEM
70CM	RB01	433.0250	434.6250	GB3HN	103.5	Ludlow	M0HOY
70CM	RB01	433.0250	434.6250	GB3DV	71.9	Maltby	G0EPX
70CM	RB01	433.0250	434.6250	GB3JS	94.8	Great Yarmouth	M0JGX
70CM	RB01	433.0250	434.6250	GB3BV	82.5	Hemel Hempstead	G3YXZ
70CM	RB02	433.0500	434.6500	GB7DZ	103.5	Congleton	M0XDR
70CM	RB02	433.0500	434.6500	GB3AV	118.8	Aylesbury	G8GQJ
70CM	RB02	433.0500	434.6500	GB3HK	118.8	Selkirk	GM0FTJ
70CM	RB02	433.0500	434.6500	GB3LS	71.9	Lincoln	M0TEF
70CM	RB02	433.0500	434.6500	GB3LV	82.5	North London	G4DFB
70CM	RB02	433.0500	434.6500	GB3WM	67	Longbridge	M0JZT
70CM	RB02	433.0500	434.6500	GB3NN	94.8	Wells Norfolk	G0FVF
70CM	RB02	433.0500	434.6500	GB3UL	110.9	Belfast N.I.	MI1VOX
70CM	RB02	433.0500	434.6500	GB7ZE	103.5	Hastings	M0JBR
70CM	RB03	433.0750	434.6750	GB3ER	110.9	Danbury	G6JYB
70CM	RB03	433.0750	434.6750	GB3VS	94.8	Taunton	G4UVZ
70CM	RB03	433.0750	434.6750	GB3KU	82.5	Ashton-U-Lyne	M0NCZ
70CM	RB04	433.1000	434.7000	GB3KL	94.8	Kings Lynn	G0IJU
70CM	RB04	433.1000	434.7000	GB3CW	103.5	Newtown Powys	GW4IQP
70CM	RB04	433.1000	434.7000	GB3IH	110.9	Ipswich	G7CIY
70CM	RB04	433.1000	434.7000	GB3VE	77	Dufton	G7ITT
70CM	RB04	433.1000	434.7000	GB3LE	77	Markfield	M1NAS
70CM	RB04	433.1000	434.7000	GB3NK	103.5	Erith	G4EGU
70CM	RB04	433.1000	434.7000	GB3UB	118.8	Bath Avon	G4KVI
70CM	RB04	433.1000	434.7000	GB3SP	94.8	Pembroke	GW3XJQ
70CM	RB05	433.1250	434.7250	GB3GH	118.8	Gloucester	G3LVP
70CM	RB05	433.1250	434.7250	GB3IM	110.9	Douglas Iom	GD4HOZ
70CM	RB05	433.1250	434.7250	GB3OV	94.8	St Neots	G6OHM
70CM	RB05	433.1250	434.7250	GB3IC	67	Wolverhampton	G7CFC
70CM	RB06	433.1500	434.7500	GB3HC	118.8	Hereford	G0JWJ
70CM	RB06	433.1500	434.7500	GB3CR	110.9	Caergwrle	M0OBW
70CM	RB06	433.1500	434.7500	GB3SK	103.5	Canterbury	G6DIK
70CM	RB06	433.1500	434.7500	GB3DI	118.8	Didcot	G8CUL

Voice Repeaters (by output frequency)

Band	CH	Out (MHz)	In (MHz)	Callsign	CTCSS	Location	Keeper
70CM	RB06	433.1500	434.7500	GB3SY	71.9	Barnsley	G8POK
70CM	RB06	433.1500	434.7500	GB3ME	67	Rugby	G7BQM
70CM	RB07	433.1750	434.7750	GB3DE	110.9	Ipswich	G1NRL
70CM	RB07	433.1750	434.7750	GB3TS	118.8	Sunderland	G7MFN
70CM	RB07	433.1750	434.7750	GB7SJ	103.5	Northwich	M0WTX
70CM	RB07	433.1750	434.7750	GB3AU	82.5	Amersham	G0WTZ
70CM	RB08	433.2000	434.8000	GB3RB	71.9	Bolsover	G1SLE
70CM	RB08	433.2000	434.8000	GB3TF	103.5	Telford	M0JZH
70CM	RB08	433.2000	434.8000	GB3AN	110.9	Amlwch	MW0GCT
70CM	RB08	433.2000	434.8000	GB3SU	71.9	Southampton	G4MYS
70CM	RB09	433.2250	434.8250	GB3CL	67	Clacton	G0MBA
70CM	RB09	433.2250	434.8250	GB3ST	103.5	Stoke On Trent	G8NSS
70CM	RB09	433.2250	434.8250	GB3HD	82.5	Huddersfield	G0ISX
70CM	RB09	433.2250	434.8250	GB3IW	71.9	Ventnor	G4IKI
70CM	RB09	433.2250	434.8250	GB3TU	77	Tring Herts	G0WTZ
70CM	RB10	433.2500	434.8500	GB3MW	67	Leamington Spa	G6FEO
70CM	RB10	433.2500	434.8500	GB3AW	71.9	Newbury	G8DOR
70CM	RB10	433.2500	434.8500	GB3DY	71.9	Wirksworth	G3ZYC
70CM	RB10	433.2500	434.8500	GB3LI	82.5	Liverpool	G3WIC
70CM	RB10	433.2500	434.8500	GB3DD	94.8	Dundee	GM4UGF
70CM	RB10	433.2500	434.8500	GB3LT	77	Luton	G6OUA
70CM	RB11	433.2750	434.8750	GB3GY	88.5	Cleethorpes	M0KWK
70CM	RB11	433.2750	434.8750	GB3AH	94.8	East Dereham	G0LGJ
70CM	RB11	433.2750	434.8750	GB3HT	77	Hinckley	M0RVD
70CM	RB11	433.2750	434.8750	GB3RE	103.5	Maidstone	G6RVS
70CM	RB11	433.2750	434.8750	GB3RH	94.8	Axminster	G6WWY
70CM	RB11	433.2750	434.8750	GB3RU	118.8	Reading	G8DOR
70CM	RB11	433.2750	434.8750	GB3MY	82.5	Atherton	G4YYB
70CM	RB11	433.2750	434.8750	GB7SV	82.5	Hitchin	G6YIQ
70CM	RB12	433.3000	434.9000	GB3TJ	118.8	Corbridge	G7UUR
70CM	RB12	433.3000	434.9000	GB3DX	110.9	Derry/Londonderry	GI4YWT
70CM	RB12	433.3000	434.9000	GB3GF	88.5	Guildford	G4EML
70CM	RB12	433.3000	434.9000	GB3WB	94.8	Weston-S-Mare	G1VSX
70CM	RB12	433.3000	434.9000	GB3EE	71.9	Mansfield	G0RRZ
70CM	RB12	433.3000	434.9000	GB3GS	94.8	Downham Market	G0FHM
70CM	RB12	433.3000	434.9000	GB3PF	82.5	Accrington	M0NFI
70CM	RB13	433.3250	434.9250	GB3CA	77	Carlisle	G1XSZ
70CM	RB13	433.3250	434.9250	GB3DS	71.9	Worksop	G3XXN
70CM	RB13	433.3250	434.9250	GB3GU	71.9	Guernsey	GU6EFB
70CM	RB13	433.3250	434.9250	GB3SM	103.5	Stoke On Trent	G4SCY
70CM	RB13	433.3250	434.9250	GB3VH	82.5	Welwyn Garden City	G4THF
70CM	RB13	433.3250	434.9250	GB3BA	71.9	Basingstoke	G8GTZ
70CM	RB13	433.3250	434.9250	GB3HW	110.9	Gidea Park	G4GBW
70CM	RB14	433.3500	434.9500	GB3ED	94.8	Edinburgh	GM4GZW
70CM	RB14	433.3500	434.9500	GB3CB	67	Birmingham	G8NDT
70CM	RB14	433.3500	434.9500	GB3CE	67	Colchester	G0MBA
70CM	RB14	433.3500	434.9500	GB3HR	82.5	Harrow	G3YXZ
70CM	RB14	433.3500	434.9500	GB3YL	94.8	Lowestoft	G4RKP
70CM	RB14	433.3500	434.9500	GB3LF	110.9	Kendal	G3VVT
70CM	RB14	433.3500	434.9500	GB3MR	82.5	Stockport	G8LZO
70CM	RB14	433.3500	434.9500	GB3SD	71.9	Weymouth	G0ECX
70CM	RB14	433.3500	434.9500	GB3WF	82.5	Otley	M0SNW
70CM	RB15	433.3750	434.9750	GB3TH	67	Tamworth	G8YUQ
70CM	RB15	433.3750	434.9750	GB3MB	94.8	Merthyr	GW0UXJ
70CM	RB15	433.3750	434.9750	GB3LH	103.5	Shrewsbury	G8DIR
70CM	RB15	433.3750	434.9750	GB3WI	94.8	Wisbech	M0DUQ
70CM	RB15	433.3750	434.9750	GB3AB	82.5	Sheffield	M0GAV
70CM	RB15	433.3750	434.9750	GB3HB	77	Roche	G4WVD
70CM	RB15	433.3750	434.9750	GB3PP	82.5	Preston	M0NED
70CM	RB15	433.3750	434.9750	GB3FN	82.5	Farnham	G4EPX
70CM	UR63	438.3875	430.7875	GB3WN	67	Wolverhampton	G8NDT
70CM	UR63	438.3875	430.7875	GB3TD	118.8	Swindon	G4XUT
70CM	UR67	438.4375	430.8375	GB3EU	77	Cold Ashby	M1FJB
70CM	DVU13	439.1625	430.1625	GB7TD	CC1	Wakefield	G1XCC
70CM	DVU13	439.1625	430.1625	GB7NS	CC3	Caterham	G0OLX
70CM	DVU13	439.1625	430.1625	GB7BS	CC3	Bristol	G4SDR
70CM	DMU26	439.3250	430.3250	GB3NU	71.9	Nottingham	M0VUB
70CM	DMU27	439.3375	430.3375	GB3MC	110.9	Leasowe	G4YWD
70CM	DMU27	439.3375	430.3375	GB3GB	67	Sutton Coldfield	G8NDT
70CM	DMU30	439.3750	430.3750	GB7EA	CC1	Erskine	MM0PAZ
70CM	DMU30	439.3750	430.3750	GB7WH	110.9	Wrexham	M0XDR
70CM	DVU32	439.4000	430.4000	GB7MD		Blandford Forum	M0RHS
70CM	DVU32	439.4000	430.4000	GB7LN	CC1	Lincoln	G0RZR
70CM	DVU32	439.4000	430.4000	GB7LP	CC1	Liverpool	M1SWB
70CM	DVU32	439.4000	430.4000	GB7CW	CC3	Bridgend	MW0IBF
70CM	DVU32	439.4000	430.4000	GB7EP	CC3	Epsom	G0OXZ
70CM	DVU32	439.4000	430.4000	GB7AL	CC2	Ipswich	M1NIZ
70CM	DVU32	439.4000	430.4000	GB7BM	CC3	Birmingham	G8VIQ
70CM	DVU32	439.4000	430.4000	GB7LS		Galashiels	GM7LUN
70CM	DVU33	439.4125	430.4125	GB7WI	CC1	Hull	M0TTL
70CM	DVU33	439.4125	430.4125	GB7BZ		Bozeat	M5WOB
70CM	DVU33	439.4125	430.4125	GB7ND	CC1	Great Ellingham	G0LGJ
70CM	DVU33	439.4125	430.4125	GB7MN	CC2	Stockport	G8NSS
70CM	DVU33	439.4125	430.4125	GB7RT	CC3	Sunningdale	M0HBK
70CM	DVU33	439.4125	430.4125	GB7SD	CC1	Weymouth	G3VPF
70CM	DVU33	439.4125	430.4125	GB7DN		Dungiven	GI0AZB
70CM	DVU33	439.4125	430.4125	GB7GT	CC13	Clee Hill	G3PWJ
70CM	DVU33	439.4125	430.4125	GB7JD	CC1	Jedburgh	GM4UPX
70CM	DVU34	439.4250	430.4250	GB7AS	CC3	Ashford	M1CMN
70CM	DVU34	439.4250	430.4250	GB7HA		Harpenden	G7HMV
70CM	DVU34	439.4250	430.4250	GB7PN	CC1	Prestatyn	G4NOY
70CM	DVU34	439.4250	430.4250	GB7DR	CC5	Poole	G7ICH
70CM	DVU34	439.4250	430.4250	GB7HS	CC2	Heptonstall	G1XCC
70CM	DVU34	439.4250	430.4250	GB7XY	CC10	Newcastle	M0DMP
70CM	DVU34	439.4250	430.4250	GB7DS	CC1	Norwich	M0ZAH
70CM	DVU34	439.4250	430.4250	GB7AE	CC3	Marlborough	M1CJE
70CM	DVU35	439.4375	430.4375	GB7LD	CC15	Leyland	M0XRS
70CM	DVU35	439.4375	430.4375	GB7RK	CC3	Belfast	MI5DAW
70CM	DVU35	439.4375	430.4375	GB7PM		Cheddar	G4RKY
70CM	DVU35	439.4375	430.4375	GB7AV	CC3	Aylesbury	G0RAS
70CM	DVU35	439.4375	430.4375	GB7ES		Eastbourne	M0LRE
70CM	DVU35	439.4375	430.4375	GB7NR	CC1	Nottingham	M0VUB
70CM	DVU35	439.4375	430.4375	GB7PP	CC1	Ipswich	G0FEA
70CM	DVU35	439.4375	430.4375	GB7GC	CC2	Grimsby	G7EOG
70CM	DVU35	439.4375	430.4375	GB7KN	CC1	Edinburgh	GM7HHB
70CM	DVU35	439.4375	430.4375	GB7BX	CC11	Cleobury North	G4VZO
70CM	DVU36	439.4500	430.4500	GB7AY	CC1	Kilmarnock	GM0ONX
70CM	DVU36	439.4500	430.4500	GB7BP		Milton Keynes	M1ACB
70CM	DVU36	439.4500	430.4500	GB7IC		Herne Bay	G4TKR
70CM	DVU36	439.4500	430.4500	GB7II	CC3	Inverness	GM6GRE
70CM	DVU36	439.4500	430.4500	GB7GS	CC3	Downham Market	G0FHM
70CM	DVU36	439.4500	430.4500	GB7GR	CC5	Grantham	G6SSN
70CM	DVU36	439.4500	430.4500	GB7NE	CC3	Ashington	G0UDZ
70CM	DVU36	439.4500	430.4500	GB7ST	CC1	Stoke On Trent	G8NSS
70CM	DVU36	439.4500	430.4500	GB7NA-B		Barnsley	G4LUE
70CM	DVU36	439.4500	430.4500	GB7HR	CC3	Heathrow	G0OXZ
70CM	DVU36	439.4500	430.4500	GB7SU	CC8	Southampton	G4WFR
70CM	DVU37	439.4625	430.4625	GB7NL		Exeter	M0ZZT
70CM	DVU37	439.4625	430.4625	GB7SK	CC1	Leicester	M1FJB
70CM	DVU37	439.4625	430.4625	GB7XX	CC10	Felling	G4MSF
70CM	DVU37	439.4625	430.4625	GB7JB	CC1	Wincanton	G3ZXX
70CM	DVU37	439.4625	430.4625	GB7AN		Accrington	M0NFI
70CM	DVU37	439.4625	430.4625	GB7WR		Great Malvern	M0XZS
70CM	DVU37	439.4625	430.4625	GB7WL	CC3	Amersham	G0WTZ
70CM	DVU38	439.4750	430.4750	GB7SE	CC3	Thurrock	M0PFX
70CM	DVU38	439.4750	430.4750	GB7FE	CC1	Stirling	GM0MZB
70CM	DVU38	439.4750	430.4750	GB7SC	CC3	Bognor Regis	G0AFN
70CM	DVU38	439.4750	430.4750	GB7LR	CC1	Leicester	M1FJB
70CM	DVU38	439.4750	430.4750	GB7YZ		Mold	M0WTX
70CM	DVU38	439.4750	430.4750	GB7PL	CC8	Plymouth	G2JP
70CM	DVU39	439.4875	430.4875	GB7ZP		Chelmsford	G0URK
70CM	DVU39	439.4875	430.4875	GB7HU		South Cave	G0VRM
70CM	DVU39	439.4875	430.4875	GB7WB	CC2	Weston-S-Mare	G4SZM
70CM	DVU39	439.4875	430.4875	GB7WC		Warrington	G4VSS
70CM	DVU39	439.4875	430.4875	GB7CO	CC3	Fareham	G3ZDF
70CM	DVU39	439.4875	430.4875	GB7NF	CC1	Newhaven	G0TJH
70CM	DVU39	439.4875	430.4875	GB7MW	CC15	Carrickfergus	GI6DKQ
70CM	DVU39	439.4875	430.4875	GB7WF	CC13	Bewdley	G8OXG
70CM	DVU40	439.4969	430.4969	GB7LL	CC12	Llangollen	GW0NIS
70CM	DVU40	439.5000	430.5000	GB7KT	CC1	Andover	G8FHI
70CM	DVU40	439.5000	430.5000	GB7CZ	CC10	Crook	G0OCB
70CM	DVU40	439.5000	430.5000	GB7PE	CC3	Peterborough	M0ZPU
70CM	DVU40	439.5000	430.5000	GB7FR	CC4	Worthing	G7RZU
70CM	DVU40	439.5000	430.5000	GB7FC		Blackpool	M0LJH
70CM	DVU40	439.5031	430.5031	GB7TT		Ludlow	G1MAW
70CM	DVU41	439.5125	430.5125	GB7WX	CC3	Tarvin	G7NEH
70CM	DVU41	439.5125	430.5125	GB7EY	CC1	Selkirk	GM4AUP
70CM	DVU41	439.5125	430.5125	GB7IT	CC1	Weston-S-Mare	G4SZM
70CM	DVU41	439.5125	430.5125	GB7YD		Barnsley	G4LUE
70CM	DVU41	439.5125	430.5125	GB7LO	CC3	Beckenham	G1HIG
70CM	DVU41	439.5125	430.5125	GB7WK	CC15	Woking	M0MXC
70CM	DVU42	439.5250	430.5250	GB7TC	CC2	Swindon	G8VRI
70CM	DVU42	439.5250	430.5250	GB7SI	CC1	Stoke-On-Trent	G8NSS
70CM	DVU42	439.5250	430.5250	GB7SO	CC1	Gosport	G7CHO
70CM	DVU42	439.5250	430.5250	GB7UL	CC1	Carrickfergus	GI6DKQ
70CM	DVU42	439.5250	430.5250	GB7AR	CC15	New Ash Green	G6VBJ
70CM	DVU42	439.5250	430.5250	GB7KU	CC5	Bourne	G6SSN
70CM	DVU42	439.5250	430.5250	GB7EW	CC3	Exeter	G6ATJ
70CM	DVU42	439.5250	430.5250	GB7HC	CC5	Hereford	G3PWJ
70CM	DVU42	439.5250	430.5250	GB7YR	CC2	Doncaster	M1DAH
70CM	DVU43	439.5375	430.5375	GB7MC		St Austell	M1DNS
70CM	DVU43	439.5375	430.5375	GB7LF		Lancaster	G3VVT
70CM	DVU43	439.5375	430.5375	GB7HE		Hastings	M0HOW
70CM	DVU43	439.5375	430.5375	GB7RJ	CC5	Chirk	G7MHF
70CM	DVU43	439.5375	430.5375	GB7IK	CC3	Rochester	G6CKK
70CM	DVU43	439.5375	430.5375	GB7IS		Weston-S-Mare	G4SZM
70CM	DVU46	439.5750	430.5750	GB7DY	CC11	Dudley	G4VZO
70CM	DVU46	439.5750	430.5750	GB7HX	CC1	Huddersfield	G0ISX
70CM	DVU46	439.5750	430.5750	GB7EB	CC2	Beccles	M0JGX
70CM	DVU46	439.5750	430.5750	GB7YI	CC10	Hexham	G1HZI
70CM	DVU46	439.5750	430.5750	GB7BH	CC4	Brighton	G7TXU
70CM	DVU46	439.5750	430.5750	GB7KP	CC1	Comber	GI0VKP
70CM	DVU47	439.5875	430.5875	GB7HZ	CC5	Strabane	MI0HOZ
70CM	DVU47	439.5875	430.5875	GB7YJ	CC3	Leamington Spa	G4USP
70CM	DVU47	439.5875	430.5875	GB7SM	103.5	St Monans	MM0DXE
70CM	DVU47	439.5875	430.5875	GB7JM	CC3	Alford	M0AQC
70CM	DVU47	439.5875	430.5875	GB7DP	CC3	Treffgarne	MW0XDN
70CM	DVU47	439.5875	430.5875	GB7DH		Northwich	M0XDR
70CM	DVU47	439.5875	430.5875	GB7BN	CC3	Bognor Regis	G0TJH
70CM	DVU47	439.5875	430.5875	GB7SS	CC1	Elgin	GM7LSI
70CM	DVU48	439.6000	430.6000	GB7WT	CC1	Omagh	GI3NVW
70CM	DVU48	439.6000	430.6000	GB7MK	CC13	Ipswich	M1NIZ
70CM	DVU49	439.6125	430.6125	GB7DV		St. Helens	G1DVA
70CM	DVU49	439.6125	430.6125	GB7RR	CC1	Nottingham	G0LCG
70CM	DVU49	439.6125	430.6125	GB7TY	CC10	Hexham	G1HZI
70CM	DVU49	439.6125	430.6125	GB7CQ	CC4	Worthing	G4WTV
70CM	DVU50	439.6250	430.6250	GB7HB	CC1	Tandragee	MI0IRZ
70CM	DVU50	439.6250	430.6250	GB7GG	CC1	Airdrie	GM4AUP
70CM	DVU50	439.6250	430.6250	GB7RV	CC2	Blackburn	M0NWI
70CM	DVU50	439.6250	430.6250	GB7CY		Crook	G0OCB
70CM	DVU51	439.6375	430.6375	GB7MJ	CC5	Romsey	G3KYG
70CM	DVU51	439.6375	430.6375	GB7GW	103.5	Washington	G1LBU
70CM	DVU51	439.6375	430.6375	GB7JF		Northampton	G1IRG
70CM	DVU51	439.6375	430.6375	GB7SR	CC2	Sheffield	M0GAV
70CM	DVU51	439.6375	430.6375	GB7HM	CC1	Caergwrle	G1SYG
70CM	DVU51	439.6375	430.6375	GB7MH	CC2	East Grinstead	G3NZP
70CM	DVU51	439.6375	430.6375	GB7CL	CC3	Frinton On Sea	G0MBA
70CM	DVU52	439.6500	430.6500	GB7IN	CC1	Alfreton	G0LCG
70CM	DVU52	439.6500	430.6500	GB7GP	CC5	High Wycombe	M0GUY
70CM	DVU52	439.6500	430.6500	GB7DK		Stranraer	GM0HPK

Voice Repeaters (by output frequency)

Band	CH	Out (MHz)	In (MHz)	Callsign	CTCSS	Location	Keeper
70CM	DVU52	439.6500	430.6500	GB7BC	CC3	Brighton	G7TXU
70CM	DVU52	439.6500	430.6500	GB7RA	CC5	Neston	M0ORA
70CM	DVU52	439.6500	430.6500	GB7KY	103.5	Kirkcaldy	GM0MMN
70CM	DVU53	439.6625	430.6625	GB7AB		Aberdeen	GM7KBK
70CM	DVU53	439.6625	430.6625	GB7DD	CC1	Dundee	MM0DUN
70CM	DVU53	439.6625	430.6625	GB7BI	CC1	Muir Of Ord	GM0UDL
70CM	DVU53	439.6625	430.6625	GB7BE		Beccles	M0JGX
70CM	DVU53	439.6625	430.6625	GB7LY	CC1	Derry/Londonderry	GI4YWT
70CM	DVU53	439.6625	430.6625	GB7DJ	CC3	Northwich	M0WTX
70CM	DVU53	439.6625	430.6625	GB7MT		Southampton	G4MYS
70CM	DVU53	439.6625	430.6625	GB7DB	CC3	Woburn Sands	G0WTZ
70CM	DVU53	439.6625	430.6625	GB7SB	CC5	Birmingham	M0UEM
70CM	DVU53	439.6625	430.6625	GB7LE	CC2	Leeds	G1XCC
70CM	DVU54	439.6750	430.6750	GB3NS	82.5	Caterham	G0OLX
70CM	DVU54	439.6750	430.6750	GB7MF	CC3	Magherafelt	MI0JPD
70CM	DVU54	439.6750	430.6750	GB7AA	CC1	Bristol	G4CJZ
70CM	DVU54	439.6750	430.6750	GB7SN	CC1	Sheffield	M1ERS
70CM	DVU54	439.6750	430.6750	GB7ED	CC5	Exeter	M0ZZT
70CM	DVU54	439.6750	430.6750	GB7DL		Skegness	G4HFG
70CM	DVU54	439.6750	430.6750	GB7BF	CC5	Blackpool	M0DKR
70CM	DVU54	439.6750	430.6750	GB7SQ	CC1	Dalkeith	MM1BJO
70CM	DVU54	439.6750	430.6750	GB7FK-B		Folkestone	M1CMN
70CM	DVU55	439.6875	430.6875	GB7NY	CC1	Bessbrook	MI0PYN
70CM	DVU55	439.6875	430.6875	GB3AD		Bristol	G7NSY
70CM	DVU55	439.6875	430.6875	GB7WM		Astley	G4WLI
70CM	DVU55	439.6875	430.6875	GB7NB		Norwich	G0LGJ
70CM	DVU55	439.6875	430.6875	GB7GF	CC3	Guildford	G4EML
70CM	DVU55	439.6875	430.6875	GB7WV	CC2	Market Harborough	G1IVG
70CM	DVU55	439.6875	430.6875	GB7RD	CC3	Yelverton	G1ZKJ
70CM	DVU55	439.6875	430.6875	GB7TP	CC1	Shipley	M0IRK
70CM	DVU55	439.6875	430.6875	GB7KD	CC3	Hailsham	G1ERJ
70CM	DVU55	439.6875	430.6875	GB7GY	CC5	St. Peter Port	GU7DAI
70CM	DVU56	439.7000	430.7000	GB7DX	CC3	New Romney	G0GCQ
70CM	DVU56	439.7000	430.7000	GB7CD		Cardiff	GW6CUR
70CM	DVU56	439.7000	430.7000	GB7AU		Amersham	G0WTZ
70CM	DVU56	439.7000	430.7000	GB7MB		Morecambe	G0VGS
70CM	DVU56	439.7000	430.7000	GB7DO	CC2	Doncaster	G0WDA
70CM	DVU56	439.7000	430.7000	GB7VO		Tenbury Wells	G8XYJ
70CM	DVU57	439.7125	430.7125	GB7RE	CC1	Retford	G0EAK
70CM	DVU57	439.7125	430.7125	GB7FT	CC3	Portsmouth	M0PPR
70CM	DVU57	439.7125	430.7125	GB7EX	CC3	Southend On Sea	G8YPK
70CM	DVU57	439.7125	430.7125	GB7WD	CC11	Sedgley	G4VZO
70CM	DVU57	439.7125	430.7125	GB7EL	CC2	Nelson	G4BLH
70CM	DVU57	439.7125	430.7125	GB7BB	CC1	Banff	MM0BUH
70CM	DVU57	439.7125	430.7125	GB7PT		Royston	M0ZPU
70CM	DVU58	439.7250	430.7250	GB7HL	CC3	Hull	M0ORH
70CM	DVU58	439.7250	430.7250	GB7FD		Fleetwood	M0RGV
70CM	DVU58	439.7250	430.7250	GB7ME	CC3	Rugby	M0IJS
70CM	DVU58	439.7250	430.7250	GB7NW		Swindon	G4XIB
70CM	DVU59	439.7375	430.7375	GB7PO		Portsmouth	M0VMX
70CM	DVU59	439.7375	430.7375	GB7CK	CC3	Folkestone	M1CMN
70CM	DVU59	439.7375	430.7375	GB7TM		Eye	G0JSV
70CM	DVU59	439.7375	430.7375	GB7BK	CC3	Reading	G8DOR
70CM	DVU59	439.7375	430.7375	GB7BJ	CC13	Malvern	G3PWJ
70CM	DVU59	439.7375	430.7375	GB7HI	CC1	Lisburn	GI6DKQ
70CM	DVU59	439.7375	430.7375	GB7MR	CC2	Bury	G1XCC
70CM	DVU59	439.7375	430.7375	GB7HT	10	Ashington	G4NAB
70CM	DVU60	439.7500	430.7500	GB7HH	CC3	Romford	M0UPM
70CM	DVU60	439.7500	430.7500	GB7SP	CC3	Salisbury	G8PCB
70CM	DVU60	439.7500	430.7500	GB7NN		Worksop	M0GET
70CM	DVU60	439.7500	430.7500	GB7MM	CC5	Nairn	GM0RML
70CM	DVU60	439.7500	430.7500	GB7WN	CC8	Wolverhampton	G8NDT
70CM	DVU60	439.7500	430.7500	GB7MP	CC3	Heysham	G6CRV
70CM	DVU60	439.7500	430.7500	GB7DM	103.5	Monifieth	MM0DRA
70CM	DVU61	439.7625	430.7625	GB7WP		Birkenhead	G4BKF
70CM	DVU61	439.7625	430.7625	GB7AD		Bristol	G4CJZ
70CM	DVU61	439.7625	430.7625	GB7PI		Royston	M0ZPU
70CM	DVU61	439.7625	430.7625	GB7GB	CC8	Sutton Coldfield	G8NDT
70CM	DVU61	439.7625	430.7625	GB7EG	CC2	East Grinstead	G7KBR
70CM	DVU61	439.7625	430.7625	GB7BA	CC13	Bradford	G1XCC
70CM	DVU61	439.7625	430.7625	GB7WW	CC1	Melksham	G0RMA
70CM	DVU62	439.7750	430.7750	GB7SX	88.5	Bognor Regis	G0AFN
70CM	DVU62	439.7750	430.7750	GB7EZ	CC14	Erdington	M0SNR
70CM	DVU62	439.7750	430.7750	GB7PX	CC10	Royston	M0ZPU
70CM	DVU62	439.7750	430.7750	GB7CS		Glasgow	GM7GDE
23CM		1,290.6500	1,270.6500	GB7IC-A		Herne Bay	G4TKR
23CM	RM0	1,297.0000	1,291.0000	GB3NO	94.8	Norwich	G8VLL
23CM	RM1	1,297.0250	1,291.0250	GB3XZ	77	Daventry	G8KHF
23CM	RM2	1,297.0500	1,291.0500	GB3FM	100	Farnham	G4EPX
23CM	RM3	1,297.0750	1,291.0750	GB3PS	77	Royston	M0ZPU
23CM	RM3	1,297.0750	1,291.0750	GB3SE	103.5	Stoke On Trent	G8NSS
23CM	RM4	1,297.1000	1,291.1000	GB3EJ	77	Cambridge	G8MLA
23CM	RM6	1,297.1500	1,291.1500	GB3MM	67	Wolverhampton	G8NDT
23CM	RM12	1,297.3000	1,291.3000	GB7NA-A		Barnsley	G4LUE
23CM	RM14X	1,297.3500	1,277.3500	GB3AK	94.8	Bristol	G4CJZ
23CM	RM15	1,297.3750	1,291.3750	GB3WC	82.5	Wakefield	G1XCC

All Repeaters (by Callsign)

MODES: A=ANALOGUE, M=DMR, D=D-STAR, F=FUSION, N=NXDN, T=TV, E=TETRA, X=PACKET, P=P25, 7=M17

Repeater	Band	Channel	TX	RX	Modes	QTHR	Where	CTCSS/CC	DMR CC
GB3AA	2M	RV53	145.6625	145.0625	A	IO81RO	BRISTOL	94.8	
GB3AB	70CM	RB15	433.3750	434.9750	A	IO93FK	SHEFFIELD	82.5	
GB3AC	70CM	RU78	430.9750	438.5750	A	IO81TS	BRISTOL	94.8	
GB3AG	2M	RV58	145.7250	145.1250	AF	IO86ON	FORFAR	94.8	
GB3AH	70CM	RB11	433.2750	434.8750	A	JO02KP	EAST DEREHAM	94.8	
GB3AK	23CM	RM14X	1297.3500	1277.3500	A	IO81RO	BRISTOL	94.8	
GB3AL	2M	RV59	145.7375	145.1375	AF	IO91QP	AMERSHAM	77.0	
GB3AM	6M	R50-13	50.8400	51.3400	A	IO91QP	AMERSHAM	77.0	
GB3AN	70CM	RB08	433.2000	434.8000	AF	IO73UJ	AMLWCH	110.9	
GB3AR	2M	RV56	145.7000	145.1000	AF	IO73VC	CAERNARFON	110.9	
GB3AS	2M	RV48	145.6000	145.0000	A	IO85MC	LANGHOLM	77.0	
GB3AU	70CM	RB07	433.1750	434.7750	A	IO91QP	AMERSHAM	82.5	
GB3AV	70CM	RB02	433.0500	434.6500	A	IO91OT	AYLESBURY	118.8	
GB3AW	70CM	RB10	433.2500	434.8500	A	IO91GH	NEWBURY	71.9	
GB3AY	2M	RV52	145.6500	145.0500	A	IO75OR	DALRY	103.5	
GB3BA	70CM	RB13	433.3250	434.9250	A	IO91	BASINGSTOKE	71.9	
GB3BB	2M	RV56	145.7000	145.1000	A	IO81	BRECON	94.8	
GB3BC	2M	RV60	145.7500	145.1500	A	IO81IO	PONTYPRIDD	94.8	
GB3BE	70CM	RU69	430.8625	438.4625	AF	IO85	DUNS	118.8	
GB3BF	2M	RV63	145.7875	145.1875	A	IO92	BEDFORD	77.0	
GB3BH	70CM	RU79	430.9875	438.5875	A	JO00FU	BEXHILL-ON-SEA	77.0	
GB3BI	2M	RV58	145.7250	145.1250	A	IO77RM	MUIR OF ORD	67.0	
GB3BK	70CM	RU72	430.9000	438.5000	A	JO01AK	BROMLEY	103.5	
GB3BL	70CM	RU70	430.8750	438.4750	A	IO92SD	BEDFORD	77.0	
GB3BM	2M	RV57	145.7125	145.1125	A	IO82	DUDLEY	67.0	
GB3BM-B	2M	RV57	145.7125	145.1125	A	IO92	BIRMINGHAM	67.0	
GB3BN	70CM	RB0	433.0000	434.6000	A	IO91OJ	BRACKNELL	118.8	
GB3BR	70CM	RU65	430.8125	438.4125	A	IO90	HOVE	88.5	
GB3BS	70CM	RU68	430.8500	438.4500	A	IO81TK	BRISTOL	118.8	
GB3BT	2M	RV56	145.7000	145.1000	A	IO85WT	BERWICK ON TWEED	118.8	
GB3BV	70CM	RB01	433.0250	434.6250	AF	IO91SR	HEMEL HEMPSTEAD	82.5	
GB3BW	70CM	RU65	430.8125	438.4125	A	IO93HO	WAKEFIELD	82.5	
GB3BX	2M	RV54	145.6750	145.0750	M	IO82	CLEOBURY NORTH		13
GB3BY	70CM	RU78	430.9750	438.5750	A	IO83QP	CHORLEY	82.5	
GB3BZ	70CM	RU68	430.8500	438.4500	AM	JO01GW	BRAINTREE		3
GB3CA	70CM	RB13	433.3250	434.9250	A	IO84OT	CARLISLE	77.0	
GB3CB	70CM	RB14	433.3500	434.9500	A	IO92BL	BIRMINGHAM	67.0	
GB3CC	70CM	RU77	430.9625	438.5625	A	IO90	CHICHESTER	88.5	
GB3CD	2M	RV55	145.6875	145.0875	AF	IO94DR	CROOK	118.8	
GB3CE	70CM	RB14	433.3500	434.9500	A	JO01KV	COLCHESTER	67.0	
GB3CF	2M	RV48	145.6000	145.0000	ADFN	IO92IQ	MARKFIELD	77.0	
GB3CG	2M	RV58	145.7250	145.1250	A	IO81VU	GLOUCESTER	118.8	
GB3CH	70CM	RU70	430.8750	438.4750	A	IO70SM	LISKEARD	77.0	

All Repeaters (by Callsign)

Repeater	Band	Channel	TX	RX	Modes	QTHR	Where	CTCSS/CC	DMR CC
GB3CI	70CM	RU66	430.8250	438.4250	A	IO92PM	CORBY	77.0	
GB3CJ	10M	10M	29.6400	29.5400	A	IO92QF	BOZEAT	77.0	
GB3CK	70CM	RB0	433.0000	434.6000	A	JO01JF	CHARING KENT	103.5	
GB3CL	70CM	RB09	433.2250	434.8250	A	JO01OT	CLACTON	67.0	
GB3CM	70CM	RU69	430.8625	438.4625	AF	IO71VW	CARMARTHEN	94.8	
GB3CN	2M	RV55	145.6875	145.0875	A	IO71TX	NEWCASTLE EMLYN	94.8	
GB3CO	2M	RV53	145.6625	145.0625	A	IO92QM	CORBY	77.0	
GB3CP	2M	RV59	145.7375	145.1375	A	IO64IG	FERMANAGH	110.9	
GB3CQ	10M	10M	29.6300	29.5300	A	IO90	WORTHING	88.5	
GB3CR	70CM	RB06	433.1500	434.7500	A	IO83	CAERGWRLE	110.9	
GB3CS	2M	RV60	145.7500	145.1500	A	IO85	MOTHERWELL	103.5	
GB3CV	70CM	DMU29	439.3625	430.3625	A	IO92FJ	COVENTRY	67.0	
GB3CW	70CM	RB04	433.1000	434.7000	A	IO82HL	NEWTOWN POWYS	103.5	
GB3CY	70CM	RU69	430.8625	438.4625	A	IO93KW	YORK	88.5	
GB3DA	2M	RV58	145.7250	145.1250	AF	JO01GR	DANBURY	110.9	
GB3DB	6M	R50-06	50.7700	51.2700	AF	JO01HR	DANBURY	110.9	
GB3DC	2M	RV55	145.6875	145.0875	A	IO92GW	DERBY	71.9	
GB3DD	70CM	RB10	433.2500	434.8500	AF	IO86MM	DUNDEE	94.8	
GB3DE	70CM	RB07	433.1750	434.7750	A	JO02NF	IPSWICH	110.9	
GB3DG	2M	RV62	145.7750	145.1750	AF	IO74UV	NEWTON STEWART	103.5	
GB3DI	70CM	RB06	433.1500	434.7500	A	IO91	DIDCOT	118.8	
GB3DM	70CM	RU74	430.9250	438.5250	AF	IO75QX	DUMBARTON	103.5	
GB3DN	2M	RV51	145.6375	145.0375	AF	IO70UW	STIBB CROSS	77.0	
GB3DP	70CM	RU69	430.8625	438.4625	A	IO92TR	BASTON	88.5	
GB3DQ	70CM	RU78	430.9750	438.5750	A	IO70RI	POLPERRO	77.0	
GB3DR	2M	RV59	145.7375	145.1375	AF	IO80SN	WEYMOUTH		
GB3DS	70CM	RB13	433.3250	434.9250	A	IO93KH	WORKSOP	71.9	
GB3DT	70CM	RB0	433.0000	434.6000	AF	IO80WU	BLANDFORD	71.9	
GB3DU	2M	RV61	145.7625	145.1625	ADMF	IO85	DUNS	1.0	
GB3DV	70CM	RB01	433.0250	434.6250	A	IO93JK	MALTBY	71.9	
GB3DW	2M	RV53	145.6625	145.0625	AF	IO72WT	HARLECH	110.9	
GB3DX	70CM	RB12	433.3000	434.9000	A	IO65HA	DERRY/LONDONDERRY	110.9	
GB3DY	70CM	RB10	433.2500	434.8500	AM	IO93FB	WIRKSWORTH	71.9	
GB3DZ	70CM	RU79	430.9875	438.5875	A	IO92	DERBY	71.9	
GB3EA	2M	RV55	145.6875	145.0875	A	JO02GE	WICKHAMBROOK	110.9	
GB3EB	2M	RV63	145.7875	145.1875	ADM	JO00AT	NEWHAVEN	88.5	
GB3ED	70CM	RB14	433.3500	434.9500	AD	IO85JW	EDINBURGH	94.8	
GB3EE	70CM	RB12	433.3000	434.9000	A	IO93	MANSFIELD	71.9	
GB3EF	6M	R50-01	50.7200	51.2200	A	JO02NF	STOWMARKET	110.9	
GB3EG	70CM	RU73	430.9125	438.5125	A	IO83QN	WIGAN	82.5	
GB3EH	70CM	RU77	430.9625	438.5625	A	IO92GC	BANBURY	67.0	
GB3EI	2M	RV49	145.6125	145.0125	ADMF	IO67IN	LOCHMADDY	88.5	
GB3EJ	23CM	RM04	1297.1000	1291.1000	A	JO02	CAMBRIDGE	77.0	
GB3EK	70CM	RU71	430.8875	438.4875	A	JO01QJ	MARGATE	103.5	
GB3EL	2M	RV48	145.6000	145.0000	A	JO01AM	LONDON	82.5	
GB3EM	2M	RV61	145.7625	145.1625	A	IO92LJ	COLD ASHBY	77.0	
GB3ER	70CM	RB03	433.0750	434.6750	AF	JO01GR	DANBURY	110.9	
GB3ES	2M	RV54	145.6750	145.0750	A	JO00	HASTINGS	103.5	
GB3EU	70CM	UR67	438.4375	430.8375	A	IO92LJ	COLD ASHBY	77.0	
GB3EV	2M	RV56	145.7000	145.1000	A	IO84SQ	DUFTON	77.0	
GB3EW	2M	RV49	145.6125	145.0125	A	IO80	EXETER	77.0	
GB3EX	70CM	RU76	430.9500	438.5500	A	IO80GT	SILVERTON	77.0	
GB3EZ	70CM	RU78	430.9750	438.5750	A	JO02GE	WICKHAMBROOK	110.9	
GB3FA	2M	RV57	145.7125	145.1125	A	IO93HI	SHEFFIELD	82.5	
GB3FC	70CM	RU68	430.8500	438.4500	A	IO83LU	BLACKPOOL	82.5	
GB3FE	2M	RV53	145.6625	145.0625	A	IO86BC	STIRLING	103.5	
GB3FF	2M	RV48	145.6000	145.0000	A	IO86	UPPER LARGO	103.5	
GB3FG	2M	RV57	145.7125	145.1125	AF	IO71	CARMARTHEN	94.8	
GB3FH	6M	R50-06	50.7700	51.2700	A	IO81OH	SOMERSET	77.0	
GB3FI	70CM	RU74	430.9250	438.5250	A	IO81OH	CHEDDAR	77.0	
GB3FJ	70CM	RU76	430.9500	438.5500	A	IO93XE	ASGARBY	71.9	
GB3FK	2M	RV60	145.7500	145.1500	AF	JO01	FOLKESTONE	103.5	
GB3FM	23CM	RM2	1297.0500	1291.0500	A	IO91OF	FARNHAM	100.0	
GB3FN	70CM	RB15	433.3750	434.9750	A	IO91OF	FARNHAM	82.5	
GB3FR	2M	RV62	145.7750	145.1750	A	JO03AE	SPILSBY LINCS.	71.9	
GB3FU	2M	RV60	145.7500	145.1500	A	IO92	NOTTINGHAM	71.9	
GB3FV	70CM	RU65	430.8125	438.4125	A	IO86	STIRLING	88.5	
GB3FX	6M	R50-10	50.8100	51.3100	A	IO91OF	FARNHAM	82.5	
GB3GB	70CM	DMU27	439.3375	430.3375	A	IO92BM	SUTTON COLDFIELD	67.0	
GB3GC	6M	R50-02	50.7300	51.2300	A	IO70TL	LISKEARD	77.0	
GB3GD	2M	RV50	145.6250	145.0250	A	IO74SG	SNAEFELL IOM	110.9	
GB3GF	70CM	RB12	433.3000	434.9000	A	IO91RF	GUILDFORD	88.5	
GB3GH	70CM	RB05	433.1250	434.7250	A	IO81VU	GLOUCESTER	118.8	
GB3GJ	2M	RV51	145.6375	145.0375	A	IN89WE	ST HELIER	71.9	
GB3GL	70CM	RU72	430.9000	438.5000	AF	IO75WU	GLASGOW	103.5	
GB3GM	2M	RV50	145.6250	145.0250	F	IO82UC	GREAT MALVERN	67.0	
GB3GN	2M	RV62	145.7750	145.1750	A	IO87TA	BANCHORY	67.0	
GB3GO	2M	RV63	145.7875	145.1875	AF	IO83BH	LLANDUDNO	110.9	
GB3GR	70CM	RU68	430.8500	438.4500	A	IO92QV	GRANTHAM	71.9	
GB3GU	70CM	RB13	433.3250	434.9250	A	IN89RK	GUERNSEY	71.9	
GB3HA	2M	RV60	145.7500	145.1500	AF	IO84	HEXHAM	118.8	
GB3HB	70CM	RB15	433.3750	434.9750	AF	IO70OJ	ROCHE	77.0	
GB3HC	70CM	RB06	433.1500	434.7500	AF	IO82PB	HEREFORD	118.8	
GB3HD	70CM	RB09	433.2250	434.8250	AF	IO93BP	HUDDERSFIELD	82.5	
GB3HE	70CM	RU74	430.9250	438.5250	AF	JO00HV	HASTINGS	103.5	
GB3HF	10M	10M	29.6900	29.5900	A	IO82	LLANGOLLEN	103.5	
GB3HG	2M	RV50	145.6250	145.0250	AF	IO94JF	THIRSK	88.5	
GB3HH	2M	RV56	145.7000	145.1000	A	IO93BF	BUXTON	71.9	
GB3HI	2M	RV56	145.7000	145.1000	A	IO76DK	ISLE OF MULL	103.5	
GB3HJ	70CM	RB01	433.0250	434.6250	A	IO94EB	HARROGATE	118.8	
GB3HK	70CM	RB02	433.0500	434.6500	A	IO85	SELKIRK	118.8	
GB3HM	6M	R50-11	50.8200	51.3200	A	IO93GA	BELPER	71.9	
GB3HN	70CM	RB01	433.0250	434.6250	A	IO82	LUDLOW	103.5	
GB3HO	70CM	RU71	430.8875	438.4875	A	IO91TB	HORSHAM	88.5	
GB3HR	70CM	RB14	433.3500	434.9500	A	IO91TO	HARROW	82.5	
GB3HS	2M	RV52	145.6500	145.0500	ADF	IO93RS	HULL	88.5	
GB3HT	70CM	RB11	433.2750	434.8750	A	IO92HN	HINCKLEY	77.0	
GB3HW	70CM	RB13	433.3250	434.9250	A	JO01CN	GIDEA PARK	110.9	
GB3HY	70CM	RU72	430.9000	438.5000	A	IO90WX	HAYWARDS HEATH	88.5	

All Repeaters (by Callsign)

Repeater	Band	Channel	TX	RX	Modes	QTHR	Where	CTCSS/CC	DMR CC
GB3IC	70CM	RB05	433.1250	434.7250	A	IO82	WOLVERHAMPTON	67.0	
GB3IE	70CM	RU68	430.8500	438.4500	D	IO70XJ	PLYMOUTH	77.0	
GB3IG	2M	RV62	145.7750	145.1750	A	IO68QE	STORNOWAY	88.5	
GB3IH	70CM	RB04	433.1000	434.7000	A	JO02OB	IPSWICH	110.9	
GB3IK	2M	RV61	145.7625	145.1625	A	JO01FJ	ROCHESTER	103.5	
GB3IM-C	70CM	RU66	430.8250	438.4250	A	IO74SD	DOUGLAS IOM	110.9	
GB3IM-P	70CM	RU70	430.8750	438.4750	A	IO74PF	PEEL	110.9	
GB3IM-R	70CM	RU66	430.8250	438.4250	A	IO74TI	RAMSEY IOM	71.9	
GB3IM-S	70CM	RB05	433.1250	434.7250	A	IO74SG	DOUGLAS IOM	110.9	
GB3IN	2M	RV51	145.6375	145.0375	ADMFPN	IO93GD	ALFRETON	71.9	
GB3IP	70CM	RU78	430.9750	438.5750	A	IO91WF	SOUTH NUTFIELD	82.5	
GB3IR	2M	RV61	145.7625	145.1625	A	IO94DJ	RICHMOND YORKS	88.5	
GB3IS	70CM	RU71	430.8875	438.4875	ADMF	IO67HC	ISLE OF SOUTH UIST	94.8	
GB3IW	70CM	RB09	433.2250	434.8250	A	IO90JO	EAST COWES	71.9	
GB3JB	2M	RV63	145.7875	145.1875	AM	IO81VC	MERE WILTSHIRE		5
GB3JL	2M	RV63	145.7875	145.1875	A	IO70SM	LISKEARD	77.0	
GB3JM	70CM	RU70	430.8750	438.4750	A	IO82	LEOMINISTER	118.8	
GB3JR	70CM	RU78	430.9750	438.5750	A	IO93HB	MANSFIELD	71.9	
GB3JS	70CM	RB01	433.0250	434.6250	A	JO02UO	GREAT YARMOUTH	94.8	
GB3KA	70CM	RU71	430.8875	438.4875	A	IO82	KIDDERMINSTER	118.8	
GB3KC	70CM	RU72	430.9000	438.5000	A	IO82WK	STOURBRIDGE	67.0	
GB3KE	2M	RV55	145.6875	145.0875	AF	IO75UV	GLASGOW	103.5	
GB3KI	2M	RV53	145.6625	145.0625	AD	JO01NI	HERNE BAY	103.5	
GB3KK	70CM	RU78	430.9750	438.5750	A	IO65VE	BALLYCASTLE	110.9	
GB3KL	70CM	RB04	433.1000	434.7000	AF	JO02FR	KINGS LYNN	94.8	
GB3KN	2M	RV56	145.7000	145.1000	A	JO01HH	MAIDSTONE	103.5	
GB3KR	70CM	RU65	430.8125	438.4125	A	IO82UI	STOURPORT ON SEVERN	118.8	
GB3KS	2M	RV50	145.6250	145.0250	AF	JO01PA	DOVER	103.5	
GB3KT	70CM	RU68	430.8500	438.4500	A	IO93AU	KEIGHLEY	82.5	
GB3KU	70CM	RB03	433.0750	434.6750	ADMFPN	IO83XM	ASHTON-UNDER-LYNE	82.5	
GB3KV	70CM	RU78	430.9750	438.5750	A	IO75XX	KILSYTH	103.5	
GB3KW	70CM	RU66	430.8250	438.4250	AF	IO75UV	GLASGOW	103.5	
GB3KY	2M	RV57	145.7125	145.1125	A	JO02FS	KINGS LYNN	94.8	
GB3LA	2M	RV57	145.7125	145.1125	A	IO85CJ	SANQUHAR	103.5	
GB3LB	2M	RV63	145.7875	145.1875	A	IO85	LAUDER	118.8	
GB3LC	70CM	RU75	430.9375	438.5375	A	IO93WH	LOUTH	71.9	
GB3LD	2M	RV54	145.6750	145.0750	A	IO84OA	LANCASTER	110.9	
GB3LE	70CM	RB04	433.1000	434.7000	AF	IO92IQ	MARKFIELD	77.0	
GB3LF	70CM	RB14	433.3500	434.9500	A	IO84PH	KENDAL	110.9	
GB3LG	6M	R50-12	50.8300	51.3300	A	IO82	LLANGOLLEN		
GB3LH	70CM	RB15	433.3750	434.9750	AF	IO82OP	SHREWSBURY	103.5	
GB3LI	70CM	RB10	433.2500	434.8500	A	IO83LL	LIVERPOOL	82.5	
GB3LL	70CM	RB0	433.0000	434.6000	A	IO82	LLANGOLLEN	94.8	
GB3LM	2M	RV58	145.7250	145.1250	AF	IO93RF	LINCOLN	71.9	
GB3LN	70CM	RU74	430.9250	438.5250	A	IO92JO	LEICESTER	118.8	
GB3LP	6M	R50-04	50.7500	51.2500	A	IO83MK	LIVERPOOL	77.0	
GB3LR	70CM	RU69	430.8625	438.4625	A	JO00AS	NEWHAVEN	88.5	
GB3LS	70CM	RB02	433.0500	434.6500	A	IO93RF	LINCOLN	71.9	
GB3LT	70CM	RB10	433.2500	434.8500	AF	IO91SV	LUTON	77.0	
GB3LU	2M	RV54	145.6750	145.0750	A	IP90JD	LERWICK	77.0	
GB3LV	70CM	RB02	433.0500	434.6500	A	IO91XP	NORTH LONDON	82.5	
GB3LY	2M	RV48	145.6000	145.0000	A	IO65NC	LIMAVADY	110.9	
GB3LZ	2M	RV53	145.6625	145.0625	AMF	IO80CV	LAPFORD	77.0	
GB3MA	6M	R50-01	50.7200	51.2200	A	IO83SM	ATHERTON	82.5	
GB3MB	70CM	RB15	433.3750	434.9750	AF	IO81HR	MERTHYR	94.8	
GB3MC	70CM	DMU27	439.3375	430.3375	AF	IO83KJ	LEASOWE	110.9	
GB3MD	70CM	RU77	430.9625	438.5625	AF	IO82GN	LLANFAIR CAEREINION	103.5	
GB3ME	70CM	RB06	433.1500	434.7500	A	IO92JJ	RUGBY	67.0	
GB3MG	70CM	RU72	430.9000	438.5000	F	IO81KP	PONTYPRIDD		
GB3MH	2M	RV50	145.6250	145.0250	A	IO91WC	EAST GRINSTEAD	88.5	
GB3MI	2M	RV57	145.7125	145.1125	AD	IO83	MANCHESTER	82.5	
GB3MK	70CM	RB0	433.0000	434.6000	AF	IO92OB	MILTON KEYNES	77.0	
GB3ML	70CM	RU73	430.9125	438.5125	ADMF	IO91RK	STAINES UPON THAMES		3
GB3MM	23CM	RM06	1297.1500	1291.1500	A	IO82XP	WOLVERHAMPTON	67.0	
GB3MN	2M	RV52	145.6500	145.0500	AF	IO83XH	DISLEY	82.5	
GB3MO	70CM	UR63	438.3875	430.7875	A	IO84	MORECAMBE	110.9	
GB3MP	2M	RV60	145.7500	145.1500	A	IO83	PRESTATYN	110.9	
GB3MR	70CM	RB14	433.3500	434.9500	A	IO83XH	STOCKPORT	82.5	
GB3MW	70CM	RB10	433.2500	434.8500	A	IO92FH	LEAMINGTON SPA	67.0	
GB3MY	70CM	RB11	433.2750	434.8750	AF	IO83SM	ATHERTON	82.5	
GB3NA	2M	RV49	145.6125	145.0125	ADF	IO93HN	BARNSLEY	71.9	
GB3NB	2M	RV50	145.6250	145.0250	A	JO02PN	NORWICH	94.8	
GB3NC	2M	RV58	145.7250	145.1250	AF	IO70OJ	ROCHE	77.0	
GB3ND	70CM	DVU59	439.7375	430.7375	M	IO70	HOLSWORTHY		1
GB3NE	2M	RV61	145.7625	145.1625	A	IO91HJ	NEWBURY	118.8	
GB3NF	2M	RV50	145.6250	145.0250	AMF	IO92	NOTTINGHAM	77.0	
GB3NG	2M	RV50	145.6250	145.0250	A	IO87XO	FRASERBURGH	67.0	
GB3NH	70CM	RU76	430.9500	438.5500	A	IO92NF	NORTHAMPTON	77.0	
GB3NI	2M	RV58	145.7250	145.1250	A	IO74CO	BELFAST N.I.	110.9	
GB3NK	70CM	RB04	433.1000	434.7000	A	JO01BL	ERITH	103.5	
GB3NL	2M	RV62	145.7750	145.1750	A	IO91XP	NORTH LONDON		
GB3NM	70CM	RB07	433.1750	434.7750	AMFN	IO92	NOTTINGHAM	71.9	
GB3NN	70CM	RB02	433.0500	434.6500	A	JO02JV	WELLS NORFOLK	94.8	
GB3NO	23CM	RM0	1297.0000	1291.0000	A	JO02PP	NORWICH	94.8	
GB3NP	70CM	RU71	430.8875	438.4875	AF	IO92LD	TOWCESTER	77.0	
GB3NR	70CM	RB0	433.0000	434.6000	A	JO02PP	NORWICH	94.8	
GB3NS	70CM	DVU54	439.6750	430.6750	A	IO91WG	CATERHAM	82.5	
GB3NT	70CM	RB0	433.0000	434.6000	AF	IO94FW	NEWCASTLE UPON TYNE	118.8	
GB3NU	70CM	DMU26	439.3250	430.3250	ADMFPN7	IO92	NOTTINGHAM	71.9	
GB3NY	70CM	RU67	430.8375	438.4375	AF	IO93RT	HULL	67.0	
GB3NZ	2M	RV48	145.6000	145.0000	A	JO02NN	WYMONDHAM	94.8	
GB3OA	2M	RV49	145.6125	145.0125	AF	IO83LP	SOUTHPORT	82.5	
GB3OC	2M	RV52	145.6500	145.0500	A	IO88LX	KIRKWALL	77.0	
GB3OH	70CM	RU76	430.9500	438.5500	ADMF	IO85EX	LINLITHGOW	94.8	
GB3OM	70CM	RU76	430.9500	438.5500	AM	IO64JQ	OMAGH		1
GB3OR	70CM	RU67	430.8375	438.4375	AF	IO83BH	LLANDUDNO	103.5	
GB3OV	70CM	RB05	433.1250	434.7250	A	IO92VG	ST NEOTS	94.8	
GB3OY	70CM	RU76	430.9500	438.5500	A	JO01AO	BUCKHURST HILL	82.5	
GB3PA	2M	RV50	145.6250	145.0250	A	IO75QV	PAISLEY	103.5	

All Repeaters (by Callsign)

Repeater	Band	Channel	TX	RX	Modes	QTHR	Where	CTCSS/CC	DMR CC
GB3PB	2M	RV50	145.6250	145.0250	A	IO90AT	WIMBORNE	71.9	
GB3PD	70CM	RU68	430.8500	438.4500	AF	IO80SN	WEYMOUTH		
GB3PE	2M	RV54	145.6750	145.0750	A	IO92TL	PETERBOROUGH	94.8	
GB3PF	70CM	RB12	433.3000	434.9000	A	IO83TR	ACCRINGTON	82.5	
GB3PH	2M	RV48	145.6000	145.0000	A	IO72SX	PWLLHELI	118.8	
GB3PI	2M	RV60	145.7500	145.1500	A	IO92XA	ROYSTON	77.0	
GB3PK	2M	RV53	145.6625	145.0625	A	IO65VE	BALLYCASTLE	110.9	
GB3PL	2M	RV61	145.7625	145.1625	A	IO70	PLYMPTON	77.0	
GB3PM	70CM	RU72	430.9000	438.5000	A	IO93	SHEFFIELD	71.9	
GB3PN	70CM	RU76	430.9500	438.5500	A	IO83PS	PRESTON	82.5	
GB3PO	2M	RV52	145.6500	145.0500	A	JO02OB	IPSWICH	110.9	
GB3PP	70CM	RB15	433.3750	434.9750	AMF	IO83	PRESTON	82.5	
GB3PR	2M	RV54	145.6750	145.0750	AF	IO86GI	PERTH	94.8	
GB3PS	23CM	RM03	1297.0750	1291.0750	A	IO92XA	ROYSTON	77.0	
GB3PT	70CM	RU69	430.8625	438.4625	A	IO83	PRESCOT	94.8	
GB3PU	70CM	RB0	433.0000	434.6000	AF	IO86GI	PERTH	94.8	
GB3PW	2M	RV62	145.7750	145.1750	A	IO82HL	NEWTOWN	103.5	
GB3PX	6M	R50-07	50.7800	51.2800	A	IO92XA	ROYSTON	77.0	
GB3PY	70CM	RU72	430.9000	438.5000	A	JO02AF	CAMBRIDGE	77.0	
GB3PZ	70CM	RU72	430.9000	438.5000	A	IO83XL	DUKINFIELD	82.5	
GB3RB	70CM	RB08	433.2000	434.8000	A	IO93IF	BOLSOVER	71.9	
GB3RC	70CM	RU69	430.8625	438.4625	A	IO81	CHELTENHAM	103.5	
GB3RD	2M	RV54	145.6750	145.0750	A	IO91JM	READING	118.8	
GB3RE	70CM	RB11	433.2750	434.8750	A	JO01HH	MAIDSTONE	103.5	
GB3RF	2M	RV62	145.7750	145.1750	ADMF	IO83TR	ACCRINGTON	82.5	
GB3RH	70CM	RB11	433.2750	434.8750	A	IO80MS	AXMINSTER	94.8	
GB3RO	70CM	RU66	430.8250	438.4250	A	IO83UQ	HASLINGDEN	82.5	
GB3RR	70CM	RU68	430.8500	438.4500	A	IO82	DUDLEY	67.0	
GB3RU	70CM	RB11	433.2750	434.8750	A	IO91LK	READING	118.8	
GB3RW	2M	RV48	145.6000	145.0000	A	IO90	WORTHING	88.5	
GB3RX	70CM	RU65	430.8125	438.4125	AF	IO83	PRESCOT	77.0	
GB3SA	70CM	RU77	430.9625	438.5625	A	IO92RP	STAMFORD	82.5	
GB3SB	2M	RV52	145.6500	145.0500	A	IO85	SELKIRK	118.8	
GB3SC	70CM	RU69	430.8625	438.4625	A	IO92ES	BURTON-UPON-TRENT	67.0	
GB3SD	70CM	RB14	433.3500	434.9500	A	IO80SQ	WEYMOUTH	71.9	
GB3SE	23CM	RM03	1297.0750	1291.0750	A	IO83	STOKE ON TRENT	103.5	
GB3SF	2M	RV50	145.6250	145.0250	F	IO86	PERTH		
GB3SG	70CM	RU73	430.9125	438.5125	A	IO81	STROUD	118.8	
GB3SH	2M	RV53	145.6625	145.0625	A	IO90IV	SOUTHAMPTON	71.9	
GB3SI	2M	RV50	145.6250	145.0250	A	IO70JB	HELSTON		
GB3SJ	2M	RV55	145.6875	145.0875	A	IO83RF	NORTHWICH	103.5	
GB3SK	70CM	RB06	433.1500	434.7500	A	JO01MH	CANTERBURY	103.5	
GB3SL	6M	R50-02	50.7300	51.2300	A	IO75XX	KILSYTH	103.5	
GB3SM	70CM	RB13	433.3250	434.9250	A	IO93BA	STOKE ON TRENT	103.5	
GB3SN	2M	RV58	145.7250	145.1250	A	IO91LC	ALTON HANTS	71.9	
GB3SO	70CM	RB0	433.0000	434.6000	ADMF	JO03AR	WITHERNSEA	71.9	
GB3SP	70CM	RB04	433.1000	434.7000	AF	IO71NQ	PEMBROKE	94.8	
GB3SR	70CM	RU73	430.9125	438.5125	A	JO03CB	SKEGNESS	71.9	
GB3SS	2M	RV48	145.6000	145.0000	AF	IO87	ELGIN MORAY	67.0	
GB3ST	70CM	RB09	433.2250	434.8250	A	IO83	STOKE ON TRENT	103.5	
GB3SU	70CM	RB08	433.2000	434.8000	A	IO90IV	SOUTHAMPTON	71.9	
GB3SV	2M	RV49	145.6125	145.0125	A	IO82WS	STAFFORD	94.8	
GB3SW	2M	RV57	145.7125	145.1125	A	IO80JQ	SIDMOUTH	77.0	
GB3SX	6M	R50-08	50.7900	51.2900	A	IO93BA	STOKE ON TRENT	103.5	
GB3SY	70CM	RB06	433.1500	434.7500	A	IO93	BARNSLEY	71.9	
GB3TA	2M	RV59	145.7375	145.1375	A	IO92DP	TAMWORTH	67.0	
GB3TD	70CM	UR63	438.3875	430.7875	A	IO91DL	SWINDON	118.8	
GB3TE	2M	RV49	145.6125	145.0125	A	JO01OU	FRINTON ON SEA	67.0	
GB3TF	70CM	RB08	433.2000	434.8000	AF	IO82RP	TELFORD	103.5	
GB3TH	70CM	RB15	433.3750	434.9750	AF	IO92	TAMWORTH	67.0	
GB3TJ	70CM	RB12	433.3000	434.9000	AF	IO85XA	CORBRIDGE	118.8	
GB3TP	2M	RV58	145.7250	145.1250	A	IO93CT	SHIPLEY	82.5	
GB3TR	2M	RV52	145.6500	145.0500	A	IO80FM	TORQUAY	94.8	
GB3TS	70CM	RB07	433.1750	434.7750	AF	IO94GX	SUNDERLAND	118.8	
GB3TU	70CM	RB09	433.2250	434.8250	A	IO91PS	TRING	77.0	
GB3TW	2M	RV58	145.7250	145.1250	AF	IO94DW	GATESHEAD	118.8	
GB3TX	10M	10M	29.6500	29.5500	A	JO02QP	NORWICH	94.8	
GB3UB	70CM	RB04	433.1000	434.7000	A	IO81UJ	BATH AVON	118.8	
GB3UK	2M	RV49	145.6125	145.0125	ADMF	IO94FV	WASHINGTON	10.0	
GB3UL	70CM	RB02	433.0500	434.6500	A	IO74CO	BELFAST N.I.	110.9	
GB3UM	6M	R50-03	50.7400	51.2400	A	IO92IQ	MARKFIELD	77.0	
GB3UO	70CM	RU76	430.9500	438.5500	AF	IO83LA	WREXHAM	110.9	
GB3UP	70CM	RU73	430.9125	438.5125	A	IO71MX	FISHGUARD	94.8	
GB3US	70CM	RB0	433.0000	434.6000	A	IO93	SHEFFIELD	103.5	
GB3VA	2M	RV56	145.7000	145.1000	A	IO91LT	BRILL	118.8	
GB3VE	70CM	RB04	433.1000	434.7000	A	IO84SQ	DUFTON	77.0	
GB3VH	70CM	RB13	433.3250	434.9250	A	IO91VT	WELWYN GARDEN CITY	82.5	
GB3VI	6M	R50-15	50.8600	51.3600	A	IO82	DUDLEY	67.0	
GB3VM	2M	RV49	145.6125	145.0125	AF	IO82QJ	TENBURY WELLS	103.5	
GB3VO	2M	RV51	145.6375	145.0375	AF	IO83LA	WREXHAM	110.9	
GB3VR	70CM	RU65	430.8125	438.4125	A	IO91TU	HARPENDEN	67.0	
GB3VS	70CM	RB03	433.0750	434.6750	A	IO80LV	TRULL	94.8	
GB3VT	2M	RV58	145.7250	145.1250	D	IO83	STOKE ON TRENT	103.5	
GB3VW	70CM	RU65	430.8125	438.4125	A	JO02NN	WYMONDHAM	82.5	
GB3WA	70CM	RU70	430.8750	438.4750	A	IO75OR	DALRY	103.5	
GB3WB	70CM	RB12	433.3000	434.9000	A	IO81MH	WESTON-SUPER-MARE	94.8	
GB3WC	23CM	RM15	1297.3750	1291.3750	A	IO93EP	WAKEFIELD	82.5	
GB3WD	2M	RV56	145.7000	145.1000	A	IO80	IVYBRIDGE	77.0	
GB3WE	2M	RV55	145.6875	145.0875	AD	IO81MH	WESTON-SUPER-MARE	94.8	
GB3WF	70CM	RB14	433.3500	434.9500	AF	IO93DV	OTLEY	82.5	
GB3WG	70CM	RU70	430.8750	438.4750	AF	IO81CP	PORT TALBOT	94.8	
GB3WH	2M	RV52	145.6500	145.0500	A	IO91EM	SWINDON	118.8	
GB3WI	70CM	RB15	433.3750	434.9750	AF	JO02	WISBECH	94.8	
GB3WK	2M	RV62	145.7750	145.1750	A	IO92FH	LEAMINGTON SPA	67.0	
GB3WL	70CM	RU76	430.9500	438.5500	ADF	IO92BJ	WYTHALL	67.0	
GB3WM	70CM	RB02	433.0500	434.6500	AF	IO92AJ	LONGBRIDGE	67.0	
GB3WN	70CM	UR63	438.3875	430.7875	A	IO82XP	WOLVERHAMPTON	67.0	
GB3WO	70CM	RU78	430.9750	438.5750	A	IO90TT	WORTHING	88.5	
GB3WP	10M	10M	29.6600	29.6500	A	IO93HP	WAKEFIELD	82.5	

All Repeaters (by Callsign)

Repeater	Band	Channel	TX	RX	Modes	QTHR	Where	CTCSS/CC	DMR CC
GB3WQ	70CM	RU71	430.8875	438.4875	A	IO84	WHITEHAVEN	77.0	
GB3WR	2M	RV48	145.6000	145.0000	A	IO81PH	CHEDDAR	94.8	
GB3WS	2M	RV60	145.7500	145.1500	A	IO91WB	HORSHAM	88.5	
GB3WT	2M	RV62	145.7750	145.1750	A	IO64JQ	OMAGH	110.9	
GB3WU	70CM	RU66	430.8250	438.4250	A	IO82VE	WORCESTER	118.8	
GB3WW	2M	RV62	145.7750	145.1750	AF	IO81CP	PORT TALBOT	94.8	
GB3WX	6M	R50-12	50.8300	51.3300	A	JO02NN	WYMONDHAM	94.8	
GB3WY	6M	R50-09	50.8000	51.3000	A	IO93EP	WAKEFIELD	82.5	
GB3XD	6M	R50-02	50.7300	51.2300	A	IO93WH	LOUTH	71.9	
GB3XL	70CM	RU71	430.8875	438.4875	AM	IO93CT	SHIPLEY	82.5	
GB3XN	70CM	RU74	430.9250	438.5250	A	IO93KJ	WORKSOP	71.9	
GB3XP	2M	RV55	145.6875	145.0875	AF	IO91VJ	MORDEN	82.5	
GB3XS	6M	R50-14	50.8500	51.3500	A	IO92MG	NORTHAMPTON	77.0	
GB3XV	70CM	DVU-R20	430.2500	439.2500	F	IO92KG	DAVENTRY		
GB3XX	70CM	RU73	430.9125	438.5125	AF	IO92KG	DAVENTRY	77.0	
GB3XZ	23CM	RM1	1297.0250	1291.0250	AD	IO92KG	DAVENTRY	77.0	
GB3YA	2M	RV57	145.7125	145.1125	A	IO81LP	CWMBRAN	94.8	
GB3YC	2M	RV60	145.7500	145.1500	A	IO94SE	FILEY	88.5	
GB3YL	70CM	RB14	433.3500	434.9500	A	JO02UL	LOWESTOFT	94.8	
GB3YW	2M	RV63	145.7875	145.1875	A	IO93EP	WAKEFIELD	82.5	
GB3ZA	2M	RV59	145.7375	145.1375	AF	IO82PB	HEREFORD	118.8	
GB3ZB	70CM	RU66	430.8250	438.4250	A	IO81QJ	BRISTOL	77.0	
GB3ZI	70CM	RU78	430.9750	438.5750	AF	IO82	STAFFORD	103.5	
GB3ZN	70CM	RU71	430.8875	438.4875	A	IO94GX	HEBBURN	118.8	
GB3ZW	6M	R50-10	50.8100	51.3100	A	IO82HL	NEWTOWN POWYS	103.5	
GB3ZX	70CM	RU66	430.8250	438.4250	A	JO00	EASTBOURNE	88.5	
GB3ZY	6M	R50-09	50.8000	51.3000	A	IO81QJ	BRISTOL	77.0	
GB7AA	70CM	DVU54	439.6750	430.6750	M	IO81RO	BRISTOL		1
GB7AB	70CM	DVU53	439.6625	430.6625	F	IO87VD	ABERDEEN		
GB7AC	70CM	RU76	430.9500	438.5500	AM	IO87	LARGS		3
GB7AD	70CM	DVU61	439.7625	430.7625	D	IO81RO	BRISTOL		
GB7AE	70CM	DVU34	439.4250	430.4250	D	IO91CJ	MARLBOROUGH		3
GB7AG	70CM	RU72	430.9000	438.5000	ADMF	IO86	FORFAR	94.8	
GB7AI	70CM	DMU27	439.3375	430.3375	AM	IO85	HAWICK		3
GB7AL	70CM	DVU32	439.4000	430.4000	M	JO02RD	IPSWICH		2
GB7AM	2M	RV53	145.6625	145.0625	DFP	IO83WI	HIGHER POYNTON		
GB7AN	70CM	DVU37	439.4625	430.4625	F	IO83TS	ACCRINGTON		
GB7AR	70CM	DVU42	439.5250	430.5250	DMFPN	JO01	NEW ASH GREEN		
GB7AS	70CM	DVU34	439.4250	430.4250	M	JO01KD	ASHFORD		3
GB7AU	70CM	DVU56	439.7000	430.7000	DF	IO91GP	AMERSHAM		
GB7AV	70CM	DVU35	439.4375	430.4375	M	IO91OT	AYLESBURY		3
GB7AY	70CM	DVU36	439.4500	430.4500	DMF	IO75TQ	KILMARNOCK		1
GB7BA	70CM	DVU61	439.7625	430.7625	M	IO93CS	BRADFORD		13
GB7BB	70CM	DVU57	439.7125	430.7125	DMF	IO87QP	BANFF		1
GB7BC	70CM	DVU52	439.6500	430.6500	DMFPN	IO90	BRIGHTON		3
GB7BD	70CM	RU76	430.9500	438.5500	M	IO81QJ	BRISTOL		3
GB7BE	70CM	DVU53	439.6625	430.6625	D	JO02TK	BECCLES		
GB7BF	70CM	DVU-R24	430.3000	439.3000	DMF	IO83LU	BLACKPOOL		5
GB7BH	70CM	DVU46	439.5750	430.5750	DMFP	IO90WU	BRIGHTON		4
GB7BI	70CM	DVU53	439.6625	430.6625	M	IO77RM	MUIR OF ORD		1
GB7BJ	70CM	DVU59	439.7375	430.7375	M	IO82	MALVERN		13
GB7BK	70CM	DVU59	439.7375	430.7375	M	IO91JM	READING		3
GB7BL	70CM	DVU54	439.6750	430.6750	ADMFPN	IO84VV	ALLENDALE		5
GB7BM	70CM	DVU32	439.4000	430.4000	MP	IO92BL	BIRMINGHAM		8
GB7BN	70CM	DVU47	439.5875	430.5875	DM	IO90PT	BOGNOR REGIS		3
GB7BP	70CM	DVU36	439.4500	430.4500	D	IO91PX	MILTON KEYNES		
GB7BR	70CM	RU74	430.9250	438.5250	M	IO74TI	RAMSEY IOM		3
GB7BS	70CM	DVU13	439.1625	430.1625	M	IO81TK	BRISTOL		3
GB7BT	70CM	DVU51	439.6375	430.6375	DMFP	IO81SK	BRISTOL		3
GB7BW	70CM	DVU58	439.7250	430.7250	DMF	IO93HP	WAKEFIELD		2
GB7BX	70CM	DVU35	439.4375	430.4375	M	IO82	CLEOBURY NORTH		11
GB7BZ	70CM	DVU33	439.4125	430.4125	D	IO92QF	BOZEAT		
GB7CA	70CM	RU74	430.9250	438.5250	M	IO74SD	DOUGLAS		2
GB7CB	70CM	DVU57	439.7125	430.7125	M	IO85	BATHGATE		4
GB7CC	70CM	DVU-R22	430.2750	439.2750	M	IO81	CHELTENHAM		2
GB7CD	70CM	DVU56	439.7000	430.7000	D	IO81JL	CARDIFF		
GB7CE	70CM	DVU49	439.6125	430.6125	F	JO01NT	GREAT CLACTON		
GB7CF	70CM	DVU50	439.6250	430.6250	M	IO90LW	CLANFIELD		3
GB7CG	2M	RV57	145.7125	145.1125	AF	IO92QA	WOBURN SANDS	77.0	
GB7CH	70CM	RU66	430.8250	438.4250	ADMF	IO85	DUNS		3
GB7CK	70CM	DVU59	439.7375	430.7375	M	JO01	FOLKESTONE		3
GB7CL	70CM	DVU51	439.6375	430.6375	M	JO01OU	FRINTON ON SEA		3
GB7CM	2M	RV56	145.7000	145.1000	AF	IO80UT	MILTON ABBAS	71.9	
GB7CO	70CM	DVU39	439.4875	430.4875	DMFPN	IO90JU	FAREHAM		3
GB7CP	70CM	RU66	430.8250	438.4250	F	IO91RO	SLOUGH	77.0	
GB7CQ	70CM	DVU49	439.6125	430.6125	DMF	IO90	WORTHING		4
GB7CS	70CM	DVU62	439.7750	430.7750	F	IO75VU	GLASGOW		
GB7CT	2M	RV51	145.6375	145.0375	M	IO91PS	TRING		3
GB7CU	70CM	DVU61	439.7625	430.7625	F	IO84GQ	MARYPORT		
GB7CV	70CM	DVU-R23	430.2875	439.2875	M	IO92GK	COVENTRY		5
GB7CX	70CM	RU69	430.8625	438.4625	DMF	IO65OD	COLERAINE		12
GB7CY	70CM	DVU50	439.6250	430.6250	F	IO94DR	CROOK		
GB7CZ	70CM	DVU40	439.5000	430.5000	M	IO94DR	CROOK		10
GB7DA	2M	RV62	145.7750	145.1750	DMF	IO85AV	AIRDRIE		1
GB7DB	70CM	DVU53	439.6625	430.6625	M	IO92QA	WOBURN SANDS		3
GB7DC	70CM	RU70	430.8750	438.4750	ADMFPN	IO92	DERBY	71.9	
GB7DD	70CM	DVU53	439.6625	430.6625	MN	IO86	DUNDEE		1
GB7DE	2M	RV51	145.6375	145.0375	AM	IO85	KENNOWAY FIFE	94.8	
GB7DH	70CM	DVU47	439.5875	430.5875	DF	IO83	NORTHWICH		
GB7DJ	70CM	DVU53	439.6625	430.6625	M	IO83RF	NORTHWICH		3
GB7DK	2M	RV55	145.6875	145.0875	AF	IO74LV	STRANRAER	103.5	
GB7DK-B	70CM	DVU52	439.6500	430.6500	D	IO74LV	STRANRAER		
GB7DL	70CM	DVU54	439.6750	430.6750	D	JO03DD	SKEGNESS		
GB7DM	70CM	DVU60	439.7500	430.7500	ADMF	IO86NL	MONIFIETH		1
GB7DN	70CM	DVU33	439.4125	430.4125	D	IO64MW	DUNGIVEN		
GB7DP	70CM	DVU47	439.5875	430.5875	M	IO71MV	TREFFGARNE		3
GB7DR	70CM	DVU34	439.4250	430.4250	M	IO90AR	POOLE		5
GB7DS	70CM	DVU34	439.4250	430.4250	ADMFP7N	JO02PP	NORWICH	94.8	
GB7DW	2M	RV50	145.6250	145.0250	AM	IO80JU	DUNKERSWELL		7
GB7DX	70CM	DVU56	439.7000	430.7000	M	JO00LW	NEW ROMNEY		3

All Repeaters (by Callsign)

Repeater	Band	Channel	TX	RX	Modes	QTHR	Where	CTCSS/CC	DMR CC
GB7DY	70CM	DVU46	439.5750	430.5750	M	IO82	DUDLEY		11
GB7DZ	70CM	RB02	433.0500	434.6500	AF	IO83	CONGLETON	103.5	
GB7EA	70CM	DMU30	439.3750	430.3750	AF	IO75SV	ERSKINE		1
GB7EB	70CM	DVU46	439.5750	430.5750	M	JO02TK	BECCLES		2
GB7EC	70CM	RU74	430.9250	438.5250	M	IO87	ABERDEEN		11
GB7ED	70CM	DVU54	439.6750	430.6750	M	IO80GR	EXETER		5
GB7EE	70CM	RU74	430.9250	438.5250	M	IO85JW	EDINBURGH		1
GB7EG	70CM	DVU61	439.7625	430.7625	M	IO91	EAST GRINSTEAD		2
GB7EH	70CM	RU75	430.9375	438.5375	M	IO92GC	BANBURY		8
GB7EI	70CM	RU68	430.8500	438.4500	ADMF	IO67IN	LOCHMADDY	88.5	
GB7EL	70CM	DVU57	439.7125	430.7125	M	IO83	NELSON		2
GB7EP	70CM	DVU32	439.4000	430.4000	M	IO91UI	EPSOM		3
GB7ER	2M	RV62	145.7750	145.1750	AF	IO80GR	EXETER	77.0	
GB7ES	70CM	DVU35	439.4375	430.4375	D	JO00DT	EASTBOURNE		
GB7EU	70CM	RU67	430.8375	438.4375	AN	IO85JW	EDINBURGH	94.8	
GB7EV	70CM	U94	431.1750	431.1750	M	IO85JW	EDINBURGH		11
GB7EW	70CM	DVU42	439.5250	430.5250	M	IO80	EXETER		3
GB7EX	70CM	DVU57	439.7125	430.7125	M	JO01GN	SOUTHEND ON SEA		3
GB7EY	70CM	DVU41	439.5125	430.5125	DMFPN	IO85	SELKIRK		1
GB7EZ	70CM	DVU62	439.7750	430.7750	M	IO92CM	ERDINGTON		14
GB7FB	70CM	DVU38	439.4750	430.4750	DMF	IO71VA	BIDEFORD		5
GB7FC	70CM	DVU40	439.5000	430.5000	D	IO83LU	CLEVELEYS		
GB7FD	70CM	DVU58	439.7250	430.7250	F	IO83LV	FLEETWOOD		
GB7FE	70CM	DVU38	439.4750	430.4750	DMF	IO86BC	STIRLING		1
GB7FG	70CM	RU76	430.9500	438.5500	AF	IO71WW	CARMARTHEN	94.8	
GB7FI	70CM	RU71	430.8875	438.4875	M	IO81OH	AXBRIDGE		3
GB7FK-B	70CM	DVU54	439.6750	430.6750	D	JO01OC	FOLKESTONE		
GB7FL	70CM	DVU61	439.7625	430.7625	DF	IO83	FRECKLETON		
GB7FO	70CM	DVU-R22	430.2750	439.2750	M	IO83LT	BLACKPOOL		1
GB7FR	70CM	DVU40	439.5000	430.5000	M	IO90	WORTHING		4
GB7FT	70CM	DVU57	439.7125	430.7125	M	IO90	PORTSMOUTH		3
GB7FU	70CM	RU71	430.8875	438.4875	AF	IO93	NOTTINGHAM		
GB7FW	70CM	DVU54	439.6750	430.6750	D	IO90LU	PORTSMOUTH		
GB7GA	70CM	RU72	430.9000	438.5000	ADMF	IO85	GALASHIELS		1
GB7GB	70CM	DVU61	439.7625	430.7625	M	IO92BM	SUTTON COLDFIELD		8
GB7GD	2M	RV55	145.6875	145.0875	DMF	IO87WE	ABERDEEN		1
GB7GF	70CM	DVU55	439.6875	430.6875	M	IO91RF	GUILDFORD		3
GB7GG	70CM	DVU50	439.6250	430.6250	DMF	IO85AV	AIRDRIE		1
GB7GJ	70CM	RU72	430.9000	438.5000	AM	IN89WE	ST HELIER	88.5	
GB7GL	70CM	RU68	430.8500	438.4500	M	IO75	GLASGOW		7
GB7GM	70CM	RU78	430.9750	438.5750	ADMF	IO85	DUMFRIES		3
GB7GP	70CM	DVU52	439.6500	430.6500	MF	IO91	HIGH WYCOMBE		5
GB7GR	70CM	DVU36	439.4500	430.4500	DMF	IO92QW	GRANTHAM		5
GB7GS	70CM	DVU36	439.4500	430.4500	MF	JO02EO	DOWNHAM MARKET		3
GB7GT	70CM	DVU33	439.4125	430.4125	M	IO82	CLEE HILL		13
GB7GW	70CM	RU65	430.8125	438.4125	ADMF	IO94FV	WASHINGTON		10
GB7GX	2M	RV59	145.7375	145.1375	ADMFPN	IO86	GLENROTHES		1
GB7GY	70CM	DVU55	439.6875	430.6875	M	IN89RL	ST. PETER PORT		5
GB7HA	70CM	DVU-R23	430.2875	439.2875	F	IO91TU	HARPENDEN		
GB7HB	70CM	DVU50	439.6250	430.6250	DMF	IO64	TANDRAGEE		1
GB7HC	70CM	DVU42	439.5250	430.5250	M	IO82	HEREFORD		5
GB7HD	2M	RV55	145.6875	145.0875	M	IO70	TAVISTOCK		3
GB7HH	70CM	DVU60	439.7500	430.7500	DMF	JO01CN	ROMFORD		3
GB7HI	70CM	DVU59	439.7375	430.7375	DMF	IO64	LISBURN		1
GB7HK	2M	RV50	145.6250	145.0250	AM	IO85	HAWICK		3
GB7HM	70CM	DVU51	439.6375	430.6375	M	IO83	CAERGWRLE		1
GB7HN	70CM	RU70	430.8750	438.4750	A	IO83RL	LEIGH	82.5	
GB7HR	70CM	DVU36	439.4500	430.4500	M	IO91SL	HEATHROW		3
GB7HS	70CM	DVU34	439.4250	430.4250	M	IO83XR	HEPTONSTALL		2
GB7HT	70CM	DVU59	439.7375	430.7375	M	IO95	ASHINGTON		10
GB7HU	70CM	DVU39	439.4875	430.4875	D	IO93RS	SOUTH CAVE		
GB7HX	70CM	DVU46	439.5750	430.5750	DM	IO93BP	HUDDERSFIELD		1
GB7IC-A	23CM	23CM	1290.6500	1270.6500	D	JO01NI	HERNE BAY		
GB7IC-B	70CM	DVU36	439.4500	430.4500	D	JO01NI	HERNE BAY		
GB7IC-C	2M	RV53	145.6625	145.0625	AD	JO01NI	HERNE BAY		
GB7IE	2M	RV54	145.6750	145.0750	AF	IO70WJ	PLYMOUTH	77.0	
GB7II	70CM	DVU36	439.4500	430.4500	M	IO77UL	INVERNESS		3
GB7IK	70CM	DVU43	439.5375	430.5375	M	JO01FJ	ROCHESTER		3
GB7IL	2M	RV53	145.6625	145.0625	M	IO77UL	INVERNESS		5
GB7IN	70CM	DVU52	439.6500	430.6500	M	IO93GD	ALFRETON		1
GB7IP	2M	RV52	145.6500	145.0500	M	IO92KP	LEICESTER		1
GB7IS	70CM	DVU43	439.5375	430.5375	F	IO81MH	WESTON-SUPER-MARE		
GB7IT	70CM	DVU41	439.5125	430.5125	M	IO81MH	WESTON-SUPER-MARE		1
GB7IV	2M	RV62	145.7750	145.1750	F	IO90IV	SOUTHAMPTON		
GB7JB	70CM	DVU37	439.4625	430.4625	M	IO81TB	WINCANTON		1
GB7JD	70CM	DVU33	439.4125	430.4125	DMN	IO85RL	JEDBURGH		1
GB7JF	70CM	DVU51	439.6375	430.6375	D	IO92NE	NORTHAMPTON		
GB7JI	70CM	RU68	430.8500	438.4500	D	IN89WE	ST HELIER		2
GB7JL	70CM	RU77	430.9625	438.5625	AM	IO83QL	WIGAN	82.5	
GB7JM	70CM	DVU47	439.5875	430.5875	DMF	JO03BI	ALFORD		3
GB7KB	2M	RV52	145.6500	145.0500	M	JO00CX	HEATHFIELD		3
GB7KD	70CM	DVU55	439.6875	430.6875	M	JO00DV	HAILSHAM		3
GB7KE	70CM	DVU47	439.5875	430.5875	MF	JO01GJ	GILLINGHAM		3
GB7KK	70CM	RU74	430.9250	438.5250	DMF	IO65VE	BALLYCASTLE		1
GB7KL	70CM	RU70	430.8750	438.4750	AF	IO84RD	KIRKBY LONSDALE	110.9	
GB7KN	70CM	DVU35	439.4375	430.4375	DMF	IO85	EDINBURGH		1
GB7KP	70CM	DVU46	439.5750	430.5750	DMF	IO74	COMBER		1
GB7KR	70CM	DVU-R24	430.3000	439.3000	M	IO82	KIDDERMINSTER		8
GB7KS	70CM	DVU-R20	430.2500	439.2500	M	IO81	STROUD		2
GB7KT	70CM	DVU40	439.5000	430.5000	M	IO91GE	ANDOVER		1
GB7KU	70CM	DVU42	439.5250	430.5250	DMF	IO92TT	BOURNE		5
GB7KY	70CM	DVU52	439.6500	430.6500	ADMFPN	IO86	KIRKCALDY		1
GB7LD	70CM	DVU35	439.4375	430.4375	M	IO83	LEYLAND		15
GB7LE	70CM	DVU53	439.6625	430.6625	M	IO93FU	LEEDS		2
GB7LF	70CM	DVU43	439.5375	430.5375	D	IO84OA	LANCASTER		
GB7LK	70CM	RU65	430.8125	438.4125	M	IO70SM	LISKEARD		5
GB7LL	70CM	DVU40	439.4969	430.4969	MN	IO82	LLANGOLLEN		12
GB7LN	70CM	DVU32	439.4000	430.4000	M	IO93RF	LINCOLN		1
GB7LO	70CM	DVU41	439.5125	430.5125	M	IO91	BECKENHAM		3
GB7LP	70CM	DVU32	439.4000	430.4000	M	IO83MK	LIVERPOOL		1

All Repeaters (by Callsign)

Repeater	Band	Channel	TX	RX	Modes	QTHR	Where	CTCSS/CC	DMR CC
GB7LR	70CM	DVU38	439.4750	430.4750	M	IO92IQ	LEICESTER		
GB7LS	70CM	DVU32	439.4000	430.4000	DMF	IO85	GALASHIELS		
GB7LT	70CM	U324	434.0500	434.0500	E	IO92	BOURNE		
GB7LV	2M	RV49	145.6125	145.0125	ADMF	IO85FV	LIVINGSTON		1
GB7LZ	70CM	RU73	430.9125	438.5125	AM	IO80CV	LAPFORD	77.0	
GB7MB	70CM	DVU56	439.7000	430.7000	F	IO84	MORECAMBE		
GB7MC	70CM	DVU43	439.5375	430.5375	DF	IO70	ST AUSTELL		
GB7MD	70CM	DVU32	439.4000	430.4000	F	IO80UT	BLANDFORD FORUM		
GB7ME	70CM	DVU58	439.7250	430.7250	D	IO92JJ	RUGBY		3
GB7MF	70CM	DVU54	439.6750	430.6750	DMF	IO64QR	MAGHERAFELT		3
GB7MH	70CM	DVU51	439.6375	430.6375	DMF	IO91WC	EAST GRINSTEAD		2
GB7MJ	70CM	DVU51	439.6375	430.6375	M	IO91GA	ROMSEY		5
GB7MK	70CM	DVU48	439.6000	430.6000	M	JO02OB	IPSWICH		13
GB7MM	70CM	DVU60	439.7500	430.7500	DMF	IO87BN	NAIRN		5
GB7MN	70CM	DVU33	439.4125	430.4125	M	IO83	STOCKPORT		2
GB7MP	70CM	DVU60	439.7500	430.7500	MP	IO84NB	HEYSHAM		3
GB7MR	70CM	DVU59	439.7375	430.7375	M	IO83	BURY		2
GB7MS	2M	RV62	145.7750	145.1750	M	IO94EX	NEWCASTLE UPON TYNE		10
GB7MT	70CM	DVU53	439.6625	430.6625	F	IO90IV	SOUTHAMPTON		
GB7NA-A	23CM	RM12	1297.3000	1291.3000	D	IO93HN	BARNSLEY		
GB7NA-B	70CM	DVU36	439.4500	430.4500	D	IO93HN	BARNSLEY		
GB7NA-C	2M	RV54	145.6750	145.0750	D	IO93HN	BARNSLEY		
GB7NB	70CM	DVU55	439.6875	430.6875	D	JO02PN	NORWICH		
GB7NC	70CM	DVU58	439.7250	430.7250	DMF	JO02OL	LONG STRATTON		3
GB7ND	70CM	DVU33	439.4125	430.4125	M	JO02LM	GREAT ELLINGHAM		1
GB7NE	70CM	DVU36	439.4500	430.4500	DMF	IO95FE	ASHINGTON		3
GB7NF	70CM	DVU39	439.4875	430.4875	M	JO00AS	NEWHAVEN		1
GB7NH	70CM	DVU50	439.6250	430.6250	F	IO91OF	ALDERSHOT		
GB7NL	70CM	DVU37	439.4625	430.4625	D	IO80FR	EXETER		
GB7NM	2M	RV59	145.7375	145.1375	DF	JO02LM	ATTLEBOURGH		
GB7NN	70CM	DVU60	439.7500	430.7500	F	IO93	WORKSOP		
GB7NR	70CM	DVU35	439.4375	430.4375	DM	IO92KX	NOTTINGHAM		1
GB7NS	70CM	DVU13	439.1625	430.1625	M	IO91WG	CATERHAM		3
GB7NT	70CM	RU74	430.9250	438.5250	AF	IO91IJ	NEWBURY	118.8	
GB7NV	70CM	RU66	430.8250	438.4250	ADM	IO93	HUCKNALL		1
GB7NW	70CM	DVU58	439.7250	430.7250	F	IO91DN	SWINDON		
GB7NY	70CM	DVU55	439.6875	430.6875	DMF	IO64	BESSBROOK		1
GB7OK	2M	RV57	145.7125	145.1125	DMF	IO91	BECKENHAM		3
GB7OR	70CM	DVU37	439.4625	430.4625	M	IO88LX	KIRKWALL		1
GB7PB	2M	RV53	145.6625	145.0625	D	IO94FR	DURHAM		
GB7PD	2M	RV59	145.7375	145.1375	ADF	IO71OV	MAENCLOCHOG	94.8	
GB7PE	70CM	DVU40	439.5000	430.5000	M	IO92TL	PETERBOROUGH		3
GB7PF	70CM	DVU51	439.6375	430.6375	M	IO80AN	PRINCETOWN		3
GB7PH	70CM	DVU37	439.4625	430.4625	DM	JO00	HAILSHAM		3
GB7PI	70CM	DVU61	439.7625	430.7625	D	IO92XA	ROYSTON		
GB7PM	70CM	DVU35	439.4375	430.4375	F	IO81PH	CHEDDAR		
GB7PN	70CM	DVU34	439.4250	430.4250	M	IO83HH	PRESTATYN		1
GB7PO	70CM	DVU59	439.7375	430.7375	F	IO90KT	PORTSMOUTH		
GB7PR	70CM	U324	434.0500	434.0500	E	IO83PS	PRESTON		
GB7PS	2M	RV55	145.6875	145.0875	M	IO90	PORTSMOUTH		3
GB7PT	70CM	DVU57	439.7125	430.7125	DMF	IO92XA	ROYSTON		
GB7PX	70CM	DVU62	439.7750	430.7750	DMP	IO92XA	ROYSTON		10
GB7PY	70CM	RU74	430.9250	438.5250	M	JO02AF	CAMBRIDGE		3
GB7RA	70CM	DVU52	439.6500	430.6500	DMF	IO83LH	NESTON		5
GB7RB-C	2M	RV54	145.6750	145.0750	D	IO81GK	COWBRIDGE		
GB7RC	70CM	DVU-R21	430.2625	439.2625	D	IO81	CHELTENHAM		
GB7RD	70CM	DVU55	439.6875	430.6875	M	IO70	YELVERTON		3
GB7RE	70CM	DVU57	439.7125	430.7125	M	IO93MH	RETFORD		1
GB7RJ	70CM	DVU43	439.5375	430.5375	DM	IO82LW	CHIRK		5
GB7RK	70CM	DVU35	439.4375	430.4375	DMFPN	IO74AN	BELFAST		3
GB7RL	70CM	DVU62	439.7750	430.7750	DMFPN	IO83RU	LONGRIDGE		10
GB7RM	70CM	U324	434.0500	434.0500	E	JO01CO	ROMFORD		
GB7RN	2M	RV51	145.6375	145.0375	D	IO90JT	FAREHAM		
GB7RO	70CM	DVU-R21	430.2625	439.2625	F	IO83UQ	HASLINGDEN		
GB7RR	70CM	DVU49	439.6125	430.6125	M	IO92	NOTTINGHAM		1
GB7RS	70CM	DVU49	439.6125	430.6125	M	IO91PX	MILTON KEYNES		4
GB7RT	70CM	DVU33	439.4125	430.4125	M	IO91	SUNNINGDALE		3
GB7RV	70CM	DVU50	439.6250	430.6250	M	IO83	BLACKBURN		2
GB7RW	2M	RV48	145.6000	145.0000	AD	IO94SO	WHITBY	88.5	
GB7RX	2M	RV62	145.7750	145.1750	AM	JO00GU	HASTINGS		3
GB7RY	70CM	RU76	430.9500	438.5500	AF	JO00HW	RYE	103.5	
GB7SA	70CM	DVU-R24	430.3000	439.3000	M	IO81	PORT TALBOT		5
GB7SB	70CM	DVU53	439.6625	430.6625	M	IO92	BIRMINGHAM		5
GB7SD	70CM	DVU33	439.4125	430.4125	M	IO80SQ	WEYMOUTH		1
GB7SE	70CM	DVU38	439.4750	430.4750	M	JO01DM	THURROCK		3
GB7SF-C	2M	RV59	145.7375	145.1375	D	IO93GK	SHEFFIELD		
GB7SG	2M	RV61	145.7625	145.1625	M	IO83MK	LIVERPOOL		1
GB7SH	70CM	RU77	430.9625	438.5625	AF	IO93GK	SHEFFIELD	71.9	
GB7SI	70CM	DVU61	439.7625	430.7625	M	IO83	LEEK		1
GB7SJ	70CM	RB07	433.1750	434.7750	A	IO83RF	NORTHWICH	103.5	
GB7SK	70CM	DVU37	439.4625	430.4625	M	IO92NO	LEICESTER		1
GB7SL	70CM	DVU-R21	430.2625	439.2625	DMFPN	IO92TR	BASTON		8
GB7SM	70CM	DVU47	439.5875	430.5875	ADMFPN	IO86	ST MONANS		1
GB7SN	70CM	DVU54	439.6750	430.6750	M	IO93GH	SHEFFIELD		1
GB7SO	70CM	DVU42	439.5250	430.5250	DM	IO90KT	GOSPORT		1
GB7SP	70CM	DVU60	439.7500	430.7500	M	IO91CB	SALISBURY		3
GB7SQ	70CM	DVU54	439.6750	430.6750	MN	IO85	DALKEITH		1
GB7SS	70CM	DVU47	439.5875	430.5875	M	IO87IP	ELGIN		1
GB7ST	70CM	DVU36	439.4500	430.4500	M	IO83	STOKE ON TRENT		1
GB7SU	70CM	DVU36	439.4500	430.4500	M	IO90IV	SOUTHAMPTON		8
GB7SV	70CM	RB11	433.2750	434.8750	AF	IO91UW	HITCHIN	82.5	
GB7SW	2M	RV51	145.6375	145.0375	D	IO81VL	CHIPPENHAM		2
GB7SZ	70CM	RU75	430.9375	438.5375	A	IO90AR	POOLE	71.9	
GB7TB	70CM		439.2188	430.2188	E	IO92	MORTON		
GB7TC	70CM	DVU42	439.5250	430.5250	M	IO91DL	SWINDON		2
GB7TD	70CM	DVU13	439.1625	430.1625	M	IO93EP	WAKEFIELD		1
GB7TE	2M	RV62	145.7750	145.1750	D	JO01OT	CLACTON ON SEA		
GB7TF	70CM	DVU39	439.4875	430.4875	F	IO91QT	TRING		
GB7TH	70CM	DVU-R23	430.2875	439.2875	M	JO01QJ	MARGATE		5
GB7TM	70CM	DVU59	439.7375	430.7375	F	JO02NH	EYE		

All Repeaters (by Callsign)

Repeater	Band	Channel	TX	RX	Modes	QTHR	Where	CTCSS/CC	DMR CC
GB7TP	70CM	DVU55	439.6875	430.6875	DMF	IO93CT	SHIPLEY		1
GB7TQ	70CM	RU71	430.8875	438.4875	AF	IO80FM	TORQUAY	94.8	
GB7TT	70CM	DVU40	439.5031	430.5031	N	IO82	LUDLOW		
GB7TX	70CM	U321A	434.0188	434.0188	E	JO02TK	BECCLES		
GB7TY	70CM	DVU49	439.6125	430.6125	DM	IO84XX	HEXHAM		10
GB7UL	70CM	DVU42	439.5250	430.5250	M	IO74	CARRICKFERGUS		1
GB7VO	70CM	DVU56	439.7000	430.7000	DMFN	IO82QJ	TENBURY WELLS		1
GB7VT	70CM	U94	431.1750	431.1750	M	IO83	STOKE-ON-TRENT		5
GB7WA	2M	RV55	145.6875	145.0875	DM	IO93WM	GRIMSBY		2
GB7WB	70CM	DVU39	439.4875	430.4875	D	IO81MH	WESTON-SUPER-MARE		2
GB7WC	70CM	DVU39	439.4875	430.4875	D	IO83QI	WARRINGTON		
GB7WD	70CM	DVU57	439.7125	430.7125	M	IO82	SEDGLEY		11
GB7WF	70CM	DVU39	439.4875	430.4875	DMFN	IO82	BEWDLEY		13
GB7WH	70CM	DMU30	439.3750	430.3750	AF	IO83	WREXHAM	110.9	
GB7WI	70CM	DVU33	439.4125	430.4125	M	IO93TR	HULL		1
GB7WK	70CM	DVU41	439.5125	430.5125	DMFPN	IO91	WOKING		15
GB7WL	70CM	DVU37	439.4625	430.4625	M	IO91QP	AMERSHAM		3
GB7WM	70CM	DVU55	439.6875	430.6875	F	IO83SM	ASTLEY		
GB7WN	70CM	DVU60	439.7500	430.7500	M	IO82XP	WOLVERHAMPTON		8
GB7WR	70CM	DVU37	439.4625	430.4625	D	IO82UB	GREAT MALVERN		
GB7WS	70CM	DVU34	439.4250	430.4250	M	IO91SA	BILLINGSHURST		3
GB7WT	70CM	DVU48	439.6000	430.6000	DMF	IO64HQ	OMAGH		1
GB7WV	70CM	DVU55	439.6875	430.6875	M	IO92NL	MARKET HARBOROUGH		2
GB7WW	70CM	DVU38	439.4750	430.4750	M	IO81	KINGTON LANGLEY		7
GB7WX	70CM	DVU41	439.5125	430.5125	DMF	IO83OE	TARVIN		3
GB7WY	70CM	DVU41	439.5125	430.5125	D	IO93FR	WAKEFIELD		
GB7XG	70CM	RU70	430.8750	438.4750	ADMFPN	IO86JF	GLENROTHES		1
GB7XX	70CM	DVU37	439.4625	430.4625	M	IO94	FELLING		10
GB7XY	70CM	DVU34	439.4250	430.4250	M	IO94EX	NEWCASTLE UPON TYNE		10
GB7YD	70CM	DVU55	439.6875	430.6875	M	IO80PW	YEOVIL		5
GB7YF	2M	RV61	145.7625	145.1625	AF	IO80PW	YEOVIL TOWN CENTER	94.8	
GB7YI	70CM	DVU46	439.5750	430.5750	M	IO84WX	HEXHAM		10
GB7YJ	70CM	DVU47	439.5875	430.5875	M	IO92FH	LEAMINGTON SPA		3
GB7YL	2M	RV63	145.7875	145.1875	M	JO02UL	LOWESTOFT		1
GB7YR	70CM	DVU42	439.5250	430.5250	M	IO93LM	DONCASTER		2
GB7YZ	70CM	DVU38	439.4750	430.4750	MF	IO83JE	MOLD		
GB7ZE	70CM	RB02	433.0500	434.6500	AM	JO00GU	HASTINGS		3
GB7ZP	70CM	DVU39	439.4875	430.4875	D	JO01FS	CHELMSFORD		

Echolink and IRLP Linking (by Callsign)

Call	Chan	Output MHz	In MHz	Mode	QTHR	Location	Keeper	ECHO	IRLP
GB3AG	RV58	145.7250	145.1250	AV	IO86ON	Forfar	MM0XET	117931	
GB3AM	R50-13	50.8400	51.3400	AV	IO91QP	Amersham	G0WTZ	4125	
GB3AR	RV56	145.7000	145.1000	AV	IO73VC	Caernarfon	GW4KAZ	206003	
GB3AS	RV48	145.6000	145.0000	AV	IO85MC	Langholm	GM6LJE	412685	
GB3BC	RV60	145.7500	145.1500	AV	IO81IO	Pontypridd	GW0UXJ	39300	
GB3BN	RB0	433.0000	434.6000	AV	IO91OJ	Bracknell	G8DOR	1938	
GB3BW	RU65	430.8125	438.4125	AV	IO93HO	Wakefield	M0YDG		5775
GB3CA	RB13	433.3250	434.9250	AV	IO84OT	Carlisle	G1XSZ	412685	5280
GB3DC	RV55	145.6875	145.0875	AV	IO92GW	Derby	G7NPW	92369	
GB3DQ	RU78	430.9750	438.5750	AV	IO70RI	Polperro	G1YDQ	418341	5612
GB3DV	RB01	433.0250	434.6250	AV	IO93JK	Maltby	G0EPX	120618	5130
GB3DX	RB12	433.3000	434.9000	AV	IO65HA	Derry/Londonderry	GI4YWT	7125	
GB3EK	RU71	430.8875	438.4875	AV	JO01QJ	Margate	M0LMK	48360	
GB3FH	R50-06	50.7700	51.2700	AV	IO81OH	Somerset	G4RKY	228585	5361
GB3HH	RV56	145.7000	145.1000	AV	IO93BF	Buxton	G7EKY	97616	
GB3HT	RB11	433.2750	434.8750	AV	IO92HN	Hinckley	M0RVD	662746	
GB3HX	R50-05	50.7600	51.2600	AV	JO00HV	Hastings	M0HOW	71066	
GB3IE	RU68	430.8500	438.4500	DV	IO70XJ	Plymouth	G7DQC	27871	
GB3IK	RV61	145.7625	145.1625	AV	JO01FJ	Rochester	G6CKK	263025	
GB3IM-C	RU66	430.8250	438.4250	AV	IO74SD	Douglas Iom	GD4HOZ	464453	
GB3IM-R	RU66	430.8250	438.4250	AV	IO74TI	Ramsey Iom	GD4HOZ	464453	
GB3IM-S	RB05	433.1250	434.7250	AV	IO74SG	Douglas Iom	GD4HOZ	464453	
GB3IN	RV51	145.6375	145.0375	DM	IO93GD	Alfreton	G4TSN	98258	
GB3IR	RV61	145.7625	145.1625	AV	IO94DJ	Richmond Yorks	G4FZN	1353	5562
GB3KC	RU72	430.9000	438.5000	AV	IO82WK	Stourbridge	G0EWH	430900	
GB3KE	RV55	145.6875	145.0875	DM	IO75UV	Glasgow	GM7SVK	5411	5410
GB3KL	RB04	433.1000	434.7000	DM	JO02FR	Kings Lynn	G0IJU 77266		
GB3KS	RV50	145.6250	145.0250	DM	JO01PA	Dover	M1CMN	346463	
GB3LF	RB14	433.3500	434.9500	AV	IO84PH	Kendal	G3VVT	184457	5140
GB3LR	RU69	430.8625	438.4625	AV	JO00AS	Newhaven	G0TJH	494669	
GB3LS	RB02	433.0500	434.6500	AV	IO93RF	Lincoln	M0TEF	268511	
GB3LV	RB02	433.0500	434.6500	AV	IO91XP	North London	G4DFB	155403	5600
GB3MH	RV50	145.6250	145.0250	AV	IO91WC	East Grinstead	G3NZP	453929	5569
GB3MI	RV57	145.7125	145.1125	DM	IO83	Manchester	G0TOG	197681	
GB3MW	RB10	433.2500	434.8500	AV	IO92FH	Leamington Spa	G6FEO	749161	
GB3NK	RB04	433.1000	434.7000	AV	JO01BL	Erith	G4EGU	54760	
GB3OA	RV49	145.6125	145.0125	AV	IO83LP	Southport	G4EID 5302	5302	
GB3PA	RV50	145.6250	145.0250	AV	IO75QV	Paisley	GM7OAW	116678	
GB3PY	RU72	430.9000	438.5000	AV	JO02AF	Cambridge	M0ZPU	222303	
GB3PZ	RU72	430.9000	438.5000	AV	IO83XL	Dukinfield	G4ZPZ	2591	5400
GB3SB	RV52	145.6500	145.0500	AV	IO85	Selkirk	GM0FTJ	116678	
GB3SD	RB14	433.3500	434.9500	AV	IO80SQ	Weymouth	G0ECX	112689	
GB3TR	RV52	145.6500	145.0500	AV	IO80FM	Torquay	G8XST		5582
GB3UB	RB04	433.1000	434.7000	AV	IO81UJ	Bath Avon	G4KVI 201135		
GB3WK	RV62	145.7750	145.1750	AV	IO92FH	Leamington Spa	G6FEO	749154	
GB3XN	RU74	430.9250	438.5250	AV	IO93KJ	Worksop	G3XXN	153126	5708
GB3ZB	RU66	430.8250	438.4250	AV	IO81QJ	Bristol	G4RKY		5429
GB7RW	RV48	145.6000	145.0000	DM	IO94SO	Whitby	G4EQS	921135	
GB7SJ	RB07	433.1750	434.7750	DM	IO83RF	Northwich	M0WTX	455339	41360
MB7AJS		430.0750	430.0750	AG	IO92	Burntwood	G6OTZ		5269
MB7ILH		144.9625	144.9625	AG	IO92GE	Leamington Spa	G4USP	250973	
MB7ISC		145.2875	145.2875	AG	IO86AC	Stirling	GM0WUR	459689	

Amateur TV Repeaters (by Callsign)

Call	Tx MHz	Rx MHz	QTHR	Location
GB3CT	3404	2435	IO93UM	Caistor
GB3CZ	2432	2346.5	JO01OT	Clacton
GB3DO	1310	1248	IO74DP	Bangor
GB3EN	1312	1249	IO91XP	North London
GB3ET	1322	1249	IO92GC	Edgehill nr Banbury
GB3EY	1308	1265	IO93RS	South Cave
GB3FT	1315	1249/1255/1265/1278	IO83LU	Blackpool
GB3FW	3406	2430	IO90LU	Portsmouth
GB3GG	1310	1280	IO93XN	Grimsby
GB3GV	1318.5	1249	IO92IQ	Markfield
GB3HV	1308	1248	IO91LD	Alton
GB3JT	1318	1249	JO00CS	Eastbourne
GB3JV	3404	2440	JO01AJ	Petts Wood
GB3KM	3406/10065/2440/1304	5665/2328/1280/10315	IO94EQ	Spennymoor
GB3LO	1316	1249	JO02UL	Lowestoft
GB3LX	10240	10425	IO93RF	Lincoln
GB3MV	1316	1249	IO92NF	Northampton
GB3NQ	1316	1249	IO70OJ	Roche
GB3NV	3406	2440	JO02PP	Norwich
GB3PV	1316	1249	JO02AF	Cambridge
GB3SQ	1304	1244	IO90BR	Bournemouth
GB3TB	1316	1249	IO80FM	Torquay
GB3TG	10240	10425	IO91PX	Milton Keynes
GB3TM	1316	1249	IO73UJ	Amlwch
GB3TN	1316	1249	JO02KS	Fakenham
GB3TT	1310	1278	IO93GK	Sheffield
GB3TV	1318.5	1249	IO91RU	Dunstable
GB3TZ	2440/2326	2397	IO91SV	Luton Beds
GB3UD	1318.5	1249	IO83VC	Stoke On Trent
GB3UT	1311.5	1249	IO81UJ	Bath
GB3VL	1310	1270	IO93RF	Lincoln
GB3VX	1310	1249	JO00	Eastbourne
GB3XY	10065	10315	IO93RS	Hull
GB3YT	1316	1276	IO93DP	Mirfield
GB3ZZ	1316	1249	IO81RM	Bristol City Centre

*Please check on: *https://ukrepeater.net/tvrepeaterlist.htm* for additional frequencies

As new applications are being made on a regular basis please visit the RSGB website and you will find more information on repeaters at: *www.ukrepeater.net*

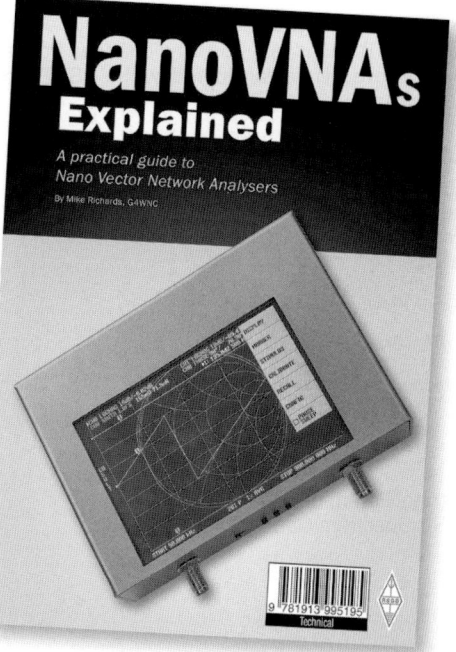

RSGB Band Plan

EFFECTIVE FROM 1ST JANUARY 2023 UNLESS OTHERWISE SHOWN

In this edition of the Yearbook are the 2023 Band Plans. These are based on a combination of IARU Region-1 recommendations, as well as UK-specific usage and feedback by RSGB Spectrum Forum members.

They are also available on the RSGB website in various formats, where more detailed change notes can also be found to guide users in the Excel master.

Whilst several changes were implemented soon after the 2020 Region-1 Conference, there have been relatively few since. This edition thus has no significant updates, but that does not reflect various ongoing developments that may result in changes in 2024 to account for:

a) The November 2023 IARU Region-1 Conference which is expected to update the HF band plans to accommodate growing digital usage

b) Implications of WRC-23 – the World Radio Conference in December 2023

c) Making accommodation for the Ofcom Licence review, notably in the VHF/UHF bands

The IARU HF changes needs agreement by IARU Member Societies in all three Regions. Any resulting changes from this and the other factors above will be considered by the RSGB Spectrum Forum in due course.

Further Information

In addition to RadCom, the latest band plans can be found on the Operating section of the RSGB website at rsgb.org/bandplans - and, if you are unsure, by all means contact either the relevant spectrum manager:

hf.manager@rsgb.org.uk

vhf.manager@rsgb.org.uk

microwave.manager@rsgb.org.uk

Murray Niman, G6JYB,
RSGB Spectrum Chair

136kHz	NECESSARY BANDWIDTH	UK USAGE
135.7-137.8kHz	200Hz	CW, QRSS and Narrowband Digital Modes

Licence Notes: Amateur Service – Secondary User. 1 watt (0dBW) ERP.
R.R. 5.67B. The use of the band 135.7-137.8kHz in Algeria, Egypt, Iran (Islamic Republic of), Iraq, Lebanon, Syrian Arab Republic Sudan, South Sudan and Tunisia is limited to fixed and maritime mobile services. The amateur service shall not be used in the above-mentioned countries in the band 135.7-137.8kHz, and this should be taken into account by the countries authorising such use. (WRC-19).

472kHz (600m)	NECESSARY BANDWIDTH	UK USAGE
IARU Region 1 does not have a formal band plan for this allocation but has a usage recommendation (Note 1).		
472-479kHz	500Hz	CW, QRSS and Narrowband Digital Modes

Note 1: Usage recommendation – 472-475kHz CW only 200Hz maximum bandwidth, 475-479kHz CW and Digimodes.
Note 2: It should be emphasised that this band is available on a non-interference basis to existing services. UK amateurs should be aware that some overseas stations may be restricted in terms of transmit frequency in order to avoid interference to nearby radio navigation service Non-Directional Beacons.
Licence Notes: Amateur Service – Secondary User. Full Licensees only, **5 watts EIRP maximum**. Note that conditions regarding this band are specified by the Licence Schedule notes.
R.R. 5.80B. The use of the frequency band 472-479kHz in Algeria, Saudi Arabia, Azerbaijan, Bahrain, Belarus, China, Comoros, Djibouti, Egypt, United Arab Emirates, the Russian Federation, Iraq, Jordan, Kazakhstan, Kuwait, Lebanon, Libya, Mauritania, Oman, Uzbekistan, Qatar, Syrian Arab Republic, Kyrgyzstan, Somalia, Sudan, Tunisia and Yemen is limited to the maritime mobile and aeronautical radionavigation services. The amateur service shall not be used in the above-mentioned countries in this frequency band, and this should be taken into account by the countries authorising such use. (WRC 12)

1.8MHz (160m)	NECESSARY BANDWIDTH	UK USAGE
1,810-1,838kHz	200Hz	Telegraphy
1,838-1,840	500Hz	Narrowband Modes
1,840-1,843	2.7kHz	All Modes
1,843-2,000	2.7kHz	Telephony (Note 1), Telegraphy 1,836kHz – QRP (low power) Centre of Activity 1,960kHz – DF Contest Beacons (14dBW)

Note 1: Lowest LSB carrier frequency (dial setting) should be 1,843kHz. AX25 packet should not be used on the 1.8MHz band.
Licence Notes: 1,810-1,850kHz – Primary User: 1,810-1,830kHz on a non-interference basis to stations outside of the UK. 1,850-2,000kHz – Secondary User. 32W (15dBW) maximum.
Notes to the Band Plan: As on page 171.

3.5MHz (80m)	NECESSARY BANDWIDTH	UK USAGE
3,500-3,510kHz	200Hz	Telegraphy – Priority for Inter-Continental Operation
3,510-3,560	200Hz	Telegraphy – Contest Preferred. 3,555kHz – QRS (slow telegraphy) Centre of Activity
3,560-3,570	200Hz	Telegraphy 3,560kHz – QRP (low power) Centre of Activity
3,570-3,580	200Hz	Narrowband Modes
3,580-3,590	500Hz	Narrowband Modes
3,590-3,600	500Hz	Narrowband Modes – Automatically Controlled Data Stations (unattended)
3,600-3,620	2.7kHz	All Modes – Automatically Controlled Data Stations (unattended), (Note 1)
3,600-3,650	2.7kHz	All Modes – Phone Contest Preferred, (Note 1). 3,630kHz – Digital Voice Centre of Activity
3,650-3,700	2.7kHz	All Modes – Telephony, Telegraphy 3,663kHz May Be Used For UK Emergency Comms Traffic 3,690kHz SSB QRP (low power) Centre of Activity
3,700-3,775	2.7kHz	All Modes – Phone Contest Preferred 3,735kHz – Image Mode Centre of Activity 3,760kHz – IARU Region 1 Emergency Centre of Activity
3,775-3,800	2.7kHz	All modes - Phone contest preferred Priority for Inter-Continental Telephony (SSB) Operation

Note 1. Lowest LSB carrier frequency (dial setting) should be 3,603kHz.
Licence Notes: Primary User: Shared with other user services.
Notes to the Band Plan: As on page 171.

5MHz (60m)	AVAILABLE WIDTH	UK USAGE
5,258.5-5,264kHz	5.5kHz	5,262kHz – CW QRP Centre of Activity
5,276-5,284	8kHz	5,278.5kHz – May be used for UK Emergency Comms
5,288.5-5,292	3.5kHz	Beacons on 5290kHz (Note 2),
5,298-5,307	9kHz	
5,313-5,323	10kHz	5,317kHz – AM 6kHz maximum bandwidth
5,333-5,338	5kHz	
5,354-5,358	4kHz	Within WRC-15 Band
5,362-5,374.5	12.5kHz	Partly within WRC-15 band, WSPR
5,378-5,382	4kHz	
5,395-5,401.5	6.5kHz	
5,403.5-5,406.5	3kHz	

Unless indicated, usage is All Modes (necessary bandwidth to be within channel limits).
Note 1: Upper Sideband is recommended for SSB activity.
Note 2: Activity should avoid interference to the experimental beacons on 5290kHz.
Note 3: Amplitude Modulation is permitted with a maximum bandwidth of 6kHz, on frequencies with at least 6kHz available width.
Note 4: Contacts within the UK should avoid the WRC-15 band (5351.5 - 5366.5 kHz) if possible
Licence Notes: Full Licensees only, **Secondary User, 100 watts maximum**. Note that conditions on transmission bandwidth, power and antennas are specified in the Licence. For the latest current guidance, refer to the RSGB website
Notes to the Band Plan. As on page 171.

7MHz (40m)	NECESSARY BANDWIDTH	UK USAGE
7,000-7,040kHz	200Hz	Telegraphy – 7,030kHz QRP (low power) Centre of Activity
7,040-7,047	500Hz	Narrowband Modes (Note 2)
7,047-7,050	500Hz	Narrowband Modes, Automatically Controlled Data Stations (unattended)
7,050-7,053	2.7kHz	All Modes, Automatically Controlled Data Stations (unattended), (Note 1)
7,053-7,060	2.7kHz	All Modes, Digimodes
7,060-7,100	2.7kHz	All Modes, SSB Contest Preferred Segment Digital Voice 7,070kHz; SSB QRP Centre of Activity 7,090kHz
7,100-7,130	2.7kHz	All Modes, 7,110kHz – Region 1 Emergency Centre of Activity
7,130-7,200	2.7kHz	All Modes, SSB Contest Preferred Segment; 7,165kHz – Image Centre of Activity
7,175-7,200	2.7kHz	All Modes, Priority For Inter-Continental Operation

Note 1: Lowest LSB carrier frequency (dial setting) should be 7,053kHz.
Note 2: PSK31 activity starts from 7,040kHz. Since 2009, the narrowband modes segment starts at 7,040kHz.
Licence Notes: 7,000-7,100kHz Amateur and Amateur Satellite Service – Primary User. 7,100-7,200kHz Amateur Service – Primary User.
Notes to the Band Plan: As on page 171.

10MHz (30m)	NECESSARY BANDWIDTH	UK USAGE
10,100-10,130kHz	200Hz	Telegraphy (CW) 10,116kHz – QRP (low power) Centre of Activity
10,130-10,150	500Hz	Narrowband Modes Automatically Controlled Data Stations (unattended) should avoid the use of the 10MHz band

Licence Notes: Amateur Service – Secondary User.
Notes to the Band Plan: As on page 171.
The 10MHz band is allocated to the amateur service only on a secondary basis. The IARU has agreed that only CW and other narrow bandwidth modes are to be used on this band. Likewise the band is not to be used for contests and bulletins. SSB may be used on the 10MHz band during emergencies involving the immediate safety of life and property, and only by stations actually involved with the handling of emergency traffic. The band segment 10,120-10,140kHz may only be used for SSB transmissions in the area of Africa south of the equator during local daylight hours.

14MHz (20m)	NECESSARY BANDWIDTH	UK USAGE
14,000-14,060kHz	200Hz	Telegraphy – Contest Preferred 14,055kHz – QRS (slow telegraphy) Centre of Activity
14,060-14,070	200Hz	Telegraphy 14,060kHz – QRP (low power) Centre of Activity
14,070-14,089	500Hz	Narrowband Modes
14,089-14,099	500Hz	Narrowband Modes – Automatically Controlled Data Stations (unattended)
14,099-14,101		IBP – Reserved Exclusively for Beacons
14,101-14,112	2.7kHz	All Modes – Automatically Controlled Data Stations (unattended)
14,112-14,125	2.7kHz	All Modes (excluding digimodes)
14,125-14,300	2.7kHz	All Modes – SSB Contest Preferred Segment 14,130kHz – Digital Voice Centre of Activity 14,195 ±5kHz – Priority for DXpeditions 14,230kHz – Image Centre of Activity 14,285kHz – QRP Centre of Activity
14,300-14,350	2.7kHz	All Modes 14,300kHz – Global Emergency Centre of Activity

Licence Notes: Amateur Service – Primary User. 14,000-14,250kHz Amateur Satellite Service – Primary User.
Notes to the Band Plan: As on page 171.

18MHz (17m)	NECESSARY BANDWIDTH	UK USAGE
18,068-18,095kHz	200Hz	Telegraphy – 18,086kHz QRP (low power) Centre of Activity
18,095-18,105	500Hz	Narrowband Modes
18,105-18,109	500Hz	Narrowband Modes – Automatically Controlled Data Stations (unattended)
18,109-18,111		IBP – Reserved Exclusively for Beacons
18,111-18,120	2.7kHz	All Modes – Automatically Controlled Data Stations (unattended)
18,120-18,168	2.7kHz	All Modes, 18,130kHz – SSB QRP Centre of Activity 18,150kHz – Digital Voice Centre of Activity 18,160kHz – Global Emergency Centre of Activity

Licence Notes: Amateur and Amateur Satellite Service – Primary User. The band is not to be used for contests or bulletins.
Notes to the Band Plan: As on page 171.

21MHz (15m)	NECESSARY BANDWIDTH	UK USAGE
21,000-21,070kHz	200Hz	Telegraphy 21,055kHz – QRS (slow telegraphy) Centre of Activity 21,060kHz – QRP (low power) Centre of Activity
21,070-21,090	500Hz	Narrowband Modes
21,090-21,110	500Hz	Narrowband Modes – Automatically Controlled Data Stations (unattended)
21,110-21,120	2.7kHz	All Modes (excluding SSB) – Automatically Controlled Data Stations (unattended)
21,120-21,149	500Hz	Narrowband Modes
21,149-21,151		IBP – Reserved Exclusively For Beacons
21,151-21,450	2.7kHz	All Modes 21,180kHz – Digital Voice Centre of Activity 21,285kHz – QRP Centre of Activity 21,340kHz – Image Centre of Activity 21,360kHz – Global Emergency Centre of Activity

Note 1: 21,125-21,245 is also designated for use by amateur satellites
Licence Notes: Amateur and Amateur Satellite Service – Primary User.
Notes to the Band Plan: As on page 171.

24MHz (12m)	NECESSARY BANDWIDTH	UK USAGE
24,890-24,915kHz	200Hz	Telegraphy 24,906kHz – QRP (low power) Centre of Activity
24,915-24,925	500Hz	Narrowband Modes
24,925-24,929	500Hz	Narrowband Modes – Automatically Controlled Data Stations (unattended)
24.929-24.931		IBP – Reserved Exclusively For Beacons
24,931-24,940	2.7kHz	All Modes – Automatically Controlled Data Stations (unattended)
24,940-24,990	2.7kHz	All Modes, 24,950kHz – SSB QRP Centre of Activity 24,960kHz – Digital Voice Centre of Activity

Licence Notes: Amateur and Amateur Satellite Service – Primary User. The band is not to be used for contests or bulletins.
Notes to the Band Plan: As on page 171.

28MHz (10m)	NECESSARY BANDWIDTH	UK USAGE
28,000-28,070kHz	200Hz	Telegraphy 28,055kHz – QRS (slow telegraphy) Centre of Activity 28,060kHz – QRP (low power) Centre of Activity
28,070-28,120	500Hz	Narrowband Modes
28,120-28,150	500Hz	Narrowband Modes – Automatically Controlled Data Stations (unattended)
28,150-28,190	500Hz	Narrowband Modes
28,190-28,199		IBP – Regional Time Shared Beacons
28,199-28,201		IBP – World Wide Time Shared Beacons
28,201-28,225		IBP – Continuous-Duty Beacons
28,225-28,300	2.7kHz	All Modes – Beacons
28,300-28,320	2.7kHz	All Modes – Automatically Controlled Data Stations (unattended)
28,320-29,000	2.7kHz	All modes 28,330kHz – Digital Voice Centre of Activity 28,360kHz – QRP Centre of Activity 28,680kHz – Image Centre of Activity
29,000-29,100		All Modes – See Note 1 regarding 29,000-29,510kHz
29,100-29,200		All Modes – FM Simplex – 10kHz Channels
29,200-29,300		All Modes – Automatically Controlled Data Stations (unattended) 29,270kHz – Internet Gateways Channel 29,280kHz – UK Internet Voice Gateway (unattended) 29,290kHz – UK Internet Voice Gateway (unattended)
29,300-29,510		Satellite Links
29,510-29,520	Guard Channel	
29,520-29,590		All Modes – FM Repeater Inputs (RH1-RH8)
29,600		All Modes – FM Calling Channel
29,610		All Modes – FM Simplex Repeater (parrot) – input and output
29,620-29,700		All Modes – FM Repeater Outputs (RH1-RH8)

Note 1: Experimental wide bandwidth operation within 29,000 - 29,510 must be on a non-interference basis to other stations, including the amateur satellite service segment at 29300 - 29510 kHz.
Licence Notes: Amateur and Amateur Satellite Service – Primary User: 26dBW permitted. Beacons may be established for DF competitions except within 50km of NGR SK985640 (Waddington).
Notes to the Band Plan: As on page 171.

50MHz (6m)	NECESSARY BANDWIDTH	UK USAGE
50.000-50.100MHz	500Hz	Telegraphy Only (except for Beacon Project) (Note 2)
		50.000-50.030MHz reserved for future Synchronised Beacon Project (Note 2)
		Region 1: 50.000-50.010; Region 2: 50.010-50.020; Region 3: 50.020-50.030
		50.050MHz – Future International Centre of Activity
		50.090MHz – Inter-Continental DX Centre of Activity (Note 1)
50.100-50.200	2.7kHz	SSB/Telegraphy – International Preferred
		50.100-50.130MHz – Inter-Continental DX Telegraphy & SSB (Note 1)
		50.110MHz – Inter-Continental DX Centre of Activity
		50.130-50.200MHz – General International Telegraphy & SSB
		50.150MHz – International Centre of Activity
50.200-50.300	2.7kHz	SSB/Telegraphy – General Usage
		50.285MHz – Crossband Centre of Activity
50.300-50.400	2.7kHz	MGM/Narrowband/Telegraphy
		50.305MHz – PSK Centre of Activity
		50.310-50.320MHz – EME
		50.320-50.380MHz – MS
50.400-50.500		Propagation Beacons only
50.500-50.700		All Modes
		50.520MHz – FM/DV Internet Voice Gateway
		50.530MHz – FM/DV Internet Voice Gateway
		50.540MHz – FM/DV Internet Voice Gateway
		50.600-50.700MHz – Digital communications
		50.630MHz – Digital Voice (DV) calling
50.700-50.900	12kHz	50.710-50.890MHz – FM/DV Repeater Outputs (10kHz channel spacing)
50.900-51.200		All Modes
51.200-51.400	12kHz	51.210-51.390MHz – FM/DV Repeater Inputs (10kHz channel spacing) (Note 4)
51.400-52.000		All Modes
		51.410-51.590MHz – FM/DV Simplex (Note 3) (Note 4)
		51.510MHz – FM Calling Frequency
		51.530MHz – GB2RS News Broadcast and Slow Morse
		51.650 & 51.750MHz – See Note 5 (25kHz aligned)
		51.970 & 51.990MHz – See Note 5

Note 1: Only to be used between stations in different continents (not for intra-European QSOs).
Note 2: 50.0-50.1MHz is currently shared with Propagation Beacons. These are due to be migrated to 50.4-50.5MHz, to create more space for Telegraphy and a new Synchronised Beacon Project.
Note 3: 20kHz channel spacing. Channel centre frequencies start at 51.430MHz.
Note 4: Embedded data traffic is allowed with digital voice (DV).
Note 5: May be used for Emergency Communications and Community Events.
Note 6: Digital experiments to support innovation may occur at 50.6, 51.0 or 51.7MHz with maximum bandwidths of 50, 200 and 400kHz respectively on a non-interference basis.
Licence Notes: Amateur Service 50.0-51.0MHz – Primary User. Amateur Service 51.0-52.0MHz – Secondary User. 100W (20dBW) maximum. Available on the basis on non-interference to other services (inside or outside the UK).
Notes to the Band Plan: As on page 171.

70MHz (4m)	NECESSARY BANDWIDTH	UK USAGE (NOTE 1)
70.000-70.090MHz	1kHz	Propagation Beacons Only
70.090-70.100	1kHz	Personal Beacons
70.100-70.250	2.7kHz	Narrowband Modes
		70.185MHz – Cross-band Activity Centre
		70.200MHz – CW/SSB Centre
		70.250MHz – MS Centre
70.250-70.294	12kHz	All Modes
		70.260MHz – AM/FM Calling
		70.270MHz MGM Centre of Activity
70.294-70.500	12kHz	All Modes Channelised Operations Using 12.5kHz Spacing
		70.3000MHz
		70.3125MHz – Digital Modes
		70.3250MHz – DX Cluster
		70.3375MHz – Digital Modes
		70.3500MHz – Internet Voice Gateway (Note 2)
		70.3625MHz – Internet Voice Gateway
		70.3750MHz – See Note 2
		70.3875MHz – Internet Voice Gateway
		70.4000MHz – See Note 2
		70.4125MHz – Internet Voice Gateway
		70.4250MHz – FM Simplex – used by GB2RS news broadcast
		70.4375MHz – Digital Modes (special projects)
		70.4500MHz – FM Calling
		70.4625MHz – Digital Modes
		70.4750MHz
		70.4875MHz – Digital Modes

Note 1: Usage by operators in other countries may be influenced by restrictions in their national allocations.
Note 2: May be used for Emergency Communications and Community Events.
Licence Notes: Amateur Service 70.0-70.5MHz – Secondary User: 160W (22dBW) maximum. Available on the basis of non-interference to other services (inside or outside the UK).
Notes to the Band Plan: As on page 171.

144MHz (2m)	NECESSARY BANDWIDTH	UK USAGE
144.000-144.025MHz	2700Hz	All Modes – including Satellite Downlinks
144.025-144.100	500Hz	Telegraphy (including EME CW)
		144.050MHz – Telegraphy Centre of Activity
		144.100MHz – Random MS Telegraphy Calling, (Note 1)
144.110-144.150	500Hz	Telegraphy and MGM
		EME MGM Activity
144.150-144.400	2700Hz	Telegraphy, MGM and SSB
		144.250MHz – GB2RS News Broadcast and Slow Morse
		144.260MHz – See Note 10
		144.300MHz – SSB Centre of Activity
		144.370MHz – MGM MS Calling
		Propagation Beacons only
144.490-144.500		Beacon guard band
		144.491-144.493 Personal Weak Signal MGM Beacons (BW: 500Hz max)
144.500-144.794	20 kHz	All Modes (Note-8)
		144.500 MHz Image Modes (SSTV, Fax etc)
		144.600 MHz Data Centre of Activity (MGM, RTTY etc)
		144.6125 MHz UK Digital Voice (DV) calling (Note 9)
		144.625-144.675 MHz See Note 10
		144.750 MHz ATV Talk-back
		144.775-144.794 MHz See Note 10
144.794-144.990	12 kHz	MGM / Digital Communications
		144.800-144.9875 MHz Digital modes (including unattended)
		144.8000 MHz Unconnected nets - APRS, UiView etc (Note 14)
		144.8125 MHz DV Internet voice gateway
		144.8250 MHz DV Internet voice gateway
		144.8375 MHz DV Internet voice gateway
		144.8500 MHz DV Internet voice gateway
		144.8625 MHz DV Internet voice gateway
		144.8750 - 144.9125 MHz - Internet Gateways
		144.9250 MHz Digital usage
		144.9375 MHz Digital usage
		144.9500 MHz Digital usage
		144.9625 MHz FM Internet voice gateway
		144.9750, 144.9875 MHz tbd (Note 11)
144.990-145.1935	12 kHz	FM/DV RV48 - RV63 Repeater input exclusive (Note 2) (Note 5)
145.200	12 kHz	FM/DV Space communications (e.g. I.S.S.) - Earth-to-Space
		145.2000 MHz (Note 4) & (Note 10)
145.200-145.5935	12kHz	FM/DV V16-V47 – FM/DV Simplex (Note 3, 5 & 6)
		145.2250MHz – See Note 10
		145.2375MHz – FM Internet Voice Gateway (IARU common channel)
		145.2500MHz – Used for Slow Morse Transmissions
		145.2875MHz – FM Internet Voice Gateway (IARU common channel)
		145.3375MHz – FM Internet Voice Gateway (IARU common channel)
		145.5000MHz – FM Calling (Note 12)
		145.5250MHz – Used for GB2RS News Broadcast.
		145.5500MHz – Used for Rally/exhibition Talk-in
		145.5750MHz, 145.5875MHz (Note 11)
145.5935-145.7935	12kHz	FM/DV RV48-RV63 – Repeater Output (Note 2)
145.800	12kHz	FM/DV Space Communications (eg ISS) – Space-Earth
145.806-146.000	12kHz	All Modes – Satellite Exclusive

Note 1: Meteor scatter operation can take place up to 26kHz higher than the reference frequency.
Note 2: 12.5kHz channels numbered RV48-RV63. RV48 input = 145.000MHz, output = 145.600MHz.
Note 3: 12.5kHz simplex channels numbered V16-V47. V16 = 145.200MHz.
Note 4: Emergency Communications Groups utilising this frequency should take steps to avoid interference to ISS operations in non-emergency situations.
Note 5: Embedded data traffic is allowed with digital voice (DV).
Note 6: Simplex use only – no DV gateways.
Note 7: Not Used.
Note 8: Amplitude Modulation (AM) is acceptable within the All Modes segment. AM usage is typically found on 144.550MHz. Users should consider adjacent channel activity when selecting operating frequencies.
Note 9: In other countries IARU Region 1 recommends 145.375MHz.
Note 10: May be used for Emergency Communications and Community Events.
Note 11: May be used for repeaters in other IARU Region 1 countries.
Note 12: DV users are asked not to use this channel, and use 144.6125MHz for calling.
Note 13: Not used.
Note 14: 144.800MHz use should be NBFM to avoid interference to 144.8125MHz DV Gateways.
Licence Notes: Amateur Service and Amateur Satellite Service – Primary User. Beacons may be established for DF competitions except within 50km of TA 012869 (Scarborough).
Notes to the Band Plan: As on page 171.

146MHz	NECESSARY BANDWIDTH	UK USAGE
146.000-146.900MHz	500kHz	Wideband Digital Modes (High speed data, DATV etc) 146.500MHz Centre frequency for wideband modes (Note 1)
146.900-147.000MHz	12kHz	Narrowband Digital Modes including Digital Voice 146.900 146.9125 146.925 146.9375 Not available in/near Scotland (see Licence Notes & NoV terms) 146.9500 146.9625 146.9750 146.9875

Note 1: Users of wideband modes must ensure their spectral emissions are contained with the band limits.
Licence Notes: Full Licensees only, with NoV, 50W ERP max – not available in the Isle of Man or Channel Isles. Note that additional restrictions on geographic location, antenna height and upper frequency limit are specified by the NoV terms.
It should be emphasised that this band is UK-specific and is available on a non-interference basis to existing services. Upper Band limit 147.000MHz (or 146.93750 where applicable) are absolute limits and not centre frequencies. The absolute band frequency limit in or within 40km of Scotland is 146.93750MHz – see NoV schedule
Notes to the Band Plan: As on page 171.

430MHz (70cm) IARU Recommendation	NECESSARY BANDWIDTH	UK USAGE
430.0000-431.9810MHz		430.0125-430.0750MHz – FM Internet Voice Gateways (Notes 7, 8)
All Modes		
		430.4000-430.7750 – UK DV 9MHz Split Repeaters – inputs
Digital Repeaters		430.8250-430.9750MHz – RU66-RU78 7.6MHz Split Repeaters – outputs See Licence Exclusion Note; 431-432MHz 430.9900-431.9000MHz – Digital Communications 431.0750-431.1750MHz – DV Internet Voice Gateways (Note 8)
Telegraphy, MGM		432.0500MHz – Telegraphy Centre of Activity
432.1000-432.4000	2700Hz	
		432.2000MHz – SSB Centre of Activity
SSB, Telegraphy		432.3500MHz – Microwave Talk-back (Europe)
MGM		432.3700MHz – Meteor Scatter Calling
432.4000-432.4900	500Hz	Propagation Beacons only
432.4900-432.9940	25kHz	432.491-432.493 MHz Personal Weak Signal MGM Beacons (BW: 500 Hz max)
		432.5000MHz – Narrowband SSTV Activity Centre
All Modes Non-channelised	(Note 11)	432.6250-432.6750MHz Digital Communications (25kHz channels)
		432.7750MHz 1.6MHz Talk-through – Base TX (Note 10)
432.9940-433.3810	25kHz	433.0000-433.3750MHz (RB0-RB15) – RU240-RU270
FM repeater outputs in UK only (Note 1)	(Note 11)	FM/DV Repeater Outputs (25kHz channels) in UK Only
433.3940-433.5810	25kHz	433.4000MHz U272 – IARU Region 1 SSTV (FM/AFSK)
	(Note 11)	433.4250MHz U274
FM/DV (Notes 12, 13)		433.450MHz U276 (Note 5)
Simplex Channels		433.4750MHz U278 433.5000MHz U280 – FM Calling Channel 433.5250MHz U282 433.5500MHz U284 – Used for Rally/Exhibition Talk-in 433.5750MHz U286
433.6000-434.0000 All Modes		433.6250-6750MHz – Digital Communications (25kHz channels)
433.8000MHz for APRS where 144.8000MHz cannot be used		433.7000MHz-433.7750MHz (Note 10) 433.8000-434.2500 MHz Digital communications & Experiments 434.0000 Low Power Non-NoV Personal Hot-Spot usage
434.0000-434.5940	25kHz	433.9500-434.0500MHz – Internet Voice Gateways (Note 8)
	(Note 11)	434.3750MHz 1.6MHz Talk-through – Mobile TX (Note 10)
		434.4750-434.5250MHz DV Internet voice gateways (Note 8)
434.5940-434.9810	25kHz	434.6000-434.9750MHz (RB0-RB15) RU240-RU270
FM repeater inputs in UK only	(Note 11)	FM/DV Repeater Inputs (25kHz channels) in UK Only (Note 12)
435.0000-436.0000		Satellites
436.0000-438.0000		Satellites and Experimental DATV/Data 437.0000 Experimental DATV Centre of Activity (Note 14)
438.0000-440.0000		438.8000 Low Power Non-NoV Personal Hot-Spot usage 438.0250-438.1750MHz – IARU Region 1 Digital Communications
All Modes		438.2000-439.4250MHz (Note 1) 438.4000MHz – 7.6MHz Talk-through (Note 10) 438.4250-438.5750MHz RU66-RU78 – 7.6MHz Split Repeaters – inputs 438.6125MHz – UK DV calling (Note 12) (Note 13) 438.8000 Low Power Non-NoV Personal Hot-Spot usage 439.2500-439.3000MHz UK DV 9MHz reverse-split repeaters – Inputs
430MHz (70cm) IARU Recommendation	NECESSARY BANDWIDTH	UK USAGE Contd
		439.6000-440.0000MHz – Digital Communications 439.4000-439.7750MHz – UK DV 9MHz split repeaters - Outputs

Note 1: In Switzerland, Germany and Austria, repeater inputs are 431.0500-431.8250MHz with 25kHz spacing and outputs 438.6500-439.4250MHz. In Belgium, France and the Netherlands repeater outputs are 430.0250-430.3750MHz with 12.5kHz spacing and inputs at 431.6250-431.9750MHz. In other European countries repeater inputs are 433.0000-433.3750MHz with 25kHz spacing and outputs at 434.6000-434.9750MHz, ie the reverse of the UK allocation.
Note 2: 430-440 MHz FM/DV maximum bandwidths are 12.5 or 25 kHz as appropriate.
Note 4: Not used.
Note 5: In other countries IARU Region 1 recommends 433.4500MHz for DV calling.
Note 7: Users must accept interference from repeater output channels in France and the Netherlands at 430.0250-430.5750MHz. Users with sites that allow propagation to other countries (notably France and the Netherlands) must survey the proposed frequency before use to ensure that they will not cause interference to users in those countries.
Note 8: All internet voice gateways: 12.5kHz channels, maximum deviation ±2.4kHz, maximum effective radiated power 5W (7dBW), attended only operation in the presence of the NoV holder.
Note 10: May be used for Emergency Communications and Community Events.
Note 11: IARU Region 1 recommended maximum bandwidths are 12.5 or 20kHz.
Note 12: Embedded data traffic is allowed with digital voice (DV).
Note 13: Simplex use only - no DV gateways.
Note 14: QPSK 2 Mega-symbols/second maximum recommended.
Licence Notes: Amateur Service – Secondary User. Amateur Satellite Service: 435-438MHz – Secondary User. Exclusion: 431-432MHz not available within 100km radius of Charing Cross, London. Power Restriction 430-432MHz is 40 watts effective radiated power maximum.
Notes to the Band Plan: As on page 171.

1.3GHz (23cm)	NECESSARY BANDWIDTH	UK USAGE
1240.000-1240.500MHz	2700Hz	Alternative Narrowband Segment – see Note 7 – 1240.00-1240.750MHz
1240.500-1240.750		Alternative Propagation Beacon Segment
1240.750-1241.000	20kHz	FM/DV Repeater Inputs
1241.000-1241.750	150kHz	DD High Speed Digital Data – 5 x 150kHz channels
All Modes		1241.075, 1241.225, 1241.375, 1241.525, 1241.675MHz (±75kHz)
1241.750-1242.000	20kHz	25kHz Channels available for FM/DV use
All Modes		1241.775-1241.975MHz
1242.000-1249.000		TV Repeaters (Note 9)
ATV		New DATV Repeater Inputs Original ATV Repeater Inputs: 1248, 1249
1249.000-1249.250	20kHz	FM/DV Repeater Outputs, 25kHz Channels (Note 9) 1249.025-1249.225MHz
1250.00		In order to prevent interference to Primary Users, caution must be exercised prior to using 1250-1290MHz in the UK
1260.000-1270.000		Amateur Satellite Service – Earth to Space Uplinks Only
Satellites		
1290.000		
1290.994-1291.481	20kHz	FM/DV Repeater Inputs (Note 5) 1291.000-1291.375MHz (RM0-RM15) 25kHz spacing
1291.494-1296.000	All Modes	
All Modes		Preferred Narrowband segment
1296.000-1296.150	500Hz	1296.000-1296.025MHz – Moonbounce
Telegraphy, MGM		
1296.150-1296.800	2700Hz	1296.200MHz – Narrowband Centre of Activity
Telegraphy, SSB & MGM		1296.400-1296.600MHz – Linear Transponder Input
(Note 1)		1296.500MHz – Image Mode Centre of Activity (SSTV, FAX etc) 1296.600MHz – Narrowband Data Centre of Activity (MGM, RTTY etc) 1296.600-1296.700MHz – Linear Transponder Output 1296.741-1296.743MHz Personal Weak Signal MGM Beacons
1296.800-1296.994		1296.750-1296.800MHz – Local Beacons, 10W ERP max 1296.800-1296.990MHz – Propagation Beacons only Beacons exclusive
1296.994-1297.481	20kHz	FM/DV Repeater Outputs (Note 5) 1297.000-1297.375MHz (RM0-RM15)
1297.494-1297.981	20kHz	FM/DV Simplex ((Notes 2, 5 & 6)) 25kHz spacing 1297.500-1297.750MHz (SM20-SM30)
FM/DV simplex (Notes 2, 5, 6)		1297.725MHz – Digital Voice (DV) Calling (IARU recommended) 1297.900-1297.975MHz – FM Internet Voice Gateways (IARU common channels, 25kHz)
1298.000-1299.000	20kHz	All Modes
All Modes		General mixed analogue or digital use in channels 1298.025-1298.975MHz (RS1-RS39)
1299.000-1299.750	150kHz	DD High Speed Digital Data – 5 x 150kHz channels
All Modes		

1.3GHz (23cm) IARU Recommendation	NECESSARY BANDWIDTH	UK USAGE Contd
		1299.075, 1299.225, 1299.375, 1299.525, 1299.675MHz (±75kHz)
1299.750-1300.000	20kHz	25kHz Channels Available for FM/DV use
All Modes		1299.775-1299.975MHz
1300.000-1325.000		TV Repeaters (UK only) (Note 9)
ATV		New DATV Repeater Outputs
		Original ATV Repeater Outputs: 1308.0, 1310.0, 1311.5, 1312.0, 1316.0, 1318.5MHz

Note 1: Local traffic using narrowband modes should operate between 1296.500-1296.800MHz during contests and band openings.
Note 2: Stations in countries that do not have access to 1298-1300MHz may also use the FM simplex segment for digital communications.
Note 3, Note 4: Not used.
Note 5: Embedded data traffic is allowed with digital voice (DV).
Note 6: Simplex use only – no DV gateways.
Note 7: 1240.000-1240.750 has been designated by IARU as an alternative centre for narrowband activity and beacons. Operations in this range should be on a flexible basis to enable coordinated activation of this alternate usage.
Note 8: The band 1240-1300MHz is subject to major replanning. Contact the Microwave Manager for further information.
Note 9: Repeaters and Migration to DATV, inc option for new DATV simplex are subject to further development and coordination.
Note 10: QPSK 4 Mega-symbols/second maximum recommended.
Licence Notes: Amateur Service – Secondary User. Amateur Satellite Service: 1,260-1,270MHz – Secondary User Earth to Space only. In the sub-band 1,298-1,300MHz unattended operation is not allowed within 50km of SS206127 (Bude), SE202577 (Harrogate), or in Northern Ireland.
Notes to the Band Plan: As on page 171.

2.3-2.302GHz IARU Recommendation	NECESSARY BANDWIDTH	UK USAGE

Access to this band requires an appropriate NoV, which is available to Full licensees only. Please note that the current NoVs last for up to three years prior to expiry.

2300.000-2300.400MHz	2.7kHz	Narrowband Modes (including CW, SSB, MGM) 2300.350-2300.400MHz Attended Beacons
2300.400-2301.800MHz	500kHz	Wideband Modes (NBFM, DV, Data, DATV, etc) Note 1
2301.800-2302.000MHz	2.7kHz	Narrowband modes (including CW, SSB, MGM) EME Usage

Note 1: Users of wideband modes must ensure their spectral emissions are contained within the band limits.
Note 2: Full licensees only with NoV, 400 watts maximum, not available in the Isle of Man or Channel isles. Note additional restrictions on usage are specified by the NoV terms. It should be emphasised that this is UK-specific and is available on a non interference basis to exisiting services.
Notes to the Band Plan: As on page 171.

2.3GHz (13cm) IARU Recommendation	NECESSARY BANDWIDTH	UK USAGE
2,310.000-2,320.000MHz		
(National band plans)	200kHz	2,310.000-2,310.500MHz – Repeater links
		2,311.000-2,315.000MHz – High speed data
		Preferred Narrowband Segment
2,320.000-2,320.150	2.7kHz	2,320.000-2,320.025MHz – Moonbounce
2,320.150-2,320.800	2.7kHz	2,320.200MHz – SSB Centre of Activity
		2,320.750-2,320.800MHz – Local Beacons, 10W ERP max
2,320.800-2,321.000		2,320.800-2,320.990MHz – Propagation Beacons Only
Beacons exclusive		
2321.000-2322.000	20kHz	FM/DV. See also Note 1
2,322.000-2,350.000		Wideband Modes including Data, ATV
2,390.000-2,400.000		All Modes
2,400.000-2,450.000MHz		2,435.000MHz ATV Repeater Outputs
Satellites		2,440.000MHz ATV Repeater Outputs

Note 1: Stations in countries which do not have access to the All Modes section 2,322-2,390MHz, use the simplex and repeater segment 2,320-2,322MHz for data transmission.
Note 2: Stations in countries that do not have access to the narrowband segment 2,320-2,322MHz, use the alternative narrowband segment 2,304-2,306MHz and 2,308-2,310MHz.
Note 3: The segment 2,433-2,443MHz may be used for ATV if no satellite is using the segment.
Licence Notes: Amateur Service – Secondary User. Users must accept interference from ISM users. Amateur Satellite Service: 2,400-2,450MHz – Secondary User. Users must accept interference from ISM users. Operation in 2310-2350 and 2390-2400 MHz are subject to specific conditions and guidance In the sub-bands 2,310.000-2,310.4125 and 2,392-2,450MHz unattended operation is not allowed within 50km of SS206127 (Bude) or SE202577 (Harrogate). ISM = Industrial, scientific and medical.
Notes to the Band Plan: As on page 171.

3.4GHz (9cm) IARU Recommendation	NECESSARY BANDWIDTH	UK USAGE
3,400.000-3,400.800MHz	2.7kHz	Narrowband Modes (including CW, SSB, MGM, EME)
		3,400.100MHz – Centre of Activity (Note 1)
		3,400.750-3,400.800MHz – Local Beacons, 10W ERP max
3,400.800-3,400.995		3,400.800-3,400.995MHz – Propagation Beacons Only
3,400.000-3,401.000MHz	200kHz	3,401.000-3,402.000MHz Data, Remote Control
3,402.000-3,410.000		Wideband Modes including DATV Repeater Outputs
All Modes (Notes 2, 3)		

Note 1: EME has migrated from 3456MHz to 3400MHz to promote harmonised usage and activity.
Note 2: Stations in many European countries have access to 3400-3410MHz as permitted by CEPT ECA table.
Note 3: Amateur Satellite downlinks planned.
Licence Notes: Amateur Service – Secondary User. Subject to specific conditions and guidance.
Notes to the Band Plan: As on page 171.

5.7GHz (6cm) IARU Recommendation	NECESSARY BANDWIDTH	UK USAGE
5,650.000-5,668.000MHz		All Modes
Satellite Uplinks		Amateur Satellite Service – Earth to Space Only
5,668.000-5,670.000	2.7kHz	5,668.200MHz – Alternative Narrowband Centre
5,670.000-5,680.000		All Modes
5,755.000-5,760.000		All Modes
5,760.000-5,762.000	2.7kHz	Narrowband Modes (including CW, SSB, MGM, EME)
		5,760.100MHz – Preferred Centre of Activity
		5,760.750-5,760.800MHz – Local Beacons, 10W ERP max
5760.800-5760.995		5,760.800-5,760.995MHz – Propagation Beacons Only
Propagation Beacons		
5,762.000-5,765.000		All Modes
5,820.000-5,830.000		All Modes
5,830.000-5,850.000		All Modes
Satellite Downlinks		Amateur Satellite Service – Space to Earth Only

Licence Notes: Amateur Service: 5,650-5,680MHz – Secondary User. 5,755-5,765 and 5,820-5,850MHz – Secondary User. Users must accept interference from ISM users. Amateur Satellite Service: 5,650-5,670MHz and 5,830-5,850MHz – Secondary User. Users must accept interference from ISM users. Unattended operation is permitted for remote control, digital modes and beacons, except in the sub-bands 5,670-5,680MHz within 50km of SS206127 (Bude) and SE202577 (Harrogate). ISM = Industrial, scientific and medical.
Notes to the Band Plan As on page 171.

10GHz (3cm) IARU Recommendation	NECESSARY BANDWIDTH	UK USAGE
10,000.000-10,125.000MHz		Note 4
All Modes		10,065MHz ATV Repeater Outputs
10,225.000-10,250.000		
All Modes		10,240MHz ATV Repeaters
10,250.000-10,350.000		
Digital Modes		
10,350.000-10,368.000		10,352.5-10,368MHz Wideband Modes (Note 2)
All Modes		
10,368-10,370MHz	2.7kHz	10,368-10,370 Narrowband Modes (Note 3)
Narrowband Telegraphy		10,368.1MHz Centre of Activity
EME/SSB		
		10,368.750-10,368.800MHz – Local Beacons, 10W ERP max
10,368.800-10,368.995		10,368.800-10,368.995MHz – Propagation Beacons Only
Propagation Beacons		
10,370.000-10,450.000		10,371MHz Voice Repeaters Rx
All Modes		10,425 ATV Repeaters
10,450.000-10,475.000		10,400-10,475MHz Unattended Operation
All Modes & Satellites		10,450-10,452MHz Alternative Narrowband Segment (Note 3)
		10,471MHz Voice Repeaters Tx
10,475.000-10,500.000		
All Modes and satellites		Amateur Satellite Service ONLY

Note 1: Deleted.
Note 2: Wideband FM is preferred between 10,350-10,400MHz to encourage compatibility between narrowband systems.
Note 3: 10,450MHz is used as an alternative narrowband segment in countires where 10,368MHz is not available.
Note 4: 10,000-10,125MHz is subject to increased Primary user utilisation and NoV restrictions.
Note 5: 10,475-10,500MHz is allocated ONLY to the Amateur Satellite Service and NOT to the Amateur Service.
Licence Notes: Amateur Service – Secondary User. Foundation licensees 1 watt maximum. Amateur Satellite Service: 10,450-10,500MHz – Secondary User. Unattended operation is permitted for remote control, digital modes and beacons except in the sub-bands 10,000-10,125MHz within 50km of SO916223 (Cheltenham), SS206127 (Bude), SK985640 (Waddington) and SE202577 (Harrogate).
Notes to the Band Plan: As on page 171.

24GHz (12mm) IARU Recommendation	UK USAGE
24,000.000-24,050.000MHz Satellites	24,025MHz Preferred Operating Frequency for Wideband Equipment 24,048.2MHz – Narrowband Centre of Activity 24,048.750-24,048.800MHz – Local Beacons, 10W ERP max
24,048.800-24,048.995 Propagation Beacons	24,048.800-24,048.995MHz – Propagation Beacons Only
24,050.000-24,250.000 All Modes	

Licence Notes: Amateur Service: 24,000-24,050MHz – Primary User: Users must accept interference from ISM users. 24,050-24,150MHz – Secondary User. May only be used with the written permission of Ofcom. Users must accept interference from ISM users. 24,150-24,250MHz – Secondary User. Users must accept interference from ISM users. Amateur Satellite Service: 24,000-24,050MHz – Primary User: Users must accept interference from ISM users. Unattended operation is permitted for remote control, digital modes and beacons, except in the sub-bands 24,000-24,050MHz within 50km of SK985640 (Waddington) and SE202577 (Harrogate).

ISM = Industrial, scientific and medical.

Notes to the Band Plan: As on page 171.

47GHz (6mm) IARU Recommendation	UK USAGE
47,000.000-47,200.000MHz	47,088.2MHz – Centre of Narrowband Activity
47,088.000-47,090.000 Narrowband Segment	47,088.8-47,089.0MHz – Propagation Beacons Only

Licence Notes: Amateur Service and Amateur Satellite Service – Primary User. Unattended operation is permitted for remote control, digital modes and beacons, except within 50km of SK985640 (Waddington) and SE202577 (Harrogate).
Notes to the Band Plan: As on page 171.

76GHz (4mm) IARU Recommendation	UK USAGE
75,500-76,000MHz All Modes (preferred)	75,976.200MHz – IARU Region 1 Preferred Centre of Activity
76,000.000-77,500.000 All Modes	
77,500-78,000 All Modes (preferred)	77,500.200MHz – Alternative IARU Recommended Narrowband Segment
78,000-81,000 All Modes	

Licence Notes:
75,500-75,875MHz Amateur Service and Amateur Satellite Service – Secondary User.
75,875-76,000MHz Amateur Service and Amateur Satellite Service – Primary User.
76,000-77,500MHz Amateur Service and Amateur Satellite Service – Secondary User.
77,500-78,000MHz Amateur Service and Amateur Satellite Service – Primary User.
78,000-81,000MHz Amateur service and Amateur Satellite Service – Secondary User.
Unattended operation is permitted for remote control, digital modes and beacons, except within 50km of SK985640 (Waddington) and SE202577 (Harrogate).
Notes to the Band Plan: As on page 171.

134GHz (2mm) IARU Recommendation	UK USAGE
134,000-134,928MHz All Modes	
134,928 -134,930 Narrowband Modes	IARU Region 1 Preferred Centre of Activity 134,928.800-134,928.990 – Propagation Beacons Only
134,930 -136,000 All Modes	

Licence Notes: 134,000-136,000MHz Amateur Service and Amateur Satellite Service – Primary User. Unattended operation is permitted for remote control, digital modes and beacons, except within 50km of SK985640 (Waddington) and SE202577 (Harrogate).

THE FOLLOWING BANDS ARE ALSO ALLOCATED TO THE AMATEUR SERVICE AND THE AMATEUR SATELLITE SERVICE

122,250-123,000MHz – Amateur Service only, Secondary User
136,000-141,000MHz – Secondary User
241,000-248,000MHz – Secondary User
248,000-250,000MHz – Primary User
Note 1: Access to frequencies >275 GHz by Full Licensees is also possible by NoV
Notes to the Band Plan: As on page 171.

NOTES TO THE BAND PLAN

ITU-R radio regulation RR 1.152 and Recommendation SM.328 (extract):

Necessary bandwidth: For a given class of emission, the width of the frequency band which is just sufficient to ensure the transmission of information at the rate and with the quality required under specified conditions.

Foundation and Intermediate Licence holders are advised to check their Licences for the permitted power limits and conditions applicable to their class of Licence.

All Modes: CW, SSB and those modes listed as Centres of Activity, plus AM. Consideration should be given to adjacent channel users.

Image Modes: Any analogue or digital image modes within the appropriate bandwidth, for example SSTV and FAX.

Narrowband Modes: All modes using up to 500Hz bandwidth, including CW, RTTY, PSK, etc.

Digimodes: Any digital mode used within the appropriate bandwidth, for example RTTY, PSK, MT63, etc.

Sideband usage: Below 10MHz use lower sideband (LSB), above 10MHz use upper sideband (USB). Note the lowest dial settings for LSB Voice modes are 1843, 3603 and 7053kHz on 160, 80 and 40m. Note that on (5MHz) USB is used.

Amplitude Modulation (AM): AM with a bandwidth greater than 2.7kHz is acceptable in the All Modes segments provided users consider adjacent channel activity when selecting operating frequencies (Davos 2005).

Extended SSB (eSSB): Extended SSB (eSSB) is only acceptable in the All Modes segments provided users consider adjacent channel activity when selecting operating frequencies.

Digital Voice (DV): Users of Digital Voice (DV) should check that the channel is not in use by other modes (CT08_C5_Rec20).

FM Repeater & Gateway Access: CTCSS Access is recommended. Toneburst access is being withdrawn in line with IARU-R1 recommendations.

Beacons Propagation Beacon Sub-bands are highlighted – please avoid transmitting in them!

MGM: Machine Generated Modes indicates those transmission modes relying fully on computer processing such as RTTY, AMTOR, PSK31, JTxx, FSK441 and the like. This does not include Digital Voice (DV) or Digital Data (DD).

WSPR: Above 30MHz, WSPR frequencies in the band plan are the centre of the transmitted frequency (not the suppressed carrier frequency or the VFO dial setting).

Transmitter setup and Linearity: Close attention should be given to power amplifier linearity to control the final transmitted bandwidth and avoid spectral regrowth affecting adjacent users. In particular this can be a major issue when operating digital modes. It is recommended that operators do not use more power than is necessary, and that care is taken to ensure sound cards, interfaces, and other equipment are properly set up so as to minimise the potential for interference.

CW QSOs are accepted across all bands, except within beacon segments (Recommendation DV05_C4_Rec_13).

Contest activity shall not take place on the 10, 18 and 24MHz (30, 17 and 12m) bands.

Non-contesting radio amateurs are recommended to use the contest-free HF bands (30, 17 and 12m) during the largest international contests (DV05_C4_Rev_07).

The term 'automatically controlled data stations' include Store and Forward stations.

Transmitting Frequencies: The announced frequencies in the band plan are understood as 'transmitted frequencies' (not those of the suppressed carrier!).

Centre of Activity (CoA): A guide to where users of a particular mode or activity tend to operate. The bandplan does not give such users precedence over other modes or activities.

Unmanned transmitting stations: IARU member societies are requested to limit this activity on the HF bands. It is recommended that any unmanned transmitting stations on HF shall only be activated under operator control except for beacons agreed with the IARU Region 1 Beacon Coordinator, or specially licensed experimental stations.

472-479kHz: Access is available to Full licensees only - see licence schedule for additional conditions.

1.8MHz: Radio amateurs in countries that have a SSB allocation ONLY below 1840kHz, may continue to use it, but the National Societies in those countries are requested to take all necessary steps with their licence administrations to adjust phone allocations in accordance with the Region 1 Band Plan (UBA – Davos 2005).

3.5MHz: Inter-Continental operations should be given priority in the segments 3500-3510kHz and 3775- 3800kHz. Where no DX traffic is involved, the contest segments should not include 3500-3510kHz or 3775-3800kHz. Member societies will be permitted to set other (lower) limits for national contests (within these limits). 3510-3600kHz may be used for unmanned ARDF beacons (CW, A1A) (Recommendation DV05_C4_Rec_12).

5MHz: Access is available to Full licensees only- see licence schedule for additional conditions.

7MHz: The band segment 7040-7060kHz may be used for automatic controlled data stations (unattended) traffic in the areas of Africa south from the equator during local daylight hours. Where no DX traffic is involved, the contest segment should not include 7,175-7,200kHz.

10MHz: SSB may be used during emergencies involving the immediate safety of life and property and only by stations actually involved in the handling of emergency traffic.

The band segment 10120kHz to 10140kHz may be used for SSB transmissions in the area of Africa south of the equator during local daylight hours.

News bulletins on any mode should not be transmitted on the 10MHz band.

28MHz: Operators should not transmit on frequencies between 29.3 and 29.51MHz to avoid interference to amateur satellite downlinks.

Experimentation with NBFM Packet Radio on 29MHz band: Preferred operating frequencies on each 10kHz from 29.210 to 29.290MHz inclusive should be used. A deviation of ±2.5kHz being used with 2.5kHz as maximum modulation frequency.

1.3GHz
The band is subject to re-planning. It is also shared with air traffic radar.

2.3GHz (2310-2350 & 2390-2400MHz)
Operation is subject to specific licence conditions and guidance - see also the Ofcom PSSR statement.

3.4GHz (3400-3410MHz)
Operation is subject to specific licence conditions and guidance - see also the Ofcom PSSR statement.

Innovation Bands: 70.5-71.5 MHz, 146-147 MHz, 2300-2302 MHz and >275 GHz
Access to these bands requires an appropriate NoV, which is available to Full licensees only.

The latest band plan information, including the master Excel files, can be found in the Operating section of the RSGB website.
Please ensure you only refer or link to the current Band Plans. Remove / delete any older versions you have locally or online.

Prefix List

Callsigns for the world's nations are determined by the International Telecommunications Union (ITU). This is the United Nations agency that co-ordinates radio activity for all spectrum users. The prefixes used by a country for both commercial and amateur radio purposes are determined from one or more ITU allocation blocks issued to that country. The amateur radio callsigns in use for a particular country might use one or a number of combinations derived from the authorised ITU allocation(s) for that country. The following list shows callsign prefixes currently in use. Most are derived from the callsign blocks allocated to administrations by the ITU for use within the countries, territories and dependencies for which a country is responsible. Also shown are some unauthorised prefixes which may be heard and which may or may not be recognised as a DXCC entity, eg 1A0 (SMOM). 1B (the Turkish area of North Cyprus) and 1Z (Karea State - Myanmar) are unofficial and are not recognised for DXCC purposes, so these are not shown.

Full information on prefixes is contained in the ***RSGB Prefix Guide***.

Prefix	Entity	Cont.	ITU	CQ	1.8	3.5	7.0	10.1	14	18	21	24	28	50	144	oth.
1A	Sov. Mil. Order of Malta	EU	28	15												
3A	Monaco	EU	27	14												
3B6, 7	Agalega & St. Brandon Is.	AF	53	39												
3B8	Mauritius	AF	53	39												
3B9	Rodriguez I.	AF	53	39												
3C	Equatorial Guinea	AF	47	36												
3C0	Annobon I.	AF	52	36												
3D2	Fiji	OC	56	32												
3D2	Conway Reef	OC	56	32												
3D2	Rotuma I.	OC	56	32												
3DA	Swaziland	AF	57	38												
3V	Tunisia	AF	37	33												
3W, XV	Vietnam	AS	49	26												
3X	Guinea	AF	46	35												
3Y	Bouvet	AF	67	38												
3Y	Peter 1 I.	AN	72	12												
4J, 4K	Azerbaijan	AS	29	21												
4L	Georgia	AS	29	21												
4O	Montenegro	EU	28	15												
4S	Sri Lanka	AS	41	22												
4U_ITU	ITU HQ	EU	28	14												
4U_UN	United Nations HQ	NA	08	05												
4W	Timor - Leste	OC	54	28												
4X, 4Z	Israel	AS	39	20												
5A	Libya	AF	38	34												
5B, C4, P3	Cyprus	AS	39	20												
5H-5I	Tanzania	AF	53	37												
5N	Nigeria	AF	46	35												
5R	Madagascar	AF	53	39												
5T	Mauritania	AF	46	35												
5U	Niger	AF	46	35												
5V	Togo	AF	46	35												
5W	Samoa	OC	62	32												
5X	Uganda	AF	48	37												
5Y-5Z	Kenya	AF	48	37												
6V-6W	Senegal	AF	46	35												
6Y	Jamaica	NA	11	08												
7O5	Yemen	AS	39	21												
7P	Lesotho	AF	57	38												
7Q	Malawi	AF	53	37												
7T-7Y	Algeria	AF	37	33												
8P	Barbados	NA	11	08												
8Q	Maldives	AS/AF	41	22												
8R	Guyana	SA	12	09												
9A	Croatia	EU	28	15												
9G	Ghana	AF	46	35												
9H	Malta	EU	28	15												
9I-9J	Zambia	AF	53	36												
9K	Kuwait	AS	39	21												
9L	Sierra Leone	AF	46	35												
9M2, 4	West Malaysia	AS	54	28												
9M6, 8	East Malaysia	OC	54	28												
9N	Nepal	AS	42	22												
9Q-9T	Dem. Rep. of Congo	AF	52	36												

Key to abbreviations of continents:
AF = Africa, **AN** = Antarctica, **AS** = Asia, **EU** = Europe,
NA = North America, **OC** = Oceania, **SA** = South America

Prefix	Entity	Cont.	ITU	CQ	1.8	3.5	7.0	10.1	14	18	21	24	28	50	144	oth.
9U	Burundi	AF	52	36												
9V	Singapore	AS	54	28												
9X	Rwanda	AF	52	36												
9Y-9Z	Trinidad & Tobago	SA	11	09												
A2	Botswana	AF	57	38												
A3	Tonga	OC	62	32												
A4	Oman	AS	39	21												
A5	Bhutan	AS	41	22												
A6	United Arab Emirates	AS	39	21												
A7	Qatar	AS	39	21												
A9	Bahrain	AS	39	21												
AP	Pakistan	AS	41	21												
B	China	AS	33, 42-44	23, 24												
BS7	Scarborough Reef	AS	50	27												
BU-BX	Taiwan	AS	44	24												
BV9P	Pratas I.	AS	44	24												
C2	Nauru	OC	65	31												
C3	Andorra	EU	27	14												
C5	The Gambia	AF	46	35												
C6	Bahamas	NA	11	08												
C8-9	Mozambique	AF	53	37												
CA-CE	Chile	SA	14,16	12												
CE0	Easter I.	SA	63	12												
CE0	Juan Fernandez Is.	SA	14	12												
CE0	San Felix & San Ambrosio	SA	14	12												
CE9/KC4	Antarctica	AN	67,69-74	S												
CM, CO	Cuba	NA	11	08												
CN	Morocco	AF	37	33												
CP	Bolivia	SA	12,14	10												
CT	Portugal	EU	37	14												
CT3	Madeira Is.	AF	36	33												
CU	Azores	EU	36	14												
CV-CX	Uruguay	SA	14	13												
CY0	Sable I.	NA	09	05												
CY9	St. Paul I.	NA	09	05												
D2-3	Angola	AF	52	36												
D4	Cape Verde	AF	46	35												
D6	Comoros	AF	53	39												
DA-DR	Fed. Rep. of Germany	EU	28	14												
DU-DZ	Philippines	OC	50	27												
E3	Eritrea	AF	48	37												
E4	Palestine	AS	39	20												
E5	N. Cook Is.	OC	62	32												
E5	S. Cook Is.	OC	62	32												
E6	Niue	OC	62	32												
E7	Bosnia-Herzegovina	EU	28	15												
EA-EH	Spain	EU	37	14												
EA6-EH6	Balearic Is.	EU	37	14												
EA8-EH8	Canary Is.	AF	36	33												
EA9-EH9	Ceuta & Melilla	AF	37	33												
EI-EJ	Ireland	EU	27	14												
EK	Armenia	AS	29	21												
EL	Liberia	AF	46	35												
EP-EQ	Iran	AS	40	21												
ER	Moldova	EU	29	16												
ES	Estonia	EU	29	15												
ET	Ethiopia	AF	48	37												
EU-EW	Belarus	EU	29	16												
EX	Kyrgyzstan	AS	30, 31	17												
EY	Tajikistan	AS	30	17												
EZ	Turkmenistan	AS	30	17												
F	France	EU	27	14												
FG, TO	Guadeloupe	NA	11	08												
FH, TO	Mayotte	AF	53	39												
FJ, TO	Saint Barthelemy	NA	11	08												
FK, TX	New Caledonia	OC	56	32												
FK, TX	Chesterfield Is.	OC	56	30												
FM, TO	Martinique	NA	11	08												
FO, TX	Austral I.	OC	63	32												
FO, TX	Clipperton I.	NA	10	07												
FO, TX	French Polynesia	OC	63	32												
FO, TX	Marquesas Is.	OC	63	31												

The letter 'S' against an ITU or CQ Zone indicates that the entity is split across several.

Prefix	Entity	Cont.	ITU	CQ	1.8	3.5	7.0	10.1	14	18	21	24	28	50	144	oth.
FP	St. Pierre & Miquelon	NA	09	05												
FR, TO	Reunion I.	AF	53	39												
FT/G, TO	Glorioso Is.	AF	53	39												
FT/J,E, TO	Juan de Nova, Europa	AF	53	39												
FT/T, TO	Tromelin I.	AF	53	39												
FS, TO	Saint Martin	NA	11	08												
FT/W	Crozet I.	AF	68	39												
FT/X	Kerguelen Is.	AF	68	39												
FT/Z	Amsterdam & St. Paul Is.	AF	68	39												
FW	Wallis & Futuna Is.	OC	62	32												
FY	French Guiana	SA	12	09												
G, GX, M, MX, 2E	England	EU	27	14												
GD, GT, MD, MT, 2D	Isle of Man	EU	27	14												
GI, GN, MI, MN, 2I	Northern Ireland	EU	27	14												
GJ, GH, MJ, MH, 2J	Jersey	EU	27	14												
GM, GS, MM, MS, 2M	Scotland	EU	27	14												
GU, GP, MU, MP, 2U	Guernsey	EU	27	14												
GW, GC, MW, MC, 2W	Wales	EU	27	14												
H4	Solomon Is.	OC	51	28												
H40	Temotu Province	OC	51	32												
HA, HG	Hungary	EU	28	15												
HB	Switzerland	EU	28	14												
HB0	Liechtenstein	EU	28	14												
HC-HD	Ecuador	SA	12	10												
HC8-HD8	Galapagos Is.	SA	12	10												
HH	Haiti	NA	11	08												
HI	Dominican Republic	NA	11	08												
HJ-HK, 5J-5K	Colombia	SA	12	09												
HK0	Malpelo I.	SA	12	09												
HK0	San Andres & Providencia	NA	11	07												
HL, 6K-6N	Republic of Korea	AS	44	25												
HO-HP	Panama	NA	11	07												
HQ-HR	Honduras	NA	11	07												
HS, E2	Thailand	AS	49	26												
HV	Vatican	EU	28	15												
HZ	Saudi Arabia	AS	39	21												
I	Italy	EU	28	15,33												
IS0, IM0	Sardinia	EU	28	15												
J2	Djibouti	AF	48	37												
J3	Grenada	NA	11	08												
J5	Guinea-Bissau	AF	46	35												
J6	St. Lucia	NA	11	08												
J7	Dominica	NA	11	08												
J8	St. Vincent	NA	11	08												
JA-JS, 7J-7N	Japan	AS	45	25												
JD	Minami Torishima	OC	90	27												
JD	Ogasawara	AS	45	27												
JT-JV	Mongolia	AS	32,33	23												
JW	Svalbard	EU	18	40												
JX	Jan Mayen	EU	18	40												
JY	Jordan	AS	39	20												
K, W, N, AA-AK	United States of America	NA	6,7,8	3,4,5												
KG4	Guantanamo Bay	NA	11	08												
KH0	Mariana Is.	OC	64	27												
KH1	Baker & Howland Is.	OC	61	31												
KH2	Guam	OC	64	27												
KH3	Johnston I.	OC	61	31												
KH4	Midway I.	OC	61	31												
KH5	Palmyra & Jarvis Is.	OC	61, 62	31												
KH5K	Kingman Reef	OC	61	31												
KH6,7	Hawaii	OC	61	31												
KH7K	Kure I.	OC	61	31												
KH8	American Samoa	OC	62	32												
KH8	Swains I.	OC	62	32												
KH9	Wake I.	OC	65	31												
KL, AL, NL, WL	Alaska	NA	1, 2	1												
KP1	Navassa I.	NA	11	08												
KP2	Virgin Is.	NA	11	08												
KP3, 4	Puerto Rico	NA	11	08												
KP5	Desecheo I.	NA	11	08												
LA-LN	Norway	EU	18	14												
LO-LW	Argentina	SA	14,16	13												

There are numerous instances of entities that do not count for DXCC before (or after) certain dates. Check the ARRL website for details: *www.arrl.org/country-lists-prefixes*

Prefix	Entity	Cont.	ITU	CQ	1.8	3.5	7.0	10.1	14	18	21	24	28	50	144	oth.
LX	Luxembourg	EU	27	14												
LY	Lithuania	EU	29	15												
LZ	Bulgaria	EU	28	20												
OA-OC	Peru	SA	12	10												
OD	Lebanon	AS	39	20												
OE	Austria	EU	28	15												
OF-OI	Finland	EU	18	15												
OH0	Aland Is.	EU	18	15												
OJ0	Market Reef	EU	18	15												
OK-OL	Czech Republic	EU	28	15												
OM	Slovak Republic	EU	28	15												
ON-OT	Belgium	EU	27	14												
OU-OW, OZ	Denmark	EU	18	14												
OX	Greenland	NA	5, 75	40												
OY	Faroe Is.	EU	18	14												
P2	Papua New Guinea	OC	51	28												
P4	Aruba	SA	11	09												
P5	DPR of Korea	AS	44	25												
PA-PI	Netherlands	EU	27	14												
PJ2	Curacao	SA	11	09												
PJ4	Bonaire	SA	11	09												
PJ5, 6	Saba & St. Eustatius	NA	11	08												
PJ7	St Maarten	NA	11	08												
PP-PY, ZV-ZZ	Brazil	SA	12,13,15	11												
PP0-PY0F	Fernando de Noronha	SA	13	11												
PP0-PY0S	St. Peter & St. Paul Rocks	SA	13	11												
PP0-PY0T	Trindade & Martim Vaz Is.	SA	15	11												
PZ	Suriname	SA	12	09												
R1/F	Franz Josef Land	EU	75	40												
S0	Western Sahara	AF	46	33												
S2	Bangladesh	AS	41	22												
S5	Slovenia	EU	28	15												
S7	Seychelles	AF	53	39												
S9	Sao Tome & Principe	AF	47	36												
SA-SM, 7S-8S	Sweden	EU	18	14												
SN-SR	Poland	EU	28	15												
ST	Sudan	AF	47, 48	34												
SU	Egypt	AF	38	34												
SV-SZ, J4	Greece	EU	28	20												
SV/A	Mount Athos	EU	28	20												
SV5, J45	Dodecanese	EU	28	20												
SV9, J49	Crete	EU	28	20												
T2	Tuvalu	OC	65	31												
T30	W. Kiribati (Gilbert Is.)	OC	65	31												
T31	C. Kiribati (British Phoenix Is)	OC	62	31												
T32	E. Kiribati (Line Is.)	OC	61, 63	31												
T33	Banaba I. (Ocean I.)	OC	65	31												
T5, 6O	Somalia	AF	48	37												
T7	San Marino	EU	28	15												
T8	Palau	OC	64	27												
TA-TC	Turkey	EU/AS	39	20												
TF	Iceland	EU	17	40												
TG, TD	Guatemala	NA	12	07												
TI, TE	Costa Rica	NA	11	07												
TI9	Cocos I.	NA	12	07												
TJ	Cameroon	AF	47	36												
TK	Corsica	EU	28	15												
TL	Central Africa	AF	47	36												
TN	Congo (Republic of the)	AF	52	36												
TR	Gabon	AF	52	36												
TT	Chad	AF	47	36												
TU	Cote d'Ivoire	AF	46	35												
TY	Benin	AF	46	35												
TZ	Mali	AF	46	35												
UA-UI1-7, RA-RZ	European Russia	EU	S	16												
UA2, RA2	Kaliningrad	EU	29	15												
UA-UI8, 9, 0, RA-RZ	Asiatic Russia	AS	S	S												
UJ-UM	Uzbekistan	AS	30	17												
UN-UQ	Kazakhstan	AS	29-31	17												
UR-UZ, EM-EO	Ukraine	EU	29	16												
V2	Antigua & Barbuda	NA	11	08												
V3	Belize	NA	11	07												

Prefix	Entity	Cont.	ITU	CQ	1.8	3.5	7.0	10.1	14	18	21	24	28	50	144	oth.
V4	St. Kitts & Nevis	NA	11	08												
V5	Namibia	AF	57	38												
V6	Micronesia	OC	65	27												
V7	Marshall Is.	OC	65	31												
V8	Brunei Darussalam	OC	54	28												
VA-VG, VO,VY	Canada	NA	2-4, 9, 75	1-5												
VK, AX	Australia	OC	55, 58-59	29,30												
VK0	Heard I.	AF	68	39												
VK0	Macquarie I.	OC	60	30												
VK9C	Cocos (Keeling) Is.	OC	54	29												
VK9L	Lord Howe I.	OC	60	30												
VK9M	Mellish Reef	OC	56	30												
VK9N	Norfolk I.	OC	60	32												
VK9W	Willis I.	OC	55	30												
VK9X	Christmas I.	OC	54	29												
VP2E	Anguilla	NA	11	08												
VP2M	Montserrat	NA	11	08												
VP2V	British Virgin Is.	NA	11	08												
VP5	Turks & Caicos Is.	NA	11	08												
VP6	Pitcairn I.	OC	63	32												
VP646	Ducie I.	OC	63	32												
VP8	Falkland Is.	SA	16	13												
VP8, LU	South Georgia I.	SA	73	13												
VP8, LU	South Orkney Is.	SA	73	13												
VP8, LU	South Sandwich Is.	SA	73	13												
VP8, LU, CE9, HF0, 4K1	South Shetland Is.	SA	73	13												
VP9	Bermuda	NA	11	05												
VQ9	Chagos Is.	AF	41	39												
VR	Hong Kong	AS	44	24												
VU	India	AS	41	22												
VU4	Andaman & Nicobar Is.	AS	49	26												
VU7	Lakshadweep Is.	AS	41	22												
XA-XI	Mexico	NA	10	06												
XA4-XI4	Revillagigedo	NA	10	06												
XT	Burkina Faso	AF	46	35												
XU	Cambodia	AS	49	26												
XW	Laos	AS	49	26												
XX9	Macao	AS	44	24												
XY-XZ	Myanmar	AS	49	26												
YA, T6	Afghanistan	AS	40	21												
YB-YH	Indonesia	OC	51,54	28												
YI	Iraq	AS	39	21												
YJ	Vanuatu	OC	56	32												
YK	Syria	AS	39	20												
YL	Latvia	EU	29	15												
YN,H6-7,HT	Nicaragua	NA	11	07												
YO-YR	Romania	EU	28	20												
YS, HU	El Salvador	NA	11	07												
YT-YU	Serbia	EU	28	15												
YV-YY, 4M	Venezuela	SA	12	09												
YV0	Aves I.	NA	11	08												
Z2	Zimbabwe	AF	53	38												
Z3	Macedonia	EU	28	15												
Z8	South Sudan (Rep of)	AF	48	34												
ZA	Albania	EU	28	15												
ZB2	Gibraltar	EU	37	14												
ZC4	UK Sov. Base Areas on Cyprus		AS	3 9												
20																
ZD7	St. Helena	AF	66	36												
ZD8	Ascension I.	AF	66	36												
ZD9	Tristan da Cunha & Gough I.	AF	66	38												
ZF	Cayman Is.	NA	11	08												
ZK3	Tokelau Is.	OC	62	31												
ZL-ZM	New Zealand	OC	60	32												
ZL7	Chatham Is.	OC	60	32												
ZL8	Kermadec Is.	OC	60	32												
ZL9	Auckland & Campbell Is.	OC	60	32												
ZP	Paraguay	SA	14	11												
ZR-ZU	South Africa	AF	57	38												
ZS8	Prince Edward & Marion Is.	AF	57	38												

Callsign Listings

Approximate Issue Dates for UK Callsigns

Full Licence		G3QAA	not issued	G0JAA	1988	G8OAA	1977-78	M1CAA	1997
Two letters		G3RAA	1962-63	G0LAA	1989	G8PAA	1978	M1DAA	1998-99
G2AA	1920-39	G3SAA	1963-64	G0MAA	1990	G8QAA	not issued	M1EAA	1999-00
G3AA	1937-38	G3TAA	1964-65	G0NAA	1991	G8TAA	1979	**Intermediate**	
G4AA	1938-39	G3UAA	1965-66	G0SAA	1992	G8ZAA	1981	2E0AAA	1991
G5AA	1921-39	G3VAA	1966-67	G0TAA	1993	G6CAA	1981	2E1AAA	1991
G6AA	1921-39	G3WAA	1967	G0VAA	1994	G6QAA	not issued	2E1BAA	1992
G8AA	1936-37	G3XAA	1967-68	G0WAA	1995	G6RAA	1982	2E1CAA	1993
		G3YAA	1968-69	M0AAA	1996	G1AAA	1983	2E1DAA	1994
Three letters		G3ZAA	1969-71	M0BAA	1997-98	G1DAA	1984	2E1EAA	1995-97
G2AAA	Pre-war	G4AAA	1971-72	M0CAA	1998-00	G1LAA	1985	2E1GAA	1997-99
G3AAA	1946	G4BAA	1972-73	M5AAA	1999-00	G1QAA	not issued	2E1HAA	1999-00
G3CAA	1947	G4DAA	1974-75	G8AAA	1964-67	G1SAA	1986	**Foundation**	
G3EAA	1948	G4EAA	1975-76	G8BAA	1967-68	G1XAA	1987	M3AAA	2002
G3GAA	1949-50	G4GAA	1977	G8CAA	1968-69	G7AAA	1988	M6AAA	2008
G3HAA	1950-51	G4IAA	1979	G8DAA	1969-70	G7EAA	1989	M7AAA	2018
G3IAA	1951-52	G4MAA	1981	G8EAA	1970-71	G7FAA	1990	**Note:** *From April 2000, out-of-sequence callsigns could be requested, so calls in later series may be heard.*	
G3JAA	1952-54	G4QAA	not issued	G8FAA	1971-72	G7HAA	1991		
G3KAA	1954-56	G4RAA	1982	G8HAA	1973	G7MAA	1992		
G3LAA	1956-57	G4SAA	1983	G8IAA	1973-74	G7OAA	1993		
G3MAA	1957-58	G4WAA	1984	G8JAA	1974-75	G7SAA	1994		
G3NAA	1958-60	G0AAA	1985	G8KAA	1975	G7TAA	1995		
G3OAA	1960-61	G0EAA	1986	G8MAA	1976-77	G7WAA	1996		
G3PAA	1961-62	G0HAA	1987	G8NAA	1977	M1AAA	1996		

Please Read important Notices and Frequently Asked Questions on page 181

Callsign Formatting

Each year the RSGB requests callsign data from Ofcom for the production of this book. This data was output in the middle of June 2023 and therefore any callsign changes made after this date will not appear in the following pages. The RSGB makes no amendments to the data supplied by Ofcom other than formatting. This includes removing upper case from town names and text. First names and most titles (Mr, Mrs, Ms, etc.) some honorific titles (Sir, Lord, Dr, etc.) are though retained. This produces a consistent format presentation of the data.

Callsigns that are included in the 'Details withheld' section of the Yearbook are only formatted for appearance. In these instances the information about the holder is not passed to the RSGB so we are not able to release any further information. We are also not able to move individuals from one list to the other on request, as the data reproduced here is a reflection of the data held by Ofcom not the RSGB. Should you want to change this information this will need to be done via Ofcom and any change will then be reflected in future editions of the Yearbook.

It can be noted that we do add Regional Secondary Locators (RSLs) which identify locations within the UK. This does not form part of the licence information so is no longer supplied by Ofcom to the RSGB. The RSGB uses UK postcodes to identify the station locations and the details of the changes are laid out in the table below. Where appropriate these appear as the suffix to the first character of the callsign. We only incorporate suffixes for individuals although we do recognise that in practice some clubs may use the appropriate suffix to identify themselves as such. We recognise that this process may contain a degree of inaccuracy and therefore apologise to anyone if their callsign is displayed incorrectly.

RSL Postcode Changes

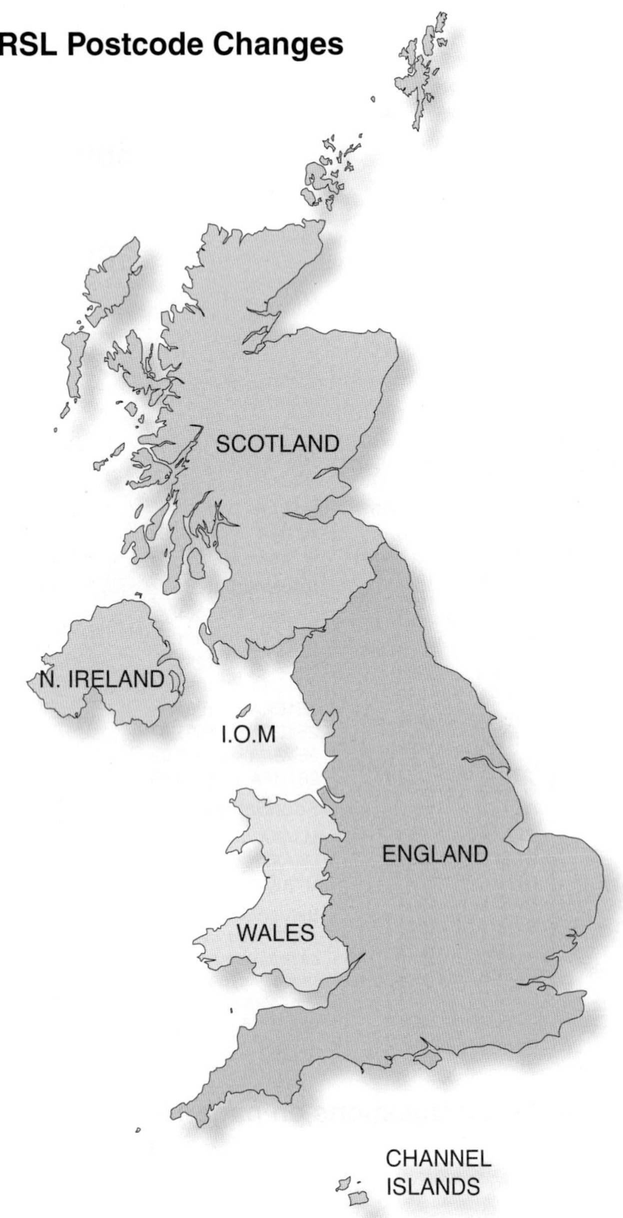

Postcode	Suffix	Country	Notes
AB	M	Scotland	
BT	I	N. Ireland	
CF	W	Wales	
CH	W	Wales	Only CH5, CH6, CH7 & CH8
DD	M	Scotland	
DG	M	Scotland	
EH	M	Scotland	
FK	M	Scotland	
G	M	Scotland	
GY	U	Gurnsey	
HS	M	Scotland	
IM	D	Isle of Man	
IV	M	Scotland	
JE	J	Jersey	
KA	M	Scotland	
KW	M	Scotland	
KY	M	Scotland	
LD	W	Wales	
LL	W	Wales	
ML	M	Scotland	
NP	W	Wales	
PA	M	Scotland	
PH	M	Scotland	
SA	W	Wales	
SY	W	Wales	Only SY10, SY11, SY16, SY18, SY19, SY20, SY21, SY22 & SY23
TD	M	Scotland	Not: Berwick TD15
ZE	M	Scotland	

Frequently Asked Questions

Throughout the year the RSGB receives a number of standard queries about an individual's entry in the *Yearbook*. Here are the answers given:

Q: I have just gained my new licence, please include my new details in the next Yearbook.

A: *Your details will be passed on to us by Ofcom from the data you supply on your licence application form. There is no need to contact the Society directly.*

Q: I am an RSGB member. Thank you for sending *RadCom* to my new address, but why didn't you change my *Yearbook* entry?

A: *The Yearbook lists all UK amateurs, not just RSGB members, and the records are entirely separate. Your callbook address details come from Ofcom. Did you re-validate your licence with your up-to-date details or let them know you moved? (***read Important Note below***)*

Q: My entry shows me as being 'details withheld', but I want my full address to be listed.

A: *Please contact Ofcom Licensing and ask them to release your details.* **The Society cannot accept direct input from licensees**.

Q: You have published my address, but I would like it withheld.

A: *As above, but ask Ofcom Licensing to withhold it from* ***all*** *callbook publishers*.

Q: My callsign is not shown at all, please include it.

A: *If a callsign is not shown in the Yearbook it was not licensed at the time the data was supplied to the RSGB. Sometimes, through an administrative error or misunderstanding, an amateur believes that he or she is licensed but the licensing records show that the licence has lapsed. If this is the case, contact Ofcom Licensing and discuss the matter with them as it is important that you re-validate your licence if you intend using it.*

Important Note

We do not make amendments to the basic callsign data. Should the details that appear in the callsign section of this *Yearbook* need amending, you should contact the licensing authority direct. Your submission should be prior to the **1st June** to allow sufficient time for the amendment to be processed.

If you have not validated your licence, which you must do every 5 years, or if any of your details have changed, ie change of address, then your licence may be void and will not appear in the Callsign listings section of this *Yearbook*, and you may not be licensed to operate.

To validate your licence, log in to *https://services.ofcom.org.uk/* and amend any details as soon as possible, this will automatically validate your licence.

Ofcom Callsign Data Disclaimer

Please note that the use of this callsign data is entirely at your own risk. While every effort is made to ensure that the information provided to you is accurate, no guarantees for the currency or accuracy of information are made.

The callsign data is provided 'as is'. It is provided without any representation or endorsement made and without warranty of any kind, whether express or implied, including but not limited to the implied warranties of satisfactory quality, fitness for a particular purpose, non-infringement, compatibility, security and accuracy.

Ofcom does not accept any responsibility for any loss, disruption or damage to your data or your computer system which may occur whilst using the data provided by Ofcom.

In no event will Ofcom be liable for any loss or damage including, without limitation, indirect or consequential loss or damage, or any loss or damages whatsoever arising from use of loss of use of, data or profits arising out of or in connection with the use of information provided by Ofcom.

UK Callsigns

2#0

2E0 AAC G. Lockett, 10 Lunesdale Street, Houghton le Spring, DH50DB
2E0 AAF S. Rhenius, Baythorne Cottage Baythorne End, Halstead, CO9 4AB
2E0 AAI D. Simmons, 23 Fairey Street, Cofton Hackett, Birmingham, B45 8GU
2E0 AAJ R. Yarrow, 27 Staplers Road, Newport, PO30 2DB
2E0 AAK R. Berridge, Bracklyn, St. Clare Road, Deal, CT14 7QB
2E0 AAN M. Keilty, 25 Lathom Avenue Wallasey, Wirral, CH44 5UH
2E0 AAO S. Thirlwall, 2 Crossfield Avenue Blythe Bridge, Stoke-on-Trent, ST11 9PL
2M0 AAQ W. Ferguson, 1D Macphail Drive, Kilmarnock, KA3 7EJ
2E0 AAX E. Wills, 1 Scott Close, High Street, Marlborough, SN8 3AF
2E0 AAZ W. Elston, 3 Ennerdale Close, Northampton, NN3 6BL
2E0 ABD R. Mold, 134 Kipling Avenue, Brighton, BN2 6UE
2E0 ABL F. Hayes, 88 Johns Road, Fareham, PO16 0RX
2E0 ABT L. Simons, Westwood, Faris Lane, Addlestone, KT15 3DJ
2E0 ABU K. Williams, 26 Charles Bassett Close, Helston, TR13 8BG
2E0 ACA I. Craig, Ferndown, Tilley Lane, Hailsham, BN27 4UT
2E0 ACE A. Backhouse, 6 Millfield, Brampton, CA8 1TT
2E0 ACQ H. Barnes, 24 Burleigh Place, Oakley, Bedford, MK43 7SG
2E0 ACR H. Smallwood, 14 Shaftesbury Avenue, Bedford, MK40 3SA
2E0 ACV E. Barnes, 3 Layton Road, Ashton-On-Ribble, Preston, PR2 1PB
2E0 ADA R. Gaskell, 18 Woodcroft Kennington, Oxford, OX1 5NH
2E0 ADL J. Wresdell, 4 Fairfield Close, Nafferton, Nafferton, YO25 4JH
2E0 ADN M. Holmes, 55 Fitzpain Road, West Parley, Ferndown, BH22 8RZ
2E0 ADR S. Watson, 8 Ventnor Road, Middlesbrough, TS5 6DU
2E0 ADY A. Jones, 67 Mandara Grove, Abbeydale, Gloucester, GL4 5XT
2E0 AED A. Sumner, 78 Woodlands Way, Southwater, West Sussex, RH13 9HZ
2E0 AES B. Hyde, 54 The Byway, Darlington, DL1 1EQ
2E0 AFL M. Coxhead, Barronsway, Parker Lane, Whitestake, Preston, PR4 4JX
2E0 AFP R. Forrest-Webb, 1 Trelasdee Farm Cottages, St. Weonards, Hereford, HR2 8PU
2E0 AFT A. Rowe, 27 Silver Street, Cheddar, BS27 3LE
2E0 AGB A. Boocock, 25 Smallwood Road, Dewsbury, WF12 7RU
2E0 AGI Dr E. Mclusky, 11 Ripon Road, Killinghall, Harrogate, HG3 2DG
2E0 AGQ B. Read, 10 Shamblers Road, Cowes, PO31 7HF
2M0 AGS G. Robbins, 39 Locheil Gardens, Glenrothes, KY7 6YL
2E0 AHB A. Beveridge, Kernick Cottage, Sparry Bottom, Carharrack, Redruth, TR16 5SH
2M0 AIE M. Hooks, Braehead Mains, Main Street, Braehead Forth, Lanark, ML11 8HA
2E0 AIK M. Footring, 26 Ernest Road, Wivenhoe, Colchester, CO7 9LG
2E0 AIL G. Harper, 48 Norlands Lane, Widnes, WA8 5AS
2E0 AIM J. Cowley, 39 Alpine Way, Tow Law, Bishop Auckland, DL13 4DS
2E0 AIT K. Lloyd, 2 Bishopstone Drive, Beltinge, Herne Bay, CT6 6RE
2E0 AJG A. Golding, 44 Blendon Drive, Andover, SP10 3NG
2E0 AJK D. Forsyth, 20 Chapel View, Rowlands Gill, NE39 2PN
2E0 AJP L. Humphrey, 133 Hocombe Road, Chandler'S Ford, Eastleigh, SO53 5QD
2E0 AJQ A. Nelson, 29 Coxford Road, Southampton, SO16 5FG
2E0 AJT W. Atkins, 55 Park Grange Croft, Norfolk Park, Sheffield, S2 3QJ
2E0 AJX C. Overson, Studio Flat, 6 Grenville Street, Bideford, EX39 2EA
2M0 AKI D. Hague, 13 North Dell, Ness, Isle of Lewis, HS2 0SW
2E0 AKJ S. Davin, 40 Theynes Croft, Long Ashton, Bristol, BS41 9NA
2E0 AKK E. Jones, 43 Wesley Road, Wimborne, BH21 2QB
2E0 AKL R. Mcdermott, 2 Monument Close, Wellington, TA21 9AL
2E0 AKN P. Newton, 61 Ashbourne Crescent, Taunton, TA1 2RA
2E0 AKO M. Newton, 43 Hayfield Road, Minehead, TA246AD
2E0 AKQ J. Paul, East Park, Church Road, Cowes, PO31 8HA
2M0 AKS Dr A. Curlis, 94 Kirkhill Road, Aberdeen, AB11 8FX
2E0 AKU R. Brisley, 15 Elm Fields, Old Romney, Romney Marsh, TN29 9SN
2E0 ALA N. Kind, 18 Cunningham Road, Bentley, Walsall, WS2 0AY
2E0 ALB Dr M. Ali, 4 The Crescent, Great Horkesley, Colchester, CO6 4EH
2E0 ALD P. Girling, The Gorse, Leiston Road, Aldeburgh, IP15 5QE
2E0 ALF A. Taylor, 113 Queensway, Grantham, NG31 9RG
2E0 ALH J. Side, Railway Crossing Cottage, Ash Road, Sandwich, CT13 9JB
2E0 ALJ G. Brierley, 35 Ochrewell Avenue, Deighton, Huddersfield, HD2 1LL
2E0 ALL A. Sawyer, 26 Dallas Brett Crescent, Folkestone, CT19 6NE
2M0 ALS M. Gill, Easter Templand, Fortrose, IV10 8RA
2E0 ALW R. Hatcher, 61 Holland Road, Oxted, RH8 9AU
2W0 ALZ D. Gunning, 19 Gordon Avenue, Prestatyn, LL19 8RU
2W0 AMB J. Lloyd-Owen, 34 Maes Y Castell, Llandudno, LL30 1NG
2M0 AMD A. Dawson, Colt House, Kinlochbervie, Lairg, IV27 4RP
2W0 AMN C. Weaver, 2 Ty Cerrig, Llanddoged, Llanrwst, LL26 0TY
2E0 AMS A. Saville, 1 Bellingham Close, Shaw, Oldham, OL2 7UU
2E0 AMW G. Fowle, 12 Lytham Road, Broadstone, BH18 8JS
2E0 AND A. Miller, East Lyn, Lazonby, Penrith, CA10 1BX
2M0 ANE G. Steele, 1 James Street, Bannockburn, Whins Of Milton, Stirling, FK7 0NQ
2E0 AOK P. Godolphin, Shepherds Cottage, Flakebridge, Appleby-in-Westmorland, CA16 6JZ
2E0 AOL T. Money, 119 Twyford Way, Canford Heath, Poole, BH17 8SR
2E0 AOS E. Hedley, 17 Chowdene Bank, Gateshead, NE9 6JJ
2E0 AOU S. Tilly, 24 Whinham Way, Morpeth, NE61 2TF
2E0 AOZ I. Marshall, 44 Cromwell Crescent, Pontefract, WF8 2EJ
2M0 APB J. Macdonald, 63 Perceval Road South, Stornoway, HS1 2TL
2E0 APG G. Wilson, 9 Newcomen Way, Telford, TF7 5UB
2E0 APJ M. France, 10 Grampian Avenue, Wakefield, WF2 8JZ
2E0 APN T. Humble, 43 Whiteley Crescent, Bletchley, Milton Keynes, MK3 5DQ
2M0 APX S. Barbour, 40 Mannerston Holdings, Linlithgow, EH49 7ND
2E0 APY J. Rayson, 18 Station Street, South Wigston, Wigston, LE18 4TH
2E0 APZ M. Meyer, 22 Orchard Grove, Newton Abbot, TQ12 1FZ
2E0 AQE J. Brown, 55 Barrington Road, Rubery, Birmingham, B45 9EU
2E0 AQI D. Black, 8 Cornwood Close, Finchley, London, N2 0HP
2E0 AQU P. Webb, 41 Lancaster Gardens, Wolverhampton, WV4 4DN
2E0 ARA J. Griffin, 35 Cottage Street, Kingswinford, DY6 7QE
2E0 ARC R. Evans, 14 Doncaster Drive, Upton, Wirral, CH494NX
2E0 ARJ R. Birch, 15 Chester Street, Cirencester, GL7 1HF
2E0 ARV I. Strachan, 1 Gantley Avenue, Billinge, Wigan, WN5 7AF
2E0 AST L. Compton, 1 Southhaven, Beach Road, Bacton, Norwich, NR12 0EP
2E0 ASU J. Bell, 122 Howard Drive, Letchworth Garden City, SG6 2DE
2E0 ASW S. Pearson, 8 The Pastures, Edlesborough, Dunstable, LU6 2HL
2E0 ASX R. Bannister, 60 St. Johns Avenue, Bridlington, YO16 4NL
2E0 ATB L. Coneley, 4 Primrose Close, Gosport, PO13 0WP
2E0 ATT N. Watson, 37 Riber Avenue, Barnsley, S71 3PU
2E0 ATY R. Browne, 2 Martham Close, London, SE28 8NF
2E0 ATZ R. Macdonald, 4 Throwley Drive, Herne Bay, CT6 8LP
2W0 AUC R. Henderson, 150 Heol Maes Eglwys, Pantlasau, Swansea, SA6 6NW
2E0 AUI D. Tyson, 21 Providence Place, Filey, YO14 9DU
2E0 AUM A. Peach, 14 Northfield Avenue Rocester, Uttoxeter, ST14 5LE
2E0 AUU P. Christopoulos, Cosenda, Tye Green Village, Harlow, CM18 6QY
2E0 AUV C. Brice, 10 Swan Close, Weston-Super-Mare, BS22 8XR
2E0 AVA T. Milne, 3 Birch Road, New Ollerton, Newark, NG22 9PU
2E0 AVB M. Orr, 42 Mayfield Road, London, W12 9LU
2E0 AVD W. Jukes, 22 Hazelmere Road, Creswell, Worksop, S80 4HS
2E0 AVK R. Harrall, 47 Hawksmoor Road, Stafford, ST17 9DS
2E0 AVQ J. Paul, 67 Fleet Avenue, Upminster, RM14 1PZ
2E0 AVS S. Panczel, 11 Chauncy Road, Manchester, M40 3GG
2E0 AVW R. Smith, Renee . Shardlow Marina. London Road, Shardlow., Derby, DE72 2GL
2W0 AVZ N. Heyne, New House, Hope, Welshpool, SY21 8JD
2E0 AWC K. Necchi, 59 Winters Way, Waltham Abbey, EN9 3HP
2E0 AWE J. Sanders, 219 Station Road, Drayton, Portsmouth, PO6 1PY
2E0 AWG S. Rope, 12 Edrich Close, Norwich, NR134JD
2E0 AWI R. Desbois, 12 Dove Walk, Hornchurch, RM12 5HH
2E0 AWK J. Jackson, 35 Coppingford Close, Rochdale, OL12 7PR
2E0 AWR J. Gibson, 14 Douglas Bank Drive, Wigan, WN6 7NH
2E0 AWS A. Storer, 16 Eastfields, Braunston, Daventry, NN11 7JN
2E0 AWT G. Griffiths, 16 Back Lane, Winteringham, Scunthorpe, DN15 9NW
2W0 AWW R. Evans, 4 Green Terrace, Deiniolen, Caernarfon, LL55 3LE
2M0 AWY J. Fulton, 35 Heatherbank, Livingston, EH54 6EE
2E0 AWZ R. Turton, 31 Second Avenue, Woodlands, Doncaster, DN6 7QQ
2E0 AXA K. Kerridge, 31 Second Avenue, Woodlands, Doncaster, DN6 7QQ
2E0 AXB M. Costello, 25 Campden Green, Solihull, B92 8HQ
2E0 AXN S. Jenner, 4 Christie Close, Chatham, ME5 7NG
2E0 AXT A. Smith, 186 Longmead Drive, Nottingham, NG5 6DJ
2E0 AXZ C. Cook, 112 Waterford Road, Ipswich, IP1 5NJ
2E0 AYI K. Williams, 36 Castle St., Tiverton, EX16 6RG
2E0 AYK M. Longbottom, 32 Ann'S Hill Road, Gosport, PO12 3JY
2E0 AYQ M. Walsh, 25 Felstead Road, Orpington, BR6 9AA
2E0 AYT R. Cole, 2 Station Road, Sharpthorne, RH19 4NY
2M0 AYU J. Bruce, 33 Kintail Place, Dingwall, IV15 9RL
2E0 AYW M. Pedley, Comfort Cottage, Goadsbarrow, Ulverston, LA12 0RE
2E0 AYX C. Penfold, 14 Romney Road, Tetbury, GL8 8JU
2E0 AYY P. Rainer, 6 Highland Close, Folkestone, CT20 3SA
2M0 AYZ C. Gajewski, 7/2 Brunswick Road, Edinburgh, EH7 5NG
2E0 AZA P. Livsey, 33 Aldingham Walk, Morecambe, LA4 4EW
2E0 AZB M. Livsey, 26 Aldingham Walk, Morecambe, LA4 4EW
2E0 AZD F. Russell, 157 Durnford St., Plymouth, PL1 3QR
2E0 AZF D. Whitworth, 570 Dereham Road, Norwich, NR5 8TE
2E0 AZJ V. Hocking, 80 Barton Tors, Bideford, EX39 4HA
2E0 AZK A. Cartwright, 17 Greenheath Way, Wirral, CH46 3RX
2E0 AZL L. Murphy, 22 Shenley Fields Drive, Birmingham, B31 1XH
2W0 AZM R. Harper, 17 Conway Drive, Wrexham, LL13 9HR
2W0 AZP D. Edwards, 9 Ffos Y Cerridden, Nelson, Treharris, CF46 6HQ
2E0 AZU A. Groat, 23 Knightwake Road, New Mills, High Peak, SK22 3DQ
2M0 AZW T. Galt, 119 Cleeves Road, Glasgow, G53 6NQ
2E0 AZZ A. Dalzell, 9 Pyms Lane, Crewe, CW1 3PJ
2E0 BAB G. Gragon, 2 Greenacres Grove, Shelf, Halifax, HX3 7RN
2I0 BAC C. Mcconnell, 41 Moyra Road, Doagh, Ballyclare, BT39 0SQ
2I0 BAD A. Hayes, 35 Ballyhamage, Doagh, BT39 0PZ
2E0 BAF D. Davies, 35 Orchard Court Carlton, Nottingham, NG4 1BD
2E0 BAH P. Mannering, 101 Burford, Brookside, Telford, TF3 1LJ
2E0 BAI R. Froud, 48 Campbell Road, Maidstone, ME15 6QB
2E0 BAJ F. Cappleman, 8 The Woodlands, Lilleshall, Newport, TF10 9EN
2E0 BAK M. Verrall, Flat 4, 84 Briar Way, Skegness, PE25 3PU
2W0 BAO T. Mulcuck, 61 Hillside View, Graigwen, Pontypridd, CF37 2LG
2E0 BAP S. Savage, 450 Locking Road, Milton, Weston-Super-Mare, BS22 8PS
2E0 BAQ D. Wills, 37 Hawden Road, Bournemouth, bh118rp
2E0 BAU C. Young, 15 Shelton Avenue, East Ayton, Scarborough, YO13 9HB
2E0 BAV A. Yorke, 33 Avon Crescent, Stratford-upon-Avon, CV37 7EX
2E0 BAX S. Amesbury, 222 Bond Street, Macclesfield, SK11 6RG
2E0 BAY A. Harris, 32 King Edward Road, Gillingham, ME7 2RE
2E0 BBA I. Wengraf, 45 School Lane, Higham, Rochester, ME3 7JR
2E0 BBG D. Skinner, 77 Rolleston Avenue, Petts Wood, Orpington, BR5 1AL
2E0 BBI A. Hayward, 23 Greenbank Close, Grampound Road, Truro, TR2 4TD
2E0 BBL C. Lishman, 6 Clarence Road, Accrington, BB5 0NA
2E0 BBM S. Beales, 49 Greenacres, Old Newton, Stowmarket, IP14 4EJ
2E0 BBN S. Woodcock, 3 Lower House Road, Leyland, PR25 1HT
2W0 BBO L. Powell, 2 Gelliderw Pontardawe, Swansea, SA8 4NB
2E0 BBQ R. Gash, 2 The Marsh, Benington, Boston, PE22 0DH
2E0 BBS M. Janes, 1 Cheswell Gardens, Church Crookham, Fleet, GU51 5NJ
2W0 BBT S. Christie-Powell, 2 Gelliderw Pontardawe, Swansea, SA8 4NB
2E0 BBX J. Akinin, 70 Valley Road, West Bridgford, Nottingham, NG2 6HQ
2E0 BBY F. Coles, 8 Moore Close, Church Crookham, Fleet, GU52 6JD
2E0 BBZ M. Thyer, 8 Roman Road, Taunton, TA1 2BD

2E0 BCB D. Holland, 13 Linley Drive, Boston, PE21 7EJ
2E0 BCD B. Grice, 17 Albion Field Drive, West Bromwich, B71 4HN
2E0 BCE J. Ferris, 39 Gladstone St., Beeston, Nottingham, NG9 1EU
2E0 BCF Dr H. Donnelly, 6 Famet Walk, Purley, CR8 2DY
2E0 BCG Rev. W. Walker, 9 Malthouse Lane, Dorchester-On-Thames, Wallingford, OX10 7LF
2E0 BCJ D. Neville, 70 Ruskin Road, Crewe, CW27JS
2E0 BCK Dr T. Gale, 38 Lantree Crescent, Trumpington, Cambridge, CB2 9NJ
2M0 BCL T. Mccall, 119 Claremont, Alloa, FK10 2ER
2E0 BCM B. Marston, 38 Bulstrode Road, Ipswich, IP2 8HA
2E0 BCO S. Norman, 27 Ashburton Road, Ickburgh, Thetford, IP26 5JA
2E0 BCQ R. Britt, Thoroughfare House, South Burlingham Road, Norwich, NR13 4FA
2D0 BCR R. Cunningham, 3 Kellets Cottage, Lhergy Cripperty, Union Mills, Isle Of Man, IM4 4NF
2E0 BCS Dr J. Schofield, 6 Robin Royd Avenue, Mirfield, WF14 0LF
2E0 BCW S. Bywater, Birch Wood, Norwich Road, Cromer, NR27 0HG
2E0 BCX J. Allen, 20 Spa Hill, Kirton Lindsey, Gainsborough, DN21 4BA
2E0 BDB A. Grosvenor, 10 Neves Close, Lingwood, Norwich, NR13 4AW
2E0 BDD A. Morrison, 70 The Drive, Northampton, NN1 4SP
2E0 BDI S. Moakes, 46 Parsonage St., Stockport, SK4 1HZ
2E0 BDJ D. Anthony, 20 Parkfield Close, Leyland, PR26 7XJ
2M0 BDN J. Walker, 49 Great King St., Edinburgh, EH3 6RP
2E0 BDO S. Grainger, 15 Carr House Lane, Wirral, CH46 6EN
2E0 BDP M. Saunders, 70 Underwood Lane, Crewe, CW1 3LE
2E0 BDQ W. Taylor, 99 St. Marys Close, Littlehampton, BN17 5QQ
2M0 BDR B. Keiller, Da Cro, Branchiclate, Burra Isle, ZE2 9LA
2E0 BDS B. Derbin-Sykes, 1, Lentons Lane, Friskney, Boston, PE22 8RR
2M0 BDT A. Halcrow, Da Cro, Branchiclate, Burra Isle, ZE2 9LA
2E0 BDV J. Goulding, 79 Dalston Drive, Manchester, M20 5LQ
2E0 BEA J. Rushton, 25 Garth Meadow, Catterick, Richmond, DL10 7RT
2M0 BEB G. Bourhill, 30C Salters Road, Wallyford, Musselburgh, EH21 8AA
2M0 BEC M. Mitchell, Easter Kilwhiss Farm, Ladybank, Cupar, KY15 7UR
2W0 BED C. Rosser, 16 Thomas St., Penygraig, Tonypandy, CF40 1EU
2E0 BEE A. Bell, 2 Croft Foot, Sandwith, Whitehaven, CA28 9UG
2E0 BEF M. Staton, 52 School Road, Newborough, Peterborough, PE6 7RG
2E0 BEG M. Davidson, 19 Mason Street, Workington, CA14 3EH
2E0 BEH C. Loughran, 8 Douglas Road, Dover, CT17 0BD
2E0 BEI C. Brett, 35 Gossilin Street, Whitstable, CT5 4LQ
2M0 BEL P. Mccluskey, 119 Tower Drive, Gourock, PA19 1SG
2E0 BEP A. Holmes, 10 Doe Park, York, YO30 4UQ
2E0 BEQ D. Whitehead, 89, Cowpes Close, Sutton-in-Ashfield, NG17 2BU
2M0 BET D. Boden, 42 Kirkwynd, Maybole, KA19 7AE
2E0 BEV I. Lowe, 1 Hazelby Road, Creswell, Worksop, S80 4BB
2E0 BEW D. Powis, Fircroft, Pound Lane, Woodbridge, IP13 0LN
2E0 BFA J. Mullarkey, 41 Foyle Avenue, Chaddesden, Derby, DE21 6TZ
2W0 BFC M. Williams, 30 Elm Drive, Risca, Newport, NP11 6HJ
2W0 BFD W. Corbett, 27 Waunfawr Gardens. Crosskeys, Newport, NP11 7AJ
2E0 BFF A. Bateman, 154 Arnold Road Mangotsfield, Bristol, BS16 9LB
2E0 BFG P. Davies, 53 Lammas Road, Cheddington, Leighton Buzzard, LU7 0RY
2E0 BFJ G. Swain, 3 Flaxfield Drive, Crewkerne, TA18 8DF
2E0 BFM A. Knights, 81 Green Lane, Barnard Castle, DL12 8LF
2E0 BFN W. Carty, 49 Princess Gardens, Blackburn, bb25ej
2E0 BFS B. Townshend, 9 Norfolk Place, Boston, PE21 9JJ
2E0 BFT D. Doroba, Flat 3, 305A London Road South, Lowestoft, NR33 0DX
2W0 BFV W. Harries, 18 Bro Teify, Alltyblacca, Llanybydder, SA40 9SR
2E0 BFZ P. Hinde, 19 Tadcaster Road, Sheffield, S8 0RA
2E0 BGC N. Speight, Flat 14, Cranbrook, London, NW1 0LJ
2U0 BGE J. Bligh, The Bounty, Salines Lane, St Sampson, Guernsey, GY2 4FL
2W0 BGI M. Lewis, 4 Coldwell Terrace, Pembroke, SA71 4QL
2E0 BGJ M. Whitaker, 5 Horns Drove, Rownhams, Southampton, SO16 8AH
2E0 BGL B. Lloyd, 104 Wootton St., Bedworth, CV12 9DZ
2E0 BGM S. Russon, 165 Billington Avenue, Newton-le-Willows, WA12 0AU
2E0 BGO J. Woodruff, 10 Bailey Close, Blackburn, BB2 4FT
2E0 BGP E. Neil, 5 Winsford Hill, Furzton, Milton Keynes, MK4 1BJ
2W0 BGQ I. Canterbury, Brynllethryd Bungalow, Senghenydd, Caerphilly, CF83 4HJ
2W0 BGS J. Brydges, 9 Twynygarreg, Treharris, CF46 5RL
2E0 BGV R. Johnson, 90 Regent Street Church Gresley, Swadlincote, DE11 9PJ
2E0 BGX P. Sarll, 81 Austendyke Road, Weston Hills, Spalding, PE12 6BX
2E0 BGZ R. Dale, 17 Spencer Gardens, Brackley, NN13 6AQ
2E0 BHA P. Bennett, 1 Queens Road, Carterton, OX18 3YB
2E0 BHB R. Fallows, Loughrigg, Cranmore Avenue, Yarmouth, PO41 0XS
2E0 BHC L. Heenan, 1 Howard Close, Daventry, NN11 4TD
2E0 BHH R. Bullen, 2 Redlands Cottages, East Coker, Yeovil, BA22 9HF
2E0 BHJ J. List, 41 Westbury Crescent, Dover, CT17 9QQ
2E0 BHN J. Mccoll, 6 Grenville Close, Bodmin, PL31 2FB
2E0 BHP G. Suter, 1 Laburnham Lodge, Worthing, BN13 3dn
2E0 BHQ D. George, 9 Winscombe Court, Frome, BA11 2DZ
2E0 BHS N. Tideswell, 19 Wish Court, Ingram Crescent West, Hove, BN3 5NY
2I0 BHT S. Quigg, 100 Whispering Pines, Limavady, BT49 0UF
2E0 BHU J. Hickman, Ardoch, Harlestone Road, Northampton, NN6 8AW
2E0 BHY S. Preston, 14 Hulles Way, North Baddesley, Southampton, SO52 9NS
2E0 BHZ T. Mildenhall, 58 Montrose Avenue, Datchet, Slough, SL3 9NJ
2E0 BIB R. Orton, 38 Whitehill Avenue, Barnsley, S70 6PP
2E0 BIC T. Humphries, 10 Cropthorne Avenue, Leicester, LE5 4QL
2I0 BID A. Jamison, 11 Richmond Gardens, Newtownabbey, BT36 5LA
2E0 BII R. Hathaway, 38 Windsor Walk, Lindford, Bordon, GU35 0SG
2M0 BIL W. Mccue, 188 Redburn, Alexandria, G83 9BU
2E0 BIM K. Hawke, 111 Dorchester Avenue, Plymouth, PL5 4AZ
2M0 BIN F. Coombes, 44 Lochfield Road, Paisley, PA2 7RL
2E0 BIO R. Aston, 8 Parliament Road, Thame, OX9 3TE
2I0 BIQ K. Blake, Chestnut Lodge Care Home, 47 Carrickaness Road, Dungannon, BT71 7NH
2I0 BIR J. Smyth, 37 Ardfreelin, Newry, BT34 1JG
2E0 BIS J. Cosson, 25 Fox Brook, St Neots, PE19 6AL
2E0 BIU P. Rushby, 16 Foxhill Lane, Selby, YO8 9AR
2E0 BIY P. Lewis, 16 Valley Road, St Albans, AL3 6LR
2E0 BJA B. Johnson, 15 Oak Avenue, Willington, Crook, DL15 0BJ
2E0 BJB A. Chaplin, 33 The Crofts, Little Wakering, Southend-on-Sea, SS3 0JS
2E0 BJD R. Rowe, 16 Orchard Road, Plymouth, PL2 2QY
2W0 BJE S. Merrifield, 37 South View Drive Rumney, Cardiff, CF3 3LX
2E0 BJK R. Robertson, 2 Dunsley Close, Middlesbrough, TS3 7DR
2E0 BJL M. Placidi, 3 Eleanor Avenue, Epsom, KT19 9HD
2E0 BJM B. Maggs, 44 Coldharbour Road, Hungerford, RG17 0AZ
2E0 BJP P. Norman, 5 Stirling Close West Row, Bury St Edmunds, IP28 8QD
2E0 BJQ M. Roebuck, 13H Elizabeth Street, Elland, HX5 0JH
2W0 BJR R. Johnson, 25 Lon Tyrhaul Llansamlet, Swansea, SA7 9SF
2E0 BJS N. Roberts, 40 Armour Road, Tilehurst, Reading, RG31 6HN
2E0 BJT R. Carter, 43 Sheldon Avenue, Standish, Wigan, WN6 0LW
2M0 BJU G. Craig, 1 Butt Avenue, Helensburgh, G84 9DA
2E0 BJV T. Gabriel, 57 West Down Road, Delabole, PL33 9DT
2E0 BJW B. White, 9 Springfield Close, Wirral, CH49 7NJ
2E0 BJX G. Dyson, 111 Chester Road, Whitby, Ellesmere Port, CH65 6SB
2E0 BJY S. Billingham, Kewell House, Wombourne Road, Swindon, Dudley, DY3 4NF
2W0 BKA M. Mainwaring, 36 Oak Street, Gilfach Goch, Porth, CF39 8UG
2E0 BKB B. Hall, 6 Marshall Close, Parkgate, Rotherham, S62 6DB
2E0 BKD G. Flack, 20 The Pastures, Hardwick, Cambridge, CB23 7XA
2E0 BKE A. Anderson, 89A Malmesbury Park Road, Bournemouth, BH8 8PS
2I0 BKI E. Kyle, 2 Wattstown Crescent, Coleraine, BT52 1SP
2E0 BKJ G. Douch, 63 Greenaways Ebley, Stroud, GL5 4UN
2M0 BKL S. Fradley, 30 Polmont Park, Polmont, Falkirk, FK2 0XT
2W0 BKM E. Thomas, 29 Maes Y Wern, Carway, Kidwelly, SA17 4HF
2W0 BKN M. Stokes, 23 Goetre Fawr Road Killay, Swansea, SA2 7QS
2E0 BKO A. Oxlade, 27 Spenfield Court, Northampton, NN3 8LZ
2E0 BKP S. Martin, The Shieling, Bolton Low Houses, Wigton, CA7 8PF
2E0 BKQ D. Griffin, 101 Kingsway, Duxford, Cambridge, CB22 4QN
2E0 BKT G. Milsom, 31 Chichester Close, Bowerdean Road, High Wycombe, HP13 6AU
2E0 BKU T. Coles, 88C Dursley Road, Trowbridge Ba14 0Ns, Trowbridge, BA14 0NS
2E0 BKV S. Bird, 9 Almery Drive, Carlisle, CA2 4EX
2E0 BKY F. Goodall, 1 Parkfield Grove, Leeds, LS11 7LS
2E0 BKZ W. Owen, 8 Sandhurst Avenue, Lytham St Annes, FY8 2DA
2W0 BLA J. Jones, Bronydd, Blaenffos, Boncath, SA37 0HZ
2E0 BLC J. Farina, 9 Mallards Close, Alveley, Bridgnorth, WV15 6JL
2E0 BLD V. Parton, 51 Marston Grove, Stoke-on-Trent, ST1 6EF
2E0 BLF A. Spaxman, 70 Park View, Shafton, Barnsley, S72 8PY
2W0 BLG R. Williams, Plaen Cottage, Bodfari, Denbigh, LL16 4BS
2E0 BLI Dr J. Skittrall, 14 Tamarin Gardens, Cambridge, CB1 9GH
2E0 BLJ A. Marks, Grosvenor Hotel, 51 Grosvenor Road, Scarborough, YO112LZ
2E0 BLK S. Heaton, 13A Roche Avenue Bilton, Harrogate, Hg1 4es
2E0 BLL M. Green, 103 Preston Street, Kirkham, Preston, PR4 2XA
2E0 BLN D. Blake, 9 Malling Avenue, Eastfield, Scarborough, YO11 3FA
2E0 BLQ J. Brewster, 44 Beaulieu Close, Hounslow, TW4 5EW
2E0 BLR R. Payne, 74 Churchill Avenue, Newmarket, CB8 0BY
2E0 BLS L. Turner, 16 Woodland Place, Scarborough, YO12 6EP
2E0 BLT J. Owen, 90 Granville Drive, Kingswinford, DY6 8LW
2E0 BLW R. Silcox, 103 Oakdale Road, Downend, Bristol, BS16 6EG
2E0 BLX G. Thorpe, 81 Knoll Drive, Coventry, CV3 5PJ
2E0 BLY A. Bailey, 58 Billy Buns Lane, Wombourne, Wolverhampton, WV5 9BP
2E0 BLZ D. Forster, 23 Field Street, Padiham, Burnley, BB12 7AU
2E0 BMB A. Bolla, 11 Shelley Crescent, Blyth, NE24 5RH
2E0 BMD J. Turner, 17 Beechwood Road, Dronfield, S18 1PW
2E0 BME C. King, 21 Lowdham, Wilnecote, Tamworth, B77 4LX
2M0 BMF F. Thomson, 11 Carmichael Place, Irvine, KA12 0XH
2E0 BMG W. Mcgill, 49 Anthony Close, Colchester, CO4 0LD
2E0 BMH B. Hawes, 3 Orchard Close, Cassington, Witney, OX29 4BU
2E0 BMJ R. Burrows, 62 Fletcher Road, Burbage, Hinckley, LE10 2PS
2M0 BMK J. Cosgrove, The Cottages, Kirkinch, Blairgowrie, PH12 8SL
2W0 BMM N. Hughes, Mountain Farm Cottage, Clynderwen, Llandissilio, SA66 7PX
2M0 BMN C. Lewis, 9 Cessnock Road, Troon, KA10 6NJ
2W0 BMO R. Rimmer, Dwyfor, Heol Las, Llantrisant, Pontyclun, CF72 8EG
2E0 BMP J. Redfern, 19 Stanton Green, Shrewsbury, SY1 4PL
2W0 BMR S. Williams, Flat 28, Llys Celyn Cedar Crescent, Tonteg, Pontypridd, CF38 1LF
2E0 BMT Rev. B. Topham, 2 Highgrove Gardens, Stamford, PE9 2GR
2E0 BMU P. Wright, 34 Coles Lane, West Bromwich, B71 2QJ
2E0 BMW A. Rowe, Southern Point, Grange View, Houghton le Spring, DH4 4HU
2E0 BMY A. Farrar, 8 Wensley Street, Thurnscoe, Rotherham, S63 0PX
2E0 BNA N. Anderson, 3 Phelipps Road, Corfe Mullen, Wimborne, BH21 3NN
2E0 BNC C. Murray, 7 Pilmoor Drive, Richmond, DL10 5BJ
2E0 BND L. Woollard, 46 Woodhill Lane, Morecambe, LA4 4NN
2E0 BNE S. Sanderson, 65 Holm Flatt Street, Parkgate, Rotherham, S62 6HJ
2E0 BNF A. Webb, 47 Granville Street, Gloucester, GL1 5HL
2E0 BNH J. Hawkes, 53 Mill Hill, Derby, DE24 5AF
2E0 BNI G. Marshall, 12 Arthur Avenue, Caister-On-Sea, Great Yarmouth, NR30 5PQ
2E0 BNJ S. Scott, Watercrook Bungalow, Natland, Kendal, LA9 7QB
2E0 BNK J. Stoppard, 15 South Lodge Court, Old Road, Chesterfield, S40 3QG
2E0 BNO S. Pearson, 370 Queen Elizabeth Road, Nuneaton, CV10 9GS
2E0 BNT B. Trayhurn, 15 Wight Drive, Caister-On-Sea, Great Yarmouth, NR30 5UN
2E0 BNV K. Peel, 123 Cunningham Road, Tamerton Foliot, Plymouth, PL5 4PU
2E0 BNW P. Ashton, 14 Poppy Close, Boston, PE21 7TJ
2E0 BNZ W. Whitcher, 17 Watermead, Stratton St. Margaret, Swindon, SN3 4WE
2E0 BOB R. Hastings, 2 Boleyn Way, Boreham, Chelmsford, CM3 3JJ
2W0 BOC O. Williams, 39 Camden Road, Maes-Y-Coed, Brecon, LD3 7RT
2E0 BOD J. Tusler, Kilima, Batts Corner, Farnham, GU10 4EX
2E0 BOF P. Callaghan, 41 Higher Ash Road, Talke, Stoke-on-Trent, ST7 1JN
2E0 BOI R. Massimino, 115 Trelowarren Street, Camborne, TR14 8AW
2E0 BOJ M. Anthony, Magpie Bungalow, Goongumpas, St. Day, Crofthandy, Redruth, TR16 5JL
2W0 BOK A. Budding, 54 Wern Isaf, Dowlais, Merthyr Tydfil, CF48 3NY
2E0 BON J. Ball, Manor House, Tolgus Hill, Redruth, TR15 1AX
2E0 BOR R. Gee, Flat 1D, Quarmby Road, Huddersfield, HD3 4HQ
2M0 BOS S. Mccurdy, 5 Kestrel Place, Greenock, PA16 7BL
2E0 BOT A. Canning, The Mount, Birmingham Road, Alcester, B49 5EG
2E0 BOV T. Liu, 52 Tenison Road, Cambridge, CB1 2DW
2W0 BOX C. Davis, 132 Steynton Road, Steynton, Milford Haven, SA73 1AN
2M0 BOY S. Kirkpatrick, Sandyhill House Dunbae Farm, Stranraer, DG9 8LX

2E0 BOZ D. Tidswell, 1 Cherrytree Grove, Spalding, PE11 2NA
2E0 BPE J. Freeman, 38 City Road, Cambridge, CB1 1DP
2E0 BPF F. Banner, 14 Dale Road, Barnard Castle, DL12 8LQ
2E0 BPG B. Dixon, 21 Pankhurst Road, Hoo, Rochester, ME3 9DF
2E0 BPI A. Schuler, 6 Tatham Court, Taunton, TA1 5QZ
2W0 BPJ R. Jones, 5 Heol Llwyn Gollen, Merthyr Tydfil, CF48 1LR
2E0 BPL D. Golding, Windrush Cottage 84-85 Bradenstoke, Chippenham, SN15 4EL
2M0 BPM M. Dickeson, 44 Mossmill Park, Mosstodloch, Fochabers, IV32 7JY
2E0 BPN J. Barker, 26 Ardley Road, Fewcott, Bicester, OX27 7PA
2I0 BPO J. Rice, 42 The Crescent, Ballymoney, BT53 6ES
2E0 BPP C. Bell, 20 Kingfisher Close, Blackburn, BB1 8NS
2E0 BPS J. Murray, 2 The Cuttings Hampstead Norreys, Thatcham, RG18 0RR
2E0 BPT L. Clark, 16 Kibblewhite Crescent, Twyford, Reading, RG10 9AX
2E0 BPU N. Phillips, First Floor Flat, 116 Lodge Road, Croydon, CR0 2PF
2M0 BPV N. Davidson, 25 Hopetoun Court, Bucksburn, Aberdeen, AB21 9QS
2E0 BPW J. Hall, 1 Nash Close, Earley, Reading, RG6 5SL
2E0 BPX D. Adshead, 16 Moat Way, Swavesey, Cambridge, CB24 4TR
2E0 BPY M. Rogers, 41 Barton Hill Drive, Minster On Sea, Sheerness, ME12 3NF
2E0 BQB T. Horsoo, 5 Kelmarsh Court, Great Holm, Milton Keynes, MK8 9EN
2E0 BQC K. Bindley, 56 Iona Close, Beaumont Leys, Leicester, LE4 0QY
2E0 BQE J. Callis, 51 Pipistrelle Way, Oadby, Leicester, LE2 4QA
2E0 BQF D. Gunn, 40 The Pastures, Oadby, Leicester, LE2 4QD
2E0 BQG S. Turner, 12 Park Street, Morecambe, LA4 6BN
2E0 BQH M. Bailey, 17 Sparrowhawk Way, Hartford, Huntingdon, PE29 1XE
2E0 BQJ D. Trudgian, 18 Hart Close, Wootton Bassett, Swindon, SN4 7FN
2E0 BQK M. Carr, 25 Malvern Avenue, Fareham, PO14 1QF
2E0 BQL B. Hall, 64 Synehurst Crescent, Badsey, Evesham, WR11 7XX
2E0 BQM N. Jewitt, 10 Gorse Lane, Oadby, Leicester, LE2 4RQ
2E0 BQN R. Suffling, 9 Tamerton Square, Woking, GU22 7SZ
2E0 BQO D. Green, 12 Nostell Road, Ashton-In-Makerfield, Wigan, WN4 9XD
2E0 BQP K. Mills, 6A Deacons Walk, Schofield Street, Mexborough, S64 9NH
2E0 BQQ J. Addy, 12 Wortley Avenue, Swinton, Mexborough, S64 8PT
2E0 BQR S. James, 94 North Road, Withernsea, HU19 2AY
2E0 BQU C. Langmaid, Flat 4, Woodlawn High Street, Partridge Green West Sussex, RH13 8HR
2E0 BQV J. Parrett, 6 Shelley Road, East Grinstead, RH19 1TA
2E0 BQW A. Chapman, 24 Eaton Grange Drive, Long Eaton, Nottingham, NG10 3QE
2E0 BQX N. Groat, 138 Freedom Road, Sheffield, S6 2XE
2E0 BQY A. Clinchant, 14 Taylor Close Norton Fitzwarren, Taunton, TA2 6TA
2E0 BQZ K. Puttock, 12 Beechfields, School Lane, Petworth, GU28 9DH
2E0 BRC E. Smith, 13 Eagle Avenue, Waterlooville, PO8 9UB
2M0 BRD B. Donnelly, 19 Douglas Drive, Dunfermline, KY12 9YG
2E0 BRF C. Dawson, 9 Mulberry Close, Poringland, Norwich, NR14 7WF
2M0 BRH C. Jones, Croy Lodge, Shandon, Helensburgh, G84 8NN
2E0 BRI B. Bott, 15 Lansdowne Crescent, Darton, Barnsley, S75 5PW
2E0 BRJ P. Shirlaw, 32 West Street, Faversham, ME13 7JG
2E0 BRK B. Kemp, 4 Creek View, Basildon, SS16 4RU
2E0 BRL R. Dewis, 6 St Nicolas Close, Pevensey, BN245LB
2E0 BRP G. Armitage, Windmill Cottage, Greens Gardens, Nottingham, NG2 4QD
2E0 BRQ K. Shaw, 2 Montrose Avenue, Montrose Street, Hull, HU8 7RY
2E0 BRS B. Simmonds, 55 Pepys Road, St Neots, PE19 2EN
2E0 BRT P. Allen, 21 Chase Vale, Burntwood, WS7 3GD
2E0 BRU B. Pickering, 7 Front Street Grindale, Bridlington, YO16 4XU
2E0 BRY M. Edmond, 12 Yeoman Close, Worksop, S80 2RR
2I0 BSA D. Cooke, 7 Killyclooney Road, Dunamanagh, Strabane, BT82 0LZ
2E0 BSB T. Wooldridge, 12 Redwood Avenue, Leyland, PR25 1RN
2M0 BSE R. Ewing, 7 Middlemas Drive, Kilmarnock, KA1 3DZ
2E0 BSF B. Streeter, Fairway, West Chiltington Road, Pulborough, RH20 2EE
2E0 BSG B. Holland, 11 Silverlands Park, Buxton, SK17 6QX
2I0 BSH R. Vage, 80 Chinauley Park, Banbridge, BT32 4JL
2E0 BSI Dr C. Ferguson, Royd Moor, Royd Moor Lane, Badsworth, Pontefract, WF9 1AZ
2E0 BSJ M. Croxford Simmons, 37 Queens Road, Askern, Doncaster, DN6 0LU
2E0 BSK S. Everson, 41 Westminster Lane, Newport, PO30 5ZF
2E0 BSM T. Hall, 18 Common Lane New Haw 18 Comman Lane New Haw, Addlestone, kt153lh
2E0 BSN M. Blagg, 17 Flint Avenue, Forest Town, Mansfield, NG19 0DS
2E0 BSQ D. Gray, 68 Endeavour Way, Hythe Marina Village, Southampton, SO45 6LA
2E0 BSR R. Gay, 47 Egerton Street, Flat C, Chester, CH1 3ND
2E0 BSS Lady C. Windsor, 44 Paragon Place, Norwich, NR2 4BL
2E0 BST S. Clay, Akers Lodge, 6 Penn Way, Rickmansworth, WD3 5HQ
2E0 BSU J. Grosvenor, 10 Neves Close, Lingwood, Norwich, NR13 4AW
2E0 BSW C. Godwin, 6, Red Earl Lane, Malvern, WR142ST
2E0 BSX I. Colvin, 80 Silvester Road, Cowplain, Waterlooville, PO8 8TS
2E0 BTA C. Wynne, 43 Lansdown Road, Broughton, Chester, CH4 0NZ
2E0 BTB B. Bowen, 2 Veronica Road, Manchester, M20 6SU
2E0 BTD J. Wynne, 43 Lansdown Road, Broughton, Chester, CH4 0NZ
2W0 BTE J. Davies, 122 Heol Frank, Penlan, Swansea, SA5 7EG
2W0 BTF T. Banks, 18 Leicester Road, Newport, NP19 7ER
2E0 BTO A. King, 6 Dunsfold Close, Crawley, RH11 8EY
2W0 BTP B. Page, 9 De Braose Close, Cardiff, CF5 2DH
2W0 BTQ T. Rogers, 59 Park Place, Newport, NP11 6BN
2E0 BTR A. Passey, 3 The Yard, Bayton, Kidderminster, DY14 9LH
2E0 BTS G. Tyler, Crofton, Stoney Ley, Worcester, WR6 5NG
2I0 BTT R. Laverty, 23 Hyacinth Avenue, Ballykelly, Limavady, BT49 9HT
2E0 BTU J. Katz, 8 Astor Drive, Birmingham, B13 9QR
2E0 BTV J. Barbieri, 20 Gilbard Court, Chineham, Basingstoke, RG24 8RG
2E0 BTW D. Richards, 73 Greenfields Avenue, Alton, GU34 2EW
2E0 BTX P. Chamberlain, 22 Stanedge Grove, Wigan, WN3 5PL
2W0 BTZ D. Shipton, Hillside, Tintern, Chepstow, NP16 6TF
2E0 BUA J. Watson, 20 St. Marys Gardens, Hilperton Marsh, Trowbridge, BA14 7PG
2E0 BUD A. Hanson, Pilgrim Cottage, South Road, Truro, TR3 7AD
2E0 BUE S. Mather, 35 Neargates Charnock Richard, Chorley, PR7 5EY
2E0 BUF N. Ham, 4 Heighes Drive, Alton, GU34 2fj
2E0 BUI S. Chuter, 32 Hill Top Drive, Harrogate, HG1 3BU
2E0 BUJ C. Cherry, 12 Scarisbrick New Road, Southport, PR8 6PY
2E0 BUK M. Buckland, 7 Heath Close, Newport, PO30 1HN
2E0 BUN L. Knight, Flat 4, Barrington House, Portsmouth, PO2 7DD
2E0 BUO B. Jenson, 10 Tintern Close, Paulsgrove, Portsmouth, PO6 4LS
2E0 BUP G. Knight, 131 Washington Road, Portsmouth, PO2 7DF
2W0 BUQ S. Gibbon, 39 Pen-Y-Groes, Penyrheol, Caerphilly, CF83 2JL
2M0 BUT F. Pudsey, 21/2 Bathfield, Edinburgh, EH6 4DU
2E0 BUV A. Markettos, 98 Foster Road, Trumpington, Cambridge, CB2 9JR
2M0 BUX A. Young, 4/4 Prestonfield Terrace, Edinburgh, eh165ee
2M0 BUY Dr J. Henderson, 7 Rowanhill Close, Port Seton, Prestonpans, EH32 0SY
2E0 BUZ D. Burrows, 19 Fleming Avenue, Bottesford, Nottingham, NG13 0ED
2E0 BVB J. Roberts, Worlds Wonder, Warehorne, Ashford, TN26 2LU
2E0 BVD A. Burns, 76, 76, Morprth, NE65 0TF
2E0 BVE J. Mcbride, 65 Dunston Road, Hull, HU5 5ES
2E0 BVG J. Thomas, 42 Brant Road, Lincoln, Ln5 8sj
2E0 BVH M. Norris, 35 Sudbrooke Road, London, SW12 8TQ
2E0 BVJ W. Tunstall, 89 Lever Street, Little Lever, Bolton, BL3 1BA
2W0 BVK A. Bennett, Erwenni, Llanbedrog, Pwllheli, LL53 7PA
2M0 BVN I. Hepworth, Bronte Cottage, Inverugie, Peterhead, AB42 3DN
2E0 BVO A. Little, Wisteria Cottage, 7 Shawfield Road, Havant, PO9 2SY
2E0 BVQ G. Turner, 28 Chapel Close, Needingworth, St Ives, PE27 4SH
2E0 BVR N. Breckons, 5 Berrybut Way, Stamford, PE9 1DS
2W0 BVS J. Smith, 15 Harvey Crescent, Aberavon, Port Talbot, SA12 6DF
2E0 BVT M. Elkington, 32 The Knoll, Kingswinford, DY68JT
2E0 BVU T. Loker, 24 St. Albans Hill, Hemel Hempstead, HP3 9NG
2W0 BVV T. Leaworthy, 7 Maesderwen Rise, Stafford Road, Pontypool, NP4 5SS
2E0 BVZ J. Harris, Flat 3 49 Enys Road, Eastbourne, BN21 2DN
2E0 BWA L. Garnett, 6 Tremaine Close, Norwich, NR6 5EL
2E0 BWD E. Mehmet, 8 Hailsham Road, Tooting, London, SW17 9EN
2E0 BWH B. Harrison, 24 Alderton Road, Nottingham, NG5 6DX
2E0 BWI A. Tring, 12 Ainsdale Close, Orpington, BR6 8DJ
2E0 BWJ J. Swift, 56 Leymoor Road, Huddersfield, HD3 4SW
2E0 BWK K. Bushell, 4 Birch Grove, Harrogate, HG1 4HR
2E0 BWL R. Lane, 9 Hartoft Road, Hull, HU5 4JZ
2E0 BWN C. Mason, Apartment 23, Burnside House, Carleton Road, Skipton, BD23 2BE
2E0 BWP D. Le Mare, The Sycamore, Church Bank, Barnard Castle, DL12 0AH
2E0 BWQ B. North, 54 Parklands, Mablethorpe, LN12 1BY
2E0 BWT A. Cockett, 10 San Marcos Drive, Chafford Hundred, Grays, RM16 6LT
2E0 BWU P. Moule, 30 Hillview Road, Chelmsford, CM1 7RX
2E0 BWV D. Jones, 77 Brinkburn Grove, Banbury, OX16 3WX
2E0 BWX T. Ward, 20 Ollerton Road, Edwinstowe, Mansfield, NG21 9QG
2E0 BWY A. Ferenc, 2A Rosedene Avenue, London, SW16 2LT
2E0 BXA J. Oakley, 59 Bewsey Street, Warrington, WA2 7JQ
2E0 BXC M. Wilkins, 31 Stratton Audley Road, Fringford, Bicester, OX27 8ED
2E0 BXD J. Bell, 8 Firsleigh Park, Roche, St Austell, PL26 8JN
2E0 BXE R. Steele, 175 Vale Road, Seaford, BN25 3HH
2E0 BXF P. Freeman, 24 Roe Green Close, Hatfield, AL10 9PE
2E0 BXG H. Hughes, 27 The Holt, Hailsham, BN27 3ND
2I0 BXJ J. Steele, 46 Circular Road, Newtownards, BT23 4BN
2E0 BXK R. Hope, 32 Winstanley Place, Rugeley, WS15 2QB
2E0 BXM K. Foster, 10 Bleaswood Road Oxenholme, Kendal, LA9 7EY
2M0 BXN G. Moir, 3 Three Wells Steading, Inverbervie, DD10 0PH
2E0 BXQ L. Denham, 92 Windermere Avenue, Southampton, SO16 9GF
2E0 BXS B. Sims, 4 New Cottages, Cranwich Road, Thetford, IP26 5EQ
2E0 BXV M. Browne, 60 Lindsay Avenue, Abington, Northampton, NN3 2LP
2E0 BXW S. Johnston, 67 Eversfield Road, Horsham, RH13 5JS
2E0 BXX J. Searle, 2 Tukes Avenue, Gosport, PO13 0SE
2M0 BXY W. Taylor, Garth Wood, Fishers Brae, Eyemouth, TD14 5NJ
2E0 BXZ J. Reynolds, 15 Chestnut Mead, Oxford Road, Redhill, RH1 1DR
2E0 BYA D. Passey, 5 The Croftings, Felton Close, Ludlow, SY8 1DS
2E0 BYC M. Matthews, 28 Kempson Drive, Great Cornard, Sudbury, CO10 0YE
2E0 BYF J. Milne, 9 Roman Road, Colchester, CO1 1UR
2E0 BYG P. Bailey, 103 Jarden, Letchworth Garden City, SG6 2NZ
2E0 BYH S. Leadbetter, 11 Cogos Park, Mylor Bridge, Falmouth, TR11 5SF
2M0 BYI C. Williamson, 31 Medrox Gardens, Cumbernauld, Glasgow, G67 4AJ
2E0 BYJ E. Isaac, 162B Hitchin Road, Stotfold, Hitchin, SG5 4JE
2M0 BYK S. Troscheit, 20 James Street, St Andrews, KY16 8YA
2I0 BYL B. Crozier, 33 Cullentragh Road, Poyntzpass, Newry, BT35 6SD
2E0 BYM M. Brough, 12 Longfield Court, Barnoldswick, BB18 5LP
2E0 BYN I. Patterson, 63 Orchard Road, South Ockendon, RM15 6HP
2E0 BYO J. Morris, 96 Bradford Crescent, Durham, DH1 1HW
2E0 BYQ J. Summerhill, 43 Rangers Walk, Bristol, BS15 3PW
2M0 BYT J. Cairney, 5 James Street, Bannockburn, Stirling, FK7 0NQ
2E0 BYW M. Bradley, 55 Queensway, Penwortham, Preston, PR1 0DT
2E0 BZA K. Moulder, 51A Aston Cantlow Road, Wilmcote, Stratford-upon-Avon, CV37 9XN
2M0 BZB S. Campbell, 11 Convener Street, New Elgin, Elgin, IV30 6BP
2E0 BZC P. Davies, 68 Sidmouth Avenue, Stafford, ST17 0HF
2E0 BZE T. Munro, 71 Zig Zag Road, Liverpool, L12 9EQ
2E0 BZG C. Warwick, 104 Church Road, Formby, Liverpool, L37 3NH
2E0 BZH R. Hopkins, 17 Springside Court, Josephs Road, Guildford, GU1 1BT
2E0 BZI J. Donnelly, 23 Pitts Croft, Neston, Corsham, SN13 9ST
2E0 BZJ B. Cooke, 2 Harvey Place, Andover, SP10 2BU
2M0 BZL C. Watkinson, The Hillock Farmhouse, Lumphanan, Banchory, AB31 4QL
2E0 BZM M. Tew, Willowell, Spring Valley Lane, Colchester, CO7 7SD
2E0 BZO D. Chatzikos, 53 Benbow Court, Shenley Church End, Milton Keynes, MK5 6JE
2E0 BZS P. Burt, 56 Winslade Road, Sidmouth, EX10 9EX
2E0 BZT P. Burgess, Tally Ho Cottage, High Street, Swindon, SN4 0AE
2E0 BZU P. Fenney, 4 Frampton Crescent, Bristol, BS16 4JA
2E0 BZV I. Govan, 9 Willowbank, Sandwich, CT13 9QA
2E0 BZX R. Crockford, 17 Tadcroft Walk, Calcot, Reading, RG31 7JR
2M0 BZZ W. Anderson, 4 Brackendene, Houston, PA6 7DE
2E0 CAA C. Orange, 79 Heath Avenue, Werrington, Stoke-on-Trent, ST9 0HU
2E0 CAD C. Dodgson, 16 Jefferson St., Goole, DN14 6SH
2E0 CAE C. Frizzell, 85 Gibbon Road, Newhaven, BN9 9ER
2E0 CAJ C. Travis, 4 Kingsdale, Worksop, S81 0XJ
2E0 CAK A. Martyn, 54 North Side, Hepthorne Lane North Wingfield, Chesterfield, S42 5hy
2E0 CAL C. Bowley, 2 Cottage, Middle Battenhall Farm, Worcester, WR5 2JL

2E0 CAO M. Castanheira, 7C Palace Road, London, N8 8QH
2E0 CAP J. Williams, 41 Overton Lane, Hammerwich, Burntwood, WS7 0LQ
2E0 CAQ D. Arnold, The Chase, Rectory Road, Penzance, TR19 6BB
2E0 CAR C. Goulding, 9 Lune Drive, Leyland, PR25 5SX
2E0 CAS C. Spence, 30 Chestnut Drive, Shirebrook, Mansfield, NG20 8NH
2E0 CAT E. Taylor, 39 Gill St., Newcastle upon Tyne, NE4 8BH
2E0 CAU D. Holmes, 5 The Cottages, Low Road, North Tuddenham, Dereham, NR20 3DG
2E0 CAV C. Vernon, 29 Alice St., Deane, Bolton, BL3 5PJ
2E0 CAW S. Court, 16 Worcester Road, Woodthorpe, Nottingham, NG5 4HY
2E0 CBA P. Humphreys, 30 The Chestnuts, Hinstock, Market Drayton, TF9 2SX
2E0 CBB B. Clements, 23 Croft Terrace, Egremont, CA22 2AT
2M0 CBC C. Bryson, 29 Roull Road, Edinburgh, EH12 7JW
2M0 CBE C. Ellison, 1 Newton Road, St. Fergus, Peterhead, AB42 3DD
2E0 CBF A. Cunningham, 18 Bradway, Whitwell, Hitchin, SG4 8BE
2E0 CBG C. Gough, 104 Canley Road, Coventry, CV5 6AR
2E0 CBH J. Wooldridge, 7 Heather Gardens, Belton, Great Yarmouth, NR31 9PP
2W0 CBJ V. Holden, 2 Anglesey Close, Tonteg, Pontypridd, CF38 1LY
2E0 CBL I. Botham, 12 Lairgill, Bentham, Lancaster, LA2 7JZ
2E0 CBO P. Rogers, 16 Begonia Close, Basingstoke, RG22 5RA
2E0 CBQ R. Smith, 5 Elizabeth Place, 13 Heath Road, Haywards Heath, RH16 3AX
2E0 CBT M. Hersom, Room 303, 95-98 Talbot Street, Dublin, DO1 WR94
2E0 CBU J. Horry, 5 Donington Road, Bicker, Boston, PE20 3EF
2I0 CBV A. Shilliday, 26 Iskymeadow Road, Armagh, BT60 3JS
2E0 CBX I. Connors, 3 Wheatfield Way, Chelmsford, CM1 2QZ
2W0 CBZ C. Nicholls, 26 Maes Geraint, Pentraeth, LL75 8UR
2E0 CCA C. Chambers, 21 Sullington Way, Shoreham-by-Sea, BN43 6PJ
2E0 CCB S. Connolly, 82 Cheswood Drive, Minworth, Sutton Coldfield, B76 1YE
2E0 CCC J. Restall, 1 Johndory, Dosthill, Tamworth, B77 1NY
2E0 CCD C. Raynerd, 43 Astley Hall Drive, Ramsbottom, BL0 9DF
2E0 CCF G. Wright, 2 Hillcrest Drive, Castleford, WF103QN
2W0 CCG G. Brierley, 15 First Avenue, Flint, CH6 5LP
2E0 CCJ C. Jones, 6 Cleeve Park Cottages, Icknield Road, Reading, RG8 0DJ
2W0 CCK M. Buxton, 25 Pen Y Bryn, Sychdyn, Mold, CH7 6EE
2E0 CCL S. Gillard, 1 Chevening Close, Stoke Gifford, Bristol, BS34 8NJ
2E0 CCQ T. Panesar, 135A Hargate Way, Hampton Hargate, Peterborough, PE7 8FL
2E0 CCR P. Alborough, 7 Edrich Square, Andover, SP10 5BS
2E0 CCW M. Shoyer, 22 St. Andrews Road, Whitehill, Bordon, GU35 9QN
2E0 CCY F. Gear, 251 Abington Avenue, Northampton, NN3 2BU
2E0 CDF B. Myler-Cook, 11A York Street, Boston, PE21 6JN
2E0 CDI S. Lee-Ray, The Paddock, Sutton Road, Alford, LN13 9RL
2W0 CDJ C. Josey, 726 Llangyfelach Road, Treboeth, Swansea, SA5 9EL
2E0 CDK D. Barnett, 20 Middlemead Road, Great Bookham, Bookham, Leatherhead, KT23 3DA
2E0 CDL C. Lote, 8 Warren Place, Walsall, WS8 6BY
2E0 CDM C. Mcnulty, 91 Barn Hey Crescent, Meols, CH47 9RW
2M0 CDO D. Fleming, 4 Oxenrig Farm Cottage, Coldstream, TD12 4EY
2E0 CDQ W. Haddock, 5 Bradley Close, Middlewich, CW10 0PF
2E0 CDR C. Russell, 255 Leeds Road, Shipley, BD18 1EH
2E0 CDS C. Small, Riddings Barn, Hope Bagot, Ludlow, SY8 3AE
2E0 CDT C. Taylor, 1 Jasmine Gardens, Warrington, WA5 1GU
2E0 CDU M. Hall, 29 The Spinney, Finchampstead, Wokingham, RG40 4UN
2E0 CDV A. Drew, 51 Hobart Road, Cambridge, CB1 3PT
2E0 CDW C. Wade, 31 Melton Green, Wath-Upon-Dearne, Rotherham, S63 6AA
2W0 CDZ P. Gough, 92 Pendwyallt Road, Whitchurch, Cardiff, CF14 7EH
2E0 CEA M. Anderson, 27 Laing Road, Colchester, CO4 3UT
2E0 CEB B. Smithers, 4 Bidmead Court Kent Way, Surbiton, KT6 7SX
2W0 CED C. Davies, 3 Penhydd Houses, Oakwood Avenue, Pontrhydyfen, Port Talbot, SA12 9SE
2M0 CEE R. Renshaw, Smithy House, Scotscalder, Halkirk, KW12 6XJ
2W0 CEF F. Price, 2 Bryniau Duon Estate, Llandegfan, Menai Bridge, LL59 5PP
2E0 CEG I. Greathead, 3 Helmington Terrace, Hunwick, Crook, DL15 0LQ
2E0 CEH P. Blyth, 20 Common Lane, Beccles, NR34 9RH
2I0 CEI B. Craney, 8A Drumhoy Drive, Carrickfergus, BT38 8NN
2E0 CEJ R. Parrish, 5 Kestrel Lane, Cheadle, Stoke-on-Trent, ST10 1RU
2E0 CEK D. Mcauslan, Casa Arco Iris, Via Variante Nascente, 8005-491, Portugal, SANTA BARBARA DE NEX
2E0 CEM M. Miller, Barn Cottage, Wingfield Hall, Manor Road, Alfreton, DE55 7NH
2E0 CEN D. Mctaggart, 59 Gainsborough Road, Richmond, TW9 2DZ
2W0 CEO C. Gozzard, Craig Dulas, Rhydyfoel Road, Llanddulas, Abergele, LL22 8EG
2E0 CEP Dr J. Pelham, 5 The Crescent, Shortstown, MK42 0UJ
2E0 CER M. Everitt, 10 Morris Close, Hatherleigh, Okehampton, EX20 3NX
2E0 CES J. Brawn, 5 Downs Cote View, Westbury On Trym, Bristol, BS9 3TU
2E0 CEU Dr N. Hoare, 5 Kelsey Head, Port Solent, Portsmouth, PO6 4TA
2M0 CEX P. Rice, 255 Eskhill, Penicuik, EH26 8DF
2E0 CEY T. Edwards, 17 The Green, Woodbastwick, Norwich, NR13 6HH
2M0 CFA R. Mannifield, 2 Plewlands Avenue, Edinburgh, EH10 5JY
2M0 CFB I. Watson, 10 Christie Place, Elgin, IV30 4HX
2E0 CFD A. Mcneil, 2 Palmerston Crescent, Liverpool, L19 1RB
2E0 CFE T. Carpenter, 11 Castle Road, Southwick, Fareham, PO17 6EY
2E0 CFG F. Fang, 49 Calthrope Road, Norwich, NR5 8RN
2E0 CFH A. Andrew, 80 Hamble Drive, Abingdon, OX14 3TE
2E0 CFI M. Summers, 21 Quantock Avenue, Caversham, Reading, RG4 6PY
2E0 CFK P. Rimmington, 28 Skipton Road, Swallownest, Sheffield, S26 4NQ
2E0 CFL A. Norville, 137 Foster Road, Trumpington, Cambridge, CB2 9JW
2E0 CFM E. Vaughan, 10 Evans Close, Manchester, M20 2SQ
2E0 CFN A. Cornelius, 16 Crown House, North Street, Nailsea, Bristol, BS48 4SX
2E0 CFP J. Hitchens, 57 Batchelor Way, Downton, Salisbury, SP5 3FN
2E0 CFQ J. Hunt, 14 Nevill Close, Hanslope, Milton Keynes, MK19 7NY
2E0 CFT A. Wells, 186 Manford Way, Chigwell, IG7 4DG
2E0 CFV E. Beever, 160 Granby Road, Buxton, SK17 7TA
2I0 CFW D. Mcglone, 10 O'Neill Terrace, Dromore, Omagh, BT78 3AW
2E0 CFY A. Davis, 101 Kenn Road, Clevedon, BS21 6JE
2D0 CFZ C. Ingles, 1 Hillberry View, Onchan, Isle Of Man, IM3 3GB
2E0 CGB G. Grimshaw, 1 Hardy Close, Pinner, HA5 1NL
2M0 CGE A. Macdonald, 1 Edinmore Cottage, Rothesay, Isle of Bute, PA20 0QT
2E0 CGG S. Brown, 4 Dorado Gardens, Orpington, BR6 7TD
2E0 CGH W. Durrant, 19 Rydal Rd, Gosport, PO12 4ES
2E0 CGI J. Garrard, 40 Wright Crescent, Bridlington, yo164rg
2E0 CGJ D. Ramsell, 36 West Street, Burton-on-Trent, DE15 0BW
2E0 CGK N. Allen, 59 Sherborne Road, Chichester, PO19 3AN
2E0 CGL C. Lewis, 3 Sovereign Way, Calcot, Reading, RG31 4US
2W0 CGM N. Wells, 52 Lowerdale Drive, Llantrisant, Pontyclun, CF72 8DY
2E0 CGP I. Duffie, Trebeighan Farm, Saltash, PL12 5AE
2E0 CGR M. Cashman, Flat 3, Linden Court Linden Road, Romsey, SO51 8BR
2E0 CGS J. Goodyear, 30 Ashburton Road, Alresford, SO24 9HH
2E0 CGT J. Russell, 81 Chapman Street, Loughborough, LE11 1DD
2E0 CGU I. Albrighton, 12 Clewley Road, Branston, Burton-on-Trent, DE14 3JE
2E0 CGV D. Ingrey, 1 Ponders Road, Fordham, Colchester, CO6 3LX
2E0 CGW N. Bown, 14 Parsons Mead, Abingdon, OX14 1LS
2E0 CGX J. Shufflebotham, 316 Stockport Road, Hyde, SK14 5RU
2I0 CGZ D. Mudd, 14 Bloomfield Road, Belfast, BT5 5LT
2E0 CHA C. Nakajima, 22 Royds Crescent, Rhodesia, Worksop, S80 3HF
2W0 CHH C. Hill, 9 Oliver Road, Newport, NP19 0HU
2E0 CHK A. Wild, 181 The Strand, Goring-By-Sea, Worthing, BN12 6DY
2E0 CHL T. Chapman, 1 East Dean Road, Lockerley, Romsey, SO51 0JL
2E0 CHN P. Gale, 37 Hazlebury Road, Poole, BH17 7AX
2E0 CHQ S. Drake-Brockman, 13 St. Johns Place, Bury St Edmunds, IP33 1SW
2E0 CHT Dr M. Cook, 14 Speyside Close, Carterton, OX18 1TT
2E0 CHU G. Hatt, 4H Colman House, Earlham Road, Norwich, NR4 7TJ
2W0 CHV S. Taylor, 43 Toronnen, Bangor, LL57 4TG
2E0 CHW C. West, 1 Willetts Mews, Hoddesdon, EN11 9DX
2E0 CHY G. Lyon, 1 Eckersley Street, Wigan, WN1 3PP
2E0 CHZ C. Jewell, 3 Marsh Gate, Clee St. Margaret, Craven Arms, SY7 9DU
2E0 CIA D. Campanario, 3 Foxearth Hall, Leek Road, Stoke on Trent, ST9 0DG
2E0 CID K. Percy, 55 Buxton Avenue, Heanor, DE75 7UN
2E0 CIG P. Holland, 30 Knighton Park Road, London, SE26 5RJ
2E0 CIH M. Verrechia, 20 The Wyvern Grafham, Huntingdon, PE28 0GG
2E0 CII T. Pearsall, 16 Langdale Road, Leyland, PR25 3AR
2E0 CIJ D. Simpson, 50 Castle Hill, Berkhamsted, HP4 1HF
2E0 CIK C. Storr, 40 Weelsby Way, Hessle, HU13 0JW
2E0 CIM S. Preece, 14 Bettespol Meadows, Redbourn, St Albans, AL3 7EW
2E0 CIN A. Mawson, 14 Pontop View, Delves Lane, Consett, DH8 7JB
2E0 CIQ J. Chapman, South View, Mill End Rushden, Buntingford, SG9 0SU
2E0 CIR I. Sapstead, 7 Shrubbery Grove, Royston, SG8 9LJ
2E0 CIS D. Woodbine, 29 Compass Tower, Munnings Road, Norwich, NR7 9TW
2E0 CIT J. O'Brian, 83 Bramdean Crescent, London, SE12 0UJ
2W0 CIV M. Ireland, Pen Y Gadlas, Ffordd Bryniau, Prestatyn, LL19 8RD
2E0 CIX J. Rowsell, 55 Scarlet Oaks, Camberley, GU15 1RD
2E0 CJA L. Mcgaughey, 1 Marsh Cottage, Keyingham Marsh, Keyingham, Hull, HU12 9JZ
2E0 CJB C. Beresford, 13 Chaseside Avenue, Twyford, Reading, RG10 9BT
2E0 CJD C. Eyre, 23 Nelson Street, Congleton, CW12 4BS
2E0 CJF K. Lambert, 38 Whittleford Road, Nuneaton, CV10 9HU
2W0 CJG J. Loughlin, 453 Heol-Y-Waun, Penrhys, Ferndale, CF43 3NW
2W0 CJJ R. Squires, Hillcrest, Pontfadog, Llangollen, LL20 7AS
2E0 CJK P. Mullen, 14 Anderson Road, Hemswell Cliff, Gainsborough, DN21 5XP
2E0 CJM K. Nicholson, 11 Lancaster Way, Skellingthorpe, Lincoln, LN6 5UF
2E0 CJO M. Reynolds, 24 Burton Close, Corringham, Stanford-le-Hope, SS17 7SB
2E0 CJP C. Price, 10 St. James Park, Lower Milkwall, Coleford, GL16 7LG
2E0 CJQ T. Willis, 143 Clarence House, Leeds, LS10 1LH
2E0 CJS P. Hanman, 7 Tremenheere Road, Penzance, TR18 2AH
2E0 CJV A. Sharam, 30 Heywood Avenue, Maidenhead, SL6 3JA
2E0 CJW C. Wright, Home Farm, Bretford, CV230LB
2W0 CJZ J. Smith, 38 Marshfield Street, Newport, NP19 0GX
2I0 CKB C. Kelly, 7A Bancran Road, Draperstown, Magherafelt, BT45 7DT
2E0 CKC A. Bradshaw, 130 Low Lane, Morecambe, LA4 6PS
2E0 CKE A. Cooke, Cooke Towers, 9 Nyetimber Crescent, Pagham, PO213NN
2E0 CKI A. Pickles, 87A Laburnum Road, Waterlooville, PO7 7EW
2E0 CKJ P. Wright, 16 Hainault Avenue, Giffard Park, Milton Keynes, MK14 5PA
2E0 CKL M. Atherton, 4 Bakers Park Saltney, Chester, CH4 8FB
2E0 CKM A. Pawlak, 8 Healey Close, Crewe, CW1 4RS
2I0 CKN M. Allen, 48 Kevlin Gardens, Omagh, BT78 1QS
2E0 CKO Dr B. Le Page, Apple Tree House, Reading Road North, Fleet, GU51 4AG
2E0 CKP C. Prior, 38 Windmill Road, Wombwell, Barnsley, S73 8PP
2E0 CKQ J. Legrain, 17 Route De La Cote, St Laurent Sur Gorre, France, 87310
2E0 CKR B. Cox, 7 Wolsey Avenue, London, E6 6HG
2E0 CKS A. Hickey, 144 Gisburn Road, Barnoldswick, BB18 5LQ
2E0 CKT B. Cunningham, 33 Barry Street, Burnley, BB12 6DT
2W0 CKV E. Price, 1 Brynderi, Pontyates, Llanelli, SA15 5SU
2E0 CKW R. Pringle, 14 Marjorie Street, Cramlington, NE23 6XQ
2E0 CKX Dr L. Schuy, 32 Sudeley Street, Brighton, BN2 1HE
2E0 CKY A. Barrett, 4 Wood Cottages, Cummings Cross, Liverton, Newton Abbot, TQ12 6HJ
2E0 CLD M. Shopland, 128 Whitewood Park, Liverpool, L9 7LG
2E0 CLE C. Edgson, 59 Gilmour Crescent, Worcester, WR3 7PJ
2M0 CLF C. Forsyth, 14B Osborne Terrace, Edinburgh, EH12 5HG
2E0 CLH S. Saunders, 5 Park Court, Woking, GU22 7NW
2E0 CLI M. Randle, 2 Oak Gardens, White Farm Road, Sutton Coldfield B74 4Lq, Birmingham, B74 4LQ
2W0 CLJ C. Jones, 33 Graig Ebbw, Rassau, Ebbw Vale, NP23 5SF
2E0 CLL C. Lee, 46 Little Lane, Huthwaite, Sutton-in-Ashfield, NG17 2RA
2M0 CLN C. Cosgrove, 62 Cowgate, Tayport, DD6 9DT
2E0 CLP R. Smith, Five Elms, Lullington Road Edingale, Tamworth, B79 9JA
2I0 CLS G. O'Neill, 46 Ashgrove Road, Newtownabbey, BT36 6LJ
2W0 CLT G. Williams, 36 Park Street, Taibach, Port Talbot, SA13 1TD
2M0 CLU R. Adamson, 6 Camdean Crescent, Rosyth, Dunfermline, KY11 2TJ
2E0 CLW D. Clewer, 45 Ashfield Road, Andover, SP10 3PE
2E0 CLY A. Page, 207 Brooklyn Road, Cheltenham, GL51 8DZ
2E0 CLZ I. Jones, 21 Kennet Green, Worcester, WR5 1JQ
2M0 CMA A. Campbell, 1B Craig Road, Troon, KA10 6DA
2E0 CMC A. Applegate, 13 Deacons Close, Kings Stanley, Stonehouse, GL10 3GA
2E0 CMD B. Eames, 22 Ashgrove Close, Hardwicke, Gloucester, GL2 4RT
2E0 CME R. Boardman, 12 St. Margarets Road, Alderton, Tewkesbury, GL20 8NN
2E0 CMF S. Mason, 8 Barrowby Gate., Grantham, NG31 7LT
2E0 CMH A. Brown, 3 Alston Road, New Hartley, Whitley Bay, NE25 0ST

2E0 CMK C. Norris, 115 Sutton Road, Walpole Cross Keys, PE344HE
2E0 CMN N. Curtis, 102 Henley Meadows, Tenterden, TN30 6EQ
2E0 CMO C. Overton, 99 Hope Avenue, Goldthorpe, Rotherham, S63 9DZ
2E0 CMP C. Pegrum, 3 Bretland Road, Tunbridge Wells, TN4 8PS
2E0 CMQ P. Chester, 33 Salehurst Road, London, SE4 1AS
2E0 CMT S. Metcalfe, 55 Coventry Close, Corfe Mullen, Wimborne, BH21 3UW
2E0 CMY F. Southgate, 52 Jeffrey Lane, Belton, Doncaster, DN9 1LT
2E0 CMZ C. Rawlin, 5 Japonica Hill, Immingham, DN40 1LT
2E0 CNA P. Davies, 2 Lynfords Drive, Runwell, Wickford, SS11 7PP
2E0 CNB Dr D. James, Bramble Cottage, Tray Lane, Atherington, Umberleigh, EX37 9HY
2E0 CNC T. Ward, 1 Darrismere Villas, Edinburgh Street, Hull, HU3 5AS
2E0 CND S. Bradley, 6 Downing Street, South Normanton, Alfreton, DE55 2HE
2E0 CNE M. Shepherd, North Waver Cottage, Bells Road Belchamp Walter, Sudbury, CO10 7AR
2E0 CNG C. Smith, 44 Brooksfield, Bildeston, Ipswich, IP7 7EJ
2E0 CNH N. Sinclair, 8 Kareen Avenue, Scarborough, YO12 4LD
2E0 CNL C. Lockyear, 26 Wentworth Gardens, Exeter, EX4 1NH
2E0 CNM C. Hunt, 105 Worlds End Lane, Weston Turville, Aylesbury, HP22 5RX
2E0 CNN D. Turnbull, 63 Brecklands, Mundford, Thetford, IP26 5EG
2E0 CNP F. Hatfull, 16B Church Street, Easton On The Hill, Stamford, PE9 3LL
2E0 CNQ C. Greenwood, 21 Valley Drive, Thornhill Dewsbury, wf120he
2M0 CNS M. Hatfull, Flat 2 1 Northumberland Place Lane, Edinburgh, EH3 6LD
2E0 CNU T. Cocks, 9 Mountfield Way, Westgate-on-Sea, CT8 8HR
2E0 CNV T. Pelham, 172A Gloucester Road, Patchway, Bristol, BS34 5BG
2E0 CNW A. Palmer, 18 Windsor Avenue, Great Yarmouth, NR30 4EA
2E0 CNX I. Buckton, 67 Tennyson Avenue, Middlesbrough, TS6 7ND
2W0 CNY G. Fryer, 9 Church Road, Chepstow, NP16 5HP
2M0 CNZ J. Rayne, 8 Bankton Grove, Livingston, EH54 9DW
2E0 COA L. Layland, 3 Thirlmere Road, Golborne, Warrington, WA3 3HH
2E0 COB J. Cobbold, 2 The Green, Blencogo, Wigton, CA7 0DF
2E0 COD G. Glasgow, 4 Beech Avenue, Culcheth, Warrington, WA3 4JF
2E0 COF Dr J. Murray, 69 Helsby Road, Lincoln, LN5 8SN
2E0 COI P. Bailey, 81A Kings Parade Holland-On-Sea, Clacton-on-Sea, CO15 5JF
2E0 COJ K. Cartwright, 53 Sedgley Road, Dudley, DY1 4NE
2E0 COL Dr S. Britt-Hazard, 9 Serbin Close, London, E10 6JL
2E0 COM R. Hammond, Bye Road Cottage Pot Kiln Chase Co93Bh, Gestingthorpe, Co93bh
2E0 CON J. Middleton, 16 Kyme Road, Boston, PE21 8NQ
2J0 COQ L. Langlois, Brookfield, La Rue D'Empierre, Trinity, Jersey, JE3 5QF
2M0 COT F. Gordon, Crofts Of Torrancroy, Strathdon, AB36 8UJ
2E0 COV V. Hopkins, 6 Daimler Road, Coventry, CV6 3GD
2E0 COX A. Cox, 11 Windmill Drive, Audlem, Crewe, CW3 0BE
2E0 COY A. Pointon, 9 Parkwood Avenue, Stoke-on-Trent, ST4 8PD
2E0 CPB C. Bond, Tryfan, Vicarage Lane, Neston, CH64 5TJ
2W0 CPD R. Briant, Talarvor, Llanon, SY23 5HG
2E0 CPE M. Phillips, 59 Bradeley Road Haslington Crewe, Crewe, CW1 5PX
2E0 CPF G. Evans, 13 Lydgate Road, Sale, M33 3LW
2E0 CPG C. Leviston, 13 Pryors Walk, Askam-in-Furness, LA16 7JG
2E0 CPK K. Jackson, 4 Milfoil Close, Marton-In-Cleveland, Middlesbrough, TS7 8SE
2E0 CPL C. Luckett, 257 Folkestone Road, Dover, CT17 9LL
2M0 CPN A. Woodford, Nordkette, Evnabrek, Levenwick, Shetland, ZE2 9GY
2E0 CPP A. Collins, 4 The Avenue, London, W4 1HT
2E0 CPR R. Packman, 62 Amsterdam Road, London, E14 3JB
2E0 CPS W. Coburn, 42 Hinton Wood Avenue, Christchurch, BH23 5AH
2M0 CPV J. Hutchinson, Hawthorn Cottage, Muirhall, West Calder, EH55 8NL
2E0 CPX P. Strickland, 100 Spitfire Road, Castle Donington, Derby, DE74 2AU
2E0 CQB I. Mckean, 14 Maltings Close, Cranfield, MK43 0EL
2E0 CQC D. Clavey, 32 Apollo Close, Dunstable, LU5 4AQ
2E0 CQD A. Kennington, 63 Emmanuel Court, Scunthorpe, DN16 2LR
2E0 CQG P. Jones, Tallonhouse, Mill Lane, Pulham St. Mary, Diss, IP21 4QY
2E0 CQH S. Baynton, 50 Briton Way, Wymondham, NR18 0TT
2M0 CQI J. Browne, 33 Pilgrims Hill, Linlithgow, EH49 7LN
2E0 CQJ N. Parry, 125 Lawsons Road, Thornton-Cleveleys, FY5 4PL
2E0 CQL K. Jeffery, 9 Gordon Road, Tunbridge Wells, TN4 9BL
2E0 CQM C. Harding, 27 Eston Avenue, Malvern, WR14 2SR
2E0 CQN B. Mcglynn, 22 Bracken Bank Way, Keighley, BD22 7AB
2E0 CQO M. Mutkin, 13 The Grove, Radlett, WD7 7NF
2E0 CQQ I. Pryke, 9 Charles Avenue, Grundisburgh, Woodbridge, IP13 6TH
2E0 CQR P. Moye, 13 Post Mill Gardens, Grundisburgh, Woodbridge, IP13 6UP
2E0 CQS P. Elsey, 62B Coleraine Road, London, SE3 7PE
2E0 CQT Dr F. Agboma, Flat 6, Mayfair Court, Edgware, HA8 7UH
2E0 CQX A. Davis, Old Malt Kiln House, Barden, Leyburn, DL8 5JS
2E0 CQZ J. Street, 22 Roman Acre, Wick, Littlehampton, BN17 7HN
2W0 CRB K. Dutfield-Cooke, Tan Yr Efail, Segurinside, Llandudno Junction, LL31 9QE
2E0 CRC A. Bhakoo, 4 Bryden Cottages, High Street, Uxbridge, UB8 2NY
2E0 CRD C. Densham, 27 Lloyds Crescent, Exeter, EX1 3JQ
2E0 CRH P. Bull, 87 Braemor Road, Calne, SN11 9DU
2E0 CRI J. Parris, 10, Wharfedale Grange, Ben Rhydding Drive, Ilkley, West Yorkshire, LS29 8AR
2E0 CRM C. Murly, C/O 1 Mount Pleasant, Middleton, Leeds, LS10 3TB
2E0 CRN J. Cranston, 7 Cowen Gardens, Gateshead, NE9 7TY
2M0 CRQ C. Welsh, 28 Peacock Wynd, Motherwell, ML1 4ZL
2M0 CRR C. Rodger, 23 Harrysmuir Road, Pumpherston, Livingston, EH53 0NT
2E0 CRS C. Marsh, 9 Westland Road Westwoodside, Doncaster, DN9 2PE
2E0 CRU C. Redmond, 6 Apsley Road, Southsea, PO4 8RH
2E0 CRV S. Hodder, 19 Kingsclere, Huntington, York, YO32 9SF
2E0 CRX S. Cross, 31 Parkfields, Abram, Wigan, WN2 5XR
2E0 CSD C. Davies, 3 King Alfreds Green, Leeds, LS6 4PZ
2E0 CSE T. Chapman, 12 Greenways Chilcompton, Radstock, BA3 4HT
2E0 CSF C. Finnis, 44 Disraeli Road, Christchurch, BH23 3NB
2E0 CSG D. Pollard, 8 Drammen Avenue, Burnley, BB11 5EA
2E0 CSH G. Webster, 15 Bridge Road, Chichester, PO19 7NW
2E0 CSJ M. Catchpole, Woodcote, Five Oaks Road, Horsham, RH13 0RQ
2E0 CSK D. Rudling, Rose Cottage, Ludwells Lane, Southampton, SO32 2NP
2E0 CSL C. Lester, 21 Barwell Way, Witham, CM8 2TY
2E0 CSN G. Flinn, 38 Fir Grove, Whitehill, Bordon, GU35 9ED
2E0 CSO C. Opie, 354 Beaumont Road, Plymouth, PL4 9EN
2E0 CSQ N. Saunders, 128 St Michaels Gardens, South Petherton, TA135BQ
2E0 CSU G. Wilman, 8 Oldfield Drive, Mobberley, Knutsford, WA16 7HB
2E0 CSV A. Parker, 9 Milecastle Court, Newcastle upon Tyne, NE5 2PA
2E0 CSW B. Wilson, 119 Fountains Close, Washington, NE38 7TQ
2M0 CSX A. Thomson, 5 Gib Grove, Dunfermline, KY11 8DH
2E0 CSY S. Bradley, 9 Crofton Road Crofton Road, Southsea, PO4 8NX
2M0 CSZ S. Leighton, 4 Earn Court, Alloa, FK10 1PT
2E0 CTA A. Cross, 12 Appleby Drive, Langdon Hills, Basildon, SS16 6NU
2M0 CTB W. Scott, 11 The Marches Armadale, Bathgate, EH48 2PG
2E0 CTC C. Castle, 32 Inglefield Road, Ilkeston, DE7 5AP
2E0 CTD F. Davison, 137 Hellis Wartha, Helston, TR13 8WF
2E0 CTE C. Etchells, 7 Woodlands Drive, Sandford, Wareham, BH20 7QA
2E0 CTF P. Evans, 67 Grenville Street, Stokport, SK3 9ER
2M0 CTI M. Jamieson, 5 Straid Bheag, Barremman, Helensburgh, G84 0QX
2E0 CTJ C. Johnson, 29 Linden Road Creswell, Worksop, S80 4JT
2E0 CTK J. Schleswick, 9 Wick House Close Saltford, Bristol, BS31 3BZ
2E0 CTL K. Baker, 27 St. Matthews Close, Cherry Willingham, Lincoln, LN3 4LS
2E0 CTM C. Meakin, 102 Ryknield Road, Kilburn, Belper, DE56 0PF
2M0 CTN R. Tait, 9 Fifth Avenue Claythorn, Glasgow, G12 0AS
2E0 CTO J. Killman, 19 Moorland Avenue, Walkeringham, Doncaster, DN10 4LG
2E0 CTQ T. Crew, Lower Creedy, Upton Hellions, Crediton, EX17 4AE
2E0 CTR T. Wright, 11 Ash Close, Daventry, NN11 0XH
2E0 CTT M. Ward, 15 Northfield Crescent, Wells-Next-the-Sea, NR23 1LP
2E0 CTU D. Foley, 1 Hill Rise Close, Harrogate, HG2 0DQ
2E0 CTV I. Ftaiha, 8 Parkside, London, NW7 2LH
2E0 CTW J. Hobbs, 82 Perry'S Lane, Wroughton, Swindon, SN4 9AP
2E0 CTZ M. Carr, 51 Langton Road, Holton-Le-Clay, Grimsby, DN36 5BH
2E0 CUA J. Valle Espin, 203 Broadway, Horsforth, Leeds, LS18 4HL
2E0 CUB A. Clements, 28 Durham Way Wyton, Huntingdon, PE28 2EQ
2E0 CUC A. Hunter, 22 Lindsay Court, Whitburn, Sunderland, SR6 7LN
2E0 CUE R. Weaver, 15 Sharps Field, Headcorn, Ashford, TN27 9UF
2E0 CUH J. Farr, Bunkabin 1, Treyew Road Tr13Ay, Truro, TR13AY
2W0 CUJ J. Barry, 7 Rockfield Rise, Undy, Caldicot, NP26 3FG
2W0 CUK I. Troughton, Rhiwbina, Pentre Lane, Cwmbran, NP44 3AP
2E0 CUL K. Richards, 207 Beaulieu Gardens, Blackwater, Camberley, GU17 0LG
2E0 CUO H. Hope, 51 Margravine Gardens, London, W6 8RN
2E0 CUS A. Cussen, The Poplars, Mill End, Southminster, CM0 7HJ
2E0 CUU M. Boon, 45 Carlton Road, Wickford, SS11 7ND
2E0 CUV J. Stewart, 19 Salisbury Road, Dover, CT16 1EX
2E0 CUW Dr R. Tofts, Elmcroft, Redhill Road, Ross-on-Wye, HR9 5AU
2E0 CUY S. Bache, 62 Whittingham Road, Halesowen, B63 3TP
2E0 CVC A. Chaplin, 10 St. Leonards Road, Malinslee, Telford, TF4 2EB
2E0 CVD C. Herd, 10 Amethyst Close, Rainworth, Mansfield, NG21 0GH
2W0 CVE C. Wilkinson, High Breck Dolwyd, Colwyn Bay, LL28 5HS
2E0 CVF A. Davies, 4 Capella Path, Hailsham, BN27 2JY
2E0 CVG C. Green, 4 Lyme Grove, Knott End-On-Sea, Poulton-le-Fylde, FY6 0AJ
2E0 CVJ R. Fidler, 44 Windermere Avenue, Ramsgate, CT11 0PF
2M0 CVK S. Muir, Cairnside, Burnhead, Dundee, DD3 0QN
2W0 CVL N. Edwards, 17 Queensway, Garnlydan, Ebbw Vale, NP23 5EE
2W0 CVM S. Broderick, 179 Malpas Road, Newport, NP20 5PP
2E0 CVN S. Treacher, 9 Noelle Drive, Newton Abbot, TQ12 1PS
2W0 CVO J. Pauline, 54 Laurel Road, Bassaleg, Newport, NP10 8NY
2E0 CVP J. Mason, 77 Albutts Road, Walsall, WS8 7ND
2E0 CVQ S. Knott, 15 Meadowlands Drive, Haslemere, GU27 2FD
2I0 CVR J. Allen, 192 Joanmount Gardens, Belfast, BT14 6PA
2E0 CVU P. Swingewood, 9 Goodall Grove, Great Barr, Birmingham, B43 7PQ
2E0 CVV C. Sayles, 11 Malton Close, Monkston, Milton Keynes, MK10 9HR
2E0 CVW D. Levy, Flat 36, Claydon House, London, NW4 1LS
2E0 CVX G. Spicer, 44 Cowley Lane, Chapeltown, Sheffield, S35 1SY
2E0 CVY R. Smith, 15 Hollybush Road, North Walsham, NR28 9XT
2E0 CVZ P. Herron, 102 Garden City Villas, Ashington, NE63 0EU
2E0 CWB M. Fletcher, 7 Richard Street, Bacup, OL113 8QJ
2E0 CWC C. Lewis, 9 Chatsworth Gardens, Sydenham, Leamington Spa, CV31 1WA
2E0 CWI R. Last, 30 Abbot Road, Bury St Edmunds, IP33 3UB
2E0 CWJ J. Taylor, 90 Village Road, Gosport, PO12 2LG
2E0 CWK J. Deery, Flat 6, 33-34 Philbeach Gardens, London, SW5 9EB
2W0 CWL K. Saltmarsh, 15 Colbourne Road, Beddau, Pontypridd, CF38 2LN
2E0 CWM G. Moore, 40 Main Street, South Rauceby, Sleaford, NG34 8QG
2E0 CWO B. Ledson, 16 Caton Close, Southport, PR9 9XF
2E0 CWQ J. Clarke, 160 Hall Lane Estate, Willington, Crook, DL15 0PP
2E0 CWR C. Ralphson, 20 Monsal Grove, Buxton, SK17 7TF
2E0 CWS K. Roberts, 4 Quartz Close, Tolvaddon, Camborne, TR14 0FT
2M0 CWV N. Ford, 11 Kenmore Way, Coatbridge, ML5 4FN
2E0 CWW A. Rowland-Stuart, 86 Wiltshire House, Lavender Street, Brighton, BN2 1LE
2E0 CWZ K. Legg, Bennetts, High Street, Thorpe-Le-Soken, Clacton-on-Sea, CO16 0EG
2E0 CXA D. Storer, 13 The Square, Lower Burraton, Saltash, PL12 4SH
2E0 CXD J. Johnson, 226 Preston New Road, Southport, PR9 8NY
2E0 CXE P. Naik, 82 Misbourne Road, Uxbridge, UB10 0HW
2M0 CXG A. Donald, 10 Fraser Road, Burghead, Elgin, IV30 5YN
2E0 CXH P. Arnold, 20 Upper Seagry, Chippenham, SN15 5EX
2M0 CXI D. Plummer, 39 St. Nicholas Drive, Banchory, AB31 5YG
2E0 CXJ C. Norman, 15 Maple Close, Sedbergh, LA10 5JE
2E0 CXK I. Nicholls, 34 West Close, Bath, BA2 1PY
2E0 CXL R. Chandler, 12 Lyndhurst Drive, Hale, Altrincham, WA15 8EA
2E0 CXM B. Cullen, 25 Nea Close, Christchurch, BH23 4QQ
2E0 CXN J. Bligh-Wall, 3 George Street, Elworth, Sandbach, CW11 3BL
2E0 CXO K. Ralph, 11 Burrard Road, London, E16 3QL
2E0 CXP R. Gill, 45 Biggin Lane, Ramsey, Huntingdon, PE26 1NB
2E0 CXQ T. Pettis, 11 Curlew Drive, Hythe, Southampton, SO45 3GB
2E0 CXU C. Ashworth, 40 Fairholme, Sedbergh, LA10 5AY
2W0 CXV Dr D. Morgan, Ty Bettws, Kilgwrrwg, Chepstow, NP16 6PN
2E0 CXW C. Wilson, 87 Levensgarth Avenue, Fulwood, Preston, PR2 9FP
2E0 CYC N. Marley, Penstemons, Chapel Lane Pen Selwood, Wincanton, BA9 8LY
2W0 CYE J. Jones, Isfryn Bungalow, Glan-Y-Nant, Llanidloes, SY18 6PQ
2E0 CYL R. Horvath, 3 Back Knowl Road, Mirfield, WF14 9SA
2W0 CYM A. Rowlands, Lluest Wen, Penygarth, Caernarfon, LL55 1EY
2E0 CYO P. Gregory, 2 Lennox Close, Hunmanby, Filey, YO14 0PY

2E0 CYP S. Challis, 73 Rivenhall Way, Hoo, Rochester, ME3 9GF
2E0 CYR M. Woodruff, 5 Regency Heights, Caversham, Reading, RG4 7RH
2E0 CYS P. Martin, 108 Headlands Grove, Swindon, SN2 7HP
2E0 CYT L. Starrett, 50 Danes Road, Bicester, OX26 2LP
2E0 CYU J. Powell, 23 Park Road, Norton, Malton, YO17 9DZ
2I0 CYW C. Wang, 32 Broadlands, Carrickfergus, BT38 7BL
2W0 CYX S. Rogers, 30 Coed Celynen Drive, Abercarn, Newport, NP11 5AU
2W0 CYY C. Smith, 29 Heol Cwarrel Clark, Caerphilly, CF83 2NE
2E0 CYZ M. Tarrant, Wayside Cottage, Gabber Lane, Plymouth, PL9 0AW
2E0 CZA G. Youll, 4 Shaftsbury Court, Barnstaple Road, Scunthorpe, DN17 1YB
2E0 CZC S. Ingledew, 34 Sunningbrook Road, Tiverton, EX16 6EB
2J0 CZD J. Crill, Keukenhof, La Route Des Cotils, Grouville, Jersey, JE3 9AP
2E0 CZG S. Sissens, 20 Fallow Drive, Eaton Socon, St Neots, PE19 8QL
2E0 CZI P. Patterson, 3 Barnes Close, Southampton, SO18 5FE
2E0 CZJ J. Chalmers, 19 Brettenham Crescent, Ipswich, IP4 2UB
2E0 CZK R. Silcock, 18 Saxon Road, Southampton, SO15 1JJ
2E0 CZN D. Endean, 11 Forrester Drive, Brackley, NN13 6NE
2W0 CZP A. Jones, 31 Russell Terrace, Carmarthen, SA31 1SZ
2E0 CZR D. Rogers, 5 Semple Gardens, Chatham, ME4 6QD
2E0 CZS P. Handley, 97 Applegarth Avenue, Guildford, GU2 8LX
2E0 CZT R. Morgan, 14 Ash Road, Ashurst, Southampton, SO40 7AT
2W0 CZU N. Adam, Tan Ffordd, Mynydd Llandygai, Bangor, LL57 4LX
2E0 CZW G. Street, 105 Jeals Lane, Sandown, PO36 9NS
2W0 CZZ P. Hampton, Caretakers Flat, T.A. Centre, Newport, NP20 5XE
2W0 DAA D. Wilson, 94 Lon Hedydd, Llanfairpwllgwyngyll, LL61 5JY
2E0 DAB D. Bambrook, 18 Vervain Close, Bicester, OX26 3SR
2M0 DAC D. Campbell, 10 Balgate Mill, Kiltarlity, Beauly, IV4 7GL
2E0 DAH D. Horner, 21 Ainsworth Road, Little Lever, Bolton, BL3 1RG
2E0 DAI D. Holman, 20 Green Drive, Wolverhampton, WV10 6DW
2E0 DAJ J. Bowley, 2 Cottage, Middle Battenhall Farm, Worcester, WR5 2JL
2E0 DAL D. Stringer, 18 Townfield Close, Ravenglass, CA18 1SL
2E0 DAO D. Oddie, 5 The Bridleway, Forest Town, Mansfield, NG19 0QJ
2W0 DAP T. Woodley, 2 Parc Onen, Neath, SA10 6AA
2E0 DAQ G. Clarke, 11 Blackfordby Lane, Moira, Swadlincote, DE12 6EX
2E0 DAR D. Robertson, 53 Moor Lane, Weston-Super-Mare, BS22 6RA
2E0 DAT D. Toyne, 19 Poachers Rest, Welton, Lincoln, LN2 3TR
2E0 DAW D. Williams, 18 Lower Greave Road, Meltham, Holmfirth, HD9 4DY
2E0 DAX D. Sobey, Flat 2 73 Park Road, Blackpool, FY1 4JQ
2E0 DAY S. Darby, 4 Whately Mews, Whately Road, Lymington, SO41 0XS
2E0 DBA D. Barnes, 11 Yewside, Gosport, PO13 0ZD
2E0 DBB D. Walker, Flat 5, Seward Court, 380-396 Lymington Road, Christchurch, BH23 5HD
2E0 DBH D. Hoare, 47 High Street, Chalgrove, Oxford, OX44 7SJ
2E0 DBI P. Marchant, 16 Melrose Drive, Peterborough, PE2 9DN
2I0 DBK D. O'Hale, 6 Cochron Road, Newry, BT35 6DD
2E0 DBL P. Kimberlee, 24 Jacey Road, Shirley, Solihull, B90 3LJ
2E0 DBM D. Mellor, 28 Winster Road, Staveley, Chesterfield, S43 3NJ
2E0 DBN D. Booth, 5 Denham Close, Skegness, PE25 3PH
2E0 DBO J. Ledger, 39Eascroft Drive, Sheffield, S20 8JG
2E0 DBP D. Payne, 31 Cockering Road, Canterbury, CT1 3UP
2E0 DBQ D. Moger, 23 Elmsleigh Road, Paignton, TQ4 5AX
2E0 DBS D. Baines, 21 Vera Road, Norwich, NR6 5HU
2E0 DBW D. Whyatt, 11 The Perrings, Nailsea, Bristol, BS48 4YD
2E0 DBY C. Riley, 2 Shottisham Hall Cottages, Alderton Road, Shottisham, Woodbridge, IP12 3EP
2E0 DBZ D. Teasdale, 43 Easington Road, Stockton-on-Tees, TS19 8ES
2E0 DCA C. Price, 9 Arlington Avenue, Aston, Sheffield, S26 2AA
2E0 DCD B. Savage, Rufford, Barnes Lane, Milford On Sea, Lymington, SO41 0RR
2E0 DCF C. Lonie Jr, 41 De La Hay Avenue, Plymouth, PL3 4HS
2E0 DCH D. Hannington, 3 Canadian Avenue, Gillingham, ME7 2DN
2E0 DCJ D. Bishop, 62 Brindley Crescent, Hednesford, Cannock, WS12 4DS
2W0 DCK R. Powell, Gerdd Y Don, Llansantffraed, Llanon, SY23 5HS
2E0 DCL D. Clark, 34 Magdalene Road, Owlsmoor, Sandhurst, GU47 0UT
2E0 DCM D. Martin, 14 Freeston Terrace, St. Georges, Telford, TF2 9HD
2E0 DCN D. Filby, 21 Knightlands Road, Irthlingborough, Wellingborough, NN9 5SU
2E0 DCP D. Sproston, 22 Oakland Avenue, Haslington, Crewe, CW1 5PB
2E0 DCS D. Sharpen, 52 Woodsend Road, Urmston, Manchester, M41 8QT
2E0 DCV J. Hurlbutt, 55 Prospect Avenue, Seaton Delaval, Whitley Bay, NE25 0EL
2E0 DCX D. Cook, 44 Statfold Lane, Fradley, Lichfield, WS13 8NY
2E0 DCZ D. Mooney, 107 Tedder Road, South Croydon, CR2 8AR
2E0 DDB C. Massey, 23 Ladymead Lane, Langford, Bristol, BS40 5EG
2E0 DDC D. Coe, 199 Newark Road, North Hykeham, Lincoln, LN6 8QS
2E0 DDD R. Hirst, 106-108 Washerwall Lane, Werrington, Stoke-on-Trent, ST9 0LR
2E0 DDF D. Pennison, 69 Caneland Court, Waltham Abbey, EN9 3DS
2E0 DDH D. De La Haye, 4 Nicola Mews, Ilford, IG6 2QE
2W0 DDJ D. Jones, 11 Alma Place, Sebastopol, Pontypool, NP4 5EA
2E0 DDL D. Curtis, 7 Neale Close, Aylsham, Norwich, NR11 6DJ
2E0 DDN B. Southern, 25 Chilgrove Avenue, Blackrod, Bolton, BL6 5TR
2M0 DDO D. Lyons, 2 Goswick Farm Cottages, Berwick-upon-Tweed, TD15 2RW
2W0 DDP R. Carpenter, 2 Greenfield Terrace, Trinant, Newport, NP11 3LJ
2E0 DDQ G. Mutch, 94 Abbotswood Road, Brockworth, Gloucester, GL3 4PF
2E0 DDR D. Randles, 12 Wain Court. Rakeway, Saughall. Chester, CH16BF
2M0 DDS D. Scott, Farewell, Arnage, Auchnagatt, Ellon, AB41 8UW
2E0 DDU S. Borrell, Rose Cottage, Colchester Main Road, Colchester, CO7 8DD
2M0 DDX D. Cunningham, 3 Dallerie, Crieff, PH7 4JH
2W0 DDZ D. May, 12 Marl Crescent, Llandudno Junction, LL31 9HS
2E0 DEC D. Chebsey, 21 Shortlands Lane, Walsall, WS3 4AG
2E0 DEE D. Nicholson, 24 Barnmead, Haywards Heath, RH16 1UZ
2E0 DEG A. Titmus, The Old Police House, Arundel Road, Fontwell, Arundel, BN18 0SX
2E0 DEH L. Mathlin, 29 Wagtail Drive, Stowmarket, IP14 5GH
2E0 DEI G. Watson, 88 Avenue Road, Sandown, PO36 8BE
2E0 DEK D. Gibson, 25 Middleham Close, Ouston, Chester le Street, DH2 1TA
2E0 DEM C. Wright, 24 Charlemont Crescent, West Bromwich, B71 3DA
2E0 DEO C. Phillips, 14 Laburnum Way, Hatfield Peverel, Chelmsford, CM3 2LP
2E0 DEP C. Garner, 30 Pendula Road, Wisbech, PE13 3RR
2E0 DEQ D. Pritchard, Flat 23 Egan Court Price Street, Birkenhead, CH416JA
2W0 DER D. Williams, 10 Bronllys, Gaerwen, LL60 6JN
2M0 DES G. Taylor, 15 Ronaldsvoe, Kirkwall, KW15 1XE
2E0 DEU M. Hanbuerger, Arboris, New Road Hill, Reading, RG7 5RY
2E0 DEV R. Barter, 8 Orchard Close, Newton Abbot, TQ12 3DF
2E0 DEX D. Rigby, 32 Springs Road, Chorley, PR6 7AN
2E0 DEZ D. Turner, 29 Balmoral Road, Castle Bromwich, Birmingham, B36 0JT
2E0 DFA B. Geall, 129 Jewell Road, Bournemouth, BH8 0JP
2E0 DFB D. Grundy, 44 Heathend Road, Alsager, Stoke-on-Trent, ST7 2SH
2E0 DFF M. Bookham, 116 Clare Gardens, Petersfield, GU31 4EU
2E0 DFG J. Tarrant, 70 Sunnymead, Midsomer Norton, Radstock, BA3 2SD
2E0 DFI M. Philpott, Garmisch, Hazel Road, Aldershot, GU12 6HP
2E0 DFJ D. Jacobs, 7 Coppice Close Ravenstone, Coalville, LE67 2NS
2E0 DFL D. Humm, 15 Sherborne Road, Farnborough, GU14 6JS
2W0 DFM R. Russell, 1 Horeb Cottages, Rhiw Road, Colwyn Bay, LL29 7TL
2W0 DFN G. Edwards, 17 Glan Y Mor Road, Penrhyn Bay, Llandudno, LL30 3NL
2E0 DFO P. Heiney, 12 Church Lane, Walberswick, IP186UZ
2E0 DFP D. Parker, 53 Brisbane Way, Cannock, WS12 2GR
2E0 DFQ P. Love, 30 Salisbury Road, Canterbury, CT2 7HH
2E0 DFS D. Smith, 186 Weekes Drive, Slough, SL1 2YR
2E0 DFT D. Fisher, 34 Orange Croft, Tickhill, Doncaster, DN11 9EW
2E0 DFV R. Rigby, 11 Mallow Close, Thornbury, Bristol, BS35 1UE
2E0 DGA P. Wilson, 45 Newquay Close, Hartlepool, TS26 0XG
2M0 DGB D. Baillie, 126 Main St., Fauldhouse, Bathgate, EH47 9BW
2E0 DGC D. Cowling, 11 Shakespeare Avenue, Scunthorpe, DN17 1SA
2E0 DGD D. Bailey, 2B Queens Road, Enfield, EN1 1NE
2E0 DGG M. Tinsell-Stanton, 38 Comberton Road, Kidderminster, DY10 3DT
2E0 DGH D. Humphrey, 42 Ratcliffe Road Sileby, Loughborough, LE12 7PZ
2M0 DGI P. Riddle, Carngeal, Pitlochry, PH16 5JL
2M0 DGJ Z. Bak, 62/6 North Gyle Loan, Edinburgh, EH12 8LD
2E0 DGL D. Lock1, Flat 33, Harrison Court, Harrison Close, Hitchin, SG4 9SG
2E0 DGM S. Hopkins, 3B Tolfa House Wellington Terrace, Truro, TR1 3JA
2E0 DGQ P. Flanagan, 71 Fellway, Pelton Fell, Chester le Street, DH2 2BY
2E0 DGR A. Norton, Flat 9, Brownhill Court, Southampton, SO16 9LB
2E0 DGS D. Smith, 48 Shirley Gardens, Tunbridge Wells, TN4 8TH
2E0 DGT R. Fripp, 41 Sweyns Lease, East Boldre, Brockenhurst, SO42 7WQ
2E0 DGU W. Alexander, 81 Cherry Lane, Lymm, WA13 0SY
2E0 DGV P. Hodson, 21 Green Hill, London Road, Worcester, WR5 2AA
2E0 DHA D. Atkins, 20 Nappsbury Road, Luton, LU4 9AL
2E0 DHB D. Bisson, 48 Elmsfield Avenue, Rochdale, OL11 5XN
2I0 DHC P. White, 46 Pine Cross, Dunmurry, Belfast, BT17 9QY
2E0 DHE R. Barnard, 3 Heaths Close, Enfield, EN1 3UP
2E0 DHF J. Jackson, 49 Leafield Rise, Two Mile Ash, Milton Keynes, MK8 8BX
2E0 DHG R. Holmes, Coach House, Cadlington House Estate, Waterlooville, PO8 0AA
2M0 DHI R. Buchan, 5 Fairview Terrace, Danestone, Aberdeen, AB22 8ZH
2E0 DHJ T. Thompson, 14 Queen Street, Northwich, CW9 5JL
2E0 DHK B. Davies, 12 Scalebor Gardens, Burley In Wharfedale, Ilkley, LS29 7BX
2E0 DHO P. Chadwick, 112 Sandy Lane, Warrington, WA2 9JA
2E0 DHQ D. Rimmer, 41 Ashburton Road, Wallasey, CH44 5XB
2I0 DHR D. Richards, 70 Cherryhill Avenue, Dundonald, Belfast, BT16 1JD
2E0 DHS D. Sherwin, 5 North Road, Buxton, SK17 7EA
2E0 DHT J. Berry, 245A Eaves Lane, Chorley, PR6 0AG
2E0 DHV P. Walton, 11 Parkfield Road, Northwich, CW9 7AR
2E0 DHW A. Stabler, 11 Lincolns Avenue, Gedney Hill, Spalding, PE12 0PQ
2E0 DHX M. Watkins, 108 Honiton Road, Exeter, EX1 3EQ
2E0 DHY D. Ball, 27 Bramble Close Aston, Birmingham, B6 5HW
2E0 DHZ A. Hitchcott, 121 Oakhurst Road, Acocks Green, Birmingham, B27 7PB
2M0 DIB C. Morris, 23 Sedgebank, Livingston, EH54 6HE
2E0 DID D. Foyston, 22 Ronaldsway Drive, Newcastle under Lyme, ST5 9HE
2M0 DIF I. Smith, 32 Kaimes Avenue, Kirknewton, EH27 8AU
2E0 DIG A. Taylor, 130A Hazelwood Avenue, Eastbourne, BN22 0UX
2E0 DIH E. Coles, 41 Venn Court Brixton, Plymouth, PL82AX
2E0 DII I. Phillips, 124 Brookwood Drive, Stoke-on-Trent, ST3 6LP
2E0 DIJ D. Yates, 16 Sunnyfield Road, Prestwich, Manchester, M25 2RD
2E0 DIL D. Yakub, 42 Swift Close, Blackburn, BB1 6LF
2E0 DIM D. Vainas, 51 Magister Road, Bowerhill, Melksham, SN12 6FD
2E0 DIP D. Webb, 52 Simpkin Close, Eaton Socon, St Neots, PE19 8PD
2E0 DIQ A. Collins, 14 Double Corner, Mendlesham Road, Cotton, Stowmarket, IP14 4RF
2E0 DIS A. George, 15 Ely Road, Croydon, CR0 2LW
2E0 DIT N. Baulf, 1 Lower Chart Cottages, Brasted Chart, Westerham, TN16 1LS
2E0 DIU I. Jones, 90 Preston, Cirencester, GL7 5PR
2W0 DIV P. Williams, 63 Trem Eryri, Llanfairpwllgwyngyll, LL61 5JF
2E0 DIX B. Smith, 7 Kestrel Avenue, Bransholme, Hull, HU7 4ST
2W0 DJC D. Cole, 14 Inner Loop Road, Beachley, Chepstow, NP16 7HF
2E0 DJF D. Fagg, 62 Hawkins Road, Folkestone, CT19 4JA
2E0 DJH D. Harris, 13 Horsecroft, Ewyas Harold, Hereford, HR2 0EQ
2E0 DJI D. Oliver, 20 Five Oaks Close, Malvern, WR14 2SW
2E0 DJJ G. Mccaffery, 7 Cliffe Court, Sunderland, SR6 9NT
2I0 DJM J. Mcbride, 22 Birchwood, Omagh, BT79 7RA
2E0 DJQ M. Barnaby, 8 Callowood Croft, Purleigh, Chelmsford, CM3 6NZ
2E0 DJR J. Rudd, 5 St Andrews Close, Blofield, NR13 4JX
2W0 DJS D. James, 10 Hafan Deg, Pencoed, Bridgend, CF35 6YG
2M0 DJT D. Rodger, 25 Wilson Road, Banchory, AB31 5UY
2E0 DJU T. Bannister, 6 Tanners Road, North Baddesley, Southampton, SO52 9FD
2E0 DJV A. Wheeler, 8 Elsworth Grove, Birmingham, B25 8EJ
2E0 DJW D. Creech, 29 Lake Road, Poole, BH15 4LE
2E0 DJX S. Scott, 13 Silver Close, Harrow, HA3 6JT
2E0 DJY D. Bennett, 31 Park Road North, Urmston, Manchester, M41 5AT
2E0 DJZ Dr B. Issac, 9B Poplar Grove, Stockport, SK2 7JD
2E0 DKA D. Carmichael, 22 California Close Great Sankey, Warrington, WA5 8WU
2E0 DKB D. Banks, 41 East Road, Rotherham, S65 2UX
2E0 DKD D. Saxon, 53 Westmorland Road, South Shields, NE34 7JJ
2E0 DKE C. Boyle, 8 Westlees Close, North Holmwood, Dorking, RH5 4TN
2E0 DKF A. Gow, Flat 203, Viotti Heights Sandy Hill Road, London, SE18 6PA
2E0 DKG E. Gomez Lozano, 2 Annesley Road, Oxford, OX4 4JQ
2E0 DKI D. Iveson, 11 Newport Road, North Cave, Brough, HU15 2NU
2E0 DKM W. Molloy, 32 Millers Barn Road, Jaywick, Clacton-on-Sea, CO15 2QB
2E0 DKO P. Taylor, 32 Heliers Road, Liverpool, L13 4DH
2E0 DKP A. Lutley, Springfield, Rookery Hill, Ashtead, KT21 1HY

2I0 DKQ G. Gardiner, 60 Limestone Meadows, Moira, Craigavon, BT67 0UT
2E0 DKR D. Reeves, 67 The Cliff, Bryanston, Blandford Forum, DT11 0PP
2E0 DKS C. Wilkes, 2 Kings Crescent, Edlington, Doncaster, DN12 1BD
2M0 DKU S. Boyd, 269 Cloch Caravans Cloch Road, Gourock, PA19 1AZ
2M0 DKV J. Flannigan, 21 Kirkbean Avenue Rutherglen, Glasgow, G73 4EA
2E0 DKW C. Nicholl, 36 Eylewood Road, London, SE27 9NA
2M0 DKX P. Bacon, 12 The Greens, Maddiston, Falkirk, FK2 0FN
2E0 DKY B. Cross, 22 Park Avenue Washingborough, Lincolnshire, LN 4 1 DB
2E0 DKZ D. Hyde, 136 Station Road, Woodmancote, Cheltenham, GL52 9HN
2E0 DLA P. Booth, 7 Handley Crescent, East Rainton, Houghton le Spring, DH5 9QX
2E0 DLC B. Tufnell, 1 Moorlands Court Wath-Upon-Dearne, Rotherham, S63 6DD
2E0 DLD D. Dewsbury, 62 Yew Tree Drive, Leicester, LE3 6PL
2W0 DLE T. Edwards, Broadfield House, Vicarage Road, Tonypandy, CF40 1HP
2E0 DLF A. Cook, 84 Clent View Road, Birmingham, B32 4LW
2E0 DLJ D. Johnstone, 14 Carr Hey, Wirral, CH46 6EL
2E0 DLK L. Sargent, 18 Lyndhurst, Maghull, l316dy
2E0 DLO S. Strange, 94 Digby Avenue, Nottingham, NG3 6DY
2E0 DLP D. Ion, 78 Blackmore Street, Derby, DE23 8AX
2E0 DLR D. Taylor, Flat 27, Laurel Court, 24 Stanley Road, Folkestone, CT19 4RL
2E0 DLT M. Tucker, 182 Salisbury Road, Amesbury, Salisbury, SP4 7HW
2E0 DLV C. Lombao, 82 Cirrus Drive, Shinfield, Reading, RG2 9FL
2E0 DLZ D. Dunne, 1 Burton Gardens, Brierfield, Nelson, BB9 5DR
2E0 DMA D. Aldridge, 5 Alpine Close, Paulton, BS39 7SE
2E0 DMB D. Browne, 27 Lewis Road, Emsworth, PO10 7RP
2I0 DMC D. Mccloskey, 1 Dernaflaw Cottages Dernaflaw Road, Dungiven, Londonderry, BT47 4PP
2E0 DMD D. Mcarthur, 7 Gore Avenue, Salford, M5 5LF
2E0 DME D. Priestley, 8 Cokefield Avenue, Nuthall, Nottingham, NG16 1AU
2W0 DMG D. Griffiths, 8 Heol Cynwyd, Llangynwyd, Maesteg, CF34 9TB
2E0 DMI A. Yates, 1 Roberts Court, Whitwell, Hitchin, SG4 8AF
2E0 DMJ D. Gegg, 84 Aberconway Crescent, New Rossington, Doncaster, DN11 0JP
2E0 DMM D. Moran, 37 Collingwood Road, Long Eaton, Nottingham, NG10 1DR
2E0 DMN D. Zubrzycki, 11 Bishopdale Holme, Bradford, BD62AB
2E0 DMP D. Powell, 58, Lessingham Avenue, Wigan, WN12HX
2E0 DMQ K. Henney, 6 Peel House, Rusham Road, Egham, TW20 9LP
2E0 DMS D. Stewart, 79 Eastfield Road, Driffield, YO25 5EZ
2E0 DMU D. Pearson, 37 Elmridge, Leigh, WN7 1HN
2E0 DMV G. Weston, 131 Ringwood Road, Eastbourne, BN22 8TQ
2E0 DMW R. Weir, 130 Alexander Square, Eastleigh, SO50 4BX
2E0 DMX D. Moffat, 27 Cinque Ports Way, Seaford, BN25 3UE
2E0 DMY D. Humphrey, 11 Colborne Close, Poole, BH15 1UR
2E0 DNB N. Brown, 241 Bury Road, Tottington, Bury, BL8 3DY
2E0 DNC D. Swift, 31 Meadow Lane, Westbury, BA13 3AE
2E0 DND D. Driver, 34 Potter Avenue, Wakefield, WF2 8HE
2E0 DNE B. Rajagopal, 4 Balliol Road, Caversham, Reading, RG4 7DT
2E0 DNF D. Featherby, 14 Station Road, Sutton, Ely, CB6 2RL
2E0 DNH R. Finch, Garth Cottage North Cowton, Northallerton, DL7 0HL
2W0 DNI W. Oliver, Pwllmeyric, Chepstow, NP16 6LE
2E0 DNJ D. Jarvice, 15 Meden Avenue, Warsop, Mansfield, NG20 0PS
2E0 DNL D. Neill, 7 Ashbrow Road, Northampton, NN4 8ST
2M0 DNM D. Mackenzie, 4 Nabhar Laxay, Isle of Lewis, HS2 9PJ
2E0 DNO D. Close, 22 Station Road, Dodworth, Barnsley, S75 3JE
2W0 DNR J. Hughes, Midfield Farm, Midfield Caravan Site, Aberystwyth, SY23 4DX
2E0 DNS J. Cummins, 55 Rowley Street, Walsall, WS1 2AZ
2E0 DNU G. Cater, 7 Seymour Street, Chelmsford, CM2 0RX
2W0 DNV D. Hooper, Nant Y Dryslwyn Cottage, Ty Mawr, Llanybydder, SA40 9RD
2E0 DNW L. Stamper, 22 Douglas Road, Workington, CA14 2QY
2E0 DNX D. Smith, 3 Cain Street, Bigrigg, Egremont, CA22 2TP
2E0 DOD A. Dodd, 14 Davies Street, Macclesfield, Sk10 1GE
2W0 DOE C. Taylor, 23 Heol Derw, Brynmawr, Ebbw Vale, NP23 4TT
2E0 DOF S. Black, 7 Harwood Close, Gosport, PO13 0TY
2E0 DOG R. Scholefield, 4 Minnie Street, Haworth, Keighley, BD22 8PR
2M0 DOI D. Speirs, 45 Elmbank Crescent, Arbroath, DD11 4EZ
2E0 DOJ D. Jenkins, 1 Green End Road, Sawtry, Huntingdon, PE28 5UX
2M0 DOL J. Marsh, 8 Hazelton Way, Broughty Ferry, Dundee, DD5 3BT
2E0 DOM D. Maddison, 9 Rowley Way, Sunnyside, Rotherham, S66 3ZY
2E0 DOP A. Seelig, 9 Warren House Court 17 St Peters Avenue Caversham, Reading, RG4 7RG
2E0 DOQ M. Dinally, 208 Ley Hill Farm Road, Birmingham, B31 1UQ
2E0 DOW B. Lewin, 68 Brackley Square, Woodford Green, IG8 7LS
2E0 DOX D. Richards, Flat 40, Leander Court, Teignmouth, TQ14 8AQ
2E0 DOZ D. Logan, Cedar House Reading Road North, Fleet, GU51 4AQ
2E0 DPD B. Scannell, 60 Burnside Road, Dagenham, RM8 1XD
2E0 DPH P. Hughes, 111 Wisbech Road, Littleport, Ely, CB6 1JJ
2W0 DPI A. Studdart, 24 Wepre Park, Connah'S Quay, Deeside, CH5 4HN
2E0 DPL C. Glass, The Old Homestead, Havikil Lane, Knaresborough, HG5 9HN
2E0 DPO D. Greenland, 1 Hilltop, Tuesley Lane, Godalming, GU7 1SB
2E0 DPR Dr D. Richardson, 89A Bean Oak Road, Wokingham, RG40 1RJ
2E0 DPS D. Seedhouse, 8 Levett Road, Tamworth, B77 4AB
2E0 DPT C. Appleton, 140 St. James Road, Orrell, Wigan, WN5 7AA
2E0 DPW D. Williams, 6 Wellhouse Avenue, West Mersea, Colchester, CO5 8GF
2E0 DPX W. Currie, 24 Mill Lane, Walton on the Naze, CO14 8PE
2E0 DPY Dr D. Pye, 151 Smallbrook Lane, Leigh, WN7 5PZ
2E0 DPZ R. Lyddall, 102 Chapel Road, Brightlingsea, Colchester, CO7 0HE
2E0 DQB A. Smith, 116 Pilling Lane, Preesall, Poulton-le-Fylde, FY6 0HG
2E0 DQD P. Driver, 68 Ripon Road, Dewsbury, WF12 7LG
2E0 DQH A. Robnett, 38B Woodmere Avenue, Watford, WD24 7LN
2E0 DQJ H. Woodfin, 8 Bank Hall Close, Bury, BL8 2UL
2E0 DQL D. Coles, 36 York Hill, Loughton, IG10 1HT
2M0 DQN G. Mcleod, 75 Grange Avenue, Wishaw, ML2 0AH
2E0 DQO J. Marks, Chantry End, Oak Hill, Epsom, KT18 7BU
2E0 DQP Rev. A. Lewis, Four Winds Cottage, Main Street, Brough, HU15 1RJ
2E0 DQQ S. Mcloughlin, 7 Wilmots Way, Pill, Bristol, BS20 0JT
2W0 DQT A. Williams, 27 Hick Street, Llanelli, SA15 1AR
2E0 DQU I. Lawton, 11 Goosewell Terrace, Plymstock, Plymouth, PL9 9HW
2E0 DQX J. Saunders, 123 Medway Road, Ferndown, BH22 8UR
2M0 DQY J. Dow, 58 Beatty Crescent, Kirkcaldy, KY1 2HS
2E0 DQZ T. Brooks, 200 Kingsway, College Estate, Hereford, HR1 1HE
2E0 DRA M. Draper, 160 Chanctonbury Road, Burgess Hill, RH15 9HA
2E0 DRB D. Barraclough, Flat 18 89 Park Road, London, SW19 2BD
2E0 DRE D. Dean, 12 Abbeydale Road South, Sheffield, S7 2QN
2D0 DRG C. Schofield, Rockside, Dreemskerry Road, Maughold, Isle Of Man, IM7 1BL
2E0 DRI K. Abel, 7 Foldgate View, Ludlow, SY8 1NB
2W0 DRK D. Machon, 22 Albert Street, Caerau, Maesteg, CF34 0UF
2E0 DRN D. Rayne, 159A Arbury Road, Nuneaton, cv107nh
2M0 DRO R. Drummond, 11 Firwood Drive, Bo'ness, EH51 0NX
2E0 DRQ D. Roberts, 3 Heather Avenue, Melksham, SN12 6FX
2E0 DRT C. Brock, 30 Cromer Road, Norwich, NR6 6LZ
2E0 DRW Dr W. Wightman, 36 Holyoake Avenue, Woking, GU21 4PW
2M0 DRY D. Drysder, 37 Farburn Drive, Stonehaven, AB39 2BZ
2E0 DRZ A. Potts, 103 Etherstone Street, Leigh, WN7 4HY
2E0 DSB D. Slade, 22 Oaklands Road, Mangotsfield, Bristol, BS16 9EY
2M0 DSG D. Gartshore, 85 Springhill Street, Douglas, Lanark, ML11 0NZ
2E0 DSH D. Hind, 19 Ellington Road, Arnold, Nottingham, NG5 8SJ
2E0 DSI D. Stocker, 113 St. Marys Road, Bodmin, PL31 1NH
2W0 DSJ D. Jenkins, 16 Celtic Road, Whitchurch, Cardiff, CF14 1EG
2E0 DSK D. Manning, 153 Pavilion Road, Worthing, BN14 7EG
2M0 DSL D. Latto, 8 Aspen Avenue, Glenrothes, KY7 5TA
2W0 DSO C. Summerfield, 11 Woodland Park, Penderyn, Aberdare, CF44 9TX
2W0 DSP A. Elias, 31 Banc Y Gors, Upper Tumble, Llanelli, SA146BR
2E0 DSQ P. Taylor, 104 Winstanley Drive, Leicester, LE3 1PA
2E0 DSS W. Stewart, 43 Newlands Drive, Halesowen, B62 9DX
2M0 DSU R. Murray, 18 Braids Road, Kirkcaldy, KY2 6JE
2E0 DSV R. Nicholson, 1 Isis Close, Aylesbury, HP21 9LY
2E0 DSW D. Weight, 19 Shakespeare Drive, Upper Caldecote, Biggleswade, SG18 9DD
2M0 DSY S. Chacko, 25 Wemyss Court, Rosyth, Dunfermline, KY11 2LL
2E0 DSZ V. Perovic, Staudenbühlstrasse 126, Zurich, Switzerland, 8052
2E0 DTB D. Bradley, 45 Fourth Avenue, Ketley Bank, Telford, TF2 0AS
2E0 DTC D. Beck, 94 Shaldon Crescent, Plymouth, PL5 3RB
2E0 DTD T. Davies, 7 Crescent Road, Warley, Brentwood, CM14 5JR
2I0 DTE D. Best, 13 Cranley Green, Bangor, BT19 7FE
2E0 DTF M. Bostock, 86 Beauvale Drive, Ilkest on, DE7 8SJ
2E0 DTG D. Griffiths, 44 Lowlands Road, Bolton Le Sands, Carnforth, LA5 8HE
2E0 DTH D. Thompson, 17 Sandpiper Close, Blyth, NE24 3QN
2M0 DTJ G. Lewin, Larch Cottage, Lein Road, Fochabers, IV32 7NW
2E0 DTL C. Brink, 138 Brookside, Burbage, Hinckley, LE10 2TN
2W0 DTM N. Sugg, 12 Caerleon Grove Castle Park Cf48 1Jh, Merthyr Tydfil, CF48 1JH
2E0 DTN D. Niggemann, 35 Holm Court, Twycross Road, Godalming, GU7 2QT
2E0 DTO E. Bray, 28, Henshall Avenue, Latchford, Warrington, WA4 1PY
2M0 DTP T. Burnett, 45 The Murrays Brae, Edinburgh, EH17 8UF
2E0 DTQ T. Kilfeather, Flat 1, 57 Chalk Hill, Watford, WD19 4DA
2W0 DTR D. Rosser, 42 Cobden Street, Aberaman, Aberdare, CF44 6EN
2E0 DTS B. Oldfield, 60 Alexandra Road, Capel-Le-Ferne, Folkestone, CT18 7LS
2E0 DTV M. King, 6 Aylesbury Close, Hockley Heath, Solihull, B94 6PA
2E0 DTW D. Williams, 96 Bingley Court, Canterbury, CT1 2SX
2E0 DTX L. Cook, 43 Midge Hall Drive, Rochdale, OL11 4AX
2E0 DTY M. Pesendorfer, 13 Blake Road, London, N11 2AD
2M0 DTZ J. Hogg, 31 Woodlea Court, Crosshouse, Kilmarnock, KA2 0ES
2E0 DUA K. Lynch, Medindie, Woodside, Ryton, NE40 4SY
2E0 DUB T. Price, 40 East Street, Kidderminster, DY10 1SE
2E0 DUD J. Storey, 3 Woodside Road, Poole, BH14 9JH
2E0 DUE I. Warnecke, 12 Caxton Road, Margate, CT9 5NP
2E0 DUF S. Lo, Fieldgate, The Avenue, Claverton Down, Bath, BA2 7AX
2E0 DUH J. Shears, 161 Park Road, Keynsham, Bristol, BS31 1AS
2E0 DUI R. Sutton, 80 Fishbourne Lane, Ryde, PO33 4EU
2E0 DUJ S. Bluff, 2 Astor Mews, High Street, Tidworth, SP9 7TR
2W0 DUL D. Davies, 2 Hendre Ddu, Manod, Blaenau Ffestiniog, LL41 4BH
2E0 DUM I. Clarke, 140 Cashmere Drive, Andover, SP11 6SS
2W0 DUN T. Dungey, 93 Boverton Road, Llantwit Major, CF61 1YA
2E0 DUO A. Richards, 16 Fruiterers Arms Caravan Park, Uphampton Lane, Ombersley, Droitwich, WR9 0JW
2E0 DUP V. Downes, 55 Ashfield Road Bromborough, Wirral, CH62 7EE
2E0 DUQ A. Ali, 42 Blease Close, Staverton, Trowbridge, BA14 8WD
2I0 DUR A. Savage, 469 Old Belfast Road, Bangor, BT19 1RQ
2E0 DUS J. Fenton, 4 Forest Hills, Newport, PO30 5NG
2E0 DUU I. Holdford, 46 Hildreth Road, Prestwood, Great Missenden, HP16 0LY
2E0 DUW L. Jones, 44 Althorpe Drive, Loughborough, LE11 4QU
2E0 DUZ T. Hope, 59 Chatsworth Crescent, Walsall, WS4 1QU
2E0 DVD A. Swan, 47 Warren Close, Whitehill, Bordon, GU35 9EX
2E0 DVF L. Lemmon, 4 Honington Close, Wickford, SS11 8XB
2E0 DVH D. Hilton, 6 Stow Villas The Street, Stow Maries, Chelmsford, CM3 6RX
2E0 DVI G. Ridley, 12 Garforth Avenue, Steeton, Keighley, BD20 6SP
2E0 DVK R. Steele, 244A Uttoxeter Road, Blythe Bridge, Stoke-on-Trent, ST11 9LY
2E0 DVM D. Vincelli, 90 Broadbottom Road, Mottram, SK14 6JA
2E0 DVN J. Sanders, 76 Fullerton Road, Plymouth, PL2 3AX
2E0 DVO K. Wells, 45 Laburnum Avenue, Yaxley, Peterborough, PE7 3YQ
2W0 DVP D. Price, 11 Cefn Melindwr, Capel Bangor, Aberystwyth, SY23 3LS
2E0 DVU M. Hughes, 183 Station Road, Hednesford, Cannock, WS12 4DP
2E0 DVV J. Allen, 17 Inglemere Gardens Arnside, Carnforth, LA5 0BX
2E0 DVW M. Roberts, 5 Cliff Cottages, Cliff Road, Hessle, HU13 0HB
2E0 DVX N. Malyon, 50 Swanage Road, Southend-on-Sea, SS2 5HY
2E0 DVY J. Poriyath, 48 Howard Close, Cambridge, CB5 8QU
2E0 DWA Dr A. Dickson, 1 Roebuck Drive Baldwins Gate, Newcastle-under-Lyme, ST5 5FE
2E0 DWB D. Belton, 4 Sandown Road, Toton, Nottingham, NG9 6GN
2M0 DWC D. Cowie, 69 Broomfield Park, Portlethen, Aberdeen, AB12 4XT
2U0 DWD A. Prosser, Woodlands, La Vassalerie, St Andrew, Guernsey, GY6 8XL
2E0 DWE D. Toller, Field Cottage, Ash Lane, Etwall, Derby, DE65 6HT
2E0 DWF D. Leyland, 20 Newton Heath, Middlewich, CW10 9HL
2E0 DWG D. Gough, 29 Belvedere Road, Biggin Hill, Westerham, TN16 3HX
2E0 DWJ D. Johnston, 34 Coxley Crescent, Netherton, Wakefield, WF4 4LR
2E0 DWK C. Duckworth, 15 St Johns Court, Burnley, BB12 6QE
2E0 DWM D. Mardlin, 13 Churchill Crescent Sonning Common, Reading, RG4 9RU
2E0 DWP D. Prout, 2 Pine Crest Way, Bream, Lydney, GL15 6HG
2W0 DWR D. Rich, 41 Ronald Road, Newport, NP19 7GF

2E0 DWS D. Sewell, 19 St. Leonards Way, Ashley Heath, Ringwood, BH24 2HS
2E0 DWT D. Wells, 40 Barnham Broom Road, Wymondham, NR18 0DF
2E0 DWU A. Edmonds, 20 Tomline Road, Ipswich, IP3 8BZ
2E0 DWV E. Preda, 4 Playsteds Lane, Great Cambourne, Cambridge, CB23 6GA
2E0 DWW K. Gorringe, 6 Buttercup Park, Pevensey Bay, BN24 6BE
2E0 DWZ M. Davey, Romea, Long Street, Attleborough, NR17 1LW
2M0 DXC A. Brown, 37 Sherwood Loan, Bonnyrigg, EH19 3NF
2E0 DXJ D. Barrett, 208 Doncaster Road, Rotherham, S65 2UE
2E0 DXK J. Whitworth, 31 Shirley Close, Chesterfield, S40 4RJ
2E0 DXL I. Newton, 16 Cross Close, Newquay, TR7 3LB
2E0 DXO P. Higginson, 9 St. Mildreds Way, Heysham, Morecambe, LA3 2QJ
2E0 DXW M. Roper, 26 Malpas Close, Bransholme, Hull, HU7 4HH
2E0 DXZ L. Brearley, Ash Tree Lodge, Snaith Road, Goole, DN14 0AT
2E0 DYB D. Brough, 38 Tynedale Avenue, Crewe, CW2 7NY
2E0 DYD G. Lloyd, 1 Holmside Terrace, Stanley, DH9 6ET
2E0 DYG D. Young, Thring Lodge, Granley Gardens, Cheltenham, GL51 6LQ
2E0 DYJ M. Bentley, 5 Stokewell Road, Wath-Upon-Dearne, Rotherham, S63 6EL
2E0 DYM P. Cottam, Eastleigh, Kings Nympton, Umberleigh, EX37 9ST
2E0 DYN D. Jones, Drove Farm, Sheepdrove, Lambourn, Hungerford, RG17 7UN
2E0 DYP W. Hartley, 30 Coltman Avenue, Beverley, HU17 0EY
2E0 DYQ P. Sykes, 16 Hill Fold, South Elmsall, Pontefract, WF9 2BZ
2E0 DYU B. Healey, 14 Orchard Close, Ferring, Worthing, BN12 6QP
2E0 DYV P. Collier, Flat 9, Henry House, Wyvil Road, London, SW8 2TF
2E0 DYX T. Carpenter, 1 Terriers Lane, Hayling Island, PO11 0FF
2E0 DYY L. Hopgood, 62 Briarwood Drive, Blackpool, FY2 0EB
2E0 DZA H. Harclerode, The Bungalow, Silverlace Green, Parham, Woodbridge, IP13 9AD
2E0 DZC L. Davison, 58 Priestley Court, South Shields, NE34 9NQ
2E0 DZE M. White, 9 Hawksworth Close, Rotherham, S65 3JX
2E0 DZF P. Ridgers, 231 The Greenway, Epsom, KT18 7JE
2E0 DZJ G. Allen, 14 The Parsonage, Sixpenny Handley, Salisbury, SP5 5QJ
2E0 DZM R. Williams, 10 Bramley Close, Twickenham, TW2 7EU
2E0 DZN M. Gosi, 49 Elms Drive, Marston, Oxford, OX3 0NW
2E0 DZO P. Sargeant, 6 Meldon Way, Winlaton, Blaydon-on-Tyne, NE21 6HJ
2E0 DZQ J. Ostapiuk, 33 Kent Terrace, Haswell, DH62EL
2E0 DZR P. Pain, 12 Maple Drive, Bamber Bridge, Preston, PR5 6RA
2E0 DZS Dr Z. Derzsi, 217 Bensham Road, Bensham, Gateshead, NE8 1US
2E0 DZT J. Emery, Mulberry Cottage, Quarry Lane, Combe St Nicholas, Chard, TA20 3PH
2E0 DZV N. Harris, 10 Pentland Drive, North Hykeham, Lincoln, LN6 9TG
2M0 DZX A. Mccormick, Flat 2 16 Marine Drive, Edinburgh, EH5 1FD
2E0 DZY D. Atkins, 32 Braybrook, Orton Goldhay, Peterborough, PE2 5SH
2M0 DZZ D. Taylor, 1 Mayfield Farm Cottages, Reston, Eyemouth, TD14 5LG
2E0 EAA B. Matthews, 30 Oaklands Drive, Brandon, IP27 0NR
2M0 EAC J. Woods, 12 Westbank Terrace, Macmerry, Tranent, EH33 1QE
2W0 EAD D. Cook, 19 Almond Avenue Risca, Newport, NP11 6PF
2E0 EAF R. Bond, 21 Coleridge Close, Bletchley, Milton Keynes, MK3 5AF
2E0 EAI S. Bide, 9 Greenway, Watchet, TA23 0BP
2E0 EAL J. Oldman, High Waters, Bentfield Green, Stansted, CM24 8HX
2E0 EAN P. Fulbrook, 167 Droitwich Road, Fernhill Heath, WR3 7TZ
2E0 EAO M. Harrison, 91 Rye Road, Hastings, TN35 5DH
2I0 EAR E. Rainey, 22 Cherry Gardens, Ballymoney, BT53 7AS
2I0 EAS D. Elliott, 15 Derrychara Park, Enniskillen, BT74 6JP
2E0 EAU E. Aksamit, 14 Popplewell Gardens, Gateshead, NE9 6TU
2E0 EAV D. Blake, Pound Farm, Swan Lane, Leigh, Swindon, SN6 6RD
2E0 EAW E. Cross, 12B Oakridge, Three Rivers Country Park, Clitheroe, BB7 3JW
2E0 EAY A. Driver, 12 Almond Tree Avenue, Malton, YO17 7DF
2E0 EAZ E. Heath, 63 Meadway, Dunstable, LU6 3JT
2E0 EBA R. Aldridge, 37 Vincent Road, Luton, LU4 9AN
2I0 EBB A. Connolly, 68 Willowbank Gardens, Belfast, BT15 5AJ
2E0 EBD B. Clayton, 26 Wood Walk, Mexborough, S64 9SG
2E0 EBJ F. Taylor, 55 Leicester Avenue, Horwich, Bolton, BL6 5QX
2E0 EBK D. Levett, 11 Love Lane, London, SE25 4NG
2E0 EBL C. Haynes, 25 Barnards Hill Lane, Seaton, EX12 2EQ
2E0 EBN P. Singleton, 9 Sherbourne Road, Middleton, Manchester, M24 6FF
2E0 EBP T. Stokes, 1 Hunters Reach, Bradwell, Milton Keynes, MK13 9BT
2E0 EBQ M. Priest, 35 Albert Road, Chaddesden, Derby, DE21 6SJ
2E0 EBR A. Buckland, 21 Malton Close, Monkston, Milton Keynes, MK10 9HR
2I0 EBS E. Mcknight, 14 Marlacoo Beg Road, Portadown, Craigavon, BT62 3TF
2M0 EBU R. Clow, 25 Scott Street, Newcastleton, TD9 0QQ
2E0 EBV D. Cutts, 38 Berkeley Drive, Hornchurch, RM11 3PY
2M0 EBW D. Gemmell, 36 Church St., Dumfries, DG2 7AS
2E0 EBX K. Carter, 50 Elliman Avenue Bottom Flat, Slough, SL2 5BG
2E0 EBZ A. Chance, 24 Doddsfield Road, Slough, SL2 2AD
2E0 ECC E. Wright, 2 Wulfrath Way, Ware, SG12 0DN
2E0 ECD C. Dennis, 1 West Villa, Crathorne, Yarm, TS15 0BA
2E0 ECG T. Anthony, 27 Evelyn Avenue, Doncaster, DN2 6LN
2E0 ECI W. Canavan, 9 The Ridings, Deanshanger, Milton Keynes, MK19 6JD
2E0 ECJ H. Mcevoy, 18 Brookfield Gardens, West Kirby, CH48 4EL
2M0 ECK A. Falconer, 61 Mountcastle Drive North, Edinburgh, EH8 7SP
2E0 ECO A. Hawksworth, 87 Bradeley Road, Haslington, Crewe, CW1 5PX
2E0 ECP S. Andrews, 24 Beesley Road, Banbury, OX16 0HL
2E0 ECQ J. O'Donnell, 24 Foxhill, Watford, WD24 6SY
2E0 ECU P. Escott, 84 Salisbury Avenue, Bootle, L30 1PZ
2E0 ECV A. Jovanovic, 33 Seward Road, London, W7 2JS
2E0 ECW E. Williams, 19 Raglans, Exeter, EX2 8XN
2E0 ECY G. Bryant, 11 Cadwallon Road, London, SE9 3PX
2W0 ECZ M. Douglas, 486 Malpas Road, Newport, NP20 6NB
2E0 EDA J. Sim, 11 Haven Close Istead Rise, Gravesend, DA13 9JR
2E0 EDC D. Lavell, 51 Kingfisher Close, Newport, PO30 5XS
2E0 EDF T. Talbot-Humphries, 22 Vicar Street, Wednesbury, WS10 9HF
2E0 EDG N. Chamberlain, 8 Southfields, Binbrook, Market Rasen, LN8 6DX
2E0 EDI E. Stormes, 1 Meadowbank, Belton, Doncaster, DN9 1NW
2E0 EDL D. Simmons, 8 Lower Grange, Huddersfield, HD2 1RU
2E0 EDM E. Moore, 44 Bridge Street, Oxford, OX2 0BB
2E0 EDN V. Lucock, 34 Wentworth Drive, Ipswich, IP8 3RX
2E0 EDP E. Palo, 13 Welwyn Close, St Helens, WA9 5HL
2D0 EDQ E. Quinney, 69 Clagh Vane, Ballasalla, Isle Of Man, IM9 2HF
2E0 EDR D. Lock, 20 Jasmine Close Trimley St. Martin, Felixstowe, IP11 0UY
2E0 EDS E. Kaye, 119 St. Bernards Avenue, Louth, LN11 8AS
2M0 EDV E. Stuart, 6D Dundee Street, Letham, DD8 2PQ
2E0 EDX I. Taylor, 37 Wood Green Drive, Thornton-Cleveleys, FY5 3DH
2M0 EDY E. Higgins, 44A Mossvale St., Paisley, PA3 2LR
2E0 EEB J. Cameron, 29 Webster Road, Stanford-le-Hope, SS17 0BE
2E0 EEF D. Finlay, 23 Glen Way, Oadby, Leicester, LE2 5YF
2E0 EEI M. Rose, 149 Claremont Road, Blackpool, FY1 2QJ
2E0 EEJ T. Scott, Fiddlers Den, Ellingstring, Ripon, HG4 4PW
2E0 EEM N. Bristow, Flat 204, Gilbert House Barbican, London, EC2Y 8BD
2E0 EEO Dr S. Leask, 1 Collington Street, Beeston, Nottingham, NG9 1FJ
2M0 EEQ A. Robson, 100 Dawson Avenue, East Kilbride, Glasgow, G75 8LH
2E0 EER J. Salter, 20 Burrow Road, Chigwell, IG7 4HQ
2E0 EES S. Rattley, 2 Burnt Cottages Beanacre, Melksham, SN12 7PT
2E0 EET P. Woodburn, 21 The Row, Silverdale, Carnforth, LA5 0UG
2E0 EEU A. Bullard, 15 Rowan Drive, Lutterworth, LE17 4SP
2M0 EEV W. Mcbain, 56 Scotstoun Park, South Queensferry, EH30 9PQ
2E0 EEW E. Thresher, 18 Sandy Lane, Preesall, Poulton-le-Fylde, FY6 0EH
2E0 EEZ P. Day, 1 Pine Close, Lutterworth, LE17 4UT
2E0 EFA M. Elbourn, 1 Downside, Gosport, PO13 0JS
2E0 EFC D. Owings, 11 Thingwall Road East, Thingwall, Wirral, CH61 3UY
2M0 EFD A. Keogh, 251 Main Street Plains, Airdrie, ML6 7JH
2E0 EFG S. Ruddy, 27 Grove Park Walk, Harrogate, HG1 4BP
2E0 EFH T. Hull, 12 Durley Road, Gosport, PO12 4RT
2M0 EFI F. Wenseth, 2 Sunnybank Cottage, Logie Coldstone, Aboyne, AB34 5PQ
2E0 EFK T. Bell, 28 Inwood Drive, Coleford, GL168EZ
2E0 EFN E. Almas, 10, Rindal, Norway, 6657
2E0 EFO M. Luper, 49 Russell Road, Buckhurst Hill, IG9 5QF
2E0 EFP S. Nelson, 25 Waterloo Place, North Shields, NE29 0NA
2U0 EFR D. Robert, Nos Treis 7 Liberation Drive, Route Des Clos Landais, St. Saviour, Guernsey, GY7 9PH
2E0 EFS R. Hubbard, Southbroom School House, Estcourt Street, Devizes, SN10 1LW
2E0 EFU J. Leeson, 2 Hawthorn Road, Radstock, BA3 3NW
2E0 EFW H. Talbot, Flat 11, Sedley Court Malta Road, Cambridge, CB1 3LW
2E0 EFY J. Goodman, 70 Bradford Road Eccles, Manchester, M30 9FT
2E0 EFZ A. Haylor, 33 Crimp Hill Road, Old Windsor, SL4 2QY
2M0 EGE R. Bisset, 8/1 North Bughtlin Brae, Edinburgh, EH12 8XH
2M0 EGI E. Ireland, The Steading, Blairmains, Shotts, ML7 5TJ
2W0 EGK K. Martin, 12 Heol Fargoed, Bargoed, CF81 8PP
2W0 EGL M. Lima Barbosa, 65 Precelly Place, Milford Haven, SA73 2BW
2E0 EGM D. Chapman, 27 Cuff Crescent, London, SE9 5RF
2I0 EGN R. Nelson, 32 Rallagh Road, Dungiven, Londonderry, BT47 4TT
2E0 EGO M. Read, 3613 Clary Ave, Fort Worth, USA, 76111
2E0 EGP J. Churchill, West Winds, Brandheath Lane, New End, Astwood Bank, Redditch, B96 6NG
2E0 EGQ D. Rowe, 18 Burman Road, Wath-Upon-Dearne, Rotherham, S63 7ND
2E0 EGR W. Smith, 81 Hazelgrove Residential Park Milton Street, Saltburn-by-the-Sea, TS12 1FE
2E0 EGS J. Booth, 27 Moorlands Scholes, Holmfirth, HD9 1SW
2E0 EHA H. Redington, Galeholm, Whitecroft, Gosforth, Seascale, CA20 1AY
2E0 EHB N. Livingstone, 2 Mickleton, Wilnecote, Tamworth, B77 4QY
2E0 EHD E. Delasalle, 31 West Hill Road, Hoddesdon, EN11 9DL
2E0 EHH N. Reeve, 124 Greenhills Road, Eastwood, Nottingham, NG16 3FR
2E0 EHJ S. Marsh, 6 Sparsholt Road, Southampton, SO19 9NJ
2E0 EHK J. Oldham, 5 Amersham Rise, Nottingham, NG8 5QG
2E0 EHN P. Norman, 455 Willerby Road, Kingston upon Hull, HU55JD
2E0 EHO D. Turford, 51 Moorfield Avenue, Bolsover, Chesterfield, S44 6EJ
2E0 EHP C. Hughes, 41 Rotherham Road, Dinnington, Sheffield, S25 3RG
2E0 EHQ N. Crudgington, Appledore Blackness Lane, Keston, BR2 6HL
2E0 EHR A. Lamont, 89 Newlands, Whitfield, Dover, CT16 3ND
2E0 EHS H. Butcher, 12 Bath Road, Willesborough, Ashford, TN24 0BJ
2E0 EHT G. Clark, 65 Chyvelah Vale, Gloweth, Truro, TR1 3YJ
2E0 EHV A. Macdonald, Woodside Cottage, Horton Way, Verwood, BH31 6JJ
2I0 EHW Dr P. Donaghy, 41 Greenvale Manor Antrim, Antrim, BT41 1SB
2E0 EHX C. Ogidih, 89 West Road, Birmingham, B43 5PG
2I0 EIB M. Rush, 122 Cullaville Road Crossmaglen, Newry, BT35 9AQ
2E0 EID D. Goodchild, Gravel Lane, Ringwood, Bh24 1xy
2E0 EIE A. Goodchild, Gravel Lane, Ringwood, Bh24 1ll
2I0 EIG J. Mcgoldrick, 45 Stewarts Road, Dromara, Dromore, BT25 2AN
2I0 EIU A. Dowling, 74 Ashmount Gardens, Lisburn, BT27 5DA
2E0 EIX R. Clark, 67 Seymour Street, Chorley, PR6 0RR
2E0 EIZ T. Baker, 92 Conway Avenue, Derby, DE723GR
2E0 EJA E. Cole, 24 Patrick'S Orchard Uffington, Faringdon, SN7 7RL
2E0 EJC M. Dominguez, 52 St. Leonards Road, Amersham, HP6 6DR
2E0 EJJ E. Potter, 3 Thomson Court Chadwick Close, Crawley, RH11 9LH
2E0 EJK R. Keast, 7 The Finches, Newport, PO30 5GU
2U0 EJL J. Littlewood, Wayland Les Martin, L'islet, GY2 4XW
2E0 EJM E. Marsh, 16 Laurel Close, North Warnborough, Hook, RG29 1BH
2E0 EJO J. Wooldridge, 6 Heskin Road, Lydiate, Liverpool, L31 0BS
2W0 EJP Dr P. Messenger, 9 South Parade, Maesteg, CF34 0AB
2E0 EJQ M. Lisle, 16 Collegiate Crescent, Sheffield, S10 2BA
2E0 EJR R. Edwards, 23 Queens Walk, Ruislip, HA4 0LX
2I0 EJT J. Bingham, 27 Carrickdale Gardens, Portadown, Craigavon, BT62 3BN
2E0 EJV A. Jarvis, 10 West Park, Wadebridge, PL27 6AN
2E0 EJW E. Woodward, 309 Hartfields Manor, Hartfields, Hartlepool, TS26 0NW
2E0 EJZ J. Berrio, 29 Scalborough Close, Countesthorpe, Leicester, LE8 5XH
2E0 EKA J. Parton, 6 Windmill Road, Atherstone, CV9 1HP
2E0 EKB G. Patrick, Athena, 121 Ringmer Road, Worthing, BN13 1DX
2E0 EKC R. Fulcher, 1 Edwards Close Hutton, Brentwood, CM13 1BU
2E0 EKD T. Patatu, 20 Compair Crescent, Ipswich, IP2 0EH
2E0 EKI C. Abbott, 38 Foxcover, Linton Colliery, Morpeth, NE61 5SR
2E0 EKJ T. Smith, Chy Crowshensy, Clifton Road, Redruth, TR15 3UD
2E0 EKK G. Cliffe, 5 Laurel Cottages, Ongar Hill Road, King's Lynn, PE34 4JB
2E0 EKM P. Bray, 24 Eldon Terrace, Bristol, BS3 4NZ
2I0 EKN W. Martin, 81 Camgart Road, Tempo, Tempo, BT94 3EQ
2E0 EKP S. Crane, 1 Nicol Mere Drive Ashton-In-Makerfield, Wigan, WN4 8DQ
2E0 EKR N. Harris, 45 Sleigh Road, Sturry, Canterbury, ct2 0ht
2E0 EKS D. Storton, 50 Cromwell Road, Blackpool, FY1 2RG

2E0 EKT M. Evans, 9 Hollis Close, Long Ashton, Bristol, BS41 9AZ
2E0 EKU J. Woods, Invicta Cottage, Carbrooke Road, Thetford, IP25 6SD
2E0 EKW R. Macdonald, 31 Addison Drive, Stratford upon Avon, CV377PL
2E0 EKX A. Trueman, 10 Mountbatten Ave, Dukinfield, SK165BU
2E0 ELA P. Milton, 7 North Crescent, Steeple Bumpstead, Haverhill, CB9 7DL
2E0 ELD E. Dalton, 120 Goodway Road, Birmingham, B44 8RG
2E0 ELE A. Jepson, Edmonton Road, Mansfield, NG21 9ah
2E0 ELI G. Hayers, 87 Bradleigh Avenue, Grays, RM17 5RH
2E0 ELK Lt. A. Hawkins, 5 Ranworth Road, Great Sankey, Warrington, WA5 3EH
2E0 ELO D. Smith, 7 Kestrel Avenue Bransholme, Hull, HU7 4ST
2M0 ELP C. Maxwell, 29 Ambleside Rise, Hamilton, ML3 7HJ
2J0 ELQ M. Thorpe, Dolphin Cottage, Union Road, Grouville, Jersey, JE3 9ER
2I0 ELR C. Adjey, 1 Foyle Park, Portstewart, BT55 7DL
2E0 ELT W. Foster, 55 Drake Avenue Minster On Sea, Sheerness, ME12 3SA
2E0 EMB L. Lee, 8 William Avenue, Margate, CT9 3XT
2E0 EME P. Smith, 5 Olivers Hill, Cherhill, Calne, SN11 8UR
2E0 EMF W. Johnson, 10 Archdale Road, Nottingham, NG5 6EB
2E0 EMG E. Macgurk, 10 Elmore Road, Lee on Solent, PO139DU
2E0 EMK M. Robino, 61 Mill Hill Little Hulton, Manchester, M38 9TN
2E0 EML D. Polley, 6 Coneygear Road, Hartford, Huntingdon, PE29 1QL
2M0 EMM E. Munro, 55 Abergeldie Road, Aberdeen, AB10 6ED
2E0 EMN C. Young, 21A Union Crescent, Margate, CT9 1NS
2E0 EMP C. Keszei, 2 Blackmore Hill Farm Cottages, Calvert Road, Buckingham, MK18 2HA
2E0 EMQ J. Arnold, 17 Larch Road Roby, Liverpool, l369ty
2E0 EMX Dr A. Holt, 36 The Maltings, Malmesbury, SN16 0RN
2E0 EMZ M. Brown, 99 Apprentice Drive, Colchester, CO45SE
2E0 ENB G. Wall, Flat 2, Coniston House Holyoake Road, Worsley, Manchester, M28 3DH
2E0 ENC E. Woolfenden, 20 Belvedere Avenue Atherton, Manchester, M46 9LQ
2W0 END M. Townsend, 71 Elm Court, Newbridge, Newport, NP11 5LU
2E0 ENE E. Smethurst, 4 Lower New Row, Worsley, Manchester, M28 1BE
2E0 ENF S. Klee, 4 Abbots Road, Pershore, WR10 1LL
2E0 ENG M. Winch, 2 Cranleigh Gardens, Cowes, PO31 8AS
2E0 ENI G. Howard, 8 Paddock Road, Woodford, Kettering, NN14 4FL
2W0 ENJ J. Turner, 4 Imble Street, Uk, SA726QL
2E0 ENM N. Winfield, 1 Southview, School Lane, Stoke Row, Henley-on-Thames, RG9 5QX
2E0 ENN S. Burton, 20 Flowerdown Avenue, Cranwell, Sleaford, NG34 8HZ
2E0 ENP K. Matthews, St. Helens Cottage, Flimby, Maryport, CA15 8RX
2W0 ENQ M. Mckenna, 25 Heaton Place Norton Road, Rhos On Sea, Colwyn Bay, LL284TL
2W0 ENV C. Hurley, 14 Magazine Street, Maesteg, CF34 0TG
2E0 ENW E. Wilson, 60 Seathorne, Withernsea, HU19 2BB
2E0 ENZ P. Joynson, 10 Rothesay Gardens, Prenton Hall Road, Prenton, CH43 3DW
2E0 EOC L. Guy, 34 Brindley Avenue, Wolverhampton, WV11 2PB
2E0 EOD A. Billingham, 6 Kemble Close, Lincoln, LN6 0NR
2E0 EOF C. Westcott, 50 Bentinck Street, Sutton-in-Ashfield, NG17 4AZ
2E0 EOH C. Shipman, 19 Westerkirk Drive, Madeley, Telford, TF7 5RJ
2E0 EOI D. Rolfe, 49 Hillrise Avenue, Sompting, Lancing, BN15 0LU
2E0 EOL Rev. D. Palmer, 14 Walcot Parade, Bath, BA1 5NF
2E0 EOM I. Garrard, 33 Uplands Road, Hockley, SS5 4DL
2E0 EON D. Hendy, 12 Rumsam Gardens, Barnstaple, EX32 9EY
2E0 EOP N. Haigh, 10 Moor Park Gardens, Dewsbury, WF12 7AS
2I0 EOS J. Kelso, 32 Old Park Manor, Ballymena, BT42 1RW
2E0 EOU J. Hinds, 83 Feckenham Road, Astwood Bank, Redditch, B96 6DE
2E0 EOV A. Blackwell, 27 Prince Charles Crescent, Farnborough, GU14 8DJ
2E0 EOX R. Waterhouse, 19 Honeysuckle Close, St Leonards-on-Sea, TN37 7LX
2E0 EOZ J. Green, 26 Foxhill Road Burton Joyce, Nottingham, NG14 5DB
2E0 EPA A. Bailey, 52 Berkeley Road, Shirley, Solihull, B90 2HT
2E0 EPB S. Duncan, 5 Spring Vale Bilton, Hull, HU11 4DN
2I0 EPC T. Crawford, 21 Ardranny Drive, Newtownabbey, BT36 6BD
2W0 EPE P. Plummer, 26 Hill Road, Neath Abbey, Neath, SA10 7NR
2E0 EPF A. Lane, 26 Astral Gardens Sutton-On-Hull, Hull, HU7 4YS
2I0 EPG I. Cairns, 18 Molyneaux Avenue, Larne, BT40 2TU
2E0 EPH C. Grant, 27 Bulrush Close, Chatham, ME5 9BN
2E0 EPJ Dr M. Heng, 1 St. Giles Croft, Beverley, HU17 8LA
2M0 EPK J. Wilson, 16 Big Brigs Way, Newtongrange, EH22 4DG
2E0 EPM J. Caswell, 10 Beech Close, Scole, Diss, IP21 4EH
2I0 EPN R. Hetherington, 112 Screeby Road, Fivemiletown, BT75 0LG
2E0 EPP P. Petersen, 15 Kent Gardens, Birchington, CT7 9RS
2E0 EPR E. Phiri, 26 Sedge Road, Andover, SP11 6RL
2E0 EPT R. Paris, 14 Jarden, Letchworth Garden City, SG6 2NP
2E0 EPU L. Pitman, 14 Walton Village, Liverpool, L4 6TJ
2E0 EPZ D. Taylor, 8 Highfields, Mill Croft Close, Norwich, NR5 0RU
2E0 EQB D. Baker, 22 Cleveland Road, Plymouth, PL4 9DF
2I0 EQC M. Mccourt, 52A Moira Road, Crumlin, BT29 4JL
2E0 EQD N. Peppe, 5 Cherford Road, Bournemouth, BH11 8SU
2E0 EQE C. Howell, 47 Birch Park Coalway, Coleford, GL16 7RU
2E0 EQG M. Barrett, 57 Marlborough Avenue, Hornsea, HU18 1UA
2E0 EQH P. Catterall, 117 Beech Hill Lane, Wigan, WN6 8PJ
2E0 EQI P. Wright, 25, Paynes Meadow, Whitminster, Gloucester U. K., GL2 7PS
2E0 EQJ Dr B. Minnis, 16 Dene Tye, Crawley, RH10 7TS
2E0 EQK Dr F. Harvey, 14 Wilton Road, Hornsea, HU18 1QU
2E0 EQL S. Daniels, 48 Gason Hill Road, Tidworth, SP9 7JX
2E0 EQN D. Ellens, 150 Lumley Avenue, South Shields, NE34 7DJ
2E0 EQP G. White, 19 Abbots Hall Avenue Clock Face, St Helens, WA9 4UX
2W0 EQQ L. Pickering, 14 Bryn Wyndham Terrace, Cardiff, CF42 5NG
2I0 EQR D. Mcdonnell, 52 Moira Road, Glenavy, Crumlin, BT29 4JL
2I0 EQS T. Mcdonnell, 52 Moira Road, Glenavy, Crumlin, BT29 4JL
2E0 EQT T. Canning, Whitegates, Mayfield Avenue New Haw, Addlestone, KT15 3AG
2E0 EQU P. Hill, 83 Gladeside, Croydon, CR0 7RW
2E0 EQW J. Hope, 4 Beaumont Way, Prudhoe, NE42 6RA
2E0 EQZ B. Davies, 16 Pearmains Close, Orwell, Royston, SG8 5QY
2E0 ERB J. Neal, 40 Channel View Road, Portland, DT5 2AY
2E0 ERD R. Overy, 62 Dykelands Road, Sunderland, SR6 8ER
2E0 ERE M. Bruce, 28 Pheasants Way, Rickmansworth, WD3 7ES
2E0 ERF G. Sheppard, 32 Bramble Drive, Hailsham, BN27 3EG
2E0 ERG R. Harriman, 9 Millers Close, Rushden, NN10 9RP
2E0 ERJ D. Cassidy, 186 Kingsman Drive, Boorley Green, Botley, SO32 2TG
2E0 ERK E. Kissin, 115 Aarons Hill, Godalming, GU7 2LJ
2E0 ERL R. Carpenter Junior, 218 Mansel Road West, Southampton, SO16 9LR
2E0 ERM E. Milner, 16 Spring Valley Court, Bramley, Leeds, LS13 4TT
2E0 ERO A. Cattell, 40 Collyweston Road, Northampton, NN3 5ET
2E0 ERP A. Burton, 12 Munden Grove, Watford, WD24 7EE
2E0 ERQ M. Pittas, 89 Seddon Road, Morden, SM4 6ED
2E0 ERS R. Springall, 18 Westbourne Park, Scarborough, YO12 4AT
2E0 ERT R. Smith, 21 Canal Road Crossflatts, Bingley, BD16 2SR
2E0 ERU M. Curtis, 9 Oaklands Crescent, Holt, NR25 6UD
2E0 ERV S. Slack, 10 Micheldever Road, Whitchurch, RG28 7JG
2E0 ERX M. Hales, 23 Raeburn Road, Sidcup, da158rd
2E0 ERZ E. Vrentzos, 12 Floyer Close, Queens Road, Richmond, TW10 6HS
2I0 ESA J. Mcgoldrick, 45 Stewarts Road, Dromara, Dromore, BT25 2AN
2E0 ESB A. Volkov, Flat 8, Despard House, 43 Palace Road, London, SW2 3EW
2E0 ESC M. Bennett-Blacklock, 46 Friern Road, London, SE22 0AX
2E0 ESI A. Ahuja, 32 Riverview Gardens, London, SW13 8QY
2E0 ESJ N. Evans, 38 Cockster Road, Longton, Stoke-on-Trent, ST3 2EG
2M0 ESL G. Robinson, Asgard, 12 Upper Waston Road, Burray, KW172TT
2E0 ESO M. Bull, 14 Ermin Walk, Thatcham, RG19 3SD
2E0 ESS E. Slevin, Woodcock Hall, Cobbs Brow Lane, Newburgh, WN8 7NB
2E0 ESU M. Eade, Roemah-Kita, Strathcona Avenue, Leatherhead, KT23 4HP
2E0 ESX J. Hawthorn, 1 Tudor Close, Leigh-on-Sea, SS9 5AR
2E0 ESY M. Clitheroe, Green End, North Street, Castle Acre, King's Lynn, PE32 2BA
2E0 ETA M. Shasby, 19 Crawshaw Grange, Crawshawbooth, Rossendale, BB4 8LY
2I0 ETB W. Curry, 7 Ballyversal Road, Coleraine, BT52 2ND
2E0 ETC L. Kay, 20 Mytton Lane, Shawbury, Shrewsbury, SY4 4JE
2E0 ETD R. Sindall, 16 Chantrell Road, Wirral, CH48 9XP
2E0 ETE A. Beardsley, 10 Moreton Close, Church Crookham, Fleet, GU52 8NS
2E0 ETF P. Molloy, 4 Tilt Meadow, Cobham, KT11 3AJ
2E0 ETG M. Landon, 29 Portland Road, Hucknall, Nottingham, NG15 7SL
2E0 ETH C. Marshall, 51 Hedgerow Close, Redditch, B98 7QF
2E0 ETI A. Thompson, 25 Ardingly Road, Cuckfield, Haywards Heath, RH17 5HD
2M0 ETJ G. Cartwright, Redhouses Outertown, Annan, DG12 5LN
2M0 ETL I. Ferguson, 5 Grahamsfield Court Kirkpatrick Fleming, Lockerbie, DG7 3BD
2E0 ETN R. Bradshaw, 272 Councillor Lane, Cheadle Hulme, Cheadle, SK8 5PN
2E0 ETP S. Cook, 68, Barnsley, S71 4ry
2E0 ETQ G. Hinson, The Leys, Brierley Hill, DY5 3UJ
2E0 ETT T. Horsten, Kastelsvej 4, 2.Tv, Copenhagen E, Denmark, 2100
2E0 ETU E. Dutson, 19 Maythorn Drive, Cheltenham, GL51 0QH
2E0 ETV K. Bianchini, 10 St. Leonards Road, Headington, Oxford, OX3 8AA
2I0 ETW P. Moore, 32 Kinnegar Rocks, Donaghadee, BT21 0EZ
2E0 ETY C. Wrobel, 33 Harsnett Road, Colchester, CO1 2HS
2W0 ETZ W. Cooper, 1 Bedw Street, Maesteg, CF34 0TF
2E0 EUB C. Wooldridge, 26 Grieg Close, Basingstoke, RG22 4DU
2E0 EUD B. Clark, 46 Fraser Close, Laindon, Basildon, SS15 6SU
2E0 EUF F. Boyce, 277 Manor View, Par, PL24 2EP
2E0 EUH P. Davies, 2 Lynfords Drive, Runwell, Wickford, SS11 7PP
2E0 EUI D. Nolan, Flat 7, Fonthill Court, London, SE23 3SJ
2E0 EUJ C. Cromie, 140 Whalley Road Wilpshire, Blackburn, BB1 9LJ
2E0 EUM T. Mann, 6 Kenley Close, Wickford, SS11 8XL
2E0 EUN E. Underhill, 61 Goldthorne Avenue, Sheldon, Birmingham, B26 3LA
2W0 EUO A. Jones, 75 Hollybush Road, Cardiff, CF23 6SZ
2E0 EUP J. Foxall, 2 Millers Walk Pelsall, Walsall, WS3 4QS
2E0 EUR S. Milner, Pavilion House, School Lane, Ormskirk, L40 3TG
2I0 EUV R. Skelton, 16 Demiville Avenue, Lisburn, BT27 5RE
2E0 EUW M. Augustus, 3 Heathend Cottages Heathend, Wotton-under-Edge, GL12 8AS
2W0 EUY P. Squire, 18 Rayon Road Greenfield, Holywell, CH8 7EQ
2E0 EUZ A. Fisher, 63 Soloman Drive, Bideford, EX39 5XY
2E0 EVA M. Knowles, 12 Dalestorth Avenue, Mansfield, NG19 6NT
2E0 EVC Dr C. Hall, 6 Browning Road Church Crookham, Fleet, GU52 0YJ
2E0 EVE D. Cattermole, Blaxhall Hall Crossing, Little Glemham, Woodbridge, IP13 0BP
2E0 EVF D. Redmayne, 10 The Square, Kington, HR5 3BA
2E0 EVI G. Rodriguez, 32 Mount Pleasant, Prestwich, Manchester, M25 2SD
2E0 EVM E. Scott, 64 Smallcombe Road, Paignton, TQ3 3RX
2E0 EVP M. Brasher, 48 Eldertree Road Thorpe Hesley, Rotherham, S61 2TQ
2E0 EVQ G. Singleton, 2 Rome Avenue, Burnley, bb115lq
2E0 EVR G. Round, 128 Leicester Road, Shepshed, Loughborough, LE12 9DH
2E0 EVU D. Mallinson, 7 Abbots Way, Yeovil, BA21 3HX
2E0 EVV T. Roberts, 16 Malfort Road, London, SE5 8DQ
2E0 EVW B. Skelton, 1 Summer Hill Road, Bexhill-on-Sea, TN39 4LN
2E0 EVX S. Hall, 12 Lady Jane Grey Road, King's Lynn, PE30 2NW
2E0 EWB J. Holley, Lookers, Blenheim Road Littlestone, New Romney, TN28 8PR
2E0 EWF S. Totham, Cavaliers, Dane Court Manor, School Road, Tilmanstone, Deal, CT14 0JL
2E0 EWH M. Lovering, 19 Chilton Avenue, Sittingbourne, ME10 4TB
2E0 EWL J. Keefe, 64 Heath Lane, Blackfordby, Swadlincote, DE11 8AA
2E0 EWM E. Melman, 177 Grantham Road, London, E12 5NB
2E0 EWQ N. Booth, 10 Games Walk, Wythenshawe, Manchester, M22 1SN
2E0 EWS E. Sherwin, 3 Warren Place, Barnsley, S70 4LR
2E0 EWT C. Barker, 11 Long Meadows, Chorley, PR7 2YA
2E0 EWX J. Dowson, 4 Thorntree Close, Goole, DN14 6HJ
2M0 EWY A. Kinnersley, 5A Regent Terrace, Dunshalt, Cupar, KY14 7HB
2E0 EXA D. O Brien, 5 Orson Leys, Rugby, CV22 5RG
2E0 EXB R. Fearnley, 31 Radburn Court, Dunstable, LU6 1HW
2E0 EXC C. Wilson, 21 New Road, Hythe, Southampton, SO45 6BN
2E0 EXG R. Low, 4A Bellevue Road, Romford, RM5 3AY
2E0 EXH A. Cooke, 17 St. Aldwyn Road, Seaham, SR7 0AN
2E0 EXI C. Smith, 37 Mace House, Union Lane, Isleworth, TW7 6GP
2E0 EXJ S. Baldwin, 143 Oxford Road, Swindon, SN3 4JA
2E0 EXK M. Constantine, 82 The Oval, Brough, HU15 1DD
2E0 EXL P. Morris, 10 Haslam Avenue, Sutton, SM3 9ND
2E0 EXM H. Baker, 300 Lowerhouse Lane, Burnley, BB12 6LZ
2E0 EXN T. Jones, 5 Broomfields Road, Appleton, Warrington, WA4 3AE
2E0 EXO T. Crocker, 32 Godmanston Close, Poole, BH17 8BU

2I0 EXP A. Boyd, 27 The Meadows, Dungannon, BT71 6PW
2E0 EXQ A. Parkin, 22 Glorney Mead, Badshot Lea, Farnham, GU9 9NL
2E0 EXU R. Coles, Bay Cottage, St. Catherines Road, Ventnor, PO38 2NE
2E0 EXV J. Kelly, Flat 4, The Corner Place, 1 North Road, Harborne, Birmingham, B17 9PA
2E0 EXW R. Whitehead, 1 Smithy Site, Farnborough, Wantage, OX12 8NS
2E0 EXX B. Smith, 8 Mill Field Close, South Kilworth, Lutterworth, LE17 6FE
2E0 EXY J. Sergeant, 16 Green Park, Whalley, Clitheroe, BB7 9TJ
2W0 EXZ D. Morgan, 28 Harbour Village, Goodwick, SA64 0DY
2E0 EYB M. Jenkins, 5 Arthurs Hill, Shanklin, PO37 6EW
2E0 EYC C. Revell, Westview, 1 Chapel Lane, Hull, HU12 0US
2E0 EYE E. Ross, Foundry Cottage, Crowders Lane, Battle, TN33 9LP
2E0 EYF A. Kazmi, 39 Herga Road, Harrow, HA3 5AX
2E0 EYG J. Miller, 9 St. Nicholas Road Tillingham, Chelmsford, CM0 7SQ
2E0 EYH W. Rudge, 33 Wyrley Rd, Wolverhampton, WV11 3NY
2E0 EYI R. Wheeler, 21 Abbey Drive Houghton Le Spring, Tyne&Wear, DH45JZ
2D0 EYK N. Smith, 4 Cooil Farrane, Douglas, IM2 1NX
2E0 EYP J. Read, 31 Merebrook Road, Macclesfield, SK11 8RH
2E0 EYS R. Bewick, 357 Franklands Village, Haywards Heath, RH16 3RP
2E0 EYU S. Kembrey, 101 Yew Tree Drive, Bristol, BS15 4UF
2E0 EYY M. Murray, 31 Feeny Street Sutton Manor, St Helens, WA9 4BJ
2M0 EZA R. Cantwell, 78 New Street, Musselburgh, EH21 6JQ
2E0 EZC A. De Mora, 36 West Park, Minehead, TA24 8AN
2E0 EZF G. Smith, 92 Brighton Road, Banstead, SM7 1BU
2E0 EZG M. Lewis, 1 Kingsmead Stretton, Burton on Trent, DE13 0FQ
2E0 EZL T. Newton, 1 Brimley Park, Bovey Tracey, Newton Abbot, TQ13 9DE
2E0 EZO M. O'Connor, 5 Kesbrook Drive Ashwood Park, Overseal, DE126NS
2W0 EZS P. Collier, 42 Derwen Road Alltwen, Pontardawe, Swansea, SA8 3AU
2E0 EZV G. Adam, 33A Claremont Square, London, N1 9LS
2E0 EZX W. Taylor, 5 Council Bungalows, Churchtown, Belton, Doncaster, DN9 1PD
2E0 EZY H. Foster, 6 Lakeway, Blackpool, FY3 8PF
2E0 EZZ G. Hutchings, 2 Beaufort Road Yate, Bristol, BS37 5DS
2E0 FAA D. Mcglone, 32 Shipley Mill Close, Kingsnorth, Ashford, TN23 3NR
2E0 FAB J. Whalley, Flat 9, 37 Church Street, Southport, PR9 0QT
2E0 FAC P. Burke, 38 Bosworth Square, Rochdale, OL11 3QG
2E0 FAE A. Foote, Flat One, Kimber'S Close Kennet Road, Newbury, RG14 5JF
2E0 FAH P. Harris, Flat 33, Buckingham Court Shrubbs Drive, Bognor Regis, PO22 7SE
2E0 FAJ S. Kingstone, 3 Roman Way, Tamworth, B79 8NF
2E0 FAM C. Horridge, 6 Back Sreet, East Stockwith, Gainsborough, DN21 3DL
2E0 FAN P. Pearce, 41 Tennyson Avenue, Boldon Colliery, NE35 9EP
2W0 FAP A. Phillips, 3 Pen Y Llys, Rhyl, LL18 4EH
2E0 FAQ G. Austin, 15K Hardie Avenue, Rugeley, Ws15 1nt
2W0 FAR A. Burgess, 18 Fairmeadows, Maesteg, CF34 9JL
2E0 FAS N. Duke, 5 Hannay Close, Barrow-in-Furness, LA14 1SZ
2E0 FAU B. Cairns, 4 Spence Court, Great Ayton, Middlesbrough, TS9 6DW
2E0 FAV A. Barter, 4 Blackberry Way Kingsteignton, Newton Abbot, TQ12 3QX
2E0 FAY N. Fahey, 5 Hillside, Felmingham, North Walsham, NR28 0LE
2E0 FBA M. Mcphee, 63 Mumford Close, West Bergholt, Colchester, CO6 3HY
2E0 FBC F. Foy, 4 The Square, East Rounton, Northallerton, DL6 2LB
2E0 FBD D. Smith, Heath Farm, Heath Road, Woolpit, Bury St Edmunds, IP30 9RL
2E0 FBE M. Goodman, 80 Clay Street, Soham, Ely, CB7 5HL
2E0 FBF R. Boyes, 63 Larch Road, New Ollerton, Newark, NG22 9SX
2E0 FBH Z. Yao, Room 4C, Unit 12-2, Zhonghaibanshanxigu Garden, No.15 Zhongqing Rd., Yantian District, Shenzhen, China, 518000
2E0 FBJ A. Tomczynski, 66 Commercial Street, London, E1 6LT
2E0 FBL F. Baker, 275 Bye Pass Road, Beeston, Nottingham, NG9 5HS
2E0 FBN T. Weston, 4 The Pightle, Peasemore, Newbury, RG20 7JS
2E0 FBO J. Raybould, 33 Lincoln Road, Dorrington, Lincoln, LN4 3PT
2E0 FBQ R. Copus, 58 Colchester Rd, Holland on Sea, CO15 5DG
2E0 FBR S. Davis, 74 The Driveway, Canvey Island, SS8 0AD
2M0 FBU K. Govan, 30 Glenwell Avenue, Stranraer, Dg9 7ba
2I0 FBY I. Gillespie, 32 Maghaberry Manor Moira, Craigavon, BT67 0JZ
2E0 FCB Dr R. Hopkins, 26 Seymour Road, Bordon, GU35 8JX
2E0 FCE B. Clark, 8 Langdale Close, Farnborough, GU14 0LQ
2W0 FCF D. Clark, 137 Llanedeyrn Road, Penylan, Cardiff, CF23 9DW
2E0 FCG C. Staples, 32 Browns Lane, Netherton, Bootle, L30 5RW
2E0 FCH A. Finch, 6 Clover Way, Thetford, IP24 1LQ
2E0 FCJ C. Pendlebury, 73 Pool Street, Wigan, WN3 5BT
2W0 FCM L. Paschalis, 45 Pencisely Road, Cardiff, CF5 1DH
2E0 FCO M. Killoran, 4 Victoria Road, Pudsey Leeds, LS28 7SR
2M0 FCQ J. Degnan, 6B Manse Road, Whitburn, EH470QA
2E0 FCS C. Spencer, 18 Coatsby Road Kimberley, Nottingham, NG16 2TH
2M0 FCT S. Mulligan, 17 Crawfurd Gardens, Rutherglen, Glasgow, G73 4JP
2E0 FCZ C. Cooper, 25 Waterside Close, Loughborough, LE11 1LP
2E0 FDB R. Goldup, 57 Partridge Way, Old Sarum, Salisbury, SP4 6PX
2W0 FDC S. Quinn, 1334 Carmarthen Road Fforestfach, Swansea, SA5 4BR
2E0 FDD B. Evans, 2 Hastings Road, Eccles, Manchester, M30 8JR
2E0 FDE R. James, 159 Orton Avenue, Birmingham, B76 1JN
2E0 FDG B. Pettit, 35 Lakeside Rise, Blundeston, Lowestoft, NR32 5BE
2E0 FDH C. Ward, 1 Granary Close, Codford, Warminster, BA12 0PR
2M0 FDI D. Robertson, Grindhus, Gonfirth, Voe, ZE2 9PY
2E0 FDS K. Bowyer, 84 Old Chester Road Helsby, Frodsham, WA6 9PG
2M0 FDT S. Doonan, West Clanfin Farm Waterside, Kilmarnock, KA3 6JQ
2E0 FDU D. Knowles, 12 Forest Grove, Harrogate, HG2-7JU
2E0 FDW F. Quinn, Harley Cottage Beech Grove Gardens, Carlisle, CA30LR
2M0 FDZ S. Mclaughlin, 21 Shirrel Road, Motherwell, ML1 4RD
2E0 FEB S. Scotching, 26 Newton Way, Leighton Buzzard, LU7 4YU
2E0 FEC G. Eycott, 1 Ham Road, Wanborough, Swindon, SN4 0DF
2E0 FEE W. Byers, 37 Windermere Drive, Bletchley, Milton Keynes, MK2 3NR
2E0 FEJ I. Davison, 35 Newhouse Avenue, Esh Winning, Durham, DH7 9JH
2E0 FEM S. Simpson, 48 Weatherhill Road Smallfield, Horley, RH6 9LY
2E0 FEO R. Hunter, 3 Sandy Way, Croyde, Braunton, EX33 1PP
2E0 FEP J. Harkins, 22 Croomes Hill, Bristol, BS16 5EH
2E0 FEQ K. Polston, 10 Marsh Farm Road, South Woodham Ferrers, Chelmsford, CM3 5WP
2E0 FEV D. Walton, 106 Cumberworth Lane, Lower Cumberworth, Huddersfield, HD8 8PG
2I0 FEX D. Rantin, 8 Buchanans Road, Newry, BT35 6NS
2W0 FEY A. Fey, 28 Bryn Rhedyn, Caerphilly, CF83 3BT
2E0 FEZ A. Waller, 155 Bridgemary Road, Gosport, PO13 0UT
2E0 FFB R. Evered, Ivy Cottage, Old Bristol Road, Wells, BA5 3AL
2E0 FFC D. Berry, 27 Harcourt Terrace, Headington, Oxford, OX3 7QF
2W0 FFI D. Evans, 1 Heol Glyndwr, Fishguard, SA65 9LN
2E0 FFJ G. Saunders, Hastings Road, Battle, TN330TA
2E0 FFK S. Hagan, 5 High Street, Brampton, Huntingdonshire, PE28 4TG
2W0 FFL Dr F. Labrosse, 72 Ger Y Llan Penrhyncoch, Aberystwyth, SY23 3HQ
2E0 FFM J. O'Reilly, 116 Coleridge Way, Crewe, CW1 5LF
2W0 FFN A. Prime, 106 Snowden Road, Cardiff, CF5 4PS
2E0 FFO M. Norfield, 122 Huntingdon Road, Upwood, Ramsey, Huntingdon, PE26 2QQ
2E0 FFP G. Peters, Curlew Court, Guys Head Road, Sutton Bridge, Spalding, PE12 9QQ
2E0 FFS R. Baines, 2 Lower Lune Street, Fleetwood, FY7 6DA
2E0 FFV A. Hawkes, 9 Ferneley Avenue, Hinckley, LE100FE
2E0 FFW P. Drake, Flat 1, Richmond Court, Eagle Close, Yeovil, BA22 8JY
2M0 FFY C. Northcott, 4/6 Castleview House, 2 Craigour Place, Edinburgh, EH17 7Rt
2E0 FGA P. Meanwell, 20 Crow Park Avenue, Sutton-On-Trent, Newark, NG23 6QG
2E0 FGH N. Speller, 2 Hurst Rise Road, Oxford, OX2 9HQ
2E0 FGI I. Kendrick, 10 Bankhouse Drive, Congleton, CW12 2BH
2E0 FGM S. Haigh, 17 Glebe Street, Swadlincote, DE11 9BW
2E0 FGO P. Offord, 1 Adare Close, Dunmow, CM6 2GR
2E0 FGQ N. Bennett, 35 West Shepton, Shepton Mallet, BA4 5UD
2E0 FGR M. Newman, 274 Long Drive, Ruislip, HA4 0HY
2E0 FGT M. Jones, 93 Barrs Road, Cradley Heath, B64 7HH
2E0 FGV G. Brindle, 25 Chedworth Drive, Witney, OX28 5FS
2E0 FGW D. Crouch, 7 Tresco Road, Berkhamsted, HP4 3JZ
2E0 FGY B. Bestwick, 185 Ashbourne Road, Turnditch, Belper, DE56 2LH
2E0 FHA M. Sprague, 11 The Orchards, Witcham, Ely, CB6 2LR
2E0 FHB N. Connolly, 32 St. Oswald Road, Bridlington, YO16 7SD
2E0 FHC R. Gaskell, Cottonwood California, Baldock, SG7 6NU
2E0 FHD L. Rodgers, 27 Arden Houses Normanby-By-Spital, Market Rasen, LN8 2HE
2E0 FHE I. Gould, 2 Harps Avenue Minster, Sheerness, ME12 3PF
2E0 FHH C. Wilkinson, Flat 5, Hometide House Beach Road, Lee-on-the-Solent, PO13 9BP
2E0 FHJ R. Lane, 2 Dickiemoor Lane, Plymouth, PL5 3NU
2E0 FHK I. Harding, 7 Hawthorne Close, River, Dover, CT17 0NG
2E0 FHM A. Papiewski, 42 Balmoral Avenue, Spalding, PE11 2RU
2E0 FHN J. Clarke-Stanley, 6 Culpins Close, Spalding, PE11 2JL
2E0 FHQ J. Gale, 46 Ingshead Avenue Rawmarsh, Rotherham., S625BW
2E0 FHR K. Scott, 86 Holcombe Drive, Burnley, BB10 4BH
2E0 FHV A. Shaw, 8 Champion Way, Mablethorpe, LN12 2EJ
2E0 FHX M. Ross, Whitewalls Off Golf Road, Mablethorpe, ln12 1lp
2E0 FIA A. Smeed, Flatt 8 9 Peir Terrace, Lowestoft, Nr330ab
2E0 FIB S. Watling, 1 Chediston Green, Chediston, Halesworth, IP19 0BB
2E0 FIF M. Hadfield, 22 Mansfield Road, Clowne, Chesterfield, S43 4DH
2E0 FIJ Dr D. De-Cogan, 52 Gurney Road, New Costessey, Norwich, NR5 0HL
2E0 FIK A. Holmes, Lodge 7, Arrowbank Caravan Park, Leominster, HR6 9BG
2E0 FIO G. Chalklin, 18 Trinity Close, West Mersea, Colchester, CO5 8RW
2I0 FIP J. Miskimmin, 47 Shore Road, Kircubbin, Newtownards, BT22 2RP
2E0 FIR M. Firth, 209 High Street, Wickham Market, Woodbridge, IP13 0RQ
2E0 FIT R. Lynch, 2 Launceston Close, Oldham, OL8 2XE
2E0 FIU J. Smith, 54 Sneinton Hermitage, Sneinton, Nottingham, NG2 4BS
2E0 FIV J. Rymell, 9 Carter Avenue Ruddington, Nottingham, NG11 6NP
2E0 FIW D. Peacock, 41 Oxford Meadow, Sible Hedingham, Halstead, CO9 3QW
2E0 FIY G. Walmsley, 27 Camberwell Way, Hull, HU8 0RU
2E0 FJA A. Ferriroli, 142 Hillbury Road, Warlingham, CR6 9TD
2E0 FJC A. Winton, 10 Old Parsonage Court Guithavon Street, Withem, CM8 1XP
2E0 FJD K. Dobson, 1 Howarth Road, Ashton-On-Ribble, Preston, PR2 2HH
2E0 FJE A. Pyatt, 13 Summervale Road, Tunbridge Wells, TN4 8JJ
2E0 FJG Z. Zhao, Christs College, Cambridge, CB2 3BU
2E0 FJI A. Atkinson, 21 Dennington Crescent, Basildon, SS14 2FF
2E0 FJJ R. Norwood, Flat 27, Cranfield Court, Galadriel Spring, South Woodham Ferrers Essex, Cm3 7Bd., Chelmsford, CM3 7BD
2E0 FJK S. Chaney, 54 Clementine Avenue, Seaford, BN25 2XG
2E0 FJN D. Picker, 12 Crown Close, Barnsley, S70 4DB
2E0 FJP J. Park, 18 Ladgate Grange, Middlesbrough, TS3 7SL
2E0 FJQ P. Hall, 11 Middleton Court, Mansfield, NG18 3RN
2E0 FJT A. Fletcher, 74 Devonshire Road, Maltby, Rotherham, S66 7DQ
2E0 FJU J. Fletcher, 74 Devonshire Road, Maltby, Rotherham, S66 7DQ
2E0 FJV C. Walson, 30 West Crescent, Duckmanton, Chesterfield, S44 5HE
2E0 FJW J. Mcbride, 78 Sherwood Drive, Redcar, TS11 6DY
2E0 FJX D. Roake, 9 Falcondale Walk, Westbury On Trym, Bristol, BS93JG
2E0 FKB M. Edge, 15 Littlemoor Avenue, Kiveton Park, Sheffield, S26 5NZ
2E0 FKC R. Chappell, 24 Woodend Drive, Shipley, BD18 2BW
2E0 FKD J. Main, 15 Byron Road, Lydiate, Liverpool, L31 0DB
2E0 FKE D. Buchan, 40 Bradstone Road, Winterbourne, Bristol, BS36 1HQ
2M0 FKF D. Simpson, 35 Westbourne Avenue Tillicoultry Clackmannanshire Fk136Pu, Tillicoultry, Fk136pu
2E0 FKH S. Caddy, 133A Barnby Gate, Newark, NG24 1QZ
2E0 FKJ R. Robinson, 4 Limetree Court, Taverham, Norwich, NR8 6QY
2I0 FKM G. Colgan, 42 Loughview Village, Carrickfergus, BT38 7PD
2E0 FKN K. Bramley, 21 Charles Street, Sutton-in-Ashfield, NG17 4LG
2E0 FKS B. Hoare, 2 St. Peters Close, South Newington, Banbury, OX15 4JL
2E0 FKT L. Allcock, 7 Arundel Court, Chesterfield, S43 3UY
2E0 FKU C. Gibson, 11 Parkside Avenue, Queensbury, Bradford, BD13 2HQ
2W0 FKW D. Delve, 28 Cader Avenue Kinmel Bay, Rhyl, LL18 5HY
2E0 FKY P. Robbins, 2 Bramble Drive Claremont Park, Berrow, Burnham-on-Sea, TA8 2NH
2E0 FLA D. Walker, 290 Shannon Road, Hull, HU8 9RY
2E0 FLF C. Wilson, 12 Desmond Avenue, Hornsea, HU18 1AF
2M0 FLG M. Bradshaw, 32 Greycraigs, Cairneyhill, Dunfermline, KY12 8XL
2E0 FLH M. Crees, Flat 3 Fox House, Fox Lane North, Chertsey, KT16 9GY
2W0 FLI Dr E. Flikkema, 7 St. James Mews, Great Darkgate Street, Aberystwyth, SY23 1DW
2M0 FLJ J. Morris, 10 Middlemas Road, Dunbar, EH42 1GJ

2E0 FLK H. Bond, Flat 4, Athrington Court, First Avenue, Felpham, Bognor Regis, PO22 7LB
2E0 FLL B. Murtagh, 4 Sandcroft Court, 76 Garlands Road, Redhill, RH1 6GZ
2E0 FLM G. Powell, 7 Donstan Road, Highbridge, TA9 3LA
2E0 FLN J. Horn, 8 Princess Close, Watton, Thetford, IP25 6XA
2I0 FLO I. Nicholl, 7 Killyclooney Road, Dunamanagh, Strabane, BT82 0LZ
2E0 FLQ O. Phillips, 54 Marshall Road, Cambridge, CB1 7TY
2E0 FLR S. Rhenius, Baythorne Cottage, Baythorne End, Halstead, CO9 4AB
2E0 FLU D. Chivers, 12 Highburn Close, Burnham on Sea, TA8 1LU
2E0 FLV M. Osband, 22 Samian Crescent, Folkestone, CT19 4JW
2W0 FLW D. Flewin, Huntingdon Way, Swansea, SA2 9HN
2E0 FMA A. West, 33 Mundays Row, Waterlooville, PO8 0HF
2E0 FMB T. Huntriss, 1 Threefields, Ingol, Preston, PR2 7BE
2E0 FME D. Bennett, 29 Margraten Avenue, Canvey Island, SS8 7JD
2E0 FMG P. Booker, 17 Colton Copse, Chandler'S Ford, Eastleigh, SO53 4HQ
2E0 FMH D. Gaskell, 126 Higher Green Lane Astley, Tyldesley, Manchester, M29 7JB
2E0 FMI W. Terry, 11 Crescent Ave, Overhulton., Bolton, BL51EN
2E0 FMK F. Hennigan, 50 Fairford Crescent, Downhead Park, Milton Keynes, MK15 9AE
2E0 FML A. Smith, 6 Rawlinson Avenue, Caistor, Market Rasen, LN7 6NQ
2I0 FMN J. Mcniece, 35 Farran Road, Ballymoney, BT53 8HD
2E0 FMO E. Harriott, 21 Slindon Croft, Alvaston, Derby, DE24SD
2E0 FMP T. Leatherbarrow, 17 Egerton, Skelmersdale, WN8 6AA
2W0 FMQ G. Goodbourn, The Gables Cwmduad, Cwmduad, Carmarthen, SA33 6XJ
2W0 FMR C. Worka, 9 Bertha Street, Pontypridd, Cf37 1ts
2E0 FMS G. Bailey, 61 Great Ranton, Pitsea, Basildon, SS13 1JS
2E0 FMT P. Young, 11 St. Andrews Avenue, Washington, NE37 1AH
2E0 FMV V. Hayes, 68 Billingsley Road, Birmingham, B26 2EA
2E0 FMX N. Shajan, 19 Sturgess Avenue, London, NW4 3TR
2E0 FMY F. Masters, 91 Mayfair Avenue, Worcester Park, KT4 7SJ
2W0 FNA K. Byast, 10 Chapel Street, Amlwch Port, Amlwch, LL68 9HT
2E0 FNB F. Nuttall, 4 Kingholm Gardens, Bolton, BL1 3DJ
2E0 FNG M. Grice, 48 St. Ives Road, Coventry, CV2 5FZ
2E0 FNH D. Flood, 506 Preston Old Road, Blackburn, BB2 5LY
2M0 FNI Capt. R. Smith, Buttview Cottage, South Bragar, Isle of Lewis, HS2 9DH
2E0 FNJ C. Wilkinson, 67 Middleton Park Grove, Leeds, LS10 4BG
2E0 FNM M. Bamber, 10 Sedgeley Mews, Freckleton, Preston, PR4 1PT
2I0 FNN D. Wilson, 81 Parknasilla Way, Aghagallon, Craigavon, BT67 0AU
2E0 FNO R. Bishop, 12A Goseley Avenue, Hartshorne, Swadlincote, DE11 7EZ
2E0 FNQ J. Mason, 11 Scriven Grove, Haxby, York, YO32 3NW
2E0 FNT M. Morse, 5 Northload Terrace, Glastonbury, BA6 9JW
2E0 FNV A. Price, 87 Derricke Road, Bristol, BS14 8NH
2E0 FNY D. Chatterton, 3 Hunt Close, South Wonston, Winchester, SO21 3HY
2E0 FOD P. Dekkers, 21 Nodens Way, Lydney, GL15 5NP
2E0 FOE C. Pascoe, Treleven, Primrose Hill, Goldsithney, Penzance, TR20 9JR
2W0 FOG M. Haywood-Samuel, 38 Tanygraig Road, Llanelli, SA14 9LH
2E0 FOH C. Donachie, 22 Glasgow Street, Hull, HU3 3PR
2E0 FOI P. Pearce, Criggion Mw Radio Station, Back Lane, Criggion, Shrewsbury, SY5 9BE
2E0 FOK D. Sutherland, 78 Holmden Avenue, Wigston, LE18 2EF
2E0 FOL W. Bartle, 5 Crosskeys Row Chapel Milton, Chapel En le Frith, SK230QQ
2E0 FOP J. Bovey, 17 North Bank Road, Bingley, BD16 1UH
2E0 FOR I. Forester, 35 Thackeray Street, Sinfin, Derby, DE24 9GY
2E0 FOT S. Reason, 49 Highfield Road, Pudsey, LS28 7JW
2E0 FOX B. Hopkins, 60 Hales Gardens, Birmingham, B23 5DF
2E0 FPA R. Etchells, 6 Woodbank Court, Canterbury Road, Manchester, M41 7DY
2I0 FPB D. Neill, 8 Castle Meadows Carrowdore, Newtownards, BT22 2TZ
2E0 FPF M. Waite, 108A Chantry Gardens, Southwick, Trowbridge, BA14 9QS
2M0 FPI M. Drever, 7 Abbotswell Crescent, Aberdeen, AB12 5AQ
2E0 FPJ J. Phillips, 1 Wath Cottages, Cundall, York, YO61 2RL
2E0 FPO B. Bruce, 1 Stone Cottages, Chilmington Green, Great Chart, Ashford, TN23 3DW
2E0 FPQ E. Turner, 2 Shepherds Close, Fen Ditton, Cambridge, CB5 8XJ
2I0 FPT D. Hawthorne, 58 Seagahan Road, Collone, Armagh, BT60 2BH
2E0 FPW W. Garvey, 254 Bury Road, Tottington, Bury, BL8 3DT
2E0 FPX M. Howard, Flat 10, Oakdale, 6 Westgate Road, Beckenham, BR3 5DY
2E0 FPY A. Capitan, 41 Cunningham Drive, Runcorn, WA7 4DL
2E0 FQA L. Scully, 17 Fenwick Close, Woking, GU21 3BY
2E0 FQC J. Roberts, 41 Mcneill Avenue, Crewe, CW1 3NW
2E0 FQE A. Brook, Inkerman House 113 Clovelly Road, Bideford, EX39 3BY
2E0 FQG K. Lambert, Flat 6, Vernons Court, Vernons Lane, Nuneaton, CV10 8BB
2E0 FQI S. Wilson-King, Maple House, Newton Reighny, ca11 0ay
2E0 FQJ R. Goldston, 30 Vincent Road, Liverpool, L21 7NX
2W0 FQM B. Davies, 10 Long Acre Court, Bishopston, Swansea, SA3 3AY
2E0 FQP D. Daniel, 38 Netherthorpe Lane, Killamarsh, Sheffield, S21 1DA
2E0 FQR A. Coote, 148 Clarendon Street, Dover, CT17 9RB
2E0 FQS P. Dyke, 10 Elmtree Close, Ashurst, Southampton, SO40 7FD
2E0 FQT J. Mumby, 68 East Common Lane, Scunthorpe, DN16 1QH
2E0 FQU A. Thomas, Penberthy Road, Portreath, Tr164lu
2I0 FQW M. Alsallal, Flat 3 Tollgate House Bradbury Place, Belfast, BT7 1PH
2M0 FRA C. Fraser-Hopewell, 2/1 70 Albert Road, Glasgow, G42 8DW
2E0 FRB D. Munday, 29 Coombe Park, Wroxall, Ventnor, PO38 3PH
2E0 FRC F. Clements, 40 Ellison Fold Terrace, Darwen, BB3 3EB
2E0 FRD D. Collins-Cubitt, 75 Anthony Drive, Norwich, NR3 4EW
2E0 FRF A. Church, The Willows, Warboys Road, Huntingdon, PE28 3AH
2E0 FRG F. Noble, 1045, 45Th Street Apartment A, California, USA, 94608
2M0 FRJ R. Fair, 6 Fairways, Stewarton, Kilmarnock, KA3 5DA
2E0 FRK D. Church, The Willows, Warboys Road, Huntingdon, PE28 3AH
2E0 FRL Rev. L. Williams, 36 Royd Court, Mirfield, WF14 9DJ
2E0 FRO P. Froggatt, 11 Goldsmith Road, Walsall, WS3 1DL
2E0 FRP K. Hiscock, 45 Gloucester Road, Reading, RG30 2TH
2E0 FRQ R. Withers, 50 Coneygear Road, Hartford, Huntingdon, PE29 1QL
2M0 FRR A. Macintyre, 2 Memorial Square, Main Street, Castletown ?, Thurso, KW14 8TU
2E0 FRS R. Haughton, 5 Fairfield Way Wesham, Preston, PR4 3EP
2E0 FRU C. Andrew, 85 Priory Street, Corsham, SN13 0BA
2E0 FRY C. Fryer, 14 Perks Road, Wolverhampton, WV11 2ND
2E0 FSA G. Walker, Flat 74, Riversley Court, 205 Wensley Road, Reading, RG1 6ED
2M0 FSB D. Kelly, 21 Dhailling Road, Dunoon, PA23 8EA
2E0 FSC C. Bugariu, 17 Foxon Lane, Caterham, CR3 5SG
2E0 FSD D. Park, 29 Tresco Close, Blackburn, bb2 4rt
2E0 FSE L. Beaney, Brookside, Walkers Lane, Shorwell, Newport, PO30 3JZ
2E0 FSG B. Jones, 9 St. James Close, Hanslope, Milton Keynes, MK19 7LF
2E0 FSH C. Winfield, 11 Lowe Avenue, Smalley, DE7 8PW
2E0 FSI D. Woolger, 17 Wyvern Close, Tangmere, Chichester, PO20 2GQ
2E0 FSK G. Hardill, 107 Leicester Road Whitwick, Coalville, LE67 5GN
2E0 FSM M. Bruce, 27 Blaenant, Emmer Green, Reading, RG4 8PH
2E0 FSN H. Hamilton, Flat B, 9 Cambridge Drive, London, SE12 8AG
2M0 FSP S. Paterson, Springbank, 2 Mains Of Cuffurach, Clochan, Buckie, AB56 5HP
2E0 FSQ D. Smith, 7 Oakley Grove, Wolverhampton, WV4 4LN
2E0 FSS D. Booth, 19 Oak Avenue, Elloughton, Brough, HU15 1LA
2M0 FSV K. Winkler, 108 Gilmore Place, Edinburgh, EH3 9PL
2E0 FSX F. Riches, 4 Priory Close, Chelmsford, CM1 2SY
2M0 FTA L. Pinkowski, 69 Nelson Avenue, Livingston, EH54 6BZ
2E0 FTB A. Price, Barn Owl Roost, Astwith, Chesterfield, S45 8AN
2E0 FTC J. Forbes, Weald Barkfold Farm, Plaistow, Billingshurst, RH14 0PJ
2E0 FTD T. Raymond, 12 Mill Race Wolsingham, Bishop Auckland, DL13 3BW
2M0 FTH S. Macdonald, 110 High Street, Cuminestown, AB53 5YH
2M0 FTI D. Yeaman, 7 Brimmond Crescent, Westhill, AB32 6RD
2E0 FTL M. Dorrington, 19 Shaftesbury Drive, Wardle, Rochdale, OL12 9LT
2E0 FTM D. Fletcher, 97 Wallace Crescent, Carshalton, SM5 3SU
2E0 FTO R. Parker, 21 Yates Way, Ketley Bank, Telford, TF2 0AZ
2E0 FTP N. Busley, Busley, South Drove, Spalding, PE11 3BD
2E0 FTQ A. Street, 110 Magdalen Street, Colchester, CO1 2LF
2E0 FTS P. Johnson, 12A Beechroyd, Pudsey, LS28 8BH
2E0 FTT M. Thompson, 10 Kilchurn, Consett, Dh88tq
2E0 FTU R. Dutton, 3 Kilkenny Road, Guisborough, TS14 7LE
2E0 FTV S. Barber, 31 Aysgarth Avenue, Crewe, CW1 4QE
2E0 FTW J. Northall, 5 West Winds Road, Winterton, Scunthorpe, DN15 9RU
2E0 FTX J. Krol, 40 Hampton Gardens, Southend-on-Sea, SS2 6RW
2E0 FTZ B. Louder, Flat 1, 30 Mill Street, Bideford, EX39 2JJ
2E0 FUA Dr R. Pediani, Old School House Great Coxwell, Faringdon, SN7 7NB
2E0 FUD Dr R. Blackwell, Vikings Hall, Baylham, Ipswich, IP6 8JS
2E0 FUE Maj. A. Taylor-Roberts, 38 Old Coach Road, Bulford, Salisbury, SP4 9DA
2E0 FUF G. Huff, 135 Westminster Road Toothill, Swindon, SN5 8JE
2E0 FUG R. Yarrow, 7 Pitts Close, Binfield, Bracknell, RG42 4ES
2E0 FUH Dr C. Pomfrett, 17 Manifold Close, Sandbach, CW11 1XP
2E0 FUJ M. Webster, Summerfield Cottage, Walker Lane Wadsworth, Hebden Bridge, HX7 8SJ
2E0 FUN S. Southern, 37 Conway Road, Calcot, Reading, RG31 4XP
2E0 FUO D. Shaw, 81 Chesterfield Road Tibshelf, Alfreton, DE55 5NJ
2M0 FUP G. Crawford, 4 Windram Road Chirnside, Duns, TD11 3UT
2E0 FUQ J. Matthews, 2 Farm Close, Bungay, NR35 1JG
2W0 FUR J. Pattinson, 79 Waterloo Road, Penygroes, Llanelli, SA14 7PN
2M0 FUS S. Mason, 31 Clifton Road, Lossiemouth, IV31 6DJ
2I0 FUT M. Crozier, 33 Cullentragh Road, Poyntzpass, Newry, BT35 6SD
2E0 FUZ D. King, 215 Hartland Road, Reading, RG2 8DN
2E0 FVB K. Smith, 30 Azalea Drive, Swanley, BR8 8HZ
2E0 FVC G. Gee, 17 Portherras Villas, Pendeen, Penzance, TR19 7TJ
2E0 FVF J. Dooley, 29 The Drive, Alsagers Bank, Stoke-on-Trent, ST7 8BB
2M0 FVJ M. Butterworth, 97A Dunearn Drive, Kirkcaldy, KY2 6AL
2E0 FVK J. Wright, 149 Wroslyn Road, Freeland, Witney, OX29 8HR
2E0 FVL P. Penycate, 8 Campbell Road, Tangmere, Chichester, PO20 2HX
2E0 FVO M. Burge, 24 Oakdale Road, North Anston, Sheffield, S25 4EY
2E0 FVR R. Farrington, Sunny Brook, Broadway Rd, Evesham, WR11 7RN
2E0 FVS S. Horne, Beaucroft, Keswick Road, Benfleet, SS7 3HU
2E0 FVT G. Smith, 148 Firhill Road, London, SE6 3SQ
2E0 FVV M. Roberts, Flat 6, 463 Brighton Road, Lancing, BN15 8LF
2I0 FVX J. Robinson, 15 Union Court, Lurgan, Craigavon, BT66 8EE
2E0 FVZ A. Rawson, 64 Dukes Mead, Fleet, GU51 4HE
2E0 FWB R. Saddler, 41 Clapham Close, Swindon, SN2 2FL
2E0 FWC F. Clark, 14 Warwick Road, Bude, EX23 8EU
2E0 FWD C. Board, Pinmoor, Moretonhampstead, Newton Abbot, TQ13 8QA
2E0 FWK N. Woodruffe, 139 North Home Road, Cirencester, GL7 1DY
2E0 FWN E. Hunter, 38 Blewitt Street, Hednesford, Cannock, WS12 4BD
2E0 FWR F. Waller, 19 Wortley Avenue S738Sb, Wombwell, s738sb
2E0 FWS M. Le-Petit, 3 Stanley Drive, Hatfield, AL10 8XX
2E0 FWT T. Gore, 22 Stoppard Road, Burnham-on-Sea, TA8 1QB
2E0 FWW J. Haslam, 25 Lulworth Road, Eccles, Manchester, M30 8WP
2E0 FWY D. Smith, 7 Oakley Grove, Wolverhampton, WV4 4LN
2M0 FWZ A. Rowlands, 35 Dublin Croft, Great Sutton, Ellesmere Port, CH66 2TD
2E0 FXA P. Richardson, 91 Park Road, Blackpool, FY1 4JE
2E0 FXE C. Troughton, 20 Oakleigh Road, Uxbridge, UB10 9EL
2W0 FXF P. Brind, Golwg Y Bryn, Tyn Y Morfa, Gwespyr, Holywell, CH8 9JN
2E0 FXH A. Williamson, 32 Beech Close Eastfield, Scarborough, YO11 3QZ
2E0 FXJ D. Parkes, 4 Round Saw Croft, Rubery, Birmingham, B45 9TT
2M0 FXL F. Lane, 23 Mayfield Avenue, Tillicoultry, FK13 6HB
2E0 FXM Dr C. Fletcher, 7 Highfield Crescent, Baildon, Shipley, BD17 5NR
2E0 FXO D. Kerr, 19 Park Lane, Sutton Bonington, Loughborough, LE12 5NQ
2E0 FXP S. Westley, 76 Rockingham Close, Birchwood, Warrington, WA3 6UY
2E0 FXU A. Campbell, 21 Sherbrook Gardens, London, N21 2NX
2E0 FXV J. Keates, Pickersleigh Court, North End Lane, Malvern, WR14 2ET
2E0 FXW G. Davies, 11, Liverpool, L12 3HS
2M0 FXX M. Mcgrorty, 59 Craighall Street, Stirling, FK8 1TA
2E0 FXY Rev. W. Hackman, Kynance, Barden Road, Speldhurst, Tunbridge Wells, TN3 0QB
2E0 FYA M. Champion, 155 Walton Road, Walton on the Naze, CO14 8NF
2E0 FYC D. Jarman, 62 Combe Drive, Dunstable, LU6 2AE
2E0 FYD M. Harper, 67 Ludsden Grove, Thame, OX9 3BY
2E0 FYE R. Fye, 201 North Wing The Residence, Kershaw Drive, Lancaster, LA1 3SY
2M0 FYF J. Fyfe, 53A Ware Road, Glasgow, G34 9AR
2M0 FYG J. Rodger, 7 Heathryfold Circle, Aberdeen, AB16 7DQ
2E0 FYL P. Catton, 97 High Street, South Hiendley, Barnsley, S72 9AN
2E0 FYO K. Hedges, 28 Hill Park, Congresbury, Bristol, BS49 5BT
2W0 FYP G. Stewart, 62 Norton Road Penygroes, Llanelli, SA14 7RS
2E0 FYR G. Hepworth, 12 Fourlands Gardens, Bradford, BD10 9SP

2E0 FYS S. Smith, 133 Radcliffe New Road, Whitefield, Manchester, M45 7RP
2E0 FYT A. Bent, Lime Garth, Sherburn Road, Durham, DH1 2JR
2E0 FYU I. Lowbridge, 4802 East Ray Road #23-513, Phoenix, USA, 85044
2E0 FYW A. Shakesby, 14 Dawnay Road, Bilton, Hull, HU11 4HB
2M0 FYX J. Burton, 6 Kilfinnan Lodges, Spean Bridge, PH34 4EB
2E0 FYZ F. Islam, 88 Ladies Grove, St Albans, AL3 5UB
2M0 FZA A. Chambers, 35 Echline Grove, South Queensferry, EH30 9RU
2E0 FZB C. Sjostedt, 14 New Road, Marlow, SL7 3NG
2W0 FZC J. Williams, 12 Pantycelyn, Fishguard, SA65 9EH
2M0 FZD S. Boyd, 6 Schaw Road, Prestonpans, EH32 9HA
2E0 FZE G. Ashcroft, 8 Launceston Close, Winsford, CW7 1LY
2E0 FZF F. Cairns, 20 St. Davids Close, Maidenhead, SL6 3BB
2E0 FZG Dr J. Tromp, St. Mary'S Street Practice Ltd, 63 St. Mary Street, Chippenham, SN15 3JF
2E0 FZJ J. Foster, Westfield House, 23 High Street, Cumnor, OX2 9PE
2E0 FZK A. Mcbirnie, 25 Ulverston Road, Swarthmoor, Ulverston, LA12 0JB
2E0 FZM C. Ramsdale, 87 Mill Lane, Kirk Ella, Hull, HU10 7JN
2I0 FZO A. Mckenzie, 32 Abbot Gardens, Newtownards, BT23 8UL
2E0 FZS S. Carro, 149 Ruston Rd, London, SE18 5QY
2E0 FZT A. Craven, 14 Deverel Road, Charlton Down, Dorchester, DT2 9UD
2E0 FZU J. Davidson, 78 Old Heath, Shrewsbury, SY1 4SE
2M0 FZW M. Love, 17 Lindsay Road, East Kilbride, Glasgow, G74 4HZ
2E0 FZY S. Whiting, 56 Station Road, Branston, Lincoln, LN4 1LH
2E0 FZZ D. Sandland, 54 Bishopdale Drive, Rainhill, Prescot, L35 4QH
2E0 GAC N. Bexon, 60 Whitwell Road, Nottingham, NG8 6JT
2E0 GAF G. Sole, 16 Beech Crescent, Hythe, Southampton, SO45 3QG
2E0 GAH G. Hudson, 26 Griffins Brook Lane, Birmingham, B30 1PU
2E0 GAL R. Hughes, 19 Pendine Crescent, North Hykeham, Lincoln, LN6 8UW
2M0 GAN G. Prior, 41 Beechwood, Linlithgow, EH49 6SD
2E0 GAO G. Bridge, 16 Victoria Street, Ramsbottom, Bury, BL0 9ED
2E0 GAP A. Pilkington, 26 Ryelands Close, Market Harborough, LE16 7XE
2W0 GAQ I. Williams, 5 Fron Goch, Llanberis, Caernarfon, LL55 4LE
2E0 GAR G. Farrar, 174 Houghton Road, Thurnscoe, Rotherham, S63 0SA
2E0 GAU G. Cooper, Holmfield, Chelmorton, Buxton, SK17 9SG
2E0 GAV C. Holmes, 8 Byron Way, Caister-On-Sea, Great Yarmouth, NR30 5RW
2E0 GAW G. Webster, 84 Sparrows Herne, Basildon, SS16 5EN
2W0 GAY A. Ferguson, The Mount Stables, Salem, Llandeilo, SA19 7HD
2E0 GBA I. Stevenson, 79 Lunedale Road, Darlington, DL3 9AT
2E0 GBB G. Foster, 22 Bradley Cottages, Consett, DH8 6JZ
2E0 GBE G. Bell, 83 Coopers Green, Bicester, OX26 4XJ
2E0 GBF P. Dawes, 49 Altofts Lodge Drive, Altofts, Normanton, WF6 2LB
2E0 GBG D. Gillingham, 5 Hillfield, St. Marks, Cheltenham, GL51 7BQ
2E0 GBH M. Brinnen, 82 Victoria Road, Mablethorpe, LN12 2AJ
2E0 GBI L. Catterall, 14 Dunham Drive, Whittle-Le-Woods, Chorley, PR6 7DN
2E0 GBJ B. Cave, 2 Beaufort Close, Newcastle upon Tyne, NE5 3XL
2E0 GBK Dr J. Bell, 6 Highfields, Fetcham, Leatherhead, KT22 9XA
2D0 GBM P. Birchall, 7 Richmond Close, Douglas, Isle Of Man, IM2 6HR
2E0 GBN G. Barton, 9 Tees Crescent, Stanley, DH9 6HX
2E0 GBO M. King, 7 Battismore Road, Morecambe, LA4 4QG
2E0 GBP P. Lindley, 17 Swallow Lane, Aston, Sheffield, S26 2GR
2E0 GBT F. Woods, 30 Hurst Close, Chandler'S Ford, Eastleigh, SO53 3PA
2E0 GBU D. Torrance, 30 St. Norbert Drive, Ilkeston, DE7 4EH
2E0 GBV G. Brotherhood, 17 Baldwin Close, Forest Town, Mansfield, NG19 0LR
2E0 GCB G. Buxton, 18 Savernake Close Rubery, Rednal, Birmingham, B45 0DD
2I0 GCC G. O'Reilly, 20 Lower Clonard Street, Belfast, BT12 4NH
2E0 GCD G. Bosnyak, 10 Station Road, Stannington Station, Morpeth, NE61 6DS
2E0 GCE G. Elsworthy, 40 Moorfield Way, Wilberfoss, York, YO41 5PL
2M0 GCF J. Brown, 78 Egilsay St., Glasgow, G22 7RG
2E0 GCG H. Vecenans, 155 Upper Dale Road, Derby, DE23 8BP
2E0 GCH S. Shreeves, 20 Selly Oak Road, Jordanthorpe, Sheffield, S8 8DU
2E0 GCI A. Clarke, 57 Welland Avenue, Grimsby, DN34 5JP
2E0 GCJ G. Jacks, 2 Corve View, Fishmore, Ludlow, SY8 2QD
2E0 GCL A. Finn, 202 Northgate Road, Stockport, SK3 9NJ
2E0 GCM N. Baker, 56 Chalklands, Bourne End, SL8 5TJ
2I0 GCN S. Morrow, 769 Farranseer Park, Macosquin, Coleraine, BT51 4NB
2E0 GCO S. Lake, 85 Clarkson Road, Norwich, NR5 8ED
2E0 GCP G. Piddington, 45 Pleasant View Road, Crowborough, TN6 2UU
2M0 GCS G. Cochrane, 33 Portland Road, Galston, KA4 8EA
2E0 GCW B. Barker, 44 Falcon Crescent, Bilston, WV14 9BE
2E0 GCY G. Cornish, 78 Kerry Avenue, Ipswich, IP1 5LD
2W0 GDA R. Davison, 2 Marlow Terrace, Mold, CH7 1HH
2W0 GDB G. Brookes, Hafan Deg, Caergeiliog, LL653yd
2E0 GDF M. Joynson-Ellis, 20 Morland Court, Skaters Way, Peterborough, PE4 6GW
2E0 GDG Dr G. Di Genova, 7 Marina Way, Abingdon, OX14 5TN
2E0 GDH S. Hunter, 9 Gelt Burn, Didcot, OX11 7TZ
2E0 GDL G. Ludlow, 13 Laburnum Walk, Gilberdyke, Brough, HU15 2TU
2E0 GDM G. Martin, Flat 16, Sorrel House, Birmingham, B24 0TQ
2E0 GDN K. Young, 51 Haven Road, Barton-upon-Humber, DN18 5BS
2E0 GDO G. Cochrane, 133 Cotman Fields, Norwich, NR1 4EP
2E0 GDT A. Shaw, 20 Hillcrest Close, Thrapston, Kettering, NN14 4TB
2E0 GDV Dr J. Hunt, 10 Couzens Close, Chippenham, SN15 1US
2E0 GDY V. Dealey, 11 Ashcombe Close, Witney, OX28 6NL
2E0 GEB S. Marr, 49 Gallows Hill, Ripon, HG4 1RG
2E0 GEC C. Whatmough, 11 Blackchapel Drive, Rochdale, OL16 4QU
2E0 GEE N. Patel, 78 Wesley Close, South Harrow, Harrow, HA2 0QE
2E0 GEF G. Winterbottom, 35 Abingdon View, Worksop, S81 7RT
2E0 GEG G. Bramham, 1 Watson Avenue, Dewsbury, WF12 8PZ
2M0 GEJ G. Jamieson, 6 Maryville Park, Aberdeen, AB15 6DU
2M0 GEK J. Wright, 43 Spey Court, Stirling, FK7 7QZ
2E0 GEL J. Willetts, 102 Welch Road, Cheltenham, GL51 0EG
2W0 GEM P. Murray, 45 Commercial Street, Risca, Newport, NP11 6AW
2E0 GEN A. Askam, 8 The Pastures, Weston-On-Trent, Derby, DE72 2DQ
2W0 GER T. Doak, 27 Hill St., Gilfach Goch, Porth, CF39 8TW
2E0 GES L. Tarlow, 59 Crouch Hall Road, London, N8 8HD
2E0 GET D. Baker, 25 Hitherspring, Corsham, SN13 9UT
2E0 GEV A. Sherman, 31 Peartree Avenue, Kingsbury, Tamworth, B78 2LG
2E0 GEX G. Evans, 19 Windsor Street, Thurnscoe, S63 0HB
2W0 GEZ N. Shepherd, Prospect, Newchapel, Llanidloes, SY18 6JY
2E0 GFB A. Durrant, 22 Supple Close, Norwich, NR1 4PP
2M0 GFC P. Davis, 4 Daisy Park, Baltasound, Unst, Shetland, ZE2 9EA
2E0 GFE G. Teale, Flat 3, 89-91 Barrack Road, Christchurch, BH23 2AJ
2E0 GFF A. Banks, 2 Holt Close, Farnborough, GU14 8DG
2E0 GFJ G. Frost, The Old Chestnut, 51 Knightcott Gardens, Banwell, BS29 6HD
2E0 GFK I. Graham, 49 Eagle Close, Leighton Buzzard, LU7 4AT
2I0 GFO G. Craig, 103 Moyle Parade, Larne, BT40 1ET
2E0 GFQ R. Haynes, 28 Ridgeway View, Montgomery, SY15 6BF
2I0 GFR G. Ramsey, 6 Creamery Park, Lisbellaw, BT94 5BU
2M0 GFT Dr V. Vyshemirsky, 2103 Great Western Road, Glasgow, G13 2XX
2E0 GFU C. Bourdiec, 30 The Ridgeway, South Shields, NE34 8BB
2E0 GFV J. Butcher, 14 Park Road, Lowestoft, NR32 1SW
2E0 GFW G. Watson, The Dell, Nova Scotia Road, Great Yarmouth, NR29 3QD
2W0 GFX J. Duffain, 37 Mount Pleasant Street, Dowlais, Merthyr Tydfil, CF48 3AF
2E0 GFY D. Horne, Alias, Lowthorpe, Southrey, Lincoln, LN3 5TD
2E0 GFZ B. Gentry, 5 Whitlock Drive, Great Yeldham, Halstead, CO9 4EE
2E0 GGA G. Amos, Willow Tree House, Deers Green, Clavering, Saffron Walden, CB11 4PX
2W0 GGG M. Brady, Ty Mawr Uchaf, Dulas, LL70 9DQ
2E0 GGI R. Gleave, 52 Cranborne Avenue, Warrington, WA4 6DE
2E0 GGJ R. Edminson, 22 Southwold Place, Cramlington, NE23 8HE
2E0 GGM A. Mansfield, 20 High Street, Broughton, Kettering, NN14 1NG
2E0 GGN N. Bentall, 15 Maple Close, Oxford, OX2 9DZ
2E0 GGO G. Jones, 109 Montgomery Avenue, Lowestoft, NR32 4DU
2E0 GGP G. Peters, Flat 6, Park Court, 46 North Park Road, Harrogate, HG1 5AD
2E0 GGQ G. Wilson, 2 Hill Row, Braunston, NN117HU
2E0 GGT G. Townsend, 23 Lodgefield Park, Stafford, ST17 0YE
2E0 GGU C. Braiden, Flat 5, Lansdowne House, 12 Twickenham Close, Swindon, SN3 3FQ
2E0 GGV C. Leek, 46 The Hollies, Holbeach, Spalding, PE12 7JQ
2E0 GGW G. Willard, 4 Varrier Jones Place, Papworth Everard, Cambridge, CB23 3XP
2M0 GGY A. Espie, 70 Everard Rise, Livingston, EH54 6JD
2E0 GGZ G. Talbot, 1 Leechwell Street, Totnes, TQ9 5SX
2E0 GHA G. Hill-Adams, 6 Broadleaze Way, Winscombe, BS25 1JX
2E0 GHB G. Bourne, 72 Cornish Way, Royton, Oldham, OL2 6JY
2E0 GHC Dr T. Hoban, 28 Cranley Road, Guildford, GU1 2JS
2W0 GHD A. Burleton, 9 Waghausel Close, Caldecott, Np26 4qr
2E0 GHE J. Merrick, 8 Maldowers Lane, Bristol, BS5 7QT
2M0 GHF G. Fleming, 169 Oldtown Road, Inverness, Scotland, IV2 4QD
2E0 GHG Dr G. Turner, 8 Scarborough Terrace, York, YO30 7AW
2E0 GHH C. West, 12 St. Georges Road, Wallington, SM6 0AS
2E0 GHJ Dr D. Pegler, September Cottage, East Bank, Winster, DE42DT
2E0 GHK D. Hindle, 18 Haig Street, Selby, YO8 4BY
2E0 GHP J. Phillips, The Manor, Blackwoods, York, YO61 3ER
2E0 GHR R. Gill, 84 Leypark Road, Exeter, EX1 3NT
2E0 GHS S. Stanhope, 61 Heathfield St., Manchester, M40 1LF
2E0 GHT P. Carter, 16 Alexandra Park Paulton, Bristol, BS39 7QS
2W0 GHV G. Thomas, 4 Blenheim Court, Picton Road, Neyland, Milford Haven, SA73 1QR
2E0 GHX S. Aucoin, 296 Turkey Road, Bexhill-on-Sea, TN39 5HY
2I0 GHY I. Gibb, 1 Shankill Road, Garvary, Enniskillen, BT94 3DB
2E0 GHZ J. Wakefield, Oakhurst, Lower Common Road, Romsey, SO51 6BT
2W0 GIA D. Owen, Tanrallt, Blaenpennal, Aberystwyth, SY23 4TP
2E0 GIE R. Harper, 19 Tennyson Avenue, King's Lynn, PE30 2QG
2I0 GIF G. Clarke, 12 Church Green, Dromore, BT25 1LL
2E0 GIH E. Peck, 11 Blake Road, Stapleford, Nottingham, NG9 7HN
2W0 GIW G. Williams, 99 Maes Llwyn, Amlwch, LL68 9BG
2W0 GIX A. Hodgson, Browns Holiday Park Towyn Road, Conwy, LL22 9HD
2E0 GIY T. Orzechowski, 15 Rothesay Terrace, Northampton, NN2 7ER
2E0 GJE G. Groves, 5 Beech Road, Ashurst, Southampton, SO40 7AY
2M0 GJG R. Hudson, 128 Neilston Road 2/2, Paisley, PA2 6EP
2M0 GJJ G. Johnson, Speur Mor, Gifford Road, Longformacus, Duns, TD11 3NZ
2E0 GJN M. Stokes, 207 Sunderland Road, South Shields, NE34 6AQ
2E0 GJP B. Coley, 17 Livingstone Road Handsworth, Birmingham, B20 3LS
2E0 GJQ N. Napiorkowski, 74 Wigley Road, London, TW13 5HE
2W0 GJR G. Reason, 454 Cowbridge Road West, Cardiff, CF5 5BZ
2E0 GJT N. Cooper, Romer Cottage, Long Reach, Ockham, Woking, GU23 6PF
2E0 GKA P. Houghton, 6 Olivers Court, Calne, SN11 0FL
2I0 GKB G. Black, 45 Meeting House Lane, Lisburn, BT27 5BY
2M0 GKD A. Mccreadie, 37 Beddie Crescent, Wigtown, Newton Stewart, DG8 9HX
2E0 GKF M. Feakins, 29 Whitesfield Road Nailsea, Bristol, BS48 2DY
2E0 GKG N. Triantafyllou, Flat 50 45 Provost Street, London, N1 7NW
2E0 GKL K. Lord, 21 Norfolk Road, Littlehampton, BN17 5PW
2E0 GKM R. Ley, 23 Heronbridge Close, Westlea, Swindon, SN5 7DR
2E0 GKN J. Duck, 24 Shearwater Grove, Innsworth, Gloucester, GL3 1DB
2E0 GKQ S. Dean, 154 Broad Lane, Walsall, WS3 2TQ
2E0 GKR L. Bullen11, 5 West View, Long Sutton, TA10 9LT
2E0 GKT K. Taylor, 56 Gibraltar Lane, Haughton Green Denton, Manchester, M34 7GG
2E0 GKU Dr A. Baker, Victoria Cottage, Swan Lane, Aughton, Ormskirk, L39 6SU
2E0 GKX R. Newell, 44 New Street, Chagford, Newton Abbot, TQ13 8BB
2E0 GLA M. Gladders, 2 Albion Mansions, Saltburn-by-the-Sea, TS12 1JP
2I0 GLC G. Crabbe, 39 Arran Avenue, Ballymena, BT42 4AP
2E0 GLD A. Goold, 6 The Elms, Kempston, Bedford, MK42 7JN
2M0 GLI G. Irvine, 120A Shore Road, Innellan, Dunoon, PA23 7SS
2E0 GLJ G. Jenkin, 24 Coronation Avenue, Camborne, TR14 7PE
2E0 GLL D. Firth, 5 Mowhay Gardens, Hatherleigh, EX20 3FE
2E0 GLM G. Parr, 50 Broadmeadow Close, Birmingham, B30 3NG
2E0 GLR E. Rimmer, 26 Kenmore Road, Prenton, CH43 3AS
2E0 GLS G. Stevens, 17 Manston Close, Ernesettle, Plymouth, PL5 2SN
2E0 GLT G. Cheeran, 37 Farnol Road, Dartford, DA1 5NG
2W0 GLV C. Tanner, Pen Y Gogarth Llaneilian, Amlwch, LL68 9NH
2E0 GLW G. Whittle, 22 Warwick Street, Leigh, WN7 2NH
2I0 GLY N. Sands, 6 The Granary, Waringstown, Craigavon, BT66 7TG
2E0 GMA G. Marsden, 38 Sandhill Road, Rawmarsh, Rotherham, S62 5NT
2M0 GMB I. Birse, North Milton Of Corsindae, Midmar, Inverurie, AB51 7QP
2E0 GMD M. Drury, 19 Cuffley Avenue, Watford, WD25 9RB
2E0 GMF S. Dearne, 26 Dilly Lane Barton On Sea, New Milton, BH25 7DQ
2E0 GMG Viscount A. Andover, Bishoper Farmhouse, Brokenborough, Malmesbury,

SN16 9SR
2E0 GMM J. Cook, 42 Pampas Close, Colchester, CO4 9ST
2E0 GMN J. Rufes, Flat 1 & 3-8 12 Smyrna Road, London, NW6 4LY
2E0 GMS G. Brooks, 14 Chalton Crescent, Havant, PO9 4PT
2E0 GMU G. Murch, 79 Alderson Crescent Formby, Liverpool, L37 3LY
2E0 GMW S. Coombs, 34 Mast Drive, Hull, HU9 1ST
2W0 GMZ H. Hughes, 21 Maes Geraint, Pentraeth, LL75 8UR
2E0 GNC B. Chandler, 1 Rambridge Farm Cottages, Weyhill, Andover, SP11 0QF
2W0 GNG P. Smith, 3 Islington Road, Bridgend, CF31 4QY
2E0 GNH G. Spiers, 2 Sponnes Road, Towcester, NN12 6ED
2E0 GNI L. Mercer, 5 Brinchcombe Mews, Plymouth, PL9 7FB
2E0 GNJ W. Neill, 21 Geelong Close, Weymouth, DT3 6RE
2E0 GNL S. Bayntun, 15 West Croft, Addingham, Ilkley, LS29 0SP
2E0 GNN A. Glover, 103A Latimer Street, Liverpool, L5 2RF
2E0 GNO W. Walton, 1 West Hall, Yeadon, Leeds, LS19 7AJ
2E0 GNQ M. Hughes, 16 Marlborough Road, Urmston, Manchester, M41 5QG
2E0 GNS G. Sandell, 20 Kirkby View, Sheffield, S12 2NB
2E0 GNU E. Brook, 30 Pitchstone Court, Farnley, Leeds, LS12 5SZ
2E0 GNV M. Wanless, 3 Bromlow Hall Barns, Bromlow, Minsterley, Shrewsbury, SY5 0DX
2E0 GNW A. Land, 12 Beverley Close Holton Le Clay, Grimsby, DN36 5HG
2M0 GOE T. Macdonald, Main Road Farm, Balephuil, Isle of Tiree, PA77 6UE
2E0 GOG B. Garrison, 121 Bromford Lane Erdington, Birmingham, B24 8JP
2I0 GOK S. Lannie, 81 Huntingdale Green, Ballyclare, BT39 9FL
2E0 GOL A. Goldsmith, 61 Fengate Drove, Weeting, Brandon, IP27 0PW
2E0 GOM C. Blackburn, 158 Dyas Road, Great Barr, Birmingham, B44 8SW
2E0 GON P. Gonczarow, 25 Ribchester Avenue, Burnley, BB10 4PD
2E0 GOO J. Barker, Pearl Bungalow, Killerby Cliff, Cayton Bay, Scarborough, YO11 3NR
2E0 GOP S. Mansfield, The Old Piggery, Ham Lane, Compton Dundon, Somerton, TA11 6PQ
2E0 GOQ R. Gibbs, 7 Thornhill, Eastfield, Scarborough, YO11 3LY
2E0 GOS N. Gostling, 49 Roundhouse Road, Dudley, DY3 2AX
2E0 GOW C. Gowing, 5 Curson Road, Tasburgh, Norwich, NR15 1NH
2M0 GOY A. Mcdonald, Kettlehills, Cupar, KY15 7TW
2E0 GOZ R. Cichocki, 12 Crossland Crescent, Wolverhampton, WV6 9JY
2E0 GPA M. Phillips, 36 Hyde Heath Court, Crawley, RH10 3UQ
2E0 GPB G. Beacher, 22 Trowbridge Gardens, Luton, LU2 7JY
2E0 GPC G. Coleman, 15 Redwood Drive, Ormskirk, L39 3NS
2E0 GPD P. Dimes, 5 Meadowbrook, Oxted, RH8 9LT
2E0 GPE G. Pearson, 41 Myrica Grove, Hoole, Chester, CH2 3EW
2E0 GPF G. Fleming, 17 Greenfield Close, Kippax, Leeds, LS25 7PX
2E0 GPG G. Bates, 230 Brook Street, Erith, DA8 1DZ
2E0 GPH G. Hart, 11 Sadlers Ride, West Molesey, KT8 1SU
2E0 GPI R. Mackay, 7 Darwin Close, Lee-on-the-Solent, PO13 8LS
2E0 GPJ A. Allan, The Oxford Health Co Ltd, Unit 4, Longlands Road, Bicester, OX26 5AH
2E0 GPK G. Kendall, 6 Kershaw Road, Walsden, Todmorden, OL14 7QF
2E0 GPL T. Arrow, Crystalwood, Stonemans Hill, Newton Abbot, TQ12 5PZ
2I0 GPQ D. Boyd, 11 Abbey Gardens, Belfast, BT5 7HL
2E0 GPS G. Perkins, Gamekeepers Cottage, Snarehill, Thetford, IP24 2QA
2E0 GPT R. Hampson, 12 Oakhays, South Molton, EX36 4DB
2E0 GPU A. Taylor, 16 Bellmans Road Whittlesey, Peterborough, PE7 1TY
2E0 GPV A. Alty, 20 The Hawthorns Eccleston, Chorley, PR75QW
2E0 GPX G. Matthews, 81 Kipling Avenue, Goring-By-Sea, Worthing, BN12 6LH
2E0 GPY N. Shepherd, 12 Barnfield Close, Radcliffe, Manchester, M26 3UA
2I0 GQA R. Bestek, 111 Cloughwater Road, Ballymena, BT43 6SZ
2E0 GQC D. Hughes, 14 Holts Lane, Clayton, Bradford, BD14 6BL
2E0 GQD J. Reynolds, 3 Ardleigh, Basildon, SS16 5RA
2E0 GQE L. Walsh, 4 Musbury Crescent, Rossendale, BB4 6AY
2M0 GQH Dr Z. Yang, 48 Foxglove Road, Newton Mearns, Glasgow, G77 6FP
2E0 GQI A. Hicks, 15 West Road, Ruskington, Sleaford, NG34 9AL
2E0 GQK C. Schroth, Flat 3, 15B Cavendish Road, Bournemouth, BH1 1QX
2E0 GQL R. Abbott, The Bungalow, Wrangaton, South Brent, TQ10 9HH
2E0 GQM R. Laidler, Swallow Cottage, Wiverton, Plympton, Plymouth, PL7 5AA
2E0 GQN D. Shingleton, 6 Newsham Walk, Manchester, M12 5QB
2E0 GQO S. Dudley, 3 Lake Croft Drive Meir Heath, Stoke on Trent, ST3 7SS
2E0 GQP I. Thompson, 26 Countrymans Way, Shepshed, Loughborough, LE12 9RB
2I0 GQR G. English, 15 Murrays Hollows, Ballyroney, Banbridge, BT32 5ES
2M0 GQS J. Rae, 4 Hillside Crescent, Langholm, DG13 0EE
2E0 GQT I. Alderman, 107 Manton Drive, Luton, LU2 7DL
2E0 GQU A. Rudgley, The New House, Plymouth Road, Buckfastleigh, TQ11 0DB
2E0 GQW E. Tart, Sunnybank Farm, Wattlesborough Heath, Shrewsbury, SY5 9EG
2I0 GQZ C. Kelly, Bt457Dt, Magherafelt, BT45 7DT
2E0 GRA G. Hayward, 129 Nipsells Chase, Mayland, Chelmsford, CM3 6EJ
2M0 GRE G. Lailvaux, 4 Oxenfoord Avenue, Pathhead, EH37 5QD
2E0 GRF G. Hewis, 10 Albert Road, New Malden, KT3 6BS
2E0 GRH S. Walters, 87 Fairbourne Close, Bransholme, Hull, HU7 5DH
2E0 GRI G. Reywer, 1 Tiverton Close, Houghton le Spring, DH4 4XR
2M0 GRK J. Black, 13 Dunlop Street, Greenock, PA16 9BG
2E0 GRL L. Spear, 57 Station Road, Melbourne, Derby, DE73 8EB
2E0 GRM N. Hallwood, 32 Hawthorne Avenue, Ripley, DE5 3PJ
2E0 GRN A. Young, 48 Sussex Street, Cleethorpes, DN35 7NP
2E0 GRP G. Priestley, 53 Millfield Gardens, Crowland, Peterborough, PE6 0HA
2E0 GRQ A. Waters, 21 Kings Road, Lee-on-the-Solent, PO13 9NU
2E0 GRR C. Hebden, 1 Ringwood Avenue, Newbold, Chesterfield, S41 8RA
2E0 GRS G. Street, Flat 9, Weavers Cottages, Congleton, CW12 1AG
2E0 GRV S. Whittaker, 115 Grange Close, Hoveton, Norwich, NR12 8EB
2E0 GRW M. Harvey, 129 Goldthorn Hill, Wolverhampton, WV2 4PS
2E0 GRX G. Kennedy, 4 Calder Crescent, Whitefield, Manchester, M45 8LH
2E0 GRY G. Collis, 16 Hill Grove, Barrow Hill, Chesterfield, S43 2NW
2E0 GRZ G. Green, 90 Princes Way, Fleetwood, FY78DX
2E0 GSA G. Smith, 2 Hawthorn Rise, Groby, Leicester, LE6 0EX
2E0 GSB G. Bertola, 17 Caraway Drive Branston, Burton-on-Trent, DE14 3FQ
2E0 GSC G. Chaffey, 63 Underwood Road, Eastleigh, SO50 6FX
2W0 GSE R. Smith, Hendafarn, Sarnau, Llanymynech, SY22 6QJ
2E0 GSF T. Gadd, 7 Dolina Road, Swindon, SN25 1TL
2I0 GSG G. Gregg, 30 Claremont Avenue, Moira, Craigavon, BT67 0SS
2E0 GSH J. Haigh, 1 Smithy Site, Farnborough, Wantage, OX12 8NS
2E0 GSJ K. Andrews, 80 Hollywell Road, Lincoln, LN5 9DA
2E0 GSK M. Silver, 52 Park Crescent, Elstree, Borehamwood, WD6 3PU
2E0 GSL Rev. L. Clark, 226 Philip Lane Tottenham, London, N15 4HH
2E0 GSN T. Savidis, 32 Peace Close Rosedale, Cheshunt Waltham Cross, EN7 5EQ
2E0 GSO T. Allen, 11 Church View Highworth, Swindon, SN6 7ER
2E0 GSP M. Carter, 32 Victoria Avenue, Brighouse, HD6 1QT
2E0 GSR G. Iredale, Ship Cottage, Main Street, Maryport, CA15 7DX
2E0 GST G. Starling, 4 Three Corner Drive, Norwich, NR6 7HA
2E0 GSW T. Rowlands, 7 Northfield Crescent, Beeston, Nottingham, NG9 5GR
2E0 GTA A. Holbrook, 6 Birch Tree Way, Maidstone, ME15 7RR
2E0 GTB P. Rigden, 11 Railway Cottages, Station Road, Whitstable, CT5 1JZ
2E0 GTE G. Cockburn, 20 Hexham Avenue, Hebburn, NE31 2HN
2E0 GTJ G. Mylonas, 68 Lancaster Gate, Cambourne, CB236AT
2E0 GTL G. Taylor, 31 Ashfurlong Crescent, Sutton Coldfield, B75 6EN
2E0 GTM G. Moon, 1 Brankenwall, Muncaster, Ravenglass, CA18 1RG
2E0 GTN G. Norbury, 3 Sherard Croft, Birmingham, B36 0LS
2I0 GTO G. Shaw, 49 Cloughey Road, Portaferry, Newtownards, BT22 1NQ
2M0 GTR S. Higgins, 24 Centre Street, Kelty, KY4 0EQ
2E0 GTT M. Smith, 24 Fifth Avenue, Portsmouth, PO6 3PE
2E0 GTU J. Haskell, 60 Blenheim Drive, Witney, OX28 5LJ
2E0 GTZ A. Blamire, 21 The Laurels, Banstead, SM7 2HG
2E0 GUA J. Hammond, 8 Rowntree Way, Saffron Walden, CB11 4DG
2E0 GUC M. Vowles, Calle Mar Serena, 29, Torre de Banagalbon, Malaga, Spain, 29738
2E0 GUF A. Bennett, 112 Vicarage Crescent, Redditch, B97 4RP
2E0 GUH D. Hughes, Flat 2, 13 Thorgam Court, Grimsby, DN31 2EU
2E0 GUJ S. Noller, 3 Thor Road, Norwich, NR7 0JS
2M0 GUL J. Hume, 8/11 Leslie Place, Edinburgh, EH4 1NH
2E0 GUN A. Price, 67 Mansfield Road, Glapwell, Chesterfield, S44 5QA
2E0 GUR G. Urban, 33 High Meadow, Hathern, Loughborough, LE12 5HW
2E0 GUT G. Taljaard, Flat 140, 105 London Street, Reading, RG1 4QD
2E0 GUU P. Edwards, 4 Stodart Road, London, SE20 8ET
2E0 GUV P. Brown, 1 Octavian Close, Hatch Warren, Basingstoke, RG22 4TY
2E0 GUY S. Mellor, 11 Bolton Meadow, Leyland, PR26 7AJ
2E0 GVC G. Clayton, The Forge, High Street, Moreton-in-Marsh, GL56 0LL
2E0 GVI S. Dickson, 11 Benfield, Grasmere, Ambleside, LA22 9RD
2W0 GVK J. Doyle, 40 Llanbeblig Road, Carrnarfon, LL55 2LW
2W0 GVN J. Bradley, Ty Newydd, Tregarth, Bangor, LL57 4AF
2W0 GVO G. Owen, Flat 3, Osborne House, Little Lane, Beaumaris, LL58 8DB
2E0 GVP T. Hunt, 36 Alverton Avenue, Poole, BH15 2QG
2E0 GVR V. Hodge, 37 Monkswood Crescent, Tadley, RG26 3UE
2E0 GVW S. Gadd, 62 Erica Way, Copthorne, RH10 3XQ
2E0 GVZ J. Earye, 28 Halls Drift, Kesgrave, Ipswich, IP5 2DE
2I0 GWA A. Cummings, 19 Bachelors Walk, Keady, Armagh, BT60 2NA
2E0 GWB G. Bunting, 31 Hardwick Avenue, Allestree, Derby, DE22 2LN
2E0 GWC Dr A. Colman, 5 Burn Heads Road, Hebburn, NE31 2TB
2E0 GWD N. Reeves, Flat 2, Delamore, Ivybridge, PL21 9QT
2E0 GWE G. Watson, 20 Windermere Drive, West Auckland, Bishop Auckland, DL149LF
2E0 GWF B. Whitemore, 24 Rectory Close, Wraxall, Bristol, BS48 1LT
2E0 GWI N. Davies, 27 Grafton Road, Ellesmere Port, CH65 2BD
2M0 GWJ A. Stanley, 22 Braid Mount, Edinburgh, EH10 6JJ
2W0 GWK K. Jones, Gorswen Brynrefail, Caernarfon, LL55 3NT
2W0 GWM M. Martin, 1 Y Gorlan, Bryn Street, Newtown, SY16 2HN
2W0 GWO C. Marrs, 49 Abraham Court, Lutton Close, Oswestry, SY11 2TH
2E0 GWP G. Prescott, 3 View Fields, Station Road, Doncaster, DN9 3AE
2E0 GWR B. Bosson, 80 White Horse Road, Marlborough, SN8 2FE
2E0 GWS G. Salter, 9 Spring Gardens, Malvern Link, Malvern, WR14 1AP
2M0 GWU I. Colborn, Gardeners Cottage, Craighouse, Isle of Jura, PA60 7XG
2E0 GWV N. Parker, 18 Sandown Road Bishops Cleeve, Cheltenham, GL52 8BZ
2E0 GWZ C. Aldous, 342 Feltham Hill Road, Ashford, TW151lW
2E0 GXB G. Beaver, 106 Dobede Way Soham, Ely, CB7 5fn
2E0 GXC G. Hardman, 12 Fernleigh Chorley New Road, Horwich, Bolton, BL6 6HD
2E0 GXD J. Marchant, 129 Highbury Grove, Clapham, Bedford, MK41 6DU
2E0 GXE B. Thomson, 40 Northgate Road, Stockport, SK3 0LQ
2E0 GXF G. Fernando, 1 Rosemary Avenue, West Molesey, KT8 1QF
2E0 GXG G. Eaton, 16 Holly Walk, Nuneaton, CV11 6UU
2W0 GXI D. Burt, 2 Cae Masarn, Pentre Halkyn, Holywell, CH8 8JY
2E0 GXK P. Blagden, 26 Duncroft Road, Hucclecote, Gloucester, GL3 3AS
2E0 GXM S. Elton, 43 Glenville Road, Bournemouth, BH10 5DD
2E0 GXP P. Brennan, 2 Ethel Road, Birmingham, B17 0EL
2E0 GXQ S. Lovell, 20 Courtenay Walk, Weston-Super-Mare, BS22 7TQ
2E0 GXT D. Killingley, 17 Colbert Drive, Leicester, LE3 2JB
2E0 GXX I. Bardell, 17 Stanton Avenue, Bradville, Milton Keynes, MK13 7AR
2E0 GXY M. Ferenc, 2A Rosedene Avenue, London, SW16 2LT
2M0 GXZ G. Sinclair, 33 Keptie Road, Arbroath, DD11 3EF
2W0 GYB L. Brown, 13 Station Road Loughor, Swansea, Sa46tr
2E0 GYC R. Mcknight, Ardralla, Church Cross, Skibbereen, Ireland, P81 RK12
2E0 GYF D. Hudson, 30 Sarmatian Fold, Ribchester, Preston, PR3 3YG
2E0 GYH P. Bates, 34 Lamerton Road, Reading, RG2 8AS
2E0 GYJ G. Booton, 69 The Street, Deal, CT14 0AJ
2E0 GYK J. Coates, 2 Holstein Drive, Scunthorpe, DN16 3TT
2I0 GYL G. Mccormick, 24 Warren Park Drive, Lisburn, BT28 1HF
2M0 GYM J. Branson, East Bank, South Road, Fochabers, IV32 7LU
2M0 GYN A. Connell, 39 Glebe Crescent, Maybole, KA19 7HZ
2E0 GYP J. Wells, 18 Roewood Road, Holbury, Southampton, SO45 2JH
2W0 GYQ B. Williams, Bryn Celyn, Hendre Road, Conwy, LL32 8RJ
2E0 GYV G. Johnson, 4 Delta Park Drive, Hesketh Bank, Preston, PR4 6SE
2E0 GYW T. Clark, 23 Shelley Grove, Bradford, BD8 0JZ
2E0 GYX J. Margarson, 26 David Street, Grimsby, DN32 9NL
2E0 GYY L. Monshall, 70 Oakley Road, Harwich, CO12 4QU
2E0 GYZ W. Rittman, 70 Market Street, Chapel-En-Le-Frith, High Peak, SK23 0HY
2M0 GZA S. Hargreaves, 4 Oxenfoord Avenue, Pathhead, EH37 5QD
2E0 GZD J. Norris, 55 Main Road, Great Leighs, CM3 1ND
2E0 GZF D. Higton, 1 Allendale Road, Caister-on-Sea, NR30 5ES
2E0 GZH Dr G. Howling, Sysonby Knoll Hotel, Asfordby Road, Melton Mowbray, LE13 0HP
2E0 GZJ C. Little, 28 Cecil Avenue, Warmsworth, Doncaster, DN4 9QW

2E0 GZM L. Wardle, 35 Woodland Close, Barnstaple, EX32 0EG
2W0 GZN D. Williams, Long Nursery Coppice Walton, Presteigne, LD82RE
2E0 GZP M. Stockdale, 3 Manor Close, Sproatley, Hull, HU11 4PY
2M0 GZQ T. Goodenough, 83 Craufurdland Road, Kilmarnock, KA3 2HU
2E0 GZT T. Marshall, 63A Newport Road, Ventnor, PO38 1BD
2M0 GZU M. Gallon, 43/18 Viewcraig Gardens, Edinburgh, EH8 9UW
2E0 GZW D. Hamilton, 16 Conifer Crest, Newbury, RG14 6RT
2I0 GZX P. Brennan, 23 Ardchrois, Donaghmore, Dungannon, BT70 3LB
2E0 HAB H. Atifeh, 57 Lincoln Drive, Rugby, CV23 1BS
2W0 HAC C. Thomas, 2 Ffordd Donaldson, Copper Quarter, Swansea, SA1 7FJ
2E0 HAE A. Hodgeon, 30 Rock Bank, Buxton, SK17 9JF
2E0 HAF J. Raehse Felstead, 10 Rubens Close, Aylesbury, HP19 8SW
2E0 HAG C. Hall, 28 Tidebrook Place, Stoke-on-Trent, ST6 6XF
2E0 HAH M. Mckenna, 54 Whickham Road, Hebburn, NE31 1QU
2E0 HAJ S. Cordner, 29 Buxton Road, Aylsham, Norwich, NR11 6JD
2W0 HAK K. Vaughan, 26 Mount Pleasant, Bargoed, CF81 8UU
2E0 HAL G. Smith, 7 Kestrel Avenue, Bransholme, Hull, HU7 4ST
2E0 HAN H. Hopkins, 3 Colegrave Road, Bloxham, Banbury, OX15 4NT
2E0 HAP A. Craven, 45 Benhams Drive, Horley, RH6 8QT
2W0 HAS R. Williams, 34 Maendu Terrace, Brecon, LD3 9HH
2E0 HAT R. Hatton, 18 Tangier Road, Guildford, GU1 2DF
2E0 HAV S. James, 35 Prospect Road, Dronfield, S18 2EA
2E0 HAW C. Cross, 131 Arnold Lane, Gedling, Nottingham, NG4 4HF
2E0 HAZ G. Hazlewood, 102 Throne Road, Rowley Regis, B65 9JX
2E0 HBB H. Russell, 4 Dearnsdale Close, Stafford, ST16 1SD
2E0 HBD D. Hardy, Flat 10, Bridport House, Hillwood Road, Birmingham, B31 1DN
2E0 HBE D. Newman, 78 Clapham Court, Gloucester, GL1 3DE
2E0 HBJ D. Howard, 31 White Mullein Drive, Redlodge, ip288xp
2E0 HBK P. Jones, 2 Whitley Place, Stoneley Park, Crewe, CW1 4GH
2I0 HBL J. Coles, 20 Upper Cairncastle Road, Larne, BT40 2DT
2I0 HBO K. Mikicki, 17 Glenhoy Drive, Belfast, BT5 5LB
2E0 HBP M. Gorrill, 30 Higher Dunscar Egerton, Bolton, BL7 9TF
2E0 HBQ S. Benniman, 5 Round Hill Lane, Shrewsbury, SY1 2NE
2E0 HBT T. Richley, 30 Chicheley Road, Harrow, HA3 6QL
2E0 HBV J. Haynes, 16 Mountsfield, Frome, BA11 5AR
2W0 HBW J. Turobin-Harrington, Michaelmas Barn, Sawmills, Kerry, Newtown, SY16 4LL
2E0 HBX H. Gruen, Allfrey House, Herstmonceux, Hailsham, BN27 4RS
2E0 HBY A. Foster, 62 Spa Road, Atherton, Manchester, M46 9NQ
2E0 HCC G. Belgium, 590 Wells Road, Bristol, BS14 9BD
2M0 HCF Dr C. Moir, 3 Farmfield Terrace, West Kilbride, KA23 9ED
2E0 HCG C. Loud, 24 Harrington Avenue, Lowestoft, NR32 4JU
2E0 HCL G. Holland, 6 Moorfield Road, Widnes, WA8 3JE
2E0 HCM W. Rowland, 5 Gwinear Downs, Leedstown, Hayle, TR27 6DJ
2E0 HCO D. Shelsher, Gables, Colchester Road, Ardleigh, Colchester, CO7 7PQ
2E0 HCS C. Salisbury, 9 Oakville Road, Heysham, Morecambe, LA3 2TB
2E0 HCV M. Britton, Butterwick Low, Hales Street Tivetshall St. Margaret, Norwich, NR15 2EE
2E0 HCW C. Haynes, 4 Thorn Close, Rugby, CV21 1JN
2E0 HCY S. O'Neill, 33 Norlands Park, Widnes, WA8 5BH
2E0 HCZ P. Kilby, 1 Home Farm Barns, South End, Milton Bryan, Milton Keynes, MK17 9HS
2W0 HDB H. Bancroft, Stop And Call, Goodwick, SA64 0EX
2W0 HDC N. Orchard, 152 Garden Suburbs, Trimsaran, sa174af
2E0 HDD M. Davies, 40 Lake Crescent, Daventry, NN11 9EB
2E0 HDE D. Edmondson, 21 Hawthorne Close, Heathfield, TN21 8HP
2E0 HDF A. Brain, 20 South St., Spennymoor, DL16 7TU
2E0 HDI J. Birkinshaw, 57 Walnut Tree Avenue, Hereford, HR2 7JU
2M0 HDL J. Howes, 22 Criffel Drive, Lincluden, Dumfries, DG2 0PE
2E0 HDM P. Allin, 8 Kiln Close, Dove Holes, SK178FQ
2W0 HDP M. Waldman, 9 David Street, Cwmbwrla, Swansea, SA5 8NX
2E0 HDQ J. Warburton, 82 Hampton Drive, Newport, TF10 7RF
2E0 HDU P. Loomes, 107 Main Street, Sedgeberrow, Evesham, WR11 7UE
2E0 HDW M. Langham, 12 Cornflower Drive, Chelmsford, CM1 6XY
2E0 HDX S. Hewick, 56 Hemswell Avenue, Hull, HU9 5JZ
2E0 HDY A. Hardy, 54 Trueway Drive Shepshed, Loughborough, LE12 9HG
2E0 HDZ I. Hammond, 88 Great Innings North, Watton at Stone, SG14 3TD
2E0 HEA J. Heagren, 14 Pepperbox Rise, Whaddon, Salisbury, SP5 3BF
2E0 HEC P. Mellers, 34 Heasman Place, Southwater, Horsham, RH13 9FT
2E0 HEF D. Robinson, Height End Farm, Kirk Hill Road, Haslingden, Rossendale, BB4 8TZ
2E0 HEG P. Sellick, 1, The Gallery, Northwick Park, Blockley, GL56 9RJ
2E0 HEI T. Bartholomew, 6 Kilmaine Road, Harwich, CO12 4UZ
2E0 HEM R. Milton, 66 Beech Road, Hoo Marina Park, Hoo, ME39TG
2M0 HEO L. Davis-Edmonds, 6 Barlockhart Park, Glenluce, Newton Stewart, DG8 0JQ
2E0 HEP J. Hobbs, 2 Eccles Road, Wittering, Peterborough, PE8 6AU
2E0 HEQ D. Barras, 36 Carew Close, Yarm, TS15 9TJ
2E0 HES R. Heslop, 7 Fieldfare Close, Clanfield, Waterlooville, PO8 0NQ
2E0 HET H. Taylor, 25 Northolme Avenue, Nottingham, NG6 9AP
2E0 HEU M. Gardiner, 21 Bell Chase, Aldershot, GU11 3GY
2E0 HEX J. Ash, 47 Stein Road, Emsworth, PO10 8LB
2I0 HEZ P. Robinson, 119 Avenue Road, Lurgan, Craigavon, BT66 7BD
2E0 HFA J. Wilson, Flat 5, Blake House, London, SE1 7DX
2E0 HFE A. Royds, 3A Fairfield Avenue, Rossendale, BB4 9TG
2E0 HFG L. Wale, 4 Essex Gardens, Market Harborough, LE16 9JS
2E0 HFJ D. Sampson, 116 South Mossley Hill Road, Liverpool, L19 9BJ
2E0 HFK C. Steele, 40 Landor Road, Whitnash, Leamington Spa, CV31 2JX
2E0 HFN A. Toal, 29 Highland Drive, Oakley, Basingstoke, RG23 7LF
2E0 HFO S. Bracegirdle, Flat 28, West Fryerne, Parkside Road, Reading, RG30 2BY
2E0 HFS H. Starling, 23 Lathom Road, Manchester, M20 4NX
2E0 HFT D. Merridale, The Granary, Falledge Lane, Upper Denby, HD8 8YH
2E0 HGA R. Bergmore, 28 Broadway, Ilkeston, DE7 8TD
2E0 HGD J. Thirlwell, 6A Brook Street, Warminster, BA12 8DN
2E0 HGE A. Hitchens, 16 Harrisons Place, Northwich, CW8 1HX
2I0 HGF T. Smyth, 139 Tullyreagh Road, Gorteen, Tempo, Enniskillen, BT94 3PH
2E0 HGG D. Rennie, 27 Orrell Road, Liverpool, L21 8NQ
2I0 HGI A. Glasgow, 17B Loy Street, Cookstown, BT80 8PZ
2E0 HGK A. Dykes, 16 Paterson Drive, Stafford, ST16 1WH
2E0 HGM G. Heath, 2 Lower Drake Fold, Westhoughton, Bolton, BL5 2RE
2E0 HGO P. Niewiadomski, Flat1 79A Dartmouth Road, London, SE23 3HT
2E0 HGP G. Hawes, 59 Hartlands, Bedlington, NE22 6JG
2E0 HGQ R. Morris, 45 St. Kildas Road, Bath, BA2 3QL
2E0 HGR S. Rayner, 109 Peckover Drive, Pudsey, LS28 8EQ
2E0 HGT A. Todd, 27 Woodleigh Crescent, Ackworth, Pontefract, WF7 7JG
2E0 HGU P. Culmer, 56 Main Road Washingborough, Lincoln, LN4 1AU
2E0 HGW M. Spreafico, 5A Bradfords Close, Bottisham, Cambridge, CB25 9DW
2E0 HGX D. Clark, 12 Wilson Crescent, Lostock Gralam, Northwich, CW9 7QH
2E0 HHA P. Freeman, Grove Orchard, Knapp Lane Coaley, Dursley, GL11 5AR
2E0 HHB D. Matulewicz-Boyle, 11 The Birches, Farnborough, GU14 9RP
2E0 HHC R. Duff, 55 Bigland Drive, Ulverston, LA12 9PD
2E0 HHE J. Cryan, 12 Stamford Avenue, Sunderland, SR34AT
2E0 HHF A. Stevens, Flat 19 Aylesbury House, London, SE15 1RW
2E0 HHG D. Rudgley, The New House, Plymouth Road, Buckfastleigh, TQ11 0DB
2E0 HHJ A. Prestwich, Highfield, Exminster, Exeter, EX6 8AT
2E0 HHK S. Collins, 5 Fernleigh Gardens, Stafford, ST16 1HA
2W0 HHN D. Evans, Bleddfa, Maes Meyrick, Heolgerrig, Merthyr Tydfil, CF48 1RZ
2E0 HHP J. Cumming, Farley Court, 100 Homer Close, Gosport, PO13 9TL
2E0 HHT R. Flevill, 62 Grosvenor Way, Horwich, Bolton, BL6 6DJ
2E0 HHU J. Newham, 6 Belsfield Gardens, Jarrow, NE32 5QB
2E0 HHW C. Fox, 209 Sulgrave Road, Washington, NE37 3DE
2I0 HHX S. Gibson, 22 Station Road, Bangor, BT19 1HD
2M0 HIB D. Hibberd, 18 Whitestripes Path, Bridge Of Don, Aberdeen, AB22 8WF
2M0 HIC Dr L. Campbell, 21 Fordyce Way, Auchterarder, PH3 1BE
2M0 HIE S. Scott, 32 Dixon Terrace, Whitburn, EH47 0LH
2E0 HIG B. Higgins, 2 Bishops Yard, High Street, Huntingdon, PE28 3JB
2E0 HIJ I. Halliday, 23 Park Avenue, Keyworth, NG12 5JY
2E0 HIK J. Hawbrook, 7 Birkdale, Norwich, NR4 6AF
2E0 HIL D. Taylor, 62 Floyds Lane, Walsall, WS4 1LE
2E0 HIO Dr A. Ward, 22 Whitecross Drive, Weymouth, DT4 9PA
2E0 HIP D. Slater, 13 Longford Close, Rainham, Gillingham, ME8 8EW
2E0 HIQ M. Pope, Drove Lodge 39 The Drove Barroway Drove, Downham Market, PE38 0AJ
2E0 HIR M. Jenkin, 13 Fore Street, St Columb Major, tr96rh
2E0 HIS G. Fenton, 40 High Street, Easington Lane, Houghton le Spring, DH5 0JN
2E0 HIT I. Petrie, 88 Vicarage Road, Henley-on-Thames, RG9 1JT
2E0 HIW C. Wilson, Collingwood Avenue, Tolworth, KT59PT
2E0 HIY A. Zahid Begum, 3 Mercury House, 4-8 Cheam Road, Epsom, KT17 1SN
2E0 HIZ J. Easdown, 38 North Street, Barming, Maidstone, ME16 9HF
2E0 HJB B. Halliwell, 9 Tennyson Rd, Rothwell, Kettering, NN14 6jh
2E0 HJC J. Howarth, 67A Crawford Ave, Manchester, M298ET
2E0 HJE K. Earwicker, Greenbank, Cripplestyle, Fordingbridge, SP6 3DU
2M0 HJG J. Burns, 191 Grahams Road, Falkirk, FK2 7BU
2E0 HJI E. Bray, 56 The Links, Gosport, PO13 0DX
2E0 HJJ H. Jackson, Wingfield Farm, Cublington Road, Leighton Buzzard, LU7 0LB
2E0 HJK A. Scott, 19 Estuary Drive, Felixstowe, IP11 9TL
2E0 HJN E. Buret, 22 Russell Close Laindon, Basildon, SS15 6AS
2M0 HJP H. Phillips, Maplebank, Leithen Road, Innerleithen, EH446NJ
2M0 HJS J. Smith, 10 High Street, Portknockie, Buckie, AB56 4LD
2E0 HJV J. Moore, 65 Hamsey Green Gardens, Warlingham, CR6 9RT
2E0 HJY P. Latham, 135 Ashgate Road, Chesterfield, S40 4AN
2E0 HJZ M. Pridmore, 61 Gillards, Bishops Hull, Taunton, TA1 5HH
2E0 HKC A. Chambers, 5 Blandford Road, Shepton Mallet, BA4 4FB
2E0 HKE P. Wightman, 11 Sloway Lane, West Huntspill, Highbridge, TA9 3RJ
2E0 HKG C. Cheung, 23 Mosley Road, Timperley, Altrincham, WA15 7TF
2U0 HKI R. Price, Les Vaurioufs, St Martins, Guernsey, GY4 6TE
2E0 HKK P. Buick, Poundgate Farm, Uckfield Road, Crowborough, TN6 3TA
2E0 HKL H. Klettke, 13 Hastings Close, Wythall, Birmingham, B47 6AW
2E0 HKP J. Coles, 4 Carloggas Close, St Mawgan. Newquay, Truro, TR8 4HJ
2E0 HKR A. Hill, 11 Pelham Drive, Hull, HU9 2AS
2I0 HKW R. Mcelholm, 19 Summerhill, Prehen, Derry, BT47 2PL
2E0 HKZ J. Beer, 92 Ashby Road, Hull, HU4 7JT
2E0 HLC T. Winter, 8 Thorpe Street, Hartlepool, TS24 0DX
2E0 HLF J. Taylor, 41 Waters View, Yarwell Mill, Yarwell, Peterborough, PE8 6EU
2E0 HLG A. Allmark, 44 Butterwick Fields, Horwich, Bolton, BL6 5GZ
2E0 HLH H. Hall, 7 Front Street Grindale, Bridlington, YO16 4XU
2E0 HLM K. Schmidt, Church, Corner, Mareham-le-Fen, PE22 7RA
2E0 HLO A. Rowan, 14 Craven Lea, Liverpool, L12 0NF
2E0 HLP P. Hallas, 37 Oakfield Road, Bromborough, Wirral, CH62 7BA
2E0 HLR L. Richardson, 35 Vidgeon Avenue, Hoo, Rochester, ME3 9DE
2E0 HMA C. Cowley, 103 Acre Road, Kingston upon Thames, KT2 6ES
2E0 HMC Dr H. Coghlan, 1Bell View Cross Houses, Shrewsbury, SY5 6JJ
2E0 HMG H. Partridge, 19 Dickens Drive, Melton Mowbray, LE13 1HZ
2E0 HML P. Venables, 7 Lawton Road, Rushden, NN10 0DX
2E0 HMM D. Lawson, 2 The Blossoms, Fulwood, Preston, PR2 9RF
2E0 HMP R. Scott, 34 Moorfield Road, Birmingham, B34 6QY
2E0 HMQ D. Allen, 19 Brooklands Close, Uttoxeter, ST148UH
2E0 HMR W. Andes, 141 Sunningdale Avenue, Hanworth, Feltham, TW13 5JS
2W0 HMS E. Barnes, 2 Trem Y Garnedd, Bangor, LL57 1NA
2M0 HMU G. Mcara, 24 Balfour Street, Alloa, FK10 1RU
2M0 HMV R. Rothon, 112 Ravenswood Rise, Livingston, EH54 6PG
2E0 HNC H. Clark, 36 Market Oak Lane, Hemel Hempstead, HP3 8JL
2E0 HNF D. Glover, 24 Cadeby Road, Sprotbrough, Doncaster, DN5 7SD
2E0 HNG M. Bell, 11 Greaves Road, Sheffield, S5 9DB
2E0 HNH A. Clark, 62 New Road, London, SE2 0QG
2E0 HNJ P. Harris, 33 Kirklea Road, Houghton le Spring, DH5 8DP
2E0 HNK H. Baird, 35 St. Peters Road, Wolvercote, Oxford, OX2 8AX
2M0 HNM M. Cowie, Larachmohr, Hawkhill, Keiss, Wick, KW1 4XF
2E0 HNP Dr D. Potts, 3A Westfield, Blean, Canterbury, CT2 9ER
2M0 HNQ J. Devlin, 200 Second Avenue, Clydebank, Glasgow, G81 3LE
2E0 HNR T. Bickerstaff, 43 Riverway, Durrington, Salisbury, SP48ES
2E0 HNX C. Shearan, 59 Arnold Crescent, Mexborough, S64 9JX
2E0 HOB J. Sparrow, 12 St. Huberts Close, Gerrards Cross, SL9 7EN
2E0 HOG M. Stroud, 1 Sefton Court, Welwyn Garden City, AL8 6WW
2W0 HOH P. Gostelow, 6 Tree Field Caerau Farm, Llanidloes, SY18 6ll
2E0 HOK K. Hough, 15 Moorside Road Endmoor, Kendal, LA8 0EN
2E0 HOL K. Craner, 46 Meadowhill Crescent, Redditch, B98 8HT

2E0 HOM A. Banwell, Hazel Cottage, Chapel Lane, Hinton, Chippenham, SN14 8HD
2E0 HOO H. Coram, 20 Eton Walk, Exeter, EX4 1FD
2E0 HOQ N. Lambert, 17 Starcross Road, Weston-Super-Mare, BS22 6NY
2M0 HOS J. Wallace, 323 High Street, Dalbeattie, DG5 4DX
2W0 HOT K. Jones, 1 Alway Crescent, Newport, NP19 9SX
2E0 HOU S. Houssart, Flat 3, Virginia Court, London, SE16 6PU
2E0 HOV K. Taylor, 44 Main Street, Willoughby, Rugby, CV23 8BH
2E0 HOX S. Hogg, 38 The Gables, Widdrington, Morpeth, NE61 5RA
2E0 HOY S. Hoyle, 7 Horderns Road, Chapel-En-Le-Frith, High Peak, SK23 9ST
2E0 HPB P. Brundrett, 114 Ack Lane East, Bramhall, Stockport, SK7 2AB
2E0 HPD R. Hodson, 99 Alcester Road, Hollywood, Birmingham, B47 5NR
2E0 HPE J. Blakemore, 16 Brentwood Avenue, Newbiggin-by-the-Sea, NE64 6JH
2E0 HPF J. Scannell, 53 Morley Croft, Farington Moss, Leyland, PR26 6QS
2E0 HPI C. Gorse, Apartment 11, 30 Stockton Road, Hartlepool, TS25 1RY
2E0 HPJ J. Whiteside, The Old Antique Shop, Bank Street Pulham Market, Diss, IP21 4TG
2W0 HPK R. Hopkins, 132 Laurel Road Bassaleg, Newport, NP10 8PT
2W0 HPL P. Evans, Flat 1 Red Cow Annex, Lloyds Terrace, Adpar, Newcastle Emlyn, SA38 9EH
2W0 HPM M. Melody, 7 Tegfan, Pontyclun, CF72 9BP
2E0 HPO A. Riches, 84 Elgar Drive, Shefford, SG17 5RA
2E0 HPR I. Hooper, 25 Honey Lane, Buntingford, SG9 9BQ
2W0 HQA C. Cowling, 14 Shelley Drive, Bridgend, CF31 4QA
2E0 HQB D. Dobbie, 24 Harrow Road, Leighton Buzzard, LU7 4UQ
2W0 HQD R. Wilkes, 9 Brynmawr Road, Ebbw Vale, NP23 5FF
2E0 HQF R. Jones, St. Fillans, The Warren, East Horsley, Leatherhead, KT24 5RH
2E0 HQJ Rev. C. Sherwood, 1 Savile Road, Elland, HX5 0LA
2E0 HQO R. Olive, Lorien, The Ridge, Thatcham, RG18 9HZ
2E0 HQQ J. Kelly, 8 Greyhound Lane, Overton, Basingstoke, RG25 3LE
2E0 HQU L. Russell, 11 Hartbushes Station Town, Wingate, TS28 5GA
2E0 HQV S. Ranger, 137 Warren Avenue, Southampton, SO16 6AF
2E0 HQY T. Jones, 7 Sedum Close, Huntington, Chester, CH3 6BL
2E0 HRB E. O'Neill, 13 Goodwood Close, Market Harborough, LE168JF
2E0 HRD V. Greatwood, 11 The Green, Long Preston, Long Preston, Skipton, BD23 4PQ
2E0 HRE S. Holroyd, 31A Cross Street, Tenbury Wells, WR15 8EF
2W0 HRG P. Sherwood, High Croft Jeffreyston, Kilgetty, SA68 0RG
2W0 HRH M. Bloore, Halfway House, Hyfrydle Road, Talysarn, Caernarfon, LL54 6HG
2E0 HRI R. Suchocki, 8 High Street, Bluntisham, Huntingdon, PE28 3LD
2E0 HRJ R. Huelin, 15 Hill Chase, Walderslade, Chatham, ME5 9HE
2W0 HRL H. Leonard, 11 Newton Road Grangetown, Cardiff, CF11 8AJ
2I0 HRM J. Mercer, 32 Templemore Avenue, Belfast, BT5 4FT
2M0 HRN R. Maccormack, 90 Wardlaw Crescent, East Kilbride, Glasgow, G75 0PY
2M0 HRP I. Morrison, Coach House, Olivers Brae, Stornoway, HS12SX
2E0 HRR A. Hannen, 9 Noble Road, Hednesford, Cannock, WS12 4RW
2E0 HRS T. Dunne, 23 Warstone Lane, Birmingham, B18 6JQ
2E0 HRT A. Hartwell, 41 Orchard Grove, Brixham, TQ5 9RH
2I0 HRV T. Darrah, 42 Pinewood Avenue, Carrickfergus, BT38 8EW
2E0 HRY H. Roxbrough, 17 Stanwell Close, Sheffield, S9 1PZ
2E0 HSB D. Murphy, 23 Lowndes Close, Stockport, SK2 6DW
2E0 HSC R. Coombes, Rose Cottage, Choice Hill Farm, Choice Hill Road, Over Norton, Chipping Norton, OX7 5PZ
2E0 HSE M. Ding, 130 Windermere Avenue, Warrington, WA2 0NE
2E0 HSG K. Dillow, 101 Martins Lane Hardingstone, Northampton, NN4 6DJ
2I0 HSL N. Davis, 19 Toberhewny Hall, Lurgan, BT66 8JZ
2E0 HSO N. Hawkins, Deganwy Hardwick Road, King's Lynn, PE30 5BB
2E0 HSP C. Bolton, 201 Lime Tree Avenue, Crewe, CW1 4HZ
2W0 HSU C. Vugts, Coed Coch, Llangammarch Wells, LD4 4BS
2E0 HSV N. Silveston, Caravan C, East End House, Oak Lane, Minster On Sea, Sheerness, ME12 3QR
2E0 HSW H. Scott Whittle, 7 Skyline House, Dickens Yard, Longfield Avenue, London, W5 2BJ
2E0 HSY J. Housley, Lesede Cottage, The Town, Carsington, Matlock, DE4 4PX
2E0 HTB H. Banasiak, 16 St Christophers Close, Bath, BA2 6RG
2E0 HTC G. Carter, 19 Brathay Crescent, Barrow-in-Furness, LA14 2BG
2E0 HTG C. Gouveia, 86A Spareacre Lane, Eynsham, Witney, OX29 4NP
2E0 HTI A. Ariyapala, 87 Lord Avenue, Ilford, IG5 0HN
2E0 HTJ T. Hare, 2 Primula Drive, Norwich, NR4 7LZ
2E0 HTK A. Dowley, 8 Paulet Place, Old Basing, Basingstoke, RG24 7LA
2E0 HTM A. Taylor, 14 The Lawns, Hinckley, LE10 1DY
2E0 HTN D. Palmer, 12 Grange Road, Southampton, SO16 6UH
2E0 HTO A. Hine, Tall Trees 13, Wilton Crescent, Alderley Edge, SK9 7RE
2E0 HTR H. Moore, 52 Limefield Street, Accrington, BB52AF
2E0 HTS S. Davison, 5 Denby Drive, Baildon, Shipley, BD17 7PQ
2E0 HTV G. Radulescu, 41 Sherard Road, London, SE9 6EX
2E0 HTW H. Whiteley-Boocock, 25 The Arches, Claremount Road, Halifax, HX3 6LD
2E0 HTX A. Meredith, 47 Northwick Road, South Oxhey, Watford, WD19 6NE
2M0 HUD A. Allan, 117 Bruce Gardens, Inverness, IV3 5BD
2E0 HUH K. Pugh, 4 Salt Boxes, Pinvin, Pershore, WR10 2LB
2M0 HUJ A. Hood, 26 Annan Avenue, East Killbride, G75 8XT
2E0 HUK M. Pratchett, Adastra Cottage, Letcombe Regis, Wantage, OX12 9JP
2W0 HUL K. Hulme, 13 Lime Street, Gorseinon, Swansea, SA4 4AD
2E0 HUM S. Crabb, 22 Mary Warner Road Ardleigh, Colchester, CO7 7RP
2E0 HUN A. Instone, 63 Larch Road, New Ollerton, Newark, NG22 9SX
2E0 HUR J. Jefferies, Millfield Cottage, 1 Bolnhurst Road, Bedford, MK44 2LF
2E0 HUT B. Hudson, 12 Elmfield Road, Hebburn, NE31 2DY
2W0 HUU P. Harper, 6 Mor Awel, Abergele, LL22 7ND
2W0 HUX M. Ladd, 50 Brynmelyn Avenue, Llanelli, SA15 3RT
2E0 HUZ H. Hughes, 24 Broadley Green, Windlesham, GU20 6AL
2E0 HVB T. Evans, 68 Wildbrook Road, Little Hulton, Manchester, M38 0FU
2E0 HVD P. Dutton, 10 River Lane, Partington, Manchester, M31 4DB
2E0 HVE M. Simonsohn, 5 Pitt Close, Blandford St. Mary, Blandford Forum, DT11 9PS
2E0 HVF S. Pearson, 3 Berkeley Road, Shirley, Solihull, B90 2HS
2M0 HVG C. Coutts, 1 Springfield Avenue, Duns, TD11 3BF
2E0 HVH C. Sidaway, 6 Rookery Chase, Deepcar, Sheffield, S36 2NF
2E0 HVL H. Ly, 11 Nodes Drive, Stevenage, SG2 8AL
2E0 HVM P. Ashton, 32 Sycamore Road New Ollerton, Nr Newark, NG22 9PS
2E0 HVQ G. Tant, 61 Western Way, Sandy, SG19 1DU
2E0 HVT R. Lewis, 46 Graydon Avenue, Chichester, PO19 8RG
2E0 HVU P. Westall, 1 Manor Gardens Morrison Road, Swanage, BH19 1JT
2E0 HVV M. Dutton, 10 Braemore Close Shaw Ol2 7Na, Oldham, OL2 7NA
2E0 HVW S. Downe, 9 Danesway, Exeter, EX4 9ES
2E0 HVY A. Riddle, 19 Cottey Crescent, Exeter, EX4 9DT
2E0 HVZ J. Godfrey, 4 Glenfield Road, Bideford, EX39 2LU
2E0 HWC H. Cheesman, 49 Front Street, Chirton, North Shields, NE29 7QN
2E0 HWE J. Emery, Flat 3, 12 Buxton Road, Ashbourne, DE6 1EX
2M0 HWG C. Haws, 1/1 4 Fenella Street, Glasgow, G32 7JT
2M0 HWI R. Rae, 1 Jura Drive, Tweedbank, Galashiels, TD1 3ST
2E0 HWJ B. Kemble, 3 Red House Close, Chudleigh Knighton, Chudleigh, Newton Abbot, TQ13 0RH
2E0 HWK A. Brand, 6 Walnut Close, Milton, Cambridge, CB24 6ET
2E0 HWN D. Wardman, 45 The Grainger, North West Side, Gateshead, NE8 2BG
2E0 HWO Dr H. Orridge, Stonecross Cottage, Wadworth Hall Lane, Wadworth, Doncaster, DN11 9BH
2W0 HWQ P. Rogers, 109 North Road, Pontywaun, Cross Keys, Newport, NP11 7FS
2E0 HWS I. Hemmens, 26 Grove Gate, Staplegrove Taunton, TA26DF
2I0 HWW P. Ford, 25 Carnhill, Londonderry, BT48 8BA
2E0 HWY T. Kerswill, 11A Vernon Drive, Prestwich, Manchester, M25 9RA
2E0 HXC H. Castree, 31 Fairview Thickwood, Colerne, Chippenham, SN14 8BS
2E0 HXF C. Shepherdson, 149 Scarborough Road, Norton, Malton, YO17 8AD
2E0 HXG P. Stokes, 26 Ashford Road, Hastings, TN34 2HA
2E0 HXH R. Spooner, 20 Hazledine Way, Bridgnorth, WV16 5AE
2E0 HXJ S. Skelton, 7 Sycamore Avenue, The Elms, Torksey, Lincoln, LN1 2NJ
2D0 HXL D. Wilson, 23 Snugborough Avenue, Union Mills, Braddan, Isle of Man, IM4 4LT
2E0 HXN M. Harrington, Tanglewood House, Station Road, Tilbrook, PE28 0JY
2E0 HXP M. Savage, 5 Mere Hall Barns, Mere Lane, Enville, Stourbridge, DY7 5JL
2M0 HXQ A. Sloan, 36 Paterson Avenue, Irvine, KA12 9JJ
2E0 HXU D. Sutton, 18 Knights Hill Severn Stoke, Worcester, WR8 9JD
2E0 HXW S. Miller, Foxglove Mead, North Common, Sherfield English, Romsey, SO51 6JT
2E0 HXY G. Jones, 158 Withington Lane, Aspull, Wigan, WN2 1JE
2E0 HYC T. Rutt, Granthorpe, Hull Road, Hull, HU11 5RN
2E0 HYD D. Conner, 20 Birch Avenue, Rochdale, OL12 9QH
2E0 HYE T. Byers, 1 Hazelwood Avenue, Sunderland, SR5 5AH
2E0 HYF J. Mc Elhinney, 16 Chaucer Road, London, E17 4BE
2E0 HYG P. Haygarth, 5 Forth Close, Peterlee, SR8 1DG
2E0 HYH D. Livingstone, 4 Hatfield Road, Southport, PR8 2PE
2E0 HYI M. Webb, 10 Sausthorpe Street, Lincoln, LN5 7XW
2E0 HYJ W. Lockley-Gardiner, 193 Westbourne, Telford, TF7 5QP
2E0 HYK T. Ward, 10 Limefield, Oakham, LE15 6ND
2E0 HYM A. Mears, Scotland Yard, Priors Leaze Lane, Hambrook, Chichester, PO18 8RQ
2E0 HYO Prof. R. Houlston, 51 Adelaide Road, Surbiton, KT6 4SR
2E0 HYP S. Bailey, 23 Maple Avenue, Tolladine, Worcester, WR4 9RD
2E0 HYS Z. Gong, Room 1111, Scape Wembley, Fulton Rd, Wembley Park, Wembley, London, HA90TF
2E0 HYU A. Hitchcott, 19 Corwen Road, Liverpool, L4 7TL
2W0 HYW H. Williams, 174 Squirrel Walk, Pontarddulais, SA4 0UG
2E0 HYY K. Mills, 3 The Seasons, Summerway, Exeter, EX4 8DQ
2M0 HYZ S. Magee, 30 Burnfield Drive, Mansewood, Glasgow, G43 1BW
2E0 HZA A. Fox, 8 Giles Road, Swindon, SN25 1QD
2E0 HZB S. Harvey, 110 Vicarage Road, Wednesfield, Wolverhampton, WV11 1SF
2W0 HZC R. Davis, 19 Ael-Y-Bryn, Caerphilly, CF83 2QX
2E0 HZE F. Stepien, 11A Wirral Gardens, Wirral, CH63 3BD
2E0 HZH S. O'Connor, 4 Charlton Grove, Bradford, BD20 0QG
2E0 HZK M. Ratcliffe, 30 Newton Cross Road, Newton In Furness, Barrow-in-Furness, LA13 0NB
2M0 HZL H. Mckay, 3 Flemington Gardens, Whitburn, Bathgate, EH47 0NS
2M0 HZO A. Moir, 2/1 7 Stewartville Street, Glasgow, G11 5PE
2E0 HZP D. Westwood, 28 Weybridge Mead, Yateley, GU46 7UY
2E0 HZQ B. Capewell, 18 Westminster Road, Kidderminster, DY11 6HG
2M0 HZR N. Mansfield, Blencathra House, Township Road, Auckengill, KW14xp
2E0 HZS H. Smith, 11 Kettlebrook Road, Tamworth, B77 1AB
2E0 HZU J. Howarth, 19 Farnham Croft, Leeds, LS14 2HR
2E0 HZW C. Norris, Heather View, Forest Front, Hythe, Southampton, SO45 3RJ
2E0 IAF S. Pryke, 12 Seaward Avenue, Leiston, IP16 4BB
2E0 IAG I. Marsh, 56B Oliver Crescent, Farningham, Dartford, DA4 0BE
2M0 IAH I. Macdonald, 20 Newbigging Terrace, Auchtertool, Kirkcaldy, KY2 5XL
2E0 IAJ I. Lockyer, 11 Lorina Road, Ramsgate, CT12 6DD
2E0 IAK H. Jeffery, Albany, Cranbrook, TN17 3JR
2E0 IAL I. Macdonald, Broomhill Mill Lane, Worthing, BN13 3DH
2E0 IAN I. Campbell, 19B Elphinstone Road, Southsea, PO5 3HP
2W0 IAO A. Oconnell, 31 North Avenue, Tredegar, NP22 3HF
2M0 IAQ J. Campbell, 113 Brent Field Circle, Ellon, AB41 9DB
2E0 IAS E. Alexander, 25 Dunelm Road, Hetton-Le-Hole, Houghton le Spring, DH5 9LB
2E0 IAZ D. Goldthorpe, 6 West End Grove Haydock, St Helens, WA11 0AP
2E0 IBA S. Jones, 11 Croft Close, Rowton, Chester, CH3 7QQ
2E0 IBB B. Beard, 8 Monks Close, Newcastle-under-Lyme, ST5 3QU
2E0 IBI A. Douglas, 3 Beech Avenue, Bilsborrow, Preston, PR3 0RH
2W0 IBM A. Mcelroy, 36 Bod Offa Drive, Buckley, CH7 2PB
2E0 IBN L. Thacker, Hunters Moon, Reigate Road, Horley, RH6 0HU
2E0 IBP N. Whittaker, 3 Westwood Dr Hellesdon, Norwich, NR6 5DE
2E0 IBR M. Perescu, 8 Marauder Road, Norwich, NR66HD
2E0 IBT J. Ferrol, 29 Westlands, Haltwhistle, NE49 9BS
2E0 IBU A. Buckley, 25 Queensway, Pilsley, Chesterfield, S45 8EJ
2M0 IBW I. Macdonald, The Cottage, High Craigton, Glasgow, G62 7HA
2W0 IBY L. Rodgers, 16 Pembrey Gardens Pontllanfraith, Blackwood, NP12 2LR
2M0 ICB I. Buchner, 33 Blakewell Gardens Tweedmouth, Berwick-upon-Tweed, TD15 2HJ
2E0 ICC I. Chilton, 47 Mayfield Crescent, Eaglescliffe, Stockton-on-Tees, TS16 0NH
2W0 ICD I. Davies, 3 Keteringham Close, Sully, Penarth, CF64 5JW
2E0 ICE P. Hallas, 53 Thornleigh Avenue Eastham, Wirral, CH62 9AZ
2W0 ICF K. Earnshaw, 5 Castle Mews George Street, Pontypool, NP4 6BU
2E0 ICI I. Lippett, 41 Springvale, Gillingham, ME8 0JG

2E0 ICK A. Morris, 71 Lurdin Lane, Standish, Wigan, WN6 0AQ
2E0 ICN H. Lee, Pantos Logistics, Attn50924, 776 Buckingham Avenue, Slough, SL1 4NL
2E0 ICP I. Pass, 69 Cotswold Road, Bath, BA2 2DL
2E0 ICR D. Burton, 16 Cage Lane, Great Staughton, St Neots, PE19 5DB
2E0 ICT M. Gascoyne, 31 Dale View, Hemsworth, Pontefract, WF9 4TA
2E0 ICU D. Marsh, 16 Laurel Close, North Warnborough, Hook, RG29 1BH
2E0 ICX D. Lucock, 34 Wentworth Drive, Ipswich, IP8 3RX
2E0 ICY E. Moore, 33 Avon Drive, Congleton, CW12 3RQ
2E0 IDA I. Sharples, 2 Moor Road Croston, Leyland, PR26 9HQ
2E0 IDC I. Cosham, 54 Hawkins Crescent, Shoreham-by-Sea, BN43 6TP
2E0 IDE B. Smith, Bees Corner, Ideford, Chudleigh, Newton Abbot, TQ13 0AZ
2E0 IDF I. Firth, 124 Viking Road, Bridlington, YO16 6TB
2I0 IDJ K. Mclaverty, 123A Castle Road, Antrim, BT41 4NG
2E0 IDK I. King, 7, Greenacres Avenue, Blythe Bridge, Stoke on Trent, ST11 9HU
2E0 IDL D. Garner, 1 Rowland Avenue, Field Street, Hull, HU9 1HR
2E0 IDM D. Melbourne, 160 Markfield Courtwood Lane, London, cr0 9hq
2E0 IDN I. Norfolk, Arwelfa, High Street, Uckfield, TN22 3LP
2E0 IDO Y. Wang, 50 Sleaford Street, Cambridge, CB1 2PU
2E0 IDR I. Reeve, 36 Stone Pippin Orchard, Badsey, Evesham, WR11 7AA
2W0 IDT I. Booth, 18 Clos Y Wiwer, Pentre Cwrt, Llantwit Major, CF61 2SG
2E0 IDY N. Downes, 17 Knightswood Close, Rosliston, Swadlincote, DE12 8JJ
2E0 IEA E. Aspden, 4 Lilac Close, Newcastle, ST5 7DH
2E0 IEB I. Beales, 6 Edge Well Rise, Sheffield, S6 1FB
2M0 IEC E. Cohen, 234 Allison Street, Glasgow, G42 8RT
2E0 IED A. Wedge, 30 Primrose Way Locks Heath, Southampton, SO31 6WX
2E0 IEE S. Spice, Ebor Lodge, Udimore Road, Rye, TN31 6BX
2E0 IEI R. Adams, 40 Lichfield Road, Gloucester, GL4 3AL
2E0 IEL N. Raptopoulos, 12 Wolversdene Road, Andover, SP10 2AX
2E0 IEM J. Paine, 11 Ferndale Park Fifield Road, Bray, Maidenhead, SL6 2DZ
2E0 IEO M. Sanderson, 2 East Crescent, Canvey Island, SS8 9HL
2W0 IEP P. Roberts, 18 Maes Mawr, Llanrwst, Ll260hw
2E0 IEQ J. Law, 58 Westfields, Zeals, BA126PW
2E0 IER B. Brooks, 54 New Hills Road, Doncaster, DN2 5TR
2E0 IET C. Blount, 55 Silverthorne Drive, Caversham, Reading, RG4 7NR
2E0 IEW J. Milner, 30 Rowena Drive, Thurcroft, Rotherham, S66 9HT
2E0 IEX T. Barris, Vogelgartenstr. 41/1, Eislingen/Fils, Germany, 73054
2E0 IFA C. Bolton, 1A Salterns Terrace, Bideford, EX39 4AG
2E0 IFB A. Wood, Berrybrook Northall Road, Eaton Bray, LU6 2DQ
2E0 IFC N. Burnet, 27 Mackenzie Way, Tiverton, EX16 4AW
2E0 IFF P. Webster, 15 Napier Street, Workington, CA14 2PT
2E0 IFG A. Evans, 56 Meadow Way, Bradley Stoke, Bristol, BS32 8BP
2E0 IFK D. Foxcroft, 8 Cowlard Close, Launceston, PL15 7EQ
2E0 IFN Dr D. Rodriguez, 23 Thorpe Way, Cambridge, cb5 8uj
2E0 IFO N. Cranston, 42 Curling Vale, Guildford, GU27PH
2E0 IFP D. Greenwood, 7 Royal Gardens, Ramsbottom, Bury, BL0 9SB
2E0 IFS I. Stephens, 14 Vardon Close Kingston Hill, Stafford, ST16 3YW
2E0 IFT G. Saunders, 140 Highbridge Road, Burnham-on-Sea, TA8 1LW
2E0 IFV P. Kearney, 22 Kingston Drive, Cheltenham, GL51 0UB
2E0 IFW E. Evans, 313 Parkgate Road, Chester, CH1 4BE
2E0 IFZ I. Huggett, 104 Hinchcliffe Orton Goldhay, Peterborough, PE2 5SS
2I0 IGA G. Castorio, 1A Marlborough Park North, Belfast, BT9 6HJ
2W0 IGC I. Curnock, 62 Heol Y Banc, Bancffosfelen, Llanelli, SA15 5DL
2E0 IGE A. Insarov, Broadway Chambers, 20 Hammersmith Broadway, London, W6 7AF
2E0 IGH P. Frankowski, Flat 7, Swift House, Chigwell Road, London, E18 1TP
2E0 IGL P. Penn-Bixby, 11 Iveston Road, Consett, DH8 7HS
2E0 IGM G. Benford, 34 Victoria Gardens, Colchester, CO4 9YD
2W0 IGN R. Buchan-Terrey, Godre'R Coed, Aberhosan, Machynlleth, SY20 8RA
2E0 IGO R. Fountain, 4 Den Hill, Eastbourne, BN20 8SY
2E0 IGP Dr J. Harmer, 1 Wynford Rise, Leeds, LS16 6HX
2E0 IGQ I. Jameson, 9 Pine Hey, Neston, CH64 3TJ
2E0 IGT A. Williamson, 4 Garden End, Melbourn, Royston, SG8 6HD
2E0 IGW G. Taylor, 39 Gill St., Newcastle upon Tyne, NE4 8BH
2I0 IGX I. Kondrashenkov, 44 Forge Manor, Magheralin, Craigavon, BT67 0XP
2E0 IHB Dr R. Bleaney, 40 Broadstone Road, Harpenden, AL5 1RF
2E0 IHE D. Loveridge, 2 Abbots Mead, Cholsey, OX10 9RJ
2E0 IHH I. Hutchinson, Bridgend, Mill Lane, North Hykeham, Lincoln, LN6 9PA
2E0 IHI C. Colless, 128 Ditton Lane, Fen Ditton, Cambridge, CB5 8SS
2E0 IHL A. Prime, 68 Langley Drive, Crewe, CW2 8LN
2E0 IHM S. Nelson-Smith, Welbeck, Wickham Road, Fareham, PO17 5BU
2E0 IHN N. Gooding, 41 The Crescent, Wolverhampton, WV6 8LA
2E0 IHO K. Brown, 143 Princes Road, Ellesmere Port, CH65 8EP
2E0 IHR O. Williams, 6 Millfield Drive, Bristol, BS30 5NR
2E0 IHS G. Sangwell, 34 Gatcombe Close, Calcot, Reading, RG31 4XQ
2E0 IHT C. Scouller, 72 Brookmans Avenue Brookmans Park, Hatfield, AL9 7QQ
2E0 IHU I. Leyland, 61 Stacey Avenue, Milton Keynes, MK12 5DN
2E0 IHV S. Jones, 36 Woodbury, Lambourn, Hungerford, RG17 7LT
2E0 IHW M. Rogers, The Blue Mushroom, L E E, Ilfracombe, EX34 8LR
2E0 IID I. Dodds, 54 Philip Road, Newark, NG24 4PD
2E0 IIF G. Gribbin, 96 Old Church Lane, Stanmore, HA7 2RR
2M0 IIH J. Jarvie, Berryhill Farm, Tak-Ma-Doon Road, Kilsyth, Glasgow, G65 0RY
2E0 IIJ C. Bassett, 2 Culverden Square, Tunbridge Wells, TN4 9NS
2E0 IIP E. Lewis, 49 Ridgebourne Road, Shrewsbury, SY3 9AB
2I0 IIQ A. Bandeja, 16, Craigavon Crescent, Dungannon, BT71 7BD
2E0 IIT G. Coleman, 398 Chemin De Meynot, la Sauvetat Du Dropt, France, 47800
2W0 IIU P. Harding, 6 Ridgeway View, Newport, NP205AW
2E0 IIV S. Lilley, 1 Ormonde Road, Chester, CH2 2AH
2E0 IIY A. Wade, 6 Elliott Grove, Brixham, TQ5 8RT
2E0 IIZ S. Martin, 6 Cherry Tree Drive, Sedgefield, Stockton-on-Tees, TS21 3DN
2W0 IJD T. Hughes, 1 Craig Y Don, Amlwch, LL68 9DN
2E0 IJG C. Nutu, 37 Brading Road, London, SW2 2AP
2E0 IJH I. Hope, 5 The Crescent, Northfleet, Gravesend, DA11 7EB
2W0 IJL K. Smith, 62 Waterloo Road, Talywain, Pontypool, NP4 7HJ
2E0 IJP C. Simon, 55 Shirecroft Road, Weymouth, DT4 0NH
2E0 IJQ R. Bedford, 29 Kent Road, Brookenby, Market Rasen, LN8 6EW
2E0 IJT R. Darby, 39 Leonard Road Wollaston, Stourbridge, DY8 3LU
2E0 IJW G. Williamson, 26 Portland Mews, Bridlington, YO16 4EH
2E0 IJX D. Wilderspin, 14 Tannery Court, North Street, Crewkerne, TA18 7AY
2E0 IJZ K. Cullen, 29 Colman Road, London, E16 3JY
2E0 IKB B. Wilkes, 2 Kings Crescent, Edlington, Doncaster, DN12 1BD
2E0 IKC E. Cooper, 51 Lakeside Hightown, Ringwood, BH24 3DX
2E0 IKG R. Butt, Waterloo Cottage, Barrack Hill, Little Birch, Hereford, HR2 8AX
2E0 IKH T. Anderson, 45 Dudley Road, Clacton-on-Sea, CO15 3DW
2E0 IKM M. Moffat, 7 Lingfell Avenue, Cockermouth, CA13 9BE
2E0 IKN C. Burls, 86 All Saints Road, Kings Heath, Birmingham, B14 7LN
2E0 IKO M. Tarrant, Holly Cottage Nanstallon, Bodmin, PL30 5JZ
2E0 IKS J. Mccosh, Ch611Dl, Merseyside, Ch611dl
2E0 IKT C. Chisholm, Hill Top Kellah, Haltwhistle, NE49 0JL
2E0 IKU Dr S. Mitchell, 56 Mill Green Road, Amesbury, Salisbury, SP4 7RE
2E0 IKV S. Rush, 88 Mountview, Borden, Sittingbourne, ME9 8JZ
2E0 IKW G. Cannon, 30 Main Street, Flixton, Scarborough, YO11 3UB
2E0 IKX P. Scarratt, 4 Sandy Lonning, Maryport, CA157LW
2E0 ILC C. Ng, 9A Romsey Road, Southampton, SO16 4BY
2E0 ILD R. Abeykoon, 20, Holmewood Crescent, Nottingham, NG5 5JH
2E0 ILE M. Percival, 10 Binder Close, Higham Ferrers, NN108PH
2E0 ILF B. Campbell, 9 Granams Croft, Liverpool, l30 0ph
2E0 ILJ Rev. P. Campbell, 116 Kingsway Park, Urmston, Manchester, M41 7FH
2E0 ILL J. Robinson, 96 White Lodge Park, Shawbury, Shrewsbury, SY4 4NU
2E0 ILN D. Bee, 25 Blatcher Close, Minster On Sea, Sheerness, ME12 3PG
2E0 ILO M. Noblet, 1 Lingdale Road, Wirral, CH48 5DG
2W0 ILQ W. Stevenson, 310 Argosy Way, Newport, NP19 0LP
2E0 ILX M. Bennett, 33 Charles Street, Redditch, B97 5AA
2M0 ILZ M. Greig, 22 Rowanhill Close, Port Seton, Prestonpans, EH32 0SY
2I0 IMB I. Barr, 64 Owenreagh Drive, Strabane, BT82 9DT
2E0 IMC I. Coggon, 45, Ansten Crescent, Doncaster, DN4 6EZ
2E0 IMD H. Ilie, Sos. Iancului Nr. 33 Bloc 105A Scara B Apt 63, Bucharest, Romania, 21717
2E0 IMG I. Gough, Holythorn Cottage, Moreton Morrell, CV359AR
2E0 IMH M. Williams, 1 Montague Road Broughton Astley, Leicester, LE9 6RL
2E0 IMJ P. Coppin, 3 Firtree Close, Rough Common, Canterbury, CT2 9DB
2W0 IMM C. Milne, 18 Chapelfield, Deganwy, LL31 9BF
2D0 IMN T. Hardwick, 3 Poplar Terrace, Douglas, Isle of Man, IM2 4AR
2I0 IMO T. Mcnaughter, 36 Elms Park, Coleraine, BT52 2QE
2M0 IMP S. Robertson, 20 Knockard Place, Pitlochry, PH16 5JF
2E0 IMS S. Edwards, 68 Hampshire Road, Droylsden, Manchester, M43 7PL
2E0 IMV D. Pelling, 8 Fowler Close Earley, Reading, RG6 7SS
2E0 IMW I. Walker, 24 Hawthorn Road, Norwich, NR5 0LP
2W0 IMX P. Lowe, 9 Felin Uchaf, Dolgellau, LL40 1NS
2E0 IMZ I. Ross, 48 Henry Drive, Leigh-on-Sea, SS9 3QF
2I0 INA A. Mcguigan, 30 Cogry Hill, Ballyclare, BT39 0RY
2E0 INC C. Jenkins, 25 Longmeadow Grove West Heath, Birmingham, B31 4SU
2E0 IND P. Ind, 30 Thompson Road, Stroud, GL5 1SY
2M0 INE M. Scott, 28B Highfield Place, Birkhill, Dundee, DD2 5PZ
2E0 INK J. Everatt, 50 Barnsley Road, Thorpe Hesley, Rotherham, S61 2RR
2M0 INS C. Ralph, 37 Seaview Terrace, Edinburgh, EH15 2HE
2E0 INT P. Kerton, 15 Barncroft Way, Havant, PO9 3AA
2E0 INX Prof. I. Neal, 8 Rushey Gill, Brandon, Durham, DH7 8BL
2W0 INY A. Noakes, Westra Holt, Westra, Dinas Powys, CF64 4HA
2M0 IOB R. Kennedy, 45 Rodney Road, Gourock, PA19 1XG
2E0 IOG D. Mathewson, 33 Thornton Road, Bootle, L20 5AN
2I0 IOI P. Gibson, 118 Coleraine Rd, Portstewart, bt55 7hs
2M0 IOK S. Webb, Fawn View, Kennethmont, Huntly, AB54 4PF
2E0 IOL R. Morley, 7 Eliot Road, St Austell, PL25 4NL
2D0 IOM D. Allcote, 18 Maynrys, Castletown, Isle Of Man, IM9 1HR
2E0 ION A. Chlebikova, St. Catharine'S College, Cambridge, CB2 1RL
2E0 IOS R. Smith, 62 Norwich Avenue, Plymouth, PL5 4JQ
2E0 IOU D. Savage, 72 Pitmore Road, Eastleigh, SO50 4LW
2E0 IOZ A. Nikitits, 270A The Ridgeway, St Albans, AL4 9XQ
2E0 IPA D. Bradley, 37 Leamoor Avenue, Somercotes, Alfreton, DE55 1RL
2I0 IPB A. Kincaid, 428 Cushendall Road, Ballymena, BT43 6QE
2E0 IPC W. Catney, 92 Second Avenue, Liversedge, WF15 8JW
2E0 IPG N. Head-Jenner, 5 Ponders Road, Colchester, CO63LX
2E0 IPK T. Bone, 104 Rowdowns Road, Dagenham, RM9 6NH
2E0 IPL C. Panaitescu, 131 Stafford Road, Croydon, CR0 4NN
2M0 IPO P. O'Hara, 13/3 135 Kirkton Avenue, Glasgow, G13 3EP
2E0 IPW P. White, 13 Field Court, Oxted, RH8 0PD
2E0 IPX P. Ashby, 45 Seaton Road, Felixstowe, IP11 9BS
2E0 IQO M. Vardy, 1 Sutton Middle Lane, Kirkby-In-Ashfield, Nottingham, NG17 8FX
2M0 IQR H. Chen, Iq Fountainbridge Room 118E 114 Dundee Street, Edinburgh, EH11 1AD
2E0 IQT S. Magrys, 13 Norfolk Road, Kidsgrove, Stoke-on-Trent, ST7 1EZ
2M0 IQU W. Curry, 36 Banklands Newburgh, Cupar, KY146DN
2E0 IQW D. Parker, 51 Parsonage Road, Henfield, BN5 9HZ
2E0 IQX A. Emmerson, 8 Weston Close, Heath Hayes, Cannock, WS11 7YX
2M0 IRC J. Livingstone, 17 Livingstone Drive, Bo'ness, EH51 0BQ
2E0 IRE M. O'Donovan, Wyllsden House, Stroud, GL52PA
2M0 IRG Dr M. Sutcliffe, 11 Low Borland Way, Eaglesham, Glasgow, G76 0BP
2E0 IRJ I. Jones, 33 Cobham Avenue, Liverpool, L9 3BP
2E0 IRN J. Glicklich, 86 Ainsdale Road, Bolton, BL3 3ER
2E0 IRO G. Clarke, 2 Gort Lomie, Clonlara, Ireland, V94 H96H
2E0 IRX G. Harman, 58 Laurence Avenue, Witham, CM8 1JB
2M0 ISA I. Watson, 64 Anstruther Street Law, Carluke, ML8 5JG
2E0 ISB S. Brown, 18 Goring Ave. Gorton., Manchester, M18 8WW
2E0 ISK K. Hyde, 8 Pennant Close, Birchwood, Warrington, WA3 6RR
2E0 ISN S. Narwan, 14 Fermoy Road, Greenford, London, UB6 9HX
2E0 ISO I. Butler, 23 Owen Way, Basingstoke, RG24 9GH
2E0 ISQ R. Lilley, 3 Coultshead Avenue, Billinge, Wigan, WN5 7HS
2E0 ISS P. Dann, 4 Middlefields Court, Middlefields, Letchworth Garden City, SG6 4NQ
2E0 ISU M. Dunstan, 5 Polgooth Close, Redruth, TR15 1QL
2M0 ISY I. Phillips, 16 Seton Court, Port Seton, Prestonpans, EH32 0TU
2W0 ISZ R. Priamo, 58 Ffordd Glyn, Coed-Y-Glyn, Wrexham, LL13 7QW
2E0 ITC A. Checketts, 77 Shenstone Avenue, Stourbridge, DY8 3EH
2E0 ITF C. Owen, 3 The Flexton, Ottery St Mary, EX11 1DJ
2W0 ITH M. Findlay, 26 Swan Street, Llantrisant, Pontyclun, CF72 8ED

2M0	ITI	Dr J. Wills, 2 Old Dalmore Gardens Auchendinny, Penicuik, EH26 0RR
2E0	ITJ	T. Johnson, 15 Tennyson Road, Creswell, Worksop, S80 4DW
2W0	ITM	I. Mitchell, 9 Rhiw Grange, Colwyn Bay, LL29 7TT
2E0	ITN	R. Goodall, 76 Beaconfield Road, Plymouth, PL2 3LF
2M0	ITO	R. Bell, 10 Lorraine Drive, Cupar, KY155DY
2E0	ITQ	J. Swales, 90 Earlswood Road, Dorridge, Solihull, B93 8RN
2E0	ITU	P. James, Flat 2, Marigold House, 2 Ironbridge Road, Twigworth, Gloucester, GL2 9GS
2E0	ITV	G. Moorhouse, 1 Woodlands Avenue, Spilsby, PE23 5EP
2E0	ITX	D. Gibbeson, 33 Treefield Walk, Barnstaple, EX32 8PE
2I0	ITY	D. Given, 15 Middle Road, Lisburn, BT27 6UU
2M0	IUE	A. Young, 2 Dunvegan Place, Ellon, , AB41 9TF
2E0	IUH	A. Morris, 22 Dixon Avenue, Newton-le-Willows, WA12 0NE
2E0	IUI	D. Marshall, 7 Lancaster Close, Newton-le-Willows, WA12 9EY
2E0	IUK	D. Grayson, 79 Errington Avenue, Sheffield, S2 2EA
2W0	IUN	I. Jones, 21 Albert Street, Maesteg, CF34 0UF
2E0	IVO	W. Gissing, 2 Yeo Moor, Clevedon, BS21 6UQ
2E0	IVR	I. Ross, 15 Earlswood Drive, Madeley, Telford, TF7 5SF
2M0	IVS	G. Brass, 114 Torbrex Rd Cumbernauld, Glasgow, G67 2JS
2E0	IVY	D. Slee, Turnpike, Brearton, Harrogate, HG3 3BX
2W0	IVZ	R. Vials, 46-48 Park Place, Gilfach, Bargoed, CF8 8NA
2E0	IWB	I. Bunting, 14 Mill Pightle, Aylsham, Norwich, NR11 6LX
2M0	IWD	I. Davidson, 3 Hillcrest Avenue, Kirkcaldy, KY2 5TU
2E0	IWF	I. Francis, 34 Furlong Road, Bourne End, SL85AA
2E0	IWI	D. Gilbertson, 6 Lewens Lane, Wimborne, BH21 1LE
2W0	IWM	I. Miles, 40 Seymour Street, Caegarw, Mountain Ash, CF45 4BL
2E0	IWS	I. Smith, 8 Tregarth Road, Chichester, PO19 5QU
2E0	IWT	M. Champness, 10 Isaac Square, Great Baddow, Chelmsford, CM2 7PP
2E0	IWZ	A. Hanna, 35 Orchard Drive, Mayland, Chelmsford, CM3 6EP
2E0	IXC	D. Watson, 10 Gimson Close, Tuffley, Gloucester, GL4 0YQ
2E0	IXH	R. Maas, 15 Pine Court, Attleborough, NR17 2HU
2E0	IXI	R. Forss, Ginger House, High Street, Blockley, Moreton-in-Marsh, GL56 9EX
2E0	IXM	Dr D. Buckley, 7 Clary Meadow, Northwich, CW8 4XG
2E0	IXS	I. Smith, 50 Aylton Road, Liverpool, L36 2LX
2M0	IXT	S. Kelso, 98 Highmains Avenue, Dumbarton, G82 2QB
2E0	IXX	D. Cattermole, 39 Moor Lea, Braunton, EX33 2PF
2I0	IYH	A. Wilson, 108A Salia Avenue, Carrickfergus, BT38 8NE
2E0	IYY	H. Burnett, 10 Galton Avenue, Christchurch, BH23 1JU
2E0	IZB	P. Lewis, 37 Speedwell Close, Melksham, SN12 7TE
2I0	IZI	A. Ismay, 21 Hillsborough Drive, Belfast, BT6 9DS
2E0	IZJ	A. Dunn, 185 Waterloo Road Yardley, Birmingham, B25 8LH
2E0	IZP	I. Pipe, 9 Sherlock Hoy Close, Broseley, Telford, Tf125jb
2E0	IZR	J. Bridson, 10 Clegg Street, Astley, Tyldesley, Manchester, M29 7DB
2E0	IZW	I. Whiteley, 2 The Meade, Manchester, M21 8FA
2E0	IZZ	P. Smith, 14 Highfield Crescent, Kettering, NN15 6JS
2E0	JAA	J. Ahmed, 59 Ramsgate, Lofthouse, Wakefield, WF3 3PX
2E0	JAB	J. Bradbury, 11 Ravensbourne House, Arlington Road, Twickenham, TW1 2AX
2E0	JAC	J. Crank, 38 Harley Avenue, Harwood, Bolton, BL2 4NU
2E0	JAF	J. King, 22 Latchmere Gardens, Leeds, LS16 5DN
2W0	JAI	J. Griffiths, 85 Tudor Estate, Maesteg, CF34 0SW
2E0	JAJ	J. Johnson, 1 Stretton Lodge Gordon Road, London, W13 8PR
2M0	JAL	C. Maclean, 16 Glamis Avenue, Elderslie, Johnstone, PA5 9NR
2E0	JAM	J. Dodds, 8 York Road, Rowley Regis, B65 0RR
2E0	JAN	G. Martin, 12 Poolside, Phase 1, St Joseph, Trinidad And Tobago,
2E0	JAO	J. Lockyear, Berrow Bank, Bromsberrow, Ledbury, HR8 1SG
2I0	JAP	E. Rantin, 8A Buchanans Road, Newry, BT35 6NS
2E0	JAQ	J. Bosworth, 10 Aston Street, Leeds, LS13 2BJ
2E0	JAR	J. Rance, 11 Orchard Lane, Longton, Preston, PR4 5AX
2M0	JAT	A. Murray, 119 Carnarc Crescent, Inverness, IV3 8SJ
2E0	JAW	J. Phipps, 42 Cavell Avenue, Peacehaven, BN10 7NS
2E0	JAX	J. Burdett, 5 Winston Drive, Wainscott, Rochester, ME2 4LJ
2E0	JAZ	J. Sadler, 10 Spindle Warren, Havant, PO9 2PU
2E0	JBA	J. Woods, 3 Ingle Avenue Morley, Leeds, LS27 9NP
2E0	JBC	D. Croft, 33 Roughaw Road, Skipton, BD23 2PY
2E0	JBD	J. Dukes, 79 Jubilee Avenue, Boston, PE21 9LE
2D0	JBE	J. Mccartney, 5 Riverside, Ramsey, Isle of Man, IM8 3DA
2E0	JBF	G. Brown, 51 Arncliffe Drive, Knottingley, WF11 8RH
2E0	JBI	J. Bethell, 19 Kiln Cottages, The Brickfields, Stowmarket, IP14 1RY
2W0	JBJ	J. Jenkins, 13 Birch Hill, Newport, NP20 6JD
2E0	JBK	S. Glass, 16 Norman Way, Colchester, CO3 4PS
2E0	JBL	J. Chatterton, 6 Bayliss Road, Wargrave, Reading, RG10 8DR
2E0	JBM	J. Moppett, 59 Piccadilly, Tamworth, B78 2ER
2E0	JBQ	D. Robinson, 27 Abbeylea Drive, Westhoughton, Bolton, bl5 3zd
2E0	JBS	J. Byrne, 45 Jenkins Drive, Sheffield, S9 1AR
2E0	JBW	S. Pascoe, Trewyn, Carnmenellis, Redruth, TR16 6PG
2E0	JBX	J. Barratt, 17 Main Road, Collyweston, Stamford, PE9 3PF
2E0	JBY	J. Bibby, 12 James Street, Burton-on-Trent, DE14 3SB
2M0	JBZ	J. Coubrough, 41 Bridge Court, Alexandria, G83 0BZ
2E0	JCA	J. Aubury, 27 Gravel Walk, Tewkesbury, GL20 5NH
2E0	JCB	J. Waring, 12 Mary Street, Farnhill, Keighley, BD20 9AU
2E0	JCC	J. Whitton, 11 Dursley Road, Bristol, BS11 9XB
2E0	JCD	J. Dalgliesh, 61 Clonners Field, Stapeley, Nantwich, CW5 7GU
2E0	JCE	J. Tompkins, 3 Hartwell Road, Portsmouth, PO3 5TN
2M0	JCF	J. Forsyth, 14B Osborne Terrace, Edinburgh, EH12 5HG
2M0	JCG	J. Rankin, 17 Dippin Place, Saltcoats, KA21 6AB
2I0	JCH	Dr J. Henderson, 1 Brook Lodge Ballinderry Lower, Lisburn, BT28 2GZ
2E0	JCK	E. Beechill, Belleroyd Farm Blackshaw Head, Hebden Bridge, HX7 7JP
2E0	JCL	J. Swift, Raf Holmpton Rysome Lane, Holmpton, HU19 2RG
2E0	JCM	J. Clark-Mcintyre, 6 Belvedere Road, Blackburn, BB1 9NS
2W0	JCN	J. Pritchard, 1 Tan Y Coed, Maesgeirchen, Bangor, LL57 1LU
2E0	JCO	J. Crewe, 22 Myrtle Tree Crescent Sandbay, Weston-Super-Mare, BS22 9UL
2W0	JCP	J. Percival, Blue Cedars, Gresford, Wrexham, LL12 8RN
2E0	JCQ	J. Stevens, 51 Cheddington Road, Pitstone, Leighton Buzzard, LU7 9AQ
2E0	JCR	J. Seear, Old Gossington Chapel, Gossington, Slimbridge, Gloucester, GL2 7DN
2E0	JCS	J. Sanderson, 54 Kelvedon Close, Chelmsford, CM1 4DG
2E0	JCT	J. Griffiths, 83 Golborne Road Ashton-In-Makerfield, Wigan, WN4 8XA
2E0	JCU	J. Underwood, 12 Forsythia Close, Bicester, OX26 3GA
2W0	JCV	J. Baldwin, 94 King Street Abertridwr, Caerphilly, CF83 4BG
2E0	JCW	J. Wright, 19 Halstead Close, Woodley, Reading, RG5 4LD
2E0	JCX	S. Jackson, 5 Duchess Park Close, Shaw, Oldham, OL2 7YN
2E0	JCY	J. Connolly, 2 Waring Avenue, St Helens, WA9 2QG
2E0	JCZ	J. Robb, 37 Wroxham Road, Woodley, Reading, RG5 3AX
2E0	JDA	J. Dale, Corydon, Church Street, Sevenoaks, TN14 7SW
2E0	JDC	L. Short, 10 Carville Crescent, Brentford, TW8 9RD
2E0	JDD	J. Delves, 14 Stanthorne Avenue, Crewe, CW2 8NH
2E0	JDE	J. Hardingham, 11 All Saints Close, Weybourne, Holt, NR25 7HH
2E0	JDF	A. Powell, Crosstrees, Main Road, Theberton, Leiston, IP16 4RX
2E0	JDH	D. Harbron, 6 West View, Penshaw, Houghton le Spring, DH4 7HP
2E0	JDI	J. Redhead, 28 Sandfields, Frodsham, WA6 6PT
2E0	JDK	J. Hazeltine, 21 Hassock Way, Wimblington, March, PE15 0PJ
2W0	JDL	G. Wilkins, 7 Byron Road, Newport, NP20 3HJ
2E0	JDM	J. Matheson, 24 Grange Road, Dacre Banks, Harrogate, HG3 4HA
2E0	JDO	J. Harriott, Flat 10, 401 High Street, Cheltenham, GL50 3FJ
2E0	JDP	N. Pipkin, 46 Charles Avenue, Albrighton, Wolverhampton, WV7 3LF
2E0	JDY	J. Boslem, 14 Morrins Close, Great Wakering, SS3 0DY
2D0	JEA	J. Hill, 54 Wybourn Drive, Onchan, Isle Of Man, IM3 4AT
2E0	JED	G. Maguire, 39 Cooper St., Stretford, Manchester, M32 8NA
2E0	JEE	Dr J. Egleton, 81 Pinchbeck Road, Spalding, PE11 1QF
2E0	JEH	J. Hewart, 14 Kestrel Close, Marple, Stockport, SK6 7JS
2E0	JEK	J. Kay, 36 Winnington Road. Marple, Stockport, SK6 6PT
2E0	JEM	J. Carvill, The Lodge, Oldbury Road, Worcester, WR2 6AA
2E0	JEN	P. Halpin, 50 Celtic Road, Deal, CT14 9EF
2E0	JES	C. Ember, 35 Mattock Lane, Ealing, London, W5 5BH
2E0	JET	R. Cunliffe, 89 Colyers Lane, Erith, DA8 3NG
2W0	JEV	E. Jones, 11 Alma Place Sebastopol, Pontypool, NP45EA
2E0	JEZ	J. Powell, 46 Woodmancote, Yate, Bristol, BS37 4LL
2E0	JFK	J. Kirk, 3 Knold Park, Margate, CT9 3BH
2E0	JFL	J. Lugsden, 21 Overhill Way, Beckenham, BR3 6SN
2I0	JFO	W. Forde, 35 Torr Gardens, Larne, BT40 2JH
2E0	JFW	J. Witchell, 7 Watercombe Lane, Yeovil, BA20 2ED
2E0	JFY	Dr C. Holmes, Old Vicarage Farmhouse, Course Lane, Wigan, WN8 7LA
2E0	JGD	J. Dinsbier, Little Downs, Sandgate Lane, Pulborough, RH20 3HJ
2E0	JGE	J. Glover, 12 Willow Street, London, E4 7EG
2E0	JGG	J. Gibbons, 28 Deveron Gardens, South Ockendon, RM15 5ET
2E0	JGH	J. Hunt, 15 Greenway, London, SW20 9BQ
2E0	JGM	I. Mccorquodale, 4 Wise Grove, Warwick, CV34 5JW
2E0	JGP	J. Paradi, 168 Castle Road, Northolt, UB5 4SG
2E0	JGS	J. Seaton, 88 Stanley Road, Cambridge, CB5 8LB
2E0	JGW	J. Whitehead, 12 Polkerris Road, Carharrack, Redruth, TR16 5RJ
2E0	JHC	J. Clark, 27 The Gabriels, Newbury, RG14 6PZ
2E0	JHE	L. Batten, Y Felin Barn, Llawr-Y-Glyn, Caersws, SY17 5RH
2E0	JHF	P. Attwater, 42 Danescourt Crescent, Sutton, SM1 3EA
2E0	JHG	J. Ginever, 66 London Road, Maidstone, ME16 8QU
2I0	JHH	J. Hannon, 75 Suffolk Road, Belfast, BT11 9PU
2M0	JHN	J. Brown, 60 Laburnum Lea, Hamilton, ML3 7LZ
2E0	JHO	B. Hodgson, 28 Grove Drive, Woodhall Spa, LN10 6RT
2E0	JHP	D. Porter, 105 Shore Road, Littleborough, OL15 9LJ
2E0	JHT	J. Hill, 45 Venus Street, Congresbury, Bristol, BS49 5HA
2E0	JIA	I. Boddy, 5 Boverton Avenue, Brockworth, Gloucester, GL3 4ER
2E0	JIB	C. Tacon, 2 Pavillion Court, 74-76 Northlands Road, Southampton, SO15 2NN
2I0	JIE	J. Evans, 404 Foreglen Road, Dungiven, Londonderry, BT47 4PN
2E0	JIG	D. Harris, 35 Itchenor Road, Hayling Island, PO11 9SN
2E0	JIL	G. Whitehead, 29 Coulsons Road, Bristol, BS14 0NN
2E0	JIP	J. Pownall, 75 Park Barn Drive, Guildford, GU2 8ER
2E0	JIR	S. Buckley, 35 Marlborough Road, Irlam, Manchester, M44 6HH
2E0	JIS	S. Smart, 26 Wycote Road, Gosport, PO13 0TG
2E0	JIT	D. Hook, 6 Cliveden Close, Allington, Maidstone, ME16 0PQ
2E0	JIW	J. Biggin, Galadean, Farriers Way, Newport, PO30 3JP
2E0	JIX	J. Packer, 20 Shipman Road, Market Weighton, Market Weighton, YO43 3RB
2E0	JIZ	S. Phillimore, 21 First Avenue, West Molesey, KT8 2QJ
2E0	JJA	J. Adler, 1 Searles Meadow, Dry Drayton, Cambridge, CB23 8BW
2E0	JJB	J. Barton, 93 Cardigan Road, Bridlington, YO15 3JU
2E0	JJK	J. King, 18 Ross Road, Wallington, SM6 8QB
2E0	JJN	J. Nicholls, 6 Marasca End, Holt Drive, Colchester, CO2 0DL
2E0	JJP	J. Walker, Flat A 14 Elswick Road, London, SE13 7SR
2E0	JJR	R. Ellery, 12 Ssentry Close, St Issey Wadebridge, pl27 7qd
2E0	JJS	J. Siddall, 16 Alandale Avenue, Langwith Junction, Mansfield, NG20 9RU
2W0	JJW	R. Woodland, Flat 14, Mays Court, Windsor Road, Neath, SA11 1NG
2W0	JJY	J. Young, Rhos Owen, Llangristiolus, Bodorgan, LL62 5RD
2E0	JJZ	J. Prince, Wellfield Avenue, Luton, LU33AT
2E0	JKB	S. Lucas, 31 Lilian Close, Norwich, NR6 6RZ
2E0	JKD	K. Winward, 123 Fulbeck Road, Middlesbrough, TS3 0RL
2E0	JKG	J. Gaskin, Badgers Barn, Canterbury Road, Folkestone, CT18 8DF
2W0	JKN	E. Jenkins, 41 Park Street, Bridgend, CF31 4AX
2E0	JKP	J. Poole, 61 Lower Vickers Street, Miles Platting, Manchester, M40 7LX
2E0	JKR	R. Whatmough, 150 Whelley, Wigan, WN1 3UE
2E0	JKS	J. Slyne, 3 Heaths Close, Enfield, EN1 3UP
2E0	JKT	J. Turner, 34 Vaughan Road, Stotfold, Hitchin, SG5 4EH
2D0	JKW	Dr J. Wardle, Cooyrt Vane, Ballamodha Straight, Ballamodha, Isle of Man, IM9 3AY
2E0	JKY	K. Young, 48 Sussex Street, Cleethorpes, DN35 7NP
2E0	JLB	J. Bookham, 116 Clare Gardens, Petersfield, GU31 4EU
2E0	JLD	J. Drinkell, 11 Valley Walk, Kettering, NN16 0LY
2E0	JLK	K. Robinson, 103 Recreation Street, Mansfield, NG18 2HP
2E0	JLM	J. Meek, Rose Tree Cottage, Charlbury, OX7 3RX
2E0	JLR	B. Taylor, 1 Whinchat Close, Stockport, SK2 5UU
2M0	JLS	J. Leitch, 25 Lime Street, Grangemouth, FK3 8LZ
2E0	JLW	J. Williamson, 7 Dale Close, Wrecclesham, Farnham, GU10 4PQ
2E0	JLX	J. Landless, 2 Aspen Way, Banstead, SM7 1LE
2E0	JMB	J. Banham, Timandra, Mill Road, Hardwick, Norwich, NR15 2ST
2E0	JMC	J. Mcinnes-Boylan, 17 Hindsford Bridge Mews, Atherton, Manchester, M46 9QZ
2E0	JMD	J. Davies, 34 Station Road, Benton, Newcastle upon Tyne, NE129NQ

2E0 JME J. Smith, 10 Wayside Road, Bridlington, YO16 4BA
2E0 JMF M. Hardwick, 34 The Pastures, Long Bennington, Newark, NG23 5EG
2W0 JMH J. Hewitt, 1 Highfield, Gloucester Road, Chepstow, NP16 7DF
2M0 JMI J. Mclaren, 33 Foulford Street, Cowdenbeath, KY4 9NB
2E0 JMJ J. Jeffries, 62 Rowans, Welwyn Garden City, AL7 1PD
2W0 JMK M. Dean, 4 Crud Yr Awel, Efailwen, Clynderwen, SA66 7UX
2E0 JMR J. Rodley, 268 Grovehill Road, Beverley, HU17 0HP
2E0 JMW J. Walker, 194A Mount Vale, Mount Vale, York, YO24 1DL
2W0 JMX B. Preece, 4 Waltham Close, Morriston, Swansea, SA6 7PH
2E0 JMY J. Bickers, 3 The Old Brickyard, West Haddon, Northampton, NN6 7GP
2E0 JNE C. Arthur, 25 Pump Hollow Lane, Mansfield, NG18 3DU
2E0 JNH J. Betteridge, 57 Wood Road, Chaddesden, Derby, DE21 4LY
2E0 JNM J. Morris, Parsonage Farm, Croscombe, Wells, BA5 3QN
2E0 JNR A. Davies, 4 Erpingham Road, Poole, BH12 1EX
2E0 JNU J. Nuttall, 73 Severn Drive, Walton-Le-Dale, Preston, PR5 4TD
2E0 JNY J. Warman, 6 Down View Road, Denbury, Newton Abbot, TQ12 6ER
2E0 JOA J. Adler, 1 Searles Meadow Dry Drayton, Cambridge, CB23 8BW
2E0 JOE J. Fletcher, 32 Chapel Lane, Barwick In Elmet, Leeds, LS15 4EJ
2E0 JOF J. Miles, 11 Enborne Gate, Newbury, RG14 6AZ
2E0 JOG G. Smith, 36 Palma Park Homes, Shelly Street, Loughborough, LE11 5LB
2E0 JOH J. Hatton, 49 Buxton Street, Morecambe, LA4 5SR
2E0 JOI W. Pickles, 31 Longfield Road, South Woodham Ferrers, CM3 5JL
2M0 JOK J. Stewart, 1 Barns Park, Dalgety Bay, Dunfermline, KY11 9XX
2E0 JOP J. Priestman, 198 Felmongers, Harlow, CM20 3DW
2I0 JOS J. Millar, 3, Ahoghill, BT42 1JN
2E0 JOV A. Bogg, 13, Emerald Grove, Hull, HU35AE
2E0 JOX J. Jones, 15 Kinnaird Road, Sheffield, S5 0NN
2E0 JPA J. Wake, 60 Cloverville Approach, Bradford, BD6 1ET
2E0 JPC J. Clark, 1 Brooklime Road, Liverpool, L11 2YH
2E0 JPD R. Young, 26 Silent Woman Park Coldharbour, Wareham, BH20 7PE
2W0 JPE J. Elsmore, 8 Clos Aberconway, Prestatyn, LL19 9HU
2W0 JPF J. Freelove, 12 Honeyborough Road, Neyland, Milford Haven, SA73 1RE
2E0 JPH J. Hazell, 90 Lichfield Road, St. Annes, Bristol, BS4 4BN
2E0 JPM C. Monahan, 48 Church Road, Earley, Reading, RG6 1HS
2E0 JPO J. Owen, 21 Marlborough Road, Luton, LU3 1EF
2I0 JPP J. Page, 259 Bridge Street, Portadown, Craigavon, BT63 5AR
2E0 JPR P. Gaur, 34 Queensberry Avenue, Copford, Colchester, CO6 1YN
2E0 JPS J. Smith, 38 New Waverley Road, Basildon, SS15 4BH
2E0 JPT J. Taylor, 6 Hawks Close, Walsall, WS6 7LE
2E0 JPU J. Patient, 4 Bucklebury Heath, South Woodham Ferrers, Chelmsford, CM3 5ZU
2E0 JPW J. Websdale, Blacksmith Cottage, Black Street, Lowestoft, NR33 8EG
2E0 JPX J. Parkes, 24 Kenilworth Road, Lichfield, WS14 9DP
2E0 JQB J. Kowalski, 126 Harland Avenue, Sidcup, DA15 7PA
2E0 JQF S. Walrond, 17 Madam Lane, Weston-Super-Mare, BS22 6PW
2E0 JQG D. Start, The Rise, Valley Lane Swaby, Alford, LN13 0BH
2E0 JQI E. Lightbown, 18 Formby Close, Blackburn, BB2 3JZ
2E0 JQK P. Blackie, 30 Queens Avenue, Ilfracombe, EX34 9LS
2E0 JQW J. Wang, Flat 31, 74 Arlington Avenue, London, N1 7AY
2E0 JRA A. Jessop, 4 Katherine Street, Thurcroft, Rotherham, S66 9LG
2E0 JRB J. Barrow, 45 Windermere Road, Seaham, SR7 8HW
2E0 JRD J. Dowdeswell, 18 Lechlade Gardens, Fareham, PO15 6HF
2E0 JRE V. Vaznais, 35 Lynwood Drive, London, KT4 7AA
2W0 JRJ J. Jillings, Cane Garden, Dolau, Llandrindod Wells, LD1 5TE
2E0 JRL J. Lambert, 82 22Nd Avenue, Hull, HU6 9LS
2E0 JRN J. Lynn, 72 Badger Close, Guildford, GU2 9WA
2E0 JRP J. Preston, 5 Bodiam Crescent, Eastbourne, BN22 9HQ
2E0 JRR J. Riley, The Thatched House, 18 Bond Street, Norwich, NR9 4HA
2E0 JRS J. Stallard, 6 Richmond Crescent, Leominster, HR6 8RX
2E0 JRT N. Wing, 39 Whittington Road, Hutton, Brentwood, CM13 1JX
2E0 JRU C. Denton, 100 Lincoln Road Deeping Gate, Peterborough, PE6 9BA
2E0 JRW M. Williams, 48 Thackeray Drive, Tamworth, B79 8HZ
2E0 JRZ B. Handley, 68 Northfield Avenue, Rothwell, Leeds, Ls260sw
2M0 JSB J. Bence, 5 Braeside Gardens, Hamilton, ML3 7PN
2E0 JSE E. Riddle, 37B Stubbs Lane, Braintree, CM7 3NR
2M0 JSF J. French, 19 Woodside Avenue, Bridge of Weir, PA11 3PQ
2E0 JSG J. Godfrey, 29 Ridgewood Drive, Harpenden, AL5 3LJ
2E0 JSH J. Harris, 23 Winchester Avenue, Chatham, ME5 9AR
2E0 JSJ T. Peck, 2 Primrose Lane, Miami Beach, Sutton on Sea, LN12 2JZ
2E0 JSK S. Killian, Flat 4, Buttermere Court, Windermere Close, Exeter, EX4 2QD
2E0 JSM J. Preston, 3 Essex Drive, Kidsgrove, ST71HE
2E0 JSN J. Slattery, 64 Rydal Avenue, Chadderton, Oldham, OL9 0QX
2E0 JSO J. O'Shea, 56 Crummock Gardens, London, NW9 0DJ
2I0 JSQ J. Quinn, 5 Woodhill Heights, Lurgan, Craigavon, BT66 7DJ
2E0 JSR Dr Z. Rathore, 7 Ashdale Avenue, Bolton, BL3 4PH
2E0 JSS J. Symonds, 14 Phillips Crescent Needham Market Ip6 8Tf, Ipswich, IP6 8TF
2M0 JST J. Stuart, 120 Wellbeck Crescent, Troon, KA10 6AW
2E0 JSX J. Kelly, Braeview, Buzzacott Lane, Combe Martin, Ilfracombe, EX34 0NL
2E0 JSY B. Josyfon, 25 Norrice Lea, London, N2 0RD
2E0 JTB D. Baker, 65 Madison Street, Tunstall, St6 5hs
2E0 JTG A. Gilbert, 31 Elizabeth Road, Leamington Spa, CV31 3LJ
2E0 JTH J. Thornhill, 47 Hopton Lane, Mirfield, WF14 8JP
2E0 JTM D. Bache, 62 Whittingham Road, Halesowen, B63 3TP
2E0 JTN M. Chivers, 4 Hunters Lodge, Fareham, PO15 5NF
2E0 JTO J. O'Connell, 23 Halstead Road, Gosfield, CO9 1PG
2E0 JTQ J. Cook, 28 Wenny Estate, Chatteris, PE16 6UX
2E0 JTS J. Walker-Wilson, Rest Harrow Southside, Scorton, Richmond, DL10 6DN
2E0 JTT J. Thompson, 11 Foxbury Drive, Orpington, BR66EJ
2E0 JTW J. Johnson, 4 Wallace Close, Hullbridge, Hockley, SS5 6NE
2E0 JTY J. Hewlett, 21 Stedman Close Ickenham, Uxbridge, UB10 8DY
2E0 JUD J. Boutros, 15 Navigation House, Whiting Way, London, SE16 7EG
2M0 JUH J. Heywood, 2 Broaddykes Drive Kingswells, Aberdeen, AB15 8UE
2E0 JUL J. Hardy, Lambda House, Seanor Lane, Chesterfield, S45 8DH
2M0 JUM C. Jump, 15 The Lade, Bonhill, Alexandria, G83 9JR
2E0 JUX A. Brown, 17 Quail Ridge, Ford, Shrewsbury, SY5 9LF
2E0 JVA A. Vasarhelyi, 1 Eldon Close, Langley Park, Durham, DH7 9FR
2E0 JVC N. Cochrane, 3 Roman Bank Moulton Seas End, Spalding, PE12 6LG
2E0 JVD J. Boyd, 62 West Road, Shoeburyness, Southend-on-Sea, SS3 9DP
2E0 JVM A. Mori, 33 Valerian Road, Hedge End, Southampton, SO30 0GR
2E0 JVP M. Mason, 7 Langland Close, Malvern, WR14 2UY
2M0 JVR J. Robertson, 3 Richmond Place, Fochabers, IV32 7HF
2E0 JVS J. Swanbrow, 7 Manor Crescent, Rookley, Ventnor, PO38 3NS
2E0 JVT S. Black, 55 North Road, Hertford, SG14 1NE
2E0 JVY P. Smythe, 15 Falcon Drive, Trowbridge, BA14 7GE
2E0 JWC J. Cater, 5 Shady Grove, Hilton, Derby, DE65 5FX
2E0 JWE J. Elsmore, 142 St. Marks Road, Chester, CH4 8DH
2E0 JWG J. Guess, Flat 257, Helen Gladstone House Nelson Square, London, SE1 0QB
2E0 JWH J. Hope, 40 Birch Lane, Rugeley, WS15 1EJ
2E0 JWJ D. Shields, 42 Studland Park, Westbury, BA13 3HL
2E0 JWP J. Wishart, 19 Chepstow Close, Chippenham, SN14 0XP
2E0 JWS C. Shaw, 15 Rosamund Avenue, Wimborne, BH21 1TE
2E0 JWW J. Webster, 72 Grosvenor Street, Derby, de248at
2E0 JWY J. Willby, 10 Sunbury Road, Birmingham, B31 4LJ
2I0 JXA C. Matchett, 28 Glendale Avenue East, Belfast, BT8 6LF
2E0 JXB J. Bischoff, 6 Harton Close, Bromley, BR1 2UD
2M0 JXC J. Cockburn, 5 Carribber Avenue, Whitecross, Linlithgow, EH49 6JS
2E0 JXF D. Wells, 96 Tennyson Avenue, Rugby, CV22 6JF
2E0 JXG C. Millar, 195 Worsley Road, Frimley, Camberley, GU16 9BH
2M0 JXK J. Webster, 6 Livingston Crescent, Winchburgh, Broxburn, EH52 6FX
2E0 JXN P. Jackson, Langsmead Barn, Eastbourne Road, Blindley Heath, Lingfield, RH7 6JX
2I0 JXO D. Burke, 7 Edinburgh Villas, Omagh, BT79 0DW
2E0 JXT J. Townsend, 16 Tower Gardens, Hythe, CT21 6DG
2E0 JXU R. Sansom, 72 Wannock Lane, Eastbourne, BN20 9SQ
2E0 JXX J. Creaser, 8 Millwood Road, Hounslow, TW3 2HH
2E0 JYA J. Young, 1 Pen Tye, Gwinear, Hayle, TR27 5HL
2W0 JYB J. Byast, Ty Arbennig Bull Bay Road, Amlwch, LL68 9EA
2W0 JYC M. Coleman, Felin Newydd Ciliau Aeron Lampeter Sa48 7Px, Lampeter, SA48 7PX
2E0 JYE E. Brereton, 21 Sydney Street, Colchester, CO2 8UP
2W0 JYI Dr J. Blaxland, 7 Maes Y Sarn, Pentyrch, Cardiff, CF15 9QQ
2E0 JYM J. Downes, 78 Astral Way, Kingston upon Hull, Hu74xz
2W0 JYN S. Bobby, 56 Ffordd Offa, Rhosllanerchrugog, Wrexham, LL14 2EY
2E0 JYP J. Capstick, 186 Forest Lane, Harrogate, HG2 7EE
2E0 JZC J. Clark, 26 Heron Way, Sandbach, CW11 3AU
2E0 JZK J. Howlett, 29 Little London, Heytesbury, Warminster, BA12 0ES
2E0 JZU P. Holmes, 82 Moore Avenue, Norwich, NR6 7LG
2E0 KAB K. Bull, 12 Hinksford Mobile Home Park, Kingswinford, DY6 0BG
2E0 KAC D. Carr, 19 Kingsmead Walk, Speedwell, Bristol, BS5 7RL
2E0 KAF R. Dyson, 4 Royston Lane, Royston, Barnsley, S71 4NL
2E0 KAG M. Hughes, 19 Pendine Crescent, North Hykeham, Lincoln, LN6 8UW
2E0 KAK R. Hambly, 144 Station Road, Irchester, Wellingborough, NN29 7EW
2E0 KAL K. Lott, 6 Centurion Close, Sandhurst, GU470HH
2W0 KAP K. Gamwasam, 01, Churton Drive, Wrexham, LL13 8RU
2E0 KAS K. Sharpe, 18 Dudhill Road, Rowley Regis, B65 8HT
2M0 KAU M. Krawczyk, 19 Wishart Archway, Dundee, DD1 2JA
2E0 KAX N. Griffiths, 67 Warstones Drive, Wolverhampton, WV4 4PF
2E0 KBA J. Fletcher, 13 Lloyd George Grove, Cannock, WS11 7GY
2I0 KBB J. Elliott, 12 Drumbeg Drive, Lisburn, BT28 1NY
2E0 KBD P. Smiths, 12 Lambton Road, Stockton-on-Tees, TS19 0ER
2E0 KBE J. Hatton, 37 St. Cuthberts Way, Holystone, Newcastle upon Tyne, NE27 0UZ
2E0 KBF B. Wiggins, The Wigwam 13 Hastings Road, Bromsgrove, B60 3NX
2E0 KBG K. Glaysher, 66 Talbot Road, Farnham, GU9 8RR
2E0 KBJ S. Inman, 9 Colbert Avenue, Ilkley, LS29 8LU
2E0 KBL S. Kneeshaw, 39 Cherrytree Walk East Ardsley, Wakefield, wf32hs
2E0 KBN K. Burness, 4 Fenwick Street, Boldon Colliery, NE35 9HU
2E0 KBP B. Al-Rawi, Flat 17, Harrow Lodge, London, NW8 8HR
2I0 KBS K. Boyle, 764 Springfield Road, Belfast, BT12 7JD
2E0 KBX K. Buxey, 17 Gort Crescent, Southampton, SO19 8LH
2E0 KBZ S. Stretton, 9 Kilton Road, Worksop, S80 2EG
2E0 KCB K. Smart, 33 East Street, Littlehampton, BN17 6AU
2E0 KCF A. Hankins, 16 Eastwick Barton, Nomansland, Tiverton, EX16 8PP
2E0 KCL J. Jopson, 12 Charing Close, Ringwood, BH24 1FA
2E0 KCN B. Somerville Roberts, 21 Regency Way, Ponteland, Newcastle upon Tyne, NE20 9AU
2E0 KCO K. Cornmell, 19 Forest Road, Chandler'S Ford, Eastleigh, SO53 1NA
2E0 KCP R. Hutton, 22A Victoria Road, Maldon, CM9 5HF
2E0 KCV K. Conlon, 136 Chart Downs, Dorking, RH5 4DG
2E0 KCW P. Walsh, 181 Hermes Close, Hull, HU9 4DR
2E0 KCZ R. Robinson, 22 Riddings Court, Timperley, Altrincham, WA15 6BG
2E0 KDB D. Barnett, 12 Craig Walk Alsager, Stoke-on-Trent, ST7 2Rj
2E0 KDE A. Lee, 14 Bernice Avenue, Chadderton, Oldham, OL9 8QJ
2E0 KDF A. Hamilton, 56 Wyvern, Telford, TF7 5QH
2E0 KDG C. Harman, 58 Laurence Avenue, Witham, CM8 1JB
2E0 KDH K. Hall, 18 Brooklands Drive, Littleover, Derby, DE23 1DN
2E0 KDI L. Azzopardi, 11 Ilynton Avenue, Firsdown, Salisbury, SP5 1SH
2W0 KDR M. Kidner, 4 Tonypistyll Road, Newbridge, Newport, NP11 4HJ
2E0 KDS K. Selvaratnam, 15 Mark Close, Southall, UB1 3QJ
2E0 KDT R. Simpson, 48 Weatherhill Road, Horley, RH6 9LY
2E0 KDV D. Craven, 69 Markham Avenue, Rawdon, Leeds, LS19 6NE
2E0 KEA T. Lupton, 81 Home Farm Lane, Bury St Edmunds, IP33 2QL
2W0 KED S. Kedward, 9 Lawrence Avenue, Aberdare, CF44 9EW
2E0 KEG K. Stone, 34 Daventry Grove, Birmingham, B32 1JA
2E0 KEI K. Hastings, 161 Cottingley Road, Allerton, Bradford, BD15 9LD
2E0 KEJ S. Jordan, 8 Averham Close, Swadlincote, DE11 9SG
2E0 KEK J. Russell, Flat 1, Knapp Cottage, Roadwater, Watchet, TA23 0QY
2E0 KEP T. Keep, Coombe Cottage, Coombe Lane, Cradley, Malvern, WR13 5JF
2W0 KEQ K. Mogford, 49 Cefn Road, Rogerstone, Newport, NP10 9AQ
2E0 KES R. Noakes, 3 Firtree Road, Hastings, TN34 3TR
2I0 KEW K. Mcdonald, 37 Ardgarvan Cottages, Limavady, BT49 0NF
2E0 KEY D. Ferguson, 39 Fouracres, Maghull, Liverpool, L317BP
2E0 KEZ D. Brough, 57 Francis Road, Ashford, TN23 7UP
2E0 KFA A. Walker, 50 Parkstone Crescent, Hellaby, Rotherham, S66 8HD
2E0 KFB Dr F. Kuttikkate, 34 Shetland Crescent, Rochford, SS4 3FJ
2I0 KFD K. Dorman, 25 Blackthorn Road, Newtownabbey, BT37 0GH

2E0 KFG C. Freebody, 94 Fingringhoe Road, Colchester, CO2 8EE
2E0 KFH J. Rowe, 22 Treaty Road Glenfield, Leicester, LE3 8LU
2E0 KFI S. Gilbert, 34 Ramsey Road, St Ives, PE27 5RD
2E0 KFJ K. Jeffrey, 26 Sandringham Drive, Paignton, TQ3 1HY
2E0 KFK K. Kozlowski, Woking Homes / Flat 2, Oriental Road, Woking, GU22 7BE
2E0 KFL K. Ballard, 34 Orange Street South Wigston, Leicester, LE184QB
2E0 KFM A. Burfield, 4 Eastern Crescent, Chelmsford, CM1 4JQ
2E0 KFO S. Howard, 11 Jolly Gardeners Court, Norwich, NR3 3HD
2E0 KFR C. Pearcey, 17 Peppercorn Close, Christchurch, BH23 3BL
2E0 KFT J. Ferrol, 29 Westlands, Haltwhistle, NE49 9BS
2E0 KFV K. O'Connell, 63 Hazelton Road, Colchester, co4 3ds
2E0 KFX M. Hill, 4 Farmland Way, Hailsham, BN27 1SP
2E0 KFY C. Day, 14 Windsor Drive, Ramsey Forty Foot, Ramsey, Huntingdon, PE26 2XX
2E0 KGA K. Gracey, Gwendoline Close, Merseyside, Ch611dl
2E0 KGB S. Moissejev, 41 Queens Road, Caversham, Reading, RG4 8DN
2E0 KGC K. Cossey, 34 Pinewood Road, Hordle, Lymington, SO41 0GP
2E0 KGD M. Tabberer, 29 Chase Vale, Chasetown, Burntwood, WS7 3GD
2E0 KGJ K. Green, 12 Merton Place, Grays, RM16 4HL
2E0 KGM A. Fielding, 4 Poot Hall, Rochdale, OL120AS
2E0 KGN J. Badham, Comwillgur House Ross Road, Longhope, GL17 0LP
2W0 KGP Dr P. Kelly, Arosfa, Westminster Road, Wrexham, LL11 6DN
2W0 KGQ J. Kelly, Arosfa, Westminster Road, Wrexham, LL11 6DN
2E0 KGT G. Tagg, Tinkers Cottage, Nevendon Road, Wickford, SS12 0QB
2E0 KGV C. Moulding, 28 Queens Avenue, Highworth, Swindon, SN6 7BA
2E0 KGY R. King, 169 Morley Road, Oakwood, Derby, DE21 4QY
2E0 KHA A. Hughes, 8 Bullens Green Lane Colney Heath, St Albans, AL4 0QS
2E0 KHG K. Gibbs, 30 George Road, Water Orton, Birmingham, B46 1PE
2E0 KHI K. Hale, 27 Heyes Street, Liverpool, L5 6SE
2W0 KHK S. Rosser, 25 Clos Tir-Y-Pwll, Newbridge, Pantside, Newport, NP11 5GE
2E0 KHM K. Maddy, 56 Coachwell Close, Telford, TF3 2JB
2E0 KHO D. Pluright, 21 Buscot Drive, Abingdon, OX14 2BJ
2E0 KHW K. White, 4 Top Birches, St Neots, PE19 6BD
2E0 KHX K. Hewson, 48 Ruskin Road, Belvedere, DA17 5BB
2M0 KIA M. West, Prieshach, Station Brae, Macduff, AB44 1UL
2E0 KIL T. Lee, 68 Wharton Drive, Springfield, Chelmsford, CM1 6BF
2E0 KIM K. Matthews, 4 George Croft, Gayton Le Marsh, Alford, LN13 0NP
2W0 KIN A. Hawkins, 5 Keats Road, Caldicot, NP26 4LH
2E0 KIO D. Bloomfield, 30 Wye Road, Clayton, Newcastle, ST5 4AZ
2E0 KIS M. Mcnamara, 50 Portsdown Road, Portsmouth, PO6 4QH
2E0 KIT A. Gladman, 19 Colchester Road, Wymering, Portsmouth, PO6 3RH
2E0 KJD K. Hale, 44 Churchville, Micklefield Micklefield, Leeds, ls254ap
2E0 KJG K. Gallyot, 18, High Pines, St. Georges Close, Christchurch, BH23 4LN
2E0 KJI M. Bidwell, 3 Walsingham Place, London, SW4 9RR
2E0 KJJ J. Kent, Meadow Bank, Rye Lane, Sevenoaks, TN14 5JF
2E0 KJK K. Kolesnik, 15 Steer Road, Swanage, BH19 2RU
2W0 KJO J. Orchard, The Burrows, Spring Gardens, Whitland, SA34 0HL
2E0 KJT K. Toohey, 197 Broad Oak Road, St Helens, WA9 2AQ
2E0 KKC E. Brown, Flat 8, Jacobs Court, High Street, Colchester, CO7 0AQ
2E0 KKF F. Fitton, 6 Leaford Close, Denton, Manchester, M34 3QH
2E0 KKG A. Fraley, 28 Riverside Park Colehouse Lane, Clevedon, BS21 6TQ
2E0 KKH J. Fairbrother, 6, The Dene, Bethersdon, Ashford, TN26 3AR
2E0 KKJ K. Juszczak, 68 College Road, Sandy, SG19 1RH
2E0 KKL Dr H. Chen, Flat 37, Castel Mill, Rodger Dudman Way, Oxford, OX1 1AD
2E0 KKM K. Minett, Rosedene, Honey Hill, Wimbotsham, King's Lynn, PE34 3QD
2E0 KKO T. Stack, 31A Chester Road South, Kidderminster, DY10 1XJ
2E0 KKQ W. Dunk, 27 Trevor Drive, Maidstone, ME16 0QW
2W0 KKR T. Rowan, 12 Pencoed Avenue The Common, Pontypridd, CF37 4AN
2E0 KKT P. Hodgson, 32 Main Road, Seaton, Workington, CA14 1HS
2E0 KKY R. Pas, 12 Monument Way, Bodmin, PL31 1NZ
2E0 KKZ D. Kirk, 57 Sandringham Avenue, London, SW20 8JY
2E0 KLB S. Arundale, 5 Bowdon Street, Stockport, SK3 9EA
2E0 KLD R. Delrosa, 18 Rangers Avenue, Dursley, GL11 4AS
2E0 KLF K. Francis, 203 Colchester Road Lawford, Manningtree, CO11 2BU
2E0 KLI J. Heng, 1 St. Giles Croft, Beverley, HU17 8LA
2M0 KLL R. Bertram, 46 Main Street Main Street, Pathhead, EH37 5QB
2M0 KLN K. Nevins, 4 Ubbanford, Norham, Berwick-upon-Tweed, TD15 2LA
2E0 KLR A. Henshall, 37 Marlow Drive, Branston, Burton-on-Trent, DE14 3TX
2E0 KLS P. Elstub, 4 Granville Lodge, Church Street, Telford, TF2 9LX
2E0 KLT C. Taylor, 9 Pendeen Crescent, Threemilestone, Truro, TR3 6SP
2E0 KLV K. Brown, 41 Church Street, Swinton, Mexborough, S64 8EF
2E0 KLW L. Woods, 193 Wimberley Street, Blackburn, BB1 8HU
2E0 KLY K. Young, 14 Beechwood Avenue, Chatham, ME5 7HH
2E0 KMA K. Machen, 18 Peveril Close, Whitefield, Manchester, M45 6NR
2E0 KMB K. Bailey, 58 Billy Buns Lane, Wombourne, Wolverhampton, WV5 9BP
2E0 KMD K. Deans, 31 Northcroft, Sandy, SG19 1JJ
2E0 KMF K. Moysey, 109 Langbrook Cottages, Langbrook, Ivybridge, PL21 9JX
2E0 KMI K. Mills, 6 West Coombe, Bristol, BS9 2BA
2E0 KMN M. Wickens, Haven Lea, Queens Drive, Windermere, LA23 2EL
2E0 KMP A. Adams, 45 Four Oaks Road Tedburn St. Mary, Exeter, EX6 6AP
2M0 KMQ W. Donnelly, 10 Sentry Knowe, Selkirk, TD7 4BG
2E0 KMS K. Shenton, 2 The Croft, Stramshall, Uttoxeter, ST14 5AG
2E0 KMT K. Murphy, 120 Elmway, Chester le Street, DH2 2LQ
2W0 KMU G. Cattle, 12 Claerwen Gelligaer, Hengoed, CF828EW
2E0 KMY K. Murray, 5 Princes Crescent, Basingstoke, RG22 6DP
2E0 KMZ K. Missenden, 47 Roseacre Drive, Elswick, Preston, PR4 3UQ
2E0 KNA T. Clarke, 111 Telford Way, High Wycombe, HP13 5SZ
2E0 KNB M. Ronan, 49 Dorset Street, Nottingham, NG8 1PU
2E0 KNC S. Henson, 297 Underwood Lane, Crewe, CW1 3SG
2E0 KNE J. Dale, 37 Bussey Road, Norwich, NR6 6JF
2E0 KNF K. Sumner, 18 Grange Road, Fleetwood, FY7 8BH
2E0 KNH K. Holman, 39 Trellech Court, Yeovil, BA21 3TE
2E0 KNK K. Kariraman, 26 Tythe Barn Lane, Shirley, Solihull, B90 1RW
2E0 KNL G. Knowles, 29 Shepherds Cote Drive, Hepscott Park, Stannington, Morpeth, NE61 6FN
2E0 KNM M. Hill, 109 Kitchener Street, St Helens, WA10 4LU
2E0 KNT K. Royce, 11 Church Lane, Stibbington, Peterborough, PE8 6LP
2E0 KNU R. Weightman, 2 Bannister Grove, Winsford, CW7 1RJ
2E0 KOA P. Whitehead, 29 Cleveland, Bradville, Milton Keynes, MK13 7AZ
2E0 KOD A. Adams, 27 George Street Stockton, Southam, CV47 8JS
2E0 KOI H. Walker, 9 Humphries Close, Leicester, LE5 4LU
2E0 KOM L. Wilson, 6 Marrick Road, Middlesbrough, TS3 7RX
2W0 KOP A. Martin, 117 Fforchaman Road, Cwmaman, Aberdare, CF44 6NL
2E0 KOR L. Ross, 133 Petersmith Drive, New Ollerton, Newark, NG22 9SG
2E0 KOS A. Hollings, 39 Rendham Road, Saxmundham, IP17 1EA
2I0 KPA S. Frazer, 2 Cavanballaghy Road, Killylea, Armagh, BT60 4NZ
2E0 KPB P. Bridgwater, 18 Angelica, Amington, Tamworth, B77 3JZ
2E0 KPC P. Gagliardi, 7 Saxon Way, Jarrow, NE32 3QA
2E0 KPD P. Davies, 60 Southerndown Road, Sedgley, Dudley, DY3 3NA
2M0 KPE K. Page, 43 Lothian Court, Glenrothes, KY6 1LZ
2W0 KPF K. Foster, Prince Of Wales House Short Bridge Street Sy18 6Ad, Llanidloes, SY18 6AD
2W0 KPH K. House, Liddington, Dehewydd Lane Llantwit Fardre, Pontypridd, CF38 2EN
2E0 KPI N. Gregg, 84 Aberconway Crescent, New Rossington, Doncaster, DN11 0JP
2E0 KPL L. Kiddell, 1 Sparham Hill, Sparham, Norwich, NR9 5QT
2E0 KPM D. Roberts, 19 East Avenue, Warrington, WA2 8AD
2W0 KPN A. Thomas, 10 Chapel Street, Gorseinon, Swansea, SA4 4DT
2E0 KPO S. Warren, 7 Crich Way, Newhall, Swadlincote, DE11 0UU
2E0 KPP D. Fellows, 147 Olive Lane, Halesowen, B62 8LR
2E0 KPR G. Andrews, 151 Great Gregorie, Basildon, SS16 5QQ
2E0 KPS K. Parow-Souchon, 87 Saxton Road, Abingdon, OX14 5JD
2E0 KPT B. Harris, 1 Newbridge Cottages, Midgehole, Hebden Bridge, HX7 7AL
2E0 KPU C. Weight, 166 Prince Henry Road, London, SE7 8PJ
2E0 KPX D. Stephenson, Flat 8, Thornwood Court 84-88 Hudson Road, Leigh-on-Sea, SS9 5NF
2M0 KPZ N. Stewart, 35 Newbattle Gardens, Dalkeith, EH22 3DR
2E0 KQR A. Kinsella, Apartment 28 241 Liverpool Road, Widnes, WA8 7HL
2E0 KQV D. Harrison, 64 Crosby Road, Grimsby, DN33 1LU
2E0 KRB A. Cowan, 217 South Park Road, Wimbledon, London, SW19 8RY
2E0 KRD K. Dukes, 127 Carlton Road, Boston, PE21 8LL
2E0 KRE K. Knight, 11 Sweetbriar Lane Holcombe, Dawlish, EX7 0JZ
2M0 KRG C. Mcintyre, 70 Oldwood Place, Livingston, EH54 6US
2E0 KRH C. Pierce, 42 Staplehurst Road, Reigate, RH2 7PY
2M0 KRI K. Waitz, C/O Waitz-Rainey, 16 Inverkeithing Road, Aberdour, Burntisland, KY3 0RS
2E0 KRK J. Birch, 3 Partridge Way, High Wycombe, HP13 5JX
2E0 KRN K. Pidwell, 2 Welford Road, Chapel Brampton, Northampton, NN6 8AF
2E0 KRR R. Rao, 3 Weller Mews, Enfield, EN2 8FG
2E0 KRS K. Tharanee, 525 Burton Road, Littleover, Derby, DE23 6FT
2E0 KRT K. Taylor, 3 The Drive, Lichfield, WS14 9QT
2E0 KRX K. Rosema, Apartment 801, 25 Goswell Road, London, EC1M 7AJ
2W0 KSA P. Terrell, 82 Baglan Street, Treherbert, Treorchy, CF42 5AR
2E0 KSC C. Moss, 19 Tozer Close, Wallisdown, Bournemouth, BH11 8RB
2E0 KSG K. Stevens, 61A Main Road, Hoo, Rochester, ME3 9AA
2E0 KSH K. Haywood, 126 Derby Street, Sheffield, S2 3NF
2E0 KSO A. Jackson, 5 Woodside Lane, London, N12 8RB
2E0 KSP D. Volkov, 2 Robert Close, Potters Bar, EN6 2DH
2E0 KSR R. Brimfield, 6 Radways Close, Fairford, GL7 4FZ
2E0 KSW K. Sewell, 12 Haylands Square, South Shields, NE34 0JB
2E0 KSX B. Elms-Lester, Ferndale House Kerry'S Gate, Hereford, HR2 0AH
2E0 KTD K. Davidson, 5 Hanover Parc, Indian Queens, St Columb, TR9 6ER
2E0 KTG K. Gribben, 33 Strawberry Close, Birchwood, Warrington, WA3 7NT
2D0 KTH Dr J. Daniels, 24 King Orry Road Glen Vine, Isle of Man, IM4 4ES
2E0 KTK D. Van Dijk, 76 High Street, Tetsworth, Thame, OX9 7AE
2M0 KTL K. Lane, 23 Mayfield Avenue, Tillicoultry, FK13 6HB
2E0 KTQ K. Welch, Jacobs Leys, Falcon Lane, Ledbury, HR8 2JS
2E0 KTV B. Chauhan, 45 Burnham Drive, Whetstone, Leicester, LE8 6HY
2E0 KTW K. Wheeler, 3 Praze Road, Porthleven, Helston, TR13 9LR
2E0 KTX K. Browne, 56 Moorhouse Avenue, Wakefield, WF2 9QG
2E0 KUC D. Holman, 38 Polyear Close, Polgooth, St Austell, PL26 7BH
2E0 KUE I. Vernon, 7 Seaton Road Wick, Littlehampton, BN17 7LG
2E0 KUF J. Inwood, Flat 13, Kelvestone House, 47 Park Road, Cannock, WS11 1NZ
2E0 KUH S. Steinhoefel, 222 Stretford Road, Urmston, Manchester, M41 9NT
2I0 KUJ T. Calka, 71 Willowfield Street, Belfast, BT6 9AW
2E0 KUK A. Mallinson, 50 Cleveland Road, Huddersfield, HD1 4PW
2I0 KUN A. Davidson, 8 Rashee Court, Ballyclare, BT39 9SE
2E0 KUR P. Fletcher, 14 Orchard Avenue, Aylesford, ME20 7LY
2E0 KVA A. Kossick, 33 Verney Road, Winslow, Buckingham, MK18 3BN
2E0 KVB M. Wall, 227 Wayfield Road, Chatham, ME5 0HJ
2E0 KVE K. Waterhouse, 74 Clifford Road, West Bromwich, B70 8JY
2E0 KVF K. Emery-Ford, 48 Welham Grove, Retford, DN22 6TS
2M0 KVI K. Vielhaber, Nether Littlefold, Crieff, PH7 3NY
2E0 KVJ D. Boyes, The Shute Fluxton, Ottery St Mary, EX11 1RL
2E0 KVK K. Sim, 49 St. Julians Wells, Kirk Ella, Hull, HU10 7AF
2M0 KVM K. Mair, 92 Graham Street, Wishaw, ML2 8HR
2E0 KVR A. King, 23 Tower Crescent, Lincoln, LN2 5QF
2E0 KVS J. Fletcher, 3 Moorend Glade Charlton Kings, Cheltenham, GL53 9AT
2E0 KVV A. Taylor, 60 Wood Ride, Petts Wood, Orpington, BR5 1PY
2E0 KWA A. Miller, Ashtree Farm, Rugby Road, Princethorpe, Rugby, CV23 9PN
2E0 KWB K. Wilkinson, 9 Mersey Rd, Redcar, ts101ls
2E0 KWC K. Comben, 9 West Lane, North Baddesley, Southampton, SO52 9GB
2E0 KWE S. Coppins, 33 Honywood Road, Lenham, Maidstone, ME17 2HH
2E0 KWG R. Knight, Thrift, Ashwells Road, Brentwood, CM15 9SG
2E0 KWL R. Barker, 6 Trenoweth Cressent, Penzance, tr184ry
2E0 KWM D. Wright, 203 Winn Street, Lincoln, LN2 5EY
2M0 KWP K. Rae, 6 Dovecot Way, Pencaitland, EH34 5HA
2E0 KWW K. Willson, Ludpit Cottage, Ludpit Lane, Etchingham, TN19 7DB
2E0 KXD C. Collins, 2 Kew Crescent, Sheffield, S12 3LP
2E0 KXI K. Dowling, 8 Chathill Close, Morpeth, NE61 2TH
2E0 KXL R. Bamber, 28 Market Street, Cheltenham, GL50 3NH
2I0 KXM K. Mitchell, 80 Markethill Road, Collone, Armagh, BT60 1LE
2W0 KXN K. Nicholls, 35 Partridge Road, Cardiff, CF24 3QW
2E0 KXR K. Golloghly, 7 Jenner Close, Bungay, NR35 1QR
2E0 KXV A. Jordan, 19 Arrow Lane, Newhaven, BN9 0FG
2E0 KXX B. Cook, 11 Arbrook Lane, Esher, KT10 9EG

2E0 KYB M. Tielemans, 2 Parr Close, Grange Park, Swindon, SN5 6JY
2E0 KYD R. Harwood, 73 Rose Way, Cirencester, GL7 1PS
2W0 KYH D. Phillips, 9 Baldwin Street, Newport, NP20 2LT
2E0 KYI K. Armstrong, 29 Thorntree Avenue, Crofton, Wakefield, WF4 1NU
2E0 KYT S. Bishop, 5 Mulberry Court, 266 Goring Road, Goring-By-Sea, Worthing, BN12 4PF
2E0 KYX A. Tandler, 6 Field View Cottages, Brimfield, Ludlow, SY8 4LB
2W0 KYZ Dr A. Rykala, 20 Mount Pleasant, Blaina, Abertillery, NP13 3DD
2E0 KZB C. Kyle-Davidson, Block C, Ingram Court, Garrowby Way, Heslington, York, YO10 5DL
2E0 KZC M. Clack, 42A Provost Street, Fordingbridge, SP6 1AY
2E0 KZH D. Taylor, 49 Boggart Hill Gardens, Seacroft, Leeds, LS14 1LJ
2M0 KZI A. Broom, 140 Green Road, Paisley, PA2 9AJ
2E0 KZJ P. Lamb, 13 Pool End, St Helens, WA9 3RE
2E0 KZK A. Whybrow, 64 Church Road Sevington, Ashford, TN24 0LF
2E0 KZL I. Hallatt, 11 Cheshire St., Audlem, Crewe, CW3 0AH
2E0 KZM M. Osborne, 9 Sunningdale Court, Jupps Lane, Goring-By-Sea, Worthing, BN12 4TU
2E0 KZN J. Spence, Officers Mess, Royal Air Force, Brize Norton, Carterton, OX18 3LX
2E0 KZQ M. Pott, 29 Lilliebrooke Crescent, Maidenhead, SL6 3XJ
2E0 KZT L. Kinzett, 52 Thistle Drive, Peterborough, PE2 8HX
2E0 KZU M. Macrae, 91 Chosen Way Hucclecote, Gloucester, GL3 3BX
2E0 KZV P. Mckie, 65 White Cross, Hexham, NE46 1JH
2E0 LAB L. Bain, 45 Larpool Crescent, Whitby, YO22 4JD
2W0 LAC S. Llewellyn, 53 Cripps Avenue, Cefn Golau, Tredegar, NP22 3PF
2E0 LAD M. Leech, 11 Westlake Close, Torpoint, PL11 2BZ
2E0 LAH B. Beckett, 21 Horseshoes Lane, Langley, Maidstone, ME17 1SR
2E0 LAI L. Potter, 18 Elizabeth Gardens, Wakefield, WF1 3SZ
2E0 LAL L. Larkins, 34 Guycroft, Otley, LS21 3DS
2M0 LAO S. Ramsay, 6 Cross Road, Peebles, EH45 8DH
2E0 LAR A. Davies, 6 Gribble Road, Liverpool, L10 7NF
2M0 LAS F. Davidson, 27 Gordon Way, Livingston, EH54 8JG
2E0 LAV A. Loyd, Maple House, Pangbourne Road, Reading, RG8 8LN
2M0 LAW A. Mccaig, 46 Patterson Drive, Law, Carluke, ML8 5LT
2E0 LAX A. Coombes, 3 Marshall Close, Purley On Thames, Reading, RG8 8DQ
2E0 LAY S. Lay, 7 Hunt Street, Swindon, SN1 3HW
2E0 LBA A. Polakovs, 76 Sandringham Crescent, Leeds, LS17 8DF
2E0 LBB L. Brown, 20 Pinfold Lane, Mirfield, WF14 9HZ
2W0 LBE M. Ryall, 27 Clos Afon Twyi, Blackwood, NP12 3FX
2E0 LBG J. Shatford, 31 Pinner Park Avenue, Harrow, HA2 6LG
2M0 LBH A. Hunsley, 42 Waverley Place, Edinburgh, EH7 5SA
2E0 LBI L. Bell, 42 Ocean Road, Walney, Barrow-in-Furness, LA14 3DX
2E0 LBJ L. Pinkney, 18 Bridlington Road, Driffield, YO25 5HZ
2E0 LBK L. Karthauser, 17 Manor Close Abbotts Ann, Andover, SP11 7BJ
2E0 LBL L. Lay, 17 Herbert Road, Hornchurch, RM11 3LD
2W0 LBN L. Au-Yeung, 104 Fleet Street, Swansea, SA1 3UX
2I0 LBS L. O'Sullivan, 24 Swifts Quay, Carrickfergus, BT38 8BQ
2E0 LBW R. Walker, 159 Cuckfield Road, Hurstpierpoint, Hassocks, BN6 9RT
2E0 LBY T. Littlebury, 71 Hampden Drive, Kidlington, OX5 2LT
2E0 LBZ D. Taylor, 143 Sandhurst Road, London, SE6 1UR
2E0 LCE I. Pilton, Caleril Barn, Pool Foot Farm Haverthwaite, Ulverston, LA12 8AA
2W0 LCJ L. Jones, Ty'R Ysgol, Holland Street, Ebbw Vale, NP23 6HT
2E0 LCK C. Cutting, 5A Clifton Mansions, Clifton Road, Folkestone, CT20 2EJ
2E0 LCM L. Micallef, 1 Brockholme Mews, Great Cambourne, Cambourne, CB23 6GU
2E0 LCN T. Cass, April Cottage, South Eau Bank, Spalding, PE12 0QR
2E0 LCR L. Rimington, 3 Amicombe, Wilnecote, Tamworth, B77 4JJ
2E0 LCW T. Wright, 2 Lilyville Road, London, SW6 5DW
2I0 LDC P. Floyd, 25 Glenside, Omagh, BT79 7GL
2E0 LDE G. Matts, 34 Barry Road, Leicester, LE5 1FA
2E0 LDF R. Irving, 2 Wasdale Close, Cockermouth, CA13 9JD
2E0 LDG A. Mcdonald, 11 Micklegate Murdishaw, Runcorn, WA76HT
2E0 LDJ L. Jepson, 143 Walnut Avenue Weaverham, Northwich, CW8 3DX
2E0 LDL D. Loftus, 5 Knowle Mount, Burley, Leeds, LS4 2PP
2E0 LDM L. Mason, 9 Trenethick Avenue, Helston, TR13 8LU
2E0 LDN K. Poulton, 21 East View, London, E4 9JA
2E0 LDQ L. Dobinson, 20 Newholme Crescent, Evenwood, Bishop Auckland, DL14 9RY
2E0 LDR L. Reynolds, 12 Providence Crescent, Boundary Way, Hull, HU4 6EF
2E0 LDV D. Butterfield, 57 Holmes Road, Retford, DN22 6QU
2W0 LDX S. Jones, 14 Lower Cross Road, Llanelli, SA15 1NQ
2E0 LDY D. Hayes, 60 Shelby Road, Worthing, BN13 2TT
2W0 LDZ T. Clapp, Crunns Farm. Coxhill, Narberth, SA67 8EH
2E0 LEA D. Anderson, 35 Sycamore Road, East Leake, Loughborough, LE12 6PP
2E0 LEE L. Tunstall, 8 York Road, Rowley Regis, B65 0RR
2E0 LEF N. Briggs, 20 Broad Lane, Pelsall, Walsall, WS4 1AP
2E0 LEG J. Landless, 2 Aspen Way, Banstead, SM7 1LE
2E0 LEL J. Jackson, 7 Burford Crescent, Wilmslow, SK9 6BL
2E0 LEM J. Francis, 22 Acre Lane, Carshalton, SM5 3AB
2W0 LEN L. Hayes, 56 Snowden Road, Cardiff, CF5 4PR
2E0 LEO L. Steer, 51 Kings Chase, East Molesey, KT8 9DG
2E0 LET R. Dunnaker, 12 Dagger Lane, West Bromwich, B71 4BA
2M0 LEW M. Strachan, 62 Charleston Drive, Dundee, DD2 2EZ
2M0 LEX A. Jenkins, 19 Hogg Avenue, Johnstone, PA5 0EZ
2E0 LEY L. Medley, 9 Polyplatt Lane, Scampton, Lincoln, LN1 2TL
2E0 LEZ L. Trend, 140 Ardleigh, Basildon, SS16 5RW
2W0 LFE L. Ansell, 114 Bowleaze, Greenmeadow, Cwmbran, NP44 4LG
2E0 LFI A. Birch, 22 Ullswater Road, Burnley, BB10 4HX
2E0 LFK L. Brown, 71 Chiltern Way, Nottingham, NG5 5NP
2E0 LFM P. Dimmick, Keir Hardie Court, Woodlea, Newbiggin-by-the-Sea, NE64 6LH
2E0 LFR L. Rofix, Birds Hill Cottage, Clopton, Woodbridge, IP13 6SE
2M0 LFS A. Miles, 9 Buchanan Drive, Lenzie, Kirkintilloch, Glasgow, G66 5HS
2E0 LFT A. Norden, 10 School Lane, Watton At Stone, Hertford, SG14 3SF
2E0 LFW F. Simon, 7 Yew Tree Court, Hemel Hempstead, HP11NE
2E0 LFX J. Lamb, 9A Matlock Road, Canvey Island, SS8 0EW
2W0 LFY A. Jones, 10 Dan Y Bryn, Caerau, Maesteg, CF34 0UW
2E0 LGB G. Benson, 2 Guisborough Road, Nunthorpe, Middlesbrough, TS7 0LB
2W0 LGE R. Samphire, Courtlands, Newport Road, Magor, Caldicot, NP26 3BZ
2W0 LGG L. Martin, 2 Ael Y Glyn, Nant Road, Harlech, LL46 2UJ
2E0 LGH S. Hallam, 18A Market Street, Hoylake, Wirral, CH47 2AE
2E0 LGL M. Rusu, 1 Turnbull Road, March, PE15 9RX
2E0 LGO P. Frost, 26 Hollies Court, Britannia Road, Banbury, OX16 5DR
2E0 LGR L. Rickman, 42 Sycamore Close, Poole, BH17 7YJ
2E0 LGS P. Schoenmaker, 24 Greenheys Drive, London, E18 2HB
2E0 LGT D. Page, 17 Hedge End Walk, Havant, PO9 5LS
2E0 LGV A. Meek, 25 Somerset Close Oswaldtwistle, Accrington, BB5 3AU
2E0 LGW A. O'Connell, 5 Westend Terrace, Gloucester, GL1 2RX
2E0 LGZ G. Cash, 60 Vittoria Court, Birkenhead, CH41 3LF
2E0 LHD L. Hoddinott, 30 Deans Mead, Bristol, BS11 0QX
2E0 LHE L. Horne, 1268 Evesham Road Astwood Bank, Redditch, B96 6AX
2E0 LHF K. Gibson, 44 Churchville, Micklefield, Leeds, LS25 4AP
2E0 LHI C. Smith, 17 Essex Street Wash Common, Newbury, RG14 6QJ
2E0 LHR S. Kapadia, 7 Elms Lane, Wembley, HA0 2NX
2E0 LHS G. Carver, 4 Andrews Road, Farnborough, GU14 9RY
2E0 LHY S. Raneses Grospe, 8 Lincombe Road, Manchester, M22 1GA
2E0 LIT C. Martin, 20 Hall Green Road, West Bromwich, B71 3LA
2E0 LIV A. Nicholson, 24 Barnmead, Haywards Heath, RH16 1UZ
2E0 LIW W. Sawyer, 20 Park Terrace Willington, Crook, DL15 0QL
2D0 LIX M. Behrman, 60 Scarlett Road, Castletown, Isle Of Man, IM9 1PN
2E0 LIZ E. Greatorex, 22 Marlborough Way, Uttoxeter, ST14 7HL
2E0 LJB L. Bedford, 29 Kent Road, Brookenby, Market Rasen, LN8 6EW
2W0 LJC C. Jones, 19 Crud Y Castell, Denbigh, LL16 4PQ
2W0 LJD J. Dyer, Abergwawr House, Belmont Terrace, Aberdare, CF44 6UW
2E0 LJG L. Goldsmith, Hunters Cottage, 61 Fengate Drove, Weeting, Brandon, IP27 0PW
2E0 LJH L. Halloway, 82 Northwall Road, Deal, CT14 6PP
2E0 LJJ L. Johnson, 33 Yetholm Place, Newcastle upon Tyne, NE5 4ED
2E0 LJK L. Marriott, 94 Lyndhurst Road, Worthing, BN11 2DW
2E0 LJL L. Lewis, 12 Hall Drive, Middleton, Morecambe, LA3 3LF
2I0 LJQ W. Phair, 34 Sketrick Island Park, Newtownards, BT23 7BN
2E0 LJR A. Siddall, 12 Russell Gardens, Sipson, West Drayton, UB7 0LS
2E0 LJS M. Fearon, 70 George Street, Heywood, OL10 4PW
2E0 LJT A. Tranter, 122 Summerhill Road, Bristol, BS5 8JU
2E0 LKC M. Balshaw, 16 East Avenue, Heald Green, Cheadle, SK8 3DL
2E0 LKE L. Kett, 52 Northgate, Hornsea, HU18 1EU
2E0 LKH K. Hull, 3 Enas Crescent, Ena Street, Hull, HU3 2TL
2E0 LKM L. Mcdonnell, 108 Long Lane, Garston, Liverpool, L19 6PQ
2E0 LKR R. King, 12 South Road, Marden, TN12 9EN
2E0 LKS L. Spriggs, 19 Mackenzie Square, Shephall, Stevenage, SG2 9TT
2W0 LKX W. Dabrowski, 2 Underhill Crescent, Knighton, LD7 1DG
2W0 LLA A. Jones, 2 Erw Terrace, Bethel, Caernarfon, LL55 1YT
2E0 LLC L. Addison, 45 Fir Terraces Esh Winning, Durham, DH7 9JQ
2E0 LLD E. Daniels, 2 Garstons Close, Titchfield Fareham, PO14 4EN
2E0 LLE R. Redmond, 28 Common Lane, Polesworth, Tamworth, B78 1LS
2I0 LLG S. Horner, 10 Meadow Court, Bushmills, BT57 8SD
2E0 LLI A. Foley, 23 Church Lane, Wymington, Rushden, NN10 9LW
2W0 LLJ J. Henley, Rhewin Glas, Glynarthen, Llandysul, SA44 6PR
2E0 LLK Dr D. Roberts, Beggarwood House Ravensworth Park Estate, Gateshead, NE11 0HQ
2W0 LLL C. Wadsworth, Tyn Llwyn, Talybont, LL43 2AN
2E0 LLM J. Proudman, 61 Iffley Road, Oxford, OX4 1EB
2E0 LLO J. Lovelock, Sea Spray, Lighthouse Road The Lizard, Helston, TR12 7NU
2W0 LLT L. Thomas, 15 Blaenwern, Newcastle Emlyn, SA38 9BE
2M0 LLU B. Gaudie, Sunnyside, Harray, Orkney, KW17 2JS
2E0 LLW P. Sampson, 16 Rutland Place, Cirencester, GL7 1PR
2E0 LLX J. Wright, 2 Regent Road, Church, Accrington, BB5 4AR
2W0 LLY P. Kyte, 1 Dan Y Bryn, Caerau, Maesteg, CF34 0UW
2E0 LLZ L. Andrews, 72 Grange Road, Alresford, SO24 9HF
2E0 LMD A. Bate, 16 East Avenue, Heald Green, Cheadle, SK8 3DL
2E0 LME G. Beardmore, 9 Ashmore Drive, Gnosall, Stafford, ST20 0RP
2E0 LMG C. Murray, 6 Maple Grove, Stocksbridge, S36 1ED
2E0 LMH L. Hudson, 68 Eleanor Road, Harrogate, HG2 7AJ
2E0 LMI N. Dorling, 51 Haygarth Close, Cirencester, GL7 1WY
2E0 LMK S. Wills, 8 Amherst Road, Newcastle upon Tyne, NE3 2QQ
2W0 LMM P. Jones, 115 Wordsworth Gardens, Pontypridd, CF37 5HH
2E0 LMO M. Broyd, 16 George Downing House, Miles Mitchell Avenue, Plymouth, PL6 5XJ
2E0 LMQ R. Lewis, 20 Hillary Road, Rugby, CV22 6EU
2E0 LMR G. Southall, 28 Manor Road, Woodford Halse, Daventry, NN11 3QP
2M0 LMY L. Mullaney, 8 School Road, Wellbank, Broughty Ferry, Dundee, DD5 3PL
2E0 LNC D. Lancaster, Linkhill View, Frith Common, Eardiston, Tenbury Wells, WR15 8JX
2M0 LNF I. Mcgurk, 16 Smith Crescent, Alexandria, G83 8BL
2E0 LNG R. Rider, 25 Kimber Close, Lancing, BN15 8QD
2M0 LNR T. Couper, 10 Sclandersburn Road, Denny, FK6 5LP
2E0 LNU M. Durban, 62 Westfield Way, Charlton, Wantage, OX12 7EP
2E0 LNX N. Stone, Flat, 97-99 Stoke Road, Gosport, PO12 1LR
2I0 LNZ L. Craig, 170 Donaghadee Road, Bangor, BT20 4PP
2E0 LOA Capt. J. Pearce, Clematis Cottage, 52 Cheselbourne, Dorchester, DT2 7NP
2E0 LOE N. Chaplin, 5 Maxwell Street, Bury, BL9 7QA
2E0 LOG D. Whelan, 431 Leeds Road, Huddersfield, HD2 1XT
2E0 LOH P. Rasmussen, 7 Portal Drive North, Upper Heyford, Bicester, OX25 5TH
2E0 LOL L. Woolley, 4 Robert Street, Warrington, WA5 1TQ
2E0 LON I. Lonsdale, 23 Hunts Field Clayton-Le-Woods, Chorley, PR6 7TT
2I0 LOP Rev. M. Donald, Station Road, Garvagh, BT51 5LA
2I0 LOR Dr A. Bell, 4 Mount Pleasant View, Newtownabbey, BT37 0ZY
2E0 LOW M. Duchar, 13 Thirlmere Avenue, Chester le Street, DH2 3ED
2E0 LOX S. Cilliers, 61 Heathwood Gardens, London, SE7 8ET
2E0 LPD D. Lester, 171 Glenavon Road, Birmingham, B14 5BT
2E0 LPE A. Monaghan, Briar Patch 117 Elm High Rd, Wisbech, pe14 0dn
2I0 LPG B. Bruce, 19 Ashvale Heights, Hillsborough, BT26 6DJ
2E0 LPJ A. Pomfrey-Jones, 46 Hampton Road, Erdington, Birmingham, B23 7JJ
2I0 LPO J. Mcerlean, 24 Mccorley Road, Toomebridge, Antrim, BT41 3NH
2E0 LPP P. Kenderes, 9 Waverton Road, London, SW18 3BY
2E0 LPR M. Price, 9 Herbarth Close, Liverpool, L9 1JZ
2E0 LPW L. Walker, 7 Stroudley Close, Ashford, TN240TY

2M0 LPX M. Macfarlane, 9 Dreghorn Park, Colinton, Edinburgh, EH13 9PH
2E0 LPZ G. Harmath, 101 Whitton Avenue East, Greenford, UB6 0QE
2M0 LQF J. Bennet, 24/11 Greenpark, Edinburgh, EH177TA
2E0 LQR T. Longmore, 3 Dairy Farm Cottages, Northlands Road, Gainsborough, DN21 5DN
2E0 LQW R. Edwards, 46 Lavers Oak, Martock, TA12 6HG
2E0 LQY I. Gilmore, 19 Green Sward Lane, Redditch, B98 0EN
2E0 LQZ L. Woodward, Oaklands, Ashford Lane, Hockley Heath, Solihull, B94 6RH
2E0 LRA L. Ayre, 30 Tithe Lane, Calverton, Nottingham, NG14 6HY
2U0 LRB L. Bichard, Brise De Mer, Les Rouvets De Bas, Guernsey, GY7 9QF
2E0 LRD K. Bainbridge, 29 Bluebell Grove, Calne, SN11 9QH
2E0 LRG M. Parker, Ridgeways, Mill Common, Westhall, Halesworth, IP19 8RQ
2E0 LRJ A. Smith, 101 Chaucer Drive, Lincoln, LN2 4LT
2E0 LRK L. Kelsey, 111-113 George Street, Mablethorpe, LN12 2BS
2E0 LRL M. Russell, 65 Court Farm Road, Bristol, BS14 0EF
2I0 LRN K. Bell, 3 Alexandra Crescent, Larne, BT40 1NE
2M0 LRO J. Barton, 58 Aberdour Road, Dunfermline, KY11 4PE
2E0 LRP R. Hughes, 7 Willow Place, Darlington, DL1 5LX
2E0 LRR L. Rhodes, 91 Bedonwell Road, Bexleyheath, DA7 5PS
2E0 LRV L. Emanuel, 20 Wychwood Drive, Redditch, B97 5NW
2E0 LRZ P. Jones, Sussex Cottage, The Limes, Felbridge, East Grinstead, RH19 2QY
2E0 LSB A. Shepherd, 39 Minehead Road, Dudley, DY1 2NZ
2E0 LSC S. Deakin, 20 Riccat Lane, Stevenage, SG1 3XY
2E0 LSE G. Bystryakov, 20 Elmhurst Gardens, Leeds, LS17 8BG
2M0 LSG S. Gray, 17 Pine Grove Cumbernauld, Glasgow, G67 3AX
2E0 LSI N. Highfield, 298 Mersea Road, Colchester, CO2 8QY
2E0 LSL A. Wood, 4 St. Andrews, The Common, Cranleigh, GU6 8NX
2E0 LSO G. Coltman, The Oaks Rushden, Buntingford, SG9 0SN
2E0 LSR K. Titmarsh, 8 Bainbridge Close, North Walsham, NR28 9UP
2E0 LSS J. Lusted, 17 St James Road, East Grinstead, RH19 1DL
2E0 LST L. Timmins, 83 Loxdale Sidings, Bilston, WV14 0TN
2E0 LSV R. Derham, Netherwood, Copse Lane, Hook, RG29 1SX
2E0 LSW L. Wellington, 5 Pasture Road, London, SE6 1JF
2E0 LSX A. Dale, 37 Bussey Road, Norwich, NR6 6JF
2E0 LTC C. Anderson, 191 Waveney Road, Hull, HU8 9NA
2E0 LTD S. Boyles, 5 Penclose Road, Fleckney, LE8 8TE
2E0 LTF L. Farrell, 1 Meadway, Ince, Wigan, WN2 2BZ
2E0 LTH L. Thornton, 11 Polruan Road, Truro, TR1 1QR
2E0 LTJ M. Brett, Lindon, Bunkers Hill, Wisbech, PE13 4SQ
2E0 LTL D. Cadden, 24 Lindenbrook Vale, Stafford, ST17 4QN
2E0 LTM P. Bone, 11 Fox Hill Drive, Stalybridge, SK15 2RP
2E0 LTP C. Brennan, 19 The Furrow, Littleport, Ely, CB6 1GL
2E0 LTR N. Rice, 30 Oveton Way, Bookham, Leatherhead, KT23 4ND
2E0 LTT M. Lovatt, 3 Withington Close, Atherton, Manchester, M46 0EZ
2E0 LTU A. Jakstas, 36 Boxridge Avenue, Purley, CR8 3AQ
2E0 LTX D. Akerman, The Brick Barn, Coppice Farm, Ross-on-Wye, HR9 7QW
2E0 LUD P. Lloyd, 8 Maydor Avenue, Saltney Ferry, Chester, CH4 0AH
2E0 LUF A. Clack, 3 Darwin Close, Swindon, SN3 3NF
2M0 LUG L. Affleck, 1 Fank Brae, Mallaig, PH41 4RQ
2E0 LUJ S. Cottam, 14 Barnard Close, Rednal, Birmingham, B45 9SZ
2E0 LUK C. Cavalcante Pinheiro Filho, Flat 7, Hollybush House, Hollybush Gardens, London, E2 9QT
2E0 LUL L. Pearson, 4 Brentwood Close, Thorpe Audlin, Pontefract, WF8 3ES
2E0 LUN N. Evetts, 35 Wood End Road, Kempston, Bedford, MK43 9BB
2E0 LUT D. Murdoch, 48 Bracklesham Gardens, Luton, LU2 8QL
2E0 LUY L. Isaac Sneath, 21 Garrick Close, Lincoln, LN5 8TG
2E0 LVC D. Tofield, 6 Church Grove, Barnstaple, EX32 9DJ
2W0 LVE A. Moody, Perthiteg, Cwmhiraeth, Llandysul, SA44 5XJ
2E0 LVG M. Brooks, 34 Roundmead, Stevenage, SG2 9PH
2E0 LVL M. Bentley, 5 Grendon Green, Stoke-on-Trent, ST2 0EH
2E0 LVP I. Ahmed, 93 Tavistock Road, Sheffield, S7 1GF
2E0 LVR Dr O. De Peyer, Flat 5, Molasses House, Clove Hitch Quay, London, SW11 3TN
2E0 LWA D. Mccorrie, 145 Mildenhall Road, Fordham, Ely, CB7 5NW
2M0 LWB L. Bradley, Amon Sul, Kiltarlity, Beauly, IV4 7HT
2M0 LWD G. Ledgerwood, 28 Cannerton Park, Milton Of Campsie, Glasgow, G66 8HR
2E0 LWL P. Gill, 38 Claygate, Thorneywood, Nottingham, NG36JX
2E0 LWQ G. Ford, The Lodge, Home Farm Lane, Rimpton, Yeovil, BA22 8AS
2E0 LWR A. Mundy, 27 Yorke Road, Croxley Green, Rickmansworth, WD3 3DW
2E0 LWT A. Teed, 57 Lymington Road, Torquay, TQ1 4BG
2E0 LXA A. Lowery, 102 Brompton Park, Brompton On Swale, Richmond, DL10 7JP
2E0 LXB R. Thompson, Croft Michael Farm, Croft Mitchell, Troon, Camborne, TR14 9JJ
2E0 LXC G. Goddard, 13 Rosslyn Avenue, Coventry, CV6 1GL
2E0 LXD D. Collins, 30 Upham Road, Swindon, SN3 1DN
2E0 LXF D. Shuttleworth, 27 Union St., Egerton, Bolton, BL7 9SP
2E0 LXI E. Pestano, 38 Third Avenue, Bexhill-on-Sea, TN40 2PA
2E0 LXQ J. Ratty, 12 Challoners Close, East Molesey, KT8 0DW
2E0 LXR L. Rowlands, 6 St. Michaels Avenue, Clevedon, BS21 6LL
2I0 LXS A. Logan, 15 Park Lane, Saintfield, Ballynahinch, BT24 7PR
2W0 LXT L. Thomas, 64 Min Y Llan, Letterston, Haverfordwest, SA62 5SP
2I0 LXW M. Lewis, 7 Liester Park, Ballyrobert, Ballyclare, BT39 9RZ
2M0 LXX C. Sturgeon, 7 Chalmers Avenue, Ayr, KA7 2NF
2E0 LXY D. Loxley, 33 Longwood Road, Tingley, Wakefield, WF3 1UG
2E0 LYB P. Lyba, 6 Ambassadors Way, North Shields, NE29 8ST
2E0 LYD B. Vile, 24 Hudson Close, Dover, CT16 2SG
2E0 LYF K. Fletcher, Beverley Hotel, 55 Old Brumby Street, Scunthorpe, DN16 2AJ
2E0 LYK A. Harrison, 19 Marlborough Avenue Haxey, Doncaster, DN9 2HL
2E0 LYM S. Owens, 18 Harvester Way, Lymington, SO41 8YD
2E0 LYR T. Martin, 46 Hayes Crescent, Frodsham, WA6 7PG
2E0 LYY I. Lloyd, 30 Church View Gardens, Kinver, Stourbridge, DY7 6EE
2E0 LYZ N. Goult, 5 Colbourne Close, Bransgore, Christchurch, BH23 8BW
2E0 LZB A. Highfield, 38 Brunswick Gardens, Garforth, Leeds, LS25 1HF
2E0 LZD G. Kelley, 31 Cherry Park, Brandon, Durham, DH7 8TN
2E0 LZE E. Stone, Oakley, Main Road, Salisbury, SP4 6EE
2E0 LZG J. Kenny, 29 Siskin Close, Bishops Waltham, Southampton, SO32 1RP
2E0 LZI I. Duxbury, 334 Linnet Drive, Chelmsford, CM2 8AL
2E0 LZM J. Roberts, 51 Bradfield Road, Broxtowe, Nottingham, NG8 6GP
2E0 LZN J. Marshall, 6 Foster Walk, Sherburn in Elmet, LS25 6EU
2E0 LZQ L. Hollingworth, 43 Wingfield Road, Hull, HU9 4PR
2E0 LZT D. Young, 20 Summerhouse, Tickenham, Clevedon, BS21 6SN
2W0 LZU P. Matthews, The Chateau, Wynnstay Hall Estate, Ruabon, Wrexham, LL14 6LA
2E0 LZX J. Mutter, 27 Snowdonia Way, Huntingdon, PE29 6XP
2M0 LZY D. Clark, 106 Braes Avenue, Clydebank, G81 1DP
2E0 MAA M. Milne, Flambards, Manor Road, Dunmow, CM6 2JR
2E0 MAB M. Bridgeland, 17 Oldfield Lane, Wisbech, PE13 2RJ
2W0 MAC C. Mccarthy, 7 Aneurin Avenue, Crumlin, Newport, NP11 5HN
2E0 MAD M. Davidson, 26 Hurford Drive, Thatcham, RG19 4WA
2E0 MAF M. Beckett, 59 Broadacre, Caton, Lancaster, LA2 9NH
2E0 MAH M. Holbrook, 9 Beechwood Mount, Hemsworth, Pontefract, WF9 4ES
2E0 MAJ M. Jones, 20 Chelsea Drive, Sutton Coldfield, B74 4UG
2E0 MAL M. Frame, 23 Greenside Court, Sunderland, SR3 4HS
2E0 MAN R. Lomax, 11 Sherbourne Drive, Heywood, OL10 4ST
2W0 MAO M. Cowhey, 11 Aspen Way, Newport, NP20 6LB
2E0 MAP M. Woolley, 84 Bowthorpe Road, Norwich, NR2 3TP
2E0 MAQ M. Finn, Atzenbach 38, Idar-Oberstein, Germany, 54473
2M0 MAV J. Cattigan, Lunan Home Farm Cottage, Lunan Bay, Arbroath, DD11 5ST
2E0 MAX M. Mcsherry, 5 Briery Croft, Stainburn, Workington, CA14 1XJ
2E0 MAY D. Maydew, 128 Thorne Road, Willenhall, WV13 1AW
2E0 MAZ M. Stevenson, 127 Walton Road, Chesterfield, S40 3BX
2E0 MBA H. Anderton, 69 Sycamore Grove, Lancaster, LA1 5RS
2E0 MBB M. Bartley, 19 South Avenue, Shadforth, Durham, DH6 1LB
2E0 MBD N. Draper, 107 Arkwrights, Harlow, CM20 3LY
2M0 MBE C. Hebenton, 43 East Avenue, Uddingston, Glasgow, G71 6LG
2E0 MBG J. Girard, 49 Beech Crescent, Hythe, Southampton, SO45 3QF
2I0 MBI B. Mcdonald, 20 Aughan Park, Poyntzpass, Newry, BT35 6TW
2E0 MBK M. Lewis, 73 Addenbrooke Street, Wednesbury, WS108HJ
2E0 MBO M. Hughes, 58 Grange Lane North, Scunthorpe, DN16 1RW
2E0 MBQ A. Blamires, 2 Foldings Grove, Scholes, Cleckheaton, BD19 6DQ
2E0 MBR P. Threakall, 83 Gregory Avenue, Birmingham, B29 5DG
2E0 MBS M. Strange, 101 Southbroom Road, Devizes, SN10 1LY
2E0 MBT M. Buchanan, 36 Church Lane, Manby, Louth, LN11 8HL
2E0 MBV M. Mchugh, 51 Rutland Street, Hyde, SK14 4SY
2E0 MBW M. Rickaby, 57 Hylton Road, Jarrow, NE32 5DN
2E0 MBX M. Bowell, 28 Jubilee Close, Byfield, Daventry, NN11 6UZ
2E0 MCA M. Addison, 319 Long Lane, London, N2 8JW
2W0 MCB M. Luxton, 3 The Paddocks, Newgate Street, Brecon, LD3 8DJ
2E0 MCC Dr H. Shekhdar, Manora Lodge, Sea Bank Road, Skegness, PE24 5QU
2M0 MCD M. Mcdonald, 17 Ramsay Mews, Strathaven, ML10 6GN
2E0 MCG J. Mcgill, 7 Willow Crescent, Wistaston, Crewe, CW2 8RL
2E0 MCH M. Bailey, 34 Jephson Drive, Birmingham, B26 2HW
2E0 MCJ M. Jeffrey, 9 Stoney Lands, Plymouth, Pl125df
2E0 MCK M. Bridgehouse, 43 Age Croft, Oldham, OL8 2HG
2E0 MCL M. Carney, 2 Lilac Meadows, Lawley Village, Telford, TF4 2NX
2E0 MCM B. Cameron, 9 Finchale Road, Hebburn, NE31 2HR
2E0 MCN M. Bridger, 11, Beecham Close, Newcastle upon Tyne, NE15 6LG
2W0 MCQ M. Ellis, Heddwch Brithdir, Dolgellau, LL40 2SF
2E0 MCS L. Allcock, 26 Castleton Grove, Inkersall, Chesterfield, S43 3HU
2W0 MCT A. Mctaggart, Brick Hall, Hundleton, Pembroke, SA71 5QX
2E0 MCW M. Carlin, 44 Sileby Road, Barrow Upon Soar, Loughborough, LE12 8LR
2E0 MD M. Davies, 2 Ellins Terrace, Normanton, WF6 1BL
2E0 MDC M. Crawley, 16 The Meadows, Herne Bay, CT6 7XF
2E0 MDE C. Cundall, 43 High Street, Great Gonerby, Grantham, NG31 8JR
2W0 MDG M. Griffiths, Mandalay, Bromfield Street, Wrexham, LL14 1NF
2E0 MDH M. Walters, 65 Bannawell Street, Tavistock, PL19 0DP
2E0 MDK A. Currie, 31 Launceston Road, Bodmin, SO50 6AY
2E0 MDN M. Bray, 13 Rosebay Close, Hartlepool, TS26 0ZL
2E0 MDR M. Bradley, 13 Elizabeth Avenue, Bilston, WV14 8EA
2E0 MDT B. Hiley, 9 Pinfold Lane, Harby, Melton Mowbray, LE14 4BU
2E0 MDU N. Alders, 14 Forest Rise, Crowborough, TN6 2ES
2E0 MDZ M. Smith, 31 Atlantic Crescent, Sheffield, S8 7FW
2E0 MED A. Medhurst, 44 Battle Road, Hailsham, BN27 1DS
2E0 MEG S. Ridley, 123 Lanercost Drive, Newcastle upon Tyne, NE5 2DL
2E0 MEH D. Howarth, 32 Cotswold Drive Rothwell, Leeds, LS26 0QZ
2E0 MEI M. Bennetts, 2 Chywoone Terrace, Newlyn, Penzance, TR18 5NR
2E0 MEK M. Prentice, Cherry Ash, Shaftesbury Road, Gillingham, SP8 4LL
2E0 MEL M. Mcgoldrick, 7 Walnut Drive, Tiverton, EX16 6HE
2E0 MEO A. Bond, 74 Long Meadows, Chorley, PR7 2YB
2E0 MEQ M. Rea, 15 Wensleydale Close, Royton, Oldham, OL2 5TQ
2E0 MES M. Skinner, 5 Sycamore Avenue, Upminster, RM14 2HR
2E0 MET G. Lewis, 5 Framfield Road, Uckfield, TN22 5AG
2E0 MEU A. Sharif, 6 Buckle Rise, Seaford, BN25 2QN
2E0 MEV M. Clarke, 54 Stafford Grove, Shenley Church End, Milton Keynes, MK5 6AZ
2W0 MEX R. Hicks, 14 Carn Celyn Beddau, Pontypridd, cf38 2tf
2E0 MEY M. Sadler, 14 Woodlands Avenue, Water Orton, Birmingham, B46 1SA
2E0 MEZ M. Marsh, 25 Southdown Road, Seaford, BN25 4PD
2E0 MFA F. Alfrey, 16 Walls Road, Bembridge, PO35 5RA
2E0 MFC L. Ross, 2 Bedford Street, Blackburn, BB2 4EU
2W0 MFD T. Heath, 16 Beacons Park, Brecon, LD3 9BR
2E0 MFF M. Clarke, 4 Mill Lane, Brant Broughton, Lincoln, LN5 0RP
2E0 MFG M. Mcgowan, 48 Alderley Road, Thelwall, Warrington, WA4 2JA
2E0 MFI M. Berrisford, 5 Branwell Drive, Haworth, Keighley, BD22 8HG
2I0 MFJ B. Traynor, 94 Markville, Portadown, Armagh, BT63 5SZ
2M0 MFK M. Cook, 6 Kirkstead Drive, Dundee, DD2 2FB
2E0 MFN P. Roberts, 17 Cannon Hill, Prenton, CH43 4XR
2E0 MFS M. Feast, 30 Peveril Drive, Riddings, Alfreton, DE55 4AP
2E0 MFT A. Hill, 1 Rochester Close, Mountsorrel, Loughborough, LE12 7UH
2E0 MFX Dr M. Fox, 11 Park End, London, NW3 2SE
2E0 MFZ M. Fairchild, Brooklyn Caravan Park 21 Almond Brow Gravel Lane Banks, Southport, PR98BU
2E0 MGA M. Greensmith, 14 Fountain Road, Draycott-In-The-Clay, Ashbourne, DE6 5HP
2E0 MGC M. Chanter, 7 Woodford Crescent, Plymouth, PL7 4QY
2E0 MGI M. Isbell, 20 Woodland Crescent, Wolverhampton, WV3 8AS
2E0 MGL M. Talbot, 26 Chevalier Grove, Crownhill, Milton Keynes, MK8 0EJ

2M0 MGM M. Geldart, 13B Greystone Place, Newtonhill, Stonehaven, AB39 3UL
2E0 MGP G. Champion, 34 Greenfields, Edenside, Kirby Cross., Frinton-on-Sea, CO13 0SW
2E0 MGR M. Reeks, 33 Madresfield Village, Madresfield, Malvern, WR13 5AA
2E0 MGT J. Gardiner, 31 Rodway Road Tilehurst, Reading, RG30 6EH
2E0 MGU M. Guidolin, Flat 3, 3 Grimston Gardens, Folkestone, CT20 2PT
2E0 MGV L. Copeland, 78 Penderyn Crescent Ingleby Barwick, Stockton-on-Tees, TS17 5HQ
2E0 MGW M. Whitticombe, 18 Fairclough Street, Burtonwood, Warrington, WA5 4HJ
2E0 MGX M. Deeley, Unit 4 Beechwood Business Park, Cannock, WS11 7GB
2M0 MGY J. Boag, 182 St. Fillans Road, Dundee, DD3 9LH
2E0 MHC G. Cardoso, 11 Coppice Way, Aylesbury, HP20 1XG
2E0 MHE S. Snelson, 212 Dickson Road, Blackpool, FY1 2JS
2E0 MHG J. Gray, 7 Ruskin Avenue, Melksham, SN12 7NG
2E0 MHJ I. Jabegu, 51 South Crescent, Blandford Camp, Blandford Forum, DT11 8AJ
2E0 MHL M. Lacey, 82 Bowerings Road, Bridgwater, TA6 6HF
2E0 MHM M. Byard, 1 Fieldside, Long Wittenham, Abingdon, OX14 4QB
2M0 MHN N. Morris, 23 Sedgebank, Sedgebank, Livingston, EH54 6HE
2E0 MHT M. Thompson, 4 Jubilee Court Ravenscroft, Holmes Chapel, Crewe, CW4 7HA
2E0 MIB V. Ball, 24 Carr Lane, Warsop, Mansfield, NG20 0BN
2E0 MID P. Staite, Chestnut Farm, Eastville, Boston, PE22 8LX
2M0 MIF D. Mifsud, 25 Priory Road, Linlithgow, EH49 6BP
2E0 MIG P. Cattermole, Blaxhall Hall Crossing, Little Glemham, Woodbridge, IP13 0BP
2E0 MIH M. Humphries, 5 Coppice Mead, Stotfold, Hitchin, SG5 4JX
2E0 MIL I. Millman, 3 Oyster Mews, 1-3 Forest Road, Poole, BH13 6EN
2E0 MIS D. Smout, Sunrays, Warbage Lane, Bromsgrove, B61 9BH
2E0 MIU J. Marsh, 14 Eyam Road, Hazel Grove, Stockport, SK7 6HP
2E0 MIV B. Davies, 60 Queensway, Blackburn, BB2 4QT
2E0 MIX D. Edge, 18 Sandringham Avenue, Whitehaven, CA28 6XL
2E0 MIY P. Billingham, 393 Landseer Road, Ipswich, IP3 9LT
2E0 MIZ A. Bartlett, 62 Kewstoke Road, Bath, BA2 5PU
2W0 MJA A. Sneddon, 3 Marigold Close, Gurnos, Merthyr Tydfil, CF47 9DA
2W0 MJC M. Churcher, 71 Twyn Road, Abercarn, Newport, NP11 5JY
2E0 MJE M. Parker, 1 Ham Road, Wanborough, Swindon, SN4 0DF
2E0 MJG Rev. M. Gillingham, 14 Nethergreen Gardens, Killamarsh, S21 1FX
2E0 MJH M. Holroyd, 9 Coniston Green, Aylesbury, HP20 2AJ
2E0 MJJ J. Jones, 19 Southbank Street, Leek, ST13 5LS
2E0 MJL M. Lee, Up To Date House, Shore Road, Boston, PE22 0NA
2E0 MJM M. Middleditch, 8 Royal Close, Yeovil, BA21 4NX
2E0 MJO S. Marcomini, 18 Chadwick Place, Long Ditton, Surbiton, KT6 5RE
2E0 MJP M. Marsh, 19 First Avenue, South Kirkby, Pontefract, WF9 3EP
2E0 MJS S. Mcmurtrie, 5 Hill Road, Carshalton, SM5 3RA
2E0 MJT M. Troth, 3 Laburnum Grove, Bromsgrove, B61 8NB
2E0 MJX M. Cresswell, 44 The Lea, Birmingham, B33 8JP
2M0 MJY M. Yarrow, Lomond Villa, Downies Village, Aberdeen, AB12 4QX
2E0 MJZ D. Cook, 15 Kendricks Fold, Rainhill, Prescot, L35 9LX
2E0 MKB M. Ballard, 41 Middlefield Ave, Halesowen, B62 9QJ
2E0 MKE M. Gregory, 65 Nursery Crescent North Anston, Sheffield, S25 4BR
2E0 MKF M. Amphlett, Highbanks, Charnes Road, Market Drayton, TF9 4LQ
2W0 MKG M. Gray, 15 The Circle, Cwmbran, NP44 7JP
2E0 MKH M. Heaton-Bentley, 65 Brookfield Road, Thornton-Cleveleys, FY5 4DR
2E0 MKI T. Palmer, 29 Field End, Maresfield, Uckfield, TN22 2DJ
2E0 MKJ M. Johnson, 7 Norfolk Wing, Tortington Manor, Ford Road, Tortington, Arundel, BN18 0FD
2E0 MKK M. Kilkenny, 23 Hazelhurst Road, Stalybridge, SK15 1HD
2E0 MKT T. Walker, 11 Banburies Close, Bletchley, Milton Keynes, MK3 6JP
2E0 MKV M. Vardy, 60 Hucklow Avenue, North Wingfield, Chesterfield, S42 5PU
2E0 MKW M. Wiggins, 2 Cherry Tree Close, Halstead, CO9 2UA
2E0 MKX M. Keyte, 3 Lower High St., Mow Cop, Stoke on Trent, ST7 3PB
2M0 MKZ M. Devlin, 4 County Place, Forgandenny, Perth, PH2 9EP
2E0 MLA A. Highfield, 29 Blewitt Street, Brierley Hill, DY5 4AW
2E0 MLC S. Sedgwick, Flat 2/A, St. Georges Court 44 Thorne Road, Doncaster, DN1 2JA
2E0 MLD M. Dixon, 37 Carlton Close, Parkgate, Neston, CH64 6RB
2E0 MLE D. Pilkington, 197 Saltings Road, Snodland, ME6 5HP
2E0 MLF M. Raynor, 21 Teversal Avenue, Pleasley, Mansfield, NG19 7QQ
2W0 MLG S. Gordon, 8 Maesteg, Cymau, Wrexham, LL11 5EP
2E0 MLH M. Howse, 12 Queens Road Moretonhampstead, Newton Abbot, TQ13 8LP
2E0 MLJ M. Jones, 11 Lower Glen Park Pensilva, Liskeard, PL14 5PP
2E0 MLK M. Kipling, 12 Jolly Brows, Harwood, Bolton, BL2 4LZ
2E0 MLL A. Mccall, 95 Newton Drive, Blackpool, FY3 8LX
2E0 MLS M. Heywood, 16 Edinburgh Drive, Hindley Green, Wigan, WN2 4HL
2E0 MLV M. Grantham, 7 Goodwin Close, Sandiacre, Nottingham, NG10 5FF
2E0 MLX A. Boyes, 12 Leyburn Grove, Stockton-on-Tees, TS18 5NH
2M0 MMB M. Cleland, 85 Carfin St., Motherwell, ML1 4JL
2W0 MMD D. Holloway, 14 Woodbrook Terrace, Burry Port, SA16 0NF
2E0 MME R. Shulver, 63 Hill Farm Way, Southwick, Brighton, BN42 4YG
2M0 MMF R. Tripney, 7 Sunnyside St., Camelon, Falkirk, FK1 4BJ
2M0 MMG M. Gourlay, 14 Holmes Holdings, Broxburn, EH52 5NS
2E0 MMH M. Hall, 67 Darlinghurst Grove, Leigh-on-Sea, SS9 3LF
2E0 MMJ M. Majhail, 3 Poynders Hill, Hemel Hempstead, HP2 4PQ
2E0 MML M. Thompson, 22 Churchfield Terrace, Barnsley, S72 8JT
2M0 MMM M. Greig, 7 St. Ronans Road, Forres, IV36 1BQ
2M0 MMO M. Overthrow, 63 Primrose Avenue, Larkhall, ML9 1JX
2E0 MMP M. Porter, Michael Porter C O 219 Burnley Road, Colne, BB88JD
2I0 MMT M. Torley, 4 Yew Tree Park, Newry, BT34 2QP
2E0 MMU A. Moss, Winstons, Mayfield Lane Durgates, Wadhurst, TN5 6DG
2E0 MMX R. Hewson, 10 Miriam Grove, Leigh, WN7 3EX
2W0 MNA M. Mcdonald, Falcondale, Pleasant Valley, Stepaside, Wisemans Bridge, Narberth, SA67 8NT
2E0 MNC M. Craner, 46 Meadowhill Crescent, Redditch, B98 8HT
2E0 MND A. Harrop, 35 Langdale Crescent, Dalton-in-Furness, LA15 8NR
2E0 MNG N. Giuliano, 13 Walton Drive, Littleover, Derby, DE23 1GN
2E0 MNH M. Harvey, May Tree Cottage, Kelvedon Road, Tiptree, Colchester, CO5 0LJ
2W0 MNJ M. Kenny, 27 Brangwyn Crescent, Newport, NP19 7QY
2E0 MNL S. Manley, The Cottage, Crow Ash Road, Berry Hill, Coleford, GL16 7RB
2E0 MNP M. Pomfret, 5 Malvern Crescent, Ince, Wigan, WN3 4QA
2E0 MNU P. Richardson, 14 Portland Street, Worksop, S80 1RZ
2E0 MNX C. Leece, 101 Wellstead Way, Hedge End, Southampton, SO30 2BH
2E0 MNY R. Parker, 53 Tunstall Road, Canterbury, CT2 7BX
2E0 MNZ L. Moyle, 14 St. Lukes Close, Kettering, NN15 5HD
2E0 MOB C. Larner, 98 Allandale, Hemel Hempstead, HP2 5AT
2M0 MOF T. Moffat, 16 Drumellan Road, Ayr, KA7 4XQ
2M0 MOK W. Fulton, 15 Staffa Avenue, Port Glasgow, PA14 6DT
2E0 MOL R. Moles, 14 Dorsett Road, Stourport-on-Severn, DY13 8EL
2W0 MON A. West, Bryn Goleu, Gwalchmai, Holyhead, LL65 4SW
2E0 MOR A. Willmore, 31 Oaklands, Bugbrooke, Northampton, NN7 3QU
2I0 MOU G. Moucka, 5 Glebe Gardens, Moira, Craigavon, BT67 0TU
2E0 MOW C. Naylor, 16 Boston Avenue, Norbreck Blackpool, fy29bz
2E0 MOY R. Moys, 12A Palmerston Avenue, Fareham, PO16 7DP
2E0 MOZ M. Meadowcroft, 8 Lamlash Road, Blackburn, BB1 2AS
2E0 MPA M. Ashworth, 123 Forest Road, Liss, GU33 7BP
2E0 MPB R. Johnson, 24 Fairfields, Upper Denby, Huddersfield, HD8 8UB
2E0 MPC M. Carter, 113 Old Road Tintwistle, Glossop, SK13 1JZ
2E0 MPE R. Eaton, 31 Pinfold Lane Ruskington, Sleaford, NG34 9EU
2E0 MPF C. Le Marchant, Bow House, Green Lane, Prestwood, Great Missenden, HP16 0QE
2E0 MPG P. Mcgrath, 24 Broadoak Drive, Lanchester, Durham, DH7 0QA
2E0 MPJ J. Neal, 75 Park Lane, Castle Donington, Derby, DE74 2JG
2E0 MPN M. Nolan, Bath Road Post Office, Post Restante. Bath Road, Devizes, SN10 1QG
2E0 MPO K. O'Hara, 41 Exeter Street, Blackburn, BB2 4AU
2E0 MPR M. Rolls, 3 Gleneagles Crescent, New Holland, Barrow-upon-Humber, DN19 7TL
2E0 MPX P. Matthew, 24 Jubilee Close, Pamber Heath, Tadley, RG26 3HP
2W0 MPY P. Man, 17A Cradock Street, Swansea, SA1 3HE
2E0 MQA D. Rogers, 9 Prospect Place, Stafford, ST17 4HZ
2E0 MQB M. Boldero, 1 Grasleigh Avenue, Allerton, Bradford, BD15 9AR
2E0 MQC M. Le Moine, 115 Rothesay Road, Blackburn, BB1 2ER
2E0 MQT F. Walker, 54 Burnage Lane, Burnage, Manchester, M19 2NL
2E0 MRA A. Armson, 2 Windmill Gardens, St Helens, WA9 1EN
2E0 MRC D. Letton, 21 Westfield, Bradninch, Exeter, EX5 4QU
2E0 MRD D. Millard, 112 Avenue Road, Sandown, PO36 8DZ
2E0 MRF M. Keal, 81 St. Catherines Grove, Lincoln, LN5 8ND
2E0 MRG G. Whittle, 22 St. Oswalds Close, Finningley, Doncaster, DN9 3ED
2E0 MRJ M. Jarrett, 17 Greenhill Gardens, Minster, Ramsgate, CT12 4EP
2E0 MRM M. Tetley, 27 Cunningham Hill Road, St Albans, AL1 5BX
2M0 MRO M. Reid, 2 Pinkie Gardens, Newmachar, Aberdeen, AB21 0QF
2E0 MRP M. Peters, 25 Windsor Court, Falmouth, TR11 3DZ
2E0 MRQ P. Gibbs, 19 Lupin Close, Etherley Dene, Bishop Auckland, DL14 0TP
2I0 MRY M. Ruddy, C/O 6 Iveagh Park, Greysteel, Londonderry, BT47 3DD
2E0 MRZ M. Roberts, 13 Stanley Road, Portslade, Brighton, BN41 1SW
2E0 MSA M. Statham, Broad Oak Bungalow, Manston, Sturminster Newton, DT10 1EZ
2M0 MSB S. Brown, 21 Whiteford Avenue, Dumbarton, G82 3JU
2E0 MSE M. Edmonds, 60 Shenstone Road, Maypole, Birmingham, B14 4TJ
2E0 MSI M. Sims, 5 Sandy Leaze, Bradford-on-Avon, BA15 1LX
2W0 MSL N. Berrall, 41 Nantgarw Road, Caerphilly, CF83 3FB
2E0 MSM M. Mohammed Shafi, 575 Wood Lane, Dagenham, London, RM8 1DR
2E0 MSN M. Boisriveau-Mitchell, 16 Pochard Close, Quedgeley, Gloucester, GL2 4LL
2I0 MSO P. Hosey, 13 Glenelly Gardens, Omagh, BT79 7XG
2E0 MSR M. Ramsay, 28A Strathmore Drive, Charvil, Reading, RG10 9QT
2E0 MSS M. Smith, The Lawns Tylers Green, High Wycombe, HP108BH
2M0 MSU R. Sutherland, 25 Burns Wynd, Maybole, KA19 8FG
2E0 MSZ M. Smith, 12 Lincoln Drive Waddington, Lincoln, LN5 9NH
2E0 MTA T. Allman, 46 Belmont Road, Rugby, CV22 5NZ
2E0 MTB T. Beckett, 95 Warrens Hall Road, Dudley, DY2 8DH
2E0 MTC C. Mathewson, 33 Thornton Road, Bootle, L20 5AN
2W0 MTD M. Davies, 11 High Street, Malltraeth, Bodorgan, LL62 5AS
2W0 MTE E. Thomas, 1 Cambrian Gardens Y Drenewydd, Newtown, SY16 2AW
2E0 MTH M. Knowles, 11 Thorneycroft Avenue, Birkenhead, CH41 8HJ
2E0 MTL M. Leach, 64 Grove Street, Wantage, OX12 7BG
2E0 MTM T. Mloduchowski, Flat 4, Gwynne House, London, E1 2AG
2E0 MTN M. Newell, 55 Station Road Brimington, Chesterfield, S43 1JU
2M0 MTO R. Foulds, 83 Croftfoot Road, Glasgow, G44 5JU
2E0 MTQ D. Hartley, 102 Moss Lane, Sale, M33 5BE
2E0 MTR M. Reilly, 26 Roman Way, Folkestone, CT19 4JT
2E0 MTT Dr M. Nassau, 4A London Road, Liphook, GU30 7AN
2E0 MTX N. Challis, 48 Brunsfield Close, Wirral, CH46 6HE
2E0 MTY D. Raine, 91 Lulworth Avenue, Jarrow, NE32 3SB
2E0 MUA J. Anderson, 121 Barton Road, Stretford, Manchester, M32 9AF
2E0 MUD S. Sparks, 25 Wilwick Lane, Macclesfield, SK11 8RS
2E0 MUI M. Tsun, Dyson'S Farm, Long Row, Tibenham, NR16 1PD
2E0 MUN A. Munford, 4 Chatburn Park Avenue, Brierfield, Nelson, BB9 5QB
2M0 MUR G. Murray, The Barn House, Springfield Farm, Carluke, ML8 4QZ
2E0 MUS A. Sutton, 3 Grotes Buildings, London, SE3 0QG
2E0 MUT J. Merritt, 41 Great Grove, Bushey, WD23 3BQ
2E0 MUW J. Blaylock, 23 Sunnyway, Blakelaw, Newcastle upon Tyne, NE5 3QB
2E0 MUZ M. Colpman, 20 Rochford Road, Basingstoke, RG21 7TQ
2E0 MVD M. Denham, 2 Shorts Corner, Frithville, Boston, PE22 7EA
2E0 MVH S. Smith, 65 Furness Avenue, St Helens, WA10 6QF
2M0 MVI M. Vaci, Davaar, Lochawe, Dalmally, PA33 1AQ
2E0 MVN D. Mills, 31 Gorseway, Hatfield, AL10 9GS
2E0 MVT D. Harris, 27 Ashley Road, Poole, BH14 9BS
2M0 MVX D. Wilson, Rivendell Lodge, Glenkindie, Alford, AB33 8RN
2E0 MWA M. Austin, 14 The Green, Brown Edge, Stoke-on-Trent, ST6 8RN
2E0 MWB M. Bryant, 284 Brantingham Road Chorlton Cum Hardy, Manchester, M21 0QU
2E0 MWD M. Day, 14 Windsor Drive, Ramsey Forty Foot, Ramsey, Huntingdon, PE26 2XX
2E0 MWH M. Hetherington, 18 Wesley Street, Low Fell, Gateshead, NE9 5YN
2E0 MWJ M. Willis, 51 Barnsdale Close, Loughborough, LE11 5AN
2E0 MWN M. Singer, 1 Bentley Road, Slough, SL1 5BB
2E0 MWT R. Woolley, 10 Hazelmoor Fold, Blackley, Elland, HX5 0DR
2E0 MWW M. Wheal, 7 Ryecroft Drive, Withernsea, HU19 2LP
2E0 MXA M. Amos, Willow Tree House, Deers Green, Clavering, Saffron Walden, CB11

4PX
2E0 MXC M. Craven, 78 Connaught Road, Brookwood, Woking, GU24 0HF
2E0 MXG M. Glen, 3 Mill Bungalows Catterick, Richmond, DL10 7LY
2E0 MXL M. Lee, 2 Rupert Road Chaddesden, Derby, de214nd
2E0 MXM M. Meehan, 14 Grosvenor Road, Walton, Liverpool, L4 5RB
2E0 MXN R. Lovell, Formby, Formby, Livepool, L37 4BP
2E0 MXP M. Palmer, New Haven, Stoneraise, Durdar, Carlisle, CA5 7AX
2E0 MXR A. Wilson, 28 Langham Road, Bristol, BS4 2LJ
2E0 MXW D. Platt, 50 Poplars Road, Stalybridge, SK15 3EN
2E0 MYA D. Mycroft, 11 Paisley Walk, Church Gresley, Swadlincote, DE11 9FF
2E0 MYB H. Ibbitson, Tor View, Whitstone, Holsworthy, EX22 6TB
2E0 MYD E. Chauvelaine, 292 Mount Pleasant, Redditch, B97 4JL
2E0 MYE D. Sykes, 2 The Street, Claxton, Norwich, NR14 7AS
2E0 MYG M. Young, 72 Goddard Way, Saffron Walden, CB10 2EB
2E0 MYH D. Morgan, 87 Pool Hayes Lane, Willenhall, WV12 4PX
2E0 MYK M. Knowles, 86 West Shore Road, Walney, Barrow-in-Furness, LA14 3UD
2E0 MYL J. Swann, 5 Lanark Close, Hazel Grove, Stockport, SK7 4RU
2E0 MYS M. Broad, 29 Oakdale Close, Clay Cross, Chesterfield, S45 9RY
2E0 MYT M. Corrigan, 33 Westbourne Road, Knott End-On-Sea, Poulton-le-Fylde, FY6 0BS
2E0 MYX S. Elliott, 79 Somerton Road, Bolton, BL2 6LN
2E0 MYZ J. Hughes, 22 Marklay Drive, South Woodham Ferrers, Chelmsford, CM3 5NP
2E0 MZB B. Gutteridge, 54 Malthouse Road, Southgate, Crawley, RH10 6BG
2E0 MZC G. Fox, 23 The Driveway, Canvey Island, SS8 0AB
2E0 MZD B. Greenwood, 13 Mayflower Street, Blackburn, BB2 2RX
2E0 MZE M. Hartshorn, 21 Bidford Road, Leicester, LE3 3AE
2E0 MZK M. Keay, 8 Coronation Road, Stoke-on-Trent, ST4 6BH
2E0 MZL G. Johnson, 22 Beechwood Close, Blythe Bridge, Stoke-on-Trent, ST11 9RH
2W0 MZS M. Simons, 68 Harbour Village, Goodwick, SA64 0DZ
2E0 MZU C. Smith, 105 Netherton Road, Worksop, S80 2SA
2E0 MZZ E. Reeve, 12 Sime Street, Worksop, S80 1TD
2W0 NAD O. Bross, 8 Queens Drive, Buckley, CH7 2LJ
2E0 NAE M. Ridgewell, 75 Hillshaw Park Way, Ripon, HG4 1JU
2E0 NAF N. Foster, 18 Austen Ave, Sawley, Nottingham, NG103GG
2E0 NAG T. Bown, 16 Sandringham Court, Queen Elizabeth Road, Nuneaton, CV10 9AR
2E0 NAI P. Turner, 35 Gattington Park, Hawthorn Hill, Dogdyke, Coningsby, Lincoln, LN4 4XA
2E0 NAM N. Carey, 28 Tremayne Road, St Austell, PL25 4NE
2M0 NAN S. Ram, 28 Craigievar Gardens, Kirkcaldy, KY2 5SD
2E0 NAP N. Deery, 25 Ribblesdale Place, Preston, PR1 3NA
2E0 NAQ N. Barnard, 10 Whites Lane Kessingland, Lowestoft, NR33 7TF
2E0 NAR R. Nagy, 40 Oakhampton Road, London, NW7 1NH
2E0 NAS N. Inglis, 74 Runswick Avenue, Whitby, YO21 3UE
2I0 NAT C. Mooney, 12 Curragh Walk, Derry City, BT48 8HX
2D0 NAU D. Smith, Alwyn, Four Roads Port St. Mary, Isle of Man, IM9 5LH
2M0 NAX A. Anderson, 18 Selkirk Street, Wishaw, ML2 8RA
2W0 NAY M. Williams, 10 Clettwr Terrace Pontsian, Llandysul, SA44 4TU
2E0 NAZ N. Azizoff, 7 Spencer Close, London, N3 3TX
2E0 NBC D. Waters, Flat 1, Pastors Hill House, Pastors Hill, Lydney, GL15 6NA
2E0 NBE N. Irvine, 100 Cavendish Road, Sunbury-on-Thames, TW16 7PL
2E0 NBG N. Newman, 1 Ockendon Road, North Ockendon, Upminster, RM14 3PT
2E0 NBI N. Beresford, 2 Meadow View, Great Addington, Kettering, NN14 4BN
2E0 NBK N. Brooks, 105B Upper Woodcote Road, Caversham, Reading, RG4 7JZ
2E0 NBM N. Modi, 20 Hereford Road, Basingstoke, RG23 8QL
2E0 NBR S. Warren, 7 Crich Way, Swadlincote, DE11 0UU
2E0 NBX N. Cohen, 8 Henry Gepp Close, Adderbury, Banbury, OX17 3FE
2E0 NBZ N. Birnie, 61 Pipers Croft, Dunstable, LU6 3JZ
2W0 NCA N. Alward, 22 Laugharne Court, Caldy Close, Barry, CF62 9DW
2E0 NCB D. Lawson, 30 Meadowcroft, St Helens, WA9 3XQ
2E0 NCC N. Croft, 22 King Edward Crescent, Leeds, LS18 4BE
2E0 NCD N. Chandler, 4 Nursery Cottages, Woodgreen, Fordingbridge, SP6 2AL
2E0 NCE D. Stanley, 58 Wells Gardens, Basildon, SS14 3QS
2E0 NCG M. Dumpleton, 3 Gooch Close Bacton, Norwich, NR12 0FA
2E0 NCI S. Ldr. B. Dowley, 120 Capel Street, Capel-Le-Ferne, Folkestone, CT18 7HB
2E0 NCJ C. Nicholson, 97 Station Road, Burgess Hill, RH15 9ED
2E0 NCK N. Taylor, 212 Plantation Hill, Worksop, S81 0HD
2M0 NCM N. Cunningham, 11 Glendoune Street, Girvan, KA26 0AA
2E0 NCN K. Clarke, 15 Grig Place Alsager, Stoke-on-Trent, ST7 2SU
2E0 NCO K. Tonge, 25 Southcote Grove, Birmingham, B38 8ED
2E0 NCR G. Paton, 17 Blakeney Road, Stevenage, SG1 2LH
2E0 NCS N. Sunley, 1 East Lea View, Cayton, Scarborough, YO11 3TN
2E0 NCT M. Steeples, 44 Trunch Road, Mundesley, Norwich, NR11 8JX
2E0 NCV G. Killpack, 20 Fisher Close, Banbury, OX16 3ZW
2E0 NCY J. Mooneapillay, 354 Upper Elmers End Road, Beckenham, BR3 3HG
2E0 NDA N. Ayre, 58 Burford Avenue, Swindon, SN3 1BN
2E0 NDG N. Graven, 33 Sheldrake Road, Broadheath, Altrincham, WA14 5LJ
2E0 NDH N. Hewitt, 36 Kenilworth Road, Doncaster, DN4 0UD
2I0 NDJ N. Jameson, 15A Ednagee Road, Castlederg, BT81 7QF
2M0 NDO D. Morrison, Osbourne Cottage, Benderloch, Oban, PA37 1QP
2E0 NDP N. Plunkett, 11 Stoneleigh Gardens, Grappenhall, Warrington, WA4 3LE
2E0 NDR N. Nash, Roann, Bedmond Road, Pimlico, Hemel Hempstead, HP3 8SH
2E0 NDT R. Thatcher, 83 Westfield Drive, North Greetwell, Lincoln, LN2 4RE
2E0 NDW P. Mackay, 30 Main Road, Austrey, Atherstone, CV9 3EH
2E0 NDY A. Williams, 12 St. Wilfrids Crescent, Brayton, Selby, YO8 9EU
2E0 NDZ A. Humphriss, 44 Bishops Close, Stratford-upon-Avon, CV37 9ED
2E0 NEC C. Humphries, 44 Linksway, Folkestone, CT19 5LS
2W0 NED K. Nedin, 28 Llys-Y-Coed Birchgrove, Swansea, SA7 9PR
2E0 NEI N. Yorke, 30 Bramdene Avenue, Nuneaton, CV10 0DH
2I0 NEJ D. Mulligan, 10 Seaview, Ardglass, Downpatrick, BT30 7SQ
2E0 NEL C. Nelson, 14 Windy Harbour Road, Southport, PR8 3DU
2E0 NEM P. Sanders, 6 Primrose Hill, Warwick, CV34 5HW
2E0 NEN B. Daniels, Broomhill, Holmrook, CA19 1UL
2M0 NEO N. Thomson, Four Winds, Holland Bush Hightae, Lockerbie, DG11 1JL
2E0 NEQ J. Clarke, 24 Telford Court, East Howdon, NE280JH
2E0 NER M. Straughan, 71 Silcoates Lane, Wrenthorpe, Wakefield, WF2 0PA
2E0 NEV N. Griffin, 54 Edinburgh Road, Newmarket, CB8 0QD
2E0 NEY P. Millard1, Weavern House, Hartham Lane, Chippenham, SN14 7EA
2E0 NFB N. Bisiker, 31 Lansdowne Avenue, Waterlooville, PO7 5BL
2E0 NFC A. Cockburn, 52 Devon Road, Hebburn, NE31 2DW
2E0 NFI N. Mooney, 60 Rhyddings Street, Oswaldtwistle, Accrington, BB5 3EY
2E0 NFS N. Stephens, 3 Spinney House, College Road, Windermere, LA23 1PX
2E0 NFX N. Friend, 4 Bartholomew Street, Dover, CT16 2LH
2E0 NGB N. Bland, 63 Swindon Road, Wroughton, Swindon, SN4 9AG
2E0 NGC L. Akred, 25 Kitchener Street, Walney, Barrow-in-Furness, LA14 3QW
2E0 NGG N. Clare, 123 Cunningham Road, Tamerton Foliot, Plymouth, PL5 4PU
2I0 NGK J. Allen, 3 Malwood Close, Belfast, BT9 6QX
2E0 NGL N. Green, 44 Rushyford Drive, Chilton, Ferryhill, DL17 0EQ
2I0 NGM A. Mckay, 17 Thorn Hill Road, Banbridge, BT32 3TL
2E0 NGN G. Nelson, 4 Garnet Field, Yateley, GU46 6FN
2M0 NGO J. Nattress, 44 Broadlands, Carnoustie, DD7 6JY
2E0 NGR N. Grey, 131 Links Avenue, Hellesdon, Norwich, NR6 5PQ
2E0 NGT N. Tooby, The Gin Case, Whiteholme, Roweltown, CA6 6LJ
2E0 NGZ S. Lawrance, 69 Athelstan Gardens, Wickford, SS11 7EF
2E0 NHB N. Barker, 17 Pippin Walk, Hardwick, Cambridge, CB23 7QD
2E0 NHJ N. Heywood, 38 Thurne Rise, Martham, Great Yarmouth, NR29 4PU
2E0 NHM N. Meakin, 60 Canberra Way, Warton, Preston, PR4 1XY
2E0 NHR S. Sawyer, 19 Malvern Close, Ashington, NE630TD
2E0 NHS J. Kelly, 12 Park Road, Milford On Sea, Lymington, SO41 0QU
2M0 NIA N. Hague, 11 Auchriny Circle, Bucksburn, Aberdeen, AB21 9JJ
2E0 NIB N. Bennett, 44 Glenmoor Road, Buxton, SK17 7DD
2I0 NIE C. Morton, 29 Lackaboy View, Enniskillen, BT74 4DY
2E0 NIF G. Calder, 41 Wood End Way, Chandler'S Ford, Eastleigh, SO53 4LN
2I0 NIO J. Tipping, 16 The Oaks, Portadown, Craigavon, BT62 4HX
2M0 NIT D. De Freitas, 14 York Street, Clydebank, G81 2PH
2M0 NIX N. Robertson, Ladyburn, Port William, Newton Stewart, DG8 9QN
2E0 NJC N. Long, 25 Blendworth Lane, Southampton, SO18 5GY
2E0 NJE N. Gonzales, 46 Whitton View, Rothbury, Morpeth, NE65 7QN
2E0 NJI N. Isherwood, 41 Livingstone Road, Blackburn, BB2 6NE
2E0 NJJ D. Wharlley, 15 Crampton Court, Grosvenor Road, Broadstairs, CT10 2XU
2E0 NJK N. Kendall, 19 Clowance Lane, Mount Wise, Plymouth, PL1 4HU
2M0 NJM S. Spencer, 55 Blackwell Court, Inverness, IV2 7AR
2E0 NJO N. Jones, 5 Montgomery Crescent, Quarry Bank, Brierley Hill, DY5 2HB
2M0 NJS N. Sheridan, Cemetery Lodge, Lochmaben, Lockerbie, DG11 1RL
2E0 NJT N. Jones, 63 Bernwell Road, London, E4 6HX
2E0 NJV D. Villa, 33 North Street, Tywardreath, Par, PL24 2PW
2E0 NKC D. Ansell, 30 Curzon Avenue, Horsham, RH12 2LB
2E0 NKE N. Rotherham, 4 Spenser Road, Cheltenham, GL51 7EA
2E0 NKI N. Crabb, 1 Council Houses, Hall Lane, Crostwick, Norwich, NR12 7BB
2E0 NKK M. Remplakowski, 11 De Montfort Road Speen, Newbury, rg14 1ta
2E0 NKM N. Morse, 33 Tower Close, Bassingbourn, Royston, SG8 5JX
2E0 NKN E. Coomer, Torquay Road, Exeter, TQ12 5EZ
2E0 NKP N. Palin, 21 Ford Lane, Crewe, CW1 3EQ
2E0 NKR M. Moss, 27 Dunn Side, Chelmsford, CM1 1DL
2W0 NKS M. Moyse, 10 Clifton Rise, Abergele, LL22 7DN
2E0 NKT N. Kent, Flat 1, Manor House, Redruth, TR15 1AX
2M0 NLA A. Cunningham, 36 Station Brae Gardens, Dreghorn, Irvine, KA11 4FB
2E0 NLB N. Brown, 9 Devonshire Avenue, Wigston, LE18 4LP
2E0 NLE M. Pack, 59 West End Falls, Nafferton, YO25 4QA
2E0 NLK N. Lake, 64 Womersley Road, Norwich, NR1 4QB
2E0 NLM P. Maybin, 16 Appleby Road, London, E16 1LQ
2E0 NLP J. Watts, 70 Castleway North, Leasowe, Wirral, CH46 1RW
2E0 NLT P. Austin, Bush Farmhouse Clee St. Margaret, Craven Arms, SY7 9DT
2E0 NLW S. Jones, 30 Crown Fields Close, Newton-le-Willows, WA12 0JW
2E0 NLX P. Cooke, 26 Welby Way, Coxhoe, Durham, DH6 4BT
2E0 NLY B. Plackett, 36 Dartmouth Crescent, Brinnington, Stockport, SK5 8BG
2E0 NMA M. Baynes, 92 Belgrave Drive, Hull, HU4 6DW
2E0 NMC N. Mcintyre, 27 Chapel Close, St Ann'S Chapel, Gunnislake, PL18 9JB
2M0 NMD T. Ormiston, 22 St. Ronans Road, Innerleithen, EH44 6LZ
2E0 NMK S. Bateson, 2 Green Crescent, Coxhoe, Durham, DH6 4BE
2E0 NNB A. Hopper, 7 Holmesdale Villas, Swallow Lane, Dorking, RH5 4EY
2E0 NNE D. Hanwell, 28 Chipperfield Road, Norwich, NR7 9RR
2E0 NNF N. Ferenc, 2A Rosedene Ave, London, sw162lt
2E0 NNG R. Turner, 38 Cotman Road, Clacton-on-Sea, CO16 8YB
2E0 NNH G. Bansil, 15 Abington Close, Crewe, Cw13tl
2E0 NNI M. Bailey, 54 Springfields Road, Stoke-on-Trent, ST4 6RZ
2E0 NNP J. Pinto, 21 Steed Crescent, Colchester, CO2 7SJ
2E0 NNQ J. Blamey, 46 First Avenue, Canvey Island, SS8 9LP
2E0 NNX D. Austin, 1002 Marsden House Marsden Road, Bolton, BL1 2JX
2E0 NNZ J. Doughty, 18 Kirk Place, Chelmsford, CM2 6TN
2E0 NOC C. Arbon, 8 Orchard Avenue, Ashford, TW15 2JB
2E0 NOD N. Lightfoot, 4 Prospect Close, Hatfield Peverel, Chelmsford, CM3 2JE
2E0 NOK C. Smith, 8 Pitts Street, Bradford, BD4 9JJ
2E0 NOM R. Paterson, 21 Beechgrove, Little Oakley, Harwich, CO12 5NN
2E0 NON G. Fielding, Chapel Court, Chapel Lane, Malvern, WR13 5HX
2M0 NOP N. Price, 5 Haltree Cottage, Heriot, EH38 5YD
2E0 NOS W. Warren, 36 Devizes Road, Swindon, SN1 4BG
2D0 NOT D. Ali, 25, Sunnydale Avenue, Port Erin, Isle Of Man, IM9 6EU
2E0 NOU L. Mccarthy, 34 Shawley Way, Epsom, KT18 5PB
2E0 NOW M. Clarke, 40 Fingringhoe Road, Langenhoe, CO5 5AD
2E0 NOZ J. Norrington, 32 Fulfen Way, Saffron Walden, CB11 4DW
2E0 NPE J. Perfect, 62 Warwick Close Holmwood, Dorking, RH5 4NL
2E0 NPF R. Mcmanus, 125 Whitaker Road, Derby, DE23 6AQ
2E0 NPL A. Lyman, 12 Chicheley Street, Newport Pagnell, MK16 9AR
2E0 NPN N. Swift, 59 Milton Avenue, Malton, YO17 7LB
2E0 NPP P. Hayward, 14 Micklewright Avenue, Crewe, CW1 4DF
2E0 NPR N. Prater, 100 Pitfold Road, London, SE12 9HY
2E0 NPS C. Kenyon, The Farmhouse, 10, Watermill Lane, Spilsby, PE23 5AG
2U0 NPT N. Thomas, 6 Tunstall Terrace, Gibauderie, St Peter Port, Guernsey, GY1 1XJ
2E0 NPX N. Paxman, 128 Coggeshall Road, Braintree, CM7 9ES
2E0 NQA R. Clark, 12 Ash Drive Haughton, Stafford, ST18 9EU
2W0 NQE S. Walmsley, 29 Shelley Court Machen, Caerphilly, CF83 8TT
2E0 NQF J. Rookyard, 3 Drapers House St. Johns Road, Banbury, OX16 5BE
2M0 NQT M. Simon, 2/11 Robertson Gait, Edinburgh, EH111HJ
2E0 NQU E. Wagner, 3 Sarre Road, London, NW2 3SN

2W0 NRA M. Chell, Mesen Fach, Llanybydder, SA40 9TY
2E0 NRB M. Beckett, 4 Sandcross Close, Orrell, Wigan, WN5 7AH
2W0 NRE N. Holding, Tir Nanog Crosswell, Crymych, SA41 3UF
2E0 NRH N. Hickson, 27 Cressing Road, Witham, CM8 2NP
2E0 NRJ N. Johnson, Belair, Western Road, Crediton, EX17 3NB
2E0 NRL E. Tattersall, 17 Malt Kiln Way, Sandbach, CW11 1JL
2E0 NRN L. Lewis, 653 Main Road, Dovercourt, Harwich, CO12 4NF
2E0 NRS N. Roberts, 34 Copeland Drive, Standish, Wigan, WN6 0XR
2E0 NRW N. Waters, 9 Shirley Road, Droitwich, WR9 8NR
2E0 NRX S. Cook, Deganwy, Hardwick Road, King's Lynn, PE30 5BB
2M0 NSA A. Custura, 14 Deansloch Terrace, Aberdeen, AB16 5SN
2E0 NSC N. Smith, 40 Fairdale Drive, Newthorpe, Nottingham, NG16 2FG
2E0 NSG N. Gregson, 4 Pollard House, Maldwyn Avenue, Bolton, BL3 3RB
2E0 NSQ S. Beedham, 27 Malpas Close Bransholme, Hull, HU7 4HH
2E0 NSR E. Parrish, 89 Delamere Drive, Macclesfield, SK10 2PS
2E0 NSS M. Price, 25 School Crescent, Lydney, GL15 5TA
2M0 NSW N. White, 2 Appleby Cottages, Whithorn, Newton Stewart, DG8 8DQ
2E0 NSY S. O'Riordan, 46 Grange Road, London, HA20LW
2E0 NTC G. Bull, 9 Kilburn Place, Dudley, DY2 8HP
2I0 NTH C. Mccormick, Flat 4, Legacorry House, Main Street, Armagh, BT61 9RW
2E0 NTJ N. Jones, 1 Olaf Close, Andover, SP10 5NJ
2E0 NTM C. Hoult, 11 Queen Street, Alnwick, NE66 1RD
2I0 NTP N. Prentice, 26 Claranagh Road, Claranagh, Enniskillen, BT94 3FJ
2E0 NTR A. Butcher, 31 Wittonwood Road, Frinton-on-Sea, CO13 9JZ
2W0 NTS N. Hewelt, 17 Saint Marys Road, Llandudno, LL30 2UB
2E0 NTV Rev. N. Wood, 12 Spring Close, Verwood, BH31 6LB
2E0 NTW C. Northwood, Apartment 50, 2 Munday Street, Manchester, M4 7BB
2M0 NTY C. Mcclymont, 115 Glenavon Road, Flat13/1, Glasgow, G20 0HT
2W0 NUC W. Groves, 3 Tetbury Close, Newport, NP20 5HX
2E0 NUD N. Cull, 8 Eaton Road, Norwich, NR4 6PY
2E0 NUG M. Wells, 23 Eastmead, Bognor Regis, PO21 4QT
2E0 NUL J. Unwin, 39 Whinfell Drive, Normanby, Middlesbrough, TS6 0BG
2M0 NUO Dr C. Brown, 4 Damselfly View, Edinburgh, EH17 8XH
2E0 NUQ M. Dickenson, 6 The Pavilions, Blandford Forum, DT11 7GF
2E0 NVB N. Betts, 12 Sandy Lane, Worksop, S80 1SW
2E0 NVK L. Bolton, 9 Nab Crescent, Meltham, Holmfirth, HD9 5LT
2E0 NVL A. Wake, Wardens Bungalow, Wareham Forest Tourist Park, North Trigon, Wareham, BH20 7NZ
2E0 NVP M. Weir, 153 Tyndale Crescent, Birmingham, B43 7HX
2E0 NVT N. Tennant, 22 The Lizard, Wymondham, NR18 9BH
2E0 NWA N. Wong, Montefiore House, Wessex Lane, Southampton, SO18 2NU
2E0 NWB J. Benbow, 44 Copthorne Park, Shrewsbury, SY3 8TJ
2E0 NWE O. Price, Queens Arms House 1 King Street Odiham, Hook, RG29 1NN
2W0 NWJ Dr N. Jones, 54 Glanrhyd, Coed Eva, Cwmbran, NP44 6TY
2I0 NWO D. Adams, 65 Rose Park, Limavady, BT49 0BF
2E0 NWR W. Westlake, 2 Chegwin Court, Newquay, TR7 2DE
2E0 NWT N. Topping, 7 Beckstone Close, Harrington, Workington, CA14 5QR
2E0 NWY S. Newhouse, 28 Hillmorton Lane, Lilbourne, Rugby, CV23 0SS
2E0 NXB N. Beck, 2 Killerton View, Wyndham Road, Silverton, Exeter, EX5 4JZ
2E0 NXE P. Symon, 8 Glebe Close, St Columb, TR9 6TA
2E0 NXL P. Mcgee, 2 Fishers Close, Norwich, NR5 0QH
2E0 NXM N. Masters, Wyngrove Vicarage Road, Wrawby, Brigg, DN208RR
2E0 NXP Y. Weng, 20 Kendal Grove, Leeds, LS3 1NS
2W0 NXV B. Williams, The Grange Llanddewi, Llandrindod Wells, LD1 6SF
2E0 NYC S. Vzor, 40 Henlow Road, Birmingham, B14 5DS
2E0 NYE N. Whittaker, 1 Edendale, Hull, HU7 4BX
2E0 NYF D. Lamble, 4 Laburnum Road, Chorley, PR6 7BG
2E0 NYG N. Cox, 182 North Tenth Street, Milton Keynes, MK9 3AY
2M0 NYK I. Nicolson, 1 Gullane Place, Dundee, DD2 3BF
2I0 NYL L. Elliott, 19 Gosford Road, Collone, Armagh, BT60 1LQ
2E0 NYM C. Matthews, 18 Tennyson Gardens, Darlington, DL1 5BJ
2E0 NYP J. Dyson, 77 Grantham Road, Southport, PR8 4LT
2E0 NYT J. Woodcock, 31 Moorway Lane Littleover, Derby, DE23 2FR
2E0 NZA G. Lund, 1 Thrush Close, Gloucester, GL4 4WZ
2E0 NZD M. Phillips, 44 Hilderic Crescent, Dudley, DY1 2ET
2E0 NZT T. Reseigh, 10 Higher Croft Parc, The Lizard, Helston, TR12 7RL
2E0 OAA C. King, 8A Barton Road, Bedford, MK42 0NA
2M0 OAB S. Milne, 5 Moriston Court, Grangemouth, FK3 0JJ
2E0 OAC G. Dray, 2 Mulberry Drive, Malvern, WR14 4AT
2W0 OAG A. Graham, 2 Heol Undeb, Beddau, Pontypridd, CF38 2LB
2E0 OAH K. Johnson, 32 Redmire Close, Bransholme, Hull, HU7 5AQ
2E0 OAI D. Saunders, 17 Sandy Lane, Prestwich, Manchester, M25 9RU
2E0 OAJ A. Johnson, 19 Meadow Vale, Bristol, BS5 7RG
2E0 OAK J. Burnett, Hobart House, 16 Church Lane, Reepham, Lincoln, LN3 4DQ
2E0 OAM A. Mullinex, 84 Danesbury Crescent, Birmingham, B44 0QS
2E0 OAO A. Al-Shakarchi, 17 Fairfax Place, London, NW6 4EJ
2E0 OAP M. Deary, 7 Newbold Avenue, Sunderland, SR5 1LG
2E0 OAR D. Edwards, 212, Eastern Avenue North Kingsthorpe, Northampton, NN2 7AT
2E0 OAS C. Sole, 2 Shepley Street, Manchester, M359dy
2E0 OAU L. Boylan, 30 Pembrey Way, Liverpool, L25 9SN
2I0 OAZ N. Armstrong, 1 Diamond Cottages, Ardmore Road, Crumlin, BT29 4QU
2E0 OBB O. Boar, 19 Blyford Road, Lowestoft, NR32 4PZ
2E0 OBC Dr C. Bridges, 23 Bramley Vale, Cranleigh, GU6 7FY
2E0 OBD M. Hopkins, 27 Girtford Crescent, Sandy, SG19 1HR
2E0 OBI P. Sherratt, 39 Vimy Road, Leighton Buzzard, LU7 1FQ
2E0 OBK D. Cull, 6 Compass Way, Bromsgrove, B60 3GP
2E0 OBL M. Orbell, 21 Reedings Road, Barrowby, Grantham, NG32 1AU
2E0 OBM D. Barwick, 2 Sutton Close, Bury St Edmunds, Ip327ep
2E0 OBO R. Blackman, 32 Kingfisher Road, Sprowston, NR7 8GX
2E0 OBP P. Percival, 2 St. Catherine Street, Ventnor, PO38 1HG
2E0 OBS B. Heath, 108 Cow Lane, Bramcote, Nottingham, NG9 3BB
2E0 OBZ D. Thomas, 51 Sandringham Avenue, Vicars Cross, Chester, CH3 5JF
2E0 OCB O. Carpenter-Beale, 6 Betherinden Cottages, Bodiam Road, Cranbrook, TN18 5LW
2M0 OCC N. Nuttall, 94 Newpark Road, Stirling, FK7 0QD
2W0 OCF N. Smith, 23 Pennyroyal Close, St. Mellons, Cardiff, CF3 0NB
2E0 OCG O. Giles, Holly Cottage, Main Road, Crewe, CW4 8LL
2E0 OCH C. Howard, 1 Beale Road, Cheltenham, GL51 0JN
2E0 OCL L. Hendry, 109 Grove Avenue, New Costessey, Norwich, NR5 0HZ
2E0 OCM I. Johnson, 9 Brook Road, Pontesbury, Shrewsbury, SY5 0QZ
2E0 OCN H. Peberdy, 53 Far Lane, Normanton On Soar, Loughborough, LE12 5HA
2E0 OCP D. Drynski, Flat 22, Nicholas Court, Corney Reach Way, Chiswick, London, W4 2TS
2E0 OCS R. Wells, 31 Bracklesham Road, Hayling Island, PO11 9SJ
2E0 OCT A. Lawrence, 5 Heighton Close, Bexhill-on-Sea, TN39 3UP
2E0 OCW W. Joyce, 2 Palmers Cottage, Main Street, Oakham, LE15 8DH
2E0 OCY P. Gibbons, 2A Gipsy Castle Estate, Hay-On-Wye, Hereford, HR3 5EG
2E0 ODB D. Bambrough, 7 Barnwell View, Herrington Burn, Houghton le Spring, DH4 7FB
2E0 ODF J. Lashley, 33 Goodes Avenue, Syston, Leicester, LE7 2JH
2E0 ODJ D. Jones, 20 Nightingale Road, Guisborough, TS14 8HA
2E0 ODL K. Bedford, 29 Kent Road Brookenby, Binbrook, Market Rasen, LN8 6EW
2E0 ODN A. Swallow, 16 Quarry Lane, Chesterfield, S40 3AS
2E0 ODO E. White, 3 Davy Drive, Maltby, Rotherham, S66 7EN
2M0 ODP D. Pounder, Birchbank, Lindean, Galashiels, TD1 3PA
2W0 ODS D. Robins, 25 Clos Mancheldowne, Barry, CF62 5AB
2E0 ODT B. Hendry, 109 Grove Avenue, New Costessey, Norwich, NR5 0HZ
2J0 ODX P. Ahier, Les Trois Carres, La Rue D'Aval, Jersey, JE3 6ER
2E0 ODZ D. Ziaja, 46 Merlin Crescent, Edgware, HA8 6JE
2E0 OEE J. Hendry, 109 Grove Avenue, New Costessey, Norwich, NR5 0HZ
2E0 OEK C. Taylor, 51 Barnes Close, Sturminster Newton, DT10 1BN
2E0 OEL I. Robinson, 12 Osprey Road, Flitwick, MK45 1RU
2E0 OEM J. Summers, 5 The Meadow, Bosvigo Lane, Truro, TR1 3NG
2E0 OES R. Barnes, 1 Moira Close, Chaddesden, Derby, DE21 4RL
2E0 OET B. Mead, 8 Wordsworth Road, Kettering, NN16 9LB
2E0 OEV M. Cuff, 1 Ivy Close, St. Leonards, Ringwood, BH24 2QZ
2E0 OEZ I. Beresford, 16A Holbeck Hill, Scarborough, YO11 2XD
2E0 OFF G. Kenyon, 2 Langdale Terrace, Stalybridge, Sk151ex
2E0 OFM P. Joyce, 2 Harold Road, Cuxton, Rochester, ME2 1EE
2E0 OFT J. Saunders, Homestead, Chapel Street Stapleton, Leicester, LE9 8JH
2E0 OGA C. Ingram, 17 Stackfield, Harlow, CM202LA
2E0 OGB R. Evans, 364 Aldermans Green Road, Coventry, CV2 1NN
2E0 OGK G. Kendall, 39 Foundry Gate, Wombwell, Barnsley, S73 0LF
2E0 OGR M. Coombes, 33 Woodside Road North Baddesley, Southampton, SO52 9NB
2W0 OGT T. Cording, 17 Maes Y Derwen, Llanrhaeadr Ym Mochnant, Oswestry, SY10 0LE
2W0 OGY C. Hodgetts, 16 Myrtle Drive, Rogerstone, Newport, NP10 9EA
2E0 OGZ G. Wilkinson, 50 Sherburn Road, Durham, DH1 2JR
2E0 OHC M. Hill, Buzon 169, Avd De La Condomina 53, Local 3, Alicante, Spain, 3540
2I0 OHE E. Paterson, 1 Sycamore Grove, Belfast, BT4 2RB
2E0 OHF K. Charlton, 14 Rubens Close, Aylesbury, HP19 8SW
2E0 OHN N. Holden, Plum Cottage, Avon Dassett, Southam, CV47 2AP
2E0 OIK N. Titmus, 68 Hobart Road, Cambridge, CB1 3PT
2E0 OIL M. O'Connor, 28 Cardigan Road, Southport, PR8 4SF
2E0 OIN H. List, 41 Westbury Crescent, Dover, CT17 9QQ
2E0 OIR A. Birkett, 21 Cedar Drive Wyke, Bradford, BD12 9HL
2E0 OJB J. Bayliss, 39 Elms Avenue, Littleover, Derby, DE23 6FB
2E0 OJD K. Winton, 130 George V Avenue, Worthing, BN11 5RX
2E0 OJE A. Forbes, Flat 10, Denby House, Paignton, TQ4 6ES
2E0 OJI J. Isaacs, 20 Vine Street, Worcester, WR3 7DY
2I0 OJK J. Kavanagh, 11 Pinewood Crescent, Claudy, BT47 4AB
2E0 OJN P. Field, 63 Hartford Road, Davenham, Northwich, CW9 8JE
2E0 OJS O. Squire, 91 Victoria Road, London, N22 7XG
2E0 OKB K. Barnett, 22 Highclere Road, Southampton, SO16 7AW
2E0 OKC A. Londors, 112 Kingston Hill Avenue, Romford, RM6 5QL
2E0 OKG K. George, Wylye, Auberrow, Hereford, HR4 8AN
2E0 OKH O. Hopkins, Apartment 17, White Croft Works, Sheffield, S3 7AH
2E0 OKK I. Day, 137 Tuffley Lane, Tuffley, Gloucester, GL4 0NZ
2E0 OKL A. Gates, 46 Gloucester Place, Littlehampton, BN17 7AL
2M0 OKO K. Biegun, 27 School Walk, Aberdeen, AB24 1XX
2E0 OKP S. Walsh, 12 The Lawns Broadley Avenue Anlaby, Hull, HU10 7HD
2E0 OKQ M. Broum, 137 Culvers Avenue, Carshalton, SM5 2BA
2E0 OKS M. Biadon, 57 Fern Hill Road, Oxford, OX4 2JW
2E0 OKY R. Eglington, 33 Bradley Lane, Bilston, WV14 8EW
2E0 OKZ A. Cammish, 6 West Vale, Filey, YO14 9AY
2E0 OLE O. Rofix, 17 Potters Crescent, Great Moulton, NR15 2HL
2E0 OLF M. Mills, 17 Hornby Street, Plymouth, PL2 1JD
2E0 OLG D. Rugen, 19 Jacksons Close, Haskayne, Ormskirk, L39 7LD
2M0 OLK O. Keast, 6 Prospecthill Place, Greenock, PA15 4DW
2E0 OLM G. Dewey, West Riding Five Ash Down, Uckfield, TN22 3AP
2W0 OLT O. Thomas, Garth Celyn, St. Davids Road, Aberystwyth, SY23 1EU
2I0 OMA J. Martin, 23 Winters Gardens, Omagh, BT79 0DZ
2E0 OME R. Hauk, 6 Hall Road, Stowmarket, IP14 1TN
2E0 OMG M. Robinson, 10 Bramley Gardens, Poulton-le-Fylde, FY6 7RD
2E0 OMI H. Kassier, 26 Higher Port View, Saltash, PL12 4BX
2D0 OMN R. Corrin, 11 Cronk Y Berry, Douglas, Isle of Man, IM2 6EY
2E0 OMR B. Withers, 29 Yeoman Way, Trowbridge, BA14 0QL
2M0 OMS M. Scullion, 24 Langmuir Road, Kirkintilloch, Glasgow, G66 2QE
2E0 OMT T. Baggley, 16 Seaton Road, Seaton, Workington, CA14 1DT
2E0 OMV J. Barton, 37 Lytton Road, Sheffield, S5 8AX
2E0 ONC G. Stephens, 8 New Molinnis, Bugle, Saint Austell, PL26 8QL
2E0 ONE P. Greenwood, 1 The Garth, Whitby, YO21 3PD
2E0 ONH A. Brown, 51 Towncroft, Chelmsford, CM1 4JX
2E0 ONI P. Cannings, 30 Graham Gardens, Luton, LU3 1NQ
2E0 ONM M. Kulakowski, Flat 4 Clement Mellish House East Stockwell St., Colchester, CO1 1GJ
2M0 ONS D. Anderson, Dail Darach, Monydrain Road, Lochgilphead, PA31 8LG
2E0 ONV J. Bonar, Flat, 5A Friday Street, Minehead, TA24 6EE
2M0 ONW K. Harper, 35 Amlaird Road, Kilmarnock, KA3 2EU
2E0 OOC B. Cooper, 71 High Street, Birstall, Batley, WF17 9RG
2E0 OOH D. Meakin, 27 Spencer Road, Long Buckby, Northampton, NN6 7YP
2E0 OOM M. Buist, 23 St. Chads Drive, Gravesend, DA12 4EL
2E0 OON C. May, 16 Trelawn Road, London, E10 5QD
2E0 OOO R. Clayton, 9 Green Island, Irton, Scarborough, YO12 4RN
2M0 OOR S. Boal, 20 Fairinsfell, Broxburn, EH526AL

2M0 OOT J. Ferrans, 77 Knockinlaw Road, Kilmarnock, KA3 2AS
2E0 OOU G. Cameron, 33 Railway Terrace, York, YO24 4BN
2E0 OOV S. Phillips, 38 Corbet Ride, Leighton Buzzard, LU7 2SJ
2E0 OOW P. Turner, 77 High Street St. Lawrence, Ramsgate, CT11 0QR
2E0 OOX A. Tunley, 5 Camborne Close, Lower Earley, Reading, RG6 4EN
2E0 OOY P. Sullivan, 25 Cavendish Drive, Birkenhead, CH42 6RG
2E0 OPB O. Blackburn, 128 High St., Crigglestone, Wakefield, WF4 3EF
2E0 OPC O. Campbell, 3 Hillside Close, Helsby, Frodsham, WA6 9LB
2E0 OPM D. Whitehouse, 6 Larch Close, Heathfield, TN21 8YW
2E0 OPO O. Silva, Flat 71, Long Acre House, Pettacre Close, London, SE28 0PB
2E0 OPS D. Lapham-Crozier, 109 Aylesbury Crescent, Plymouth, PL5 4HX
2E0 OPT J. Nihill, 14 Hereford Avenue, Clayton, Newcastle, ST5 3ED
2E0 OPU K. Morris, 80 Bridge Street, Chatteris, PE16 6RN
2E0 OQD L. Sproule, Dragonpits North Perrott, Crewkerne, TA18 7TH
2E0 OQH D. Cooper, 52 Meadow Lane, Birkenhead, CH42 3YE
2M0 OQR P. Taylor, 2 Laurel Grove, Aberdeen, AB22 8YJ
2E0 OQV N. Connor, 28 Church Street, Hungerford, RG170JE
2E0 OQZ P. Hall, 13 Sheard Avenue, Ashton-under-Lyne, OL6 8DS
2E0 ORD R. Drage, 4 Bruce'S Close Conington, Peterborough, PE73QW
2E0 ORE C. Dodds, 22 Cambridge Street, Wolverton, Milton Keynes, MK12 5AJ
2W0 ORH O. Hopkin, The Forge, Rock Road, St Athan, CF62 4PG
2E0 ORI D. Dart, Ticklebelly Cottage Lower Charlton Trading Estate, Shepton Mallet, BA4 5QE
2M0 ORK M. Herridge, The Hollies, Petticoat Lane, Orkney, KW17 2RP
2E0 ORM T. Rowlands, 7 Swan Delph, Aughton, Ormskirk, L39 5QG
2E0 ORP M. Orpen, Daymer, Tey Road, Colchester, CO6 3RY
2E0 ORS P. Lewis, 9 The Hill, Glapwell, Chesterfield, S44 5LX
2W0 ORT B. Roberts, 6 Trem Y Moelwyn, Tanygrisiau, Blaenau Ffestiniog, LL41 3SS
2E0 ORX T. Hedger, 1 Berry Terrace, Acton Square, Sudbury, CO10 1HT
2M0 OSC J. Hanley, 44 Waverley Crescent, Livingston, EH54 8JN
2E0 OSE P. Pope, 2 Hale Villas, Honiton, EX14 9TQ
2W0 OSG J. Bellis, 32 Broughton Road, Lodge, Wrexham, LL11 5NG
2W0 OSH S. Owen, 8 Old Tanymanod Terrace, Blaenau Ffestiniog, LL41 4BU
2M0 OSK A. Twort, 17 Balallan, Isle of Lewis, HS2 9PN
2E0 OSS S. Fraser, Old Post Office, Mill Road Barton St. David, Somerton, TA11 6DF
2E0 OST F. Limbert, Knowlecroft Little Ribston, Wetherby, LS22 4ET
2E0 OSX A. Logan, 23 Cherry Tree Rise, Walkern, Stevenage, SG2 7JL
2E0 OSY D. Coppenhall, 55 Vicarage Lane, Elworth, Sandbach, CW11 3BU
2E0 OTA S. Jenkinson, 17A Britannia Road Burbage, Hinckley, LE10 2HE
2E0 OTB P. Hateley, 44 Painters Croft, Coseley, Bilston, WV14 8AP
2I0 OTC B. Emerson1, 67 Castlemore Avenue, Belfast, BT6 9RH
2E0 OTG S. Ball, 18 Lighthurst Avenue, Chorley, PR7 3HY
2E0 OTI D. Scotcher, 17 St. Dominics Square, Luton, LU4 0UN
2E0 OTM O. Morris, 1 Crawford Avenue, Peterlee, SR8 5EG
2E0 OTP M. Anostalgia, 136 Avenue Road Extension, Leicester, LE2 3EH
2E0 OTR T. Rood, 5 Sunny View, Rise, Hull, HU11 5BW
2E0 OTT J. Slobin, 45 Dale Edge, Eastfield, Scarborough, YO11 3EP
2I0 OTW P. Fallon, 18 Church View, Killough, Downpatrick, BT30 7RJ
2E0 OTY P. Lane, 5 Swan Court, Middle Watch, Swavesey, Cambridge, CB24 4AG
2E0 OTZ M. Wright, 8 St. Wilfrids Road, Oundle, Peterborough, PE8 4NX
2E0 OUH P. Loxton, 32 Parkhill Crescent, Wakefield, WF1 4EZ
2I0 OUI M. Doogan, 54 Birchdale Manor, Lurgan, Craigavon, BT66 7SY
2E0 OUK I. Hrynkiewicz, 21 York Road, Cannock, WS11 8ES
2E0 OUR D. Mills, 91 Harp Road, London, W7 1JQ
2M0 OUU C. Mcconnochie, 72 Duddingston Avenue, Kilwinning, KA13 6RS
2E0 OVB R. Gowers, 43 Tungstone Way, Market Harborough, LE16 9GA
2M0 OVD D. Adamson, 5 Central Quadrant, Ardrossan, KA22 7DY
2E0 OVF S. Hedgecock, 37 Tennyson Road, Maldon, CM9 6BE
2E0 OVI O. Popa, 5 Lanark Close, Horsham, RH13 5RY
2E0 OVL J. Hirst, 57 Newgate Street, Doddington, PE15 0SR
2E0 OVR F. Farrer, 16 High Street, Eagle, Lincoln, LN6 9DH
2W0 OVT J. Jones, 40 Ffordd Coed Marion, Caernarfon, LL55 2EF
2M0 OVV S. Monaghan, 13 Ballyhennan Crescent, Tarbet, Arrochar, G83 7DB
2E0 OVW G. Sandell, 1 St. Margaret Road, Ludlow, SY8 1XN
2E0 OVX M. Walton, 358 Old Heath Road, Colchester, CO2 8DD
2E0 OWC S. Iles, Bigbury Bay Holiday Park, Challaborough, Kingsbridge, TQ7 4HS
2E0 OWE D. Fincham, 2 Glebe Close, St Columb, TR9 6TA
2E0 OWG D. Burgfeld, 20 Wilson Row, Crowthorne, RG45 6WE
2E0 OWH T. Harrison, 58 Ascot Drive, Cannock, WS11 1PE
2E0 OWL S. Hurley, 11 Beresford Avenue, Wirral, CH63 7LR
2M0 OWT Z. Mckinnon, 8 Rowanlea Avenue, Paisley, PA2 0RP
2E0 OXF A. Comerford, 21 New Cross Road, Headington, Oxford, OX3 8LP
2E0 OXO D. Harden, 59 Violet Avenue, Edlington, Doncaster, DN12 1NW
2M0 OXQ S. Mckinnon, 8 Rowanlea Avenue, Paisley, PA2 0RP
2E0 OXT R. Ingram, 17 Stackfield, Harlow, CM20 2LA
2M0 OXX A. Berry, 8 Hill Street Striling Fk7 0Dh, Striling, FK7 0DH
2E0 OYG B. Nuttall, 22 Countess Crescent, Bispham, Blackpool, FY2 9LQ
2W0 OYL A. Owen, Bro Dawel, Brynrefail, Caernarfon, LL55 3NR
2E0 OYN R. Watson, 60 Beresford Avenue, Surbiton, KT5 9LJ
2E0 OYQ A. Gibbons, 305 Monks Road, Lincoln, LN2 5LB
2E0 OYR S. Roberts, 77 Lambwath Road, Hull, HU8 0HB
2M0 OYS R. Thomson, Meadow Steading Tornaveen, Torphins, Banchory, AB31 4PJ
2E0 OYY N. Weiner, 3 Bluebell Court Lower Mardyke Avenue, Rainham, RM13 8GF
2E0 OZE M. Crockford, Centre Cottage Kelk, Kelk, YO258HL
2E0 OZG K. Randle, 36 All Saints Road, Sittingbourne, ME10 3PB
2E0 OZI S. Carpenter, 52 Mewstone Avenue, Wembury, Plymouth, PL9 0JZ
2E0 OZK M. Kaszewski, 87 Blenheim Road, Northolt, UB5 4TS
2E0 OZM M. Collantine, Little Mill Farm, Little Mill, Egremont, CA22 2NN
2W0 OZO S. Hayward, 22 Dewsland Street, Milford Haven, SA73 2AU
2E0 OZQ A. Sherer, 28 Baroness Road, Grimsby, DN34 4DP
2E0 OZW K. Ozwell, 109 Abbey Road, Grimsby, DN32 0HN
2E0 OZX C. Russell, 9 Sandfields Road, St Neots, PE19 1PF
2W0 OZY C. Osborne, Gwinwydden, Tremont Road, Llandrindod Wells, LD1 5BH
2E0 PAA C. Jago, 20 Glanville Road, Tavistock, PL19 0EA
2E0 PAB P. Stone, 60 Acorn Avenue, Braintree, CM7 2LR
2I0 PAC P. Dallas, 12 Glendun Crescent, Coleraine, BT52 1UJ
2E0 PAD J. Stainton, 40-41 Dyke End Golcar, Huddersfield, HD7 4LA
2E0 PAE P. Illidge, 55 East Park Road, Spofforth, Harrogate, HG3 1BH
2E0 PAF P. Taylor, 47 Pickhurst Park, Bromley, BR2 0TN
2E0 PAH P. Hunt, 33 Drakes Close, Bridgwater, TA6 3TD
2E0 PAJ P. Kiernan, 4 Bradley Street, Southport, PR9 9HW
2E0 PAK P. Watson, 10 Whitelands Crescent, Baildon, Shipley, BD17 6NN
2W0 PAN A. Paffey, 1 St. Vincent Road, Newport, NP19 0AN
2E0 PAO D. Pike, 46 Haymans Close, Cullompton, EX15 1EH
2E0 PAP P. Woodyard, 65 Raglan Street, Lowestoft, NR32 2JS
2E0 PAS A. Spinks, 10 Foxley Close, Norwich, Nr58dq
2E0 PAT G. Dobson, 4 Durley Gardens, Orpington, BR6 9LL
2E0 PAU P. Dossett, 92 Dale Valley Road, Southampton, SO16 6QU
2E0 PAV S. Richards, 18 Lowfields Staxton, Scarborough, YO12 4SR
2E0 PAX E. Goodwin, 55 Twickenham Road, Sunderland, SR3 4JN
2E0 PBB W. Ramsell, 8 Didcot Drive, Marchington, Uttoxeter, ST14 8LT
2M0 PBC C. Montague, 74 Holmbyre Road, Glasgow, G45 9QD
2E0 PBE P. Edwards, 791 Windmill Lane, Denton, Manchester, M34 2ER
2E0 PBH A. Billings, 46 Thorley Drive, Cheadle, Stoke-on-Trent, ST10 1SA
2E0 PBJ P. Mansfield, 15 Earl Avenue, New Waltham, Grimsby, DN36 4NE
2E0 PBK P. Parkin, 29 Robinia Close, Lutterworth, LE17 4FS
2E0 PBL P. Ball, 101 Chelwood Drive, Bath, BA2 2PS
2I0 PBM P. Bingham, 45 Gowanvale Drive, Banbridge, Bt323gd
2E0 PBN P. Dent, 25 Clyde Avenue, Hebburn, NE31 2JN
2E0 PBO P. Kay, 30 Broadway, Grange Park, St Helens, WA10 3RX
2E0 PBP P. Jones, 50 Clay Lane, Doncaster, DN2 4RJ
2E0 PBR P. Alley, 58 Osprey Close, Watford, WD25 9AR
2E0 PBS P. Sycamore, 17 Markham Avenue, Weymouth, DT4 0QL
2E0 PBT P. Burgess, 61 Grosvenor Avenue, Torquay, TQ2 7JX
2W0 PBU Dr W. Dickson, The Rowans, Pwllmeyric, Chepstow, NP16 6LA
2E0 PBV M. James, 42 Doone Way, Ilfracombe, EX34 8HS
2E0 PBW I. Yovchev, 11 Beverley Drive, Edgware, HA8 5NQ
2E0 PBY P. Barker, 27 Sandwood Road, Sandwich, CT13 0AQ
2I0 PBZ P. Bell, 3 Alexandra Crescent, Larne, BT40 1NE
2E0 PCC P. Garraway, The Poplars, Crowell Road, Chinnor, OX39 4HP
2W0 PCD P. Day, 15-16 Troedrhiw-Trwyn, Pontypridd, CF37 2SE
2W0 PCE P. Iles, 150 Pen-Y-Bryn, Caerphilly, CF83 2LA
2E0 PCF P. Faulkner, 32 Manvers Road, Beighton, Sheffield, S20 1AY
2E0 PCG P. Green, 8 Grassthorpe Road, Sheffield, S12 2JH
2E0 PCI P. Collins, 16 Fern Grove, Haverhill, CB9 9ND
2E0 PCL G. Owen, 38 Trentham Drive, Bridlington, YO16 6ES
2W0 PCN P. Nash, 110 Aberporth Road, Cardiff, CF14 2RY
2E0 PCO P. Coombes, 2 Bissoe Cottages, Bissoe, Truro, TR4 8SU
2E0 PCQ P. Morris, 14 Marina Road, Darlington, DL3 0AL
2E0 PCR C. Vincent, 81 Trethannas Gardens, Praze, Camborne, TR14 0LL
2W0 PCT S. Gau, Disgwylfa, The Downs, Cardiff, CF5 6SB
2E0 PCU P. Roberts, 1 Ballard Road, Wirral, CH48 9XU
2E0 PCV D. Whiting, 133 Belfield Road, Accrington, BB5 2JD
2M0 PCW Prof. S. Skerratt, 8/11 Leslie Place, Edinburgh, EH4 1NH
2E0 PCX P. James, 44 Narbonne Avenue Ellesmere Park Eccles, Manchester, M30 9DL
2E0 PCZ P. Colyer, 23 Florida Road, Torquay, TQ1 1JY
2E0 PDB P. Phipps, Meakers Cottage, Long Load, Langport, TA10 9JX
2E0 PDG E. Aitken, 20 Plover Drive, Bury, BL9 6JH
2E0 PDH C. Macleod, 21 Halsetown, St Ives, TR26 3LY
2E0 PDJ J. Sandon, 461 Archer Road, Pin Green, Stevenage, SG1 5QP
2E0 PDL M. Garry, 34 Conway Road, Paignton, TQ4 5LH
2E0 PDM P. March, 46 Christchurch Road, Tilbury, RM18 8XP
2E0 PDO A. Dingwall, 48 Village Farm Caravan Site, Bilton Lane, Harrogate, HG1 4DL
2E0 PDP J. Clarkson, 56 Edward Bailey Close, Binley, Coventry, CV3 2LZ
2E0 PDQ D. Carpenter, 34B Carey Park, Killigarth, Looe, PL13 2JP
2E0 PDU L. Fuller, Rosemar Lodge Westford, Wellington, TA21 0DX
2E0 PDX G. Swindells, 15 Benedict Close, Salford, M7 2GB
2E0 PDZ P. Harper, 36 Barrow Close, Marlborough, SN8 2BD
2E0 PEC B. Clayton, 1 Maude Crescent Sowerby Bridge, Halifax, HX6 1lb
2W0 PEE N. Tanner, 3 Maes Y Tyra, Resolven, Neath, SA11 4NN
2E0 PEF P. Freeman, 57 Ruffa Lane, Pickering, YO18 7HN
2W0 PEG J. Reason, 158 Caerau Lane, Cardiff, CF5 5JS
2W0 PEH B. Sellers, 86 St. John Street, Ogmore Vale, Bridgend, CF32 7BB
2E0 PEI I. Daraban, 2 Bassenthwaite, Huntingdon, PE29 6UL
2E0 PEL S. Peel, 21 Fairfield Avenue, Ormesby, Middlesbrough, TS7 9BB
2E0 PEM P. Metters, 137 Nevinson Avenue, South Shields, NE34 8NE
2E0 PEP S. Hill, 35 Longs Way, Wokingham, RG40 1QW
2E0 PET P. Munson, 8 Longley Lane, Spondon, Derby, DE21 7AT
2E0 PEW P. Woolley, 84 Bowthorpe Road, Norwich, NR2 3TP
2E0 PEX D. Munn, 36 Moor Lea, Braunton, EX33 2PE
2D0 PEY G. Wilby, Byways, Glenlough Circle, Glen Vine, Isle of Man, IM4 4AX
2E0 PFA P. Fernie, 39 North Parade, Falmouth, TR112TE
2E0 PFB P. Browne, 151 North Road, St. Andrews, Bristol, BS6 5AH
2W0 PFD P. Devlin, Brynteg, Fron Bache, Llangollen, LL20 7BP
2E0 PFF S. Debenham, 10 Elizabeth Close, Wellingborough, NN8 2JA
2E0 PFG P. Goddard, 62 Woodlands Drive, Thetford, IP24 1JJ
2E0 PFH P. Holmes, 18 Raleigh Avenue, Whiston, Prescot, L35 3PL
2E0 PFL P. Leng, The Barn, Gildersleets, Settle, BD24 0AH
2E0 PFO P. Noble, 14 Park Street, Swallownest, Sheffield, S26 4UP
2E0 PFR P. Ratcliffe, 2 Newlands Avenue, Whitby, YO21 3DX
2E0 PFT A. Perfect, 3 Chelmarsh Close, Chellaston, Derby, DE73 6PB
2E0 PFY R. Ball, 7 Cliff Closes Road, Scunthorpe, DN15 7HT
2E0 PGC P. Challans, Flat 4, Sandringham Court, 2 Chandos Square, Broadstairs, CT10 1QN
2E0 PGH P. Hill, 14 Drovers Way, Woodlands, Ivybridge, PL21 9XA
2E0 PGI G. Hartless, 32 Long Acre, Mablethorpe, LN12 1JF
2E0 PGJ P. Johnston, 35 Staunton Ave, Hayling Island, PO11 0EW
2E0 PGL P. Lewis, 154 Meadow Head, Sheffield, S8 7UF
2E0 PGM P. Mcfadden, Maple Cottage, Leighton Buzzard, LU7 9DZ
2M0 PGO P. Greig, 22 Rowanhill Close, Port Seton, Prestonpans, EH32 0SY
2E0 PGP W. Dover, Windcrest, Fox Lane, Basingstoke, RG23 7BB
2E0 PGR P. Grainger, 36 Orchard Road, Wigton, CA7 9JL
2E0 PGS P. Stevenson, 6 Dighton Gate, Stoke Gifford, Bristol, BS34 8XA
2E0 PGT P. Thompson, 25 Pitclose Road, Birmingham, B31 3HU
2I0 PHA S. Wright, 3 Harryville, Portstewart, BT55 7AU

2E0 PHB P. Haslam-Brunt, 488 Klightwood Road, Lightwood, Stoke on Trent, ST3 7EW
2E0 PHE D. Thomas, 8 Cedar Avenue, Weston-Super-Mare, BS22 8HL
2E0 PHK P. Bentley, The Vauce Farm, Langley-On-Tyne, Hexham, NE47 5NA
2E0 PHL P. Probst, 37 Devonshire Street, Skipton, BD23 2ET
2E0 PHM P. Meerman, 24 Horseshoe Crescent, Burghfield Common, Reading, RG7 3XW
2M0 PHO C. Hunter, 1 North Gate Lodge, Erines, Tarbert, PA29 6YL
2W0 PHP C. Maggs, 15 Stuart Street, Treorchy, CF42 6SN
2E0 PHS P. Sladen, 25 Linden Grove, Beeston, Nottingham, NG9 2AD
2E0 PHU P. Uttley, 55 Dunce Park Close, Elland, HX5 0PF
2E0 PHX P. Hunter, 160 Pembroke Road, Northampton, NN5 7ER
2E0 PIA M. Hill, 56 Moorhouse Avenue, Wakefield, WF2 9QG
2M0 PID T. Hamilton, 57/6 North Street, Bo'ness, EH51 0AE
2M0 PIJ G. Wilson, Flat 7 29 Second Avenue, Clydebank, g813ab
2E0 PIK B. Pike, 19 Cardigan Gardens, Reading, RG1 5QP
2E0 PIO T. Nakagawa, 7 Milton Street, Barrowford, Nelson, BB9 6HE
2E0 PIP P. Marsh, 16 Laurel Close, North Warnborough, Hook, RG29 1BH
2E0 PIT C. Fox, Millstone Cottage, Prior Wath Road, Scarborough, YO13 0AZ
2E0 PIU P. Lounton, 107 Browning Hill Coxhoe, Durham, DH6 4SA
2E0 PIW R. Metcalfe, 33 Midland Terrace, Hellifield, Skipton, BD23 4HJ
2E0 PIX B. Wolff, 7 Church Terrace, Reading, RG1 6AS
2E0 PJC P. Carne, 1 Curlew Close, Letchworth Garden City, SG6 4TG
2E0 PJD P. Dawson, 88 Urmson Road, Wallasey, CH45 7LQ
2E0 PJE P. Elmore, 8 Gray Street, Elsecar, Barnsley, S74 8JR
2E0 PJH P. Holmes, 1 Leonards Place, Bingley., BD16 1AD
2W0 PJJ P. Jones, 23 Pinecroft Avenue, Aberdare, CF44 0HY
2W0 PJM P. Mclaren, 10 Haulfryn, Ruthin, LL15 1HB
2E0 PJN P. Northover, 181 Mullway, Letchworth Garden City, SG6 4BD
2E0 PJR P. Radford, 43 Bells Lane, Nottingham, NG8 6EX
2E0 PJS P. Spilman, 28 Staines Way, Louth, LN11 0DF
2E0 PJT P. Tomlinson, 11 Haynes Close, Clifton, Nottingham, NG11 8JN
2E0 PJY P. Jay, 6 Sportsmans Way, Longwick, HP27 9FZ
2M0 PKA P. Akula, 270 Springhill Road, Aberdeen, AB16 7SL
2E0 PKB P. Beier, 20 Markham Avenue, Armthorpe, Doncaster, DN3 2AZ
2E0 PKF A. Goose, 12 Brown Street, Rainham, Gillingham, ME8 7JN
2E0 PKH P. Holman, Brindle Lodge, Wilney Green Bressingham, Diss, IP22 2AJ
2E0 PKK P. Knight, 73 Bramley Crescent, Southampton, SO19 9LJ
2E0 PKL D. Cadet, 2 Paddockside, Middleton, Ludlow, SY8 3EB
2E0 PKM M. Brookes, 73 Radwinter Road, Saffron Walden, CB11 3HU
2E0 PKR S. Parker, 57 Queen Street, Horncastle, LN9 6BH
2E0 PKS R. Harlow, 28 Dovecliff Crescent, Stretton, Burton-on-Trent, DE13 0JH
2E0 PKU D. Winstanley, 43 Florence Street, St Helens, WA9 5NA
2E0 PLA P. Brown, 11 Booth Crescent, Rossendale, BB4 9BT
2M0 PLB P. Bromley, Broadwood Treovis Upton Cross, Liskeard, G68 9JY
2E0 PLC Dr T. Kyriacou, 54 Sutton Avenue, Silverdale, Newcastle-under-Lyme, ST5 6TB
2E0 PLE G. Dennis, 21 Rydal Crescent, Scarborough, YO12 4JJ
2E0 PLH P. Holmes, 28 Ackworth Drive, Manchester, M23 1LD
2E0 PLJ N. Bull, Eldoret, Castle Street Bampton, Tiverton, EX16 9NS
2E0 PLK A. Jedryka, 71 West Royd Drive, Shipley, BD18 1HL
2E0 PLR P. Tolcher, 15 Langstone Close, Torquay, TQ1 3TX
2E0 PLS J. Walczak, 18 Heathfield, Chippenham, SN15 1BQ
2E0 PLU T. Palmer, Edison House, Bow Street, Great Ellingham, NR17 1JB
2E0 PLV P. Le Vallois, 14 London Row, Arlesey, SG15 6RX
2E0 PLX P. Levy, 43 Conroy Drive, Dawley, Telford, TF4 2RW
2E0 PLY D. Hensman, 130 Clittaford Road, Plymouth, PL6 6DW
2E0 PMA N. Perkins, 39 Ladychapel Road, Abbeymead, Gloucester, GL4 5FQ
2E0 PMB P. Browne, Ham Cottage, Hammingden Lane, Haywards Heath, RH17 6SR
2E0 PMC C. Bowman, 26 Albany Hill, Tunbridge Wells, TN2 3RX
2E0 PMD P. Dowling, 22 Chelkar Way, York, YO30 5ZH
2E0 PME P. Martin, 5 Shropshire Drive, Wilpshire, Blackburn., BB1 9NF
2E0 PMI P. Mccormick, Fieldview, Crown East Lane, Lower Broadheath, Worcester, WR2 6RH
2E0 PML C. Suddell, Lynhurst, Littleworth Lane Partridge Green, Horsham, RH13 8JX
2E0 PMM P. Mather, 11 Odette Court, Bingley, BD16 3QN
2E0 PMO P. Brindley, 70 Atherfield Road, Reigate, RH2 7PS
2E0 PMP P. Punjabi, 62 Cleveland Road, London, W13 8AJ
2M0 PMR A. Graham, 27 Crichton Road, Pathhead, EH37 5RA
2E0 PMU P. Muston, 59 The Pastures Narborough, Leicester, LE19 3DY
2E0 PMV P. Mansfield, 27 Popplechurch Drive, Swindon, SN3 5DE
2E0 PMX A. Marlow, Yeomans Barn, Kingsbridge, TQ7 3BH
2W0 PMZ P. Eckersley, 14 Bronantfer, Gwaun Cae Gurwen, Ammanford, SA18 1EN
2E0 PNA N. Hine, 13 Wilton Crescent, Alderley Edge, SK9 7RE
2E0 PNB P. Bozikis, 336 Higham Hill Road, London, E17 5RG
2E0 PNC G. Billington, 47 Smithy Leisure Park, Cabus Nook Lane, Preston, PR3 1AA
2I0 PND P. Doherty, 69 Fivemile Straight, Draperstown, BT457ht
2E0 PNG P. Green, 34 Drydens Close, Titchmarsh, Kettering, NN14 3DD
2E0 PNK C. Taylor, 212 Plantation Hill, Worksop, S81 0HD
2E0 PNN P. Bowen, 12 Powell Place, Newport, TF10 7BS
2E0 PNP R. Bartha, 6 Chappell Close, Aylesbury, HP19 9QA
2E0 PNR A. Ault, 124 High Street, Aylesbury, HP20 1RB
2E0 PNW P. Winkley, 8 Thelusson Court, Woodfield Road, Radlett, WD7 8JF
2E0 PNZ P. Norman, Four Acres Bungalow Farm, Winwick Gated Road West Haddon, Northampton, NN6 7BH
2E0 POB A. Carden, Hazelgrove, South Allington, Kingsbridge, TQ7 2NB
2I0 POD A. Hunter, 38 Robinson Road, Bangor, BT19 6NJ
2E0 POE Lord J. Williams, 7 Southrop Road Kingsway, Quedgeley, Gloucester, GL2 2HN
2E0 POQ R. Finch, 19B Kiln Road, Newbury, RG14 2LS
2E0 POU P. Daubaris, 32 Chalcombe Road, Abbey Wood, London, SE2 9QS
2E0 POZ P. Stilwell, 3 Ridgeway Cottages, Foxhill, Swindon, SN4 0DU
2E0 PPA P. Ridley, 218 Lichfield Road, Rushall, Walsall, WS4 1SA
2E0 PPB P. Perkins, 52 Ashley Piece, Ramsbury, Marlborough, SN8 2QE
2E0 PPD P. Bengey, 3 Millmead Road, Bath, BA2 3JW
2E0 PPF S. Harding, Barley Hill Cottage, Combe St. Nicholas, Chard, TA20 3HJ
2E0 PPH P. Huband, Flat 26, Tintern House, Selcroft Avenue, Birmingham, B32 2BS
2E0 PPJ P. Joannou, 5 Crowhurst Mead, Godstone, RH9 8BF
2E0 PPK N. Green, 11 Wythburn Way, Rugby, CV21 1PZ
2W0 PPL A. Dighton, 84 Trefelin, Aberdare, CF44 8LF
2E0 PPM I. Barnes, 35 Copley Road, Stanmore, HA7 4PF
2E0 PPO J. Grint, 9 Mountbatten Drive, Leverington, PE13 5AF
2E0 PPR P. Rickwood, 8 Bealeys Avenue, Wolverhampton, WV11 1EG
2I0 PPW J. Macfarlane, 1 Main Street, Uttony, Magheraveely, Enniskillen, BT92 6NB
2E0 PPY S. Evans, 51 St. Georges Road, Dudley, DY2 8EY
2E0 PPZ L. Westwood, 28 Ash Crescent, Kingswinford, DY6 8DJ
2E0 PQA P. Casado Arias, 24 Oldridge Road, London, SW12 8PJ
2E0 PQR P. Nathan, 7 Pitt Drive, Seaford, BN25 3JB
2W0 PQU E. Maher, 25 Ffordd Cadfan, Bridgend, CF31 2DP
2E0 PRA P. Randerson, 7 Roman Crescent, Swindon, SN1 4HH
2E0 PRC P. Craig, 4 Poolside, Burston, Stafford, ST18 0DR
2E0 PRD P. Denham, Flat 1, 10 Prince Alfred Avenue, Skegness, PE25 2UH
2E0 PRG P. Garland, Flat 4, Amante Court, 190 Southwood Road, Hayling Island, PO119QL
2I0 PRL P. Reid, 1 Nettlehill Mews, Lisburn, BT28 3HN
2I0 PRM E. Simpson, 10 Woodview Park, Tandragee, Craigavon, BT62 2DD
2E0 PRO P. Robinson, 16 Bartlett Close, Liverpool, L31 8BZ
2E0 PRS P. Shaw, 32 Hardwick Road East, Worksop, S80 2NT
2E0 PRV P. Bleasdale, 12 Malvern Ave, Padiham, bb127dt
2M0 PSA P. Smith, 13 Newmills Grove, Balerno, EH14 5SY
2E0 PSC P. Croxford, 1 Meteor Close, Bicester, OX26 4YA
2E0 PSD J. Eames, 6 The Oaklands, Cold Meece, Stone, ST15 0QH
2E0 PSH S. Storey, 10 Amble Way, Trimdon Station, TS29 6DZ
2E0 PSK S. Pierce, 117 Victoria Avenue, Hastings, TN35 5BS
2E0 PSM P. Smart, 142 Finch Road, Chipping Sodbury, Bristol, BS37 6JB
2E0 PSN G. James, 28 Redcar Road, Romford, RM3 9PT
2E0 PSO P. Sheffield, 13 St. Winifred Road, Wallasey, CH45 5EJ
2E0 PSP J. Anderson, 57 Chapel Lane, Hadfield, Glossop, SK13 1NX
2E0 PSR Dr P. Shaw, 25 Headcorn Road, Platts Heath, Maidstone, ME17 2NH
2E0 PSW J. Godfrey, 4 Cherry Close, Houghton Conquest, Bedford, MK45 3LQ
2E0 PSZ S. Macdonald, Woodside Cottage, Horton Way, Verwood, BH31 6JJ
2E0 PTA P. Chambers, 257 Kings Acre Road, Hereford, HR4 0SR
2U0 PTB P. Brucher, No 2 Mon Desir, Les Monmains, Vale., Guernsey, GY3 5TZ
2M0 PTE P. Pirie, Willowbank, Kirkton Of Tough, Alford, AB33 8ER
2E0 PTF A. Norton, 9 The Common, West Tytherley, Salisbury, SP5 1NS
2E0 PTG P. Gavin, 11 Campbell Close, Yateley, GU46 6GZ
2E0 PTI B. Parton, 51 Marston Grove, Stoke-on-Trent, ST1 6EF
2E0 PTM P. Millington, 125 Telford Way, High Wycombe, HP13 5SZ
2E0 PTS P. Boultwood, 32 Makepiece Road, Bracknell, RG42 2HJ
2E0 PTY S. Brace, 2 Greenfields Cottages, Brockhampton Estate, Bringsty, WR6 5TB
2E0 PTZ Prof. P. Curtis, Cotswold, Salisbury Road, Abbotts Ann, Andover, SP11 7NX
2E0 PUB A. Fulton, 8 Priest Hill Gardens, Wetherby, LS22 7UD
2E0 PUE A. Hannon, 8 Circular Road West, Liverpool, L11 1AZ
2E0 PUF K. Simkin, 19 Summercourt, Hailsham, BN273AW
2E0 PUG D. Cobbold, 65 St. Olaves Road, Bury St Edmunds, IP32 6RR
2E0 PUH Dr M. Foster, 58 New Terrace, Staverton, Trowbridge, BA14 6NY
2E0 PUL D. Pullen, 5 Weldon Close, Shotton Colliery, Durham, DH6 2YJ
2E0 PUN J. Rideout, 35 Colmead Court, Northampton, NN38QE
2E0 PUS P. Ellis, 40 Grasmere Road Royton, Oldham, OL2 6SR
2E0 PUT S. Croston, 12 Sefton Street, London, SW15 1LZ
2E0 PUZ B. Titmus, 68 Hobart Road, Cambridge, CB1 3PT
2E0 PVK R. Banks, 1 Holly Meadows, Ashford, TN23 3QR
2E0 PVN P. Nicholls, 23 Bishops Gate, Birmingham, B31 4AJ
2E0 PVQ E. Rhodes, The Old Forge, Stoke Gabriel, Totnes, TQ9 6RL
2E0 PVU R. Coomer, 30 Torquay Road, Kingskerswell, Newton Abbot, TQ12 5EZ
2E0 PVW P. Armstrong, 10 Shirdley Avenue, Liverpool, L32 7QG
2E0 PWC P. Castle, 3 Wye Road, Brockworth, Gloucester, GL3 4PP
2E0 PWD J. Shaw, 71 Rowntree Lodge New Earswick, York, YO32 4AA
2E0 PWF C. Cousins, 43 Avon Close, Little Dawley, Telford, TF4 3HP
2E0 PWG P. Green, 15 Dickenson Road, Chesterfield, S41 0RX
2E0 PWI P. Warwick, 24 Chiltern Close, Berinsfield, Wallingford, OX10 7PZ
2E0 PWJ P. Williamson, 22 Earls Road, Shavington, CW2 5ez
2E0 PWK S. Platts, 59 Sea View Road, Drayton, Portsmouth, PO6 1EW
2E0 PWL M. Mynn, The Town House, Parsons Field, St. Mary'S, Hugh Town, TR21 0JJ
2E0 PWM P. Mitchell, 13 Ashorne Close, Matchborough, Redditch, B98 0EY
2W0 PWO P. Oseland, 6 Oaklands Close, Bridgend, CF31 4SJ
2E0 PWP Dr P. Thompson, 3 Floyers Field, West Stafford, Dorchester, DT2 8FJ
2W0 PWR D. Riley-Kydd, 26 Talwrn Road, Wrexham, LL11 3PG
2E0 PXD P. Donaghy, 67 Brockenhurst Way, Bicknacre, Chelmsford, CM3 4XN
2M0 PXH Dr P. Holmes, Maraval, Doune Road, Dunblane, FK15 9AT
2E0 PXI L. Willis, 12 Robartes Road, St. Dennis, St Austell, PL26 8DS
2E0 PXP J. Maudsley, Knight Stainforth Hall, Little Stainforth, Settle, BD24 0DP
2E0 PXW B. Smith, 19 Alexandra Square, Winsford, CW7 2YR
2E0 PXY C. Fox, 45 Park Road, Wivenhoe, Colchester, CO7 9LS
2E0 PXZ Dr C. Fox, 45 Park Road, Wivenhoe, Colchester, CO7 9LS
2E0 PYA W. Allen, 109 Barston Road, Oldbury, B68 0PU
2E0 PYC R. Watson, 8 Bourne Close, Warminster, BA12 9PT
2W0 PYL C. Thorley, Helston, The Mountain, Holyhead, LL65 1YR
2E0 PYM A. Nutt, 77 Exeter Close, Stevenage, SG1 4PW
2E0 PYN L. Scott, 28 Cavendish Place, New Silksworth, Sunderland, SR3 1JW
2E0 PYR P. Robinson, 15 Cornelius Drive, Wirral, CH61 9PY
2E0 PZK P. Kirby, 11 Bembridge Court, Crowthorne, RG45 6BN
2E0 PZM T. Menzies, 4 Meadow Road, Muxton, Telford, TF2 8JH
2E0 RAA R. Keeley, 17 Pembroke Avenue, Wirral, CH46 0TP
2W0 RAD R. Miles, 63 Phillip Street, Mountain Ash, CF45 4BG
2E0 RAF A. Woodrup, 440 New Hall Lane, Preston, PR1 4TA
2E0 RAG A. Green, 18 Harold Avenue, Ashton-In-Makerfield, Wigan, WN4 9UZ
2E0 RAH R. Haynes, 47 Alder Drive, Alderholt, Fordingbridge, SP6 3EP
2E0 RAI R. Trim, 23 Coleman Road, Bournemouth, BH11 8EQ
2E0 RAJ P. Penfold, 2 The Leas, Essenden Road, St Leonards-on-Sea, TN38 0PU
2E0 RAK P. Graham, 28 Newburgh House, Highworth, Swindon, SN6 7DW
2E0 RAL A. Clewes, 20 Linden Drive, Crewe, CW1 6HN
2E0 RAM R. Mason, 73 Edinburgh Road, Chatham, ME4 5BZ
2E0 RAS R. Shippey, 43 Westbury Street, Bradford, BD4 8PB
2E0 RBA Dr R. Bowman, 48 Eliot Drive St. Germans, Saltash, PL12 5NL
2E0 RBC D. Judge, 18 Shepherd Street, Bacup, OL13 8BH
2E0 RBG R. Rigden, 36A Atherston, Bristol, BS30 8YB

2E0 RBH M. Clifford, 32 Tiverton Way, Chessington, KT92QS
2E0 RBI R. Gilbert, 61 Coltstead, New Ash Green, Longfield, DA3 8LN
2E0 RBK B. Kelly, 21 Hogarth Walk, Bristol, BS7 9XS
2E0 RBN V. Steele, 175 Vale Road, Seaford, BN25 3HH
2E0 RBO R. Coleman, 5 Meeting Lane, Burton Latimer, Kettering, NN15 5LS
2E0 RBP R. Cobb, 57 Adams Drive, Willesborough, Ashford, TN24 0FX
2M0 RBQ R. Titmarsh, Caberfeidh Clachan Na Luib, North Uist, HS6 5HD
2E0 RBR C. Dunstan, 67 Knights Way, Mount Ambrose, Redruth, TR15 1PA
2E0 RBU R. Booth, 142 Heath Lane, Earl Shilton, LE9 7PD
2I0 RBV R. Montgomery, 85 Rockfield Heights, Connor, Ballymena, BT42 3GH
2W0 RBW R. Williams, Bardsville, Porthdafarch Road, Holyhead, LL65 2LL
2W0 RBX R. Jones, 25 Whiteway Drive, Gresford, Wrexham, LL12 8HW
2E0 RBZ R. Booker, 6 Kipling Road, Dursley, GL11 4QB
2E0 RCA B. Whiteley, 2A Beechfield Close, Thorpe Willoughby, Selby, YO8 9QJ
2E0 RCB R. Brown, 9 Bayleaf Crescent, Oakwood, Derby, DE21 2UG
2E0 RCC R. Chadwick, 4 Gleneagles Drive, Haydock, St Helens, WA11 0YS
2M0 RCD S. Mckenzie, 0/2 69 Glenkirk Drive, Glasgow, G15 6AU
2E0 RCF R. Goody, 113 Kenneth Road, Basildon, SS13 2BH
2E0 RCI S. Lofthouse, 30 Broughton Grove, Skipton, BD23 1TL
2E0 RCL R. Buckland, 34 Beechwood Drive, Meopham, Gravesend, DA13 0TX
2E0 RCM R. Medland, 5 Bay Tree Cottages, Hospital Road, Bude, EX23 9BP
2E0 RCN R. Northway, 8 Dean Close, Wick, Littlehampton, BN17 7ND
2E0 RCO B. Walker, Pilgrims Gore Road, Burnham-on-Sea, TA8 2HL
2E0 RCR R. Rawson, 30 Harty Road, Haydock, St Helens, WA11 0YY
2E0 RCT M. Russell, 107 Cambridge Road, Hitchin, SG4 0JH
2W0 RCU R. Gripp, 23 Edmond Locard Court, Chepstow, NP16 6FA
2E0 RCV R. Treacher, 93 Elibank Road, London, SE9 1QJ
2E0 RCW C. Wilson, 31 Violet Road, South Woodford, London, E18 1DG
2M0 RCZ A. Conlon, Kilrae, Barrpath, Glasgow, G65 0EX
2E0 RDA M. Salt, 1 Chantry Close, Harrow, HA3 9QZ
2W0 RDD R. Cotterell, 49 Graham Court, Caerphilly, CF83 1RF
2E0 RDE R. Seeley, 2 Church Road, Folkestone, CT20 3LH
2E0 RDF J. Bailey, 22 Wilford Drive, Ely, CB6 1TL
2M0 RDG R. Rogerson, 93 Auchencrieff Road, Locharbriggs, Dumfries, DG1 1UZ
2M0 RDH R. Hutton, 2 Watson Place, Dunfermline, KY12 0DR
2E0 RDI R. Topley, 85 Stuart Road, Aylsham, Norwich, NR11 6HW
2W0 RDJ R. Cole, 14 Inner Loop Road, Beachley, Chepstow, NP16 7HF
2E0 RDN R. Newton, 38 Bedford Road, Denton, Northampton, NN7 1DR
2E0 RDO R. Owen, 23 Bevan Close, Stockton-on-Tees, TS19 8RF
2E0 RDP M. Payne, The Devonhurst, 13 Eastern Esplanade, Broadstairs, CT10 1DR
2E0 RDQ R. Owen, 4 Aldersleigh Drive, Stafford, ST17 4RY
2I0 RDR Rev. R. Rowe, 31 Main Street, Brookeborough, BT94 4EZ
2M0 RDT R. Tourish, 8 Linnpark Gardens, Johnstone, PA5 8LH
2E0 RDU J. Crosby, 65 Bradwell Avenue, Stretford, Manchester, M32 9RT
2E0 RDW R. Wyatt, 297 Weston Road Weston Coyney, Stoke-on-Trent, ST3 6HA
2E0 RDX B. Wilkes, 9 Barnsley Avenue, Conisbrough, Doncaster, DN12 3LB
2W0 RDZ R. Shipman, 1 Lledfair Place, Heol Pentrerhedyn, Machynlleth, SY20 8DL
2E0 REB R. Beardsley, 10 Moreton Close, Church Crookham, Fleet, GU52 8NS
2E0 REC M. Barker, 18 Nickleby Road, Waterlooville, PO8 0RH
2M0 REH R. Hay, Roddach Cottage East, Cummingston, Burghead, Elgin, IV30 5XY
2W0 REJ J. Richardson, 15 Calland Street, Plasmarl, Swansea, SA6 8LE
2E0 REL R. Allen, 44 Overing Avenue, Great Waldingfield, Sudbury, CO10 0RJ
2E0 REM R. Whalley, 188 Astley Street, Astley, Tyldesley, Manchester, M29 7AX
2E0 REN A. Pendle, 17 Norrington Grove, Birmingham, B31 5NY
2E0 RER R. Ridge, Roskellan House, Maenlay, Helston, TR12 7QR
2E0 RES R. Strong, 2 Dean Avenue, Thornbury, Bristol, BS35 1JJ
2E0 REU B. Robinson, 11 Wimbledon Drive, Stockport, SK3 9RZ
2W0 REX C. Moreton, 20 Millbrook Court, Little Mill, Pontypool, NP4 0HT
2E0 REY R. Milton, 66 Hoo Marina Park, Vicarage Lane, Rochester, ME3 9TG
2E0 REZ M. Ricketts, 45 Jesmond Road, Grays, RM16 2QS
2E0 RFD R. Dell, 18 Greenacres, Fulwood, Preston, PR2 7DA
2E0 RFE R. Parkinson, 27 Hopton Avenue, Mirfield, WF14 8JW
2E0 RFF M. Petchey, 74 Avondale, Ellesmere Port, CH65 6RW
2E0 RFG R. Gray, Upper Bisterne Farmhouse, Bisterne, Ringwood, BH24 3BP
2E0 RFH R. Henderson, 9 Green Mead, South Woodham Ferrers, Chelmsford, CM3 5NL
2E0 RFI A. Driscoll, 82 Station Road, Langford, Biggleswade, SG18 9PQ
2E0 RFK Rev. R. Eardley, Bridge Cottage, Martin, Fordingbridge, SP6 3LD
2E0 RFL R. Tongs, 9 Woodland Drive, Winterslow, Salisbury, SP5 1SZ
2E0 RFM R. Corney, Lavender Cottage, Worlds End, Hambledon, Waterlooville, PO7 4QU
2E0 RFN P. Collins, 7 Fitzmaurice Square, Calne, SN11 8NL
2W0 RFT S. Beer, 49 Central Street, Pwllypant, Caerphilly, CF83 2NJ
2E0 RFU J. Byrne, 316 Turncroft Lane, Stockport, SK1 4BP
2W0 RGA R. Anderson, 156 Cockett Road, Cockett, Swansea, SA2 0FQ
2E0 RGB F. Birtwistle, 27 Southwell Road, Wisbech, PE13 3LF
2E0 RGC R. Charbonneau, 22 Redgrave Drive, Crawley, RH10 7WF
2I0 RGD B. Gilliland, 28 Baird Avenue, Donaghcloney, Craigavon, BT66 7LP
2E0 RGH R. Hillman, 28 Alpha Street, Exeter, EX1 2SP
2E0 RGK K. Harley, Care Of: Mr K Harley 9 Amyas Way, Northam, Bideford, EX39 1UT
2W0 RGL G. Lewis, Bryn Cottage, Clydach, Abergavenny, NP7 0LL
2I0 RGM R. Murphy, 40 Stoneypath, Londonderry, BT47 2AF
2E0 RGO M. White, 7 Overthwart Crescent, Worcester, WR4 0JW
2E0 RGR R. Hopwood, 1 Shakespeare Grove, Cannock, WS11 4BQ
2E0 RGS R. Simpson, 22 Kenworthy Road, Stocksbridge, Sheffield, S36 1BZ
2I0 RGT R. Todd, 3 Granville Manor, Kells, Ballymena, BT42 3JE
2M0 RGU D. Small, 30 Caledonia Crescent, Ardrossan, KA22 8LW
2D0 RGW D. Corkish, 10 Vicarage Close, Ballabeg, Castletown, Isle Of Man, IM9 4LQ
2E0 RGY P. Haydon, 101 Blandford Street, Ashton-under-Lyne, OL6 7HG
2E0 RHA P. Costall, 3 Gaynsford Place, Little Canfield, Dunmow, CM6 1WB
2E0 RHC R. Cook, 16 First Square, Stainforth, Doncaster, DN7 5RH
2E0 RHE R. East, 6 Ashley Road, Worcester, WR5 3AY
2E0 RHH R. Hawkins, 3 Fairways Drive, Harrogate, HG2 7ES
2W0 RHI A. Johnston, 44 Cradoc Road, Brecon, LD3 9LH
2E0 RHL R. Landragin, 101 Linden Gardens, Enfield, EN1 4DY
2E0 RHM R. Murphy, 23 Lowndes Close, Stockport, SK2 6DW
2I0 RHN R. Dunwoody, 16 Dernalea Road, Milford, Armagh, BT60 4DZ
2E0 RHO R. Celyn, 207 Beacon One, Beacon Court, Dublin, Ireland, D18AP94
2E0 RHP R. Hanson, 27 Almond Rise, Forest Town, Mansfield, NG19 0NA
2I0 RHQ H. Selfridge, 147 Culcrum Road, Dunloy, Ballymena, BT44 9DT
2E0 RHS S. Hawkins, Forest Edge, Deer Park, Milton Abbas, Blandford Forum, DT11 0AY
2M0 RHT Dr R. Harkness, Fernwood, 4 Cassalands, Dumfries, DG2 7NS
2E0 RIA M. Landragin, 101 Linden Gardens, Enfield, EN1 4DY
2E0 RIH R. Hart, Jays, South Street, Kington Magna, Gillingham, SP8 5ET
2E0 RIL J. Riley, 67 Moss Bank Road, St Helens, WA11 7DE
2E0 RIN R. Horner, 21 Ainsworth Road Little Lever, Bolton, BL3 1RG
2E0 RIO S. Rutt, Granthorpe, Hull Road, Hull, HU11 5RN
2E0 RIQ R. Back, 79 Aspin Close Wellington Home, Somerset, TA21 9EG
2I0 RIR E. Stevenson, 25 Woodlands Manor, Portadown, Craigavon, BT62 4JP
2E0 RIS J. Edwards, 45 Bramshaw Gardens, Bournemouth, BH8 0BT
2E0 RIT R. Royds, 1 Castle Croft, Bolton, BL2 3QT
2E0 RIW R. Cruse Howse, 12 Queens Road Moretonhampstead, Newton Abbot, TQ13 8LP
2E0 RIZ C. Chilton, 5 Braithwaite Ave, Keighley, BD22 6ET
2E0 RJA R. Ashman, 61 Fairfield Road, Burgess Hill, RH15 8NP
2E0 RJB B. Roaf, 8 Weare Close, Portland, DT5 1JP
2E0 RJD J. Dixon, 23 Dee Way, Winsford, CW7 3JB
2E0 RJE R. Earnshaw, 53 Blue Waters Drive, Paignton, TQ4 6JF
2E0 RJH R. Harrison, 18-20 Hall Lane, Kirkburton, Huddersfield, HD8 0QW
2M0 RJJ R. James, The Garret, Alyth, Blairgowrie, PH11 8HQ
2W0 RJL R. Lovesey, 33 Ty Isaf Park Avenue, Risca, Newport, NP11 6NB
2E0 RJM R. Millington, Quaintways, The Avenue, Tarporley, CW6 0BA
2E0 RJP R. Pounder, 65 Stubsmead, Swindon, SN3 3TB
2E0 RJR R. Radley, 20 Thorntondale Drive, Bridlington, YO16 6GW
2E0 RJU A. Foster, 7 Moira Dale, Castle Donington, Derby, DE74 2PG
2E0 RJX R. Miller, 23 Clarendon Road, Sevenoaks, TN13 1EU
2E0 RJY R. Kumar, 1 Stilton Close, Aylesbury, HP19 8JH
2E0 RJZ R. Duthie, 14 Kettles Close, Oakington, Cambridge, CB24 3XA
2E0 RKB A. Turner, 140 Thorndon Avenue, West Horndon, Brentwood, CM13 3TR
2E0 RKD R. Dainton, 1 The Woodlands, Stroud, GL5 1QE
2E0 RKE R. Elger, 3 Alexander Square, Eastleigh, SO50 4BW
2W0 RKF W. Lewis, 53 Leyshon Road, Gwaun Cae Gurwen, Ammanford, SA18 1EN
2E0 RKK E. Whiten, 17 Scott Close, Ashby-de-la-Zouch, LE65 1HT
2W0 RKM M. Beasley, Ffynnon Wen, Bontnewydd, Aberystwyth, SY23 4JJ
2E0 RKO N. Jacobs, 43 The Pines, Yapton, Arundel, BN18 0EG
2E0 RKR G. Hirst, 94 Upper Brighton Road Sompting, Lancing, BN15 0LB
2E0 RKS R. Styles, 4 Coningsby Close, Gainsborough, DN21 1SS
2E0 RKV A. Gallop, 75 Shearmans, Fullers Slade, Milton Keynes, MK11 2BQ
2E0 RKX M. Hemming, 11 Blackberry Way, Evesham, WR11 2AH
2E0 RKZ F. Holmes, 2 Station Lane, Hartlepool, TS25 1AX
2W0 RLB M. Bannister, 45 Queens Drive, Llantwit Fardre, Pontypridd, CF38 2NT
2E0 RLE J. Earle, 24 Brook Vale, Charlton Kings, Cheltenham, GL52 6JD
2E0 RLG J. Berrisford, 18 Trowels Ln, Derby, DE22 3LS
2E0 RLJ R. Johnson, 23 Friars Dene Road, Gateshead, NE10 0DR
2E0 RLP L. Paston, 138A Forest View Road, Manor Park, London, E12 5HX
2E0 RLR R. Hanson, 664 Leeds Road, Huddersfield, HD2 1UB
2E0 RLV R. Oliver, 3 Histon Road Cottenham, Cambridge, CB24 8UF
2E0 RLW R. Wood, 7 Wishart Green, Old Farm Park, Milton Keynes, MK7 8QB
2E0 RLX K. Lewis, 15 Gilbert Scott Way, Kidderminster, DY10 2EZ
2E0 RLY A. Rock, 29 Marley Road, Kingswinford, DY6 8RQ
2E0 RMC Dr R. Campbell, 2 Hesketh Bank, York, YO10 5HH
2M0 RMD R. Nelson, 8 South Street Cambus Fk102Pa, Stirling, FK102PA
2E0 RMF B. Hudson, 19 The Ropery, Whitby, Yo224ey
2M0 RMH R. Hay, 12 Mitchell Brae, Balmedie, Aberdeen, AB23 8PW
2E0 RMI J. Salmon, 25 Helston Road, Chelmsford, CM1 6JF
2E0 RMJ J. Mitchell, 11 Brookside Drive, Oadby, Leicester, LE2 4PB
2I0 RMK A. Mackenzie, 30 Dalriada Gardens, Ballycastle, BT54 6DZ
2E0 RML R. Lunson, 130 Pilgrims Way, Bedford, MK42 9TZ
2E0 RMM J. Hodson, 77 Waltham Road, Woodford Green, IG8 8DW
2M0 RMN R. Nicoll, 15 Redford Walk, Edinburgh, EH13 0AF
2W0 RMO R. Orchard, The Burrows, Spring Gardens, Whitland, SA34 0HL
2M0 RMP J. Mclelland, 24 Hyslop Street, Airdrie, ML6 0ES
2W0 RMR M. Manley, 25 Warwick Road, Brynmawr, Ebbw Vale, NP23 4HW
2E0 RMS R. Stevenson, 97 Queen Street, Crewe, CW1 4AL
2E0 RMT M. Callow, 4 The Firs, Canvey Island, SS8 9TW
2M0 RMV J. Vine, Seaview, Half Of 4 Kilvaxter, Kilmuir, IV51 9YR
2E0 RMW R. Ambler, 21 Whitley Spring Road, Ossett, WF5 0QA
2W0 RMY A. Lewis, 3 Aster View, Port Talbot, SA12 7ED
2E0 RMZ R. Mansfield, 8 Haysoms Drive, Greenham, Thatcham, RG19 8EY
2E0 RNA P. Bannon, 73 London Road, Worcester, WR5 2DU
2E0 RNB R. Ballard, 11 Thurston, Skelmersdale, WN8 8QU
2M0 RND M. Gerrard, 10 Whinhill Gardens, Aberdeen, AB11 7WD
2E0 RNE R. Emery, 67 Victoria Street, Gillingham, ME7 1EW
2E0 RNF M. Faulkner, 6 Stanley Avenue, Queenborough, ME11 5DT
2E0 RNH I. Justice, 4 Saxon Street, Droylsden, Manchester, M43 7FR
2E0 RNI D. Taylor, 10 Church Rd, Northwich, CW9 5NT
2E0 RNJ Dr J. Reynolds, 38 Spring Lane, Hockley Heath, Solihull, B94 6QY
2E0 RNM M. James, 7 Pixey Place, Oxford, OX2 8BB
2E0 RNO J. Franks, 14 The Hamlet, Slades Hill, Templecombe, BA8 0HJ
2E0 RNP R. Penaluna, 113 Brocklesby Road, Scunthorpe, DN17 2LW
2E0 RNR J. Shettler, 504 Leeds Road, Huddersfield, HD2 1YW
2E0 RNS P. Rainey, 27 School Road, Silver End, Witham, CM8 3RZ
2E0 RNT R. Taylor, 22 Shakespeare Court, Chaucer Way, Hoddesdon, EN11 9QS
2E0 RNU M. Nutt, 110 Birkinstyle Lane, Shirland, Alfreton, DE55 6BT
2E0 RNW N. Rapson, 15 School Close, Bampton, Tiverton, EX16 9NN
2I0 ROC A. Mccann, 6 Bowens Meadow, Lurgan, Craigavon, BT66 7UT
2E0 ROD R. Burton, 23 Freston, Paston, Peterborough, PE4 7EN
2E0 ROI W. Chorlton, 25 Ash Grove, Orrell, Wigan, WN5 8NG
2E0 ROJ A. Hodson, 22 Walmley Ash Road, Sutton Coldfield, B76 1HY
2E0 ROM R. Pasika, 192 Longfield Lane, Cheshunt, Waltham Cross, EN7 6AQ
2E0 ROO P. Purnell, 89 St. Aubyns, Goldsithney, Penzance, TR20 9LS
2E0 ROP P. Bolton, 1 Acorn Rise, Hollesley, Woodbridge, IP12 3JT
2E0 ROR A. Threlfall, 148 Windlehurst Road High Lane, Stockport, SK6 8AG
2E0 ROS C. Ross, 27 The Meadows, Skegness, PE25 2JA
2E0 ROV R. Van-Der-Wijst, 6 Willow Street, Romford, RM7 7LJ

2E0 ROX M. Johnson, 27 Tyndall Walk, Birmingham, B32 3UN
2E0 ROY R. Trzeciak-Hicks, 24 Wolston Meadow, Middleton, Milton Keynes, MK10 9AY
2E0 RPA R. Ashley, 15 Wimbourne Drive, Gillingham, ME8 9EN
2E0 RPC R. Cockayne, 20 The Shrubbery, Rugeley, WS15 1JJ
2E0 RPD I. Handley, Rosedale, Chapman Street, Market Rasen, LN8 3DS
2E0 RPE Dr A. Wallman, Oakwood, 30 Elmsway, Bramhall, Stockport, SK7 2AE
2E0 RPF R. Fullagar, 6 Locke Way, Stafford, ST16 3RE
2E0 RPH R. Hann, Flat 4, Broadview Haye Down, Tavistock, PL19 0NN
2E0 RPL R. Lyons, 103A Oxney Road, Peterborough, PE1 5NG
2I0 RPM R. Gault, 7 Gardenmore Place, Larne, BT40 1SE
2E0 RPN P. Whiterod, 9 Brewhouse Lane, Soham, Ely, CB7 5JD
2E0 RPO W. Donnelly, 9 Old Laundry Mews, Laundry Lane, Ingleton, Carnforth, LA6 3GH
2E0 RPR M. Roper, 13 St. Cuthbert Street, Worksop, S80 2HN
2E0 RPU C. Pallett, 110 St. Leonards View, Polesworth, Tamworth, B78 1JY
2E0 RPW R. Webb, Norbury, Terrills Lane, Tenbury Wells, WR15 8DD
2E0 RPY H. Kennedy, 11 Green Road, High Wycombe, HP13 5BD
2W0 RQC R. Chegwin, 17 Cyncoed Crescent, Cardiff, CF23 6SW
2E0 RQD S. Cownley, 5 Pumphouse Lane, East Cowes, PO32 6FJ
2E0 RQK B. Dare, 1 St. Johns Villas, Sivell Place, Exeter, EX2 5ES
2E0 RQN J. Foster, 23 High Street, Cumnor, Oxford, OX2 9PE
2E0 RQQ Dr R. Baldwin, Red Barn, Elder Lane, Grimston, King's Lynn, PE32 1BJ
2E0 RRA M. Hall, Ward 2 Berth 15, Royal Hospital, Chelsea Royal Hospital Road, London, SW3 4SR
2E0 RRC R. Wilson, 84 Sir Thomas Whites Road, Coventry, CV5 8DR
2I0 RRE R. Rantin, 8A Buchanans Road, Newry, BT35 6NS
2E0 RRF D. White, 2 Birchwood Avenue, Breaston, Derby, DE72 3AQ
2E0 RRL K. Woodhams, 83 Langdale Place, Newton Aycliffe, DL5 7DY
2M0 RRO C. Duncan, 131 Croftend Avenue, Glasgow, G44 5PF
2W0 RRY P. Smith, 19 Grandison Street, Neath, SA11 2PG
2E0 RSB R. Blandford, 60 Benomley Road, Almondbury, Huddersfield, HD5 8LS
2E0 RSC J. Lawrence, 42 Alma Street, Weston-Super-Mare, BS23 1RD
2E0 RSD S. Ray, 28 Stenbury View, Wroxall, Ventnor, PO38 3DB
2E0 RSG S. Cowgill, 17 Tooley Street, Boston, PE21 6DP
2E0 RSH Dr R. Hodgkinson, 39 Oxford Road, Carlton-In-Lindrick, Worksop, S81 9BD
2E0 RSI R. Simms, 3 The Byeway, London, SW14 7NL
2E0 RSM M. Shortreed, 13 Marshfield Road, Settle, BD24 9DA
2W0 RSV G. Jones, 31 Liverpool Road, Buckley, CH7 3LH
2M0 RTA R. Turpie, 11 Ashkirk Place, Dundee, DD4 0TN
2E0 RTC C. Ring, 29 Shelley Close, Newport Pagnell, MK16 8JB
2M0 RTD D. Robertson, 17 Keswick Drive, Hamilton, ML3 7HN
2E0 RTE P. Ballington, 7 Links Close, Sinfin, Derby, DE24 9PF
2E0 RTG I. Sharpe, 4 Low Dowfold, Crook, DL15 9AE
2E0 RTH R. Hunter, Poplars, March Road Guyhirn, Wisbech, PE13 4DA
2M0 RTI I. Swanston, 6 Roberts Grove, Galashiels, TD1 2BJ
2W0 RTJ R. Johns, 39 Tyla Coch, Llanharry, Pontyclun, CF72 9LT
2E0 RTK Dr R. Tempo, 35 Warminster Road, Bath, BA2 6XG
2E0 RTM S. Key, 159 Launcelot Road, Bromley, BR1 5EA
2E0 RTN M. Symmonds, 24 Woodville Grove, Stockport, SK5 7HU
2M0 RTO M. Smith, 2 Richmond Court, Dundee, DD2 1BF
2E0 RTP R. Paster, 8 Rachaels Lake View, Warfield, Bracknell, RG42 3XU
2E0 RTU N. Sargent, 21 St. Michaels Road, Claverdon, Warwick, CV35 8NT
2E0 RTW R. Whitfield, 47 Denchworth Road, Wantage, OX12 9AY
2E0 RTY A. Loukes, 14 Batchwood View, St Albans, AL3 5TD
2W0 RTZ T. Rowlands, Caer Gog Farm, Bodffordd, LL77 7BX
2W0 RUA E. Jones, 8 Ty Newydd Court, High Street, Ruabon, Wrexham, LL14 6BF
2E0 RUD R. Rudd, 11 Woodlands Way, Lepton, Huddersfield, HD8 0JA
2E0 RUG J. Fry, Deal Cottage, Ipswich Road, Long Stratton, Norwich, NR15 2TF
2E0 RUI R. Lima Matos, 12C Crouch End Hill, London, N8 8AA
2M0 RUP R. Hart, Rosalis, Piperhill, Nairn, IV12 5SD
2E0 RUS R. Garland, 113 The Drive, Feltham, TW14 0AH
2E0 RUT R. Rutland, 53 Downs Avenue, Whitstable, CT5 1RR
2E0 RUU N. Cooksley, 10 Honeywick Close, Bristol, BS3 5ND
2E0 RUX A. Rowland, 5, Gwinear Downs, Leedstown, Hayle, TR27 6DJ
2E0 RUZ R. Brierley, 39 Hatfield Road, Alvaston, Derby, DE24 0BU
2E0 RVE R. Harvey, 1 Bickley Moss, Whitchurch, SY13 4JF
2M0 RVF S. Wilson, 8 Robertson Drive, Bellshill, ML4 2EQ
2I0 RVH T. Nelson, 25 Monaghan Road Annashanco Rosslea, Belfast, bt927pt
2E0 RVI R. Miranda, 33 Hill Rise, Luton, LU3 3EA
2M0 RVM A. Johnstone, 10 Earlspark Avenue, Bieldside, Aberdeen, AB15 9BU
2I0 RVT R. Todd, 58 Kilrea Road, Portglenone, Ballymena, BT44 8JB
2E0 RVV D. Kemp, 2 Darwin Close, Elston, Newark, NG23 5PQ
2E0 RWB R. Whalley, 65 Stanley Street, Nelson Lancashire, BB9 7ET
2E0 RWC R. Clarke, 110 Fairmile Road, Halesowen, B63 3QD
2E0 RWE W. Eustace, 34 Hertford Avenue, London, SW14 8EQ
2W0 RWF T. Mitchell, 9 Rhiw Grange, Colwyn Bay, LL29 7TT
2I0 RWG R. Gilmore, 11 Abbots Gardens, Newtownabbey, Belfast, BT379QZ
2M0 RWH R. Humphrey, 11 Pearce Grove, Edinburgh, EH12 8SP
2E0 RWJ R. Jones, 31 Mary Mead, Warfield, Bracknell, RG42 3SZ
2E0 RWK R. Westrup, 31 Robert Balding Road, Dersingham, King's Lynn, PE31 6UR
2E0 RWN R. Nock, 83 Coles Lane, West Bromwich, B71 2QW
2E0 RWP R. Penny, 93 Mirfield Grove, Hull, HU9 4QR
2E0 RWR W. Rees, 69 Pewley Way, Guildford, gu13pz
2E0 RWT A. White, 3 Ipswich Place, Thornton-Cleveleys, FY5 1SP
2E0 RWW M. White, 130 Main Street Walton, Street, BA16 9QX
2E0 RWX R. Williamson, 23 Harrowdyke, Barton-upon-Humber, DN18 5LN
2E0 RXC R. Codrai, 27 Howard Avenue West Wittering, Chichester, PO20 8EX
2E0 RXG M. Robson, 270 Calder Road, Lincoln, LN5 9TL
2E0 RXR R. Rich, 54 Paddocks Way, Ferndown, BH22 9FW
2E0 RXT M. Harrison, 2 Rosemount Court, Holly Bank Road, York, YO24 4EG
2E0 RXW R. Wells, The Turkey House, Park Farm, Tudeley, Tonbridge, TN11 0NL
2E0 RXX G. Acton, 39 Craig Road, Macclesfield, SK11 7YH
2E0 RYD P. Ryder, 5 Weyhouse Close, Stroud, GL5 2JJ
2M0 RYG R. Young, 22 Station Road Armadale, Bathgate, EH48 3LN
2M0 RYL R. Haynes, 29 Invercauld Road, Aberdeen, AB16 5RP
2E0 RYN D. Renshaw, 25 Ashley Road, Worksop, S81 7JS
2E0 RYP C. Barrett, 1 Mead Road, Padgate, Warrington, WA1 3TN
2E0 RYR R. Brown, 2 Westgate, Leominster, HR68SA
2E0 RYS R. Sayre, 8 Lorne Road, Richmond, TW10 6DS
2E0 RYX Dr I. Van Der Linde, 77 Port Vale, Hertford, SG14 3AF
2J0 RZD R. Luscombe, Flat 3, 1 Rouge Bouillon, St. Helier, Jersey, JE2 3ZA
2M0 RZE G. Smith, 40 Pirleyhill Drive, Shieldhill, Falkirk, FK1 2EA
2W0 RZL R. Higgins, 44 Maeshyfryd Road, Holyhead, LL65 2AL
2E0 RZM G. Kingstone, 17 Ullswater Drive, Leighton Buzzard, LU7 2QR
2E0 RZS R. Stone, 63 Sands Lane, Lowestoft, NR32 3ER
2I0 RZT J. Thompson, 119 Rathkyle, Antrim, BT41 1LN
2E0 RZX B. Forrest, 32 Idonia Road, Wolverhampton, WV6 7NQ
2E0 SAA S. Garrett, 44 Wardle Crescent, Leek, ST13 5PW
2E0 SAF S. Finch, 25 Bluebell Avenue, Wigan, WN6 8NS
2I0 SAI S. Barnes, 191 Marlacoo Road, Portadown, Craigavon, BT62 3TD
2W0 SAK S. Poyser, Glandwr Snowdon Street, Porthmadog, LL49 9DF
2E0 SAN S. Neachell, 59 Gilmour Crescent, Worcester, WR3 7PJ
2E0 SAT M. Ward, Flat 1, The Old Chapel, Chapel Street, Holsworthy, EX22 6AY
2E0 SAW S. Williams, Flat 35, Winterton House, London, E1 2QR
2M0 SAX G. Sproul, 25 Mulben Place, Glasgow, G53 7UP
2E0 SAY M. Baker, 20 Centurion Rise, Hastings, TN34 2UL
2E0 SAZ S. Waller, 17 Vere Road, Peterborough, PE1 3DZ
2E0 SBB A. Southwell, 56 Lambrook Road, Taunton, TA1 2AF
2E0 SBC K. Smith, 7 Rosebery Avenue, Morecambe, LA4 5RU
2E0 SBD S. Shailes, 9 Ingham Street, Padiham, Burnley, BB12 8DR
2E0 SBH S. Ray, 18 Crescent Way, Cholsey, Wallingford, OX10 9NE
2E0 SBJ C. Didcott, 9 Maid Marion Avenue, Selston, Nottingham, NG16 6QH
2E0 SBK S. Harvey, 40 Littlemoor Lane, Newton, Alfreton, DE55 5TY
2E0 SBL S. York, 1 The Cottage, Dogdyke Bank, Lincoln, LN4 4JQ
2E0 SBM M. Burrows, 4 Melton Street, Earl Shilton, Leicester, LE9 7FP
2E0 SBN H. Buckley, 10 Lower Hey Lane, Mossley, Ashton-under-Lyne, OL5 9DE
2E0 SBO R. Raine, 110 Stirling Avenue, Jarrow, NE32 4HS
2M0 SBP K. Verrall, 7 Roshven View, Arisaig, PH39 4NX
2E0 SBS G. Hamilton, 11 The Spinney, West Lavington, Devizes, SN10 4HP
2E0 SBW S. Jeffery, 79 Greenbank Road, Watford, WD17 4FJ
2E0 SBX S. Elliott, 74 Preston Avenue, Alfreton, DE55 7JX
2E0 SBZ D. Smith, 105 Princes Street, Dunstable, LU6 3AS
2E0 SCA S. Waudby, 31Bloomfield Avenue Hu5 5Nh, Hull, HU5 5NH
2W0 SCB S. Elias, 20 Attlee Way, Cefn Golau, Tredegar, NP22 3TA
2E0 SCC D. Riman, 22 Princess Road, Hinckley, LE10 1EB
2E0 SCD J. Lord, 5 Langworthy Avenue Little Hulton, Manchester, M38 9GQ
2E0 SCE D. Hagan, 8 Charles Close, Westcliff-on-Sea, SS0 0EU
2E0 SCF S. Faulkner, Mount Pleasant, Elkstones, Buxton, SK17 0LU
2M0 SCG S. Greenland, Flat 12, Weymouth Court, 201 Weymouth Drive, Glasgow, G12 0ER
2E0 SCH P. Hayes, 4 London Road, Roade, Northampton, NN7 2NL
2E0 SCJ C. Short, 35 Whitley Willows Lepton, Huddersfield, HD8 0GD
2E0 SCK M. Stillman, 58 Highfield Road, Bognor Regis, PO22 8PH
2W0 SCL A. Davies, 25 Llanfair Road, Tonypandy, CF40 1TA
2E0 SCM J. Stephenson, 4 Carlow Drive West Sleekburn, Choppington, NE62 5UT
2E0 SCN J. Porter, Flat 2, 14 Trafalgar Square, Scarborough, YO12 7PY
2E0 SCO J. Hepburn, 32 Green Croft, Ashington, NE63 8EF
2W0 SCP S. Peel, 28 Dan Yr Allt, Llanelli, SA14 8AT
2E0 SCQ S. Chandler, 7 Clinton Road, Redruth, TR15 2LL
2E0 SCR C. Petrie, 14 Rotherfield Avenue, Eastbourne, BN23 8JQ
2E0 SCS A. Mcallister-Bowditch, 24 Morse Close, Chippenham, SN15 3FY
2E0 SCV S. Shaw, 3 Wellington Terrace, Littleborough, OL15 9DA
2E0 SCW S. Whittaker, 25 Cleveleys Road, Blackburn, BB2 3JS
2E0 SCX S. Hassall, 21 Bridgnorth Grove, Chesterton, Newcastle under Lyme, ST5 7QP
2E0 SDA A. Sanderson, 65 Holm Flatt Street, Parkgate, Rotherham, S62 6HJ
2E0 SDC S. Cornwell, 16 Chesterfield Way, Eynesbury, St Neots, PE19 2JY
2E0 SDD S. David, 34 Hardwicke Walk, Kings Heath, Birmingham, B14 5XX
2E0 SDJ D. Wild, Flat 10, Merchants Place, Rendezvous Street, Folkestone, CT20 1ET
2E0 SDK S. Allington, 137 Marshall Lane, Northwich, CW8 1LA
2E0 SDM C. Reid, 28 Albion Road, London, N16 9PH
2W0 SDO M. Johns, 151 Somerset Street, Abertillery, NP13 1DR
2E0 SDP R. De Vries, Corner Cottage, Hillcrest Close, Sturminster Newton, DT10 2DL
2E0 SDQ P. Weaver, 1 Madeley Street, Newcastle, ST5 6LS
2E0 SDR A. Lane, 36 The Crescent, Eastbourne Estate, Coleford, GL16 8BS
2E0 SDT S. Theaker, 10 Grange Fields Mount, Leeds, LS10 4QN
2E0 SDV J. Williams, 41 Overton Lane, Hammerwich, Burntwood, WS7 0LQ
2E0 SDW S. Whitehead, 55 Crombie Road, Sidcup, DA15 8AT
2E0 SDY J. Popple, 11, Chapel Close, Waterbeach, Cambridge, CB25 9JW
2E0 SDZ E. Ayre, 1 Spring Gardens, Broadmayne, Dorchester, DT2 8PP
2E0 SEA M. Bown, 47 Ullswater Crescent, Weymouth, DT3 5HF
2E0 SEB S. Gordon, 6 Aspinall Grove, Hailsham, BN27 3GP
2I0 SEC S. Carlin, 9 Mullandra Park, Kilcoo, Newry, BT34 5LS
2E0 SEE A. Seedig, 21 Ambleside Close, Mytchett, GU16 6DG
2M0 SEF S. Fleming, 20 Park Road, Invergowrie, Dundee, DD2 5AH
2E0 SEJ S. Johnson, 2 North Square, Edlington, Doncaster, DN12 1ED
2I0 SEK R. Thomson, 1 Litchfield Park, Coleraine, BT51 3TN
2E0 SEO S. Bates, 6 Foxdell, Northwood, HA6 2BU
2E0 SEP A. Howard, 24 Ladybower Lane, Poulton-le-Fylde, FY6 7FY
2E0 SER M. Edmonds, 20 Tomline Road, Ipswich, IP3 8BZ
2E0 SES Dr M. Cianni, 121 Springfield Park Avenue, Chelmsford, CM2 6EW
2E0 SET P. Setter, 199 Southbourne Grove, Westcliff-on-Sea, SS0 0AN
2E0 SEW G. Ferguson, 31 Barton Court Road, New Milton, BH25 6NW
2E0 SEY J. Averill, 14 Shannon Close, Saltney, Chester, CH4 8PJ
2W0 SEZ S. Ezard, 59 Station Farm, Croesyceiliog, Cwmbran, NP44 2JW
2I0 SFA N. Jenkinson, 35 Scarvagh Heights, Scarva, Craigavon, BT63 6LY
2W0 SFB P. Latham, 20 Kenyon Avenue, Wrexham, LL11 2ST
2E0 SFC R. Bates, 61 Park View, Crowmarsh Gifford, Wallingford, OX10 8BN
2E0 SFP S. Finlayson, 41 Low Catton Road, Stamford Bridge, York, YO41 1DZ
2E0 SFQ Dr S. Pearce, 15 Hillfield Court Road, Gloucester, GL1 3QS
2E0 SFS M. Heenan, 15 Woodacre Green, Bardsey, Leeds, LS17 9AB
2E0 SFT M. Shelton, 92 Timberley Drive, Grimsby, DN37 9QZ
2E0 SFV Dr S. Favell, Lodge Farm, Moor Lane, Reepham, Lincoln, LN3 4EE
2E0 SFZ S. Driscoll, 21 Hall Drive, Middleton, Morecambe, LA3 3LF
2E0 SGG S. Mallows, 74 St Alban'S Close, Gillingham, me7 1tx

UK Callsigns

2E0 SGH S. Holt, 14 Fir Street, Cadishead, Manchester, M44 5AU
2E0 SGI A. Parker, 43 Springfield Avenue, Shirebrook, Mansfield, NG20 8LF
2I0 SGK J. Hunter, 160 Ballynure Road, Ballyclare, BT39 9AJ
2E0 SGL P. Hilton, Shankly Cottage 161 Highgate, Jennings Yard, Kendal, LA9 4EN
2I0 SGM S. Murray, 117, Knockview Drive, Tandragee, BT622BL
2M0 SGO S. O'Neill, Flat 5-8 460 Sauchiehall Street, Glasgow, G2 3JW
2M0 SGQ S. Gill, 5 Ramornie Place, Kingskettle, Cupar, KY15 7PT
2E0 SGS S. Smith, 2 Burnsall Avenue, Blackpool, FY3 7LQ
2E0 SGY S. Gregory, 9 Croftlands Road, Wythenshawe, M22 9YE
2E0 SHA J. Ridley, 73 The Markhams, New Ollerton, Newark, NG22 9QY
2E0 SHB S. Burrows, 78A Coronation Road, Earl Shilton, Leicester, LE9 7HJ
2E0 SHG D. Grindrod, 6 Croft Way, Market Drayton, TF9 3UB
2E0 SHH R. Devos, 76 North Parade, Sleaford, NG348AW
2E0 SHK S. Helm, 10 St. Annes Avenue, Middlewich, CW10 0AE
2I0 SHM S. Montgomery, 2 Woodland Gardens, Lisburn, BT27 4PL
2E0 SHN S. Pallister, 32 Greensnook Lane, Bacup, OL13 9DQ
2E0 SHO S. Hodgkiss, 14 Dales Close, Biddulph Moor, Stoke-on-Trent, ST8 7LZ
2E0 SHP A. Sharp, 30 Burbidge Close, Calcot, Reading, RG31 7ZU
2E0 SHR R. Shepherd, 19 Elford Avenue, Newcastle upon Tyne, NE13 9AP
2E0 SHV A. Sidhu-Brar, White Gates, Main Road, Northampton, NN7 3NA
2E0 SHW A. Dransfield, 34 Ernest Kirwood Close, Hull, hu5 5xx
2W0 SHX J. Nicholas, Reservoir House, St. Lythan'S, Wenvoe, CF5 6BQ
2E0 SHY W. Shelley, 91 Canterbury House, Stratfield Road, Borehamwood, WD6 1NT
2E0 SHZ S. Lewis, 58 Ocean Close, Fareham, PO15 6QP
2E0 SIA S. Skirving, 1 Hallington Close, Bolton, BL3 6YH
2E0 SIB D. Sibley, 85 Brick Crescent, Stewartby, MK43 9GG
2E0 SID S. Frampton, 20 Winslow Close, Boldon Colliery, NE35 9LR
2E0 SIF S. Frost, 4 Banister Way, Shipston-on-Stour, CV36 4JU
2E0 SIH S. Hammond, Ellsworth, Thrigby Road, Filby, Great Yarmouth, NR29 3HJ
2E0 SIJ J. Simkins, 37 St. Andrews Meadow, Harlow, CM18 6BL
2E0 SIK N. Butler, 75 Rutland Street, Derby, DE23 8PR
2E0 SIM S. Matley, 67 Alexandra Road, Chandlers Ford, SO53 2BP
2E0 SIS A. Tanseli, 157 Warwick Road, Rayleigh, SS6 8SG
2E0 SIX S. Hannah, 4 Station Road, Minsterley, Shrewsbury, SY5 0BG
2E0 SIY S. Shaul, 31 Chatterton Avenue Ermine West, Lincoln, LN1 3SZ
2E0 SIZ S. Riley, 51 Kenilworth Avenue, Reading, RG30 3DL
2E0 SJA A. Spears, 26 Aragon Close, Ashford, TN23 5DF
2E0 SJE S. Etty, 31 Mill Drive Leven, Beverley, HU17 5NR
2E0 SJF S. Farnell, 16 Lily Way, Lowestoft, NR33 8NN
2E0 SJG S. Goodridge, Trelane, Pelynt, Looe, PL13 2LF
2E0 SJH S. Humphreys, Flat 1, Winslow Court 100 Fordwych Road, London, NW2 3NN
2E0 SJI S. Ibbotson, 17 Marley Combe Road, Haslemere, GU27 3SN
2E0 SJK S. Kendrick, 29 Waterside Silsden, Keighley, BD20 0LQ
2E0 SJM S. Moore, 33 Church Street, Heavitree, Exeter, EX2 5EP
2E0 SJN S. Nash, 3 Brightside, Waterlooville, PO7 7BA
2E0 SJP S. Parker, 36 Eton Close, Lincoln, LN6 0YF
2E0 SJQ S. Carpenter, Field View, Old Lyndhurst Road, Southampton, SO40 2NL
2E0 SJR S. Roberts, 7 Alberta Grove, Prescot, L34 1PX
2E0 SJS S. Spencer, 55 Witton Lane, West Bromwich, B71 2AA
2W0 SJT S. Tweddle, 3 Bron Ffinan, Pentraeth, LL75 8UT
2I0 SJV D. Parkinson, 16 Beechwood Gardens, Moira, Moira, BT67 0LB
2E0 SKA G. Brown, 3 Willow Lane, Goostrey, Crewe, CW4 8PP
2E0 SKE M. Swan, 35 Colston Close, Plymouth, PL6 6AY
2W0 SKG W. Lloyd, 29 Duffryn Street, Mountain Ash, CF45 3NU
2E0 SKI A. Skarzynski, 1 River View Moorings, Bridge Road, Stoke Ferry, King's Lynn, PE33 9TS
2E0 SKK A. Greig, 3 Fir Grange Avenue, Weybridge, KT13 9AR
2E0 SKL C. Hare, 25 Southend Place, Sheffield, S2 5FQ
2E0 SKP R. Ibbotson, 33 St. Peters Avenue, Caversham, Reading, RG4 7DH
2M0 SKR C. Aitken, Windybraes, Upper Gills, Canisbay, Caithness, Canisbay, KW1 4YB
2E0 SKZ J. Parfitt, 5 Sheridan Road, Frimley, Camberley, GU16 7DU
2W0 SLD D. Dash, 36 Rockvilla Close, Varteg, Pontypool, NP4 7QF
2E0 SLG M. Jordan-Reed, Pump Corner, High Street Green, Sible Hedingham, Halstead, CO9 3LG
2E0 SLH S. Hill, 108 Hitchin Close, Romford, RM3 7EQ
2E0 SLJ S. Legg, 98 Shenstone Valley Road, Halesowen, B62 9TF
2E0 SLK J. Blezard, 10 North Row, Barrow-in-Furness, LA13 0HE
2E0 SLM J. Stringer, 31 Pipit Lane, Birchwood, Warrington, WA3 6NY
2M0 SLN R. Nixon, 7 Shellacres Farm Cottage, Cornhill-on-Tweed, TD12 4XB
2E0 SLO D. Rolph, 17 Moorlands Park Ashby Road, Sinope, Le67 3bd
2W0 SLP S. Williams, 63 Trem Eryri, Llanfairpwllgwyngyll, LL61 5JF
2E0 SLR B. Robins, 99 Uplands Road, Dudley, DY2 8BB
2E0 SLS K. Holloway, 6 Britons Lane Close, Beeston Regis, Sheringham, NR26 8SH
2E0 SLT S. Thompson, 80 Mountain Road, Dewsbury, WF12 0BP
2E0 SLV B. Gimbert, 49 Ratcliffe Road, Loughborough, LE11 1LF
2E0 SMA S. Harcourt, 71 Ingleby Road, Long Eaton, Nottingham, NG10 3DG
2I0 SMD Dr S. Davey, 16 Maritime Drive, Carrickfergus, BT38 8GQ
2E0 SMF C. Smith, 21 Mill House Drive, Cheltenham, GL50 4RG
2E0 SMK S. Kendrick, 103A Latimer Street, Liverpool, L5 2RF
2E0 SML S. Weightman, 131 Leeds Road, Birstall, Batley, WF17 0JZ
2M0 SMN S. Nicoll, 15 Redford Walk, Edinburgh, EH13 0AF
2E0 SMO S. O'Neill, 1 Avenue Cottages, Winchester Road, Fareham, PO17 5EX
2E0 SMS S. Stratford, 23 The Fairway, Banbury, OX16 0RR
2E0 SMX S. Hogg, 57 The Grange, Burton-on-Trent, DE14 2EX
2I0 SMY S. Dallas, 101 Coagh Road Stewartstown, Dungannon, BT71 5JL
2E0 SMZ S. Edwards, 30 Morrison Road, Tipton, DY4 7PU
2E0 SNA P. Harvey, 35 Isaac Street, Liverpool, L8 4TH
2E0 SNB S. Bodsworth, 51 Berry Way, Andover, SP10 3RZ
2E0 SNC M. Hemmings, 62 Spencer Way, Stevenage, SG2 8GD
2E0 SND A. Williamson, 25 Manor Road, Rugby, CV21 2SZ
2E0 SNE M. Bryan, 13 Elmwood Avenue, Sunderland, SR5 5AW
2I0 SNG S. Gilmour, 14G Malcolm Road, Lurgan, Craigavon, BT66 8DF
2E0 SNJ S. Worger, 6 Glendale Terrace, Mornington Road, Whitehill Bordon, GU35 9AJ
2E0 SNM S. Hindmarsh, 7 George Street Murton, Murton, SR7 9BN
2E0 SNP S. Powell, 3 Tadgedale Avenue, Loggerheads, Market Drayton, TF9 4DD
2E0 SNS S. Button, 6 Farfield, Retford, DN22 7TL
2E0 SNU E. Collins, 38 Meynell Road, Sheffield, S5 8GN
2W0 SNW S. Williams, Hillsboro Aberkenfig, Bridgend, CF32 0EW
2E0 SNZ M. Ross, 11 Queens Place, Otley, LS21 3HY
2M0 SOE R. Baxter, 7 Clegg Gardens, Troon, KA10 7GZ
2E0 SOF W. Soffe, 96 Urban Road, Doncaster, DN4 0EP
2E0 SOJ J. Sykes, 79 Owler Park Road, Ilkley, LS29 0BG
2M0 SOP A. Gordon, 26 East Millicent Avenue, Golspie, KW10 6TL
2E0 SOR S. Orchard, 30 Wilkes Court, Ipswich, IP5 2EQ
2E0 SOT S. Gregory, 11 Ribblesdale Avenue, Congleton, CW12 2BS
2M0 SOV D. Kinnear, 55 Wilmington Drive, Glenrothes, KY7 6US
2E0 SOX R. Cook, 46 Wheatsheaf Road, Tividale, Oldbury, B69 1SW
2E0 SOZ R. Mather, 4 Vimy Road, Wednesbury, WS10 9BQ
2E0 SPA S. Etheridge, 50 Pond Road, Horsford, Norwich, NR10 3SW
2E0 SPB D. Evans, 21 Quilter Close, Bilston, WV14 9AX
2E0 SPD S. Denman, 12 Dyke Vale Road, Sheffield, S12 4ER
2E0 SPE G. Bradley, Greentree Cottage, Town End, Broadclyst, Exeter, EX5 3HW
2E0 SPG M. Hennessey, 3 Northgate Cottage, Falmer Road, Rottingdean, Brighton, BN2 7DT
2E0 SPH D. Janowicz, 20 Salisbury Road, St Leonards-on-Sea, TN37 6RX
2M0 SPL S. Ling, Leadburnlea Leadburn, West Linton, EH46 7BE
2E0 SPM S. Morris, 23 De Courtenai Close, Bournemouth, BH11 9PG
2E0 SPN J. Daniels, 27 Hammerwater Drive, Warsop, Mansfield, NG20 0DJ
2E0 SPR S. Randall, 23 Onslow Road, Plymouth, PL2 3QG
2E0 SPS J. Bogdaniec, 3 Cavalry Chase, Okehampton, EX20 1GR
2E0 SPT K. Tremain, 26 Longbeech Park, Canterbury Road, Ashford, TN27 0HA
2E0 SPU P. Saunders, 62 Parkfield Avenue, Eastbourne, BN22 9SF
2E0 SPW S. Woodmore, 66 Imperial Way, Chislehurst, BR7 6JR
2E0 SPZ R. Hinde, 3 Baunhill Close, Northampton, NN3 3EQ
2E0 SQB S. Stevens, 40 Heath Road, Exeter, EX2 5JX
2E0 SQK P. Henry, 22 Huddleston Close, Wirral, CH49 8JP
2M0 SQL P. Goodhall, 12 Templand Road, Lhanbryde, Elgin, IV30 8PP
2E0 SQN I. Franklin, 23 Ingle Drive, Ashby-de-la-Zouch, LE65 2LW
2E0 SRB S. Blaikie, 22 Juno Close, Goring-By-Sea, Worthing, BN12 4UB
2E0 SRC D. Barker, 12 The Weavers, Denstone, Uttoxeter, ST14 5DP
2W0 SRD D. Bray, 24 Lon Gwesyn, Birchgrove, Swansea, SA7 9LD
2M0 SRF R. Farrer, 23 Upper Craigour, Edinburgh, EH17 7SE
2E0 SRH S. Robinson, 7 Higney Road, Hampton Vale, Peterborough, PE7 8LZ
2E0 SRJ J. Statham, Oakwoods, School Lane Upper Basildon, Reading, RG8 8LT
2E0 SRO S. Rowland, 6 Peach Hall, Tonbridge, TN10 3HD
2E0 SRP P. Skidmore, 36 Princes Drive, Harrow, HA1 1XH
2E0 SRT C. Royle, Eaton Bank House 3 Valerian Close, Buxton, SK17 6PJ
2E0 SRV S. Vickers, 22 Thistle Green, Birmingham, B38 9TT
2M0 SRX S. Russell, 3 Rankin Road, Wishaw, ML2 8PG
2M0 SRY S. Young, 6 Ramsey Cottages, Bonnyrigg, EH19 3JG
2E0 SRZ S. Ahmed, Flat 10 Jubilee Court 51 Eaton Road, London, SM2 5AQ
2E0 SSA A. Dossa, 24 Warwick Drive, Cheshunt, Waltham Cross, EN8 0BW
2E0 SSC S. Charles, 29 Woolford Close, Winchester, SO22 4DN
2E0 SSD A. Gillard, 4 Horton Avenue, Thame, OX9 3NJ
2E0 SSE S. Evans, 78 High Brooms Road, Tunbridge Wells, TN4 9BN
2E0 SSF S. Farnsworth, 26 Burrows Close, Headington, Oxford, OX3 8AN
2E0 SSG A. Butler, 12 South Bank Cottages South Stoke, Reading, RG8 0HX
2E0 SSI S. Irwin, 18 Manse Way, Swanley, BR8 8DD
2E0 SSJ S. Shortland, 10 Highclere Court, Knaphill, Woking, GU21 2QP
2E0 SSK S. Kamal, 184 Aycliffe Road, Borehamwood, WD6 4EG
2E0 SSL G. Sawyer, 432 Rowood Drive, Solihull, B92 9LQ
2E0 SSM S. Mcbain, 13A St. Lukes Close, Cherry Willingham, Lincoln, LN3 4LY
2E0 SSN B. Woods, 28 Delph Drive, Burscough, Ormskirk, L40 5BE
2M0 SSO C. Haldane, 6A Earls Gate, Bothwell, Glasgow, G71 8BP
2E0 SST S. Slapper, 1 Standards Keep, Standards Road, Bridgwater, TA7 0EZ
2E0 SSX K. Baker, 64 Pendle Drive, Basildon, SS14 3LZ
2E0 SSY M. Hossell, 80 Murray Road, Sheffield, S11 7GG
2E0 SSZ S. Ziya, 32 Latchford Road, Wirral, CH60 3RW
2E0 STA S. Collis, 21 Holme Hall Crescent, Chesterfield, S40 4PQ
2M0 STB I. Learmonth, 14 Deansloch Terrace, Aberdeen, AB16 5SN
2E0 STH S. Holt, 65 Brown Royd Avenue, Huddersfield, HD5 9QA
2E0 STI S. Pearce, 20 Barcote Walk, Plymouth, PL6 5QE
2M0 STK T. Storkey, Flat E, 29 Herbert Street, Glasgow, G20 6NB
2D0 STL S. Leslie, 16 Little Meddow, Andreas, Isle Of Man, IM7 4HY
2I0 STN S. Nash, 5 Drumard Cottages, Dans Road, Ballymena, BT42 2PX
2E0 STP B. Freeman, 45 Pryor Road, Baldock, SG7 6LH
2E0 STQ E. St Quinton, Mill Cottage, The Thorofare, Woodbridge, IP13 8BB
2E0 STT A. Burgess, 1 Laurel Crescent, Long Eaton, Nottingham, NG10 3NL
2M0 STV S. Mccormick, 4 Poplar Avenue, Johnstone, PA50EF
2E0 STX S. Jackson, 64 Main Road, Moulton, Northwich, CW9 8PB
2I0 SUB C. Williams, 92 Whitehouse Park, Newtownabbey, BT37 9sh
2E0 SUC C. Stokes, 45 Chaucer Close, Basingstoke, RG24 9DW
2E0 SUD A. Greenhalgh, 61 Long Meadows, Chorley, PR7 2YB
2E0 SUE S. Turford, 1 Portland Crescent Bolsover, Chesterfield, s446eg
2E0 SUF S. Batley, 2 Boulge Road, Hasketon, Woodbridge, IP13 6LA
2E0 SUH S. Halewood, 12 Silver Street Riccall, York, YO19 6PB
2E0 SUK A. Buck, 10 Northfield Park, Mansfield Woodhouse, Mansfield, NG19 8PA
2E0 SUS S. Millard, 112 Avenue Road, Sandown, PO36 8DZ
2E0 SUT R. Sutton, Yew Tree Farm, Paddol Green, Shrewsbury, SY4 5QZ
2E0 SUU W. Malcolm-Brown, Flat 11, Chiltern Court, Harpenden, AL5 5LY
2E0 SUX A. Suttle, 61 Albert Street, Shildon, DL4 2DN
2E0 SUY N. Jones, 8 Regent Court Belvedere Close, Guildford, GU2 9GA
2E0 SUZ S. Coombes, 33 Clarence Park Road, Bournemouth, BH7 6LF
2E0 SVB J. Ratcliffe, 7 De Havilland Drive, Hazlemere, HP15 7FP
2E0 SVG J. Savage, 44 Hastings Road, Maidstone, ME15 7SP
2E0 SVK L. Mikolka, Flat, 1 Scotney Court, Romney Marsh, TN29 9JP
2E0 SVP V. Sterea, 8 Hamilton Street, Stalybridge, SK15 1LL
2E0 SVT L. Tayler, 22 Wheatley Road, Leicester, LE4 2HN
2E0 SVV A. Kemp, 2 Darwin Close, Elston, Newark, NG23 5PQ
2W0 SVW D. Wilkinson, Bryn Y Mor, North Road, Caernarfon, LL55 1BE
2E0 SVZ A. Abson, 117 Lysander Road, Rubery, Birmingham, B45 0EN
2M0 SWA S. Anderson, 33 Dryden Avenue, Loanhead, EH20 9JT
2E0 SWB S. Bennett, 155 Warstones Road Penn, Wolverhampton, WV4 4LG

2E0 SWE O. Hall, 2 Beverley Lodge, Paradise Road, Richmond, TW9 1LL
2E0 SWF R. Jenkinson, Esperance, West End Road, Doncaster, DN9 1LB
2E0 SWK M. Sedgwick, 44 Cundall Road, Hartlepool, TS26 8LG
2M0 SWM P. Holmes, 15A Ochlochy Park, Dunblane, FK15 0DU
2E0 SWN S. Neale, 43 Crompton Road Pleasley, Mansfield, NG19 7RG
2W0 SWO S. Owen, 500 Cowbridge Road West, Cardiff, CF5 5DA
2E0 SWP S. Probert, 10 The Green, Church Lawton, St73ed
2E0 SWS Rev. S. Scotson, 16 Merryfield Road, Dudley, DY1 2PD
2E0 SWT T. Akay, Caddebostan Mah. Plaj Yolu Sok. No25/15 Kad?Köy, ?Stanbul (Asya), Turkey, 34710
2E0 SXC S. Ward, 22 St. Margarets Close, Horstead, Norwich, NR12 7ER
2E0 SXF S. Levsen, 8 Craig Close Broughton, Brigg, DN20 0SE
2E0 SXI S. Harriss, 30 Chatsworth Place, Harrogate, HG1 5HR
2M0 SXJ P. Duckles, 45 Redhouse Place Blackburn, Bathgate, EH47 7QB
2I0 SXM G. Hutton, 13 Meadowbank, Sepatrick, Banbridge, BT32 4PZ
2E0 SXP S. Perring, 16 Salford Road, Bolton, BL5 1BL
2E0 SXS D. Truscott, 37 Langley, Chulmleigh, EX18 7BQ
2M0 SXT S. Thorogood, 38 Forres Drive, Glenrothes, KY6 2JU
2E0 SXY S. Lord, 34 Alsop Street, Leek, ST13 5NZ
2E0 SYA S. Yem, 8 Beechwood Avenue, Wallasey, CH45 8NX
2E0 SYB J. Redgrave, 24 Burnham Close Trimley St. Mary, Felixstowe, IP11 0XG
2E0 SYD K. Hunt, 13 Beaumaris Court, Spondon, Derby, DE21 7RG
2E0 SYE S. Greenheart, 168 Beech Hill Lane, Wigan, WN6 8PL
2M0 SYH S. Hancox, 45 Park Road, Brechin, DD9 7AE
2E0 SYI S. De Koster, 21 Normoor Road, Burghfield Common, Reading, RG7 3QG
2M0 SYL A. Haynes, 29 Invercauld Road, Aberdeen, AB16 5RP
2E0 SYM C. Symons, 159 Middlecroft Road South Staveley, Chesterfield, S43 3NF
2E0 SYN M. Hickey, 100 Chester Road, Poynton, Stockport, SK12 1HG
2E0 SYO B. Hender, 11 Cambridge Road, Walton-on-Thames, KT12 2DP
2E0 SYS M. Styles, 46 Daffodil Drive, Lydney, GL15 5RE
2W0 SYT M. Burnell, Ty Talwyn Farm, Cefn Cribwr, Bridgend, CF32 0BP
2E0 SYW S. Ward, Wellbeck, Wheel Road, Alpington., Norwich, NR14 7NH
2E0 SYX S. Young, 26 Keel Close, Carlton Colville, Lowestoft, NR33 8GT
2E0 SYY C. Gibson, 3 Conway Drive, Billinge, Wigan, WN5 7LH
2M0 SZC S. Kirkbride, 18 North Roundall, Limekilns, Dunfermline, KY11 3JY
2E0 SZG S. Gray, The Pigsty, Cleverton, Chippenham, SN15 5BT
2E0 SZH S. Hegarty, 6 Wymbush Crescent, Bristol, BS13 0BB
2J0 SZI M. Brown, 77 Andium Court, Langtry Gardens, St. Saviours Hill, St. Saviour, Jersey, JE2 7AH
2E0 SZL R. Williams, 4 Bluebell Close, Barlborough, Barlborough, S43 4WT
2E0 SZT Z. Szot, 45 Ealing Park Gardens, London, W5 4EX
2I0 TAA T. Boyd, 40 Walnut Park, Larne, BT40 2WF
2E0 TAB T. Balls, Rostan 99 Front Road, Murrow, Wisbech, PE13 4JQ
2E0 TAC A. Paxton, 20F Green End, Granborough, Buckingham, MK18 3NT
2E0 TAD G. Aldridge, Greenridge, Fore Street Bishopsteignton, Teignmouth, TQ14 9QR
2E0 TAG J. Bingham, 31 Wyre Close, Paignton, TQ4 7RU
2W0 TAI T. Evans, 27 Addison Road, Neath, SA11 2AY
2E0 TAJ A. Turner, 174 Preston Road, Standish, Wigan, WN6 0NP
2E0 TAK D. Heathcote, 154 High Street, Harriseahead, Stoke-on-Trent, ST7 4JX
2E0 TAL A. Walton, 65 Broadway East, Rotherham, S65 2XA
2E0 TAM A. Dodds, 19 Westgate, Oldbury, B69 1BA
2I0 TAN A. Kelly, 19 Union Street Mews, Coleraine, BT52 1EN
2E0 TAO G. Jones, 57 Oxford Road, Banbury, OX16 9AJ
2E0 TAQ R. Dicker, 38 Inkerman Road, Southampton, SO19 9DA
2W0 TAR K. Parry, 80 Cripps Avenue, Cefn Golau, Tredegar, NP22 3PB
2M0 TAS D. Branson, Derelochy, Kingsteps, Nairn, IV12 5LF
2E0 TAT A. Horton, 11 Hilton Road, Tividale, Oldbury, B69 1JU
2E0 TAU A. Smith, 46 Mulberry Close, Goldthorpe, Rotherham, S63 9LB
2E0 TAV D. Rajanayagam, 87 Riffel Road, London, NW2 4PG
2E0 TAW T. Williamson, 286 Glynswood, Chard, TA20 1BX
2W0 TAX T. Mcintyre, Coach House, Commercial Street, Griffithstown, Pontypool, NP4 5JF
2E0 TAY A. Taylor, 15 Woodford Glebe, Welford, Northampton, NN6 6AF
2E0 TAZ A. Barron, 80 Primrose Crescent, Norwich, NR7 0SF
2E0 TBD A. Marshall, 13 The Markhams, New Ollerton, Newark, NG22 9QX
2E0 TBE E. Whitehouse, 16 Rue Gaston De Caillavet, Paris, France, 75015
2E0 TBF D. Bateson, 19 Rothesay Road, Heysham, Morecambe, LA3 2UR
2E0 TBH J. Smith, 10, Seacroft Road Plymouth Pl51Ph, Plymouth, PL51PH
2E0 TBI J. Lynch, Beechway, Raddel Lane, Warrington, WA4 4EE
2E0 TBL J. Wilson, 35 Lawson Avenue, Jarrow, NE32 5UF
2E0 TBQ A. Wong, 43 Northern Road, Aylesbury, HP19 9QT
2E0 TBR H. Hylton, 214 School Road, Hall Green, Birmingham, B28 8PF
2E0 TBS C. Burnham, 93 Colliers Break, Emersons Green, Bristol, BS16 7EB
2E0 TBV A. Hodgkinson, 41 Allott Crescent, Jump, Barnsley, S74 0LB
2E0 TBW B. Fitzgerald-O'Connor, 24 Routh Street, London, E6 5XX
2E0 TBX C. Pulford, 17 Canada Road, Cromer, NR27 9AH
2E0 TBZ Dr C. Cowen, Rosita, White Street Green, Sudbury, CO10 5JN
2E0 TCB A. Bruton, 29 Helyers Green, Wick, Littlehampton, BN17 7HB
2E0 TCC A. Fleming, 39 Urswick Green, Barrow-in-Furness, LA13 0BH
2E0 TCF T. Guy, 16 Cogdeane Road, Poole, BH17 9AS
2E0 TCG A. Gilberts, 22 Granby Road, Buxton, SK17 7TW
2E0 TCI A. Ashton, 46 Kingsland, Harlow, CM18 6XL
2I0 TCJ T. Mckee, 4 Earlford Heights, Newtownabbey, BT36 5WZ
2E0 TCK P. Pritchard, 11 Beacon Avenue, Dunstable, LU6 2AD
2W0 TCM M. Digby, 40 Waterloo Road, Ammanford, SA18 3SF
2E0 TCN C. Lyne, 4 Bridge Close, Catterick Garrison, DL9 4PG
2E0 TCO T. Corcoran, 191 Queensway, Rochdale, OL11 2NA
2E0 TCQ M. Nicholas, 56 Morrell Street. Maltby., Rotherham, S667LJ
2E0 TCU B. Neal, 6 Canterbury Street, Chaddesden, Derby, DE21 4LG
2E0 TCV T. Curnow, 5 Bosleake Row, Bosleake, Redruth, TR15 3YG
2E0 TCX A. Wilson, 38 Cotleigh Drive, Sheffield, S12 4HU
2E0 TCY C. Maddex, 154 Tintern Avenue, Westcliff, Westcliff-on-Sea, SS0 9QF
2E0 TCZ A. Cudworth, 30 Compass Tower, Munnings Road, Norwich, NR7 9TW
2W0 TDD T. Dixon, Troedyrhiw, Abercych, Boncath, SA37 0EY
2E0 TDE A. Dilworth, 808 Liverpool Road, Southport, PR8 3QF
2W0 TDF W. Welch, Kenilworth, School Lane, Gobowen, Oswestry, SY11 3LD
2E0 TDH P. Thorley, 57 Riverside Drive, Hambleton, Poulton-le-Fylde, FY6 9EH
2E0 TDI D. Davies, 32 Newlyn Crescent, Puriton, Bridgwater, TA7 8BS
2E0 TDJ T. Kelly, 50 Ivanhoe Road, Herne Bay, CT6 6EQ
2I0 TDL T. Browne, 7 Hawthorn Park Greysteel, Londonderry, BT47 3YE
2E0 TDN S. Friend, 1 Orford Road, Tunstall, Woodbridge, IP12 2JH
2E0 TDO Dr T. Digman, 74 Baddlesmere Road, Whitstable, CT5 2LA
2E0 TDP A. Pickett, 4 Trembel Road, Mullion, Helston, TR12 7DY
2E0 TDR J. Mcmullen, 281 The Broadway, Cullercoats, North Shields, NE30 3LH
2E0 TDT A. Maclean, 10 Elizabeth Close, West Hallam, Ilkeston, DE7 6LW
2E0 TDV T. Millichamp, 9 Wells Close, Bridgnorth, WV16 5JQ
2E0 TED G. Cahill, 81 Albemarle Road, Willesborough, Ashford, TN24 0HJ
2E0 TEE A. Chapman, 6 Chaffer Lane, Birdham, Chichester, PO20 7EZ
2E0 TEF R. Ashman, 85A Fakenham Road, Great Ryburgh, Fakenham, NR21 7AQ
2E0 TEH T. Harris, 12 Maple Close, Stourport-on-Severn, DY13 8TA
2E0 TEI M. Bell, 36 Schneider Road, Barrow-in-Furness, LA14 5DW
2E0 TEJ A. Bell, 36 Schneider Road, Barrow-in-Furness, LA14 5DW
2E0 TEK G. Tecklenberg, 33 Tilman Drive, Peterborough, PE7 0LU
2E0 TEN A. Riddick, 30 Britannia Road, Banbury, OX16 5DW
2E0 TEU Dr S. Hill, 36 The Woodlands, Market Harborough, LE16 7BW
2E0 TEW J. Woodland, 14 Kelham Green, Nottingham, NG3 2LP
2E0 TEZ T. Kemp, 30 Tawny Sedge, King's Lynn, PE30 3PW
2E0 TFD T. Daniel, 28 Polefield Circle Prestwich, Manchester, M25 2WP
2E0 TFE J. King, Plum Tree Cottage, Royston Place, Barton On Sea, New Milton, BH25 7AJ
2M0 TFF T. Feltus, 5/5 Moat House, Moat Drive, Edinburgh, EH14 1NS
2E0 TFH T. Harrison, 1 Tall Trees, Colchester, CO4 5DU
2E0 TFI B. May, 10 Farrier Place, Downs Barn, Milton Keynes, MK14 7PJ
2E0 TFK J. Anthony, 21 Belgrave Street, Denton, Manchester, M34 3WP
2E0 TFM I. Whiteley, 29 Harvey Avenue, Wirral, CH49 1RT
2E0 TFN C. Tompkins, 9 Billbrook Road, Hucclecote, Gloucester, GL3 3QS
2E0 TFO R. Styles, 52 Vernham Grove, Bath, BA2 2TB
2E0 TFT S. Mayor, 12 Yealand Avenue, Heysham, Morecambe, LA3 2LT
2E0 TFX T. Fisk, 2 Hall Farm Cottage, Caston Road, Caston, Attleborough, NR17 1BW
2E0 TGA Dr M. Depardieu, 4 Belvedere Fff, Bath, BA1 5ED
2E0 TGB W. Millington, 93 Feiashill Road, Trysull, Wolverhampton, WV5 7HT
2E0 TGC G. Cooper, 21 Thistle Bridge Road, Chivenor, Barnstaple, EX31 4FL
2M0 TGD I. Currie, 4 Greendyke Cottage, Falkirk, FK2 8PP
2E0 TGF A. Mcgoff, 55 Knights End Road, March, PE15 9QA
2E0 TGG T. Sutton, Yew Tree Farm, Paddol Green, Shrewsbury, SY4 5QZ
2E0 TGI T. Griffiths, 6 Witney Lane, Edge, Malpas, SY148JJ
2E0 TGJ T. Jones, 27 Chamberlain Grove, Fareham, PO14 1HH
2E0 TGK R. Willis, 12 Robartes Road St. Dennis, St Austell, PL26 8DS
2E0 TGL A. Wallace, 17 Dennis Road, Liskeard, PL14 3NS
2M0 TGM T. Mcconnell, 2 Stewartgill Place, Ashgill, Larkhall, ML9 3BB
2M0 TGN A. Barclay, 21 Netherlea, Scone, Perth, PH2 6QA
2E0 TGO T. Gentry, 175 Huddersfield Road, Halifax, HX3 0AS
2E0 TGQ T. Garcia-Quismondo, 11 Half Moon Lane, Worthing, BN13 2EN
2E0 TGV D. Cook, 35 Elmwood Drive, Breadsall, Derby, DE21 4GA
2E0 THE S. Griffiths, 37 Stourton Close, Knowle, Solihull, B93 9NP
2E0 THG L. Bennett, 19 Campion Crescent, Cranbrook, TN17 3QJ
2E0 THP A. Thorpe, 8 Syke Avenue, Tingley, Wakefield, WF3 1LU
2E0 THR J. Summers, 263 Stroud Road, Gloucester, GL1 5JZ
2E0 THS E. Graham, 25 South End Road, Ottringham, Hull, HU12 0DP
2E0 THT A. Hoyte, 43 Orchard Drive Mayland, Chelmsford, CM3 6EP
2E0 THZ M. Hopewell, 4 Cotes Crescent Bicton Heath, Shrewsbury, SY3 5AS
2W0 TID S. Cook, 114 Caerphilly Road, Bassaleg, Newport, NP10 8LJ
2E0 TIF C. Townsend, 64 Burnbridge Road Old Whittington, Chesterfield, S41 9LR
2E0 TIL P. Athersmith, 5 Aqueduct Lane Stirchley, Telford, TF3 1BW
2E0 TIN D. Tinn, 28 South Road, Kirkby Stephen, CA174sn
2W0 TIR G. Williams, 1 Pont Y Berllan, Talycafn Road, Llanrwst, LL26 0EF
2E0 TIV S. Bryan, 7 Authers Heights, Tiverton, EX16 5PE
2E0 TIZ A. Tyrrell, Flat 1, 61 Vicarage Road, Eastbourne, BN20 8AH
2E0 TJC T. Catton, 97 High St., South Hiendley, Barnsley, S72 9AN
2E0 TJD T. Dix, Willow Cottage 31 London Road, Woolmer Green, SG3 6JE
2E0 TJE T. Ellis, 10 Digby Crescent, Water Orton, Birmingham, B46 1NP
2I0 TJF T. Ferguson, 3 Wheatfield Park, Ballybogy, Ballymoney, BT53 6NT
2E0 TJG J. Wong, St Edmund'S College, Mount Pleasant, Cambridgeshire, CB3 0BN
2E0 TJI S. Guest, 19 Ellesmere Avenue, Ashton-under-Lyne, OL6 8UT
2E0 TJJ T. Jinkerson, 104 Foxcote, Finchampstead, Wokingham, RG40 3PE
2I0 TJK J. Mackenzie, 30 Dalriada Gardens, Ballycastle, BT54 6DZ
2I0 TJM T. Mulholland, 215 Finaghy Road North, Belfast, BT11 9ED
2E0 TJN T. Newton, 4 Manor Close, Bradford Abbas, Sherborne, DT9 6RN
2M0 TJO T. Johnston, The Old Schoolhouse, Luggate Burn, Haddington, EH41 4QA
2E0 TJP P. Neal, 14 Hilltop Close, Desborough, Kettering, NN14 2LQ
2E0 TJQ D. Roberts, 61 Teign Bank Road, Hinckley, LE10 0ED
2I0 TJR T. Ruddell, 30 Ballynacor Meadows, Portadown, Craigavon, BT63 5UU
2E0 TJS A. Steer, 51 Kings Chase, East Molesey, KT8 9DG
2E0 TJT J. Thompson, 32 Church Street, Warnham, Horsham, RH12 3QR
2E0 TJU E. Duffield, 92 Crosby Street, Stockport, SK2 6SP
2E0 TJV T. Humphreys, 93 Cornwall Crescent Yate, Bristol, BS37 7RU
2E0 TJW T. Ward, 25 Chislehurst Road, Carlton Colville, Lowestoft, NR33 8BY
2E0 TJX T. Jones, 15 Kinnaird Road, Sheffield, S50NN
2E0 TJY J. Yeo, 10, Bradgate Close, Leicester, LE79NP
2E0 TKD S. Stebbings, 4 Coltsfoot Lane, Bull'S Green, Knebworth, SG3 6SB
2M0 TKE R. Mckie, 16 Silver Street, Creetown, Newton Stewart, DG8 7HU
2E0 TKF D. Long, 25 St. Matthias Road, Deepcar, Sheffield, S36 2SG
2W0 TKS S. Davies, 5 Maldwyn Street, Cardiff, CF11 9JR
2E0 TKV T. King, 24 Royston Avenue, Basildon, SS15 4EW
2E0 TKX A. Sood, Parima, Sewardstone Road, London, E4 7RA
2E0 TKY S. Potter, 93 Church Lane, Bocking, Braintree, CM7 5SD
2E0 TLB F. Smith, 13 Heather Walk, Crowborough, TN6 2HA
2E0 TLC C. Parry, 27 Tynedale Close, Stockport, SK5 7NA
2E0 TLD N. Sinclair, 11 Primrose Close, Warton, Preston, PR4 1EN
2M0 TLE K. Quillien, 1/2 81 Laurel Street, Glasgow, G11 7QX
2W0 TLG D. James, Coombe House, Coombe, Presteigne, LD8 2HL
2E0 TLJ R. Mcwilliam, 3 Fountains Close, Riccall, York, YO19 6QN
2E0 TLM I. Williams, 36 Telford Road, Tamworth, B79 8EY
2I0 TLT W. Thompson, 25 Darby Road, Carrickfergus, BT38 7XU

2E0 TLW C. Conghos, Flat 30, The Dutch Quarter Ap. West Stockwell Street, Colchester, CO11FQ
2E0 TLX D. Burdsall, 37 Fulmar Walk, Whitburn, Sunderland, SR6 7BW
2E0 TLY I. Bain, 45 Larpool Crescent, Whitby, YO22 4JD
2E0 TMA T. Ahern, 39 Essex Road, Romford, RM7 8BE
2W0 TMB M. Woodington, 44 Glas Y Gors, Aberdare, CF44 0BQ
2E0 TMC S. Oram, 7 West Bank, Main Street, Old Weston, Huntingdon, PE28 5LJ
2E0 TMD P. Kirby, 36 Durham Road, London, E12 5AX
2I0 TME T. Evans, 30 Seagoe Park, Portadown, Craigavon, BT63 5HR
2E0 TMF A. Taylor, 11 Fillingfir Drive, Leeds, LS16 5EG
2I0 TMH T. Mcelwee, Orchard Bank, 1A Dunover Road, Ballywalter, Newtownards, BT22 2LE
2E0 TMM J. Woodcock, 1 Main Street Brandesburton, Driffield, YO258RL
2E0 TMN A. Macnauton, 27A Lincoln Road, Poole, BH12 2HT
2E0 TMO T. Williams, Moor Farm, Moor Lane, Lincoln, LN3 4EG
2E0 TMS T. Mackenzie, 2 Newcastle Street, Carlisle, CA2 5UH
2E0 TMX A. Lee, 21 Percheron Place, Westbury, BA13 2GR
2E0 TMY T. Haley, 3 Akeman Rise, Ramsden, Chipping Norton, OX7 3BJ
2W0 TNB N. Williams, 60 Denbigh Close, Wrexham, LL12 7TW
2E0 TNC T. Humphries, 32 Bonds Meadow, Oulton Broad, NR32 3QL
2E0 TNE D. Camp, Kidbrooke Lodge, Lewes Road, Forest Row, RH18 5AF
2M0 TNM P. Rae, Tigh-Na-Mara, East Kilbride, Isle of South Uist, HS8 5TS
2E0 TNN M. Tinnion, 3 Hillhead Road, Newcastle upon Tyne, NE5 5AP
2E0 TNO R. Naden, 10 Suffolk Close, Holland On Sea, Clacton-on-Sea, CO15 5SQ
2E0 TNV M. Clough, 8 Skeldyke Road, Kirton, Boston, PE20 1LR
2E0 TNZ S. Swallow, 16 Quarry Lane, Chesterfield, S40 3AS
2E0 TOF J. Morgan, Glas Y Dorlan, Pontrhydfendigaid, Ystrad Meurig, SY25 6EJ
2E0 TOG B. Whelan, 147 Lawsons Road, Thornton-Cleveleys, FY5 4PL
2M0 TOK J. Mcgurk, Pf2, 18 Warriston Road, Edinburgh, EH7 4HN
2E0 TOL S. Tomlinson, 8 Levett Road, Stanford-le-Hope, SS17 0BB
2E0 TOM T. Wilcox, 17 Westminster Avenue, Royton, Oldham, OL2 5XY
2E0 TOP R. Styles, 16 Hatton Park, Bromyard, HR7 4EY
2M0 TOR S. Gibson, 41 Kane Place, Stonehouse, ML9 3NR
2E0 TOT S. Dix, 8 Beaumont Road, Longlevens, Gloucester, GL2 0EJ
2E0 TOX P. Tivey, 4 Appleby Glade, Castle Gresley, Swadlincote, DE11 9EJ
2E0 TOY L. Bodnar, 47 Alchester Court, Towcester, NN12 6RL
2E0 TPA A. Patrick, 2 Beacon Grange, Malvern, WR14 3EU
2I0 TPC T. Crozier, 6 Garden Of Eden, Carrickfergus, BT38 7ls
2E0 TPD F. Harwood, 1, South Highall Cottage, Lincolnshire, LN10 6UR
2M0 TPE B. Angus, Springfield Church Street, Dufftown, AB55 4AR
2E0 TPG A. Gravell, 21 Wickridge Close, Stroud, GL5 1ST
2E0 TPH T. Hazel, 84 Rodwell Avenue, Weymouth, DT4 8SQ
2E0 TPI T. Idiculla, 3 Stanhope Road, Slough, SL1 6JR
2E0 TPL A. Capon, 1 Windermere Way, North Common, BS305XN
2E0 TPN G. Grace, 194 Lillechurch Road, Dagenham, RM8 2EW
2E0 TPP G. Finney, 4 Cressida Court, Braunstone Lane, Leicester, LE3 3AP
2E0 TPR T. Prince, Flat24 Killbridge Close, Redcar, TS118DT
2E0 TPY T. Pollett, 10 Bridport Road, Poole, BH12 4BS
2E0 TQB D. Holmes, 4 Council House, Nidds Lane, Boston, PE20 1LZ
2E0 TQF C. Bailey, 55 Bridgend Park Brewery Road, Wooler, NE71 6QG
2E0 TQS A. Raeper, 11 Ashbourne Drive, Stoke-on-Trent, St56rl
2M0 TRA T. Mussell, Dunelm, Thornhill Road, Cuminestown, Turriff, AB53 5WH
2W0 TRD R. George, 18 Bryndedwyddfa, Penygroes, Llanelli, SA14 7PR
2E0 TRE T. Ellis, 84 Revelstoke Road, London, SW18 5PB
2E0 TRH A. Hargreaves, 27 Meadow Head Close, Blackburn, BB2 4TY
2E0 TRI N. Rostant, 14 Gas Street, Leamington Spa, CV31 3BY
2E0 TRJ R. Nicholson, 8 Meadway, Maghull, Liverpool, L31 8AX
2E0 TRK A. Cooper, 23 Ash St., Manchester, M9 5XY
2D0 TRL T. Leece, Thie Sy Cheyll Ballastrooan Colby, Isle of Man, IM9 4NR
2I0 TRM S. Davison, 60 Cornation Place, Craigavon, BT66 7AN
2E0 TRO M. Tromans, 10 Crofters View, Little Wenlock, Telford, TF6 5AU
2E0 TRP T. Parsons, 5 Blackmoor Road, Aubourn, Lincoln, LN5 9SX
2W0 TRR T. Rees, 4 Maes Y Glo, Llanelli, SA149PZ
2W0 TRS D. Humphreys, 48 High Street, Abergwynfi, Port Talbot, SA13 3YW
2E0 TRU G. Truman, 3 Mulberry Road, Rotherham, S254BH
2E0 TRW C. Jacobs, Flat 33, The Lodge, Lavender Road, Waterlooville, PO7 8BX
2E0 TRX R. Moldoveanu, 17 Lynchford Road, Farnborough, GU14 6AR
2E0 TRY H. Chawdhry, Trinity College, Cambridge, CB2 1TQ
2E0 TSA T. Stamp, 41 Glamorgan Close, Mitcham, CR4 1XG
2E0 TSC M. Hemmings, 212 High St., Pensnett, Brierley Hill, DY5 4JF
2D0 TSE A. Elliott, Round Table House, Ronague, Castletown, Isle of Man, IM9 4HJ
2E0 TSI S. Gilham, 32 Whitby Road, Lytham St Annes, FY8 3HA
2W0 TSJ S. Trott, 6 Mounton Drive, Chepstow, NP16 5EH
2E0 TSO I. Macfarlane, 70 Ashby Drive, Rushden, NN10 9HH
2E0 TSP R. Connolly, 29 Gayer Street, Coventry, CV6 7EU
2M0 TSR D. Murphy, 38 Lothian Road, Stewarton, Kilmarnock, KA3 3BT
2E0 TSU S. Watson, 5 Birchwood Avenue, Whickham, Newcastle upon Tyne, NE16 5QS
2E0 TSY I. Oxenham, 16 Chester Long Court, Vaughan Road, Exeter, EX1 3WU
2E0 TTA T. Arscott, 29 Parsonsfield Road, Banstead, SM7 1JW
2E0 TTB T. Bate, 87 Dunsheath, Telford, TF3 2BY
2E0 TTE C. Macrae, 16 Handfield Road, Liverpool, L22 onx
2M0 TTF J. Mcmorland, 382 Maryhill Road, Glasgow, G20 7YQ
2E0 TTG T. Hoyle, 60 Greenbank Crescent, Marple, Stockport, SK6 7PB
2E0 TTH P. Troth, Beuna Vista, Hawford Wood, Droitwich, WR9 0EZ
2E0 TTM T. Moncaster, 24 Minster Yard, Lincoln, LN2 1PY
2E0 TTN M. Groves, 15 Plains Lane, Littleport, Ely, CB6 1RJ
2E0 TTP S. Bailey, 39 Aintree Drive, Balby, Doncaster, DN4 8TU
2E0 TTS D. Brook, 140 Dearne Hall Road, Barugh Green, Barnsley, S75 1LX
2E0 TTT A. Rosenschein, 101 Christchurch Road, London, SW14 7AT
2E0 TTW R. Glynn, 106 Fairway Avenue, West Drayton, UB7 7AP
2E0 TTY A. Smith, 93 Sheriffs Highway, Gateshead, NE9 6QN
2E0 TUB C. Austen, Fenbank House, Roman Bank, Holbeach Clough, Spalding, PE12 8DH
2E0 TUC B. Tucker, 12 Alpha Place, Appledore, Bideford, EX39 1QY
2E0 TUD R. Bolton, 5 Beacon Hill Avenue, Harwich, CO12 3NR
2E0 TUE N. Wadsworth, Haygarth, Docker, Kendal, LA8 0DF
2E0 TUF J. Caswell, 3 Pavilion Court, Roydon, Diss, IP22 5SP
2E0 TUG A. Barrett-Sprot, 1 Malting End, Wickhambrook, Newmarket, CB8 8YH
2E0 TUH R. Taylor, Penhawger Park, Liskeard, PL14 3LW
2I0 TUI C. Stockdale, 3 Hightown Drive, Newtownabbey, BT36 7TG
2E0 TUK T. Goddard, 217 Speedwell Road, Bristol, BS5 7SP
2E0 TUM T. Mcgoun, 64 Buttfield Lane, Howden, Goole, DN14 7DS
2E0 TUN A. Tunney, 79 Scott Street, Burnley, BB12 6NJ
2E0 TUR T. Baines, 10 Croydon Avenue, Leigh, WN7 1TP
2E0 TUS R. Austen, Fenbank House, Roman Bank, Holbeach Clough, Spalding, PE12 8DH
2E0 TUT C. Howard, 75 Gordon Road, Herne Bay, CT6 5QX
2I0 TUV C. Bailie, 26 Moatview Park, Dundonald, Belfast, BT16 2BE
2E0 TUW R. Harris, 7 Fosse Lane, Shepton Mallet, BA4 4PS
2E0 TUX N. Trangmar, 8 Maxstoke Close, Meriden, Coventry, CV7 7NB
2E0 TVD S. Pettit, 24 Bickington Lodge Estate, Barnstaple, EX31 2LH
2E0 TVG T. Gunnoo, 76 Katherine Drive, Dunstable, LU5 4NU
2E0 TVM T. Nott, 87 Powney Road, Maidenhead, SL6 6EG
2E0 TVR S. Martin, 35 Hermitage Green, Hermitage, Thatcham, RG18 9SL
2E0 TVS N. Simmonds, 3 Noneley Hall Barns, Noneley, Wem, Shrewsbury, SY4 5SL
2E0 TVW A. Wilson, 11 Headland Way, Alton, Stoke-on-Trent, ST10 4AN
2E0 TVZ D. Darby, 116 Middle Road, Southampton, SO19 8FS
2E0 TWA A. Weatherall, The Old Telephone Exchange, The Street, Canterbury, CT3 1ED
2E0 TWD P. Roche, 8 Fold Mews Hazel Grove, Stockport, SK7 4NU
2E0 TWI C. Wild, 203 High Street, Saltney, Chester, CH4 8SJ
2E0 TWK A. Champion, 4 Oldcastle Croft, Tattenhoe, Milton Keynes, MK4 3EN
2E0 TWO K. Monaghan, The Bulstone Hotel, Branscombe, Seaton, EX12 3BL
2E0 TWP P. Miller, 46 Great Brooms Road, Tunbridge Wells, TN4 9DH
2E0 TWQ J. Attwood, 13 John Winter Court, Euston Road, Great Yarmouth, NR30 1DU
2E0 TWS T. Wood, 44 Wincobank Lane, Sheffield, S4 8AA
2E0 TWT J. Colbourne, 43, Westfield Road, Bilston, WV14 6EW
2E0 TWW T. Larman, 861B London Road, Westcliff on Sea, SS0 9SZ
2E0 TWX S. Waterson, 7 West Farm Road, Newcastle upon Tyne, NE6 4JA
2E0 TWZ N. Houghton, 100 Ellerburn Avenue, Hull, HU6 9RW
2M0 TXA T. Dorricott, 11 Dakota Way, Renfrew, PA4 0NP
2I0 TXB J. Morrison, 9 Coral Cottages, Kilkeel, BT34 4FT
2E0 TXE D. Parker, 51 Ruskin Road, Congleton, CW12 4EA
2E0 TXG T. Ruddick, Hazel Gill, Croglin, Carlisle, CA4 9RR
2E0 TXH T. Holden, 132 Sutherland Street, Barrow in Furness, LA14 2BJ
2E0 TXI P. Deeprose, 2 Denehurst Gardens, Hastings, TN35 4PB
2E0 TXJ J. Hopkins, 53 Sprules Road, London, SE4 2NL
2M0 TXK Prof. T. Kerby, 1 St. Mark'S Lane, Edinburgh, EH15 2PX
2E0 TXL P. Dunnicliffe, 19 Woodland Road, Chelmsford, CM1 2AT
2I0 TXM A. Mcgarvey, 66A Scaddy Road, Downpatrick, BT30 9BS
2E0 TXP P. Cassells, 49 Dodds Lane, Maghull, Liverpool, L31 0BD
2E0 TXQ B. Roberts, 20 Mitchell Road, Enderby, LE19 4NX
2M0 TXR A. Mcneill, 13 Spinkhill, Laurieston, Falkirk, FK2 9JR
2E0 TXU M. Holmes, Lower Farm, Stony Moor, Newton-On-Rawcliffe, Pickering, YO18 8QJ
2M0 TXY A. Caldwell, 180 Pappert, Alexandria, G83 9LG
2E0 TYC T. Corcoran, 50 Grange Road, Bracebridge Heath, Lincoln, LN4 2PW
2W0 TYE M. Armstrong, Tan Yr Efail, Segurinside, Llandudno Junction, LL31 9QE
2W0 TYG C. Morris, 17 Percy Road, Wrexham, LL13 7EA
2E0 TYH T. Hill, 14 Hunters Mead, Motcombe, Shaftesbury, SP7 9QG
2E0 TYL M. Tyler, 40 Bullards Lane, Woodbridge, IP12 4HE
2E0 TYT C. Gibson, 17 Clyde Court, Grantham, NG31 7RB
2E0 TYY T. Bradshaw, 12 Whitegate Hill, Caistor, Market Rasen, LN7 6SW
2M0 TZB M. Lawson, 23 Kirkfield View, Livingston Village, Livingston, EH54 7BP
2E0 TZD A. Maiden, 79 Green End Road, Manchester, M19 1LE
2E0 TZE F. Simpson, 36 Lyndhurst Ave, Blackpool, FY4 3AX
2E0 TZG T. Grabiec, 16 Jubilee Crescent, Clowne, Chesterfield, S43 4ND
2E0 TZM A. Laister, 23 Berry Street, Greenfield, Oldham, OL3 7EF
2E0 TZR T. Tozer, 9 Bainbridge Court, Plymouth, PL7 4HH
2E0 TZY S. Crabb, 1 Council Houses, Hall Lane, Crostwick, Norwich, NR12 7BB
2E0 TZZ P. Moore, 24 Plough Road, Dormansland, Lingfield, RH7 6PS
2W0 UAA M. Uphill, 1 Brynview Avenue, Ystrad Mynach, Hengoed, CF82 7DB
2E0 UAB C. Kent, 19 Coppice Rise, Harrogate, HG1 2DP
2E0 UAC M. Timms, 29 Sutherland Avenue, Coventry, CV5 7ND
2I0 UAD A. Pritchard, 16 Ballymaconnell Road South, Bangor, BT19 6DQ
2E0 UAE J. Firth, 36 Howley Grange Road, Halesowen, B62 0HW
2E0 UAK A. Vaile, 66 Grasmere Point, Old Kent Road, London, SE15 1DU
2M0 UAL K. Mackenzie, Alderwood, Braes, Ullapool, IV26 2TB
2E0 UAM C. Byrne, 31 Graham Drive, Castleford, WF10 3EY
2E0 UAO M. Lewis, 6 Remembrance Road, Newbury, RG14 6BA
2E0 UAP P. Richards, 22 Waterpump Court, Northampton, NN3 8US
2W0 UAR P. Barnes, 95 Marcroft Road, Port Tennant, Swansea, SA1 8PN
2W0 UAS C. Walker, Perthi, Llaneilian, Amlwch, LL68 9LY
2E0 UAV T. Ward, Flat 26, Bassett Court, Bassett Avenue, Southampton, SO16 7DR
2E0 UAW A. Wragg, 14 Grizedale Avenue Sothall, Sheffield, S20 2DL
2E0 UAY D. Levey, Heriots Wood, The Common, Stanmore, HA7 3HT
2I0 UBE L. Calderwood, 43 Rathview Park, Mullybritt, Enniskillen, BT94 5EW
2E0 UBH R. Tolman, 10 Woodcote Way, Abingdon, OX14 5NE
2E0 UBM B. Mcdowell, 10 Vineyard Lane, Kingswood, Wotton-under-Edge, GL12 8SB
2E0 UBN B. Shephard, 13 Forest Street, Annesley Woodhouse, Kirkby In Ashfield, Nottingham, NG17 9HE
2E0 UBT C. Bowes, 11 Burghwallis Lane, Sutton, Doncaster, DN69JU
2E0 UBU D. Hampton, 6 St. Georges Lane South, Worcester, WR1 1QZ
2E0 UBW J. Cardwell, 35 Bush Lane, Freckleton, Preston, PR4 1SB
2E0 UCB C. Barber, 32 Ashcroft Road, Ipswich, IP1 6AB
2E0 UCD N. Booth, Spook Hollow, Spook Hill, North Holmwood, Dorking, RH5 4JP
2E0 UCE S. Roderick, 47 Cuckmans Drive, St Albans, AL2 3AY
2E0 UCH W. Toher, The Chapel, Station Road, Darlington, DL2 1JG
2W0 UCL K. Ucele, 30 Britannia Apartments, Phoebe Road, Copper Quarter, Pentrechwyth, Swansea, SA1 7FG
2E0 UCR C. Rogers, 1 Tregurthen Close, Camborne, TR14 7EB
2I0 UCS M. Edwards, 15 Highgrove Road, Carrickfergus, BT38 9AG
2E0 UCT C. Thorne, 67 Devon Road, Cadishead, Manchester, M44 5HB
2E0 UCV T. Kilroy, 55 Summerfield Crescent, Brimington, Chesterfield, S43 1HB
2I0 UCY L. Stock, 15 Mahon Drive, Portadown, Craigavon, BT62 3JB

2E0 UDA A. Cattell, 2 St. James Close, Ruscombe, Reading, RG10 9LJ
2E0 UDB D. Beard, 1 Bond Close, Leonard Stanley, Stonehouse, GL10 3GQ
2E0 UDE R. Brown, 24 Malthouse Court, Wellington, Telford, TF1 1QJ
2E0 UDM D. Meehan, 47 Clinton Road, Shirley, Solihull, B90 4RN
2I0 UDR J. Mcclarence, 10 Cloyfin Park, Coleraine, BT52 2BL
2M0 UEA E. Ewing, Arisaig, Priestland, Darvel, KA17 0LP
2E0 UED E. Cook, 10 Firwood Drive, Gloucester, GL4 0AB
2E0 UEE R. Goldsack, 5 Parc Dellen, Croft Farm Park, Luxulyan, Bodmin, PL30 5EW
2E0 UEH Dr S. Smith, 557, Riverside Island Marina, Isleham, Ely, CB7 5SL.
2E0 UEL M. Tointon, 13 Ridgeway, Broadstone, BH18 8DY
2E0 UER S. Mauer, Flat 18, Channing Court, Osborne Road, London, W3 8SY
2E0 UET R. Taylor, 68 Charter Road, Chippenham, SN15 2RA
2E0 UFM A. Parkhouse, 3 St. Margarets Avenue, Ashford, TW15 1DR
2M0 UFO S. Mcdougall, 6F Watchmeal Crescent, Clydebank, G81 5EA
2E0 UGE R. Evans, 11 Swane Road Stockwood, Bristol, bs14 8nq
2E0 UGF M. Sims, 133 Canterbury Road, Hawkinge, Folkestone, CT18 7BS
2M0 UGL N. Rogers, 108 Beechwood Road, Cumbernauld, Glasgow, G67 2NP
2E0 UGM G. Mountain, 34 Albert Road, Warlingham, CR6 9EP
2E0 UGO D. Hill, 19 Farren Road, Birmingham, B31 5HH
2E0 UGP G. Palin, 104 Nelson Street, Crewe, CW2 7LN
2E0 UHF N. Booth, Greenfield, Westmancote, Tewkesbury, GL20 7EP
2E0 UHJ P. Hopkinson, 28 Stockdove Way, Thornton-Cleveleys, FY5 2AR
2E0 UHL M. Simpson, 32 Underhill Lane, Wolverhampton, WV10 8NS
2E0 UHS I. Phillpott, 14 Buttercup Close, Paddock Wood, Tonbridge, TN12 6BG
2E0 UID U. Decruz, Victoria Road, Lowestoft, NR33 9LR
2E0 UIP D. Woodhouse, 1 Low Metham Cottages, Metham, DN14 7YA
2E0 UIT A. Rackett, 151 Stoke Road, Gosport, PO12 1SE
2E0 UJD K. Colman, 10 South Rise, North Walsham, NR28 0EE
2E0 UJM R. Banks, 3 Parkhayes, Woodbury Salterton, Exeter, EX5 1QS
2E0 UJR J. Risby, 112 Stratton Heights, Cirencester, GL7 2RL
2E0 UKA A. Reay, 12 Victoria Avenue, South Hylton, Sunderland, SR4 0QZ
2E0 UKB J. English, 1 Niton Cottage Pound Lane, Meonstoke, Southampton, SO32 3NP
2E0 UKD J. Parfrey, 47 Ford Lane, Rainham, RM13 7AS
2E0 UKK G. Radulescu, 41 Sherard Road, London, SE9 6EX
2E0 UKM M. Qassim, Winchester Road, Kings Somborne, Stockbridge, SO20 6NY
2E0 UKT N. Radulescu, 41 Sherard Road, London, SE9 6EX
2E0 UKU S. Crowther, 4 Brigsteer Close, Clayton Le Moors, Accrington, BB5 5GE
2E0 UKW D. Watts, 3 Witney Road, Crawley, RH10 6GJ
2E0 UKX A. Fletcher, 2 Brow Crescent, Halfway, Sheffield, S20 4GB
2E0 UKY G. Smith, 47 Percy Road, Carlisle, CA2 6ER
2E0 ULC P. Power, 16 Mainstone Close, Redditch, B98 0PP
2M0 ULD N. Chalmers, 4 Hercules Place, Arbroath, DD11 4HT
2E0 ULH D. Butchart, Flat 26, Seldon House, London, SW8 4DP
2E0 ULY A. Ulyatt, 7 Southend Mews, Tatworth Road, Chard, TA20 2DP
2E0 UMD A. Slee, Foal Cottage Mare Hill Common, Pulborough, RH20 2DX
2E0 UMG M. Ribbands, Dyson'S Farm, Long Row, Tibenham, NR16 1PD
2M0 UMH L. Mitchell Hynd, Smithy House, Bruichladdich, Isle of Islay, PA49 7UN
2E0 UMO K. Laing, 16 Cherrywood Drive Gonerby Hill Foot, Grantham, NG31 8QL
2E0 UMP I. Gilbert, 84 Carlyle Road, London, W5 4BJ
2E0 UMR U. Munir, Flat 6, Horton House, Field Road, London, W6 8HW
2E0 UNI G. Rigby, Gas House Farm, Shavington Park, Market Drayton, TF9 3SY
2W0 UNY L. Betts, 12A Maesgwyn, Pontnewydd, Cwmbran, NP44 1BQ
2E0 UOJ K. Barry, 25 Delabole Road, Merstham, Redhill, RH1 3PB
2E0 UOK A. Abraham, Flat 2, 41 Francis Road, Birmingham, B33 8SL
2E0 UOM S. Johnson, 43 Terry Gardens, Kesgrave, Ipswich, IP5 2EP
2E0 UPA J. Van Der Elsen, 6 Kent Close, Churchdown, Gloucester, GL3 2HQ
2W0 UPH A. Williams, 8 Old Tanymanod Terrace, Blaenau Ffestiniog, LL41 4BU
2E0 UPK D. Harmer, 98 King Georges Avenue, Coventry, CV6 6FF
2E0 UPM P. Mynors, 109 Walton Road, Frinton-on-Sea, CO13 0AB
2W0 UPR S. Stupple, 7 Glan Preseli, Llanddewi Velfrey, Narberth, SA67 7PG
2E0 UPS A. Clarke, Little Acre, Pound Lane, Hardwicke, Gloucester, GL2 4RJ
2E0 UPT S. Thomas, 103 Liverpool Road, Upton, Chester, CH2 1BB
2E0 UPU A. Stirk, 59 West Avenue, Lightcliffe, Halifax, HX3 8TJ
2E0 URA R. Scott, 157 Walton Back Lane, Walton, Chesterfield, S42 7LT
2E0 URD J. Ratcliffe, 63 Dickens Lane, Poynton, Stockport, SK12 1NN
2E0 URF R. Freeman, 9 Bramley Road, Wisbech, PE13 3PA
2E0 URJ Capt. R. Jordan, Pheasants Walk, Copyhold Lane, Haslemere, GU27 3DZ
2E0 URK M. Johnson, 54 Birchwood Drive, Ulverston, LA12 9PN
2M0 URP S. Gray-Jones, Flat C, 7 Nelson Street, Aberdeen, AB24 5EP
2M0 URT H. Knox, Shalderha Holm, Orkney, KW17 2SA
2E0 USA A. Jepson, 24 Shortbrook Close, Westfield, Sheffield, S20 8LE
2E0 USC S. Coleman, 32 Southwell Road, Wisbech, PE13 3LQ
2E0 USD S. Dean, 39 Low Grange View, Leeds, LS10 3DT
2E0 USG S. Green, 11 Lavender Walk, Beverley, HU17 8WE
2W0 USK T. Keenan, 54 Burrium Gate, Usk, NP15 1TN
2E0 USM J. Wright, 32 Carlton Road, Nottingham, NG10 3LF
2W0 USN J. Barry, Flat 3, 31 Ely Road, Cardiff, CF5 2JF
2E0 USS L. Sanduly, 1 Archford Croft, Emerson Valley, Milton Keynes, MK4 2EZ
2E0 USV D. Soames, 40 Woodland Drive, North Anston, Sheffield, S25 4EP
2E0 UTB M. Mcdonald, 55 Bournemouth Avenue, Middlesbrough, TS3 0NN
2E0 UTC J. Alincastre, 90 York Crescent, Durham, DH1 5PT
2E0 UTD A. Mclean, 47 Tarn Drive, Bury, BL9 9QB
2M0 UTH P. Dower, 1670 Maryhill Road, Glasgow, G20 0HJ
2E0 UTL A. Salt, 1 Chantry Close, Harrow, HA3 9QZ
2W0 UTT B. Bull, Swan Cottage, Swan Road, Welshpool, SY21 0RH
2E0 UTX A. Emmerson, 31 Culver Road, Stockport, SK3 8PG
2M0 UTZ C. Mackie, 4/36 Gillsland Road, Edinburgh, EH10 5BW
2W0 UUA I. Ward, 37 Maes Gwydryn, Abersoch, Pwllheli, LL53 7ED
2E0 UUW J. Douch, 63 Greenaways, Ebley, Stroud, GL5 4UN
2E0 UVO A. Ruocco, 16 Conyers Avenue, Grimsby, DN33 2BY
2E0 UVP D. Bisbey, 17 Benson Close, Lichfield, WS13 6DA
2E0 UVW A. Randle, 2B Truman Street, Kimberley, Nottingham, NG16 2HA
2E0 UVZ K. Such, 38 Hornby Grove, Hull, HU9 4PG
2E0 UWI R. Wilmot, 41 Milton Brow, Weston-Super-Mare, BS22 8DD
2E0 UWK J. Hadley, 75 Glendower Avenue, Coventry, CV5 8BD
2E0 UWT A. Chapman, 10 Derwent Road, Seaton Sluice, Whitley Bay, NE26 4JH
2E0 UXC J. Mckie, 59 Leaholme Terrace, Blackhall Colliery, Hartlepool, TS27 4AB
2E0 UXM K. Taylor, 9 Whitcliffe Grange, Richmond, DL10 4ES
2E0 UXS C. Dutton, Twillingate Farm, Tiptoe, Lymington, SO41 6EJ
2E0 UXV I. Renton, 6 Parham Road, Bournemouth, BH10 4BB
2E0 UYB J. Smith, 9 Trafalgar Road, Newport, PO30 1QD
2E0 UYF E. Martin, 28 Mountford Close, Wellesbourne, Warwick, CV35 9QQ
2E0 UZK A. Sweet, 3 Beechwood Grove, Blackpool, FY2 0DZ
2W0 UZO D. White, 222 St. Fagans Road, Cardiff, CF5 3EW
2E0 VAA K. Britain, Blenheim Cottage, Falkenham, Ipswich, IP10 0QU
2W0 VAC C. Rayment, Brambles, Alltami Road, Mold, CH7 6RT
2E0 VAF J. Edmunds, 17 Stephens Road, Liskeard, PL143SX
2E0 VAG S. Smith, 12 Stoneleigh Avenue, Hordle, SO41 0GS
2E0 VAM M. Higham, 30 Broome Road, Southport, PR8 4EQ
2E0 VAN D. Stinson, 1 The Croft, Earls Colne, Colchester, CO6 2NH
2E0 VAO S. Pankhurst, 57 Barley Croft, Harlow, CM18 7QZ
2E0 VAS V. Papanikolaou, 104 West Drive Gardens, Soham, Ely, CB7 5EX
2E0 VAT M. Raynor, 68 Cambridge Street, South Elmsall, Pontefract, WF9 2AR
2E0 VAU S. Fairbourn, 17 Perry'S Lane, Wroughton, Swindon, SN4 9AX
2E0 VAV A. Burnett-Provan, 6 The Park-Dodwell Park-Evesham Road, Stratford-upon-Avon, CV37 9SR
2W0 VAW I. Kane, 44 Hafod Arthen Estate, Brynithel, Abertillery, NP13 2HY
2E0 VBB S. Coulson, 17 Charlton St., York, YO23 1JN
2I0 VBH V. Hazelton, 12 The Elms Bush, Dungannon, BT71 6UE
2E0 VBJ A. Barrett, 2 Friars Close, Clacton-on-Sea, CO15 4EU
2E0 VBK Dr B. Kalogerakis, Inglewood, Madingley Road, Cambridge, CB23 7PH
2E0 VBL D. Wood, 3 Ripley Close, Wakefield, WF3 2FG
2E0 VBM A. Bairstow, 12 Danesfield Avenue, Waltham, Grimsby, DN37 0QE
2E0 VBN K. Sloan, Woodland Halt, Old Station Road, Winchester, SO21 1BA
2E0 VBR J. Baughan, Chestnut Farm, Eastville, Boston, PE22 8LX
2E0 VBS K. Florence, 30 Lancaster Gardens, Ealing, London, W13 9JY
2W0 VBT M. Kveksas, 25 Barnard Way, Church Village, Pontypridd, CF38 1DQ
2E0 VBW B. Whall, 3 Farrow Close, Great Moulton, Norwich, NR15 2HR
2E0 VBX P. Otterwell, 50 Hythe Road, Staines-upon-Thames, TW18 3EE
2E0 VBY D. Potter, 30 Mersham Gardens, Goring-By-Sea, Worthing, BN12 4TQ
2E0 VCA A. Hill, 24 Keelham Drive, Leeds, LS19 6SG
2E0 VCB B. Ashall, 22 Bloomer Wood View, Sutton-in-Ashfield, NG17 1HA
2E0 VCC D. Jacobs, 5 Tor View, Tregadillett, Launceston, PL15 7HB
2E0 VCD D. Jones, 102 Bryce Road, Brierley Hill, DY5 4ND
2E0 VCE B. Mcguirk, 9 Almond Crescent Standish, Wigan, WN6 0AZ
2E0 VCG P. Robins, 20 Saffron Close, Chineham, Basingstoke, RG24 8XQ
2E0 VCK K. Vickers, 11 Kendal Drive, Rainhill, Prescot, L35 9JQ
2E0 VCM C. Mallory, 11 Baymead Meadow, North Petherton, North Petherton, ta6 6qw
2E0 VCP C. Pinder, 70 Highfield Road, Beverley, HU17 9QR
2E0 VCU P. Pritchard, 12 Easton Crescent, Billingshurst, RH14 9TU
2E0 VCW C. Gray, The Pigsty, Cleverton, Chippenham, SN15 5BT
2E0 VCY G. Shakespeare, 6 Waterworks Cottages, Clough Road, Hull, HU6 7QB
2E0 VCZ Dr C. Jenkins, 52 Warden Abbey, Bedford, MK41 0SN
2E0 VDC N. Fairbairn, 15 Hewitt Road, Dover, CT16 1TH
2E0 VDE R. Ferguson, 31 Barton Court Road, New Milton, BH25 6NW
2E0 VDL D. Lodwig, 15 Frithwood Park, Brownshill, Stroud, GL6 8AB
2E0 VDQ A. Bolster, 45 Headlands Drive, Hessle, HU13 0JP
2E0 VDV C. Lemin, 44 Barton Road, Berrow, Burnham-on-Sea, TA8 2LT
2W0 VDW N. Thomas, 57 Brynhyfryd Street, Treorchy, CF42 6DT
2E0 VEF P. Johnson, 90 North Road, Southport, PR9 8QR
2W0 VEH D. Thomas, 57 Brynhyfryd Street, Treorchy, CF42 6DT
2E0 VEK K. Lowcock, 43 Larch Street, Nelson, BB9 9RH
2D0 VES C. Glaister, Balleigh Villa, Jurby Road, Isle of Man, IM8 3NZ
2E0 VET S. Froggatt, 140 Greenlea Court, Huddersfield, HD5 8QB
2E0 VEW V. Wheatley, 22 Woodlands Avenue, Shelton Lock, Derby, DE24 9FQ
2E0 VEX A. Crawford, 4 Trimpley Drive, Kidderminster, DY11 5LB
2E0 VEZ P. Barrows, 5A Magdalen Road, Willoughby, Rugby, CV23 8BJ
2E0 VFA A. Colcombe, 217 Church Drive, Quedgeley, Gloucester, GL2 4US
2E0 VFD S. Harriott, 188 Tean Road, Cheadle, Stoke-on-Trent, ST10 1NQ
2E0 VFN B. Bewick, 2 Grange Close, Hoveton, Norwich, NR12 8EA
2I0 VFO J. Sills, 145 Ballycolman Estate, Strabane, BT82 9AJ
2E0 VFT A. Middleton, 89 Crediton Road, Okehampton, EX20 1NU
2M0 VFV S. Young, 103 Feorlin Way, Garelochhead, Helensburgh, G84 0EB
2E0 VFX S. Bint, 5 Chapel Corner, Hullavington, Chippenham, SN14 6RT
2E0 VGA S. Unsworth, Heath Terrace, Towcester, NN128UP
2E0 VGB V. Greenway-Brown, 207 Lowe Avenue, Wednesbury, WS10 8NS
2E0 VGC D. Carter, 18 Silent Woman Park, Coldharbour, Wareham, BH20 7PE
2E0 VGF P. Williamson, 17 Shaw Street Biddulph, Stoke-on-Trent, ST8 6JE
2E0 VGK B. Lunn, 204A Main Street, Horsley Woodhouse, Ilkeston, DE7 6AX
2M0 VGT E. Smith, 15 School Wynd, Quarrier's Village, PA11 3NL
2E0 VGV G. Venugopalan, 3 Southwater Close, London, E14 7TE
2M0 VGY E. Blakeway, 25 Allanton Grove, Wishaw, ML2 7LL
2E0 VHA R. Parker, 29 Hill Lea Gardens, Cheddar, BS27 3JH
2E0 VHC M. Sands, Room 3, 8 Upperton Gardens, Eastbourne, BN21 2AH
2E0 VHF C. Craswell, 49 Alexandria Walk, Cheltenham, GL525LG
2E0 VHV L. Kelly, 9 Ham Lane, Farrington Gurney, Bristol, BS39 6TW
2E0 VHZ J. Telfer, 50 Agraria Road, Guildford, GU2 4LF
2E0 VIA Dr L. Kirkcaldy, 17 Central Avenue, Exeter, EX4 8NG
2E0 VIB W. Pilling, 12 Brooke Close, Accrington, BB5 2QX
2E0 VID J. Brewer, 1 Bentley Road, Forncett St. Peter, Norwich, NR16 1LH
2E0 VIS A. Oliveira, 13 A Lakefield Road, London, N22 6RR
2E0 VIT A. Vitiello, 8 Pegasus Road, Leighton Buzzard, LU7 3NJ
2E0 VJB V. Bowkett, 9 Gwealmayowe Park, Helston, TR13 0PE
2E0 VJH R. Heming, Milepost Cottage, Benenden Road, Ashford, TN27 8BY
2D0 VJK R. Smith, 3 Rheast Barrule, Castletown, IM9 1HW
2W0 VJL V. Lea, 30 Cardiff Road, Pwllheli, LL53 5NU
2E0 VJO J. Sawyer, 27 Croft Road, Wallingford, OX10 0HN
2M0 VJS J. Scott, 81 Kinellar Drive, Glasgow, G14 0EU
2E0 VJX B. Walker, 255 Packington Avenue, Birmingham, B34 7RU
2W0 VKA A. Vincent, 88 Lake Street, Ferndale, CF43 4HE
2E0 VKB K. Davies, 23 Egmanton Road, Meden Vale, Mansfield, NG20 9QN
2E0 VKG A. Smith, 305, 1414 5Th St Sw, Calgary, Canada, T2R 0Y8
2E0 VKK R. Cresswell, Meadow View, Hulver Road, Beccles, NR34 7UW
2E0 VKM S. Latimer, 40 Petersham Road, Long Eaton, Nottingham, NG10 4DD
2E0 VKN I. Astley, 6 Shay Court, Crofton, Wakefield, WF4 1SL

2M0 VKO S. Macdonald, 366 Millcroft Road Cumbernauld, Glasgow, G67 2QW
2E0 VKQ R. Vickerstaff, 16 Sewell Wontner Close, Kesgrave, Ipswich, IP5 2GB
2E0 VKR V. Rotaru, 15 Leicester Drive, Glossop, SK13 8SH
2E0 VKS D. Vickers, 178 Bakewell Road, Matlock, DE4 3BA
2E0 VKW V. Williams, Moor Farm, Moor Lane, Lincoln, LN3 4EG
2E0 VKY V. Fleming, 17 Greenfield Close, Kippax, Leeds, LS25 7PX
2E0 VKZ M. Hillman, Flat 5, 32 South Terrace, Littlehampton, BN17 5NU
2E0 VLB Dr V. Boev, Flat 4, Camborne House, Sutton, SM2 6RL
2E0 VLD S. Sawyer, 85 Beechwood Road, Sheffield, S6 4LQ
2M0 VLF I. Mcglynn, 25 Fairhill Avenue, Hamilton, ML3 8JS
2E0 VLL L. Burbidge, 33 Burcote Fields, Towcester, NN12 6TH
2W0 VLO B. Henley, Rhewin Glas, Glynarthen, Llandysul, SA44 6PR
2E0 VLT P. Honey, 3 Peterswood, Harlow, CM18 7RJ
2E0 VMA C. Skupski, 57 Three Nooks, Bamber Bridge, Preston, PR5 8EN
2W0 VMC J. Argent, 7 Lloyds Hill, Buckley, CH7 3ER
2E0 VMD H. Melhuish, 22 Mayflower Close, Glossop, SK13 8UD
2E0 VMG S. Brett, 51 West End, Wirksworth, DE44EG
2D0 VMN V. Matthewman, Monte Rosa, 7 Ballaughton Close, Isle of Man, IM2 1JE
2E0 VMV R. Vale, 611 College Road, Birmingham, B44 0AY
2E0 VNE S. Swinton, 4 Fallow Road, Newton Aycliffe, DL5 4SU
2E0 VNL M. Harris, 5 Lynmore Close, Northampton, NN4 9QU
2E0 VNN V. Nikolaidis, 35-46 Ernst Chain Road, Manor Park, Guildford, GU2 7YW
2E0 VNO D. Harwood, 36 Seaview Drive, Great Wakering, Southend-on-Sea, SS3 0BE
2E0 VNT V. Sheppard, Flat 8, Riverside Court Cambridge Road, Harlow, CM20 2AD
2E0 VNV D. Nicholls, 12 Northview, Tufnell Park Road, London, N7 0QB
2M0 VNW A. Sim, 44 Hillmoss, Kilmaurs, Kilmarnock, KA3 2RS
2E0 VNX A. Wright, 32 Temple Grove, Leeds, LS15 0HT
2E0 VNY V. Brindle, 185 Brunshaw Road, Burnley, BB10 4DL
2M0 VOB D. Johnson, 8 Sandmartin Grove, Lenzie, Kirkintilloch, Glasgow, G66 3WF
2W0 VOC A. Parsons, 21 Rectory Drive, St. Athan, Barry, CF62 4PD
2E0 VOD N. Mclean, 21 Matlock Avenue, Wigston, LE18 4NA
2I0 VOF P. Mcfadden, 35 West Wind Terrace, Hillsborough, BT26 6BS
2W0 VOG S. Cawsey, 8 Hickman Road, Penarth, CF64 2AJ
2E0 VOK R. Remnant, 172 Burnham Road, Highbridge, TA9 3EH
2I0 VOQ D. Sinton, 34 West Link, Holywood, BT18 9NX
2M0 VOZ C. Docherty, 23 The Maltings, Haddington, EH41 4EF
2E0 VPA P. Larcombe, 55 Forest Drive, Weston-Super-Mare, BS23 2UG
2E0 VPG J. Jackson, 80A Clarence Road, Leighton Buzzard, LU7 3EL
2E0 VPH P. Haywood, 1 Scott Close Bidford-On-Avon, Alcester, B50 4HY
2E0 VPI V. Prystaj, 69 Kingsdale Crescent, Bradford, BD2 4DP
2E0 VPJ C. Rainbow, 13 Audley Road, Talke Pits, Stoke-on-Trent, ST7 1UG
2E0 VPL J. Ross-Monclus, 2 Putland Place Harrietsham, Maidstone, ME17 1SZ
2M0 VPM A. Cowan, 32 Esk Valley Terrace, Dalkeith, EH22 3FT
2E0 VPN D. Forbes, Flat 3, Arundel Court 1 Cherrywood Drive, London, SW15 6DS
2E0 VPO R. Dean, 15 Gorge Road, Dudley, DY3 1LF
2W0 VPT G. Taylorr, 17, Llygad Yr Haul, Glynneath, SA11 5RL
2M0 VPU J. Smith, 82 Overton Road, Netherburn, Larkhall, ML9 3BT
2E0 VRA K. Cope, 9 Amber Heights, Ripley, DE5 3SP
2E0 VRB A. Morehen, 20 Castleton Grove, Inkersall, Chesterfield, S43 3HU
2E0 VRC A. Newbould, 9 Laburnum Road, Rudheath, Northwich, CW9 7JT
2E0 VRD V. Bowen, 4 Crossley Gardens, Halifax, HX1 5PU
2E0 VRE D. Easden, 20 Brunel Way, Calne, SN11 9FN
2E0 VRI J. Groves, 350 Middle Deal Road, Deal, CT14 9SN
2E0 VRO G. Bryant, 15 The Clock Inn Park Lydeway, Devizes, SN10 3PP
2E0 VRP I. Bell, 53 Shurland Avenue, Sittingbourne, ME10 4QT
2E0 VRR R. Oxley, 17 Hardhurst Road, Alvaston, Derby, DE24 0LF
2E0 VRT J. Garwood, 4 Ryedale, Carlton Colville, Lowestoft, NR33 8TB
2E0 VRX C. Bradley, 47 Long Meadow, Skipton, BD23 1BP
2E0 VRZ A. Dean, 17, The Grange, 51 Gwendolyn Drive, Coventry, CV3 1QU
2E0 VSE Dr S. Vellaichamy, 37 Briarswood, Chelmsford, CM1 6UH
2W0 VSW V. Wallace, 10 Maes Llydan, Benllech, Tyn-Y-Gongl, LL74 8RD
2E0 VSZ F. Zanchi, Flat 2 4 Helios Road, London, SM6 7BZ
2M0 VTB D. Hamilton, 57/6 North Street, Bo'ness, EH51 0AE
2E0 VTC J. White, 10 Meaux Road, Wawne, Hull, HU7 5XD
2E0 VTF S. Campbell, 50 Northwood Road, Whitstable, CT5 2ES
2W0 VTK J. Martin, 78 Llwyn Ynn, Talybont, LL43 2AG
2W0 VTP M. Davies, 26 Ty Nant, Caerphilly, CF83 2RA
2E0 VTR N. Burton, 11 Weldon Avenue, Stoke-on-Trent, ST3 6PN
2E0 VTS P. Wilkes, 8 Cloverdale, Stafford, ST17 4QJ
2E0 VTT J. Owen, 8 Highridge Crescent, Bristol, BS13 8HN
2E0 VTV I. Roberts, 15 Broadcroft, Hemel Hempstead, HP2 5YX
2I0 VTZ S. Gore, 5 Rosebrook, Dungiven, Londonderry, BT47 4GA
2E0 VUB S. Daley, 1 North Green, Calverton, Nottingham, NG14 6NT
2E0 VUK B. Lewis, 68 Irwin Avenue, Rednal, Birmingham, B45 8QU
2E0 VUL K. Anand, 19 Tempest Road, Upper Cambourne, Cambridge, CB23 6HW
2E0 VUM D. Bailey, 1 The Magpies, Maulden, MK45 2EG
2M0 VUS C. Doherty, 22 Castlelaw Street, Glasgow, G32 0NF
2M0 VUV R. Fraser, 72 Ferguson Drive, Denny, FK6 5AG
2E0 VVA A. Amos, Willow Tree House, Deers Green, Clavering, Saffron Walden, CB11 4PX
2E0 VVB J. Hunter, 53 Grove Road, Tiptree, Colchester, CO5 0JJ
2E0 VVE K. Macmanus, 23 Mount Pleasant Residential Park, Bloomhill Road, Doncaster, DN8 4ST
2E0 VVF A. Blackburn, Anvil House Strait Lane, Hurworth, Darlington, DL2 2AH
2E0 VVJ T. Sandham, 96 South Road, Morecambe, LA4 6JS
2E0 VVK P. Dickson, 23 Balmoral Avenue, Huddersfield, HD4 5LL
2W0 VVO S. Barry, 1 Pearson Cottages, St. Brides, Haverfordwest, SA62 3BN
2M0 VVS I. Lindsay, 265 Stirling Street, Denny, FK6 6QJ
2E0 VVX J. Platt, 12 Tawny Grove Four Marks, Alton, GU34 5DU
2E0 VWF D. Smith, 3 Skipton Close, Corby, NN18 0NS
2E0 VWG M. Cross, 11 Polyplatt Lane, Scampton, Lincoln, LN1 2TL
2E0 VWK M. Poole, 15 Roberts Place, Dorchester, DT1 2JJ
2E0 VWL J. Campion, 25 Suffield Road, Liverpool, L4 1UL
2E0 VWM M. Bridge-Wilkinson, 14 Ascot Close, Bedworth, CV12 8TB
2E0 VWT C. Poole, 15 Devon Close, Macclesfield, SK10 3HB
2E0 VWW S. Hegarty, 31 Beaconsfield Road, Deal, CT14 7DA
2E0 VWX T. Mcbride, 53 Blackdown Grove, St Helens, WA9 2BD
2M0 VXB M. Al Saeed, 9 Appin Place, Edinburgh, EH14 1NJ
2E0 VXI D. Holden, 24 Penny Gate Close, Hindley, Wigan, WN2 3DP
2M0 VXL J. May, 12 Clochbar Gardens, Milngavie, Glasgow, G62 7JP
2E0 VXR A. Murdoch, 128 Whalley Road, Langho, Blackburn, BB6 8DD
2E0 VXT A. Calvert, 5 Pond Cottages, Butts Pond, Sturminster Newton, DT10 1BE
2E0 VXX T. Quiney, 20 Britannia Gardens, Stourport-on-Severn, DY13 9NZ
2E0 VYN M. Roberts, 11 Oakleigh Road, Pinner, HA5 4HB
2E0 VYW A. Willsher, 1 Tolputt Court Gladstone Road, Folkestone, CT19 5NE
2E0 VZL S. Haycock, 51 South Crescent, Southend-on-Sea, SS2 6TB
2M0 VZX D. Mcintosh, Crunes Way, Greenock, PA15 2WH
2E0 WAA O. Prin, 19 The Colliers, Heybridge Basin, Maldon, CM9 4SE
2E0 WAE A. Ward, 29 Mainwaring Road, Wallasey, CH44 9DN
2E0 WAF H. Burch, 46 School Lane, Horton Kirby, Dartford, DA4 9DQ
2E0 WAG D. Wagstaff, 68 Braziers Quay, South Street, Bishop's Stortford, CM23 3YW
2I0 WAH T. Quin, 165 Marlacoo Road, Portadown, Craigavon, BT62 3TD
2I0 WAI M. Mcerlean, 38A Culbane Road Portglenone, Portglenone, BT448NZ
2E0 WAJ W. Johnson, 145 Netherton Road, Worksop, S80 2SA
2E0 WAK P. Holton, 66 Mill Road, Gillingham, ME7 1JB
2E0 WAP A. Woodhouse, 4 Grafton Close, St Albans, AL4 0EX
2I0 WAS W. Campbell, 9 Rochester Court, Coleraine, BT52 2JL
2E0 WAT A. Watmough, 37 Heath Park Road, Buxton, SK17 6NY
2E0 WAV A. Snelson, 6 Rayleigh Close, Braintree, CM7 9TX
2E0 WAY W. Davies, 17 Oakdale Avenue, Harrogate, HG1 2JN
2I0 WBD W. Mcdonald, 14 Edenmore Park, Limavady, BT49 0RG
2E0 WBE D. Buckley, 99 Kings Drive, Bradwell, Great Yarmouth, NR31 8TF
2I0 WBF W. Turkington, 8A Drummullan Road, Moneymore, Magherafelt, BT45 7XS
2E0 WBG W. Bennison, 21 Ashdene Close, Chadderton, Oldham, OL1 2QG
2E0 WBH W. Howie, 152 Norwood Road, Birkby, Huddersfield, HD2 2YD
2E0 WBI G. Birch, 23 Hanson Street, Great Harwood, Blackburn, BB6 7LP
2M0 WBJ W. Jackson, 3 Annick Road, Dreghorn, Irvine, KA11 4EY
2E0 WBL A. Riley, 35 Ross Avenue, Wirral, CH46 2SA
2E0 WBO W. Jones, 50 Bridge Place, Croydon, CR0 2BB
2E0 WBQ D. Hodgson, 11 Harmony Place, Mountain, Queensbury, Bradford, BD13 1LD
2E0 WBR W. Reeves, 33 Pond Bank, Blisworth, Northampton, NN7 3EL
2E0 WBS A. Whadcoat, 38 Edwin Panks Road, Hadleigh, Ipswich, IP7 5JL
2E0 WBT B. Weston, 10 Clement Drive, Peterborough, PE2 9RQ
2I0 WBU W. Bradley, 16 Mullaghanagh Lane, Dungannon, BT71 7NY
2E0 WBW R. Howard, 13 Top Common, East Runton, Cromer, NR27 9PW
2E0 WBX A. Coats, 57 Mill Hill, Boulton Moor, Derby, DE24 5AF
2E0 WBZ K. Hunt, 11 De Marnham Close, West Bromwich, B70 6RJ
2E0 WCB A. Williams, 74 Broadfield Road, Bristol, BS4 2UW
2E0 WCC C. Calvert, 1 Moorsholme Avenue, Manchester, M40 9BW
2E0 WCE M. Hall, 259 Lambourn Drive, Allestree, Derby, DE22 2ur
2E0 WCG S. Marsh, 31A Broad Street, Stamford, PE9 1PJ
2E0 WCL O. Fallon, 26 Central Avenue, Corfe Mullen, Wimborne, BH21 3JD
2E0 WCM D. Neumunn, 92 Miner Street, Walsall, WS2 8QL
2E0 WCO P. Austen, 20 Victor Close, Seaford, Bn25 2jq
2W0 WCP Dr E. Harries, Ty Traeth, Caerwedros, Llandysul, SA44 6BS
2W0 WCQ E. Edwards, 1 Brynhyfryd, Sarn, Bridgend, CF32 9UR
2E0 WCX W. Dix, 21 Pine Vale Crescent, Bournemouth, BH10 6BG
2M0 WDA W. Shaw, Shaws Farm, Selkirk, TD7 4PR
2E0 WDB W. Bull, 117 Walton Road, Wednesbury, WS10 0EU
2I0 WDD D. Milligan, 30 Belgrano Ahoghill, Ballymena, BT42 2QQ
2M0 WDG W. Goodfellow, 1 Yester Place., Haddington, EH41 3BE
2E0 WDH Dr W. Henderson, 14 Highfield Road, Newcastle upon Tyne, NE5 5HS
2E0 WDI I. Woollen, 33 The Oaks, Taunton, TA1 2QX
2E0 WDM S. Thompson, 64 Church Road, Fordham, Colchester, CO6 3NJ
2E0 WDN S. Gibson, Radio Licence, 122 Bridge Street, Whaddon, Royston, SG8 5SN
2M0 WDP D. Park, 54 Coblecrook Gardens, Alva, FK12 5BL
2E0 WDR S. Lacey, 2 Purbeck Cottages, Acton, Swanage, BH19 3LU
2E0 WDS W. Sargeant, 2 Church Mews, Judith Road, Kettering, NN16 0QR
2E0 WDY C. Johnson, 45 Gordon Road, Chelmsford, CM2 9LN
2E0 WEC C. Newton, 7 Moss Close, Bridgwater, TA6 4NA
2E0 WED D. Richards, 58 Holm Lane, Oxton, Prenton, CH43 2HS
2E0 WEE N. Froggatt, 6 Beech Grove, New Malden, KT3 3HR
2E0 WEF I. Chambers, 2 Belford Road, Borehamwood, WD6 4HY
2E0 WEG W. Gray, 39 Guest Avenue, Poole, Bh121ja
2E0 WEI B. Fitchett, 1 Hilly Fields Mews, Parsonage Estate, Rogate, Petersfield, GU31 5BF
2E0 WEJ W. Jefferies, 26 Norcutt Road, Twickenham, TW2 6SR
2E0 WEK K. Alabaster, 16 Butlers Road, Horsham, RH13 6AJ
2E0 WEL W. Easdown, 11 Mulcaster Avenue, Kidlington, OX5 2HG
2E0 WEO S. Kiel, 32 Weavers Avenue, Frizington, CA26 3AT
2E0 WES S. Weston, 73 Priory Road, Ashton-In-Makerfield, Wigan, WN4 9UP
2E0 WET A. Forrest, 1 Errington Bungalows, Sacriston, Durham, DH7 6NE
2M0 WEV G. Weir, 95 White Street, Whitburn, Bathgate, EH47 0BH
2E0 WEZ J. Weston, 25 Cambridge Road, Orrell, Wigan, WN5 8PL
2W0 WFB L. Bowman, Chanrick, Penderyn Road, Aberdare, CF44 9RU
2E0 WFC J. Paradas, 1A Brocks Ghyll, Eastbourne, BN20 9RQ
2E0 WFD P. Blundell, 6 Pennwell Lawn, Leeds, LS14 5NR
2E0 WFG W. Griffiths, 68 Altcar Lane, Formby, Liverpool, L37 6AY
2D0 WFH W. Hogg, Medhamstead, Lhergydhoo, Isle of Man, IM5 2AE
2D0 WFK R. Kissack, 6 Falcon Cliff Court, Douglas, Isle of Man, IM2 4AQ
2M0 WFN W. Noon, 0/1 445 Royston Road, Glasgow, G21 2DE
2E0 WFR G. Kirby, East Road, West Mersea, CO5 8EB
2E0 WFS R. Wormald, 10 Erica Court, Woking, GU22 0JB
2E0 WFW W. Fletcher-Wells, 46 London Road, Buxton, SK17 9NU
2E0 WFY K. Payne, Eastern Esplanade, Broadstairs, CT10 1DR
2E0 WGB G. Beale, 34 Teville Road, Worthing, BN11 1UG
2E0 WGC C. Watkins, 25 Citadilla Close, Gatherley Road, Richmond, DL10 7JE
2E0 WGD D. Warriner, 1 St. Johns Avenue, North Hykeham, Lincoln, LN6 8QR
2U0 WGE R. Batiste, Asile De Paix, Clos Des Sablons, Sandy Lane, Guernsey, GY2 4RN
2E0 WGF E. Lewis, 105 Wards Hill Road, Minster On Sea, Sheerness, ME12 2LH
2E0 WGG S. Biggs, 81 St Abbs Drive, Bradford, BD6 1EJ
2E0 WGI S. Sugihara, Southfield, Park Lane, Wokingham, RG40 4PY
2I0 WGL W. Leonard, 57 Old Coach Road Mullanavehy, Enniskillen, BT92 2EW
2I0 WGM G. Mccusker, 10 Birchdale, Lurgan, Craigavon, BT66 7TR
2W0 WGN C. Morris, Hideaway Bettws Cedewain, Newtown, Powys, SY16 3DS

2E0 WGO I. Paterson, 11 Ocho Rios Mews, Eastbourne, BN23 5UB
2E0 WGP W. Power, 111 Woodlands Road, Ditton, Aylesford, Maidstone, ME206EF
2E0 WGT B. Withington, 20 Bond Way, Hednesford, Cannock, WS12 4SN
2E0 WHA W. Armes, 11 Rutland Road, Broadheath, Altrincham, WA14 4HW
2E0 WHB B. Marks, 167 Linnet Drive, Chelmsford, CM2 8AH
2E0 WHD J. Gardner, Silverdale, Vicarage Lane, Ormskirk, L40 6HQ
2E0 WHF D. Cracknell, 120 Woodhill, London, SE18 5JL
2E0 WHH J. Cook, 20 Huntingdon Close, Totton, Southampton, SO40 3NX
2E0 WHO M. Wells, 42 Eggesford Road, Stenson Fields, Derby, DE24 3BH
2E0 WHP D. Cooper, 58 Serpentine Road, Widley, Waterlooville, PO7 5EF
2E0 WHT K. Snipe, 5 Draycott Road, Chiseldon, Swindon, SN4 0LT
2E0 WHU A. Hodgson, 515 Ashingdon Road, Rochford, SS4 3HE
2M0 WHX D. Strachan, 30 Belhaven Park, Muirhead, G69 9FB
2M0 WIC R. Mackay, 12 Robertson Square, Wick, KW1 5NF
2E0 WIE L. Shasby, 19 Crawshaw Grange, Crawshawbooth, Rossendale, BB4 8LY
2E0 WIG J. Shaw, 54 Dicconson Street, Wigan, WN1 2AT
2E0 WIL W. Whyatt, 686 Whitchurch Lane, Whitchurch, Bristol, BS14 0EJ
2E0 WIO D. Owen, 8 Crag Bank Crescent, Carnforth, LA5 9EQ
2E0 WIS D. Wiskow, 15 Ferndale Close, Sandbach, CW11 4HZ
2E0 WIV P. Sanders, Alresford Road, Wivenhoe, CO7 9JX
2E0 WIZ S. Keen, 13 Ivy Road, Kettering, NN16 9TG
2E0 WJB W. Bradley, 4 Forest View Avenue, London, E10 6DX
2E0 WJC W. Cromack, 45 Southroyd Park, Pudsey, LS28 8AX
2E0 WJE W. Ellis, 16 Furlong Drive, Tean, Stoke-on-Trent, ST10 4LD
2E0 WJF W. Furnell, 4 The Gap, Canterbury, CT13NN
2E0 WJH M. Hooper, 2 Captains Parade, East Cowes, PO32 6GT
2E0 WJI J. Dale, 47 Mungo Park Road, Rainham, RM13 7PD
2E0 WJL W. Lloyd, 14 Lewis Grove, Wolverhampton, WV11 3HR
2I0 WJM W. Mcclean, 6 Alveston Drive Carryduff, Belfast, BT8 8RL
2M0 WJP W. Paterson, 1 Burnside Terrace, Stranraer, DG9 8HH
2E0 WJR D. Howard, 3 Ashdown Drive, Crawley, RH10 5AB
2M0 WJS S. Wilson, 2 Kinnear Court, Guardbridge, St Andrews, KY16 0UE
2E0 WJT W. Twemlow, Flat 6, 27 Marmion Road, Liverpool, L17 8TT
2E0 WJW J. Withington, 20 Bond Way, Hednesford, Cannock, WS12 4SN
2I0 WKE J. Wilkinson, 11 Fairview Park, Dromore, BT25 1PN
2E0 WKG A. Cole, 104 Newport Road, Cowes, PO31 7PS
2E0 WKM C. Wise, 32 Commercial Street, Willington, DL15 0AD
2E0 WKT G. Williams, 18 Elmsleigh Road, Farnborough, GU14 0ET
2E0 WKV K. Dale, 26 Warwick Place, Langdon Hills, Basildon, SS16 6DU
2E0 WKW E. Stammers, 40 Tillingbourne Road, Shalford, Guildford, GU4 8EY
2I0 WKY P. Ffitch, 1 Lisburn Road, Ballynahinch, BT24 8BL
2E0 WKZ B. Drury, 6 Ellen Grove, Harrogate, HG1 4RH
2M0 WLA W. Lawson, 60 Inglis Avenue, Port Seton, EH32 0AQ
2E0 WLD W. Daley, 27 Rosebery Street, Manchester, M14 4UR
2E0 WLK R. Readman, 1 Millside Close, Kilham, Driffield, YO25 4SF
2E0 WLN J. Wilson, 125 Langroyd Road, Colne, BB8 9ED
2E0 WLQ J. Clark, 4 Exeter Street, North Tawton, EX20 2HB
2M0 WLX B. Ewart, 94 Kirkness Street, Airdrie, ML6 6ET
2E0 WLY M. Walters, 39 Portland Place, Coseley, Bilston, WV14 9TB
2W0 WMB A. Gibbs, 105 Oak Place, Bargoed, CF81 8NT
2I0 WMC W. Mccormick, 6 Church Street, Rosslea, Enniskillen, BT92 7DD
2E0 WMD M. Peters, 9 Evelyn Close, Twickenham, TW2 7BL
2E0 WMG K. Pugh, Col Bern, Church Rd, Colchester, CO7 8HS
2I0 WMH W. Hawkes, 12 Meadow Court, Newtownards, BT23 8YE
2M0 WMJ W. Mackenzie, 7 Urquhart Grove, Elgin, IV30 8TB
2M0 WML W. Little, Burnside, Main Street, Lochans, Stranraer, DG9 9AW
2E0 WMM J. Wilkinson, 8 Hunters Point, Chinnor, OX39 4TG
2I0 WMN W. Mcmullen, 69 Lissize Avenue Rathfriland, Newry, BT34 5DE
2E0 WMP M. Weaver, 16 Avocet Drive, Kidderminster, DY10 4JT
2E0 WMT R. Tew, 61 Magna Road, Bournemouth, BH11 9ND
2M0 WMU Dr M. Marino, 35 Niddrie House Park, Edinburgh, EH16 4UH
2E0 WMY C. Walmsley, 6 Holly Close, Brighton, BN1 6RZ
2M0 WNB W. Bennett, 8 Laxay, Lochs, Isle of Lewis, HS2 9PJ
2E0 WNI R. Karpinski, 55 Cambridge Avenue, New Malden, KT3 4LD
2E0 WNL S. Adams, 1 Byford Way, Winslow, Buckingham, MK18 3RJ
2E0 WNM W. Mccoo, Ivy House, West Drove South, Walpole Highway, Wisbech, PE14 7RA
2E0 WNO R. Springall, 7 The Spinney Grange Park, Northampton, NN4 5BT
2E0 WNS C. Pryke, 50 Raglan Gardens, Watford, WD19 4LL
2E0 WNT T. Corker, North Side, Wingerworth Hall Estate, Chesterfield, S42 6PL
2E0 WNW A. Walker, 4 Pretymen Crescent, New Waltham, Grimsby, DN36 4NS
2D0 WNY C. Larkham, Monte Rosa, 7 Ballaughton Close, Douglas, IM2 1JE
2E0 WOB R. Landragin, 101 Linden Gardens Enfield En1 4Dy United Kingd101, England, EN1 4DY
2W0 WOD W. Davies, Foelallt, North Road, Aberystwyth, SY23 2EL
2E0 WOH B. Pyatt, 16 Colin Blythe Road, Tonbridge, TN10 4LB
2E0 WOL W. Walther, 139 East Street, Epsom, KT17 1EJ
2E0 WOS W. Barnes, Cushendall, Lyngate Road, North Walsham, NR28 0DH
2E0 WOW D. Martin, Kiln Close, Main Road, Lincoln, LN4 4QH
2W0 WOY C. Wood, 50 Heather Court, Cwmbran, NP446JR
2E0 WOZ H. Greenhalgh, 61 Long Meadows, Chorley, PR7 2YB
2E0 WPB W. Bruen, 25 Carlton Avenue Upholland, Skelmersdale, WN8 0AE
2E0 WPD D. Woodhams, 83 Langdale Place, Newton Aycliffe, DL5 7DY
2E0 WPE M. Shaw, 10 Beechwood Avenue, Shevington, Wigan, WN6 8EH
2E0 WPH P. Holmquest, 6 Rhyme Hall Mews, Fawley, Southampton, SO45 1FX
2E0 WPI T. Hobson-Smith, 15 Henconner Lane, Chapel Allerton, Leeds, LS7 3NX
2E0 WPJ P. Joyner, 3 Barton Road, Canterbury, CT1 1YG
2E0 WPN A. Hunt, 200 Chorley Old Road Whittle-Le-Woods, Chorley, PR6 7NA
2E0 WPS W. Phillips, 36 Beeches Road, Great Barr, Birmingham, B42 2HF
2E0 WPT M. Thompson, 35 Princes Avenue, Desborough, Kettering, NN14 2RQ
2E0 WPZ D. Chilvers, Flint Cottage, Cherrytree Road, Plumstead, Norwich, NR11 7LQ
2E0 WQK D. Blackie, 8 Kingswood Road, Manchester, M14 6SB
2E0 WRF W. Fuller, 28 Birch Road, Normanton, WF6 1LB
2E0 WRI P. Wright, 4A Alma Street, Melbourne, Derby, DE73 8GA
2E0 WRK J. Erinjeri, 63 Butts Green, Westbrook, Warrington, WA5 7XT
2E0 WRL B. Rickard, Corminnow Lanjeth High Street, St Austell, PL26 7TE
2E0 WRO R. Woodford, 7 Steer Road, Swanage, BH19 2RU
2E0 WRQ S. Dawe, Whiterocks House, St Anns Chapel, PL18 9HN
2E0 WRS S. Webber, 59 Mincinglake Road, Stoke Hill, Exeter, EX4 7DY
2E0 WRT C. Smith, 2 Burley Gardens, Street, BA16 0SN
2M0 WRX K. Glacken, 14 Hailes Avenue, Edinburgh, EH13 0NA
2E0 WRY L. Curtis, 39 Mount Stewart Street, Seaham, SR7 7NG
2E0 WSC W. Cuddeford, 5 Rosevalley Threemilestone, Truro, TR3 6BH
2E0 WSG D. Baker, 17 Woodroyd Gardens, Luddendenfoot, Halifax, HX2 6BG
2I0 WSH W. Hamilton, 9 Susan Street, Belfast, BT5 4FE
2E0 WSJ J. Woodroof, 37 Danefield Road, Northampton, NN3 2LT
2M0 WSK J. Muchowski, 71 The Braes, Tullibody, Alloa, FK10 2TT
2E0 WSM J. Claydon, 17 Canterbury Close, Weston-Super-Mare, BS22 7TS
2E0 WSN D. Clempson, 27 Spalding Road, Hartlepool, TS25 2LD
2E0 WSP S. Peare, 15 Clydesdale Gardens, Bognor Regis, PO22 9BE
2E0 WSR A. Rigler, 10 The Ball, Dunster, Minehead, TA24 6SD
2E0 WSS L. Shand, 52 Ten Acre Way, Rainham, Gillingham, ME8 8TL
2E0 WST R. West, 557 East Bank Road, Sheffield, S2 2AG
2E0 WSW S. Thorne, 2 Ellfield Close, Bristol, BS13 8EF
2E0 WSX A. Holmes, 49 Elm Grove South, Barnham, Bognor Regis, PO22 0EJ
2E0 WSZ J. Hocking, 26 Musket Road, Heathfield, Newton Abbot, TQ12 6SB
2E0 WTA T. Wood, 61 Berry Avenue, Watford, WD24 6RU
2E0 WTD K. Metcalfe, 33 Corsican Drive, Hednesford, Hednesford, WS12 4SS
2M0 WTE P. Jackson, 4 Wester Tarbat House, Kildary, Invergordon, IV18 0GF
2E0 WTG D. Cooper, Little Heath, Bradfield Common, North Walsham, NR28 0QR
2E0 WTH P. Newth, 3 Mulberry Court, Mulberry Street, Stratford-upon-Avon, CV37 6RT
2M0 WTN A. Ross, 29 East Banks, Wick, KW1 5NL
2E0 WTQ T. Walsh, 6 Brass Thill, Durham, DH1 4DS
2E0 WTR B. Shackleton, 7 Erringden Street, Todmorden, OL14 6AW
2M0 WTT S. Paterson, Free Church Manse, Church Street, Golspie, KW10 6TT
2E0 WTU A. Kolesnyk, 3 Hilary Close, Fulham Road, London, SW6 1EA
2E0 WTY R. Clare, Kimberley, Boston Road, Bicker, Boston, PE20 3AP
2E0 WTZ D. Scott, 198 Slade Green Road, Erith, DA8 2JG
2E0 WUF A. Butkus, 73A Hudson Road, Bexleyheath, DA7 4PQ
2M0 WUI P. Bingham, 129 Livingstone Terrace, Irvine, KA12 9ER
2E0 WUK R. Mallinson, 20 Moorside Court, Moorends, Doncaster, DN8 4SL
2M0 WUL W. Murdoch, 64 Cotton Street, Castle Douglas, DG7 1AH
2E0 WUN R. Lester, 17 Clarence Road, Capel-Le-Ferne, Folkestone, CT18 7LW
2E0 WVD M. Bradshaw, 118 Queens Road, Vicars Cross, Chester, CH3 5HE
2E0 WVE M. Huggett, 12 West View Cottages Lewes Road, Lindfield, Haywards Heath, RH16 2LJ
2E0 WVG D. Millward, 77A Meadowcroft, St Albans, AL1 1UG
2E0 WVL C. Lacey, 2 Purbeck Cottages Acton, Langton Matravers, Swanage, BH19 3LU
2E0 WVM M. Edwards, Rouse Farm, Normans Lane, Warrington, WA4 4PY
2E0 WVS W. Symons, Pammel-House, 4 Trevassack Court, Hayle, TR27 4NA
2E0 WVW C. Ring, Acorn Cottage Prospect Place, Helston, TR13 8RU
2E0 WWA A. Kerr, 9 Martindale Way, Sawston, Cambridge, CB22 3BT
2I0 WWB W. Bradley, 14 Ardmore Grange, Ballygowan, Newtownards, BT23 5TZ
2M0 WWC W. Stevenson, 28 Lightburn Road Cambuslang, Glasgow, G72 8UE
2E0 WWD R. Dick, 15 Havenwood, Arundel, BN18 0AH
2E0 WWE Dr G. Geh, 7 River Gardens, Carshalton, SM5 2NH
2E0 WWF E. Field, Sunnyside Farm, Shapwick Road, Westhay, Glastonbury, BA6 9TT
2E0 WWG A. Higgins, 7 Waterloo Terrace, Bideford, EX39 3DJ
2E0 WWJ T. Evans, 36A Swanmore Road, Ryde, PO33 2TQ
2E0 WWK J. Sell, 3 Powter Close, Elsenham, Cm226ut
2E0 WWL S. Wood, 18 Eller Drive, West Winch, King's Lynn, PE33 0NN
2M0 WWM R. Jowett, Fearnoch Ardentallen, Oban, PA34 4SF
2E0 WWN W. Northover, 13 Dagenham Avenue, Dagenham, RM9 6LD
2E0 WWP L. Rowe, 4 Allandale View, Lincoln, LN1 3RD
2W0 WWR P. Abram, 2 Blackthorn Close Marford, Wrexham, LL12 8LB
2E0 WWS D. Cox, 9 Northbrook Copse, Bracknell, RG12 0UA
2E0 WWT A. Walls, 7 Waveney Grove, York, YO30 6EQ
2E0 WWV S. Forshaw, 32 Fox Way, Eastfield, YO11 3PH
2M0 WWX A. Prentice, 24 Victoria Road, Grangemouth, FK3 9JN
2E0 WWY W. Hind, 6 Whinfield Avenue, Shotton Colliery, Co Durham, DH6 2HE
2E0 WWZ R. Alexander, 14 Ashfield Terrace, Appley Bridge, Wigan, WN6 9AG
2E0 WXD B. Wild, 1 Sunnymount, Midsomer Norton, Radstock, BA3 2AS
2E0 WXK P. Marsh, 30 Mount Pleasant, Aylesford, ME20 7BE
2W0 WXM E. Lake, 28 Hampsons Grove Ruabon, Wrexham, LL14 6AN
2M0 WXS A. Mccall, 1 Finlayson Drive, Airdrie, ML6 8LU
2E0 WXT T. Nimash, 6 Wallingford Road, Bristol, BS4 1SL
2E0 WXU S. Reed, 101 Milton Road, Weston Super Mare, Bs23 2ux
2E0 WXY J. Freeman, 46 Wargrove Drive, Sandhurst, GU470DU
2E0 WXZ B. Harding, 76 Grasmere Road, Bare, Morecambe, LA4 6EN
2E0 WYE A. Ayres, Bryn Hyffryd, Phocle Green, Ross-on-Wye, HR9 7TW
2E0 WYG A. Wood, 85 Love Lane, Rayleigh, SS6 7DX
2E0 WYH Q. Wang, 14, Marquis House, 45 Beadon Road, London, W6 0BT
2E0 WYT T. Webster, 1 Fen Close, Newton, Alfreton, DE55 5TD
2E0 WYZ S. Melton, 2A The Orchard, Bishopthorpe, York, YO23 2RX
2E0 WZA N. Lee, Whiteoak, Ashurst Drive, Tadworth, KT20 7LN
2E0 WZR I. Bird, 2 Church Street, Wiveliscombe, Taunton, TA4 2LR
2U0 WZY S. Kirkpatrick, Ste Helene Manor, St Andrew, Guernsey, GY6 8XN
2W0 XAA D. Pollard, 110 Rowan Way, Malpas, Newport, NP20 6JN
2E0 XAE A. Oxborrow, 24 Ickworth Crescent, Rushmere St. Andrew, Ipswich, IP4 5PQ
2E0 XAG D. Robinson, 1 Common Piece, Swinefleet, Goole, DN14 8DE
2E0 XAH A. Harden, 16 Shining Cliff Court, Bawtry, Doncaster, DN10 6SW
2E0 XAI H. Parfitt, 5 Sheridan Road, Frimley, Camberley, GU16 7DU
2E0 XAL G. Davies, 6 Bayleys Close, Empingham, Oakham, LE15 8PJ
2E0 XAM A. Morgan, 18 Keysworth Drive, Wareham, BH20 7BD
2I0 XAN W. Nelson, 44 Lacky Road, Tattynageeragh, Rosslea, Enniskillen, BT92 7GA
2E0 XAO J. Cooke, 9 Nyetimber Crescent, Pagham, Bognor Regis, PO21 3NN
2E0 XAR S. Halliday, 8 Newby Farm Road, Scarborough, YO12 6UN
2E0 XAS J. Saxon, 134 Sherwood Drive, Wigan, Wn5 9rs
2E0 XAV J. Martin, 20 Hall Green Road, West Bromwich, B71 3LA
2E0 XAW A. Winkley, 77 Lechlade Road, Birmingham, B43 5ND
2E0 XAY S. Greaves, The Grange Farmhouse, Leicester Forest East, LE3 3GA
2E0 XAZ J. Deacon, 42A Fairfield, Christchurch, BH23 1QX
2E0 XBA G. Harvey, 55 Trelawney Road Hainault, Ilford, IG6 2NJ
2E0 XBB R. Boland, 10 Kenilworth Drive, Kidderminster, DY10 1YD
2W0 XBC C. Powell, 1 Llwyn-Onn, Penderyn, Aberdare, CF44 9YJ

2M0 XBD S. Boyd, 1 St. Marks Lane, Edinburgh, EH15 2PX
2W0 XBE S. Best, 38 Greensway Abertysswg, Rhymney, Tredegar, NP22 5AR
2E0 XBG J. Walker, 6 Wellington Terrace, Islip, Kettering, NN14 3LJ
2E0 XBM C. Atkinson, 7 Hamilton Road, Grantham, NG31 9QG
2E0 XBN B. Johnson, 6 Trevor Road, Swinton, Manchester, M27 0YH
2E0 XBT S. Pryer, 16 Wayside Avenue, Worthing, BN13 3JU
2E0 XBW B. Woollett, 23 Kinglake House, Southall, UB24FZ
2E0 XBX M. Lee, 46 Little Lane, Huthwaite, Sutton-in-Ashfield, NG17 2RA
2E0 XBZ M. Stead, 38 Park Road, Bracknell, RG12 2LU
2E0 XCA C. Archer, 31 Stoney Bank Drive Kiveton Park, Sheffield, S26 6SJ
2E0 XCB D. Beech, 23 Holding Crescent, Halmer End, Stoke-on-Trent, ST7 8AS
2E0 XCC J. Blackwell, 68 Scarborough Drive, Minster On Sea, Sheerness, ME12 2NQ
2E0 XCD G. Coldham, 27 Welsby Road, Leyland, PR25 1JA
2E0 XCF I. Titchener, 18 King Edgar Close, Ely, CB6 1DP
2E0 XCH C. Hayes, 7 Hadstock Close, Sandiacre, Nottingham, NG10 5LQ
2E0 XCM R. Styles, Padcroft, Weir, OL13 8QL
2E0 XCN C. Norton, 34 The Grove, Little Aston, Sutton Coldfield, B74 3UD
2E0 XCO M. Macdonell, 54 Cinque Foil, Peacehaven, BN10 8DZ
2E0 XCP C. Parker, The Grange, Watercombe Cornwood, Nr. Ivybridge, PL21 9RB
2M0 XCS C. Sharp, 12 Manse Place, Inverkeithing, KY111AZ
2M0 XCT E. Whitaker, Breal House, Drumindorsair, Beauly, IV4 7AH
2E0 XCV M. Abberley, 10 Cranesbill Close, Featherstone, Wolverhampton, WV10 7TY
2E0 XCZ M. Hartley, 24 Burnham Avenue, Bognor Regis, PO21 2JU
2E0 XDA C. Ashley, 22 Pasture Close, Lower Earley, Reading, RG6 4UY
2E0 XDC D. Copsey, Fairview, Mill Lane, Hook End, Brentwood, CM15 0PP
2E0 XDD D. Baseden Butt, 29 Shearwater Way, Stowmarket, IP14 5UG
2E0 XDF D. Ferrington, 20 Innings Lane, Warfield, Bracknell, RG42 3TR
2E0 XDG Dr G. Welch, Amazonas, Sandy Lane, Hightown, Liverpool, L38 3RP
2E0 XDH D. Hudson, 34 Upton Gardens, Upton upon Severn, WR8 0NU
2E0 XDI J. Peain, 29 Wild Flower Way, Ditchingham, Bungay, NR35 2SF
2E0 XDM D. Cole, 39 Hillside Road, Southminster, CM0 7AL
2I0 XDR R. Cross, 15 Ballyfore Road, Larne, BT40 3NF
2M0 XDS D. Suttie, 37 Ullapool Crescent, Dundee, DD2 4TT
2W0 XDT R. Snape, Bodlondeb, North Road, Whitland, SA34 0AX
2M0 XDX A. Lark, 20 Lawfield, Coldingham, Eyemouth, TD14 5PB
2E0 XDY C. Bass, 51 Dane Park Road, Ramsgate, CT11 7LP
2E0 XDZ G. Parsons, 2 The Close, East Grinstead, rh191dq
2E0 XEA A. Welch, 18 Monk Close, Tipton, DY4 7TP
2E0 XEB R. Macgregor, 125 Spring Lane, Birmingham, B24 9BY
2E0 XEE A. King, 25 Nash Road, Dibden Purlieu, Southampton, SO45 4RS
2E0 XEM S. Christie, 124 Bickershaw Lane, Abram, Wigan, WN2 5PP
2E0 XEN J. Fautley, 71 Pullman Lane, Godalming, GU7 1YB
2E0 XES R. Vickery, 17 Plain-An-Gwarry, Redruth, TR15 1JB
2E0 XET J. Daws, 1157 Evesham Road, Astwood Bank, Redditch, B96 6DY
2E0 XEV N. Wright, 10 Olden Mead, Letchworth Garden City, SG6 2SP
2E0 XEW A. Diaz, 29, Parkside Gardens, Widdrington, Morpeth, NE61 5RP
2E0 XFD L. Austin, 2 List Meadows, Littlebourne, Canterbury, CT3 1XW
2E0 XFF M. Chamberlain, 30 Roxton Rd, Great Barford, MK44 3 LR
2W0 XFG S. Gair, Forest View House, Woodcroft, Chepstow, NP16 7PZ
2E0 XFH M. Ashton, Lodge Farm Bungalow, Wattisham Road, Ipswich, IP7 7LU
2M0 XFM B. Burrows, 27 Bughtknowes Drive, Bathgate, EH48 4DP
2E0 XFR J. Stamford, 12 Springhead, Tunbridge Wells, TN2 3NY
2E0 XFX J. Walters-Pennell, Clopton Grange, Clopton, Woodbridge, IP13 6QR
2E0 XGA G. White, 89 Kings Drive, Thingwall, Wirral, CH61 9QA
2E0 XGB M. Carpenter, The Retreat, High Lane Manaccan, Helston, TR12 6HT
2E0 XGO R. Mcgowan, 67 Loop Road North, Whitehaven, CA28 6LS
2E0 XGS G. Stanley, 95 Old Vicarage, Westhoughton, Bolton, BL5 2EG
2E0 XGW G. Whall, 10 Hillcrest Court, Ipswich Road, Diss, IP21 4YJ
2E0 XGX J. Parry, 12 Kerrysdale Close Sutton, St Helens, WA9 3WA
2E0 XGY G. Young, 6 The Maypole Thaxted, Dunmow, CM6 2QZ
2E0 XHB B. Hobbs, 2 Miller Court, Bedford, MK42 9PB
2E0 XHG M. Bolton, 11 Silvia Way, Fleetwood, FY7 7JF
2E0 XHL P. Snook, 7 Sandhurst Avenue, Kwazulu Natal, South Africa, 3610
2E0 XIA N. Pilling, 12 Brooke Close, Baxenden, BB5 2QX
2E0 XIG D. Cassidy, 172 Lyde Road, Yeovil, BA21 5PN
2E0 XII B. Wood, 111 Beech Crescent, Castleford, WF10 3RN
2E0 XIK J. Lambert Hurley, 64 Henry Road, West Bridgford, Nottingham, NG2 7ND
2E0 XIP I. Prior, 81 Ladymeade, Ilminster, TA19 0EA
2E0 XIS M. Morton, 26 Elderberry Gardens, Witham, CM8 2PT
2E0 XIT D. Fincham, 2 Glebe Close, St Columb Major, TR9 6TA
2E0 XJJ J. Jones, 2 Lavender Gardens, Warrington, WA5 1BQ
2E0 XJL J. Welch, 49 Walshs Manor, Stantonbury, Milton Keynes, MK14 6BU
2E0 XJM J. Meredith, 26 Hallam Drive, Shrewsbury, SY1 4YE
2E0 XJN J. Neal, 5 Shelley Close, Huntingdon, PE29 1NF
2E0 XJP C. Dennis, Hillsdene, Plex Lane, Ormskirk, L39 7JY
2W0 XJQ D. Smith, Tyddyn Bach, Bethel, Caernarfon, LL55 1YD
2E0 XJR J. Raffill, 34 Almond Road, Kettering, NN16 9PF
2E0 XJT S. Ramsden, 76 Brigg Lane, Camblesforth, Selby, YO8 8HD
2E0 XJW J. Wood, 3 Onslow Mews, Cranleigh, GU6 8FD
2E0 XKC J. Cunningham, Boleyn House, Erwarton, Ipswich, IP9 1LL
2W0 XKL R. Jones, Flat 2, Tan Y Geraint, 33 Princess Street, Llangollen, LL20 8RD
2E0 XKM M. Koster, Stalworthy Manor Farm, Suton Lane, Suton, Wymondham, NR18 9JG
2E0 XKO P. Goodridge, 22 Horefield, Porton, Salisbury, SP4 0LE
2E0 XKS K. Spowage, 3 Arcadia Avenue, Mansfield, NG20 8JS
2E0 XKT K. Todman, 12 Winscombe, Bracknell, RG12 8UD
2E0 XKX S. Lyon, 10 Sycamore Close, Preston, Hull, HU12 8TZ
2E0 XLG D. Moore, 2 Queens Garth, Thornton In Craven, Skipton, BD23 3TH
2E0 XLH A. Rutson-Edwards, 89 Bemerton Gardens, Kirby Cross, Frinton-on-Sea, CO13 0LQ
2E0 XLI R. Messen, 45 Church Lane North Bradley, Trowbridge, BA14 0TE
2E0 XLJ L. Justin, Garth, Park View Road, Pinner, HA5 3YF
2E0 XLM Dr S. Wing, 107 Highlands Boulevard, Leigh-on-Sea, SS9 3TH
2E0 XLO P. Ffitch, 22 Beeching Close, Halwill Junction, EX21 5XY
2E0 XLR L. Mcintyre, 35 Coley Hill, Reading, RG1 6AE
2E0 XLS O. Hutley, 1 John Ray Street, Braintree, CM7 9DZ
2W0 XLT W. Murphy, 148 Caergynydd Road, Waunarlwydd, Swansea, SA5 4RE
2E0 XLX J. Gascoigne, 1 Mill Meadow, Aylesbury, HP19 8GW
2E0 XLY M. Oxley, 49 Dalton Crescent, Shildon, DL4 2LE
2E0 XMC M. Callis, 1 Webb Close, Letchworth Garden City, SG6 2TY
2E0 XMF P. Foster, 100 Howe Road, Norton, Malton, YO17 9BL
2W0 XMG P. Provis, Dingle Gardens, Croesbychan, Aberdare, CF44 0EJ
2E0 XMK M. Rose, 19 Hawthorn Street, Peterlee, SR8 3LY
2E0 XMO E. Martin, 61 Uffington Avenue, Lincoln, LN6 0AG
2E0 XMP M. Pearce, 1 Hillside Close, Helsby, Frodsham, WA6 9LB
2M0 XMQ L. Anderson, 6 St. Keiran Crescent, Stonehaven, AB39 2GQ
2E0 XMS M. Street, Flat 6, Derwent Court, Solihull, B92 7BU
2E0 XMT T. Wrenn, 2 Wood Street, Grays, RM17 6EQ
2W0 XMW P. Hoath, 8 Liverpool Terrace, Llithfaen, Pwllheli, LL53 6NN
2E0 XNC R. Jeffery, 126 Woodham Lane, New Haw, Addlestone, KT15 3NQ
2M0 XND R. Biggart, Lodgebush Cottage, Craigie, Kilmarnock, KA1 5NA
2E0 XNF N. Ferrington, 20 Innings Lane, Warfield, Bracknell, RG42 3TR
2E0 XNL G. Molendijk, 47 Lodge Road, Scunthorpe, DN15 7EN
2M0 XOA Dr A. Onken, 19/7 Damside, Edinburgh, EH4 3BB
2E0 XOD I. Donnelly, 17 Jessop Close, Horncastle, LN9 6RR
2E0 XOE D. Hart, 7 Penrose Road, Ferndown, BH22 9JF
2E0 XOH V. Fox, 48 Grenham Avenue, Manchester, M15 4HD
2E0 XOJ M. Benson, 11 Hield Grove, Aston By Budworth, Northwich, CW9 6LN
2E0 XOK D. Mitchell, 3 Ivy Cottage, Main Road, Theberton, Leiston, IP16 4RX
2E0 XOL T. Brownen, 43 Great Rea Road., Brixham., TQ5 9SW
2E0 XON H. Tranter, Flat 6 Oak Apple Court 25 Acorn Road Catshill, Bromsgrove, B61 0TR
2E0 XOR M. Hauser, 23 Prince William Way Sawston, Cambridge, CB22 3SZ
2W0 XOT J. Messenger, 34 Goylands Close, Llandrindod Wells, LD1 5RB
2E0 XPC P. Cole, 28 Norfolk Gardens, Newcastle upon Tyne, NE28 7HP
2E0 XPD P. Douglas, 76 Woodside Avenue, Benfleet, SS7 4NY
2E0 XPH P. Hennessey, 11 Monmouth Drive, Eaglescliffe, Stockton-on-Tees, TS16 9HU
2E0 XPJ J. Parfitt, 12 Jodrell Place, Selsey, Chichester, PO20 0FQ
2E0 XPK C. Park, 197 Occupation Road, Albert Village, Swadlincote, DE11 8HD
2E0 XPM P. Mullen, 12 Poplar Grove, Conisbrough, Doncaster, DN12 2JG
2E0 XPP P. Jarvis, 24 St. Peters Gardens, Leeds, LS13 3EH
2D0 XPS D. Heaton, Rushen Vicarage Barracks Rd, Port St Mary, Isle Of Man, IM9 5LP
2E0 XPT T. Parfitt, 5 Sheridan Road, Frimley, Camberley, GU16 7DU
2E0 XPY P. Elliott, 1 Speldhurst Gardens Cliftonville, Margate, CT9 3HJ
2E0 XQA M. Ashford, 11 Pond Close, Felixstowe, IP112JW
2E0 XQK D. Goodfellow, 60 Pickering Green, Gateshead, NE9 7DX
2E0 XQX A. Walrond, Leigh Hill Cottage, Lowton, Taunton, TA3 7SU
2E0 XRD C. Darby, 2 Lindsey Court, Alfred Street, Lincoln, LN5 7PZ
2E0 XRG G. Duffy, 34 Twentyfifth Avenue, Blyth, NE24 2QW
2E0 XRL S. Bennett, 7 Holme Park, High Bentham, LA2 7ND
2E0 XRM D. Bingham, 33 Sheffield Road, Creswell, Worksop, S80 4HN
2E0 XRO A. Dainty, 6 St Nicholas Way Wygate Park Spalding, Spalding, PE11 3GF
2E0 XRQ T. Jones, 3 Langford Road, Liverpool, L19 3RA
2E0 XRS R. Stevens, 53 Keeble Way, Braintree, CM7 3JX
2M0 XRV J. Caldwell, 35, Clunie Drive, Larbert, FK5 4UA
2M0 XRX A. Wright, 17A James Court, 493 Lawnmarket, Edinburgh, eh1 2pb
2M0 XRZ M. Nicholls, Grahams Onsett Farm, Newcastleton, TD9 0TT
2E0 XSB S. Bolton, The Conifers, Methwold Road Methwold Hythe, Thetford, IP26 4QW
2E0 XSD C. Catlin, 27 Main Street, Frizington, CA26 3SA
2E0 XSG C. Braisby, 4 Langmans Way, Woking, GU21 3QY
2E0 XSJ S. Jordan, 74 Soane Gardens, South Shields, Ne34 8nn
2E0 XSL S. Looker, 165 Mollison Drive, Wallington, SM6 9GX
2E0 XSW S. Whall, 17 Vicarage Road, Deopham, Deopham, NR18 9DR
2E0 XSZ K. Edwards, 24 Abbey Road, Halesowen, B63 2HE
2E0 XTA J. Moore, Flat 4, 219 Holland Road, Holland On Sea, Clacton-on-Sea, CO15 6NL
2E0 XTB A. Blews, 57 Highfield Grove, Stafford, ST17 9RA
2E0 XTC K. Haworth, 11 Petersfield Close, Bootle, L30 1SG
2M0 XTH T. Henderson, 139 Crewe Crescent, Edinburgh, EH5 2JN
2E0 XTL M. Porter, 20, Southfield Road, Much Wenlock, TF13 6AX
2E0 XTM J. Maguire, 14 Botha Road, St. Eval, Wadebridge, PL27 7TS
2W0 XTP D. Willis, 51 Fforchaman Road, Cwmaman, Aberdare, CF44 6NG
2M0 XTS C. Robertson, 5 Broomlands Place, Irvine, KA12 0DU
2E0 XTV T. Benson, 83 Glovers Road, Birmingham, B10 0LE
2E0 XUA Dr M. Sinclair, 40 Grotto Road, South Shields, NE34 7AH
2E0 XUH Dr G. Thomas, 3 The Croft, Wilton, Egremont, CA22 2PW
2E0 XUI J. Ribbands, Dyson'S Farm, Long Row, Tibenham, NR16 1PD
2E0 XUK B. Chadwick, 44 Glendale Drive, Mellor, Blackburn, BB2 7HD
2W0 XUL M. Brooks, Garth, Cemmaes, Machynlleth, SY20 9PR
2E0 XUM W. Webb, 84 Bruce Street, Swindon, SN2 2EN
2E0 XUN A. Southern, 89 Purlewent Drive Weston, Bath, BA1 4BD
2E0 XUU R. Gopan, 84 Hilmanton, Lower Earley, Reading, RG6 4HN
2E0 XUZ N. Hanson-Collins, 92 Howbury Lane, Slade Green, Erith, DA8 2DR
2E0 XVD S. Jones, 19 Runshaw Lane, Euxton, Chorley, PR7 6AU
2E0 XVF J. Smith, 8 Mayfields, Spennymoor, DL16 6RN
2E0 XVK D. Pounder, 15 Eldon Grove, Hartlepool, TS26 9LY
2W0 XVT C. Williams, 1 South View, Freeholdland Road, Pontypool, NP4 8LL
2E0 XVX M. Lawrence, 16 Timson Close, Market Harborough, LE16 7UU
2E0 XVZ A. Sibley, 27 Sherwood Road, Tetbury, GL8 8BU
2W0 XWD R. Hawkins, Nook Cottage, Common-Y-Coed, Caldicot, NP26 3AX
2E0 XWX D. Wilkins, Malt Hill Cottage, Malt Hill, Warfield, Bracknell, RG42 6JG
2E0 XXB S. Bunce, 15 Downs View Road, Bembridge, PO35 5QS
2E0 XXC A. Carter, 36 Marriotts Close, Ramsey Mereside, Ramsey, Huntingdon, PE26 2TX
2E0 XXI P. Wright, 59 Vera Crescent, Rainworth, Mansfield, NG21 0EU
2E0 XXK K. Mitchell, 1 Denstroude Cottages, Denstroude Lane, Canterbury, CT2 9JX
2E0 XXM M. Jennings, Springfield Farm, The Causeway, Stow Bridge, King's Lynn, PE34 3PP
2E0 XXO S. Nesling, 64 Ruskin Avenue, Lincoln, LN2 4BT
2M0 XXP A. Pitkethley, 99 Margaretvale Drive, Larkhall, ML9 1EH
2E0 XXT T. Archer, 1 Banks Road, Ashford, TN234NR
2E0 XXX M. Davey, 27 Earls Drive, Newcastle, ST5 3QR
2E0 XYA P. Hodges, 191 Broadstone Road, Stockport, SK4 5HP

2E0 XYM M. Leonard, 75 Skillings Lane, Brough, HU15 1BA
2E0 XYT E. Wood, 13 Rosedale, Welwyn Garden City, AL7 1DW
2E0 XYX A. Austin, 4 Cornwall Avenue, Oldbury, B68 0SW
2E0 XYZ C. Edgar, 9 Winchester Avenue, Morecambe, LA4 6DX
2M0 XZA M. Saunders, 8D Springhill Road, Port Glasgow, PA14 5QP
2E0 XZI J. Ball, Ponsharden Boatyard, Falmouth Road Ponsharden, Penryn, TR10 8AB
2E0 XZM A. Ifrim, 46 Northdown Road, Solihull, B91 3NB
2E0 XZW C. Set, Hughes Hall, Wollaston Road, Cambridge, CB1 2EW
2M0 XZX G. Queen, G/R 31 Provost Road, Dundee, DD3 8AF
2E0 XZZ S. Rouse, 7 Cranbrook Road, Thurnby, Leicester, LE7 9UA
2W0 YAB S. Morgan, 38 Ffordd Cadfan, Bridgend, CF31 2DP
2W0 YAD R. Williams, 92 Bowleaze, Greenmeadow, Cwmbran, NP44 4LF
2W0 YAE G. Thatcher-Sharp, 20 Dilys Street, Blaencwm, Treorchy, CF42 5DT
2M0 YAF C. Tait, 127 Bonnyton Road, Kilmarnock, KA1 2NU
2E0 YAH S. Abdullah, 24 The Grove, Walsall, WS5 4BX
2E0 YAL D. Parker, 16 Aldborough Road, Dagenham, RM10 8AS
2E0 YAO L. Jex, 26 Springdale Crescent, Brundall, Norwich, NR135RA
2E0 YAP A. Kissin, 115 Aarons Hill, Godalming, GU7 2LJ
2E0 YAQ L. Iakovlev, 20A Brownlow Road, London, N3 1NA
2E0 YAR F. Lees, 5 St. Winifred Road, Rainhill, Prescot, L35 8PY
2E0 YAS J. Stacey, 12 Kimbridge Road, East Wittering, Chichester, PO20 8PE
2E0 YAV W. Jones, 8 Oakbrook Close, Ewyas Harold, Hereford, HR2 0NX
2E0 YAW M. Driscoll, 14B Pretoria Road Hedge End, Southampton, SO30 0BS
2E0 YAX A. Garn, 5 Bassett Street, Walsall, WS2 9PZ
2J0 YAY J. Bryant, 5 Louiseberg Court Queen'S Road, St Helier, Jersey, JE2 3GQ
2E0 YBB B. Brown, 3 Swaledale, Worksop, S81 0UY
2I0 YBH J. Mccourt, 70A Sessiagh Scott Road Rock, Dungannon, BT70 3JU
2E0 YBL L. Jones, 8 Oakbrook Close, Ewyas Harold, Hereford, HR2 0NX
2M0 YBR G. Fordyce, 2 Church Street, East End, Earlston, TD4 6HS
2E0 YBS B. Scroggs, Thatchways, High Street, Banbury, OX15 5HW
2W0 YBZ P. Smith, 29 Heol Cwarrel Clark, Caerphilly, CF83 2NE
2E0 YCA E. Cooper, 20 Manor Close, Baston, Peterborough, PE6 9PH
2E0 YCD D. Ashton-Hilton, 14 Weetwood Road, Congresbury, Bristol, BS49 5BN
2M0 YCG C. Graham, 64 Forgewood Road, Motherwell, ML1 3TH
2M0 YCJ Prof. C. Jones, 11B Ettrick Road, Edinburgh, EH10 5BJ
2E0 YDA A. Bedford, 1 Carder Crescent, Bilston, WV14 0JT
2E0 YDB D. Bower, 4 Winsford Road, Sheffield, S6 1HT
2I0 YDF D. Foley, 14 Chestnut Hall Court, Moira, Maghaberry, BT67 0GJ
2E0 YDJ D. Jackson, 3 Laburnum Road, Cadishead, Manchester, M44 5AS
2W0 YDK D. Edwards, 25 Bryn Coed, Gwersyllt, Wrexham, LL11 4UE
2E0 YDM C. Preece, 14 Dock Street, Widnes, WA8 0QX
2E0 YDT A. Carney, 9 Hart Square, Sunderland, SR4 8BS
2E0 YDX J. Leighton, 27 The Pastures, Cayton, Scarborough, YO11 3UU
2E0 YEA E. Raine, Stable Cottage, St. Martins, Richmond, DL10 4SJ
2E0 YEK D. Noyek, 34 The Spinney, Sidcup, DA14 5NF
2E0 YEO R. Froggatt, The Hawthorns, 2 Laurels Drive, Barton St David, TA11 6AT
2E0 YEP S. Quinn, 7 Poppleton Court, Tingley, Wakefield, WF3 1UY
2M0 YEQ G. Pearce, 2 Kirkriggs, Forfar, DD8 2AT
2E0 YES M. Casey, 7 Cobham Avenue, Manchester, M40 5QW
2E0 YEW A. Mcewen, 4 The Pantyles, Nightingale Lane, Sevenoaks, TN14 6BX
2E0 YEY G. Matthews, 255 Coach Road Estate, Washington, NE37 2EU
2E0 YEZ A. Atkinson, 2E Bagridge Road, Wolverhampton, WV3 8HW
2W0 YFC J. Evans, 11 Dew Crescent, Cardiff, CF5 5PB
2E0 YFG F. Grande, Restrup Engvej 26, Aalborg, Denmark, 9000
2E0 YFR D. Cockburn, 30 Queensberry Road, Burnley, BB11 4LH
2E0 YFZ J. Blower, 4 Lamorna Close, Luton, LU3 2TH
2E0 YGB A. Birch, 3 Partridge Way, High Wycombe, HP13 5JX
2E0 YGC A. Cottell, 3 Honeylight View, Swindon, SN25 4XS
2E0 YGH F. Armstrong, 38 Dovecote Drive, Haydock, St Helens, WA11 0SD
2E0 YGM M. Clayton, Swallows 5 The Waldrons East Garston, Hungerford, RG17 7JB
2E0 YGR G. Robertson, 12 Chester Close Ince, Wigan, WN3 4JP
2E0 YGS G. Moss, 125 Lavender Avenue, Mitcham, CR4 3RS
2E0 YGT S. Mucklow, Blithbury House, Blithbury Road, Rugeley, WS15 3HR
2E0 YHF J. Marvel, Dean House Farm, Nordan, Leominster, HR6 0AW
2E0 YHI R. Bragg, 50 Moorclose Road Harrington, Workington, CA14 5LB
2E0 YHN J. Penniston, 1 Line Cottage The Causeway, Thorney, Peterborough, PE6 0QJ
2E0 YHW R. Lloyd, 18 Newbury Grove Blurton, Stoke-on-Trent, ST3 3DD
2E0 YHZ K. Nelson, 185 Headlands, Fenstanton, Huntingdon, PE28 9LP
2E0 YIB R. Kershaw, 40 The Butts, Frome, BA11 4AA
2W0 YIF D. Williams, 371 Coed-Y-Gores, Llanedeyrn, Cardiff, CF23 9NR
2E0 YIN D. Smith, 60 Stocken Close, Gloucester, GL3 3UL
2M0 YIO B. Fullerton, 55 Alexander Avenue, Stevenston, KA20 4BG
2E0 YIT C. Peters, 15 Jasmine Crescent Trimdon Village, Trimdon Station, TS29 6QE
2E0 YJB J. Blackall, 2 Ryson Avenue, Blackpool, FY4 4DN
2E0 YJF D. Moran, 23 Abbotsfield Crescent, Tavistock, PL19 8EY
2E0 YJL J. Cairns, 17 Alfred Avenue, Worsley, Manchester, M28 2TX
2E0 YJW J. Williams, 66 Oakfield Avenue, Hitchin, SG4 9JD
2E0 YJY M. Kealey, 24 Ben Nevis Road, Birkenhead, CH42 6QY
2E0 YKC D. Forshaw, 14 Hope Carr Road, Leigh, WN7 3ET
2E0 YKX Dr J. Dunn, 6 Webbs Way, Ashley Heath, Ringwood, BH24 2DU
2E0 YLD L. Davis, 9 High Street, Chapel-En-Le-Frith, High Peak, SK23 0HD
2W0 YLE A. Doyle, 54 Bro Syr Ifor, Tregarth, Bangor, LL57 4AS
2E0 YLH J. Haystead, 11 Lumley Close, Maltby, Rotherham, S66 7SG
2W0 YLL R. Bowen, 25 Maendu Terrace, Brecon, LD3 9HH
2E0 YLP R. Campbell-Black, 10 Wren Close, Towcester, NN12 6RD
2E0 YLR J. Raynerd, 3 Brooksbottoms Close, Ramsbottom, Bury, BL0 9YP
2I0 YLT S. Mccormick, 46 Lany Road, Moira, Craigavon, BT67 0NZ
2D0 YLX D. Cain, 7 Cronk Y Berry Mews, Douglas, Isle Of Man, IM2 6HQ
2E0 YLZ N. Bacala, Flat 3, 232A Seven Sisters Road, London, N4 3NX
2M0 YMA Dr A. Brasier, 10 Wellgrove Crescent, Westhill, AB32 6TH
2E0 YME M. Smith, 2 Tullig, Cahirciveen, County Kerry, Ireland, V23 V348
2I0 YMF M. Foley, 44 Gallows Street, Dromore, BT25 1BD
2W0 YMG M. Graham, 60 Heol Seward Beddau, Pontypridd, CF38 2SR
2E0 YMM J. Griffiths, 9 Cauldale Close, Middleton, Manchester, M24 5SU
2W0 YMP S. Keeble, Mynachlog, Tyn Y Gongl, LL74 8SG
2E0 YMR L. Humphreys, Flat 5 Sabie Court, 21 Argyll Road, Bournemouth, BH5 1EB
2W0 YMS M. Stevenson, 64 Leslie Terrace, Porth, CF39 9TE
2E0 YMT P. Horrox, 39 Wilton Grove, Heywood, OL10 1AS
2E0 YMV M. Vaites, 19 Campion Drive, Sheffield, S21 1TG
2E0 YMW M. Williamson, 5 John F Kennedy Walk, Tipton, DY4 0SF
2E0 YND B. Higgins, 41 Lower Meadow, Harlow, CM18 7RE
2E0 YNI S. Widdowson, 19 Waysidebrimington, Chesterfield, S43 1BQ
2M0 YNK N. Kirtley, 10 Millcroft Road, Auldearn, Nairn, IV12 5TW
2E0 YNT A. Ashby, 5 New Street, Osbournby, Sleaford, NG34 0DL
2E0 YNY N. Oldrid, 4 Bar Lane Mapplewell, Barnsley, S75 6DQ
2E0 YOI I. Eliade, 44 Brookside Road, Stratford-upon-Avon, CV37 9PH
2E0 YOK C. Bietz, 7 Harry Watson Court, Norwich, NR3 3SX
2E0 YOM J. Thresher, Quarry Grange, Nuneaton Road Over Whitacre, Coleshill, Birmingham, B46 2NH
2E0 YOP T. Court, Eastgate Cottage, Perrys Lane, Norwich, NR10 4HJ
2M0 YOY J. Moir, 41 Brisbane Terrace, East Kilbride, Glasgow, G75 8DL
2E0 YOZ J. Wilson, 5 Queens Road Hoylake, Wirral, CH47 2AG
2E0 YPG R. Irwin, 21 Penn St., Belper, DE56 1GH
2E0 YPJ P. Kirby, 30 New Street, Eccleston, Chorley, PR7 5TW
2E0 YPK E. Maddex, 58 Bohemia Chase, Leigh-on-Sea, SS9 4PP
2E0 YPU T. Galloway, 1 Farley Close Shadoxhurst, Ashford, TN26 1NB
2E0 YPW Dr P. Woodfin, Laurel Cottage, Barrow Street, Much Wenlock, TF13 6EN
2E0 YQC G. Hope, 27 Clearmount Drive Charing, Ashford, TN27 0LH
2E0 YQT J. Best, 24 Suggitts Lane, Cleethorpes, DN35 7JJ
2M0 YRB R. Rimmer, Glenalty Cottage, Barrhill, KA26 0QT
2E0 YRF Dr A. Saje, 72 Bedworth Road, Bulkington, Bedworth, CV12 9LL
2E0 YRM M. Radulov, 14 Grove Road, Chatham, ME4 5HS
2E0 YRT C. Williams, 16 Mill Meadow, Tenbury Wells, WR15 8HX
2E0 YRU K. Michael, 55 St Olaves Road, Norwich, NR3 4QB
2E0 YRW J. Woods, 21 Appleyard Crescent, Norwich, NR3 2QN
2E0 YSB P. Bunting, 29 Marion Avenue, Alverthorpe, Wakefield, WF20BJ
2E0 YSF L. Metcalfe, 40 St. Anns Court, Hartlepool, TS24 7HY
2E0 YSK A. Bragg, Buena Vista Low Moresby, Whitehaven, CA28 6RR
2E0 YSO M. Mayson, Bell Cottage, School Road, Broughton, Huntingdon, PE28 3AT
2E0 YSP A. Gobey, Nut Tree Cottage, Valley Rd, Ipswich, IP8 4LR
2M0 YSR C. Phillips, 8 The Square, Newtongrange, Dalkeith, EH22 4QD
2E0 YSU G. Cummings, 10 Perth Close, Skegness, PE25 2HY
2M0 YTA J. Anderson, The Grange, Leslie Road, Scotlandwell, Kinross, KY13 9JE
2E0 YTB R. Thomas, 9 Spa View Terrace, Sheffield, S12 4HG
2M0 YTD T. Davidson, 141 Huron Avenue, Livingston, EH54 6LQ
2E0 YTF G. Garman, 11 Rye Close, Norwich, NR3 2LF
2M0 YTN J. Keymer, 2 Crunklaw Farm Cottage, Duns, TD11 3RA
2E0 YTT R. Spooner, 45 Shaftesbury Avenue, Southport, PR8 4NH
2E0 YTZ R. Price, 7 Keymer Way, Colchester, CO3 9XJ
2E0 YUD S. Nutt, 77 Exeter Close, Stevenage, SG1 4PW
2E0 YUN Y. Li Song, 8 Birkin Court, Welwyn Garden City, AL7 3FA
2M0 YUP S. Duff, 1 Castle Moffat Cottages, Garvald, Haddington, EH41 4LW
2E0 YVR L. Palir, 116 Carville Crescent, Brentford, TW8 9RD
2E0 YVT P. Van Staveren, 14 Fortune Green Road Flat 3, London, NW6 1UE
2W0 YVY C. Ibarra-Rivadeneira, 79 Heol Frank, Penlan, Swansea, SA5 7AH
2E0 YWA I. Stanley, 11 Reading Road, Burghfield Common, RG7 3PY
2E0 YWN R. Cook, Flat 2, 1 Duke Street, Salford, M7 1PR
2E0 YWO M. Bruyneel, 14 Riversmead, St Neots, PE19 1HA
2E0 YWP D. Spooner, 30 Clover Road, Norwich, NR7 8TF
2E0 YXB R. Barrett, 18 Bullstake Close, Oxford, OX2 0HN
2E0 YXO M. Bilverstone, 12 Westlea Road, Sywell, Northampton, NN6 0BY
2E0 YXX K. Ramminger, 54 White Hedge Drive, St Albans, AL3 5TX
2E0 YXZ S. Christofi, 19 Kingsland Avenue, Northampton, NN2 7PP
2E0 YYA D. Crane, 3 Middlemead Close West Hanningfield, Chelmsford, CM2 8UR
2E0 YYB C. Norwood, 94 Waverley, Woodside, Telford, TF7 5LU
2E0 YYD P. Dumpleton, 20 Cambridge Road North, Mablethorpe, LN12 1QR
2E0 YYG D. Harwood, 32 John Reid Road, South Shields, NE34 9EB
2E0 YYK S. Hoyle, 20 Brandsby Grove, Huntington, York, YO31 9HL
2E0 YYM P. Boulding, 16 Higher Croft Road, Lower Darwen, Darwen, BB3 0QR
2W0 YYP P. Griffey, 61 Cottesmore Way, Cross Inn, Pontyclun, CF72 8BG
2E0 YYT G. Finney, 121 School Lane, Caverswall, Stoke-on-Trent, ST11 9EN
2M0 YYU A. Anderson, 16 Walker Court, Glasgow, G11 6QP
2E0 YYY M. Hunter, 126 Turner Street, Stoke-on-Trent, ST1 2NE
2E0 YYZ S. Wall, 26 Wallace Lane, Whelley, Wigan, WN1 3XT
2E0 YZA B. Wilson, 69 Rotherwood Road, Sheffield, S21 2dt
2E0 YZC D. Matheson, 21 Warren Hill Road, Woodbridge, IP12 4DU
2E0 YZI B. Gray, 5 Oldfield Grove, London, SE16 2NA
2E0 YZM G. Brown, 134 Skipper Way, Lee-on-the-Solent, PO13 8HD
2E0 YZQ A. Kent, 4 Sellerdale Drive, Wyke, Bradford, BD12 9DA
2M0 YZT P. Connon, 4 Highfield Court, Stonehaven, AB39 2PL
2E0 YZW M. Wandby, 44 Windrush Road, Hollywood, Birmingham, B47 5QA
2E0 YZX M. Galloway, 2 Edendale Terrace, Horden, Peterlee, SR8 4RD
2E0 YZY I. Hodgkiss, 190 Ulverley Green Road, Solihull, B92 8AD
2E0 YZZ P. Collingham, 1 Wychwood Drive, Trowell, Nottingham, NG9 3RB
2W0 ZAA S. Tozer, 110 Glanffornwg, Wildmill, Bridgend, CF31 1RL
2E0 ZAC M. Cotton, 18 St. Oswalds Crescent, Brereton, Sandbach, CW11 1RW
2W0 ZAE P. Mason, 20 Coronation Road, Six Bells, Abertillery, NP13 2PJ
2E0 ZAF R. Amos, 6 Eccles Road, Wittering, Peterborough, PE8 6AU
2E0 ZAH L. Almond, 26 Ashbourne Drive, Desborough, Kettering, NN14 2XG
2E0 ZAI B. Hardy, 10 Spring Farm Road, Burton-on-Trent, DE15 9BN
2E0 ZAJ S. Rafter, 30 Monmouth Grove, St Helens, WA9 1QB
2E0 ZAL J. Moore, Moorelake Lodge, Barholm Road, Stamford, PE9 4RJ
2E0 ZAP R. Bird, 78 Arden Road, Hockley, Tamworth, B77 5JE
2E0 ZAU M. Southgate, 107 Englands Lane, Loughton, IG10 2QL
2E0 ZBB Dr M. Palmer, 116 Claverham Road, Yatton, Bristol, BS49 4LE
2W0 ZBC G. Jones, 12 Wilson Place Ely, Cardiff, cf54ln
2E0 ZBD J. Wells, 15 Phillips Crescent, Needham Market, Ipswich, IP6 8TF
2E0 ZBE J. Ellery, 7 Midanbury Crescent, Southampton, SO18 4FN
2M0 ZBF G. Irving, 55 Gillbank Avenue, Carluke, ML8 5UW
2M0 ZBH P. Mclaren, 1 Morayvale, Aberdour, Burntisland, KY3 0XE
2E0 ZBW R. Weaver, 116 Carville Crescent, Brentford, TW8 9RD
2E0 ZBZ M. Carvell, 10 Burns Close, Stevenage, SG2 0JN
2E0 ZCB C. Button, 59 Lindsey Road, Harworth, Doncaster, DN11 8QJ
2E0 ZCG C. Gregory, 81 Fiskerton Way, Oakwood, Derby, DE21 2HY

2I0 ZCM A. Birkhead, 21 Carson Villas, Upperlands Maghera, BT46 5SH
2E0 ZCP C. Palawinna, 3 Stirling Court Road, Burgess Hill, RH15 0PS
2E0 ZDA D. Phillips, 7 Broughton Road, Banbury, OX16 9QB
2E0 ZDB D. Brownsea, 47 Southill Road, Bournemouth, BH9 1SH
2E0 ZDC Dr D. Cooke, Apartment 9, 27 Sheldon Square, London, W2 6DW
2E0 ZDE D. Kirkden, 57 Crow Hill Road, Margate, CT9 5PF
2E0 ZDH D. Hardwick, 30 Halfcot Avenue, Stourbridge, DY9 0YB
2E0 ZDJ L. Macrides, 5 Apple Farm Lane, Weston-Super-Mare, BS24 7TJ
2E0 ZDM D. Mason, 94 Guessburn, Stocksfield, NE43 7QR
2E0 ZDW D. Whitley, 10 Kenmore Drive, Cleckheaton, BD19 3EJ
2E0 ZDX P. Jones, Stonehead Farm Over Wyresdale, Lancaster, LA2 9DL
2M0 ZEB G. Barrie, 24 Mauldeth Road, Broxburn, EH52 6FB
2E0 ZED A. Henderson, Clocktower Lodge, Cragside, Rothbury, Morpeth, NE65 7PU
2M0 ZEE P. Russell, 21 St. Andrews Drive, Law, Carluke, ML8 5GB
2E0 ZEH S. Hendy, Flat 2, 33 Kingston Road, Leatherhead, KT22 7SL
2E0 ZEN H. Haydon, 101 Blandford Street, Ashton-under-Lyne, OL6 7HG
2M0 ZET H. Dally, 3 Gremmasgaet, Lerwick, Shetland, ZE1 0NE
2E0 ZEV B. Dexter, 237 Wordsworth Avenue, Sheffield, S5 8NE
2M0 ZFG S. Street, Tangaroa, Fairfield Gardens, Kilcreggan, Kilcreggan, G84 0HS
2E0 ZFV P. Whiteley, 1 Newton Close, Fareham, PO14 3LF
2I0 ZFZ R. Mckay, 31 Squires Hill Crescent, Belfast, BT14 8RE
2E0 ZGA G. Campbell, 10 Welbeck Road, Rochdale, OL16 4XP
2E0 ZGL A. Lunn, 57 Greets Green Road, West Bromwich, B70 9ES
2E0 ZGS Z. Sznober, 9 Moor Road Dawley, Telford, TF4 2AR
2E0 ZGX P. Beltrami, 15 Woodroffe Square, Calne, SN11 8PW
2E0 ZHG C. Barnes, 23 South Street, Crewe, CW2 6HN
2E0 ZHN E. Mady, 130 Staveley Gardens, London, W4 2SF
2E0 ZIP S. Kiley, 178 Kingfisher Drive, Woodley, Reading, RG5 3LQ
2E0 ZIV I. Vickers, 3 Nesbit Road, St. Marys Bay, Romney Marsh, TN29 0SF
2W0 ZJA D. Bowen, 25 Maendu Terrace, Brecon, LD3 9HH
2E0 ZJB M. Collier, 32 London Road, Dereham, NR19 1AW
2E0 ZJO J. Rawlinson, Westfield Farm, Risden Lane, Cranbrook, TN18 5DU
2E0 ZJQ A. Rawlinson, Westfield Farm, Risden Lane, Hawkhurst, Sandhurst, Cranbrook, TN18 5DU
2E0 ZKT P. Stone, 2 Endeavour Close, Lower Stondon, Bracknell, SG166JR
2E0 ZLA A. Zeller, Flat 1, 57 Chalk Hill, Watford, WD19 4DA
2E0 ZLD Z. Dunne, 1 Burton Gardens, Brierfield, Nelson, BB9 5DR
2E0 ZLM L. Milburn, 17 Hammingden Court, Crawley, RH10 3FR
2E0 ZLO M. Holmes, 6 Wells Court, Saxilby, Lincoln, LN1 2GY
2E0 ZMB M. Breslin, 15 Acorn Gardens, East Cowes, PO32 6TD
2E0 ZMI T. Kelly, 2 Weaver House, Chester Road, Runcorn, WA7 3EG
2E0 ZML J. Marr, Touchstone, Heathfield Road, Bembridge, PO35 5UW
2E0 ZMM D. Mainwaring, 1 Buckingham Close, Didcot, OX11 8TX
2E0 ZMO A. Maguire, 132 Wigan Road, Ormskirk, L39 2BA
2E0 ZMR I. Nicholson, 2 Broom Close, Leyland, PR25 5RQ
2E0 ZMS M. Strickland, Ancoats, Piercy End, York, YO62 6DQ
2E0 ZMT M. Thompson, 133 Redford Avenue, Horsham, RH12 2HH
2E0 ZMZ M. Coleman, 3 Tummon Road, Sheffield, S2 5FD
2M0 ZNQ W. Beaton, 4 Moorfield Gardens, Springfield, Cupar, KY15 5SH
2E0 ZNZ Dr G. Richardson, Berwick Cottage, Bailes Lane, Guildford, GU3 2AX
2E0 ZOM J. Skinner, 36 Milton Road, Waterloo, Liverpool, L22 4RF
2E0 ZOR A. Mccrystal, 15 Amory'S Holt Way, Maltby, Rotherham, S66 8RF
2E0 ZOT M. Toher, The Chapel, Station Road, Darlington, DL2 1JG
2E0 ZOZ A. Hunter, 9 Gelt Burn, Didcot, OX11 7TZ
2E0 ZPA P. Archer, 31 Stoney Bank Drive, Kiveton Park, Sheffield, S26 6SJ
2E0 ZPN L. Gibbs, 37 Oxford Road, Fulwood, Preston, PR2 3JL
2E0 ZPT M. Atfield, 42 Pauls Croft Cricklade, Swindon, SN6 6AJ
2E0 ZPY R. Pyner, 1 Avon Court, 63 Shakespeare Road, Bedford, MK40 2DS
2E0 ZRB R. Brown, 194 Wymersley Road, Hull, HU5 5LN
2E0 ZRG R. Greaves, 7 Eller Brook Close, Heath Charnock, Chorley, PR6 9NQ
2E0 ZRL A. Briggs, 3 Swallow Avenue, Leeds, LS12 4RD
2E0 ZRM D. Morgan, 2 Raymond Way, Plymouth, PL7 4EG
2E0 ZRQ G. Crane, 33A Carlisle Gardens, Horncastle, LN9 5LP
2E0 ZRT T. Cooper, 9 Websters Close, Shepshed, Loughborough, LE12 9AT
2E0 ZRX C. Waterworth, 4 Mossdale Road, Ashton-In-Makerfield, Wigan, WN4 0EQ
2E0 ZSA S. Airs, 6 The Willows, Culham, Abingdon, OX14 4NN
2E0 ZSB S. Bannister, 162 Dobcroft Road, Sheffield, S11 9LH
2E0 ZSE P. Holmes, 53 Bishops Hull Road, Bishops Hull, Taunton, TA1 5EP
2E0 ZSH S. Hampson, 12 Flying Fields Drive, Macclesfield, SK11 7GE
2E0 ZSJ J. Gibson, 22 Woodburn Drive Chapeltown, Sheffield, S351YS
2E0 ZSK S. Kneller, 12A Richard Street, Crewe, CW1 3AF
2E0 ZSR S. Robottom-Scott, 73 St. Bernards Road, Solihull, B92 7DF
2E0 ZST S. Harris, 61 Monks Park Avenue, Bristol, BS7 0UA
2E0 ZSU Capt. P. Westwell, Roden House, Dobsons Bridge, Whitchurch, SY13 2QL
2E0 ZSY T. Symons, Southgate, The Commons, Mullion, TR12 7HZ
2E0 ZTD G. Berry, 5 Oakholme Rise, Worksop, S81 7LJ
2E0 ZTE S. Broom, 128 Springhill Road, Wolverhampton, WV11 3AQ
2E0 ZTG A. Hill, 5 Park Road, Thurnscoe, Rotherham, S63 0TG
2E0 ZTL C. Ingamells, 2 St. Mary'S Drive, Sutterton, Boston, PE20 2LU
2E0 ZTM T. Moore, 16 Warwick Drive, Earby, Barnoldswick, BB18 6LX
2E0 ZUX A. Remnant, 172 Burnham Road, Highbridge, TA9 3EH
2E0 ZVG I. Browne, 85 White Eagle Road, Haydon Leigh, Swindon, SN25 1PY
2E0 ZVL V. Lynch, 16 Okehampton Crescent, Sale, M33 5HR
2E0 ZVR B. Bateman, 27 Imperial Avenue, Kidderminster, DY10 2RA
2E0 ZWA D. Tordoff, 49 Dale Edge, Eastfield, Scarborough, YO11 3EP
2W0 ZWR G. Spicer, 6 Cromwell Road, Neath, SA10 8DR
2E0 ZWW Dr W. Warwicker, 28 Porters Wood, Petteridge Lane, Matfield, Tonbridge, TN12 7LR
2I0 ZXD J. Baker, 324 Clonmeen, Drumgor, Craigavon, BT65 4AT
2E0 ZXG G. Carless, Silver Cottage, Silver Street, South Petherton, TA13 5BY
2E0 ZXJ J. Wildsmith, Flat 34, Romsley Hill Grange, Farley Lane, Romsley, Halesowen, B62 0LN
2I0 ZXM M. Meagher, 42 Mourne View Park, C.Down, Newry, BT35 6BZ
2E0 ZXQ I. Talbot, 41 Elmwood Close, Cannock, WS11 6LX
2E0 ZXR M. Holbrook-Bull, 66 Wayman Road, Corfe Mullen, Wimborne, BH21 3PN
2E0 ZXV A. Booth, 11 Kinnaird Close, Elland, HX5 9JF
2E0 ZYG C. Richardson, Heathercroft, Kirkby Mills, Kirkbymoorside, York, YO62 6NN
2E0 ZYK M. Edge, 19 Burton Ave, Rushall, WS41NH
2E0 ZYL S. Allen, Milverton, Mill Road, Pulborough, RH20 2PZ
2E0 ZYX S. Entwisle, 30 Arden Mhor, Pinner, HA5 2HR
2E0 ZZC S. Alexander, 13 Padgate, Thorpe End, Norwich, NR13 5DG
2W0 ZZF D. Jones, 1 Brig Y Nant, Llangefni, LL77 7QD
2E0 ZZT N. Payne, 19 Sid Park Road, Sidmouth, EX10 9BW
2W0 ZZU E. Jones, 39 Ger-Y-Llan, Velindre, Llandysul, SA44 5YB
2E0 ZZZ J. Earnshaw, Dunelm, Ayton Road, Irton, Scarborough, YO12 4RQ

2#1

2E1 ABE H. Forder, 4 Jackson Drive, Kennington, Oxford, OX1 5LL
2E1 ABN J. Morrison, 52 Kimberley Close, Dover, CT16 2JW
2E1 ABQ N. Harman, 7 Maple Avenue, Torpoint, PL11 2NE
2E1 ABW S. Minnock, 32 Sandwood Road, Sandwich, CT13 0AQ
2E1 ABY M. O'Brien, 14 Westdean Close, Dover, CT17 0NP
2E1 ACG V. Hammonds, 22 The Croft Meriden, Coventry, CV7 7NQ
2E1 ACK D. Nye, 5 Charles Road, Deal, CT14 9AT
2W1 ACM D. Young, 1 Hawthorn Road Llanharry, Pontyclun, CF72 9JD
2E1 ACS N. Roberts, 37A Rockley Avenue, Birdwell, Barnsley, S70 5QY
2E1 ACT D. Thomas, 4 Greenwood, Bamber Bridge, Preston, PR5 8JS
2E1 ACW C. Pooler, 18 Johnstone Close, Wrockwardine Wood, Telford, TF2 7DA
2E1 ACZ R. Stanley, 219 Fartown, Pudsey, LS28 8NH
2E1 ADJ J. Bridgman, 5 Drayton Avenue, Mackworth, Derby, DE22 4JU
2M1 ADM A. Mair, 8 Cockburn Crescent Whitecross, Linlithgow, EH49 6JT
2W1 ADO K. Jenkins, 79 Beaufort Road, Newport, NP19 7PB
2E1 ADP T. Thompson, 19 Park End, Summer Lane Caravan Park, Banwell, BS29 6JD
2E1 ADQ C. Hammett, 63 Treffry Road, Truro, TR1 1WL
2E1 ADR D. Palmer, 133 Victoria Road East, Thornton-Cleveleys, FY5 5HH
2E1 ADT K. Barbery, 17 Polbreen Avenue., St Agnes, TR50TR
2E1 AEC C. Vincent, 134 Wolds Drive, Keyworth, Nottingham, NG12 5DA
2E1 AEJ E. Jones, 19 Foxhollow Bar Hill, Cambridge, CB23 8EP
2E1 AEQ V. Parrish, 89 Delamere Drive, Macclesfield, SK10 2PS
2E1 AEU A. Prescott, 43 Coombe Road, Southminster, CM0 7AH
2E1 AFA J. Davis, 5 James Way, Camberley, GU15 2RQ
2E1 AFC J. Rossiter, 7 Valley View, Bodmin, PL31 1BE
2E1 AFI J. Charnley, 30 Dunkirk Avenue, Fulwood, Preston, PR2 3RY
2E1 AFN L. Swindale, 17 Crofton Close, Bracknell, RG12 0UR
2E1 AFR F. Batty, 26 Kingsmead Park, Elstead, Godalming, GU8 6DZ
2E1 AFS L. Jenkins, 49 Harts Grove, Chiddingfold, Godalming, GU8 4RG
2E1 AGE J. Prince, Field House, 25 Chiltern Road, Slough, SL1 7NF
2E1 AGO K. Beech, 44 Ashdene Close Willerby, Hull, HU10 6LW
2E1 AGQ J. Collins, 61 Albemarle Road, Gorleston, Great Yarmouth, NR31 7AS
2E1 AGV E. Muircroft, 84 Longley Avenue West, Sheffield, S5 8WF
2E1 AHK J. Robinson, 35 Vegal Crescent, Halifax, HX3 5PA
2E1 AHU T. Hassall, 5 Ashworth Street, Bacup, OL13 9LS
2W1 AID S. Williams, 27 Barnardo Street, Maesteg, CF34 0HT
2E1 AII D. Swann, 89 Leazes View, Rowlands Gill, NE39 2JT
2E1 AIT J. Tonks, 295 Quinton Road West, Quinton, Birmingham, B32 1PG
2E1 AIY B. Whalley, 46 Wayside, Woodside, Telford, TF7 5NG
2E1 AJB A. Burgess, 5 Wilkie Road, Birchington, CT7 9HE
2E1 AKK T. Goodwin, 37 Purdy Meadow, Long Eaton, Nottingham, NG10 3DJ
2E1 AKW M. Bradley, Flat 20, Crown Court, Portsmouth, PO1 1QN
2E1 ALC A. Coward, 11 Oaklands, Ardingly, Haywards Heath, RH17 6UE
2I1 ALE D. Auld, 37 Castlewellan Road, Rathfriland, Newry, BT34 5EL
2E1 ALX A. Wilson, 196 Spring Lane Lambley, Nottingham, NG4 4PE
2E1 AMB A. Collins, 141 Downside Avenue, Findon Valley, Worthing, BN14 0EY
2E1 AMW D. Evans, 9 Robin Close, Farnworth, Bolton, BL4 0RG
2E1 ANG P. Jackson, 55 Bomers Field, Rednal, Birmingham, B45 8TQ
2E1 ANH C. Jackson, 55 Bomers Field, Rednal, Birmingham, B45 8TQ
2E1 ANN M. Kearney, 18 Wayside Mews, Maidenhead, SL6 7EJ
2E1 ANQ A. Bell, 8 Silk Mill Green, Leeds, LS16 6DU
2M1 ANY L. Waterall, 3 Wavell Street, Grangemouth, FK3 8TG
2E1 AOF G. Tutt, 46 Heathcroft Avenue, Sunbury-on-Thames, TW16 7TL
2E1 AOG J. Menday, 3 Ash Grove, Guildford, GU2 8UT
2E1 AOK R. Gill, 45 Biggin Lane, Ramsey, Huntingdon, PE26 1NB
2M1 AOL G. Sibbald, 1 Ormiston Drive East Calder, Livingston, EH53 0RN
2E1 APW D. Jenkinson, 9 Dalton Street, Cockermouth, CA13 0AR
2E1 APX D. Johnson, 3 Plantation Avenue, Swalwell, Newcastle upon Tyne, NE16 3JN
2E1 AQH S. Allgood, 53 The Avenue, Leighton Bromswold, Huntingdon, PE28 5AW
2E1 AQT R. Scott, 39A Highgate Lane Goldthorpe, Rotherham, S63 9BA
2E1 ARG M. Trigg, 41 Veasey Road, Hartford, Huntingdon, PE29 1TA
2E1 ARS W. Hornby, Lindenstrasse 9, Allschwil, Switzerland, 4123
2E1 ASF T. Stevens, 20 The Butts, Crudwell, Malmesbury, SN16 9HF
2E1 ATH J. Minnock, 32 Sandwood Road, Sandwich, CT13 0AQ
2E1 ATV R. Corden, Konrad Cottage, Welburn, York, YO60 7DX
2E1 AUN D. Austin, 66 Homewood Avenue, Sittingbourne, ME10 1XJ
2E1 AUQ E. Harding, 17 Summerfield Close, Wokingham, RG41 1PH
2W1 AVM S. Peacock, 16 Banalog Terrace, Hollybush, Blackwood, NP12 0SF
2E1 AVT J. Hoggan, 39 The Glade, Crawley, RH10 6JL
2M1 AVZ N. Sinclair, 16 Sycamore Glade, Livingston, EH54 9JG
2E1 AWZ V. Hilton, 232 Hurst Rise, Matlock, DE4 3EW
2E1 AXD A. Richmond, 57 The Fairway, Daventry, NN11 4NW
2E1 AXE L. Richmond, 57 The Fairway, Daventry, NN11 4NW
2I1 AXH K. Bird, 115 Halftown Road, Lisburn, BT27 5RF
2E1 AXI F. Preece, 8 Gregory Road, Hedgerley, Slough, SL2 3XL
2E1 AXL W. Rose, 7 Harby Close, Grantham, NG31 7XA
2W1 AYO E. Phillips, 2 Oak St., Newport, NP9 7HW
2E1 AYS P. Cartwright, 41 Sandgate Drive Kippax, Leeds, LS257EX
2E1 AZA E. Williams, 37 Danesby Crescent, Denby, Ripley, DE5 8RF
2E1 AZK C. Morris, 10 Hempits Grove, Acton Trussell, Stafford, ST17 0SL
2E1 AZQ J. Perkins, Highfield House, Newtown, Buxton, SK17 0NF
2W1 AZU D. Williams, 75 Queens Avenue, Maesgeirchen, Bangor, LL57 1NH
2E1 AZW R. Roberts, Rose Cottage, Castle Hill, Leyburn, DL8 4QN
2E1 BAD B. Rowland, Deacons Cottage, Bridleway, Croft, LE9 6EE
2E1 BAE T. Ladley, Marisdene, London Road, Faversham, ME13 9LF
2M1 BBY J. Duncan, 4 Lady Moss, Tweedbank, Galashiels, TD1 3SB
2E1 BCC N. Kluger-Langer, 23 Vernon Walk, Northampton, NN1 5ST
2E1 BCF J. Brown, 9 South Street High Spen, Rowlands Gill, NE39 2HF

2E1 BDB P. Hudson, 1 Dean Moore Close, St Albans, AL1 1DW
2E1 BDC P. Kennedy, Spinney Bungalow, Thearne Lane, Woodmansey, Beverley, HU17 0SA
2E1 BDV P. Mclusky, 11 Ripon Road, Killinghall, Harrogate, HG3 2DG
2E1 BEB J. Wright, 4 Sweeters Field Road, Alfold, GU6 8UD
2E1 BEV M. Norman, 41 Avon Grove, Bletchley, Milton Keynes, MK3 7BP
2E1 BFH M. Knight, 359 Shelley Road, Wellingborough, NN8 3EW
2E1 BFP J. Philpot, 7 Providence Place, Ilkeston, DE7 8AL
2E1 BFW P. Hyde, 10 Highfield Crescent, Taunton, TA1 5JH
2E1 BFX C. Mann, 11 North River Road, Runham Vauxhall, Great Yarmouth, NR30 1JY
2E1 BGN A. Rollitt-Smith, 9 St Helens Road, Doncaster, DN4 5EQ
2E1 BGQ M. Wynn, 22 Matthews Drive, Wickersley, Rotherham, S66 1NN
2E1 BHB A. Comis, 178 Lordswood Road, Birmingham, B17 8QH
2E1 BHC P. Comis, 65 Montague Way Chellaston, Derby, DE73 5as
2E1 BHF J. Clifford, 16 Park View Road, Birmingham, B31 5AU
2E1 BHU P. Gosling, 10 Prospect Road, Carlton, Nottingham, NG4 1LY
2E1 BIM C. Faulkner, 1 Westland, Martlesham Heath, Ipswich, IP5 3SU
2E1 BIT C. Swain, 2 Cinder Road Somercotes, Alfreton, DE55 4JY
2E1 BJD R. Saunders, 31 Greenwood Road, High Green, Sheffield, S35 3GU
2E1 BJG S. Benson, 45 Maple Way, Selston, Nottingham, NG16 6FA
2E1 BKF G. Muircroft, 62 New Road, Rotherham, S61 2DU
2E1 BKK C. Berry, Roseneath, Walcote Road, Lutterworth, LE17 6EQ
2E1 BKP S. Martin, The Shieling, Bolton Low Houses, Wigton, CA7 8PF
2E1 BKT D. Jump, 15 Swallowfield, Chorley New Road, Horwich, Bolton, BL6 6HN
2E1 BLG M. Levinson-Withall, The Bungalows, 20 Moor End Avenue, Salford, M7 3NX
2E1 BLP B. Mulley, 8 Drinkstone Road, Gedding, Bury St Edmunds, IP30 0QB
2E1 BMB B. Porthouse, 10 Pecklewell Terrace, Maryport, CA15 7QJ
2E1 BME Dr M. Riley, 2 Keeble Drive, Washingborough, Lincoln, LN4 1DZ
2E1 BMF D. Carskake, 38 Loppets Road, Crawley, RH10 5DW
2E1 BMJ D. Peters, 7 Gravel Lane, Drayton, Abingdon, OX14 4HY
2E1 BMV B. Mycock, 69 Bentley Road, Uttoxeter, ST14 7EN
2W1 BOG I. Skinner, 4 St. Marys Crescent, Rogiet, Caldicot, NP26 3TB
2E1 BOM M. Love, 72A Hart Plain Avenue, Waterlooville, PO8 8RX
2E1 BOO M. Constantine, 18 Hillbeck, Halifax, HX3 5LU
2E1 BOZ S. Razzaq, 15 Wild Herons, Hook, RG27 9SF
2E1 BPN A. Collins, Flat 2, 49 Dukes Head Street, Lowestoft, NR32 1JY
2E1 BPV D. Roberts, 98 Pinkneys Road, Maidenhead, SL6 5DN
2E1 BRA M. Scotton, 15 Grove Road, Aston, Stone, ST15 0DW
2E1 BRC G. Thornsby, 25 Kipling Way, Stowmarket, IP14 1TS
2E1 BRD M. Larcombe, 52 Orchard Road, Burgess Hill, RH15 9PL
2E1 BRG C. Sanderson, 14 Hazelwood Avenue, York, YO10 3PD
2E1 BSC C. Castle, 26 Chestnut Walk, Pulborough, RH20 1AW
2W1 BST S. Long, 10 St. George Road, Bulwark, Chepstow, NP16 5LA
2E1 BTG M. Joyner, Brimar, Nelson Park Road, Dover, CT15 6HL
2E1 BTK K. Creamer, Ellipsis. 35 Salisbury Road, St Margaret'S Bay. Ct15 6Dl, Dover, Ct15 6dl
2E1 BUJ P. Stott, 12 Castle View, Ovingham, Prudhoe, NE42 6AT
2E1 BUM F. Stone, 7 Cherry Tree Close, Hilton, Derby, DE65 5FD
2E1 BUR J. Izzard, Sunnyside, West End, Hailsham, BN27 4NH
2E1 BUV P. Fletcher, 18 Woodside View, Holmesfield, Dronfield, S18 7WX
2E1 BVJ E. Constantine, 18 Hillbeck, Halifax, HX3 5LU
2E1 BVQ N. Brown, 55 Shakespeare Terrace, Chorley, PR6 7AQ
2E1 BVY D. Fox, 36A Northmere Road, Poole, BH12 4DY
2E1 BWT A. Ross, 42 New Heritage Way, North Chailey, Lewes, BN8 4GD
2E1 BYH M. Hearn, Flat 8, Ascot House, 26-30 Tontine Street, Folkestone, CT20 1JU
2E1 BYI S. Piccavey, 611 Manchester Road, Linthwaite, Huddersfield, HD7 5QX
2W1 BYK A. Sellors, 12 Morfa View, Bodelwyddan, Rhyl, LL18 5TT
2M1 BYW T. Conlan, 12 Rowantree Road, Mayfield, Dalkeith, EH22 5ER
2E1 BYY W. Kennedy, 1 Lynton Road, Hindley, Wigan, WN2 4EH
2E1 BZB D. Peter, 41 Coleswood Road, Harpenden, AL5 1EF
2E1 BZH G. Low, 30 Laburnum Grove, Runcorn, WA7 5EL
2E1 BZI D. Low, 6 Pale Manor Close, Malvern, WR14 1SZ
2E1 CAF S. Kirkpatrick, 29 Barnfields, Gloucester, GL4 6WE
2E1 CAH R. Richards, 39 North Holme Court, Northampton, NN3 8UX
2E1 CAJ J. Godding, 70 Rodway Road, Tilehurst, Reading, RG30 6DT
2E1 CAQ M. Porter, 7 Dunkirks Mews, Hertford, SG13 8BA
2E1 CAT L. O'Ryan, 12 Minton Close, Congleton, CW12 3TD
2E1 CAU A. Burt, 20 Medina Breeze Walk, Binfield, Newport, PO30 2GS
2E1 CAW R. Marshall, 15 Whisby Court, Holton-Le-Clay, Grimsby, DN36 5BG
2E1 CBH S. Stretch, 5 Ledwych Road, Droitwich, WR9 9LA
2E1 CBU C. Richards, 11 Purvis Road, Rushden, NN10 9QA
2E1 CCF P. Izzard, 7 Yardley Drive, Northampton, NN2 8PE
2E1 CCG R. Winship, 32 Lytes Cary Road, Keynsham, Bristol, BS31 1XD
2E1 CCI A. Murphy, 34 Hawkenbury Way, Lewes, BN7 1LT
2E1 CCN D. Hill, 28 Pendarves Flats, St. Clare Street, Penzance, TR18 2PL
2E1 CDK G. Langdon, 43 Daniel St., Ryde, PO33 2BH
2E1 CDS B. Allen, 78 Bargates, Christchurch, BH23 1QL
2E1 CDZ A. Woods, 40 Windsor Way, Sandy, SG19 1JL
2W1 CEE D. Whish, 62 Marion Road, Prestatyn, LL19 7DF
2E1 CEQ M. Sayers, Flat 12, Quay Court, Loring Road, Christchurch, BH23 2AU
2E1 CEU J. Columbine, West Lodge, 166 Tollerton Lane, Nottingham, NG12 4FW
2E1 CEZ J. Brown, Kingsdown Cottage, Fron, Montgomery, SY15 6SB
2E1 CFB D. Williams, 100 Hills Lane Drive, Madeley, Telford, TF7 4BX
2E1 CGS M. Smith, 27 Oldway Lane, Slough, SL1 5LA
2E1 CHX M. Reavill, 11 Clarence Road, Beeston, Nottingham, NG9 5HY
2E1 CIK D. Gillatt, 1 City Mills, Skeldergate, York, YO1 6DB
2E1 CIO C. Brooks, 1 Gloucester Road, Guisborough, TS14 7DZ
2W1 CIP R. Bufton, 7 Laburnum Close, Rassau, Ebbw Vale, NP23 5TS
2E1 CIR R. Bush, Church View, Overcross, Banham, Norwich, NR16 2BY
2E1 CIT J. Watkins-Field, Sharlions, 27 Bosvigo Road, Truro, TR1 3DG
2E1 CIY B. Harratt, 8 Aaron Wilkinson Court, South Kirkby, Pontefract, WF9 3JT
2E1 CJB K. Jordan, 11 Sandringham Place, Hucknall, Nottingham, NG15 8EU
2E1 CJC R. Richardson, 3 Cautley Drive, Killinghall, Harrogate, HG3 2DJ
2E1 CJD P. Taylor, 13 Mackenzie Crescent, Burncross, Sheffield, S35 1UR
2E1 CJF S. Curtis, 354 St. Helens Road, Leigh, WN7 3PQ
2E1 CJJ S. Hull, 1 Occupation Lane, New Bolingbroke, Boston, PE22 7LW
2M1 CJL A. Jones, 9 Firs Street, Falkirk, FK2 7AY
2E1 CJN H. Hughes, 46 The Boundary, Oldbrook, Milton Keynes, MK6 2HT
2E1 CJZ Z. Hodges, 12 Linwal Avenue, Houghton-On-The-Hill, Leicester, LE7 9HD
2E1 CKH K. Riley, 16 King St., Westhoughton, Bolton, BL5 3AX
2E1 CKQ E. Swain, 11 Blackdown, Fullers Slade, Milton Keynes, MK11 2AA
2W1 CLC M. Holmes, 77 Harlech Drive, Merthyr Tydfil, CF48 1JU
2E1 CLG G. Harvey, 3 Mulberry Way, Sittingbourne, ME10 3TG
2E1 CLM E. Woolley, 82 Pennycroft Road, Uttoxeter, ST14 7ET
2E1 CML B. Norris, 20 Laburnum Close, Guildford, GU1 1NA
2E1 CMZ D. Bracher, 29 Bungalow Park, Holders Road, Amesbury, Salisbury, SP4 7PJ
2E1 CNM A. Raxworthy, 32 St. Marys Avenue, Alverstoke, Gosport, PO12 2HX
2W1 CNN A. Gray, 36 Heol Pentre Felen, Morriston, Swansea, SA6 6BY
2E1 CNO L. Call, 3 Southfield, Bramhope, Leeds, LS16 9DR
2E1 COD B. Call, 6 Harecroft Road. Otley, Leeds, LS212BQ
2E1 COG D. Oakes, 2 Hillcrest, Scotton, Catterick Garrison, DL9 3NJ
2E1 COM M. Holt, 20 Lingfield Mount, Leeds, LS17 7EP
2E1 COV I. Cockshoot, 72 Princess Margaret Avenue, Cliftonville, Margate, CT9 3EF
2E1 CPB A. Cadey, Wakeley House, High Street, Charing, Ashford, TN27 0LS
2E1 CPC M. Sheppard, 11 Parvian Road, Leicester, LE2 6TS
2E1 CPF R. Kensall, 16 Parkwood Close, Broadstairs, CT102XN
2E1 CPI B. Rowley, 7 Hall Farm Close, Castle Donington, Derby, DE74 2NG
2E1 CPJ P. Fisk, 38 Bedingfield Crescent, Halesworth, IP19 8EE
2E1 CPP N. Newman, 89 Sea Place, Goring-By-Sea, Worthing, BN12 4BH
2E1 CPQ P. Goodwin, 60 Dale Crescent, Congleton, CW12 3EP
2W1 CPS D. Probert, 32 Heol Penlan Longford, Neath, SA10 7LB
2E1 CPV P. Porter, 7 Long Road, Framingham Earl, Norwich, NR14 7RY
2E1 CQM I. Hurst, 14 Longhill Rise, Kirkby in Ashfield, NG17 9FL
2E1 CQP V. Brightwell, 40 Streete Court Road, Westgate-on-Sea, CT8 8BX
2E1 CQQ E. Whelan, 54 Boroughbridge Road, Northallerton, DL7 8BN
2E1 CRA M. Lewis, 15 Highcliffe Avenue, Chester, CH1 5DP
2E1 CRI J. Mosby, 1 School Road, Golcar, Huddersfield, HD7 4NU
2E1 CSD A. Smith, 30 Lime Grove, Grantham, NG31 9JD
2E1 CTU S. Crawshaw, 1 Wardley Close, Burnley, BB12 6ET
2M1 CUS R. Higgins, 30 Wallfield Crescent, Aberdeen, AB25 2JX
2E1 CVE R. Trow, 5 Cranberry Drive, Stourport-on-Severn, DY13 8TH
2E1 CWE M. Skewes, 47 Pentrevah Road, Penwithick, St Austell, PL26 8UA
2J1 CWG J. Totty, Flat 4, Beech Court, Woodlands Apartments, La Route Des Cotils, Grouville, Jersey, JE3 9AY
2J1 CWH C. Totty, Flat 4, Beech Court, Woodlands Apartments, La Rue Des Cotils, Grouville, Jersey, JE3 9AY
2E1 CWJ D. Thatcher, 6 Ivel View, Sandy, SG19 1AU
2E1 CWN P. Kemble, 88 Mayfield Road, Ipswich, IP4 3NG
2E1 CWP J. Parker, 19 Mayfair Close, Dukinfield, SK16 5HR
2E1 CWQ P. Millward, 28 Olive Grove, Burton Joyce, Nottingham, NG14 5FG
2E1 CWX E. Parker, 19 Mayfair Close, Dukinfield, SK16 5HR
2E1 CXE J. Mortimer, 14 Oakfield Road, Bourne End, SL8 5QN
2E1 CXF A. Whittaker, 62 Ingham Street, Padiham, Burnley, BB12 8DR
2E1 CXI R. Maunder, 12 Hamble Springs, Bishops Waltham, Southampton, SO32 1SG
2E1 CXP E. Bradshaw, 41 Sherwood Road, Woodley, Stockport, SK6 1LH
2W1 CYC D. Tiltman, 16 St. Georges Road, Heath, Cardiff, CF14 4AQ
2E1 CYD G. Reilly, 10 Gloucester Road, Huyton, Liverpool, L36 1XX
2E1 CYE R. Stenhouse, High Park, Common Road, Norwich, NR16 1HH
2E1 CYI D. Pomery, 3 Mayfair Park, Minorca Lane, St Austell, PL26 8QN
2E1 CYM N. Griffiths, 85 Foljambe Road, Chesterfield, S40 1NJ
2E1 CYP J. Bick, 45 Gloucester Road, Almondsbury, Bristol, BS32 4HH
2E1 CYS I. Limbert, 9 Lyme Grove, Liverpool, L36 8BN
2E1 CYT S. Gallagher-Willmer, 14 Bellhurst Cottages, Bellhurst Cottages, Bellhurst Road, Robertsbridge, TN32 5DN
2E1 CYU S. Barker, 26 Rye Court Helmsley, York, YO62 5DY
2E1 CYZ M. Baxter, 5 Farnborough Street, Farnborough, GU14 8AG
2E1 CZB B. Fido, 7 Claires Walk, Parklands Mobile Hom, Scunthorpe, DN17 1SW
2E1 CZF A. Pink, 31 The Fairway, Daventry, NN11 4NW
2E1 CZJ J. Bosworth, 57 Livingstone Road, Derby, DE23 6PS
2E1 CZO B. Coombs, 10 Horseshoe Walk, Widcombe, Bath, BA2 6DE
2E1 CZS S. Backhouse, 49 Western Avenue, Felixstowe, IP11 9SL
2E1 DAK C. Wilderspin, 3 Ferndale, Eaglestone, Milton Keynes, MK6 5AE
2W1 DAO J. Patrick, 52 Huntsmans Corner, Wrexham, LL12 7UH
2E1 DAP D. Pope, 24 Wanstead Crescent, Blackpool, FY4 4AR
2E1 DAR A. Cottle, 66 Elmhurst Estate, Batheaston, Bath, BA1 7NX
2E1 DAT D. Taylor, 24 Quince, Tamworth, B77 4EN
2E1 DBP A. Gener, 3E Dartmouth Terrace, Greenwich, London, SE10 8AX
2E1 DBQ S. Walker, 33 Parkside Somercotes, Alfreton, DE55 4LA
2E1 DBS K. Budd, 20 Marshal Road, Poole, BH17 7HA
2E1 DBT Dr R. Verma, 43 Farley Road, Derby, DE23 6BW
2E1 DBZ S. Issatt, 7 Birch Road, Doncaster, DN4 6PD
2E1 DCV B. Shields, 20 Gresley Court, Grantham, NG31 7RH
2E1 DDJ J. Gallagher, 71 Castle Hill, Beccles, NR34 7BJ
2E1 DDZ S. Arter, 18 Essex Road, Westgate-on-Sea, CT8 8AP
2W1 DEA P. Smith, 23 Gainsborough Close, Llantarnam, Cwmbran, NP44 3BX
2E1 DEM N. Humphreys, 4 Rose Cottages, Annscroft, Shrewsbury, SY5 8AU
2E1 DEN P. Hart, 18 Harewood Road, Rochdale, OL11 5TG
2E1 DEP L. Darton, 8 Foster Grove, Sandy, SG19 1HP
2E1 DET A. Whyman, 8 Staplers Close, Great Totham, Maldon, CM9 8UN
2E1 DFB B. Williams, Hillside, Wigmore Lane Eythorne, Dover, CT15 4AW
2E1 DFE R. Scott, 315 Ormskirk Road, Wigan, WN5 9DL
2E1 DFS R. Bearcroft, 45 Broad Marston Road, Pebworth, Stratford-upon-Avon, CV37 8XT
2E1 DFZ M. Axon, 48 Cowslip Road, Broadstone, BH18 9QZ
2E1 DGL P. Lewis, 166 Euston Road, Morecambe, LA4 5LE
2W1 DGM J. Foster, 56 Hillrise Park, Clydach, Swansea, SA6 5DX
2E1 DHC M. Axworthy, 3Shepherds Court, 1 Shepherds Street, St Leonard's on Sea, TN38 0ET
2M1 DHG G. Russell, 76 Duffus Crescent, Elgin, IV30 5PY
2E1 DHJ D. Jones, 11 Kylemilne Way, Stourport-on-Severn, DY13 9NA
2E1 DHX D. Walker, 17 Pinehurst Avenue, Christchurch, BH23 3NS
2E1 DIA J. Laffin, 154 Blenheim Drive, Allestree, Derby, DE22 2GN
2E1 DIH J. Bentley, 4 Highway, Crowthorne, RG45 6HE
2E1 DIJ T. Willard, 4 Chapel Cottages, Cowfold Road, Bolney, Haywards Heath, RH17

5QU
2E1 DKU A. Whittle, 9 Dale View, Littleborough, OL15 0BP
2E1 DLA P. Craig, 25 Harts Green Road, Birmingham, B17 9TZ
2E1 DLD E. Harrison, 55 Hudson Close, Worcester, WR2 4DP
2E1 DLM D. Wilson, 76 Cheadle Road, Uttoxeter, ST14 7BY
2E1 DLO B. Bush, Church View, Overcross, Banham, Norwich, NR16 2BY
2E1 DLR R. Diaper, 30 Holmcroft Road, Kidderminster, DY10 3AG
2E1 DLR D. Carr, 33 Livingstone St., Leek, ST13 5JU
2E1 DLS J. Field, 27 Lovelace Road, Barnet, EN4 8EA
2E1 DLT P. Lilley, 12 Trueman Gardens, Arnold, Nottingham, NG5 6QT
2E1 DLX P. Thackray, 20 Darfield St., Leeds, LS8 5DB
2E1 DMH M. Rippin, Gaverne, Welford Road, Stratford-upon-Avon, CV37 8RA
2E1 DMI R. Dixon, 14 Blackdown Close, Peterlee, SR8 2JW
2E1 DMU S. Wilson, 218 Bredhurst Road, Gillingham, ME8 0RD
2E1 DMZ R. Laverick, 12 Greenlands Road, Redcar, TS10 2DG
2E1 DNB P. Lomas, 42 Lane End, Pudsey, LS28 9AD
2E1 DNC M. Harris, Flat 22, Alexandra Court, The Royal Seabathing, Margate, CT9 5NT
2E1 DNF J. Foy, 2 Lark Rise, Northampton, NN3 8QT
2W1 DNK A. Waller, 4 Rose Court, Ty Canol, Cwmbran, NP44 6JH
2E1 DNX M. Mackenzie, 73 Newstead Road, Weymouth, DT4 0AS
2E1 DOA P. Dalby, Windfall, 11 Greensward Lane, Hockley, SS5 5HD
2E1 DOG H. Godzisz, 1 Jutland Place, Egham, TW20 8ET
2E1 DOZ S. Brenchley, 89 Thicket Mead, Midsomer Norton, Radstock, BA3 2SL
2E1 DPK S. Hill, 22B Strait Lane, Hurworth On Tees, Darlington, DL2 2AL
2E1 DPQ R. Tattersall, 70 Selwyn Street, Hillstown, Bolsover, Chesterfield, S44 6LR
2E1 DQM N. Edwards, 609 Upper Richmond Road West, Richmond, TW10 5DU
2E1 DQQ D. Horsley1, 1 Mead Close, Swanley, BR8 8DQ
2E1 DQT A. Charles, 6 Bridewell St., Wymondham, NR18 0AR
2E1 DQZ C. Houlden, 29 Court Barton, Portland, DT5 2HJ
2W1 DRB P. Waller, 4 Rose Court, Ty Canol, Cwmbran, NP44 6JH
2E1 DRC P. Hatcher, 32 Slough Road, Iver, SL0 0DT
2E1 DRU K. Mccann, Treverven, Back Lane, Selby, YO8 6QP
2E1 DRV J. Wohlgemuth, 37 Broadcoombe, South Croydon, CR2 8HR
2E1 DRX R. Hauxwell, 65 Harleston Way, Heworth, Gateshead, NE10 9BQ
2E1 DRY G. Symonds, Flat 13, Bradbury House, Norwich, NR2 3PT
2E1 DSU J. Hewitt, 49 Calder Drive, Sutton Coldfield, B76 1YR
2E1 DSX C. Haddon, 1 Victoria Place, Weston, Portland, DT5 2AA
2E1 DTE Dr C. Folkerd, 4 Beechwood, Shaw, Oldham, OL2 8LP
2E1 DTF D. Folkerd, The Old Manse, London, W5 5QT
2E1 DTG V. Thomas, 4 Greenwood, Bamber Bridge, Preston, PR5 8JS
2E1 DTK R. Cleary, 5 Gregson Road, Widnes, WA8 0BX
2E1 DTN J. Bennett, Rectory Cottage, Broad Street, Bristol, BS40 5LD
2E1 DUI G. Tunley, 36 Hamilton Drive, Chippenham, SN14 0XW
2E1 DUW M. Goldby, Waylands Gate, St. Johns Road, New Milton, BH25 5SD
2M1 DWK D. Keay, Parkhill, Cromwell Park, Almondbank, Perth, PH1 3LW
2E1 DWM J. Bush, Church View, Overcross, Banham, Norwich, NR16 2BY
2E1 DWP B. Hanson, 35 Market Lane, Wolverhampton, WV4 4UL
2E1 DWW D. Chapman, 38 Cranfleet Way, Long Eaton, Nottingham, NG10 3RJ
2E1 DXB I. Humberstone, 20 Kingswood Road, Colchester, CO4 5JX
2W1 DXV A. Powers, 9 Courtybella Gardens, Newport, NP20 2GN
2E1 DYL M. Reid, 24 Mitchel Way, Madeley, Telford, TF7 5SN
2E1 DYT C. Block, 13 Beatrice Road, Capel-Le-Ferne, Folkestone, CT18 7LH
2E1 DZH J. Richards, 1 Stocks Mead, Washington, Pulborough, RH20 4AU
2E1 DZJ R. Morris, 7 Chapmans Close, Stirchley, Telford, TF3 1ED
2E1 DZL R. Neville, 4 Danson Gardens, Blackpool, FY2 0XH
2E1 DZP A. Cannon, 20 Gladwyn Street, Stoke-on-Trent, ST2 8JZ
2M1 DZS C. Clark, 3 Old Cottages, Seton Mains, Longniddry, EH32 0PG
2E1 DZV J. Keegan, The Cottage, 11 Condor Grove, Lytham St Annes, FY8 2HE
2M1 DZW B. Waugh, 93 Denholm Road, Musselburgh, EH21 6TU
2M1 DZX R. Stuart, Tigh Na Coille, Bishop Kinkell, Conon Bridge, Dingwall, IV7 8AW
2E1 EAG J. Oxley, 97 Defoe Crescent, Colchester, CO4 5LQ
2E1 EAK Y. Wood, 67 Bay View Road, Duporth, St Austell, PL26 6BN
2W1 EAN G. Taylor, 55 Haulfryn, Tregynwr, Carmarthen, SA31 2DT
2E1 EAV I. Duggan, 21 Chetwynd Road, Toton, Nottingham, NG9 6FW
2E1 EAW M. Barnett, 23 Francis Close, Penkridge, Stafford, ST19 5HP
2E1 EAX R. Edwards, Flat 10, Longford Court, 60 London Road, Sevenoaks, TN13 2UG
2E1 EAZ D. Finch, 174 Park Road, Chesterfield, S40 2LL
2M1 EBJ D. Martin, Inverleod, Avoch, IV9 8PR
2E1 EBL K. Dignall, 11 Mottershead Road, Widnes, WA8 7LD
2E1 EBN S. Valvona, 123 Binstead Lodge Road, Ryde, PO33 3UB
2E1 EBR C. Smith, 16 Meadowbrook, Ancaster, Grantham, NG32 3RR
2E1 EBX L. Collinson, 12 Victoria Avenue, Hunstanton, PE36 6BX
2E1 EBZ M. Watkins, 2 Griffin Close, Billingshurst, RH14 9GS
2M1 ECF J. Mackenzie, Rainbows End, Lochside, Lairg, IV27 4EG
2E1 ECG A. Topping, 30 St. Pauls Avenue, Nottingham, NG7 5EB
2E1 ECL C. Gray, 19 Marsh View, Newton, Preston, PR4 3SX
2E1 ECM P. Fletcher, 11 The Banks, Long Buckby, Northampton, NN6 7QQ
2E1 ECN M. Chapman, 15 Norwood Road, Somersham, Huntingdon, PE28 3EY
2E1 ECV P. Elsey, 129 Kingsway, Chandler'S Ford, Eastleigh, SO53 5BX
2E1 EDA E. Kurtz, 1 Leonard Medler Way, Hevingham, Norwich, NR10 5LE
2E1 EDB M. Bradwell, 6 Moorfoot Gardens, Gateshead, NE11 9LA
2E1 EDD S. Lansdell, 42 Marylebone Crescent, Derby, DE22 4JX
2M1 EDM D. Martin, 45 Tiree Place, Newton Mearns, Glasgow, G77 6UJ
2J1 EDR A. Price, Wheatlands, La Rue Des Longchamps, St Brelade, Jersey, JE3 8BN
2M1 EDT J. Wilson, 4F Langside Street, Clydebank, G81 5HJ
2E1 EDV J. Blanche, 11 Woodside, Southminster, CM0 7RD
2E1 EDW J. Blanche, 11 Woodside, Southminster, CM0 7RD
2E1 EEK S. Paffett, 15 Centaury Gardens Horton Heath, Eastleigh, so507ny
2W1 EEP J. Jones, Glanrafon Garage, Pontfadog, Llangollen, LL20 7AR
2E1 EET L. Edwards, 39 Foskitt Court, Northampton, NN3 9AX
2E1 EFG S. Philpot, 93 Princess Drive, Grantham, NG31 9QA
2E1 EFQ D. Bushby, 66 Sandy Road, Everton, Sandy, SG19 2JU
2E1 EFT L. Froggatt11, 255 Rushton Road, Desborough, Kettering, NN14 2QB
2E1 EGI G. Durrant, 51 Raglan Avenue, Waltham Cross, EN8 8DA
2E1 EGU W. Lupton, Eaton Cottage, Shorts Green Lane, Shaftesbury, SP7 9PA
2E1 EGV M. Townson, 42 Osborne Grove, Shavington, CW2 5BY
2E1 EHB I. Greenall, 356 Warrington Road, Abram, Wigan, WN2 5XA
2E1 EHF Rev. J. Goodman, The Vicarage, 4 Austenway, Chalfont St Peter, SL9 8NW
2E1 EHM D. Whittaker, 68 Querns Road, Canterbury, CT1 1PZ
2E1 EHP D. Moses, 121 Badger Avenue, Crewe, CW1 3JN
2E1 EHY N. Waters, 23 Harold Road, Birchington, CT7 9NA
2W1 EID R. Owens, 11 Elm Grove, Newport, NP20 6JF
2W1 EIN C. Bodley, 38 Bryn Hawddgar, Clydach, Swansea, SA6 5LA
2E1 EIO A. Godolphin, Shepherds Cottage Flakebridge, Appleby-in-Westmorland, CA16 6JZ
2E1 EIU D. Meddings, 42A Argyle Road, Poulton-le-Fylde, FY6 7EW
2E1 EIV A. Eyre, St. Michael Mead, The Common, Norwich, NR12 8BA
2E1 EIX A. Chruscinski, 39 Sherwood Rise, Mansfield Woodhouse, Mansfield, NG19 7NP
2E1 EJC T. Bain, 23 Salisbury Crescent, Blandford Forum, DT11 7LX
2E1 EJD W. Ling, Valley Farm Equestrian Centre, Wickham Market, Woodbridge, IP13 0ND
2U1 EJF A. Scheffer, Foveat, Rue De Jardins, Les Prins, Guernsey, GY6 8EZ
2E1 EJU D. Lumley, 19 Bramley Avenue, Needingworth, St Ives, PE27 4UD
2E1 EJX M. Tullett, The Lodge, York Road, Knaresborough, HG5 0SW
2U1 EKE C. Ayres, Rousay, Bailiffs Cross Road, St Andrew, Guernsey, GY6 8RY
2U1 EKH K. Johnson, 9 Clos Spurway, Victoria Avenue, Guernsey, GY2 4AH
2M1 EKI K. Hughes, 49 Marmion Drive, Kirkintilloch, Glasgow, G66 2BH
2E1 EKM D. Baldwin, 51 Queens Road, Broadstairs, CT10 1PG
2E1 EKQ M. Eades, 19 Egham Court, Horsa Street, Bolton, BL2 2DE
2W1 EKR D. Pollard, 19 The Drive, Bargoed, CF81 8JX
2E1 ELE P. Williams, 20 Elm Close, Great Haywood, Stafford, ST18 0SP
2M1 ELU I. Sinclair, Airdaniar, Kilchrenan, Taynuilt, PA35 1HG
2E1 EMH V. Newton, 60 The Lynch, Winscombe, BS25 1AR
2E1 EMI G. Paterson, 4 Rowallan Drive, Bedford, MK41 8AW
2E1 EMK J. Roff, 4 Needham House, Victoria Road, Devizes, SN10 1FA
2E1 EMN J. Thompson, 97 King Lane, Leeds, LS17 5AX
2M1 ENI D. Paterson, Leuchlands Croft, Whitecairns, Aberdeen, AB23 8UT
2M1 ENK D. Paterson, 29 Arnothill Gardens, Falkirk, FK1 5BQ
2E1 ENN G. Cattle, 50 Oakland Avenue, York, YO31 1DF
2E1 ENR S. Taylor, 65 Cross Bank Road, Batley, WF17 8PN
2E1 ENZ R. Baxter, 20 Thorpe St., Thorpe Hesley, Rotherham, S61 2RP
2E1 EOD N. Bridges, Acomb, Station Road, Pershore, WR10 3BB
2E1 EOI J. Allum, 122 Long Chaulden, Hemel Hempstead, HP1 2HY
2E1 EOK C. Blackman, 7 Deanery Close, Ripley, DE5 3TR
2E1 EOO D. Webb, 11 Alfriston Road, Worthing, BN14 7QU
2E1 EOQ D. Horton, Glen View, New Road, Bude, EX23 9LE
2E1 EOR K. Horton, The Old School House, Kelly, Lifton, PL16 0HJ
2M1 EOV M. Macleod, 4 Portnaguran, Isle of Lewis, HS2 0HD
2E1 EOZ R. Cannon, 31 Moretons Mews, Basildon, SS13 3NB
2E1 EPA P. Allaker, 3 Eden Cottages, Watling Street, Consett, DH8 6HZ
2E1 EPD M. Freedman, Rivermeade, Irwell Vale, Bury, BL0 0QA
2E1 EPE P. Freedman, Rivermeade, Irwell Vale, Bury, BL0 0QA
2W1 EPL G. Morris, Flat 12, Windsor Court Crescent Road, Rhyl, LL18 1TF
2W1 EPO F. Hodge, Flat 7, Windsor Court, Rhyl, LL18 1TF
2E1 EPQ T. Fanning, 26 Mandeville Close, Tilehurst, Reading, RG30 4JT
2M1 EPV D. Stewart, 19 Arthur Street, Blairgowrie, PH10 6PF
2E1 EQE D. Evans, 5 Compton Drive, Streetly, Sutton Coldfield, B74 2DA
2E1 EQI V. Collins, 30 Upham Road, Swindon, SN3 1DN
2E1 EQK P. Gibson, 73 Marlborough Road Hadley, Telford, TF1 5LN
2E1 EQQ K. Wood, 52 Ashfield Avenue, Beeston, Nottingham, NG9 1PY
2E1 EQR J. Jones, 17 Dunkeld Drive, Shrewsbury, SY2 5UZ
2E1 ERJ D. Lusty, 104 Polstain Road Threemilestone, Truro, TR3 6DB
2E1 ERU A. Ainslie, 16 Ryeburn Close, Kessingland, Lowestoft, NR33 7UH
2E1 ESK L. Hodge, 11 Glebelands, Bampton, OX18 2LH
2E1 ESM K. James, 36 Lemon Hill, Mylor Bridge, Falmouth, TR11 5NA
2E1 ESN C. Mclean, 18 Chatfield Road, Gosport, PO13 0TN
2E1 ESQ A. Mould, 95 Stanton Road, Southampton, SO15 4HU
2E1 ESW D. Wilkinson, 139 Church Road, Jackfield, Telford, TF8 7ND
2E1 ETB R. Moore, Thorpefield Farm, 91 Thorpe Street, Rotherham, S61 2RP
2E1 ETJ A. Willis, Kilncroft, Broadlayings, Newbury, RG20 9TS
2M1 ETM S. Mackie, 25 Carlaverock Drive, Tranent, EH33 2EE
2W1 ETN D. Jorgensen, Unit 11, Llandow Trading Estate, Llandow, Cowbridge, CF71 7PB
2W1 ETW R. Thomas, 24 Heol Innes, Llanelli, SA15 4LA
2E1 EUE E. Hunt, 5 Freshfield Square, Southampton, SO15 8QU
2W1 EUR A. Joynes, 25 St. Annes Gardens, Maesycwmmer, Hengoed, CF82 7QQ
2M1 EUV E. Clark, 3 Old Cottages, Seton Mains, Longniddry, EH32 0PG
2E1 EUY W. Partridge, 15 Cranbourne Avenue, Wolverhampton, WV4 6RJ
2E1 EVH F. Stevenson, 56 Barrows Hill Lane, Westwood, Nottingham, NG16 5HJ
2E1 EVJ R. Chipperfield, 5 Lullingstone Close, Hempstead, Gillingham, ME7 3TS
2E1 EVK R. Chipperfield, 5 Clayton Avenue, Upminster, RM14 2EZ
2E1 EVM J. Merrick, The Birches, Dunn Street Bredhurst, Gillingham, ME7 3NA
2E1 EWK B. Ashman, 108 Eastwood Drive, Highwoods, Colchester, CO4 9SL
2E1 EWN B. Storkey, 9 Snatchup, Redbourn, St Albans, AL3 7HD
2E1 EXA P. Johnson, 110 Rachel Clarke Close, Stanford le Hope, SS17 7SX
2E1 EXI D. Taylor, 188 Walstead Road, Walsall, WS5 4DN
2E1 EXK J. Wilkes, 47 Greenwood Park, Hednesford, Cannock, WS12 4DQ
2E1 EXP C. Staerck, 8 Cresswell Road, Worksop, S80 1SU
2I1 EXU P. Robinson, 20 Harwood Park, Carrickfergus, BT38 7LZ
2E1 EYC M. Davis, 15 Farmcroft Road Mansfield Woodhouse, Mansfield, NG19 8QU
2E1 EYF E. Hedley, 17 Chowdene Bank, Gateshead, NE9 6JJ
2E1 EYL M. Harris, 1 Brampton Court, Bowerhill, Melksham, SN12 6TH
2E1 EYS B. Egglestone, 4 Lancaster Close, Etherley Dene, Bishop Auckland, DL14 0RP
2W1 EYZ S. Gray, 36 Heol Pentre Felen, Morriston, Swansea, SA6 6BY
2M1 EZA J. Boyle, Flat 3/2, 33 St. Mungo Avenue, Glasgow, G4 0PH
2E1 FAN K. Blanchard, 17 Stephens Way, Sleaford, NG34 7JN
2E1 FAT S. Hallsworth, 27 Westfield Avenue, Heanor, DE75 7BN
2E1 FBA R. Locke, 12A Kennedy Court, Walesby, Newark, NG22 9PQ
2E1 FBK C. Strange, 12 Cricketts Lane, Chippenham, SN15 3EF
2E1 FBS D. Seabridge, 31 Charlestown Drive, Allestree, Derby, DE22 2HA
2E1 FBY I. Roper, 8 Mulhalls Mill Wharf Street, Sowerby Bridge, HX6 2EF
2E1 FCC M. Williams, Cornacres, Dodwell, Stratford upon Avon, CV37 9ST

2E1 FCD M. Williams, Cornacres, Dodwell, Stratford upon Avon, CV37 9ST
2E1 FCE J. Lindsay, 10 New Station Road, Swinton, Mexborough, S64 8AH
2E1 FCO I. Brady, 6 Bristow Close, Bletchley, Milton Keynes, MK2 2XP
2E1 FDC L. Peck, 17 Mill Lane, Barton Le Clay, Bedford, MK45 4LN
2E1 FDD M. Warren, 145 Shirehall Road, Sheffield, S5 0JL
2E1 FDF G. Bilson, 34 Bramlyn Close, Clowne, Chesterfield, S43 4QP
2E1 FDJ K. Newnam, 1 Wheatlands Close, Maulden, Bedford, MK45 2AQ
2E1 FDK L. Rule, 46 Meadowsweet Road, Poole, BH17 7XT
2E1 FDM J. Mew, 1 Council Houses, Norwich Road, Norwich, NR9 4NY
2E1 FDP G. Westwood, 6 Monkton Road, Borough Green, Sevenoaks, TN15 8SD
2E1 FDT P. Brown, 19 Huxley Close, Nottingham, NG8 4PU
2E1 FDU M. Darton, 8 Foster Grove, Sandy, SG19 1HP
2E1 FDY K. Rankin, 31 Gorsey Bank Road, Hockley, Tamworth, B77 5JD
2E1 FEC F. Dunmore, 36 Dove Rise, Oadby, Leicester, LE2 4NY
2E1 FEF K. Rhodes, 30 Priory Road, Louth, LN11 9AL
2E1 FET N. Moore, 84 Franklynn Road, Haywards Heath, RH16 4DH
2E1 FFJ R. Denniss, Chapel House, Farlesthorpe, Alford, LN13 9PH
2E1 FFL P. Harbinson, 19 Pennine Road, Dewsbury, WF12 7AW
2E1 FFX P. Fryer, 19A Wintringham Way, Purley On Thames, Reading, RG8 8BH
2E1 FFZ D. Fawcett, 6 Wand Hill, Boosbeck, Saltburn-by-the-Sea, TS12 3AW
2E1 FGB N. Ash, 35 Fairford Road, Tilehurst, Reading, RG31 6PY
2W1 FGR J. Clark, 22 Heol Yr Wylan, Cwmrhydyceirw, Swansea, SA6 6TB
2E1 FHO P. Sawyers, 15 Park Terrace, Whitby, YO21 1PN
2E1 FHQ C. Ward, 18 Aspen Grove, Aldershot, GU12 4EU
2E1 FHZ D. Hebb, 45 Marklew Avenue, Grimsby, DN34 4AD
2E1 FIE M. Robertson, 98 Hawthorn Road, Bognor Regis, PO21 2DG
2E1 FIQ H. Pook, Beverley Friory, Friars Lane, Beverley, HU17 0DF
2E1 FIV D. Layton, 69 Elm Drive, Crewe, CW1 4EL
2E1 FIX J. Shaw, 27 St. Davids Avenue Romiley, Stockport, SK6 3JT
2W1 FJG A. Heigh, Janneen, Henry Street, Wrexham, LL14 4DA
2E1 FJL M. Gray, 19 Marsh View Newton, Preston, PR4 3SX
2W1 FJN I. Pearson, Warren Cottage, Pontfadog, Llangollen, LL20 7AT
2E1 FJP B. Melling, 68 Westfield Drive, Ribbleton, Preston, PR2 6TH
2E1 FJV I. Page, 84 Beaulieu Close, Toothill, Swindon, SN5 8AH
2W1 FJZ Prof. P. Edwards, Cam O'R Afon, Dolywern, Llangollen, LL20 7AD
2E1 FKD E. Merrington, Cartref, Ball Lane, Frodsham, WA6 8HP
2E1 FKJ J. Ewing, 130 Uttoxeter Road, Hill Ridware, Rugeley, WS15 3QX
2E1 FKM C. Andrews, Flat 2, 1St Floor, 153 Thanet St., Chesterfield, S45 9JT
2E1 FKT M. Boyes, 19 Rowe Ashe Way, Locks Heath, Southampton, SO31 7EY
2E1 FKZ A. Woodward, 19 Hazel Grove, Winchester, SO22 4PQ
2E1 FLD P. Whittaker, 34 New Road, Newhall, Swadlincote, DE11 0SP
2E1 FLN G. Mcmillan, 10 Thornton Court, Girton, Cambridge, CB3 0NS
2E1 FLW S. Burling, Ongar Cottage, 28 Main Road, Macclesfield, SK11 0BU
2E1 FMC R. Sanderson, 92 Edge Lane, Dewsbury, WF12 0HB
2E1 FMW A. Oughton, 176 South Lodge Drive, Southgate, London, N14 4XN
2E1 FNB F. Morgan, 171 Town Road, London, N9 0HJ
2E1 FNJ L. Carr, 10 Bonds Road, Hemblington, Norwich, NR13 4QF
2U1 FNQ J. Le Page, Heathwick, Les Martins, St Martin, Guernsey, GY4 6QJ
2E1 FNR D. Rowland, 32 Treloweth Way, Pool, Redruth, TR15 3TT
2E1 FNX J. Meredith, 2 Hamilton Road, Dawley, Telford, TF4 3NG
2E1 FNY C. Warren, Clifden Farm, Quenchwell Road, Truro, TR3 6LN
2E1 FON S. Simpson, 20 Staveley Grove, Keighley, BD22 7DH
2E1 FOU K. Perry, 238 Sherwood Street, Warsop, Mansfield, NG20 0HJ
2E1 FOW J. Thacker, 15 Riverside Mews, Hall Yard, Stoke-on-Trent, ST10 4FE
2E1 FOX J. Camp, 1 Higher Tresillian Cottages, Tresillian, Newquay, TR8 4PL
2W1 FPK A. Cartwright, 7 Pen Parc, Malltraeth, Bodorgan, LL62 5BG
2E1 FPM L. Noel, 58 Easenhall Lane, Redditch, B98 0BJ
2E1 FPP D. Paddon, 21 Oak Park Drive, Havant, PO9 2XE
2E1 FPU J. Overland, 73 Butchers Lane, Walton on the Naze, CO14 8UE
2E1 FPV J. Stringer, The Cottage, High Street, Radstock, BA3 5AL
2E1 FPW J. Green, 2 Broadmeadow, Kingswinford, DY6 7HG
2E1 FQB P. Elcombe, 16 Blenheim Avenue, Martham, Great Yarmouth, NR29 4TW
2E1 FQE C. Carter, 331A Ordnance Road, Enfield, EN3 6HE
2M1 FQI G. Robinson, 12 Hannahston Avenue, Drongan, Ayr, KA6 7AU
2E1 FQO M. Dunthorne, 52 New Street, Oakengates, Telford, TF2 6ES
2E1 FQY O. James, 24 Fryer Avenue, Leamington Spa, CV32 6HY
2E1 FRC H. Doyle, Hurst House, Stratford Road, Henley-in-Arden, B95 6AB
2E1 FRE D. Franklin, The Old Vicarage, Westville Road, Boston, PE22 7HJ
2E1 FRI J. Wood, 3 Harold Collins Place, Colchester, CO1 2GQ
2E1 FRQ A. Storry, 99 Swineshead Road, Wyberton Fen, Boston, PE21 7JG
2E1 FRW T. Depledge, 16 Pennington Walk, Retford, DN22 6LR
2E1 FRY E. Young, 7 The Quadrant, Fordingbridge, SP6 1BW
2E1 FRZ T. Jennings, 23 De Lacy Court, New Ollerton, Newark, NG22 9RN
2E1 FSF J. Kinch, 7 Fox Lane, Oakley, Basingstoke, RG23 7BB
2E1 FSG R. Kinch, 7 Fox Lane, Oakley, Basingstoke, RG23 7BB
2E1 FSH C. Forber, 32 Larch Avenue, Newton-le-Willows, WA12 8JF
2E1 FSR G. Hammond, 50 Fernhill Close, Woodbridge, IP12 1LB
2E1 FSV D. Feetenby, 32 Hawkins Close, Daventry, NN11 4JQ
2E1 FSX D. Charlton, 20 Bailey Crescent, South Elmsall, Pontefract, WF9 2TL
2E1 FSY M. Fey, 37 Winnards Close, West Parley, Ferndown, BH22 8PA
2E1 FSZ T. Bashford, 198 Uplands Road, West Moors, Ferndown, BH22 0EY
2E1 FTA S. Goodall, Red Roofs, 4 Chapel Street, Stapleton, LE9 8JH
2E1 FTE C. Bingham, 56 Newgate Lane, Mansfield, NG18 2LQ
2E1 FTF E. Sheppard, 4 Lindrick Avenue, Swinton, Mexborough, S64 8TE
2E1 FTH R. Megone, 16 Mercer Close, Basingstoke, RG22 6NZ
2E1 FTI M. Burrows, 40 Fairmile Road, Christchurch, BH23 2LL
2E1 FTV J. Bailey, 13 Newark Road, Mexborough, S64 9EZ
2W1 FUD D. Green, 17 Glyn-Y-Mel Pencoed, Bridgend, CF35 6YA
2E1 FUH D. Hartley, 2 Thirlmere Avenue, Burnley, BB10 1HU
2E1 FUJ J. Hartley, 23 Broomfield Road, Fleetwood, FY7 7HA
2E1 FUQ J. Quartermaine, 3 Markham Close, Duston, Northampton, NN5 6TW
2W1 FVH R. Hallett, Llamedos, 151 Trealaw Road, Tonypandy, CF40 2NX
2E1 FVJ R. Halford, 104 Gladstone Avenue, London, N22 6LH
2E1 FVK T. Kiely, 192 Morley Avenue, London, N22 6NT
2E1 FVS M. Hotchin, 122 Buckingham Avenue, Scunthorpe, DN15 8NS
2E1 FVY D. Ripley, 5 Rope Walk, Cranbrook, TN17 3DZ
2E1 FWA C. Walker, 12 Bradshaw Crescent, Honley, C/O Mr C Walker, Holmfirth, HD9 6EG
2E1 FWD C. Booth, 8 Heathfield Mews, Martlesham Heath, Ipswich, IP5 3UF
2E1 FWM M. Foister, 38 Nine Acres, Kennington, Ashford, TN24 9JW
2E1 FWX G. Sessions, 5 Luxton Court, Cullompton, EX15 1FJ
2E1 FXN D. Boland, 29 Broom House, Baneberry Road, Gloucester, GL4 6UY
2E1 FYC D. Martyr, 52 Parklawn Avenue, Epsom, KT18 7SL
2E1 FYZ T. Wilkie, Bramcote, Grange Road, Sutton on Sea, LN12 2RE
2E1 FZC J. Charter, 36 Northumberland Avenue, London, E12 5HD
2E1 FZH M. Saunders, 105 Raynham Road, Bury St Edmunds, IP32 6ED
2E1 FZU D. Peters, 25 Corndon Crescent, Shrewsbury, SY1 4LD
2E1 FZY J. Doyle, Orchard Corner, Piddington Road, Ludgershall, Aylesbury, Hp189pj
2W1 GAC W. Hicks, 2 Second Avenue, Clase, Swansea, SA6 7LN
2M1 GBG M. Thomson, 194 East Main Street, Broxburn, EH52 5HQ
2E1 GBM J. Lake, 25 Erlensee Way, Biggleswade, SG18 8GG
2E1 GBN A. Vincent, 12 Broad Park Avenue, Ilfracombe, EX34 8DZ
2E1 GCB E. Muxlow, 14 Maidwell Way, Grimsby, DN34 5UP
2D1 GCC B. Smith, 82 Royal Park, Ramsey, Isle of Man, IM8 3UH
2E1 GCF C. Horn, 12 Melbourne Road, Chichester, PO19 7NE
2W1 GCU J. Bennett, 28 Neyland Path Fairwater, Cwmbran, NP44 4PX
2E1 GDA C. Norris, 8 Sutton Lane, Middlewich, CW10 9AU
2E1 GDB M. Jeffery, 14 Holly Mount Shavington, Crewe, CW2 5AZ
2E1 GDD D. Townson, 4 Crawford Street, Bradford, BD4 7JJ
2E1 GDF S. Okeefe, 9 Ivernia Close, Derby, DE23 1XF
2E1 GDG D. Baker, 78 Station Road, Whittlesey, Peterborough, PE7 1UE
2E1 GDK V. Potts, 14 Topcliffe Mead, Morley, Leeds, LS27 8UH
2E1 GDM M. Banner, 23 Astral Grove, Hucknall, Nottingham, NG15 6FY
2E1 GDO R. Hacker, Flat 7, Dove Court Packers Lane, Ramsgate, CT11 8QA
2W1 GDY A. Palmer, 4A Clomendy Road, Cwmbran, NP44 3LS
2E1 GES C. Knowlson, 23 Hawthorne Avenue, Shipley, BD18 2JB
2M1 GEZ R. Scott, Kirklands, Craigend Road, Galashiels, TD1 2RJ
2M1 GFG R. Dempster, 42 Kirkhill Road, Edinburgh, EH16 5DD
2E1 GFS S. Thomas, 24 Jenwood Road, Dunkeswell, Honiton, EX14 4UZ
2E1 GFW M. Jones, 68 Hampton Park Road, Hereford, HR1 1TJ
2E1 GGL E. Mills, 70 Crescent Road, Rochdale, OL11 3LG
2E1 GGT A. Walker, 12 Bladen Close, Cheadle Hulme, Cheadle, SK8 5RU
2E1 GHE A. Carter, 50 Harberd Tye, Chelmsford, CM2 9GJ
2E1 GHF A. Richardson, 42 Gypsey Road, Bridlington, YO16 4AZ
2E1 GHI E. Mills, 80 Bransdale Road, Nottingham, NG11 9JB
2E1 GHX C. Lewin, The Hawthorns, Hawthorne Drive, Stafford, ST19 9NQ
2E1 GHZ P. Mather, 18 Watcombe Cottages, Richmond, TW9 3BD
2I1 GIH L. Costford, Aughakeerin, Derrygonnelly, Co Fermanagh, BT93 6FR
2E1 GIK A. Craven, 4 Amanda Drive, Louth, LN11 0AZ
2E1 GIZ E. Christieson, September Cottage, Rushlake Green, Heathfield, TN21 9PP
2E1 GJC A. Brown, Hurst Farm, Ashworth Road, Rochdale, OL11 5UP
2E1 GJD S. Burgess, 14 Shrubcote, Tenterden, TN30 7BA
2E1 GJE P. Lee, 2 Dennis St., Worksop, S80 2LL
2E1 GJG H. Kennedy, 19 High Street, East Hoathly, Lewes, BN8 6DR
2E1 GJJ A. Noon, 12 Stoney Fold, Telford, TF3 5GQ
2E1 GJL P. Lovesey, 20 Lindsey Way, Louth, LN11 8RP
2E1 GJO S. Higgs, 239 Whalley Drive, Bletchley, Milton Keynes, MK3 6PL
2E1 GJP K. Turner, 2 Bungalow, Dunston Fen, Lincoln, LN4 3AP
2E1 GJT C. Wojcik, 153 Netherton Road, Worksop, S80 2SD
2E1 GJY S. Brazier, 14 Pynes Lane, Bideford, EX39 3EB
2E1 GKB M. Brearley, 136 Elmsfield Avenue, Rochdale, OL11 5XA
2E1 GKE D. Shephard, 107 Withywood Drive, Telford, TF3 2HX
2E1 GKF N. Salter, 10 Jarvis Place St. Michaels, Tenterden, TN30 6DQ
2W1 GKJ M. Moore, 56 Morfa Street, Bridgend, CF31 1HD
2E1 GKP C. Dix, 141A Jerningham Road, London, SE14 5NJ
2E1 GKY A. Reed, 32 Hollis Gardens, Cheltenham, GL51 6JQ
2M1 GLD E. Clark, 3 Old Cottages, Seton Mains, Longniddry, EH32 0PG
2E1 GLQ J. Lutener, 22 Heron Park, Basingstoke, RG24 8UJ
2E1 GLR J. Halsall, 83 Poole Road, Leeds, LS15 7HD
2E1 GLS D. Brown, 22 Hillcrest Road, Castleford, WF10 3QX
2E1 GLT M. Simpson, 20 Mount Pleasant Residential Park, Bloomhill Road, Moorends, Doncaster, DN84ST
2W1 GLY C. Mills, 26 Goossens Close, Ringland, Newport, NP19 9JN
2E1 GMA S. Jordan, 31 Rathmore Road, Birkenhead, CH43 2 HE
2W1 GMM B. Richards, 8 East Avenue, Griffithstown, Pontypool, NP4 5AB
2E1 GMO M. Hastry, 56 Kilsyth Close, Fearnhead, Warrington, WA2 0SQ
2E1 GMQ J. Barrett, 114 William Street, Long Eaton, Nottingham, NG10 4GD
2E1 GMT J. Williams, Hillsborough Lodge, Lower St. German'S Road, Exeter, EX4 4PW
2E1 GMV S. Fletcher, 5 Hayeswood Road, Stanley Common, Ilkeston, DE7 6GB
2E1 GNE R. Wright, 3 Ednall Lane, Bromsgrove, B60 2DB
2E1 GNK G. Smith, 106 Broadoak Road, Manchester, M22 9PL
2E1 GNN P. Saben, Tredinneck Moor, Newmill, Penzance, TR20 8XT
2E1 GNR A. Taylor, 16 Penny Lane, Collins Green, Warrington, WA5 4DS
2E1 GNU A. Watson, 7 Branksome Drive, Morecambe, LA4 5UJ
2E1 GNZ D. Hoyle, 31 Rochester Avenue, Bolton, BL2 5ED
2E1 GOC S. Bedell, 1 Pheasant Field Drive, Spondon, Derby, DE21 7LR
2E1 GOE C. Hopkins, 28 Pitcairn Road, Mitcham, CR4 3LL
2E1 GOK K. Phillips, 30 Allendale, Ilkeston, DE7 4LE
2E1 GOM Dr J. Constable, 1900 Olden Glen, de Pere, USA, 54115
2E1 GOP B. Cartwright, 14 Zealand Close, Hinckley, LE10 1TJ
2E1 GOZ I. Press, 19 Banwell Close, Keynsham, Bristol, BS31 1JX
2E1 GPE M. Ruff, Brambledown, Camberlot Road, Hailsham, BN27 3QG
2E1 GPG D. Howells, Flat 1-2, 130 Essex Road, London, N1 8LX
2E1 GPO R. Dinnage, 108 Chaddock Lane Boothstown, Salford, M28 1DF
2E1 GPV M. Allott, 1 Charles Street, Lancaster, LA1 4UU
2E1 GQB G. Meek, 443 Springfield Road, Chelmsford, CM2 6AP
2E1 GQD A. Smith, 31 Haywards Place, Easterton, Devizes, SN10 4PP
2E1 GQN T. Stevens, 25 Avenue Road, New Milton, BH25 5JP
2E1 GQU M. Forster, 33 Deer Valley Road, Holsworthy, EX22 6DA
2E1 GRA A. Merrill, 66 Royal Oak Drive, Selston, Nottingham, NG16 6RJ
2E1 GRG D. Brady, 6 Bristow Close, Bletchley, Milton Keynes, MK2 2XP
2E1 GRT M. Keeling, 3 Waltham Hall Cottages, Norwich Road, Little Stonham, IP14 5LX
2E1 GSC P. Sherratt, 23 Paulina Avenue, Nottingham, NG15 8JA
2E1 GSM D. Ridgway, 145 Bodmin Road, Astley, Manchester, M29 7PE

2E1 GSW J. Harratt, 8 The Runcie Building, Ripon College, Oxford, OX44 9EX
2E1 GSX H. Mustoe, 5 Dartmouth Park Avenue, London, NW5 1JL
2E1 GTB C. Barnes, 533 Maidstone Road, Blue Bell Hill, Chatham, ME5 9QP
2E1 GTD S. Keene, 17 Chaffers Close Long Sutton, Hook, RG29 1SY
2E1 GTE S. Hall, 122 Norwich Road, New Costessey, Norwich, NR5 0EH
2E1 GTF S. Silk, 89 St. Barts Road, Sandwich, CT13 0AS
2E1 GTI A. Cadier, 28 Romney Avenue, Folkestone, CT20 3QJ
2E1 GTQ A. Smith, 5 Langton Avenue, Chelmsford, CM1 2BW
2E1 GTT A. Davies, Little Platt, Bodle Street Green, Hailsham, BN27 4RA
2E1 GTW F. James, 1 The Gorseway, Bexhill-on-Sea, TN39 4PP
2E1 GUC A. Ibrahim, 14 Dunvegan Road, London, SE9 1SA
2E1 GUN S. Kirby, 2 Kneeton Park, Middleton Tyas, Richmond, DL10 6SB
2E1 GVB E. Artis, 40 Nottingham Way, Great Yarmouth, NR30 2SA
2E1 GVC G. Carlin, 38 Balfour St., East Bowling, Bradford, BD4 7JT
2E1 GVD G. Carlin, 82 Kingsway, Drighlington, Bradford, BD11 1ET
2E1 GVJ J. Millington, 6 Fentonhouse Lane, Wheaton Aston, Stafford, ST19 9NU
2E1 GVS G. Gallacher, 112 Central Drive, Stoke-on-Trent, ST3 2AJ
2E1 GW Rev. G. Wellington, 94 Arlington Road, London, N14 5AT
2E1 GWX R. Shepherd, 91 Saxon Road, Hastings, TN35 5HH
2E1 GXE M. Rogers, 45 Church Road, Westoning, Bedford, MK45 5LP
2E1 GXH D. Webb, 51 Garden Road, Walton on the Naze, CO14 8RR
2E1 GXI J. Adams, 28 Greenside, Stoke Prior, Bromsgrove, B60 4EB
2E1 GXL P. Travis, 42 Trafalgar Road, Wallasey, CH44 0EB
2E1 GXQ P. Walker, 78 Kirkby Road, Desford, Leicester, LE9 9JG
2E1 GXS R. Tomlin, 6 Gaviots Green, Gerrards Cross, SL9 7EB
2E1 GXU R. Steven, 23 Collingwood Road, Woodbridge, IP12 1JL
2E1 GXV M. Rogers, 2 Tudor Close, Framlingham, Woodbridge, IP13 9SL
2M1 GXX G. Scott, Kirklands, Craigend Road, Galashiels, TD1 2RJ
2E1 GXY P. Batty, 134 Plymouth Road, Scunthorpe, DN17 1TS
2E1 GYB M. Peterson, 29 Warwick Close, Saxilby, LN1 2FT
2E1 GYC S. Jenkins, 45 Dorchester Road, Solihull, B91 1LN
2E1 GYD M. Hall, 45 Dorchester Road, Solihull, B91 1LN
2E1 GYG D. Abbott, 5 Heathcote Gardens, Rudheath, Northwich, CW9 7JB
2E1 GYH S. Pearson, 8 The Pastures, Edlesborough, Dunstable, LU6 2HL
2E1 GYO C. Stanley, 9 Pyramid Caravan Site, Beeston, Nottingham, NG9 1NS
2E1 GYR H. Webster, 67 Greenways, Over Kellet, Carnforth, LA6 1DE
2M1 GYX R. Macleod, 13 Wyvern Park, Edinburgh, EH9 2JY
2E1 GYZ L. Rowley, 20 Long Leasow, Selly Oak, Birmingham, B29 4LT
2E1 GZF J. Walker, 1 Riverside Court, Louth, LN11 7AG
2E1 GZV A. Ager, 5 Matthews Close, Bedhampton, Havant, PO9 3NJ
2E1 GZY A. Wilson, 22 Ormesby Road, Raf Coltishall, Norwich, NR10 5JY
2E1 GZZ M. Coles, 29 Sydney Road, Exeter, EX2 9AH
2E1 HAC M. Lucas, 48 Sycamore Drive, Ash Vale, Aldershot, GU12 5PR
2E1 HAL M. Farraway, 35 Wingbourne Walk, Nottingham, NG6 8DT
2E1 HAM D. Langmead, 38 Milton Grove, London, N11 1AX
2E1 HAQ K. Graffham, 15 Hayes Road, Clacton-on-Sea, CO15 1TX
2E1 HAS B. Heirene, 9 Ryecroft Crescent, Barnet, EN5 3BP
2E1 HAU M. Durrant, 8 Drake Avenue, Great Yarmouth, NR30 4BS
2E1 HAW J. Beith, 18 Avenue Road, New Milton, BH25 5JP
2E1 HBF L. Buggs, 2 Archway Cottages, Valley Road, Leiston, IP16 4AR
2E1 HBJ C. Masters, 85 Petersham Road, Creekmoor, Poole, BH17 7DW
2E1 HBS D. Treacher, 6 Beech View, Whitwell, York, YO60 7JW
2E1 HBT E. Marlow, 17 Fellows Court, Weymouth Terrace, London, E2 8LP
2E1 HCB D. Broom, 114 Gammons Lane, Watford, WD2 5HY
2E1 HCG A. Docherty, The Flat, Winton House, Stoke on Trent, ST4 2RQ
2M1 HCP A. Smith, 4 The Terrace, Lhanbryde, Elgin, IV30 8NY
2M1 HCQ A. Smith, 4 The Terrace, Lhanbryde, Elgin, IV30 8NY
2E1 HCT R. Cowlishaw, 23 Aldrich Drive, Willen, Milton Keynes, MK15 9HP
2E1 HDB M. Dimambro, 26 Fetcham Court, Bank Top, Newcastle upon Tyne, NE3 2UL
2E1 HDC P. Dimambro, 26 Fetcham Court, Bank Top, Newcastle upon Tyne, NE3 2UL
2E1 HDE A. Morris, 71 Lurdin Lane, Standish, Wigan, WN6 0AQ
2E1 HDF R. Harrison, 36 Windermere Road, Farnworth, Bolton, BL40QH
2E1 HDZ T. Lockett, 14 Tildsley Crescent, Weston, Runcorn, WA7 4RN
2E1 HEE H. Merrington, Cartref, Ball Lane, Frodsham, WA6 8HP
2E1 HEF J. Hastry, 56 Kilsyth Close, Fearnhead, Warrington, WA2 0SQ
2E1 HEG E. Mather, 72 Cranleigh Road, Worthing, BN14 7QW
2E1 HEK L. Taylor, 127 Dundee Close, Fearnhead, Warrington, WA2 0UJ
2E1 HEM C. Kulikovsky, 42 Highlands Grove, Bradford, BD7 4BG
2E1 HEO A. Austin, Flat 4, De Cham Court, 33 De Cham Road, St Leonards-on-Sea, TN37 6JA
2E1 HER D. Webster, 28 Longmeadow Lane Heysham, Morecambe, LA32FH
2E1 HES M. Davenport, 24 Willow Grove, Belper, DE56 1LX
2E1 HEV S. Brandon, 2 Moss Bank, Winsford, CW7 2ED
2E1 HFA S. Emerton, 18 Avenue Road, New Milton, BH25 5JP
2M1 HFE C. Budas, 20 Oak Avenue, Bearsden, Glasgow, G61 3HD
2E1 HFH S. Overall, Flat 74, Douglas Buildings Marshalsea Road, London, SE1 1EL
2E1 HFN J. Newell, 4 Honeyfield Drive, Ripley, DE5 3JL
2E1 HFS P. Dunne, 2 All Saints Road, Wyke Regis, Weymouth, DT4 9EZ
2E1 HFV H. Shackleton, Woodroyd, 66 Spring Ave, Keighley, BD21 4TA
2E1 HFW M. Hemstock, 4 Tavistock Avenue, Perivale, Greenford, UB6 8AJ
2E1 HFX K. Taylor, 46 Hunters Field, Stanford In The Vale, Faringdon, SN7 8LX
2W1 HFZ L. Jones, 52 George Street, Aberdare, CF44 6SH
2E1 HGA T. Winwood, 2 The Warren, Abingdon, OX14 3XB
2E1 HGE A. Higgs, 26 Avon Avenue, Ringwood, BH24 2BH
2E1 HGF P. Crookes, 11 Degens Way, Hugglescote, Coalville, LE67 2XD
2E1 HGG P. Mead, 9 Abraham Drive, Silver End, Witham, CM8 3SP
2E1 HGJ J. Jordan, 4 Tuckers Close, Loughborough, LE11 2PG
2E1 HGM K. Cronin, East Cottage Westviile Rd, Thornton le Fen, LN4 4YJ
2E1 HGR D. Edwards, Whitehaven, High Street, Buxted, Uckfield, TN22 4JU
2E1 HGT A. Ruaux, 85 St. Catherines Road, Crawley, RH10 3TB
2E1 HGU T. Lawrence, 85 West Avenue, Clacton-on-Sea, CO15 1HB
2E1 HGY M. Henson, 35 Westbrook Drive Rainworth, Mansfield, NG21 0FB
2E1 HHA A. Gittens, Elyria, 22 Charles Lovell Way, Scunthorpe, DN17 1YL
2E1 HHB D. Botterell, 12 Selsey Avenue, Clacton-on-Sea, CO15 1NQ
2E1 HHE Dr T. Carlson, 15 White Hedge Drive, St Albans, AL3 5TU
2E1 HHG J. Bradbury, 192 Greenwood Road, Bakersfield, Nottingham, NG3 7FY
2E1 HHJ S. Brown, 3 Marsden Drive, Scunthorpe, DN15 8AD
2E1 HHK M. Tyldesley, 24 Trinity Court Broughton Brigg N Lincolnshire Dn200Sj, Broughton, DN200SJ
2E1 HHL S. Williams, Alwent Farm, Staindrop, Darlington, DL2 3NS
2E1 HHY S. Day, 10 Second Avenue, Wolverhampton, WV10 9PP
2E1 HID D. Smith, 23 Marlborough Road, Long Eaton, Nottingham, NG10 2BS
2E1 HIL I. Finch, 29 Sherwood Road, Grimsby, DN34 5TG
2M1 HIN J. Mcphee, 101 Birkenside, Gorebridge, EH23 4JF
2E1 HIO W. Roberts, 30 Park Boulevard, Clacton-on-Sea, CO15 5RH
2E1 HIQ A. Mayes, 126 Walesby Lane, New Ollerton, Newark, NG22 9UU
2E1 HJA K. Mieske, 10 Cowdrey Road, London, SW19 8TU
2E1 HJE C. Clifton, 35 Farrowdene Road, Reading, RG2 8SD
2E1 HJO M. Hyde, 10 Devonshire Drive, Barnsley, S75 1EE
2E1 HJS R. Burns, 130 Kingsway, Mapplewell, Barnsley, S75 6EU
2E1 HJW L. Carey, 5 Park Avenue, Harlow, CM17 9NL
2M1 HKA S. Mcintyre-Stewart, 4 Howie Crescent, Rosneath, Helensburgh, G84 0RL
2E1 HKB K. Brunning, 45 Dover Road, Ipswich, IP3 8JQ
2E1 HKC K. Cullum, 22 Orwell View Road, Shotley, Ipswich, IP9 1NP
2E1 HKE R. Wilson, Newstead Farm, Clay Lane, Norwich, NR10 4PP
2E1 HKS S. Hall, 1 George Place, Wellington, Telford, TF1 2AJ
2E1 HKY D. Green, 89 Upper Ratton Drive, Eastbourne, BN20 9DJ
2E1 HLA J. Moran, 8 Doffcocker Lane, Bolton., BL1 5RG
2M1 HLE G. Stephen, 9 Carse Court Iv3 8Te, Inverness, Iv3 8te
2E1 HLF E. Carder, 45 Chalklands, Linton, Cambridge, CB21 4JQ
2E1 HLH E. Gardner, New House, Birdbush Avenue, Saffron Walden, CB11 4DJ
2E1 HLL S. Batchelor, 2 Belmont Avenue, Atherton, Manchester, M46 9RR
2E1 HLO C. Mitchell-Watson, 144 Shakespeare Crescent, Dronfield, S18 1ND
2E1 HLP L. Lawrence, 85 West Avenue, Clacton-on-Sea, CO15 1HB
2E1 HLS J. Preen, 12 Isaac Walk, Worcester, WR2 5EQ
2E1 HLU T. Kimm, 99 Midland Road, Bramhall, Stockport, SK7 3DT
2E1 HLW M. Mcneany, 29 Grenfell Road, Manchester, M20 6TG
2E1 HMB T. Emmett, Meadow Sweet, Ramsden Lane, Offwell, EX149RY
2E1 HMQ K. Vance, 5 Riversdale Close, Birstall, Leicester, LE4 4EH
2E1 HNB K. Hughes, High Lane Cottage, Congleton Road, Macclesfield, SK11 9RR
2E1 HNF J. Li, 13 Tothill Street, Minster, Ramsgate, CT12 4AG
2W1 HNH A. Chalk, 42 Erskine Road, Colwyn Bay, LL29 8EU
2E1 HNN T. Bennellick, 18 Bailie Close, Abingdon, OX14 5RF
2E1 HNS T. Middleton, 31 Coltman Avenue, Long Crendon, Aylesbury, HP18 9DP
2I1 HNZ M. Mccrum, 2 New Road, Donaghadee, BT21 0DR
2E1 HOF J. Johnson, 2 Meadow Drive, Canon Pyon, Hereford, HR4 8NT
2E1 HOK S. Langley, 100 Pogmoor Road, Barnsley, S75 2EF
2E1 HOO D. Anstie, 20 Keyes Road, Norwich, NR1 2JX
2E1 HOS T. Keating, 65 Shorncliffe Avenue, Norwich, NR3 2HT
2E1 HOT J. Tosh, 65 Shorncliffe Avenue, Norwich, NR3 2HT
2E1 HOU S. Clarke, 14 Findley Drive, Wirral, CH46 3SG
2E1 HPC S. Carr, 10 Bonds Road, Hemblington, Norwich, NR13 4QF
2E1 HPD D. Carr, 10 Bonds Road, Hemblington, Norwich, NR13 4QF
2E1 HPM L. Hall, 9 Stone Court, South Hiendley, Barnsley, S72 9DL
2E1 HPS J. Vandervord, 13 Granville Avenue, Ramsgate, CT12 6DX
2E1 HPT A. Spain, 60 The Maples, Broadstairs, CT10 2PE
2E1 HPZ W. Bellis, Cliffe Bungalow, Barnsley Road, Barnsley, S72 9JX
2E1 HQA F. Kennedy, 19 High Street, East Hoathly, Lewes, BN8 6DR
2E1 HQH J. Garner, 30 Birds Avenue, Garlinge, Margate, CT9 5NE
2E1 HQP M. Newth, 13 Okement Close, West End, Southampton, SO18 3PP
2E1 HQQ R. Woodruff, 224A Spendmore Lane, 2Coppull, Chorley, PR7 5BZ
2E1 HQV B. Cummings, 8 Spey House, Criterion Street, Stockport, SK5 6TD
2E1 HQW S. Hackney, 8 Spey House, Criterion Street, Stockport, SK5 6TD
2E1 HQY P. Procter, 1 Brow Hey, Bamber Bridge, Preston, PR5 8DS
2E1 HQZ A. Webb, 52 Stanfield Road, Stoke-on-Trent, ST6 1AT
2E1 HRB S. Dykes, 15 Sunningdale Road, Chelmsford, CM1 2NH
2E1 HRJ A. Hollis, 89 Longfield Lane, Cheshunt, Waltham Cross, EN7 6AN
2E1 HRM D. Morgan, 171 Town Road, London, N9 0HJ
2E1 HRN A. Fower, 17 Oakwood Road, Leek, ST13 8LW
2M1 HRS R. Duncan, 90 Faulds Gate, Aberdeen, AB12 5QT
2E1 HRY P. Odle, 24 Longfellow Road, Gillingham, ME7 5QG
2E1 HSA T. Newman, Sometimes (The Workshop), South Pew, Dorchester, DT2 9HZ
2E1 HSB C. Price, 162 Stamshaw Road, Portsmouth, PO2 8LX
2E1 HSD B. Manning, Barton Farm, Barton Road, Wisbech, PE13 4TL
2M1 HSG M. Douglas, 195 Dumbuck Road, Dumbarton, G82 3NU
2E1 HSJ M. Tulk, 8 Cleves Close, Weymouth, DT4 9JU
2E1 HSL T. Brodie, 8 Meadow Way, Plymouth, PL7 4JB
2E1 HSP A. Saville, 4 Shannon Court, Downs Barn, Milton Keynes, MK14 7PP
2E1 HSR M. Rogers, 47 Tregarrian Road, Tolvaddon, Camborne, TR14 0HD
2E1 HTE C. Hall, 2 Brambling Lane, Wath-Upon-Dearne, Rotherham, S63 7GT
2W1 HTK E. Bateman, 32 Park Avenue, Bodelwyddan, Rhyl, LL18 5TB
2E1 HTM J. Brice, 10 Swan Close, Weston-Super-Mare, BS22 8XR
2M1 HTR A. Pollard, 12 Royfold Crescent, Aberdeen, AB15 6BH
2E1 HTU R. Corbett, 11 Old Office Close, Dawley Bank, Telford, TF4 2QA
2E1 HTV C. Sargent, 8 Gilwell Grove, Priorslee, Telford, TF2 9SR
2E1 HTY A. Semple, Linden Lea, Lydham, Bishops Castle, SY9 5HB
2E1 HUB R. Semple, The Inn On The Green, Wentnor, Bishops Castle, SY9 5EF
2E1 HUC K. Jones, 24 Brooke Avenue, Margate, CT9 5NG
2E1 HUE D. Taylor, 2 Delph Cottages, Barkisland, Halifax, HX4 0BW
2E1 HUJ P. Adams, 20 Grosvenor Road, London, W4 4EH
2E1 HUQ K. Gerrard, 8 Windsor Crescent, Little Houghton, Barnsley, S72 0HG
2W1 HUX H. Huxley, Hen Berllan, Nant Mawr Road, Buckley, CH7 2BS
2E1 HVB A. Scothern, 42 Griceson Close, Ollerton, Newark, NG22 9BD
2E1 HVI G. Reilly, B I Z Ltd, Millmarsh Lane, Enfield, EN3 7QA
2E1 HVL G. Batsman, 66A Merivale Road, Harrow, HA1 4BH
2E1 HVM C. Jackson, 76 Margards Lane, Verwood, BH31 6JP
2E1 HVN P. Harris, 17 Seymour Avenue, Great Yarmouth, NR30 4BB
2M1 HVR C. Hextall, 4 Hawthornbank, Cockenzie, Prestonpans, EH32 0HZ
2E1 HVT R. Hicks, 31 Arundel Road, Great Yarmouth, NR30 4LD
2E1 HVU S. Hicks, 50 Garfield Road, Great Yarmouth, NR30 4JU
2E1 HVZ C. Stevens, 82 Bembridge Drive, Alvaston, Derby, DE24 0UQ
2E1 HWI C. Zdziech, 200 Kensington Street, Rochdale, OL11 1QS
2E1 HWJ T. Winch, Whitehall Barn, Stowmarket Road, Stowmarket, IP14 6BU
2E1 HWQ S. Cannon, 22 Maid Marion Rise Warsop, Mansfield, NG20 0LD
2E1 HWU R. Noakes, 14 Sackville Road, Immingham, DN40 1EE
2E1 HWV M. Brett, 30 Belmont Avenue, London, N13 4HD

2E1 HXB B. Wheat, 62 Havenwood Rise, Nottingham, NG11 9HE
2E1 HXC A. Rodger, 4 Burns Nurseries, Wootton Road, King's Lynn, PE30 3BG
2E1 HXD A. Brett, 1C Old Park Road, Palmers Green, London, N13 4RG
2E1 HXM L. Tunstall, 8 York Road, Rowley Regis, B65 0RR
2E1 HXN J. Dodds, 8 York Road, Rowley Regis, B65 0RR
2E1 HXP M. Walsh, 23 Moss Fold Road, Darwen, BB3 0AQ
2E1 HXR E. Lacey, 6 Weetshaw Close, Shafton, Barnsley, S72 8PZ
2W1 HXT K. Jones, 11 St. Davids Close, Gobowen, Oswestry, SY11 3JF
2E1 HXY R. Ling, 8 Spa Hill, Kirton Lindsey, Gainsborough, DN21 4NE
2E1 HXZ P. Ling, 8 Spa Hill, Kirton Lindsey, Gainsborough, DN21 4NE
2E1 HYD A. Wilson, 135 Britannia Road Morley, Leeds, LS27 0DS
2E1 HYE D. Bruce, 6 Princes Way, King's Lynn, PE30 2QL
2E1 HYI P. Redfern, 42 Newton Street, Retford, DN22 7AD
2E1 HYX M. Hammond, 10 Collingwood Close, Braintree, CM7 9UG
2E1 HZI E. Sinclair, 21 Longport Avenue, Manchester, M20 1EN
2E1 HZM M. Fleming, 54 Madison Avenue, Bradford, BD4 0JJ
2E1 HZV E. Colley, 14 Hawthorne Close, Tyldesley, Manchester, M29 8PH
2E1 HZY P. Hartley, 14 Medway Walk, Wigan, WN5 9NQ
2E1 IAB J. Jackson, 40 Lightbounds Road, Bolton, BL1 5UN
2E1 IAC M. Dix, 10 College Hill, Godalming, GU7 1YA
2E1 IAE M. Hedges, 14 Weavers Mill Way, Holmfirth, HD9 7FB
2E1 IAI N. Revell, York House, The Street, Great Saling, Braintree, CM7 5FS
2E1 IAS S. Loane, 30 St Wilfrids Road, Burgess Hill, RH15 8BD
2E1 IAT B. Randell, 32 Windsor Close, Rubery, Birmingham, B45 0DA
2E1 IAX M. Frankland, 90 Kensington Road, Coventry, CV5 6GH
2E1 IAY J. Healey, 28 Thatchers Place, Westlands, Droitwich, WR9 9ED
2E1 IAZ P. Healey, 28 Thatchers Place, Westlands, Droitwich, WR9 9ED
2M1 IBE P. Woods, 92 Preston Crescent, Prestonpans, EH32 9RD
2M1 IBH G. Noonan, 8 Johnston Terrace, Port Seton, Prestonpans, EH32 0BB
2E1 IBJ D. Marshall, 143 Middleton Road, Banbury, OX16 3QS
2W1 IBN M. Tucker, 21 Pen Y Fan Close, Pentwyn Crumlin, Newport, NP11 3JQ
2E1 IBP J. Siddall, 17 Dalebrook Road, Sale, M33 3LD
2E1 IBS I. Grahame, 24 Bushby Avenue, Broxbourne, EN10 6QE
2E1 IBT J. Jordan, 25 Morris Close, Loughborough, LE11 1PU
2M1 IBX J. Millar, 18A Dougall St., Tayport, DD6 9JD
2E1 ICI C. Rengifo, 61 Clarendon Road, Sale, M33 2DY
2E1 ICM L. Kerswill, 18C Jewell Road, Bournemouth, BH8 0JQ
2E1 ICO A. Ryan, 18 Belmount Avenue, Newcastle upon Tyne, NE3 5QD
2E1 ICT G. Mears, 4 Uplands Road, Woodford Green, IG8 8JN
2E1 ICU R. Ropinski, 38 The Leys, Little Eaton, Derby, DE21 5AR
2E1 ICW A. Spiers, 2 Adeline Cottages, Jacobs Well Road, Guildford, GU4 7PD
2E1 IDC C. Hillier, 15 Dabbs Hill Lane, Northolt Park, Northolt, UB5 4AQ
2E1 IDE G. Swain, 33 Saville St., Blidworth, Mansfield, NG21 0RW
2E1 IDG D. Greaves, The White House, 25 London Road, Leicester, LE8 9GF
2E1 IDH K. Reynolds, 3 Lilac Close, Chelmsford, CM2 9NY
2E1 IDI C. Welsh, 56 Longacres, St Albans, AL4 0DR
2E1 IDM D. Cheetham, 405 Jenkin Road, Sheffield, S9 1AY
2E1 IEK E. Howie, 102 Rushdean Road, Rochester, ME2 2QB
2E1 IEM J. Cutler, 11 Margaret Ashton Close, Manchester, M9 4PZ
2E1 IFA F. Maddison, 87 Godolphin Road, London, W12 8JN
2E1 IFL C. Wynn, 45 Hillcrest Road, Berry Hill, Coleford, GL16 7RG
2E1 IFM A. Gow, 11 Rodley Square, Lydney, GL15 5AZ
2E1 IFN K. Gow, 11 Rodley Square, Lydney, GL15 5AZ
2E1 IFW D. Hyde, The Grove, 7 Mill Lane, Kidderminster, DY10 3ND
2E1 IGA E. Crane, 161 Heath Way, Horsham, RH12 5XX
2E1 IGG K. Horsley, 6 Flatford Place, Kidlington, OX5 1TH
2E1 IGI D. Willimot, 5 Green Lane, Upton, Huntingdon, PE28 5YE
2E1 IGJ Capt. H. Schnaar, 18 Witham Lodge, Witham, CM8 1HG
2E1 IGK C. Smith, 21 Earl Spencer Court, Peterborough, PE2 9PQ
2M1 IGO V. Gray, 55 Prestongrange Road, Prestonpans, EH32 9DD
2E1 IGP A. Wolstencroft, 201 Walton Road, Sale, M33 4ER
2M1 IGQ C. Bond, Strathllan House, Doune Road, Dunblane, FK15 9AR
2E1 IGU R. Ransom, 97 Park Square West, Jaywick, Clacton-on-Sea, CO15 2NU
2E1 IGY J. Kerr, 14 Seafield Close, Barton On Sea, New Milton, BH25 7HR
2E1 IHB W. Clarke, 41 Upton Road, Atherton, Manchester, M46 9RQ
2E1 IHE D. Siviter, 72 Sandfield Road, West Bromwich, B71 3NE
2E1 IHF L. Parrish, 5 Kestrel Lane, Cheadle, Stoke-on-Trent, ST10 1RU
2E1 IHJ D. Seeby, 59 Dallamoor, Telford, TF3 2EE
2E1 IHK J. Smith, 54 Hillside Avenue, Bridgnorth, WV15 6BU
2E1 IHO J. Willingham, 3 Cherry Road, Nailsea, Bristol, BS48 2EE
2E1 IHS A. Hardman, 47 Oatlands Road, Manchester, M22 1AH
2E1 IHT G. Large, 11 Ranworth, King's Lynn, PE30 4XD
2E1 IHW J. Eaton, 184 Gore Road, New Milton, BH25 5NQ
2E1 IHY C. Sneap, 14 Calver Close, Nottingham, NG8 1AT
2E1 IIA C. Lee, 154 Grangeway, Runcorn, WA7 5JA
2E1 IIC G. Foreman, 41 Winnards Park, Sarisbury Green, Southampton, SO31 7BX
2E1 IID G. Gudgeon, 28 Park Close, Stevenage, SG2 8PX
2E1 IIE D. Jones, 31 Summerhill Drive, Liverpool, L31 3DN
2E1 IIG H. Holmes De Wyvill Sinclair, 27 Mount Crescent, Bridlington, YO16 7HR
2E1 IIJ G. Colclough, Little Hallands, Norton, Seaford, BN25 2UN
2E1 IIL T. Rodgers, 57 Knowles Hill, Rolleston-On-Dove, Burton-on-Trent, DE13 9DY
2E1 IIM A. Bagg, 1 Stone Road, Burnham-on-Sea, TA8 1JU
2E1 IIP R. Mullen, 58 Jasmin Avenue, Newcastle upon Tyne, NE5 1TL
2E1 IIV G. Guinan, 5A Temple Lane, Silver End, Witham, CM8 3QY
2M1 IIW M. Smith, 25 Charleston Crescent, Cove, Aberdeen, AB12 3DZ
2E1 IJD C. Wilcockson, Conybeare House, Willowbrook, Windsor, SL4 6HL
2E1 IJE A. Wilcockson, Conybeare House, Willowbrook, Windsor, SL4 6HL
2E1 IJF J. Grubb, Waterloo Farm, Foston-On-The-Wolds, Driffield, YO25 8BH
2E1 IJK D. Butler, Church Cottage, Church Road, Badminton, GL9 1HT
2E1 IJL B. Gow, 11 Rodley Square, Lydney, GL15 5AZ
2E1 IJM E. Lawless, 99 Ribbleton Avenue, Ribbleton, Preston, PR2 6DA
2E1 IJN P. Nevard, Millinder House, Westerdale, Whitby, YO21 2DE
2E1 IJO B. Blackham, 5 Reedham Drive, Bramley, Rotherham, S66 2SW
2E1 IJQ J. Mallichan, 17 Napier Road, Gillingham, ME7 4HB
2E1 IJR B. Rendell, 43 Springmead, Chard, TA20 2EW
2E1 IJS A. Reed, 139 Wigmore Road, Gillingham, ME8 0TH
2E1 IJU E. Heagren, 156 Eastbrooks, Pitsea, Basildon, SS13 3QH
2E1 IJX J. Underwood, 12 Forge Lane, Gillingham, ME7 1UG
2E1 IJY A. Brown, 24 Greenfields, Langley Mill, Nottingham, NG16 4GJ
2E1 IKA M. Flanagan, 33 Ullswater Road, Chorley, PR7 2JB
2E1 IKB B. Randall, 5 Percival Road, Eastbourne, BN22 9JL
2I1 IKL D. Mcmichael, 33 Shelton Road, Armoy, Ballymoney, BT53 8YH
2E1 IKM K. Ralph, 15 Hansell Road, Norwich, NR7 0LY
2E1 ILH D. Ledger, 12 Reedling Drive, Southsea, PO4 8UF
2E1 ILM Dr A. Chandoo, 59 Stanton Road, Birmingham, B43 5HH
2E1 INC R. Davenport, 10 Woodend Lane, Hyde, SK14 1DT
2E1 INT D. Kenyon, 88A Knutsford Road, Wilmslow, SK9 6JD
2E1 INW I. Williamson, 196 Bruntwood Lane, Heald Green, Cheadle, SK8 3AS
2E1 IOS H. Handrick, 13 Ormonde Way, New Rossington, Doncaster, DN11 0SB
2E1 ITE G. Cahill, 81 Albemarle Road, Willesborough, Ashford, TN24 0HJ
2W1 ITI D. Quick, Rosegarth, Woodbine Road, Blackwood, NP12 1QH
2E1 IVT F. Handley, 155 Heathcote Street, Longton, Stoke-on-Trent, ST3 1AD
2E1 IWD D. Brooks, 61 Carisbrooke High St., Newport, PO30 1NR
2E1 IWG W. Goodwin, 4 Southfields Close, Bishops Waltham, Southampton, SO32 1EY
2E1 JAA J. Ahmed, 59 Ramsgate, Lofthouse, Wakefield, WF3 3PX
2E1 JAC J. Marter, 4 Meadow Way, Seaford, BN25 4QT
2E1 JBJ A. Berry, 4 Newlands Park Way, Newick, Lewes, BN8 4PG
2E1 JCM J. Matthews, 9 Clive Green, Shrewsbury, SY2 5QL
2E1 JEF G. Hensby, Flat 12, Edel Quinn House, Wirral, CH49 6PN
2E1 JEH J. Haimes, 15 The Avenue, Hersden, Canterbury, CT3 4HL
2E1 JGD J. Green, 42 Longfield Ave, Newbarn, DA3 7LA
2E1 JGM J. Moody, 44 Frensham, Cheshunt, Waltham Cross, EN7 6HB
2E1 JGW J. Wright, 2 St. Leonards Park, East Grinstead, RH19 1EE
2E1 JIM J. Deacon, Spring Valley, Churt Road, Farnham, GU10 2QU
2E1 JJM J. Martin, 1 Collins Lane, West Harting, Petersfield, GU31 5NZ
2E1 JJN J. Nicholson, 6 Mill Gardens, West End, Southampton, SO18 3AG
2M1 JKG J. Grieve, 39 Kenmount Drive, Kennoway, Leven, KY8 5HA
2E1 JKL J. Loader, Furlong Farm, Henley, Langport, TA10 9AX
2E1 JKP J. Peggram, Cherry Trees, Broad Lane, Bracknell, RG12 9BY
2E1 JLC J. Chaundy, Ambleside, Barnes Lane, Lymington, SO41 0RL
2I1 JMC J. Mccaw, 62 High Street, Ballymena, BT43 6DT
2E1 JME E. Marshall, 60 Dudsbury Road, West Parley, Ferndown, BH22 8RG
2E1 JMG J. Greaves, The White House, 25 London Road, Leicester, LE8 9GF
2E1 JMW J. Williams, 73 Telford Road, Tamworth, B79 8EY
2E1 JOD J. Antimano, 23 Nupton Drive, Barnet, EN5 2QU
2W1 JOL T. Philpott, 6 Hillside, Fochriw, Bargoed, CF81 9LQ
2E1 JON J. Sturgeon, Windyridge, Linkside West, Hindhead, GU26 6PA
2E1 JOY E. Nye, 11 Barnhill Close, Marlow, SL7 3HA
2E1 JPH J. Holden, 47 Copse Hill, London, SW20 0NJ
2E1 JRB J. Bancroft, 7 Cordwainer Grove, Sedgefield, Stockton-on-Tees, TS21 2JY
2E1 JRS J. Stew, 19 Salisbury Close, Sittingbourne, ME10 3BL
2E1 KAJ K. Johnson, 22 Prior Road, Greatstone, New Romney, TN28 8SB
2E1 KCC C. Coates, 104 Orion Way, Willesborough, Ashford, TN24 0DZ
2E1 KFS D. O'Hagan, 44 Mulberry Close, Paignton, TQ3 3GD
2E1 KID R. Gow, 11 Rodley Square, Lydney, GL15 5AZ
2E1 KIP J. Barker, 5 Severn Avenue, Weston-Super-Mare, BS23 4DH
2E1 KJB K. Blanch, Sticelett Farm, Rolls Hill, Cowes, PO31 8NE
2E1 KLT G. Clark, 28 Manor Road, Woolton, Liverpool, L25 8QG
2M1 KOJ C. Smith, 68 Craigmore Street, Dundee, DD3 0EA
2W1 KWK P. Stagg, 8 Canal Terrace, Ystalyfera, SA9 2LP
2E1 KYQ L. Fisher, Annstuyvonne, Baghill Green, Wakefield, WF3 1DL
2E1 LAM A. Lam, Partridge Cottage, Redpale, Heathfield, TN21 9NR
2W1 LCO M. Kinsey, Hyfrydle, Cyffylliog, Ruthin, LL15 2DW
2E1 LEC C. Cattel, 21 School Hill, Chickerell, Weymouth, DT3 4BA
2E1 LED F. Chorlton, 25 Ash Grove, Orrell, Wigan, WN5 8NG
2E1 LEN L. Kinley, 100 Withington Lane, Aspull, Wigan, WN2 1JE
2E1 LES L. Tomkins, 46 The Boulevard, Great Sutton, Ellesmere Port, CH65 7DZ
2E1 LEW S. Freestone, 47 Salisbury Street, Gainsborough, DN21 2RS
2E1 LEX A. Champkin, 17 Blundell Place, Bedford, MK42 9XB
2E1 LGA A. Edwards, 23 Brittany Avenue, Ashby-de-la-Zouch, LE65 2QY
2E1 LGE L. Edwards, 8 St. Andrews Close, High Ham, Langport, TA10 9DD
2E1 LGJ L. Gaston-Johnston, 23A Nashleigh Hill, Chesham, HP5 3JQ
2E1 LGV L. Verghese, 19 Old Mansion Close, Eastbourne, BN20 9DP
2E1 LIS E. Buckland, 11 Veronica Close, Basingstoke, RG22 5NW
2E1 LIZ E. Greatorex, 22 Marlborough Way, Uttoxeter, ST14 7HL
2E1 LJL L. Lewis, 28 Brow Hey, Bamber Bridge, Preston, PR5 8DS
2E1 LJW L. Walker, 125 Devereux Road, West Bromwich, B70 6RQ
2E1 LME M. Hulme, 13 Cherry Tree Court, Diss, IP22 4QW
2E1 LOZ L. Dodman, 30 Cambridge St., Rugby, CV21 3NQ
2M1 LPT C. Macnab, 18 Lochport, North Uist, Western Isles, HS6 5EU
2E1 LSI Dr L. Soares Indrusiak, 6 Trafalgar House, Piccadilly, York, YO1 9QP
2E1 MAR M. Davies, Newton Lodge, Newton, Ellesmere, SY12 0PF
2E1 MAZ M. Adlington, 21 Newstead Road, Stoke-on-Trent, ST2 8HU
2E1 MDC C. Mackay, 665A Edenfield Road, Rochdale, OL11 5XE
2E1 MEL P. Cosens, 34 Waterloo Road, Salisbury, SP1 2JX
2E1 MEP M. Pearce, 137 Westwood Road, Salisbury, SP2 9HN
2E1 MFC M. Clews, 16 Chestnut Street, Worcester, WR1 1PA
2E1 MGB H. Knight, 10 Welford Road, Barton, Alcester, B50 4NP
2E1 MHB M. Balyuzi, 48 Cleveland Gardens, London, NW2 1DY
2E1 MIB D. Clegg, 1 Green Croft, Hereford, HR2 7NT
2M1 MIC M. Budas, 20 Oak Avenue, Bearsden, Glasgow, G61 3HD
2E1 MIN M. Dawson, 140A Healey Road, Scunthorpe, DN16 1HT
2E1 MJF M. Faulkner, 49 Oakfield Road, Shrewsbury, SY3 8AD
2E1 MJH M. Hickford, 3 Ashen Road, Clare, Sudbury, CO10 8LQ
2E1 MJM M. Marter, 4 Meadow Way, Seaford, BN25 4QT
2E1 MPB M. Bates, Apartment 801, Imperial Point, Salford, M50 3RB
2E1 MPN M. Nurse, 81 Lexden Drive, Seaford, BN25 3JF
2E1 MPQ Dr A. Gair-Harris, Osterley, White Lackington, Piddletrenthide, Dorchester, DT2 7QU
2W1 MRK M. Morgan, 30 Hafan Werdd, Mornington Meadows, Caerphilly, CF83 3BU
2W1 MSC M. Cook, 9 Drenewydd, Park Hall, Oswestry, SY11 4AH
2W1 MWS M. Southall, 23 Ffordd Elias, Old Colwyn, Colwyn Bay, LL29 9LA
2E1 NAC N. Cosens, 1 Bake Farm Cottages, Salisbury Road, Salisbury, SP5 4JT
2E1 NEW A. Fleck, 108 Hillsview Avenue, Newcastle upon Tyne, NE3 3LA
2E1 NFD N. De-Thabrew, 12 Balfour Road, Dover, CT16 2NQ

2E1 NII C. Sims, 7 Ainthorpe Lane, Ainthorpe, Whitby, YO21 2JN
2E1 NJC N. Cook, Idle Shores Springfield Road, Woolacombe, EX34 7BX
2E1 NPH A. Arnold, 2 Duck Lane, Haddenham, Ely, CB6 3UE
2E1 NRL N. Loveridge, 26 Haylands, Portland, DT5 2JZ
2E1 NRQ J. Dean, 9 School Hill, Chickerell, Weymouth, DT3 4BA
2E1 NVK E. Ridoutt, The Bungalow, Main Road, Porchfield, Newport, PO30 4LP
2E1 OBI B. Atkins, 30 Rishworth Rise, Shaw, Oldham, OL2 7QA
2E1 ODG D. Goodall, 19 Rossefield Avenue, Leeds, LS13 3SG
2E1 OLI O. Bradley, 19 Lincoln Close, Eastbourne, BN20 7TZ
2E1 ORT M. Hickford, Conifers, 3 Ashen Road, Sudbury, CO10 8LQ
2E1 OUJ J. Walsh, 155 Hunter Drive, Bletchley, Milton Keynes, MK2 3NG
2E1 OZO D. Day, Blakeney, Arbor Lane, Lowestoft, NR33 7BQ
2E1 OZY O. Morris, 44 Leamington Road, Weymouth, DT4 0EZ
2E1 PAL A. Barnes, 78 Greenhurst Drive, East Grinstead, RH19 3NE
2E1 PAW P. Woodward, 5 Lupin Way, Clacton, CO167DX
2E1 PDH P. Hasney, 22 Newburn Court, South Shields, NE33 4HR
2E1 PDM P. Marney, 137 Thistle Grove, Welwyn Garden City, AL7 4AG
2E1 PDQ A. Pennington, 16 Invicta Road, Margate, CT9 3SL
2E1 PEC P. Cosens, 1 Bake Farm Cottage, Salisbury Road, Salisbury, SP5 4JT
2E1 PEW A. Pewsey, Pineholm, High Close, Bovey Tracey, Newton Abbot, TQ13 9EX
2E1 PGA P. Graham, 556 Mather Avenue, Liverpool, L19 4UG
2E1 PGB D. Brierley, 639 Borough Road, Birkenhead, CH42 9QA
2W1 PGL P. Lloyd, Pentrip, Wynnstay Yard, Wrexham, LL14 6DP
2E1 PHW P. Wade, 94 Steyne Road, Bembridge, PO35 5SL
2E1 PJJ J. Miller, Flat 1 Block 2, St. Phillips Place, Eastbourne, BN22 8LW
2E1 PMT J. Green, 10 Holme Dene, Haxey, Doncaster, DN9 2JX
2E1 PPK A. Boag, 53 Castlewood Road, London, N16 6DJ
2E1 PPM C. Martin, 4 Chaloner Place, Aylesbury, HP21 8NW
2E1 RAD J. Elliott, 13 Ormonde Way, New Rossington, Doncaster, DN11 0SB
2E1 RAF R. Walker, 35 Romany Close, Letchworth Garden City, SG6 4LA
2E1 RAO N. Brent, 53 Middlewich Street, Crewe, CW1 4DA
2E1 RBA A. Russell - Bishop, 227 Ardleigh Green Road, Hornchurch, RM11 2ST
2E1 RBH C. Hodges, Marine Cottage, 1A Clements Lane, Portland, DT5 1AS
2E1 RFS R. Starling, 10 School Lane, Lawford, Manningtree, CO11 2HZ
2E1 RIO R. Odle, 24 Longfellow Road, Gillingham, ME7 5QG
2E1 RJH Sir R. Heygate, 34 Abercrombie Street, London, SW11 2JD
2E1 RJS R. Spevack, 87 Albany Road, Hersham, Walton-on-Thames, KT12 5QG
2E1 RMD B. Debenham, 80 Stewart Road, Chelmsford, CM2 9BD
2E1 RMS R. Stevenson, 97 Queen Street, Crewe, CW1 4AL
2E1 RON C. Robinson, 11 Poplar Way, Leeds, LS13 4SU
2E1 ROO M. Ratcliffe, 76 Churchill Road, Stone, ST15 0DY
2E1 RSB R. Mather, 76 Stavordale Road, Moreton, Wirral, CH46 9PS
2E1 RSH R. Davies, Lamb Cottage 3 Manor Barns, Snowshill, Broadway, WR12 7JR
2W1 RSS R. Hark, 5 Victoria Park, Bagillt, CH6 6JS
2E1 RWC R. Cornwall, 9 Bishop Close, Dunholme, Lincoln, LN2 3US
2E1 RWN N. Rowan, 27 Crieff Road, London, SW18 2EB
2E1 SAZ S. Greatorex, 54 Lilac Grove, Glapwell, Chesterfield, S44 5NG
2E1 SBF C. Noon, 24 Sunset Walk Bush Estate, Eccles-On-Sea, Norwich, NR12 0SX
2M1 SCO R. Jeffrey, Burnbrae, Crocketford, Dumfries, DG2 8QR
2E1 SCR S. Ride, 61 Pepper Street, Sutton-in-Ashfield, NG17 5GD
2E1 SDI G. Baines, 60 Parkdale Road, Thurmaston, Leicester, LE4 8JP
2W1 SDR A. Rose, 4 Ramsey Road, Barry, CF62 9DF
2E1 SGK S. Knott, 24 John Street, Leek, ST13 8BL
2E1 SHE S. Brown, 11 Wordsworth Avenue, Tamworth, B79 8BZ
2E1 SIS L. Downes, 6 Greenland Crescent, Beeston, Nottingham, NG9 5LB
2M1 SJB S. Budas, 20 Oak Avenue, Bearsden, Glasgow, G61 3HD
2E1 SJG S. Grant, 47 Coneyford Road, Shard End, Birmingham, B34 7AY
2E1 SKA W. Pitt, 237 Broadway, Dunscroft, Doncaster, DN7 4HS
2E1 SKR A. Skrzypecki, The Chestnuts, Birdcage Lane, Halifax, HX3 0JQ
2E1 SKY P. Staerck, 42 Plantation Hill, Worksop, S81 0RJ
2E1 SOB A. Weatherall, 1 Dean Place, Stoke-on-Trent, ST1 3HS
2E1 SOX A. Hughes, 4 Cobden Court, Birkenhead, CH42 3YH
2E1 SPH S. Harris, Lackington Drove, Dorchester, DT2 7QU
2E1 SPY S. Palmer, 21 Ibbett Close, Kempston, Bedford, MK43 9BT
2W1 SRB S. Bowen, 41 Bro Dawel, Merthyr Tydfil, CF47 0YU
2E1 SRI S. Edwards, 70 Summer Field Court Altona Close Stone, Staffs, ST15 8AR
2E1 STK L. Meek, 3 St. Johns Grove, Kirk Hammerton, York, YO26 8DE
2E1 STO G. Stone, 40 Friars Road, Stoke-on-Trent, ST2 8DS
2E1 STU S. Green, Flat 12, Canute House, Strand Street, Poole, BH15 1EJ
2E1 SUE S. Macpherson, 18 Mountbatten Avenue, Dukinfield, SK16 5BU
2W1 SWB E. Jenkins, 8 Ffordd Elias, Old Colwyn, Colwyn Bay, LL29 9LA
2I1 SWD S. Mcauley, Layde View, 19 Rathlin Avenue, Ballycastle, BT54 6DQ
2E1 SWS S. Walters-Smith, 83 Chesterfield Road, Tibshelf, Alfreton, DE55 5NJ
2E1 TAB T. Brierley, 6 Bridle Avenue, Wallasey, CH44 7BJ
2E1 TAG D. Taggart, 26 Realmwood Close, Canterbury, CT1 1GY
2E1 TAP A. Weaver, 14 Ferry View, Thorngumbald, Hull, HU12 9GB
2W1 TBD T. Davies, 72 Eversley Road, Sketty, Swansea, SA2 9DF
2E1 TBW A. Walker, 76 Greenway, Birmingham, B20 1EQ
2E1 TCP T. Clayton, 14 Medway Walk, Wigan, WN5 9NQ
2W1 TDM T. Meredith, 18 Hyde Place, Llanhilleth, Abertillery, NP13 2RT
2E1 TIM T. Wightman, Laithbutts Farm, Cowan Bridge, Carnforth, LA6 2JL
2E1 TKD D. Hensby, 28 Moorland Crescent, Whitworth, Rochdale, OL12 8SU
2E1 TMB T. Bolderstone, 20 Wellington Crescent, Sculthorpe, Fakenham, NR21 7PU
2E1 TNE A. Millard, 6 Connaught Road, Weymouth, DT4 0SA
2E1 TOM T. Lake, 77 Grafton Road, King's Lynn, PE30 3EX
2E1 TON A. Steele, 15 Roman Meadow, Downton, Salisbury, SP5 3LB
2E1 TSO G. Smith, 37 The Crescent, Bracebridge Heath, Lincoln, LN4 2NP
2E1 TWB A. Brown, 7 Brookfield Road, Wooburn Green, High Wycombe, HP10 0PZ
2E1 UJE A. Dalzell, 9 Pyms Lane, Crewe, CW1 3PJ
2E1 UKT G. Eason, Whitegates, Parsonage Road, Takeley, Bishop's Stortford, CM22 6QX
2E1 UTD R. Smith, 14 Oakfield Cottages, Brockton, Shrewsbury, SY5 9JA
2E1 VAR R. Preece, 5 Pavey Run, Ottery St Mary, EX11 1FQ
2M1 VFO M. Bartlett, 93 Lumsden Crescent, Almondbank, PH1 3UA
2W1 VMR M. Sides, 17 Mayville Avenue, Llay, Wrexham, LL12 0PW
2M1 VXB C. Andrew, 23 Shore Street, Inverallochy, Fraserburgh, AB43 8WA
2I1 WBC E. Paulikas, 33 Clarefield, Dungannon, BT71 6TQ
2E1 WCD W. Denny, 86 Lloyds Avenue, Kessingland, Lowestoft, NR33 7TR
2E1 WEB C. Webb, 1 The Square, Eltisley Road, Sandy, SG19 3BT
2E1 WGB G. Barber, 35 Lower Park Crescent, Bishop's Stortford, CM23 3PU
2E1 WIN C. Wingfield, 35 Causey Farm Road, Hayley Green, Halesowen, B63 1EQ
2E1 WJB B. Walsh, 20 Edge Fold Crescent, Worsley, Manchester, M28 7EX
2E1 WNA I. Jones, 151 Atherton Road, Hindley, Wigan, WN2 3EE
2E1 WPW A. Newton, 12 Hewett Street, Warsop Vale, Mansfield, NG20 8XN
2E1 WRC W. Chorlton, 25 Ash Grove, Orrell, Wigan, WN5 8NG
2E1 WVF D. Fowler, Millhouse, 8 Church Road, Stockport, SK6 5PR
2E1 WWD E. Durkin, 30 Douglas Road West, Stafford, ST16 3NX
2E1 XDJ K. Senior, 20A Union Street, Hemsworth, Pontefract, WF9 4AP
2E1 XGX P. Hughes, 17 Worcester Park, Bath, BA1 6QU
2E1 XRM J. Thompson, 3 Fern Avenue, Staveley, Chesterfield, S43 3RH
2E1 XXX J. Day, 98 Steynburg Street, Hull, HU9 2PF
2W1 YEG J. Thorne, 11 Dowland Road, Penarth, CF64 3QX
2E1 YES J. Glover, 31 West Drive, Lancaster, LA1 5BY
2E1 YHZ J. Hiltz, 35 Highview Road, London, W13 0HA
2E1 YRK C. Wright, 55 Booth Street, Denton, Manchester, M34 3HU
2E1 ZPR W. Toomer, 10 Northfield, Tarrant Hinton, Blandford Forum, DT11 8JD

G0

G0 AAA Three A's Contest Group, c/o K. Pritchard, 9 Golf Close, Pyrford, Woking, GU22 8PE
G0 AAG W. Furnival, 24 Moor Street, Hereford, HR4 9LA
G0 AAM G. Willetts, 2 Underlane, Boyton, Launceston, PL15 9RR
G0 AAN Prof. L. Schnurr, 42 Basin Road, Heybridge Basin, Maldon, CM9 4RQ
G0 AAT P. Wheatley, 44 Primrose Crescent, Worcester, WR5 3HT
G0 AAU J. Blades, 42 Ellesmere, Burnmoor, Houghton le Spring, DH4 6EA
G0 AAW A. Whitehouse, Forest Green, Cotchford Lane, Hartfield, TN7 4DN
GM0 AAX G. Anthoney, 10 Cedar Road, Kilmarnock, KA1 2HP
G0 AAY I. Brooks, The Lodge, Bilsborrow Lane, Bilsborrow, Preston, PR3 0RP
G0 ABB M. Honeywell, 23 Deverell Place, Waterlooville, PO7 5ED
GW0 ABE P. Hughes, 59 Jeffreys Road, Wrexham, LL12 7PD
G0 ABI P. Green, Camellia Cottage, The Challices, Chulmleigh, EX18 7QX
G0 ABM T. Johnson, 143 Queens Road, Tunbridge Wells, TN4 9JY
G0 ABN A. Samuels, 45 Mermaid Close, Chatham, ME5 7PT
G0 ABP D. Orgill, 32 Upland Avenue, Chesham, HP5 2EB
GW0 ABT T. Thomas, Tymawr Farm, Llanwern, Brecon, LD3 7UW
G0 ABV D. Wood, 18 Bankhouse Road, Nelson, BB9 7RA
G0 ABW J. Harding, Fen End Farm, High Street, Huntingdon, PE19 4UE
G0 ACA J. Kliffen, 8 West Park, Minehead, TA24 8AW
G0 ACD M. Amos, 41 Jocelyn Road, Richmond, TW9 2TJ
GW0 ACH J. Cooper, 157 Bryn Road, Brynmenyn, Bridgend, CF32 9LU
G0 ACK D. Lamb, 339 Victoria Road, Ruislip, HA4 0DS
G0 ACQ M. Godden, 20 Channel View Road, Portland, DT5 2AY
G0 ACZ A. Hope, Knab Hall Bungalow, Knab Hall Lane, Tansley Matlock, DE4 5FS
G0 ADA N. Law, 43 Canonsfield, Peterborough, PE4 5AQ
G0 ADB C. Willis, 5 Gower Drive, Biddenham, Bedford, MK40 4PZ
GW0 ADC K. Shepherd, 15 Gronant Road, Prestatyn, LL199DT
GI0 ADD OBO Armagh and Dungannon District A.R.C., c/o J. Ashe, 49 Deans Walk, Richhill, Armagh, BT61 9LD
GM0 ADF D. Mackinnon, 60 Mount Stuart Drive, Wemyss Bay, PA18 6DX
G0 ADG R. Bristeir, 94 Burnthwaite Road, Fulham, London, SW6 5BG
G0 ADH R. Razey, 2 Park Farm Cottage, 26 St. Georges Road, Wallingford, OX10 8HP
G0 ADJ D. Elsworth, 34 Seal Road, Bramhall, Stockport, SK7 2JR
G0 ADK J. Saueressig, 8 The Ridgeway, River, Dover, CT17 0NX
G0 ADL B. Barlow, 134 Bury Road, Radcliffe, Manchester, M26 2UX
G0 ADO A. Hodgson, Arla Burn Farm, Middleton in Teesdale, DL12 0QU
G0 ADP A. Wither, 30 Mersey Road, Aigburth, Liverpool, L17 6AD
GW0 ADS J. Jenkins, Derwen Las, Llanwnnen, Lampeter, SA48 7LG
G0 ADT I. Andrew, 28 Beechnut Drive Blackwater, Camberley, GU17 0DJ
G0 ADU B. Gibson, 55 Ledward Street, Winsford, CW7 3EN
G0 ADW P. Radford, 42 Ashbury Drive, Weston-Super-Mare, BS22 9QS
GM0 ADX Kilmarnock and Loudoun ARC, c/o A. Mckay, 2 Osprey Drive Sparrow Plantation Kilmarnock Ka13 Lq, Kilmarnock, KA1 3LQ
G0 ADZ D. Price, Pippins, High St., Newmarket, CB8 9DQ
G0 AED H. Gerard, 18 Hunstanton Road, Dersingham, King's Lynn, PE31 6HQ
GM0 AEG D. Greatorex, 40 Robertson Road Lhanbryde, Elgin, IV30 8PR
G0 AEL K. Horton, 16 Linden Close West Parley, Ferndown, BH22 8RS
G0 AEN S. Webb, 7 Holbear Grange, Forton Road, Chard, TA20 2ED
G0 AEP G. Roper, 17 Slepe Crescent, Poole, BH12 4DH
G0 AEU P. Tietz, 5 Chevin Road, Belper, DE56 2UW
G0 AEV Dr S. Reed, Bridlands, Middle Common, Chippenham, SN15 5NN
G0 AEW D. Arlette, 12 Polmear, Par, PL24 2AT
G0 AEX E. Hannaby, 170A Weston Drive Otley, Leeds, LS21 2DT
GM0 AEY D. Palmer, 36 Kilsyth Road, Haggs, Bonnybridge, FK4 1HE
GW0 AEZ J. Howarth, 7A Liddell Drive, Llandudno, LL30 1UH
G0 AFH I. Burns, Little Delmar Farm, Leywood Road, Meopham, DA13 0UD
G0 AFJ A. Brown, 33 Marion Road, Haydock, St Helens, WA11 0PY
G0 AFN P. Howard, 1 Avon Close, Bognor Regis, PO22 6BX
G0 AFP D. Rayner, 69 Lovelace Drive, Woking, GU22 8QZ
G0 AFQ J. Cook, 71 Richmond Avenue, Burscough, Ormskirk, L40 7RB
G0 AFR C. Mears, 11 Aberford Close, Reading North, Reading, RG30 2NX
G0 AFT C. Bovey, 12 The Mead, Beaconsfield, HP9 1AW
G0 AFU M. Bousfield, 49 Armstrong Street, Ridsdale, Hexham, NE48 2TN
G0 AFY R. Norman, 98 Foxwell Drive, Headington, Oxford, OX3 9QF
G0 AFZ C. Wilson, 9-10 Daventry Street, Southam, CV47 1PH
G0 AGB R. Ellis, 22 Fern Road, Storrington, Pulborough, RH20 4LW
G0 AGC A. Core, 1 Partridge Ride, Loggerheads, Market Drayton, TF9 2QX
G0 AGD P. Shortland, 69 South Parade, Worksop, S81 0BS
GW0 AGL E. Williams, Caerbergam Cottage, Llanbedr, LL45 2HT
GM0 AGN G. Speirs, 43 Sheuchan View, Stranraer, DG9 7TA
G0 AGO R. White, 137 Fennells, Harlow, CM19 4RR
G0 AGR Aylesbury Raynet Grp, c/o R. Needs, 13 Greenway, Great Horwood, Milton Keynes, MK17 0QR
G0 AGU B. Humphries, 22 Leander Close, Burntwood, WS7 1PW
GM0 AGV A. Rollings, 24 Millburn Court, Sheuchan St., Stranraer, DG9 0DX

GW0 AGZ J. Squire, Dyffryn, Llanymynech, SY22 6EW
G0 AHA A. Pannell, 3 Nethercourt Gardens, Ramsgate, CT11 0RY
G0 AHB A. Bennett, 2 Portland Place, Hertford Heath, Hertford, SG13 7RR
G0 AHC B. Hewitt, 32 Pinehurst Drive, Kings Norton, Birmingham, B38 8TH
G0 AHD C. Richardson, 122 Elmton Road, Creswell, Worksop, S80 4DE
G0 AHE P. Moss, Vivenda, Wick Road, Langham, CO4 5PE
G0 AHI T. Deacon, 9 Mulberry Close, Woodley, Reading, RG5 3LR
G0 AHJ J. Johnson, 14 Park House Lane, Prestbury, Macclesfield, SK10 4HZ
G0 AHK D. Layne, 5 Howe Close, Christchurch, BH23 3JA
G0 AHL R. Sears, 19 Shepherds Grove Park, Stanton, Bury St Edmunds, IP31 2AY
G0 AHM P. Gregg, 5 Rosevear Road Bugle, St Austell, PL26 8PH
G0 AHO C. Grant, 20 Muriel Kenny Court, Hethersett, Norwich, NR9 3EZ
G0 AHR P. Cuthbert, 115 Tintern Avenue, Whitefield, Manchester, M45 8WY
G0 AHU J. Tolson, 1 Old Mill Court Station Road, Plympton, Plymouth, PL7 2AJ
G0 AHV R. Gilling, 24 Bellerby Road, Skellow, Doncaster, DN6 8PD
G0 AIG M. Sutherland, 28 Sycamore Way, Littlethorpe, Leicester, LE19 2HT
G0 AIH R. Baker, 38 The Front, Middleton One Row, Darlington, DL2 1AU
GI0 AIJ I. Greenwood, Deers Leap, 24 Tullyrusk Road, Crumlin, BT29 4JQ
G0 AIL D. Penny, 30 Belvedere Road, Yeovil, BA21 5JB
G0 AIM S. Robinson, 164 Leigh Road, Westhoughton, Bolton, BL5 2LE
G0 AIN S. Withnell, 85 Headroomgate Road, Lytham St Annes, FY8 3BG
G0 AIO A. Owens, 69 Locomotion Way, North Shields, NE29 6XE
GI0 AIQ F. Holland, 413 Ballyoran Park, Portadown, Craigavon, BT62 1JX
GM0 AIR D. Parker, Devon Cottage, Main Street, Cupar, KY15 7QX
G0 AIS J. Borg, 94 Coldershaw Road, London, W13 9DT
G0 AIX D. Westlake, Chyvellin, Newmill, Penzance, TR20 8XW
GW0 AIY R. Gibbons, Cefnysgwyn, Capel Isaac, Llandeilo, SA19 7UA
G0 AIZ W. Chesterton, Homelea Farm, Fosse Way, Coventry, CV7 9LR
G0 AJA A. Astley, 16 Cedar Avenue, Ellesmere, SY12 9PA
G0 AJB A. Botherway, 4 Brodrick Drive, Ilkley, LS29 9SN
G0 AJF S. Harrison, 25 High Spring Road, Keighley, BD21 4TF
G0 AJH J. Hornsby, 15 Coronation Drive, Hornchurch, RM12 5BL
G0 AJJ L. Leavold, 8 Wilkinson Way, North Walsham, NR28 9BB
G0 AJL B. Bromsgrove, 34 Boundary Drive, Hunts Cross, Liverpool, L25 0QD
GM0 AJT D. Rollings, 2 Challoch Crescent, Leswalt, Stranraer, DG9 0LN
GW0 AJU A. Underwood, Rock Hill, Cefn Abbey, Llanarthne, Carmarthen, Carmarthen, Uk, sa32 8lj
G0 AJW J. Welsby, 47 Links Road, Knott End-On-Sea, Poulton-le-Fylde, FY6 0DF
G0 AJX M. Coles, 13 Dawnay Road, Bilton, Hull, HU11 4HB
G0 AJZ B. Park, 147 Castle Road, Ings Farm, Redcar, TS10 2LT
G0 AKC T. Chamberlain, 455 Norwich Road, Ipswich, IP1 5DR
G0 AKF K. Farrance, Clarewood, Tabley Road, Knutsford, WA16 0NE
G0 AKH K. Hale, 58 St. Stephens Road, Saltash, PL12 4BJ
GM0 AKJ P. Seaton, 51 Leachkin Avenue, Inverness, IV3 8LH
G0 AKK J. Chivers, 4 Laurel Drive, Bognor Regis, PO21 3ND
G0 AKL R. King, 4 Pinewood Drive, Horning, Norwich, NR12 8LZ
G0 AKM K. Muchamore, 3 Belfont Walk, London, N7 0SN
G0 AKO R. Cleveland, 2 Morse Close, Brundall, Norwich, NR13 5LG
G0 AKR A. Poynter, Hill Top Farm, Warren Road, Chatham, ME5 9RD
G0 AKS D. Newell, 7 Edward Road West, Clevedon, BS21 7DY
G0 AKU R. Reanney, 9 Stapleford Court, Ellesmere Port, CH66 1RW
GW0 AKV J. Anderson, 173 Gate Road, Penygroes, Llanelli, SA14 7RW
G0 ALA D. Whitehouse, 40 Fernleigh Crescent, Up Hatherley, Cheltenham, GL51 3QL
G0 ALB C. Matcham, 28 Buckingham Drive, Luton, LU2 9RA
G0 ALC J. Richards, 77 Poxon Road, Walsall Wood, Walsall, WS9 9JR
G0 ALE Tatsfield Arts, c/o P. Madagan, 40 Lagham Park, South Godstone, Godstone, RH9 8ER
G0 ALI A. Soars, 8 Nomis Park Congresbury, Bristol, BS49 5HB
G0 ALJ R. Prior, 274 West View Lodge, Canterbury Road, Herne, , CT6 7HB
G0 ALQ J. Amer, 1 Kingfisher Way, Watton, Thetford, IP25 6SR
G0 ALR A. Wagenaar, La Villette, Folles, France, 87250
GM0 ALS F. Roe, 74 Willow Grove, Livingston, EH54 5NA
GM0 ALW G. Chalmers, 38 Grove Hill, Kelso, TD5 7AS
GM0 ALX M. Chalmers, 38 Grovehill, Kelso, TD5 7AS
GD0 AMD A. Dorman, 1 Sprucewood Rise, Foxdale, Douglas, Isle Of Man, IM4 3JP
G0 AMO M. Adams, 9 Brancaster Avenue, Charlton, Andover, SP10 4EN
G0 AMP R. Senft, 11 Maltravers Drive, Littlehampton, BN17 5EY
G0 AMS A. Sinclair, 4 Blackbrook Park Avenue, Fareham, PO15 5JJ
G0 AMU L. Parrott, 3 Fox Gardens, Lymm, WA13 9EY
G0 AMW C. Dunn, 90 Mushroom Green, Dudley, DY2 0EE
G0 AMX P. Appleyard, 28 Romany Close, Letchworth Garden City, SG6 4JZ
G0 AMY J. Sheppeck, 25 Kingsleigh Road, Heaton Mersey, Stockport, SK4 3QF
G0 AMZ K. Fay, 2 Chapel Row, Hartley Wintney, Hook, RG27 8NJ
GW0 ANA G. Jones, Nirvana 2 Castle Precinct, Llandough, Cowbridge, CF71 7LX
G0 ANE W. Burrows, 35 Bamford Avenue, Barnsley, S71 3SJ
GM0 ANG J. Fish, Senekal, Alma Road, Fort William, PH33 6HB
G0 ANH J. Wright, 5 Abbeville Avenue, Whitby, YO21 1JD
G0 ANK I. Smallwood, 27 Cormorant Way, Herne Bay, CT6 6HG
G0 ANL P. Ellis, 8 Shropshire Close, Woolston, Warrington, WA1 4DY
G0 ANM J. Mooney, 2 Madford Lane, Launceston, PL15 9EB
G0 ANN M. Viney, 12 Palgrave House, Sherwell Road, Norwich, NR6 6PU
G0 ANO D. Lawton, Grenehurst, Pinewood Road, High Wycombe, HP12 4DD
G0 ANP D. Guy, 7 Park Avenue, Castle Cary, BA7 7HE
G0 ANT Eden Valley Radio Society, c/o D. Shaw, Cotehouse, Bleatarn, Appleby-in-Westmorland, CA16 6PX
G0 ANW J. Hockley, 6 King Edward Road, Birchington, CT7 0EL
G0 AOA T. Marten, 10 Chieveley Drive, Tunbridge Wells, TN2 5HG
G0 AOB M. Beal, 45 Castle Meadows, Launceston, PL15 7DZ
G0 AOC S. Colledge, 32 Wakeman St., Worcester, WR3 8BQ
G0 AOD D. Heathcote, 8 Ferrers Avenue, Tutbury, Burton-on-Trent, DE13 9JR
G0 AOE N. Evans, 16 Humbledon Park, Sunderland, SR3 4AA
GM0 AOF R. Wallace, 1 Holding West Kincardine, Crieff, PH7 3RP
G0 AOH J. Abbruscato, 22199 Pine Tree Ln, Hockley, USA, 77447
G0 AOJ F. Fenwick, 6 School Lane, Bobby, Brigg, DN20 0PP
G0 AOK D. Gall, 2 Norham Close, Wideopen, Newcastle upon Tyne, NE13 7HS
G0 AOL G. Parsons, 248 Filey Road, Scarborough, YO11 3AQ
G0 AOM R. Day, 3 Railway Cottages, Newby Bridge, Ulverston, LA12 8AW
G0 AOO J. Butterwick, 45 Fox Howe Coulby Newham, Middlesbrough, TS8 0RU
G0 AOP P. Warriner, 36 Eskdaleside, Sleights, Whitby, YO22 5EP
G0 AOQ S. Boyd, 17 Ropery Walk, Pocklington, York, YO42 2BF
G0 AOS M. Reynolds, Willhay Cottage, Willhay Lane, Axminster, EX13 5RW
G0 AOW S. Sands, 33 High Street, Kinver, Stourbridge, DY7 6HF
G0 AOX A. Sands, Gabledown, Bridgnorth Road, Stourbridge, DY7 6RW
G0 AOY T. Rudd, Grasmere, Burgh, Woodbridge, IP13 6SU
G0 AOZ R. Powell, Town Pond Cottage, Town Pond Lane, Southmoor, Abingdon, OX13 5HS
G0 APB P. Buckley, 7 Callams Close, Rainham, Gillingham, ME8 9ES
G0 API J. Fell, 14 Rectory Avenue, Corfe Mullen, Wimborne, BH21 3EZ
GM0 APN J. Loveday, Crombiebrae, Inverurie, AB51 0JT
G0 APP A. Pamment, 5 New Captains Road, West Mersea, Colchester, CO5 8QP
G0 APV V. Covell-London, 15 St. Nicholas Way, Potter Heigham, Great Yarmouth, NR29 5LG
G0 APY G. Flood, 4 Campbell Crescent, Great Sankey, Warrington, WA5 3DA
G0 AQA S. Baynes, 1 Reeves Paddock, Townsend, Priddy, Wells, BA5 3FG
G0 AQB J. Ireland, The Grange, Grange Lane, Worcester, WR2 6RW
GI0 AQD D. Burn, 135 Main Road Portavogie, Newtownards, BT22 1EL
G0 AQF D. Dolphin, 16 Golden Cross Lane, Catshill, Bromsgrove, B61 0LQ
G0 AQH G. Griffin, 23 St. Giles Close, Shoreham-by-Sea, BN43 6GR
G0 AQI E. Greenhalgh, 19 Rooks Nest Lane, Therfield, Royston, SG8 9QX
GW0 AQR J. Richards, 21 Cae Gwyn, Caernarfon, LL55 1LL
G0 AQS P. Goldthorpe, 29 Broadoak Road, Ashton-under-Lyne, OL6 8QN
G0 AQT V. Taylor, 5 St. Matthews Drive, St Leonards-on-Sea, TN38 0TR
G0 AQU R. Mason, 28 Vandyke, Great Hollands, Bracknell, RG12 8UP
G0 AQZ D. Mcdonald, 24 Wenning Street, Nelson, BB9 0LE
GW0 ARA Aberystwyth & District Radio Society, c/o R. Clews, Maesygaer, Ciliau Aeron, Lampeter, SA48 7SG
GM0 ARD J. Hoey, 152 Muirhouse Avenue, Motherwell, ML1 2LB
G0 ARF R. Canning, Green Lane Cottage, Eardisland, Leominster, HR6 9BN
GM0 ARH A. Haxton, Lanimar, Dickson Avenue, Montrose, DD10 9EJ
GW0 ARK K. Hudspeth, 67 Bloomfield Road, Blackwood, NP12 1LX
G0 ARL P. Shorland, Jen-Lee, Tolvaddon, Camborne, TR14 0EQ
G0 ARP A. Price, Brook House, Drury Lane, Shrewsbury, SY4 1DT
GM0 ART A. Gray, 191 Greengairs Road, Greengairs, Airdrie, ML6 7SZ
G0 ARU J. Lumb, 2 Briarwood Avenue, Bury St Edmunds, IP33 3QF
G0 ARV A. Gates, 278 Higher Road, Liverpool, L26 9uf
GM0 ARY N. Hamilton, Glenwood Cottage, Enzie, Buckie, AB56 5BW
G0 ARZ A. Everard, 3 St. Hild Close, Darlington, DL3 8LD
G0 ASG P. Fautley, The Old Reading Room Ashwater, Beaworthy, EX21 5EF
G0 ASH J. Coupe, 65 Irongate, Bamber Bridge, Preston, PR5 6UY
G0 ASK S. Gray, 29 Verity Walk, Stourbridge, DY8 4XS
G0 ASL J. Gray, 29 Verity Walk, Stourbridge, DY8 4XS
G0 ASM N. Marston, 14 Greystoke Avenue, Sunderland, SR2 9DX
G0 ASN S. Hendry, 5 Harvey Road, Great Totham, Maldon, CM9 8QA
G0 ASP E. Mason, 28 Pendil Close Wellington, Telford, TF1 2PQ
G0 ASQ L. Williams, 24 Alston Drive, Bare, Morecambe, LA4 6QR
G0 ASX N. Ward, 104 Kingsley Road, Bishops Tachbrook, Leamington Spa, CV33 9RZ
G0 ASZ E. Gamble, 87 Silverdale Drive, Waterlooville, PO7 6DP
GM0 ATA Dr R. Mulholland, 1 Larch Grove, Milton Of Campsie, Glasgow, G66 8HG
G0 ATB V. Herbert, 98 Blithdale Road, Abbey Wood, London, SE2 9HL
G0 ATC Rafac ARC, c/o D. Taylor, 76 Heworth Village, York, YO31 1AL
G0 ATD D. Welch, 51 Verbena Way, Worle, Weston-Super-Mare, BS22 6RL
G0 ATE A. Lovell, Bogardesgatan 5, Goteborg, Sweden, 41654
G0 ATG T. Goodyer, 11 Upper Bere Wood, Waterlooville, PO7 7HX
G0 ATK G. Atkins, 97 South Street, Tillingham, Southminster, CM0 7TH
GM0 ATL P. Smith, 29 Rowan Drive, Bearsden, Glasgow, G61 3HQ
G0 ATO Dr N. Gregorian, 449 E Providencia Ave, Burbank, California, USA, 91501 2916
G0 ATP C. Abela, 14 Warren Road, Barkingside, Ilford, IG6 1BJ
GM0 ATQ L. Morgan, 12 Bayview Road, Gourock, PA19 1XE
G0 ATS E. Green, Chylean, Tintagel, PL34 0HH
G0 ATW J. Ferrier, 30 Grimsby Road, Laceby, Grimsby, DN37 7DB
G0 ATZ C. Best, 34 Julius Hill, Warfield, Bracknell, RG42 3UN
G0 AUB F. Groves, 14 Edenfield Road, Mobberley, Knutsford, WA16 7HE
G0 AUE A. Knowles, 3 The Avenue, Wighill Park, Tadcaster, LS24 8BS
G0 AUF R. Griffiths, 26 Hamilton Road, Morecambe, LA4 6QG
G0 AUG T. Hopkinson, Whitbarrow Hall, Caravan Park, Penrith, CA11 0XB
G0 AUH M. Hopkinson, Whitbarrow Hall, Caravan Park, Penrith, CA11 0XB
G0 AUI C. Stiller, 6 Barn Cottage Lane, Haywards Heath, RH16 3QW
G0 AUJ M. O'Dell, 19 Redwing Rise, Royston, SG8 7XU
G0 AUK Amsat-UK, c/o J. Heck, Pickles Orchard, Memorial Road Great Hampden, Great Missenden, HP16 9RE
GM0 AUL R. Mckenzie, 26 Gladstone Place, Woodside, Aberdeen, AB24 2RP
G0 AUN D. Taylor, 24 Kingshill Drive, Hoo, Rochester, ME3 9JP
G0 AUR A. Hogg, 1 Champions Way, South Woodham Ferrers, Chelmsford, CM3 5NJ
G0 AUT A. Utting, 9 Sydney Road, Spixworth, Norwich, NR10 3PG
G0 AUV A. Johnson, 125 Charles Street Sileby, Loughborough, LE12 7SH
G0 AUW R. Philpott, 4 Dukeswood, Chestfield, Whitstable, CT5 3PJ
G0 AUX K. Daly, Granarogue, Carrickmacross, Ireland, A81 XD62
GM0 AVB K. Graham, 98 Dalswinton Avenue, Dumfries, DG2 9NR
GW0 AVD P. Richards, 11 Seabourne Road, Holyhead, LL65 1AL
G0 AVE D. Brannon, 10 Rochester Crescent, Crewe, CW1 5YF
G0 AVH T. Mcguire, 33 Sandy Lane, Hindley, Wigan, WN2 4EJ
G0 AVJ M. Searley, 49 Hollymount Close, Exmouth, EX8 5PQ
G0 AVP P. Raxworthy, 32 St. Marys Avenue, Alverstoke, Gosport, PO12 2HX
G0 AVU G. Logan, Fenton Hill Farm, Wooler, NE71 6JL
G0 AWG W. Mcnamara, 22 Cissbury Avenue, Peacehaven, BN10 8TJ
G0 AWH N. Bergstrom-Allen, Garden Cottage, Thicket Priory, Thorganby, York, YO19 6DE
GI0 AWK D. Payne, 10 Grovemount Court, Altnagelvin, Londonderry, BT47 5JP
G0 AWM C. Warr, 29 Barton Road, Lancaster, LA1 4ER
G0 AWR D. Proud, Willow Cottage, Tresevern, Truro, TR3 7AT
GW0 AWT S. Richardson, Glasfryn, Porthyrhyd, Llanwrda, SA19 8DF
G0 AWW D. Elliott, 188 Seaview Road, Wallasey, CH45 5HB
G0 AWY R. Titmuss, 70 Mallards Rise, Church Langley, Harlow, CM17 9PL

G0 AWZ D. Richardson, Beckside, Stockton Lane, York, YO32 9UA
G0 AXB J. Moore, 53 Thatchers Court, Westlands, Droitwich, WR9 9EG
G0 AXC D. Lee, 188 Manstone Avenue, Sidmouth, EX10 9TJ
G0 AXD Dr S. Waters, 27 Mill Lane, Shepherdswell, Dover, CT15 7LJ
G0 AXE M. Davenport, 119 Gravel Lane, Wilmslow, SK9 6EG
G0 AXI Dr N. Davies, 15 Fyfield Road, Oxford, OX2 6QE
G0 AXJ R. Mccluskey, 29 Hotspur Avenue, Bedlington, NE22 5TD
GM0 AXM J. Thomson, 16 Ravelstone Terrace, Edinburgh, EH4 3TP
G0 AXO R. Bainbridge, 9 Hamilton Crescent, North Shields, NE29 8DW
G0 AXQ P. Austin, 28 Britannia Close, Sittingbourne, ME10 2JF
G0 AXS S. Birch, 29 Manners Road, Southsea, PO4 0BA
GM0 AXX F. Gilhooly, 26 Arnott Gardens, Edinburgh, EH14 2LB
GM0 AXY E. Dons, 37 Ashley Drive, Edinburgh, EH11 1RP
G0 AXZ W. Johnson, 29 Wentworth Park, Allendale, Hexham, NE47 9DR
G0 AYA M. Chown, 15 Hambleden Walk, Maidenhead, SL6 7UH
GI0 AYB J. Throne, Fascadail, 12 Mason Road, Londonderry, BT47 2RY
G0 AYC C. Porter, 10 Cotefield Drive, Leighton Buzzard, LU7 3DS
G0 AYD D. Dixon, 3 Towns End, Wylye, Warminster, BA12 0RN
G0 AYF W. Collier, 8 Douglas Street, Hindley, Wigan, WN2 3HP
GI0 AYG M. Evans, 12 Tullymore Park, Ballymena, BT42 2AU
G0 AYI B. Spencer, 161A West Lane, Hayling Island, PO11 0JW
G0 AYM K. Luxton, 2 Trinity Court, Westward Ho, Bideford, EX39 1LT
GW0 AYP D. Graves, 185 Rhyl Coast Road, Rhyl, LL18 3US
GW0 AYQ R. Smith, 4 Glan Ysgethin, Talybont, LL43 2BB
GM0 AYT A. Mcdougall, Ceol Na Mara, 3 Bowfield Road, West Kilbride, KA23 9LB
G0 AYX P. Towell, 25 Cedar Close, Grafham, Huntingdon, PE28 0DZ
G0 AYY A. Perry, Chala, Woodhouse Lane Hill, Lyme Regis, DT7 3SX
GI0 AZA Dr E. Harper, 404 Foreglen Road Dungiven, Londonderry, BT47 4PN
GI0 AZB H. Evans, Oville House 404 Foreglen Road, Dungiven, Londonderry, BT47 4PN
GM0 AZC J. Sherry, 26 Grahamshill Terrace, Fankerton, Denny, FK6 5HX
G0 AZD G. Hayter, 19 Austin Road, Cirencester, GL7 1BT
G0 AZE L. Owen, 68 Clevedon Road, Tickenham, Clevedon, BS21 6RD
G0 AZG T. Wharton, Onanole, Clitheroe Road, Clitheroe, BB7 3DA
G0 AZH J. Wharton, 66 Hayhurst Street, Clitheroe, BB7 1ND
G0 AZM M. Walton, 5 Home Farm Court, Home Farm Close, Leicester, LE4 0SU
G0 AZP S. Tricker, 1 Drewitt Court, 75 Godstow Road, Oxford, OX2 8PE
G0 AZQ A. Pearce, 21 Cherry Way, Nafferton, Driffield, YO25 4PA
G0 AZR J. Norman, The End Peg, The Street, Norwich, NR11 7AQ
GM0 AZU J. Aiken, 48 Kirkwall Avenue, Blantyre, Glasgow, G72 9NX
GM0 AZV C. Norton, 3 Nether Balfour Cottages, Durris, Banchory, AB31 6BL
GW0 AZW R. Dooley, 98 Gelli Aur, Treboeth, Swansea, SA5 9DG
G0 AZX R. Pritchard, 10 Dolphin Crescent, Paignton, TQ3 1AE
G0 BAA North Cheshire Radio Club, c/o G. Gourley, 6A Longsight Lane, Cheadle Hulme, Cheadle, SK8 6PW
G0 BAF N. Quinn, 15 Newham Lane, Steyning, BN44 3LR
G0 BAG R. Cox, 2 Yardlea Close, Rowland's Castle, PO9 6DQ
GW0 BAH P. Bateman, 48 Ogmore Drive, Nottage, Porthcawl, CF36 3HR
G0 BAI P. Goodger, 125 Mill Hill Wood Way, Ibstock, LE67 6QD
G0 BAJ R. Edinborough, Flat 5 Fairlawn Apartments, 10 Elmsleigh Park, Paignton, TQ4 5AT
G0 BAK W. Schofield, 24 Meltham Road, Honley, Holmfirth, HD9 6HX
G0 BAM E. Smith, 166 Tudor Way, Dines Green, Worcester, WR2 5QY
G0 BAN J. Emerson, 26 Cardwell Street, Roker, Sunderland, SR6 0JP
G0 BAO M. Heath, Brambles, 1 Bestwall Road, Wareham, BH20 4HY
G0 BAP R. Harman, 42 Newlyn Close, Bransholme, Hull, HU7 4PQ
G0 BAQ I. Irving, Fourwinds, Woodburn Drive, Leyburn, DL8 5HU
G0 BAT W. Mcdonagh, 35 Manor Way Deeping St James, Peterborough, PE68PS
G0 BAU S. Craggs, 79 Silverdale Road, Cramlington, NE23 3LW
G0 BAW L. Haynes, 9 Heather Gardens, Belton, Great Yarmouth, NR31 9PP
G0 BAX A. Harrison, 10 Knoll Road, Sidcup, DA14 4QU
G0 BAY S. Parker, 85 Highfield Road, Glossop, SK13 8NZ
G0 BBB U. Grunewald, Nuptown Orchard, Nuptown, Warfield, Bracknell, RG42 6HU
GW0 BBC D. Thomas, 88 Cefn Graig, Rhiwbina, Cardiff, CF14 6JZ
G0 BBE L. Conlon, 4 Hill Crest Drive, Slack Head, Milnthorpe, LA7 7BB
G0 BBJ D. Fry, 9 Brook Gardens, Emsworth, PO10 7JY
G0 BBK P. Marshall, 35 Rosewood Close, Burnham-on-Sea, TA8 1HG
G0 BBN K. Hendry, 23 Briscoe Way, Lakenheath, Brandon, IP27 9SA
G0 BBO J. Stanbury, 6 Waterside Apartments, Weech Road, Dawlish, EX7 9FA
G0 BBR W. Kemp, 4 Blacksmiths Field Crowhurst, Battle, TN33 9AX
G0 BBT E. Snape, 1 Stephen Crescent, Humberston, Grimsby, DN36 4DS
G0 BBV H. Watts, 44 Laurel Road, Norwich, NR7 9LL
G0 BCF R. Pickering, 14 Dalestorth Gardens, Skegby, Sutton-in-Ashfield, NG17 3FT
G0 BCH P. Guppy, 202 Exeter Road, Kingsteignton, TQ12 3NJ
GD0 BCJ P. Mcgrath, 69 Clagh Vane, Ballasalla, Isle Of Man, IM9 2HF
GW0 BCL R. Johnson, 55 Maes Yr Haf, Llanelli, SA15 3NF
GD0 BCN H. Glaister, 42 Barrule Drive, Onchan, Douglas, Isle Of Man, IM3 4NR
G0 BCO A. Duffield, 32 Mount Close, Honiton, EX14 1QZ
GI0 BCP M. Jones, 2 Pine Ridge, Donaghadee, BT21 0QR
GW0 BCR J. Watts, 39 Turnberry Drive, Abergele, LL22 7UD
G0 BCS P. Rose, 53 South St., Pennington, Lymington, SO41 8DY
G0 BCT R. Sayer, Vignouse, Paimpont, France, 35380
G0 BCU D. Charnock, 44 Bramshill Close, Birchwood, Warrington, WA3 6TZ
G0 BCW D. Grevett, 45 East Bridge Road, South Woodham Ferrers, Chelmsford, CM3 5SB
G0 BCX G. Eden, 83 Windle Hall Drive, St Helens, WA10 6QG
GM0 BCY A. Carslaw, 51 Stonefield Drive, Paisley, PA2 7QY
G0 BCZ R. Johnson, 89 Arlington Gardens, Romford, RM3 0EB
G0 BDB P. Stephens, 259 Beaumont Road, Plymouth, PL4 9EL
GU0 BDI D. Prosser, Rustlings, Les Friquets, St Andrew, Guernsey, GY6 8SJ
G0 BDJ K. Dower, 16 Mills Drive, Wellington, TA21 9ED
G0 BDK B. Steele, 82 Margetts Road, Kempston, Bedford, MK42 8DT
G0 BDM D. Briggs, 72 Showell Grove, Droitwich, WR9 8UD
G0 BDN A. Slater, 1 Linglongs Avenue, Whaley Bridge, High Peak, SK23 7DT
G0 BDP R. Wills, 8 Owlswood, Ridingsmead, Salisbury, SP2 8DN
G0 BDR M. Gee, 100 Plantation Hill, Worksop, S81 0QN
G0 BDS B. Scroggs, 10 Lyon Close, Chelmsford, CM2 8NY
GI0 BDU V. Fortune, 10C Ards Drive, Newtownabbey, BT37 0JN
GW0 BDW R. Kelsall, 1+2 Tyn Rhos, Llanddona, Beaumaris, LL58 8YG
GI0 BDZ D. Pavis, 269 Lower Braniel Road, Belfast, BT5 7NR
GI0 BEB T. Magee, 10 Abernethy Park, Newtownabbey, BT36 6QQ
G0 BEC S. Christmas, Hitherto, Moats Tye Combs, Stowmarket, IP14 2EY
G0 BEE S. Bloom, 8 South Riding, Bricket Wood, St Albans, AL2 3ND
G0 BEJ P. Mallett, Woodrow, Chalk Lane, Spalding, PE12 9YF
GM0 BEL G. Lindsay, 6 Netherhouse Avenue, Lenzie, Glasgow, G66 5NG
G0 BEN D. Pykett, 20 Rochester Drive, Lincoln, LN6 0XQ
G0 BEP D. Boalch, 90 Belle Vue Road, Wivenhoe, Colchester, CO7 9EH
G0 BES S. Illsley, 88 Arnold Road, Eastleigh, SO50 5RR
G0 BEV Dr M. Hill, Windrush, Jesmond Gardens, Newcastle upon Tyne, NE2 2JN
G0 BEX A. Penfold, 115 Applegarth Park, Seasalter Lane, Whitstable, CT5 4BZ
GI0 BEY N. Turkington, 16 Glenshesk Park, Bangor, BT20 4US
GU0 BEZ V. Stamps, Mill Cottage, Sark, Guernsey, GY10 1SA
GI0 BFA R. Mckersie, 4 Burnside Park, Belfast, BT8 6HU
G0 BFC C. Simkin, Flat 33, Weymouth House Balfour, Tamworth, B79 7BE
GI0 BFD A. Magee, 70 Hillview Park, Enniskillen, BT74 6EU
G0 BFJ J. Stocks, 96 North Street, Lockwood, Huddersfield, HD1 3SL
G0 BFK K. James, Yew Tree Cottage, Whiston, Stafford, ST19 5QH
G0 BFM A. Anderson, 23 Aldred Road, Sheffield, S10 1PD
GD0 BFN J. Kneale, 51 Maple Avenue, Onchan, Douglas, Isle Of Man, IM3 3GA
GM0 BFW A. Mcclelland, 4 Walkerston Avenue, Largs, KA30 8ER
G0 BFZ A. Nock, 57 Mushroom Green, Dudley, DY2 0EE
G0 BGA G. Allan, 24 Leadbetter Drive, Bromsgrove, B61 7JG
G0 BGB M. Davis, 1 Westwood View, Crawcrook, Ryton, NE40 4HR
G0 BGH I. Bamford, 60 Coast Drive, Greatstone, New Romney, TN28 8NX
G0 BGI S. Knight, 14A Manor Road, Upton Lovel, Warminster, BA12 0JW
G0 BGR R. Jones, Glenroy, 5 School Road, Blackpool, FY4 5DS
G0 BGV H. Eastwood, 3 The Brambles, Thorpe Willoughby, Selby, YO8 9LL
G0 BGX O. Pauley, 235 Roughton Road, Cromer, NR27 9LQ
G0 BGY J. Moult, New Bungalow, Bar Bridge Lane Swineshead., Boston, PE20 3PG
G0 BHA P. White, 42 Abbey Road, Medstead, Medstead, Alton, GU34 5PB
G0 BHH P. Gregory, 20 Heyes Grove, Rainford, St Helens, WA11 8BW
G0 BHK E. Stiles, 16 Henry Avenue, Rustington, Littlehampton, BN16 2NY
G0 BHP P. Fambely, 126 Ashton Lane, Sale, M33 5QJ
G0 BHR D. Egan, 44 Wrights Lane, Warley, Cradley Heath, B64 6QX
G0 BHS D. Angell, 12 Harrod Drive, Market Harborough, LE16 7EH
G0 BHT W. Blake, Tylers Farm, Grantham Road, Grantham, NG33 5HG
G0 BHU M. Jamieson, 3 Rowan Way, Bourne, PE10 9SB
G0 BIA R. Armstrong, 64 Churchill Drive, Marske by the Sea, TS11 6BE
G0 BIE D. Lucas, 9 Newborough Road, Alvaston, Derby, DE24 0LH
G0 BIN E. Ashley, 2 School Road, Bulkington, Bedworth, CV12 9JB
G0 BIQ R. Bellamy, Emathu, No. 48 28Th April Street 1688, Xahhra, Malta, XRA1033
G0 BIR A. Skinner, Halfway Lock Cottage, Upper Gambolds Lane, Stoke Prior, Bromsgrove, B60 3HB
G0 BIV J. Young, 3 Kensington Close, Kings Sutton, Banbury, OX17 3XB
G0 BIW M. Smith, 394 Longbridge Road, Barking, IG11 9EE
G0 BIX T. Dansey, C/O 14 Stirling Park, Rochester, ME1 3QR
G0 BJA G. Vincent-Squibb, 7 Chudleigh Road, Henley Green, Coventry, CV2 1AF
G0 BJD S. Wood, 55 Megdale, Wolds, Matlock, DE4 3TE
G0 BJI D. Peacock, 15 Farmfield Road, Banbury, OX16 9AP
G0 BJK D. Thomas, 33 Chatsworth Road, Stretford, Manchester, M32 9QF
G0 BJL D. Oxley, 21 Ringwood Avenue, Chesterfield, S41 8RA
G0 BJP T. Fox, 206 Hollinsend Road, Sheffield, S12 2EJ
G0 BJR G. Oliver, 158 High Barn St., Royton, Oldham, OL2 6RW
G0 BKA J. Leedham, 27 St. Andrews Crescent, Stratford-upon-Avon, CV37 9QL
G0 BKB V. Leedham, 27 St. Andrews Crescent, Stratford-upon-Avon, CV37 9QL
G0 BKC P. Glasper, 2 Iris Close, Stockton, TS18 1ax
G0 BKD B. Mawn, 13 Micklethwaite Grove, Moorends, Doncaster, DN8 4NU
G0 BKE F. Barker, 17 Walders Avenue, Sheffield, S6 4AY
G0 BKH B. Minton, 8 Rosebank Walk, Barnton, Northwich, CW8 4PU
GW0 BKJ M. Glover, 4 New Hospital Villa, Hospital Road, Brecon, LD3 0DU
G0 BKL E. Smith, 40 Grays End Close, Grays, RM17 5QR
G0 BKN I. Mciver, 3 Asgard Drive, Bedford, MK41 0UP
G0 BKP J. Dadswell, Ivy House, Chivery, Tring, HP23 6LE
G0 BKQ P. Rowe, 45 Springhill Avenue, Wolverhampton, WV4 4ST
G0 BKR Dr G. Clark, 2 Roundheads End Forty Green, Beaconsfield, HP9 1YB
GM0 BKS G. Christison, 13/5 West Winnelstrae, Edinburgh, EH5 2ET
G0 BKU S. Coles, 88C Dursley Road, Trowbridge, BA14 0NS
G0 BKW K. Weston, 14 Auchinleck Court Burleigh Way, Crawley Down, Crawley, RH10 4UP
GM0 BKX T. Stewart, 104 Barrhill Road, Cumnock, KA18 1PU
G0 BKZ D. Hyde, 75 Bury Street, Stockport, SK5 7RE
G0 BLB R. Baker, Homelea, Upper Bristol Road, Clutton, Bristol, BS39 5RJ
G0 BLM P. Mathews, 25 Shore Mount, Littleborough, OL15 8EN
G0 BLO B. Osborn, 4 Highfield Close, Collier Row, Romford, RM5 3RX
G0 BLQ H. Cooper, 26 Badgers Way, Buckingham, MK18 7EQ
G0 BLS A. Wilkie, 7 Willow Drive, Droitwich, WR9 7QE
G0 BLT G. Blackmoor, 4 St. Godwalds Crescent, Aston Fields, Bromsgrove, B60 2EB
G0 BLU E. Mustard, 108 Allandale, Hemel Hempstead, HP2 5AT
G0 BLV M. Puncer, 17 St. Michaels Walk, Eye, Peterborough, PE6 7XG
G0 BLW A. Crimlisk, 14 Long Lane, Aughton, Ormskirk, L39 5AT
G0 BMG X. Green, 65 Rosamond Road, Bedford, MK40 3UG
G0 BMH J. Mclean, 24 Durham Drive, Oswaldtwistle, BB5 3AT
G0 BML T. Garvey, 162 Birchfields Road, Manchester, M14 6PE
G0 BMM R. Howarth, 95 Riverside Drive, Radcliffe, Manchester, M26 1HY
G0 BMN K. Hutchins, 23 Salisbury Road, Tunbridge Wells, TN4 9DJ
G0 BMP L. Gyurgyak, 7 Lakeside Close, New Park, Newton Abbot, TQ13 9FE
G0 BMQ M. Armstrong, 4 Medway Drive Preston, Weymouth, DT3 6LF
G0 BMS I. Cooper, Flat 1, Philip Howard Court, Glynne Street, Farnworth, Bolton, BL4 7DQ
G0 BMT J. Walkley, 10 Exton, Dunster Crescent, Weston-Super-Mare, BS24 9EH
G0 BMU T. Drewitt, 6 Copse View Cottages, Ascot Road, Maidenhead, SL6 3JY
G0 BMZ B. Lindgren, Berzeliigatan 26, Goteborg, Sweden, SE-41 53
G0 BNE Dr A. Knell, 13 Northumberland Road, Leamington Spa, CV32 6HE
G0 BNF J. Holmes, 63 Grange Avenue, Street, BA16 9PF
G0 BNG L. Britton, 36 Frampton Court, Trowbridge, BA14 9HL
G0 BNJ B. Northway, 3 Kingston Close, Kingskerswell, Newton Abbot, TQ12 5EW

G0 BNK Wearside Electronics At ARS, c/o I. Douglas, 13 Castlereagh Street, New Silksworth, Sunderland, SR3 1HJ
GW0 BNN J. Bulpin, 12 Waungron Close, Treboeth, Swansea, SA5 7DH
GW0 BNO D. Lindley, 29 Belvedere Close, Kittle, Swansea, SA3 3LA
GM0 BNQ D. Macdonald, Greenbrae Cottage, Auchterless, Turriff, AB53 8HD
G0 BNR N. Keightley, Wavendon, Daintree Road, Ramsey St. Marys, Ramsey, Huntingdon, PE26 2TF
G0 BNU D. Wright, Dovetail Cottage, South Petherwin, Launceston, PL15 7LQ
G0 BNW W. Wheeler, 201 Topsham Road, Exeter, EX2 6AN
G0 BNY C. Lee, 29 Meadow Dale, Chilton, Ferryhill, DL17 0RW
G0 BNZ J. Hewitt, 35 Birmingham Road, Alvechurch, Birmingham, B48 7TB
G0 BOC E. Pitman, 35 Brackley Way, Totton, Southampton, SO40 3HP
GW0 BOE Dr W. Piotrowski, Ty Nant, Abbeycwmhir, LD1 6PH
G0 BOH D. Bowman, 6 Linksfield, Denton, Manchester, M34 3TE
G0 BOM A. Samouelle, 4 Fox Road, Bourn, Cambridge, CB23 2TU
G0 BON I. Rogers, 33 Sandstone Road, Swindon, SN25 2FE
G0 BOO B. Gilbert, Silver Micha, Hunts Corner, Norwich, NR16 2HL
G0 BOR J. Goodall, 1 Purn Lane, Weston-Super-Mare, BS24 9JG
G0 BOT D. Ashby, 36 Mersey Grove, Birmingham, B38 9LA
G0 BPA B. Lyford, 48 Wilverley Place, Blackfield, Southampton, SO45 1XW
GM0 BPF W. Johnstone, Byre Cottage, Kilmichael Glassary, Lochgilphead, PA31 8QL
G0 BPK N. Ferguson, Royd Moor, Royd Moor Lane, Badsworth, Pontefract, WF9 1AZ
G0 BPL D. Farnham, 24 Downham Road, Watlington, King's Lynn, PE33 0HS
G0 BPM D. Coffey, 121 Worksop Road, Swallownest, Sheffield, S26 4WB
G0 BPQ J. Jones, 16 Laurel Avenue, Darwen, BB3 3AG
G0 BPR S. Whitnear, 55 Brier Crescent, Nelson, BB9 0QD
G0 BPS R. Pascoe, 12 Oak Rise, Terlingham Gardens, Hawkinge, CT18 7FU
GM0 BPT A. Murray, 1 Gordon Road, Edinburgh, EH12 6NB
G0 BPU M. Johnson, 23 Camden Road, Ipswich, IP3 8JW
G0 BPX D. Norridge, 125 Ferry Road, Marston, Oxford, OX3 0EX
G0 BPZ P. Maple, Beech House, Derby Road, Ashbourne, DE6 1LZ
G0 BQB P. Smith, 10 Denby Lane, Grange Moor, Wakefield, WF4 4ED
G0 BQC W. Scrivener, Folly Hall Farm, Kings Causeway, Nelson, BB9 0EZ
G0 BQE J. Chown, 15 Hambleden Walk, Maidenhead, SL6 7UH
G0 BQG R. Jefferies, 35 Lambrok Close, Trowbridge, BA14 9HH
G0 BQI M. Chapman, 6A Rees Street, Islington, London, N1 7AR
G0 BQK M. Hall, 31 Winchester Close, Stratton, Swindon, SN3 4HB
G0 BQO J. Rawson, 3 Cooks Close, Kesgrave, Ipswich, IP5 2YT
G0 BQP Dr J. Simpson, 18 Southdean Drive, Hemlington, Middlesbrough, TS8 9HH
GM0 BQQ I. Laczko, 34 Airds Drive, Dumfries, DG1 4EW
G0 BQV M. Ashdown, 42 Alpine Avenue, Tolworth, Surbiton, KT5 9RJ
G0 BQW S. Robinson, 114 Hopefield Avenue, Sheffield, S12 4XE
GI0 BQX D. Maguire, 16 Kilmacormick Drive, Enniskillen, BT74 6EP
G0 BQZ S. Smith, 18 Stratford Drive, Eynsham, Witney, OX29 4QJ
G0 BRA Banbury Amateur R, c/o F. Humphris, 169 Bloxham Road, Banbury, OX16 9JU
G0 BRH C. Waters, 468 Buckfield Road, Leominster, HR6 8SD
GM0 BRJ D. Wilson, Four Winds, High Barrwood Road, Kilsyth, G65 0EE
G0 BRL M. Bevan, 22 Spring Crescent, Brown Edge, Stoke-on-Trent, ST6 8QH
G0 BRM N. Entwistle, Sonunda House, Church Lane, Freckenham, Bury St Edmunds, IP28 8JF
GI0 BRO R. Wilson, 68 Kensington Road, Belfast, BT5 6NG
G0 BRQ A. Morris, 108 Lytchett Drive, Broadstone, BH18 9NR
GM0 BRS Border ARS, c/o A. Scott, 20 Treaty Park, Birgham, Coldstream, TD12 4NG
G0 BRW B. Whatling, 6 Rock Road, Dursley, GL11 6LF
G0 BRX S. Shirras, 72 Sefton Avenue, Poulton-le-Fylde, FY6 8BL
G0 BRZ L. Rimmer, 7 Dorfold Close, Sandbach, CW11 1EB
G0 BSA M. Embling, 23 Sinodun Road, Didcot, OX11 8HP
G0 BSD D. Tringham, 47 The Knoll Palacefields, Runcorn, WA7 2UH
G0 BSF H. Rolfe, 46 Great Gardens Road, Hornchurch, RM11 2BA
G0 BSH R. Harman, 22 Ridgebrook Road, Kidbrooke, London, SE3 9QN
G0 BSJ E. New, 6 Witchampton Close West Leigh, Havant, PO9 5RY
G0 BSK K. Bryant, 9 Cunningham Park, Mabe Burnthouse, Penryn, TR10 9HB
G0 BSN R. Vincent-Squibb, 1 Alexander Avenue, Earl Shilton, Leicester, LE9 7AF
G0 BSP R. Snow, 73 Boxtree Road, Harrow Weald, Harrow, HA3 6TN
G0 BST J. I'Anson-Holton, Lake View, Brookside Avenue, Telford, TF3 1LA
G0 BSX Dr P. Meiring, 18 Slayleigh Lane, Sheffield, S10 3RF
G0 BTA D. Round, 21 Bunting Drive, Bradford, BD6 3XE
GW0 BTB B. Botham, The Cherries, Anglesey, LL74 8SR
G0 BTD G. Owen, 7 East View, Crossgates, Leeds, LS15 8AY
G0 BTH P. Clark, 180 Roselands Drive, Paignton, TQ4 7RW
GM0 BTK Dr W. Rossmann, Strathvale, Milton Of Ogilvie, Forfar, DD8 1UN
G0 BTQ N. Burge, 43 Bourn Rise, Pinhoe, Exeter, EX4 8QD
G0 BTT E. Clay, 23 New Street, Sleaford, NG34 7HG
G0 BTU K. Tupman, Magpies, 6 Larcombe Road, Petersfield, GU32 3LS
G0 BTV S. Coleman, 42 Regent Street, Sutton-in-Ashfield, NG17 2EH
G0 BUB M. Sharpe, New House, Highfield Farm, Grantham, NG32 3SJ
G0 BUC K. Simpson, 5 Cedar Close, The Elms, Lincoln, LN1 2NH
G0 BUD J. Spearing, 19 Elizabeth Square, London, SE16 5XN
GM0 BUE A. Stark, 30 Kelvin Way, Kilsyth, Glasgow, G65 9UL
GM0 BUI G. Williamson, 2 Laburnum Grove, Burntisland, KY3 9EU
G0 BUJ R. Pelling, The Orchard, Shirwell, Barnstaple, EX31 4JR
G0 BUK I. Mitchell, Cornerfield, Five Ash Down, Uckfield, TN22 3AP
G0 BUV J. Howard, 130 Coventry Road, Coleshill, Birmingham, B46 3EH
G0 BUW P. Martin, 39A The Grove, Bearsted, Maidstone, ME14 4JB
G0 BUX R. Clinton, Appletrees, Alexandra Road, Mayfield, TN20 6UD
G0 BUZ E. Piper, 44 Parsonage Estate, Rogate, Petersfield, GU31 5HJ
G0 BVA S. Fawcett, 34 Wantsume Lees, Sandwich, CT13 9JF
G0 BVC G. Broadhurst, Summerhayes, 10 Ottervale Close, Honiton, EX14 9TA
G0 BVD P. Oakley, Elmsleigh House, New Street, Great Torrington, EX388BY
GM0 BVG J. Graham, Clintpark, Lockerbie, DG11 3JH
G0 BVK E. Evers, 13 Grasmere Close, Penistone, Sheffield, S36 8HP
G0 BVM J. Speers, 187 Worsley Road, Eccles, Manchester, M30 8BP
G0 BVO J. Teasdale, 1 Newtown Bungalows, Newtown, Spennymoor, DL16 7QS
G0 BVQ C. Hawkes, 12 Summer Hall Ing, Wyke, Bradford, BD12 8DN
G0 BVS M. Arthur, 8 Hanbury Road, Bedworth, CV12 9BX
G0 BVT A. Tonge, 30 Cardigan Avenue, Morley, Leeds, LS27 0DP
G0 BVU D. Davies, 30 Bullpit Road, Balderton, Newark, NG24 3LY
G0 BVV D. Clench, 70 Shalbourne Crescent, Bracklesham Bay, Chichester, PO20 8RG
G0 BVW Q. Curzon, 154 Clophill Road, Maulden, Bedford, MK45 2AE
G0 BWB J. Chappell, 49 Midway, South Crosland, Huddersfield, HD4 7DA
G0 BWC Bolton Wireless Club, c/o R. Wilkinson, 84 Park Road, Bolton, BL1 4RQ
GW0 BWE D. Allen, 48 Castle Road, Crickhowell, NP8 1AP
G0 BWG J. Raynes, 115 Deerlands Avenue, Sheffield, S5 7WU
G0 BWJ K. Miller, 8 Horsham Gardens, Sunderland, SR3 1UJ
G0 BWK R. Lee, 57 Hart Lane, Luton, LU2 0JF
G0 BWO D. Wilkinson, 139 Grosvenor Road, Dalton, Huddersfield, HD5 9HX
G0 BWP B. Passmore, 364 Franklin Road, Kings Norton, Birmingham, B30 1NG
G0 BWQ K. Kitson, 278 Cowcliffe Hill Road, Huddersfield, HD2 2NE
G0 BWV J. Puttock, Sutton & Cheam Rs, 53 Alexandra Avenue, Sutton, SM1 2PA
G0 BWY G. Fearnside, 16 Lee Court, Thwaites Brow, Keighley, BD21 4TL
G0 BXC P. Hughes, 123 Garth Road, Morden, SM4 4LF
G0 BXD A. Wootton, Dower House, Hilton, Bridgnorth, WV15 5PB
G0 BXG R. Dring, 22 Castle Street, Eastwood, Nottingham, NG16 3GW
G0 BXH B. Hansell, 7 Fry Road, Stevenage, SG2 0QG
G0 BXJ A. Peachey, 60 Upwell Road, March, PE15 9EA
G0 BXL T. White, 79 Elmbridge, Harlow, CM17 0JY
G0 BXM M. Smyth, Sunset Cottage, Cloonfree, Strokestown, Ireland, F42 YX92
GM0 BXR T. Holland, 7 Burnside, Flotta, Stromness, KW16 3NP
G0 BXS R. Rennolds, 6 Roman Way, Bourton-On-The-Water, Cheltenham, GL54 2EW
G0 BXU A. Collick, 39 Beech Rise, Sleaford, NG34 8BJ
G0 BXV D. Dixon, Flat 19, Maple Court 3A Staunton Avenue, Hayling Island, PO11 0EF
G0 BYA S. Houlding, 90 Wordsworth Avenue, Stafford, ST17 9UE
G0 BYF R. Aucote, 12 Craddock Court, Craddock Drive, Nuneaton, CV10 9EL
G0 BYH G. Stokes, 23 Maynard Close, Bradwell, Milton Keynes, MK13 9HS
G0 BYK M. Jackson, Stockswood, Stocks Lane, Southwold, IP18 6UJ
G0 BYL J. Luxton, 2 Trinity Court, Westward Ho, Bideford, EX39 1LT
G0 BYQ S. Ratcliffe, 173 Whinney Lane, New Ollerton, Newark, NG22 9TJ
G0 BYU F. Cooper, 23 York Close, Gillow Heath, Stoke-on-Trent, ST8 6SE
G0 BYX C. Pavier, 12A Friend Lane, Edwinstowe, Mansfield, NG21 9QZ
GW0 BYZ R. Jones, 9 Dennithorne Close, Merthyr Tydfil, CF48 3HE
GW0 BZA K. Duckfield, Ellesmere, 29 Esplanade Avenue, Porthcawl, CF36 3YS
G0 BZB A. Volpe, 7 Oakwell Terrace, Haltwhistle, NE49 9LR
G0 BZC G. Smith, 9 Blackett Avenue, Stockton-on-Tees, TS20 2EX
G0 BZF D. Reid, Nicolaas Beetsstraat 29, Hengelo Ov, Netherlands, 7552HW
G0 BZH G. Hodgson, 16 Dockroyd, Oakworth, Keighley, BD22 7RH
GW0 BZJ J. Newton, 21 Village Court, Penrhiw Avenue, Blackwood, NP12 0LU
GI0 BZM C. Colhoun, 85 Whitehill Park, Limavady, BT49 0QF
G0 BZN C. Johnson, 23 Medlar Street, Weston Turville, Aylesbury, HP22 5YQ
G0 BZP J. Bates, 28 Westbourne Road, West Bromwich, B70 8LD
GM0 BZS A. Ince, Burnside, Braefield, Glen Urquhart, Inverness, IV63 6TN
G0 BZT R. Targonski, 4 Woodville Gardens, Dudley, DY3 1LB
G0 BZU K. Barlow, 105 Buller Street, Bury, BL8 2BQ
G0 BZV E. Trett, 11 Langton Avenue, Bierley, Bradford, BD4 6BY
G0 BZW A. Porter, 1 Bloomfield Terrace, Weston, Portland, DT5 2AB
G0 BZX P. Smith, 8 Avalon Close, Orpington, BR6 9BS
GM0 CAD D. Hewitt, Willowburn, Ardross, Alness, IV17 0XN
G0 CAE K. Walsh, 13 Weston Park Homes, Weston Road, Portland, DT5 2DE
G0 CAG D. Talaber, 54 Southfield Park, North Harrow, Harrow, HA2 6HE
GI0 CAH P. Leonard, 4 Clonurson Road, Enniskillen, BT92 3BU
G0 CAJ T. Morgan, 73 Winnards Park, Sarisbury Green, Southampton, SO31 7BX
G0 CAK G. Russell, 15 Warblington Close, Tadley, RG26 3YW
G0 CAL R. Faulkner, 10 Fell Wilson St., Warsop, Mansfield, NG20 0PT
G0 CAM C. Chislett, Woodview, Crofthandy, Redruth, TR16 5PT
G0 CAS N. Clarke, Lyme View, 3 East Cliff House, Dawlish, EX7 9JB
G0 CAX I. Moore, 25 Granby Close, Winyates, Redditch, B98 0PJ
G0 CAY K. Arnold, 14 Ford Close, St.Ive, Cornwall, PL14 3FN
GM0 CBA B. Aitken, 48 Kenilworth Rise, Livingston, EH54 6JJ
G0 CBB M. Phillips, 52 Rivington Drive, Burscough, Ormskirk, L40 7RP
GM0 CBC J. Johnstone, 12 Castle Acre, Ecclefechan, Lockerbie, DG11 3DU
G0 CBD G. Roby, 40 Lulworth Drive, Hindley Green, Wigan, WN2 4QS
G0 CBI S. Mcarthur, 36 Ingham Close, Bradshaw, Halifax, HX2 9PQ
G0 CBJ A. Whittingham, 61 Hillcroft Road, Altrincham, WA14 4JE
G0 CBK N. Priestley, 13 Orchard Close, Charfield, Wotton-under-Edge, GL12 8TJ
GW0 CBL J. Follant, 76 Wern Road, Skewen, Neath, SA10 6DL
G0 CBM C. Wilkie, Bramcote, Grange Road, Sandilands, Mablethorpe, LN12 2RE
G0 CBN P. Forster, 2 Rockingham Close, Birchwood, Warrington, WA3 6UY
G0 CBO R. Hoffman, 26 Penn Road Taverham, Norwich, NR8 6NN
G0 CBP R. Mcmahon, 34 Denmark Road, Poole, BH15 2DB
G0 CBT S. Peat, 64 Grange Road, Romford, RM3 7DX
G0 CBU W. Drea, 146 Slewins Lane, Hornchurch, RM11 2BS
G0 CBW M. Bentley, Nickers Hill Farm, Falhouse Lane, Dewsbury, WF12 0NL
G0 CCA G. Coles, Walnut Lodge, Staunton Lane, Bristol, BS14 0QG
G0 CCB C. Gill, 52 Southfield Road, Nailsea, Bristol, BS48 1JD
G0 CCC Caversham Con G, c/o C. Young, 18 Wincroft Road, Caversham, Reading, RG4 7HH
G0 CCF J. Broome, 35 Claygate Road, Cannock, WS12 2RN
G0 CCG H. Toh, 3 Priory Road, Dover, CT17 9RQ
G0 CCJ J. Weinstock, 24 Hamilton Road, Tiddington, Stratford-upon-Avon, CV37 7DD
G0 CCL Cambridge Consultants ARC, c/o L. Laprade, 22 Langley Way, Hemingford Grey, Huntingdon, PE28 9DB
G0 CCM M. Hurst, 2 Poplar Road, Oughtibridge, Sheffield, S35 0HR
G0 CCN P. Hurst, 23 Cantilupe Crescent, Aston, Sheffield, S26 2AS
GW0 CCO L. Holderness, 8 Jubilee Gardens Templeton, Narberth, SA67 8ST
G0 CCQ J. Smith, Dobie Lodge, Rochford, Tenbury Wells, WR15 8SR
G0 CCS J. Zissler, 8 Norman Drive, Whittington, King's Lynn, PE33 9TQ
G0 CCU L. Whitelegg, 30 Chatsworth Road, Arnos Vale, Bristol, BS4 3EY
G0 CCV J. Gaut, 18 First Avenue, Clipstone Village, Mansfield, NG21 9EA
G0 CCX A. Gilbert, 6 Tarring Close, South Heighton, Newhaven, BN9 0QU
G0 CDA M. Ryder, 9 Lincoln Close, Woolston, Warrington, WA1 4LU
G0 CDB J. May, 6 Hodson Close, Paignton, TQ3 3NU
GI0 CDM J. Neill, 16 Rathlin Street, Belfast, BT13 3DZ
G0 CDO J. Faulkner-Court, Yew Tree Cottage, Avon Dassett, Southam, CV47 2AT

G0 CDQ M. Murphy, 10 Bayham Road, Sevenoaks, TN13 3XA
G0 CDR J. Warwick, 29 Hay Brow Crescent, Scalby, Scarborough, YO13 0SG
G0 CDS C. Brind, 8 Pezenas Drive, Market Drayton, TF9 3UJ
GM0 CDV R. Evans, 9 Courthill Farm Cottages, Kelso, TD5 7RU
GM0 CDW A. Thomson, 24 Craigmount Gardens, Edinburgh, EH12 8EA
G0 CDY A. Hulme, 71 Victoria Gardens, Ferndown, BH22 9JQ
G0 CDZ E. Hicks-Arnold, Ingleside, Junction Road, Salisbury, SP5 3AZ
GM0 CEA R. Robertson, St. Madoes Cottage, St. Madoes, Perth, PH2 7NF
G0 CEB E. Hutchinson, 58 Avon Rise, Retford, DN22 6QH
G0 CEC P. Coenraats11, 54 Falstaff Avenue, Earley, Reading, RG6 5TG
G0 CEF A. Sheard, 8 Hazel Beck, Bingley, BD16 1LZ
G0 CEG P. Worsdale, 10 Manton Road, Lincoln, LN2 2JL
G0 CEI P. Olliffe, 4 Orpwood Paddock, School Road, Ardington, Wantage, OX12 8RB
G0 CEJ C. Baguley, 44 Royds Crescent, Rhodesia, Worksop, S80 3HG
G0 CEL N. Evely, 11 St.Margarets New Road, Teignmouth, TQ14 8UE
G0 CEM T. Barry, 26 Gatcombe Road, Hartcliffe, Bristol, BS13 9RB
G0 CEN F. Field, 19 The Maples, Nailsea, Bristol, BS48 4RT
G0 CEO M. Ohta, House No-0120 Marfori Ext. Puroku-01 Barangay Masiit, Calauan-City, Philippines, 4012
G0 CEP D. Craker, Brynamlwg, Llanddewi Brefi, Tregaron, SY25 6PE
G0 CEQ D. Griffiths, 297 Shurland Avenue, Barnet, EN4 8DQ
G0 CER D. Harris, 9 Garden City, Tern Hill, Market Drayton, TF9 3QB
GW0 CES M. Smith, 8 Ridgeway Avenue, Marford, Wrexham, LL12 8ST
G0 CEU C. Hawkins, 3 Offord Close, Tottenham, London, N17 0TE
G0 CEV K. Towers, 7 Copeland Road, Hucknall, Nottingham, NG15 8EB
G0 CEW P. Allanson, 16 Woodhouse Lane, Kirkhamgate, Wakefield, WF2 0SE
G0 CEY G. Cadey, 45 St. Mildreds Road, Westgate-on-Sea, CT8 8RJ
G0 CFB R. Tyler, The Firs, Laundry Lane Huntingfield, Halesworth, IP19 0PY
GM0 CFC G. Percival, 5 Murrell Terrace, Aberdour, Burntisland, KY3 0XH
G0 CFD F. Dimmock, The Cottage, Lutton Garnsgate, Long Sutton, PE12 9JP
G0 CFI R. Deans, 23 Aldringham Park, Aldringham, Leiston, IP16 4QZ
GM0 CFK C. Knight, 8 Ednam Drive, Glenrothes, KY6 1NA
G0 CFM B. Robbins, 46 Purton Close, Kingswood, Bristol, BS15 9ZE
G0 CFN T. Hastings, 1 Pottle Close, Botley, Oxford, OX2 9SN
G0 CFT T. Feaviour, 57 Elizabeth Way, Felixstowe, IP11 2PQ
GM0 CFW K. Moffat, 11 Russell Court, Lochgelly, KY5 9EU
G0 CGA W. Roberts, 10 The Meadows, Station Road, Hodnet, TF9 3QF
G0 CGD C. O'Neill, 4 Bronte Walk, Backford, Chester, CH1 6QJ
G0 CGE B. Griffin, 75 Greenham Wood, Bracknell, RG12 7WH
G0 CGH R. Hurley, 57 Cannons Close, Bishop's Stortford, CM23 2BQ
G0 CGI B. Redfern, The Old Blacksmiths Forge, Old Newton, Craven Arms, SY7 9PG
G0 CGM A. Cutcliffe, 9 Dixon Close, Paignton, TQ3 3NA
G0 CGQ R. Hazlewood, 9 The Brambles, Haslington, Crewe, CW1 5RA
G0 CGS D. Morris, 2 Willow Close, Brinsworth, Rotherham, S60 5JU
G0 CGT B. Birch, 59 Shepperson Road, Sheffield, S6 4FG
G0 CGW G. Ashcroft, 49 Tantallon, Birtley, Chester le Street, DH3 2JG
G0 CGZ G. Allison, 24 Southfield Road, Scartho, Grimsby, DN33 2PL
G0 CHB B. Small, 3 Katherine Crescent, Skegness, PE25 3LF
G0 CHC M. Smith, 12 High Street, West Wickham, Cambridge, CB21 4RY
G0 CHE K. Piper, Flat 5, Marine Court, 4 Marine Drive West, Bognor Regis, PO21 2QA
G0 CHG R. Moate, Garth House, Redbrook Street, Ashford, TN26 3QS
G0 CHJ R. Poulter, 19 Homestead Way, Winscombe, BS25 1HL
G0 CHK R. Kerby, Blackboy Lane, Chichester, PO18 8BE
G0 CHL K. Dewhurst, Rock Cottage, 82 New Street, Stoke-on-Trent, ST8 7NW
GM0 CHM J. Stephen, 26 Douglas Avenue, Elderslie, Johnstone, PA5 9NE
G0 CHN J. Sutton, 40 Dane Valley Road, Margate, CT9 3RX
G0 CHO C. Ousbey, 30 Hawthorn Way, Shipston-on-Stour, CV36 4FD
G0 CHP R. Angus, 25 Jellicoe House, Capstan Road, Hull, HU6 7AS
G0 CHQ J. Pepper, 7 East Towers, Pinner, HA5 1TN
G0 CHR A. Steward, 94 Whittington Avenue, Hayes, UB4 0AE
G0 CHV T. Sherriff, 5 Pembroke Avenue, Morecambe, LA4 6EJ
G0 CHY P. Moulton, 4 The Ridge, Withyham Road, Tunbridge Wells, TN3 9QU
G0 CIG A. Aldridge, 15 Doubletrees, St. Blazey, Par, PL24 2LD
GM0 CII W. Rogers, 170 Boswall Parkway, Edinburgh, EH5 2JJ
G0 CIM S. Dodd, 61 Church Road, Hove, BN3 2BP
G0 CIR R. Jasper, 84 Rose Green Road, Bognor Regis, PO21 3EQ
G0 CIT G. Davies, 18 Cheltenham Place, Brighton, BN1 4AB
G0 CIX R. Wilton, 30 Barrington Crescent, Birchington, CT7 9DF
G0 CJA L. Harper, 23 Beech Avenue, Hazel Grove, Stockport, SK7 4QP
G0 CJD K. Spratley, 92 Plantation Hill, Worksop, S81 0QN
G0 CJG D. Slatter, 13 Hill Burn, Henleaze, Bristol, BS9 4RH
G0 CJO R. Nelson, Flat 1, 17 Ashburnham Road, Hastings, TN35 5JN
G0 CJQ R. Saw, 33 Rectory Lane, Southoe, St Neots, PE19 5YA
G0 CJV C. Scholey, 11A Guildford St., Grimsby, DN32 7PL
G0 CJX J. Stott, 12A Henley Close, Saxmundham, IP17 1EY
G0 CJZ R. Waller, 29 Valley Drive, Withdean, Brighton, BN1 5FA
G0 CKA G. Waller, The Pines, 18 Old London Road, Brighton, BN1 8XQ
G0 CKD F. Wall, 37 Newton Way, St. Osyth, Clacton-on-Sea, CO16 8RQ
G0 CKE J. Centanni, 5 Mickle Meadow, Water Orton, Birmingham, B46 1SN
G0 CKF B. Woodward, 10 Darley Close, Kilham, YO25 4UA
G0 CKH A. Affolter, 12 Newfound Drive, Norwich, NR4 7RY
G0 CKI K. Tarbett, 20 Leeholme, Houghton le Spring, DH5 8HR
GW0 CKK D. Robinson, Island View Caravan Park, Beach Road, Penarth, CF64 5UG
GW0 CKL I. Gulyas, 2 Eglwysilan Way, Abertridwr, Caerphilly, CF83 4EQ
G0 CKM C. Gee, 6 Canterbury Close, Dukinfield, SK16 5RT
G0 CKV O. Lundberg, Rowan House, Cavendish Road, Weybridge, KT13 0JW
G0 CLC E. Rogers, Room 21, Bodmeyrick Residential Home, Holsworthy, EX22 6HB
G0 CLD H. Davey, 6 Cambridge Grove, Otley, LS21 1DH
G0 CLG S. Haynes, 9 Heather Gardens, Belton, Great Yarmouth, NR31 9PP
G0 CLH D. Lingard, 17 Feltwell Road, Methwold Hythe, Thetford, IP26 4QJ
G0 CLJ S. Banks, 15 Hunters Way, Saffron Walden, CB11 4DE
G0 CLM A. Greig, 98 Appletree Lane, Redditch, B97 6TS
G0 CLR R. Hunt, 1 Avon View, Cotswold Grange Country Park, Twyning, Tewkesbury, GL20 6DL
G0 CLT P. Hodgson, 14 Catherine Howard Close, Thetford, IP24 1TQ
G0 CLV S. Barr, 11 Mallard Way, Wirral, CH46 7SJ
G0 CLX G. Glotham, 89 Mellish Road, Walsall, WS4 2DF
G0 CMB R. Burling, 28 Croydon Road, Arrington, Royston, SG8 0DJ
GW0 CMI K. Voller, Machlud Haul, Tanygroes, Cardigan, SA43 2HR
G0 CMK N. Sherwood, The Orchards, Barton Road, Brigg, DN20 8SH
G0 CMM J. Bell, Le Magnou, Romagne, France, 86700
GM0 CMO M. Murray, 1 Gordon Road, Edinburgh, EH12 6NB
G0 CMP V. Warren, 129 Market Road, Thrapston, Kettering, NN14 4JT
G0 CMR R. Melia, 14 Friar Park Road, Wednesbury, WS10 0TB
G0 CMT E. Gadeberg, Hojmarksvej 35, Po Box 56, Horsens, Denmark, 8700
G0 CMU C. Beecham, Moorview, 2 Endor Crescent, Ilkley, LS29 7QH
G0 CMW J. Groom, Windyridge, High Road, Maidenhead, SL6 9JF
G0 CNA M. Wyatt, 32 Stafford Road, Bridgwater, TA6 5PH
G0 CND I. Sharrott, 31 The Fleet, Stoney Stanton, Leicester, LE9 4DZ
G0 CNG C. Roberts, 72 Nairn Road, Walsall, WS3 3XB
GW0 CNJ S. Edwards, 31 Eagleswell Road, Boverton, Llantwit Major, CF61 2UG
GW0 CNK G. Orchard, 8 Glyn Avenue, Prestatyn, LL19 9NN
G0 CNL C. Rogers, 34 Martin Court, Werrington, Peterborough, PE4 6JS
GM0 CNP J. Mullen, 46 Templars Crescent, Kinghorn, Burntisland, KY3 9XS
G0 CNU J. Mckenzie, 4 Hadrian Court, Humshaugh, Hexham, NE46 4DE
G0 CNV T. Mcmanus, 84 Beverley Road, Hessle, HU13 9BP
GM0 CNW H. Taylor, 20 Woodhaven Avenue, Wormit, Newport-on-Tay, DD6 8LF
G0 COA G. Coates, 1 Ash Brow, Flockton, Wakefield, WF4 4TE
G0 COC R. Vallis, Bartley Villa, Southampton Road, Southampton, SO40 2NA
G0 COE E. Coe, 20 Bitterne Way, Lymington, SO41 3PB
G0 COG K. Stringer, 33 Brookes Road, Flitwick, Bedford, MK45 1BU
GW0 COH J. Rixon, Bro Afallon, Rhoslefain, Tywyn, LL36 9LY
G0 COI J. Atkinson, 90 Priors Road, Cheltenham, GL52 5AN
G0 COJ B. Ellery, 384 Sutton Way, Great Sutton, Ellesmere Port, CH66 3LL
G0 COL C. Shepherd, 9 Wrea Head Close Scalby, Scarborough, YO13 0RX
G0 COQ W. Kelsey-Stead, 65/4 Moo 2, Rawai, Phuket, Thailand, 83130
GW0 COU S. Jones, 635 Clydach Road, Ynystawe, Swansea, SA6 5AX
G0 COY T. Frearson, 31 Paradise Street, Rugby, CV21 3SZ
G0 COZ J. Wayman, 1 Waterloo Close, Bredon, Tewkesbury, GL20 7WL
G0 CPA A. Prichard, 1 Poltondale, Swindon, SN3 5BN
G0 CPD G. Forster, 2 Rockingham Close, Birchwood, Warrington, WA3 6UY
G0 CPF I. Whyte, 36 Chestnut Road, Ashford, TW15 1DG
G0 CPJ P. Walker, 46 Ribble Avenue, Freckleton, Preston, PR4 1RX
G0 CPN I. Gurton, 28 Bloomfield Road, Harpenden, AL5 4DB
G0 CPO S Norm Alfreton, c/o G. Childe, 20 Glenmore Drive, Stenson Fields, Derby, DE24 3HE
G0 CPP J. Linfoot, Flat, 10 Pembroke Court, Oxford, OX4 1BY
G0 CPR J. Stewart, 45 Dawn Crescent, Upper Beeding, Steyning, BN44 3WH
G0 CPT A. Fielder, 46 Route De Pontivy, 22570 Plelauff, Plelauff, France, 22570
G0 CPU M. Cracknell, 17 Windmill Fields, Harlow, CM17 0LQ
G0 CPV S. Pocock, 14572 West, 152Nd Place, Kansas, USA, 66062
G0 CPZ B. Adams, 85 Copperfields, Lydd, Romney Marsh, TN29 9UU
G0 CQB G. Cant, 4 The Mount, Docking, King's Lynn, PE31 8LN
G0 CQC Prof. N. Wilding, Lyncombe Ridge Lyncombe Vale Road, Bath, BA2 4LP
G0 CQD A. Cooper, 10 Buckthorne Court, East Ardsley, Wakefield, WF3 2DD
G0 CQH T. Neal, 34 Dene View, Ashington, NE63 8JF
G0 CQI G. Baker, 26 Gardeners Road, Halstead, CO9 2TB
G0 CQJ J. Kilmister, 9 Wheal Jane Meadows, Threemilestone, Truro, TR3 6EN
G0 CQK J. Coombes, 22 Chollerford Close, Gosforth, Newcastle upon Tyne, NE3 4RN
GM0 CQL P. Rudd, 41 Broadford Terrace, Broughty Ferry, Dundee, DD5 3EF
G0 CQO K. Clark, Flat 9, Middlewood Hall, Doncaster Road, Barnsley, S73 9HQ
G0 CQP Dr R. Berrisford, 19 Moorlands Drive, Mayfield, Ashbourne, DE6 2LP
GM0 CQQ A. Herring, Mountpleasant, 8 Linlithgow Road, Bo'ness, EH51 0DD
G0 CQR P. Smith, 93 Nottingham Road, Long Eaton, Nottingham, NG10 2BY
G0 CQS S. Faulkner, 96 Ashby Road East, Bretby, Burton-on-Trent, DE15 0PT
G0 CQT J. Aulsebrook, 38 South Road, Beeston, Nottingham, NG9 1LY
G0 CQU M. Dean, 3 Buxton Road West, Disley, Stockport, SK12 2AE
GM0 CQV B. Hynes, 1 Hillside Court, Hillside Place, Peterculter, AB14 0TU
G0 CQY V. Tharp, 14 Cumberland Road, Congleton, CW12 4PH
G0 CQZ N. Gardner, 9 Curbridge Road, Witney, OX28 5JT
G0 CRB A. Paddock, 15 Castle Close, Henley-in-Arden, B95 5LR
G0 CRD M. Wallis, Quernmore, Hammer Lane, Hailsham, BN27 4JL
G0 CRE A. Green, 59 London Road, Sleaford, NG34 7LQ
G0 CRF T. Bailey, 65 Edge Lane, Chorlton Cum Hardy, Manchester, M21 9JU
G0 CRJ A. Reader, 8 Daddyhole Road, Torquay, TQ1 2ED
G0 CRK D. Bowles, 9 The Meadows, Breachwood Green, Hitchin, SG4 8PR
G0 CRL K. Moate, 32 Bournewood, Hamstreet, Ashford, TN26 2HL
G0 CRN B. Hopper, 189 Western Road, Mickleover, Derby, DE3 9GT
G0 CRO R. Crow, 20 Victoria Grove, Wombourne, Wolverhampton, WV5 9AJ
G0 CRT N. Jelley, 64 Leicester Road, Broughton Astley, Leicester, LE9 6QE
G0 CRU K. Tham, Flat 4, Warwick Court, 4 Lansdowne Road, London, SW20 8AP
G0 CRX K. Pearson, Kalmia, Bishampton Road, Worcester, WR7 4BT
G0 CRY T. Sparrey, 16 Rosemary Road, Parkstone, Poole, BH12 3HB
G0 CSK P. Senior, 13 St. Michaels Avenue, Swinton, Mexborough, S64 8NX
GM0 CSN R. Trussler, 19 Royellen Avenue, Hamilton, ML3 8QH
G0 CSS H. Shillitto, 25 Commonside, Selston, Nottingham, NG16 6FN
G0 CSU T. Chadwick, 9 Ernest Street, Prestwich, Manchester, M25 3HZ
G0 CSV J. Capindale, 2 Rivan Grove, Grimsby, DN33 3BL
G0 CSW W. Hudson, 9 Nethergate, Dudley, DY3 1XW
G0 CSY W. Dunstan, 5 Twemlow Lane, Holmes Chapel, Crewe, CW4 8DT
G0 CSZ F. Dinger, 5A Knockroe, Delgany, Ireland, A63HK03
G0 CTC E. Finnesey, 17 Gilbert Close Formby, Liverpool, L37 6FA
G0 CTD A. Shipperley, 72 Hithercroft Road, Downley, High Wycombe, HP13 5RH
G0 CTF W. Sargent, Likoma, 32 Seaton Down Road, Seaton, EX12 2SB
GW0 CTG G. Darrell, 10 Clatter Brune Estate, Presteigne, LD8 2LB
G0 CTH H. Churchman, Westcroft, Church Road, Chester, CH4 9NG
GI0 CTI M. Murphy, 97 Longfield Road, Mullaghbawn, Newry, BT35 9TX
G0 CTP J. Smart, 4, Sycamore Close, Holmes Chapel, CW4 7BT
G0 CTQ I. Whiffin, 42 Canute Road, Birchington, CT7 9QH
G0 CTR P. Martin, 32 Warkworth Court, Ellesmere Port, CH65 9EN
G0 CTS R. Gumb, 17 Castle Lane Bolsover, Chesterfield, S44 6PS
G0 CTZ E. Gray, 18 Shepherds Court, Sheep House, Farnham, GU9 8LF
G0 CUA M. Dissanayake, 9 Sweyn Place, Blackheath, London, SE3 0EZ
G0 CUB G. Smith, 55 Countess Way, Euxton, Chorley, PR7 6PT
G0 CUH S. Crane, 70 Highertown, Truro, TR1 3QD

G0 CUI L. Hobson, 25 Bevan Close, Elsecar, Barnsley, S74 8DR
G0 CUL B. Chilvers, 99 Links Avenue, Hellesdon, Norwich, NR6 5PQ
G0 CUN P. Connolly, 21 Hartwood Green, Hartwood, Chorley, PR6 7BJ
G0 CUO J. Hewitt, Peel House, Sacriston Lane, Durham, DH7 6TF
G0 CUX R. Bedford, 26 Devon Avenue, Fleetwood, FY7 7EA
G0 CUZ C. Morris, 12 Turners Hill Road, Lower Gornal, Dudley, DY32JU
G0 CVA C. Andrews, 10 Hartford Road Hartley Wintney, Hook, RG27 8QW
G0 CVB K. Albon, 65 Belmont Road, Kirkby-In-Ashfield, Nottingham, NG17 9DY
G0 CVC E. Buckley, 34 Newstead Terrace, Halifax, HX1 4TA
GM0 CVD F. Nicholl, Trees, 7 Holmisdale, Isle of Skye, IV55 8WS
G0 CVH V. Hughes, 27 Billy Lane, Clifton, Manchester, M27 8FS
G0 CVI S. Khandro, 107 Castle Hill Gardens, Torrington, EX38 8EX
G0 CVM W. Auty, 22 Hurley Close, Great Sankey, Warrington, WA5 1XG
G0 CVN I. Howsham, West Street, North Kelsey, Lincoln, LN7 6EL
GM0 CVP OBO Rhosllannerchrugog Group, c/o C. Phillips, Lonnie Sanday, Orkney, KW17 2BA
GW0 CVY D. Edwards, 26 Russell Terrace, Carmarthen, SA31 1SY
G0 CWA N. Strong, 46 Malpas Drive, Great Sankey, Warrington, WA5 1HN
G0 CWB M. Starkey, 23 Arthur Street, Cannock, WS11 5HD
G0 CWD I. Donachie, 72 Gresham Road, Norwich, NR3 2NQ
G0 CWF M. Warr, 17 Moray Close, Peterlee, SR8 1DQ
G0 CWH A. Carden, Hazelgrove, South Allington, Kingsbridge, TQ7 2NB
G0 CWL D. Chapaton, 1 Rue De L'Angile, Lyon, France, 69005
G0 CWO L. Gough, 7 Congreve Road, Stoke-on-Trent, ST3 2HA
G0 CWP K. Stather, 5 St. Margarets Road, Bolton Le Sands, Carnforth, LA5 8EN
G0 CWQ A. Nesbitt, Amberley, Stokeinteignhead, Newton Abbot, TQ12 4QS
GM0 CWR W. Pentland, Cambir, Faskally, Pitlochry, PH16 5LA
G0 CWS A. Hattersley, The Bungalow, Top Lane, Buxton, SK17 8LP
G0 CWU F. Humphreys, 31 Springfield Way, Oakham, LE15 6QA
G0 CWW R. Gooden, 39 Heath Road, Ipswich, IP4 5RZ
G0 CWX M. Shotter, Peverley, Newport Road, Cowes, PO31 8PE
GW0 CWZ T. Cattley, Yew Tree Cottage, Ifton Heath, Oswestry, SY11 3DH
G0 CXD T. Moore, 13 Moss Lane, Hulland Ward, Ashbourne, DE6 3FB
G0 CXJ A. Beasley, 2 Ilmington Road, Blackwell, Shipston-on-Stour, CV36 4PG
GW0 CXK F. Johns, Manteg, Penslade, Fishguard, SA65 9PB
G0 CXO S. Wellings, 11 Matlock Road, Walsall, WS3 3QD
G0 CXU R. Pitter, 57 Greenhill Way, Farnham, GU9 8TA
G0 CXV R. Clark, Woodlands, Islet Road, Maidenhead, SL6 8HT
G0 CXW T. Pearce, 16 Beech Lodge, Rosewoodlane, Shoeburyness, SS3 9FA
G0 CXX J. Jopling, 54 Redesdale Gardens, Gateshead, NE11 9XH
GM0 CXY C. Monteith, 46 Lochryan Street, Stranraer, DG97HR
G0 CYB P. Kelsall, 200 Town Street, Middleton, Leeds, LS10 3TJ
G0 CYC R. Curtis, 125 Handside Lane, Welwyn Garden City, AL8 6TA
G0 CYD W. Snow, Willbeard Farm, Greenditch Street, Bristol, BS35 4HJ
GW0 CYG D. Davies, 40A Furzeland Drive, Neath, SA10 7UG
G0 CYI J. Knight, Urbanizacao Quinta Da Torre, Edificio Perola, Armacao de Pera, Portugal, 8365-184
GW0 CYK S. Radford, 11 South Parade, Maesteg, CF34 0AB
G0 CYN R. Blackmoore, 4 Haycocks Close Dothill Wellington, Telford, TF13NN
G0 CYO G. Beddow, 12 Wulfruna Gardens, Finchfield, Wolverhampton, WV3 9HZ
G0 CYR D. Bramley, 10 Thirlmere Close, Huncoat, Accrington, BB5 6JQ
G0 CYX J. Faulkner, 11 Valley View, South Elmsall, Pontefract, WF9 2DD
G0 CZD M. Kinder, 13 Oak Bank Close Willaston, Nantwich, CW5 7JA
G0 CZR K. Laws, 6 Crabbe'S Close Feltwell, Thetford, IP26 4BD
G0 CZU H. Richardson, 14 Melbreak Close, Whitehaven, CA28 9TG
G0 CZY P. Heath, The Cottage, Great Staughton Road, Bedford, MK44 2BA
G0 DAB D. Buik, 54A Buckshaft Road, Cinderford, GL14 3DZ
G0 DAC D. Cowley, 81 Ashtree Road, Walsall, WS3 4LS
G0 DAE C. Tidwell, 86 Powerscourt Road, North End, Portsmouth, PO2 7JW
G0 DAF J. Randall, 26 Marian Road, Boston, PE21 9HA
G0 DAH B. Smith, 1 High Street Over, Cambridge, CB24 5NB
G0 DAI D. Isom, 36 Deerfold, Astley Village, Chorley, PR7 1UH
G0 DAM R. Clayton, 171 Warning Tongue Lane, Cantley, Doncaster, DN4 6TU
G0 DAU M. Saunders, 22 Humphreys Close, St. Cleer, Liskeard, PL14 5DP
G0 DAV D. Paine, Woodland View, St. Mellion, Saltash, PL12 6RH
G0 DAX D. Burt, 19B Midhurst Road, Eastbourne, BN22 9HP
G0 DAY K. Banks, 52 Hunter Avenue, Burntwood, WS7 9AQ
G0 DAZ C. Mister, Woodbine Cottage, Hanley Childe, Tenbury Wells, WR15 8QY
G0 DBC J. Hudson, 1 Linnet Way, Biddulph, Stoke-on-Trent, ST8 7UF
G0 DBD J. West, Stonecroft, 4 Trevella Road, Bude, EX23 8NA
G0 DBE L. Marsland, 154, Moss Lane, Litherland., Liverpool, L21 7NN
G0 DBI K. Danks, 28 Warnes Lane, Burley, Ringwood, BH24 4EL
G0 DBJ G. Willetts, Waterside, 48 Stourton Crescent, Stourbridge, DY7 6RR
GM0 DBK D. Kerr, 3 Glaisnock View, Cumnock, KA18 3GA
G0 DBM S. Lovesey, 20 Ferry Gardens, Quedgeley, Gloucester, GL2 4PB
G0 DBP S. Hender, 23 Broadacres, Honley, Holmfirth, HD9 6ND
G0 DBS P. Le Feuvre, 56 Greenfields Avenue, Alton, GU34 2EE
GM0 DBW M. Bolton, 11 Covenanters Drive, Corston, Aberdeen, AB12 5AB
G0 DBX D. Beale, 17 Rue Du Passolis, Montseret, France, 11200
G0 DBY P. Dodge, 425 Sutton Road, Maidstone, ME15 8RA
G0 DCF R. Green, Kenville, West Lane, Sheffield, S26 3XS
G0 DCG M. Brown, 31 Victoria Road, Littlestone, New Romney, TN28 8NL
G0 DCI A. Merrylees, 90 Grangehill Road, Eltham, London, SE9 1SE
G0 DCJ J. Warren, The Old Barn, Scotgate Close, Thetford, IP24 1PF
G0 DCO M. Beirne, 14 Swiss Cottage, Bollinbrook Road, Macclesfield, SK10 3DJ
G0 DCP P. Holdaway, 8 Beaufort Avenue, Market Deeping, Peterborough, PE6 8JD
G0 DCR J. Frost, 36 York Gardens, Braintree, CM7 9NF
G0 DCS P. Ashton, 27 Dunsby Road, Luton, LU3 2UA
G0 DCU J. Faithfull, 54 Cardiff Place, Bassingbourn, Royston, SG8 5LR
G0 DCW A. Corallini, 8 Britannia, Puckeridge, Ware, SG11 1SY
G0 DCZ P. Richards, 16 Charles Eaton Court, Bedworth, CV12 0AX
G0 DDA P. White, 11 Elms Road, Fareham, PO16 0SQ
G0 DDE B. Dignum, 16 Stirling Court Road, Burgess Hill, RH15 0PT
G0 DDF D. Fairchild, Fairhaven, 2 Linacre Road, Torquay, TQ2 8LE
G0 DDJ A. King, 4 Tyne Close, Wellingborough, NN8 5WT
GW0 DDK E. Down, Silver Hill, Pen Y Cwm, Haverfordwest, SA62 6JZ
GW0 DDL G. Jones, 9 George St., New Quay, SA45 9QR
G0 DDT J. Barnes, 262 King Henrys Drive, New Addington, Croydon, CR0 0AA
G0 DDU J. Norris, 15 Liverpool Road North, Burscough, Ormskirk, L40 5TN
G0 DDV N. Hewett, 49 Harrow Way, Carpenders Park, Watford, WD19 5EH
G0 DDW M. Hillier, 28 Meadow Walk, Bridgemary, Gosport, PO13 0YN
G0 DDY P. Piper, 5 Goodwood Close, Midhurst, GU29 9JG
G0 DDZ M. Eastman, 23 Haughgate Close, Woodbridge, IP12 1LQ
G0 DEB D. Bannister, 60 St. Johns Avenue, Bridlington, YO16 4NL
G0 DEC D. Willicombe, 26 Falkland Court, Braintree, CM7 9LL
G0 DEE R. Spilling, 20 Saxonfields, Poringland, Norwich, NR14 7JE
G0 DEF M. Mutton, 39 Martin Road, Kettering, NN15 6HF
G0 DEH P. Finbow, 6 Down Road, Teddington, TW11 9HA
G0 DEJ W. Rutt, 24 Coopers Lane, Verwood, BH31 7PG
G0 DEK F. Underwood, Hobletts, Fen Lane, Grays, RM16 3LT
GM0 DEM A. Jones, 9 Firs Street, Falkirk, FK2 7AY
G0 DEO W. Batey, 13 Cassiobury Avenue, Feltham, TW14 9JE
G0 DEP D. Jarrard, 26 Lingmell Court, Tolladine, Worcester, WR4 9YU
GM0 DEQ R. Alexander, 9 Weston Place, Prestwick, KA9 2ED
G0 DER D. Gillmore, 4 Holly Ridge, Fenns Lane, Woking, GU24 9QE
G0 DEU J. Tournant, 47 High St., Linton, Cambridge, CB1 6HS
GM0 DEX A. Goldie, 87 Ardrossan Road, Seamill, West Kilbride, KA23 9NF
G0 DEZ D. Watson, 1 Castle House Drive, Stafford, ST16 1DS
G0 DFA D. Allsopp, 10 Chalfont Close, Middleton-On-Sea, Bognor Regis, PO22 7SL
G0 DFC L. Cropley, San Ferryann, Brundish Road, Wilby, Eye, IP21 5LS
GI0 DFD R. Mcalister, 78 Cairn Road, Carrickfergus, BT38 9AP
G0 DFE J. Stone, 12 Main Road, Hawkwell, Hockley, SS5 4JN
G0 DFF J. Sidnell, 17 Barlings Road, Harpenden, AL5 2AL
G0 DFI D. Oakley, 6 Staplehurst Gardens, Cliftonville, Margate, CT9 3JB
G0 DFO J. Tomlinson, 33 Belgrave St., Nelson, BB9 9HR
G0 DFT J. Maw, 10 Shamrock Close, Newcastle upon Tyne, NE15 8TW
G0 DFV R. Cadd, 27 Grove Road, Brafield On The Green, Northampton, NN7 1BW
GW0 DFY R. Anthony, 2 Barrfield Road, Rhuddlan, Rhyl, LL18 2RY
G0 DGA D. Gullick, Greenleas, Courthay Orchard, Langport, TA10 9AE
G0 DGB L. Connell, 24 Finchale Road, Framwellgate Moor, Durham, DH1 5JN
G0 DGE J. Gullick, Greenleas, Courthay Orchard, Langport, TA10 9AE
G0 DGF M. Beakhust, 63 Chadacre Road, Epsom, KT17 2HD
G0 DGH G. Daniels, 81 London Road, Clacton-on-Sea, CO15 3SR
GW0 DGJ C. Riddle, 18 Windsor Mews, Adamsdown Square, Cardiff, CF24 0HS
GM0 DGK A. Donaldson, 30 Jeanfield Crescent, Forfar, DD8 1JR
G0 DGQ G. Dudley, 95 Alfreton Road, South Normanton, Alfreton, DE55 2BJ
G0 DGU P. Brown, 17 Freyden Way, Frettenham, Norwich, NR12 7NB
G0 DGW D. Barham, 64 Gorran Avenue, Rowner, Gosport, PO13 0NF
GW0 DHA R. Waller, 4 Rose Court, Ty Canol, Cwmbran, NP44 6JH
G0 DHB R. Evans, 20 Pulley Avenue, Eaton Bishop, Hereford, HR2 9QN
GM0 DHD A. Lymer, 16 Gerson Park, Greendykes Road, Broxburn, EH52 6PL
GW0 DHG B. Ristic, 168 Heather Road, Newport, NP19 7QW
G0 DHI A. Rutherford, 19 Briar Bank, Carlisle, CA3 9SN
G0 DHJ J. Wraight, 59 Sandy Lane, Walton, Liverpool, L9 9AY
G0 DHL M. Mason, 7 Clayhill Copse Peatmoor, Swindon, SN5 5AL
G0 DHM D. Moore, Stoke Hall Farm, Stoke On Tern, Market Drayton, TF9 2DU
G0 DHR Dr H. Rutt, 3 Russell Place, Highfield, Southampton, SO17 1NU
G0 DHS A. Kitching, 1 Borrowdale, Albany, Washington, NE37 1QD
G0 DHT K. Smith, Lower Carniggey Farm, Greenbottom, Truro, TR4 8QL
GI0 DHW C. Cleland, 19 Sheskin Way, Belfast, BT6 0ER
G0 DHZ H. Johnsen, 47 Bondfields Cres, Havant, PO9 5ER
G0 DIA A. Holland, 156 Perry Rise, London, SE23 2QP
G0 DIG P. Johnson, 5 Brook Bank, Whitehaven, CA28 8PZ
G0 DIH P. Strong, 2 Jasper Cottages, Cornworthy, Totnes, TQ9 7EY
G0 DIM A. Latham, 49 Tithe Barn Road, Wootton, Bedford, MK43 9EZ
G0 DIP J. Huggins, 7 Coniston Drive, Jarrow, NE32 4AE
GW0 DIQ M. Smith, 7 Clos Gorsfawr, Grovesend, Swansea, SA4 4GZ
G0 DIR S. Caslake, Bishopwood Cottage, Wistow Common, Selby, YO8 3RD
G0 DIS P. Pearce, 42 Sinclair Garth, Wakefield, WF2 6RE
G0 DIU T. Rumble, 1 Victoria St., Brighouse, HD6 1HH
GW0 DIV R. Griffiths, 5 Heol-Y-Sarn, Llantrisant, Pontyclun, CF72 8DA
GW0 DIX R. Rees, 22 The Complex, Tan Y Bryn, Burry Port, SA16 0HP
G0 DIY R. Allen, 65, Fitzroy Drive, Leeds, LS8 4AG
G0 DIZ R. Quaintance, 18 Queens Avenue, Ilfracombe, EX34 9LN
G0 DJA D. Ackrill, 59 Moor Lane, Bolsover, Chesterfield, S44 6EW
G0 DJC D. Collins, 71 Trench Road, Tonbridge, TN10 3HG
GM0 DJG J. Walker, Mo Bhothan, Lochlibo Road, Uplawmoor, Glasgow, G78 4AA
G0 DJK D. Keates, 13 Willow Rise, Witham, CM8 2LL
G0 DJL C. Blount, 42 Penmere Drive, Newquay, TR7 1QQ
G0 DJM M. Rawson, 4 Fleet Lane, Tockwith, Yo26 7qd
G0 DJO A. Robson, 19 Barnard Close, Bedlington, NE22 6NE
G0 DJQ R. Salt, Croft View, Little Cubley Cubley, Ashbourne, DE6 2FB
G0 DJS H. Stemp, 5 Depot Road, Horsham, RH13 5HB
G0 DJT P. Odegaard, Flat 4, 31 Birdhurst Road, South Croydon, CR2 7EF
GW0 DJU D. Lee, 7 Redwood Close, Neath, SA10 7US
G0 DJV B. Joy, 1 Riverbourne Road, Salisbury, SP1 1NU
GW0 DJX A. Cullen, 30 Llys Cyncoed, Oakdale, Blackwood, NP12 0NQ
GW0 DKF R. Thomas, 48 Maryport Street, Usk, NP15 1AD
GW0 DKG R. Malpas, 26 St. Davids Avenue, Whitland, SA34 0AF
GM0 DKK A. Mackenzie, 33 Castle Crescent, Denny, FK6 6PN
G0 DKM S. Daniels, 3 Warren Close, Hutton, Weston-Super-Mare, BS24 9QX
G0 DKN A. Mcclelland, 12 Grove Heath North, Ripley, Woking, GU23 6EN
G0 DKO G. Maskort, 133 Borstal St., Rochester, ME1 3JU
G0 DKR M. Cole, 54 Ribble Road, Coventry, CV3 1AU
G0 DKS R. Barnes, Pentwyn, Graeme Road, Yarmouth, PO41 0RX
G0 DKV E. Peberdy, 18 Arden Road, Kenilworth, CV8 2DU
G0 DKX G. Birk, 30 Maple Drive, Alvaston, Derby, DE24 0FT
G0 DKY J. Bass, 8 Ann Close, Hassocks, BN6 8NB
G0 DKZ B. Yates, 9 Cloister Walk, Whittington, Lichfield, WS14 9LN
GW0 DLA E. Stuckey, 32 Stanley Road, Gelli, Pentre, CF41 7NJ
G0 DLB R. Burdett, 11 Fisher Avenue, Rugby, CV22 5HN
G0 DLF A. Keon, 72 Rochelle Way, Duston, Northampton, NN5 6YW
G0 DLL W. Jackson, 22 Cliff Gardens, Scunthorpe, DN15 7PJ
G0 DLP P. Lee, 14 Downs Court Road, Purley, CR8 1BB
G0 DLS D. Sparey, 21 Buxton Road, Ashbourne, DE6 1EX
G0 DLT J. Aizlewood, 7 Scott Avenue, Simonstone, Burnley, BB12 7HY

GW0 DLW J. Goldsmith, Woodlands, Candy, Oswestry, SY10 9AZ
G0 DMA J. Slaney, 50 Laburnum Road, Langold, Worksop, S81 9RR
G0 DMB C. Cade, 6 Court Close, Kirby Muxloe, Leicester, LE9 2DD
G0 DME J. Hallam, 34 Danethorpe Vale, Nottingham, NG5 3DA
G0 DMH B. Cox, 21 Shelthorpe Road, Loughborough, LE11 2PB
G0 DMJ N. Beck, 36 Grove Street, Great Hale, Sleaford, NG34 9JZ
G0 DMK D. King, 94 Western Road, Mickleover, Derby, DE3 9GQ
G0 DMN S. Langham, 43 Greenwood Drive, Kirkby-In-Ashfield, Nottingham, NG17 8JT
G0 DMP D. Potter, 102 Normandy Avenue, Beverley, HU17 8PF
G0 DMS F. Spencer, 35 Askew Grove, Repton, Derby, DE65 6GR
G0 DMU I. Morison, 4 Arley Close, Macclesfield, SK11 8QP
G0 DMV R. Parrish, 89 Delamere Drive, Macclesfield, SK10 2PS
G0 DMW F. Cosgrove, Denton Park Middle School, Linhope Road, Newcastle upon Tyne, NE5 2NW
G0 DND N. Downs, 10 Oak Street, Northwich, CW9 5LJ
G0 DNF D. Fisher, 69 Priors Orchard, Southbourne, Emsworth, PO10 8GE
GM0 DNG G. Wallace, 21 The Grange, Perceton, Irvine, KA11 2EU
GM0 DNH P. Moore, 105 Fintry Drive, Dundee, DD4 9HQ
G0 DNI G. Wann, Manor Barn, Shotatton, Shrewsbury, SY4 1JH
G0 DNQ C. Anderton, 19 Berrylands Close, Wirral, CH46 7UT
G0 DNV R. Hammett, 47 Sowden Park, Barnstaple, EX32 8EJ
G0 DNY B. Whysker, 21 Heyland Road, Manchester, M23 1HF
G0 DOA C. Dale, 11 Roman Close Newby, Scarborough, YO12 5RG
G0 DOB R. Gray, 3 Perth Way, Immingham, DN40 1PW
G0 DOC H. Kiff, Rock Cottage, Cloudside, Congleton, CW12 3QG
G0 DOE T. Purcell, 38A Moor Lane, Chessington, KT9 1BW
G0 DOG C. Purcell, 76 Wensleydale Road, Great Barr, Birmingham, B42 1PL
G0 DOK R. Cornish, 18 Rooksbury Croft, Havant, PO9 5HU
G0 DOM D. Oskis, 10 Moultrie Way, Cranham, Upminster, RM14 1NB
G0 DOR P. Davidson, 5 Derby Grove, Maghull, Liverpool, L31 5JJ
G0 DOU H. Grandfield, 2 Bolshaw Road, Heald Green, Cheadle, SK8 3PJ
G0 DOZ Dr J. Rozday, 130 Walton Park, Pannal, Harrogate, HG3 1RJ
G0 DPC J. Cook, 88 South Avenue, Southend on Sea, SS2 4HU
G0 DPE B. Aldersey, 4 Salterbeck Terrace, Salterbeck, Workington, CA14 5HP
G0 DPG D. Ganner, 58 Kilngate, Lostock Hall, Preston, PR5 5UW
G0 DPI D. Barkley, 39 Fulbeck Avenue, Wigan, WN3 5QN
G0 DPJ K. Ellis, 7 Beech Close, Dudley, DY3 1NG
G0 DPK P. Mccaldon, 47 Merritt Road, Didcot, OX11 7DF
G0 DPO K. Glazebrook, 86 Deveraux Drive, Wallasey, CH44 4DL
G0 DPQ A. Henning, 24 Garfield Avenue, Draycott, Derby, DE72 3NP
G0 DPS J. Fyrth, 2 Merton Gardens, Farsley, Pudsey, LS28 5DZ
G0 DPT M. Smith, 32 Amesbury Drive, London, E4 7PZ
GI0 DPV J. Mangan, 141 Glen Road, Andersonstown, Belfast, BT11 8BP
G0 DPW D. Welham, 1 Torridge Road Keynsham, Bristol, BS31 1QG
G0 DPX J. Brown, 72 Whitcliffe Road, Cleckheaton, BD19 3BY
G0 DPY A. Garner, 7 Danes Court, Grimoldby, Louth, LN11 8TA
G0 DQB R. White, 27 Windsor Walk Scawsby, Doncaster, DN5 8NQ
GM0 DQC H. Meikle, 20 Muirsland Place, Lesmahagow, Lanark, ML11 0FF
G0 DQH J. Colwill, 18 Collingbourne Drive, Chandler'S Ford, Eastleigh, SO53 4SW
G0 DQI D. Harding, High Peak, Hillcrest Road, Deal, CT14 8EB
GI0 DQJ D. Livingstone, 16 Stronge Court, Portadown, Craigavon, BT62 3QX
G0 DQM J. Williams, Maes-Yr-Awel, Suttonfield Road, Doncaster, DN6 9JX
G0 DQO W. Cartwright, 50 Kings Road, Walsall, WS4 1JB
G0 DQQ S. Power, 10 Beach Road, Hartford, Northwich, CW8 4BA
G0 DQS M. Glen, 10 Field Lane, Dursley, GL11 6JE
GW0 DQT A. Duck, 15 Ambryn Road, New Inn, Pontypool, NP4 0NJ
GM0 DQV G. George, 13 Balmoral Terrace, Elgin, IV30 4JH
GW0 DQW C. Hughes, Cym Lane, Rogerstone, Newport, NP10 9EN
GW0 DQY G. Hughes, 39 Thornhill Close, Upper Cwmbran, Cwmbran, NP44 5TQ
G0 DRA D. Love, 4 St. Chads Road, Lichfield, WS13 7LZ
G0 DRD S. Carvin, 43 Brackenstown Village, Dublin, Ireland,
G0 DRE J. Webster, 3 Badby Road West, Daventry, NN11 4HJ
G0 DRH B. Harris, 9 Woodlands Close, Rayleigh, SS6 7RG
GW0 DRI J. Smith, 7 Clos Gorsfawr, Grovesend, Swansea, SA4 4GZ
G0 DRJ J. Connell, 20 Finchale Road, Framwellgate Moor, Durham, DH1 5JN
G0 DRK K. Fox, 54 Tuskar Street, London, SE10 9UZ
G0 DRL G. Howarth, 15C Shaftesbury Road, Southsea, PO5 3JA
G0 DRM D. Cookson, 70 Rope Lane, Wistaston, Crewe, CW2 6RD
G0 DRN M. Jenkin, 3A Westminster Street, Crewe, CW2 7LQ
G0 DRO D. Roberts, Flat 1, 129 Prestbury Road, Macclesfield, SK10 3DA
G0 DRQ R. Hope, 3 Farm Crescent, Sittingbourne, ME10 4QD
G0 DRR O. Perry, 9 Home Park Close, Bramley, Guildford, GU5 0JP
GW0 DRS R. Ford, 11 Lincoln Road, Ewloe, Deeside, CH5 3RW
G0 DRT P. Quested, Nethercroft, Southsea Avenue, Minster On Sea, Minster on Sea Sheerness, ME12 2NH
GM0 DRU I. Maclennan, 70 Kenneth St., Stornoway, HS1 2DS
G0 DRV W. Hoyle, 10 Picton Gardens, Rayleigh, SS6 7LB
G0 DRW T. Wright, 73 West Street, Ryde, PO33 2QQ
G0 DRX P. Hill, 34 Church Road, Whitchurch, Bristol, BS14 0PP
G0 DSB T. Winship, 32 Lytes Cary Road, Keynsham, Bristol, BS31 1XD
GI0 DSG W. Mckeever, 17 The Hawthornes, Londonderry, BT48 8TH
GW0 DSJ E. Shipton, 51 Maes Stanley, Bodelwyddan, Rhyl, LL18 5TL
G0 DSN L. Nash, Four Furlongs, Wells Road, Wells-Next-the-Sea, NR23 1QE
G0 DSO R. Calvert, Shortley Close, Robin Hoods Bay, Nr Whitby, YO22 4PB
GW0 DSP M. Lamb, 4 Hadfield Close, Connah'S Quay, Deeside, CH5 4JP
G0 DSQ Dr K. Myint, 1 Belton Road, Camberley, GU15 2DE
G0 DSR D. Jones, 20 Marsh Green, Wigan, WN5 0PU
G0 DSX N. Hanking, 7 Clayside House, Kenton Court, South Shields, NE33 4HP
G0 DTC Dr D. Coxon, 13 Gate Farm Road, Shotley Gate, Ipswich, IP9 1QH
G0 DTI K. Sumner, 7 Largs Road, Shadsworth, Blackburn, BB1 2JQ
G0 DTP L. Quantrill, Innisfree, 1 Ironwell Lane, Hockley, SS5 4JY
G0 DTQ W. Goldstraw, 5 Council Houses, Wantage Road, Hungerford, RG17 7DG
G0 DTT R. Mycock, Beresford Crescent, Stockport, SK56NU
G0 DTW S. Bolton, 40 Claytonwood Road, Stoke-on-Trent, ST4 6LD
G0 DUA S. Linden, 4 Downing Drive, Great Barton, Bury St Edmunds, IP31 2RP
G0 DUB F. Mossop, 4 Brookdale Way, Waverton, Chester, CH3 7NT
G0 DUE A. Braybrook, 4 Bodinar Road, Penryn, TR10 8JD
G0 DUF R. Hoare, 7 Springfield Close, Watlington, OX49 5RF
G0 DUG D. Marsden, 127 Morley Crescent, Kelloe, Durham, DH6 4NP
G0 DUH P. Murrell, 10 Irving Close, Braunton, EX33 1DH
G0 DUI P. Smith, 20 Deanscroft Way, Stoke-on-Trent, ST3 5XW
G0 DUK K. Bennett, 78 Rectory Road, Upper Deal, Deal, CT14 9NB
G0 DUM D. Gibbons, 17 Della Avenue, Barnsley, S70 6LG
G0 DUN D. Wakeford, 2 Rooley House Cottage, Rooley Lane, Sowerby Bridge, HX6 1NS
GI0 DUP R. Miskimmin, 15 Abbeydale Avenue, Newtownards, BT23 8RT
G0 DUQ R. Wilkes, 47 Greenwood Park, Hednesford, Cannock, WS12 4DQ
G0 DUS M. Marlow, 56 Harvest House, Cobbold Road, Felixstowe, IP11 7SP
GM0 DUX R. Mcgowan, 4 The Cottages Ashfield, Dunblane, FK15 0JS
G0 DVB J. Morris, 18 Ellingdon Road, Wroughton, Swindon, SN4 9HY
G0 DVC P. Robinson, 256 Victoria Road, Ruislip, HA4 0DW
G0 DVE S. Hutchings, 5 Dales Close, Wimborne, BH21 2JU
G0 DVG T. Callaghan, 27 Thealby Lane, Thealby, Scunthorpe, DN15 9AG
GM0 DVH N. Mcnulty, 6 Main Road, Crookedholm, Kilmarnock, KA3 6JT
G0 DVJ J. Mitchener, Cabins, Wenham Road, Ipswich, IP8 3EY
G0 DVL R. Nash, 28 Squires Way, Wilmington, Dartford, DA2 7NW
GM0 DVO A. Gemmell, 23 Busby Road, Carmunnock, Glasgow, G76 9BN
G0 DVP G. Gulliford, 29 Windsor Road, Seaham, SR7 8DG
G0 DVQ G. De-Wilton Homes, 81 Oriel Avenue, Gorleston, Great Yarmouth, NR31 7JJ
G0 DVS E. Holding, 31 Popple Street, Sheffield, S4 8JH
G0 DVT J. Brindle, 1 Holywell Close, Bury St Edmunds, IP33 2LS
GI0 DVU J. Henry, 3 Kirkwoods Park, Lisburn, BT28 3RR
G0 DVY R. Ibbotson, Fern Lea, Alford, LN13 0JP
G0 DWB D. Wathen-Blower, 61 Dykes End, Collingham, Newark, NG23 7LD
G0 DWC S. Beadle, 18 The Shrubberies, Cliffe, Selby, YO8 6PW
G0 DWD J. Hawes, Cherry Lodge, Woodwaye, Reading, RG5 3HA
G0 DWE C. Brown, 12 Forest Close, Newport, PO30 5SF
G0 DWF D. Fouche, 17 Burlington Gardens, Rainham, Gillingham, ME8 8TA
GM0 DWH J. Cobley, 15 Fintry Terrace, Bourtreehill South, Irvine, KA11 1JD
G0 DWJ N. Hall, 10 Newnham Road, Leamington Spa, CV32 7SN
G0 DWM C. Pearsons, 12 Abbey Way, Farnborough, GU14 7DA
GI0 DWN M. Dougan, 97 Redrock Road, Collone, Armagh, BT60 2BN
G0 DWO P. Labron, 22 Fourth Avenue, Morpeth, NE61 2HJ
GW0 DWQ S. Outen, 2 Heol Vaughan, Burry Port, SA16 0HF
G0 DWR P. Bell, 5 Sages Lane, Privett, Alton, GU34 3NP
G0 DWS J. Tubbs, 19 Greenhill Road, Northfleet, Gravesend, DA11 7EZ
G0 DWV C. Danby, Fir Trees, Hall Road, Norwich, NR10 3LX
GM0 DWY R. Price, 80 Eastern Avenue, Largs, KA30 9EQ
G0 DWZ M. Sharp, 50 Milton Drive, Southwick, Brighton, BN42 4NE
G0 DX Burnham Beeches Radio Club, c/o U. Grunewald, Nuptown Orchard, Nuptown, Warfield, Bracknell, RG42 6HU
GM0 DXB W. Mcgill, 112 West Main Street, Armadale, Bathgate, EH48 3JB
GM0 DXE H. Quin, Flat/Edinbane Shop, Edinbane Shop, Portree, IV51 9PW
GW0 DXG Cwmbran Contest & DX Group, c/o D. Stoole, Brookside Farm, Baltic Terrace, Cwmbran, NP44 7AH
G0 DXH D X Hunter, c/o C. Mortlock, 27 Baldwin Road, Greatstone, New Romney, TN28 8SY
GM0 DXI H. Lakhaney, 5 Snowberry Fields, Thankerton, Biggar, ML12 6RJ
G0 DXK M. Bedford, 12 Winchester Drive, Mablethorpe, LN12 2AY
GW0 DXO D. Oates, 86 Queens Avenue, Maesgeirchen, Bangor, LL57 1NG
G0 DXT T. Pearson, 7 Lathkill Grove, Tibshelf, Alfreton, DE55 5LU
GU0 DXX P. Guilbert, Chinq, La Ferme Es Frases, Rue Des Issues, Guernsey, GY7 9FS
GW0 DXZ G. Stephens, Ty Coch, Rhydwen Place, Clydach, SA6 5RN
GM0 DYD D. Young, 4 Primrose Avenue, Rosyth, Dunfermline, KY11 2SS
G0 DYG D. Doyle, 40 Howson St., Rock Ferry, Birkenhead, L42 2BR
GW0 DYH S. Fergusson, 37 Station Road, Old Colwyn, Colwyn Bay, LL29 9EL
G0 DYL C. Jones, 63 Hockenhull Avenue, Tarvin, Chester, CH3 8LR
G0 DYM M. Rivers, 5 Ann Carter Close, Hereford, HR2 7LS
GM0 DYU M. Mcwhinnie, 35 Morrison Place Cruden Bay, Peterhead, AB423HZ
G0 DYW I. Dowse, 57 Palmer Crescent, Leighton Buzzard, LU7 4HY
G0 DZA P. Wentworth, 46 Woodside Avenue, Cinderford, GL14 2DW
G0 DZB P. Onion, 18 The Maples, Bedford, MK42 7JX
G0 DZC C. Cosgrif, 53 Lower Manor Lane, Burnley, BB12 0EF
G0 DZH D. Gough, 20 Lawn Close, Ruislip, HA4 6ED
G0 DZI C. Kuss, 20 Windermere Road, Haydock, St Helens, WA11 0ES
GW0 DZL J. Davies, 5 Talbot St., Llanelli, SA15 1DG
G0 DZM P. Gainey, Prencott, Harley Wood, Stroud, GL6 0LD
G0 DZQ B. Stevens, 24 Waverley Crescent, Wickford, SS11 7LN
GW0 DZU P. Barker, 2 The Uplands Pontrhydyfen Port Talbot, Port Talbot, SA12 9TG
G0 DZV K. Rose, 57 Cheriton Road, Winchester, SO22 5AX
GM0 DZW R. Webster, Tigh-Na-Darroch, Old Line Road, Aberdeenshire, AB35 5UT
G0 DZX J. Huggins, 12 Willow Chase, North Anston, Sheffield, S25 4DQ
G0 DZY B. Sparrow, 11B Croft Place, Mildenhall, Bury St Edmunds, IP28 7LN
G0 DZZ J. Hall, 4 Dorking Crescent, Clacton-on-Sea, CO16 8FQ
G0 EAE M. Taylor, 26 St. Marys Road, Bozeat, Wellingborough, NN29 7JU
G0 EAG A. Sammons, 35 Fernlea Avenue, Herne Bay, CT6 8UQ
GM0 EAH A. Mcdougall, 16 Rotherwood Avenue, Glasgow, G13 2RJ
G0 EAM A. Moore, 69 Renfrew Avenue, St Helens, WA11 9RW
G0 EAN C. Bibb, 54 Dorsett Road, Wednesbury, WS10 0JF
G0 EAT S. Anderson, 65 Sands Lane, Holme-On-Spalding-Moor, York, YO43 4HJ
G0 EAU A. Surooprajally, 26 Walton Avenue, North Cheam, Sutton, SM3 9UB
GW0 EAW J. Humphries, 19 Tai Newydd, Llanfaelog, Ty Croes, LL63 5TW
G0 EBD M. Element, 9 Longbridge Close, Shrewsbury, SY2 5YD
G0 EBF L. Taylor, 18 Manifold Road, Eastbourne, BN22 8EH
G0 EBG R. Fuller, 41 Burnham Road, Hullbridge, Hockley, SS5 6BG
G0 EBI J. Speller, 43 Castle Hill Park London Road, Clacton-on-Sea, CO16 9QP
G0 EBL K. Mayes, The Stone House, Goathland, Whitby, YO22 5AN
G0 EBP A. Bowmaker, 1 Hestham Drive, Morecambe, LA4 4QD
G0 EBQ N. Flatman, 2 Deben Valley Drive, Kesgrave, Ipswich, IP5 2FB
G0 EBS S. Artus, 14 Bramley Road, East Peckham, Tonbridge, TN12 5BW
G0 EBW D. Agar, 122 Salisbury Road, Moseley, Birmingham, B13 8JZ
G0 EBZ M. Whitfield, Camp Farm, Elberton, Bristol, BS35 4AQ
G0 ECB M. Hayhurst, 3 Burton Gardens, Brierfield, Nelson, BB9 5DR
G0 ECG M. Higgin, 24 Tiverton Drive, Briercliffe, Burnley, BB10 2JT

G0 ECI R. Matthews, Sunrise, Back Bank, Whaplode Drove, Spalding, PE12 0TT
G0 ECJ H. Hughes, Asham, Walton Hill, Gloucester, GL19 4BT
G0 ECK P. Jenkins, 49 Ewell Park Way, Ewell, Epsom, KT17 2NW
G0 ECL K. Clift, 28 Redgate Road, Girton, Cambridge, CB3 0PP
G0 ECM M. Bell, 18 Linnet Close, Patchway, Bristol, BS34 5RN
G0 ECN K. Watkinson, Sunnyview, Beacon Way, Skegness, PE25 1HL
G0 ECQ C. Quinnin, 10 Willow Avenue, Blyth, NE24 1PG
G0 ECS G. Westaby, 2 Goodwood, Bottesford, Scunthorpe, DN17 2TP
GM0 ECU R. Low, 56 George Street, Whithorn, Newton Stewart, DG8 8NZ
G0 ECW E. Wilson, 20 Wivelsfield Road, Saltdean, Brighton, BN2 8FQ
G0 ECX R. Mott, 2 Dennis Road, Weymouth, DT4 0NJ
G0 ECZ B. Clarke, 4 Prospect Road, Langford, Biggleswade, SG18 9NY
G0 EDC B. Hall, 7 Ferndale Close, Penyffordd, Chester, CH4 0NH
G0 EDE D. Proctor, 7 Main Avenue, Westhill, Torquay, TQ1 4HZ
G0 EDF G. Lunt, 45 Malvern Road, Liverpool, L6 6BN
GM0 EDJ P. Temple, 23 Ramsay Place, Johnstone, PA5 0EX
G0 EDK A. Lee, 44 Lynn Road, North Shields, NE29 8HS
G0 EDO R. Ormond, 75 Desford Road, Newbold Verdon, Leicester, LE9 9LG
GM0 EDQ J. Shaw, 28 Drumcross Road, Bathgate, EH48 4HG
GM0 EDR J. Bell, 52 Turnberry Road, Glasgow, G11 5AP
G0 EDS Eastbourne District Scouts ARC, c/o A. Seabrook, 63 St. Annes Road Willingdon, Eastbourne, BN20 9NJ
G0 EDT J. Hopwood, 53 St. Marys Road, Stratford-upon-Avon, CV37 6XG
G0 EDU P. Martin, 20 Eastbourne Terrace, Westward Ho, Bideford, EX39 1HG
G0 EDY P. Gould, 53 Green Road, Kidlington, OX5 2EU
G0 EEA G. Reffell, 26 Barnwood Road, Gloucester, GL2 0RX
G0 EEF G. Legg, 12 Churchill Road, Wimborne, BH21 2AU
GM0 EEG D. Mctaggart, 65 Oronsay Road, Airdrie, ML6 8FX
GM0 EEH J. Hunter, 2 Hagen Drive, Motherwell, ML1 5RZ
G0 EEI A. Jones, 18 Manor Road North, Nantwich, CW5 5NW
G0 EEJ A. Mitchell, 18 Burnards Court, Berrycombe Road, Bodmin, PL31 2NU
GI0 EEO J. Ryan, 11 Carnesure Heights, Comber, Newtownards, BT23 5RN
G0 EET B. Henderson, 34 Paget House Grove Place Upton Lane Nursling, Southampton, SO16 0AQ
GM0 EEY D. Smith, 23 Torness, Kirkwall, KW15 1UU
G0 EEZ C. Wright, 60 Grove Crescent, Hanworth, Feltham, TW13 6LZ
G0 EFA C. Mansfield, 10 Priory Drive, Abbey Wood, London, SE2 0PP
GM0 EFC G. Perry, 14B Meadowfoot Road, West Kilbride, KA23 9BX
GM0 EFD C. Perry, 14B Meadowfoot Road, West Kilbride, KA23 9BX
G0 EFG B. Balmer, 13 Chillingham Crescent, Ashington, NE63 8BQ
GM0 EFH A. Buchan, Flat 7, 1 Castlebank Court, Glasgow, G13 2LA
G0 EFI V. Newman, 14 Hilltop Close, Rayleigh, SS6 7TD
G0 EFL L. Cohen, 6 Branksome Walk Manor Branksomewood Road, Fleet, GU51 4SW
G0 EFN H. Dunne, 19 Bute Brae, Bletchley, Milton Keynes, MK3 7TA
G0 EFO M. Shortland, 4 Hillier Road, Guildford, GU1 2JQ
G0 EFP N. Hughes, 43A Wellhouse Road, Beech, Alton, GU34 4AQ
GM0 EFQ H. Fisher, 1 Millhill Lane, Musselburgh, EH21 7RD
G0 EFR L. Wheeler, 29 Belmont Park, Pensilva, Liskeard, PL14 5QT
G0 EFS P. Hancock, 7 Carlton Avenue, Hayes, UB3 4AD
GM0 EFT R. Neilson, 54 Macdonald Smith Drive, Carnoustie, DD7 7TB
GI0 EFW J. O'Hara, 284 Foreglen Road, Dungiven, Londonderry, BT47 4PJ
G0 EFY B. Warman, 177 Scotter Road, Scunthorpe, DN15 8AU
G0 EFZ I. Gerrard, 7 High Street, Wootton, Wootton, Ulceby, DN39 6SG
G0 EGC W. Stormont, 3 Bridge Cottages, Greenham, Crewkerne, TA18 8QE
G0 EGE W. Trotter, Bungalow, Stoupe Cross Farm, Whitby, YO22 4JU
G0 EGG D. Jackson, 41 Colman Avenue, Wolverhampton, WV11 3RT
GW0 EGH D. Taylor, 81 Goldfinch Close, Caldicot, NP26 5BW
GM0 EGI B. Devlin, 112 Benview, Bannockburn, Stirling, FK7 0HJ
G0 EGP D. Williamson, Wrybourne Lodge, 200 Bushbury Road, Wolverhampton, WV10 0NA
GW0 EGQ G. Peters, 5 Roman Way, Buckley, CH7 2EQ
G0 EGR C. Davis, 49 Brackendale Road, Bournemouth, BH8 9HY
G0 EGT G. Pitts, 119 Rusper Road, Crawley, RH11 0HW
G0 EGW V. Readhead, White Lodge, Rendham Road, Saxmundham, IP17 2AA
GW0 EHA I. Pemberton, 26 Stanley Grove, Ruabon, Wrexham, LL14 6AH
G0 EHE B. Evans, 27 Mulso Road, Finedon, Wellingborough, NN9 5DP
G0 EHK G. Cheetham, 172A Hesketh Lane, Tarleton, Preston, PR4 6AT
GM0 EHL B. Armstrong, 31 Old Abbey Road, North Berwick, EH39 4BP
G0 EHO R. Mellor11, 1 The Square, Lybury Lane, Redbourn, St Albans, AL3 7JB
G0 EHQ F. Skinner, Halfway Lock Cottage Upper Gambolds Lane Stoke Prior, Bromsgrove, B60 3HB
G0 EHR M. Clark, 60A Clatterford Road, Newport, PO30 1PA
GW0 EHS F. Fennah, 7 Y Ddol, Llanbrynmair, SY19 7DJ
G0 EHT J. Tommey, The Birches, Upton Bishop, Ross-on-Wye, HR9 7UF
G0 EHV E. Ashburner, 8 Shellbark, Houghton le Spring, DH4 7TD
G0 EHW B. Coles, 32 Victoria Street, Lostock Hall, Preston, PR5 5RA
G0 EHX T. Elliott, 18 Hollinside Square, Sunderland, SR4 8AU
G0 EIB L. Allen, 28 Clarence Place, Maltby, Rotherham, S66 7HA
G0 EIF P. Massheder, 5 Hazel Close, Penwortham, Preston, PR1 0YE
G0 EIG M. Holtham, 37 Higher Efford Road, Plymouth, PL3 6LD
G0 EIH T. Briley, 9 Wheatfield Way, Cranbrook, TN17 3LS
G0 EIM M. Heyes, 11 Beech Close, Isleham, Ely, CB7 5UU
G0 EIQ M. Richards, 3 Derwent Crescent, Whetstone, London, N20 0QN
G0 EIR G. Kemp, 27 Shady Grove, Alsager, Stoke-on-Trent, ST7 2NQ
GM0 EIT A. George, 7 Mid St., Keith, AB55 5AG
G0 EIY S. Pryce, 40 Gains Avenue, Bicton Heath, Shrewsbury, SY3 5AN
G0 EIZ W. Kenyon, Flat 21 House 4, Copper Place, Manchester, M14 7FZ
G0 EJD M. Moss, 1 Orchard Rise, Beckingham, DN10 4NG
GW0 EJE Pembs Radio Society, c/o E. Hollowell, 54 Portfield, Haverfordwest, SA61 1BW
G0 EJI L. Garden, 9 Gateway Avenue, Smithville, Canada, L0R 2A0
G0 EJO G. Nock, 20 Chigwell Road, Bournemouth, BH8 9HW
G0 EJR A. Love, 15 Mountain Ash, Weston Park, Bath, BA1 2UU
GI0 EJT S. Rafferty, 81 Mullaghmore Drive, Omagh, BT79 7PQ
GI0 EJU V. Hutchinson, 10 Golan Road, Knockmoyle, Omagh, BT79 7TJ
G0 EJV L. Hodges, 34 Wiseholme Road, Skellingthorpe, Lincoln, LN6 5TF
G0 EKH K. Harper, 2 Vale Road, Decoy, Newton Abbot, TQ12 1DZ
G0 EKK H. Wright, 61D Clapgun Street, Castle Donington, Derby, DE74 2LF
GM0 EKM C. Duncan, Roadside Cottage, Hoswick, Shetland, ZE2 9HL
G0 ELB B. Bray, 34 Newlands Drive, Forest Town, Mansfield, NG19 0HZ
G0 ELC P. Hall, 28 Maria Drive, Stockton-on-Tees, TS19 7JL
G0 ELJ D. Dawson, Moonstone, Rottenstone Lane Scratby, Great Yarmouth, NR29 3QT
G0 ELK E. Hausler, 5 Balaton Place, Snailwell Road, Newmarket, CB8 7YP
GM0 ELL N. Elliot, Flat 1, Tarfside, Ascog, Isle of Bute, PA20 9EU
G0 ELM P. Green, 14 Beech Avenue, Parbold, Wigan, WN8 7NS
G0 ELN S. Macdonald, 61 Pavilion Road, Worthing, BN14 7EE
G0 ELO J. Ellis, 21 Coxway, Clevedon, BS21 5AQ
GM0 ELP D. Maxwell, 29 Ambleside Rise, Hamilton, ML3 7HJ
G0 ELR East Lancashire Raynet, c/o N. Isherwood, 41 Livingstone Road, Blackburn, BB2 6NE
G0 ELU K. Kyriacou, 11 Mead Way, Bromley, BR2 9EN
G0 ELX P. Wilson, 39 Tintern Grove, Stockport, SK1 4DS
GD0 ELY J. Brown, Cleckheaton, Ballaragh, Laxey, Isle Of Man, IM4 7PW
G0 ELZ W. Cross, 31 Joshua Close, Liverpool, L5 0TD
GW0 EMB H. Blore, 17 Kendal Way, Wrexham, LL12 8AF
GM0 EMC K. Mclaren, 3 Bracany Gardens, Fogwatt, Dalriada, Elgin, IV30 8SY
G0 EMF D. Brown, 21 Scoular Drive, North Seaton, Ashington, NE63 9SE
G0 EMK M. Kendall, 88 Coldnailhurst Avenue, Braintree, CM7 5PY
G0 EML R. Bullock, 40 Little Harlescott Lane, Shrewsbury, SY13PY
G0 EMM K. Dockray, 54 Kelsick Park, Seaton, Workington, CA14 1PY
GM0 EMQ J. Vinton, 2 Luncarty Place, Turriff, AB53 4UD
G0 EMR P. Page, 144 Cody Road, Farnborough, GU14 0DD
G0 EMS H. Brown, Priors Lea, Besford Road, Worcester, WR8 9AN
G0 EMT M. Chapman, 102 Fangrove Park, Lyne, Chertsey, KT16 0BP
G0 EMU Hillbillies Contest Group, c/o N. Evely, 11 St.Margarets New Road, Teignmouth, TQ14 8UE
G0 EMV W. Van Aswegen, 16 Spencer Road, Southampton, SO19 6QX
G0 EMX E. Parr, 74 Stanley Road, Coventry, CV5 6FF
G0 ENA R. Procter, 41 Bracken Road, Ferndown, BH22 9PD
G0 ENB R. Felton, 16 Quidenham Road, East Harling, Norwich, NR16 2JD
G0 END R. Mcgarvie, Croftlands, Thornthwaite, Keswick, CA12 5SA
G0 ENF G. Crawshaw, 51 Templeway West, Lydney, GL15 5JD
G0 ENJ D. Buckingham, 208 Bannings Vale, Saltdean, Brighton, BN2 8DJ
G0 ENM D. James, 22 Gretton Road, Walsall, WS9 0DT
G0 ENN B. Bowden, 49 Springfield Drive, Westcliff-on-Sea, SS0 0RA
G0 ENO K. Dempster, 2 Hillside House, 64 Graham Road, Malvern, WR14 2HU
GM0 ENQ W. Smith, 10 Woodlands Place, Inverbervie, Montrose, DD10 0SL
GW0 ENT J. Comerford, Bod Elen, Bontnewydd, Caernarfon, LL54 7YE
GW0 ENU D. Walters, 132 Tan Y Bryn, Valley, Holyhead, LL65 3ES
G0 ENV A. Wood, 262 Egmanton Road, Meden Vale, Mansfield, NG20 9PY
G0 ENW G. Fingerhut, Wild Rose, Behind Hayes, Templecombe, BA8 0BP
G0 ENY R. Ford, 27 Albert Road, Millisons Wood, Coventry, CV5 9AS
G0 ENZ T. Trudgeon, 1 Bessy Beneath Cottages, Ruan High Lanes, Truro, TR2 5JX
G0 EOF G. Potter, 88 Highlands Close, Kidderminster, DY11 6JU
G0 EOG J. Jennings, 81 Newgate St., Burntwood, WS7 8TX
G0 EOH D. Bristow, Herniss Bungalow, Nr Penryn, TR10 9DT
G0 EOI T. Froggatt, Diestseweg 115, Geel, Belgium, 2440
G0 EOJ D. Shore, 9 Hawthorn Close, Clowne, Chesterfield, S43 4SX
G0 EOK P. Gibson, 44 Martindale Road, Hemel Hempstead, HP1 2QR
G0 EOL W. Prater, 44 Alundale Road, Winsford, CW7 2QD
G0 EOM A. Deakin, 36 The Ridgway, Romiley, Stockport, SK6 3EY
G0 EON F. Cloke, 9 Mill Close, East Coker, Yeovil, BA22 9LF
G0 EOP L. Herf, Old Chapel, Fore Street, South Molton, EX36 3HL
G0 EOS P. Smith, 35 Tanglewood Close, Birmingham, B34 7QX
G0 EOX I. Hunton, 123 Huddersfield Road, Diggle, Oldham, OL3 5NU
G0 EOY S. Outterside, 21 Coquet Grove, Throckley, Newcastle upon Tyne, NE15 9JU
G0 EOZ P. Kerton, Mooraless, 11 North Filham Cot, Ivybridge, PL21 9DH
G0 EPA G. Whitham, 55 Bisley Grove, Bransholme, Hull, HU7 4PY
G0 EPC P. Eggleton, 33 Elsham Way, Swindon, SN25 4TJ
G0 EPE D. Cargill, 41 Grosvenor Road, Skegness, PE25 2DD
G0 EPL J. Love, 191 High Street, Henley-in-Arden, B95 5BA
GM0 EPO J. Shades, 15 Balminnoch Park, Doonfoot, Ayr, KA7 4EQ
G0 EPP A. Przybyla, 18 Cherwell, Washington, NE37 3LA
G0 EPR P. Wardale, 104 Rectory Place, Woolwich, London, SE18 5BY
G0 EPU C. Crosby, 37 Malwood Way, Maltby, Rotherham, S66 7HF
G0 EPV J. Collins, 49 Alspath Road, Meriden, Coventry, CV7 7LU
G0 EPY C. Hirst, Sunnyside, Main Street Hellifield, Skipton, BD23 4HX
G0 EQB K. Curson, 20 Penelope Grove, Peterborough, PE2 8XP
G0 EQC I. Williamson, 4 Edgewell Road West, Prudhoe, NE42 6JP
G0 EQD J. Archer, 29 Easby Close, Bishop Auckland, DL14 0RX
G0 EQE D. Cunningham, 2 Fairmead Road Moreton, Wirral, CH46 8TX
G0 EQH E. Shackleton, 2 Culcheth Avenue, Marple, Stockport, SK6 6NA
G0 EQI R. Mallinson, 3 Captain Cooks Crescent, Whitby, YO22 4HL
GM0 EQS Dr S. Palmer, Fintry Schoolhouse, Turriff, AB53 5RN
G0 EQV D. Buckley, Cl/ Juan Ramon Jimenez 23, Formentera Del Segura, Alicante, Spain, 1379
GM0 EQW T. Olsen, Ard Chuan, Taynuilt, PA35 1HY
GM0 ERB N. Calder, 16 Camesky Road, Caol, Fort William, PH33 7ER
G0 ERF K. Watkins, 29 Saddlers Close, Billingshurst, RH14 9GL
G0 ERL W. Marbus, Elm Farm House, Debenham Road, Stowmarket, IP14 5LP
G0 ERS R. Smith, 79 Froxfield Road, Havant, PO9 5PW
GM0 ERT R. Tannahill, 62 Caroline Park, Mid Calder, Livingston, EH53 0SJ
GM0 ERV S. Mclennan, 6 Mull Terrace, Oban, PA34 4YB
G0 ERW W. Spencer, 46 Saxon Way, Bourne, PE10 9QY
G0 ERY R. Saunders, 24322 Augustin Street, Mission Viejo, USA, 92691
G0 ESA W. Wilkinson, 10 Chemin Des Mardeilles, Chirac, France, 16150
G0 ESD T. Roberts, Alcana 58.2 Bajo, La Romana, Alicante, Spain, 3669
G0 ESF J. Southgate, 36 Porch Way, Whetstone, London, N20 0DS
G0 ESH G. Cooper, The Bumbles, Shatterford, Bewdley, DY12 1TR
G0 ESI D. Williams, Miramare, Egremont Road, St Bees, CA27 0AS
GW0 ESK C. Williams, 3 Lodge Orchard, Mona Street, Amlwch, LL68 9RX
G0 ESL M. Musgrave, 11 Hillside Drive, Yealmpton, Plymouth, PL8 2NT
G0 ESO S. Llewellyn, Eastfield Cottage, Mavis Enderby, Spilsby, PE23 4EJ
GW0 ESU W. Lee, 8 Bronheulog, Bodffordd, LL77 7SU

G0 ESW R. Graham, 454 Lobley Hill Road, Lobley Hill, Gateshead, NE11 0BS
G0 ESY M. Perrett, 2 Barn Park Ashwater, Beaworthy, EX21 5eu
G0 ETA G. Luhman, 31 Flexmore Way, Langford, Biggleswade, SG18 9PT
GW0 ETF S. Rolfe, Tynlon Minffordd, Bangor, LL57 4DR
G0 ETI J. Bedford, 41A Arden Road, Herne Bay, CT6 7UW
G0 ETJ J. Bathurst, 7 Princethorpe Road, Birmingham, B29 5PU
G0 ETL G. Tatterson, 2 Eden Road, Leeds, LS4 2TT
GW0 ETM J. Davies, 1 Mount View, Plas Road, Blackwood, NP12 3RH
G0 ETP T. Howe, 76 Birch Trees Road, Great Shelford, Cambridge, CB22 5AW
G0 ETQ J. Curnow, 4 Penmere Court, Falmouth, TR11 2RN
GW0 ETU E. Cabban, Garmonfa, Capel Garmon, Llanrwst, LL26 0RG
G0 ETV G. Mcquire, 43 Elmridge Crescent, Blackpool, FY2 0NQ
G0 ETZ S. Gurney, Crimond The Common, Exmouth, EX8 5EE
G0 EUC Capt. R. Burnet, 41 Douglas Crescent, Southampton, SO19 5JP
G0 EUD C. Jackson, Old Stable Cottage, Hall Road, Walpole Highway, Wisbech, PE14 7QD
GI0 EUG E. Hagan, 34 Coolshinney Road, Magherafelt, BT45 5JF
G0 EUJ K. Neville, 5 Coleville Avenue, Fawley, Southampton, SO45 1DA
GM0 EUL J. Estibeiro, The Joiners House Preston, Duns, TD11 3TQ
GM0 EUM J. Mackinnon, 60 Mount Stuart Drive, Wemyss Bay, PA18 6DX
G0 EUN J. Nichol, 58 Benson Crescent, Doddington Park, Lincoln, LN6 3NU
G0 EUP M. Rigg, 64 Hathaway, Blackpool, FY4 4AB
G0 EUR B. Barber, 1 Shore Place, Trowbridge, BA14 9TB
G0 EUV R. Jones, Greens Cottage, Luton Road, Offley, Hitchin, SG5 3DR
G0 EUZ R. Jones, 90 High Street, Cottenham, Cambridge, CB24 8SD
G0 EVA D. Evans, 113 Denby Dale Road, Wakefield, WF2 8EB
G0 EVB J. Barnes, 7 Parkfield Mews, Little Parkfield Road, Liverpool, L17 8UD
G0 EVD J. Noble, 22 Hadleigh, Letchworth Garden City, SG6 2LU
GW0 EVE V. Berry, 40 Yerburgh Avenue, Colwyn Bay, LL29 7NB
G0 EVF J. Mowbray, 44 Monkdale Avenue Cowpen Estate, Blyth, NE24 4EB
GW0 EVG N. Berry, 40 Yerburgh Avenue, Colwyn Bay, LL29 7NB
G0 EVH A. Ferneyhough, 30 Bedford Drive, Sutton Coldfield, B75 6AU
G0 EVI S. Monk, 310 Hinckley Road, Leicester, LE3 0TN
G0 EVJ S. Evans, 181 Curborough Road, Lichfield, WS13 7PW
G0 EVM P. Stewardson, Hyland, St. Kenelms Road, Halesowen, B62 0NE
G0 EVN S. Parkin, 13 Queens Drive, Nuthall, Nottingham, NG16 1EG
G0 EVO F. Robinson, 42 The Paddock, York, YO26 6AW
G0 EVP K. Griffiths, 44 Curzon Road, Poynton, Stockport, SK12 1YE
G0 EVQ A. Sewell, 53 Soame Close Aylsham, Norwich, NR11 6JF
G0 EVR A. Kittrick, 28 Jubilee Street, Hall Green, Wakefield, WF4 3JZ
G0 EVS H. Angus, 111 Great Elms Road, Hemel Hempstead, HP3 9UQ
G0 EVT J. Hoban, 3 Lake Lock Grove, Stanley, Wakefield, WF3 4JJ
G0 EVU F. Samet, 4 Pembroke Grove, Glinton, Peterborough, PE6 7LG
G0 EVV D. Stansfield, 22 Low Stobhill, Morpeth, NE61 2SG
G0 EVW G. Watts, 3 Maple Grove Knightsdale Road, Weymouth, DT4 0FE
G0 EVX A. Cater, 2 Turkdean Road, Cheltenham, GL51 6AL
G0 EVY D. Strobel, Dell Cottage, Copyholt Lane Stoke Pound, Bromsgrove, B60 3AY
G0 EVZ S. Males, 6 Lammas Path, Stevenage, SG2 9RN
G0 EWD P. Kennedy, 262 Green Road, Springvale, Sheffield, S36 6BH
GI0 EWE P. Holland, 35 Ashfield Road, Clogher, BT76 0HJ
GM0 EWF D. Pettigrew, 112 South Street, Armadale, Bathgate, EH48 3JU
G0 EWH R. Newton, 74 Walker Avenue, Stourbridge, DY9 9EL
G0 EWI J. Daramy, 6 Boulton Close, Chesterfield, S40 4XJ
GI0 EWP J. Hartin, 2 Berryhill Close, Dunamanagh, Strabane, BT82 0GZ
G0 EWR D. Wale, 33 Westground Way, Tintagel, PL34 0BH
G0 EWT A. Coates, 11 Canterbury Road, Brotton, Saltburn-by-the-Sea, TS12 2XG
GM0 EWU C. Craig, Knipoch Hotel, Knipoch, Oban, PA34 4QT
G0 EWV T. Buckle, 15 Gleaves Avenue, Harwood, Bolton, BL2 4ET
GM0 EWW J. Moore, 19 Mansefield Tyndrum, Crianlarich, FK20 8RQ
GM0 EWX C. Macpherson, 6 Borve, Skeabost Bridge, Portree, IV51 9PE
GW0 EWY W. Woods, 40 Ger-Y-Llan, Velindre, Llandysul, SA44 5YB
G0 EWZ I. Mason, 56 Evenlode Crescent, Coventry, CV6 1BY
G0 EXA A. Bendall, 3 St Michaels Gate, Brimfield, Ludlow, SY8 4NE
G0 EXB P. Langdon, Dahlia Cottage, Kidderminster, DY14 9HP
GW0 EXD Dr C. Challinor, Bryn Tirion, Lower Frankton, Oswestry, SY11 4PA
G0 EXN J. Eden, 4 Halescourt, Church Lane, Shifnal, TF11 8RD
G0 EXU A. Davies-Jones, 10 Ponsford Road, Knowle, Bristol, BS4 2UP
G0 EYA J. Morris, 31 Beldham Road, Farnham, GU9 8TW
G0 EYE A. Lunn, 45 St. Anthonys Avenue, Eastbourne, BN23 6LN
G0 EYF B. Ford, 15 Derby Road, Barnstaple, EX32 7HW
G0 EYG C. Tite, 13 Potter Way, Bedford, MK42 9RG
GW0 EYH R. Dawson, 74 Dylan Avenue Cefn Fforest, Blackwood, NP12 3NG
G0 EYL W. Mcadam, 2 Manor Orchards, Knaresborough, HG5 0BW
G0 EYM R. Rowlett, No 1 Bungalow, Main Road, Wisbech, PE14 9JR
G0 EYO C. Pettitt, 12 Hennals Avenue, Redditch, B97 5RX
G0 EYP R. Francis, 42 Carmarthen Road, Cheltenham, GL51 3LA
G0 EYR P. Robinson, 17 Highfield, Taunton, TA1 5JE
G0 EYT M. Fox, 49 Manor Drive, Esher, KT10 0AZ
G0 EYU C. Talbot, 59 Heywood Avenue, Austerlands, Oldham, OL4 4AZ
G0 EYW D. Jones, 4 Granville Crest, Kidderminster, DY10 3QS
G0 EYX D. Southey, 253 Sandon Road, Stafford, ST16 3HQ
G0 EYZ P. Orchard, 5 Vicarage Road, Bletchley, Milton Keynes, MK2 2EZ
GW0 EZB S. Cowie, 37 Rockfield Drive, Llandudno, LL30 1PF
G0 EZI J. Pitfield, 42 Spinney Green, Eccleston, St Helens, WA10 5AH
G0 EZJ D. Drake, 60 Jessopp Avenue, Bridport, DT6 4ES
G0 EZL D. Hodgkinson, 16 Fortescue Avenue, Twickenham, TW2 5LS
GW0 EZQ Llanelli ARC, c/o P. Cavallucci, 23 Pier Street, Swansea, SA1 1RY
GM0 EZR P. Newton, 115 Napier Road, Glenrothes, KY6 1DU
G0 EZT E. Humphries, 37 Grove Meadow, Cleobury Mortimer, Kidderminster, DY14 8AG
G0 EZU A. Davies, Flat 16, Beechcroft Salisbury Terrace, Teignmouth, TQ14 8JA
G0 EZX C. Wood, 34 Rosemary Lane, Stourbridge, DY8 3EP
G0 EZY T. Jeacock, 9 Parkwood Rise, Barnby Dun, Doncaster, DN3 1LY
G0 FAB H. Kay, 51 Colin Crescent, Colindale, London, NW9 6EU
G0 FAD J. Bowers, 22 Kilmiston Drive, Fareham, PO16 8DY
G0 FAE R. Beer, 65 Bridgefield Road, Whitstable, CT5 2PH
G0 FAH W. Wright, 46 Homestall Road, East Dulwich, London, SE22 0SB
G0 FAJ L. Barnes, 22 Holton Heath Park, Wareham Road, Poole, BH16 6JS
G0 FAS G. West, 33 Dorcis Avenue, Bexleyheath, DA7 4RL
G0 FAU A. Doyle, 2 Ferndown Way, Weston, Crewe, CW2 5GS
G0 FAW H. Jenkinson, Flat 2, 3 Kassima, Kissonerga/Paphos, Cyprus, 8574
G0 FBB Meopham Parish Radio Club, c/o I. Burns, Little Delmar Farm, Leywood Road, Meopham, DA13 0UD
G0 FBC C. Hyatt, 44 Barnes Lane, Sarisbury Green, Southampton, SO31 7BZ
G0 FBG G. Hands, 74 Berrington Road, Nuneaton, CV10 0LB
G0 FBL B. Sharman, 64 Collingwood Drive, Shiney Row, Houghton le Spring, DH4 7LP
G0 FBM D. Lawson, 52 Ryefield Road, Eastfield, Scarborough, YO11 3DR
G0 FBO C. Roberts, 86 Peake Road, Brownhills, Walsall, WS8 7BZ
G0 FBQ W. Wootton, 94 Dyas Avenue, Great Barr, Birmingham, B42 1HF
G0 FBS G. Nicolson, 34 Chalbury Close, Weymouth, DT3 6LE
GW0 FBT C. Hughes, Llanhennock Cheshire Home, Caerleon, Newport, NP18 1LT
G0 FBW A. Armstrong, 1 Montfalcon Close, Peterlee, SR8 1DD
G0 FBX A. Leigh, Roleystone, 7 Fieldfare, Gloucester, GL4 4WH
G0 FCA I. Groom, 12 Billington Avenue, Rossendale, BB4 8UW
G0 FCB C. Tatlow, Frenton Farm, Whitemoor, St Austell, PL26 7XQ
G0 FCG P. O'Neill, 36 Grantley Gardens, Mannamead, Plymouth, PL3 5BS
G0 FCH R. Last, 39 Upham Road, Swindon, SN3 1DJ
GM0 FCI P. Reid, 129 Mckinlay Crescent, Irvine, KA12 8DR
G0 FCJ M. Lawson, 1A Hunters Close, Stroud, GL5 4UW
G0 FCM I. Sircombe, 4 Long Eights, Northway, Tewkesbury, GL20 8QY
G0 FCO B. Cooke, 49 Shepherds Croft, Slade, Stroud, GL5 1US
G0 FCQ D. James, 15 Kington Gardens, Birmingham, B37 5HX
G0 FCT I. Pawson, 3 Orion, Bracknell, RG12 7YX
G0 FCU S. Kennedy, Laurel Cottage, Pond Lane, Peaslake, Guildford, GU5 9RS
G0 FCV A. Woods, 8 Wareham Road, Lytchett Matravers, Poole, BH16 6DP
G0 FCX A. Traynor, 2 Mansfield Road, Mossley, Ashton-under-Lyne, OL5 9JN
G0 FCZ L. Standley, 26 Ullswater Drive, Middleton, Manchester, M24 5RL
G0 FDA R. Dresser, 6 Acacia Avenue, Fencehouses, Houghton le Spring, DH4 6JG
G0 FDD C. Shalley, 4 Almond Walk, Lydney, GL15 5LP
G0 FDE M. Jackson, 2 Sunnybank, Watledge, Stroud, GL6 0AP
G0 FDH R. Wishart, 15 Plumer Avenue, Tang Hall, York, YO31 0PX
G0 FDJ K. Castley, Zone 2, (Opposite High School), Cagayan de Oro City, Philippines, 9000
G0 FDP F. Parradine, 87 Oakways Eltham, London, SE9 2NZ
G0 FDS J. Johnson, 150 Lenthall Avenue, Grays, RM17 5AB
G0 FDT M. Flatman, 8 The Pines, Cringleford, Norwich, NR4 7LT
G0 FDV D. Jardine, 11 Gorse Hill, Broad Oak, Heathfield, TN21 8TW
G0 FDX Central Lancs A R C, c/o J. Lawson, 14 Kentmere Avenue, Farington, Leyland, PR25 3UH
G0 FDZ C. Whitmarsh, 35 Dorchester Avenue, Bexley, DA5 3AH
G0 FEH D. Thickett, 23 Helmsdale Road, Leamington Spa, CV32 7DN
G0 FEI V. Ward, Romayne, St. Johns Road, Great Yarmouth, NR31 9JT
G0 FEJ G. Marshall, Birchlands, 3 Longbridge Close, Hook, RG27 0DQ
G0 FEK R. Wilson, 34 Belfairs Drive, Chadwell Heath, Romford, RM6 4EB
GW0 FEM G. Felton, 10 Penbodeistedd, Llanfechell, Gwynedd, LL68 0RE
G0 FEO A. Quy, 17 Fircroft, Kingsbury, Tamworth, B78 2JU
G0 FEP Dr H. Maclean, 15 Keystone Cres, Kew East, Australia, 3102
G0 FEQ K. Howard, 11 Station Road, Ulceby, DN39 6UQ
GW0 FEU D. Davies, Coedfryn, Halkyn, Holywell, CH8 8ES
G0 FEV E. Kittrick, 28 Jubilee Street, Hall Green, Wakefield, WF4 3JZ
G0 FEY S. Goodwin, 5 St. Wenefredes Green Bickley, Whitchurch, SY13 4EB
G0 FEZ K. Wragg, 11A Fall Road, Heanor, DE75 7PQ
G0 FFB S. Maughan, 17 Upper Dane, Desborough, Kettering, NN14 2LB
G0 FFF R. Ford, 15 Holloway, Pershore, WR10 1HW
G0 FFK R. Mckenzie, 40 Fairway Avenue, West Drayton, UB7 7AN
G0 FFL R. O'Keeffe, 40 Edinburgh Road, Maidenhead, SL6 7SH
G0 FFN E. Fisher, 19 Keats Way, West Drayton, UB7 9DR
G0 FFQ R. Baldock, 1A Thorneywood Road, Long Eaton, Nottingham, NG10 2DZ
G0 FGA A. Walker, 2 Chelwood Drive, Sandhurst, GU47 8HT
G0 FGC J. Biggs, 40 Packmore St., Warwick, CV34 5BX
G0 FGE B. Bolt, 106 Barley Farm Road, Exeter, EX4 1NJ
G0 FGG J. Thompson, 17 Fryer Crescent Haughton Le Skerne, Darlington, DL1 2DX
G0 FGI H. Cromack, 6 West Park, South Molton, EX36 4HJ
G0 FGJ D. Joyce, 44 St. Marys Close, Marston Moretaine, Bedford, MK43 0QZ
G0 FGK B. Whitehouse, Flat 53, Peel House, Tamworth, B79 7BQ
GW0 FGO W. Waldron, 100 Porthmawr Road, Cwmbran, NP44 1NB
G0 FGP R. Bradwell, Summer Fields School, Mayfield Road, Oxford, OX2 7EN
G0 FGS Dr M. Bosley, Crossroads Palmerston Road, Ross-on-Wye, HR9 5PN
G0 FGW R. Clements, 28 Willow Grove, Chippenham, SN15 1AR
G0 FGX R. Mccreadie, 45 Gwealhellis Warren, Helston, TR13 8PQ
G0 FGZ Dr C. Newton, Peartree Cottage, Little London, Longhope, GL17 0PH
G0 FHC B. Trimmer, Sydney Cottage, Salisbury Road, Romsey, SO51 6EE
GM0 FHD E. Mottart, 1 Muirake Cottages, Cornhill, Banff, AB45 2BQ
GM0 FHF C. Ashdown, Oliver Cottage 7 Fernilea, Carbost, Isle of Skye, IV47 8SJ
G0 FHH P. Beatty, 21 Avery Hill, Kingsteignton, TQ123LB
GM0 FHJ R. Menzies, 105 Yoker Mill Road, Glasgow, G13 4HL
G0 FHK R. Peart, 33 Fieldfare, Abbeydale, Gloucester, GL4 4WH
GW0 FHL B. Thomas, Plastirion, Padeswood Road, Buckley, CH7 2JL
G0 FHO G. Taylor, 93 Fengate Mobile Home Park, Peterborough, PE1 5XE
G0 FHT G. Bate, 7 Albany Court, Redruth, TR15 2NY
G0 FHW T. Opie, 3 Colborne Avenue Illogan, Redruth, TR16 4EB
G0 FHX A. Hocking, 79 Cornish Crescent, Truro, TR1 3PE
G0 FHY J. Hocking, 79 Cornish Crescent, Truro, TR1 3PE
G0 FIC K. Tarry, 38 Tresithney Road, Carharrack Tr165Qz, Reduth, TR165QZ
G0 FIG A. Trusler, 42 Mill Hill, Shoreham-by-Sea, BN43 5TH
G0 FIJ C. Jones, 46 Wilmington Close, Woodley, Reading, RG5 4LR
G0 FIN A. Findlay, 218 Lower Hillmorton Road, Rugby, CV21 3TS
G0 FIP E. Tugwell, 14 Martinique Way, Eastbourne, BN23 5TH
GM0 FIQ H. Bohan, 12 Loch Way, Kemnay, Inverurie, AB51 5QZ
G0 FIT K. Menzel, 8 Higher Bockhampton, Dorchester, DT2 8QJ
G0 FIU R. Edwards, Stanmore Cottage, Brockley Corner, Culford, Bury St Edmunds, IP28 6UA
G0 FIW I. Osborne, Alacoo, Tan Lane, Clacton-on-Sea, CO16 9PS
G0 FJA B. Samuels, 63 Mill Road, Okehampton, EX20 1PR
G0 FJB J. Bailey, Powney Cottage, Powney Street, Ipswich, IP7 7AL

G0 FJD D. Tyers, 18 Gardeners Close, Kidderminster, DY11 5DW
GW0 FJE R. Hill, 13 Maesglas Grove, Newport, NP20 3DJ
GW0 FJH C. Lonsdale, 6 Oak Tree Close, New Inn, Pontypool, NP4 0DG
G0 FJJ A. Winkler, 10 Havenside, Shoreham-by-Sea, BN43 5LN
GW0 FJP D. Stanley, 9 Haywain Court, Bridgend, CF31 2ED
GW0 FJQ A. Marshall, 26 Avondale Road, Gelli, Pentre, CF41 7TW
G0 FJR D. Paynter, 6 Blacksmiths Close, Ramsey Forty Foot, Huntingdon, PE26 2YW
G0 FJS P. Copeland, 6 Waverley Road, Northampton, NN2 7DA
G0 FJZ G. Bradley, 59 Main Road, Watnall, Nottingham, NG16 1HE
G0 FKF J. Smith, Erw Las, The Cross Roads, Redruth, TR16 5PN
G0 FKG C. Skillings, 11 Curtis Road, Norwich, NR6 6RB
G0 FKI A. Brown, Panorama, Highway Lane, Redruth, TR15 1SE
G0 FKJ C. Currey, 1 Newport Close, Portishead, Bristol, BS20 8DD
G0 FKK R. Parrish, 11 Pitt Road, Maidstone, ME16 8PA
GM0 FKP J. Tohill, 71 Campsie Road, Kilmarnock, KA1 3RY
G0 FKS K. Stancliffe, 3 Upper Lambricks, Rayleigh, SS6 8BP
G0 FKW A. Timms, 63 High Street, Astcote, Towcester, NN12 8NW
G0 FKX D. Warren, 1 Ruby Terrace, Porkellis, Helston, TR13 0LD
G0 FKY J. Merifield, 84 Wareham Road Corfe Mullen, Wimborne, BH21 3LG
G0 FLA W. Frewen, Tantalus, Main Street, Rye, TN31 6NB
G0 FLD K. Pitman, Pump Cottage, High Street, Bridlington, YO15 1JT
G0 FLG B. Parker, 30 Caistor Road, Market Rasen, LN8 3JA
G0 FLI N. Penistone, 114 Long Lane, Worrall, Sheffield, S35 0AF
G0 FLP J. Hammond, Ashmond House, Queens Street, March, PE15 8SN
G0 FLQ R. Cheetham, 65 Avondale Avenue, Hazel Grove, Stockport, SK7 4QE
G0 FLT A. Auker-Howlett, 7 Caxton End, Eltisley, St Neots, PE19 6TJ
G0 FLU M. Crane, 40 Dukes Way, Newquay, TR7 2RW
G0 FLV R. Pennock, 4 Millers Way, Heckington, Sleaford, NG34 9JG
G0 FLW L. Williams, 56 Meriden Avenue, Stourbridge, DY8 4QS
G0 FLX D. Gray, Jaig House, Mill Lane Hemingbrough, Selby, YO8 6QX
G0 FMB B. Collins, 28 Marlborough Road, South Woodford, London, E18 1AP
G0 FMG J. Voss, 4 Chaucer Avenue, Mablethorpe, LN12 1DA
G0 FMI R. Friston, Savmar, 72 Bradenham Road, Thetford, IP25 7PJ
G0 FMJ R. Bennett, 86 Westons Hill Drive, Emersons Green, Bristol, BS16 7DN
G0 FMN A. Mcquarrie, 47 Ramsons Avenue, Conniburrow, Milton Keynes, MK14 7BB
G0 FMP I. Hodgkiss, 41 Buckingham Rise, Worksop, S81 7ED
G0 FMT D. Unwin, 11 Carlton Rise, Melbourn, Royston, SG8 6BZ
G0 FMU A. Turner, 17 The Dell, Great Warley, Brentwood, CM13 3AL
GM0 FMW D. Enderby, 45 Oxgangs Park, Edinburgh, EH13 9LF
G0 FMX A. Roberts, Am Weinberg 5, Bergen, Germany, 29303
G0 FNA C. Halliday, 42 Denshaw, Upholland, Skelmersdale, WN8 0AY
G0 FNB S. Mudd, 91 Chalkwell Avenue, Westcliff-on-Sea, SS0 8NL
G0 FND M. Cousins, Wiccan Lodge, Mumbys Drove, Wisbech, PE14 9JT
GM0 FNE T. Wilson, 20 Mace Court, Stirling, FK7 7XA
G0 FNF I. Gilbert, 82 Abbotswood Road, Brockworth, Gloucester, GL3 4PF
G0 FNH J. Toon, 2 Home Farm Park, Burton-on-Trent, DE13 9BJ
G0 FNJ D. Earnshaw, 43 Bank Parade, Burnley, BB11 1UG
G0 FNM E. Walker, 216 Milnrow Road, Rochdale, OL16 5BB
G0 FNP P. Radcliffe, Hill View, Wilton, Pickering, YO18 7LE
G0 FNS J. Brayshaw, 26 Ashfield Avenue, Malton, YO17 7LE
G0 FNV N. Moult, 16 Selwyn Close, Nottingham, NG6 0EY
G0 FOB N. Robinson, 15 Hollins Bank, Sowerby Bridge, HX6 2RU
G0 FOC M. Clowes, 55 Landswood Close, Birmingham, B44 0LF
G0 FOE P. Hume-Spry, 23 Appledore Avenue, Wollaton, Nottingham, NG8 2RE
G0 FOG C. Singleton, 1 Abbey Close, Aslockton, Nottingham, NG13 9AF
G0 FOH M. Holdsworth, Merrill, Ringwood Road, Southampton, SO40 7GY
G0 FOI J. Wilde, 2 Bottoms Lane, Birkenshaw, Bradford, BD11 2NN
G0 FOK R. Cox123, 12 East Street, Thame, OX9 3JS
GW0 FOL J. Brazier, Flat 5, Llys Madryn Caernarvon Road, Pwllheli, LL53 5LF
G0 FOT R. Gibbons, 3 Fairfield, Gamlingay, Sandy, SG19 3LG
G0 FOU G. Binns, 21, Rydal Close, Winsford, CW7 2SE
G0 FOY G. Grigg, 12 Townfield Road, Mobberley, Knutsford, WA16 7HF
G0 FPI J. Spence, 60 Railey Road, Crawley, RH10 8BZ
G0 FPM P. Fennell, 45 Badby Road West, Daventry, NN11 4HJ
G0 FPN D. Waller, Apartment 23, 3 Woodbrooke Grove, Birmingham, B31 2FG
G0 FPO W. Coulthard, 1 Lambton St., Eccles, Manchester, M30 8DD
G0 FPT A. Pettigrew, 12 Pensford Close, Crowthorne, RG45 6QR
G0 FPU M. Densham, 69 Mortimer Way, Leicester, LE3 1GR
GW0 FPY J. Bortowski, 4 Bryn Deiniol, Valley Road, Llanfairfechan, LL33 0SR
G0 FPZ R. Wileman, 3 Primrose Way, Stamford, PE9 4BU
G0 FQA A. Wallis, 14 Varley Close, Wellingborough, NN8 4UZ
G0 FQC P. Wood, 26 Church Road, Bamber Bridge, Preston, PR5 6EP
G0 FQD R. Harvey, 42 Groomsland Drive, Billingshurst, RH14 9HB
G0 FQF J. Rae, Sunnybank House, Burnley Road East, Rossendale, BB4 9PX
G0 FQI D. Breen, 53 Swift Close, Grange Park, Northampton, NN4 5AZ
G0 FQO D. Berry, The Bungalow, Basil Road, Kings Lynn, PE33 9RP
G0 FQP G. Owens, 73 Edinburgh Road, Widnes, WA8 8BG
GM0 FQQ B. Campbell, 93 Treeswoodhead Road, Kilmarnock, KA1 4PB
GM0 FQS A. Smith, 60 Gordon Avenue, Bonnyrigg, EH19 2PQ
G0 FQU A. Date, 23 Gilpin Way, Olney, MK46 4DN
GM0 FQV J. Black, Solway View, Carlisle Road, Annan, DG12 6QX
GW0 FQZ C. Emblen, 12 Gefnan, Mynydd Llandygai, Bangor, LL57 4DJ
G0 FRB S. Gresty, 4 Palace Road, Sale, M33 6WU
GM0 FRC Falkirk & District ARS, c/o P. Howson, 1 Howetown Fishcross, Alloa, FK10 3AW
G0 FRD S. Baldwin, 18 Derwent Road, Leighton Buzzard, LU7 2QW
G0 FRL R. Lowe, 12 Cavenham Grove, Bolton, BL1 4UA
G0 FRM H. White, 51 New Close, Knebworth, SG3 6NU
G0 FRN D. Oakes, 56 Middle Way, Chinnor, OX39 4TP
G0 FRO A. Medcalf, 1 The Old School, School Lane, Didcot, OX11 0ES
G0 FRR Flight Refuelling ARS, c/o A. Baker, Highleaze, Deans Drove, Poole, BH16 6EQ
G0 FRS Farnborough Contest Group, c/o R. Konowicz, 12 Ambleside Crescent, Farnham, GU9 0RZ
GM0 FRT Funny Contest G, c/o A. Duncan, Barrhill House, Peterculter, AB14 0LN
G0 FRU C. Osbourn, Bourn Bungalow, Back Lane, Newmarket, CB8 9NB
G0 FRV S. Adams, 63 Turnbull Drive, Leicester, LE3 2JU
G0 FRX G. Cowling, Laissez Faire, Reedness, Goole, DN14 8ET
G0 FRY J. Walker, Wildersley, Wildersley Road, Belper, DE56 1PD
G0 FRZ R. Swinney, 27 Auckland Close, Houghton le Spring, DH4 6GG
G0 FSA M. Hughes, 2 Harrington Road, Desborough, Kettering, NN14 2NH
G0 FSB S. Barton, 154 The Hill Cromford, Matlock, DE4 3QU
G0 FSD P. Robinson, 198 Westfield, Plymouth, PL7 2EJ
G0 FSF D. Cawser, 26 Queen Street, Burton-on-Trent, DE14 3LR
G0 FSG C. Easton, Dallmore, Lockwood Beck Road, Saltburn, TS12 3LE
GM0 FSH W. Mackenzie, Nambrac, Cairnmount, Jedburgh, TD8 6SA
G0 FSJ D. Hutchinson, 6 Birdsall Avenue, Nottingham, NG8 2EH
G0 FSL T. Fleet, 51 The Crescent, Walsall, WS1 2DA
G0 FSM J. Bent, 32 Cross Waters Close, Wootton, Northampton, NN4 6AL
G0 FSP J. Pears, 19 Lichfield Close, Grantham, NG31 8RS
GM0 FSV J. Mcgowan, 26 Wallace Gardens, Stirling, FK9 5LS
GM0 FSW N. Mcallister, 36 Kinneff Crescent, Dundee, DD3 9RG
GM0 FSY D. Macdonald, 4A Brue, Isle of Lewis, HS2 0QW
GM0 FSZ E. Sandilands, Eric Sandilands, 12 Kerr Court, Girvan, KA26 0BP
GM0 FTG R. Hill, 9 Chambers Drive, Carron, Falkirk, FK2 8DX
GM0 FTH H. Livingston, Monthouse, Parkhead Road, Linlithgow, EH49 7BS
G0 FTI R. Hughes, 46 The Boundary, Oldbrook, Milton Keynes, MK6 2HT
GM0 FTJ N. Mitchinson, 85 Forest Road, Selkirk, TD7 5DD
GM0 FTK W. Kirk, 59 Silverbuthall Road, Hawick, TD9 7BH
G0 FTN A. Lautman, 3 Windsor Close, Leigh-on-Sea, SS9 4EA
G0 FTO R. Morton, 44 Cromer Drive, Atherton, Manchester, M46 0QE
G0 FTR D. Gill, 80 Bramwell Street, Sheffield, S3 7PB
G0 FTU C. Jones, 2 Ladycross Cottages, Dormansland, RH76PB
GM0 FTX I. Birkett, 25 Darnhall Crescent Craigend, Perth, PH2 0HH
G0 FUE D. Denton, 7 Uplands Avenue, East Ayton, Scarborough, YO13 9EU
G0 FUH R. Douglas, 5 Portnalls Road, Coulsdon, CR5 3DD
G0 FUI P. Abbott, 170 Hangleton Valley Drive, Hove, BN3 8FE
G0 FUN Apau Contest Group, c/o W. Somerville, Glendella, Wycombe Road Stokenchurch, High Wycombe, HP14 3RP
G0 FUO D. Harrop, 7 Haythorne Way Swinton, Mexborough, S64 8SQ
G0 FUR D. Buckle, 63 Ashley Drive South, Ashley Heath, Ringwood, BH24 2JP
G0 FUS P. Fry, Flat 2, National Westminster Bank, North Street, Wiveliscombe, Taunton, TA4 2JY
G0 FUU P. Firmin, 25 The Heights, Hastings, TN35 5EP
G0 FUV J. Mills, Smiths Hill Petrockstow, Okehampton, EX20 3EZ
G0 FUW S. Hartley, 5 Sydenham Buildings, Bath, BA2 3BS
G0 FUY A. Firth, 10 Holroyd Hill, Wibsey, Bradford, BD6 1PQ
G0 FUZ C. Muten, 20 Middlewich Road, Nantwich, CW5 6HL
G0 FVB G. Stokes, 87A Wimborne Road, Southend on Sea, SS2 4JR
GW0 FVC J. Newman, 57 Heritage Park, St. Mellons, Cardiff, CF3 0DQ
G0 FVD R. Nichol, 32 Greenwood Avenue, Harworth, Doncaster, DN11 8HT
G0 FVF A. Howman, 32 Dereham Road, Pudding Norton, Fakenham, NR21 7NA
G0 FVH D. Dolling, 41 Bullfinch Close, Poole, BH17 7UP
G0 FVI Dr A. Gilfillan, 57 Brant Road, Lincoln, LN5 8RX
GM0 FVJ L. Martin, 44 Glenearn Court, Pittenzie Street, Crieff, PH7 3LE
G0 FVM P. Irons, 55A Main Street, Cayton, Scarborough, YO11 3RS
G0 FVN P. Johnston, Woburn House, 38 Chatsworth Avenue, Pontefract, WF8 2UP
G0 FVO J. Leach, 19 Fyfe Crescent, Baildon, Shipley, BD17 6DR
G0 FVS J. Jackson, 165 Hall St., Briston, Melton Constable, NR24 2LQ
G0 FVT D. Lisney, 14 Clarendon Drive, Martham, Great Yarmouth, NR29 4TD
G0 FVU M. Smith, 31 Cromford Road, Crich, Matlock, DE4 5DJ
G0 FWA P. Waddington, 19 Olivers Drive, Witham, CM8 1QJ
G0 FWD J. Adcock, 102 Richmond Way, Newport Pagnell, MK16 0LH
G0 FWF D. Stanton, 53 Chester Road, London, N17 6EH
G0 FWP J. Purvess, 389 Otley Old Road, Leeds, LS16 6BX
GM0 FWY P. Gray, 21 Simpson Street, Glasgow, G20 6XZ
GW0 FXC R. Rowland, 60 Ombersley Road, Newport, NP20 3EE
G0 FXD D. Harrison, 41 North End Lane, Malvern, WR14 2NG
G0 FXI R. Oram, 4 Hardy Avenue, Bristol, BS3 2BP
G0 FXK J. Paxton, 27 Holborn View, Leeds, LS6 2RD
G0 FXL D. Garratt, 238 Hockley Road, Hockley, Tamworth, B77 5EY
G0 FXM S. Vickery, 17 Trenowah Road, St Austell, PL25 3EB
G0 FXQ I. Drury, 263 Waxwing Lane, Strasburg, USA, VA 22657
G0 FXR R. Clark, 9 Kensington Avenue, Normanby, Middlesbrough, TS6 0QQ
G0 FXS K. Rutter, 6 Chetney Close, Stafford, ST16 1XA
G0 FXT C. Richardson, 47 Leighton Close, Crossgates, Scarborough, YO12 4LA
G0 FXY M. Smith, 19 Legion St., South Milford, Leeds, LS25 5AY
GJ0 FYB C. Le Jehan, Sundora, 4 St. Marys Village, St Mary, Jersey, JE3 3BQ
G0 FYD I. Mccabe, 99 Edgeway Road, Blackpool, FY4 3NH
G0 FYE B. Moss, 22 Battersby St., Ince, Wigan, WN2 2NA
G0 FYH R. Butterworth, 12 Strickland Drive, Morecambe, LA4 6TB
G0 FYL J. Samuels, 63 Mill Road, Okehampton, EX20 1PR
GW0 FYO A. Williams, 106 Garrod Avenue Dunvant, Swansea, SA2 7XQ
G0 FYP C. Holland, 44 Brightstowe Road, Burnham-on-Sea, TA8 2HP
G0 FYU T. Liggins, 55 Kirkland Street, Pocklington, York, YO42 2BX
G0 FYW B. Morrin, Flat 19 Elms Hall, Elms Rd, Morecombe, LA4 6DD
G0 FYX S. Swain, 40, Parkside, Havant, PO9 3PL
G0 FZA N. Dickinson, 14 Chelsea Mews, Lancaster, LA1 2AS
G0 FZB J. Atherfold, 42 Mansell Road, Shoreham-by-Sea, BN43 6GP
G0 FZC G. Palmer, 29 Lindsay Road, Leicester, LE3 2EJ
G0 FZD C. Nichols, 7 Royston Avenue, Owlthorpe, Sheffield, S20 6SG
G0 FZE B. Gould, 2 Parkdale, Ibstock, LE67 6JW
G0 FZF C. Guinan, 20 Putney Close, Oldham, OL1 2JS
G0 FZG H. Eddleston, 40 High Street Corby Glen, Grantham, NG33 4LX
G0 FZH D. Moore, 12 Newtown Park, Langport, TA10 9TF
G0 FZM K. Blabey, 19 Lud Lane, Tamworth, B79 7EW
G0 FZO C. Lee, 21 Sandholme, Market Weighton, York, YO43 3ND
GI0 FZT P. Frend, 41 Brunswick Manor Abbey Street, Bangor, BT20 4JD
G0 FZU J. Tinsley, 21 Peckforton View, Kidsgrove, Stoke-on-Trent, ST7 4TA
GW0 FZY Dr J. Woolgar, Glan-Yr-Afon, Cwmcerdinen, Felindre, Swansea, SA5 7PU
G0 FZZ J. Foster, 23 Shrewsbury Street, Hartlepool, TS25 5RQ
G0 GAG M. Cowley, 46 Mapletoft Avenue, Mansfield Woodhouse, Mansfield, NG19 8HT
G0 GAJ F. Bunce, 45 Hailes Road, Gloucester, GL4 4RB
G0 GAL E. Howells, 5 Bowland Close, Overdale, Telford, TF3 5HE
G0 GAP W. Meecham, 38 Douglas Road West, Stafford, ST16 3NX

G0 GAQ B. Johnson, 67 Nursery Lane, Northampton, NN2 7PT
G0 GAR A. Robinson, 28 Briarsleigh, Wildwood, Stafford, ST17 4QP
GM0 GAT A. Thomasson, 1 Eastside Green, Westhill, AB32 6XY
GM0 GAV G. Taylor, South Lodge, Fingask, Errol, Perth, PH2 7SS
G0 GBC P. Dempster, Flat 218A, Peachfield Road, Malvern, WR14 4AP
G0 GBE C. Thompson, 135 Stafford Road, Bloxwich, Walsall, WS3 3PG
G0 GBG B. Wilkinson, 3 Friarage Avenue, Northallerton, DL6 1DZ
GM0 GBH P. Young, 4 Primrose Avenue, Rosyth, Dunfermline, KY11 2SS
G0 GBI G. Loake, 81 Duchess Road, Bedford, MK42 0SE
G0 GBL D. Taylor, 38 Seward Road, Badsey, Evesham, WR11 7HQ
G0 GBN J. Henshaw, 7 Gorsefield Close, Wirral, CH62 6BU
G0 GBP D. Porter, 10 Broomholme, Shevington, Wigan, WN6 8DT
G0 GBQ H. Whitfield, 81 Tenter Balk Lane, Adwick-Le-Street, Doncaster, DN6 7EE
G0 GBR S. Mattinson, 76 Fairway Avenue, Tilehurst, Reading, RG30 4QB
G0 GBU M. Mccallum, 65 Whalley Road, Altham West, Accrington, BB5 5DH
GM0 GBV D. Hackett, 117 Nethanvale, Auchlochan Lesmahagow, Lanark, ML11 0FX
G0 GBW C. Oswald, 3 Belvedere Drive, Bilton, Hull, HU11 4AX
G0 GBY M. Francis, 50 Edinburgh Avenue, Leigh-on-Sea, SS9 3SG
G0 GCA M. Collins, 22 Southfield Close, Woolavington, Bridgwater, TA7 8HJ
G0 GCJ R. Mellor, 2 Taxal View, Fernilee, High Peak, SK23 7HD
G0 GCK M. Johnson, 7 Ash Grove, Northallerton, DL6 1RQ
GM0 GCO B. Carson, 46 Tweed Drive, Bearsden, Glasgow, G61 1EJ
G0 GCQ J. Rivers, 16, Longshore Grove, New Romney, TN28 8FP
GM0 GDD A. Mcmillan, 10 Lyon Road, Erskine, PA8 6HG
GI0 GDF E. Hooks, 9 Curtis Walk, Lisburn, BT28 1HE
GW0 GDI N. Erskine, 302 Caerphilly Road, Cardiff, CF14 4NS
G0 GDJ J. Brooks, Bream, The Fitches, Saxmundham, IP17 1UX
G0 GDL M. Rodgers, 21 Dovedale Rise, Allestree, Derby, DE22 2RE
G0 GDS J. Hyde, 7 Wandells View, Brantingham, Brough, HU15 1QL
G0 GDV J. Strickland, 11 Chilworth Gardens, Waterlooville, PO8 0LD
G0 GEB R. East, Bramleys, 34 Boyd Avenue, Dereham, NR19 1LU
GM0 GEE L. Stapleton, 22 Ashie Road, Inverness, IV2 4EN
G0 GEF P. Chipman, 2 Cornforth Close, Trinity Road, Tamworth, B78 2LA
G0 GEH L. Howarth, 12 Thomas Street, Hemsworth, Pontefract, WF9 4AY
GW0 GEI S. Jones, Blaenpant, Dihewyd, Lampeter, SA48 7PJ
G0 GEL J. Pruden, 77 Browning Crescent, Ford Houses, Wolverhampton, WV10 6BQ
G0 GEP G. Perry, 123 Green Lanes, Wylde Green, Sutton Coldfield, B73 5LT
G0 GEQ J. Rogerson, 44 Romney Close, Clacton-on-Sea, CO16 8YE
G0 GER R. Knighton, 262 Victoria Road West, Thornton-Cleveleys, FY5 3QB
G0 GEU S. Holt, 22 Sulby Grove, Morecambe, LA4 6HD
GW0 GEV T. Hurst, Woodside, Parc Seymour, Caldicot, NP26 3AB
G0 GEZ D. Creek, The Coach House, Basketts Lane, Yarmouth, PO41 0PY
G0 GFA C. Armitage, 19 Park Road, Barlow, Selby, YO8 8ES
G0 GFC R. Johnson, 3 Lance Drive, Chase Terrace, Burntwood, WS7 1FA
G0 GFD K. Venn, 28 Streamleaze, Titchfield Common, Fareham, PO14 4NP
G0 GFE B. Keechan, 24 Tolley Road, Kidderminster, DY11 7EW
G0 GFI C. Burrows, 29 Hampden Road, Malvern Link, Malvern, WR14 1NB
G0 GFK J. Denford, 10 Churchill Road, Bideford, EX39 4HG
GM0 GFL A. Marriott, Parkview, Dunrossness, ZE2 9JG
GW0 GFN D. Anderson, Penrheol Farm, Meidrim, Carmarthen, SA33 5NX
G0 GFP A. Fowler, 43 Eastbourne Heights, Oak Tree Lane, Eastbourne, BN23 8FB
G0 GFQ K. Martin, 21 All Saints Close, Weybourne, Holt, NR25 7HH
G0 GFR B. Holloway, 28 Elmsdale Road, Ledbury, HR8 2EG
GD0 GFV J. Angiolini, Elm Lodge Patrick Road, St Johns, Isle Of Man, IM4 3BP
G0 GFY C. James, 180 Mitcham Road, Croydon, CR0 3JF
G0 GFZ P. Taylor, 18 Redriff Road, Romford, RM7 8HD
G0 GGA R. Parkins, 5 Ferndale Grove, Hinckley, LE100PH
G0 GGB A. Norman, 13 Market Place, Great Yarmouth, NR30 1LY
G0 GGE P. Burrow, 9 Minsmere Road, Belton, Great Yarmouth, NR31 9NX
G0 GGG N. Rogers, 34 Broadway, Warminster, BA12 8EB
G0 GGH F. Sell, 17 Auriel Avenue, Dagenham, RM10 8BS
G0 GGL D. Ellins, 52 Littlewood, Stokenchurch, High Wycombe, HP14 3TF
G0 GGM B. Fitzsimmons, 64 Belle Vue Road Wivenhoe, Colchester, CO7 9LD
G0 GGN R. Samways, 7 St. Michaels Way, Steeple Claydon, Buckingham, MK18 2QD
G0 GGQ G. Redding, 50 Great Hill, Shefford, SG17 5EA
GW0 GGW A. Patrick, 1 Sunningdale Grove, Colwyn Bay, LL29 6DG
GI0 GGY J. Porter, 24 Cooleen Park, Londonderry, BT48 8AQ
G0 GHB J. Graham, Caxtonian, Brimbelow Road, Norwich, NR12 8UJ
G0 GHD N. Houghton, 1 Alma Street, Alfreton, DE55 7HX
G0 GHE A. Goldspink, Danehaven, Themelthorpe, Dereham, NR20 5PS
GW0 GHF B. Williams, 10 Pantycelyn Road, Llandough, Penarth, CF64 2PG
GW0 GHG D. Roberts, 16 Min Y Mor, Aberffraw, Ty Croes, LL63 5PQ
G0 GHH P. Cross, Balls Farm Cottage, Musbury Road, Axminster, EX13 8TT
G0 GHK Finningley ARS, c/o M. Hotchin, 122 Buckingham Avenue, Scunthorpe, DN15 8NS
G0 GHL T. Reeves, Rowney Farm, Newcastle Road, Loggerheads, Market Drayton, TF9 2QG
G0 GHM D. Coxon, 7 Kingston Way, Nailsea, Bristol, BS48 4RA
GM0 GHN T. Taylor, 10 Woodside Drive, Forres, IV36 2UF
G0 GHO W. Small, 24 Barrowdale Close, Exmouth, EX8 5PN
G0 GHT R. Pearce, 52 Pearse Close, Hatherleigh, Okehampton, EX20 3QW
G0 GHW G. Whitehead, 27 Cheyney Walk, Westbury, BA13 3UH
G0 GIA R. Keeley-Osgood, 2A St. Anns Crescent, Gosport, PO12 3JJ
GM0 GIB M. Gibb, Am Fasgadh, Drumtian, Glasgow, G63 0NP
G0 GIE D. Adams, Penlon, Kingsley Gardens, Codsall, Wolverhampton, WV8 2AJ
G0 GIF J. Catterson, St. Sebastians Presbytery, Gerald Road, Salford, M6 6DW
GW0 GIH J. Thomas, 41 The Uplands, Brecon, LD3 9HT
G0 GII B. Jackson, 94 Nether Court, Halstead, CO9 2HF
G0 GIL J. Carter, 112 Landor Road, Whitnash, Leamington Spa, CV31 2JZ
G0 GIN K. Mills, 19 Thomas Bassett Drive, Colyford, Colyton, EX24 6PN
G0 GIR P. Martin, 47 Bryant Road, Rochester, ME2 3EP
G0 GIT M. Jinks, Caixa Postal 003, Campina Grande De Sul, Parana, Brazil,
G0 GJA D. Wright, 110 Mancroft Road, Caddington, Luton, LU1 4EN
G0 GJC M. Hole, 110 North Boundary Road, Brixham, TQ5 8JT
G0 GJD R. Radcliffe, 23 Rose Vale, Heald Green, Cheadle, SK8 3RN
G0 GJE R. Cleverley, 22 The Tinings, Monkton Park, Chippenham, SN15 3LX
G0 GJG B. Phillips, 6 Greenways, Penkridge, Stafford, ST19 5HD
G0 GJH G. Hughes, 51 Kingsmead, Seaford, BN25 2HA
G0 GJL S. Moyses, 10 Jones Close, March, PE15 9RZ
G0 GJM J. Rattigan, 4 Grosvenor Street, Barrow-in-Furness, LA14 4AH
G0 GJN T. Roberts, 14 Glen Park Gardens, Bristol, BS5 7NE
G0 GJR A. Beard1, 9 Whitebeam Close, Basingstoke, RG22 5FH
G0 GJV M. Goodey, 62 Rose Hill, Binfield, Bracknell, RG42 5LG
G0 GJW A. Sale, 5 Kingswood Road, Gillingham, ME7 1DZ
G0 GJX F. Norton, 147 Wells Road, Glastonbury, BA6 9AN
G0 GKH D. Birch, 32 Union Street, Trowbridge, BA14 8RY
G0 GKI C. Crawford, 70 Westmoreland Avenue, Welling, DA16 2QD
G0 GKK A. Pickering, 5 West Farm Court, Medomsley, Consett, DH8 6TL
G0 GKL M. Stevens, 33 Langham Road, Hastings, TN34 2JE
G0 GKN J. Dowse, 46 Nantwich Road, Middlewich, CW10 9HG
G0 GKO G. Lumsden, Spencer Buildings, Front Street, Hartlepool, TS27 4RT
G0 GKP J. Bonner, 40 Lyles Road, Cottenham, Cambridge, CB24 8QR
GM0 GKR J. Stapleton, Ravenscourt, 23 Strathkinness High Road, St Andrews, KY16 9UA
G0 GLA M. Hoppe, 67 Belmont Road, Maidenhead, SL6 6LG
G0 GLG H. Torunski, 33 Brickhill Way, Calvert, Buckingham, MK18 2FS
G0 GLH T. Mitchell, 18 Park Avenue, Bedlington, NE22 7EJ
GW0 GLI L. Edwards, 23 Tan Y Bryn, Llanbedr Dc, Ruthin, LL15 1AQ
G0 GLJ H. Robertson, 13 St. Martins Road, Chatteris, PE16 6JF
G0 GLQ P. Davenport, 1 Boobery, Sampford Peverell, Tiverton, EX16 7BS
G0 GLU M. Fry, 1 Hebden Avenue, Warwick, CV34 5XD
G0 GLW G. White, 3 Kent Road, Gosport, PO13 0SP
GW0 GLX M. Mcfarland, Min Afon, Garreg Fawr Road, Caernarvon, LL54 7ED
G0 GLZ Dr R. Sugden, 7 Westbourne Grove, Goole, DN14 6NA
G0 GMA P. Sweeting, 35 The Ridings, Market Rasen, LN8 3EE
G0 GMB M. Baker, 25 Pentlands, Fullers Slade, Milton Keynes, MK11 2AF
G0 GMC C. Cook, 4 Woodlands Close Rustington, Littlehampton, BN16 3ET
GM0 GMD T. Astbury, 8 Auchinlay Holdings, Auchinlay, Dunblane, FK15 9NA
G0 GME R. Turner, 17 Wrose Brow Road, Shipley, BD18 2NT
GM0 GMI J. Bavin, Garvan, 10 Grampian Way, Glasgow, G78 2DH
G0 GMJ J. Bowles, 38 Rydal Grove, Liversedge, WF15 7DN
G0 GML B. Ogden, 14 Fermandy Lane, Crawley Down, Crawley, RH10 4UB
GM0 GMN J. Bertram, 20 Kyles View, Largs, KA30 9ET
GM0 GMO G. Wylie, 9 Friar Avenue, Bishopbriggs, Glasgow, G64 2HP
G0 GMS A. Read, 7 Grange Court, Hixon, Stafford, ST18 0GQ
G0 GMY P. Whitelock, 2 Shippards Road, Brighstone, Newport, PO30 4BG
G0 GNA J. Weller, Pytchley, Chichester Close, Dorking, RH4 1LP
G0 GNE R. Maddison, Tom Butt, Hope Street, Elstead, Godalming, GU8 6DE
G0 GNF G. Frykman, 8 Orchard Close, Bishops Itchington, Southam, CV47 2QS
G0 GNI M. Anderson, 17 Orchard Road, Seaview, PO34 5JE
GM0 GNK Inverclyde ARG, c/o A. Givens, 5 Langhouse Place, Inverkip, Greenock, PA16 0EW
G0 GNP B. Ackerman, 31 Melton Mill Lane, High Melton, Doncaster, DN5 7TE
G0 GNQ L. West, 22 Lyndhurst Avenue, Margate, CT9 2PS
G0 GNU J. Norton, 7 Maudsley St., Bradford, BD3 9JT
G0 GNV M. Mundy, The Homestead, Homestead Lane, Burgess Hill, RH15 0RQ
G0 GNW G. Goddard, 113 Linden Walk, Louth, LN11 9HT
GM0 GNY L. Graupner, 5 Carden Close, Alves, Elgin, IV30 8FE
G0 GOB R. Drage, 13 Manor Ride, Brent Knoll, Highbridge, TA9 4DY
G0 GOE P. Bozac, La Vigne, Motemboeuf, France, 16310
G0 GOH R. Cornell, 81 Mercel Avenue, Armthorpe, Doncaster, DN3 3HS
G0 GOI J. Macham, 9 Bankfield Grove, Scot Hay, Newcastle, ST5 6AR
GM0 GON A. Harrison, Moneen, High Askomil, Campbeltown, PA28 6EN
G0 GOO B. Evans, 17 Clarence Place, Maltby, Rotherham, S66 7HA
GM0 GOV R. Dinning, 1 South Brae Aiket Road, Dunlop, Kilmarnock, KA3 4BP
G0 GOX F. Goodger, 66 Selkirk Close, Wimborne, BH21 1TP
G0 GOZ G. Reay, 53 Tithe Barn Road, Stafford, ST16 3PL
G0 GPB K. Love, 16 Tenscore Avenue, Walsall, WS6 7BX
G0 GPE D. Wells, Browtop, Old Lane, Crowborough, TN6 2AD
G0 GPF P. Crowley, 45 Beeches Road Great Barr, Birmingham, B42 2HJ
GI0 GPG D. Mckee, 38 Nursery Road, Armagh, BT60 4BL
G0 GPH P. Butterworth, 1 The Avenue, Bury, BL9 5DQ
G0 GPK J. Burrows, Applefields, Wilmingham Lane, Yarmouth, PO41 0SL
GW0 GPQ R. Rees, 5 Golwg Y Gaer, Salem, Llandeilo, SA19 7PA
G0 GPR R. Wordsworth, 59 Highgate Lane, Goldthorpe, Rotherham, S63 9BA
G0 GPS I. Jones, 107 Wolverley Road, Kidderminster, DY11 5JN
G0 GPT S. Carter, 8 St. Crispins Way, Ottershaw, Chertsey, KT16 0RE
G0 GPV J. Barnes, 24 Burleigh Place, Oakley, Bedford, MK43 7SG
G0 GPX M. Wise, 440 Tuahiwi Road Rd1, Kaiapoi, New Zealand, 7691
GW0 GQC B. Morgan, 5 Brynmawr, Bettws, Bridgend, CF32 8SD
GI0 GQG J. Mills, 60 Loughmacrory Road, Omagh, BT79 0PH
G0 GQH A. Jenner, 53 The Leys, Woburn Sands, Milton Keynes, MK17 8QG
G0 GQI A. Orton, 2 The Grove, Brampton Abbotts, Ross-on-Wye, HR9 7JH
G0 GQJ T. Waters, 42 Tregundy Road, Perranporth, TR6 0EF
G0 GQK M. Evans, St. Bega, Shay Lane, Newport, TF10 8DA
G0 GQO S. Taylor, 24 Catstree, Stirchley, Telford, TF3 1XZ
G0 GQP D. Jackson, 38 Chestnut Crescent, Bletchley, Milton Keynes, MK2 2LA
G0 GQT M. Bishop, 52 Lingley Drive, Wainscott, Rochester, ME2 4NE
G0 GQV P. Leech, Holly Tree Cottage, 1 Goat Lane, Norwich, NR13 4NF
G0 GQW H. Jones, 21 Sandiways Road, Wallasey, CH45 3HJ
G0 GQX M. Chappell, 43 Aigburth Hall Avenue, Liverpool, L19 9EA
G0 GQY D. Smith, 62 Beresford Road, Dorking, RH4 2DG
G0 GRB P. Attew, 23 Kingsleigh Close, Trunch, North Walsham, NR28 0QU
G0 GRC Grantham Rad Cl, c/o F. Seddon, 20 Pinfold Lane, Bottesford, Nottingham, NG13 0AR
GM0 GRD R. Kelly, Overmill John Allan Drive, Cumnock, KA18 3AG
G0 GRI I. Carter, 17 Collingham Close, Templecombe, BA8 0LR
GM0 GRL D. Moore, Parkview, Claredon Place, Dunblane, FK15 9HB
G0 GRM G. Meanley, Hemplands, 8 Bennett Drive, Warwick, CV34 6QJ
G0 GRO B. Phillips, 12 Fairview Avenue, Weston, Crewe, CW2 5LX
GW0 GRQ J. Mitchell, 44 Crossways St., Barry, CF63 4PQ
G0 GRS G. Sawford, 17 Church Road, Pytchley, Kettering, NN14 1EL
G0 GRU R. Foster, 30 Wimberley Way, South Witham, Grantham, NG33 5PU
G0 GRV Y. Katoh, 4-6-3 Minami-Aoyama, Minato-Ku, Tokyo, Japan, 107-0062
GM0 GRW R. Young, 8 Nursery Lane, Mauchline, KA5 6EH
G0 GRX Bolton Raynet G, c/o C. Ashlin, 13 Brantfell Grove, Bolton, BL2 5LY

G0 GRZ F. Pounder, 29 Read Avenue, Beeston, Nottingham, NG9 2FJ
G0 GSA R. Hall, 47 Main Street, Stretton Under Fosse, Rugby, CV23 0PE
G0 GSF Dr B. Austin, 110 Frankby Road, West Kirby, Wirral, CH48 9UX
GM0 GSG H. Cameron, 36 Lynn Crescent, Kirkwall, KW15 1FF
G0 GSH C. Summers, 18 Hays Lane, Hinckley, LE10 0LA
G0 GSJ D. Howie, 22 Jason Street, Walney, Barrow-in-Furness, LA14 3EJ
G0 GSK W. Blay, 50 Fir St., Cadishead, Manchester, M44 5AU
G0 GSL T. Ritchie, 6 High Road East, Felixstowe, IP11 9JT
G0 GSM T. Pritchard, 21 Newlyn Drive, Bredbury, Stockport, SK6 1EF
G0 GSN N. Pope, 1 Knowsley Road West, Clayton Le Dale, Blackburn, BB1 9PW
G0 GSR F. Johnson, 9 Manor Close, Tavistock, PL19 0PN
GW0 GST A. Williams, 34 Gwydyr Road, Llandudno, LL30 1HQ
G0 GSU A. Dixon, 33A Valley Road. Thornhill, Dewsbury, WF12 0HY
G0 GSX C. Guerrero, 183 Edinburgh House, Queensway, Gibraltar,
G0 GSY B. Thomsen, 8 Richmond Road, Cleethorpes, DN35 8PD
G0 GSZ P. Hunter, 28 Hanover Court, Canterbury Way, Thetford, IP24 1BZ
G0 GTI A. Dickinson, 6 Church Lane, Bessacarr, Doncaster, DN4 6QB
GM0 GTL T. Lorimer, 443 Delgatie Court, Pitteuchar, Glenrothes, KY7 4RW
G0 GTN J. Bumford, 10 St. Alkmonds Square, Shrewsbury, SY1 1UH
G0 GTV S. Moore, 2 Sheppards Close, Heighington, Lincoln, LN4 1TU
GW0 GTW G. Williams, 1 Pont Y Berllan, Talycafn Road, Llanrwst, LL26 0EF
GM0 GTY F. Stirling, 42 Mcbain Place, Kinross, KY13 8QZ
GW0 GUA J. Waddell, 24 Gower View, Llanelli, SA15 3SN
G0 GUC A. Russon, 92 St. Georges Road, Dudley, DY2 8ER
G0 GUD D. Law, 10 Derwent, Tamworth, B77 2LD
G0 GUE M. Tyou, 19 Trinity Square, Margate, CT9 1HU
G0 GUF J. Youde, 4 Greenacre, Hixon, Stafford, ST18 0QE
G0 GUG M. Grove, 9 Foxcote Lane, Cradley Forge, Halesowen, B63 2JJ
GM0 GUJ J. Cumming, 24 Parkhill Wynd, Leven, KY8 4LH
G0 GUN C. Critchley, 3 Beaconsfield View, Robert Road, Slough, SL2 3XT
G0 GUO J. Rosindale, Treworder Farm, Ruan Minor, Helston, TR12 7JL
G0 GUS S. Fellick, Po Box 337, Kelmscott WA, Australia, 6995
G0 GUT N. Reed, 30 Wrey Avenue, Liskeard, PL14 3HX
GM0 GUU J. Stirling, 42 Mcbain Place, Kinross, KY13 8QZ
G0 GUV R. Murkin, 59 Teagues Crescent, Trench, Telford, TF2 6RF
G0 GUW B. Standen, 43 Westover Gardens, St. Peters, Broadstairs, CT10 3EY
GU0 GUX J. Scheffer, Route De Carteret, Cobo, Castel, Guernsey, GY5 7YS
GW0 GUY E. Hollowell, 54 Portfield, Haverfordwest, SA61 1BW
G0 GVA J. Lawson, 14 Kentmere Avenue, Farington, Leyland, PR25 3UH
G0 GVB S. Balme, 97 Backhold Drive, Halifax, HX3 9DT
G0 GVE S. Thomas, 14 Goodley, Oakworth, Keighley, BD22 7PD
G0 GVF M. Davidson, Cristina Bungalows, 26 Ave Del Pacific, Benelmadena Malaga, Spain,
G0 GVN C. Wackett, 7 Newell Road, Hemel Hempstead, HP3 9PD
G0 GVS N. Clayton, 220 Milnrow Road, Rochdale, OL16 5BB
G0 GVT J. Gibb, 37 St. James Road, Melton, North Ferriby, HU14 3HZ
G0 GVX B. Sherwood, 363 Old Laira Road, Laira, Plymouth, PL3 6DH
G0 GVZ A. Grimes, 37 Cavendish Avenue, Cambridge, CB1 7UR
G0 GWA Dr S. Browne, 108 Whirley Road, Macclesfield, SK10 3JL
G0 GWC D. Roberts, 179 Southfield Avenue, Preston, Paignton, TQ3 1JX
G0 GWD A. Brown, 6 Laing Square, Wingate, TS28 5JE
GW0 GWE R. Edwards, 11 Trem Y Eglwys, Coed-Y-Glyn, Wrexham, LL13 7QE
G0 GWG C. Partridge, 44 Pine Close, South Wonston, Winchester, SO21 3EB
G0 GWH A. Clampitt, 139 Smiths Rd, Emerald Beach, Australia, 2456
G0 GWI P. Bell, 6 Dorchester Close, Hale, Altrincham, WA15 8PW
G0 GWL D. Baker, 2 Hall Lea Sedgefield, Stockton-on-Tees, TS21 2AN
G0 GWM R. Sawkins, 48 South Road, Saffron Walden, CB11 3DN
G0 GWN J. Faulconbridge, 32 Beridge Road, Halstead, CO9 1LB
G0 GWP A. Horton, 184 Mount Pleasant, Keyworth, Nottingham, NG12 5ET
G0 GWS W. Duff, Highfield, Maundown, Taunton, TA4 2BU
G0 GWY G. Birch, 33 Kenilworth Road, Scunthorpe, DN16 1EY
G0 GXF P. Hirst, 57 Etherington Drive, Hull, HU6 7JT
G0 GXH P. Rogers, 20 Clayfield Close, Nottingham, NG6 8DG
G0 GXI S. Suter, Fair View, Station Road, York, YO62 4DG
G0 GXO H. Swaddle, 12 Belmont Gardens, Haydon Bridge, Hexham, NE47 6HG
GW0 GXQ J. Williams, 91 Mold Road, Buckley, CH7 2JA
G0 GXS R. Wareham, 8 Meeting Lane Needingworth, St Ives, PE27 4SN
G0 GXT D. Mellor, The Village Stores, Cleobury North, Bridgnorth, WV166RP
G0 GXU M. Reeve, 25 Chiltern Road, Hitchin, SG4 9PJ
G0 GXX M. Smallwood, 12 Rowan Walk, Hornsea, HU18 1TT
G0 GXZ M. Butler, 1A Springhead Avenue, Hull, HU5 5HZ
G0 GYA C. Taylor, 37 Manor Park Avenue, Allerton Bywater, Castleford, WF10 2DN
G0 GYH D. Goodall, 5 Coach Drive, Eastwood, Nottingham, NG16 3DR
G0 GYI R. Griffith, Wayside, Colchester Main Road, Colchester, CO7 8DH
G0 GYJ M. Dawson, 11 Eastholme Drive, York, YO30 5SU
GM0 GYM T. Quinn, 40 Drumry Road, Clydebank, G81 2LL
GM0 GYN P. Gibson, 7 Rogerhill Drive, Kirkmuirhill, Lanark, ML11 9XS
G0 GYO M. Mcpherson, 4 Highfield Place, Wideopen, Newcastle upon Tyne, NE13 7HW
G0 GYP C. Fiedler, 51 Fleet Lane, Tockwith, York, YO26 7QD
GM0 GYQ Dr H. Garmany, Woodside Cottage, Shielhill, Dundee, DD4 0PW
GM0 GYT J. Ritchie, 24 Kirkton Crescent, Knightswood, Glasgow, G13 3AQ
G0 GYU P. Walters, Stonecroft, Main Street, Knaresborough, HG5 9LD
G0 GYY P. Markham, 15 Victoria Road, Walton on the Naze, CO14 8BU
G0 GZB J. Bartlett, 44 Beverington Road, Eastbourne, BN21 2SD
G0 GZE P. Wylie, 15 Semley Road, Hassocks, BN6 8PD
G0 GZF A. Pierce, 17B Alderford Street, Sible Hedingham, Halstead, CO9 3HX
G0 GZI R. Jeffery, 7 Corfe Way, Winsford, CW7 1LU
G0 GZL R. Daniels, 8 Dun Cow Close, Brinklow, Rugby, CV23 0NZ
G0 GZM M. Johns, 3 Carew Close, Coulsdon, CR5 1QS
G0 GZN L. Edmunds, 27A Sea View Road, Parkstone, Poole, BH12 3LP
G0 GZO K. Walters, 18 Leabrooks Avenue, Sutton-in-Ashfield, NG17 5HU
G0 GZP B. Lee, 18 Boston Close, Eastbourne, BN23 5RA
GW0 GZR M. Heel, 27 Englefield Drive Oakenholt, Flint, CH6 5SB
G0 GZU N. Harris, 16 Gibbs Field, Bishop's Stortford, CM23 4EY
G0 GZV K. Bailey, 35 Edgehill Road, Chislehurst, BR7 6LA
G0 HAE R. Isaac, 12 Abbeyfields Close, Netley Abbey, Southampton, SO31 5GR
G0 HAK P. Owens, Flat 1, 45 The High Street, Enfield, EN3 4EF
G0 HAL A. Paterson, 1 Birch Grove, Timperley, Altrincham, WA15 7YH
GM0 HAN K. Monaghan, 10 Sauchiewood Cottages, Mintlaw, Peterhead, AB42 5LR
G0 HAS A. Jordan, 30 Spring Meadows, Trowbridge, BA14 0HD
G0 HAU E. Hazell, 12 Fulford Close, Bideford, EX39 4DX
G0 HAW C. Blackmoor, 4 St. Godwalds Crescent, Aston Fields, Bromsgrove, B60 2EB
G0 HBA R. Curzon, 24 Edwards Drive, Wellingborough, NN8 3JJ
G0 HBB D. Brown, 15 Osborne Avenue, Tuffley, Gloucester, GL4 0QN
GW0 HBD A. Kings, 13 Fairfield Road, Bulwark, Chepstow, NP16 5JP
GM0 HBF C. Fraser, Rockside, Locheport, Isle of North Uist, HS6 5EU
G0 HBJ R. Millett, 8 Sidestrand Road, Newbury, RG14 6HP
GM0 HBK C. Robertson, 3 Sasaig, Teangue, Isle of Skye, IV44 8RD
G0 HBL K. Alderman, How Green Farm, Baldock Road, Buntingford, SG9 9RH
G0 HBN W. Riley, 18 Leyland Close, Trawden, Colne, BB8 8TB
G0 HBO S. Holmes, 17 Portland Gardens, Low Fell, Gateshead, NE9 6UX
G0 HBS R. Darlington, 7 Binbrook Place, Chorley, PR7 2QU
G0 HBU A. Dwyer, 10 Shaftway Close, Haydock, St Helens, WA11 0YQ
GM0 HBV T. Sandilands, 8 Cheviot View, Lowick, Berwick-upon-Tweed, TD152TY
G0 HBW R. Fitzgerald, 93 Finland Road, Brockley, London, SE4 2JQ
G0 HBX J. Turbefield, 126 Preston Road, Chorley, PR6 7AU
GW0 HBZ D. Wright, Drws-Y-Nant, St. Asaph Avenue, Kinmel Bay, Rhyl, LL18 5EY
GW0 HCB P. Matheson, 34 Pentwyn, Radyr, Cardiff, CF15 8RE
G0 HCC Herts County C/, c/o T. Groves, 31 Tunnel Wood Close, Watford, WD17 4SW
G0 HCD T. Carroll, 2 West Durham Cottages, Roddymoor, Crook, DL15 9QX
G0 HCE M. Wilkinson, 98 Whitestiles, High Seaton, Workington, CA14 1LL
G0 HCI M. Bodle, 10 Watts Road, Hedge End, SO304EZ
GW0 HCK A. Morgan, 3 Gelli Newydd, Golden Grove, Carmarthen, SA32 8LP
GW0 HCN T. Hayden, 7 Attlee Close, Garnlydan, Ebbw Vale, NP23 5ES
G0 HCP Rev. D. Goode, 12 Hills Lane, Ely, CB6 1AY
GM0 HCQ M. Gloistein, 27 Stormont Way, Scone, Perth, PH2 6SP
G0 HCR S. Turner, 71 Valley Road Lillington, Leamington Spa, CV32 7RX
G0 HCY G. Vine, 56 Colchester Road, St. Osyth, Clacton-on-Sea, CO16 8HB
G0 HDB A. Davies, 11 Gravel Pits Close, Bredon, Tewkesbury, GL20 7QL
G0 HDC G. Evans, 34 Coronation Drive, Donnington, Telford, TF2 8HY
G0 HDD A. Adey, 37 Cranmere Avenue, Wolverhampton, WV6 8TR
G0 HDF R. Bartlam, 34 Quarry Road, Selly Oak, Birmingham, B29 5NX
G0 HDG A. Edwards, 29 Larch Road, Maltby, Rotherham, S66 8AZ
G0 HDH C. Worsfold, 7 West View Road, Cowes, PO31 8NR
G0 HDI B. Walker, Lantilla, Elmfield Lane, Calshot, Southampton, SO45 1BJ
G0 HDJ A. Douglas, Threave House, Blind Lane, Barton St David, TA11 6BW
GI0 HDO M. Deehan, 9 Farland Way, Londonderry, BT48 0RS
G0 HDP B. Wainwright, 19 Durkar Low Lane, Durkar, Wakefield, WF4 3BL
G0 HDS R. Hurt, 7 Atwood Close, Immingham, DN40 2DQ
GW0 HDY R. Finnis, 6 The Crwscent, Cwmbran, NP44 7JG
G0 HDZ I. Rose, 82 Little Brays, Harlow, CM18 6ES
G0 HEA B. Griffin, 26 Hamer Street, Radcliffe, Manchester, M26 2RS
G0 HEE A. Salt, New Meg, Greetwell Lane Nettleham, Lincoln, LN2 2NQ
G0 HEF R. Mckeever, 38 Brookfield Avenue, Runcorn, WA7 5RF
G0 HEJ E. Lloyd, 43 Queensway, Upton, Chester, CH2 1PF
G0 HEL A. Nicholls, 19B Dark Lane South, Steeple Ashton, Trowbridge, BA14 6EZ
G0 HEM G. Gardner, New House, Birdbush Avenue, Saffron Walden, CB11 4DJ
G0 HEN G. Henstock, 36 Cornwallis Road, Oxford, OX4 3NW
G0 HER K. Kibblewhite, 53 Woodcote, Bedford, MK41 8EL
G0 HET P. Nutkins, 31 Higher Spence Cottage, Bridport, DT6 6DF
G0 HEU P. Stott, 70 Wansbeck, Washington, NE38 9EG
G0 HEV P. Brindley, 6 Chaucer Close, Stowmarket, IP14 1GH
G0 HEW J. Mchale, 21 Bonython Road, Newquay, TR7 3AW
G0 HEX D. Cheetham, 4 Battersbay Grove, Hazel Grove, Stockport, SK7 4QW
G0 HFA Dr J. Lansdowne, 8 Nansloe Close, Helston, TR13 8BP
G0 HFC F. Chadwick, 59 Beech Avenue, Greenfield, Oldham, OL3 7AW
G0 HFE T. Cadman, 28 Denbigh Court, Ellesmere Port, CH65 5DX
G0 HFK R. Fuller, Flat 13, Samuel Lewis Trust Dwellings, London, SW6 1BS
G0 HFL N. Major, 10 Heather Close, Branston, DE14 3FL
G0 HFO D. Hirst, 10 The Rogers, Shanklin, PO37 7HH
G0 HFX C. Parnell, 29 Southfield, Southwick, Trowbridge, BA14 9PW
GW0 HGC B. Roberts, Ty Clyd, Ffordd Mela, Pwllheli, LL53 5AP
G0 HGG S. Entwisle, 30 Arden Mhor, Pinner, HA5 2HR
G0 HGH J. Scott, 3 Westminster Drive, Spalding, PE11 2UW
G0 HGI B. Boden, 71 Park Head Road, Sheffield, S11 9RA
G0 HGM J. Jenkinson, 4 Greenways, Ilfracombe, EX34 8DT
GW0 HGN T. Jones, 11 Lon Ogwen, Bangor, LL57 2UD
G0 HGO D. Cady, 45 The Hill, Wheathampstead, St Albans, AL4 8PR
GW0 HGP Dr L. Magfhogartai, Talaharian, Abergwili, Carmarthen, SA31 2JL
G0 HGV D. Burgess, 243 Lichfield Avenue, Torquay, TQ2 8AJ
G0 HHA P. Dixon, 68 Chelsea Road, Sheffield, S11 9BR
G0 HHC A. Gorton, 7 Sterling Close, Colchester, CO3 9DP
GW0 HHD G. Williams, Gwastad Annas, Barmouth, LL42 1DX
GI0 HHE J. Dowey, 19B The Bridges, Newtownabbey, BT37 0TD
G0 HHN P. Le-Brun, The Granary, Henton, Chinnor, OX39 4AE
GI0 HHV S. Johnstone, 6 Gilbert Crescent, Bangor, BT20 4PE
GW0 HHW B. Hayward, Tyn Y Gerddi, Deiniolen, Caernarfon, LL55 3ND
GI0 HHZ R. Fitzsimons, 9 Ingledene Park, Newtownards, BT23 8QT
G0 HIC J. East, 35 Preachers Vale, Coleford, Radstock, BA3 5PT
G0 HID D. Paine, 35 Marfield Close, West Midlands, B76 1YD
G0 HIF A. Edmonds, 44 Blea Tarn Road, Kendal, LA9 7NA
GM0 HIG C. Cameron, 36 Lynn Crescent, Kirkwall, KW15 1FF
G0 HIJ W. Roberts, 13 Roseacre Road, Elswick, Preston, PR4 3UD
G0 HIK N. Gregory, Town End, Kirkby Road, Askam-in-Furness, LA16 7EY
GM0 HIM J. Pert, 56 Lochiel Drive, Milton Of Campsie, Glasgow, G66 8ET
GW0 HIR A. Edwards, Flat 6, Oaktree Court, Fields Road, Cwmbran, NP44 3AZ
G0 HIU M. Valentine, 34 Meadow Avenue, Preesall, Poulton-le-Fylde, FY6 0HA
G0 HIW D. Ross, 60 Kingsway, South Molton, EX36 4AL
G0 HIZ D. Hughes, 27 Thornfields, Crewe, CW1 4TY
G0 HJB F. Little, 26 Hoghton Road, Longridge, Preston, PR3 3UA
G0 HJD J. Ford, 42 The Grove, Walton-on-Thames, KT12 2HS
G0 HJK J. Everett, 92 Thackeray Road, Ipswich, IP1 6JB
G0 HJL J. Levesley, 96 Brookside Road, Bransgore, Christchurch, BH23 8NA
G0 HJM R. Smith, 100 Braemar Road, Billingham, TS23 2AN
G0 HJR R. Filby, 10 Malvern Avenue, Burton-on-Trent, DE15 9EB

GM0 HJU J. Mackenzie, 17 Maple Avenue, Milton Of Campsie, Glasgow, G66 8BB
GM0 HJV C. Mcclure, 27 St. Andrews Drive, Gourock, PA19 1HY
G0 HJW P. Webber, Palm Court Hotel, 1 Lansdowne Road, Falmouth, TR11 4BE
G0 HJX P. Devine, 41 Carodoc Road, Wingate, TS28 5BT
G0 HJZ J. Durrell, 8 Woodcote Cottages, Graffham, Petworth, GU28 0NY
G0 HKB R. Conneely, Harford House, Wells Road, Radstock, BA3 4EX
G0 HKC E. Chambers, 19, Courville Close, Alveston, BS35 3RR
G0 HKE J. Boyton, 3 Wenny Estate, Chatteris, PE16 6UX
G0 HKF A. Roberts, 149 Cannock Road, Burntwood, WS7 0BB
G0 HKN R. Leigh, 19 Richmond Court, Worthing, BN11 4JB
GW0 HKQ D. Suddes, Villa Dobochet, Holway Road, Holywell, CH8 7DR
G0 HKW D. Spencer, 48 Ingleborough Way, Leyland, PR25 4ZR
G0 HKZ S. Liptrott, 8 Fox Bank Close, Widnes, WA8 9DP
G0 HLA T. Barker, Peartree Cottage, Ash Hill Common, Romsey, SO51 6FU
G0 HLB Dr R. Evans, 191 St. Leonards Road East, Lytham St Annes, FY8 2HW
G0 HLI A. Dudley, 7 St. Michaels Close, Willington, Derby, DE65 6EB
G0 HLJ I. Barber, 17 Copley Crescent, Scawsby, Doncaster, DN5 8QW
GM0 HLK M. Borthwick, 5/11 Dalgety Avenue, Edinburgh, EH7 5UF
G0 HLL R. Steel, 43 Westfield Avenue, Wigston, LE18 1HY
G0 HLS S. Deacon, 32 Westfield Way, Charlton, Wantage, OX12 7EW
G0 HLU A. Mckay, 48 Holdenby Close, Retford, DN22 6UB
GM0 HLV D. Gill, Slowbend Cottage, Loth, Helmsdale, KW8 6HP
G0 HLW M. Neale, 126 Brookvale Road, Solihull, B92 7JB
G0 HMD M. Mills, 44 East Acridge, Barton-upon-Humber, DN18 5HH
G0 HME C. Statham, 24 St. Johns Close, Heather, Coalville, LE67 2QL
G0 HMF J. Tite, 40 Dinglederry, Olney, MK46 5ES
G0 HMG S. Finch, Combe Shorney Farm, Brompton Ralph, Taunton, TA4 2SB
G0 HMK T. Angier, 29 Sunnyhill Road, Herne Bay, CT6 8LT
GM0 HMM F. Robertson-Mudie, 18 Portnaguran, Isle of Lewis, HS2 0HD
G0 HMO E. Cooke, 190 Newark Crescent, Nottingham, NG2 4NY
G0 HMX M. Wragg, 27 Rosedale, Worksop, S81 0TB
G0 HND A. Simmonds1, Fern Way, Purssells Meadow, Naphill, High Wycombe, HP14 4SG
GW0 HNE K. Luke, 19 Heol Y Gors, Cwmgors, Ammanford, SA18 1PE
G0 HNG B. Horton, 3 Cromer Road, Finedon, Wellingborough, NN9 5LP
G0 HNI J. Buckler, 181 Glen Road, Oadby, Leicester, LE2 4RJ
GM0 HNJ J. Cowan, 20A Harbour View, Invergordon, IV18 0EY
G0 HNL T. Cannon, 10 Badger Close, Guildford, GU2 9PJ
G0 HNO T. Callanan, 39 Greenlands Way, Henbury, Bristol, BS10 7PH
GM0 HNP E. Bottomley, 33 Duke Street, Coldstream, TD12 4BS
G0 HNQ H. Brammeld, School House, Rosley, Wigton, CA7 8AU
GW0 HNS S. Yates, 46 Y Berllan, Dunvant, Swansea, SA2 7RW
GW0 HNT W. South, Kimberlee, 45 Dan Y Bryn, Neath, SA11 3PJ
GM0 HNV P. Mackenzie, 16 Cedar Drive, Milton Of Campsie, Glasgow, G66 8AY
G0 HNW P. Widger, Notre Revie, Cinderhills Road, Holmfirth, HD9 1EH
G0 HNZ R. Disney, 25 Davos Way, Skegness, PE25 1EL
G0 HOB J. Mc Glynn, 1 Primrose Crescent, Leeds, LS15 7QW
G0 HOC M. Harris, Po Box 841, Hobart, Tasmania, Australia, 7001
G0 HOD N. Colbourn, 7 Brookwood Road, Farnborough, GU14 7HH
G0 HOF K. Barnett, 5 Morborne Road, Folksworth, Peterborough, PE7 3SS
G0 HOJ J. Hoyland, 67 Coronation Road, Wroughton, Swindon, SN4 9AT
G0 HOP G. Evans, 12 Marlowe Close, Stevenage, SG2 0JJ
G0 HOQ M. Piper, Applegarth Milford Road, Barton on Sea, BH25 5PW
G0 HOS B. Tufnail, 197 Portland Road, Wyke Regis, Weymouth, DT4 9BH
G0 HOT M. Reeds, 6 Leach Way, Riddlesden, Keighley, BD20 5DB
G0 HOV A. Howes, 11 Stretton Road, Wolston, Coventry, CV8 3FR
G0 HOX I. Atkins, 64 Twin Hill Lane, Stafford, Virginia, USA,
G0 HPA G. Smith, 3 Park Lane, Featherstone, Pontefract, WF7 6BL
GW0 HPC G. Evans, 49 Pen Yr Ally Avenue, Neath, SA10 6DS
G0 HPG L. Brookes, 177 Charnwood Close, Rubery, Birmingham, B45 0JY
G0 HPH P. Brookes, 177 Charnwood Close, Rubery, Birmingham, B45 0JY
GM0 HPK A. Gaston, 9 Lochans Mill Avenue, Lochans, Stranraer, DG9 9BZ
GM0 HPL T. Mccutcheon, 25 Millburn Court, Sheuchan St., Stranraer, DG9 0DX
G0 HPM R. Waddingham, 59 Second Avenue, Frinton-on-Sea, CO13 9LY
G0 HPN L. Denyer, 7 Long Close, Chippenham, SN15 3JY
G0 HPQ P. Delaney, 27 Grasmere Road, Redcar, TS10 1JA
G0 HPS G. Bukin, Granary Cottage, Uffculme, Cullompton, EX15 3DN
G0 HPV K. Hardy, 12 Liquorpond Street, Boston, PE21 8UF
GM0 HQF E. Kolonko, West Manse, School Road, Kilbirnie, KA25 7LB
GM0 HQG G. Flett, Stenadale, Orphir, Orkney, KW17 2RF
G0 HQH S. Mayer, 7 Wright Avenue, Chesterton, Newcastle under Lyme, ST5 7PB
G0 HQK J. Jones, 28 Clares Lane Close, The Rock, Telford, TF3 5DA
G0 HQN C. Wilson, 34 Lyndhurst St., Salford, M6 5YB
GM0 HQT B. Bell, 4 Broadlee Bank, Tweedbank, Galashiels, TD1 3RF
G0 HQU S. Broadhead, 39 Haw Avenue, Yeadon, Leeds, LS19 7XE
G0 HQX R. Rea, 11 Wissage Lane, Lichfield, WS13 6DQ
G0 HRD L. Lawrence, 119 Ladywood Road, Lane End, Dartford, DA2 7LP
G0 HRF B. Barwick, 100 Westwood, Golcar, Huddersfield, HD7 4JY
GW0 HRG Halkyn Radio Group, c/o P. Clark, 14 Lincoln Road, Ewloe, Deeside, CH5 3RW
G0 HRH G. Vaughton, Higher Woodhayne, Whitford, Axminster, EX13 7PB
G0 HRJ C. Mason, 22 Eskdale Avenue, Halifax, HX3 7NH
G0 HRK R. Eselgroth, 15 Hedgerley Gardens, Greenford, UB6 9NT
G0 HRL G. Ward, 67 Sebright Road, Wolverley, Kidderminster, DY11 5UA
G0 HRO J. Bland, 5 King Street, Mansfield, NG18 2PX
G0 HRR K. Mott, 191 Joyners Field, Harlow, CM18 7QD
G0 HRS Hilderstone Radio and Electronic Society, c/o I. Lowe, 54 College Road, Margate, CT9 2SW
G0 HRT R. Harwood, 4 Bartholomew Close, Walton Park, Milton Keynes, MK7 7HH
G0 HRW Wigan & Dist AR, c/o D. Barkley, 39 Fulbeck Avenue, Wigan, WN3 5QN
G0 HRX C. Deakin, Wainscot, Lanreath, Looe, PL13 2NX
G0 HSA A. Bennett, 21 Hunstone Avenue, Norton, Sheffield, S8 8GE
GI0 HSB W. Dickson, 8 Drumglass Avenue, Bangor, BT20 3HA
GM0 HSC H. Cumming, 42 Hadrian Way, Bo'ness, EH51 9QN
G0 HSD A. Shaw, 15 Austerby, Bourne, PE10 9JJ
G0 HSH D. Denford, 20 Shaxton Crescent, New Addington, Croydon, CR0 0NU
G0 HSK J. Hakes, 86 Station Crescent, Rayleigh, SS6 8AR
G0 HSN P. Broadley, 8 Langley Gardens, Lowestoft, NR33 9JE
G0 HSR Huntingdon ARS, c/o A. Dunham, 28 Kingfisher Close, Chatteris, PE16 6TP
G0 HSV S. Jacob, 3 Broadfield Close, Gomeldon, Salisbury, SP4 6LX
G0 HSW B. Statham, 11 Old Woods Hill, Torquay, TQ2 7NR
G0 HSX J. Harvey, 38 Honey Close, Dagenham, RM10 8TE
G0 HTD H. Todd, 105 Brownhill Road, Blackburn, BB1 9QY
G0 HTG P. Hague, 14 Camellia Close, Driffield, YO25 6QT
GM0 HTH J. Grieve, Langamo Harray, Kirkwall, KW17 2JU
G0 HTK T. St John-Murphy, Sherbourne, Sherbourne Drive, Windsor, SL4 4AE
G0 HTL B. Sargent, 25 Jordans Way, Bricket Wood, St Albans, AL2 3SJ
G0 HTM Dr R. Bushell, 121 Rickmansworth Road, Watford, WD18 7JD
G0 HTO V. Morris, 16 Wentworth Close, Longlevens, Gloucester, GL2 9RB
G0 HTS S. Alder, 56 Sparrowbill Way, Patchway, Bristol, BS34 5AU
GM0 HTT A. Flett, Shannon Dounby, Orkney, KW17 2HR
G0 HTX A. Coyle, 8 Farm Place, Kensington, London, W8 7SX
G0 HUD D. Withers, 141 Broadway, Walsall, WS1 3HB
G0 HUF P. Peterson, 13 Orchard Road, Smallfield, Horley, RH6 9QP
G0 HUG P. Hemphill, Springhill, George Lane, Glemsford, Sudbury, CO10 7SB
G0 HUH J. O'Dowd, 139A Town Lane, Denton, M342DJ
G0 HUK G. Roberts, 24 Cornwall Road, Bingley, BD16 4RN
GW0 HUN M. Roberts, 19 Bro Geirionydd, Trefriw, LL27 0JE
G0 HUQ R. Monk, 43 Lichfield Drive, Bury, BL8 1BJ
G0 HUT T. Hutton, 4 Victoria Gardens, Farnborough, GU14 9UH
G0 HUW A. Dyson, 24 Newborough Close, Austrey, Atherstone, CV9 3EX
G0 HUX D. Griffin, 62 Buckshaft Road, Cinderford, GL14 3AX
G0 HVA J. Curtis, Glebe Farm, West Knighton, Dorchester, DT2 8PE
G0 HVB D. Clark, 3 West Well Lane, Theale, Wedmore, BS28 4SW
G0 HVC W. Hutchins, 25 Manor Road, Herne Bay, CT6 6RF
GI0 HVJ R. Cunliffe, 82 Strabane Road, Newtownstewart, Omagh, BT78 4JZ
G0 HVN D. Cottam, 14 Barnard Close, Rednal, Birmingham, B45 9SZ
G0 HVO R. Minter, 15 Harold Close, Pevensey Bay, Pevensey, BN24 6SL
G0 HVP D. Cooper, 11 Downland Court, Magazine Road, Ashford, TN24 8NF
GM0 HVS D. Kearns, Mo Dhachaidh, Portnacroish, Appin, PA38 4bl
G0 HVX R. Mcquillan, 9 Sandpit Road, Welwyn Garden City, AL7 3TW
GD0 HWA C. Howard, 5 Ballure Grove, Ramsey, Isle Of Man, IM8 1NF
GM0 HWB P. Lafferty, 18 Chesters Crescent, Motherwell, ML1 3QU
G0 HWC P. Young, 14 Carisbrooke Avenue, Clacton-on-Sea, CO15 4RZ
G0 HWI P. Mardle, 55 Chelmsford Avenue, Southend-on-Sea, SS2 6JG
G0 HWK M. Drew, 10 Marina Drive, Dunstable, LU6 2AH
GI0 HWO J. Crawford-Baker, 131 Gobbins Road, Islandmagee, Larne, BT40 3TX
G0 HWP A. Wilson, 41 Bentley Drive, Walsall, WS2 8RX
G0 HWQ P. Wood, 2 Kingsbridge Close, Braintree, CM7 5NB
G0 HWS P. Dowsett, Furze Cottage, West Chiltington Road, Pulborough, RH20 2PR
G0 HWT P. Littlechild, 13 Castle Hill, Daventry, NN11 4AQ
G0 HWU A. Edwards, Higher Tregiddle Farm, Gunwalloe, Helston, TR12 7QW
G0 HWV B. Lawton, 9 Byron Court, Kidsgrove, Stoke-on-Trent, ST7 4JF
G0 HWY P. Adams, 229 Upper Selsdon Road, South Croydon, CR2 0DZ
G0 HXC F. Pegg, 20 Swaledale Avenue, Cowpen Estate, Blyth, NE24 4DT
G0 HXD P. Halsall, 22 Northway, Northwich, CW8 4DF
G0 HXF A. Etheridge, Wyngra, Haywards Heath Road, Haywards Heath, RH17 6NJ
GI0 HXH A. Mccaldin, 7 Mount Ida Road, Banbridge, BT32 4HF
G0 HXL E. Calthorpe, 49 Cross Coates Road, Grimsby, DN34 4QH
G0 HXM A. Delves, 11 Willoughby Road, Langley, Slough, SL3 8JH
G0 HXR G. Martin, 34 Parson Drove, West Pinchbeck, Spalding, PE11 3QW
GW0 HXS T. Williams, Bryndewi, Llanarth, SA47 0QN
G0 HXU S. Dawson, Hamlet Cottage, West End, Tadcaster, LS24 9DL
G0 HYG R. Taylor, 79 South Drive, Harwood, Bolton, BL2 3NS
GW0 HYH V. Grayson, Willow Lodge, Croeslan, Llandysul, SA44 4SJ
GW0 HYL V. Underwood, Rock Hill, Llanarthney, Carmarthen, SA32 8LJ
GD0 HYM M. Dunning, 55 Station Park, Colby, Isle Of Man, IM9 4NL
G0 HYN D. Robertson, 5 Chandlers, Orton Brimbles, Peterborough, PE2 5YW
G0 HYP K. Ferguson, 4 Honister Road, Whitehaven, CA28 8HS
G0 HYR R. Deakin, 5 Burton Road, Oakthorpe, Swadlincote, DE12 7QU
G0 HYS S. Hutchinson, 55 Richmond Avenue, Sheffield, S13 8TH
G0 HYT P. Gray, 2 Bryan Close, Sunbury Oppn Thames, TW16 7UA
GW0 HYU K. Richards, 7 Capel Eyedyrn, Pontprennau, Cardiff, cf238xj
G0 HZA M. Rolph, 60 Queen Street, Swaffham, PE37 7BT
G0 HZB M. Roberts, 4 Elba Close, Paignton, TQ4 7LW
G0 HZC Dr G. Hutt, 21 Bentley Crescent, Fareham, PO16 7LU
G0 HZD W. Brown, Longstone Cottage, Old Pound, St Austell, PL26 7XS
G0 HZE R. Howell, 161 Coneygree Road, Stanground, Peterborough, PE2 8LH
G0 HZG P. Sturgess, 45 Queensmead Close, Groby, Leicester, LE6 0YP
GM0 HZI N. Mclaren, 10 Newton Avenue, Skinflats, Falkirk, FK2 8NP
G0 HZK R. Muggleton, 70 Front Street East, Wingate, TS28 5AG
G0 HZL M. Hawkins, Flat 8, Library Mews, Hawes Side Lane, Blackpool, FY4 5RF
GM0 HZO T. Leckie, 1 Dykehead, Port Of Menteith, Stirling, FK8 3JY
G0 HZQ R. Eldrett, 20 Hill Rise, Horspath, Oxford, OX33 1TJ
G0 HZY H. Harding, 29 Brighton Avenue, Elson, Gosport, PO12 4BU
G0 IAA H. Singer, Neptune Gap, Kent, CT5 1EL
G0 IAC C. Caspell, 28 Peel Terrace, Stafford, ST16 3HD
G0 IAD A. Dobbyn, Roadways, Selwick Drive, Bridlington, YO15 1AP
G0 IAE D. Yeo, 356 Radcliffe Road, Fleetwood, FY7 7NH
G0 IAG A. King, 2 Ebenezer Cottages, Thorney Road, Peterborough, PE6 7UB
G0 IAH T. Preece, White House, Bredwardine, Hereford, HR3 6BY
G0 IAI P. Dean, Down Farm, Lovaton, Yelverton, PL20 6PT
G0 IAK M. Watts, Hainault, Coronation Avenue, Bradford-on-Avon, BA15 1AX
G0 IAL R. Scott, 46 St. Albans Road, Hemel Hempstead, HP2 4BA
G0 IAP P. Allen, 25 Wayside Avenue, Hornchurch, RM12 4LL
G0 IAS A. Hickman, The Conifers, High Street, Retford, DN22 8AJ
G0 IAX R. Bailey, 3 Charlotte Close, Birstall, Batley, WF17 9BX
G0 IAY L. Wiltshire, 7 Burleaze, Chippenham, SN15 2AY
GI0 IBC Radio Amateur Invalid and Blind Club, c/o P. Hallam, 95 Belfast Road, Carrickfergus, BT38 8BY
G0 IBE R. Higgs, 60 Lichfield Avenue, Evesham, WR11 3EA
G0 IBG B. Armstrong, 65 Asquith Road, Bentley, Doncaster, DN5 0NT
G0 IBI R. Clarke, 14 Cranwell Avenue, Cranwell, Sleaford, NG34 8HG
G0 IBJ R. Bishop, 25 Hennings Park Road, Poole, BH15 3QU
G0 IBN A. Kersey, 35 Sceptre Close Tollesbury, Maldon, CM9 8XB
G0 IBR S. Dalley, 5 Anstey Mill Close, Alton, GU34 2QT

G0 IBS K. Brown, Fernbank, Foreside Lane, Bradford, BD13 4EY
G0 IBT H. Percy, 13 Cherry Tree Walk, Astley Cross, Stourport-on-Severn, DY13 0JT
G0 IBW D. Jones, 5 Luccombe Close, Ingleby Barwick, Stockton-on-Tees, TS17 0NL
G0 IBY F. Hubbard, 187 Standhill Road, Carlton, Nottingham, NG4 1LE
G0 IBZ M. Hackford, Snuggles, Rockalls Road, Colchester, CO6 5AR
G0 ICB J. Howell, 21 Peaslands Road, Saffron Walden, CB11 3ED
G0 ICC R. Shirvington, 3 Main Street, Preston Bissett, Buckingham, MK18 4LH
G0 ICD D. Matthews, 32 Stewards Avenue, Widnes, WA8 7BN
G0 ICE M. Knott, 24 Walsingham Way, Ely, CB6 3AL
G0 ICG Sir T. Lees, Post Green, Lytchett Minster, Dorset, BH16 6AP
G0 ICJ D. Dawkes, 95 Houndsfield Lane, Wythall, Birmingham, B47 6LX
G0 ICK Sth Tottenhm AR, c/o I. Kraven, 55 Cranfield Crescent, Cuffley, Potters Bar, EN6 4DZ
G0 ICP Dr J. Ker, 1 The Willows, Ulcombe Road, Ashford, TN27 9QR
G0 ICW M. Bagnall, 6 Silver Fir Close, Hednesford, Cannock, WS12 4SU
G0 IDB S. Jeffreys, 26 Meadway, Esher, KT10 9HF
G0 IDD D. Clarke, 60 Dunedin Crescent, Burton-on-Trent, DE15 0EJ
G0 IDE D. Dobson, 79 Wood Green, Leyland, PR25 2YL
G0 IDF D. Fletcher, 12 Drayton Road, Dorchester-On-Thames, Wallingford, OX10 7PJ
G0 IDH M. Frost, Kernyk, Trevanion Terrace, Carn Brea Village, Redruth, TR15 3BP
GM0 IDJ J. Low, 56 George St., Whithorn, Newton Stewart, DG8 8NZ
G0 IDL M. Bedell, 1 Pheasant Field Drive, Spondon, Derby, DE21 7LR
G0 IDP D. Ford, 2 Bedelands Close, Burgess Hill, RH15 8DL
G0 IDS S. Mcmaster, 12 Gilpin Road, Newton Aycliffe, DL5 5EQ
GD0 IDU N. Bayliss, Teal, Tromode Road, Douglas, Isle Of Man, IM2 5EH
GM0 IDV A. Price, Upper Arsdale, Evie, Orkney, KW17 2NN
G0 IDZ L. Bailey, 22 Coventry Grove, Wheatley, Doncaster, DN2 4QA
G0 IEB R. Neal, 6 Wheatcroft Avenue, Scarborough, YO11 3BN
G0 IEE K. Harris123, 14 Dunstall Close, St. Marys Bay, Romney Marsh, TN29 0QX
G0 IEH T. Lister, 6 Fordlands Crescent, Fulford, York, YO19 4QQ
G0 IEN Dr R. Wharton, 1407 Briar Bayou Dr, Houston, USA, 77077
G0 IEO R. Woodberry, 35 Whybridge Close, Rainham, RM13 8BB
G0 IEQ G. Kennedy, 1 Hornbeam Avenue, Great Sutton, Ellesmere Port, CH66 2US
G0 IER B. Smith, 73 Devon Street, Hull, HU4 6PL
G0 IES D. Johnson, 32 Middleton Road, North Reddish, Stockport, SK5 6SH
GM0 IET M. Shell, 15 Lundin View, Leven, KY8 5TL
G0 IEW J. Rose, 1 Nelson Place, Whiston, Prescot, L35 3PP
G0 IEY S. Tribe, 6 Privett Road, Waterlooville, PO7 5HJ
G0 IFA K. Keitch, 10 Sycamore Close, Willand, Cullompton, EX15 2SH
G0 IFC A. Burnett, 2 Courtney Road, Tiverton, EX16 6EE
G0 IFD T. Street, Flat 18, 58 Mapledene Road, London, E8 3LE
G0 IFF G. Spinney, 7 Nightingale Gardens Nailsea, Bristol, BS48 2BH
G0 IFL R. Finch, 6 Clover Way, Thetford, IP24 1LQ
G0 IFN R. Moriarty, 46 Oak Avenue, Morecambe, LA4 6HS
G0 IFQ P. Lait, Glenside, 8 Kingston Lane, Shoreham-by-Sea, BN43 6YB
G0 IFS N. Clayton, Ringinglow, Fair Lawn, Whitstable, CT5 3JZ
G0 IFT P. Nurse, 67 Grasleigh Way, Allerton, Bradford, BD15 9BD
GD0 IFU W. Corkish, 23 Bollan Drive, Ballagarey, Glen Vine, Isle of Man, IM4 4FE
G0 IFW M. Mcnamara, 7 Moray Road, Chadderton, Oldham, OL9 8AE
G0 IFX B. Galloway, 2 Summers Close, Knutsford, WA16 9AW
G0 IGA R. Clark, 9 Windsor Close, Hove, BN3 6WQ
G0 IGB V. Elliott, 22 Kirkstead Road, Cheadle Hulme, Cheadle, SK8 7PZ
G0 IGC J. Golightly, 123 Littlefield Lane, Grimsby, DN34 4PN
G0 IGH G. Golightly, 123 Littlefield Lane, Grimsby, DN34 4PN
GM0 IGJ J. Dickson, Eilrig, Roberton, Hawick, TD9 7PR
G0 IGK G. Knox, 33 St. Lukes Road, Aller, Newton Abbot, TQ12 4NE
G0 IGM M. Manning, 31 Harcourt Crescent, Shrewsbury, SY2 5LQ
G0 IGT H. Terry, 7 Marlow Drive, Irlam, Manchester, M44 6LR
G0 IGU G. Morton, 23 Bembridge Road, Eastbourne, BN23 8DX
G0 IHA G. Goodwin, Thumpers, 16 St. Catherines Road, Winchester, SO23 0PP
G0 IHC M. Nash, 49 Oakfield Way, Sharpness, Berkeley, GL13 9UT
G0 IHE Q. Reed, 3 Carre Gardens, Worle, Weston-Super-Mare, BS22 7YB
G0 IHF W. Hough, 19 Farnham Close, Appleton, Warrington, WA4 3BG
G0 IHI M. Eyers, 190 Greenhill Road, Herne Bay, CT6 7RS
G0 IHK I. Tough, 15 Headlands Way Whittlesey, Peterborough, PE7 1RL
G0 IHO R. Priestley, Le Haut Courtigne, Parcay Les Pins, Maine Et Loire, France, 49390
G0 IHU C. Smith, 66 Bronte Farm Road, Shirley, Solihull, B90 3DF
G0 IIA J. Stokes, The Beehive, Debenham Road, Ipswich, IP6 9TD
G0 IID J. Batley, 3 Folldon Avenue, Sunderland, SR6 9HP
G0 IIE J. Colliton, Po Box 17, Ramsgate 2217, Sydney, Australia, 2219
G0 IIF J. Green, 1 Huntley Grove, Sheffield, S11 7LX
G0 IIG G. Fleming, 168 Blythway, Welwyn Garden City, AL7 1DU
G0 IIK N. Ackland, 69 Great South West Road, Hounslow, TW4 7NH
GM0 IIO G. Berrich, Juno Lodge The Woods Diverswell Farm Fishcross, Alloa, FK10 3AN
G0 IIP R. Chambers, 12 Dorchester Close, Maidenhead, SL6 6RX
G0 IIQ D. Pykett, 35 Harneis Crescent, Laceby, Grimsby, DN37 7BA
GI0 IJB J. Stevenson, 22 Knockfergus Park, Greenisland, Carrickfergus, BT38 8SN
GM0 IJD P. Taggerty, 44 Ravenswood Drive, Glenrothes, KY6 2PA
G0 IJI A. Muir, 8 Whitley Drive, Halifax, HX2 9SJ
G0 IJK B. Puncher, Danbys Oast, Coldbridge Lane, Maidstone, ME17 2AX
G0 IJN A. Slade, Skerries, Summerhill Althorne, Chelmsford, CM3 6BY
GM0 IJR I. Ross, 14 Kilmundy Drive, Burntisland, KY3 0JW
G0 IJU P. Elms, 12 Suffield Way, King's Lynn, PE30 3DE
GW0 IJY I. Roberts, 7 Heol Bradwen, Four Mile Bridge, Holyhead, LL65 2NF
G0 IJZ Dr M. Walden, 181 Coleridge Road, Cambridge, CB1 3PW
G0 IKB K. Brown, 1 Deerwood Close, Macclesfield, SK10 3RE
G0 IKC R. Hole, 3 Holywell Park, Halwill, Beaworthy, EX21 5UD
G0 IKD A. Galvin, 27 Hill Top Lane, Tingley, Wakefield, WF3 1HT
G0 IKE L. Dodson, 1 Limmer Lane, Booker, High Wycombe, HP12 4QR
G0 IKI M. Holbrough, 21 Malyns Close, Chinnor, OX39 4EW
G0 IKN S. Mcdonald, 152 Stony Lane, Burton, Christchurch, BH23 7LD
G0 IKP T. Clements, 72 Stanbridge Road, Haddenham, Aylesbury, HP17 8HN
G0 IKQ A. Cook, 17 Lothersdale, Wilnecote, Tamworth, B77 4HT
G0 IKR M. Best, 21 Hubble Road, Corby, NN17 1JD
GM0 IKY A. Diamond, 51 Marchlands Avenue, Bo'ness, EH51 9ER
G0 IKZ J. Pugh, 1 The Lawns, Everton, Sandy, SG19 2LB
G0 ILA N. Smith, 165 Upper Deacon Road, Southampton, SO19 5LN
GM0 ILB I. Brown, Vadill, Brae, Shetland, ZE2 9QN
G0 ILC J. Price, 30 Pottery Close, Whiston, Prescot, L35 3RW
G0 ILD B. Harrison, 61 Foyle Road, Blackheath, London, SE3 7RQ
G0 ILH J. Collins, 14 Balmain Crescent, Wolverhampton, WV11 1BG
G0 ILI G. Pomroy, 17 Rock Close, Pengegon, Camborne, TR14 7TT
G0 ILK B. Loram, 12 The Finches, Castleham, St Leonards-on-Sea, TN38 9LQ
G0 ILN R. Putnam, 95 Martyns Way, Bexhill-on-Sea, TN40 2SH
G0 ILO Dr P. Taylor, 16 Petrel Close, Herne Bay, CT6 6NT
GM0 ILQ I. Ferguson, C/O Mr Jb Ferguson, 8 Cleveden Crescent, Glasgow, G12 0PB
G0 ILT T. Drummond, 10 Delamere Road, Earley, Reading, RG6 1AP
G0 ILV K. Betts, 6, The Grove, Sunderland, SR5 3EG
G0 ILZ M. Hanraads, 6 Oak Hill, Hollesley, Woodbridge, IP12 3JY
G0 IMA P. Pearce, 7 Otter Road, Clevedon, BS21 6LQ
G0 IMB R. Battersby, 4 Gorsey Brow, Urmston, Manchester, M41 9QE
G0 IMD R. Clamp, 276 The Parade, Greatstone, New Romney, TN28 8UL
G0 IMG M. Gotch, 44 Audley Road, Saffron Walden, CB11 3HD
GM0 IMH C. Taylor, The Grange Smithy, Errol, Perth, PH2 7TB
G0 IMK N. Sparrey, The Ashes, Mamble Road, Kidderminster, DY14 9HX
G0 IMP N. Wheeldon, 28 Constance Avenue, Lincoln, LN6 8SN
G0 IMQ N. Cummings, 14 Cemetery Road, Gloucester, GL4 6PB
G0 IMU D. Spearman, 7 Farman Close, Salhouse, Norwich, NR13 6QD
G0 IMV R. Hill, Marclecote, Ledbury Road, Ross-on-Wye, HR9 7BE
GM0 IMW A. Mcbride, 4 Clova Place, Uddingston, Glasgow, G71 7BQ
G0 IMX C. Lacey, 69 Field Lane, Pelsall, Walsall, WS4 1DQ
GM0 IMZ I. Mcewan, Granary Cottage, Vogrie Grange, Gorebridge, EH23 4NT
G0 INA E. Cairns, 2 Stockhill Circus, Stockhill, Nottingham, NG6 0LS
GM0 INC C. Connor, 2 Clyde View, Girvan, KA26 9DJ
G0 ING Nthants Expd Gr, c/o L. Parker, 128 Northampton Road, Wellingborough, NN8 3PJ
G0 INJ P. Shuttlewood, 10 Church Lane, Costock, Loughborough, LE12 6UZ
G0 INK S. Taylor, 37 Crestfield Drive, Pye Nest, Halifax, HX2 7HG
GW0 INN P. Garner, 44 Tycoch Road, Sketty, Swansea, SA2 9EQ
G0 INO D. Elvin, 56 Cock Bank, Whittlesey, Peterborough, PE7 2HN
G0 INQ C. Hannell, Toat Lodge, Pulborough, RH20 1BZ
G0 INT J. Lee, 2 York Terrace, Birchington, CT7 9AZ
G0 INV N. Wileman, 3 Primrose Way, Stamford, PE9 4BU
G0 INZ R. Ball, 24 Healey Close, Abingdon, OX14 5RL
GM0 IOA S. Aitken, 52 Craigellachie Court, Glenrothes, KY7 6XE
G0 IOE B. Haynes, 6 Epping Walk, Furnace Green, Crawley, RH10 6LX
G0 IOF G. Moore, 64 Walmer Road, Seaford, BN25 3TN
G0 IOH P. Thomas, 1 Jervis Close, Daventry, NN11 4LL
G0 IOI T. Maunder, 19 St. Monica Road, Southampton, SO19 8FF
G0 IOK G. Markeson, 23 Chantry Lane, Tideswell, Buxton, SK17 8NP
GD0 IOM R. Ferguson, Moaney Moar House, Corlea Road, Ballasalla, Isle Of Man, IM9 3BA
G0 IOO A. Willis, 46 Maryland Road, Thornton Heath, CR7 8DF
G0 IOP Dr D. Melville, Little Buckden, Milberry Lane, Stoughton, Chichester, PO18 9JJ
G0 IOQ Denton Park Middle School Radio Club, c/o F. Cosgrove, Denton Park Middle School, Linhope Road, Newcastle upon Tyne, NE5 2NW
G0 IOR M. Robertson, 12B Southfield Road, Grimsby, DN33 2PL
GI0 IOT M. Rabbett, 41 Richill Park, Londonderry, BT47 5QY
G0 IOU B. Gleed, 19 Silver Birch Caravan Site, Walters Ash, High Wycombe, HP14 4UY
GM0 IOY E. Watt, 43 Larkfield Road, Gourock, PA19 1YA
G0 IOZ R. Southerington, 4 Reculver Close, Sunnyhill, Derby, DE23 1WN
G0 IPB L. Collins, 32 Parkstone Road, Syston, Leicester, LE7 1LY
G0 IPC J. Hilton, 99 Kirby Road, Stone, Dartford, DA2 6HD
G0 IPH D. Sutton, 10 Normanhurst Road, Borough Green, Sevenoaks, TN15 8HT
G0 IPJ C. Whitaker, Cross Acres, Pottersheath Road, Welwyn, AL6 9SZ
G0 IPK J. Marsh, 44 Richmond Gardens, Harrow Weald, Harrow, HA3 6AJ
G0 IPN A. Greenwood, 4 Roach Place, Rochdale, OL16 2DD
G0 IPO E. Landor, Silverden, Silverden Lane, Cranbrook, TN18 5LX
GM0 IPV B. Stephenson, 11 Bishop Forbes Crescent, Blackburn, Aberdeen, AB21 0TW
GM0 IPW Dr R. Dickie, Taynish, 11 Churchill Drive, Stornoway, HS1 2NP
G0 IPX Fists/International Morse Pre 'Soc', c/o J. Griffin, 35 Cottage Street, Kingswinford, DY6 7QE
GI0 IQA Dr S. Ruff, 9 Cooleen Park, Newtownabbey, BT37 0RR
GW0 IQC S. Steed, 3 Claremont Road, Llandudno, LL30 2UF
GM0 IQD D. Ross, 24 Eriskay Road, Inverness, IV2 3LX
G0 IQH F. Manning, 180 Priestley Terrace, Wibsey, Bradford, BD6 1QU
GM0 IQI P. James, 6 Jenny Moore Road, St Boswells Melrose, TD6 0AL
G0 IQK W. Chewter, 93 Eton Road, Ilford, IG1 2UF
G0 IQM S. Atkinson, 26 Skipton Road, Trawden, Colne, BB8 8QS
G0 IQN C. Jones, 189 Moor Lane, Cranham, Upminster, RM14 1HN
G0 IQP P. Williams, 28 Cross Likey, Church Stoke, Montgomery, SY15 6AL
GW0 IQZ W. Williams, 31 Stad Ty Croes, Llanfairpwllgwyngyll, LL61 5JR
GW0 IRC D. Knibbs, 24 Corbett Grove, Caerphilly, CF83 1SZ
G0 IRH D. Pond, 31 Quintilis, Bracknell, RG12 7QQ
G0 IRI J. Wade, 12 Kendal Road, Harlescott, Shrewsbury, SY1 4ER
G0 IRJ A. Wilson, 42 Bacheler St., Hull, HU3 2TZ
G0 IRK N. Porter, 23 Calder Court, 7 Britannia Road, Surbiton, KT5 8TS
G0 IRM N. Chambers, 78 Durley Avenue, Pinner, HA5 1JH
GW0 IRP C. Evans, 19 Bryn Terrace, Caerau, Maesteg, CF34 0UR
G0 IRQ F. Wagner, 4 Hamilton Close, South Walsham, Norwich, NR13 6DP
GW0 IRT P. Williams, 15 Rhymney Close, Rassau, Ebbw Vale, NP23 5TF
G0 IRY W. Gallacher, 143 Acre Street, Huddersfield, HD3 3EJ
GM0 ISA D. Arcari, 184 Fintry Drive, Fintry, Dundee, DD4 9LP
G0 ISC R. Slyfield, 359 Ringwood Road, Poole, BH12 4LT
G0 ISE G. Carter, 12 Somerford Road, Broughton, Chester, CH4 0SZ
G0 ISG R. Hale, 5 Land Oak Drive, Kidderminster, DY10 2ST
G0 ISH R. Pownall, Beechgrove, Field Road Whiteshill, Stroud, GL6 6AG
G0 ISI R. Christopher, 33 Grange Park, Albrighton, Wolverhampton, WV7 3EN
G0 ISJ G. Parkin, Mil-Rune, Marsh Gate, Cornwall, PL32 9YN
G0 ISK M. Glover, 50 Broadway, Brinsworth, Rotherham, S60 5ES

G0 ISL Rev. B. Shersby, 4 Blenheim Gardens, Chichester, PO19 7XE
G0 ISM R. Faversham, 19 Pelwood Road, Camber, Rye, TN31 7RU
G0 ISO M. Edmunds, 27A Sea View Road, Parkstone, Poole, BH12 3LP
G0 ISP M. Harrington, 56 Harris Croft, Wem, Shrewsbury, SY4 5DU
GI0 ISQ D. Christie, 8 Ballytober Road Portballintrae, Bushmills, BT57 8UX
GM0 IST J. Purtell, 31 Daleally Crescent, Errol, Perth, PH2 7QA
G0 ISX M. Williams, 22 New Avenue, Huddersfield, HD5 0JD
G0 ISY J. Davies, Winter Cottage, Church Road, St Ives, TR26 3LE
GI0 ITJ Dr D. Linton, 4 Elmwood, Cullybackey, Ballymena, BT43 5PY
G0 ITL J. Dunkley, 34 Sparrow Close, Ilkeston, DE7 4PW
G0 ITM J. Dutton, 16 Briarfield, Washington, NE38 8RX
G0 ITO A. O'Shaughnessy, Southby, Buckland, Faringdon, SN7 8QR
G0 ITU M. Wood, 78 Haycliffe Road, Bradford, BD5 9HB
G0 ITZ E. Pieroni, 395 Huddersfield Road, Millbrook, Stalybridge, SK15 3HU
G0 IUA S. Ratcliffe, 425 Manchester Road Westhoughton, Bolton, Bl53js
G0 IUD A. Wakeman, 133 Stanshawe Crescent, Yate, Bristol, BS37 4EG
G0 IUH P. Battershill, 45 Winkworth Road, Banstead, SM7 2QJ
G0 IUK E. Pickerill, 6 Pitmore Walk, Moston, Manchester, M40 0GB
G0 IUN C. Stain, 6 Sutton Close, Sutton-in-Ashfield, NG17 3DP
GI0 IUP G. Lyle, 40 Enagh Crescent, Maydown, Londonderry, BT47 6UG
G0 IUV K. Knowles, 18 Croft Close, Pinxton, Nottingham, NG16 6RF
G0 IUW R. Hill, 7 Berkeley Close, Cashe'S Green, Stroud, GL5 4SA
G0 IUY J. Tribe, 6 Privett Road, Waterlooville, PO7 5HJ
G0 IVB M. Llewellyn, 28 North St., Clay Cross, Chesterfield, S45 9PL
G0 IVD P. Mcgarvey, 125 Esme Road, Sparkhill, Birmingham, B11 4NJ
GW0 IVG W. Jones, 30 Maesglas Pontyates, Llanelli, SA15 5SG
G0 IVI K. Edwards, 39 King Edward Street, Sandiacre, Nottingham, NG10 5BS
GI0 IVJ J. Mccausland, 17 Dunavon Heights, Dungannon, BT71 6TN
G0 IVO B. Singleton, Langdales, Broadbury, Okehampton, EX20 4LL
G0 IVP S. Parrish, Ambleside, Carlton Avenue, Hornsea, HU18 1JG
GM0 IVQ G. Mckinlay, 68 Hillend Road, Clarkston, Glasgow, G76 7XT
G0 IVR Itchen Valley ARC, c/o K. Hastie, 3 The Woodlands, Kings Worthy, Winchester, SO23 7QQ
GW0 IVT T. Jones, Heathbrook, Maesmawr Close, Brecon, LD3 7JF
G0 IVV J. Sheldrake, 66 Bibbys Way, Framlingham, Woodbridge, IP13 9FD
G0 IVX H. Harrison, 8 North Leigh, Tanfield Lea, Stanley, DH9 9PA
G0 IVZ J. Fisher, Farland, Rillaton, Callington, PL17 7PA
G0 IWB J. Wells, 12 Church Road, Grafham, Huntingdon, PE28 0BB
GW0 IWD A. Aston, Ty Newydd, Y Ffor, Pwllheli, LL53 6UY
G0 IWF F. Oakton, 180 Wragley Way, Stenson Fields, Derby, DE24 3DZ
G0 IWI Dr D. Ranson, 27 Kruses Road, North Warrandyte, Victoria, Australia, 3113
G0 IWN D. Greenhalgh, 8 Pleasant Road, Milton, Southsea, PO4 8JU
GM0 IWX T. Lorimer, 9 Orchard House, Orchard Grove, Leven, KY8 5XA
G0 IWZ A. White, 13 Woodcote Drive, Poole, BH16 5RA
GW0 IXK R. Griffiths, 71 Elder Grove, Llangunnor, Carmarthen, SA31 2LH
GW0 IXM O. Williams, 11 Hafod Road, Tycroes, Ammanford, SA18 3QL
GM0 IXO G. Michie, 2 Moncur St., Townhill, Dunfermline, KY12 0HN
GW0 IXQ D. Graham, 1 Maestir, Llanelli, SA15 3NS
G0 IXS P. Turner-Hicks, 64 Sharpley Avenue, Coalville, LE67 4DT
G0 IXT G. Lambert, Church House, Chapel Lane, Christchurch, BH23 6BE
G0 IXV P. Biscombe, Keverine, Victoria Road, Folkestone, CT18 7JS
G0 IXZ D. Fleetwood, Lynton House, Station Road Bolsover, Chesterfield, S44 6BH
GM0 IYA W. Brown, 29 Bankhead Crescent, Dennyloanhead, Bonnybridge, FK4 1RY
G0 IYD P. Bates, 29 Juler Close, North Walsham, NR28 0SY
G0 IYE D. Chalmers, 42 Thornbury Drive, Uphill, Weston-Super-Mare, BS23 4YH
G0 IYK M. Harrison, 12 Richmond Rise, Reepham, Norwich, NR10 4LS
G0 IYM M. Linton, 134 Eccles Old Road, Salford, M6 8QQ
G0 IYO K. Stanmore, 48 St. Michaels Avenue, Bishops Cleeve, Cheltenham, GL52 8NX
GM0 IYP Sutherland and District ARC, c/o C. O'Hennessy, Savalbeg, Challenger Estate, Lairg, IV27 4ED
G0 IYQ N. Mitchelson, Trevena, Winskill, Penrith, CA10 1PD
G0 IYS P. Willis, 5 Binbrook Walk, Corby, NN18 9HH
G0 IYT C. Savin, Jenalri, Union Lane, Preston, PR3 6SS
G0 IYU P. Kirk, 38 Carleton Street, Morecambe, LA4 4NY
G0 IYV T. Jones, 159 Cobden View Road, Crookes, Sheffield, S10 1HT
G0 IYW G. Farndon, 28 Willow Close, Collycroft, Bedworth, CV12 8BE
G0 IYX Capt. I. Roberts, 2 Samuel Fold, Pendlebury Lane, Wigan, WN2 1LT
G0 IYY M. Lindsay, 5 Eatonhill, Norwich, NR4 7PY
G0 IZC J. Carver, 131 Rutland Avenue, High Wycombe, HP12 3JQ
G0 IZE R. Bates, 36 Maple Crescent, Alveley, Bridgnorth, WV15 6LT
G0 IZG G. Bennett, 21 Porlock Close Platt Bridge, Wigan, WN2 5HY
G0 IZI A. Warwick, 58 Longworth Avenue, Tilehurst, Reading, RG31 5JY
G0 IZJ E. Whittaker, 17 Packer St., Bolton, BL1 3LD
G0 IZK M. Dennehy, 45 Vine Road, Tiptree, Colchester, CO5 0LR
G0 IZL N. Procter, 19 Manitoba Way, Selston, Nottingham, NG16 6FP
G0 IZN A. Halliday, 12 Fowley Common Lane, Glazebury, Warrington, WA3 5JJ
G0 IZP R. Mcdonald, 4 The Paddocks, Baunton, Cirencester, GL7 7DL
G0 IZQ J. Bradnock, 5 Milverton Road, Knowle, Solihull, B93 0HX
G0 IZR J. Bridson, 10 Clegg Street, Astley, Tyldesley, Manchester, M29 7DB
G0 IZV D. Gowers, 32 Silver Fox Crescent, Woodley, Reading, RG5 3JA
G0 IZY D. Hicks, Woodlands, Crawley Ridge, Camberley, GU15 2AJ
G0 JAA T. Watson, 89 Addison Road, Wednesbury, WS100LW
G0 JAC M. Hill, 9 Longacre, Woodthorpe, Nottingham, NG5 4JS
G0 JAF J. Foy, 23 Lee Road, Nelson, BB9 8SD
G0 JAG R. Glyn, 171 Bull Close Road, Norwich, NR3 1NY
GW0 JAI T. Jones, 26 Treowain, Machynlleth, SY20 8EJ
G0 JAJ G. West, 39 Court Farm Road, Eltham, London, SE9 4JL
G0 JAL W. Bell, 244 Westbourne, Woodside, Telford, TF7 5QR
G0 JAM J. Morrison, 107 Crown Meadow, Colnbrook, Slough, SL3 0LJ
G0 JAN J. Ilston, 6 Dovedale, Canvey Island, SS8 8HX
G0 JAO L. White, 56 Grange Road, Leigh-on-Sea, SS9 2HT
G0 JAP C. Collins, 26 Nicholsons Wharf, Mather Road, Newark, NG24 1FN
G0 JAQ G. Blackwood, 63 Illingworth Avenue, Halifax, HX2 9JH
G0 JAR R. Offord, 116 Townsend Road, Snodland, ME6 5RL
G0 JBA P. Boorman, Gladstone House, Marshborough Road, Sandwich, CT13 0PE
G0 JBC C. Collins, 3 Hollin Hall Farm, Long Causeway, Denholme, Bradford, BD13 4DX
G0 JBH Dr A. Bingham, 5 Burns Rd, Southampton, SO19 6QT
G0 JBJ A. Tungate, 171 Marlborough Gardens, Faringdon, SN7 7DG
G0 JBM L. Humphreys, 19 Clinch Green Avenue, Bexhill-on-Sea, TN39 5HN
G0 JBO B. Lees, Preston Hall, Preston, Telford, TF6 6DH
G0 JBP R. Lipscomb, Redmoor, Bickley Road, Bromley, BR1 2NF
G0 JBR E. Thorley, 7 Drake Street, St Helens, WA10 4JG
G0 JBS M. Notman, 3 Pilling Avenue, Lytham St Annes, FY8 3QF
G0 JBV A. Mcintosh, 17 The Chase, Abbeydale, Gloucester, GL4 4WP
G0 JBY J. Youd, 25 Hanson Road, Andover, SP10 3HL
G0 JBZ R. Stevens, 51 Beacon Park Crescent, Poole, BH16 5PB
G0 JCA J. Andress, 46 Bridwell Road, Plymouth, PL5 1AB
GW0 JCB C. Jones, 8 West Walk, Barry, CF62 8BY
G0 JCC A. Lancaster, Little Edwards Barn, Newton St Margarets, HR2 0QG
G0 JCD P. Busby, 7 Rimmer Green, Scarisbrick, Southport, PR8 5LP
G0 JCF D. Smith, 14 College Drive, Ruislip, HA4 8SB
G0 JCG A. Butler, 97 Hermitage Road, Saughall, Chester, CH1 6AQ
G0 JCK W. Pattinson, 2 The Green, Ticknall, Derby, DE73 7GY
G0 JCN M. Lovatt, 37 Hartland Avenue, Bilston, WV14 9AN
GM0 JCP D. Grey, 7 Cemetery Lane, Tweedmouth, Berwick-upon-Tweed, TD15 2BS
G0 JCQ B. Rimmer, 8A Mallee Avenue, Southport, PR9 8NL
GW0 JCT P. Jones, Bronallt, Cenarth, Newcastle Emlyn, SA38 9JS
G0 JCY P. Stuart, Skylark Corner, Seaborough Hill, Crewkerne, TA18 8PL
G0 JCZ M. Martin, 2821 Bissonnet St, Houston, USA, 77005-4014
GM0 JDB J. Ratter, Foulawick Wethersta, Brae, ZE2 9QS
G0 JDC J. Gwynn, 117 Main Street, Goldthorpe, Rotherham, S63 9JW
G0 JDD I. Douce, 67 Glenbervie Drive, Leigh-on-Sea, SS9 3JT
G0 JDE R. Doyle, 61 St. Peters Road, West Mersea, Colchester, CO5 8LN
G0 JDG A. Purseglove, 122 Chesterfield Road, Huthwaite, Sutton-in-Ashfield, NG17 2QF
G0 JDL J. Clarke, 17-18 Sotherton Corner, Sotherton, Beccles, NR34 8AP
G0 JDM T. Kewell, The Old Skittle Alley Baker Street, Frome, BA11 3BL
G0 JDO T. Thomas, 26 Corfe Crescent, Torquay, TQ2 7QX
G0 JDQ M. Cozens, Lot 321, 1001 Starkey Rd, Largo, USA, 33771
GW0 JDS J. Stonehouse, 54 Port Tennant Road, Port Tennant, Swansea, SA1 8JF
GW0 JDW D. Willis, 5 Dan Lan Rd, Llanelli, Dyfed, SA16 0NF
GW0 JDY M. Lewis, 12 Fern Rise, Neyland, Milford Haven, SA73 1RA
G0 JEA R. Kaye, 63 Coronation Drive, Birdwell, Barnsley, S70 5RL
G0 JEC D. Naylor, 6 Front St., Kirk Merrington, Spennymoor, DL16 7HZ
G0 JEE B. Greer, Willow Farm, Main Road, Burton-on-Trent, DE13 9QD
GM0 JEF J. Fysh, 7 Chestnut Place, Ellon, AB41 9HF
G0 JEH S. Rosbottom, 26 Wellington Street, Preston, PR1 8TP
G0 JEK C. Kelland, 11 The Meads, West Hanney, Wantage, OX12 0LJ
GW0 JEQ R. Wicks, Gooseberry Cottage, Llangunllo, Knighton, LD7 1SW
G0 JEU S. Dunsmore, 33 Church Street, Messingham, Scunthorpe, DN17 3SB
GI0 JEV H. Kernohan, 40 Lisnafillon Road, Gracehill, Ballymena, BT42 1JA
G0 JEW P. Wells, 12 Shelley Drive, Lutterworth, LE17 4XF
G0 JEZ B. Wells, 37 Elder Road, Denvilles, Havant, PO9 2UW
G0 JFA A. Jones, Highercombe West, Highercombe, Dulverton, TA22 9PT
G0 JFC J. Aithison, 14 Claymore Rise, Silsden, Keighley, BD20 0QQ
G0 JFD G. Butcher, 15 Leander Drive, Gosport, PO12 4GG
G0 JFE P. Sixsmith, 10 Wisbeck Road, Bolton, BL2 2TA
GI0 JFF W. Mcbride, 5 Aylesbury Road, Newtownabbey, BT36 7YP
GM0 JFH J. Mair, 43 Todhill Avenue, Onthank, Kilmarnock, KA3 2EQ
GM0 JFK C. Harper, Glencoul Cottage, Cullicudden, Dingwall, IV7 8LL
GM0 JFL B. Harper, 16 Brae Park, Munlochy, IV8 8PJ
G0 JFM S. Nicholls, Fieldway, The Street, Eyke, Woodbridge, IP12 2QG
G0 JFP J. Scott, 23 Botesworth Green, Milnrow, Rochdale, OL16 3PJ
GW0 JFQ N. Williams, Flat 4, 234-237 Chapmans High St., Swansea, SA1 1NZ
G0 JGB J. Barnett, 43 Westsprink Crescent, Stoke-on-Trent, ST3 5JD
G0 JGF S. Cox, 60 Leawood Road, Trent Vale, Stoke-on-Trent, ST4 6LA
G0 JGI C. Davison, 28 Ashford Crescent, Hythe, Southampton, SO45 6EU
G0 JGV K. Bewley, 38 Great Innings South, Watton At Stone, Hertford, SG14 3TF
GD0 JGX D. Ginsberg, 26 Keeill Pharick Park, Glen Vine, Isle Of Man, IM4 4EW
G0 JHD R. Jennings, 54A Pensbury Street, Darlington, DL1 5LH
GM0 JHE K. Hunter, 2/2 206 Skirsa Street, Glasgow, G23 5DJ
G0 JHG C. Holmes, 17 Wenning Court, Morecambe, LA3 3SH
GW0 JHH L. Ireland, 109 Dan-Y-Cribyn, Ynysybwl, Pontypridd, CF37 3EU
G0 JHJ W. Fowler, 1 East Orchard, Sileby, Loughborough, LE12 7SX
G0 JHK M. Hedges, 24 Fletcher Avenue, St Leonards-on-Sea, TN37 7QX
G0 JHL R. Wilmot, 43 A 2, Pã„Ã„Skylã„Ntie, JÇ?msÇ?, Finland, 42100
G0 JHQ Dr A. Cotton, Coburg Cottage, Barton Estate, East Cowes, PO32 6NT
GI0 JHR H. Oxtoby, 13 Castle Court, Cookstown, BT80 8QJ
GI0 JHS N. Greer, 11 Ardtrea Road, Stewartstown, Dungannon, BT71 5LY
G0 JHT D. Talbot, Southways, Tichborne Down, Alresford, SO24 9PL
G0 JHU M. Stephenson, 38 St. Helens Crescent, Low Fell, Gateshead, NE9 6DH
G0 JHW J. Waterhouse, 81 Barkham Road, Wokingham, RG41 2RJ
G0 JIA A. Sharp, 9 Higher Park, East Prawle, Kingsbridge, TQ7 2DB
G0 JIB P. Moran, 12 Sapphire Drive, Kirkby, Liverpool, L33 1UW
G0 JIF A. Fennell, 12 Vale Road, Ramsgate, CT11 9LU
G0 JII C. Davis, 10 Marnhull Road, Poole, BH15 2EX
G0 JIL G. Hampson, Flat 4, Priory Lodge, Stony Lane South, Christchurch, BH23 1FA
G0 JIM J. King, 4 Glenhurst Avenue, Ruislip, HA4 7LZ
G0 JIR A. Potter, 8 Oaklands, The Street, Ashford, TN25 6NE
G0 JIS A. Myland, 8 Burseldon Court, East Cliff Road, Devon, EX7 0BP
G0 JIT R. Atherton, Frensham, Grange Lane, Northwich, CW8 2BQ
G0 JIW M. Edwards, 98 Cornwallis Drive, Eaton Socon, St Neots, PE19 8TZ
G0 JJD G. Thorne, 4 Barronwood Court, Tarleton, Preston, PR4 6TR
G0 JJE H. Spratt, 9 Kennedy Close, Halesworth, IP19 8EG
GW0 JJF I. Mccormick, The Old Railway Inn, Station Road, Gilwern, NP7 0BY
G0 JJG J. Butt, 29 Shearwater Way, Stowmarket, IP14 5UG
G0 JJI P. Forshaw, 73 Galloway Road, Hamworthy, Poole, BH15 4JS
G0 JJK T. Atkins, 46 Fallowfield, Ampthill, Bedford, MK45 2TP
G0 JJM A. Eves, 136 Thistle Grove, Welwyn Garden City, AL7 4AQ
G0 JJO M. Wheatley, 25 Sheringham Drive, Etchinghill, Rugeley, WS15 2YG
G0 JJP R. Fowler, 6 Salts Croft Hawksyard, Rugeley, WS15 1SR
G0 JJQ W. Dillon, 49 Goring Way, Greenford, UB6 9NN
G0 JJR D. Briggs, 130 Beresford Avenue, Skegness, PE25 3JN

G0 JJS J. Smith, 48 Oakleigh Gardens, Oldland Common, Bristol, BS30 6RH
G0 JJV S. Hill, 26 Harborough Way, Sheffield, S2 1RG
G0 JJW J. Walsh, Flints House, Coates, Peterborough, PE7 2DD
G0 JJY I. Bowen, 169 Clopton Road, Stratford-upon-Avon, CV37 6TF
GD0 JKA A. Brook, 9 Stonecrop Grove, Douglas, Isle Of Man, IM2 7DX
G0 JKC K. Clarke, 55 Compton Avenue, Aston-On-Trent, Derby, DE72 2AU
G0 JKE S. Ward, 27 Greenock Street, Sheffield, S6 4NB
G0 JKG F. Shinn, 23 Cygnet Close, Brampton Bierlow, Rotherham, S63 6EY
G0 JKH D. Hinson, 6 Nethergate, Stannington, Sheffield, S6 6DJ
G0 JKI M. Edmunds, 2 Jubilee Road, Bungay, NR35 1RE
G0 JKJ S. Marshall, 31 Postbridge Road, Coventry, CV3 5AG
G0 JKL J. Hallendorff, 8 Third Avenue, Kidsgrove, Stoke-on-Trent, ST7 1BZ
G0 JKM P. Renvoize, Flat 14 Block M, Peabody Buildings, Dufferin Street, London, EC1Y 8NL
G0 JKP A. Whibley, 8 Ticehurst Road, Brighton, BN2 5PU
G0 JKU A. Bowering, 137A Knole Lane, Brentry, Bristol, BS10 6JN
G0 JKY B. Hadley, 60 Chapel St., Pensnett, Brierley Hill, DY5 4EF
G0 JLE R. Adams, 18 Dundridge Gardens, Bristol, BS5 8SZ
G0 JLF J. Flowers, 4 Ashleigh Crescent, Yatton, Bristol, BS49 4DF
G0 JLI T. Davies, 31 Burnbush Close, Bristol, BS14 8LQ
GM0 JLJ E. Mottart, 1 Muirake Cottages, Cornhill, Banff, AB45 2BQ
G0 JLL N. Sheen, 26 Springvale Rise, Hemsworth, Pontefract, WF9 5HY
G0 JLP M. Bray, 205 Woodlands Road, Gillingham, ME7 2SW
G0 JLR I. Sabey, 21 Althorpe Street, Radford, Nottingham, NG7 3GN
G0 JLS M. Brown, 4 River Gardens, Shawbury, Shrewsbury, SY4 4LA
G0 JLU E. Wood, 65 Walford Road, Rolleston-On-Dove, Burton-on-Trent, DE13 9AR
G0 JLV D. Vaughan, 23 Beckmeadow Way, Mundesley, Norwich, NR11 8LP
GW0 JLX A. Digby, 6 Melin Y Coed, Cilgerran, SA43 2AQ
G0 JMD J. Davis, 38 Mountbatten Close, Yate, BS37 5TE
G0 JME J. Pither, 74 Bucklands Road, Teddington, TW11 9QS
G0 JMI M. Parkin, 17 Bolle Road, Alton, GU34 1PW
GW0 JMJ W. Williams, 16 Chapel Close, Elim Way, Blackwood, NP12 2AD
G0 JMK M. Kemble, 74 Teg Down Meads, Winchester, SO22 5ND
G0 JML M. Pivac, 72 Old Mill Road, Saffron Walden, CB11 3ER
G0 JMN H. Marshall, 23 Cranbourne Drive, Chorley, PR6 0LJ
GM0 JMO J. Bell, 5 Louisa Drive, Girvan, KA26 9AH
G0 JMR D. Williams, 27 Grindlestone Hirst, Colne, BB8 8BF
G0 JMS M. Standen, 11 Hazel Gardens, Sonning Common, Reading, RG4 9TF
G0 JMW M. Williams, 51 Crackley Hill, Coventry Road, Kenilworth, CV8 2EE
G0 JMZ P. Farrar, 2 Ancaster Avenue, Chapel St. Leonards, Skegness, PE24 5SL
G0 JNA R. Janes, 37 Valley View, Market Drayton, TF9 1EA
G0 JNE T. Royle, 35 Patrons Drive, Sandbach, CW11 3AS
G0 JNG A. Stothard, Lanshaw Farm, Otley Road, Harrogate, HG3 1QX
G0 JNJ A. Denny, 85 Delamere Drive, Macclesfield, SK10 2PS
G0 JNK A. Powers, 42 Newbridge Road Ambergate, Belper, DE56 2GS
G0 JNQ S. Mitchell, 78 Bellasize Park, Gilberdyke, Brough, HU15 2XU
G0 JNR S. Doveton, 2 Red Scar Drive, Scarborough, YO12 5RQ
G0 JNT L. Keeton, 66 Worlaby Road, Scartho Top, Grimsby, DN33 3JP
G0 JNY H. Potter, 720 Old Norwich Road, Ipswich, IP1 6LB
G0 JNZ T. Bray, 135 Fort Austin Avenue, Crownhill, Plymouth, PL6 5NR
G0 JOC J. Lassemillante, 25 Rissington Walk, Thornaby, Stockton-on-Tees, TS17 9QJ
G0 JOD R. Degg, 28 The Spinneys, Welton, Lincoln, LN2 3TU
G0 JOG B. Wilson, 41 Palmerston Close, Ramsbottom, Bury, BL0 9YN
GM0 JOL Rev. J. Lincoln, 59 Obsdale Park, Alness, IV17 0TR
G0 JOM J. Goddard, 65 New Street, North Wingfield, Chesterfield, S42 5JP
G0 JON J. Swain, 1 Ganstead Way, Low Grange, Billingham, TS23 3SY
G0 JOP J. Bryder, 110 Georgelands, Ripley, Woking, GU23 6DQ
G0 JOS E. Christmas, 15 Norton Avenue, Surbiton, KT5 9DX
GM0 JOV A. Farquhar, 57 Woodcroft Avenue, Bridge Of Don, Aberdeen, AB22 8WY
G0 JOX D. Sykes, 449 Westdale Lane, Mapperley, Nottingham, NG3 6DH
G0 JPC K. Dale, 31 Cadshaw Close, Birchwood, Warrington, WA3 7LR
G0 JPE P. Roberts, 61 Abbey Lane, Sheffield, S8 0BN
G0 JPF S. Lee, 15 Wilson Way, Earls Barton, Northampton, NN6 0NZ
GM0 JPG D. Arnold, 1 Knockenhair Road, Dunbar, EH42 1BA
G0 JPH J. Holden, 47 Copse Hill, London, SW20 0NJ
G0 JPI S. Martin, 34 Parson Drove, West Pinchbeck, Spalding, PE11 3QW
G0 JPJ P. Smith, S.V Kiwiroa, New Zealand Reg. Ship, New Zealand, ON-876019
G0 JPL D. Smith, 16 Leander Close, Nottingham, NG11 7BE
G0 JPM J. Mitchell, 8 Eldred Drive, Orpington, BR5 4PF
G0 JPQ C. Barber, Charity Farm House, Mill Lane, Skegness, PE24 5NN
G0 JPT C. Hamlet, 9 Covert Road, Manchester, M22 4QS
G0 JPU Dr M. Ferguson, 15 Squires Leaze, Thornbury, Bristol, BS35 1TB
G0 JPY D. Polley, 42 Lindfield Road, Eastbourne, BN22 0AJ
G0 JPZ M. Kirk, 5 The Paddock, Kirkby-In-Ashfield, Nottingham, NG17 8BT
G0 JQA Hambleton ARC, c/o B. Alderson, 43 Brompton Road, Northallerton, DL6 1ED
GM0 JQE I. Templeton, 39 Cairngorm Court, Irvine, KA11 1PN
G0 JQK J. Hughes, 18 Monmouth Road, Wallasey, CH44 3ED
G0 JQP G. Bradley, 7 Copeland Row, Evenwood, Bishop Auckland, DL14 9PY
GI0 JQQ E. Butler, 59 Ballinlea Road, Maghernahar, Ballycastle, BT54 6JL
G0 JQR D. Allison, 12 Goosander Close, Kings Lynn, PE31 7AF
G0 JQS A. Downing, 1 Raglans, Alphington, Exeter, EX2 8XN
GW0 JQT S. Lloyd, 10 Park Crescent, Llanelli, SA15 3AE
G0 JQX C. Ditchfield, 8 Meerbrook Way, Quedgeley, Gloucester, GL2 4QE
G0 JQZ S. Chamberlain, 54 Henray Avenue, Glen Parva, Leicester, LE2 9QJ
G0 JRB J. Reed, 290 Messingham Road, Bottesford, Scunthorpe, DN17 2QY
G0 JRC J. Clayton, 49 Bramble Lane, Mansfield, NG18 3NP
GI0 JRD B. Mcanespie, 23 Ashton Park, Belfast, BT10 0JQ
G0 JRE I. Perks, 9 Atherton Close, Shalford, Guildford, GU4 8HZ
GW0 JRF F. Rees, Caerleon, Picton Road, Tenby, SA70 7DP
G0 JRH P. Scott-Dickinson, Ninicsu, 18 Pennington Drive, Weybridge, KT13 9RU
GI0 JRI K. Murray, 67 Sicily Park, Belfast, BT10 0AN
G0 JRM C. Brown, 8 The Elms, Horringer, Bury St Edmunds, IP29 5SE
G0 JRN A. Hansley, 238 Milton Road, Cowplain, Waterlooville, PO8 8SE
GM0 JRQ B. Baker, The Ridge, Peat Inn, Cupar, KY15 5LH
G0 JRR C. Curtis, 1 Westover Drive, Burton-Upon-Stather, Scunthorpe, DN15 9HH
G0 JRT T. Hanratty, 12 Clarendon Street, Consett, DH8 5LS
G0 JRV J. Broomfield, 14 Woodfen Crescent, Leominster, HR6 8SS
G0 JRX T. Cave, 71 Cambo Drive, Cramlington, NE23 6TW
G0 JRY C. Mulvany, 25 Redwing Close, Bicester, OX26 6SR
G0 JRZ T. Brookes, Jemora, Littleham, Bideford, EX39 5HN
G0 JSA R. Taylor, 11 Yeadon Close, Accrington, BB5 0FN
G0 JSC J. Wheatley, 8 Winchester Close, Feniton, Honiton, EX14 3EX
G0 JSE J. Edwards, 49 The Fleet, Stoney Stanton, Leicester, LE9 4DZ
G0 JSF B. Halmshaw, 7 Gerrard St., Rochdale, OL11 2EB
G0 JSG P. Holden, 19 Briar Close, Lowestoft, NR32 4SU
G0 JSJ R. Jackett, Bourne House, 105 Moor Road, Leyland, PR26 9HP
G0 JSK A. Wakefield, Kuleana, Barnrigg, Carnforth, LA6 2LJ
G0 JSL G. Brown, 9 Western Drive, Leyland, PR25 1YB
G0 JSM J. Brown, 9 Western Drive, Leyland, PR25 1YB
G0 JSO R. Wynne, South Graceholme, High Lorton, Cockermouth, CA13 9UQ
G0 JSP P. Fauchon, 114 Petersfield Avenue, Staines-upon-Thames, TW18 1DJ
G0 JSR S. Rickman, 35 Cedar Way, Basingstoke, RG23 8NG
G0 JST OBO Jubilee Sailing Trust (ARS), c/o J. Wheatley, 8 Winchester Close, Feniton, Honiton, EX14 3EX
G0 JSU A. Collett, 5 Park View Drive, Lydiard Millicent, Swindon, SN5 3LX
GW0 JSX R. Davies, 8 Princes Park, Rhuddlan, Rhyl, LL18 5RW
GJ0 JSY S. Smith-Gauvin, 31 Le Jardin A Pommiers, La Rue De Patier, St Saviour, Jersey, JE2 7LT
G0 JSZ J. Crellin, 89 Wapshare Road, West Derby, Liverpool, L11 8LR
G0 JTA D. Ellison, Blackburn Hall, Grinton, Richmond, DL11 6HH
G0 JTD R. Lyne, 32 Davenwood Upper Stratton, Swindon, SN2 7LL
GW0 JTE L. Horne, 29 Station Terrace, Dowlais, Merthyr Tydfil, CF48 3PU
GW0 JTF J. Hosking, 14 School Terrace, Cwm, Ebbw Vale, NP23 7QY
GW0 JTJ T. Watkins, Ty Unig, Forest Road, Treharris, CF46 5HG
G0 JTL J. Hinchliffe, 19 The Terrace, Honley, Holmfirth, HD9 6DS
G0 JTM P. Smith, 999 Manchester Road, Linthwaite, Huddersfield, HD7 5LS
G0 JTN C. Smith, 35 Allendale Road, Earley, Reading, RG6 7PD
G0 JTP A. Hill, 159 Sandford Road, Bradford, BD3 9NU
G0 JTR B. Thatcher, 18 Harescombe, Yate, Bristol, BS37 8UA
G0 JTT S. Hemsworth, 4 Spoonhill Road, Stannington, Sheffield, S6 5PA
GW0 JTU A. Lewis, 33 Heol Helig, Brynmawr, Ebbw Vale, NP23 4TY
G0 JUA J. Hardcastle, 37 Caithness Road, Liverpool, L18 9SJ
G0 JUE T. Cruse, Watch Tower House, The Ridgeway, London, NW7 1RS
G0 JUI E. Gibson, 107 Church Avenue, Meanwood, Leeds, LS6 4JT
G0 JUK N. Mayes, Cliffe Cottage, Rotherham Road, Barnsley, S71 5QX
G0 JUL G. Hogben, Calle El Arado 7, Las Brenas, Spain, 35570
G0 JUM S. Barker, 11 Pennington Close, Copplestone, Crediton, EX17 5NA
G0 JUN R. Softley, 14 Topps Drive, Bedworth, CV12 0DE
G0 JUQ M. Eborall, 26 Bishopton Lane, Stratford-upon-Avon, CV37 9JN
G0 JUR N. Barnett, 60 Commercial Road, Spalding, PE11 2HE
G0 JUT D. Horne, 24 Ringwood Drive, Leeds, LS14 1AP
G0 JUV A. Phillips, 2 New Zealand Terrace, Bridport, DT6 3PW
G0 JUY P. Barden, 38 Silver Close, Tonbridge, TN9 2UY
G0 JVB M. Jackson, 114 Norman Street, Ilkeston, DE7 8NL
GM0 JVC Dr M. Grant, Monikie, Gryffe Road, Kilmacolm, PA13 4BB
G0 JVF D. Cleaver, 4 Wyvern Close, Devizes, SN10 2UE
G0 JVH A. Puffett, 142 Cheltenham Road East, Gloucester, GL3 1AA
G0 JVI A. Mawson, 38 Springbank Road, Gildersome, Leeds, LS27 7DJ
G0 JVK R. Cook, 15 Hucklow Court, Mansfield, NG18 3QP
G0 JVL A. Sclater, 6 Balmoral Close, Alton, GU34 1QY
G0 JVN R. Horne, 7 Alexander Road, Bentley, Walsall, WS2 0HJ
G0 JVU N. Pattinson, 4 Carlisle Road, Brampton, CA8 1SR
GM0 JVV J. Stevenson, 52 Fernbrae Avenue, Rutherglen, Glasgow, G73 4AE
G0 JVW A. Thornton, 1 Primula Close, Clifton, Nottingham, NG11 8SL
GW0 JWC C. Jones, 31 Tynewdd Nantybwch, Tredegar, NP223SQ
G0 JWD J. Brambley, Trail View, Biggin, Buxton, SK17 0DH
GW0 JWF B. Matthews, 25 Manor Park, Newbridge, Newport, NP11 4RS
G0 JWG J. Gaunt, Fenton House, Church Road, King's Lynn, PE33 0HE
G0 JWJ T. Bridgland-Taylor, 1 Overbury Court, Hereford, HR1-1DG
G0 JWL K. Lindsay, 11A Pyrford Close, Waterlooville, PO7 6BT
G0 JWM J. Mcmullen, 281 The Broadway, Cullercoats, North Shields, NE30 3LH
G0 JWO T. Forbes, 63 Wardle Drive, Annitsford, Cramlington, NE23 7DE
GD0 JWR H. Richardson, Pitcairn, Quarterbridge Road, Douglas, Isle Of Man, IM2 3RQ
G0 JWV P. Bryant, Crugsillick Cottage, Ruan High Lanes, Truro, TR2 5JP
G0 JWY J. Curtis, Conway, 27 Southgate, Hornsea, HU18 1RE
G0 JXF A. Hancox, 5 East Glade Close, Sheffield, S12 4QL
GW0 JXG M. Price, 4 Vale View, Woodfieldside, Blackwood, NP12 0DB
G0 JXI M. Brown, 11 Miles Close, Birchwood, Warrington, WA3 6QD
G0 JXJ D. Copeland, 2B Rose Road, Canvey Island, SS8 0BP
G0 JXO R. Smith, 24 Kirkstead Road, Carlisle, CA2 7RD
G0 JXP K. Pile, 14 Semper Close, Knaphill, Woking, GU21 2NG
G0 JXQ D. Davis, 17 Welbourne Close, Raunds, Wellingborough, NN9 6HE
G0 JXR P. Keasley, 55 Hillside, Hoddesdon, EN11 8RW
G0 JXX M. Hoddy, 52 Hayling Rise, High Salvington, Worthing, BN13 3AG
G0 JXY T. Bartholomew, 52 Lauderdale Avenue, Wallsend, NE28 9HU
G0 JYC P. Hodgson, Uhland Str 4, Neunkirchen, Seelscheid, Germany, 53819
G0 JYE P. Foss, 37 Ling Crescent, Ruddington, Nottingham, NG11 6GG
G0 JYF S. Deakin, Brook House, Ivy Lane, Lower South Wraxall, BA152RZ
G0 JYH H. Ryan, Fairview, Imperial Avenue, Sheerness, ME12 2HG
G0 JYI A. Street, 11 Leigh Gardens, Leigh-on-Sea, SS9 2PX
G0 JYJ C. Hodgson, 113 Roman Road, East Ham, London, E6 3RY
G0 JYK J. Sharp, 22 Boat Lane, Irlam, Manchester, M44 6EN
G0 JYL J. Bartram, 2 Reeves Piece, Bratton, Westbury, BA13 4TH
G0 JYN S. Ashcroft, 90 Kestrel Close, Chipping Sodbury, Bristol, BS37 6XA
G0 JYQ M. Gregory, 21 Jacaranda Close, Fareham, PO15 5LG
G0 JYS A. Jepson, 45 Cotefield Road, Manchester, M22 1UR
G0 JYU D. Smith, 104 Hanley Road, London, N4 3DW
G0 JYV J. Rowlands, 67 Woodside, Gosport, PO13 0YX
G0 JYX T. Cloke, 14 Bickley Close, Hanham Green, Bristol, BS15 3TB
G0 JYZ I. Broomhall, 49 Funtley Hill, Fareham, PO16 7XA
G0 JZA N. Cox, Flat 9, Fontigary, Harts Close, Teignmouth, TQ14 9HG
G0 JZE A. Mcfadyen, 26 Lewis Road, Chipping Norton, OX7 5JS
G0 JZF N. Rogers, 15 Templar Road Yate, Bristol, BS375TF
G0 JZH R. Morris, 96 Chandag Road, Keynsham, Bristol, BS31 1QE
G0 JZJ F. Russell, 37 Overpool Road, Ellesmere Port, CH66 1JW

G0 JZL G. Galley, 1 St. James Avenue, South Anston, Sheffield, S25 5DR
G0 JZS G. Corbett, 359 London Road, Stoke-on-Trent, ST4 5AN
G0 JZT J. Chappell, 2 Wayside, Knott End-On-Sea, Poulton-le-Fylde, FY6 0DD
G0 JZU W. Etherington, 15 East Bank, North End Road, Arundel, BN18 0DJ
GM0 JZV J. Warden, Westlea, Little Brechin, Brechin, DD9 6RQ
G0 JZW A. Elford, 10 Meadowlands, Lymington, SO41 9LB
G0 KAB A. Pilkington, Flat 17, Blackshaw House, Bolton, BL3 5NU
G0 KAK I. Walker, 6 Granary Court, Northampton, NN4 0XX
GW0 KAM G. Jones, 14 Plantation Drive, Croesyceiliog, Cwmbran, NP44 2AN
G0 KAQ E. Walker, 2 Newtown Road, Uppingham, Oakham, LE15 9TS
G0 KAS M. Stevens, 20 Melton Place, Epsom, KT19 9EE
G0 KAT V. Chapman, 20 St. Chad, Barrow-upon-Humber, DN19 7AU
G0 KAU R. Crocker, 10 Westhall Close Carlton-Le-Moorland, Lincoln, LN5 9JD
GW0 KAX P. Owen, 13 Highland Close, Sarn, Bridgend, CF32 9SB
G0 KAY T. Davies, Netaerial.Com, Europa House, Barcroft Street, Bury, BL9 5BT
GM0 KAZ A. White, 65 Orchard Street, Galston, KA4 8EJ
G0 KBA I. Langtree, 243 Devonshire Road, Atherton, Manchester, M46 9QB
G0 KBJ E. Burndred, 52 Everest Road, Kidsgrove, Stoke-on-Trent, ST7 4DY
G0 KBK R. Sleigh, 14 Brook House Flats, Chetwynd End, Newport, TF10 7JD
G0 KBL S. Rudcenko, 39 The Avenue, Sutton, SM2 7QA
G0 KBM D. Manning, 2 Sluice Farm Cottages, Kirton, Ipswich, IP10 0QF
G0 KBN G. Beech, 2 Whitely Avenue, Ilkeston, DE7 8WB
G0 KBO V. Kravchenko, Flat 16, Birchfield House, London, E14 8EY
G0 KBP M. Dearing, 1 Woodbine Villas, New Village Road, Cottingham, HU16 4NF
GM0 KBR J. Mcfadyen, 8 Ramsay Crescent, Bathgate, EH48 1DD
G0 KBS L. Kay, 2 Childwall Crescent Childwall, Liverpool, L16 7PQ
G0 KBZ K. Harrison, 6 Staveley Road, Alford, LN13 0PN
G0 KCA I. Walder-Davis, 93 Church St., St. Peters, Broadstairs, CT10 2TX
G0 KCB D. Beckley, Fen Hill, Hall Road, Great Yarmouth, NR29 5NU
G0 KCC Dr W. Randolph, 13 Links Road, Poole, BH14 9QP
G0 KCD M. Leech, 20 Walton Road, Frinton-on-Sea, CO13 0AQ
G0 KCE V. Harding, 17 St. Anns Road South, Heald Green, Cheadle, SK8 3DZ
G0 KCF C. Fosbrook, 4A Yew Tree Road, Hayling Island, PO11 0QE
G0 KCG D. Hall, 47 Sunningdale Road, Fareham, PO16 9PA
G0 KCH M. Mccarthy, 75 Taynton Drive, Merstham, Redhill, RH1 3PX
G0 KCL Kings College ARS, c/o J. Greenberg, 12 Broadhurst Avenue, Edgware, HA8 8TR
GM0 KCN D. Smith, 12 Cannon Street, Selkirk, TD7 5BP
GM0 KCY D. Michael, 84 Bourtreehall, Girvan, KA26 9EL
G0 KCZ W. Bowles, Willow Grove, Little Common North Bradley, Trowbridge, BA14 0TX
G0 KDA P. Cooper, 6 Norwich Close, Scalby, Scarborough, YO13 0PP
G0 KDB D. Greenhalgh, Hillcroft, Colby, Appleby-in-Westmorland, CA16 6BD
GM0 KDC R. Smith, 21 Glen View Crescent, Gorebridge, EH23 4BT
G0 KDD B. Woodward, 58 Marine Drive, Bishopstone, Seaford, BN25 2RU
GM0 KDF R. Thomson, 25 Cheviot Road, Silvertonhill, Hamilton, ML3 7HB
G0 KDG R. Simpson, 29, Hampsfell Grange, Hampsfell Road, Grange-over-Sands, LA11 6AZ
GI0 KDH A. Brown, 3 Gargrim Road, Fintona, Omagh, BT78 2EH
G0 KDI R. Steans, 302 Walton Road, West Molesey, KT8 2HY
G0 KDL W. Cooper, 24 Ambleside Road, Lightwater, GU18 5TA
GM0 KDO G. Kirkland, 34 Langhouse Green, Crail, Anstruther, KY10 3UD
GM0 KDP I. Dunbar, Mabruk, 25 Kinord Drive, Aboyne, AB34 5JZ
G0 KDQ M. Ward, 2 Hollin Gate, Otley, LS21 2DP
G0 KDR R. Lintott, Upper Grove Farm, Rendham, Saxmundham, IP17 2AS
G0 KDS S. Lindsay, 27 Bagnell Road, Bristol, BS14 8PZ
G0 KDT P. Cracknell, 54 Yannon Drive, Teignmouth, TQ14 9JP
G0 KDV Darenth Valley Radio Society, c/o P. Bourke, 26 Craylands Lane, Swanscombe, DA10 0LP
G0 KDW G. Burrett, 10 Prospect Walk, Lower Burraton, Saltash, PL12 4RG
G0 KDX B. Ashton, Squirrel Wood, Anderton Mill, Chorley, PR7 5PY
G0 KDY A. Spry, Newlands Farm, Bradworthy, Holsworthy, EX22 7RN
G0 KEB C. Frost, 61 Selbourne Avenue, Surbiton, KT6 7NR
G0 KEC H. Opitz, 26 Holme Court, Lower Warberry Road, Torquay, TQ1 1QR
G0 KED J. Bower, Linwood, Stain Lane, Mablethorpe, LN12 1QB
G0 KEE C. Simons, 51 Moorville Drive South, Carlisle, CA3 0AW
G0 KEI D. Kennard, 37 Shelley Square, Southend-on-Sea, SS2 5JP
G0 KEK B. Curtis, Beggars Roost, Rea Barn Road, Brixham, TQ5 9EE
GD0 KEO A. Birchenough, 20 St. Stephens Meadow, Waterloo Road, Sulby, Ramsey, Isle Of Man, IM7 3DA
GM0 KEQ R. Crawford, Glengarry, East Terrace, Kingussie, PH21 1JS
G0 KEV K. Gallagher, 8 Holme Grove, Burley In Wharfedale, Ilkley, LS29 7QB
G0 KEX A. O'Hara, 26 Thompson Avenue, Ainsworth, Bolton, BL2 5RJ
G0 KEY S. Cole, 160 New Haw Road, Addlestone, KT15 2DN
G0 KFD P. Bailey, 273 Humberston Avenue, Humberston, Grimsby, DN36 4JA
G0 KFF K. Field, 50 Madrona, Amington, Tamworth, B77 4EJ
GW0 KFL R. Rees, 15 Taliesin Close, Pencoed, Bridgend, CF35 6JR
G0 KFM J. Collins, 19 Brookside Park Homes, Waterloo Road, Wimborne, BH21 3SP
G0 KFQ B. Wilson, 20 Peacock Way, Littleport, Ely, CB6 1AB
G0 KFS A. Purcell, 33 Fishley Close, Bloxwich, Walsall, WS3 3QA
G0 KFT C. Dickerson, 1 Park Farm Lane, Nuthampstead, Royston, SG8 8LT
G0 KFV M. Evans, 40 Park Square East, Jaywick, Clacton-on-Sea, CO15 2NN
G0 KFW W. Cole, 5A Park Lane, Kemsing, Sevenoaks, TN15 6NU
G0 KFY P. Elliot-West, 135 Tunstall Road, Sunderland, SR2 9BB
G0 KGA A. Danby, Cornerstone, Foulbridge Lane, Snainton, Scarborough, YO13 9AY
GW0 KGD Prof. V. Zakharov, Ty-Brith, Cloddiau, Welshpool, SY21 9JE
G0 KGE H. Johnson, 2 Thirlmere, Kennington, Ashford, TN24 9BD
G0 KGI J. Coleman, 80 Ormston Avenue, Horwich, Bolton, BL6 7ED
G0 KGL G. Lindsay, 66 Jubilee Crescent, Mangotsfield, Bristol, BS16 9AZ
G0 KGR R. Beadle, 2 Edward Cottages, Great Munden, Ware, SG11 1HT
G0 KGT B. Williams, 8 Grimbald Road, Knaresborough, HG5 8HD
G0 KHA K. Seddon, 17 Dunmail Drive, Kendal, LA9 7JG
G0 KHF P. Witley, 18 Seagate Road, Hunstanton, PE36 5BD
G0 KHH A. Rogers, 3 Ripley Drive, Wigan, WN3 6AJ
G0 KHJ J. Warburton, 92 Worsley Road Farnworth, Bolton, BL4 9LX
G0 KHK P. Shaw, 15 Greenfield Avenue, Marlbrook, Bromsgrove, B60 1HE
G0 KHQ P. Hughes, 4 Millards Close, Hilperton Marsh, Trowbridge, BA14 7UN
G0 KHR E. Forsyth, 11 Brooklyn Road, Stockport, SK2 6BX
G0 KHY J. Jenkins, 3 Gosslan Close, St Ives, PE27 3YZ
G0 KHZ M. Crane, Drewton House, Back Lane, Goole, DN14 7HD
G0 KIA R. Harris, 19 Old Bath Road, Sonning, Reading, RG4 6SZ
G0 KIC B. Hayward, 22 Waldron Street, Bishop Auckland, DL14 7DS
GW0 KIG K. O'Reilly, 14 Catherine Close, Abercanaid, Merthyr Tydfil, CF48 1YY
G0 KIK S. Berry, 85 Lake View Close, West Park, Plymouth, PL5 4LT
G0 KIM J. West, 242 Grane Road, Haslingden, Rossendale, BB4 4PB
G0 KIN T. Harper, 72 School Road, Salford Priors, Evesham, WR11 8XN
GW0 KIR K. Jones, 1 Heyope Road, Heyope, Knighton, LD7 1PT
G0 KIY N. Brook, 2 Back Regent Place, Harrogate, HG1 4QR
G0 KJC J. Clayton, 49 Bramble Lane, Mansfield, NG18 3NP
G0 KJF R. Warner, Barley Hill Farm, Combe St. Nicholas, Chard, TA20 3HJ
G0 KJG B. Bevington, 12 Buckingham Road, Rowley Regis, B65 9JN
G0 KJJ W. Ritchie, 16 Avenue Mezidon Canon, Honiton, EX14 2TT
G0 KJK K. Ranger, Flat 12, Dulverton Hall Esplanade, Scarborough, YO11 2AR
G0 KJM J. Richards, 14 Southwood Drive East, Bristol, BS9 2QP
G0 KJN J. Windebank, 9 Townsend Place, St. Ippolyts, Hitchin, SG4 7RQ
G0 KJP D. Scott, 19 The Fillybrooks, Aston, Stone, ST15 0DH
G0 KJR K. Robbertze, 19 Shapton Close, Holbury, Southampton, SO45 2QJ
GW0 KJT T. Lewis, 2 Railway Terrace, Pontyberem, Llanelli, SA15 5HN
G0 KJU J. Robertson, 28 Frith Road, Bognor Regis, PO21 5LL
GW0 KJZ J. Jones, 64 Cleviston Park, Llangennech, Llanelli, SA14 9UP
G0 KKC D. Browning, 81 Bishop Road, Bishopston, Bristol, BS7 8LU
G0 KKD A. Parker, Old Rectory, 1 Church Lane, Matlock, DE4 2GL
GM0 KKE I. Coulson, 11 Redcliffs, Kingoodie, Dundee, DD2 5DL
G0 KKF B. Hough, 54 Woodbourne Road, Sale, M33 3TN
G0 KKH J. Henderson, Hodgkins Farm, Norton Heath Road, Willingale, Willingale, CM5 0QG
G0 KKL P. Mayer, Flat 7 Broomrigg, 5 Belle Vue Road, Poole, BH14 8UE
G0 KKQ E. Birch, 6 Totterton, Lydbury North, SY7 8AN
G0 KKR B. Chapman, Millbrooke Cottage, Covenham St. Bartholomew, Louth, LN11 0PB
G0 KKS A. Nance, 33 Oak Close, Copthorne, Crawley, RH10 3QT
G0 KKT I. Osborne, 19 Lumber Leys, Walton on the Naze, CO14 8SS
G0 KKU J. Howard, 111 Heath Road, Penketh, Warrington, WA5 2DB
G0 KKV M. Lowe, 11 Priory Walk Mancetter, Atherstone, CV91QA
G0 KLA C. Thompson, 425 Mortlake Ave, St Lambert, Canada, J4P 3C7
G0 KLD C. Wren, 38 Green Street, Hyde, SK14 1QX
G0 KLF N. Anderton11, 29 Cliftonville Drive, Swinton, Manchester, M27 5NA
G0 KLG R. Hodds, 17 Oaklands Drive, Willerby, Hull, HU10 6BJ
G0 KLJ J. Leader, 9 Southerwicks, Corsham, SN13 9NH
G0 KLK A. James, 14 Randle Drive, Sutton Coldfield, B75 5LH
G0 KLN C. Ryalls, 22 Carr Lanebd205Hn, Keighley, BD205HN
GM0 KLO C. Grossart, 11 Woodlands Drive, Brightons, Falkirk, FK2 0TF
GM0 KLP J. Pentland, 2 Glenniston Cottages, Auchtertool, Fife, KY5 0AX
G0 KLQ D. Cross, 15 Fernside Road, West Moors, Ferndown, BH22 0EE
G0 KLT D. Rogers Jones, 20 Birchwood, Leyland, PR26 7QJ
G0 KLU P. Fairhurst, 161 Daniells, Welwyn Garden City, AL7 1QP
GW0 KLY R. Jones, 20 Newfoundland Way, Blackwood, NP12 1FS
GM0 KMA Dr M. Rainey, 2 Shields Holdings, Lochwinnoch, PA12 4HL
G0 KMB K. Bowdler, 18 Cavendish St., Leigh, WN7 1SG
G0 KMC A. Slaughter, 42 Goss Avenue, Waddesdon, Aylesbury, HP18 0LY
G0 KMF R. Holmshaw, 142 Oakleigh Park Drive, Leigh-on-Sea, SS9 1RU
GM0 KMJ P. Johnstone, 26 Lomond Crescent, Stenhousemuir, Larbert, FK5 4LT
G0 KMK M. Aslam, 38 Grey St., Burnley, BB10 1BA
G0 KML B. Hawes, 201 Ridgeway, Plympton, Plymouth, PL7 2HP
G0 KMN S. Hepworth, 9 College View, Ackworth, Pontefract, WF7 7LA
G0 KMP G. Aungiers, 6 Woodlands Crescent, Barton, Preston, PR3 5HB
G0 KMV H. King, Que Lindo, Church Lane, Bristol, BS39 5UP
G0 KMW H. Shepherd, Whydown, 3 White House Close, Abingdon, OX13 6LP
G0 KNJ R. Bygrave, 69 Albert Gardens, Harlow, CM17 9QG
G0 KNL W. Christlo, 6 Nether Ley Gardens, Chapeltown, Sheffield, S35 1AH
G0 KNN M. Gregg, 22 Mayfields, Spennymoor, DL16 6RN
GM0 KNT Dr A. Bates, Caberfeidh, Balnageith, Forres, IV36 2SG
G0 KNW C. Wiles, Everest, Mile Road, Morpeth, NE61 5QW
G0 KNX G. Allen, 3 Ryton Close, Coventry, CV4 8HF
G0 KNY K. Heaton, 5 Perriams Place, Budleigh Salterton, EX9 6LY
G0 KOC A. Kinson, 6 Uplands Park, Broad Oak, Heathfield, TN21 8SJ
G0 KOE T. Foxton, Dunbar, Dam Lane, Malton, YO17 9SJ
G0 KOF D. Henretty, 13 Siskin Chase, Cullompton, EX15 1UD
G0 KOI M. Cooper, 15 Woodleigh Avenue, Harborne, Birmingham, B17 0NW
G0 KOJ B. Thomas, Harpley House, Police Road, Walpole St. Andrew, Wisbech, PE14 7NN
G0 KOK P. Love, 2 Meadway, Dover, CT17 0PS
G0 KOM A. Mcgonigle, 5 Whitehouse Crescent, Chelmsford, CM2 7LP
G0 KOO A. Mcdowell, Fern Cottage, The Gride, Boston, PE22 9LS
G0 KOU B. Arrowsmith, 25 Watchouse Road, Chelmsford, CM2 8PT
GI0 KOW R. Cummings, 19 Bachelors Walk, Keady, Armagh, BT60 2NA
G0 KOY B. Clues, 8 Acland Avenue, Colchester, CO3 3RS
GW0 KPD J. James, 1 Pellau Road, Margam, Port Talbot, SA13 2LF
G0 KPE C. Mcgowan, Belvedere, Tylers Lane, Reading, RG7 6TN
GI0 KPF B. Mccausland, 5 Hollyfields, Dungannon, BT71 7BH
G0 KPG R. Moore, 34 Fishponds Road, Kenilworth, CV8 1EZ
G0 KPH P. Keighley, 15 Stuart Court, Warwick Terrace, Leamington Spa, CV32 5NU
GD0 KPN J. Mcloughlin, 8 Governors Hill, Douglas, Isle Of Man, IM2 7AW
G0 KPQ J. Morgan, 9 Consort Close, Plymouth, PL3 5TX
GW0 KPU R. Harper, 4 Gresford Road, Llay, Wrexham, LL12 0NW
GW0 KPV G. Owen, 2 Ffordd Beibio, Holyhead, LL65 2EF
G0 KPY A. Baker-Munton, 66 Stanway Road, Headington, Oxford, OX3 8HX
G0 KPZ D. Portch, 148 Brixham Road, Welling, DA16 1EJ
G0 KQA P. Davies, Silver Birches, Orchard Road, Basingstoke, RG22 6NU
GM0 KQB R. Kemp, 1 Grendon Court, Stirling, FK8 2JX
GD0 KQE K. Jordan, Engadine, Little Switzerland, Isle of Man, IM2 6AG
G0 KQH B. O'Donoghue, 30 Lake Drive, Hamworthy, Poole, BH15 4LT
G0 KQI L. Painter, 185 Albion St., St Helens, WA10 2HA
G0 KQK T. Chambers, Autumn, Water Lane, Castle Bytham, Grantham, NG33 4RT
G0 KQO G. Elliott, 32 Chapel Street, Newport, PO30 1PZ
G0 KQP J. Carroll, 5 Montagu View, Leeds, LS8 2RH

G0 KQR R. Bundell, 24 Sylvan Avenue, East Cowes, PO32 6PS
G0 KQS G. Griffiths, Sherwood House, Buggen Lane, Neston, CH64 6QB
G0 KQT A. Holdway, 18 The Quantocks, Thatcham, RG19 3SF
GW0 KQU G. Slatter, 6 Glannant St., Penygraig, Tonypandy, CF40 1JT
GW0 KQV D. Clark, Martinique, Wolfscastle, Haverfordwest, SA62 5DY
GW0 KQX W. Cook, Fronoleu, Bryngwy, Rhayader, LD6 5BN
G0 KQY D. Murrell, 25 Waverley Road, Hoylake, CH47 3DD
G0 KRB G. Phillips, 57 Hollytrees, Bar Hill, Cambridge, CB23 8SF
G0 KRD D. Downes, 7 Sandy Lane, Fakenham, NR21 9ES
G0 KRG Keighley Raynet Group, c/o L. Conlon, 4 Hill Crest Drive, Slack Head, Milnthorpe, LA7 7BB
G0 KRH C. Tarrant, 91 Dunes Road, Greatstone, New Romney, TN28 8SW
G0 KRK B. Cockfield, 47 Aston Road, Willenhall, WV13 3DG
GW0 KRL I. Capon, Pentre Garreg Bach, Marianglas, LL73 8PP
GW0 KRQ J. Cartwright, 20 Castlefield Place, Cardiff, CF14 3DU
G0 KRR J. Hough, 1 Rock Lane, Linslade, Leighton Buzzard, LU7 2QQ
G0 KRS Keighley ARS, c/o K. Conlon, 4 Hill Crest Drive, Slack Head, Milnthorpe, LA7 7BB
G0 KRT E. Masters, 91 Mayfair Avenue, Worcester Park, KT4 7SJ
G0 KRU A. Wright, Cherry Tree Cottage, High Road, Beighton, Norwich, NR13 3LA
G0 KRX P. Ruder, 34 Chelmsford Road, South Woodford, London, E18 2PL
G0 KRY D. Sanders, 149 Sutton Road, Walsall, WS5 3AW
G0 KSC J. Johnson, 20 Sanders Road, Canvey Island, SS8 9NY
G0 KSD R. Allgood, 7 The Chase, Blofield, Norwich, NR13 4LZ
G0 KSJ J. Graves, 172 Hall Lane, Upminster, RM14 1AT
G0 KSL R. Torr, 68 East Towers, Pinner, HA5 1TL
G0 KSN H. Dabhi, 23 Shalgrove Field, Fulwood, Preston, PR2 3SX
G0 KSS K. Nobbs, 49 St. James Drive, Burton, Carnforth, LA6 1HY
G0 KTC C. Ayres, 219 Ashingdon Road, Rochford, SS4 1RS
G0 KTD A. Bonney, 6 Mitchell Road, St Austell, PL25 3AU
GM0 KTH D. Wakefield, Millfield, Burray, KW17 2SU
GW0 KTL G. Edmunds, 14 Avon Close, Bettws, Newport, NP20 7BZ
G0 KTN T. Smithers, 14 Georgian View, Bath, BA2 2LZ
GM0 KTO J. Power, 0/2 20 Eastercraigs, Glasgow, G31 3LJ
G0 KTP R. Cockbill, 45 Mills Road, Melksham, SN12 7DT
G0 KTR R. Farnley, 67 Barons Court Failsworth, Manchester, M35 0LH
G0 KTS J. Wright, 65 Groombridge Close, Welling, DA16 2BP
G0 KTT P. Riley, 11 Pinewood Court, South Downs Road, Altrincham, WA14 3HY
G0 KTU A. Miller, 11 Blackbrook Avenue, Paignton, TQ4 7ND
G0 KTV J. Edgington, 83 Woking Road, Guildford, GU1 1QL
G0 KTW J. Moss, 1 Millers View, Much Hadham, SG10 6BN
G0 KTX A. Hornsby, 328 Pelham Road, Immingham, DN40 1PT
G0 KTY G. Salisbury, Nythfa, 1 Stad-Y-Garnedd, Anglesey, L60 6BB
G0 KUA V. Kathuria, 2 Bevan Road, Lovedean, Waterlooville, PO8 9QH
G0 KUC D. Bloomfield, 14 Horsham Close, Luton, LU2 8JH
G0 KUD P. Haith, 17 Lime Tree Avenue, Grimsby, DN33 2BB
G0 KUE P. Webb, 119 Chipstead Valley Road, Coulsdon, CR5 3BP
G0 KUF N. Buchanan, Meadowside, Jacobs Well Road, Guildford, GU4 7PD
GI0 KUH J. Mccabe, 121 Garvaghy Road, Craigavon, BT62 1EH
G0 KUI J. Wood, 6 West Terrace, Stakeford, Choppington, NE62 5UL
GM0 KUJ J. Mcgifford, 52 Gartons Road, Glasgow, G21 3HY
GM0 KUP F. Mann, 12 Greenbank Court, Falkirk, FK1 5DS
G0 KUQ F. Hills, 112 Boxfield Green, Stevenage, SG2 7DS
G0 KUU F. Gillham, 260 Summerhouse Drive, Wilmington, Dartford, DA2 7PB
G0 KUW C. Bennett, 67 King St., Clowne, Chesterfield, S43 4BS
G0 KUX P. Kay, 97 Avenue Road, London, N14 4DH
G0 KUY S. Crane, 13 Kirkstone Drive, Royton, Oldham, OL2 6TP
G0 KUZ B. Long, 81 Easthorpe Street, Ruddington, Nottingham, NG11 6LB
G0 KVA A. Sargent, 25 Jordans Way, Bricket Wood, St Albans, AL2 3SJ
G0 KVC H. Crouch, 21A Victoria Gardens, Horsforth, Leeds, LS18 4PJ
GM0 KVD C. Mackay, 5 Cromer Gardens, Glasgow, G20 9JQ
GM0 KVE A. Dickson, 17 Junction Road, Kinross, KY13 8TA
G0 KVF R. Croucher, 26 Edith Avenue, Peacehaven, BN10 8JB
G0 KVG R. Neal, 14 Saxton Close, Worksop, S80 3DE
G0 KVJ Peterlee ARC, c/o A. Pennell, 99 Westheath Avenue, Sunderland, SR2 9LQ
G0 KVK G. Cooper, 33 Lawnswood Road, Wordsley, Stourbridge, DY8 5PH
G0 KVM W. Gravenor, 3 Foxhill Grove, Queensbury, Bradford, BD13 2JN
G0 KVO D. Pallister, 9 Curtis Hayward Drive, Quedgeley, Gloucester, GL2 4WJ
GI0 KVQ G. Millar, 1 Mullybrannon Road, Dungannon, BT71 7ER
G0 KVR C. Mayo, 118 Burden Road, Beverley, HU17 9LH
G0 KVS Prof. M. Gill, 21 Priory Terrace, London, NW6 4DG
G0 KVU H. Sheratte, Redcar Villa, 382 Buxton Road, Macclesfield, SK11 7ES
GW0 KWA D. Clark, 37 Rotherslade Road, Langland, Swansea, SA3 4QW
G0 KWC R. Waite, 95 Westlea Road, Leamington Spa, CV31 3JE
G0 KWD K. Dyer, 79 Station Road, Woolton, Liverpool, L25 3PY
G0 KWE J. Knowles, 10 Grove Hill, Hessle, HU13 0RT
G0 KWF W. Taylor, 14 Rossiters Lane, St. George, Bristol, BS5 8TW
G0 KWG S. Lodge, 7 Primrose Drive, Milkwall, Coleford, GL16 7PU
GM0 KWL B. Mulleady, 9 Elizabeth Crescent, Camelon, Falkirk, FK1 4JF
GD0 KWM G. Brown, 10 Albert Street, Ramsey, Isle Of Man, IM8 1JF
GW0 KWO K. Williams, 39 Lewis Drive, Caerphilly, CF83 3FT
G0 KWQ P. Bates, 46 Kingsley Avenue, Redditch, B98 8PL
GM0 KWW J. Alexander, Shore Cottage, Girvan, KA26 9JH
G0 KXD H. Worden, 3 Tower View, Darwen, BB3 3GZ
G0 KXG J. Nicholls, 93 Swan Road, Hanworth, Feltham, TW13 6PE
G0 KXL S. Maton, 117 Woodchurch Road, Birkenhead, CH42 9LJ
G0 KXV G. Meredith, Hedgerow, Watton Road, Larling, Norwich, NR16 2AJ
G0 KXW A. Fitzmaurice, 14 Welwyn Close, Thelwall, Warrington, WA4 2HE
G0 KXY P. Wroe, 44 Hillberry Crescent, Warrington, WA4 6AF
G0 KXZ A. Sockett, 35 Whernside Road, Woodthorpe, Nottingham, NG5 4LB
G0 KYA S. Nichols, 61B Norwich Common, Wymondham, NR18 0SW
G0 KYB M. Kinder, 58 Longridge Avenue, Stalybridge, SK15 1HL
G0 KYD A. Hull, 1 Occupation Lane, New Bolingbroke, Boston, PE22 7LW
G0 KYE L. Landricombe, 19 Crackston Close, Eggbuckland, Plymouth, PL6 5SN
G0 KYG P. Willetts, 197 Norwich Road, Fakenham, NR21 8LR
G0 KYH J. Elliott, Tregerrick, Martinstown, Dorchester, DT2 9JN
G0 KYJ G. Liversidge, 65 Rowelfield, Luton, LU2 9HL
G0 KYK R. Beardsmore, 2 Fitzmaurice Road, Wednesfield, Wolverhampton, WV11 3EG
G0 KYL J. Lawrence, 8 Murray Terrace, Dipton, Stanley, DH9 9HB
G0 KYM W. Lowder, 24 Plantation Lane, Bearsted, Maidstone, ME14 4BH
G0 KYN R. Markham, 25 Burndell Way, Hayes, UB4 9YF
G0 KYR D. Cooper, 7 Kendal Rise, Bedlington, NE22 6PB
G0 KYS R. Edgar, 45 Exeter Road, Dawlish, EX7 0AB
GM0 KYU J. Robertson, 143 Rankin Court, Greenock, PA16 9AZ
G0 KYX Rev. P. Morgan, 4 Cromwell Mews, Burgess Hill, RH15 8QF
GW0 KYY M. Warner, 76 Rhodfar Eos, Cwmrhydyceirw, Swansea, SA6 6SW
GJ0 KYZ Dr P. Mahrer, 2 Oakley, La Rue Parcqthee, St Lawrence, Jersey, JE3 1FR
G0 KZA E. Bishop, 21 Mandalay St., Basford, Nottingham, NG6 0BH
G0 KZD M. Withey, 9 Marnhull Road, Longfleet, Poole, BH15 2EX
GW0 KZE Dr C. Dublon, Tyn-Y-Waun, Dare Road, Aberdare, CF44 8UB
GW0 KZG A. Adams, 1 Nant Y Ffynnon, Letterston, Haverfordwest, SA62 5SX
G0 KZH E. Clayton, 220 Milnrow Road, Rochdale, OL16 5BB
G0 KZI J. Williams, 18 St. Andrews Close, Holme Hale, Thetford, IP25 7EH
G0 KZM D. Egan, 56 Walker Avenue, Wollescote, Stourbridge, DY9 9EL
G0 KZN A. Sargeant, 27 Sandygate Crescent, Old Leake, Boston, PE22 9RA
G0 KZO E. Lomas, 2 Linney Road, Bramhall, Stockport, SK7 3JW
G0 KZT A. Briers, 33 Deans Walk, Coulsdon, CR5 1HR
GW0 KZW W. Jones, Tanglewood, 2 Bryntirion Avenue, Prestatyn, LL19 9PB
GM0 KZX B. Spink, 9 St. Andrews Crescent, Dumbarton, G82 3ER
G0 LAA E. Martin, 90 Grand Drive, Herne Bay, CT6 8LS
G0 LAD J. Parfett, 65 Brompton Lane, Rochester, ME2 3BA
G0 LAG J. Penney, 2A St John St., Wainfleet, Skegness, PE24 4DL
G0 LAK J. Rogers, 186 Beavers Lane, Birleywood, Skelmersdale, WN8 9BP
GI0 LAM G. Lamb, 31 Dromara Road, Ballyward, Castlewellan, BT31 9SJ
G0 LAN A. Taylor, 24 Marlborough Drive, Mablethorpe, LN12 2 BA
G0 LAP S. Jeffery, 3 Cromwell Close Walcote, Lutterworth, LE17 4JJ
G0 LAU J. Barber, Jasmine Cottage, Spend Lane, Ashbourne, DE6 2AS
G0 LAX A. Duggan, 28 Higher Rads End, Eversholt, Milton Keynes, MK17 9ED
G0 LAZ J. Pryer, Apesford Crossing Cottage, Apesford, Leek, ST13 7EX
GW0 LBA A. Hughes, Derwen Las, Valley Road, Llanfairfechan, LL33 0SS
G0 LBB R. Batty, 31 Spring Lane, Balderton, Newark, NG24 3NZ
G0 LBE S. Watkinson, 40 Wharfedale, Westhoughton, Bolton, BL5 3DP
GW0 LBI L. Smart, Wordsley, Gwerthonor Road, Bargoed, CF81 8JS
GM0 LBN Dr J. Clark, 35 Jedburgh Avenue, Rutherglen, Glasgow, G73 3EN
G0 LBO J. Ross, The Gables, Jack Lane, Northwich, CW9 8QA
G0 LBQ R. Perks, 120 Cranes Park Road, Sheldon, Birmingham, B26 3ST
GM0 LBR B. Gourlay, Muirhead House, Chryston, Glasgow, G69 9ND
G0 LBT K. Tromans, 7 Heathfield, Heath Charnock, Chorley, PR6 9LA
G0 LBZ P. Cahill, 56 Dene Road, Headington, Oxford, OX3 7EE
G0 LCB A. Cleaver, 19 Newlands Drive, Grove, Wantage, OX12 0NY
G0 LCC H. Grinter, 16 Gladiolus Road, Langport, TA109TA
G0 LCD P. Chinnock, 30 Trelissick Road, Paignton, TQ3 3GW
G0 LCE K. Robinson, 33 Mirlaw Road, Cramlington, ne23 6ub
G0 LCG S. Sorockyj, 8 Bowden Avenue, Bestwood Village, Nottingham, NG6 8XN
G0 LCH M. Nash, 22 Northleigh Close, Loose, Maidstone, ME15 9RP
G0 LCJ B. Lucock, 15 Mayfield Road, Newquay, TR7 2DG
G0 LCN D. Prendiville, 40 Caerleon Drive, Southampton, SO19 5LF
G0 LCO R. Foster, 18 Stokesay Way, Sutton Hill, Telford, TF7 4QE
G0 LCS K. Rochester, 22 Langford Road, Cockfosters, Barnet, EN4 9DS
G0 LCT G. Moss, 15 Coppice Avenue, Hatfield, Doncaster, DN7 6AH
G0 LCU B. Walker, 70 King George Road, Loughborough, LE11 2PA
G0 LCV J. Fidoe, 85 Sedgemoor Road, Bridgwater, TA6 5NS
G0 LCX D. Weatherill, Northend Cottage, North End Road, Yatton, Bristol, BS49 4AS
G0 LDB M. Mallinson, 25 The Fairway, Banbury, OX16 0RR
GI0 LDI D. Keys, 71 Madison Avenue, Eglinton, Londonderry, BT47 3PW
G0 LDJ D. Cansfield, 1 Brook Walk, Calmore, Southampton, SO40 2UY
G0 LDO R. Summerfield, 64 Station Road, Broughton Astley, Leicester, LE9 6PT
G0 LDP K. Starkey, 13A Cardigan Road, Bedworth, CV12 0LY
G0 LDR J. Marlow, 21 Thames Rise, Kettering, NN16 9JL
G0 LDU K. Allies, 6 Alston Close, Hazel Grove, Stockport, SK7 5LR
G0 LDY K. Jenkinson, 2 Madeira Avenue, Codsall, Wolverhampton, WV8 2DS
GW0 LDZ B. Garland, 20 Bryn Avenue, Upper Brynamman, Ammanford, SA18 1BD
GI0 LEC Lough Erne ARC, c/o A. Duffy, 81A Arney Road, Bellanaleck, Enniskillen, BT92 2DL
G0 LEE R. Lee, 7 Long Meadow, Little Hoole, Preston, PR4 4RQ
G0 LEF T. Bell, 16 North Seaton Road, Newbiggin-by-the-Sea, NE64 6XT
G0 LEH G. Chatfield, 1A Sheringham Way Orton Longueville, Peterborough, PE2 7AH
G0 LEI E. Hughes, 1 Leith Gardens Tanfield Lea, Stanley, DH9 9LZ
G0 LEJ M. Huggett, Rosslyn, Station Road, Brampton, CA8 1EX
G0 LEL F. Sunley, 39 Winton Road, Northallerton, DL6 1QQ
G0 LEN West Lincs Ry G, c/o P. Worsdale, Emergency Planning Department, Fire Brigade Headquarters, Lincoln, LN5 8EL
G0 LEP D. Stewart, Buckskin, 16 Prescelly Close, Basingstoke, RG22 5DN
G0 LES E. Simpson, Laneside, Cliff Lane, Bridlington, YO15 1JF
G0 LEU P. Johnson, 5 The Hawthorns, Broadstairs, CT10 2NG
G0 LEV D. Painter, Troutbeck, Mary Tavy, Tavistock, PL19 9PR
G0 LEY R. Smith, 63 Windsor Road, Wellingborough, NN8 2ND
G0 LFA N. Swallow, 178 Barcroft St., Cleethorpes, DN35 7DX
G0 LFE Rev. K. Gray, 25 The Pastures, Blyth, ne24 3ha
G0 LFF R. Hide, Flat 22, Church Court Church Road, Haywards Heath, RH16 3UE
G0 LFH P. Mustchin, 6 Spinney North, Pulborough, RH20 2AT
G0 LFI F. Cotton, 49 Cornwall Road, Fratton, Portsmouth, PO1 5AR
G0 LFM V. Sancto, Meadowbank, 15A Spratling Street, Ramsgate, CT12 5AW
G0 LFN S. Southwell, Sullys, 12 Somerset Road, Southsea, PO5 2NL
G0 LFP S. Courtney-Crowe, 28 Brymore Close, Prestbury, Cheltenham, GL52 3DY
G0 LFQ P. Mason, Penrose House, Tavistock Road, Launceston, PL15 9LE
G0 LFV P. Fisher, Chevalier, Marks Corner, Newport, PO30 5UH
G0 LFX M. Harrison, 8 Browns Lane, Uckfield, TN22 1RT
G0 LFY D. Recardo, 1 Heronfield Close, Redditch, B98 8QL
G0 LFZ A. Recardo, 1 Heronfield Close, Redditch, B98 8QL
G0 LGA R. Letts, 28 Catlin Crescent, Shepperton, TW17 8EU
G0 LGB J. Walker, 22 Temperance Field, Wyke, Bradford, BD12 9NR
G0 LGC L. Culshaw, 15 Naunton Avenue, Leigh, WN7 4SX
G0 LGE M. Brooman, 25 Knockholt Road, Cliftonville, Margate, CT9 3HL

G0 LGF T. Evennett, The Homestead, Pound Green Lane Shipdham, Thetford, IP25 7LS
G0 LGG N. Challacombe, 17 Tanners Lane, Chalkhouse Green, Reading, RG4 9AD
G0 LGJ M. Taylor, 6 Welden Road, Scarning, Dereham, NR19 2UB
G0 LGK E. Wall, Shrubbery Cottage, Felderland Lane, Deal, CT14 0BT
G0 LGO A. Turton, 58 Highfield Lane, Quinton, Birmingham, B32 1QT
GI0 LGV H. Magill, 51 Ballybracken Road, Doagh, Ballyclare, BT39 0TQ
G0 LGW R. Caine, 148 Dumpton Park Drive, Broadstairs, CT10 1RP
G0 LGZ B. Grimes, Flat 12, Clyde House, Ventnor, PO38 1QL
G0 LHB A. Okubo, 1427-9-608, Yamazaki-Cho Machida City, Tokyo, Japan, 195-0074
G0 LHD R. Caton, 13 Goss Barton, Nailsea, Bristol, BS48 2XD
G0 LHE R. Wilmot, Elm Tres, Drayson Lane, Northampton, NN6 7SR
G0 LHG A. Rayner, 24 Syers Lane, Whittlesey, Peterborough, PE7 1AT
G0 LHL M. Humphreys, 25 Dalestorth Close, Sutton-in-Ashfield, NG17 4EH
G0 LHM B. Tuffrey, 53 Sheffield Road, Warmsworth, Doncaster, DN4 9QR
G0 LHN J. Butterworth, 38 Stuart Avenue, Moreton, Wirral, CH46 9PF
G0 LHR L. Robinson, 82 Grassholme, Wilnecote, Tamworth, B77 4BZ
G0 LHU J. Lawton, 37 Southway, Horsforth, Leeds, LS18 5RN
G0 LHV R. Kay, 24 Chapel Garth, Gilberdyke, Brough, HU15 2UH
G0 LHX H. Passmore, 1 Knights Close, Westonzoyland, Bridgwater, TA7 0AZ
G0 LHZ J. Carter, 22 Orchard Coombe, Whitchurch Hill, Reading, RG8 7QL
G0 LIA R. James, 77 Charlotte Close Mount Hawke, Nr Truro., TR4 8TT
G0 LIB R. Weston, 38 Church Road, Peasedown St. John, Bath, BA2 8AF
G0 LII S. Hodgson, 4 Nikolaou Michael Street, Dasaki Achnas, Cyprus, 5523
GW0 LIK C. Raymond, 23 Castle Pill Crescent, Steynton, Milford Haven, SA73 1HD
GM0 LIM J. Duffy, 39 Kylerhea Road, Thornliebank, Glasgow, G46 8AB
G0 LIN C. Smith, 2 Ha'Penny Drive, Holbrook, Ipswich, IP9 2TT
G0 LIQ J. Cunningham, 219 Alfreton Road, Underwood, Nottingham, NG16 5GX
GM0 LIR P. Woods, 29 Yarrow Crescent, Wishaw, ML2 7JX
GW0 LIS A. Wright, 8 Bryn Mor Terrace, Holyhead, LL65 1EU
G0 LIW P. Hopkins, 2641 Suncoast Lakes Blvd, Port Charlotte, USA, 33980
G0 LIY P. Smit, 18 Owlwood Lane, Dunnington, York, YO19 5PH
G0 LIZ E. Saunders, 40 Walkley Street, Sheffield, S6 3RG
G0 LJB P. Williams, 44 Meadow Road, Mirehouse, Whitehaven, CA28 8EP
G0 LJD B. Howard, 8 St. Margarets Close, Seasalter, Whitstable, CT5 4ST
G0 LJF M. Binks, 24 Mill Lane, Stockport, SK5 6UU
G0 LJG D. Green, The Archways, St. Georges Road, Trowbridge, BA14 6JQ
G0 LJH C. Holmes, Manacor, 4 Dovefields, Uttoxeter, ST14 5LT
G0 LJI G. Evans, 241 St. Johns Road, Newbold Moor, Chesterfield, S41 8PE
G0 LJJ D. Mackenny, 21 Chilton Way, Hungerford, RG17 0JR
G0 LJK W. Dancock, 11 St. Davids Close, Stourport-on-Severn, DY13 8RZ
G0 LJM D. Roebuck, 8 Runnymede Court, Bradford, BD10 9JW
G0 LJP P. Mcleod, 4 Caple Avenue, Kings Caple, Hereford, HR1 4UL
G0 LJU D. Koveos, 235 Poynters Road, Dunstable, LU5 4SH
G0 LJV S. Swinbourne, 11 Stapleton Road, Warmsworth, Doncaster, DN4 9LA
GW0 LJW D. Goodwin, 25 Bevan Crescent, Blackwood, NP12 1EW
G0 LKA C. Cruddas, 81 Church Walk, Atherstone, CV9 1PS
G0 LKI W. Cockerell, 3 Churchford Road, Knowle, Braunton, EX33 2LT
GW0 LKJ W. Halliwell, 20 Llwynon Road, Oakdale, Blackwood, NP2 0LX
GM0 LKS E. Mcgreevy, 47 Fairfield Drive, Renfrew, PA4 0EG
GM0 LKT A. Ferris, 60 Appin Crescent, Kirkcaldy, KY2 6ES
G0 LKY G. Civil, Whitehouse Farm, Magpie Lane, Brentwood, CM13 3DZ
G0 LLB R. Smith, 34 Churchill Rise, Chelmsford, CM1 6FD
G0 LLC M. Bridges, 7 Sun Road, Woodland, Bishop Auckland, DL13 5NF
GW0 LLD H. Jones, Dolau Bran, Cynghordy, Llandovery, SA20 0LD
G0 LLE P. Ferris, 116 Capel Road, London, E7 0JS
G0 LLG D. Davies, 4 Whitendale, Lancaster, LA1 5JD
GM0 LLJ B. Borrows, 27 Craigdimas Grove, Dalgety Bay, Dunfermline, KY11 9XR
G0 LLL J. Roberts, 69 Barnoldswick Road, Barrowford, Nelson, BB9 6BQ
G0 LLP L. Proud, 26 Drayton Court, The Green, Nuneaton, CV10 0SL
G0 LLU A. Harrison, 4 Hardwick Close, High Lane, Stockport, SK6 8DG
G0 LLX A. Bassett, 125 Stonyhill Avenue, South Shore, Blackpool, FY4 1PW
G0 LLY S. Munro, 10 Aykroft, Bourne, PE10 0QX
G0 LMA S. Crooks, 10 Mere Close, Mountsorrel, Loughborough, LE12 7BP
G0 LMD M. Butler, 44 East Stratton, Winchester, SO21 3DU
G0 LMJ E. Garrott, Lynden, Clappers Lane, Chichester, PO20 7JJ
GI0 LMR W. Redmond, 6 Hazelwood Crescent, Craigywarren, Ballymena, BT43 6TA
G0 LMX V. Denecker, Kernanderry, Faringdon Road, Abingdon, OX13 6QJ
G0 LNA R. Henderson, 65 Rowarth Road, Manchester, M23 2UL
G0 LNB G. Goodwin, 16 Hucklow Avenue, Newall Green, Manchester, M23 2YX
G0 LNE T. Stokes, 22 Armada Close, Erdington, Birmingham, B23 7PB
G0 LNI J. Stringer, 2 West End, Marston Magna, Yeovil, BA22 8BW
G0 LNK P. Bower, 103 Henson Park, Chard, TA20 1NJ
GW0 LNM P. Pentecost, Brynhyfryd, Maes Y Bont Road, Llanelli, SA14 7NA
G0 LNN D. Draycott, 3 Sycamore Gardens, Dymchurch, Romney Marsh, TN29 0LA
GW0 LNO Dr S. Feeney, Tardd Y Dwr, Star Crossing Road Cilcain, Mold, CH7 5NU
G0 LNS G. Robinson, 9 Greenlands Court, Seaton Delaval, Whitley Bay, NE25 0BU
G0 LNT M. Millward, 50 Barnsley Road, Moorends, Doncaster, DN8 4QT
G0 LNV Dr T. Appleyard, 78 Chelsea Road, Sheffield, S11 9BR
G0 LNW T. Horabin, 69 Birchwood Avenue, North Gosforth, Newcastle upon Tyne, NE13 6QB
G0 LNX I. Davison, 20 Littlegreen Gardens Compton, Chichester, PO18 9NP
G0 LOC T. Loraine, Fieldgate, Coltstaple Lane, Horsham, RH13 9BB
GM0 LOD G. Collier, 64 Hadfast Road, Cousland, Dalkeith, EH22 2NZ
G0 LOE S. Phillips, 26 Belvedere Drive, Dukinfield, SK16 5NW
G0 LOF F. James, 6 Pinewood Close, East Preston, Littlehampton, BN16 1HF
G0 LOH K. Dutson, 3 The Barracks, Wynford Eagle, Dorchester, DT2 0ER
G0 LOJ Dr C. Budd, 18 Rossendale Close, Worle, Weston-Super-Mare, BS229HA
GM0 LOK J. Leggat, Ailach, St. Aethans Road, Elgin, IV30 2YR
G0 LOL C. Kidger, 25 Elmham Road Cantley, Doncaster, DN4 6LF
GM0 LOO H. Hunter, 25 Braehead Road, Kirkcaldy, KY2 6XP
G0 LOP G. Tweedy, 8 Greencliffe Drive, York, YO30 6NA
GM0 LOT R. Clasper, 32 Murieston Park, Livingston, EH54 9DT
G0 LOU A. Hordle, 152 Evering Avenue, Poole, BH12 4JH
G0 LOW Shortwave Shop, c/o D. Kemp, Zeacombe House Caravan Park East Anstey, Tiverton, EX16 9JU
G0 LOZ I. Tomson, 13 Valley View, Bewdley, DY12 2JX
GM0 LPB J. Gault, 25 Beech Brae, Bishopmill, Elgin, IV30 4NS
G0 LPF R. Willkins, 20 Fairholme Drive, Yapton, Arundel, BN18 0JH
G0 LPG B. Gaunt, Po Box 50, Guildford, GU1 2FJ
G0 LPN B. Alperowicz, 20 Chemin Du Cabanis, Meynes, France, 30840
G0 LPP C. Galea, 36 Godwit Close, Gosport, PO12 4JF
G0 LPQ J. Hagen, 7 Oak Close, Whiston, Prescot, L35 2YG
G0 LPT G. Wegg, 23 Kerdane, Hull, HU6 9EB
G0 LPU A. Newton, 10 Rowan Court, Greasby, Wirral, CH49 3QH
G0 LPV B. Chappell, 49 Midway, South Crosland, Huddersfield, HD4 7DA
G0 LPX B. Garbutt, 34 The Green, Tockwith, York, YO26 7RA
G0 LQC D. Briggs, 57 Charlton Drive, High Green, Sheffield, S35 3PA
G0 LQD P. Valleley, 9 Lavender Road, Basingstoke, RG22 5NN
G0 LQI M. Murphy, 133 Preston Road, Preston, Weymouth, DT3 6BG
G0 LQK M. Hinchliffe, 2 Ash Grove, New Longton, Preston, PR4 4XJ
GD0 LQL S. Kewley, 16 Hillcrest Grove, Onchan, Isle of Man, IM3 3HY
G0 LQM J. Hipwell, 5 Dolphin Crescent, Paignton, TQ3 1AE
G0 LQN L. Fish, 44 Maycroft Avenue, Poulton-le-Fylde, FY6 7NE
G0 LQO R. Taylor, 17 York Close, Clayton Le Moors, Accrington, BB5 5RB
G0 LQT H. Smith, 66 The Avenue, Clacton-on-Sea, CO15 4ND
G0 LQU P. Fardell, 90 Beechwood Avenue, St Albans, AL1 4XZ
G0 LQV M. Fordham, 24A Main Street, Prickwillow, Ely, CB7 4UN
G0 LQW P. Macolive, 6 Pembroke Way, Hayes, UB3 1PZ
G0 LQX T. Newstead, 17 Aspland Road, Norwich, NR1 1SH
G0 LQZ C. Walkup, 1 Darley Hall, Luton, LU2 8PP
GM0 LRA Lorn Radio Amateur Club, c/o S. Mciver, 9 Balvicar Road, Oban, PA34 4RP
GI0 LRB P. Strawbridge, 98 Moyola Drive, Londonderry, BT48 8EF
G0 LRE J. Norman, 9A St. Johns Grove, Heysham, Morecambe, LA3 1ET
G0 LRI S. Kennedy, Colmans Farm, Elmstone-Hardwicke, Cheltenham, GL51 9TG
G0 LRJ P. Daymond, 14 Philip Close, Plymouth, PL9 8QZ
G0 LRK K. Wall, 27 Broomfield Road, Fleetwood, FY7 7HA
G0 LRM A. Littler, 365 Westhorne Avenue, London, SE12 9AB
G0 LRO D. Watmough, 41 West Crayke, Bridlington, YO16 6XR
G0 LRP P. Waters, Unit3, Site2 Sandpitlane, Nr Beccles, NR34 7TH
G0 LRR Rssdale Rayt Gr, c/o S. Greenwood, Carter Place Farm, Hall Park, Rossendale, BB4 5BQ
G0 LRU F. Alderson, Old School House, Tattersett, King's Lynn, PE31 8RS
G0 LRW M. Simmons, 6 The Crescent, Bletchley, Milton Keynes, MK2 2QD
GI0 LRZ Dr N. Mitchell, 6 Brae Road, Newry, BT34 1NZ
G0 LSA Fyle Coast R.G., c/o R. Knighton, 262 Victoria Road West, Thornton-Cleveleys, FY5 3QB
G0 LSE P. Harris, 7 Rowan Avenue, Egham, TW20 8AN
G0 LSI D. Peachey, Thornbury Cottage, Ashmill, Beaworthy, EX21 5HA
G0 LSJ C. Jones, 179 Blandford Road, Efford, Plymouth, PL3 6JZ
G0 LSK D. Taylor, 22 Meon Road, Mickleton, Chipping Campden, GL55 6TD
G0 LSP L. Pawlik, 2 Woodcock Close, Bamford, Rochdale, OL11 5QA
G0 LSQ D. Williams, 41 Ravensgate Road, Charlton Kings, Cheltenham, GL53 8NS
G0 LSU J. Hair, 84 Oxford Street, Barrow-in-Furness, LA14 5QQ
G0 LSV I. Limbert, 9 Lyme Grove, Liverpool, L36 8BN
G0 LSX D. Barber, 179 Rye Hills, Bignall End, Stoke-on-Trent, ST7 8LP
G0 LTB A. Veal, 64 Hither Bath Bridge, Bristol, BS4 5DJ
GW0 LTC G. Edwards, 5 Lower Farm Court, Rhoose, Barry, CF62 3HQ
G0 LTD B. Tugwell, 14 Martinique Way, Eastbourne, BN23 5TH
G0 LTE D. Prout, 8 Ferenberge Close, Farmborough, Bath, BA2 0DH
GI0 LTF H. Irwin, 9 Edward Street, Armagh, BT61 7QU
G0 LTO R. Summers, 18B Rose Road, Canvey Island, SS8 0BP
G0 LTP D. Freeman, 71 Longleaze, Wootton Bassett, Swindon, SN4 8AS
G0 LTR Tamworth & Lichfield Raynet, c/o R. Williams, 76 Quince, Amington, Tamworth, B77 4EU
GI0 LTT S. Beattie, 28 Millers Lane, Newtownards, BT23 7AR
G0 LTV R. Pearce, Talara, Kent Street, Battle, TN33 0SF
G0 LTX S. Painting, Claytons, Inkpen, Newbury, RG17 9QE
GW0 LUA D. Jewell, 3 Grove Road, Colwyn Bay, LL29 8ER
G0 LUB A. Nicholls, 235 Thorpe Road, Melton Mowbray, LE13 1SH
G0 LUC P. Dyke, 5328 Malcolm Street, Oceanside, USA, 92056
G0 LUD R. Spacey, 18 Longdale Avenue, Ravenshead, Nottingham, NG15 9EA
GM0 LUF T. Traill, 30 Strathesk Road, Penicuik, EH26 8EF
G0 LUH D. Goodison, 33 Witham Road, Isleworth, TW7 4AJ
G0 LUI P. Draper, 265 Nottingham Road, Ilkeston, DE7 5AT
G0 LUK D. Palmer, Braidwood, Enborne Row, Newbury, RG20 0LY
G0 LUL P. Clune, 50 St. Marks Road, Mitcham, CR4 2LF
G0 LUM W. Mitchell-Watson, 144 Shakespeare Crescent, Dronfield, S18 1ND
G0 LUN C. Stayt, 100 Cromwell Way, Oddington, Kidlington, OX5 2LL
G0 LUP K. Chambers, 9 Village Farm Road, Preston, Hull, HU12 8QH
G0 LUQ T. Vale, Grange Farm Flat, Station Road, Bicester, OX26 5DX
G0 LUU K. Blackburn, 63 Robsons Drive, Huddersfield, HD5 9JW
G0 LUY G. Oneill, 16 Aldam Street, Darlington, DL1 2HY
G0 LVF F. Talmage, 54 Rodwell Avenue, Weymouth, DT4 8SG
G0 LVG P. Nilan, 15 Broomhall Road Pendlebury, Swinton, Manchester, M27 8XP
GW0 LVH J. Wimpenny, Gwili House, The Ropewalk, Milford Haven, SA73 3LW
GM0 LVI D. Warburton, Lawvista, High Street, Perth, PH2 7QQ
G0 LVJ B. Bradshaw, 18 Burley Avenue, Lowton, Warrington, WA3 2ES
GM0 LVK L. Alexander, 97 Land Street, Keith, AB55 5AP
GM0 LVL C. Mcewan, 42 Marionville Crescent, Edinburgh, EH7 6AU
G0 LVR J. Russell, 9 Pear Tree Lane, Rowledge, Farnham, GU10 4DW
G0 LVT B. Thornber, 78 Skipton Road, Silsden, Keighley, BD20 9LL
G0 LVX T. Burns, 52 Somerset Drive, Bury, BL9 9DQ
G0 LVY J. Littler, 39 Wigan Road, Golborne, Warrington, WA3 3TZ
G0 LWC P. Timlett, 10 Reynolds Gardens, Moulton, Spalding, PE12 6PT
GM0 LWD L. Mcwilliams, 38 Churchill Street, Alloa, FK10 2JG
G0 LWE W. Short, 6 Kensington Avenue, Normanby, Middlesbrough, TS6 0QQ
G0 LWG N. Sheriden, 10 Grimsby Road, Louth, LN11 0DY
G0 LWI F. Butler, 8 Bradwell Road, Buckhurst Hill, IG9 6BY
G0 LWL D. Spooner, 7 East Avenue, Althorne, Chelmsford, CM3 6DD
G0 LWM A. Mothew, 7 Ashfields, Loughton, IG10 1SB
G0 LWN D. Parsons, 107 Larkswood Road, Chingford, London, E4 9DU
GI0 LWO Dr S. Magill, 40 Gardners Road, Lisburn, BT27 5PD
G0 LWU A. Scarr, Kerrera, 33 Chapel Lane, Overton, Morecambe, LA3 3JA
G0 LXB C. Tapping, 49 Coventry Gardens, Herne Bay, CT6 6SB
G0 LXC J. Searle, 232 Park Lane, Frampton Cotterell, Bristol, BS36 2EN

G0 LXF R. Turner, 5 Darenth Court, Quilter Road, Orpington, BR5 4NS
G0 LXG B. Clulee, 25 Cloister Crofts, Leamington Spa, CV32 6QG
G0 LXI C. Sidney, 25 John Mcguire Crescent, Binley, Coventry, CV3 2QG
G0 LXL G. Truckel, 26 Elm Close, Chipping Sodbury, Bristol, BS37 6HE
GI0 LXN W. Black, 14 Killyliss Road, Fintona, Omagh, BT78 2DL
G0 LXP R. Gant, 25 Worcester Avenue, Garstang, Preston, PR3 1FJ
G0 LXR B. Morgan, 208 Main Street, Burley In Wharfedal, Ilkey, LS29 7HS
G0 LXV M. Lee, 23 Lyndale Road, Redhill, RH1 2HA
G0 LXW A. Pollard, 16 Bellenger Way, Kidlington, OX5 1TR
G0 LXX J. Mayfield, 9 Middlefell Way, Clifton, Nottingham, NG11 9JN
G0 LXY J. Scarr, Betula House, Barford Road Bloxham, Banbury, OX15 4EZ
G0 LYC P. Hindle, Three Ways, Oakwood, Hexham, NE46 4LE
GW0 LYF D. Hobbs, 9 Llwynypia Terrace, Llwynypia, Rhondda, CF40 2JD
G0 LYG F. Aris, 5 Horley Road, Mottingham, London, SE9 4LF
GM0 LYH H. Cochrane, 69 Balgray Avenue, Kilmarnock, KA1 4QT
G0 LYI W. Stevens, 97 Kelvin Grove, Portchester, Fareham, PO16 8LF
G0 LYJ W. Hughes, 26 Cambridge Cottages, Richmond, TW9 3AY
GW0 LYK S. Parker, Maesycoed, Blaenycoed, Carmarthen, SA33 6ES
G0 LYN L. Roper, 57 Burnt Hills, Cromer, NR27 9LW
GM0 LYO J. Fletcher, 1 Silverwood Farm Cottage, Kilmarnock, KA3 6HJ
G0 LYQ L. Selman, 156 Bradley Drive, Santa Cruz, USA, 95060
G0 LYR P. Spencer, 4 Jubilee Close, Duloe, Liskeard, PL14 4PA
GM0 LYT A. Fegen, Acharn, Losset Road, Blairgowrie, PH11 8BU
G0 LYX D. Brown, 67 Croft Road, Benfleet, SS7 5RL
G0 LYZ C. Knaggs, 29 Wansford Road, Driffield, YO25 5NB
G0 LZD R. Wood, 4 Burns Road, Royston, SG8 5PT
GM0 LZE D. Morrison, 27B Benside, Newmarket, Stornoway, HS2 0DZ
G0 LZF J. Rivers, 211 Upper Wickham Lane, Welling, DA16 3AW
G0 LZG W. Miles-Williams, 13 Cavendish Road, Chesham, HP5 1RW
G0 LZI M. Tudor, 32 Ringwood, Oxton, Prenton, CH43 2LZ
G0 LZL D. Bates, 92 Thirlmere Road, Partington, Manchester, M31 4PT
G0 LZS T. Wright, 8 Glentham Close, Lincoln, LN6 8BX
G0 LZV C. Nicholas, 19 Spring Crofts, Bushey, WD23 3AR
G0 LZW D. Riddick, 289 Hatfield Road, St Albans, AL4 0DH
G0 LZX R. Knowles, 9 Malham Close, Southport, PR8 6UP
G0 LZY R. Smith, Pleasanton, Church Street, Halstead, CO9 3AZ
G0 MAA S. Rawson, The Cabin, Tully, Four Mile House, Roscommon, Ireland,
GM0 MAC A. Macfarlane, Breadalbane, Ferrindonald, Isle of Skye, IV44 8RF
G0 MAD Mapperley & District ARC, c/o K. Moody, 5 Moore Road, Mapperley, Nottingham, NG3 6EF
G0 MAF M. Plaskitt, 4 Church Lane, Immingham, DN402EU
G0 MAH G. Humphrey, 57 Haig Avenue, Leyland, PR25 2DD
G0 MAL W. Bowden, 43 Burlish Close, Stourport-on-Severn, DY13 8XW
GD0 MAN Manx Sthrn DX G, c/o M. Dunning, 55 Station Park, Colby, Isle Of Man, IM9 4NL
G0 MAR N. Buchan, Acorns, 37 Forge Rise, Uckfield, TN22 5BU
G0 MAS A. Sedgbeer, 7 Leofric Road, Pinnex Moor, Tiverton, EX16 6JU
G0 MAT R. Johnson, 30 Wheatlands, Titchfield Common, Fareham, PO14 4SL
G0 MAY P. Holmes, 11 Bingham Road, Cotgrave, Nottingham, NG12 3JS
G0 MAZ M. Gurr, Elan, Sandown Road, Sandwich, CT13 9NY
G0 MBA A. Horsman, 15 Hanwell Close, Clacton-on-Sea, CO16 7HF
G0 MBB A. Cutter, Hundred House, Pink Road Lacey Green, Princes Risborough, HP27 0PG
G0 MBD J. Curzon, 17 Bullfinch Way, Cottenham, Cambridge, CB24 8AW
G0 MBG B. Drew, 59 Coventry St., Kidderminster, DY10 2BZ
G0 MBI A. Roxburgh, Flat 1, Tudor Court, Sutton Coldfield, B74 4AN
G0 MBK M. Kerr, 83 Greenwood Crescent, Sheffield, S9 4HE
G0 MBL A. Plant, 99 Pegwell Road, Ramsgate, CT110ND
GW0 MBN P. Salt, Acorns, Llwyncelyn, Cardigan, SA43 2PE
G0 MBP B. Vanson, 25 St. Helens Road, Westcliff-on-Sea, SS0 7LA
G0 MBQ S. Withers, 14 Rushes Road, Petersfield, GU32 3BW
G0 MBR Mid-Beds Raynet Group, c/o I. Mciver, 3 Asgard Drive, Bedford, MK41 0UP
G0 MBS B. Sinclair, 48 East Crescent, Duckmanton, Chesterfield, S44 5ET
G0 MBU G. Buckwell, 24 Ings Road, Redcar, TS10 2DL
G0 MBV R. Buckwell, 31 West Green Stokesley, Stokesley, TS95BE
GW0 MBW M. Watkins, Llwyn Onn, Wainfelin Road, Pontypool, NP4 6DF
G0 MBY P. Eaton, 3 Thirslet Drive, Heybridge, Maldon, CM9 4YN
G0 MBZ M. Phillips, 14 Kingsclere Drive, Bishops Cleeve, Cheltenham, GL52 8TG
G0 MCE R. Daw, Flat 11, Tong Court, Boscobel Crescent, Wolverhampton, WV1 1QQ
GM0 MCJ H. Munro, Flat 6, Charlie Devine Court Bridge Of Don, Aberdeen, AB22 8WG
G0 MCM M. Sniezko-Blocki, 18 Westwoods Hollow, Burntwood, WS7 9AT
G0 MCO D. Belcher, 7 Bower End, Chalgrove, Oxford, OX44 7YN
G0 MCP K. Scott, 16 Hawton Road, Newark, NG24 4QB
G0 MCQ M. Quicke, 53 Newfield Avenue, Farnborough, GU14 9PJ
G0 MCT R. Craig, 24 Avondale, Sunderland, SR4 0LZ
G0 MCV S. Morley, Mill Lane, Sileby, Loughborough, LE127UX
GM0 MDD J. Clough, Obo Largs & District Ars, Redbank, Skelmorlie, PA17 5DX
G0 MDJ K. Smithyes, Roseville, Childs Ercall, Market Drayton, TF9 2DG
G0 MDK C. Hobson, 1 Martindale Avenue, Wimborne, BH21 2LE
G0 MDM R. Robbins, 3 North Approach, Watford, WD25 0EH
G0 MDN B. Millward, 5 Regency Close, Weddington, Nuneaton, CV10 0DF
G0 MDO D. Ward, 53 The Drive, Bingley, BD16 2EY
GW0 MDQ P. Firmstone, Bod Awen, Ffordd Top Y Rhos Treuddyn, Mold, CH7 4NE
G0 MDR F. Lupton, 51 Bullens Green Lane, Colney Heath, St Albans, AL4 0QR
G0 MDV M. Bellas, 3 Elm Terrace, Penrith, CA11 7JY
GM0 MDX W. Dempster, 124 Chatelherault Crescent, Hamilton, ML3 7PW
G0 MEA P. Smith, 7 Prospect Drive, Keighley, BD22 6DD
G0 MEC M. Spurgeon, 11 Homestead Road, Bodicote Chase, Banbury, OX16 9TW
G0 MEE B. Shelton, 12 Meadowlands, Blundeston, Lowestoft, NR32 5AS
G0 MEF M. Frear, 18 Boulsworth Road, Preston Grange, North Shields, NE29 9EN
G0 MEN A. Fitzgerald, 39 Rue Marcel Miquel, Issy-Les-Moulineaux, France, 92130
G0 MEO R. Davis, 17 Welbourne Close, Raunds, Wellingborough, NN9 6HE
G0 MEQ H. Rigby, 33 Herne Rise, Ilminster, TA19 0HH
G0 MEV J. Thorndyke, 23 Fordhams Close, Stanton, Bury St Edmunds, IP31 2EE
G0 MEW L. Whiteside, 9 Nutfield Gardens, Ilford, IG3 9TB
G0 MEX H. Horne, 410 Bacup Road, Waterfoot, Rossendale, BB4 7JA
G0 MEY M. Coulter, 52 Pine Close, Brant Road, Lincoln, LN5 9UT
G0 MEZ C. Thorndyke, 23 Fordhams Close, Stanton, Bury St Edmunds, IP31 2EE
G0 MFH P. Lee, 56 Cockshead Road, Liverpool, L25 2RB
G0 MFQ Capt. G. Dunster, 21 Brunel Quays, Great Western Village, Lostwithiel, PL22 0JB
G0 MFR G. Ayre, Hintocks, Elm Hill, Motcombe, Shaftesbury, SP7 9HL
G0 MFT G. Drake, The Bungalow, Church Lane, Tydd St. Giles, Wisbech, PE13 5LG
G0 MFY L. Choong, 15 Hounsfield Lodge, 5 Chambers Park Hill, Wimbledon, London, SW20 0QE
G0 MGC G. Clark, Holly Cottage, New Road, Bristol, BS35 4DX
G0 MGG S. Smith, 60 Grange Road Tuffley, Gloucester, GL4 0PG
G0 MGH A. Strevens, 14 Larchfield Way, Horndean, Waterlooville, PO8 9HE
G0 MGI M. Goodall, 2 Meadow Court, Littleport, Ely, CB6 1JW
G0 MGJ K. Hancock, 12 Westmorland Close, Stoke-on-Trent, ST6 6UR
G0 MGL G. Lees, 68 Green Lane, Oldham, OL8 3BA
G0 MGM R. Dunne, 8 Telston Close, Bourne End, SL8 5TY
G0 MGN J. Brand, 38 Canterbury Gardens, Hadleigh, Ipswich, IP7 5BS
GM0 MGO D. Macdonald, 22 Christie Place, Elgin, IV30 4HX
GW0 MGQ G. Budge, 47 Rhyd Y Defaid Drive, Sketty, Swansea, SA2 8al
G0 MGT D. Carruthers, 19 Creek View Avenue, Hullbridge, Hockley, SS5 6LU
G0 MGU C. Brown, The Cottage, Tylers Road, Harlow, CM19 5LJ
G0 MGX M. Jones, 8 Stanton Avenue, Belper, DE56 1EE
G0 MGZ D. Chaloner, 16 Farriers Road Middle Barton, Chipping Norton, OX7 7EU
G0 MHA P. Bakrania, 31 South Priors Court, Northampton, NN3 8LD
GI0 MHB P. Mckee, 168 Ballynamoney Road, Lurgan, Craigavon, BT66 6LD
G0 MHC G. Ford, Thornley Road, Trimdon Station, TS29 6DA
GM0 MHD P. Overton, Cluanie, Cairnballoch, Alford, AB33 8HQ
GM0 MHE B. Hyde, The Cottage, Killiechronan, Isle of Mull, PA72 6JU
G0 MHF J. Bisson, 14 Howbeck Drive, Oxton, Prenton, CH43 6UY
GW0 MHK P. Lee, 22 Bron Y Graig, Bodedern, Holyhead, LL65 3SY
G0 MHN D. Tebay, 19 St. Johns Road, Newport, PO30 1LN
GM0 MHS D. Rendall, Aranthrue, 17 Scapa Crescent, Kirkwall, KW15 1RL
G0 MHY R. Jones, 339 Glenashton Drive, Oakville, Canada, L6H 4W2
G0 MHZ R. Pardoe, 138 Fowler Road, Aylesbury, HP19 7QJ
G0 MIA C. Murray, Brookfield, Collaroy Road, Thatcham, RG18 9PB
G0 MIB D. Hussey, 15 The Ridings, Telscombe Cliffs, Peacehaven, BN10 7EF
G0 MID R. Jeffery, 3 New Road, Paddock Wood, Tonbridge, TN12 6HP
G0 MIE R. Ebbs, 25 Foxtail Close, Gloucester, GL4 6DW
G0 MIF I. Buckle, 28 Leybourne Road, Rochester, ME2 3QG
G0 MIG N. May, Spring Barn, Eastwood Park, Wotton-under-Edge, GL12 8DA
G0 MIH P. Swift, Solvig, Broadbridge Lane, Horley, RH6 9RF
G0 MIJ C. Nolan, 95 Strodes Crescent, Staines, TW18 1DG
G0 MIK M. Ritson, 24 Chapel Road, Pawlett, Bridgwater, TA6 4SH
GM0 MIS Dr S. Buchanan, 7 Eilean Rise, Ellon, AB41 9NF
G0 MIT G. Bromfield, 63 Herondale Road, Mossley Hill, Liverpool, L18 1JZ
GM0 MIW A. Mcnicol, The Glebe House, Arbirlot, Arbroath, DD11 2NX
G0 MIX M. Jones, 15 Quadrant Close, Murdishaw, Runcorn, WA7 6DW
G0 MIZ J. Munro, Flat 3/4, 24 The Strand, Ryde, PO33 1JD
G0 MJA G. Taylor, 32 Main Road, Great Holland, Frinton-on-Sea, CO13 0JL
G0 MJB R. Daynes, 25 Redwood Close, Keighley, BD21 4YG
G0 MJC A. Keeble, 5 Thistledown Road, Horsford, Norwich, NR10 3ST
G0 MJF M. Weaver, 91 Mantle Street, Wellington, TA21 8BB
G0 MJG S. Cartlidge, 19 Thornfield Road Crosby, Liverpool, L23 9XY
G0 MJJ L. Challis, 30 London Road, Kirton, Boston, PE20 1JA
G0 MJK D. Linnell, 19 Beech Avenue, Northampton, NN3 2HE
G0 MJO G. Lucas, Flat 3, The Gateway, 2 Wilderton Road West, Poole, BH13 6EF
G0 MJP R. Davies, 59 Gaunts Way, Letchworth Garden City, SG6 4PL
GM0 MJR E. Baviello, 18 Glaskhill Terrace, Penicuik, EH26 0EL
G0 MJT M. Tandy, 10 Palace Close, Rowley Regis, B65 9LG
G0 MJV A. Williams, 16 Hoy Crescent, Seaham, SR7 0JT
G0 MJX J. Harper, 109 Baxter Avenue, Kidderminster, DY10 2HB
G0 MJY D. Gourley, 86 Upton Road, Kidderminster, DY10 2YB
G0 MJZ J. Edwards, 42 Tenterfield Road, Ossett, WF5 0RU
G0 MKA T. Chapman, 17 Trevor Road, Swinton, Manchester, M27 0YH
G0 MKC W. Dunn, 25 Magdalene Court, Seaham, SR7 7DJ
G0 MKD K. Davenport, 6 Malpas Road, Runcorn, WA7 4AD
G0 MKK M. Stockdale, First Floor, 22 Commercial St., Harrogate, HG1 1TY
G0 MKL R. Chell, 3 Elderberry Close, Stourport-on-Severn, DY13 8TF
G0 MKN K. Brady, 17B Furzefield Road, Welwyn Garden City, AL7 3RL
G0 MKP N. Grice, 7 Brecon Avenue, Huddersfield, HD3 3QF
G0 MKU G. Langford, 11 Nearhill Road, Kings Norton, Birmingham, B38 8LB
G0 MKW A. Jones, 26 Clarendon Street, Bloxwich, Walsall, WS3 2HT
G0 MKY Dr J. Herries, Elmfold, Witney, OX8 6PZ
G0 MKZ T. Pougher, 8 Wensleydale, Hull, HU7 6DE
G0 MLB K. Walters, 14 Varo Terrace, Stockton on Tees, TS18 1JY
G0 MLC W. Lowe, 54 St. Lesmo Road, Edgeley, Stockport, SK3 0TX
G0 MLE D. Sabin, 1 West Nolands, Nolands Road, Calne, SN11 8YD
G0 MLF P. Marshall, 45 Haylings Road, Leiston, IP16 4DJ
G0 MLJ M. Jones, Mark Jones Eye Care, 21 Fisherton Street, Salisbury, SP2 7SU
G0 MLL J. Lyons, 40 Waddington Avenue, Burnley, BB10 4LB
G0 MLM T. Leeman, 5 Serlby Rise, Nottingham, NG3 2LS
GW0 MLN E. Jones, 55 Blackoak Road, Cyncoed, Cardiff, CF23 6QU
G0 MLO K. Packard, The Haven, Howe Green Road, Purleigh, Chelmsford, CM3 6PZ
G0 MLQ D. Pearson, Yewtree Cottage, Worcester Road, Chipping Norton, OX7 5XX
G0 MLY C. Rolinson, 534 Haslucks Green Road, Shirley, Solihull, B90 1DS
G0 MMA K. Plumridge, Flat 23, Barton Court, Tewkesbury, GL20 5RL
GW0 MMB P. Evans, 138 Pont Adam Crescent, Ruabon, Wrexham, LL14 6EG
G0 MMC J. Cuthill, 17 Elmwood Drive, Keighley, BD22 7DN
G0 MMH P. Walker, 11 Flixton Drive, Crewe, CW2 8AP
G0 MMI C. Underhill, 5 Grove Way, Waddesdon, Aylesbury, HP18 0LH
G0 MMJ D. Wilkins, 18 Garendon Road, Loughborough, LE11 4QD
G0 MMO N. Laud, 3 Woodlands, Wirksworth, Matlock, DE4 4PG
G0 MMQ H. Dadak, 3 Cadogan Close, Holyport, Maidenhead, SL6 2JS
G0 MMT L. Catherall, New Haven, Peckforton Hall Lane, Tarporley, CW6 9TF
G0 MMW L. Roberts, 12 Deveron Close, Plymouth, PL7 2YF
G0 MMX D. Hebden, 7 Whitecroft Ave, Shaw, Oldham, OL2 8HY
GW0 MMY W. Ellis, Broad Oak Cottage, Llyndir Lane, Burton, Wrexham, LL12 0AU
G0 MNA A. Munir, 39 Gulberg V, Lahore, Pakistan,
G0 MNC J. Williams, 34 Brassington Street, Clay Cross, Chesterfield, S45 9NH
G0 MND T. Rogers, 40 Rowell Way, Sawtry, Huntingdon, PE28 5WB

G0 MNH M. Brown, 15 Hamilton Row, Waterhouses, Durham, DH7 9AU
G0 MNI A. Carlile, Top Flat 13B Mill Road, Cleethorpes, DN35 8HZ
G0 MNN M. Franklin, 7 Auburn Close, Bridlington, YO16 7PN
GW0 MNO N. Bufton, 7 Laburnum Close, Rassau, Ebbw Vale, NP23 5TS
GW0 MNP M. Butler, 1 Green Meadow, Cefn Cribwr, Bridgend, CF32 0BJ
GM0 MNV R. Gandy, 102 The Henge, Glenrothes, KY7 6XX
GM0 MNW K. Carmichael, 8G Colonsay Terrace, Soroba, Oban, PA34 4YL
G0 MNY A. Dagnall, 10 Rosebury Avenue, Leigh, WN7 1JZ
GW0 MOF G. Greenhalgh, 6 Clifton Grove, Rhyl, LL18 4AF
G0 MOH R. Greaves, Paradise Meadow, Church Street, Okehampton, EX20 1JF
G0 MOK R. Hamer, 4 Maldon Road, Standish, WN6 0EX
G0 MOM S. Kendall, 220 Marsh St., Barrow in Furness, LA14 1BQ
GW0 MOQ N. Brush, 25 Heol Y Ffynnon, Efail Isaf, Pontypridd, CF38 1AU
G0 MOR Dr S. Morrey, 22 Wellpond Close, Sharnbrook, Bedford, MK44 1PL
G0 MOU R. Clark, 20 Oakcroft Gardens, Littlehampton, BN17 6LT
GW0 MOW R. Harris, 25 Twynyffald Road, Blackwood, NP12 1HQ
G0 MOX D. Gleek, 10 Castlereagh House, Lady Aylesford Avenue, Stanmore, HA7 4FP
G0 MPA B. Coram, 18A Lake Green Road, Sandown, PO36 9HW
G0 MPI S. Sutcliffe, 142 Sandy Lane, Farnborough, GU14 9JQ
G0 MPJ B. Osborne, 12 Arminers Close, Gosport, PO12 2HB
G0 MPK D. Knights, 11 King Edward St., Kirton Lindsey, Gainsborough, DN21 4NF
G0 MPM J. Claughton, 14 Witch Close, East Stour, Gillingham, SP8 5LB
G0 MPO A. Neenan, 50 Middleton Road, Brownhills, Walsall, WS8 6JF
G0 MPP J. Anderson, Hazel Wood Lee Lane, Bingley, BD16 1UF
G0 MPQ J. Wood, Garthmere, 4 Hunters Lane, Lincoln, LN4 4PB
G0 MPR F. Gibbons, 26 Tenbury Close, Bentley, Walsall, WS2 0NH
G0 MPT E. Roddy, 1546 West St, Stoughton, USA, 2072
G0 MPW J. Woods, 26 Compton Road, Southport, PR8 4HA
G0 MQC P. Capewell, 191 Monyhull Hall Road, Birmingham, B30 3QN
G0 MQD R. Field, 10 Somerville Close, Waddington, Lincoln, LN5 9QR
G0 MQE J. Peirce, 600 Highland Avenue, Ottawa, Ontario, Canada, K2A 2K3
G0 MQH D. Robinson, 88 Cotton Lane Halton Lodge, Runcorn, WA7 5JB
G0 MQI R. Ingle, 232 North Park Avenue, Norwich, NR4 7ED
G0 MQJ P. Robinson, 92 Greasby Road, Greasby, Wirral, CH49 3NG
G0 MQK V. Murton, 4 Cross Park Road, Wembury, Plymouth, PL9 0EU
G0 MQL D. Franklin, 50 The Elms, Chatteris, PE16 6JN
G0 MQM M. Hillier, 5 Sinodun Road, Didcot, OX11 8HP
GI0 MQN R. Browning, 53 Caulside Park, Antrim, BT41 2DR
G0 MQR I. Tuson, 6 Buffs Lane, Heswall, Wirral, CH60 2SG
GW0 MQU P. Smyth, 19 North Avenue, Tredegar, NP22 3HE
G0 MQV N. Cook, 17 Moorside Road, Richmond, DL10 5DJ
G0 MQW C. Mcwhinnie, 32 The Horse Close, Emmer Green, Reading, RG4 8TT
G0 MQX R. Bowers, 54 Buxton Road, Dawley, Telford, TF4 2EW
G0 MRA E. Southon, 20 Edinburgh Crescent, Kirton, Boston, PE20 1JT
G0 MRB R. Broughton, 6 Lumley Place, Lincoln, LN5 7UT
G0 MRD D. Gordon, 152 Oldham Road, Ashton-under-Lyne, OL7 9AN
G0 MRF D. Bowman, 11 Crane Way, Twickenham, TW2 7NH
GM0 MRJ M. Johnston, 27 Denholm Court, Glenrothes, KY6 1JP
G0 MRK J. Kelly, 14 Arden Walk, Sale, M33 5NY
G0 MRL L. Bradshaw, 342 Manchester Road, Blackrod, Bolton, BL6 5BG
G0 MRM E. Caligari, 209 Ormskirk Road, Upholland, Skelmersdale, WN8 0AA
G0 MRP D. Pidgeon, 87 Suckling Green Lane, Codsall, Wolverhampton, WV8 2BY
G0 MRR C. Denton-Powell, 4 Korresia Walk, Bridgwater, TA5 2GT
G0 MRY M. Hazzledine, 52 Springfield Road, Repton, Derby, DE65 6GP
G0 MRZ B. Rowell, 73 Halsteads Road, Torquay, TQ2 8HB
G0 MSA A. Hagland, 11 Coppice View, Heathfield, TN21 8YS
G0 MSF G. Obey, 51 Chichester Close, Murdishaw, Runcorn, WA7 6DQ
GI0 MSG T. Mc Geown, 1 Drumcairn Road, Armagh, BT61 7SA
GI0 MSH D. Mcelroy, 81 Keady Road, Armagh, BT60 3AA
GI0 MSI E. Nesbitt, 47 Mossfield, Glenanne, Armagh, BT60 2JF
GI0 MSK H. Rattray, 20 Charlemont Gardens, Armagh, BT61 9BB
G0 MSO A. Webb, 12 Forthlin Road, Allerton, Liverpool, L18 9TN
G0 MSR S. Rutt, 3 Russell Place, Highfield, Southampton, SO17 1NU
G0 MSS J. Taft, 8 Dresden Close, Mickleover, Derby, DE3 0RD
GM0 MST J. Scotter, 3 George Street, Halkirk, KW12 6YE
GW0 MSW E. Goodwin, Tremayne, 11 Duchess Road, Monmouth, NP25 3HT
GW0 MSY H. Duggan, 41 Maesglas Road, Newport, NP20 3DE
G0 MSZ J. Lycett, 24 Milbank Court, Darlington, DL3 9PF
G0 MTA Rev. F. Bligh, 60 Hoole Road, Chester, CH2 3NL
G0 MTB P. Pearson, 53 Station Road, Dersingham, King's Lynn, PE31 6PR
G0 MTD S. Topping, 7 Beckstone Close, Harrington, Workington, CA14 5QR
GI0 MTE P. Robinson, 8 Annaboe Road, Kilmore, Armagh, BT61 8NP
G0 MTF G. Sanders, 18 Impey Close, Thorpe Astley, Leicester, LE3 3SW
GW0 MTI M. White, 52 James Street Trethomas, Caerphilly, CF83 8FY
G0 MTJ J. Boothroyd, Quince Cottage, Church Lane, Ashford, TN26 1LS
G0 MTK I. Chapman, 12 Guernsey Farm Lane, Bognor Regis, PO22 6BU
G0 MTN L. Volante, Richmond House, Icknield Street, Birmingham, B38 0EP
G0 MTP A. Owen, 26 Gresham Street, Coventry, CV2 4EU
G0 MTQ J. Baker, Moffat House, Church Road Broughton Moor, Maryport, CA15 7SS
G0 MTT R. Williamson, 47 Ochre Dike Walk, Rotherham, S61 4DL
G0 MTV D. Wright, Blakey Ridge, 2 Abbey Gardens, Wimborne, BH21 2EA
G0 MTW F. Puffett, 1 Far Sandfield Churchdown, Gloucester, GL3 2JS
G0 MTY P. Stunden, 81 Treloweth Way, Pool, Redruth, TR15 3TS
G0 MUC S. Markwick, Flat 64, Lakeside Court, 35 Dallington Road, Eastbourne, BN22 9EJ
G0 MUD Christchurch ARS, c/o D. Layne, 5 Howe Close, Christchurch, BH23 3JA
G0 MUH S. Riley, 7 Crow Wood Avenue, Burnley, BB12 0JG
G0 MUJ S. Spink, 12 Chaucer Court, Ewelme, Wallingford, OX10 6HW
G0 MUK R. Sanders, 1A Lychgate Drive, Waterlooville, PO8 9QE
G0 MUN A. Collins, Flat 7, Haydon Court, Newton Abbot, TQ12 1GQ
G0 MUR R. Garrett, 27 Victoria Park Road, Buxton, SK17 7PU
G0 MUZ J. Lockyer, Flat 1, 11 Birch Hill Court, Birchington, CT7 9UQ
G0 MVC C. Neil, 208 Stonelow Road, Dronfield, S18 2ER
G0 MVE M. Storkey, 9 Waterman Court, Acomb, York, YO24 3FB
G0 MVM A. Frost, 15 Church Street Lacock, Chippenham, SN15 2LB
G0 MVP M. Isted, 62 Chippers Road, Worthing, BN13 1DG
G0 MVR M. Bentley, 1 Cotswold Road, Lupset, Wakefield, WF2 8EL
G0 MVT W. Brindley, 41 Boon Hill Road Bignall End, Stoke-on-Trent, ST7 8LA
G0 MVV C. Howes, 8 Alder Way, Hazel Slade, Cannock, WS12 0SX
G0 MVW W. Barker, Fieldhead, School Lane, Thame, OX9 2NE
G0 MVX J. Reeves, 5 Arrows Crescent, Boroughbridge, York, YO51 9LP
G0 MVY Dr J. Donnett, 51 Beaumont Avenue, St Albans, AL1 4TT
G0 MWE R. Woodward, 22 Maryport Road, Dearham, Maryport, CA15 7EG
G0 MWH R. Atkins, 24 Hill House Road, Norwich, NR1 4BE
GM0 MWJ D. Robertson, 73 Kettilstoun Mains, Linlithgow, EH49 6SH
GD0 MWL A. Crowther, 3 Lime Street, Port St. Mary, Isle of Man, IM9 5ED
G0 MWM E. Bailey, 8 Blackthorn Close, Thornton-Cleveleys, FY5 2ZA
GW0 MWN Dr D. Harries, Rhydiau, Pencader, SA39 9BY
G0 MWS S. Mcphee, 19 Lyttelton Road, Stourbridge, DY8 3RP
G0 MWT Chelmsford ARS, c/o P. Tittensor, 47 St. Johns Road, Chelmsford, CM2 0TY
G0 MWU B. Pebody, 30 High St., Oakfield, Ryde, PO33 1EL
G0 MWV D. Campbell, Bychance, Higher Gardens, Corfe Castle, Wareham, BH20 5ES
G0 MWW C. Murt, 17 Drake Road, Padstow, PL28 8ES
G0 MWZ R. Rutherford, 26 St.Golder Road, Newlyn Coombe, Penzance, TR18 5QW
G0 MXB D. Burnett, Cloonmaghaura, Williamstown, Ireland, 907
G0 MXD D. Wroe, Tylers House, Coton, Whitchurch, SY13 3LT
G0 MXE F. Jennings, 2 Hickman Close, London, E16 3TA
GW0 MXG P. Taylor, 2 Pen-Y-Dre, Caerphilly, CF83 2NZ
G0 MXH D. Kay, 8 Meadowbrook Close, Lostock, Bolton, BL6 4HX
GM0 MXP R. Park, The Loft, Front Street, Dunblane, FK15 9PX
G0 MXU S. Smith, 99 Greenwood, Bamber Bridge, Preston, PR5 8JX
G0 MXW D. Houghton, 127 Melwood Drive, Liverpool, L12 8RN
G0 MXX Prof. B. Clarke, Linden Cottage, School Lane, Nottingham, NG12 3FD
G0 MXY C. Erratt, 60 Allen Court, Ridding Lane, Greenford, UB6 0JZ
GM0 MXZ T. Goody, Lambs' Park, Forgandenny, Scotland, PH2 9HS
G0 MYA A. Gray, 57 Dominie Cross Road, Retford, DN22 6NH
G0 MYC R. Clifton, Heathwood, Thrigby Road, Great Yarmouth, NR29 3HJ
G0 MYD G. Hughes, 292 Mount Pleasant, Southcrest, Redditch, B97 4JL
G0 MYH J. Foster, 5 Jacobs Close, Glastonbury, BA6 8EJ
G0 MYJ D. Sutherland, Devonia, Hamble Lane, Southampton, SO31 8EL
G0 MYL Sir H. Pigott, Brook Farm, Shobley, Ringwood, BH24 3HT
G0 MYM P. Harrison, 16 Bodiham Hill, Garforth, Leeds, LS25 2LF
G0 MYN W. Hughes, 60 Pineways, Appleton, Warrington, WA4 5EJ
GM0 MYQ J. Frati, 10 Benbecula Road, Aberdeen, AB16 6FU
G0 MYR R. Worsley, Omaru, Higher Pennance, Redruth, TR16 5TQ
GM0 MZD A. Coutts, 37 West High Street, Bishopmill, Elgin, IV30 4DJ
G0 MZF Dr B. Nicholson, 349 City Road, London, EC1V 1LR
GM0 MZH R. Wallace, 80 Tourhill Road, Kilmarnock, KA3 2DA
G0 MZJ C. Earp, 9 The City, Edington, Westbury, BA13 4QQ
G0 MZK R. Little, Mythe House, 129 Slad Road, Stroud, GL5 1RD
G0 MZN J. Nunn, 20 Somerton Gardens, Earley, Reading, RG6 5XG
G0 MZP R. Kay, 7 Alderson Road, Worksop, S80 1UZ
G0 MZQ W. Greed, 5 West View, Creech St. Michael, Taunton, TA3 5QP
G0 MZY F. Woodhall, 13 Whitegate Drive, Clifton, Manchester, M27 8RE
G0 MZZ A. Benson, 1 Oxford Close, Gomersal, Cleckheaton, BD19 4RU
G0 NAA A. Leake, Thorpe Garth, East Newton, Hull, HU11 4SD
G0 NAD R. Naden, 10 Suffolk Close, Holland On Sea, Clacton-on-Sea, CO15 5SQ
GM0 NAE J. Carlin, 24 Hillcrest Avenue, Paisley, PA2 8QW
GM0 NAI J. Fisher, High Birches, Culbokie, Dingwall, IV7 8JS
G0 NAJ J. Neary, 266 Yew Tree Lane, Dukinfield, SK16 5DN
G0 NAP P. Howell, 29 South View Park, Plymouth, PL7 4JE
GM0 NAQ G. Furmage, 25 Craigton Crescent, Alva, FK12 5DS
G0 NAR C. Fenton, Oakwell, Newtons Hill, Hartfield, TN7 4DH
G0 NAS L. Aykroyd, 3 Bank Cottages, Orton Road, Penrith, CA10 3TW
G0 NAU M. Best, 81 Maybury Road, Hull, HU9 3LB
G0 NAX D. Edmonds, 1 Ashtree Close, Chelmsford, CM1 2RR
GM0 NAZ A. Heggie, 75 Doon Walk, Craigshill, Livingston, EH54 5AD
GM0 NBA T. Adam, Burnbank, Cairnbaan, Lochgilphead, PA31 8SQ
G0 NBB M. Watkins, 7 Sand End, Whitstable, CT5 4TH
G0 NBC L. Steenvoorden, 1 Thornbury Road, Immingham, DN40 1HH
G0 NBD A. Brown, 20 Sheen Road, Wallasey, CH45 1HA
G0 NBE R. Allen, 5 Crompton Grove, Stoke-on-Trent, ST4 8UZ
GM0 NBG J. Mcvittie, 19 Beech Way, Girvan, KA26 0BX
G0 NBH J. Goodwin, Hankelow Court, Hall Lane, Crewe, CW3 0JB
G0 NBI G. Coomber, 3 Dolly Drove, Chard, TA20 1PF
G0 NBJ N. Foster, 20A Pear Tree Road, Ashford, TW15 1PW
G0 NBP A. Stevens, Gate Farm Barns, Earthcott Green, Bristol, BS35 3TA
G0 NBW B. Nolan, 4 Shetland Road, Blackpool, FY1 6LP
GI0 NCA R. Pinkerton, 9 Cloghole Road Campsie, Londonderry, BT47 3JW
G0 NCE O. Wheeler, 56 Rochester Gate High Street, Rochester, me1 1jg
G0 NCH R. Magri, 13 Roebuck Road, Chessington, KT9 1JY
G0 NCL R. Harris, 5 Campbell Close, High Wycombe, HP13 5XY
G0 NCO D. Murray, 11 Blandy Road, Henley-on-Thames, RG9 1PH
G0 NCQ M. Hemmings, 2 Holly Walk, Nuneaton, CV11 6UU
G0 NCS C. Healey, 22 Stirling Road, St. Budeaux, Plymouth, PL5 1PD
G0 NCT F. Norman, 101 Central Avenue, Canvey Island, SS8 9QP
GW0 NCU S. David, 142 Robert Street, Manselton, Swansea, SA5 9NH
G0 NCW A. Judge, 106 Bicknor Road, Park Wood, Maidstone, ME15 9PD
G0 NCX R. Hughesdon, 3 Lyndhurst Road, Gosport, PO12 3QY
G0 NCY R. Hartley, 23 Broomfield Road, Fleetwood, FY7 7HA
GW0 NDA S. Frost, Fron Hyfryd, Nebo, Caernarfon, LL54 6EW
G0 NDB R. Evans, 51532 Range Road 224, Sherwood Park, Canada, T8C 1H5
G0 NDC R. Little, Maranatha, Higher Moresk Road, Truro, TR1 1BW
G0 NDD J. Jackson, 26 Wadham Close, Peterlee, SR8 2NN
G0 NDF C. Peters, 347 Mile Oak Road, Portslade, Brighton, BN41 2RD
G0 NDS Northampton ARG, c/o J. Johnson, 30 Millside Close, Kingsley, Northampton, NN2 7TR
G0 NDU J. Dykes, 33 Mill House Drive, Cheltenham, GL50 4RG
G0 NDV P. Timmins, Flat 2 34 Lumley Avenue, Skegness, PE25 2TH
G0 NDY R. Wayne, Colkirk House, Manor House Street, Horncastle, LN9 5HF
GW0 NDZ G. Davies, 4 Crichton Street, Treorchy, CF42 6DF
G0 NEB J. Dorning, 51 O'Sullivan Crescent, St Helens, WA11 9RE
GW0 NEC V. Fletcher, Brig Y Don, Pendre Road Penrhynside, Llandudno, LL30 3BY
G0 NED E. Dudley, 4 Lake Croft Drive, Stoke-on-Trent, ST3 7SS
G0 NEE M. Stott, Wellview, 12 Castle View, Prudhoe, NE42 6AT

G0 NEF A. Chapman, 13 Clayton Grove, Bracknell, RG12 2PT
G0 NEM M. Purser, 17 Firecrest Road, Chelmsford, CM2 9SN
G0 NEN G. Lewin, 9 Westgate, Bridgnorth, WV16 5BL
G0 NEO J. Boland, 28 Vicarage Road, Orrell, Wigan, WN5 7AX
G0 NEP P. Whitling, 17 Balcomb Crescent, Margate, CT9 3XJ
G0 NEQ Aldridge and Barr Beacon ARC, c/o L. Horton, 4 Summer Lane, Walsall, WS4 1DS
G0 NER S. Stones, 38 Mill View, Ferrybridge, Knottingley, WF11 8SR
G0 NES D. Bryant, 35 Truemans Heath Lane, Hollywood, Birmingham, B47 5QE
GM0 NET Ayrshire Raynet Group, c/o T. Stewart, 104 Barrhill Road, Cumnock, KA18 1PU
G0 NEU A. Dawson, 182 Ladysmith Road, Enfield, EN1 3AE
G0 NEV M. Carter, 1 Mill Lane Preston, Weymouth, DT3 6DE
G0 NFA D. Gilbert, 2 Greenfield Cottages, Bentley, Farnham, GU10 5HZ
G0 NFB L. Raynor, 19 West View, Doncaster Road, Worksop, S81 9RA
G0 NFE R. Ransome, 66 Spencer Way, Stowmarket, IP14 1UQ
G0 NFG D. Hopper, 28 Western Avenue, Herne Bay, CT6 8TU
G0 NFH J. Acton, 63 Bevington Close, Patchway, Bristol, BS34 5NP
G0 NFI P. Edwards, 59 Treffry Road, Truro, TR1 1WL
G0 NFL M. George, 2 Jubilee Terrace, Isham, Kettering, NN14 1HG
GD0 NFN J. Butler, 15 Church Close, Lonan, Laxey, Isle Of Man, IM4 7JY
G0 NFO R. Charteris, 7 Kennedy Close, Kidderminster, DY10 1LR
G0 NFR R. Glynn, 118 Pelham Road, Birmingham, B8 2PD
G0 NFV A. Hunt, 1 Avon View Cotswold Grange Country Park, Downfield Lane, Twyning, Tewkesbury, GL20 6DL
G0 NFY A. Cordwell, 71 Tadcaster Road, Norton Woodseats, Sheffield, S8 0RA
G0 NFZ R. Hodges, 82 Frankholmes Drive, Shirley, Solihull, B90 4YB
G0 NGA P. Golder, 282 Noak Hill Road, Basildon, SS15 4DE
G0 NGD C. Stenbacka, 11 Mount View, Billericay, CM11 1HB
G0 NGE W. Clarke, 20 Langdale Road, Leyland, PR25 3AR
G0 NGG R. Brown, 20 King Edward Road, Stanford-le-Hope, SS17 0EF
G0 NGI D. Johnson, 16 Woodcote House. 188 Brookwood Farm Drive . Knaphill, Woking, GU21 2fz
GM0 NGJ A. Caldwell, 7 Gladstone Terrace, New Deer, Turriff, AB53 6TE
G0 NGK P. Walmsley, Valley View, Longworth Ave, Chorley, PR7 4PJ
G0 NGN C. Jordan, Green Lane Cottage, Leintwardine, Craven Arms, SY7 0NB
G0 NGP P. Wilson, Jansil, Worthing Road, Littlehampton, BN17 6JN
G0 NGQ J. Gibbs, 3 Holts Green, Great Brickhill, Milton Keynes, MK17 9AJ
G0 NGW R. Ramplin, 17 Cross St., Langold, Worksop, S81 9SL
G0 NHB Dr W. Stallard, 28 Wheatfield Road, Stanway, Colchester, CO3 0YJ
GU0 NHD K. Benton, Keukenhof, Route De Carteret, Castel, Guernsey, GY5 7YS
GW0 NHE A. Smith, 7 Clos Gorsfawr, Grovesend, Swansea, SA4 4GZ
G0 NHG J. Godwin, 22 Stonebeck Avenue, Harrogate, HG1 2BW
G0 NHJ J. Robson, 35 Melling Road, Cramlington, NE23 6AS
G0 NHK K. Robson, 35 Melling Road, Cramlington, NE23 6AS
GM0 NHL C. Waldron, 24/2 Vennel Street, Dalry, KA24 4AF
G0 NHM N. Robertshaw, Cherry Tree House, Church Street, York, YO26 8DD
G0 NHO K. Crookes, 64 Heron Drive, Audenshaw, Manchester, M34 5QX
G0 NHP T. Maguire, 1 Gosford St., Balsall Heath, Birmingham, B12 9ER
G0 NHR Nunsfield House ARG, c/o K. Clarke, 55 Compton Avenue, Aston-On-Trent, Derby, DE72 2AU
GM0 NHT H. Cherrie, 6 Milking Hill, Tong, Isle of Lewis, HS2 0HU
G0 NHZ A. Pollard, 2 Forest Close, Cowplain, Waterlooville, PO8 8JE
G0 NID E. Page, Flat 29, Westfields, 212 Hall Lane, Manchester, M23 1LP
G0 NIF S. Wilkins, 15 Roundway, Egham, TW20 8Bs
G0 NIG N. Smith, 45 The Gills, Otley, LS21 2BY
G0 NIK P. Gell, 25 Westland, Martlesham Heath, Ipswich, IP5 3SU
G0 NIL D. Woods, 30 Longridge Avenue, Stalybridge, SK15 1HG
G0 NIN N. Beer, 2 Marcelle Court, School Road, Hindhead, GU26 6LR
G0 NIQ J. Naylor, 6 Mallard Close, Christchurch, BH23 4DD
G0 NIX W. Jones, 13 Kilrush Terrace, Woking, GU21 5EG
GW0 NIY R. Milton, 49 Heol Y Deri, Rhiwbina, Cardiff, CF4 6HD
G0 NJD R. Mallett, 71 Olivet Road, Woodseats, Sheffield, S8 8QR
G0 NJG K. Binns, 31A Dellands, Overton, Basingstoke, RG25 3LD
G0 NJJ N. Jones, 19 Foxhollow, Bar Hill, Cambridge, CB23 8EP
GM0 NJL R. Watson, 24 Hillock Avenue, Redding, Falkirk, FK2 9UT
G0 NJO J. Howard, 37 Carisbrooke Road, St Leonards-on-Sea, TN38 0JN
G0 NJP M. Murakami, 5-8 Takamidai, Takatsuki, Osaka, Japan, 5691020
G0 NJQ P. Schlatter, Churchgate House, Sutton Road, Maidenhead, SL6 9SN
G0 NJS M. Davie, 101 Upperfield Road, Welwyn Garden City, AL7 3LR
G0 NJT J. Perkins, Flat 2Block 122, Flat 2. Leachgreen Lane Rednal, Birmingham, B45 8EH
G0 NJZ T. Dodds, 33 Westgate, Warley, Oldbury, B69 1BA
G0 NKC M. Connolly, 3 Port Mer Close, Exmouth, EX8 5RF
GW0 NKG M. York, 9 Fox Hollows, Brackla, Bridgend, CF31 2NE
GW0 NKH R. Hearne, Mora, Rhydypandy Road, Morriston, SA6 6NX
GW0 NKJ D. Davies, 6 Dulais Fach Road, Tonna, Neath, SA11 3JW
G0 NKK A. Dipper, 31 Stratton Heights, Cirencester, GL7 2RH
G0 NKM H. Monks, 100 Crossefield Road, Cheadle Hulme, Cheadle, SK8 5PF
G0 NKQ V. Nunns, 8 Trevithick Road, Tregurra, Truro, TR1 1RU
G0 NKU S. Fowler, 58 Buxton Road, High Lane, Stockport, SK6 8BH
G0 NKZ K. Everard, Woodside, Staple Lane, Taunton, TA4 4DE
G0 NLA R. Bryan, 23 Quarry Lane, Halesowen, B63 4PB
GW0 NLB W. Rees, 51 Heol Capel Ifan, Pontyberem, Llanelli, SA15 5HF
G0 NLG R. Chapman, 49 Walden Way, Frinton-on-Sea, CO13 0BH
G0 NLJ S. Oliver, Chalk Lodge, Peters Lane, Princes Risborough, HP27 0LG
G0 NLL J. Bowers, 35 Peveril Gardens, Newtown, Stockport, SK12 2RG
G0 NLM C. Ridley, 20 Victoria Gardens, Ferndown, Dorset, BH22 9JH
G0 NLN D. Ashworth, 59A Normanby Road, London, NW10 1BU
G0 NLQ P. Dunn, 10 Endsleigh Close, Upton, Chester, CH2 1LX
G0 NLT D. Wentworth, 7 Gilbeys Close, Stourbridge, DY8 4XU
GM0 NLU N. Harvey, The Shieling, Tealing, Dundee, DD4 0QU
G0 NLV F. Fairman, 26 Marina Gardens, Cheshunt, Waltham Cross, EN8 9QY
G0 NLX S. Pearson, 18 New Road, Amersham, HP6 6LD
G0 NMB A. Haberman, 4 Allendale Drive, Copford, Colchester, CO6 1BP
G0 NMC T. Neal, Sunnymead, Ewyas Harold, Hereford, HR2 0JA
G0 NMD Rev. L. Austin, 7 Kennedy Close, Chester, CH2 2PL
G0 NMH B. Markey, 164 Bourn View Road, Netherton, Huddersfield, HD4 7JS
G0 NMJ J. Denniss, 61 Checkstone Avenue, Bessacarr, Doncaster, DN4 7JY
G0 NMP P. Chapman, 1 Bader Close, Watton, Thetford, IP25 6FF
G0 NMS J. Howes11, 39 Pound Hill, Bacton, Stowmarket, IP14 4LP
G0 NMY M. Longson, 54 Beresford Street, Shelton, Stoke-on-Trent, ST4 2EX
GW0 NNE R. Hart, River View, Kilkewydd, Welshpool, SY21 8RT
G0 NNG R. Westley, 91 Lincoln Way, Daventry, NN11 4SU
GI0 NNK J. Cairns, Fridge Air, 10 Dunsilly Road, Antrim, BT41 2JH
G0 NNN W. Forbes, 29 High St., Maryport, CA15 6BQ
G0 NNO M. Shore, 12 Boscoppa Road, St Austell, PL25 3DR
G0 NNR B. Thomas, Creekside, Greenbank Road, Truro, TR3 6PQ
G0 NNS Norwich North Scouts Fellowship ARC, c/o C. Hendry, 109 Grove Avenue, New Costessey, Norwich, NR5 0HZ
G0 NNT Prof. V. Martinelli, 23 St. Angelo Street, Sliema, Malta, SLM1334
G0 NNU L. Payne, 147 Upper Marehay Road, Marehay, Ripley, DE5 8JG
G0 NNZ J. Belfield, 17 Burtondale Road, Crossgates, Scarborough, YO12 4JR
G0 NOB L. Leek, 21 Riverside Drive, Solihull, B91 3HH
G0 NOH J. Reed, 23 Morehall Avenue, Folkestone, CT19 4EQ
G0 NON T. Behan, Maytree Cottage, Marley Lane, Haslemere, GU27 3RG
GW0 NOO S. Coburn, 54 Queensway, Hope, Wrexham, LL12 9PE
GW0 NOP P. Coburn, 54 Queensway, Hope, Wrexham, LL12 9PE
G0 NOU Dr W. Ayers, Peach Lodge, Foolow, Hope Valley, S32 5QB
GI0 NOX S. Mcateer, 33 Knocknamuckley Lane, Portadown, BT63 5PF
G0 NPA R. Stephens, 50 Windrush Way, Abingdon, OX14 3SX
G0 NPC G. Hill, 1 Gleneagles Court, Edwalton, Nottingham, NG12 4DN
G0 NPE P. Hulme, 92 London Road, Cowplain, Waterlooville, PO8 8EW
G0 NPF D. Delacassa, The Tree House, Easthams Road, Crewkerne, TA18 7AQ
G0 NPG K. Heaviside, 58 Arundel Drive, Ranskill, Retford, DN22 8PQ
G0 NPI J. Podvoiskis, 3 Barnview Drive, Irlam, Manchester, M44 6WY
G0 NPJ L. Jackson, 60 East Park Avenue, Darwen, BB3 2SQ
G0 NPK D. Goulbourne, Widowscroft Farm, Hollingworth, Hyde, SK14 8LE
GW0 NPL S. Instone, 61 Llanfach Road, Abercarn, Newport, NP11 5LA
GW0 NPM H. Thomas, 34 Upland Road, Pontllanfraith, Blackwood, NP12 2ND
G0 NPN B. Harris, 23 Pound Road, Highworth, Swindon, SN6 7LA
G0 NPO I. Brown, Egremont, Arterial Road, Basildon, SS14 3JN
G0 NPP T. Watson, 20 Ivanhoe View, Gateshead, NE9 7TR
G0 NPQ H. Carruthers, 55 Inskip Terrace, Gateshead, NE8 4AJ
G0 NPV B. Ham, 37 Lower Moor, Barnstaple, EX32 8NW
G0 NPW M. Ambach, Karwendelstrabe 7, Tyrol, Austria, A-6130
G0 NPY P. Yates, 31 Wallpark Close, Brixham, TQ5 9UN
G0 NQA A. Gurbutt, 16 Crabtree Lane, Sutton-On-Sea, Mablethorpe, LN12 2RT
GI0 NQC A. Dornford-Smith, 10 Carmorn Road, Toomebridge, Antrim, BT41 3NX
G0 NQE C. Wilkinson, 8 Westfield Avenue, Knottingley, WF11 0JH
G0 NQG S. Latham, 27 Rockside Gardens, Frampton Cotterell, Bristol, BS36 2HL
G0 NQI J. Shepherd, Yew Tree Cottage, The Stenders, Mitcheldean, GL17 0JE
G0 NQJ D. Scaplehorn, 9 Stockwell Avenue, Mangotsfield, Bristol, BS16 9DR
G0 NQK R. Edwards, Manor Cottage, Manor Road, Ipswich, IP7 6PN
G0 NQN T. Fricker, 8 Folly View Necton, Swaffham, PE37 8LU
G0 NQV D. Litchfield, 37 Graeme Road, Enfield, EN1 3UU
G0 NQW D. Marshall, 15 Whisby Court, Holton-Le-Clay, Grimsby, DN36 5BG
G0 NQY S. Seggar, 145 Mount View Road, Norton, Sheffield, S8 8PJ
G0 NRA G. Lowe, 25 Manor House Court, Kirkby-In-Ashfield, Nottingham, NG17 8LH
G0 NRB R. Bellamy, 4 Wimbourne Walk, Corby, NN18 0BN
G0 NRF G. Stilgoe, 47 Chesterton Close, Redditch, B97 5XS
G0 NRI W. Hilton, 5 North Street, Williton, Taunton, TA4 4SL
G0 NRJ R. Croucher, 17 Sundridge Road, Woking, GU22 9AU
G0 NRK J. Butler, 14 Fairfield Road, Barnard Castle, DL12 8EB
G0 NRM R. Stout, 7 Thornbridge Drive, Sheffield, S12 4YF
G0 NRN G. Harrison, 14 Hardy Avenue, South Ruislip, Ruislip, HA4 6SX
GM0 NRT W. Cardno, 52 Salisbury Terrace, Aberdeen, AB10 6QH
GW0 NRW S. Chadwick, 27 Refail Farm Estate, Four Mile Bridge, Holyhead, LL65 2EX
G0 NRX S. Godbold, 13 Dawn Crescent, Upper Beeding, Steyning, BN44 3WH
G0 NRZ A. Pill, 5 St. Leonards Close, Upton St. Leonards, Gloucester, GL4 8AL
G0 NSA T. Brown, 5 St. Valentines Close, Kettering, NN15 5EG
G0 NSC W. Thornton, 46 Lavender Court, Croft Road, Barnsley, S70 3FG
G0 NSG P. Crespel, Via Leopardi 6, Cavaion Veronese, Italy, 37010
G0 NSH B. Piggott, 33 Lawrence Close, Hertford, SG14 2HH
G0 NSI J. Man, 13 Cheriton Close, Barnet, EN4 9TX
G0 NSK A. Pennell, 99 Westheath Avenue, Sunderland, SR2 9LQ
G0 NSL C. Russell, 163 Halton Road, Runcorn, WA7 5RJ
G0 NSO T. Barfield, 91 Ollerton Road, New Southgate, London, N11 2JY
G0 NSP B. Teasdale, 18 Valley Forge, Washington, NE38 7JN
GW0 NSZ J. Swinden, 22 Heol Awel, Abergele, LL22 7UQ
G0 NTA A. Jarvis, Willowmead, Nugents Park, Pinner, HA5 4RA
G0 NTB B. Jarvis, Willowmead, Nugents Park, Middlesex, HA54RA
GJ0 NTD G. Blake, 29 Pied Du Cotil, St. Andrews Road, St Helier, Jersey, JE2 3JF
G0 NTG D. Chawner, 49 St. Anns Road, Middlewich, CW10 9BY
G0 NTH A. Gardner, Flat 6, Eastcliff Court, Shanklin, PO37 6EJ
GM0 NTI L. Grieve, 1 Orchard Way, Inchture, Perth, PH14 9QB
G0 NTJ A. Williams, 23 Lancaster Gardens, Aylsham, Norwich, NR11 6LB
GM0 NTL R. Fraser, Hopefield Cottage, Gladsmuir, Tranent, EH33 2AL
GM0 NTR J. Harrison, 17B High Street, Oban, PA34 4BG
G0 NTT L. Lloyd, 8 Coastal Rise Hest Bank, Lancaster, LA2 6HJ
GM0 NTY D. Rankin, 25 Tinto Avenue Bellfield, Kilmarnock, KA1 3SJ
G0 NUA K. Franklin, 7 Auburn Close, Bridlington, YO16 7PN
G0 NUD B. Bell, 74 Henderson Road, Carlisle, CA2 4PZ
G0 NUH M. Darling, 1 Roman Way, Highworth, Swindon, SN6 7BU
GM0 NUI R. Honeyman, 81 Glen Avenue, Largs, KA30 8RH
G0 NUL D. Forster, 33 Wigeon Lane, Walton Cardiff, Tewkesbury, GL20 7RS
G0 NUN R. Barker, 5 Wickridge Close, Uplands, Stroud, GL5 1ST
G0 NUO G. Du Feu, 17 Oak Road, Tavistock, PL19 9LJ
G0 NUP K. Prince, 59 Chantry Road, East Ayton, Scarborough, YO13 9ER
GM0 NUQ R. Handyside, 113 Stockiemuir Avenue, Bearsden, Glasgow, G61 3LX
G0 NUR A. Bushell, 121 Rickmansworth Road, Watford, WD18 7JD
GW0 NUS G. Dyer, 15 Park Road, Newbridge, Newport, NP11 4RE
G0 NUT D. Mckay, 43 Mordales Drive, Marske-By-The-Sea, Redcar, TS11 7HT
G0 NUU R. Browne, Fox Cottage, Bukehorn Road, Peterborough, PE6 0QG
G0 NUZ L. Wildman, 22 Berrys Wood, Newton Abbot, TQ12 1UP
G0 NVA F. Stainsby, 11 Stonehouse Park, Thursby, Carlisle, CA5 6NS

G0 NVD J. Nothard, Ashmount, Fockerby, Scunthorpe, DN17 4RZ
G0 NVJ S. Winter, 509 Stockwood Road, Brislington, Bristol, BS4 5ES
G0 NVM J. Chandler, 14 Highfield Road, Chelmsford, CM1 2NQ
G0 NVO P. Oldham, 59 Wellspring Dale, Stapleford, Nottingham, NG9 7ET
G0 NVS M. Fletcher, 31 Woodthorpe Court, Sherwood, Nottingham, NG5 4DY
G0 NVT P. Boyle, 99 Heath Road, Penketh, Warrington, WA5 2BY
G0 NVV G. Price, 58 Hollowfields Close, Redditch, B98 7NR
G0 NVX C. Watts, 41 Salter Street, Berkeley, GL13 9BU
G0 NVY P. Hanson, 10 Parkfield Road, Ruskington, Sleaford, NG34 9HS
G0 NWC H. Cooper, 24 Queens Road, Haydock, St Helens, WA11 0RH
G0 NWE G. Egan, 11 Shepherds Row, Castlefields, Runcorn, WA7 2LG
G0 NWF S. Webster, 31 Park Estate, Shavington, Crewe, CW2 5AW
GI0 NWG A. Williamson, 23 Iskymeadow Road, Armagh, BT60 3JS
G0 NWH North West Hampshire Raynet, c/o J. Long, 1 Tangway, Chineham, Basingstoke, RG24 8SU
GM0 NWI A. Cunningham, 33 Broom Court, Stirling, FK7 7UL
G0 NWJ G. Blomeley, 13 Edale Grove, Sale, M33 4RG
G0 NWL H. Jordan, 33 Earlsbourne, Church Crookham, Fleet, GU52 8XG
G0 NWM Tynemouth ARC, c/o G. Errington, 22 Willoughby Drive, Whitley Bay, NE26 3DY
GI0 NWN M. Coyle, 67 Glen Road, Londonderry, BT48 0BY
GW0 NWR N.W.R.R.C., c/o E. Shipton, 51 Maes Stanley, Bodelwyddan, Rhyl, LL18 5TL
G0 NWS A. Edwards, 130 Bedowan Meadows, Tretherras, Newquay, TR7 2TB
G0 NWT North Norfolk ARG, c/o L. Nash, Four Furlongs, Wells Road, Wells-Next-the-Sea, NR23 1QE
G0 NWV D. Brown, 65 Warstones Drive, Penn, Wolverhampton, WV4 4PF
G0 NWY I. Peters, 62 Kingston Avenue, Seaham, SR7 8NL
G0 NXA G. Herbert, Savory Cottage, 1 Dingle Lane, Nr Tewkesbury, GL20 6DW
G0 NXC R. Sillito, 25 Naisbett Avenue, Peterlee, SR8 4BW
G0 NXD P. Brazenall, Flat 80, Clent Court, Dudley, DY1 2AZ
G0 NXE F. Rogers, 5 Station Gardens, Eckington, Pershore, WR10 3EZ
G0 NXF D. Robinson, 5 Hazel Grove, Welton, Lincoln, LN2 3JX
G0 NXH P. Cunningham, 2 The Park, Mistley, Manningtree, CO11 2AL
G0 NXI L. Edgecumbe, 51 Aller Park Road, Newton Abbot, TQ12 4NH
G0 NXL B. Ewald, Sto.Nino 2, Lower Casili Consolacion, Cebu, Philippines, 6001
G0 NXM R. Najman, 9 Bevin House, Alfred St., London, E3 2BB
G0 NXN B. Mitchell, 2 Mariners Court, Great Wakering, Southend-on-Sea, SS3 0DR
GM0 NXO G. Fyall, 105 St. Kilda Crescent, Kirkcaldy, KY2 6DR
G0 NXQ W. Love, 2 Longmead Cottages, Milborne St. Andrew, Blandford Forum, DT11 0HU
G0 NXR A. Moreton, 5 Gorse Avenue, Thornton-Cleveleys, FY5 2PH
G0 NXS A. Ellis, High Nentsberry, Alston, CA9 3LZ
G0 NXT S. Platts, 15 Holywell Avenue, Smisby, Ashby-de-la-Zouch, LE65 2HL
GW0 NXW G. Scanlin, 4 Eifion Close, Barry, CF63 1RQ
G0 NXX J. Lynch, 14 The Pastures, Cayton, Scarborough, YO11 3UU
G0 NXY P. Martin, 48 Mill Lane, Fazeley, Tamworth, B78 3QD
GM0 NYD J. Burrow, Bongate, Jedburgh, TD8 6DU
G0 NYE J. Dixon, 17 Marlowe Close, East Hunsbury, Northampton, NN4 0QQ
G0 NYH J. Moseley, 42 Burford Road, Chipping Norton, OX7 5DZ
GI0 NYI J. Benson, 18 Alexander Avenue, Armagh, BT61 7JD
G0 NYJ S. Au, Flat 2, 1St Floor, Block C, Greenland Garden, Tuen Mun, N.T., Hong Kong,
G0 NYK J. Hoose, 91 Brevere Road, Hedon, Hull, HU12 8LX
G0 NYM M. Borer, 37 Broadway, Ripley, DE5 3LJ
GM0 NYP N. Purtell, 31 Daleally Crescent, Errol, Perth, PH2 7QA
G0 NYQ J. Pape, 12A High Wiend, Appleby-in-Westmorland, CA16 6RD
G0 NYR R. Cheetham, 8 Fairway, Huyton, Liverpool, L36 1UD
G0 NYS D. Cox, 10 Calder Close, Bollington, Macclesfield, SK10 5LJ
G0 NYY S. Nicholas, 39 Cubitts Close, Welwyn, AL6 0DZ
G0 NYZ S. Maloney, 34 Keswick Road, Normanby, Middlesbrough, TS6 0BN
G0 NZA M. Lowe, 25 Manor House Court, Kirkby-In-Ashfield, Nottingham, NG17 8LH
G0 NZE A. Benfield, 12 St. Marys Court, Weald, Bampton, OX18 2HX
G0 NZI C. Peake, 3 Marigold Walk, Bermuda Park, Nuneaton, CV10 7SW
G0 NZJ Rev. P. Forbes, 18 Francis Road, Hinxworth, Baldock, SG7 5HL
GW0 NZN J. Smith, 6 Cherry Grove, Croespenmaen, Newport, NP11 3DF
G0 NZR D. Catterall, 86 Broomfield Road, Swanscombe, DA10 0LT
G0 NZT D. Nash, 27 Sandling Avenue, Horfield, Bristol, BS7 0HS
G0 NZU R. Blanning, 38 Northville Road, Northville, Bristol, BS7 0RG
GM0 OAA M. Wigg, 1/1 2 Glencairn Drive, Glasgow, G41 4QN
G0 OAB D. Griffith, 5 Upthorpe Drive, Wantage, OX12 7DF
GW0 OAJ J. Morrice, 184 Rowan Way, Newport, NP20 6JT
G0 OAS N. Goddard, 15 Canada Road, Cobham, KT11 2BB
G0 OAT R. Petri, Tarnwood, Denesway, Gravesend, DA13 0EA
G0 OAW W. Waldron, Redstone Farm, Germans Week, Beaworthy, EX21 5BQ
G0 OAZ J. Fox, L'Auziziere De St.Marsault, la Foret Sur Serve, France, 79380
GW0 OBB W. Evans, Brynawel, Cross Inn, Llanon, SY23 5NB
G0 OBE J. Clarke, 16 Silver Birch Avenue, Bedworth, CV12 0AZ
G0 OBH H. Cox, Windrush, Malthouse Lane, Great Yarmouth, NR29 5QL
G0 OBJ G. Pratley, 34 Druce Way, Thatcham, RG19 3PF
G0 OBK K. Warnes, 3 Blue Bell Close, Underwood, Nottingham, NG16 5FN
G0 OBN C. Hodgson, 25 Pembroke Court, Sunderland, SR5 4DF
G0 OBO K. Weeks, 11 Sandwich Road, Preston Grange, North Shields, NE29 9HT
G0 OBP K. Hudson, 2 Colwill Walk, Plymouth, PL6 8XF
G0 OBQ G. Lang, 63 Grosvenor Drive, Whitley Bay, NE26 2JR
G0 OBT S. Fortt, 59 Coombe Dale, Sea Mills, Bristol, BS9 2JF
G0 OBV M. Roberts, The Bourne, The Avenue, Reading, RG7 6NN
G0 OCB Dr R. Dingle, 29 Castle View, Witton Le Wear, Bishop Auckland, DL14 0DH
G0 OCC G. Allen, 2 Haworth Drive, Bootle, L20 6EJ
G0 OCF R. Mccoye, 26 Hansby Close, Oldham, OL1 2UA
G0 OCK B. Pilkington, 219 Brownhill Avenue, Burnley, BB10 4QH
G0 OCL F. Pilkington, 219 Brownhill Avenue, Burnley, BB10 4QH
G0 OCR F. Batkin, 24 Sandyfields Road, Sedgley, Dudley, DY3 3LB
G0 OCS M. Stabbins, Primrose Cottage, Carlidnack Lane, Falmouth, TR11 5HE
G0 OCT L. Stabbins, Primrose Cottage, Carlidnack Lane, Falmouth, TR11 5HE
G0 OCW P. Brazier, 1 Ravenshore Cottages, Holcombe Road, Rossendale, BB4 4AN
G0 OCY P. Beeston, 100 Suffield Road, High Wycombe, HP11 2JL
G0 ODA G. Hill, 163 Parsonage Rd, Castle Hill, Australia, 2154
GM0 ODB J. Kane, 21 Sersley Drive, Kilbirnie, KA25 6EY
G0 ODE D. Williams, 212 Birchanger Lane, Birchanger, Bishop's Stortford, CM23 5QH
G0 ODG P. Duce, Curlew Cottage, Weston, Hr6 9Je, Hereford, HR6 9je
G0 ODH D. Hibberd, Northview, The Common, Dilhorne, Stoke on Trent, St10 2pa
G0 ODI R. Sutton, 87 Downs Valley Road, Woodingdean, Brighton, BN2 6RG
G0 ODK W. Everett, 120 Wantage Road, Reading, RG30 2SF
G0 ODM J. Chomer, 14 Holly Park Gardens, Finchley, London, N3 3NJ
G0 ODN M. Hall, 31 Meendhurst Road, Cinderford, GL14 2EF
G0 ODP P. Warman, 107 Hillside Road, Corfe Mullen, Wimborne, BH21 3SB
G0 ODQ J. Hall, Two Chestnuts, Emmington, Chinnor, OX39 4AA
G0 ODR C. Hendry, 109 Grove Avenue, New Costessey, Norwich, NR5 0HZ
G0 ODS M. Treacher, 6 Beech View, Whitwell, York, YO60 7JW
G0 ODU K. Petherick, 17 Castle Close, Totternhoe, Dunstable, LU6 1QJ
G0 ODX D. Hughes, Balshult, Eriksmala, Sweden, 361 94
G0 ODY G. Fleming, 27 Crawthorne Crescent, Huddersfield, HD2 1LB
G0 OEA A. Moggridge, Outer Bailey, Kingsland, Leominster, HR6 9QN
G0 OEB C. Donald, 8 Greenway, Walsall, WS9 8XE
G0 OED A. Mardo, 10 Meadow View, Uffculme, Cullompton, EX15 3DS
GI0 OEH K. Patterson, 8 Beechwood Gardens, Moira, Craigavon, BT67 0LB
G0 OEI M. Hopkins, 30 Commonside, Brownhills, Walsall, WS8 7AY
G0 OEJ M. Garbutt, 92 Owlet Road, Windhill, Shipley, BD18 2LT
G0 OEK D. Spooner, 60 St. Pauls Road, Staines, TW18 3HH
G0 OEM G. Cunningham, 7 Wykin Lane, Stoke Golding, Nuneaton, CV13 6HN
G0 OEQ J. Blichfeldt, 2 Duck Cottages, Rolvenden Road, Cranbrook, TN17 4BT
G0 OER P. Roberts, 1 Ardern Close, Bristol, BS9 2QT
G0 OES D. Owen, 5 Vicarage Walk, Rosliston, Swadlincote, DE12 8LB
G0 OEW D. Rooke, The Grange, 107 Wybunbury Road, Nantwich, CW5 7ER
G0 OEY A. Kerrison, 63 Stour Road, Harwich, CO12 3HS
G0 OFA R. Dennis, 8 Newton Hall, Coach Road, Newton Abbot, TQ12 1ER
G0 OFB J. Jones, 59 East Street, Long Buckby, Northampton, NN6 7RB
G0 OFD J. Gilbert, 6 Mill Hill, Brancaster, King's Lynn, PE31 8AQ
G0 OFE J. Smith, 38 Wilson Road, Bournemouth, BH1 4PH
G0 OFF S. Hipkin, 62 Woodberry Way, Walton on the Naze, CO14 8EW
GW0 OFH S. Williams, 5 Brynmelyn Avenue, Llanelli, SA15 3RU
GM0 OFL J. Wilkie, Hope Cottage, 4 Main Street, Cupar, KY15 4SS
GM0 OFM J. Park, Rameldry Mill Cottage, Rameldry Mill Road, Rameldry, Kingskettle, Cupar, KY15 7TY
G0 OFN I. Clabon, 14 Melrose Avenue, Twickenham, TW2 7JE
G0 OFR G. Borrowdale, 30 Barton View, Penrith, CA11 8AX
G0 OFT S. Duncan, 10 Huntingdon Rise, Bradford-on-Avon, BA15 1RJ
G0 OFW P. Lightfoot, 18 Fields Close, Alsager, Stoke-on-Trent, ST7 2ND
G0 OFX F. Rawlins, 12 Arundel Road, Eastleigh, SO50 4PQ
G0 OFY J. Wane, 25 Holmdale Avenue, Crossens, Southport, PR9 8PS
G0 OFZ A. Grace, 7 Sandringham Heights, St Leonards-on-Sea, TN38 9UA
G0 OGB J. Wallis, 10 Middlewood Road, Lanchester, Durham, DH7 0HL
G0 OGE M. Free, 2 St Pauls Court, Princess Street, Maidenhead, SL6 1NX
GW0 OGI D. Keely, 15 Ffordd Cerrig Mawr, Caergeiliog, Holyhead, LL65 3LU
G0 OGJ K. Marshall, 8 Porter Way, Northwich, CW9 7JA
GW0 OGL S. Glanville, Fron Haulog, Llanelian, Colwyn Bay, LL29 8UY
G0 OGM S. Bowerman, 24 Bingham Close, Emerson Valley, Milton Keynes, MK4 2AU
GM0 OGN R. Hall, 13 Cleat, Castlebay, Isle of Barra, HS9 5XX
G0 OGP Y. Powell, 18 Carrington Road, Stockport, SK1 2QE
G0 OGS S. Malpass, 21 Tollhouse Way, Wombourne, Wolverhampton, WV5 8AF
G0 OGW C. Douglas, 29 Barnes Close West Wellow, Romsey, SO51 6ET
G0 OGX J. Pennington, 1 Chisel Close, Hereford, HR4 9XF
GM0 OGZ R. Goodall, 3 Croftcrunie Cottages, Tore, Muir of Ord, IV6 7SB
G0 OHA A. White, 11 Garden Close, Consett, DH8 5PA
GM0 OHD M. Holding, Craigvar Main Street, Lochfoot, DG2 8NR
G0 OHF F. Wilson, 3 Foundry Mews, Burgh Le Marsh, Skegness, PE24 5HQ
GI0 OHG E. Bennett, 53 Condiere Avenue, Connor, Ballymena, BT42 3LD
GW0 OHJ D. Workman, 4 Rhuddlan Road, Buckley, CH7 3QA
G0 OHK N. King, 7 Fountains Close, Biddick Village, Washington, NE38 7TA
G0 OHQ R. Bunyan, 11 Kenhill Close, Snettisham, King's Lynn, PE31 7PA
G0 OHR J. Armitage, 17 Worsbrough Road, Birdwell, Barnsley, S70 5QR
GI0 OHT W. Stanley, 95 Bangor Road, Newtownards, BT23 7BZ
GI0 OHU R. King, 28 Moss Road, Waringstown, Craigavon, BT66 7QY
G0 OHW J. Vasek, 20 West Hall Road, Richmond, TW9 4EE
G0 OID F. Tett, 2 Church Park, Bradenstoke, Chippenham, SN15 4ER
G0 OIE M. Gathergood, 54 Robin Lane, Bentham, Lancaster, LA2 7AG
G0 OIF D. Read, 17 Lumber Leys, Walton on the Naze, CO14 8SS
G0 OII R. Pullen, 1 Ridings Court, 5 Crown Crescent, Scarborough, YO11 2BJ
G0 OIK P. King, 96 Mancroft Road, Caddington, Luton, LU1 4EN
G0 OIM B. Turnbull, 56 Bryans Close Road, Calne, SN11 9AB
G0 OIN A. Fairey, 1 Imbert Close, New Romney, TN28 8XP
G0 OIO J. Fuller, 1 The Courtyard, Snape, Saxmundham, IP17 1FB
G0 OIQ A. Welland, 6 Chartwell Road, Stafford, ST17 0AJ
G0 OIR G. Wicks, 28 Old School Lane, Milton, Cambridge, CB24 6BS
G0 OIS T. Smith, 33 Kites Nest Lane, Lightpill, Stroud, GL5 3PJ
G0 OIU A. Smith Jones, 16 Armley Road, Liverpool, L4 2UN
G0 OIV A. Sait, 124 Dicksons Drive, Newton, Chester, CH2 2BX
G0 OIW M. Palmer, 28 Westfield Road, Caversham, Reading, RG4 8HH
G0 OIX D. Wright, 132 Longmoor Lane, Liverpool, L9 9BZ
G0 OIY J. Smith, 59 Charlecote Drive, Dudley, DY1 2GG
G0 OJB L. Chadwick, 18 Pakenham Place, Haverhill, CB9 0JA
G0 OJC M. Crowly, 3 Eleanor Street, Ellesmere Port, CH65 4BB
G0 OJF R. Shireby, Norwood, Rookery Lane, Sudbrook, Grantham, NG32 3RU
G0 OJG J. Omalley, 8 Hawks Court, Hallwood Park, Runcorn, WA7 2FR
G0 OJJ A. Green1, 6 Goulds Close Palgrave, Diss, IP22 1AR
G0 OJP R. Melton, 4 Ashleigh Avenue, Maiden Newton, Dorchester, DT2 0BP
G0 OJR B. Fox, 6 Bury Gardens, Elmdon, Saffron Walden, CB11 4LX
G0 OJS S. John, 40 Elizabeth Avenue, Brixham, TQ5 0AY
G0 OJT W. Jenkinson, 7 Moortown Road, Watford, WD19 6JH
G0 OJW J. Taylor, C/O 16 Caroline Place, Plymouth, PL1 3PS
G0 OJX P. Williams, 3 Nutwell Cottages, Exmouth Road, Exmouth, EX8 5AP
G0 OJY A. Hurt, The Manse, High Oak Road, Ware, SG12 7PD
G0 OKA D. Martin, 67 Mill Street, Torrington, EX38 8AL
G0 OKD R. Bradley, 4 Paddocks Close, Pinxton, Nottingham, NG16 6JR
G0 OKF S. Bolam, 100 Bushfield Road, Scunthorpe, DN16 1NA
G0 OKI R. Morris, 4 Greenway Gardens, Kings Norton, Birmingham, B38 9RY

GM0 OKJ J. Fraser, 2 Barra Place, Stenhousemuir, Larbert, FK5 4UF
G0 OKK B. Crowe-Haylett, 13 Lynton Close, Ely, Cb61dj
G0 OKL J. Collins, 3 Burford Grove, Bristol, BS11 9RT
GI0 OKM G. Nesbitt, 205 Longstone Street, Lisburn, BT28 1TY
G0 OKN R. Maloney, Rosewell, Jacobstow, Bude, EX23 0BN
G0 OKT C. Parker, 1 Richmond Drive, Perton, Wolverhampton, WV6 7RR
G0 OKV K. Cowell, 4 St. Georges Close, Colne, BB8 8DP
G0 OKX K. Gardner, 13 Beanshaw, Eltham, London, SE9 3HL
G0 OKZ J. Thorpe, Four Jays, 46A High Street, Doncaster, DN10 4BU
G0 OLD D. Mclaren, 11 St. Matthew Close, Uxbridge, UB8 3SR
G0 OLE Boothferry ARS, c/o K. Mccann, Treverven, Back Lane, Hemingbrough, Selby, YO8 6QP
G0 OLL E. Platts, 38 Swanbourne Road, Sheffield, S5 7TL
G0 OLO D. Collinson, 20 Carlisle Crescent, Penshaw, Houghton le Spring, DH4 7RD
G0 OLR L. Roberts, Rose Cottage, Castle Hill, Leyburn, DL8 4QN
G0 OLS T. Humphries, 23 Sycamore Drive, Lutterworth, LE17 4TR
G0 OLT L. Tringale, 19 Lysander Road, Kings Hill, West Malling, ME19 4TT
G0 OLX D. Stanton, 106 Scrapsgate Road, Minster On Sea, Sheerness, ME12 2DJ
GW0 OLZ G. Smith, 23 Gainsborough Close, Llantarnam, Cwmbran, NP44 3BX
G0 OMB B. Walker, 15 Infirmary Road, Workington, CA14 2UG
GM0 OMC C. Cook, Briarwood, 95 Old Edinburgh Road, Inverness, IV2 3HT
G0 OMD A. Gilbert, 19 Farrs Avenue, Andover, SP10 2AH
G0 OMF D. Hupton, 90 Warwick Road, Atherton, Manchester, M46 9PQ
G0 OMH P. Burbeck, 5 Wouldham Terrace, Saxville Road, St Paul's Cray, BR5 3AT
G0 OMM S. Adams, 13 Bells Drove, Sutton St. James, Spalding, PE12 0JG
G0 OMN G. Charman, 4 Hornton Grove, Hatton Park, Warwick, CV35 7UA
G0 OMZ R. Lomas, 7 Chaunterell Way, Abingdon, OX14 5PP
G0 ONA P. Nicholls, 5 Dingle Road, Ashford, TW15 1HF
G0 ONB CSMT Group, c/o Dr N. Shaxted, 5 Arnold Road, Chartham, Canterbury, CT4 7QL
GI0 OND J. Lappin, 46 Grange Road, Kilmore, Armagh, BT61 8NX
G0 ONF V. Szendzielarz, 5 Granville Road, Urmston, Manchester, M41 0XY
G0 ONG J. Mobbs, 5 Distaff Road, Poynton, Stockport, SK12 1HN
G0 ONH B. Fellows, 36 Balmoral Road, Stourbridge, DY8 5HR
GM0 ONN I. Barnetson, 38 Woodlands Drive, Lhanbryde, Elgin, IV30 8JU
G0 ONS J. Chinnery, 31 Kingsway, Northampton, NN2 8HD
G0 ONT R. Broom, Staging Post Abbotskerswell, Newton Abbot, TQ12 5NX
GW0 ONU D. Harris, 2 Sheppard St., Pwllgwaun, Pontypridd, CF37 1HT
GM0 ONX L. Paget, 40 Davaar Drive, Kilmarnock, KA3 2JG
GW0 ONY J. Edwards, 3 High Street, Bryngwran, Holyhead, LL65 3PL
G0 OOB D. Walpole, 12 Damgate Lane, Acle, Norwich, NR13 3DH
G0 OOD T. Chapman, 21, Links Close., Norwich.Norfolk., N6 5PJ
G0 OOF R. Williams, Dyffryn Coed, Union Road, Coleford, GL16 7QB
G0 OOI W. Humphries, 76 Mortlake Road, Richmond, TW9 4AS
G0 OON P. Healey, 10 Wroxham Road, Great Sankey, Warrington, WA5 3EE
G0 OOO Scarborough Seg, c/o R. Clayton, 9 Green Island, Irton, Scarborough, YO12 4RN
G0 OOQ S. Whitehurst, 28 Severn Way Cressage, Shrewsbury, SY5 6DS
G0 OOR A. Jex, 26 Springdale Crescent, Brundall, Norwich, NR135RA
G0 OOS L. Marobin, Flat 60, Tudor Court, King Henrys Walk, London, N1 4NU
G0 OOU R. Field, 34 Piltdown Close, Hastings, TN34 1UU
G0 OPA J. Lee, Holly Lodge, Carrhouse Road, Doncaster, DN9 1PG
G0 OPC M. Marriott, 188 Leverington Common, Leverington, Wisbech, PE13 5BP
G0 OPG C. Knowlson, 28 Hill Drive, Handforth, Wilmslow, SK9 3AR
G0 OPI A. Bennett, 32 Gainsborough Road, Bournemouth, BH7 7BD
G0 OPL W. Cowell, 72A The Malting, Ramsey, Huntingdon, PE26 1LZ
G0 OPM G. Melia, Sunnyside, Little Asby, Appleby-in-Westmorland, CA16 6QE
GW0 OPP R. Owens, 62 Ty Llwyd Parc Estate, Quakers Yard, Treharris, CF46 5LB
G0 OPT P. Tennant, 128 Devonshire Street, Keighley, BD21 2QJ
G0 OPV R. Heatley, 68 Jeckyll Road, Wymondham, NR18 0WQ
GM0 OPX D. Mcferran, Ardlair, Milltimber, AB13 0ER
G0 OQE F. Porter, Kinross, 12 Brooklands Road, Milton Keynes, MK2 2RN
G0 OQI K. Zak, 5 The Rookery, Sandy, SG19 2UR
G0 OQK N. Garrod, 121 Totteridge Lane, High Wycombe, HP13 7PH
G0 OQP A. Caton, 20 Lower Oxford Road, Newcastle, ST5 0PB
G0 OQQ B. Wood, 52 Ashfield Avenue, Beeston, Nottingham, NG9 1PY
G0 OQR A. Glen, 70 Moscow Road East, Stockport, SK3 9QL
G0 OQT M. Jones, 6 Eastleigh Close, Burnham-on-Sea, TA8 2EW
G0 OQX J. East, 30 Auckland Road, Scunthorpe, DN15 7BT
G0 OQZ H. Dawson, 6 Maer Top Way, Barnstaple, EX31 1RZ
G0 ORC V. Shirley, 160 Over Lane, Belper, DE56 0HN
G0 ORD Dr E. Chantler, Hilltop Gardens, High Beech Road, The Pludds, Ruardean, GL17 9UD
G0 ORE N. Reddish, 15 Drakes Close, Redditch, B97 5NG
G0 ORG N. Robertson, Clayhill Cottage, The Street, Aldham, Ipswich, IP7 6NN
G0 ORJ J. Bamford, 39 Skelldale View, Ripon, HG4 1UJ
G0 ORK S. Humberstone, 4 Rowcroft Road, Paignton, TQ3 2RE
G0 ORL C. Rowley, 31 Keepers Croft East Goscote, Leicester, LE7 3ZJ
G0 ORM D. Birch, 31 Grasmere Terrace, Maryport, CA15 7QN
G0 ORO D. Martin, The Shieling, Bolton Low Houses, Wigton, CA7 8PF
G0 ORP M. Simpson, 3 Front Street, Barnby, Newark, NG24 2SA
G0 ORT D. Leonard, Three Ashes Cottage, 442 Outwood Common Road, Billericay, CM11 1ET
G0 ORV V. Wilton, Fairthorn Trotts Ln, Pooks Green, Southampton, SO4 4WQ
G0 ORX J. Melton, 4 Charlwood Close, Copthorne, Crawley, RH10 3TG
G0 ORY A. Moss, 10 Shakespeare Drive, Leicester, LE3 2SP
G0 OSA C. Wilkinson, 10 Chemin Des Mardeilles, Chirac, France, 16150
GW0 OSB I. Price, 16 Carmarthen Court, Caerphilly, CF83 2TX
G0 OSC G. Mason, 18 Nithsdale Road, Liverpool, L15 5AX
G0 OSD G. Alexander, 15 Brackley Way, Totton, Southampton, SO40 3HP
G0 OSG R. Brazier, 9 Wheelers Walk, Blackfield, Southampton, SO45 1WX
G0 OSI K. Pallant, 7 Council Bungalows, Church Lane, Braintree, CM7 5SH
GM0 OSJ W. Legge, 57 High Street Archiestown, Aberlour, AB38 7QZ
G0 OSK C. Saggers, 49 Revels Road, Hertford, SG14 3JU
G0 OSO P. Markham, Moor Farm, Moor Lane, Lincoln, LN3 4EG
G0 OSU J. Collier, 27 Birdham Close, Bognor Regis, PO21 5TD
G0 OSW R. Sainsbury, Salem Park Farm, Southampton Road Landford, Salisbury, SP5 2BE
G0 OSX N. Shackley, 20A Pear Tree Road, Ashford, TW15 1PW
GM0 OTB R. Pugh, 28 Pladda Road, Saltcoats, KA21 6AQ
GI0 OTC T. Doherty, 37 Magheramenagh Drive, Portrush, BT56 8SP
G0 OTE E. Bowell, 7 Bede House Bank, Bourne, PE10 9JX
G0 OTF G. George, 211 Bromford Road, Birmingham, B36 8HA
G0 OTH R. Topliss, 12 Dorothy Avenue, Skegness, PE25 2BP
GM0 OTI Dr J. Grieve, 1 Orchard Way, Inchture, Perth, PH14 9QB
G0 OTJ J. Cummins, Tarr House, Lumb Lane, Matlock, DE4 2HP
GM0 OTS W. Mcintosh, 14 East Road, Hopeman, Elgin, IV30 5SU
G0 OTT D. Mcdonald, 118 Torrington Avenue, Tile Hill, Coventry, CV4 9AA
GM0 OTU A. King, 31 Pendreich Grove, Bonnyrigg, EH19 2EH
GW0 OTY W. Cooper, 50 Tennyson Road, Penarth, CF64 2SA
G0 OUC R. Rodgerson, 8 Dearne Street, Darton, Barnsley, S75 5HL
GD0 OUD S. Hill, 54 Wybourn Drive Onchan, Isle of Man, IM3 4AT
G0 OUG I. Hole, 50 Westcroft Drive, Westfield, Sheffield, S20 8EF
GW0 OUH H. Griffiths, 45 Jubilee Road, Godreaman, Aberdare, CF44 6DD
G0 OUK J. Hinton, Wayside, Beauchief Drive, Sheffield, S17 4RJ
GI0 OUM R. Ferris, 3 Kingsland Drive, Belfast, BT5 7EY
G0 OUN P. Bingham, 12 Sandpiper Road, Thorpe Hesley, Rotherham, S61 2UN
G0 OUO S. Palk, 10 Springfield Close, Andover, SP10 2QT
G0 OUR Open University ARC, c/o A. Rawlings, 57 High Street, Nash, Milton Keynes, MK17 0EP
GW0 OUV M. Williams, 7 Heol Isaf, Nelson, Treharris, CF46 6NS
GI0 OUZ B. Prunty, 16 Old Portadown Road, Lurgan, Craigavon, BT66 8RH
G0 OVA P. Crake, 1 Ashdown Close, Bracknell, RG12 2SE
G0 OVC B. Godfrey, 291 Collier Row Lane, Romford, RM5 3ND
GM0 OVD R. Darroch, 36 Tweed Street, Dunfermline, KY11 4NA
G0 OVE K. Mohammed, 63 Shirley Gardens, Barking, IG11 9XB
G0 OVK R. Mansell1, 2 Ambrose Close, Willenhall, WV13 3DQ
G0 OVQ A. Bannister, 34 Morningside Drive, East Didsbury, Manchester, M20 5PL
G0 OVT B. Navier, 12 Brooklyn Avenue, Brooklyn Street, Hull, HU5 1ND
G0 OVV M. Bolton, 85 Oak Park Road, Wordsley, Stourbridge, DY8 5YJ
G0 OVY P. Maggs, 85 Helsby Road, Sale, M33 2XF
G0 OWA J. Wright, 10 Whalley Road, Heskin, Chorley, PR7 5NY
G0 OWC P. Bush, 52 Asker Lane, Matlock, DE4 5LA
G0 OWE D. Matthews, 54 The Wynding, Bedlington, NE22 6HW
G0 OWH J. Dobbs, 9 Highlands, Littleborough, OL15 0DS
G0 OWI A. Hawkridge, Thorntrees, 109 Allerton Road, Bradford, BD8 0AA
G0 OWJ A. Cooper, 28 Belmont Road, Pensnett, Brierley Hill, DY5 4EX
G0 OWK S. Searle, 14 Edison Gardens, Colchester, CO4 0AJ
GM0 OWM Orkney Wireless Museum, c/o E. Holt, Ashwell, St. Ola, Kirkwall, KW15 1SX
G0 OWP D. Edwards, 9 Mark Road, Hightown, Liverpool, L38 0BG
G0 OWR C. Howard, 75 Westbury Park, Wootton Bassett, Swindon, SN4 7DN
G0 OWU R. Wilkes, 39 Hillside Road, Dudley, DY1 3LE
G0 OWV J. Harbottle, 42 Littlemede, Eltham, London, SE9 3EB
G0 OXA G. Landen-Turner, 59 Mill Road, Higher Bebington, Wirral, CH63 5PA
G0 OXB P. Draycott, 41 Ashleigh Avenue, Bridgwater, TA6 6AX
G0 OXE Morse Club, c/o C. Tapping, 49 Coventry Gardens, Herne Bay, CT6 6SB
GI0 OXK D. Taggart, 106 Moorfield Road, Dromore, Omagh, BT78 3LR
G0 OXL Dr C. Robinson, 33 Windsor Rd, Wellesley, Ma, USA, 2481
G0 OXP L. Matthews, 6 Spotland Tops, Cutgate, Rochdale, OL12 7NX
GM0 OXS M. Beith, 30 Raith Road, Fenwick, Kilmarnock, KA3 6DB
G0 OXT P. Hutchinson, Rosebank Cottage, Marcombe Road, Torquay, TQ2 6LL
G0 OXV K. Mahood, 1A Heskin Lane, Ormskirk, L39 1LR
G0 OXW V. Soutter, 2 Hyde Barton, Churchill Way, Bideford, EX39 1NX
G0 OXX Dr J. Berridge, Bracklyn, St. Clare Road, Deal, CT14 7QB
G0 OXY M. Gray, 142 Harrowden Road, Bedford, MK42 0SJ
G0 OXZ M. Stone, 29 Chesterfield Road, Epsom, KT19 9QR
G0 OYA M. Clapperton, 99 Bath Road, Bridgwater, TA6 4PN
G0 OYC K. Saunders, 1 Chesham Way, Watford, WD18 6NX
G0 OYF S. Harvey, 68 Stuart Road, Rowley Regis, B65 9HZ
G0 OYI G. Holden, The House On The Green, Linstock, Carlisle, CA6 4PZ
G0 OYJ T. Gonsalves, 30 Cunnington Street, Chiswick, London, W4 5EN
G0 OYL W. Waring, 1 Innerhaugh Mews, Haydon Bridge, Hexham, NE47 6DE
G0 OYM M. Trahearn, 16 Grange Lane, Lichfield, WS13 7ED
G0 OYN D. Hedley, 42 Liphook Road, Lindford, Bordon, GU35 0PP
G0 OYO D. James, 7 Abbotts Road, Plymouth, PL3 4PD
G0 OYP B. Barber, 3 Catherine Avenue, Mansfield Woodhouse, Mansfield, NG19 9AZ
G0 OYQ S. Lowe, 14 Kensington Avenue, Kingswood, Hull, HU7 3AF
G0 OYR N. Ashfield, 167 Greville Road, Warwick, CV34 5PU
G0 OYS D. Temple, 4 Cameron Avenue, Abingdon, OX14 3SR
GM0 OYU M. Chesters, Blackhill, Blackhill Road, Kirkwall, KW15 1FP
G0 OYX D. Medley, 9 Northolme Crescent, Hessle, HU13 9HU
G0 OYY G. Mantle, 26 Graham Road, Wordsley, Stourbridge, DY8 5PU
G0 OYZ R. Bray, 10 Upwell Road, March, PE15 9DT
GW0 OZB A. Gardner, 28 Usk Court, Thornhill, Cwmbran, NP44 5UN
G0 OZG D. Turner, 27 Aylesbury Avenue, Langney Point, Eastbourne, BN23 6AB
G0 OZJ G. Gourley, 6A Longsight Lane, Cheadle Hulme, Cheadle, SK8 6PW
G0 OZL B. Smith, 19 Fieldstone Court, Howick, Northpark, New Zealand, 1705
G0 OZM C. Rapson, Kaloma, Northiam Road, Rye, TN31 6EP
G0 OZO J. Harris, 31 Grasby Road, Limber, Grimsby, DN37 8LB
G0 OZP B. Salt, 9 Ashville Gardens, Pellon, Halifax, HX2 0PJ
GI0 OZQ D. Gillespie, 81 Lisfannon Park, Londonderry, BT48 9DU
G0 OZR G. Markham, 2 Edwin Avenue, Woodbridge, IP12 1JS
G0 OZS I. Moffat, The Hatchets, The Street, Brockford, Stowmarket, IP14 5PE
G0 PAB P. Betts, 14 Saltergate Road, Messingham, Scunthorpe, DN17 3SZ
GM0 PAC G. Macaulay, 22 Glencairn Drive, Glasgow, G41 4PW
G0 PAD A. Jacobs, 16 Clwyd Walk, Corby, NN17 2LN
G0 PAE C. Hewitt, 28 Amersham Avenue, Langdon Hills, Basildon, SS16 6SJ
G0 PAG N. Page, 54 Queensway, Old Dalby, Melton Mowbray, LE14 3QH
G0 PAI I. Leitch, 70 Hanover Road, Rowley Regis, B65 9DZ
G0 PAN D. Elkington, 45 Heathfield, Leeds, LS16 7AB
G0 PAO C. Muddimer, 7 Tots Gardens, Acton, Sudbury, CO10 0DJ
G0 PAR D. How, 25 Lovelace Road, London, SE21 8JY
G0 PAS M. Lord, 5 Wasdale Green, Cottingham, HU16 4HN
G0 PAZ D. Utley, 30 Station Road, Ackworth, Pontefract, WF7 7NA
G0 PBB Forest of Dean ARG, c/o W. Bonser, 24 Meend Garden Terrace, Cinderford, GL14 2EB

G0 PBE D. Yates, 101 Coach House Drive, Shevington, Wigan, WN6 8AU
G0 PBF J. Brown, 71 Piccadilly Road, Swinton, Mexborough, S64 8LF
G0 PBH D. Mason, 11 Bryony Close, Killamarsh, Sheffield, S21 1TF
GW0 PBJ L. Wright, Cedar House, Old Aston Hill Ewloe, Deeside, CH5 3AL
G0 PBL P. Davies, 85 Church Road, Byfleet, West Byfleet, KT14 7NG
G0 PBM A. Razzell, 96 Weston Road, Aston-On-Trent, Derby, DE72 2BA
G0 PBN A. Moulder, 10 Parsonage Road, Rainham, RM13 9LW
G0 PBO A. Coleman, 16 Cowley Close, Swineshead, Boston, PE20 3ES
G0 PBP A. Evans, 24 Oakleigh Avenue, Glen Parva, Leicester, LE2 9TH
G0 PBQ D. Hodge, 2 Leyland Close, Southport, PR9 8AT
G0 PBR R. Clark, 4 Haigh St., Cleethorpes, DN35 8QN
G0 PBS D. Webber, Lowenva, Shripple Lane, Winterslow, Salisbury, SP5 1PW
G0 PBU D. Bradley, 2A Mitchell Street, Kettering, NN16 9HA
G0 PBV N. Plumb, 35 Foamcourt Waye, Ferring, Worthing, BN12 5RD
G0 PBW R. Brown, 26 Lynnes Close, Blidworth, Mansfield, NG21 0TU
G0 PBY R. Freer, 15 Fosse Close Enderby, Leicester, LE192AW
G0 PCA K. Godwin, 11 St. Lukes Way, Allhallows, Rochester, ME3 9PR
G0 PCB E. Godwin, 11 St. Lukes Way, Allhallows, Rochester, ME3 9PR
G0 PCD S. Farrow, 7 Bakewell Close, Hull, HU9 5LH
G0 PCE R. Barnes, Flat 113, Queens Quay, 58 Upper Thames Street, London, EC4V 3EJ
G0 PCF B. Foxall, 11 Cranley Gardens, Shoeburyness, Southend-on-Sea, SS3 9JP
GW0 PCJ C. Watson, 4 Brookland Close, Maesycwmmer, Hengoed, CF82 7RH
G0 PCK Rev. A. Lord, 47 Nottingham Road, Trowell, Nottingham, NG9 3PF
G0 PCM I. Calvert, 16 Nab Wood Drive, Shipley, BD18 4EJ
G0 PCP R. Baldock, 19 Ferndale Close, Burntwood, WS7 4US
G0 PCQ I. Yeo, Chyventon, Smithams Hill, Bristol, BS40 6BZ
G0 PCT D. Hambly, Culver Park, Rattery, South Brent, TQ10 9LL
GI0 PCU A. Stewart, 1 Lislaynan, Ballycarry, Carrickfergus, BT38 9GZ
G0 PCW J. Budden, Fieldgate, Durnstown, Lymington, SO41 6AL
G0 PCY J. Radford, 93 Hook Road, Surbiton, KT6 5AF
G0 PCZ B. Lody, 41 Galsworthy Road, Chertsey, KT16 8EP
G0 PDA W. Cole, Y Marian, Bow Street, SY24 5BE
GW0 PDB G. Griffiths, Dolcoed, Llandysul, SA44 4RJ
G0 PDE D. Livingstone, 68 Brimley, Leonard Stanley, Stonehouse, GL10 3NA
G0 PDH D. Hyde, 9 Empress Avenue, Marple, Stockport, SK6 7BG
GJ0 PDJ M. Turner, 4 Le Clos Sara, St Lawrence, Jersey, JE3 1GT
G0 PDK W. Marsden, 8 Albert Road, Eston, Middlesbrough, TS6 9QW
G0 PDM M. Glover, 22 Fern Street, Sutton-in-Ashfield, NG17 2DW
GD0 PDN D. Beedan, Ashmawr, Mount Rule Road, Douglas, Isle Of Man, IM4 4QZ
G0 PDP A. Farmer, 76 Wood Lane, Kingsnorth, Ashford, TN23 3AG
GM0 PDQ M. Kusin, East Overhill Farm, Stewarton, Kilmarnock, KA3 5JT
G0 PDV R. Netherway, 2 Avon Court, Lawn Road, Bristol, BS16 5BL
G0 PDZ I. Lowe, 54 College Road, Margate, CT9 2SW
G0 PEB R. Williams, 10 Barton Close, Whippingham, East Cowes, PO32 6LS
G0 PEC I. Tutt, 1 Castle Road, Hadleigh, Ipswich, IP7 6JH
G0 PEF I. Williams, 6 Newport Road, Godshill, Ventnor, PO38 3HR
G0 PEG J. Jenner, 1 Bellflower Rise, Ashford, TN24 0GS
G0 PEH A. Lifton, 70 Scrapsgate Road, Minster On Sea, Sheerness, ME12 2DJ
GM0 PEI A. Pollock, 113 Gartmorn Road, Sauchie, Alloa, FK10 3PD
G0 PEJ G. Ford, 5 Rosslyn Close, Hockley, SS5 5BP
G0 PEK K. Richardson, 35 Vidgeon Avenue, Hoo, Rochester, ME3 9DE
G0 PEP Waters & Stanton Electronics, c/o P. Waters, 9 Tudor Way, Hawkwell, Hockley, SS5 4EY
G0 PEQ P. Cook, 88 Sprowston Road, Norwich, NR3 4QW
G0 PER K. Kreuchen, 211 Creek Road, March, PE15 8RY
G0 PEV R. Dawson, 6 Oxton Lane, Tadcaster, LS24 8AG
G0 PEW J. Lyne, 157 Westwick Road, Sheffield, S8 7BW
GM0 PEX P. Bendermacher, 1 Cedar Drive, Milton Of Campsie, Glasgow, G66 8AY
G0 PEY R. Pearson, 26 Ammonite Drive, Needham Market, Ipswich, IP6 8FJ
G0 PFA M. Sole, 44 Chestnut Avenue, Ewell, Epsom, KT19 0SZ
G0 PFD A. Davison, 45 Cheyne Garth, Hornsea, HU18 1BF
G0 PFE R. Lees, Lyndric, 23 New Queen St., Scarborough, YO12 7HL
GM0 PFH G. Spurr, 6 The Granary, Glebe Street, Dumfries, DG1 2LU
G0 PFI E. Ball, 57 Cherry Tree Road, Sheffield, S11 9AA
G0 PFJ F. Poynter, 7 Howards Way, Cawston, Norwich, NR10 4AZ
GI0 PFL S. Mcclean, 22 Whiteways, Newtownards, BT23 4UW
G0 PFM E. Ashworth, 88 Hawthorn Avenue, Colchester, CO4 3JP
G0 PFN D. Catchpole, 43 Welsford Road, Norwich, NR4 6QB
G0 PFO D. Butler, 1901 Dean Avenue, Michigan, USA, 48842
G0 PFQ S. Streluk, 11 Ninefoot Lane, Belgrave, Tamworth, B77 2NA
G0 PFT M. Farrell, Hobberley House, Hobberley Lane, Leeds, LS17 8LX
G0 PFU K. Wignall, 4 Weavers Fold, Bretherton, Leyland, PR26 9AP
G0 PFY R. Marshall, 66 Oakwood Hill, Loughton, IG10 3EP
GW0 PFZ A. Powell, Rich Lyn, Carmel Road, Holywell, CH8 7DF
G0 PGA C. Smith, 5 Northfield Drive, Mansfield, NG18 3DD
G0 PGB C. Hosking, 32 Queen St., Penzance, TR18 4BH
GI0 PGC J. Forsythe, 1 Coulson Avenue, Lisburn, BT28 1YJ
GM0 PGD A. Paterson, 21 Kirkwood Avenue, Redding, Falkirk, FK2 9UF
G0 PGI Dr D. Beckly, Knighton, Buckland Monachorum, Yelverton, PL20 7LH
G0 PGJ G. Smith, 16 Weeth Lane, Camborne, TR14 7JN
G0 PGK D. Lawrence, 7 Richmond Road, Appledore, Bideford, EX39 1PE
G0 PGL D. Blight, 73 Stoke Road, Taunton, TA1 3EL
G0 PGQ M. Molloy, 20 The Lawn, Whittlesford, Cambridge, CB22 4NG
G0 PGS P. Slater, 1 Greyhound Road, Glemsford, Sudbury, CO10 7SJ
G0 PGT J. Newman, Sometimes (The Workshop), South Pew, Dorchester, DT2 9HZ
G0 PGW G. Dunn, 6 Rosewood Avenue, Haslingden, Rossendale, BB4 5NG
G0 PGX S. Thomas, Creekside, Greenbank Road, Truro, TR3 6PQ
G0 PGY J. Underwood, 56 Bassenhally Road, Whittlesey, Peterborough, PE7 1RR
G0 PGZ B. Hill, 48 Lackford Avenue, Totton, Southampton, SO40 9BT
G0 PHC G. Woodhouse, 12 Matthew Street, Alvaston, Derby, DE24 0ER
G0 PHD C. Whitehead, 27-28 St. Nicholas St., Scarborough, YO11 2HF
G0 PHE P. Long, 40D Curborough Road, Lichfield, WS13 7NQ
GM0 PHG D. Mclaughlin, 96 Craighlaw Avenue, Eaglesham, Glasgow, G76 0HA
G0 PHI P. Hirst, 4 Brook House, Brook House Lane, Huddersfield, HD8 8LX
G0 PHO C. Wilson, 448 Hythe Road Willesborough, Ashford, TN24 0JH
G0 PHP K. Green, 39 Fleetgate, Barton-upon-Humber, DN18 5QA
G0 PHR M. Andrews, 9 Irving Road, Solihull, B92 9DQ
GM0 PHW M. Whitehead, 185 Allanton Road, Allanton, Shotts, ML7 5AX
G0 PHY O. Williams, 30 Franklin Road, Biggleswade, SG18 8DX
G0 PIA J. Brown, 14 St. Georges Avenue, Hornchurch, RM11 3PD
G0 PIB K. Simmonds, 9 Packman Drive, Ruddington, Nottingham, NG11 6GF
G0 PID B. Thomas, 112 Pen Park Road, Bristol, BS10 6BP
G0 PIK A. Clements, 37 Sun St., Isleham, Ely, CB7 5RU
G0 PIL T. O'Brien, 45 Rossall Promenade, Thornton-Cleveleys, FY5 1LP
G0 PIN A. Pinnock, 1 Rutland Gardens, Ealing, London, W13 0ED
G0 PIS J. Bird, 12 Beresford Gardens, Romford, RM6 6RX
G0 PIT A. Freeman, 3 Greenleas Road, Wallasey, CH45 8LR
G0 PIU G. Papadopoulos, 1 Darenth Road, London, N16 6EP
GM0 PIV M. Black, Drumtochty, 37 Clepington Road, Dundee, DD4 7EL
GW0 PIX Pheonix ARC Cmyru, c/o C. Hughes, Cym Lane, Rogerstone, Newport, NP10 9EN
G0 PIY C. Pollock, Flat 5, 93 Priory Grove, London, SW8 2PD
GW0 PJA P. Baston, 27 Higher Common Road, Buckley, CH7 3NG
G0 PJC A. Jones, 35 Orchard Way, Letchworth Garden City, SG6 4RZ
GM0 PJD Dr P. Dobie, Cairnview Cottage, Quothquan, Biggar, ML12 6NB
G0 PJG J. Geraghty, 61 Bridle Lane, Streetly, Sutton Coldfield, B74 3QE
GI0 PJH W. Stewart, 23 Sandy Grove, Magherafelt, BT45 6PU
G0 PJI P. Wood, 2 Central Crescent, Hethersett Nr93Ep, Norwich, NR9 3EP
G0 PJM M. Hughes, The Cottage, Astley Burf, Stourport-on-Severn, DY13 0RX
G0 PJO M. Waller, Olive Cottage, 6 Church Road, Ipswich, IP9 1HS
G0 PJR P. Ruffle, 55 Nailers Drive, Burntwood, WS7 0ES
G0 PJS P. Spicer, 86 Main St., Wilsford, Grantham, NG32 3NR
G0 PJU J. Brown, 20 Stamford Avenue, Seaton Delaval, Whitley Bay, NE25 0PA
G0 PJW C. Wormald, 22 Tulworth Road, Poynton, Stockport, SK12 1BL
G0 PJY P. Graham, 11 Raby Court, Ellesmere Port, CH65 9DZ
G0 PJZ R. Dorling, Aletheia, St. Marys Road, Colchester, CO7 8NN
GM0 PKF P. French, 7 Knockothie Hill, Ellon, AB41 8BA
G0 PKJ D. Stallon, 8 Hidcote Close, Eastcombe, Stroud, GL6 7EF
G0 PKN T. Finneran, 23 Longdales Road, Lincoln, LN2 2JR
GM0 PKP W. Carroll, 20 Pinewood Road, Mayfield, Dalkeith, EH22 5HX
GM0 PKQ F. Grant, Silverknowes, Arbeadie Road, Banchory, AB31 5XA
G0 PKR K. Ritson, 14 Dunsdale Road, Holywell, Whitley Bay, NE25 0NG
G0 PKT C L P K, c/o A. Horsman, 15 Hanwell Close, Clacton-on-Sea, CO16 7HF
G0 PKV E. Bennett, 17 Bixhead Walk, Broadwell, Coleford, GL16 7EB
G0 PKW O. Fairgrieve, 8 Aird, Point, Isle of Lewis, H52 0EU
GM0 PKX E. Michael, 8 Castlepark Grove, Kintore, Inverurie, AB51 0SN
G0 PLA T. Reddish, 72 Edgmond Close, Redditch, B98 0JQ
G0 PLB K. Murray, Viamory, Wistanswick, Market Drayton, TF9 2BD
G0 PLC P. Gosnell, 230 Rowley Gardens, London, N4 1HN
G0 PLD T. Pogson, 64 New North Road, Slaithwaite, Huddersfield, HD7 5BW
GM0 PLH W. Chan, 5 Lansdowne Drive, Cumbernauld, Glasgow, G68 0JB
G0 PLK T. Kennedy, Woodgreen, Williamstown, Ireland,
GD0 PLQ J. Mitchell, 5 Westminster Drive, Douglas, Isle Of Man, IM1 4EG
GD0 PLR W. Smith, 1 High View Road, Douglas, Isle Of Man, IM2 5BQ
G0 PLS I. Wallis, 20 Gerard Avenue, Bishop's Stortford, CM23 4DU
G0 PLX J. Parker, 24 Egmont St., Salford, M6 7LA
G0 PLZ D. Lindsay, 33 Varna Road, Bordon, GU35 0DG
G0 PMB G. Banks, 10 Gregory Road, Glass Houghton, Castleford, WF10 4PH
G0 PMF G. Dellbridge, 19 Cleeve Close, Astley Cross, Stourport-on-Severn, DY13 0NY
G0 PMG R. Dellbridge, 4 Woodford Way, Wombourne, WV5 8HB
G0 PMI R. Spencer, 4 Barstow Avenue, York, YO10 3HE
GI0 PML M. Mcpeake, 28D Greenview Avenue, Antrim, BT41 4EQ
G0 PMM D. Carrott, 5 Raeburn House, 42 Brighton Road, Sutton, SM2 5JH
GM0 PMO A. Fawcett, Glennairn, Stromness, KW16 3EX
G0 PMP M. Overend, 58 Church Road, Liversedge, WF15 7LP
G0 PMS R. Sweeney, 33 Traherne Close, Lugwardine, Hereford, HR1 4AF
G0 PMU R. Nolson, 50 Shelf Hall Lane, Shelf, Halifax, HX3 7NA
GM0 PMW A. Renwick, 21 Scroggie Meadow, Annan, DG12 6dy
G0 PMX J. Garnham, 20 Deans Walk, Durham, DH1 1HA
G0 PMZ I. Brydon, 12 Pearce Road, Maidenhead, SL6 7LF
G0 PNA M. Cranwell, 21 Cockhaven Road, Bishopsteignton, Teignmouth, TQ14 9RF
G0 PNB R. Hope, 7 Irwell Green, Taunton, TA1 2TA
GW0 PNC H. Hartwell, Heulwen, Llanfair Clydogau, Lampeter, SA48 8LH
GW0 PNE D. Hutson, Sandalwood, 60 Glyndwr Road, Colwyn Bay, LL29 8TA
G0 PNF W. Warren, 38 Stoneyhurst Drive, Curry Rivel, Langport, TA10 0JH
G0 PNG G. Buckley, 46 King Street, Portland, DT5 1NH
GW0 PNI Dr D. Pitkin, Highbury Aberporth, Cardigan, SA43 2BZ
G0 PNM P. Sobye, 2 Willowbank, Fraddon, St Columb, TR9 6TW
G0 PNO P. Studdart, 656 Rayleigh Road, Hutton, Brentwood, CM13 1SJ
GI0 PNP R. Pritchard, 79 Harbour Road, Ballyhalbert, Newtownards, BT22 1BW
G0 PNQ A. Varley, 37 Forest Road, Cambridge, CB1 9JA
G0 PNR G. Mcilroy, 1 Belmont Walk, Worcester, WR3 7HY
G0 PNS Radio Club of Pabay, c/o J. Harris, 37 Long Orchard Way, Martock, TA12 6FA
G0 PNT S. Poulter, 119 Aragon Road, Morden, SM4 4QG
GW0 POA M. Hale, 5 Marchwood Close, Rumney, Cardiff, CF3 3LZ
GI0 POB G. Eldridge, 59 Beechwood Gardens, Bangor, BT20 3JD
G0 POC Dr P. Elwood, 55 Madan Road, Westerham, TN16 1DX
GM0 POD W. Mccallum, St Brides Way, Colyton, Ayr, KA6 6QG
GW0 POG C. Gavin, Hafod Wen, Bagillt Road, Bagillt, CH6 6JE
G0 POK D. Quinnear, 5 Heath Drive, Chelmsford, CM2 9HA
G0 POM P. Harris, 44 Boston Road, Heckington, Sleaford, NG34 9JE
G0 POQ D. Kemp, 7 St. Nicholas Avenue, Hull, HU4 7AH
G0 POT M. Sansom, 19 Baily Avenue, Thatcham, RG18 3EG
G0 POU J. Crosby, 9 Hermitage Close, North Mundham, Chichester, PO20 1JZ
G0 POY A. Eskelson, 90 Charlton Crescent, Barking, IG11 0NL
GW0 POZ D. Morgan, Coedybryn, Synod Inn, Dyfed, SA44 6JE
G0 PPH W. Blythe, 4 Beresford Road, Stubbington, Fareham, PO14 2QX
G0 PPI D. Chenery, 25 Aldreth Road, Haddenham, Ely, CB6 3PW
G0 PPJ P. Johnson, 20 Bearmore Road, Warley, Cradley Heath, B64 6DU
G0 PPK W. Gill, 21 Flockton Avenue, Standish Lower Ground, Wigan, WN6 8LH
G0 PPL G. Lattka, 9 The Row, Sutton, Ely, CB6 2PD
G0 PPM K. Powell, 86 Nortonwood, Forest Green, Nailsworth, Stroud, GL6 0TB
G0 PPQ P. Jackson, 8 Buttree Court, South Kirkby, Pontefract, WF9 3NB
G0 PPR G. Whaling, 3 Bell Close, Little Snoring, Fakenham, NR21 0HX

G0 PPS Prudential ARS, c/o D. Dyer, 57 Garrison Lane, Felixstowe, IP11 7RR
G0 PPU H. Bennett, 32 Sculthorpe Road, Blakedown, Kidderminster, DY10 3JR
G0 PPX J. Omara, 18 Tarrant Grove, Quinton, Birmingham, B32 2NW
G0 PPY N. Turner, 31 Shamrock Avenue, Whitstable, CT5 4EL
G0 PQB S. Slater, 118 Danziger Way, Borehamwood, WD6 5DG
G0 PQD K. Skuse, 4 Barton Close, Berrow, Burnham-on-Sea, TA8 2NN
G0 PQF A. Judge, 44 Thorley Lane, Bishop's Stortford, CM23 4AD
G0 PQG A. Harper, 81 High Street, Great Houghton, Barnsley, S72 0AU
G0 PQO Dr K. Martin, 8 Taylors Close, Meppershall, Shefford, SG17 5NH
G0 PQR C. Wardle, P O Box N 3189, Nassau, Bahamas
GM0 PQV J. Maguire, 64 High Street, Loanhead, EH20 9RR
G0 PQW P. Bartholomew, 29 Beatrice Avenue, East Cowes, PO32 6HR
G0 PQX S. Shipley, 102 Jackson Street, Goole, DN14 6DH
G0 PQY A. Langford, 53 Cambridge Avenue, Bottesford, Scunthorpe, DN16 3PH
G0 PRF J. Goodwin, 146 Grimescar Road, Ainley Top, Huddersfield, HD2 2EB
GM0 PRG Perth Repeater Group, c/o D. Morris, Ash Cottage, Perth Road, Perth, PH2 9LW
G0 PRH M. Grassi, Little Ash, Sleight Lane, Wimborne, BH21 3HL
G0 PRI L. Ward, 20 The Green, Newby, Scarborough, YO12 5JA
G0 PRK R. Weller, 15 Richmond Avenue Highams Park, London, E4 9RR
GW0 PRM B. Goodier, 14 Meadowbank, Old Colwyn, LL29 8EX
GM0 PRO P. Greenway, 5 Java Place, Craignure, Isle of Mull, PA65 6BG
GM0 PRQ M. Shield, Castleshield, Fiscavaig, Isle of Skye, IV47 8SN
G0 PRS Poole Radio Scouts (Prs), c/o C. Baverstock, 43 Tatnam Road, Poole, BH15 2DW
G0 PRU Prudential ARS, c/o D. Dyer, 57 Garrison Lane, Felixstowe, IP11 7RR
G0 PRY D. Mcnab, 10 Rainham Gardens, Alvaston, Derby, DE24 0DJ
G0 PSD P. Hayward, 6 Greenock Close, Westlands, Newcastle-under-Lyme, ST5 2LG
G0 PSF P. Yeatman, 73 Roundway, Waterlooville, PO7 7QB
G0 PSG R. Carvell, 26 Greenfield Avenue, Kettering, NN15 7LL
G0 PSH A. Goldstraw, 59 Lansbury Grove, Stoke-on-Trent, ST3 6JY
G0 PSI J. Wood, 18 Kennedy Avenue, Long Eaton, Nottingham, NG10 3GF
G0 PSJ S. Jacques, Torr Garth, 38 Cheyne Walk, Hornsea, HU18 1BX
G0 PSK G. Hawkins, 8 Broughton Road, West Ayton, Scarborough, YO13 9JW
G0 PSL P. Daddy, 52 Seafield Avenue, Hull, HU9 3JQ
G0 PSO P. O'Nion, 11 Capitol Close, Swindon, SN3 4AB
GU0 PSP M. Dowding, L'Ancrage, Les Marais, Guernsey, GY7 9LD
GW0 PSV G. Wardman, 5 High Street, Trelewis, Treharris, CF46 6AB
G0 PSY S. Brodie, Waterloo Cottage, Tanners Green, Norwich, NR9 4QS
G0 PSZ L. Banaszak, 17 Stoney Piece Close Bozeat, Wellingborough, NN29 7NS
G0 PTA R. Attwood, 2 Elizabeth Road, Basingstoke, RG22 6AX
G0 PTD A. Washington, 22 Elm Tree Drive, Bignall End, Stoke-on-Trent, ST7 8NG
G0 PTE P. Davidson, 28 Daneswell Drive, Wirral, CH46 1QH
G0 PTG J. Mattison, 21 Maynard House, Dunmow Road, Dunmow, CM6 2DL
G0 PTI H. Aigeldinger, 14 Peregrine Avenue, Morley, Leeds, LS27 8TD
G0 PTK D. Dunford, 25 Northfields Lane, Brixham, TQ5 8RS
G0 PTL D. Caley, 5 Crosswood Close, Bransholme, Hull, HU7 5BU
G0 PTM A. Baird, 65 Waterpump Court, Thorplands, Northampton, NN3 8UR
GI0 PTQ P. Keenan, Drumbadreeuagh, Belleek, Enniskillen, BT93 3FT
G0 PTR A. Ryland, 19 Redwood Court, Northway, Tewkesbury, GL20 8SN
G0 PTT K. Caunce, 7 Trevanions Way, Totland Bay, PO39 0JL
G0 PTU J. Davies, 8668 Ne Orchard Loop Road, Leland, USA, 28451
GM0 PTY A. Higgins, 26 Waterton Road, Bucksburn, Aberdeen, AB21 9HS
G0 PUD D. Shaw, 27 St. Davids Avenue, Romiley, Stockport, SK6 3JT
G0 PUK A. Johnson, 3 Plantation Avenue, Swalwell, Newcastle upon Tyne, NE16 3JN
GW0 PUM D. Jenkins, Gwalia House, 143A Priory Street, Carmarthen, SA31 1LR
GM0 PUN H. Heritage, 6 Newton Place, Rosyth, Dunfermline, KY11 2LX
GW0 PUP G. Brown, 17 High Street, Senghenydd, Caerphilly, CF83 4GG
G0 PUQ H. O'Hare, 39 Crichton Road, Carshalton, SM5 3LS
G0 PUW G. Taylor, 33 Heol Aberwennol, Borth, SY24 5NP
G0 PUY C. Duckworth, 121 Mill Gate, Newark, NG24 4UA
G0 PVB B. Sketcher, 147 Moorside Road, Bradford, BD2 3HD
G0 PVE K. Greaves, 10 Chatsworth Drive, Syston, Leicester, LE7 1HX
G0 PVF P. Benson, 21 Farleigh Road, New Haw, Addlestone, KT15 3HS
GI0 PVG T. Lyons, 3 Clanbrassil Gardens, Portadown, Craigavon, BT63 5YD
G0 PVJ E. Hewitt, 8 Embleton Road, Headley Down, Bordon, GU35 8AJ
G0 PVN C. Fleet, 14 Fairwood Road, Penleigh, Westbury, BA13 4EA
G0 PVO L. Hewitt, Sunny Nook, Grains Road, Shaw, Oldham, OL2 8JF
G0 PVP C. Duffy, 590 Chorley Old Road, Bolton, BL1 6AA
G0 PVQ P. Fuller, 19 Greenwood Court, Webb Close, Crawley, RH11 9JH
G0 PVR J. Davies, 13 The Close, Stalybridge, SK15 1HU
G0 PVT D. Henderson, 7 Love Avenue, Dudley, Cramlington, NE23 7BH
G0 PVU Rev. J. Roberts, 31 Seaton Way, Marshside, Southport, PR9 9GJ
G0 PVW R. Hamer, 2 Back Lane Pontesford, Pontesbury, Shrewsbury, SY5 0UD
G0 PVY A. Heward, 22 Ross Avenue, Leasowe, Wirral, CH46 2SB
G0 PWA D. Williams, 31 Piper Hill Avenue, Manchester, M22 4DZ
G0 PWC B. Dawe, 6 Ullswater Avenue, Stourport-on-Severn, DY13 8QP
G0 PWH P. Hughes, 21A Erua Road, Waiheke Island, New Zealand, 1081
G0 PWK S. Alder, 5 Inglesgarth Court, Spennymoor, DL16 7UG
G0 PWL S. Wright, 33 Virginia Avenue, Lydiate, Liverpool, L31 2NN
G0 PWO A. Boyes, 7 Thornwood Covert, Foxwood, York, YO24 3LF
G0 PWQ W. Tonks, 295 Quinton Road West, Quinton, Birmingham, B32 1PG
GM0 PWS N. Doherty, Cairdeas, Carrbridge, PH23 3AA
G0 PWU G. Brown, 21 Armada Drive, Teignmouth, TQ14 9NF
G0 PWV H. Chorley, 19 Cleeve Road, Priorswood, Taunton, TA2 8DX
G0 PWW Rev. T. Edwards, Kenneggy Lodge, Polperro Road, Looe, PL13 2JS
G0 PWX G. Richards, 87 Woodlands Road, Ditton, Aylesford, ME20 6EF
G0 PXA G. Petri, Mount Holly, Castledon Road, Billericay, CM11 1LH
G0 PXB C. Marsh, White Rose Robin Hoods Walk, Boston, PE21 9LW
G0 PXD A. Harrison, 25 Lansbury Avenue, New Rossington, Doncaster, DN11 0AA
G0 PXE M. Cook, 16 Ascot Drive, Doncaster, DN5 8QA
G0 PXF K. Linsley, 132 Rein Road, Tingley, Wakefield, WF3 1JB
G0 PXG M. Hardman, 47 Oatlands Road, Manchester, M22 1AH
G0 PXH B. Wilkinson, 22 Portree Crescent, Blackburn, BB1 2HB
G0 PXI P. Rigby, 41 St. Huberts Road, Great Harwood, Blackburn, BB6 7AS
G0 PXK N. Pratt, 23 Hall Lane, Whitwick, Coalville, LE67 5FD
G0 PXL D. Martland, 6 Omega Way, Trentham, Stoke-on-Trent, ST4 8TF
G0 PXM G. Kirby, Tralee, Main Road Brighstone, Newport, PO30 4DJ
G0 PXO J. Morgan, 5 Sealy Close, Wirral, CH63 9LP
G0 PXP T. Cox, 60 Seven Oaks Crescent, Bramcote, Nottingham, NG9 3FP
G0 PXQ C. Bell, 17 Jubilee Square, South Hetton, Durham, DH6 2TR
GI0 PXS J. Madden, The Cottage, 53 Clarendon Street, Londonderry, BT48 7ER
G0 PXT E. Denman, 15 Clare Way, Bexleyheath, DA7 5JU
GM0 PXV P. Barclay, 15 Craigmount Avenue North, Edinburgh, EH12 8DH
G0 PXX E. Mason, 36 Gattison Lane, New Rossington, Doncaster, DN11 0NQ
G0 PXY D. Smith, 27 Hanbury Close, Cheshunt, Waltham Cross, EN8 9BZ
G0 PXZ G. Walker, 54 Burnage Lane, Burnage, Manchester, M19 2NL
G0 PYE A. Arnold, 2 Duck Lane, Haddenham, Ely, CB6 3UE
G0 PYF A. De Buriatte, Tanglewood, East End, North Leigh, OX8 6PZ
G0 PYI G. Bodaly, 41 Robert Street, Northampton, NN1 3BL
G0 PYJ J. Swatton, 30 Squires Close, Crawley Down, Crawley, RH10 4JQ
G0 PYL R. Harriss, 38 Portland Drive Whittleford, Nuneaton, CV10 9HY
GM0 PYM Paisley ARC, c/o S. Mckinnon, 8 Rowanlea Avenue, Paisley, PA2 0RP
G0 PYS D. Rose, 99 Blackfriars, Rushden, NN10 9PF
GW0 PYU H. Clarke, 3 Tanyrallt Avenue, Bridgend, CF31 1PQ
G0 PYV M. Hainesborough, 39 Princes Close, North Weald, Epping, CM16 6EW
G0 PYW A. Haworth, Clayfoot, Collins Green, Knightwick, Worcester, WR6 5PS
G0 PZB W. Hattrick, 38 Nithsdale Road, Weston-Super-Mare, BS23 4JR
G0 PZC D. Flitterman, Flat 7, 1 Rutland Gate, London, SW7 1BL
G0 PZD G. Holmes, 6 Darleydale Drive, Eastham, Wirral, CH62 8EX
G0 PZF J. O'Connell, Apartment 5, Roxboro House, Bailick Road, Co Cork, Ireland,
G0 PZJ Duxford Radio Limited, c/o D. Featherby, 14 Station Road, Sutton, Ely, CB6 2RL
G0 PZM N. Byron, 2 St. Aidans View, Boosbeck, Saltburn-by-the-Sea, TS12 3LS
G0 PZO C. Jordan, 31 Rathmore Road, Birkenhead, CH43 2HE
G0 PZP W. Rabbitt, 21 Barnfield Road, Woolston, Warrington, WA1 4NW
G0 PZR Penzance Radio Club, c/o O. Prosser, 2 Caroline Close, Ventonleague, Hayle, TR27 4EX
GW0 PZS T. Edwards, 6 Cottage Home, Newborough, Llanfairpwllgwyngyll, LL61 6SY
GW0 PZT E. Allely, Dwyfor, Rhiw, Pwllheli, LL53 8AE
GW0 PZU A. Ward, 158 Mold Road, Mynydd Isa, Mold, CH7 6TF
G0 PZW Dr J. Birch, 32 Poplar Grove, Scotter, Gainsborough, DN21 3TZ
G0 PZX A. Dennis, 44 Larksfield Road, Faversham, ME13 7ES
GW0 PZZ M. Owen, 90 Shakespeare Avenue, Penarth, CF64 2RX
GM0 QKá K. Traill, 31 Sherwood Place, Bonnyrigg, EH19 3JY
GW0 RAD J. Lewis, 189 Heol Y Gors Cwmgors, Ammanford, SA18 1RF
G0 RAE R. Walker, 12 Hill Drive, Ackworth, Pontefract, WF7 7LQ
G0 RAF Raf Waddington ARC, c/o R. Pickles, Bramcote Lorne, Rectory Lane Gamston, Retford, DN22 0QQ
GU0 RAG K. De La Haye, Flat 4, Forest Lodge Flats, Forest Lane, Guernsey, GY1 1WJ
G0 RAL P. Vallis, Gryphon, Dirtham Lane, Leatherhead, KT24 5SD
G0 RAM M. King, 65 Chepstow Close, Stevenage, SG1 5TT
G0 RAN M. Jamil, 29 Harrow Close, Bury, BL9 9UD
GM0 RAO A. Williamson, Cairn Cottage, Durris, Banchory, AB31 6DT
G0 RAR A. Walton, Flat 25 Albert Weedall Centre, 23 Gravelly Hill North, Birmingham, B23 6BT
G0 RAS V. Maddex, 3 The Vines, Shabbington, Aylesbury, HP18 9HH
G0 RAT W. Barnes, 17 Saxon Way, Bradley Stoke, Bristol, BS32 9AR
G0 RAU D. Woodnutt, 17 Hill Farm Road, Chalfont St. Peter, Gerrards Cross, SL9 0DD
G0 RAV R. Ravenscroft, 4 The Paddock, Lidlington, Bedford, MK43 0RW
G0 RAX R. Preston, 45 Long Meadow, Skipton, BD23 1BP
G0 RBA E. Bannister, 59 Home Farm Park Lee Green Lane, Church Minshull, Nantwich, CW5 6ED
G0 RBB M. Batchelor, 16 Clementi Avenue Holmer Green, High Wycombe, HP15 6TN
GI0 RBC J. Thompson, 3 Strandburn Park, Sydenham, Belfast, BT4 1ND
G0 RBD D. Kiely, 45 Redland, Chippenham, SN14 0JB
GW0 RBH Dr R. Hughes, 17 Pentrosfa Road, Llandrindod Wells, LD1 5NL
G0 RBI S. Ward, Oaklands, Burtonwood Road, Warrington, WA5 3AN
G0 RBJ P. Evans, Flat 7, 150 Booker Avenue, Liverpool, L18 9TB
G0 RBM C. Boland, 13 Rushfield Crescent, Brookvale, Runcorn, WA7 6BN
GI0 RBO J. Kernohan, 17 Tullygrawley Road, Teeshan, Ballymena, BT43 5NP
G0 RBQ R. Gibbs, Lime Tree House, Top Road, Slindon, Arundel, BN18 0RP
GI0 RBS Rev. R. Rainey, 4 Crossnadonnell Road, Limavady, BT49 0BD
G0 RBV P. Brunton, Flat 26, Absalom Court, Wright Close, Gillingham, ME8 6XP
G0 RBW T. Jones, 11 Coppice Close, Willaston, Nantwich, CW5 6NL
G0 RCF E. Carrington, 4 Lancaster Drive, East Grinstead, RH19 3XF
G0 RCH T. Cullup, 13 London Street, Whittlesey, Peterborough, PE7 1BP
G0 RCI A. Gibson, 1 Oakleigh Road, Grantham, NG31 7NN
G0 RCJ J. Topham, 23 St. Nicholas View, West Boldon, East Boldon, NE36 0RF
G0 RCL O. Baldwin, 23 Cherry Tree Walk, Tadcaster, LS24 9HS
G0 RCN N. Allen, 78 Bargates, Christchurch, BH23 1QL
G0 RCP P. Mellors, 64 Pinewood Way, North Colerne, Chippenham, SN14 8QU
G0 RCS Royal Signals ARS, c/o I. Mcgowan, Meld House, Hawthorn Road, Shrewsbury, SY3 7NB
G0 RCU R. Thomas, 164 Kings Head Lane, Bristol, BS13 7BW
G0 RCW W Cheshire Ray, c/o F. Mossop, 4 Brookdale Way, Waverton, Chester, CH3 7NT
G0 RCX C. Garbett, 27 Burghley Drive, West Bromwich, B71 3LX
G0 RCY M. Crimes, 27 Dunmore Road, Little Sutton, Ellesmere Port, CH66 4PD
GM0 RDA G. Adamson, 10 Rossend Terrace, Burntisland, KY3 0DQ
G0 RDB C. Fernie, 2 Hopkins Close, Cambridge, CB4 1FD
GM0 RDC C. Robinson, Greenhouse Farm, Lilliesleaf, Melrose, TD6 9EP
G0 RDD M. Prendergast, 1 Olaman Walk, Peterlee, SR8 2EA
G0 RDF L. Wolstenholme, The Hollies, Avondale Road, Chesterfield, S40 4TF
G0 RDG G. Kowalski, 47 Graveney Place, Springfield, Milton Keynes, MK6 3LU
G0 RDH B. Watson, 7 Branksome Drive, Morecambe, LA4 5UJ
GI0 RDJ I. Mcmullan, 35 Howard Place, Lisburn, BT28 1EX
G0 RDK C. Wiseman, 42 Merlin Way, Kidsgrove, Stoke-on-Trent, ST7 4YL
GI0 RDM K. Mcguckin, 20 Lisnahull Park, Dungannon, BT70 1UH
G0 RDN G. Johnston, 11 Granville Street, Deal, CT14 7EZ
G0 RDO J. Snell, 5 Waverley Road, Newton Abbot, TQ12 2ND
G0 RDP D. Peat, 24 Brookland Ave., Mansfield, NG18 5NB
G0 RDR B. Lowe, 18 Lowther Drive, Swillington, Leeds, LS26 8QG
G0 RDS A. Williams, 30 Swan Close, Talke, Stoke-on-Trent, ST7 1TA
G0 RDT D. Treen, 13 Peveril Road, Duston, Northampton, NN5 6JW
G0 RDU S. Emms, 33 Whitworth Avenue, Stoke Aldermoor, Coventry, CV3 1EQ

G0 RDV L. Davies, 3 Rydalside, Kettering, NN15 7DR
G0 RDX P. Walker, Moze Cross Cottage, Beaumont Road, Harwich, CO12 5BQ
G0 RDY G. Steel, Long Close, 82 Whatton Road, Derby, DE74 2DT
GM0 RDZ S. Smith, 12 Home Avenue, Duns, TD11 3HQ
G0 REA R. James, Woodpeckers, Freshwater Lane, Truro, TR2 5AR
G0 REB C. Salmon, 14 Surrey Drive, Congleton, CW12 1NU
GM0 RED East Dunbartonshire Raynet Group, c/o J. Pert, 56 Lochiel Drive, Milton Of Campsie, Glasgow, G66 8ET
G0 REE D. Jones, 120 Heathfield Road, Keston, BR2 6BF
G0 REF Epping Forest Raynet Group, c/o J. Andrews, 85 Little Cattins, Harlow, CM19 5RN
G0 REL D. Gaskell, 18 Woodcroft, Kennington, Oxford, OX1 5NH
G0 REN C. Wienrich, 94 Sandling Lane, Penenden Heath, Maidstone, ME14 2EA
G0 REO P. Hill, 16 Robins Way, Nuneaton, CV10 8PA
G0 REP A. Blackburn, 2 Blackthorn Road, Stratford-upon-Avon, CV37 6TD
G0 REQ D. Hibberd, 25 Manor Road, Rugby, CV21 2SZ
G0 REU T. Lam, 53 Beaufort Road, Upper Cambourne, Cambridge, CB23 6FP
G0 REV A. Bowmaker, Post Cottage, Ardley Road, Bicester, OX25 6LP
GM0 REZ A. Dailey, 82 Don Drive, Livingston, EH54 5LP
G0 RFA S. Garczynski, 19 Thornhill Croft, Leeds, LS12 4JX
G0 RFE A. Moore, 139 Argyle Street, Heywood, OL10 3RS
G0 RFF C. Bourne, Essams, 11 The Grove, Hailsham, BN27 3HU
G0 RFG E. Hyde, 63 Newlyn Drive, Sale, M33 3LH
G0 RFI S. Brackley, 12 Farnborough Grove, Liverpool, L26 6LW
G0 RFL T. Pooley, 133 Hardie Road, Dagenham, RM10 7BT
G0 RFM J. Copplestone, 25 Bruche Avenue, Paddington, Warrington, WA1 3HX
G0 RFN J. Taylor, 121 Garesfield Gardens, Burnopfield, Newcastle upon Tyne, NE16 6LQ
G0 RFQ G. Buck, 3 Church View, Trawden, Colne, BB8 8SA
G0 RFS Dr C. Bradley, 71A Bagley Wood Road, Kennington, Oxford, OX1 5LY
G0 RFT R. Lagar, 25 Neville Avenue, Warrington, WA2 9BQ
G0 RFV K. Goodworth, 12 Ewood Drive, Doncaster, DN4 6AU
G0 RFX P. Walford, Suite 184, 2 Old Brompton Road, London, SW7 3DQ
G0 RFY D. Horton, 21 St. James Street, Waterfoot, Rossendale, BB4 7HN
G0 RGC J. Bridge, Little House, Castle House Yard, Langport, TA10 9PR
G0 RGE M. Jenkinson, 25 Porchester Close, Hucknall, Nottingham, NG15 7UB
G0 RGG J. Hubbard, 4 Avondale, Ellesmere Port, CH65 6RW
G0 RGH HARIG, c/o J. Mitchener, Cabins, Wenham Road, Ipswich, IP8 3EY
G0 RGJ R. Provins, 42 Forest View Road, Tuffley, Gloucester, GL4 0BX
G0 RGL D. Edmondson, 64 Raleigh Avenue, Hayes, UB4 0EF
G0 RGM J. Trice, 71 Deerswood Road, Crawley, RH11 7JP
G0 RGN B. Woodhead, 16 Dow St., Hyde, SK14 4BS
G0 RGO Rev. J. Drummond, 14 Bulls Head Cottages, Turton, Bolton, BL7 0HS
G0 RGP A. Gibbs, 17 Manor Bend, Galmpton, Brixham, TQ5 0PB
G0 RGU J. O'Gorman, 141 Chesterfield Road, Huthwaite, Sutton-in-Ashfield, NG17 2QF
G0 RGW R. Slatter, Ashwell House, Stratford Road, Oversley Green, Alcester, B49 6PG
G0 RGX J. Sandys, 28 The Maultway, Camberley, GU15 1PS
G0 RHB L. Mulford, 55 Mill Farm Crescent, Hounslow, TW4 5PF
GW0 RHC K. Dyer, 34 Lundy Drive, West Cross, Swansea, SA3 5QL
GW0 RHE S. Williams, 5 Llys Yr Orsaf, Llanelli, SA15 2LB
G0 RHF P. Ellwood, Coire Cas, Marsh Lane, Poulton-le-Fylde, FY6 9AW
G0 RHG D. Stewart, 14 The Dell, East Grinstead, RH19 3XP
G0 RHI B. Dooks, 7 Manor Drive, Kirby Hill, Boroughbridge, York, YO51 9DY
G0 RHJ R. Judson, 27 Newcombe Drive, Arnold, NG5 6RX
G0 RHK P. Ford, 19 Swan Bank, Hay-On-Wye, Hereford, HR3 5DW
G0 RHO J. Belling, 77 Chantry Road, Marden, Tonbridge, TN12 9JD
GM0 RHP D. Crooke, 2 Main Street, Carnock, Dunfermline, KY12 9JQ
G0 RHV J. Parish, 83 Harold Road, Stubbington, Fareham, PO14 2QS
G0 RIB A. Shaw, 38 Longmead Gardens, Havant, PO9 1RR
G0 RIC R. Cannell, 284 Archway Road, Highgate, London, N6 5AU
G0 RIE D. Reilly, 15 Shutewater Close, Bishops Hull, Taunton, TA1 5EH
G0 RIF D. Barnes, 11 Back Lane, Whittington, Lichfield, WS14 9NH
G0 RII J. Spacey, 43 Woodlands Road, Allestree, Derby, DE22 2HG
G0 RIJ W. Sykes, Summerfield, Second Avenue, Ross-on-Wye, HR9 7HT
G0 RIK N. Stockwell, 12 Weavers Mead, Great Cheverell, Devizes, SN10 5TP
G0 RIP J. Austwick, 22 Shurmer Street, Bolton, BL3 4BW
G0 RIQ D. Wisbey, 22 Rutland Drive, Hornchurch, RM11 3EN
G0 RIR W. Lewis, 88A Clifton Road, Grimsby, DN34 4QN
G0 RIU P. Davis, 21 Newton Way, St. Osyth, Clacton-on-Sea, CO16 8RQ
G0 RIX B. Cook, 7 Rosewood Gardens, New Milton, BH25 5NA
G0 RIY G. Watson, 52 Elmwood Park, Loddiswell, Kingsbridge, TQ7 4SD
G0 RIZ B. Body, 12A Elm Court Gardens, Truro, TR1 1DS
G0 RJA K. Jones, 10 Dale Terrace, Lingdale, Saltburn-by-the-Sea, TS12 3EE
G0 RJC V. Fletcher, 7 Highfield Crescent, Baildon, Shipley, BD17 5NR
G0 RJE R. Enright, 17 Ripston Road, Ashford, TW15 1PQ
GM0 RJG E. Kelly, Durness, Newbridge, Dumfries, DG2 0QX
G0 RJI N. Rapson, 21 Ashley Close, Penwithick, St Austell, PL26 8UB
G0 RJJ E. Foord, 65 Dane Court Gardens, Broadstairs, CT10 2SD
G0 RJL J. Hilton, 177 Wilmot Road, Dartford, DA1 3BP
G0 RJM J. Marchant, 129 Highbury Grove, Clapham, Bedford, MK41 6DU
G0 RJN H. Vicary, The Brambles, Wrotham Road, Gravesend, DA13 0QA
GI0 RJO L. Douglas, 15 Bramhall Crescent, Londonderry, BT47 5HE
G0 RJT Rev. H. Leak, 15 Sutherland Road, Tittensor, Stoke-on-Trent, ST12 9JQ
G0 RJV G. Rogers, Maes Gwersyll, Garthmyl, Montgomery, SY15 6RS
G0 RJX E. Gaskell, 18 Woodcroft Kennington, Oxford, OX1 5NH
G0 RKB D. Roberts, 20 Beech Grove, Trowbridge, BA14 0HG
G0 RKC A. Alecio, Flat 4, 19 The Beacon, Exmouth, EX8 2AF
GW0 RKD S. Gray, Whispers, Front Street, Rosemarket, Milford Haven, SA73 1JT
G0 RKE C. Burgess, 12 Middleway, Grotton, Oldham, OL4 5SH
G0 RKG R. Gaskell, 18 Woodcroft Kennington, Oxford, OX1 5NH
G0 RKN H. Burn, 60 Uplands Croft, Stoke-on-Trent, ST9 0LF
G0 RKP J. Aubin, 46 Kenilworth Drive, Clitheroe, BB7 2QN
G0 RKQ R. Plumtree, 80 Dewsbury Avenue, Scunthorpe, DN15 8BP
G0 RKS G. Goss, Little Ashcroft, Parkgate Road, Dorking, RH5 5DZ
G0 RKT D. Dukesell, Mayfield, Ashbourne Road, Buxton, SK17 9RY
GM0 RKU P. Craft, 2 Luke Place, Broughty Ferry, Dundee, DD5 3BN
G0 RKV V. Webley, 2 Octavian Drive, Bancroft, Milton Keynes, MK13 0PN
G0 RLA P. Harvey, Rowlands Barn, Dunbridge Lane, Awbridge, Romsey, SO51 0GQ
G0 RLB B. Stoneley, 44 Ilthorpe, Hull, HU6 9ER
G0 RLF G. Low, 61 Fenwick Lane, Halton Lodge, Runcorn, WA7 5YU
G0 RLH E. Miles, 31 Winnipeg Road, Bentley, Doncaster, DN5 0ED
G0 RLI J. Thomas, 204 Watchouse Road, Galleywood, Chelmsford, CM2 8NF
G0 RLJ P. Tyson, 44 Windmill Avenue, Kilburn, Belper, DE56 0PQ
G0 RLL T. Dyson, 4 Lyspitt Common, Meppershall, Shefford, SG17 5GZ
G0 RLN K. Taylor, 29 School Road, Pontefract, WF8 2AJ
G0 RLO K. Conlon, 4 Hill Crest Drive, Slack Head, Milnthorpe, LA7 7BB
GW0 RLQ J. Ellwood, 5 Smallwood Road, Baglan, Port Talbot, SA12 8AP
G0 RLS P. Ashcroft, 38A Wood End, Bluntisham, Huntingdon, PE28 3LE
G0 RLT R. Taylor, Flat 4, 3 St. Pauls Square, Southport, PR8 1NQ
G0 RLV E. Jones, 16 Fisher Avenue, Rugby, CV22 5HN
G0 RLY J. Karkoszka, 5 Wood Street, Haworth, Haworth/Keighley, BD22 8BJ
GM0 RLZ C. Brown, 41 Russell Avenue, Kingseat, Dunfermline, KY12 0YX
GW0 RMB S. Ferris, Temorfa, Y Ffor, Pwllheli, LL53 6UB
G0 RMC M. Charlton, 53 Dunstone View, Plymouth, PL9 8TW
G0 RMD P. Calter, 8 Exeter Road, Scunthorpe, DN15 7AT
G0 RMG R. Jones, 6 Wychwood Drive Hunt End, Redditch, B97 5NW
G0 RMJ S. Hogg, 38A High St., Ventnor, Isle of Wight, PO38 1RZ
GM0 RML A. Smart, 6 Alton Bank, Nairn, IV12 5PJ
G0 RMN A. Younger, 4 Esk Hause Close, West Bridgford, Nottingham, NG2 6SG
G0 RMO M. Miller, 8 Pilton Walk, Newcastle upon Tyne, NE5 4PQ
G0 RMP R. Seal, 2 Shaftesbury Road, Bridlington, YO15 3NP
G0 RMR C. Rabey, 23 Thorn Lane, Four Marks, Alton, GU34 5BX
GM0 RMT G. Wilkie, 25 Barn Rd., Stirling, FK8 1EP
G0 RMU R. Clover, Teffont, 42 Warren Road, Addlestone, KT15 3UA
GM0 RMV M. Verity, 19 Vivian Terrace, Edinburgh, EH4 5AW
G0 RMX D. Esdale, The Bell Inn, Central Lydbrook, Lydbrook, GL17 9SB
G0 RNA T. Rawlinson, 330 Blackpool Old Road, Poulton-le-Fylde, FY6 7QY
G0 RNB N. Brooks, 57 Mansel Crescent, Parson Cross, Sheffield, S5 9QR
G0 RNC A. Pritchard, 27 Walkley Crescent Road, Walkley, Sheffield, S6 5BA
G0 RNF I. Hunnisett, 69 Cornwall Road, Ruislip, HA4 6AJ
G0 RNH M. Ahmed, 75 Drove Road, Swindon, SN1 3AE
G0 RNI Luton Rep Grp, c/o D. Thorpe, 70 Willow Way, Ampthill, MK45 2SP
GW0 RNK K. Williams, 8 Trinity Place, Pontarddulais, Swansea, SA4 8RD
G0 RNP D. Eves, 64 Hillingdon Road, Gravesend, DA11 7LG
G0 RNQ B. Willson, 4 Caldew Grove, Sittingbourne, ME10 4SL
G0 RNS J. White, 24 Malines Avenue, Peacehaven, BN10 7PS
G0 RNV B. Sherriff, 27 Magellan Way, Spalding, PE11 2FG
G0 RNX S. Onions, 18 St. Cuthberts Crescent, Albrighton, Wolverhampton, WV7 3HW
G0 RNY A. Attle, 2 Watson Park, Spennymoor, DL16 6NB
G0 ROA H. Seidner, 7411 Morocca Lake Drive, Delray Beach, USA, 33446
G0 ROC Rochdale & Dsictrict ARS, c/o P. Hewitt, 11 Thetford Close, Bury, BL8 1XB
G0 ROD C. Reaney, 81A Bargate Road, Belper, DE56 1NE
G0 RON R. Mcneil, 3 Thorncliffe Gardens, Auckley, Doncaster, DN9 3PE
G0 ROO Dover Construction Club, c/o I. Keyser, Rosemount, Church Whitfield Road, Whitfield, Dover, CT16 3HZ
G0 ROS R. Kent, 40 Waxes Close, Abingdon, OX14 2NG
G0 ROT M. Davis, 7 Walter Close, Chickerell, Weymouth, DT3 4GU
GM0 ROU A. Butcher, 224 Laird Street, Dundee, DD3 9PL
G0 ROW A. Gurnhill, 53 Millbrook Avenue, Denton, Manchester, M34 2DQ
G0 ROX D. Lee, 131 Abbotsbury Road, Weymouth, DT4 0JX
G0 ROY R. Biddle, 21 Kingsway West, Newton, Chester, CH2 2LA
G0 ROZ Dorset Police AR, c/o C. Hardy, 40 Beresford Road, Poole, BH12 2HE
G0 RPA I. Mcavoy, 74 Parkstone Heights, Poole, BH14 0RZ
G0 RPD J. Barton, 183 Windy Arbor Road, Whiston, Prescot, L35 3SF
G0 RPF L. Smith, 28 Chester Road, Stockton Heath, Warrington, WA4 2RX
G0 RPG J. Riley, 1 Chatsworth Avenue, Culcheth, Warrington, WA3 4LD
G0 RPJ D. Wesil, 8 Camber Way, Pevensey Bay, Pevensey, BN24 6RW
G0 RPL N. Alison, 9 South Drive, Burgess Hill, RH15 9PY
G0 RPM N. Williams, 11 Berkeley Gardens, London, N21 2BE
G0 RPO R. Dowd, Belgrano, 1 Watson Avenue, Warrington, WA3 3QX
G0 RPU J. Symonds, La Cumbre, 35 Byward Drive, Scarborough, YO12 4JE
G0 RPV W. Till, 97 Haslar Crescent, Waterlooville, PO7 6DD
G0 RPW D. Wilson, 39 The Wintles, Bishops Castle, SY9 5ES
G0 RPY C. Button, 8 Heywood Road, Diss, IP22 4DJ
GW0 RQC R. Chegwin, 17 Cyncoed Crescent, Cardiff, CF23 6SW
G0 RQF K. Hales, 3 New Barnfields, Hereford, HR4 7AZ
G0 RQG J. Gill, 24 Greenfields Court, Bridgnorth, WV16 4JS
G0 RQH D. Hughes, 31 Sussex Drive, Pagham, Bognor Regis, PO21 4RN
G0 RQI S. Spragg, 4 Valley Road Arleston, Telford, TF1 2JP
G0 RQL D. Roomes, View Field, Milton Damerel, Holsworthy, EX22 7NY
G0 RQN P. Robertson, 1 Yaffle Mews, Great Cambourne, Cambridge, CB23 5HY
G0 RQO D. Hillyer, 32A Belbroughton Road, Blakedown, Kidderminster, DY10 3JG
GW0 RQP G. Ashford, 26 Laura Street Treforest, Pontypridd, CF37 1NW
G0 RQQ K. Ballinger, 3 Cliff Court, Burton Road, Lincoln, LN1 3NN
GW0 RQS L. Pritchard, 86 Bryn Road, Markham, Blackwood, NP12 0QE
G0 RQX D. Townend, 38 Kingston Drive, Shrewsbury, SY2 6SJ
G0 RQZ R. Lawrence, 74 Principal Rise, Dringhouses, York, YO24 1UF
G0 RRI I. Burden, 2 Essex Road Flat 2, Lowestoft, NR32 2HH
G0 RRL R. Leigh, 33 Beddington Road, Orpington, BR5 2TF
G0 RRM P. Brumby, 69 Gilbert Walk, Nether Stowe, Lichfield, WS13 6AU
G0 RRO J. Breingan, 44 Farmstead Road, Corby, NN18 0LG
G0 RRR J. Marsden, 11 Firethorn Drive, Hyde, SK14 3SN
G0 RRV A. Tomson, 5 Fordham Close, Ashwell, Baldock, SG7 5LJ
G0 RRZ R. Carrington, 45 Crompton Road, Pleasley, Mansfield, NG19 7RG
G0 RSA J. King, 39 Nursery Gardens, St Ives, PE27 3NL
GM0 RSE Glenrothes & District ARC, c/o T. Brown, 11 Approach Row, East Wemyss, Kirkcaldy, KY1 4LB
G0 RSG 1st Ringmer Scout Group, c/o T. Mcconnell, 51 Langney Road, Eastbourne, BN21 3QD
GM0 RSI J. Ritchie, 36 James Mitchell Place, Mintlaw, Peterhead, AB42 5ES
G0 RSL K. White, 25 Curson Rise, Kendal, LA9 7PN
G0 RSR Reading Scouts Radio, c/o S. French, 22 Amity Street, Reading, RG13LP
G0 RSS R. Simmonds, 4 Corys Close, Kirby Road, Norwich, NR14 7DP
G0 RSU G. Weston, 2 Whitburn Road, Toton, Nottingham, NG9 6HP

G0 RSV W. Webster, 21 Quince Tree Way, Hook, RG27 9SG
G0 RSW R. Waters, 17 Wilson Road, Southend-on-Sea, SS1 1HG
G0 RSY A. Gibbs, Orchard Court, Woodside Road, Wootton Bridge, Ryde, PO33 4JR
G0 RTA T. Arakawa, 2-974-8-1502, Sayama, Osakasayama, Japan, 589-0005
G0 RTC T. Chisholm, 316 Birchfield Road East, Northampton, NN3 2SY
G0 RTF I. Slaney, 6 Little Shardeloes, High Street, Amersham, HP7 0EF
G0 RTH A. Elcoate, 9 Parsonage Lane, Laindon, Basildon, SS15 5YN
G0 RTI S. Harriss, 6 Redland Road, Leamington Spa, CV31 2PB
G0 RTM P. Mcknight, 39 Dunmail Drive, Kendal, LA9 7JG
G0 RTN G. Lynch, 52 Queens Road, Devizes, SN10 5HW
GW0 RTP C. Llewellyn, 16 Garth Street, Kenfig Hill, Bridgend, CF33 6EU
G0 RTQ D. Lawrence, 11 Pembroke Court St. Johns Road, Newbold, Chesterfield, S41 8NX
GW0 RTR R. Rees, 9 Langland Road, Mumbles, Swansea, SA3 4ND
G0 RTU P. Kirkup, 337 Wheatley Lane Road, Fence, Burnley, BB12 9QA
GM0 RTY D. Inns, 57 Craiglomond Gardens, Balloch., Alexandria., G83 8RP
G0 RTZ G. Hurst, 35A Trewsbury Road, London, SE26 5DP
GI0 RUC R. Kerr, 194 Shore Road, Greenisland, Carrickfergus, BT38 8TX
GW0 RUD P. Marriott, 16 Heol Morlais, Llannon, Llanelli, SA14 6BD
G0 RUF N. Taylor, 30 Leonard St., Hull, HU3 1SA
G0 RUH Dr M. Roberts, 82 Glover Road, Scunthorpe, DN17 1AS
G0 RUR P. Simpson, Amber Lodge Nursing Home, 684-686 Osmaston Road, Derby, DE24 8GT
G0 RUS R. Lenthall, 182 Chelmsford Avenue, Grimsby, DN34 5DB
G0 RUT R. Russell, 4 Hinton Road, Newport, PO30 5QZ
G0 RUV M. Gent, 111 Portland Street Clowne, Chesterfield, S43 4SA
GM0 RUW J. Coughtrie, 61 Bells Burn Avenue, Linlithgow, EH49 7LD
G0 RUX W. Taylor, 21 Summerdale Road, Cudworth, Barnsley, S72 8XG
G0 RUY A. Pritchard, 41 Borough Close, Kings Stanley, Kings Stanley, GL10 3LJ
G0 RUZ C. Farlow, 4 Nether Road, Silkstone, Barnsley, S75 4NN
G0 RVE A. Pierce, 34 Church Close, Shawbury, Shrewsbury, SY4 4JX
G0 RVH K. Dailey, 55 Chesterton Avenue, Harpenden, AL5 5SU
G0 RVI J. Davis, 4 Stockbridge Close, Canford Heath, Poole, BH17 8SU
G0 RVK M. Fogg, 15 Elm Grove, Bisley, Woking, GU24 9DG
G0 RVM A. Gawthrope, 62 Meadow Way, Bradley Stoke, Bristol, BS32 8BP
GW0 RVR R. Goodall, 8 Heol Penderyn, Brackla, Bridgend, CF31 2EA
G0 RVS B. Roff, 1 Kennel Cottages, Arlington, Barnstaple, EX31 4LP
G0 RWA B. Chorley, 19 Cleeve Road, Priorswood, Taunton, TA2 8DX
G0 RWI E. Johns, 3 The Rowans, Portishead, Bristol, BS20 6SR
G0 RWJ D. King, 78 Andersey Way, Abingdon, OX14 5NW
G0 RWL B. Mackenzie, 73 Newstead Road, Weymouth, DT4 0AS
G0 RWM R. Martin, 82 Woodlands Avenue, West Byfleet, KT14 6AP
GI0 RWO B. Madden, 1 Skegoneill Drive, Belfast, BT15 3FY
G0 RWQ N. Monument, C/Isla Cabrera 14.1.10, Regia Roig Blq 2, Orihuela, Spain, 3189
G0 RWS W. Scott, Hunters Lodge, Broadmore Green, Worcester, WR2 5TE
G0 RWT R. Pine, Rhodanna, Tennis Court Road, Bristol, BS39 7LU
GM0 RWU Dr J. Ponton, Old Cottage Gardens, Legerwood, Earlston, TD4 6AS
G0 RWW M. Barrass, 11 Flintham Court, Mansfield, NG18 4NB
G0 RWY Dr D. Ramsay, 2 Old Church Road, Colwall, Malvern, WR13 6ET
G0 RXA N. Roscoe, 35 Kenilworth Road, Cheadle Heath, Stockport, SK3 0QL
G0 RXQ F. Lockey, The Dormers, Cirencester Road, Tetbury, GL8 8HA
G0 RXU F. Nethercott, 6 Laking Avenue, Broadstairs, CT10 3NE
GM0 RYA G. Roberts, 8 Parkview, Lhanbryde, Elgin, IV30 8JZ
GM0 RYD J. Van Dyke, 112 Alexander Avenue, Largs, KA30 9EX
GI0 RYK R. White, 1 Woodland Park, Lisburn, BT28 1LD
G0 RYL R. Hodges, 1A Clements Lane, Portland, DT5 1AS
G0 RYM S. Goodwin, 14 Greenhill, Alveston, Bristol, BS35 2QX
G0 RYP C. Martin, 7 St Lawrence Street, B'kara, Malta, BKR1521
G0 RYQ P. Irwin, 11 Cowdale Cottages, Cowdale, Buxton, SK17 9SE
G0 RYR T. Ballinger, 9 Somerville Court, Cirencester, GL7 1TG
G0 RYS Richmond School Amateur, c/o M. Vann, Richmond School, Darlington Road, Richmond, DL10 7BQ
GW0 RYT R. Pitman, 7 Cleveland Drive, Risca, Newport, NP11 6RD
GI0 RYU B. Millar, 312 Churchill Park, Portadown, Craigavon, BT62 1EY
G0 RYW S. Preston, The Chapel, Robson St., Shildon, DL4 1EB
G0 RZB D. Mcdonnell, Glencoe, The Ridge, Salisbury, SP5 2LN
G0 RZG R. Hayward, Old School Farm, Wickham Market, Suffolk, IP13 0HE
G0 RZI B. Easdon, 20 Winder Gate, Frizington, CA26 3QS
G0 RZM B. Judd, 24 Haywood Way, Reading, RG30 4QP
G0 SAC Sutton Area Contest Group, c/o A. Cross, 31 Mountcombe Close, Surbiton, KT6 6LJ
GW0 SAJ Dr H. Jones, 6 Westfa Road, Uplands, Swansea, SA2 0PR
G0 SAR South Anglia Raynet, c/o D. Sparrow, 23 Tranmere Grove, Ipswich, IP1 6DU
G0 SAY C. Thorpe, 78 Bowland Rd, Baguley, Manchester, M23 1JX
G0 SBA D. Sant, Marjar Marden, Hereford, HR1 3EP
G0 SBB D. Barton, Manuka, West Hill, Worthing, BN13 3BZ
G0 SBC R. Harris, 142 St. Nicolas Park Drive, Nuneaton, CV11 6EE
G0 SBH T. Wernham, 6 The Hill, Wangford, Beccles, NR34 8AT
G0 SBK M. Jenkins, 9 Tothill Road, Swaffham Prior, Cambridge, CB25 0JX
G0 SBM South Devon Raynet Group, c/o C. Coker, 46 Clarendon Road, Ipplepen, Newton Abbot, TQ12 5QS
G0 SBO E. Hodgson, 21 Royd Avenue, Mapplewell, Barnsley, S75 6HH
G0 SBP F. Parkinson, 28 Tillage Green, Darlington, DL2 2GL
G0 SBU B. Wedgwood, 40 Ford Street, Delves Lane, Consett, DH8 7AE
G0 SBV R. Talbot, 11 Whitefield Road, Holbury, Southampton, SO45 2HP
G0 SBX E. Barclay, 58 Stockton Road, Hartlepool, TS25 1RW
G0 SBY J. Thompson, 4 Ridgemont, Fulwood, Preston, PR2 3FQ
G0 SBZ W. Sandle, 507B Harrogate Road, Leeds, LS17 7DU
GM0 SCA S. Edwards, The Old Police House, Broughton, Biggar, ML12 6HQ
G0 SCG A. Leavey, 14 Cherry Close, Ealing, London, W5 4JW
G0 SCI B. Young, 11 Gainsborough Avenue, Washington, NE38 7EF
G0 SCK D. Britton, 31 Clay Bottom, Bristol, BS5 7EJ
G0 SCL S. Lawrence, 4 Dale Park Rise, Leeds, LS16 7PP
G0 SCM F. Binnington, 7 Webbs Close, Combs, Stowmarket, IP14 2NZ
G0 SCO Scottish Office ARC, c/o J. Lefever, 30 Holland Road, Melton Mowbray, LE13 0LU
G0 SCQ D. Brusch, 7 Tyrell Close, Stanford In The Vale, Faringdon, SN7 8EY
G0 SCR Caterham Radio Group, c/o P. Lewis, 20 Annes Walk, Caterham, CR3 5EL
G0 SCT R. Bricknell, 82 Hills Road, Saham Hills, Thetford, IP25 7EZ
G0 SCU F. Taylor, 6 Shelley Close, Bolton Le Sands, Carnforth, LA5 8HQ
G0 SCV G. Belt, 3 Prospect Hill, Whitby, YO21 1QE
GM0 SCW R. Anderson, 10 Cyril Crescent, Paisley, PA1 1GT
G0 SCX G. Hobbs, 37 Winnards Park, Sarisbury Green, Southampton, SO31 7BX
G0 SCY W. Best, 61 Gainsborough, Hanworth, Bracknell, RG12 7WL
G0 SDC Southern DX Club, c/o P. Robinson, 11 The Avenue, Hambrook, Chichester, PO18 8TZ
G0 SDD C. James, 4 Hill Top, Bream, Lydney, GL15 6JQ
G0 SDE B. Jupp, 25 Briscoe Way, Lakenheath, Brandon, IP27 9SA
G0 SDF J. Atkins, 30 Bransby Road, Chessington, KT9 2LA
G0 SDG Rev. L. Wilkes, Springfield, Sunnyfield Lane, Up Hatherley, Cheltenham, GL51 6JE
G0 SDJ A. Mcmullon, Carwood House, Hothersall Lane, Preston, PR3 2XB
G0 SDL J. Wilson, Appletrees, Combeinteignhead, Newton Abbot, TQ12 4RE
G0 SDM P. Robinson, 12 Maple Way, Donington, PE11 4XL
G0 SDR S. Harriss, 30 Chatsworth Place, Harrogate, HG1 5HR
GM0 SDS B. Wills, 24 Hopes Avenue, Dalmellington, Ayr, KA6 7RN
G0 SDT J. Sparkes, 18 Hermes Avenue, St. Erme, Truro, TR4 9FW
G0 SDW M. Pattman, 4 Branscombe Road, Bristol, BS9 1SN
G0 SDX Willpower Contest Group, c/o J. Faulkner-Court, Yew Tree Cottage, Avon Dassett, Southam, CV47 2AT
G0 SEB J. Shepherd, 25 Station Road, St. Helens, Ryde, PO33 1YF
G0 SEC J. Curtis, 24 Brisbane Road, Weymouth, DT3 6RD
GM0 SEF D. Stolting, 3 Eden Park, Cupar, KY15 4HS
GM0 SEI R. Vennard, 4 Braehead, Girdle Toll, Irvine, KA11 1BD
GM0 SEP Strathclyde Emergency Planning Unit, c/o R. Cowan, 85 Eastwoodmains Road, Clarkston, Glasgow, G76 7HG
G0 SET H. Pearson, 110 The Gateway, Dover, CT16 1LH
G0 SEU L. Payas, C/O Jo-Anne Maclaren, 7801 Hibiscus Court, Gibraltar, Gibraltar,
G0 SEW K. Green, 13 Knowle Road, Sheffield, S5 9GA
G0 SEY E. Russell, 60 Icknield Way, Tring, HP23 4HZ
G0 SFA B. Hyde, 108 St. Bedes Crescent, Cambridge, CB1 3UB
G0 SFE K. Spring, 18 Greenway, Woodmancote, Cheltenham, GL52 9HU
G0 SFG T. Coneley, 4 Beaconsfield Road Fareham, Fareham, po16 0qb
GD0 SFI B. Hull, Uplands, Ballavitchel Road, Crosby, Isle of Man, IM4 2DN
G0 SFJ A. Thomas, 21 Great Bowden Road, Market Harborough, LE16 7DE
GW0 SFP B. Rish, 15 Bryn Marl, Deganwy, Llandudno Junction, LL31 9BZ
GM0 SFQ J. Stirling, 86 Obsdale Park, Alness, IV17 0TR
GI0 SFT P. Mcdonald, 13 Heathfield, Culmore, Londonderry, BT48 8JD
G0 SFV D. Burton, 3 Norwood, Carden Hill, Brighton, BN1 8AH
G0 SGF J. Barrett, 12 Trent View Gardens, Radcliffe-On-Trent, Nottingham, NG12 1AY
GM0 SGH M. Brown, 3 Arnott Road, Blackford, Auchterarder, PH4 1QE
G0 SGI J. Sankey, 56 Gorsey Lane, Mawdesley, Ormskirk, L40 3TF
G0 SGP A. Danton, 3 Cliffe Close, Ruskington, Sleaford, NG34 9AT
G0 SGR S. Rice, 94 St. Johns Avenue, Bridlington, YO16 4NL
G0 SGT D. Huddleston, 162 Manor Road, Newton St. Faith, Norwich, NR10 3LG
G0 SGV J. Allen, 57 Watford Road, Kings Langley, WD4 8DY
G0 SGX F. Dingwall, 20 Whitehills Road, Loughton, IG10 1TS
G0 SHC M. Lane, Cherry Tree House, Pipwell Gate Saracens Head, Holbeach, Spalding, PE12 8BA
GM0 SHD G. Balfour, 6 Kirkden Street, Friockheim, Arbroath, DD11 4SX
G0 SHJ R. Harrison, 22 East Anglian Way, Gorleston, Great Yarmouth, NR31 6QY
G0 SHM B. Coates, 74 Colescliffe Road, Scarborough, YO12 6SB
G0 SHN G. Jacot, Boucle De L'Observatoire, Le Grand Revard, Pugny-Chatenod, France, 73100
G0 SHO B. Lawrence, 70 Beacon Road, Rolleston-On-Dove, Burton-on-Trent, DE13 9EG
G0 SHP A. Pratt, 4 Chestnut Close, Braunton, EX33 2EH
G0 SHT S. Rossi, 21 Rattigan Gardens, Whiteley, Fareham, PO15 7EA
G0 SHU G. Bennett, 57 Princess Way, Euxton, Chorley, PR7 6PL
G0 SHY M. Bamber, 1 Penair Crescent, Truro, TR1 1YS
GM0 SIA P. Brooks, 3 Jamiesons Court, Kelso, TD5 7EU
G0 SIE A. Swingler, 9 Princess Drive, Wistaston Green, Crewe, CW2 8HP
G0 SIG Signallers Interest Group, c/o K. Prince, 59 Chantry Road, East Ayton, Scarborough, YO13 9ER
G0 SII T. Richards, 142 Princes Mews, Royston, SG8 9BN
GM0 SIM I. Simpson, 1 West Abercromby Street, Helensburgh, G84 9LL
G0 SIQ Rev. R. Myerscough, Hamer, The Street, Holt, NR25 6NW
GW0 SIS K. Barrett, Gorbio House, 47 West Road, Bridgend, CF31 4HD
G0 SIU B. Durrant, 3 Parklands, Shoreham-by-Sea, BN43 6NN
G0 SIW B. Ellison, 6 Eskdale Road, Ashton-In-Makerfield, Wigan, WN4 8QT
G0 SIY A. Hopkinson, 55 Nordale Park Norden, Rochdale, OL12 7RT
G0 SJB S. Barraclough, 67 Sude Hill, New Mill, Holmfirth, HD9 7ER
G0 SJG S. Gunning, 99 Mile Oak Road, Portslade, Brighton, BN41 2PJ
G0 SJH S. Harris, 19 Mundays Boro Road, Puttenham, Guildford, GU3 1AZ
G0 SJP M. Windle, 1A Prices Avenue, Margate, CT9 2NS
G0 SJR R. Brand, Foxgrove, 19A Mill End Close, Dunstable, LU6 2FH
G0 SJU T. Bousfield, 8 Harpington View, Mordon, Stockton-on-Tees, TS21 2EZ
G0 SJV P. Gostick, 25 Cashmere Lane, Cashmere, Queenslands, Australia, 4500
G0 SKA C. Mitchell, Nevada, Slad Lane, Lacey Green, Princes Risborough, HP27 0PW
GW0 SKC B. Foote, Red Roofs, 5 Woodland Avenue, Colwyn Bay, LL29 9NL
G0 SKD T. Ward, 4 Burrows Grove, Wombwell, Barnsley, S73 8PS
G0 SKI M. Foy, 335 South Avenue, Southend-on-Sea, SS2 4HR
G0 SKJ K. Cockburn, 11 Highlands Avenue, Barrow-in-Furness, LA13 0AU
G0 SKK D. Chadwick, 386 Tamworth Road, Amington, Tamworth, B77 4AQ
G0 SKM M. Tunstall, 24 Barbrook Avenue, Stoke-on-Trent, ST3 5UG
G0 SKN P. Hartley, 9 Weston Road, Wimborne, BH21 2SF
G0 SKQ C. Haines, 29 Woodlands Close, Aston, Stone, ST15 0DX
G0 SKR J. Goodall, Red Roofs, 4 Chapel Street, Stapleton, LE9 8JH
G0 SKW K. Walker, 20 Thornhill Close, Bramcote, Nottingham, NG9 3FS
G0 SLB C. Mattison, 14A Buckingham Drive, Colchester, CO4 3YH
GW0 SLC R. Thomas, 6 Grovers Close, Glyncoch, Pontypridd, CF37 3DF
G0 SLD P. Westripp, 2 Ridgeway, Horns Road, Cranbrook, TN18 4RA
G0 SLH C. Shoesmith, 2 Caravelle Gardens, Northolt, UB5 6EU
G0 SLI T. Day, 21 Mowbray Road, Ham, Richmond, TW10 7NQ

G0 SLJ Dr D. Pepper, 17 Cliffe House, Radnor Cliff, Folkestone, CT20 2TY
G0 SLK E. Patterson, 45 Sandhurst Road, Rainhill, Prescot, L35 8NE
G0 SLL R. Petrie, Royal Hospital Chelsea, Royal Hospital Road, London, SW3 4SR
G0 SLN P. Grainger, 11 Smith Grove, Ryhope, Sunderland, SR2 0JU
G0 SLP M. Coultas, 35 Monteigne Drive, Bowburn, Durham, DH6 5QB
G0 SLQ S. Quinn, 48 Aldsworth Close, Springwell Village, Gateshead, NE9 7PG
G0 SLR R. Lisle, 21 Porlock Close, Penketh, Warrington, WA5 2QE
G0 SLU C. Barr, 17 Knighton Road, Otford, Sevenoaks, TN14 5LD
G0 SLW J. Waite, 28 Overdown Rise, Portslade, Brighton, BN41 2YG
G0 SLY C. Kratzer, 9900 Dale Ridge Ct, Vienna, USA, VA 22181
G0 SLZ A. Matthews, 28 Sherwin Road, Stapleford, Nottingham, NG9 8PQ
G0 SMH B. Marchant, 20 Wrench Road, Norwich, NR5 8AS
G0 SMJ M. Jackson, Sunny Cot, 44 Dulwich Road, Holland-On-Sea, Clacton-on-Sea, CO15 5NA
G0 SMM J. O'Nion, 7 Ettington Close, Cheltenham, GL51 0NY
G0 SMN A. Mckenzie, 311 Weston Road, Weston Coyney, Stoke-on-Trent, ST3 6HA
G0 SMO C. Cash, 27 Robert Wynd, Bilston, WV14 9SE
G0 SMP S. Pountain, 21 Hayfield Road, Chapel-En-Le-Frith, High Peak, SK23 0JF
G0 SMR J. Ballard, 7 Chapelcroft Court, Liverpool, L12 9GY
G0 SMS D. Barnham, 35 Post Office Road, Frettenham, Norwich, NR12 7AB
GI0 SMU A. Hanna, 39 Dalton Crescent, Comber, Newtownards, BT23 5HE
G0 SMZ R. Clews, 99 Kilbury Drive, Worcester, WR5 2NG
G0 SNB W. Bonser, 24 Meend Garden Terrace, Cinderford, GL14 2EB
G0 SNF J. Culling, 4 Ash Road, Princes Risborough, HP27 0BQ
G0 SNG D. Hill, 3 Morcar Road, Stamford Bridge, York, YO41 1PR
G0 SNK A. Gill, Bradgate, Kings Lane, Lymington, SO41 6BQ
G0 SNM K. Killick, 15 Popplechurch Drive, Swindon, SN3 5DE
G0 SNO D. Lauder, 20 Sutherland Close, Barnet, EN5 2JL
G0 SNP J. Du Heaume, 10 Water Lane, Pill, Bristol, BS20 0EQ
G0 SNQ H. Davis, 44 Kenyon Street, Ashton under Lyne, OL6 7DU
G0 SNS B. Harrison, 8 Elm Park, Pontefract, WF8 4LG
GM0 SNT A. Carpenter, 9 Glenbervie Road, Kirkcaldy, KY2 6HR
G0 SNU I. Gray, 27 Meadow Close, Lavenham, Sudbury, CO10 9RU
G0 SNV J. Worsnop, 1217 Thornton Road, Thornton, Bradford, BD13 3BE
G0 SNW P. Rogers, 126 Bradford Road, Otley, LS21 3LE
G0 SNX N. Johnson, 12 Bleach Mill Lane, Menston, Ilkley, LS29 6HE
G0 SNZ A. Flood, 3 Ongar Walk, Blackley, Manchester, M9 8JD
G0 SOA Statford-Upon-Avon & District Radio Society, c/o C. Ousbey, 30 Hawthorn Way, Shipston-on-Stour, CV36 4FD
G0 SOF G. Hedley, 260E 100 Sts, Raymond, Alberta, Canada, TOK 250
G0 SOG F. Mellings, 4 Kiln Lane, Horley, RH6 8JG
G0 SOK P. Mayer, 248 Dimsdale Parade West, Newcastle, ST5 8EA
G0 SON R. Peard, 28 New Road, Shoreham-by-Sea, BN43 6RA
G0 SOU J. Pendleton, 17A Langley Drive, Kegworth, Derby, DE74 2DN
G0 SOX P. Chapman, 4 Churchill Close, Brightlingsea, Colchester, CO7 0RS
G0 SOY A. Croft, 34 Fourth Avenue, Wolverhampton, WV10 9LZ
G0 SPA P. Benson, 7 Crofton Close Attenborough, Nottingham, NG9 5HX
G0 SPB G. Rusby, 12 Park Meadow, Princes Risborough, HP27 0EB
G0 SPC M. Chawner, Timbertops, Churchfield Lane, Benson, Wallingford, OX10 6SH
G0 SPF T. Hill, 7 Broadlands Road, Paignton, TQ4 5NY
G0 SPH K. Brooks, 25 Lagos Grove, Winsford, CW7 2BJ
G0 SPK D. Neal, 490 Aureole Walk, Newmarket, CB8 7BQ
G0 SPQ I. Wilson, 3 Caring Lane, Bearsted, Maidstone, ME14 4NJ
G0 SPS M. Forder, 157 Kennington Road, Kennington, Oxford, OX1 5PE
G0 SPX J. Sparks, 34 Green Park Avenue, Skircoat Green, Halifax, HX3 0SR
GW0 SPY J. Thorley, 8 Bryn Gwyn, Abergele, LL22 8JA
G0 SPZ M. Rickard-Worth, 2 Harwood Road, Littlehampton, BN17 7AT
G0 SQE E. Young, 14 Hanover Court, Gateshead, NE9 6TZ
G0 SQF J. Bubez, 4 Southway, Burgess Hill, RH15 9ST
G0 SQH D. Higbee, 12 Shelley Close, Ashley Heath, Ringwood, BH24 2JA
G0 SQI N. Blythe, 14 The Green, South Creake, Fakenham, NR21 9PD
G0 SQK M. Stuckey, 212 Roughton Road, Cromer, NR27 9LQ
G0 SQL R. Bishop, 73 Broomgrove Gardens, Edgware, HA8 5RJ
G0 SQP F. Ott, 16 Hornbeam Close, Chelmsford, CM2 9LW
G0 SQS M. Hewitt, 1 Harpswell Hill Park, Hemswell, Gainsborough, DN21 5UT
GW0 SQT A. Davis, 8 Cook Road, Barry, CF62 9HD
G0 SQX T. Donley, 21 Elmridge, Leigh, WN7 1HN
GW0 SQY S. Morgan, Oakfield House, Barry Road, Pontypridd, CF37 1HY
G0 SRC South Derbyshire & Ashby Woulds ARG, c/o V. Stocker, 25 Davies Drive, Uttoxeter, ST14 7EQ
GW0 SRE D. Price, 1 Rhas Cottages, Pontyates, Llanelli, SA15 5SF
GW0 SRF D. Daniels, 73 Bethania Road, Upper Tumble, Llanelli, SA14 6DT
G0 SRG Sunderland Raynet Group, c/o S. Green, 133 Sevenoaks Drive, Sunderland, SR4 9NQ
GI0 SRL A. Harbison, 26 Ballymartin Road, Templepatrick, Ballyclare, BT39 0BW
GM0 SRO I. Maclean, 7 Thirlestane Crescent, Lauder, TD2 6TT
GI0 SRP N. Averill, 3 Edmund Court, Tobermore, Magherafelt, BT45 5QA
GM0 SRQ D. Beacher, 17 Cairn Grove, Crossford, Dunfermline, KY12 8YD
G0 SRR M. Baugh, 97 Wilson Avenue, Deal, CT14 9NJ
G0 SRY J. Smart, 60 Blaze Park, Wall Heath, Kingswinford, DY6 0LN
G0 SRZ E. Hart, 47 Northfield Crescent, Wells-Next-the-Sea, NR23 1LR
GI0 SSA J. Stevenson, 27 Kinnegar Rocks, Donaghadee, BT21 0EZ
G0 SSC G. Mills, 49 Priestley Close, Doncaster, DN4 9DQ
G0 SSE N. Morton, Cami Terrapico 54, Roquetes, Spain, 43520
G0 SSG R. Andre, 54 Covertside, Wirral, CH48 9UL
G0 SSJ A. Powell, 61 Albert Road, Grappenhall, Warrington, WA4 2PF
G0 SSK G. Johnson, 503 Holden Road, Leigh, WN7 2JJ
G0 SSL A. Mcevoy, 12 Fountains Avenue, Haydock, St Helens, WA11 0RS
G0 SSN C. Ireland, 14 Castlefields, Istead Rise, Gravesend, DA13 9EJ
GM0 SSQ A. Winchester, 23 Craigmount, Avenue North, Edinburgh, EH12 8DL
G0 SSV G. Kendall, 3 Brayton Avenue, Sale, M33 5HF
G0 SSX P. Ellis, 4 Rostwold Way, Norwich, NR3 3NN
G0 SSY D. Webb, 3 Cams Hill Lane, Hambledon, Waterlooville, PO7 4SP
G0 SSZ R. Stanley, 219 Fartown, Pudsey, LS28 8NH
GM0 STB Scottish Tourist Board Radio, c/o R. Aitkenhead, 11 Elm Court, Quarter, Hamilton, ML3 7FB
GI0 STC P. Dellett, 14 Fox Park, Omagh, BT79 0JX
G0 STF T. Clements, 29 Bonchurch Drive, Wavertree, Liverpool, L15 4PW
G0 STH St Helens & District ARC, c/o R. Vaughan, 6 Dellside Grove, St Helens, WA9 5AR
G0 STK A. Hardcastle, 25 Aire Crescent, Cross Hills, Keighley, BD20 7RW
GI0 STM K. Murray, 17 Glebe Court, Dungannon, BT70 3PU
G0 STR W. Shaw, 161 Springwood Crescent, Edgware, HA8 8SH
GI0 STS R. Todd, 73 Lakeview Park, Drumgor, Craigavon, BT65 4AL
G0 STW C. Kendrick, 18 Ainger Road, Upper Dovercourt, Harwich, CO12 4TS
G0 SUA A. Edwards, 49 Griggs Meadow, Dunsfold, Godalming, GU8 4ND
G0 SUB G. Thomas, 83 Hollingthorpe Road, Hall Green, Wakefield, WF4 3NW
G0 SUC B. Smith, 89 Farndale Drive, Guisborough, TS14 8JX
GM0 SUF H. Butcher, 14 Newbattle Road, Dalkeith, EH22 3DB
GM0 SUH J. Montgomery, Woods 2, Tralee Bay Holidays, Benderloch, Oban, PA37 1QR
G0 SUI G. Scarlett, 10 Laythorpe Terrace, East Morton, Keighley, BD20 5TL
G0 SUL J. Ward, 8 Wilderness Road, Guildford, GU2 7QN
GU0 SUP P. Cooper, 1 Clos Au Pre, La Route De La Hougue Du Pommier, Castel, Guernsey, GY5 7FQ
G0 SUQ I. Johnson, 181 Broad Street, Bromsgrove, B61 8NQ
G0 SUT J. Burdett, 1 Main Street, Egginton, Derby, DE65 6HL
G0 SUU J. Ogier, 37 Hill Park Road, Torquay, TQ1 4LD
GM0 SUY C. Auld, 148 Echline Drive, South Queensferry, EH30 9XG
G0 SVA D. Nock, 431 Locking Road, Weston-Super-Mare, BS22 8QN
G0 SVB P. Herrmann, 5992 Royal Court, Lockport, USA,
G0 SVH E. Wright, 26 Walmsley Close Church, Accrington, BB5 4HL
G0 SVJ M. Creswick, 5 Wheatlands Drive, Easington, Saltburn-by-the-Sea, TS13 4PB
G0 SVK J. Hosfield, 29 Whitecroft, Gosforth, Seascale, CA20 1AY
G0 SVN N. Savin, 138A Lillibrooke Crescent, Maidenhead, SL6 3XH
G0 SVP G. Broadhurst, 9 Sharples Street, Accrington, BB5 0HQ
G0 SVQ A. Haydock, 8 Corbridge Close, Blackpool, FY4 5EZ
GM0 SVS Dr M. Whiteley, 9 Pathfoot Avenue, Bridge Of Allan, Stirling, FK9 4SA
G0 SVU R. Morris, 57 Marten Drive, Huddersfield, HD4 7JX
G0 SVX I. Roebuck, 3 Lodge Hill Drive, Kiveton Park, Sheffield, S26 5RU
G0 SVY D. Beardsley, 1 Amber Villas, Sutton St. Nicholas, Hereford, HR1 3DF
G0 SVZ M. Cooper, 20 Bankfield Road, Widnes, WA8 7UW
G0 SWB R. Atkinson, 57 Jobling Avenue, Blaydon-on-Tyne, NE21 4RR
G0 SWC R. Eeles, 50 Nightingale Road, Guildford, GU1 1EP
G0 SWE S. Whitbourn, 50 Nightingale Road, Guildford, GU1 1EP
G0 SWF J. Durrant, 16 Bugdens Lane, Verwood, BH31 6EY
G0 SWH B. Fletcher, 58 Broomfield Avenue, Worthing, BN14 7SB
G0 SWL C. Niles, 23 Randsfield Avenue, Brough, HU15 1BE
G0 SWN Gilwell Park Scout Radio Club, c/o S. Barber, Homedale, St. Monicas Road, Tadworth, KT20 6ET
G0 SWO B. Atkinson, 165 Alliance Avenue, Hull, HU3 6QY
G0 SWS T. Stow, 38 The Strand, Mablethorpe, LN12 1BQ
G0 SWU P. Broad, 78 Lenham Road, Sutton, SM1 4BG
G0 SWV Short Wave Dial Group, c/o S. Down, 1 Dove Close, Honiton, EX14 2GP
G0 SWW M. Stevens, Autumn Cottage, Silver Street, Horncastle, LN9 5NH
G0 SWY M. Humphrey, 5 Ventnor Court, Southampton, SO16 3EB
G0 SXA W. Daly, 85 Lordens Road, Huyton, Liverpool, L14 9PA
GW0 SXE J. Williams, 11 Courbet Drive, Connah'S Quay, Deeside, CH5 4WP
G0 SXG K. Stammers, 102 Eaton Road, Appleton, Abingdon, OX13 5JJ
G0 SXK A. Dodd, 20 Braemar Avenue, Chelmsford, CM2 9PW
G0 SXM D. Smith, River Meadow, Harlyn Bay, Padstow, PL28 8SB
G0 SXN B. Watts, 9 Weavers Walk, Cullompton, EX15 1SS
GM0 SXO A. Segar, 36 Kestrel Avenue, Dunfermline, KY11 8JL
GM0 SXP Dr R. Cliff, 32 Lochardil Road, Inverness, IV2 4LD
GM0 SXQ M. Hepburn, The Toll House, Corse, Lumphanan, Banchory, AB31 4RY
GW0 SXS K. Wheeler, The Glen, Dreenhill, Haverfordwest, SA62 3XH
G0 SXU C. Bocock, 20 Court Avenue, Stoke Gifford, Bristol, BS34 8PJ
G0 SXW P. Cressey, Skidby Hill Farm, Beverley Road, Cottingham, HU16 5TF
G0 SXY P. Davies, 46 Wagstaff Way, Olney, MK46 5FB
GW0 SXZ J. Green, 23 Litchard Park, Bridgend, CF31 1PF
G0 SYF R. Pearce, 14 Shepherds Leaze, Wotton-under-Edge, GL12 7LQ
GW0 SYG Cleddau ARS, c/o R. Richards, 77 Church Road, Llanstadwell, Milford Haven, SA73 1EA
G0 SYI K. Treasure, 30 Grace Park Road, Brislington, Bristol, BS4 5JA
G0 SYP C. Steinhoefel, 222 Stretford Road, Urmston, Manchester, M41 9NT
G0 SYQ M. Wood, 4 Gordon Road, Hastings, TN34 3JN
G0 SYR B. Petifer, 14 Wood Lane, Caterham, CR3 5RT
G0 SYS D. Hall, 72 Mansfield Road, Edwinstowe, Mansfield, NG21 9NH
G0 SYT D. Firks, Bryn Garth Cottage, Hereford, HR2 8HJ
GM0 SYV J. Kelly, 2 Ryeland Street, Strathaven, ML10 6DL
GM0 SZA I. Stones, 8 Cloverfield Place, Bucksburn, Aberdeen, AB21 9RH
G0 SZE C. Andrews, 33 Blackheath Road, Barnsley, S71 3RH
G0 SZG J. Greene, 308 Cedar Road, Nuneaton, CV10 9DY
GI0 SZH I. Mcewen, 234 Legahory Court, Legahory, Craigavon, BT65 5DH
G0 SZI D. Green, 81A Long Lane, Holbury, Southampton, SO45 2ND
G0 SZJ S. Fletcher, Apartado 39, Ourique, Portugal, 7670
G0 SZK L. Joyce, 106A Victoria Drive, Bognor Regis, PO21 2EJ
GW0 SZN M. Lawrence, 1 Greenwood Cottages, Gelligroes, Blackwood, NP12 2JB
G0 SZO J. Everard, Woodside, Staple Lane, Taunton, TA4 4DE
G0 SZT D. Allibone, Virginia, North Street, Langport, TA10 9RH
GW0 SZU B. Osborne, 163 Park Street, Bridgend, CF31 4BB
G0 TAA W. Chandler, 38 Falkland Road, Chandler'S Ford, Eastleigh, SO53 3GD
GM0 TAE C. Porter, 20 Baird Avenue, Kilwinning, KA13 7AR
G0 TAG D. Beane, 12 Strafford Court, Pondcroft Road, Knebworth, SG3 6DF
G0 TAH K. Aldus, 77 Springvale Road, Kings Worthy, Winchester, SO23 7ND
G0 TAI I. Hawkins, 1 Millhayes, Great Linford, Milton Keynes, MK14 5EP
G0 TAK R. Walker, 35 Romany Close, Letchworth Garden City, SG6 4LA
G0 TAL S. Walsworth, 4 The Homestead, Heckmondwike, WF16 9JL
G0 TAM A. Farrow, 18 The Green Trimingham, Norwich, NR11 8ED
G0 TAN S. Venner, 36 Broadwood Avenue, Ruislip, HA4 7XR
G0 TAO R. Lokuge, 11 Porchester Close, Southwater, Horsham, RH13 9XR
G0 TAR B. Lucas, 8 Gilbert Close, Hempstead, Gillingham, ME7 3QQ
G0 TAS J. Taylor, 19 Castle Close, Leconfield, Beverley, HU17 7NX
G0 TAT M. Wilmot, Southview, Roman Road, Weston-Super-Mare, BS24 0AB
G0 TAX T. Welch, 63 Vicarage Close, New Silksworth, Sunderland, SR3 1JF
GM0 TAY OBO Tayside Raynet, c/o I. Strachan, 238 Coupar Angus Road, Muirhead, Dundee, DD2 5QN

G0 TAZ J. Burgess, Combeside House, Symonsburrow, Cullompton, EX15 3XA
G0 TBC S. Lawson, 27 Broadlands Avenue, Eastleigh, SO50 4PP
GM0 TBH J. Mcmaster, 96 Cunningham Crescent, Ayr, KA7 3JB
G0 TBI S. Mckinnon, 145 Enville Road, Kinver, Stourbridge, DY7 6BN
GW0 TBM J. Goulden, Wenffrwd Cottage, Llangollen Road, Llangollen, LL20 7UH
G0 TBO P. Clark, 30 Alicia Avenue, Garlinge, Margate, CT9 5JZ
G0 TBS W. Lucas, 67 Tower Ride, Uckfield, TN22 1NU
G0 TBU R. Brightwell, 40 Streete Court Road, Westgate-on-Sea, CT8 8BX
G0 TBW G. Davis, 32 Medlock Close, Farnworth, Bolton, BL4 9QW
G0 TCA R. Seaward, 13 Blythe Close, Catford, London, SE6 4UW
GM0 TCC K. Walker, 174 South Seton Park, Port Seton, Prestonpans, EH32 0BP
G0 TCE Dr P. Taylor, 67 Rectory Park, South Croydon, CR2 9JR
G0 TCF P. Loch, 400 Loughborough Road, West Bridgford, Nottingham, NG2 7FD
G0 TCH T. Noszkay, 41 Brue Close, Weston-Super-Mare, BS23 3BX
G0 TCI B. Hughes, 29 East View Close, Radwinter, Saffron Walden, CB10 2TZ
G0 TCJ T. Worrall, 9 Barnstaple Close, Wigston, LE18 2QX
GW0 TCL D. Parsons, Aston Hall Res.Home, Lower Aston Hall Lane, Deeside, CH5 3EX
G0 TCO M. Roper, 1 The Cottages, Norwich Road, Norwich, NR9 5BY
G0 TCP P. Maynard, Seaton House, Lower Road, Westerfield, Ipswich, IP6 9AR
G0 TCQ A. Cudlip, The Oaks, 16 Bilberry Close, Southampton, SO31 6XX
GM0 TCU C. Pirie, 10 Annesley Park, Torphins, Banchory, AB31 4HG
GW0 TCV A. Thomas, 4 Gordon Terrace, King Street, Mold, CH7 1LD
G0 TCW A. Furmston, 46 Twydall Lane, Gillingham, ME8 6JE
G0 TCY M. Templeman, 44 Wisbeck Road, Bolton, BL2 2TA
GW0 TDA P. Price, 1 Brynderi, Pontyates, Llanelli, SA15 5SU
G0 TDC S. Tinsley, 48 Smithy Lane, Croft, Warrington, WA3 7JG
G0 TDE R. Barrett, Brookfield, Hobbacott Lane, Bude, EX23 0ES
G0 TDG N. Hotson, 1 Holly Court 24 Hastings Road, London, E16 1GJ
G0 TDJ S. Smith, Westgarth Flat 4, 145-146 Marina, StLeonards-on-Sea, TN38 0BT
G0 TDM J. Sutton, 15 Lowther Street, Penrith, CA11 7UW
G0 TDN S. Maclennan, Shalam, Rising Sun, Callington, PL17 8JE
GI0 TDP J. Driscoll, 67 Whinney Hill, Holywood, BT18 0HG
G0 TDQ P. Luscombe, 33 Rea Barn Road, Brixham, TQ5 9ED
G0 TDR T. Round, 92 Church View Gardens, Kinver, Stourbridge, DY7 6EE
G0 TDV R. King, 10 Lansdown Road, Kingswood, Bristol, BS15 1XB
G0 TDX Y. Kimoto, 523 Yukinaga, Maizuru-City, Kyoto, Japan, 625-0052
GM0 TEA A. Cherry, 39 Clark Road, Edinburgh, EH5 3AR
G0 TEB C. Sexton, Lapswater, Marsh, Honiton, EX14 9AL
G0 TED E. Thane, 19 Churchill Road, Walton, Stone, ST15 0EB
G0 TEE T. Speight, 1 Lyndene Avenue, Worsley, Manchester, M28 2RJ
G0 TEI R. Sparks, Radio Licence Centre, Po Box 885, Bristol, BS99 5LG
G0 TEL R. Bellenot, 22 Roderick Avenue, Peacehaven, BN10 8JT
G0 TEM A. Merrix, 53 Pear Tree Road, Great Barr, Birmingham, B43 6HX
G0 TEO L. Baker, Flat 6, Francis Snary Lodge 12 Chesterton Close, London, SW18 1SD
GD0 TEP A. Kissack, 30 High View Road, Douglas, Isle Of Man, IM2 5BH
G0 TES R. Dicks, 10 Westfield Avenue, Raunds, Wellingborough, NN9 6DQ
G0 TFB Wallen Antennae Radio Club, c/o L. Wallen, Lambda Works, 45A Whitehall Road, Ramsgate, CT12 6DE
G0 TFC A. Stanley, 145 Tribune Drive, Houghton, Carlisle, CA3 0LF
G0 TFD D. Eyre, 29 Old Acre Lane, Brocton, Stafford, ST17 0TW
GM0 TFE C. Stewart, 185 Newbattle Abbey Crescent, Dalkeith, EH22 3LT
GM0 TFF A. Mcghie, 16 Boyach Crescent, Isle Of Whithorn, Newton Stewart, DG8 8LD
GD0 TFG J. Dowling, 2 Ballabridson Park, Ballasalla, Isle Of Man, IM9 2ES
G0 TFI D. Roberts, 9 The Pound, Westoning, Bedford, MK45 5JN
G0 TFK J. Sutcliffe, 4 Lancaster Gate, Nelson, BB9 0AP
G0 TFL J. Williams, 107 Clay Lane, Rochdale, OL11 5QW
GD0 TFO J. Bellis, Jandakot, Old Castletown Road Port Soderick, Isle of Man, IM4 1BB
G0 TFP J. Brett, 11 Manor Road, Astley, Manchester, M29 7PH
GM0 TFQ H. Wignall, 7 Windyedge, Inverurie, AB51 3WJ
G0 TFR A. Vining, The Cedars, Thorney Road, Peterborough, PE6 0LH
G0 TFT A. Gibson, 28 Finchale Terrace, Jarrow, NE32 3TX
G0 TFU M. Wilson, 6 The Chase, Calcot, Reading, RG31 7DN
G0 TFV T. Carvell, 18 Park View, School Lane, Rye, TN31 6UR
G0 TFX S. Roberts, 20 Beech Grove, Trowbridge, BA14 0HG
G0 TGB T. Blore, Glendhoon, Laneham Street, Retford, DN22 0JX
GM0 TGE I. Ross, Idlewilde, Kintore, AB51 0XA
GM0 TGG L. Mackenzie, 90 Tay St., Newport on Tay, DD6 8AP
G0 TGH S. Leak, 12 Kentmere Approach, Leeds, LS14 1JP
G0 TGM S. Fowler, 38 Meadowcroft, Aylesbury, HP19 9LN
G0 TGP W. Ford, Flat 2, Hillyard Court, Wareham, BH20 4QX
G0 TGQ O. Cubitt, 97 Sutton Lane, Langley, Slough, SL3 8AU
G0 TGR D. Greenacre, 38 Toon Crescent, Bury, BL8 1JB
G0 TGU W. Collier, 20 Wainers Croft, Greenleys, Milton Keynes, MK12 6AL
G0 TGX P. Ireland, 2 The Ridings, Hull, HU5 5HW
G0 THD P. Hart, 39 Barley Drive, Burgess Hill, RH15 9XG
G0 THF K. Greatorex, 54 Lilac Grove, Glapwell, Chesterfield, S44 5NG
G0 THH H. Hudders, 7 Cedar Crescent, Thame, OX9 2AX
G0 THI W. Davey, 15 Park Avenue, Histon, Cambridge, CB24 9JU
G0 THJ J. Stowell, 6 Westage Lane, Great Budworth, Nr Northwich, CW9 6HJ
GI0 THO W. Edmondson, 7 Rathcavan Drive, Ballymena, BT42 2QH
G0 THQ R. King, 6 Mayon Green Crescent, Sennen, Penzance, TR19 7BS
GI0 THR J. Mcdonald, 54 Bettys Hill Road, Newry, BT34 2ND
G0 THS S. Graham, 25 South End Road, Ottringham, Hull, HU12 0DP
G0 THV J. Coote, 8 St. Francis Chase, Bexhill-on-Sea, TN39 4HZ
G0 THW A. Brentnall, Sandy Rise, Mill Lane, Woodbridge, IP12 3LL
G0 THY M. Preston, 15 Poplar Close, Kidlington, OX5 1HH
G0 TIA P. Lodge, 79, Via Circonvallazione, 79, Milano, Italy, 20090
G0 TID C. Alexander, 25 Diamedes Avenue, Stanwell, Staines-upon-Thames, TW19 7JE
G0 TIG H. Janes, 91 Thorpe Bay Gardens, Southend-on-Sea, SS1 3NW
G0 TII S. Graham, 4 Oakland Avenue, Ellenborough, Maryport, CA15 7BU
G0 TIJ D. Page, 16 Arlington Close, Yeovil, BA21 3TB
G0 TIL J. Parmenter, 48 Honey Way, Royston, SG8 7EU
G0 TIP G. Thorne, 19 Lapwing Lane, Brinnington, Stockport, SK5 8JY
G0 TIS P. Jones, 61 North Road, Bourne, PE10 9AU
G0 TIW T. Parker, The Bungalow, 178 Green End Lane, Hemel Hempstead, HP1 2BQ
G0 TIX P. Wilson, 2 Staley Close, Stalybridge, SK15 3HJ
G0 TIZ E. Tometzki, 11 Southey Close, Enderby, Leicester, LE19 4QZ
G0 TJC L. Taylor, 35 Leafield Road, Darlington, DL1 5DF
G0 TJD A. Wedgwood, 10 Milner Place, London, N1 1TN
G0 TJE S. Sullivan, 7 Gosfield Road, Dagenham, RM8 1JY
G0 TJG M. Spafford, Old School House, 11 Old School Lane, Chesterfield, S44 5UE
G0 TJH I. Foord, 25 The Rose Walk, Newhaven, BN9 9NH
G0 TJI G. Woods, 126 Luddenham Close, Ashford, TN23 5SA
G0 TJN T. Newland, 80 Burnway, Hornchurch, RM11 3SG
G0 TJP A. Person, 78 Green Street, Ston Easton, Radstock, BA3 4BZ
G0 TJQ C. Dowell, 19 Field Top, Bailiff Bridge, Brighouse, HD6 4EQ
G0 TJR L. Ellams, 131 Broadway, Dunscroft, Doncaster, DN7 4HB
G0 TJT R. Francis, 661 Osmaston Road, Derby, DE24 8NF
GI0 TJV A. Gibson, 58 Cairnmore Park, Lisburn, BT28 2DN
G0 TJY Capt. T. Herd, The Old Dairy, High Street, Ipswich, IP8 3AP
GM0 TKB L. Thomas, Greengeo, Scarfskerry, Thurso, KW14 8XN
GM0 TKC T. Maxwell, 28 Cedar Grove, Dunfermline, KY11 8BH
GM0 TKE J. Mcluckie, 118 Lady Nairn Avenue, Kirkcaldy, KY1 2AT
G0 TKF W. Stuart, 3 Rookery Vale, Deepcar, Sheffield, S36 2NP
G0 TKG D. Mawson, 14 Windermere Close, Worksop, S81 7QE
G0 TKJ T. Clayton, 40 Morrison Road, Darfield, Barnsley, S73 9ED
G0 TKL E. Roberts, 43 Ashbourne Crescent, Sale, M33 3LQ
G0 TKR E. Martin, 20 Easters Grove, Stoke-on-Trent, ST2 7PF
GW0 TKX A. Mason, 101 Aneurin Bevan Avenue, Gelligaer, Hengoed, CF82 8ET
G0 TKZ R. Hamblin, 5 Streatfield, Edenbridge, TN8 5DF
G0 TLA R. Lythall, 6 Belmont Crescent, Little Houghton, Barnsley, S72 0HT
G0 TLI C. Vince, 8 Kent Road, Swindon, SN1 3NJ
GW0 TLJ A. Mathias, 75 Coombs Drive, Milford Haven, SA73 2NU
G0 TLN G. Sim, 24 Fernmoor Drive, Irthlingborough, Wellingborough, NN9 5TL
G0 TLP A. Whitwam, 9 Oak Apple Close, Stourport on Severn, DY13 0JR
G0 TLQ R. Parsons, 10 Waterside, Isleham, Ely, CB7 5SH
GW0 TLS R. Upton, 17 Bryn Gannock Deganwy, Conwy, LL31 9UG
G0 TLT D. Davies, 14 Hammerwater Drive, Warsop, Mansfield, NG20 0DJ
G0 TLU P. Thompson, Flat 4, 14 West End Way, Lancing, BN15 8RL
G0 TLZ J. Trefry, 67 Axminster Close, Cramlington, NE23 2UE
G0 TMA R. Griffiths, Flat 1, Buckhurst Court, 29 Buckhurst Road, Bexhill-on-Sea, TN40 1QE
G0 TME B. Park, 69 Lea Villa Residential Park, Lea, Ross-on-Wye, HR9 7GP
G0 TMF M. Fleetwood, 9 Reynolds Close, Swindon, Dudley, DY3 4NQ
G0 TMH C. Neary, 3 Wordsworth Close, Torquay, TQ2 6EA
G0 TMJ E. Jones, 1 Ivel View, Sandy, SG19 1AU
G0 TMK W. Howell, 41 Chestnut Avenue, Shavington, Crewe, CW2 5BJ
G0 TML A. Rowley, 32 Spring Lane, Flore, Northampton, NN7 4LS
GI0 TMS M. Smyth, 41 Coolaghy Road, Newtownstewart, Omagh, BT78 4LG
G0 TMT M. Tuttle, 7 Mill Lane Horsford, Norwich, NR10 3ES
GW0 TMU D. Edwards, 68 Heol Y Meinciau, Pontyates, Llanelli, SA15 5RT
GW0 TMV T. Vlismas, Maes Yr Awel, Newport Rd, Crymych, SA41 3RR
G0 TMW G. Boswell, 7 Chestnut Avenue, Wootton, Northampton, NN4 6LA
G0 TMZ N. Lilley, 18 Beechwood Avenue, New Milton, BH25 5NB
G0 TNC G. Stephenson, 54 Rock Road, Sittingbourne, ME10 1JF
G0 TNF M. Wills, Chapel View, Cliburn, Penrith, CA10 3AL
G0 TNG P. Hayward, 63 Devereaux Crescent, Ebley, Stroud, GL5 4PX
G0 TNH L. Crow, 181 Foxlydiate Crescent, Batchley Estate, Redditch, B97 6NS
GM0 TNK D. Mackenzie, 32 Miller Gardens, Inverness, IV2 3DT
G0 TNL A. Beattie, Pine Croft, Blitterlees, Wigton, CA7 4JJ
G0 TNM R. Guppy, 12 Highfield Road, Caterham, CR3 6QX
G0 TNO C. Billington, 5 Lamers Road, Luton, LU2 9BL
G0 TNP P. Unstead, 9 Buckingham Road, Swindon, SN3 1HZ
G0 TNQ A. Lewis, 8 Newts Way, St Leonards-on-Sea, TN38 9TH
G0 TNS H. Hauton, 8 St. Catherines Crescent, Scunthorpe, DN16 3LQ
G0 TNY A. Rose, Heronsgate, River Gardens, Maidenhead, SL6 2BJ
G0 TOB J. Horsfall, 10 Derwent Close, Hebden Bridge, HX7 7ED
G0 TOC M. Litchman, 26 Oak Tree Close, Loughton, IG10 2RE
G0 TOD T. Northover, 13 Dagenham Avenue, Dagenham, RM9 6LD
G0 TOE A. Gallagher, 1A Wynsome Street, Southwick, Trowbridge, BA14 9RB
GM0 TOF M. Farnworth, 30E Roseangle, Dundee, DD1 4LY
GW0 TOI A. Lipian, 10 Field Street Trelewis, Treharris, Treharris, CF46 6AW
G0 TOK B. Dixon, 97 Sunny Blunts, Peterlee, SR8 1LN
GW0 TOM T. Beedle, 2 Chestnut Grove, Maesteg, CF34 0NT
G0 TOO C. Richmond, 11 Harewood Avenue, Morecambe, LA3 1JH
G0 TOQ S. Harrison, 7 Harvey Street, Kingstone, Barnsley, S70 6JT
G0 TOS B. Harper, 51 Cross Lane, Scarborough, YO12 6DQ
G0 TOT R. Claxton, Purbeck View, 2 Hoburne Road, Swanage, BH19 2SL
G0 TOX C. Wheeler, 190 Mount Pleasant, Redditch, B97 4JL
G0 TOY K. Li, 16 Garthland Drive, Arkley, Barnet, EN5 3BB
G0 TPA A. Taylor, The Grey House, Chipping Campden, GL55 6XP
G0 TPB B. Smith, 34 Wheatlands, Fareham, PO14 4SL
G0 TPD J. Gayther, 33 Greenways, Winchcombe, Cheltenham, GL54 5LQ
G0 TPE A. Davis, 320 Preston Old Road, Blackburn, BB2 2TX
GW0 TPF N. Pugh, 28 Hill View Road, Llanrhos, Llandudno, LL30 1SL
G0 TPG B. Taylor, 3 Stonepits Lane, Hunt End, Redditch, B97 5LX
G0 TPH A. Horne, 54 Manor Road, Desford, Leicester, LE9 9JR
GM0 TPI D. Harris, 42 Shira Terrace, East Kilbride, Glasgow, G74 2HU
GW0 TPL D. Blundell, 30 Heol Ffynnon Wen, Cardiff, CF14 7TP
G0 TPM P. Mercer, 10 Holmcliffe Avenue, Huddersfield, HD4 7RJ
G0 TPN M. Padgett, 97 Larks Hill, Pontefract, WF8 4RP
G0 TPO M. Cook, 11 Atherton Close, Shurdington, Cheltenham, GL51 4SB
G0 TPP P. Pimblett, 5 Edgeside, Great Harwood, Blackburn, BB6 7JS
GW0 TPR B. Carter, Rhyd Y Mwyn, Cilgwyn Street, Llanerchymedd, LL71 8ED
G0 TPY J. Brunt, 1 Dane Grove, Cheadle, Stoke on Trent, ST10 1QS
GM0 TQB M. De Vries, Old Post Office House Knockbain, Munlochy, IV8 8PG
G0 TQC K. Sharman, 7 Watkins Way, Paignton, TQ3 3JJ
GI0 TQD J. Gough, 50 Culmore Point, Londonderry, BT48 8JW
G0 TQJ C. Vernon, 13 Greenlands Close, Tarporley, CW6 0DA
G0 TQP J. Smith, High Tree, Radford Lane, Wolverhampton, WV3 8JT
G0 TQR A. Baker, 23 Trematon Drive, Ivybridge, PL21 0HT
G0 TQS C. Forsyth, 8 Oriole Drive, Exeter, EX4 4SJ
G0 TQT J. Joll, 16 Jephson Road, St. Judes, Plymouth, PL4 9ET

G0 TQV K. Taylor, 1 Chapel Close, Reepham, Lincoln, LN3 4EJ
G0 TQZ M. Emm, Highwood Cottage, Daggons Road, Fordingbridge, SP6 3DJ
G0 TRB R. Betts, 15 Cleasby, Wilnecote, Tamworth, B77 4JL
G0 TRD T. Thorman, Tir Na Nog, Coombe Ridings, Kingston upon Thames, KT2 7JT
G0 TRE N. Dimbleby, 4 Rossetti Place, Holmer Green, High Wycombe, HP15 6XA
G0 TRG Thames ARG, c/o A. Atkinson, 21 Dennington Crescent, Basildon, SS14 2FF
G0 TRH D. Brooke, 19 Albion Court, Anlaby Common, Hull, HU4 7PL
G0 TRI L. Pritchard, The Granary, Greenways Farm, Ross-on-Wye, HR9 6DH
G0 TRJ R. Makepeace, 8 Lethlean Close, Phillack, Hayle, TR27 5AN
G0 TRK J. Ogden, 65 Elm St., Middleton, Manchester, M24 2EQ
G0 TRM C. Page, 1 The Leeway Danbury, Chelmsford, CM3 4PS
G0 TRN J. De Frece, 15 Kent Close, Highfield Road, Chesterfield, S41 7HA
G0 TRU A. Irvine, 21 Rutland Road, Partington, Manchester, M31 4NP
G0 TRW R. Cross, 7 The Island, Anthorn, Wigton, CA7 5AN
G0 TRY G. Lyon, 33 Barlborough Road, Pemberton, Wigan, WN5 9HZ
GI0 TSA D. Moore, 3 Knightsbridge, Londonderry, BT47 6FE
G0 TSB B. Snell, 30 Queens Crescent, Brixham, TQ5 9PJ
GW0 TSE L. Owen, Cartref, Llangain, Carmarthen, SA33 5AH
G0 TSG P. Ryder, 4 Edgeway, Nottingham, NG8 6LY
G0 TSH K. Hutt, Fenwick Crossing House, Fenwick Lane, Fenwick, Doncaster, DN6 0EZ
G0 TSJ S. Ruud, 5 Wood Street, Haworth, Keighley, BD22 8BJ
G0 TSK G. Wilkins, 156 Buckingham Crescent, Bicester, OX26 4HB
GW0 TSL H. Chapman, Flat 6, Archer Court, Phyllis Street, Barry, CF62 5US
G0 TSQ C. De Lacy, Monday Cottage, Hammerwood, East Grinstead, RH19 3QE
G0 TSR Dr N. Depledge, 29 Scargill Drive, Spennymoor, DL16 6LY
GI0 TSS C. Tait, 116 Aughnaskeagh Road, Dromara, BT25 2NT
G0 TST R. Hawtree, 9 Stonechat Road, Billericay, CM11 2NX
G0 TSU D. Michael, 5 Evelyn Close, Twickenham, TW2 7BL
G0 TTE S. Sutherland, 11 Brentwood Road, Leicester, LE2 6AD
GW0 TTF W. Price, Tramore, 67 High Street, Bridgend, CF32 0HL
G0 TTG M. Warriner, 34 Cabrera Avenue, Virginia Water, GU25 4EZ
G0 TTI R. Rayment, 145 Feeches Road, Southend-on-Sea, SS2 6TF
G0 TTL R. Booth, Old School House, Old School Lane, Doncaster, DN11 9BW
G0 TTM A. Radley, 16 Kingsley Lane, Thundersley, Benfleet, SS7 3TU
GW0 TTN P. Brzenczek, Flat 18, Ty Newydd House, Ty Newydd Court, Cwmbran, NP44 1LH
G0 TTO L. Chadwick, 2 Auden Place, Longton, Stoke-on-Trent, ST3 1SJ
G0 TTQ B. Trivett, 712 East Myrtle Ave, Foley, USA, 36535
G0 TTR Dr A. Bartle, 10 Holme Dene, Haxey, Doncaster, DN9 2JX
G0 TTS P. Bulmer, 61 Middleham Avenue, York, YO31 9BD
G0 TTW K. Yeates, Newlands, Ashby Lane, Lutterworth, LE17 4SQ
GM0 TTY W. Mcburney, 6 Hill Street, Tillicoultry, FK13 6HF
G0 TUC D. Wilkes, 85 Moss Lane, Hesketh Bank, Preston, PR4 6AA
G0 TUE R. Gilchrist, 132 Ramsden Street, Barrow in Furness, Cumbria, LA14 2BU
G0 TUI B. Hudson, 5 Rylands Road, Southend-on-Sea, SS2 4LW
G0 TUJ W. Spencer, 111 Rosmead Street, Hull, HU9 2TE
G0 TUL R. Woollard, 68 Trunk Furlong, Aspley Guise, Milton Keynes, MK17 8HX
G0 TUM B. Cooper, 13 Highbury Road, Leeds, LS6 4EX
G0 TUN A. Powney, 16 Westbrook Way, Wombourne, Wolverhampton, WV5 0EA
G0 TUO M. Brough, 24 St. Georges Road, Bletchley, Milton Keynes, MK3 5EN
G0 TUP N. Callow, 3 Maunleigh, Forest Town, Mansfield, NG19 0PP
GM0 TUS R. Young, 50 Berrywell Drive, Duns, TD11 3HG
G0 TUU R. Hudson, Norton House, 27 Torne View, Doncaster, DN9 3PQ
G0 TUV L. Kennedy, 28 Murton Garth, Murton, York, YO19 5UL
G0 TUW G. Woolfenden, 11 Chesshire Close, Areley Kings, Stourport-on-Severn, DY13 0EB
G0 TUX J. Fossey, 12 Hitchin Road, Arlesey, SG15 6RP
G0 TUZ W. Donovan, 28 Nevill Road, Snodland, ME6 5HX
G0 TVB P. Rigg, 1 Stones Hey Gate, Widdop Road, Hebden Bridge, HX7 7HD
G0 TVC A. Lanham, 7 College Lane, Stratford-upon-Avon, CV37 6DD
G0 TVD M. Smith, 9 Oak Green Way, Abbots Langley, WD5 0PJ
G0 TVL S. Birkenshaw, 19A Vale Head Grove, Knottingley, WF11 8JL
G0 TVM A. Bashir, 70 Smith Lane, Bradford, BD9 6DQ
G0 TVO B. Hewson, 6 Sanworth St., Todmorden, OL14 5BU
G0 TVR C. Binnell, 146 Hales Crescent, Warley, Smethwick, B67 6QX
G0 TVS C. Saunders, 17 Bure Road, Friars Cliff, Christchurch, BH23 4ED
GM0 TVT R. Hemmings, 2A Caversta, Isle of Lewis, HS2 9QE
G0 TVU M. Binns, 49 Fairview, Pontefract, WF8 3NT
G0 TVV E. Graham, Ibis Styles London, 543 Lea Bridge Road Reception Desk, London, E10 7EB
G0 TVW D. Rodman, Flat 4, Heatherfield Court, 197 Baslow Road, Totley Rise, Sheffield, S17 4DT
GW0 TVX R. Elms, 7 Manor Daf Gardens, St. Clears, Carmarthen, SA33 4ES
GM0 TWB I. Lindsay, Fallady Cottage, Angus, DD8 2SP
G0 TWD OBO Wigan Deanery High School, c/o Rev. J. Drummond, 14 Bulls Head Cottages, Turton, Bolton, BL7 0HS
G0 TWE Rev. T. Walker, 1 Neville Turner Way, Waltham, Grimsby, DN37 0YJ
GW0 TWF M. Patterson, 6 Devonshire Place, Port Talbot, SA13 1SG
G0 TWH M. Jewkes, 11 Maple Grove, Crewe, CW1 4DY
GW0 TWI S. Foote, Red Roofs, 5 Woodland Avenue, Colwyn Bay, LL29 9NL
GW0 TWL M. Lewis, 75 Lon Maesycoed, Newtown, SY16 1QQ
GW0 TWO P. Murdoch, 19 Penhelyg Road, Aberdovey, LL35 0PT
GW0 TWR C. Harrison, 28 Brynau Wood, Cimla, Neath, SA11 3YQ
G0 TWT B. Lee, Ridge Hill, Ledbury Road, Ledbury, HR8 1ND
G0 TWV K. Stewart, 97 Chase Meadows, Blyth, NE24 4LB
GI0 TWX Dr I. Chin, 25 Meadowbrook, Islandmagee, Larne, BT40 3UG
G0 TXA B. Carter, 16 Hollow Street, Canterbury, CT34DS
GM0 TXJ N. Service, 13 Garden Terrace, Falkirk, FK1 1RL
G0 TXL P. Elliott, 32 Crichton Avenue, Wallington, SM6 8HL
G0 TXN M. Thoyts, 17 Solent Avenue, Lymington, SO41 3SD
G0 TXO D. Whitaker, 19 Bradwell Fold, Glossop, SK13 6HX
GW0 TXP A. Smith, Courtlands, Llysonnen Road, Carmarthen, SA33 5DR
G0 TXU G. Redmond, 21 Grosvenor Gardens, Bognor Regis, PO21 3EZ
G0 TXY A. Hicks, 29 Oak Tree Close Strensall, York, YO32 5TE
G0 TYM T. Allison12, 2 Westlands, Stokesley, Middlesbrough, TS9 5BU
G0 TYN M. Holmes, 48 Woodpecker Close, Wirral, CH49 4QP
G0 TYQ G. Clark, 2 Keith Road, Swanton Morley, Dereham, NR20 4NQ
G0 TYS S. Allanson, 5 Kingsley Avenue Crofton, Wakefield, WF4 1RN
G0 TYW P. Cocker, 52 Heathlee Road, London, SE3 9HP
G0 TYZ T. Turley, 6 Rowan Grove, Oxford, OX4 7FD
G0 TZC P. Sables, 45 Carr Head Lane, Bolton-Upon-Dearne, Rotherham, S63 8DA
G0 TZD R. Leah, 64 Valley Road, Chatham, Canada, N7L 5G3
G0 TZH M. Bushnell, Rose Cottage, Street Ashton, Rugby, CV23 0PH
G0 TZM K. Winfield, Russettwalls, 58 Bretby Lane, Burton-on-Trent, DE15 0QW
G0 TZO R. Nelson, 61 Broken Cross, Charminster, Dorchester, DT2 9QB
G0 TZP A. Seals, 94A Snakes Lane, Southend-on-Sea, SS2 6UA
G0 TZR J. Macknish, 21A Knoll Rise, Orpington, BR6 0EJ
G0 TZV G. Bryant, 54 Drew Road, Stourbridge, DY9 0UP
G0 TZY A. Davies, 20 Peel Park Close, Accrington, BB5 6PL
G0 TZZ C. Soames, Stud Farm Bungalow, 3 The Street, Sporle, PE32 2EA
G0 UAA I. Fallows, 47 Melrose Avenue, Burnley, BB11 4DN
G0 UAC C. Martin, 17 Chambers Grove, Chapeltown, Sheffield, S35 2TD
G0 UAD G. Rowe, 8 Dove Close, Bolton-Upon-Dearne, Rotherham, S63 8JL
GI0 UAG R. Anderson, Derry Lodge, 2 Tullymally Road, Newtownards, BT22 1JX
G0 UAI S. Marshall, 68 Parkfield Avenue, Hampden Park, Eastbourne, BN22 9SF
G0 UAK R. Thompson, 51 Rydal Avenue, Ramsgate, CT11 0PX
G0 UAO J. Smith, 124 Parkside Avenue, Barnehurst, Bexleyheath, DA7 6NL
G0 UAP P. Westbury, 5 Smithfield Place, Bournemouth, BH9 2QJ
G0 UAS A. Smith, 3 Woodcourt Close, Sittingbourne, ME10 1QT
G0 UAV P. Tristram, 26 St. Andrews Road, Paignton, TQ4 6HA
G0 UAY E. Fletcher, 5 Butt Hill Court, Bury New Road, Manchester, M25 9NT
G0 UBA R. Gibbs, 357 Downham Way Bromley, Kent, London, BR1 5EW
G0 UBG D. Endean, 6 Higher Westonfields, Totnes, TQ9 5QY
G0 UBJ K. Beach, Taylor Cove, Harwich Road, Clacton-on-Sea, CO16 0AX
G0 UBK C. Carvell, 6 Field Close, Whitby, YO21 3LR
G0 UBL M. Stracey, 9 Boundary Drive, Hutton, Brentwood, CM13 1RH
G0 UBM J. Horsfield, 8 St. Edmunds Road, Ipswich, IP1 3QZ
G0 UBO E. Martin, 9 The Valley Green, Welwyn Garden City, AL8 7DQ
G0 UBX A. Quince, 9 Biscay Close, Irchester, Wellingborough, NN29 7FD
G0 UBY L. Duffill, 14 Leonard Street, Hull, HU3 1SA
GW0 UCA Dr T. Ogden, Tyn Llidiart, Brithdir, Dolgellau, LL40 2RP
G0 UCC M. Sayegh, 1 Hoylake Road, East Acton, London, W3 7NP
G0 UCD M. West, 12 Jenny Gill Crescent, Skipton, BD23 2RR
G0 UCE J. Swartz, 15076 New Salem Bluff Road, Petersburg, Il, USA, 62675
G0 UCF Dr G. Knox, 117 Old Shoreham Road, Hove, BN3 7AQ
G0 UCH C. Hawes, 64 Whitmore Court, Whitmore Way, Basildon, SS14 2TN
G0 UCI R. Jones, 29 Avon Dale, Newport, TF10 7LS
G0 UCK B. Linehan, 28 Hurlstone Grove Furzton, Milton Keynes, MK4 1EF
G0 UCN P. Turner, 176 Crescent Road, Hadley, Telford, TF1 5LF
G0 UCP Dr J. Seager, 2 Waterford Road, Oxton, Prenton, CH43 6UT
G0 UCS W. Dingley, 63 Hurst Green, Mawdesley, Ormskirk, L40 2QS
G0 UCT B. Obrien, 59 Riddlesdown Avenue, Purley, CR8 1JL
G0 UCX T. Taylor, 1 Cooke Gardens, Branksome, Poole, BH12 1QE
G0 UDB D. Birch, 4 Godolphin Road, Helston, TR13 8PY
G0 UDG K. Deegan, 7 Oldcott Crescent, Kidsgrove, Stoke-on-Trent, ST7 4HF
GW0 UDH R. Pritchard, 1 Limetree Court, Station Road, Abergavenny, NP7 5JA
G0 UDI J. Murphy, 8 Spencer Avenue, Wribbenhall, Bewdley, DY12 1DB
GW0 UDJ D. Jones, Bryn Awelon, Brynsannan, Holywell, CH8 8AX
GM0 UDL A. Cowan, House Of Shannon, Wester Templands, Fortrose, IV10 8RA
G0 UDO R. Dodd, 33 Dogcroft Road, Stoke-on-Trent, ST6 6PE
G0 UDP K. Fairbotham, 32 Northolme Drive, York, YO30 5RP
GM0 UDY A. Hyslop, 9 Shalloch Square, Girvan, KA26 0EA
G0 UDZ M. Baister, 7B Hepple Road, Spital Estate, Newbiggin-by-the-Sea, NE64 6ST
G0 UEA J. Heald, 94 Haylings Road, Leiston, IP16 4DT
G0 UEB R. Fisher, 14 Colindeep Lane, Sprowston, Norwich, NR7 8EG
G0 UEC J. Bird, 56 Garsdale, Birtley, Chester le Street, DH3 2EY
G0 UED I. Harkness, 2 Trevor Drive, Maidstone, ME16 0QP
GI0 UEG Radio Amateur Special Event Group, c/o R. Wilson, 68 Kensington Road, Belfast, BT5 6NG
G0 UEH R. Hislop, 79 Norwood Avenue, Hasland, Chesterfield, S41 0NJ
G0 UEK S. Payne, 55 Binstead Lodge Road, Binstead, Ryde, PO33 3TL
GW0 UEO J. Ewan, 71 Kingston Drive, Connah's Quay, Deeside, CH5 4TN
GM0 UET R. Henderson, 22 Bowmont Place, East Kilbride, Glasgow, G75 8YG
G0 UEU A. Jackson, 7 Rushfield, Sawbridgeworth, CM21 9NF
G0 UEW St Christophers School ARC, c/o R. Jones, Greens Cottage, Luton Road, Offley, Hitchin, SG5 3DR
G0 UFB P. Short, 23 Barn Close, Hartford, Huntingdon, PE29 1XF
G0 UFC Dr D. Meacock, The Limes, Davids Lane, Boston, PE22 0BZ
G0 UFD T. Reynolds, 37 Clarendon Street, Rochdale, OL16 4UB
G0 UFE S. Bird, 15 Ludlow Drive, Stirchley, Telford, TF3 1EG
G0 UFF R. Reynolds, 5 Dymond Court, Bodmin, PL31 2FP
G0 UFI G. Brady, Thirn Grange, Thirn, Ripon, HG4 4AU
G0 UFJ S. Glover, 46 Merton Close, Oldbury, B68 8NG
G0 UFL T. Reid, Menwith Hill Station, Po Box 985, Harrogate, HG3 2RF
G0 UFN P. Dean, 80 Escallond Drive, Dalton-Le-Dale, Seaham, SR7 8JZ
G0 UFP C. Beesley-Reynolds, Kaos Roams, Palmerston Close, Leicester, LE8 0JJ
G0 UFU C. Jameson, 35 Bilberry Grove, Taunton, TA1 3XN
G0 UFV P. Shaw, 15 Moorfield Road, St. Giles-On-The-Heath, Launceston, PL15 9SY
G0 UFW J. Marvill, 242 Hillmorton Road, Rugby, CV22 5BG
G0 UFY J. Brown, 3 Slipper Mill, Slipper Road, Emsworth, PO10 8XD
G0 UFZ Dr J. Howard, 32 Eastgate, North Newbald, York, YO43 4SD
G0 UGA B. Moody, 38 Bromwich Road, Willerby, Hull, HU10 6SF
G0 UGD N. Boyd, 16 Edensor Road, Eastbourne, BN20 7XR
GM0 UGG A. Aird, 3 Graystones, Kilwinning, KA13 7DT
GM0 UGH M. Westland, 142 Claremont, Alloa, FK10 2EG
G0 UGI H. Argument, 9 Oxley Close, Shepshed, Loughborough, LE12 9LS
G0 UGJ D. Basford, 91 Hollins Spring Avenue, Dronfield, S18 1RP
G0 UGM E. Peasey, Greenfield Cottage, Greenfield Road, Colne, BB8 9PE
GW0 UGQ M. Webb, 75 Bolingbroke Heights, Flint, CH6 5AN
G0 UGR M. Chaloner, Barnhay House, Newport, Berkeley, GL13 9PY
G0 UGS F. Taberner, 51 Canford View Drive, Colehill, Wimborne, BH21 2UW
G0 UGW Lt. P. Grant, 37 Glenmore Park, Dundalk, County Louth, Ireland,
G0 UGX J. Kinsella, 85 Lowther St., Coventry, CV2 4GL
G0 UGY S. Jackson, 32 Sherwell Drive, Alcester, B49 5HA
GM0 UHC I. Ropper, 2 Deerhill, Dechmont, Broxburn, EH52 6LY

G0 UHD K. Hore, 15 Heriot Way, Great Totham, Maldon, CM9 8BW
G0 UHF R. Darwent, 139 The Oval, Sheffield, S5 6SQ
G0 UHG R. Warne, 32 Chance Street, Tewkesbury, GL20 5RF
G0 UHI D. Norton, 52 Letchworth Road, Leicester, LE3 6FG
GW0 UHJ W. Griffiths, 147 High Street, Tonyrefail, Porth, CF39 8PL
G0 UHK Prof. M. Peiperl, 45 High St., Harrow, HA1 3HT
G0 UHM L. Ruddock, 2 Cross Lane, Waterlooville, PO8 9TJ
GW0 UHO P. Sage, 4 Gladstone Terrace, Miskin, Mountain Ash, CF45 3BS
G0 UHQ M. Minihane, 60 Wolsey Drive, Walton-on-Thames, KT12 3BA
G0 UHS R. Hatch, 99-101 Hornby Road, Blackpool, FY1 4QP
G0 UHU G. Bloyce, 8 Olivers Court Olivers Close, Clacton-on-Sea, CO15 3QX
GW0 UHX S. Jones, 64 Springfield Gardens, Hirwaun, Aberdare, CF44 9LY
G0 UID J. Parker, 1 Church End, Syresham, Brackley, NN13 5HU
G0 UIF G. Fairbrass, 230 Kirkby Road, Barwell, Leicester, LE9 8FS
GM0 UIG C. Cowan, 85 Eastwoodmains Road, Clarkston, Glasgow, G76 7HG
G0 UIH S. Lawman, 44 Barnwell, Peterborough, PE8 5PS
G0 UIL D. Brice, 4 Bishop Fox Drive, Taunton, TA1 3HQ
GW0 UIP N. Wallace, Tan Y Bryn, Bryn Road, Flint, CH6 5HU
G0 UIQ W. Furze, 2 Lynewood Road, Cromer, NR27 0EE
G0 UIW A. Jones, 26 Grosvenor Road, Harrogate, HG1 4EG
G0 UIX A. Lambert, 10 The Green, Cirencester, GL7 1AU
GW0 UIZ B. Galsworthy, 30 Pen Yr Yrfa, Morriston, Swansea, SA6 6BA
G0 UJD M. Bartle, 14 Litton Avenue, Skegby, Sutton-in-Ashfield, NG17 3AB
GI0 UJG R. Stinson, 51 Cloncarrish Road, Portadown, Craigavon, BT62 1RN
G0 UJI D. Sparkes, 31 Lockyers Drive, Ferndown, BH22 8AL
G0 UJP J. Fleetwood, 11 Chichester Road, Southampton, SO18 6BB
G0 UJT J. Deal, 19 Coniston Road, Gloucester, GL2 0NA
G0 UJU D. Barlow, 7 Prospect Terrace, Canal Road, Taunton, TA1 1PH
G0 UKA J. Black, 8 Cornwood Close, Finchley, London, N2 0HP
GW0 UKC M. Price, 50 Llangorse Road, Cwmbach, Aberdare, CF44 0HR
GM0 UKD S. Munro, 76 John Neilson Avenue, Paisley, PA1 2SX
GW0 UKF J. Griffith, Craig Artro, Llanbedr, LL45 2LU
GW0 UKG A. Powell, 80 St. Andrews Crescent, Abergavenny, NP7 6HN
G0 UKK K. Stanyer, 15 Wilbrahams Way, Alsager, Stoke-on-Trent, ST7 2NR
G0 UKL P. Charlton, 6 Maiden Road Shirebrook, Mansfield, NG20 8GA
G0 UKM M. Russell, 10 Baytree Grove, Ramsbottom, Bury, BL0 9UF
G0 UKO R. Theakston, 130 Greenshaw Drive Haxby, York, YO32 2DG
G0 UKP B. Jopson, 21 Richmond Street, Southend-on-Sea, SS2 4NW
G0 UKS R. Towler, 77 Glebe Road, Hull, HU7 0DU
GW0 UKT W. Waldron, Torfaen Scouts Arc, 23 Forest Close, Cwmbran, NP44 4TE
GM0 UKZ C. Gibson, 45 Tiree Place, Newton Mearns, Glasgow, G77 6UJ
G0 ULA N. Foster, 68 Brookfield Way, Lower Cambourne, Cambridge, CB23 5ED
G0 ULG K. Dewing, 124 Proctor Road, Norwich, NR6 7PH
G0 ULH L. Harris, 183A Painswick Road, Gloucester, GL4 4AG
G0 ULI M. Kaliski, 132 Woodland Road, Hellesdon, Norwich, NR6 5RQ
G0 ULL E. Williams, 50 Broad Lawn, New Eltham, London, SE9 3XD
G0 ULM P. Wilson, 56 Highfield Road, Blacon, Chester, CH1 5AZ
G0 ULN L. Fuller, 30 Linthorpe Grove, Willerby, Hull, HU10 6SA
G0 ULO P. Ravenscroft, 32 Tennyson Road, Wolverhampton, WV10 8NG
GW0 ULP L. Parsons, 105 Victoria Road West, Prestatyn, LL19 7DS
G0 ULQ W. Bucknell, 119 Fossway, York, YO31 8SQ
G0 ULS F. Mortimer, 115 Dell Road, Lowestoft, NR33 9NX
G0 UMI E. Latter, 76 Crossway, Plympton, Plymouth, PL7 4HY
GM0 UMJ Dr S. Heerma Van Voss, Blarachaorachan, Fort William, PH33 6SZ
G0 UMK R. Adams, 9 Chestnut Garth, Roos, Brooklands, Hull, HU12 0LE
G0 UML N. Watling, 36 All Saints Walk, Mattishall, Dereham, NR20 3RF
G0 UMM N. Roberson, 6 Long Lane, West Winch, King's Lynn, PE33 0PG
G0 UMP M. Parker, 102 Cavendish Road, Patchway, Bristol, BS34 5HH
G0 UMS R. Petrie, 11 St. James'S Close, Yeovil, BA21 3AH
G0 UMV P. Johnson, 52 Evesham Road, Cookhill, Alcester, B49 5LJ
G0 UMY C. Lowe, 37 Parsonage Brow, Upholland, Skelmersdale, WN8 0JG
G0 UNB J. Jeffers, 11 Polywell, Appledore, Bideford, EX39 1SG
G0 UNC G. Hancock, 3C Richmond Street, Hull, HU5 3JY
G0 UND D. Scargill, 10 Mendip Avenue, North Hykeham, Lincoln, LN6 9SZ
G0 UNE J. Spencer, 6 Redcar Road, Sunderland, SR5 5QA
G0 UNF G. Taylor, Flat 67A, Bramley Grange Hotel Flats, Guildford, GU5 0BL
G0 UNG B. Massey, 40 East St., Ashton In Makerfield, Wigan, WN4 8ST
G0 UNK T. Wright, 2 Regent Road Church, Accrington, BB5 4AR
G0 UNW R. Martin, 44 Threadneedle Cres, Willowdale, Ontario, Canada,
G0 UNY M. Lindley, 23 Townend Lane, Deepcar, Sheffield, S36 2TN
G0 UOB A. Chilinski, 54, Foxes Meadow, King's Lynn, Norfolk, PE32 2AS
G0 UOD E. Sheather, Clare Cottage, North End, Shaftesbury, SP7 9HX
G0 UOI R. Fox, 20 Levett Road, Polegate, BN26 6NE
G0 UOK R. Dutton, 3 Kilkenny Road, Guisborough, TS14 7LE
GI0 UOL D. Hards, 22 Balloo Walk, Bangor, BT19 7HL
G0 UOM N. Taylor, 16 Josephine Road, Rotherham, S61 1BJ
G0 UOP OBO University of Plymouth A.R.S., c/o R. Linford, Department Of Communication, And Electronic Engineering, Drake Circus Plymouth, PL4 8AA
G0 UOQ P. Creissen, 143 Hawthorn Bank, Spalding, PE11 2UN
G0 UOS J. Butterworth, 12 Ingswell Drive Notton, Wakefield, WF4 2NF
GM0 UOU C. Muir, 1/1 132 Falside Road, Paisley, PA2 6JT
G0 UOV S. Porter, 20 Newbridge Road, Ambergate, Belper, DE56 2GR
G0 UOZ A. Morris, 4 Pleasant Terrace, Lincoln, LN5 8DA
G0 UPD R. Brinkley, 70 Leopold Road, Felixstowe, IP11 7NR
GM0 UPE Dr G. Sutherland, 22 Montrose Drive., Bearsden, G61 3LG
G0 UPG J. Dunne, 40 Egmont Road, Poole, BH16 5BZ
G0 UPK D. Harmer, 98 King Georges Avenue, Coventry, CV6 6FF
G0 UPL H. Summers, 2 Hillview, Highgate Road, Forest Row, RH18 5AZ
G0 UPO M. May, Flat 4, Hyde Bank Court, Hyde Bank Road, New Mills, High Peak, SK22 4NE
G0 UPP S. Bennett, 102 Laflouder Fields, Mullion, Helston, TR12 7EJ
G0 UPS P. Zimmermann, 85 Wimborne Road West, Wimborne, BH21 2DH
G0 UPV T. Berrisford, 126 Star & Garter Road, Stoke-on-Trent, ST3 7HN
G0 UPY M. Rogers, 3 Wheelwright Gardens, Long Compton, Shipston-on-Stour, CV36 5LN
G0 UQB D. Brittain, 9 Highfield Road, Cookley, Kidderminster, DY10 3UB
G0 UQC R. Birkett, 14 Greta Street, Keswick, CA12 4HS
G0 UQE R. Weaver, 107 Patch Lane, Redditch, B98 7XE
G0 UQF G. Merrils, 2 East St., Darfield, Barnsley, S73 9AE
GW0 UQH S. Provan, 12 Trewarren Road, St. Ishmaels, Haverfordwest, SA62 3SZ
G0 UQI L. Snowden, 25 Brixham Road, Paignton, TQ4 7HG
G0 UQJ N. Smith, 36 Maple Road, Sutton Coldfield, B72 1JP
GI0 UQK P. Sinclair, 37 Willow Avenue, Banbridge, BT32 4RE
G0 UQO G. Rekers, 18 The Ridge, Purley, CR8 3PE
G0 UQP F. Waters, 96 Stockley Road, Barmston Village, Washington, NE38 8DR
G0 UQQ N. Baskerville, 10 Park Avenue, Sprotbrough, Doncaster, DN5 7LW
G0 UQT S. Nash, 26 Lyndhurst Crescent, Wembdon, Bridgwater, TA6 7QG
G0 UQU J. Squires, 44 St. Marys Road, Doncaster, DN1 2NP
G0 UQV M. Parkin, Chenet, 32 The Nooking, Doncaster, DN9 2JQ
G0 UQY P. Cox, 17 Hyde Lane, Upper Beeding, Steyning, BN44 3WJ
G0 UQZ D. Eyre, 41 Lindsay Road, Sheffield, S5 7WE
G0 URB J. Morse, 14 Strathmore Road, Bournemouth, BH9 3NS
G0 URC P. Ball, 12 Warren Way, Digswell, Welwyn, AL6 0DH
GM0 URD I. Thomson, 33 Aytoun Grove, Dunfermline, KY12 9YA
G0 URF M. Dodsworth, 359 Upper Town Street, Bramley, Leeds, LS13 3JX
GI0 URI S. Robinson, 19 Pattonville, Dungannon, BT71 6DD
G0 URK J. Alderman, 56 Edward Harvey Link, Beaulieu Park, Chelmsford, CM1 6BU
GI0 URN Royal Naval (Ulster) ARC, c/o N. Mckee, 54 Castlemore Park, Belfast, BT6 9RP
G0 URO D. Fosh, 8 The Pines, Horsham, RH12 4UF
G0 URT S. Bradbury, 39 Grosvenor Road, Hyde, SK14 5AB
GM0 URU Dr J. Cartlidge, 14 Davidson Street, Broughty Ferry, Dundee, DD5 3AT
G0 URW A. Paone, Viale Dei Quattro, Venti 128 C/O Int6, Roma, Italy, 00 152
G0 USA L. Civita, 53 Rockhurst Drive, Eastbourne, BN20 8XD
GI0 USC J. Smith, 5 Old Turn, Carrickfergus, BT38 7EH
G0 USE J. Davy-Jones, 1 Wensley Gardens, Emsworth, PO10 7RA
G0 USF M. James, 8 Swan Quay Bath Lane, Fareham, PO16 0DX
GM0 USI A. Dimmick, 02 120 Shakespeare Street, Glasgow, G20 8LF
G0 USJ S. Johnson, 36 Langar Woods, Langar, Nottingham, NG13 9HZ
G0 USK P. Perera, 13 Dalcross Road, Hounslow, TW4 7RA
G0 USM B. Parker, 24 Mayfield Road, Chorley, PR6 0DG
GI0 USQ P. Fox-Roberts, 8 Lynwood Park, Holywood, BT18 9EU
GI0 USS K. Chambers, 59 Ravenswood, Banbridge, BT32 3RD
GI0 USW P. Mcdonald, 13 Serpintine Road, Newtownabbey, BT36 7HA
G0 UTA C. Gurney, 9 Snowdrop Mews, Exeter, EX4 2PN
G0 UTB Dr P. Cordrey, Redline C.A Edif. Fannellis 5 Y 6, Av. Ricaurte, Cojedes, Venezuela, 2201
GW0 UTC J. Lomas, 5 Gwelfor, Rhos On Sea, Colwyn Bay, LL28 4AJ
GM0 UTD H. Urquhart, The Cless, Peebles, EH45 8NU
GI0 UTE D. Auld, 37 Castlewellan Road, Rathfriland, Newry, BT34 5EL
G0 UTM B. Watson, 6 Shakespeare Avenue, Scunthorpe, DN17 1SA
G0 UTN S. Harrison, 22 Dales Avenue, Sutton-in-Ashfield, NG17 4BY
G0 UTP R. Walker, 1 Farmcote Court, Hemlington, Middlesbrough, TS8 9LJ
G0 UTR D. Coleman, 1 Kerstin Close, Cheltenham, GL50 4SA
G0 UTT Dengie Hundred ARS, c/o A. Slade, Skerries, Summerhill Althorne, Chelmsford, CM3 6BY
G0 UTU P. Humphreys, 47 Crescent Road, Locks Heath, Southampton, SO31 6PE
GI0 UTV I. Ross, 312 Castlereagh Road, Belfast, BT5 6AD
G0 UTX P. Curtis, 31 Dorking Close Ings Estate, Hull, HU8 9DG
G0 UTZ R. Davies, 36 Woodside Avenue, Brown Edge, Stoke-on-Trent, ST6 8RX
G0 UUA Dr W. Hutchings, Bitternsdale Farm, Bustomley Lane, Stoke-on-Trent, ST10 4PE
GM0 UUB G. Matthews, 88 Nevis Crescent, Alloa, FK10 2BN
G0 UUC W. Slater, 44 Hope St., Chesterfield, S40 1DG
G0 UUF S. Errington, 23 Pinewood Drive, Bletchley, Milton Keynes, MK2 2HT
G0 UUI R. Newport, 17 College Crescent, Oakley, Aylesbury, HP18 9QZ
G0 UUM J. Lewis, Burley Cottage, Shortwood, Stafford, ST21 6RG
G0 UUN J. Lewis, Burley Cottage, Shortwood, Stafford, ST21 6RG
G0 UUP M. Stevens, 24 Oakroyd Close, Burgess Hill, RH15 0QN
G0 UUR C. Branch, 10 Queens Road, Thame, OX9 3NQ
G0 UUS G. Hanson, 34 Old Garden Close, Locks Heath, Southampton, SO31 6RN
G0 UUT I. Paim, 8 Sparrow Close, Attleborough, NR17 1GP
G0 UUU P. Earnshaw, 7 Hampton Road, Scarborough, YO12 5PU
G0 UUZ A. Breeze, Drakelow Cottage, Drakelow Lane, Wolverley, Kidderminster, DY11 5RU
GI0 UVD W. Bustard, 66 Hertford Crescent, Lisburn, BT28 1SQ
G0 UVE I. Reynolds, Five Bars, Hereford Road Weobley, Hereford, HR4 8SW
G0 UVG P. Ellis, 9 Matilda Gardens, Shenley Church End, Milton Keynes, MK5 6HT
GU0 UVH T. Bosher, Highlea, Clos Des Mouriaux Les Mouriaux, Alderney, Guernsey, GY9 3UY
G0 UVL L. Cartwright, 14 Norley Hall Avenue, Wigan, WN5 9TG
G0 UVN B. Goolding, 10 Oakwell Close, Stevenage, SG2 8UG
G0 UVR G. Akse, Drive Cottage, Ebberston, Scarborough, YO13 9PA
G0 UVT J. Goldie, The Coachmans Cottage, Pulford Lane, Chester, CH4 9NN
G0 UVX G. Valentine, 19 Kingsway Stotfold, Hitchin, SG5 4EL
G0 UWA E. Shanklin, The Coachmans Cottage, Pulford Lane, Chester, CH4 9NN
G0 UWB R. Moyle, Aitchill House, Lower Brailes, Banbury, OX15 5AP
GW0 UWD J. Matthews, 42 Wexham Street, Beaumaris, LL58 8HW
G0 UWF S. Bowles, 37 Manor Road, Paignton, TQ3 2HZ
G0 UWI H. Chipper, 36 Newbridge Way, Truro, TR1 3LX
G0 UWK I. Goodier, 2 Chatterley Drive Kidsgrove, Stoke-on-Trent, ST7 4HW
G0 UWO N. Winfield, Oaklea, Chyvogue Meadow, Truro, TR3 7JP
G0 UWS A. Sharman, 3 Deben Crescent, Swindon, SN25 3QB
G0 UWU M. Claridge, 105 Barnwood Avenue, Gloucester, GL4 3AG
GM0 UWV J. Reynolds, The Paddock, Leswalt, Stranraer, DG9 0LJ
G0 UWX D. Pollard, 191 High Road, Halton, Lancaster, LA2 6QB
GI0 UXD D. Burns, 207 Rathfriland Road, Dromara, Dromore, BT25 2EQ
G0 UXF C. Whittaker, 3A Oak Avenue, Horwich, Bolton, BL6 6JE
G0 UXG P. Blunt, 17 Offens Drive, Staplehurst, Tonbridge, TN12 0LR
G0 UXH A. Mawson, 93 Glenridding Drive, Barrow-in-Furness, LA14 4PA
G0 UXI M. Whitehead, 7 Avalon Drive, Manchester, M20 5WN
GW0 UXJ A. Burns, 34 Lakeside Gardens, Merthyr Tydfil, CF48 1EN
G0 UXO B. Hillman, 2 Holmes Chapel Road, Congleton, CW12 4NE
G0 UXR T. Gilmore, 48 Ash Lane, Hale, Altrincham, WA15 8PD
GW0 UXX D. Cumiskey, 16 Delapoer Drive, Haverfordwest, SA611HJ

G0 UXZ A. Walmsley, Runaways Cottage, Ring Street Stalbridge, Sturminster Newton, DT10 2LZ
G0 UYA S. Chamberlin, 15 Bull Close, East Tuddenham, Dereham, NR20 3LX
G0 UYC D. Rolph, 3 Bell Close, Bawdeswell, Dereham, NR20 4SL
G0 UYE A. Colton, 9 Pineway, Bridgnorth, WV15 5DS
G0 UYF P. Moss, 34 Almond Tree Avenue, Goole, DN14 9QR
G0 UYG A. Forster, 9 Pinewood Walk, Stokesley, Middlesbrough, TS9 5HU
G0 UYH A. Smart, Nine Hamelin Street, Pye Green Road, Cannock, WS11 2SE
G0 UYM OBO Thames Amateur Radio, c/o J. Hall, Pump Farm Cottage, Pump Lane South, Marlow, SL7 3RB
G0 UYP R. Arkell, 76, Wilnecote Lane, Tamworth, B77 2JA
G0 UYQ M. Melbourne, 42 Pasture Road, Stapleford, Nottingham, NG9 8GL
GM0 UYS S. Jeffrey, 33 The Rowans, Insch, AB52 6ZD
G0 UYT J. Hemming, 62 Beaumont Road Bournville, Birmingham, B30 2DY
G0 UYV B. Smith, 80A Bramcote Lane, Beeston, Nottingham, NG9 4ES
GI0 UYY L. O'Flaherty, 1 Ravensdale Villas, Newry, BT34 2PG
GM0 UYZ J. Wheeler, 54 Wittet Drive, Elgin, IV30 1TB
G0 UZD A. Gisby, 18 Westmead Gardens, West Avenue, Worthing, BN11 5LP
G0 UZE K. Brookes, Kohima, Spout Lane, Stoke-on-Trent, ST2 7LR
G0 UZF I. Riley, 102 Harrison Road, Chorley, PR7 3HS
G0 UZK A. Rushton, Rosemary Cottage, The Green, East End, North Leigh, Witney, ox296py
GM0 UZV W. Cargill, 23 Ceres Road, Craigrothie, Cupar, KY15 5QB
G0 UZW N. Donald, 20 Parkhill Road, Barnby Dun, Doncaster, DN3 1DP
GW0 UZX F. Thompson, 15 Aneurin Crescent, Twynyrodyn, Merthyr Tydfil, CF47 0TB
G0 UZY J. Shotter, Elthorne, Hawkswood Road, Hailsham, BN27 1UN
GI0 VAB P. Moore, 59 Belmont Avenue, Belfast, BT4 3DE
G0 VAD J. Whitehall, 29 Melrose Terrace, Newbiggin-by-the-Sea, NE64 6XN
G0 VAE M. Clarke, 21 Sycamore Road, Greenstead Estate, Colchester, CO4 3NF
G0 VAG P. Mason, 8 Westbourne Park, Scarborough, YO12 4AT
G0 VAH A. Heyworth, 3 The Knowl Churton, Chester, CH3 6NE
G0 VAI R. Frow, Valley Yard, Skeete Road, Folkestone, CT18 8DS
G0 VAJ S. Hobden, 10 Turton Close, Brighton, BN2 5DA
G0 VAL D. Wyatt, 96 Woodlands Close, Clacton-on-Sea, CO15 4RU
G0 VAM A. Witter, 44 Regent St., Newton le Willows, WA12 9LS
G0 VAR C. Wilson, 15 Biddenden Close, Bearsted, Maidstone, ME15 8JP
G0 VAS V. Ikonomou, 18 Canhams Road, Great Cornard, Sudbury, CO10 0EP
G0 VAU T. James, 114 Broadway, Loughborough, LE11 2JG
G0 VAV J. Farrington, 6 The Gravel, Mere Brow, Preston, PR4 6JX
G0 VAX B. Bowers, 31 Gresford Avenue, Wirral, CH48 6DA
G0 VAY G. Gower, 14 Beck Garth, Hedon, Hull, HU12 8LH
G0 VAZ R. Barker, 56 Southend Road, Weston-Super-Mare, BS23 4JZ
GM0 VBE B. Higton, The Straith, Priestland, Darvel, KA17 0LP
G0 VBG W. Chandler, 6 St. Thomas Close, Hinton Waldrist, Faringdon, SN7 8RP
G0 VBK M. Blackmore, Flat 20 Hill View House, Bristol, BS15 1TA
G0 VBM R. Caddy, 4 Londesborough Park, Seamer, Scarborough, YO12 4QT
G0 VBN J. Cressey, 32 Ballifield Road, Sheffield, S13 9HX
G0 VBP F. Madely, 138 Coningsby Drive, Kidderminster, DY11 5LZ
G0 VBQ F. Webster, 14 Redbank Avenue, Erdington, Birmingham, B23 7JR
G0 VBR M. Dunstan, 3 Coronation Road, Rawmarsh, Rotherham, S62 5LW
G0 VBT G. Dawson, 22 High St., Tean, Stoke on Trent, ST10 4DZ
G0 VBX J. Collingwood, 31 Corn Close, South Normanton, Alfreton, DE55 2JD
G0 VBZ D. Stinton, 43 The Meadows, Bidford-On-Avon, Alcester, B50 4AP
G0 VCD L. Browne, 7 Mornington Drive, Cheltenham, GL53 0BH
G0 VCJ K. Emblen, 17 Larch Close, Bognor Regis, PO22 9LA
GM0 VCN W. Long, 52 Stirling Road, Milnathort, Kinross, KY13 9XG
G0 VCV J. Partridge, 27 Leigh Road, Penhill, Swindon, SN2 5DE
G0 VCW R. Evans, 4 Cedar Drive, Carlton Manor Park, Chapel Road, Carlton Colville, Lowestoft, NR33 8BL
G0 VDE Dr W. Rothwell, 30 Wellbrook Way, Girton, Cambridge, CB3 0GP
G0 VDJ E. Webster, 33 Cherry Gardens, Bitton, Bristol, BS30 6JA
G0 VDN Dr B. Logan, 22 Chiltern House, Hillcrest Road, London, W5 1HL
G0 VDO C. Pond, 316 St. Faiths Road, Old Catton, Norwich, NR6 7BL
G0 VDP K. Hutley, Three Ways, 1 Walden House Road, Maldon, CM9 8PJ
G0 VDQ A. Ramsden, 7 Florence Road, Pakefield, Lowestoft, NR33 7BX
G0 VDR L. Goffin, The Hollies, Belaugh Green Lane, Norwich, NR12 7AJ
G0 VDT P. Clark, 21 Uplands, Welwyn Garden City, AL8 7EN
G0 VDU J. Newman, Shangrila, Treverbyn Road, Stenalees, St Austell, PL26 8TL
G0 VDV D. Steele, Tanglewood, Beckingham Street, Maldon, CM9 8LL
G0 VDZ N. Newby, 167 Watersplash Road, Shepperton, TW17 0EN
G0 VEC B. Hoare, Kilclare, Carrick On Shannon, Drumaleague House, Co Leitrim, Ireland,
G0 VEH J. Mulye, 83 Forest Drive East, London, E11 1JX
G0 VEI B. Davison, Pond House, Moores Lane, Colchester, CO7 6RF
GM0 VEK P. Davie, 62 Monkland Avenue, Kirkintilloch, Glasgow, G66 3BW
GW0 VEM A. Mccleverty, 3 The Fold, Upper Thornton, Milford Haven, SA73 3UE
G0 VEO R. Myatt, 48 Adaston Avenue, Eastham, Wirral, CH62 8BS
G0 VEP P. Steed, 1 Falcon Green, Portsmouth, PO6 1LW
G0 VEQ T. Arkadiusz, Shio^I-Chq 5- 66, Yokosuka, Japan, 238-0042
G0 VET P. Burnand, 5 Northgate, Hornsea, HU18 1ES
GW0 VEU G. Chantry, Summerfield, The Avenue, Oswestry, SY11 4LF
GW0 VEW D. Davies, 27 Twyniago, Pontarddulais, Swansea, SA4 8HX
G0 VEX P. Dickinson, 34 Marshall Avenue, Bridlington, YO15 2DS
G0 VFB D. Banks, 6 Kirkstall Close, Walsall, WS3 2SS
GM0 VFD A. Adam, 10 Greenmount Road North, Burntisland, KY3 9JQ
GW0 VFF S. Emanuel, 98 Moorland Road, Cimla. Neath, SA11 1JL
G0 VFL K. Gardner, Str Matei Vasilescu 82, Drobeth Turnu Sevferin 1500, Mehedinti, Romania,
G0 VFS R. Bailey, 13 Whiteland Rise, Westbury, BA13 3HP
G0 VFU F. Cokayne, 101 Neston Drive, Bulwell, Nottingham, NG6 8QY
G0 VFV G. Marley, 41 Scalby Road, Burniston, Scarborough, YO13 0HN
G0 VFW T. Thirlwell, 58 Chesham Road, Bovingdon, Hemel Hempstead, HP3 0EA
GM0 VFY G. Stuart, Easter Ardoe Cottage, Ardoe, Aberdeen, AB12 5XT
G0 VFZ R. Barnes, 18 Battle Road, Tewkesbury, GL20 5TZ
G0 VGB D. London, 113 Westbrooke Avenue, Hartlepool, TS25 5HZ
G0 VGC B. Jenkins, 27 Glendale, South Woodham Ferrers, Chelmsford, CM3 5TS
G0 VGD J. Constance, 4 Hopgarden Road, Tonbridge, TN10 4QS
GM0 VGI G. Anderson, 21 Bydand Gardens, Inverurie, AB51 4FL
G0 VGJ Prof. D. Graham, Parkburn, Colby, Appleby-in-Westmorland, CA16 6BD
G0 VGK P. Jenkinson, 5 Inglemere Drive, Stafford, st17 4qx
GI0 VGL G. Warnock, 98 Skerriff Road, Altnamachin, Newry, BT35 0PJ
G0 VGN A. Fawcett, 26 Merlin Court, Oswaldtwistle, Accrington, BB5 3TA
G0 VGP B. Davies, 18 Wyresdale Gardens, Lancaster, LA1 3FA
G0 VGR P. Tamplin, 74 Asquith Road, Gillingham, ME8 0JD
G0 VGT A. Trent, 18 Castle Drive, Reigate, RH2 8DQ
GI0 VGV A. Wright, 29 Wynfort Lodge, Moira, Craigavon, BT67 0QT
G0 VGY P. Keen, 89 Raleigh Avenue, Hayes, UB4 0EF
G0 VHF Colchester Contest Group, c/o J. Lemay, Carlton House, White Hart Lane, Colchester, CO6 3DB
GI0 VHG P. Hughes, 17 Cardinal Dalton Park, Keady, Armagh, BT60 3TS
G0 VHH G. Mann, 10 Earsham Drive, King's Lynn, PE30 3UZ
G0 VHK D. Godding, 20 Southwood Gardens Burghfield Common, Reading, RG7 3HY
G0 VHL P. Jones, 28 Helena Road, Capel-Le-Ferne, Folkestone, CT18 7LQ
G0 VHO Dr D. Ross, Rosehill, Leyland Lane, Leyland, PR26 8LB
G0 VHQ A. Harding, Po Box 10620 Apo, Grand Cayman, Cayman Islands,
GM0 VHR C. Cook, 2 Nortonhall Cottage Eildon, Melrose, TD6 9HB
G0 VHT P. Morrison, Primrose Bank, Holme Lacy, HR26LJ
G0 VHY R. Wilson, 7 Scarratt Close, Forsbrook, Stoke-on-Trent, ST11 9AP
G0 VIA C. Cannon, 1 Long Walk, Northstead, Scarborough, YO12 6BQ
GI0 VIB A. Smith, 42 Ballycullen Road, Moy, Dungannon, BT71 7HT
G0 VID D. Morris, 117 Lonsdale Avenue, Doncaster, DN2 6HF
GI0 VIF D. Canning, 10 Cotswold Close, Saintfield, Ballynahinch, BT24 7FQ
G0 VIG L. Merrick, 4 Berryfield Glade, Churchdown, Gloucester, GL3 2BT
G0 VII G. Boundey, 25 Ivy Place, Tantobie, Stanley, DH9 9PT
G0 VIJ J. Johnson, 1 Rosa Vella Drive, Dereham, NR20 3SB
GD0 VIK D. Wood, The Hawthorns, Droghadfayle Road, Isle of Man, IM9 6EL
G0 VIM M. Rivers, Snagsmount, Lambden Road, Ashford, TN27 0RB
G0 VIQ E. Sully, 10 The Paddock, Pound Hill, Crawley, RH10 7RQ
GM0 VIT W. Henderson, Strone View, Bridge Of Cally, Blairgowrie, PH10 7JL
G0 VIX M. Rutland, Roselea, 28 Eastfield Avenue, Fareham, PO14 1EG
GM0 VIY Dr J. Oates, 14 Craighlaw Avenue, Eaglesham, Glasgow, G76 0EU
G0 VJB B. Vaughan, 43 Bankfield Road, Shipley, BD18 4AW
G0 VJC C. Smith, 19 Wiscombe Avenue, Penkridge, Stafford, ST19 5EH
GI0 VJE C. Hannigan, 4 Silverhill Road, Strabane, BT82 0AE
G0 VJH J. Herrington, 84 Glenn Road, Poringland, Norwich, NR14 7LU
G0 VJI R. Clay, 38 Hubbards Road, Chorleywood, Rickmansworth, WD3 5JJ
G0 VJJ S. Gould, 87 Wentworth Drive, Bedford, MK41 8QD
G0 VJK Dr M. Stanton, 11 Eldean Road, Duston, Northampton, NN5 6RF
G0 VJM A. Howell, Beach Lawns Care Home, 67 Beach Road, Weston-Super-Mare, BS23 4BG
G0 VJN P. Andres, Flat 80, Northfield House, Bristol, BS3 1XB
GJ0 VJP N. Collier Webb, Chatelet, Les Marais Avenue, La Route De La Haule, Jersey, Jersey, JE3 1LE
G0 VJR R. Henshall, 9 Murrayfield Drive, Willaston, Nantwich, CW5 6QF
G0 VJY R. Welbourn, 30 Bower Road, Swinton, Mexborough, S648NU
G0 VKC K. Cunningham, Claerwern Cottage, Three Ashes, Hereford, HR2 8NA
G0 VKE A. Willis, 5 Robin Hill, Shoppenhangers Road, Maidenhead, SL6 2GZ
G0 VKF M. George, 3 Oak Crescent, Cherry Willingham, Lincoln, LN3 4AX
GM0 VKG OBO Largs & District ARS, c/o J. Clough, Obo Largs & District Ars, Redbank, Skelmorlie, PA17 5DX
G0 VKH H. Venus, 45 St. Albans Road, Seven Kings, Ilford, IG3 8NN
G0 VKI A. Hopkinson, 34 Welby St., Fenton, Stoke on Trent, ST4 4PL
G0 VKL R. Butler, 34 Commissioners Road, Strood, Rochester, ME2 4EB
G0 VKS H. Klein, 30 Odenwaldstrasse, Frankfurt, Germany, 60528
G0 VKX Dr R. Smith, 16 Southbrook Road, Langstone, Havant, PO9 1RN
G0 VKY D. Russell, 1 Debden Close, Ernesettle, Plymouth, PL5 2DB
G0 VLC Dr A. Betts, 42 Goring Road, Steyning, BN44 3GF
GI0 VLE W. Dalton, 16 Junction Road, Randalstown, Antrim, BT41 4NP
G0 VLF R. Mcaleer, 44 Harvey Avenue, Durham, DH1 5ZG
G0 VLI D. Taylor, 118 Portsmouth Road, Lee-on-the-Solent, PO13 9AF
G0 VLJ Prof. G. Heald, 2 Holyrood Terrace, Weymouth, DT4 0BE
G0 VLK A. Palfreeman, 29 Boulby Road, Redcar, TS10 5EB
G0 VLQ B. Close, 74 Heston Avenue, Hounslow, TW5 9EX
G0 VLR Leicester Raynet Group, c/o A. Holmes, 5 Launde Park, Market Harborough, LE16 8BH
G0 VLV D. Hardman, 15 Wordsworth Avenue, Bolton Le Sands, Carnforth, LA5 8HJ
G0 VMA G. Skupski, 57 Three Nooks, Bamber Bridge, Preston, PR5 8EN
G0 VMC J. Williams, 62 Hollows Close, Salisbury, SP2 8JX
GW0 VMD K. Peacey, 2 Robin Close, Cardiff, CF23 7HN
G0 VME M. Macdonald, 44 Hillesden Avenue, Elstow, Bedford, MK42 9YX
G0 VMF P. Quirk, 70 Sands Lane, Oulton, Lowestoft, NR32 3HS
G0 VMK P. Nicholls, 53 Fastolff Avenue Gorleston, Great Yarmouth, NR31 7ND
G0 VMN D. Turton, 68 Bartholomew Street, Wombwell, Barnsley, S73 8LD
G0 VMP J. Roze, 9 Ralfland View, Shap, Penrith, CA10 3PF
G0 VMQ P. Ellis, 104 Gravesend Road, Strood, Rochester, ME2 3PN
GW0 VMR P. Smith, Bron Awel, Brynisa Road, Wrexham, LL11 6NS
GW0 VMS L. Beedle, 2 Chestnut Grove, Maesteg, CF34 0NT
G0 VMT Dr T. Moorhead, 53 Childwall Priory Road, Liverpool, L16 7PA
GM0 VMV E. Kennedy, 1 Greenbank Gardens, Edinburgh, EH10 5SL
GW0 VMW R. Price, 8 Tanllwyfan, Old Colwyn, Colwyn Bay, LL29 9LQ
GW0 VMZ J. Davies, 34 Penlan View, Ynysfach, Merthyr Tydfil, CF47 8NJ
G0 VNA G. Kent, Plum Tree Cottage, Colesbrook Lane, Gillingham, SP8 4HH
GW0 VND M. Goodridge, 17 Charles St., Neyland, Milford Haven, SA73 1SA
G0 VNE H. Stokes, 9 Causeway Glade, Dore, Sheffield, S17 3EZ
G0 VNH C. Langdon, 652 Hotham Road South, Hull, HU5 5LE
G0 VNI S. Williams, The Croft, Ringwood Road, Southampton, SO40 7LA
G0 VNJ M. Guy, 38 Sandy Lane, Charlton Kings, Cheltenham, GL53 9DQ
G0 VNK J. Harders, Kalckreuthweg 17, Hamburg, Germany, 22607
G0 VNO D. Johns, 8 Hill Fold, Dawley Bank, Telford, TF4 2QE
G0 VNQ W. Askam, 10 Staunton Close, Castle Donington, Derby, DE74 2XA
G0 VNW J. Carrington, 74 Ogle Street, Hucknall, Nottingham, NG15 7FR
G0 VNY J. Crawford, 2C Papillon House, Balkerne Gardens, Colchester, CO1 1PR
G0 VOB K. Fuller, 28 Bradshaw Way, Irchester, Wellingborough, NN29 7DP
G0 VOE C. Iles, Moon Cottage, 2 Mountview Terrace, Pawlett, Bridgwater, TA6 4SL
G0 VOF M. Walmsley, 121 Roe Lee Park, Blackburn, BB1 9SA

GW0	VOG	D. Roberts, 16 Pentre Isaf, Old Colwyn, Colwyn Bay, LL29 8UT
G0	VOJ	S. Williams, 6 Oak Road, Clanfield, Waterlooville, PO8 0LJ
G0	VOK	N. Reilly, 22 Lee Drive, Northwich, CW8 1BW
GM0	VOL	V. Nerurkar, 26 Fothringham Drive, Monifieth, Dundee, DD5 4SW
GM0	VOU	P. Scott, 30 Main St., Newmills, Dunfermline, KY12 8SS
G0	VOV	D. Hartwell, 57 Wyatts Covert, Denham, Uxbridge, UB9 5DJ
GU0	VPA	R. Peeters, 17 Grosse Hougue, Saltpans Road, St. Sampson, Guernsey, Guernsey, GY2 4NS
G0	VPC	M. Davies, 27 Bedford Road, Orpington, BR6 0QJ
G0	VPE	Reading and West Berkshire Raynet, c/o D. Pibworth, 20 Marathon Close, Woodley, Reading, RG5 4UN
GM0	VPG	N. Thackrey, Rhumore, Kilmun, Dunoon, PA23 8SB
G0	VPH	M. Austin, 107 Spicer Close, London, SW9 7UE
G0	VPJ	J. Stacey, 16 Crane Drive, Verwood, BH31 6QB
G0	VPO	A. Robinson, 2 Fort Street, Sandown, PO36 8BA
G0	VPS	D. Bennett, 3 Sivilla Road, Kilnhurst, Mexborough, S64 5TY
G0	VPT	L. Smith, Hillside, Kings Mill Lane, Stroud, GL6 6SA
G0	VPU	Dr M. Burbidge, 3 Kirklands, Hest Bank, Lancaster, LA2 6ER
G0	VPV	G. Pesarini, 53 Llanvanor Road, London, NW2 2AR
G0	VPW	M. Rhodes, 102 Malvern Crescent, Little Dawley, Telford, TF4 3JF
G0	VPX	M. Worsfold, 9 Montacute Road, Lewes, BN7 1EN
G0	VPY	E. Weston, 33 William Street, Tunbridge Wells, TN4 9RP
G0	VPZ	J. Greenfield, 36 Barttelot Road, Horsham, RH12 1DQ
G0	VQA	J. Groves, 24 Gimble Way, Pembury, Tunbridge Wells, TN2 4BX
G0	VQB	M. Grainger, 174 Woodlands Road, Gillingham, ME7 2SX
G0	VQD	P. Newberry, 2 Gatewycke Terrace, Tanyard Lane, Steyning, BN44 3RL
G0	VQG	M. Thomas, 10 Finchale Crescent, Darlington, DL3 9SA
G0	VQH	J. Bailey, 9 Little Ditton Woodditton, Newmarket, CB8 9SA
G0	VQJ	J. Metcalfe, 158 Barrowford Road, Colne, BB8 9QR
G0	VQK	S. Haigh, 2 Locker Avenue, Warrington, WA2 9PS
G0	VQL	G. Guild, 15 Canalside Cottages, Chester Road, Runcorn, WA7 3AQ
G0	VQM	C. Reid, 54 Montacute Road, New Addington, Croydon, CR0 0JE
G0	VQO	N. Haynes, 139 Hull Road Anlaby, Hull, HU10 6ST
G0	VQR	T. Cannon, 35 Loddon Bridge Road, Woodley, Reading, RG5 4AP
G0	VQS	P. Beck, 6 Holly Close, Little Bealings, Woodbridge, IP13 6PL
G0	VQT	F. Seabourne, 11 Heyford Avenue, London, SW20 9JT
G0	VQW	A. Jack, 1 Dockle Way, Upper Stratton, Swindon, SN2 7LQ
G0	VQX	D. Thomas, 11 Fordwells Drive, Bracknell, RG12 9YL
G0	VQY	P. Wooding, 31 Douglas Avenue, Brixham, TQ5 9EL
GW0	VQZ	N. Jones, 59 Woodfield Terrace, Penrhiwceiber, Mountain Ash, CF45 3YA
G0	VRE	A. Challis, 12 Moorland Crescent, Boultham Moor, Lincoln, LN6 7NL
G0	VRF	M. Waples, 42 Butts Road, Wellingborough, NN8 2PU
G0	VRH	D. Bower, 29 Worksop Road, Mastin Moor, Chesterfield, S43 3DH
G0	VRK	C. Seabridge, 13 Hillside Avenue Forsbrook, Stoke-on-Trent, ST11 9BH
GW0	VRL	C. Saunders, 14 Portway, Bishopston, Swansea, SA3 3JR
G0	VRM	A. Russell, 3 St. Nicholas Close, North Newbald, York, YO43 4TT
GM0	VRP	R. Phillips, 18 Broomridge Road St. Ninians, Stirling, FK7 0DT
G0	VRQ	R. Rendall, 633 Moston Lane, Manchester, M40 5QD
G0	VRS	K. Gillen, 33 Norwich Close, Ashington, NE63 9RY
G0	VRT	M. Ritson, 14 Dunsdale Road, Holywell, Whitley Bay, NE25 0NG
G0	VRU	B. Gee, 11 Spitfire Avenue, Grimoldby, Louth, LN11 8UJ
G0	VRV	S. Philipps, 24 Acres End, Amersham, HP7 9DZ
G0	VRW	P. Wadhams, Brickwood, Bigbury Road, Canterbury, CT4 7ND
G0	VRX	C. Jenkins, 31 Ashbrook Crescent, Rochdale, OL12 9AJ
G0	VRY	P. Maccormick, 94 Hillmead, Norwich, NR3 3PF
G0	VRZ	K. Tuer, Broad Ing, Penrith, CA10 2LL
G0	VSB	A. Gibbs, 18 Grange Close, Ludham, Great Yarmouth, NR29 5PZ
G0	VSG	I. Smyth, Fuchsia Cottage Holywell Lake, Wellington, TA21 0EL
GU0	VSH	M. Wright, 3 St Anthony, Melleiha, Malta,
G0	VSJ	P. Taylor, 133 Beech Drive, Shifnal, TF11 8HZ
G0	VSK	A. Thomas, 3 Barnfield Way, Cannock, WS12 0PR
G0	VSL	K. Hill, 18 Baker Close, Chasetown, Burntwood, WS7 4GU
G0	VSM	T. Day, Box 204, The Postroom, Calle San Jaime 5, Benijofar, Spain, 3178
GW0	VSO	P. Burchell, 5 Brentwood Place, Ebbw Vale, NP23 6JR
G0	VSS	R. King, 19 Greenhayes, Cheddar, BS27 3HZ
GW0	VST	P. Sandham, 6 Skomer Close, Nottage, Porthcawl, CF36 3QH
GW0	VSW	J. Mason, 2 Golwg-Y-Bryn, Off Woodland Road, Skewen, Neath, SA10 6SP
G0	VSY	G. Forster, 3 Foxmoor, Bishops Cleeve, Cheltenham, GL52 8SS
G0	VSZ	D. Forster, 3 Foxmoor, Bishops Cleeve, Cheltenham, GL52 8SS
G0	VTA	R. Collins, 30 Upham Road, Swindon, SN3 1DN
G0	VTC	L. King, 4 Glenhurst Avenue, Ruislip, HA4 7LZ
G0	VTD	S. Towler, 77 Glebe Road, Hull, HU7 0DU
G0	VTI	T. Ibbitson, 36 Knoll Park, East Ardsley, Wakefield, WF3 2AX
G0	VTJ	A. Jameson, 9 White Laithe Green, Leeds, LS14 2EP
G0	VTL	H. Bradfield, Glebe Farm, Wharf Street, Leicester, LE4 8AY
G0	VTM	J. Pearson, 23 Glebe Avenue, Mitcham, CR4 3DZ
GI0	VTS	R. Boyle, 53 Portaferry Road, Cloughey, Newtownards, BT22 1HP
G0	VTV	K. Groom, 2 Ruins Barn Road, Tunstall, Sittingbourne, ME10 4HS
G0	VUC	C. Boughton, 79 Hawfield Lane, Burton-on-Trent, DE15 0BY
G0	VUH	A. France, 31 Broadway Swinton, Mexborough, S64 8HD
G0	VUL	A. Hardie, Tana-Merah, Church Lane, Retford, DN22 9NQ
G0	VUM	P. Cooper, 16 Mortomley Close, High Green, Sheffield, S35 3HZ
G0	VUN	T. Wooding, 101 Park Farm Road, Ryarsh, West Malling, ME19 5JX
G0	VUT	C. Turner, 28 Reading Street, Broadstairs, CT10 3AZ
G0	VUX	Bolton School ARC, c/o C. Walker, Bolton School Ltd, Chorley New Road, Bolton, BL1 4PA
GM0	VUY	R. Turnbull, 6 Letham Gait, Dalgety Bay, Dunfermline, KY11 9GT
G0	VVA	M. Newbold, 10 Shaw St., Derby, DE22 3AS
GI0	VVC	J. Serridge, 21 Lassara Heights, Warrenpoint, Newry, BT34 3PG
G0	VVF	D. Turner, 2 Toulmin Drive, Swadlincote, DE11 0BH
G0	VVG	A. Lomas, 14 Ingledene Caravan Site, Lawsons Road, Thornton-Cleveleys, FY5 4DL
G0	VVK	P. Neale, 15 Kenmore Walk, Wibsey, Bradford, BD6 3JQ
G0	VVP	R. Hatton, Plot 2, Sutton Road, Hull, HU7 5YY
G0	VVQ	J. Mahon, 2 Rob Lane, Newton-le-Willows, WA12 0DR
G0	VVR	C. Leigh, 4 Corrick Close, Draycott, Cheddar, BS27 3UB
G0	VVT	E. Murphy, 21 Standard Street, Stoke-on-Trent, ST4 4NG
G0	VVX	Croham Callers, c/o M. Samuel, 71 Brighton Road, South Croydon, CR2 6EE
G0	VVZ	Rev. D. Matthiae, 142 South Wing, Fairfield Hall, Kingsley Avenue, Fairfield, Hitchin, SG5 4FY
G0	VWB	M. Davies, 23 Star Lane, Folkestone, CT19 4QH
GW0	VWD	A. Ball, 20 Hillcrest, Brynna, Pontyclun, CF72 9SJ
G0	VWE	P. Whitfield, 55 Greenways, Sutton, Woodbridge, IP12 3TP
G0	VWF	C. Howell, 24 Bellring Close, Belvedere, DA17 6LP
G0	VWH	M. Wright, 27 Ellesmere Close Hucclecote, Gloucester, GL3 3DH
G0	VWP	T. Sayner, 59 Horner St., York, YO30 6DZ
G0	VWQ	D. Cockburn, Spindleberry, Weare Giffard, Bideford, EX39 4QR
G0	VWT	T. Holland, 135 Newcastle Road, Stone, ST15 8LF
GI0	VWU	A. Mccabe, 40 Rydalmere St., Belfast, BT12 6GF
G0	VWV	R. Giles, 9 Bower Green, Lords Wood, Chatham, ME5 8TN
G0	VWW	T. Robson, 58 Burton Road, Lincoln, LN1 3LB
G0	VWX	E. Collinson, 40 Cock Robin Lane, Catterall, Preston, PR3 1YL
GM0	VWZ	S. Lawrie, 4 Glenavon Drive, Airdrie, ML6 8QG
GM0	VXA	P. Marriott, Craiglea Cottage, Omoa Road Cleland, Motherwell, ML1 5LQ
G0	VXB	S. Cockburn, 3 Inglebrook Heights, Westward Ho!, Bideford, EX39 1GU
G0	VXC	M. Coles, 133 Highthorn Road, Kilnhurst, Mexborough, S64 5UU
G0	VXD	G. Clayton, 27 Simpsons Lane, Knottingley, WF11 0HG
G0	VXE	D. Herbert, 50 St. Leonards Crescent, Scarborough, YO12 6SP
G0	VXG	R. Wilkinson, 139 Church Road, Jackfield, Telford, TF8 7ND
G0	VXJ	M. Finch, 73 Cordingley Way, Donnington, Telford, TF2 7LJ
G0	VXK	J. Davies, 26 Beverley Close, Wylde Green, Sutton Coldfield, B72 1YF
G0	VXM	G. Simmons, 90 Pollards Fields, Knottingley, WF11 8TD
GM0	VXQ	P. Mulheron, 78 South Commonhead Avenue, Airdrie, ML6 6PA
G0	VXS	C. Newman, 89 Goose Cote Lane, Oakworth, Keighley, BD22 7NQ
G0	VXV	P. Nicholls, 21 Woodlea Close Yeadon, Leeds, LS19 7NL
G0	VXW	M. Rendall, 633 Moston Lane, Manchester, M40 5QD
G0	VXX	A. Fry, 96 Westcroft Gardens, Morden, SM4 4DL
G0	VXY	G. Forde, 22 Alloa Road, London, SE8 5AJ
GM0	VXZ	I. Terris, 1 Linden Avenue, Wishaw, ML2 8SE
GM0	VYB	T. Cleghorn, 9 Bridge Of Aldouran, Leswalt, Stranraer, DG9 0LW
G0	VYC	M. Dawson, 11 Owls Retreat, Colchester, CO4 3FE
GW0	VYF	P. Simmons, 153 St. Davids Road, Letterston, Haverfordwest, SA62 5SS
GW0	VYG	GB7ADX CSG/Cymru Contest Group, c/o T. Jones, Penrhiw Bach, Bryngwran, Holyhead, LL65 3RD
G0	VYK	R. Heesom, 41 Ridgeway, Pembury, Tunbridge Wells, TN2 4ER
GM0	VYL	P. Maver, 69 Mayfield Crescent, Musselburgh, EH21 6EX
G0	VYN	M. Guest, 53 Ringwood Close, Furnace Green, Crawley, RH10 6HQ
G0	VYP	W. Southworth, 58 Moyse Avenue, Walshaw, Bury, BL8 3BL
G0	VYQ	T. Mcinerney, 41 Newton Way, Tongham, Farnham, GU10 1BY
G0	VYT	M. Garry, 14 Adrians Close, Mansfield, NG18 4HG
G0	VYU	R. Raynor-Smith, 10 Marsh Road, Trowbridge, BA14 7PR
G0	VYV	E. Smith, Magnolia House, Main Road, Highbridge, TA9 3QZ
G0	VYX	P. Rumsam, 24 Hamer Avenue, Rossendale, BB4 8QH
GM0	VYY	D. Macconnell, Mine Cottage, Tashieburn Road, Lanark, ML11 8ES
G0	VZA	B. Howlett, 156 Lanarch Road, Waverley, Dunedin, New Zealand,
G0	VZB	R. Ferris, Polmarth House, Carnmenellis, Redruth, TR16 6NT
G0	VZE	E. Cook, 8 Calvert Grove, Newcastle, ST5 8QA
G0	VZI	P. Robinson, 5 Coppice Close, Haxby, York, YO32 3RR
G0	VZK	C. Sampson, 1 Warton Lane, Austrey, Atherstone, CV9 3EJ
G0	VZL	E. Levring, 3 Evelyn Croft, Wylde Green, Sutton Coldfield, B73 5LF
G0	VZN	K. Voller, 20 Browns Lane, Uckfield, tn221ry
G0	VZO	J. Koops, 64 Winchester Avenue, Nuneaton, CV10 0DW
G0	VZT	S. Hopton, 5 Wellington Close Marske-By-The-Sea, Redcar, TS11 6NW
G0	VZV	D. A'Bear, 7 Meadow Close, Bembridge, PO35 5YJ
G0	VZX	P. Dart, 208 Elburton Road, Plymouth, PL9 8HU
G0	WAB	W. Neale, 5 Gibbs Court, Dane Close, Wirral, CH61 3XS
G0	WAC	K. Wells, 42 Eggesford Road, Stenson Fields, Derby, DE24 3BH
G0	WAD	C. Dyson, 21 Highmoor, Kirkhill, Morpeth, NE61 2AS
G0	WAE	D. Phillips, 14 Seymour Road, Newton Abbot, TQ12 2PU
GI0	WAH	W. Hutchman, 35 Carlingford Park, Newry, BT34 2NY
G0	WAL	W. Reed, 10 Ashmeads Way, Wimborne, BH21 2NZ
G0	WAM	S. Stevens, 25 Busticle Lane, Sompting, Lancing, BN15 0DJ
G0	WAN	M. Gill, 23 Walmers Avenue, Higham, Rochester, ME3 7EH
G0	WAS	A. Smith, 9 Moor Lane Maulden, Bedford, MK45 2DJ
G0	WAT	P. Brice-Stevens, 31 Lodgefield, Welwyn Garden City, AL7 1SD
G0	WAW	I. Oura, The Quoins, Gloucester Road, Bath, BA1 8AD
G0	WAX	L. Merrin, 10 Hawkesbury Road, Canvey Island, SS8 0EX
G0	WAY	R. Slimmon, 22A Southbourne Place, Cannock, WS11 4SA
G0	WBA	K. Fradgley, 84 Church Road, Wordsley, Stourbridge, DY8 5AU
G0	WBC	Dr C. Mortimer, 6 Honeycomb Close, Narborough, Leicester, LE19 3PS
G0	WBL	S. Hughes, 43 The Cloisters, Rickmansworth, WD3 1HL
G0	WBR	T. Johnson, 7 Southover Way, Hunston, Chichester, PO20 1NY
G0	WBS	Monmouth School Amateur School Society, c/o P. Wentworth, 46 Woodside Avenue, Cinderford, GL14 2DW
G0	WBT	J. Woodcock, 54 Longworth Road, Horwich, Bolton, BL6 7BE
G0	WBV	R. Burchell, 1 Broad Forstal Farm Cottages, Tilden Lane, Tonbridge, TN12 9AX
G0	WCB	A. Bathurst, 81 Heatherstone Avenue, Dibden Purlieu, Southampton, SO45 4LE
GI0	WCE	D. Cromie, 11 Cherryvalley Park West, Belfast, BT5 6PU
G0	WCH	B. Prestage, 17 Moorgate Road, Hindringham, Fakenham, NR21 0PT
G0	WCI	M. Hand, 24 Ettingshall Road, Bilston, WV14 9UJ
G0	WCJ	J. Slater, 25 Croft Lane, Diss, IP22 4NA
G0	WCK	D. Hornby, 7 Milton Street, West Bromwich, B71 1NJ
G0	WCO	P. Anness, 18 Plaisir Place, Thurston Road, Lowestoft, NR32 1RY
G0	WCR	M. Knott, 76 New Barns Avenue, Mitcham, CR4 1LF
G0	WCS	S. Ormerod, 6 New Wokingham Road, Crowthorne, RG45 7NR
G0	WCU	R. Roberts, 9 Stones Close, Hogsthorpe, Skegness, PE24 5NZ
G0	WCZ	G. Sutherland, 8 Dukes Avenue, London, N3 2DD
G0	WDC	D. Clark, Meadowcroft Bungalow, Ugthorpe, Whitby, YO21 2BL
GM0	WDF	J. Dunlop, West Dougliehill Farm, Dougliehill Road, Port Glasgow, PA14 5XF
G0	WDG	L. Morgan, Flat 30, Raleigh Court, Sherborne, DT9 3EQ
G0	WDK	D. Hackett, 131 Station Road South, Walpole St. Andrew, Wisbech, PE14 7LZ
G0	WDQ	S. Collins, 30 Upham Road, Swindon, SN3 1DN
G0	WDT	R. Emery, 10 Penarth Place, Newcastle, ST5 2JL
G0	WDU	M. Round, 2 Bell Mead, Studley, B80 7SH

GW0 WDV S. Steddy, 1 Springfield Close, Wenvoe, Cardiff, CF5 6DA
G0 WDW W. Williamson, Monfa Walford Heath, Shrewsbury, SY4 2HT
G0 WDX World DX Radio Club, c/o R. Morgan, 56 The Meadows, Hull, HU7 6EE
G0 WEA R. Foster, 60 Sandford Close Bransholme, Hull, HU7 4HN
G0 WEB R. Webb, 25 Greenfield Croft, Bilston, WV14 8XD
GM0 WED E. Holt, Ashwell, St. Ola, Kirkwall, KW15 1SX
G0 WEF N. Kenworthy, 19 Parkside Close, Radcliffe, Manchester, M26 2QS
G0 WEO D. Waterfield, 20 St. Andrew Road, Evesham, WR11 2NR
GW0 WER P. Moran, 1 Plas Issa, Brook Street, Wrexham, LL14 3EE
G0 WET R. Wright, Ul Dordoy 32/34 Kb 52, Bishkek, Kyrgyzstan, 720082
G0 WEV S. Mcmullen, 70 Sylvan Avenue, Timperley, Altrincham, WA15 6AB
G0 WEX S. Elliott, 1 The Gables, Oddfellows Road, Hope Valley, S32 1DU
GW0 WEY J. Doores, 14 Parc Tyddyn, Red Wharf Bay, Pentraeth, LL75 8NQ
GM0 WEZ Dr P. Ewing, Kildonan House, Caerlaverock Farm, Crieff, PH5 2BD
GM0 WFA T. Clark, 23 Letham Place, St Andrews, KY16 8RB
GM0 WFB J. Keenan, 53 Clermiston Crescent, Edinburgh, EH4 7DF
G0 WFD M. Farrell, 24 Lindsay Street, Stalybridge, SK15 2LT
G0 WFE S. Dean, 42 North Street, Wareham, BH20 4AQ
G0 WFF C. Whelan, 50 Garrick Close, Ings Road Estate, Hull, HU8 0ST
G0 WFG A. Vinters, 106 Halifax Road, Ripponden, Sowerby Bridge, HX6 4AG
G0 WFH C. Gresswell, 11 Dandy Dinmont Caravan Park Blackford, Carlisle, CA6 4EA
G0 WFK H. Bamford, Upper Twynings Farm, Pumphouse Lane, Droitwich, WR9 7EB
G0 WFL M. Street, 262 Carter Knowle Road, Sheffield, S7 2EB
G0 WFM P. Wallace, 8 Wemmick Close, Rochester, ME1 2DL
G0 WFO D. Tarry, 2 Kestrel Heights, Codnor Park, Nottingham, NG16 5PW
G0 WFP W. Painz, 8 Warminger Court, Ber Street, Norwich, NR1 3ED
G0 WFQ M. Troy, 22 Jackie Wigg Gardens Totton, Totton, Southampton, so409lz
G0 WFT D. Saunders, 14 Shelton Avenue, Toddington, Dunstable, LU5 6EL
G0 WFV A. Corbett, Lan Y Llyn, 10 Rutland Avenue, Waddington, LN5 9FW
G0 WFX S. Hayes, 94 Kingshurst Way, Birmingham, B37 6JQ
G0 WGA M. Merrin, 10 Hawkesbury Road, Canvey Island, SS8 0EX
G0 WGB J. Atkins, 6 Carolina Gardens, Plymouth, PL2 2ER
GW0 WGE D. Thomas, 85 Tan Y Bryn, Burry Port, SA16 0LD
G0 WGH J. Edwards, 21 Ridgeway, Ottery St Mary, EX11 1DT
G0 WGI K. Sherwin, 1 Nursery Close, Wroughton, Swindon, SN4 9DR
G0 WGJ G. Jarvis, 38 West Cliff Road, Dawlish, EX7 9DY
G0 WGL P. Furness, 6 Westbrook Square, Manchester, M12 5PU
GW0 WGM T. Jones, 26 Heol Maenofferen, Blaenau Ffestiniog, LL41 3DL
GW0 WGN J. Lyons, 4 London Road, Pembroke Dock, SA72 6DU
G0 WGP R. Glover, 5 The Vineries, Burgess Hill, RH15 0ND
G0 WGV I. French, Penmellyn, Tarrandean Lane, Truro, TR3 7NW
GW0 WGW B. James, 72 Park Place, Bargoed, CF81 8NB
G0 WHC W. Chesterton, 61 Butt Lane, Blackfordby, Swadlincote, DE11 8BG
G0 WHD G. Hassall, 3 Sunny Bank, Cark In Cartmel, Grange-over-Sands, LA11 7PF
G0 WHL E. Barnes, 10 Cranbourne Road, Rochdale, OL11 5JD
G0 WHN J. Edwardes, Pippins, 1 Horse Lane Orchard, Ledbury, HR8 1PP
G0 WHO R. Clutson, 151 Stepney Road, Scarborough, YO12 5NJ
G0 WHQ K. Wilson, 11 Harbour Close, Murdishaw, Runcorn, WA7 6EH
G0 WHV R. Bessell, 6 Bayford Lodge, Wellington Road, Pinner, HA5 4NJ
G0 WHY Rev. P. White, 11A Dowse Road, Devizes, SN10 3FN
G0 WHZ T. Tipping, Flat 24, The Moorings, Kingsbridge, TQ7 1LP
GW0 WIB M. Mcdermott, Brynmeudwy, Llwyndrain, Llanfyrnach, SA35 0AT
G0 WIC E. Clark, 8 Rose Cottages, Shotton Colliery, Durham, DH6 2NF
G0 WIE P. Swan, The Old Rectory, Sampford Brett, Taunton, TA4 4LA
G0 WIG P. Ellison, 28 Kingscote Road East, Cheltenham, GL51 6JS
G0 WIL M. Williams, The Croft, Ringwood Road, Southampton, SO40 7LA
G0 WIS G. Woodbury, 4 Henley Drive, Droitwich, WR9 7RX
G0 WIT M. Perry, 52 Somerset Avenue, Chessington, KT9 1PN
G0 WIW N. Appleby, Westholme, Asterby Lane Asterby, Louth, LN11 9UE
G0 WIX A. Beeching, 174 Grove Road, Rayleigh, SS6 8UA
G0 WIY W. King, 3 Grove Court, The Waterloo, Cirencester, GL7 2PZ
GM0 WIZ I. Waugh, 2 Gilloch Avenue, Dumfries, DG1 4DN
G0 WJA L. Coleman, 1 Kerstin Close, Cheltenham, GL50 4SA
G0 WJC M. Briscoe, 34 Winterton Drive, Low Moor, Bradford, BD12 0UX
G0 WJD D. Jackson, 10 Wisdoms Green, Coggeshall, Colchester, CO6 1SG
G0 WJH J. Kemp, 85 St. Andrews Way, Church Aston, Newport, TF10 9JQ
GI0 WJI R. Bicker, 62 Spa Road, Ballynahinch, BT24 8PT
G0 WJJ D. Mullaney, 62 Darby Road, Wednesbury, WS10 0PN
G0 WJK A. Cobb, 37 Nordham, North Cave, Brough, HU15 2LT
G0 WJN P. Howland, 73 Timberdine Avenue, Worcester, WR5 2BG
G0 WJS D. Mellings, 36 Hillwood Drive, Glossop, SK13 8RJ
G0 WJU R. Crewe, 11 Osbert Close, Norwich, NR1 2NL
G0 WJV C. Page, 73 Two Saints Close, Hoveton, Norwich, NR12 8QR
G0 WJX R. Davies, 84 Hob Hey Lane, Culcheth, Warrington, WA3 4NW
G0 WJZ I. Donachie, 72 Gresham Road, Norwich, NR3 2NQ
G0 WKA M. Drever, 66 Milton Road, Branton, Doncaster, DN3 3PB
G0 WKH M. Thomas, 63 Corfe Way, Broadstone, BH18 9ND
G0 WKI H. Yearl, 191 Tamworth Road, Kettlebrook, Tamworth, B77 1BT
G0 WKJ C. Towle, 32 Charlwood Road, Luton, LU4 0BU
G0 WKM N. Purchon, Mitchells Elm House, Wanstrow, Shepton Mallet, BA4 4SN
G0 WKN K. Whitmore, Amosford, Lutton Gowts, Spalding, PE12 9LQ
G0 WKQ South Tyneside ARS, c/o D. Harbron, 48 Sheridan Road, South Shields, NE34 9JJ
G0 WKT M. Pugh, 28 Marsh Road, Wilmcote, Stratford-upon-Avon, CV37 9XR
G0 WKU P. Yea, 89 Laxton Road, Taunton, TA1 2XF
G0 WKW V. Pleshkevich, Rya Ängaväg 53, RydebŽCk, Sweden, 25730
GM0 WKZ B. Pybus, 43 Kinacres Grove Carriden, Bo'ness, EH51 9LT
G0 WLC B. Cole, 499 Lightwood Road, Stoke-on-Trent, ST3 7EN
G0 WLD M. Russell, 23A St Ann'S Road, Barnes, London, SW13 9LH
G0 WLF W. Storace-Rutter, 46 Norbury Court Road, London, SW16 4HT
G0 WLG P. Blizzard, 12 Hilton Road, Malvern, WR14 3NP
GW0 WLI J. Ruddle, Flat 58, Thomas Court, Cardiff, CF23 5EZ
GW0 WLN C. Purcell, 31 Brookdale Court, Church Village, Pontypridd, CF38 1RP
GW0 WLQ R. Evans, 4 Llantrisant Road, Tonyrefail, Porth, CF39 8PP
G0 WLR Dr M. Pettigrew, 26 Victoria Court, Sheffield, S11 9DR
G0 WLX A. Suckling, 54 Lake Hill, Sandown, PO36 9HF
G0 WMB A. Houghton, 2 Beanhill Crescent, Alveston, Bristol, BS35 3JG
G0 WMC P. Norman, 53 Saddlers Park, Eynsford, Dartford, DA4 0HA
G0 WMD M. De Silva, 31 Rosemary Avenue, Hounslow, TW4 7JQ
G0 WMG G. Studd, 34 The Broadway, Lancing, BN15 8NY
GM0 WMH A. Hughes, 7 Eden Grove, Kirkpatrick Fleming, Lockerbie, DG11 3AT
G0 WMJ J. Walker, 29 Akenside Court, New Fort Way, Bootle, L20 4UU
G0 WMN W. Mcnab, 74 Parkhouse Road, Minehead, TA24 8AF
G0 WMQ E. Williams, 16 Birch Grove, Wallasey, CH45 1JG
GW0 WMT S. Robinson, 23 Thornhill Road, Cardiff, CF14 6PE
G0 WMU R. Davis, 49 Goodway Road, Great Barr, Birmingham, B44 8RL
G0 WMW D. Roberts, 17 Edgecombe Avenue, Weston-Super-Mare, BS22 9AY
G0 WMX G. Giuliani, 32 Davison Street, Newburn, Newcastle upon Tyne, NE15 8NB
G0 WMY P. Hanmer, 182 Tollemache Road, Prenton, CH43 7SE
G0 WMZ R. Duckworth, 39 George Lane, Bredbury, Stockport, SK6 1AS
GW0 WNB S. Blumson, 15 Bryn Colwyn, Colwyn Bay, LL29 9LJ
G0 WND T. Sunouchi, 200-20 Morooka-Cho, Kouhoku-Ku, Yokohama, Japan, 222-0002
G0 WNF E. Clark, Flat B, 145 Church Street, Whitby, YO22 4DE
G0 WNJ M. Bottomley, 21 Priory Park, Grosmont, Whitby, YO22 5QQ
GM0 WNR A. Campbell, 17 Arran Road, Motherwell, ML1 3NA
GM0 WNS I. Calder, Cut The Wind Cottage, Arbroath, DD11 4RH
G0 WOA M. Lyon, 128 Lydyett Lane, Barnton, Northwich, CW8 4JU
G0 WOC J. Doxey, 41 Shady Grove, Hilton, Derby, DE65 5FX
G0 WOI Avon Scouts ARC, c/o R. Laney, 7 Downfield Close, Alveston, Bristol, BS35 3NJ
G0 WOM T. Renshaw, 1 Basford View, Cheddleton, Leek, ST13 7HJ
G0 WON M. Kelly, 9 Thetford Road, Brandon, IP27 0BS
G0 WOP P. Dabell, 21 Tatton Road North, Heaton Moor, Stockport, SK4 4RL
G0 WOU D. Atkinson, 596 Wolseley Road, St. Budeaux, Plymouth, PL5 1UX
GI0 WOW R. Cooper, Rockmount, 10 Carmorn Road, Toomebridge, Antrim, BT41 3NX
G0 WPC N. Crossley, 55 Scholey Avenue Woodsetts, Worksop, S81 8SF
G0 WPF J. Towle, 36 Old Orchard, Haxby, York, YO32 3DT
GM0 WPI A. Lister, Heathfield Easter Hardmuir, Nairn, IV12 5QG
G0 WPL J. Wozniak, 40 Cockhill, Trowbridge, BA14 9BQ
G0 WPM N. Gilboy, 6 Talisman Close, Sherburn Village, Durham, DH6 1RJ
G0 WPO N. Griffiths, Orchard Marina (Nb T'Dreme) Robbo Farm School Road Rudheath, Northwich, CW9 7RG
GM0 WPU K. Faloon, Moss-Side Croft, 6 Rothiemay, Huntly, AB54 5NY
GI0 WPV O. Price, 18 Hill Crest Walk, Bangor, BT20 4DF
GM0 WPW T. Halligan, 4 Trainers Brae, North Berwick, EH39 4NR
G0 WPX Wpx Contest Group, c/o D. Beattie, Hares Cottage, Woolston, Church Stretton, SY6 6QD
G0 WQA J. Ward, 49 Woodhall Drive, Lincoln, LN2 2AE
G0 WQC J. Keeling, 31 Tudor Drive, Otford, Sevenoaks, TN14 5QP
GW0 WQP D. Taylor, 18 Oakfield Avenue, Chepstow, NP16 5NE
G0 WQQ D. Bennett, Shrove Furlong Longwick Road, Princes Risborough, HP27 9HE
G0 WQW J. Thompson, 78 Lowestoft Road, Worlingham, Beccles, NR34 7RD
G0 WQY L. Mansfield, 25 Carlton Road, Derby, DE23 6HB
G0 WRC Wythall Contest Group, c/o L. Volante, Richmond House, Icknield Street, Birmingham, B38 0EP
G0 WRE P. Scarratt, 339 Utting Avenue East, Norris Green, Liverpool, L11 1DF
GM0 WRH E. Castle, 38 Davieland Road, Giffnock, Glasgow, G46 7LU
GW0 WRI R. Lewis, 26 Bryn Road, Upper Brynamman, Ammanford, SA18 1AU
G0 WRK P. Taylor, Old Acres, Priory Road, Yeovil, BA22 8NY
G0 WRL D. Webb, 16 Burrowfield, Bruton, BA10 0HR
G0 WRM Lichfield Raynet Group, c/o M. White, 3 High Street, Clay Cross, Chesterfield, S45 9DX
G0 WRN J. Hodges, 48 Beach Road, Severn Beach, Bristol, BS35 4PF
G0 WRQ O. Baxter, 5 Church Path, Bridgwater, TA6 7AJ
GM0 WRR J. Scott, 70 Montford Avenue, Glasgow, G44 4PA
G0 WRS Warrington ARC, c/o M. Isherwood, 32 Franklin Close Old Hall, Warrington, WA5 8QL
G0 WRT P. Winfield, 150 Tinshill Road, Leeds, LS16 7PN
GM0 WRU R. Holmes, 3 Buchan Street, Wishaw, ML2 7HG
GM0 WRV R. Spink, 44 West Park, Inverbervie, Montrose, DD10 0TT
G0 WSA J. Proctor, Barfords Farm, Swineshead Road, Kirton Holme, Boston, PE20 1SQ
G0 WSB M. Brimley, 42 Grange Road, Netley Abbey, Southampton, SO31 5FE
G0 WSC R. Connett, 15 Channels Lane, Horton, Ilminster, TA19 9QL
G0 WSD G. Swann, 1 Beaver Close, Chichester, PO19 3QU
G0 WSH R. Munt, Box 166, Se-177 23 Jarfalla, Sweden, Sweden
G0 WSI Dr W. Warburton, Marlins, Water Lane, Dibden Purlieu, Southampton, SO45 4SB
G0 WSJ S. Jelly, 10 Whitemarsh, Mere, Warminster, BA12 6BP
GD0 WSK S. Kempner, Sonnish-Ny-Marrey, St. Mary'S Avenue, Port St Mary, Isle Of Man, IM9 5ET
G0 WSP P. Croft, 82 Granby Road, Buxton, SK17 7TJ
GM0 WSR Strathclyde Regional Raynet Groups, c/o D. Mackinnon, 60 Mount Stuart Drive, Wemyss Bay, PA18 6DX
G0 WSS Horsham Scout ARG, c/o P. Head, 36A Ashacre Lane, Worthing, BN13 2DH
G0 WSY J. Adams, 53 Princess Road, Rochdale, OL16 4AY
G0 WTA N. Midworth, 1 Highfields Drive, Loughborough, LE11 3JS
G0 WTB C. Hicks, 59 Elmsfield Avenue, Norden, Rochdale, OL11 5XW
G0 WTC W. Clark, 41 Brook Close, Jarvis Brook, Crowborough, TN6 2ET
G0 WTD T. Stokes, 33 Talbot Drive, Euxton, Chorley, PR7 6PD
G0 WTF A. Wright, 4 Wilshere Road, Welwyn, AL6 9PX
G0 WTG D. Way, 25 Kinson Avenue, Poole, BH15 3PH
G0 WTI G. Lightfoot, 1 Trewyn Road, Holsworthy, EX22 6HX
G0 WTK R. Jenkins, 11 Westfield Drive, Worksop, S81 0JS
G0 WTL G. Fisher, 6 Totternhoe Road, Dunstable, LU6 2AG
G0 WTM D. Sutton, 32 Queensway, Euxton, Chorley, PR7 6PW
G0 WTO E. Birch, 4 Kynnesworth Gardens, Higham Ferrers, Rushden, NN10 8NH
GM0 WTP J. Whitecross, 7 Northfield Crescent, Edinburgh, EH8 7PU
G0 WTR M. Robertson, Clayhill Cottage, The Street, Aldham, Ipswich, IP7 6NN
G0 WTW G. Thompson, Lyngrove, Seaton Lane, Seaham, SR7 0LP
G0 WUA P. Harris, 1 Newlands, Landkey, Barnstaple, EX32 0NJ
G0 WUG G. Bishop, 8 Bulstrode Place, Kegworth, Derby, DE74 2DS
G0 WUH H. White, 111 Beacon Glade, South Shields, NE34 7QU
G0 WUI W. Cassidy, 17 Catcheside Close, Whickham, Newcastle upon Tyne, NE16 5RX

GW0 WUL C. Minard, 3 Riverside Close, Aberfan, Merthyr Tydfil, CF48 4RN
GW0 WUM E. Roobottom, Puffin View, Abergwyngregyn, Llanfairfechan, LL33 0LL
G0 WUO R. Berkeley, 4 Raleigh Road, Leasowe, Wirral, CH46 2QZ
GM0 WUP W. Steele, 35 Devlin Court, Whins Of Milton, Stirling, FK7 0NP
GM0 WUR G. Steele, 1 James Street, Bannockburn, Whins Of Milton, Stirling, FK7 0NQ
G0 WUS C. Rayns, 28 Hopyard Close, Leicester, LE2 9GY
G0 WUU J. Purcell, 2 Windsor Close, Cawood, YO8 3WE
G0 WUV J. Dickinson, 112 Stoneleigh Avenue, Longbenton, Newcastle upon Tyne, NE12 8XQ
G0 WUW V. Claridge, 105 Barnwood Avenue, Gloucester, GL4 3AG
GM0 WUX J. Donnan, 41 Annick Drive, Dreghorn, Irvine, KA11 4ER
G0 WUY A. Williamson, Millfield Lodge, 151 Hull Road, York, YO10 3JX
G0 WVA D. Townsend, Flat 32, Weymouth House, Eaton Place, Margate, CT9 1FE
G0 WVD T. Lishman, 13 Meadow Way, Sandown, PO36 8QE
G0 WVE B. Richardson, 12 Stoney Lane, Barrow, Bury St Edmunds, IP29 5DD
G0 WVM M. Brooker, 4 May Close, Sidlesham, Chichester, PO20 7RR
G0 WVT W. Carwood, 57 Upton Road, Kidderminster, DY10 2YB
G0 WVV D. Jones, 4 Back Bower Lane, Gee Cross, Hyde, SK14 5NS
G0 WVW D. Smith, 16 Browning Drive, Great Sutton, Ellesmere Port, CH65 7BW
G0 WVY F. Holt, 22 First Avenue, Tottington, Bury, BL8 3JA
G0 WWA D. Nicholas, 41 Grayling Road, Stourbridge, DY9 7AZ
G0 WWD D. Weston, 25 Ley Lane, Kingsteignton, Newton Abbot, TQ12 3JE
G0 WWE S. Fitton, 7 Blackburn Way, West Wick, Weston-Super-Mare, BS24 7GT
G0 WWF P. Baron, 55 Church View, Brompton, Northallerton, DL6 2RD
G0 WWH A. Houghton, 3 Billinge Close, Bolton, BL1 2JP
G0 WWL A. Babbage, 10 Heather Close, Honiton, EX14 2YP
G0 WWM Prof. T. Yukawa, 5349 Route 12, Birch Hill, Richmond, Canada, C0B 1Y0
G0 WWO C. Sawyer, 4 Padley Close, Ripley, DE5 3FG
G0 WWP S. Griffith, 12 Beech Grove, Cliffsend, Ramsgate, CT12 5LD
GW0 WWQ H. Burton, Vale View, Porth Y Waen Bodfari, Denbigh, LL16 4BU
G0 WWR R. Prior, 20 Churchfield Road, Walton on the Naze, CO14 8BL
G0 WWT J. Smith, 54 Greenfield Avenue, Kettering, NN15 7LL
G0 WWU R. Pratt, 4 King John Avenue Gaywood, King's Lynn, PE30 4QA
GM0 WWX R. Johnstone, 8 Harris Court, Alloa, FK10 1DD
G0 WWZ M. Charlesworth, Po Box 841, 9506, Korondal City, South Cotabato, Philippines,
G0 WXA S. Everitt, 125 Victoria Road, Warley, Oldbury, B68 9UL
G0 WXC R. Needham, 1 The Leas, Sedgefield, Stockton-on-Tees, TS21 2DS
G0 WXD P. Atkinson, 27 Ranworth Road, Bramley, Rotherham, S66 2SP
G0 WXE A. Kelleher, 144 Alibon Road, Dagenham, RM10 8DE
G0 WXF P. Wright, 60 Farnborough Road, Clifton, Nottingham, NG11 8GF
G0 WXG J. Lamb, 10 Lenhurst Avenue, Leeds, LS12 2RE
G0 WXH S. Croot, 58 Dixie Street, Jacksdale, Nottingham, NG16 5JZ
G0 WXJ P. Badham, 201 York Avenue, East Cowes, PO32 6BH
G0 WXL E. Harper, 20 Bar Meadows, Malpas, Truro, TR1 1SS
GW0 WXO A. Davies, 81 Brynifor, Mountain Ash, CF45 3AB
G0 WXP S. Parkes, 23 Curlew Close, Whittlesey, Peterborough, PE7 1XQ
G0 WXZ D. Milne, 22 Jessica Avenue, Verwood, BH31 6LH
G0 WYA N. Taylor, 61 Oldbury Road, Rowley Regis, B65 0NP
GI0 WYB J. Simpson, 66 Ballyportery Road, Dunloy, Ballymena, BT44 9BN
G0 WYD R. Coleman, 2 Chestnut Nepaul Road, Tidworth, SP9 7EU
G0 WYF C. Rudge, 1 Mill Lane, Alfington, Ottery St Mary, EX11 1PF
G0 WYG D. Biginton, 67 Capstone Road, Bromley, BR1 5NA
G0 WYI C. Sharpe, 35 Fairmead Close, Nottingham, NG3 3EQ
GI0 WYK W. Carress, 12 Ashbourne Park, Newtownards, BT23 7RE
G0 WYM A. Shields, 8 Thames Drive, Melton Mowbray, LE13 0DS
G0 WYN A. Godwin, 27 Melbourne Avenue, Dronfield Woodhouse, Dronfield, S18 8YW
GI0 WYO R. Kilgore, 3 Summer Meadows Manor, Londonderry, BT47 6SE
G0 WYP S. Barker, C/O, 14 Cordova Avenue, Manchester, M34 2WP
G0 WYQ J. Richardson, 24 Brockhall Road, Kingsley, Northampton, NN2 7RY
G0 WYR Dr C. Fox, 36 Haig Drive, Slough, SL1 9HB
G0 WYT I. Seabright, 78 Lazy Hill, Birmingham, B38 9PA
G0 WYU E. Martin, 2 Bospowis, St. Martins Crescent, Camborne, TR14 7HN
G0 WYV P. Little, 24 Pickwick Crescent, Rochester, ME1 2HZ
G0 WYY R. Bell, 41 Ravenshill Road, West Denton, Newcastle upon Tyne, NE5 5EA
G0 WYZ S. Gillen, 33 Norwich Close, Ashington, NE63 9RY
G0 WZA D. Wood, 2 Buckingham Mews Flitwick, Bedford, MK45 1TB
G0 WZB B. Burdis, 10 Johnston Avenue, Hebburn, NE31 2LJ
G0 WZC S. Cooper, 27 Polweath Road, Penzance, TR18 3PW
G0 WZD Dr J. Paloschi, 1 Lander Close, Milton, Cambridge, CB24 6EB
G0 WZG K. Norris, 15 East View, Choppington, NE62 5UF
G0 WZH S. Frankum, 39 Leighton Road, Wingrave, Aylesbury, HP22 4PA
G0 WZJ J. Pickering, 11 Smithson Court, Malton, YO17 7BQ
G0 WZK N. Collis Bird, 4 Manwell Drive, Swanage, BH19 2RB
GW0 WZL J. Rushton, 2 Fron Felen, Clogwyn Melyn, Penygroes, Caernarfon, LL54 6PT
G0 WZM I. Kitchen, 102 Riverview Road, Epsom, kt19 0jp
GM0 WZO I. Finlayson, 10B Flesherin, Isle of Lewis, HS2 0HE
G0 WZV K. Aston, 10 Browning Close, Lexden, Colchester, CO3 4JJ
GI0 WZW R. Pollock, 5 Brooke Grove, Banbridge, BT32 3YA
G0 WZX B. Stevens, 172 Gordon Avenue, Camberley, GU15 2NT
G0 WZY M. Davies, 45 Elm Grove, Swainswick, Bath, BA1 7BA
GW0 WZZ M. Bobby, Hafan, Church Street, Penycae, LL14 2RL
G0 XAA Ansty Contest Club, c/o K. Evans, Littlefield House, Bolney Road, Haywards Heath, RH17 5AW
G0 XAB R. Dawson, 28 Calf Close, Haxby, York, YO32 3NS
GI0 XAC Dr A. Chin, 25 Meadowbrook, Islandmagee, Larne, BT40 3UG
G0 XAD B. Wright, 96 Ellenborough Close, Thorley, Bishop's Stortford, CM23 4HU
G0 XAE M. Buckland, The Homestead, Bollow, Westbury-on-Severn, GL14 1QX
G0 XAF P. Coles, 10 Springfield Close, Mangotsfield, Bristol, BS16 9BZ
G0 XAG I. White-Sharman, 62 Timberleys, Littlehampton, BN17 6QB
G0 XAH D. Tecklenberg, 1 Proby Close, Yaxley, Peterborough, PE7 3ZF
G0 XAI S. Bennett, Maggies Meadow, Hoe Lane, Flansham, PO228NS
G0 XAK S. Curtis, 389 Portway, Shirehampton, Bristol, BS11 9UF
G0 XAM L. Camber, Sundown, Strawberry Gardens, Newick, BN8 4QX
G0 XAN G. Aylward, 53 Overdown Rise, Portslade, Brighton, BN41 2YF
G0 XAO P. Cole, 18 Mundys Field, Ruan Minor, Helston, TR12 7LF
GW0 XAP B. Blake, 39 Heol Sant Gattwg, Llanspyddid, Brecon, LD3 8PD
G0 XAR S. Farthing, 21 Cavell Close Swardeston, Norwich, NR14 8DH
G0 XAS J. Gavin, 22 Rotherwick Way, Cambridge, CB1 8RX
G0 XAT R. Wallbank, 32 Truro Place, Cannock, WS12 3YJ
G0 XAU C. Martin, 7 Mayfair Close, Dukinfield, SK16 5HR
GM0 XAV A. Main, 1 Border Avenue, Saltcoats, KA21 5NH
G0 XAW A. Santillo, 34 Wearde Road, Saltash, PL12 4PP
G0 XAY R. Elford, Prospects, Tormarton Road, Badminton, GL9 1HP
G0 XAZ H. Simmons, 96 Porlock Road, Southampton, SO16 9JF
G0 XBA A. Hill, 13 Sycamore Way, Winklebury, Basingstoke, RG23 8AD
G0 XBC M. Soane, 24 Nurseries Road, Wheathampstead, St Albans, AL4 8TP
G0 XBG A. Marinho, 13 Pipkin Way, Oxford, OX4 4AR
GI0 XBH K. Morgan, 40 St. Marys Gardens, Belfast, BT12 7LG
G0 XBL H. Watts, 7 Hartwood Road, Liverpool, L32 7QH
G0 XBO E. Smith, 8 Nene Road, Hunstanton, PE36 5BZ
G0 XBQ C. Levingston, 44 Lewis Road, South Australia, Glynde, Australia, 5070
G0 XBV A. Nottage, 99 Fermor Way, Crowborough, TN6 3BH
G0 XCF C. Foley, 12 Cross Street, Northam, Bideford, EX39 1BS
G0 XDI M. Cabban, 2 Sandycroft Road, Amersham, HP6 6QL
G0 XDL G. Edwards, 82 Regent Drive, Skipton, BD23 1BB
G0 XDX G. Kendall, 39 Foundry Gate, Wombwell, Barnsley, S73 0LF
G0 XEG D. Riches, 21 Brinkley Way, Felixstowe, IP11 9TX
GM0 XFK J. Neary, 17 Harkins Avenue, Blantyre, Glasgow, G72 0RQ
G0 XGL G. Lawrence, 20 Branewick Close, Fareham, PO15 5RS
G0 XGM R. Burgess, 22 Lee Close, Kidlington, OX5 2XZ
G0 XIT B. Davis, Westfield House, Wood End Road, Bedford, MK43 9BB
G0 XJS J. Sliman, 4 Coleridge Close, Exmouth, EX8 5SP
G0 XKK K. Keenan, 25 Harris Road, Harpur Hill, Buxton, SK17 9JS
G0 XOX P. Thorndike, 56 Durham Road, Southend-on-Sea, SS2 4LU
G0 XPD A. Sutton, Karena Gweek, Helston, TR12 6UB
G0 XRC Exmouth ARC, c/o A. Howell-Jones, 11 Staffick Close, Kenton, Exeter, EX6 8NS
G0 XTA R. Skells, 95 Sutton Road, Leverington, Wisbech, PE13 5DR
G0 XTL T. Layphries, 11 Crossfield, Fernhurst, Haslemere, GU27 3JL
G0 XTM Dr N. Marshall, 3 The Green, Dorking Road, Tadworth, KT20 5SQ
G0 XVC R. Nixon, 31 Ashfield, Shotley Bridge, Consett, DH8 0RF
G0 XVL D. Veale, 5 Heathfield Close, Dronfield, S18 1RJ
G0 XXX 5xx Group Daventry, c/o D. Knowler, Apartedo 1009, 8670 - 999, Aljezur, Portugal,
GW0 XYL J. Hockley, 44 Brookfields, Crickhowell, NP8 1DJ
GI0 XYZ Co-Antrim A.R.S DX Group, c/o J. Hoey, 66 Woodburn Road, Carrickfergus, BT38 8ps
G0 XZT J. Gilbert, 17 Phoenix Way, Portishead, Bristol, BS20 7FG
G0 YBU Dr D. Cuff, 7 Parsons Pool, Shaftesbury, SP7 8AL
G0 YCE F. Ricci, 4 Plymouth Street, Oldham, OL8 1PP
G0 YDX R. Bates, Apartment 608, 465 West Dominion Drive, Wood Dale, USA, 60191-2309
G0 YKC G. Clampin, Inverkeris, Tydd Road, Spalding, PE12 0HP
G0 YKK K. Kelsall, 56 Glenwood Avenue, Baildon, Shipley, BD17 5RS
G0 YLO Wincanton Ladies Contest Group, c/o C. Monksummers, 29 Cloverfields, Peacemarsh, Gillingham, SP8 4UP
G0 YOU C. Haye, 15 Byron Close, Yateley, GU46 6YW
G0 YRT B. Howarth, 23 Yew Tree Road, Denton, Manchester, M34 6JY
G0 YSS A. Jones, 122 Slater St., Latchford, Warrington, WA4 1DW
G0 YYH K. Yeung, Flat 1, 2 Brunswick Hill, Reading, RG1 7YT
G0 YYY R. Konowicz, 12 Ambleside Crescent, Farnham, GU9 0RZ
GI0 ZAK F. Tanner, 4 Leopold Gardens, Belfast, BT13 3XN
GM0 ZAM J. Glennon, 68 Carronshore Road, Carron, Falkirk, FK2 8EE
G0 ZAP R. Crozier, 930-932 Burnley Road, Loveclough, Rossendale, BB4 8QL
G0 ZAT W. Skipper, 18 Central Avenue, South Shields, NE34 6AZ
G0 ZDL D. Lister, 7 The Fairways, Condover, Shrewsbury, SY5 7BW
G0 ZEE C. Monksummers, 29 Cloverfields, Peacemarsh, Gillingham, SP8 4UP
G0 ZEP R. Carter, 12 Glebe Close, Abbotsbury, Weymouth, DT3 4LD
GI0 ZER E. Robinson, 32 Corrycroar Road, Pomeroy, Dungannon, BT70 3DY
G0 ZGN M. Geernaert-Davies, Bampton, Nettlestone Green, Seaview, PO34 5DY
G0 ZHP Polish Scout ARC - London, c/o K. Jasinski, 35 Friars Place Lane, London, W3 7AQ
G0 ZIP F. Marston, 1 Weaver Road, Leicester, LE5 2RL
G0 ZMC M. Conlon, 3 Selside, Brownsover, Rugby, CV21 1PG
G0 ZMH M. Howell, Orchard House, Blennerhasset, Wigton, CA7 3QX
G0 ZPV J. Roberts, Long Meadow, Kidderton Lane, Nantwich, CW5 8JD

G1

G1 AAC P. Watterson, 25 Church Lane, Mablethorpe, LN12 2NU
G1 AAD N. Youd, 8 Forest Road, Piddington, Northampton, NN7 2DA
G1 AAG A. Withers, 23 Fernie Road, Guisborough, TS14 7LZ
G1 AAH P. Worledge, 8 Forest Edge Road, Sandford, Wareham, BH20 7BX
G1 AAK P. Webster, 30 Belvoir Road, Widnes, WA8 6HR
G1 AAL A. Woodward, 40 Berwood Farm Road, Wylde Green, Sutton Coldfield, B72 1AG
G1 AAP M. Wilson, 18 Briars Close, Southwood, Farnborough, GU14 0PB
G1 AAQ E. Writer, 78 Henley Way, Ely, CB7 4YJ
G1 AAR M. West, The Lair, Lewes Road, Newhaven, BN9 9AH
G1 AAV M. Young, 325 Blair Avenue, Friday Harbor, USA, 98250
G1 ABA D. Lambert, 72 Johnson Drive, Barrs Court, Bristol, BS30 7BS
G1 ABJ A. Norton, 29 Long Lane, Chapel-En-Le-Frith, High Peak, SK23 0TA
G1 ABM I. Macdonald, 23 Cymberline Way, Warwick Gates, Warwick, CV34 6FQ
G1 ABQ F. Macdonald, Boothlands Farm, Newdigate, Dorking, RH5 5BS
G1 ABW B. Webber, 5 Sheldon Way, Berkhamsted, HP4 1FG
GW1 ABX C. Passey, 119 Ffordd Y Mileniwm, Barry, CF62 5BD
G1 ACA J. Garner, Cobwebs, Lewes Road, Haywards Heath, RH17 7PG
G1 ACB G. Gifford, 42 Green Park, Brinkley, Newmarket, CB8 0SQ
G1 ACD P. Hughes, 7 Dellfield Lane, Liverpool, L31 6AS
GI1 ACN T. Hutton, 23 Enniscrone Park, Portadown, Craigavon, BT63 5DQ
GW1 ACV W. Harrison, Top Flat, Craiglwyd Hall, Penmaenmawr, LL34 6ER
G1 ACY M. Holley, 95 Lyes Green, Corsley, Warminster, BA12 7PA
G1 ADB M. Hunt, 46 Cunningham Drive, Bury, BL9 8PD

G1 ADE H. Kirk, 34 Tilworth Road, Hull, HU8 9BN
GM1 ADI R. Stevens, 10 Tiel Path, Glenrothes, KY7 5AX
GW1 ADY L. Rees, 40 High St., Abergwili, Carmarthen, SA31 2JB
G1 AEA A. Read, 16 Western Close, Penton Park, Chertsey, KT16 8QB
G1 AEB J. Savage, 30 Green Meadows, Caravan Park, Cheltenham, GL51 6SN
G1 AEF R. Smith, 130 Winchester Road, Ford Houses, Wolverhampton, WV10 6EZ
G1 AEI P. Stevenson, 2 Lazonby Hall Cottages, Lazonby, Penrith, CA10 1BA
G1 AEJ L. Smith, 4 Penhale Road, Braunstone, Leicester, LE3 2UU
G1 AEQ D. Lewis, 10 Addington Road, Bolton, BL3 4QZ
G1 AET F. Lawson, 10 Avebury Close, Tuffley, Gloucester, GL4 0TS
G1 AEU A. Lott, 27 Queens Crescent, Brixham, TQ5 9PJ
G1 AEX G. Muggeridge, Gribble House, Wey Street, Ashford, TN26 2QH
G1 AFI D. Mills, 25 Lower Park Crescent, Poynton, Stockport, SK12 1EF
G1 AFJ P. Keyte, 11 Woodward Road, Pershore, WR10 1LW
G1 AFK B. Kelsey, 18 Parkside Avenue, Littlehampton, BN17 6BG
G1 AFW R. Harris, 15 Rodmer Close, Minster On Sea, Sheerness, ME12 2BS
G1 AGA A. Gee, 74 New Street, Milnsbridge, Huddersfield, HD3 4LD
G1 AGB D. Gee, 74 New Street, Milnsbridge, Huddersfield, HD3 4LD
G1 AGK D. Woodruffe, 7 Orchard Close, Melton, Woodbridge, IP12 1LD
G1 AGM G. Williams, 22 Moor Tarn Lane, Walney, Barrow-in-Furness, LA14 3LP
G1 AGW D. Yeatman, 302 Canterbury Road, Herne Bay, CT6 7HD
GM1 AHF A. Mccormack, 18 Harris Court, North Muirton, Perth, PH1 3DD
GM1 AHG N. Mccormack, 18 Harris Court, North Muirton, Perth, PH1 3DD
G1 AHM A. Martland, Knowleswood, Wrennals Lane, Chorley, PR7 5PW
G1 AHQ D. Mobley, 17 Butts Close, Aynho, Banbury, OX17 3AE
G1 AHS A. Mitchell, 7 Cross Park Street Horbury, Wakefield, WF4 6AE
G1 AHT C. Wager-Bradley, Swallowdale, Longhoughton Road, Alnwick, NE66 3AT
G1 AIA E. Oliver, 9 Taylor Terrace, West Allotment, Newcastle upon Tyne, NE27 0EF
GW1 AIB E. Robinson, 35 Bolgoed Road, Pontarddulais, Swansea, SA4 8JF
G1 AIF R. Robinson, 9 Prospect Terrace, New Kyo, Stanley, DH9 7TR
G1 AIG S. Rimell, 1 Francis Way, Bridgeyate, BS30 5WJ
G1 AII J. Rushton, 8 Keats Road, Stonebroom, Alfreton, DE55 6JG
G1 AIO A. Peters, 28 Drake Avenue, Didcot, OX11 0AD
G1 AJC T. Howe, 4 Willow Crescent, Broughton Gifford, Melksham, SN12 8NB
G1 AJD M. Jacobsen, 3 Green Lane, Tickton, Beverley, HU17 9RH
G1 AJE M. Chambers, 1 Green Lane, Tickton, Beverley, HU17 9RH
G1 AJK D. Neale, 161 Antrobus Road, Birmingham, B21 9NU
G1 AJN G. Overy, Flat2 37 Magdalene Rd, St Leonards on Sea, TN376ET
G1 AJQ D. Padfield, 20 Gainsborough Crescent, Chelmsford, CM2 6DJ
G1 AJS N. Parker, 7 The Hollies, Clee Hill, Ludlow, SY8 3NZ
G1 AJT C. Pickering, 16 Ashworth Way, Newport, TF10 7EG
G1 AJU J. Parsons, 40 Tynings Close, Kidderminster, DY11 5JP
G1 AJV E. Roberts, 4 Willow Way, Redditch, B97 6PH
G1 AJY A. Young, 4 Woodlea, Leybourne, West Malling, ME19 5QY
G1 AJZ P. Rayson, 1 Grange Gardens, Taunton, TA2 7EN
G1 AKA H. Rowley, Apartment 27, The Bridges, Buxton Road, Macclesfield, SK10 1FW
G1 AKB S. Ryder, Corner Cottage, Crown Lane, Defford, Worcester, WR8 9BE
G1 AKD A. Rideout, 7 Beech Road, Martock, TA12 6DT
G1 AKE P. Rowland, 7 Maxwell Street, Bury, BL9 7QA
G1 AKV T. Alexander, 8 Greenway, Eastbourne, BN20 8UG
G1 ALA J. Bird, 17 Sherrards Way, Barnet, EN5 2BW
G1 ALD A. Brown, 40 Sutherland Road, Edmonton, London, N9 7QG
G1 ALK R. Batchelor, 20 Mallard Way, Lower Stoke, Rochester, ME3 9ST
G1 ALL M. Bond, 8 Alfred Street, Irchester, Wellingborough, NN29 7DR
G1 ALR A. Stafford, 24 Bourne Street, Croydon, CR0 1XL
G1 ALU K. Spiers, Robins Post, North Heath Lane, Horsham, RH12 5PJ
GW1 ALV A. Shaw, Derlwyn, Efailwen, Clynderwen, SA66 7JP
G1 AML I. Tidey, April Cottage, Cansiron Lane, East Grinstead, RH19 3SE
G1 AMN N. Webb, 3 Allens Lane, Norwich, NR2 2JB
G1 AMS J. Winter, Flat 23, Knightlow Lodge Knightlow Avenue, Coventry, CV3 3HH
G1 ANA J. Coles, 10 Westgate Hill Street, Bradford, BD4 0SJ
G1 AND I. Dodds, 54 Philip Road, Newark, NG24 4PD
G1 ANF T. Crowe, 15 Lambert Road, Kendray, Barnsley, S70 3AA
GI1 ANG G. Carvill, 18 Hospital Road, Newry, BT35 8PW
G1 ANI M. Cooper, 33 Park View, Royston, Barnsley, S71 4AA
G1 ANK R. Bond, 21 Dinglebank Close, Lymm, WA13 0QR
G1 ANQ P. Fincher, 11 Verney Mews, Reading, RG30 2NT
G1 ANS K. Elvin, 97 Jeans Way, Dunstable, LU5 4PR
G1 ANV R. Edgar, 19 Butt Hedge, Long Marston, York, YO26 7LW
GW1 ANW D. Evans, 6 Oakfield Terrace, Ammanford, SA18 2NG
G1 ANZ P. Thirst1, The Haywain, Thirsts Farm, Happisburgh, Norwich, NR12 0RU
G1 AOC R. Doughty, 7 High Street, Garlinge, CT9 5LN
G1 AOE J. Darley, 16 Ivydene, Knaphill, Woking, GU21 2TA
G1 AOF R. Dean, 10 Livingstone Road, Ellesmere Port, CH65 2BE
G1 AOQ I. Gorsuch, Elmstone Farm, Fosten Green, Ashford, TN27 8ER
G1 AOR P. Grayshon, 90 Park Lea Bradley, Huddersfield, HD2 1QP
G1 AOZ J. Henderson, The Bungalow, Appleby Grammar School, Appleby-in-Westmorland, CA16 6XU
G1 APA C. Hibbert, 19 Fern Road, Maia, Dunedin, New Zealand, 9022
G1 APL R. Hopkins, 15 Oak Meadow Shipdham, Thetford, IP25 7FD
G1 APQ A. Howells, 16 Oakley Wood Road, Bishops Tachbrook, Leamington Spa, CV33 9RN
GW1 APU K. Pierson, Cefn Glaniwrch Cottage, Llanrhaeadr Ym, SY10 0DR
G1 AQF A. Matthews, 44 Essex Close, Dines Green, Worcester, WR2 5RW
G1 AQI J. Aldersey, 36 Walls Road, Salterbeck, Workington, CA14 5JA
G1 AQP A. Bell, 1 Purbeck Drive, Lostock, Bolton, BL6 4JF
G1 AQV J. Blackman, 30 Parklands Way, Penrith, CA11 8SD
G1 AQX K. Armstrong, 30 Cobholm Place, Cambridge, CB4 2UN
GD1 AQY S. Broad, Ballaberragh, Farm, Ramsey, Isle Of Man, IM73EB
G1 ARD G. Ashton, 101 Wickenby Garth, Bransholme, Hull, HU7 4RF
G1 ARF J. Makin, 6 Cambridge House, Courtfield Gardens, London, W13 0HP
G1 ARH R. Lupton, 19 Avenue Close, Harrogate, HG2 7LJ
G1 ARL C. Mandall, 11A Hazel Road, Park Street, St Albans, AL2 2AH
G1 ARM R. Preece, 5 Pavey Run, Ottery St Mary, EX11 1FQ
G1 ARU A. Judge, 44 Thorley Lane, Bishop's Stortford, CM23 4AD
GM1 ASA H. Kilpatrick, 80 Livingstone Terrace, Irvine, KA12 9DN
G1 ASD K. Knight, 80 Winton Road, Reading, RG2 8HJ
G1 ASG T. Stokes, 21 Guildford View, Sheffield, S2 2NZ
G1 ASN J. Stansfield, Flat 12, Harty House Church Street, Eccles, Manchester, M30 0LT
G1 ASR D. Thornton, 8 Chestnut Court, Toft Hill, Bishop Auckland, DL14 0TG
G1 ASU A. Clews, 47 Gaydon Road, Solihull, B92 9BJ
GM1 ASY J. Christie, 99 Meadow Crescent, Elgin, IV30 6ER
G1 ATA D. Cotton, 24 Stirling Rise, Stretton, Burton-on-Trent, DE13 0JP
G1 ATC Air Cadet ARS, c/o V. Tuff, 8 Millcroft Court, Blyth, NE24 3JG
G1 ATG C. Clarke, 29 Huyton Hey Road, Huyton, Liverpool, L36 5SF
G1 ATL K. Bishop, 8 Sandbanks Grove, Hailsham, BN27 3LS
G1 ATQ H. Rogers, 74 Front Road, Murrow, Wisbech, PE13 4HU
G1 ATU G. Krause, 23 Daniel Crescent, Lincoln, LN4 1QT
GM1 ATW E. King, Marionville, Donibristle, Cowdenbeath, KY4 8EU
G1 ATY B. Jarratt, 4 Jazz Road, Aylesbury, HP18 0EZ
GW1 ATZ G. Morris, 18 Grosvenor Road, Shotton, CH5 1NU
G1 AUH R. Holland, 17 Eaton Square, Leeds, LS10 4SN
G1 AUI C. Hayes, 37 St. Radigunds Street, Canterbury, CT1 2AA
G1 AUM G. Gumbrell, 24 Tattershall Drive, Market Deeping, Peterborough, PE6 8BS
G1 AUR H. Graham, 4 Eisenhower Road, Basildon, SS15 6JR
G1 AUU S. Goan, 15 Winchester Avenue, Chorley, PR7 4AQ
G1 AUY R. Curzon, 24 Edwards Drive, Wellingborough, NN8 3JJ
GM1 AUZ G. Crockford, 13 Forvie Street, Bridge Of Don, Aberdeen, AB22 8TP
G1 AVA B. Carter, 10 Opal Street, Keighley, BD22 7BP
G1 AVB R. Davidson, 3 Eastridge Drive Bishopsworth, Bristol, BS13 8HQ
G1 AVC J. Dean, 62 Melwood Drive, Liverpool, L12 8RW
G1 AVF B. Eastick, 30 Farmcote Road, Aldermans Green, Coventry, CV2 1SA
G1 AVW H. Sterry, 23 Eardisley Close, Matchborough, Redditch, B98 0BX
G1 AVZ J. Toolan, 12 Stillington Road, Huby, York, YO61 1HW
G1 AW A. Wragg, 14 Grizedale Avenue Sothall, Sheffield, S20 2DL
G1 AWD T. Wells, 2 Stephens Close, Mortimer Common, Reading, RG7 3TL
G1 AWF A. Wyspianski, 53 Alington Crescent, Kingsbury, London, NW9 8JL
GW1 AWH I. Robinson, 13 Clas Ty Wern, Cardiff, CF14 4SB
G1 AWJ Dr J. Moyle, Amberley, Cotswold Close, Shipston-on-Stour, CV36 4NR
G1 AWK D. Moulson, 15 Bramble Way, Leavenheath, Colchester, CO6 4UN
G1 AWU A. Pickles, 49 Hermitage Street, Crewkerne, TA18 8ET
GM1 AXI T. Ball, 3 Kersland Place, Glengarnock, Beith, KA14 3BQ
GW1 AXU J. Cook, 22 Northlands Park, Bishopston, Swansea, SA3 3JW
G1 AXW C. Dann, 113 Belle Vue Road, Cinderford, GL14 3BL
G1 AYH M. Newell, 15 The Grove, Luton, LU1 5PE
G1 AYI T. Smith, 19 Leaf Road, Houghton Regis, Dunstable, LU5 5JG
G1 AYP K. Lawton, Meadowbank, Sutton St. Nicholas, Hereford, HR1 3BJ
G1 AYU C. Chambers, 1 Saunders Close, Pound Hill, Crawley, RH10 7AE
G1 AZA J. Cook, 72 Valebridge Road, Burgess Hill, RH15 0RP
G1 AZC M. Cannings, 7 Whinlatter Place, Newton Aycliffe, DL5 7DR.
G1 AZD A. Drage, 51 Greenbank Avenue, Kettering, NN15 7EF
G1 AZE B. Davies, 22 Hillside Road, Four Oaks, Sutton Coldfield, B74 4DQ
G1 AZJ G. Taylor, 31 Ashfurlong Crescent, Sutton Coldfield, B75 6EN
G1 BAA F. Whittaker, 14 Mill Lane, Wombourne, Wolverhampton, WV5 0LG
G1 BAB R. Wilson, Barley Corners, Arundel Road, Seaford, BN25 4LZ
GW1 BAI K. Lowe, 18 Aberdovey Close, Dinas Powys, CF64 4PS
G1 BAL J. Mahoney, 27 Linby Drive, Bircotes, Harworth, Doncaster, DN11 8FP
GM1 BAN D. Morrison, 4 West Murkle, Murkle, Thurso, KW14 8YT
G1 BAQ A. Miller, 44 Spring Gardens, Newport Pagnell, MK16 0EE
G1 BAR B. Norris, 1 Earleswood, Benfleet, SS7 1DN
G1 BAX J. Peters, Ferndale Cottage, Brea, Camborne, TR14 9AT
G1 BBA T. Tilley, Mizpah, Carpalla Foxhole, St Austell, PL26 7TY
G1 BBC J. Duxbury, 7 Osprey Close, Blackburn, BB1 8LP
GW1 BBH J. Sharkey, 33 Ffordd Morfa, Llandudno, LL30 1ES
G1 BBI K. Ford, 2 Ford Close, Ferndown, Bh22 8AA
G1 BBK B. Artingstall, 19 Town Lane, Denton, Manchester, M34 6AF
G1 BBT J. Dackham, 4 Overbury Close, Weymouth, DT4 9UE
G1 BBY W. Brown, 53 Drummonds Close, Longhorsley, Morpeth, NE65 8UR
G1 BCB A. Blake, 58 Greenacres, Bath, BA1 4NR
G1 BCE B. Brough, 6 Higgs Road, Wednesfield, Wolverhampton, WV11 2PD
G1 BCG P. Edwards, 5 Playden Close, Brighton, BN2 5GH
GW1 BCI A. Gray, 69 Tyn Y Parc Road, Cardiff, CF14 6BJ
G1 BCN A. Hopkinson, 104 Everill Gate Lane, Wombwell, Barnsley, S73 0YJ
G1 BCU R. Tagg, 38 Salhouse Road, Rackheath, Norwich, NR13 6QH
GW1 BDF K. Jones, 10 Trinity Road, Tonypandy, CF40 1DQ
GW1 BDG O. Jones, 10 Trinity Road, Tonypandy, CF40 1DQ
GW1 BDH B. Jones, 8 Walton Crescent, Llandudno Junction, LL31 9ER
G1 BDI B. Jones, 56 Mount Grace Road, Luton, LU2 8EP
G1 BDP M. Broad, 7 Steventon Road, Drayton, Abingdon, OX14 4JX
G1 BDQ A. Bell, Doddington Mill, Mill Lane, Nantwich, CW5 7NN
G1 BDU A. Bradshaw, Lyndale, 145 Alder Lane, Wigan, WN2 4ET
G1 BDY M. Bragg, 33 Mosley Street, Barnoldswick, BB18 5BS
G1 BEB R. Cocking, 22 Dunbeath Avenue, Rainhill, Prescot, L35 0QH
G1 BEG J. Ellsmore, 15 Greenbush Drive, Halesowen, B63 3TJ
G1 BEJ I. Dixon, 60A Woodlands Road, Allestree, Derby, DE22 2HF
G1 BEK G. Death, 105 Belvedere Road, Ipswich, IP4 4AD
G1 BES J. Gibbard, 2 Almond Court, Liverpool, L19 2QZ
G1 BET A. Waters, 12 Anvil Court, Whittonstall, Consett, DH8 9JU
GI1 BEU L. Gough, 76-78 Culmore Point, Londonderry, BT48 8JW
GW1 BFB A. Frayne, 20 Springfield Avenue, Upper Killay, Swansea, SA2 7HW
G1 BFF T. Fishlock, 62 Red Barn Road, Brightlingsea, Colchester, CO7 0SJ
G1 BFG A. Harper, 144 Ashfield Road, Bispham, Blackpool, FY2 0EN
G1 BFK R. Else, 134 Market Street, South Normanton, Alfreton, DE55 2EJ
G1 BFS P. Sainsbury, 103A Pilot Road, Hastings, TN34 2AU
G1 BFV S. Harman, Ivy Cottage, Grove Road, Friston, Saxmundham, IP17 1PP
G1 BGC S. Javes, 95 West Way, Lancing, BN15 8LZ
G1 BGF D. Mclachlan, 1 North Holme Court, Northampton, NN3 8UX
G1 BGH P. Nicholls, 40 Sedgefield Close, Worth, Crawley, RH10 7XG
G1 BGJ E. Morgan, 79 Mayland Avenue, Canvey Island, SS8 0BU
G1 BGK G. Lang, Belford Barn, Ashburton, Newton Abbot, TQ13 7HT
G1 BGM P. Leslie, 5 Maple Croft, Netherton, Huddersfield, HD4 7HS
G1 BGO D. Lee, 36 Westwick, Hedon, Hull, HU12 8HQ
G1 BGQ P. Sampson, 23 Westfield Road, Mirfield, WF14 9PW
G1 BHB S. Matthews, 222 Widney Lane, Solihull, B91 3JY

G1 BHF S. Oldfield, The Sycamores, Fulford Road, Stoke-on-Trent, ST11 9QT
G1 BHG B. Osborne, 12 Arminers Close, Gosport, PO12 2HB
G1 BHO B. Palmer, Flat 5, Brook Court Burcot Lane, Bromsgrove, B60 1AD
G1 BHQ R. Pritchard, 41 Greenland Avenue, Maltby, Rotherham, S66 7EU
G1 BHR I. Rabbitt, 66 Parkfield Avenue, Delapre, Northampton, NN4 8QB
G1 BHV R. Green, 20 Haygate Drive, Wellington, Telford, TF1 2BY
G1 BHW C. Vaughan, 11 Fremantle Road, Aylesbury, HP21 8EH
G1 BIA H. Wilmshurst, Langholm Lodge, Raydaleside, Darlington, DL3 7SJ
G1 BIF T. Whelan, 43 Martin Avenue, Little Lever, Bolton, BL3 1NX
G1 BIM R. Ross, 46 Arbour Close, Rugby, CV22 6EH
G1 BIN B. Talbot, 27 Shuttleworth Road, Clifton Upon Dunsmore, Rugby, CV23 0DB
G1 BIU C. Utting, 22 Yew Tree Drive, Bromsgrove, B60 1AL
G1 BJE A. Stimpson, 2 Church Avenue, Kings Sutton, Banbury, OX17 3RJ
G1 BJK P. Lough, 87 Finchley Road, Kingstanding, Birmingham, B44 0LB
G1 BJN C. Stroud, 6 Church Road Ideford, Chudleigh, Newton Abbot, TQ13 0BB
G1 BJZ G. Garlick, 1 Shannon Way, Burton Latimer, Kettering, NN15 5SX
G1 BKB S. Stokes, 52 Brantley Avenue, Wolverhampton, WV3 9AR
G1 BKI M. York, 38 Rannoch Way, Corby, NN17 2LH
G1 BKJ J. Prior, 6 Emfield Grove, Grimsby, DN33 3BS
G1 BKL D. Ambler, Corrig, 4 Old Main Road, Bridgwater, TA6 4RY
GM1 BKR J. Rankin, 3 Spalding Drive, Largs, KA30 9BZ
G1 BKU R. Lawn, 39 Wetherby Road, Knaresborough, HG5 8LH
G1 BKZ W. Moore, 10 Progress St., Darwen, BB3 2DT
G1 BLB P. Thurman, 11 Copperfield Drive, Langley, Maidstone, ME17 1SX
G1 BLJ S. Lovell, 12 The Holloway, Swindon, Dudley, DY3 4NT
G1 BLK C. Ridley, 14 Painswick Road, Hall Green, Birmingham, B28 0HH
G1 BLO N. Swan, 8 Tyrrells Court Bransgore, Bransgore, BH238BU
G1 BLQ P. Lloyd, 35 Westfield Road, Hertford, SG14 3DL
G1 BLV S. Davies, Swallow Cottage, The Chantry, Leyburn, DL8 4NA
GM1 BLX I. Dewar, 11 Abbotshall Road, Kirkcaldy, KY2 5PH
G1 BMB G. Fallows, 66 Ulverston Road, Swarthmoor, Ulverston, LA12 0JF
G1 BMN N. Lamb, 106 St. Davids Road, Leyland, PR25 4XY
G1 BMP S. Norminton, 63 Candish Drive, Plymouth, PL9 8DB
G1 BMT P. Tuthill, 12 Herbert Road, Salisbury, SP2 9LF
G1 BMW P. Winterton, 35 Paynesfield Road, Tatsfield, Westerham, TN16 2AT
G1 BMZ S. Wilkins, Garden House Westdown Farm, Exmouth, EX8 5BU
GM1 BNA R. Main, 14 Hunters Grove, East Kilbride, G74 3HZ
G1 BNE A. Perkins, 7 High Street, Wollaston, Wellingborough, NN29 7QE
G1 BNG S. Marsh, 28 Orcheston Road, Bournemouth, BH8 8SR
G1 BNN S. Tilly, 24 Whinham Way, Morpeth, NE61 2TF
GM1 BNP R. Holt, Whitlam Farmhouse, Newmachar, Aberdeen, AB21 0RS
GM1 BNS W. Graham, 6 Braemar View, Clydebank, G81 3RR
G1 BNV S. Henderson, 7 Havering, Castlehaven Road, London, NW1 8TH
G1 BNX R. Huxley, 83 Gleneagles Road, Wyken, Coventry, CV2 3BH
G1 BOB C. Balsdon, 4 Queens Hayes, Willey Lane, Okehampton, EX20 2NG
G1 BOO F. Crompton, 24 Alcester Road, Sale, M33 3QP
G1 BOX M. Platten, 48 Brier Road, Sittingbourne, ME10 1YL
G1 BPD I. Attridge, 12 Ascot Road, Orpington, BR5 2JF
G1 BPE M. Barwick, 32 St. Georges Road, Harrogate, HG2 9BS
G1 BPS M. Rowell, 1 Willow Street, Haslingden, Rossendale, BB4 5NA
G1 BPU L. Staal, 5 Hunt Court, 236 Chase Side, London, N14 4PG
G1 BPV A. Stalker, 1 Ganymede Lane, Brackley, NN13 6RA
G1 BQG A. Bright, 15 Cross Road, Maldon, CM9 5EE
G1 BQH J. Murphy, 34 Knights Hill, Walsall, WS9 0TG
G1 BQI G. Smith, 92 Lime Road, Accrington, BB5 6BJ
GM1 BQP W. Macrobbie, 44 Moray Park Terrace, Culloden, Inverness, IV2 7RW
G1 BQQ J. Sowerbutts, 22 Worsley St., Accrington, BB5 2PA
G1 BQR M. Spinks, 26 Church Hill, Royston, Barnsley, S71 4NH
G1 BQV M. Sunderland, 36 Moorlands Avenue, Leeds, LS19 6AD
G1 BRB A. Patton, 72 Sanctuary Way, Grimsby, DN37 9RZ
G1 BRD J. Smith, 127 Wolverhampton Road, Cannock, WS11 1AR
G1 BRF P. Williams, 292 Hagley Road, Hasbury, Halesowen, B63 4QG
G1 BRP K. Crawley, 19 Park Mount, Harpenden, AL5 3AS
G1 BRS Bournemouth Radio Society, c/o M. Stevens, 16 Golf Links Road, Ferndown, BH22 8BY
GM1 BSG J. Ross, 16 Myreton Drive, Bannockburn, Stirling, FK7 8PX
GI1 BSJ J. Cunningham, 4 Garvaghy Road, Portglenone, Ballymena, BT44 8EF
G1 BSY K. Morris, 3 Moravian Close, Dukinfield, SK16 4EW
G1 BSZ R. Nash, Roann, Bedmond Road, Hemel Hempstead, HP3 8SH
G1 BTF A. Hardy, 14 Parsonage Road, Rainham, RM13 9LW
G1 BTI K. Forrest, 61 Woodbury Road, Bridgwater, TA6 7LJ
GM1 BTL W. Erskine, 30 Market Road, Kirkintilloch, Glasgow, G66 3JL
G1 BTN I. Evans, 37 Lyndale Avenue, Lostock Hall, Preston, PR5 5UU
G1 BTV D. Holt, 2 London Heights, Dudley, DY1 2QZ
G1 BUJ A. Bates, 17 Walkers Heath Road, Kings Norton, Birmingham, B38 0AB
G1 BUQ J. Spink, 38 Hemlingford Road, Sutton Coldfield, B76 1JQ
G1 BUV M. Osborne, 5 Wells Road, Riseley, Bedford, MK44 1DY
GM1 BUY C. Rose, 14 Spoutwells Drive, Scone, Perth, PH2 6RR
G1 BUZ R. Scott, 27 Woodford Av, Henderson, Auckland, New Zealand, 610
GM1 BVA L. Nieto, 71 Clyde Street Camelon, Falkirk, FK1 4ED
GM1 BVT W. Pettett, 15 Strude Howe, Alva, FK12 5JU
G1 BVV G. Pemberton, 8 Hotchin Road, Sutton-On-Sea, Mablethorpe, LN12 2NP
G1 BWG R. Dodd, 5 Halesworth Road, Wolverhampton, WV9 5PH
G1 BWH P. Eaton, 8 Chester Road, Barnwood, Gloucester, GL4 3AX
G1 BWI J. Eastham, 81 Park Lee Road, Blackburn, BB2 3NZ
G1 BWP W. Webb, The New Bungalow, Tram Road, Coleford, GL16 8DN
GU1 BWW D. Ash, Trigale House, La Trigale, St Anne, Guernsey, GY9 3TX
G1 BWX P. Caton, 39 Farmerie Road, Hundon, Sudbury, CO10 8HA
G1 BWZ N. Fieldsend, 47 Hollycroft, Barmston, Driffield, YO25 8PP
GM1 BXI R. Clark, 5 Abbotsfield Terrace, Auchterarder, PH3 1DD
G1 BXQ J. Smith, Noonhill Farm, Grove Road, Coventry, CV7 9JE
G1 BXT I. Thomas, Myrtle Cottage, Swan Lane, Sellindge, Ashford, TN25 6EB
GW1 BXX A. Gillard, 10 Parc Pendre, Brecon, LD3 9ES
G1 BYI J. Moore, 11 Sherborne Road, Wallasey, CH44 2EY
G1 BYJ G. Noon, 48 Sanderling Close, Letchworth Garden City, SG6 4HY
G1 BYO S. Slaughter, 96 Adys Road, London, SE15 4DZ
G1 BYP P. Snitch, 79 Albion Avenue, York, YO26 5QZ
G1 BYQ D. Hatton, 34 Avocet Way, Bicester, OX26 6YP
G1 BYS A. Kempton, 14 Lower Gravel Road, Bromley, BR2 8LT
G1 BYT N. Kinselley, 5 Helford Close, Bedford, MK41 7TU
GM1 BZD P. Ward, Burnside, Flotta Stromness, KW163NP
G1 BZE C. Weeds, Flat 1, Bank House, Campbell Street, Belper, DE56 1AP
G1 BZM P. Endean, 11 Forrester Drive, Brackley, NN13 6NE
GM1 BZR D. Cameron, 14 Queen St., Castle Douglas, DG7 1HX
GI1 BZT Dr L. Gornall, 14 Ballymoghan Lane, Magherafelt, BT45 6HW
G1 BZU Royal Naval ARS, c/o J. Kirk, 111 Stockbridge Road, Chichester, PO19 8QR
G1 BZW R. Kimber, 38 Greenmere, Brightwell-Cum-Sotwell, Wallingford, OX10 0QG
GI1 CAI J. Mcbride, Dunaree, 20 Oldcastle Road, Omagh, BT78 4HX
G1 CAN P. Shonfield, 242 Chickerell Road, Weymouth, DT4 0QY
G1 CAR Mastcar ARC, c/o J. Murray, 20 East End, Cholsey, Wallingford, OX10 9RT
G1 CAY D. Shea, 38 Ranworth Avenue, Hoddesdon, EN11 9NR
G1 CBB L. Walker, The Biel, Furze Vale Road, Bordon, GU35 8EP
G1 CBK S. Mole, 53 Parkfield Road, Rainham, Gillingham, ME8 7TA
G1 CBL L. Mason, Reflow, Unit 2 Spring Lane North, Malvern, WR14 1BU
G1 CBS J. Hatt, 77 Pentland Close, Basingstoke, RG22 5BQ
G1 CBY S. Fountaine, 142 Elvaston Road, North Wingfield, Chesterfield, S42 5GA
G1 CCD M. Williamson, 15 Nook Fields Harwood, Bolton, BL2 4LN
GM1 CCI C. Watson, 11 Ladybridge Houses, Banff, AB45 2JR
G1 CCL D. Livsey, 12 Aldingham Walk, Morecambe, LA4 4EW
G1 CCM T. Mcmillan, 47 Sandsend Road, Eston, Middlesbrough, TS6 8AF
GM1 CCN C. Orr, Easter Cowden Farm, Dalkeith, EH22 2NS
G1 CCW F. Haselden, 7 Chestnut Avenue, Gosfield, Halstead, CO9 1TD
G1 CCX P. Kennedy, 24 Leadhall Drive, Harrogate, HG2 9NL
GW1 CDH D. Davies, 10 Bryn Castell, Abergele, LL22 8QA
G1 CDN D. Wormall, 20 Greenfield Road, Hemsworth, Pontefract, WF9 4RL
G1 CDO R. Wormall, 17 Newstead Grove, Fitzwilliam, Pontefract, WF9 5DS
G1 CDQ A. Sturman, 3 Windward House, 73, Lytham St Annes, FY8 1LZ
GW1 CDY D. Forth, 1 Upper Cwrt, Cwrt, Pennal, Machynlleth, SY20 9LA
G1 CEI P. Hirons, 27 Ashley Close, Crondall, GU10 5RD
GM1 CEJ R. Stout, 16 Ardoch Park, Balgeddie, Glenrothes, KY6 3PJ
G1 CEO R. Day, 17 Barry Avenue, Bicester, OX26 2DZ
GI1 CET J. Barr, 2 Willowvale Close, Islandmagee, Larne, BT40 3SD
G1 CEU J. Clarke, Timbers, Wayford Road, Norwich, NR12 9LH
G1 CFA P. Middleton, 36 Station Court, Railway Street, Hornsea, HU18 1QD
G1 CFB K. Rhodes, 34 Bannister Drive, Hull, HU9 1EJ
G1 CFE A. Wyatt, 32 Wensleydale Avenue, Blackpool, FY3 7RS
G1 CFG C. Rigby, 4 Humber Street, Longridge, Preston, PR3 3WD
G1 CFJ G. Gardner, 165 Brookhouse Road, Brookhouse, Lancaster, LA2 9NY
G1 CFK D. Bowles, Fiddlers Nook, Thurston Road, Bury St Edmunds, IP31 2PL
GW1 CFM Dr N. Bristow, Bryn Gwydion, Pontllyfni, Caernarfon, LL54 5EY
GW1 CGD S. Oliver, 21 Hillside Court, Holywell, CH8 7PJ
G1 CGH A. Rawlins, 4 Low Park, West Woodburn, Hexham, NE48 2SQ
G1 CGJ M. Davies, Kuling, Bridgwater Road, Winscombe, BS25 1NB
G1 CGP A. Gillard, 28 Moor Tarn Lane, Walney, Barrow-in-Furness, LA14 3LP
G1 CGU G. Fitzpatrick, 12 Dunster Close, Minehead, TA24 6BY
G1 CHE D. Ogarr, 10 Ellastone Grove, Stoke-on-Trent, ST4 5EE
G1 CHM C. Milburn, Field House, Copper Hill, Hayle, TR27 4LY
G1 CHN A. James, The Red House, Gandish Road, Colchester, CO7 6TP
G1 CHQ M. Wells, 7 Vint Rise, Idle, Bradford, BD10 8PU
GW1 CHS J. Hughes, Reservoir House, St. Lythan'S, Cardiff, CF5 6BQ
GM1 CHT A. Hyde, 19 Drum Brae Gardens, Edinburgh, EH12 8SY
G1 CHV C. Compton, 21 Vange Riverview Centre, Vange, Basildon, SS16 4NE
G1 CIA M. Ferentiuk, 74 Fallowfield Drive, Rochdale, OL12 6LZ
G1 CIM S. Hancock, Monrad, Back Street, Gainsborough, DN21 3DL
G1 CIT M. Whalley, 6 Rookery Walk, Clifton, Shefford, SG17 5HW
G1 CIV D. Owen, 23 Munnings Drive, Hinckley, LE10 0LG
GW1 CIY G. Evans, Maes Yr Haf, Beulah, Newcastle Emlyn, SA38 9QB
G1 CJC L. Gilbert, Holmefield Cottage, Oker, Matlock, DE4 2JJ
G1 CJI P. Arnold, 36 Gopsall Road, Hinckley, LE10 0DY
GW1 CJJ P. Williams, 6 Parc Ffynnon, Llysfaen, Colwyn Bay, LL29 8SA
G1 CJK R. Baiey, 318 Plumstead Common Road, London, SE18 2RT
G1 CJL B. Hallybone, 38 Anvil House, Champion Way, Bedford, MK42 9EH
G1 CKF J. Darby, 97 Littlehaven Lane, Horsham, RH12 4JE
G1 CKJ T. Martin, 201 Gloucester Road, Kidsgrove, Stoke-on-Trent, ST7 4DQ
G1 CKR T. Miller, 31 Hampden Road, Malvern, WR141NB
G1 CKT R. Nelson, 11 Meadow Way, Plymouth, PL7 4JB
GI1 CKU T. Gardiner, 17 Grange Valley Gardens, Ballyclare, BT39 9HE
G1 CKV N. Derbyshire, 54 Windy Arbor Road, Whiston, Prescot, L35 3SG
G1 CKY P. Turner, 16 Pendragon Way, Leicester Forest East, Leicester, LE3 3EY
G1 CLD S. Patterson, Dunedin, Little Ness, Shrewsbury, SY4 2LG
G1 CLJ P. Kennedy, 12 Newbroke Road, Rowner, Gosport, PO13 9UJ
G1 CLT R. Bokor, 54 Granwood Road, Middlesbrough, TS6 9HX
G1 CMC A. Lickley, 19 Sandy Rise, Selby, YO8 9DW
GM1 CMF P. Carnegie, 29 Dalgetty Court, Muirhead, Dundee, DD2 5QJ
G1 CMH J. Norwood, Flat 28 The Manor, Church Road, Gloucester, GL3 2HT
G1 CMZ S. Lewkowicz, 7 The Mart, Locking Road, Weston-Super-Mare, BS23 3DE
GM1 CNH N. Stewart, 160 Carrick Knowe Drive, Edinburgh, EH12 7EW
G1 CNI S. Dwyer, Po Box 44, Tahmoor, Australia, 2573
G1 CNN P. Beeson, Flat 6, Oxford Court, London, W3 0HH
G1 CNV T. Thornton, Orchard Walk, 23 Crookham Road, Fleet, GU51 5DP
G1 CNZ R. Reid, 34 Wellesley Street, Taunton, TA2 7DT
G1 COD P. Stellings, 10 Thornbrook Road Chapel-En-Le-Frith, High Peak, SK23 0LX
G1 COE C. Howard, 30 Torquay Road Kingskerswell, Newton Abbot, TQ12 5EZ
GM1 COF P. Mcgowan, 38 Mckenzie Crescent, Lochgelly, KY5 9LT
G1 COV Coventry Raynet Group, c/o D. Green, 67 Coombe Park Road, Binley, Coventry, CV3 2NW
G1 COW R. Penfold, 1 Padworth Road, Burghfield Common, Reading, RG7 3QE
G1 COX A. Berkeley, 42 Ringley Drive, Whitefield, Manchester, M45 7LR
G1 COY C. Robson, 43 Longdyke Drive, Carlisle, CA1 3HT
G1 CPA P. Nairne, 137 Barden Road, Tonbridge, TN9 1UX
G1 CPC J. Arthur, St. Aubin, Plomer Green Lane, High Wycombe, HP13 5XN
G1 CPD G. Ghetti, 7 Rue De Provence, Paris, France, 75009
G1 CPM E. Rose, 26 Lavender Way, Bourne, PE10 9TT
G1 CPO C. Haygarth, 3 Rew Close, Ventnor, PO38 1BH
G1 CPU G. Milligan, 100 Churchfield Road, Gateacre, Liverpool, L25 3SE
G1 CPX I. Clarke, 19 Welbeck, Bracknell, RG12 8UQ

G1 CQA R. Chaney, 55 Bartlow Road, Linton, Cambridge, CB21 4LY
GM1 CQC H. Smith, 601 Ferry Road, Edinburgh, EH4 2TT
G1 CQG D. Perry, 5 Beech Hill, Wellington, TA21 8ER
G1 CQK G. Knott, 7 Nunney Close, Cheltenham, GL51 0TU
G1 CQR D. Fuller, 26 Longfields, Ely, CB6 3DN
G1 CQT P. Turley, 35 Alwinton Avenue, Stockport, SK4 3PU
G1 CRH P. Everitt, 10411 Tula Lane, Cupertino, USA, 95014
G1 CRN W. Murray, 91 Chaucer Avenue, Hounslow, TW4 6NA
G1 CRT W. Cambridge, 36 Selwyn Avenue, Richmond, TW9 2HA
G1 CSA J. Walton, 23 Keighley Avenue, Sunderland, SR5 4BU
G1 CSN R. Beasley, 26 Retford Close, Harold Hill, Romford, RM3 9NA
G1 CSO J. Dent, 18 New Street, St Neots, PE19 1AE
G1 CSR Civil Service ARS, c/o N. Sanderson, 54 Kelvedon Close, Chelmsford, CM1 4DG
G1 CSS M. Wilson, 210 London Road, Worcester, WR5 2JT
G1 CSY E. Hartley, 2 Lamberts Close, Weasenham, King's Lynn, PE32 2TE
G1 CTF R. Fenton, 37 Martingale Chase, Newbury, RG14 2EN
GW1 CTO D. Powell, 88 Church View, Chirk, Wrexham, LL14 5PF
G1 CTQ N. Losardo, 14 Arnside Close, Clayton Le Moors, Accrington, BB5 5GG
GM1 CUC H. Mattinson, 11 Riverside Park, Canonbie, DG14 0UY
G1 CUG D. Laughton, 2 Stamford Road, Careby, Stamford, PE9 4EB
G1 CUM J. Rawlings, Castle House, Barrow Haven, Barrow-upon-Humber, DN19 7EY
GW1 CUQ N. Paull, 6 Llys Caradog, Creigiau, Cardiff, CF15 9JP
G1 CUZ S. Seal, Crantock, Bellingdon, Chesham, HP5 2XW
G1 CWI M. Kemp, Casa Lucia, Vale Formosilho, S. Marcos Da Serra, Portugal, 8375-210
G1 CWJ A. Burton, 303 Heneage Road, Grimsby, DN32 9NW
G1 CWP R. Hide, Flat 22, Church Court Church Road, Haywards Heath, RH16 3UE
G1 CWQ B. Wyatt, 3 Shipley Close, Blackpool, FY3 7UJ
G1 CWW G. Stone, 37 Canterbury Drive, Ashby-de-la-Zouch, LE65 2QQ
G1 CWZ D. Penrose, 7 Two Ashes, Bayston Hill, Shrewsbury, SY3 0QF
G1 CXQ C. Roberts, 11 Adel Wood Drive, Leeds, LS16 8JQ
GW1 CXT P. Lodge, 1 Brookland Terrace, Nantymoel, Bridgend, CF32 7SY
G1 CYQ B. Wheeldon, 27 Lawrence Walk, Newport Pagnell, MK16 8RF
G1 CYY T. Brien, 54 Central Avenue, Fartown, Huddersfield, HD2 1DA
GM1 CZE J. Jindra, Rtrap 5 South Charlotte St Edinburgh, Edinburgh, EH2 4AN
G1 CZH P. Harkins, 27 Hillfoot Green, Liverpool, L25 7UH
G1 CZN P. Burrows, 7 Eton Terrace, Ince, Wigan, WN3 4NS
G1 CZU M. Abram, 28 Langport Drive, Vicars Cross, Chester, CH3 5LY
G1 CZW R. Silcocks, 69 Kennaway Road, Clevedon, BS21 6JJ
G1 DAE I. Rusby, 12 Park Meadow, Princes Risborough, HP27 0EB
G1 DAK S. Felton, 8 Clancutt Lane, Coppull, Chorley, PR7 4NS
G1 DAT P. Burnett, 14 Hollywalk Drive, Middlesbrough, TS6 0PL
G1 DAV D. Forsey, 3 Northwood Drive, Newbury, RG14 2HB
G1 DAX P. Costigan, 10 The Paddock, Clevedon, BS21 6JU
G1 DAZ S. Burchell, 31 Thornton Road, Girton, Cambridge, CB3 0NP
G1 DBH B. Cobb, 28 Sandringham Road, Newton Abbot, TQ12 4HA
G1 DBI G. Doig, 78 Plane Tree Drive, Crewe, CW1 4ES
G1 DBL L. Owen, 27 Coniston Drive, Holmes Chapel, Crewe, CW4 7LA
G1 DBR D. Ross, 113 Nun House Drive, Winsford, CW7 3LE
G1 DBZ R. Cooper, 31 Erskine Crescent, Sheffield, S2 3LQ
GM1 DCB M. Senior, The Raw, Bridgend, Isle of Islay, PA44 7PZ
G1 DCI J. Griffith, My Home, Highworth Road, Swindon, SN3 4SF
G1 DCU P. Gardner, 38Apley Rd Dy84Pa, Stourbridge, Dy84pa
G1 DCX M. Race, 76 Lonsdale Road, Stamford, PE9 2SG
G1 DCY S. Richmond, 1042 Evesham Road, Astwood Bank, Redditch, B96 6ED
G1 DCZ J. Sandall, 7 St. Road, Compton Dundon, Somerton, TA11 6PX
G1 DDA F. Wood, 7 Yew Tree Park, The Rowe, Newcastle, ST5 4EN
G1 DDF F. Griffin, 77 Widmore Drive, Hemel Hempstead, HP2 5JL
G1 DDK M. Abraham, Skywave Marine Services, Unit 1, The Arcade, Falmouth, TR11 2TD
G1 DDR R. Oakley, 20 Halton Lane, Wendover, Aylesbury, HP22 6AR
G1 DDS D. Seccombe, 14 Millfield, Bedlington, NE22 5DZ
G1 DEN P. Edinburgh, 77 Westerley Lane, Shelley, Huddersfield, HD8 8HP
G1 DEO B. Davies, 12 Woodbine Close, Newport, PO30 1AF
G1 DEP J. Dunhill, 8 Brentwood Avenue, Thornton-Cleveleys, FY5 3QR
G1 DEQ D. Gilbey, 7 Victory Way, Cottenham, Cambridge, CB24 8TG
G1 DER J. Hacker, 4 Foxglove Close, Bamber Bridge, Preston, PR5 6XR
G1 DES D. Smith, 14 College Drive, Ruislip, HA4 8SB
G1 DEU D. Hagger, 11 Recreation Walk, Great Cornard, Sudbury, CO10 0HH
G1 DEX H. Irvin, 30 Bank View, Earlsheaton, Dewsbury, WF12 8HH
G1 DEY C. Jacob, 10 Wynchgate, Southgate, London, N14 6RR
G1 DEZ P. Baxter, 27 Manor Crescent, Brinsworth, Rotherham, S60 5HG
G1 DFF M. Smith, 22 Cedars Avenue, Wombourne, Wolverhampton, WV5 0JX
G1 DFI J. Swift-Hook, 12 Warwick Drive, Newbury, RG14 7TT
G1 DFM A. Westlake, 47 Quarry Road, Kingswood, Bristol, BS15 8NZ
G1 DFN F. Wright, 1 Old Engine Houses, Brusselton, Shildon, DL4 1QA
G1 DFP G. Fielding, 35 Amos Avenue, Litherland, Liverpool, L21 7QH
G1 DFR A. Gemmill, The Steps, Bliss Gate Road, Bliss Gate, Rock, Kidderminster, DY14 9XT
G1 DFT I. Hampson, 293 Sandbrook Road, Southport, PR8 3RP
G1 DFW D. Hoare, 51 Hartington Road, Dronfield, S18 2LE
G1 DFZ R. Jobbins, 8 Newark Road, Hartlepool, TS25 2LA
G1 DGL R. Simpson, 51 Ramleaze Drive, Salisbury, SP2 9PA
G1 DGW I. Johnson, 24 York Road, Maghull, Liverpool, L31 5NL
G1 DGY A. Koch, 65 Collier Lane, Ockbrook, Derby, DE72 3RP
G1 DHB R. Fagence, 5 Balmoral Close, Billericay, CM11 2LL
G1 DHM G. Miller, 32 Belbroughton Close, Lodge Park, Redditch, B98 7NH
G1 DHQ D. Palmer, 60 Heathcote Drive, Sileby, Loughborough, LE12 7ND
G1 DHY N. Roe, 41 Highfield Lane, Chaddesden, Derby, DE21 6PH
G1 DIA P. Rowe, 5 Bramble Close, Great Boughton, Chester, CH3 5XN
G1 DIF Devon Data Group, c/o D. Roomes, View Field, Milton Damerel, Holsworthy, EX22 7NY
G1 DIG S. Cadman, 71 Gayfield Avenue, Withymoor, Brierley Hill, DY5 2BU
G1 DIK A. Smith, Windycross, Newbourne Road, Woodbridge, IP12 4PT
G1 DIL A. Witts, Langdale 3 Forton Glade, Newport, TF10 8BP
G1 DIM C. Smith, 37 Ivory Close, Tuffley, Gloucester, GL4 0QY
G1 DIO A. Heath-Anderson, 12 The Medway, Daventry, NN11 4QU
G1 DIR A. Wilkinson, 15 St. Margarets Grove, Leeds, LS8 1RZ
G1 DJI J. Short, 7 Bushfields, Loughton, IG10 3JT
G1 DJQ N. Lofthouse, Cambridge Park, 8 Abbott Clough Avenue, Blackburn, BB1 3LP
G1 DJU C. Whitby, 7 Wentworth Way, Stoke Bruerne, Towcester, NN12 7SA
G1 DKB K. Ball, 74 North Street, Okehampton, EX20 1BD
G1 DKE M. Spry, 71 High Street, Topsham, Exeter, EX3 0DY
G1 DKI M. Lindenbergh, 26 Manston Drive, Perton, Wolverhampton, WV6 7LX
G1 DKV G. Charlton, 20 Bailey Crescent, South Elmsall, Pontefract, WF9 2TL
G1 DKX A. Thomas, 92 Singleton Crescent, Goring-By-Sea, Worthing, BN12 5DJ
G1 DKY J. Miller, 40 Central Avenue, Herne Bay, CT6 8RX
G1 DLA R. Deacon, 22 Islip Gardens, Northolt, UB5 5BX
G1 DLB J. Desborough, 106 Grand Avenue, Lancing, BN15 9QD
G1 DLH M. Ogle, 22 Warwick St., Daventry, NN11 4AL
G1 DLJ G. Hope, 3 Farm Crescent, Sittingbourne, ME10 4QD
GW1 DLP W. Jones, 160 Christchurch Road, Newport, NP19 7SA
GM1 DLS C. Barry, 32 Prospect Drive, Ashgill, Larkhall, ML9 3AJ
G1 DMH L. Lees, 3 Ockbrook Court, Muskham Avenue, Ilkeston, DE7 8EY
G1 DMN P. Snow, 14 Beechwood Avenue, Darlington, DL3 7HP
G1 DMR R. Manser, 53 Downs Barn Boulevard, Downs Barn, Milton Keynes, MK14 7LL
G1 DMS D. Segal, Flat 1, Masons House, London, NW9 9NG
G1 DMW F. Latham, Higher Lane, Parbold, Wigan, WN8 7RA
G1 DNA J. Birkmyre, Swarland, Morpeth, NE65 9JW
G1 DNI W. Darling, 2 Strathaird Avenue, Walney, Barrow-in-Furness, LA14 3DE
G1 DNK B. Cunningham, 14 Leeson Drive, Ferndown, BH22 9QQ
G1 DNO C. Birtchnell, Linnetts Roost End, Sturmer, Haverhill, CB9 7XW
G1 DNP R. Collins, 12 Bean Oak Road, Wokingham, RG40 1RL
G1 DNT M. Cole, 52 Lower Meadow, Quedgeley, Gloucester, GL2 4YY
G1 DNY R. Clay, 38 Hubbards Road, Chorleywood, Rickmansworth, WD3 5JJ
G1 DNZ G. Clarke, 150 Minver Crescent, Nottingham, NG8 5PN
G1 DOA K. Chappell, 17 Linton Close, Winyates, Redditch, B98 0NA
G1 DOG I. Cheeseman, 445 Uttoxeter Road, Blythe Bridge, Stoke-on-Trent, ST11 9NT
G1 DOJ T. Brodrick, 16 Wallenge Drive, Paulton, Bristol, BS39 7PX
G1 DOL R. Breakspear, 7 Woodside, North Leigh, Witney, OX29 6SQ
G1 DON D. Macnamara, 56 Macdonald Street, Orrell, Wigan, WN5 0AJ
G1 DOT D. Barker, 60 Rolvenden Road, Wainscott, Rochester, ME2 4PG
G1 DOX J. Acton, 63 Bevington Close, Patchway, Bristol, BS34 5NP
G1 DPI A. Barratt, 23 Wilberforce Road, South Anston, Sheffield, S25 5EG
G1 DPJ C. Beasley, 12 East Leys Court, Moulton, Northampton, NN3 7TX
GW1 DPL M. Beer, 67 Killan Road, Dunvant, Swansea, SA2 7TH
G1 DPN D. Bettany, 10 Redbrook Crescent, Melton Mowbray, LE13 0EU
G1 DPT T. Cairney, Dolgoch, Hall Lane, Lutterworth, LE17 5RP
GW1 DPV M. Carter, 14 Cerdin Avenue, Pontyclun, CF72 9ER
G1 DPW S. Cmoch, 25 Monro Place, Epsom, KT19 7LD
G1 DPX R. Colley, 12 Glenfield Road, Banstead, SM7 2DG
G1 DQD C. Anderson, 11 Swallowfield Drive, Hull, HU4 6UG
G1 DQF P. Buckmaster, 7 Yew Tree Close, New Ollerton, Newark, NG22 9UP
G1 DQL R. Bradley, 1 Audley Place, Sutton, SM2 6RW
G1 DQQ D. Dwight, 19 The Highway, Stanmore, HA7 3PL
G1 DQU M. Elliott, 52 Wellfield Road, Alrewas, Burton-on-Trent, DE13 7EZ
GW1 DQV B. Emary, 2 The Paddocks, Penarth, CF64 5BW
G1 DRG G. Foster, 19 Asquith Avenue, Burnholme, York, YO31 0PZ
G1 DRI K. Gill, 33 Hazel Croft, Werrington, Peterborough, PE4 5BJ
G1 DRR N. Hamilton, 8 Sudbury Court, Mansfield, NG18 3RZ
G1 DRW D. Hart, 71 Breinton Road, Hereford, HR4 0JY
G1 DRY S. Cox, 25 Church Close, Stoke St. Gregory, Taunton, TA3 6HA
G1 DSA J. Critchley, 3 Beaconsfield View, Robert Road, Slough, SL2 3XT
G1 DSB J. Darling, 145 Hartlands, Bedlington, NE22 6JJ
G1 DSF A. Daw, 19 Rowan Close, Yarnfield, Stone, ST15 0EP
G1 DSG M. Degerdon, 25 Rosslyn Road, Billericay, CM12 9JN
G1 DSJ P. Morgan, 29 Brisbane Road, Reading, RG30 2PE
GM1 DSK D. Keay, Parkhill, Cromwell Park Almondbank, Perth, PH1 3LW
G1 DSM A. Hicks, 5 Restwell Avenue, Cranleigh, GU6 8PQ
G1 DSP Spalding and District ARS, c/o A. Hensman, 24 Belchmire Lane, Gosberton, Spalding, PE11 4HG
G1 DSZ J. Phillips, 20 The Meadows, Broomfield, Herne Bay, CT6 7XF
GW1 DTA M. Pilot, 92 Llanllienwen Road, Cwmrhydyceirw, Swansea, SA6 6LU
G1 DTE W. Merz, 38 Lime Avenue, Colchester, CO4 3NL
G1 DTF A. Middleton, 2 Beccles Way, Bramley, Rotherham, S66 2SJ
G1 DTS E. Kier, 9 Newbridge Way, Truro, TR1 3LX
G1 DUI P. Norman, 3 Church View, Witchford, Ely, CB6 2HH
G1 DUJ B. Oakley, 6 Staplehurst Gardens, Cliftonville, Margate, CT9 3JB
G1 DUO P. Richards, 16 Fruiterers Arms Caravan Park, Uphampton Lane, Ombersley, Droitwich, WR9 0JW
G1 DUS D. Roberts, Westpark, 296 Westleigh Lane, Leigh, WN7 5PW
G1 DUT J. Robertson, 4 Pembroke Road, Macclesfield, SK11 8RT
G1 DVD B. Marshall, 1 Anglers Way, Chesterton, Cambridge, CB4 1TZ
G1 DVH J. Knighton, 90 Sherwood Crescent, Market Drayton, TF9 1NP
GM1 DVO C. Hepworth, 20 Station Avenue, Duns, TD11 3HW
G1 DVU J. Green, 788 The Ridge, St Leonards-on-Sea, TN37 7PS
G1 DWC Dr R. Everett, 73 Fordwych Road, London, NW2 3TL
GU1 DWO A. Smith, La Cambrette, La Rue Des Reines, Forest, Guernsey, GY8 0JB
G1 DWT J. Dwight, 59 Highfield Road, Bramley, Leeds, LS13 2BX
G1 DWU A. Swales, 90 Earlswood Road, Dorridge, Solihull, B93 8RN
G1 DXD P. Darke, 18 Colchester Close, Southend-on-Sea, SS2 6HR
G1 DXH R. Crissell, 1 Medlar Drive, South Ockendon, RM15 6TS
G1 DXM D. Walling, 37 Ulverston Road, Swarthmoor, Ulverston, LA12 0JB
G1 DXN R. Walters, 16 Lune Drive, Morecambe, LA3 3RZ
G1 DXQ P. Postle, 20 Courtenay Close, Norwich, NR5 9LB
G1 DYC D. Winkley, Southall Cottage, Hadley, Droitwich, WR9 0AU
G1 DYL K. Tysoe, Valeside, School Road, West Hanney, Wantage, OX12 0LB
G1 DYN J. Snowling, 5 Verbena Close, Beechwood, Runcorn, WA7 3JA
G1 DYQ N. Prosser, 35 Holmfirth Close, Belmont, Hereford, HR2 7UG
G1 DYR R. Munday, 12 Glisson Road, Hillingdon, Uxbridge, UB10 0HH
G1 DZB N. Babbage, 248 Molesey Avenue, West Molesey, KT8 2ET
G1 DZC G. Baldwin, 29 Ramshead Approach, Leeds, LS14 1HH
G1 DZD L. Ball, 14 St. Wilfrids Road, Burgess Hill, RH15 8BD
G1 DZY K. Bricknall, 21 Uplands Way, Springwell Village, Gateshead, NE9 7NQ

G1 DZZ K. Bridle, 8 Hardy Avenue, Dorchester, DT1 1LL
G1 EAB A. Bolton, 5 Willow Crescent, Gedling, Nottingham, NG4 4BL
G1 EAE A. Broughton, Cobwebs, The Fleet, Pulborough, RH20 1HS
GM1 EAH W. Buchanan, 38 Kenmount Place, Kennoway, Leven, KY8 5LT
G1 EAJ M. Bunting, 22 Ling Close Coltishall, Norwich, NR12 7HZ
G1 EAM A. Bush, 30 New Road, Smallfield, Horley, RH6 9QN
G1 EAN A. Butler, Ty Ni, Hall Lane, Leamington Spa, CV33 9HG
GW1 EAV S. Davies, Laburnum House Guilsfield, Welshpool, SY21 9PX
G1 EAX R. Dawkins, 17 Dacer Close, Stirchley, Birmingham, B30 3BZ
G1 EBB A. Di Duca, 15 Moray Close, Halesowen, B62 9PP
G1 EBP C. Jermany, 5 Lexington Close, Hemsby, Great Yarmouth, NR29 4ES
G1 EBT A. Jones, 34 Benbow Quay, Coton Hill, Shrewsbury, SY1 2DL
G1 EBV S. Dawswell, 66 Priory Walk, Leicester Forest East, Leicester, LE3 3PP
G1 EBW P. Challen, 20 Drummond Road, Cawston, Rugby, CV22 7TN
G1 EBX S. Challen, The Vicarage, Chantry Lane, Towcester, NN12 6YY
G1 EBZ R. Charlton, Meer Booth Rd, Boston, PE22 7AB
G1 ECC D. Chippendale, 19 East Park Avenue, Darwen, BB3 2SQ
G1 ECE B. Clark, 9 Conigre, Chinnor, OX39 4JY
G1 ECI J. Christy, 1 Edinburgh Drive, Hindley Green, Wigan, WN2 4HL
G1 ECK N. Preval, 63 Dudley Avenue, Leicester, LE5 2EF
G1 ECS C. Frettsome, 16 Botany Avenue, Mansfield, NG18 5NG
G1 ECV J. Gardener, 32 Beckington Crescent, Chard, TA20 2BU
G1 ECY G. Giles, 74 The Larches, Uxbridge, UB10 0DN
G1 EDA R. Goff, 21 Findon Road, Elson, Gosport, PO12 4EP
G1 EDH T. Hacker, 179A Churchill Avenue, Chatham, ME5 0DQ
G1 EDK P. Hammond, 160 Westlands Caravan Park, Herne Bay, CT6 7LE
G1 EDM J. Hargreaves, 5 Nuttall Avenue, Little Lever, Bolton, BL3 1PW
G1 EDP M. Hazell, 15 Lords Hill, Coleford, GL16 8BG
G1 EDT W. Hewitt, 99 Derrydown Road, Perry Barr, Birmingham, B42 1RY
G1 EDU A. Hicks, 22 Manor Park, Mirfield, WF14 0EW
G1 EDX M. Holtam, 16 Cowley Close, Cheltenham, GL51 6NP
G1 EEA S. Howcroft, 23 Alderley Avenue, Blackpool, FY4 1QG
G1 EEO M. Kirby, Church Cottage, Burrington, Umberleigh, EX37 9JG
G1 EEZ J. Lewis, 516 Wellsway, Bath, BA2 2UD
G1 EFF A. Marriott, 75 St. Johns Road, Cudworth, Barnsley, S72 8DE
G1 EFG C. Mcara, 6 Winniford Close, Chideock, Bridport, DT6 6SA
G1 EFK G. Means, Ferry Farm, Witham Bank, Lincoln, LN4 4QA
G1 EFL M. Medcalf, 47 Paddock Drive, Chelmsford, CM1 6UX
G1 EFO P. Hyde, 24 Grassam Close, Preston, Hull, HU12 8XF
G1 EFP A. Jarrett, 4 Langstone Close, Horwich, Bolton, BL65SZ
G1 EFS S. Newell, 7 Edward Road West Walton Park, Clevedon, BS21 7DY
G1 EFT P. Nicholson, 20 Rowley Road, Torquay, TQ1 4PX
G1 EFU A. Nixon, 14 Carlton Road, Lowton, Warrington, WA3 2EP
G1 EFX C. Nutkins, Higher Spence, Bridport, DT6 6DF
G1 EGB G. Page, 23 Maskelyne Close, Battersea, London, SW11 4AA
G1 EGE K. Pay, 1 Swallow Walk Biddulph, Stoke-on-Trent, ST86TY
G1 EGI D. Phillips, Bethune, Rame Cross, Penryn, TR10 9DZ
G1 EGK Dr J. Preece, The Grange, Harewood Road, Wetherby, LS22 5BL
G1 EGL R. Preston, 45 Gaynor Close, Wymondham, NR18 0EA
G1 EGR G. Roberts, 32 Ancaster Court, Horncastle, LN9 6HG
G1 EGZ A. Adams, Radnor, Shorts Road, Carshalton, SM5 2PB
G1 EHB P. Allcock, 1 Fredrick Dunford Close, Marlborough, SN8 4YS
G1 EHE M. Appleton, Flat 2, Black Swan Buildings, Winchester, SO23 9DT
G1 EHF D. Austen, Tudorlands, Silchester Road, Bramley, Tadley, RG26 5DG
GW1 EHI R. Davies, 1 Mount View, Plas Road, Blackwood, NP12 3RH
G1 EHK R. Birch, 17 White Wood Road, Eastry, Sandwich, CT13 0JZ
G1 EHM P. Bird, 4 Parkside Avenue, Tilbury, RM18 8DT
G1 EHS B. Brodribb, 1 Ponswood Road, St Leonards-on-Sea, TN38 9BU
G1 EHU M. Hostekens, 1 Ponswood Road, St Leonards-on-Sea, TN38 9BU
G1 EHX C. Cameron, Rose Cottage Orchard Way Berry Hill, Coleford, GL16 7AQ
G1 EIB N. Purkins, 16 Nunburnholme Avenue, North Ferriby, HU14 3AN
G1 EIG J. Ryan, 71B Gunterstone Road, London, W14 9BS
G1 EIH D. Samber, 102 Midsummer Avenue, Hounslow, TW4 5BB
G1 EIO B. Smith, 43 Oak Avenue, Hindley Green, Wigan, WN2 4LZ
G1 EIP G. Smith, 52 Penhill Crescent, St Johns, Worcester, WR2 5PX
G1 EIR R. Smith, 29 Windmill Lane, Henbury, Bristol, BS10 7XE
G1 EIV S. Stanley, 11 Mandeen Grove, Mansfield, NG18 4FA
G1 EIX H. Stephens, 16 Addison Drive, Stratford-upon-Avon, CV37 7PL
G1 EIZ M. Stewart, 2 Patmore Link Road, Hemel Hempstead, HP2 4PX
G1 EJA R. Stone, 1 Poplar Close, Ashford, TN23 3DY
G1 EJK G. Tomlinson, 4 Werneth Close, Denton, Manchester, M34 6LR
G1 EJQ A. Walker, Gymru Fach, 51 The Crescent, Consett, DH8 5JF
GW1 EKC M. Davis, Minffordd, Oakeley Square, Blaenau Ffestiniog, LL41 3PU
G1 EKM S. Evans, 41 Clocktower Drive, Liverpool, L9 1AG
G1 EKP G. Fowler, 18 Lossie Drive, Iver, SL0 0JS
G1 EKU N. Kernahan, 15 Howgill Lane, Sedbergh, LA10 5DE
G1 ELE R. Watt, 88 Graham Crescent, Portslade, Brighton, BN41 2YB
G1 ELF P. Wood, 3 Elf Meadow, Poulton, Cirencester, GL7 5HQ
G1 ELJ R. Williams, 53 Springhill Park, Wolverhampton, WV4 4TR
G1 ELK P. Wilson, Barkstan Lodge, Quadring Bank, Spalding, PE11 4RF
GI1 ELP D. Allen, 12 Scriggan Road, Limavady, BT49 0DH
G1 ELQ B. Bentley, Sandy Ridge, Church Street, Stoke-on-Trent, ST7 4RS
G1 ELX Prof. A. Challinor, 24 West End Rise, Horsforth, Leeds, LS18 5JL
G1 ELZ M. Cook, Well Cottage, Old Road, Alderbury, Salisbury, SP5 3AR
G1 EME OBO The Worcester Moonbounce Society, c/o W. Day, 4 Queenswood Drive, Worcester, WR5 3SZ
G1 EMF Emf Hams, c/o J. Murray, 20 East End, Cholsey, Wallingford, OX10 9RT
G1 EML H. Hill, 5 Wentworth Gardens, Alton, GU34 2BJ
G1 EMM K. Hill, 73 Yellow Birch Drive, Kitchener, Canada, N2N 2M3
G1 EMW R. Dearsley, Prince William Farm, Lynn Road, King's Lynn, PE33 9BD
G1 ENA G. Edwards, 22 Whalley Lane, Uplyme, Lyme Regis, DT7 3UR
G1 END Dr M. Friedman, Flat 28, Hertford Mews, Potters Bar, EN6 1XW
GW1 ENG P. Gibson, The Nook, Trimsaran Road, Kidwelly, SA17 4EB
GW1 ENP P. Hewett, Tarn Hows, 7 Lakeside Drive, Presteigne, LD8 2EG
G1 ENR A. Deakin, The Farmhouse, New House Farm, Tenbury Wells, WR15 8TW
GM1 EOA J. Minaudo, Meadowside, Newbridge, Dumfries, DG2 0QX
G1 EOH S. Braybrooke, 6 Tubbenden Lane, Orpington, BR6 9PN
GW1 EOI G. John, 31 Gellifawr Road Morriston, Swansea, SA6 7PN
G1 EOJ M. Kay, 59 Palmer Crescent, Leighton Buzzard, LU7 4HY
G1 EOK E. Keeble, 17 Moat Avenue, Green Lane, Coventry, CV3 6BT
G1 EOM H. Kinghorn, 29 Meadowview Road, Sompting, Lancing, BN15 0HU
GI1 EOS P. Leitch, 212 Belfast Road, Muckamore, Antrim, BT41 2EY
G1 EPD D. Hathaway, 46 Blackwell Avenue, Newcastle upon Tyne, NE6 4DR
G1 EPF L. Marshall, 75 Acacia Crescent, Wigan, WN6 8NJ
G1 EPL R. O'Callaghan, 47 Seabrook Drive, Thornton-Cleveleys, FY5 3SE
G1 EPO J. Pragnell, Sundale, Northampton Road, Brackley, NN13 7TY
GW1 EPR R. Rees, 16 Railway View, Caldicot, NP26 5GB
G1 EPS M. Rhodes, 155 High Park Road, Southport, PR9 7BY
G1 EQF P. Uttridge, Springers Rest, Beck Lane, Hull, HU12 9RG
G1 EQJ M. Whittle, Churchfield Cottage, West Road, Wareham, BH20 5RY
G1 EQL P. Wootton, 20 Oakhill Road, Dronfield, S18 2EJ
G1 EQM R. Agacy, 23 Highgate Lane, Bolton-Upon-Dearne, Rotherham, S63 8HR
G1 EQU R. Percival, 23 Plumtree Road, Thorngumbald, Hull, HU12 9QG
GW1 ERA A. Price, 2 Ger Y Coed, Brackla, Bridgend, CF31 2LA
G1 ERF S. Rogers, 31 Morgan Road, Southsea, PO4 8JS
G1 ERM D. Salter, 94 Clifton Street, Swindon, SN1 3QA
G1 ERQ R. Stevens, 172 Branksome Avenue, Stanford-le-Hope, SS17 8DE
G1 ERS D. Strange, 15 Truman Road, Bournemouth, BH119BP
G1 ERU S. Mole, 17A Marlborough, Seaham, SR7 7SA
G1 ERY A. Moseley, 15 Gillsway, Northampton, NN2 8HT
G1 ERZ A. Moules, 5 Hill Road, Borstal, Rochester, ME1 3NJ
G1 ESC C. Mountain, 16 Temples Court, Helpston, Peterborough, PE6 7EU
G1 ESW K. Laughton, 33 Tiverton Close, Radcliffe, Manchester, M26 3UJ
G1 ESX K. Love, 63 Buxton Road, Spixworth, Norwich, NR10 3PP
G1 ETD Dr J. Newland, 84 Waterman Way, London, E1W 2QW
G1 ETQ B. Sweeney, 51 Tristram Avenue, Hartlepool, TS25 5PA
G1 ETZ C. Walker, 1 Shepherds Close, Shepshed, Loughborough, LE12 9SQ
G1 EUA B. Wall, 3 Grenville Avenue, Teignmouth, TQ14 9NJ
G1 EUD D. Wiles, 62 Taylor St., Tunbridge Wells, TN4 0DX
G1 EUF R. Wilson, Street Farm, Henny Street, Sudbury, CO10 7LS
G1 EUG D. Wolfe, 48 Wilby Lane, Great Doddington, Wellingborough, NN29 7TP
G1 EUH J. Woods, 5 Sand Martin Avenue, Wesham, Preston, PR4 3FE
G1 EUI M. Wright, 71 Oakridge Road, High Wycombe, HP11 2PL
G1 EUM S. Foote, Harroway, South Hanningfield Road, Wickford, SS11 7PF
G1 EUN A. Friend, 43 Gildale, Peterborough, PE4 6QY
G1 EUQ A. Gammon, 5 Sommerville Close, Faversham, ME13 8HP
G1 EUU M. Gibson, 1 Oakleigh Road, Grantham, NG31 7NN
G1 EVA M. Hattersley, 190 Elmton Road, Creswell, Worksop, S80 4DY
G1 EVI B. Green, 49 Brockman Crescent, Dymchurch, Romney Marsh, TN29 0UA
G1 EVR P. Lowe, 155 Long Lane, Bolton, BL2 6EU
G1 EVV C. Mylchreest, 21 Bexhill Gardens, St Helens, WA9 5FQ
G1 EWC A. Webster, 49 Uplands Croft, Werrington, Stoke-on-Trent, ST9 0LF
G1 EWE T. Williams, 20 Sandringham Drive, Dartford, DA2 7WB
G1 EWH S. Bell, The Haven, Low Street, Retford, DN22 0LN
G1 EWM T. Conlin, 5 Morland Drive, Rochester, ME2 3LW
GW1 EWW K. Edwards, 25 Woodland Road, Neath, SA11 3AL
GW1 EWY B. Owen, Llys Helen, Croesor, Penrhyndeudraeth, LL48 6SR
G1 EXG J. Hare, 1 The Copse, 50-52 Princes Road, Brighton, BN2 3RH
G1 EXK S. Bussey, 7 Ilderton Crescent, Seaton Delaval, Whitley Bay, NE25 0FH
G1 EXM C. Bussey, 6 Ray Court, Wimblington, March, PE15 0FE
G1 EXR W. Cosgrove, 62 Twyford Avenue, Great Wakering, Southend-on-Sea, SS3 0EX
G1 EXU R. Cloke, Woodlands, Roxton Lane, Keelby, DN41 8JB
G1 EXV D. Cooper, 75 Merevale Avenue, Nuneaton, CV11 5LU
G1 EYD K. Rutter, 12 Berwick Terrace, North Shields, NE29 7AW
G1 EYG G. Shipperley, 72 Hithercroft Road, Downley, High Wycombe, HP13 5RH
G1 EYJ Rev. I. Smith, 21 Gorsehill Road, Wallasey, CH45 9JA
G1 EYS P. Jorquera, 21 Highlands Road, Orpington, BR5 4JP
G1 EYT P. Vickers, 21 Blackwood Drive, Sutton Coldfield, B74 3QP
G1 EYW S. Warren, 43 Glebe Rise Kings Sutton, Banbury, OX17 3PH
G1 EYY D. Whincup, 172 Appleton Road, Hull, HU5 4PF
G1 EYZ S. White, 67 Wingfield Road, Lakenheath, Brandon, IP27 9HR
G1 EZF M. Allmark, 11 Potternewton Crescent, Leeds, LS7 2DY
G1 EZI S. Armstrong, 5 Dashpers, Brixham, TQ5 9LJ
G1 EZJ C. Barker, 52 Spode Street, Stoke-on-Trent, ST4 4DY
G1 EZU D. Harpham, 16 Scotts Way, Kirkby-In-Ashfield, Nottingham, NG17 9DN
G1 FAA S. Jeffery, 35 Lynton Avenue, Orpington, BR5 2EH
G1 FAD T. Kenney, 7 Hickin Close, Charlton, London, SE7 8SH
GM1 FAF J. Marshall, Drummorlie, Wallyford Toll, Musselburgh, EH21 8JT
GM1 FAI A. Miller, 21 Merker Terrace, Linlithgow, EH49 6DD
G1 FBE D. Telford, 9 Central Avenue, Carlisle, CA1 3QB
G1 FBI J. Hughes, 47 Stambourne Way, Upper Norwood, London, SE19 2PY
GW1 FBL T. Boorman, 43 Ffordd Taliesin, Killay, Swansea, SA2 7DF
GM1 FBM B. Borland, Beechwood Cottage, Muirhall Road, Perth, PH2 7LL
G1 FBQ Dr S. Fraser, Walnut Tree Cottage, Main Road, Abingdon, OX13 5LN
GW1 FBU K. Goodchild, 3 Mill Leat Lane, Gorseinon Swansea, SA4 4QE
G1 FBW T. Howchen, 1 Ash Road, Canvey Island, SS8 7EA
G1 FBZ W. Hicks, 7 Meadow Close, Thundersley, Benfleet, SS7 3RJ
G1 FCN M. Nutton, Rose Cottage, Henley Street, Luddesdown, Gravesend, DA13 0XB
G1 FCU S. Reed, 20 Mead Crescent, Bookham, Leatherhead, KT23 3DU
G1 FCW Essex CW ARC, c/o P. Tittensor, 47 St. Johns Road, Chelmsford, CM2 0TY
G1 FCX Dr N. Coleman, The Old Croft Reeth, Richmond, DL11 6TE
G1 FDD N. Groeber, 113 Kings Road, Kings Heath, Birmingham, B14 6TN
G1 FDL V. Kelk, 7 Rowan Place, Garforth, Leeds, LS25 2JR
G1 FDN W. Kenyon, 22 Barons Way, Lower Darwen, Darwen, BB3 0RG
G1 FDO T. King, 32 Bagnall Avenue, Arnold, Nottingham, NG5 6FT
G1 FEF C. Smith, 27 Wye Road, Wooburn Green, High Wycombe, HP10 0DU
GM1 FEM I. Smith, 150 Eden Park, Clayton Caravan Park, St Andrews, KY16 9YB
G1 FEO G. Jones, 8 Kenilworth Road Lighthorne Heath, Leamington Spa, CV33 9TH
G1 FEP D. Twidale, 18 Kinnaird Road, Wallasey, CH45 5HN
G1 FET P. Taylor, 29 Dunstall Road, Halesowen, B63 1BB
G1 FEX J. Allsop, 15 Woodland Grove, Mansfield Woodhouse, Mansfield, NG19 8AZ
G1 FFH T. Firth, 126 Tombridge Crescent, Kinsley, Pontefract, WF9 5HE
G1 FFO P. Thomas, 9 Awefield Crescent, Smethwick, W Midlands, B67 6PR
G1 FFR S. Tucker, 28 Peregrine Road, Stockport, SK2 5UR

G1 FFU D. Woolmer, 2 Muccleshell Close, Havant, PO9 2HR
G1 FGC P. Chalkley, 10 Preston Gardens, Luton, LU2 7NL
G1 FGE N. Collier, 104 Grosvenor Street, Kearsley, Bolton, BL4 8DW
G1 FGI P. Escreet, Wortley Cottage, Husthwaite, York, YO61 4Py
G1 FGK W. Grint, 15 Ivythorn Road, Street, BA16 0TE
GM1 FGN R. Hussey, 21 Maidenfield, Mossbank, Shetland, ZE2 9TD
G1 FHH P. Johnson, 30 Copplestone Grove, Longton, Stoke-on-Trent, ST3 5UD
G1 FHI S. Murray, 51 Huddersfield Road, Newhey, Rochdale, OL16 3QZ
G1 FHK C. Peart, 33 Fieldfare, Abbeydale, Gloucester, GL4 4WH
G1 FHR R. Roots, 14 Sussex Close, London Road, Sevenoaks, TN15 6BB
G1 FHY S. Wise, 23 Wordsworth Drive, Eastbourne, BN23 7QP
G1 FIG J. Clark, 58 Kelvin Grove, Fareham, PO16 8LE
G1 FIM P. Mcdonnell, 55 Lodge Hall, Harlow, CM18 7SY
G1 FIP S. Rowlandson, 48 Greville Road, Warwick, CV34 5PB
G1 FJD A. Wood, 2 Towning Close, Deeping St. James, Peterborough, PE6 8HR
G1 FJF A. Bawden, 67 Silo Drive, Farncombe, Godalming, GU7 3NZ
G1 FJH P. Bruce, 26 Queens Road, Wilbarston, Market Harborough, LE16 8QJ
GW1 FJI R. Bullock, 32 Tinmans Green, Redbrook, Monmouth, NP25 4NB
G1 FJJ M. Cammish, 20 Chantry Avenue, Hartley, Longfield, DA3 8DD
G1 FJS K. Davis, 16 West Field Close, Taunton, TA1 5JU
G1 FKJ G. Portlock, 1 Windmill Cottage, Oxford Street, Marlborough, SN8 2DH
GW1 FKL G. Howells, 70 Meadow Street, Treforest, Pontypridd, CF37 1SS
G1 FKM J. Kendall, 6 Wellington Street, Allerton, Bradford, BD15 7QZ
G1 FKP D. Le Vine, Anglecroft, Borough Road, Westerham, TN16 2LA
G1 FKS C. Macliesh, 33 River Way, Twickenham, TW2 5JP
G1 FKT J. Mansley, 2 Beech Tree Close, Cuerden Residential Park, Leyland, PR25 5PA
GW1 FKY K. Eaton, 21 Westminster Way, Bridgend, CF31 4QX
G1 FLI C. Franklin, 3 Park Road, Rugby, CV21 2QU
G1 FLV M. Parr, 5 Suffolk Grove, Leigh, WN7 4TA
G1 FLW P. Partridge, 18 Chaucers Drive, St Peters Field, Nuneaton, CV10 9SD
G1 FLX S. Pickstone, 48 Oak Tree Drive, London, N20 8QH
GW1 FLY A. Rosier, 2 Watkin Drive, Oswestry, SY11 1SQ
G1 FMA A. Robinson, 5 Alford Road, Heaton Chapel, Stockport, SK4 5AW
G1 FMT C. Hardy, 110 Jubilee Road, Waterlooville, PO7 7RG
G1 FMU K. Harris, 8 Trelawney Rise, Callington, PL17 7PT
GM1 FMV G. Hind, 135 Pilton Avenue, Edinburgh, EH5 2HP
G1 FMW D. Lewis, 4 Westwood Grove, Solihull, B91 1QB
GM1 FMX W. Mccandlish, Lingdowey, Stoneykirk Road, Stranraer, DG9 7BX
G1 FNA L. Phillips, 2 Stratton Green, Bedgrove, Aylesbury, HP21 7EP
G1 FND N. Stephens, 7 Quarry Road, Alveston, Bristol, BS35 3JL
G1 FNF B. Walker, 47 Coppice Avenue, Eastbourne, BN20 9QJ
G1 FNN W. Ball, 43 Fairfax Drive, Herne Bay, CT6 6QZ
G1 FNP M. Bellas, 3 Elm Terrace, Penrith, CA11 7JY
G1 FNS B. Cutts, 7 Lych Gate Close, Sandhurst, GU47 8JH
G1 FNU J. Dodd, 38 The Quadrant, North Shields, NE29 7HP
GM1 FNX P. Ewing, Arisaig, Priestland, Darvel, KA17 0LP
G1 FOA P. Franklin, 14 St. Erth Hill, St. Erth, Hayle, TR27 6EX
G1 FOE P. Holt, 27 Sandown Close, Blackwater, Camberley, GU17 0EN
G1 FOF M. James, 9 Denham Crescent, Mitcham, CR4 4LZ
G1 FOM P. Lyttle, 3 Woodlands, East Ardsley, Wakefield, WF3 2JG
G1 FON M. Mangan, Schuetzenstr. 54, Kaiserslautern, Germany, 67659
G1 FOW K. Worsley, 102 Cabul Close, Warrington, WA2 7SE
G1 FOZ M. Slattery, 9 Barns Close, Walsall, WS9 9BD
G1 FPC A. Sands, 7 Kimberley Avenue, Seymour Street, Hull, HU3 5PP
GM1 FPD D. King, 18 Ford Spence Court, Benderloch, Oban, PA37 1PY
G1 FPK R. Kerridge, 80 Melton Road, Wymondham, NR18 0DE
G1 FPP N. Sirkett, Flat 2, Barfield House, Ryde, PO33 2JP
G1 FPY A. Watson, 9 Linthurst Newtown, Blackwell, Bromsgrove, B60 1BP
G1 FPZ A. Wilcox, 2 Dawkins Road, Poole, BH15 4JD
G1 FQD S. Eldredge, 2 Chelmsford Drive, Worcester, WR5 1QX
G1 FQI D. Jackson, 28A Wawne Road Sutton-On-Hull, Hull, HU7 4YE
G1 FQX J. Lamb, 205 Springfield Road, Sutton Coldfield, B76 2SY
G1 FRD G. Bennett, 14 Thessaly Road, Stratton, Cirencester, GL7 2NG
GM1 FRG T. Watson, Iolaire, Mannachie Grove, Forres, IV36 2WE
G1 FRJ P. Bearne, 59 Foxhole Road, Paignton, TQ3 3TD
G1 FRL G. Mott, 191 Joyners Field, Harlow, CM18 7QD
G1 FRM R. Doughty, 4 Trinity Road, Wisbech, PE13 3UN
G1 FRS Farnborough Contest Group, c/o B. Dawson, Isca, 12 Lestock Way, Fleet, GU51 3EB
G1 FSE K. Stokes, 33 The Crescent, Burntwood, WS7 2PA
G1 FSF J. Williams, 52B Pensford Drive, Eastbourne, BN23 7NY
GI1 FSJ N. Colgan, 11 St. Johns Park, Moira, Craigavon, BT67 0NL
GM1 FSU I. Menzies, 33 Lochside Drive, Bridge Of Don, Aberdeen, AB23 8EH
G1 FSW R. Consolante, 19 Chestnut Gardens, Stamford, PE9 2JY
G1 FSX J. Nash, 21 St. Marys Close, Peterborough, PE1 4DR
GM1 FSZ K. Hall, 12 Brockhill Rise, Inverurie, AB51 5RH
G1 FTD J. Hobbs, Fetchalls, The Green, Bury St Edmunds, IP30 9AF
GM1 FTG M. Lonnen, Hunt Hall, Glendevon, Dollar, FK14 7JZ
G1 FTH A. Marston, 92 Sorrell Road, Nuneaton, CV10 7AW
G1 FTJ S. Tonks, Milestones Watling Street, Cannock, WS11 1SH
G1 FTK N. Apps, 71 De Cham Road, St Leonards-on-Sea, TN37 6HF
G1 FTU J. Pearson, Largo, Hemming Green, Chesterfield, S42 7JQ
G1 FTV A. Bryant, 42 Kemerton Walk, Swindon, SN3 2EA
G1 FTX E. Demeza, 2 Adam Close, St Leonards-on-Sea, TN38 9QW
GM1 FTZ H. Simpson, Clachan Farm Cottages, Rosneath, Helensburgh, G84 0QR
G1 FUG A. Sagar, 28 Rangoon Road, Solihull, B92 9DB
G1 FUJ B. Jones, 4 Caddick Close, Kingswood, Bristol, BS15 4RT
G1 FVA K. Irons, 14 Beech Grove, Houghton, Carlisle, CA3 0NU
G1 FVC I. Batten, 17 Cornfield Road, Birmingham, B31 2EB
G1 FVE D. Ward, 36 Croxby Avenue, Scartho, Grimsby, DN33 2NW
G1 FVH M. Jordan, 160 Beta Road, Farnborough, GU14 8PH
G1 FVP R. Slone, 55 Chilton Way, Hungerford, RG17 0JR
G1 FVS P. Dean, 10 Moor Terrace, Bradford, BD2 4SG
G1 FVU K. Bloomfield, 14 Manners Road, Fornham St. Martin, Bury St Edmunds, IP31 1TE
GW1 FWC W. Williams, Llwynprenteg, Llanafan, Ceredigion, SY23 4BQ
GW1 FWE J. Duggan-Keen, Bodlondeb, Chapel Street Caerwys, Mold, CH7 5AE
G1 FWF H. Morgan, 2 Mayfield Park South, Fishponds, Bristol, BS16 3NG
G1 FWR P. Harman, 35 Point Clear Road, St. Osyth, Clacton-on-Sea, CO16 8EP
G1 FWS F. Swaine, 2 Norwich Close, Stevenage, SG1 4NU
G1 FWU T. Norbury, 19 Charles Cope Road, Orton Waterville, Peterborough, PE2 5ER
G1 FWY C. Pitt, 31 D'Arcy Way, Tolleshunt D'Arcy, Maldon, CM9 8UD
G1 FWZ B. Lakey, 3 Simons Close, Worle, Weston-Super-Mare, BS22 6DJ
G1 FXB A. Turquand, 63 Sundown Avenue, Dunstable, LU5 4AL
G1 FXC R. Lambourne, Rosedale, Townsend, Bicester, OX27 0EY
G1 FXD O. Rogers, The Barn, Millways Farm, Rosenannon, Bodmin, PL30 5PJ
GW1 FXL S. Annetts, Hengwm, Rhayader, LD6 5LD
G1 FXM T. Young, Elm Field Lodge, Tollesbury Road, Tolleshunt D'Arcy, Maldon, CM9 8UA
G1 FXS P. Striplin, 64 Ebrington Road, Malvern, WR14 4NL
G1 FXT R. Robinson, 5 Lilac Close, Newton Longville, Milton Keynes, MK17 0DQ
G1 FXX R. Tunbridge, 35 Coworth Close, Ascot, SL5 0NR
G1 FYE C. Sterland, 103 Main St., Distington, Workington, CA14 5UJ
G1 FYF C. Bradley, 7 Coltsfoot, Biggleswade, SG18 8SR
G1 FYQ Pontefract and District ARS, c/o K. Taylor, 29 School Road, Pontefract, WF8 2AJ
G1 FYS K. Boothroyd, 16 Kelvin Avenue Dalton, Huddersfield, HD5 9HG
G1 FYU M. Blockley, 7 Stoke Lane, Stoke Bardolph, Burton Joyce, Nottingham, NG14 5HR
G1 FZL P. Dyer, 18 Christopher Close, Yeovil, BA20 2EH
G1 FZR R. Delve, 18 Thame Road, Piddington, Bicester, OX25 1PX
G1 FZS R. Fleet, 17 Crown Road, Portslade, Brighton, BN41 1SJ
G1 FZV A. Ogden, 5 Lower Bristol Road, Clutton, Bristol, BS39 5PB
G1 GAD F. Mcloughlin, 21 Darwin Crescent, Newcastle upon Tyne, NE3 4TT
G1 GAN P. Cartwright, 1 Railway Cottages, Sutton Bingham, Yeovil, BA22 9QW
G1 GAR M. Kipping, 46 Old Hardenwaye, Totteridge, High Wycombe, HP13 6TJ
G1 GAS C. Kelley, Dunkeld, Bridge Street Fenny Compton, Southam, CV47 2XY
G1 GAT R. Fernihough, 3 Sandpiper Close, Quedgeley, Gloucester, GL2 4LZ
GW1 GAU M. Sullivan, 14 Moorview Road, Gendros, Swansea, SA5 8BU
G1 GAW R. Smith, 18 Curlew, Wilnecote, Tamworth, B77 5PL
G1 GBC W. Boucher, 12 Highfield Terrace, Ilfracombe, EX34 9LG
G1 GBF J. Delaney, 31 Roose Road, Barrow-in-Furness, LA13 9RG
GW1 GBH S. Parkins, 100 Welsh Road, Garden City, CH52HX
G1 GBI M. Pearson, 34 Downside Road, Sutton, SM2 5HP
G1 GBR A. Ziemacki, 3 Wheatcroft Road, Rawmarsh, Rotherham, S62 5JR
G1 GBV D. Evans, 59 Watlington Road, Benfleet, SS7 5DT
G1 GBX D. Vanbeck, 101 Upper St., Islington, London, N1 1QN
GM1 GCB S. Raisey-Skeats, 20 Gordon Street, Boddam, Peterhead, AB42 3AY
G1 GCF F. Clough, 2 Hudson Close, Tadcaster, LS24 8JD
G1 GCJ G. Morris, 21 Orchard Place, Deer Park, Ledbury, HR8 2XD
G1 GCY G. Gowland, 7 Canewdon Hall Close, Canewdon, Rochford, SS4 3PY
G1 GDA M. Austin, 10 Simon Place, Wideopen, (Division Of Tyneside Motor Sport, Newcastle upon Tyne, NE13 7HT
G1 GDB D. Thwaytes, 3 California Terrace, Bothel, Wigton, CA7 2JF
G1 GDJ C. Godward, 3 Court Close, Brighton, BN1 8YG
G1 GDM J. Coombes, 22 Chollerford Close, Gosforth, Newcastle upon Tyne, NE3 4RN
GM1 GDO J. Morton, 6 Deanpark Place, Balerno, EH14 7ED
G1 GDR B. Smith, 4 Planetree Close, Bromsgrove, B60 1AW
G1 GDT C. Groom, Woodstock Cottage, 2 Woodstock Terrace, Dursley, GL11 5SW
GM1 GEQ T. Menzies, 239 Eskhill, Penicuik, EH26 8DF
G1 GER A. Goodings, 2 Mulberry Grove, Bradwell, Great Yarmouth, NR31 8QJ
GM1 GES W. Barbour, 27 Drove Road, Langholm, DG13 0JW
G1 GET F. Wood, 40 Mill Lane, Spalding, Pe11 4tl
G1 GEV K. Argyle, 62 Yew Tree Drive, Leicester, LE3 6PL
G1 GEY D. Stoker, Headlands, Front Street, Aldborough, Boroughbridge, York, YO51 9ES
G1 GFA N. Pitt, 37 Shelley Drive, Four Oaks, Sutton Coldfield, B74 4YD
G1 GFC S. Bradley, 75 New Road, Sawston, Cambridge, CB22 3BN
G1 GFD A. Crook, 54 Somerset Way, Paulton, Bristol, BS39 7YX
G1 GFF V. Thomas, Flat 38, Lazonby Court, St Leonards-on-Sea, TN38 0QP
GW1 GFO P. Carlisle, 5 Pengry Road, Loughor, Swansea, SA4 6PH
G1 GFW J. Jacobs, 11 Delamere Close, Castle Bromwich, Birmingham, B36 9TW
G1 GFZ R. Pelling, Spring Cottage, Main Road, Hastings, TN35 4SL
G1 GGB R. Phillips, 58 Baranscraig Avenue, Patcham, Brighton, BN1 8RE
G1 GGI S. Hargreaves, 10 Reedham Crescent, Cliffe Woods, Rochester, ME3 8HT
G1 GGK F. Coldham, 5 Church Lane, Towersey, Thame, OX9 3QL
G1 GGN R. Barkley, 9 Eagle Close, Erpingham, Norwich, NR11 7AW
G1 GGT Dr R. Sharp, Arosa, The Park, Harwell, OX11 0HB
G1 GHG K. Knibbs, 8 Ferguson Way, Huntington, York, YO32 9YG
GD1 GHK W. Corlett, 9 Kerrocruin, Kirk Michael, Isle of Man, IM6 1AF
G1 GHU T. Smith, 19 Higher Holcombe Road, Teignmouth, TQ14 8RJ
G1 GHY A. Laszkiewicz, 38 Langley Lane, Ifield, Crawley, RH11 0NA
GM1 GHZ Backpackers Radio Activity Group, c/o P. Thompson, 31 St. Marys Drive, Perth, PH2 7BY
G1 GIA I. Sinclair, 40 Holders Hill Gardens, London, NW4 1NP
G1 GID G. Watt, 48 Southdown Road, Portslade, Brighton, BN41 2HN
G1 GIE R. Buckley, 6280 Hawkes Bluff Avenue, Davie, Fort Lauderdale, USA, 33331-3419
G1 GIJ D. Hadjidakis, 19 Eastfield Road, Royston, SG8 7ED
G1 GJD S. Corson, 9 St. Marys Way, Weedon, Northampton, NN7 4QL
G1 GJT P. Jackson, 10 Claremont Road, Nottingham, NG5 1BH
G1 GKA R. Mason, 32 Linden Drive, Evington, Leicester, LE5 6AH
G1 GKF R. Mann, Little Chysauster, Penzance, TR20 8XA
G1 GKH G. Hinds, 35 Lime Grove, Burntwood, WS7 0HA
GI1 GKI T. Campbell, 265 Ballynahinch Road, Lisburn, BT27 5LS
G1 GKK S. Brooke, 14 Saxton Avenue, Heanor, DE75 7PZ
G1 GKN A. Tyler, West Hanningfield Road Great Baddow, Chelmsford, CM2 8HN
G1 GKR A. Barlow, 454 Shaw Road, Royton, Oldham, OL2 6PG
GW1 GKV P. Jones, Pen Y Galchen Farm, Pwlldu, Pontypool, NP4 9SS
G1 GKW E. Perryman, Flat 4, Garth House, Bognor Regis, PO21 1HQ
G1 GLG S. Padgham, 4 Hollamby Park, Hailsham, BN27 2LX
G1 GLN D. Beasley, 40 Susannah Street, London, E14 6LS
G1 GLS R. Lees, 28 Lyngarth House, Grosvenor Road, Altrincham, WA14 1LH
G1 GLZ B. Start, 9 Front Street, Corbridge, NE455AP
GI1 GME A. Mcilwee, 48 Carnduff Drive, Ballymena, BT43 7ap

G1 GMF R. Sievert, 5 Sandmoor Road, New Marske, TS11 8BP
G1 GMG S. Gainswin, 1 Buckfast Road, Buckfast, Buckfastleigh, TQ11 0EA
G1 GMH D. Peat, 23 Hill Bottom Close, Whitchurch Hill, Reading, RG8 7PX
G1 GMM H. Barczynski, 64 Kings Acre, Coggeshall, Colchester, CO6 1NY
G1 GMQ A. Bolton, 8 Turners Walk, Chesham, HP5 3BT
G1 GMV A. Brewer, 25 Ackerman Road, Dorchester, DT1 1NZ
G1 GMX J. Mold, Sycamore Lodge, Butt Lane, Goulceby, Louth, LN11 9UP
G1 GNP D. Roe, 9 The Orchard, Fairfield Road, Ilkeston, DE7 6DD
G1 GNQ G. Nurse, Orchard Bungalow, Priors Green, Stisted, Braintree, CM77 8BP
G1 GNX G. Leonard, 11 St. Leonards Drive, Timperley, Altrincham, WA15 7RS
G1 GOP A. Abbott, The Shieling, West Road, Weaverham, Northwich, CW8 3HH
G1 GOQ S. Abbott, 5 Heathcote Gardens, Rudheath, Northwich, CW9 7JB
G1 GOY P. Miles, 37 Central Avenue, Northampton, NN2 8EA
G1 GPE D. Murray, 8 Tweed Crescent, Rushden, NN10 0GS
G1 GPL L. Pointon, 10 Lovers Walk, Dunstable, LU5 4BG
G1 GPM M. Stevenson, 6 Charnock Crescent, Sheffield, S12 3HB
G1 GPT A. Parsons, 157 Deighton Road, Huddersfield, HD2 1JT
G1 GQB J. Bagshaw, 7 Queen Street, Gomersal, Cleckheaton, BD19 4LG
G1 GQJ C. Clark, 9 Conigre, Chinnor, OX39 4JY
G1 GQQ M. Rowbotham, 37 Crawford Rise, Arnold, Nottingham, NG5 8QF
G1 GQY A. Armstrong, 18 Flaxfield Way, Kirkham, Preston, PR4 2AY
G1 GQZ C. Armstrong, 18 Flaxfield Way, Kirkham, Preston, PR4 2AY
G1 GRB R. Arnold, 26 Pinehurst Park, West Moors, Ferndown, BH22 0BW
G1 GRM B. Sinclair, 97 Lear Drive, Wistaston Green, Crewe, CW2 8DS
G1 GRN L. Skorupinski, 49 Pool Lane, Winterley, Sandbach, CW11 4RZ
G1 GRT G. Thomas, 83 Hollingthorpe Road, Hall Green, Wakefield, WF4 3NW
G1 GRU W. Smith, Flat 48, Winehala Court, 50A Sandbeds Road, Willenhall, WV12 4GA
G1 GRZ W. Baker, 26 Gardeners Road, Halstead, CO9 2TB
G1 GSB P. Standley, Bligh House, 1 Norwich Road, Norwich, NR16 1DJ
G1 GSG M. Taylor, 7 Marshall Road, Cropwell Bishop, Nottingham, NG12 3DP
G1 GSJ W. Mclaren, Ingleside, Waterloo, Whitchurch, SY13 2PX
G1 GSK M. Baylis, 45 Florence Avenue, Hove, BN3 7GX
G1 GSN P. Bradfield, 118 East Road, Langford, Biggleswade, SG18 9QP
G1 GST J. Thomas, 59 Cross Lane, Dudley, DY3 1PD
G1 GSY G. Bridle, 43 Cornflower Close, Locks Heath, Southampton, SO31 6SP
G1 GTA P. Butler, 25 Harringdale Road, High Harrington, Workington, CA14 4NU
G1 GTF E. Chilton, 33 Kersall Court, Nottingham, NG6 9DT
G1 GTH C. Clark, 24 Daisy Royd, Huddersfield, HD4 6RA
G1 GTK D. Towers, 50 Westbeech Road, Pattingham, Wolverhampton, WV6 7AQ
G1 GTM A. Turner, Carnbrae, Woodhouse Hill, Lyme Regis, DT7 3SL
G1 GTP B. Warnaby, 69 Caledonian Road, Hartlepool, TS25 5LB
G1 GTQ A. Clarke, 18 Waterloo Road, Brighouse, HD6 2AT
G1 GTR P. Clarke, 13 Mitchell Street, Brighouse, HD6 2AY
G1 GTS R. Clarke, 47 Peartree Road, Enfield, EN1 3DE
G1 GTX P. Cork, 95 Barmston Way Barmston Village, Washington, NE38 8DD
G1 GUI S. Watts, 16 Northampton Close, Bracknell, RG12 9EF
G1 GVJ P. Allen, 17 Winfield Road, Sedbergh, LA10 5AZ
GW1 GVM D. Gale, 5 Gadlys Terrace, Glyncorrwg, sa13 3bp
G1 GVP J. Gibbon, 18 Eagle Street, Penn Fields, Wolverhampton, WV3 7DN
G1 GWE A. Friel, 10 Marvejols Park, Cockermouth, CA13 0QR
G1 GWF B. Maxwell, Hillcrest, Castle View, Egremont, CA22 2NA
G1 GWJ A. Gillon, 94 Pelham St., Ashton under Lyne, OL7 0DU
G1 GWO J. Green, 32 Elizabeth Close, Highwoods, Colchester, CO4 9YU
G1 GWS J. Mossop, 14 Websters Lane, Great Sutton, Ellesmere Port, CH66 2LH
G1 GWX R. Patrick, 9 Brant Avenue, Illingworth, Halifax, HX2 8DL
G1 GXB K. Ray, 4 Elm Road Bishops Waltham, Southampton, SO32 1JR
G1 GXC S. Ray, 75 The Meads, Edgware, HA8 9HE
G1 GXF T. Scott, 9 Walker Drive, Leigh-on-Sea, SS9 3QS
GM1 GXH I. Sinclair, Clan Sinclair House, Nosshead Lighthouse, Caithness, KW1 4QT
GW1 GXQ R. Tulk, Home Farm Lodge, Pen Y Lan, Wrexham, LL14 6HS
G1 GXW B. Woodhouse, 5 Filby Drive, Durham, DH1 1LT
G1 GXX A. Mayes, 31 Holsey Lane, Bletchley, Milton Keynes, MK2 3FH
G1 GYC M. Hallsworth, 87 Talbot Street, Hazel Grove, Stockport, SK7 4BJ
G1 GYF D. Harvey, 264 Rangefield Road, Bromley, BR1 4QY
G1 GYH J. Hay, 23 Manor Close, Wilmslow, SK9 5PX
G1 GYJ F. Mallows, 31 Booth Road, Hartford, Northwich, CW8 1RD
G1 GYM P. Mcewen, 7 Springfield, Longhoughton, Alnwick, NE66 3NT
G1 GYQ A. Hayward, 1 Cleveland Road, Basildon, SS14 1NF
G1 GYT T. Down, 1 Park View, East Tytherley Road, Romsey, SO51 0LW
G1 GZG M. Newport, 9 Highbury Park, Exmouth, EX8 3EJ
G1 GZI A. Farmar, Hawkes Place, Horslett Hill, Holsworthy, EX22 6RS
G1 GZK K. Feay, 19 Dorset Avenue, Diggle, Oldham, OL3 5PL
G1 GZM I. Ford, 97 Green Rock Lane, Walsall, WS3 1NQ
G1 GZT N. O'Connor, 39 Strickland Drive, Morecambe, LA4 6TD
G1 HAB H. Birkmyre, Swarland, Morpeth, NE65 9JW
G1 HAC J. Hilton, 32 Dowry St., Fitton Hill, Oldham, OL8 2LP
G1 HAH B. Hodgson, 40 Trentham Drive, Bridlington, YO16 6ES
GW1 HAX N. Bevan, Mountain View, Whip Lane, Oswestry, SY10 8HU
G1 HBC T. Hopkins, 58 Broom Grove, Knebworth, SG3 6BQ
G1 HBD A. Hornby, 2 Maple Close, Winnersh, Wokingham, RG41 5PE
G1 HBE A. Howlett, 43 Cheetham Hill Road, Dukinfield, SK16 5JL
G1 HBF M. Hughes, 2 Chaldon Road, Canford Heath, Poole, BH17 8DB
G1 HBR K. Jackaman, 14A Cloverdale Gardens, Sidcup, DA15 8QL
G1 HBV E. Jones, 37 Sluice Road, Denver, Downham Market, PE38 0DY
G1 HBW F. Jones, 184 Harwich Road, Little Clacton, Clacton-on-Sea, CO16 9PU
G1 HCC E. Kent, 100 Waskerley Road, Washington, NE38 8DS
G1 HCI R. De Ste Croix, 49 Oxford Street, Grimsby, DN32 7JE
G1 HCJ G. De Ste Croix, 49 Oxford Street, Grimsby, DN32 7JE
G1 HCM F. Dawson, 33 Oakwood Road, Ryde, PO33 3JU
G1 HCU G. Gratton, 5 Nursery Avenue, Ovenden, Halifax, HX3 5SZ
G1 HDG P. Greed, 12 Bailey Close, Windsor, SL4 3RD
G1 HDK W. Akhurst, 20 Newton Road, Faversham, ME13 8DZ
G1 HDO A. Appleton, Flat 9, 19-21 West Cliff Road, Bournemouth, BH4 8AT
G1 HDR R. Stanford, 1 South End, Bassingbourn, Royston, SG8 5NG
G1 HDX M. Robertson, 12 James Park Homes, Egremont, CA22 2QQ
G1 HEA Dr A. Steele, 18 Lace Crescent, Tiverton, EX16 5FH
G1 HEJ J. Alexander, 1 Locarno Road, Swanage, BH19 1HY
G1 HEN D. Coates, 2 Penfold Drive, Countesthorpe, Leicester, LE8 5TP
G1 HEP C. Heptonstall, Badger Cottage, 27 Bolster Moor Road, Huddersfield, HD7 4JU
G1 HEQ K. Tucker, 507 New North Road, Ilford, IG6 3TF
G1 HER G. Dasilva-Hill, 12 St. Stephens Crescent, Thornton Heath, CR7 7NP
G1 HEU G. Tybora, 37 Nunsfield Drive, Alvaston, Derby, DE24 0GH
GW1 HEV D. Thomas, 3 Oaklands Terrace, Wiston, Haverfordwest, SA62 4PR
G1 HEW P. Travers, 49 West Bank Drive, South Anston, Sheffield, S25 5JG
G1 HEX J. Dunn, 8 Ettrick Terrace North, Craghead, Stanley, DH9 6BE
G1 HEY J. Todd, Atlast, 7 Marine Avenue West, Mablethorpe, LN12 2TX
G1 HFA R. Winder, 176 Ambleside Road, Lancaster, LA1 3ND
G1 HFE S. Wood, 18 Rosemellin, Camborne, TR14 8QF
G1 HFH D. Ward, 10 Fulshaw Avenue, Wilmslow, SK9 5JA
G1 HFK W. Willoughby, 27 Foxwood Grove, Sheffield, S12 2FN
G1 HFS K. Burgess, 32 Hendon Street, Leigh, WN7 1TS
G1 HFT E. Bennett, 20 Cromford Road, Clay Cross, Chesterfield, S45 9RE
G1 HFY B. Watson, 20 St. Marys Gardens, Hilperton Marsh, Trowbridge, BA14 7PG
G1 HGA K. Yates, 3 Flaxland Crescent, Sileby, Loughborough, LE12 7SB
G1 HGB J. Neville, 44 Thorpe House Avenue, Sheffield, S8 9NG
G1 HGC B. Newton, 42 Heath Road, Widnes, WA8 7NQ
G1 HGD M. Newell, 189 Humber Road, Coventry, CV3 1NZ
G1 HGF T. Standring, 52 Beechcroft Road, Ipswich, IP1 6BD
G1 HGT A. Berkerey, 36 Erlesmere Gardens, London, W13 9TY
G1 HGY Rev. P. Parry, Forge House, Church Road, Wellingborough, NN9 6BQ
G1 HHB C. Brown, 12 Forest Close, Newport, PO30 5SF
G1 HHC D. Bolt, C/O Old School Hse, Maristow Roborough, Plymouth, PL6 7BY
G1 HHD E. Bolt, Old School House, Roborough, Plymouth, PL6 7BY
G1 HHG J. Pilfold-Bagwell, 16 Anselm Close, Sittingbourne, ME10 1EY
GW1 HHM K. Roberts, Gwenallt, Lon Crecrist, Holyhead, LL65 2AZ
G1 HHO G. Reeve, 10 Badgers Copse, New Milton, BH25 5PE
G1 HHQ J. Brown, 14 The Green, Winscombe, BS25 1AL
G1 HHS A. Burrows, 1 Browns Avenue, Runwell, Wickford, SS11 7PT
G1 HHT A. Benstock, 10 Wike Ridge Avenue, Leeds, LS17 9NL
G1 HHU N. Ball, 140 Albert Avenue, Prestwich, Manchester, M25 0HE
G1 HHW W. Curtis, Rio Taibilla 11, San Miguel De Salinas, Alicante, Spain, 3193
GD1 HIA P. Smith, 98 Silverburn Crescent, Ballasalla, Isle Of Man, IM9 2ED
G1 HIB M. Standing, 7 Oxcliffe New Farm Caravan Park, Oxcliffe Road, Morecambe, LA3 3EF
G1 HIG T. Ravelini, 85 Glanfield Road, Beckenham, BR3 3JT
G1 HIJ R. Dimmock, 67 Meadway, Dunstable, LU6 3JT
GW1 HIN R. Ellwood-Thompson, 15 Skinner Street, Aberystwyth, SY23 2JU
G1 HIO M. Horsfield, 59 Queens Drive, Newton-le-Willows, WA12 0LY
G1 HIP K. Horsfield, 59 Queens Drive, Newton-le-Willows, WA12 0LY
G1 HIU J. Clarke, 3 Shelley Priory Cottages, Shelley, Ipswich, IP7 5RQ
G1 HJD B. Taylor, 111 High St., Warboys, Huntingdon, PE17 2TB
GM1 HJL I. Copland, 6 Dunadd View Kilmichael Glassary, Lochgilphead, PA31 8QA
G1 HJO J. Cornall, Fern Holme, Taylors Lane, Preston, PR3 6AB
G1 HJP T. Carter, 84 Colvile Road, Wisbech, PE13 2EL
G1 HJS M. Todd, 38 The Churchlands, New Romney, TN28 8LB
GM1 HJX M. Williams, 33/9 Marlborough Street, Edinburgh, EH15 2BD
G1 HKF C. Maclennan, 72 Sandsfield Lane, Gainsborough, DN21 1DD
G1 HKM F. Woods, 275 Scotter Road, Scunthorpe, DN15 7EH
G1 HKR T. Whittam, 27 Dimples Lane, Garstang, Preston, PR3 1RD
G1 HKS D. Wilson, 34 Belfairs Drive, Chadwell Heath, Romford, RM6 4EB
G1 HKU C. Weatherley, Flat 27, Briary Court, Egypt Esplanade, Cowes, PO31 8BT
G1 HLP D. Plant, 15 Heathcombe Road, Bridgwater, TA6 7PD
G1 HLQ M. Edwards, 11 Lightwood, Crown Wood, Bracknell, RG12 0TR
G1 HLS W. Etherton-Scott, 62 Spencer Road, Walthamstow, London, E17 4BD
G1 HLT I. Fay, 7 Oakridge Close, Forest Town, Mansfield, NG19 0EY
G1 HLV J. Lee, Deighton Manor, Deighton, Northallerton, DL6 2SN
G1 HLY R. Powell, 55 Lumley Road, Horley, RH6 7JF
G1 HMI T. Rock, 4 Hunters Gate, Much Wenlock, TF13 6BW
G1 HML H. Willard, M V Irma, Paglesham Boatyard, Paglesham Eastend, SS4 2ER
G1 HMT G. Gray, Home Farm, Furlong Drove, Ely, CB6 2EQ
G1 HMW W. Gardner, 25 Prospect Place, Wing, Leighton Buzzard, LU7 0NT
G1 HMY P. Batty, 14 Woodville Road, Penwortham, Preston, PR1 9DR
G1 HMZ K. Breedon, 17 Emmanuel Avenue, Arnold, Nottingham, NG5 9QN
G1 HND M. Burling, 28 Croydon Road, Arrington, Royston, SG8 0DJ
GW1 HNF N. Bufton, 7 Laburnum Close, Rassau, Ebbw Vale, NP23 5TS
GW1 HNG P. Beesley, 12 Bryngolwg, Aberdare, CF44 0ER
G1 HNH P. Bannister, 40 Regent St., Stowmarket, IP14 1RJ
G1 HNN R. Cockman, 31 Kensington Road, Southend-on-Sea, SS1 2SX
G1 HNU M. Gray, 28 The Close, Bradwell, Great Yarmouth, NR31 8DR
GM1 HNZ A. Simmers, Loanside, Crossroads, Keith, AB55 6LP
G1 HOD A. Smith, 14 Bridge Street, Shepshed, Loughborough, LE12 9AD
G1 HOI N. Ballard, 185 Nw Harwood, Spc #45, Prineville, USA, 97754
G1 HOJ B. Baylis, 118 Eastgate, Deeping St. James, Peterborough, PE6 8RD
G1 HOL S. Cook, 20 Gaymore Road, Kidderminster, DY10 3TU
G1 HOP R. Chaston, 157 Winston Avenue, Coventry, cv2 1dl
G1 HOU A. Kirby, 185 High Street, Dunsville, Doncaster, DN7 4BU
G1 HPB R. Wearing, 163 Birmingham Road, Stratford-upon-Avon, CV37 0AP
G1 HPS T. Jones, 25 Foxcotte Road, Charlton, Andover, SP10 4AR
G1 HPU P. James, 8 Pipers Wood Cottages, Little Missenden, Amersham, HP7 0RQ
G1 HPV T. Jones, 175 New Road, Great Wakering, Southend-on-Sea, SS3 0AR
G1 HPZ I. Russell, 10 Launceston Road, Bristol, BS15 1EP
G1 HQE C. Close, 3 Hay Green, Therfield, Royston, SG8 9QL
G1 HQG A. Coley, 5 Arundel Way Highcliffe, Christchurch, BH23 5DX
G1 HQH I. Church, 8 Keats Drive Harwell, Didcot, OX11 6FA
G1 HQJ D. Robinson, 16 Green Lane, Platts Heath, Maidstone, ME17 2NS
G1 HQK I. Richardson, 1 Cedar Drive, Lowestoft, NR33 9HA
G1 HQN J. Rattenbury, Compton Lodge, High Ham, Langport, TA10 9DH
G1 HQO Dr G. Spaven, Spout House Farm, Macclesfield Road, High Peak, SK23 7QU
G1 HQQ F. Jensen, 79 The Drakes, Shoeburyness, Southend-on-Sea, SS3 9NY
G1 HQW J. Kierman, 26 Popes Lane, Gorefield Road, Wisbech, PE13 5BD
G1 HRA D. Lloyd, 35 Charles Close, Abbotts Barton, Winchester, SO23 7HT
G1 HRD V. Allen, 155, Reigate Avenue, Sutton, SM13RJ
G1 HRH M. Gregory, 45 Larksfield Avenue, Bournemouth, BH9 3LW
G1 HRJ P. Deakes, 108 Glaisdale Drive East, Nottingham, NG8 4LZ

G1 HRL E. Dillow, 18 Laburnum Grove, Warwick, CV34 5TG
G1 HRM T. Davenport, 36 Rydale Road, Nottingham, NG5 3GS
G1 HRQ B. Madore, 66A West Street, Ryde, PO33 2QF
G1 HRU S. Hill, 26 Crescent Road, Dudley, DY20NW
G1 HRV K. Higgins, 22 Thatchers Lane, Cliffe, Rochester, ME3 7TN
GM1 HRY A. Davis, 3 High Shore, Banff, AB45 1DB
G1 HSA S. Arnold, 30 Pine Avenue, Newton-le-Willows, WA12 8JE
G1 HSF M. Evans, Fernlea, Pit Hill Lane, Bridgwater, TA7 9BT
G1 HSG N. Evans, 25 Chetwyn Avenue, Bromley Cross, Bolton, BL7 9BN
G1 HSH M. Ellerby, 3 Gilwern Court, Ingleby Barwick, Stockton-on-Tees, TS17 5DJ
G1 HSI M. Glazier, 19 West Place Brookland, Romney Marsh, TN29 9RG
G1 HSJ L. Godden, The Conifers, 14 Pirehill Lane, Stone, ST15 0JN
G1 HSL J. Girt, 22 Medway Road, Ipswich, IP3 0QH
G1 HSM L. Heller, 1 Princes Road, St Leonards-on-Sea, TN37 6EL
G1 HSO M. Hoey, 37 Newhouse Road, Blackpool, FY4 4JJ
G1 HSP C. Hunt, 700 Western Boulevard, Nottingham, NG8 5FH
G1 HSX P. Kimber, 16 Sycamore Close, Lydd, Romney Marsh, TN29 9LF
G1 HTF H. Heron, 43 Cheetham Hill Road, Dukinfield, SK16 5JL
G1 HTL J. Foster, 25 Hunter Road, Arnold, Nottingham, NG5 6QZ
G1 HTM S. Froggatt, 17 Queensway, Saxilby, Lincoln, LN1 2QB
G1 HTN P. Farrow, 10 St. Thomas Close, Chilworth, Guildford, GU4 8LQ
G1 HTO R. Fortescue, 7 Bodkin Lane, Weymouth, DT3 6QL
G1 HTR A. Gee, 4 Malkin Avenue, Radcliffe on Trent, NG12 1DP
GU1 HTY B. Ayres, Rousay, Bailiffs Cross Road, St Andrew, Guernsey, GY6 8RY
G1 HUM R. Beech, 6 Law Cliff Road, Birmingham, B42 1LP
G1 HVL P. Howarth, 7A Fox Court, Durkar, Wakefield, WF4 3BH
G1 HVW J. Craft, 3 Coltsfoot Drive, Royston, SG8 9EU
G1 HWA Dr K. Harris, 27 Middle Field Road, Rotherham, S60 3JJ
G1 HWJ P. Milner, 3 Larne Avenue, Cheadle Heath, Stockport, SK3 0UJ
G1 HWK T. Mccarthy, 25 Henley Avenue, North Cheam, Sutton, SM3 9SG
G1 HWO T. Miller, 27 Richmond Way, Oadby, Leicester, LE2 5TR
G1 HWP R. Davies, 7 The Nook, Tupsley, Hereford, HR1 1NH
G1 HWR L. Mills, 54 Petters Road, Ashtead, KT21 1NE
G1 HWY M. Jupp, 54 Shooting Field, Steyning, BN44 3RQ
G1 HXN G. King, 1 Sudan Cottage, Frogge Lane, Coltishall, Norwich, NR12 7JU
G1 HXP J. Lennard, 10 Orston Road East, West Bridgford, Nottingham, NG2 5FU
G1 HXR R. Davis, 6 Fairway Drive, Northmoor, Wareham, BH20 4SG
G1 HXT G. Eden, Heathend Cottage, Cromhall, Wotton under Edge, GL12 8AS
G1 HXZ C. Cave, 20 Meadow View, Banbury, OX16 9SR
G1 HYA N. Cramp, 3 Sowood Court, Ossett, WF5 0TJ
G1 HYC D. Curson, 25 Colbert Park, Swindon, SN25 4YJ
G1 HYG B. Crowther, 104 John St., Beamish, Stanley, DH9 0QP
G1 HYM P. Bailey, 10 Bell Drive Hednesford, Cannock, WS12 4RA
GU1 HYN B. Bolderston, 12 Hartlebury Estate, Steam Mill Lane, St Martin, Guernsey, GY4 6NH
G1 HYO M. Green, 26 Hunters Field, Stanford In The Vale, Faringdon, SN7 8LR
G1 HYQ D. Chenoweth, 20 Churchlands Road, Bedminster, Bristol, BS3 3PW
G1 HYU K. Church, 31 Riversway, King's Lynn, PE30 2ED
G1 HYX A. Chance, 18 Egdon Glen, Crossways, Dorchester, DT2 8BQ
G1 HZD C. Legate, 101 Butchers Lane, Walton on the Naze, CO14 8UD
G1 HZI I. Dodd, Gardeners Cottage, Sandhoe, Hexham, NE46 4LU
G1 HZJ M. Devine, 7 South Parade, Seascale, CA20 1PZ
GM1 HZL M. Donaldson, 43 Stair Street, Drummore, DG9 9PT
G1 HZN C. Dadd, 60 Rosaire Place, Scartho, Grimsby, DN33 2JS
G1 HZR K. Farrar, 8 Ascot Avenue Cantley, Doncaster, DN4 6HE
G1 IAB S. Matthews, 66 West End Road Epworth, Doncaster, DN9 1LB
G1 IAD D. Morton, 12 Pennygate Drive, Lowestoft, NR33 9HL
G1 IAG P. Morris, 18 Greenway Close, London, NW9 5AZ
G1 IAL A. Plant, 148 Chatsworth Road, Halesowen, B62 8TH
G1 IAQ D. Memory, 22 Marstown Avenue, South Wigston, Leicester, LE18 4UH
G1 IAV P. Costello, Newgrange, Poplar Road, New Milton, BH25 5XP
GW1 IAW I. Woodward, Corlander, Middle Road, Wrexham, LL11 3TW
G1 IBF G. Tullock, 16 Ward Lea Nafferton, Driffield, YO25 4JZ
G1 IBJ C. Diaper, 163 Edwin Road, Gillingham, ME8 0AQ
G1 IBO D. Buss, Rlc, Po Box 885, Bristol, BS99 5LG
G1 IBP A. Heaysman, 325 Broomfield Road, Chelmsford, CM14DU
G1 IBS W. Chadwick, 102 Feltham Road, Ashford, TW15 1DP
G1 IBX C. Drayton, The Lindens, Main Street, Kelfield, York, YO19 6RG
G1 ICA D. Keable, 90 King Edward Road, Rugby, CV21 2TE
G1 ICH M. Adams, 61 Monks Park Road, Northampton, NN1 4LU
G1 ICI M. Thorpe, 18 Sherrier Way, Lutterworth, LE17 4NW
G1 ICK D. Winton, 16 Lord Avenue, Clayhall, Ilford, IG5 0HP
G1 ICQ A. Orgee, 54 Riverview Close, Hallow, Worcester, WR2 6DA
G1 ICX D. Palmer, 18 Newfields, Sporle, King's Lynn, PE32 2UA
G1 IDE S. Roy, 28 Kingston Rise, New Haw, Addlestone, KT15 3EY
G1 IDF A. Edwards, 51 Redrock Road, Rotherham, S60 3JN
G1 IDJ G. Perry, 61 Ollands Road, Reepham, Norwich, NR10 4EL
G1 IDQ G. Arnold, 36 Market Street, Rugeley, WS15 2JL
G1 IDR Gliding Centre ARS, c/o A. Gilmore, Ashfields, Naseby Road Clipston, Market Harborough, LE16 9RZ
G1 IDV G. Bain, 99 Longford Lane, Gloucester, GL2 9HB
G1 IDZ D. Young, 70 North Malvern Road, Malvern, WR14 4LX
GW1 IEB L. Tatham, Hebron Stores, Llangwnadl, Pwllheli, LL53 8NW
G1 IEC P. Walton, 2 Albert Road, Bromsgrove, B61 7BE
GM1 IEL J. Bruce, 24 South Green Drive, Airth, Falkirk, FK2 8JP
G1 IEO J. Turner, 2 Hilberry Road, Canvey Island, SS8 7EL
G1 IEP G. Tomkins, The Close, Broomfield Clayton, Bradford, BD14 6PJ
G1 IEX D. Broughton, 33 Queens Park Flats, Queens Park Close, Mablethorpe, LN12 2AS
G1 IEY S. Reigate, 9 Effingham Road, Croydon, CR0 3NF
G1 IFF A. Rose, 15 Elderwood Way Tuffley, Gloucester, GL4 0RA
G1 IFH G. Reading, 34A Harding Avenue, Rawmarsh, Rotherham, S62 7ED
G1 IFV N. Ginger, Barnlea, Fairwarp, Uckfield, TN22 3DT
G1 IFW W. Gain, 14 Clarence Road, St Leonards-on-Sea, TN37 6SD
G1 IFX D. Garratt, 3 Fort Road, Mountsorrel, Loughborough, LE12 7HB
G1 IGA C. Brooks, 14 The Furlong, King Street, Tring, HP23 6BX
G1 IGC S. Brookes, 52 Larch Grove, Kendal, LA9 6AU
G1 IGN G. Scroggs, 52 Eastern Road, Burnham-on-Crouch, CM0 8BT
G1 IGP G. Spinks, 89 Uplands Road, Oadby, Leicester, LE2 4NT
G1 IGW D. Cliff, 23 Grey Towers Drive, Nunthorpe, Middlesbrough, TS7 0LT
G1 IHA R. Stravens, 75 Telford Road, London, N11 2RL
G1 IHE J. Smith, 65 Woods Avenue, Hatfield, AL10 8QF
G1 IHI M. Godsave, 35 Furlong Close, Midsomer Norton, Radstock, BA3 2PR
G1 IHJ A. Homer, 6 Ensall Drive, Wordsley, Stourbridge, DY8 4XX
G1 IHL S. Hopkins, 98 Court Road, Kingswood, Bristol, BS15 9QP
G1 IHS C. Currie, 33 Ashridge Drive, Bricket Wood, St Albans, AL2 3SR
G1 IHY J. Shilson, 3 Hereford Close, Desborough, Kettering, NN14 2XA
G1 III C. Smith, 199A Richardshaw Lane, Stanningley, Pudsey, LS28 6AA
G1 IIO B. Thornton, 21 Valley Road, Banbury, OX16 9BQ
GU1 IIW R. Loveridge, Shamley, Route De Portinfer, Vale, Guernsey, GY6 8LN
G1 IIX B. Lee, 58 Shaw Avenue, Normanton, WF6 2TT
G1 IIY D. Turner, 154 Rowlett Road, Corby, NN17 2BS
GW1 IIZ J. Underwood, Rock Hill, Llanarthney, Carmarthen, SA32 8LJ
G1 IJC M. Williamson, 2 Lancaster Close, Fakenham, NR21 8DW
G1 IJJ J. Lainchbury, 17 Pearmain Avenue, Wellingborough, NN8 4SF
G1 IJM M. Shoosmith, 18 Pottery Close, Aylesbury, HP19 7FY
G1 IJQ D. Martin, 27 St. Andrews Road, Stratton, Bude, EX23 9AG
G1 IJY R. Wise, Flat 1, Capelia House, 18-21 West Parade, Worthing, BN11 3RB
G1 IKF R. Blakemore, 31 Millstone Rise, Liversedge, WF15 7BW
G1 IKG P. Mcgahon, 520 Chessington Road, West Ewell, Epsom, KT19 9HH
G1 IKH I. Mclaughlin, 28 Jarvis Avenue, Nottingham, NG3 7BH
G1 IKL P. O'Sullivan, Japonica Cottage, Ardens Grafton, Alcester, B49 6DR
G1 IKT S. Elliott, Manor House, Bewholme, Driffield, YO25 8DX
G1 IKV B. Austin, 16 Heathlands, Westfield, Hastings, TN35 4QZ
G1 IKW G. Winn, 16 Highbury Place, Bramley, Leeds, LS13 4PW
G1 ILC S. Colley, 8 Tennyson Road, Maltby, Rotherham, S66 7LU
G1 ILF P. Ellis, 48 Willington Road Skellow, Doncaster South Yorkshire, DN6 8JE
G1 ILG B. Evans, 12 The Mead Thaxted, Dunmow, CM6 2PU
G1 ILH G. Farr, 18 The Loont, Winsford, CW7 1EU
G1 ILJ C. Wood, Potter Brompton Wold Farm, Potter Brompton, Scarborough, YO12 4PH
G1 ILO R. Bell, 80 West Avenue, Lightcliffe, Halifax, HX3 8TJ
G1 ILY C. Sims, 226 Exeter Road, Exmouth, EX8 3NB
G1 IMD M. Hall, 6 Poplar Avenue, New Mills, High Peak, SK22 4HR
G1 IME N. Hopkins, 22 Cornfield Way, Ashton-Under-Hill, Evesham, WR11 7TA
G1 IMI C. Foreman, Thornham Farm, Wansford, Driffield, YO25 8JJ
G1 IMM A. Gee, 24 Granhams Close, Great Shelford, Cambridge, CB22 5LG
G1 IMS I. Stewart, 34 Newgate Street Village, Hertford, SG13 8RB
G1 IMY R. Laycock, 24 Farmcroft Road, Mansfield Woodhouse, Mansfield, NG19 8QT
G1 INA A. Lowe, 47 Springfield Park Road, Chelmsford, CM2 6EB
G1 INB P. Laker, 207 Columbia Road, Ensbury Park, Bournemouth, BH10 4EE
G1 IND V. Lowe, 35 Elm Place, Armthorpe, Doncaster, DN3 2DE
G1 INI B. Ginsburg, 27 Park Crescent, Elstree, Borehamwood, WD6 3PT
G1 INJ Dr R. Ginsburg, 3 Basing Hill, London, NW11 8TE
G1 INK S. Green, 48 Queen Elizabeth Road Humberston, Grimsby, DN36 4DE
GM1 INS B. Skakle, 190 West Road, Fraserburgh, AB43 9NL
G1 INU M. Sweet, 67 Swinley Road, Wigan, WN1 2DL
GD1 IOM IOM ARS(Iomars), c/o A. Morgan, Thal'Loo Glass, Nassau Road, Mwyljyn Moddey, Isle Of Man, IM7 4AQ
G1 IOO D. Camac, 6 Wisbeck Road, Tonge Fold, Bolton, BL2 2TA
G1 IOP A. Cheer, 15 Stibbs Way, Bransgore, Christchurch, BH23 8HG
GW1 IOT H. Harrison, 2 Hendre, Newtown, Ebbw Vale, NP23 5FE
G1 IOU G. Holland, 16 Hancox Street, Oldbury, B68 9LQ
G1 IPD D. Mobbs, 64 Cranford Road, Kingsley, Northampton, NN2 7QX
G1 IPE A. Medcalf, 23 Allesborough Drive, Pershore, WR10 1JH
G1 IPI R. Taylor, Lower Manaton, South Hill Road, Callington, PL17 7LW
GW1 IPJ I. Parry-Jones, 21 Laurels Avenue, Bangor-On-Dee, Wrexham, LL13 0BQ
G1 IPP M. Allen, 23 Waterloo Crescent, Countesthorpe, Leicester, LE8 5SU
G1 IPU G. Coote, 10 Curlew Close, Clacton-on-Sea, CO169EN
G1 IPY J. Rowlands, 2 Wellfield, Longton, Preston, PR4 5BX
G1 IQA G. Adkins, 117 Connolly Drive, Rothwell, Kettering, NN14 6TN
G1 IQE F. Angwin, 171 Windsor Road, Wellingborough, NN8 2LZ
G1 IQF M. Ames, 7 Northgate, Leyland, PR25 3NR
G1 IQG A. Bloodworth, 79 Hands Road, Heanor, DE75 7HB
G1 IQK B. Shaw, 23 Lodge Drive, Culcheth, Warrington, WA3 4ES
G1 IQN J. Spicer, 5 Berries Mount, Bude, EX23 8AP
GW1 IQS I. Jones, 7 The Oaks, Quakers Yard, Treharris, CF46 5HQ
G1 IQU D. Jolley, 212 Eastern Esplanade, Southend-on-Sea, SS1 3AD
G1 IRG S. Manning, 11 Broomhill Crescent, Southfields, Northampton, NN3 5BH
G1 IRQ N. Tansley, 11 Juniper Close, Lutterworth, LE17 4US
GM1 ISJ D. Barker, 7 The Keys, Kildonan, Isle of Arran, KA27 8SA
GW1 ISK F. Davies, Hendref, Red Wharf Bay, Pentraeth, LL75 8YG
G1 ISN T. Evans, 22 Malthouse Lane, Ashover, Chesterfield, S45 0AL
G1 ISP B. Etherington, 24 Broomcroft Road, Ossett, WF5 8LH
GW1 ISR Dr P. Kenington, Trap Farm, Devauden, Chepstow, NP16 6PE
G1 ISS B. Lyons, 51 Wade Reach, Walton on the Naze, CO14 8RE
G1 ISX C. Hall, 9 Moneyhill Court, Dellwood, Rickmansworth, WD3 7DY
G1 ISY N. Morris, 15 Turners Close, Highnam, Gloucester, GL2 8EH
G1 ITE P. Hayler, 27 Birch Way, Heathfield, TN21 8BB
G1 ITJ K. Edmett, Solstice, Youngs Paddock, Salisbury, SP5 1RS
G1 ITL D. Gilbey, 34 Farnhurst Road, Barnham, Bognor Regis, PO22 0JN
G1 ITS T. Williams, 86 Hillcrest Road, Rochdale, OL11 2QB
G1 ITV K. Ward, 5 Clarence Court, Bolton, BL1 2XP
G1 IUA N. Harris, Sunnyside Lodge, Mongeham Road, Deal, CT14 8JW
G1 IUD C. Sermons, 17 Wellside, Marks Tey, Colchester, CO6 1XG
G1 IUF M. Kilkenny, 138 Stanbury Road, Hull, HU6 7BW
G1 IUL B. Jackson, 10 Wood End Croft, Coventry, CV4 9RN
G1 IUT T. Christmas, 3 Mount Road, Cosby, Leicester, LE9 1SX
G1 IUW G. Diaper, 89 East St., Sudbury, CO10 2TP
G1 IUZ S. Groves, 135 Ring Road, Crossgates, Leeds, LS15 7QE
G1 IVF D. Lowe, 21 Farndon Road, Market Harborough, LE16 9NW
G1 IVG C. Lowe, 22 Ryelands Close, Market Harborough, LE16 7XE
GM1 IVI M. Grey, 7 Cemetery Lane, Tweedmouth, Berwick-upon-Tweed, TD15 2BS
G1 IVK T. Garnham, 45 Buscot Drive, Abingdon, OX14 2BL
G1 IVL R. Hudson, 80 Drake Avenue, Worcester, WR2 5RR

G1 IVO L. Ladner, 7 Polventon Close, Heamoor, Penzance, TR18 3LD
G1 IVP J. Lamb, 5 Honeycroft, Loughton, IG10 3PR
G1 IVV G. Merrington, Cartref, Ball Lane, Frodsham, WA6 8HP
G1 IWE T. Coombs, 114 Talbot Street, Whitwick, Coalville, LE67 5AZ
G1 IWT R. Moore, 9 Rowland St., Allenton, Derby, DE24 9BT
G1 IXE V. Green, 50 Alcove Road, Fishponds, Bristol, BS16 3DR
G1 IXF I. Green, 50 Alcove Road, Bristol, BS16 3DR
G1 IXV C. Haver, 18 Church Lane, Edenham, Bourne, PE10 0LS
G1 IYA A. Greenwood, 21 Ovenden Crescent, Halifax, HX3 5PE
G1 IYB B. Haines, 66 North Drive, Grove, Wantage, OX12 7PN
G1 IYE I. Hawes, 129 Manor Road, Ash, Aldershot, GU12 6QB
G1 IYF C. Parfitt, 12 Marigold Close, Basingstoke, RG22 5RG
G1 IYO R. Reed, 482 Baring Road, London, SE12 0EG
G1 IZA D. Lamb, 33 Cherston Road, Loughton, IG10 3PL
G1 IZB F. Smith, 6 Mill Close, Marshchapel, Grimsby, DN36 5TP
G1 IZH A. Trueman, 3 Higher Road. Woolavington, Bridgwater, TA7 8EA
G1 IZN R. Mitchell, 45 Kent Close, Mitcham, CR4 1XN
G1 JAA R. Lees-Oakes, 6 Tabley Street, Mossley, Ashton-under-Lyne, OL5 9PD
G1 JAB J. Burke, 48 Medina Road, Portsmouth, PO6 3HD
G1 JAG K. Powell, 89 Avenue Road, Leicester, LE2 3EA
G1 JAH J. Hagues, 7 Eastern Green Park Two, Eastern Green, Penzance, TR18 3BA
G1 JAL P. Westbury, 6 Bradford Road, Rode, Frome, BA11 6PR
G1 JBB L. Richards, 14 St. Julitta, Luxulyan, Bodmin, PL30 5ED
G1 JBC N. Thompson, 24 Braemor Road, Calne, SN11 9DT
G1 JBE D. Blackburn, 24 Reservoir St., Darwen, BB3 1LQ
G1 JBG J. Beacon, 14 Siskin Close, Bishops Waltham, Southampton, SO32 1RQ
G1 JBJ S. Bartlett, Lieu Dit La Gaule, Nantheuil, 24800
G1 JBM M. Clark, 28 Compton Crescent, West Moors, Ferndown, BH22 0BZ
G1 JBW B. Ellison, 931 Burnley Road, Todmorden, OL14 7ET
G1 JBZ I. Halsey, Cowtrott, Back Lane, Great Yarmouth, NR29 5ED
G1 JCC I. Jefferson, 19 Orchard Way, Flitwick, Bedford, MK45 1LF
G1 JCL M. Munn, 11 Foxley Road, Queenborough, ME11 5AW
G1 JCP J. Pasfield, Fairlands, White Lodge Crescent, Clacton-on-Sea, CO16 0HT
G1 JCT S. Farrant, The Bungalow, Brewery Yard, Stroud, GL5 4JW
G1 JCW A. Duffy, 14 Garden Street, Padiham, Burnley, BB12 8NP
G1 JDE O. Graffham, 106 Barford Road, Edgbaston, Birmingham, B16 0EF
G1 JDF D. Gray, 11 Field Close, Bollington, Macclesfield, SK10 5JG
GM1 JDJ L. Mcleman, 14 Flures Place, Erskine, PA8 7DH
G1 JDO P. Oliver, 3 Savile Walk, Brierley, Barnsley, S72 9HJ
G1 JDQ P. Paterson, Oak Lea, 11A Fletsand Road, Wilmslow, SK9 2AD
G1 JDT G. Palmer, 5 Dunstar Avenue, Audenshaw, Manchester, M34 5LJ
G1 JDV J. Vasey, 22 Rickleton Village Centre, Washington, NE38 9ET
G1 JEA D. Hart, 10 Marina Drive, March, PE15 0AU
G1 JEH K. Schneider, 65 Alpha Road, Birchington, CT7 9ED
G1 JER Dr J. Johnson, 5 Hunters Ride, Appleton Wiske, Northallerton, DL6 2BD
G1 JEZ S. Taylor, 72 Molyneux Drive, Wallasey, CH45 1JT
GM1 JFF A. Weddell, 10 High Street, Eyemouth, TD14 5EU
G1 JFL M. Woolridge, 23 Marina Drive, May Bank, Newcastle, ST5 9NL
GW1 JFT R. Beaugie, 32 Court Gardens, Rogerstone, Newport, NP10 9FU
G1 JFU W. Cmd. D. Bryant, 22 Highfield Park, Heaton Mersey, Stockport, SK4 3HD
G1 JGD N. Cullis, 39 Gilbert Drive, Langdon Hills, Basildon, SS16 6SP
G1 JGE M. Colley, 118 Devon Crescent, Birtley, Chester le Street, DH3 1HP
G1 JGF L. Cox, 7 Timberdine Avenue, Worcester, WR5 2BD
G1 JGM B. Easey, 4 Ash Trees, East Brent, Highbridge, TA9 4DQ
G1 JGR C. Fortnum, 11 Ayr Close, Stamford, PE9 2TS
G1 JGS M. Garland, 4A St Andrews Way, Freshwater, PO40 9NH
G1 JGT J. Giller, 9 Alberta Crescent, Huntingdon, PE29 1TL
G1 JGY H. Ketley, 24 Farmcroft Road, Mansfield Woodhouse, Mansfield, NG19 8QT
G1 JHB C. Lawrence, Flat 22, St. Pauls Court, Salford, M7 3NZ
G1 JHD M. Plant, 7 Kendale, Hemel Hempstead, HP3 8NN
G1 JHG T. Mccormick, 33 Bryanston Road, Aigburth, Liverpool, L17 7AL
GM1 JHM A. Harding, Bandrum House, Roadside Of Catterline, Catterline, Stonehaven, AB39 2UA
G1 JHN F. Harris, 4 Parc An Ithan, The Lizard, Helston, TR12 7PA
G1 JHP H. Hamer, 126 Mellor Brow, Mellor, Blackburn, BB2 7PN
GI1 JHQ J. Selfridge, 6 Belmont Place, Coleraine, BT52 1QH
G1 JHS S. Barrett-Jolley, 61 Eyston Drive, Weybridge, KT13 0XE
GM1 JHU J. Adams, 3A Glenpatrick Road, Elderslie, Johnstone, PA5 9BH
G1 JHX A. Bennett, 32 Park Road, Stretford, Manchester, M32 8DQ
G1 JHY A. Potter, 25 Robinsons Meadow, Ledbury, HR8 1SU
G1 JHZ S. Potter, 25 Robinsons Meadow, Ledbury, HR8 1SU
GW1 JIE K. Robertson, Tyn Y Pwll, Fachwen, Caernarfon, LL55 3HD
G1 JIG S. Ridgard, 5 Orchard Way, Luton, LU4 9LT
G1 JIH E. Rowell, 80 Kings Delph, Whittlesey, Peterborough, PE7 2PD
G1 JIJ J. Passfield, 2 Parker Road, Chelmsford, CM2 0ES
G1 JIR D. Robinson, 4 Mayorlowe Avenue, Stockport, SK5 8DB
G1 JIW P. Spooner, 62 Chester Crescent, Newcastle, ST5 3RW
G1 JJA S. Bessent, 347 Birchfield Road, Redditch, B97 4NE
G1 JJE N. Banks, 6 Bylands Place, Newcastle, ST5 3PQ
G1 JJK P. Naylor, 14 Wrockwardine Road, Wellington, Telford, TF1 3DB
G1 JJQ J. Schulz, 4 Collinge Close, East Malling, West Malling, ME19 6QS
G1 JJR V. Smith, 99 Dewhurst Road, Fartown, Huddersfield, HD2 1BN
G1 JJT N. Stubbs, 8 West Avenue, Hilton, Derby, DE65 5FY
G1 JJZ S. Willey, 34 Kernick Road, Penryn, TR10 8NT
G1 JKE N. Knapton, 4 Crabmill Lane, Easingwold, York, YO61 3DE
G1 JKF N. Leaney, 31 Saxon Way, Willingham, Cambridge, CB24 5UR
GM1 JKJ A. Britton, 15 Glenbrook, Balerno, EH14 7JE
G1 JKL A. Crouch, 107 Waterleat Avenue, Paignton, TQ3 3UD
G1 JKN P. Cooper, 5 Lower Leys Way, Leominster, HR6 0SS
G1 JKO G. Cooper, Hazeldene, Fleet Coy, Spalding, PE12 0RU
GW1 JKP R. Coleman, Rockwood, Clive Road, Fishguard, SA65 9DB
G1 JKV T. Whittaker, Flat 2, Paul Vanson Court, New Berry Lane, Walton-on-Thames, KT12 4HQ
G1 JKX J. West, 4 Coronation Terrace, Longhorsley, Morpeth, NE65 8UN
G1 JLB M. Denison, 9 Derwent Road, Harrogate, HG1 4SG
G1 JLG B. Giddings, 71 Tyrone Road, Southend-on-Sea, SS1 3HD
G1 JLM S. Brosnan, 4 Black Rod Close, Hayes, UB3 4QJ
GM1 JLP K. Robson, 13 Woodstock Avenue, Galashiels, TD1 2EE
G1 JLQ J. Yarnall, 4 Parklands, Evesham, WR11 2QJ
G1 JLX N. Povey, 33 Church Lane, Fradley, Lichfield, WS13 8NJ
G1 JMC B. Harris, 55 Valiant Way, Melton Mowbray, LE13 0GE
G1 JMD P. Hall, 64 Synehurst Crescent, Badsey, Evesham, WR11 7XX
G1 JMF A. Hooper, 5 Nine Elms Road, Longlevens, Gloucester, GL2 0HA
G1 JMH J. Hickey, 36 Station Road, Alderholt, Fordingbridge, SP6 3RB
G1 JMK M. Justice, 6 Stanley Terrace, Devizes, SN10 5AJ
G1 JMN R. Andrews, Owls Rest, Park Lane, Worcester, WR2 6PQ
G1 JMP R. Ainsworth, 95 Heysham Close, Murdishaw, Runcorn, WA7 6DT
G1 JMS J. Stoddart, 24 Vicarage Close, Platt Bridge, Wigan, WN2 5DW
G1 JMV P. Slark, 11 Hillfield Walk, Bolton, BL2 2UR
G1 JMW W. Smith, 37 Peake Road, Brownhills, Walsall, WS8 7BZ
G1 JMY F. Taylor, Wold Lodge, Pocklington Road, York, YO42 1YJ
GD1 JNB P. Clarke, 5 Sumark Avenue Douglas, Isle of Man, IM2 2AD
GM1 JNC A. Campbell, 17 Moulin Circus, Cardonald, Glasgow, G52 3JY
G1 JNG G. Eccles, 1 Bridge Place, Amersham, HP6 6JF
GW1 JNH W. Francis, 8 George Street, Treherbert, Treorchy, CF42 5AH
GW1 JNI A. Fennah, 7 Y Ddol, Llanbrynmair, SY19 7DJ
G1 JNQ P. Auld, 80 Milestone Road, Stone, Dartford, DA2 6DN
GW1 JNR P. Adcock, Bleak House, Cefn Coch, Welshpool, SY21 0AE
GM1 JNS G. Bartram, 6 Craigewan Crescent, Peterhead, AB42 1HL
G1 JNX S. Langston, 4 Lagwood Close, Hassocks, BN6 8HZ
G1 JNY A. Larkin, 16 Thetford Close, Corby, NN18 9PH
G1 JOA B. Marsh, 96 Coopers Lane, Clacton-on-Sea, CO15 2DA
G1 JOD R. Norton, Middleton House, Bleathwood, Ludlow, SY8 4LX
G1 JOJ B. Smith, 146 Battram Road, Ellistown, Coalville, LE67 1GB
G1 JOL B. Shane, 7 Oakwood Glade, Holbeach, Spalding, PE12 7JS
G1 JON J. Storey, 34 Austin Rise, Longbridge, Birmingham, B31 4QN
G1 JOO R. Seymour, 5 Clifton Place, Easton, Bristol, BS5 0SE
G1 JOR J. Ormsby-Rymer, 109 Goldcrest Road, Chipping Sodbury, Bristol, BS37 6XJ
GW1 JOV T. Moore, Dan Bryn Coch, Llandyfan, Ammanford, SA18 2TY
G1 JOW C. Oates-Miller, 32 Burnlee Road, Holmfirth, HD9 2PS
G1 JPC M. Ward, Four Winds, 13 Westfield Avenue, Wellingborough, NN9 6DQ
G1 JPI M. Taylor, 26 Monks Close, Lancing, BN15 9DD
GM1 JPJ R. Jamieson, 6A Mary Street, Stonehaven, AB39 2AD
G1 JPK T. Jefferies, 98 St. Johns Road, Frome, BA11 2BD
G1 JPP A. Hawes, 25 Folly Close, Fleet, GU52 7LN
G1 JPT B. Gleave, 1 Fearnley Way, Newton-le-Willows, WA12 8SQ
G1 JQK S. Gibbs, 43 Reddish Vale Road, Stockport, SK5 7EU
GI1 JQP J. Innes, 22 Ashley Lodge, Dunmurry, Belfast, BT17 0AF
G1 JQR D. Sell, 17 Auriel Avenue, Dagenham, RM10 8BS
G1 JRD J. Barber, 7 Ash Green, Canewdon, Rochford, SS4 3QN
G1 JRF D. Bruckshaw, 18 Old Moat Drive, Northfield, Birmingham, B31 2LY
G1 JRL H. Cook, 31 Butley Road, Felixstowe, IP11 2NY
G1 JRP C. Davis, 5 Redwing Avenue, Chippenham, SN14 6XJ
G1 JRR R. Chalker, 182 Bridge Road, Chessington, KT9 2EY
G1 JRU D. Evans, 63 Malwood Road West, Hythe, Southampton, SO45 5DL
G1 JRW D. Gilchrist, 204 Great West Road, Heston, Hounslow, TW5 9AW
G1 JRX Dr M. Girgis, Rozel, Wilson Road, Kidderminster, DY11 7XU
G1 JRZ G. Hobbs, 3 Glebe Cottage, Bremhill, Calne, SN11 9LD
G1 JSK P. Lees, 2 Russet Close, Braintree, CM7 1DR
G1 JSP J. Marston, 29 Ise Road, Kettering, NN15 7DU
G1 JST P. Johnson, 156 Norby Estate, Norby, Thirsk, YO7 1BQ
G1 JTC S. Baskerville, Shalimar, Grove Lane, Leeds, LS6 2AP
GM1 JTK A. Doig, 18 Gotterstone Drive, Broughty Ferry, Dundee, DD5 1QW
G1 JTM M. Ferris, 22 Route De Matha, Aigre, France, 16140
G1 JTX L. Sharman, 14 Northlands Avenue, Orpington, BR6 9LY
G1 JTZ B. Linn, 35 New Street Carcroft, Doncaster, DN6 8EH
G1 JUD R. Robinson, 24 Affleck Avenue, Radcliffe, Manchester, M26 1HN
G1 JUI M. Lister, Beaconfield, Middle Road, Poole, BH16 6HJ
G1 JUO I. Penney, 11 Eclipse Drive, Sittingbourne, ME10 2HR
G1 JUP M. Baker, 9A, Manor Road, Alton, GU34 2PF
GW1 JVB G. Evans, 2 Old Village Road, Barry, CF62 6RA
G1 JVF G. Hannan, 20 Arlington Drive, Stockport, SK2 7EB
G1 JVG D. Jewsbury, 68 Grainger Close, Basingstoke, RG22 4EA
GW1 JVH C. Kirkman, The Nant, Nantmawr, Oswestry, SY10 9HN
G1 JVL J. Leggett, 10 Home Park Road, Nuneaton, CV11 5UB
G1 JVM D. Turton, 8 Lightwoods Road, Warley, Smethwick, B67 5AY
G1 JVN F. Ursell, 110 Watt Lane, Sheffield, S10 5RE
G1 JVO C. Ursell, 110 Watt Lane, Sheffield, S10 5RE
GM1 JVU R. Aitken, 81 Rashgill, Locharbriggs, Dumfries, DG1 1QN
G1 JVY A. Smith, Woodlands, Old School Lane, Biggleswade, SG18 9JL
G1 JWD R. Osborne, 24 Brockington Road, Bodenham, Hereford, HR1 3LR
G1 JWG D. Mckay, 15 Wellington Crescent, Baughurst, Tadley, RG26 5PJ
GM1 JWJ R. Male, 13 Briar Grove, Forfar, DD8 1DQ
G1 JWL M. Warren, Flat 15, Fulton Lodge, Harrogate, HG3 2UT
G1 JWO A. Stone, 32 Berrynarbor Park, Sterridge Valley, Ilfracombe, EX34 9TA
G1 JWY T. Yorke, 12 Shanklin Drive, Weddington, Nuneaton, CV10 0BA
G1 JXA R. Whatley, 7 Okefield Road, Crediton, EX17 2DN
GI1 JXE G. Murray, Ashgrove, 61 Monteith Road, Banbridge, BT32 5RD
G1 JXG A. Mcmillan, 183 Forest Road, Clipstone Village, Mansfield, NG21 9DS
G1 JXL M. Phillips, 71 Juniper Square, Havant, PO9 1HZ
G1 JXP I. Wilson, Meadow Lodge, Kilhallon, Par, PL24 2RL
G1 JXS C. Bunkum, 7 Goose Green Close, Wolvercote, Oxford, OX2 8QT
G1 JXX H. Williams, 24 Vaughan Close, Four Oaks, Sutton Coldfield, B74 4XR
G1 JYB B. Cartledge, Oysterber Farm, Burton Road, Lower Bentham, Lancaster, LA2 7ET
G1 JYH M. Cherry, 36 Meads Avenue, Hove, BN3 8EE
G1 JYK S. Langdale, Bramley House, Dishforth Road, Sharow, Ripon, HG4 5BU
G1 JYR D. Fraley, 1334 Warwick Road, Knowle, Solihull, B93 9LQ
G1 JYZ A. Philpott, 2 Ocean View Road, Ventnor, PO38 1AA
G1 JZG M. Wilmshurst, Chalklands, Main Street, Horncastle, LN9 5PT
G1 JZL S. Beckett, 15 Peaks Avenue, New Waltham, Grimsby, DN36 4LJ
GM1 JZM D. Johnstone, 7 Gleneagles Avenue, Glenrothes, KY6 2QA
G1 JZN J. Jacklin, 26 Rockmill End, Willingham, Cambridge, CB24 5HY
G1 JZT R. Everitt, 55 Risborough Road, Bedford, MK41 9QR
G1 JZU D. Goulden, 26 Derwent Walk, Greenacres, Oldham, OL4 2DJ
G1 JZX J. Hesketh, 735 Manchester Road, Over Hulton, Bolton, BL5 1BA

G1 JZY T. Mitchell, 22 Grundy Avenue, Prestwich, Manchester, M25 9TG
G1 JZZ D. Porter, 2 Flour Mill Close, Burscough, Ormskirk, L40 5TL
G1 KAG A. Watson, 60 Beresford Avenue, Surbiton, KT5 9LJ
G1 KAK C. Buttery, Yew Tree Cottage, Chapel Lane, Newport, PO30 3DD
G1 KAO A. Hawxby, 73 Anthea Drive, Huntington, York, YO31 9DB
G1 KAR Southdown ARS, c/o A. Seabrook, 63 St. Annes Road Willingdon, Eastbourne, BN20 9NJ
G1 KAS M. Hughes, 9. Greenacres, Kirkby-in-Ashfield, NG17 7GE
G1 KAT C. Lawrence, 23 Brutus Drive, Coleshill, Birmingham, B46 1UF
G1 KBC S. Barrington, Fawley Cottage, Butt Lane, Loughborough, LE12 5EE
G1 KBE T. Bradley, 32 Laurel Gardens, Marlowe Road, Hartlepool, TS25 4NZ
G1 KBF C. Halls, 16 Stoats Close, South Molton, EX36 4JU
G1 KBG D. Arthur, Durnaford, Callington, PL17 7HP
G1 KBH P. Bloy, 29 Mannington Place, South Wootton, King's Lynn, PE30 3UD
GM1 KBJ A. Wagstaff, 2 Birnock Water, Moffat, DG10 9DY
G1 KBL M. Rack, 212 Willingham Street, Grimsby, DN32 9PY
GM1 KBZ S. Crockford11, 101 Cromwell Road, Aberdeen, AB15 4UE
G1 KCA D. Thomas, 2 Reynards Meadow, Sutton Hill, Telford, TF7 4NQ
GM1 KCH W. Curran, 10 The Laurels, Dundee, DD4 0AD
G1 KCR J. Smith, 125 De Montfort Way, Coventry, CV4 7DU
G1 KCS A. Scrutton, Ashleigh, Butt Hill, Southam, CV47 8NE
G1 KCU J. Warrington, 204 High St., Feltham, TW13 4HX
G1 KCV C. Shepherd, 18 Nichols Street Desborough, Kettering, NN14 2QU
G1 KCW H. Elleray, 34 Trent Close, Plymouth, PL3 6PB
G1 KDO D. Cattell, 22 Budmouth Avenue, Weymouth, DT3 6JW
GI1 KDS K. Lewis, 763 Antrim Road, Belfast, BT15 4EP
G1 KEB R. Farmer, 72 Bradleys Lane, Wallbrook, Bilston, WV14 8YW
G1 KEI P. Smith, 47 Bostock Road, Abingdon, OX14 1DW
G1 KEV H. Denton, 27 Melrose Gardens, Hersham, Walton-on-Thames, KT12 5HF
G1 KEW N. Coote, 36 Summerfield Road, Manchester, M22 1AF
G1 KFB T. Jesson, The Hawthorns, The Outwoods, Burbage, Hinckley, LE10 2UD
G1 KFG J. Feay, 19 Dorset Avenue, Diggle, Oldham, OL3 5PL
G1 KFH J. Richmond, 11 Elm Avenue, Pennington, Lymington, SO41 8BD
G1 KFQ P. King, 10 Hockley Lane, Eastern Green, Coventry, CV5 7FR
G1 KGA O. Himmo, 227B Caterham Drive, Coulsdon, CR5 1JS
G1 KGC P. Simpson, 100 Pitchford Avenue, Maddington, Australia, WA 6109
G1 KGE M. Phelps, Windermere, Wyson, Ludlow, SY8 4NQ
G1 KGL S. Dorrington, Rosewood, 74 Hangleton Way, Hove, BN3 8EQ
G1 KGO S. Coben, 106 Fleming Mead, Mitcham, CR4 3LW
G1 KGQ C. Buxton, 3 Goodwood Way, Mansfield, NG18 3BY
G1 KGU J. Amos, Mite View, Ravenglass, CA18 1SW
G1 KGV C. Ashcroft, Wood End House, Wood End, Huntington, PE28 3LE
GW1 KGW C. Walker, Perthi, Llaneilian, Amlwch, LL68 9LY
GI1 KGZ P. Knott, 47 Pretoria Street, Belfast, BT9 5AQ
GI1 KHF R. Maternaghan, 1 Pinegrove Crescent, Ballymena, BT43 6TL
GW1 KHH H. Mosley, 17 Cadwgan Road, Old Colwyn, Colwyn Bay, LL29 9PY
G1 KHM K. Morgan, 157 Headlands, Fenstanton, Huntingdon, PE28 9LP
G1 KHS D. Tucker, 5 Uplands Close, Hawkwell, Hockley, SS5 4DN
GM1 KHU C. Wall, 27 Golf Terrace, Insch, AB52 6JY
G1 KHY R. Staszewski, Green Rigg West Woodburn, Hexham, NE48 2SG
G1 KIB J. Martin, The Old Bakehouse, Shotteswell, Banbury, OX17 1JA
G1 KII D. Beale, 88 Long Innage, Halesowen, B63 2UY
G1 KIJ D. Marsters, 21 Cow Lane, Rampton, Cambridge, CB24 8QG
G1 KIT J. Fisher, 63 Rogers Avenue, Creswell, Worksop, S80 4JR
G1 KIW J. Moss, 42 Chantry Lane, Necton, Swaffham, PE37 8ET
G1 KIZ T. Head, 36A Ashacre Lane, Worthing, BN13 2DH
G1 KJG C. Robinson, Steinhalden 3, Muehlau, Switzerland, 5642
G1 KJH J. Beach, 8 Harvey Lane, Norwich, NR7 0BQ
G1 KJQ M. Haymes, 30 Holly Road, Blackpool, FY1 2SF
G1 KJX B. Hobbs, 33 Hawthorn Drive, Heswall, Wirral, CH61 6UP
G1 KKA P. Montgomery, 7 Birchwood Close, Tavistock, PL19 8DR
G1 KKD T. Elcock, Little Grange, 33 Cromford Drive, Derby, DE3 9JT
G1 KKE D. Rose, 6 The Holt, Mollington, Banbury, OX17 1BE
G1 KKF A. Lawes, 15 Leybourne Road, Brighton, BN2 4LT
G1 KKH P. Cunliffe, 37 Rectory Road, Worthing, BN14 7PE
GM1 KKI K. Johnston, Innisfree, Gulberwick, Shetland, ZE2 9JX
GW1 KKJ K. Taylor, 23 Vardre Avenue, Deganwy, Conwy, LL31 9UT
G1 KKS I. Gott, Tayman House, The Street, Badminton, GL9 1HH
G1 KLI M. Smith, Sycamore Farm, 6 Station Road, Spalding, PE12 0NP
G1 KLK A. Dutton, 111 St. Michaels Road, Crosby, Liverpool, L23 7UL
G1 KLP A. Jackson, 81 Suffield Way, King's Lynn, PE30 3DX
G1 KLW P. Golding, 80 Birdbrook Road, London, SE3 9QP
G1 KLZ D. Ellershaw, 38 Lakeber Avenue, Bentham, Lancaster, LA2 7JN
G1 KMJ J. Couzins, 30 Camden Road, St. Peters, Broadstairs, CT10 3DR
G1 KMN N. Thompson, 10 Belmont Crescent, Swindon, SN1 4EY
G1 KMS I. Millar, The Grange, 105 High Street, Northampton, NN3 3JX
G1 KNA J. Burdett, Glencorse, 13 Fairfax Avenue, Selby, YO8 4AZ
G1 KNI S. Martin, 6 Prinsted Walk, Fareham, PO14 3AD
G1 KNK G. Mellors, 21 Church Close, Stoke St. Gregory, Taunton, TA3 6HA
G1 KNQ G. Roberts, 22 Grove St., New Balderton, Newark, NG24 3AZ
G1 KNU P. Sharp, Purbrook Cottage, Lyme Road, Axminster, EX13 5BL
G1 KNX S. Jones, 31 Church Street, Tewkesbury, GL20 5PD
G1 KNZ J. Washby, 2 Olivier Court, Council Avenue, Hull, HU4 6RW
G1 KOD J. Rodgers, 5 Bridge Avenue, Latchford, Warrington, WA4 1RJ
G1 KOG E. Beir, 17 Deansway, Hemel Hempstead, HP3 9UE
G1 KOH G. Hall, 22 Waterfield Close, Leicester, LE5 4EN
G1 KON L. Mccoy, 56 Curate Road, Anfield, Liverpool, L6 0BZ
G1 KOP Liverpool Raynet Group, c/o J. Gibbard, 2 Almond Court, Liverpool, L19 2QZ
G1 KOR A. Waddoups, 20 Stevenson Way, Wickford, SS12 9DY
G1 KOT M. Lynn, 52 Vowler Road, Langdon Hills, Basildon, SS16 6AQ
G1 KOX T. Williams, 24 Elm Tree Close, Northolt, UB5 6AR
G1 KPI T. Houghton, 24A Studley Road, Torquay, TQ1 3JN
G1 KPU D. Skilton, 137 Coast Drive, Lydd On Sea, Romney Marsh, TN29 9NS
G1 KPV A. Rayner, 147 Ramuz Drive, Westcliff-on-Sea, SS0 9JN
G1 KPZ M. Gillott, 2 Firthwood Avenue, Coal Aston, Dronfield, S18 3BQ
G1 KQD S. Bennett, Mumford Cottage, Wick Lane, Lower Apperley, Gloucester, GL19 4DS
G1 KQE H. Sweet, 5 Dence Close, Herne Bay, CT6 6BH
G1 KQH S. Wigg, 45 Cambrian Lane, Rugeley, WS15 2XH
G1 KQN A. Bowyer, 80 Holme Fen, Holme, Peterborough, PE7 3PR
G1 KQP S. Price, Whitton Paddocks, Pulley, Shrewsbury, SY3 0AG
G1 KQU G. Gray, 10A Albert Close, Rayleigh, SS6 8HP
GW1 KQV C. Caudy, 43 Graham Avenue, Pen-Y-Fai, Bridgend, CF31 4NR
GW1 KQY B. Francis, 4 Heol Tir Coch, Efail Isaf, Pontypridd, CF38 1BW
G1 KQZ A. Quinn, 581 Chorley Old Road, Bolton, BL1 6BL
G1 KRU A. Graves, 49 Robin Lane, Edgmond, Newport, TF10 8JL
G1 KRX B. Piper, 26 Hare Law Gardens, Stanley, DH9 8DG
G1 KSC S. Harrison, 44 Rosslyn Road, Whitwick, Coalville, LE67 5PT
G1 KSE A. Robinson, 2 Frome Close, Marchwood, Southampton, SO40 4SL
G1 KSH P. Sherwood, 43 Brighton Street, Penkhull, Stoke-on-Trent, ST4 7HH
G1 KSI S. Bennington, The Oaks, Lynn Road, Wisbech, PE14 7DF
G1 KSK E. Mullin, 26 Fearnhead Lane, Fearnhead, Warrington, WA2 0BE
G1 KSN V. Tankard, 55 Filching Road, Eastbourne, BN20 8SD
G1 KST J. Almond, 49 The Promenade, Withernsea, HU19 2DW
G1 KSW E. Giacani, 21 Barton Terrace, Leeds, LS11 8TP
G1 KTF D. Webb, Fairway, Barkham Road, Wokingham, RG41 4DH
G1 KTS G. Goodridge, 110 Quarrendon Road, Amersham, HP7 9EP
GW1 KTW C. Thomas, 18 Acrefield Avenue, Guilsfield, Welshpool, SY21 9PN
G1 KTY B. Rider, Rose Cottage, Coley Road, Bristol, BS40 6AP
G1 KTZ D. Robins, Ayala, Higher Road, Liskeard, PL14 5NQ
G1 KUG L. Preece, Oakdean, Crow Ash Road, Berry Hill, Coleford, GL16 7RB
GM1 KUI G. Smith, 38 Crown Cottages, Stuartfield, Peterhead, AB42 5HR
G1 KUN T. Sexton, 41 St. Bedes Gardens, Cambridge, CB1 3UF
G1 KUO G. Edwards, 40 Dent Street, Hartlepool, TS26 8AY
G1 KUQ P. Andrews, 37 Lonsdale Road, Harborne, Birmingham, B17 9QX
G1 KVC J. Norris, 3 St. Pauls Close, Adlington, Chorley, PR6 9RS
GW1 KVI D. Bedson, Crud Yr Awel, Old Llanfair Road, Harlech, LL46 2SS
G1 KVO M. Haydon, 1 Glencrofts, Hockley, SS5 4GN
G1 KVP B. Froggatt, 59 Queens Road, Rushall, Walsall, WS4 1HP
G1 KVQ A. Mallin, 88 Highbridge Road, Burnham-on-Sea, TA8 1LN
G1 KVR M. Wood, Kviabol, Sheep Pen Lane, Seaford, BN25 4QR
G1 KVW I. Wilson, 45 Meadway, Halstead, Sevenoaks, TN14 7EY
GM1 KWA B. Simpson, 2 Cowden Way, Comrie, Crieff, PH6 2NW
G1 KWF S. Watson, 1 Owl Way, Hartford, Huntingdon, PE29 1YZ
GM1 KWG J. Winterbourne, Birkenbush, Clochan, Buckie, AB56 5AL
G1 KWK M. Duerden, 12 Masefield Avenue, Bradford, BD9 6EX
GM1 KWX C. Carter, Annachmor House, Clynder, Helensburgh, G84 0QD
G1 KXJ A. Poole, 17 Adelaide Street, Stonehouse, Plymouth, PL1 3JF
G1 KXP S. Lindsay-Smith, 47 Shaftesbury Avenue, Timperley, Altrincham, WA15 7NP
G1 KXQ M. Bloxham, 34 Northcote Road, Farnborough, GU14 9EA
G1 KXX A. Shaw, 70 Field Gardens Steventon, Abingdon, OX13 6TF
G1 KXZ A. Moore, 103 Park Grove, Barnsley, S70 1QE
G1 KYK M. Partridge, Flat 2, Lymington Court Station Road, Sutton Coldfield, B73 5JY
G1 KYN D. Jackson, 37/39 Cloverhill, Sunniside, NE16 5PT
G1 KYV S. Youngs, Glenlovat, Oakley Road, Cheltenham, GL52 6NZ
G1 KZA E. Kilner, 3 Ruskin Close, Wath-Upon-Dearne, Rotherham, S63 6NU
G1 KZD T. Asker, 34 Post Office Road, Frettenham, Norwich, NR12 7AB
GM1 KZG J. Southworth, 2 School Street, New Pitsligo, Fraserburgh, AB43 6NE
G1 KZI N. Lansley, 126 The Promenade, Peacehaven, BN10 7JA
G1 LAN D. Roberts, 192 Boothferry Road, Hull, HU4 6EW
G1 LAO J. Flower, 12 Balmoral Close, Alton, GU34 1QY
G1 LAP J. Beecham, Newholme, Whitemill Lane, Stone, ST15 0EG
G1 LAR L. Rogers, 37 Gilbey Road, Tooting, London, SW17 0QQ
G1 LAT S. Kirkwood, 1 Nether View Lodge Lane, Wennington, Lancaster, LA2 8NP
G1 LAW E. Boyce, 214 London Road, Benfleet, SS7 5SJ
G1 LBH P. Tomkins, 64 Glenwood Gardens, Bedworth, CV12 8DA
GI1 LBI H. Budina, 46 Dunderg Road, Macosquin, Coleraine, BT51 4NE
G1 LBK M. Kirk, Badgers Rise, Dunley Gardens, Stourport-on-Severn, DY13 0LL
G1 LBM C. Taylor, 28 Grace Court, Dial Lane, Bristol, BS16 5UP
G1 LBU R. Parry, 84 Sulgrave Road, Washington, NE37 3BZ
G1 LCC J. Edwards, 1 Herons Way, Runcorn, WA7 1UH
G1 LCE S. Turner, Rainow Villa, Under Rainow Road, Congleton, CW12 3PL
G1 LCN G. Clegg, 16 The Pastures, Lower Westwood, Bradford-on-Avon, BA15 2BH
G1 LCR Leicester Raynet Grp, c/o D. Harrison, 7 Shirley Close, Castle Donington, Derby, DE74 2XB
G1 LCS J. Burrows, 1 Browns Avenue, Runwell, Wickford, SS11 7PT
G1 LCY R. Brown, 3 Penmare Close, Hayle, TR27 4PJ
G1 LCZ P. Milston, 47 Wellington Road, Todmorden, OL14 5EQ
G1 LDC P. Gibson, 18 Vicarage Walk, Northwich, CW9 5PS
G1 LDJ M. Ward, 23, Tricketts Lane, Ferndown, BH22 8AT
G1 LDN J. Burnet, 41 Douglas Crescent, Southampton, SO19 5JP
G1 LDY A. Smith, 7 Gladstone Road, Broughton, Chester, CH4 0RN
G1 LEC R. Gugi, 47 Witton Crescent, Darlington, DL3 0JQ
G1 LED K. Pinkard, 2 Lonsdale Court, Lache Lane, Chester, CH4 7LZ
G1 LEH M. Nicholson, 7 Ellerbeck Close, Workington, CA14 4HY
GW1 LEL A. Rowland, 86 Dee Road, Connah'S Quay, Deeside, CH5 4PA
G1 LEN L. Taylor, 35 Leafield Road, Darlington, DL1 5DF
G1 LEO J. Brittain, 26 Saxby Close, Eastbourne, BN23 7BH
G1 LES J. Buckle, 14 Alder Close, Mapplewell, Barnsley, S75 6JA
G1 LEX J. Blything, Blythwood, 319 Great Brickkiln Street, Wolverhampton, WV3 0PY
G1 LFD V. Middleton, 5 Fieldhouse, Holmfirth, HD9 1EN
G1 LFI G. Hepworth, 3 College View, Ackworth, Pontefract, WF7 7LA
G1 LFM D. Mawdsley, 3 Chapel Lane, Cronton, Widnes, WA8 4NT
GW1 LFN D. Rees, 17 Cwm Mwyn, Gorslas, Llanelli, SA14 7HY
G1 LFR D. Love, 17 Longstaff Avenue, Rawnsley, Cannock, WS12 0QE
GW1 LFX M. Lamb, 4 Hadfield Close, Connah'S Quay, Deeside, CH5 4JP
G1 LGB G. Gundry, 181 Stoneleigh Avenue, Worcester Park, KT4 8YA
G1 LGJ S. Stephens, 41 Elliotts Lane, Codsall, Wolverhampton, WV8 1PG
GI1 LGM C. Dowdall, 24 Glasmullen Road, Glenariffe, Ballymena, BT44 0QZ
G1 LGQ P. White, 25 Witton Road, Ferryhill, DL17 8QE
G1 LGY B. Bishop, 26 Robin Gardens, Totton, Southampton, SO40 8US
G1 LHD M. Goodes, 17 Ashmead Close, Lords Wood, Chatham, ME5 8NY
G1 LHE G. Courtney, 24 Greenwood Close, Bognor Regis, PO22 9DG
G1 LHL B. Mayson, 19 Hudson Close, Sturry, Canterbury, CT2 0HX
GW1 LHT G. Seldon, 28 Monterey Street, Manselton, Swansea, SA5 9PE

GW1 LHV J. O'Nions, Pant-Glas, Gegin Lane, Wrexham, LL11 3YT
G1 LIG M. Coker, 5 Penling Close, Cookham, Maidenhead, SL6 9NF
G1 LIK S. Church, 24 Lovel End, Chalfont St. Peter, Gerrards Cross, SL9 9PA
G1 LJL B. Duncan, 13 Westwick Grove, Sheffield, S8 7DP
GM1 LKD J. Craib, 10 Cameron Road, Bridge Of Don, Aberdeen, AB23 8QN
GW1 LKG Dr R. Cannon, 43 Plas St. Pol De Leon, Portway Marina, Penarth, CF64 1TR
G1 LKH R. Hilton, 8 Hogshill Lane, Cobham, KT11 2AQ
G1 LKJ P. Manning, 1 Waverley Gardens, Ash Vale, Aldershot, GU12 5JP
G1 LKK C. Wardle, 16 Tedworth Avenue, Stenson Fields, Derby, DE24 3BS
G1 LKL J. Wilkinson, 160 Staines Road, Feltham, TW14 9ED
G1 LLA C. Davis, Fourwinds, Ringwood Road, Christchurch, BH23 7BE
G1 LLI D. Hutchings, 3 Willoway Grove, Braunton, EX33 1AT
G1 LLQ P. Dollimore, 96 Hayes Bridge Ct, Uxbridge Road, Hayes, UB4 0JH
G1 LLU I. Olver, 10 Celtic Road, Deal, CT14 9EE
G1 LLW D. Rogers, 36 Guessens Road, Welwyn Garden City, AL8 6RH
G1 LLZ M. Jackson, 146 Sea Road, Chapel St. Leonards, Skegness, PE24 5RY
G1 LMC J. Trainer, 86 Plessey Road, Blyth, NE24 3HX
G1 LML J. Thornton, 7 Queens Road, Vicars Cross, Chester, CH3 5HA
G1 LMN S. Froggatt, 255 Rushton Road, Desborough, NN14 2QB
G1 LMQ G. Donachie, 36 Eastfields Narborough, King's Lynn, PE32 1SS
G1 LMS J. Honeyball, 3 Mill End Close, Warboys, Warboys, PE28 2FP
G1 LMT S. Langson, 7 John Street, Knutton, Newcastle, ST5 6DT
G1 LMU S. Hodgetts, 4 Stoke Park Court, Bishops Cleeve, Cheltenham, GL52 8US
G1 LMW D. Partridge, 44 Trumpet Terrace, Cleator, CA23 3DY
G1 LMZ G. Clennell, 69 Seventh Row, Ashington, NE63 8HX
G1 LNA D. Wood, 18 Rosemellin, Camborne, TR14 8QF
G1 LNQ P. Bury, 2 Manor Rise, Thornton In Craven, Skipton, BD23 3TP
G1 LNR L. Button, 37 Abbots Way, Preston Farm, North Shields, NE29 8LU
G1 LOE K. Gosling, 485B Blandford Road, Plymouth, PL3 6JF
G1 LOK P. Blyth, 12 Beulah Street, King's Lynn, PE30 4DN
G1 LOL G. Davidson, 18 Gotham Lane, Bunny, Nottingham, NG11 6QJ
G1 LOU L. Vaisey, Esperanza, 5 Hudson Close, Ringwood, BH24 1XL
G1 LOV P. Foster, 6 Croxteth Road, Bootle, L20 5EA
G1 LOW P. Collick, 39 Beech Rise, Sleaford, NG34 8BJ
GM1 LOZ L. Hird, 19 Alexandra Road, Keith, AB55 5BX
G1 LPQ D. Thorpe, 167 Southwell Road West, Mansfield, NG18 4HD
G1 LPS T. Roxby, 3 Coulton Terrace, Kirk Merrington, Spennymoor, DL16 7HN
G1 LQB L. Clarke, 10 Stonecliff Park Prebend Lane, Welton, Lincoln, LN2 3JS
G1 LQC P. Chambers, 26 Drummond Gardens, Christ Church Mount, Epsom, KT19 8RP
G1 LQE A. Dickinson, 5 Stone Croft, Barrowford, Nelson, BB9 6BL
G1 LQH M. Lewis, 8 Wetherdown, Herne Farm, Petersfield, GU31 4PN
G1 LQM C. Costello, The Coach House, The Green, Norwich, NR12 9PZ
G1 LQP P. Gautrey, 17 Heath Road, Market Bosworth, Nuneaton, CV13 0NT
G1 LQT C. Matthews, The Jasmine, School Hill, High Street, St Austell, PL26 7TP
G1 LQV G. Austin, 10 St. Peters Close, Ruislip, HA4 9JT
G1 LQX R. Saunders, 3 Lancaster Road, Cressex Industrial Estat, High Wycombe, HP12 3NN
G1 LRK C. Bassett, 11 Redcastle Road, Thetford, IP24 3NF
G1 LRM R. Cartmell, 16 Churchfield Drive, Wigginton, York, YO32 2FL
G1 LRU R. Jones, 8 Mullen Avenue, Downs Barn, Milton Keynes, MK14 7LU
G1 LRV Dr T. Edgar, Unit 5, 15 Church Street, Kelmscott, Australia, 6111
G1 LSB P. Brockett, 146 Winsover Road, Spalding, PE11 1HQ
G1 LSK I. Wiseman, Flat 4 Granville House, Glebe Avenue, Hunstanton, PE36 6BS
G1 LSN M. Felton, 1 Barnwell Close, Wistaston, Crewe, CW2 6TG
G1 LSX J. Humphreys, 44 Grosmont Grove, Hereford, HR2 7EG
G1 LSZ P. Lambert, 22 Cullingworth Avenue, Hull, HU6 7DD
G1 LTC E. Rodd, 32 Cranfield, Plympton, Plymouth, PL7 4PF
G1 LTE G. Eades, 117 Booths Farm Road, Birmingham, B42 2NU
G1 LTG G. Perry, 123 Green Lanes, Wylde Green, Sutton Coldfield, B73 5LT
G1 LTH D. Williams, 5 West Road, Ormesby, Great Yarmouth, NR29 3RJ
G1 LTI P. Smith, 12 Cleeves Court, Cleeves Way, Rustington, Littlehampton, BN16 3TS
G1 LTK A. Sturgess, 11 Keats Close, Earl Shilton, Goonhilly, Leicester, LE9 7DU
G1 LTL Dr D. Gardner, New House, Birdbush Avenue, Saffron Walden, CB11 4DJ
GM1 LTM U. Wallace, No 1 Holding Wester, Kincardie, Tayside, PH7 3RP
G1 LUC N. Spring, Old Orchard House, Copyhold Lane, Dorchester, DT2 9LT
G1 LUF J. Wright, 298 Field Road, Bloxwich, Walsall, WS3 3NB
G1 LUN P. Baker, 651D Puketona Road, Paihia, New Zealand, 204
G1 LUX C. Deacon, 12 Russet Way, Burnham-on-Crouch, CM0 8RB
GM1 LUZ C. Campbell, 18 Parkview Avenue, Falkirk, FK1 5JX
G1 LVH D. Barnett11, Steepholme, Front Street, South Clifton, Newark, NG23 7AA
G1 LVR P. Cousins, 38 Braunston Drive, Hayes, UB4 9RB
G1 LVV A. Critchlow, 51B West Road, Buxton, SK17 6HQ
G1 LVW P. Parker, 43 Meadow Close, Farmoor, Oxford, OX2 9PA
GD1 LVY J. Dorman, 1 Sprucewood Rise, Foxdale, Douglas, Isle Of Man, IM4 3JP
G1 LVZ F. West, 14 Ashley Drive, Twickenham, TW2 6HW
G1 LWE J. Etchells, 34 Link Avenue, Urmston, Manchester, M41 9NJ
G1 LWF T. Finch, 9 Halstead Road, Southampton, SO18 2PQ
G1 LWH B. Whitehouse, 105 Quarry Road, Birmingham, B29 5LE
G1 LWL M. Patrick, 39 Poplar Road, Healing, Grimsby, DN41 7RE
G1 LWX M. Berry, 133 Rectory Road, Ashton-In-Makerfield, Wigan, WN4 0QF
G1 LWY K. Slater, 56 Berners Road, Sheffield, S2 2GB
GM1 LXA S. Sellick, 1 Hatton Home Farm Cottages, Turriff, AB53 8ED
G1 LXK P. Branston, 19 Highfield Duddington, Stamford, PE9 3QD
GM1 LXM S. Mcintier, 2A Glenacre Drive, Largs, KA30 9BH
G1 LYX N. Bristow, 21 Dudley Street, Leighton Buzzard, LU7 1SE
G1 LZF B. Dickinson, 178 Wycliffe Gardens, Shipley, BD18 3JB
G1 LZH P. Taylor, 10 Pickenham Road, Birmingham, B14 4TG
G1 LZL D. Atkinson, 336 Moss Bay Road, Workington, CA14 5AF
G1 LZS S. Stallworthy, 6 Kenwood Close, Hastings, TN34 2AT
G1 LZZ D. Dennett, Redhill Cottage, 14 Main Road, Wareham, BH20 5RN
G1 MAC M. Macbeth, 58 The Common, Abberley, Worcester, WR6 6AY
G1 MAD Moorlands and District ARS, c/o C. Beesley, 15 Byron Close, Cheadle, Stoke-on-Trent, ST10 1XB
G1 MAL D. Simmons, 47 Lower Street, Haslemere, GU27 2NY
G1 MAR Midland ARS, c/o N. Gutteridge, 68 Max Road, Quinton, Birmingham, B32 1LB
G1 MAS M. Crooks, 19 The Nook Crookesmoor, Sheffield, S10 1EJ
G1 MAV G. Seaman, 22A Mount Road, Bexleyheath, DA6 8JS
GW1 MAX R. Pullen, 8 Carmarthen Court, Caerphilly, CF83 2TX
G1 MBE C. Batty, 32 The Warings, Heskin, Chorley, PR7 5NZ
G1 MBM M. Kitson, 54 Hollins Lane, Sowerby Bridge, HX6 2RP
G1 MBN P. Doyle, 11 Clifford Avenue, Longton, Preston, PR4 5BH
GW1 MBV J. Follett, 136 Westbourne Road, Penarth, CF64 3HH
G1 MBW M. Smith, 32 Amesbury Drive, London, E4 7PZ
G1 MCG D. Minton, 8 Rosebank Walk, Barnton, Northwich, CW8 4PU
G1 MCI A. Rawson, 43 County Road North, Hull, HU5 4HN
GM1 MCN Canon C. Stanley, Nazareth House, 34 Claremont Street, Aberdeen, AB10 6RA
G1 MCT G. Williams, 29 Coleridge Road, Barnby Dun, Doncaster, DN3 1AN
G1 MCW W. Higgins, 15 Redburn Close, Liverpool, L8 4XR
G1 MCY C. Toogood, 16 Penlea Avenue, Bridgwater, TA6 6JU
G1 MDC D. Tommey, 99 Fairfield Park Road, Bath, BA1 6JR
G1 MDE D. Harrison, 19 Oakwood Road East, Rotherham, S60 3ER
G1 MDG Chesham & District ARS, c/o M. Appleby, 6 Mandeville Road, Prestwood, Great Missenden, HP16 9DS
G1 MDJ J. Murch, Downings, Prinsted Lane, Emsworth, PO10 8HS
GM1 MDO J. Stewart, 104 Barrhill Road, Cumnock, KA18 1PU
G1 MDQ M. Mitchell, 310 Parlaunt Road, Slough, SL3 8AX
G1 MDS M. Lewis, Westbank, 46 Weyside Road, Guildford, GU1 1HX
G1 MET Dr R. Heywood, 22 Catterall Close, Blackpool, FY1 3RB
G1 MFK M. Morris, 1 Fruitlands, Malvern, WR14 4AH
G1 MGF A. Hall, 172 Aldershot Road, Guildford, GU2 8BL
GW1 MGI L. Long, 18 Pentre Poeth Road, Bassaleg, Newport, NP10 8LL
G1 MGN T. Jones, 35 Manta Road, Dosthill, Tamworth, B77 1PE
G1 MGU J. Studd, 64 Moggs Mead, Herne Farm, Petersfield, GU31 4NX
G1 MGZ M. Brophy, 78 Foley Road West, Streetly, Sutton Coldfield, B74 3NP
G1 MHA C. Corner, 100 Monkseaton Drive, Whitley Bay, NE26 3DJ
G1 MHB G. Patten, 31 Sea View Road, Skegness, PE25 1BN
G1 MHF G. Fleming, 25 Waverton Avenue, Prenton, CH43 0XB
G1 MHM W. Taylor, Corsend Farmhouse, Corsend Road, Gloucester, GL19 3BP
G1 MHN M. Dawson, Fairview, Wilkins Road, Wisbech, PE14 8DQ
G1 MHP D. Gilbert, 10 Pigeon Grove, Bracknell, RG12 8AP
G1 MIE K. Martin, 21 All Saints Close, Weybourne, Holt, NR25 7HH
GD1 MIP A. Morgan, Thal'Loo Glass, Nassau Road, Mwyljyn Moddey, Isle Of Man, IM7 4AQ
G1 MIY R. Lunnon, 9 Hennerton Way, High Wycombe, HP13 7UE
G1 MJA P. Griffin, 147 Kingshayes Road, Aldridge, Walsall, WS9 8SN
G1 MJI D. Fry, 23 South Lea, Braunton, EX33 2HN
GI1 MJJ F. Gilliland, 48 Malone Heights, Belfast, BT9 5PG
G1 MJN M. Newbold, Sawubona, Vicarage Lane, Skegness, PE24 4JJ
G1 MJO A. Hemming, 64 Haslucks Green Road, Shirley, Solihull, B90 2EJ
G1 MJT R. Naylor, 6 Alden Close, Morley, Leeds, LS27 0SG
G1 MJV N. Liddiard, Orchard End, Dunwich Lane, Peasenhall, Saxmundham, IP17 2JP
GM1 MKC C. Strong, Little Couchercairn St Katherines, Inverurie, AB51 8TQ
G1 MKE R. Cox, Carlingford, Brimley Drive, Teignmouth, TQ14 8LE
G1 MKP M. Pauley, 12 Coxs End, Over, Cambridge, CB24 5TZ
G1 MKR Milton Keynes Raynet Group, c/o J. Breen, 68 Honeysuckle Way, Bedford, MK41 0TF
G1 MKS P. Healy, 43 Brockley View, London, SE23 1SL
GW1 MKV A. Coe, Woodbrook Cottage, Paddock Row, Clwyd, LL14 6DD
G1 MKY J. Hargreaves, 14 Newlands Close Blackfield, Southampton, SO45 1WH
G1 MLC H. Taha, 53 Barbers Hill, Werrington, Peterborough, PE4 5ED
GM1 MLS A. Pearson, 1 Bridgefield, Inverbervie, Montrose, DD10 0SR
G1 MLV P. Kelsey, 573 Stannington Road, Stannington, Sheffield, S6 6AB
GM1 MLW C. Lindsay, 51 Perrays Crescent, Dumbarton, G82 5HP
GM1 MLY M. Bull, 20 Grenitote, Lochmaddy, Isle of North Uist, HS6 5BP
G1 MMA S. Butcher, 155 Crow Lane East, Newton-le-Willows, WA12 9UD
G1 MMD K. Morris, 16 Goldcrest Road, Forest Town, Mansfield, NG19 0GP
G1 MMI T. Ryder, 117 Cotefield Drive, Leighton Buzzard, LU7 3DN
GM1 MMK K. Cupples, 16 Glebe Crescent, Airdrie, ML6 7DH
G1 MMN C. Lovett, 49 Tame St. East, Walsall, WS1 3LB
G1 MMT Dr S. Quick, 21 The Glade, Ashley Heath, Ringwood, BH24 2HR
G1 MMZ A. Roffey, 32 Hertford Road, Digswell, Welwyn, AL6 0DB
GW1 MNC A. Dykes, 16 The Mercies, Porthcawl, CF36 5HN
GW1 MNU I. Jukes, 14 Beechwood Place, Narberth, SA67 7EE
G1 MNX P. Lane, 1 St. Davids Close, Lower Willingdon, Eastbourne, BN22 0UZ
G1 MNY R. Smith, 3 Florence Farm Mobile Home Park, London Road, Sevenoaks, TN15 6BP
G1 MOB Dr N. Booth, Greencotes, Warden, Hexham, NE46 4SS
G1 MOK T. Dommett, 9 Causeway Close, Woolavington, TA7 8DW
GM1 MON J. Mcqueen, Rowan Cottage Redcastle, Lunanbay, Arbroath, DD11 5SS
G1 MOS H. Whitbread, Foresters, Main Road, Martlesham, IP12 4SL
G1 MOV M. Ballantyne, 248 Calshot Road, Great Barr, Birmingham, B42 2BX
G1 MOW D. Mallin, 60 Arundel Road, Littlehampton, BN17 7DF
G1 MOZ J. Nicholson, 11 West View Rise, Huddersfield, HD1 4UR
G1 MPC M. Human, 28 Lincoln Drive, Croxley Green, Rickmansworth, WD3 3NH
G1 MPD M. Champion, 14 Meadow Court Road, Earl Shilton, Leicester, LE9 7FF
G1 MPG C. Alefs, 27 Millfields, Beckermet, CA21 2YY
G1 MPI M. Hillman, 28 Murndal Dr, Donvale, Australia, 3111
G1 MPL C. Hitcham, 27 Kirby Cane Walk, Lowestoft, NR32 3EL
G1 MPP A. Baxter, 94 Abbeyfield Drive, Fareham, PO15 5PF
G1 MPT D. Bowden, 4 Cornmill Close, Bardsey, Leeds, LS17 9EG
G1 MPU P. Brolan, 2 Mount Road, Barnet, EN4 9RL
G1 MPW S. Cooke, 21 Wealdon Close, Southwater, Horsham, RH13 9HP
GM1 MQA R. Cooke, Taigh Na Greine, Lower Bayble, Isle of Lewis, HS2 0QB
G1 MQB P. Barker, 7 Verbena Close, Nottingham, NG3 4PZ
G1 MQC Devon County Council Emergency Plng, c/o M. Newport, 9 Highbury Park, Exmouth, EX8 3EJ
G1 MQQ G. Stainton, 168 Slades Road, Golcar, Huddersfield, HD7 4JR
G1 MRC M. Johnson, 24 Stjames'S Park, Wakefield, WF1 4EU
G1 MRE P. Langford, 33 Briscoe Road, Hoddesdon, EN11 9DG
G1 MRI P. Gardiner, 12 Weston Road, Aston Clinton, Aylesbury, HP22 5EG
GM1 MRS L. Alexander, 97 Land Street, Keith, AB55 5AP

G1 MRX P. Young, 30 Badminton Road, Maidenhead, SL6 4QT
GM1 MRY W. Andrew, 11 Eddington Gardens, Chryston, Glasgow, G69 0JW
G1 MSA A. Taylor, Apartment 125, Earlsdon Park Retirement Village, Albany Road, Coventry, Cv5 6lf
G1 MSB E. Turner, 14 Lauderdale Gardens, Bushbury, Wolverhampton, WV10 8AY
G1 MSD R. Newsome, 6 Woodhouse Grove, Fartown, Huddersfield, HD2 1AS
G1 MSG M. White, 11 Beck Way, Loddon, Norwich, NR14 6UZ
G1 MSK J. Riley, Hillcrest, Norwich Road, Chedgrave, Norwich, NR14 6BQ
GM1 MSN W. Clark, 66 Winstanley Wynd, Kilwinning, KA13 6EB
GM1 MSO D. Clark, 50 Dalry Road, Kilwinning, KA13 7HE
G1 MSR J. Cockroft, 8 Harris Road, Standish, Wigan, WN6 0QR
GM1 MSS Dr I. Coates, 55 Whitehaugh Park, Peebles, EH45 9DB
G1 MSY K. Ludgate, 138 Halton Road, Runcorn, WA7 5RW
G1 MTA R. Grey, 23 Chapel Rise, Ringwood, BH24 2BL
G1 MTB S. White, 12 Morgan Close, Bexhill-on-Sea, TN39 5EQ
GW1 MTH H. Williams, 10, Moriah Street, Merthyr Tydfil, CF47 8LU
G1 MTJ S. Tickle, 14 Rothesay Drive, Crosby, Liverpool, L23 0RF
G1 MTP N. Mcneil, 3 Grosvenor Court, Water Lane, York, YO30 6PX
G1 MTU D. Ward, 27 Penzer St., Kingswinford, DY6 7AA
G1 MUC C. Shingles, 20 Spencer Close, Lingwood, Norwich, NR13 4BB
G1 MUM H. Phillips, 6 Peaks Down, Peatmoor, Swindon, SN5 5BH
GU1 MUP P. Rudd, Val Des Arquets, Les Arquets, St Pierre Du Bois, Guernsey, GY7 9HE
G1 MUQ P. Hillier, Bythan, Avenbury Lane, Bromyard, HR7 4LB
G1 MUT P. Lawrence, 2 Chapel Terrace, Station Street, Ashbourne, DE6 1DF
GM1 MUY L. Coxon, 40 Hamilton Street, Broughty Ferry, Dundee, DD5 2RE
G1 MVE P. Tither, 32 Manor Avenue, Marston, Northwich, CW9 6DS
G1 MVF S. Rasmussen, 10 The Pightle, Grafham, Huntingdon, PE28 0UU
G1 MVG Dr F. Marshall, Hartwell, Newgrounds, Fordingbridge, SP6 2LJ
G1 MVI P. Bowe, 197 Gloucester Avenue, Chelmsford, CM2 9DX
G1 MVQ S. Hamilton-Cooper, 4 Wren Close, Appleby Magna, Swadlincote, DE12 7BD
G1 MVT D. Driscoll, 25 Broom Close, Dawlish, EX7 0RP
GW1 MVZ D. Petrie, 48A Lower Quay Road, Hook, Haverfordwest, SA62 4LR
GM1 MWK J. Grieve, 10 Jubilee Court, Kirkwall, KW15 1XR
G1 MWS R. Bell, 92 Dean Drive, Wilmslow, SK9 2EY
G1 MWT M. Clancy, 34 High Meadows, Greetland, Halifax, HX4 8QF
G1 MXC K. Sampson, 40 Crisp Road, Lewes, BN7 2TX
G1 MXD G. Waldron, 55 Sheringham Road, Poole, BH12 1NS
GM1 MXE G. Schafers, Dahlsteven, Orkney, KW17 2RD
G1 MXM I. Hunt, Four Seasons, Westmarsh, Canterbury, CT3 2LP
GM1 MYF R. Jones, 46B Forest Road, Aberdeen, AB15 4BP
G1 MYM E. Emons, 18 Haig Road, Stanmore, HA7 4EP
G1 MYO F. Ross, 2 Mount Pleasant, Steeple Claydon, Buckingham, MK18 2QS
G1 MYQ P. England, Moonstones, Down Ampney, Cirencester, GL7 5QS
GM1 MYR R. Cook, 95 Old Edinburgh Road, Inverness, IV2 3HT
G1 MZD D. Barlow, 34 Mays Way, Potterspury, Towcester, NN12 7PP
G1 MZG S. Banks, 18 Sheerstock, Haddenham, Aylesbury, HP17 8EU
G1 MZH E. Schamp, 21 Beechwood Avenue, Melton Mowbray, LE13 1RT
GD1 MZJ J. Sutherland, Archallagan Park, Marown, Isle Of Man, IO9 9SU
G1 MZM M. Bignell, 53 Rosebay Avenue, Birmingham, B38 9QT
G1 MZP A. Froggatt, 16 Seagull Close Hull, Hull, HU4 6SN
G1 MZT T. Tipper, 114 Paddock Lane, Redditch, B98 7XT
G1 MZW I. Gillson, 13 Beech Green, Southcourt, Aylesbury, HP21 8JG
GM1 MZZ J. Frearson, 27 Miltonbank Crescent, Guardbridge, St Andrews, KY16 0XE
G1 NAA M. Crabtree, 23 Ava Crescent, Richmond Hill, Ontario, Canada, L4B 2X1
G1 NAB G. Rainy Brown, Old Stores Cottage, Newbury, RG20 8SE
G1 NAN A. Gateley, 2 Langmere Road, Watton, Thetford, IP25 6LG
G1 NAP J. Hopkins, 3 De Havilland Road, Upper Rissington, Cheltenham, GL54 2NZ
G1 NAQ A. Ashton, 6 Lansdowne Crescent, Darton, Barnsley, S75 5PW
G1 NAT D. Mcgowan, 7 Eccles Close, Henley Green, Coventry, CV2 1EF
G1 NAU J. Cawsey, 134 Goddard Avenue, Swindon, SN1 4HX
G1 NBK W. Winning, Plump House, Terrington, York, YO60 6QB
G1 NBO R. Mewis, 52 Princess St., Burton on Trent, DE14 2NP
G1 NBP Dr P. Allan, 24 Farm Piece, Stanford In The Vale, Faringdon, SN7 8FA
G1 NBT A. Bradley, 59 Main Road, Watnall, Nottingham, NG16 1HE
G1 NBU L. Wellbeloved, 8 Orchard Close, South Wonston, Winchester, SO21 3EY
GW1 NBW W. Morris, 17 Fairway, Port Talbot, SA12 7HG
G1 NBY Rev. R. Roeschlaub, 20 Pannatt Hill, Millom, LA18 5DB
G1 NCD P. Moss, 54 Merton Drive, Westminster Park, Chester, CH4 7PQ
G1 NCG K. Powell, 43 Mallard Close, Swindon, SN3 5JG
G1 NCK D. Bates, 71 Nicholas Crescent, Fareham, PO15 5AJ
G1 NCL C. Holmes, 16 Industrial St., Pelton, Chester le Street, DH2 1NR
G1 NCM A. Urquhart, 75 Springvale Road, Winchester, SO23 7ND
G1 NCN H. Jones, 8 Warren Close, Old Catton, Norwich, NR6 7NL
G1 NCO P. Robinson, 12 Mountain Ash Avenue, Leigh-on-Sea, SS9 4SZ
G1 NCR North Cheshire Radio Club, c/o G. Gourley, 6A Longsight Lane, Cheadle Hulme, Cheadle, SK8 6PW
G1 NDK K. Dunn, Marylands, Maidstone Road, Tonbridge, TN12 0RH
G1 NDL K. Harrison, 20 Springfield Avenue, Ashbourne, DE6 1BJ
G1 NDQ M. Taylor, 41 Hill View Gardens, Halifax, HX3 7BT
G1 NDV R. Airey, 30 White Horse Crescent, Grove, Wantage, OX12 0PY
GW1 NED H. Anderson, Penrheol Farm, Meidrim, Carmarthen, SA33 5NX
G1 NEG P. Mcgarry, 10 Douglas Avenue, Soothill, Batley, WF17 6HG
G1 NEN A. Hefford, 31 High Street, Rushton, Kettering, NN14 1RQ
GM1 NET Strathclyde R G, c/o R. Campbell, 32 Harvie Avenue Newton Mearns, Glasgow, G77 6LQ
G1 NEV D. Dawson, 4 Hawksworth Lane, Guiseley, Leeds, LS20 8HA
GM1 NEW J. Kerins, 30 Beech Avenue, Newton Mearns, Glasgow, G77 5PP
G1 NEZ B. Stiff, 4 Timberlaine Road, Pevensey Bay, Pevensey, BN24 6DE
G1 NFB D. Bagley, 38 Ashchurch Road, Tewkesbury, GL20 8BT
G1 NFE C. Duffy, 25 Redcar Avenue, Thornton-Cleveleys, FY5 2LG
G1 NFN R. Hammond, 124 Maney Hill Road, Sutton Coldfield, B72 1JU
G1 NFO B. Webb, 63 Rother Road, Rotherham, S60 2UZ
G1 NFQ R. Vivian, Flat 1, 26 Beer Road, Seaton, EX12 2PD
G1 NGE R. Nelson, Woodlands, 35 Cromer Road, Hevingham, Norwich, NR10 5QX
G1 NGI A. Jones, Rotherhithe New Road Tissington Court Flat 5, London, SE16 2AG
G1 NGL K. Roberts, 4631 Chatham Se St, Salem, USA, 97302
GW1 NGN P. Johansson, 63 Grange Road, Rhyl, LL18 4AD
G1 NGR R. Sharman, Flat 1 11 Sherbourne Road, Blackpool, fy12pw
G1 NHG Dr M. Hausler, 5 Balaton Place, Snailwell Road, Newmarket, CB8 7YP
G1 NHX P. Severn, 310 Worlds End Lane, Birmingham, B32 2SB
G1 NIC N. James, America Lodge, Ridgeway Road, Torquay, TQ1 2EU
G1 NIM R. Hill, Cama De Mingot 3 La Alcoraia, Alicante, Spain, 3699
G1 NIT M. Virtue, 50 Borthwick Park, Orton Wistow, Peterborough, PE2 6YY
G1 NIV J. Young, The Estate Office, Granary Court, Chadwell Heath, RM6 6PY
G1 NJG B. Asker, 34 Post Office Road, Frettenham, Norwich, NR12 7AB
G1 NJI R. Foster, 35 Colin Road, Barnwood, Gloucester, GL4 3JL
G1 NJV M. Wing, 27 Hill Street, Hunstanton, PE36 5BS
G1 NKF I. Spindler, Birkett House, Weeton Village, Weeton, Preston, PR4 3NB
G1 NKN Dr A. Mason, 51 Benett Drive, Hove, BN3 6UQ
G1 NKT Lt. P. Grant, 37 Glenmore Park, Dundalk, County Louth, Ireland,
G1 NKV F. Smith, 18-26 Hendon Rise, Thorneywood, Nottingham, NG3 3AN
G1 NLQ D. Arter, 18 Essex Road, Westgate-on-Sea, CT8 8AP
G1 NLS G. Borrett, 45 Yarwells Headland, Whittlesey, Peterborough, PE7 1RF
G1 NLZ R. Brown, 15 Johnson Road, Great Baddow, Chelmsford, CM2 7JL
G1 NMI C. Whitehead, 31 Makeney Road, Holbrook, Belper, DE56 0TZ
G1 NML J. Hyde, 6 Crown Green, Coventry, CV6 6FA
G1 NMN M. Durey, 71 Orchard Road, Maldon, CM9 6EW
G1 NMP N. Fenner, 22 Gowers Field, Aylesbury, HP20 2QT
G1 NMQ L. Gibbs, 45 Woolavington Hill, Woolavington, Bridgwater, TA7 8HQ
G1 NMW A. Stone, 5 Bridge Street, Cheltenham, GL51 9DQ
G1 NNA B. Lloyd, 17 Brooklands Road, Brantham, Manningtree, CO11 1RN
G1 NNB G. Lloyd, 9 Hornbeam Walk, Witham, CM8 2SZ
G1 NNF R. Kenny, 35 Broom Leys Road, Coalville, LE67 4DD
G1 NNN S. Moore, 104 Gloucester Avenue, Chelmsford, CM2 9LF
G1 NNR D. Newman, 78 Vale Road, Poole, BH14 9AU
G1 NOO I. Allgood, 53 The Avenue, Leighton Bromswold, Huntingdon, PE28 5AW
G1 NOR G. Galbraith, 44 Parker Road, Grays, RM17 5YN
G1 NOS K. Brookes, 20 School Avenue, Guide Post, Choppington, NE62 5DN
G1 NPA P. Allan, 214 Westwood Road, Sutton Coldfield, B73 6UQ
G1 NPC G. Hammond, 31 Earlsway, Macclesfield, SK11 8RJ
G1 NPI C. Badcock, 7 Heathfield Road, Chandler'S Ford, Eastleigh, SO53 5RP
G1 NPJ K. Hyslop, Smallgains Marina Prout Industrial Estate, Point Road, SS8 7TJ
G1 NPN K. Mcdougal, 51 Argyll Avenue, Wirral, CH62 8EB
G1 NPX H. Nurse, 46 Kelvedon Road Coggeshall, Colchester, CO6 1RQ
G1 NQB C. Carpenter, 10 D'Arcy Road, St. Osyth, Clacton-on-Sea, CO16 8QE
G1 NQH M. Cook, Brooksdie Cottage, Brook Lane, Market Harborough, LE16 8SJ
G1 NQN K. Benfold, 56 Cornwall Avenue, Blackpool, FY2 9QW
G1 NQO J. Jacques, 65 Daggers Hall Lane, Marton, Blackpool, FY4 4AX
G1 NQU M. Peacock, 19 Ashfield Terrace, Haworth, Keighley, BD22 8PL
G1 NRE D. Paul, 99 Wilkinson Road, Bedford, MK42 7FR
G1 NRF M. Halloway, 41 Trenoweth Estate North Country, North Country, Redruth, TR164AQ
G1 NRG Northants Raynet Group, c/o S. Manning, 11 Broomhill Crescent, Southfields, Northampton, NN3 5BH
G1 NRK P. Slater, 12A Apsley Close, Bishop's Stortford, CM23 3PX
G1 NRM A. Harrison, 34 Marsh Lane, Mill Hill, London, NW7 4QP
G1 NRN K. Johnson, 98 Wroxham Road, Great Sankey, Warrington, WA5 3NU
G1 NRX P. Hill, 13 Onslow Road, Newent, GL18 1TL
G1 NRY I. Leach, 36 Harrowden, Bradville, Milton Keynes, MK13 7DA
G1 NSB R. Winterburn, Flat 15, Elms Farm, Mather Avenue, Manchester, M45 8NT
G1 NSD D. Young, 13 Crawshaw Park, Pudsey, LS28 7EP
G1 NSG H. Goodwin, 91 Grange Lane, Sutton Coldfield, B75 5LD
G1 NSK F. Godfrey, 50 Eskdale Drive, Worksop, S81 7QB
G1 NSQ C. Joyce, 70 Campbell Road, Twickenham, TW2 5BY
G1 NST S. Dodd, Chedburgh House, Hall Lane, North Walsham, NR28 0RZ
G1 NSV D. Kemplen, 2 Vicarage Close, Menheniot, Liskeard, PL14 3QG
G1 NTI B. Longstaff, 23 Chester Road Estate, Stanley, DH9 0QD
G1 NTJ P. Worden, 86 Buckwood Road Markyate, St Albans, AL3 8JB
G1 NTK P. Hull, Hazelwood, 2 Cheats Road, Taunton, TA3 5JW
G1 NTL J. Mcshane, 12 Virginia Gardens, Middlesbrough, TS5 8BT
G1 NTN A. Edwards, 9 Lincoln Close, Woodley, Romsey, SO51 7TJ
G1 NTP B. Haden, 72 Charlton Road, Blackheath, London, SE3 8TT
G1 NTR S. Mayer, 41 Lowe Street, Macclesfield, SK11 7NJ
G1 NTV D. Merry, 18 Tremabe Park, Dobwalls, Liskeard, PL14 6JS
G1 NTX S. Taylor, 5 Collingwood, Farnborough, GU14 6LX
G1 NUH R. Hardwick, 5 Seaview, Oakmere Park, Little Neston, CH64 0XP
G1 NUO E. Oram, 31 Nathaniel Walk, Tring, HP23 5DG
G1 NUS J. Thornley, 270 Hurdsfield Road, Macclesfield, SK10 2PN
G1 NVE V. Wood, 175 Windleshaw Road Dentons Green, St Helens, WA10 6TP
G1 NVL G. Officer, Flat 5, 8 Charlton Drive, Sale, M33 2BJ
G1 NVN A. Parkin, 1 Dunelm Walk, Leadgate, Consett, DH8 7QT
G1 NVO B. Kimber, 27 Court Road, Brockworth, Gloucester, GL3 4ES
G1 NVS R. Pennington, 5 Park Close, Northway, Tewkesbury, GL20 8RB
G1 NVV M. Timlett, Ashley Lodge, Church Lane, Strubby, Alford, LN13 0LR
G1 NVY K. Peers, 47 Walpole Avenue, Whiston, Prescot, L35 2XX
G1 NWA C. Rickerby, Brownsea, 113 Cliftonville Road, Woolston, Warrington, WA1 4BJ
GW1 NWF R. Gray, 36 Heol Pentre Felen, Morriston, Swansea, SA6 6BY
G1 NWG C. Scates, 17 Trecastle Way, Carleton Road, London, N7 0EL
G1 NWH H. Walker, 24 Castleton Avenue, Riddings, Alfreton, DE55 4AG
G1 NWM J. Westwood, 9 Landbeach Road, Milton, Cambridge, CB24 6DA
G1 NWO J. Sharp, Hunrosa, Crowlas, Penzance, TR20 8DS
G1 NWT B. Worviell, Cliddesden, The St., Shaftesbury, SP7 9PF
G1 NWZ M. Spacey, 4 Hitchman Court, Copenhagen Close, Luton, LU3 3TW
G1 NXB G. Fardoe, 3 Park Avenue, Wallasey, CH44 9DZ
G1 NXI C. Foster, 203 Hempshill Lane, Nottingham, NG6 8PF
G1 NXR V. Barrett, 4 Alexandra Street, Heywood, OL10 2AU
G1 NXS B. Martlew, 15 Dunscar Close, Birchwood, Warrington, WA3 7LS
G1 NXT K. Rushton, 9 Laburnum Avenue, Woolston, Warrington, WA1 4NY
G1 NYI T. Rozier, 26 Watersmeet Way, London, SE28 8PU
G1 NYJ D. Hopton, 32 Braemar Avenue, Urmston, Manchester, M41 6HP
G1 NYP M. Fleet, 152 Bridge Road, Chessington, KT9 2EY
GW1 NYS R. Parkhurst, 13 Tasker Way, Haverfordwest, SA611FB
G1 NYZ D. Robinson, 4 Rushden Drive, Reading, RG2 8LJ
G1 NZD P. Bates, 132 Brownings Avenue, Chelmsford, CM1 4HJ

GW1 NZF R. Stuckey, 8 Gelli Crossing, Gelli, Pentre, CF41 7UD
G1 NZH J. Edgecock, 13 Holmsdale Close, Durgates, Wadhurst, TN5 6UT
G1 NZK A. Davis, 73A Milton Road, Taunton, TA1 2JQ
G1 NZL C. Elliott, 7 Elizabeth Diamond Gardens, South Shields, NE33 5HX
G1 NZN J. Rolley, 12 Ravenscroft Drive, Chaddesden, Derby, DE21 6NX
G1 NZP E. Elliott, 7 Red House Road, Hebburn, NE31 2XS
G1 NZQ R. Finch, 12 Simcox Street, Hednesford, Cannock, WS12 1BG
G1 NZZ R. Nicol, 37 Thicknall Drive, Stourbridge, DY9 0YH
G1 OAE R. Steel11, 7 Derwent Bank, Seaton, Workington, CA14 1EE
G1 OAM T. Wilson, 15 Whipperley Way, Luton, LU1 5LB
G1 OAR P. Wallace, 7 Trinity View, Ketley Bank, Telford, TF2 0DX
G1 OAU V. English, 29E High Street, Eye, Peterborough, PE6 7UP
G1 OAW J. Smith, 62 Elson Lane, Elson, Gosport, PO12 4EU
G1 OAX L. Pugh, 15 Didcott Way, Appleby Magna, Swadlincote, DE12 7AS
G1 OAZ G. Sugden, 247 Yorkland Avenue, Welling, DA16 2LH
G1 OBA I. Bpophy, 78 Foley Road West, Streetly, Sutton Coldfield, B74 3NP
G1 OBC I. Evans, 6 Park End, Lichfield, WS14 9US
G1 OBM J. Miller, 141 Discovery Road Mount Wise, Plymouth, PL1 4PR
G1 OCH C. Hillman, Mayjon, Crow, Ringwood, BH24 3ER
G1 OCK R. Liepziger, 21 Third Avenue, Woodside Park, Poulton-le-Fylde, FY6 0PW
G1 OCL D. Potter, 9 Beachcroft Place, Lancing, BN15 8JN
GM1 OCR N. Law, Kilbrannan, Benderloch, Oban, PA37 1QU
G1 OCS A. Heap, 56 Moorside Road, Eccleshill, Bradford, BD2 3RB
G1 OCY K. Minihane, 34 Wavell Road, Gosport, PO13 0XR
G1 ODB D. Price, 21 Orchard Road, Nailsea, Bristol, BS48 2DZ
G1 ODD B. Westlake, 47 Quarry Road, Kingswood, Bristol, BS15 8NZ
G1 ODE G. Wheeler, 14 Mina Road, Bristol, BS2 9TB
G1 ODJ T. Boycott, 17 Brook Street, Whitley Bay, NE26 1AF
G1 ODK G. Cole, 87 Chichester Road, Ramsgate, CT12 6NZ
G1 ODN J. Thompson, 14 Redwing Close, Horsham, RH13 5PE
G1 ODQ C. Bond, 6 Copse Close, Hugglescote, Coalville, LE67 2GL
G1 ODT A. Pargeter, Acres Cottage, Smallburn Road, Longhorsley, Morpeth, NE65 8QH
G1 ODZ G. Fountain, Tinkers Lane, Wigginton, Tring, HP23 6JB
G1 OEB A. Orchard, Flat No 3, 35 The High Street, Hemel Hempstead, HP3 0HG
G1 OEF Dr E. Earland, 7 Paxford House Square, Ottery St Mary, EX11 1BX
G1 OEM T. Webb, 95 Devereaux Crescent, Ebley, Stroud, GL5 4PX
G1 OEP J. Harris, 109 Hook Rise South, Surbiton, KT6 7NA
G1 OEQ D. Tribute, 'Pathey', Lower Polstain Road, Truro, TR3 6BQ
G1 OER A. Parkin, 4 Waverley Road, Farnborough, GU14 7EY
G1 OET K. Doswell, 15A Queen Street, Desborough, Kettering, NN14 2RE
G1 OFG P. Howard, Cork Farm, Ruthern Bridge, Bodmin, PL30 5LU
G1 OFL R. Hall-Osman, 67 Livingstone Road, Gravesend, DA12 5DN
G1 OFW S. Gadsby, 30 Woodside Close, Knaphill, Woking, GU21 2DD
G1 OFX W. Coates, 3 Graysmead, Sible Hedingham, Halstead, CO9 3NX
G1 OFY Capt. P. Hendy, 4 The Pack, Burgh-By-Sands, Carlisle, CA5 6BE
G1 OGB P. Campion, 2 Woodside, Plymouth, PL4 8QE
G1 OGC J. Kelly, 18 Mount Pleasant, Riddings, Alfreton, DE55 4BL
G1 OGE A. Wilson, Moor Cottage, Ellastone Road, Stoke-on-Trent, ST10 3ER
G1 OGH K. Atkinson, 62 Fines Park, Stanley, DH9 8QY
G1 OGR R. Welch, 2 Broadlands Avenue, Waterlooville, PO7 7JE
G1 OGV R. Bicknell-Thompson, 4 Linden Court, Greenfrith Drive, Tonbridge, TN10 3LW
G1 OGY D. Gilligan, Two Worlds, Tan Lane, Clacton-on-Sea, CO16 9PS
GM1 OGZ D. Madden, Flat 2/2, 24 Collier Street, Johnstone, PA5 8AR
G1 OHD B. Beswick, 2 Ferndale Road, Peak Dale, Buxton, SK17 8AY
G1 OHH S. Griffin, 6 Raygill Place, Lancaster, LA1 2UQ
G1 OHL Dr W. Morden, Apartment 3, Fernside House 49 Hollington Park Road, St Leonards-on-Sea, TN38 0SE
G1 OHU J. Freeman, 81 West Hill, Kimberworth, Rotherham, S61 2EX
G1 OHV R. De Havilland, 11 Morvale Street, Stourbridge, DY9 8DE
G1 OHX V. Pears, 10 Fremantle Road, South Shields, NE34 7RF
GW1 OIB R. Jones, 10 Ferndale Crescent, Gobowen, Oswestry, SY11 3PJ
GW1 OII S. Lloyd, 10 Park Crescent, Llanelli, SA15 3AE
GW1 OIK W. Jaggard, 2 Aled Drive, Rhos On Sea, Colwyn Bay, LL28 4UU
G1 OIO W. Benton, 2 Regents Close, Seaford, BN25 2EB
G1 OIS N. Ellis, 1A Northcote Road, Croydon, CR0 2HX
G1 OIZ R. Young, 1 Croft Walk, Whitwell, Worksop, S80 4UD
G1 OJB P. Critchley, 4 Shandon Avenue, Northenden, Manchester, M22 4DP
G1 OJD D. Facer, 7 Lowry Close, Bedworth, CV12 8DG
G1 OJL S. Evenden, 11 Chapel Street, Tavistock, PL19 8DX
G1 OJO D. Pearson, 16 Hilldown Road, Hayes, Bromley, BR2 7HX
G1 OJQ P. Brooks, 7 Ashbourne Road, Underwood, Nottingham, NG16 5EH
G1 OJS Dr A. Robinson, 3 Clifton Mews, Fareham, PO168TY
G1 OJT R. Barnish, 64 Braithwell Road, Maltby, Rotherham, S66 8JU
G1 OKB A. Ibbotson, 62 Crag View Crescent, Oughtibridge, Sheffield, S35 0GD
G1 OKF D. Cannon, 44 Grange Bottom, Royston, SG8 9UQ
G1 OKI D. Hyde, 108 St. Bedes Crescent, Cambridge, CB1 3UB
G1 OKP R. Lloyd, 52 Roman Way, Ross-on-Wye, HR9 5RL
G1 OKV G. Hendricks, 105 Hillcrest Park, Wilbury Hills Road, Letchworth Garden City, SG6 4LF
GW1 OKY J. Cottrell, Bryn Dewi, Llanallgo, Moelfre, LL72 8HB
G1 OLE D. Bates, 10 Upton Gardens, Worthing, BN13 1DA
G1 OLM J. Hesketh, 87 Condor Grove, Blackpool, FY1 5NA
G1 OLQ W. Dacey, 69 Freshfield Gardens, Allerton, Bradford, BD15 7PR
G1 OLT C. Sawyer, 4 Padley Close, Ripley, DE5 3FG
G1 OLY O. Brooks, 34 Jubilee Road, Stokenchurch, High Wycombe, HP14 3SJ
GI1 OMD T. Mcquaid, 5 Edenamohill, Drumkeen, Enniskillen, BT93 0FQ
G1 OMI J. Canning, 130 Main Road, Duston, Northampton, NN5 6RA
G1 OMX P. Cumiskey, 1 York Terrace, Gateshead, NE10 9NB
G1 OMY D. Ainscough, 11 Tressel Drive, Sutton Manor, St Helens, WA9 4BS
G1 ONC P. Maitland, 7 Spinners Court, Stalham, Norwich, NR12 9EQ
G1 OND B. Hastry, 56 Kilsyth Close, Fearnhead, Warrington, WA2 0SQ
G1 ONE Bolton Wireless Club, c/o D. Lewis, 10 Addington Road, Bolton, BL3 4QZ
G1 ONH F. Slater, 32 Winthorpe Avenue, Morecambe, LA4 4RE
G1 ONJ D. Tennant, 128 Devonshire St., Keighley, BD21 2QJ
G1 ONK L. Boston, Lissa Park Sitesi No 16 (86/R), 965 Sokak Mustafa Kemal Bulvari, Calis, Turkey, FETHIYE 48300
G1 ONQ F. Pearce, 1B Council Street, Bozeat, Wellingborough, NN29 7LS
G1 ONV R. Bonar, 6 Harepark, Allerford, Minehead, TA24 8HL
G1 OOB J. Clark, The Bungalow, Sutton Lane, Walton, Street, BA16 9RJ
G1 OOG A. Cooper, Riverfield, Creek View Avenue, Hockley, SS5 6LU
G1 OOJ M. Watson, 15 Ellis Park, St. Georges, Weston-Super-Mare, BS22 7FA
G1 OOM B. Woodcock, 27 Main Street, Cosby, Leicester, LE9 1UW
G1 OOS E. Rowthorn, 4 Woburn Court, Rushden, NN10 9HL
G1 OOU J. Pearce, 19 Cawthorne Road, Kettlethorpe, Wakefield, WF2 7HW
G1 OOW D. Streeter, 78 Stockfield Road, Acocks Green, Birmingham, B27 6BB
G1 OOZ D. Parsons, 1 Carlyle Road, Rowley Regis, B65 9BQ
G1 OPA D. Smith, 15 Billington Close, Coventry, CV2 5NQ
G1 OPD P. Elsom, 7B Church Lane, Keelby, Grimsby, DN41 8ED
G1 OPG K. Chappell, 21 Victoria Street, Long Eaton, Nottingham, NG10 3EW
G1 OPJ A. Golding Brown, 17 Main Street, Withybrook, Coventry, CV7 9LT
GM1 OPO G. Askew, 49 Kittlegairy Road, Peebles, EH45 9LX
G1 OPT P. Harvey, 4 Linden Grove, Teddington, TW11 8LT
G1 OPV P. Drew, 20 Russell St., Accrington, BB5 2NF
G1 OPW F. Cox, 44 Mountain Wood, Bathford, Bath, BA1 7SB
G1 OQB R. Tams, 7 Hermitage Road, Abingdon, OX14 5RN
G1 OQF C. Caines, 6 Abel Smith Gardens, Branston, Lincoln, LN4 1NN
G1 OQG D. Fryer, 16 Elston Place, Aldershot, GU124HY
G1 OQI C. Kill, 169 Spring Road, Southampton, SO19 2NU
G1 OQM M. Palmer, 250 Kinson Road, East Howe, Bournemouth, BH10 5EP
G1 OQO M. Williamson, Greenfields Farm, Plumley Moor Road, Knutsford, WA16 9SB
GM1 OQT J. Watson, 64 Anstruther Street Law, Carluke, ML8 5JG
G1 OQU A. Windsor, 23 Pear Tree Way, Oakridge, Basingstoke, RG21 5QE
G1 OQV J. Connor, 28 Church Street, Hungerford, RG17 0JE
G1 OQW A. Boot, 63 Hunters Way, Stoke-on-Trent, ST4 5EF
G1 OQX M. Dunham, 5 King Street, Wimblington, March, PE15 0QF
G1 ORB D. Tomsett, 20 North Avenue, Bognor Regis, PO22 6HG
G1 ORC Oldham Am Rad C, c/o G. Oliver, 158 High Barn St., Royton, Oldham, OL2 6RW
G1 ORG D. Stanley, 25 Kingsley Crescent, Bulkington, Bedworth, CV12 9PS
G1 ORK D. Spicer, 35 Strood Road, St Leonards-on-Sea, TN37 6PN
G1 ORL C. Barlow, 16 Fosseway South, Midsomer Norton, Radstock, BA3 4AN
G1 ORN F. Daniels, 6 Middlemead, Stratton-On-The-Fosse, Radstock, BA3 4QH
GW1 ORP J. Marlow, West Bulthy, Bulthy, Welshpool, SY21 8ER
G1 ORS B. Williams, 3 Welton Close, Wilmslow, SK9 6HD
G1 ORT Rev. S. Smale, The Old Vicarage, 68 Cardigan Road, Bridlington, YO15 3JT
G1 OSA A. Sleigh, 2 Rock Terrace, Buxton, SK17 6HN
G1 OSE R. Howlett, 37 Waveney Drive, Hoveton, Norwich, NR12 8DP
G1 OSG W. Beilby, 119 Beaconsfield, Withernsea, Nth Humberside, HU19 2EW
G1 OSH G. Slater, 12A Apsley Close, Bishop's Stortford, CM23 3PX
G1 OSI J. Nicholson, 117 Lower Meadow, Harlow, CM18 7RF
G1 OSJ M. Howard, 8 Abbotts Crescent, St. Ives, Huntingdon, PE17 6YB
G1 OSL G. Hunter, 2 Dilloway Street, St Helens, WA10 4LN
G1 OSO A. Durbridge, 16 Nightingale Drive, Mytchett, Camberley, GU16 6BZ
G1 OSP J. Woollons, 28 Columbus Ravine, Scarborough, YO12 7JT
GM1 OST T. Ferguson, 40 Dallowie Road, Patna, Ayr, KA6 7ND
G1 OTA D. Lee, 25 Elm View, Steeton, Keighley, BD20 6SZ
GW1 OTI K. Nickson, 25 Burntwood Road, Buckley, CH7 3EL
G1 OTN R. Weight, 1 Crowland Road, Thornton Heath, CR7 8RP
G1 OTZ J. Macdonald, 42 Lion Lane, Haslemere, GU27 1JD
G1 OUA H. Saunders, 8 Norfolk Road, Luton, LU2 0RE
G1 OUG K. Anderson, 9 Bradford Park Drive, Bolton, BL2 1PA
GW1 OUP D. George, 24 Ty Fry Close, Brynmenyn, Bridgend, CF32 8YB
G1 OUX P. Bolderson, 113 Kirkdale Crescent, Leeds, LS12 6AY
G1 OUY T. Willans, 3 Highfield, Hatton Park, Warwick, CV35 7TQ
G1 OVG T. Powell, 11 Wymering Lane, Portsmouth, PO6 3QT
G1 OVH N. Waud, 32 Wellsfield, Huntingdon, PE29 1LW
G1 OVK P. Beard, 33 Sanctuary Close, Worcester, WR2 5PY
G1 OVO R. Cox, 60 Prospect Crescent, Whitton, Twickenham, TW2 7EA
GM1 OVW R. Hetherington, 37 Brockwood Avenue, Penicuik, EH26 9AN
G1 OVY P. Mcclelland, 30 Bowyer Road, Abingdon, OX14 2EP
G1 OWD M. Forsyth, 13 Hillside Close, Paulton, Bristol, BS39 7PN
G1 OWI M. Branch, 38 Kynaston Road, Didcot, OX11 8HD
G1 OWJ A. Copsey, 13 Monro Avenue, Crownhill, Milton Keynes, MK8 0BB
G1 OWK G. Garner, 8 Lansdowne Road, Swadlincote, DE11 9DZ
G1 OWM G. Woodley, 16 Albert St., St. Barnabas, Oxford, OX2 6AY
G1 OWZ J. Scott, 10 Beethoven Close Old Farm Park, Milton Keynes, MK7 8PL
GM1 OXB D. Owen, Ordiga, Wellheads, Clochan, Buckie, AB56 5HB
G1 OXF R. Lewis, 17 Hollow Lane, Hayling Island, PO11 9AA
G1 OXH A. Price, 10 Low Meadow, Whaley Bridge, High Peak, SK23 7AY
GW1 OXJ I. Jones, 15 Victoria Road, Penygroes, Caernarfon, LL54 6HD
G1 OXO J. Ilston, 6 Dovedale, Canvey Island, SS8 8HX
GM1 OXQ I. Mckune, 16 Queensberry Court, Dumfries, DG1 1BT
G1 OXT R. Faulkner, 54 Clegg Hill Drive, Sutton-in-Ashfield, NG17 2QA
G1 OYF V. Shirley, 18 Crotch Crescent, Marston, Oxford, OX3 0JJ
G1 OYG D. Crowe, 18 Bengairn Avenue, Patcham, Brighton, BN1 8RH
G1 OYH H. Grinter, 16 Gladiolus Road, Langport, TA109TA
G1 OYM A. Ingram, 78 Kenwood Gardens, Gants Hill, Ilford, IG2 6YG
G1 OYU B. Toon, 2 Marstonlane Park, Rolleston On Dove, Staffs, DE13 9BJ
G1 OYZ P. Smith, 189 Rolleston Road, Burton-on-Trent, DE13 0LD
G1 OZB A. Saunders, 4, Severn Way, Bewdley, DY12 2JQ
G1 OZD Dr J. Anderson, 179 Rolleston Road, Burton-on-Trent, DE13 0LD
G1 OZG D. Lewis, 768 Rochdale Road Middleton, Manchester, M24 2RF
G1 OZR I. Thomson, 2 Casaubon Close, Dereham, NR19 1EG
G1 OZV M. Jolly, The Oaks, 6 Gwealhellis Warren, Helston, TR13 8PQ
GW1 OZW B. Donovan, Henysgol, Drope Road, Cardiff, CF5 6EP
G1 PAF P. Foster, 3 The Greenway, Ickenham, Uxbridge, UB10 8LS
G1 PAK C. Parker, 6 Chilham Close, Hemel Hempstead, HP2 4UG
G1 PAT P. Chapman, 24 Broad Lane, Moulton, Spalding, PE12 6PN
GW1 PAV P. Grey, 4 Lon Carreg Bica, Birchgrove, Swansea, SA7 9QH
G1 PBB D. Smith, 4 Field Rose Court Adlington, Chorley, PR6 9SS
G1 PBF D. Pearce, 32 Marshall Road, Willenhall, WV13 3PB
G1 PBX E. Churchill, 87 Bradley Crescent, Shirehampton, Bristol, BS11 9SR
G1 PBY A. Parrott, 54 Dockin Hill Road, Doncaster, DN1 2QU
G1 PCA R. Blandford, 16B Sherwood Road, Keynsham, Bristol, BS31 1DB

GW1 PCD P. Dicken, Trosgol, Deiniolen, Caernarfon, LL55 3LU
G1 PCG J. Dunwell, 8 Violet Grove, Thatcham, RG18 4DQ
G1 PCN E. Benzie, 6 Priors Park, Emerson Valley, Milton Keynes, MK4 2BT
G1 PCQ A. Popplewell, 38 Welbeck Street, Hull, HU5 3SQ
G1 PCR C. Carter, 80 Cranbrook Drive, Maidenhead, SL6 6SS
G1 PCU C. Mather, 5 Knolles Road, Cowley, Oxford, OX4 3HT
G1 PDA Dr E. Evans, 4 Zig Zag Road, Wallasey, CH45 7NZ
G1 PDS S. Cliffe, 24 Dalehurst Road, Bexhill-on-Sea, TN39 4BN
G1 PEE S. Jackson, 71 Slyne Road, Bolton Le Sands, Carnforth, LA5 8AQ
G1 PEI L. Berridge, 33 Wesley Drive, Weston-Super-Mare, BS22 7TJ
G1 PEK M. Sutton, 17 Barton Close, Witchford, Ely, CB6 2HS
GM1 PEL B. Taynton, 32 Broomhall Road, Edinburgh, EH12 7PD
G1 PER M. Payne, 14 Linacres Drive, Chellaston, Derby, DE73 6XH
G1 PEU G. Gibbons, 43 Buckland Avenue, Basingstoke, RG22 6JA
GW1 PFK A. Romano, The Glen, Glen Road, Swansea, SA3 5QJ
GM1 PFU C. Hewlett, 6 Glenturret Terrace, Perth, PH2 0AR
G1 PFY J. Gold, 6 Woodland Avenue, Bournemouth, BH5 2DJ
G1 PFZ W. Gill, 2 Rufford Court, Rufford Avenue, Leeds, LS19 7ED
G1 PGD K. Biggs, 30 Elder Lane, Burntwood, WS7 9BT
G1 PGH J. Baker, 2882 Rikkard Dr, Thousand Oaks, USA, 91362
G1 PGI C. Smith, The Laurels, Maple Court, Rodmersham, Sittingbourne, ME9 0LR
G1 PGJ D. Castle, 8 Woodhall Court, Welwyn Garden City, AL7 3TD
G1 PGN D. Brooks, 10 Elmtree Road, Ruskington, Sleaford, NG34 9BT
GM1 PGP R. Barbour, 25/27 Drove Road, Langholm, DG13 0JW
G1 PGQ R. Pennycook, 28 Marine Court, Southsea, PO4 9QU
G1 PGS D. Hands, 45 Croft Avenue, West Wickham, BR4 0QH
G1 PGV J. Davidson, 5 Hanover Parc, Indian Queens, St Columb, TR9 6ER
G1 PGX M. Bingham, 6 Bittern Close, Hull, HU4 6SQ
G1 PHA G. Day, 102 Meadlands Drive, Ham, Richmond, TW10 7ED
GM1 PHD Dr N. Muir, 25 Drylaw House Gardens, Edinburgh, EH4 2UE
G1 PHJ H. Johnson, 27 Ridgeway Avenue, Gravesend, DA12 5BD
G1 PHK R. Baines, 319 Pontefract Road, Featherstone, Pontefract, WF7 5AB
G1 PHN M. Morley, 8 The Becks, Alvechurch, Birmingham, B48 7NE
G1 PHS P. Street, 17 Roxby Gardens Thornton-Le-Dale, Pickering, YO18 7SR
G1 PHU B. Burton, Natson, Tedburn St. Mary, Exeter, EX6 6ET
G1 PHV C. Marsh, 85 Cromwell Crescent, Market Harborough, Le169jw
G1 PIF D. Rogers, 20 Chapel Close, Acomb, Hexham, NE46 4RX
GW1 PIH H. Owen, Llys Gwynedd, Bethel, Caernarfon, LL55 1YB
G1 PII G. Cooper, 39 Church Road, Harlington, Dunstable, LU5 6LE
G1 PIX R. Bibby, 40 Morval Crescent, Runcorn, WA7 2QS
G1 PIY F. Cholerton, 17 Stringer Crescent, Warrington, WA4 1QN
G1 PJB T. Henderson, 8 Sandford House, Sandford, BH20 7DH
G1 PJC G. Siarey, 23 Celsus Grove, Swindon, SN1 4GE
G1 PJI S. Scott, 11A Lodge Crescent, Orpington, BR6 0QE
G1 PJJ S. Turvey, 5 Ingestre Street, Harwich, CO12 3JA
G1 PJK J. Foster, 15 Parklands Way, Liverpool, L22 3YX
GW1 PJL Dr D. Brookfield, The Barns, Milwr, Holywell, CH8 8HE
G1 PJM P. Mitchell, 11 Wingle Tye Road, Burgess Hill, RH15 9HR
G1 PJO K. Thompson, 13 Kirby Walk, Bretton, Peterborough, PE3 9UD
GW1 PJP P. Probert, 7 Albany Road, Blackwood, NP12 1DZ
G1 PJR J. Garrett, 2 Wantsume Lees, Sandwich, CT13 9JF
G1 PJT R. Wood, Lynwood, Halley Road, Heathfield, TN21 8TG
G1 PJV N. Saunders, 24 Gateland Close, Haxby, York, YO32 2ZZ
G1 PJZ J. Rogers, 55 York Road, Driffield, YO25 5AY
GM1 PKB W. Graham, 98 Dalswinton Avenue, Dumfries, DG2 9NR
G1 PKG A. Stockton, 190 Sommerfield Road, Woodgate, Birmingham, B32 3TA
GW1 PKM P. Miller, Ddaugae Farm, Gwrhyd Road, Swansea, SA9 2RY
GM1 PKN D. Gillies, 10 Killeonan, Campbeltown, PA28 6PL
G1 PKO C. Jones, 52 The Drive, Bury, BL9 5DL
G1 PKP M. Lloyd, 243 Stand Lane, Radcliffe, Manchester, M26 1JA
G1 PKR D. Bannister, 11 Keats Drive, Swadlincote, DE11 0DS
G1 PKV J. Sennitt, 44 Pear Tree Avenue, Newhall, Swadlincote, DE11 0NB
GW1 PKW P. Kingsley-Williams, Banhadlen Uchaf, Back Road, Mold, CH7 4QD
GW1 PLJ J. Morgan, Holly Cottage, Old Racecourse, Oswestry, SY10 7PQ
G1 PLU S. Goy, 352 Chanterlands Avenue, Hull, HU5 4ED
G1 PLV R. Cilia, 18 London Fields House, Kensington Road, Crawley, RH11 9NS
G1 PMA D. Harding, 37 Junction Cottages, London Road, Hardham, Pulborough, RH20 1LA
G1 PMF P. Felton, 13 New St., Sudbury, CO10 6JB
G1 PMJ D. Loon, 18 Stourcliffe Road, Wallasey, CH44 3AF
G1 PMK A. Mills, 12 Sydney Street, Kimberley, Nottingham, NG16 2LQ
G1 PML C. Roberts, 21 Dryden Close, Grantham, NG31 9QS
G1 PMU L. Ellis, 4 Hunt Avenue Heybridge, Maldon, CM9 7TY
G1 PNB S. Garwell, 8 Shakespeare Road Prestwich, Manchester, M25 9GW
G1 PNC T. Collins, 1 Artillery Place, Hollyhedge Road, Manchester, M22 4GG
GW1 PND K. Hassall, 2 Hilton House, Steynton Road, Milford Haven, . Sa731Bd, Milford Haven, SA73 1BD
G1 PNL D. Johnson, 27 Ridgeway Avenue, Gravesend, DA12 5BD
G1 PNX D. Staples, 2 Bulcote Road, Clifton, Nottingham, NG11 8FD
GM1 POA J. Jamieson, 11 Binns Road, Glasgow, G33 5HU
G1 POC C. Elsom, 8 King Avenue, Maltby, Rotherham, S66 7HX
G1 POD E. Mitchell, 11 Wingle Tye Road, Burgess Hill, RH15 9HR
G1 POJ K. Phillips, Flat 18, Edith Ramsay House, 134 Duckett Street, London, E1 4TD
G1 POM E. Baker, 19 Ramsey Road, Thornton Heath, CR7 6BX
G1 POR A. Porter, 1125 Yardley Wood Road, Warstock, Birmingham, B14 4LS
G1 PPB T. Roots, 11 Windermere Avenue, Eastern Green, Coventry, CV5 7GP
G1 PPD A. Shons, 108 Southdown Road, Catherington, Waterlooville, PO8 0NF
G1 PPG G. Joyner, Valle De Los Nogales 609, Fraccionamiento Real Del Valle, Nuevo Leon, Mexico, ZP 66350
G1 PPK J. Knight, 183 Northumberland Avenue, Thornton-Cleveleys, FY5 2JS
G1 PPO A. Wade, 40 Throxenby Lane, Scarborough, YO12 5HW
G1 PPQ S. Walker, 22 Ward Close, Aylestone, Leicester, LE2 8NJ
G1 PPU D. Walker, 115 Kilby Road, Fleckney, Leicester, LE8 8BP
G1 PPX G. Nicholls, 2 Leybrook Croft, Hemsworth, Pontefract, WF9 4JA
G1 PPZ A. Parton, 17 Causey Farm Road, Halesowen, B63 1EQ
G1 PQJ P. Wilson, Orchard Cottage, Rectory Road, Norwich, NR14 8HT
G1 PQK P. Harrison, 20 Priory Road, Stanford-le-Hope, SS17 7EW
G1 PQO S. Stevens, 49 The Beeches, Upton-Upon-Severn, Worcester, WR8 0QQ
G1 PQR J. Haynes, 33 St. Vincents Avenue, Kettering, NN15 5DR
G1 PQT J. Mayes, 44 Foxwarren, Claygate, Esher, KT10 0JZ
G1 PQX R. Moat, 27 Pioneer Road, Dover, CT16 2AR
G1 PQY G. Jones, 25 Myvod Road, Wednesbury, WS10 9BT
G1 PRF S. Haden, 33 Poplar Avenue, Chelmsley Wood, Birmingham, B37 7RD
G1 PRH M. Watkins, 7 Sand End, Whitstable, CT5 4TH
G1 PRL R. Williams, 54 Windways, Little Sutton, Ellesmere Port, CH66 1JF
G1 PRM A. Webber, 37 Rasen Road, Tealby, Market Rasen, LN8 3XL
G1 PRP A. Moore, Wyndrush, Northend Lane, Southampton, SO32 3QN
G1 PRS C. Ladley, 25 Laburnum Crescent, Louth, LN11 8SG
G1 PRW A. Swift, 38 Knightsbridge Way, Stretton, Burton-on-Trent, DE13 0WJ
G1 PRZ B. Saich, 65 Orchard Rise West, Sidcup, DA15 8TA
G1 PSH L. Walton, 2 Church Road, Colmworth, Bedford, MK44 2JX
G1 PSL C. Sparrow, 112 Hill Cot Road, Bolton, BL1 8RW
G1 PSS A. Laszkiewicz, 13 Darwall Drive, Ascot, SL5 8NB
GM1 PST P. Stanhope, The Roundal, Alva, FK12 5HU
GM1 PSU I. Manson, 25 Etna Court, Armadale, Bathgate, EH48 2TD
GW1 PSW A. Bunting, 11 Lon Y Gaer, Deganwy, Conwy, LL31 9RG
GM1 PSZ D. Liddle, 9 Rullion Road, Penicuik, EH26 9HS
G1 PUK D. Jones, 100 Cop Lane, Penwortham, Preston, PR1 0UR
G1 PUO D. West, 30 Farm Avenue, Swanley, BR8 7JA
G1 PUQ C. Tripp, Kingshill House, Church Street, Somerton, TA11 6ER
GM1 PUR D. Wood, 50 Riverside Road, Eaglesham, Glasgow, G76 0DG
G1 PUU S. Pantall, 27 Woodlands Drive, Foston, Derby, DE65 5DL
G1 PUV C. Wiseman, 42 Merlin Way, Kidsgrove, Stoke-on-Trent, ST7 4YL
G1 PUY J. De Renzi, Bankside, South Newington, Banbury, OX15 4JE
G1 PUZ S. Box, 103 Ilkeston Road, Bramcote, Nottingham, NG9 3JT
G1 PVA G. Kemp, 5 Gosselin Street, Whitstable, CT5 4LA
G1 PVD H. Murray, 93 Burridge Road, Burridge, Southampton, SO31 1BY
GI1 PVE R. Law, 147 Lone Moor Road, Londonderry, BT48 9LA
GW1 PVN P. Jones, 72 Lon Maesycoed, Newtown, SY16 1QQ
G1 PVR R. Kelly, Hogbrook Farm, Banbury Road, Leamington Spa, CV33 9QL
G1 PVT D. Broad, 14 Albion Road, Westcliff-on-Sea, SS0 7DR
G1 PVZ J. Vincent, Brookfield, North Street, Crewkerne, TA18 7AX
G1 PWF D. King, 79 Wootton Drive, Hemel Hempstead, HP2 6LA
G1 PWH "South Normanton, Alfreton & District ARC", c/o A. Jones, 16 Collumbell Avenue, Ockbrook, Derby, DE72 3TF
GM1 PWL K. Robertson, 7 Meadows Crescent, Lochgilphead, PA31 8AG
G1 PWM J. Bulman, 3 South View, Littlethorpe, Ripon, HG4 3LL
G1 PWO P. Owens, Flat 1, 45 The High Street, Enfield, EN3 4EF
G1 PWS R. Manson, Smavollen 11, Stavanger, Norway, 4017
G1 PWU D. Dwyer, 24 Alder Way, Melksham, SN12 6UL
G1 PWY P. Gardner, 2 South Road, Morecambe, LA4 5RA
GW1 PXM Dr R. Blakeway, Ty Nantglyn, Glascwm, Llandrindod Wells, LD1 5SE
G1 PXQ A. Scivetti, 10 Rippleside, Basildon, SS14 1UA
G1 PXW P. Hollands, 16 Hazel Crescent, Thornbury, Bristol, BS35 2LX
GI1 PXX S. Argue, 2 Lisnisky Walk, Portadown, Craigavon, BT63 5FY
GW1 PYY A. Thomas, 22 Sea Road, Abergele, LL22 7BU
G1 PZA W. Grech-Cini, Byram Garnge, Great North Road, Byram, WF11 9PA
G1 PZD M. Pattison, 13 Mixes Hill Road, Luton, LU2 7TX
G1 PZP C. Collett, Yaffle, 26 Hertford Road, Hoddesdon, EN11 9JR
G1 PZS J. Alltimes, 9 Wrexham Road, Romford, RM3 7YX
GM1 PZT G. Hardacre, 242 Sutherland Way, Knightsridge, Livingston, EH54 8JB
GI1 RAA T. Hourican, 43 Burren Road, Warrenpoint, Newry, BT34 3SA
G1 RAE M. Buckley-Brown, 2 Brothertoft Road, Boston, PE21 8HD
G1 RAF Royal Air Force Halton Radio Society, c/o A. Mockford, 58 Wendover Heights, Wendover, HP22 6PH
G1 RAG J. Jones, 10 Huntington Close, Redditch, B98 0NF
G1 RAO K. Hughes, 20 Pickering Close, Bury, BL8 1UE
G1 RAP R. Prosser, 27 Dorset Gardens, Rochford, SS4 3AH
G1 RAX D. Bendall, Brambles, 17 Berryfield Road, Lymington, SO41 0HQ
G1 RBA A. Wheatley, 3 Woodsbank Terrace, Wednesbury, WS10 7RQ
G1 RBH K. Dunn, 65 Lime Street, Sutton-in-Ashfield, NG17 4GA
GI1 RBI W. Mckeown, 15 Laragh Lee, Ballycassidy, Enniskillen, BT94 2JA
G1 RBO A. Thompson, 33 Hunters Way, Saffron Walden, CB11 4DE
GM1 RBQ R. Johnson, Wester Balquhandy House, Dunning, Perth, PH2 0RB
G1 RBX R. Baker, 3 Hazelton Close, Solihull, B91 3GA
G1 RBY P. Hurp, 55 Brooklyn Grove, Coseley, Bilston, WV14 8YH
G1 RBZ G. Norris, 26 Westwood Road, Leyland, PR25 3NS
G1 RCE A. Hills, 12 Heathway, Chaldon, Caterham, CR3 5DL
G1 RCI P. Mason, 34 Central Park Avenue, Wallasey, CH44 0AQ
G1 RCN P. Wilson, 146 Wilkinson Street, Nottingham, NG8 5FJ
G1 RCV Cray Valley Radio Society, c/o A. Styles, 6 Hill Brow, Crayford, Dartford, DA1 3NX
G1 RCW J. Chetwynd, 35 Cordelia Close, Dibden, Southampton, SO45 5UD
G1 RCX P. Tweney, 9 Dovehouse Close, Eynsham, Witney, OX29 4EW
GM1 RDG J. Horsburgh, Donvilla, 66 Harlaw Road, Inverurie, AB51 4TB
G1 RDJ C. Price, 44 Poplar Road, Stourbridge, DY8 3BD
G1 RDU P. Fanning, 9 Fishermans Walk, Shoreham-by-Sea, BN43 5LW
G1 RDX S. Newbold, 7 Rookery Meadow Holmer Green, High Wycombe, HP15 6XF
G1 REO J. Telford, 85 Medway, Great Lumley, Chester le Street, DH3 4HU
G1 RET R. Taylor, 3 Solent View, Calshot, Southampton, SO45 1BH
G1 RFB R. Palmer, 55 Edinburgh Road, Freshwater, PO40 9DL
G1 RFC R. Armitage, 15 Northolmby St., Howden, Goole, DN14 7JL
G1 RFI A. Venables, 10 Tilehouse, Redditch, B97 4PL
G1 RFQ D. Jenks, Tre-Vorgan, Courtenay Road, Tavistock, PL19 0EE
G1 RFS T. Niner, 281 Nightingale Road, Edmonton, London, N9 8QL
G1 RFX K. Tonner, Millstream Cottage, Golden Valley, Malvern, WR13 6AA
G1 RGG P. King, 124 Henley Grove Road, Rotherham, S61 1RY
GM1 RGM D. Keddie, 26 Daleally Crescent, Errol, Perth, PH2 7QA
G1 RGT G. Williams, 76 Eastern Avenue, Pinner, HA5 1NJ
G1 RHB K. Sears, Lodge 7 Kelsey Wood Country Park, Monksthorpe Spilsby, PE23 5PP
G1 RHE W. Berry, The Bungalow, Basil Road, Kings Lynn, PE33 9RP
G1 RHW T. Ibbitson, 36 Knoll Park, East Ardsley, Wakefield, WF3 2AX
GI1 RIB B. Rafferty, 81 Mullaghmore Drive, Omagh, BT79 7PQ
GM1 RIG I. Mcgowan, Feddal Lodge, Braco, Dunblane, FK15 9RA

G1 RIR L. Shears, 7 Lower Furlongs, Brading, Sandown, PO36 0DX
G1 RIV T. Dyson, 4 Lyspitt Common, Meppershall, Shefford, SG17 5GZ
G1 RIX Dr I. Steele, 10 Manor Road, Irby, Wirral, CH61 4UA
G1 RIY M. Gray, Brook House, Brandside, Buxton, SK17 0SG
G1 RJA M. Johnson, 28 Bittles Green, Motcombe, Shaftesbury, SP7 9NX
G1 RJD W. Blower, 129 Kingsway, Kirkby-In-Ashfield, Nottingham, NG17 7FH
G1 RJN S. Hall, Little Dene, Eastwick Road, Leatherhead, KT23 4BJ
GM1 RJS J. Hambrook, 32 Blackdales Avenue, Largs, KA30 8HU
G1 RJW R. Wicks, 32 Shelley Close, Northcourt, Abingdon, OX14 1PR
G1 RKD B. Allport, 1 Percy Drive, Swarland, Morpeth, NE65 9JN
GM1 RKI A. Morris, 39 Old Town, Peebles, EH45 8JE
G1 RKJ R. Wilson, 107 Hamilton Avenue, Uttoxeter, ST14 7FE
G1 RKR R. Luker, Grantham Cottage, Haywards Heath Road, Lewes, BN8 4DS
G1 RLA R. Hall, Bliss Lodge, Worcester Road, Chipping Norton, OX7 5XS
G1 RLB R. Lawson, 28 Hallett Way, Bude, EX23 8PG
G1 RLD D. Beeton, 50 Hanson Avenue, Shipston-on-Stour, CV36 4HS
G1 RLF R. Walter, 10 Birch Meadow, Clehonger, Hereford, HR2 9RH
G1 RLI P. Webb, 41 Lancaster Gardens, Wolverhampton, WV4 4DN
G1 RLK B. Ian, 35 Stanley Road, Heysham, Morecambe, LA3 1UR
G1 RLR P. Pedley, 24 Appledore Road, Walsall, WS5 3DT
G1 RLT P. Vipond, The Old Forge, Nentsbury, Alston, CA9 3LH
GM1 RLV D. Mackay, Burnlea, Harrapool Broadford, Broadford, IV49 9AQ
G1 RMC SW London Raynet, c/o I. Jackson, 5 Vivien Close, Chessington, KT9 2DE
G1 RMN M. Richards, 20 Tas Combe Way, Willingdon, Eastbourne, BN20 9JA
G1 RNL M. Kinsella, The Nook, Eaudyke Road, Boston, PE22 8RU
G1 RNV B. Purse, 28 Holford Road, Guildford, GU1 2QF
G1 RNY P. Roe, Kalmia, Sheffield Park, Uckfield, TN22 3RA
G1 RNZ G. Saville, 4 Shannon Court, Downs Barn, Milton Keynes, MK14 7PP
GM1 ROB R. Rimmer, Glenalty Cottage, Barrhill, KA26 0QT
G1 ROD I. Lupton, 19 Avenue Close, Harrogate, HG2 7LJ
GW1 ROE I. Roe, Tyddyn Berth, Chwilog, Pwllheli, LL53 6RQ
G1 ROH D. Gower, 68 Wood Common, Hatfield, AL10 0UB
G1 ROK P. Court, Hamara, Shortlands Grove, Bromley, BR2 0LS
GM1 ROL S. O'Connor, 2 Latch Farm Cottages, Off Leyden Road, Kirknewton, EH27 8DQ
GM1 ROM D. O'Connor, 2 Latch Farm Cottages Off Leyden Road, Kirknewton, EH27 8DQ
G1 RON F. Donnachie, 2 The Mall, Patrington Haven Leisure Park, Patrington, HU12 0PT
GM1 ROX A. Donald, South Sandlaw House, Alvah, Banff, AB45 3UD
G1 RPE I. Smith, 66 Aire Road, C/O Beech Croft, Wetherby, LS22 7UE
G1 RPO C. Reed, Colins, Throcking Road, Cottered, Buntingford, SG9 9RA
G1 RPP C. Broadbent, 7 Wharfe Park, Addingham, Ilkley, LS29 0QZ
G1 RPT M. Tribe, 11 Heathlands Close, Crossways, Dorchester, DT2 8TS
G1 RPV R. Hardiman, 27 Staithe Road, Martham, Great Yarmouth, NR29 4PT
GM1 RQD D. Marwick, 17 Laverock Road, Kirkwall, KW15 1EE
G1 RQI P. Quirk, 75 Harcourt Road, Folkestone, CT19 4AF
G1 RQK S. Bland, 6 Bryden Close, Northallerton, DL6 1SF
GW1 RQM M. Aquilina, 3 Aldergrove Close, Port Talbot, SA12 8EY
G1 RRE J. Eckersley, 88 New Heys Way, Bradshaw, Bolton, BL2 4AQ
G1 RRG H. Johnstone, 16 Riverside Crescent, Otley, LS21 2RS
GM1 RRJ B. Elliott, 195 Braehead Road, Cumbernauld, Glasgow, G67 2BL
G1 RRR K. Bareham, 19 Northfield Road, Ringwood, BH24 1LS
G1 RRU P. Atkinson, 19 Haggar Street, Wolverhampton, WV2 3ET
G1 RRW A. Henderson, Eastbury Farm House, Tarrant Gunville, Blandford Forum, DT11 8JQ
G1 RSC S. Hall, 17 Nevill Road, Rottingdean, Brighton, BN2 7HH
G1 RSE J. Rod, 42 Westwood Avenue, Ferndown, BH22 9HN
G1 RSF A. Dean, Les Monneries, Combieres, Charente, France,
G1 RSK C. Broughton, 65 Manby Road, Immingham, DN40 2SG
GI1 RSR C. Fogarty, 46 Knockview Drive, Tandragee, Craigavon, BT62 2BH
G1 RTW G. White, 101 London Road, Hailsham, BN27 3AH
G1 RTX I. Poole, 8 Bates Close, Higham Ferrers, Rushden, NN10 8HF
G1 RUE D. John, 28 Cliff Road Winteringham, Scunthorpe, DN15 9NQ
G1 RUG A. Kay, Pear Tree Cottage, Hale House Lane, Farnham, GU10 2JG
G1 RUL A. Cake, 8 Carrick Close, Dorchester, DT1 2SB
G1 RUZ Dr E. Byrne, 25 South Road, Grassendale Park, Liverpool, L19 0LS
GW1 RVC R. Baker, 24 Nant Road, Connah's Quay, CH5 4AL
G1 RVF B. Dempster, 5 Church Walk, Bozeat, Wellingborough, NN29 7ND
G1 RVH C. Stancer, 20 Overton Avenue, Willerby, Hull, HU10 6AR
G1 RVK I. Brooks, 10 Foxgloves, Deeping St. James, Peterborough, PE6 8SH
GW1 RVP P. Scott, 22 Powys Road, Llandudno, LL30 1HZ
G1 RVT B. Kavanagh, 73 Esh Wood View, Ushaw Moor, Durham, DH7 7FD
G1 RWR S. Lee, 7 Ridge Way Close, Rotherham, S65 3NH
G1 RWT K. Whitton, 11 Dursley Road, Shirehampton, Bristol, BS11 9XB
G1 RWX K. Ikin, 15 Broadway, Farnworth, Bolton, BL4 0HQ
GI1 RXL C. O'Connell, 15 Grange Road, Coleraine, BT52 1NG
GI1 RXM A. Murphy, 3 Church Lane, Crossgar, Downpatrick, BT30 9PX
G1 RXV S. Mugele, 19 Ambassador, Bracknell, RG12 8XP
G1 RYF M. O'Callaghan, 79 Kingsfield Avenue, Harrow, HA2 6AQ
G1 RYM R. Knox, 13 Grosvenor Road, Billingham, TS22 5HA
G1 RYQ S. Marshall, 15 Moulton Close, Belper, DE56 0EA
G1 RYS I. Mcculloch11, 5 Knighthead Point, The Quarterdeck, London, E14 8SR
G1 RYY R. Howse, 37 Great Eastern Road, Hockley, SS5 4BX
GW1 RZE P. Morgan, Flat 24, Ynysderw House, Swansea Road, Swansea, SA8 4AA
G1 RZJ L. Ward, 19 Spring Close View, Sheffield, S14 1RJ
G1 RZZ R. Wood, 40 Ashville Gardens, Pellon, Halifax, HX2 0PL
G1 SAK M. Weatherley, 95 Cambalt Road, Putney, London, SW15 6EX
GW1 SAM A. Hodgkinson, 64 Rhodfa Wen, Llysfaen, Colwyn Bay, LL29 8LE
G1 SAR South Anglia Raynet, c/o K. Gaunt, 21 Abbey Close, Rendlesham, Woodbridge, IP12 2UD
G1 SAT Inmarsat ARC, c/o R. Smith, 32 Wolseley Gardens, London, W4 3LR
GM1 SBD M. Angiolini, Innis Chonain, Mill Road, Stirling, FK7 9LP
G1 SBK J. Spink, Highfields, Church Lane, Leeds, LS16 8DE
G1 SBN J. Davison, 29 Glenfield Avenue, Wetherby, LS22 6RN
G1 SBW I. Case, 4 Portside, Preston Brook, Runcorn, WA7 3LE
G1 SBZ T. Lumley, 32 Downland Road, Woodingdean, Brighton, BN2 6DJ
G1 SCA S. Allen, 28 Neville Road, Luton, LU3 2JJ
G1 SCB K. Gray, Donkleywood House, Donkleywood, Hexham, NE48 1AQ
G1 SCL N. Stackhouse, 16 Tintern Avenue, Urmston, Manchester, M41 6FJ
G1 SCN C. Taylor, 4 Tunnel Road, Beaminster, DT8 3BQ
G1 SCO R. Brown, 52 Challenger Drive, Sprotbrough, Doncaster, DN5 7RY
G1 SCQ K. Kent, 5 Jubilee Road, Heacham, King's Lynn, PE31 7AR
G1 SCR Shropshire Raynet, c/o M. Jones, 35 Pendle Way, Meole Brace, Shrewsbury, SY3 9QS
G1 SCT M. Lane, Cherry Tree House, Pipwell Gate Saracens Head, Holbeach, Spalding, PE12 8BA
G1 SCV A. Faulkner, Northwood, Cranham, Gloucester, GL4 8HB
G1 SCY F. West, 9 Tregarland Close Coads Green, Launceston, PL15 7NE
G1 SDJ C. Cooper, Tapshays Cottage, Burton Street, Sturminster Newton, DT10 1PS
G1 SDK S. Karpasitis, Riverdene Blythe Road, Hoddesdon, EN11 0BB
G1 SDN R. Mordue, 29 Sycamore Close, Witham, CM8 2PE
G1 SDX G. Taylor, 3 Erica Drive, Torquay, TQ2 8LP
G1 SEA D. Tucker, 12 Chatsworth Way, Heanor, DE75 7TJ
G1 SEF A. Fearnley, 1 Dover Road, London, E12 5DZ
G1 SEH South East Hampshire Raynet, c/o J. Woonton, 44 Jubilee Road, Southsea, PO4 0JE
G1 SEO M. Lines, 158 Nine Mile Ride, Finchampstead, Wokingham, RG40 4JA
G1 SES M. Halden, 19 Fenwick Lane Halton Lodge, Runcorn, WA7 5YU
G1 SEW A. Goatman, 150 Merlin Park, Portishead, Bristol, BS20 8RW
G1 SFU D. Coffey, 14 Shawbridge, Harlow, CM19 4NJ
G1 SGA R. Blunt, 8 The Crescent, Wolverhampton, WV6 8LA
GW1 SGE A. Morgan, 3 Gelli Newydd, Golden Grove, Carmarthen, SA32 8LP
GW1 SGG M. Jones, 48 Maes Alltwen, Dwygyfylchi, Penmaenmawr, LL34 6UA
GW1 SGH R. Thorne, 6 Cromwell Avenue, Rhyddings, Neath, SA10 8DW
G1 SGM R. Parkin, Craigside, 15 Holly Drive, Leeds, LS16 6EF
G1 SGP N. Barnes, 44 Cromford Road, Wirksworth, Matlock, DE4 4FR
G1 SGR N. Marsh, 16 Daytona Quay, Eastbourne, BN23 5BN
G1 SGS B. Kimber, 4 Nautilus Drive, Minster On Sea, Sheerness, ME12 3NJ
G1 SGZ P. Gamble, 9 Windmill Close, Ockbrook, Derby, DE72 3TE
G1 SHH A. Compton, Fairlight, 25 Framfield Road, Uckfield, TN22 5AH
G1 SHI M. Kuik, 196 Prestbury Road, Macclesfield, SK10 3BS
G1 SHN G. Richardson, 12 Northenhay Walk, Morden, SM4 4BS
G1 SHT D. Wooster, 34 New Road, Penn, High Wycombe, HP10 8DL
G1 SHU N. Carr, 7 Pear Tree Drive, Sedgeberrow, Evesham, WR11 7GQ
G1 SID C. Siddons, 423 London Road, Grays, RM20 4AB
G1 SIG B. Scholte, 266 Hednesford Road, Heath Hayes, Cannock, WS12 3DS
G1 SIM S. Hallam, 46 Holte Road, Atherstone, CV9 1HN
G1 SIO J. Robson, Ealands, The Stanners, Corbridge, NE45 5BA
G1 SIP R. Webb, 54 Ashby Avenue, Chessington, KT9 2BU
G1 SIU A. Bingley, 51 Kirkby Folly Road, Sutton-in-Ashfield, NG17 5HP
G1 SIX R. Foster, 70 Mansfield Road, Edwinstowe, Mansfield, NG21 9NH
G1 SJB C. Thompson, 135 Stafford Road, Bloxwich, Walsall, WS3 3PG
G1 SJD Lord N. Peirce, 73 Knights Mead, Chudleigh Knighton, Chudleigh, Newton Abbot, TQ13 0RF
G1 SJG E. Boydon, 56 Oliver Leese Court, Ten Butts Crescent, Stafford, ST17 9HP
G1 SJO A. Banthorpe, 32 Long Close, Station Road, Henlow, SG16 6JS
G1 SJT A. Long, 23 Beech Road, Sutton Weaver, Runcorn, WA7 3ER
G1 SJU D. Maciver, 176 Burges Road, London, E6 2BS
G1 SJZ P. Randall, 7 Eastbourne Avenue, Featherstone, Pontefract, WF7 6LQ
G1 SKE D. Turner, 26 Carlton Way, Cleckheaton, BD19 3DG
G1 SKI K. Weaver, 26 Southdown Close, Haywards Heath, RH16 4JR
G1 SKQ J. Forster, 4 Rydal Road, Lemington, Newcastle upon Tyne, NE15 7LR
G1 SKV B. Abell, 5 Aldborough House Brook Street, York, Yo317qq
G1 SKW R. Sidwell, 81 Oakengates Road, Donnington, Telford, TF2 7LQ
G1 SLA C. Baker, 17 Dawlish Avenue, Chadderton, Oldham, OL9 0RF
G1 SLE R. Drabble, 37 Barton Street, Clowne, Chesterfield, S43 4RS
G1 SLG M. Butcher, 4 Sotheby Rise, Ecton Brook, Northampton, NN3 5AD
G1 SLI B. Elliott, 51 Allerhope, Hall Close Grange, Cramlington, NE23 6SX
G1 SLO K. Turner, 155 Bure Lane, Christchurch, BH23 4HB
G1 SLP E. Methven, 10 Woodbine Road, Durham, DH1 5DR
G1 SLU K. Ford, 123 Stockwood Lane, Bristol, BS14 8SZ
G1 SMB M. Chitty, Timbercroft, Faris Lane, Addlestone, KT15 3DL
G1 SMC S. Mccloy, 28 Ferndown Drive, Godmanchester, Huntingdon, PE29 2LU
GW1 SMG A. Walker, Blaenlluest Oakford, Llanath, SA470RT
GW1 SMJ F. Beavan, Uplands, Bronllys, Brecon, LD3 0HN
G1 SMP P. Devlin, 42 Trafalgar Way, Lichfield, WS14 9FD
GW1 SMT C. Manning, Walker Road, Cardiff, CF24 2EL
G1 SMY S. Fisher, 19 Sandown Road, Sandown, PO36 9JL
G1 SNI P. Smith, 27 Briar Close, Gillingham, SP8 4SS
G1 SNO D. Tubb, 42 Hill Farm Road, Marlow, SL7 3LU
G1 SNQ I. Boss, 11 Penkridge Road, Church Gresley, Swadlincote, DE11 9FH
G1 SNU D. Tunbridge, 12 Burnham Road, Latchingdon, Chelmsford, CM3 6EU
G1 SOB T. Yetton, 7 Warwick Close, Canvey Island, SS8 9YB
G1 SOG R. Stearn, 18 Kings Avenue, Chippenham, SN14 0UJ
G1 SOM Taunton & Somerset Raynet Group, c/o D. Smith, 47 Laburnum St., Taunton, TA1 1LB
G1 SOX W. Stennett, 26 Moorfield Road, St. Giles-On-The-Heath, Launceston, PL15 9SY
G1 SOY J. Moseley, 72 Wisden Road, Stevenage, SG1 5JA
G1 SPA D. Harrison, 7 Shirley Close, Castle Donington, Derby, DE74 2XB
G1 SPJ A. Gibbs, 86 Broadmark Road, Slough, SL2 5PN
G1 SPM A. Hughes, 8 Rigby Grove, Little Hulton, Manchester, M38 0FQ
G1 SPT S. Tatem, 55 Chelwood Road, Chellaston, Derby, DE73 5SJ
G1 SPU A. Burnett, 16 Shielding Way, Stafford, ST16 3WG
GW1 SPW B. Saunders, Bishops Mill, Llanwrda, SA19 8AD
G1 SPX K. Shires, 19 Prince Charles Avenue, Sittingbourne, ME10 4NA
GM1 SQA J. Yates, 67 Closeburn, Thornhill, DG3 5HR
G1 SQC K. Boote, 51 Sunnyfield Oval, Stoke-on-Trent, ST2 7PA
G1 SQG P. Dennis, Fuchsia House, 18 West View Road, Yelverton, PL20 7DD
G1 SQI J. Bewley, 21 Duloe Gardens, Pennycross, Plymouth, PL2 3RS
G1 SQW R. Rawson, 43 County Road North, Hull, HU5 4HN
GM1 SQZ G. Pocock, 1 Pitcairn Grove, East Kilbride, Glasgow, G75 8TN
G1 SRA T. Binns, Cross Farm, 1. Cross Lane Oxenhope, Keighley, BD22 9LE
GW1 SRB K. Gough, 2 Church Road, Abertridwr, Caerphilly, CF8 4DL
G1 SRD P. Foulds, 7 Bridge Road, Little Sutton, Spalding, PE12 9EG

GM1 SRP J. Mcculloch, Wester Curr Cottage, Dulnain Bridge, Grantown-on-Spey, PH26 3LX
GM1 SRR M. Christmas, Lindens, Smithy Loan, Dunblane, FK15 0HQ
G1 SSL M. Belcher, 52 Kynaston Road, Didcot, OX11 8HD
G1 SSS Capt. J. Banfield, 2 Laleham Close, Eastbourne, East Sussex, BN21 2LQ
G1 SSZ F. Bowhill, 78 East Gomeldon Road, Gomeldon, Salisbury, SP4 6NB
G1 STK A. Goddard, 65 Langley Hall Road, Solihull, B92 7HE
G1 STP V. Trolan, Hilbre, East Taphouse, Liskeard, PL14 4NJ
G1 STQ J. Taylor, C/O 22 Welland Grove, Newcastle, ST5 4EP
G1 STW R. Gerrard, 7 Wisteria Drive, Healing, Grimsby, DN41 7JB
G1 SUH J. Speakman, 130 Dicconson Street, Wigan, WN1 2BA
GW1 SUK A. Dimmock, Gwyndy, Llandegfan, Menai Bridge, LL59 5PW
G1 SUM T. Cull, 25 Queensway, Ponteland, Newcastle upon Tyne, NE20 9RZ
GJ1 SUP T. Stone, Les Quatres Saisons La Route De La Porte, St John, Jersey, JE34DE
G1 SVD R. Roper, 57 Burnt Hills, Cromer, NR27 9LW
G1 SVI T. Metcalfe, 38 Station Road, Branston, Lincoln, LN4 1LH
G1 SVJ C. Murphy, 13 Northfield Road, Ringwood, BH24 1LS
G1 SVL K. Stocker, 22 Hadlow Down Close, Luton, LU3 2PY
G1 SVN M. Churchman, Westcroft, Church Road, Chester, CH4 9NG
G1 SVP B. Howard, 15 Four Acres, Bideford, EX39 3RW
GM1 SVQ C. Pringle, 12 Atkinson Road, Dumfries, DG2 7DH
G1 SVR Severn Valley Rd, c/o E. Churchyard, 11 Greenfields Drive, Bridgnorth, WV16 4JW
GW1 SVV D. Osborne, 22 Springfield Gardens, Hirwaun, Aberdare, CF44 9LY
G1 SWE M. Foster, 15 Parklands Way, Liverpool, L22 3YX
G1 SWF S. Roberts, 43 Lawn Close, Ruislip, HA4 6ED
G1 SWH G. Schoof, 4-5 Canal Row, Haigh, Wigan, WN2 1NA
G1 SWI B. Gillett, 18 Rookery Close, Fenny Drayton, Nuneaton, CV13 6BB
G1 SWK T. King, 10 Berkeley Close, Ipswich, IP4 2TS
G1 SWR Warwickshire Avon Raynet Group, c/o C. Ousbey, 30 Hawthorn Way, Shipston-on-Stour, CV36 4FD
G1 SWS S. Walters-Smith, 83 Chesterfield Road, Tibshelf, Alfreton, DE55 5NJ
G1 SWU G. Denham, 36 Redstone Farm Road, Hall Green, Birmingham, B28 9NT
G1 SWX C. Harrap, 10 Newhaven Place, Portishead, Bristol, BS20 8EG
G1 SWZ R. Wright, 61 Quarry Road, Hurtmore, Godalming, GU7 2RW
G1 SXB G. Whetstone, 60 Worple Road, Staines, TW18 1EE
GM1 SXJ P. Duckles, 45 Redhouse Place Blackburn, Bathgate, EH47 7QB
GW1 SXT M. Kerry, 40 Oaklands Road, Sebastopol, Pontypool, NP4 5BZ
GW1 SXU P. Janes, 19 Fair View, Chepstow, NP16 5BX
GM1 SXX A. Copland, 74 Whitehaugh Avenue, Paisley, PA1 3SR
G1 SXY Y. Entwistle, Sonunda House, Church Lane, Freckenham, Bury St Edmunds, IP28 8JF
GM1 SYC W. Graham, 7 Brunt Place, Dunbar, EH42 1RT
GI1 SYM G. Thompson, 57 Rosepark, Donaghadee, BT21 0BN
G1 SYP M. Thornton, 58 The Green View, Shafton, Barnsley, S72 8PW
G1 SYU M. Oubridge, 54 Cantle Avenue, Downs Barn, Milton Keynes, MK14 7QS
G1 SYV B. Goodier, 42 Manchester Road, Clifton, Swinton, Manchester, M27 6WY
G1 SYZ D. Setterfield, 6 Murrayfield Close, Plymouth, PL2 3FB
GI1 SZC Dr D. Mcmanus, 38 Deanfield, Bangor, BT19 6NX
G1 SZD T. Trengove, 8 Kemp Close, Truro, TR1 1EF
G1 SZK R. Frost, 68 Wessex Road, Didcot, OX11 8BP
GM1 SZM W. Robertson, 28 Dewars Avenue, Kelty, KY4 0BG
G1 SZT W. Dowkes, Woodlea, Gillamoor Road, York, YO62 6EL
G1 TAI P. Gabel, 4 Blacksmiths Green, Shutlanger, Towcester, NN12 7RS
G1 TAR C. Anderton, 5 Leyland Avenue, Hindley, Wigan, WN2 3SB
G1 TAU J. Drewry, 10 Drayton Drive, Heald Green, Cheadle, SK8 3LF
G1 TAY A. Shearer, 101 Millside, Stalham, Norwich, NR12 9PB
G1 TAZ F. Perkin, 5 Highgrove, Trevadlock Hall Park, Launceston, PL15 7PW
G1 TBE J. Kelday, 20 Lowfield, Eastfield, Scarborough, YO11 3LQ
G1 TBI W. Monk, Brook House, River View, Buxton, SK17 8SW
G1 TBK D. Watson, 72 Dawes Avenue, West Bromwich, B70 7LS
G1 TBN A. Malhi, 517 Clifton Drive North, Lytham St Annes, FY8 2QX
G1 TBT D. Taylor, Top Wath Laer, Top Wath Road, Harrogate, HG3 5PG
GM1 TBW A. Napier, Jehrada Cottage, Longhaven, Peterhead, AB42 0NY
G1 TBX G. Williams, 16 Coppice Road, Talke, Stoke-on-Trent, ST7 1UB
G1 TCH C. Hinton, 5 Vale Rise, Matlock, DE4 3SN
G1 TCK D. Adams, 28 Greenside, Stoke Prior, Bromsgrove, B60 4EB
GM1 TCN T. Curran, Miltonbank, East Cottage, Forfar, DD8 3TU
GM1 TCP J. Campbell, 16 Barony Road, Auchinleck, Cumnock, KA18 2LL
G1 TDL M. Mundy, The Homestead, Homestead Lane, Burgess Hill, RH15 0RQ
G1 TDN T. Collinson, 26 Westway Avenue, Hull, HU6 9SA
G1 TDO F. Sanchez-Garci, 74 Gorthorpe, Hull, HU6 9EZ
G1 TDP J. Spry, 21 Christchurch Gardens, Waterlooville, PO7 5BT
GM1 TDT D. Robertson, 131 Foxbar Road, Paisley, PA2 0BD
GM1 TDU J. Rooney, Rob Roy, Kinneff, Montrose, DD10 0TG
GW1 TDV A. Potts, 4 Bloomfield Close, Newport, NP19 9ET
G1 TEX N. Swann, 9 Alexandra Road, Parkstone, Poole, BH14 9EL
GW1 TFB A. Hughes, 38 Llys Dyffryn, St Asaph, LL17 0SX
GI1 TFC R. Mcmaster, 43 Craigs Road, Carrickfergus, BT38 9RL
GM1 TFF I. Hynd, 48 Ben Ledi Crescent, Cumbernauld, Glasgow, G68 9NG
GW1 TFL J. Robson, 16 Dunraven Road, Sketty, Swansea, SA2 9LG
G1 TFM J. Bishop, 93 The Vale, Feltham, TW14 0JY
G1 TFY D. Mackinnon, 20 Saxon Grange, Sheep Street, Chipping Campden, GL55 6BY
GM1 TFZ S. Bavin, Garvan, 10 Grampian Way, Glasgow, G78 2DH
GM1 TGY C. Christie, Firlands, Spey Valley Drive, Aberlour, AB38 9NU
G1 TGZ R. Mclintock, 10 The Close, Riverhead, Sevenoaks, TN13 2HE
G1 THA V. Collins, Flat 2, Marsh Mead, Glebe Road, Petersfield, GU31 5SB
G1 THD A. Simmons, 22 Willow Way, Princes Risborough, HP27 9AY
G1 THF The Ham Fellowship, c/o A. Nixon, 14 Carlton Road, Lowton, Warrington, WA3 2EP
G1 THG D. Moore, East View, The Common, Gillingham, SP8 5NB
G1 THP M. Fudge, 7 Shepherds Close, Winchester, SO22 4HU
GM1 THR N. Harrison, 127 Bruntsfield Place, Flat 1F3, Edinburgh, EH10 4EQ
GM1 THS G. Slessor, 27 Scurdie Ness, Aberdeen, AB12 3NG
G1 THW A. Taylor, 5 Brookside, Beare Green, Dorking, RH5 4QH
G1 TIF J. Batey, The Hemmel, Barrasford, Hexham, NE48 4BD
G1 TIH F. Bell, 143 Peter St., Blackpool, FY1 3NN
G1 TIJ P. Seaman, 18 Earlsford Road, Mellis, Eye, IP23 8DY
G1 TIK D. Waters, Gardeners Cottage, Sandhoe, Hexham, NE46 4LU
G1 TIQ S. Crellin, 89 Wapshare Road, West Derby, Liverpool, L11 8LR
G1 TJH C. Barfoot11, 6 Maldon Close Bishopstoke, Eastleigh, SO50 6BD
GW1 TJK A. Evans, Maes Yr Onnen, 134 Waterloo Road, Ammanford, SA18 3RY
GJ1 TJP J. Poole, Jardin Du Puits, La Longue Rue, St Martin, Jersey, JE3 6ED
G1 TJR R. Bromley, 33 Bromley Road, Lytham St Annes, FY8 1PQ
G1 TJT W. Adam, 9 Maple Drive, South Ockendon, RM15 6XE
G1 TJW B. Smith, 25 The Ferns, Tetbury, GL8 8JE
G1 TKE A. Gibson, 9 Fishers Mead, Dulverton, TA22 9EN
G1 TKQ I. Burdon, 72 Greenway Road, Taunton, TA2 6LE
G1 TKY P. Draper, 4 Woodmans Croft, Hatton, Derby, DE65 5QQ
G1 TLA B. Pattenden, Inshallah, Abbeystrewery, Skibbereen, Ireland,
G1 TLC A. Myers, 7 Hillside, Chelveston, Wellingborough, NN9 6AQ
G1 TLE M. Rowe, 11 Parkfield, Stillington, York, YO61 1JR
G1 TLH D. Koopman, 11 Tufts Field, Midhurst, GU29 9BU
G1 TLW A. Lloyd, 96 Fairdene Road, Coulsdon, CR5 1RF
G1 TMF J. Firth, 29 Curzon Street, Newcastle, ST5 0PD
G1 TML P. Brooks, 7 Beechcombe Close, Pershore, WR10 1PW
G1 TMW A. Caspersz, 25 Cheltenham Place, Harrow, HA3 9NB
G1 TNK J. Flattley, 53 The Drive, Bredbury, Stockport, SK6 2ED
G1 TNP A. Karande, Flat 1, 6 St. Dominics Close, Torquay, TQ1 4UN
G1 TNR D. Jordan, 21 Rosewood Park, Walsall, WS6 7HD
G1 TOB T. Hall, 3 Saville Road, Twickenham, TW1 4BQ
G1 TOL S. Talbot, 8 Thornford Drive, Swindon, SN5 7BB
G1 TPC M. Bellamy, 2 Nelson Drive, Rothwell, Kettering, NN14 6DZ
G1 TPN R. Whateley, 14 Eastfield Road, Delapre, Northampton, NN4 8PE
G1 TPO R. Steel, 15 Thornbury Avenue, Seghill, Cramlington, NE23 7RT
G1 TPQ S. Codman, 4 Farmland Road, New Costessey, NR5 0HX
G1 TPV D. Juett, 10 Leys Road, Cambridge, CB4 2AU
G1 TQH K. Scroggins, 44 Hillcroft Road, Herne, Herne Bay, CT6 7EW
G1 TQN A. Hobbs, The Sail Loft, 604 Blandford Road, Poole, BH16 5EQ
G1 TQR J. Harris, 48 Beech Close, Corby, NN17 2AF
G1 TQT M. Costello, Flat 24, Wisley House, London, SW1V 2QS
G1 TQU C. Chambers, 8 Dagtail Lane, Redditch, B97 5QT
G1 TQY P. Bishop, 2 Spruce Avenue, Whitehill, Bordon, GU35 9TA
G1 TRI C. Curtis, Westbury, Walkers Lane North Blackfield, Southampton, SO45 1YE
G1 TRL J. Deacon, 28 Dollicott, Haddenham, Aylesbury, HP17 8JG
GI1 TRZ W. Hamilton-Sturdy, 243 The Woods, Larne, BT40 1BD
G1 TST P. Jackman, 1 Palmer Road, Trowbridge, BA14 8QP
G1 TSV S. Corrigan, 163 Blackburn Road, Heapey, Chorley, PR6 8EJ
G1 TTB G. Birkby, 44 Lady Bay Road, West Bridgford, Nottingham, NG2 5DS
G1 TTC K. Howard, 73 Challacombe, Furzton, Milton Keynes, MK4 1DP
G1 TTG J. Brickwood, Datemachi House #301, 3-33-5, Tokyo, Japan, 150
G1 TTH E. Roughton, 18 Church Close, Braybrooke, Market Harborough, LE16 8LD
G1 TTK G. Lewis, 57 Edgecumbe Road, Roche, St Austell, PL26 8JH
G1 TTL F. Moss, 64 Birch Avenue, Cuerden Residential Park, Leyland, PR25 5PD
G1 TTX A. Docherty, Sunnybrae, Wickhurst Road, Sevenoaks, TN14 6LY
G1 TUI J. Tracy, 18 Preston St., Kirkham, Preston, PR4 2ZA
G1 TUL S. Neale, 28 Needham Drive, Sutton St. James, Spalding, PE12 0EG
G1 TUS R. Rodley, Meadow Cottage, Cold Ashby Road, Northampton, NN6 8QP
G1 TUU M. Dixon, 118 Kings Ash Road, Paignton, TQ3 3TU
G1 TUZ P. Nelson, 42 York Avenue, East Cowes, PO32 6RU
G1 TVW J. Halliday, 24 Duncan Avenue, Otley, LS21 3LN
G1 TWH S. Greenfield, Byways, Brightlingsea Road, Colchester, CO7 8JH
G1 TWS M. Dench, 110 Eastwood Road, Rayleigh, SS6 7JR
G1 TWT A. Scott, 21 Hexham, Oxclose, Washington, NE38 0NR
G1 TWW R. Reeves, 40 Kennett Road, Romsey, SO51 5PQ
G1 TWY B. Panton, Lavers, Preston Road, Lavenham, Sudbury, CO10 9QD
G1 TXO M. Riches, 32 Wyncham Avenue, Sidcup, DA15 8ER
G1 TYP E. Summers, 262 Huddersfield Road, Stalybridge, SK15 3DZ
G1 TYU R. Ward, 1 Kirkcroft Close Thorpe Hesley, Rotherham, S61 2UH
G1 TZZ M. Foster, 7 Orion Way, Braintree, CM7 9UR
G1 UAF A. Webster, 7 Castlehythe, Ely, CB7 4BU
G1 UAL C. Bagwell, 1 Waldegrave Court, Movers Lane, Barking, IG11 7UW
G1 UAY K. Varnals, Regent Studio, Skidden Hill, St Ives, TR26 2DU
G1 UAZ C. Musson, 14 Alfreton Road, South Normanton, Alfreton, DE55 2AS
G1 UBC J. Pedley, 24 Appledore Road, Walsall, WS5 3DT
G1 UBH M. Howell, 4 Chattisham Close, Stowmarket, IP14 2RE
G1 UBL A. Camm, 24 Croyde Avenue, Greenford, UB6 9LS
G1 UBN S. Knight, 46 Hollybank Road, Hythe, Southampton, SO45 5FQ
G1 UBT E. Dunn, 118 James Turner St., Birmingham, B18 4NE
G1 UBV N. Brigden, 176 Hulverston Close, Sutton, SM2 6UA
G1 UCC D. Boot, 13 Westland Street, Stoke-on-Trent, ST4 7HE
G1 UCG W. Roberts, 13 Roseacre Road, Elswick, Preston, PR4 3UD
G1 UCI D. Roberts, 56 Meadow Lane, Ainsdale, Southport, PR8 3RS
G1 UCN M. Davis, Flat 3, South Court, 29 Second Avenue, Bridlington, YO15 2LW
G1 UCO D. Ralph, 55 St. Marys Road, Warley, Smethwick, B67 5DH
G1 UCR R. Irving, 52 Holsworthy Square, London, WC1X 0BG
G1 UCT D. Pye, 95 Lansdowne Way, High Wycombe, HP11 1UB
G1 UCZ D. Dewar, 224 Seaside Road, Aldbrough, Hull, HU11 4RY
G1 UDB G. Wardle, 53 Braine Road, Wetherby, LS22 6NP
G1 UDE D. Grayson, 28 Chesterfield Road, Swallownest, Sheffield, S26 4TL
G1 UDR J. Scott, 47 Corinthian Road, Chandler'S Ford, Eastleigh, SO53 2AY
G1 UDS P. Spence, 17 Springvale Rise, Parkside, Stafford, ST16 1TE
G1 UDT M. Boydon, 56 Oliver Leese Court, Ten Butts Crescent, Stafford, ST17 9HP
G1 UDW D. Upton, 14 Ipley Way, Hythe, Southampton, SO45 3LJ
G1 UDX M. Stasuik, 30 Ramsey Drive, Arnold, Nottingham, NG5 6QL
G1 UEA D. Mills, 56 Canterbury Road, Birchington, CT7 9AS
G1 UEO J. Aldred, 1E North End Road, Steeple Claydon, Buckingham, MK18 2PF
G1 UEQ J. Sims, 345 Blandford Road, Hamworthy, Poole, BH15 4HP
G1 UEV S. Brown, 10 Walkers Lane South, Blackfield, Southampton, SO45 1YN
G1 UFA B. Bailey, 10 Milton Road, Waterlooville, PO7 6AA
G1 UFH J. Coyne, 20 Hawthorn Hill, Letchworth Garden City, SG6 4HG
G1 UFJ R. Davis, 10 Greenhill Gardens, Minster, Ramsgate, CT12 4EW
G1 UFL K. Whistance, 20 Solent Way, Milford On Sea, Lymington, SO41 0TE
G1 UFM S. Pugh, 18 Styal Avenue, Stretford, Manchester, M32 9SJ

G1 UFS N. Chandler, 7 Sherlock Avenue, Parklands, Chichester, PO19 3AE
G1 UFT P. Zara, 53 Castleton Road, Wigston, LE18 1FQ
G1 UFX A. Hern, Flat 9, Block P, Peabody Estate, London, SE1 8DU
G1 UGB J. Slater, 57 Freshbrook Road, Lancing, BN15 8DE
G1 UGG K. Blagg, 2 Geldof Drive, Blackpool, FY1 2AQ
G1 UGH T. Chaplin, 21 Shillitoe Close, Bury St Edmunds, IP33 3DU
G1 UGJ B. Lowe, 5 Kingfisher Drive, Necton, Swaffham, PE37 8NN
G1 UGL P. Beardshaw, 12 Halesworth Close, Chesterfield, S40 3LW
G1 UGO S. Lexton, 70 Halsbury Road, Westbury Park, Bristol, BS6 7SU
G1 UGV D. Bodman, 34 Churchlands, North Bradley, Trowbridge, BA14 0TD
G1 UGX G. Hodgkins, 20 Broadstone Close, Barnwood, Gloucester, GL4 3TX
G1 UHB S. Tomkins, The Close, Broomfield, Bradford, BD14 6PJ
GW1 UHF M. Perrett, Penlan, Whitland, SA34 0QX
G1 UHO H. Court, 10 Dorset Road, West Kirby, Wirral, CH48 6DJ
G1 UIB R. Moxon, 16 Kielder Oval, Harrogate, HG2 7HQ
G1 UID H. Kentfield, 1 Torquay Avenue, Gosport, PO12 4NS
G1 UIO M. Mcvittie, 19 Alder Crescent, Poole, BH12 4BD
GM1 UIR A. Perks, The Lodge, Cemetery Drive, Dumbarton, G82 5HD
G1 UJX J. Huddlestone, 8 Wilmot Avenue, Chaddesden, Derby, DE21 6PL
G1 UK A. Buck, 10 Northfield Park, Mansfield Woodhouse, Mansfield, NG19 8PA
G1 UKA M. Wilkie, 1 Celandine Close, Billericay, CM12 0SU
G1 UKH R. Rodgers, 88 Norwich Road, Watton, Thetford, IP25 6DW
G1 UKS D. Simmons, 54 Rydal Court, Morecambe, LA4 5LT
G1 UKW C. Crane, 3 Hawkshead Drive, Royton, Oldham, OL2 6TW
G1 UKZ P. Hall, 50 Sunningdale Drive, Irlam, M44 6WH
G1 ULB G. Walker, 20 Clough Drive, Prestwich, Manchester, M25 3JL
G1 ULG D. Porter, 44 Weir Road, London, SW12 0NA
G1 ULP T. Garner, 8 Brookside Park, Station Road, Hugglescote, Coalville, LE67 2GB
G1 ULQ P. Rowsell, Thomley Hall Farm, Worminghall, Bucks, HP18 9JZ
G1 ULR D. Ross, 2 Jemmetts Close, Dorchester-On-Thames, Wallingford, OX10 7RA
G1 UMS J. Preece, White House, Bredwardine, Hereford, HR3 6BY
G1 UMY R. Wiltshire, Danesboro, Stonehill Road, Chertsey, KT16 0ER
G1 UNB P. Durrant, Glenmore, 20 Linersh Wood Close, Bramley, Guildford, GU50EG
G1 UNN P. Coldicott, 45 Orrian Close, Stratford-upon-Avon, CV37 0TT
G1 UNQ R. Davis, Lowlands, Station Road, Evesham, WR11 7QG
G1 UNU K. Etwell, Hawthorn Cottage, Old Worcester Road, Kidderminster, DY11 7XS
G1 UOD P. Farmer, 22 Nortune Close, Birmingham, B38 8AJ
G1 UOJ M. Lichtaowicz, 219 Hamilton Drive West, York, YO24 4PL
G1 UOR W. Rodgers, 9 Hillcrest, Skelmersdale, WN8 9JZ
GW1 UOV A. Ham, Tynffordd, Blaencwrt, Lampeter, SA39 9AZ
GW1 UOY W. Bagley, Glan Severn, Trefeglwys Road, Llanidloes, SY18 6HZ
G1 UPP M. Maude, 6 Malin Parade, Portishead, Portishead, BS20 7FW
G1 UPT J. Ravelini, 15 Clarendon Green, Orpington, BR5 2NY
G1 UPX T. Giles, 108 Queensway, Didcot, OX11 8SW
G1 UQC Capt. D. Rusbridge, 1 Ray Bond Way, Aylsham, Norwich, NR11 6UT
G1 UQF R. Davies, 33 Melbourne House, Melbourne Road, Northampton, NN5 5LW
G1 UQK N. Lewis, Woodstock, Bridge Street, Pershore, WR10 2PL
G1 UQT S. Alliott, 32 Broad Gates, Silkstone, Barnsley, S75 4HD
GW1 URD R. Ameson, 9 Coed-Y-Fronallt Estate, Dolgellau, LL40 2YG
GW1 URF Dr A. Jones, Hafandeg, Southgate, Aberystwyth, SY23 1RY
G1 URH S. Woodley, Stables Edge, Lower Road, Chilton, Didcot, OX11 0RR
G1 URJ N. Capon, 13 West Croft, Berinsfield, Wallingford, OX10 7NL
G1 URQ D. Traynor, 5 Mount Street, Widnes, WA8 6TL
G1 URR C. Gough, 67 Pickmere Lane, Wincham, Northwich, CW9 6EB
G1 URW J. Carman, 5 Melbourne Road, Blacon, Chester, CH1 5JQ
G1 URZ T. Bennett, 16 Montgomery Avenue, Hemel Hempstead, HP2 4HE
G1 USF P. Napp, 23 Harriot Drive, Newcastle upon Tyne, NE12 7EU
GD1 USI A. Tawney, Croym Dty Chione, The Howe, Isle of Man, IM9 5PR
G1 USK R. Firth, 40 Ashfield Road, Chippenham, SN15 1QQ
GM1 USN J. Challis, Bay Villa, Strachur, Cairndow, PA27 8DE
G1 USV R. York, 10 Severn View Road, Thornbury, Bristol, BS35 1AY
G1 USW P. Daniels, 29 Station Road, Wickwar, Wotton-under-Edge, GL12 8NB
G1 USZ M. Trim, 10 Oldends Lane, Stonehouse, GL10 2DG
G1 UTC C. Thomas, Hazel Mount, Lockhams Road, Curdridge, Southampton, SO32 2BD
G1 UTF R. Beer, 8 Littlestone Court, Grand Parade, New Romney, TN28 8NF
G1 UTJ A. Lees, Timbercroft, Elliotts Orchard, Warwick, CV35 8ED
G1 UTM P. Yearsley, 25 Dinmor Road, Manchester, M22 1NN
G1 UTN C. King, 7 Hillcrest, Hyde, SK14 5LJ
G1 UTP S. Darlington, 17 Eleanor Road, Royton, Oldham, OL2 6BH
G1 UTS G. May, 95 Moorfield Avenue, Denton, Manchester, M34 7TX
G1 UTZ A. Peters, 10 Hill View Close, Grantham, NG31 7PH
G1 UUF I. Grounsell, 23 Loughbrow Park, Hexham, NE46 2QD
G1 UUJ D. Eyre, 29 Old Acre Lane, Brocton, Stafford, ST17 0TW
G1 UUK G. Perry, 6 Morgan Close, Arley, Coventry, CV7 8PR
G1 UUL S. Brough, 9 Beech Tree Lane, Cannock, WS11 1AZ
G1 UUO R. Hanson, 1 Ashmore Road, Coventry, CV6 1LH
G1 UUP A. Robins, 38 Eastbourne Road, Willingdon, Eastbourne, BN20 9NS
G1 UUS M. Carter, 4 Asbury Road, Balsall Common, Coventry, CV7 7QN
G1 UUT R. Bygate, 91 Woodcote Avenue, Kenilworth, CV8 1BE
G1 UUV R. Fairholm, 63 Rugby Road, Clifton Upon Dunsmore, Rugby, CV23 0DE
G1 UUZ M. Davis, 20 Pavilion Avenue, Warley, Smethwick, B67 6LA
G1 UVD R. Rothery, 2 Highcroft, Mount Pleasant, Batley, WF17 7NT
G1 UVE A. Shackleton, 31 Ashton Street, Leeds, LS8 5BY
G1 UVI M. Stanley, 35 Moorgate Road, Kippax, Leeds, LS25 7ET
G1 UVJ D. Rogers, 11 Beech Crescent, Mexborough, S64 9EH
G1 UVK A. Linfoot, 19 Vicarage Close, Bubwith, Selby, YO8 6LN
GW1 UVN J. Jones, Silversprings, Llanelly Church Road, Abergavenny, NP7 0EL
G1 UWD D. Peachey, 28 Broad St., Truro, TR1 1JD
GM1 UWE M. Peachey, 8/7 Durar Drive, Edinburgh, EH4 7HN
G1 UWQ J. Naughton, 86 Brookford Avenue, Coventry, CV6 2GQ
G1 UWV B. Dixon, Terridene, Park Road, New Milton, BH25 6QE
GW1 UXW M. Lewis, Gorse Cottage, Graig Road, Cwmbran, NP44 5AS
G1 UXZ Gravesend and Bean DX Group, c/o J. Sage, 8 Foxwood Road, Bean, Dartford, DA2 8BH
G1 UYT S. Tomkinson, 3 Heysham Close, Weston Coyney, Stoke-on-Trent, ST3 6RG
GW1 UYW K. Wallis, Cartref Newydd, Llanarmon D.C., LL20 X
G1 UYZ D. Payne, 8 Gascoigne Drive, Spondon, Derby, DE21 7GL
G1 UZC P. Jarrett, 17 Wolmers Hey, Great Waltham, Chelmsford, CM3 1DA
G1 UZD J. Yates, 8 Holt Drive, Wickham Bishops, Witham, CM8 3JR
G1 UZS E. Lees, 11A Edale Avenue, Mickleover, Derby, DE3 9FY
G1 UZW N. Boag, 60 Harebell, Amington, Tamworth, B77 4NA
G1 VAA N. Wills-Browne, 2 Parkfield Road, Aigburth, Liverpool, L17 8UH
G1 VAB D. Goodwill, 94 Palmerston Street, Derby, DE23 6PF
GM1 VAD S. Scanlain, Crossraguel, Old Newton, Nairn, IV12 5RA
G1 VAG A. Grant, 26 Fountains Avenue, Boston Spa, Wetherby, LS23 6PX
G1 VAJ A. Leach, 8 Eskdale, Brownsover, Rugby, CV21 1NJ
G1 VAL M. Pearson, 5 Craven Court, Warwick Drive, Barnoldswick, BB18 6WA
G1 VAN P. Van Falier, 572 Stafford Road, Ford Houses, Wolverhampton, WV10 6NN
G1 VAO A. Andrews, 12 Kings Lea, Ossett, WF5 8RY
GW1 VAW B. Williams, 10 Tynybedw Terrace, Treorchy, CF42 6RL
G1 VAY D. Hearn, 90 Princes Drive, Valley Dip, Seaford, BN25 2TX
GI1 VAZ G. Richardson, 6 Cedarhurst Rise, Belfast, BT8 7RJ
G1 VBA P. Buckle, 63 Ashley Drive South, Ashley Heath, Ringwood, BH24 2JP
G1 VBB R. Bedwell, 24 Tiger Moth Drive, Southam, CV47 1AS
GM1 VBD R. Scott, Enzie Slackhead, Buckie, Moray, AB5 2BJ
GM1 VBE Dr S. Clink, Southsyde, Woodhead Avenue, Glasgow, G71 8AR
G1 VBL R. Kelsall, 11 Manor Road, Ducklington, Witney, OX29 7YD
G1 VBO R. Piper, Tatnam Farm, St. Mary'S Road, Romney Marsh, TN29 0PW
G1 VBP P. Smith, The Rufford Care Centre, Room 1, Gateford Road, S81 7BH
G1 VBQ P. Wright, 81 High St., Syston, Leicester, LE7 1GQ
G1 VBY K. Moreton, 29 Tiber Drive, Newcastle, ST5 7QD
G1 VCU A. Smithson, 118 High Street, Linton, Cambridge, CB21 4JT
G1 VCZ A. Hewes, 89 Fronks Road, Dovercourt, Harwich, CO12 4EQ
G1 VDE D. Taylor, 21 Munday Close, Bussage, Stroud, GL6 8DG
G1 VDO S. Preston, 50 Milton Avenue, Malton, YO17 7LB
G1 VDP C. Colclough, 53 St Marys Road, Nuneaton, CV11 5AT
GW1 VDT P. Whatley, Fairacre, Bishton Lane, Chepstow, NP16 7LG
GW1 VDW L. Marquardt, 2 Pembroke Terrace, Varteg, Pontypool, NP4 7UJ
GM1 VDZ G. Neil, 7 Dalfarson Avenue, Dalmellington, Ayr, KA6 7TX
G1 VFH M. Staley, 128 Pinewood Crescent, Stoke-on-Trent, ST3 6HZ
GM1 VFQ J. Murdoch, 8 Primpton Avenue, Dalrymple, Ayr, KA6 6EL
GM1 VFR H. Mcdonald, 43 Southfield Road Cumbernauld, Glasgow, G68 9DZ
G1 VFW I. Beeby, 32 Edditch Grove, Bolton, BL2 6BJ
G1 VGA K. Crane, 92 Dimond Road, Southampton, SO18 1JS
G1 VGI K. Waterson, 20 Cadogan Road, Bury St Edmunds, IP33 3QJ
G1 VGK M. Whittington, 253 Kings Drive, Eastbourne, BN21 2UR
G1 VGM S. Ball, 16 Stonewood Gate, Field Lane, St. Helens, Ryde, PO33 1FY
G1 VGO D. Holdsworth, 28 Moorbank Close, Wombwell, Barnsley, S73 8RX
G1 VGP G. Anderson, 1 White Rose Mead, Garforth, Leeds, LS25 2EG
G1 VHC P. Ward, Parkhill House Parkhill, Toddington, lu5 6aw
G1 VHN D. Barnes, 46 Lawnswood Avenue, Wordsley, Stourbridge, DY8 5LR
G1 VHW W. Killeen, 12 Meriden Grove, Lostock, Bolton, BL6 4RQ
G1 VHY D. Symonds, 79 Kingsway, Kirkby-In-Ashfield, Nottingham, NG17 7EH
G1 VID T. Howe, The Old School House, 3 Bairds Hill, Broadstairs, CT10 3AA
G1 VIF C. Morphett, 138 Healds Road, Dewsbury, WF13 4HT
G1 VIG D. Carrick, 5 Kings Close, Market Overton, Oakham, LE15 7PS
G1 VII N. Pooley, 64 Lynwood Grove, Orpington, BR6 0BH
G1 VIN C. Kirkland, 29 Shelley Road, Enderby, Leicester, LE19 4QX
G1 VIO A. Eades, 41 Woodhall Gate, Pinner, HA5 4TX
G1 VIP M. Walker, 43 Wimborne Road Cogdean, Corfe Mullen, BH21 3DS
GW1 VIR I. Berry, 1-2 Pottery Cottages, Trefonen, Oswestry, SY10 9DF
G1 VIS S. Grimes, 73 Ryston Road, Denver, Downham Market, PE38 0DP
G1 VIT N. Cliff, 12 New Church Road, Wellington, Telford, TF1 1JH
G1 VIW R. Paterson, 27 Copt Heath Drive, Knowle, Solihull, B93 9PA
G1 VIY T. Depledge, 13 Peel Drive, Astbury, Congleton, CW12 4RF
G1 VIZ A. Hemenway, 69 Sixth Avenue, Heworth, York, YO31 0UR
GW1 VJB B. Flounders, 76 Springfield Road, Sebastopol, Pontypool Torfaen, NP4 5BX
GM1 VJD R. Terras, 8 Haddington Gardens, Lawthorn, Irvine, KA11 2EB
G1 VJE M. Sayers, Sgs, Trenchard Building, Jssu (Cy), BFPO59
G1 VJG C. Rhenius, 8 Rutland Close, Cambridge, CB4 2HT
G1 VJJ C. Silvey, 106 Southchurch Avenue, Southend-on-Sea, SS1 2RP
G1 VJN B. Jones, 73 Tonge Road, Murston, Sittingbourne, ME10 3NR
G1 VJQ T. Monaghan, 15 Mulgrave Road, Worsley, Manchester, M28 2RW
G1 VJY A. Birch, Maximilien, Les Ecovets, Chesieres, Switzerland, 1885
G1 VKB A. Morris, 140 Astwood Road, Worcester, WR3 8EZ
G1 VKC D. Close, 60 Lead Lane, Ripon, HG4 2LN
G1 VKG J. Howarth, 10 Poplar Place, Penrith, CA11 9HN
GI1 VKJ B. Duffy, 45 Boulevard Green, Newcastle, BT33 0FA
G1 VKN S. Matthews, Matilda - Kings Marina, Mather Road, Newark, NG24 1FW
G1 VKT E. Dale, 29 Hulme Road, Leigh, WN7 5BT
GM1 VLA A. Lee, Sandana, Kirkpatrick Fleming, Lockerbie, DG11 3BA
G1 VLD A. Burkitt, 6 Chewells Close, Haddenham, Ely, CB6 3XE
G1 VLS G. Turner, 19 Hector Road Darwen, Blackburn With Darwen, BB3 0AY
G1 VLU D. Green, 5 New Mill Road, Holmfirth, HD9 7SG
GW1 VMA M. Hills, Ty Newydd, Rhos, Llandyssul, SA44 5HE
GI1 VMF J. Oliver, 29 Callan Bridge Park, Armagh, BT60 4BU
G1 VMX E. Driver, 39 Witham Road, Woodhall Spa, LN10 6RW
G1 VNB D. Neeves, 18 Beechwood Road, Bedworth, CV12 9AG
G1 VNE S. Nocera, Strada Provinciale, Mulazzano 88, Parma, Italy, 43010
G1 VNH V. Palmer, 57 Old Tiverton Road, Exeter, EX4 6NG
G1 VNL A. Smith, 112 Manor Lane, Charfield, Wotton-under-Edge, GL12 8TN
G1 VNM S. Eyers, 190 Greenhill Road, Herne Bay, CT6 7RS
G1 VNS K. Baum, 11 St. Ives Road, Wigston, LE18 2JB
G1 VNU S. Exell, 22 Woodside Road, Beare Green, Dorking, RH5 4RH
G1 VNV M. Gold, 14 Brewers Lane, Badsey, Evesham, WR11 7EU
G1 VNZ A. Atkins, 28 Third Avenue, Pebsham, Bexhill-on-Sea, TN40 2PG
G1 VOB M. Prior, 36 Bassnage Road, Halesowen, B63 4HQ
G1 VOC A. James, 70 Martin Croft, Silkstone, Barnsley, S75 4JS
G1 VOJ A. Duce, 16 Gillmans Road, Orpington, BR5 4LA
G1 VON S. Howard, 95 Greenbarn Way, Blackrod, Bolton, BL6 5TE
G1 VOP Dr D. Hettiarratchi, 2 Carham Close, Gosforth, Newcastle upon Tyne, NE3 5DX
G1 VOQ P. Tandy, Old Channel Hill Farm, North End, Fordingbridge, SP6 3HA
G1 VOR S. Twigg, 16 Merlin Avenue, Nuneaton, CV10 9JZ
G1 VOY R. Ward, Overdale, Egton, Whitby, YO21 1UE

GI1 VPA D. Smythe, 57 Ballymacormick Avenue, Bangor, BT19 6AY
G1 VPC M. Wigley, 36 Wivelsfield Road, Haywards Heath, RH16 4EW
G1 VPE R. Wilcockson, Oaks Farm, Markham Road, Duckmanton, Chesterfield, S44 5HP
G1 VPS M. Wright, 27 Willow Rope Road, Kettering, NN15 7BA
G1 VQB G. Doughty, 95 Buxton Road, Chaddesden, Derby, DE21 4JL
G1 VQG G. Hannaford, 8 Porthmellon Gardens, Callington, PL17 7QL
G1 VQH G. King, Mill Cottage, 48 Mill Street, Swadlincote, DE12 8ES
G1 VQI C. Coates, Ecalox Ltd, Hammonds Farm, Stapleford Road, Stapleford Tawney, Stapleford Abbotts, Romford, RM4 1RR
G1 VQK E. Wright, 94 Bachelor Gardens, Harrogate, HG1 3EA
G1 VQV S. Gent, 66 Apperley Way, Cradley Forge, Halesowen, B63 2PY
G1 VRA E. Jones, Bramley Lodge, Back Lane, Royston, SG8 6DD
G1 VRC T. Nicholson, 8 East Street, High Spen, Rowlands Gill, NE39 2HD
G1 VRJ J. Ager, 20 Kirktonhill Road, Westlea, Swindon, SN5 7AF
GW1 VRR J. Williams, 31 Syr Davids Avenue, Cardiff, CF5 1GH
GW1 VRW C. King, 27 Gadlys Road West, Barry, CF62 7HX
G1 VSD W. Bennett, 90 Garwood Road, Yardley, Birmingham, B26 2AW
G1 VSH G. Pettit, 7 Dunster Crescent, Hornchurch, RM11 3QD
G1 VSK A. Heyes, 41 Coronation Drive, Penketh, Warrington, WA5 2DD
G1 VSM D. Pratt, 17 Worcester Gardens, Greenford, UB6 0BH
G1 VSO C. Champ, 31 Nobles Close, Oxford, OX2 9DN
GM1 VSR A. Bain, 4 Bawdley Head, Fraserburgh, Aberdeenshire, AB4 5SE
G1 VSX G. Buck, 1 Hobart Road, Weston-Super-Mare, BS23 4QQ
G1 VTE M. Joynson, 90 Fairhope Avenue, Bare, Morecambe, LA4 6LA
G1 VTK R. Swan, 13 Mendip Road, Torquay, TQ2 6UQ
G1 VTN C. Peacock, 1 Furnace Lane, Madeley, Crewe, CW3 9EU
G1 VTO B. Kemp, 193 Cavalry Park, March, PE15 9DL
G1 VTP J. Bridgehouse, 12 Castle Hall Close, Stalybridge, SK15 2HR
G1 VTQ B. Gorman, 40 Maudland Bank, Preston, PR1 2YL
G1 VTS J. Smith, 9 Birchway, Hayes, UB3 3PA
G1 VUG M. Cooper, Osborne House, Main Street, Louth, LN11 0XF
G1 VUK S. Hunt, 1 Lucknow Cottages, Northbridge Street, Robertsbridge, TN32 5NP
G1 VUP A. Cheeseman, Flat 5 Dubarry House, Hove Park Villas Hove Park Villas, Hove, BN3 6HP
G1 VUY D. White, 14 Waggoners Way, Bugbrooke, Northampton, NN7 3QT
G1 VVB S. Patrick, 9 Brant Avenue, Illingworth, Halifax, HX2 8DL
G1 VVE G. Bindon, 74 Cashford Gate, Taunton, TA2 8QB
G1 VVH P. Fisher, Flat 46 Morris House, Fairchild Close, London, SW11 2SU
G1 VVL A. Proctor, 448 Tuttle Hill, Nuneaton, CV10 0HR
G1 VVM G. Brett, 47 Windermere Avenue, Huncoat, Accrington, BB5 6JG
G1 VVT H. Mawson, 118 Byron Street, Loughborough, LE11 5JW
G1 VVU J. Stephens, 34 King Street, Seahouses, NE68 7XR
G1 VVX I. Andronov, 53 Broad Street, Ludlow, SY8 1NH
G1 VVY J. Waters, 2 Tea Caddy Cottages, Worthing Road, West Grinstead, Horsham, RH13 8LG
GM1 VWA J. Gilruth, 88 Fintry Crescent, Dundee, DD4 9EX
G1 VWC G. Matthews, 9 Sadlers Way, Hemingford Grey, Huntingdon, PE28 9EW
G1 VWL R. Knowles, 9 Malham Close, Southport, PR8 6UP
G1 VWP S. Goodwin, 75 Farleigh Hill, Tovil, Maidstone, ME15 6AA
G1 VWU A. Driver, Grace Barn, Pencarrow, Advent, Camelford, PL32 9RZ
G1 VWZ R. Huntley, 49 Main Street, Wetwang, Driffield, YO25 9XL
G1 VXD R. Roche, 98 Grange Road, Cheddleton, Leek, ST13 7NP
GM1 VXE A. Rennie, 5 Barbieston Cottage, Drongan, Ayr, KA6 7EF
G1 VXS S. Shirvington, 22 Bassett Road, Northleach, Cheltenham, GL54 3QJ
G1 VXX A. Powell, 9A Shaftesbury Close, Bracknell, RG12 9PX
G1 VXY T. Tomkins, 40 Diksmuide Drive, Ellesmere, SY12 9QA
G1 VYA H. Seddon, 25 Thistledown Close, Wigan, WN6 7PA
G1 VYB T. Smith, 11 River Terrace, Wisbech, PE13 1PZ
GM1 VYF P. Letters, 23 East Lennox Drive, Helensburgh, G84 9JD
G1 VYG J. Mcdonald, 4 Braeside Bowfield Road, Howwood, Johnstone, PA9 1BP
G1 VYM D. Eveleigh, 7 Malin Road, Littlehampton, BN17 6NN
G1 VYS A. Walker, 49 Selworthy Road, Stoke-on-Trent, ST6 8PL
G1 VZB C. Lillis, 6 Whitelake View, Urmston, Manchester, M41 8UT
GM1 VZG T. Gilmour, Fiold, Rope Walk, Kirkwall, KW15 1XJ
G1 VZT B. Rayner, 44 Foxhall Fields, East Bergholt, Colchester, CO7 6QY
G1 VZW B. Hitchen, 40 Methuen Avenue, Fulwood, Preston, PR2 9QX
G1 WAB Derbys Workd All Britain G.ARC, c/o J. Wainwright, 8 Common Lane, Cutthorpe, Chesterfield, S42 7AN
G1 WAC Wythall Radio Club, c/o M. Pugh, 44 Simms Lane, Hollywood, Birmingham, B47 5HY
G1 WAE C. Rogers, 63 Greenwell Road Haydock, St Helens, WA11 0SQ
G1 WAP B. Stott, 35 Sheridan Road, Laneshawbridge, Colne, BB8 7HW
G1 WAS P. Delaney, 61 Lyndale Avenue, Eastham, Wirral, CH62 8DG
G1 WAW Wessex Aw Club, c/o P. Mitchell, 3 High Howe Close, Bournemouth, BH11 8NN
G1 WCY Ripon & District ARS, c/o J. Reeves, 5 Arrows Crescent, Boroughbridge, York, YO51 9LP
G1 WDQ T. Hutton, 23 Dines Close, Wilstead, Bedford, MK45 3BU
G1 WEF R. Cornwell, 13 Milford Road, Thurrock, Grays, RM16 2QL
G1 WEV I. Henderson, 1 Prestbury Road, Pennywell, Sunderland, SR4 9DW
G1 WFA Dr C. Ward, 7 Coates Close, Heybridge, Maldon, CM9 4PB
G1 WFG S. Lake, 42 Haling Park Road, South Croydon, CR2 6NE
G1 WFJ A. Woodhouse, 5 Dudley Road, Kingswinford, DY6 8BT
G1 WFO P. Burden, 110 Westbury Leigh, Westbury, BA13 3SH
GI1 WFP N. Mcloughlin, 44 Kilbroney Rd, Rostrevor, BT34 3BL
G1 WFS S. Rogers, 19 Stoke Street, Hull, HU2 9BL
G1 WFU R. Dickson, 49 Ashgrove, Peasedown St. John, Bath, BA2 8EF
GI1 WGK W. Steele, 19 William Street, Donaghadee, BT21 0HL
G1 WGL S. Morris, 9 Starling Close, Burgess Hill, RH15 9XR
G1 WGM K. Perry, 25 Hillary Drive, Crowthorne, RG45 6QF
G1 WGO A. Smith, Dray Cottage, Main Street Bishop Wilton, Bishop Wilton, YO421RX
GW1 WGR West Glamorgan Raynet, c/o M. Rowles, 7 Gelli Deg, Bryncoch, Neath, SA10 7PL
G1 WHT M. Watts, 11 Bywood Place, Grimsby, DN37 9RH
G1 WHU D. Baines-Jones, 22-24 Grove Royd, Halifax, HX3 5QU
G1 WHY J. Lowe, 23 Hoylake Drive, Tividale, Oldbury, B69 1QA
G1 WID J. Hartridge, 15 Hundred Acres, Wickham, Fareham, PO17 6JB
G1 WIS D. Mountain, 45 Westway Gardens, Redhill, RH1 2JB
G1 WIW R. Dowdeswell, 5 Croft Close, Barwell, Leicester, LE9 8EW
GU1 WJA W. Ayres, Rousay, Bailiffs Cross Road, St Andrew, Guernsey, GY6 8RY
G1 WJG J. Coates, The Old Timbers, 23 Yoells Lane, Waterlooville, PO8 9SG
G1 WJK G. Richards, 3 Pleasant Close, Kingswinford, DY6 9TQ
G1 WJO A. Blackwell, 1 Gladstone Terrace, Hinckley, LE10 1HE
G1 WJR W. Rollins, 5 Little Clacton Road, Great Holland, Frinton-on-Sea, CO13 0ET
GM1 WKH N. Graves, The Lythe, 8 Tree Road, Tarves, Ellon, AB41 7JY
G1 WKK J. Arnott, 27 Main Road, Tadley, RG26 3NJ
G1 WKO R. Reichmann, 9 Rue Du Croteau, la Neuville Les Wasigny, France, 8270
G1 WKS West Kent ARS, c/o L. Featherstone, Prices Wood Bungalow Leigh, Tonbridge, TN11 8HP
G1 WKZ G. Evans, 4 The Mallards, Fareham, PO16 7XR
G1 WLD S. Evans, 4 The Mallards, Fareham, PO16 7XR
GI1 WLJ G. Mccutcheon, 73 Tullynagardy Road, Newtownards, BT23 4TB
G1 WLN N. Roskruge, 2 Lanner Green Terrace, Lanner, Redruth, TR16 6DQ
G1 WLO M. Head, 46 Ridgewood Gardens, Bexhill-on-Sea, TN40 1TS
G1 WLU S. Brookes, Ivy Cottage, Haselor, Alcester, B49 6LX
G1 WLW A. Fordyce, 41 Benscliffe Drive, Loughborough, LE11 3JP
G1 WLX J. Goacher, 41 Clay Hill, Two Mile Ash, Milton Keynes, MK8 8AY
G1 WMK J. Mayo, 10 Church Close, Fringford, Bicester, OX27 8DR
G1 WMN K. Harvey, 61 Westfield Road, Northchurch, Berkhamsted, HP4 3PW
G1 WMS R. Cadwallader, Rambla Grande, Los Reyes, Urcal, Spain, 4691
GM1 WMU S. Webster, 15 Forrest Place, Armadale, Bathgate, EH48 2GZ
G1 WMV B. Catchpoole, 8 Buckland Avenue, Basingstoke, RG22 6JL
G1 WNL F. Thomas, 38 Partridge Avenue, Yateley, GU46 6PB
G1 WNZ G. Sollazzo, 7 Miles Road, London, N8 7SJ
G1 WOR Worthing & Dist ARC, c/o P. Godbold, 13 Dawn Crescent, Upper Beeding, Steyning, BN44 3WH
G1 WPG B. Groome, The Old Smithy, High Street, Dorchester, DT2 8JW
G1 WPH L. Eden, 23 Elm Green Close, Worcester, WR5 3HD
G1 WPL R. Balkwell, 2 Franklyn Road, Droylsden, Manchester, M43 6DS
G1 WPR T. Bromley, 7 Brookside, Desborough, Kettering, NN14 2UD
G1 WQC R. Pratt, 11 Park Road, Ryde, PO33 2BG
G1 WQH R. Booth, 66 Fairburn Crescent, Pelsall, Walsall, WS3 4PU
G1 WQL P. Kenyon-Brodie, 17 Potterdale Drive, Little Weighton, Cottingham, HU20 3UU
G1 WQN S. Mangan, 48 Emblett Drive, Newton Abbot, TQ12 1YJ
G1 WQU T. Gregg, 27 Somerleaze Close, Wells, BA5 1UD
G1 WQX A. Peek, 7 De Havilland Road Upper Rissington, Cheltenham, GL54 2NZ
G1 WQY K. Webber, 2 Henniker Road, Ipswich, IP1 5HD
G1 WRC Wisbech AR & Electronics Club, c/o J. Balls, 70 Risegate Road, Gosberton, Spalding, PE11 4EY
G1 WRD M. Simpson, Orchard House, Todwick Grange, Sheffield, S26 1JQ
G1 WRE R. Constantine, 34 Lynden Close, Ripon, HG4 1US
G1 WRF A. Jolly, 27 Murrayfield Drive, Brandon, Durham, DH7 8TG
G1 WRH G. Marshall, 1 Portland Close, Braintree, CM7 9NJ
G1 WRN North Warks Raynet Group, c/o T. Yorke, 12 Shanklin Drive, Weddington, Nuneaton, CV10 0BA
G1 WRO M. Smith, 8 Milldale Road, Farnsfield, Newark, NG22 8DQ
G1 WRS Wakefield and District Radio Society, c/o D. Burden, 16 Milnthorpe Lane, Wakefield, WF2 7DE
G1 WRU J. Jinks, 27 Taryn Drive, Darlaston, Wednesbury, WS10 8XY
GW1 WRV R. Marston, 34 Kevin Ryan Court, Georgetown, Merthyr Tydfil, CF48 1EE
G1 WRY A. White, 34 Pain'S Way, Amesbury, Salisbury, SP4 7RG
G1 WSC J. Bland, 9 Earl Street, Grimsby, DN31 2NB
G1 WSD D. Garratt, 87 Garden Road, Eastwood, Nottingham, NG16 3FY
G1 WSE J. Frizell, 17 St. Johns Terrace, Lewes, BN7 2DL
G1 WSF D. Pettican, 52 Shepherds Way, Saffron Walden, CB10 2AH
G1 WSN J. Spillett, Mockbeggar Cottage, Mockbeggar, Ringwood, BH24 3NQ
G1 WSW M. Flewitt, 38 Laburnum Avenue, Newbold Verdon, Leicester, LE9 9LQ
G1 WSZ V. Tosney, 126 Norburn Park Witton Gilbert, Durham, DH7 6SQ
G1 WTB E. Musson, 110 Marples Avenue, Mansfield Woodhouse, Mansfield, NG19 9DW
G1 WTH C. Piddock, 118 Howley Grange Road, Halesowen, B62 0HU
GW1 WTL J. Beachey, 24 Trem-Y-Mynydd Court, Blaenavon, Pontypool, NP4 9LX
G1 WTN R. Wroe, 13 Silverdale Drive, Barnsley, S71 2PP
G1 WTS M. Roper, 19 Normay Rise, Newbury, RG14 6RY
G1 WTW Prof. C. Underwood, 4 Hawthorn Road, Godalming, GU7 2NE
G1 WTX P. Egan, 13 Beechcroft Drive, Guildford, GU2 7SA
G1 WTY B. Parkes, 11 Hampton Grove, Cheadle Hulme, Cheadle, SK8 6DG
GW1 WTZ C. Green, 11 Brookfield Close, Gorseinon, Swansea, SA4 4GW
G1 WUC M. Garner, 40 Studley Road, Harrogate, HG1 5JU
G1 WUH K. Carr, 41 Surrey Road, Dagenham, RM10 8ES
G1 WUM R. Miles, Haseley Lodge, Birmingham Road, Warwick, CV35 7HF
G1 WUU J. Neate, 23 Crossley Moor Road, Kingsteignton, Newton Abbot, TQ12 3LE
G1 WUY J. Wilkins, 21 Stocks Loke, Cawston, Norwich, NR10 4BS
G1 WVD M. Shrago, 12 Oakwood Road, Bricket Wood, St Albans, AL2 3PU
G1 WVK J. Power, 45 Grace Gardens, Cheltenham, GL51 6QE
G1 WVM R. Vowles, 47 Tyndale Avenue, Yate, Bristol, BS37 5EX
G1 WVP P. Marsh, 182 Oldbrook Boulevard, Oldbrook, Milton Keynes, MK6 2HG
G1 WVR Welland Valley ARS, c/o D. Lowe, 21 Farndon Road, Market Harborough, LE16 9NW
G1 WVS P. Gibson, 17 Nene Side Close, Badby, Daventry, NN11 3AD
G1 WVV R. Sutton, 28 Shrubbery Gardens, Wem, Shrewsbury, SY4 5BX
G1 WVW R. Jones, 8 Downing Avenue, Newcastle, ST5 0JY
G1 WVZ R. Mccutcheon, 13 The Beeches, Rugeley, WS15 2QY
G1 WWA R. Williams, Coombe Farm Cottage, Stottesdon, Kidderminster, DY14 8LS
G1 WWB R. Eeles, 23 Elgin Avenue, Ashford, TW15 1QE
GW1 WWE J. Peake, Winley, 70 Higher Lane, Swansea, SA3 4PD
G1 WWH A. Benn, Burneston, Bedale, DL8 2HT
G1 WWI M. Dronfield, White Lodge Farm, High Bradfield, Sheffield, S6 6LJ
G1 WWP J. Sharpe, 10 Stocking Green Close, Hanslope, Milton Keynes, MK19 7NH
GW1 WWW Lord W. Edmondson, Glaslyn, Penysarn, LL69 9YB
G1 WWY L. Donald, 53 Andrews Way, Raunds, Wellingborough, NN9 6RD
G1 WXC D. Blackman-Wells, 15 Purbeck Place, Littlehampton, BN17 5DP
G1 WXF J. Pearson, 17 Hebden Avenue, Woodloes Park, Warwick, CV34 5XD
G1 WXK M. Bell, 151 Towngate, Ossett, WF5 0PP

G1 WXS P. Springall, 31 The Orchards, Epping, CM16 7BB
G1 WXT M. Moorecroft, 4 St. Davids Road, Locks Heath, Southampton, SO31 6EP
G1 WXU G. Parsons, 73 Worthing Avenue, Elson, Gosport, PO12 4DB
G1 WXW P. Prescott, 13 The Boltons, Waterlooville, PO7 5QR
G1 WYA R. Webb, 1, London Road, Oldham, OL1 4BJ
G1 WYB Lord E. Coupe, Killidina, 28 Wellington Road, Blackburn, BB2 2NQ
G1 WYC S. Smith, 82 Wignals Gate, Holbeach, Spalding, PE12 7HR
G1 WYD A. Darlington, 15 Kestrel Close, Carterton, OX18 3LS
G1 WYG D. Biginton, 67 Capstone Road, Bromley, BR1 5NA
G1 WYM Dr M. Cheema, Lower Ebford Barton, Ebford, Exeter, EX3 0RA
G1 WYN F. Wilkinson, 218 Thorney Leys, Witney, OX28 5NZ
G1 WYP H. Milsom, 1 Wyld Court, Blunsdon, Swindon, SN25 2EE
GM1 WYV A. Henderson, 30 Pentland Crescent, Larkhall, ML9 1UR
GI1 WYZ R. Kennedy, 3 St. Annes Crescent, Newtownabbey, BT36 5JZ
G1 WZB I. Ross, 46 Fordbridge Road, Ashford, TW15 2SJ
G1 WZG J. Endicott, 16 Packs Close, Harbertonford, Totnes, TQ9 7TL
GW1 WZI J. Cartwright, 20 Castlefield Place, Cardiff, CF14 3DU
G1 WZK D. Heaton, 39 Bridgwater Road, Romford, RM3 7UB
G1 WZM C. Turner, 2 Martins Mews, Haverhill, CB9 7FU
G1 WZO L. Leach, Leyland, The Street, Gloucester, GL2 7ED
G1 WZQ A. Utting11, 20 Davenport Road, Leicester, LE5 6SA
G1 XAA H. Jacklin, 26 Rockmill End, Willingham, Cambridge, CB24 5HY
G1 XAJ J. Franklin, 16 Mountbatten Drive, Colchester, CO2 8BH
G1 XAL A. Dangerfield, Brookside, High Street, Gloucester, GL2 7LW
G1 XAM J. Bryant, 12 Dale Tree Road, Barrow, Bury St Edmunds, IP29 5AD
G1 XAP P. Whittingham, 28 Wedge Avenue, Haydock, St Helens, WA11 0DY
GW1 XAS J. Hume, 2 Llain Wen, Pentrefelin, Amlwch, LL68 9PD
G1 XBE T. Beecher, 77 Grime Lane, Sharlston Common, Wakefield, WF4 1EH
GW1 XBG P. Smith, 35 Terrace Road, Swansea, SA1 6HN
GM1 XBK K. Mcclure, 9 Cumnock Road, Mauchline, KA5 5AE
G1 XBL B. Darby, Pippins, Green Street, Worcester, WR5 3QB
G1 XBR S. Loney, 4 Mendip Road, Southampton, SO16 4BN
G1 XCB P. Davies, 91 Station Road, Hadfield, Glossop, SK13 1AR
G1 XCK W. Potter, Flat, 1 Tabernacle Walk, Blandford Forum, DT11 7DL
G1 XCY E. Knott, 24 Walsingham Way, Ely, CB6 3AL
G1 XDJ H. Opitz, 26 Holme Court, Lower Warberry Road, Torquay, TQ1 1QR
G1 XDK D. Rouse, 23 Montgomery Way, King's Lynn, PE30 4YH
G1 XDS J. Fyson, One Redlands Estate, Ibstock, LE6 1HT
G1 XDV T. Gale, 1A Aldridge Road, Streetly, Sutton Coldfield, B74 3TU
GM1 XEA P. Thomson, 13 Westwood Drive, Westhill, AB32 6WW
GM1 XEB M. Mcculloch, 6 Learmont Place, Milngavie, Glasgow, G62 7DT
G1 XEH R. Burton, 18 Churchfield, Harpenden, AL5 1LL
G1 XEP L. Lambert, 124 Frankland Road, Croxley Green, Rickmansworth, WD3 3AU
G1 XES E. Turner, 1104 Wimborne Road, Bournemouth, BH10 7AA
G1 XET C. Burton, 14 Fotherley Road, Mill End, Rickmansworth, WD3 8QG
GW1 XFB D. Evans, Bwthyn Bach, 2 Old Village Road, Barry, CF62 6RA
G1 XFE J. King, 40 Galway Avenue, Chaddesden, Derby, DE21 6TR
G1 XFL K. Lanham, 22 Ascot Close, Ladywood, Birmingham, B16 9EY
G1 XFM J. Shaw, 33 Park Farm Close, Horsham, RH12 5EU
G1 XFO M. Price, 67 Broadway, Oldbury, B68 9DP
G1 XFR B. Bance, 36 Leafy Oak Road, Grove Park, London, SE12 9RS
G1 XGE D. Hutchinson, 32 The Causeway, Kingswood, Hull, HU7 3AL
G1 XGM M. De-Wynter, 8 Eldon Place Cutler Heights, Bradford, BD4 9JH
G1 XGN M. Brady, 67 Boardman Fold Road Middleton, Middleton, M24 1QD
G1 XGP S. Blinkhorn, 12 Eloura Lane, New South Wales, Australia, 2577
G1 XGW A. Gray, 12 Peak Close, Oldham, OL4 2TH
G1 XGZ D. Richards, Flat 8, Thackeray Court, London, SW3 3LB
G1 XHA J. De Bank, 5 Horn Hill View, Beaminster, DT8 3PJ
G1 XHO T. Williams, 145 Bulwell Lane, Old Basford, Nottingham, NG6 0BS
G1 XHR E. Goodwin, Hankelow Court, Hall Lane, Crewe, CW3 0JB
GM1 XHZ T. Valentine, 4 Angus Cottages, Friockheim, Arbroath, DD11 4SR
GI1 XIB J. Wilkinson, 67 Glenwood, Ahoghill, Ballymena, BT42 1GW
G1 XIE R. Dyer, 21 Allden Avenue, Aldershot, GU12 4AG
G1 XIH L. Taylor, 76 Sidney Road, Blackley, Manchester, M9 8AT
G1 XII T. Lovatt, 5 Acre Rise, Willenhall, WV12 4SL
GM1 XIN W. Allan, Corse Farm, Kininmonth, Peterhead, AB42 4JU
G1 XIO A. Faram, 4 Wellington Road, Gillingham, ME7 4NN
G1 XIV G. Reynolds, The Thatched Cottage, St. Thomas Drive, Bognor Regis, PO21 4TN
G1 XIY Dr M. Nottingham, 11 Taverners Drive, Ramsey, Huntingdon, PE26 1SF
GM1 XJE Dr J. Hopkins, 9 Pathfoot Avenue, Bridge Of Allan, Stirling, FK9 4SA
GW1 XJJ H. Worgan, 29 Mayfield Avenue, Laleston, Bridgend, CF32 0LH
G1 XJK F. Tilley, 37B Fant Lane, Maidstone, ME16 8NP
G1 XJM R. Davidson, 17 Willows Avenue, Alfreton, DE55 7ER
G1 XJN D. Jones, 12 Brockhill Close, Kettering, NN15 7DS
G1 XJO N. Snowden, 11 Marion Drive, Shipley, BD18 2EY
G1 XJT A. Nichols, Amherst Harnham Lane, Withington, GL54 4DD
G1 XJZ D. Layton, 7 Turton Street, Kidderminster, DY10 2TH
G1 XKB N. Bowen, 2 Thorncliffe Road Great Barr, Birmingham, B44 9DB
G1 XKD G. Lawton, 23 Fiske Court, Cavendish Road, Sutton, SM2 5ER
G1 XKJ K. Higlett, 3 Clover Way, Killinghall, Harrogate, HG3 2WE
G1 XKL B. Smith, 1 Hirsts Cottages, Spa Lane, Ormskirk, L40 6JG
G1 XKN A. Chambers, 34 Haunchwood Drive, Sutton Coldfield, B76 1JR
G1 XKQ B. Neate, 30 Berry Avenue, Paignton, TQ3 3QN
G1 XKY E. Marsh, 15 Beacon Close, Rubery, Birmingham, B45 9DA
G1 XLE P. Bryan, 2 Regency Court, Armitage Road, Brereton, Rugeley, WS15 1PE
G1 XLG C. Proctor, 24 Orchard Way, Southam, CV47 1EX
GM1 XLH C. Cullingworth, Lochmoss, Ythanwells, Huntly, AB54 6HA
G1 XLL P. Greetham, Flat 10, Hillman House, Coventry, CV1 1FZ
G1 XLN J. Banks, 2 Birchlands Road, Stoke-on-Trent, ST1 6TW
G1 XLT R. Perrat, 18 Petts Hill, Northolt, UB5 4NL
G1 XLW A. Harrington, 44 Fairburn Crescent, Pelsall, Walsall, WS3 4PU
GD1 XMA M. Haley, Yn Croit, Ballamanagh Road, Sulby, Isle of Man, IM7 2HB
G1 XMH R. Fisher, White House, Slough Road, Manningtree, CO11 1NS
G1 XMI J. Brown, 78 Park Way, St Austell, PL25 4HR
G1 XMP C. Brookes, 58 Brookwood Drive, Stoke-on-Trent, ST3 6HY
G1 XNC R. Gearing, 6 Boughton Close, Gillingham, ME8 6ND
G1 XNG C. Whitehead, 6 Welbeck Street, Sutton-in-Ashfield, NG17 4AY
G1 XNI N. Dingle, 29 Castle View, Witton Le Wear, Bishop Auckland, DL14 0DH
G1 XNK R. Rafter, 8 Bishops Walk, Ilchester, Yeovil, BA22 8NS
G1 XNN R. Harding, Highview, High Road, Wallingford, OX10 0QT
G1 XNX M. Cowell, 105 Belgrave Road, Darwen, BB3 2SF
G1 XOG I. Wilson, 131 Weatherly Road, Torbay, Auckland, New Zealand, 630
GM1 XOI Mid Lanark ARS, c/o C. Welsh, 28 Peacock Wynd, Motherwell, ML1 4ZL
G1 XOT B. Blake, Ty Capel Tynygraig, Ystrad Meurig, SY25 6AE
G1 XOW S. Wragge, Treetops, Priory Road, Thurgarton, Nottingham, NG14 7GW
G1 XOZ C. Harding, 24 Bryer Close, Bridgwater, TA6 6UR
G1 XPD L. Wheatley, 25 Hobbis House, Redditch Road, Birmingham, B38 8LS
GM1 XPE J. Graham, Lodge, Stronsay, Orkney, KW17 2AN
G1 XPF A. Duell, 3 Jail Lane, Biggin Hill, Westerham, TN16 3SA
G1 XPI R. Wilson, 10 Ringway Garforth, Leeds, LS25 1BN
GI1 XPV K. Nesbitt, 2 Church Road Gracehill, Ballymena, BT42 2NL
G1 XPW M. King, 104 Green Lane, Vicars Cross, Chester, CH3 5LE
G1 XQI K. Bates, Flat 2, 26 Leinster Road West, Harolds Cross, Ireland, DUBLIN 6
G1 XQP J. Jackson, Greengable Upcott, Bishops Hull, Taunton, TA4 1AQ
G1 XRE S. Staton, 6 Greenhowsyke Lane, Northallerton, DL6 1HP
G1 XRF B. Rogers, 5 Springfield Road, Ruskington, Sleaford, NG34 9HG
G1 XRJ S. Hancock, 8 Elanor Road, Sandbach, CW11 3FZ
G1 XRM M. Braybrook, 12 Bossington Close, Rownhams, Southampton, SO16 8DW
G1 XRO C. Frost, 110 Spring Hill, Weston-Super-Mare, BS22 9BD
G1 XRQ J. Koenig, 216 Bretch Hill, Banbury, OX16 0LU
G1 XRT R. Taylor, 24 Hoestock Road, Sawbridgeworth, CM21 0DZ
G1 XSA S. Cattle, 5 Highworth Drive, Newcastle upon Tyne, NE7 7FB
G1 XSM P. Elwell, Woodcroft, Vann Lake Road, Ockley, Dorking, RH5 5JB
G1 XSQ C. Nightingale, 3 Cedar Rise Crookham Common, Thatcham, RG19 8DY
G1 XST S. Davies, 64 Oakfields, Worth, Crawley, RH10 7FL
G1 XSV M. Scarr, 15 Biddesden Lane, Ludgershall, Andover, SP11 9PG
G1 XTA S. Hodson, Flagstones, 12 Duns Tew, Bicester, OX6 4JR
G1 XTD I. Clark, Flat, Redhill Farm, Penrith, CA11 0DT
GI1 XTK B. Braniff, 5 Cintons Park, Downpatrick, BT30 6NS
GW1 XUD R. Andrews, 270 Barry Road, Barry, CF62 8BJ
G1 XUE G. Henne, 57 Heaf Gardens, Bentley Close, Aylesford, ME20 7SF
G1 XUH M. Thornton, 46 Lavender Court, Croft Road, Barnsley, S70 3FG
G1 XUU S. Bishop, 22 John St., Brightlingsea, Colchester, CO7 0NA
G1 XUW D. Austin, 17 Patricia Avenue, Horstead, Norwich, NR12 7EW
GW1 XVC S. Ward, Beech Cottage, Saron Road, Penperlleni, Pontypool, NP4 0BN
G1 XVD C. Snow, 77 Oxford Drive, Hadleigh, Ipswich, IP7 6AW
G1 XVF T. Pottage, 18 Pennine Close, Huthwaite, Sutton-in-Ashfield, NG17 2QD
G1 XVL B. Perry, 152 Stanborough Avenue, Borehamwood, WD6 5LR
GW1 XVM J. Duggan, 112 Gaer Park Drive, Newport, NP20 3NR
G1 XVR D. Briggs, 17 The Lonnen, South Shields, NE34 8EJ
G1 XVW A. Pogorzelski, 28C Mosslea Road, London, SE20 7BW
G1 XVY R. Sacharewicz, 15 Milford Close, Walkwood, Redditch, B97 5PZ
G1 XWD A. Rhodes, 2 Kent Avenue, Theddlethorpe, Mablethorpe, LN12 1QE
G1 XWK A. Rich, The Court House, Wadborough Road, Worcester, WR7 4RF
G1 XWM M. Cox, 17 Tybalt Close, Heathcote, Warwick, CV34 6XB
G1 XWN G. Andrews, 22 Arnhem Grove, Braintree, CM7 5UQ
G1 XWO F. Williams, 15 Hartsbourne Way, Stafford, ST17 4NR
G1 XWS H. Seatory, Ivydene, The Street, Woodbridge, IP12 3QU
G1 XWZ F. Millbank, Room 216 Kate House, Pitchill House Nursing Home, Evesham, WR11 8SN
G1 XXE P. Yeates, 9 Arlington Road, St. Annes, Bristol, BS4 4AF
G1 XXF J. Ellison, 68 Rocket Way, Forest Hall, Newcastle upon Tyne, NE12 9RL
G1 XXH R. Tapp, 26A Main Road, Grendon, Northampton, NN7 1JW
G1 XXR S. Austin, 5 Mercia Road, Baldock, SG7 6RZ
G1 XXV T. Blackmore, 56 Fraser Close, Shoeburyness, Southend-on-Sea, SS3 9YS
G1 XXW P. King, 2 Ebenezer Cottages, Thorney Road, Peterborough, PE6 7UB
G1 XYD A. Bates, 29 Juler Close, North Walsham, NR28 0SY
G1 XYF C. Hudson, 8 College Road, Bredon, Tewkesbury, GL20 7EH
G1 XYG R. Butler, 15 Bracknell Crescent, Nottingham, NG8 5EU
G1 XYN I. Pritchard, 8 Hoon Avenue, Newcastle, ST5 9NY
G1 XYO S. Glazzard, 109 Highfields Road, Chasetown, Burntwood, WS7 4QS
G1 XYR D. Searle, 33 Claypool Road, Kingswood, Bristol, BS15 9QJ
G1 XYS A. Brown, 5 Somersby Drive, Kenton, Newcastle upon Tyne, NE3 3TN
G1 XYV M. Minshull, 12 Dunnett Close, Attleborough, NR17 2NG
G1 XYZ Kings Lynn ARC, c/o E. Haskett, 23 Gloucester Road, King's Lynn, PE30 4AB
G1 XZA I. Rawlingson, 1 Wadham Street, Stoke-on-Trent, ST4 7HF
G1 XZB J. Rawlingson, 1 Wadham St., Penkhull, Stoke on Trent, ST4 7HF
G1 XZG D. Collins, 5 Elmwood Close, Lincoln, LN6 0LZ
GW1 XZI R. Magwood, 13 Inverness Place, Cardiff, CF24 4RU
G1 XZQ R. Carville, 66 Ludlow Road, Paulsgrove, Portsmouth, PO6 4AE
G1 XZV J. Heys, 2 Oakenhill Walk, Bristol, BS4 4LP
G1 XZW R. Hudson, 7 Grange Avenue, Luton, LU4 9AS
G1 XZX B. Strutt, 10 Park Cottages, Lower Somersham, Ipswich, IP8 4PP
G1 YAB C. Rogers, 221 Dales Road, Ipswich, IP1 4JY
G1 YAE B. Thompson, 1 Littlehoughton Farm Cottages, Littlehoughton, Alnwick, NE66 3JZ
G1 YAF A. Tyler, 16 Harridge Road, Leigh-on-Sea, SS9 4HA
G1 YAH J. Mcsoley, 88 Rodings Avenue, Stanford-le-Hope, SS17 8DT
G1 YAS D. Elliott, 48A Great Lane, Reach, Cambridge, CB5 0JF
G1 YBA I. Hardaker, 57 Windermere Road, Nottingham, NG7 6HL
G1 YBB S. Clements, 46 Brampton Road, Newton Farm, Hereford, HR2 7DF
GW1 YBF L. Ward, 11 Verlands Way, Pencoed, Bridgend, CF35 6TY
G1 YBG Dr J. Ballance, Orchid Bank, Woolhope, Hereford, HR1 4RQ
G1 YBI A. Jones, 43 Oakleigh Road, Droitwich, WR9 0RP
G1 YBK J. Fyson, 1 Redlands Estate, Ibstock, Leicester, LE6 1HT
G1 YBM J. Pedley, 92 Ashfield Drive, Moira, Swadlincote, DE12 6HQ
G1 YBT J. Bagshaw, 2 Boulton Court, Robin Hood Road, Skegness, PE25 3QU
G1 YCK M. Travis, 10 Victoria Road, Kearsley, Bolton, BL4 8NR
G1 YCM A. Laughlan, 33 Park House, Gorseyfields, Manchester, M43 6DX
G1 YCN D. Lewis, 81 Ashton Avenue, Rainhill, Prescot, L35 0QR
G1 YCR R. Lawrence, 82 Moseley St., Southend on Sea, SS2 4NN
G1 YDA M. Davies, 10 Rue Alphonse Delaveau, Pouzauges, France, 85700
G1 YDD A. Forster, 56 Tantobie Road, Denton Burn, Newcastle upon Tyne, NE15 7DQ
G1 YDG A. Miles, Yew Tree House, Main Street, Wantage, OX12 0HT

G1 YDI C. Lambeth, 36 Mill Lane, Oxford, OX3 0QA
G1 YDJ T. Polley, 9 Otter Road, Clevedon, BS21 6LQ
G1 YDQ J. Carpenter, 34B Carey Park, Killigarth, Looe, PL13 2JP
GI1 YEA L. O'Flaherty, 1 Ravensdale Villas, Newry, BT34 2PG
G1 YED W. Ross-Fraser, 47 Lichford Road, Sheffield, S2 3LB
G1 YEH J. Davis, 38 Mountbatten Close, Yate, BS37 5TE
G1 YEP F. Russell, 7 Glenmore Avenue, Liverpool, L18 4QE
G1 YES B. Underhay, 24 Rutland Road, Southall, UB1 2UP
G1 YEU M. Eales, 32 Selston Drive, Nottingham, NG8 1DE
G1 YEV Dr D. Martin, 73 Summerfields Way, Ilkeston, DE7 9HE
G1 YEW R. Hill, Grange Barn, Funtington, Chichester, PO18 9LN
G1 YEZ A. Lord, 66 Salcombe Drive, Glenfield, Leicester, LE3 8AF
G1 YFA J. Rymsza, 24 Green Lane, Studley, B80 7HD
G1 YFC P. Neades, 57 Bullingham Lane, Hereford, HR2 6RU
G1 YFD S. Lycett, 27 Ropewalk, Alcester, B49 5DD
G1 YFE J. Dent, 90 Eastwood, Chatteris, PE16 6RX
G1 YFG R. Szemeti, The Stables, Hoarstone Court Trimpley Lane, Bewdley, DY12 1RB
G1 YFI J. Simmonds, 94 Gravel Hill, Tile Hill, Coventry, CV4 9JH
GM1 YFO P. Mirtle, 11 Humbie Road, Kirkliston, EH29 9AN
GW1 YFP M. Hearne, Mora, Rhydypandy Road, Morriston, SA6 6NX
G1 YFQ D. Scothern, 5 Wilkinson Drive Middle Rasen, Market Rasen, LN8 3LD
G1 YFT R. Allsopp, 271 Wigston Lane, Aylestone, Leicester, LE2 8DL
G1 YGP S. Jarman, 55 The Meadows, Todwick, Sheffield, S26 1JG
GM1 YGV R. Johnstone, 10 Lundy Road, Inverlochy, Fort William, PH33 6NX
GM1 YGW Dr G. Craib, 1C Cherry Bank, Dunfermline, KY12 7RG
G1 YGY C. Weaver, 11 Thirlmere, Swindon, SN3 6LA
GW1 YHA P. George, 24 Ty Fry Close, Brynmenyn, Bridgend, CF32 8YB
G1 YHB J. Moggeridge, 22 St. Michaels Court, Faircross Avenue, Weymouth, DT4 0DS
G1 YHE D. Coate, 74 Wimborne Road, Poole, BH15 2BZ
G1 YHG M. Kennedy, Milestones, Blandford Forum, DT11 9DW
G1 YHI K. Davies, 2 Orchard Close, Lytchett Minster, Poole, BH16 6JH
G1 YHJ G. Williams, 2 Cotton Close, Broadstone, BH18 9AJ
GW1 YHL D. Crawshaw, 12 Glanmor Crescent, Uplands, Swansea, SA2 0PJ
G1 YHN S. Rhodes, 221 Ormonds Close, Bradley Stoke, Bristol, BS32 0DW
G1 YHP J. Hogg, 1 Deepdale, Guisborough, TS14 8JY
G1 YHV S. Briscoe, 8B Corfe View Road, Corfe Mullen, Wimborne, BH21 3LZ
G1 YIL A. Kaye, 2 Church Place 135 Edward Road, Balsall Heath, Birmingham, B12 9JQ
G1 YIQ J. Stafford, 6 Gardners Drive, Hullavington, Chippenham, SN14 6EL
G1 YIZ A. Ward, 42 Felstead Crescent, Sunderland, SR4 0AB
G1 YJB G. Evans, 16 Kynaston Drive, Wem, Shrewsbury, SY4 5DE
G1 YJF G. Moulds, Mountain Ash, 2 Fremantle Close, Chelmsford, CM3 5TY
G1 YJH M. Blackman, Clough Head, Hollinsclough, Buxton, SK17 0RG
G1 YJI P. Kay, 97 Avenue Road, London, N14 4DH
G1 YJJ R. Colman, 197 Coppins Road, Clacton-on-Sea, CO15 3LA
G1 YJL W. Pond, The Wheatlands, Calais Street, Sudbury, CO10 5JA
G1 YJQ J. Duffy, Flat 14, The Sycamores, Newcastle upon Tyne, NE4 7ER
G1 YJR J. Davies, 70 Ash Road, Sandiway, Northwich, CW8 2PB
G1 YJY P. Sengupta, 48 Badger Close, Guildford, GU2 9WA
GM1 YKE J. Campbell, 16 Barony Road, Auchinleck, Cumnock, KA18 2LL
G1 YKI J. Heathfield, 82 Auriel Avenue, Dagenham, RM10 8BT
G1 YKK S. O'Connor, 32 Whitfield Cross, Glossop, SK13 8NW
G1 YKL J. Brackenridge, 21 St. Mark Road, Deepcar, Sheffield, S36 2TF
GW1 YKT W. John, 4 Heol Y Bryn, Rhiwbina, Cardiff, CF4 6HY
G1 YKX M. Rouse, 105 Great Spenders, Basildon, SS14 2NS
GW1 YKY S. Jones, 14 Plantation Drive, Croesyceiliog, Cwmbran, NP44 2AN
G1 YKZ R. Burt, 11 Long Common, Heybridge, Maldon, CM9 4US
G1 YLB S. Doyle, 2 The Greenways, Paddock Wood, Tonbridge, TN12 6LS
G1 YLE P. Adams, 25 Main Road, Kesgrave, Ipswich, IP5 1AQ
G1 YLG A. Hodkin, 18 Habershon Drive, Chapeltown, Sheffield, S35 2ZT
G1 YLJ A. Hunt, 14 Sandalwood Close, Willenhall, WV12 5YJ
G1 YLM E. Bradshaw, 38 Whiteford Drive, Kettering, NN15 6HH
G1 YLN M. Swetman, 11 Outer Circle, Taunton, TA1 2BS
G1 YLV M. Bayliss, 2 Plattens Court, Wroxham, Norwich, NR12 8SQ
G1 YMA W. Scoles, 26 The Close, Brancaster Staithe, King's Lynn, PE31 8BS
G1 YMC S. Fitzpatrick, 19 Claremont Falls, Killigarth, Looe, PL13 2HT
GM1 YME J. Hein, 78 Montgomery Street, Edinburgh, EH7 5JA
G1 YMH M. Homer, 86 Victoria Road, Brierley Hill, DY5 1DB
G1 YMJ A. Smith, 6 Norton Crescent, Towcester, NN12 6DN
G1 YMP S. Ellis, 6 Newark Road, Hindley, Wigan, WN2 3HR
G1 YMR P. Webster, 117 Warley Road, Blackpool, FY1 2RW
G1 YMV H. Johnson, 2 Greenbank Avenue, Storth, Milnthorpe, LA7 7JP
G1 YMY A. Hussain, 4 Riverside, Chadderton, Oldham, OL1 2TX
G1 YNH C. Arundel, 54 Broadmead, Castleford, WF10 4SE
G1 YNJ M. Rocke, Orchard House, 55 Tarvin Road, Chester, CH3 5DY
G1 YNO B. Surtees, 5 Haweswater Grove, West Auckland, Bishop Auckland, DL14 9LQ
G1 YNQ J. Crow, 71 Stockshill Road, Ashby, Scunthorpe, DN16 2LQ
G1 YOA G. Hawkins, 11523 Sun Ray Court, San Diego, USA, 92131
G1 YOF A. Lockwood, 9 Hartley Road, Exmouth, EX8 2SG
G1 YOS M. Avenell, Lime House, Worlds End, Newbury, RG20 8SD
GJ1 YOT N. Paisnel, 11 Bon Air Apartments, La Grande Route De La Cote, St Clement, Jersey, JE2 6SE
G1 YOU S. Nicholls, Fieldway, The Street, Eyke, Woodbridge, IP12 2QG
G1 YOY J. Bowen, 16 Cotham Lawn Road, Bristol, BS6 6DU
G1 YPH A. Roberts, 18 Surtees Grove, Stoke-on-Trent, ST4 3HH
GM1 YPJ L. Davies, 24 Ardgour Road, Caol, Fort William, PH33 7PQ
G1 YPM R. Northcott, 32 Lichfield Road, Exwick, Exeter, EX4 2EU
G1 YPR G. Dearden, 125 Campsall Field Road, Wath-Upon-Dearne, Rotherham, S63 7ST
G1 YPT G. Hartshorn, 11 Lime Avenue, Ripley, DE5 3HD
G1 YPU E. Rowberry, 69 Alpha Terrace, Trumpington, Cambridge, CB2 9HS
G1 YPZ L. Pritchett, Flat 75, Dalehead, London, NW1 2JL
G1 YQI P. Bennett, 7 Woburn Avenue, Firwood Ind Est, Bolton, BL2 3AY
G1 YQL C. Stagg, 559 Dividy Road, Stoke-on-Trent, ST2 0BX
GW1 YQM R. Evans, Maesyronnen, Sarnau, Llanymynech, SY22 6QL
G1 YQN P. Mcshea, Heathercot, Cross Drive, Maidstone, ME17 3NP
G1 YQP M. West, 27 Nidderdale Road, The Meadows, Wigston, LE18 3XW
G1 YQU J. Stapleford, 7 Garfield Road, Hugglescote, Coalville, LE67 2HU
G1 YQY W. Oakes, 2 Hillcrest Scotton, Catterick Garrison, DL9 3NJ
G1 YRC York ARC, c/o A. Palfrey, 5 Ings View, York, YO30 5XE
GM1 YRD J. Jones, Kirkland Of Glencairn, Kirkland Moniaive, Thornhill, DG3 4HD
G1 YRE S. Piper, Willow End, The Street, Ipswich, IP6 9HG
G1 YRF D. Buggs, 2 Archway Cottages Valley Road, Leiston, IP16 4AR
G1 YRJ M. Stott, 8 Kingfisher Way, Stowmarket, IP14 5BB
G1 YRM C. Mead, 32 Sandy Road, Potton, Sandy, SG19 2QQ
G1 YRQ R. Nock, 43 Delph Drive, Brierley Hill, DY5 2LQ
G1 YRR W. Fry, 227 London Road North, Merstham, Redhill, RH1 3BN
G1 YRY S. Roberts, 23 Deal Court, Haldane Road, Southall, UB1 3NT
G1 YSA M. Crick, 85 Ashurst Road, London, N12 9AU
GI1 YSG P. Kennedy, 29A Barnfield Road, Lisburn, BT28 3TQ
G1 YSX D. Taylor, 103 Southend, Garsington, Oxford, OX44 9DL
G1 YTG A. Thackray, 39 Totnes Close, Corby, NN18 8DB
G1 YTL H. Vyvyan, 13 Shearwater Close, Peel Common, Gosport, PO13 0RB
G1 YTO K. Wenman, 2 Hythe Road, Sittingbourne, ME10 2LR
G1 YTV S. Fisher, Farlands, Lower Rillaton, Callington, Lower Rillaton, PL17 7PF
G1 YTX T. Clayton, 40 Morrison Road, Darfield, Barnsley, S73 9ED
G1 YUB B. Harrison, 15 Helmington Terrace, Hunwick, Crook, DL15 0LQ
G1 YUL J. Robson, 28 Eastfield Street, Sunderland, SR4 7SA
G1 YUN K. Hetherington, 29 Broomridge Avenue, Newcastle upon Tyne, NE15 6QN
G1 YUS A. Lees, 692 Walmersley Road, Bury, BL9 6RN
G1 YUU A. Bagworth, 127 Barnsley Road, Darfield, Barnsley, S73 9PE
G1 YUX J. Garnett, 21 Vicarage Close, Mossley Hill, Mossley Hill, Liverpool, L18 7HU
G1 YVI K. Biddlecombe, 29 Stone Close, Worthing, BN13 2AU
G1 YVS C. Newby-Robson, 1 Bramley Drive, Offord D'Arcy, St Neots, PE19 5SF
G1 YVV I. Coleman, 69 Glebelands, West Molesey, KT8 2PY
G1 YVZ A. Bell, 159 Hounslow Road, Hanworth, Feltham, TW13 6PX
G1 YWI A. Williams, 26 Matlock Road, Bloxwich, Walsall, WS3 3QD
G1 YWN A. Whitworth, 183 Logan St., Bulwell, Nottingham, NG6 9FX
G1 YWY M. Jones, 16 Cumnock Road, Castle Cary, BA7 7FE
G1 YXA B. Dixon, 16 Dyrham Parade, Patchway, Bristol, BS34 6EF
G1 YXH R. Harrison, 14 St. Leonards Avenue, Chatham, ME4 6HL
G1 YXJ A. Palmer, 14 Garibaldi Road, Redhill, RH1 6PB
GW1 YXR N. Williams, 31 Syr Davids Avenue, Cardiff, CF5 1GH
G1 YXT C. Wise, 28 Southlands, East Grinstead, RH19 4BZ
G1 YXY L. White, The Garden Flat, 47 Hamerton Road, Gravesend, DA11 9DX
G1 YYC F. Karlinski, 100 Lindsay Avenue, Wakefield, WF2 8AS
G1 YYD E. Brown, 25 Cork Road, Lancaster, LA1 4BD
G1 YYH J. Heaton, 85 Morris Green Lane, Bolton, BL3 3JD
G1 YYL T. Mulloy, 49 Barons Close Kirby Muxloe, Leicester, LE9 2BW
G1 YYP M. Illston, 4 The Sett, Oxhill, Warwick, CV35 0RE
G1 YYU W. Fludgate, Holly Lodge, Thorpe Bank, Little Steeping, Spilsby, PE23 5BB
G1 YYY Braintree Raynet Group, c/o D. Willicombe, 26 Falkland Court, Braintree, CM7 9LL
GW1 YZF D. Edwards, 240 Berthin, Greenmeadow, Cwmbran, NP44 4LB
G1 YZH R. Baxter, 4 Kendal Gardens, Woodley, Stockport, SK6 1BL
G1 YZJ R. Rennison, Foxhall Cottage, Kelshall, Royston, SG8 9SE
G1 YZT E. Fenlon, 17 Hawes Avenue, Ramsgate, CT11 0RN
G1 ZAA A. Tice, 20 Manor Road, Middle Littleton, Evesham, WR11 8LL
G1 ZAK A. Powney, 16 Westbrook Way, Wombourne, Wolverhampton, WV5 0EA
G1 ZAR S. Tyler, 43 Wharf Road, Pinxton, Nottingham, NG16 6LH
G1 ZAW M. Smoker, 41 Queens Gardens, Dartford, DA2 6HZ
G1 ZAY A. Robinson, 31 Heathview Road, Socketts Heath, Grays, RM16 2RS
G1 ZBB S. Brown, 6 Pathfinder Way, Ramsey, Huntingdon, PE26 1LX
G1 ZBG M. Bason, 52 Wroslyn Road, Freeland, Witney, OX29 8HH
G1 ZBH G. Brock, 148 Lonsdale Drive, Rainham, Gillingham, ME8 9HX
G1 ZBJ A. Manning, 12 Clifford Drive, Heathfield, Newton Abbot, TQ12 6GX
G1 ZBL N. Marsh, 16 Laurel Close, North Warnborough, Hook, RG29 1BH
G1 ZBO Dr E. Mclusky, 11 Ripon Road, Killinghall, Harrogate, HG3 2DG
G1 ZBP A. Whipp, 114 Lower Manor Lane, Burnley, BB12 0EF
G1 ZBU G. Newby, 77 Darby Road, Garston, Liverpool, L19 9AN
G1 ZBW W. Baker, El-Granaro, Brampton, Appleby-in-Westmoreland, CA16 6JS
G1 ZBY S. Parker, 8 Greenbank Drive, Lincoln, LN6 7LQ
G1 ZCC C. Feather, 10 Thruffle Way, Bar Hill, Cambridge, CB23 8TR
G1 ZCS E. Davis, 10 Fairfield Drive, Lowestoft, NR33 8QG
G1 ZDG P. Whittaker, 156 Stoughton Road, Guildford, GU2 9PG
G1 ZDR J. Angus, 8 Gravel Road, Bromley, BR2 8PF
G1 ZDT A. Gregory, 9 Fordbridge Road, Ashford, TW15 2TD
G1 ZDU B. Rowles, 4 Milton Road, Aston Clinton, Aylesbury, HP22 5LA
G1 ZDX M. Bodecott, 1 The Park, St. Pegas Road, Peakirk, Peterborough, PE6 7NG
G1 ZDY C. Tipp, 27 Lakeland Avenue, Bognor Regis, PO21 5FA
G1 ZEA P. Jones, 21 Hill Top Rise, Harrogate, HG1 3BW
G1 ZEC G. Stevens, 25 Avenue Road, New Milton, BH25 5JP
G1 ZED B. Haworth, 139 Manchester Road, Accrington, BB5 2NY
G1 ZEI J. Wyatt, Ciampia, 10 St. Georges Hill, Perranporth, TR6 0DZ
G1 ZEK D. Ault, 68 Moira Dale Castle Donington, Derby, DE74 2PJ
G1 ZEU S. Aspey, 7 Foresters Path, School Aycliffe, Newton Aycliffe, DL5 6TA
G1 ZEW D. Pentin, 35 Seafore Close, Liverpool, L31 2JS
G1 ZEX R. Davies, 71 Higher Croft Road, Lower Darwen, Darwen, BB3 0QT
G1 ZFB K. Barton, 67, Grange Rd Flat3, Ramsgate, CT11 9LP
G1 ZFD J. Davies, 71 Higher Croft Road, Lower Darwen, Darwen, BB3 0QT
G1 ZFF T. Voisey, 26 Gorlands Road, Chipping Sodbury, Bristol, BS37 6LA
G1 ZFG J. Stephenson, 5 Hunstrete, Pensford, Bristol, BS39 4NT
G1 ZFS N. Woolard, 159 Medway Road, Worcester, WR5 1LL
GW1 ZFX J. Milosevic, 38 Thornhill Close, Upper Cwmbran, Cwmbran, NP44 5TQ
G1 ZGF R. Jackson, 37 Carisbrooke Road, Harpenden, AL5 5QS
G1 ZGH J. Sharpe, 204A Featherstone Lane, Featherstone, Pontefract, WF7 6AH
G1 ZHD A. Gilmore, Ashfields, Naseby Road Clipston, Market Harborough, LE16 9RZ
GW1 ZHI C. Owens, 7 Frondeg, Southsea, Wrexham, LL11 6RH
G1 ZHL M. Dharas, 225 Redmile Walk, Peterborough, PE1 4UR
G1 ZHN M. Griffiths, 2 Muirway, Benfleet, SS7 4LS
G1 ZHZ J. Hall, 27 Quarry Hill Road, Ilkeston, DE7 4DA
G1 ZIM S. Agnew, Rose Mount, The Hill, Millom, LA18 5HE
GM1 ZIV J. Large, 9 Maitland Terrace, Kildrochat, Stranraer, DG9 9EX

G1 ZJK D. Ellard, 35 Edgehill Drive, Daventry, NN11 0GR
G1 ZJP R. Offer, Chapel Yard Cottage, Quadring Eaudyke, Spalding, PE11 4QB
G1 ZJQ D. Smith, 44 Yarmouth Drive, Cramlington, NE23 1TS
GW1 ZKE M. Grindle, 57 Islwyn Street, Cwmfelinfach, Newport, NP11 7HY
GW1 ZKN R. Ogden, Plas Yn Bonwm Farm, Holyhead Road, Corwen, LL21 9EG
G1 ZKZ G. Kenealy, 20 Penny Lane, Haydock, St Helens, WA11 0QS
G1 ZLA M. Roberts, 18 Craster Drive, Nottingham, NG6 7FJ
G1 ZLB J. Kynaston, Smithy Cottage, Main Street, Nottingham, NG12 5PY
G1 ZLC P. Ashby, 12 Treeford Close, Solihull, B91 3PW
G1 ZLD M. Bignell, 57 Ramsey Road, Halstead, CO9 1AS
GW1 ZLL D. Ball, 38 Heol Sirhwi, Barry, CF62 7TG
G1 ZLY N. Cooper, 56 Kingfisher Road, Mansfield, NG19 6EG
G1 ZME R. Coatman, 3 Harold Avenue, Blackpool, FY4 5HG
G1 ZMG R. Hoad, Broad Lea, Amsbury Road, Maidstone, ME17 4DN
G1 ZMJ Rev. D. Roberts, 31 Seaton Way, Marshside, Southport, PR9 9GJ
G1 ZMW C. Tubey, 2 Rowley Close, Swadlincote, DE11 8LX
GW1 ZNC S. Elworthy, 70 Maple Drive, Brackla, Bridgend, CF31 2PF
G1 ZND A. Soble, 6 The Glebe, Hildersley, Ross-on-Wye, HR9 5BL
G1 ZNK A. Edwards, 68 Middlemarch Road, Coventry, CV6 3GF
GM1 ZNR V. Roberts, 4 Ladieside, Brae, Shetland, ZE2 9SX
G1 ZNT W. Barton, 27 Hornby Crescent, Clock Face, St Helens, WA9 4RY
G1 ZNV R. Charteris, 7 Kennedy Close, Kidderminster, DY10 1LR
G1 ZNX R. Agnew, 156 Goswell End Road, Harlington, Dunstable, LU5 6NT
G1 ZNZ A. Steele, 72 Park Lane, Knypersley, Stoke-on-Trent, ST8 7AS
G1 ZOB R. Brown, 28 Albertus Road, Hayle, TR27 4JQ
G1 ZOQ A. Mather, 330 Lever Street, Radcliffe, Manchester, M26 4PT
G1 ZOS C. Wood, Wurzerstr. 180, Bonn, Germany, 53175
GM1 ZOX N. Senior, 36 Lathro Park, Kinross, KY13 8RU
G1 ZOY M. Knowles, 17 Stainmore Close, Birchwood, Warrington, WA3 6TP
G1 ZPA N. Johanssen, 10 Waverley Court, Verulam Place, St Leonards-on-Sea, TN37 6QR
G1 ZPC C. Rule, 1 Park En Venton, Mullion, Helston, TR12 7JH
G1 ZPJ P. Read, 58 Godolphin Road, Helston, TR13 8QJ
G1 ZPO A. Brookes, 212 Pontefract Road, Featherstone, Pontefract, WF7 5AG
G1 ZPQ J. Pitchford, 7 Firecrest Drive, Leegomery, Telford, TF1 6FZ
G1 ZPU R. Compton, 18 Drove Road, Gamlingay, SG193NY
G1 ZQE D. Marsden, 94 Blackford Road, Shirley, Solihull, B90 4BX
GM1 ZQF G. Milne, 6 Alexandra Street, Alyth, Blairgowrie, PH11 8AS
G1 ZQG P. Huntley, 5 Beacon Avenue, Barton-upon-Humber, DN18 5DP
G1 ZQN P. Gibson, 60 Raglan Road, Bromley, BR2 9NW
G1 ZQO C. Stokes, 21 Deerswood Lane, Bexhill-on-Sea, TN39 4LT
G1 ZQR R. Twyman, Farmend, Halls Lane, Reading, RG10 0JB
G1 ZQV M. Lowe, 34 Woodbank Road, Groby, Leicester, LE6 0BN
G1 ZRE R. Ellis, 7 Bromley Close, Blackpool, FY2 0SD
G1 ZRP M. Crook, 21 Treyew Road, Truro, TR1 2BY
G1 ZRQ D. Barrett, 10 Trelawny Road, Menheniot, Liskeard, PL14 3TS
G1 ZRR N. Youngman-Smith, 12 Timber Way, Chinnor, OX39 4EU
G1 ZRS J. Cantwell, 10 Cathedral Drive, Fairfield, Stockton-on-Tees, TS19 7JT
G1 ZRT C. Guymer, 74 West Common Lane, Scunthorpe, DN17 1DU
G1 ZSE K. Crocker, 32 Godmanston Close, Poole, BH17 8BU
G1 ZSF A. Lewis, 76 Reading Road, Finchampstead, Wokingham, RG40 4RA
G1 ZSG C. Bell, 41A Handel Road, Canvey Island, SS8 7HL
G1 ZSK T. Adams, 26 Hillside Avenue, Plymouth, PL4 6PR
G1 ZST L. Sherwood, 50 Thornton Road, Manchester, M14 7WT
G1 ZSV R. Mercer, 23 Larne Road, Bilton Grange, Hull, HU9 4UE
G1 ZSY A. Hughes, 37 Brisbane Road, Reading, RG30 2PE
G1 ZSZ H. Hossle, Flat 91, Castlemeads Court, 143 Westgate Street, Gloucester, GL1 2PB
GM1 ZTB W. Bell, 77 Bongate, Jedburgh, TD8 6DU
G1 ZTG P. Wilsdon, 64 Chestnut Avenue, Euxton, Chorley, PR7 6BS
G1 ZTJ G. Bailey, 34 Newton Road, Bideford, EX39 2LL
G1 ZTK W. Causer, 47 Sandringham Road, Wombourne, Wolverhampton, WV5 8EF
G1 ZTM A. Corp, 158 Somerton Road, Street, BA16 0SA
G1 ZTN P. Harvey, 64 Privett Road, Gosport, PO12 3SX
G1 ZUB P. Thompson, Berry Brow, Wetherby Road, Leeds, LS14 3AU
G1 ZUC Dr R. Johnson, Mile House, Lansdown Road, Bath, BA1 5SY
G1 ZUH M. Pomroy, 21 Nook Farm Avenue, Syke, Rochdale, OL12 0SH
G1 ZUS N. Watts, The Vista, Churchill Way, Bideford, EX39 1PA
G1 ZUU Avon Valley ARS, c/o S. Brookes, Ivy Cottage, Haselor, Alcester, B49 6LX
G1 ZUZ J. Rowling, 11 Barncroft, Norton, Runcorn, WA7 6RJ
G1 ZVC S. Beith, 18 Avenue Road, New Milton, BH25 5JP
G1 ZVE H. Barugh, Westwinds, 40 Ruden Way, Epsom, KT17 3LN
GM1 ZVJ J. Hilton, 25 Alford Way, Dunfermline, KY11 8BF
G1 ZVO W. Scott, 16 Sweetbriar Lane, Holcombe, Dawlish, EX7 0JZ
G1 ZVZ K. Fish, 11 Little Meadow Way, Bideford, EX39 3QZ
G1 ZWB B. Ward, Parkhill House. Parkhill, Toddington, LU5 6AW
G1 ZWH T. Sanders, Sandalwood, 41 Tinney Drive, Truro, TR1 1AT
G1 ZWQ J. Bowker, 9 Scarthwood Close, Bolton, BL2 4DU
G1 ZWY C. Roe, 17 Hanbury Close, Chesterfield, S40 4SQ
G1 ZXC J. Westgate, 19 Granville Street, Gloucester, GL1 5HL
G1 ZXD M. Williams, 43 Hamble Road, Poole, BH15 3NJ
G1 ZYJ A. Ainger, 16 Hillside Road, Harpenden, AL5 4BT
G1 ZYN N. Suffolk, 2 Tamerton Road, Leicester, LE2 9DD
G1 ZYS F. Woodland, 12 Toll House Way, Chard, TA20 1FH
G1 ZZA P. Harvey, 10 Barnfield Close, Wirral, CH47 7DA
G1 ZZC P. Golds, 7 Selsey Close, Worthing, BN13 1LQ
G1 ZZG B. Thomas, 4 Gilbert Close, Torquay, TQ2 6BS
G1 ZZL M. Low, 23 Larch Crescent, Tonbridge, TN10 3NN

G2

GW2 ABJ G. Edwards, 2 Heol Y Glo, Tonna, Neath, SA11 3NJ
G2 ABR C. Mayman, Greenacre, Stones Green Road, Harwich, CO12 5BS
G2 AD East of England DX Group, c/o A. Mcgonigle, 5 Whitehouse Crescent, Chelmsford, CM2 7LP
G2 AIW M. Lambeth, 11 Ellerman Avenue, Twickenham, TW2 6AA
GM2 AJW J. Jack, Malindella, Main Road, Dumfries, DG1 1RZ
G2 ALM R. Wilkins, 36 Offington Gardens, Worthing, BN14 9AU
G2 ALN L. Taylor, 76 Sidney Road, Blackley, Manchester, M9 8AT
G2 ALX A. Wilson, 196 Spring Lane Lambley, Nottingham, NG4 4PE
G2 AMG H. Mitchell, Stone Cottage, Yeovil Road, Yeovil, BA22 9RR
G2 ANC J. Bromiley, 28 Clive Road, Westhoughton, Bolton, BL5 2HR
G2 API H. Batty, 64 North St., Scalby, Scarborough, YO13 0RU
G2 AQJ R. Collins, 17 Archers Court, Salisbury, SP1 3WE
G2 ART F. Cawson, 43 Trafalgar Road, Southport, PR8 2HF
G2 ARU R. Loveland, Apartment 8, Royal Bay Court, 86A Barrack Lane, Bognor Regis, PO21 4DY
G2 ARV R. Bennett, 16 Emily Street, St Helens, WA9 5LZ
G2 ARY G. Lee, 16 Phoenix Chase, North Shields, NE29 8SS
G2 AS Sheffield HF DX Group, c/o P. Day, 38 Broomhill Road, Old Whittington, Chesterfield, S41 9DA
G2 ASF Coventry ARS, c/o J. Beech, 124 Belgrave Road, Coventry, CV2 5BH
G2 AXO W. Purser, Bethel And Bethesda Residential Home, Equity Road East, Leicester, LE9 7FY
G2 AZM E. Oakley, 67 South Road, Northfield, Birmingham, B31 2QZ
G2 BAR B. Hill, 38 Westons Brake, Emersons Green, Bristol, BS16 7BP
G2 BBC Ariel Rad Group, c/o D. Pick, 178 Alcester Road South, Kings Heath, Birmingham, B14 6DE
G2 BBI L. Steel, 1B Trinity Avenue, Westcliff-on-Sea, SS0 7PU
G2 BGG J. Garner, Barbon, Aigburth Hall Road, Liverpool, L19 9DG
G2 BHG G. Harrison, 13 High View Park, Cromer, NR27 0HQ
G2 BHY A. Bonner, 57 Downsview, Heathfield, TN21 8PF
G2 BJK G. Brown, 25 The Cloisters, South Street, Wells, BA5 1SA
G2 BKZ R. Mctait, 20 Rowland Road, Stevenage, SG1 1TE
G2 BQP P. Gully, 23 Lawrence Grove, Henleaze, Bristol, BS9 4EL
G2 BQY Trowbridge and District ARC, c/o D. Birch, 32 Union Street, Trowbridge, BA14 8RY
G2 BRS Bournemouth Radio Society, c/o M. Stevens, 16 Golf Links Road, Ferndown, BH22 8BY
G2 BSJ R. Biltcliffe, 3 Church View, Steeple Claydon, Buckingham, MK18 2QR
G2 BSW R. Ward, Serendipity, 17 Marlpit Lane, Seaton, EX12 2HH
G2 BTZ E. Moreman, 5 Sheridan Way, Longwell Green, Bristol, BS30 9UE
G2 BUJ S. Greenwood, 29 The Elms, Nine Elms, Swindon, SN5 5XA
GM2 BWW A. Barrett, Mains Of Glasclune Farm Middleton Road, Blairgowrie, PH10 6SF
GI2 BX City of Belfast Radio Amateur Society, c/o F. Hunter, Flat 9 50 Edenvale Crescent, Belfast, BT4 2BH
G2 BXH A. Perkins, 3 Greenway Close, Radcliffe-On-Trent, Nottingham, NG12 2BU
G2 BXP M. Prestidge, 48 Parkfield Road, Warley, Oldbury, B68 8PT
G2 BZR R. Bassford, 59 Watling St., Dordon, Tamworth, B78 1SY
G2 CD R. Matthews, 7 Coolgardie Avenue, Chigwell, IG7 5AU
G2 CFC G. Fretwell, 17 Cross Lane, Stocksbridge, Sheffield, S36 1AY
GW2 CGF S. Griffiths, 1 Nicholl Court, Mumbles, Swansea, SA3 4LZ
G2 CHI W. Bailey, 25 Lenham Road East, Saltdean, Brighton, BN2 8AF
G2 CIW J. Moseley, 33 Cathedral Court, London Road, Gloucester, GL1 3QE
G2 CJK A. Clarkson, 6 Mather Avenue, Accrington, BB5 5AU
G2 CKR M. Garfitt, 90 Wedderburn Road, Malvern, WR14 2DQ
G2 CNN S. Ball, 16 Stonewood Gate, Field Lane, St. Helens, Ryde, PO33 1FY
G2 CO F. Cooknell, 65 Coombe Valley Road, Preston, Weymouth, DT3 6NL
G2 CP Scarborough ARC, c/o D. Herbert, 50 St. Leonards Crescent, Scarborough, YO12 6SP
G2 CQX V. Pugh, 8 Beech Close, Hanwood, Shrewsbury, SY5 8RA
G2 CR Cambridgeshire Raynet, c/o D. Burkin, 26 Rampton Road, Cottenham, Cambridge, CB24 8UL
G2 DAN S. Whiteley, 142 Brisbane Road, Mickleover, Derby, DE3 9JW
G2 DBH G. Dodd, St Nicholas Cottage, 14 Bury Fields, Guildford, GU2 4AZ
G2 DD L. James, Pinecroft, Green Drive, Wokingham, RG40 2HT
G2 DGB A. Short, 12 Grosvenor Crescent, Dorchester, DT1 2BA
GW2 DHM W. Andrews, 69 Fairwater Grove West, Cardiff, CF5 2JN
G2 DJ Derby & District ARS, c/o R. Buckby, 22 Woodstead, Embleton, Alnwick, NE66 3XY
G2 DJM Dr N. Chilton, 38 Kingswood Avenue, Newcastle upon Tyne, NE2 3NS
G2 DLX D. Mitchell, 1 Denstroude Cottages, Denstroude Lane, Canterbury, CT2 9JX
G2 DML J. Crossfield, Forest Lodge, Chopwell Wood, Rowlands Gill, NE39 1LT
GW2 DNJ N. Brierley, Minera, 6 Trinity Crescent, Llandudno, LL30 2PQ
G2 DP Tingley Moonbounce Society, c/o D. Parker, 50 Rein Road, Tingley, Wakefield, WF3 1HZ
G2 DPA M. Brashill, 42 Bannister Street, Withernsea, HU19 2DT
G2 DPY D. Silverson, 63 Downside, Shoreham-by-Sea, BN43 6HF
G2 DRM D. Mobey, 21 Langdale Court 151 Windermere Drive, Wellingborough, NN83XA
G2 DT Furness ARS, c/o M. Bell, 36 Schneider Road, Barrow-in-Furness, LA14 5DW
G2 DWB N. Webster, 1 Gratton Dale, Carlton Colville, Lowestoft, NR33 8WP
G2 DWC Dubmire ARC, c/o B. Wheeler, 2 Rose Street, Houghton le Spring, DH4 5BB
G2 DXU A. Wyatt, Flat 3, Soper House, 9 Dart View Road, Galmpton, Brixham, TQ5 0BQ
G2 DZH N. Talbot, 105 Westwood Lane, Welling, DA16 2HJ
G2 EC Blandford Garrison ARC, c/o R. Carter, 12 Glebe Close, Abbotsbury, Weymouth, DT3 4LD
G2 FA Folkestone & District ARS, c/o Dr D. Pepper, 17 Cliffe House, Radnor Cliff, Folkestone, CT20 2TY
G2 FCP F. Varley, 39 Nettleton Road, Mirfield, WF14 9AW
G2 FFD D. Skipworth, Melrose, West End Road, Boston, PE22 0BU
G2 FGT R. Rogers, 67 Kingswell Avenue, Arnold, Nottingham, NG5 6SY
G2 FHF J. Illsley, Fancys Farm, Glacis, Portland, DT5 1FR
G2 FJA MARTS, c/o K. Earl, 210 Churchill Avenue, Chatham, ME5 0JS
G2 FKO Appledore and District ARC, c/o J. Lovell, Kowloon, Slade, Bideford, EX39 3LZ
G2 FKZ Radio Society of Great Britain, c/o S. Thomas, 2 Myrtle Cottages, Sandy Lane, Saxmundham, IP17 1HR
G2 FLW M. Clarkson, Causeway Cottage, Sawley Road, Clitheroe, BB7 4RS
G2 FM Flaxton Moor Contest Group, c/o C. Quarton, Flaxton Gatehouse, Flaxton, York, YO60 7QT
G2 FMW E. Baker, 86 Osborne Gardens, Herne Bay, CT6 6SE
GW2 FOF Rhondda ARS, c/o J. Howells, 13 Vicarage Road, Penygraig, Tonypandy, CF40 1HR
G2 FQZ R. Day, Resting Oak Cottage, Resting Oak Hill, Lewes, BN8 4PS

G2 FSH B. Weeden, 24 Berkeley Close, Rochester, ME1 2UA
G2 FSJ K. Levitt, 1 Charnwood Close, Andover, SP10 2RB
G2 FSR J. Hunt, 4 Warmdene Road, Brighton, BN1 8NL
G2 FVL L. Carrick-Smith, Highfields House, Sheffield Road, Clowne, Chesterfield, S43 4AP
G2 FXJ S. Moisy, 15 Charles Street, Redditch, B97 5AA
G2 FXQ S. Saddington, South Ridding, Sibson Road, Atherstone, CV9 3RE
G2 FXV M. Middleton, Dolphin View Nursing And, Residential Home, Harbour Road, Morpeth, NE65 0AP
G2 FXZ J. Hodgetts, 59 Woodland Road, Halesowen, B62 8JS
G2 FYO H. Terraneau, 2653 Nutmeg Circle, Simi Valley, USA, 93065-1327
GW2 HCA L. Sanders, 2 Cae Neuadd, Penybontfawr, Oswestry, SY10 0NS
G2 HCG B. Sykes, Flat 7, Solent Pines Whitby Road, Milford On Sea, Lymington, SO41 0UX
G2 HDF Midland Contest Group, c/o M. Waldron, 32 Windmill Street, Upper Gornal, Dudley, DY3 2DQ
G2 HFP S. Trudgill, 267 Clifton Drive South, Lytham St Annes, FY8 1HW
GW2 HFR J. Kelly, Arosfa, Westminster Road, Wrexham, LL11 6DN
G2 HHH T. Bayliss, 55 Foxlydiate Crescent, Redditch, B97 6NJ
G2 HIX Dr D. Craig, Pear Tree Cottage, Cripps Corner, Staplecross, TN32 5QS
G2 HKQ A. Knight, 17 Moorland Crescent, Upton, Poole, BH16 5LA
G2 HKS R. Udall, Longfield, 20 Upper Way, Rugeley, WS15 1QA
G2 HKU E. Trowell, 316 Minster Road, Minster On Sea, Sheerness, ME12 3NR
G2 HLB C. Maltby, The Willows Farm, Stallingborough Road, Grimsby, DN40 1NR
G2 HLP D. Hearsum, 1225 Duckview Court, Centerville, USA, 45458-2784
G2 HMK T. Brown, 99 Brinkburn Drive, Darlington, DL3 0JY
G2 HNA J. Weaver, 7 Cramer St., Stafford, ST17 4BX
G2 HNI L. Hewitt, 60 Shaftesbury Avenue, Southampton, SO17 1SD
G2 HR Silverthorn Radio Club, c/o L. Butterfields, 22 Horsley Road, London, E4 7HX
G2 HS Echelford DX Group, c/o Dr P. Miller Tate, 19 Esher Avenue, Walton-on-Thames, KT12 2SZ
G2 HW South Manchester Radio Club, c/o R. Smith, 16 Coniston Avenue, Sale, M33 3GT
G2 HX Gloucester Amateur Radio & Electronics Society, c/o L. Harris, 183A Painswick Road, Gloucester, GL4 4AG
G2 IF W. Setterfield, 54 Hallam Road, Nelson, BB9 8AB
G2 JL T. Mortimer, 10 Harold Road, Hayling Island, PO11 9LT
G2 KF T. Harris, Summerfield, Coombe Road, Lanjeth, High Street, St Austell, PL26 7TL
G2 KG C. Hill, 47 Belswains Lane, Hemel Hempstead, HP3 9PW
G2 KQ B. Hawes, 3 Orchard Close, Cassington, Witney, OX29 4BU
G2 KS K. Matthews, St. Helens Cottage, Flimby, Maryport, CA15 8RX
G2 LK L. Piggott, 37 Moss Lane Worsley, Manchester, M28 3WD
G2 LW Crystal Palace Radio & Electronics Club, c/o R. Burns, 84 Portnalls Road, Coulsdon, CR5 3DE
G2 MN J. Walker-Wilson, Rest Harrow Southside, Scorton, Richmond, DL10 6DN
GM2 MP North of Scotland Contest Group, c/o Prof. K. Kerr, East Loanhead, Auchnagatt, Ellon, AB41 8YH
G2 NF A. Canning, 261 Loddon Bridge Road, Woodley, Reading, RG5 4BL
G2 OA Southport & District ARC (SADARC), c/o C. Staples, 32 Browns Lane, Netherton, Bootle, L30 5RW
GW2 OG J. Hogg, Bwthyn Y Briallu, Ynys Ferw Bach, Gaerwen, LL60 6NW
GW2 OP Pembrokeshire Contest Group, c/o M. Shelley, Sunray, Pendine, Carmarthen, SA33 4PD
G2 OU Farmors School Radio Club, c/o D. Tatlow, Mulberry House, Bettys Grave, Cirencester, GL7 5ST
G2 PA P. Dyke, 5328 Malcolm Street, Oceanside, USA, 92056
G2 PB P. Baron, 55 Church View, Brompton, Northallerton, DL6 2RD
G2 PK J. Ellison, Jowsers, Northfield Lane, Wells-Next-the-Sea, NR23 1JZ
G2 RE R. Evered, Ivy Cottage, Old Bristol Road, Wells, BA5 3AL
GU2 RS R. Robilliard, Moss Bank, La Mare, St Andrew, Guernsey, GY6 8XX
G2 RSA P. King, 32 Millstream Way, Leegomery, Telford, TF1 6QR
G2 SH J. Shearme, Chevin, Penn Street, Amersham, HP7 0PY
G2 SR Surrey Raynet, c/o T. Dabbs, 4 Caverleigh, Cadogan Road, Surbiton, KT6 4DH
G2 SU Northern Hgts Rd, c/o A. Robinson, 9 Illingworth Close, Illingworth, Halifax, HX2 9JQ
G2 SZ D. Goyder, 8 Bloomsbury Walk, Southampton, SO19 9GB
G2 TO Bury St Edmunds ARS, c/o M. Green, 4 Boundary Cottages, Great Finborough, Stowmarket, IP14 3AG
G2 UG Halifax & District ARS, c/o D. Baker, 17 Woodroyd Gardens, Luddendenfoot, Halifax, HX2 6BG
G2 UH D. Hayward, Hope, Churchwell Street, Sherborne, DT9 6RG
G2 UT K. Reid, 4 Harles Acres, Hickling, Le14 3af
GM2 UWX P. Riggs, 106 Braes Avenue, Clydebank, G81 1DP
G2 VS R. Barrett, 76 Westgate Park, Sleaford, NG34 7QP
G2 XG E. Davie, 7 Cranworth Crescent, Chingford, London, E4 7HN
G2 XP Sutton & Cheam, c/o J. Puttock, Sutton & Cheam Rs, 53 Alexandra Avenue, Sutton, SM1 2PA
G2 YC R. Mcknight, Ardralla, Church Cross, Skibbereen, Ireland, P81 RK12
G2 YL S. Quarton, Flaxton Gatehouse, Flaxton, York, YO60 7QT
G2 YT P. Fox, Hillside House, Almshoebury, St. Ippolyts, Hitchin, SG4 7NT

G3

G3 AAF K. Avery, 4 Whiphill Close, Doncaster, DN4 6DX
G3 AAS M. Glynn, 39 Moor Allerton Drive, Leeds, LS17 6RY
G3 AB A. Chadwick, 5 Thorpe Chase, Ripon, HG4 1UA
G3 ABG WAB Awards Group, c/o J. Brooks, 28 Avon Vale Road, Loughborough, LE11 2AA
G3 ACQ H. Harmsworth, 43 Cornelian Avenue, Scarborough, YO11 3AN
G3 ADZ K. Gaunt, 21 Abbey Close, Rendlesham, Woodbridge, IP12 2UD
GM3 AEI J. Rosselle, 140 Main Street, Neilston, G78 3JX
G3 AER G. Wright, 70 Gunton Drive, Lowestoft, NR32 4QB
G3 AFB D. Tait, 34 Mount St., Dorking, RH4 3HX
G3 AGC W. Curphey, 8 Emily Davison Avenue, Morpeth, NE61 2PL
G3 AGF R. Edginton, 9 Churchill Road, Seaford, BN25 2UL
G3 AHE R. James, 40 Barrack Road, Hounslow, TW4 6AG
GM3 AHR A. Thomson, Meadowrise, 4 Law View Gardens, Leven, KY8 5SW
G3 AJD T. Moore, 6 Old Parsonage Court, Otterbourne, Winchester, SO21 2EP
G3 AJK R. Earland, 7 Trews Weir Court, Exeter, EX2 4JS
G3 AKF Reading & District ARC, c/o V. Robinson, 4 Hilltop Road, Caversham, Reading, RG4 7HR
G3 AKI F. Knowles, 1 Mayfield Close, Bishops Cleeve, Cheltenham, GL52 8NA
G3 AKJ A. Wheele, 4 Mannings Way, Barnstaple, EX31 1QF
GM3 ALF A. Megson, 2 Garden Cottage, Glendelvine, Murthly, PH1 4JN
G3 ALG G. Starling, 207 Shirley Road, Croydon, CR0 8SB
G3 ALK E. Holmes, 7 Castle Drive, Ilford, IG4 5AE
GM3 ALZ F. Gordon, Crofts Of Torrancroy, Strathdon, AB36 8UJ
GJ3 AME P. Landor, Lauge, Rue Des Raisies, St Martin, Jersey, JE3 6AT
G3 AMH H. Green, 9 Robert Avenue, Cundy Cross, Barnsley, S71 5RB
G3 AMK B. Littleproud, 25 Fern Avenue, Lowestoft, NR32 3JF
G3 AMW Hull and District ARS, c/o B. Atkinson, 165 Alliance Avenue, Hull, HU3 6QY
GI3 AMY J. Collett, 10 Cronstown Road, Newtownards, BT23 8QS
G3 APL J. Russon, 59 Ridge Road, Kingswinford, DY6 9RE
G3 APS L. Shergold, 8 The Moors, Lydiard Millicent, Swindon, SN5 3LE
G3 APU J. Andrews, 44 Eastridge View, East The Water, Bideford, EX39 4RS
G3 AQB W. Stephenson, 20 Chapel Court, Chapel Row, Seahouses, NE68 7TD
G3 AQF A. Kearns, 8 Pennyfathers Lane, Welwyn, AL6 0EN
G3 ARE F. Chubb, 2 Brook Close, Plympton, Plymouth, PL7 1JR
GW3 ARS J. Sagar, 75 Hookland Road, Newton, Porthcawl, CF36 5SG
G3 ASG R. Fautley, 7 Kingfisher Road, Downham Market, PE38 9RQ
G3 ASR Edgware & District Radio Society, c/o H. Haria, 34 Larkfield Avenue, Harrow, HA3 8NF
G3 AST J. Plowman, 17 Orchardleigh, East Chinnock, Yeovil, BA22 9EN
G3 ASV G. Pope, 5 Penn Crescent, Haywards Heath, RH16 3HW
GW3 ASW Aberdare ARS, c/o B. Werrell, 26 Glynhafod Street, Cwmaman, Aberdare, CF44 6LD
G3 ASX D. Paine, 43 Wilton Road, Muswell Hill, London, N10 1LX
G3 ATC OBO Air Cadet Radio Society, c/o W. Green, 2 Irkdale Avenue, Enfield, EN1 4BD
G3 ATI A. Williams, 74 Broadfield Road, Bristol, BS4 2UW
G3 ATX A. Perry, The Cottage, The Green, Bristol, BS48 3BG
GW3 ATZ G. Morris, 18 Grosvenor Road, Shotton, CH5 1NU
GM3 AUE A. Mcghie, 1 Boyach Crescent, Isle Of Whithorn, Newton Stewart, DG8 8LD
G3 AVE F. Flanner, 1 Ludford Close, Sutton Coldfield, B75 6DW
G3 AVL R. Reynolds, 12 Eastham Rake, Wirral, CH62 9AA
G3 AVN P. Parker, Flat 15, The Rise Care Home, Dawlish, EX7 0QL
G3 AWK N. Gough, 3 Sycamore Close, Cherry Willingham, Lincoln, LN3 4BJ
G3 AWP P. Gifford, 21 Bengal Road, Bournemouth, BH9 2ND
GM3 AXX A. Fraser, 58 Rigghead, Stewarton, Kilmarnock, KA3 3DQ
G3 AYL North Yorkshire YI Group, c/o R. Duffield, 4 Crabmill Lane, Easingwold, York, YO61 3DE
G3 AZI A. Mccann, 105 Todd Lane North, Lostock Hall, Preston, PR5 5UP
G3 AZW A. Bates, 68 Hill St., Hilperton, Trowbridge, BA14 7RS
G3 BAC R. Bastow, 2A New Road, Meopham, Gravesend, DA13 0LS
G3 BAR K. Brenchley, 6 Windwards Close Lanreath, Looe, PL132WP
G3 BBK J. Orrin, Greenacres, Church Street, Heathfield, TN21 9AL
G3 BBX D. Holloway, 10 Spencers Orchard, Bradford-on-Avon, BA15 1TJ
G3 BCE D. Nichols, Marsh Farm, Camp Road, Templecombe, BA8 0TH
G3 BDQ J. Heys, White Friars, Friars Hill, Hastings, TN35 4EP
G3 BDT A. Searle, 30 Hawthorne Grove, Poulton-le-Fylde, FY6 7PN
G3 BEX W. Short, Highland Light, 26 Howard Crescent, Beaconsfield, HP9 2XP
G3 BFL H. Siebert, 3 Greenlands Road, Kingsclere, Newbury, RG20 5RJ
G3 BGF R. Winkworth, 1 Collingwood Drive, Mundesley, Norwich, NR11 8JB
G3 BHA N. Taylor, 8 Aragon Way, Bournemouth, BH9 3SB
G3 BHM H. Kempson, 8 Hounds Way, Hayes, Wimborne, BH21 2LD
G3 BII A. Clark, 19 Lakes Lane, Beaconsfield, HP9 2LA
G3 BIK E. Chicken, Ivy Thorn Cottage, Morpeth, NE61 6LQ
G3 BJ D. Beattie, Hares Cottage, Woolston, Church Stretton, SY6 6QD
G3 BJD J. Maxwell, 10 Castle View, Egremont, CA22 2NA
G3 BKJ H. Alderson, 31 Rumbold Road, Edgerton, Huddersfield, HD3 3DB
G3 BLS D. Walker, 32 South St., Osney, Oxford, OX2 0BE
G3 BMO H. Speed, 45 Willow Glade, Huntington, York, YO32 9NJ
G3 BMQ G. Humphrey, 56A Park Lane, Wallington, SM6 0TN
G3 BNE G. Alderman, 35 Eynswood Drive, Sidcup, DA14 6JQ
G3 BNF A. Embleton, 34 Riverdale Park, Bent Lane, Chesterfield, S43 3UH
G3 BNW J. Bailey, 13 Heywood Road, Alderley Edge, SK9 7PN
G3 BOK S. Rutt, Granthorpe, Hull Road, Hull, HU11 5RN
G3 BPF A. Painter, Cold Green, Rochford, Tenbury Wells, WR15 8SP
G3 BPK Douglas Vall AR, c/o D. Snape, 30 Culcross Avenue, Wigan, WN3 6AA
G3 BPP R. Hampton, 11 Greenlands, Hutton Rudby, Yarm, TS15 0JQ
G3 BPQ E. Smith, 23 The Ladysmith, Ashton-under-Lyne, OL6 9AP
G3 BQE R. Fussey, 9 Alicia Gardens, Harrow, HA3 8JB
G3 BQT E. Hulme, 21 Brookside Crescent, Greenmount, Bury, BL8 4BG
G3 BRQ K. Tackley, 1 Greenways, Fleet, GU52 7UG
G3 BRS Bury Radio Society, c/o P. Stocks, 12 Bredbury Drive Farnworth, Bolton, BL4 7QD
G3 BSN P. Stanley, 1 Thames View, Cliffe Woods, Rochester, ME3 8LR
GM3 BSQ Aberdeen ARS, c/o I. Munro, 57 Craigiebuckler Avenue, Aberdeen, AB15 8SF
GM3 BST J. Tuke, 2/23 Hawthorn Gardens, Loanhead, EH20 9EE
GW3 BV Q. Cruse, Glas Y Dorlan, Llanfihangel-Y-Creuddyn, Aberystwyth, SY23 4LA
G3 BVA E. Digman, 75 Ramsden Road, Orpington, BR5 4LU
G3 BVB D. Adair, 3 Belmont Close, Shaftesbury, SP7 8NF
G3 BWI W. Timms, 22 Padway, Penwortham, Preston, PR1 9EL
G3 BYG N. Williams, Chapel Lake Halwill, Beaworthy, EX21 5UF
G3 BZB R. Cunliffe, 5 Silk Mill Lane, Tutbury, Burton-on-Trent, DE13 9LE
G3 BZU Royal Naval ARS, c/o J. Kirk, 111 Stockbridge Road, Chichester, PO19 8QR
G3 CAJ R. Prince, 52 Mafeking Road, Southsea, PO4 9BG
G3 CAZ J. Shaw, 128 Perth Road, Ilford, IG2 6AS
GW3 CBA H. Kellaway, 34 Winston Road, Barry, CF62 9SW
G3 CDM I. Gardner, 30 Pierremont Crescent, Darlington, DL3 9PB
G3 CEI Dr C. Brown, Downlands, Off Hackwood Lane, Basingstoke, RG25 2NH
GW3 CF Prestatyn ARC, c/o K. Shepherd, 15 Gronant Road, Prestatyn, LL199DT
GI3 CFH Nth West Irelan, c/o D. Fulton, 120 Dunnalong Road, Bready, Strabane, BT82 0DP

G3 CFR J. Jowett, Ashleigh, Kilmington, Axminster, EX13 7ST
G3 CGD J. Yeend, 30 St. Lukes Road, Cheltenham, GL53 7JJ
G3 CGE R. Gardner, 62 Rosewall Road, Southampton, SO16 5DW
G3 CIK H. Romer, 96 Mortlake Road, Richmond, TW9 4AS
G3 CIL M. Holley, 6586 196Th Street, Langley, Canada, BC V2Y 1R3
G3 CIM S. Denney, 52A Intwood Road, Cringleford, Norwich, NR4 6AA
GM3 CIO RSARS - Royal Signals Amatuer Radio Society, c/o C. Hall, 21 Peterculter Retirement Park Peterculter, Aberdeen, AB14 0AB
G3 CKE S. Mason, 46 Frankton Close, Redditch, B98 0HJ
G3 CKR M. Ryder, 9 Lincoln Close, Woolston, Warrington, WA1 4LU
G3 CLW L. Hutton, 46 Penwill Way, Paignton, TQ4 5JQ
G3 CMH Yeovil ARC, c/o D. Bowden, 58 Southville, Yeovil, BA21 4JF
G3 CMU H. Meyers, Cornerways, 2 Old Mill Lane, Polegate, BN26 5NS
G3 CNO Fort Purbrook ARC, c/o M. Ponsford, 83 Grant Road, Farlington, Portsmouth, PO6 1DU
G3 CO Colchester Radio Amateurs Club, c/o H. Yeldham, 19 Wade Reach, Walton on the Naze, CO14 8RG
GM3 COB J. Paterson, 10 Hathaway Drive, Giffnock, Glasgow, G46 7AE
G3 CON L. Crabbe, 6 Node Hill, Studley, B80 7RR
GM3 COQ D. Oswald, 8 Redfield Road, Montrose, DD10 8TW
G3 CPG L. Damon, 18 Scafell Court, Dewsbury, WF12 7PD
GW3 CPM C. Vugts, Coed Coch, Llangammarch Wells, LD4 4BS
G3 CPN M. Stevens, 16 Golf Links Road, Ferndown, BH22 8BY
G3 CPT D. Capp, 46 Stoke Road, Bletchley, Milton Keynes, MK2 3AD
G3 CQL M. Clarke, 3 Shelley Priory Cottages, Shelley, Ipswich, IP7 5RQ
G3 CQU K. Raffield, 113 Waddington Avenue, Coulsdon, CR5 1QP
GW3 CR R. Richards, 77 Church Road, Llanstadwell, Milford Haven, SA73 1EA
G3 CRC Clacton Radio Club, c/o C. Day, 35 Rochford Road St. Osyth, Clacton-on-Sea, CO16 8PH
G3 CRH H. Sanders, Little Orchard, 68A Park Road, Burton-on-Trent, DE13 7AJ
G3 CRS Royal Naval ARS, c/o J. Kirk, 111 Stockbridge Road, Chichester, PO19 8QR
G3 CSA Ellesmre P&D AR, c/o T. Saggerson, 18 Ploughmans Way, Great Sutton, Ellesmere Port, CH66 2YJ
G3 CSR Civil Service ARS, c/o N. Sanderson, 54 Kelvedon Close, Chelmsford, CM1 4DG
G3 CSY K. Hill, 30 Hestham Avenue, Morecambe, LA4 4PZ
G3 CTP J. Swift, 20 Leighlands, Crawley, RH10 3DW
G3 CTQ H. Westwell, 224 Dickson Road, Blackpool, FY1 2JS
G3 CTZ A. Jones, 17 Oaklea Way, Old Tupton, Chesterfield, S42 6JD
G3 CUF H. Ashworth, 97 Winchcombe Road, Sedgeberrow, Evesham, WR11 7UZ
G3 CUR R. Collette, 8A Woolwich Road, Belvedere, DA17 5EW
G3 CUY E. Paul, 91 Windmill Drive, Brighton, BN1 5HH
G3 CVK P. Bolton, 50 Meadow Road, West Malvern, Malvern, WR14 2SD
G3 CWD J. Robinson, 4 Phoenix Court The Mount, Taunton, TA1 3NR
G3 CWH R. Rogers, 107 Rotherham Road, Coventry, CV6 4FH
G3 CWI R. Newstead, 89 Victoria Road, Macclesfield, SK10 3JA
G3 CXP R. Gill, 45 Biggin Lane, Ramsey, Huntingdon, PE26 1NB
G3 CYU J. Wilson, 1 Beeches Farm Road, Crowborough, TN6 2NY
G3 CYX P. Lambert, 11 Marlborough Close, Musbury, Axminster, EX13 8AP
G3 CZL R. Buckman, Heathfield, Hang Hill Road, Lydney, GL15 6LQ
G3 CZU Dorking & District Radio Society, c/o T. Ellinor, 53 Hillside, Banstead, SM7 1HG
G3 DAE C. Bland, 84 Milton Road, Grimsby, DN33 1DE
G3 DAQ R. Braithwaite, 32 Rupert Crescent, Queniborough, Leicester, LE7 3TU
G3 DAT D. Taylor, 24 Quince, Tamworth, B77 4EN
G3 DAV J. Waller, 17 Spencer Close, Marske-By-The-Sea, Redcar, TS11 6BD
G3 DBJ D. Buggs, 2 Archway Cottages Valley Road, Leiston, IP16 4AR
G3 DBV S. Hedges, 25 Rudland Close, Thatcham, RG19 3XW
G3 DCE F. Humphries, Little Hayes, 1 Meadway, Sidmouth, EX10 9JA
G3 DCO B. Coyne, 58 Osborne Road, New Milton, BH25 6AB
G3 DCT Duddon Contest Team, c/o C. Leviston, 13 Pryors Walk, Askam-in-Furness, LA16 7JG
G3 DCV A. Watson, 93 St. Dunstans Drive, Gravesend, DA12 4BJ
GM3 DDL J. Jackson, 74 Cairngorm Crescent, Paisley, PA2 8AW
G3 DEJ T. Wiseman, 70 Dove House Lane, Solihull, B91 2EG
G3 DEN R. Lea, 90 Wroxham Gardens, Potters Bar, EN6 3DL
G3 DEY E. Ford, 177 Latters Orchard, Old Road, Maidstone, ME18 5PR
G3 DFY N. Devine, 46 Tytton Lane West, Wyberton, Boston, PE21 7HL
G3 DID J. Doyle, 16 Park Hall Crescent, Birmingham, B36 9SN
GM3 DIE T. Dickson, 91 Milton Road West, Edinburgh, EH15 1RA
GM3 DIN A. Clark, 11 Regent Park Square, Glasgow, G41 2AF
G3 DIT Prtsmth & Dars, c/o T. Mortimer, 10 Harold Road, Hayling Island, PO11 9LT
G3 DMO C. Earnshaw, 35 Rogersfield, Langho, Blackburn, BB6 8HB
G3 DNN G. Saville, 2 Gaskell Close, Littleborough, OL15 8EB
G3 DNS N. King, 31 Great Norwood St., Cheltenham, GL50 2AW
GM3 DOD A. Murray, 50 Castlepark Drive, Fairlie, Largs, KA29 0DG
G3 DOV D. Dove, 3 Walnut Grove, Watton, Thetford, IP25 6EY
G3 DPM D. Cooknell, 23 The Hyde, Winchcombe, Cheltenham, GL54 5QR
G3 DQQ D. Winterburn, 47 Hilda Avenue, Tottington, Bury, BL8 3JE
G3 DQT J. Ayres, 8 Cornfield Road, Seaford, BN25 1SW
G3 DQW Peterborough Radio and Electronic Soc., c/o B. Vaughan, 7 Oundle Road, Chesterton, Peterborough, PE7 3UA
G3 DR RSGB Contest Club, c/o N. Totterdell, Moscar Cross House, Hollow Meadows, Sheffield, S6 6GL
G3 DRN E. Allen, 30 Bodnant Gardens, Wimbledon, London, SW20 0UD
G3 DSZ A. Kent, 23 Pagehall Close, Scartho, Grimsby, DN33 2HF
G3 DT L. Boorman, 2 Bull Lane Cottages, Bull Lane, Ashford, TN26 3HA
G3 DTP A. Jackson, Flat 6, St. Albans Court, Rochdale, OL11 4HW
G3 DTU C. Prior, 36 Bassnage Road, Halesowen, B63 4HQ
G3 DTX I. Duck, Chenies, Loudhams Wood Lane, Chalfont St Giles, HP8 4AR
G3 DUW R. Hodgson, The Shealing, Forest Moor Drive, Knaresborough, HG5 8JT
G3 DVF G. Cain, 23 Wiltshire Avenue, Crowthorne, RG45 6NR
G3 DWI G. Lusty, Sundial House, High St., Chipping Campden, GL55 6AG
G3 DXD D. Rolph, 2 Victoria Court, Victoria Road, Marlow, SL7 1DR
G3 DXZ C. Fletcher, 12 Park Crescent, Retford, DN22 6UF
G3 DYO N. Alder, Greenwoods, Eastfield Road, Ross on Wye, HR9 5JY
GW3 DZJ F. Pardy, 5 Y Bryn, Glan Conwy, Colwyn Bay, LL28 5NJ
G3 EAE G. Billington, 75 Mount Vernon Road, Barnsley, S70 4DW
G3 EAR East Ardsley Radio Society, c/o P. Driver, 68 Ripon Road, Dewsbury, WF12 7LG
G3 EBP M. Courcoux, 116 Ameysford Road, Ferndown, BH22 9QE
G3 EBV S. Squire, Leafield, 4 Little Green Lane, Rickmansworth, WD3 3JQ
GJ3 ECC R. Taylor, 21 Samares Avenue La Grande Route De St. Clement, St Clement, Jersey, JE2 6NY
G3 ECM P. Bowles, 29 Coleman Avenue, Hove, BN3 5ND
G3 ECP J. Brown, Manor Cottage, 2 The Maltings, Huntingdon, PE28 4DZ
GI3 ECQ G. Mcgarry, 18 Marna Brae Park, Lisburn, BT28 3PD
G3 EDM M. Marris, 6 High Street, Wye, Ashford, TN25 5AL
G3 EEH Dr J. Watkinson, The Moorings, 63 Ruffa Lane, Pickering, YO18 7HN
G3 EEO Nunsfield House - ARG, c/o A. Price, 10 Low Meadow, Whaley Bridge, High Peak, SK23 7AY
GD3 EFD M. Thompson, Whitehouse Cottage, St. Marks, Ballasalla, Isle Of Man, IM9 3AH
G3 EFL W. Preston, 8 Pencraig View, Greytree, Ross-on-Wye, HR9 7JR
G3 EFS W. Borland, Sloane Nursing Home, 28 Southend Road, Beckenham, BR3 5AA
G3 EFX Rad Soc Harrow, c/o C. Friel, 102A Sharps Lane, Ruislip, HA4 7JB
G3 EGF T. Kellett, Braville, St. Ives Road, Consett, DH8 7SJ
G3 EGV R. Staniforth, 26 Winslow Road, Preston, Weymouth, DT3 6NE
G3 EHQ H. Bone, 2 Waterville Gardens, Orton Waterville, Peterborough, PE2 5LG
G3 EHW J. Watkins, 19 Barrow Grove, Sittingbourne, ME10 1LB
GW3 EIZ C. Lyon, Ardraeth, The Drive, Bodorgan, LL62 5AW
G3 EJH W. Peatman, 110 Cator Lane, Beeston, Nottingham, NG9 4BB
GW3 EJR J. Armstrong, Mirianog, 1 Bryn Bedw, Cardigan, SA43 2NY
G3 EKE L. Stockley, C/O Glebe Cottage, Baylham, Ipswich, IP6 8JS
G3 EKJ H. Mattacks, Fieldfare, Eastbourne Road, Lewes, BN8 6PS
G3 EKL Colburn & Richmondshire District ARS, c/o M. Vann, 3 Mile Planting, Richmond, DL10 5DB
G3 EKW ARC of Nottm, c/o S. Williams, Haywood Community Centre, 46 Haywood Road, Nottingham, NG3 6AD
G3 ELS B. Rudd, Orchard Bungalow, 14 Walmer Close, Colchester, CO7 0PE
G3 ELV Royal Air Force Henlow Radio & Electronics Club, c/o R. Walker, 35 Romany Close, Letchworth Garden City, SG6 4LA
G3 ENO R. Green, 8B The Beck, Elford, Tamworth, B79 9BP
GM3 EOB C. Merrilees, 6 Spoutwells Drive, Scone, Perth, PH2 6RR
G3 EOO J. Hamlett, 23 Riddings Road, Timperley, Altrincham, WA15 6BW
G3 ERD Derby and District ARS, c/o C. Gent, 16 Coronation Avenue, Alvaston, Derby, DE24 0LQ
G3 ESY P. Jones, Fieldfarm House, Residential Home, Hampton Bishop, Hereford, HR1 4JP
G3 ETP P. Woodyard, 65 Raglan Street, Lowestoft, NR32 2JS
G3 EUE E. Jones, White Lodge, The Street, Steyning, BN44 3WE
G3 EVA S. Ayers, Old Catton, Old Catton, Norwich, NR6 7HF
G3 EVT R. Mutton, Summer Hayes, Mill Lane, Alcester, B49 6LF
G3 EWF A. Harris, 5 Wickham Court, Stapleton, Bristol, BS16 1DQ
G3 EWM P. Green, 23 Tilton Road, Borough Green, Sevenoaks, TN15 8RS
G3 EWT C. Tamkin, 4 Stanmer Villas, Brighton, BN1 7HP
G3 EXL D. Derham, 3 Riverbank Cottages, Old Ferry Road, Saltash, PL12 6BJ
G3 EZB J. Rackett, Little Vectis, Folgate Lane, Norwich, NR8 5DP
G3 FBT S. Briggs, 20 Bluebell Close, Newton Aycliffe, DL5 7LN
G3 FBU W. Brown, 79 Mill Hill, Deal, CT14 9EW
G3 FCM A. Cowley, 13 Steward Close, Stuntney, Ely, CB7 5TW
G3 FDW A. Gibbings, 16 Turnberry Avenue, Eaglescliffe, Stockton-on-Tees, TS16 9EH
GW3 FDZ D. Whitehead, Tyddyn Bach, Dyffryn Ardudwy, LL44 2RQ
G3 FEW E. Rule, 15 Norwich Road, Lenwade, Norwich, NR9 5SH
GI3 FFF Ballymena ARC, c/o J. Clarke, 154 Galgorm Road, Ballymena, BT42 1DE
G3 FGP R. Brooks, 10 The Oval, Longfield, DA3 7HD
G3 FHG M. Hopkins, Hylton Cottage, Grafton, Tewkesbury, GL20 7AT
G3 FHN E. Aldworth, Glenaire, 15 Heather Way, Hastings, TN35 4BL
G3 FHT A. Lewis, 8 Lutyens Fold, Milton Abbot, Tavistock, PL19 0NR
G3 FIA A. Lowden, 3 Boscobel Road, Great Barr, Birmingham, B43 6BB
G3 FIC J. Glover, 53 Swanpool Lane, Aughton, Ormskirk, L39 5AY
G3 FJE Shefford & District ARS, c/o D. Ross, 3 Little Lane, Clophill, Bedford, MK45 4BG
GW3 FJI E. Jones, 8 Merllyn Road, Rhyl, LL18 4HH
G3 FJL J. Hall, 250 Scraptoft Lane, Leicester, LE5 1PA
G3 FJO A. Ellefsen, 121 The Furlongs, Ingatestone, CM4 0AL
GI3 FJX J. Davidson, 7 Keel Point, Dundrum, Newcastle, BT33 0NQ
G3 FKI E. Lambert, 6 Abercorn Gardens, Kenton, Harrow, HA3 0PB
GD3 FLH IOM ARS, c/o A. Sinclair, 1, Marathon Drive, Douglas, Isle Of Man, IM4 2BP
G3 FLV L. Keighley, 24 St. Annes Road, Headingley, Leeds, LS6 3NX
G3 FMO G. Elliott, Oatlands, Southend Road, Chelmsford, CM2 7TD
GW3 FMR C. Dwyer, Ystrad, 29 The Oval, Llandudno, LL30 2BU
G3 FMU D. Mcdiarmid, 102 Shalloak Road, Broad Oak, Canterbury, CT2 0QH
G3 FMW J. Stockley, 22 Manor Gardens, Killinghall, Harrogate, HG3 2DS
G3 FNL R. Grubb, 7762 Brockway Drive, Boulder, USA, 80303
G3 FNO G. Morgan, 27 Kestrel Close, Downley, High Wycombe, HP13 5JN
G3 FNZ J. Lambert, 49 Rede Court Road, Strood, Rochester, ME2 3SP
GW3 FPH J. Hayes, 4 St. Marys Drive, Northop Hall, Mold, CH7 6JF
G3 FPY J. Dew, 62 Monks Park Avenue, Horfield, Bristol, BS7 0UH
G3 FRE W. Frith, 56 Ringleas, Cotgrave, Nottingham, NG12 3NE
GM3 FRU D. Wark, Flat 37A, Northwood House, Edinburgh, EH9 2EL
G3 FRV R. Vaughan, 1 Langstone Close, Maidenbower, West Sussex, RH10 7JR
G3 FSA A. Davis, Willow Cottage, Hedging, Bridgwater, TA7 0DE
GW3 FSP L. Davies, Glanmor, Brynna Road, Pencoed, Bridgend, CF35 6PD
G3 FSX R. Ellis, Laura House, 79 Sunte Avenue, Haywards Heath, RH16 2AB
G3 FTK L. Gray, 109 Foxholes Road, Poole, BH15 3NE
GI3 FTT W. Brennan, 10 Dunhugh Park, Londonderry, BT47 2NL
G3 FUJ W. Scott, 10 Pavilion Road, Littleover, Derby, DE23 6XL
G3 FVA Sth Manchestr Rd, c/o D. Armitage, 12 Loughborough Close, Sale, M33 5UF
G3 FVR R. Bannister, 22 Manton Road, Hitchin, SG4 9NW
G3 FWD B. Purchase, 126 Renton Road, Wolverhampton, WV10 6XH
G3 FWI W. Sutton, Pendle, 6 Cuperham Close, Romsey, SO51 7LH
G3 FWU L. Richardson, Belmont Cottage, Christys Lane, Shaftesbury, SP7 8NQ
G3 FXI P. Cardwell, 3 Old Talbot, Llanwnog, Caersws, SY17 5JG
GD3 FXN A. Radcliffe, 3 Cronk Drine, Union Mills, Douglas, Isle Of Man, IM4 4NG

G3 FYF P. Acke, Kinghurst Farm, Holne, Newton Abbot, TQ13 7RU
G3 FYQ Pontefract & District ARS, c/o N. Ferguson, Royd Moor, Royd Moor Lane, Badsworth, Pontefract, WF9 1AZ
G3 FYX R. Emery, 30 Station Road Winterbourne Down, Bristol, BS36 1EP
GW3 FZV R. Lewis, 1 Victoria Avenue, Penarth, CF64 3EN
G3 GAA W. Jeans, 36 Pimms Grove, High Wycombe, HP13 7EF
G3 GAF Dr C. Dollery, 101 Corringham Road, London, NW11 7DL
G3 GAH D. Johnson, 31 Coniston Avenue, Penketh, Warrington, WA5 2QY
G3 GAQ D. Bottomley, 24 Midhope Road, Woking, GU22 7UE
G3 GBD S. Hancock, 53 Friary Grange Park, Winterbourne, Bristol, BS36 1NA
GD3 GBG A. Moore, 114 Ballabrooie Drive, Douglas, Isle Of Man, IM1 4HQ
G3 GBN S. Feldman, Flat 5 Maitland Joseph House, 35 Marlowes, Hemel Hempstead, HP1 1LB
G3 GBS M. Sandoz, Edelweiss, Broad Lane, Solihull, B94 5DP
G3 GBU Stoke on Trent ARS, c/o A. Allen, 3 Wayfield Grove, Harpfields, Stoke on Trent, ST46DB
GM3 GBZ OBO Strathmore ARC, c/o G. Balfour, 6 Kirkden Street, Friockheim, Arbroath, DD11 4SX
G3 GCU R. Adams, 13 Birch Rise, Ashley Heath, Market Drayton, TF9 4PZ
G3 GCW B. Jones, 44 Winner Hill Road, Paignton, TQ3 3BT
G3 GDB G. Bird, 16 Simnel Road, London, SE12 9BG
G3 GDH D. Silveston, 192 Rosemary Avenue, Minster On Sea, Sheerness, ME12 3HX
G3 GDZ R. Hooten, 17 Clarence Road, Teddington, TW11 0BQ
G3 GEF J. Andrews, 45 Sandes Court, Sandes Avenue, Kendal, LA9 4LN
G3 GEG E. Cooper, Ciren, 19 Ventnor Road, Sandown, PO36 0JT
G3 GEI Solihull ARS, c/o Dr R. Hancock, 80 Ulleries Road, Solihull, B92 8EE
G3 GEJ L. Airey, 32 Brookside Close, Bedale, DL8 2DR
G3 GEX P. Burton, 18 Tankerfield Place, Romeland Hill, St Albans, AL3 4HH
G3 GGG R. Bishop, 31 Blenheim Close, Didcot, OX11 7JQ
G3 GGH P. Horn, Darfield, 50 Barrack Road, Bexhill-on-Sea, TN40 2AZ
G3 GGI A. Laurence, 70 Firs Avenue, London, N11 3NQ
G3 GGK P. Simpson, 109 Highfields Road, Highfields Caldecote, Cambridge, CB23 7NX
G3 GGL D. Wormald, Long Acre, Mamble, Kidderminster, DY14 9JY
G3 GGN D. Shute, 100 Wick Street, Wick, Littlehampton, BN17 7JS
G3 GGR J. Sykes, 49 Chapel St., Pelsall, Walsall, WS3 4LW
G3 GGS W. Waring, 51 Church Road, Leyland, PR25 3AA
G3 GGU G. Smith, Greenacres, Top Road, Chesterfield, S44 5AE
G3 GHN Clifton ARS, c/o S. Fletcher, 90 Westcombe Park Road, Blackheath, London, SE3 7QS
G3 GIB A. Wake, 42 Charles Avenue, Watton, Thetford, IP25 6BZ
G3 GIH J. Bird, The Old Stackyard, Daisy Green, Bury St Edmunds, IP31 3HX
G3 GIZ Chester & District Radio Society, c/o P. Holland, Chatterton, Chapel Lane, Threapwood, Malpas, SY14 7AX
G3 GJA C. Reynolds, 49 Westborough Way, Anlaby Common, Hull, HU4 7SW
G3 GJJ P. Watson, 5 High Garth, Winston, Darlington, DL2 3RY
G3 GJL Worcester Radio Amateur Association, c/o R. Moles, 14 Dorsett Road, Stourport-on-Severn, DY13 8EL
GW3 GJQ S. Ldr. R. Handley, Flat 2, 11 Trinity Square, Llandudno, LL30 2RA
G3 GJW T. Lundegard, Saxby, Botsom Lane, Sevenoaks, TN15 6BL
GM3 GKJ A. Gordon, The Paddock, Greenhead Farm, West Saltoun, EH34 5EH
G3 GKS R. Christian, 27 Howey Rise, Frodsham, WA6 6DN
G3 GLA B. Mase, 18 Norton Drive, Norwich, NR4 6JD
G3 GLL T. Green, 6 Woodrolfe Road, Tollesbury, Maldon, CM9 8SB
G3 GLW P. Willis, 26 Snellgrove Close, Calmore, Southampton, SO40 2WD
G3 GLX J. Simmonds, 99 Foljambe Avenue, Chesterfield, S40 3EY
G3 GMC P. Mcvey, 18 Worlebury Hill Road, Weston-Super-Mare, BS22 9SP
G3 GML F. Murray, 3 Rosemary Close, Tiptree, Colchester, CO5 0QD
G3 GMM E. Mcfarland, 60 Sutton Oaks, London Road, Crewe, CW4 7AS
G3 GMS M. Thayne, 14 Tynedale Avenue, Monkseaton, Whitley Bay, NE26 3BA
G3 GMW L. Nichols, 5 Middle Pasture, Peterborough, PE4 5AU
G3 GMY F. Green, 5 Silvercliffe Gardens, New Barnet, Barnet, EN4 9QT
G3 GNA D. Macmillan, Brook Farm, Broadwas, Worcester, WR6 5NE
G3 GOS P. Peach, The Firs, Goldsmith Lane, Axminster, EX13 7LU
G3 GQC OBO Manfield Am Radio Society, c/o D. Riley, 9 Century Avenue, Mansfield, NG18 5EE
G3 GQK J. Wall, P.O Box 631, Nambucca Heads, Australia, 2448
GM3 GRG D. Rollo, 25 Beaufort Drive, Kirkintilloch, Glasgow, G66 1AX
G3 GRL S. Houlton, 97 Mansfield Road, Alfreton, DE55 7JP
G3 GRQ C. Hebden, 129 Millers Way, Honiton, EX14 1JB
G3 GRS Gravesend ARS, c/o D. Lawley, 5 The Limes, Buckland, Buntingford, SG9 0PW
G3 GRV G. Halse, 10 Charnock Close, Hordle, Lymington, SO41 0GU
G3 GRY F. Wiseman, 14 Parkway, Crowthorne, RG45 6EN
G3 GTA J. Shute, 32 Woodborough Drive, Winscombe, BS25 1HB
G3 GTF B. Harris, 6 The Priory, Lewes Road, Cross In Hand, Heathfield, TN21 0FE
GM3 GTQ A. Mcphedran, 3 Argyll Road, Bearsden, Glasgow, G61 3JX
GI3 GTR R. Mckinty, 3 Rhanbuoy Road, Craigavad, Holywood, BT18 0DY
G3 GUE A. Dowling, Church Cottage, Frittenden, Cranbrook, TN17 2DD
G3 GUR J. Scully, 1 Wyde Feld, Bognor Regis, PO21 3DH
GW3 GUX J. Brimecombe, Llwyn Onn, Llangoed, Beaumaris, LL58 8PH
G3 GVM F. Robins, 59 Titchfield Road, Stubbington, Fareham, PO14 2JF
G3 GWB Northampton Radio Club, c/o J. Cockrill, 28 Northampton Road, Harpole, Northampton, NN7 4DD
G3 GWC E. Ramsdale, 8 May Cottages, Monkswell Lane, Coulsdon, CR5 3SX
G3 GWE A. Daum, 100 Shawbridge, Harlow, CM19 4NW
G3 GXG C. Lee, 7 Lilac Avenue, Lower Quinton, Stratford-upon-Avon, CV37 8US
G3 GXI Eccles and District ARS, c/o Prof. C. Harrison, 11 Ringley Park, Whitefield, Manchester, M45 7NT
G3 GXQ W. Roberts, 24 Leeds Road, Barwick In Elmet, Leeds, LS15 4JD
G3 GYQ C. Spackman, 10 Norton Drive, Warwick, CV345FE
G3 GZT R. Moores, 117 Horton Road, Brighton, BN1 7EG
GW3 GZX A. Bladon, 6 Quarry Bank, Mold Road, Denbigh, LL16 4DT
G3 GZZ A. Bevan, 14 Parsonage Road, Berrow, Burnham-on-Sea, TA8 2NL
G3 HAA J. Morgan, 10 Bamber Gardens, Southport, PR9 7PQ
G3 HAL R. Parrott, 3 Ash Grove, Chard, TA20 1BZ
GM3 HAM Lothians Radio Society, c/o P. Bates, 10 Swanston Avenue, Edinburgh, EH10 7BU
G3 HAN M. Hitchman, 12 Briar Walk, Oadby, Leicester, LE2 5UE
G3 HCO G. Errock, 307 Main Road, Emsworth, PO10 8JG
G3 HCS H. Stratton, 26 Marjorie Road, Chaddesden, Derby, DE21 4HQ
G3 HCT J. Bazley, C/O Mr P Chadwick, Three Oaks, Swindon, SN5 0AD
G3 HCZ B. Edmondson, 1 Harbour Lane, Turton, Bolton, BL7 0PA
GW3 HDF K. Groves, 6 Overleigh Drive, Buckley, CH7 2PA
G3 HDM S. Campbell, Carrer De Ses Sevines, 1 Bajo, Mallorca, Spain,
GM3 HDT J. Graham, Lodge, Stronsay, KW17 2AN
G3 HEH E. Parker, 39 Hellath Wen, Nantwich, CW5 7BB
G3 HEJ D. Stanners, Tanglewood Samarkand Close, Camberley, GU15 1DG
GM3 HEN A. White, Byeways Whiting Bay, Isle of Arran, KA27 8QH
GW3 HEU D. Rickers, 4 St. Marks Terrace, Wrexham, LL13 0PQ
GD3 HFC F. Arrowsmith, The Evergreens, South Cape, Laxey, Isle Of Man, IM4 7JB
G3 HFM A. Vickers, Foxcroft, 4 Woodlands End, Macclesfield, SK11 9BF
GU3 HFN Guernsey AR, c/o P. Cooper, 1 Clos Au Pre, La Route De La Hougue Du Pommier, Castel, Guernsey, GY5 7FQ
GM3 HGA J. Mccall, 1 Pinewood Place, Aberdeen, AB15 8LT
G3 HGD V. Best, 3 Old Auction Mart, Kirkby Lonsdale, Carnforth, LA6 2AF
G3 HGE T. Withers, Woodpeckers, West Stow, Bury St Edmunds, IP28 6ER
G3 HHD T. Hayward, Skirt Bank, Nether Silton, North Yorkshire, YO7 2LL
G3 HHU Dr J. Ickringill, 28 Deena Close, Queens Drive, London, W3 0HR
G3 HIU Milton Keynes ARS, c/o D. White, 1 Whaddon Road, Shenley Brook End, Milton Keynes, MK5 7AF
GI3 HJH R. Mcburney, 8 Main Road, Ballymartin, Newry, BT34 4NU
G3 HJP G. Cooper, 25 Plantation Avenue, Shadwell, Leeds, LS17 8TB
G3 HKA C. Booth, 88 Green Drive, Thornton-Cleveleys, FY5 1JD
G3 HKD D. Money, 125 Wroxham Road, Norwich, NR7 8AD
G3 HKF B. Ferris, 5 Guildway, Todwick, Sheffield, S26 1JN
G3 HKH M. Harrison, 3 Stert Street, Abingdon, OX14 3JF
G3 HKT A. Partner, 10 The Tanners, Titchfield Common, Fareham, PO14 4BH
G3 HLG D. Johnson, Robins, 4 Station Road, Newark, NG23 7RA
G3 HLI M. Bradford, 101 Oxendon Way, Binley, Coventry, CV3 2HA
G3 HLN P. Woods, 145 Hollybush Lane, Welwyn Garden City, AL7 4JT
G3 HMB I. Elliot, Grange House, Manningtree Road, Ipswich, IP9 2SW
G3 HMG A. Macgregor, 14 Quantock Grove, Williton, Taunton, TA4 4PD
G3 HMO J. Osborne, 141 Chadwick Road, London, SE15 4PY
G3 HMQ J. Robson, 32 St. Stephens Road, Cold Norton, Chelmsford, CM3 6JE
G3 HMR G. Moser, 30 Blackhall Croft, Blackhall Road, Kendal, LA9 4UU
G3 HMV N. Bolton, 2 Selborne Villas, Clayton, Bradford, BD14 6JZ
G3 HNC B. Dyer, 30 Smithson Avenue, Castleford, WF10 3HN
GM3 HNE G. Campbell, 17 Roseburn Terrace, Edinburgh, EH12 5NG
GI3 HNM C. Davies, 121 Comber Road, Toye, Downpatrick, BT30 9PD
GM3 HOM J. Reilly, 30 Park Crescent, Bishopbriggs, Glasgow, G64 2NS
G3 HPB F. Tooley, 70 Langbury Lane, Ferring, Worthing, BN12 6QA
G3 HPD F. Dews, 341 Crossley Lane, Mirfield, WF14 0NR
G3 HQG G. Atkins, 20 Mansfield Road, Killamarsh, Sheffield, S21 2BX
G3 HQS C. Baker, Roffensis, 16 Boulderside Close, Norwich, NR7 0JJ
G3 HQT P. Ball, 68 Brook Lane, Warsash, Southampton, SO31 9FG
G3 HQX J. Brodzky, 3 Ropewalk House, Hyde Abbey Road, Winchester, SO23 7XH
G3 HRE F. Watson, 54 Tavistock Road, Cambridge, CB4 3ND
G3 HRH R. Hills, 2 The Dell, Otterbourne Road, Winchester, SO21 2DE
G3 HRK D. Willies, 17 Campion Way, Sheringham, NR26 8UN
G3 HRX J. Hilling, 24 Gloucester Road, Gaywood, King's Lynn, PE30 4AB
GM3 HSO J. Yang, Flat 6, 14 Bothwell Street, Edinburgh, EH7 5PS
G3 HST G. Allen, Moor Farm Cottage, Salcombe, TQ8 8PW
G3 HSV D. Alesbury, 23 Cullerne Road, Swindon, SN3 4HU
G3 HTA J. Forward, Sunrays, Barnstaple Cross, Crediton, EX17 2EP
G3 HTB M. Squance, Church Lane, Cubbington, Leamington Spa, CV32 7JT
G3 HTC C. Storey, 12 Vereker Drive, Sunbury-on-Thames, TW16 6HF
G3 HTJ W. Walker, 53 Wolfridge Ride, Alveston, Bristol, BS35 3PR
G3 HTO R. Dolton, 43 Jubilee Meadow, St Austell, PL25 3EX
G3 HTT W. Cheeseworth, 10 Barton Mill Court, Station Road West, Canterbury, CT2 7JZ
G3 HUB M. Harrison, Rolling Hills, Brandy Lane, Lostwithiel, PL22 0QH
G3 HUD M. Brown, 10 Park House Mews, Congleton Road, Sandbach, CW11 4SP
G3 HUK M. Morrissey, 1 Hamilton Road, Church Crookham, Fleet, GU52 6AS
G3 HUO K. Young, 80 Darbys Lane, Oakdale, Poole, BH15 3ET
G3 HUR D. Brough, 18 Lark Hall Road, Macclesfield, SK10 1QP
G3 HUX J. Matthews, 4 Berrington Grove, Ashton-In-Makerfield, Wigan, WN4 9LD
G3 HVA D. Pinnock, 2 Oak Close, Oakley, Basingstoke, RG23 7DD
G3 HVJ A. Chappell, 22206 Del Valle St, Woodland Hills, USA, 91364-1515
GM3 HVK J. Craig, 147 Avon Road, Larkhall, ML9 1RA
G3 HWF South and West Yorkshire Wing ATC, c/o D. Taylor, 76 Heworth Village, York, YO31 1AL
G3 HWM J. Cowling, 19 The Drive, Hullbridge, Hockley, SS5 6LZ
G3 HWW York ARS, c/o C. Rouse, 86 Melton Avenue, Clifton, York, YO30 5QG
G3 HXK P. Nethercot, Ronhill, Stoodleigh, Tiverton, EX16 9PJ
G3 HXN J. Crisp, 371 Stroud Road, Tuffley, Gloucester, GL4 0DA
G3 HYG D. Topping, Bentley, Middle Street, Waltham Abbey, EN9 2LB
G3 HYH S. Hay, 27 Acres Road, Leicester Forest East, Leicester, LE3 3HB
GM3 HYX C. Rattray, 58 Aberdour Road, Dunfermline, KY11 4PE
G3 HZP H. James, 10 Playsted Lane, Cambourne, Cambridge, CB3 6GA
G3 HZT P. Fraser, 45 The Martlet, Hove, BN3 6NT
G3 IAR M. Crowther-Watson, The Snicket, 14 The Avenue, Sevenoaks, TN15 8EA
G3 IAZ A. Wickham, Apartment 6, Panama Reach, Eastbourne, BN23 5PL
G3 IBI P. Scutt, 62 Old Street, Fareham, PO14 3HW
G3 IBQ K. Holt, 61 Millford Avenue, Nepean, Ontario, Canada, K2G-1C4
GM3 IBU 59 Degrees North ARG, c/o E. Holt, Ashwell, St. Ola, Kirkwall, KW15 1SX
G3 IBY Dr T. Wilmshurst, 4 Eastern Road, West End, Southampton, SO30 3EQ
G3 ICA G. Adams, Sue Marey, Selsley Hill, Stroud, GL5 5JS
G3 ICB A. Bull, 91 Lower Way, Thatcham, RG19 3RS
G3 ICC I. Chilton, 47 Mayfield Crescent, Eaglescliffe, Stockton-on-Tees, TS16 0NH
G3 ICG K. Mcfarlane, Clifton, 18 Needham Road, Harleston, IP20 9JY
G3 ICN D. Collado Castells, 37 Herbert Street, London, NW5 4HB
G3 ICZ W. Clowes, 144 Norton Lane, Norton-In-The-Moors, Stoke-on-Trent, ST6 8BZ
G3 IDB A. Brooks, 45 Northfield Road, Townhill Park, Southampton, SO18 2QE
G3 IDW R. Reynolds, 6 Church Way, Stratton, Swindon, SN3 4NF
G3 IDY R. Robson, 66 Tilstock Crescent, Shrewsbury, SY2 6HQ

G3 IEJ S. Watson, 6 Hope Street, Lytham St Annes, FY8 3SL
G3 IFX A. Cooke, 9 Lee Crescent, Ilkeston, DE7 5EF
G3 IGC A. Garforth, 110 Foxdenton Lane, Chadderton, Oldham, OL9 9QR
G3 IGU K. Coates, 76 Copley Crescent, Scawsby, Doncaster, DN5 8QP
G3 IGV J. Birkbeck, 4 Tregullan View, Bodmin, PL31 1BH
G3 IGZ D. Bruce, 22 Brownspring Drive, New Eltham, London, SE9 3JX
G3 IHX N. Bond, 333 Hillandale Drive, Charlotte, USA, 28270
G3 IIN M. Griffin, Michaelmas, Southdown Road, Freshwater, PO40 9UA
G3 IIO D. Harriott, 23 Hamsey Crescent, Lewes, BN7 1NP
G3 IIV A. Davies, Paarl, 129 Cotwall End Road, Dudley, DY3 3YQ
G3 IIW M. Sands, Beech Lea, St. Marks Road, Tunbridge Wells, TN2 5LU
G3 IJA J. Allan, 5 Terrington Court, Strensall, York, YO32 5PA
G3 IJL A. Sephton, 16 Bloemfontein Avenue, Shepherds Bush, London, W12 7BL
G3 IJS J. Stratfull, 55 Craigweil Lane, Aldwick, Bognor Regis, PO21 4XN
G3 IJU E. Briggs, 32 Lethbridge Road, Wells, BA5 2FN
G3 IJV R. Harvey, 16 Gatesgarth Close, Hartlepool, TS24 8RB
G3 IKB D. Giddens, 89 Pollards Oak Road, Oxted, RH8 0JE
G3 IKL R. Craxton, 103 Clifton Road, Rugby, CV21 3QH
G3 IKQ R. Chilton, 80 Plantation Road, Hextable, Swanley, BR8 7SB
G3 ILE E. Marsh, 63 Willows Lane, Accrington, BB5 0SQ
G3 ILO S. Spencer, 9 Vaisey Field, Whitminster, GL2 7pt
G3 IMW S. Whitfield, 7 Sir Alex Walk, Topsham, Exeter, EX3 0LG
G3 IMX E. Jolliffe, 96 Cowes Road, Newport, PO30 5TP
G3 INP G. Stanway, Bramble Edge Cottage, Bates Lane, Frodsham, WA6 9LL
G3 INQ Capt. B. Podmore, 6 Alfred Court, Furlong Road, Bourne End, SL8 5AZ
G3 INR P. Buchan, 79 Cavendish Avenue, Cambridge, CB1 7UR
G3 INU R. Appleby, 14 Truro Court, Canterbury Way, Stevenage, SG1 4LF
G3 INY E. Tudor, Mowhills House, 133 High Street, Bedford, MK43 7ED
G3 INZ J. Tournier, Avalon 13 Greenlands Flackwell Heath, High Wycombe, HP10 9PL
G3 IOB P. Revell, 54 Lytham Road, Perton, Wolverhampton, WV6 7YY
G3 IOI N. Pascoe, 36 Kilbirnie Road, Bristol, BS14 0HS
G3 IOJ B. Rixon, 1 Carde Close, Hertford, SG14 2EU
G3 IOM R. Chidzey, 8 Dormans Close, Dormansland, Lingfield, RH7 6RL
G3 IOR P. Gowen, 17 Heath Crescent, Norwich, NR6 6XD
G3 IPD C. Oakley, 4 Cross Keys Lane, Low Fell, Gateshead, NE9 6DA
G3 IPG G. Phipps, 12 Mill Close, Pulham Market, Diss, IP21 4TQ
G3 IPL R. Winters, 43 Manor Close, Harpole, Northampton, NN7 4BX
G3 IPP M. Dance, Golf Cottage, 8 St. Johns Road, Crawley, RH11 7BD
G3 IQF R. Fowler, 49 Westhorpe Park, Westhorpe, Marlow, SL7 3RH
G3 IQX E. Popplewell, 71 Thornbury Road, Southbourne, Bournemouth, BH6 4HU
G3 IQY A. Rees, 59 Hillside Gardens, Barnet, EN5 2NQ
G3 IRA J. Wren, 29 Carisbrook Terrace, Chiseldon, Swindon, SN4 0LW
G3 IRQ P. Rackham, Upyonda, Otley Bottom, Ipswich, IP6 9NG
G3 ISB C. Brock, 24 Glebelands Road, Knutsford, WA16 9DZ
G3 ISD E. Hatch, 147 Borden Lane, Sittingbourne, ME10 1BY
G3 IST S. Turner, 8001 Bayshore Drive, Seminole, USA, 34646
G3 ISX C. Leal, 61 Light Oaks Avenue, Light Oaks, Stoke-on-Trent, ST2 7NF
GJ3 IT Jersey ARS, c/o M. Turner, 4 Le Clos Sara, St Lawrence, Jersey, JE3 1GT
G3 ITB T. Bartlett, 19 Hardley Street Hardley, Norwich, NR14 6BY
G3 ITF B. Freeman, 47 Gorham Avenue, Rottingdean, Brighton, BN2 7DP
G3 ITH R. Franklin, 2 Berkeley Drive, Kingswinford, DY6 9DX
G3 ITL J. Humpoletz, 76 Marlborough Road, Braintree, CM7 9LR
GM3 ITN L. Hamilton, Halls Land, Cochno Road, Clydebank, G81 6NR
G3 IUB Birmingham A R S, c/o D. Cottam, 14 Barnard Close, Rednal, Birmingham, B45 9SZ
G3 IUC R. Mcmillan, East Orchard, Almeley, Hereford, HR3 6LF
G3 IUE M. Newell, 35 Ingleside Crescent, Lancing, BN15 8EN
G3 IUJ R. Rogerson, 19 Martins Road, Shortlands, Bromley, BR2 0EE
G3 IUO G. Allen, 157 Lynton Road, Bedminster, Bristol, BS3 5LN
G3 IUV G. Loveday, 2 St. Aldwyns Close, Bristol, BS7 0UQ
G3 IUW L. Pritchard, Green Horizons, Send Hill, Woking, GU23 7HR
G3 IUY J. Presland, 6 Pippin Close, Sutton Ely, CB6 2RX
G3 IUZ Rev. H. Davis, 6 St. Thomas Terrace, Wells, BA5 2XG
G3 IVC A. Sycamore, Fir Tree Cottage, Compton Valence, Dorchester, DT2 9ES
GW3 IVK D. Evans, 11 Hill View, Bryn-Y-Baal, Mold, CH7 6SL
G3 IVP A. Page, The Farmhouse, Budges Shop, Trerulefoot, Saltash, PL12 5DA
G3 IW British Aerospace ARS, c/o W. Wilkie, 14 Horseshoe Close, Northwood, Cowes, PO31 8PZ
G3 IWE A. Wyse, 29 Tregainlands Park, Washaway, Bodmin, PL30 3AU
G3 IWH I. Hall, 46 Bushmead Road, Luton, LU2 7EU
GW3 IWM M. Holland, 7 Willans Court, Willans Drive, Newtown, SY16 4DB
G3 IWV J. Parker, 472-474 Castle Lane West, Bournemouth, BH8 9UD
G3 IWW R. Hopkins, 34 Shelley Close, Abingdon, OX14 1PR
GM3 IWX W. Ritchie, 8 Cheviot Place, Grangemouth, FK3 0DE
G3 IXI K. Landon, 1 The Laurels, Leedons Park, Broadway, WR12 7HB
G3 IXN M. Lovejoy, 73 Stoneham Lane, Swaythling, Southampton, SO16 2NZ
G3 IXZ R. Bowden, 41 Brockington Road, Bodenham, Hereford, HR1 3LP
G3 IYF D. Baker, Long Haul, 3 Chapel Lane, Lincoln, LN6 9EX
G3 IZA D. Allison, 71 South Hill Road, Bromley, BR2 0RW
G3 IZD I. Davies, 13 Thurlow Way, Barrow-in-Furness, LA14 5XP
G3 IZF D. Taylor, 24 Woodville Avenue, Crosby, Liverpool, L23 3BZ
G3 IZM J. Harper Bill, 1 Shepherds Close, Staple Hill, Bristol, BS16 5LE
G3 IZQ H. Hyman, 19 Black Horse Drive, Acton, USA, 1720
G3 JAL R. Taylor, 304 Brigstock Road, Thornton Heath, CR7 7JE
G3 JAU C. Davies, 107 Talbot Road, Bournemouth, BH9 2JE
G3 JBF L. Brown, Ladygate, St. Michaels Road, Stafford, ST19 5AH
GW3 JBJ F. Mathers, 17 Penlon, Menai Bridge, LL59 5LR
GW3 JBZ J. Brace, 12 Heol Gwili, Gorseinon, Swansea, SA4 4GE
G3 JCK F. Chilvers, 5 Low Common Close, Foulsham, Dereham, NR20 5TW
G3 JCM D. Bolwell, 3 Mildmays, Danbury, Chelmsford, CM3 4DP
G3 JCR K. Smith, 20 Manor House Gardens, Abbots Langley, WD5 0DH
G3 JDD R. Dobson, 16 Howden Road, Fulham, Australia, 5024
G3 JDO H. Martin, 7 Nairn Street, Jarrow, NE32 4HX
G3 JDT B. Read, Glenside, 4 Hatton Lane, Warrington, WA4 4BY
G3 JDY OBO Royal Air Force ARS E Riding Area, c/o B. Atkinson, 165 Alliance Avenue, Hull, HU3 6QY
G3 JFD B. Brown, 130 Ashland Road West, Sutton-in-Ashfield, NG17 2HS
GM3 JFG Stornoway Repeater Group, c/o N. Doherty, Cairdeas, Carrbridge, PH23 3AA
G3 JFR N. Cottrell, 28 Colley Wood, Kennington, Oxford, OX1 5NF
G3 JFS P. Cole, 25 Wardlow Gardens, Plymouth, PL6 5PU
G3 JFT B. Dare, 128 Sancroft Road, Spondon, Derby, DE21 7ES
G3 JFW P. Beevers, Hill Farm Granary, Lower Somersham, Ipswich, IP8 4PU
GW3 JGA J. Lawrence, 40 Aberconway Road, Prestatyn, LL19 9HL
GW3 JGE V. Owen, 17 Knowles Avenue, Prestatyn, LL19 8SG
G3 JGP E. Robinson, 16 Shaw Green Storth, Milnthorpe, LA7 7JB
G3 JHH S. Burgess, 34 Redcliffe Road, London, SW10 9NJ
G3 JHI R. Hathaway, 30 Berkeley Drive, Hornchurch, RM11 3PY
G3 JHP E. Allen, 11 Newlands Close, Horley, RH6 8JR
G3 JHU C. Pavey, 3 Field Close, Chatham, ME5 9TD
G3 JIE D. Youngs, 12 Fox Grove, East Harling, Norwich, NR16 2PS
GM3 JIG K. Hodge, 66 Ardrossan Road, Seamill, West Kilbride, KA23 9LX
GM3 JIJ Stornoway Repeater Group, c/o N. Doherty, Cairdeas, Carrbridge, PH23 3AA
G3 JIP J. Hill, Calle El Palomar 13, El Romeral, Malaga, Spain, 29130
G3 JIR J. Hardcastle, 8 Norwood Grove, Rainford, St Helens, WA11 8AT
G3 JIS R. Heaton, 20 Tewkesbury Avenue, Urmston, Manchester, M41 0RJ
GD3 JIU M. Thompson, 3 Close Cam, Port Erin, Isle Of Man, IM9 6NB
G3 JIX Dr K. Smith, Staple Farm House, Durlock Road, Canterbury, CT3 1JX
GM3 JJQ D. Millar, 51 Tiree Crescent, Polmont, Falkirk, FK2 0UX
G3 JJR J. Rickwood, 44A The Bridle Path, Madeley, Crewe, CW3 9EL
G3 JJT C. Kempson, 8 Arle Gardens, Cheltenham, GL51 8HR
G3 JKB D. Simmonds, 73 Tor-O-Moor, Woodhall Spa, LN10 6SD
GM3 JKC C. Cooper, 28 Kippford St., Glasgow, G32 9BW
G3 JKE G. Thomas, 13 Essex Drive, Taunton, TA1 4JX
G3 JKF K. Franklin, 4 Princes Close, Seaford, BN25 2EW
G3 JKL J. Lovell, Kowloon, Slade, Bideford, EX39 3LZ
G3 JKM D. Buckland, 29, Longfields, Swaffham, Swaffham, PE37 7RH
GM3 JKS F. Claytonsmith, 16 Templand, Crossmichael, Castle Douglas, DG7 3BF
G3 JKX M. Street, 12 Ullswater Close, Priorslee, Telford, TF2 9RB
G3 JLK C. Jeffery, 33 Thirlmere Road, Weston-Super-Mare, BS23 3UY
G3 JLN F. Blain, High Ridge, Howgate Lane, Bembridge, PO35 5QW
G3 JLZ V. Ludlow, 6 Raleigh Crescent, Stevenage, SG2 0EQ
G3 JMJ D. Nunn, Oak Lea, Crouch House Road, Edenbridge, TN8 5EL
GM3 JMM J. Murdoch, 4 Cedar Drive, Milton Of Campsie, Glasgow, G66 8AY
G3 JMZ J. Hilton, Windsor House, Preston Road, Charnock Richard, Chorley, PR7 5HH
G3 JNB V. Brand, 8 Greenway, Campton, Shefford, SG17 5BN
G3 JNJ D. Platt, 22 Charcroft Gardens, Enfield, EN3 7HA
G3 JNM T. Whittaker, 16 Acresdale, Lostock, Bolton, BL6 4PJ
GM3 JOB G. Bryce, 3 West Bowhouse Way, Girdle Toll, Irvine, KA11 1NJ
G3 JOE J. Brown, 10 Park House Mews, Congleton Road, Sandbach, CW11 4SP
G3 JOR V. Capell, Endways, 15 Copse Road, Bexhill-on-Sea, TN39 3UA
G3 JOT F. Whatley, 1 Mill Close, Wroughton, Swindon, SN4 9AR
G3 JOX A. Greaves, Jacobs Well, Woodhill Road, Chelmsford, CM2 7SF
G3 JPB C. Noden, Brownhills Cottage Farm, Brownhills, Market Drayton, TF9 4BE
G3 JPG R. Parker, 6 Cambridge Road, Chingford, London, E4 7BP
G3 JPJ J. Peerless, 101 Greenside, Borehamwood, WD6 4JD
G3 JPM B. Grainge, 4 Maltings Close, Chevington, Bury St Edmunds, IP29 5RP
G3 JPO M. Fielding, 68 Mitford Road, South Shields, NE34 0EQ
G3 JPU D. Plant, Briarfields, Raby Crescent, Shropshire, SY3 7JN
G3 JPZ I. Denney, 5 Howard Close, Harleston, IP20 9HY
G3 JQ A. Webster, 5 Brookside Court, 142 Prestbury Road, Macclesfield, SK10 3BR
G3 JQC G. Hawksworth, 16 Birkhead Street, Heckmondwike, WF16 0BE
G3 JQK E. Jones, Appleton Thorne, Lower Broad Lane, Redruth, TR15 3HJ
G3 JQL J. Haggart, 22 Alnwick Road, Newton Hall, Durham, DH1 5NL
G3 JQS J. Guttridge, Victoria Cottage, The Common, Cambridge, CB21 5LR
G3 JRD R. Dancy, 1 Ladds Corner, Eastcourt Lane, Gillingham, ME7 2UW
G3 JRE F. Thomas, 99 Eastfield Road, Wollaston, Wellingborough, NN29 7RS
G3 JRH P. Horne, The Annexe, Burntwood Farm, Winchester, SO21 1AF
G3 JRK J. Knight, 10 Lynton Drive, Burnage, Manchester, M19 2LQ
G3 JRL F. Armstrong, 4 Medway Drive Preston, Weymouth, DT3 6LF
G3 JRS A. Kidd, 35 Hollands Way, Kegworth, Derby, DE74 2GQ
G3 JRY S. Auty, 3 Rochford Crescent, Boston, PE21 9AE
G3 JSA D. Wilcox, 13 Richards Drive, Dartmouth, Canada, NS B3A 2P1
GW3 JSG J. Gunn, Flat 18, Bro Llewelyn, Penrhyndeudraeth, LL48 6AL
G3 JSK D. Dean, 8 Bradford Road, Corsham, SN13 0QR
G3 JSU L. Sampson, 107 South St., Lancing, BN15 8AS
GW3 JSV D. Holmes, Fair Oaks, Berriew, Welshpool, SY21 8AU
G3 JTJ J. Jones, Westerland, 1 Roborough Close, Plymouth, PL6 6AH
G3 JTK G. Allen, 119 Haymoor Road, Poole, BH15 3NR
G3 JTO F. Gell, 93 Pasture Road, Stapleford, Nottingham, NG9 8HR
G3 JTQ R. Griffiths, 7 Dever Way, Oakley, Basingstoke, RG23 7AQ
G3 JTT P. Thompson, 30 Farnol Road, Birmingham, B26 2AF
G3 JUU D. Adams, 23 Arlington Gardens, Attleborough, NR17 2NH
G3 JUW C. Lovell, 5 Montpelier Road, Ilfracombe, EX34 9HP
G3 JUX J. Mcfarlane, 141 Tyler Grove, Aston, Stone, ST15 0JA
G3 JVC J. Cleeve, 44 Ditton Hill Road, Long Ditton, Surbiton, KT6 5JD
G3 JVL M. Walters, 26 Fernhurst Close, Hayling Island, PO11 0DT
G3 JVM R. Medcraft, 134 Dulverton Road, Ruislip, HA4 9AG
G3 JVN D. Keen, 14 Penina Avenue, Newquay, TR7 2LE
G3 JVP V. Purdy, 99 Belmont Road, Uxbridge, UB8 1QX
G3 JVR D. Nokes, 16 Salisbury Grove, Giffard Park, Milton Keynes, MK14 5QA
G3 JWB J. Boote, 1 Shacklock Close, Arnold, Nottingham, NG5 9QE
G3 JWI R. Page-Jones, 34 Edwards Way, Hutton, Brentwood, CM13 1BT
G3 JWN F. Walker, 2 Croft Place, Brighouse, HD6 4AP
G3 JWQ B. Maycock, Hill House, Bullock Lane, Alfreton, DE55 4BP
G3 JXC C. Gregory, 51 Calgary Avenue, Blackburn, BB2 7DS
G3 JYG J. Kirby, 14 Grovelands Road, Hailsham, BN27 3BZ
G3 JYS R. Finch, 8 Chalfont Close, Allesley, Coventry, CV5 9HL
G3 JZF J. Smith, 17A, Sutton Coldfield, B74 2QA
G3 JZL W. Montford, 3 The Close Brandon, Coventry, CV8 3JF
G3 JZT R. Cheetham, 7 Parkway, Stockport, SK3 0PX
G3 KAE J. Rowley, 41 Main Street, East Ayton, Scarborough, YO13 9HL
G3 KAF J. France, 34 Ladythorn Road Bramhall, Stockport, SK7 2ER
G3 KAG A. Parker, Hillside Main Street, Roston, Ashbourne, DE6 2EH
G3 KAN A. Shrewsbury, 1 Dardis Close Kingsley Northampton, Northampton, Nn2

7dn
G3 KAP R. Taylor, 2 Brenchley Mews, Charing, Ashford, TN27 0JQ
G3 KAR D. Hammond, Christen Mares, Willersey Hill, Broadway, WR12 7PF
G3 KAU L. Laszkiewicz, 38 Langley Lane, Ifield, Crawley, RH11 0NA
GW3 KAX G. Mackrell, Preseli Newchapel, Boncath, SA37 0EH
G3 KBH Dr M. Hughes, Northdean, Brimstone Lane, Gravesend, DA13 0BW
G3 KBI T. Waller, 12 Skelton Road, Brotton, Saltburn-by-the-Sea, TS12 2TJ
GM3 KBP A. Kerr, 47 Hillpark Avenue, Edinburgh, EH4 7AH
GM3 KC Montrose Amateur Radio Station, c/o B. Murray, Sherwood Cottage, Farnell, Brechin, DD9 6UH
G3 KCB B. Green, 18 Kenilworth Road, Sale, M33 5FB
G3 KCD P. Bedwell, Narrowgate Rose Cottage, Court Road Rollesby, Great Yarmouth, NR29 5HQ
G3 KCG D. Tyerman, 20 Grace Gardens, Bishop's Stortford, CM23 3EX
GW3 KCQ J. Williams, Y-Fedw, Cwmann, Lampeter, SA48 8DT
G3 KCT D. Blythe, 6 Penn House, Mallory Street, London, NW8 8SX
G3 KCV J. Saunders, 7 Stone Lane, Yeovil, BA21 4NN
GM3 KCY G. Buchanan, 30 Gilmour Avenue, Clydebank, G81 6AW
G3 KCZ W. Siertsema, 21 Rowles Close, Kennington, Oxford, OX1 5LX
GW3 KDB P. Miles, Y Gorlan, Cross Inn, Llandysul, SA44 6NP
G3 KDD V. Barrett, 2 Carlisle Close, Sandy, SG19 1TX
G3 KDP A. Bounds, 32 Tregwary Road, St Ives, TR26 1BL
G3 KDU M. Crawford, 95 Victoria Avenue, Princes Avenue, Hull, HU5 3DW
G3 KDW H. Turnbull, 13 Linden Court, Wessex Road, Southampton, SO18 3RB
G3 KDY R. Folgate, Stile Cottage, Wilkinson Drive, Market Rasen, LN8 3LD
G3 KEG C. Rogers, 100 Sparth Road, Clayton Le Moors, Accrington, BB5 5QD
G3 KEK G. Carr, 88 Woodrow Crescent, Knowle, Solihull, B93 9EQ
G3 KEL R. Bray, Croft House, Blencogo, Wigton, CA7 0BZ
G3 KEP D. Pratt, 11 Moorleigh Close, Kippax, Leeds, LS25 7PB
G3 KEQ J. Mitchell, Chellow Dene, Viewlands Avenue, Westerham, TN16 2JE
G3 KEV M. Hamilton, 2 Wordsworth Close, Scalby, Scarborough, YO13 0SN
GM3 KEZ J. Little, 33 Manor Court, Forfar, DD8 1AD
G3 KFB N. Parkinson, 16 Collinson Avenue, Scunthorpe, DN15 8AB
G3 KFG H. Taylor, 17 Rose Acre Road, Littlebourne, Canterbury, CT3 1SY
G3 KFP A. Olds, 43 Fourth Avenue, Teignmouth, TQ14 9DT
G3 KFU P. Barry, 21 Old Pasture Road, Frimley, Camberley, GU16 8SA
GW3 KGI M. Bowen, 24 Parklands View, Sketty, Swansea, SA2 8LX
G3 KGP M. Palmer, Fairways, 8 Gwealdues, Helston, TR13 8JZ
G3 KGT J. Nicolson, 24 Pottersfield Road, Woodmancote, Cheltenham, GL52 9PY
GW3 KGV K. Bates, 5 Ffordd Nant Goch, Llangadfan, Welshpool, SY21 0PW
GM3 KHH W. Cecil, Innes House Oran, Buckie, AB56 5EP
G3 KHK D. Connolly, Jonquil, 6 Sanderson Mews, Colchester, CO5 7HF
G3 KHQ A. Langley, 58 Dumbarton Road, Brixton Hill, London, SW2 5LU
G3 KHR J. Fox, 25 Langdale Crescent, Bexleyheath, DA7 5DZ
G3 KHU R. Gabbitas, 12 Thornyville Drive, Oreston, Plymouth, PL9 7LF
G3 KHZ D. Cox, 18 Station Road, Castle Bytham, Grantham, NG33 4SB
G3 KII G. Lively, 9 Wilson Road, Shurdington, Cheltenham, GL51 4SN
G3 KIJ E. Lugmayer, 17 Borough End, Beccles, NR34 9YW
G3 KIL R. Messer, The Shambles, Swinbrook Road, Oxford, OX18 1DX
G3 KIP K. Grover, 1 Powdermill Close, Tunbridge Wells, TN4 9DR
G3 KIQ J. Elliot, 2 Pennine Close, Blackley, Manchester, M9 6HR
G3 KIW G. Jenner, Pogles Wood Cottage, Paradise Lane, Reading, RG7 6NU
G3 KJC R. Church, Three Birches, Sandy Close, Thatcham, RG18 9QP
GM3 KJE J. Scott, 5 Garthdee Terrace, Aberdeen, AB10 7JE
G3 KJK L. Wilkes, Parc Crane, Penmenner Road, Helston, TR12 7NN
GW3 KJN I. Winter, 5 Uwch Y Nant, Mynydd Isa, Mold, CH7 6YP
G3 KJO Huddersfield Technical College ARS, c/o M. Lupton, 44, Carr Road, Deepcar, , Sheffield, S36 2NR
G3 KJS W. Smith, 32 Lumley Road, Chester, CH2 2AQ
GW3 KJW P. Allely, Dwyfor, Rhiw, Pwllheli, LL53 8AE
G3 KJX B. Alderson, 43 Brompton Road, Northallerton, DL6 1ED
G3 KJY J. York, 48 Browhead Court, Shackleton Street, Burnley, BB10 3DS
GM3 KJZ G. Paterson, 3 Ferry Barns Court, North Queensferry, Inverkeithing, KY11 1ET
G3 KKC A. Rumbelow, 7 Hoof Close, Littleport, Ely, CB6 1HU
G3 KKJ A. Shannon, 23 Glebeland Close, West Stafford, Dorchester, DT2 8AE
G3 KKP J. Burgess, Moorend, Main St., Leeds, LS20 8NX
G3 KKZ P. Champion, 7 All Saints Drive, South Croydon, CR2 9ES
G3 KLC J. Bennett, Koivula, Station Road, Boston, PE20 3QT
G3 KLD R. Russell, 43 Ingestre Road, Hall Green, Birmingham, B28 9EQ
G3 KLF I. Crowther, 3 Glenelg, Fareham, PO15 6JU
G3 KLH D. Alexander, 20 Deans Court, Milford On Sea, Lymington, SO41 0SG
G3 KLK B. Page, 7 Marconi Way, Southall, UB1 3JP
G3 KLN N. Whittaker, 111 Burnley Road, Colne, BB8 8DT
G3 KLP J. Young, Woodglades, 34 The Demesne, Ashington, NE63 9TP
G3 KLV G. Vine, 4 Tollgate Close, Northampton, NN2 6RP
G3 KLZ D. Enoch, 7A Mount Road, Evesham, WR11 3HE
G3 KMA R. Balister, La Quinta, Mimbridge, Woking, GU24 8AR
G3 KMD J. Bass, 3 Tennyson Avenue, Grays, RM17 5RG
G3 KME L. Pennell, 182 Northampton Road, Wellingborough, NN8 3PJ
GM3 KMF G. Robbins, 39 Locheil Gardens, Glenrothes, KY7 6YL
G3 KMG D. Plumridge, Rose Cottage, Castleside, Consett, DH8 9AP
G3 KMI Southampton University Wireless Society, c/o P. Crump, 41 Vernon Way, Guildford, GU2 8DE
GM3 KML R. Whitfield, 42 Greenwood, Tweedmouth, Berwick-upon-Tweed, TD15 2EB
G3 KMM J. Crowther, 15 Chemin De Bausses, Villelongue D'Aude, Limoux, France, 11300
G3 KMO M. Birch, 12 The Heath, Hevingham, Norwich, NR10 5QW
G3 KMQ R. Heslop, Fairways, Meadow Drive, Bude, EX23 8HZ
G3 KMS D. Swain, 3 Nevy Fold Avenue, Horwich, Bolton, BL6 6QG
G3 KMV R. Birchall, Willow Tree House, Poole, Nantwich, CW5 6AL
G3 KND J. Hardy, Vogelenzang, 1B Roberts Road, Aldershot, GU12 4RD
G3 KNG A. Embrey, 59 Oaken Lanes, Codsall, Wolverhampton, WV8 2AW
G3 KNJ J. Otter, 7 Longacre Road, Dronfield, S18 1UQ
G3 KNU P. Jackson, 7 Ferriby Road, Scunthorpe, DN17 2EQ
GW3 KNZ A. Eccles, 78 Uplands Avenue, Connah'S Quay, Deeside, CH5 4LG
G3 KOA T. Robinson, 32 Campbell Crescent, East Grinstead, RH19 1JR
G3 KOB R. Goodman, 8 Decouttere Close, Church Crookham, Fleet, GU52 0UR
G3 KOD P. Kay, 7 St. Regis Close, London, N10 2DE
G3 KOJ R. Ezra, 39 Buckland Close, Waterlooville, PO7 6ED
G3 KOM F. Foulkes, 27 Aspian Drive, Coxheath, Maidstone, ME17 4JZ
G3 KOQ B. Parker, 9 Yewdale Avenue, Heysham, Morecambe, LA3 2LR
G3 KOS B. Faithfull, 68 Lampton Road, Long Ashton, Bristol, BS41 9AQ
G3 KOX N. Waite, 7 Lanercost Close, Welwyn, AL6 0RW
G3 KOZ W. Henderson, 9 Chiselbury Grove, Salisbury, SP2 8EP
G3 KPO Rosemary ARG, c/o A. Thornton, 78 Wellington Road, Ryde, PO33 3QJ
G3 KPU E. Prince, 9 Alwyn Road, Thorne, Doncaster, DN8 5JG
G3 KPV J. Killeen, 10 Den Brook Close, Torquay, TQ1 3TP
G3 KQB Raf Digby ARC, c/o P. Hanson, 10 Parkfield Road, Ruskington, Sleaford, NG34 9HS
G3 KQG Dr E. James, The Meadows, Kennford, Exeter, EX6 7TZ
G3 KQQ C. Mattacks, 68 Middlesex Drive, Bletchley, Milton Keynes, MK3 7EU
G3 KQV J. Ryley, 30 St. Helens Drive, Leicester, LE4 0GS
G3 KQY R. Disley, 6 St. Margarets Road, Farington, Leyland, PR25 4XT
G3 KRT G. Hodges, 102 Torrington Road, Ruislip, HA4 0AU
G3 KRW K. Whelan, Killiney, Longsplatt, Corsham, SN13 8DF
G3 KRX W. Addy, 14 Cresttor Road, Liverpool, L25 6DW
G3 KRZ J. Greenwood, Lea Cottage, Meadow Close Grimoldby, Louth, LN11 8HY
G3 KSF R. Harper, 21 Howard Oliver House, Harvey Gardens, Southampton, SO45 3LS
G3 KSP P. Hooper, 1 Victoria Mews, Morecambe, LA4 5QD
G3 KTA P. Munt, 130 Chipstead Way, Woodmansterne, Banstead, SM7 3JR
G3 KTH M. Darkin, 3 Adrian Close, Shell, Droitwich, WR9 7AY
G3 KTI M. Rees, Blue Pillars, 6 Grove Crescent, Coleford, GL16 8AZ
G3 KTM R. Atthill, Flat 23, Folland Court 70 Hamble Lane, Hamble, Southampton, SO31 4JS
G3 KTP D. West, 84 High Street, Castle Donington, Derby, DE74 2PQ
G3 KTR A. Rock, Licensee Address, Not Applicable, Resides in Usa, USA, .
G3 KTT M. Gallon, 41 Dene Gardens, Newcastle upon Tyne, NE15 8RL
G3 KTZ C. Lindsay, 38 St. Vincents Close, Littlebourne, Canterbury, CT3 1TZ
G3 KUD J. Duncan, 9 Springhill Close, Westlea, Swindon, SN5 7BG
G3 KUE Preston ARS, c/o A. Mcphail, 300 Fletcher Road, Preston, PR1 5HJ
GI3 KVD D. Jones, 5 Whitehill Park, Limavady, BT49 0QF
G3 KVJ S. Tomlinson, 31 The Quarry, Alwoodley, Leeds, LS17 7NH
G3 KVP D. Kitchen, Folkingham Place, Market Place, Sleaford, NG34 0SE
G3 KVR S. Davis, 3 Coronation Road, Banwell, BS29 6AZ
G3 KVT A. Smith, Winston House, Felthorpe Road, Norwich, NR9 5TF
GW3 KWB R. Neville, 35 Beechcroft Road, Newport, NP19 8AG
G3 KWJ N. Valentine, The White House, Dene Road, Ashtead, KT21 1EB
G3 KWK R. Nolan, 6 Plymouth Close, Redditch, B97 4NP
G3 KWN Lt. Col. W. Nicoll, Yonder, Milldown Road, Blandford Forum, DT11 7DE
G3 KWO K. Dawson, 44 Avondale Road, Darwen, BB3 1NS
G3 KWT I. Shaw, The Hawthorns, Woodlands Drive, Leeds, LS19 6JX
G3 KWW Dr R. Wilkinson, 83 Palewell Park, London, SW14 8JJ
G3 KWY A. Swain, 5 Hilderstone Close Alvaston, Derby, DE24 0SA
G3 KXB D. Pantony, 71 South Street, Whitstable, CT5 3EJ
G3 KXE E. Bettles, 15 St. Francis Avenue, Southampton, SO18 5QL
G3 KXF D. Wallis, 17 Upper Belgrave Road, Seaford, BN25 3AD
G3 KXI D. Keeler, 16 Honeysuckle Way, Witham, CM8 2XG
GM3 KXQ S. Floyd, 3 Crarae Place, Newton Mearns, Glasgow, G77 6XX
G3 KXS H. Perry, 688 Durham Road, Madison, USA, 6443
G3 KXV V. Johnston, 9 Holbeck Avenue, Middlesbrough, TS5 8DR
GW3 KXX R. Weaver, 59 Broad St., Leckwith, Cardiff, CF1 8BZ
G3 KYE J. Orr, 102 Manor House Lane, Yardley, Birmingham, B26 1PR
G3 KYF K. Sullivan, 14 Wigston Road, Blaby, Leicester, LE8 4FU
G3 KYM H. Stamper, The Bungalow, School Hill, St Austell, PL26 7TP
G3 KYZ D. Clarke, Primrose Mount, Old Neighbourhood, Stroud Glos, GL6 8AA
G3 KZB M. Ward, Flat 14, Meadrow Court, Godalming, GU7 3HG
G3 KZC R. Harknett, 28 Woodyleaze Drive, Hanham, Bristol, BS15 3BY
G3 KZE J. Davies, 45 Dahn Drive, Ludlow, SY8 1XZ
G3 KZG A. Bills, Brooklands 2 The Acre, Stourbridge, DY7 6HW
GW3 KZO M. Dennis, 11 Maes Yr Ysgol, Templeton, Narberth, SA67 8TZ
G3 KZR I. Davies, Lusty Hill Farm, Lusty Gardens, Bruton, BA10 0BS
GW3 KZT A. James, 143 Gaer Park Drive, Newport, NP20 3NS
G3 KZU M. Dolan, 15 Ringwood Road, Headington, Oxford, OX3 8JB
G3 KZX L. Loveland, 21 Roseland Close, Keyworth, Nottingham, NG12 5LQ
G3 KZZ D. Forster, 281 Mortimer Road, South Shields, NE34 0DR
G3 LAA A. Sedman, 69 Beechwood Avenue, Locking, Weston-Super-Mare, BS24 8DS
G3 LAG H. Gow, 43 Ringstead Crescent, Weymouth, DT3 6PT
G3 LAI G. Livingston, 24 Duncannon Drive, Falmouth, TR11 4AQ
G3 LAS J. Butcher, The Stables, Priory Farm, Thorpe Tilney, Lincoln, LN4 3SL
G3 LAU F. Adkin, 5 Cosway Mansions, Shroton St., London, NW1 6UE
GM3 LAW W. Walker, 45 Watts Gardens, Cupar, KY15 4UG
G3 LAZ R. Gerrard, Rnib Wavertree House, 211 Somerhill Road, Hove, BN3 1RN
G3 LBM A. Mulcahy, 8 Old Barn Close, Winkleigh, North Devon, EX19 8JX
G3 LBS Dr G. Cleeton, 24 Severn Drive, Newcastle, ST5 4BH
G3 LCF P. Baldwin, 49 King George Vi Mansions, Court Farm Road, Hove, BN3 7QX
G3 LCH M. Pharaoh, 1 Madeira Road, Mitcham, CR4 4HD
G3 LCI H. Young, 23 Willow Grove, Wirral, CH46 0TU
G3 LCL S. Ldr. A. Baylis, Queen Oak Inn, Bourton, Gillingham, SP8 5AL
GW3 LCQ M. Williams, Dwyros, 12 Penrhos Avenue, Llandudno Junction, LL31 9EL
G3 LCY J. Tamlin, 53 Hele Gardens, Plympton, Plymouth, PL7 1JY
GW3 LDC J. Phillips, 9 Trelawny Close, Usk, NP15 1SP
G3 LDG B. Gee, Daisy Bank, Carlton Road, Bedford, MK43 7JL
G3 LDI R. Cooke, The Old Nursery, The Drift, Swardeston, Norwich, NR14 8LQ
G3 LDJ K. Day, 45 Thick Hollins Drive, Meltham, Holmfirth, HD9 4DR
GI3 LEG D. Wilson, 189 Cregagh St., Belfast, BT6 8NL
G3 LEK L. Kitching, Woodsyde Lower Road, Harmer Hill, Shropshire, SY4 3QX
G3 LEO G. Brigham, Waterside House, West Tanfield, Ripon, HG4 5LF
G3 LET P. Hobbs, Honeysuckle Cottage, Stairbridge Lane Bolney, Haywards Heath, RH17 5PA
GW3 LEW G. Weale, Winfield, Templeton, Narberth, SA67 8SP
G3 LFD R. Widders, 82 Azalea Walk, Eastcote, Pinner, HA5 2EH
GJ3 LFJ H. Mesny, La Trigale, Route De L'Eglise, St Lawrence, Jersey, JE3 1LA
G3 LFR M. Everett, Thrupp Wharf, Cosgrove, Milton Keynes, MK19 7BE

G3 LFV R. Manser, 39 Long Meadow, Markyate, St Albans, AL3 8JN
G3 LFX Dr D. Pedder, 37 Hersham Road, Walton-on-Thames, KT12 1LE
G3 LGA M. Hayward, Brindle, Romsey Road, Stockbridge, SO20 8DB
G3 LGF G. Falding, 10 Angel Court, Shaftesbury, SP7 8HX
G3 LGK B. Sandall, Amber Croft, Main Road Higham, Alfreton, DE55 6EH
G3 LGQ P. Marsden, 49 Southfield Park, North Harrow, Harrow, HA2 6HF
G3 LGR M. Hooles, 114 Cassiobury Drive, Watford, WD17 3AQ
G3 LGT J. Tate, Pine Holt, 34 Queens Road, Fleet, GU52 7LE
GM3 LGU R. Pryde, Room 5, Abbeyfield, Paterson Court, Haddington, EH41 3DU
G3 LGW D. Spencer, Paladin, 89 Watling Street, Tamworth, B78 3DE
G3 LHG E. Smith, 3 Meadow Avenue, Wetley Rocks, Stoke-on-Trent, ST9 0BD
G3 LHJ D. Webber, 43 Lime Tree Walk, Milber, Newton Abbot, TQ12 4LF
GW3 LHK G. Griffiths, Glyndwr, Lampeter Road, Aberaeron, SA46 0ED
G3 LHN R. Muir, 19 Eastwick Drive, Bookham, Leatherhead, KT23 3PY
G3 LHS L. Matthews, 14 Mayforth Gardens, Ramsgate, CT11 0LL
G3 LHU M. Dixon, 27 Planets Lane, Cheltenham, GL51 6GR
G3 LHZ Prof. M. Underhill, Hatchgate, Tandridge Lane, Lingfield, RH7 6LL
G3 LIK M. Puttick, 21 Sandyfield Crescent Cowplain, Waterlooville, PO8 8SQ
G3 LIO J. Gibbs, 13 Bromley Road, Macclesfield, SK10 3LN
G3 LIV J. Melvin, 2 Salters Court, Newcastle upon Tyne, NE3 5BH
GM3 LIW A. Wood, 97A Fort St. Broughty Ferry, Dundee, Dd51dy
G3 LJD J. Davies, 57 Madeira Court, Knightstone Road, Weston Super Mare, BS23 2BH
GM3 LJR T. Saxton, 38 William Street, Dalbeattie, DG5 4EN
GW3 LJS Dr T. Bloxam, 15 Cleveland Avenue, Mumbles, Swansea, SA3 4JD
G3 LKV D. Locke, 4 Glebe Close, Doveridge, Ashbourne, DE6 5NY
G3 LKW D. Wiltshire, 71 Ferndale, Waterlooville, PO7 7PG
G3 LLD S. Collier, 64 Slonk Hill Road, Shoreham-by-Sea, BN43 6HY
G3 LLE K. Webster, 25 Carlin Gate, Blackpool, FY2 9QT
G3 LLG R. Loveday, 42 Bridle Path, Woodcote, Reading, RG8 0SE
G3 LLK J. Gale, 66 Burys Bank Road, Crookham Common, Thatcham, RG19 8DD
GM3 LLP B. Watson, 4 Caldwell Road, West Kilbride, KA23 9LE
G3 LLV J. Mcelvenney, 10 Bignor Place, Sheffield, S6 1JE
G3 LLZ D. Goacher, 27 Glevum Road, Swindon, SN3 4AA
G3 LME K. Taylor, 7 Bowen Close, Cheltenham, GL52 5EG
G3 LMH R. Wellbeloved, 8 Orchard Close, South Wonston, Winchester, SO21 3EY
G3 LMQ J. Hamer, 7 Arundel Road, Coventry, CV3 5JT
G3 LMR J. Eley, 25 Peckleton View, Desford, Leicester, LE9 9QF
G3 LMX T. Mitchell, 27 Hanmer Road, Simpson, Milton Keynes, MK6 3AY
G3 LNL P. Lovelady, 14 Maunders Court, Liverpool, L23 9YU
G3 LNM R. Scrivens, 6 Highland Close, Cantley, Norwich, NR13 3SW
G3 LNN J. Symes, 19 Boundary Close, Kirkby-In-Ashfield, Nottingham, NG17 8RS
G3 LNP A. Preedy, 2 Hunters Gate, Much Wenlock, TF13 6BW
GW3 LNR A. Gwynne, 77 Edward St., Pant, Merthyr Tydfil, CF48 2BB
G3 LNS G. Beasley, Po Box 1344, Paphos, Cyprus
G3 LNW J. Mcguire, 5 Primrose Way, Trevadlock Hall Park, Launceston, PL15 7PW
G3 LOD D. Rowse, 48 Oatlands Avenue, Bar Hill, Cambridge, CB23 8EQ
G3 LOE W. Roberts, 13 Brean Road, Stafford, ST17 0PA
G3 LOF Dr G. Peskett, 13 Warneford Road, Oxford, OX4 1LT
G3 LOV M. Francis, Cherry Tree Cottage, Atlantic Close, Tintagel, PL34 0EL
G3 LPC Weald ARS, c/o C. Desborough, 22 Westland Road, Faringdon, SN7 7EY
G3 LPL P. Sherdley, 2 Stable Yard, Taylors Lane, Preston, PR3 6AP
G3 LPN J. Hunt, 28 Robins Bow, Camberley, GU15 3NR
G3 LPT G. Woods, Bamburgh House, Hunston, Bury St Edmunds, IP31 3EN
G3 LPU E. Burrell, 20 The Avenue Richnond Wood Norton Eveham Road, Wood Norton Worcs. Wr11 4Ty, Evesham, WR11 4TY
GU3 LPV T. Catts, Po Box 1029, Alderney, GY9 3JD
G3 LPY R. Nye, Beech Cottage, Gorelands Lane, Chalfont St Giles, HP8 4HQ
G3 LQB K. Bishop, Friedrich-Ebert Strasse 23, Osterode, Germany, 37520
GW3 LQE A. Ernest, 2 Osborne 7 Clive Crescent, Penarth, CF64 1WW
G3 LQJ R. Cox, 12A Kelling Close, Holt, NR25 6RU
G3 LQO E. Harris, 10 Girdle Road, Walsworth, Hitchin, SG4 0AN
G3 LQP R. Brown, 262 Fir Tree Road, Epsom, KT17 3NL
G3 LQR S. Freeman, West Farm, Cransford, Woodbridge, IP13 9PQ
G3 LQS Raf Coningsby ARC, c/o D. Bloomfield, 14 Horsham Close, Luton, LU2 8JH
G3 LQW K. Wallace, 55 Lamborne Road, Leicester, LE2 6HQ
G3 LQX M. Nicholls, 6 Lyme Bay Road, Teignmouth, TQ14 8RS
GI3 LQY J. Stronach, 20 Monaville Drive, Lisburn, BT28 2DR
G3 LRA C. Eley, 1 Hilldale View, Gaisgill, Penrith, CA10 3UE
GM3 LRG J. Gray, 47 South Street, Greenock, PA16 8QG
G3 LRH G. Frampton, 1 Ludlow Road, Church Stretton, SY6 6DD
G3 LRI J. Blakey, 10 Wilson Terrace, Newcastle upon Tyne, NE12 7JP
G3 LRL R. Bowell, 16 Margarite Way, Wickford, SS12 0ER
G3 LRQ M. Humphries, 2 South View Close, Twyford, Reading, RG10 9AY
G3 LRS Leicester Radio Society, c/o A. Moss, 10 Shakespeare Drive, Leicester, LE3 2SP
G3 LRU J. Miller, 57 Clarendon Villas, Hove, BN3 3RE
G3 LRX R. Durell, Middleton Farm, Hubbards Hill, Lenham, ME17 2EJ
G3 LSA D. Moore, 5 Seahaven Springs Estate, Seaholme Road, Mablethorpe, LN12 2QS
GD3 LSF E. Ellis, Ballahams, 3 Glen Road, Laxey, Isle Of Man, IM4 7AP
G3 LSJ C. Gerrard, 6 Bridle Close, Sleaford, NG34 7TD
G3 LSQ P. Aitchison, Upper Weston House, Cot Lane, Chichester, PO18 8SU
G3 LST P. Clarke, Half Moon House, Church Street, Colchester, CO6 4QH
G3 LSX G. Townsend, 21 Grange Avenue, East Barnet, Barnet, EN4 8NJ
G3 LTF P. Blair, Woodleigh, Upper Wyke, Andover, SP11 6EA
G3 LTM B. Moyler, 1 Bay Walk, Aldwick, Bognor Regis, PO21 4ET
GW3 LTX R. Savage, Plas Gwyntog, Rhoslefain, Tywyn, LL36 9ND
G3 LUA A. Knowles, 73 Kingslea Road, Solihull, B91 1TJ
G3 LUC E. Bate, 5 Elm Road, Shildon, DL4 1BH
G3 LUH K. Reader, 21 Broadwater Avenue, Poole, BH14 8QY
G3 LUK 59 Squadron Air, c/o I. Haliwell, 61 Cliffe Road, Shepley, Huddersfield, HD8 8AG
G3 LUN Special Communications (TA) Association, c/o D. Smith, The Old Forge, High Street, Brinkley, Newmarket, CB8 0SE
G3 LUO C. Evans, Peas Gill House, Gawthrop, Sedbergh, LA10 5QB
G3 LUW B. Whittaker, Woodlands, Newton Down, Lifton, PL16 0AS
G3 LUZ F. Machin, 70 Poors Lane, Hadleigh, Benfleet, SS7 2LN
GM3 LVA D. Simpson, Larchwood, Tomatin, Inverness, IV13 7YR
G3 LVB G. Brooks, 11 Prince Charles Way, Seaton, EX12 2TU
G3 LVL B. Ash, 5 Church Close, Wickham Bishops, Witham, CM8 3LN
G3 LVP K. Eastty, 7 The Grange, The Reddings, Cheltenham, GL51 6RL
G3 LVW R. Smith, 40 Highwoods Drive, Marlow Bottom, Marlow, SL7 3PY
G3 LWD P. Stone, Bramley, Stone Street, Hythe, CT21 4JP
G3 LWF L. Franklin, 1 Woodwell Cottages Woodwell Road, Bristol, BS11 9UP
G3 LWJ C. Way, 8 Stratford Place, Eaton Socon, St Neots, PE19 8HY
G3 LWM J. Harris, 37 Long Orchard Way, Martock, TA12 6FA
G3 LWR J. Evans, 18 Mandeville Road, Isleworth, TW7 6AD
G3 LWT P. Buck, 17 Sanden Close, Hungerford, RG17 0LA
GW3 LWU G. Brisbar, 97 Chambers Lane, Mynydd Isa, Mold, CH7 6UZ
G3 LXB S. Jones, 43 New St., Chase Terrace, Walsall, WS7 8BT
GW3 LXE J. Boden, Plas Heulwen, Llanfair Road, Newtown, SY16 3JY
G3 LXJ F. Fisher, 7 Greenbank, Halesworth, IP19 8RP
G3 LXQ D. Gallop, 4 Volunteer Road, Theale, Reading, RG7 5DN
GU3 LYC T. De Putron, Shieling Cottage, La Rue Marquand, St Andrew, Guernsey, GY6 8RB
G3 LYD Dr E. Henderson, The Homestead, High Street, Ventnor, PO38 3HZ
G3 LYG Dr A. Macgregor, 10 Balroy Court Forest Hall, Newcastle upon Tyne, NE12 9AW
G3 LYP Dr M. Scott, The Magnolias, Marlow Road, High Wycombe, HP14 3JW
GW3 LYU D. Price, Llansilin, Oswestry, SY10 7QB
G3 LYZ B. Currey, 42 Westfield Avenue, Goole, DN14 6JX
G3 LZC A. Stirland, 98 Aldreds Lane, Heanor, DE75 7HG
G3 LZI J. Oates, Cherry Tree Cottage, Green Moor Wortley, Sheffield, S35 7DQ
G3 LZM M. Bush, 5 Quay Close, Hereford, HR1 2RQ
G3 LZN G. Ellison, Little Flushing, St. Peters Road, Falmouth, TR11 5TJ
G3 LZO P. Thomas, 20 Bleasdale Court, Longridge, Preston, PR3 3TX
G3 LZR E. Speller, 78 Chelmsford Road, Holland-On-Sea, Clacton-on-Sea, CO15 5DJ
G3 LZZ A. Pomfret, Flat 2, Ingwell House, Grange-over-Sands, LA11 6DP
G3 MAE Dr A. Wilson, 8 The Paddock Appleton Wiske, Northallerton, DL6 2BE
G3 MAI R. Stevens, 10 Grange Drive, Stratton, Swindon, SN3 4LA
G3 MAJ E. Holden, 10 Rowan Tree Close, Greasby, Wirral, CH49 3AW
G3 MAR Midland ARS, c/o N. Gutteridge, 68 Max Road, Quinton, Birmingham, B32 1LB
GM3 MAS A. Pringle, 1 Falloch Road, Milngavie, Glasgow, G62 7RR
G3 MAU J. Wardle, 17 Frederick Neal Avenue, Coventry, CV5 7EH
G3 MAV J. Bradley, 17 Talboys Walk, Tetbury, GL8 8YU
G3 MAZ H. Bell, Downside, North Street Mere, Warminster, BA12 6HH
GI3 MBB A. Mcmurtry, 20 Towerview Crescent, Bangor, BT19 6BA
GD3 MBC R. Wernham, Fair Isle, Lhoobs Road, Douglas, Isle Of Man, IM4 3JB
G3 MBD H. Dannatt-Brader, 20 Shire Place, Northampton, NN3 8DE
G3 MBK D. Underdown, 26 Birch Road, Farncombe, Godalming, GU7 3NT
G3 MBM J. Masters, 8 Purbeck Terrace Road, Swanage, BH19 2DE
G3 MBN B. Gibbs, 15 Moor Barton, Neston, Corsham, SN13 9SH
G3 MBO B. Aspinwall, 33 Clipstone Crescent, Leighton Buzzard, LU7 3LU
G3 MBU M. Standige, 7 Hill Crest Avenue, Burnley, BB10 4JA
G3 MCA D. Owen, 1 Mosslea Road, Orpington, BR6 8HP
G3 MCB A. Williams, 1 Wyvern Road, Sutton Coldfield, B74 2PS
G3 MCC K. Worrall, 21 Northwood Avenue, Middlewich, CW10 0HR
G3 MCD K. Holland, Ravendale St. Lawrence, Bodmin, PL30 5JL
G3 MCE L. Lee, 34 Westby Way, Poulton-le-Fylde, FY6 8AD
G3 MCK G. Stancey, 22 Peterborough Avenue, Oakham, LE15 6EB
G3 MCL C. Simpkins, 6 Compton Way, Olivers Battery, Winchester, SO22 4EY
G3 MCP P. Goadby, 535 Welford Road, Leicester, LE2 6FN
G3 MCV B. Vaughan, 17 Richmond Close, West Town, Hayling Island, PO11 0ER
G3 MCX J. Kennedy, 22 Croham Park Avenue, South Croydon, CR2 7HH
G3 MD T. Drew, Quinta Bela, Po De Asnos, Pastor, Portugal, 3230-248
G3 MDD B. Mudge, 9 Crossmead, Woolavington, Bridgwater, TA7 8ER
G3 MDG Chesham & District ARS, c/o M. Appleby, 6 Mandeville Road, Prestwood, Great Missenden, HP16 9DS
G3 MDI M. Plummer, Kembali, 14 Turnberry Drive, Woodhall Spa, LN10 6UE
G3 MDM G. Mcgee, 2 Ilynton Avenue, Firsdown, Salisbury, SP5 1SH
G3 MDR M. Hallet, 33 Latimer Street, Romsey, SO51 8DF
G3 MEC J. Pearce, 86 Sopers Lane, Poole, BH17 7EU
G3 MED F. Griffiths, 105 Hillcroft Crescent, South Oxhey, Watford, WD19 4PA
G3 MEH R. Piper, 8 Osborne Way, Wigginton, Tring, HP23 6EN
G3 MEV C. Cory, Tekelex, Chapel Lane, Thatcham, RG19 8BE
G3 MEY J. Lawrence, 16 Waverley Court, Corsham, SN13 9NN
G3 MFG D. Close, 27 High St., Collyweston, Stamford, PE9 3PW
G3 MFH G. Dale, 20 Blythe Avenue, Stoke-on-Trent, ST3 7JY
G3 MFJ G. Firth, 13 Wynmore Drive, Bramhope, Leeds, LS16 9DQ
G3 MFK M. Camp, 82 Leicester Road, Hinckley, LE10 1LT
G3 MFL A. Russell, Pear Tree Cottage, Savernake Road, Marlborough, SN8 3AS
G3 MFO P. Elliot, 3 Shickle Place, Hopton, Diss, IP22 2QR
G3 MFW H. Woodhouse, 143 Bodmin Road, Truro, TR1 1RA
G3 MGL A. Davis, 22 Yarmouth Close, Crawley, RH10 6TH
G3 MGQ P. Parkman, 2A Frank Woolley Road, Tonbridge, TN10 4LE
G3 MGS C. Stephens, 12 Berkshire Road, Bristol, BS7 8EX
G3 MGU A. Dodson, 53 Simons Lane, Wokingham, RG41 3HG
G3 MGW R. Wheeler, 51 Seaview Road, Brightlingsea, Colchester, CO7 0PR
G3 MGX J. Tomlinson, 34 Bentley Road, Tacolneston, Norwich, NR16 1DL
G3 MGY Dr M. Uotome, 3-10-14 Kujayarna Sugimami-Ku, Tokyo, Japan, 168-0082
G3 MHD A. Williams, 9 Charlotte Cove Road Charlotte Cove, Tasmania, Australia, 7112
G3 MHF M. Ockenden, 16 Ripley Chase, 17 The Goffs, Eastbourne, BN21 1HB
G3 MHT E. Landon, 14 The Blackthorns, Broughton, Brigg, DN20 0BB
G3 MHV Dr T. Langdon, 58 Upper Marsh Road, Warminster, BA12 9PN
G3 MHX M. Tate, 48 Crossgates Bedwell Plash, Stevenage, SG1 1LS
G3 MIP S. Heilbron, 8 Beechwood Drive, Formby, Liverpool, L37 2DG
G3 MJM A. Marshall, 2 Westwood Way, Beverley, HU17 8GE
G3 MJW C. Edmunds, 23 Dorset Gardens, Northampton, NN2 7PX
G3 MKE W. Smith, 12 Benscliffe Drive, Loughborough, LE11 3JP
GW3 MKT M. Hooks, 1 Llwyn Castan, Pentwyn, Cardiff, CF23 7DA
G3 MKU A. Bower, 82 Anson Road, Shepshed, Loughborough, LE12 9PU
G3 MKV C. Curtis, 24 Rodney Road, Hartford, Huntingdon, PE29 1RZ

G3 MLO P. Weatherall, Woodside, Stone Street Stelling Minnis, Canterbury, CT4 6DN

G3 MLQ S. Blundell, 12 Brookfield St., Melton Mowbray, LE13 0NB

G3 MLS D. Nappin, New Edge Farm, Heptonstall, Hebden Bridge, HX7 7PG

G3 MMA D. Mayes, Flat 15, Hanover Court, Blackman Way, Witham, CM8 1JZ

G3 MME P. Whitford, Three Pieces, Vernon Lane, Kelstedge, Chesterfield, S45 0EA

GI3 MMF W. Mcaleer, 90 Gortin Park, Belfast, BT5 7EQ

GI3 MMG D. Noon, 34 Rodney Park, Bangor, BT19 6FN

G3 MMJ G. Browne, 39A Cromwell Road, Canterbury, CT1 3LD

G3 MMN B. Newman, 101 Tally Ho Road, Shadoxhurst, Ashford, TN26 1HW

G3 MMS G. Whiting, 25 Obthorpe Lane Thurlby, Bourne, PE10 0ES

G3 MMX E. Lawley, 3 Barnicott Close, Newton Ferrers, Plymouth, PL8 1BP

G3 MNB H. Benjamin, 21 Sheephouse Green, Wotton, Dorking, RH5 6QW

G3 MNJ J. Yates, Trinder Cottage Filkins, Lechlade, GL7 3JG

G3 MNS I. Swan, Flat 6 Mason Court, Alford Road, Sutton-On-Sea, Mablethorpe, LN122GY

G3 MNV P. Darragh, 48 Goodwood Park Road, Northam, Bideford, EX39 2RR

G3 MOA J. Ruff, 17 Harts Close, Teignmouth, TQ14 9HG

G3 MOL J. Lixenberg, Orchard House, 77A Pembroke Crescent, Hove, BN3 5DF

G3 MON D. Gent, 12 Field Road, Billinghay, Lincoln, LN4 4EA

GM3 MOR Dr R. Webster, Meric, 7 Woodmuir Crescent, Newport-on-Tay, DD6 8HL

G3 MOT J. Lambert Hurley, 64 Henry Road, West Bridgford, Nottingham, NG2 7ND

GW3 MOV Dr C. Smith, 38 Plas Taliesin, Penarth, CF64 1TN

G3 MPB A. Smith, 10 Goodwood Road, Redhill, RH1 2HH

G3 MPD OBO Poldhu ARC, c/o L. Jones, Treharne Cottage, Meaver Road, Mullion, Helston, TR12 7DN

G3 MPF C. Smith, 29 Cloisters, Tarleton, Preston, PR4 6UL

G3 MPN D. Johnson, 54 Norwich Road, Wymondham, NR18 0NT

GW3 MPP G. Price, 17 Celtic Close, Undy, Caldicot, NP26 3PB

G3 MPW A. Walker, 14 St. Joans Drive, Scawby, Brigg, DN20 9BE

G3 MQD P. Greed, The Bungalow, Townsend, Devizes, SN10 4RR

G3 MQI March Army Cadet Force, c/o R. Gill, 45 Biggin Lane, Ramsey, Huntingdon, PE26 1NB

GM3 MQO G. Olesen, 8 Rowallan Crescent, Prestwick, KA9 2HE

G3 MQR D. Robinson, 32 Bullock Wood Close, Colchester, CO4 0HX

G3 MRQ D. Byne, Storm Bay, Church Street Charwelton, Daventry, NN11 3YT

G3 MRT R. Strafford, Chy Lowarth, Sparnock, Truro, TR3 6EB

GM3 MRV G. Carrick, 4 Kingfisher Lane, Gretna, DG16 5JS

G3 MRX Dr P. Robinson, 9 Barton Close, Cambridge, CB3 9LQ

G3 MRZ M. Crutchley, 40 Ufton Crescent, Shirley, Solihull, B90 3SA

G3 MSL R. Ives, 11 Coombe Drive, Fleet, GU51 3DY

G3 MSO E. Tunstall, 11 The Broadway, Charlton On Otmoor, Kidlington, OX5 2UB

G3 MSW K. Ashcroft, Fendley Corner, Common Lane, Harpenden, AL5 5DW

G3 MTD B. Kissack, 13 Church Street, The Old Saddlers, Braunton, EX33 2EL

G3 MTG R. Prior, 35 Hanson Drive, Fowey, PL23 1ET

G3 MTJ R. Skoyles, 2 Hay Close Great Oakley, Corby, NN18 8HX

G3 MTP P. Gadsden, Rose Cottage, Salwayash, Bridport, DT6 5HX

G3 MTR B. Wolfe, 24 Marchbank Drive, Cheadle, SK8 1QY

GM3 MTW C. Wolstencroft, 29 Fasach, Glendale, Isle of Skye, IV55 8WP

GM3 MUA P. Lawlor, Woodside, North Kessock, Inverness, IV1 3XG

G3 MUO G. Gott, 10 Churchill Crescent, Marple, Stockport, SK6 6HJ

G3 MUX C. Benson, Orchard Croft, 85 Runcorn Road, Warrington, WA4 6UA

G3 MVE A. Bullimore, 8 St. Georges Road, Felixstowe, IP11 9PL

G3 MVM P. Pierson, 7 Beehive Road, Goffs Oak, Waltham Cross, EN7 5NL

G3 MVV N. Miller, Avon, Gardiners Lane North, Billericay, CM11 2XA

G3 MVX J. Burke, 120 Seabourne Road, Bexhill-on-Sea, TN40 2SD

G3 MVZ F. Garrett, 18 Wolfe Close, Chichester, PO19 6BY

G3 MWM Dr D. Murden, Po Box 06, Curitiba, Parana, Brazil, 80011 970

G3 MWO D. Beales, 2 Wood Close, Tostock, Bury St Edmunds, IP30 9PX

GM3 MWX A. Winton, 2 Castlehill Cottages, Brisbane Glen Road, Largs, KA30 8SN

G3 MWZ J. Casling, 19 Orchard Close, Tavistock, PL19 8HA

G3 MXA B. Collins, 2 Pilgrims Way, Ely, CB6 3DL

G3 MXF P. Cutler, 14 Verulam Road, Poole, BH14 0PP

G3 MXH T. Downing, 8 Auction Yard, Haughley, Stowmarket, IP14 3GA

G3 MXJ D. Andrews, Coupelle, Levignac de Guyenne, France, 47120

GM3 MXN T. Sorbie, 9 Lynn Court, Larkhall, ML9 1QT

G3 MXP J. Palfrey, Caprice, 4 Laverstock Park West, Salisbury, SP1 1QL

G3 MXV H. Pierson, 65 Station Road, Countesthorpe, Leicester, LE8 5TB

G3 MYA A. Martindale, 16 Charles Miller Court, Leiston, IP16 4BY

G3 MYC C. Cheatle, 56 Ashfurlong Crescent, Sutton Coldfield, B75 6EN

G3 MYG R. Inman, 60 Abercorn Road, Mill Hill, London, NW7 1JL

G3 MYI J. Lewis, 50 Robin Gardens, Waterlooville, PO8 9XF

G3 MYM R. Micklewright, 5 Sandringham Road, Yeovil, BA21 5JE

G3 MYY S. Boston, Beavers, Mill Lane, Ipswich, IP8 4AU

G3 MYZ P. Nicholson, 3 Welborn Court Main Street, Flixton, Scarborough, YO11 3XA

G3 MZA E. Hamblen, 64 Tollers Lane, Coulsdon, CR5 1BB

G3 MZC C. Sutcliffe, 1 Tollgate Road, Culham, Abingdon, OX14 4NL

G3 MZI J. Hood, 89 Freemens Way, Deal, CT14 9DQ

G3 MZN R. Lightfoot, 28 Wheal Gorland Road, St. Day, Redruth, TR16 5LT

GM3 MZX M. Pedreschi, Clary Lodge, Carse Of Clary, Newton Stewart, DG8 6BH

G3 MZZ A. Kightley, 29 The Parkway, Gosport, PO13 0PT

G3 NAE C. Richardson, 10 Fielders Way, East Wellow, Romsey, SO51 6EX

G3 NAI R. Norman, 19 Hughes Croft, Bletchley, Milton Keynes, MK3 5HA

G3 NAK G. Mallinson, 145 Huddersfield Road, Meltham, Holmfirth, HD9 4AJ

G3 NAN R. Henderson, 48 Cartwright Crescent, Brackley, NN13 6HA

G3 NAP B. Sowter, 56 Alderminster Road, Coventry, CV5 7JU

G3 NAQ Dr G. Grayer, Bagatelle, 3 Southend, Newbury, RG20 7BE

G3 NAT London Raynet G, c/o A. Brooker, 18 Honeybourne Way, Petts Wood, Orpington, BR5 1EZ

G3 NAV E. Cook, Edward R.Cook, 152 O St. Spc.51, Lincoln, USA, 95648

G3 NAW J. Ryan, 4 Ferry Lane, Bath, BA2 4HS

G3 NAY S. Whithorn, 53 Torbay Road, Allesley, Coventry, CV5 9JY

G3 NBL Dr J. Larson, Nyhem, Whitton Village, Stockton-on-Tees, TS21 1LQ

G3 NBN R. Weaving, 7 Fairway Gardens Sparkwell, Plymouth, PL7 5FE

G3 NBQ P. Burt, 3335 Mountain Highway, North Vancouver, Canada, V7K 2H4

G3 NBS A. Bairstow, 27 Williams Way, Longwick, Princes Risborough, HP27 9RP

G3 NBY H. Murray, 36 Sterndale Road, Davenport, Stockport, SK3 8QU

G3 NBZ K. Thorne, 3 Cherry Gardens, Abstacle Hill, Tring, HP23 4EA

G3 NCN J. Ellerton, 7 Cotterell Close, Bracknell, RG42 2HL

GM3 NCO A. Mustard, Tigh Ard, Knockhouse Hill, Crossford, Dunfermline, KY12 8PT

GW3 NCT R. Lord, 8 Llys Steffan, Llantwit Major, CF61 2UF

GW3 NDB G. Wyatt, 3 Creidiol Road, Mayhill, Swansea, SA1 6TZ

G3 NDC C. Deamer, Gatehouse, Warren Lane, Stanmore, HA7 4LD

G3 NDK R. Webb, 142 Penrose Avenue, Carpenders Park, Watford, WD19 5AA

G3 NDM B. Mahony, 12 Orchard Green Marden, Hereford, HR1 3ED

G3 NDN D. Newey, 15 Clent View Road, Stourbridge, DY8 3JE

G3 NDS R. Oliver, Flat 21, Exeter Court, 52 Wharncliffe Road, Highcliffe, Christchurch, BH23 5DF

GU3 NDX Castel Contest Club, c/o R. Beebe, San Grato, Les Houguette, Castel, Guernsey, Guernsey, GY5 7DZ

G3 NEH J. Isles, 3 Drovers Croft, Greenleys, Milton Keynes, MK12 6AN

G3 NEO P. Bagshaw, 48 Kiveton Lane, Todwick, Sheffield, S26 1HL

G3 NEP C. Wager-Bradley, Swallowdale, Longhoughton Road, Alnwick, NE66 3AT

GM3 NEQ A. Finlay, 19 Fraser Avenue, Newton Mearns, Glasgow, G77 6HP

G3 NFB J. Leviston, 9 Barnes Avenue, Fearnhead, Warrington, WA2 0BL

G3 NFC Burton & District Radio Society, c/o G. Newstead, 97 Hawthorn Crescent, Burton-on-Trent, DE15 9QN

G3 NFJ M. Coward, High Bank, 51 High Street, Warminster, BA12 7AP

GI3 NFM K. Mcelhatton, 2A Orpheus Drive, Dungannon, BT71 6DR

G3 NFP L. Beckwith, Westgate Burghill, Hereford, HR4 7RW

G3 NFV R. Sykes, 16 The Ridgeway, Fetcham, Leatherhead, KT22 9AZ

G3 NFW J. Carroll, White Lodge, Hunston, Chichester, PO20 1PA

G3 NFY B. Twist, 11 Church Street, Minehead, TA24 5JU

G3 NGJ W. Epton, 2 Eastcliff Road, Lincoln, LN2 5RU

G3 NGK D. Chapman, 6 Pickhurst Green, Hayes, Bromley, BR2 7QT

GM3 NGW W. Webb, 5 Thornlea Drive Giffnock, Glasgow, G46 6DB

G3 NGX H. Hogg, Crossways, Ferry Road, Reading, RG8 0JL

G3 NGZ Pelican Radio Group Little Rissington, c/o M. Grierson, 1 Blenheim Close, Upper Rissington, Cheltenham, GL54 2QX

G3 NHB Dr D. Bowyer, 41A High Street, Trumpington, Cambridge, CB2 9HR

G3 NHE M. Dann, 61 Alms Hill Road, Parkhead, Sheffield, S11 9RR

G3 NHF J. Noble, 27 Chestnut Avenue, Donington, Spalding, PE11 4XH

G3 NHL C. Lewis, The Anchorage, Quay Road, Devoran, Truro, TR3 6PW

G3 NHP G. Peacock, Hallowsgate House, Flat Lane, Tarporley, CW6 0PU

GM3 NHQ T. Harrison, 7 Cults Gardens, Broughty Ferry, Dundee, DD5 1QT

G3 NHR H. Rogers, Aughavore, Church Walk, Louth, LN11 8LJ

G3 NHS J. Carp, 45 Rochester Road, Lowestoft, NR33 0JR

G3 NHV D. Hare, White Lodge, Mount Gabriel, Schull, Ireland,

G3 NHX G. Quarterman, 2 Milton Avenue, Sutton, SM1 3QB

G3 NIC K. Plant, Rose Cottage, Lincoln Road, Lincoln, LN2 2NE

G3 NID I. Douglas, 6 Ansley Road, Houghton, Huntingdon, PE28 2DQ

GM3 NIG D. Cram, 61 Gailes Road, Troon, KA10 6TB

G3 NII R. Porter, 6 Clifton Road, Shefford, SG17 5AA

G3 NIJ B. Barker, 4 Glantlees, West Denton, Newcastle upon Tyne, NE5 2PJ

G3 NIL G. Munden, 126 Stanley Green Road, Poole, BH15 3AQ

GW3 NIN J. Brogan, 38 Graig Park Circle, Newport, NP20 6HE

G3 NIQ R. Gorton, 2 Clyde Court, Clyde Close, Redhill, RH1 4AY

G3 NIW P. Ives, Allt Na Crioch, Ockham Road South, Leatherhead, KT24 6QJ

G3 NJA Torbay ARS, c/o D. Webber, 43 Lime Tree Walk, Milber, Newton Abbot, TQ12 4LF

G3 NJB WACRL(World Association of Christian Ras), c/o P. Jackson, 8 Buttree Court, South Kirkby, Pontefract, WF9 3NB

G3 NJG T. George, 8 Lanehays Road, Hythe, Southampton, SO45 5ER

G3 NJV P. Randall, Myresyke, Ruan Minor, Helston, TR12 7LU

G3 NJX R. Geeson, The Grove, Main Road, Ripley, DE5 3RE

G3 NJY Dr M. Bibby, 47 Whitney Tavern Road, Weston, USA, 2493

G3 NKC D. Sharred, 4 Rufford Close, Wistaston, Crewe, CW2 6XP

GM3 NKG A. Campbell, 22 Saltire Crescent, Larkhall, ML9 2LG

G3 NKH R. Dowling, Orchard House, Oughtrington Lane, Lymm, WA13 0RD

G3 NKJ R. Gill, 45 Biggin Lane, Ramsey, Huntingdon, PE26 1NB

G3 NKL R. Jones, 12 Crumpax Meadows Longridge, Preston, PR3 3JG

GW3 NKM C. Jones, 77 Margam Road, Port Talbot, SA13 2LB

G3 NKQ C. Burchell, 4 Bakers Way, Perry, Huntingdon, PE28 0BS

G3 NKS D. Thom, 78 Farmfield Road, Cheltenham, GL51 3RA

G3 NKW H. White, 16 Turnberry Close, Lymm, WA13 9LY

GM3 NLB Dr F. Inglis, 3 Fleming Road, Bishopton, PA7 5HW

G3 NLY R. Smethers, 46 Church Road, Burntwood, WS7 9EA

G3 NMD Houghton ARC, c/o I. Laidler, 5 South Street, West Rainton, Houghton le Spring, DH4 6PA

G3 NMH H. Perkins, 31 Dorchester Road, Weybridge, KT13 8PE

G3 NML M. Slater, 46 Ladywood, Eastleigh, SO50 4RW

GM3 NMN R. Dunlop, 39 Braid Drive, Glenrothes, KY7 4ES

G3 NMW T. Whateley, 285 Harborne Road, Birmingham, B15 3JB

G3 NMX D. Wills, 4100 Jackson Ave, Apt 530, Austin, USA, 78731

G3 NMZ G. Bath, 11 Heron Way, Hickling, Norwich, NR12 0YQ

G3 NN C. Bolt, 147 Swan Avenue, Bingley, BD16 3PL

G3 NNA M. Codd, 1 Shaftesbury Place, Lancaster, LA1 4PZ

GW3 NNB R. Evans, Cemlyn, Ffordd Dewi Sant, Pwllheli, LL53 6EG

G3 NNG C. Desborough, 22 Westland Road, Faringdon, SN7 7EY

G3 NNN P. Mason, 1 Morley Lane, Stanley, Ilkeston, DE7 6EZ

G3 NNO M. George-Powell, Old Church Lane Cottage, Pateley Bridge, Harrogate, HG3 5LY

G3 NNT S. Pilkington, The Quarries, Quarry Drive, Ormskirk, L39 5BG

G3 NNV P. Swanson, 11 Grassmoor Close, Wirral, CH62 7JY

G3 NNW K. Taylor, 34 Shore Road, Warsash, Southampton, SO31 9FU

GM3 NNZ Dr B. East, 26 Hyndford Road, Lanark, ML11 9AE

G3 NOA P. Reynolds, Brook Bushes, Bramshaw, Lyndhurst, SO43 7JB

G3 NOC A. Waldie, Gwyn Lyn, 85 Park Road, Coleford, GL16 7AG

G3 NOI R. Cumming, 21 Britannia Way, Woodmancote, Cheltenham, GL52 9QP

G3 NOP D. Peacock, Robin Hill, Cottingham, HU16 5JG

G3 NOX J. Royle, Keepers Cottage, Duddenhoe End, Saffron Walden, CB11 4UU

G3 NPC Dr J. Swanson, 23 Oatlands Road, Tadworth, KT20 6BS

G3 NPM A. Macdonald, 5 Arlington Close, Swindon, SN3 3NB

GI3 NPP R. Gibson, 109 Bush Road, Dungannon, BT71 6QG

G3 NPS B. Harrad, 32 Woodfield Avenue, Northfleet, Gravesend, DA11 7QG

G3 NPT G. Bell, 9 Humber View, Hessle, HU13 0PY

G3 NPY J. Joslin, 150 Roman Bank, Skegness, PE25 1SE

G3	NPZ	T. Griffiths, 18 Lulworth Road, Lee-on-the-Solent, PO13 9HU
G3	NQA	S. Hall, 76 Cheltenham Drive, Bromford, Birmingham, B36 8QG
G3	NQF	R. Fenton, Harmins Green, France Lynch, Stroud, GL6 8LZ
G3	NQK	J. Beddows, 17 Rue Francois Mitterand, Pleven, France, 22130
G3	NQN	M. Hartung, 31 Ellenbrook Lane, Hatfield, AL10 9RW
G3	NQT	R. Levi, 24 Stanmore Way, Loughton, IG10 2SA
G3	NQX	W. Brown, 73 Church Avenue, Preston, PR1 4UD
G3	NQZ	Dr G. Lockhart, 179 Poolbrook Road, Malvern, WR14 3JZ
G3	NR	A. Birt, 36 Queens Road, Swanage, BH19 2ET
G3	NRD	J. Packer, The Hayloft, Butts Bank Farm, Gulval, Penzance, TR18 3BB
G3	NRH	B. Perrin, Apartment 37, Hardy Lodge, Coppice Street, Shaftesbury, SP7 8GY
G3	NRM	M. Moore, 127 Adel Lane, Leeds, LS16 8BL
G3	NRQ	C. Higgins, Billdoro, Mill Lane, Saltfleet, Louth, LN11 7SA
G3	NRU	D. Brook-Foster, 246 St. Margarets Banks, High Street, Rochester, ME1 1HY
G3	NRW	I. Wade, 7 Daubeney Close, Harlington, Dunstable, LU5 6NF
G3	NRX	R. Murphy, 3 Lady Leasow, Shrewsbury, SY3 6AB
G3	NRZ	C. Hogg, 7 Elm Grove, Erith, DA8 3BL
G3	NSD	Dr B. Styles, York House, Bluntisham Road, Colne, Huntingdon, PE28 3LY
G3	NSF	T. Simpson, 41 Benyon Grove, Orton Malborne, Peterborough, PE2 5XS
G3	NSL	I. Whitter, The Old Hall, Hall Lane, Lincoln, LN3 4HT
G3	NSO	G. Brookes, 27 Pineside Avenue, Cannock Wood, Rugeley, WS15 4RG
G3	NSP	Dr J. Lennox, Kestrel, School Lane, Bicester, OX25 4AW
G3	NSS	T. Spain, Manor View, Shotatton, Shrewsbury, SY4 1JD
G3	NSW	R. Kay, 7 Lea Drive, Blackley, Manchester, M9 7AR
G3	NTD	A. Marsden, 15 Northfield Way, Retford, DN22 7LJ
G3	NTF	I. Neary, 65 Vicarage Road, Ashton-under-Lyne, OL7 9QY
G3	NTI	R. Blain, 11 Mill Bank, Ness, Neston, CH64 4BJ
G3	NTM	W. Brown, 18 Georgian Close, Staines, TW18 4NR
G3	NUA	J. Hogg, 16 Moorston Close, Naisberry Park, Hartlepool, TS26 0PJ
G3	NUB	M. Bursnall, Panorama, Church Lane, Bishops Castle, SY9 5AF
G3	NUL	V. Johnston, 119 High St., Cheveley, Newmarket, CB8 9DG
G3	NUN	A. Langford-Brown, 9 Orchard Lane, Corfe Mullen, Wimborne, BH21 3SU
GW3	NUO	Dr P. Williams, Crud Y Gwynt, 27 Mynydd Garnllwyd Road, Swansea, SA6 7PB
G3	NUQ	I. Macarthur, 2 Bramley Close, Bramhall, Stockport, SK7 2DT
GM3	NUU	Dr J. Reid, Rochelle, Findon, Aberdeen, AB12 3RL
G3	NVB	A. Bryant, 1 Downlands Road, Winchester, SO22 4ET
G3	NVL	R. Allen, 692 Hitchin Road, Luton, LU2 7UH
G3	NVM	D. Arigho, 81 Crookham Road Fleet Gu51 5Np Hampshire, Fleet, GU51 5NP
G3	NVP	B. Mapp, 33 Cotswold Drive, Redcar, TS10 4AG
GM3	NVQ	G. Martin, 39 St. Johns Drive, Dunfermline, KY12 7TB
GI3	NVW	W. Pollock, J R Pollock & Co, 155 Doogary Road, Omagh, BT79 0HF
G3	NVX	R. Davison, 76 Poplars Way, Beverley, HU17 8PU
G3	NWG	D. Stevens, 8 Dane Road, Chelmsford, CM1 2SS
G3	NWH	A. Collis, C/O 510 Lowther Road, Dunstable, LU6 3LJ
G3	NWL	A. Lock, 7 Heather Close, St. Leonards, Ringwood, BH24 2QJ
G3	NWR	Wirral ARS, c/o W. Davies, Davies Electrical Services (N W) Ltd, 104 Bromborough Village Road, Bromborough, Wirral, CH62 7EX
GW3	NWS	F. Clare, Glen View, Newport Road, Caldicot, NP26 3BZ
G3	NWW	M. Wakely, Chyandour, 3 Ganges Close, Mylor Harbour, Falmouth, TR11 5UG
G3	NWX	K. Morgan, 97 Elmwood, Sawbridgeworth, CM21 9NN
G3	NWY	D. Forster, 79 Westbrooke Avenue, Hartlepool, TS25 5HX
G3	NXC	A. Plant, 178 Clay Lane, Yardley, Birmingham, B26 1DY
G3	NXK	O. Diplock, North Lodge, Messing Park, Colchester, CO5 9TD
G3	NXL	P. Lamming, 25 Leconfield Garth, Follifoot, Harrogate, HG3 1NF
G3	NXN	F. Wickens, 32 Kenilworth Avenue, Wimbledon, London, SW19 7LW
G3	NXO	A. Watt, Little Owls, Singleton, Chichester, PO18 0EX
GW3	NXR	T. Miles, West Uplands Lodge, Upland Arms, Dyfed, SA32 8DX
G3	NXS	F. Shaw, 69 Finedon Road, Irthlingborough, Wellingborough, NN9 5TY
G3	NXT	W. Fletcher, 3 Orchard Close, Metheringham, Lincoln, LN4 3DT
G3	NXX	I. Miller, 11 Lynton Drive, High Lane, Stockport, SK6 8JE
G3	NXZ	J. Howe, 18 Laburnum Grove, Conisbrough, Doncaster, DN12 2JW
G3	NY	Easingwold & District ARS, c/o N. Knapton, 4 Crabmill Lane, Easingwold, York, YO61 3DE
G3	NYB	W. Bingham, 7 Bolton Hill Road, Doncaster, DN4 6DQ
G3	NYD	D. Coles, 113 Berrow Road, Burnham-on-Sea, TA8 2PH
G3	NYE	A. Taylor, 25 Burnside Road, Gatley, Cheadle, SK8 4NA
GM3	NYG	J. Fish, 31 Oaklands Avenue, Irvine, KA12 0SE
GI3	NYJ	S. Currie, 122 Belfast Road, Comber, Newtownards, BT23 5QP
G3	NYK	A. Melia, 67A Deben Avenue, Martlesham Heath, Ipswich, IP5 3QR
GI3	NYL	L. Elliott, 19 Gosford Road, Collone, Armagh, BT60 1LQ
G3	NYM	North Yorkshire Contest Group, c/o A. Duffield, 4 Crabmill Lane, Easingwold, York, YO61 3DE
G3	NYR	D. Rayner, 42 Canford Drive, Allerton, Bradford, BD15 7AU
G3	NYS	C. Whiteley, 30 Lynch Hill Park, Whitchurch, RG28 7NF
G3	NYX	J. Heaviside, 110A Cuckfield Road Hurstpierpoint, Hassocks, BN6 9RZ
G3	NYZ	A. Stafford, Blakefield, Jawbone Lane, Derby, DE73 1BW
G3	NZL	H. Chapman, 57 Athelstan Road, Southampton, SO19 4DE
G3	NZP	M. Harman, 19 Hill House Close, Turners Hill, Crawley, RH10 4YY
G3	NZR	W. Young, 5 Grasmere Grove, Frindsbury, Rochester, ME2 4PN
G3	NZS	H. Parkes, 35 Dovey Road, Tividale, Oldbury, B69 1NT
G3	NZV	A. Park, Waterside Cottage, Bowden Lane, High Peak, SK23 0QF
G3	NZW	S. James, Beresford, Latchmoor Ave, Gerrards Cross, SL9 8LJ
G3	NZY	R. Shelley, 4 Fairview Court, St. Martins Avenue, Scarborough, YO11 2DA
G3	OAD	T. Haydu Jones, 1 Beggars Roost, Golf Course Road, Stroud, GL6 6TJ
G3	OAF	W. Jeffs, Silver Jay, Colehill Lane, Wimborne, BH21 7AN
G3	OAH	Dr P. Whittlestone, Turangi, 14A Croft Bank, Malvern, WR14 4DW
GW3	OAJ	C. Davies, 11 Beaumaris Way Grove Park, Blackwood, NP12 1DF
G3	OAL	E. Lincoln, Lynholme, Millbank, Newton Aycliffe, DL5 6RF
G3	OAR	G. Greenwood, 1 Maltkiln Lane, Castleford, WF10 4LF
G3	OAZ	J. Randall, 243 Paddock Road, Basingstoke, RG22 6QP
GM3	OBC	R. Thomson, 1 Knowehead, Star, Glenrothes, KY7 6LA
GM3	OBG	P. Bridges, 29 Kirkbank, Auchmithie, Arbroath, DD11 5SY
G3	OBL	J. Tyrrell, 2 Briar Close, Yeovil, BA21 5XA
GI3	OBO	D. Waugh, 16 Seaview Avenue, Millisle, Newtownards, BT22 2BN
G3	OBV	P. Harris, 15 Ratliffe Road, Rugby, CV22 6HB
G3	OBZ	M. Birkett, Hazelwood, Cromwell Ave, Woodhall Spa, LN10 6TH
G3	OCA	K. Frankcom, 1 Chesterton Road, Spondon, Derby, DE21 7EN
G3	OCB	C. Bowden, Tregwyn, Tregonning Road, Stithians, Truro, TR3 7FG
G3	OCH	J. Hulett, 21 Exmoor Avenue, Leicester, LE4 0BJ
G3	OCP	D. Wallace, 11 Station Road, Haddenham, Aylesbury, HP17 8AN
G3	OCR	S. Nutt, 23A Hesketh Drive, Southport, PR9 7JX
GW3	ODB	A. Pritchard, 41 Maes Cantaba, Ruthin, LL15 1YP
G3	ODC	D. Martin, 7 Seaview Avenue, Eastham, Wirral, CH62 0BD
G3	ODD	E. Stables, Manor Croft, Water Lane, Selby, YO8 6QL
G3	ODO	W. Buckett, 2725, Halcomb Bridge Road, Alpharetta, USA, 30022
GM3	ODP	Dr T. Salvesen, Easter Catter, Croftamie, Glasgow, G63 0EX
G3	ODX	S. Clarke, 18 Dunedin Drive, Caterham, CR3 6BA
G3	OEB	R. Downs, 23 Old London Road, Benson, Wallingford, OX10 6RR
G3	OEC	Prof. C. Isham, 2 Lime Grove, Ruislip, HA4 8RY
G3	OEQ	D. Bunn, Heliophilia Gardens 1, Agias Annis St. 17, Kato Pafos, Cyprus, 8036
G3	OFI	B. Bisley, 132-1919 St Andrews Place, Courtenay, British Columbia, Canada, V9N 9J4
G3	OFP	G. Cunnah, 225 Springwell Lane, Balby, Doncaster, DN4 9AJ
GM3	OFT	P. Bower, An Cluain, Ballplay Road Dg10 9Ju, Moffat, DG10 9JU
G3	OFW	H. Blake, 19 Segsbury Grove, Harmans Water, Bracknell, RG12 9JL
G3	OFX	R. Welch, 112 Copsewood Road, Bitterne, Southampton, SO18 1QR
G3	OGE	J. Rose, 1 Westgate House, 22 Westgate, Hornsea, HU18 1BP
G3	OGH	A. Brooker-Carey, 29 Byron St., Amble, Morpeth, NE65 0ER
G3	OGK	Dr G. Kennedy, Thayers, Edwyn Ralph, Bromyard, HR7 4LY
G3	OGP	R. Powell, Garlands Farm, The Haven, Billingshurst, RH14 9BH
G3	OGX	J. Allsop, 17 Hambro Hill, Rayleigh, SS6 8BN
G3	OGZ	M. Beer, 24 Byron Court, Beech Grove, Harrogate, HG2 0LL
G3	OHC	G. Badger, 3 Hesketh Close, Cranleigh, GU6 7JB
G3	OHH	R. Hargreaves, 46 Castle Road, Mow Cop, Stoke-on-Trent, ST7 3PH
G3	OHL	D. White, Holme Fell Cottage, Hallbankgate, Brampton, CA8 2NJ
G3	OHM	South Birmingham Radio Society, c/o J. Storey, 34 Austin Rise, Longbridge, Birmingham, B31 4QN
G3	OHN	K. Whitehouse, 27A Howdles Lane, Brownhills, Walsall, WS8 7PL
G3	OHP	M. Winter, 9 Higham Road, Cliffe, Rochester, ME3 7SH
G3	OHS	J. Perry, 517 Longbridge Road, Barking, IG11 9DD
G3	OHX	I. Jackson, Brattle House, Manor Road, Beaconsfield, HP9 2QU
GM3	OIB	K. Younger, 183 Main St., Pathhead, EH37 5SQ
G3	OIC	I. Croxford, 16 Chesterwood, Hollywood, Birmingham, B47 5EN
G3	OIF	P. Squires, 191 Station Road, Knowle, Solihull, B93 0PT
G3	OIH	B. Shields, 24 Churchfield, Fulwood, Preston, PR2 8GT
G3	OIL	M. Wills, 23 Falcons Way, Salisbury, SP2 8NR
GW3	OIN	J. Nicholas, 28 Hardy Avenue, Rhyl, LL18 3BG
G3	OIP	P. Holker, 9 Limetree Grove, Braunton, EX33 1HE
GM3	OIV	W. Anderson, 6 Winchburgh Road, Winchburgh, Broxburn, EH52 6QB
G3	OJ	J. Hobin, 14 St. Martins Green, Trimley St. Martin, Felixstowe, IP11 0UU
G3	OJG	Dr P. Gale, Garden Cottage, Sacombe Green, Ware, SG12 0JQ
G3	OJI	J. Sleight, Orchard House, School Hill, Napton, Southam, CV47 8NN
G3	OJK	J. Bates, 8 Spaxton Road, Durleigh, Bridgwater, TA5 2AP
G3	OJL	M. Plaster, Combe House, Milton Lane, Wookey Hole, Wells, BA5 1DG
G3	OJS	H. Braham, 10 Glebe Way, Frinton-on-Sea, CO13 9HR
G3	OJV	P. Waters, 9 Tudor Way, Hawkwell, Hockley, SS5 4EY
G3	OJZ	B. Todd-White, 3 Alexandra Road, Capel-Le-Ferne, Folkestone, CT18 7LB
G3	OKA	Dr J. Share, 82 Birkenhead Road, Meols, Wirral, CH47 0LB
G3	OKB	M. Ireson, 15 Digby Drive, North Luffenham, Oakham, LE15 8JS
G3	OKD	Z. Nilski, The Poplars, Wistanswick, Market Drayton, TF9 2BA
G3	OKH	G. Hillman, 504 Chester Road, Kingshurst, Birmingham, B36 0LG
G3	OKS	S. Smithies, Moorcroft, Fernhill., Horley, RH6 9SY
GW3	OKT	Dr J. Thompson, The Old Place, Old Racecourse, Oswestry, SY10 7HL
G3	OKU	M. Cross, 39 Westfield, The Marld, Ashtead, KT21 1RH
G3	OLB	T. Boucher, Hedgerows, Sheldon, Honiton, EX14 4QS
G3	OLH	A. Remsbury, Nodali, 16 Little Green Lane, Chertsey, KT16 9PH
G3	OLP	B. Wadsworth, 5 Birch Avenue, Todmorden, OL14 5NX
G3	OLU	J. Saunders, Apartamento 6306, Forum Mare Nostrum, Camino De Pinxo 2, Alfaz Del Pi, Spain, 3580
G3	OLW	J. Burnett, Wenrisc, Chapel Lane, Tewkesbury, GL20 8HS
G3	OLX	J. Parker, Palfreys, Picquets Way, Banstead, SM7 1AJ
G3	OMA	Dr S. Kay, 5 Chevalier Close, Swindon, SN5 5TS
G3	OMB	R. Spurgeon, 57 Laburnum Crescent, Kirby Cross, Frinton-on-Sea, CO13 0QH
G3	OMD	A. Callegari, Danebridge Nursery, Much Hadham, SG10 6JG
G3	OMJ	P. Judkins, 18 St. Johns Square, Wakefield, WF1 2RA
G3	OMK	T. Kirk, 54 Highfields Drive, Loughborough, LE11 3JT
GW3	OMN	M. Jenkins, 25Stepneyroad, Swansea, SA2 0FZ
G3	OMR	M. Russoff, Flat 3 Hartsbourne Court, Hartsbourne Road, Bushey Heath, WD23 1PZ
G3	OMS	Dr R. Simpson, 23 Larkhill, Rushden, NN10 6BG
G3	OMT	A. Russell, 5 Little Close, Swadlincote, DE11 0EB
G3	OMY	D. Hancock, 17 Forestlake Avenue, Ringwood, BH24 1QU
G3	OMZ	D. Lee, 19 Sarum Lodge, Three Swans Chequer, Salisbury, SP1 1AL
G3	OND	J. Denman, 167 Minnis Road, Birchington, CT7 9QD
GI3	ONF	R. Sinton, 35 The Rose Garden, Tandragee, Craigavon, BT62 2NJ
G3	ONI	D. Woods, Flat 26, Chapel Court, Wilmslow, SK9 5EN
GU3	ONJ	A. Richmond, The Cedars, 3 Holly Drive, Braye Road, St Sampson, Guernsey, GY2 4EF
G3	ONL	P. Brodribb, 18 Ipswich Road, Debenham, Stowmarket, IP14 6LB
G3	ONR	B. Reynolds, 17 Cresswells Mead, Holyport, Maidenhead, SL6 2YP
G3	ONU	D. Barry, 2 Catherine Close, Shrivenham, Swindon, SN6 8ER
G3	ONV	J. Verity, Tall Pine, Station Road Kirby Muxloe, Leicester, LE9 2EN
G3	OOH	G. Lander, 132 Chemin De Saule, Bernex, Switzerland, 1233
G3	OOK	J. Plenderleith, 3D Deluxe Court, Jalan Pahlawan Kepayan, Kota Kinabalu, Malaysia, 88200
G3	OOL	J. Hatch, 628-707 Esquimalt Road, Victoria, Canada, BC V9A 3L7
G3	OOP	Dr B. Havenhand, 15 Sandiway, Chesterfield, S40 3HG
G3	OOU	R. Burns, 84 Portnalls Road, Coulsdon, CR5 3DE
G3	OOW	M. Docker, Apartment 219, Clarence Park, 415 Worcester Road, Malvern, WR14 1FU
G3	OPB	M. Bues, 7A Alice Parkins Close, Hadleigh, Ipswich, IP7 6FE
GW3	OPC	N. Ward, 17 Heol Nant, Llanelli, SA14 8EL

G3 OPG R. Tingay, 18 Grove Road, Newbury, RG14 1UH
G3 OPH R. Atkinson, Lake Walk, Adderbury, OX17 3PF
G3 OPJ C. Harrisson, 129 Granville Way, Sherborne, DT9 4AT
G3 OPW J. Cook, Upwood Park, Black Moor Road, Keighley, BD22 9SS
G3 OPX C. Green, Gothic, Plymouth Road, Totnes, TQ9 5LH
G3 OQC J. Woods, 1 Dean Road, Cosham, Portsmouth, PO6 3DG
G3 OQD M. Emmerson, 6 Mounthurst Road, Hayes, Bromley, BR2 7QN
G3 OQF R. Kay, 7 Chemin Des Grands-Champs, Bogis-Bossey, Switzerland, CH 1279
GM3 OQI J. Ramsay, 150 City Road, Dundee, DD2 2PW
GW3 OQK A. Fairgrieve, 3 Pleasant Road, Gorseinon, Swansea, SA4 9WH
G3 OQO D. Henley, 36 Main Street, Newbold, Rugby, CV21 1HW
GI3 OQR D. Gibson, 93 Cavan Road, Dungannon, BT71 6QN
G3 OQT R. Mclachlan, Oak Trees, Park Lane, Rodsley, Ashbourne, DE6 3AJ
G3 ORG I. Taylor, 10 Westfield Road, Henlow, SG16 6BN
G3 ORI J. Vickers, 45 Willow Park Drive, Stourbridge, DY8 2HL
G3 ORK R. Talbot, 9 Bracebridge Drive, Southport, PR8 6XH
GW3 ORL D. Williams, 14 Seymour Avenue, Parc Seymour, Caldicot, NP26 3AG
G3 ORN W. Thomas, 20 Vinnicombes Road, Stoke Canon, Exeter, EX5 4BB
G3 ORP P. Pickering, 21 Palmar Road, Maidstone, ME16 0DL
G3 ORV M. Saunders, 40 Archfield Road, Cotham, Bristol, BS6 6BE
G3 ORY R. Titterington, Wyclif House, St. Marys Road, Lutterworth, LE17 4PS
G3 OS A. Boor, 3 Croft Cottages, Beltoft, Doncaster, DN9 1NA
G3 OSI D. Swanson, 48 Moscow Drive, Liverpool, L13 7DJ
G3 OSP S. Plumtree, Flat 18, Oliver Leese Court, Ten Butts Crescent, Stafford, ST17 9HP
G3 OSQ D. Beakhust, Nonsuch Lodge, Morgans Vale Road, Redlynch, Salisbury, SP5 2HU
G3 OSR P. Hughes, Flat 12, Highcliffe, 32 Albemarle Road, Beckenham, BR3 5HJ
G3 OST D. Wilson, Chemin D'Arques, Ambrumesnil, Offranville, France, 76550
G3 OTH C. Cook, Swiss Cottage, Netherton Lane, Bedlington, NE22 6DR
G3 OTK R. Harris, 4 Alford Court, Hambridge, Langport, TA10 0BS
G3 OTN P. Seaman, 5 Berkeley Close, Maidenhead, SL6 5JP
G3 OTR M. Beckley, Mallards, Albury Road, Ware, SG11 2DN
GI3 OTU A. Burge, 38 Bayview Road, Bangor, BT19 6AR
G3 OTV P. O'Kane, 36 Coolkill, Sandyford, Dublin, Ireland, D18 P7F4
G3 OTW W. Miller, 418 Old Chester Road, Birkenhead, CH42 4PD
G3 OTY Capt. R. Cogzell, Flat 242, Clydesdale Tower Holloway Head, Birmingham, B1 1UJ
G3 OUA D. Tarr, 17 Allendale Avenue, Findon Valley, Worthing, BN14 0AH
G3 OUC P. Painting, 15 Turnpike Road, Shaw, Newbury, RG14 2ND
G3 OUI I. Dickinson, 64A Richmond St, College Park, Australia, 5069
G3 OUT A. Walker, High Beacon Farm, Fulletby, Horncastle, LN9 6LB
GM3 OUU G. Rennie, 60 Woodend Place, Aberdeen, AB15 6AN
G3 OUV P. Perkins, 47 Priory Avenue, High Wycombe, HP13 6SN
G3 OVE Dr M. Brown, 25 Carpenters Lane, West Kirby, Wirral, CH48 7EX
G3 OVH A. Abbey, 1 The Fairway, Kirby Muxloe, Leicester, LE9 2EU
G3 OVL M. Hubbard, 7 Creake Road Syderstone, King's Lynn, PE31 8SF
G3 OVX H. Hammett, 27 Courtman Road, Tottenham, London, N17 7HT
G3 OWB J. Holland Carter, 37 Highfield Avenue, Cambridge, CB4 2AJ
G3 OWE D. Saunders, 4A Ullswater Crescent, Radipole, Weymouth, DT3 5HE
G3 OWJ P. Jarvis, 44 Torrin Drive, Shrewsbury, SY3 6AW
G3 OWQ J. Clarke, 29 Long Brackland, Bury St Edmunds, IP33 1JH
GM3 OWU V. Stewart, 9 Baberton Park, Juniper Green, EH14 5DW
G3 OWX J. Greany, Flat 3 Crete Hill House, Cote House Lane, Bristol, BS9 3UW
GM3 OXA A. Fosters, 16 Reid Crescent Milnathort, Kinross, KY13 9TB
G3 OXG D. Thompson, 34 Sandy Road, Potton, Sandy, SG19 2QQ
GM3 OXK J. Carson, 23 Whinny Rig, Heathhall, Dumfries, DG1 3RJ
G3 OXL D. Westbury, Rose Cottage, Cruise Hill, Ham Green, Redditch, B97 5UA
G3 OXN D. Swainson, 4 Grasmere Avenue Spondon, Derby, DE21 7JZ
G3 OXR P. Garthwaite, 16 Newtown Avenue, Royston, Barnsley, S71 4HF
GM3 OXX G. Burt, Clunie Lodge, Netherdale, Turriff, AB53 4GN
G3 OYB W. Waters, 4 Calartha Road, Pendeen, Penzance, TR19 7DZ
GI3 OYG J. Semple, 5 Tullaghgore Road, Ballymoney, BT53 6QF
G3 OYL D. Gilbert, 348 Willington Road, Kirton End, Boston, PE20 1NU
G3 OYN G. Saunders, 17 Chester Street Caversham, Reading, RG4 8JH
G3 OYT G. Clinton, 2 Greenways, Abbots Langley, WD5 0EU
G3 OYX M. Rignall, Ashdown, Nupend, Stroud, GL6 0PY
GM3 OZB A. Mckay, 2 Osprey Drive Sparrow Plantation Kilmarnock Ka13 Lq, Kilmarnock, KA1 3LQ
G3 OZC J. Holstead, 72 Woodlands Avenue, Feniscowles, Blackburn, BB2 5NN
G3 OZD P. Cross, 5 Lings Lane, Hatfield, Doncaster, DN7 6AB
G3 OZE J. Grainger, 6 Fulford Cross, Fulford, York, YO10 4PB
GM3 OZJ I. Morgan, 43 Dalgety Gardens, Dalgety Bay, Dunfermline, KY11 9LF
G3 OZK M. James, 11 Shortborough Avenue, Princes Risborough, HP27 9HU
G3 OZL Dr A. Jeavons, Wadsley Grove, Worrall Road, Sheffield, S6 4BE
G3 OZN E. Badger, 20 Tennyson Drive, Worksop, S81 0EE
G3 OZP P. Smith, 39 Sherborne Avenue, North Shields, NE29 8NT
G3 OZT R. German, 10 Beverley Road, Dibden Purlieu, Southampton, SO45 4HS
GI3 OZW P. Dynes, 1 Rossin View, Donaghmore, Dungannon, BT70 1SZ
G3 PAG J. Davies, Cedar Croft, School Lane, West Malling, ME19 5EH
G3 PAI J. Rabson, 55 Severn Road, Ipswich, IP3 0PU
GM3 PAK Dr M. Senior, The Raw, Bridgend, Isle of Islay, PA44 7PZ
G3 PAQ J. Davis, 76 Allfarthing Lane, London, SW18 2AJ
G3 PAX J. Barker, 2 Barons Hall Lane, Fakenham, NR21 8HB
G3 PBF J. Orford, 63 Flowerhill Way, Istead Rise, Gravesend, DA13 9DS
G3 PBI A. Davies, 69 Sycamore Road, Chalfont St Giles, HP8 4LG
G3 PBR A. Green, 6 Shipley Close, Woodley, Reading, RG5 4RT
G3 PBT R. Hilsley, 1 Chelmerton Avenue, Great Baddow, Chelmsford, CM2 9RE
G3 PCG D. Askew, Lapthorne, Adsborough, Taunton, TA2 8RP
G3 PCJ T. Walford, Vedal House, Langport, TA10 9FB
G3 PCL D. Shaw, 3 Randolph Close, Cheltenham, GL53 7RT
G3 PCT P. Hurst, Anchorage House, Upper Wood Lane, Dartmouth, TQ6 0DQ
G3 PCW M. Watling, 8 Preetz Way, Blandford Forum, DT11 7XG
G3 PCX B. Dodge, 34 Downs Road, Penenden Heath, Maidstone, ME14 2JN
G3 PCY J. Wilson, 5 Huntsham Court Stables, Huntsham, Tiverton, EX16 7NA
G3 PDC R. Curwen, 53 Karslake Road, Liverpool, L18 1EY
G3 PDD J. Dolby, Oaklea, School Lane, Belper, DE56 2AL
G3 PDH M. Prestwood, Salatiga, Bell Lane, Salhouse, Norwich, NR13 6RR
GI3 PDN R. Harbison, 26 Ballymartin Road, Templepatrick, Ballyclare, BT39 0BW
G3 PDP A. Ralls, 12 Oakhill Close, Bursledon, Southampton, SO31 1AP
GM3 PDX J. Barker, 44 Priory Road, Linlithgow, EH49 6BS
G3 PEJ P. Watson, 37 Chestnut Bank, Scarborough, YO12 5QJ
G3 PEK B. Simpson, 20 Monterey Street, St Ives, Australia, NSW 2075
G3 PEM C. Thomson, 109 Hillside Grove, Chelmsford, CM2 9DD
G3 PET A. Widdowson, 34 Highfields Road, Chasetown, Burntwood, WS7 4QU
G3 PEW J. Hudson, 68 Lower Street, Stansted, CM24 8LR
G3 PEZ J. Gutteridge, 66 Croft Drive, Moreton, Wirral, CH46 0QT
G3 PFE G. Spriggs, Brookbank Cottage, Newcastle Road, Nantwich, CW5 7EJ
G3 PFH M. Blunden, 24 Mill View Close, Woodbridge, IP12 4HR
G3 PFJ J. Harris, 3 Chimney Mills, West Stow, Bury St Edmunds, IP28 6ES
G3 PFM A. Baker, Highleaze, Deans Drove, Poole, BH16 6EQ
G3 PFO C. Barr, Riders Way, Collum Green Road, Stoke Poges, SL2 4AX
G3 PFT A. Heeley, 108 Valley Lane, Lichfield, WS13 6ST
GW3 PFV K. Robbins, 1 Rhiw Parc Road, Abertillery, NP13 1BS
G3 PFX C. Small, Overlangs, Kingston, Kingsbridge, TQ7 4PF
G3 PGA A. Hammond, 23 St. Andrews Road, Fremington, Barnstaple, EX31 3BS
G3 PGC R. Armstrong, 6 Barnstaple Road, North Shields, NE29 8QA
G3 PGJ R. Bashford, Popplestone, 33 Penair View, Truro, TR1 1XR
G3 PGK C. Pearless, 26 Church Road, Preston, Weymouth, DT3 6RP
G3 PGN H. Buckenham, Tweed Cottage, Tilbury Road, Great Yeldham, Halstead, CO9 4JG
G3 PGQ D. Yates, 26 Lowestoft Road, Carlton Colville, Lowestoft, NR33 8JD
GM3 PGY A. Mc Ewen, 4 Reef Terrace, Crossapol, Isle of Tiree, PA77 6UT
G3 PHD I. Gardiner, 189 Brennan Road, Tilbury, RM18 8BA
G3 PHG A. Gibbs, 223 Crimea Street, Noranda, Australia, WA 6062
G3 PHJ J. Johnston, 9 Appleby Glade, Haxby, York, YO32 3YW
G3 PHL B. Davies, 17 Linksway, Leigh-on-Sea, SS9 4QY
G3 PHO P. Day, 38 Broomhill Road, Old Whittington, Chesterfield, S41 9DA
G3 PIA Harwell ARS, c/o C. Desborough, 22 Westland Road, Faringdon, SN7 7EY
G3 PID P. Chandler, 528 Goffs Lane, Goffs Oak, Waltham Cross, EN7 5EW
G3 PIJ P. Mellett, 16 Tutton Hill Colerne, Chippenham, SN14 8DN
GM3 PIL R. Munro, 20 County Cottages, Piperhill, Nairn, IV12 5SE
G3 PIN J. Patten, 8 Leacroft Road, Penkridge, Stafford, ST19 5BX
G3 PIY C. Isaacs, Holme View, Brick Lane, Christchurch, BH23 8DU
G3 PIZ T. Watts, 26 Woodger Close, Guildford, GU4 7XR
G3 PJC C. Arnold, 47 Peartree Lane, Danbury, Chelmsford, CM3 4LS
G3 PJQ A. Aldridge, 1 Mary Grove, Highnam, Gloucester, GL2 8NH
G3 PJT Dr R. Whelan, 36 Green End, Comberton, Cambridge, CB23 7DY
G3 PJV P. Walsh, 23 Moss Fold Road, Darwen, BB3 0AQ
G3 PJW R. Unsworth, 8 Coleridge Road, Billinge, Wigan, WN5 7EB
G3 PJY R. Millman, 103 The Crescent, Walsall, WS1 2DA
G3 PKC J. Tinker, 72 Jackson Avenue, Leeds, LS8 1NS
G3 PKD R. Sharples, 40 Greetham Road, Cottesmore, Oakham, LE15 7DB
G3 PKL C. Fox, 2 Mill Cottages, Wareham Road, Poole, BH16 6ET
G3 PKQ J. Holmes, 36 Hillside Gardens, Walthamstow, London, E17 3RJ
G3 PKR K. Parker, 263 High St., Hayes, UB3 5ET
G3 PKY Rev. P. Okelly, The Ravel, School Lane, Drogheda, Ireland,
GW3 PLB R. Howe, Brooklands, Caeffynnon, Kidwelly, SA17 5EJ
G3 PLE D. Barlow, Pine, Churchtown, Helston, TR12 7BW
G3 PLJ P. Fairrington, 30 Orchard St., Weston Super Mare, BS23 1RQ
GI3 PLL FL R. Moore, 818 Seacoast Road, Castlerock, Coleraine, BT51 4SD
G3 PLN G. Smith, 7 Coniston Avenue, Grimsby, DN33 3EE
GM3 PLO J. Gray, Norland, South End, Stromness, KW16 3DJ
G3 PLP R. Cox, 30 Brooks Road, Sutton Coldfield, B72 1HP
G3 PLR D. Skye, 16 Lulworth Avenue, Poole, BH15 4DQ
G3 PLT G. Lawes, 7 Tormynton Road, Weston-Super-Mare, BS22 9HU
G3 PLW J. Norton, 32 Fismes Way, Wem, Shrewsbury, SY4 5YD
G3 PLX J. Martinez, High Blakebank Farm Underbarrow, Kendal, LA8 8HP
G3 PLY G. Mcneil, 168 Chobham Road, Ascot, SL5 0HU
GM3 PMB W. Miller, 15 Glenalla Crescent, Ayr, KA7 4DA
G3 PMH "OBO March & Dist ARS, British Legion Club, Rookswood Road", c/o V. Cracknell, 106 High St., Upwood, Huntingdon, PE26 2QE
G3 PMJ S. Revell, 11 Mere Fold, Worsley, Manchester, M28 0SX
GD3 PML Dr D. Smith, Cooilbane Cottage, Main Road, Sulby, Isle Of Man, IM7 2HR
G3 PMR A. Jubb, Psathi Village, Pafos, Cyprus, 8749
G3 PMV A. Feist, 1 Lowry Drive, Marple Bridge, Stockport, SK6 5BR
G3 PMW K. Dews, 14 Baddow Place Avenue, Great Baddow, Chelmsford, CM2 7JN
G3 PND S. Appleyard, Plumtree House, Mill Lane, Cromer, NR27 9PH
G3 PNO I. Hawkins, Victoria House, Victoria Street, Totnes, TQ9 5EF
G3 PNP J. Ward, 131 Monarch Road, Eaton Socon, St Neots, PE19 8GU
G3 PNQ A. Floyd, 27 Beechfield Parbold, Wigan, WN8 7AR
G3 PNT C. Durell, 17 Ryders Avenue, Westgate-on-Sea, CT8 8LW
G3 PNU E. Clark, 1 Station Road, Drigg, Holmrook, CA19 1XH
G3 POG D. Mawdsley, 20 Cable Street, Formby, Liverpool, L37 3LX
GM3 POI C. Penna, North Windbreck, Deerness, Orkney, KW17 2QL
G3 POM G. Morgan, 7 Quantock Grove, Williton, Taunton, TA4 4PD
G3 POQ P. Hayes, 16 Melton Drive, Storrington, Pulborough, RH20 4LU
GI3 POS G. Smyth, 91A Gilford Road Lurgan, Craigavon, BT66 7EB
GM3 POT J. Walford, Chorcaill, Reay, Thurso, KW14 7RG
G3 PPB P. Perkins, 52 Ashley Piece, Ramsbury, Marlborough, SN8 2QE
G3 PPC D. Taylerson, 18 The Grove, Teddington, TW11 8AS
GM3 PPE Dr M. Eccles, Newtonlees Bungalow, Kelso, TD5 7SZ
G3 PPO L. Hook, 79 Whiteley Crescent, Bletchley, Milton Keynes, MK3 5DQ
G3 PPR Dr J. Beavon, 24 Cromer Road Mundesley, Norwich, NR11 8BE
G3 PPT L. Sear, 4 Mount Pleasant Road, Threemilestone, Truro, TR3 6BB
G3 PPU P. Smith, 56 Alphington Avenue, Frimley, Camberley, GU16 8LR
G3 PQA J. Rogers, Dromore, Strande Lane Cookham, Maidenhead, SL6 9DN
G3 PQB S. Harbour, 43 Warbon Avenue, Peterborough, PE1 3DS
G3 PQC P. Turk, 13 The Crescent, Farnborough, GU14 7AR
G3 PQD D. St John, 26 Henry Street Rainham, Gillingham, ME8 8HE
G3 PQF D. Dell, 7 Blunden Road, Cove, Farnborough, GU14 8QJ
G3 PQJ B. Cole, 17 Coburg Court, East Cowes, PO32 6SS
G3 PQM M. Thorp, Cecil, Thorrington Road, Clacton-on-Sea, CO16 9ES
G3 PQP T. Foster, 136 Sladepool Farm Road Kings Heath, Birmingham, B14 5EF
G3 PQY J. Lawrence, 2A Hall Road, Hull, HU6 8SA
G3 PRC Plymouth Radio Community, c/o P. Connor, 20 Longfield, Lutton, Ivybridge,

PL21 9SN
G3 PRE W. Armstrong, 24 Newbury St., South Shields, NE33 4UE
G3 PRH M. Coward, 51 Farleigh Road, Backwell, Bristol, BS48 3PB
G3 PRI D. Quigley, 1A Elizabeth Road, Bishop's Stortford, CM23 3RJ
G3 PRK A. Yilmaz, 7 Kerdistone Close, Potters Bar, EN6 1LG
GW3 PRL D. Snow, Rhwngyddwy Dre, Brynsiencyn, Llanfairpwllgwyngyll, LL61 6TZ
G3 PRQ E. Wooden, Mullins, Windsor Road, Alton, GU34 5EF
G3 PRR Rev. I. Partridge, 4 Thames Street, Louth, LN11 7AD
G3 PRU J. Nicholas-Letch, 53 Hayden Road, Rushden, NN10 0JH
GW3 PRW J. Dolan, 24 Pen Derwydd, Llangefni, LL77 7QE
G3 PS A. Mccann, 5 Arrowsmith Drive, Hoghton, Preston, PR5 0DT
G3 PSB P. Bottomley, 47 Stonelea, Barkisland, Halifax, HX4 0HD
G3 PSC J. Holton, 1204 Greenford Road, Greenford, UB6 0HQ
G3 PSG North Riding Rafars, c/o R. Armstrong, 64 Churchill Drive, Marske by the Sea, TS11 6BE
G3 PSM C. Thomas, 16 Fordlands, Thorpe Willoughby, Selby, YO8 9PD
GM3 PSP Dr A. Masson, 20 Frogston Avenue, Edinburgh, EH10 7AQ
GI3 PSQ C. Bristow, 58 Bristow Park, Belfast, BT9 6TJ
G3 PSR M. Gibbs, 62 Abinger Drive, Chatham, ME5 8UL
G3 PSS M. Kent, 99 London Road, Newington, Sittingbourne, ME9 7RH
G3 PSU P. Martin, 15 St. Lukes Close, Cannock, WS11 1BB
G3 PSV D. Park, 18 Widworthy Drive, Broadstone, BH18 9BD
G3 PSZ K. Jones, 24 Station Road, Okehampton, EX20 1EA
G3 PTB A. Tomalin, Chapel Street, Barford, Norwich, NR9 4AB
G3 PTG R. Gealy, 14 Wivelsfield, Eaton Bray, Dunstable, LU6 2JQ
G3 PTI K. Atter, 60 Hough Road, Barkston, Grantham, NG32 2NS
G3 PTQ T. Chapman, 5 Maple Close, Bottisham, Cambridge, CB25 9BQ
G3 PTS Dr G. Holt, 7 Beech Close, Olivers Battery, Winchester, SO22 4JY
G3 PTX L. Buckley, 188 Compstall Road, Romiley, Stockport, SK6 4JF
G3 PTZ A. Bensley, 13 Lime Grove, Cherry Willingham, Lincoln, LN3 4BE
G3 PUO L. Rooks, 17 The Close, Clayton Le Moors, Accrington, BB5 5RX
G3 PUQ N. Semmens, 4 South Park, Redruth, TR15 3AW
G3 PUR R. Tarr, 37 Warwick Avenue, Coventry, CV5 6DJ
G3 PUX I. Champion, Mill Bungalow, Billinghurst, RH14 0DY
GM3 PUY I. Forsyth, 68 Drumover Drive, Glasgow, G31 5RP
G3 PUZ D. Hogan, 17 Buckingham Mansions, Bath Road, Bournemouth, BH1 2PG
G3 PVG J. Bennett, 11 Enderby Road, Thurlaston, Leicester, LE9 7TF
G3 PVH D. Sumner, 64 Kelsey Avenue, Emsworth, PO10 8NQ
G3 PVJ Dr H. Coltman, 68 Cressex Road, High Wycombe, HP12 4TY
G3 PVU J. Hunt, 28 Harris Road, Lincoln, LN6 7PN
G3 PWB I. Dufour, 3 Western Close, Rushmere St. Andrew, Ipswich, IP4 5UU
G3 PWJ R. Fisher, 34 Doctors Hill, Stourbridge, DY9 0YE
G3 PWN G. Grimshaw, 1 Sandsacre Drive, Bridlington, YO16 6UA
G3 PWS R. Dalton, 23 Muswell Road, Mackworth, Derby, DE22 4HN
G3 PWX A. Boyd, 15 Northend, Batheaston, Bath, BA1 7EE
G3 PWY D. Gresswell, 10 Cherrywood Gardens, Flackwell Heath, High Wycombe, HP10 9AX
G3 PXH M. Bartlett, 3 Jessopp Avenue, Bridport, DT6 4AN
G3 PXI A. Evans, Apartado 286, Luz Lagos, Portugal, 8601-929 LUZ GS
GM3 PXK Mid Lanark ARS, c/o M. Overthrow, 63 Primrose Avenue, Larkhall, ML9 1JX
G3 PXL A. Hickin, 14 Churston Broadway, Dartmouth Road, Paignton, TQ4 6LE
G3 PXU G. Grove, 11 Croft Close, Warwick, CV34 6QY
G3 PXV R. Wiseman, 3 Springfield Road, Ruskington, Sleaford, NG34 9HG
GW3 PYD D. Stephens, 1 Awelfryn Terrace, Merthyr Tydfil, CF47 9YP
G3 PYE Cambridgeshire Repeater Group, c/o P. Nice, 31 Elizabeth Drive, Chapel St. Leonards, Skegness, PE24 5RS
G3 PYF J. Green, 68 Magdalen Lane Wingfield, Trowbridge, BA14 9LQ
G3 PYH A. Broadbent, 52 Norman Street, Failsworth, Manchester, M35 9EJ
G3 PYI D. Coy, 26 Hardy Road, Bishops Cleeve, Cheltenham, GL52 8BN
G3 PYL D. Justice, 4 Birley Moor Avenue, Sheffield, S12 3AQ
G3 PYO J. Dann, 1 Ffinch Close, Ditton, Aylesford, ME20 6ET
G3 PYW Rev. A. Speight, Glebe Cottage, Hollow Lane, Woodbridge, IP13 8LZ
GW3 PYX J. Chetcuti, 3 Beechwood Drive, Penarth, CF64 3RB
G3 PYZ RSARS - Royal Signals ARS, c/o M. Foster, 7 Church Street, Fenstanton, Huntingdon, PE28 9JL
G3 PZB A. Ash, 34 Coronation Avenue, Northwood, Cowes, PO31 8PN
G3 PZE C. Burkitt, The Old Wheelwright, 17 Oxford Road, Hitchin, SG4 8NP
G3 PZF G. Dale, 16 Palfrey Close, St Albans, AL3 5RE
G3 PZL P. Brown, Little Langford, Newlands Lane, Henley-on-Thames, RG9 5PS
G3 PZN C. Wood, 24 Talveneth, Pendeen, Penzance, TR19 7UT
G3 PZU B. Brown, 138 First Avenue, Sudbury, CO10 1YU
G3 PZV P. Greed, The Bungalow, Townsend, Devizes, SN10 4RR
G3 PZX A. Ward, 20 Tower Close, Costessey, Norwich, NR8 5AU
G3 PZZ P. Smith, 38 Leasway, Wickford, SS12 0HE
G3 QI Flaxton Moor Contest Group, c/o C. Quarton, Flaxton Gatehouse, Flaxton, York, YO60 7QT
G3 RAC Thales ARC, c/o D. Waterworth, 116 Reading Road, Woodley, Reading, RG5 3AD
G3 RAF Royal Air Force ARS, c/o M. Garrett, 489 Dorchester Road, Weymouth, DT3 5BP
G3 RAL Loughborough and District ARC, c/o A. Harrison, 44 Rosslyn Road, Whitwick, Coalville, LE67 5PT
G3 RAM C. Langmaid, Flat 4, Woodlawn High Street, Partridge Green West Sussex, RH13 8HR
G3 RAU D. Moffatt, Mill House, Middle Street, Glentworth, Gainsborough, DN21 5BZ
G3 RBD F. Hanson, 207 Grant Road, Liverpool, L14 0LG
G3 RBJ A. Payne, Laurel Bank, Sand Road, Wedmore, BS28 4BZ
G3 RBP R. Parsons, Netherhall Barn, Hallmoor Road, Darley Dale, Matlock, DE4 2HF
G3 RBY A. Stagles, 8 Goodwood Close, Cowplain, Waterlooville, PO8 8BG
G3 RCB N. Kingsley, 1 Wensleydale Gardens, Hampton, TW12 2LU
G3 RCD C. Brockbank, 31 Park Hill, Church Crookham, Fleet, GU52 6PW
G3 RCE R. Allbright, 50 Portsdown Road, Portsmouth, PO6 4QH
G3 RCM Sheffield ARC, c/o D. Littlewood, 50 Industry Road, Sheffield, S9 5FQ
G3 RCQ D. Cole, Amber Lights, Market Lane, Walpole St. Andrew, Wisbech, PE14 7LT
G3 RCV Cray Valley Radio Society, c/o A. Styles, 6 Hill Brow, Crayford, Dartford, DA1 3NX
G3 RCW Worksop Amatuer Radio Society, c/o A. Bostock, 26 Ingham Road, Bawtry, Doncaster, DN10 6NW
G3 RCX L. Gibson, 7 Heycroft Road, Eastwood, Leigh-on-Sea, SS9 5SW
G3 RCZ G. Thompson, 22 Warton Avenue, Heysham, Morecambe, LA3 2LX
G3 RDA S. Whitehead, 98 Oak Road, Fareham, PO15 5HP
GW3 RDB Hoover ARC, c/o T. George, 80 Yew Street, Troedyrhiw, Merthyr Tydfil, CF48 4EE
G3 RDC A. Wood, 14 Little Paradise, Marden, Hereford, HR1 3DR
G3 RDF J. Jeffrey, Old Church Cottage, Ipsden, Wallingford, OX10 6AE
G3 RDG K. Michaelson, 40 The Vale, Golders Green, London, NW11 8SG
G3 RDH J. Barnes, 4 Deepdene Drive, Dorking, RH5 4AD
G3 RDP H. Cutts, 50 Cropton Road, Hull, HU5 4LP
G3 RDQ D. Griffiths, Upcote Cottage, Chilbolton, Stockbridge, SO20 6BA
G3 RDR P. Rudwick, 29 Fuller Street, London, NW4 4RR
G3 RDZ J. Walker, 94 Keys Park, Parnwell Way, Peterborough, PE1 4SN
G3 RE S. Dixon, 33 Medhurst Crescent, Gravesend, DA12 4HJ
G3 REB R. Cole, Lyndale, Brimscombe Lane, Brimscombe, Stroud, GL5 2RF
G3 RED D. Sylvester, 10 Ivy Grove Gunthorpe, Peterborough, PE4 7TW
G3 REH H. Neale, Thornlea Fishergate, Spalding, PE12 0EZ
G3 REL B. Woodfield, 49 Oakfield Road, Blackwater, Camberley, GU17 9DZ
G3 REP R. Parkes, 2 Saxon Road, Steyning, BN44 3FP
G3 REU G. Hearn, 70 Cranmer St., Long Eaton, Nottingham, NG10 1NL
G3 REV R. Pulling, 410 Leach Lane, Sutton Leach, St Helens, WA9 4NA
G3 REW D. Morris, 66 Windmill Close, Brixham, TQ5 9SQ
GM3 RFA D. Garrington, 3 Sutherland Avenue, Fort William, PH33 6JS
G3 RFH K. Randall, 25 Kingsway, Thornton-Cleveleys, FY5 1DL
GD3 RFK D. Dodd, Ellan Geay, Ballayockey Lane, Regaby, Isle of Man, IM7 3HP
G3 RFN G. Wild, 17 New Church Close, Clayton Le Moors, Accrington, BB5 5GH
G3 RGB A. Moon, 6 Troon Close, Saltersgill, Middlesbrough, TS4 3HX
G3 RGC T. Matthews, 38 Foxhill, Grimsby, DN37 9QL
G3 RGD R. Dobdinson, 73 Watwood Road, Hall Green, Birmingham, B28 0TW
G3 RGE K. King, Blandford Garrison A.R.C., Cole Block, Blandford Forum, DT11 8RH
G3 RGJ R. Weston, 43 Pearce Avenue, Parkstone, Poole, BH14 8EG
G3 RGM D. Mullins, Flat 23, Kennington Palace Court, Sancroft Street, London, SE11 5UL
G3 RGN L. Binns, Leamar, 707 Halifax Road, Cleckheaton, BD19 6LJ
G3 RGP R. Pratt, 1 Colebrooke Avenue, Ealing, London, W13 8JZ
G3 RGS D. Thomson, Skippers Down, Old Coach Road, Wrotham, Sevenoaks, TN15 7NR
GM3 RGU J. Connelly, 9 Glenhead Crescent Hardgate, Clydebank, G81 6LW
G3 RHP J. Garrett, 5 The Courtyard, Sudbourne Park, Sudbourne, Woodbridge, IP12 2AJ
G3 RHQ K. Vickers, Hillview, Barton-upon-Humber, DN18 5DZ
G3 RHR K. Drinkwater, Brearton Lodge, Brearton, Harrogate, HG3 3BX
G3 RHU M. Stanbridge, 183 Charlton Park Midsomer Norton, Radstock, BA3 4BR
G3 RHW C. Cushion, 3 The Copse, Bridgwater, TA6 4DW
G3 RHZ A. Wilkinson, 41 Church Road, Nailstone, CV13 0QH
GW3 RIB W. Huxley, No2 Bungalow, Nant Mawr Rd, Clwyd, CH7 2BS
G3 RID D. Nancarrow, 6 Trythogga Road, Gulval, Penzance, TR18 3NA
GW3 RIH W. Elton, 15 Main Avenue, Peterston-Super-Ely, Cardiff, CF5 6LQ
G3 RIK D. Carden, 9 Wood Hey Grove, Rochdale, OL12 9TY
G3 RIM T. Emeney, 10 Kilnside, Claygate, Esher, KT10 0HS
G3 RIR N. Ackerley, 24 Macaulay Road, Lutterworth, LE17 4XB
G3 RIX M. Tetley, 87 Main Street, Irton, Scarborough, YO12 4RJ
GW3 RIY A. Chapman, 14 Birch Walk, Porthcawl, CF36 5AN
G3 RJE J. Hunt, 33 Rainhill Road, Rainhill, Prescot, L35 4PA
G3 RJF I. Walker, 28 Norrington Road, Maidstone, ME15 9RA
G3 RJH Dr R. Harding, High Trees, Arrowsmith Road, Wimborne, BH21 3BG
G3 RJI A. Paul, 3 Brunswick Avenue, Upminster, RM14 1NA
G3 RJM R. Cutts, 60 Holmpton Road, Withernsea, HU19 2QD
G3 RJS P. Barry, 235 Manor Way, Aldwick, Bognor Regis, PO21 4HT
G3 RJT C. Garland, 48 Underbank End Road, Holmfirth, HD9 1ES
G3 RJV Rev. G. Dobbs, 9 Highlands, Littleborough, OL15 0DS
G3 RKF T. Roeves, 33 York Crescent, Wilmslow, SK9 2BB
G3 RKH Rev. J. Marshall, 166 Calton Road, Gloucester, GL1 5ER
G3 RKJ N. Summers, 126 Kestrel House, 1 Alma Road, Enfield, EN3 4QE
G3 RKK Dr A. Shepherd, 59 Lime Avenue, Camberley, GU15 2BH
G3 RKL Dr A. Whitaker, 160 Derbyshire Lane, Sheffield, S8 8SE
G3 RKM J. Meaker, 11 Woodend View, Mossley, Ashton-under-Lyne, OL5 0SN
G3 RKQ A. Balmforth, Leam Brink, Lutton Gowts Lutton, Long Sutton, Spalding, PE12 9LQ
GW3 RKV R. Volck, Maes-Y-Bryn, Rosebush, Clynderwen, SA66 7QS
G3 RKZ B. Tibbert, 99A Main Street Horsley Woodhouse, Ilkeston, DE7 6AW
G3 RLA C. Phillips, Bella Vista, The Moorings, Wirral, CH60 9JT
G3 RLD R. Ramshaw, 132 Main Road, Duston, Northampton, NN5 6RA
G3 RLE B. Turner, 56 Bamford Way, Rochdale, OL11 5NB
G3 RLF D. Price, 8 Newland Road, Droitwich, WR9 7AF
G3 RLJ J. Harper, East View, 35 John Street, Sutton in Ashfield, NG17 4EN
G3 RLL D. Grindell, 23 Park Hall Avenue, Walton, Chesterfield, S42 7LR
G3 RLO D. Cadman, 32 Breedon Hill Road, Derby, DE23 6TG
G3 RLT W. Stewart, 4 Denmark Road, Kingston upon Thames, KT1 2RU
G3 RLV M. Vann, 3 Mile Planting, Richmond, DL10 5DB
G3 RLX N. Taylor, 146 Morledge, Matlock, DE4 3SD
G3 RMD F. Regan, 7 Hilltop Road, Cheltenham, Gl504NW
G3 RMF B. Magill, 14 Barry Street, Worcester, WR1 1NR
GW3 RMJ P. Jennings, Myrtle Cottage, Harpers Lane, Presteigne, LD8 2AN
G3 RMK R. Ruaux, Park View, Wallage Lane, Crawley, RH10 4NG
G3 RMN M. Smith, 121 Shirley Way, Croydon, CR0 8PN
G3 RMQ J. Ingham, High Croft, 1 Layton Crescent, Leeds, LS19 6RJ
G3 RMX W. Hall, 52 Barley Gate, Leven, Beverley, HU17 5NU
G3 RMY J. Andrews, 12 Gerald Close, Burgess Hill, RH15 0NB
G3 RMZ A. Pink, 37 Shute Park Road, Plymouth, PL9 8RB
GI3 RNO P. Greenan, 9 Ashville Park, Antrim, BT41 1HH
G3 RNP J. Price, 125 Oakfield Road, Malvern, WR14 1DT
G3 RNX W. Walker, 44 South Road, Weston-Super-Mare, BS23 2HE
G3 ROC R. Collins, Thorn Acacia, Rye Road Northiam, Rye, TN31 6NJ
G3 ROD R. Davenport, 7 Nether Close, Duffield, Belper, DE56 4DR
G3 ROG G. Morgan, 22 Monks Road, Winchester, SO23 7EQ
G3 ROM B. Sweetman, Cortijasa Los Perez 220, Benajarafe, Malaga, Spain, 29790
G3 ROO I. Keyser, Rosemount, Church Whitfield Road, Whitfield, Dover, CT16 3HZ

G3	ROP	M. Goodrick, 18 Milford Street, Cambridge, CB1 2LP
G3	ROQ	R. Gill, 45 Biggin Lane, Ramsey, Huntingdon, PE26 1NB
G3	ROS	H. Williams, Roslyn, Whalley Road, Burnley, BB12 7HT
G3	ROW	S. Smith, Po Box 2738, Silver City, USA, 88062
G3	RPA	J. Knowles, Springhill, Gilpins Ride, Berkhamstead, HP4 2PD
G3	RPB	K. Spicer, Grove Cottage, Dallinghoo, Woodbridge, IP13 0LR
G3	RPD	G. Clinch, 2 Storrs Close, Bovey Tracey, Newton Abbot, TQ13 9HR
G3	RPL	T. Neyland, 22 Pax Hill, Bedford, MK41 8BT
GM3	RPM	J. Mcavoy, 120 Donaldswood Road, Paisley, PA2 8EB
G3	RPO	F. Seddon, 23 Countessway, Euxton, Chorley, PR7 6PT
G3	RPV	T. Venn, 22 Eaton Close Hartford, Huntingdon, PE29 1SR
G3	RPZ	H. Trunley, Bijou House, 75 Belgrave Road, Leigh-on-Sea, SS9 5EL
G3	RQF	D. Keith, 108 Lower Northam Road Hedge End, Southampton, SO30 4FT
GM3	RQQ	H. Robertson, 102 Orchy Crescent, Bearsden, Glasgow, G61 1RE
G3	RQR	N. Kirtley, 14 Byron Avenue, Winchester, SO22 5AT
G3	RQS	R. Rimmer, 25 Haig Court, Chesterton, Cambridge, CB4 1TT
GI3	RQU	Dr S. Laverty, 572 Antrim Road, Belfast, BT15 5GL
G3	RQX	P. Lewis, 20 Osborne Road, Penn, Wolverhampton, WV4 4AY
G3	RQZ	P. Madagan, 40 Lagham Park, South Godstone, Godstone, RH9 8ER
G3	RR	Hucknall Rolls Royce ARC, c/o S. Sorockyj, 8 Bowden Avenue, Bestwood Village, Nottingham, NG6 8XN
G3	RRG	P. Taylor, 44 Leegate Road, Stockport, SK4 4AX
G3	RRI	R. Wilmot, Swinside, Branthwaite Lane, Workington, CA14 1HE
G3	RRM	J. Hughes, 41 Highfield Avenue, Great Sankey, Warrington, WA5 2TW
G3	RRN	Dr K. Jones, Field House, Wragby Road, Lincoln, LN2 2QU
G3	RRP	R. Pine, 21 Hatherden Avenue, Poole, BH14 0PJ
G3	RRS	Rutherford Appleton Laboratory ARC, c/o J. Wright, 2 Barnfield, Charney Bassett, Wantage, OX12 0HA
G3	RRW	J. Francis, 5 Central Park Avenue, Plymouth, PL4 6NW
G3	RSB	R. Scaife, 7 Woodgates Close, North Ferriby, HU14 3JS
G3	RSC	Sutton Coldfield Radio Society, c/o B. Adkins, 4 Orion Close, Ward End, Birmingham, B8 2AU
G3	RSE	C. Cheney, 35 Metcalfe Road, Cambridge, CB4 2DB
G3	RSF	A. Notschild, 8 Hillpark, Buckland Brewer, Bideford, EX39 5HY
G3	RSI	F. Mckeracher, Wickets, 1 Marshal Close, Alton, GU34 1RA
G3	RSM	F. Burnett, 4 Woodlands Drive Fulwood, Preston, PR2 9SQ
G3	RST	R. Southern, 30 Barnfield, Crowborough, TN6 2RY
G3	RSU	D. Bindon, Forth House, Water Street, Langport, TA10 0HH
G3	RSV	R. Dowsett, 23 South Wootton Lane, King's Lynn, PE30 3BS
G3	RSW	W. Mullarkey, The Barn, Skull House Lane, Wigan, WN6 9DJ
G3	RTB	R. Bell, 14 Wacker Field Road, Rendlesham, Woodbridge, IP12 2UT
G3	RTD	J. Gailer, Shelleys, King Stag, Sturminster Newton, DT10 2BE
G3	RTE	G. Kellaway, 55 Ladbrooke Drive, Potters Bar, EN6 1QW
G3	RTM	A. Chaddock, 4 Liddiard Close Kennington, Oxford, OX1 5RY
G3	RTO	N. Pratt, 87A Dovecote Road, Newthorpe, Nottingham, NG16 3QL
G3	RTP	J. Pennington, Brambling, Forest Road, Waterlooville, PO7 6UE
G3	RTY	H. Meers, 10 Lawnswood Avenue, Chasetown, Burntwood, WS7 4YD
G3	RUD	E. Workman, Sunset, 2 Burnham Drive, Weston-Super-Mare, BS24 9LW
GW3	RUE	E. Edwards, Ceris, Ruthin Road, Denbigh, LL16 3EU
G3	RUG	G. Twiss, 9 Brae Head, Eaglescliffe, Stockton-on-Tees, TS16 9HP
G3	RUH	J. Miller, 3 Bennys Way, Coton, CB23 7PS
GM3	RUI	R. Furness, 43 Glebe Street, Leven, KY8 4QN
G3	RUJ	R. Powell, 13 Bridges Drive, Bristol, BS16 2UB
G3	RUO	W. Williamson, 84 Atfield Drive, Whetstone, Leicester, LE8 3NE
GM3	RUP	C. Morton, 295 Byres Road, Hillhead, Glasgow, G12 8TL
G3	RUV	A. James, 4 The Chestnuts, Aylesbeare, Exeter, EX5 2BY
G3	RUZ	D. Martin, 4 Chilbolton Mews, 19 Chilbolton Avenue, Winchester, SO22 5HU
G3	RVA	R. Crowe, 37 Huccaby Close, Brixham, TQ5 0RJ
G3	RVC	Prof. P. Cochrane, Willow Barn, East Lane, Woodbridge, IP13 6EB
GW3	RVG	S. Sedgebeer, 50 Minffrwd Road Pencoed, Bridgend, CF35 6SD
G3	RVI	J. Walch, 52 Marsh House Road, Sheffield, S11 9SP
GM3	RVL	Dr H. Brash, 5 Hillview Drive, Edinburgh, EH12 8QW
G3	RVM	C. Trusson, 27A Roman Way, Thatcham, RG18 3BP
G3	RVS	G. Haynes, Littleton House Cottage Blandford St. Mary, Blandford Forum, DT11 9NB
G3	RVX	J. Colegate, 1 Oldmere Cottages High Street, Bathford, Bath, BA1 7TJ
G3	RVY	P. Colegate, 65 Forest Road, Melksham, SN12 7AB
G3	RWE	T. Yates, 3 Sycamore Crescent, Macclesfield, SK11 8LL
G3	RWF	P. Henwood, Conifers, Church Road, Littlebourne, Canterbury, CT3 1UA
G3	RWI	Dr P. Cross, Home Farm House, Icomb, Cheltenham, GL54 1JD
G3	RWL	R. Limebear, 60 Willow Road, Enfield, EN1 3NQ
G3	RWV	M. Sanders, 7 Netherby Close, Tring, HP23 5PJ
G3	RWW	G. Southern, 27 Eldred Road, Liverpool, L16 8NZ
GW3	RWX	D. Thomas, 88 Cefn Graig, Rhiwbina, Cardiff, CF14 6JZ
G3	RXA	J. Thomas, Blair House, Market Place, Norwich, NR16 2AN
G3	RXG	R. Burgess, 11 Beech Road, Shipham, Winscombe, BS25 1SA
G3	RXI	E. Blundell, 29 Garden Close, Hook, RG27 9QZ
G3	RXO	R. Brown, Lower Dicker, Hailsham, BN27 4BG
G3	RXP	D. Mason, 5 Spa Top, Caistor, Market Rasen, LN7 6RB
G3	RXS	W. Scarlett, 14 Warren Drive, Bingley, BD16 3BX
GM3	RXU	Prof. I. Macpherson, 1 Broomie Dell, Earlston, TD4 6BN
GI3	RXV	N. Graham, 3 Shilgrove Place, Castledawson, Magherafelt, BT45 8AL
GM3	RXZ	R. Marshall, 52 Lumsdaine Drive, Dalgety Bay, Dunfermline, KY11 9YU
GW3	RYE	J. Harris, Treweryn, Llwyndafydd Road, Llandysul, SA44 6BT
G3	RYH	J. Brodie, Huntlands Farm, Gaines Road, Worcester, WR6 5RD
G3	RYK	I. Grayson, 156 Little Brays, Harlow, CM18 6EY
G3	RYP	D. Craggs, New House, Dacre Banks, Harrogate, HG3 4EW
GW3	RYR	Dr C. Morgan, 33 West Grove, Merthyr Tydfil, CF47 8HJ
G3	RYW	D. Wardlaw, 21 Tormey Street, Balwyn North, Australia, VIC 3104
G3	RYZ	M. Byrne, 16 Downham Gardens Tamerton Foliot, Plymouth, PL5 4QE
G3	RZC	R. Pellett, La Biochere, Aizenay, France, 85190
G3	RZF	D. Horton, 26 The Crescent, Slough, SL1 2LQ
G3	RZG	M. Box, 18 Stottingway Street, Weymouth, DT3 5QA
G3	RZI	M. Moss, 1082 Evesham Road, Astwood Bank, Redditch, B96 6ED
G3	RZJ	G. Hall, 185 Dialstone Lane, Stockport, SK2 7LQ
G3	RZP	P. Chadwick, Three Oaks, Braydon, Swindon, SN5 0AD
G3	RZV	A. Lawrance, 97 Dorchester Road, Oakdale, Poole, BH15 3QZ
G3	RZY	C. Abrey, 31 Yew Tree Lane, Leeds, LS15 9JD
G3	SAD	Stevenage & District A.R.S, c/o R. Mctait, 20 Rowland Road, Stevenage, SG1 1TE
GM3	SAE	R. Mcmillan, 54 Birchwood, Invergordon, IV18 0BG
G3	SAH	R. Matthews, 2, Newton Close, Oakenshaw South, Redditch, B98 7YR
GM3	SAN	S. Weir, 19 Ellismuir Road, Baillieston, Glasgow, G69 7HW
G3	SAO	Dr J. Midgley, 3 Chipping Fold, Milnrow, Rochdale, OL16 4YD
G3	SAR	R. Warner, Cubs Wood, Rycroft Lane, Sevenoaks, TN14 6HT
GM3	SBC	E. Murphy, 65 Silverknowes Crescent, Edinburgh, EH4 5JA
G3	SBF	S. Eames, 4 Dabey Close, Markfield, LE67 9UJ
G3	SBL	Stafford & Districts ARS, c/o G. Reay, 53 Tithe Barn Road, Stafford, ST16 3PL
G3	SBM	D. Turner, 50 Hardings, Chalgrove, Oxford, OX44 7TJ
G3	SBP	R. Gynn, Honeywood Belvidere Road, Exeter, EX4 4RR
G3	SBT	W. Turnbull, 15 Marshallsay Road, Chickerell, Weymouth, DT3 4BB
G3	SCB	Rev. R. Hinder, 18 Mapledale Avenue, Croydon, CR0 5TB
G3	SCD	D. Dunn, 5 Oaks, Harrington, Spilsby, PE23 4NH
G3	SCJ	D. Power, 47 Marlborough Street, Gainsborough, DN21 1BT
G3	SCL	R. Houghton, Hans-Miederer-Str. 10B, Schliersee, Germany, 83727
GI3	SCM	T. Mccullough, 16 Mccormack Gardens, Lurgan, Craigavon, BT66 8LE
G3	SCT	Thurrock Sea Cadet Corp - Tilbury, c/o N. Wilkinson, 12 Woodlands Close, Grays, RM16 2GB
G3	SCV	Rev. G. Stanton, 8 Kennett Close, Norwich, NR4 7JA
G3	SCZ	R. Brown, 22 Lordswood Silchester, Reading, RG7 2PZ
G3	SDC	De Montfort University ARS, c/o R. Titterington, Wyclif House, St. Marys Road, Lutterworth, LE17 4PS
G3	SDG	J. Bottom, 48 Chesterton Avenue, Harpenden, AL5 5SU
G3	SDH	P. Kelly, Martyndale, The Street, Compton Martin, Bristol, BS40 6JE
G3	SDL	D. Court, Connogue, River Lane, Shankill, Ireland, D18 W2R4
G3	SDO	K. Heathfield, 2 Georgian Close, Broadway, Weymouth, DT3 5PF
G3	SDS	South Dorset Rs, c/o G. Watts, 3 Maple Grove Knightsdale Road, Weymouth, DT4 0FE
G3	SDT	D. Allen, Chelsea Cottage, The Turnpike, Carleton Rode, Norwich, NR16 1RS
G3	SDW	K. Underwood, Apartment 7, Imperial Court, Palermo Road, Torquay, TQ1 3NW
G3	SDY	G. Edinburgh, 77 Westerley Lane, Shelley, Huddersfield, HD8 8HP
G3	SEA	P. Perretta, 1511 Punahou St., Apt 208, Honolulu, USA, 96822
G3	SED	E. Devereux, 191 Botley Road, Burridge, Southampton, SO31 1BJ
G3	SEF	R. Frew, Sawley House, 82 Wormholt Road, London, W12 0LP
G3	SEG	W. Gordon, 55 Trajan Avenue, South Shields, NE33 2AN
G3	SEJ	E. John, Obo St. Dunstans Ars, 52 Broadway Avenue, Wallasey, CH45 6TD
GM3	SEK	Dr I. White, 2 Appleby Cottages, Whithorn, Newton Stewart, DG8 8DQ
G3	SEM	P. Cort-Wright, 11-13 Hardingham Street Hingham, Norwich, NR9 4JB
G3	SEN	R. Dawes, 18 Sutherland Road, Nottingham, NG3 7AP
G3	SEQ	J. Crossfield, Forest Lodge, Chopwell Wood, Rowlands Gill, NE39 1LT
GM3	SER	H. Bremner, 2 Rowan Crescent, Lenzie, Glasgow, G66 4RE
G3	SES	P. Stevens, 20 Abbots Park, Chester, CH1 4AN
G3	SET	G. Aram, 5 Lancaster Green, Hemswell Cliff, Gainsborough, DN21 5TQ
G3	SEY	R. Mackey, 44 South Street, Ossett, WF5 8LF
G3	SEZ	P. Luft, Swan House, Livesey Road, Ludlow, SY8 1EY
G3	SFB	C. Hale, 16 Windmill Court, East Wittering, Chichester, PO20 8RJ
GW3	SFC	A. Richards, 30 Well Place, Aberdare, CF44 0PB
G3	SFE	P. Everett, 58 Greenwood Avenue, Bognor Regis, PO22 9EX
G3	SFG	Southgate ARC, c/o D. Berry, 4 Holly Hill, Winchmore Hill, London, N21 1NP
G3	SFK	P. Kerry, 251 Upper Rainham Road, Hornchurch, RM12 4EY
GW3	SFQ	R. Mugford, 27 Highfield Close, Dinas Powys, CF64 4LR
G3	SFU	P. Woodfield, 49 Oakfield Road, Blackwater, Camberley, GU17 9DZ
G3	SFV	E. Meachen, 46 Rainsborough Gardens, Market Harborough, LE16 9LW
GI3	SG	J. Patty, 3C Finwood Park, Belfast, BT9 6QR
G3	SGA	A. Jones, Po Box 355, Crestholme 3652, Natal, South Africa,
G3	SGC	G. Morris, Norcrest, Beach Road, Norwich, NR12 0AL
G3	SGF	P. Casemore, 9 Wellcroft Cottages, Church Lane, Albourne, Hassocks, BN6 9BZ
GW3	SGK	Dr B. King, Ty Derwen Vinegar Hill, Undy, Caldicot, NP26 3EJ
G3	SGL	A. Isaacs, Holme View, Brick Lane, Christchurch, BH23 8DU
G3	SGR	Dr J. Craig, Ferndown, Tilley Lane, Hailsham, BN27 4UT
G3	SGV	J. Fallon, 8 Tretower Close, Plymouth, PL6 6BH
G3	SGX	R. Bona, 1 Maxwell Road, Broadstone, BH18 9JG
G3	SGY	A. Nesbitt, 28 Fairfax Road, Middleton St. George, Darlington, DL2 1HF
G3	SGZ	T. Chapple, 39 Maynards Park, Bere Alston, Yelverton, PL20 7AR
G3	SHD	L. Dray, 1 Chalfont Close, Bradville, Milton Keynes, MK13 7HS
G3	SHF	B. Naylor, 47 Chester Road Poynton, Stockport, SK12 1HA
G3	SHK	R. Pett, 5 Kingford Close, Woodfalls, Salisbury, SP5 2NQ
G3	SHL	J. Harlow, 3 The Fairways, Sherford, Taunton, TA1 3PA
GM3	SHR	J. Coster, 17 Glamis Place, Dalgety Bay, Dunfermline, KY11 9UA
G3	SHX	R. West, 15 St. Andrews Close, Margate, CT9 4HA
G3	SHY	R. Cottrell, 157 Ridge Lane, Watford, WD17 4SU
G3	SHZ	Dr J. Whittington, Twyford Manor, Bicester Road, Buckingham, MK18 4EL
G3	SIA	B. Keyte, 9 Swanns Meadow, Bookham, Leatherhead, KT23 4JX
G3	SIG	RSARS - Royal Signals Amatuer Radio Society, c/o A. Watt, 5 Brambling Road, Horsham, RH13 6AX
GW3	SIK	K. Pugh, Tanybanc, Blaenporth, Cardigan, SA43 2BD
G3	SIR	D. Durham, 29 Waverley Road, Stratton St Margaret, Swindon, SN3 4AY
G3	SIT	R. Kressman, 12 School Lane, Fenstanton, Huntingdon, PE28 9JR
G3	SIU	P. Hearson, 14 Osgood Gardens, Orpington, BR6 6JU
G3	SJH	Dr C. Eyles, 9 St. Peters Road, Harborne, Birmingham, B17 0AT
G3	SJI	M. Batt, 9 Grange Park, Westbury-On-Trym, Bristol, BS9 4BU
G3	SJJ	C. Burbanks, 16 Cotgrave Road, Plumtree, Nottingham, NG12 5NX
G3	SJK	S. Cherry, 4 West Hill Road South, South Wonston, Winchester, SO21 3HP
G3	SJR	W. Tynan, 22 Belchmire Lane, Gosberton, Spalding, PE11 4HG
G3	SJW	S. Haigh, 29 West Street, Chichester, PO19 1QS
G3	SJX	P. Hart, The Willows, Paice Lane, Medstead, Alton, GU34 5PR
GM3	SJY	C. Lawrenson, Hollyburn, West Port, Cupar, KY15 7BW
G3	SKI	R. Bravery, 19 Lindum Road, Worthing, BN13 1LX
G3	SKN	D. Naylor, 52 Rue Du Port, Pontorson, France, 50170
G3	SKR	A. Gold, 60 Wynnstay Gardens, London, W8 6UU
G3	SKV	S. Hobday, 31 Sackville Crescent, Harold Wood, Romford, RM3 0EJ
G3	SKY	IOW Rad Soc, c/o A. Ash, 34 Coronation Avenue, Northwood, Cowes, PO31 8PN

GD3 SKZ K. Manktelow, Tramman House, Ballabeg, Isle Of Man, IM9 4HA
G3 SLI A. Osborne, 18A Cumnor Road, Boars Hill, Oxford, OX1 5JP
G3 SLJ D. Parsons, Am Dorfplatz 12, Winden Am Aign, Germany, 85084
G3 SLK R. Pickering, 147 Windermere Avenue, Nuneaton, CV11 6HN
G3 SLL H. Tyreman, U1 11Jenaya Place, Labrador, Australia, 4215
G3 SLS A. Hancock, 6 The Fairway, Mablethorpe, LN12 1LL
G3 SLT D. Ormerod, 21 Valletta Close, Chelmsford, CM1 2PT
G3 SLX J. Smith, 256 Stone Road, Stoke-on-Trent, ST4 8NJ
G3 SMD R. Turner, 7 Paddocks Lane, Cheltenham, GL50 4NU
G3 SMF I. Hamill, 74 Lampits Hill, Corringham, Stanford-le-Hope, SS17 9AJ
G3 SMN R. Forster, 28 Springbridge Road, Manchester, M16 8PW
GW3 SMT P. Torry, Pen-Y-Rhos Old Racecourse, Oswestry, SY10 7HP
G3 SMV J. Smith, 18 Hounslow Road, Mackworth, Derby, DE22 4BW
G3 SMZ R. Hill, 68 Chestnut Street, Chadderton, Oldham, OL9 8HH
G3 SNA S. Andrew, Berry Brow House, Berry Brow, Greenfield, Oldham, OL3 7EJ
GJ3 SND B. Walster, Le Ponterrin Cottage, La Rue Du Ponterrin, St Saviour, Jersey, JE2 7HP
G3 SNG A. Ambler, 12 Oakdene Road, Marple, Stockport, SK6 6PJ
G3 SNH W. Harrison, 44 Briar Road, Thornton-Cleveleys, FY5 4NB
G3 SNN A. Woolford, 39 Apple Orchard, Prestbury, Cheltenham, GL52 3EH
G3 SNO G. Smith, Stoneycroft, Godsons Lane, Napton, Southam, CV47 8LX
G3 SNP M. Pitcher, Sandycot, Cadsden Road, Princes Risborough, HP27 0NB
G3 SNR G. Morgan, Eaton House, Eaton Bank, Belper, DE56 4BH
G3 SNT R. Dixon, Copper Beeches, Witton Gilbert, Durham, DH7 6TW
G3 SNU K. Selleck, Westphalia, Dartington, Totnes, TQ9 6DJ
G3 SOA W. Mccartney, Lychgate House, Uffington, Shrewsbury, SY4 4SN
G3 SOE R. Jennings, 31 Copper Beech Drive, Wombourne, Wolverhampton, WV5 0LH
GI3 SOO M. Foley, 5 Woodland Drive, Cookstown, BT80 8PL
G3 SOU Southampton ARC, c/o M. Troy, 22 Jackie Wigg Gardens Totton, Totton, Southampton, so409lz
GW3 SPA R. Alban, 73 Plymouth Road, Penarth, CF64 3DD
G3 SPI I. Dawe, 10 Selsden Close, Elburton, Plymouth, PL9 8UR
G3 SPJ C. Wooff, 55 Bostall Hill, Abbey Wood, London, SE2 0QX
G3 SPL P. Lee, 10 Antony Gardner Crescent, Whitnash, Leamington Spa, CV31 2TQ
G3 SPN N. Collins, Flat 1, Vista Mare West, 44 West Parade, Worthing, BN11 5EF
G3 SPO P. Oneill, Recreation Cottage, Slad Road, Stroud, GL6 7QA
G3 SPP A. Minett, 45 Patterdale Drive, Worcester, WR4 9HS
GM3 SPT G. Mckay, 152 Inveresk St., Greenfield, Glasgow, G32 6TA
G3 SPV K. Richardson, Brookfield Grove, The Dukes Drive, Bakewell, DE45 1QQ
G3 SPY Gpt(Coventry)ARS, c/o R. Harris, Clevelands, Tamworth Road, Coventry, CV7 8JJ
G3 SQN J. Grant, 8 Thornhill Way, Mannamead, Plymouth, PL3 5NP
G3 SQO D. Best, Tanglewood, Showley Road, Blackburn, BB1 9DP
G3 SQQ J. Franks, 11 Thoresby Avenue, Kirkby-In-Ashfield, Nottingham, NG17 7LY
G3 SQU C. Clarke, 14 Woodlea Gardens, Newcastle upon Tyne, NE3 5BY
G3 SQX E. Taylor, 115 St. Albans Avenue, London, W4 5JS
G3 SRC Surrey Ra Con C, c/o M. Fagg, 113 Bute Road, Wallington, SM6 8AE
GW3 SRF D. Woolen, Rose Cottage, Newcastle Hill, Bridgend, CF31 4EY
GW3 SRG A. Peake, 70 Higher Lane, Langland, Swansea, SA3 4PD
G3 SRJ G. Carlisle, 3 Grimms Meadow, Walters Ash, High Wycombe, HP14 4UH
GW3 SRM S. Hulme, 64 Salem Street, Amlwch, LL68 9BT
G3 SRN G. Bray, 4 Ledway Drive, Wembley, HA9 9TQ
G3 SRQ R. Bisseker, 24 Millgate, High Wycombe, HP11 1GL
G3 SRR M. Rees, 83 Salisbury Road, Farnborough, GU14 7AE
G3 SRT Salop A.R.S., c/o R. Golding, 7 Belvidere Avenue, Shrewsbury, SY2 5PF
GM3 SRV R. Tatton, 17 Paties Road, Edinburgh, EH14 1EF
G3 SRX N. Down, 23 Christopher Close, Heckington, Sleaford, NG34 9SA
GW3 SSK J. Williams, 5A Derllwyn Close, Tondu, Bridgend, CF32 9DH
G3 SSN J. Brand, 133 Hatfield Avenue, Fleetwood, FY7 7DU
G3 SSW Dr S. Erents, 50 Blandy Avenue, Southmoor, Abingdon, OX13 5DB
G3 SSZ L. Lavelle, 49 Jones Road Goffs Oak, Waltham Cross, EN7 5JT
G3 STF P. Sandiford, 11 Calle Menendez Pelayo, San Javier, Spain, 30730
G3 STJ B. Riley, 2 Watson Street, Swinton, Manchester, M27 6AQ
GM3 STM S. Mitchell, 97 Barbieston Road, Auchinleck, KA18 2ED
G3 STP P. La Pierre, 42 Berry Vale, South Woodham Ferrers, Chelmsford, CM3 5GY
G3 STT W. Haynes, 37 Hawthorn Grove, Southport, PR9 7AA
G3 STZ C. Thorn, 4 Riveredge, Framilode, Gloucester, GL2 7LH
G3 SUA E. Winstanley, 1 Drews Court Churchdown, Gloucester, GL3 2LD
GW3 SUH K. Hughes, 2 Graig Terrace, Ferndale, CF43 4EU
G3 SUI J. Burrows, 68 Grosvenor Road, Sale, M33 6NW
G3 SUK M. Baker, 8 Wynton Rise, Stowmarket, IP14 2AB
G3 SUL D. Waller, 66 Wallace Drive, Dunstable, LU6 2DF
G3 SUN G. Hodgkinson, Cfone Communications, 9 Adler Industrial Estate, Betam Road, Hayes, UB3 1ST
G3 SUS S. Jacobs, 16 Mayfield Park, Thorley, Bishop's Stortford, CM23 4JL
G3 SUX D. Bradshaw, 25 Meare Close, Tadworth, KT20 5RZ
GM3 SUZ D. Mclean, Whitecroft Farm, Barrs Brae, Port Glasgow, PA14 5QG
G3 SVC Spen Valley ARS, c/o J. Wilde, 2 Bottoms Lane, Birkenshaw, Bradford, BD11 2NN
G3 SVD A. Hewitt Mbe, Redwood House, Adbury Holt, Newbury, RG20 9BW
G3 SVI D. Davis, 188 Eastwood Old Road, Leigh-on-Sea, SS9 4RY
G3 SVJ Luton VHF Group, c/o A. Barter, 503 Northdown Road, Margate, CT9 3HD
G3 SVK F. Curtis, 32 Elgin Avenue, Harold Wood, Romford, RM3 0YT
G3 SVQ A. Yallop, Whitehill, 16 High Street, Bedford, MK43 7JX
G3 SVR Severn Valley Rd, c/o E. Churchyard, 11 Greenfields Drive, Bridgnorth, WV16 4JW
G3 SVW R. Smith, 16 Coniston Avenue, Sale, M33 3GT
G3 SVZ P. Laxton, 52 Reddington Road, Plymouth, PL3 6PT
G3 SWC B. Tinton, 1 Bridge Road, Rudgwick, Horsham, RH12 3HD
G3 SWH P. Whitchurch, 21 Dickensons Grove, Congresbury, Bristol, BS49 5HQ
GM3 SWK A. Shearer, 12 Coolin Drive, Portree, IV51 9DN
G3 SWU T. Heeley, 34 Worlaby Road, Scartho Top, Grimsby, DN33 3JT
G3 SWW H. Cooper, 9 Fortyfoot, Bridlington, YO16 7SA
G3 SXA J. Croft, 14 Stanstead Road, Forest Hill, London, SE23 1BW
G3 SXC A. Critchley, 39 Westcliffe, Great Harwood, Blackburn, BB6 7PH
G3 SXE L. Lethbridge, 24 Furze Road, High Salvington, Worthing, BN13 3BH
G3 SXH A. Henderson, 50 Sylvan Road, Exeter, EX4 6EY
G3 SXI D. Ashmore, Flat 8, Carter Bench House, Clarence Road, Macclesfield, SK10 5JZ
G3 SXP Capt. J. Redford, Woodgates Harling Road, Gt Hockham, IP24 1NP
G3 SXQ E. Rockett, Devoran, The Causeway, Highbridge, TA9 4QT
G3 SXR A. Read, Readymoney Cove, Fowey, PL23 1JH
G3 SXT Dr P. Sweeny, 1A Market Street, Eckington, S21 4EG
G3 SXV B. Vincent, 18 Rowanhayes Close, Ipswich, IP2 9SX
G3 SXW R. Western, 7 Field Close, Chessington, KT9 2QD
G3 SYA D. Ashworth, 31 Belmont Avenue, Ribbleton, Preston, PR2 6DH
G3 SYB H. Barker, Azul Avion, Main Road, Alford, LN13 0JP
G3 SYC Pontefract + District ARS, c/o N. Ferguson, Royd Moor, Royd Moor Lane, Badsworth, Pontefract, WF9 1AZ
G3 SYD S. Beauchamp, 1 Gosden Close, Furnace Green, Crawley, RH10 6SE
G3 SYM D. Coltart, The Sycamores St. Clether, Launceston, PL15 8PP
G3 SYS Dr D. Emerson, 4269 N.Soldier Trail, Tucson, USA, 85749
G3 SYZ A. Rogers, Draycott, Primrose Hill, Hastings, TN35 4DN
G3 SZ King's Lynn ARC, c/o E. Haskett, 23 Gloucester Road, King's Lynn, PE30 4AB
G3 SZE J. Evrall, 38 Eastlang Road Fillongley, Coventry, CV7 8ER
G3 SZF R. Frost, 24 Mount Pleasant, Hertford Heath, Hertford, SG13 7QU
G3 SZG J. Wright, Flat 4 39A Sion Hill, Kidderminster, DY10 2XT
G3 SZJ M. Shardlow, 19 Portreath Drive, Allestree, Derby, DE22 2BJ
G3 SZM J. Wuille, 45 Keymer Crescent, Goring-By-Sea, Worthing, BN12 4LD
G3 SZR C. Davis, 148 Birkbeck Road, Beckenham, BR3 4SS
G3 SZU K. Radford, 30 Whitendale Drive, Bolton le Sands, LA5 8LY
G3 SZV B. Ward, 138 County Road, Ormskirk, L39 1NN
G3 SZY G. Douglas, 169 High Street, Cheveley, Newmarket, CB8 9DG
G3 TA C. Lambert, Stonecroft Notch Rd Winstone, Cirencester, GL7 7JU
G3 TAA K. Jessop, 15 Courtenay Gardens, Newton Abbot, TQ12 1HS
GI3 TAC D. Campbell, 56 The Quay Killyleagh, Downpatrick, BT30 9GB
G3 TAF D. Cassere, 9 St. Marys Garth, East Keswick, Leeds, LS17 9ER
G3 TAG R. Gouldstone, 11 School Lane, Toft School Lane Toft, Cambridge, CB23 2RE
G3 TAI C. Ward, 50 Lakeside, Bracknell, RG42 2LE
G3 TAJ R. Marchant, Cascade, The Street, Canterbury, CT3 1LN
GM3 TAL M. Hamilton, 3 Charles Court, Limekilns, Dunfermline, KY11 3LG
G3 TAO W. Eaton, 8E St. Aubyns Road, London, SE19 3AD
G3 TAQ N. Bullock, 29 St. Marys Road, Stowmarket, IP14 1LP
G3 TAW C. Wood, 22 Habberley Road, Kidderminster, DY11 6AA
G3 TAX J. Boydell, 13 Lynch Road, Farnham, GU9 8BZ
G3 TAY A. Yarker, 6 Moor Top Road, Halifax, HX2 0NP
G3 TAZ R. Davies, 69 Stopsley Way, Luton, LU2 7UU
G3 TBF H. Wilkins, 17 Bathleaze, Kings Stanley, Stonehouse, GL10 3JN
G3 TBG G. Goulbourn, 41 Rutland Road, Stamford, PE9 1UP
G3 TBJ C. Webster, Wayside, Rosenithon St. Keverne, Helston, TR12 6QR
G3 TBK J. Cree, 24 Old Lincoln Road, Caythorpe, Grantham, NG32 3EJ
G3 TBL R. Ashman, 85A Fakenham Road, Great Ryburgh, Fakenham, NR21 7AQ
G3 TBW T. Westbury, 6 Ellerdene Close, Redditch, B98 7PW
G3 TCG M. Trundle, 20 Denehurst Gardens, Hastings, TN35 4PB
G3 TCI A. Bye, 7 Larkfield Avenue, Gillingham, ME7 2LN
G3 TCL M. Dawson, 11 St. Georges Close, Brampton, Huntingdon, PE28 4US
G3 TCO Dr A. Preece, 12 South Dene, Bristol, BS9 2BW
G3 TCQ Raf South Yorkshire Area-ARC, c/o R. Clayton, 171 Warning Tongue Lane, Cantley, Doncaster, DN4 6TU
G3 TCT G. Kimbell, Eastfield Farmhouse, Fair Place West Lydford, Somerton, TA11 7DN
G3 TCU P. Guttridge, 33 Franklyn Road, Godalming, GU7 2LD
GW3 TCV J. Edwards, Pen Y Maes, Trehelig, Welshpool, SY21 8SG
GM3 TCW J. Kelly, 144A Manse Road, Newmains, Wishaw, ML2 9BL
G3 TCY J. Lewis, 10 Sheringham Drive, Etchinghill, Rugeley, WS15 2YG
G3 TCZ R. Freeman, 3 Hoffmann Gardens, South Croydon, CR2 7GE
G3 TDC J. Yates, 20 Ferncliffe, Stourbridge, DY7 6DX
G3 TDH R. Stevens, 19 Canberra Road, Bramhall, Stockport, SK7 1LG
G3 TDL R. Davis, 105 Eldred Avenue, Brighton, BN1 5EL
G3 TDM R. Mason, 19 Lawrence Road, St Agnes, TR5 0XQ
G3 TDT I. Hollingsbee, 89 Swift Road, Abbeydale, Gloucester, GL4 4XJ
G3 TEB G. Addis, 1 Smallridge, Newbury, RG20 0LH
G3 TEC T. Rutherford, 17 Rosedale Avenue, Sunderland, SR6 8BD
G3 TEE F. Stork, 20 Gay Meadows, Stockton On The Forest, York, YO32 9UJ
G3 TEH A. Storey, 23 Foster Street, Barnsley, S70 3EW
G3 TEI Thorn Emi ARC, c/o J. Gaffney, 77 South Street, Pennington, Lymington, SO41 8DY
G3 TEL P. Mcpherson, 2 Osborne Place, Lower Street, Merriott, TA16 5NP
G3 TEP B. Atkinson, 20 King Street, Seahouses, NE68 7XP
G3 TEU A. Sherer, 35 Beverley Road, Willerby, Hull, HU10 6AW
G3 TEV M. Mills, Shepton, 3 Tylers Way, Stroud, GL6 8ND
G3 TEX P. Painter, 1 Linden House, Barkleys Hill, Bristol, BS16 1FB
G3 TFA G. Whenham, Hogs Hollow, Welsh Road East, Southam, CV47 1NF
G3 TFF G. Fuller, 99 Stanbury, Keighley, BD22 0HA
G3 TFL G. Rogers, 19 Manor Road, Henley-on-Thames, RG9 1LT
G3 TFO J. Auty, 64 Ainley Road, Huddersfield, HD3 3QX
G3 TFR J. Hardstone, 17 Whitefield, Stockport, SK4 2PE
G3 TFV E. Tokley, 14 Maple Way, Earl Shilton, LE9 7HW
G3 TFX R. Fusniak, 35 High Street, Burwell, CB25 0HD
GM3 TFY D. Guest, 31 Newmills Crescent, Balerno, EH14 5SX
G3 TGB B. Ely, 375 Cressing Road, Braintree, CM7 3PE
G3 TGD M. Allenson, 4 The Orchard, Powick, Worcester, WR2 4SE
G3 TGE D. Cahill, Hillcrest, 1 The Greenyard, Northampton, NN7 1EQ
G3 TGF C. Bonner, 57 Downsview, Heathfield, TN21 8PF
GM3 TGG T. Gratton, 23 Culhorn Road, Stranraer, DG9 8DB
G3 TGL A. Fantham, 52 Calverley Road, Kings Norton, Birmingham, B38 8PW
G3 TGN Dr P. Collar, 5 Oak Tree Lane, Tavistock, PL19 9DA
G3 TGO B. Vaughan, 7 Oundle Road, Chesterton, Peterborough, PE7 3UA
G3 THC D. Stimson, 94 Casterton Road, Stamford, PE9 2UB
G3 THF B. Mchugh, 283 Coppice Road, Poynton, Stockport, SK12 1SP
G3 THG W. Cmd. A. Kent, The Coach House, Dipford, Taunton, TA3 7NR
GM3 THI Dr R. Harkess, Friarton Bank Rhynd Road, Perth, PH2 8PT
G3 THQ B. Greenaway, 5 Lansdowne Grove, Neasden, London, NW10 1PL

G3 THS P. Last, 66 Hare Close, Buckingham, MK18 7EW
G3 THT E. Bennett, 20 Cromford Road, Clay Cross, Chesterfield, S45 9RE
G3 THV G. Swindells, 4 Fitzhenry Mews, Norwich, NR5 9BH
G3 THW P. Walters, 22 Windmill Rise, Woodhouse Eaves, Loughborough, LE12 8SG
G3 TIE A. Dutton, 130 Wades Hill, Winchmore Hill, London, N21 1EH
G3 TIG P. Turner, 44 Bestwall Road, Wareham, BH20 4JA
GI3 TIJ F. Eccles, 31 Ballydawley Road, Moneymore, Magherafelt, BT45 7NU
G3 TIK D. French, 37 Warner Road, Ware, SG12 9JN
G3 TIN B. Taylor, Perry House, 188 Walstead Road, Walsall, WS5 4DN
G3 TIR D. Stewart, Apt 347, 8601-902, Luz-Lagos, Algarve, Portugal,
G3 TIX R. Hardy, 47, Maple Mews Cressy Road, Alfreton, DE55 7PL
G3 TJA R. Street, 11 Royal Close, Rugeley, WS15 2DD
G3 TJE P. Smith, 7 Tower Walk, Weston-Super-Mare, BS23 2JR
G3 TJH W. Bickham, 22 Ash Crescent, Galmington, Taunton, TA1 5PW
G3 TJI G. Roff, 47 Penshurst Rise, Frimley, Camberley, GU16 8XX
GI3 TJJ J. Boyce, 19 Dunvale Park, Londonderry, BT48 0AU
GI3 TJM R. Miller, 47A Newtownards Road, Donaghadee, BT21 0PY
G3 TJP D. Lankshear, 28 Monmouth Place, Newcastle-under-Lyme, ST5 3DF
G3 TJS P. Goodenough, Llys Aderyn, 11 Guildford Road, Lightwater, GU18 5RZ
G3 TJU L. Grant, 4 Berry Close, Purdis Farm, Ipswich, IP3 8SP
G3 TJX G. Tillson, 95 Kelverlow Street, Oldham, OL4 1LX
G3 TKA P. Duncan, 18 Pickering Road, Hull, HU4 6TL
G3 TKB J. Foster, 4 Hookergate Lane, Rowlands Gill, NE39 2AD
G3 TKF R. Thompson, Walnut House, Chestnut Walk Saltford, Bristol, BS31 3BG
GW3 TKG D. Locke, 201 Tyn Y Tower, Baglan, Port Talbot, SA12 8YE
GW3 TKH K. Winnard, 208 Heol Hir, Thornhill, Cardiff, CF14 9LA
G3 TKK Dr P. Doughty, Mallows, Ballam Road, Preston, PR4 3PN
G3 TKN V. Lear, 53 Chaplains Avenue, Cowplain, Waterlooville, PO8 8QH
G3 TKS J. Sanderson, 28 Finmere, Hanworth, Bracknell, RG12 7WF
GM3 TLA Dr D. Pearson, 23 Binghill Road West, Milltimber, AB13 0JB
G3 TLD M. Selwyn, 50 Tufthorn Avenue, Coleford, GL16 8PT
G3 TLH I. Brown, 15 Juniper Close, Exeter, EX4 9JT
G3 TLK A. Endacott, Redacres Market Garden St. Marychurch Road Coffinswell, Newton Abbot, TQ12 4SB
GW3 TLP I. Jones, Tyddyn Brith, Star, Gaerwen, LL60 6AL
G3 TLU J. Serlin, 8/34 Ehud Manor, Netanya, Israel, 4265935
G3 TLY S. Alexander, Pinetrees Wilmslow Avenue, Woodbridge, IP12 4HW
G3 TMA I. Buffham, 62/70 Soi Sukhumuit 13, Sukhumvit Rd, Klongtoey Nua, Bangkok, Thailand, 10110
G3 TMB J. Baker, 29 Garstang Road, Southport, PR9 9XW
G3 TMD E. Parsons, 22 Colins Walk, Scotter Dn21 3Sr, Gainsborough, DN21 3SR
GI3 TME R. Hargan, 13 Drumlerry, Londonderry, BT48 8GQ
GW3 TMJ A. Taylor, 24 Emroch St., Goytre, Port Talbot, SA13 2YE
GW3 TMP J. Jones, Haulfryn Stryt-Cae-Rhedyn Leeswood, Mold, CH7 4SS
G3 TMQ R. Harrison, 57 Rue Des Bouviers, Mansle, France, 16230
G3 TMR D. Emmett, 22 Syljon, 114 Villiers Road, Walmer, South Africa, 6070
GW3 TMS D. Smith, 2 Glan Yr Afon Gardens, Sketty, Swansea, SA2 9HY
G3 TMU C. Neale, 63 Rosemary Gardens, Blackwater, Camberley, GU17 0NJ
G3 TMX S. Bennett, 12 Angel Lane, Bury St Edmunds, IP33 1RF
G3 TNE C. Wantling, 28 Moss Green, Welwyn Garden City, AL7 3TE
G3 TNI J. Clingan, 41 Cranham Close, Headless Cross, Redditch, B97 5AY
GI3 TNK S. Dornan, 9 Clonallon Gardens, Belfast, BT4 2BY
G3 TNN N. Sinclair, 11 Primrose Close, Warton, Preston, PR4 1EN
G3 TNQ C. Davis, 963 Manchester Road, Bury, BL9 8DN
GD3 TNS A. Sinclair, 1, Marathon Drive, Douglas, Isle Of Man, IM4 2BP
G3 TNX V. Allison, 24 Colston Gate, Cotgrave, Nottingham, NG12 3JY
G3 TOA B. Otter, Po Box 31191, Lusaka, Zambia, 3000
GW3 TOB A. Coughlin, 37 Parc Y Felin, Creigiau, Cardiff, CF15 9PB
G3 TOJ G. Steel, 10 Rossmere Avenue, Rochdale, OL11 4BT
G3 TON A. Fentham, 106 Elm Road, New Malden, KT3 3HP
G3 TOP A. Peperell11, 47 Glade Road, Marlow, SL7 1DQ
G3 TOQ N. Taylor, Ministro Raul Fernandes, 180 Apt 1805, Bothfogo, Rio de Janeiro, Brazil, 22260-040
G3 TOV G. Miles, 200 Ladybank Road, Mickleover, Derby, DE3 0RR
GW3 TOW A. Hirst, 17 Beech Hollows Lavister, Rossett, Wrexham, LL12 0DA
G3 TOY R. Wright, High View Cottage, Tatenhill Common, Burton-on-Trent, DE13 9RT
G3 TPB Dr J. Knight, 2120 North Pantops Drive, Charlottesville, USA, 22901
G3 TPH R. Henville, 67 Salisbury Road, Blandford Forum, DT11 7LW
G3 TPI T. Wager, 1 Sundown Close, New Mills, High Peak, SK22 3DH
G3 TPJ O. Tillett, 27 Cranbrook Drive, Gidea Park, Romford, RM2 6AP
G3 TPO C. Ockendon, 29 Garlies Road, Forest Hill, London, SE23 2RU
G3 TPP B. Eyre, 56A Ounsdale Road Wombourne, Wolverhampton, WV5 8BH
G3 TPQ G. Harris, 12 Highridge Close, Purton, Swindon, SN5 4BS
G3 TPV F. Robinson, 28 Homer Park, West Common, Soton, SO45 1XP
G3 TPW S. Webb, 1 The Green, Swinton, Malton, YO17 6SY
G3 TQA A. Robinson, 9 Illingworth Close, Illingworth, Halifax, HX2 9JQ
G3 TQC J. Sunderland, 7 Beavers Close, Guildford, GU3 3BX
G3 TQF G. Findon, 3 The Paddock, Newton, Rugby, CV23 0EE
G3 TQL J. Jones, 9 Stonehouse Avenue, Willenhall, WV13 1AP
G3 TQQ J. Bottomley, 32 Ruffa Lane, Pickering, YO18 7HN
G3 TQX G. Grimshaw, 50 Rembrandt Way, Bury St Edmunds, IP33 2LT
G3 TQY M. Knights, Springside Farm, Tismans Common, Horsham, RH12 3DU
G3 TQZ R. Allan, Longfield, Upper Wick Lane, Rushwick, Worcester, WR2 5SU
G3 TRB T. Barber, 48 Newland Road, Droitwich, WR9 7AZ
G3 TRC R. Collins, 8 Sylvan Way, Redhill, RH1 4DE
G3 TRD J. Bellamy, 90 Colneis Road, Felixstowe, IP11 9LG
G3 TRG R. Green, 2 Ragley Walk, Rowley Regis, B65 9NT
G3 TRH R. Farrance, 63 Salisbury Close, Rayleigh, SS6 9UH
G3 TRK D. Kitson, 11 Deerstone Road, Nelson, BB9 9LN
G3 TRL A. Green, Rembrandt Stud, Clotton Common, Tarporley, CW6 0HJ
G3 TRR A. Mills, 207 Sutherland Drive, Wirral, CH62 8EQ
G3 TRX C. Bailey, 15 Seymour Avenue, Margate, CT9 5HT
G3 TRY Mid-Thames Radio Direction-Finding Club, c/o W. Pechey, Jays Lodge, Crays Pond, Reading, RG8 7QG
G3 TSA J. Denby, 107 Station Road, Fenay Bridge, Huddersfield, HD8 0DE
G3 TSC Trinity School Radio Club, c/o R. Evans, 7 Westland Drive, Bromley, BR2 7HE
G3 TSE D. Brealy, 1 Lydford Close, Ivybridge, PL21 0YW
G3 TSF E. Glasscott, 26 Columbus Circle, Bluffton, USA, 29909
G3 TSM V. Mallows, 13 Greatfield Way, Rowland's Castle, PO9 6AG
G3 TSO M. Grierson, 1 Blenheim Close, Upper Rissington, Cheltenham, GL54 2QX
G3 TSR Col. P. Reader, 42 Chiltley Way, Liphook, GU30 7HG
G3 TSS C. Waters, 1 Chantry Estate, Corbridge, NE45 5JH
G3 TSV T. Clay, 132 Underdale Road, Shrewsbury, SY2 5EF
G3 TSZ A. Macwalter, 142 Altrincham Road, Wilmslow, SK9 5NQ
G3 TTB P. Clegg, 6 Ricketts Drive, Billericay, CM12 0HH
G3 TTC K. Orchard, 32 Myton Crescent, Warwick, CV34 6QA
G3 TTG V. Batchelor, 31Bsupakarn Condo, 1057 Charoen Nakorn Road, Bangkok, Thailand, 10600
G3 TTI L. Meikle, 3 Hillcrest, West Woodburn, Hexham, NE48 2RZ
G3 TTJ J. Barber, 33 Midford Lane, Limpley Stoke, Bath, BA2 7GR
G3 TTP B. Horsey, Nethercotts, Gurney Street, Bridgwater, TA5 2HW
G3 TTU R. Holt, 50 Alverley Lane, Doncaster, DN4 9AR
G3 TTY B. Field, Greenleaves, 2 Duke Street, Southport, PR8 1RS
G3 TUF F. Long, 37 St. Catherines Road, Bitterne, Southampton, SO18 1LS
G3 TUL J. Copson, 16 Fothergill Way, Wem, Shrewsbury, SY4 5NX
G3 TUU C. Keeble, 86 Kirby Road, Walton on the Naze, CO14 8RL
G3 TUW P. Moore, 54 Herbert Ave, Palmerston North, New Zealand, 4412
GU3 TUX C. Rees, 2 Rue De La Saline, Alderney, Guernsey, GY9 3XD
G3 TUY M. Bruce, 3 Redlands Place, Wokingham, RG41 4ED
G3 TVC L. Rice, Beechwood, 11 Barnoldby Road, Grimsby, DN37 0JR
G3 TVD J. Shersby, 29 Vale Square, Ramsgate, CT11 9DE
G3 TVH J. Harknett, 60 Windmill Drive, Croxley Green, Rickmansworth, WD3 3FE
G3 TVI R. Stevens, 64 Ferndale, Waterlooville, PO7 7PB
G3 TVL P. Hunt, 14 Walnut Close, Epsom, KT18 5JL
G3 TVM H. Fletcher, 20 Westfield Road, Great Shelford, Cambridge, CB22 5JW
G3 TVN R. Williams, 23A Acacia Avenue, Liverpool, L36 5TN
G3 TVR E. Churchyard, 11 Greenfields Drive, Bridgnorth, WV16 4JW
G3 TVT I. Fraser, 18 Savick Avenue, Bolton, BL2 6JJ
G3 TVU I. Brown, 63 Peak View Drive, Ashbourne, DE6 1BR
G3 TVV A. Coates, 35 Mogg St., St. Werburghs, Bristol, BS2 9UB
G3 TVX D. Ashwood, Apartment 43, Dane Court 21 Mill Green, Congleton, CW12 1FS
G3 TVY J. Sutton, 3 Sunrise Avenue, Nottingham, NG5 1NH
G3 TWB R. Ballard, 31 South Devon Avenue, Nottingham, NG3 6FT
G3 TWJ M. Roach, 35 Hartley Old Road, Purley, CR8 4HH
GW3 TWN F. Mason, Awel Mon, Bodffordd, Llangefni, LL77 7LJ
G3 TWX D. Woodhouse, 38 Jenny Road, Spixworth, Norwich, NR10 3QW
G3 TWY G. Mills, 11 Milton Street, Narborough, LE19 3EZ
G3 TXC N. Harris, April Cottage, Sheepcote Green, Saffron Walden, CB11 4SJ
G3 TXE A. Parker, 7 St. Peters Court, Claydon, Ipswich, IP6 0HZ
G3 TXF N. Cawthorne, Dormers, The Green, Hinton Charterhouse, Bath, BA2 7TJ
G3 TXH B. Levett, 18 Forge Road, Little Sutton, Ellesmere Port, CH66 3SQ
G3 TXK C. Moss, 2 Sutton Lane Adlington, Chorley, PR6 9PA
G3 TXL A. Graham, Woodtown, Sampford Spiney, Yelverton, PL20 6LJ
G3 TXZ C. Tucker, Flat 35, Martlets Court, Crowborough, TN6 1JF
G3 TYA J. Grant, Tanyanga, Wheal Leisure, Perranporth, TR6 0EY
G3 TYG B. Winslow, 10 Almond Walk, Hazlemere, High Wycombe, HP15 7RE
GW3 TYI D. West, 44 Glanmor Park Road, Sketty, Swansea, SA2 0QE
G3 TYO J. Stringer, 1 Hazel Road, Tavistock, PL19 9DN
G3 TYP I. Jackson, 20 Daventry Road, Barby, Rugby, CV23 8TR
G3 TZA J. Riley, 38 West Broadway, Bristol, BS9 4TB
GI3 TZB W. Mckinney, 33 Heatherstone Road, Bangor, BT19 6AE
G3 TZD R. Mansell, 354 Allen Road, Salt Point, USA, 12578
G3 TZE R. Armitage, 3 Holst Mead, Stowmarket, IP14 1TD
G3 TZG J. Glanville, 3 Seneschal Road, Cheylesmore, Coventry, CV3 5LF
G3 TZL P. Bowen, White House, Durleigh Marsh, Petersfield, GU31 5AX
G3 TZM W. Mahoney, 61 Starbold Crescent, Knowle, Solihull, B93 9LA
G3 TZO P. Holland, Chatterton, Chapel Lane, Threapwood, Malpas, SY14 7AX
G3 TZQ S. Ridgway, 12 The Mead, Plymouth, PL7 4HS
G3 TZT M. Mead, 2 Market Court, Market Place, Wincanton, BA9 9PB
G3 TZU J. Harding, 5 Salisbury Road, Whitchurch, SY13 1RQ
GI3 TZX W. Nesbitt, 101 Belfast Road, Bangor, BT20 3PP
GM3 UA A. Pairman, Seabank, Largiebeg, Brodick, KA27 8RL
G3 UAA D. Ramsey, The Orchard, Carmen Grove, Leicester, LE6 0BA
G3 UAE J. Gill, 22 Maddever Crescent, Liskeard, PL14 3PT
G3 UAF Dr M. Smith, 138 Market St., Clay Cross, Chesterfield, S45 9LY
GM3 UAG J. Davidson, Cairntoul, Ellon, AB41 8QS
G3 UAP P. Parker, Avenue Kersbeek 116, 1190 Brussels, Belgium,
G3 UAS T. Morgan, 2 Park View, Hatch End, Pinner, HA5 4LN
G3 UAX R. Stansfield, 22 Reeds Avenue, Earley, Reading, RG6 5SR
GW3 UAY C. Butters, 39 Parc Y Ffynnon, Ferryside, SA17 5TQ
GI3 UBA R. Reid, 21 Ballymaconnell Road, Bangor, BT20 5PN
G3 UBB D. Fill, 2 Brook Close, Packington, Ashby-de-la-Zouch, LE65 1WA
G3 UBD G. Higgins, Lower Laithe Farm, Providence Lane, Keighley, BD22 7QS
GW3 UBH J. Pugh, 5 Pen Y Maes, Llanfechain, SY22 6XL
G3 UBI M. Fisher, Bank Top Farm, Cropton, Pickering, YO18 8HH
GM3 UBJ Dr W. Hossack, Kincrig, 39 Skene Street, Macduff, AB44 1RP
G3 UBL C. Ledger, Kinrara, Sandhills Road, Salcombe, TQ8 8JP
G3 UBP C. Riches, 28 Saxondale Avenue, Burnham-on-Sea, TA8 2PS
G3 UBS B. Speakman, Merrydown, Burley Lane, Derby, DE6 4JS
G3 UBV D. Roberts, 8 Churnet Close, Bedford, MK41 7ST
G3 UBX P. Burden, 68 Coalway Road, Wolverhampton, WV3 7LZ
G3 UBY A. Clark, Sans Souci, Fairmead Road, Saltash, PL12 4JH
G3 UCA P. Sinclair, 32 Barn Meadow, Bamber Bridge, Preston, PR5 8DU
G3 UCD R. Pescod, 7 Brian Close, Chelmsford, CM2 9DZ
G3 UCF A. Passmore, 16 Chaffinch Close, Basingstoke, RG22 5QD
GM3 UCH W. Wright, 460 Main Street, Stenhousemuir, Larbert, FK5 3JU
GM3 UCI G. Mccallum, 15 Quarry Road, Law, Carluke, ML8 5HB
GW3 UCJ M. Evans, 31 Cilmaengwyn Road, Pontardawe, Swansea, SA8 4QL
G3 UCK G. Downs, 2 Dyehouse, Wilsden, Bradford, BD15 0BE
G3 UCL University College London ARS, c/o Dr G. Smart, 30 Cornmills Road, Soham, CB7 5AT
GM3 UCN F. Hetherington, 4 Rosebery Place, Livingston, EH54 6RP
G3 UCQ J. Farrar, 2 Marsh Lane, Hayle, TR27 4PS
G3 UCT M. Taylor, Orchard House, Leigh, Sherborne, DT9 6HL
G3 UCW M. Pettit, 3C Clive Court, Grand Parade, Eastbourne, BN21 3DD

G3 UD G. Bloor, 26 Leveson Road, Hanford, Stoke-on-Trent, ST4 4QP
G3 UDA K. Linney, Sunnybank, Oak Lane, Bicton Heath, Shrewsbury, SY3 5BW
G3 UDD Dr S. Chandler, Malt House, Box, Stroud, GL6 9HF
G3 UDH P. Butcher, 55 Offington Lane, Worthing, BN14 9RJ
G3 UDI Dr R. Butcher, Temple Lodge, Six Mile Bottom Road, Cambridge, CB21 5LD
GM3 UDK Dr C. Oliver, 40 Charles Way, Limekilns, Dunfermline, KY11 3LH
G3 UDN Mid Warwickshire ARS, c/o Q. Wright, 9 Browning Avenue, Warwick, CV34 6JQ
G3 UDP M. Brown, 4 Boyfields, Quadring, Spalding, PE11 4QQ
G3 UDV P. Lindsley, Oak Lodge, Cromer Road, Cromer, NR27 9QT
G3 UED J. Jones, 12 Francis Groves Close, Bedford, MK41 7DH
G3 UEE D. Diamond, 36 Darbys Lane, Oakdale, Poole, BH15 3ET
G3 UEG D. Gould, 2 Mayfield Close, Harlow, CM17 0LH
G3 UEK J. Whitehouse, P.O. Box 364, Guntersville, Alabama, USA, 35976
GW3 UEP R. Plimmer, Fron Haul, Tregroes, Llandysul Sa44 4Ne, Llandysul, SA44 4NE
G3 UEQ A. Hearn, 53 Twyford Gardens, Salvington, Worthing, BN13 2NT
G3 UES Echelford ARS, c/o S. Roy, 28 Kingston Rise, New Haw, Addlestone, KT15 3EY
G3 UEU J. Holmes, 23 School Lane Berry Brow Hd4 7Ra, Huddersfield, HD4 7RA
GI3 UEX R. Thompson, 94 Orangefield Crescent, Belfast, BT6 9GJ
G3 UEY D. Browning11, 13 Beechcombe Close, Pershore, WR10 1PW
G3 UEZ R. Gilbert, 18 Peckham Avenue, New Milton, BH25 6SL
G3 UFB N. Brinkworth, 11 Haycroft Road, Stevenage, SG1 3JL
G3 UFF P. Rodway, 37 Neville Avenue, Portchester, Fareham, PO16 9NR
G3 UFI P. Conway, 1 The Woodlands, Hastings, TN34 2SF
G3 UFJ L. Symons, 31 Springfield Way, Threemilestone, Truro, TR3 6BJ
G3 UFQ D. Eckley, 27 Apsley Grove, Dorridge, Solihull, B93 8QP
G3 UFS C. Smith, 50 Grand Avenue, Lancing, BN15 9PZ
G3 UFV P. Crawshaw, 35 Bishopton Avenue, Stockton-on-Tees, TS19 0RA
G3 UFX H. Julian, Brigantine, Lower Market Street, Penryn, TR10 8BH
G3 UFY S. Knowles, 77 Bensham Manor Road, Thornton Heath, CR7 7AF
G3 UGC J. Smethurst, 81 Springside Road, Bury, BL9 5JG
G3 UGF R. Constantine, 18 Hillbeck, Halifax, HX3 5LU
G3 UGJ D. Smith, 33 Rippington Drive, Marston, Oxford, OX3 0RJ
G3 UGX R. Heaton, Flat 5, 73 Belsize Park Gardens, London, NW3 4JP
G3 UHF Sth Manchester, c/o C. Ward, 2 Arlington Drive, Stockport, SK2 7EB
G3 UHJ R. Gordon, 77 Alwyn Road, Darlington, DL3 0AH
G3 UHK J. Baldwin, 19 Lutyens Close, Stapleton, Bristol, BS16 1WL
G3 UHN P. Neale, 98 Meadway, Harpenden, AL5 1JQ
G3 UHS C. Houltby, 9 Bayard St., Gainsborough, DN21 2JZ
GM3 UHT W. Garner, Sarkshields Cottage, Eaglesfield, Lockerbie, DG11 3AE
G3 UHU D. Hampton, 9 Portwey Close, Weymouth, DT4 8RF
G3 UHV C. Sutton, Braehead, Old Lane, Stoke-on-Trent, ST6 8TG
G3 UHW H. Tomlinson, 32 Manor Road, Farnborough, GU14 7EU
G3 UHX A. Thorpe, 12 Newnham Lane, Ryde, PO33 4ED
G3 UI L. Cobb, 27 Moorlands Crescent, Halifax, HX2 8AA
G3 UIB C. Hearn, 8 The Poles, Upchurch, Sittingbourne, ME9 7EX
G3 UID K. Baldock, 284 Rocky Mountain High, Camano Island, USA, WA 98282
G3 UIF G. Thorne, Flagstaff House, Main Street, Hull, HU12 0RY
GI3 UIH W. Aylward, 37 Stewartstown Avenue, Belfast, BT11 9GF
G3 UIK J. Young, Shirley Lodge, 45 Graham Road, Malvern, WR14 2HU
GM3 UIN M. Mcleman, 6 Stenton Road, West Barns, Dunbar, EH42 1UG
G3 UIS A. Stone, West Lodge, The Downs, Poulton-le-Fylde, FY6 7EG
G3 UIT B. Seedle, 54 Normoss Road, Blackpool, FY3 0AL
G3 UJA B. Mcclory, 12 The Crescent, Mottram St. Andrew, Macclesfield, SK10 4QW
G3 UJB B. Davis, 2 Rawden Close, Harwich, CO12 4BW
G3 UJE B. Gale, Tall Trees Farm, Noah'S Ark Lane, Great Warford, WA16 7AX
G3 UJI S. Turner, 51 Hilton Road, Stoke-on-Trent, ST4 6QZ
G3 UJM R. Banks, 3 Parkhayes, Woodbury Salterton, Exeter, EX5 1QS
G3 UJO P. Bradley, 60 Weyland Road, Headington, Oxford, OX3 8PD
G3 UJV R. Heath, 26 Lancaster Avenue, Hadley Wood, Barnet, EN4 0EX
G3 UJZ J. Mcnaught, Ryton House, Glos, GL7 3AR
G3 UK J. Whittaker, Riverside, 48 Baunton, Cirencester, GL7 7BB
G3 UKB R. Cowdery, 80 Caxton End, Eltisley, St Neots, PE19 6TJ
G3 UKC University of Kent, c/o F. Barnes, 4 Pound Close, Ducklington, Witney, OX29 7TH
G3 UKD A. Golding, 40 Unicorn Lane, Eastern Green, Coventry, CV5 7LJ
G3 UKE P. Adams, 34 Mount Pleasant Close, Lightwater, GU18 5TP
GM3 UKG G. Grant, 35 Inward Road, Buckie, AB56 1DD
G3 UKH P. Hopwood, 58 Bolbec Road, Newcastle upon Tyne, NE4 9EP
G3 UKI B. Curnow, Flat 3A, Olivers Wharf, 64 Wapping High Street, London, E1W 2PJ
G3 UKL M. Bennett, Shireley, Munns Lane, Sittingbourne, ME9 7SY
G3 UKM M. Leighton, 85 Kemps Green Road, Balsall Common, Coventry, CV7 7QF
G3 UKV M. Vincent, 9 Sleapford, Long Lane, Telford, TF6 6HQ
G3 UKW M. Newton, 11 Chestnut Close, Rushmere St. Andrew, Ipswich, IP5 1ED
G3 ULD G. Cawkwell, 50 Station Road, Patrington, Hull, HU12 0NE
G3 ULN M. Hibbitt, 123 Stanborough Road, Plymstock, Plymouth, PL9 8PJ
G3 ULO I. Spencer, Fichtenweg 10C, Much, Germany, 53804
G3 ULT Reading & District ARC, c/o J. Carter, 22 Orchard Coombe, Whitchurch Hill, Reading, RG8 7QL
GW3 UMD N. Maxwell, 1 Nant Fawr Crescent, Cardiff, CF2 6JN
G3 UMF Dr A. Simpson, Forest Farm, Old Road, Shotover Hill, Oxford, OX3 8TA
G3 UML L. Margolis, 52 Park View Gardens, Hendon, London, NW4 2PN
G3 UMM P. Hudson, 105 Southlands, Weston, Bath, BA1 4DZ
G3 UMT B. Turvey, 90 Jenkinson Road, Towcester, NN12 6AW
G3 UMV P. Johnson, 52 Evesham Road, Cookhill, Alcester, B49 5LJ
GU3 UMX D. Ozanne, Eturs Lodge, Les Eturs, Castel, Guernsey, GY5 7DT
G3 UNA D. Cutter, David Cutter Engineering, 34 Greengate Lane, Knaresborough, HG5 9EL
G3 UNI T. Wood, 4 Musgrave Road, Chinnor, OX39 4PL
G3 UNK C. Hebden, 1 Ringwood Avenue, Newbold, Chesterfield, S41 8RA
G3 UNM A. Matthews, Winsford, The Common, Stoke-on-Trent, ST10 2PA
G3 UNS T. Mills, 22 The Dingle, Crawley, RH11 7JE
G3 UOA University of Aston Radio Society, c/o Dr P. Best, 21 Greening Drive, Edgbaston, Birmingham, B15 2XA
G3 UOC D. Brown, Rexfield, Alcester Road, Henley-in-Arden, B95 6BH
G3 UOD M. Spencer, Cleeve House, Melton Road, Melton Mowbray, LE14 3QG
G3 UOI J. Firby, 19 Cliffe Avenue, Harden, Bingley, BD16 1LN
G3 UOJ J. Hartwell, Fulling Mill Oast, Caring Lane, Maidstone, ME17 1TJ
G3 UOM D. Horsburgh, 11 Delamare Way, Oxford, OX2 9HZ
G3 UON D. Geere, Tinos Premier Marinas Ltd, Western Concourse, Brighton, BN2 5UP
GW3 UOO D. Rogers, Green Tops, 69 Megs Lane, Buckley, CH7 2AG
GU3 UOQ P. Le Boutillier, Room 10 Maison L'Aumone, Castel, Guernsey, GY5 7RT
G3 UOS Dr A. Whitaker, Univer Of Sheffield, Dept Of Elec Eng, Sheffield, S1 3JD
G3 UPA M. Foden, 10 Maud Road, Water Orton, Birmingham, B46 1PD
G3 UPD H. Del Monte, 5 Scotts Close, Colden Common, Winchester, SO21 1US
GI3 UPG R. Mckimm, 227 Millisle Road, Donaghadee, BT21 0LN
G3 UPI T. Codling, 21 Willow Close, Saxilby, Lincoln, LN1 2QL
G3 UPJ D. Trainer, 153 High Street, Cherry Hinton, Cambridge, CB1 9LN
G3 UPM T. Burke, 12 Worthing Road, Laindon, Basildon, SS15 6AL
G3 UPN K. Snape, Delamere, Ryston End, Norfolk, PE38 9AX
G3 UPS R. Keyte, 18 Mclean Drive, Kessingland, Lowestoft, NR33 7TY
G3 UPW P. Smith, 9 Ash Road, Shepperton, TW17 0DN
G3 UPY D. Houghton, 119 Welsby Road, Leyland, PR25 1JD
G3 UPZ H. James, 18 Clydesdale Road, Whiteley, Whiteley, PO15 7BD
G3 UQD R. Whittington, 65 King Edward Avenue, Worthing, BN14 8DG
G3 UQL M. Baker, 10 Catchpole Close, Greenleys, Milton Keynes, MK12 6LR
G3 UQR D. Robinson, 3 Marriott Close, Irthlingborough, Wellingborough, NN9 5RB
G3 UQW A. Ball, 3 Orchard Lea, Sherfield-On-Loddon, Sherfield on Loddon, RG27 0ES
G3 URA R. Whittering, Vrouhas, Crete, GR - 72053
G3 URE J. Thexton, 78 Greenfield Road, Newcastle upon Tyne, NE3 5TQ
G3 URI Newbury Vintage Wireless Society, c/o Dr M. Franks, 13 Fifth Road, Newbury, RG14 6DN
G3 URJ A. Moss, 17 Surrey Drive, Finchfield, Wolverhampton, WV3 9LW
G3 URK I. Campbell, 27 Lewis Close, Adlington, Chorley, PR7 4JU
G3 URL C. Adams, 25 Avon Road, Cannock, WS11 1LJ
G3 URN Dr M. Jolley, 34469 N Circle Drive, Round Lake, USA, 60073
G3 URQ J. Letts, Bridgeways, Snows Lane, Leicester, LE7 9JS
G3 URU Dr R. Edworthy, 44 Middleton Avenue, Littleover, Derby, DE23 6DL
G3 URV F. Stevens, 38 Endhill Road, Birmingham, B44 9RR
G3 URX J. Speake, 211 Milton Road, Cambridge, CB4 1XG
G3 URZ Dr B. Ewen-Smith, 1 Kinnersley, Severn Stoke, Worcester, WR8 9JR
G3 USA C. Taylor, 39 School Road, Great Alne, Alcester, B49 6HQ
G3 USC M. Hall, Redthorn Bungalow, Upton, Langport, TA10 9NL
G3 USD D. Mason, 2A Devon Road, Bedford, MK40 3DF
G3 USE S. Down, 1 Dove Close, Honiton, EX14 2GP
GI3 USK H. Kernaghan, 1 Elizabeth Road, Holywood, BT18 0PL
GM3 USL Cunninghame & District Amt Radio Club, c/o J. Walker, Mo Bhothan, Lochlibo Road, Uplawmoor, Glasgow, G78 4AA
G3 USO C. Walker, St. Jude, Stoney Lane, Wilmslow, SK9 6LG
G3 USR G. Rolland, 3B Reeves Lane, Wing, Oakham, LE15 8SD
G3 UST J. Turner, Flat 4, Barnett Janner House, Leicester, LE4 0UR
G3 USW W. Clough, 32 Jackson Crescent, Rawmarsh, Rotherham, S62 7EN
G3 USX M. Robertson, 1 Lindvale Horsell Rise, Woking, GU21 4BG
G3 UTA K. Smyth, 154 Scrub Lane, Benfleet, SS7 2JP
G3 UTC G. Farr, 26 Burstead Drive, Billericay, CM11 2QN
G3 UTE N. Wright-Williams, Trinco, 9 Orpine Close, Bicester, OX26 3ZJ
GW3 UTG A. Antley, 12 Fairfield Avenue, Rhyl, LL18 3EE
GW3 UTL R. Barker, 51 Rockfield Drive, Llandudno, LL30 1PF
G3 UTS T. Belshaw, 20 Greencroft Road, Delves Lane Industrial Estate, Consett, DH8 7DY
G3 UUB N. Bateman, 10 Telford Crescent, Woodley, Reading, RG5 4QT
G3 UUC J. Nurse, 25 Dobson Road, Crawley, RH11 7UH
G3 UUF J. Hansom, 12 Torquay Avenue, Hartlepool, TS25 3DP
G3 UUG E. Nightingale, 61 The Cockpit, Marden, Tonbridge, TN12 9TQ
G3 UUI M. Mapson, 253 Central Avenue, Southend-on-Sea, SS2 4ED
G3 UUL T. Jones, 32 Oakwood Drive Hucclecote, Gloucester, GL3 3JF
G3 UUM A. Page, 22 Tower Road, Feniscowles, Blackburn, BB2 5LE
G3 UUQ A. Clelland, Rieschbogen 7, Hohenkirchen, Germany, 85635
G3 UUR Dr D. Gordon-Smith, The Chalet, Bell Road, Rockland St Peter, Attleborough, NR17 1UL
G3 UUT J. Wilson, 20B High Green, Great Shelford, Cambridge, CB22 5EG
G3 UUU L. Newman, Eastholme, Mill Street, Newton Abbot, TQ13 8AR
G3 UUV R. Frost, Chez Nous, Coldharbour Yard, Swindon Road, Kington Langley, Chippenham, SN15 5LY
G3 UUY D. Wright, St. Julians, 55 Old Road, Harlow, CM17 0HD
GW3 UVA D. Knowles, The Clappers, Spon Green, Buckley, CH7 3BL
G3 UVB D. Barnes, 27 Royal Court, Worksop, S80 2DL
G3 UVM M. Simpson, 36 Rectory Close, Newbury, RG14 6DD
G3 UVQ N. Mercer, 19 Sycamore Road, Brookhouse, Lancaster, LA2 9PB
G3 UVR D. Jones, 39 Pensby Road Heswall, Wirral, CH60 7RA
G3 UVU J. Curry, Clonlea, New Ridley, Stocksfield, NE43 7RQ
G3 UVW Coventry Tech ARC, c/o R. Harris, Clevelands, Tamworth Road, Coventry, CV7 8JJ
G3 UVY L. Parkin, 8 Smithfield Close, Ripon, HG4 2PG
G3 UWE R. Simpson, 30 Heath Lawns, Fareham, PO15 5QB
G3 UWH Cmdr. J. Endicott, The Mill House, Halse, Taunton, TA4 3AQ
GW3 UWL Sir S. Grant, The Court, 19 Marine Parade, Penarth, CF64 3BE
G3 UWM P. Marchant, 12 Laurel Way, Ickleford, Hitchin, SG5 3UP
G3 UWP R. Pickering, 41 Maiden Greve, Malton, YO17 7BE
G3 UWR C. Bonsall, Parkside, Lodge Road, Doncaster, DN6 8EB
GW3 UWS Uws Radio Society, c/o Dr T. Davies, 44 Carnglas Road, Sketty, Swansea, SA2 9BW
G3 UWT P. Myers, 22 High Street Barnby Dun, Doncaster, DN3 1DS
GM3 UWX J. Stirling, 25 Maxwell Road, Bishopton, PA7 5HE
G3 UWZ M. Newman, 26 Highbank, Westdene, Brighton, BN1 5GB
G3 UXH P. Carey, 44 Monteney Gardens, Sheffield, S5 9DY
G3 UXM J. Greaves, 23 Woodhouse Road, Intake, Sheffield, S12 2AY
G3 UXO Dr A. Eardley, Flat E, 113 Sutherland Avenue, London, W9 2QH
G3 UXR N. Goddard, 1 Aston Mead, St. Catherine'S Hill, Christchurch, BH23 2SP
G3 UXY A. Baker, 1 Napier Road, Maidenhead, SL6 5AR
G3 UYB M. Shaw, Beech Farm Cottage, Hawkhurst Road, Battle, TN33 0QS
G3 UYC J. Peirson, Ashfield Farm, Ulting, Maldon, CM9 6QP

G3 UYD E. Clarke, 65 Oakmount Road, Chandler'S Ford, Eastleigh, SO53 2LJ
G3 UYE M. Richer, 1 Station Road, Surfleet, Spalding, PE11 4DA
G3 UYG J. Clegg, 11 South Park Road, Gatley, Cheadle, SK8 4AL
G3 UYK P. Kemble, 74 Teg Down Meads, Winchester, SO22 5ND
G3 UYL D. Knott, 22 Linden Close, Prestbury, Cheltenham, GL52 3DU
G3 UYN C. Malcolm, Glen Mor, Trenance, St Keverne, Helston, TR12 6QL
GM3 UYR Dr P. Gamble, 21 St. Marys Drive, Perth, PH2 7BY
G3 UYX J. Ball, 8 Withybed Close, Alvechurch, B48 7PL
G3 UYY E. Bradley, 3 Windrush, Wargrave Road, Henley-on-Thames, RG9 2LX
G3 UZB J. Shewan, 42 Stirling Road, Redcar, TS10 2JZ
GI3 UZJ D. Singleton, 38A Cloughey Road, Portaferry, Newtownards, BT22 1NQ
G3 UZK D. Bloomfield, 22 Laurel Road Locks Heath, Southampton, SO31 6QG
G3 UZM C. Haddock, 26 Featherbed Lane, Exmouth, EX8 3NE
GW3 UZS J. Diplock, Cartref, 98 Pendwyallt Road, Cardiff, CF14 7EH
G3 UZW R. Andrews, 10 Hilltop Rise, Bookham, Leatherhead, KT23 4DB
G3 UZX F. Mitchell, 158 Cobham Road, Fetcham, Leatherhead, KT22 9JR
GI3 VAF R. Best, 6 Knightsbridge Court, Bangor, BT19 6SD
GM3 VAJ I. Gray, 1 Greenside Avenue, Berwick-upon-Tweed, TD15 1BZ
G3 VAK M. Sutcliffe, 26 Weald Road, Burgess Hill, RH15 9SP
GM3 VAL G. Talbot, Whitethorn House, Milnathort, KY13 9XU
G3 VAO M. Farmer, Horton Brook Cottage, Horton Wem, Shrewsbury, SY4 5NB
GM3 VAP C. Weston, 18 Kirkbrae Mews, Cults, Aberdeen, AB15 9QF
GI3 VAW R. Sherrard, 39 Shanreagh Park, Limavady, BT49 0SF
G3 VBA K. Hatton, 38 Doric Avenue, Frodsham, WA6 6QQ
G3 VBE F. Miles, 65 Montgomery Street, Hove, BN3 5BE
G3 VBG B. Morris, 88 Newcastle Road, Leek, ST13 7AA
G3 VBL C. Pedder, Thorncliffe, 5 Royalty Lane, Preston, PR4 4JD
G3 VBQ D. Wright, 5 Padin Close, Chalford, Stroud, GL6 8FB
GM3 VBT T. Logan, 137 Buccleuch St., Garnethill, Glasgow, G3 6QN
G3 VBU J. Lynch, 11 Rosenthorpe Road, London, SE15 3EG
G3 VBV S. Boyce, 58 Woodbury Road, Halesowen, B62 9AW
GM3 VBY F. Hindley, The White House, 17 Main Road, Elgin, IV30 8UR
G3 VC M. Bridge, 5 Garrett'S Place, Donington, Spalding, PE11 4YL
G3 VCA R. Pickles, Bramcote Lorne, Rectory Lane Gamston, Retford, DN22 0QQ
G3 VCG D. Wilks, 36 Greenways, Chelmsford, CM1 4EF
GI3 VCI M. Mcfadden, 121 Greystown Avenue, Belfast, BT9 6UH
G3 VCK J. Fenwick, 78 Loveridge Road, London, NW6 2DT
G3 VCL B. Clark, 60 Somerset Avenue, Harefield, Southampton, SO18 5FS
G3 VCM I. Anderson-Mochrie, 10214 Hunt Club Lane, Palm Beach Gardens, USA, 33418
G3 VCN P. Kalas, 110A Underlane, Plympton, Plymouth, PL7 1QZ
G3 VCP N. Kail, 1 Siemons Street, One Mile, Queensland, Australia, 4503
G3 VCQ C. Wilson, 2 Frankham Close, Dinnington, Sheffield, S25 3QG
G3 VCR C. Rooney, 129 Drift Road Clanfield, Waterlooville, PO8 0PD
G3 VCT R. Hemmings, Wood View, Cryers Hill Road, Cryers Hill, High Wycombe, HP15 6JR
G3 VCV D. Prout, 7 Chemin Des Estimeurs Nord, Plan De La Dame, Valreas, France, 84600
G3 VCX D. Bridgen, 22 Maple Grove, Immingham, DN40 2JH
G3 VCY Dr C. Clayton, Wildfield, West Flexford Lane, Guildford, GU3 2JW
G3 VDB J. Evans, 7 Barncroft Close, Chelford, Macclesfield, SK11 9SW
G3 VDE J. Sellers, Blacktoft Grange, Blacktoft Grange Road, Sandholme, Brough, HU15 2ZU
G3 VDF H. Gregory, 44 Mowlands Close, Sutton-in-Ashfield, NG17 5GH
G3 VDH R. Godwin, Hopworthy Moor Cottage, Pyworthy, Holsworthy, EX22 6XX
G3 VDK S. Bailey, 6 Minnie Street, Keighley, BD21 1HY
G3 VDL J. St Leger, Warmbrook, Throwleigh, Okehampton, EX20 2JF
G3 VDO I. Hacking, 1 Pine Crescent, Poulton-le-Fylde, FY6 8EB
G3 VDS R. Higham, 34 Ashford Road, Wirral, CH47 5AW
G3 VDU P. Bennett, 56 Winchester Avenue, Weddington, Nuneaton, CV10 0DW
G3 VDV N. Brinnen, 134 Victoria Road, Mablethorpe, LN12 2AJ
G3 VDZ T. Richardson, Flat 11, The Moorings, 21 Albert Way, East Cowes, PO32 6GA
G3 VEB R. Bridson, 14 Zig Zag Road, Wallasey, CH45 7NZ
G3 VEF Fareham and District ARC, c/o C. Jenkins-Powell, 43 Cambridge Road, Lee-on-the-Solent, PO13 9DH
G3 VEH C. Morcom, 15 Markson Road, South Wonston, Winchester, SO21 3EZ
GM3 VEI I. Sheffield, 37 Bellevue Court, Queens Road, Dunbar, EH42 1YR
G3 VEK S. Holden, Garden Flat 1 Emmanuel Road, Hastings, TN34 3LB
G3 VER Verulam ARC, c/o P. King, 96 Mancroft Road, Caddington, Luton, LU1 4EN
G3 VES H. Martin, 1 Houghton Park Cottages, Ampthill, MK45 2EY
G3 VET M. Langwade, 19 South Wootton Lane, King's Lynn, PE30 3BS
G3 VEV R. Butterfield, 7 Kipling Close, Lincoln, LN2 4EW
GM3 VEY F. Baxter, 8 Northcote Park, Aberdeen, AB15 7SX
G3 VFB A. Matthews, 45 Kings Square, Taunton, TA1 3FN
G3 VFC T. Chipperfield, 5 Lullingstone Close, Hempstead, Gillingham, ME7 3TS
G3 VFD C. Westwood, Uplands, The Hillside, Orpington, BR6 7SD
G3 VFF D. Hine, Whirlwind, Chesboule Lane Gosberton Risegate, Spalding, PE11 4EU
G3 VFH L. Moore, 15 Elmete Drive, Roundhay, Leeds, LS8 2LA
GW3 VFL A. Lightly, 9 The Kymin, Monmouth, NP25 3SD
G3 VFO T. Hart, The Hawthorns, 163 Hastings Road, Battle, TN33 0TP
G3 VFX D. Davison, 28 Treve Avenue, Harrow, HA1 4AJ
GW3 VFZ M. Hughes, Cefn Dinas, Bangor, LL57 4DP
G3 VG J. Wood, 7 Sherring Close, Bracknell, RG42 2LD
G3 VGD D. Jones, 31 Meadow Road, Windermere, LA23 2EU
G3 VGE M. Hickman, 75 Carlton Road, Redhill, RH1 2BZ
G3 VGH Dr B. Hutchinson, 78 Strensall Road, Huntington, York, YO32 9SH
G3 VGK K. Blackburn, 57 Hope Street, Leigh, WN7 1NB
G3 VGR D. Aldridge, 62 Roding View, Buckhurst Hill, IG9 6AQ
G3 VGW R. Buckby, 22 Woodstead, Embleton, Alnwick, NE66 3XY
G3 VGX Dr R. Orton, 15 Middleton Close, Cambridge, CB4 1DG
G3 VGY R. Ricketts, 30 Water Lane, Tiverton, EX16 6RB
G3 VGZ B. Duffell, 7 Potto Close, Yarm, TS15 9RZ
G3 VHE R. Evans, 23 Hardwell Close, Grove, Wantage, OX12 0BN
G3 VHF M. Eavis, 2 Burrsholt, Cople, Bedford, MK44 3UJ
G3 VHH J. Delves, 11 Willoughby Road, Langley, Slough, SL3 8JH
G3 VHI G. Boultbee, 6 Laxton Close, Heckington, Sleaford, NG34 9TS
G3 VHK Dr J. Robinson, 8 Lorraine Park, Harrow, HA3 6BX
G3 VHL H. Buttress, 132 Elan Avenue, Stourport-on-Severn, DY13 8LR
GI3 VHM V. Addidle, 23 Church Lodge, Moneyrea, Newtownards, BT23 6ES
G3 VHN J. Burge, 14 Robinson Place, Brant Broughton, Lincoln, LN5 0SJ
G3 VHS J. Cobb, Middle Cottage, Abingdon, OX13 5LR
G3 VHU M. Herring, Flat 6, 1 Royal Crescent, Bridlington, YO15 2PG
G3 VHW N. Humphrey, 10 Pembroke Close, Eastleigh, SO50 4QY
G3 VHZ B. Neary, 30 Laneham Close, Doncaster, DN4 7HU
G3 VID T. Howe, 33 Devon Gardens, Birchington, CT7 9SR
G3 VIP G. Wood, 47 Church Lane, Holton-Le-Clay, Grimsby, DN36 5AQ
G3 VIR R. Brade, 9 Magness Road, Deal, CT14 9JF
G3 VIX T. Stevens, 97 Broad Acres, Hatfield, AL10 9LE
G3 VIY R. Vasper, 31 Oakland Road, Forest Town, Mansfield, NG19 0EJ
G3 VJE H. Cole, 3 Canberra Crescent, Grantham, NG31 9RD
G3 VJG M. Deutsch, 1 Hodge Court, Kettering, NN15 7EZ
G3 VJI J. Steel, 4 Broom Close, Kendal, LA9 6BN
G3 VJM A. Wood, Danehill, Brookhill Road, Crawley, RH10 3PS
G3 VJN A. Ryan, 10 Skasmata, Koili, Cyprus, 8543
G3 VJR J. Longstaff, 23 Harlington Road, Adwick-Upon-Dearne, Mexborough, S64 0NL
G3 VJV C. Hartley, 16 Cyril Bell Close, Lymm, WA13 0JS
G3 VJX M. Gill, Upper Bean Hall, Church Road, Redditch, B69 6RN
GM3 VJY J. Evans, 64 Craigmount Avenue North, Edinburgh, EH12 8DL
G3 VKB J. Orr, 9 Chemin Des Postes, Villers Carbonnel, France, 80200
G3 VKF K. Kelly, 2 Longden Lane, Macclesfield, SK11 7EN
G3 VKI F. Turner-Smith, 26 Ash Church Road, Ash, Aldershot, GU12 6LX
G3 VKK Chesterfield & District ARS, c/o J. Otter, 7 Longacre Road, Dronfield, S18 1UQ
GW3 VKL Barry ARS, c/o P. King, 11 Lord Street, Penarth, CF64 1DD
G3 VKM R. Basford, Newgate, Thorpe Road Haddiscoe, Norwich, NR14 6PP
GM3 VKN P. Mansell, Broad Meadows, Fort Augustus, PH32 4DW
G3 VKQ C. Mcewen, 37 Malvern Way, Twyford, Reading, RG10 9PY
G3 VKT R. Smith, 32 Wolseley Gardens, London, W4 3LR
G3 VKU D. Hollingsworth, 4 Cairn View, Longframlington, Morpeth, NE65 8JT
G3 VKV G. Jones, 32 The Grove, Hales Road, Cheltenham, GL52 6SX
G3 VKW K. Evans, Littlefield House, Bolney Road, Haywards Heath, RH17 5AW
G3 VLC C. Hawkins, 2 Benett Drive, Hove, BN3 6PL
G3 VLD T. Denney, Spindrift-East-, Terrace, Walton on Naze, CO14 8PX
G3 VLF Dr T. Beamond, Park View, Middle Lane, Whatstandwell, Matlock, DE4 5EG
G3 VLG Hinckley Amateur Radio and Electronics Society, c/o C. Colclough, 53 St Marys Road, Nuneaton, CV11 5AT
G3 VLH J. Longhurst, 13 Hophurst Drive, Crawley Down, Crawley, RH10 4XA
G3 VLJ Dr A. Hansen, 1829 Francisco St, Berkeley, USA, 94703
G3 VLL G. Gauntlett, 7 Riverside Drive, Sprotbrough, Doncaster, DN5 7LH
G3 VLN J. Allin, 57 Burleigh Road, West Bridgford, Nottingham, NG2 6FQ
G3 VLO Dr J. Owen-Jones, 16 Cotswold Close, Torquay, TQ2 6UB
G3 VLR B. Rispin, 37 Ferry Road South Cave, Brough, HU15 2JG
G3 VLW P. Martin, Orchard Rise, Littlemoor Road, Highbridge, TA9 4NG
G3 VM D. Grace, 107 Bush Avenue, Little Stoke, Bristol, BS34 8NG
G3 VMI D. Pike, 11 Cavalry Drive, March, PE15 9EQ
G3 VMK N. Chadwick, 1 St. Francis Meadow Mitchell, Newquay, TR8 5DB
G3 VMP B. Mills, Highlands Cottage, Crow Lane, Clacton on Sea, CO16 9AN
G3 VMR R. Redding, 53 Cadwell Drive, Maidenhead, SL6 3YS
G3 VMT T. Poole, 64 Humber Close, Thatcham, RG18 3DT
G3 VMU C. Davis, 23 Vernon Walk, Northampton, NN1 5ST
G3 VMV C. Whiting, 5 Carlton, Elloughton, Brough, HU15 1FF
G3 VMW S. Wilson, 3 Crag Gardens, Bramham, Wetherby, LS23 6RP
G3 VMY Dr E. Searle, 203 Church Road, Earley, Reading, RG6 1HW
G3 VMZ D. Nicholls, 26 Highfield Close, Semington, Trowbridge, BA14 6JZ
G3 VNB R. Thomas, 7 Lane Gardens, Bushey Heath, Bushey, WD23 1PE
G3 VNG D. Hind, 4 Thornyville Villas, Plymouth, PL9 7LA
G3 VNH P. Hardy, Lambda House, Seanor Lane, Chesterfield, S45 8DH
G3 VNI S. Cammies, 5 Sheringham Close, Allington, Maidstone, ME16 0NF
G3 VNP P. Dowles, 1A Queen Street, Maldon, CM9 5DP
G3 VNQ M. Pritchard, 9 Tamarack Drive, Cortlandt Manor, USA, 10567
G3 VNT L. Pearson, Hatherly, The Street Ashfield, Stowmarket, IP14 6LX
G3 VNU J. Finch, 286 Sea Front, Hayling Island, PO11 0AZ
G3 VNY I. Walker, 45 Terry Drive, Walmley, Sutton Coldfield, B76 2PT
GW3 VNZ D. Jacklin, 40 Westbourne Road, Penarth, CF64 3HF
G3 VOB D. Vivian, Belle View Cottage, Blandford Road North, Poole, BH16 5PP
G3 VOF M. Foster, 1 Clavering Court Lincombe Drive, Torquay, TQ1 2HH
GW3 VOL J. Phillips, 96 Maes Y Sarn, Pentyrch, Cardiff, CF15 9QR
G3 VOM D. Lane, 2 Eden Close, Wilmslow, SK9 6BG
G3 VOO M. Barnett-Bone, 7 Dorchester Hill, Milborne St. Andrew, Blandford Forum, DT11 0JG
G3 VOS R. Cottrell, Larkhill, 47 Bullsland Lane, Rickmansworth, WD3 5BD
G3 VOT G. Webster, Red House Farm, Ashford Lane, Bakewell, DE45 1NJ
G3 VOU J. Barlow, 68 Willow Avenue, Cheadle Hulme, Cheadle, SK8 6AX
G3 VOV M. Lane, 56 Main Street, Bushby, Leicester, LE7 9PP
G3 VOW M. Fereday, Spindlewood, Stoney Lane, Thatcham, RG18 9HQ
G3 VPA M. Rose, 59 Park Drive, Sittingbourne, ME10 1RD
G3 VPE H. Pinchin, Birchmere Retirement & Care Home 1270 Warwick Road, Knowle, Solihull, B93 9LQ
G3 VPF E. Harland, 5 Bramdon Lane, Portesham, Weymouth, DT3 4HG
G3 VPG C. Jacob, 18 Compton Way, Olivers Battery, Winchester, SO22 4HS
G3 VPH J. Mayall, 10 Manor Close, Droitwich, WR9 8HG
G3 VPK W. Mcclintock, 1 Small Horse Farm Close, Freshwater, PO40 9FY
GM3 VPN J. Gardner, Taringa, Edentown, Cupar, KY15 7UH
G3 VPQ I. Westwood, 14 Staplegrove Road, Taunton, TA1 1DQ
G3 VPR R. Harrison, 512 Broadgate, Weston Hills, Spalding, PE12 6DA
G3 VPS P. Lennard, 5 Parkside, East Grinstead, RH19 1JG
G3 VPT P. Burgess, 26 William Peck Road, Spixworth, Norwich, NR10 3QB
GI3 VPV R. Aughey, 30 Glen Road, Hillsborough, BT26 6ES
G3 VPW J. Wright, 2 Barnfield, Charney Bassett, Wantage, OX12 0HA
G3 VPX I. Sumner, 132 Barrs Road, Cradley Heath, B64 7EZ
G3 VQF J. Moorhouse, 185 Aldermoor Road, Southampton, SO16 5NQ
G3 VQG R. Beadle, 5 Badgeney Road, March, PE15 9AP
G3 VQO L. Allwood, 9 Gorse End, Horsham, RH12 5XW

G3 VQQ M. Hall, 5 Kings Lea, Kingsway, WF5 8RY
G3 VQR A. Henshaw, 3 Lewens Close, Wimborne, BH21 1JJ
G3 VQS R. Kirby, 197 Longfield, Falmouth, TR11 4SR
G3 VQW B. Fawkes, 6 Oak Avenue, Worcester, WR4 9UG
G3 VQY J. Cumming, Camelot, Cheltenham Road, Hockley, SS5 5HJ
G3 VRB J. Nias, 49 St. Margarets Road, Bishopstoke, Eastleigh, SO50 6DG
G3 VRE Chippenham and District ARC, c/o B. Tanner, 2 Doveys Cottage, Kington Langley, Chippenham, SN15 5NT
G3 VRF J. Charlton, 57 Victoria Road, Bidford-On-Avon, Alcester, B50 4AR
G3 VRU P. Ford, 15 Doles Lane, Whitwell, Worksop, S80 4SN
G3 VRV M. Huish, Becketts, Woodbury, Exeter, EX5 1JD
G3 VRW P. Lamb, 5 The Templars, Bridge End, Warwick, CV34 6PF
G3 VRY J. Pitt, 30 Hillcroft Road, Chesham, HP5 3DJ
G3 VSB G. Jones, Braemar, Alton Road, Uttoxeter, ST14 5DH
G3 VSE K. Thompson, 3 Parkside, Morecambe, LA4 4TJ
G3 VSH D. Freedman, 102 Collingwood Road, Sutton, SM1 2RB
G3 VSI N. Prince, 96 Foxglove Way, Springfield, Chelmsford, CM1 6QR
G3 VSJ D. Chaloner, 38 Barnfield Close, Hoddesdon, EN11 9EP
G3 VSK T. Mccurry, 148 Moorgate Road, Rotherham, S603AZ
G3 VSL J. Arscott, 122 Woodlands Road, Ashurst, Southampton, SO40 7AL
G3 VSQ R. West, 10 Hawkshill Drive, Hemel Hempstead, HP3 0BS
G3 VSR T. Barraclough, 27 Kestrel Park, Skelmersdale, WN8 6TA
G3 VST F. Moore, Causeway House, Risbury, Leominster, HR6 0NG
G3 VSU A. Moore, 48 Cransley Rise, Mawsley, Kettering, NN14 1TB
G3 VSV D. Middleton, 8 Fulmar Close, Bradwell, Great Yarmouth, NR31 8JG
GM3 VTB V. Budas, 20 Oak Avenue, Bearsden, Glasgow, G61 3HD
G3 VTD R. Price, 36, Hadleigh Rise, Pontefract, WF8 4SJ
G3 VTE R. Swetmore, 18 Tideswell Road, Stoke-on-Trent, ST3 5EG
GM3 VTH D. Coutts, 29 Barons Hill Avenue, Linlithgow, EH49 7JU
G3 VTL J. Levett, 56 St. Nicholas Avenue, Kenilworth, CV8 1JW
G3 VTO M. Coombs, 10 Horseshoe Walk, Widcombe, Bath, BA2 6DE
G3 VTR A. Davis, Fieldings, Bury Road, Bury St Edmunds, IP29 4PL
G3 VTS C. Walker, 2 Georgian Close, Abbeydale, Gloucester, GL4 5DG
G3 VTT C. Turner, 84 Gravel Hill Way Dovercourt, Harwich, CO12 4XN
G3 VUD P. Bentley, 12 West Terrace Seaton Sluice, Whitley Bay, NE26 4RE
G3 VUE T. Mowbray, Elmhirst House, Lincoln Road, Horncastle, LN9 5AW
G3 VUH M. Blackwell, Room 2, Mulroy House Peaker Park, Market Harborough, LE16 7FP
G3 VUI M. Harris, 3 Ross Road East, Stanley PO Box 226, Falkland Islands (Malvinas), FIQQ 1ZZ
G3 VUK R. Knight, 8 Narromine Drive, Calcot, Reading, RG31 7ZL
G3 VUL Dr J. Lotz, 29 Burton Manor Road, Stafford, ST17 9QJ
G3 VUN G. Ackerley, The Paddock, Stoney Lane, Tarporley, CW6 0SX
G3 VUO J. Mills, 9 Sandpiper Walk, Chelmsford, CM2 8XJ
G3 VUR K. Evans, 68 Downs Road, Hastings, TN34 2DZ
G3 VUS D. Latimer, Braefoot, Lanercost Road, Brampton, CA8 1EN
G3 VUY D. Bradley, 4 Felthorpe Close, Upton, Wirral, CH49 4GY
GW3 VVC J. Parry, 'Ar Allt' Lon Hedydd Llanfairpwll Anglesey, North Wales, LL61 5JY
G3 VVE H. Robinson, 4 Cross Street, Mansfield Woodhouse, Mansfield, NG19 9NA
GM3 VVF A. Ross, 17 Tarvit Green, Glenrothes, KY7 4SJ
G3 VVL K. Lax, 17 Malt Rise, Crew Green, Shrewsbury, SY5 9EU
G3 VVR J. Grace, Woodside, Easthorpe, Malton, YO17 6QX
G3 VVT R. Wilkinson, 18 Green Road, Kendal, LA9 4QR
G3 VWA C. Marflow, 13 Walthew Green, Roby Mill, Skelmersdale, WN8 0QT
G3 VWC A. Marriott, 28 Horseshoe Walk, Bath, BA2 6DF
G3 VWD C. Bean, 11 Nightingale Lane, Coventry, CV5 6AY
G3 VWH B. Wilde, 34 Grangefields Road, Shrewsbury, SY3 9DB
G3 VWJ G. Westwood, 133 Torrisholme Road, Lancaster, LA1 2TZ
G3 VWK A. Hammett, (Hammett), Ladock, Truro, TR2 4PQ
G3 VWQ P. Forster, 59 Woodland View, Stratton Strawless, Norwich, NR10 5LT
G3 VWX E. Perks, The Oaklands, Bromfield Road, Ludlow, SY8 1DW
GM3 VWY I. Malcolm, 2 Morton Crescent, St Andrews, KY16 8RA
G3 VXA M. Harrold, 26 Leys Close, Harefield, Uxbridge, UB9 6QB
G3 VXE G. Brindle, 8 Peckover Drive, Pudsey, LS28 8EF
G3 VXF B. Ellis, 15A The Street, Stedham, Midhurst, GU29 0NQ
G3 VXH R. Huffadine, 19 Cumberland Street, Worcester, WR1 1QE
G3 VXJ R. Rylatt, 16 First Avenue, Worthing, BN14 9NJ
G3 VXK R. Porter, 16 Millcroft, Crosby, Liverpool, L23 9XJ
G3 VXS D. Peach, Flat 35, Homeshire House, 36 Sandbach Road South, Stoke-on-Trent, ST7 2LP
G3 VXY Prof. B. Cotton, 12 Tower Gardens, Bassett, Southampton, SO16 7EL
G3 VYA B. Atkiss, 47 Russell Road, Partington, Manchester, M31 4DY
G3 VYD J. Bourne, Tyndalls, 8 Kelvedon Road, Witham, CM8 3LZ
G3 VYF M. Lee, 11 Sturrocks, Vange, Basildon, SS16 4PQ
G3 VYG R. Walpole, 7 Springfield Road, Taverham, Norwich, NR8 6QU
G3 VYI M. Franklin, 6 Tor Road, Farnham, GU9 7BX
GM3 VYJ T. Jameson, 8 River View, Dalgety Bay, Dunfermline, KY11 9YE
G3 VYK P. Frost, 164 Newthorpe Common, Newthorpe, Nottingham, NG16 2EN
G3 VYN M. Turner, Plumtree Cottage, Spring Lane, Norwich, NR15 2NT
G3 VYS A. Nell, Teaselwood, Parkham, Bideford, EX39 5PL
G3 VYU R. Chamberlain, 1 Thornemead Werrington Meadows, Peterborough, PE4 7ZD
G3 VYW D. Carter, 13 Sturdy Close, Hythe, CT21 6AG
G3 VYX C. Burr, The Old Rectory House, Radipole Lane, Weymouth, DT4 9RN
GI3 VYY B. Hamilton, 4 Castleton Court, 16 Osborne Park, Belfast, BT9 6HA
G3 VYZ L. Thompson, 44 Tillmouth Avenue, Holywell, Whitley Bay, NE25 0NP
G3 VZE D. Kennedy, 79 High Street, Dunsville, Doncaster, DN7 4BS
G3 VZF J. Adams, Chilterns, Bellingdon, Chesham, HP5 2XL
G3 VZG R. Golding, 7 Belvidere Avenue, Shrewsbury, SY2 5PF
G3 VZH Dr C. Doran, 16 Wordsworth Road, Penge, London, SE20 7JG
G3 VZJ A. Clemmetsen, 6 Lord Louis Crescent, Plymouth, PL9 9SH
G3 VZL R. Newman, 20 Glapthorn Road Oundle, Peterborough, PE8 4JQ
G3 VZM F. Houghton, 14 Windfield Gardens Little Sutton, Ellesmere Port, CH66 1JJ
G3 VZO V. Hartshorn, 61 Fulmerton Crescent, Redcar, TS10 4NJ
G3 VZR E. Thompson, Meadowside, Bromsberrow Heath, Ledbury, HR8 1NX
G3 VZU W. Mooney, 538 Liverpool Road, Great Sankey, Warrington, WA5 3LU
G3 VZV G. Shirville, Birdwood, Heath Lane, Woburn Sands Heath Lane, Woburn Sands, Milton Keynes, MK17 8TN
G3 WAB P. Harrison, 8 Buxtons Lane, Guilden Morden, Royston, SG8 0JU
G3 WAE I. Harris, Orchard Cottage, The Street, Devizes, SN10 2LD
G3 WAG D. Gillett, 20 Redcar Avenue, Hereford, HR4 9TJ
G3 WAH N. Hodgson, 42 Tofts Grove, Rastrick, Brighouse, HD6 3NP
G3 WAL J. Barker, 76 Halebrose Court, Seafield Road, Bournemouth, BH6 3DU
G3 WAM M. Taplin, 146 Ashover Road, Old Tupton, Chesterfield, S42 6HG
GM3 WAP A. Philp, Philp House, High Street, Blairgowrie, PH11 8DW
G3 WAS Lichfield A.R.S, c/o R. Smethers, 46 Church Road, Burntwood, WS7 9EA
G3 WBA I. Currell, 47 Highdale Avenue, Clevedon, BS21 7LU
G3 WBB E. Avery, 2 Blythe Avenue, Thornton-Cleveleys, FY5 2LL
G3 WBC R. Bryant, 12 Laburnum Grove, Luton, LU3 2DW
G3 WBG H. Hindle, 6 Windsor Road, Conisbrough, Doncaster, DN12 3DF
G3 WBI P. Lewis, 15 Norwood Road, Lytham St Annes, FY8 2QN
G3 WBK Dr P. Tofts, 48 Rugby Road, Brighton, BN1 6EB
G3 WBL K. Weller, Charbury House, Bayton, Kidderminster, DY14 9LJ
G3 WBN A. Thurlow, Chesnet House, Croydon, CR0 5BA
G3 WBP J. Broadley, 13 Portland Close, Bedford, MK41 9NE
G3 WBQ T. Brook, 22 Downside Road, Guildford, GU4 8PH
GI3 WBR R. Mccrea, 1 Killynoogan Terrace, Killynoogan, Enniskillen, BT93 8DF
G3 WBS D. Thomson, 2A The Landway, Kemsing, Sevenoaks, TN15 6TG
GW3 WBU B. Vodden, 22 Heath Avenue, Penarth, CF64 2QZ
GW3 WCA P. Dunbar, Pengwern Fach, Penrherber, Newcastle Emlyn, SA38 9RL
G3 WCB D. John, 27 Churchfields, Dartmouth, TQ6 9HJ
G3 WCD C. Dillon, 63 High Street, Toseland, St Neots, PE19 6RX
G3 WCE B. Edwards, Elder Cottage, Norwich, NR10 5BB
G3 WCJ P. Hackett, Po Box 1 Wooroloo Wa 6558, Wooroloo, Australia, 6558
G3 WCL J. Croker, 29 Alexandra Road, Bedminster Down, Bristol, BS13 7DF
G3 WCM F. Chidlow, 64 Mitchell Avenue, Northside, Workington, CA14 1AA
G3 WCQ R. Bailey, 43 Earlsdon Avenue South, Coventry, CV5 6DR
G3 WCU J. Pealing, 93 Fernside Road, Poole, BH15 2JQ
GW3 WCV D. Howell, 6 Douglas Close, Cardiff, CF5 2QT
G3 WCY B. Smith, 26 Sandhurst Lane, Blackwater, Camberley, GU17 0DH
G3 WDD T. Horrobin, 29 Ambleside Road Maghull, Liverpool, L31 6BY
G3 WDE P. Ford, 11 Brook Lane, Felixstowe, IP11 7EG
G3 WDG Dr C. Suckling, 314A Newton Road, Rushden, NN10 0SY
G3 WDM C. Care, 127 Brooklands Crescent, Fulwood, Sheffield, S10 4GF
G3 WDN E. Fielding, The Birches, 3 Sneath Road, Norwich, NR15 2DS
G3 WDS D. Spooner, 45 Otterburn Avenue, Whitley Bay, NE25 9QR
G3 WDU I. Peterkin, 243 Hampton Street, Hampton, Australia, 3188
G3 WDX P. Hickey, 11 Bembridge Place, Linden Lea, Leavesden, Watford, WD25 7DN
G3 WEA A. Cross, 34 Pinewood Drive, Potters Bar, EN6 2BD
GM3 WED A. Rose, Craiglea, Schoolcroft, Dingwall, IV7 8LB
G3 WEF A. Beazley, 24 Tealsbrook, Covingham, Swindon, SN3 5AU
G3 WEG P. Webster, 22 Whincroft Drive, Ferndown, BH22 9LJ
G3 WEI D. Turner, Birchwood, Heath Top, Market Drayton, TF9 4QR
G3 WEJ S. Bradshaw, 11 Meadow Park, Dawlish, EX7 9BS
GI3 WEL R. Knox, 91 Banbridge Road, Waringstown, Craigavon, BT66 7RU
GI3 WEM V. Gracey, 23 Cascum Road, Banbridge, BT32 4LF
GW3 WEQ C. Collins, 21 Bron Wern Llanddulas, Abergele, LL22 8JD
G3 WEU K. Gregory, 67 Clowne Road, Barlborough, Chesterfield, S43 4EH
G3 WEW Dr R. Wood, 8305 El Matador Drive, Gilroy, USA, 95020
G3 WEY A. Nelson, 4 Bell Close, Farmborough, Bath, BA2 0AP
GW3 WEZ J. Lawrence, 3 Siskin Crescent, Rogiet, Caldicot, NP26 3UW
G3 WF D. Cockings, Elettra, 207A Birchfield Road, Redditch, B97 4LX
G3 WFF B. Tew, 96 Mill Lane, Sawston, CB223HZ
G3 WFH D. Morris, 27 Albert Square, Bowdon, Altrincham, WA14 2ND
GI3 WFP P. Mcalpine, 20 Gransha Road South, Bangor, BT19 7QB
G3 WFT D. Holland, 32 Woodville Drive, Sale, M33 6NF
G3 WFW K. Hampson, 11 Gladstone Grove, Stockport, SK4 4BX
G3 WGE E. Law, Brunanburgh, 1B Ponds Road, Galleywood., Chelmsford, CM2 8QP
G3 WGH M. Reeve, 18-20 Radford Road, Nottingham, NG7 5FS
G3 WGK B. Wormwell, 26 Windsor Avenue, Longridge, Preston, PR3 3EL
G3 WGN D. Aslin, Old Smithy Cornworthy, Totnes, TQ9 7HH
G3 WGV J. Linford, Pennine View, Sleagill, Penrith, CA10 3HD
G3 WGY H. Ashford, 56 Guarlford Road, Malvern, WR14 3QP
G3 WGZ G. Sowden, Villa Clare, The Lizard, Helston, TR12 7NU
GI3 WHA L. Hanna, Igrangeville Park, Newtownards, BT23 8TE
G3 WHB S. Christie, 4 Dairy Court Holyport, Maidenhead, SL6 2US
G3 WHG M. Key, 12 Great Melton Road, Hethersett, Norwich, NR9 3AB
G3 WHJ A. Johnson, 49 Tennyson Drive, Malvern, WR14 2UL
GM3 WHT M. Smith, Vakterlee, Cumliewick, Shetland, ZE2 9HH
G3 WI M. Hulme, 44 Thirlmere Avenue, Ashton-under-Lyne, OL7 9HN
G3 WIA R. Ottley, 15 Orchard Way, Thrapston, Kettering, NN14 4RE
G3 WII F. Clarke, 8 Tristram Close, Chandler'S Ford, Eastleigh, SO53 4TT
GM3 WIJ N. Mackenzie, 57 Countesswells Terrace, Aberdeen, AB15 8LQ
G3 WIK M. Shorland, Baxhill Bungalow, Chase Road, Upper Colwall, Malvern, WR13 6DL
G3 WIM Wimbledon & District ARS, c/o S. Pelling, 10 Narrow Way, Bromley, BR2 8JB
G3 WIN Windscale ARS, c/o R. Wood, Abbots Croft, Abbey Road, St Bees, CA27 0EG
G3 WIO E. Obrien, Tanglewood, Anthonys Way, Wirral, CH60 0BP
G3 WIP Dr G. Bulger, Flat C/21, Herbal Hill Gardens, 9 Herbal Hill, London, EC1R 5XB
G3 WIS B. Day, 54 South Avenue, Hope Carr, Leigh, WN7 3BU
G3 WIU W. Bekenn, 35 Blackdown Avenue, Rushmere St. Andrew, Ipswich, IP5 1AY
G3 WIW A. Leach, 199 Braemor Road, Calne, SN11 9EA
GM3 WJE J. Thom, 50 Dickson Avenue, Dundee, DD2 4EG
G3 WJG G. Lean, 54 Blacketts Wood Drive, Chorleywood, Rickmansworth, WD3 5QH
G3 WJH W. Wilkinson, Chiriqui, 15 Camerton Road, Workington, CA14 1LP
G3 WJI P. White, Linden House, Willisham, Ipswich, IP8 4SP
G3 WJJ D. Finnemore, 4 Purbeck Gardens, Felton Road, Poole, BH14 0QS
G3 WJM B. Schoth, 3 Solent Drive, Hythe, Southampton, SO45 5FP
G3 WJN R. Hassell Bennett, 30 Greenlands Avenue, Redditch, B98 7QA
G3 WJS J. Starling, 16 Queenscliffe Road, Ipswich, IP2 9AS
G3 WKA N. Bardell, 4 Church End, Arlesey, SG15 6UY
GM3 WKB D. Topham, Dairy Cottage, Kilmany, Cupar, KY15 4PT
G3 WKE S. Braidwood, Flat 229, Helen Gladstone House, Nelson Square, London, SE1 0QB

G3 WKF M. Richards, Wayside Cottage, Penwithick Road, St Austell, PL26 8UH
G3 WKH R. Martin, 130 Main Avenue S Apartment 208, Renton, USA, WA 98057
G3 WKI M. Hewins, 37 Ringwood Close, Furnace Green, Crawley, RH10 6HQ
G3 WKL Dr J. Gould, 116 Wolverton Road, Newport Pagnell, MK16 8JG
G3 WKP P. King, Nirvana, Comprigney Hill, Truro, TR1 3TX
G3 WKR M. Goodwin, 6 Hobbs Hill, Rothwell, Kettering, NN14 6YG
G3 WKS West Kent ARS, c/o D. Green, St. Annes Poundfield Road, Crowborough, TN6 2BG
G3 WKW R. Thornton, 26 Florence Road, Fleet, GU52 6LQ
G3 WKX Maidenhead & District ARC, c/o M. Palmer, 28 Westfield Road, Caversham, Reading, RG4 8HH
G3 WLA A. Macpherson, 15A Monkstone Drive, Berrow, Burnham-on-Sea, TA8 2NW
G3 WLD J. Hall, 22 Haverhill Road, Stapleford, Cambridge, CB22 5BX
G3 WLG Dr M. Griffiths, The Oaklands, Hollybush Lane, Worcester, WR6 6HQ
G3 WLH Dr C. Pell, 1 Glenville Gardens, Hindhead, GU26 6SX
G3 WLM R. Joyce, 20 Barking Close, Luton, LU4 9HG
GW3 WLN Dr J. Pritchard, 13 Cefn Graig, Rhiwbina, Cardiff, CF14 6SW
G3 WLO E. Denton, 11 Highland Road, Amersham, HP7 9AU
G3 WLT Dr D. Firth, 3 School Lane, Shaldon, Teignmouth, TQ14 0DG
G3 WLV J. Bushby, 14 Clayton Drive, Thurnscoe, Rotherham, S63 0RZ
G3 WLW R. Millar, 1229 Leeds Road, Bradley, Huddersfield, HD2 1UY
G3 WLY J. Harwood, 12 Longwood Avenue, Cowplain, Waterlooville, PO8 8HX
G3 WMA W. Shepperd, 28 Tyne Road, Oakham, LE15 6SJ
G3 WMD J. Whomes, 44 Russell Close, Steeple Morden, Royston, SG8 0NE
G3 WME M. Groom, 409 Finchampstead Road, Finchampstead, Wokingham, RG40 3RL
G3 WMJ G. Jillings, 16 Greenwich Road, Diep River, South Africa, 7800
GW3 WMP Dr J. Hopton, 9 Bryneuraidd, Ammanford, SA18 3TG
G3 WMQ M. Watson, Chant House, Dark Lane, Nailsworth, GL6 0DR
G3 WMS I. Vance, Larkfield, Debden Road, Saffron Walden, CB11 3RU
G3 WMT R. Dowling, 80 Elmfield Way, Sanderstead, South Croydon, CR2 0EF
G3 WMU Amberley Radio Group, c/o E. Spicer, 3 Golden Avenue Close East Preston, Littlehampton, BN16 1QS
G3 WMX C. Knott, 7 Manor View, Crewkerne, TA18 8JT
G3 WMY S. Smith, Five Oaks, Sandy Lane, Henfield, BN5 9UX
GM3 WNB J. Ohare, 208 Gilmartin Road, Linwood, Paisley, PA3 3ST
G3 WNC R. Todd, 17 Tudor Road, West Bridgford, Nottingham, NG2 6EB
G3 WND R. Aston, 16 St. Johns Road, Mortimer Common, Reading, RG7 3TR
G3 WNP T. Baker, 54 Hamilton Road, Reading, RG1 5RD
G3 WNQ E. Lingard, Tedulf, Rotten Row, Theddlethorpe, Mablethorpe, LN12 1NX
G3 WNR K. Grey, 15 Woodbourne Avenue, Leeds, LS17 5PQ
G3 WNS A. Willson, Hilltop, Cryers Hill Road, Cryers Hill, High Wycombe, HP15 6LJ
G3 WNV D. Field, Rackhay, Prescott, Cullompton, EX15 3BA
G3 WNW D. Bailey, 31 Antrobus Street, Congleton, CW12 1HE
G3 WOA J. Goodman, 15 Highway, East Taphouse, Liskeard, PL14 4NW
G3 WOD J. Welford, 303 Scalby Road, Scarborough, YO12 6TF
G3 WOE M. White, 76 Birch Row, Bromley, BR2 8FG
G3 WOH E. Grossmith, 4 Lincoln Way, Rainhill, Prescot, L35 6PJ
GM3 WOJ C. Tran, Achnacoille, Lamington, Invergordon, IV18 0PE
G3 WOK D. Clifton, 59 Grantham Road, Bracebridge Heath, Lincoln, LN4 2LE
G3 WOM M. Muir, 6 Broadstairs Court, Sunderland, SR4 8NP
G3 WOO R. Brace, 11 Cedar Close, Sawbridgeworth, CM21 9NT
G3 WOR Worthing and DARC, c/o A. Cheeseman, Flat 5 Dubarry House, Hove Park Villas Hove Park Villas, Hove, BN3 6HP
G3 WOS C. Gare, Old White Lodge, 183 Sycamore Road, Farnborough, GU14 6RF
G3 WOT M. Meads, 12 Burlington Way, Hemingford Grey, Huntingdon, PE28 9BS
G3 WOV G. Macnaught, 30 West End Falls, Nafferton, Driffield, YO25 4QA
GU3 WOW P. Hancock, La Breloque, Les Grandes Rues, Les Buttes, Guernsey, GY79EL
GM3 WPA S. Hutchinson, 4 Wiston Place, Dundee, DD2 3JR
G3 WPB P. Smith, 180 Victoria Road, Ferndown, BH22 9JE
G3 WPD A. Smith, 118 Bois Moor Road, Chesham, HP5 1SS
G3 WPF R. Unsworth, Spurs Lodge, Sagars Road, Styal, Wilmslow, SK9 4HE
G3 WPG D. Dye, 10 Headington Close, Bradwell, Great Yarmouth, NR31 8DN
G3 WPH Dr M. Chamberlain, 10 Clifton Rise, Wargrave, Reading, RG10 8BN
G3 WPN V. Bennellick, The Wolverns, Castle Frome, Ledbury, HR8 1HG
G3 WPP D. Minett, Melrose Cottage, South Road, Truro, TR3 7AD
G3 WPQ M. Kaye, Pucknell Lodge, Hollywell Lane Bayton Common, Kidderminster, DY14 9NR
G3 WPR C. Richmond, 1 Grangeway Gardens, Ilford, IG4 5HN
G3 WPT R. Brown, 65 Staining Rise, Staining, Blackpool, FY3 0BU
G3 WPV D. Lamont, 6B Route De Mailhac, Bize Minervois, France, 11120
G3 WQG D. Chalmers, 25 Willow Close, Flackwell Heath, High Wycombe, HP10 9LH
G3 WQK Southdown ARS, c/o A. Seabrook, 63 St. Annes Road Willingdon, Eastbourne, BN20 9NJ
G3 WQL A. Conway, 17 Mountcastle Road Leicester, Leicester, LE3 2BW
G3 WQU P. Mckay, C/O Unifil, Po Box 5852, New York, USA, 10163-5852
G3 WQY T. Codrai, Sealand, Coast Road, Norwich, NR12 0PD
G3 WRA S. Powell, 9 Belgravia Gardens, Hereford, HR1 1RB
G3 WRD R. Richardson, Common Crest, Drapery Common, Sudbury, CO10 7RW
GW3 WRE B. Jones, 6 Pentyla, Maesteg, CF34 0BB
G3 WRI Dr P. Brown, 30 Applerigg, Kendal, LA9 6EA
G3 WRJ R. Bacon, Gyffen, Gosmore Road, Hitchin, SG4 9AN
G3 WRK G. Oakes, 13 Tidnock Avenue, Congleton, CW12 2HN
G3 WRL E. Northwood, 8 Derwood Grove, Werrington, Peterborough, PE4 5DD
G3 WRO K. Haynes, 34 Pear Tree Mead, Harlow, CM18 7BY
G3 WRR Q. Collier, 19 Grangecliffe Gardens, South Norwood, London, SE25 6SY
G3 WRT Dr I. Dilworth, Ashpound Cottage, Pound Lane, Ipswich, IP9 2JB
G3 WSB Dr K. Band, 11 Denewood Close, Watford, WD17 4SZ
G3 WSC Crawley ARC, c/o J. Pitty, 12 St. Leonards Road, Horsham, RH13 6EJ
G3 WSD A. Fisher, 63 Spencer Close, Potton, Sandy, SG19 2QR
G3 WSM B. Storry, 508 Arleston Lane, Stenson Fields, Derby, DE24 3AA
GM3 WSR V. Clark, 6 Parkhill Circle, Dyce, Aberdeen, AB21 7FN
GW3 WSU C. Beynon, 16 Hardy Close, Barry, CF62 9HJ
G3 WSV J. Lawson, 3 Pearmains, Great Leighs, Chelmsford, CM3 1QS
G3 WSW J. Holmes, 37 Redwood Avenue, Leyland, PR25 1RN
G3 WTB R. Baxter, 10 Windsor Court, Oxford Road, Southport, PR8 2JJ
G3 WTD J. Davis, 71 Broughton Road, Croft, Leicester, LE9 3EB
GI3 WTG G. Thompson, 36 Crewcatt Road, Richhill, Armagh, BT61 8QN
G3 WTN R. Limehouse, 56 Lincoln Way, Daventry, NN11 4SX
G3 WTO Dr J. Spencer, 76 Durranhill Road, Carlisle, CA1 2SZ
G3 WTP Bedford and District ARC, c/o H. Ehm, 17 Stuart Road, Kempston, Bedford, MK42 8HS
G3 WTQ P. Angold, 10 Hartford Avenue, Wilmslow, SK9 6LP
G3 WTR D. Wright, 8 Calverley Park, Tunbridge Wells, TN1 2SH
G3 WTS J. Smith, Windycross, Newbourne Road, Woodbridge, IP12 4PT
G3 WTT C. Goodwin, 1 School Lane, Canwick, Lincoln, LN4 2RP
G3 WTV K. Baker, 33 Reading Road, Woodley, Reading, RG5 3DA
G3 WTY P. Hodgkiss, 28 Beaumont Rise, Worksop, S80 1YA
GW3 WTZ M. Jones, 55 Rowan Way, Malpas, Newport, NP20 6JN
G3 WUA B. Lindop, 56 Marina Court, 9-19 Mount Wise, Newquay, TR7 2EJ
G3 WUB Dr P. Rice, 23 Christchurch Square, Homerton, London, E9 7HU
G3 WUG I. Elvins, 6 Bay Road, Unit 23, Newmarket, USA, NH 03857
G3 WUH W. Dufton, 22 Windsor Road, Bexhill-on-Sea, TN39 3PB
G3 WUI G. Spink, 60 Woodhouse Hill, Huddersfield, HD2 1DH
G3 WUK J. Spencer Chapman, Apartado De Correos 156, Mojacar 04638, Almeria, ZZ2 3TP
G3 WUL R. Whillier, 9 Tudor Drive, Yateley, GU46 6BX
G3 WUN D. Holden, 99 Sheerstock, Haddenham, Aylesbury, HP17 8EY
GI3 WUO Dr L. Waring, 16 Belfast Road, Holywood, BT18 9EL
G3 WUW A. Papworth, 3570 Corey Road, Malabar, USA, 32950
GM3 WUX T. Robinson, 82 Albert Road, Glasgow, G42 8DR
G3 WVG K. Pritchard, 9 Golf Close, Pyrford, Woking, GU22 8PE
G3 WVM C. Loosemore, 24 Myrtlebury Way, Exeter, EX1 3GA
G3 WVQ J. Barratt, 26 Johnstone Road, Newent, GL18 1PZ
G3 WVR J. Green, Honeysuckle Cottage, New Green, Braintree, CM7 5EG
GW3 WVV R. Barker, Henllys, Maenygroes, New Quay, SA45 9RL
G3 WWG J. Ross, 24 Raby Road, Stockton-on-Tees, TS18 4JA
G3 WWH R. Taylor, Shambles, Davids Lane, Benington, Boston, PE22 0BZ
G3 WWI R. Oxley, 1 Elm Grove, Maidstone, ME15 7RT
G3 WWL B. Tipper, 271 Blackberry Lane, Four Oaks, Sutton Coldfield, B74 4JS
G3 WWS M. Southall, 61 Grange Close, Horam, Heathfield, TN21 0EF
G3 WWT J. Teed, 47 West Cliff Road, Dawlish, EX7 9DZ
GI3 WWY M. Anderson, 8 Loughbrickland Road, Gilford, Craigavon, BT63 6BH
GW3 WXA J. Gough, Traleen, Rhydlewis, Llandysul, SA44 5PN
G3 WXC P. Brooker, 28 Uplands Road, Northwood, Cowes, PO31 8AL
G3 WXD Dr C. Zammit, 9 Sandbanks Drive, Hatch Warren, Basingstoke, RG22 4UL
G3 WXG I. Habens, 48 Carden Avenue, Brighton, BN1 8NE
G3 WXH J. Arnold, 6 The Spinney, Weston-Super-Mare, BS24 9LH
G3 WXI A. Strong, 3 Ellorslie Drive, Stocksbridge, Sheffield, S36 2BB
G3 WXM M. Smith, Linden, 86 Grove Road, Tring, HP23 5PB
G3 WXN L. Mckown, Flat D, 310 Oldham Road, Royton, Oldham, OL2 5AS
G3 WXU S. Allbutt, 8 Langton Close, Vinters Park, Maidstone, ME14 5PG
G3 WXW C. Traveller, 13 Cosy Corner, North Walsham, NR28 0EN
G3 WYB A. Tring, 1 Crownbourne Court, St. Nicholas Way, Sutton, SM1 1JE
G3 WYD P. Patmore, 141 Cannons Close, Bishop's Stortford, CM23 2BL
G3 WYH R. Hutton, 6 The Sidings, Ruskington, Sleaford, NG34 9GA
G3 WYK P. Bysshe, Orchard House, High Road, Maidenhead, SL6 9JT
GM3 WYL A. Ritchie, 83 Larkfield Road, Lenzie, Glasgow, G66 3AS
G3 WYN J. Gibson, Four Oaks, Tylers Green, Haywards Heath, RH17 5DZ
G3 WYP D. Allan, 283 Cliffe Lane, Gomersal, Cleckheaton, BD19 4SB
G3 WYT M. Edwards, 23 Burnside, Waterlooville, PO7 7QQ
G3 WYW P. Bigwood, 18 The Martins, Thatcham, RG19 4FD
G3 WZE P. Cleary, 531 Diamond Street, San Francisco, USA, 94114
G3 WZG P. Murtha, 9 Cross Street, Southport, PR8 1HZ
G3 WZH N. Ghani, 52B Stormore, Dilton Marsh, Westbury, BA13 4BH
G3 WZI K. Reeves, 9 Tibberton Close, Solihull, B91 3UD
G3 WZJ A. Watt, 30 The Hedgerows, Collingham, NG237RL
G3 WZK S. Beal, 16 Clovelly Avenue, Warlingham, CR6 9HZ
G3 WZO J. Kyriakides, 16 Wise Lane, London, NW7 2RE
G3 WZP G. Budden, 7 Ashburton Gardens, Bournemouth, BH10 4HP
G3 WZR Dr R. Wright, 2 Jackson Close, Devizes, SN10 3AP
G3 WZS H. Williams, 7 Munn Drive, Po Box 276, Tobermory, Canada, N0H 2R0
G3 WZT J. Matthews, 46 Park Lane West Grinstead, Horsham, RH13 8LT
G3 WZW G. Laycock, 1 Campsall Cottage, Churchfield Road, Doncaster, DN6 9BY
G3 WZZ A. Huddleston, Willow Bank Cottage, Willow Bank, Keighley, BD20 5AN
G3 XAB D. Whittaker, 2 Stone Edge, Halifax Road, Burnley, BB10 3QH
G3 XAC C. Whitehead, 10 Berkeley Drive, Read, Burnley, BB12 7QG
G3 XAG J. Gibbon, The Bungalow, Manless Terrace, Saltburn-by-the-Sea, TS12 2DQ
G3 XAP A. Ashton, 2 Wickham Road, Thwaite, Eye, IP23 7EE
G3 XAQ A. Ibbetson, Katallin, Town Lane Chartham Hatch, Canterbury, CT4 7NN
G3 XAU T. Woodward, 33 Common Road, Hemsby, Great Yarmouth, NR29 4LT
G3 XAW M. Chouings, 32 Nunney Close, Keynsham, Bristol, BS31 1XG
G3 XAX A. Paley, 19 Arbour Lane, Wickham Bishops, Witham, CM8 3NS
G3 XBE H. Walton, 12 Le Page Court, Nottingham, NG8 3ES
G3 XBF Barking Radio and Electronics Society, c/o S. Peat, 64 Grange Road, Romford, RM3 7DX
G3 XBH G. Thompson, 25A Copleston Road, London, SE15 4AN
G3 XBI P. Boast, 19 Main Road, West Huntspill, Highbridge, TA9 3QU
G3 XBM R. Lapthorn, 7 Mill Close, Burwell, Cambridge, CB25 0HL
G3 XBN F. Chamberlain, 43 Old Mill Close, Patcham, Brighton, BN1 8WE
G3 XBQ A. Weseley, Loves House, Goudhurst Road, Tonbridge, TN12 9NB
G3 XBW M. Wells, 15 Rivers Reach, Frome, BA11 1AQ
G3 XBX D. Harris, The White House, High Street, Newnham, GL14 1BW
G3 XBY D. Harvey, 38 School Road, Shirley, Solihull, B90 2BB
G3 XBZ P. Ciotti, 6 Bascott Road, Bournemouth, BH11 8RH
G3 XCD H. Martin, 17 Vyner Road, Wallasey, CH45 6TE
G3 XCE E. Wells, 23 Briarfield Road, Poulton-le-Fylde, FY6 7PW
G3 XCJ W. Burden, 44 Spekehill, Eltham, London, SE9 3BW
G3 XCK J. Pegrum, 14 The Leys, Langford, Biggleswade, SG18 9RS
G3 XCO Dr D. Meldrum, 34 Graham Road, Ipswich, IP1 3QF
G3 XCS C. Squires, 5 Frith Road, Saltash, PL12 6EL
G3 XCT D. Dade, 40 Compton Avenue, Brighton, BN1 3PS
G3 XCW G. Winter, 14 Drakes Lea, Evesham, WR11 3BJ
G3 XCY K. Bristow, 34 Stanier Road, Preston, Weymouth, DT3 6PD
GI3 XCZ G. Martin, 100 Drumconnelly Road, Gortaclare, Omagh, BT79 0XS
GI3 XDD S. Crampton, 135A Ballymena Road, Doagh, Ballyclare, BT39 0TN

G3 XDK A. Maris, 140 Edward Street, Brighton, BN2 0JL
G3 XDL A. Long, 2A Hawthorndene Road, Bromley, BR2 7DY
G3 XDM A. Benson, 31 Oakhill Drive, Welwyn, AL6 9NW
G3 XDP G. Wilkinson, 509 Warrington Road, Culcheth, Warrington, WA3 5QY
G3 XDS P. Wilde, 5 Ruddington Court, Mansfield, NG18 4QD
G3 XDU K. Whitbread, 27 Duckmill Crescent, Duckmill Lane, Bedford, MK42 0AF
GI3 XDX G. Mcdowell, 13 Redford Road, Cullybackey, Ballymena, BT43 5PR
G3 XDY J. Quarmby, 12 Chestnut Close, Rushmere St. Andrew, Ipswich, IP5 1ED
G3 XDZ Z. Skrobanski, 1035 Pine Grove Pointe Drive, Roswell, USA, GA 30075-2704
G3 XEC G. Grundy, Route De L'Angouiniere, la Roche Sur Yon, France, 85000
G3 XED C. Masters, 79 Kings Head Lane, Bristol, BS13 7DB
G3 XEF M. Fleetwood, Hemmet, 235 Shingle Hill Way, Gundaroo, Australia, NSW 2620
G3 XEI J. Hooper, Long Barn House, Bolney Road, Horsham, RH13 8AZ
G3 XEN P. Mullineaux, 27 Ashfield Avenue, Lancaster, LA1 5EB
G3 XEP White Rose ARS, c/o E. Hannaby, 170A Weston Drive Otley, Leeds, LS21 2DT
GI3 XEQ J. Bailie, 7 Houston Road, Belfast, BT6 9SE
G3 XER D. Mannix, 34 Ashby Road, Ticknall, Derby, DE73 7JJ
G3 XEV J. Cooper, 34 Arcal Street, Dudley, DY3 1TG
G3 XEW G. Childs, 115 Summerhouse Drive, Bexley, DA5 2ER
G3 XEY A. Robinson, 48 Colton Road, Shrivenham, Swindon, SN6 8AZ
G3 XFD Dr R. Mannion, Flat 1, 1 Spencer Road, Bournemouth, BH1 3TE
G3 XFF E. Tuddenham, 42 Garrison Lane, Felixstowe, IP11 7RP
G3 XFU I. Hasman, Fleetway, The Spinney, Newark, NG24 2NT
G3 XG Braintree & District ARS, c/o M. Kendall, 88 Coldnailhurst Avenue, Braintree, CM7 5PY
G3 XGC Dr G. Cottrell, 36 Davenant Road, Oxford, OX2 8BY
G3 XGD G. Watson, 6 The Avenue, Lyneal, Ellesmere, SY12 0QJ
G3 XGE P. Greenhalgh, 13 Primrose Avenue, Urmston, M41 0TY
G3 XGH W. Jamison, Horseshoe Cottage, Town Fold, Stockport, SK6 5BT
G3 XGK C. Langley, Clarence Cottage, Commodore Road, Lowestoft, NR32 3NF
G3 XGU K. Hill, 42 Greenleafe Avenue, Doncaster, DN2 5RF
G3 XGV G. Fowles, Ruby House, Broad Marston, Stratford-upon-Avon, CV37 8XY
G3 XGW K. Yates, Tibblestone Lodge, Ashton Road, Beckford, Tewkesbury, GL20 7AU
G3 XGY B. Harris, 36 Holland Way, Blandford Forum, DT11 7RU
G3 XHB G. Bentley, 2 Conan Drive, Richmond, DL10 4PQ
GW3 XHD B. Walters, 16 Broomhill, Port Talbot, SA13 2US
GW3 XHG D. Griffiths, 7 Canning Street Ton Pentre, Pentre, CF41 7HF
G3 XHM A. Lewis, Bradley Villa, 41 West Street, Ryde, PO33 2UH
G3 XHW Dr J. Morris, 2 The Corniche, Sandgate, Folkestone, CT20 3TA
G3 XHY Dr C. Tinline, 56 Appletree Lane, Redditch, B97 6SE
G3 XHZ J. Farrer, Woodside Cottage, Catmere End, Saffron Walden, CB11 4XG
G3 XIA P. Bates, 28 Salvington Hill, Worthing, BN13 3AT
G3 XIB B. Johnson, 30 Tamar Way, Gunnislake, PL18 9DH
G3 XIG C. Graham, 1 Arnhem Green, Dorchester, DT1 2PS
G3 XIH W. Dixon, 17 Chestnut Bank, Scarborough, YO12 5QJ
G3 XII F. Harrison, 78 Lancaster Lane, Leyland, PR25 5SP
G3 XIP D. Aspinall, 53 Springhill Road, Fen Drayton, Cambridge, CB24 4SR
G3 XIQ K. Finch, 116 Wisbech Road, Outwell, Wisbech, PE14 8PF
G3 XIR I. Deane, 26 Callin Court Grey Friars, Chester, CH1 2NW
GW3 XIS Dr R. Belcher, 8 Bishops Grove, Sketty, Swansea, SA2 8BE
G3 XIV G. Bulleyment, 30 Brackley Avenue, Fair Oak, Eastleigh, SO50 8FL
G3 XIX J. Hobin, 14 St. Martins Green, Trimley St. Martin, Felixstowe, IP11 0UU
G3 XIY R. Hall, 22 Cumbria Close, Thornbury, Bristol, BS35 2YE
G3 XIZ C. Osborn, 116 Holme Court Avenue, Biggleswade, SG18 8PB
GW3 XJC B. Luke, 33 Maiden Street, Cwmfelin, Maesteg, CF34 9HP
GM3 XJE Dr P. Duffett-Smith, 3 Simpson Place, Carnoustie, DD7 7PJ
G3 XJI W. Wilkinson, 1 Scafell Drive, Kendal, LA9 7PE
G3 XJM J. Sawdy, 41 Ashbarn Crescent, Winchester, SO22 4QH
G3 XJN H. Duncombe, La Rochelle, Thakeham Copse, Pulborough, RH20 3JW
G3 XJP P. Rhodes, Danvers House, Wigmore, Leominster, HR6 9UF
GW3 XJQ M. Shelley, Sunray, Pendine, Carmarthen, SA33 4PD
G3 XJR A. Dickinson, 1 Pearl Bank, Apartment #33-04, Singapore, Singapore, 169016
G3 XJS P. Barville, Felucca, Pinesfield Lane, West Malling, ME19 5EN
G3 XJW L. Rix, 63 Edendale Road, Melton Mowbray, LE13 0EW
G3 XJZ C. Sykes, 15 Morpeth Close, Wirral, CH46 6HQ
GW3 XKB K. Bevan, Renhold, 25 Bryn Gannock, Conwy, LL31 9UG
G3 XKD M. King, 15 Glebe Road, Prestbury, Cheltenham, GL52 3DG
G3 XKE C. Evans, 8 Blakelands Avenue, Sydenham, Leamington Spa, CV31 1RJ
G3 XKG R. Stanton, 16 Ashwood Park, Fetcham, Leatherhead, KT22 9NT
G3 XKH B. Ward, 12 Pagets Road, Bishops Cleeve, Cheltenham, GL52 8AG
G3 XKL C. Gill, Little Acre, Plough Lane Marston, Devizes, SN10 5SR
G3 XKS H. Grattan, Grattan Grange, St. Breward, Bodmin, PL30 3PN
G3 XKU M. Rose, 71 Maryon Road, Ipswich, IP3 9NJ
G3 XKV R. Stratton, 60 Lateward Road, Brentford, TW8 0PL
G3 XKX D. Wills, 70 Hidcote Road, Oadby, Leicester, LE2 5PF
G3 XKY G. Schrager, Flat 4, 3 The Park, London, N6 4EU
G3 XLE J. Vaughan, Eastwood Lodge, Main Road, Boston, PE22 7JU
G3 XLG R. Spreadbury, Lings Farm, Blacksmiths Lane Forward Green, Stowmarket, IP14 5ET
G3 XLI P. Holland, 38 Marlin Square, Abbots Langley, WD5 0EG
G3 XLL J. Lockwood, 22 Egremont Road, Diss, IP22 4NF
G3 XLN Dr D. Russell, 29 Gold Street, Hanslope, Milton Keynes, MK19 7LU
G3 XLP I. Richardson, Brockwood, Grove Road, Ryde, PO33 3LH
G3 XLR A. Bunyan, 87 Seymer Road, Romford, RM1 4LA
G3 XLW D. Powell, Broomhill, Aveton Gifford, Kingsbridge, TQ7 4NE
G3 XLX R. Littlewood, Brewery Farm, Old Coach Road, Axbridge, BS26 2EH
G3 XLZ J. Tozer, 54 Ganges Road, Plymouth, PL2 3AZ
G3 XMB R. Richardson, 42 King Edwards Road, South Woodham Ferrers, Chelmsford, CM3 5PQ
G3 XMC D. Brain, Orchardleigh, Bristol Road, Bristol, BS40 6HF
G3 XMG M. Graham, 30 Moorlands Road, Thornton, Liverpool, L23 1US
G3 XMH P. Brumfitt, 85 Centenary Way, Copcut, Droitwich, WR9 7TD
G3 XMK A. Flather, 10 Oakleigh Court, Aston Lodge Park, Stone, ST15 8LA
G3 XMM T. Morgan, 32 Grasmere Road Longlevens, Gloucester, GL2 0NQ
G3 XMP A. Brasier, The Bears, Moor End Lane, Fakenham, NR21 0EJ
G3 XMQ P. Eggleton, 12 Newbold Road, Wellesbourne, Warwick, CV35 9NZ
GM3 XMY D. Hobden, 1 Larch Avenue, Glenrothes, KY7 5TE
G3 XNE A. Smyth, 3 Carteret Road, Bude, EX23 8DD
G3 XNN R. Jephcott, 3 Chatsworth Park, Thornbury, Bristol, BS35 1JF
GD3 XNU J. Craine, Mwyllin Squeen, Station Road, Ballaugh, Isle Of Man, IM7 5AH
G3 XNX D. Chivers, 51 Alma Road, Brixham, TQ5 8QR
G3 XOB D. Ellacott, 39 Canford Lane, Bristol, BS9 3DQ
G3 XOC M. Cooley, 21 Castle Road, Newport, PO30 1DT
G3 XOD R. Horsman, 65 Pendennis Park, Brislington, Bristol, BS4 4JL
G3 XOI A. Gordon, 20 Hawkins Crescent, Shoreham-by-Sea, BN43 6TP
GJ3 XOJ D. Gray, La Brecque, La Grande Route De La Cote, St Clement, Jersey, JE2 6FP
G3 XOK R. Kearney, 32 Springfield Road, Lower Somersham, Ipswich, IP8 4PQ
G3 XOP P. Featherstone, 2 Firs Close, Whitchurch, Aylesbury, HP22 4LH
GM3 XOQ P. Weller, Mither Tap, Bridge Road, Kemnay, Inverurie, AB51 5QT
G3 XOU D. Wright, Cross Cottage, 208 Whitchurch Road, Tavistock, PL19 9DQ
G3 XOV R. Johnson, 29 Hungary Hill, Stourbridge, DY9 7PS
G3 XPA R. Bevan, Sitio Do Laranjeiro 616F, Moncarapacho, Olhao, Portugal, 8700 077
G3 XPC R. Chapman, 22 Windsor Ride, Finchampstead, Wokingham, RG40 3LG
G3 XPD D. Smith, 5 Peel Street, Stafford, ST16 2DZ
G3 XPI B. Hallows, 3 Southdown Close, Rochdale, OL11 4PP
G3 XPJ K. George, 34 Third Avenue, Northville, Bristol, BS7 0RT
GW3 XPK J. Dore, Henfaes Isaf, Llangurig, Llanidloes, SY18 6SN
G3 XPM R. Tinson, 1 Plompton Grove, Harrogate, HG2 7DP
G3 XPQ G. Black, 24 Mount Drive, Leyburn, DL8 5JQ
G3 XPR I. Bassett-Smith, Grey Gables, Southam Road, Cheltenham, GL52 3BB
G3 XPT G. Symonds, 45 Westfield Road, Dereham, NR19 1JB
G3 XPU C. Woodley, 170 Rugby Road, Burbage, Hinckley, LE10 2ND
G3 XPW C. Moller, 344 High Street, Cottenham, Cambridge, CB24 8TX
G3 XPY A. Bagley, 11 Glamis Road, Newquay, TR7 2RY
G3 XPZ J. Appleton, 66 Bolton Road West, Ramsbottom, Bury, BL0 9ND
G3 XQJ G. Wren, 2 Netheredge Close, Knaresborough, HG5 9BZ
G3 XQM A. Finch, 15 The Brooks, Burgess Hill, RH15 8TR
GW3 XQO P. Salomon, 28 Ansell Road, Wrexham, LL13 9NQ
GM3 XQP Dr A. French, 83/16 Hopetoun St., Edinburgh, EH7 4NJ
G3 XQZ Sir P. Simpson, 8 Church Road, Egleton, Oakham, LE15 8AD
G3 XRC D. Carlsen, 57 Chignal Road, Chelmsford, CM1 2JA
G3 XRD G. Knight, 57B Oliver Road, Kirk Hallam, Ilkeston, DE7 4JY
G3 XRI P. Williams, 2 Sycamore Avenue, Newton-le-Willows, WA12 8LT
G3 XRJ S. Chappell, Lyonesse, Trebehor, St. Levan, Penzance, TR19 6LX
G3 XRK D. Griffin, 12 Charles Road, Whittlesey, Peterborough, PE7 2RG
GW3 XRM D. Dunn, 9 Mill Bank Estate, Llandegfan, Menai Bridge, LL59 5RD
G3 XRN Royal Naval Auxiliary Service ARS, c/o G. Axford, 24 Jack Branch Court, Wash Lane, Clacton-on-Sea, CO15 1EJ
GI3 XRQ Bangor and District ARS, c/o R. White, 28 Lord Warden'S Parade, Bangor, BT19 1YU
G3 XSC K. Southgate, 10 Cott Road, Lostwithiel, PL22 0ET
G3 XSD Prof. G. King, 1 Spring Gardens, North Baddesley, Southampton, SO52 9JG
G3 XSI T. Haslam, 29 Backmoor Road, Norton, Sheffield, S8 8LB
G3 XSN B. Donn, 7 Thurne Way, Liverpool, L25 4SQ
GW3 XSR A. Sutton, 3 Pendre Walk, Tywyn, LL36 0AY
G3 XSV A. Hydes, Woodcroft Bath Road, Langford, Bristol, BS40 5EB
G3 XSZ F. Mundy, 5 The Paddocks, New Haw, Addlestone, KT15 3LX
G3 XTC P. Borrett, 21 Kenley Walk, Cheam, Sutton, SM3 8ES
G3 XTH G. King, 73 Grand Avenue, Hassocks, BN6 8DD
G3 XTI J. Jarvie, 11 Guild Road Aston Cantlow, Henley-in-Arden, B95 6JA
G3 XTN R. Hough, 1 Fiddler Hall, Newby Bridge, Ulverston, LA12 8NQ
G3 XTP K. Lloyd, 10 The Verneys, Cheltenham, GL53 7DB
G3 XTQ M. Draycott, 119A High Street North, Stewkley, Leighton Buzzard, LU7 0EX
G3 XTR B. Dunn, 12 Campbell Road, Westville, South Africa, 3629
G3 XTT D. Field, Daisy Cottage, Henton, Wells, BA5 1PD
G3 XTZ G. Phillips, 27 Stanley Road, Ashford, TW15 2LP
G3 XUC A. Keohane, 6 Birchwood Fields, Tuffley, Gloucester, GL4 0AL
G3 XUD P. Kirby, 7 Lillywhite Close, Burgess Hill, RH15 8TF
G3 XUF A. Warner, 79 Kelvin Grove, Portchester, Fareham, PO16 8LF
G3 XUH R. Pearson, 8 St. Benets Close, Walton-Le-Dale, Preston, PR5 4UT
G3 XUM J. Moran, 30 Elsie Street, Farnworth Bolton, BL4 9HT
G3 XUP G. Everest, 13 Noel Rise, Burgess Hill, RH15 8BW
GM3 XUW R. Johnston, 123 Craigmount Brae, Edinburgh, EH12 8XW
G3 XUX E. Fitzgerald, 4 Southwick Road, Wickham, Fareham, PO17 6HS
G3 XVA D. Pickles, 40 Jupiter House Hindhead Knoll Walnut Tree, Milton Keynes, MK7 7FH
G3 XVB A. Vizoso, Stones Farm House, Faringdon Road, Faringdon, SN7 8NP
G3 XVC M. Collopy, New Gardens, Burnt House Lane, Dartford, DA2 7SP
G3 XVG T. Barraclough, 52 Denshaw Avenue, Denton, Manchester, M34 3NX
G3 XVH S. Franklin, 337 Hendon Way, London, NW4 3NB
G3 XVL C. Mccarthy, 23 Hornbeam Road, Stowupland, Stowmarket, IP14 4DL
G3 XVP P. Pimblott, 40 Richmondfield Lane, Barwick In Elmet, Leeds, LS15 4EZ
GW3 XVQ F. Jinks, 28 St. Anns Road, Bonnie View, Blackwood, NP12 3PG
G3 XVS R. Kitching, Ashby Down, Streetway Road, Grateley, SP11 7EH
G3 XVV M. Salmon, 54 Church Road, Rivenhall, Witham, CM8 3PH
G3 XVW K. Ball, 39 Spinney Close, Northfield, Birmingham, B31 2JG
G3 XVX H. Dearing, The Gables Bungalow, Ollerton Road Little Carlton, Newark, NG23 6BP
G3 XVY P. Coull, 40 Wear Bay Crescent, Folkestone, CT19 6BA
G3 XWA J. Ennis, 30 Hillcrest Avenue, Carlisle, CA1 2QJ
G3 XWB C. Cadogan, 8 Horncliffe Close, Rawtenstall, Rossendale, BB4 6EE
G3 XWD D. Watts, 40 Outlands Drive, Hinckley, LE10 0TW
G3 XWG M. Shen, 23 Back Lane, Whixley, York, YO26 8BG
G3 XWH R. Horton, 23 Back Lane, Whixley, York, YO26 8BG
G3 XWL J. Cripps, 3 Queens Court, Queens Road, Cranbrook, TN18 4JE
G3 XWN G. Laycock, 48 Marina Terrace, Golcar, Huddersfield, HD7 4RA
G3 XWO T. Lowe, 688 East J Street, Chula Vista, USA, 92010
G3 XWU R. Pearson, 10 Eastleigh Close, Boldon Colliery, NE35 9NG
G3 XWV C. Hamilton, The Stables Foinavon, Newsham, Richmond, DL11 7RD
GW3 XXB A. Evans, 74 Celyn Avenue, Cardiff, CF23 6EQ
G3 XXC K. Rigelsford, 14 Glebelands Avenue, South Woodford, London, E18 2AB

G3 XXE P. Williams, 4 Plantation Close, Aller, Newton Abbot, TQ12 4NS
G3 XXF C. Vine, 14 Hamilton Road, St Albans, AL1 4PZ
G3 XXG P. Sharpen, 3 Western Road, Urmston, Manchester, M41 6LE
G3 XXH Dr S. Watts, 58 Cambridge Avenue, New Malden, KT3 4LE
G3 XXM D. Richards, 836 The Ridge, St Leonards-on-Sea, TN37 7PX
G3 XXN F. Pickersgill, 3 Church St., Langold, Worksop, S81 9NW
G3 XXO E. Birks, 46 Curzon Drive, Worksop, S81 0LP
G3 XXQ L. Dixon, 24 Angerton Gardens, Newcastle upon Tyne, NE5 2JB
G3 XXR P. Higton, 13 Wilton Avenue, Bradley, Huddersfield, HD2 1RN
G3 XXX S. Bradnam, 39 Pelham Way, Cottenham, Cambridge, CB24 8TQ
G3 XYB G. Maitland, 7 Battery Road, Cowes, PO31 8DP
G3 XYC P. Crust, 16 London Lane Wymeswold, Loughborough, LE12 6UB
G3 XYD W. Gordon-Laycock, 51 Overbrook, Swindon, SN3 6AR
G3 XYE J. Clifton, Romford Cottage, Romford, Verwood, BH31 7LE
G3 XYF J. Wresdell, Bracey Bridge Farm, Harpham, Driffield, YO25 4DE
G3 XYG Dr M. George, 30 Northfield Park, Barnstaple, EX31 1QA
G3 XYH J. Hill, 35 Windmill Avenue, Marshalswick, St Albans, AL4 9SJ
G3 XYI D. Fearnley, 14 Salforal Close, Rettendon Common, Chelmsford, CM3 8EL
G3 XYJ R. Walker, 27 Archer Close, Kings Langley, WD4 9HF
G3 XYO S. Line, Cottles House, Cottles Lane, Exeter, EX5 1EE
G3 XYV I. Cooper, 118 Stagsden Road, Bromham, Bedford, MK43 8QJ
GW3 XYW D. Jones, 22 Alltiago Road Pontarddulais, Swansea, SA4 8HU
G3 XYZ Kings Lynn ARC, c/o E. Haskett, 23 Gloucester Road, King's Lynn, PE30 4AB
G3 XZB N. Edwards, 14 Churchill Close, Cowes, PO31 8HQ
G3 XZG J. Browne, 82 Cresswell Road, Chesham, HP5 1TA
G3 XZK D. Gething, 31 Lower Lodge Lane, Hazlemere, High Wycombe, HP15 7AT
GI3 XZM D. Vance, The Eaves, Reagh Island, Newtownards, BT23 6EN
G3 XZO M. Rhodes, 21 Halford Road, Ettington, Stratford-upon-Avon, CV37 7TL
G3 XZP D. Holburn, 19 Whitwell Way, Coton, Cambridge, CB23 7PW
G3 XZV J. Sonley, Ravenscliffe, Lands Lane, Knaresborough, HG5 9DE
G3 XZX J. Lowe, 24 Candish Drive, Plymouth, PL9 8DB
G3 XZY T. Garner, 122 Wainfleet Road, Skegness, PE25 3RX
GM3 YAC P. Howarth, Kilbeg House, Teangue, Isle of Skye, IV44 8RQ
G3 YAD M. Goodrich, 2 Highworth Crescent, Yate, Bristol, BS37 4EY
GW3 YAF T. Davies, Rhyd Wen, Llangyndeyrn, Kidwelly, SA17 5EN
G3 YAG W. Thompson, 2 Fern Close, Frimley, Camberley, GU16 9QU
G3 YAI T. Mills, 16 Hunts Hill, Glemsford, Sudbury, CO10 7RL
G3 YAJ D. Sellen, Prospect House, Wignall Street, Manningtree, CO11 2HX
GM3 YAO F. Offler, Ar Dachaidh, 3 King David Drive, Montrose, DD10 0SW
G3 YAR I. Gildersleve, 7 Oak Park Road, Newton Abbot, TQ12 1RQ
G3 YBA E. Cooper, Flat, 23 Westminster Crescent, Sheffield, S10 4EU
G3 YBE E. Gilbert, 2 Church Field, Stanford, Ashford, TN25 6UA
G3 YBG J. Rabjohns, Quarries Bungalow, Barley Lane, Exeter, EX4 1TA
G3 YBH P. Storey, Po Box 47060, Denman Street Postal Outlet, Vancouver, Canada, V6G 3E1
G3 YBK R. Donno, 6 Mincinglake Road, Exeter, EX4 7EA
G3 YBM R. Mitchell, 98 Marlborough Drive, Burgess Hill, RH15 0EU
GW3 YBN C. Davies, 31 Park Prospect, Graigwen, Pontypridd, CF37 2HF
G3 YBO R. Baines, 10 Chartwell Ave, Wingerworth, Chesterfield, S42 6SP
G3 YBP A. Young, Hillcrest, Graynfylde Drive, Bideford, EX39 4AP
GM3 YBQ K. Horne, 10 Blair Place, Kirkcaldy, KY2 5SQ
G3 YBR S. Cook, 6 Essex Court, Marlton, New Jersey, USA, 8053
G3 YBS R. Lindsay-Smith, 58 Chalgrove Road, London, N17 0JD
G3 YBU B. Whittle, Holmlea, Main Road, Hull, HU12 9NG
G3 YBY I. Mccarthy, 76 High Street, Purton, Swindon, SN5 4AD
GI3 YBZ J. Mccann, 61 Glengawna Road, Glengawna, Omagh, BT79 7WJ
GM3 YCB S. Riddell, 16 Lewis Drive, Old Kilpatrick, Glasgow, G60 5LE
G3 YCE J. Peck, 7 Paddock Close, Radcliffe-On-Trent, Nottingham, NG12 2BX
G3 YCH J. Sharman, 7 Watkins Way, Paignton, TQ3 3JJ
G3 YCJ P. Sheard, 52 Victoria Road, Elland, HX5 0QA
G3 YCO R. Lewis, 115 Chester Road, Whitby, Ellesmere Port, CH65 6SB
G3 YCV J. Hibbert, 5 Cliff View Road, Cliffsend, Ramsgate, CT12 5ED
G3 YCX A. Cain, 42 Wood Lane, Prescot, L34 1LW
G3 YCY R. Barrett, 47 Marshals Drive, St Albans, AL1 4RD
G3 YDD Hereford ARS, c/o T. Bridgland-Taylor, 1 Overbury Court, Hereford, HR1-1DG
G3 YDE W. Bagwell, 93 Broadley Drive, Torquay, TQ2 6UT
GI3 YDH M. Mcintyre, 36 Beechgrove Park, Belfast, BT6 0NR
GI3 YDM J. Dunlop, 34 Ballybentragh Road, Dunadry, Antrim, BT41 2HJ
GM3 YDN D. Nutt, Little Craigfin, Kilkerran, Maybole, KA19 8LR
G3 YDO W. Mcnally, 2607 The Highlands Dr., Sugar Land, Texas, USA, 77478
G3 YDT W. Patterson, 32 The Pagoda, Maidenhead, SL6 8EU
G3 YDY P. Selwood, 43 Keene Way, Galleywood, Chelmsford, CM2 8NT
G3 YEC R. Edmondson, 16 Orchard Close, Copford, Colchester, CO6 1DB
G3 YED R. Nettleton, 4 Sycamore Close, Stratford-upon-Avon, CV37 0DZ
G3 YEG N. Sears, 17 Walls Road, Bembridge, PO35 5RA
G3 YEK R. Johnston, Flat 906, Orchard Plaza, Poole, BH15 1EH
GD3 YEO R. Rimmer, 27 Manor Lane, Farmhill, Douglas, Isle Of Man, IM2 2NP
G3 YEP R. Wakeley, Fir Bank, Fell Lane, Penrith, CA11 8BJ
G3 YEQ L. Miller, 28 Arthur Road Cliftonville, Margate, CT9 2EN
G3 YER D. Lowe, Flat 14, Marklands, 37 Julian Road, Bristol, BS9 1NP
G3 YEU B. Short, 83 Rowanfield Road, Cheltenham, GL51 8AF
GM3 YEW D. Morris, Ash Cottage, Perth Road, Perth, PH2 9LW
G3 YFD W. Hewitt, 22 Derby Road, Stockport, SK4 4NE
G3 YFE J. Shaw, 57 London Road, Amesbury, Salisbury, SP4 7EE
G3 YFG S. Westell, 2 Whiteacre Lane, Barrow, Clitheroe, BB7 9BJ
G3 YFK P. Mcalister, Lower Winnington Farm, Winnington, Shrewsbury, SY5 9DJ
G3 YFL C. Wright, Holmwood, Brackley Avenue, Hook, RG27 8QX
G3 YFM M. Smyth, 21 The Paddock, Longworth, Abingdon, OX13 5BX
G3 YFO M. Bunce, 36 Burlington Road, Burnham, Slough, SL1 7BQ
G3 YFP J. Bottomley, 42 Birkdale Court, Fornham St. Martin, Bury St Edmunds, IP28 6XF
G3 YFU E. Tomlin, Indalo, Magna Mile, Market Rasen, LN8 6AJ
G3 YFV P. I'Anson, Flat 297, Latymer Court, London, W6 7LD
G3 YFW P. Rosen, 21 Hadley Heights, Hadley Road, Barnet, EN5 5QH
G3 YGA E. Warwick-Oliver, St. Madron, Throwleigh, Okehampton, EX20 2HX
G3 YGB J. Coleman, 19 Megdale, Matlock, DE4 3JW
G3 YGC E. Elliott, 18 Bear St., Lowerhouse, Burnley, BB12 6NQ
G3 YGD D. Brown, 5 Meadow Edge Barrowford, Nelson, BB9 6BT
G3 YGE J. Okas, Dipley Springs, Dipley Common, Hook, RG27 8JS
G3 YGF Dr J. Gannaway, Highview House, Winchester Road Fair Oak, Eastleigh, SO50 7HB
G3 YGG J. Kelly, 79 West Hill Avenue, Epsom, KT19 8JX
GW3 YGH A. Hughes, 267 Cyncoed Road, Cardiff, CF23 6PA
G3 YGJ D. Brierley, 4 Waterloo Terrace, Bideford, EX39 3DJ
G3 YGL F. Smith, 34 Bridle Road Eastham, Wirral, CH62 8BR
G3 YGM M. Osborne, Cheriton, Alexandra Road, St Ives, TR26 1ER
G3 YGR C. Thomas, Oakdene, School Lane, Reading, RG7 3ES
G3 YGZ A. Walsh, Green Royd, Saddleworth Road, Halifax, HX4 8NU
G3 YHB J. Baker, 109 Bermuda Road, Wirral, CH46 6AX
G3 YHC W. Hermes, 22 Mallinson Crescent, Harrogate, HG2 9HP
G3 YHF C. Skelcher, 51 Blenheim Road, Moseley, Birmingham, B13 9TY
G3 YHG D. Harding, 17 Summerfield Close, Wokingham, RG41 1PH
G3 YHH J. Froud, Summer Park, New Road, Teignmouth, TQ14 8UF
G3 YHI R. Vale, High House, Mounts Lane, Daventry, NN11 3ES
G3 YHK J. Clemence, Aroha, 67 Tomline Road, Felixstowe, IP11 7NX
G3 YHM R. Harvey, 26 Birkdale Road, Worthing, BN13 2QY
G3 YHN C. Pedley, 25 Fallowfield Road, Walsall, WS5 3DH
G3 YHO R. Yaxley, Fallow View, Swaffham Road, Dereham, NR19 2LX
G3 YHQ D. Mercer, 19 Kingsfield Drive, Didsbury, Manchester, M20 6JA
GW3 YHR C. Briscoe, Gadlas House, Ffordd Top Y Rhos, Treuddyn, Mold, CH7 4NE
GJ3 YHU D. Robinson, 16 Seagrove Court, La Rue De La Corbiere, St Brelade, Jersey, JE3 8HN
G3 YHV C. Chidgey, 46 Station Road, Shirehampton, Bristol, BS11 9TX
G3 YIA M. Harris, 100 Chapel Lane, Wymondham, NR18 0DN
G3 YIB M. Knowler, 2 Vulcan Street, Southport, PR9 0TW
G3 YIC V. Sedgley, 25 Avenue Road, Weymouth, DT4 7JH
G3 YIE E. Lusty, Stanley End Farm, Bell Lane, Stroud, GL5 5JY
G3 YIF J. Weiner, 1 Chippendayle Drive, Harrietsham, Maidstone, ME17 1AD
GW3 YIH F. Cobb, Mon Reve, Rhodfa Nant, Abergele, LL22 9ND
G3 YII N. Smith, Creekside, Nursery Lane, South Wootton, King's Lynn, PE30 3NA
G3 YIK Dr J. Morgan, Cedars, Springhill, Longworth, Abingdon, OX13 5HL
G3 YIN E. Ellery, 14 Four Acres, Bideford, EX39 3RW
G3 YIQ R. Jones, The Old Vicarage, Upper South Wraxall, Bradford-on-Avon, BA15 2SB
G3 YIR Anglo Americn Rd, c/o A. Ward, 20 Tower Close, Costessey, Norwich, NR8 5AU
G3 YIW J. Gallop, 55 Somervell Drive, Fareham, PO16 7QW
G3 YIY R. Ingram, 11 Bank Terrace, Mevagissey, St Austell, PL26 6QZ
G3 YJA B. Porter, 49 Beverley Road, Leamington Spa, CV32 6PW
G3 YJD J. Davies, 25 Harkness Close, Bletchley, Milton Keynes, MK2 3NB
G3 YJE P. Merriman, The Old Croft, Brimpsfield, Gloucester, GL4 8LD
G3 YJG G. Mason, 8 Leighton Road, Sunderland, SR2 9HQ
G3 YJN R. Hodge, 36 Binswood End, Harbury, Leamington Spa, CV33 9LN
G3 YJP D. Benham, 19 Benham Road, Otis, USA, 1253
G3 YJQ F. Bourne, 78 Normandy Way, Plymouth, PL5 1SR
G3 YJS M. Roche, 8 Northdown Close, Penenden Heath, Maidstone, ME14 2ER
G3 YJV Dr M. Hawthorne, 49 Broom Close, Teddington, TW11 9RL
G3 YJW R. Whitehouse, White Court, Camp Hill, Tonbridge, TN11 8LE
G3 YJZ A. Mitchell, 89 Queen Annes Grove, Enfield, EN1 2JU
GM3 YKA J. Wiewiorka, 47 Albert Avenue, Grangemouth, FK3 9AT
G3 YKB J. Hodgson, 18 Nascot Place, Watford, WD17 4QT
G3 YKC D. Fayers, 1 Tismeads Crescent, Swindon, SN1 4DP
G3 YKI K. Vickers, Brick Kiln House, Ditton Priors, Bridgnorth, WV16 6TW
G3 YKK C. Donne, The Hideaway, Lease Lane, East Halton, Immingham, DN40 3PT
G3 YKO D. Darwood, Briarwood Cottage, Packhorse Lane, Birmingham, B38 0DN
GM3 YKP W. Sutton, Old Police Station, Dunbeath, KW6 6EA
G3 YKS R. Butlin, 48 Roman Way, Market Harborough, LE16 7PQ
G3 YKW Capt. R. Walker, 17 Ballantyne South, Montreal West, Canada, QC H4X 2BI
GW3 YKZ M. Biddiscombe, 20 Arlington Close, Malpas, Newport, NP20 6QF
G3 YLA J. Bacon, 37 Burgh Lane, Mattishall, Dereham, NR20 3QP
GM3 YLD J. Frew, Queens Cottage, 87 Queen St., Dunoon, PA23 8AX
GJ3 YLI A. Morrissey, Flat 4, 1 Springfield Crescent, St Helier, Jersey, JE2 4GL
G3 YLJ G. Whitehead, 29 Coulsons Road, Bristol, BS14 0NN
GJ3 YLN J. Speller, Lindau, Gorey Village Main Road, Jersey, JE3 9EP
GM3 YLU W. Taylor, 4 Newbie Barns, Newbie, Annan, DG12 5QL
G3 YLV P. Jones, Oaklea, Cadney Lane, Whitchurch, SY13 2LW
G3 YLW P. Lascelles, 4 Meadowsweet Close, Snettisham, King's Lynn, PE31 7UG
G3 YLY B. Hawes, 15 Bridge Lane, Wimblington, March, PE15 0RR
G3 YMC D. Sergeant, 8 Toll Gardens, Bracknell, RG12 9EX
G3 YMD Dover ARC, c/o P. Love, 2 Meadway, Dover, CT17 0PS
G3 YMH R. Wainwright, The Olives, High Street, Uckfield, TN22 4LB
G3 YMM T. Campbell Davis, 9 Cloister Road, Acton, London, W3 0DE
G3 YMN J. Rhys, 10 Mulberry Close, Weybridge, KT13 8RA
GI3 YMT M. Higgins, 1 Cairnshill Park, Belfast, BT8 6RG
G3 YMU J. Hibberd, Barn Cottage, School Lane, Crewe, CW3 0BA
G3 YMV I. Machardie, 20 Arley Close, Swindon, SN25 4TP
G3 YMW D. Sapsworth, 16 Laxton Avenue, Hardwick, Cambridge, CB23 7XL
GI3 YMY N. Newell, 18 Kilmaine Avenue, Bangor, BT19 6DU
G3 YMZ A. Plummer, 6 Shelley Place, Kallaroo, Perth, Australia, 6025
G3 YNC C. Adams, 18 Glenavon Road, Highcliffe, Christchurch, BH23 5PN
GM3 YND I. Simpson, 3 Ravenscraig Terrace, Steelend, Dunfermline, KY12 9LU
G3 YNF B. Turner, South Lodge, Stanford Road, Fineshade, NN17 3BA
G3 YNJ C. Powell, 38 Braeside Road, St. Leonards, Ringwood, BH24 2PH
G3 YNK D. Evans, Michigan Villa, Wall Road, Hayle, TR27 5HA
G3 YNL O. Price, Queens Arms House 1 King Street Odiham, Hook, RG29 1NN
G3 YNO M. Booth, 28 Humber Road, North Ferriby, HU14 3DW
G3 YNU Dr I. Stevenson, 18 Sittingbourne Road, Wigan, WN1 2RR
G3 YOA A. Adams, The Gables, Chapel Road Trunch, North Walsham, NR28 0QG
G3 YOC R. Moore, 38 Sandygate, Wath-Upon-Dearne, Rotherham, S63 7LR
GM3 YOI Dr K. Falconer, Lumbo Farmhouse, St Andrews, KY16 8NS
G3 YOL S. Cole, Halebrook, Bridgwater Road, Winscombe, BS25 1NH
G3 YOM K. Beddoe, 30 Tamella Road, Botley, Southampton, SO30 2NY
G3 YON F. Webster, 16 Pembroke Road, Dronfield, S18 1WH
G3 YOO J. Webster, 7 Hardwick Avenue, Allestree, Derby, DE22 2LN

GM3 YOR A. Givens, 5 Langhouse Place, Inverkip, Greenock, PA16 0EW
G3 YOV T. Gammage, 23 Artizan Road, Northampton, NN1 4HU
G3 YOY H. Clark, 2 Chestnut Close, Peakirk, Peterborough, PE6 7NW
G3 YPD P. Chester, 44 Richmond Drive, Lichfield, WS14 9SZ
G3 YPE M. Greenwood, 21 Dobb Top Road, Holmbridge, Holmfirth, HD9 2PQ
G3 YPK S. Wallis, Mas La Floride - Bp61068, 250 Chemin De Magarnaud, Sommieres, France, 30250
G3 YPL D. Gray, 19 Westbury Gardens, Higher Odcombe, Yeovil, BA22 8UR
G3 YPM Dr R. Moore, 20 Ebrington Road, Malvern, WR14 4NL
G3 YPS S. Atkinson, 13 Charles Street Dn21 2Ja, Gainsborough, DN21 2JA
G3 YPT P. Tomes, 86 Hurn Road, Christchurch, BH23 2RP
G3 YPU P. Koker, 1 Cedar Court, Bridlington, YO16 6ZQ
G3 YPW P. Willingham, 49 Creek Road, Hayling Island, PO11 9RA
G3 YPY A. Head, 32 Weald View Road, Tonbridge, TN9 2NQ
G3 YPZ J. Petters, 218 New House Farm Hospital Drove, Long Sutton, PE12 9EN
G3 YQA F. Wilson, 26 Humber Road, North Ferriby, HU14 3DW
G3 YQB D. Rankin, 105 Sparrowhawk Way, Hartford, Huntingdon, PE29 1XY
G3 YQC J. Wood, 14 Little Paradise, Marden, Hereford, HR1 3DR
G3 YQF R. Linford, Department Of Communication, And Electronic Engineering, Drake Circus Plymouth, PL4 8AA
GM3 YQG N. Waylett, 3A Arabella, Tain, IV19 1QH
GW3 YQH R. Evens, 41 Heol Gwys, Upper Cwmtwrch, Swansea, SA9 2XQ
G3 YQJ P. Burnet, 166 Oundle Road, Thrapston, Kettering, NN14 4PQ
GM3 YQK J. Dillon, Steelbank Cottage, Dalgraven, Kilwinning, KA13 6PL
G3 YQL P. Murfitt, 53A Codnor Denby Lane, Codnor, Ripley, DE5 9SP
GW3 YQM Dr D. Wynford-Thomas, Coach House, The Hollies, Pentrepoeth Road, Newport, NP10 8RT
G3 YQN R. Trott, 120 Portland Road, Bromley, BR1 5AZ
G3 YQO D. Kirkwood, Greensand Lodge, High Street, Lidlington, Bedford, MK43 0QR
GW3 YQP C. Hardie, 3 Berth Glyd, Gyffin, Conwy, LL32 8NP
G3 YQV C. Railton, 14 Copford Lane, Long Ashton, Bristol, BS41 9NF
G3 YQW M. Funnell, 15 Mcindoe Road, East Grinstead, RH19 2DD
G3 YQZ M. Johnson, 36 Coventry Road, Bulkington, Bedworth, CV12 9ND
G3 YRC Yarmouth Radio Club, c/o P. Nicholls, 53 Fastolff Avenue Gorleston, Great Yarmouth, NR31 7ND
G3 YRH B. Dodds, 1 Croft View, Killingworth, Newcastle upon Tyne, NE12 6BT
GI3 YRL J. Branagh, 17 Rathmoyle Park West, Carrickfergus, BT38 7NG
GW3 YRP I. Dudley, Tynewydd, Llansantffraid, SY22 6TW
G3 YRQ I. Parkinson, 61 Cinnamon Lane, Fearnhead, Warrington, WA2 0AG
G3 YRU P. Wilby, 58 Main Street, Thorner, Leeds, LS14 3BU
G3 YRX I. Elston, 11 Knowle Drive, Exwick, Exeter, EX4 2DF
G3 YSD J. Murdoch, 32 Scalegill Road, Moor Row, CA24 3JL
G3 YSG M. Taylor, 4 Yew Tree Court, Botley Road, Southampton, SO31 1EA
GW3 YSI P. Tipping, Pen Yr Enfys Ashdale Lane, Llangwm, Haverfordwest, SA62 4NU
G3 YSK A. Button, 13 Taplings Road, Weere, Winchester, SO22 6HE
G3 YSM M. Davidson, 14 Fuchsia Walk, Wirral, CH49 3AG
G3 YSN H. Smith, Carnview, Tregender Lane, Penzance, TR20 8DJ
G3 YSQ A. Pratt, 7 The Croft, West Hanney, Wantage, OX12 0LD
G3 YSR C. Beattie, Mayerin, Churchway, Aylesbury, HP17 8RG
G3 YSW N. Thrower, 8 Upton Gardens, Worthing, BN13 1DA
G3 YSX Dr S. Bryant, 154 London Road North, Merstham, Redhill, RH1 3AA
GW3 YTC G. Rutherford, 29 Sisial Y Mor, Rhosneigr, LL64 5XB
GD3 YTE P. Gill, Hollybank, Sulby Bridge, Sulby, Isle Of Man, IM7 2AY
G3 YTG T. Blair, 56 Rue Du Golfe De Barbareu, Etaules, France, 17750
G3 YTI S. Cooper, 24 Cambridge Street, Darwen, BB3 3JH
GW3 YTL C. Lewis, 26 Ffordd Gwenllian, Llay, Wrexham, LL12 0UW
G3 YTN R. Hill, 35 Coxwold View, Wetherby, LS22 7PU
G3 YTQ C. Kidd, 118 Segensworth Road, Fareham, PO15 5EQ
G3 YTR M. Hatt, 1 Larches Way, Crawley Down, Crawley, RH10 4UJ
GM3 YTS R. Ferguson, 19 Leighton Avenue, Dunblane, FK15 0EB
G3 YTT W. Taylor, 1 Milby Drive, Nuneaton, CV11 6JR
G3 YTU C. Coward, 27 Rothley Chase, Haywards Heath, RH16 3PE
G3 YTW G. Clarke, 117 Bermuda Village, Nuneaton, CV10 7PW
G3 YTX G. Clamp, 9 Furse Close, Camberley, GU15 1BF
G3 YTY M. Edib, 84 Connaught Gardens, London, N13 5BT
G3 YTZ M. Readman, Flat 1-8, 365 Wilmslow Road, Manchester, M14 6AH
GW3 YUC D. Davies, 30 Wern Isaf, Dowlais, Merthyr Tydfil, CF48 3NY
G3 YUD P. Hawkins, 37 Alexandra Road, Dorchester, DT1 2LZ
G3 YUH R. Ayling, 25 Nash Court Road, Margate, CT9 4DH
G3 YUJ R. Steed, 53 Colchester Road, Ipswich, IP4 3BT
GD3 YUM M. Parnell, 1 Derwent Drive, Onchan, Douglas, Isle of Man, Isle Of Man, IM3 2DF
G3 YUQ E. Elsley, 25 Elmsdale Road, Wootton, Bedford, MK43 9JW
G3 YUU C. Lord, Green Sleeves, Marley Road, Maidstone, ME17 1BS
G3 YUX R. Moore, 69 Ivatt, Tamworth, B77 2HQ
G3 YUZ I. Wilson, 23 Alyth Road, Bournemouth, BH3 7DG
G3 YVA B. Edwards, 774, Calle 21 S.O., Puerto Rico, Puerto Rico, 921
G3 YVH A. Boyne, 18 Crow Lane West, Newton-le-Willows, WA12 9YG
G3 YVI R. Gilbert, Mayville, New Copse, Alton, GU34 5NP
G3 YVK H. Tabberer, 101 Broadclyst Gardens, Southend-on-Sea, SS1 3QU
GW3 YVN N. Little, Brynhyfryd Llansteffan, Carmarthen, SA33 5HA
GU3 YVV R. Outhwaite, Le Courtillet, La Route De Sausmarez, Guernsey, GY4 6SF
G3 YVW B. Blackwell, 19 Tokely Road, Frating, Colchester, CO7 7GA
GM3 YVX D. Coupar, 32 Gillies Place, Broughty Ferry, Dundee, DD5 3LE
G3 YVY D. Hanley, 5 Hallcroft Close, Billingham, TS23 1QN
G3 YVZ T. Gardner, 64 Balmoral Terrace, Heaton, Newcastle upon Tyne, NE6 5YA
G3 YWA E. Pepper, 30 Westfield Drive, Harpenden, AL5 4LP
G3 YWF P. Smith, 18 David Avenue, Cliftonville, Margate, CT9 3DU
G3 YWH F. Hill, 24 Mount St. James, Guide, Blackburn, BB1 2DR
G3 YWL C. Coverdale, 2 Lillypilly Lane, Cooranbong, Australia, 2265
G3 YWM P. Hubert, 575 Bramford Lane, Ipswich, IP1 5JX
G3 YWP Capt. P. White, Po Box 86524 Al Jazeera Po, Ras Al Khaimah, United Arab Emirates, 86524
G3 YWS J. Smith, 16 Woodlands, Winthorpe, Newark, NG24 2NL
G3 YWT P. Smith, Beechwood, Clarendon Road, Salisbury, SP5 3AT
G3 YWU S. Fisher, Arkle, 31 Frith Avenue, Northwich, CW8 2JB
G3 YWW A. Carpenter, 17 Victoria Avenue Upwey, Weymouth, DT3 5NG
G3 YWX I. Poole, 17 Glebe Road, Dorking, RH4 3DS
G3 YXH S. Marshall, The Bunglaow, Llamedos Stables, Fieldhead Lane, Bradford, BD11 1JL
G3 YXM D. Pick, 178 Alcester Road South, Kings Heath, Birmingham, B14 6DE
G3 YXN P. Whalley, Southerly, The Common, Hanworth, Norwich, NR11 7HP
G3 YXO Dr D. Watson, Norton, Gote Lane, Ringmer, Lewes, BN8 5HX
G3 YXQ Dr R. Ireland, 31 St James Street, Sackville, New Brunswick, Canada, E4L 4L7
G3 YXS D. Naylor, 7 Ruthven Court, Litherland, Liverpool, L21 2PE
G3 YXW P. Dunford, 79 Summerdown Road, Eastbourne, BN20 8DQ
GM3 YXY A. Thomson, Roselea, Lanark, ML11 7SE
G3 YYC Dr G. Sharples, 3A Green Lane Park Homes, Breinton, Hereford, HR4 7PN
G3 YYE H. Lawrence, 14 Manor Lane, Dinnington, Sheffield, S25 2SW
G3 YYG J. Bolton, 3 Fyne Drive, Linslade, Leighton Buzzard, LU7 2YG
G3 YYK R. North, 13 Nicholson Way, Havant, PO9 3AZ
G3 YYN P. Spurr, Windmill Farm Barn, High Street, Milton Keynes, MK14 5AX
G3 YYR Dr A. Parry, 17 May Pole Knap, Somerton, TA11 6HP
G3 YYW G. Wills, 137 Aldermans Drive, Peterborough, PE3 6BB
G3 YYZ J. Cuthbert, Fulbeck, 48 Mayes Lane, Harwich, CO12 5EJ
G3 YZK Dr G. West, 6 Lammas Close, Cowes, PO31 8DT
G3 YZN R. Awbery, Dashwood, Beacon View Road, Godalming, GU8 6DU
G3 YZO Dr R. Wilson, Kellers, Duck Street, Wendens Ambo, Saffron Walden, CB11 4JU
G3 YZQ P. Williams, 41 Church Street, St. Georges, Telford, TF2 9JZ
G3 YZR J. Porter, Birklands, 16 The Oval, Scarborough, YO11 3AP
G3 YZT A. Slaney, 1 Marlborough Way, Goring-By-Sea, Worthing, BN12 4HG
G3 YZV E. Woollard, 24 Griffin Close, Twyford, Banbury, OX17 3HR
G3 YZW A. Armstrong, 423 Bideford Green, Linslade, Leighton Buzzard, LU7 2TY
G3 YZZ K. Beverstock, 16 Chaucer Close, Emmer Green, Reading, RG4 8PA
G3 ZAB I. Pivac, 427 Great North Road Glendene, Auckland, New Zealand, 602
G3 ZAE S. Benstead, 15 Les Congeries, Dournazac, France, 87230
G3 ZAG B. Taylor, 27 Ridgeway, Wellingborough, NN8 4RU
G3 ZAJ D. Sutton, Deer Wood, Canterbury Road, Ashford, TN25 4DF
G3 ZAL L. Huggett, 24 Hockers Lane, Detling, Maidstone, ME14 3JN
G3 ZAU E. Lord, 41 Daven Road, Congleton, CW12 3RB
G3 ZAW K. Bird, Leys House, Main Street, Northampton, NN7 4SH
G3 ZAY M. Atherton, 41 Enniskillen Road, Cambridge, CB4 1SQ
G3 ZBB S. Jackson, Old Fir Tree Inn Peacemarsh, Gillingham, SP8 4EU
G3 ZBF M. Matthews, 13 Bursill Crescent, Ramsgate, CT12 6EZ
G3 ZBG G. Moorfield, 43 Broadmark Lane, Rustington, Littlehampton, BN16 2HH
G3 ZBI Nunsfield House, c/o K. Frankcom, 1 Chesterton Road, Spondon, Derby, DE21 7EN
G3 ZBM W. Worthington, 32 Princess Drive, Wistaston Green, Crewe, CW2 8HS
G3 ZBP M. Baker, 82 Folkestone Road, Copnor, Portsmouth, PO3 6LR
GM3 ZBR K. Melvin, 1 Charleston Village, Charleston, Forfar, DD8 1UF
G3 ZBS J. Mccall, 5 Sundew Close, Wokingham, RG40 5YB
G3 ZBU A. Watt, 5 Brambling Road, Horsham, RH13 6AX
G3 ZBZ B. Cross, 176 Outwood Road, Heald Green, Cheadle, SK8 3LL
G3 ZCA D. Lake, 9 Grafton Close, King's Lynn, PE30 3EZ
G3 ZCD R. Fogg, 24 Edinburgh Gardens, Windsor, SL4 2AN
G3 ZCG K. Young, Flat 36, Brookhurst Court, Leamington Spa, CV32 6PB
G3 ZCH D. Hill, 11 Chapeltown Road, Radcliffe, Manchester, M26 1YF
G3 ZCI R. O'Brien, 9 Holmwood Garth, Hightown, Ringwood, BH24 3DT
G3 ZCJ M. Allerton Austin, 13 Kilpin Green, North Crawley, Newport Pagnell, MK16 9LZ
GI3 ZCK G. Ward, 6B Stranmillis Road, Belfast, BT9 5AA
G3 ZCL G. Hammersley, 74 Cammel Road, West Parley, Ferndown, BH22 8SB
G3 ZCT J. Beehlar, 12 Dulverton Road, Leicester, LE3 0SA
G3 ZCX P. Fort, 1 Lowther Lane, Foulridge, Colne, BB8 7JY
G3 ZCY R. Hogg, 5 Bishop Way, Pateley Bridge, Harrogate, HG3 5LH
G3 ZCZ Dr J. Kasser, 60 Jervois Ave, Magill, Australia, 5072
G3 ZDF J. Kirk, 111 Stockbridge Road, Chichester, PO19 8QR
G3 ZDG S. Cole, 1 The Copse, Exmouth, EX8 4EY
GM3 ZDH R. Dixon, Flat 5, 10 Mavisbank Gardens, Glasgow, G51 1HG
G3 ZDK P. Given, 4 Elveden Drive, Ilkeston, DE7 9JW
G3 ZDM R. Muriel, 13 York Road, Sale, M33 6EZ
G3 ZDQ I. Flemming, Rudderhams Cottage, Blandford Hill, Winterborne Whitechurch, Blandford Forum, DT11 0AA
G3 ZDT P. Morrison, Saddlers, Upper Green Road, Tonbridge, TN11 9PL
G3 ZDU G. Marshall, Bassington, Hulne Park, Alnwick, NE66 3JE
G3 ZDW R. Hyde, 25 The Pastures, Cottesmore, Oakham, LE15 7DZ
G3 ZDY D. Palmer, Flat 3, 60 Millbank, London, SW1P 4RW
G3 ZEB B. Robinson, The Gables, The Street, Great Yarmouth, NR29 4EA
G3 ZED A. Rothwell, Brandon, Manor Brow, Keswick, CA12 4AP
G3 ZEF R. Stephen, 97 Hunters Field, Stanford In The Vale, Faringdon, SN7 8ND
G3 ZEJ R. Smith, 3 Willake Road, Kingskerswell, Newton Abbot, TQ12 5AB
G3 ZEK M. Bailey, 12 Bridgers Mill, Haywards Heath, RH16 1TF
G3 ZEM R. Henderson, Po Box 62155, Pafos, Cyprus, 8061
G3 ZEN Dr A. Glaser, 155 Little Breach, Chichester, PO19 5UA
G3 ZEO S. Wilders, Old Farm Barn, Silkstead Lane, Winchester, SO21 2LG
G3 ZEQ M. Brenig-Jones, Orchard House, Larters Lane, Stowmarket, IP14 5HB
G3 ZER I. Mercer, 28 West Way, Rickmansworth, WD3 7EN
G3 ZES A. Downing, Oaktree Bungalow, The Endway, Chelmsford, CM3 6DU
GM3 ZET Lerwick Radio Club, c/o T. Goodlad, 72 North Lochside, Lerwick, Shetland, ZE1 0PJ
GM3 ZEU C. Clarkson, 8 Moor Place, Portlethen, Aberdeen, AB12 4TF
GD3 ZEX C. Douglas, Sea Villa, The Promenade, Laxey, Isle Of Man, IM4 7DF
G3 ZEZ G. Coleman, 16 Kestrel Way, Clacton-on-Sea, CO15 4JE
G3 ZFC C. Davis, 3 Cross Road, Haslington, Crewe, CW1 5SY
G3 ZFF R. Hornbuckle, 54 Gladys Avenue, Cowplain, Waterlooville, PO8 8HS
G3 ZFP R. Penberthy, 10 Lancot Avenue, Dunstable, LU6 2AW
G3 ZFR R. Harris, Clevelands, Tamworth Road, Coventry, CV7 8JJ
G3 ZFT A. Magnus, Woodland Cottage, Linkside East, Hindhead, GU26 6NY
G3 ZFV J. Watts, Riverside, St. Georges Road, Barnstaple, EX32 7AS
G3 ZFZ G. Gibson, 174 Roose Road, Barrow-in-Furness, LA13 0EE
G3 ZGA J. Hart, Peter-Vischer-Str.9, Marktredwitz, Germany, 95615
G3 ZGC R. Jolliffe, 54 Glendale Avenue, Wash Common, Newbury, RG14 6RU
GM3 ZGH R. Yeoman, 162 Jamphlars Road, Cardenden, Lochgelly, KY5 0ND
G3 ZGI T. O'Neill, Braeside, Cookbury, Holsworthy, EX22 7YG

G3 ZGN P. Swarbrick, 1 Hill View, Charminster, Dorchester, DT2 9QX
G3 ZGP R. Cridland, 13 Clarendon Avenue, Redlands, Weymouth, DT3 5BG
G3 ZGQ L. Mead, 12 Ferniefields, High Wycombe, HP12 4SP
G3 ZGT B. Druce, 25 Boothgate Drive, Howden, Goole, DN14 7EN
G3 ZGU K. Richens, The Old Barn, Marsh Lane, Cheswardine, Market Drayton, TF9 2SF
G3 ZGY G. Paddock, 56 Clee View Road, Wombourne, Wolverhampton, WV5 0BD
G3 ZGZ D. Woodhall, 15 Cherrywood Avenue, Thornton-Cleveleys, FY5 1SU
G3 ZHA G. Gillam, 58 Downhall Road, Rayleigh, SS6 9LY
G3 ZHB A. Stuart, 207 Saunders Lane, Mayford, Woking, GU22 0NT
G3 ZHC N. Willmot, 2 Athlone Road, Walsall, WS5 3QX
G3 ZHE A. Heyes, 20 Walsingham Road, Penketh, Warrington, WA5 2AQ
G3 ZHJ P. Moss, 6 Windmill View, Houghton Conquest, MK45 3GD
G3 ZHK T. Kellow, Glenvale, St. Dominick, Saltash, PL12 6TD
G3 ZHL Dr J. Morgan, Cedars, Springhill, Abingdon, OX13 5HL
G3 ZHO R. Wilton, 23 Moorland View, Liskeard, PL14 3TQ
G3 ZHP D. Marsden, 14 Daisy Drive, Barnsley, S70 4NY
G3 ZHS R. Ray, 37 Doxey Fields, Stafford, ST16 1HJ
G3 ZHT B. Lundean, 13 Isis Close, Lympne, Hythe, CT21 4JQ
G3 ZHU G. Clark, Kenzie, Canterbury Road, Dover, CT15 7HR
G3 ZHV R. Bowler, 21 Pine Close, South Wonston, Winchester, SO21 3EB
G3 ZHZ A. Macfadyen, 19 Oldfield Road Lower Willingdon Bn20 9Qd, Eastbourne, BN20 9QD
G3 ZIB D. Tye, 21 Elmstone Drive, Tilehurst, RG315NS
G3 ZIC J. Viney, 20 South Drive, Upton, CH49 6LA
G3 ZID A. Greathead, 20 Westland, Martlesham Heath, Ipswich, IP5 3SU
G3 ZIE E. Brown, 21 Newbridge Way, Pennington, Lymington, SO41 8BG
G3 ZIF H. Wilson, 6 Risborrow Close, Etwall, Derby, DE65 6HY
G3 ZIG R. Reed, Oak Cottage, Dereham Road, Dereham, NR20 4AA
G3 ZII M. Rathbone, 25 Halsall Road, Southport, PR8 3DB
G3 ZIJ J. Stables, 9 Milbanke Close, Ouston, Chester le Street, DH2 1JJ
G3 ZIK A. Mather, 15 Claughton Avenue, Bolton, BL2 6US
G3 ZIL G. Griffiths, 14 Bassett Close, Southampton, SO16 7PE
G3 ZIM R. Wolten, 12 Well Lane, Liverpool, L16 5ET
G3 ZIN G. Spencer, 16 West Lawn, Ipswich, IP4 3LJ
G3 ZIO C. Harvey, Ham Cottage, 1A Elstan Way, Croydon, CR0 7PR
G3 ZIV K. Nolan, West End Cottage, Woodhall, Selby, YO8 6TG
G3 ZIY R. Drinkwater, 15 Woodside Crescent, Smallfield, Horley, RH6 9NA
G3 ZJF P. Broughton, 7 Old Mill Way, Wells, BA5 2JU
G3 ZJG J. Garner, 50 Thorndale, Ibstock, LE67 6JT
G3 ZJK C. Milner, The Everglades, Sawbridge Road, Grandborough, Rugby, CV23 8DN
G3 ZJO E. Bennett, 44 Central Avenue, Northampton, NN2 8DZ
G3 ZJP W. Fenton, 50 Orion Road, Rochester, ME1 2UH
G3 ZJQ Dr R. Walker, 2 Chelwood Drive, Sandhurst, GU47 8HT
GW3 ZJS J. Smith, Llainfran, New Quay, SA45 9RR
G3 ZJV M. Firth, 63 Sycamore Road, East Leake, Loughborough, LE12 6PP
G3 ZJW Dr B. Mccombe, 208 Thorpe Road, Peterborough, PE3 6LB
G3 ZJX B. Castle, 10 Oakley Drive, Bromley, BR2 8PP
G3 ZJY J. Greenwood, 91 Keyhaven Road, Milford On Sea, Lymington, SO41 0TF
G3 ZJZ J. Mason, 35 Broad Way, Hockley, SS5 5EL
G3 ZKD W. Ball, 6 Coronation Drive, Penketh, Warrington, WA5 2DD
G3 ZKG J. Riley, 41 Church Avenue, West Sleekburn, Choppington, NE62 5XF
G3 ZKH I. Bateman, Jaysville, The Strand, Pershore, WR10 3JZ
G3 ZKI A. Williams, 38 Seneca Street, Bristol, BS5 8DX
G3 ZKN D. Morgan, 38 Ryder Close, Norman Hill, Dursley, GL11 5SG
G3 ZKO P. Lee, 20 Little Haseley, Oxford, OX44 7LH
G3 ZKQ A. Walton, 3 Fox Hill Close, Selly Oak, Birmingham, B29 4AH
G3 ZKZ J. Shaw, 2 Castle Close, Felixstowe, IP11 9NN
G3 ZLD J. Barker, 57 Broom Park, Teddington, TW11 9RS
G3 ZLE D. Ward, 11 Spruce Grove Avenue, Baden, Ontario, Canada, N3A 3P7
G3 ZLF R. Nelson, 225 Walton Road, Chesterfield, S40 3BT
G3 ZLJ E. Dalton, 29 Windmill Lane, Castlecroft, Wolverhampton, WV3 8HJ
G3 ZLM R. Hook, 35 Parkwood Crescent, Hucclecote, Gloucester, GL3 3JH
G3 ZLP J. Campbell, 4 Ladygrove Cottages Preston, Hitchin, SG4 7SA
G3 ZLQ M. Adams, 41 Primrose Lane, Yeovil, BA21 5SH
G3 ZLR A. Ridley, 28 Riverbank, Laleham Road, Staines, TW18 2QE
G3 ZLS S. Craske, Shallowford Holsworthy Road Hatherleigh, Okehampton, EX20 3LE
G3 ZLX E. Jones, 94 Westbrook End, Newton Longville, Milton Keynes, MK17 0BX
GM3 ZMA J. Butler, 11 Quartalehouse, Stuartfield, Peterhead, AB42 5DE
G3 ZME Telford&Dist AR, c/o M. Vincent, 9 Sleapford, Long Lane, Telford, TF6 6HQ
G3 ZMG J. Maughan, 83 Oak Road, Peterlee, SR8 3HU
G3 ZMH D. Mcauslan, Golden Sedge, Street End, Bristol, BS40 7TL
G3 ZMK J. Askew, Po Box 4487, Linstead, St Catherine, Jamaica,
G3 ZML M. Owen, 3 Gordon Road, Mount Waverley, Victoria, Australia, 3149
G3 ZMM R. Hodgkinson11, 39 Oxford Road, Carlton-In-Lindrick, Worksop, S81 9BD
G3 ZMO J. Callum, Dales Barn Top, Town Head, Hawes, DL8 3RH
G3 ZNB P. Hannam, Bogg Hall, Oulston, York, YO61 3RE
GM3 ZNC J. Mulheron, 10 Devonview Place, Airdrie, ML6 9DF
G3 ZND J. Bartlett, Chickamauga, Tinnahinch, Clonaslee, Ireland, R32 T622
G3 ZNE N. Ingle, 81 Redmoor Close, Tavistock, PL19 0ER
G3 ZNG S. Birt, 82 Aintree Road, Thornton-Cleveleys, FY5 5HP
G3 ZNH R. Coombes, 7 Lower Grove, Whitsbury, Fordingbridge, SP6 3QA
G3 ZNK D. Hainsworth, 48 Greenacre Park, Rawdon, Leeds, LS19 6AR
G3 ZNR D. Bailey, 12 St. Philips Drive, Burley In Wharfedale, Ilkley, LS29 7EN
G3 ZNT R. Brown, 282 Luton Road, Dunstable, LU5 4LF
G3 ZNU M. Appleby, 6 Mandeville Road, Prestwood, Great Missenden, HP16 9DS
G3 ZNW R. Blasdell, 32 Fulham Close, Broadfield, Crawley, RH11 9NY
G3 ZOG A. Elliott, 15 Braemar Gardens, East Herrington, Sunderland, SR3 3PX
G3 ZOH B. George, 14 Pondfield Road, Orpington, BR6 8HJ
G3 ZOI D. Deane, 10 Stephens Road, Mortimer Common, Reading, RG7 3TU
G3 ZOL J. Powell, 156 Avon Way, Colchester, CO4 3YP
GU3 ZOM D. Pearson, Tequesta, York Avenue, Port Soif, Vale, Guernsey, GY6 8HS
G3 ZON K. Lacy, 9 Rhodes Way, Tilgate, Crawley, RH10 5DQ
G3 ZOT J. Hewitt, 8 Charles Avenue, Scotter, Gainsborough, DN21 3RR
G3 ZOW K. Clamp, 12 Cowlishaw Close, Shardlow, Derby, DE72 2GS
G3 ZOX P. Durham, 18 Maldon Road Great Totham, Maldon, CM9 8PR
G3 ZOY A. Alldrick, 23 Coxs Close, Nuneaton, CV10 7ET
G3 ZPA D. White, 1 Whaddon Road, Shenley Brook End, Milton Keynes, MK5 7AF
G3 ZPB P. Burton, 202 Coulsdon Road, Coulsdon, CR5 2LF
G3 ZPI G. Braund, 184 Faversham Road, Kennington, Ashford, TN24 9AE
G3 ZPJ M. Symons, 11 Tudor Lodge Park, Truthwall, Penzance, TR20 9BW
G3 ZPK H. Willis, 12 Combe View, Hungerford, RG17 0BZ
G3 ZPL Dr N. Richardson, 501 Forest Avenue, Palo Alto, USA, CA 94301
G3 ZPM A. Stormont, The Hawthorns, Church Lane, Louth, LN11 7JR
G3 ZPR D. Mason, 26 Upton Road, Fleetsbridge, Poole, BH17 7AH
G3 ZPS S. Shorey, 47 Stanham Road, Dartford, DA1 3AN
G3 ZPU A. Nightingale, 42 Spilsby Road, Horncastle, LN9 6AW
G3 ZPW M. Brown, 6 Castle Court, Praa Sands, Penzance, TR20 9SX
G3 ZQB A. Seabrook, 63 St. Annes Road Willingdon, Eastbourne, BN20 9NJ
G3 ZQC J. Smith, 8 Upland Rise, Westbury, BA13 3HW
G3 ZQF S. Carpenter, 4 Mount Court, West Wickham, BR4 9AH
G3 ZQH Dr D. Barrett, Linden House, Clifton Lane, Nottingham, NG11 6AA
G3 ZQI B. Downer, 9 Crabapple Road, Dereham, NR20 3GH
G3 ZQJ B. Stagg, 1 Naunton Way, Leckhampton, Cheltenham, GL53 7BQ
G3 ZQL S. Murray, Old Court, Westleigh, Tiverton, EX16 7HT
G3 ZQQ J. Peden, 51A Bewdley Road, Kidderminster, DY11 6RL
G3 ZQR B. Downton, Lickwith Cottage, Monkokehampton, Winkleigh, EX19 8SL
G3 ZQS Int. Morse Preservation Society, c/o R. Walker, Fists The International Morse Preservation Society, Po Box 6743, Tipton, DY4 4AU
G3 ZQT J. Yu, Hunterscombe, Dorking Road, Leatherhead, KT22 8JT
G3 ZQU M. Goodrum, Cedars, Church Lane, Stonham Parva, Stowmarket, IP14 5JL
G3 ZQW B. Barrington, Pinlands Cottage, Bines Road, Horsham, RH13 8EQ
G3 ZQY N. Clark, Chelsworth, Heaton Grange Road, Romford, RM2 5PP
G3 ZRA Dr R. Elliot, 1321 East Bailey Road, Naperville, USA, 60565
G3 ZRB D. Hill, 872 Oldham Road, Rochdale, OL11 2BN
G3 ZRE P. Ottewell, 30 Cumberland Avenue, Leyland, PR25 1BH
G3 ZRG I. Steward, Keeper'S Cottage, Banville Lane, Cromer, NR27 9RN
G3 ZRH A. Stokes, 34 Shenfield Crescent, Brentwood, CM15 8BW
G3 ZRJ A. Butler-Roskilly, 2 North Road, Pennymoor, Tiverton, EX16 8LQ
G3 ZRL A. Mcwatters, 22 Church Close Swillington, Leeds, LS26 8QJ
G3 ZRM M. Payne, 3 Waterside Close, Bordon, GU35 0HB
G3 ZRN D. Catherwood, 14 Hatton Lane, Hatton, Warrington, WA44BY
G3 ZRQ D. Maxfield, 40 Fegg Hayes Road, Stoke-on-Trent, ST6 6RA
G3 ZRR M. Samuel, 71 Brighton Road, South Croydon, CR2 6EE
G3 ZRS P. Rodmell, 2 Meadow Way, Walkington, Beverley, HU17 8SD
GM3 ZRT W. Strachan, Inverlaroch Albert Road, Ballachulish, PH49 4JR
G3 ZRX Raynet Association, c/o T. Lundegard, Saxby, Botsom Lane, Sevenoaks, TN15 6BL
G3 ZRY G. Stott, 8 Willow Road, Chinnor, OX39 4RA
G3 ZSB V. Poore, 216 Powder Mill Lane, Twickenham, TW2 6EJ
G3 ZSF A. Houltby, 2 Sinderson Road, Humberston, Grimsby, DN36 4UF
GM3 ZSH J. Donaldson, 4 Wallsend Court, Dunfermline, KY12 9BE
G3 ZSJ R. Troughton, 4 Owletts, Worth, Crawley, RH10 7SQ
G3 ZSQ R. Dunham, 42 Marsdale Drive, Stockingford, Nuneaton, CV10 7DE
G3 ZSS P. Bacon, 3 The Grange, Woodmancote, Emsworth, PO10 8UX
G3 ZST T. Surgey, The Little Wood, Windmill Lane, Ladbroke, Southam, CV47 2BN
G3 ZSU S. Scannell, 20 Queens Road Wilbarston, Market Harborough, LE16 8QJ
G3 ZSX Dr K. Craig, 20 Alexander Close, Abingdon, OX14 1XA
G3 ZSZ R. James, 283 High St., New Whittington, Chesterfield, S43 2AP
G3 ZTB R. Ranson, Flat 1, The Steyne, Alexandra Road, Harrogate, HG1 5JS
GW3 ZTH J. Ludlow, 44 Fox Hollows, Brackla, Bridgend, CF31 2NG
G3 ZTI K. Marshall, 2 Keepers Mill, Woodmancote, Cheltenham, GL52 9QS
G3 ZTJ C. Morgan, The Villa, The Green, Wallsend, NE28 7PH
GI3 ZTL F. Convery, 2 Coolagh Road, Maghera, BT46 5JR
GM3 ZTP S. Elwell-Sutton, 17 Lintrathen Gardens, Dundee, DD3 8EJ
G3 ZTR D. Lockwood, 25 Thorntondale Drive, Bridlington, YO16 6GW
G3 ZTT Mid Cheshire AR, c/o D. Bevan, 46 Park Lane, Hartford, Northwich, CW8 1PY
G3 ZTU J. Mace, Westgate, Bignor, Pulborough, RH20 1PQ
G3 ZTV P. Webster, 3 Templemere, Norwich, NR3 4EF
G3 ZTX P. Angell, Star Hill House, Star Hill, Forest Green, Nailsworth, Stroud, GL6 0NJ
G3 ZTY J. Yale, 15 Rectory Avenue, Corfe Mullen, Wimborne, BH21 3EZ
G3 ZTZ P. Howell, 1 Jasmine Close, Littlehampton, BN17 6UP
G3 ZUB M. Davison, 10 Springfield Close, Loughborough, LE11 3PT
G3 ZUC D. Cardell, 22 Millview Road, Heckington, Sleaford, NG34 9JP
G3 ZUE A. Nicholas, Verriotts Lane, Morecombelake, Bridport, DT6 6DU
G3 ZUI M. Johnson, Greentiles, Main Road, Maltby Le Marsh, Alford, LN13 0JW
G3 ZUK Prof. R. Whitehead, Church View, Church End, Cambridge, CB21 5PE
G3 ZUL B. Kennedy, 24 Mallards Close, Alveley, WV15 6JL
G3 ZUM B. Lonnon, 5 Mickle Meadow Water Orton, Birmingham, B46 1SN
G3 ZUN D. Sharpe, 12 Belgrave Crescent, Seaford, BN25 3AX
G3 ZUO D. Ham, Upton Manor Lodge, Upton Manor Road, Brixham, TQ5 9QZ
G3 ZUS N. Ewer, The Roost, Lyonshall, Kington, HR5 3HZ
G3 ZUT M. Thorne, Rossendale Main Road Easter Compton, Bristol, BS35 5RE
G3 ZUZ M. Fox, 41 Elgin Avenue, Ashton in Makerfield, WN4 0RH
G3 ZVC B. Comer, Corner House, High Street, Fairford, GL7 4EQ
G3 ZVH D. Bedford, 17 Tewkesbury Close Upton, Chester, CH2 1NF
G3 ZVI P. Longhurst, 18 Austen Close, Exeter, EX4 8HB
G3 ZVK J. Simons, 120 Bond Way, Hednesford, Cannock, WS12 4SN
G3 ZVM A. Greenbank, Grahamsley, Westburn, Ryton, NE40 4EU
G3 ZVN G. Peck, 4 Koonowla Close, Biggin Hill, Westerham, TN16 3BJ
G3 ZVQ J. Bridge, 8 Highfield Grove, Lostock Hall, Preston, PR5 5YB
G3 ZVS E. Park, Waterside Cottage, Bowden Lane, High Peak, SK23 0QF
G3 ZVT G. Rabstaff, Evesham, 26 Lichfield Drive, Manchester, M25 0HX
G3 ZVV J. Gellatly, 11 Archers Drive, Bilsthorpe, Notts, NG22 8SD
G3 ZVW S. White, Heatherleigh Crewkerne Road, Axminster, EX13 5SX
GI3 ZVZ Dr D. Nicholls, 2 Printshop Road, Templepatrick, Ballyclare, BT39 0HZ
G3 ZWD P. Flicos, 35 The Broadway, Northbourne, Bournemouth, BH10 7EU
G3 ZWK D. Raimbach, 12 Vineys Gardens, Tenterden, TN30 7AZ
G3 ZWM I. Morrison, The Vicarage, 1A Church Road, Sandy, SG19 2JY
G3 ZWN A. Slingsby, 10D Stanbury Road, Thruxton, Andover, SP11 8NS
G3 ZWP R. Davies, Flat 24, Hurley Court, Bracknell, RG12 9QH
G3 ZWR N. Hay, 19 Logan Road, Walkerville, Newcastle upon Tyne, NE6 4SY

GM3 ZXB A. Robertson, 77 Cobden St., Dundee, DD3 6DD
G3 ZXD M. French, Po Box 217, Leigh, New Zealand, 947
G3 ZXF D. Corner, 122 Stortford Hall Park, Bishop's Stortford, CM23 5AP
GM3 ZXG J. Higgins, 9 Waverley Street, Greenock, PA16 9DH
GW3 ZXI S. Brennan, 9 Swn Y Nant, Church Village, Pontypridd, CF38 1UE
G3 ZXM M. Brown, Curraghmore, Tullogher, Co Kilkenny, Ireland, X91 R642
G3 ZXO J. Burnie, 1 Chapel Meadow, Buckland Monachorum, Yelverton, PL20 7LR
G3 ZXV P. Veale, 13 Lawford Gardens, Kenley, CR8 5JJ
G3 ZXW J. Midmore, 9 Whiteways, Wimborne, BH21 2PQ
G3 ZXY P. Holtham, 21 Sherborne Place, Chapel Hill, Australia, QLD 4069
G3 ZXZ M. Stokes, Chaos Cottage, Otley Road, Bingley, BD16 3AY
G3 ZYC M. Sneap, Ivy Farm Bungalow, Farm Close, Pentrich, Ripley, DE5 3RR
G3 ZYD R. Alton, 23 Cemetery Road, Belper, DE56 1EJ
GM3 ZYE R. Bellerby, Glenamour, Newton Stewart, DG8 7AE
G3 ZYL Dr G. Bowhay, Windwhispers, Lewannick, Launceston, PL15 7QD
G3 ZYP A. Matheson, 1 St. Edmunds Close, Bromeswell, Woodbridge, IP12 2PL
G3 ZYQ A. Robinson, 6 The Crescent, Minster On Sea, Sheerness, ME12 3BQ
G3 ZYR A. Booer, Lower Farm Barn, Duck End Lane, Witney, OX29 5RH
G3 ZYV D. Ferigan, 191 Gillingham Road, Gillingham, ME7 4EP
G3 ZYX R. Offord, 10 Barberry Way, Blackwater, Camberley, GU17 9DX
G3 ZYY T. Day, Box 204, The Postroom, Calle San Jaime 5, Benijofar, Spain, 3178
G3 ZYZ Dr M. Joiner, 22 Sanderstead Court, Addington Road, South Croydon, CR2 8RA
GM3 ZZA P. Rose, 4 Heatherfield Glade, Livingston, EH54 9JE
G3 ZZD S. Ireland, Po Box 55, Glen Forrest, Glen Forest, Australia, 6071
G3 ZZF D. Woolley, 21 Lulworth Avenue, Wembley, HA9 8TP
G3 ZZH M. Battersby, 25 Lowther Close, Emmbrook, Wokingham, RG41 1JE
G3 ZZI G. Smith, 34 Huria Lane, Woodend, New Zealand, 7610
G3 ZZL Dr S. Keightley, Riverrun, 4, Northern Burway Laleham Reach, Chertsey, KT16 8RW
G3 ZZM M. Robinson, 7400 Old Bunch Road, Wendell, USA, 27591
GD3 ZZN M. Rickward, Dunsandle 12 Fairway Close, Port Erin, Isle of Man, IM9 6LS
G3 ZZP D. Matthews, 7 Boulsworth Avenue, Hull, HU6 7DZ
G3 ZZQ R. Ludwell, Church Lodge, Thetford Road, Thetford, IP24 2QX
G3 ZZS R. Wills, 21 Woodford Road, Glenholt Park, Plymouth, PL6 7HX
G3 ZZU C. Waldron, 22 Windermere Road, Patchway, Bristol, BS34 5PW
G3 ZZV G. Evans, 20 Creekside View, Tresillian, Truro, TR2 4BS
G3 ZZW P. Chimber, 202 Wintersdale Road, Leicester, LE5 2GP
G3 ZZX A. Evans, Ashlea, Aston Munslow, Craven Arms, SY7 9ER
G3 ZZZ J. Gibbs, 32 Gresham Avenue, Margate, CT9 5EH

G4

GM4 AAF Dundee ARC, c/o J. Wilson, 20 Ballumbie Gardens, Dundee, DD4 0NR
G4 AAH K. Lawson, 233 Southwell Road West, Mansfield, NG18 4HF
G4 AAL J. Layton, Meadow View, Martley, Worcester, WR6 6QA
G4 AAQ P. Butterfield, 29 Aire Street, Knottingley, WF11 9AT
G4 AAR Ashford ARC, c/o J. Wellard, 19 South Motto, Kingsnorth, Ashford, TN23 3NJ
G4 AAX Northumbria ARC, c/o G. Emmerson, 72 The Gables, Widdrington, Morpeth, NE61 5RB
G4 ABC Thornbury & S. Gloucestershire ARC, c/o P. Smart, 142 Finch Road, Chipping Sodbury, Bristol, BS37 6JB
G4 ABE J. Ellis, 4 Hazelmount Crescent, Warton, Carnforth, LA5 9HS
G4 ABL A. Howell, 46 Acre Court, Andover, SP10 1HH
G4 ABN T. Atkins, 55 Havenbrook Blvd, Willowdale, Ontario, Canada,
G4 ABQ J. Hudson, 46 High Street, Odell, Bedford, MK43 7BB
G4 ABW L. Willey, 7 Oaklands Road, Four Oaks, Sutton Coldfield, B74 2TB
G4 ABX Dr B. Macaulay, Old Chapel, Chapel Street, Swinford, Lutterworth, LE17 6AZ
G4 ABY D. Green, 2 Fossfield Winstone, Cirencester, GL77JY
G4 ACF ACF/CCF Int R N, c/o Capt. R. Cogzell, Flat 242, Clydesdale Tower Holloway Head, Birmingham, B1 1UJ
G4 ACI J. Blackburn, 40 Carlton Avenue, Upholland, Skelmersdale, WN8 0AE
G4 ACJ H. Reeve, 11 Heather Drive, Ferndown, BH22 9SD
G4 ACL D. Atkinson, 38 Hornbeam Road, Theydon Bois, Epping, CM16 7JX
GM4 ACM A. Miller, 38 Randolph Road, Broomhill, Glasgow, G11 7LG
G4 ACP J. Scherrer, 26 Grange Way, Willington, Bedford, MK44 3QW
G4 ACS G. Weale, 11 Heather Drive, Kinver, Stourbridge, DY7 6DR
G4 ACU M. Levy, 34428 Yucaipa Blvd, E346, Yucaipa, USA,
G4 ACW N. Roe, 10 Ramsdean Road, Stroud, Petersfield, GU32 3PJ
G4 ACY R. Ratcliffe, 173 Montague Road, Bilton Hill, Rugby, CV22 6LG
G4 ACZ R. Mutton, Summer Hayes, Mill Lane, Alcester, B49 6LF
G4 ADD W. Ricalton, 4 South Road, Longhorsley, Morpeth, NE65 8UW
G4 ADE M. Woollin, 14 St. Nicholas Drive, Hornsea, HU18 1EW
G4 ADF P. Harrison, Siedendörp 11, Fehmarn, Germany, 23769
G4 ADG P. West, 136 The Dashes, Harlow, CM20 3RU
G4 ADJ P. Hampton, 45 Mortlake Avenue, Redhill, Worcester, WR5 1QB
G4 ADK R. Cutbush, 60 Culver Way, Sandown, PO36 8QL
GW4 ADL Dr T. Davies, 44 Carnglas Road, Sketty, Swansea, SA2 9BW
G4 ADM A. Maish, 73 Edenfield Gardens, Worcester Park, KT4 7DX
G4 ADP P. Mccurrie, Lakefields, Drake Close, Southampton, SO40 4XB
G4 ADR N. Ayres, 17 Ludsden Grove, Thame, OX9 3BY
G4 ADS J. Chisman, 115 St. Lukes Avenue, Ramsgate, CT11 7HT
G4 ADV Newquay & District ARS, c/o K. Francks, 63 Parc Godrevy, Newquay, TR7 1TY
G4 AEB T. Baines, 6 Brading Avenue, Clacton-on-Sea, CO15 4PA
G4 AED B. Cator, 9 Saham Road Watton, Thetford, IP25 6EA
G4 AEE M. Bedford, 4 Holme House Lane, Oakworth, Keighley, BD22 0QY
G4 AEG I. Kemp, 21 Rednal Road, Birmingham, B38 8DT
G4 AEH J. Lee, 44 Howard Road, Nuneaton, CV10 7ES
G4 AEI G. Prater, Heathfield, Wyndham Lane, Salisbury, SP4 0BY
GM4 AEK P. Boswell, Coralach, Dunvegan, Isle of Skye, IV55 8WF
G4 AEL R. Cox, 34 Ratcliffe Drive, Stoke Gifford, Bristol, BS34 8UD
G4 AEM P. Ellis, 96 Whitelands Avenue, Chorleywood, Rickmansworth, WD3 5RG
G4 AEO P. Hunt, 93 Park Road, Coalville, LE67 3AF
G4 AEP W. Thomas, High Field Claypit Lane, Froxfield, GU32 1DD
G4 AER R. Sobey, 5 Fairway Close, Liphook, GU30 7XD
G4 AES K. Walker, 3 Glen View, Stile, Sowerby Bridge, HX6 1NL
G4 AEV R. Anderson, Trinafour, Abingdon Road, Abingdon, OX13 6NU
G4 AEY R. Twist, 1 Birchwood Drive, Rushmere St. Andrew, Ipswich, IP5 1EB
G4 AEZ B. Oughton, 326 Ware Road, Hailey, Hertford, SG13 7PG
G4 AFA Dr N. Porter, 13 Charfield Close, Winchester, SO22 4PZ
G4 AFE J. Tallentire, 4 Alston Road, Middleton-In-Teesdale, Barnard Castle, DL12 0UU
G4 AFF S. Cooper, Fairfield, Southburgh Lane, Hingham, Norwich, NR9 4PP
GI4 AFH G. Phillips, 38 Sketrick Ind Park, Newtownards, BT23 3BN
G4 AFI A. Cheetham, 39 Burns Avenue, Church Crookham, Fleet, GU52 6BN
G4 AFQ D. Warner, Treeside, School Lane, Gimingham, Norwich, NR11 8HJ
G4 AFR F. Nicholson, 5 Friars Terrace, Barrow-in-Furness, LA13 0BX
G4 AFS T. Bucknell, 7 Alexander Court Sandbeds, Keighley, BD20 5NW
G4 AFT D. Randles, Long Reach, Westerns Lane, Harrogate, HG3 3PB
G4 AFU P. Rollin, Farthings, Burneston, Bedale, DL8 2JE
G4 AFX A. Moore, Garden Wing, Copinger Hall, Stowmarket, IP14 3DJ
G4 AFY R. Perrin, 8 Granville Crest, Kidderminster, DY10 3QS
G4 AFZ V. Bott, 25 Finkle Street, Hensall, Goole, DN14 0QY
G4 AGC C. Wortham, 57 Cranleigh Drive, Swanley, BR8 8NZ
G4 AGE R. Evans, Mansfield, 1 Horsehead Lane, Chesterfield, S44 6HU
GM4 AGG W Scotland ARS, c/o A. Stewart, Three Acres, Cochno Road, Clydebank, G81 6PX
G4 AGH S. Pearson, 75 Gloucester Road, Thornbury, Bristol, BS35 1JH
GM4 AGL W. Ferguson, 72 High Parksail, Erskine, PA8 7HX
G4 AGM R. Williams, Flat 32, St. Johns Court, 59 Murray Road, Northwood, HA6 2FY
G4 AGN J. Porter, 109 Heacham Drive, Leicester, LE4 0LL
G4 AGQ J. Billingham, 14 St. Matthews Court, Sutherland Road, Brighton, BN2 2EX
GM4 AGU Glasgow University Wireless Society, c/o M. Topple, Emmaus, 30 Robshill Court, Newton Mearns, G77 6UG
G4 AGY G. Rippengill, 5 Bridge Farm Drive, Liverpool, L31 9AL
GI4 AHD Dr F. Elder, 44 Learmount Road, Claudy, Londonderry, BT47 4AQ
G4 AHG Shirehampton ARC, c/o C. Chidgey, 46 Station Road, Shirehampton, Bristol, BS11 9TX
G4 AHJ M. Downey, 11 Woodlands Drive, Lepton, Huddersfield, HD8 0JB
G4 AHK B. Palin, 11 Ashgrove Close, Marlbrook, Bromsgrove, B60 1HW
G4 AHM J. Stratton, 10 Brownshill, Maulden, Bedford, MK45 2BT
G4 AHN D. Lax, 1 Gardeners Hill Road, Wrecclesham, Farnham, GU10 4RL
G4 AHO K. Jones, 13 Upland Grove, Bromsgrove, B61 0EL
GI4 AHP T. Sloan, 13 Mount Royal, Lisburn, BT27 5BF
G4 AHT M. Niven, 16 Treewall Gardens, Bromley, BR1 5BT
G4 AHW A. Thomson, 392 Glen Ross Road, Quinte West Ontario, Canada, K0K 2C0
G4 AHZ J. Kynaston, 19 Sharples Drive, Wrea Green, Preston, PR4 2EL
G4 AIB P. Holt, 20 Hedingham Close, Ilkeston, DE7 5HR
G4 AIE W. Mackie, 23 College Park, Horncastle, LN9 6RE
G4 AIJ R. Jones, Sycamores, The Sheet, Ludlow, SY8 4JT
GI4 AIO R. Lindsay, 67 Halfpenny Gate Road, Moira, Craigavon, BT67 0HP
G4 AIR D. Bieber, Tonkins Quay House, Lanteglos-by-Fowey, PL23 1NB
G4 AIU E. Morgan, 12 Kitts, Wellington, TA21 9AX
G4 AIW A. Scarsbrook, 16 Greenbank Avenue, Uppermill, Oldham, OL3 6EB
G4 AJA C. Hoare, 16 Shrivenham Road Highworth, Swindon, SN6 7BZ
G4 AJB A. Bonwitt, 60 Wellhouse Road, Beech, Alton, GU34 4AG
G4 AJE P. Brown, 33A March Road, Wimblington, March, PE15 0RW
G4 AJG P. Perera, 43 Hillside Avenue, Woodford Green, IG8 7QU
G4 AJJ G. Smith, 17 Marshall Drive, Ruddington, NG11 6AJ
G4 AJO R. Finch, 48 Allens Lane, Sprowston, Norwich, NR7 8EJ
G4 AJQ Prof. N. Johnson, 503-97 Lawton Blvd, Toronto, Toronto, Canada, M4V 1Z6
G4 AJU I. Aldridge, 28 Robert St., Williton, Taunton, TA4 4PG
GM4 AJV M. Mackinnon, 55 Fairbrae, Edinburgh, EH11 3GZ
G4 AJW A. Wade, 139 Gilbert Road, Cambridge, CB4 3PA
G4 AJY D. Ellis, 26 Drake Close, Benfleet, SS7 3YL
G4 AKA M. Diprose, 4A Russet Close, Staines, TW19 6AX
G4 AKB M. Court, 3 Ingleby Paddocks, Enslow, Kidlington, OX5 3ET
G4 AKC D. Starkie, 5 Kidbrooke Avenue, Blackpool, FY4 1QR
G4 AKD I. Alexander, 46 Pettitts Lane, Dry Drayton, Cambridge, CB23 8BT
G4 AKE C. Gent, 16 Coronation Avenue, Alvaston, Derby, DE24 0LQ
G4 AKG P. Fry, 11 Park Road, Burgess Hill, RH15 8EU
G4 AKR G. Slack, 16 East Carr, Cayton, Scarborough, YO11 3TS
G4 AKW G. Robinson, 2 Hasketon Road, Woodbridge, IP12 4JR
GW4 AKY D. Hayes, Parth Y Barcud Blaenpennal, Aberystwyth, SY23 4TR
G4 AL J. Wood, 18 Kennedy Avenue, Long Eaton, Nottingham, NG10 3GF
G4 ALA J. Hardwick, 455 Hatton Road, Feltham, TW14 9QP
G4 ALB N. Castledine, 1 Johns Close Burbage, Hinckley, LE10 2LY
G4 ALC J. Balls, 48 Collingwood Road, Great Yarmouth, NR30 4LR
G4 ALD F. Donovan, 4 Rembrandt Drive, Northfleet, Gravesend, DA11 8NQ
G4 ALE Addiscombe ARC, c/o M. Franklin, 6 Tor Road, Farnham, GU9 7BX
G4 ALF K. Law, 93 Measham Drive, Stainforth, Doncaster, DN7 5TQ
G4 ALR M. Down, 5 Juniper Mead, Stotfold, Hitchin, SG5 4RU
G4 ALT A. Taylor, 21 Gould Avenue West, Kidderminster, DY11 7HD
G4 ALY R. Bird, 6 The Cross, St. Dominick, Saltash, PL12 6SP
G4 ALZ R. Bridgland, 20 Newling Way, Worthing, BN13 3DG
G4 AMD C. Heavens, 10815 N. Indian Wells Drive, Fountain Hills, USA, AZ 85268
G4 AMF J. Cresswell, 7 Glinton Avenue, Blackwell, Alfreton, DE55 5HD
G4 AMI M. Hearn, 63 Greswolde Road, Solihull, B91 1DX
G4 AMJ Dr D. Evans, 330 Weld County Road 16 1/2, Longmont Co, USA, 80504-9467
G4 AMN C. Wainwright, 60 Main Street, Hoby, LE14 3DT
G4 AMP B. Flack, Ave Des Hospitaliers De St Jean 7, Waterloo, Belgium, 1410
G4 AMT T. George, Stoneways, Trevescan, Sennen, Penzance, TR19 7AQ
GW4 AMX J. Barrett, Flat 5, Rhos Abbey, Rhos Promenade, Colwyn Bay, LL28 4QA
G4 AMY R. Briggs, Nickey Nook View, Lancaster New Road, Preston, PR3 1NL
GW4 AMZ P. Leach, 41 Bryn Colwyn, Colwyn Bay, LL29 9LJ
G4 ANB Dr J. Morris, 4111 Eve Road, Simi Valley, USA, 93063
G4 AND J. King, Chetwynd, Henfield Road, Steyning, BN44 3TF
G4 ANE H. Leach, 30 Taywood Road, Thornton-Cleveleys, FY5 2RT
GW4 ANK R. Davenport, 14 Milward Road, Barry, CF63 3QD
G4 ANN R. Hadfield, 45 Erica Way, Copthorne, Crawley, RH10 3XG
G4 ANP M. Valentine, 10 Thellusson Avenue, Scawsby, Doncaster, DN5 8QN
G4 ANT East Anglian Contest Group, c/o R. Reed, Oak Cottage, Dereham Road, Dereham, NR20 4AA
G4 ANU C. Columbine, 5 Thornbury Drive, Mansfield, NG19 6NB
G4 ANV P. Hudson, 3 Rowan Drive, Kilburn, Belper, DE56 0PG

G4 ANW T. Slack, 16 Woodside Avenue, Alverstone Garden Village, Sandown, PO36 0JD
G4 ANY D. Stephens, Croeso Cottage, 31 Coton, Whitchurch, SY13 2RA
G4 ANZ B. Warren, 36 Hobbiton Road, Weston-Super-Mare, BS22 7HP
G4 AOA H. Mason, 9 Chatsworth Drive, Little Eaton, Derby, DE21 5AP
G4 AOJ R. Horton, 34 Rising Lane, Knowle, Solihull, B93 0BZ
G4 AOK T. Winter, 6 Cunliffe Drive, Brooklands, Sale, M33 3WS
G4 AOL D. Harmer, 4 Somerton Gardens, Earley, Reading, RG6 5XG
G4 AOP FL D. Hibbin, 95A Thorpe Acre Road, Loughborough, LE11 4LF
GM4 AOR K. Henderson, 97 Granton Road, Edinburgh, EH5 3NH
G4 AOS Dr J. West, Horsley House, Rochester, Newcastle upon Tyne, NE19 1TA
G4 AP J. Rooke, 12 Hellings Gardens, Broadclyst, Exeter, EX5 3DX
G4 APB K. May, 53 Shearwood Crescent, Dartford, DA1 4SU
G4 APD Rugby Amateur Transmitting Society, c/o P. Wells, 12 Shelley Drive, Lutterworth, LE17 4XF
GW4 APF M. Richards, 9 Bank Road, Llangennech, Llanelli, SA14 8UB
G4 APG M. Pellatt, Old Thatch, Branscombe, Seaton, EX12 3BL
GM4 API D. Hebenton, Craigmill Cottage, 3 Craigmill Road, Dundee, DD3 0PH
G4 APJ K. Punshon, 24 Newcombe Road Ramsbottom, Bury, BL0 9UT
G4 APL P. Lewis, 20 Annes Walk, Caterham, CR3 5EL
G4 APO Dr R. Hirst, 21 Manor Farm Court, Thrybergh, Rotherham, S65 4NZ
G4 APP W. Grogan, 8 Fairway South Moor, Stanley, DH9 7HP
G4 APS D. Fiander, 2 Snowshill Close, Nuneaton, CV11 4XQ
G4 AQA P. Hall, 39 Mill Lane, Kirk Ella, Hull, HU10 7JE
G4 AQB S. Macdonald, 58A Tarbet Drive, Bolton, BL2 6LT
G4 AQE P. Saunders, Orchard Cottage, Vale Road, Broadstairs, CT10 2JG
G4 AQG University of Sussex ARS, c/o A. Maris, 140 Edward Street, Brighton, BN2 0JL
G4 AQJ K. Gordon, 96 Pear Tree Crescent, Shirley, Solihull, B90 1LF
G4 AQK D. Davis, 23 Matley Moor, Liden, Swindon, SN3 6NL
G4 AQR I. Cordingley, Orchard Cottage, Compton, Marldon, Paignton, TQ3 1TA
G4 AQS M. Bliss, 53 Rowallan Drive, Bedford, MK41 8AS
G4 AQT J. Rowbotham, 56 Longleat Crescent Beeston, Nottingham, NG9 5EU
G4 AQZ G. Axford, 24 Jack Branch Court, Wash Lane, Clacton-on-Sea, CO15 1EJ
GW4 ARC Rhyl District Amateur Radio, c/o A. Evans, 4 Elm Grove, Rhyl, LL18 3PE
G4 ARE Exeter ARS, c/o J. Rooke, 12 Hellings Gardens, Broadclyst, Exeter, EX5 3DX
G4 ARF Furness ARS, c/o M. Bell, 36 Schneider Road, Barrow-in-Furness, LA14 5DW
G4 ARI T. Raven, 15 Preston Close Stanton Under Bardon, Markfield, LE67 9TX
GM4 ARJ J. Ferguson, 26 Cleuch Avenue, Tullibody, Alloa, FK10 2RX
G4 ARN Norfolk ARC, c/o A. Hall, 122 Norwich Road, New Costessey, Norwich, NR5 0EH
G4 ARO T. Covey, 68 Wellington Close, Walton-on-Thames, KT12 1BB
G4 ARS Carlisle & Dis. ARS, c/o C. Wolf, 35A Moorhouse Road, Carlisle, CA2 7LU
GM4 ARU J. Mcintyre, 12 Johnstone Lane, Carluke, ML8 4NR
G4 ARW C. Turner, 2 Queens Way, Wotton-under-Edge, GL12 7HA
G4 ARX B. Curley, 22 Churchill Crescent, Sheringham, NR26 8NQ
G4 ARY A. Langford, 33 Briscoe Road, Hoddesdon, EN11 9DG
G4 ASF R. Mccurrach, Isa Coed, Bowden Lane, Bude, EX23 9BJ
G4 ASG P. Bayley, 9 Westbrook Green, Bromham, Half Acre, Chippenham, SN15 2EF
G4 ASH I. Roberts, 86 Federation Avenue, Desborough, Kettering, NN14 2NX
G4 ASI F. Emery, Room 10, Building 448, Westerham, TN16 3BN
G4 ASK E. Rayland, 40 Sycamore Close, Taunton, TA1 2QJ
G4 ASL S. Ayling, Kitnocks 89 Queen'S Road, Alton, GU34 1JA
G4 ASM A. Murphy, Apartment 28, Trinity Gardens, 1 Kingsmead Road South, Prenton, CH43 6TA
GU4 ASO R. Ayres, Langaller, Rue Colin, Vale, Guernsey, GY6 8LA
G4 ASP J. Holding, Old Pearmain, Eardisland, Leominster, HR6 9DN
G4 ASQ Dr M. Jordan, 4 Marchfont Close, Nuneaton, CV11 6GA
G4 ASR D. Butler, Yew Tree Cottage, Lower Maescoed, Hereford, HR2 0HP
G4 ASW M. Yorke, 8 St John Place, Port Washington, USA, NY 11050
G4 ASX O. Perry, 60 Malines Avenue, Peacehaven, BN10 7RS
G4 ASY D. Yeaman, Paddock View, Hurstbourne Priors, Whitchurch, RG28 7SE
G4 ASZ M. Hurst, 21 Bankside, Dunton Green, Sevenoaks, TN13 2UA
GM4 ATA J. Hotchin, 2 Moorfield Place, Gatehead, Kilmarnock, KA2 0AX
G4 ATB R. Shapland, 14 Charney Court, Grange-over-Sands, LA11 6DL
G4 ATG BARTG, c/o Dr A. Thomas, The Stone Barn, 1 Home Farm Close, Chesterton, OX26 1TZ
G4 ATH Thornton Cleveleys ARS (Tcars), c/o J. Rodway, 9 York Avenue, Thornton-Cleveleys, FY5 2UG
G4 ATL D. Bloomfield, 26 Preston Crowmarsh, Wallingford, OX10 6SL
G4 ATQ G. Hawkins, 18 Brook Street, Leighton Buzzard, LU7 3LH
G4 ATU S. Brown, Mullins View, 1D Turnpike Road, Ormskirk, L39 3LD
G4 AUB A. Smith, 56 Longrood Road, Rugby, CV22 7RE
G4 AUC S. Baugh, 70 Madingley, Bracknell, RG12 7TF
GW4 AUD A. Lacy, Llanoris, Llanerfyl, Welshpool, SY21 0EP
G4 AUE A. Rose, 18 Highview Gardens, St Albans, AL4 9JX
G4 AUF C. Friel, 102A Sharps Lane, Ruislip, HA4 7JB
G4 AUG R. Mortimer, 19 St. Monance Way, Colchester, CO4 0PJ
G4 AUL G. Mitchell, 10 Wealden Close, Hildenborough, Tonbridge, TN11 9HB
G4 AUN R. Collett, 70 Clifton Road, Darlington, DL1 5DX
GM4 AUP I. Suart, 37 Meldrum Mains, Glenmavis, Airdrie, ML6 0QQ
G4 AUQ F. Barker, 90 Hall Road, Hull, HU6 8SB
G4 AUR J. Mcburney, 4 Fownhope Road, Sale, M33 4RF
G4 AUV G. Wing, 105 Moore Avenue, Norwich, NR6 7LG
G4 AUY P. Sherwood, 43 Kingsland, Arleston, Telford, TF1 2LE
GW4 AVC D. Bowers, 31 Clarence Road, Wrexham, LL11 2EU
G4 AVE L. Cates, 45 Smoke Lane, Reigate, RH2 7HJ
G4 AVF A. Fletcher, 11 Little Oak Close, Lees, Oldham, OL4 3LW
G4 AVJ G. Pople, 3 Leighton Drive, Creech St. Michael, Taunton, TA3 5DW
G4 AVK S. Ripley, 62 Palewell Park, London, SW14 8JH
G4 AVL P. Newby, 238A Wherstead Road, Ipswich, IP2 8JZ
G4 AVS R. Wilson, Aerial House, 1 The Fields, Woodbridge, IP12 2HZ
G4 AVV G. Cluer, 12 Bingham Road, Addiscombe, Croydon, CR0 7EB
G4 AVX A. Newman, 101 Washbrook Road, Portsmouth, PO6 3SB
G4 AWA R. Payne, 11 Beaconsfield Road, Christchurch, BH23 1QT
GM4 AWB R. Macduff, 7 Cairngorm Crescent, Bearsden, Glasgow, G61 4EH
G4 AWF D. Wilson, 4 Caradoc Meadow, Sellack, Ross-on-Wye, HR9 6GJ
G4 AWG G. Higgs, Firtree House, Perry Wood, Faversham, ME13 9SE
G4 AWJ G. Thomas, 9 Highcroft Crescent, Heathfield, TN21 8HE
G4 AWK M. Roberts, 2 Miners Garth, Liverton Mines, Saltburn-by-the-Sea, TS13 4BU
G4 AWM D. Norfolk, 13 Oakwood Crescent, Greenford, UB6 0RF
G4 AWO R. Gray, 10 Stone Park, Broadsands, Paignton, TQ4 6HT
G4 AWU R. Lane, 8 Town Street, Lound, Retford, DN22 8RS
G4 AWW N. Shepherd, Jo Kebi, Stonehall Road, Dover, CT15 7JS
G4 AWY R. Mekka, 57 St. Johns Road, Caversham, Reading, RG4 5AL
G4 AWZ P. Matthews, 22 Rydens Road, Walton-on-Thames, KT12 3DA
G4 AXA N. Pope, Silver Hill, Norwich Road, Great Yarmouth, NR29 5PB
G4 AXC C. Burden, Cedar Croft, Hengar Lane, Bodmin, PL30 3PH
G4 AXD G. Edy, 44 Roseholme, Maidstone, ME16 8DR
G4 AXF J. Jacques, 30 Centurian Way, Bedlington, NE22 6LD
G4 AXL C. Gerrard, 22 Kelso Drive, The Priorys, North Shields, NE29 9NS
G4 AXO J. Wills, 48 Fairfield Road, Winchester, SO22 6SG
GM4 AXS P. Wilberforce, 8 Ferryfield Road, Connel, Oban, PA37 1SR
G4 AXU G. Parr, Chesil Coppice, West Bexington, Dorchester, DT2 9DD
GI4 AXV J. Doherty, 172 Dunmore Road, Ballynahinch, BT24 8QQ
G4 AXX M. Marsden, Mill Cottage, Shrowle, East Harptree, Bristol, BS40 6BJ
G4 AXY A. Mort, 86 Longfield Road, Winnall, Winchester, SO23 0NU
G4 AYB A. Kelle, Urb.Sorries 10, la Massana, Andorra, AD.400
G4 AYD T. Hodgetts, 3 Garsdale Road, Weston Super Mare, BS22 8PT
G4 AYH G. Monks, 7 Town Street, Rawdon, Leeds, LS19 6PU
G4 AYK Mid Severn Vall, c/o P. Perrins, 9 Merrick Close, Hayley Green, Halesowen, B63 1JY
G4 AYL E. Lambert, 41 Brand Hill Drive, Crofton, Wakefield, WF4 1PF
G4 AYO M. Hewitt, 10 Blacka Moor View, Sheffield, S17 3GZ
GW4 AYQ J. Durrans, 87 The Links Trevethin, Pontypool, NP4 8DQ
G4 AYR T. Greenwood, 30 Ringwood Road, Headington, Oxford, OX3 8JA
G4 AYS A. Crook, 153 Shortheath, Shortheath, Swadlincote, DE12 6BL
G4 AYU N. Kenyon, 74 Albert Road, Leyland, PR25 4YJ
G4 AZA R. Winkworth, 13 Bagley Close, Kennington, Oxford, OX1 5LS
G4 AZC P. Martin, Stoneovers, Wellow Top Road, Ningwood, Yarmouth, PO41 0TL
G4 AZD A. Edgecock, Sunnydene, Station Road, Colchester, CO7 8JA
G4 AZG Dr G. Macdonald, Pilgrims Cottage, Church Lane, Canterbury, CT4 6HX
G4 AZH M. Bushnell, Rose Cottage, Street Ashton, Rugby, CV23 0PH
GW4 AZI D. Thomas, Sunnydale, Scurlage, Swansea, SA3 1BA
GD4 AZJ R. Troughton, Flat 2, Waterfront Apartments, Mooragh Promenade, Ramsey, Isle Of Man, IM8 3AN
G4 AZL P. Justin, Garth, Park View Road, Pinner, HA5 3YF
G4 AZM C. Wilson, 17719 Phil C Peters Road, Winter Garden, USA, 34787
G4 AZS A. Bayling, 55 Shelton Road, Shrewsbury, SY3 8SU
G4 AZT T. Barker, 1 Links Road, Kennington, Oxford, OX1 5RX
G4 AZU J. Tiller, 21 Portal Road, Winchester, SO23 0PX
G4 AZX J. Robinson, 19 Sunnycroft Gardens, Cranham, Upminster, RM14 1HP
G4 BAD Bath and District ARC, c/o P. Carter, 16 Alexandra Park Paulton, Bristol, BS39 7QS
G4 BAN P. Godfrey, 5 Parkway, Southgate, London, N14 6QU
G4 BAO Dr J. Worsnop, 20 Lode Avenue, Waterbeach, Cambridge, CB25 9PX
GM4 BAP A. Beaton, 46 Balmoral Place, Aberdeen, AB10 6HP
G4 BAQ R. Chambers, 17 Exmoor Close, Worthing, BN13 2PW
G4 BAS Club of Friendship, c/o H. Ketley, 24 Farmcroft Road, Mansfield Woodhouse, Mansfield, NG19 8QT
G4 BAU R. Russell, 228 Broomhill, Downham Market, PE38 9QY
G4 BAV J. Gee, Windmill Lodge, Mill Lane Witnesham, Ipswich, IP6 9HR
G4 BBA P. Chilcott, 321 Eastfield Road, Peterborough, PE1 4RA
G4 BBD Dr M. Tooley, 4 Shelley Road, Bath, BA2 4RJ
GI4 BBE R. Bolton, Ohmvilla, 69 Newcastle Street, Newry, BT34 4AQ
G4 BBH R. Ferryman, 25 Winant Way, Dover, CT16 2AX
G4 BBI P. Nixon, 8 White Edge Close, Chesterfield, S40 4LE
G4 BBJ R. Ramsay, 1 Sapho Park, Gravesend, DA12 4NA
G4 BBL A. Thackery, 19 Pyne Point, Clevedon, BS21 7RL
G4 BBM A. Benn, 10 Roman Way, Coventry, CV3 6RD
G4 BBP Dr P. Howey, Raybarrow Farm, Nettleton Shrub Nettleton, Chippenham, SN14 7NN
G4 BBQ D. King, 62 Ansley Road, Nuneaton, CV10 8NU
G4 BBT Dr R. Hancock, 80 Ulleries Road, Solihull, B92 8EE
G4 BBU P. Whittle, 20 Marlbrook Lane Marlbrook, Bromsgrove, B60 1HN
G4 BBY R. Edwards, 27 Provis Mead, Chippenham, SN15 3UA
G4 BCA Dr D. Tunnicliffe, 4 Chesford Drive Churchdown, Gloucester, GL3 2BA
G4 BCB Dr K. Johnston, 92C Mcdowalls Road, Yugar, Australia, 4520
GJ4 BCC R. Davies, 2 Manor View Close, La Grande Route De St. Pierre S, St Peter, Jersey, JE3 7AZ
GW4 BCF R. Newman, 32A Park Avenue, Porthcawl, CF363EP
G4 BCG G. Wale, 2 The Jordans, Coventry, CV5 9JT
G4 BCH P. Burgess, Tretawn, Kite Hill, Ryde, PO33 4LG
G4 BCP L. Graves, The Beach Hut, 6 Hauxley Links, Morpeth, NE65 0JS
G4 BCT A. Gordon, 4 Victoria Road West, Thornton-Cleveleys, FY5 1BU
G4 BCV Essex Raynet, c/o N. Smith, Clare Cottage, White Ash Green, Halstead, CO9 1PD
G4 BCX A. Helm, 38 Blandford Road, Lower Compton, Plymouth, PL3 5DU
GW4 BCZ J. White, 13 Stokes Court, Ponthir, Newport, NP18 1RY
G4 BDC K. Collerton, 93 Nursery Lane, Leeds, LS17 7EE
GM4 BDJ R. Mccartney, Cairndhu Walter Street, Langholm, DG13 0AX
GI4 BDL V. Simpson, 25 Waringstown Road, Lurgan, Craigavon, BT66 7HH
G4 BDQ P. Harris, 76 Rozel Court, Southampton, SO16 9QE
GI4 BDR Dr N. Evans, 87A Oldtown Road, Castledawson, Magherafelt, BT45 8BZ
G4 BDW J. Bagley, 12 The Crescent, Caldecott, Wellingborough, NN9 6AU
G4 BDX M. Horoszko, 1 Woodgarth Cottages, Reedness, Goole, DN14 8EX
G4 BEB R. Browning, Sprackets Orchard, Curry Rivel, Langport, TA10 0PP
G4 BEH J. Eato, 77 Rutland Avenue, Nuneaton, CV10 8EG
G4 BEI J. Palmer, 124A High Street, Wyke Regis, Weymouth, DT4 9NU
G4 BEL R. Taylor, 12 The Rampart, Haddenham, Ely, CB6 3ST
G4 BEM S. Ford, 3 Hill View, Stoke-on-Trent, ST2 7AR
G4 BEO B. Hailstone, 6 Larkswood Rise, St Albans, AL4 9JU
GW4 BEQ G. Hotchkiss, 38 Arlington Road, Sully, Penarth, CF64 5TQ
G4 BEU J. Small, 20 Hastings Road, Birkdale, Southport, PR8 2LW
G4 BEV R. Taylor, 6 Churchill Crescent, Marple, Stockport, SK6 6HJ

G4 BEZ J. Phillipson, 3 Montrose Close, New Hartley, Whitley Bay, NE25 0TA
G4 BFC A. Riddell, 12 Sunrise, Malvern, WR14 2NJ
G4 BFR D. Baldwin, 112 Moorland View Road, Chesterfield, S40 3DF
G4 BFS T. Sargent, 15 Pound Lane, Blofield, Norwich, NR13 4NB
G4 BFT C. Johnson, 7 Field View, Braunston, Daventry, NN11 7JS
G4 BFV D. Sinclair, 46 Church Lane, Mablethorpe, LN12 2NU
GM4 BFX Dr A. Milne, 65 Lord Hays Grove, Aberdeen, AB24 1WT
G4 BG A. Duckworth, Ambergate, 2 Ashleigh Drive, Teignmouth, TQ14 8QX
GI4 BGB P. Kelly, 30 Cahore Road, Draperstown, Magherafelt, BT45 7LY
G4 BGD Dr R. Williams, Old Hall House, Old Hall Avenue, Littleover, Derby, DE23 6EN
G4 BGH A. Ruddell, 9 Parsonage Close, Charlton, Wantage, OX12 7HP
G4 BGM C. Zeal, 20 Hurst Park, Midhurst, GU29 0BP
G4 BGP C. Barber, 45 Cuerdale Lane Walton-Le-Dale, Preston, PR5 4BP
GM4 BGS S. Liddell, 49 Inchbrae Road, Cardonald, Glasgow, G52 3HA
G4 BGT M. Staton, 30 Shaftesbury Avenue, Chandler'S Ford, Eastleigh, SO53 3BS
G4 BGW I. Wilson, Whitethorn, Sandhurst Lane, Gloucester, GL2 9NW
G4 BHC F. Stevens, 11 Hen Wythva, Camborne, TR14 7XN
G4 BHD T. Goldsworthy, Trevarth, Atlantic Terrace, Camborne, TR14 7AW
G4 BHE B. Macklin, 4 Bramdown Heights, Basingstoke, RG22 4UB
G4 BHJ M. Fochtmann, 1 Chapmans Way, Over, Cambridge, CB24 5PZ
G4 BHL J. Firth, 10 Ridgway Avenue, Darfield, Barnsley, S73 9DU
G4 BHP G. Benwell, The Did, Tunley, Bath, BA2 0DZ
G4 BHT M. Hulands, 100 Avenue Road, Rushden, NN10 0SJ
GM4 BHU D. Aitkenhead, 37/3 Cavalry Park Drive, Edinburgh, EH15 3QG
G4 BIA R. Hood, 8 Fayre Meadow, Robertsbridge, TN32 5AU
G4 BID W. Boyd, 2 The Ramblers, Poringland, NR14 7QN
G4 BII D. Williams, 2 Main Street, Poundon, Bicester, OX27 9AZ
G4 BIK P. Mellor, 10 Greenfields, Earith, Huntingdon, PE28 3QH
G4 BIM P. Bentley, Blakes Hill, Limerstone Road, Brighstone, Newport, PO30 4AE
G4 BIN N. Long, Homedale, Bayford Hill, Wincanton, BA9 9LS
GW4 BIS A. Davies, 12 Church St., Troedyrhiw, Merthyr Tydfil, CF48 4HD
GM4 BIT R. Wilson, 5 Collins Drive, Loans, Troon, KA10 7HA
G4 BIX Dr D. Price, 34 Vanda Crescent, St Albans, AL1 5EX
G4 BIY M. Corbett, 6 Windgap Lane, Haughley, Stowmarket, IP14 3PA
G4 BIZ A. Paxton, Cleveland House, Bartley Road, Southampton, SO40 7GP
G4 BJB C. Hurst, 28 Hengistbury Road, Barton-on-Sea, BH25 7LU
G4 BJC International Shortwave League, c/o A. Kinson, 6 Uplands Park, Broad Oak, Heathfield, TN21 8SJ
G4 BJD G. Overton, 14 Aylestone Drive, Hereford, HR1 1HT
G4 BJF B. Marshall, 23 Sandgate Avenue, Birstall, Leicester, LE4 3HQ
G4 BJG P. Smith, 11 Chatsworth Avenue, Clowne, Chesterfield, S43 4SR
G4 BJJ H. Tickell, 26 Shear Brow, Blackburn, BB1 7EX
GI4 BJK K. Patterson, 1A Demesne Gate, Saintfield, Ballynahinch, BT24 7BE
G4 BJN D. Harvey, 23 Lapwing Close, Hemel Hempstead, HP2 6DS
G4 BJO B. Greeves, 65 Stowupland Road, Stowmarket, IP14 5AN
G4 BJP S. Popek, 42 Victoria Road, Polegate, BN26 6DA
G4 BJS J. Loose, Flat 30 Highbury Court, Howard Road East, Birmingham, B13 0RQ
G4 BJT Dr M. Ware, 20 Bath Road, Buxton, SK17 6HH
G4 BJX W. Whatmore, 51 The Fairways, Sherford, Taunton, TA1 3PA
G4 BKA A. Neaves, The Coach House, Cliff Hall Lane Cliff, Tamworth, B78 2DR
G4 BKB G. Jessup, 68 Danes Road, Bicester, OX26 2LR
G4 BKE D. Wright, 4 Wynne Close, Broadstone, BH18 9HQ
G4 BKF T. Howarth, 71 Ford Road, Wirral, CH49 0TD
GW4 BKG S. Emlyn-Jones, 26 Lime Tree Way, Porthcawl, CF36 5AU
G4 BKH A. Chorley, 354 Denton Lane, Chadderton, Oldham, OL9 8QD
G4 BKI P. Evans, 16 Cotgrave Road, Plumtree, Nottingham, NG12 5NX
G4 BKO J. Francis, 22 Earlswood Drive, Mickleover, Derby, DE3 9LN
G4 BKQ R. Gubbins, 29 Meadow End, Gotham, Nottingham, NG11 0HP
G4 BKR W. Taggart, Calle Zarauz 61, Urb. San Luis, Alicante, Spain, 3180
G4 BKS P. Erkiert, 129 Cannock Road, Aylesbury, HP20 2AS
G4 BLD C. Croucher, 13 Magnolia Way, Pilgrims Hatch, Brentwood, CM15 9QS
G4 BLL P. Burnett, 28 Crownest Lane, Bingley, BD16 4HL
GM4 BLO G. Milne, 65 Millburn Avenue, Clydebank, G81 1ER
G4 BLS P. Appleby, Flat 14, Maryan Court, Hailsham, BN27 3DJ
G4 BLT R. Sterry, 9 Finch Avenue, Wakefield, WF2 6SE
G4 BM T. Searle, 2 Woolfall Terrace, Seaforth, Liverpool, L21 4PJ
G4 BMC D. Barrell, 26 Yerville Gardens, Hordle, Lymington, SO41 0UL
G4 BMD M. Hayes, No 6 The Court, Dunboyne Castle, Dunboyne, Ireland, A86 FA02
G4 BMK M. Kerry, 2 Beacon Close, Seaford, BN25 2JZ
GW4 BML B. Lloyd, 1 Llys Melyn, Tregynon, Newtown, SY16 3EE
G4 BMM P. Knight, 75 Ashcroft Road, Luton, LU2 9AX
G4 BMO D. Cloke, Church Cottage, East Coker, Yeovil, BA22 9LY
G4 BMP R. Sadler, 9 Meade King Grove, Woodmancote, Cheltenham, GL52 9UD
G4 BMQ D. Harrop, 1 Edgecombe Crescent, Rowner, Gosport, PO13 9RD
GM4 BMS A. Redford, 9 Broker, Isle of Lewis, HS2 0EZ
G4 BMU S. East, 2 Linscott House, 64D Russell Road, Buckhurst Hill, IG9 5QE
G4 BMW C. Pescod, 7 Brian Close, Chelmsford, CM2 9DZ
G4 BNB R. Wynn, 48 Darnley Road, Woodford Green, IG8 9HY
GW4 BNC Dr G. Thomas, Ash Barn Penylan Road Michaelston-Y-Fedw, Cardiff, CF3 6XW
G4 BNE Dr R. Herring, 96 St. Fabians Drive, Chelmsford, CM1 2PR
GW4 BNJ D. Williams, 48 St. Hilary Drive, Killay, Swansea, SA2 7EH
G4 BNK W. Wright, 27 St. Johns Road, Farnborough, GU14 9RL
G4 BNL R. Morley, 63 Holt Park Crescent, Holt Park, Leeds, LS16 7SL
G4 BNM S. Homans, 3 Hilton Mews, Bramhope, Leeds, LS16 9LF
G4 BNO M. Ayling, 68 Littledown Avenue, Queens Park, Bournemouth, BH7 7AS
G4 BNP J. Burgess, 11 Winters Lane, Ottery St Mary, EX11 1AR
G4 BNS A. Collinson, 30 Thornton Road, Pickering, YO18 7HZ
G4 BNT G. Moore, 50 Barley Rise, Strensall, York, YO32 5AA
G4 BNW M. Knight, 18 Friary Road, Abbeymead, Gloucester, GL4 5FD
G4 BNX I. Middleton, 9 Brentnall Court, Kirk Close, Nottingham, NG9 5EZ
G4 BOB R. Ambler, 21 Whitley Spring Road, Ossett, WF5 0QA
G4 BOF P. Harry, 5 St. Michaels Avenue, Kingsland, Leominster, HR6 9QR
G4 BOH C. Cummings, Castle View, Childs Lane, Congleton, CW12 4TQ
G4 BOJ N. Greenstreet, 223 Upperthorpe, Sheffield, S6 3NG
G4 BOL R. Fineman, 4 Sherbourne Avenue, Bradley Stoke, Bristol, BS32 8BB
G4 BON J. Strutt, 163 Scalby Road, Scarborough, YO12 6TB
G4 BOO D. Rumens, 3 Flecker Close, Thatcham, RG18 3BA
G4 BOP Dr P. Berwick, Beech Croft, West Hill, Ottery St Mary, EX11 1UY
G4 BOQ J. Hall, 15 Main Street, Greetham, Oakham, LE15 7NJ
G4 BOU J. Chance-Read, 15 Garrard Way, Wheathampstead, St Albans, AL4 8PE
G4 BOV A. Horton, Martletts, 52 Lower Cookham Road, Maidenhead, SL6 8JZ
G4 BOZ A. Brock, 1 Carpenter Drive, St Leonards-on-Sea, TN38 9RX
G4 BP Scarborough ARS, c/o M. Day, 33 Ryndle Walk, Scarborough, YO12 6JT
G4 BPE A. Evans, Fairfield Main St, Claypole, Newark, NG23 5BA
G4 BPJ B. Stone, 12 Forbes Road, Newlyn, Penzance, TR18 5DQ
G4 BPN N. Kerstein, 40 Davidson Close Hythe, Southampton, SO45 6JT
G4 BPO Po Research Cnt, c/o C. Hoare, 16 Shrivenham Road Highworth, Swindon, SN6 7BZ
G4 BPV P. Barker, 2 Oriole Drive, Exeter, EX4 4SJ
G4 BQA M. Morley, 8 The Ridings, Seaford, BN25 3HW
G4 BQB J. Crocker, 4 Portland Terrace, Watchet, TA23 0DD
G4 BQC B. Makeham, 64 Benomley Road, Almondbury, Huddersfield, HD5 8LS
GM4 BQD R. Muir, 9 Craigs Court, Torphichen, Bathgate, EH48 4NU
G4 BQF M. Duce, 28 Thompson Avenue, Canvey Island, SS8 7TS
G4 BQH D. Livsey, 18 Tollards Road, Countess Wear, Exeter, EX2 6JJ
GI4 BQI W. Mccullough, 16 Ballylisk Lane, Portadown, Craigavon, BT62 3RN
G4 BQJ A. Hill, 3 Cambrai Avenue, Warrington, WA4 6QU
G4 BQN N. Marsden, 32 Chard Road, Drimpton, Beaminster, DT8 3RF
GW4 BQQ L. Dean, Tanrallt, Llangwyryfon, Aberystwyth, SY23 4SP
G4 BQR W. Carmichael, 47 Neath Drive, Ipswich, IP2 9TA
G4 BQS B. Prichard, The Gables, Wootton Lane, Canterbury, CT4 6RT
G4 BQV R. Mullard, 46 Green Lane, Clanfield, Waterlooville, PO8 0JX
G4 BQW W. Glover, Swallows Meadow Court, 33 Swallows Meadow, Solihull, B90 4PH
G4 BQY P. Aburrow, 25 Hill Crescent, Worcester Park, KT4 8NB
G4 BRA Bracknell ARC, c/o M. Goodey, 62 Rose Hill, Binfield, Bracknell, RG42 5LG
GM4 BRB A. Stewart, 121 William Street, Dalbeattie, DG5 4EE
G4 BRC Kent Raynet Group, c/o T. Lundegard, Saxby, Botsom Lane, Sevenoaks, TN15 6BL
G4 BRF R. Mickleburgh, 85 Carey Park, Killigarth, Looe, PL13 2JP
GW4 BRH J. Sniadowski, Bryn Bach Barn, Cwmdu, Crickhowell, NP8 1RT
G4 BRK N. Whiting, Forge End, Garford, Abingdon, OX13 5PF
G4 BRL A. Moore, 14 Heath Road, Ipswich, IP4 5SA
GM4 BRM A. Long, 34 Thornly Park Drive, Paisley, PA2 7RP
GM4 BRN Kingdom ARS, c/o P. Merckel, 1 Mortimer Court, Dalgety Bay, Dunfermline, KY11 9UQ
GW4 BRS Barry ARS, c/o S. Trahearn, 148 Gladstone Road, Barry, CF62 8ND
G4 BRW M. Gordon, 57 Taunton Road, Bridgwater, TA6 3LP
G4 BSA M. Draper, The Wallow, Mount Road, Bury St Edmunds, IP31 2QU
G4 BSC J. Wells, Tredworth, Sunnyfield Lane, Up Hatherley, Cheltenham, GL51 6JE
G4 BSD D. Hoose, Leonard Cheshire He, Oaklands, Garstang, PR3 1RD
G4 BSK M. Rhind-Tutt, Oldfield, Moor Road, Bridgwater, TA7 9AR
G4 BSM S. Grove, 31 Sheppard Way, Minchinhampton, Stroud, GL6 9BZ
G4 BSS J. Spence, 4 Langford Lane, Burley In Wharfedale, Ilkley, LS29 7NR
G4 BSV A. Cox, 175 Hillcrest, Weybridge, KT13 8AS
G4 BSW N. Hadley, 323 Canterbury Road, Margate, CT9 5JA
G4 BTE M. Smith, 24 Lea Bank, Wolverhampton, WV3 9HN
GI4 BTG B. Davidson, 106 Tudor Park, Newtownabbey, BT36 4WL
G4 BTI D. Case, 8 Fawley Road, Reading, RG30 3EN
G4 BTK A. Whitehouse, 690 Kingstanding Road, Kingstanding, Birmingham, B44 9SS
G4 BTN C. Brion, Passaford House, Hatherleigh, Okehampton, EX20 3LU
GW4 BTW I. Jolly, 1 Llewelyn Drive, Bryn-Y-Baal, Mold, CH7 6SW
G4 BTX N. Monument, 9 Tower Road, Felixstowe, IP11 7PR
GM4 BUA Dr T. Shepherd, 1 Spruce Gardens, Cupar Muir, Cupar, KY15 5WN
G4 BUB P. Cox, 53 Boleyn Avenue, Enfield, EN1 4HR
G4 BUE C. Page, Cherry View 28 Ellerslie Lane, Bexhill-on-Sea, TN39 4LJ
G4 BUF G. Jolley, 70 Hempstead Road, Holt, NR25 6DG
G4 BUH M. Banahan, 2 The Paddock, Ely, CB6 1TP
G4 BUI J. Simpson, 19 Greenacres, Wetheral, Carlisle, CA4 8LD
GI4 BUJ J. Sander, 696 Doagh Road, Newtownabbey, BT36 4TP
G4 BUK J. Turner, 12 Purley Bury Avenue, Purley, CR8 1JB
G4 BUO D. Lawley, 5 The Limes, Buckland, Buntingford, SG9 0PW
G4 BUP Dr P. Moss, Amalrie, Franklin Road, Chelmsford, CM3 6NF
G4 BUW K. Lamb, 10 Malthouse Gardens, Marchwood, SO40 4XY
G4 BUX Buxton Radio Amateurs, c/o D. Carson, 21 Harris Road, Harpur Hill, Buxton, SK17 9JS
GW4 BUZ J. Howells, 13 Vicarage Road, Penygraig, Tonypandy, CF40 1HR
G4 BVB R. Pridham, Treetops, Chilsworthy, Gunnislake, PL18 9PB
GM4 BVD A. Sampson, 47 Muirend Road, Perth, PH1 1JD
GW4 BVE J. Clifford, Dippers Barn Coppice Lane Pool Quay, Welshpool, SY21 9JY
G4 BVF M. Sinclair, 28 Roker Park Avenue, Ickenham, Uxbridge, UB10 8ED
G4 BVG A. Young, 90 Pine Ridge, Carshalton, SM5 4QH
G4 BVI G. Chenery, 44 Belstead Road, Ipswich, IP2 8AZ
GW4 BVJ R. Mortimore, 76 Cwmfferws Road, Tycroes, Ammanford, SA18 3UA
G4 BVK K. Stevens, 20 Coberley, Bristol, BS15 8ES
G4 BVM C. Newman, 19 Clare Road, Peterborough, PE1 3DT
G4 BVP M. Noble, Harbet, Shipley Road, Horsham, RH13 9BG
G4 BVQ P. Kennedy, 18 Rushmere Avenue, Levenshulme, Manchester, M19 3EH
G4 BVS S. Overend, Deepend Cottage, Lower Hone Lane, Bosham, Chichester, PO18 8QN
GW4 BVT R. Osborne, Plas-Y-Bryn, 1 Belle Vue Gardens, Brecon, LD3 7NY
GM4 BVU N. Macdonald, 3 Townhill Road, Hamilton, ML3 9UX
G4 BVV P. Goben, 1 Petal Close, Maltby, Rotherham, S66 7HJ
G4 BVW A. Reilly, 4 Moreton Drive, Poulton-le-Fylde, FY6 8ED
G4 BVY I. Dixon, 5 The Howsells, Lower Howsell, Malvern, WR14 1AD
GM4 BVZ J. Davidson, Rosemount, Whiting Bay, Isle of Arran, KA27 8PR
G4 BWB R. Andrews, Flat 19, 4 Salamanca Place, London, SE1 7HB
G4 BWC Bradley Wood Scout Radio Group, c/o M. Bray, 2 Camborne Drive, Fixby, Huddersfield, HD2 2NF
G4 BWE S. Price, 9 Spurcroft Road, Thatcham, RG19 3XX
G4 BWF R. Johnson, 29 Oakfield Avenue, Markfield, LE67 9WH
G4 BWG S. Marsh, 26 Station Road, Whyteleafe, CR3 0EP
G4 BWL H. Morris, 2 Brickwall Lane, Curry Rivel, Langport, TA10 0NX
GI4 BWM J. Mccullagh, 14 Parkgate Meadows, Parkgate, Ballyclare, BT39 0FA
G4 BWN P. Funnell, 6 Bolero Close, Wollaton, Nottingham, NG8 2BZ
G4 BWO D. Tyler, 5 Brentry Avenue, Bristol, BS5 0DL

G4	BWP	F. Handscombe, Sandholm, Bridge End Road, Red Lodge, Bury St Edmunds, IP28 8LQ
G4	BWR	M. Hildich, 7 Claverham Park, Claverham, Bristol, BS49 4LS
G4	BWV	A. Burchmore, 49 School Lane, Horton Kirby, Dartford, DA4 9DQ
G4	BWX	S. Egerton, 15 Hyde Road, Torrisholme, Morecambe, LA46NU
G4	BWY	P. Willcocks, 27 Manor Road, Barnet, EN5 2LE
GI4	BXB	R. Brown, Apartment 10, Anchor Watch, Donaghadee, BT21 0GA
G4	BXC	H. Pearce, 32 Marshall Road, Willenhall, WV13 3PB
G4	BXD	B. Nock, 47 Oakfield Road, Kidderminster, DY11 6PL
G4	BXH	D. Hardy, Box 52831, Dubai, United Arab Emirates,
G4	BXI	C. Godden, 84 Crescent Road, Ramsgate, CT11 9QZ
G4	BXQ	A. Pressley, 22 Springbank Avenue, Farsley, Pudsey, LS28 5LW
G4	BXS	J. Morris, Reaside, Newnham Bridge, Tenbury Wells, WR15 8LP
G4	BXY	H. Barker, 31 Briants Avenue, Caversham, Reading, RG4 5AY
G4	BXZ	Dr J. Howell, 3 Gate Farm Road, Shotley Gate, Ipswich, IP9 1QH
GW4	BYA	P. Braham, 23 Gilfach Y Gog, Penygroes, Llanelli, SA14 7RJ
G4	BYB	R. Penman, 9 Southall Avenue, Worcester, WR3 7LR
G4	BYD	A. Atkinson, 5 Ashfield Avenue, Skelmanthorpe, Huddersfield, HD8 9BW
G4	BYE	T. Miller, 4 Jessop Road, Stevenage, SG1 5NF
GM4	BYF	P. Bates, 10 Swanston Avenue, Edinburgh, EH10 7BU
G4	BYG	V. Lindgren, 143 Hull Road, Anlaby., Hull, HU106ST
G4	BYI	A. Wilson, 223 Waingaro Road, Rd 1, Ngaruawahia, New Zealand, 3793
G4	BYL	B. Smith, 27 Thorneyholme Road, Accrington, BB5 6BD
G4	BYM	B. Buzzing, 1 Westmead Close, Droitwich, WR9 9LG
G4	BYO	W. Tee, 87 Higher Blandford Road, Broadstone, BH18 9AE
G4	BYR	I. Maslen, 26 Millington Road, Wallingford, OX10 8FE
G4	BYS	Dr G. Warren, 96 Parkside Drive, Cassiobury, Watford, WD17 3BB
GM4	BYT	R. Cook, 132 Clachtoll, Lochinver, Lairg, IV27 4JD
G4	BYW	J. Lekesys, 4 Gleneagles Way, Fixby, Huddersfield, HD2 2NH
G4	BYZ	C. Mills, North Lodge, Margery Wood Lane, Tadworth, KT20 7BA
G4	BZB	D. Parsons, 27 St. Leodegars Way, Hunston, Chichester, PO20 1PE
G4	BZE	P. Bradley, Woodlands, Longdown, Exeter, EX6 7SR
G4	BZF	M. Reed, 1 The Cottages, Farm Lane, Plymouth, PL6 5RJ
G4	BZG	R. Smith11, 17 Styrrup Road, Harworth, Doncaster, DN11 8LL
G4	BZI	R. Bracey, 7 Park Estate, Shavington, Crewe, CW2 5AW
G4	BZJ	A. Mitchell, 18 Malham Fell, Bracknell, RG12 7DU
G4	BZL	D. Simpson, Ivy Cottage, Princess Street, Leeds, LS19 6BS
G4	BZM	M. Edwards, 13 Lechmere Crescent, Malvern, WR14 1TY
G4	BZP	F. Partington, 21 East Road, Wymeswold, Loughborough, LE12 6ST
G4	BZR	F. Jordan, 16 Elterwater Crescent, Barrow-in-Furness, LA14 4PH
G4	BZS	M. Pasek, 10 Prospect Place, Norwood Green, Halifax, HX3 8QF
G4	BZU	B. Beaven, 7 Glamorgan Road, Up Hatherley, Cheltenham, GL51 3JF
G4	BZV	N. Barton, 3 Plover Close, Oakham, LE15 6BE
G4	CAA	NATS & CAA Radio Society, c/o S. Rossi, 21 Rattigan Gardens, Whiteley, Fareham, PO15 7EA
GM4	CAB	S. Reynolds, 39 Panmure St., Broughty Ferry, Dundee, DD5 2EU
G4	CAF	D. Hogg, Fairview, Dordale Road, Bromsgrove, B61 9JT
G4	CAJ	M. Farr, 23 Waterfall Way, Barwell, Leicester, LE9 8EH
G4	CAK	M. Scarlett, Maple House, Westford, Wellington, TA21 0DT
GM4	CAM	D. Hamilton, 7 High Langside Holding, Craigie, Kilmarnock, KA1 5ND
GM4	CAQ	R. Miles, 15 Clark Avenue, Linlithgow, EH49 7AP
GW4	CAT	N. Schofield, Maen Llwyd-Tan Yr Alt, Llanllyfni, Caernarfon, LL54 6RT
GM4	CAU	T. Wratten, 89 Hilton Road, Aberdeen, AB24 4HX
G4	CAX	D. Borley, 95 Meadow Lane, Moulton, Northwich, CW9 8QQ
G4	CAY	C. Parker, 25 Meadow Dale, Chilton, Ferryhill, DL17 0RW
G4	CAZ	J. Lefever, 30 Holland Road, Melton Mowbray, LE13 0LU
G4	CBD	J. Swanson, 9 Park House Gardens, Twickenham, TW1 2DF
GI4	CBG	R. Smyth, 58 Gilnahirk Road, Belfast, BT5 7DH
G4	CBL	P. Tomlinson, 55 Reldene Drive, Hull, HU5 5HS
G4	CBM	G. Blakeley, Stowe House, Preston Gubbals Road, Shrewsbury, SY4 3LY
G4	CBO	D. Aiken, 16 Woodland Gardens, North Wootton, King's Lynn, PE30 3PX
G4	CBQ	P. Daniells, Holly Villa, Foxholes, Wem, Shrewsbury, SY4 5UJ
G4	CBT	H. Wall, 54 Little Harlescott Lane, Shrewsbury, SY1 3PZ
G4	CBW	A. Horsfall, 60 Talke Road, Red Street, Newcastle, ST5 7AH
G4	CBY	T. Cooper, Lincolnshire House, Brumby Wood Lane, South Humberside, DN17 1AF
G4	CBZ	A. Mepham, 32 Brambletyne Avenue Saltdean, Brighton, BN2 8EJ
GW4	CC	Gower/Gwyr Contesting Club, c/o K. Dyer, 34 Lundy Drive, West Cross, Swansea, SA3 5QL
G4	CCA	M. Fadil, 25 North Parade, Horsham, RH12 2DA
G4	CCC	C. Young, 18 Wincroft Road, Caversham, Reading, RG4 7HH
G4	CCH	H. Ling, 8 Spa Hill, Kirton Lindsey, Gainsborough, DN21 4NE
G4	CCI	J. Chapman, 7 Ravensthorpe Drive, Loughborough, LE11 4PU
GM4	CCN	T. Keats, Tigh Na Luch, Skye Of Curr Road Dulnain Bridge, Grantown-on-Spey, PH26 3PA
G4	CCQ	M. Stanton, 84 Forest Hill, Maidstone, ME15 6TH
G4	CCT	S. Hyman, 49 Southover, Woodside Park, London, N12 7JG
G4	CCY	P. Fagg, 113 Bute Road, Wallington, SM6 8AE
G4	CCZ	P. Simons, Westwood, Faris Lane, Addlestone, KT15 3DJ
G4	CDC	E. Morton, 6 Norfolk Avenue, Burton-Upon-Stather, Scunthorpe, DN15 9EW
G4	CDD	Denby Dale and District ARS, c/o J. Chappell, 49 Midway, South Crosland, Huddersfield, HD4 7DA
G4	CDF	M. Naylor, 6 Holsworthy Close, Lower Earley, Reading, RG6 3AH
G4	CDG	A. Davidson, Po Box Hm 150, Hamilton, Bermuda, HM AX
G4	CDH	J. Brade, 11 Old Farm Place, Ash Vale, Aldershot, GU12 5SF
G4	CDI	G. Boardman, 9 Byron Road, Weston-Super-Mare, BS23 3XQ
G4	CDJ	P. Jarrett, Roydan House 36 Ferndale Road, Teignmouth, TQ14 8NH
G4	CDL	F. Mepham, Avenida Robleda 16/22, San Luis, Torrevieja, Spain, 3180
G4	CDN	R. Banester, Fairfield, Church Road, Norwich, NR12 9SA
G4	CDR	C. Winstanley, 3 Peter St., Blackburn, BB1 5HQ
G4	CDU	N. Huntley, 26 Malin Parade, Portishead, Bristol, BS20 7FW
G4	CDW	G. Trickey, 3 Fairleigh Rise, Kington Langley, Chippenham, SN15 5QF
G4	CDX	P. Wheeler, 69 Waterside Road, Slyfield Green, Guildford, GU1 1RQ
G4	CDY	T. Giles, Lanherne, Meaver Road, Mullion, Helston, TR12 7DN
G4	CDZ	J. Boden, 2 The Coppice, Whaley Bridge, High Peak, SK23 7LH
G4	CEC	Dr P. Knight, 26 Meadway, Harrold, MK43 7DR
G4	CEI	M. Baker, 17 Whitehills Green, Goring, Reading, RG8 0EB
G4	CEJ	R. Moore, 17 Somme Avenue, Flookburgh, Grange-over-Sands, LA11 7LJ
G4	CEK	J. Bird, Panorama, Cobblers Lane, Swanage, BH19 2PX
G4	CEL	S. Hudson, Frekes Cottage, Moorside, Sturminster Newton, DT10 1HQ
G4	CEN	D. Davies, 35 Ruthellen Road, Chelmsford, USA, 1824
G4	CEO	M. Uotome, 5-21-6-308, Kita Karasuama, Setagaya-Ku, Tokyo, Japan, 157-0061
G4	CEP	G. Morris, 7 Manor Road, Sandy, SG19 1DT
G4	CES	Royal Air Force Cosford ARC, c/o M. Farmer, Horton Brook Cottage, Horton Wem, Shrewsbury, SY4 5NB
G4	CEU	D. Jarvis, Flat 1, Gunnery House, 2 Chapel Road, Shoeburyness, Southend-on-Sea, SS3 9SL
G4	CEX	C. Durant, 63 Ulleries Road, Solihull, B92 8DX
G4	CEY	J. Ball, 68 Swallows Court, Pool Close, Spalding, PE11 1GZ
G4	CFB	K. Henry, 80 Fernwood Rise, Westdene, Brighton, BN1 5EP
GW4	CFC	L. Gruffydd, 45 Maes Yr Hafod, Menai Bridge, LL59 5NB
G4	CFD	H. Garner, 6 Blacksmiths Close, Barrow-upon-Humber, DN19 7HG
G4	CFG	P. Arnold, 14 George Birch Close, Brinklow, Rugby, CV23 0NN
G4	CFH	J. Hill, 10 Albert Clarke Drive, Willenhall, WV12 5AU
G4	CFK	L. Smith, 36 Kinderton Park Cledford Lane, Middlewich, CW10 0JS
G4	CFP	W. Bones, 22 Rotherhead Close, Horwich, Bolton, BL6 5UG
GI4	CFQ	J. Mcsweeney, 109 Twaddell Avenue, Belfast, BT13 3LG
G4	CFS	G. Dodwell, 9 Balfour Drive, Liss, GU33 7BF
G4	CFV	R. Hall, Pinewood Lodge, 16 Tullyvarraga Hill, Co Clare, Ireland, V14 H292
G4	CFW	R. Raven, 9 Southwood Close, Ferndown, BH22 9HW
G4	CFY	A. Nailer, 12 Weatherbury Way, Dorchester, DT1 2EF
G4	CFZ	M. Stevens, 3 Rip Croft, Portland, DT5 2EE
G4	CGA	D. Sellwood, 47 Waterhall Avenue, London, E4 6NA
G4	CGB	D. Tromans, 29 Cannon Road Wombourne, Wolverhampton, WV5 9HR
G4	CGD	A. Richardson, 24 West House Close, Wimbledon, London, SW19 6QU
G4	CGE	W. Dore, 87 Thame Road, Aylesbury, HP21 8LY
G4	CGF	W. Badz, Bottom Flat, 36 Luckington Road, Bristol, BS7 0US
G4	CGG	R. I'Anson, 87 Tranby Lane, Anlaby, Hull, HU10 7DT
G4	CGH	M. Davies, 2 Manor Close, Berrow, Burnham-on-Sea, TA8 2LN
G4	CGL	J. Miller, 29 Springhill Road, Wednesfield, Wolverhampton, WV11 3AW
G4	CGM	M. Duff, Clittaford Club, Moses Close, Plymouth, PL6 6JP
G4	CGO	J. Pollock, 71 Stevenson Street, Kew, Australia, 3101
G4	CGP	P. Wright, 4 Avill Way, Wickersley, Rotherham, S66 1DL
G4	CGR	K. Davies, High View, Alcester Road Wootton Wawen, Henley-in-Arden, B95 6BH
G4	CGU	R. Taylor, 23 Ridgacre Lane, Quinton, Birmingham, B32 1EL
G4	CGV	C. Manklow, 37 Brittons Crescent, Barrow, Bury St Edmunds, IP29 5AG
G4	CGW	J. Dunglinson, Blenheim, Willow Lane, Camberley, GU17 9DL
GW4	CGZ	D. Newman, 138 Twyn Carmel, Merthyr Tydfil, CF48 1PH
G4	CHD	T. Adams, 1 Francis Drive Westward Ho, Bideford, EX39 1XE
GM4	CHF	J. Magill, 74 Garnqueen Crescent Glenboig, Coatbridge, ML5 2SX
G4	CHG	P. Ashton, 7 Conway Grove, Cheadle, Stoke-on-Trent, ST10 1QG
G4	CHI	P. Robinson, Longcroft House, Longcroft Lane, Burton-on-Trent, DE13 8NT
G4	CHJ	A. Williams, 10 Olde Hall Road, Featherstone, Wolverhampton, WV10 7BB
G4	CHL	P. Howe, 135 Rue De Pierrevert, Ste-Tulle, France, 4220
G4	CHM	R. Mcewan, Fifth Acre, Carr Lane, Alfreton, DE55 2DN
G4	CHS	Cheltenham Hackspace ARS, c/o D. Miller, 50 Sandyleaze, Gloucester, GL2 0PX
GM4	CHX	J. Kyle, 7 Fasaich, Strath, Gairloch, IV21 2DH
GU4	CHY	R. Allisette, Lilyvale House, Rue Des Houmets, Castel, Guernsey, GY5 7XZ
G4	CIA	W. Cooper, 20 Planton Way, Brightlingsea, Colchester, CO7 0LB
G4	CIB	B. Woodcock, 2 Poolhay Close Corse Lawn, Gloucester, GL19 4NY
G4	CIC	S. Edmondson, 7 Browns Road, Bradley Fold, Bolton, BL2 6RQ
GM4	CID	R. Mcclements, Eskdail, Newtown St. Boswells, Melrose, TD6 0RY
G4	CIJ	J. Chennells, 10 Lower Cippenham Lane, Slough, SL1 5DF
G4	CIO	Dr M. Phillips, Chapel House, The Cross, Stonehouse, GL10 3TU
G4	CIZ	A. Wallbank, 1 Pollards Cottages, Clanville, Andover, SP11 9JD
G4	CJJ	M. Viner, 15 St. Anthonys Drive, Hedon, Hull, HU12 8NT
G4	CJK	V. Roney, 76 Hilton Lane, Great Wyrley, Walsall, WS6 6DT
G4	CJM	J. Alcock, 1 Alma St., Fenton, Stoke on Trent, ST4 4PH
G4	CJO	A. Mountifield, 6 Sawyers Close, Teg Down, Winchester, SO22 5JX
G4	CJP	V. Duffy, 2 Moor View Close, High Harrington, Workington, CA14 4NX
G4	CJR	C. Crick, 19 The Drive, Coulsdon, CR5 2BL
G4	CJT	K. Hughes, 4 Epsom Place, Cranleigh, GU6 7ET
G4	CJV	A. Kerton, 8 Fabian Drive Stoke Gifford, Bristol, BS34 8XN
G4	CJY	B. Payne, 78 Carver Hill Road, High Wycombe, HP11 2UA
G4	CKB	J. Banester, Fairfield, Church Road, Norwich, NR12 9SA
G4	CKH	G. Jackson, 86 Lloyds Avenue, Kessingland, Lowestoft, NR33 7TR
G4	CKK	P. Atkins, 60 Wentworth Way, Harborne, Birmingham, B32 2UX
G4	CKQ	A. Horne, 1 Upper Halliford Road, Shepperton, TW17 8RX
G4	CKS	D. Fitzgerald, 36 Vardens Road, London, SW11 1RH
G4	CKT	R. Gwynne, 17 Dorrington Close, Stoke-on-Trent, ST2 7BZ
G4	CKX	S. Taylor, 5 Chiltern Avenue, Bishops Cleeve, Cheltenham, GL52 8XP
G4	CLA	P. Lindsay, The Barn, Main Street Ashby Parva, Lutterworth, LE17 5HY
G4	CLB	C. Brown, 31 Sapcote Road, Burbage, Hinckley, LE10 2AS
G4	CLC	D. Lewis, Brandywine, Westbury, BA13 4NY
G4	CLD	G. Beaver, The Gables, Reading Road, Reading, RG7 3BU
G4	CLE	T. Baker, 18 Prescott Avenue, Rufford, Ormskirk, L40 1TT
G4	CLF	J. Bryant, Hillhead Cottage, Calshot Road, Southampton, SO45 1BR
G4	CLG	S. Whittingham, 18 Northcroft, Shenley Lodge, Milton Keynes, MK5 7AJ
G4	CLI	Dr D. Sadler-Lockwood, 14 Mountain Road, Dewsbury, WF12 0BW
G4	CLJ	P. Eccles, Inghams, The Town, Dewsbury, WF12 0QX
G4	CLL	R. Goodchild, Grey Cliffe House, Owmby Cliff Road, Owmby-By-Spital, Market Rasen, LN8 2HL
G4	CLM	C. Le Marchant, Bow House, Green Lane, Prestwood, Great Missenden, HP16 0QE
G4	CLN	P. Redfern, 12, Wilbarn Road Wilbarn Road, Paignton, TQ3 2BN
G4	CLO	R. Armishaw, Drift House, Cattons Drift, The Green, Felbrigg, Norwich, NR11 8PN
G4	CLP	J. Harrison, 47 Mason Way, Padbury WA, Australia, 6025
G4	CLR	I. Hewer, 23 Thoresby Avenue, Tuffley, Gloucester, GL4 0TD
G4	CLY	N. Thompson, 6 Miena Way, Ashtead, KT21 2HU
G4	CMG	T. Milne, Lynwood, Clovelly Road, Hindhead, GU26 6RP
G4	CMH	D. Spendlove, 22 Green Bank, Harwood, Bolton, BL2 3NG
GM4	CMI	R. Campbell, 1 Gibraltar Terrace, Dalkeith, EH22 1EE

G4 CMK R. Harker, 140 Victoria Road, Beverley, HU17 8PJ
G4 CML J. Livesley, Rivendell 71B Hillfoot Road, Shillington, SG5 3NS
G4 CMM C. Pope, Silver Hill, Norwich Road, Great Yarmouth, NR29 5PB
G4 CMP P. Lennon, 53 Rycot Road, Speke, Liverpool, L24 3TH
G4 CMT Raywell Park Scout ARS, c/o A. Russell, 3 St. Nicholas Close, North Newbald, York, YO43 4TT
G4 CMU G. Brind, 9 Becket Wood, Newdigate, Dorking, RH5 5AQ
G4 CMX P. Rossiter, 36 Milton Drive Ravenshead, Nottinghamshire, NG15 9BE
G4 CMY A. Mann, 13 Rosedale Avenue, Stonehouse, GL10 2QH
G4 CMZ K. Archer, 24 Willson Road, Littleover, Derby, DE23 1BZ
G4 CNH L. Carpenter, 166 Abbey View Garsmouth Way, Watford, WD25 9DZ
G4 CNI P. Geiger, Lloyd Mount, Howard Drive, Altrincham, WA15 0LT
GW4 CNL G. Goodfield, 10 Lewis Street, Church Village, Pontypridd, CF38 1BY
G4 CNZ D. Allen, 344 Coventry Road, Hinckley, LE10 0NH
G4 COE D. Smith, 54 Warrington Road, Leigh, WN7 3EB
GW4 COJ C. Roberts, 8 Oaklands Park Drive, Rhiwderin, Newport, NP10 8RB
GW4 COL I. Braithwaite, Kew, Trefonen, Oswestry, SY10 9DH
G4 COM J. Compton, Aysgarth, Durley Brook Road, Southampton, SO32 2AR
G4 COR I. Harvey, 50 Callow Hill Way, Littleover, Derby, DE23 3RL
G4 COS J. Hansell, 87 Garratts Way, High Wycombe, HP13 5XT
G4 COT S. Brett, 8 Pinewood Grove, Hull, HU5 5YY
G4 COV C. Cardwell, 11 Manor Cottages, Heronsgate Road, Rickmansworth, WD3 5BJ
GM4 COX J. Hood, 4 Murray Road, Law, Carluke, ML8 5HR
G4 CPA G. Hanson, 11 Churchill Way, Cross Hills, Keighley, BD20 7DN
G4 CPC S. Wright, 20 Stillwell Grove, Wakefield, WF2 6RN
G4 CPD G. Knox, Glencairn, 6 Aldborough Road, York, YO51 9EA
G4 CPE A. Turner, 7 Slate Hall, Sundon, Luton, LU3 3PY
G4 CPG M. Howkins, 16 Beckett Court, Gedling, Nottingham, NG4 4GS
G4 CPI J. Housden, 5 Tothby Meadows, Alford, LN13 0EH
G4 CPL C. Mcgee, Zafra, The Knapp, Bromyard, HR7 4BD
G4 CPM A. Fielding, 95 Hillcrest, Weybridge, KT13 8AS
G4 CPN J. Bird, 166 Cowick Lane, Exeter, EX2 9JF
G4 CPQ N. Scrogie, 154 St. Albans Road, Derby, DE22 3JP
G4 CPV R. Fisk, 16 Sterry Drive, Thames Ditton, KT7 0YN
G4 CPW P. Wilson, 5 Pebble Close, Lowestoft, NR32 4DR
G4 CPY N. Grassby, 11 Eider Close, Whetstone, Leicester, LE8 6YB
G4 CQA G. Angell, 55 Golden Riddy, Leighton Buzzard, LU7 2RH
G4 CQH J. Sperry, 50 Lochinver, Hanworth, Bracknell, RG12 7LD
G4 CQI A. Lanfear, 120 Charlton Road, Kingswood, Bristol, BS15 1HF
G4 CQM D. Hilleard, Hazeldene, Bridgerule, Holsworthy, EX22 7EW
G4 CQN M. Mawby, 61 Carter Drive, Beverley, HU17 9GL
G4 CQO S. Burgess, Tretawn, Kite Hill, Ryde, PO33 4LG
G4 CQQ R. Taylor, 8 Park Avenue, Markfield, LE67 9WA
G4 CQR D. Wood, 49 Wolsey Crescent, Morden, SM4 4TD
G4 CQS A. Rowsby, 10 Echells Close, Bromsgrove, B61 7EB
GW4 CQT D. Price, Vine Cottage, Garth Road, Cwmbran, NP44 7AB
G4 CQV P. Baldwin, 26 Ashford Road, Fulshaw Park, Wilmslow, SK9 1QE
G4 CQW A. Lane, 21 Winterbourne Road, Poole, BH15 2ES
G4 CQX L. Palfrey, C/O Po Box 314, Cyprus, XX99 1AA
GW4 CQZ M. Doig, Helenfa, Ystrad Road, Denbigh, LL16 3HE
G4 CRB W. Oxley, Flat 147, Oceana Boulevard Orchard Place, Southampton, SO14 3HW
G4 CRC Cornish Radio Amateur Club, c/o S. Holland, 49 Oxland Road, Illogan, Redruth, TR16 4SH
G4 CRE D. Rush, 8 Sheaf Place, Worksop, S81 7LE
G4 CRG K. Burgin, The Pike Lock House, Eastington, Stonehouse, Glos Gl10 3rt, GL10 3RT
GW4 CRH M. Worvill, The Berwyns, Domgay Road, Four Crosses, Llanymynech, SY22 6SL
G4 CRK R. Sellman, 43 Mount Avenue, Stone, ST15 8LW
G4 CRM J. Lennon, 107 Andrew Crescent, Waterlooville, PO7 6BG
G4 CRN Dr A. Hall, Westhill, Bear Lane Longdon, Tewkesbury, GL20 6BB
G4 CRP K. Tyler, Pinfold House, 3 Pinfold Lane, Leeds, LS25 1HE
G4 CRS E.C A.R.C., c/o A. Mackay, 2 Highcliffe Grove, New Marske, Redcar, TS11 8DU
G4 CRT L. Kirby, 41 Woodville Road, Overseal, Swadlincote, DE12 6LU
G4 CRW A. Holmes, 4 Castle Avenue, Datchet, Slough, SL3 9BA
G4 CSD P. Hyde, 25 Merton Road, Basingstoke, RG21 5UA
G4 CSE M. Lewis, 10 Kenmore Drive, Bristol, BS7 0TT
G4 CSM D. Chaplin, 35 Lanes End, Totland Bay, PO39 0AL
GI4 CSO J. Mccormack, 12 Glengoland Crescent, Dunmurry, Belfast, BT17 0JG
GI4 CSP G. Robinson, 10 Ranfurly Avenue, Dungannon, BT71 6PJ
G4 CST G. Hopkins, 63 Alma Road, Brixham, TQ58QR
G4 CSV J. Jackson, 43 Ambleside, Boundary Court, Stockport, SK8 1BA
GW4 CSY B. Vickery, 6 Duffryn Close, St. Nicholas, Cardiff, CF5 6SS
G4 CSZ M. Riley, 5 Dunstarn Gardens, Leeds, LS16 8EJ
G4 CTA A. Clewer, 6 Frensham Close, Stanway, Colchester, CO3 0HP
G4 CTC T. Cann, Noahs Rough, Old Coach Road Wrotham, Sevenoaks, TN15 7NR
G4 CTD C. Vernon, 50 Copthall Road West, Ickenham, Uxbridge, UB10 8HS
G4 CTE P. Bradshaw, 43 Hill Top Road, Grenoside, Sheffield, S35 8PE
G4 CTI P. Ashcroft, 7 Kings Ripton Road, Sapley, Huntingdon, PE28 2NU
G4 CTM P. Barratt, 3 Bramshott Close, Hitchin, SG4 9EP
G4 CTT Dr T. Thirst, Thirsts Farm, Happisburgh Road, Happisburgh, NR12 0RU
G4 CTU B. Hitchins, 12 Parkland Avenue, Kidderminster, DY11 6BX
GW4 CTV S. Mee, Cysgod Y Gaer, Cwmsymlog, Aberystwyth, SY23 3EZ
G4 CTY A. Lightbody, 3 Elphicks Place, Tunbridge Wells, TN2 5NB
G4 CTZ I. Cage, 334 Stockton Lane, York, YO31 1JW
G4 CUE W. Pechey, Jays Lodge, Crays Pond, Reading, RG8 7QG
G4 CUG R. Worsell, 8 Waterworks Cottages, Old Willingdon Road, Eastbourne, BN20 0AS
G4 CUI Dr G. Cook, 1 St. Albans Road, Fulwood, Sheffield, S10 4DN
G4 CUQ B. Hughes, 30 Fuller Road, Dagenham, RM8 2TU
GI4 CUV N. Atkins, 38 Rosscoole Park, Belfast, BT14 8JX
GM4 CUX G. Winchester, 23 Craigmount, Avenue North, Edinburgh, EH12 8DL
G4 CVA Rev. J. Wardle, 27 First Avenue, Bridlington, YO15 2JW
G4 CVC J. Everist, 11 Redding Close, Dartford, DA2 6NB
G4 CVD P. Petty, 41 Hensley Road, Bath, BA2 2DR
G4 CVF B. Sheppard, 20 Lambourne Court, St. Johns Close, Uxbridge, UB8 2UL
G4 CVG W. Bullock, 14 Saxon Drive, Rillington, Malton, YO17 8LZ
G4 CVM R. Watson, 36 Abbots Close, Knowle, Solihull, B93 9PP
G4 CVN D. Williams, Lyngarth, Leatherhead Road, Leatherhead, KT23 4RR
G4 CVO W. Wyer, 11 Nether Close, Wingerworth, Chesterfield, S42 6UR
G4 CVS Dr B. Pearson, 8 The Pastures, Edlesborough, Dunstable, LU6 2HL
GW4 CVT C. Thorley, Helston, The Mountain, Holyhead, LL65 1YR
G4 CVU J. Swingewood, 5 Blaze Park, Wall Heath, Kingswinford, DY6 0LL
G4 CVX Dr R. Sims, 345 Blandford Road, Hamworthy, Poole, BH15 4HP
G4 CW North Kent Radio Society, c/o F. Connor, 134 Summerhouse Drive, Bexley, DA5 2ES
G4 CWA Rev. W. Burton, 23 Purok 5, San Pedro Li, Pampanga, Philippines,
G4 CWB D. Andrews, 100 Duchy Road, Harrogate, HG1 2HA
G4 CWC R. Barrett, Lumina, Bridegate Lane, Melton Mowbray, LE14 3QA
G4 CWE A. Humm, 32 Layton Road, Hounslow, TW3 1YH
GW4 CWG G. Crossland, 32 Long Bridge St., Llanidloes, SY18 6AR
G4 CWH Dr C. Smithers, 10 Grange Park, Bishops Stortford, CM23 2HX
G4 CWM J. Pickles, 111 Linden Avenue, Prestbury, Cheltenham, GL52 3DT
G4 CWP W. Pevy, Brambletye, Ashstead Lane, Godalming, GU7 1SY
G4 CWS S. Wood, 16 Ramley Road, Lymington, SO41 8GQ
GW4 CWU B. Heppenstall, Gwelfor, Llanrhyddlad, Holyhead, LL65 4BG
G4 CWV F. Parr, 5 Benenden Road, Wainscott, Rochester, ME2 4NU
G4 CXE P. Bolton, 93 Westfields, Narborough, King's Lynn, PE32 1SY
GM4 CXF J. Thomson, 31 Teviot Place, Troon, KA10 7EE
GW4 CXK R. Evans, 74 Alexandra Street, Ebbw Vale, NP23 6JF
G4 CXL R. Menday, Huf House, Horseshoe Ridge, Weybridge, KT13 0NR
GM4 CXP D. Dance, 18 Masons Court, Kelso, TD5 7NJ
G4 CXQ D. Dyer, 26 Locking Road, Weston-Super-Mare, BS23 3DF
G4 CXT M. Bell, Quebec Cottage, Curlew Green Kelsale, Saxmundham, IP17 2RA
G4 CXW G. Spencer, 17 Rockland Road, Bristol, BS16 2SW
G4 CXX M. Bullough, 2 Colebrook Close, London, SW15 3HZ
G4 CXZ A. Thompson, 6 Ducks Walk, Twickenham, TW1 2DD
G4 CYB F. Burnett, Herons Siege, Blundies Lane, Enville, Stourbridge, DY7 5HU
G4 CYC K. Green, Ao-Te-Aroa, 12 Hill Road, Fareham, PO16 8LB
G4 CYF C. Tully, Harmony, The Crescent, Clacton-on-Sea, CO16 0EP
G4 CYG D. Darkes, 70 Braemar Road, Lillington, Leamington Spa, CV32 7EY
G4 CYI J. Palfrey, Lower Trewince Farm, Newquay, TR8 4AW
G4 CYO K. Robinson, 3 The Woodhouses, Patshull Road, Wolverhampton, WV6 7DU
G4 CYR S. Allen, The Poplars, Wotton Underwood, HP18 0RX
G4 CYY C. Lewis, 54 Whelpley Hill Park, Whelpley Hill, Chesham, HP5 3RJ
G4 CYZ L. Large, Captains Farmhouse, Streat Lane, Hassocks, BN6 8SB
G4 CZA K. Newman, 2 Skys Wood Road, St Albans, AL4 9NZ
G4 CZB J. Cockrill, 28 Northampton Road, Harpole, Northampton, NN7 4DD
G4 CZH E. Brindley, 150A Woods Lane, Derby, DE22 3UE
G4 CZK A. Mercer, Coed Y Rhedyn, Bronydd, Clyro, HR3 5RX
GI4 CZO G. Mccomb, 1 Magheraboy Drive, Portrush, BT56 8GP
G4 CZP R. Crossley, 2 Jewel View, 31 Downside, Ventnor, PO38 1AL
G4 CZR C. Redfern, 6 Pont Croix, Mellionnec, France, 22110
G4 CZU P. Hadler, 30 Hillview Road, Whitstable, CT5 4HX
GI4 CZW C. Corderoy, 3 The Limes, Drumlyon, Enniskillen, BT74 5NQ
G4 CZX I. Godden, 163 Ringmer Road, Worthing, BN13 1DZ
G4 CZZ R. Aggus, 68 Conifer Walk, Stevenage, SG2 7QS
G4 DAC D. Squires, 91 Croham Valley Road, South Croydon, CR2 7JJ
G4 DAF G. Walker, 56 Goodwin Road, Croydon, CR0 4EG
G4 DAM R. Dence, 32 Hayeswood Road Stanley Common, Ilkeston, DE7 6GB
G4 DAP C. Ison, 19 Grays Close, Chalgrove, Oxford, OX44 7TN
G4 DAQ W. Silvester, 2 Tudor Close, Barton-Le-Clay, Bedford, MK45 4NE
G4 DAT R. Davidson, 5 St. Lucians Lane, Wallingford, OX10 9ER
GI4 DAV D. Hart, 31 Downshire Road, Carrickfergus, BT38 7QD
G4 DAX D. Smith, Red Roof, Goathland, Whitby, YO22 5AN
G4 DAY D. Sawyer, 2 Blunts Wood Road, Haywards Heath, RH16 1NB
G4 DBD A. Borland, 39 Green Lane, Willaston, Nantwich, CW5 7HY
G4 DBE J. Clark, 20 Sandy Lane, Irby, Wirral, CH61 0HD
G4 DBF Maj. D. Freeston, 20 Coningham Road, Whitley Wood, Reading, RG2 8QP
G4 DBG F. Kneale, 60 Summertrees Road, Great Sutton, Ellesmere Port, CH66 2BJ
G4 DBM B. Mcgennity, 46 St. Andrews Road, Boreham, Chelmsford, CM3 3BY
G4 DBN N. Smith, Birch Tree House Asselby, Goole, DN14 7HE
G4 DBQ B. Roberts, 7 North Square, London, NW11 7AA
G4 DBR C. Ewing, 130 Uttoxeter Road, Hill Ridware, Rugeley, WS15 3QX
G4 DBS B. Middleton, 10 Curtois Close Branston, Lincoln, LN4 1LJ
G4 DBX L. Stubbs, The Cottage, Middlewich Road, Crewe, CW1 4RA
G4 DBY P. Walker, 48 Whitefields Drive, Richmond, DL10 7DL
G4 DBZ D. Martin, 12 South Park, Redruth, TR15 3AW
G4 DCB P. Mortimer, 13A Elder Avenue, Wickford, SS12 0LP
GI4 DCC W. Chesney, 52 Taylorstown Road, Toomebridge, Antrim, BT41 3RT
G4 DCD C. Stephenson, 6 Livingstone Close, Rothwell, Kettering, NN14 6HT
G4 DCE J. Sketchley, 48 Coverdale, Whitwick, Coalville, LE67 5BP
G4 DCF M. Booth, 14 Blackstock Close, Sheffield, S14 1AE
G4 DCH C. Tucker, 29, Newton Abbot, TQ12 1US
G4 DCI P. Hopewell, 3 Hunts Orchard, Hathern, Loughborough, LE12 5HQ
G4 DCJ D. Jarrett, 15 Groveside, East Rudham, King's Lynn, PE31 8RL
G4 DCK M. Holliday, 1A Fairmile Gardens, Longford, Gloucester, GL29ED
GM4 DCL T. Main, 15 Polton Road, Lasswade, EH18 1AB
G4 DCM P. Rhodes, Parcela 1352, Calle De Zurbaran 8, Alicante, Spain, 3170
G4 DCN Dr J. Mason, Keepers Cottage, Baldon Lane, Marsh Baldon, Oxford, OX44 9LT
G4 DCP P. Hull, Seymour Cottage, Forest Road, Waterlooville, PO7 6UA
G4 DCS M. Grant, 27 Windmill Street, Whittlesey, Peterborough, PE7 1QN
G4 DCW D. Walker, 70 High Street, Cranfield, Bedford, MK43 0DF
G4 DCX E. Trickey, 53 Hollyguest Road, Hanham Green, Bristol, BS15 9NN
G4 DCY W. Dransfield, Flat 6, Heath Mount Hall, Ilkley, LS29 9JN
G4 DDB E. Connor, 8 Russell Street, Dover, CT16 1PX
G4 DDC Dunstable Downs Radio Club, c/o P. Seaford, 14 Nevis Close, Leighton Buzzard, LU7 2XD
G4 DDD R. John, 32 Hundred Acre Road, Streetly, Sutton Coldfield, B74 2LA
G4 DDI C. Guy, 7 Herrick Court, Clinton Park, Lincoln, LN4 4QU
G4 DDK S. Jewell, Blenheim Cottage, Falkenham, Ipswich, IP10 0QU
G4 DDL M. Pemberton, 37 Woodmancott Close, Forest Park, Bracknell, RG12 0XU

G4 DDM R. Finch, 1 Cherry Tree Cottage, Church Road, High Wycombe, HP10 8LN
G4 DDN G. Leonard, 65 Qualitas, Bracknell, RG12 7QG
G4 DDP R. Clark, 41 Avenue Road, Bexleyheath, DA7 4
G4 DDT A. Ives, 24 Johnson Crescent, Heacham, King's Lynn, PE31 7LQ
G4 DDV R. Bowman, 13 Wellington Road, St Albans, AL1 5NJ
G4 DDX R. Pratt, 16 Thurlow Close, Stevenage, SG1 4SD
G4 DDY M. Fagg, 113 Bute Road, Wallington, SM6 8AE
G4 DDZ N. Turner, Rose Cottage, Catshill Cross, Stafford, ST21 6LT
G4 DEA P. Dunning, Cold Harbour, Bishop Burton, Beverley, HU17 8QA
G4 DEM D. Walker, The Horseshoe Inn, 1 Horseshoe Court, Bristol, BS36 2FD
G4 DEO A. Wallis, 5 Nancevallon, Higher Brea, Camborne, TR14 9DE
GW4 DEP D. Dabinett, Pentre Isaf, Llangyniew, Welshpool, SY21 0JT
G4 DEQ A. Derrick, 4 Hillside Cottages, Barrow Street, Bristol, BS48 3RX
G4 DEU A. Fuge, 6 Haythorne Court, Staple Hill, Bristol, BS16 5QS
G4 DEV S. Newport, 18 Chacewater Crescent, Worcester, WR3 7AN
G4 DEW J. Males, 49 Gunthorpe Road, Peterborough, PE4 7TN
GM4 DEX J. Sharp, 72 Broom Road, Rimbleton, Glenrothes, KY6 2BQ
G4 DFA T. Ellinor, 53 Hillside, Banstead, SM7 1HG
G4 DFB D. Berry, 4 Holly Hill, Winchmore Hill, London, N21 1NP
G4 DFC C. Goldingay, 71 Kingham Close, Redditch, B98 0SB
G4 DFD K. Bailey, 16 Chandos Drive, Martlesham, Woodbridge, IP12 4ST
G4 DFE W. Raybould, 33 Roberts Green Road, Dudley, DY3 2BB
G4 DFG P. Gibbs, Redhill Farm Stables, Redhill, Telford, TF2 9NZ
G4 DFJ Dr J. Klein, 5 Cranley Gardens, London, N10 3AA
G4 DFN S. Widdett, 21 Lower Howsell Road, Malvern, WR14 1DX
G4 DFO S. Wainwright, 39 Ascot Road, Birmingham, B13 9EN
G4 DFP A. Morecroft, 4 Arran Close, Bolton, BL3 4PP
GW4 DFQ N. Dear, Hollybush Cottage, Candy, Oswestry, SY10 9BA
G4 DFS S. Booth, 13 Milner Avenue, Penistone, Sheffield, S36 9DB
G4 DFT R. Perrin, 131 Acacia Avenue, Ottawa, Ontario, Canada, K1M 0R2
G4 DFU F. Skillington, 6 Okehampton Crescent, Nottingham, NG3 5SE
G4 DFV D. Walters, 11 King George V Avenue, Mansfield, NG18 4ER
G4 DFX J. Taylor, 26 Courthope Road, London, NW3 2LD
G4 DFY R. Dedman, 2 Forest Villas, Long Mill Lane, Sevenoaks, TN15 8LQ
G4 DFZ K. Knight, 61 Westbourne Road, Sutton-in-Ashfield, NG17 2FB
G4 DGB T. Crute, 26 Runcorn, Sunderland, SR2 0BP
G4 DGF A. Matthews, 14 Hardy Green, Wellington College, Crowthorne, RG45 7QR
GI4 DGI Rev. D. Coyle, 16 Northland Avenue, Londonderry, BT48 7JN
G4 DGL Dr E. Mills, C/O Ac Clarke, 243 Barton Road, Cambridge, CB23 7BU
G4 DGQ J. Dussart, Seagarth, Cresswell, Northumberland, NE61 5JU
GM4 DGT W. Stirling, 16 Shire Way, Alloa, FK10 1NQ
G4 DGW A. Gagnon, 60 Woodruff Avenue, Hove, BN3 6PJ
G4 DHF D. Johnson, Dean Cottage, Dowsby Fen, Bourne, PE10 0TU
G4 DHK R. Stanleigh, Shallowpool Bungalow , Shallowpool, PL13 2ND
G4 DHL C. Durnall, 143 Green Lane, Wolverhampton, WV6 9HB
GM4 DHN N. Macleod, 54 Drum Brae South, Edinburgh, EH12 8TB
G4 DHQ A. Becket, 65 High Street, Prestwood, HP16 9EJ
G4 DHT R. Haverson, Kerri, Sunton, Marlborough, SN8 3DZ
G4 DHU D. Spender, Rose Cottage, Huntsmans Lane, Sudbury, CO10 7JX
G4 DHV C. Jones, 49 Newport Pagnell Road, Hardingstone, Northampton, NN4 6ER
G4 DHW P. Mcelroy, 2 Donohue Lane, Manchester, USA, 6040
G4 DIA B. Powell, 1 The Heights, Market Harborough, LE16 8BQ
G4 DIC R. Phipps, 4 Mill Court, Wells-Next-the-Sea, NR23 1HF
G4 DIE I. Dredge, 60 Springfield Close, Corsham, SN13 0JR
G4 DIG D. Hine, 6A Clifton Terrace, Southend-on-Sea, SS1 1DT
G4 DIH R. Coates, 57 Dalebrook Road, Burton-on-Trent, DE15 0AB
G4 DII A. Excell, 7 Kingslake Villas, Taunton Road, Bridgwater, TA6 6BW
GM4 DIJ J. Howie, 36 Clermiston Road, Edinburgh, EH12 6XB
GM4 DIN N. Burns, 24 Garioch Road, Inverurie, AB51 4RQ
G4 DIP B. Chapman, 83 Courtenay Road, Great Barr, Birmingham, B44 8JB
G4 DIS K. Mills, 7 Montgomery Close, Colchester, CO2 8SJ
G4 DIT R. Siddall, 79 The Knoll, Palacefields, Runcorn, WA7 2UH
G4 DIU A. Walker, 26 Sketchley Court, Nottingham, NG6 7DL
G4 DIV Dr L. Day, 86 Copperfield Road, Southampton, SO16 3NY
G4 DIY R. Bennett, 17 Truro Close, St Helens, WA11 9EL
GM4 DIZ H. Lydall, 27 Calder Road, Edinburgh, EH11 3PF
G4 DJB P. Roberts, 10 Tintagel Drive, Frimley, Camberley, GU16 8XQ
G4 DJC R. Baker, 42 Rushleydale, Springfield, Chelmsford, CM1 6JX
G4 DJD D. Carter, 43 Sturton Road, Sheffield, S4 7DE
G4 DJJ C. Callicott, Clare House, Hepscott, Morpeth, NE61 6LT
G4 DJK D. Corkill, 1A Hardie Crescent Braunstone, Leicester, LE3 3DQ
G4 DJP Dr J. Chivers, 33 Hazelwood Road, Duffield, Belper, DE56 4DP
G4 DJX A. Gray, 5 Meadow Close, Marshalswick, St Albans, AL4 9TG
G4 DJY C. Steeden, Parklands, Chapel Road, Blackpool, FY4 5HT
G4 DJZ A. Petrie, 3 Sharma Leas, Peterborough, PE4 6ZH
G4 DKC T. Smith, 57 St Andrew Street, Tiverton, EX16 6PL
G4 DKD E. Pascoe, 48 Bull Baulk, Middleton Cheney, Banbury, OX17 2QQ
G4 DKH K. Hastie, 3 The Woodlands, Kings Worthy, Winchester, SO23 7QQ
G4 DKM R. Lacken, 378 Wallisdown Road, Wallisdown, Bournemouth, BH11 8PS
G4 DKP J. Etheridge, 14A Areley Court, Stourport-on-Severn, DY13 0AR
G4 DKQ J. Loughlin, 24 Grassendale Rd, Liverpool, L19 0NA
G4 DKV M. Pipes, Lower Hough Park, Hulland Ward, Ashbourne, DE6 3EN
G4 DKX N. Cartwright, Little Bulmer Farm, Wiston Road, Colchester, CO6 4LT
G4 DLA L. Turner, 160 Sandbach Road, Church Lawton, Stoke-on-Trent, ST7 3RB
GW4 DLC B. Bourne, 15 Rhos Fawr, Morfa, Abergele, LL22 9YH
G4 DLD M. Garwood, 1 Orchard Terrace, Welford, NN6 6AE
GM4 DLG R. Bower, Tigh Na Bruaich, Port Logan, Stranraer, DG9 9NE
G4 DLP R. Stoddon, 34 Cromwell Road, Lancaster, LA1 5BD
G4 DLT R. Hill, Rose Lodge, 35 Colne Fields, Huntingdon, PE28 3DL
GM4 DLU A. Mccudden, 9 Dryburgh Lane, East Kilbride, Glasgow, G74 1BQ
G4 DLY P. Collister, Flat 11, Liffey Court, 165-173 London Road, Liverpool, L3 8PZ
G4 DMB W. Green, 38 Greenlands Way West, Sheringham, NR26 8XP
G4 DMC R. Cleverley, 13 The Close, Melksham, SN12 6AG
G4 DMF J. Wright, 10 Thorpes Road, Heanor, DE75 7GQ
G4 DMG D. Griffiths, 1 Shepherds Down, Alresford, SO24 9PP
G4 DMH M. Horton, 47 Checkstone Avenue, Doncaster, DN4 7JY
G4 DMI C. Armistead, 16 Rushfield, Sawbridgeworth, CM21 9NF
G4 DML G. Moore, Calvers Farm, Norwich Road, Thelveton, IP21 4NG
G4 DMM I. Stinchcombe, 2 Alexandra Drive, Bere Alston, Yelverton, PL20 7DW
G4 DMP D. Pratt, 11 Moorleigh Close, Kippax, Leeds, LS25 7PB
GM4 DMQ J. Pritchard, 36 Craigleith Hill Crescent, Edinburgh, EH4 2JU
GW4 DMR D. Bevan, 3 Trem Y Foryd, Kinmel Bay, Rhyl, LL18 5JE
G4 DMS P. Freeman, 1 Littleworth, Towcester, Northants, NN12 8AL
G4 DMT J. Southall, 4 Tye Lane, Willisham, Ipswich, IP8 4SR
G4 DND J. Kennedy, Tor View Cottage, Postbridge, Yelverton, PL20 6SY
G4 DNE G. Swaysland, 35 Keyhaven Road, Milford On Sea, Lymington, SO41 0QW
G4 DNG M. Whitaker, 332 Milton Road, Cambridge, CB4 1LW
G4 DNH J. Easteal, The Chalkers, Ermin Street, Hungerford, RG17 7TS
G4 DNJ A. Grisley, 7 Arnhill Road, Gretton, Corby, NN17 3DN
G4 DNK M. Lelliott, Well Lane Corner, Lower Froyle, Alton, GU34 4LJ
G4 DNP R. Travis, 14 Elmstead Avenue, Wembley, HA9 8NX
GI4 DNW M. Getty, 34 Magheralave Park East, Lisburn, BT28 3BT
G4 DNX D. Dyer, 57 Garrison Lane, Felixstowe, IP11 7RR
G4 DOA A. Mead, 11 Yarnton Close, Nine Elms, Swindon, SN5 5UQ
G4 DOC D. James, 76 Grove Road, Harpenden, AL5 1HD
G4 DOE J. Alford, 26 Edmunds Avenue St. Pauls Cray, Orpington, BR5 3LF
GM4 DOF R. Davidson, 3 Hillcrest Avenue, Kirkcaldy, KY2 5TU
GI4 DOH R. White, 28 Lord Warden'S Parade, Bangor, BT19 1YU
G4 DOJ N. Sanig, 22 Bruntwood Avenue, Heald Green, Cheadle, SK8 3RU
G4 DOL P. Atkins, 28 Victoria Place, Easton, Portland, DT5 2AA
GI4 DOM D. Cafolla, 87 Stockmans Lane, Belfast, BT9 7JD
GM4 DON T. Donnelly, 18 Birtwhistle Street, Gatehouse Of Fleet, Castle Douglas, DG7 2JJ
GW4 DOO A. Kenyon, 6 Abbey Road, Port Talbot, SA13 1HA
G4 DOQ J. Willis, Kilncroft, Broadlayings, Newbury, RG20 9TS
GM4 DOZ T. Findlay, 37 Adamton Road North, Prestwick, KA9 2HY
G4 DPA G. Austin, 16 Courtenay Road, Wantage, OX12 7DN
G4 DPD Capt. C. Richardson, Domaine De Calcat, Route De Cates, la Sauvetat Sur Lede, France, 47150
G4 DPF I. Ross, 37 County Rd, March, PE158ND
G4 DPH G. Jones, 7 The Avenue, Yatton, Bristol, BS49 4DA
G4 DPJ D. Wear, 84 Hulham Road, Exmouth, EX8 3LA
GD4 DPK F. Quayle, 1 Birch Hill Gardens, Onchan, Douglas, Isle Of Man, IM3 4ET
G4 DPO A. Nixon, 174 Davidson Road, Croydon, CR0 6DE
G4 DPP P. Slade, Derlee House, East Lane, Abbots Langley, WD5 0QG
G4 DPT T. Upstone, 10 Lythe Fell Avenue, Halton, Lancaster, LA2 6NH
G4 DPU J. Pilling, 223 Manchester Road, Accrington, BB5 2PF
G4 DPV S. Ford, 3 Hill View, Stoke-on-Trent, ST2 7AR
G4 DPW P. Leslie-Reed, 43 Milehouse Lane, Newcastle, ST5 9JZ
G4 DPZ D. Johnson, 96 Summerfields Avenue, Halesowen, B62 9NR
G4 DQA D. Macken, 17 Culvercroft, Binfield, Bracknell, RG42 4DF
G4 DQB G. Wallis, Ellerton Wood Farm, Little Soudley, Market Drayton, TF9 2NB
G4 DQG A. Riley, 378 Hungerford Road, Crewe, CW1 6HD
GM4 DQJ R. Grant, 31 Stormont Park, Scone, Perth, PH2 6SD
G4 DQL N. Hall, 5 Brooklyn Crescent, Cheadle, SK8 1DX
G4 DQN G. Spenceley, 168 Robin Hood Lane, Walderslade, Chatham, ME5 9LA
G4 DQP Rev. V. Lewis, Four Winds Cottage, Main Street, Broomfleet, Brough, HU15 1RJ
G4 DQQ W. Thomas, 64 West End, Silverstone, Towcester, NN12 8UY
GI4 DQT S. Taaffe, 22 Skerriff Road, Cullyhanna, Newry, BT35 0JG
G4 DQW J. Krzymuski, 3079 Aberdeen Ct, Marietta, USA, 30062
G4 DQZ W. Ellis, Gillhams House, Gillhams Lane, Haslemere, GU27 3ND
G4 DR D. Urquhart, 7 Padwell Lane, Bushby, Leicester, LE7 9PQ
G4 DRA S. Chester, 63 Hawkshead Street, Southport, PR9 9BT
G4 DRI I. Selby, 2 Ashley Close, Welwyn Garden City, AL8 7LH
G4 DRO T. Brosnan, 168 Abbots Road, Edgware, HA8 0SA
GW4 DRR G. Spencer, Tyn Cae, Llanfwrog, Anglesey, LL65 4YL
G4 DRS J. Wayman, Oak Tree Lodge, Redbridge Road, Crossways, Dorchester, DT2 8BG
G4 DRU B. Plastow, 185 Allesley Old Road, Coventry, CV5 8FL
G4 DRV J. Harris, Flat 36, Colonel Stevens Court, 10A Granville Road, Eastbourne, BN20 7HD
G4 DRX D. Mckone, 12 Hawkshead Road, Knott End-On-Sea, Poulton-le-Fylde, FY6 0QE
G4 DRZ G. Carney, 94 Combe Avenue, Portishead, Bristol, BS20 6JX
G4 DSA G. Kemp, 4 Chapter Way, Monk Bretton, Barnsley, S71 2HP
G4 DSC O. Boniface, 11 Holmefield Road, Ripon, HG4 1RZ
G4 DSD R. Woodman, 89A Western Way, Ponteland, Newcastle upon Tyne, NE20 9AW
G4 DSE P. Zollman, 92 Well Lane, Curbridge, Witney, OX29 7PA
G4 DSF S. Jones, 11 Alba Close, Middleleaze, Swindon, SN5 5TL
G4 DSI Dr I. Mcandrew, South Winds, Outrigg, St Bees, CA27 0AN
G4 DSN J. Dryden, 33 Old Station Road, Newmarket, CB8 8DT
GM4 DSO T. Hughes, 15 Boreland Road, Kirkcudbright, DG6 4HL
G4 DSP Spalding and District ARS, c/o A. Hensman, 24 Belchmire Lane, Gosberton, Spalding, PE11 4HG
G4 DSQ R. Coombe, 150 Tean Road, Cheadle, Stoke-on-Trent, ST10 1LW
G4 DSR B. Irwin, 97 Offerton Lane, Stockport, SK2 5BS
G4 DSY R. Miller, 21 Woodstock Avenue, Sutton, SM3 9EG
G4 DTB M. Bryan, 58 Grandstand Road, Hereford, HR4 9NF
G4 DTC R. Howgego, 39 Harestone Valley Road, Caterham, CR3 6HN
GM4 DTH P. Dick, Napier House, 8 Colinton Road, Edinburgh, EH10 5DS
GM4 DTJ R. Henderson, 2 Burdiehouse Avenue, Edinburgh, EH17 8AW
G4 DTL W. Young, 56 Lincoln Road, Washingborough, Lincoln, LN4 1EG
G4 DTM P. Marrable, 6 Piccadilly Close, Northampton, NN4 8RU
G4 DTP D. Pells, 6 Clarence St., Stonebroom, Alfreton, DE55 6JW
GW4 DTQ D. Gibbon, 90 Grosvenor Road, Prestatyn, LL19 7TS
G4 DTT W. Brooks, 11 Lowther Grove, Garforth, Leeds, LS25 1EN
GW4 DTU A. Roberts, Brynlludw, Van, Llanidloes, SY18 6NP
G4 DTW S. Parsons, 54 Furze Cap, Kingsteignton, Newton Abbot, TQ12 3TE
G4 DUA R. Bearne, Gap House, Over Street Stapleford, Salisbury, SP3 4LP
G4 DUB R. Harden, 2 Diamond Ridge, Barlaston, Stoke-on-Trent, ST12 9DT
G4 DUE A. Parker, 5 Geddes Close, Hawkinge, Folkestone, CT18 7QL
G4 DUF B. Phillips, Woody Nook, Petworth Road, Godalming, GU8 5TU
G4 DUI P. Wilson, 6 Hereford Road, Colne, BB8 8JX
G4 DUJ T. Morley, 6 Bakers Close, South Woodham Ferrers, Chelmsford, CM3 5JF

G4 DUL M. Coburn, 16 Chapel Close Toddington, Toddington, LU5 6AZ
G4 DUM V. Long, 2 B Pinnacle Hill, Bexleyheath, DA7 6AF
G4 DUO F. Taylor, 7 Osterley Lodge, Church Road, Isleworth, TW7 4PQ
G4 DUQ P. Keane, 45 Bramblewood Road, Worle, Weston-Super-Mare, BS22 9LW
G4 DUT D. Elliott, 3 Oakland Walk, Dawlish, EX7 9RS
G4 DUW J. Goldbey, Waylands Gate, St. Johns Road, New Milton, BH25 5SD
GM4 DUX K. Hampson, 9 North Crescent, Garlieston, Newton Stewart, DG8 8BA
G4 DVA T. Stanway, 24 Fellbrook Lane, Bucknall, Stoke-on-Trent, ST2 8AQ
GW4 DVB B. Price, 156 Parc Bryn Derwen, Llanharan, Pontyclun, CF72 9TX
G4 DVG J. Douglas, 367 Wightman Road, London, N8 0NA
G4 DVI M. Small, 15 Cannock Drive, Stockport, SK4 3JB
G4 DVJ R. Hall, 12 Britannia Gardens, Westcliff-on-Sea, SS0 8BN
G4 DVK M. Lang, 52 Gloucester Road, Burnham-on-Sea, TA8 1JA
G4 DVM M. Cartwright, Seacue, 8 Adelaide Avenue, West Bromwich, B70 0SL
G4 DVN S. Whalley, 1 Radley Way, Werrington, Stoke-on-Trent, ST9 0JN
G4 DVV J. Thomas, 57 Bourton Avenue, Stoke Lodge, Bristol, BS34 6EB
G4 DVX R. Farr, 1 Lavender Road Up Hatherley, Cheltenham, GL51 3BN
G4 DVZ T. Beaumont, 39 Meadow Road, Garforth, Leeds, LS25 2EN
G4 DWC D. Cannings, 5 Rowan Close, Brackley, NN13 6PB
G4 DWF Dr D. Faulkner, 1 Westland, Martlesham Heath, Ipswich, IP5 3SU
G4 DWM T. Hunt, 45 Front Street Frosterley, Bishop Auckland, DL13 2QR
GW4 DWN H. Richards, 61, High Street, Abergwynfi, Port Talbot, SA13 3YN
G4 DWO W. Ingham, Westfield Villa, Westfield Villas, Wakefield, WF4 6EQ
G4 DWR M. Molloy, 153 Palmdale Drive, Scarborough, Canada, MIT 1P2
G4 DWU J. Blowers, 28 Keld Close, Scarborough, YO12 6UF
GW4 DWX M. Smith, Tonn Marr, Bronybuckley, Welshpool, SY21 7NQ
GI4 DWZ P. Tucker, 1 The Courtyard, Dunadry Road, Dunadry, Antrim, BT41 4QQ
G4 DXB B. Chester, 147 Sanctuary Way, Grimsby, DN37 9RX
GI4 DXK W. Gordon, 17 Ballyheather Road, Ballymagorry, Strabane, BT82 0BD
G4 DXN G. Williams, 6 Nightingale Court, Leam Terrace, Leamington Spa, CV31 1DQ
G4 DXO P. Jones, 40 Furze Road, Worthing, BN133BH
G4 DXP C. Howells, 11 West Garth, Carlton, Stockton-on-Tees, TS21 1DZ
G4 DXT T. Shaman, 3 Padshall Park, Bideford, EX39 3NE
G4 DXW R. Smith, 29 George Street, Peterborough, PE2 9PD
G4 DXY J. Spendlove, 15 Grammer Street, Denby Village, Ripley, DE5 8PQ
G4 DYC M. Cooke, 4 Geddes Way, Mattishall, Dereham, NR20 3RE
GI4 DYE E. Macintyre, 115 Bell Doo, Strabane, BT82 9QL
G4 DYG P. Rich, 392 Doncaster Road, Stairfoot, Barnsley, S70 3RH
G4 DYH G. Dunn, Croft Cottage, Innocence Lane, Ipswich, IP10 0PL
G4 DYI C. Titheridge, 41 Church Walk, Worthing, BN11 2LT
G4 DYJ J. Cope, 4 Alexander Way, Burgh Le Marsh, Skegness, PE24 5JN
G4 DYM E. Auty, Jesla, 5 Silverstone Way, Bristol, BS49 5ES
G4 DYO B. Mccartney, 123 Reading Road, Finchampstead, Wokingham, RG40 4RD
G4 DYR R. Page, 28291 Misty Morning Lane, Beloit, Ohio, USA, 44609
G4 DYV B. Whiting, Fourwinds, Buttercake Lane, Boston, PE22 9QX
GW4 DYY R. Mander, Meadowlands, Severn Lane, Welshpool, SY21 7BB
G4 DZC Dr M. Bayes, 25 Welby Gardens, Grantham, NG31 8BN
G4 DZH H. Davies, 33 Sandown Road, Ocean Heights, Paignton, TQ4 7RL
G4 DZJ D. Paul, Flat 2, Cranfield, Alexandra Road, Penzance, TR18 4LZ
G4 DZK G. Stocker, 8 Brook Drive Astley, Manchester, M29 7HS
GM4 DZM I. Shewan, Springbank, Distillery Road, Inverurie, AB51 0ES
G4 DZS A. Watson, 59 Merdon Avenue, Chandler'S Ford, Eastleigh, SO53 1GD
G4 DZU D. Parker, 50 Rein Road, Tingley, Wakefield, WF3 1HZ
GM4 DZX R. Macleod, 6 Swanson Drive, Wick, KW1 5TF
G4 EAB J. Blackburn, 9 Pitchford Road, Albrighton, Wolverhampton, WV7 3LS
GM4 EAF Perth & Dist AR, c/o A. Hutton, 4 Linn Road, Stanley, Perth, PH1 4QS
G4 EAG S. Ruffle, 39 Nightingale Avenue, Cambridge, CB1 8SG
G4 EAK M. Betts, 19 Maracas Cove, Western Australia, Australia, 6028
G4 EAN I. Brothwell, 56 Arnot Hill Road, Arnold, Nottingham, NG5 6LQ
G4 EAQ Dr A. Churchley, 46 Birchdale Road, Appleton, Warrington, WA4 5AW
G4 EAS C. Ellery, 17 Wessex Way, Dorchester, DT1 2NR
GM4 EAU C. Murray, 43 Malleny Avenue, Balerno, EH14 7EJ
GM4 EAW J. Mathers, 36 Alexander St., Dunoon, PA23 7EW
G4 EAX J. Gell, 21 Maylands Avenue, Breaston, Derby, DE72 3EE
G4 EAZ P. Holliman, 17 Arundel Road, Tewkesbury, GL20 8AT
GD4 EBA D. Kinrade, 8 Alfred Teare Grove Douglas, Isle of Man, Isle Of Man, IM2 6EH
G4 EBE G. Harcourt, Hard Farm, Little Marsh Lane, Field Dalling, Holt, NR25 7LL
G4 EBF G. Reason, 37 Park End, Croughton, Brackley, NN13 5LX
G4 EBG B. Meredith, 20 Kestrel Avenue, Thorpe Hesley, Rotherham, S61 2TT
G4 EBI A. Hamm, 166 Sylvan Road, London, SE19 2SA
G4 EBK G. Smith, 6 Fenby Close, Grimsby, DN37 9QJ
G4 EBL R. Whitwell, 14 Green Lane, Yarpole, Leominster, HR6 0BG
G4 EBN M. Valente, Glenville, Abbey Road, Durham, DH1 5DQ
G4 EBO W. Gibbs, 25 Belvedere Road, Exmouth, EX8 1QN
G4 EBQ N. Talbot, 59 Heywood Avenue, Austerlands, Oldham, OL4 4AZ
GI4 EBS J. Mcnerlin, 5 Rosendale Avenue, Limavady, BT49 0AE
G4 EBT D. Taylor, 3 Crofters Drive, Cottingham, HU16 4SD
GM4 EBX P. Hopkinson, The Coaches, Kingussie, PH21 1NY
G4 EBY G. Head, 7 Partridge Way Downley, High Wycombe, HP13 5JX
G4 ECA M. Wrintmore, 148 Westwick Road, Sheffield, S8 7BX
G4 ECE J. Martin, 38 Parklands, Mablethorpe, LN12 1BY
G4 ECF G. Penney, 8 Drake Park, Bognor Regis, PO22 7QG
G4 ECO B. Palmer, Small Pine, Hedgerow Lane, Leicester, LE9 2BN
G4 ECS A. Wisbey, 12 Livingstone Road, Caterham, CR3 5TG
G4 EDC R. Vane-Stobbs, 2 Wood Cottages, Walford Heath, Shrewsbury, SY4 3AZ
G4 EDD J. Fletcher, 5 Hayeswood Road, Stanley Common, Ilkeston, DE7 6GB
G4 EDG S. Taylor, 80 Nadder Park Road, Exeter, EX4 1NX
G4 EDH G. Rose, 15 Pendennis Close, Winklebury, Basingstoke, RG23 8JD
G4 EDK A. Ball, Tinten House, 2 Tinten Lane, Dorchester, DT1 3WP
G4 EDM W. Concannon, 155 Walton Road, Sale, M33 4FS
G4 EDN K. Currie, 37 Golden Ridge, Freshwater, PO40 9LF
G4 EDQ R. Gulliver, 32 Lavender Close, Thornbury, Bristol, BS35 1UL
G4 EDR D. Mappin, 13 Willow Close, Filey, YO14 9NY
G4 EDW P. Eaton, Orchard House, Oxford Road, Sutton Scotney, Winchester, SO21 3JG
G4 EDX J. Fletcher, 69 Thackerays Lane, Woodthorpe, Nottingham, NG5 4HU
G4 EDY M. Grindrod, 20 Castle Mead, Kings Stanley, Stonehouse, GL10 3LD
G4 EDZ W. Russell, Blue Firs, Bower Road, Mersham, Ashford, TN25 6NW
G4 EEE A. Wood, Pezula, Brimpton, Reading, RG7 4TR
G4 EEF S. Foster, 6 Webster Close, Hornchurch, RM12 6TF
G4 EEH D. Greer, 5 Potto Close, Yarm, TS15 9RZ
G4 EEJ Dr R. Arak, 76 Halifax Road, Brighouse, HD6 2EP
G4 EEL A. Cheshire, 1 Westerby Lane, Smeeton Westerby, Leicester, LE8 0RA
G4 EEQ Rev. F. Robinson, 26 Winstanley Road, Little Neston, Neston, CH64 0UZ
G4 EES P. Smith, Forge House & Stables, Whistley Road, Devizes, SN10 5TD
G4 EET S. Greep, 5 Berkswell Close, Solihull, B91 2EH
G4 EEV D. Warwick, Orchard Cottage, Colber Lane, Bishop Thornton, Harrogate, HG3 3JR
G4 EEZ M. Bath, 146 North Road, Hertford, SG14 2BZ
G4 EFB C. Mccloud, 34 St. Stephens Road, Portsmouth, PO2 7PG
G4 EFD D. Stubbs, 63 Moss Lane, Wardley, Manchester, M27 9RD
G4 EFE M. Peters, 11 Filbert Drive, Reading, RG31 5DZ
G4 EFG D. Watton, 247 Bloxwich Road, Walsall, WS2 7BB
G4 EFH A. Johnson, Winchcombe, 3 Merrivale Lane, Ross-on-Wye, HR9 5JL
G4 EFO M. Senior, 16 Cherry Tree Close, Billingshurst, RH14 9NG
GM4 EFR J. Moar, Hansel, Stangergill Cres, Caithness, KW14 8UT
G4 EFX A. Levitt, The Coach House Strines Clough Farm, Blackshaw Head, Hebden Bridge, HX7 7JA
G4 EFY J. Hurst, 12 Dukes Mead, Fleet, GU51 4HA
G4 EGB J. Fletcher, 114 Scholes Park Road, Scarborough, YO12 6RA
GM4 EGD I. Brownlie, 16 Border Street, Greenock, PA15 2EE
G4 EGG W. Higginson, 7 Arundale, Westhoughton, Bolton, BL5 3YB
G4 EGM R. Webster, 230 Huyton Lane, Huyton, Liverpool, L36 1TH
G4 EGN V. Coles, 205 Farmers Close, Witney, OX28 1NS
G4 EGQ P. Pennington, 6 Highland Close Sandgate, Folkestone, CT20 3SA
G4 EGR D. Barwood, 41 Wingfield Road, Bristol, BS3 5EG
GW4 EGS M. Price, 19 Pencaerfenni Park, Crofty, Swansea, SA4 3SE
G4 EGU P. Wolfe, 90 Alderney Road, Erith, DA8 2JD
GM4 EGX R. Howard, 22 Kirkbrae Drive Cults, Aberdeen, AB15 9RH
G4 EGY S. Liptrott, 40 Mapperley Orchard, Arnold, Nottingham, NG5 8AG
G4 EHD W. Tait, 51 Broadley Crescent, Halifax, HX2 0RL
G4 EHG C. Bryan, 9 Brandy Hole Lane, Chichester, PO19 5RL
G4 EHJ E. Wilby, 5 Matuku Street, Heretaunga, Upper Hutt, New Zealand, 5018
G4 EHK D. Goulbourne, 8 Moor Avenue, Appley Bridge, Wigan, WN6 9JS
G4 EHN Dr J. Axe, 5 Hillgate Place, London, W8 7SL
GM4 EHP I. Petrie, Ugie Cottage, Victoria Road, Peterhead, AB42 4NL
G4 EHQ M. Holley, 76 Chatham Grove, Chatham, ME4 6LY
G4 EHR D. Kirton, 16 Silver Innage, Halesowen, B63 2PP
G4 EHT W. Watson, 7 Darwin Close, Lichfield, WS13 7ET
G4 EHW Peterborough and District ARC, c/o T. Ralph, 15 Portchester Close, Stanground, Peterborough, PE2 8UP
G4 EHX J. Fearn, 37-39 Bourne Square Breaston, Derby, DE72 3DZ
G4 EHY F. Greenough, 7 Carnforth Avenue, Hindley Green, Wigan, WN2 4LD
G4 EIA M. Wallis, 34 St. Aidans Close, Bristol, BS5 8RH
G4 EIC E. Calvert, 163 Milner Road, Heswall, Wirral, CH60 5RY
G4 EID M. Haworth, 26 Willowhey Marshside, Southport, PR9 9TW
GW4 EIE R. Francis, 18 Iscoed, Beaumaris, LL58 8HH
G4 EIG J. Vickerstaff, 5 Luddington Road, Solihull, B92 9QH
G4 EII A. Cunliffe, 35 Coultshead Avenue, Billinge, Wigan, WN5 7HT
G4 EIJ J. Rees, 17 Finch Road, Chipping Sodbury, Bristol, BS37 6JF
G4 EIK R. Currell, Brookside, Treworga, Truro, TR2 5NP
G4 EIL G. Oughtibridge, 1 Lincoln Drive, Liversedge, WF15 7NJ
G4 EIM J. Beaumont, 132 Hull Road, Woodmansey, Beverley, HU17 0TH
GW4 EIN D. Jones, Vine Tree Cottage, Mill Lane, Abergavenny, NP7 9SA
GD4 EIP Dr C. Baillie-Searle, 2 Marguerite Place Foxdale, Douglas, Foxdale, IM4 3HE
G4 EIV J. Sondhis, 47 Emlyn Road, Horley, RH6 8RX
GM4 EIW J. Dunnington, 4 Woodburn Way, Cumbernauld, Glasgow, G68 9BJ
G4 EIX D. Whalley, 1 Lees Farm Drive, Madeley, Telford, TF7 5SU
G4 EIY B. Thomas, 8 Whitehill Road, Barton-Le-Clay, Bedford, MK45 4PF
GI4 EIZ W. Stewart, 56 Ballysillan Park, Belfast, BT14 8HD
G4 EJD C. Bourne, 5 Brempton Croft, Hilderstone, Stone, ST15 8XL
G4 EJE J. Brown11, 2 Coriander Gardens, Littleover, Derby, DE23 2UB
G4 EJG I. Adams, 2 Copper Hall Close, Rustington, Littlehampton, BN16 3RZ
G4 EJH K. Middleton, 92 South Road, Portishead, Bristol, BS20 7DY
GM4 EJI G. Lucas, 20 Myreside Gardens, Kennoway, Leven, KY8 5TR
G4 EJK D. Reardon, 65 Blenheim Road, Caversham, Reading, RG4 7RP
G4 EJM M. West, 19 Park Drive, Trentham, Stoke-on-Trent, ST4 8AB
G4 EJP P. Sheppard, 220 Beckfield Lane, York, YO26 5QS
G4 EJU R. Hands, 19 Orwell Road, Walsall, WS1 2PJ
G4 EJW N. Perkins, 231 Burnham Road, Burnham-on-Sea, TA8 1LT
GM4 EJX A. Murray, 67 Carronvale Road, Larbert, FK5 3LH
G4 EKB D. Epton, 61 Cartmel Drive, Dunstable, LU6 3PT
GM4 EKC J. Mackinnon, 185 Deeside Gardens, Aberdeen, AB15 7QA
G4 EKD P. Spelman, 68 Hardwick St., Tibshelf, Alfreton, DE55 5QH
G4 EKF S. Sinclair, Wayside, Alnwick Road Lesbury., Alnwick, NE66 3PJ
G4 EKG M. Tittensor, 16 Durcott Road, Evesham, WR11 1EQ
GM4 EKI G. Marsh, Riverview, Lewiston, IV63 6UW
G4 EKJ C. Shaw, 10 St. Helens Road, Harrogate, HG2 8LB
G4 EKM S. Green, 133 Sevenoaks Drive, Sunderland, SR4 9NQ
G4 EKS R. Holtham, 27 Peyton Close, Eastbourne, BN23 6AF
G4 EKT Hornsea ARC, c/o R. I'Anson, 87 Tranby Lane, Anlaby, Hull, HU10 7DT
G4 EKV M. Lobb, 52 Ridge Park Avenue, Mutley, Plymouth, PL4 6QA
G4 EKW M. Shaw, 50 White Road, Nottingham, NG5 1JR
G4 EKZ D. Saul, 78 Ingleton Drive, Lancaster, LA1 4QZ
G4 ELA R. Dawson, 2 Bertram Drive, Wirral, CH47 0LQ
G4 ELC G. Keay, 9 Buchanan Avenue, Bournemouth, BH7 7AA
G4 ELG D. Campbell, 12 Newton Close, Newton Solney, Burton-on-Trent, DE15 0SL
G4 ELI S. Brown, Helford Lodge, The Fairway, Falmouth, TR11 5LR
G4 ELJ Dr D. Clark, 24B Heatherdale Road, Camberley, GU15 2LT
G4 ELK A. Lewis, 1 Springcroft, Parkgate, Neston, CH64 6SF
G4 ELL R. James-Robertson, 8 Whittington Road, Worcester, WR5 2JU
G4 ELM E. Jewell, 12 Patricks Copse Road, Liss, GU33 7DL
G4 ELP D. Stockley, 2 The Ridings Chestfield, Whitstable, CT5 3PE
GI4 ELQ J. Cushnahan, 34 Cornakinnegar Road, Lurgan, Craigavon, BT67 9JN
G4 ELR East Lancashire ARC, c/o N. Mooney, 60 Rhyddings Street, Oswaldtwistle, Accrington, BB5 3EY

GM4 ELV Dr D. Dhuglas, 1 Micklehouse Road, Baillieston, Glasgow, G69 6TG
G4 ELW I. Bontoft, 5 Kings Drive Westonzoyland, Bridgwater, TA7 0HJ
G4 ELY R. Panting, 124 Loddon Bridge Road, Woodley, Reading, RG5 4AW
G4 ELZ J. Pascoe, 3 Aller Brake Road, Aller, Newton Abbot, TQ12 4NJ
G4 EMA I. Welburn, 33 Bowland Way, Clifton, York, YO30 5PZ
G4 EMB N. Lockett, 18 Seagers, Great Totham, Maldon, CM9 8PB
G4 EMD R. Edge, 4 Mortimer Hill Cleobury Mortimer, Kidderminster, DY14 8QQ
G4 EMH G. Parsley, 7 Rowan Road, Martham, Great Yarmouth, NR29 4RY
G4 EMK G. Parker, 15 Burton Road, Heckington, Sleaford, NG34 9QR
G4 EML C. Durbridge, 2 Send Villas, Sandy Lane, Send, Woking, GU23 7AP
G4 EMQ J. Purchon, 19 Warburton, Emley, Huddersfield, HD8 9QP
G4 EMV P. Johnson, 4 Chapel Lane, Blackwater, Camberley, GU17 9ET
G4 EMW Prof. E. Warrington, 3 Long Meadow, Wigston, LE18 3TY
GM4 EMX C. Hall, 21 Peterculter Retirement Park Peterculter, Aberdeen, AB14 0AB
G4 ENA P. Asquith, Well Cottage, The Green, Stroud, GL5 5LN
G4 ENB C. Asquith, 36 Sunningdale, Luton, LU2 7TE
G4 ENC J. Fenton-Coopland, 14 Chevril Court, Wickersley, Rotherham, S66 2BN
GM4 ENF A. Fyffe, 39 Watts Gardens, Cupar, KY15 4UG
G4 ENH I. Hodgkiss, 190 Ulverley Green Road, Solihull, B92 8AD
G4 ENJ K. Hunter, 1 Markers Park, Payhembury, Honiton, EX14 3NL
G4 ENK P. Kelly, 14 Manville Road, Wallasey, CH45 5AY
G4 ENL P. Jewitt, Colenco Power Engineering, Tafernstrasse 26, Baden, Switzerland, CH5402
GM4 ENN A. Rae, 183 Campsie Street, Glasgow, G21 4XY
GM4 ENP Dr J. Johnston, 4 Lawhead Road West, St Andrews, KY16 9NE
G4 ENR K. Brook, 154 Ridge Nether Moor, Swindon, SN3 6NF
G4 ENS A. Morris, 6 Barrowby Gate, Grantham, NG31 7LT
G4 ENZ M. Church, 2B Meadow Way, Churchdown, Gloucester, GL3 2AU
G4 EOA Dr T. Strickland, 22A Branksome Road, St Leonards-on-Sea, TN38 0UA
G4 EOB G. Lawrance, 77 Bigland Drive, Ulverston, LA12 9PD
G4 EOC R. Grunwald, 20 Sunny Bank Road, Batley, WF17 0LJ
G4 EOE R. Everest, 2 Burley Road Parkstone, Poole, BH12 3DA
G4 EOF S. Lawrence, 34 Stanley Drive, Leicester, LE5 1EA
G4 EOG A. Heritage, 33 Peartree Lane Danbury, Chelmsford, CM3 4LS
G4 EOJ T. Ilott, 45 Parkside Snettisham, King's Lynn, PE31 7QF
GU4 EON M. Allisette, Les Amballes Lodge, Les Amballes, St. Peter Port, Guernsey, GY1 1WU
G4 EOR P. Stevens, 71 Tower Court, Westcliff Parade, Westcliff-on-Sea, SS0 7QH
G4 EOT D. Bussell, 26 Norbreck Crescent, Wigan, WN6 7RF
GM4 EOU J. Smith, 6 Rodger Street, Cellardyke, Anstruther, KY10 3HU
GW4 EOW P. Baxter, 103 Lon Conwy, Benllech, LL74 8RP
G4 EOX N. Davenport, 25 Prairie Crescent, Burnley, BB10 1EU
G4 EPA J. Pepper, 52 King Style Close, Crick, Northampton, NN6 7ST
G4 EPC J. Stevens, 16 Brindles Field, Tonbridge, TN9 2YS
G4 EPD R. Heeley, 263 Barnsley Road, Cudworth, Barnsley, S72 8JP
GW4 EPF J. Pile, 5 Western Close, Mumbles, Swansea, SA3 4HF
G4 EPH J. Splaine, 765 Wells Road, Bristol, BS14 0PB
GI4 EPK E. Coyle, 14 Colby Avenue, Culmore, Londonderry, BT48 8PF
G4 EPL L. Ward, 49 Edgewood Drive, Hucknall, Nottingham, NG15 6HY
G4 EPM N. Lewis, 97 Orsett Road, Grays, RM17 5HA
G4 EPN A. Wright, 34 Webbs Way, Stoney Stanton, Leicester, LE9 4BW
G4 EPU M. Gray, 33 Claremont Drive, Pitsea, Basildon, SS16 4TL
G4 EPW L. Goulding, 24 Lancaster Drive, Lydney, GL15 5SL
G4 EPX D. Chater-Lea, Beech Rise, Church Road, Mortimer West End, RG7 2HY
GI4 EQA E. Mooney, 33 Piney Hill, Magherafelt, BT45 6PY
G4 EQC B. Smith, 11 Tean Close, Burntwood, WS7 9JS
G4 EQD N. Smith, 1 Park View, Messingham, Scunthorpe, DN17 3TT
G4 EQE D. Smith, 7 Demesne Gardens, Martlesham Heath, Ipswich, IP5 3UA
G4 EQJ J. Lee, 12 Gainsborough Close, Folkestone, CT19 5NB
G4 EQK M. Hale, 9 Cramer Gutter, Oreton, Kidderminster, DY14 0UA
G4 EQL Dr M. Townsend, 39 Main Street, Fleckney, Leicester, LE8 8AP
G4 EQM W. Evans, 9 Edwin Road, Didcot, OX11 8LG
GI4 EQN C. Cupples, 20 Westland Avenue, Ballywalter, Newtownards, BT22 2TR
G4 EQP A. York, 8 Granville Close, Hanham, Bristol, BS15 3TJ
G4 EQS K. Dowson, Fyling Hall Lodge, Fylingdales, Whitby, YO22 4QN
G4 EQX J. Mcilroy, 17 Brownsfield Road, Yardley Gobion, Towcester, NN12 7TY
GM4 EQY J. Hately, 10 Crags Road, Paisley, PA2 6RA
G4 EQZ K. Faulkner, 18 Milton Crescent, Talke, Stoke-on-Trent, ST7 1PF
GW4 ERB B. Skidmore, Milton Oak House, Oxland Lane, Milford Haven, SA73 1LG
G4 ERD A. Hamilton, 2905 Nancy Creek Road Nw, Atlanta, USA, 30327
G4 ERF P. Jones, Woodview, 10 Barrow Hill, Barrow, Bury St Edmunds, IP29 5DX
G4 ERH J. Perry, C/O Wagons/Lits Apt168, San Antonio, Baleric Isles, Spain, ZZ9 9CO
G4 ERL E. Lawley, 23 Briar Rigg, Keswick, CA12 4NN
GI4 ERM K. Bones, 54 Derryvolgie Park, Lisburn, BT27 4DA
G4 ERO C. Leonard, 24 Lower Road, Stuntney, Ely, CB7 5TN
G4 ERP R. Marshall, 40 Evesham Road, Bishops Cleeve, Cheltenham, GL52 8SA
G4 ERQ T. Birchall, 10 Avon Court, Alsager, Stoke-on-Trent, ST7 2BA
G4 ERR J. Cummins, 5 Blenheim Orchard, Shurdington, Cheltenham, GL51 4TG
G4 ERS J. Gamblen, Illfield, High Wych, Herts, CM21 0HX
G4 ERT H. Marriott, 108 Leicester Road, Quorn, Loughborough, LE12 8BB
G4 ERV W. Coombes, 33 Clarence Park Road, Boscombe East, Bournemouth, BH7 6LF
G4 ERW D. Lurcook, 2 Drury Road, Tenterden, TN30 6QG
G4 ERX R. Elliott, No 6 Aphrodities Rock Village, 19 Lykourisson Street, Paphos, Cyprus, 8852
G4 ERY D. Tyson, 12 Melbury Road, Woodthorpe, Nottingham, NG5 4PG
G4 ERZ A. Wells, 38 Sextant Road, Hull, HU6 7BA
G4 ESG D. Neal, 2 St. Margarets Avenue, Ashford, TW15 1DR
GI4 ESI S. Mcclean, 14 Bamber Park, Ballymena, BT43 5HE
GW4 ESL P. Edwards, 14 Northfield Close, Caerleon, Newport, NP18 3EZ
G4 EST C. Cartmel, 41 Lathom Drive, Rainford, St Helens, WA11 8JR
G4 ESU C. Rouse, 86 Melton Avenue, Clifton, York, YO30 5QG
G4 ESY D. Jackson, 16 Melrose Park, Beverley, HU17 8JL
G4 ETC P. Keeble, 18 Shrubland Drive, Rushmere St. Andrew, Ipswich, IP4 5SX
G4 ETD A. Firth, 1 Wee Cottage Crook, Kendal, LA9 8LH
G4 ETG D. Humphries, 173 Herne Road, Ramsey St. Marys, Ramsey, Huntingdon, PE26 2SY
G4 ETI J. Shaw, 20 Castleton Grove, Jesmond, Newcastle upon Tyne, NE2 2HD
G4 ETK C. Bourne, 12 Sheepcoat Close, Shenley Church End, Milton Keynes, MK5 6JL
G4 ETM J. Taylor, 24 Marlborough Drive, Mablethorpe, LN12 2 BA
G4 ETN B. Smith, Cleeve Valley Farm, Chipstable, Taunton, TA4 2QF
G4 ETO J. Roach, 33 Pound Lane, Topsham, Exeter, EX3 0NA
G4 ETP T. Pinch, 1 Fernhill Close, Ivybridge, PL21 9JE
G4 ETS J. Forsey, 3 Orchard Leaze, Dursley, GL11 6HY
G4 ETW Willenhall & District ARS, c/o M. Gibbons, 117 Ettingshall Road, Bilston, WV14 9XF
G4 ETX D. Ludlow, 18 Springfield Court, Ravensbourne Place, Springfield, Milton Keynes, MK6 3JJ
G4 ETZ F. Webb, 166 Glastonbury Road, Yardley Wood, Birmingham, B14 4DS
GW4 EUA G. Smith, 13 Lapwing Close, Penarth, CF64 5GA
G4 EUC G. Mendoza, 32 The Circuit, Cheadle Hulme, Cheadle, SK8 7LG
G4 EUF G. Mayo, 28 Ring Fence, Shepshed, Loughborough, LE12 9HY
G4 EUG G. Payne, 28 Pollards Drive, Horsham, RH13 5HH
G4 EUJ R. Whiteley, 35 Wood Lane, Gedling, Nottingham, NG4 4AD
G4 EUK G. Adcock, 2 Erringham Road, Shoreham-by-Sea, BN43 5NQ
G4 EUL A. Clarke, 5 Margil Way, Richmond, New Zealand, 7020
G4 EUR M. Tout, 25 Booth Rise, Northampton, NN3 6HP
G4 EUW B. Keeling, 43 Marennes Crescent Brightlingsea, Colchester, CO7 0RU
G4 EUZ Durham and District ARS, c/o R. Mcaleer, 44 Harvey Avenue, Durham, DH1 5ZG
G4 EVA C. Roberts, Rosemary, York Road, West Byfleet, KT14 7HX
G4 EVC K. Chadwick, 1 Parklands, Southport, PR9 7HX
G4 EVD E. Parry, 60 Hunters Forstal Road, Herne Bay, CT6 7DW
G4 EVE P. Webster, The Old School, School Hill, Cirencester, GL7 2LS
G4 EVI J. Howard, 127 Goldcroft, Yeovil, BA21 4DD
G4 EVK I. Shepherd, 12 Watsons Lane, Harby, Melton Mowbray, LE14 4DD
GW4 EVL T. Hopkins, 39 Glen Road, West Cross, Swansea, SA3 5PR
G4 EVN S. Garrett, Church Farmhouse, The St., Stowmarket, IP14 6LX
G4 EVP C. Mcpartland, 55 Elliotts Lane, Codsall, Wolverhampton, WV8 1PG
G4 EVR Dr A. Davies, 10 New Street, Ludlow, SY8 2NQ
GM4 EVS D. Johnstone, Sycamore House, Kirk Loan, Perth, PH2 6TD
GW4 EVX R. Price, 19 New Brighton Road, Sychdyn, Mold, CH7 6EF
G4 EVZ M. Powrie, 31 The Grove, Billericay, CM11 1AU
G4 EWB E. Bradfield, St. Olafs, Beach Road, Bacton, Norwich, NR12 0EP
G4 EWE D. Overton, Toulouse Colwell Road, Totland Bay, PO39 0AH
G4 EWI F. Warner, 48 Brookfield Road, Walsall, WS9 8JE
G4 EWJ B. Jordan, 42 Ben Nevis Road, Birkenhead, CH42 6QY
G4 EWK D. Mellor, 18 Briar Close, Newhall, Swadlincote, DE11 0RX
GM4 EWL R. Macleod, 9 Croftcroighn Gate, Glasgow, G33 5JJ
GM4 EWM E. Mclean, 21 Milnefield Avenue, Elgin, IV30 6EJ
G4 EWT S. Mason, 15 Northfield Close, Bishops Waltham, Southampton, SO32 1EW
G4 EWV I. Mcpherson, 12 Victoria Crescent, Ashford, TN23 7HL
G4 EWW T. James, 2 The Green, Bottom Street, Southam, CV47 2FJ
G4 EWZ J. Halford, The Anchor Inn, Chesterfield Road, Alfreton, DE55 7LP
G4 EXD I. Marsh, 8 South Esk, Culgaith, Penrith, CA10 1QR
GW4 EXE B. Hope, Oriel, Moelfre, LL72 8HN
G4 EXF A. Grindrod, Mullions, Church Street, Stonehouse, GL10 3HX
GI4 EXI G. Crothers, 46 Culmore Point, Londonderry, BT48 8JW
G4 EXK P. Bradbury, Rosscairne, 42 Halfpenny Lane, Longridge, Preston, PR3 2EA
G4 EXN L. Dolman, 46 Norfolk St., Norwich, NR2 2SN
G4 EXT J. Corben, 65 Oatley Park Avenue, Oatley, New South Wales, Australia, 2223
G4 EXU D. Fisher, 1 Francolin Close, Woodhaven, Natal South Africa, ZZ1 9FR
G4 EXZ R. Fidler, 55 Sunnyvale Drive, Longwell Green, Bristol, BS30 9YQ
G4 EYA C. Evans, 64 Boyd Avenue, Toftwood, Dereham, NR19 1ND
G4 EYB D. Fernie, Shepherds Close, Reigate Road, Leatherhead, KT22 8RD
G4 EYE A. Free, Homeric, Harwich Road, Harwich, CO12 5JF
G4 EYJ D. Davies, 21 Russell Drive, Malvern, WR14 2LE
G4 EYM J. Shardlow, 19 Portreath Drive, Allestree, Derby, DE22 2BJ
G4 EYN K. Wright, 61 Albert Road, Chaddesden, Derby, DE21 6SH
GW4 EYO C. Carver, 8 Overlea Drive, Hawarden, Deeside, CH5 3HS
G4 EYT C. Williams, 35 Heath Crescent, Norwich, NR6 6XF
G4 EYV Dr P. Skolar, Apartment 12, Fircroft, Devenish Road, Ascot, SL5 9GF
G4 EYX P. Davies, 14 Saville Road, Blackpool, FY1 6JP
G4 EZC H. Spencer, 5 Carlyn Drive, Chandler'S Ford, Eastleigh, SO53 2DJ
G4 EZE J. Hinton, 7 The Glebelands, Crowborough, TN6 1TF
G4 EZG Lord M. Mac Gregor Of Stirling, Raddle Barn, South Leigh Road High Cogges, Witney, OX29 6UW
GM4 EZJ K. Glendinning, 14 Craiglockhart Avenue, Edinburgh, EH14 1HW
G4 EZM E. Green, 6 Downham Place, Blackpool, FY4 1QS
G4 EZN Dr J. Keeler, 67 Perne Avenue, Cambridge, CB1 3RY
G4 EZP I. Melville, 3 Crescent Road, Benfleet, SS7 1JL
G4 EZQ C. Doman, 6 Churnet Close, Bedford, MK41 7ST
G4 EZU Dr W. Peterson, 22 Weston Close, Potters Bar, EN6 2BQ
G4 EZX D. Titheridge, 2 The Oaks, Wilsden, Bradford, BD15 0HH
G4 FAA L. Atkinson, 56 The Spinney, Sidcup, DA14 5NF
G4 FAB S. Fox, 16 The Teasels, Bingham, Nottingham, NG13 8TY
G4 FAD R. Langford, Foxholes, Parsonage Farm, Hereford, HR4 8AJ
G4 FAE S. Hodgetts, 79 Field Lane, Alvaston, Derby, DE24 0GQ
G4 FAH D. Jones, 41 Sorrel Walk, Brierley Hill, DY5 2QG
G4 FAI A. Smith, 13 Old Library Mews, Norwich, NR1 1ET
G4 FAJ R. Sadler, East Lynne, 202 Shire Oak, Walsall, WS9 9PD
G4 FAL N. Totterdell, Moscar Cross House, Hollow Meadows, Sheffield, S6 6GL
G4 FAP R. Painting, 15 Surrey Walk Aldridge, Walsall, WS9 8JG
G4 FAQ D. Jones, 7 Camrose Gardens, Pendeford, Wolverhampton, WV9 5RN
G4 FAS G. Royle, 56 Branksome Drive, Heald Green, Cheadle, SK8 3AJ
G4 FAT N. Trollope, 21 Glenmoor, Eckington, Pershore, WR10 3BW
GM4 FAU Dr J. Walker, West Lodge, Otterstone, Dunfermline, KY11 7HZ
G4 FAV A. Bevan, 330 Stourbridge Road, Halesowen, B63 3QR
G4 FAW D. Cutts, 51 Brook Lane, Felixstowe, IP11 7LG
G4 FAX R. Macfie, 97 Chesford Road, Luton, LU2 8DP
G4 FAZ G. Brownett, Apartment 264, The Crescent, Hannover Quay, Bristol, BS1 5JR
G4 FBB D. Ellis, 17 Victoria Avenue, Yeadon, Leeds, LS19 7AS
G4 FBC R. Heron, 54 Basket Road, Kells, Whitehaven, CA28 9AH
G4 FBG D. Shone, 6 Windlehurst Road, High Lane, Stockport, SK6 8AB

G4 FBI E. Creasy, 16 Birchwood Close, Horley, RH6 9TX
G4 FBK M. Kipp, 55 Hollybrook Mews, Yate, Bristol, BS37 4GB
G4 FBN B. Neale, Badgers Sett, Harbertonford, Totnes, TQ9 7PU
GM4 FBP J. Dean, 37 Countesswells Avenue, Aberdeen, AB15 8LX
GW4 FBQ J. Hobley, Lawley The Avenue West Felton, Oswestry, SY11 4EQ
G4 FBS Horndean and District ARC, c/o S. Tribe, 6 Privett Road, Waterlooville, PO7 5HJ
GM4 FBU H. Macdougall, 17 Prospecthill St., Greenock, PA15 4HH
G4 FBV D. Vaughan, 6 Swallow Close, Felixstowe, IP11 9LR
G4 FBY B. Sorger, Courtlands, Monks Corner, Saffron Walden, CB10 2RW
G4 FBZ W. Kitching, 3 Prince Charles Crescent, Telford, TF3 2JX
G4 FCA J. Haddon, 8 Oaklands, Cradley, Malvern, WR13 5LA
G4 FCB N. Edwards, 40 Camden St., Walsall, WS1 4HF
G4 FCC G. Freeman, 12 The Haven, Beadnell, Chathill, NE67 5AW
G4 FCD R. Girling, Lee Gardens, Lower Road, Chalfont St. Peter, Gerrards Cross, SL9 8LQ
G4 FCF W. Wade, 11 St. Marys Road, Bluntisham, Huntingdon, PE28 3XA
G4 FCI A. Cullup, 201 Elm Low Road, Elm, Wisbech, PE14 0DF
G4 FCL P. Lawson, 1 Beehive Cottages, Wickham Road, Fareham, PO16 7JF
G4 FCN C. Coker, 46 Clarendon Road, Ipplepen, Newton Abbot, TQ12 5QS
G4 FCT J. Gunn, 8 College Gardens, Hornsea, HU18 1EF
G4 FCU D. Restall, 7 Medway Close, Skelton-In-Cleveland, Saltburn-by-the-Sea, TS12 2JZ
GW4 FCV R. Jones, 2 Pen-Y-Cwarel Road, Wyllie, Blackwood, NP12 2HP
G4 FCX B. Pearl, 66 Benfleet Road, Benfleet, SS7 1QB
G4 FCY I. Smith, 17031 Los Cerritos, Los Gatos, USA, 95030
G4 FCZ M. Thomas, The Old School, Church Lane, Lowestoft, NR32 5LL
G4 FDA W. Chapman, 34 Saxon Close, Oake, Taunton, TA4 1JA
G4 FDD J. Livingston, 26 Dikelands Lane, Upper Poppleton, York, YO26 6JB
G4 FDF V. Cunningham, 9 Lacon Road, Bramford, Ipswich, IP8 4HD
G4 FDG R. Taylor, Trelawn, 26A Honiton Road Cullompton, Devon, EX15 1PA
G4 FDI S. Giles, Conifers, Kington Magna, Gillingham, SP8 5EW
G4 FDK R. Palmer, 26 Silverstone Way, Congresbury, Bristol, BS49 5ES
GM4 FDM T. Wylie, 3 Kings Crescent, Elderslie, Johnstone, PA59AD
G4 FDN P. Mcguinness, 9 Farmdale Road, Carshalton Beeches, SM5 3NG
G4 FDP R. Miller, 65 West Road, Oakham, LE15 6LT
G4 FDR S. Ldr. D. Roberts, 236 Grantham Road, Sleaford, NG34 7NX
G4 FDS J. Ingram, 170 Churchill Road, Poole, BH12 2JF
GM4 FDT R. Kerr, Rosskeen Bridge, Invergordon, IV18 0PR
G4 FDU R. Mckinlay, 54 Barn Meadow Lane, Bookham, Leatherhead, KT23 3EY
G4 FDX I. Offer, Southease, Balmer Lawn Road, Brockenhurst, SO42 7TT
G4 FEA C. Beezley, 19 Beech Avenue, Claverton Down, Bath, BA2 7BA
G4 FEB D. Emery, 424 Clement Avenue, Charlotte, North Carolina, USA, 8204
GM4 FEI A. Marsden, 63 Carlogie Road, Carnoustie, DD7 6EX
G4 FEJ B. Fawcett, 75 Ark Royal, Bilton, Hull, HU11 4BN
G4 FEM P. Greatorex, 2 Briar Briggs Road, Bolsover, Chesterfield, S44 6SE
GM4 FEO J. Gaughan, 12 Fernbank Avenue, Windygates, Leven, KY8 5FA
G4 FEQ H. Stogdale, 14 Main Street, Ledston, Castleford, WF10 2AA
G4 FEU T. Southwell, 12 Chequer Lane, Upholland, Skelmersdale, WN8 0DE
G4 FEV D. Whitty, 146 Avenue Road, Rushden, NN10 0SW
G4 FF GXFF Radio Group, c/o A. Dodd, 14 Davies Street, Macclesfield, Sk10 1GE
G4 FFA R. Harris, 89 Evelyn Avenue, Ruislip, HA4 8AJ
G4 FFC M. Packer, Ricmaes Cottage, Chadwell End, Pertenhall, MK44 2AU
G4 FFE L. Marriott, 94 Lyndhurst Road, Worthing, BN11 2DW
GM4 FFF Gm Flora and Fauna, c/o J. Phunkner, 7 Plenshin Court, Glasgow, G53 6QW
GI4 FFL J. Finnegan, 15 Mossgreen, Richhill, Armagh, BT61 9JX
G4 FFM D. Bailey, 10 Manor Road, Stutton, Tadcaster, LS24 9BR
G4 FFN C. Baker, 78 Station Road, Whittlesey, Peterborough, PE7 1UE
GM4 FFP I. Campbell, 35 Radernie Place, St Andrews, KY16 8QR
G4 FFS D. Hodge, 15 Buckland Close, Peterborough, PE3 9UH
G4 FFW M. Betts, 56 Kingswood Road, Fallowfield, Manchester, M14 6RX
G4 FFX R. Clear, 33 Cedars Road, Beddington, Croydon, CR0 4PU
G4 FFY R. Howells, 16 Handel Walk, Tonbridge, TN10 4DG
G4 FGF J. Drakeley, 31 Goldstar Way, Birmingham, B33 0YP
GI4 FGH W. Tweedy, 11 Beechgrove Rise, Belfast, BT6 0NH
G4 FGJ G. Mcgowan, 6 Caldecote Green, Upper Caldecote, Biggleswade, SG18 9BX
GM4 FGL G. Williams, 1 Fife St., Keith, AB55 5EH
G4 FGM D. Lund, P.O. Box 333, 23 Vander Avenue, Blenheim, on, Canada, N0P 1A0
G4 FGO A. Oliver, Beaver Lodge, Dale Road, Brough, HU15 1HY
G4 FGR S. Porter, 138 Broad Lane, Essington, Wolverhampton, WV11 2RQ
GM4 FGS I. Douglas, 47 Meadowpark, Ayr, KA7 2LW
G4 FGW C. Hall, 5 St. Edmunds Stamford Bridge, York, YO41 1PW
G4 FGY J. Maltby, Ingle Nook, Lenton Road, Grantham, NG33 4HA
GM4 FH D. Hamilton, 57/6 North Street, Bo'ness, EH51 0AE
GI4 FHB W. Mcfaul, 9 Durham Park, Londonderry, BT47 5YD
G4 FHF J. Walker, Roseberry, Owmby Road, Barnetby, DN38 6BD
G4 FHK T. Knight, 3 Eaton Close, Rainworth, Mansfield, NG21 0AR
G4 FHN R. Lovell, 16 North View, Staple Hill, Bristol, BS16 5RU
G4 FHQ M. Hardy, 6 Apple Tree Close, Bromyard, HR7 4UL
G4 FI N. Seath, 6 Harvester Way, Sibsey, Boston, PE22 0YD
G4 FIA M. Mucklow, 7 Burns Close, Newport Pagnell, MK16 8PL
GW4 FIC D. Pearson, Hope Cottage, Parkhouse, Monmouth, NP25 4QD
G4 FIE P. Groom, 2A The Chestnuts, Countesthorpe, Leicester, LE8 5TL
G4 FIF D. Cherrington, 4 Bloomfield Close, Wombourne, Wolverhampton, WV5 8HQ
G4 FIG B. Callaway, 44 Grover Avenue, Lancing, BN15 9RQ
G4 FIH E. Fernandes, 2C Northampton Park, London, N1 2PJ
G4 FIN K. Mullaney, 5 St. Peters Way, Cogenhoe, Northampton, NN7 1NU
G4 FIQ G. Clegg, 6 Vergette Court, Towngate West, Market Deeping, Peterborough, PE6 8DJ
G4 FIT J. Chapman, 83 High St., Sutton, Ely, CB6 2NW
G4 FIV P. Morley, Ash House, Germansweek, Beaworthy, EX21 5BP
GM4 FIZ A. Murray, Woodhouse, Mount High, Balblair, IV7 8LH
G4 FJB J. Dodd, 7 Hornbrook Grove, Solihull, B92 7HH
G4 FJC R. De La Rue, Linden Lea, Balls Chase, Halstead, CO9 1NY
G4 FJF M. Thacker, 164 Mongeham Road, Great Mongeham, Deal, CT14 9LL
G4 FJH D. Powell, 24 Beaconlea, Hanham, Bristol, BS15 8NX
GD4 FJI R. Allison, 20 Droghadfayle Park, Port Erin, Isle Of Man, IM9 6EP
G4 FJJ D. Bayliss, 20 Midhill Drive, Rowley Regis, B65 9SD
G4 FJK T. Hugill, West Whitnole House, Stoodleigh, Tiverton, EX16 9QH
G4 FJP J. Perry, 108 Elm Road, New Malden, KT3 3HP
G4 FJT C. Cuthbert, 44 Towse Close, Clacton-on-Sea, CO16 8US
G4 FJV J. Barton, 5 Buttermere Grove, Willenhall, WV12 5FQ
G4 FJW C. Hook, 11 Battlesmere Road, Cliffe Woods, Rochester, ME3 8TR
G4 FJX I. Perera, 1 Francis Road, Perivale, Greenford, UB6 7AD
G4 FKA G. Plucknett, 9 Oakwood Gardens, Coalpit Heath, Bristol, BS36 2NB
G4 FKE C. White, 111 Waterbeach Road, Slough, SL1 3JU
G4 FKG G. Kirk, 124 Star Road, Peterborough, PE1 5HF
G4 FKH G. Williams, 21 Borda Close, Chelmsford, CM1 4JY
G4 FKI D. Thorpe, 70 Willow Way, Ampthill, MK45 2SP
G4 FKP B. Tarry, 6 Beech Gardens, Rainford, St Helens, WA11 8DJ
G4 FKQ M. Barnwell, 77 Elmfield Road, Peterborough, PE1 4HA
G4 FKR R. Hammond, 3 Rutledge Drive, Littleton, Winchester, SO22 6FE
G4 FKU K. Salter, Alton, 12 Perinville Road, Torquay, TQ1 3NZ
G4 FKX R. Abel, 23 Edward Gardens, Wickford, SS11 7EH
G4 FKY R. Sharpe, 1 Park Copse, Horsforth, Leeds, LS18 5UN
GI4 FLG K. Mayne, 8 Grandmere Park, Bangor, BT20 5RF
G4 FLM F. Crofts, 43 Broadlands Drive, East Ayton, Scarborough, YO13 9ET
GM4 FLP I. Strachan, 238 Coupar Angus Road, Muirhead, Dundee, DD2 5QN
G4 FLR D. Tanner, 4 Duckpitts Cottages, Bramling, Canterbury, CT3 1LY
G4 FLS A. Snow, 1A Park Avenue, Longlevens, Gloucester, GL2 0DZ
GM4 FLX A. Lovegreen, 16 Grahams Avenue, Lochwinnoch, PA12 4EG
G4 FLY G. Haynes, 39 Zinzan Street, Reading, RG1 7UG
GD4 FMB D. Taylor, Burnt Mill House, Mount William, Summer Hill, Douglas, Isle of Man, IM2 4PE
G4 FMC M. Constable, 2 The Banks, Long Buckby, Northampton, NN6 7QQ
G4 FMI F. Connor, 29 Parkdale Road, Paddington, Warrington, WA1 3EN
G4 FMJ L. Cooke, 23 Widecombe Road, Stoke-on-Trent, ST1 6SL
G4 FMM T. Walsh, 106 Westgate, Elland, HX5 0BB
G4 FMY D. Larsen, C/O Salbu (Pty) Ltd, Private Bag X 2352, Wingate Park, South Africa, 153
G4 FMZ E. Pearson, 4 Forder Walk, Salisbury, SP2 7FY
G4 FNC L. Harper, Three Oaks, Braydon, Swindon, SN5 0AD
G4 FND D. Yeates, 61 Martins Hill Lane, Burton, Christchurch, BH23 7NW
G4 FNG R. Walker, South Moor Farm, Langdale End, Scarborough, YO13 0LW
G4 FNI K. Nichols, 11 Tregonwell Road, Bournemouth, BH2 5NR
G4 FNJ P. Fuller, 4 Whitworth Road, Minehead, TA24 8EB
G4 FNK A. Jackson, 6 Blandys Hill, Kintbury, Hungerford, RG17 9UE
G4 FNL G. Bubloz, 42 Hillcrest, Westdene, Brighton, BN1 5FN
GW4 FNO G. Lloyd, 15 Budden Crescent, Caldicot, NP26 4PP
G4 FNP J. Guite, 15 Marlborough Avenue, Falmouth, TR11 2RW
G4 FNQ C. Wedgbury, 32 Cloverdale, Stoke Prior, Bromsgrove, B60 4NF
G4 FNR D. Rabone, 6 Cranwell Grove, Kesgrave, Ipswich, IP5 2YN
GI4 FNU M. Mcdowell, 50 Dunraven Parade, Belfast, BT5 6BT
G4 FNZ D. Bannister, 7 Sudeley Close, Malvern, WR14 1LP
G4 FOB J. Samuels, 8 Holm Oaks, Butleigh, Glastonbury, BA6 8UB
G4 FOH S. Foote, 14 High Street, Chrishall, Royston, SG8 8RP
GW4 FOI J. Doyle, 54 Bryncatwg, Cadoxton, Neath, SA10 8BG
G4 FOL J. Bell, 60 Queens Close, West Moors, Ferndown, BH22 0HN
GW4 FOM R. Rowles, 37 Vincent Court, Vincent Road, Cardiff, CF5 5AQ
G4 FON R. Goff, 69 Frambury Lane, Newport, CB11 3PU
G4 FOR D. Hawkes, 19 Taj Court, Ottawa, Canada, K1G 5K7
G4 FOS I. Gilmore, 4 Borton Road, Blofield, Norwich, NR13 4RU
G4 FOT H. Exley, 16 Croft Street, Horncastle, LN9 6BE
G4 FOW R. Strangeway, 88 Old Manor Way, Portsmouth, PO6 2NL
G4 FOX Melton Mowbray ARS, c/o G. Mason, 120 Scalford Road, Melton Mowbray, LE13 1JZ
G4 FOY K. Scott, 20 Tower Street, Alton, GU34 1NU
GM4 FOZ D. Moodie, 1 Lageonan Road, Grandtully, Aberfeldy, PH15 2QY
G4 FPA J. Shorthouse, 20 Boxgrove Road, Sale, M33 6QW
G4 FPB C. Roper, 52 Sandringham Avenue, Wirral, CH47 3BZ
G4 FPE G. Butterfield, 13 Windsor Walk, Batley, WF17 0JL
G4 FPG R. Hawke, Basque Close, Hastingleigh, Ashford, TN25 5JB
G4 FPI B. Wood, 193 Robin Way, Chipping Sodbury, Bristol, BS37 6JU
G4 FPM E. Keeler, 18 Clyde Road, Worthing, BN13 3LG
G4 FPO K. Wilson, 14 Stuart Grove, Eggborough, Goole, DN14 0JX
G4 FPV S. Perkins, 17 Lime Tree Avenue, Malvern, WR14 4XE
G4 FPY K. Jones, 58 Woodlands Road, Allestree, Derby, DE22 2HF
GM4 FPZ M. Lisle, 50 Lade Braes, St Andrews, KY16 9DA
GM4 FQE E. Thirkell, 20 The Glebe, Crail, Anstruther, KY10 3UJ
G4 FQF P. Herring, 34 Woodlands Road, Romford, RM1 4HD
GM4 FQG R. Mclaren, Lethendry, North Road, Dunbar, EH42 1AY
G4 FQH B. Nelmes, Birchgrove, 17 Woodfield Road, Dursley, GL11 6HB
G4 FQI M. Smith, Wilson Hall Farm, Slade Lane, Wilson, Derby, DE73 8AG
G4 FQM S. Morris, 23 Ellesmere Way, Carlisle, CA2 6LZ
G4 FQN G. Kelly, 7 Greenwood Road, Lymm, WA13 0LA
G4 FQP C. Bamford, 12 Lincoln Drive, Caistor, Market Rasen, LN7 6PA
G4 FQT R. Gregory, 24 Tilton Road, Borough Green, Sevenoaks, TN15 8RS
GW4 FQU Dr I. Jones, Rhandirmwyn, Llandygai, Bangor, LL57 4LD
G4 FQV D. Gray, 8 Foxglove Close, Wyke, Gillingham, SP8 4TW
G4 FQW B. Dunn, 17 Duke St., Clayton Le Moors, Accrington, BB5 5NQ
G4 FQZ D. Simms, 1 Old Barn Close, Little Eaton, Derby, DE21 5AX
G4 FRA V. Lane, 3 Lawnway, York, YO31 1JD
G4 FRB Dr G. Morse, Riversdale, High Street, Salisbury, SP3 5JL
G4 FRD J. Walton, 2 Billy Mill Avenue, North Shields, NE29 0QX
G4 FRF K. Johnson, 7 Bridge Croft, Clayton Le Moors, Accrington, BB5 5XP
GW4 FRH R. Dawkins, 22 Derwen Fawr, Crickhowell, NP8 1DQ
G4 FRI G. Bird, Holmwood, 101 Brookfield Road, Gloucester, GL3 2PN
G4 FRK J. Rodway, 9 York Avenue, Thornton-Cleveleys, FY5 2UG
G4 FRL N. Ambridge, 4 Staggs Road, Thame, OX9 3AG
G4 FRM P. Hill, 8 Davenport Park Road, Stockport, SK2 6JS
G4 FRO G. Orford, 29 Church Road, Stoke Bishop, Bristol, BS9 1QP
G4 FRR K. Redford, 2 Cressington Cottages, Westend, Stonehouse, GL10 3SN
G4 FRV R. Vincent, Little Poulner, White Horse Lane, Bideford, EX39 1NW
G4 FRW M. Nicholson, 10 Beechfield Road, Cheadle Hulme, Cheadle, SK8 7DS
G4 FRX J. Nelson, Bank Cottage, Crew Green, Shrewsbury, SY5 9AS
G4 FRZ A. Jarrett, 73 Abbots Road, Abbots Langley, WD5 0BJ

GM4 FSB G. Millar, 30 Albert Crescent, Newport-on-Tay, DD6 8DT
G4 FSD J. Creasey, 144 Belthorn Road, Belthorn, Blackburn, BB1 2NN
G4 FSE P. Biner, 295 Daws Heath Road, Rayleigh, SS6 7NS
GM4 FSF K. Horne, 80 New Row, Dunfermline, KY12 7EF
G4 FSG P. Murchie, 42 Catherine Road, Woodbridge, IP12 4JP
G4 FSH J. Bagnall, Rainow, Under Rainow Road, Congleton, CW12 3PL
G4 FSK P. Vallow, 1 Carolbrook Road, Ipswich, IP2 9JF
G4 FSN E. Walton, 68 Mary Street West, Horwich, Bolton, BL6 7JU
G4 FSQ J. Morley, 65 Longfield Avenue, Golcar, Huddersfield, HD7 4BT
G4 FSS D. Hocking, 10 Garfit Road, Kirby Muxloe, Leicester, LE9 2DE
G4 FSU I. Greenshields, 3 Lovers Walk, Wells, BA5 2QL
G4 FSX G. Down, 20 Thackers Way, Deeping St. James, Peterborough, PE6 8HP
G4 FTA R. Earle, 10 Crosslands, Fringford, Bicester, OX27 8DF
G4 FTG C. Norton, 2 Heathlands Drive, Maidenhead, SL6 4NF
G4 FTI A. Bowhill, 9 West Park Drive East, Roundhay, Leeds, LS8 2EE
G4 FTK N. Cridland, Alfriston, 105 Elvetham Road, Fleet, GU51 4HN
G4 FTL G. King, 59 Rookery Lane, Northampton, NN2 8BX
G4 FTN J. Grainger, Kinlet Cottage 45 Stone Lane, Kinver, DY7 6DU
G4 FTP E. Kraft, 6 The Nook, Wivenhoe, Colchester, CO7 9NH
G4 FTQ P. Clutterbuck, 19 Warwick Close, Dorking, RH5 4NN
G4 FTW Dr M. Rowland, 16 Hayter Close, West Wratting, Cambridge, CB21 5LY
G4 FTX G. Knock, 31 Northmead, Ledbury, HR8 1BE
G4 FTY R. Page, Mercury House, 19 Green Lane, Birmingham, B46 3NE
G4 FTZ B. Togwell, 26 Garraways Royal Wootton Bassett, Swindon, SN4 8LL
G4 FUA G. Cheater, 34 Robbins Close, Bristol, BS32 8AS
GI4 FUE C. Morrison, 60 Windslow Drive, Carrickfergus, BT38 9BB
G4 FUG P. Clark, 42 Shooters Hill Road, Blackheath, London, SE3 7BG
G4 FUI M. Rigby, 16 Juniper Way, Penrith, CA11 8UF
G4 FUJ G. Wright, 35 Langdale Road, Cheltenham, GL51 3LX
GI4 FUM Dr W. Hutchinson, 40 Oldstone Hill, Muckamore, Antrim, BT41 4SB
G4 FUO J. Nowell, Crofters Cottage, Back Lane, Copmanthorpe, York, YO23 3SH
G4 FUP N. Braeman, 30 Oakley Road, Wimborne, BH21 1QJ
G4 FUR Coulsdon Amateur Transmitting Society, c/o A. Briers, 33 Deans Walk, Coulsdon, CR5 1HR
G4 FUU M. Pothecary, 61 Inglewood, Pixton Way, Croydon, CR0 9LN
G4 FUY P. Bonson, 26 Shreen Way, Gillingham, SP8 4EL
G4 FUZ A. Mallows, Kilmurry House, Kilmurry Estate, Kilmurry Fermoy Cork, Ireland, 00 00
G4 FVA P. Catling, The Heights, 10 Adams Road, Cambridge, CB25 0JU
G4 FVB I. Clabon, 2 Farm View Lodge, Lodge Hill Lane, Rochester, ME3 8NE
G4 FVK D. Sewell, 11 Haddon Close, Stanground, Peterborough, PE2 8LS
G4 FVL G. Rankin, 25 The Chase, Coulsdon, CR5 2EJ
GM4 FVM J. Edgar, 7 Welltower Park, Ayton, Eyemouth, TD14 5RR
GM4 FVO C. Evans, East Cottage, Mount Melville, St Andrews, KY16 8NT
G4 FVP C. Davies, 28 Neville Road, Darlington, DL3 8HY
GM4 FVQ A. Dimmick, 02 120 Shakespeare Street, Glasgow, G20 8LF
GM4 FVS G. Cusiter, 4 Elphin Hill, Ellon, AB41 8BH
G4 FVU A. Sweetapple, Bent Oak, Axminster Road, Axminster, EX13 8AQ
G4 FVV B. Vincent, 27 Naseby Walk, Leeds, LS9 7SY
G4 FVW D. Hooper, 8 Barn Close, Crewkerne, TA18 8BL
G4 FVX R. Johns, 42 Lansdown Road Redland, Bristol, BS66NS
G4 FVZ Dr P. Gould, 152 High Street North, Stewkley, Leighton Buzzard, LU7 0EP
G4 FWA J. Beckett, 9 Gleneagles Drive, Ipswich, IP4 5SD
GM4 FWF A. Gurney, 42 Fleming Way, Invergordon, IV18 0LU
G4 FWK P. Mooney, 57 Johnstown Road, Co Dublin, Dun Laoghaire, Ireland,
G4 FWM C. Webb, 6 Chatsworth Avenue, Fleetwood, FY7 8EG
G4 FWN N. May, Sandock Nurseries, Middle Dimson, Gunnislake, PL18 9NG
GD4 FWQ C. Matthewman, 26 King Orry Road Glen Vine, Douglas, Isle of Man, IM4 4ES
G4 FWR A. Johnson, 86 Meadow Close, Thatcham, RG19 3RL
G4 FWT S. O'Shanohun, The Corner Stone, Treskinnick Cross, Bude, EX23 0DT
G4 FXA V. Arnold, 435 Manchester Road, Clifton, Manchester, M27 6WH
G4 FXE R. Lott, 83 Manor Road, Dover, CT17 9LQ
GW4 FXF G. Swan, Long Acre, New Road, Neath, SA10 8HT
G4 FXI P. Overell, 48 Bedgrove, Aylesbury, HP21 7BD
GM4 FXL A. Docherty, 10 Dumyat Road, Menstrie, FK11 7DG
G4 FXM G. Farnie, Barn End, Rughill, Wedmore, BS28 4HL
G4 FXR W. Wunderlich, 31 College Road, Bromley, BR1 3PU
G4 FXT N. Burkitt, 31 Loxwood, Earley, Reading, RG6 5QZ
G4 FXU R. Napper 1, 12 Brumell Drive, Lancaster Park, Morpeth, NE61 3RB
G4 FXY P. Staton, 52 School Road, Newborough, Peterborough, PE6 7RG
G4 FYB S. Carter, 48 Gorse Bank Road, Hale Barns, Altrincham, WA15 0AS
G4 FYE G. Coggon, 45 Ansten Crescent, Cantley, Doncaster, DN4 6EZ
G4 FYG M. Newlands, 72 Town Acres, Tonbridge, TN10 4NG
GM4 FYH C. Waddington, Wester Lathallan, Leven, KY8 5QP
G4 FYI T. Fallick, 44 Cypress Road, Newport, PO30 1HA
G4 FYJ J. Lemon, 30 Iveagh Court, Farm Hill, Exeter, EX4 2LR
G4 FYM D. Wiggs, 8 Bulbery Abbotts Ann, Andover, SP11 7BN
G4 FYO T. Foley, 16 Buckingham Road, Winslow, Buckingham, MK18 3DY
G4 FYQ M. Robins, 36 Wolverley Avenue, Wollaston, Stourbridge, DY8 3PJ
G4 FYT D. Lawrence, 23 Parkmead Road, Wyke Regis, Weymouth, DT4 9AL
G4 FZA J. Bladen, 4 St. James Close, Hanslope, Milton Keynes, MK19 7LF
G4 FZC Maj. A. Chapman, Majadilla Del Muerte 155, Malaga, Spain, 29649
GI4 FZD P. Menown, 34 Cairnburn Road, Belfast, BT4 2HS
G4 FZF A. Grinling, 32 Maybridge Square, Goring-By-Sea, Worthing, BN12 6HR
G4 FZG B. Sirignano, Eversholt, 22 Cleevelands Drive, Cheltenham, GL50 4QB
GM4 FZH Dr C. Smith, Ravenstone House Whithorn, Newton Stewart, DG8 8DU
G4 FZL L. Povoas, 9 Masons Drive, Necton, Swaffham, PE37 8EE
G4 FZM M. Davey, 37 The Common West Wratting, Cambridge, CB21 5LR
G4 FZN J. Kirby, 2 Kneeton Park, Middleton Tyas, Richmond, DL10 6SB
G4 FZP A. Drury, 31 Brook Drive, Whitefield, Manchester, M45 8FR
G4 FZR K. Dally, Ealand Grange, Ealand, Scunthorpe, DN17 4DG
G4 FZS H. Bulmer, 10 Southfield Lodge, South End Villas, Crook, DL15 8NN
G4 FZV Dr P. Redall, 106 Stowey Road, Yatton, Bristol, BS49 4EB
G4 FZZ D. Holmes, 12 Chestnut Close, Rushmere St. Andrew, Ipswich, IP5 1ED
G4 GAB R. Padbury, 8 Osbourne Drive, Holton-Le-Clay, Grimsby, DN36 5DS
GW4 GAF A. Mccann, Lower Fiddlers Green, Felindre, Knighton, LD7 1YT
G4 GAI K. Taylor, 31 Stonehill Drive, Rochdale, OL12 7JN
G4 GAK M. Sykes, 21 Croft Walk, Broxbourne, EN10 6LD
G4 GAP H. Fitzherbert, 36 Westover Road, Broadstairs, CT10 3ES
G4 GAT B. Denton, 2 Seacroft Road, Broadstairs, CT10 1TL
G4 GBA C. Brookson, Orchard View, The Street., Stonham Aspal, Stowmarket, IP14 6AJ
G4 GBC F. Orchard, 39B Breach Road Marlpool, Heanor, DE75 7NJ
G4 GBE R. Blacker, 20 Claremont Park, Lincoln Road, Sleaford, NG34 8AE
G4 GBI A. Edwards, 96 Bathurst Road, Winnersh, Wokingham, RG41 5JF
G4 GBK C. Appleton, 249 Devonshire Road, Atherton, Manchester, M46 9QB
G4 GBP C. North, Somerholme, Forest Road, Hale, Fordingbridge, SP6 2NR
G4 GBW Dr J. Wilcox, 533 Upper Brentwood Road, Gidea Park, Romford, RM2 6LD
G4 GBX W. Greed, 18 Nursteed Park, Devizes, SN10 3AN
G4 GBY J. Robson, 35 Hankin Avenue Dovercourt, Harwich, CO12 5HE
G4 GCI N. Palmer, 14 Cambria Drive, Dibden, Southampton, SO45 5UW
G4 GCJ F. Fuller, 7 Prestwick Close, Bletchley, Milton Keynes, MK3 7RQ
GM4 GCL J. Tyler, 1 Mansefield Road, Tweedmouth, Berwick-upon-Tweed, TD15 2DX
GI4 GCN R. Booth, 12 Priory Drive, Carrickfergus, BT38 8HZ
G4 GCT North Bristol ARC, c/o R. Elford, Prospects, Tormarton Road, Badminton, GL9 1HP
G4 GCU Z. Kowalczyk, 6 St Georges Crescent, Redcar, TS11 8BT
G4 GCX R. Reisch, 18 St. Margarets Grove, Leeds, LS8 1RZ
GW4 GDB Dr A. Duncan, The Return 4 Ffordd Gerwyn, Wrexham, LL13 7DX
G4 GDC S. Wiles, Conifers, Aisthorpe, Lincoln, LN1 2SG
GM4 GDF J. Cain, Carradale, Braehead, Kirkinner, Newton Stewart, DG8 9AH
G4 GDG R. Smith, 47 Windsor Road, Levenshulme, Manchester, M19 2FA
G4 GDL M. Ellis, 32 Pegholme Drive, Otley, LS21 3NZ
GW4 GDM J. Owens, Yr Hafan I Maes Gyn, An Llanarmon-Yn-Ial, Clwyd N Wales, CH7 4PY
G4 GDO F. Lamb, 13-336 Queen St. South, Mississauga, Ontario, Canada, L5M 1M2
G4 GDP J. O'Shea, 30 Sue Ryder Homes, Owning, Co Kilkenny, BN2 7HA
G4 GDR Rev. A. Heath, 227 Windrush, Highworth, Swindon, SN6 7EB
G4 GDS D. Jones, 3 Kingfisher Drive, Benfleet, SS7 5ES
G4 GDT D. Wood, 30 Semley Road, Hassocks, BN6 8PE
G4 GDU I. Hoskin, 14 Trevingey Parc, Redruth, TR15 3BZ
G4 GDX I. Smith, 25 Windrush Avenue, Brickhill, Bedford, MK41 7BS
G4 GDY M. Edwards, 9 Earls Walk, Binley Woods, Coventry, CV3 2AJ
G4 GED D. Richardson, 68 Beech Tree Road, Holmer Green, High Wycombe, HP15 6UT
G4 GEE Dr R. Nash, 135 Farren Road, Coventry, CV2 5EH
GI4 GEL R. Penn, 9 Milltown Road, Donaghcloney, Craigavon, BT66 7NE
G4 GEN A. Morriss, Pipinford Park, Millbrook Hill Nutley, East Sussex, TN22 3HX
G4 GEO C. Tomkinson, Ridgeway, Towers Road, Poynton, Stockport, SK12 1DD
G4 GEP V. Peake, 24 Holyoke Grove, Leamington Spa, CV31 2RB
G4 GET I. Jordan, 70 Hungerhill Road, Kimberworth, Rotherham, S61 3NP
G4 GEY J. Carter, 30 Braemar Road, Hazel Grove, Stockport, SK7 4QG
G4 GEZ R. Evans, 2 Greyfriars Lane, West Common, Harpenden, AL5 2QJ
G4 GFC S. Wright, 163 Croham Valley Road, South Croydon, CR2 7RE
G4 GFE D. Foulds, 12 Royal Beach Court, North Promenade, Lytham St Annes, FY8 2LT
G4 GFI M. Broadway, 91 Tattenham Grove, Epsom, KT18 5QT
G4 GFJ L. Frankham, 47 St. Marys Gardens, Hilperton Marsh, Trowbridge, BA14 7PH
GW4 GFL Abergavenny Radio Society, c/o A. Hopkins, 30 Wavell Drive, Newport, NP20 6QN
G4 GFM D. Hessom, 89 Pond Close, Overton, Basingstoke, RG25 3LZ
G4 GFN S. Dabbs, 52 Hayling Rise, Worthing, BN13 3AG
G4 GFV J. Simpson, 19 Hollinside Close, Whickham, Newcastle upon Tyne, NE16 5QZ
G4 GFY P. King, 78 Gweal Wartha, Helston, TR13 0SN
G4 GFZ S. Dunkerley, Po Box Hm 2215, Hamilton, Bermuda, ZZ9 9PO
G4 GGC M. Marsh, 21 Stour Gardens, Great Cornard, Sudbury, CO10 0JN
G4 GGE D. Nicholson, 41 Thurstons Barton, Bristol, BS5 7BQ
GM4 GGF V. Mason, 19 Sherwood Crescent, Bonnyrigg, EH19 3LQ
G4 GGH P. Ledbury, 12 Sandfield Close, Lichfield, WS13 6BF
G4 GGI R. Williamson, Burwood, Wych Hill Lane, Woking, GU22 0AA
G4 GGL T. Grainger, 34 Maple Avenue, Ripley, DE5 3PY
G4 GGR F. Gemmell, 89 Coach Road, Guiseley, Leeds, LS20 8AY
G4 GGT M. Masterson, 44 Highstone Avenue, London, E11 2PP
G4 GGX S. Randall, 66 Park Court, Harlow, CM20 2PZ
G4 GGZ J. Birch, 13 Alison Way, Aldershot, GU11 3JX
G4 GHA J. Cleaton, 1 Avon Drive, Northmoor, Wareham, BH20 4EL
G4 GHB B. Kitchen, 73 Birch Street, Ashton-under-Lyne, OL7 0JD
G4 GHI R. Crabb, 29 Horsecastles Lane, Sherborne, DT9 6BU
G4 GHK J. Donovan, 6 Manor Place, Church, Accrington, BB5 4DX
G4 GHL Prof. M. Ward, 9 Woodshears Drive, Malvern, WR14 3EA
G4 GHM J. Mills, 2 Old Vicarage Close, Chilton Polden, Bridgwater, TA7 9DY
G4 GHO S. Webb, 10, Pilch Close, Norwich, NR1 3FU
G4 GHQ P. Fisher, 95 Slaithwaite Road, Thornhill Lees, Dewsbury, WF12 9DN
G4 GHR D. Humphreys, 64 Holne Chase, Plymouth, PL6 7UB
G4 GHT M. Skyner, 15 Dart Close, Alsager, Stoke-on-Trent, ST7 2HY
G4 GHZ P. Collins, 17 Tenterden Gardens, London, NW4 1TG
GI4 GID J. Heasley, 36 Collinbridge Gardens, Newtownabbey, BT36 7SU
G4 GIG J. Mullany, Flat 3 Michelle Close, Hollybank Road, Birmingham, B13 0PR
G4 GIM B. Waters, 60 Whitewood Way, Worcester, WR5 2LN
GM4 GIO R. Marshall, 9 Belford Terrace, Edinburgh, EH4 3DQ
G4 GIR I. Frith, 50 Rowallan Drive, Bedford, MK41 8AS
G4 GIS J. Darbyshire, 7 Sandle Road, Bishop's Stortford, CM23 5HY
G4 GIX T. Kearns, 7 Flitwick Grange, Milford, Godalming, GU8 5DN
G4 GIY R. Harris, The Willows, Nanny Lane, Church Fenton, Tadcaster, LS24 9RL
GW4 GJA K. Austen, 6 Caernarvon Grove, Merthyr Tydfil, CF48 1JS
G4 GJE D. Davis, 6 Regina Drive, Walsall, WS4 2HB
GW4 GJI R. Whitley, 22 Pen Y Bryn Road, Colwyn Bay, LL29 6AF
G4 GJO D. Blampied, 113 Green Street, Enfield, EN3 7JF
G4 GJR T. Aldridge, 180 Tickford Street, Newport Pagnell, MK16 9BG
G4 GJS W. Owens, Dorfstrasse 49, Effeld, Germany, D-41849
G4 GJU P. Moxham, 233 Walsall Road, Aldridge, Walsall, WS9 0QA
G4 GJV A. Horne, 22 Hedingham Close, London, N1 8UA
G4 GJY S. Simmonds, 14 Lindsey Crescent, Kenilworth, CV8 1FL
G4 GKC C. Willoughby, 79 Liskeard Road, Walsall, WS5 3ES

GM4 GKH G. Duke, 3 Woodlands Grove, Westhill, Inverness, IV2 5DU
G4 GKK A. Hawkins, 101 Tobyfield Road, Bishops Cleeve, Cheltenham, GL52 8NZ
G4 GKT Dr F. Delaney, 6 Stour Road, Astley, Manchester, M29 7HH
G4 GKU J. Cooper, 44 Belvedere Road, Bridlington, YO15 3NA
G4 GKX J. Trevett, 12 Churchill Road, Blandford Forum, DT11 7HH
G4 GKY C. Williams, 12A Parc An Dix Lane, Phillack, Hayle, TR27 5AB
G4 GKZ R. Revill, 102 Hurst Drive, Stretton, Burton-on-Trent, DE13 0EE
G4 GLC D. Hamilton, Rome Lea, 4 Lane Ends, Settle, BD24 0AG
GM4 GLG C. Edwards, The Old Mill Ferry Road Sandbank, Dunoon, PA23 8QH
G4 GLH D. Bennett, Flat 3, Falcon Crag, Cowan Head, Kendal, LA8 9HL
G4 GLI M. King, 28 Topcliffe Way, Cambridge, CB1 8SH
G4 GLM Dr G. Manning, 63 The Drive, Edgware, HA8 8PS
G4 GLN A. Bellfield, 50 Highfield Road, Biggin Hill, Westerham, TN16 3UU
G4 GLP D. Dale-Green, 31 Robins Bow, Camberley, GU15 3NP
G4 GLQ J. Tysiorowski, 52 Meadow Croft, Penrith, CA11 8EH
GW4 GLU M. Norbury, 16 Pont Aur, Ynyscedwyn Road, Flat, Swansea, SA9 1BP
G4 GLW C. Redmayne, 20 Kings Road, Accrington, BB5 6BS
GM4 GM Clyde Valley DX Group, c/o G. Mccallum, 15 Quarry Road, Law, Carluke, ML8 5HB
G4 GMB D. Hitchins, 21 Colwell Court, Newton Aycliffe, DL5 7PS
G4 GMI J. Seddon, 8 Upper Elms Road, Aldershot, GU11 3ET
G4 GMK M. North, 10 Long Lane, Pott Shrigley, Macclesfield, SK10 5SD
G4 GMN R. Caswell, 15 Murtwell Drive, Chigwell, IG7 5ED
G4 GMS L. Hicks, 108 Northorpe, Thurlby, Bourne, PE10 0HZ
G4 GMT A. Aedy, 35 Ashlea Avenue, Brighouse, HD6 3SR
G4 GMW M. Weaver, 22 Greenhill Road, Alveston, Bristol, BS35 3LZ
G4 GMZ Prof. J. Alder, 104 Park Lane, Congleton, CW12 3DE
G4 GNA D. Townend, 442 Blackmoorfoot Road, Crosland Moor, Huddersfield, HD4 5NS
G4 GND R. Culpan, 23 Aldreth Road, Haddenham, Ely, CB6 3PP
G4 GNG C. Pemberton, 2 Henthorn St., Shaw, Oldham, OL2 7AY
GD4 GNH R. Ferguson, Moaney Moar House, Corlea Road, Ballasalla, Isle Of Man, IM9 3BA
G4 GNO J. Callaghan, Evergreen, Seale Lane, Farnham, GU10 1LE
G4 GNP S. Mcgrory, The Paddocks, High Street Hook, Goole, DN14 5NY
G4 GNQ G. Sims, 85 Surrey Street, Glossop, SK13 7AJ
GM4 GNR W. Thow, 11 St. Marys Place, Ellon, AB41 8QW
G4 GNS S. Henry, 28 Marion Avenue, Shepperton, TW17 8AY
GI4 GNT J. Taggart, Windy Brae, 5 Glasvey Drive, Limavady, BT49 9HQ
G4 GNU A. Cross, 15 Louise Road, Rayleigh, SS6 8LW
G4 GNV S. Jones, 12 Yew Tree Close, Yeovil, BA20 2PD
G4 GNW T. Hennigan, 128 Dimsdale View West, Newcastle, ST5 8EL
G4 GNX A. Baker, 11 Fairfield Close, Shoreham-by-Sea, BN43 6BH
GW4 GNY M. Davies, Laburnum House Guilsfield, Welshpool, Powys, SY21 9PX
G4 GOA J. Harris, 28 Campion Drive, Bradley Stoke, Bristol, BS32 0BH
G4 GOG T. Densham, 37 Bovingdon Park, Roman Road, Hereford, HR4 7SW
G4 GOJ G. Porter, Ye Olde Homestede, High Street, Grimsby, DN36 5PL
GI4 GOL G. Brennan, 69 Kashmir Road, Belfast, BT13 2SB
G4 GOM F. Smith, 11 Reed Field, Bamber Bridge, Preston, PR5 8HT
G4 GON Dr J. Guest, 6 The Tyning, Bath, BA2 6AL
G4 GOO M. Kimmitt, Old Oaks, Tilston Road, Malpas, SY14 7DB
G4 GOP D. Benn, 36 Church Avenue, Horsforth, Leeds, LS18 5LD
G4 GOR J. Cross, 57F Grasmere Road, Blackpool, FY1 5HP
GI4 GOS H. Sinclair, 43 Edgcumbe Gardens, Belfast, BT4 2EH
G4 GOT R. Bradbury-Harrison, 11 Derwent Drive, Goring-By-Sea, Worthing, BN12 6LA
G4 GOU M. Wilson, Reedley Marina, Barden Lane, Burnley, BB12 0DX
GI4 GOV P. Barr, 5 Rosewood Park, Belfast, BT6 9RX
GM4 GOW R. Armstrong, Lera Cottage, Charleston Village, Forfar, DD8 1UF
G4 GOX R. Pearson, 33 Livedge Hall Lane, Liversedge, West Yorkshire, WF15 7DP
G4 GOZ E. Cockerill, 6 Richmond Avenue, Barnoldswick, BB18 5JB
GI4 GPA W. Otterson, 34 Ashbourne Park, Coleraine, BT51 3RE
G4 GPB R. Cooper, 17 Cavendish Drive, Claygate, Esher, KT10 0QE
GI4 GPC J. Ferguson, 7 Lairds Road, Katesbridge, Banbridge, BT32 5NN
G4 GPD W. Horn, 9 Springwell View, Love Lane, Bodmin, PL31 2QP
G4 GPF H. Winwood, 16 Brook Lane Hackenthorpe, Sheffield, S12 4LF
G4 GPJ N. Bailey, 12 Carmarthen Close, Callands, Warrington, WA5 9UU
G4 GPL A. Fish, 32 Deacons Hill Road, Elstree, Borehamwood, WD6 3LH
GM4 GPP C. Auty, Valsgarth, Haroldswick, Shetland, ZE2 9EF
G4 GPQ T. Stockill, 26 Hunters Close, Chatteris, PE16 6BD
G4 GPR A. Mills, 116 Mays Lane, Barnet, EN5 2LS
G4 GPV A. Brown, 12 Winstone Gardens, Cirencester, GL7 1GJ
G4 GPW B. Ainsworth, 23 Cokeham Road, Sompting, Lancing, BN15 0AE
G4 GPY S. Edwards, 71 St. Leonards Road Molescroft, Beverley, HU17 7HP
G4 GQA J. Chmielewski, 2 Wolverton Avenue, Kingston upon Thames, KT2 7QD
G4 GQE N. Harris, Mere Farmhouse, Matlaske Road, Norwich, NR11 7BE
G4 GQL A. Schiffman, Flat 48, Oakside Court, Ilford, IG6 2PH
GM4 GQM G. Firmin, 75 North Road, Lerwick, Shetland, ZE1 0PQ
G4 GQP R. Foote, 22 Hippings Vale, Oswaldtwistle, Accrington, BB5 3LH
G4 GQR Brighton & Dist, c/o P. Thompson, Flat 4, 14 West End Way, Lancing, BN15 8RL
G4 GQS B. Bentley, 25 Edinburgh Drive, North Anston, Sheffield, S25 4HB
G4 GQV J. Barrett, 13 Church Bank, Church, Accrington, BB5 4JQ
G4 GQY C. Lee, 74 Ilkeston Road, Trowell, Nottingham, NG9 3PX
G4 GQZ D. Tweedie, 39 Frenchfield Way, Penrith, CA11 8TW
GM4 GRC Glenrothes & District ARC, c/o D. Francis, 2 Morlich Crescent, Dalgety Bay, Dunfermline, KY11 9UW
G4 GRG Grajon Radio Group, c/o G. Badger, 3 Hesketh Close, Cranleigh, GU6 7JB
G4 GRJ D. Gower, 2 Norview Road, Whitstable, CT5 4DN
G4 GRK P. England, 2A Firs Close, Cowes, PO31 7NF
G4 GRM L. Horton, 4 Summer Lane, Walsall, WS4 1DS
G4 GRN T. Griffiths, 75 Central Avenue, Waltham Cross, EN8 7JJ
G4 GRP G. Gardiner, 35 Westparkside, Goole, DN14 6XN
G4 GRR Dr G. Searle, Shalbourne, The Dell, Vernham Dean, Andover, SP11 0LF
G4 GRS M. Williams, Flat 2, High Point, Highgate, N6 4BA
G4 GRT D. Mounter, 36 Norwich Road, Watton, Thetford, IP25 6DB
G4 GRU D. Jones, 36 Moor Lane Woodford, Stockport, SK7 1PP
G4 GRZ R. Marsh, 54 Waverton Road, Bentilee, Stoke-on-Trent, ST2 0QY
G4 GSA P. Milsom, 214 Ormonds Close, Bradley Stoke, Bristol, BS32 0DZ
G4 GSB M. Hall, 35 Bunns Lane, Dudley, DY2 7RA
G4 GSC J. Osborne, 3 Temple Gardens, Staines, TW18 3NQ
G4 GSD A. Watkin, 41 Brockwell Lane, Chesterfield, S40 4EA
GW4 GSG E. Warner, 99 St. Peters Park, Northop, Mold, CH7 6YU
GW4 GSH M. Beynon, 16 Hardy Close, Barry, CF62 9HJ
G4 GSK P. Barnett, Dunelm House, Barley Hill, Dunbridge, Romsey, SO51 0LF
G4 GSL J. Foster, 14 Braemar Grove, Heywood, OL10 3RR
G4 GSO H. Elliott, 40 Dene House Road, Seaham, SR7 7BQ
GD4 GSR D. Roberts, 7 Knock Rushen, Castletown, IM9 1TQ
GW4 GSS R. Bennett, Penrhiw Old Rd, Bwlchgwyn, Clwyd, LL11 5UH
GI4 GST W. Johnston, 3 Glenview, Comber, Newtownards, BT23 5HR
G4 GSY M. Bainbridge, 21 Cockey Moor Road, Bury, BL8 2HD
G4 GSZ K. Court, Vereda Escorredor 138, Alicante, Spain, 3150
GW4 GTC Coleg Menai Radio Club, c/o B. Davies, Rhosyr, Llanfair Pg, LL61 5JB
G4 GTD R. Ford, 2 Jersey Avenue, St. Annes, Bristol, BS4 4RA
GW4 GTE D. Evans, Glendale, Mount Pleasant Road, Buckley, CH7 3ET
G4 GTH M. Linda, 16 Woodlinken Close, Verwood, BH31 6BS
G4 GTN P. Reeve, 2 Court Road, Tunbridge Wells, TN4 8ED
G4 GTU S. Pocock, 202-1969 Oak Bay Avenue, Victoria, Canada, V8R 1E3
GM4 GTV N. Mackenzie, 57 Countesswells Terrace, Aberdeen, AB15 8LQ
G4 GTX W. Craigen, 19 Nilverton Avenue, Sunderland, SR2 7TS
G4 GTZ M. Phillips, 12 Reydon Avenue, Wanstead, London, E11 2JD
G4 GUA Dr J. Overton, 1 Pigeon House Farm, Pigeon House Lane, Hampshire, PO7 5SF
G4 GUC D. Bailey, 12 Westbeck, Ruskington, Sleaford, NG34 9GU
G4 GUE I. Pope, P O Box 662, Durbanville, South Africa, 7551
G4 GUG M. Meadows, 8 Beeches Park Hampton Fields, Minchinhampton, Stroud, GL6 9BA
GI4 GUH J. Clarke, 1 Rathview, Banbridge, BT32 4PY
G4 GUK K. Scott-Green, 1 Pickwick, Corsham, SN13 0JD
GM4 GUL S. Macdonald, 5 Lower Glebe, Aberdour, Burntisland, KY3 0XJ
G4 GUN G. Le Good, 45 Kingsfield Crescent, Witney, OX28 2JB
G4 GUO C. Brain, 7 Elverlands Close, Ferring, Worthing, BN12 5PL
G4 GUQ E. Crawford, 28 Mccullogn Drive, Erin, Ontario, Canada, N0B 1T0
G4 GUS J. Firmin, Warren Cottage, Hill House Road, Norwich, NR14 7EE
G4 GUV Dr J. Aindow, 2 Cutlers Close, Sydling St. Nicholas, Dorchester, DT2 9RG
G4 GUW G. Baggott, 105 The Crescent, Walsall, WS1 2DA
G4 GUX J. Kuipers, 27 Shirley Street, Hove, BN3 3WJ
G4 GUY T. Eaves, 3 Barons Road, Dousland, Yelverton, PL20 6NG
G4 GVE J. Hawkings, Church Barn, Dingle Lane, Sandbach, CW11 1FY
G4 GVG V. Gormley, 24 Beech Road, Garstang, Preston, PR3 1FS
GM4 GVJ G. Marshall, Drummorlie, Wallyford Toll, Musselburgh, EH21 8JT
GM4 GVK I. Munro, 57 Craigiebuckler Avenue, Aberdeen, AB15 8SF
G4 GVQ S. Eatough, 48 Mount Marua Way, Upper Hutt, New Zealand, 5018
G4 GVR R. Mason, 3 Coronation Close, Hellesdon, Norwich, Nr65Hf, Norwich, NR6 5HF
GI4 GVS P. Hallam, 95 Belfast Road, Carrickfergus, BT38 8BY
G4 GVV S. Fox, Flat 3, Woodford House, Aldershot, GU11 3EL
G4 GVW P. Gillen, 8 Barton Hamlet, Great Barton, Bury St Edmunds, IP31 2PP
G4 GVZ D. Morris, 40E Lansdown Crescent, Cheltenham, GL50 2NG
G4 GWB I. Gibbs, 9 The Square, Choppington, NE62 5DA
G4 GWE J. Martin, 57 Crescent Road East, Palm Beach, Auckland, New Zealand, 1001
G4 GWF H. Haden, 1 Bankside Close, Marple Bridge, Stockport, SK6 5ET
G4 GWG D. Snape, 30 Culcross Avenue, Wigan, WN3 6AA
G4 GWH M. Steventon, 22 The Beeches, Uppingham, LE15 9PG
G4 GWI J. Sheehan, 1 Osierground Cottages, Agester Lane, Canterbury, CT4 6NP
G4 GWJ J. Butcher, Mount Pleasant, Trampers Lane, North Boarhunt, Fareham, PO17 6DG
G4 GWP B. Langford, Dulce Verano, 29M San Jaime, Alicante, Spain, 3720
GD4 GWQ A. Matthewman, 26 King Orry Road, Glen Vine, Douglas, Isle Of Man, IM4 4ES
G4 GWR A. Scott-Green, 58B High Street, Sutton Benger, Chippenham, SN15 4RL
G4 GWT A. Kittle, 28 Clare Crescent, Towcester, NN12 6QQ
G4 GWU T. Chapman, 11 Ash Court, Brampton, Huntingdon, PE28 4FH
G4 GWV R. Hookham, 50 Billy Mill Avenue, North Shields, NE29 0QN
G4 GWX J. Travis, 4 Merrial Close, Bakewell, DE45 1JB
G4 GWZ R. Whitehead, 14 Southgate Crescent, Rodborough, Stroud, GL5 3TS
G4 GXB P. Butcher, 52 Chandos Road, Rodborough, Stroud, GL5 3QZ
G4 GXD D. Travis, 1 Hawthorn Close, Whixall, Whitchurch, SY13 2ND
G4 GXI P. Pearson, 58 Winchester Road, Grantham, NG31 8AD
G4 GXK Saltash Dist AR, c/o K. Hale, 58 St. Stephens Road, Saltash, PL12 4BJ
G4 GXL S. Fletcher, 31 Wesley Road, North Wootton, King's Lynn, PE30 3XA
G4 GXM R. Corr, 15 Waterdell Lane, St. Ippolyts, Hitchin, SG4 7RA
G4 GXN M. Wright, 5 Woodview Park, The Donahies, Dublin, Ireland, DUBLIN 13
G4 GXO R. Taylor, 16 Chestnut Close, Culgaith, Penrith, CA10 1QX
G4 GXQ P. Swain, 5 Cromley Road High Lane, Stockport, SK6 8BP
G4 GXR J. Higginbotham, Casa Valeriana, Fuente Grande, ZéJar, Spain, 18811
G4 GXW G. Cahill, 21 Moresby Close, Westlea, Swindon, SN5 7BX
G4 GXY E. Dowlman, 4 Beald Way, Ely, CB6 3DA
G4 GXZ A. Warrilow, Gyse Lodge, Gussage All Saints, Wimborne, BH21 5ET
G4 GYA R. Williscroft, 91 Parkfield Crescent, Tamworth, B77 1HB
G4 GYF G. Hiscoe, 1 Greendale Close, Fleetwood, FY7 8BQ
G4 GYI P. Ward, 23 Ropewalk, Alcester, B49 5DD
G4 GYJ R. Littlefield, 7 Carron Mead, South Woodham Ferrers, Chelmsford, CM3 5GH
G4 GYL M. Denby, 13 Hunger Hills Avenue, Horsforth, Leeds, LS18 5JS
G4 GYN Dr R. Colson, 46 Westwood Drive, Amersham, HP6 6RJ
G4 GYP L. Ratcliff, 15 Spring Close, Biggleswade, SG18 0HL
G4 GYS J. Plested, 24 Farm Way, Bushey, WD23 3SS
G4 GZA D. Ayris, 16 Chapel Lane, Northorpe, Gainsborough, DN21 4AF
G4 GZC P. Teanby, 34 High Street, Belton, Doncaster, DN9 1LR
GM4 GZD G. Smith, Ardvourlie, Loaneckheim, Kiltarlity, IV4 7JQ
G4 GZG L. Stringer, The Spinney Lincombe Drive, Torquay, TQ1 2HH
G4 GZH D. Andrew, Little Stone House, The Crescent, Steyning, BN44 3GD
G4 GZK H. Dalton, 24 Church Lane Coven, Wolverhampton, WV9 5DE
G4 GZL D. Barker, 79 South Parade, Boston, PE21 7PN

G4 GZM A. Mcmillan, 4 Aluric Rise, Newton Abbot, TQ12 4FN
G4 GZN K. Andreang, 62 Castleton Avenue, Barnehurst, Bexleyheath, DA7 6QU
G4 GZO A. Thurbon, 37 Lealand Road, Drayton, Portsmouth, PO6 1LZ
GM4 GZQ J. Mcginty, 77 Crawford Road, Houston, Johnstone, PA6 7DA
G4 GZS K. Wallace, 11 Orson Leys, Rugby, CV22 5RG
G4 GZT P. Jensen, 7 Union Street, Mosman, New South Wales, Australia, 2088
G4 GZU R. Woodcock, 143 Berry Hill Road, Mansfield, NG18 4RT
GM4 GZW Dr E. Simon, 12 St. Colme Road, Dalgety Bay, Dunfermline, KY11 9LH
GW4 GZX J. Hunter, 245 Heathwood Road, Heath, Cardiff, CF14 4HS
G4 HAC C. Denscombe, High Holme, 4 Kendricks Bank, Shrewsbury, SY3 0EX
G4 HAG J. Long, 9 Denbrook Avenue, Bradford, BD4 0QH
G4 HAI P. Levitt, 21 Station Road, Firsby, Spilsby, PE23 5PX
G4 HAJ D. Magee, 2 Holt Park Vale, Holt Park, Leeds, LS16 7QX
G4 HAK P. Torrance, 1 Clifton Lawn, Ramsgate, CT11 9PB
GM4 HAM Edinburgh and District ARC, c/o N. Stewart, 160 Carrick Knowe Drive, Edinburgh, EH12 7EW
GM4 HAO R. Mackean, 10A Dick Place, Edinburgh, EH9 2JL
G4 HAP H. Lavin, 30 Greenslate Road, Billinge, Wigan, WN5 7BG
G4 HAS D. Buck, 4687 Bracknell Road, Burlington, Ontario, Canada, L7M 0E5
GW4 HAT P. Jones, 68 Pastoral Way Sketty, Swansea, SA2 9LY
G4 HBA S. Horne, Beaucroft, Keswick Road, Benfleet, SS7 3HU
G4 HBD P. Trepess, 3 Lawford Rise, Wimborne Road, Bournemouth, BH9 2BZ
GM4 HBG I. Robertson, 53 Carmuir Forth, Lanark, ML11 8AP
G4 HBI F. Cassidy, 55 High Bank Road, Droylsden, Manchester, M43 6FS
GW4 HBK D. Lewis, 23 Gelligroes Road, Pontllanfraith, Blackwood, NP12 2JU
G4 HBL G. Hardy, The Mill House, Thearne, Beverley, HU17 0RU
GM4 HBQ A. Taylor, 6 Bowling Green St., Methil, Leven, KY8 3DH
G4 HBR J. Mcgee, 3 Hedgelea Road, East Rainton, Houghton le Spring, DH5 9RR
GW4 HBS S. Illidge, 24 Maes Briallen, Llandudno, LL30 1JJ
G4 HBT M. Foreman, 83 Hawthorn Crescent, Yatton, bs49 4 rg
G4 HBV A. Martin, 21 Ashwood Way, Hucclecote, Gloucester, GL3 3JE
G4 HBY M. Cotton, Esterith, 113 Belvedere Road, Burton-on-Trent, DE13 0RF
GW4 HBZ B. Clowes, 7 Dukesfield Drive, Buckley, CH7 3HN
GW4 HCA C. Prentice, 1 Victoria Street, Maesteg, CF34 0YP
G4 HCB Dr J. Harrison, 36 Elmlea Avenue, Bristol, BS9 3UU
G4 HCC M. Hodgkinson, 34 Pennine Way, Brierfield, Nelson, BB9 5DT
G4 HCD A. Reed, 28 Russell Street, Sutton-in-Ashfield, NG17 4BE
GM4 HCE K. Kirkland, 11 Marchfield Park Lane, Edinburgh, EH4 5BF
G4 HCG Dr R. Gordon, Middle House, 9 Fotheringhay Road, Peterborough, PE8 5HP
G4 HCI M. Foreman, 39 Artists View Drive, Calgary, Alberta, Canada, T3Z 3N4
G4 HCK N. Wilkinson, 12 Woodlands Close, Grays, RM16 2GB
GI4 HCN J. Clarke, 154 Galgorm Road, Ballymena, BT42 1DE
GM4 HCO V. Kusin, East Overhill Farm, Stewarton, Kilmarnock, KA3 5JT
GI4 HCX I. Magill, 205 Whitechurch Road, Ballywalter, Newtownards, BT22 2LA
GM4 HCY M. Stokes, 22 Lothian Road, Jedburgh, TD8 6LA
G4 HCZ L. Fellows, 19 Grosvenor Road, Lower Gornal, Dudley, DY3 2PS
GW4 HDB M. Greatrex, 4 Lee Street, St. Thomas, Swansea, SA1 8HQ
G4 HDD S. Rose, 14 Highgate West Hill, London, N6 6JR
G4 HDE S. Green, 6 Poveys Mead, Kingsclere, Newbury, RG20 5ER
GW4 HDF V. Hill, 9 Cae Pant, Caerphilly, CF83 2UW
GI4 HDJ B. Mcgarry, 43 Umrycam Road, Feeny, Londonderry, BT47 4TJ
G4 HDL N. Sedgwick, Flat 3, Hartford Court, 33 Filey Road, Scarborough, YO11 2TP
G4 HDO A. Kirkland, 4 Laurelwood Road, Droitwich, WR9 7SE
GW4 HDR A. Evans, 4 Elm Grove, Rhyl, LL18 3PE
G4 HDS P. Unwin, Mycroft, Rochester, Newcastle upon Tyne, NE19 1RH
G4 HDU Rev. B. Keal, 46 Eastway, Liverpool, L31 6BS
G4 HDY G. Burgess, 44 Clifton Road, Winchester, SO22 5BU
GW4 HDZ D. Birch, 16 Llanharry Road, Brynsadler, Pontyclun, CF72 9DB
G4 HEB P. Tuffs, 48 Mackie Drive, Guisborough, TS14 6DJ
G4 HEC P. Stracey, 14 Portfield Road, Christchurch, BH23 2AG
G4 HEE W. Dallas, 21 Jubilee Avenue Asfordby, Melton Mowbray, LE14 3RY
G4 HEJ W. Reid, Comphurst, Comphurst Lane, Hailsham, BN27 4TX
GW4 HER S. Rogers, Green Tops, 69 Megs Lane, Buckley, CH7 2AG
G4 HEV G. Cass, 18 Rawcliffe Drive, York, YO30 6PE
G4 HEW G. Hancock, 12122-244Th Street, Maple Ridge, Canada, BC V4R 1I1
G4 HFG G. Eckersall, 40 Portland Drive, Skegness, PE25 1HF
G4 HFI M. Roberts, The Willows, Riverside, Hayle, TR27 5JD
G4 HFO M. Blythe, Trethullan Farmhouse, Sticker, Saint Austell, PL26 7EH
G4 HFQ G. Freeth, 9 South Avenue, New Milton, BH25 6EY
G4 HFS M. Davies, 4 Manor Close, Lewknor, Watlington, OX49 5BL
G4 HFU P. Spooner, The Birches, Wingrave Road, Aston Abbotts, Aylesbury, HP22 4LT
G4 HFZ S. Mccann, 6 Almond Grove, Scunthorpe, DN16 2ES
G4 HGH A. Selmes, 35 Windmill Rise, Hundon, Sudbury, CO10 8EQ
GW4 HGJ G. Carruthers, Henllys Farm, Cardigan, SA43 2HR
G4 HGK J. Davis, Hurstbourne, Westdown Road, Bexhill-on-Sea, TN39 4DY
G4 HGL Dr J. Buckley, Sandringham, Neston Road, Ness, Neston, CH64 4AT
G4 HGM M. Gregory, 1474 Profile Rd, Franconia, USA, NH03580
G4 HGN D. Hoyle, Pharmacy Cottage, Queen Street, Buxton, SK17 8JT
G4 HGR M. Baker, 39 The Cherry Orchard, Hadlow, Tonbridge, TN11 0HU
GW4 HGS G. Passmore, 18 Brickhurst Park, Johnston, SA62 3PA
G4 HGT J. Wilkinson, 7 Hilton Grange Bramhope, Leeds, LS16 9LE
G4 HGV M. Leach, 15 Beech Lea, Blunsdon, Swindon, SN26 7DE
G4 HHA K. Stalley, The Forge, Woodbridge Road, Woodbridge, IP12 2JE
GW4 HHD J. Hutchinson, 3 Erw Fawr, Henryd, Conwy, LL32 8YY
G4 HHH Maj. P. Walker, East Rigg, Fylingdales, Whitby, YO22 4QG
G4 HHJ D. Thomas, 64 Marconi Way, St Albans, AL4 0JG
G4 HHL V. Gorny, 22 Park Road, Shirehampton, Bristol, BS11 0EF
G4 HHM D. Ryder, 96 Huttoft Road, Sutton-On-Sea, Mablethorpe, LN12 2QZ
G4 HHO Rev. C. Buckley, Curraghmore, Model Farm Road, Co Cork, Ireland,
G4 HHS L. May, 20 Crescent Road, Marland, Rochdale, OL11 3LF
G4 HHX R. Edmonds, 14 Singledge Lane, Whitfield, Dover, CT16 3EJ
G4 HHZ A. Harwood, 55 Nichol Road, Chandler'S Ford, Eastleigh, SO53 5AX
G4 HIA M. Nicholls, 12 Bents Drive, Sheffield, S11 9RP
G4 HIC M. Maddison, 34 Maple Avenue, Sandiacre, Nottingham, NG10 5EF
G4 HIE M. Hammond, 53 Chiltern Road, Baldock, SG7 6LT
G4 HIF D. Mallet, 41 Kiln Close, Calvert, Buckingham, MK18 2FD
G4 HIH R. Wilson, 4 Dinmont Place, Hall Close Grange, Cramlington, NE23 6DN
G4 HIJ R. Woolley, 29 Belle Vue Road, Ashbourne, DE6 1AT
G4 HIN R. Twiggs, 31 Westlands Avenue, Slough, SL1 6AH
G4 HIQ Dr A. Sturman, 22 St. Crispins Avenue, Wellingborough, NN8 2HT
G4 HIV B. Milne, 11 Station Road Thorpe-On-The-Hill, Lincoln, LN6 9BS
G4 HIW C. Vernon, 2 Standing Butts Close, Walton-On-Trent, Swadlincote, DE12 8NJ
G4 HIX P. Duncan, 89 Felstead Crescent, Sunderland, SR4 0AE
G4 HIZ J. Easdown, 38 North Street, Barming, Maidstone, ME16 9HF
G4 HJB C. Hall, 10 Porlock Court, Cramlington, NE23 3TT
G4 HJD A. Goy, 352 Chanterlands Avenue, Hull, HU5 4ED
G4 HJE S. Small, Leydene House, 102 Crestway, Chatham, ME5 0BH
G4 HJF W. Dredge, 10 Lime Close, Locking, Weston-Super-Mare, BS24 8BH
G4 HJH M. Hardaker, 6 Caradon Place, Verwood, BH31 7PW
G4 HJI J. Bright, 18 B Rue De La Station, Aspach-le-Bas, France, 68700
GM4 HJK R. Mitchell, 9 Pine Way, Perth, PH1 1DT
G4 HJL M. Zarattini, The Pippins, Orchard Street, Derby, DE3 0DF
GM4 HJO M. Mozolowski, The Auld Manse, 8 Sandport, Kinross, KY13 8DN
GM4 HJQ D. Mackenzie, 58 High Street, East Linton, EH40 3BH
G4 HJS P. Tempest, 15 Charles Avenue, Leeds, LS9 0AE
G4 HJT Dr D. Lloyd, 39 High Street, 40 Bertrand Drive, Princeton, USA, 8540
G4 HJV D. Miller, 50 Sandyleaze, Gloucester, GL2 0PX
G4 HJW B. Wright, 39 High Street, Little Wilbraham, Cambridge, CB21 5JY
G4 HJY M. Black, 28 Cricketers Close, Chessington, KT9 1NL
G4 HKB P. Turner, 1 Longridge, Colchester, CO4 3FD
G4 HKC I. Butson, 60 Churnwood Road, Parsons Heath, Colchester, CO4 3EY
G4 HKO Thurrock Acorns ARC, c/o N. Wilkinson, 12 Woodlands Close, Grays, RM16 2GB
G4 HKP C. Turner, 37 Kabeljou Crescent, Randpark Ridge, South Africa, 2169
G4 HKQ C. Marsh, 33 Southview Road, Hockley, SS5 5DY
G4 HKR A. Reed, 85 Ringway, Garforth, Leeds, LS25 1BZ
G4 HKS M. Lynch, Wessex House, Drake Avenue, Staines-upon-Thames, TW18 2AP
GM4 HKV J. Henderson, 7 Lumsden Crescent, St Andrews, KY16 9NQ
GW4 HKX R. Rowlands, 4 Glascoed, Hermon, Bodorgan, LL62 5LF
G4 HKY L. Bower, 1 Elmfield Drive, Skelmanthorpe, Huddersfield, HD8 9BT
G4 HKZ J. Butcher, Mount Pleasant, Trampers Lane, Fareham, PO17 6DG
G4 HLA J. Sullivan, 1 Godley Hill Road, Hye, SK14 3BW
G4 HLB R. Hallam, 16 Hall Road, Haconby, Bourne, PE10 0UY
G4 HLF P. Westwell, 11 Cheshire Park, Warfield, Bracknell, RG42 3XA
G4 HLI J. Friend, 62 St. Catherines Hill, Bramley, Leeds, LS13 2LE
G4 HLL Walsall ARC, c/o C. Willoughby, 79 Liskeard Road, Walsall, WS5 3ES
G4 HLN L. Bennett, 26 Winchester Road, Burnham-on-Sea, TA8 1HY
GW4 HLO W. Davies, Erw Deg, 11 Madoc Street, Porthmadog, LL49 9BU
G4 HLT M. Eckhoff, 6 Ramsbury Drive, Earley, Reading, RG6 7RT
G4 HLW K. Turnell, 31 Greenbank Terrace, Ringstead, Kettering, NN14 4DD
G4 HLX Dr N. Taylor, 27 Alfredston Place, Wantage, OX12 8DL
G4 HLZ M. Wood, 48 High Street, Maryport, CA15 6BQ
G4 HMA M. Smith, 8A Duke Street, Cullompton, EX15 1DW
G4 HMC J. Oliver, Chalk Lodge, Peters Lane, Monks Risborough, Princes Risborough, HP27 0LG
GM4 HML S. Mcluckie, 12 Croft Place, Eliburn, Livingston, EH54 6RJ
G4 HMM B. Dearing, 44 Woodlands Way, Southwater, Horsham, RH13 9HZ
GM4 HMN A. Cumming, 18 South Covesea Terrace, Lossiemouth, IV31 6NA
GW4 HMR D. Morris, Hafodty Cottage, Tregarth, Bangor, LL57 4NS
G4 HMS RNARS London (HMS Belfast) Group, c/o C. Read, 58 Somerset Road, Chiswick, London, W4 5DN
G4 HMX J. Halliday, 16 Ennerdale Drive, Congleton, CW12 4FR
G4 HND A. Course, 14A Wood Street, Geddington, Kettering, NN14 1BG
G4 HNF D. Waterworth, 116 Reading Road, Woodley, Reading, RG5 3AD
G4 HNG G. Poulton, Tresillian Morcombelake, Bridport, DT6 6dy
GM4 HNK C. Ferguson, Leckuary, Kilmichael Glassary, Lochgilphead, PA31 8QL
G4 HNO S. Wilson, 8 Hillcrest Avenue, Stockport, SK4 3JS
G4 HNQ J. Bryden, 32 Jerusalem Road, Skellingthorpe, Lincoln, LN6 5TW
G4 HNU P. Vaughan, 26 Canterbury Road, Worthing, BN13 1AE
G4 HNW S. Walls, 11 Copperfield Close, Malton, YO17 7YN
G4 HNX E. Beal, 49 Ambersham Crescent, East Preston, Littlehampton, BN16 1AJ
G4 HNZ S. Bannister, 14 Amery Close, Worcester, WR5 2HL
G4 HOC M. Oliver, 34 Manderley Close, Coventry, CV5 7NR
G4 HOD M. Gunby, 128 Heath Road, Runcorn, WA7 4XL
G4 HOF P. Warrener, 139 Louth Road, Holton Le Clay, Grimsby, DN36 5AD
G4 HOI W. Skeels, 141 Woodward Road, Dagenham, RM9 4ST
G4 HOJ P. Hobson, High Rising, 4 Dovecote Lane, Lincoln, N5 0AD
G4 HOK J. Mckay, 2 Bransghyll Terrace, Horton-In-Ribblesdale, Settle, BD24 0HG
G4 HOL M. Holden, Avda. Jardines Del Almanzora No 62, La Alfoquia De Zurgena, Zurgena, Spain, 4661
G4 HOM F. Garratt, 90 Brushfield Road, Birmingham, B42 2QJ
G4 HON C. Ward, 2 Arlington Drive, Stockport, SK2 7EB
G4 HOP S. Fordham, 61 Cemetery Road, Dronfield, S18 1XX
G4 HOR N. Freer, 2 Welham Croft, Shirley, Solihull, B90 4UU
G4 HOU L. Anstead, 21 Tickenor Drive, Finchampstead, Wokingham, RG40 4UD
G4 HOW N. Cleaver, 18 Old Cleeve, Minehead, TA24 6HJ
G4 HOY J. Fennell, Bajamar House, Belton Road, Doncaster, DN9 1JL
GD4 HOZ D. Osborn, Kionlough House, Kionlough Lane, Bride, Isle Of Man, IM7 4AG
G4 HPA Hertfordshire Peak Assault (Scouts), c/o M. Wood, 26 Parkfield Crescent Kimpton, Hitchin, SG4 8EQ
G4 HPB R. Wilden, 48A The Crescent, Cradley Heath, B64 7JS
G4 HPD B. Constable, Dukes Pleasure Long Headland, Ombersley, Droitwich, WR9 0DX
G4 HPE S. Richards, 6 Heathfield, Royston, SG8 5BW
G4 HPH J. Littler, 363 Atherton Road, Hindley, Wigan, WN2 3XD
GM4 HPK D. Moore, Rashfield Farm By Kilmun, Dunoon, PA23 8QT
G4 HPN R. Baker, 8 The Cloisters Belmore Lane, Lymington, SO41 3QX
G4 HPS P. Barker, 11 Dipton Gardens, Sunderland, SR3 1AN
G4 HPT D. Oliver, Ashdell, Newlands Lane, Birmingham, B37 7EE
G4 HPX J. Trotter, 29 Broad Park Road, Bere Alston, Yelverton, PL20 7AH
G4 HPY R. Spragg, 3 Truro Gardens, Luton, LU3 2AP
G4 HQB P. Sandell, 1 St. Margaret Road, Ludlow, SY8 1XN
G4 HQC C. Wilcox, 42 Kentmere Close, Cheltenham, GL51 3PD
G4 HQD R. Bagley, 8 Bishop Ruzar Furrugia Street, Xaghra, Xra, Malta, 103
GM4 HQF D. Lindsay, 39 Seamount Court, Aberdeen, AB25 1DQ

G4 HQH S. Parker, 20 Swaddale Avenue, Chesterfield, S41 0SU
G4 HQM D. Waspe, 28 Wilman Way, Salisbury, SP2 8QS
GM4 HQU N. Gent, 4 Eskview Villas, Eskbank, Dalkeith, EH22 3BN
G4 HQX P. Morys, 41 Salter Street, Berkeley, GL13 9BU
GM4 HQZ A. Morrison, Block 19, 2 Sandpiper Road., Edinburgh, EH6 4TR
G4 HRB D. Taylor, 8 Fambridge Close, Maldon, CM9 6DJ
G4 HRC Havering & District ARC, c/o D. Nuttall, 92 Long Road, Lowestoft, NR33 9DH
G4 HRE D. Hollow, 8 Vermont Woods, Finchampstead, Wokingham, RG40 4PF
G4 HRG R. Denley, 50 Cranmere Avenue, Wergs, Wolverhampton, WV68TS
G4 HRH A. Allen, The Hollies Sedgeford, Whitchurch, SY13 1EX
GM4 HRJ J. Mcniff, East Cove Cottage, Main Road, Port Glasgow, PA14 6XP
GM4 HRL A. Sergeant, 24 Academy Road, Bo'ness, EH51 9QD
G4 HRS Horsham ARC, c/o J. Matthews, 46 Park Lane West Grinstead, Horsham, RH13 8LT
G4 HRU R. Profitt, 10 Taunton Vale, Hunters Hill, Guisborough, TS14 7NB
G4 HRY D. Farn, 14 Corfe Close, Coventry, CV2 2JG
G4 HS S. Hopper, 16 Stanford Avenue, Hassocks, BN6 8JL
G4 HSB P. Rovardi, 8 Cambridge Road, Linthorpe, Middlesbrough, TS5 5NQ
G4 HSC H. Hughes, 16 Dalton Drive, Goose Green, Wigan, WN3 6TQ
G4 HSD R. Smithers, 16 Derby Road, Sutton, SM1 2BL
GW4 HSH R. Williams, 114 West Cross Lane, West Cross, Swansea, SA3 5NQ
G4 HSK S. Glass, 36 Pickwick Avenue, Chelmsford, CM1 4UN
G4 HSM R. Hurrell, 97 Dovercliffe Close Se, Calgary, Alberta, Canada, T2B 1W4
G4 HSN A. Chorley, Leycot, Cornells Lane, Saffron Walden, CB11 3SP
G4 HSS P. Forshaw, 54 The Park, Penketh, Warrington, WA5 2SG
GJ4 HSW F. Le Quesne, Brookhill House, Princes Tower Road, St Saviour, Jersey, JE2 7UD
G4 HSX F. Cole, 3 Wadsworth Avenue, Todmorden, OL14 7NF
G4 HSZ P. Thacker, 23 Lulworth Avenue, Leeds, LS15 8LW
G4 HTB T. Rance, 2 Glenavon Gardens, Slough, SL3 7HN
G4 HTD L. Mason, Forest Farm, Folly Drove, Stewley, Ashill, Ilminster, TA19 9NW
G4 HTE E. Sergeant, 13 Morven Close, Potters Bar, EN6 5HE
G4 HTG A. Brunton, 409 Outwood Common Road, Billericay, CM11 1ET
G4 HTH R. Herringshaw, 35 Oxley Close, Shepshed, Loughborough, LE12 9LS
G4 HTL A. Mcculloch, 14 Harbour Close, Blouberg Sands, Cape Town, South Africa, 7441
G4 HTO I. Myford, 33A Brick Kiln Lane, Mansfield, NG18 5LA
GM4 HTU A. Langton, 71 Gray Street, Aberdeen, AB10 6JD
G4 HTV ITV West Radio Club, c/o R. Thompson, Walnut House, Chestnut Walk Saltford, Bristol, BS31 3BG
G4 HTW P. Mcveigh, The Dale, Bowns Hill, Matlock, DE4 5DG
G4 HTX R. Houghton, Elmtrees, Church End, Bedford, MK44 2RP
G4 HTY D. Stokes, Flat 6, 35-37 Gratton Road, London, W14 0JX
G4 HTZ S. Barrett, Hamstede 1 The Street, Ashen, CO10 8JN
G4 HUA T. Ellam, 3115 Carleton Street Sw, Calgary, Canada, AB T2T3L5
G4 HUD J. Bramall, 55 Wood Lane, Louth, LN11 8RY
G4 HUE A. Nehan, Danisway, Queens Road, Colmworth, Colmworth, MK44 2LA
G4 HUF P. Baguley, 16 Churchill Road, Broadheath, Altrincham, WA14 5LT
G4 HUG W. Daniels, 48 Mellanear Road, Hayle, TR27 4QT
G4 HUH P. Chapman, 1291 Los Amigos Avenue, California, USA, 93065
GM4 HUL W. Savory, 20 Broomfield, Carradale East, Campbeltown, PA28 6RZ
G4 HUM D. Hazzard, 34 Chessel Avenue, Bitterne, Southampton, SO19 4DX
G4 HUN N. Whiteside, 17 Shibleys Court, Fishers Lane, Norwich, NR2 1EE
G4 HUO M. Bennett, 9 Lavender Avenue, Blythe Bridge, Stoke-on-Trent, ST11 9RN
G4 HUQ M. Crake, 12 Bosburn Drive, Mellor Brook, Blackburn, BB2 7PA
G4 HUT D. Consitt, Saxtorpsvagen 210, Landskrona, Sweden, 26194
G4 HUW S. Faulkner, Vaarveien 8, Oslo, Norway, 1182
GM4 HUX R. Lindsay, 32A James Street, Alva, FK12 5AL
GU4 HUY R. Sarre, Le Clercs, Clos Du Murier, Rue De Bas, Guernsey, GY2 4HJ
G4 HVC A. Kiddle, 19 Old Lincoln Road, Caythorpe, Caythorpe, Grantham, NG32 3DF
G4 HVF C. Bracewell, Roseville Yoredale Avenue, Leyburn, DL8 5BH
G4 HVG J. Phipps, 5 Akeman Close, St Albans, AL3 4NJ
GI4 HVI A. Hamilton, 11 Norwell Park, Castlerock, Coleraine, BT51 4TS
GM4 HVM A. Douglas, 24 Plane Grove, Dunfermline, KY11 8RA
G4 HVO J. Fitzwater, The Olde Cottage, Babylon Lane, Tadworth, KT20 6XE
G4 HVR G. Southwell, 4A Neve Avenue, Wolverhampton, WV10 9BU
GM4 HVS Dr R. Teperek, 8 Forest Park, Stonehaven, AB39 2GF
G4 HVT N. Wilkinson, Breidablikkbakken 15, Porsgrunn, Norway, 3911
G4 HVV Haven Valley Contest Club, c/o C. Goadby, Heligan, 12 School Road, Newmarket, CB8 9RX
G4 HVW F. Moody, 87 Whitegate Walk, Rotherham, S61 4LP
G4 HWA B. Morton, Yew Tree House, 14, Baker Street, Gayton, NN7 3EZ
G4 HWC E. King, 2 Thornton Road, March, PE15 8SH
G4 HWF R. Rudd, 69 Stanford Avenue, Brighton, BN1 6FB
G4 HWH A. Jandrell, 21 Wildacres, Stourbridge, DY8 3PH
G4 HWI M. Allin, 50 Swallow Rise, Knaphill, Woking, GU21 2LH
G4 HWJ M. Dawson, Mulberry Cottage, The Hamlet, Ely, CB6 1SB
G4 HWK F. Pilling, 51 Lynton Road, Chingford, E4 9EA
G4 HWM Dr D. Jeffery, 14 Beechwood Crescent, Chandler'S Ford, Eastleigh, SO53 5PA
G4 HWN R. Heath, Flat 172, Hagley Road Retirement Village, 330 Hagley Road, Birmingham, B17 8BN
GM4 HWO C. Wright, 3 Stanedykehead, Edinburgh, EH16 6YE
G4 HWV T. Wiles, Manor Farm, Manor Close, Middlesbrough, TS9 5AG
G4 HWW R. Scott, Flat 57 Tatton Cour, 35 Derby Rd, Stockport, SK4 4NL
G4 HXC D. Edwards, 179 Pallett Drive, Nuneaton, CV11 6JA
G4 HXE A. Tilbee, 9 Moorhead Court, Southampton, SO14 3GQ
G4 HXH R. Pope, 95 Northolt Avenue, Bishop's Stortford, CM23 5DS
G4 HXK F. Rendell, 64 Rivermead, Stalham, Norwich, NR12 9PJ
G4 HXL Dr L. Manderson, 16 Archery Avenue Foulridge, Colne, BB8 7NH
G4 HXN D. Kelly, 27 Keswick Road, Bookham, Leatherhead, KT23 4BQ
GW4 HXO M. Probert, 1 Ynys Dawel, Solva, Haverfordwest, SA62 6UF
G4 HXQ G. Burlington, Podgwell Cottage Seven Leaze Lane, Edge, GL6 6NJ
G4 HXU D. Mcdermott, 6 Chiltern Grove, Thame, OX9 3NH
G4 HXX Crossways Contest Group, c/o Dr C. Dollery, 101 Corringham Road, London, NW11 7DL
G4 HXY S. Simmons, 48 Copland Road, Stanford-le-Hope, SS17 0DF
G4 HYD Capt. A. Oakley, 2 Manor Close, Beverley, HU17 7BP
G4 HYG C. Moulding, 106 Barton Road Farnworth, Bolton, BL4 9PT
GI4 HYM C. Gill, 11 Quay Street, Ardglass, Downpatrick, BT30 7SA
GM4 HYR M. Bond, 1 Saughtonhall Crescent, Edinburgh, EH12 5RF
G4 HYT Dr P. Kurian, 22A Lindisfarne Avenue, Blackburn, BB2 3EH
G4 HYW A. Wilkes, Efford Park, Milford Road, Lymington, SO41 0JD
G4 HYY T. Jackson, 33 Highcroft Road, Todmorden, OL14 5LZ
GW4 HYZ B. Green, 28 Sunnybank Road, Griffithstown, Pontypool, NP4 5LT
G4 HZE E. Hill, 14 Station Road, Saltash, PL12 4DY
G4 HZF R. Scarlett, 1 St. Martins Crescent, Grimsby, DN33 1BG
G4 HZG M. White, 3 High Street, Clay Cross, Chesterfield, S45 9DX
GW4 HZH Dr D. Doherty, 35 Ffordd Bryngwyn, Garden Village, Gorseinon, Swansea, SA4 4EB
G4 HZI W. Backhouse, 191 Wigmore Road, Gillingham, ME8 0TL
G4 HZJ L. Jackson, 1 Belvedere Avenue, Atherton, Manchester, M46 9LQ
GW4 HZM J. Styles, 5 Heol-Y-Berth, Caerphilly, CF83 1SP
G4 HZN T. Lockwood, 8 St. Nicholas Road, Thorne, Doncaster, DN8 5BS
G4 HZP A. Charlton, The Crook, Rowelton, Carlisle, CA6 6LH
G4 HZR D. Saunders, 4 Furzedene, Furze Hill, Hove, BN3 1PP
G4 HZT T. Morton, 3 Grandstand Road, Hereford, HR4 9NE
G4 HZV R. Bagwell, 30 Christmas Pie Avenue Normandy, Guildford, GU3 2EN
G4 HZW A. Usher, 14 Bucklow Avenue, Mobberley, Knutsford, WA16 7ET
G4 HZX N. Squibb, 127 Copers Cope Road, Beckenham, BR3 1NY
G4 IAB A. Bell, 10 Long Acre, Weaverham, Northwich, CW8 3PT
G4 IAD D. Crompton, The Beeches, 6 St. Johns Wood, Bolton, BL6 4FA
G4 IAG T. Court, Woodview Breach Oak Lane, Coventry, CV7 8AU
G4 IAJ T. Jefferson, Garden Flat, 5 Esplanade, Scarborough, YO11 2AF
G4 IAO A. Robertson, 7 Big Back Lane, Chedgrave, Norwich, NR14 6BH
G4 IAQ J. Brooks, 28 Avon Vale Road, Loughborough, LE11 2AA
G4 IAR D. Brooks, 28 Avon Vale Road, Loughborough, LE11 2AA
G4 IAT B. Smith, 69 Birch Hall Avenue, Darwen, BB3 0JW
G4 IAU D. Lilley, 65 Peel St., Horbury, Wakefield, WF4 5AN
G4 IAY F. Whittaker, 91 Oakdale, Worsbrough, Barnsley, S70 5NR
G4 IBC Radio Amateur Invalid and Blind Club, c/o K. Marsh, Highgrove, Creech Heathfield, Taunton, TA3 5EW
G4 IBH D. Dockery, 20 Saffron Way, Sittingbourne, ME10 2EY
GM4 IBI Dr W. Mitchell, Brownhill Of Ardo, Methlick, Ellon, AB41 7HS
G4 IBM C. Murphy, 15 Loders Close, Poole, BH17 9BF
G4 IBN K. Pointon, 65 Gypsy Lane, Castleford, WF10 3PB
G4 IBS G. Baxendale, Sarno, Granville Road, Darwen, BB3 2SS
GI4 IBV S. Johnston, 61 Ravenhill Park, Belfast, BT6 0DG
G4 IBW R. Ropinski, 38 The Leys, Little Eaton, Derby, DE21 5AR
G4 ICB B. Clarke, 59 Baden Powell Crescent, Pontefract, WF8 3QD
G4 ICC M. Gater, 17 Douglas Road, Northampton, NN5 6XX
G4 ICE A. Mitchell, 11 Poplar Lane, Cannock, WS11 1NQ
G4 ICF A. Denison, 40 Leysholme Drive, Leeds, LS12 4HQ
G4 ICH C. Wickenden, Chalfont, Little Whelnetham, Bury St Edmunds, IP30 0DG
G4 ICI R. Perks, Drayton Lodge, Drayton Manor Drive, Tamworth, B78 3TJ
G4 ICM Icom (UK) AR, c/o D. Stockley, 2 The Ridings Chestfield, Whitstable, CT5 3PE
G4 ICP R. Witney, 36 Dapifer Drive, Braintree, CM7 3LG
G4 ICU A. Jones, 15 High Street, Sedgley, Dudley, DY3 1RL
G4 ICZ B. Greatrix, 12 Swainsfield Road, Yoxall, Burton-on-Trent, DE13 8PT
GW4 IDC M. Rudge, 8 Penrallt Estate, Llanystumdwy, Criccieth, LL52 0SR
G4 IDD D. Dockar, 49 Dixon Lane, Wortley, Leeds, LS12 4RR
G4 IDF D. Hobro, 60 Linksview Crescent, Worcester, WR5 1JJ
G4 IDG G. Tonge, 6 Bickford Close, Lapley, Stafford, ST19 9JZ
G4 IDH I. Harris, 47D Tower 2 Queens Terrace, 1 Queen Street, Sheung Wan, Hong Kong, 12345
G4 IDJ D. Macgregor, 29 Terrington Hill, Marlow, SL7 2RE
G4 IDL T. Wade, 47 Rig Drive, Swinton, Mexborough, S64 8UL
G4 IDR D. Redman, 13 Halifax Road, Golcar, Huddersfield, HD7 4NS
G4 IDS I. Stanley, 11 Reading Road, Burghfield Common, RG7 3PY
G4 IDT F. Heywood, 62 Southleigh Road, Leeds, LS11 5SG
G4 IDU K. Kniveton, 32 Minster Avenue, Bude, EX23 8RY
GW4 IDV P. Brown, 3 Lon Llewelyn, Abergele, LL22 7DG
G4 IDW A. Compton, Aysgarth, Durley Brook Road, Southampton, SO32 2AR
G4 IDX D. Turner, 36 Joyes Road, Folkestone, CT19 6NX
G4 IEB C. Williamson, 72 Granville Drive, Kingswinford, DY6 8LL
G4 IEC A. Everard, 2 Oak Wood Road, Wetherby, LS22 7QY
GM4 IEF A. Hancock, Pitlair House Nursing Home, Cupar, KY15 5RF
G4 IEG C. Shearer, 2 Perigrine Close, Basildon, SS16 5HX
G4 IEH S. Lindell, 60 Lakenheath, Oakwood, London, N14 4RP
G4 IES W. Pitt, 1 Windy Ridge, James Street, Stourbridge, DY7 6ED
G4 IET J. French, 10 Sunridge Avenue, Luton, LU2 7JL
GW4 IEU W. Griffiths, 3 Garreglwyd Park, Holyhead, LL65 1NW
G4 IEV P. Gill, 48 Meeting House Lane, Balsall Common, Coventry, CV7 7FX
GW4 IEZ R. Senior, 5 Cwm Arthur, Denbigh, LL16 4BD
GW4 IFE Dr A. Strachan, 1 Cornelius Close, South Cornelly, Bridgend, CF33 4RQ
G4 IFI C. Loftus, C/O, 15 Chappell Road, Manchester, M43 7UQ
G4 IFJ M. Daniels, 8 Hathersage Drive, Glossop, SK13 8RG
G4 IFM Dr S. Petraitis, 16 Brookbank Road, Dudley, DY3 2RX
G4 IFQ A. Webb, 15 Windsor Mead, Sidford, Sidmouth, EX10 9SJ
G4 IFR P. Hanson, 42 Oak Avenue, Newport, TF10 7EF
G4 IFT D. Howorth, 11A Norwood Drive, Torrisholme, Morecambe, LA4 6LT
G4 IFX C. Deacon, Spring Valley, Churt Road, Farnham, GU10 2QU
G4 IGC L. Hall, 57 Station Hill, Swannington, Coalville, LE67 8RJ
GW4 IGF P. Higgs, Oulton, Daisy Lane, Parkside, Rossett, Wrexham, LL12 0BP
G4 IGG N. Bennett, 1 Burnham Avenue, Oxley, Wolverhampton, WV10 6DX
G4 IGK M. Wickham, 8 Verlands Close Niton, Ventnor, PO38 2BG
G4 IGL R. Coombes, 9 Beechwood Close, Evington, Leicester, LE5 6SY
GM4 IGS R. Chapman, 65 Lochgreen Avenue, Troon, KA10 6UP
G4 IGT R. Roberts, 37 Admirals Place, Gibraltar, Gibraltar, GX111AA
G4 IGU K. Blackett, 46 Lansdown, Yate, Bristol, BS37 4LR
GD4 IHC R. Furness, Breryk, Windsor Road, Ramsey, Isle Of Man, IM8 3EB
G4 IHI P. Ferrari, Maggie, Back Road, Halesworth, IP19 9DY
GW4 IHM I. Wingfield, Keyhaven, 2 Belmont Close, Abergavenny, NP7 5HW
G4 IHO D. Carson, 21 Harris Road, Harpur Hill, Buxton, SK17 9JS

G4 IHR N. Allen, 8 Shoulbard, Fleckney, Leicester, LE8 8TX
G4 IHS G. Donn, Flat 31, Rex Cohen Court, Liverpool, L17 1AB
G4 IHT R. Riddington, Beech House, Tetbury, GL8 8SN
GI4 IHY R. Clarkson, 2 Massereene Gardens, Antrim, BT41 4JQ
G4 IHZ M. Hyde, 23 Northumberland Way, Ardsley, Barnsley, S71 5DH
G4 IIA M. Stamford, The Old Wheelwrights, East Street, Leominster, HR6 9HB
G4 IIB K. Marshall, Alderbaran, Ruckcroft, Carlisle, CA4 9QR
G4 IIC C. Clifford, 11 Halfcot Avenue, Stourbridge, DY9 0YB
G4 IID C. Eastland, 40 Hillside Road, Bushey, WD23 2HA
G4 IIH P. Henson, 70 Mell Road, Tollesbury, Maldon, CM9 8SR
G4 III P. Godwin, Holgate, Selby Road, Goole, DN14 0LN
G4 IIK C. Lodge, 35 Beaumont Cottages, Kelsale, Saxmundham, IP17 2NW
G4 IIN N. Evans, 56 Homerton Road, Middlesbrough, TS3 8LX
G4 IIO P. Howe, 59 Days Road, Samford Valley, Australia, 4520
GM4 IIR A. Nelson, 5 Scarletmuir, Lanark, ML11 7PS
G4 IIX C. Wherrett, 14 Sails Drive, York, YO10 3LR
G4 IIY I. Fugler, (Fugler), Lees Hill Farm, Lees Hill, Brampton, CA8 2BB
G4 IJA B. Barnes, 28 Oaklands Park, Roughton Moor, Woodhall Spa, LN10 6UU
G4 IJB R. Butterworth, 3 Derriman Glen, Sheffield, S11 9LQ
G4 IJD J. Seddon, 38 Kemple View, Clitheroe, BB7 2QD
GU4 IJF Dr N. Roberts, Maison Du Cotil, Alderney, Guernsey, GY9 3YZ
G4 IJI M. Walker, 19 Highbury Place, Headingley, Leeds, LS6 4HD
G4 IJJ A. Spratt, 8 Pheasant Rise, Copdock, Ipswich, IP8 3LF
G4 IJM I. Arnold, 44 Elwick Avenue, Acklam, Middlesbrough, TS5 8NT
G4 IJO G. Gaunt, 7 Marine Parade, Saltburn-by-the-Sea, TS12 1DP
G4 IJR B. Moyse, 1703 Twin Pond Circle, College Station, Texas, USA, 77845-3051
G4 IJU J. Coles, 84 Mansfield Lane, Calverton, Nottingham, NG14 6HL
G4 IJV B. Dowling, Box Cottage, Box, Stroud, GL6 9HB
GI4 IKF T. Black, 147 Old Westland Road, Belfast, BT14 6TE
G4 IKI P. Gabriel, 20 Barge Lane, Ryde, po334lb
G4 IKJ P. Edwards, 34 Albion Road, Malvern Link, Malvern, WR14 1PU
G4 IKL R. Hibbin, 2 Phoenix Close, West Wickham, BR4 0TA
G4 IKQ R. Kitchener, 43 Haven Close, Swanley, BR8 7JY
G4 IKX D. Thomas, 18A Stockwell Lane, Aylburton, Lydney, GL15 6DN
G4 IKY D. Sillars, 34 Sandown Road, Stevenage, SG1 5SF
G4 ILA Rev. W. Mckae, 3 Grantham Close, Wirral, CH61 8SU
GM4 ILE J. Smy, 2 Dungavel Gardens, Hamilton, ML3 7PE
G4 ILF A. Hyde, 68 Broxburn Road, Warminster, BA12 8EZ
G4 ILH J. Acott, 2 Park Hill Road, Sidcup, DA15 7NL
G4 ILI G. Cratchley, 2 The Maples, The Reddings, Cheltenham, GL51 6RW
G4 ILM Dr M. Turnbull, Southlea, Newbury, Gillingham, SP8 4QJ
G4 ILN G. Fitt, 15 Sidegate Avenue, Ipswich, IP4 4JJ
G4 ILP C. Borkowski, 25 Stroud Road, Wimbledon, London, SW19 8DQ
G4 ILQ R. Manton, 18 Barnetts Close, Kidderminster, DY10 3DG
G4 ILR C. Howett, 38 Hillyfields, Dunstable, LU6 3NS
GM4 ILS R. Adam, 1 Woodlands Crescent, Bishopmill, Elgin, IV30 4LY
G4 ILT G. Barnacle, 58 Cotley Road, Leicester, LE4 2LH
G4 ILW J. Dingwall, Flat 3, Baltic House, London, SW2 1NQ
G4 ILX S. Sliwinski, 9 Oakhill Road, Sheffield, S7 1SJ
GI4 ILZ W. Sharpe, 22 Tweskard Park, Belfast, BT4 2JZ
G4 IMB P. Gascoigne, 108 Blandford Avenue, Castle Bromwich, Birmingham, B36 9JD
GW4 IMC T. Waters, 34 Woodlands Park, Betws, Ammanford, SA18 2HF
G4 IMH V. Tatman, 271 London Road, Bedford, MK42 0PX
G4 IML M. Giles-Holmes, 177B Babbacombe Road, Torquay, TQ1 3SU
G4 IMM R. Steele, 23 Peacocks Close, Cavendish, Sudbury, CO10 8DA
G4 IMP A. Phillpott, Southways, Stombers Lane Hawkinge, Folkestone, CT18 7AP
G4 IMS J. Roe, 5 Lawford Lane, Writtle, Chelmsford, CM1 3EA
G4 IMU K. Holley, 18 Sandford Avenue, Loughton, IG10 2AJ
G4 IMV J. Mollart, 8 Harrison St., Newcastle, ST5 1NH
G4 INA P. Grice, 48 Repington Road, Tamworth, B77 4AA
G4 INB Dr B. Dupree, 3 Hillary Road, Cheltenham, GL53 9LB
G4 INF B. Walpole, Bridge Farm, Stony Lane, Exeter, EX5 1PP
G4 ING J. Hartley, 50 Waverley Road, Hyde, SK14 5AU
G4 INI J. Church, Belle Vue, Gas Lane, Torrington, EX38 7BE
G4 INU F. Haighton, 2028 Cheviot Court, Burlington, on, Canada, L7P 1W8
G4 INX A. Harada, 3 Bazzleways Close, Milborne Port, Sherborne, DT9 5FD
G4 IOA P. Hill, 135 Village Road, Cheltenham, GL51 0AE
GM4 IOB R. Smith, Hestivald, Downies Lane, Stromness, KW16 3EP
G4 IOD W. Marshall, 92 High Street, Ossett, WF5 9RQ
G4 IOE F. Stevenson, Raakollveien 20A, Rolvsoy, Norway, N-1663
G4 IOG J. Blackett, 70 Church Lane, Newington, Sittingbourne, ME9 7JU
G4 IOJ M. Fielding, 35 Windmill Grove, Fareham, PO16 9HP
G4 IOK C. Marshall, 100 Hailey Road, Witney, OX28 1HQ
GD4 IOM Isle of Man ARS, c/o M. Webb, Coastguard House, 1 Mount Morrison, Peel, Isle Of Man, IM5 1PN
G4 ION Ionspheric P Gr, c/o Dr E. Warrington, Dept Of Engineering, University Of Leicester, Leicester, LE1 7RH
GI4 IOO R. Chambers, 32 Victoria Road, Sydenham, Belfast, BT4 1QU
GW4 IOQ A. White, Wddyn Cottage, Treflach, Oswestry, SY10 9HQ
G4 IOV P. Emmerton, 5 Portsmouth Wood Close, Lindfield, Haywards Heath, RH16 2DQ
GM4 IPA International Police Association, c/o J. Bertram, 20 Kyles View, Largs, KA30 9ET
G4 IPB P. Hodgkinson, Woodedge, Snaisgill Road, Middleton-In-Teesdale, Barnard Castle, DL12 0RP
G4 IPE R. Wilson, Brookside, Stather Road, Burton-Upon-Stather, Scunthorpe, DN15 9DH
G4 IPF L. Horseman, 55 Sackville Avenue, Hayes, Bromley, BR2 7JS
G4 IPH R. Bass, 292 Thornhills Lane, Clifton, Brighouse, HD6 4JQ
G4 IPI D. Foster, 1 Thorn Court, Four Marks, Alton, GU34 5BY
GM4 IPK A. Steven, Pangdene, Virkie, Shetland, ZE3 9JS
G4 IPL L. Winters, 58 Larkhall Lane, Harpole, Northampton, NN7 4DP
G4 IPM N. Terry, 15 Baldwins Close, Bourn, Cambridge, CB23 2TH
G4 IPN W. Flindall, 3 Meadow Drive, Gressenhall, Dereham, NR20 4LR
G4 IPR T. Jones, 130 Turkey Street, Enfield, EN1 4PS
G4 IPV G. Mayne, 228 Tutbury Road, Burton-on-Trent, DE13 0NY
G4 IPY Dr A. White, 3 Guarlford Road, Malvern, WR14 3QW
GW4 IQA R. Lloyd, Llwyn Celyn, Pandy, Gwent, NP7 8DN
GW4 IQB D. Fuller, 9 Llwyn Onn, Croesyceiliog, Cwmbran, NP44 2AL
G4 IQD N. Sivapragasam, 1 Treve Avenue, Harrow, HA1 4AL
G4 IQF S. Wilkinson, 41, Church Road, Nailstone, CV13 0QH
G4 IQJ P. Brannon, 90 Jacksmere Lane, Scarisbrick, Ormskirk, L40 9RS
G4 IQK G. Evans, 14 Beach Priory Gardens, Southport, PR8 1RT
G4 IQO C. Britton, 271 Havant Road Farlington, Portsmouth, PO6 1DB
G4 IQQ R. Phillips, 2 The Close, Dartford, DA2 7ES
G4 IQR N. Troop, 8 Fox Green, Great Bradley, Newmarket, CB8 9NR
G4 IQV G. Menzies, 40 Epsom Lane North, Epsom, KT18 5PY
G4 IQW A. Langford, 42 Amis Way, Stratford-upon-Avon, CV37 7JF
G4 IQZ J. Long, 51 Bratton Road, Westbury, BA13 3ES
G4 IRB J. Heath, 19 Anson Road Swinton, Manchester, M27 5GZ
G4 IRC Ipswich Rad Club, c/o J. Gee, Windmill Lodge, Mill Lane Witnesham, Ipswich, IP6 9HR
G4 IRD R. Richards, 11 Purvis Road, Rushden, NN10 9QA
G4 IRG E. Turner, 9 Wallingford Road, Handforth, Wilmslow, SK9 3JT
G4 IRH T. Pendleton, 17A Langley Drive, Kegworth, Derby, DE74 2DN
G4 IRP F. Boocock, 109 Northumberland Road, Harrow, HA2 7RB
G4 IRS R. Ball, 1 Mount Hindrance Close, Chard, TA20 1DZ
G4 IRU N. Ashcroft, Oaklands, 11 Greenway, Wilmslow, SK9 1LU
G4 IRV J. Hastie, 13 Thornlands, Easingwold, York, YO61 3QQ
G4 IRY R. Gladden, 145A Hampton Road, South Fremantle, Australia, WA 6162
GI4 ISH M. Fearis, 205 Dunluce Avenue, Belfast, BT9 7AX
GW4 ISJ P. Martin, 42 Leckwith Avenue, Cardiff, CF11 8HQ
G4 ISK D. Brighton, 39 Les Forets, Glenac, France, 56200
GM4 ISM M. Hughes, 6 Hawthorn Gardens, Larkhall, ML9 2TD
G4 ISN A. Holmes, 5 Launde Park, Market Harborough, LE16 8BH
G4 ISQ B. Jones, 7 Timbertree Road, Warley, Cradley Heath, B64 7LE
GI4 ISR C. Mcclurg, 4 Gracefield Lodge, Dollingstown, Craigavon, BT66 7UA
G4 ISS J. Proudfoot, Laburnum Cottage, Corby Hill, Carlisle, CA4 8PL
G4 ISU N. Whittingham, The Lilacs, 4 Ridgedale Mount, Pontefract, WF8 1SB
G4 ITB J. Stone, 35 Landseer Avenue, Chapel St. Leonards, Skegness, PE24 5QZ
G4 ITC C. Claydon, 69 Abingdon Road, Dorchester-On-Thames, Wallingford, OX10 7LB
G4 ITG B. Davey, 31 Somervell Drive, Fareham, PO16 7QL
GW4 ITJ C. Hard, 3 Longbridge, Ponthir, Newport, NP18 1GT
G4 ITP C. Owen, 334 Beaumont Leys Lane, Leicester, LE4 2BJ
GW4 ITQ B. Lindley, 4 Stryd Y Brython, Ruthin, LL15 1JA
G4 ITR K. Fisher, 51 Edge Hill, Ponteland, Newcastle upon Tyne, NE20 9RR
G4 ITV B. Dingle, 74 Fenay Lane, Almondbury, Huddersfield, HD5 8UJ
G4 ITX M. Payne, 34 Thales Drive, Arnold, Nottingham, NG5 7NF
G4 ITY D. Hardie, 42 Lagoon Road, Pagham, Bognor Regis, PO21 4TJ
G4 IUA J. Campbell, 61 Telegraph Lane, Claygate, Esher, KT10 0DT
G4 IUF M. Parker, 23 Pannal Avenue Pannal, Harrogate, HG3 1JR
G4 IUH R. Pye, 7 Meadow View, Potterspury, Towcester, NN12 7PH
G4 IUJ J. Wroe, 25 Yew Tree Lane, Poynton, Stockport, SK12 1PU
GW4 IUK H. Morley, 63 Lewis Road, Neath, SA11 1DJ
GW4 IUL D. Pullin, 32 Clinton Road, Penarth, CF64 3JD
G4 IUM G. Adams-Spink, 55 Hawthorn Drive, Harrow, HA2 7NU
GW4 IUN R. Janes, 3 Greenway Avenue, Rumney, Cardiff, CF3 3HQ
G4 IUP R. Limbert, 21 Staincliffe Drive, Keighley, BD22 6FF
GM4 IUS N. Bethune, 9 Links Gardens, Leith, Edinburgh, EH6 7JH
G4 IVC F. Wood, 20A Lynwood Avenue, Felixstowe, IP11 9HS
G4 IVD Rev. A. James, 6 Dovedale Close, Hardwicke, Gloucester, GL2 4JH
GI4 IVI A. Kerr, 29 The Rose Garden, Tandragee, Craigavon, BT62 2NJ
G4 IVL T. King, Flat 1, 159 Cheriton Road, Folkestone, CT19 5HG
G4 IVO R. Hargreaves, 23 Bracken Road, Long Eaton, Nottingham, NG10 4DA
G4 IVU A. Dixon, 98 Seaview Chalet Park, Green Lane, Kessingland, Lowestoft, NR33 7RG
G4 IVZ G. Harper, 12 Bletchley Road, Stewkley, Leighton Buzzard, LU7 0ER
G4 IWA J. Arrowsmith, 16 Mancetter Road, Mancetter, Atherstone, CV9 1NZ
G4 IWD G. Craig, 83 Pearl Road, Walthamstow, London, E17 4QY
G4 IWF G. Mason, 51 Egerton Road, Streetly, Sutton Coldfield, B74 3PG
G4 IWI J. Stocking, Bildersbrook, Grove Road, Melton Constable, NR24 2DE
G4 IWJ Capt. R. Towle, Rye Hill Humshaugh, Hexham, NE46 4BN
G4 IWN J. Andrews, 5 Chapman Avenue, Maidstone, ME15 8EG
G4 IWO N. Bradley, 6 The Old School House, Shore, Littleborough, OL15 8EZ
GI4 IWP E. Maclaine, 105 Bencran Road, Sixmilecross, Omagh, BT79 9QA
G4 IWQ D. Cannon, 57 Halswell Road, Clevedon, BS21 6LE
G4 IWR S. Berry, 40 Warrendale, Barton-upon-Humber, DN18 5NH
G4 IWS C. Caine, 10 Goodwood Close, Burghfield Common, Reading, RG7 3EZ
G4 IWU J. Scrivens, 130 Sea Lane, Rustington, Littlehampton, BN16 2RZ
G4 IWV I. Parker, 43 Longdown Road, Congleton, CW12 4QH
G4 IXB C. Tuvey, 1 Dorset Way, Heston, Hounslow, TW5 0NF
G4 IXD I. Palgrave Brown, The Abbey House, The Street, King's Lynn, PE33 9HP
G4 IXE G. Walmsley, Warwick Farm House, Cracknore Hard Lane, Southampton, SO40 4UT
G4 IXF D. Toon, 26 Reddish Avenue, Whaley Bridge, High Peak, SK23 7DP
GM4 IXH Dr J. Finlayson, 7 Abbotshall Road, Cults, Aberdeen, AB15 9JX
G4 IXQ A. Constable, Oakside, The Street, Bury St Edmunds, IP31 1NG
G4 IXT I. Jefferson, 7 Bluebell Close, Rugby, CV23 0UH
G4 IXY P. Beardsmore, 2 Spencer Place, Sandridge, St Albans, AL4 9DW
G4 IYA M. Adams, 8 Boltons Close, Brackley, NN13 6ND
G4 IYC B. Couchman, 48 Eastfields, Blewbury, Didcot, OX11 9NS
G4 IYE R. Smith, 72 Worthing Road, Patchway, Bristol, BS34 5HX
G4 IYK S. Dixon, 33 Medhurst Crescent, Gravesend, DA12 4HJ
GI4 IYO K. Burnside, 4 Cuttles Road, Comber, Newtownards, BT23 5YX
G4 IYP F. Dearden, 22 Claremont Road, Chorley, PR7 3NH
G4 IYS D. Burgess, The Beeches, Half Moon Lane, Redgrave, Diss, IP22 1RU
GM4 IYZ J. Potts, Eastwood Court, 1 Eastwoodmains Road, Glasgow, G46 6QB
G4 IZA D. Howard, 1700 3Rd Avenue West - Apt: 507, Bradenton, USA, 34205
GI4 IZF M. Weller, 58 Manse Road, Ballycarry, Carrickfergus, BT38 9LF
G4 IZH P. Robinson, 24 Haveroid Way, Crigglestone, Wakefield, WF43PG
GW4 IZJ P. Rennick, 41 Church Road, Pontnewydd, Cwmbran, NP44 1AT
GD4 IZL G. Brookes, 44 Magherchirrym, Port Erin, Isle of Man, IM9 6DB
G4 IZQ A. Scarth, 1 Beechwood Avenue, Whitley Bay, NE25 8EP
G4 IZS R. Sexton, 31 Fosters Lane, Woodley, Reading, RG5 4HH

G4 IZU D. Byers, 16 Tealby Court, Georges Road, London, N7 8HY
G4 IZX P. Beards, 3 Elm Drive Brightlingsea, Colchester, CO7 0LA
G4 IZZ M. Eggleton, 49 Gretton Road, Gotherington, Cheltenham, GL52 9QU
G4 JA P. Stenning, 20 Galba Road, Caistor, Market Rasen, LN7 6GN
G4 JAA P. Hawkins, 38 Davidson Close, Great Cornard, Sudbury, CO10 0YU
G4 JAC R. Emeny, 28 Manor House Way, Brightlingsea, Colchester, CO7 0QR
G4 JAJ B. Noble, 19 Ayrton Avenue, Blackpool, FY4 2BW
G4 JAQ M. Crofts, 43 Broadlands Drive, East Ayton, Scarborough, YO13 9ET
G4 JAR Hadrabs Cont Gr, c/o I. Melville, 3 Crescent Road, Benfleet, SS7 1JL
G4 JAV W. Bird, 42 Beaumont Way Norton Canes, Cannock, WS11 9FQ
G4 JAX A. Lunn, 11 Dibden Lodge Close, Hythe, Southampton, SO45 6AY
G4 JBA J. Alderman, 38 Greenacres, Shoreham-by-Sea, BN43 5WY
G4 JBD G. Laming, 72 Fildyke Road, Meppershall, Shefford, SG17 5LU
G4 JBE D. Lacey, 78 New Road Sutton Bridge, Spalding, PE12 9RQ
G4 JBF Dr G. Lester, Lufflands, Yettington, Budleigh Salterton, EX9 7BP
G4 JBH A. Dening, 42 Grove Avenue, Yeovil, BA20 2BD
G4 JBK A. Maude, 5 Darrowby Close, Thirsk, YO7 1FJ
G4 JBL C. White, Pegasus, Gotts Corner, Sturminster Newton, DT10 1DD
GW4 JBQ J. Cleak, Dantre, Newport Road, Cwmbran, NP44 3AE
G4 JBR P. Dixon, Hardwick House, New Road, South Molton, EX36 4BH
G4 JBW D. Barber, 3 Vestry Road, Street, BA16 0HY
G4 JBY G. Bowden, 78 Lynwood Avenue, Darwen, BB3 0HZ
G4 JCA C. How, 9 Chanctonbury Walk, Storrington, Pulborough, RH20 4LT
G4 JCF G. Hoey, Foehrer Strasse 8, Muenster, Germany, 64839
G4 JCG P. Chapman, 4 Chester Close, Garstang, Preston, PR3 1LH
G4 JCH B. Hercombe, 13 Dovecote, Shepshed, Loughborough, LE12 9RW
G4 JCJ C. Newman, 4 Winchilsea Drive, Gretton, Corby, NN17 3BT
GW4 JCK N. Warnock, 2 Sheepcourt Cottages, Bonvilston, Cardiff, CF5 6TN
G4 JCL D. Bryan, 3 New Lane, Skelmanthorpe, Huddersfield, HD8 9EH
GM4 JCM A. Glashan, 35A Lochinver Crescent, Gourdie, Dundee, DD2 4UA
G4 JCS J. Stevenson, Highfields Farm, Saltburn by the Sea, TS13 4UG
G4 JCX C. Gallacher, 1 Brislands Lane Four Marks, Alton, GU34 5AD
G4 JCY R. Thornton, Binesfield, Bines Green, Horsham, RH13 8EH
G4 JCZ A. Clifton, 87 Aubrey Road, Quinton, Birmingham, B32 2BA
G4 JDC L. Boddington, Flat 33, Sorrel House, Birmingham, B24 0TQ
GW4 JDE G. Evans, 32 Radstock Court, Abergavenny, NP7 5BQ
G4 JDF Dr P. Scovell, 69 Nursery Road, Maidenhead, SL6 0JR
G4 JDG C. Aitchison, Upper Weston House, Cot Lane, Chichester, PO18 8SU
G4 JDH R. Purbrick, 2 Oyster Cottages, Tinnocks Lane, Southminster, CM0 7NF
GM4 JDK M. Hopkinson, The Coaches, Kingussie, PH21 1NY
G4 JDO R. Tew, 4 Chetwode Close, Allesley, Coventry, CV5 9NA
G4 JDP S. Pallett, 6 Lancaster Close, Coalville, LE67 4TG
G4 JDS L. Radley, 34 Queens Road, Chelmsford, CM2 6HA
G4 JDT H. Lexton, 11 Mulberry Close, Romford, RM2 6DX
G4 JDW L. Nelson-Jones, 15 Gainsborough Road, Bournemouth, BH7 7BD
GW4 JDZ D. Samuel, 87 Llewellyn Park Drive, Morriston, Swansea, SA6 8PF
G4 JED K. Bird, 25 Knowsley Way, Hildenborough, Tonbridge, TN11 9LG
G4 JEF D. Wood, Little Burgate Farm, Markwick Lane, Godalming, GU8 4BD
G4 JEI N. Osborne, 33 Dankton Gardens, Sompting, Lancing, BN15 0DX
GM4 JEJ M. Thomson, Ravenside, Mill Road, Carnoustie, DD7 7SQ
GM4 JEM W. Redpath, 69 Ulster Crescent, Edinburgh, EH8 7JL
G4 JEO F. Kemp, 42 Baker Road, Abingdon, OX14 5LW
G4 JES M. Wells, 16 Moorside Court, North Hykeham, LN69XA
G4 JEY R. Furness, Mermaid Lodge, 68/70 Brighton Lodge, West Sussex, BN15 8LW
G4 JFC G. Hainsworth, The Annexe, 16 Rowlandson Close, Northampton, NN3 3PB
G4 JFD D. Featherstone, 6 Claremont Gardens, Tunbridge Wells, TN2 5DD
G4 JFF C. Webb, 68 Higgs Field Crescent, Warley, Cradley Heath, B64 6RB
G4 JFG J. May, Midsummers Eve, Third Cliff Walk, Bridport, DT6 4HX
GM4 JFH S. Draycott, Kinmount, Whiting Bay, Isle of Arran, KA27 8QH
G4 JFN R. Hudson, 15 Fellows Road, Farnborough, GU14 6NU
GI4 JFP D. Goodman, 60 Castlewood Avenue, Coleraine, BT52 1EW
G4 JFS J. Fitzsimons, 27 Brese Avenue, Warwick, CV34 5TS
G4 JFV R. Oldroyd, Hambledon, 197 Inner Promenade, Lytham St Annes, FY8 1DW
G4 JFX B. Mount, 4 Maplestone Road, Whitchurch, Bristol, BS14 0HH
G4 JGF J. Fitzgerald, 21 St. Aidans Avenue, Darwen, BB3 2BS
G4 JGG J. Pether, 7 Celina Close, Bletchley, Milton Keynes, MK2 3LS
G4 JGH A. Allchin, 9 Ashfield Road, Kings Heath, Birmingham, B14 7AS
G4 JGQ J. Bevan, 10 Streamdale, Abbey Wood, London, SE2 0PD
G4 JGS S. Harding, 9 Lightsfield, Oakley, Basingstoke, RG23 7BL
GW4 JGU A. Green, 9 Westbourne Grove, Sketty, Swansea, SA2 9DT
G4 JGV S. Sharred, Calle Pena De San Roque 23, Tinajo 35560, Lanzarote, Las Palmas, Birmingham, Spain, 35560
GW4 JGW K. Simpson, 59 Midland Place, Llansamlet, Swansea, SA7 9QX
G4 JGX R. Calver, Meadowbank, Lowertown, Helston, TR13 0BY
G4 JHA R. Thomas, 2 Woodlands Road, Astley, Manchester, M29 7BH
G4 JHE M. Green, 1 Morley Hill, Stanford-le-Hope, SS17 8HP
GU4 JHH R. Harvey, Courtil Masse, Les Landes, Vale, Guernsey, GY3 5JD
G4 JHI D. Miller, 10 Fair View, Horsham, RH12 2PY
G4 JHN J. Unwin, 28 Wallett Avenue, Beeston, Nottingham, NG9 2QR
G4 JHP J. Hawes, 13 Broadmead Road, Colchester, CO4 3HB
G4 JHQ Dr C. Kear, 60 Haywoods Lane, Somerset, Tasmania, Australia, 7322
G4 JHS P. Hey, 47 Hillcrest Road, Thornton, Bradford, BD13 3PQ
G4 JHU N. Fineman, Deansway, 2 The Drive, Rickmansworth, WD3 4EB
G4 JHV S. Marsh, Ravensknowle, Culgaith, Penrith, CA10 1QF
G4 JHW D. Morrison, Flat 1, 118 Anerley Park, London, SE20 8NU
GI4 JIC P. Mcauley, 68 Ballylenaghan Heights, Belfast, BT8 6WL
G4 JIG E. White, 12A Partridge Close, Great Oakley, Harwich, CO12 5DH
G4 JIH K. Adams, 12 Hawkewood Avenue, Waterlooville, PO7 6EB
G4 JII R. Green, Kingswood, Red House Lane, Doncaster, DN6 7EA
G4 JIJ I. Kraven, 55 Cranfield Crescent, Cuffley, Potters Bar, EN6 4DZ
G4 JIK D. Bird, 6 Wyebank, Bakewell, DE45 1BH
G4 JIO K. Mason, 5 Davenport Avenue, Hessle, HU13 0RL
G4 JIQ W. Barker, 69 Britten Road, Brighton Hill, Basingstoke, RG22 4HN
G4 JIR A. Rixon, 12 Vancouver St., Darlington, DL3 6HN
G4 JIU I. Mcgarrigle, 58 Langland Close, Corringham, Essex, SS17 7LB
G4 JIV C. Davies, 6 Valerie Avenue, Baulkham Hills, Australia, 2153
GI4 JIW J. Ferrin, 38 Dalewood, Newtownabbey, BT36 5WR
G4 JIX J. Bentley, 33 Lime Road, Ferryhill, DL17 8DL
GI4 JJF K. McIlroy, 69 Morston Park, Bangor, BT20 3ER
G4 JJH J. Herbert, 8 Falmouth Road, Springfield, Chelmsford, CM1 6HY
G4 JJM M. Allison, 19 Ash Grove, Kirklevington, Yarm, TS15 9NQ
G4 JJP R. Thomas, 28 Clarks Meadow, Shepton Mallet, BA4 4FD
G4 JJQ J. Wheway, 25 Mount View Avenue, Scarborough, YO12 4EW
GW4 JJR L. James, 65 Fflorens Road, Newbridge, Newport, NP11 3DW
G4 JJS S. Harrison, Seacroft Grange Care Village, The Green, Leeds, LS14 6JL
GW4 JJV M. Bell, 6 Owain Close, Cyncoed, Cardiff, CF23 6HN
GW4 JJW A. Bell, 6 Owain Close, Cyncoed, Cardiff, CF23 6HN
G4 JJX M. Grange, 6 Draysfield, Wormshill, Sittingbourne, ME9 0TY
G4 JJY J. Carline, 101 Cemetery Road, Scunthorpe, DN16 1EB
G4 JKA J. Ewen-Smith, 1 Kinnersley, Severn Stoke, Worcester, WR8 9JR
GM4 JKB J. Barnes, Capricorn, 13 Marchhill Drive, Dumfries, DG1 1PP
G4 JKC P. Howard, 72 Marlowe Way, Lexden, Colchester, CO3 4JP
G4 JKE D. King, Flat 21, Anchor Court, 2 Carey Place, London, SW1V 2RT
G4 JKF B. Hodges, Gramaur, Mucklestone Wood Lane, Market Drayton, TF9 4ED
G4 JKH J. Phillips, 57 New Sturton Lane, Garforth, Leeds, LS25 2NW
GW4 JKK A. Bexley, Pennar Fach Farm Plwmp, Llandysul, SA44 6ES
G4 JKM D. Twigg, 31 Parklands, Malmesbury, SN16 0QH
G4 JKQ T. Bowen, 40 Grange Road, Ibstock, LE67 6LF
GW4 JKR D. Wilson, 94 Lon Hedydd, Llanfairpwllgwyngyll, LL61 5JY
G4 JKS M. Claytonsmith, Hares Cottage, Woolston, Church Stretton, SY6 6QD
GM4 JKT Dr O. Thores, 5 Havens Edge, Limekilns, Dunfermline, KY11 3LJ
GW4 JKV M. Rackham, 31 Severn Road, Pontllanfraith, Blackwood, NP12 2GA
G4 JKY E. Lennox, Lowfield House, Low Street, York, YO61 4QA
GM4 JKZ K. Leggett, 3 The Steadings Swinside Townhead Farm, Jedburgh, TD8 6ND
GM4 JLD P. Woods, 12 Dalriada Place, Kilmichael Glassary, Lochgilphead, PA31 8QA
GI4 JLF R. Russell, 1 Belmont Drive, Belfast, BT4 2BL
G4 JLG Dr D. Yorke, 40 Edge Fold Road, Worsley, Manchester, M28 7QF
G4 JLJ L. Bailey, 3 Eden Close, Hutton Rudby, Yarm, TS15 0HT
G4 JLO H. Dyson, 15 Swallow Grove, Netherton, Huddersfield, HD4 7SR
G4 JLV J. Brower, 37 High Street, Steventon, Abingdon, OX13 6RZ
G4 JLX H. Braggs, 47 Manor Road, Sandown, PO36 9JA
GM4 JLZ E. Philip, 12 Pinefield, Inchmarlo., Banchory, AB31 4AF
G4 JMC J. Trickett, 86 School Road, Thurcroft, Rotherham, S66 9DL
G4 JMF D. Ollerhead, 15 Kingsley Road, Chester, CH3 5RR
G4 JMG J. Gorton, 12 Apsley Close, Harrow, HA2 6AP
G4 JMM J. Mcfadyen, Flat 28, Dunstable Court, 12 St. Johns Park, London, SE3 7TN
G4 JMO A. Oakley, The Laithe, Coal Pit Lane, Colne, BB8 8NR
G4 JMT M. Firth, 6 Eastfield Drive, Woodlesford, Leeds, LS26 8SQ
GM4 JMU K. Maxted, 33 Woodpecker Grove Symington, Kilmarnock, KA1 5SF
G4 JMY D. Liversidge, 6 Yardley Way, Grimsby, DN34 5UQ
GM4 JNB N. Baird, 23 Scorguie Avenue, Inverness, IV3 8SD
G4 JNE Dr C. Houghton, 22 Rainow Road, Macclesfield, SK10 2PF
G4 JNH R. Barker, 171 Leicester Road, New Packington, Ashby-de-la-Zouch, LE65 1TR
G4 JNK N. Kendall, 13 Oaks Drive, Cannock, WS11 1ET
G4 JNL P. Senior, 9 Seely Close, Heighington, Lincoln, LN4 1TT
G4 JNQ E. Allison, 7 Abbey Road, Flitcham, King's Lynn, PE31 6BT
GI4 JNS Dr D. Hughes, 53 Cranley Grove, Bangor, BT19 7EY
G4 JNT A. Talbot, 15 Noble Road, Hedge End, Southampton, SO30 0PH
G4 JNX N. Whyborn, Kimberlin, Southwood Road, Norwich, NR13 3AB
G4 JNZ C. Barron, The Marling Pitts, Coughton, Ross-on-Wye, HR9 5ST
G4 JOA K. Wood, Flat 98, Harbour Tower, Gosport, PO12 1HE
G4 JOB J. Barker, 6 Larkswood Close, Tilehurst, Reading, RG31 6NP
G4 JOD F. Rawlings, 14 Haddon Way, Carlyon Bay, St Austell, PL25 3QG
GW4 JOG P. Truberg, 106 Johnston Road, Llanishen, Cardiff, CF14 5HJ
G4 JOI R. Tidnam, 21 Manor Lane, Lewisham, London, SE13 5QW
G4 JOO C. Harman, 46 Chandos Crescent, Edgware, HA8 6HL
GI4 JOR J. Farrell, 36 Cumber Park, Drumaness, Ballynahinch, BT24 8GA
GW4 JOT S. Carfoot, 24 Marble Church Grove, Bodelwyddan, Rhyl, LL18 5UP
G4 JOU R. Bowden, Flat 7, 39 Anstey Road, Alton, GU34 2RD
G4 JOV J. Wedderburn, 12 Victoria Avenue, Market Harborough, LE16 7BQ
G4 JOW J. Butler, Pickstock Manor, Pickstock, Newport, TF10 8AH
G4 JPA R. Jarvis, 2135 Oak Beach Blvd, 2135 Oak Beach Blvd, Sebring FL, USA, 33875
G4 JPB Canon J. Beaumont, 9 Warren Bridge, Oundle, Peterborough, PE8 4DQ
GW4 JPC G. Woods, 178 Saron Road, Saron, Ammanford, SA18 3LN
G4 JPE B. Hatley, 9 Somerstown Court, Tilehurst Road, Reading, RG1 7TY
GM4 JPG I. Wilson, 11 Ellwyn Terrace, Galashiels, TD1 2BA
GW4 JPJ H. Genon, Dolau Cwerchyr, Penrhiwllan, Llandysul, SA44 5NZ
G4 JPK J. Pymm, Larkfield, Goxhill Road, Barrow-upon-Humber, DN19 7EE
GW4 JPP E. Jones, 1 Awel Y Mor, Cambrian Road, Tywyn, LL36 0AG
G4 JPS Bristol Raynet, c/o A. Williams, 38 Seneca Street, Bristol, BS5 8DX
G4 JPX I. Harrison, 61 Charles Street, Golborne, Warrington, WA3 3DF
GM4 JPZ C. Hall, 42 Torridon Road, Broughty Ferry, Dundee, DD5 3JG
G4 JQB R. Wickham, 35 Ashley Road, Bathford, Bath, BA1 7TT
G4 JQF M. Key, 14 Ascot Road, Wigginton, York, YO32 2QE
G4 JQJ R. Field, 12 Granson Way, Washingborough, Lincoln, LN4 1EY
G4 JQK S. Casey, 14 Harrison Close, Emersons Green, Bristol, BS16 7HB
G4 JQL S. Wayman, Oak Tree Lodge, Redbridge Road, Crossways, Dorchester, DT2 8BG
G4 JQN R. Ward, 1 Dursley Road, Heywood, Westbury, BA13 4LG
GW4 JQQ R. Henry, 7 Gronow Close Neath Abbey, Neath, SA10 7AD
G4 JQS C. Boulton, Manor Cottage, Stratton, Dorchester, DT2 9RY
G4 JQU Z. Pokusinski, 362 Long Banks, Harlow, CM18 7PG
G4 JQV C. Mee, 26 De Lisle Court, Loughborough, Leicester, LE11 4PP
G4 JQW F. Lobban, 20 Evering Avenue, Poole, BH12 4JQ
G4 JQX C. Riley, 1 Coulston, Westbury, BA13 4NX
GM4 JR A. Anderson, 232 Annan Road, Dumfries, DG1 3HE
GI4 JRA J. Harrigan, 124 Drones Road, Pharis, Ballymoney, BT53 8JT
G4 JRB M. Hahn, 21 Stanley Road South, Rainham, RM13 8AJ
G4 JRD R. De Muth, 66 Perkins Road, Ilford, IG2 7NQ
GM4 JRF H. Hamilton, 8 Ardlui Gardens, Milngavie, Glasgow, G62 7RL
G4 JRJ S. North, 2 Robey Drive, Eastwood, Nottingham, NG16 3DP
G4 JRW K. Burton, 93 Truncliffe, Bradford, BD5 8NX
G4 JRY T. Wislocki, 30 Kingston Rd, Scunthorpe, dn16 2be
G4 JS Darwen ARC, c/o W. Kenyon, Flat 21 House 4, Copper Place, Manchester,

M14 7FZ
G4 JSD J. Hamilton, 89 The Paddocks, Old Catton, Norwich, NR6 7HE
G4 JSE R. Salaman, 39 Arthur Street, Unley, Australia, SA 5061
G4 JSK L. Welch, 3 Sunnyfield Avenue, Cliviger, Burnley, BB10 4TE
G4 JSM P. Hart, 112 Shelton Avenue, Hucknall, Nottingham, NG15 7QA
G4 JSP C. Perkins, The Laurels, Higher Heath, Whitchurch, SY13 2HZ
G4 JSQ D. Piper, 102 Redhouse Lane, Walsall, WS9 0DB
G4 JSS V. Waddington, 1 Bridle Lane, Netherton, Wakefield, WF4 4HN
G4 JST F. Ogden, 11 Stocklands Close, Cuckfield, Haywards Heath, RH17 5HH
G4 JSV N. Hingley, 29 Mayfield Road, Hurst Green, Halesowen, B62 9QW
G4 JSX M. Owen, Thatched Cottage, Main Street, Rugby, CV23 0JA
G4 JSZ D. Fry, The Stocks, Lyth Bank, Shrewsbury, SY3 0BE
G4 JTC J. Bautista, 47 Valiant House, Varyl Begg Estate, Gibraltar, Gibraltar, GX11 1AA
G4 JTE Dr P. Djali, 105 Pepper Lane Standish, Wigan, WN6 0PW
GI4 JTF Dr E. Squance, 11 Ballymenoch Road, Holywood, BT18 0HH
G4 JTK J. Lee, 203 Chester Road Whitby, Ellesmere Port, CH65 6SE
G4 JTM J. Llewellyn, Pier Road, Enniscrone, Co Sligo, Ireland,
G4 JTO H. Young, 72 Perrinsfield, Lechlade, GL7 3SD
G4 JTP G. Parker, 14 Maplewood, Ashurst, Skelmersdale, WN8 6RJ
G4 JTR V. Robinson, 4 Hilltop Road, Caversham, Reading, RG4 7HR
GI4 JTS R. Macrory, 8 Manse Road, Newtownards, BT23 4TP
G4 JTX P. Simon, 19C High Street, Kilburn, Belper, DE56 0NS
GW4 JUC H. Woodward, 11 Pant-Yr-Odyn, Sketty, Swansea, SA2 9GR
G4 JUD F. Loach, 39 Park Road West, Wolverhampton, WV1 4PL
GM4 JUE J. Cormack, 16 Shore Lane, Wick, KW1 4NT
G4 JUH R. Wilkinson, 3 Anglesey Road, Dronfield, S18 1UZ
GW4 JUI D. Draper, Bryn Erin, Llangoed, Beaumaris, LL58 8SU
G4 JUK M. Neville, 103 Walsall Road, Great Wyrley, Walsall, WS6 6LD
G4 JUM B. Buller, 36 Grove Road, Ashtead, KT21 1BE
GW4 JUN V. Winton, Ty Cerrig, Rhosesmor Road, Halkyn, Holywell, CH8 8DL
G4 JUR R. Harrod, 6 Carnforth Road, Barnsley, S71 2RA
G4 JUV C. Bauers, 21 Nethergate Street, Bungay, NR35 1HE
G4 JUW W. Cole, 5 Brook Furlong, Nesscliffe, Shrewsbury, SY4 1BY
G4 JUZ N. Gabriel, 156 Clarence Avenue, New Malden, KT3 3DY
G4 JVA G. Butler, 37 Turmore Dale, Welwyn Garden City, AL8 6HT
G4 JVC I. Jones, 4 Grove Crescent South, Boston Spa, Wetherby, LS23 6AY
G4 JVD P. Hainsworth, 74 Ravensbourne Drive, Woodley, Reading, RG5 4LJ
G4 JVH G. Onions, 3 Tower Rise, Tividale, Oldbury, B69 1NP
G4 JVJ Dr R. Ashman, 44 Conan Doyle Walk, Swindon, SN3 6JB
G4 JVM F. Pearson, Coach House, The Park, Manningtree, CO11 2AL
GJ4 JVP J. Arthur, 13 Les Quennevais Park, St. Brelade, Jersey, JE3 8GB
G4 JVT G. Howell, 25 Thornhill Road, Hednesford, Cannock, WS12 4LR
G4 JVX D. Powell, 2 Curlew Close, Winsford, CW7 1SW
G4 JVZ M. Glennon, 41 Moorway, Guiseley, Leeds, LS20 8LD
G4 JWA D. Naylor, 19 Bindbarrow, Burton Bradstock, Bridport, DT6 4RG
G4 JWK L. Ball, Tree Tops, Bodiam, Robertsbridge, TN32 5UG
G4 JWL P. Woodward, Le Rosey, Rolle, Switzerland, 1180
G4 JWV N. Lyons, 114 Spring Hill, Weston-Super-Mare, BS22 9BD
GI4 JWW T. Martin, 57 Oneill Road, Newtownabbey, BT36 6UN
G4 JXC R. Butler, 6 Woodland Avenue, Dursley, GL11 4EW
G4 JXE P. King, 21 Compton Way, Olivers Battery, Winchester, SO22 4HS
G4 JXH P. Mcgivern, 83 Birdhill Avenue, Reading, RG2 7JU
G4 JXI H. Collier, 12 Coronation Drive, Leigh, WN7 2UU
G4 JXJ C. Blewitt, 12 Salton Street, Secret Harbour, Australia, 6173
G4 JXK D. Bonfield, 14 Springdale Close, Brixham, TQ5 9RL
GW4 JXN G. Roberts, 4 Frondeg, Ffordd Penmynydd, Llanfairpwllgwyngyll, LL61 5AX
GM4 JXP S. Green, 48 Barclay Park, Aboyne, AB34 5JF
G4 JXR G. Wilde, 26 Fleetham Grove, Hartburn, Stockton-on-Tees, TS18 5LH
G4 JXU K. Lee, 34 Evergreen Way, Wokingham, RG41 4BX
G4 JXZ I. Terrell, 10 Red Lion Close, Cranfield, Bedford, MK43 0JA
GM4 JYB B. Sparks, Donlyn, Lyth, Wick, KW1 4UD
G4 JYE D. Sargent, 15 Wilton Road, Balsall Common, Coventry, CV7 7QW
G4 JYF C. Golley, 10 New Molinnis, Bugle, St Austell, PL26 8QL
G4 JYG T. Watson, 59 Vincent Road, Norwich, NR1 4HQ
G4 JYH A. Curtis, 19 Donnelly Road, Bournemouth, BH6 5NW
GI4 JYJ G. Mcmaw, 26 Watch Hill Road, Ballyclare, BT39 9QW
G4 JYK P. Leather, 35 Somerset Close, Congleton, CW12 1SE
G4 JYL G. Thomas, 16 Fordlands, Thorpe Willoughby, Selby, YO8 9PD
G4 JYN Waterside ARS, c/o T. Williams, 31 Manor Road, Holbury, Southampton, SO45 2NQ
G4 JYP N. Shelley, 25 Threeways, Cuddington, Northwich, CW8 2XJ
G4 JYQ J. Tierney, 39 Daneway, Southport, PR8 2QW
G4 JYT R. Armstrong, 38 Watson Avenue, Market Harborough, LE16 9NA
G4 JYU N. Bourner, 11 Richborough Road, Sandwich, CT13 9JE
G4 JYW D. Proctor, 36 Westlands, Pickering, YO18 7HJ
G4 JZA S. Geary, Bella Vista, The Square, Truro, TR2 4DS
GM4 JZB D. Gardner, 7 Croft Road, Auchterarder, PH3 1EW
G4 JZF G. Taylor, 1 Threshers Drive, Willenhall, WV12 4AN
G4 JZL J. Adams, 1 Powell Close, Creech St. Michael, Taunton, TA3 5TE
G4 JZQ M. Noakes, 333 St. Neots Road, Hardwick, Cambridge, CB23 7QL
G4 JZR E. Williams, 7 Laurel Drive, Willaston, Neston, CH64 1TN
G4 JZV R. Bellamy, 31 Shaftesbury Avenue, Lincoln, LN6 0QN
GW4 JZY J. Price, 18 Woodland Drive, Bassaleg, Newport, NP10 8PA
G4 JZZ C. Gadd, 40 Stanley Mount, Sale, M33 4AE
G4 KAB L. Rose, 2 Westglade Court, Woodgrange Close, Harrow, HA3 0XQ
G4 KAE D. Wood, 7 Mead Close, Cheddar, BS27 3XN
G4 KAL B. Thompson, 23 South Street, Keelby, Grimsby, DN41 8HE
G4 KAM S. Greenwood, Little Oaks, Green Lane, Axminster, EX13 5TD
G4 KAR R. Jeffries, 22 Ingrams Way, Hailsham, BN27 3NP
G4 KAT M. Westwater, 3 Burns Way, Harrogate, HG1 3NA
G4 KAU Dr T. Mansfield, 2 Stratford Crescent, Cringleford, Norwich, NR4 7SF
GM4 KAV F. Bowles, 40 Craigbarnet Road, Milngavie, Glasgow, G62 7RA
G4 KAX M. Haswell, 5 Westcombe Avenue, Leeds, LS8 2BS
GW4 KAZ B. Davies, 2 Glan Llyn Terrace Bethel, Caernarfon, LL55 1YL
G4 KBA K. Boucher, 22 Emery Close, Walsall, WS1 3AL
G4 KBB B. Bristow, 13 Princes Street, Piddington, High Wycombe, HP14 3BN
G4 KBH R. Hodgson, 29456 Trailway Lane, Agoura Hills, USA, 91301
G4 KBI C. Wainman, 9 Willson Drive, Riddings, Alfreton, DE55 4AF
G4 KBK R. Fisher, 80/72 Kangan Drive, Berwick, Australia, 3806
GJ4 KBM B. Nelson, 1 La Genetiere, La Route Orange, St Brelade, Jersey, JE3 8GP
G4 KBP M. Ford, Micalma, Anslow Lane, Burton-on-Trent, DE13 9DS
G4 KBQ J. Haslam, 20 Lightfoots Avenue, Scarborough, YO12 5NS
GI4 KBW P. Henderson, 7 Clonaslea, Newtownabbey, BT37 0UL
G4 KBX C. Chapple, Woodend, Hebron, Morpeth, NE61 3LA
G4 KCC H. Holmden, 29 Cambridge Road West, Farnborough, GU14 6QA
G4 KCD B. Dean, 3 Marchant Court, Gunthorpe Road, Marlow, SL7 1UW
G4 KCF K. Sanderson, 39 Kirkland Street Pocklington, York, YO42 2BX
G4 KCM C. Sanders, 13 Meadow Court, Whiteparish, Salisbury, SP5 2SE
G4 KCN D. Salmon, The Pines, 5A Westfield Avenue, Harpenden, AL5 4HN
GI4 KCO K. Wright, 72 Elm Corner, Dunmurry, Belfast, BT17 9PY
G4 KCP D. Appleton, 28 Edgewood, Shevington, Wigan, WN6 8HR
GW4 KCQ P. Evans, 2 Cwmnantllwyd Road, Gellinudd, Swansea, SA8 3DT
G4 KCR S. Dunn, 4 St. Ronans Road, Harrogate, HG2 8LE
G4 KCT B. Firth, 8 Lyndale Avenue, Osbaldwick, York, YO10 3QB
G4 KCU D. Greatbatch, 1 Hilltop Way, Dronfield, S18 1YL
GW4 KCV Dr R. Murray-Shelley, 5 Dan Y Wern, Pwllgloyw, Brecon, LD3 9PW
G4 KCX P. Hicks, 7 North Croft, High Wycombe, HP10 0BP
GW4 KCY P. Trimmer, 15 Cypress Court, Landare, Aberdare, CF44 8YB
G4 KCZ Dr C. Conduit, 21 Shadybrook Lane, Weaverham, Northwich, CW8 3PN
G4 KDE Dr A. Lamont, 50 Hockley Road, Rayleigh, SS6 8EB
G4 KDH K. Howe, Woodlands, St. Peters Road, Hockley, SS5 6AA
GW4 KDI R. Stanton, 33 Brook Road, Shotton, CH5 1HH
G4 KDK J. Riggs, 28 Long Hill, Mere, Warminster, BA12 6LR
G4 KDL A. Seago, 50 Kimberley Road, Lowestoft, NR33 0TZ
G4 KDM J. Pearson, 110 Plover Mills, Lindley, Huddersfield, HD3 3ZF
G4 KDN J. Phaff, 28, Draycott Road, Abingdon, OX13 5BZ
G4 KDR I. Wassell, 21 Speedwell Way, Horsham, RH12 5WA
G4 KDS C. Lafferty, Apw 0414427, Addresspal, The Beacon, Mosquito Way, Hatfield, AL10 9WN
G4 KDU G. Baldwin, 31 Kilnhurst Road, Todmorden, OL14 6AX
G4 KDW I. Davidson, 24 Queenswood Drive, Hitchin, SG4 0LG
G4 KEB L. Bright, 49 Fellows Avenue, Wall Heath, Kingswinford, DY6 9ET
G4 KEC R. Cookson, 4 Wellington Gardens Selsey, Chichester, PO20 0EE
G4 KEE V. Tomkins, 58 Chancellors Way, Beacon Hill, Exeter, EX4 9DY
G4 KEG D. Fryer, 28 Hudson Road, Eastwood, Leigh-on-Sea, SS9 5NX
G4 KEI C. Gaston, Seaward, Marshlands Lane, Heathfield, TN21 8EY
G4 KEL S. Kell, 11 Streatlam Road, Darlington, DL1 4XG
G4 KEN K. Smith, 32 St. Clements Road, Harrogate, HG2 8LX
G4 KEP H. Haria, 34 Larkfield Avenue, Harrow, HA3 8NF
GI4 KEQ B. Mcmahon, 26 Ballycraigy Road, Newtownabbey, BT36 5ST
G4 KES B. Bloomer, 2 Magor Hill Cottages, Magor Hill, Camborne, TR14 0JF
G4 KEW R. Marshall, 60 Drake Road, Harrow, HA2 9EA
G4 KEX B. Hubbard, 16 Shelf Moor, Halifax, HX3 7PW
G4 KEY D. Turner, 22 Westhawe, Bretton, Peterborough, PE3 8BA
G4 KEZ B. Archer, 86 York Road, Swindon, SN1 2JU
G4 KFA T. Bearpark, 19A Humber Lane, Patrington, Hull, HU12 0PJ
G4 KFB M. Bird, 84 Penwill Way, Paignton, TQ4 5JQ
G4 KFC A. Scandrett, 45 Merryhill, Northampton, NN4 9YH
GW4 KFD B. Wilson, 1A Treetops, Llanelli, SA14 8DN
G4 KFF R. Hewson, 2 Ribchester Way, Brierfield, Nelson, BB9 0YH
G4 KFH E. Sinkinson, 24 Old Hall Park, Langthorpe, York, YO51 9BZ
GW4 KFI D. Bromfield, 3 Warwick Road, Brynmawr, Ebbw Vale, NP23 4AR
G4 KFJ C. Baker, 5 Holly Bank Rise, Dukinfield, SK16 5EG
G4 KFL R. Rowney, 58 Wychdell, Stevenage, SG2 8JD
G4 KFP J. Marshall, 92 High Street, Ossett, WF5 9RQ
G4 KFS T. Wood, 47 Marsh View, Beccles, NR34 9RT
G4 KFT M. Rothwell, 3 Chiltern Road, Prestbury, Cheltenham, GL52 5JQ
GW4 KFY J. Edwards, 15 The Meadows, Llandudno Junction, LL31 9LP
G4 KFZ R. Stanton, 50 Plymstock Road, Plymstock, Plymouth, PL9 7NU
G4 KGA M. Hattam, 11 Dukes Wood Avenue, Gerrards Cross, SL9 7LA
G4 KGC P. Suckling, 314A Newton Road, Rushden, NN10 0SY
G4 KGE J. Baldwin, 30 Petters Road, Ashtead, KT21 1NE
G4 KGF G. Brooks, 1 Highfield Close, Pembury, Tunbridge Wells, TN2 4HG
GM4 KGK N. Munro, Windyridge, Lower Bayble, Isle of Lewis, HS2 0QB
G4 KGL M. Lees, 15 Blacklock, Chelmsford, CM2 6QL
G4 KGN D. Mitchinson, 191 St. Marys Avenue Hemingbrough, Selby, YO8 6YY
G4 KGO R. Matthews, 191 Valley Road, Ipswich, IP4 3AH
G4 KGU B. Thomas, 12 Link Road, Sale, M33 4HP
G4 KGX W. Green, 3 Amos Road, Leicester, LE3 6NA
G4 KGY T. Lawford, 20 Magdalen Court Ersham Road, Canterbury, CT1 3DH
GM4 KGZ S. Low, Gartwood, 14 Dundas Avenue, North Berwick, EH39 4PS
GM4 KHE G. Phanco, 28 Park Road, Clydebank, G81 3LH
G4 KHG E. Scholes, 19 Castle Hill, Newton-le-Willows, WA12 0DU
GM4 KHI T. Ferrie, 17 Bargarron Drive, Paisley, PA3 4LL
G4 KHJ B. Geeson, 24 Rydal Avenue, Poulton-le-Fylde, FY6 7DJ
G4 KHK P. Martin, 24 Heddington Close, Trowbridge, BA14 0LH
G4 KHM J. Whitington, 18 Somerset Road, Ferring, Worthing, BN12 5QA
GW4 KHQ J. Woodland, 7 Lighthouse Park, St. Brides Wentlooge, Newport, NP10 8SL
G4 KHR R. North, 21 St. Augustine Grove, Bridlington, YO16 7DB
GM4 KHS Kelso ARS, c/o G. Chalmers, 38 Grove Hill, Kelso, TD5 7AS
G4 KHU P. Hawkins, Temple View, High Street, Templecombe, BA8 0JG
G4 KHX P. Winchester, 27A Lower Road, Milton Malsor, Northampton, NN7 3AW
G4 KIB J. Hambleton, Monte Pascoal, Cci 264 St, Sao Teotonio, Portugal, 7630-583
G4 KIF A. Sansom, 1881-9 Avenue S E, Salmon Arm, Canada, BC V1E 2J6
G4 KIH W. Bartlett, 48 Barrymore Walk, Rayleigh, SS6 8YF
G4 KII K. Harris, St. Peters Church Vicarage, Haywood Road, Birmingham, B33 0LH
G4 KIK D. Whyborn, 33 Church Road, Trull, Taunton, TA3 7LG
G4 KIL P. Williams, 4 Church Court, 130 Nevill Avenue, Hove, BN3 7NS
G4 KIM P. Newman, 16 Oakwood Road, Westlea, Swindon, SN5 7EF
G4 KIN P. Taylor, 22 Windermere Drive, Rainford, St Helens, WA11 7LD
G4 KIP J. Ball, Moss Nook Farm, Moss Nook Lane Road, St Helens, WA11 8AG
G4 KIQ A. Brooks, 10 St. James Avenue East, Stanford-le-Hope, SS17 7BQ
G4 KIR K. Chattenton, 29 Wand Hill Gardens, Boosbeck, Saltburn-by-the-Sea, TS12 3AP
G4 KIU N. Peacock, Flat 30, 55 Upper Grosvenor Road, Tunbridge Wells, TN1 2DY

GI4 KIX D. Gilmore, The Overlook, 29 Ballymaconaghy Road, Belfast, BT8 6SB
G4 KIZ D. Holmes, Lancaster House, Magna Mile, Market Rasen, LN8 6AD
G4 KJA B. Preston, 24 Nursery Close, Hucknall, Nottingham, NG15 6DQ
GI4 KJC N. Quinn, 54 Moyle Road, Newtownstewart, Omagh, BT78 4JT
G4 KJD I. Pitkin, Clover Cottage, Kenny, Ashill, Ilminster, TA19 9NH
G4 KJJ J. Smith, 30 Rookery Close, St Ives, PE27 5FX
G4 KJK D. Oliver, 15 Brixham Avenue, Cheadle Hulme, Cheadle, SK8 6JG
G4 KJP L. Jordan, Cami De Fuster No 1, Marxuquera Alta, Valencia, Spain, 46700
G4 KJS A. Gregory, 1-3 Nargate Street, Littlebourne, Canterbury, CT3 1UH
G4 KJU R. Fisher, 85 Larkway, Brickhill, Bedford, MK41 7JP
G4 KKB K. Blamey, 123 St. Edmunds Walk, Wootton Bridge, Ryde, PO33 4JJ
G4 KKG J. Taylor, 12 Glenthorne Avenue, Yeovil, BA21 4PG
G4 KKJ H. Perryman, 15 Queen Mary Crescent, Kirk Sandall, Doncaster, DN3 1JU
G4 KKN P. Roberts, 2 Samuals Fold, Pendlebury Lane, Wigan, WN2 1LT
G4 KKO J. Walton, 17 Wychperry Road, Haywards Heath, RH16 1HJ
G4 KKR R. Page, 4 Nursery Drive, March, PE15 8EQ
G4 KKS A. Morris, 4 Woodville Gardens, Wigston, LE18 1JZ
G4 KKT J. Mahoney, 18 Park Avenue, London, N22 7EX
G4 KKU A. Imianowski, 97 Bloomfield Road, Bristol, BS4 3QP
GM4 KKV P. Rucklidge, 8 Stanehead Park, Biggar, ML12 6PU
G4 KKZ K. Robinson, 13 Race Hill, Launceston, PL15 9BB
G4 KLA Dr J. Nelson, 67 Swarthmore Road, Birmingham, B29 4NH
G4 KLB C. Watts, 42 Truscott Avenue, Bournemouth, BH9 1DB
G4 KLD C. Dewhurst, 56 Collett Way, Priorslee, Telford, TF2 9SL
G4 KLE M. Foster, 7 Church Street, Fenstanton, Huntingdon, PE28 9JL
G4 KLF A. Selmes, 82 Beaufort Court, Beaufort Road, St Leonards-on-Sea, TN37 6PF
G4 KLJ D. Wellings, 41 Wroxham Drive, Wollaton, Nottingham, NG8 2QR
G4 KLM P. Raven, Wedgewood, Green Lane West, Norwich, NR13 6LT
GM4 KLN I. Moore, 7 Greenside Avenue Rosemarkie, Fortrose, IV10 8XA
GM4 KLO M. Mistofsky, 18 Troon Place, Newton Mearns, Glasgow, G77 5TQ
G4 KLT L. Jones, 52 The Drive, Bury, BL9 5DL
G4 KLX J. Naylor, 35 Old River, Denmead, PO7 6UX
G4 KMB A. Griggs, 4 Raleigh Rise, Portishead, Bristol, BS20 6LA
G4 KME J. Horley, 50 Hillswood Drive, Endon, Stoke-on-Trent, ST9 9BW
G4 KMF E. Colmer, 31 Mosyer Drive, Orpington, BR5 4PN
G4 KMH S. Cottis, 61 Oaken Grove, Maidenhead, SL6 6HN
G4 KMJ D. Edwards, 72 Parkstone Road, Hastings, TN34 2NT
G4 KMK R. Blower, 133 Almondbury Bank, Huddersfield, HD5 8EX
G4 KMM P. Northmore, Flat 23, Margaret Hill House 77 Middle Lane, Hornsey, N8 8NX
G4 KMP G. Ramsey, 21 Goldsmith Road, Eastleigh, SO50 5EN
G4 KMW R. Greenhough, 36 Churchbalk Lane, Pontefract, WF8 2QQ
G4 KMX R. Cope, 41 Hall Lane, Witherley, Atherstone, CV9 3LT
G4 KNI D. Rickard, 12 Dabryn Way, St. Stephen, St Austell, PL26 7PF
G4 KNN A. Leggett, 3 Hayes Mead, Holbury, Southampton, SO45 2JZ
G4 KNO A. Summers, Broxwood, Bury Road, Bury St Edmunds, IP29 4PH
G4 KNQ H. Smith, Grey Gables, Humphrey Gate, Buxton, SK17 9TS
G4 KNR S. Mason, 9 Bempton Close, Bridlington, YO16 7HL
G4 KNS J. Wallett, 46 Aldreth Road, Haddenham, Ely, CB6 3PW
G4 KNT I. Morton, 65 Manton Road, Hitchin, SG4 9NP
GM4 KNU A. Torrance, 306 Mearns Road, Newton Mearns, Glasgow, G77 5LS
G4 KNV Dr D. Wilkinson, Westview, Old Byland, York, YO62 5LG
G4 KNX A. Bennett, 4 Chelmarsh Close, Redditch, B98 8SQ
G4 KNZ S. Davies, 17 Haywood, Bracknell, RG12 7WG
GW4 KOE R. Lines, 19 Magnolia Close, Cardiff, CF23 7HQ
GM4 KOI S. Milne, 24 St. Ternans Road, Newtonhill, Stonehaven, AB39 3PF
G4 KOJ J. Wilson, 54 Devonshire Drive, Mickleover, Derby, DE3 9HB
G4 KOK J. Stockley, Clee View, Leys Lane, Leominster, HR6 0AZ
G4 KON L. Butt, 16A Kestrel Crescent, Oxford, OX4 6DX
GM4 KOO S. Cawthorne, 8 Captains Brae, Twynholm, Kirkcudbright, DG6 4PE
G4 KOQ G. Birkhead, 15 Crannog, Keshcarrigan, Carrick on Shannon, Ireland, N41K235
G4 KOR A. Hughes, 55 Welford Road, Shirley, Solihull, B90 3HX
G4 KOT G. Lindsay, 9 Pennine View, Durham, DH6 1QN
G4 KOU G. Martin, Flat 1, Field House, Station Road, East Preston, Littlehampton, BN16 3RU
G4 KOV H. Wright, Sandpiper Cottage, Standard Road, Wells-Next-the-Sea, NR23 1JY
G4 KOW D. Mclachlan, 48 Nursery Avenue, Bexleyheath, DA7 4JZ
G4 KOY R. Gill, 87 Penkett Road, Wallasey, CH45 7QQ
GW4 KPD Dr A. Grant, Chandlers, Welsh Street, Chepstow, NP16 5LU
G4 KPE P. Griggs, 6 Nightingale Way, Sutton Bridge, Spalding, PE12 9RG
G4 KPF T. Hart, 15 Whitefriars Meadow, Sandwich, CT13 9AS
G4 KPG W. Lam, 2 Wistaria Road, Flat 3A, Kowloon, Hong Kong,
G4 KPH D. Lewis, 4 Raymond Court, Pembroke Road, London, N10 2HS
G4 KPI J. Lorton, 14 Provis Mead, Chippenham, SN15 3UA
G4 KPL M. Young, 8 Tweed Close, Worcester, WR5 1SD
G4 KPM M. Pitt, 20 Little Halt, Portishead, Bristol, BS20 8JQ
G4 KPP C. Kelly, 115 Kingsdown Crescent, Dawlish, EX7 0HB
G4 KPS C. Saunders, 26 Henley Fields, St. Michaels, Tenterden, TN30 6EL
G4 KPU G. Taylor, 179 Bradway Road, Bradway, Sheffield, S17 4PF
G4 KPV F. Dunn, 12 Streete Court, Westgate-on-Sea, CT8 8BT
G4 KPX R. Burton, 28 Mulberry Way, Ely, CB7 4TH
G4 KPZ V. Cracknell, 106 High St., Upwood, Huntingdon, PE26 2QE
GW4 KQ D. Phillips, 37 Saint Margarets Park, Lower Ely, Cardiff, CF5 4AP
GI4 KQA T. Moffitt, 36 Greenview, Parkgate, Ballyclare, BT39 0JP
G4 KQC G. Leatherbarrow, 6 Queens Walk, Thornton-Cleveleys, FY5 1JW
G4 KQD A. Down, 5 Juniper Mead, Stotfold, Hitchin, SG5 4RU
G4 KQE A. Mead, 9 Abraham Drive, Silver End, Witham, CM8 3SP
G4 KQH D. Howes, 14 Manitoba Way, Eydon, Daventry, NN11 3PR
G4 KQK Dr C. Barnes, Glebe Farmhouse, Stafford, ST18 9DQ
G4 KQL A. Daulman, 2 Trentham Road, Hartshill, Nuneaton, CV10 0SN
G4 KQO R. Ferguson, 8 Rutland Gardens, Croydon, CR0 5ST
G4 KQP S. Jones, 114 Portland Road, Toton, Nottingham, NG9 6EW
G4 KQQ R. Jones, 2 Bubwith Walk, Wells, BA5 2EN
GM4 KQS A. Smith, 9 Woodmill, Kilwinning, KA13 7PT
G4 KQV C. Hands, 41 Coverdale Road, Solihull, B92 7NU
G4 KQY M. Pearce, 51 Grove Avenue, New Costessey, Norwich, NR5 0JB
G4 KQZ T. Thorne, 17 Pine St. South, Bury, BL9 7BU
G4 KRD M. Khalaf, 508 London Road, Thornton Heath, CR7 7HQ
G4 KRF R. Moore, Flat 15, Nelson Court, 130 Rowson Street, Wallasey, CH45 2LZ
G4 KRH R. Cane, 24 South End, Longhoughton, Alnwick, NE66 3AW
G4 KRJ E. Gaffney, 54 Dockham Road, Cinderford, GL14 2BH
G4 KRN A. Troy, 1B Lidderdale Road, Liverpool, L15 3JG
G4 KRT M. Davis, 35 Mullion Croft, Kings Norton, Birmingham, B38 8PH
G4 KRW R. Waterman, 170 Station Road, Mickleover, Derby, DE3 9FJ
G4 KSA D. Mountain, 178 Wragby Road, Lincoln, LN2 4PT
G4 KSG R. Ralph, 62 Northdown Road, Solihull, B91 3ND
GI4 KSH H. Morrow, 2 Carnhill Grove, Newtownabbey, BT36 6LS
G4 KSK R. Benyon, C/ Grecia 17, Villalbilla, Madrid, Spain, 28819
GI4 KSO D. Mawhinney, 233 Ballynahinch Road, Annahilt, Hillsborough, BT26 6BH
G4 KSQ B. Morris, 22 Burdell Avenue, Headington, Oxford, OX3 8ED
G4 KSR S. Norris, 17 Montroy Close, Bristol, BS9 4RS
G4 KST T. Hughes, 42 Western Drive, Hanslope, Milton Keynes, MK19 7LD
G4 KSU Dr K. Prettyjohns, 315 High Street, Sheerness, ME12 1UT
G4 KSY A. Street, 43 Ridgedale Road, Bolsover, Chesterfield, S44 6TX
G4 KTB T. Cottham, 4 Talisman Close, Tiptree, Colchester, CO5 0DT
G4 KTG H. Wilson, 24 Clumber Avenue, Newark, NG24 4DT
GW4 KTQ G. Davies, 56 Ffordd Cynan, Bangor, LL57 2NS
G4 KTR D. Burrell, 67 Newfield Drive, Nelson, BB9 9RR
GW4 KTT P. Valerio, The Brackens, Reynoldston, Swansea, SA3 1AE
G4 KTU K. White, 22 Ridyard Street, Wigan, WN5 9PA
G4 KTW E. Dale, The Woodlands, Cotheridge, Worcester, WR6 5LZ
G4 KTX J. Goldsmith, The Maltings, Flacks Green, Chelmsford, CM3 2QS
G4 KTZ P. Cullen, 5 Swaledale Gardens, Fleet, GU51 2TE
G4 KUC J. Goodier, 20 Poleacre Lane, Woodley, Stockport, SK6 1PG
G4 KUD B. Whittles, 12 Locksley Gardens, Birdwell, Barnsley, S70 5SU
G4 KUE C. Raspin, 35 Allesley Hall Drive, Coventry, CV5 9NS
G4 KUF A. Redman, 42 Gallows Hill Lane, Abbots Langley, WD5 0DA
G4 KUJ T. Groves, 31 Tunnel Wood Close, Watford, WD17 4SW
G4 KUL D. Hepplestone, 19 Richlans Road, Hedge End, Southampton, SO30 0HU
GI4 KUM W. Glenn, 1 Meadowside, Antrim, BT41 4HD
G4 KUQ P. Goodfellow, 10 St. Agnes Walk, Knowle, Bristol, BS4 2DL
G4 KUR S. Hammonds, 22 The Croft, Meriden, Coventry, CV7 7NQ
G4 KUX N. Peckett, Fourwinds, Woodland, Bishop Auckland, DL13 5RH
G4 KUY M. Hill, Park Villa, Park Road, Tydd St. Giles, Wisbech, PE13 5NH
GI4 KUZ W. Hamill, 47 Gracefield, Gracehill, Ballymena, BT42 2RP
G4 KVC R. Mitchell, 6 Green St., Smethwick, B67 7BX
G4 KVD J. Mcmahon, 5 Victoria Walk, Wokingham, RG40 5YL
G4 KVI C. Dunn, 71 Redfield Road, Midsomer Norton, BA3 2JH
G4 KVK P. Park, 2 Leyburn Drive, High Heaton, Newcastle upon Tyne, NE7 7AP
G4 KVL B. Tharme, 4 Longcroft Avenue, Liverpool, L19 4TB
G4 KVP A. Woodland, 45 Walsingham Road, Wallasey, CH44 9DX
G4 KVQ R. Scott, 20 Forest Hill, Carlisle, CA1 3HF
G4 KVR P. Bell, 24 Onslow Gardens, Ongar, CM5 9BG
G4 KVT J. Fairfax, 382 Wells Road, Bristol, BS4 2QP
G4 KVU M. Shearer, Appleacre, Mill Road, Haverhill, CB9 7NN
G4 KVX P. Bleiker, Waterside, 31 North Shore Road, Hayling Island, PO11 0HL
G4 KWE T. Peel, Herongate, Derwent Lane, Hope Valley, S32 1AS
G4 KWF E. Pickup, 36 Werneth Road, Glossop, SK13 6NF
G4 KWH C. Meadows, 16 Dart Road, Bedford, MK41 7BT
G4 KWJ J. Hakes, Commonbank Cottage, Lancaster, LA2 9AN
G4 KWK K. Hakes, Commonbank Cottage, Lancaster, LA2 9AN
G4 KWL T. Walter, 11 Silver Street Congresbury, Bristol, BS49 5EY
G4 KWM P. Deville, Bexton Doncaster Road, Mexborough, S64 0JD
G4 KWO G. Phillips, 20 Eastfield Drive, Solihull, B92 9ND
G4 KWQ A. Soltysik, 24 Cottage Close, Hednesford, Cannock, WS12 1BS
G4 KWT D. Pibworth, 20 Marathon Close, Woodley, Reading, RG5 4UN
G4 KWW J. Ilott, Flat 3 Berkeley Court, Scotter Road, Scunthrope, DN15 7EG
G4 KWX B. Cox, Sylvan House, Alton Road, Farnham, GU10 5EL
G4 KWY D. Gasser, 49 Pennycress, Locks Heath, Southampton, SO31 6SY
G4 KWZ G. Harris, Windmill House, Ripon Road, Kirby Hill, York, YO51 9DP
G4 KXF J. Farrar, Blanchards, The Green Saxtead, Woodbridge, IP13 9QH
G4 KXG K. Jackson, 17 Copperfield Close, Kettering, NN16 9EW
G4 KXK J. Ward, 38 Stonechat Avenue, Abbeydale, Gloucester, GL4 4XD
G4 KXL J. Redman, 488 Blair Road, Georgia, USA, 30563
G4 KXO M. Reynolds, Ilex House, Redwick Road, Bristol, BS12 3LQ
G4 KXP J. Brockett, 17 Swan Drive, Droitwich, WR9 8WA
G4 KXQ M. Wogden, 28 Magnolia Close, Barnstaple, EX32 8QH
G4 KXR A. Tipper, 10 Tithebarn Copse, Exeter, EX1 3XP
G4 KXU G. Robinson, 25 Stable Way Kingswood, Hull, HU7 3FA
G4 KXV D. Rigby1, 145 Knightlow Road, Harborne, Birmingham, B17 8PY
G4 KXW G. Redhead, 18 Paddock Way, Dronfield, S18 2FF
G4 KYE T. Carhart, C6 Tamar Park, Coxpark, Gunnislake, PL18 9BD
G4 KYH A. Waddilove, 2 Gwel Trencrom, Hayle, TR27 6PJ
G4 KYI R. Shipton, 3 Fiery Lane, Uley, Dursley, GL11 5DA
GW4 KYK J. Jones, 7 Frankwell St., Tywyn, LL36 9EP
G4 KYO G. Barber, 25 Queensway, Hayle, TR27 4NJ
GW4 KYT D. Thomas, 3 New Road, Trebanos, Swansea, SA8 4DL
G4 KYU R. Ringrose, Melford House, George Street, Ipswich, IP8 3NH
G4 KYX D. Gee, 13 Dart Road, Brickhill, Bedford, MK41 7BT
G4 KYY P. Day, 46 Beatrice Avenue, Saltash, PL12 4NG
G4 KZB P. Hazelwood, 12 Ryecroft, Stourbridge, DY9 9EH
G4 KZD J. Young, 30 Crofton Way, Enfield, EN2 8HS
G4 KZI B. Clark, 21 Church Road, Binstead, Ryde, PO33 3TA
G4 KZK R. Smith, 15 St. Anthonys Way, Brandon, IP27 0DN
G4 KZO A. Keir, Kingfisher House, Nantwich Road, Calveley, Tarporley, CW6 9JT
G4 KZQ R. Bennett, 16 Emily Street, St Helens, WA9 5LZ
G4 KZT B. Ashdown, 1 The Warren, Little Snoring, Fakenham, NR21 0JU
G4 KZU N. Rathbone, 7 Foreland Way, Keresley, Coventry, CV6 2NN
G4 KZV J. Parkin, 18 Bradnock Close, Birmingham, B13 0DL
G4 KZW S. Haydock, 60 Tong St., Bradford, BD4 9LX
G4 KZX A. Still, 17 Arundel Road, Newhaven, BN9 0ND
G4 KZZ N. Roberts, 13 Rosemoor Close, Hunmanby, Filey, YO14 0NB
G4 LAB Leicestershire Worked All Brittain Group, c/o D. Brooks, 28 Avon Vale Road, Loughborough, LE11 2AA

G4 LAD Leeds&Dist ARC, c/o M. Howes, Yarnbury Rufc, Brownberrie Lane, Leeds, LS18 5HB
G4 LAE C. Wordley, Whispering Winds, 7 Fulcher Avenue, Chelmsford, CM2 6QN
G4 LAF R. Brodrick, 16 Wallenge Drive, Paulton, Bristol, BS39 7PX
G4 LAI C. Fone, 12 Chiltern Rise, Ashby-de-la-Zouch, LE65 1EU
G4 LAJ R. Hackett, 4 Ryton Grove, Birmingham, B34 7RS
G4 LAK R. Procter, 83 Twickenham Road, Newton Abbot, TQ12 4JG
G4 LAM R. Lamberton, 28A Newtown Road, Raunds, Wellingborough, NN9 6LX
G4 LAN P. Conway, 14 Leahall Lane, Rugeley, WS15 1JE
GM4 LAO A. Waddell, 13 Auchenglen Road, Braidwood, Carluke, ML8 5PH
G4 LAU C. Stevens, 9 Newbury Avenue, Melton Mowbray, LE13 0SR
G4 LAW F. Craven, 2 Barn Owl Way, Stoke Gifford, Bristol, BS34 8RZ
G4 LAY G. Dobbs, Chaka, Grimsby Road, Market Rasen, LN8 6DH
G4 LBC P. Rusling, 20 Packman Lane, Kirk Ella, Hull, HU10 7TL
GM4 LBE A. Tait, 12 Greenwell, Gott, Shetland, ZE2 9UL
G4 LBH R. Giles, 33 Sowerby Avenue, Luton, LU2 8AF
G4 LBJ L. Gurney, Bluebridge Coach House, Colchester Road, Halstead, CO9 1QG
G4 LBM South London Raynet, c/o I. Jackson, 5 Vivien Close, Chessington, KT9 2DE
G4 LBQ J. Philipson, Clifton Farm House, Pullover Road, King's Lynn, PE34 3LS
G4 LBS Borden Gram Sch, c/o K. Groom, 2 Ruins Barn Road, Tunstall, Sittingbourne, ME10 4HS
G4 LBT R. Harmer-Knight, 3 Grendon Drive, Sutton Coldfield, B73 6QA
G4 LBU E. Kersey, 98 Campbell Road, Ipswich, IP3 9RE
G4 LBY S. Wright, 22 Crown St., Mansfield, NG18 3JL
G4 LCB Dr M. Goldman, 19 Myddelton Park, London, N20 0HT
G4 LCE N. Watson, 14 Mill Lane, Whittlesford, Cambridge, CB22 4NE
GW4 LCF G. Williams, 2 The Paddocks, Lodge Hill, Newport, NP18 3BZ
G4 LCH M. Gregory, Highleys Farm, 375 Tanworth Lane, Shirley, Solihull, B90 4DX
G4 LCL E. Beardmore, Kilaguni, The Avenue, Stoke-on-Trent, ST9 9LW
G4 LCM P. Allsopp, 32 Linden Close, Prestbury, Cheltenham, GL52 3DU
GM4 LCP J. Staruszkiewicz, 1 Nether Kirkton Way, Neilston, Glasgow, G78 3PZ
G4 LCU M. Brownlow, The Croft, 1 Byne Close, Pulborough, RH20 4BS
GW4 LDA R. Lawrence, 10 St. George Road, Bulwark, Chepstow, NP16 5LA
G4 LDB T. Kendall, 86 Rockford Close, Redditch, B98 7YL
G4 LDC A. Wallis, 4 Trevose Close, Chandler'S Ford, Eastleigh, SO53 3EB
G4 LDD P. Harling, Pimlico House, Gisburn Road, Clitheroe, BB7 4ES
G4 LDJ F. Gabell, 25 Woodland Way, Crowborough, TN6 3BQ
G4 LDL A. Bettley, 1 Dovetrees, Covingham, Swindon, SN3 5AX
GI4 LDN S. Mcquaid, Mullaghrodden, Dungannon, Co Tyrone, BT70 3LU
GW4 LDP I. Dobby, 43 Chestnut Avenue, West Cross, Swansea, SA3 5NL
G4 LDR N. Underwood, Blandings, Yarmley Lane, Winterslow, Salisbury, SP5 1RB
G4 LDS C. Baker, 14 Clarendon Road, Morecambe, LA4 4HS
G4 LDT D. Holland, 29 Lily Crescent, Sunderland, SR6 7HN
G4 LDW M. Morris, 11 Kingswood Avenue, Hampton, TW12 3AU
GM4 LDX M. Mcforsyth, Haltoun, Eddleston, Peebles, EH45 8PW
G4 LED A. Wood, 67 Bay View Road, Duporth, St Austell, PL26 6BN
G4 LEG P. Brent, 14 Stagelands, Crawley, RH11 7PE
G4 LEM E. Goodman, 83 Avondale Road, Kettering, NN16 8PL
G4 LEN A. Kendall, 18 Chivenor Way Kingsway Quedgeley, Gloucester, GL2 2BH
G4 LEP C. Jacobs, 16 Woodyard Close, Mulbarton, Norwich, NR14 8AS
GM4 LER T. Goodlad, 72 North Lochside, Lerwick, Shetland, ZE1 0PJ
G4 LES L. Macvean, 27 Babs Field, Bentley, Farnham, GU10 5LS
G4 LEV C. Veitch, 108 Racecourse Road, Rd2, Otane, New Zealand, 4277
G4 LEX G. Train, 29 Waggoners Way, Morton, Bourne, PE10 0XR
GM4 LFE R. Broom, 1 Byron Court, Banff, AB451FB
GW4 LFF Dr J. Devonshire, 19 Voss Park Drive, Llantwit Major, CF61 1YD
G4 LFG M. Davis, 53 St. Georges Avenue, South Shields, NE33 3EH
GM4 LFK L. Mclean, Lower Hatton Cottage, Dunkeld, PH8 0ET
GM4 LFL J. Rennie, 1 The Banks, Brechin, DD9 6JD
GW4 LFO Highfield ARC, c/o D. Jenkins, 16 Celtic Road, Whitchurch, Cardiff, CF14 1EG
G4 LFQ J. Holloway, Flat 1, 66 Unthank Road, Norwich, NR2 2RN
G4 LFS R. Draycott, 25 Flat Lane, Whiston, Rotherham, S60 4EF
G4 LFT A. Busby, 16A High Street, Sutton-On-Trent, Newark, NG23 6QA
GW4 LFV B. Crow, Lindisfarne, Pen Y Waun, Pentyrch, Cardiff, CF15 9SJ
GW4 LFW T. Cross, 50 Ty-Newydd, Whitchurch, Cardiff, CF14 1NQ
G4 LGB B. Graham, 19 Cannerby Lane, Sprowston, Norwich, NR7 8NQ
G4 LGH K. Garside, 191 Kenton Road, Newcastle upon Tyne, NE3 4NR
GM4 LGM J. Mcgregor, 26 Engels Street, Alexandria, G83 0RZ
G4 LGO N. Thomas, 24 Victoria Road, Teddington, TW11 0BG
GI4 LGP S. Mccracken, 29 Norwood Gardens, Belfast, BT4 2DX
G4 LGU W. Hills, Alperton, Rowhill Road, Dartford, DA2 7QQ
G4 LGX J. Hall, 30 Chatsworth Road, Harrogate, HG1 5HS
G4 LGY P. Harber, 28 Regent Road, Epping, CM16 5DL
G4 LHA G. Reoch, 350 Mavanelle Cove, Hempstead, USA, 77445
G4 LHE J. Lee, 41A Orchard Road, Seer Green, Beaconsfield, HP9 2XH
G4 LHF S. Moffat, 14 Churchill Rise, Burstwick, Hull, HU12 9HP
G4 LHI P. Rosamond, 13 Newnham Close Hartford, Huntingdon, PE29 1RP
GM4 LHJ J. Campbell, 23 Napier Avenue, Bathgate, EH48 1DF
GW4 LHL S. Edwards, 16 Maes Crugiau, Rhydyfelin, Aberystwyth, SY23 4PP
G4 LHO P. Gillen, 622 Glenwood Dr., Oxnard, USA, 93030
G4 LHP G. Griffiths, 21 Spring Lane, Olney, MK46 5HT
GM4 LHQ W. Herron, 21 Southfield Avenue, Paisley, PA2 8BY
G4 LHR E. Williams, 10 Eastbourne Close, Ingol, Preston, PR2 3YR
G4 LHT J. Mcsorley, 117 Park Avenue, Ruislip, HA4 7UL
GM4 LHW S. Burnett, 17 Crusader Drive, Roslin, EH25 9NP
G4 LIA J. Gordon, 36 Warbeck Close, Newcastle upon Tyne, NE3 2FG
G4 LIC J. Graham, 14 Ashbourne Grove, London, W4 2JH
GI4 LIF R. Goligher, Mountjoy East, County Tyrone, BT79 7JJ
G4 LIG P. Hesketh, 5 Beeches Close, Ixworth, Bury St Edmunds, IP31 2EW
G4 LIJ R. Nutt, 4 Mercers Drive, Bradville, Milton Keynes, MK13 7AY
G4 LIL C. Brown, Sandysike Cottage, Sandysike, Carlisle, CA6 5SS
G4 LIM G. Moody, 37 Pine Street, Stockton-on-Tees, TS20 2SP
G4 LIO J. Marshman, 12 Neelands Grove, Cosham, Portsmouth, PO6 4QL
G4 LIQ P. Williams, 54 High St., Yelling, St Neots, PE19 6SD
G4 LIR P. Taylor, 64 Walford Road, Rolleston-On-Dove, Burton-on-Trent, DE13 9AR
GM4 LIS D. Wilkes, 11 Trinity Crescent, Beith, KA15 2HG
G4 LIX G. Greenwood, 11 James Street, Holywell Green, Halifax, HX4 9AS
G4 LIY C. Ware, 4 Highfield Terrace, Lower Bentham, Lancaster, LA2 7EP
G4 LJB J. Wild, 6 Chestnut End, Headley, Bordon, GU35 8NA
GU4 LJC Cmdr. B. Le Lievre, Calabar Forest Road, Forest, Guernsey, GY8 0AB
G4 LJF Capt. I. Shepherd, Hutts Farm, Blagrove Lane, Wokingham, RG41 4AX
G4 LJK R. Mckee, 5 Moorcroft, Ossett, WF5 9JL
G4 LJN R. Bartlett, 37 Church Road, Ferndown, BH22 9ES
G4 LJR G. Garden, 9 Gateway Avenue, Smithville, Canada, L0R 2A0
GW4 LJS P. Harding, Harbour Light, Five Roads, Llanelli, SA15 5AQ
G4 LJT W. Hayward, 19 Woodlands Coxheath, Maidstone, ME17 4EE
G4 LJU C. Howell, 43 Copsleigh Close, Salfords, Redhill, RH1 5BJ
GW4 LJW J. Jenkins, Pantycelyn, Llanwnnen, Lampeter, SA48 7LW
G4 LJY J. Warren, Clifden Farm, Quenchwell Carnon Downs, Truro, TR3 6LN
G4 LKD J. Spurgeon, Whitgift House, Whitgift, Goole, DN14 8HL
GI4 LKG V. Tait, 30 Corby Drive, Lisburn, BT28 3HG
G4 LKM G. Clarke, 33 Mulberry Avenue, Penwortham, Preston, PR1 0LL
G4 LKP Dr K. Craven, 8 Melander Close, York, YO26 5RP
GW4 LKS W. Evans, Dan Y Craig, Craig Road, Swansea, SA7 9HS
G4 LKT P. Goodman, 34 Fullers Road, South Woodford, London, E18 2QA
G4 LKU D. Hill, 8 Lingfield Walk, Corby, NN18 9JS
G4 LKW P. Head, 36A Ashacre Lane, Worthing, BN13 2DH
G4 LKX D. Hepworth, 2 Granby Crescent, Doncaster, DN2 6AN
G4 LKZ C. Shuttleworth, 17 Stirling Close, Clitheroe, BB7 2QW
G4 LLG P. West, 5 Stonehill Close, Appleton, Warrington, WA4 5QD
G4 LLI G. Matthews, 101 Trafalgar Road, Horsham, RH12 2QL
G4 LLL Dr N. Rudgewick-Brown, 10 Windsor Park, Dereham, NR19 2SU
G4 LLM T. Barnes, 20 Mayes Close, Warlingham, CR6 9LB
G4 LLN R. Connolly, Newfane, Temple Way, Slough, SL2 3HE
G4 LLQ A. Leeming, 52 Kingfisher Drive, Pickering, YO18 8TA.
G4 LLZ A. Barr, 28 Roundway, Honley, Holmfirth, HD9 6DD
G4 LMA J. Baylis, 41 Ailesbury Way, Burbage, Marlborough, SN8 3TD
G4 LMF R. Harrison, 18 Gunners Lane, Studley, B80 7LX
GM4 LMG D. Heasman, 41 Honeyberry Drive, Rattray, Blairgowrie, PH10 7RB
G4 LMK J. Morris, 17 Overbrook Grange, Nuneaton, CV11 6BQ
G4 LML W. Turner, 11 Field View Close, Exhall, Coventry, CV7 9BJ
G4 LMM P. Stears, 127 Hughenden Avenue, High Wycombe, HP13 5SS
G4 LMN D. Piper, 45A Spielplatz, Lye Lane, Bricket Wood, St Albans, AL2 3TD
G4 LMR British Railways ARS, c/o G. Sims, 85 Surrey Street, Glossop, SK13 7AJ
G4 LMV W. Loxley, 92 Needlers End Lane, Balsall Common, Coventry, CV7 7AB
G4 LMW R. Thomson, Shire Jee Neevas, Cold Ash Hill, Cold Ash, Thatcham, RG18 9PH
G4 LMX S. Crosson Smith, The Old Pump House, Engine Road Ten Mile Bank, Downham Market, PE38 0EN
G4 LMY J. Piggott, 30 Farleigh Fields, Orton Wistow, Peterborough, PE2 6YB
G4 LNC A. Friis, 22 Garthwaite Crescent Shenley Brook End, Milton Keynes, MK5 7AX
G4 LNE N. Howorth, 42 Fairfield Avenue, Rossendale, BB4 9TQ
G4 LNG F. Hollis, 97 Manor Road, Chesterfield, S40 1HZ
G4 LNM D. Brown, 26 The Brucks, Wateringbury, Maidstone, ME18 5PX
GM4 LNN C. Foden, 4 Coastguard Houses, Cromwell Road, Kirkwall, KW15 1LN
G4 LNQ K. Marshall, 44 Rosemary Drive, Alvaston, Derby, DE24 0TA
G4 LNR L. Miles, 130 Well Lane, Willerby, Hull, HU10 6HS
G4 LNT B. Thompson, 113 Gordon Road, Stanford-le-Hope, SS17 7QZ
G4 LNY A. Thurgood, 19 Froment Way, Milton, Cambridge, CB24 6DT
G4 LNZ G. Langford, 15 Ambleside Drive, Hereford, HR4 0LP
G4 LOB A. Major, 33 Borough Road, Bridlington, YO16 4HN
GW4 LOD D. Parrott, 39 Groves Road, Newport, NP20 3SP
G4 LOE G. Tuppeny, 5 Ashlawn Crescent, Solihull, B91 1PR
G4 LOF M. Adams, 7 Finningley Drive, Allestree, Derby, DE22 2XP
G4 LOG R. Farley, Linden Lea, Close Hill, Redruth, TR15 1EW
G4 LOH T. Fern, South Boderwennack Farm, Trevenen Bal, Helston, TR13 0PR
G4 LOI B. Howell, 13 Westfield, Plympton, Plymouth, PL7 2DY
G4 LOJ C. Black, Charisma, Church Road, Norwich, NR14 7PB
G4 LOM J. Boult, Findon Lodge, Hartside, Durham, DH1 5RJ
G4 LON J. Berg, 25 Larch Close, Billinge, Wigan, WN5 7PX
G4 LOO D. Ross, 3 Little Lane, Clophill, Bedford, MK45 4BG
G4 LOP C. Hannah, 63 Chauntry Road, Alford, LN13 9HJ
G4 LOR W. Mooney, 21 Windsor Court, Poulton-le-Fylde, FY6 7UX
G4 LOV J. Sutherland, 31 Kensington Road, Sandiacre, Nottingham, NG10 5PD
G4 LOX D. Morton, 9 Metford Grove, Bristol, BS6 7LG
G4 LOY B. Carr, Spring House, Station Road, Grimsby, DN36 5QS
G4 LPD R. Mills, 3 Whitfield Close, Wilford, Nottingham, NG11 7AU
G4 LPE I. Sill, 36 Snowshill Drive, Cheswick Green, Solihull, B90 4JT
G4 LPF S. Swain, 9 Brickyard, Stanley Common, Ilkeston, DE7 6FR
GM4 LPG W. Maslen, Broomie Knowe, Skye Of Curr Road, Grantown-on-Spey, PH26 3PA
GM4 LPJ Dr G. Kolbe, Riccarton Farm, Newcastleton, TD9 0SN
G4 LPL I. Davis, 728 Ridge Rd #30, Lantana, USA, 33462
G4 LPO Dr J. Hampson, Ivy Lodge, Tor Side, Rossendale, BB4 4AJ
G4 LPP Dr P. Holt, Ellon House, Church Road Sutton, Norwich, NR12 9SG
GM4 LPT J. Hopkins, 19 Cairnport Road, Stranraer, DG9 8BQ
GW4 LPU J. Jones, 9 Aelybryn, Ceinws, Machynlleth, SY20 9EZ
G4 LPW C. Clarke, 5 The Cottages, Low Road North Tuddenham, Dereham, nr20 3dg
G4 LPY J. Carter, 147 Maidenway Road, Paignton, TQ3 2PT
G4 LPZ R. Doran, 1 Maple Drive, Chellaston, Derby, DE73 6RD
G4 LQD T. Alderman, 8 Melrose Road, Weybridge, KT13 8UP
G4 LQE N. Bishop, 33 Pollards Green, Chelmsford, CM2 6UH
G4 LQF N. Field, 14 Regent Road, Harborne, Birmingham, B17 9JU
G4 LQG C. Richardson, 25 Hookstone Drive, Harrogate, HG2 8PR
G4 LQH R. Sharpe, Owl Cottage, Royal Oak Lane, Lincoln, LN5 9DT
G4 LQI J. Cavanagh, 133 Brox Road Ottershaw, Chertsey, KT16 0LG
G4 LQL D. Lander, 1 Colby Close, Forest Town, Mansfield, NG19 0LS
G4 LQM T. Mccrimmon, The Square, Newbiggin, Heads Nook, Brampton, CA8 9DH
G4 LQP I. Plant, 6 Randol Close, Mansfield, NG18 5HY
GM4 LQR J. Reid, 80 Bellside Road, Cleland, Motherwell, ML1 5NU
G4 LQW D. Mcniel, C/O G Mcniel, Stable Cottage, Alresford, SO24 0HP
G4 LQX R. Coleman, 35 Meadowside Road, Upminster, RM14 3YT
G4 LRB K. Geen, 34 Kensington Road, Ipswich, IP1 4LD
G4 LRD D. Holt, 241 New Hey Road, Oakes, Huddersfield, HD3 4GH

G4 LRG J. West, 9 Bainbridge Court St. Helen Auckland, Bishop Auckland, DL14 9EJ
G4 LRH G. Obermaier, 9 Milton Park Avenue, Southsea, PO4 8JG
G4 LRL P. Wilkins, 12 Chadcote Way Catshill, Bromsgrove, B61 0JT
G4 LRN N. Barker, 3 Silesbourne Close, Birmingham, B36 9ST
G4 LRO R. Talbott, 33 Highfield Street, Anstey, Leicester, LE7 7DU
G4 LRP A. Boyd, 5 Walmer Close, Southwater, Horsham, RH13 9XY
G4 LRT S. Berry, Hillview, Stanford Close, Cold Ashby, Northampton, NN6 6EW
GM4 LRU T. Hood, 29 Thomson Crescent Port Seton, Prestonpans, EH32 0AN
G4 LRV N. Bundle, Whiteway Cottage, Moorside, Sturminster Newton, DT10 1HQ
G4 LSA J. Bell, Byanna Cottage, Sturbridge, Stafford, ST21 6LE
G4 LSE K. Darton, 18 Highfield Avenue, Bishop's Stortford, CM23 5LS
G4 LSG S. Smith, 1 Parkland Crescent, Norwich, NR6 7RQ
G4 LSK A. Sate, 2 Lynns Hall Close, Great Waldingfield, Sudbury, CO10 0FH
G4 LSL B. Lawrence, 42 Cross St., Crowle, Scunthorpe, DN17 4LH
G4 LSQ P. Elmer, 6 Elmers Lane, Kesgrave, Ipswich, IP5 2GW
G4 LSU A. Burnett, 72 Ightham Road, Erith, DA8 1LU
G4 LSV C. Herrett, 61 Mansfield Road, Alfreton, DE55 7JN
G4 LSX G. Pearce, 13 Walnut Close, Nailsea, Bristol, BS48 4YH
G4 LTC J. Diment, 16 Riverside Walk, Isleworth, TW7 6HW
G4 LTH J. Allan, 13 Vincent Close, Corringham, Stanford-le-Hope, SS17 7QL
G4 LTI M. Coverdale, 1A Halton Chase, Westhead, Ormskirk, L40 6JR
G4 LTK P. Hinks, 1 Richard Joy Close, Holbrooks, Coventry, CV6 4EY
GM4 LTL N. Hyde, 18 Mansefield, Methlick, Ellon, AB41 7DF
G4 LTM G. Hudsmith, 17 Greenside Close, Dukinfield, SK16 5HS
G4 LTS B. Packington, 83 Fitzroy Road, Whitstable, CT5 2LE
G4 LTT A. Willetts, 43 Galloway Avenue, Birmingham, B34 6JL
G4 LTZ J. Peake, 8 Surrey Drive, Congleton, CW12 1NU
G4 LUA R. Gathergood, 37 Hawkley Drive, Tadley, RG26 3YH
G4 LUB D. Waldron, 1 Galbraithe Close, Bilston, WV14 8HX
GM4 LUD R. Bannerman, 20 Post Box Road, Birkhill, Dundee, DD2 5PX
G4 LUE E. Bailey, 8 Hild Avenue Cudworth, Barnsley, S72 8RN
G4 LUF R. Irish, 15 Tenter Hill, Wooler, NE71 6DB
G4 LUN A. Stickland, 3 Kivernell Road, Milford On Sea, Lymington, SO41 0PP
G4 LUO C. Morgans, Merlewood, Maidstone Road, Sittingbourne, ME9 7QA
G4 LUQ M. Tust, 21 Laneside Close, Chapel En le Frith, SK23 0TS
GM4 LUS S. Smith, 80 Deanburn Park, Linlithgow, EH49 6HA
G4 LUT W. Terry, Morston, 121 Lodge Lane, Grays, RM17 5SF
G4 LUW E. Johnson, 29 Watering Lane, Collingtree, Northampton, NN4 0NJ
GW4 LUX P. Biddle, 14 Crossways Park, Howey, Llandrindod Wells, LD1 5RD
G4 LUY D. Chubb, 11 Pelham Close, Bembridge, PO35 5TS
G4 LVA A. Lucas, 4 Hewell Close, Kingswinford, DY6 7RQ
G4 LVD B. Durrant, 140 Fletcher Road, Ipswich, Ip30la
G4 LVG D. Halls, 7 Raeburn Road, Ipswich, IP3 0EW
G4 LVI A. Entwistle, 68 Sandy Lane, Stretford, Manchester, M32 9BX
G4 LVK A. Kelly, 8 Green Slade Crescent, Marlbrook, Bromsgrove, B60 1DS
G4 LVO V. Stretch, 5 Ledwych Road, Droitwich, WR9 9LA
G4 LVR Dr B. Roe, 7 Abbey Fields, Crewe, CW2 8HJ
G4 LVV A. Hanson, 1 Church Street, Kempsey, Worcester, WR5 3JG
GM4 LVW Dr M. Bowman, 5 Whinfield Gardens, Prestwick, KA9 2PW
G4 LWB P. Smith, 2A Kirby Lane, Kirby Lodge, Melton Mowbray, LE13 0BY
G4 LWC L. Collins, 44 Hollybush Lane, Penn, Wolverhampton, WV4 4JJ
GW4 LWD W. Chandler, 19 Cilhaul Terrace, Mountain Ash, CF45 3ND
G4 LWF P. Green, 1 Haddon Croft, Hayley Green, Halesowen, B63 1JQ
G4 LWG I. Lambert, 21 East View Terrace, Barnoldswick, BB18 5NW
GW4 LWL K. Edwards, 25 Gareth Close, Thornhill, Cardiff, CF14 9AF
G4 LWN R. Nock, 83 Coles Lane, West Bromwich, B71 2QW
G4 LWQ G. Simpson, Floral Cottage, 11 Summerwood Road, Street, BA16 0RL
G4 LWU R. Moore, 45 Lime Kiln Way, Salisbury, SP2 8RN
G4 LWV E. Videan, 40 Guessens Grove, Welwyn Garden City, AL8 6RF
G4 LWY J. Bryce, 6A Cawley Avenue, Culcheth, Warrington, WA3 4DF
GW4 LWZ Chepstow + District A.R.S, c/o S. Trott, 6 Mounton Drive, Chepstow, NP16 5EH
G4 LXA D. Gibson, 10 Church Close, Braybrooke, Market Harborough, LE16 8LD
G4 LXC P. Johnson, 148 Broadmead, Tunbridge Wells, TN2 5NN
G4 LXD Dr F. De Bass, Hawthorns, 10 Melville Road, Thetford, IP24 1NG
G4 LXH D. Jones, 34 Alpha Grove, Isle Of Dogs, London, E14 8LH
G4 LXJ C. Phillips, Bangla, Silchester Road, Tadley, RG26 5EP
GM4 LXM G. Low, 23 Bellfield Road, North Kessock, Inverness, IV1 3XU
GW4 LXO Dr J. Eastment, 211 Pantbach Road, Rhiwbina, Cardiff, CF14 6AE
G4 LXR R. Hooper, 88 Ninehams Road, Caterham, CR3 5LJ
G4 LXU C. Lennox, Lowfield House, Low Street, York, YO61 4QA
G4 LXV A. Rose, 40 Wilson Drive, Outwood, Wakefield, WF1 3DN
G4 LXW A. Trousdale, 65 Low Moor Side, New Farnley, Leeds, LS12 5EA
G4 LXX I. Taylor, Heatherlands, Felixstowe Road, Ipswich, IP10 0DE
G4 LXY D. Millin, Flat G12A, Elizabeth Court, Bournemouth, BH1 3DX
G4 LYB C. Kitchener, Hamrest 5, Mill Road, Cromer, NR27 0BG
G4 LYC P. Collett, 7 Saxon Rise, Earls Barton, Northampton, NN6 0NY
G4 LYD D. Palmer, 123 Bucklesham Road, Kirton, Ipswich, IP10 0PF
G4 LYE N. Pilling, 22 Templar Way, Selby, YO8 9XH
G4 LYF Dr A. Webb, 11 Crowland Road, Haverhill, CB9 9LE
G4 LYG J. Lavis, Brea Croft, St Just, TR19 7RN
G4 LYJ R. Carter, 11 Ash Close, Shrewsbury, SY2 6HU
G4 LYL H. Bonnor, 1 Christchurch Road, Winchester, SO23 9SR
G4 LYM G. Schiffeldrin, 68 The Fairway, Alwoodley, Leeds, LS17 7PD
G4 LYU B. Gauntlett, 4 Sandbanks Gardens, Hailsham, BN27 3TL
GM4 LYV W. Hattie, 47 Border Way, Kirkintilloch, Glasgow, G66 2BD
G4 LYX J. Wylie, 15 Semley Road, Hassocks, BN6 8PD
G4 LYY J. Schoolar, 140 Slades Road, Golcar, Huddersfield, HD7 4JR
G4 LZD G. Reading, 73 Mayflower Close, Dartmouth, TQ6 9JN
G4 LZE C. Lugard, 5 Woodland Gardens, South Croydon, CR2 8PH
G4 LZJ P. Garnett, Drewen Garth, Church St, Hull, HU11 4RN
G4 LZK R. Broughton, 18 Elim Court Gardens, Crowborough, TN6 1BS
GM4 LZO G. Mcdonald, Ellrigg, Ballencrieff Toll, Bathgate, EH48 4LD
GW4 LZP R. Smith, 4 Glan Ysgethin, Talybont, LL43 2BB
G4 LZQ G. Williams, 2 Whitminster Lane, Frampton On Severn, Gloucester, GL2 7HR
GI4 LZR W. Turner, 31 Thiepval Avenue, Belfast, BT6 9JF
GI4 LZS J. Smyth, 12 Cleland Park Central, Bangor, BT20 3EP
G4 LZT R. Brown, 40 Pegholme Drive, Otley, LS21 3NZ
G4 LZU E. Hayden, Firbank, 1 Watery Lane, Taunton, TA3 5BX
G4 LZV K. Brazington, 38 Tamworth Road, Amington, Tamworth, B77 3BT
G4 LZZ A. Siemieniago, 3 Skye Close, Highworth, Swindon, SN6 7HR
G4 MAB M. Barry, 19 Trinity Road, Northampton, NN3 3FA
GI4 MAC Dr M. Mckinney, 117 Downpatrick Road, Crossgar, Downpatrick, BT30 9EH
G4 MAG D. Lucas, 23 Rectory Close, Wistaston, Crewe, CW2 8HG
GM4 MAI Dr A. Mcwilliam, Lochaber, Braehead, Avoch, IV9 8QL
GI4 MAJ W. Mcclintock, 37 Belfast Road, Larne, BT40 2PH
G4 MAK J. Gregg, 11 Talbot Terrace, Rothwell, Leeds, LS26 0DR
G4 MAR C. Rowe, 29 Lucknow Road, Willenhall, WV12 4QF
G4 MAS C. Day, 59 Hoe Lane, Ware, SG12 9LS
G4 MAU Dr D. Birchall, 6 Hillmorton Road, Knowle, Solihull, B93 9JL
GI4 MAZ P. Canny, 7 Drumachose Park, Limavady, BT49 0NY
G4 MB J. Bowes, 20 Broomfield Road, Bexleyheath, DA6 7PA
G4 MBA A. Cowsill, 21 Manor Close, Bromham, Bedford, MK43 8JA
G4 MBC Mid-Beds Contest Assoc, c/o F. Handscombe, Sandholm, Bridge End Road, Red Lodge, Bury St Edmunds, IP28 8LQ
G4 MBD I. Moth, 145 Carisbrooke Road, Newport, PO30 1DG
G4 MBE R. Scargill, 17 Springfield Lane, Morley, Leeds, LS27 9PL
GM4 MBG I. Simpson, 1 Knockhall Road, Newburgh, Ellon, AB41 6BJ
G4 MBH A. Embleton, 4 Daventry Close, Mickleover, Derby, DE3 0QT
G4 MBJ R. Hyett, 18 Escley Drive, Hereford, HR2 7LU
G4 MBK J. Broadbent, Buttercross Cottage, Low Road, Gainsborough, DN21 4ER
GW4 MBL S. Elmore, Eirianfron, Llangoed, Beaumaris, LL58 8PG
GI4 MBM S. Mcconnell, 8 Carnesure Drive, Comber, Newtownards, BT23 5LP
GI4 MBQ C. Black, 5 Woodbrook Park, Warrenpoint, Newry, BT34 3HL
G4 MBZ P. Taylor, 32 Grasmere Road, Lightwater, GU18 5TJ
G4 MCA J. Hanton, 5 St. Davids Drive, Thorpe End, Norwich, NR13 5HR
G4 MCE A. Audcent, 9 Woodlands, Axbridge, BS26 2AX
G4 MCF C. Begg, 11 Lilac Road, Normanby, Middlesbrough, TS6 0BS
GI4 MCH N. Crymble, 60 Princes Drive, Newtownabbey, BT37 0AZ
G4 MCM D. Hadaway, 22 The Delph, Reading, RG6 3AN
G4 MCQ S. Bailey, 50 Quantock Close, Warmley, Bristol, BS30 8UT
GD4 MCR D. Cannon, 44 Derby Road, Peel, Isle of Man, IM5 1HP
G4 MCU G. Stow, 15 Hawthorne Gardens, Hockley, SS5 4SW
GM4 MCV A. Patterson, Wayside, 37 Abbotsford Road, Galashiels, TD1 3HW
GI4 MCW G. Edgar, 51 Kempe Stones Road, Newtownards, BT23 4SQ
G4 MD P. Howett, 2, Parkfield Road, Stourbridge, DY8 1HD
G4 MDB R. Tokley, 9 Peel Road, Springfield, Chelmsford, CM2 6AQ
G4 MDC J. Divall, 2 Brockswood Lane, Welwyn Garden City, AL8 7BG
GI4 MDD I. Gibson, 4 Ilford Avenue, Belfast, BT6 9SF
G4 MDE K. Hodkinson, 13 Clovelly Road, Edenthorpe, Doncaster, DN3 2PE
G4 MDF Mid-Thames RDF, c/o B. Bristow, Club Jays Lodge, Reading, RG8 7QG
G4 MDH G. Feary, 76 Parsons Way, Royal Wootton Bassett, Swindon, SN4 8DJ
G4 MDJ A. Smith, 48 Milltown Way, Leek, ST13 5SZ
G4 MDK G. Winfield, 328 Stone Road, Stoke-on-Trent, ST4 8NJ
G4 MDM P. Brassington, 42 Dartmouth Avenue, Newcastle, ST5 3NY
G4 MDN D. Fowler, The Dees, Cross Lanes, Gerrards Cross, SL9 0LR
GI4 MDO S. Hewitt, 23 Drumard Road, Portadown, Craigavon, BT62 4HP
G4 MDR A. Farmer, 42 Sunridge Close, Newport Pagnell, MK16 0LT
G4 MDT G. Fitton, 29 Okus Grove, Upper Stratton, Swindon, SN2 7QA
G4 MDU Dr J. Gudgeon, Shillingsworth Cottage, Leckhampstead Road, Wicken, MK19 6BY
GD4 MDY S. Keenan, Fenella Villa, Peveril Road, Peel, Isle Of Man, IM5 1PJ
G4 MDZ S. Cline, 24 Petrel Way, Hawkinge, Folkestone, CT18 7GZ
G4 MEA R. Hutchings, 16 Le Marchant Road, Frimley, Camberley, GU16 8RW
G4 MEB P. Green, 1 Haddon Croft, Hayley Green, Halesowen, B63 1JQ
G4 MEE D. Mobbs, 39 Bramwell Road, Freckleton, Preston, PR4 1SS
G4 MEF B. Rudkin, 18 Beechfield Avenue, Birstall, Leicester, LE4 4DA
G4 MEH M. Hughes, 8 Elm Beds Road, Poynton, Stockport, SK12 1TG
GW4 MEI I. Williams, 27 Y Glyn, Caernarfon, LL55 1HF
G4 MEK C. Chappell, 6 Brayside Avenue, Cowcliffe, Huddersfield, HD2 2PQ
G4 MEM M. David, Ridgeway, Grainbeck Lane, Killinghall, Harrogate, HG3 2AA
G4 MEO B. Elliott, 13 Spring Grove, Sandy, SG19 1EU
GI4 MEQ M. Kelly, 6 Beechdene Gardens, Lisburn, BT28 3JH
G4 MES J. Willis, 1 Cedar Crescent, Royston, SG8 5BP
G4 MET E. Robinson, 60 Huntsmans Drive, Hereford, HR4 0PN
G4 MEX M. Care, 12 Hallowell Road, Northwood, HA6 1DW
GM4 MFB C. Bowden, West Reidford, Drumoak, Banchory, AB31 5AU
G4 MFD W. Dean, Bowerdene, Staplehay Trull, Taunton, TA3 7HH
G4 MFE R. Kaiser, Blackcoombe Farm, Henwood, Liskeard, PL14 5BW
G4 MFI D. Roberts, 88 Woodhouse Road, Urmston, Manchester, M41 7WX
G4 MFK G. Dennick, 29 Old Post Office Lane, Badsey, Evesham, WR11 7XF
GM4 MFL Easter Ross Rc, c/o R. Kerr, Rosskeen Bridge, Invergordon, IV18 0PR
G4 MFN M. Jones, 67 Dosthill Road, Two Gates, Tamworth, B77 1JD
GM4 MFO M. Mackenzie, Ash Lodge, 8 Brookend Brae, Helensburgh, G84 0QZ
G4 MFP W. Woollen, Greensward, Townsend, Didcot, OX11 0DX
G4 MFQ R. Dunstan, 100 Trevithick Road, St Austell, PL25 4RJ
G4 MFR C. Shanks, 225 Freshfield Road, Brighton, BN2 9YE
G4 MFS M. Smith, 23 Sutcliffe Avenue, Weymouth, DT4 9SA
GI4 MFT Magherafelt ARS, c/o H. Evans, Oville House 404 Foreglen Road, Dungiven, Londonderry, BT47 4PN
G4 MFV J. Marshall, 278 Derby Road, Bramcote, Nottingham, NG9 3JN
G4 MFW B. Fletcher, 53 Onslow Gardens, London, SW7 3QF
G4 MFX G. Davis, Orchard House Peartree Avenue, Martham, Great Yarmouth, NR29 4RJ
G4 MGB H. Mayor, Lock House, Canal Bank, Preston, PR4 6HD
G4 MGD T. Sear, 14 Crowberry Drive, Scunthorpe, DN16 3DE
G4 MGG S. Esposito, 21A Spencefield Lane, Leicester, LE5 6PT
G4 MGH R. Hampson, 30 Witts Lane, Purton, Swindon, SN5 4EX
G4 MGI P. Kemmis, Coton Clanford Farm Coton Clanford, Stafford, ST18 9PE
G4 MGK W. Brown, 2 Ashworth Park, Knutsford, WA16 9DE
G4 MGN M. Norman, 7 Kingsway, Seaford, BN25 2NE
G4 MGO J. Newman, 20 Marshmoor Mobile Home Park, Wallow Lane, Ipswich, IP7 7BZ
G4 MGP H. Boddy, Greyholme, West Lane, Scarborough, YO13 9AR
G4 MGQ R. Boddy, Greyholme, West Lane, Scarborough, YO13 9AR
G4 MGR Wirral & DARS, c/o D. Jones, 39 Pensby Road Heswall, Wirral, CH60 7RA

G4 MGV R. Pass, 6 High Street, Hanslope, Milton Keynes, MK19 7LQ
G4 MGW A. Lodge, 92 Cheltenham Road, Gloucester, GL2 0LX
G4 MGX J. Freeman, 5A Beech Avenue, Briar Bank Park, Bedford, MK45 3WE
G4 MGY J. Gibbs, 27 East Hill Park, Knatts Valley, Sevenoaks, TN15 6YF
G4 MHA P. Wright, 2 Windsor Mews, Stanley, DH9 8UH
G4 MHC Malvern Hills Radio Amateurs Club, c/o D. Hobro, 60 Linksview Crescent, Worcester, WR5 1JJ
GI4 MHD G. Quaite, 4 Drakes Bridge Road, Crossgar, Downpatrick, BT30 9EW
G4 MHE R. Musto, The Thatch, Mountain Castle, Dungarvan, Ireland, X35 A073
G4 MHF J. Marshall, Chaseborough House, Village Hall Lane, Wimborne, BH21 6SG
G4 MHJ R. Hewitt, 38 Eastry Road, Erith, DA8 1NN
G4 MHK T. Fougere, 74 Longland Road, Eastbourne, BN20 8HY
G4 MHQ A. Bell, 22 Ryde Place, Lee-on-the-Solent, PO13 9AU
GW4 MHR C. Norman, 31 Cae Braenar, Holyhead, LL65 2PN
G4 MHS N. Naish, 85 Wear Bay Road, Folkestone, CT19 6PR
G4 MHX B. Smith, 6 Howbeck Crescent, Wybunbury, Nantwich, CW5 7NX
G4 MIB D. Senior, Court Barton, Bull Street, Creech St. Michael, Taunton, TA3 5PW
G4 MID E. Pratt, 65 Barton Road, Thurston, Bury St Edmunds, IP31 3PD
GM4 MIG I. Giffen, 57 Glengarry Crescent, Falkirk, FK1 5UE
G4 MIH D. Fenton, Waverley, Warrington Road, Northwich, CW8 2LW
GW4 MII P. Jenkins, 2 Gwynfi Street, Treboeth, Swansea, SA5 7DW
G4 MIJ R. Hunt, 21 Springwell, Ingleton, Darlington, DL2 3JJ
G4 MIK Dr M. Bull, Toad In The Hole, 25 Prospect Park, Tunbridge Wells, TN4 0EQ
GM4 MIM Rev. I. Morrison, 53 Eastcroft Drive, Polmont, Falkirk, FK2 0SU
G4 MIO P. Davies, 9 Place Albert 1E, la Hulpe, Belgium, 1310
GW4 MIP P. Phillips, Woodreefe, Amroth, Narberth, SA67 8NR
G4 MIS N. Allen, 17 Winfield Road, Sedbergh, LA10 5AZ
G4 MIT G. Hurst, The Hollies, Derby, DE6 6NB
G4 MIV K. Gibson, 179 Ratcliffe Road, Sileby, Loughborough, LE12 7PX
G4 MIX M. Howland, 1 Swanton Farm Cottages, Lydden, Dover, CT15 7JN
G4 MJA M. Swift, 4 Embleton Drive, Chester le Street, DH2 3JS
G4 MJC F. Jul-Christensen, 66 Bushey Lodge Cottages, Firle, Lewes, BN8 6LS
GI4 MJD M. Dunne, 26 Duncreggan Road, Londonderry, BT48 0AD
G4 MJF M. Hill, 42 Oaklands Drive, Northampton, NN3 3JL
G4 MJI T. Sturmey, 35 Lane Court, Boscobel Crescent, Wolverhampton, WV1 1QH
G4 MJT F. Harrison, 98 The Stray, South Cave, Brough, HU15 2AL
G4 MJU E. Smith, 256 Stone Road, Stoke-on-Trent, ST4 8NJ
G4 MJW S. Carey, 27 Kilmar Way, St. Cleer, Liskeard, PL14 5LU
G4 MJX G. Fisher, 87 Ethersall Road, Nelson, BB9 0RP
G4 MKD W. Kendal, 9 The Glade, Furnace Green, Crawley, RH10 6JS
G4 MKE A. Plaice, 10 Stockhill Road, Chilcompton, Radstock, BA3 4JL
G4 MKF Dr M. Franks, 13 Fifth Road, Newbury, RG14 6DN
G4 MKG P. Opie, Timbers, Stockers Hill Road Rodmersham, Sittingbourne, ME9 0PL
G4 MKI D. Bray, 180 Greenhill Road, Herne Bay, CT6 7RS
G4 MKP T. Burbidge, 11 The Drift Little Gransden, Sandy, SG19 3DX
G4 MKQ K. Barnes, 2 Zeus Lane, Waterlooville, PO7 8AG
G4 MKR R. Byford, 46 Sutton Mill Road, Potton, Sandy, SG19 2QG
G4 MKT B. Jackson, Meadow Top Farm, Edgeside Lane, Rossendale, BB4 9SD
GM4 MKU J. Flett, 40 Commerce Street, Lossiemouth, IV31 6QH
G4 MKW P. Bowden, 12 Honeywood Road, Horsham, RH13 6AE
G4 MKX C. Gericke, Pear Tree House, Water Street, Bristol, BS40 6AD
G4 MLB R. Padmore, 3 Uldale Close, Nelson, BB9 0ST
G4 MLG A. Denyer, 94 Wood Lane, Chippenham, SN15 3DZ
G4 MLI B. Mitchell, 16 Perhaver Park, Gorran Haven, St Austell, PL26 6NZ
G4 MLO G. Rae, 62 Brunel Drive, Upton, Northampton, NN5 4AJ
G4 MLQ J. Lamont, 9 Deepdale Croft, Barugh Green, Barnsley, S75 1QG
G4 MLR J. Norris, Freshfield House, Freshfield Lane, Haywards Heath, RH17 7HE
G4 MLV L. Gaunt, 31 Moat Hill, Birstall, Batley, WF17 0DX
G4 MLW Dr I. Jones, 114 Tennent Road, York, YO24 3HG
G4 MLY I. Vincent, 40 Treetops Close, London, SE2 0DN
GW4 MM Box 25 Contest Club, c/o T. Kirby, Parsonage House, St. Nicholas, Goodwick, SA64 0LG
G4 MMA K. Barnard, 89 Kings Road, Harrow, HA2 9LD
G4 MMG A. Beecher, 27 Normandale, Bexhill-on-Sea, TN39 3LU
G4 MMH M. Evans, Corners, Howbourne Lane, Uckfield, TN22 4QB
G4 MMI R. Hodge, 20 Linden Grove, Roydon, Diss, IP22 4GJ
GI4 MMJ L. Kirk, 26 Wallace Hill Road, Downpatrick, BT30 9BU
G4 MMT A. Haley, 9 Minster Avenue, Bude, EX23 8RY
G4 MNA L. Meale, 57 Chestnut Drive, Newton Abbot, TQ12 4JZ
G4 MNE J. White, 25 Fulwith Drive, Harrogate, HG2 8HW
GI4 MNF N. Foote, 4 Bushfield Road Moira, Craigavon, BT67 0JB
G4 MNI W. Loucks, 155 Brentwood Rd N, Toronto, Ontario, Canada, M8X 2C8
GI4 MNN R. Barr, 13 Fairhill Walk, Belfast, BT15 4GR
G4 MNP M. Ward, 14 Grayling Mead, Fishlake Meadows, Romsey, SO51 7RU
G4 MNT R. Calkin, 5 Bergen Court, Maldon, CM9 6UH
G4 MNX C. Leece, 101 Wellstead Way, Hedge End, Southampton, SO30 2BH
G4 MOC P. Fawkes, 118 Rhoon Road, Terrington St. Clement, King's Lynn, PE34 4HZ
G4 MOE S. Noke, 48 Hoadley Green, Salisbury, SP1 3HS
GW4 MOG C. Tombs, 14 Heol Merioneth, Boverton, Llantwit Major, CF61 2GS
G4 MOH K. Moran, 23 Dunlin Close Quedgeley, Gloucester, GL2 4GS
G4 MOI D. Bone, 69 Pick Hill, Waltham Abbey, EN9 3LD
GW4 MOK V. Cashmore, 31 Maes Y Dyffryn, Greenfield, Holywell, CH8 7QR
G4 MOL K. Highley, 3 West Hill Drive, Hythe, Southampton, SO45 6DL
G4 MOP H. Williams, 15 Hiawatha, Wellingborough, NN8 3SH
G4 MOT K. Watson, 12 Regency Park Grove, Pudsey, LS28 8QD
G4 MOV E. Durey, 71 Orchard Road, Maldon, CM9 6EW
G4 MOY D. Bristow, Flat 88, Millwood, Sycamore Avenue, Bingley, BD16 1HW
GW4 MOZ J. Upstone, 31 Broadway, Llanblethian, Cowbridge, CF71 7EX
G4 MPA G. Squibb, 36 Frognal Gardens, Teynham, Sittingbourne, ME9 9HU
GM4 MPC D. Smith, 13 Fernie Place, Dunfermline, KY12 9BX
G4 MPG P. Grace, 3 Warwick Grange, Solihull, B91 1DD
G4 MPH N. Simmonds, 77 Main Street, Long Whatton, Loughborough, LE12 5DF
G4 MPI W. Sharples, Foxgrove, Bonchurch Shute, Ventnor, PO38 1NX
G4 MPJ M. Whitfield, 26 Kingsclere Drive, Bishops Cleeve, Cheltenham, GL52 8TG
G4 MPK S. Foster, 25 Thorne Crescent, Bexhill-on-Sea, TN39 5JH
G4 MPL T. Grimbleby, 109 Downfield Avenue, Hull, HU6 7XE
G4 MPO C. Duffy, 87 Allenby Drive, Leeds, LS11 5RX
G4 MPQ K. Clark, 33 Landers Reach, Lytchett Matravers, Poole, BH16 6NB
GM4 MPR D. Miller, Old School, Ackergill, Wick, KW1 4RG
G4 MPT D. Abbott, 42 Rosebery Avenue, Blackpool, FY4 1LB
G4 MPW M. Corbett, Braemar, Heath Mill Lane, Guildford, GU3 3PR
GM4 MPY A. Bell, 48 Greenlaw Crescent, Paisley, PA1 3RT
GI4 MQA M. Bradley, 28 Church Road, Moneyrea, Newtownards, BT23 6BB
G4 MQF A. Ramsey, 51 Queens Road, Warmley, Bristol, BS30 8EJ
G4 MQG C. Winters, 45 Blackbush Spring, Harlow, CM20 3DY
G4 MQK C. Cubitt, The Flint House, Ostend Place, Flat 2, Walcott, Norwich, NR12 0NJ
G4 MQL R. Cuddington, The Barn, Vatch Lane, Stroud, GL67LE
G4 MQP P. Crowe, 22 Ringsbury Close, Purton, Swindon, SN5 4DE
G4 MQQ S. Jones, The Old Granary, Broomsmead, Lapford, EX17 6NA
G4 MQR Dr G. Blower, 30 The Glebe, Cumnor, Oxford, OX2 9QA
G4 MQS P. Elliott, 153 Glenhills Boulevard, Leicester, LE2 8UH
G4 MQV J. Sanderson, 5 Babbacombe Drive, Ferryhill, DL17 8DA
G4 MQW R. Richardson, Hazeldene, Sutton Road Fovant, Salisbury, SP3 5LF
G4 MRB J. Feeley, 177 Rock Street, Sheffield, S3 9JF
G4 MRD Dr C. Scrase, The Holt, Main Road Woolverstone, Ipswich, IP9 1AR
G4 MRK J. Veitch, 14 Dunmore Avenue, Sunderland, SR6 8ET
G4 MRL N. Thomas, 31 Gloucester Street, London, SW1V 2DB
GI4 MRN J. Mccrea, 14 Fairfield Park, Bangor, BT20 4TX
G4 MRQ R. Marchington, 78 Buxton Road, Dove Holes, Buxton, SK17 8DW
G4 MRS Martlesham Radio Society, c/o A. Cook, The Old Vicarage, High Road, Ipswich, IP6 9LP
G4 MRU P. Sables, 76 Sandyfields View, Carcroft, Doncaster, DN6 8JQ
G4 MRW R. Westmeckett, 3 Alton Grove, Portchester, Fareham, PO16 9NJ
G4 MRX J. Cooch, 6 Blackthorn Close, Newton, Preston, PR4 3TU
GI4 MRZ E. Smith, 114 Bloomfield Road South, Bangor, BT19 7HR
G4 MSA D. Price, Peacehaven House, Chalbury, Wimborne, BH21 7EZ
G4 MSE J. Sivapragasam, 1 Treve Avenue, Harrow, HA1 4AL
GW4 MSI P. Needham, Cil Y Sarn, Llanegryn, Tywyn, LL36 9SB
G4 MSJ T. Moan, 23 Laurel Grove, Sunderland, SR2 9EE
G4 MSK W. Wilkinson, 24 Greenway, Bromley, BR2 8EY
GM4 MSL G. Wallace, 21 Rosslyn Court, Rosslyn Avenue, Perth, PH2 0GY
G4 MSN R. Slator, 27 The Drive, Alwoodley, Leeds, LS17 7QB
G4 MSP P. Shaw, 1, Broadgate, Halifax, HX4 9HZ
G4 MSQ F. Watson, Syne Hurst Cottage, Kimbolton Road, Bedford, MK44 2EW
G4 MSW L. Morgan, 22 Stonelea Road, Hemel Hempstead, HP3 9JY
G4 MSY R. Naylor, 89 Pelham Avenue, Grimsby, DN33 3NG
GW4 MTD H. Mcmurray, 14 Hopkin St., Brynhyfryd, Swansea, SA5 9HN
GW4 MTE R. Smith, 6 Lavender Court, Brackla, Bridgend, CF31 2ND
G4 MTF G. Williams, 8 Blythe Close, Newport Pagnell, MK16 9DN
G4 MTG J. Sillitoe, 42 Marsham Road, Kings Heath, Birmingham, B14 5HD
G4 MTH A. Smith, 25 Lindsay Close, Stanwell, Staines, TW19 7LF
GM4 MTI D. Spence, Royal Fern, Dunollie Road, Oban, PA34 5JQ
G4 MTP J. Theodorson, 7 Kingfisher Court, Overstone Lakes, Ecton Lane, Northampton, NN6 0BD
G4 MTR D. Coulter, 9 Taylors Way, Whitehaven, CA28 9PD
G4 MTW F. Cook, 2 Burford Gardens, Sunderland, SR3 1LX
GI4 MTZ G. Downs, 19 Mullaghboy Road, Islandmagee, Larne, BT40 3TT
G4 MUA A. Kingdon, 4 Castlemead Walk, Northwich, CW9 8GP
GI4 MUE J. Gwilt, 207 Clandeboye Road, Bangor, BT19 1AA
G4 MUI D. Brown, 114 Telford Way, High Wycombe, HP13 5TA
GW4 MUJ R. Parker, Llanedw, Rhulen, Builth Wells, LD2 3UU
G4 MUL D. Mayo, 6 Leigh Avenue, Marple, Stockport, SK6 6DF
GI4 MUN A. Lennon, 5 The Drumlins, Ballynahinch, BT24 8HW
G4 MUP G. Rouse, 43 Oakwood Drive, Prenton, CH43 7NX
G4 MUQ C. Ashlin, 13 Brantfell Grove, Bolton, BL2 5LY
G4 MUS D. Elwell, 18 Padgetts Way, Hullbridge, Hockley, SS56LR
G4 MUT Dr T. Hackwill, 6 Ramsbury Drive Earley, Reading, RG6 7RT
G4 MUU F. Westall, 4 Francesca Lodge, Somerford Way, Christchurch, BH23 3QN
G4 MUV S. Nicolle, 17 Allensmore Close Matchborough, Redditch, B98 0AS
G4 MUW G. Weaver, 60 Crispin Road, Winchcombe, Cheltenham, GL54 5JX
GM4 MUZ H. Angus, South Grange Care Centre, Grange Road, Dundee, DD5 4HT
GW4 MVA G. Burhouse, 18 Leopard Moth Road, Sealand, Deeside, CH5 2FX
G4 MVB A. Berrow, 657 Old Lode Lane, Solihull, B92 8NB
G4 MVD S. Cook, 44 Ettrick Drive Sinfin, Derby, DE24 3EA
G4 MVE D. Casey, 18 Sandholme Drive, Ossett, WF5 8QP
G4 MVP J. Paskins, 190 Gore Road, New Milton, BH25 5NQ
GI4 MVQ D. Mccluney, 49 Upper Cairncastle Road, Larne, BT40 2EG
G4 MVS G. Mellett, Weston, Byways, Chichester, PO20 0HY
G4 MVX M. Gardiner, 60 Rochford Way, Walton on the Naze, CO14 8SR
GW4 MVY D. Davies, 48 Bryn Eglur Road, Morriston, Swansea, SA6 7PQ
G4 MVZ R. Pepper, 4 Marine Avenue, Skegness, PE25 3ER
GI4 MWA Dr F. Ruddell, 16 Beechfield Manor, Aghalee, Craigavon, BT67 0GB
G4 MWD I. Shaw, 33 Park Farm Close, Horsham, RH12 5EU
G4 MWF P. Wilkinson, 14 Grasmere Close, Penistone, Sheffield, S36 8HP
G4 MWG Dr I. Grant, Middle Paradise, Garsdale, Sedbergh, LA10 5PH
G4 MWH R. Blythe, 4 Ashlea Close, Selby, YO8 4NY
G4 MWJ R. Featherstone, 3 Fairfield, Coningsby, Lincoln, LN4 4SP
G4 MWL A. Keyworth, 14 Robinson Road, Sheffield, S2 5QW
G4 MWO P. Gaskell, 131 Greenfield Road, Dentons Green, St Helens, WA10 6SH
G4 MWP T. Underhill, 5 Lyndhurst Croft, Eastern Green, Coventry, CV5 7QE
G4 MWQ C. Weir, 1 Ashfield Place, Ilkley Road, Otley, LS21 3PN
G4 MWS Macclesfield & District ARS, c/o G. Acton, 39 Craig Road, Macclesfield, SK11 7YH
G4 MWW J. Mcleod, 156 Brockhurst Road, Gosport, PO12 3BB
G4 MWX L. Payne, 40 Westmorland Drive, Costhorpe, Worksop, S81 9JT
G4 MXE R. Swinnerton, 8 Maple Close, Brereton Green, Sandbach, CW11 1SQ
G4 MXF R. Wallace, 161 Alma Avenue, Hornchurch, RM12 6AT
G4 MXI K. Bruntlett, Cronk-Ny-Mona, King Street, Louth, LN11 0PN
G4 MXM W. Hawkridge, 7 Langdale Gardens, Leeds, LS6 3HB
G4 MXP D. Aunger, 2 Leyland Lodge, Morval, Looe, PL13 1PN
GI4 MXV M. Mckee, 4 Loran Parade, Larne, BT40 2DF
GI4 MXW D. Mckinney, 13 Lynden Gate, Portadown, Craigavon, BT63 5YH
G4 MXY B. Sowerby, 3 Goughs Lane, Bracknell, RG12 2JR
G4 MYB C. Barham, 10 Little Brook Road, Sale, M33 4WG
G4 MYD R. Clay, 7 Dipper Close, Kilkhampton, EX23 9RE
G4 MYE B. Chase, 9 Claremont Drive, Taunton, TA1 4JE

G4 MYN J. Thomas, 42 Allington Drive, Billingham, TS23 3UA
G4 MYQ G. Pettican, 39 Kabin Road, Norwich, NR1 1QA
G4 MYS A. Sillence, 74 Atherley Road Shirley Southampton So15 5Ds, Southampton, SO15 5DS
GI4 MYT W. Stewart, 11 Fairway Gardens, Castlereagh, Belfast, BT5 7PS
G4 MYU A. Summers, 6 Rothesay Road, Brierfield, Nelson, BB9 5RS
G4 MYW B. Mallinson, 7 Barnes Wallis Way, Churchdown, Gloucester, GL3 2TR
G4 MYY E. Ball, Ashleigh 111 Boslowick Road, Falmouth, TR11 2ER
G4 MYZ R. Roscoe, 4 Cobham Villas, Longden, Shrewsbury, SY5 8EP
GW4 MZB R. Mills, Heather Lea, Oaklands, Welshpool, SY21 8HL
G4 MZC B. Horsman, 53 Meadow Drive, Bembridge, PO35 5XU
G4 MZF D. Earnshaw, Po Box 1098, Sedona, USA, 86339
G4 MZH K. Grimshire, 8 St. Peters Close West Buckland, Barnstaple, EX32 0TX
G4 MZI G. Dunn, 20 The Grange, Wombourne, Wolverhampton, WV5 9HX
G4 MZK P. Ansell, 46 Rochford Way, Croydon, CR0 3AD
G4 MZL E. Ailsby, 15 Norman Close, Bridport, DT6 4ET
G4 MZM P. Harper, 9 The Orchard, Market Deeping, Peterborough, PE6 8JS
G4 MZQ R. Bickley, 12 Cemetery Road, Market Drayton, TF9 3BD
G4 MZU C. Harrison, 6 Woodlands Close, Chandler'S Ford, Eastleigh, SO53 5AT
G4 MZV R. Privett, 2 Stevenson Court, Eaton Ford, St Neots, PE19 7LF
G4 MZY D. Plater, 58 Mead Fields, Bridport, DT6 5RF
G4 MZZ J. Powell, 40 Kent Road, Formby, Liverpool, L37 6BQ
G4 NAC D. Bosworth, 13 Burns Road, Kettering, NN16 9LA
GI4 NAE D. Jackson, 1 Cloughey Road, Portaferry, Newtownards, BT22 1ND
G4 NAJ N. Ashdown, Cobwebs, Wilderness Lane, Uckfield, TN22 4HT
G4 NAK R. Morey, 1 Bradfield Cottages Queens Road, Freshwater, PO40 9HB
G4 NAQ C. Maby, 173 Clevedon Road Tickenham, Clevedon, BS21 6RG
G4 NAV G. Quayle, 10 Abbotsford Grove, Timperley, Altrincham, WA14 5AZ
G4 NBC M. Hoare, Chy Noweth, Seworgan, Falmouth, TR11 5QN
G4 NBF A. Topsfield, Wild Willow Cottage, Hancock Lane, Truro, TR2 5DD
G4 NBG Dr C. Budd, 12 Chedworth Close Claverton Down, Bath, BA2 7AF
G4 NBH W. Cockshaw, 14 Shropshire Road, Leicester, LE2 8HW
G4 NBI L. Everton, 18 Markham Road, Sutton Coldfield, B73 6QR
GW4 NBM M. Jones, Rhos Eithin, Brynsiencyn, Llanfairpwllgwyngyll, LL61 6TZ
G4 NBN A. Bergman, Flat 6, Grange Court, Grange Court Road, Bristol, BS9 4DW
GI4 NBO G. Barr, 51 Hillhead Road, Dundonald, Belfast, BT16 1XD
G4 NBP D. Coggins, 23 Fernleaze Coalpit Heath, Bristol, BS36 2SB
G4 NBQ R. Hassell, 2 Regnum Close, Eastbourne, BN22 0XH
G4 NBS A. Collett, 10 Quince Road, Hardwick, Cambridge, CB23 7XJ
G4 NBW J. Alford, 86 Grindleford Road, Great Barr, Birmingham, B42 2SQ
GW4 NBY K. Barrett, Gorbio House, 47 West Road, Bridgend, CF31 4HD
G4 NCA P. Cook, 38 Oak Road, Kettering, NN15 7AP
G4 NCB K. Wooffindin, Viewlands, Milford Road, Leeds, LS25 6AF
G4 NCD K. Morey, Iona, Colwell Road, Totland Bay, PO39 0AH
G4 NCI R. Smith, 25 Sisters Way, Birkenhead, CH41 4FF
G4 NCJ J. Short, Allybere, Marhamchurch, Bude, EX23 0HY
G4 NCK P. Shapero, 3 Princess Court, Leeds, LS17 8BY
G4 NCP M. Carver, 45 Harvester Way, Crowland, Peterborough, PE6 0DG
G4 NCS C. Angove, 9A Wanstead Road, Bromley, BR1 3BL
G4 NCU M. Hewitt, Hillcrest Bungalow, Middle Street, Crewkerne, TA18 8LY
G4 NCV L. Hollingworth, 55 Glenfield Avenue, Nuneaton, CV10 0DZ
G4 NCY I. Woomans, 223 Umberslade Road, Selly Oak, Birmingham, B29 7SG
G4 NCZ J. Ramsay, 79 Humphrey Lane, Urmston, Manchester, M41 9PT
G4 NDC D. Mason, 133 Bath Road, Atworth, Melksham, SN12 8LA
G4 NDD J. Lloyd, 72 Thornyville Villas, Plymouth, PL9 7LD
G4 NDL R. Davies, 2 Torrhill Cottages Godwell Lane, Ivybridge, PL21 0LT
G4 NDM R. Carter, 49 Cambridge Road, West Bridgford, Nottingham, NG2 5NA
G4 NDP J. Burnett, 42 Wentworth Drive, South Kirkby, Pontefract, WF9 3RY
G4 NDR L. Lewis, 653 Main Road, Dovercourt, Harwich, CO12 4NF
G4 NDT Dr R. Bramley, 8 Ivy Bank Park, Bath, BA2 5NF
G4 NDU Prof. A. Bramley, 8 Ivy Bank Park, Bath, BA2 5NF
GM4 NDV W. Rattray, 17 Brownside Road, Cambuslang, Glasgow, G72 8NL
G4 NEA S. Rice, 13 Wigram Way, Stevenage, SG2 9TP
G4 NEE D. Foreman, 1 Stour Valley Close, Upstreet, Canterbury, CT3 4DB
G4 NEG W. Glew, Carinya, Beltoft Belton, DN9 1MB
G4 NEH J. Hankin, Millfield Torpenhow, Wigton, CA7 1JF
GW4 NEI K. Hodge, 16A Mold Road, Mynydd Isa, Mold, CH7 6TD
G4 NEJ K. Jackson, 7533 Park Spring Circle, Orlando, USA, fl32815
G4 NEL Nel Rynt L/B Wt, c/o D. Bird, 154 Cherrydown Avenue, Chingford, London, E4 8DZ
G4 NEO C. Digby, 7 Dagnall Road, Olney, MK46 5BJ
G4 NEQ R. Welsh, Holme View, Farleton, Lancaster, LA2 9LF
G4 NER P. Lidbetter, 1 Moor Lane, Westfield, Hastings, TN35 4QU
G4 NEY J. Jarvis, 116 Balland Field, Willingham, Cambridge, CB24 5JU
G4 NFA F. Austin, 45 Southdown Crescent, Cheadle Hulme, Cheadle, SK8 6EQ
G4 NFE J. Edwards, 5 Windmill Rise, York, YO26 4TU
G4 NFF S. Frisby, 19 Woodcock Close, Norwich, NR3 3TB
GI4 NFH R. Jennings, 117 Belsize Road, Lisburn, BT27 4BS
GM4 NFI D. Leckie, 6 Galloway Place, Fort William, PH33 6UH
G4 NFL C. Peel, The Ferns, Park Wood Drive, Newcastle, ST5 5EU
G4 NFP M. Saunby, Teachmore, Jacobstowe, Okehampton, EX20 3AJ
G4 NFR R. Tyson, 49 Strathaird Avenue Walney, Barrow-in-Furness, LA14 3DE
G4 NFS N. Sharples, 47 New Fosseway Road, Bristol, BS14 9LW
G4 NFT G. Mcavoy, Flat 5, Maple Lodge, Douglas Close, Poole, BH16 5HE
G4 NFV J. Clark, 12 Ogle Avenue, Morpeth, NE61 2PN
GI4 NFW J. Hegarty, 1 Cookstown Road, Moneymore, Magherafelt, BT45 7QF
G4 NFY A. Clarke, Ravenswood, Gull Road, Wisbech, PE13 4ER
G4 NGB B. Gliddon, Apartment 1, Idle Shores, Springfield Rd, Woolacombe, EX34 7BX
G4 NGD M. Hadnum, 79 Mowbray Road, Bedford, MK42 9UX
G4 NGF M. Chapman, Millway, Dunton Lane, Lutterworth, LE17 5HX
GM4 NGJ C. Bridges, Highfield, Ballinluig, Pitlochry, PH9 0LG
G4 NGL J. Gass, 2A Orchard Way, Breachwood Green, Hitchin, SG4 8NT
GI4 NGP D. Paul, 4 Draperstown Road, Tobermore, Magherafelt, BT45 5QG
G4 NGR C. Webber, 107 Northfields Lane, Brixham, TQ5 8RN
G4 NGS Dr G. Towler, 77 Worrin Road, Shenfield, Brentwood, CM15 8JL
G4 NGV A. Whittaker, 31 Rydal Road, Haslingden, Rossendale, BB4 4EE
G4 NHA P. Hastilow, 18 Broadway Avenue, Croydon, CR0 2LP
GW4 NHB P. Boyce, 28 Alfreda Road, Cardiff, CF14 2EH
G4 NHD M. Brightman, Patriot'S Arms, 6 New Road, Swindon, SN4 0LU
G4 NHE C. Evans, 4 Fryers Copse, Wimborne, BH21 2HR
G4 NHF A. Jones, 51 Wiclif Way, Nuneaton, CV10 8NH
GW4 NHH Dr M. Buck, 23 Velindre Road, Cardiff, CF14 2TE
GM4 NHI J. Cramond, Robson'S Croft, Dunecht, Aberdeenshire, AB32 7EQ
G4 NHL T. Dixon, 30 Green Lane, Stamford, PE9 1HF
G4 NHN C. Rowsell, 31 Kingsfield Gardens, Bursledon, Southampton, SO31 8AY
G4 NHO J. Pattemore, 2 Edes Cottages Ottways Lane, Ashtead, kt21 2pg
G4 NHP S. Porter, 16 Alexander Close, Sidcup, DA15 8QY
G4 NHQ A. Mcmackin, 33 Poynder Place, Hilmarton, Calne, SN11 8SQ
G4 NHR I. Turner, 74 Diban Avenue, Elm Park Estate, Hornchurch, RM12 4YF
G4 NHT Moorlands & Dis A, c/o C. Beesley, 15 Byron Close, Cheadle, Stoke-on-Trent, ST10 1XB
G4 NHW D. Collins, 22 Stalyhill Drive, Stalybridge, SK15 2TR
GM4 NHX G. Brooks, The Old Post Office, Scotscalder, Caithness, KW12 6XJ
G4 NID G. Bromley, 46 Independent Hill, Alfreton, DE55 7DG
G4 NIF D. Lee, 22 Woodland Rise, Parkend, Lydney, GL15 4JX
G4 NIJ K. Sheldon, Whitehaven, May Tree Road, Pershore, WR10 2NY
G4 NIL R. Henshall, St. Anns, Staplehay, Taunton, TA3 7HB
G4 NIP D. Lewis, 76 Reading Road, Finchampstead, Wokingham, RG40 4RA
G4 NIU M. Sharples, 9 Lower Lune Street, Fleetwood, FY7 6DA
G4 NIV N. Rowcroft, 110 Linceslade Grove, Loughton, Milton Keynes, MK5 8BL
G4 NIX G. Cooke, 7 Lime Grove, Royston, SG8 7DJ
G4 NIY S. Cooke, 5 Honey Way, Royston, SG8 7ES
G4 NIZ R. King, 20 Woodside East, Thurlby, Bourne, PE10 0HT
G4 NJA R. Hewson, 6 Talisman Drive, Bottesford, Scunthorpe, DN16 3SW
G4 NJB K. Hepke, 25 Victoria Avenue, Willerby, Hull, HU10 6DD
G4 NJI A. Corker, 59 Foljambe Road, Rotherham, S65 2UA
G4 NJJ P. Cousins, Roman Lodge Hall Road Clenchwarton, King's Lynn, PE34 4DA
G4 NJK R. Elliott, Stables End Cottage, Fatherford Farm, Okehampton, EX20 1QQ
GW4 NJL D. Gerrard, Salem Chapel, Llanfairtalhaiarn, Abergele, LL22 8SS
G4 NJN A. Bowman, 18 Essex Avenue, Isleworth, TW7 6LF
GI4 NJQ P. Igo, 84 Glebetown Drive, Downpatrick, BT30 6PZ
G4 NJR M. Doe, 2 Summerfields, Yarnfield, Stone, ST15 0RH
G4 NJT R. Smith, 2131 - 21St Ne, Salmon Arm, Bc, V1E 2DN
G4 NJW G. Lowes, 26 Huttoft Road, Sutton-On-Sea, Mablethorpe, LN12 2QY
GI4 NKB F. Hunter, Flat 9 50 Edenvale Crescent, Belfast, BT4 2BH
G4 NKC M. Jones, Racecourse Farm, Church Lane, Bridgnorth, WV16 4NW
G4 NKI J. Shaw, 1A Greenways, The Stookes, Chesterfield, S40 3HF
GI4 NKK K. Planck, 4 Westland Drive, Ballywalter, Newtownards, BT22 2TH
G4 NKP D. Mellin, 1 Whitfield Close, Warminster, BA12 9HX
GW4 NKR G. Lewis, The Old, Post Office Cottage, Brecon, LD3 0UR
G4 NKU D. Dunn, 37 Ridgemead, Calne, SN11 9EW
G4 NKW G. Hilton, 8 Sandwich Close, St Ives, PE27 3DQ
G4 NKX P. Digby, 3 Copperview Mews, Cowplain, Waterlooville, PO8 8BW
GI4 NKY W. Campbell, 68 Richmond Court, Lisburn, BT27 4QX
G4 NLA G. Goodrich, 29 Cresswells, Corsham, SN13 9NJ
G4 NLB B. Horne, 77 Surrey Hills Residential Park, Boxhill Road, Tadworth, KT20 7LZ
G4 NLC P. Hedison, 85 Moorhouse Lane, Whiston, Rotherham, S60 4NH
GW4 NLD P. Frost, 1 Chester Street, Rhyl, LL18 3ER
G4 NLG F. Askew, 9 The Hall Spinney, Howden, Goole, DN14 7FD
G4 NLH D. Haydon, 50 Ward Close, Stratton, Bude, EX23 9BB
G4 NLI R. Scott, 10 Middle St., North Perrott, Crewkerne, TA18 7SG
GM4 NLJ J. Martin, Whitehill Foot Farm, Kelso, TD5 8LB
G4 NLK R. Southall, 7 The Willows, Brereton, Rugeley, WS15 1EP
G4 NLL D. Bell, 5 Byron Court, Dalton On Tees, Darlington, DL2 2PX
G4 NLO K. Butcher, 18 Windmill Street Whittlesey, Peterborough, PE7 1HJ
GI4 NLQ P. Burns, 41 Lambeg Road, Lambeg, Lisburn, BT27 4QA
G4 NLU R. Williams, 50 Hemerdon Heights, Plympton, Plymouth, PL7 2EY
G4 NLW J. Wilding, 8 Millbrook Way, Brierley Hill, DY5 3YY
G4 NMC D. Willis, 41 Chadbrook Crest, Richmond Hill Road, Birmingham, B15 3RL
G4 NMD Rev. G. Smith, 6 Birtley Rise, Bramley, Guildford, GU5 0HZ
G4 NMF W. Squire, 7 Essex Crescent, Seaham, SR7 8DZ
G4 NMK K. Petre, 24 Harrogate Terrace, Murton, Seaham, SR7 9PQ
G4 NMO J. Gallagher, 1 Wythburn Mews, Langdale Road, Woodlesford, Leeds, LS26 8FJ
G4 NMP B. Dudhill, Westfield, Mablethorpe, L N12
G4 NMR S. Harvey, 2 Draycote Close, Worcester, WR5 3SY
G4 NMS S. Burgess, Borne House, Romsey Road, Stockbridge, SO20 6PR
G4 NMT A. Godsiff, 59 Vinson Close, Orpington, BR6 0EQ
G4 NMU R. Jones, Flat 4, Russell Court, Southport, PR9 8NY
G4 NMV J. Baxter, 16 Avon Close, Weston-Super-Mare, BS23 4QS
G4 NMY P. Bennett, 45 Ravenbank Road, Luton, LU2 8EJ
G4 NNB G. Mackie, 8 The Avenue, Biggleswade, SG18 0PS
GM4 NNH J. Barber, 156 Jamphlars Road, Cardenden, Lochgelly, KY5 0ND
G4 NNI E. Bailey, 213 Ashby Road, Hinckley, LE10 1SJ
G4 NNJ A. Forster, 31 Par Four Lane, Lydney, GL15 5GB
GM4 NNK D. Harkness, 22 Brockwood Crescent, Blackburn, Aberdeen, AB21 0JZ
GW4 NNL M. Jones, Jodanare, 72B Princes Drive, Colwyn Bay, LL29 8PW
GI4 NNM J. Mcgillian, 48 Millfield, Ballymena, BT43 6PB
G4 NNN C. Winterflood, 12 Bourne Road, Colchester, CO2 7LQ
GW4 NNO T. Hadley, Pastures Green Llandderfel Ll23 7Rf, Bala, LL23 7RF
G4 NNP D. Andrews, 10 Brantwood Drive, Paignton, TQ4 5HZ
G4 NNS B. Coleman, Woodlands, Redenham, Andover, SP11 9AN
G4 NNX D. Ward, 18 Henders, Stony Stratford, Milton Keynes, MK11 1RB
G4 NNY D. Ransford, 52 Loughborough Road, Bunny, Nottingham, NG11 6QD
G4 NNZ G. Martorano, 81 Sapcote Drive, Melton Mowbray, LE13 1HG
G4 NOB R. Burbeck, 20 St. Johns Road, Smalley, Ilkeston, DE7 6EG
G4 NOC N. Black, 7 Woodland Crescent, Bracknell, RG42 2LH
G4 NOE M. Hollinghurst, 30 Hall Road, Cheltenham, GL53 0HE
G4 NOL B. Robinson, 9 Henderson Road Norwich Nr4 7Jw, Norwich, NR4 7JW
GW4 NOO M. Oliver, 6 Flemish Close, St. Florence, Tenby, SA70 8LT
G4 NOP R. Simpson, 14 Cumberland Way, Dibden, Southampton, SO45 5TW
G4 NOR C. Heaps, 12 Oak Tree Close, Mansfield, NG18 3EN
GW4 NOS R. Hopkins, 8 Shady Road, Gelli, Pentre, CF41 7UG
G4 NOT C. Sturgeon, Windyridge, Linkside West, Hindhead, GU26 6PA
G4 NOU R. Wigmore, Little Landguard, Whitecross Lane, Shanklin, PO37 7EJ

G4 NOX K. Campbell, 4 Orchard Close, Rowlands Gill, NE39 1EQ
G4 NOY J. Mills, 103 Irby Road, Wirral, CH61 6UZ
G4 NPA A. Abbot, 4 Nursery Drive, Birmingham, B30 1DR
G4 NPB Dr S. Abbot, 4 Nursery Drive, Birmingham, B30 1DR
GW4 NPC R. Blayney, 42 Ty Draw, Church Village, Pontypridd, CF38 1UF
G4 NPD G. Chamberlain, 13 Mayford Close, Beckenham, BR3 4XS
G4 NPG P. Duffy, 15 The Glade, Sheldon, Birmingham, B26 3PW
G4 NPH J. Arnold, 2 Duck Lane, Haddenham, Ely, CB6 3UE
G4 NPN D. Westwood, 1 Elmfield Road, Hartlebury, Kidderminster, DY11 7LA
G4 NPS J. Woolliss, Wharfdale, 245 Scartho Road, Grimsby, DN33 2EA
G4 NPU P. Williams, 122 Longhurst Lane, Mellor, Stockport, SK6 5PG
G4 NPY C. Read, 3 Wyrley Close, Lichfield, WS14 9DA
G4 NQB S. Mould, 53 Wolverley Avenue, Stourbridge, DY8 3PJ
G4 NQC J. Neal, 60 Balsdean Road, Brighton, BN2 6PF
G4 NQI R. Atterbury, Lomas De San Jose 20, la Vinuela, Spain, 29712
G4 NQJ C. Brown, Kingsdown Cottage, Fron, Montgomery, SY15 6SB
G4 NQL M. Churms1, 123 Urb Buenavista, Guadamar, Spain, 3140
G4 NQM M. Drohan, 23 Lindholme Drive, Rossington, Doncaster, DN11 0UP
G4 NQQ N. Hemmings, 3 Church Close, Shapwick, Bridgwater, TA7 9LS
G4 NQS R. Young, 79 Cradge Bank, Spalding, PE11 3AF
G4 NQW P. Perrins, 9 Merrick Close, Hayley Green, Halesowen, B63 1JY
G4 NQZ R. Riley, 103 St. Nicolas Park Drive, Nuneaton, CV11 6DZ
G4 NRA J. Tisdale, 12 Digby Road, Kingswinford, DY6 7RP
G4 NRC Raynet-UK, c/o F. Mossop, 4 Brookdale Way, Waverton, Chester, CH3 7NT
G4 NRD A. Lindsay, 21 Willow Road, Four Pools Industrial Estate, Evesham, WR11 1YW
G4 NRE W. Ward, 88 Central Road, Cromer, NR279BW
G4 NRF H. Westwood, 67 Bedford Close, Featherstone, Pontefract, WF7 5LH
G4 NRG R. Greengrass, Shanaclune, Dunhill, Ireland, X91 Y0H3
G4 NRH D. Whitehead, 50 Southey Lane, Kingskerswell, Newton Abbot, TQ12 5JG
G4 NRI M. Dhami, 118 Havelock Drive, Brampton, Ontario, Canada, L6W 4E3
G4 NRK Newborough Radio Klub, c/o P. Staton, 52 School Road, Newborough, Peterborough, PE6 7RG
G4 NRO J. Coupe, 3 Longfield Avenue Coppull, Chorley, PR7 4NT
G4 NRP J. Fish, Kennels Cottage, Manor Road, Upper Bentley, Redditch, B97 5TB
GI4 NRQ Coleraine & Dis, c/o J. Hamill, 67 Windsor Avenue, Coleraine, Ireland, BT52 2DR
G4 NRR N. Astbury-Rollason, Fern Lodge, The Parade, Newton Abbot, TQ13 0JH
G4 NRS P. Goding, 29 Caistor Drive, Hartlepool, TS25 2QG
G4 NRT Cmdr. D. Bondy, 328 Wilson Avenue, Rochester, ME1 2ST
G4 NRV A. Diplock, 24 Billings Hill Shaw, Hartley, Longfield, DA3 8EU
G4 NRX S. Mantell, 50 Coleshill St., Fazeley, Tamworth, B78 3RA
G4 NRY I. Mantell, 34 Piccadilly, Tamworth, B78 2EP
G4 NRZ K. Moody, 5 Moore Road, Mapperley, Nottingham, NG3 6EF
G4 NS J. Hudson, 22 Essex Gardens, Marsden, South Shields, NE34 7JQ
G4 NSA D. Morgan, 12 Rosalind Avenue, Bebington, Wirral, CH63 5JR
G4 NSB Dr J. Winterburn, 164 Malthouse Lane Earlswood, Solihull, B94 5SD
G4 NSC W. Weatherspoon, 12 Greenacres Close, Crawcrook, Ryton, NE40 4TD
G4 NSD I. Mitchell, Greenway Cottage, Greenway, Tatsfield, Westerham, TN16 2BT
G4 NSE J. Rank, 9 Fairfield Crescent, Scarborough, YO12 6TL
G4 NSH S. Robinson, Upper Hambleton Hill Farm, Wainstalls, Halifax, HX2 7TX
G4 NSJ G. Heffer, 35 Henty Road, Worthing, BN14 7HE
GM4 NSL G. Greenlees, 22 Hunters Grove, Hunters Quay, Dunoon, PA23 8LQ
G4 NSM A. Bruce, 5 Orchard Croft, Epworth, Doncaster, DN9 1LL
G4 NSN M. Craft, 8 Juniper Road, Farnborough, GU14 9XU
G4 NSO S. Auckland, 6 Lombard Crescent, Darfield, Barnsley, S73 9PP
GI4 NSS L. Robinson, 52 Ballymacormick Avenue, Bangor, BT19 6AY
G4 NST S. Thorpe, 11 Grove Road, Hethersett, Norwich, NR9 3JP
G4 NSW F. Lacey, 27 Howards Gardens, New Balderton, Newark, NG24 3FJ
G4 NSZ M. Stanway, 72 Sheldons Court, Winchcombe Street, Cheltenham, GL52 2NR
G4 NTA P. Allan, 2 Park View, Queensbury, Bradford, BD13 1PL
G4 NTC D. Henderson, 4 Vincent Street, Bolton, BL1 4SA
G4 NTG J. Williamson, 5 Rochester Close, Headless Cross, Redditch, B97 5FP
G4 NTJ A. Rick, 9 Sheldon Close, Loughborough, LE11 5EZ
GM4 NTL J. Mcdermott, Milking Green Gate, Eliock, Sanquhar, DG4 6LD
GD4 NTR G. Kelly, 5 Tynwald Close, Peel, Isle of Man, IM5 1JJ
G4 NTV A. Wilkes, 34 Tideswell Road, Great Barr, Birmingham, B42 2DT
G4 NTW W. Douglas, 2 Rockville, Sunderland, SR6 9EL
GM4 NTX K. Elliott, Northfield Cottage, Denny, FK6 6RB
G4 NTY J. Higson, 5 Primrose Avenue Worsley, Manchester, M28 0TP
G4 NUA E. Williams, 45 Bayford Place, Cambridge, CB4 2UF
G4 NUB M. Tyler, 27 Shakespeare Drive, Dinnington, Sheffield, S25 2RP
G4 NUF C. Muller, 118 Park Lane, Northampton, NN5 6PZ
G4 NUG R. Needs, 13 Greenway, Great Horwood, Milton Keynes, MK17 0QR
G4 NUJ K. Symonds, 30 Fairlea Crescent Northam, Bideford, EX39 1BD
G4 NUK J. Brown, 33 Balmoral Drive, Leicester, LE3 3AD
GM4 NUN G. Mackenzie, 1 Walnut Grove, Blairgowrie, PH10 6TH
G4 NUO A. Mackenzie, Ivy House, 145 High Street, Marske By The Sea, Redcar, TS11 6JX
G4 NUS A. Layland, 16 Park Road, Quarry Bank, Brierley Hill, DY5 2DA
GM4 NUU H. Park, Carndearg, Upper Steelend, Dunfermline, KY12 9LP
G4 NUV R. Richmond, 7 Bishopdale Drive, Watnall, Nottingham, NG16 1LE
G4 NUX C. Smith, Wealdon Cottage, Dunsfold Road, Billingshurst, RH14 0PJ
G4 NUY E. Wharton, Vandling, Well Bank, Well, Bedale, DL8 2QF
G4 NVA J. Dyke, 2 Brooklands Drive, Goostrey, Crewe, CW4 8JB
G4 NVH G. Boull, 80 Ascot Road, Baswich, Stafford, ST17 0AQ
GM4 NVI D. Chapman, 9 Baillieswells Terrace, Bieldside, Aberdeen, AB15 9AR
G4 NVL T. Copeman, 1 Chestnut Avenue, Welney, Wisbech, PE14 9RG
G4 NVM J. Duddridge, 19 Ridgeway, Hurst Green, Etchingham, TN19 7PJ
G4 NVN G. Harper, Pathways, 41 Somerset Close, Congleton, CW12 1SE
G4 NVP K. Kett, 24 Deancourt Drive, New Duston, Northampton, NN5 6PY
G4 NVQ S. Shirley, 93 Alfred Road, Hastings, TN35 5HZ
G4 NVT M. Musgrave, 49 Vowler Road, Langdon Hills, Basildon, SS16 6AQ
G4 NVV R. English, 124 Hillside Road, Portishead, Bristol, BS20 8LG
G4 NVY M. Juffs, 32 Brooklands Park, Longlevens, Gloucester, GL2 0DP
GM4 NWK T. Hill, 3 Swift Crescent, Glasgow, G13 4QN
G4 NWM M. Talbott, 44 Tamworth Road, Amington, Tamworth, B77 3BT
G4 NWN W. Talbott, 44 Tamworth Road, Amington, Tamworth, B77 3BT
G4 NWO K. Chaplin, 1 Beechwood Crescent, Amington, Tamworth, B77 3JH
G4 NWR Nth Wilts Rayne, c/o H. Woolrych, 20 Meadow Drive, Devizes, SN10 3BJ
G4 NWS A. Wheeler, 11 Barley Way, Rothley, Leicester, LE7 7RL
G4 NWW J. Edgeley, 12 The Glade, Horsham, RH13 6DD
G4 NXA D. Kirk, Apartment 28, Riverside, 8 Loxley Park, Sheffield, S6 4TF
G4 NXB J. Evans, 11 Columbia Close, Selston, Nottingham, NG16 6GP
GW4 NXD R. Johns, 12 Woodfield Road, New Inn, Pontypool, NP4 0PT
G4 NXG A. Birch, 6 Crescent Road, Wallasey, CH44 0BQ
G4 NXI R. Light, 72 Badger Rise, Portishead, Bristol, BS20 8AX
GI4 NXJ M. Mcfall, 4 Riverdale Close, Ballyclare, BT39 9WE
G4 NXL M. Street, 41 Shaw Lane, Holbrook, Belper, DE56 0TG
G4 NXO Sheppey Western, c/o A. Collett, 10 Quince Road, Hardwick, Cambridge, CB23 7XJ
G4 NXP D. Taylor, The Acreage, View Farm, 11 Malvern Road, Worcester, WR2 4SF
G4 NXR P. Rose, Cuptree Cottage, 17 The Cross, Colchester, CO7 9QQ
G4 NXS M. Davis, 446 Upper Wortley Road, Scholes, Rotherham, S61 2SS
GM4 NXT W. Davidson, 7A South Street, Aberchirder, Huntly, AB54 7XR
G4 NXV D. Gadsden, 37 Cambridge Street, Wymington, Rushden, NN10 9LG
G4 NXW K. Chesters, 4 Kirton Crescent, Lytham St Annes, FY8 4BJ
G4 NXX D. Harbottle, 33 Robin Crescent, Lyme Green, Macclesfield, SK11 0LJ
G4 NYA R. Hyams, 6 Gaudick Road, Eastbourne, BN20 7QE
G4 NYB R. Knighton, Merry Ways, The Green, Derby, DE72 2BJ
G4 NYC G. Jannetta, 14 Banks Lane, Heckington, Sleaford, NG34 9QY
G4 NYD I. Watling, 5 Claylands Court Bishops Waltham, Southampton, SO32 1JS
G4 NYG D. Routledge, 7 Littleton Croft, Solihull, B91 3XR
G4 NYJ C. Webb, 65 Littlebeck Drive, Darlington, DL1 2TU
G4 NYK Prof. R. Williams, 67A Sea Mills Lane, Bristol, BS9 1DR
G4 NYL S. Brown, 2 Windsor Close, Read, Burnley, BB12 7QH
GU4 NYT N. Le Page, Heathwick, Les Martins, Guernsey, GY4 6QJ
G4 NYV M. Katzmann, 6654 Barnaby Street Nw, Washington Dc, USA, 20015
G4 NYW R. Schoales, Ventura, Towngate, Hope Valley, S33 9JX
G4 NYY P. Ernster, 36 Forest End, Fleet, GU52 7XE
G4 NYZ J. Battle-Welch, 25 Moorcroft Close, Callow Hill, Redditch, B97 5WB
G4 NZB P. Neville, 66 Oak Lodge Avenue, Chigwell, IG7 5HZ
G4 NZC B. Manchett, 12 Old Parsonage Court, Otterbourne, Winchester, SO21 2EP
G4 NZE E. Third, 1 Roman Road, Corby, NN18 8FZ
G4 NZG S. Parsons, Trehunsey Cottage Quethiock, Liskeard, PL14 3SG
G4 NZK B. Laniosh, 47 Barley Mow Lane, Catshill, Bromsgrove, B61 0LU
G4 NZN K. Lowe, Springwood, Priesthorpe Road, Pudsey, LS28 5RE
G4 NZO G. Frederick, 98 Coleridge Park Drive, Winnipeg, Canada, MB R3K 0B5
G4 NZQ P. Brooks, 7 Lindford Drive, Eaton, Norwich, NR4 6LT
G4 NZU R. Wilson, 9 Greythorn Drive, West Bridgford, Nottingham, NG2 7GG
G4 NZX D. Cooper, 6 Swinburne Close, Barnby Dun, Doncaster, DN3 1BS
G4 NZY D. Coles, 91 Grove Lane, Harborne, Birmingham, B17 0QT
G4 NZZ B. Coulson, 19 St. Lukes Close, Kettering, NN15 5HD
G4 OAB M. Clutton, 8 Ash Grove, Runcorn, WA7 5LR
G4 OAE D. Crisp, 20 Crawford Close, Earley, Reading, RG6 7PE
G4 OAG A. Dymott, St. Huberts, Ashey Road, Ryde, PO33 4BB
G4 OAI H. Richter, 84 Roehampton Drive, Wigston, LE18 1HU
G4 OAK S. Richards, 1 Stocks Mead, Washington, Pulborough, RH20 4AU
G4 OAN L. Wild, The Drey, Penny Lane, Bedlington, NE22 6HD
G4 OAR N. Mclaren, Ingleside, Waterloo, Whixall, Whitchurch, SY13 2PX
GM4 OAS G. Liddle, Ashbeck, Morar, Mallaig, PH40 4PD
G4 OAU G. Austin, 38 Willow Crescent, Hatfield Peverel, Chelmsford, CM3 2LJ
G4 OAV S. Ames, 21 Common Lane, Harpenden, AL5 5BT
G4 OAX W. Joiner, 15 Laxton Grove, Great Holland, Frinton-on-Sea, CO13 0SF
GW4 OBA R. Jarvis, 3 Church View Llanwenog, Llanybydder, SA40 9UU
G4 OBB Dr D. Kaylor, 27 Fair Green, Glemsford, Sudbury, CO10 7PH
G4 OBC M. Taylor, 5 Hawford Avenue, Kidderminster, DY10 3BH
GM4 OBD G. Sangster, 36 St. Marys Drive, Ellon, AB41 9LW
G4 OBE R. Snary, 12 Borden Avenue, Enfield, EN1 2BZ
G4 OBK P. Catterall, 54 Westlands, Pickering, YO18 7HJ
G4 OBN S. Harding, 21 Abbey Road, Medstead, Alton, GU34 5PB
G4 OBT N. Day, Tanhouse Farm, Rusper Road, Dorking, RH5 5BX
G4 OBV E. Day, 42 Grosvenor Close, Bishop's Stortford, CM23 4JP
G4 OBX R. Dobson, 40 Dipton Gardens, Sunderland, SR3 1AN
GM4 OCA P. Windsor, Hillside Overbrae, Fisherie, Turriff, AB53 5QP
G4 OCC J. Johnston, 6482 Doctor Blair Crescent, North Gower, Ontario, Canada, K0A 2T0
G4 OCF D. Clayton, Flat 6 Clifton Court, Clevedon, BS21 6QS
G4 OCH K. Dickens, The Old Post Office, Cleobury Road, Ground Floor Flat, Bewdley, DY12 2QG
G4 OCJ Dr E. Mullock, 14 Cottage Walk Shawclough, Rochdale, OL12 6DZ
GI4 OCK J. Mackay, 12 Lynne Road, Bangor, BT19 1NT
GI4 OCL R. Mccurry, 82 Cumberland Road, Dundonald, Belfast, BT16 2BB
G4 OCQ I. Blackman, 69 Thorntons Close, Pelton, Chester le Street, DH2 1QH
G4 OCR Dr M. Butler, 41 Neale Road, Chorlton Cum Hardy, Manchester, M21 9DP
G4 OCU D. Gipp, 7 Edmunds Road, Cranwell Village, Sleaford, NG34 8EL
G4 OCX L. Jewell, Ashbrook Farm, Mill Hill, Bedford, MK44 2HP
G4 OCZ C. Richardson, 149 Old Fort Road, Shoreham-by-Sea, BN43 5HL
G4 ODA B. Tatnall, Poplar House, Delgate Bank, Weston Hills, Spalding, PE12 6DH
G4 ODD M. Mathers, Rose Cottage, Kirton Road Egmanton, Newark, NG22 0HF
G4 ODE Dr N. Dovaston, 53 Elmway, Chester le Street, DH2 2LX
G4 ODF D. Faulkner, Amber Croft, Dale Close, Mansfield, NG20 9EB
G4 ODG V. Cawthron, 8 Clay Hill Road, New Quarrington, Sleaford, NG34 7TF
G4 ODI A. Dyer, 34 Oakley Road, Chinnor, OX39 4HB
G4 ODM C. Mott-Gotobed, 14 Copse Road, New Milton, BH25 6ES
GW4 ODN A. Whitticombe, 160 Haven Drive, Hakin, Milford Haven, SA73 3HN
G4 ODR Enfield Contest Group, c/o J. Young, 30 Crofton Way, Enfield, EN2 8HS
GI4 ODT W. Barker, 96 Highlands Road, Limavady, BT49 9LY
GM4 ODW D. Maclean, Gramaiche Donavourd, Pitlochry, PH16 5JS
GJ4 ODX S. Langlois, L'Amarrage, La Route Orange, St Brelade, Jersey, JE3 8GP
GD4 OEA C. Gerrard, 6 Rheast Lane, Peel, Isle Of Man, IM5 1BE
G4 OEB P. Grant, 24 Dowlands Road, Bournemouth, BH10 5LG
G4 OEC E. Mcpheat, Dyche Old School House Holford, Bridgwater, TA5 1SF
G4 OED G. Perry, 12 Boydell Close, Shaw, Swindon, SN5 5QT
G4 OEF D. Bayliss, 38 Yarborough Crescent, Lincoln, LN1 3LU

G4 OEH N. Rumble, 24 Firle Road, North Lancing, Lancing, BN15 0NZ
GW4 OEJ A. England, 9 Priory Road, Milford Haven, SA73 2DS
G4 OEK R. Jones, 17B Plumpton Park Road, Doncaster, DN4 6SQ
G4 OEM G. Hooker, 42A Nether Hall Road, Doncaster, DN1 2PZ
G4 OEP Dr A. Smith, 15 Dyrham Close, Henleaze, Bristol, BS9 4TF
G4 OEQ C. Thomas, 69 Quakers Road, Downend, Bristol, BS16 6JG
G4 OER D. Warner, 28 Jameson Bridge Street, Market Rasen, LN8 3EW
GW4 OES A. Pickard, 89 Ael Y Bryn, Llanedeyrn, Cardiff, CF3 7LL
G4 OEU Q. Campbell, 8 Cookson Close, Corbridge, NE45 5HB
G4 OEX G. Jones, 3 Kings Mews, Bedford Street, Warrington, WA4 6GY
G4 OEY Dr T. Sanderson, Backershagenlaan 32, Wassenaar, Netherlands, 2243 AD
GM4 OEZ W. Taylor, 2 Jubilee Terrace, Findochty, Buckie, AB56 4QA
G4 OFA B. Millican, 24 Wellington Terrace, Bramley, Leeds, LS13 2LH
GM4 OFC J. Snelgrove, Mill House, South Bridgend, Crieff, PH7 4DH
GM4 OFI J. Robertson, Springwell House, Auckengill, Wick, KW1 4XP
G4 OFN P. Edmonds, 170 Halton Road, Sutton Coldfield, B73 6NZ
G4 OFO N. Baynes, 3 Charles Babbage Close, Chessington, KT9 2SA
G4 OFP J. Knowles, 6 Seaway Gardens, Paignton, TQ3 2PE
G4 OFU R. Burdess, 33 The Green, Dartford, DA2 6JS
G4 OGB L. Elliott, Elm Lodge, 2 Hood Croft, Doncaster, DN9 2FB
G4 OGG J. West, 2 Sanders Close, Kempston, Bedford, MK42 8RX
G4 OGL J. O'Brien, 5 Highfields Mead, East Hanningfield, Chelmsford, CM3 8XA
GM4 OGM S. Mather, Pentland View, Limekiln Road, West Linton, EH46 7BA
GW4 OGO S. Williams, 31 Kensington Park, Magor, Caldicot, NP26 3QG
GI4 OGQ W. Kernohan, 3 Camphill Park, Ballymena, BT42 2DQ
G4 OGW D. Thomas, Handley Cross Cottage, Harewood End, Hereford, HR2 8JT
G4 OGZ M. Walker, 52 Derwent Road, Harpenden, AL5 3NX
GW4 OH A. David, 45 Amanwy, Llanelli, SA14 9AH
G4 OHA L. Lux, Hyde Brae, Hyde Hill Chalford, Stroud, GL6 8NY
G4 OHB P. Taylor, Flat 11, Romsley Hill Grange, Farley Lane, Romsley, Halesowen, B62 0LN
G4 OHC R. Poore, 8 Ainsdale Close, Worthing, BN13 2QX
G4 OHF G. Bean, 141 Narborough Road, Leicester, LE3 0PB
GI4 OHH D. Cox, 32 Kilmaconnell Road, Castleroe, Coleraine, BT51 3QZ
G4 OHJ J. Porter, 77 Westholme Road, Bidford-On-Avon, Warwickshire, B50 4AN
G4 OHM 5th Bham Rnt Gr, c/o J. Parkin, 18 Bradnock Close, Birmingham, B13 0DL
G4 OHP L. Sharps, 68 Vicarage Lane, Elworth, Sandbach, CW11 3BU
G4 OHQ J. Reade, 7 Wilmar Close, Hayes, UB4 8ET
GM4 OHT T. Mitchell, 12 Dalriada Place, Kilmichael Glassary, Lochgilphead, PA31 8QA
G4 OHV C. Addison-Lees, 18 Langley Avenue, Somercotes, Alfreton, DE55 4LT
GI4 OHW N. Bell, Rocklyn, 16 Dromore Road, Omagh, BT78 1QZ
GM4 OHY R. Cameron, 88 Little Vennel, Cromarty, IV11 8XF
G4 OIA S. Hartgroves, 54 Kensey Valley Meadow, Launceston, PL15 9TJ
G4 OID J. Storry, 99 Swineshead Road, Wyberton Fen, Boston, PE21 7JG
G4 OIE R. Neale, Field House, Recreation Road, Mansfield, NG19 8TL
G4 OIG G. Peck, 45 Bentley Close, Northampton, NN3 5JS
G4 OII M. Morley, Padagi, Town Road, Grimsby, DN36 5JE
GM4 OIJ B. Robertson, Frian, Westerton, Dalcross, Inverness, IV2 7JL
G4 OIK J. Price, 4 Housman Walk, Kidderminster, DY10 3XL
G4 OIL J. Price, 26 Hales Park, Bewdley, DY12 2HT
G4 OIM P. Marchant, 29 Hilldrop Road, Bromley, BR1 4DB
G4 OIN A. Reeley, Gibraltar House, 53 Pegasus Gardens, Gloucester, GL2 4NP
G4 OIQ P. Storey, 4 Sorrel Close, Wootton Bassett, Swindon, SN4 7JG
G4 OIR N. Robinson, 14 Glendale, Orton Wistow, Peterborough, PE2 6YL
G4 OIS G. Reid, 65 Rowelfield, Luton, LU2 9HL
G4 OIV D. Mavin, 20 Rowlington Terrace, Ashington, NE63 0LZ
G4 OIW Dr C. Pine, 2 Grange Drive, Stokesley, Middlesbrough, TS9 5PQ
G4 OJD M. Guy, 73 Penn Meadows, Brixham, TQ5 9PF
G4 OJF G. Ball, 12 Kelstern Close, Northwich, CW9 5QR
G4 OJG J. Glass, 70 Canterbury Road, Lydden, Dover, CT15 7ES
G4 OJH A. Giles, 11 Stanmore Close, Clacton-on-Sea, CO16 7HQ
G4 OJI D. Schofield, 18 Berrow Walk, Bristol, BS3 5ES
G4 OJJ V. Holyoake, 281 Causeway, Green Road, West Midlands, B68 8LT
G4 OJK H. Baxendale, 72B De Villiers Avenue, Liverpool, L23 2XF
G4 OJL J. Bevan, 5 Selsdon Close, Wythall, Birmingham, B47 6HP
G4 OJN A. Semark, 11 Fir Tree Close, Thorpe Willoughby, Selby, YO8 9PF
G4 OJP R. Prosser, 17 Lloyd Street, Hereford, HR1 2HB
G4 OJQ A. Rowland, Mole Cottage, Chapel Close, Morwenstow, EX23 9JR
G4 OJR A. Stone, 67 Bluebell Way, Preston, PR5 6XQ
G4 OJS J. Rowlands, 70 Braces Lane, Marlbrook, Bromsgrove, B60 1DY
G4 OJU C. Lock, 11 Stockton Close, Bristol, BS14 0DS
G4 OJV A. Picton, 5 Tuttles Lane East, Wymondham, NR18 0EN
G4 OJW L. Szondy, 6 Stanhope Gardens, London, N6 5TS
G4 OJY A. Wright, 2 Wards End Cottages, Tow Law, Bishop Auckland, dl13 4js
G4 OKA N. Crook, 3 College Road, Reading, RG6 1QE
G4 OKB B. Bloomfield, 2 Walstead Manor Cottages, Scaynes Hill Road, Haywards Heath, RH16 2QG
G4 OKC A. Gardner, 19 Lower Rea Road, Brixham, TQ5 9UD
G4 OKD A. Forryan, 21 Blakesley Road, The Meadows, Wigston, LE18 3WD
G4 OKE M. Gould, 10 Canterbury Close, Pelsall, Walsall, WS3 4PB
GW4 OKF P. Granby, 104 Priory Road, Milford Haven, SA73 2ED
G4 OKH M. Fisher, Witham Lodge, Fen Road, Newton-In-The-Isle, Wisbech, PE13 5HT
G4 OKM M. Smith, 7 Russley Green, Wokingham, RG40 3HT
G4 OKS M. Porter, Penlee, 11 Penwithick Road, Penwithick, St Austell, PL26 8UQ
GI4 OKU T. Patton, 29 Greystone Park, Limavady, BT49 0EQ
G4 OKW C. Trayner, 2 Herisson Close, Pickering, YO18 7HB
G4 OKY R. Wilkinson, 10 Mildenhall Road, Loughborough, LE11 4SN
G4 OKZ D. Wilson1, 20 The Square, Worsthorne, Burnley, BB10 3NG
G4 OLA R. Mccubbin, 20 Wellesley Park, Wellington, TA21 8PY
GM4 OLH I. Walsh, 43 Sandyhill Road, Tayport, DD6 9NX
G4 OLK A. Mackay, 2 Highcliffe Grove, New Marske, Redcar, TS11 8DU
G4 OLL D. Wilson, 10 Winchester Drive, Stourbridge, DY8 2LH
G4 OLO G. Spencer, 5 Pitchcroft Lane, Church Aston, Newport, TF10 9AQ
G4 OLP R. Parker, 2 Laurel Road, Norwich, NR7 9LL
G4 OLS J. Lloyd, 16 Gilbanks Road, Stourbridge, DY8 4RN
G4 OLU D. Steward, 30 Riffhams Drive, Great Baddow, Chelmsford, CM2 7DD
G4 OLY C. Morgan, 316 Middle Road, Southampton, SO19 8NT
G4 OLZ I. Sirley, The Barn, Littleworth Lane, Horsham, RH13 8JF
G4 OMD D. Wilson, 1 Witchford Close, Lincoln, LN6 0SS
G4 OMG C. Prescott, 15 Sarabeth Drive, Tunley, Bath, BA2 0EA
G4 OMI Dr A. Proudler, 17 Sunnyside Road, Ketley Bank, Telford, TF2 0DT
G4 OMJ G. Yarnall, Blue Lias, Cropwell Road, Langar, Nottingham, NG13 9HD
GI4 OMK P. Murphy, 11 Danesfort Apartments, Belfast, BT9 5QL
G4 OMN D. Thorndike, 48 Cressingham Road, Reading, RG2 7JR
G4 OMP M. Nyman, 26 Silverstone Court, River Brook Drive, Birmingham, B30 2SH
G4 OMS R. Reynolds, 90 Manchester Road, Blackpool, FY3 8DP
G4 OMT M. Taylor, 320 Duffield Road, Derby, DE22 1EQ
G4 OMV D. Marlow, 53 The Lawns, Corby, NN18 0TA
G4 OMZ L. Welding, 67 Sunningdale Close, Burtonwood, Warrington, WA5 4NS
G4 ONC E. Westcott, 8 Portal Place, Ivybridge, PL21 9BT
G4 ONF P. Sergent, 6 Gurney Close, New Costessey, Norwich, NR5 0HB
G4 ONG W. Lowe, 34 Ridgeway, Lowton, Warrington, WA3 2QL
G4 ONH C. Weller, Mole End, Brent Hall Road, Braintree, CM7 4JZ
GW4 ONI D. Mears, 7 Tanydarren, Cilmaengwyn, Swansea, SA8 4QT
G4 ONJ A. Lightfoot, 13 Midhurst Close, Ifield, Crawley, RH11 0BS
G4 ONP Loughton & Epping Forest ARS, c/o J. Mulye, 83 Forest Drive East, London, E11 1JX
G4 ONS M. Slade, 5 Pedder Road, Clevedon, BS21 5HB
G4 ONV G. Parker, 49 Newlands, Dawlish, EX7 0EA
G4 ONZ P. Tebbutt, 115 Whitfield Mill, Meadow Road, Bradford, BD10 0LP
G4 OO Spalding and District ARS, c/o A. Hensman, 24 Belchmire Lane, Gosberton, Spalding, PE11 4HG
G4 OOB J. Westerman, 7 Gascoigne Court, Barwick In Elmet, Leeds, LS15 4NY
G4 OOC B. Simister, 3 Beech Tree Road, Featherstone, Pontefract, WF7 5EB
G4 OOE A. Langmead, 10 The Copse, Scarborough, YO12 5HG
G4 OOH S. Parker, Flat 4, 72 Springfield Mount, Leeds, LS18 5QE
G4 OOI C. Parker, 18 Langdale Gardens, Leeds, LS6 3HB
G4 OOJ C. Rose, 3 Harley Drive, Leeds, LS13 4QY
G4 OOK S. Stobbs, 78 Hershall Drive, Town Farm, Middlesbrough, TS3 8NX
G4 OOL Dr J. Yeandel, Fairfield Farm, Penhallow, Truro, TR4 9LT
G4 OOQ N. Ward, 8 Meadowview Road, Kempston, Bedford, MK42 7BE
GM4 OOU J. Forsyth1, 10 Rowallan Crescent, Prestwick, KA9 2HE
G4 OOX L. Harvey, 27 Guernsey Drive, Birmingham, B36 0PB
G4 OOY D. Bird, 13 Kilvington Road, Arnold, Nottingham, NG5 7HQ
G4 OPB N. Hydes, 2 Stable Court, Martlesham Heath, Ipswich, IP5 3UQ
G4 OPD A. Blissett, 26 Cherry Orchard, Holt Heath, Worcester, WR6 6ND
G4 OPE M. Hodges, 40 Ennersdale Road, Coleshill, Birmingham, B46 1EP
GI4 OPH T. Crawford, 50 Thornleigh Gardens, Bangor, BT20 4NP
G4 OPI A. Easom, 1 Station Close, West Ayton, Scarborough, YO13 9JQ
G4 OPK D. Carrett, 80 Rotherfield Way, Emmer Green, Reading, RG4 8PL
G4 OPL T. Bayliss, 4 Sycamore Close, Polgooth, St Austell, PL26 7BW
G4 OPN W. Broxup, 9 Kingsway, Hapton, Burnley, BB11 5RB
G4 OPO C. Haddrell, 9 Counterpool Road, Kingswood, Bristol, BS15 8DQ
G4 OPP I. Horsefield, 61 Lewis Court Drive, Boughton Monchelsea, Maidstone, ME17 4LG
G4 OPQ Dr M. Holli, 6 Morley Close Staple Hill, Bristol, BS16 4QE
G4 OPR R. Hayward, Sunnyfields, Lighthouse Road, Dover, CT15 6EJ
G4 OPT A. Kemsley, Newfield Lodge Rest Home, 93-99 St. Andrews Road South, Lytham St Annes, FY8 1PU
GM4 OPU J. Houston, 26 Clerk Drive, Corpach, Fort William, PH33 7LE
G4 OPV J. Jackson, 15 Jackson Crescent, Stourport-on-Severn, DY13 0EW
GW4 OPW B. Jones, 12 Ashbourne Court, Aberdare, CF44 8HA
G4 OPY S. Balmer, 101 Marsh Lane, Shepley, Huddersfield, HD8 8AP
G4 OQ G. Lidstone, 76 Thames Drive, Leigh-on-Sea, SS9 2XD
GW4 OQB A. Greatrex, Clydfan, Dinas Cross, Newport, SA42 0XS
G4 OQG Prof. M. Ayres, 3 Wicks Drive, Chippenham, SN15 3EL
G4 OQH Rev. H. Callaghan, 5 Manor Park View, Manor Park Road, Glossop, SK13 7TL
G4 OQJ I. James, 4 Lancaster Gardens, Earley, Reading, RG6 7PA
G4 OQK R. Alderton, 1 Comfrey Way, Thetford, IP24 2UU
G4 OQL Dr C. Bowley, Plum Tree House, Walk Close, Derby, DE72 3PN
G4 OQN M. Barr, 30 Hounslow Road, Twickenham, TW2 7EX
G4 OQP J. Gerrity, 14 Lostock Avenue, Hazel Grove, Stockport, SK7 5JN
G4 OQR M. Huddart, Buckminster Gliding Club Ltd Saltby Airfield Sproxton Road Skillington, Grantham, NG33 5FE
G4 OQU J. Davenport, 1 Lowfields, Staveley, Chesterfield, S43 3QB
G4 OQV R. Beecham, 7 Crummock Close, Coventry, CV6 6GY
G4 OQX G. Cooper, 61 Fallowfield Road, Hasbury, Halesowen, B63 1BZ
GI4 OQY K. Chambers, 44 Ballywillin Road, Portrush, BT56 8JN
G4 OQZ B. Dawson, Isca, 12 Lestock Way, Fleet, GU51 3EB
G4 ORB G. Busby, The Terrace, Terrace Road South, Binfield, Bracknell, RG42 4DS
G4 ORC Oldham Am Radio Club, c/o G. Oliver, 158 High Barn St., Royton, Oldham, OL2 6RW
G4 ORE A. Charles, 14 Chorleywood Bottom, Chorleywood, Rickmansworth, WD3 5JD
GI4 ORI J. Hamill, 67 Windsor Avenue, Coleraine, Ireland, BT52 2DR
G4 ORJ A. Jones, Fairview, Frenchs Road, Wisbech, PE14 7JF
G4 ORP M. Parsons, 15 Sherbourne Road, Hangleton, Hove, BN3 8BA
G4 ORQ A. Walker, 4A Winston Drive, Eston, Middlesbrough, TS6 9LY
G4 ORS W. Ragg, 14 Mocatta Way, Burgess Hill, RH15 8UR
G4 ORU G. Wadwell, 7 Barkhart Drive, Wokingham, RG40 1TW
G4 ORV D. Whatmough, Flat 3, 170 Buxton Road, Stockport, SK2 6HA
G4 ORW A. Atherley, 2 Haydock Close, Dosthill, Tamworth, B77 1QR
G4 ORX P. Baggett, 33 Foxglove Way, Thatcham, RG18 4DL
G4 ORY C. Bates, Rivieri, Coventry Road, Kingsbury, B78 2LH
G4 OSB T. Arris, 7 Rowan Road, North Hykeham, Lincoln, LN6 8LY
GI4 OSF J. Mcniece, 35 Farran Road, Ballymoney, BT53 8HD
GI4 OSG D. Robinson, 17 Dalton Glen, Comber, Newtownards, BT23 5RJ
G4 OSH A. Nevison, 10 Birch Way, Tunbridge Wells, TN2 3DA
G4 OSI D. Whitehouse, 10 Felstead Street, Stoke-on-Trent, ST2 7HJ
G4 OSJ P. Brewer, 2 Mill Close, Wing, Oakham, LE15 8RH
G4 OSK K. Hall, 21 Eardulph Avenue, Chester le Street, DH3 3PR
G4 OSO E. Binns, Fieldview, 5 Moorside, Cleckheaton, BD19 6JH
G4 OSP M. Binns, Fieldview, 5 Moorside, Cleckheaton, BD19 6JH
G4 OSR S. Roberts, 1 Lakeside Crescent, Long Eaton, Nottingham, NG10 3GH

GM4 OSS S. Campbell, 14 Hillhouse Place, Stewarton, Kilmarnock, KA3 3HT
G4 OST P. Cabban, Ivydene, Upper Tockington Rd. Tockington, Bristol, BS32 4LQ
G4 OSU M. Dixey, 50 Sandon Road, Ford Houses, Wolverhampton, WV10 6EN
GM4 OSV C. Dunn, 66 Glen Doll Road, Neilston, Glasgow, G78 3QP
G4 OSX J. Griffiths, 15 Victoria Road, Cambridge, CB4 3BW
G4 OSY D. Gascoigne, 2 Thorncliffes, Chapel Lane, Pontefract, WF9 3NJ
G4 OTB Dr N. Hailes, Yew Tree Cottage, The Hollies Common, Stafford, ST20 0JD
G4 OTC P. Gagen, 16 Melbourne Road Sidemoor, Bromsgrove, B61 8PE
G4 OTD C. Taylor, 2 Kismet Avenue, Highbury, Australia, 5089
G4 OTE M. Grayson, One Elm, 58 Kaye Lane, Huddersfield, HD5 8XU
GI4 OTG A. Mcneice, 148 Doagh Road, Newtownabbey, BT36 6BA
G4 OTI Dr P. Stockbridge, 11 Fairways, Frodsham, WA6 7RU
G4 OTJ J. Witchell, Pennyquick Cottage, Broomhill Lane, Bristol, BS39 5SA
G4 OTL M. Baker, 49 Grove Lane, Gomersal Cleckheaton, BD19 4JT
G4 OTS G. Eccleston, 24 Orton Lane, Wombourne, Wolverhampton, WV5 9AW
G4 OTU D. Fagan, 3 Oxenham Green, Torquay, TQ2 6DX
G4 OTV D. Green, St. Annes Poundfield Road, Crowborough, TN6 2BG
G4 OTX G. Hibberd, 2 Carr Bank, Oakamoor, Stoke-on-Trent, ST10 3EA
G4 OUB J. Whetstone, 5 Zetland Road, Barnard Castle, DL12 8LA
G4 OUG C. Beesley, 15 Byron Close, Cheadle, Stoke-on-Trent, ST10 1XB
G4 OUH J. Brockway, 16 Dawlish Road, Dudley, DY1 4LU
G4 OUI M. Brockway, 16 Dawlish Road, Dudley, DY1 4LU
G4 OUJ S. Carrigan, 1 Milford Crescent, Littleborough, OL15 9EF
G4 OUM T. Bolton, 25 Woodfield Drive, Lichfield, WS14 9HH
GI4 OUN D. Fulton, 120 Dunnalong Road, Bready, Strabane, BT82 0DP
GI4 OUP S. Henderson, 47 Donaghedy Road, Bready, Strabane, BT820DB
G4 OUS D. Bean, 18 Witham Court, Higham, Barnsley, S75 1PX
G4 OUT I. Cornes, 17 Chilwell Avenue, Little Haywood, Stafford, ST18 0QZ
GW4 OUU D. Grace, The Laurels, Gwbert Road, Cardigan, SA43 1AF
G4 OUZ L. Day, 3 Harris Way, Lee Mill Bridge, Ivybridge, PL21 9EU
G4 OVD G. Rugen, 24 Highgate Road, Lydiate, Liverpool, L31 0DA
GI4 OVE J. Mcelvanna, 26 Lissummon Road, Newry, BT35 6NA
G4 OVF P. Sampson, 34 Solway Road, Moresby Parks, Whitehaven, CA28 8XJ
G4 OVG J. Thompson, 29 Arun, East Tilbury, Tilbury, RM18 8SX
GW4 OVH H. Owen, 5 Arllwyn Cefn Road, Bwlchgwyn, Wrexham, LL11 5YF
G4 OVJ R. Read, 29 Imber Road, Shaftesbury, SP7 8RX
G4 OVL M. Allen, 18 Philip Garth, Wakefield, WF1 2LS
G4 OVM C. Barnes, 13 Waterworks Road, Farlington, Portsmouth, PO6 1NG
GI4 OVN S. Dawson, 31 Rock Hill Warren Road, Donaghadee, BT210FB
G4 OVO J. Featherstone, Garden Close, Greenway Road, Torquay, TQ1 4NJ
G4 OVR D. Fillingham, 6 Kings Chase, Rothwell, Leeds, LS26 0HL
G4 OVS F. Goddard, 4 St. Peters Close, Barnburgh, Doncaster, DN5 7EN
G4 OVT J. Grant, The Bungalow, Drury Lane, Horsforth, Leeds, LS18 4RL
G4 OVV N. Howard, 9 Snowshill Drive Highfield, Wigan, WN3 6AD
G4 OVW B. Jempson, Flat 1, 3 Dacre, Scarborough, YO11 2SP
G4 OVX M. Kennett, Toms Cottage, Kendal Lane, York, YO26 7QN
GI4 OWA G. Elliott, 4 Fernbrae Gardens, Londonderry, BT47 5XS
GI4 OWB J. Fallows, 22 Meadow Way, Ballygowan, Newtownards, BT23 5TQ
G4 OWH G. Gregor, 41 Stonebridge Drive, Frome, BA11 2TN
G4 OWK T. Pearsall, 6 Vernon Close, Martley, Worcester, WR6 6QX
G4 OWL C. Payne, 7 Ellis Avenue, Onslow Village, Guildford, GU2 7SR
G4 OWN A. Turner, 1 Milton Road, Flitwick, Bedford, MK45 1QA
GW4 OWQ D. Scott, Isyfoel, Morfa Bychan, Porthmadog, LL49 9YD
G4 OWS N. Cunliffe, 44 Shore Road, Hesketh Bank, Preston, PR4 6RB
G4 OWT S. Harwood, 24 Firle Crescent, Lewes, BN7 1QG
G4 OWY R. Howes, 8 Kitchener Road, Weymouth, DT4 0LN
G4 OXD T. Rose, 41 Keats Way, Hitchin, SG4 0DP
G4 OXG N. Wood, Mentmore House, Woodside Hill Chalfont St. Peter, Gerrards Cross, SL9 9TD
G4 OXK R. Waygood, 2 Brookside Close, Bransgore, Christchurch, BH23 8BT
GW4 OXL W. Smith, 11 Connacht Way, Pembroke Dock, SA72 6FB
GI4 OXO W. Fitzsimons, 83 Boghill Road, Newtownabbey, BT36 4QT
G4 OXR C. Mortimer, Meadow Bank, Dowlish Wake, Ilminster, TA19 0NZ
G4 OXU Dr D. Whan, 1 Hillclose Avenue, Darlington, DL3 8BH
GI4 OYG M. Black, 38 Town Park, Carrickfergus, BT38 8FG
G4 OYH A. Bowyer, 37 St. Mark Drive, Colchester, CO4 0LP
GI4 OYI D. Chambers, 238 Donaghadee Road, Newtownards, BT23 7QP
GI4 OYL J. Cuthbert, 19 Antrim Road, Ballymena, BT42 2BJ
GI4 OYM W. Elliott, 23 Castle View Park, Portrush, BT56 8AS
G4 OYN A. Fuller, 17 Brington Drive, Barton Seagrave, Kettering, NN15 6UW
G4 OYO A. Bramley, 13 Moorland Avenue, Stapleford, Nottingham, NG9 7FY
G4 OYP G. Savin, 19 Hailey Avenue, Loughborough, LE11 4QW
G4 OYR N. Lee, Silverstone, Alverstone Road, Apse Heath, Sandown, PO36 0LH
G4 OYT H. Moss, 101 Barnford Crescent, Oldbury, B68 8PR
G4 OYX D. Porter, 8 Stanton Drive, Ludlow, SY8 2PH
G4 OYZ M. Spencer, 29 Kliffen Place, Halifax, HX3 0AL
G4 OZC W. Smith, 15 Henbury Drive, Woodley, Stockport, SK6 1PY
G4 OZD R. Woolley, 7 Geveze Way, Broughton Astley, Leicester, LE9 6HJ
G4 OZG E. Haskett, 23 Gloucester Road, King's Lynn, PE30 4AB
GI4 OZI A. Kinghan, 14 Sunningdale Park North, Belfast, BT14 6RZ
GI4 OZJ G. Allen, 6 Lougherne Road, Annahilt, Hillsborough, BT26 6BX
G4 OZL P. Ingram, Rosehill Cottage, Mount Carmel Road, Andover, SP11 7ER
G4 OZM D. Bradberry, 6 The Close, Easton On The Hill, Stamford, PE9 3NA
G4 OZN P. Donaldson-Badger, Heckdyke Cottage, Heckdyke, Doncaster, DN10 4BE
G4 OZP S. Corrigan, 11 Pear Tree Close, Bromham, Bedford, MK43 8PR
G4 OZQ G. Fisher, 9 Shrubbery Road, Drakes Broughton, Pershore, WR10 2AX
GW4 OZU P. Hyams, Tricklewood, Pembroke Dyfed, SA71 5HY
G4 OZX C. Goble, 12 Longfield Road, Emsworth, PO10 7TR
G4 OZY A. Osborn, 35 Griston Road, Watton, Thetford, IP25 6DN
G4 PAA P. Booth, 22 Charters Lane, Brandesburton, Driffield, YO25 8QJ
G4 PAC P. Caldwell, 3 Orchard Mead, Broadwindsor, DT8 3RA
GW4 PAF J. Thomas, 2 Tudor Way, Llantwit Fardre, Pontypridd, CF38 2NH
G4 PAH Dr M. Rollason, Ash Ridge, Clint, Harrogate, HG3 3DS
G4 PAI P. Wells, 15 Apple Tree Grove, Ferndown, BH22 9LA
G4 PAS P. Searles, 63 Whalley Road, Ramsbottom, Bury, BL0 0DP
G4 PAT J. Thirsk, 47 Chestnut Avenue, Euxton, Chorley, PR7 6BP
G4 PAV G. Brutnall, 57 Wollaston Road, Irchester, Wellingborough, NN29 7DA
G4 PBC J. Kilroy, 119 Station Road Brimington, Chesterfield, S43 1LJ
G4 PBD Capt. R. Hughes, 8 Frinton Court, The Esplanade, Frinton-on-Sea, CO13 9DW
G4 PBF P. Baker, 23 Orde Close, Crawley, RH10 3NG
G4 PBJ B. Oakley, 6 Windmill Way, Haxby, York, YO32 3NL
G4 PBN J. Vivian, 3 Station Road, Gunnislake, PL18 9DX
G4 PBO D. Smith, 21 Sydney Road, Benfleet, SS7 5RD
G4 PBR D. Suttenwood, White Lodge, Mersea Road, Colchester, CO5 7LJ
GI4 PBS T. Wilson, 39 Woburn Road, Millisle, Newtownards, BT22 2HY
GI4 PBT T. Wilson, Brambly Hedge, 39 Woburn Road, Newtownards, BT22 2HY
G4 PBY B. Jones, 13 Albert Street, Cheltenham, GL50 4HS
G4 PBZ T. Ashton, 90 Secker Avenue, Warrington, WA4 2RE
G4 PCB A. Cox, 12 Merrymeet, Whitestone, Exeter, EX4 2JP
G4 PCD M. Dally, 11 Wrightson Terrace, Doncaster, DN5 9ST
G4 PCE R. Collins, 389 Lode Lane, Solihull, B92 8NN
G4 PCF P. Goodson, 46 Southwold, Bracknell, RG12 8XY
GW4 PCJ R. Belcher, Parciau, Bronwydd Arms, Carmarthen, SA33 6BN
G4 PCK B. James, Rivendell, Kingsgate Close, Torquay, TQ2 8QA
G4 PCL B. Walker, 22 Peveril Road, Tibshelf, Alfreton, DE55 5LQ
G4 PCN C. Lambert, 23 Palmars Cross Hill, Rough Common, Canterbury, CT2 9BL
GW4 PCO P. Mogford, 27 Ynysymaerdy Road, Briton Ferry, Neath, SA11 2TE
G4 PCP C. Shelton, 18 Beaconsfield Drive, Coddington, Newark, NG24 2RX
GI4 PCQ J. Quinn, 86 Knocknacarry Road, Cushendun, Ballymena, BT44 0NS
G4 PCR J. O'Hara, 12 Ray Avenue, Nantwich, CW5 6HJ
GM4 PCT A. Gordon, 2 Duchray St., Riddrie, Glasgow, G33 2DD
G4 PCW A. Tucker, 3 Eston Close, Mabe Burnthouse, Penryn, TR10 9JW
GW4 PCX R. Price, 2 Grassholm Place, Broadway, Haverfordwest, SA62 3HX
G4 PCZ D. St Quintin, 16 Cromwell Road, Sprowston, Norwich, NR7 8XH
G4 PDC P. Carter, 16 Alexandra Park Paulton, Bristol, BS39 7QS
G4 PDD F. Bibby, 14 St. Clare Terrace, Chorley New Road, Bolton, BL6 4AZ
G4 PDE R. Bradshaw, 44 Hawthorn St., Derby, DE24 8BD
G4 PDG B. Hillard, Farmlea, Hele Lane, South Petherton, TA13 5AP
G4 PDI B. Kenzie, 9 Goodliffe Avenue, Balsham, Cambridge, CB21 4AD
G4 PDK R. Davies, 11 Tamar Green, Corby, NN17 2LA
G4 PDQ J. Clayton, 217 Prestbury Road, Cheltenham, GL52 3ES
G4 PDR D. Hughes, 19 Burnsall Close, Farnborough, GU14 8NN
G4 PDU C. Carrington, 3 Jeake Drive, Rye, TN31 7FH
G4 PDY J. Brandhuber, 3 Brigham Place, Felpham, Bognor Regis, PO22 7NW
G4 PEA R. Flanders, 51 Rookwood Court, Guildford, GU2 4EL
G4 PED J. Hinde, 12A Station Parade, Ockham Road South, Leatherhead, KT24 6QN
G4 PEF Prof. W. Ingram, 141 Churchill Road Willesden Green, London, NW2 5EH
G4 PEK L. Dymond, 13 Thornlea Avenue Fremington, Barnstaple, EX31 3DA
G4 PEL W. Threapleton, Cobbs Nook Farm, Newstead Lane, Stamford, PE9 4JJ
G4 PEN R. Potts, 180 Brook Hill, Thorpe Hesley, Rotherham, S61 2PZ
G4 PEO J. Pitty, 12 St. Leonards Road, Horsham, RH13 6EJ
GI4 PES N. Robinson, 3 Moorland Drive, Lisburn, BT28 2XU
G4 PET J. Smith, Pasturefields House, Pasturefields Lane, Stafford, ST18 0RD
G4 PEU K. Smith1, Pasturefields House, Pasturefields Lane, Stafford, ST18 0RD
G4 PEW R. Wood, Abbots Croft, Abbey Road, St Bees, CA27 0EG
GW4 PEX W. Williams, 168 Mumbles Road, West Cross, Swansea, SA3 5AN
G4 PEY R. Wilmot, 1 Retreat Cottages, Church Lane, Horsham, RH12 3ND
G4 PFA P. Wheeler, 21 Browns Road, Holmer Green, High Wycombe, HP15 6SL
G4 PFE J. Laverick, 5 York Crescent, Newton Hall Estate, Durham, DH1 5PU
G4 PFF J. Potter, 8 Mill Field Close, Burton Joyce, Nottingham, NG14 5AA
G4 PFG M. Spooner, 6 Cross Road, Starston, Harleston, IP20 9NQ
G4 PFJ J. Backus, 2 Southview Villas, Dunmow Road, Takeley, Bishop's Stortford, CM22 6SW
G4 PFK G. Gifford, 184 Chantrey Crescent Great Barr, Birmingham, B43 7PG
GW4 PFL P. Freestone, 8 Harcourt Road, Llandudno, LL30 1TU
G4 PFO J. Gregory, 22 Tower View Road, Great Wyrley, Walsall, WS6 6HE
G4 PFQ Northwest Durham Raynet, c/o T. Hanratty, 12 Clarendon Street, Consett, DH8 5LS
G4 PFR J. Harding, 19 Carrington Crescent Wendover, Aylesbury, HP22 6AW
G4 PFT J. Harris, 45 Redehall Road, Smallfield, Horley, RH6 9QA
G4 PFU D. Blunt, 12 Mallard Place, East Grinstead, RH19 4TF
G4 PFW H. Palmer, 5 Hurst Close, Crawley, RH11 8LQ
G4 PFX D. Palmer, 14 Garibaldi Road, Redhill, RH1 6PB
G4 PFY Dr J. O'Hagan, 13 Chapel Road, Stanford In The Vale, Faringdon, SN7 8LE
G4 PFZ J. Aspland, 6 Trilithon Close, Hellesdon, Norwich, NR6 5EP
G4 PGA B. Gage, 2 Wellsbourne Road, Stone Cross, Pevensey, BN24 5QX
G4 PGB P. Hayward, 22 Falconers Park, Sawbridgeworth, CM21 0AU
G4 PGD R. Hall, 18 Park View, Truro, TR1 2BW
G4 PGG P. Beesley, 15 Byron Close, Cheadle, Stoke-on-Trent, ST10 1XB
GI4 PGH J. Crawford, 2 Holywood Road, Newtownards, BT23 4TQ
G4 PGJ D. Ward, 48 Moat Bank, Bretby, Burton-on-Trent, DE15 0QJ
GM4 PGM P. Brash, 4 Union Street, Lossiemouth, IV31 6BA
GI4 PGN J. Bailie, 4 Quarry Road, Greyabbey, Newtownards, BT22 2QF
G4 PGO D. Fernant, 2 Lonsboro Road, Wallasey, CH44 9BR
G4 PGQ D. Harrison, 517 Atherton Road, Hindley Green, Wigan, WN2 4QF
G4 PGS P. Clark, 23 Nova Mews, Sutton, SM3 9HY
GM4 PGV P. Lawless, 37 Oaklands Avenue, Irvine, KA12 0SE
G4 PGW N. Puttick, 33 Alder Hill Drive, Totton, Southampton, SO40 8JB
G4 PGX M. Williams, 114 Ferry Street, Burton-on-Trent, DE15 9EY
G4 PGY Dr R. White, Beech Hill, Northampton, NN7 4LL
GW4 PHB W. Vickers, Creigiau, Penrhyndeudraeth, LL48 6LS
G4 PHC G. Stearn, 31, Regents Way, Minehead, TA24 5HS
G4 PHK P. Colbeck, 76 Church Road, Winterbourne Down, Bristol, BS36 1BY
G4 PHL P. Green, Danewalk, North Road, Brotherton, Knottingley, WF11 9ED
G4 PHP D. Foster, 120 Green Lane, Cookridge, Leeds, LS16 7HF
G4 PHR T. Clough, 37 Park Avenue, Mirfield, WF14 9PB
GW4 PHT D. Dalling, 308 Townhill Road Mayhill, Swansea, SA1 6PD
G4 PHV G. Bennison, 35 Ermine Street, Thundridge, Ware, SG12 0SY
G4 PHX T. Jackson, 28 John Lee Road, Ledbury, HR8 2FE
G4 PIA L. Roberts, 18 Turret Grove, London, SW4 0ET
GI4 PID B. Little, 8 Ballynoe Road, Antrim, BT41 2QT
G4 PIE D. Tyler, 12 Bernards Way, Flackwell Heath, High Wycombe, HP10 9EQ
G4 PIJ J. Goodman, 4 Maloren Way, West Moors, Ferndown, BH22 0BQ
G4 PIP C. Bottoms, Treboro House, Ullenhall, West Midlands, B95 5NN
G4 PIQ A. Cook, The Old Vicarage, High Road, Ipswich, IP6 9LP

G4 PIR J. Child, 12 Beachill Road, Havercroft, Wakefield, WF4 2EJ
G4 PJD H. Hoare, Farvardale, The Street, Bishop's Stortford, CM22 7LT
G4 PJE R. Kershaw, 13 Silver Hill, Milnrow, Rochdale, OL16 3UJ
G4 PJJ Dr N. Garbutt, Tudor Cottage, Main Road, Gloucester, GL2 8JP
G4 PJK R. Mosedale, 21 Druids Avenue, Aldridge, Walsall, WS9 8LA
G4 PJL R. Bailey, 5 Braemar Road, Doncaster, DN2 5HN
G4 PJP M. Clay, 24 Begonia Drive, Burbage, Hinckley, LE10 2SW
GM4 PJR N. Yarrow, 10 Coxburn Brae, Bridge Of Allan, Stirling, FK9 4PS
G4 PJS P. Shields, 81 Flaxton, Skelmersdale, WN8 6PE
G4 PJT S. Schofield, 18 Ascot Close, Mexborough, S64 0JG
G4 PJZ J. Towle, 46 Querneby Road, Nottingham, NG3 5HY
G4 PKE R. Badham, Caedman, Terrace Road North, Bracknell, RG42 5JG
G4 PKF E. Wood, 68 Baswich Crest, Stafford, ST17 0HJ
GM4 PKJ D. Smith, Haremuir Bungalow, Benholm, Montrose, DD10 0HX
G4 PKK Dr S. Juden, 17A Astonville St., Southfield, London, SW18 5AN
G4 PKM J. Derrick, 37 Admiralty Street, Keyham, Plymouth, PL2 2BR
G4 PKO D. French, 14 Linden Close, Prestbury, Cheltenham, GL52 3DU
G4 PKP J. Jones, Jason Photographic, New Moss Farm, Liverpool, L37 0AH
G4 PKT D. Lewin, 14A Warwick New Road, Leamington Spa, CV32 5JG
G4 PKV D. Griffiths, 61 The Drive, North Harrow, Harrow, HA2 7EJ
G4 PKW C. Gerard, 7 Parkwood Road Sidemoor, Bromsgrove, B61 8UA
G4 PKX F. Gallimore, 3 Wilson Crescent, Lostock Gralam, Northwich, CW9 7QH
G4 PKZ R. Richardson, Manor Farm, Manor Lane, Oakham, LE15 7JL
G4 PLH R. Hughes, 73 Upland Road, Sutton, SM2 5JA
GM4 PLI J. Nellis, 64 Kirkwood Avenue, Clydebank, G81 2ST
G4 PLK S. Lewis, 189 Ashburton Road, Hugglescote, Coalville, LE67 2HE
G4 PLL I. Thomas, 15 Wakefield Road, Fitzwilliam, Pontefract, WF9 5AJ
G4 PLS A. Haigh, White Horse Cottage, Maypole, Canterbury, CT3 4LN
G4 PLU J. Bates, 63 Sunny Blunts, Peterlee, SR8 1LP
G4 PLV M. Seton, 12 Chatsworth St., Roundthorn, Oldham, OL4 5LF
G4 PLX A. Salata, 64 Wildwood Road, London, NW11 6UP
G4 PLY V. Morris, 21 Cranhill Road, Street, BA16 0BY
G4 PLZ P. Connors, Manor Cottage, Mill Road, Banningham, Norwich, NR11 7DT
G4 PMA D. Pearson, 42 Church Street, Stapleford, Nottingham, NG9 8DJ
G4 PMB F. Thompson, 28 Lordsmead, Cranfield, MK43 0HP
G4 PMG M. Green, Huntley, Chesham Road, Wigginton, Tring, HP23 6HH
GM4 PMH S. Dunn, 4 Mid Street, Rosehearty, Fraserburgh, AB43 7JS
G4 PMJ A. Santos, 17 Elm Garth, Roos, Baytree, Hull, HU12 0HH
GM4 PMK R. Blackwell, Willowbank, Pennyghael, Isle of Mull, PA70 6HB
G4 PMM R. Williams, 2 Keepers Close, Bestwood Village, Nottingham, NG6 8XE
GI4 PMP H. Smith, 1A Taylor Park, Limavady, BT49 0NT
G4 PMS P. Steele, 107 Lower Shelton Road, Marston Moretaine, Bedford, MK43 0LW
GM4 PMT A. Ross, 16 Burnside Drive, Bridge Of Don, Aberdeen, AB23 8PL
G4 PMV K. Grime, 13 Runnymede Court, Jackson Street, Bolton, BL3 5HX
G4 PMW R. Dunn, 117 All Saints Way, West Bromwich, B71 1RU
G4 PMY G. Bell, Linden Lea, Crewe Road, Sandbach, CW11 4RE
G4 PMZ P. Butcher, 9 Little Platt, Guildford, GU2 8JU
G4 PNB A. Bathurst, 64 Oakfields, Guildford, GU3 3AU
G4 PNC D. Hood, 32 Bishops Wood, Nantwich, CW5 7QD
G4 PND A. Daniel, 10 Tamarisk Close, Hatch Warren, Basingstoke, RG22 4UX
G4 PNH G. Aungiers, 17 Broadwood Drive, Fulwood, Preston, PR2 9SS
G4 PNI R. Bishop, 29 Windsor Court, Poulton-le-Fylde, FY6 7UX
G4 PNK T. Crosland, Park Farm House, Park Road, Bedford, MK43 7QF
G4 PNL A. Coe, 112 Harborough Road, Desborough, Kettering, NN14 2QY
GM4 PNM A. Wixon, Riverview Cottage, Melrose, TD6 9JB
G4 PNP B. Deak, 57 Arundel Road, Peacehaven, BN10 8RP
G4 PNQ R. Dhami, 3327 Smoke Tree Road, Mississauga, Ontario, Canada, L5N 7M5
G4 PNT A. Hellewell, 41 Woodlea Grove, Armthorpe, Doncaster, DN3 2HN
G4 PNX D. Painter, 93 Oxclose Lane Arnold, Nottingham, NG5 6FN
G4 POB T. Hutchings, 9 Little Dell, Welwyn Garden City, AL8 7HZ
GI4 POC R. Drain, 5 Ravelstone Avenue, Bangor, BT19 1EQ
G4 POD J. Gould, 30 Weymouth Avenue, Middlesbrough, TS8 9AB
G4 POF J. Hart, 1 Meadow Court, Fordingbridge, SP6 1LW
G4 POG W. Evans, 3 Coastline Village, Ostend Road, Norwich, NR12 0NE
G4 POI D. Lambert, 8 Stretton Road, Barnsley, S71 1XQ
G4 POL B. Robertson, 12 Green Lane, Woodstock, OX20 1JY
G4 POP T. Genes, 28 Hillside Road, Burnham on Crouch, CM0 8EY
G4 POR B. Banks, 30 Hospital Road, Burntwood, WS7 0ED
G4 POT D. Girling, 20 Fore Street, Praze An Beeble, Camborne, TR14 0JX
G4 POU P. Dyer, 36 Margate Road, Ipswich, IP3 9DE
G4 POW A. Owen, 60 Brighton Avenue, Elson, Gosport, PO12 4BX
G4 POY R. Kent, Talstr 4, Eriskirch, Germany, 88097
G4 PPB E. Marshall, 75 Acacia Crescent, Wigan, WN6 8NJ
G4 PPC B. Lowe, 19 Wolverhampton Road, Bloxwich, Walsall, WS3 2EZ
G4 PPE M. Bell, 55 Park Road, Hampton Hill, Hampton, TW12 1HX
G4 PPG J. O'Sullivan, 40 Sheldon Avenue, Standish, Wigan, WN6 0LW
G4 PPH A. Betts, 11 Brisbane Close Mansfield Woodhouse, Mansfield, NG19 8QZ
G4 PPJ S. Bone, 6 Manor Road, Folksworth, Peterborough, PE7 3SU
G4 PPK C. Everley, 5 Firs Close, Hazlemere, Hazlemere, HP15 7TF
G4 PPL P. Fisher, 10 Magdalene Court, Seaham, SR7 7DJ
G4 PPN D. Chapman, 22 Horsley Drive, Kingston upon Thames, KT2 5GG
G4 PPP R. Bailey, 10 Epping Close, Walsall, WS3 1TT
G4 PPR M. Spencer, 67 Holmley Lane, Dronfield, S18 2HQ
G4 PPS Dr D. Herbert, 48 Furrow Grange, Middlesbrough, TS5 8DP
GM4 PPT R. Hodge, 34 Craig View Coylton, Ayr, KA6 6LB
G4 PPU M. Roy, 17 Elgar Avenue, Tolworth, Surbiton, KT5 9JH
G4 PPW A. Keech, 2 Mountfield Road, Irthlingborough, Wellingborough, NN9 5SY
G4 PPZ J. Young, Nonsuch, Oxbridge, Bridport, DT6 3UB
G4 PQB D. Mathers, Dovedale Lodge, Bourton On The Hill, Moreton-in-Marsh, GL56 9TE
G4 PQI J. Raybould, 2 Woodland Avenue, Brierley Hill, DY5 1EQ
G4 PQM E. James, 59 Queensway, Euxton, Chorley, PR7 6PN
G4 PQP P. Malme, Newhaven, Mill Lane East Runton, Cromer, NR27 9PH
G4 PQS W. Tedbury, Tyting House, Exeter Road, Honiton, EX14 1AX
G4 PQU A. Harwood, 108 Tudor Green, Jaywick, Clacton-on-Sea, CO15 2PE
GI4 PQV T. Pollock, 33 Seahill, Donaghadee, BT21 0SH
G4 PQW M. Davis, 478 Eastern Avenue, Gants Hill, Ilford, IG2 6EQ
G4 PQX D. Taylor, 28 Main Street, Broadmayne, Dorchester, DT2 8EB
G4 PQY A. Williams, 7 Bower Hall Drive, Steeple Bumpstead, Haverhill, CB9 7ED
G4 PRB P. Ball, 21 Doonamana Road, Dun Laoghaire, Ireland, A96 W6K3
G4 PRD P. Dakin, 12 Spinney Close, Kidderminster, DY11 6DQ
G4 PRF S. Brown, 27 The Court, Anderby Creek, Skegness, PE24 5YQ
GI4 PRH D. Simpson, 31 Beech Green, Doagh, Ballyclare, BT39 0QB
G4 PRJ M. Worsfold, 5 Turner Close, Langney, Eastbourne, BN23 7PF
G4 PRL R. Hunt, 13 Westlake Rise, Heybrook Bay, Plymouth, PL9 0DS
GM4 PRO T. Oneil, 187 Main St., Chapelhall, Airdrie, ML6 8SF
GW4 PRP S. Lane, 12 Carlos St., Port Talbot, SA13 1YD
G4 PRQ I. Hooper, Flat 5, Wenlock House, 41 Stanstead Road, London, SE23 1HG
G4 PRS Poole Radio Society, c/o P. Ciotti, 6 Bascott Road, Bournemouth, BH11 8RH
G4 PRW P. Whitten, 2 Eastmead, Woking, GU21 3BP
G4 PSE M. Grime, 10 East Park Avenue, Darwen, BB3 2SQ
G4 PSI C. Franks, 11 Orchard Close, Crook, DL15 8QU
GM4 PSJ R. Stroud, 24 Cullen St., Portsoy, Banff, AB45 2PJ
GM4 PSL T. Grice, 35 Approach Row, East Wemyss, Kirkcaldy, KY1 4LB
G4 PSO A. Little, 20 Vicarage Close, Shillington, Hitchin, SG5 3LS
G4 PSP S. Gardner, 191 Charlton Park Midsomer Norton, Radstock, BA3 4BR
G4 PSR C. Sartorius, 39 Althorne Gardens, London, E18 2DA
G4 PSS S. Black, 71 Bellerby Drive, Urpeth Grange, Ouston, Chester le Street, DH2 1UF
G4 PST D. Turner, Hurdletree Bank Farm, Hurdletree Bank, Spalding, PE12 8QQ
G4 PSU A. Davidson, 5 Hanover Parc, Indian Queens, St Columb, TR9 6ER
G4 PTE Dr K. Lown, Maurice House, Callis Court Road, Broadstairs, CT10 3AH
G4 PTF C. Keeping, 12 St. Francis Avenue, Southampton, SO18 5QJ
G4 PTK G. Mason, 120 Scalford Road, Melton Mowbray, LE13 1JZ
G4 PTM J. Fyrth, 2 Merton Gardens, Farsley, Pudsey, LS28 5DZ
GM4 PTQ Rev. S. Bennie, 13A Scotland Street, Stornoway, HS1 2JN
G4 PTR G. Pinter, 29 The Mill, The Boulevard, Horsham, RH12 1GR
G4 PTU Blandford Garrison CW and DX Group, c/o R. Carter, 12 Glebe Close, Abbotsbury, Weymouth, DT3 4LD
GD4 PTV B. Brough, 4 The Bretney, Jurby, Isle Of Man, IM7 3BL
G4 PTW A. Brunning, 6 Newstead Road, Barnwood, Gloucester, GL4 3TQ
G4 PTZ J. Topley, 27 Inveraray Close, Sinfin, Derby, DE24 3JA
G4 PUB Basildon Dist Rd, c/o S. Wensley, 7 Bradshaw Close, Windsor, SL4 5PS
GW4 PUC R. Rees, 16 Brynheulog, Llanelli, SA14 8AE
G4 PUD B. Langdon, 80 Glen Rise, Birmingham, B13 0EJ
G4 PUM R. Brookes, Broadeaves, Bridgemere Lane, Hunsterson, Nantwich, CW5 7PN
G4 PUO W. Stumpf, 18 Saxhorn Road, Lane End, High Wycombe, HP14 3JN
G4 PUP B. Philipp, 2 Red Lion Park, Denbigh Road, Battle, TN33 9ET
G4 PUQ P. Mcewen, Southerly, Church Road, Halesworth, IP19 0EA
GM4 PUS Dr J. Murray, Ose Farm House, Isle of Skye, IV56 8FJ
GW4 PUX B. Gayther, Coed Park, Penisarwaun, Caernarfon, LL55 3PW
G4 PUZ N. Barrington, 18 Brambleside, Thrapston, Kettering, NN14 4PY
G4 PVC A. Smith, 21, Darwin Crescent, Morley, Australia, 6062
G4 PVM P. Tittensor, 47 St. Johns Road, Chelmsford, CM2 0TY
G4 PVN A. Taylor, 3 Mond Crescent, Billingham, TS23 1DL
G4 PVP P. Painter, 80 Willowsbrook Road, Hurst Green, Halesowen, B62 9RF
GM4 PVQ D. Ross, 11 Edinview Gardens, Stonehaven, AB39 2EG
G4 PVS J. Maude, Anthony Fold Farm, Bury Old Rd, Ramsbottom Lancs, BL0 0RY
GW4 PVU A. Wilkinson, 1 Langley Close, Penrhyn Bay, Llandudno, LL30 3LN
G4 PVX A. Daws, 9 Wellow Mead, Peasedown St. John, Bath, BA2 8SA
G4 PVY R. Limb, 3 Canford Heights, Western Road, Poole, BH13 7BE
G4 PVZ G. Loach, 39 Park Road West, Wolverhampton, WV1 4PL
G4 PWA P. Dane, Oakhill Lodge Hewelsfield, Lydney, GL15 6UN
G4 PWB G. Smith, 9A Lansdowne Drive, Rayleigh, SS6 9AL
G4 PWD M. Mchale, 41 Sheringham Drive, Etchinghill, Rugeley, WS15 2YG
G4 PWE J. Veness, 59 St. Helens Down, Hastings, TN34 2BG
G4 PWF R. Harries, The Mill, Mill Lane, Middle Rasen, LN8 3LE
G4 PWG J. Hubbard, 2 Carlton Road, Portchester, Fareham, PO16 8JW
G4 PWI P. Heredge, 118 Oxford Crescent, Didcot, OX11 7AX
G4 PWM D. East, 39 Chapel Lane Navenby, Lincoln, LN5 0ER
G4 PWP D. Blackwell, 58 Bleadon Hill, Weston Super Mare, BS24 9JW
GM4 PWQ J. Foster, 185 Sea Road, Methil, Leven, KY8 2EQ
GM4 PWR A. Coutts, Ballymeanoch Barn, Kildonan, Isle of Arran, KA27 8SF
G4 PWS S. Keen, 34 Unwin Road, Isleworth, TW7 6HX
G4 PWV D. Howton, 4 Stonepine Close, Wildwood, Stafford, ST17 4QS
G4 PWY P. Hennessy, 5 Smedley Court, Egginton, Derby, DE65 6HD
GW4 PWZ W. Evans, Windyridge Bungalow, Mount View, Merthyr Tydfil, CF47 0UX
GM4 PXB M. Gale, East Teuchan Cruden Bay, Peterhead, AB42 0PP
G4 PXC A. Lord, 16 Lark Valley Drive, Fornham St. Martin, Bury St Edmunds, IP28 6UG
G4 PXE A. Brend, 42 West Garth Road, Exeter, EX4 5AJ
G4 PXF M. Harries, 63 Oakhill Road, Dronfield, S18 2EL
GM4 PXG T. Worthington, 5 Clairmont Place, Lerwick, Shetland, ZE1 0BR
G4 PXH E. Southwell, 60 Solent Breezes, Hook Lane, Southampton, SO31 9HG
G4 PXJ J. Peet, 66 Barry Road, Northampton, NN1 5JS
GI4 PXM W. Nelson, 111 Rutherglen Street, Belfast, BT13 3LR
G4 PXN C. Sissons, 9 Mount Pleasant, Goldenbank, Falmouth, TR11 5BW
G4 PXR T. Geldart, Langdale, Coast Road, Ulverston, LA12 9QZ
G4 PXX P. Styles, 23 Mereweather Avenue, Frankstow, Victoria, Australia, 3199
G4 PXY G. Bloomfield, 15 Beaulieu Drive, Pinner, HA5 1NB
G4 PYA A. Ledger, 32 St. Augustines Crescent, Whitstable, CT5 2NW
G4 PYD C. Johnson, 51 Newstead Avenue, Holton-Le-Clay, Grimsby, DN36 5BQ
G4 PYG I. Bennett, Collins Green, School Road, Colchester, CO5 9TH
G4 PYH D. Jackson, 9 Stour Close, Altrincham, WA14 4UE
G4 PYI B. Burman, 53 Field Avenue, Hatton, Derby, DE65 5ER
GM4 PYJ J. Balfour, 36 Causewayhead Road, Stirling, FK9 5EU
G4 PYQ A. Hill, 37 Rock St., Gee Cross, Hyde, SK14 5JX
G4 PYS E. Devereux, 15 Severn Close, Paulsgrove, Portsmouth, PO6 4BB
G4 PYU S. Harding, 6 Rooks Way, Tiverton, EX16 6XJ
G4 PYV S. Parkin, 4 Acacia Close, Worksop, S80 3RD
G4 PYW M. Zubrzycki, 4 Falklands Court, Easington, Hull, HU12 0QE
G4 PZJ C. Christopher, 15 Inman Road, Earlsfield, London, SW18 3BB
G4 PZL R. Pain, Hegerston, Long Road West, Colchester, CO7 6ES
G4 PZN C. Poulson, 12 Glebe Close, Appleby, CA16 6RS
G4 PZU F. Day, 27 Prince Charles Road, Lewes, BN7 2HY

G4 PZV A. Terry, 6 Seaton Close, Stubbington, Fareham, PO14 2PX
G4 PZW R. Proctor, 6 North Street, Burwell, Cambridge, CB25 0BA
G4 PZX A. Tracey, Well 'N' Garden, Abberton Road, Colchester, CO5 7AS
G4 RAA M. Brown, 12 Mead Way, Slough, SL1 6HD
G4 RAB D. Ellis, 1 Showering Close, Bristol, BS14 8DY
G4 RAC J. Cooper, 134 Jordan Avenue, Stretton, Burton-on-Trent, DE13 0JD
G4 RAE R. Laney, 7 Downfield Close, Alveston, Bristol, BS35 3NJ
GD4 RAG J. Martin, Tradewinds, Mount Gawne Road, Port St. Mary, Isle of Man, IM9 5LX
GM4 RAH P. Robertson, 32 Crosswood Crescent, Balerno, EH14 7HS
GM4 RAI R. Shand, 12 Bexley Terrace, Wick, KW1 5HQ
G4 RAJ J. Shaw, 31 Dartmouth Avenue, Almondbury, Huddersfield, HD5 8UP
G4 RAK J. Hornby, Domains.Com Ltd, 1 The Courtyard, Southwell Business Park, Portland, DT5 2NQ
G4 RAP V. Seaman, 30 Dukes Orchard Nicholas Close, Writtle, Chelmsford, CM1 3JZ
G4 RAR P. Clemens, 18 Ladylea Road, Horsley, Derby, DE21 5BN
G4 RAV P. Evans, 2 Terence Airey Court, Harleston, IP20 9JP
G4 RAY G. Pickering, 1 Warrens Yard, Wells-Next-the-Sea, NR23 1PA
GM4 RAZ B. Smith, 8 Mosside Drive, Portlethen, Aberdeen, AB12 4NY
G4 RB Dr R. Baldwin, Red Barn, Elder Lane, Grimston, King's Lynn, PE32 1BJ
G4 RBC C. Hawkridge, 2 Windward Close, Littlehampton, BN17 6QX
G4 RBH T. Farmer, 12 Rose Avenue, Mitcham, CR4 3JS
G4 RBP R. Purdy, 4 York Road, Brookenby, Market Rasen, LN8 6EX
G4 RBQ D. Love, Highridge, South Road, Haywards Heath, RH17 7QS
G4 RBR C. Randall, 38 Kilmorey Gardens, St. Margaret'S, Twickenham, TW1 1PY
G4 RBU T. Crookes, 167 Willow Drive, Handsworth Hill, Sheffield, S9 4AU
G4 RBZ C. Dervin, 24 Willow Park Way, Aston-On-Trent, Derby, DE72 2DF
G4 RCB D. Thorp, 24 Garnstone Drive, Weobley, HR4 8TH
G4 RCC Caravan & Camping ARC, c/o A. Wright, 34 Webbs Way, Stoney Stanton, Leicester, LE9 4BW
G4 RCD M. Capstick, 186 Forest Lane, Harrogate, HG2 7EE
G4 RCE Dr M. Capstick, Gladbachstrasse 19, Zurich, Switzerland, 8006
G4 RCF J. O'Dell, 5 Further Ends Road, Freckleton, Preston, PR4 1RL
G4 RCG J. Muzyka, 2 Engine Fold Off Lindale Lane Kirkhamgate, Wakefield, WF2 0PP
G4 RCH S. Thompson, 2 Allenby Drive, Leeds, LS11 5RP
G4 RCJ D. Underwood, The Coach House, Todmorden, OL14 8EP
GI4 RCK W. Mcmillen, 26 Maymount St., Belfast, BT6 8BH
GW4 RCM W. Williams, 154 Llysfaen Road, Old Colwyn, Colwyn Bay, LL29 9HP
GM4 RCN J. Young, 13 Craig Crescent, Causewayhead, Stirling, FK9 5LR
G4 RCP C. Marriott, 19 Beechey Close, Denver, Downham Market, PE38 0DH
G4 RCR P. Starley, 40 Wordsworth Avenue, Warwick, CV34 6JD
G4 RCY A. Beglin, 3 The Mead, Shipham, Winscombe, BS25 1TR
G4 RCZ I. Dempster, 54 Fashoda Road, Selly Park, Birmingham, B29 7QJ
G4 RDC J. Gumb, 16 Raggleswood Close, Earley, Reading, RG6 7LH
G4 RDG Dr G. Murray, 176 Golfwood Drive, Hamilton, Canada, L9C 7B8
G4 RDH M. Wirthner, 51 College Road, Upper Beeding, Steyning, BN44 3TB
GM4 RDI J. White, 1 Banknowe Road, Tayport, DD6 9LG
G4 RDL Leeds Raynet Group, c/o G. Belt, 3 Prospect Hill, Whitby, YO21 1QE
G4 RDM P. Rouget, 7 Palmer Close, Wellingborough, NN8 5NX
G4 RDS B. Wood, 100 Lower Road, Hullbridge, Hockley, SS5 6DD
GW4 RDW L. Jones, 6, Norton Terrace, Glyncorrwg, Port Talbot, SA13 3AN
G4 RDY L. Harrison, 48 Bleasdale Avenue, Thornton-Cleveleys, FY5 3RQ
G4 REA A. Wilson, 101 Watmore Lane, Winnersh, Wokingham, RG41 5LG
G4 REC A. Marrows, 7 Victoria Close, Yeadon, Leeds, LS19 7AU
G4 REE P. Miller, 2 The Pavilions End, Camberley, GU15 2LD
GM4 REF W. Mclean, 1 Richmond Court, Rutherglen, Glasgow, G73 3BG
G4 REG A. Boocock, 2 Vine Garth, Clifton, Brighouse, HD6 4JZ
G4 REH R. England, 5 Weir Road, Congresbury, Bristol, BS49 5HL
GW4 REI D. May, 19 Sycamore Street, Pembroke Dock, SA72 6QN
G4 REK J. Tylee, 40 Luna Road, Thornton Heath, CR7 8NY
GM4 REN B. Strathdee, 85 Waverley Knowe Crescent, Currie, EH14 5PP
GW4 RER C. Carini, 6 Whitewell Drive, Llantwit Major, CF61 1TA
G4 REU J. Taylor, 18 Fackley Way, Stanton Hill, Sutton-in-Ashfield, NG17 3HT
GW4 REX P. Hassmann, 11 Oak Close, Bulwark, Chepstow, NP16 5RL
G4 RFA M. Tew, 25 Broad Oak Lane, Penwortham, Preston, PR1 0UX
G4 RFC S. Fletcher, 90 Westcombe Park Road, Blackheath, London, SE3 7QS
G4 RFF T. Mundiya, Apt 5, 35 Mercer Street, New York, USA, 10013
G4 RFI R. Linden, 24 Hartland Drive, Edgware, HA8 8RH
GD4 RFK M. Dodd, Ellan Geay, Ballayockey Lane, Ramsey, Isle Of Man, IM7 3HP
G4 RFN A. Robey, 54 Jarrett Avenue, Wainscott, Rochester, ME2 4NL
G4 RFO B. Wood, 11 Oakdale Avenue, Wibsey, Bradford, BD6 1RP
G4 RFP A. Goodall, 10 Beacon Close, Everton, Lymington, SO41 0LQ
G4 RFR Flight Refuelling ARS, c/o R. Hawkins, Forest Edge, Deer Park, Milton Abbas, Blandford Forum, DT11 0AY
G4 RFU D. Abbott, 21 Leckhampton Road, Cheltenham, GL53 0AZ
G4 RFV B. Adams, 14 Foxcroft Drive, Wimborne, BH21 2JZ
G4 RGA J. Dunnett, 43 Oakfield Park, Wellington, TA21 8EX
G4 RGB M. Rogers, 4 Hill Corner, Ledgemoor, Hereford, HR4 8QG
G4 RGE N. Lovely, Dolphin Cottage, Upper Green Road, Ryde, PO33 1XE
G4 RGF P. Mccall, 11 Elworthy Drive, Wellington, TA21 9AT
G4 RGH D. Mclaughlin, 1 Bosville Close, Ravenfield, Rotherham, S65 4NF
GW4 RGI W. Baker, 4 Connaught Place, Pembroke Dock, SA72 6EZ
G4 RGM Greater Manchester Raynet, c/o N. Czernuszka, 12 Durham Drive, Ashton-under-Lyne, OL6 8BP
G4 RGO J. Crocker, 8 Oakwood Avenue, Havant, PO9 3RA
G4 RGP E. Hall, 93 Sthbourne Coast, Bournemouth, BH6 4DX
GD4 RGR K. Grattan, 41 Carrick Park, Sulby, Ramsey, Isle of Man, IM7 2EY
GM4 RGS R. Smith, 8 Mosside Drive, Portlethen, Aberdeen, AB12 4NY
GM4 RGU D. Nicolson, Silver Birches, Blebo Craigs, Cupar, KY15 5UF
G4 RGY Burnham ARC, c/o M. Lang, 52 Gloucester Road, Burnham-on-Sea, TA8 1JA
G4 RHB M. Bailey, 10 Greenwood Avenue, Bolton Le Sands, Carnforth, LA5 8AW
G4 RHC M. Kellett, 1 Spa Cottages, Gilsland, Brampton, CA8 7AL
G4 RHJ P. Vickers, 2 Firbank Drive, Woking, GU21 7QT
G4 RHK L. Woodcock, 2 Poolhay Close Corse Lawn, Gloucester, GL19 4NY
G4 RHL R. Langdon, 15 St. Cuthberts Way, Sherburn Village, Durham, DH6 1RH
G4 RHR K. Backhouse, 113 Bucklesham Road, Kirton, Ipswich, IP10 0PF
G4 RHX A. Moore, 1 St. Andrews Road, New Marske, Redcar, TS11 8AU
G4 RHY M. Pratt, The Bays, Back Lane, Doncaster, DN9 3AJ
G4 RHZ B. Coupe, 9 School Lane, Auckley, Doncaster, DN9 3JR
GW4 RIB D. Stoole, Brookside Farm, Baltic Terrace, Cwmbran, NP44 7AH
G4 RIE D. Littler, 16 Lee Bank, Westhoughton, Bolton, BL5 3HQ
G4 RIH N. Thorne, 61 Horsham Avenue, London, N12 9BG
G4 RIK R. Kirkwood, 42 Porters Hill, Harpenden, AL5 5HR
G4 RIM A. Day, 3 Harris Way, Lee Mill Bridge, Ivybridge, PL21 9EU
G4 RIO S. Williams, 18 The Leas, Barkston, Grantham, NG32 2PD
G4 RIP J. Creaseyy, 8 Church Street Billingborough, Sleaford, NG34 0QG
G4 RIQ L. Rushforth, 90 Brearley Avenue, New Whittington, Chesterfield, S43 2DZ
G4 RIS B. Didmom, 45 Millstrood Road, Whitstable, CT5 1QF
G4 RIU R. Jones, 67 Plover Road, Larkfield, Aylesford, ME20 6LA
GM4 RIV Wigtownshire ARC, c/o A. Gaston, 9 Lochans Mill Avenue, Lochans, Stranraer, DG9 9BZ
G4 RJA I. Wilkinson, 24 Isis Way, Hilton, Derby, DE65 5LP
G4 RJD K. Ward, 3 Levetts Hollow, Hednesford, Cannock, WS12 2AW
GM4 RJF J. Weatherer, 20 Gilloch Crescent, Dumfries, DG1 4DW
G4 RJG I. Toon, 18 Barrowfield Road, Stroud, GL5 4DF
G4 RJM G. Heward, 4 Hillside Drive, Little Haywood, Stafford, ST18 0NN
G4 RJO B. Robertson, 28 Heath Lane, Blackfordby, Swadlincote, DE11 8AA
G4 RJQ C. Johnston, 299 Constable Avenue, Clacton-on-Sea, CO16 8YU
GM4 RJX J. Hatton, 64 Abercromby Crescent, Helensburgh, G84 9DN
G4 RJY C. Sidney, 10 Colville Close, Bampton, OX18 2NN
G4 RJZ A. Phillpott, Southways, Stombers Lane, Folkestone, CT18 7AP
G4 RKB J. Conlon, 24 Goldcrest Close, Colchester, CO4 3FN
GI4 RKC S. Jennings, 34 Palmer Avenue, Lisburn, BT28 3QB
G4 RKD T. Clarke, 19 Ratby Lane, Markfield, LE67 9RJ
G4 RKF B. Peart, 7A Grundy Close, Abingdon, OX14 3SD
G4 RKG J. Whiting, 19 Watermore Close, Frampton Cotterell, Bristol, BS36 2NQ
GM4 RKH T. Llewellyn, The Shepherds Cottage, Buckies Farm, Thurso, KW14 7XH
GW4 RKI K. Perryman, 17 Fford Maes Gwilym Ffos Las Carway Kidwelly, Carmathenshire, SA17 4AX
G4 RKK I. Welford, Mistletoe House, Watton Road, Thetford, IP24 1PB
G4 RKL W. Welford, Bowling Green House, Griffin Lane, Attleborough, NR17 2AD
GM4 RKM T. Cassidy, 34 Torr-Na-Faire, Lochaline, Oban, PA80 5XS
G4 RKN P. Wells, 24 Common Close, West Winch, King's Lynn, PE33 0LB
G4 RKO B. Cooper, 20 The Paddock, Alconbury, Huntingdon, PE28 4WS
G4 RKP R. Groom, Tryst, Rackhams Corner, Lowestoft, NR32 5LB
G4 RKR D. Geddes, 9 Rosenella Close, Northampton, NN4 8RX
G4 RKV L. Adams, 50 Selsea Avenue, Herne Bay, CT6 8SD
GW4 RKX G. Cook, 22 Northlands Park, Bishopston, Swansea, SA3 3JW
GW4 RKZ R. Cleverley, 33 Tylchawen Crescent, Tonyrefail, Porth, CF39 8AL
G4 RLA C. Butcher, 7 Lascelles Hall Road, Kirkheaton, Huddersfield, HD5 0AT
G4 RLC T. Isom, 64 Cuffling Drive, Leicester, LE3 6NF
G4 RLF M. Wright, 24 Wessex Road, Wilton, Salisbury, SP2 0LW
G4 RLL J. Woods, 4 Wheatfield Drive, Burton Latimer, Kettering, NN15 5YL
G4 RLM J. Hiscock, 62 East Borough, Wimborne, BH21 1PL
G4 RLN D. Rosevear, 37 Sharaman Close, St Austell, PL25 3DH
GW4 RLO G. Seal, Roda Villa, 14 Uchel Dre, Kerry, Newtown, SY16 4PS
GW4 RLP T. Varney, 29 Ffordd Eryri, Caernarfon, LL55 2UR
G4 RLR J. Hogan, 13 Strawberry Fields, Great Barford, Bedford, MK44 3BQ
G4 RLU P. Quickfall, Quickfall, 14 Lade Fort Crescent, Romney Marsh, TN29 9YF
GM4 RLV W. Duguid, Villach, 7 Hawthorn Place, Ballater, AB35 5QH
G4 RLX H. Cave, 3 Grace Meadow, Whitfield, Dover, CT16 3HA
GI4 RMA L. Mccullough, 9 Castleward Road Strangford, Downpatrick, BT30 7LY
G4 RMC D. Marsden, 67 Fourth Avenue, Watford, WD25 9QH
G4 RMD J. Cobley, 4 Briars Close, Hatfield, AL10 8DQ
G4 RMG E. Guy, 2 Broad Street, St Columb Major, TR9 6AS
G4 RMJ A. Cattani, Rozel, Marsh Road, Spalding, PE12 9PJ
GW4 RML D. Davies, 101 Westlands, Port Talbot, SA12 7DE
G4 RMN M. Hogan, 16 Freshfield Close, West Earlham, Norwich, NR5 8RA
G4 RMQ J. Dudley, 2 Heathcote Grove, London, E4 6RT
G4 RMT P. Johnson, 4 High Beech, Lowestoft, NR32 2RY
G4 RMV M. Buckle, 3 Tilesford Park Tilesford, Pershore, WR10 2LA
G4 RMX J. Phelps, 33 Thirlmere Drive, North Anston, Sheffield, S25 4JP
G4 RNA P. Dronfield, 1 Kestrel Rise Swallownest, Sheffield, S26 4SD
G4 RNC C. Blezard, 26 Welford Avenue, Lowton, Warrington, WA3 2RN
G4 RND C. Hawkins, 57 Links Road, Knott End-On-Sea, Poulton-le-Fylde, FY6 0DF
G4 RNF J. Handley, Flat 31, Croft Manor Mason Close, Freckleton, Preston, PR4 1RG
G4 RNI G. Tuck, 105 Johnson Estate, Wheatley Hill, Durham, DH6 3LH
G4 RNK R. Dodson, 22 Southgate Crescent Rodborough, Stroud, GL5 3TS
GI4 RNP V. Mcfarland, 108 Fernagh Road, Whiteabbey, Newtownabbey, BT37 0BE
G4 RNR J. Maunder, Boat Inn Farm, Shipley Gate Eastwood, Nottingham, NG16 3JE
G4 RNT D. Thorpe, 10 Stoke Road, Taunton, TA1 3EJ
G4 RNW M. Stewart, 29 Elstree Road, Bushey Heath, Bushey, WD23 4GH
G4 RNX A. Walker, 5 Christchurch Road, Malvern, WR14 3BH
G4 RNZ K. Page, 51 Bournville Road, Weston-Super-Mare, BS23 3RR
G4 ROA A. Chamberlain, 16 Okehampton Road, Stivichall, Coventry, CV3 5AU
G4 ROB R. Taylor, 18 Spruce Avenue, Selston, Nottingham, NG16 6DX
G4 ROC L. Odell, 30 St. Hybalds Grove, Scawby, Brigg, DN20 9DG
G4 ROH W. Smith, 10 Playford Road, Rushmere St. Andrew, Ipswich, IP4 5RH
G4 ROI S. Kiernan, 60 Riverview Road, Epsom, KT19 0LB
G4 ROJ R. Stafford, 21 Kittiwake Drive, Kidderminster, DY10 4RS
G4 ROK G. Sambrook, 73 Hayes Drive, Barnton, Northwich, CW8 4JX
G4 ROM M. Ellis, Field Cottage, Hole Lane, Farnham, GU10 5LP
G4 ROP C. White, 18 Ashton Gardens, Old Tupton, Chesterfield, S42 6JF
G4 ROR D. Harrison, 55 Hudson Close, Worcester, WR2 4DP
G4 ROU W. Maudsley, 42 Crawford St., Clock Face, St Helens, WA9 4XH
GW4 ROV P. Weaver, 24 Montclaire Avenue, Blackwood, NP12 1EE
G4 ROX A. Capel, 33 Romney Avenue, Bristol, BS7 9ST
G4 RPA D. Court, 4 Rucrofts Close, Bognor Regis, PO21 3SL
G4 RPC R. Cassling, 14 Canada Way, Lower Wick, Worcester, WR2 4DJ
G4 RPD A. Else, 77 Sherwood Street, Mansfield Woodhouse, Mansfield, NG19 9NB
GM4 RPE J. Mccabe, 109 Weirwood Avenue, Baillieston, Glasgow, G69 6LQ
G4 RPI C. Parrish, 16 Charter Close, Boston, PE21 9PD
G4 RPJ I. Flaherty, 10 Highfield Park, Heaton Mersey, Stockport, SK4 3HD
G4 RPK J. Kaine, 74 Camden Mews, London, NW1 9BX
G4 RPL D. Ingham, Auchengray, 51 Helena Street, Mexborough, S64 9PF

GM4 RPO T. Gemmell, 30 Goldie Crescent, Lochside, Dumfries, DG2 0AJ
G4 RPP G. Kyte, 25 Brasted Close, Bexleyheath, DA6 8HU
G4 RPT M. Edis, 28 High Street, Broughton, Kettering, NN14 1NG
G4 RPV K. Baker, 153 Long Nuke Road, Birmingham, B31 1DX
G4 RQA T. Mills, 14 Oxford Close, Padiham, Burnley, BB12 7DB
G4 RQF R. Langer, Elms Bungalow, Queens Road, Sheffield, S20 1AW
G4 RQG S. Baggaley, 35 Hayner Grove, Weston Coyney, Stoke-on-Trent, ST3 6PQ
G4 RQI D. Warr, 5 Monckton Drive, Castleford, WF10 3HT
G4 RQJ R. Hannan, 87 Plymouth Street, Walney, Barrow-in-Furness, LA14 3AN
G4 RQK A. Johnson, 14 Highfield, Duddington, Stamford, PE9 3QD
G4 RQL M. Wilson, The Old Chapel, Poulshot Road, Devizes, SN10 1RW
G4 RQN N. Bartzeliotis, 74B Ebrington Street, Plymouth, PL4 9AQ
G4 RQO J. Pulford, 68 York Avenue, Droitwich, WR9 7DQ
G4 RQP G. Hallett, 9 Dolcroft Road, Rookley, Ventnor, PO38 3NT
GW4 RQQ T. Jones, 19 Penlon, Menai Bridge, LL59 5LR
GW4 RQS J. Leighton, 12 Morley Avenue, Connah'S Quay, Deeside, CH5 4RE
G4 RQU D. Young, 9 Mercedes Avenue, Hunstanton, PE36 5EJ
G4 RQW A. Mcewen, Apartment 9, 10 Lismore Place, Carlisle, CA1 1LX
G4 RRA P. Pasquet, Honey Blossom Cottage Spreyton, Crediton, EX17 5AL
G4 RRD I. Reynolds, 4 Chappel Hill, Fakenham, NR21 9HW
G4 RRH J. Green, 83A High Street, Ramsey, Huntingdon, PE26 1BZ
G4 RRJ C. Pemberton, Flat 1, 14 Holmfield Road, Blackpool, FY2 9TB
G4 RRM P. Walker, 11 Flixton Drive, Crewe, CW2 8AP
G4 RRN C. Harrold, Boundary Farm, Felbrigg Road, Norwich, NR11 8PD
GM4 RRP R. Morris, The Gables, Highfield, Muir of Ord, IV6 7XN
G4 RRQ I. Wright, 39 Queen Anne Gardens, Falmouth, TR11 4SW
G4 RRR T. Bunce, Pear Tree House, Greaves Lane, Malpas, SY14 7AR
G4 RRU R. Crooks, 6 Whylands Avenue, Worthing, BN13 3HG
G4 RRX R. Saxton, 7 Huxley Road, Old Lakenham, Norwich, NR1 2JR
G4 RSC Reading School ARC, c/o T. Walter, 11 Silver Street Congresbury, Bristol, BS49 5EY
G4 RSD D. Bones, Flint Cottage, Ipswich Road, Woodbridge, IP13 7PP
G4 RSF M. Booth, 45 Park Avenue, Thackley, Bradford, BD10 0RJ
GI4 RSI K. Allen, 25 Knockgreenan Avenue, Omagh, BT79 0EB
G4 RSL C. Bagley, 47 Meadow View Road, Weymouth, DT3 5PB
G4 RSN R. Burman, Woodlands Vale, Calthorpe Road, Ryde, PO33 1PR
G4 RSP D. Sandy, The Chestnuts, Dumbs Lane, Norwich, NR10 3BH
G4 RSS J. Upton, 24 Heritage Drive, Clowne, Chesterfield, S43 4ST
G4 RST D. Martin, 111 Arkwright Road, Irchester, Wellingborough, NN29 7EE
G4 RSU P. Winnett, 148 Green Lanes, Epsom, KT19 9UL
G4 RSW A. Bairstow, 63 Barnes Road, Stafford, ST17 9RL
G4 RSX M. Dean, 117 Waltham Way, Chingford, London, E4 8HD
G4 RTA K. Mellor, 2 Clune St., Clowne, Chesterfield, S43 4NJ
G4 RTC X. Iona, 13 Vicars Close, Enfield, EN1 3DW
G4 RTH R. Hamstead, 1A The Close, North Walsham, NR28 9HS
G4 RTI E. Handy, 80 Watwood Road, Shirley, Solihull, B90 2HY
G4 RTJ C. Howe, 113 Fatfield Park, Washington, NE38 8BP
GM4 RTN T. Morton, 15 Craig Crescent, Causewayhead, Stirling, FK9 5LR
G4 RTO G. Calkin, 54 Pattermead Crescent, Ottawa, Ontario, Canada, K1V 0G2
G4 RTP Dr A. Shattock, The Stone House, Westport Road, Co Galway, Ireland
G4 RTQ I. Whitehead, 3 Botany Close, Thatcham, RG19 4GJ
G4 RTS W. Bateson, 10 Priestfield Avenue, Colne, BB8 9QJ
G4 RTV C. Tucker, 4 Kelsey Park Road, Beckenham, BR3 6LJ
G4 RTW G. Rolf, Flat 4, Hawksworth House, 73 St. Johns Road, Sandown, PO36 8HE
G4 RTX G. Kingdon, Flat 12, Courtfields, Lancing, BN15 8PA
G4 RTY R. Hayward, Alverstone, 28 Chatsworth Avenue, Shanklin, PO37 7NZ
G4 RUA R. Medcalf, 21 Greenbank, Falmouth, Cornwall, TR11 2SW
G4 RUE I. Worsdale, 10 Manton Road, Lincoln, LN2 2JL
G4 RUI A. Keeble, 9 Horsley Avenue, Shiremoor, Newcastle upon Tyne, NE27 0UF
G4 RUJ P. Evans, 706 St. Johns Road, Clacton-on-Sea, CO16 8BN
G4 RUL A. Turner, 42 Brassey Avenue, Hampden Park, Eastbourne, BN22 9QG
G4 RUN M. Beesley, 60 Ainsbury Road, Canley Gardens, Coventry, CV5 6BB
GM4 RUP Rev. J. Campbell, 96 Boghead Road, Lenzie, Glasgow, G66 4EN
G4 RUR M. Baker, 2 St. Peters Yard, Wold Street, Norton, Malton, YO17 9FH
G4 RUT H. Edwards, 19 Cameron Road, Burpengary East, Queensland, Australia, 4505
G4 RUW R. Daniel, 4 Gloucester Road, Newbury, RG14 5JP
GW4 RUX A. Jones, Forest Lodge, Glynhafod Street, Aberdare, CF44 6LD
GW4 RVA T. Nicholas, 15 Maes Llewelyn, Carmarthen, SA31 1JJ
G4 RVE C. Andrews, 29 Dell Drive, Angmering, Littlehampton, BN16 4HE
GI4 RVF J. Burke, 45 Shorelands, Greenisland, Carrickfergus, BT38 8FB
G4 RVG I. Binding, 40 Parklands, South Molton, EX36 4EW
G4 RVH D. Hird, 27 Red Beck Park, Cleator Moor, CA25 5EU
G4 RVJ D. Jones, 6 Priory Close, Pilton, Barnstaple, EX31 1QX
G4 RVK D. Bentley, 106 Pargeter Street, Walsall, WS2 8RR
G4 RVL D. Gentle, 1 Sunny Hill, Milford, Belper, DE56 0QR
G4 RVO T. Collinson, 8 Brownberrie Drive, Horsforth, Leeds, LS18 5PP
G4 RVP S. O'Donnell, 2 Derowen Drive, Hayle, TR27 4JN
GD4 RVQ J. Wornham, 64 Seafield Close, Onchan, Isle of Man, IM3 3BU
G4 RVS A. Rodgers, 278 Norton Lane, Norton, Sheffield, S8 8HE
GI4 RVT R. Jenkins, 11 Willowvale Crescent, Islandmagee, Larne, BT40 3SQ
G4 RVU P. Wigley, 7 Cavendish Close, Duffield, Belper, DE56 4DF
G4 RVV M. Stoneham, 139 Hever Avenue, West Kingsdown, Sevenoaks, TN15 6DT
G4 RVY G. Richardson, 69 O'Neill Drive, Peterlee, SR8 5UD
G4 RWD K. Cheetham, 71 Westmead Road, Barton Under Needwood, Burton-on-Trent, DE13 8JR
GM4 RWE D. Brown, Willow Crook Turin, Forfar, DD8 2UZ
G4 RWF M. Piecha, Oaklea, Gordons Close, Taunton, TA1 3DA
G4 RWG R. Guest, 67 Hanbury Road, Dorridge, Solihull, B93 8DN
G4 RWH R. Williams, 1 The Meadows, Newhall Green, Coventry, CV7 8BF
G4 RWI N. Spear, 79A Lower Icknield Way, Chinnor, OX39 4EA
G4 RWK W. Tolman, Pulland Cottage, West Down, Ilfracombe, EX34 8NH
G4 RWM P. Titherington, 5 Hayland Green, Hailsham, BN27 1SR
G4 RWN F. Rowan, 1 Massey Walk, Wythenshawe, Manchester, M22 5JY
G4 RWQ B. Wilkes, 3 Alsop Crest, Acton Trussell, Stafford, ST17 0SJ
GW4 RWR R. Thomas, Ystrad Isa, Ystrad, Denbigh, LL16 4RL
G4 RWS S. Valentine, 65 Holland Street, Bolton, BL1 8PA
G4 RWV P. Paling, 15 Longfellow Road, Banbury, OX16 9LB
G4 RWW P. Glaisher, The Firs, 279 Addiscombe Road, Croydon, CR0 7HY
G4 RXB K. Hawkings, Church Barn, Dingle Lane, Sandbach, CW11 1FW
GM4 RXD R. Gasken, 3 Hameravirin, Glendale, Isle of Skye, IV55 8WL
G4 RXF G. Bence, 10 Valley Road, Mangotsfield, Bristol, BS16 9HN
G4 RXG A. Mumford, Upper Cross Farm, Thornton Lane, Sandwich, CT13 0EU
G4 RXH H. Fowler, 164 Rectory Road, Deal, CT14 9NP
G4 RXK P. Mcmullan, 6 Deepdale Road, Blackpool, FY4 4UD
GI4 RXM T. Stitt, 51 Lakeland Road, Hillsborough, BT26 6PW
GW4 RXO P. Alexander, 23 Maengwynne, Llanelli, SA15 4NL
G4 RXQ R. Buck, 2 Talbot Cottages, Birtley, Chester le Street, DH3 1AR
G4 RXR R. Raine, 47 Buckingham Road, Peterlee, SR8 2DT
GI4 RXS R. Burnside, 19 Hilton Park, Portglenone, Ballymena, BT44 8HH
GM4 RXW N. Webster, Meric, 7 Woodmuir Crescent, Newport-on-Tay, DD6 8HL
GI4 RXX S. Tweedie, 12 Glencraig Close, Newtownabbey, BT36 5GZ
G4 RYB J. Baldwin, 31 Beech Road, Branston, Lincoln, LN4 1PG
G4 RYE D. Cocker, 34 Beechfield, Leeds, LS12 5QS
G4 RYH F. Appleby, 10 Buckingham Orchard, Chudleigh Knighton, Newton Abbot, TQ13 0EW
G4 RYI D. Ashcroft, 9 Aldermere Crescent, Urmston, Manchester, M41 8UE
GW4 RYJ A. Salisbury, Heddwch, Ash Grove, Flint, CH6 5RX
G4 RYK A. Richards, Castell Forwyn, Abermule, Powys, SY15 6JH
GI4 RYL M. Mccallan, 16 Abbey Crescent, Newtownabbey, BT37 9PD
G4 RYM I. Spalding, 2 Briery Lands, Heath End, Stratford-upon-Avon, CV37 0PP
G4 RYO P. Allan, Ivy Cottage, Lee Mill Bridge, Ivybridge, PL21 9EF
GI4 RYP J. Ferguson, Drumbee-More, Armagh, BT60 1HP
GW4 RYQ K. Edwards, 10 Bala Drive, Rogerstone, Newport, NP10 9HN
G4 RYS N. Black, 42 Stonegate Way, Leeds, LS17 6FD
G4 RYT J. Pickup, 274 Mauldeth Road West, Chorlton Cum Hardy, Manchester, M21 7TG
G4 RYV D. Rumbold, 15 Lodge Grove, Yateley, GU46 7AD
G4 RZC L. Ingerslev, 20 Stoney Run Lane, Marion, USA, 2738
G4 RZD L. Bradley, 138 Templeton Road, Birmingham, B44 9BY
GW4 RZE A. Taylor, 5 Wyebank Rise, Tutshill, Chepstow, NP16 7DS
G4 RZF G. May, 5 The Burlongs, Glebe Road, Swindon, SN4 7DR
G4 RZI M. Nagle, The Woodlands, Pilton West, Barnstaple, EX31 4JQ
G4 RZM B. Williams, Warren Cottage, Polyphant, Launceston, PL15 7PS
G4 RZN R. Leeds, 1A Clare Road, Cromer, NR27 0DD
G4 RZQ K. Russell, Courtiles, Main Road, Ventnor, PO38 3NH
G4 RZR R. Tooth, 25 Northgate, Beccles, NR34 9AS
GM4 RZW D. Taylor, 42 Craiglockhart Road, Edinburgh, EH14 1HG
G4 RZY L. Baker, The Novers Park Community Centre, Rear Of 122-124, Bristol, BS4 1RN
G4 RZZ I. Griffin, 15 Hesselyn Drive, Rainham, RM13 7EJ
G4 SAB C. Leat, 8 White Point Court, Whitby, YO21 3UR
G4 SAC S. Collings, 4 Glamis Close, Waterlooville, PO7 8JN
G4 SAJ C. Green, 76 Dibleys, Blewbury, Didcot, OX11 9PU
GI4 SAM S. Noble, 19 New Line, Dundonald, Belfast, BT16 1UU
G4 SAS R. Jones, 42 Fastmoor Oval, Birmingham, B33 0NR
G4 SAT Inmarsat ARC, c/o D. John, 27 Churchfields, Dartmouth, TQ6 9HJ
G4 SAV F. Hepworth, 5 Snydale Avenue, Normanton, WF6 1SS
G4 SAW M. Arbon, 106 The Tideway, Rochester, ME1 2NN
GI4 SBA K. Branagh, 17 Rathmoyle Park West, Carrickfergus, BT38 7NG
G4 SBB Dr C. Fay, Driftwood, Middle Road Sway, Lymington, SO41 6BB
G4 SBC I. Bateman, 29 Bradgate Croft Hasland, Chesterfield, S41 0XZ
G4 SBD G. Bax, 8 Hockeredge Gardens, Westgate-on-Sea, CT8 8AN
G4 SBE K. Bowden, 14 Pool Hey Lane, Scarisbrick, Southport, PR8 5HS
G4 SBF P. Fry, 54 Studley Avenue, Holbury, Southampton, SO45 2PP
G4 SBG H. Crossland, 107 Fairway, Normanton, WF6 1SN
G4 SBM R. Harding, 12 Keswick Avenue, Loughborough, LE11 3RL
G4 SBN J. Ayers, 3 Sovereign Way, Ryde, PO33 3DL
GM4 SBP G. Allan, 31 Jubilee Grove, Glenrothes, KY6 1HW
G4 SBQ M. Rushton, 14 Acorn Close, Leyland, PR25 3AF
G4 SBS R. Phillips, 4 Cumberland Drive, Fazeley, Tamworth, B78 3YA
G4 SBU B. Gundry, 37 Stoneham Park, Petersfield, GU32 3BT
G4 SBW T. Carberry, 10 Honeymeade Close, Stanton, Bury St Edmunds, IP31 2EF
G4 SCB M. Sargent, 19 Pine Tree Close Cowes, Isle Of Wight Po31 8Dx, Cowes, PO31 8DX
G4 SCE Dr P. Whitehead, Carrick View, Bank End, Carlisle, CA5 6QW
G4 SCG C. Curson, 3 Cranmer Road, Edgware, HA8 8UA
G4 SCJ D. Meakins, 19 Booth Lane North, Northampton, NN3 6JQ
GW4 SCK R. Hancock, 1 Nevills Close, Gowerton, Swansea, SA4 3BG
G4 SCL M. Starkey, Cutlers Forth Farm, Radley Road, Newark, NG22 8AP
G4 SCM J. Claxton, Camino Del Perpen 25, Catral, Spain, 3158
G4 SCO N. Drury, 3 Northam Close, Marshside, Southport, PR9 9GA
G4 SCS A. Fletcher, Stonehouse Stores, West Rd, Ormesby, NR29 3RJ
G4 SCV I. Gammon, The Haven, Craddock, Cullompton, EX15 3LH
G4 SCY P. Bradbury, 52 Moss Park Avenue, Werrington, Stoke-on-Trent, ST9 0EP
G4 SDI L. Footring, 26 Ernest Road, Wivenhoe, Colchester, CO7 9LG
G4 SDJ R. Freeman, Flat 2, Russett Court, 15 Kirtleton Avenue, Weymouth, DT4 7PS
G4 SDL B. Dorricott, 6 Knowsley Avenue, Urmston, Manchester, M41 7BT
GW4 SDO D. Phillips, Trem Y Fammau, Tri Thy, Tir Y Fron Lane, Mold, CH7 4TU
GW4 SDT S. Lansdown, 11 Redbrook Road, Newport, NP20 5AA
G4 SDU P. Smart, 6 Nobold Close, Baschurch, Shrewsbury, SY4 2EH
G4 SDX G. Townend, 9 Warren Park Close, Hove Edge, Brighouse, HD6 2RU
G4 SDZ M. Gayler, 39 Holmfield Avenue West Leicester Forest East, Leicester, LE3 3FF
G4 SEA R. Seabridge, 7 Heritage Avenue, Frankston South, Melbourne, Australia,
G4 SEB S. Swallow, 16 Quarry Lane, Chesterfield, S40 3AS
G4 SEF R. Jenkinson, 4 Apple Croft Skidby, Cottingham, HU16 5UG
G4 SEG A. Clayton, 448 Gisburn Road, Blacko, Nelson, BB9 6LZ
G4 SEJ B. Vane, 3 Charnwood, Chestfield, Whitstable, CT5 3QD
G4 SEK K. Aylwin, 9 Hockeredge Gardens, Westgate-on-Sea, CT8 8AN
G4 SEL G. Wilkes, 49 Charlemont Road, Walsall, WS5 3NQ
G4 SEN N. Whitham, The Cottage, Castle Gate Nancledra, Penzance, TR20 8BQ
G4 SEP C. Turner, Saxavord, Humberston Road, Grimsby, DN36 5NJ
G4 SEQ D. Vickers, 48 Bromley Road, Hanging Heaton, Batley, WF17 6EH
G4 SET C. Hall, 8 Sharps Court, Exmouth, EX8 1DT
G4 SEU J. Russell, 9 Batten Close, Christchurch, BH23 3BJ

G4	SEV	N. Le Gresley, 32 Churchill Road, Welton, NN11 2JH
G4	SEW	K. Weir, Les Ginestes Appt 231, 28 Av Dr Gerhardt, Peymeinade, France, 6530
G4	SEZ	C. Greenland, 21 Penleigh Close, Corsham, SN13 9LE
GM4	SFA	A. Keenan, Darwin, Coalhall, Ayr, KA6 6ND
G4	SFB	D. Knowler, Apartedo 1009, 8670 - 999, Aljezur, Portugal,
G4	SFD	D. Birks, Flat, 1 Pewsham House, Chippenham, SN15 3RX
GI4	SFE	J. Mccullough, 12 Bramble Grange, Newtownabbey, BT37 0XH
G4	SFG	P. O'Connor, 36 Heron Road, Oldbury, B68 8AQ
G4	SFH	N. Richardson, 22 Bramshott Drive, Hook, RG27 9EY
G4	SFJ	S. Stott, 20 Lingfield Crescent, Wigan, WN6 8QA
G4	SFN	J. Tetlow, 14 Fountains Crescent, Hebburn, NE31 2HT
G4	SFP	J. Nash, 259 Weald Drive Furnace Green, Crawley, RH10 6PN
G4	SFQ	S. Fletcher, The Bakery, Keswick Road, Norwich, NR12 0HF
G4	SFS	P. Grosjean, Garden House, West Horrington, Wells, BA5 3ED
GM4	SFT	D. Mcalonan, Glenonan House Cromlech St, Dunoon, PA23 8PQ
GM4	SFW	J. Stuart, Tigh Na Coille, Bishop Kinkell, Dingwall, IV7 8AW
G4	SFY	R. Baker, 69 Northfield Road Mundesley, Norwich, NR11 8JN
GI4	SFZ	C. Hought, 37 Oldpark Avenue, Ballymena, BT42 1AX
G4	SGA	G. Barnes, 3 Blandford Avenue, Castle Bromwich, Birmingham, B36 9HX
G4	SGD	S. Simpson, 17 Astley Way Ashby De La Zouch, Leicester, LE651LY
G4	SGE	B. Hughes, 86 Lewis Avenue, Wolverhampton, WV1 2AR
G4	SGF	Dr K. Ruiz, Flat 12, The Woodlands, 39 Shore Lane, Sheffield, S10 3BU
G4	SGG	D. Earp, 88 Linton Rise, Nottingham, NG3 7BY
G4	SGI	S. Collings, 46 St. Michaels Road, Cheltenham, GL51 3RR
G4	SGJ	J. Campbell, 9 Blackdown Close, Dibden Purlieu, Southampton, SO45 5QS
G4	SGN	P. Playle, 6 Walnut Tree Close, Cheshunt, Waltham Cross, EN8 8NH
GW4	SGQ	P. Hruza, 18 Withy Avenue, Forden, Welshpool, SY21 8NJ
GW4	SGR	S.Glam Rynt Grp, c/o R. Magwood, 13 Inverness Place, Cardiff, CF24 4RU
G4	SGU	G. Gilbertson, 6 The Stray, South Cave, Brough, HU15 2AL
G4	SGV	K. Jones, 228 Evesham Road, Headless Cross, Redditch, B97 5EP
G4	SGW	W. Prouse, 1 Springfield Cottages, Bishops Tawton, Barnstaple, EX32 0DF
G4	SGX	I. Haywood, 5 Pump Corner, Marsham, Norwich, NR10 5PW
G4	SGY	J. Winters, 94 Wharncliffe Road, Loughborough, LE11 1SN
G4	SHA	M. Webb, 9 Steele Close, Devizes, SN10 3SL
G4	SHB	P. Sheridan, 17 Boakes Drive, Stonehouse, GL10 3QW
G4	SHC	R. Bentham, 12 Tanners Way, Nantwich, CW5 7FL
GW4	SHF	S. Purser, Penbrey, Llanfair Caereinion, Welshpool, SY21 0DG
G4	SHH	P. Brooking, 49 Binstead Lodge Road, Ryde, PO33 3TL
G4	SHJ	N. Douglas, 87 Hutton Avenue, Hartlepool, TS26 9PR
G4	SHK	G. Clifford, Turnpike Road, Blunsdon, Swindon, SN26 7EA
G4	SHM	M. Baker, 92 Moy Avenue, Eastbourne, BN22 8UQ
G4	SHN	J. Pitts, 4 Tannery Close, Dagenham, RM10 7EX
G4	SHO	F. Dibden, 127 Mayola Road, Clapton, London, E5 0RG
G4	SHQ	D. West, 24 Oxfield Drive, Gorefield, Wisbech, PE13 4LX
G4	SHY	L. Afford, 3 Kem Street, Nuneaton, CV11 4LH
GM4	SID	S. Will, 53 Bishop Forbes Crescent, Blackburn, Aberdeen, AB21 0TW
G4	SIE	R. Mason, 35 Princes Gardens, Blyth, NE24 5HL
G4	SIF	R. Rowsell, 61 Barrack Road, Bexhill-on-Sea, TN40 2AZ
GW4	SII	P. Garston, 85 Wood Lane, Hawarden, Deeside, CH5 3JG
G4	SIJ	B. Hammond, 10 Grampian Close, Sleaford, NG34 7WA
G4	SIL	E. Tubman, 54 Summerfield Avenue, Whitstable, CT5 1NS
GI4	SIP	J. Mckavanagh, 28 Thompsons Grange, Carryduff, Belfast, BT8 8TG
G4	SIR	P. Rogers, Pikk 36-46, Tallinn, Estonia, 10123
G4	SIS	R. Keefe, 28 Burstead Drive, South Green, Billericay, CM11 2QN
GI4	SIW	Antrim & Dis AR, c/o Dr W. Hutchinson, 40 Oldstone Hill, Muckamore, Antrim, BT41 4SB
GI4	SIZ	T. Thompson, 135 Glenhead Road, Limavady, BT49 9LR
GM4	SJB	J. Bruce, Kinnaird, Brora, KW9 6NN
G4	SJD	S. Davis, 33 Pollard Close, Plymstock, Plymouth, PL9 9RR
G4	SJG	G. Upton, 18 Cranthorne Drive, Nottingham, NG3 7HD
G4	SJH	B. Lewis, 23 Lightwater Meadow, Lightwater, GU18 5XH
G4	SJI	A. Harris, 10 Egroms Lane, Withernsea, HU19 2LZ
G4	SJL	S. Thomson, 11 Beverley Road, London, W4 2LL
G4	SJM	J. Reeves, 5 Arrows Crescent, Boroughbridge, York, YO51 9LP
G4	SJN	B. Hunt, Tralee, Oakridge Lynch, Stroud, GL6 7NY
GW4	SJO	M. Edwards, 1 Maes Yr Efail, Penparc, Cardigan, SA43 1FB
G4	SJP	S. Prior, East Brantwood, Manor Road, Barnstaple, EX32 0JN
GI4	SJQ	G. Frazer, 20 Old Rectory Park, Portadown, Craigavon, BT62 3QH
G4	SJU	S. Browne, 38 Aldrin Road, Pennsylvania, Exeter, EX4 5DN
G4	SJV	D. Chapman, 24 Broad Lane, Moulton, Spalding, PE12 6PN
G4	SJW	Bridges Radio Club Hampshire, c/o M. Troy, 22 Jackie Wigg Gardens Totton, Totton, Southampton, so409lz
GW4	SKA	J. Barber, 49 Blackmill Road, Bryncethin, Bridgend, CF32 9YN
GM4	SKB	M. Whyatt, Backburn Cottage, Castleton Road, Auchterarder, PH3 1JS
G4	SKM	Maltby and District ARS, c/o Dr R. Cochrane, 134 Moor Lane South, Ravenfield, Rotherham, S65 4QR
G4	SKO	M. Brooke, 6 Sun Street, Stanningley, Pudsey, LS28 6DJ
GW4	SKP	A. Clark, 27 Heol Sant Bridget, St. Brides Major, Bridgend, CF32 0SL
G4	SKU	J. Blain, 1 Handley Court, Bunyan Road, Sandy, SG19 1BJ
G4	SLG	K. Oliver, 42 Minster Drive, Cherry Willingham, Lincoln, LN3 4NA
GW4	SLI	J. Plumley, 34 Graigwen Crescent, Abertridwr, Caerphilly, CF83 4BN
G4	SLL	J. Buckland, 245 Saunders Lane Mayford, Woking, GU22 ONU
GI4	SLQ	K. Boyd, 29 Benburb Road, Moy, Dungannon, BT71 7SQ
GM4	SLV	J. Pumford-Green, Greenmeadow, Clousta, Bixter, Shetland, ZE2 9LX
G4	SLW	D. Dudkowski, 7 White Street, Brighton, BN2 0JH
GM4	SLY	J. Bell, 13 Corrie Place, Troon, KA10 6TZ
G4	SMA	M. Goode, Meadowgreen, Batch Valley, Church Stretton, SY6 6JW
G4	SMB	Capt. M. Briggs, 70 Auchinleck Close Kellythorpe, Driffield, YO25 9HE
G4	SMD	D. Blackwell, 2 Courtry Cottages, Bridgehampton, Yeovil, BA22 8HF
G4	SME	Skmrsdle & D AR, c/o J. Rogers, 186 Beavers Lane, Birleywood, Skelmersdale, WN8 9BP
G4	SMK	K. Wilson, 15 Woodside Avenue, Cottingley, Bingley, BD16 1RB
GW4	SML	T. Morgan, 37 Fan Heulog Talbot Green, Pontyclun, CF72 8HQ
G4	SMM	M. Manley, Rolleston, Parkgate Road, Chester, CH1 6JS
G4	SMQ	D. Mcdonald, 3 Cloverton Drive, Bridgwater, TA6 4HQ
G4	SMT	J. Short, 42 Alt Road, Formby, Liverpool, L37 6DF
G4	SMX	J. Wilson, 168 Elms Vale Road, Dover, CT17 9PN
GI4	SNA	D. Ross, 127 Pond Park Road, Lisburn, BT28 3RE
G4	SND	M. Newey, 148 High Street, Pensford, Bristol, BS39 4BH
G4	SNI	H. Farley, 11 College Lane, Hatfield, AL10 9PB
G4	SNJ	C. Frenzel, Butlers Hall, Butlers Hall Lane, Bishop's Stortford, CM23 4BL
G4	SNL	I. Dunworth, 51 The Crest, Sawbridgeworth, CM21 0ER
G4	SNN	N. Booth-Isherwood, 65 Burnaby St., Alvaston, Derby, DE24 8RN
G4	SNO	A. Duggins, Dowles Bungalow, Dowles Road, Bewdley, DY12 3AA
G4	SNQ	T. Wadsworth, 20 Rook Wood Way, Little Kingshill, Great Missenden, HP16 0DF
G4	SNR	E. Meekers, 5 Frobisher Close, Mudeford, Christchurch, BH23 3SN
G4	SNU	R. Hardie, 12 Hopland Close, Longwell Green, Bristol, BS30 9XB
G4	SNV	G. Eastgate, 103 Western Road, Leigh-on-Sea, SS9 2PB
GM4	SNW	M. Donnelly, 23 St John Street, Creetown, Newton Stewart, DG8 7JB
G4	SOA	P. Instone, 19 Dickenson Road, Swindon, SN25 1WG
G4	SOB	W. Hammond, 27 Gainsborough Road, Colchester, CO3 4QN
GW4	SOC	V. Shaw, 130 Aberthaw Road, Ringland, Newport, NP19 9QS
G4	SOF	J. Blight, Lowbell, Handy Cross, Bideford, EX39 3ET
G4	SOG	M. Pridham, 61 Bridge Street, Kington, HR5 3DJ
G4	SOH	M. Spence, 5 St. Helens Avenue, Benson, Wallingford, OX10 6RY
G4	SOI	M. Wray, The Old Croft, Top Street, Retford, DN22 0LG
G4	SOK	R. Hollow, The Beeches, Grove Lane, Penzance, TR20 9HN
G4	SOL	I. Bale, 2B Holes Lane, Knottingley, WF11 8LH
G4	SOM	J. Spiteri, 27 Hillcote Close, Sheffield, S10 3PT
G4	SOP	K. Lillingstone, 4 Old Bakery Court, Coltishall, Norwich, NR12 7DQ
G4	SOQ	B. Lyons, 16 Ashdale Close, Sawtry, Huntingdon, PE28 5SN
G4	SOR	A. Collins, 19 Cavendish Road, Skegness, PE25 2QZ
G4	SOT	D. Goodwin, Route De Samatan, Frontignan Saves, France, 31230
GI4	SOY	M. Dornan, 25 Dromore Road, Desertmartin, Magherafelt, BT45 5JZ
G4	SOZ	D. Herd, Capricorn Cottage, Low Common, Norwich, NR14 7BU
G4	SPA	Buxton ARS, c/o R. Marchington, 78 Buxton Road, Dove Holes, Buxton, SK17 8DW
G4	SPC	B. Escreet, 198 Front Street, Sowerby, Thirsk, YO7 1JN
G4	SPD	N. Jarvis, 25 St. Augustines Close, Aldershot, GU12 4SF
G4	SPE	G. Callaghan, 1 Wessex Close, Semington, Trowbridge, BA14 6SA
GW4	SPL	P. Ace, 116 Gellionen Road, Clydach, Swansea, SA6 5HF
G4	SPR	F. Rattray, 4 Winton Manor Court, Winton, Kirkby Stephen, CA17 4HR
G4	SPS	N. Beggs, 11 Orion Close, Fareham, PO14 2SQ
G4	SPT	H. Golding, Barany Uyca 2, Hodmezovasarhely, Hungary, 6800
GI4	SPU	N. Alcock, 22 Chippendale Avenue, Bangor, BT20 4PT
G4	SPV	A. Adamson, 520 York Road, Stevenage, SG1 4EP
G4	SPW	T. Devlin, 32 Kestrel Drive, Dalton-in-Furness, LA15 8QA
G4	SPY	A. Kay, 36 Yorks Wood Drive, Birmingham, B37 6DL
G4	SPZ	P. Harris, 22 Bramley Way, Bewdley, DY12 2PU
G4	SQA	D. Yeoman, 1 Cartmell Court Bridge Street Deeping St James Pe6 8EI, Peterborough, Pe6 8el
G4	SQG	R. Nicholson, 7 Half Mile Gardens, Leeds, LS13 1BL
G4	SQI	G. Gulliford, 18 Purslane, Abingdon, OX14 3TR
G4	SQJ	G. Turner, 21 Barlow Road, Chichester, PO19 3LD
G4	SQK	G. Middleton, 212 East Markham Avenue, Durham, USA, 27701
GI4	SQL	S. Adrain, 10 Highgate Drive, Newtownabbey, BT36 4WQ
GM4	SQM	D. Anderson, 34 Culzean Crescent, Kilmarnock, KA3 7DT
GM4	SQO	R. Riddiough, 1 Cedar Road, Ayr, KA7 3PE
G4	SQQ	C. Hollister, Rosemead, 326 Passage Road, Bristol, BS10 7TE
G4	SQV	J. Hart, 88 Breckhill Road, Woodthorpe, Nottingham, NG5 4GQ
G4	SRD	R. Sealy, 10 Mallard Close, Bowerhill, Melksham, SN12 6TQ
GW4	SRE	J. Reeves, 5 Lon Y Bryn, Glynneath, Neath, SA11 5BG
G4	SRF	C. Radcliffe, 85 Brian Avenue, Cleethorpes, DN35 9DE
GW4	SRI	A. Gray, 24 Waunarlwydd Road, Cockett, Swansea, SA2 0GB
GM4	SRL	R. Cowan, 85 Eastwoodmains Road, Clarkston, Glasgow, G76 7HG
G4	SRP	G. Brierley, 6 Yeo Drive, Appledore, Bideford, EX39 1RD
GI4	SRQ	W. Mchugh, 47 Main St., Hamiltonsbawn, Armagh, BT60 1LP
G4	SRV	D. Darby, 28 Coleshill Close, Hunt End, Redditch, B97 5UN
G4	SRX	G. Sampson, 47 Netherthorpe Way, North Anston, Sheffield, S25 4FL
GM4	SSA	H. Hassel, Sumra, Eshaness, Shetland, ZE2 9RS
G4	SSC	Dr A. Taylor, 38 Summershades Lane, Grasscroft, Oldham, OL4 4ED
G4	SSD	South Devon Radio Club, c/o J. May, 6 Hodson Close, Paignton, TQ3 3NU
G4	SSE	J. Sartin, East Barn, Gubblecote, Tring, HP23 4QG
GI4	SSF	S. Craig, 6 Kingswood Park, Belfast, BT5 7EZ
G4	SSH	R. Clayton, 9 Green Island, Irton, Scarborough, YO12 4RN
G4	SSJ	R. Bolton, 83 Sandicroft Close, Birchwood, Warrington, WA3 7LY
G4	SSL	W. Lancashire, 111 Pentire Avenue, Newquay, TR7 1PF
G4	SSP	K. Waghorne, 23 Bramley Hill, Mere, Warminster, BA12 6JX
G4	SSV	S. Smith, 1 Buckfast Road, Lincoln, LN1 3JS
G4	SSW	J. Walker, 15 Hillfield Road, Bilton, Rugby, CV22 7EW
G4	SSZ	D. Fox, 49 Turnpike Hill, Hythe, CT21 4SE
G4	STB	P. Lock, 82 Rownhams Road, Throop, Bournemouth, BH8 0NL
G4	STD	B. West, 13 Tanglewood Close, Upper Shirley, Croydon, CR0 5HX
G4	STE	S. Farr, 37A Bromsgrove Road, Studley, B80 7PG
G4	STH	G. Timbrell, Crossing Cottage, Lamyatt, Shepton Mallet, BA4 6NG
G4	STI	M. Pinder, 36 West Ridge, Allesley, Coventry, CV5 9LN
G4	STK	E. Brown, 16 Springfield, Ovington, Prudhoe, NE42 6EH
G4	STO	P. Rose, Pinchbeck Farmhouse, Mill Lane, Sturton By Stow, Lincoln, LN1 2AS
G4	STP	T. Mangles, 46 Cedar Crescent, Willington, Crook, DL15 0DA
G4	STV	Hadley Wood Contest Group, c/o S. Wilson, 21 Plumian Way, Balsham, CB21 4EG
G4	STW	A. Buckley, 5 Russell Walk, Messingham, Scunthorpe, DN17 3TU
G4	STZ	T. Bromsgrove, 11 Moelwyn Drive, Ellesmere Port, CH66 1TY
G4	SUA	B. Common, 59 Thornbera Gardens, Bishop's Stortford, CM23 3NP
GW4	SUD	K. Jones, 111 Ewenny Road, Bridgend, CF31 3LN
GW4	SUE	M. Hill, 13 Maesglas Grove, Newport, NP20 3DJ
GM4	SUF	P. Gane, Ardmore Lodge, Station Road, Edderton, Tain, IV19 1LA
G4	SUK	M. Kent, 304 Reculver Road, Herne Bay, CT6 6SR
GW4	SUN	B. Le Carpentier, 43 Abernant Road, Aberdare, CF44 0PY
G4	SUO	P. Barwick, Brook Cottage, 94 Ambleside Road, Lightwater, GU18 5UJ
GM4	SUR	R. Aitken, 2 Eskdale Drive, Bonnyrigg, EH19 2LD
G4	SUS	S. Morgan, Hollaway, Northbourne Road, Deal, CT14 0LA
G4	SUU	M. Nairn, 13 Hanover Court, Lacey Street, Ipswich, IP4 2PJ

G4 SUX R. Payne, 13 Bowden Hill, Lacock, Chippenham, SN15 2PW
G4 SVA J. Appleton, 66 Bolton Road West, Ramsbottom, Bury, BL0 9ND
G4 SVB A. Gatrell, Sunnyside, Muddles Green, Lewes, BN8 6HW
G4 SVC T. Axtell, 146 Olivers Battery Road South, Winchester, SO22 4LF
GD4 SVD A. Ames, 20 Sunnybank Avenue, Onchan, Isle Of Man, IM3 3BW
G4 SVE J. Hewett, 84 Dunsgreen, Ponteland, Newcastle upon Tyne, NE20 9EJ
G4 SVG S. Wensley, 7 Bradshaw Close, Windsor, SL4 5PS
G4 SVI C. Ames, Heatherdene, 58 The Street, Norwich, NR8 6AB
G4 SVL I. Hewitt, 35 Birmingham Road, Alvechurch, Birmingham, B48 7TB
GM4 SVM G. Hudson, 17 Drylaw Crescent, Edinburgh, EH4 2AU
G4 SVQ P. Haffenden, 113 Pavilion Road, Worthing, BN14 7 EG
G4 SVR W. Baddeley, 12 Stockport Road, Altrincham, WA15 8ET
G4 SVS D. Bush, 8 Oldbury Chase, Bristol, BS30 6DY
G4 SVV D. Lees, 1, Davies Ave, Cheadle, SK8 3PF
GM4 SVW W. Mclaren, 4 Firth Crescent, Gourock, PA19 1EW
G4 SVY J. Perez, 9 Rectory Road, Shanklin, PO37 6NX
G4 SWA S. Authers, 9 Conway Avenue, Birmingham, B32 1DR
G4 SWH R. Jones, Blakeney, Fore Street Weston, Hitchin, SG4 7AS
G4 SWO S. Griffiths, 25 Hanks Close, Malmesbury, SN16 9UA
G4 SWQ R. Torence-Smith, Birch Lodge, Flax Lane, Sudbury, CO10 7RS
G4 SWR I. Hill, 7 Cosford Close, Matchborough East, Redditch, B98 0BH
G4 SWY D. Turner, 15 Brooke Close, Bushey, WD23 1FB
GW4 SXA P. Ap-Dafydd, 37 Heol Nant Cwmdare, Aberdare, CF44 8TE
G4 SXE B. Holden, 76 The Lawns, Rolleston-On-Dove, Burton-on-Trent, DE13 9DE
G4 SXG D. Fraser, 63 Vicars Hall Gardens, Worsley, Manchester, M28 1HW
G4 SXH L. Fletcher, 27 Stanford Rise, Sway, Lymington, SO41 6DW
GM4 SXJ R. Malcolm, 43 Kinghorne St., Hospitalfield, Arbroath, DD11 2LZ
G4 SXK A. Leighs, 16 Spode Close, Cheadle, Stoke-on-Trent, ST10 1DT
GU4 SXM P. Bannier, 10 Le Bouet, Longstore, St Peter Port, Guernsey, GY1 2BA
G4 SXQ D. Lempriere, Harewarren Lodge, Salisbury, SP2 0NF
G4 SXT D. Ayers, 22 Holders Road, Amesbury, Salisbury, SP4 7PP
GI4 SXV E. Barker, 39 Birchwood, Omagh, BT79 7RA
G4 SXX J. Key, 29 Scots Court, Hook, RG27 9QJ
G4 SXY G. Haines, 25 Hook Hill, South Croydon, CR2 0LB
G4 SXZ R. Hiles, 19 Station Road, Kirton Lindsey, Gainsborough, DN21 4BB
G4 SYA C. Allen, 11 Chandos Court, Martlesham, Woodbridge, IP12 4SU
G4 SYB P. Loveland, 25 White Acres Road, Mytchett, Camberley, GU16 6JJ
G4 SYC G. Lomas, 2 Linney Road, Bramhall, Stockport, SK7 3JW
G4 SYD H. Cook, 24 Front St., Sherburn Hill, Durham, DH6 1PA
G4 SYE M. Wilson, The Old Post Office, Knowl Hill, Reading, RG10 9YD
GM4 SYF N. Wallace, 2 Mansefield, Leitholm, Coldstream, TD12 4JQ
G4 SYG P. Tattersall, Sun House, The Street, Nacton, Ipswich, IP10 0EU
GW4 SYI L. Wilder, 19 Cambrian Drive, Rhos On Sea, Colwyn Bay, LL28 4SL
G4 SYL J. Frost, 68 Wessex Road, Didcot, OX11 8BP
GI4 SYM W. Donaldson, 44 Drumman Hill, Armagh, BT61 8RW
GW4 SYO H. Thomas, 1 Cambrian Terrace, Llwynypia, Tonypandy, CF40 2HN
GU4 SYQ L. Le Page, Heathwick, Les Martins, Guernsey, GY4 6QJ
G4 SYR S. Field, 4 Lyndale, Kelvedon Hatch, Brentwood, CM15 0BQ
G4 SYT D. Chambers, 26 Drummond Gardens, Christ Church Mount, Epsom, KT19 8RP
G4 SYV R. Ackroyd, 14 Isis Avenue, Bicester, OX26 2GS
G4 SYW A. Hargreaves, 13 Linworth Road, Bishops Cleeve, Cheltenham, GL52 8PA
G4 SZA I. Donaldson, 25 Alwyn Road, Maidenhead, SL6 5EG
G4 SZB E. Flannigan, 14 Westbank Avenue, Blackpool, FY4 5BT
GM4 SZG J. Freeland, 48 Elgin Place, Shawhead, Coatbridge, ML5 4JQ
G4 SZI A. Parry, 18 Spinney Lane, Rabley Heath, Welwyn, AL6 9TF
G4 SZO K. Painter, 8 Fairview Road, Broadstone, BH18 9AX
GI4 SZP N. Hughes, 32 Kinedale Park, Ballynahinch, BT24 8YS
GI4 SZQ K. Murphy, 8 Rosscolban Meadows, Kesh, Enniskillen, BT93 1UH
G4 SZS T. Harber, 27 Yarlington Close, Norton Fitzwarren, Taunton, TA2 6RR
GI4 SZU J. Mccurry, 30 Carrowdoon Road, Dunloy, Ballymena, BT44 9DL
GW4 SZV Aberporth YMCA ARC, c/o G. Carruthers, Henllys Farm, Cardigan, SA43 2HR
GI4 SZW M. Keenan, 30 Ballynabee Road Camlough, Newry, BT35 7HD
G4 SZX M. Stockton, 7 The Croft, Thorne, Doncaster, DN8 5TL
GI4 SZY R. Wilson, 57 Mill Green, Doagh, Ballyclare, BT39 0PH
G4 TAD M. Wooltorton, 4 Hall Lane, Oulton, Lowestoft, NR32 5DJ
G4 TAG T. Gammage, 31 Kennet Drive, Congleton, CW12 3RH
G4 TAH I. Conibear, 5 Brunswick Street, Redfield, Bristol, BS5 9QN
GI4 TAJ J. Bingham, 35 Rathmena Drive, Ballyclare, BT39 9HZ
G4 TAK J. Hancock, 29 Convent Close, Aughton, Ormskirk, L39 4XP
GM4 TAL A. Blyth, 73 Glassel Park Road, Longniddry, EH32 0TA
G4 TAM B. Chambers, 93 Main Road, Hoo, Rochester, ME3 9EU
G4 TAO S. Rogers, Gaythorpe, Blacketts Wood Drive, Rickmansworth, WD3 5QQ
GI4 TAP S. Mccabe, 27 Baronscourt Road, Carryduff, Belfast, BT8 8BQ
G4 TAT D. Ruth, 1 Derwent Drive, Appley, Ryde, PO33 1NT
GW4 TAU E. Davies, Gwelydon, Lon Brynteg, Ynys Mon, LL59 5UA
GI4 TAV M. Doherty, 20 Drumcairn Close, Belfast, BT8 8HQ
G4 TAW N. Perrott, 56 Park Avenue, Chatswood, Australia, 2067
G4 TAZ F. Rewaj, 5 Spring Gardens, Grizebeck, Kirkby-in-Furness, LA17 7XJ
G4 TBF E. Popham, 741 Landbase Australia, Locked Bag 25, Gosford, Australia, 2250
G4 TBG D. Smith, 34 Grays Lane Downley, High Wycombe, HP13 5TZ
G4 TBI P. Cornell, 1 Orient Drive, Winchester, SO22 6NZ
G4 TBJ R. Smith, 184 Solihull Road, Shirley, Solihull, B90 3LG
G4 TBK D. Nix, 75 Mayfield Road, Chaddesden, Derby, DE21 6FX
G4 TBM M. Tester, 6 Harvard Close, Lewes, BN7 2EJ
G4 TBN C. Anderson, 10 School Lane, Everdon, Daventry, NN11 3BW
G4 TBO C. Harris, 33 Brent Street Brent Knoll, Highbridge, TA9 4DT
G4 TBQ P. Harlow, 36 Ravens Way, Burton on Trent, DE14 2JS
G4 TCA R. Hawthorn, Weirsmeet Bungalow, Mill Lane, Derby, DE74 2EJ
G4 TCB P. Holland, 9 Garmont Road, Leeds, LS7 3LY
G4 TCC A. Hawkins, 5 Hussey Road, Norton Canes, WS11 9TP
G4 TCE P. Wade, 356 Shirehall Road, Sheffield, S5 0JP
G4 TCG R. Tams, 4 Langdale Close, Fryston, Castleford, WF10 2RB
G4 TCI M. Soars, 8 Nomis Park Congresbury, Bristol, BS49 5HB
G4 TCK N. Nicholls, 15 Millennium Way, Westward Ho, Bideford, EX39 1XN
G4 TCM W. Smith, 65 Larchwood Road, Yew Tree Estate, Walsall, WS5 4HE
G4 TCO D. Preece, Tyning House, Westrip, Stroud, GL6 6EY
G4 TCP J. Caddick, 3 Church Walk, Avonwick, South Brent, TQ10 9EJ
GI4 TCR Dr A. Jackson, 9 Cove Crescent, Groomsport, Bangor, BT19 6HW
GI4 TCS W. Jackson, Shantara, 21 Carnreagh, Hillsborough, BT26 6LJ
G4 TCT P. Johnson, 5 Moorside Drive, Drighlington, Bradford, BD11 1HE
G4 TDB D. Waterhouse, 19 Finsbury Drive, Brierley Hill, DY5 3NY
G4 TDC L. Wilson, Eastwood, Common Road, Tadcaster, LS24 9PQ
G4 TDF R. Copsey, 7 Musson Close, Marston Green, Solihull, B37 7HS
G4 TDG R. Dowson, 14 The Warren, Tuffley, Gloucester, GL4 0TT
G4 TDI D. Day, 1 Kings Paddock, Ossett, WF5 8EN
G4 TDO B. Fereday, 16 Glentworth Gardens, Wolverhampton, WV6 0SF
G4 TDP S. Bowden, 62 Manor House Road, Wednesbury, WS10 9PH
G4 TDQ L. Ashton, Bacton Wood Mill, Spa Common, North Walsham, NR28 9SH
G4 TDR C. Jay, Hill House, Badgers Orchard, Pershore, WR10 3HJ
G4 TDU B. Brandon, 8 Moor Park Avenue, Castleton, Rochdale, OL11 3JG
G4 TDV R. Stokes, Sunnybank, The Arch, Exeter, EX5 1LL
G4 TDW R. Maskew, 23 Daventry Road, Rochdale, OL11 2LN
G4 TDZ A. Jones, Riverbank, Main Street, Newark, NG22 0PP
G4 TEB P. Larbalestier, 81 Gosfield Lake Park Church Road Gosfield, Halstead, CO9 1UG
GI4 TED K. Doherty, 77 Drumflugh Road, Benburb, Dungannon, BT71 7QF
GM4 TEF A. Chalmers, Mayfield, Malcolm Road, Peterculter, AB14 0NX
G4 TEK J. Churchill, Yew Tree Cottage, Grove Lane, Salisbury, SP5 2NR
GD4 TEM J. Freestone, C/O Kololi Holiday Services, Po Box 335, Douglas, Isle Of Man, IM99 2QF
G4 TEN J. Burch, 1 South Farm Close, Tarrant Hinton, Blandford Forum, DT11 8JY
G4 TEP L. Kennedy, 69 Drayton Road, Borehamwood, WD6 2DA
GW4 TEQ J. Mattocks, Brynheulog, Bronygarth Road, Oswestry, SY10 7RQ
G4 TEU J. Burr, 4 The Fleet, Royston, SG8 5BB
G4 TEW R. Cooper, 4 Battle Close, Boroughbridge, York, YO51 9GN
G4 TEZ J. Lever, 30 Radcliffe Road, Winsford, CW7 1RE
G4 TFB B. Gray, 29 Verity Walk, Stourbridge, DY8 4XS
G4 TFC D. Gray, 29 Verity Walk, Stourbridge, DY8 4XS
G4 TFD B. Pickard, 3 Lodore Road, Bradford, BD2 4HY
G4 TFF P. Ayre, 1 Spring Gardens, Broadmayne, Dorchester, DT2 8PP
G4 TFG B. Curtis, 64 Market Avenue, Wickford, SS12 0AB
G4 TFH P. Mach, 17 Moorlands Avenue, Ossett, WF5 9PR
G4 TFI W. White, 9 Saffron Way, Tiptree, Colchester, CO5 0AY
GM4 TFJ E. Wallace, Lochiel Villa, Corpach, Fort William, PH33 7LR
GW4 TFM Rev. A. Davis, 51 Gungrog Hill, Welshpool, SY21 7UL
G4 TFO A. Coaton, 138 Ratby Road, Groby, Leicester, LE6 0BT
G4 TFP P. Herman Cranmer, 38 Barbrook Lane, Tiptree, Colchester, CO5 0EF
GW4 TFS A. Jones, 6 Gower View, Llanelli, SA15 3SN
G4 TFT C. Greenwood, 3 Moorfield Drive, Oakworth, Keighley, BD22 7EX
G4 TFU Dr A. Gerrard, 2 Dudley Road, Timperley, Altrincham, WA15 6UE
G4 TFV F. Kelly, 2 Victoria Terrace, Kirkstall, Leeds, LS5 3HX
G4 TFW H. Guy, 24 The Mall, Binstead, Ryde, PO33 3SF
GW4 TFX B. James, 9 Brangwyn Close, Morriston, Swansea, SA6 6AS
G4 TFZ D. Jarvis, 21 Ashurst Place, Stannington, Sheffield, S6 5LN
GW4 TGA S. Marvelley, 36 Muirfield Drive, Mayals, Swansea, SA3 5HS
G4 TGB D. Meadows, 39 Sylvester Street, Mansfield, NG18 5QS
G4 TGE J. Pullen, 71 Barrow Road, Barton-upon-Humber, DN18 6AE
G4 TGG G. Sifford, 25 Kingsley Court, Fraddon, St Columb, TR9 6PD
G4 TGJ R. Tomlinson, 25 Beverley Rise, Ilkley, LS29 9DB
G4 TGK W. Wimble, 87 Rolfe Lane, New Romney, TN28 8JL
GW4 TGL W. Protheroe-Thomas, Golwg Y Llan, Carmarthen, SA32 8PR
G4 TGM A. Sherratt, Anlyn, Norbury Drive, Brierley Hill, DY5 3DP
G4 TGP R. Barling, Maranello House, Pay Street, Folkestone, CT18 7DZ
GI4 TGR T. Greer, Ticino, 82 Purdysburn Hill, Belfast, BT8 8JZ
G4 TGS D. Swarbrook, 6 Westview Close, Leek, ST13 8ES
GW4 TGT T. Threlfall, 14 Clarence Street, Pembroke Dock, SA72 6JP
G4 TGV I. Soaft, 55 The Close, Thurleigh, MK44 2DT
G4 TGW D. Starmer, 20 Garners Way, Harpole, Northampton, NN7 4DN
G4 THA M. Crook, 28 Porter Street, Preston, PR1 6QN
G4 THC M. Arnison, 57 Heywood Road, Cinderford, GL14 2QU
G4 THF B. Smith, 63 Hitchin Road, Stotfold, Hitchin, SG5 4HT
G4 THI A. Robson, 5 Wetton Lane, Tibshelf, Alfreton, DE55 5NA
GW4 THK W. Moore, 2 Heol Cae Glas, Sarn, Bridgend, CF32 9UG
G4 THN M. Anthony1, Middlewood House, Blacksmiths Lane, Forward Green, Stowmarket, IP14 5ET
G4 THP D. Last, 16 South Street, Hockwold, Thetford, IP26 4JG
G4 THU J. Read, 35 Maytree Hill, Droitwich, WR9 7QU
G4 THV R. Biddlecombe, 24 West Avenue, Riverview Park, Althorne, CM3 6DF
G4 THX G. Donoughue, 1 Kings Mount, Leeds, LS17 5NS
G4 THY K. Cornes, 6 Haywood Heights, Little Haywood, Stafford, ST18 0UR
G4 TIA S. Jarvis, 1 Wakenslade Cottages, School House, Chard, TA20 4PJ
G4 TIC B. Helman, The Dingle, Redway, Minehead, TA24 8QF
G4 TID D. Hall, 6 St. Augustine Drive, Droitwich, WR9 8QR
G4 TIF M. Jones, 5 Congreve Close, Warwick, CV34 5RQ
G4 TIG W. Pope, 4 Salston Barton, Strawberry Lane, Salston, Ottery St Mary, EX11 1RG
G4 TIH C. Kay, 26 Clare Close, Elstree, Borehamwood, WD6 3NJ
G4 TIM G. Clementson, 74 Pentley Park, Welwyn Garden City, AL8 7SG
G4 TIQ D. Edwards, Rosedene, Trewint Estate, Liskeard, PL14 3RL
G4 TIV C. Roper, 12 Canford Drive, Allerton, Bradford, BD15 7AU
GW4 TIW D. Wood, 159 Liswerry Road, Newport, NP19 9QR
G4 TIX H. Woolrych, 20 Meadow Drive, Devizes, SN10 3BJ
GW4 TIZ P. Wyles, The Lawns, Halkyn Road, Holywell, CH8 7SJ
G4 TJA P. Prosser, 47 Devereux Drive, Watford, WD17 3DD
GM4 TJD M. Maclennan, Tree Tops, Ruilick, Beauly, IV4 7EY
G4 TJI R. Slim, 58 Fairways Drive, Harrogate, HG2 7ER
G4 TJK M. Porter, 17 Lancaster Avenue, Guildford, GU1 3JR
GM4 TJL J. Hebborn, Elysian Fields, Spean Bridge, PH34 4EX
G4 TJM E. Mordas, 6 Walsingham Gate, High Wycombe, HP11 1PA
GW4 TJN G. Smallwood, Rushbrit, The Spinney Old Road, Bwlchgwyn, Wrexham, LL11 5UF
GW4 TJQ J. Wallis, 17 Pantbach Place, Whitchurch, Cardiff, CF14 1UN
G4 TJS R. Dent, 4 Woodgreen Close, Callow Hill, Redditch, B975YR

G4 TJU F. Richards, 9 Dales Grove, Worsley, Manchester, M28 7JW
G4 TJY L. Barker, 75 Hills Road, Saham Hills, Thetford, IP25 7EW
G4 TKF P. Tuck, 178 St. Ediths Marsh, Bromham, Chippenham, SN15 2DJ
G4 TKO J. Sharman, 102 Commercial Road, Skelmanthorpe, Huddersfield, HD8 9DS
G4 TKP R. Peel, 3 Martins Hill Lane, Burton, Christchurch, BH23 7NJ
G4 TKS J. Clancy, 22 Audrey Needham House, Victoria Grove, Newbury, RG14 7RB
G4 TKW D. Hamilton, 38 Gosport Road, Lee-on-the-Solent, PO13 9EN
G4 TLE H. Kennard, Chestnut Cottage, Main Street, Rye, TN31 6UL
G4 TLG G. Holloway, Rose Cottage, Deblins Green, Callow End, Worcester, WR2 4UE
G4 TLL J. Jones, Amusement Depot, Station Road, Cullompton, EX15 1BQ
G4 TLM B. Jennings, 33 Queen Margarets Drive, Brotherton, Knottingley, WF11 9HR
G4 TLO P. Johnson, Flat 5, Mulberry Lodge, 26 New Brighton Road, Emsworth, PO10 7EW
G4 TLR B. Richards, 37 Salisbury Grove, Sutton Coldfield, B72 1YE
G4 TLS J. Norton, 2 Hill Rise, Great Rollright, Chipping Norton, OX7 5SW
G4 TLT A. Price, 20 Feast Field, Wooliston, NN29 7QG
G4 TLW H. Allen, 425 Broadway, Chadderton, Oldham, OL9 8AP
G4 TLY E. Holmes, 36 Corn Gastons, Malmesbury, SN16 0DR
G4 TMA P. Fielding, 22 Meadow Crescent, Poulton-le-Fylde, FY6 7QX
GI4 TMB M. Beggs, 15 Marsham Court, Cotswold Drive, Bangor, BT20 4RS
G4 TMC P. Barnett, 8 Parsonage Road, Horsham, RH12 4AR
G4 TMD L. Downes, 357 Stone Road, Stafford, ST16 1LD
G4 TMF P. Aisthorpe-Buckley, 17 Pine View Road, Verwood, BH31 6LQ
G4 TMG C. Sherwood, 14 Amberley Road, Rustington, Littlehampton, BN16 2EF
G4 TMI P. Johnson, 3A Railway Street, Tow Law, Bishop Auckland, DL13 4DU
G4 TML B. Parr, 5 Ashes Lane, Almondbury, Huddersfield, HD4 6TE
G4 TMQ J. Martin, 22 Wansbeck Court, Front Street East, Bedlington, NE22 5BU
G4 TMR S. Lacy, 26 Sterndale Drive, Newcastle, ST5 4HS
G4 TMV E. Gale, 4 Waingap Crescent, Whitworth, Rochdale, OL12 8PX
G4 TMX W. Armstrong, 121 Bede Street, Sunderland, SR6 0NT
G4 TMY N. Hounslow, 18 Crompton Place, Blackburn, BB2 6LW
G4 TMZ D. Gillott, 132 Racecommon Road, Barnsley, S70 6JY
G4 TNA K. Pope, 305 Hulton Lane, Bolton, BL3 4LF
G4 TNB P. Dollery, 22 Barley Mead, Danbury, Chelmsford, CM3 4RP
G4 TND D. Jenkins, 27 Glebelands, Chudleigh, Newton Abbot, TQ13 0GB
G4 TNE D. Horsman, 33 Chanters Hill, Barnstaple, EX32 8DN
GW4 TNF T. Jones, 9 Hinsley Drive, Wrexham, LL13 9QH
GM4 TNJ R. Milenkovic, 10 Loganbarns Road, Dumfries, DG1 4BU
GM4 TNP J. Burke, 25 Duncan Road, Auchmuty, Glenrothes, KY7 4HS
G4 TNU A. Scott, 79 Westwood Drive, Amersham, HP6 6RR
G4 TNY D. Womack, 58 Nelson Road North, Great Yarmouth, NR30 2AT
GM4 TOE B. Horning, Cemetery Lodge, Colleonard Road, Banff, AB45 1DZ
G4 TOG B. Grainger, 23 Heath Road, Hordle, Lymington, SO41 0GG
G4 TOH R. Russell, 23 Milfoil Avenue, Conniburrow, Milton Keynes, MK14 7DY
G4 TOI P. Andrews, 12 Cedarwood Grove, Sunderland, SR2 9EJ
G4 TOM T. Turbert, 200 Salisbury Terrace, York, YO26 4XP
G4 TOO M. Final, 3 Borda Close, Chelmsford, CM1 4JY
GM4 TOQ A. Stewart, Three Acres, Cochno Road, Clydebank, G81 6PX
GI4 TOR A. Kincaid, 63 Carolhill Park, Ballymena, BT42 2DG
G4 TOT R. James, Brantholme, Hasty Brow Road, Lancaster, LA2 6AG
G4 TOX J. Glover, Burrows Farm, Toot Hill Road, Ongar, CM5 9QW
G4 TOY R. Handstock, 38 Watson Close, Upavon, Upavon, SN9 6AE
G4 TOZ S. Marchini, Flat 9, Crofthill Court, Rochdale, OL12 9UX
GW4 TPG M. Evans1, 14 Heol Dewi, Hengoed, CF82 7NP
G4 TPH T. Brockman, 57 Ramsbury Drive, Hungerford, RG17 0SG
GI4 TPI Dr G. Anderson, 13 Ashley Park, Bangor, BT20 5RQ
G4 TPJ R. Mepham, 36 Bramble Close, Hildenborough, Tonbridge, TN11 9HQ
G4 TPK P. Phillips, 83 Arundel Road, Benfleet, SS7 4EE
G4 TPM A. Malcher, 68 Maryatt Avenue, South Harrow, Harrow, HA2 0SX
G4 TPO S. Mcculloch, 125 Comptons Lane, Horsham, RH13 5NZ
GM4 TPQ W. Milligan, 7 Girvan Road, Turnberry, Girvan, KA26 9LP
GM4 TPR J. Mitchell, 65 Robb Place, Castle Douglas, DG7 1LW
G4 TPS P. Stinton, 37 Harriet Close, Sutton Bridge, Spalding, PE12 9QU
G4 TPV G. Jones, 397 Fishponds Road, Eastville, Bristol, BS5 6RJ
G4 TPW H. Igglesden, Treeways, Littleworth Lane, Horsham, RH13 8ER
GM4 TPX K. Gerard, 9 Overdale Crescent, Prestwick, KA9 2DB
GI4 TPY K. Boag, 12 Plantation Road, Bangor, BT19 6AF
G4 TQB Dr P. Grannell, 6 Fermain Close, Newcastle-under-Lyme, ST5 3EF
G4 TQC M. Anson, 15 Clover Ridge, Cheslyn Hay, Walsall, WS6 7DP
GW4 TQD J. Gulley, Lawnfields, Brynhoffnant, Llandysul, SA44 6EA
G4 TQL K. Prince, Room 36, Millfield, Bury New Road, Heywood, OL10 4RQ
G4 TQO P. Fowler, 7 Ormonde Avenue, Orpington, BR6 8JP
G4 TQR R. Wilkes, 47 Richmond Drive, Glen Parva, Leicester, LE2 9TJ
G4 TQS A. Wallis, 27 Church Farm Road, Upchurch, Sittingbourne, ME9 7AG
G4 TQT I. Waller, 3 Reneville Close Moorgate, Rotherham, S60 2AT
G4 TQY M. Addison, 6 Hanley Orchard, Hanley Swan, Worcester, WR8 0DS
G4 TQZ P. Foster, 18 Stokesay Way, Sutton Hill, Telford, TF7 4QE
G4 TRA S. Redway, Hill House Grange Lane, Rodbourne, sn160es
G4 TRD R. Dafter, 49 Balmoral Road, Salisbury, SP1 3PZ
G4 TRE B. Boon, Orchards, School Lane, Woodbridge, IP13 0ES
G4 TRF I. Boon, 327 Broomfield Road, Chelmsford, CM1 4DU
G4 TRG I. Willmer, 30 Portland Road, East Grinstead, RH19 4EA
GM4 TRH A. Macdonald, Tigh Na Mara, 6 Halistra, Hallin, Isle of Skye, IV55 8GL
G4 TRI S. March, 23 Pebworth Close, Church Hill North, Redditch, B98 9JX
G4 TRM S. Burgess, Muston Farm, Winterborne Muston, Blandford Forum, DT11 9BU
G4 TRN J. Everingham, 17 Collingwood Road, Redland, Bristol, BS6 6PD
G4 TRP M. Fell, 5 Sandown Close, Goring-By-Sea, Worthing, BN12 4QA
G4 TRQ A. Westmorland, Lilly Rose, Rosevear, Bugle, St Austell, PL26 8RJ
G4 TRR P. Rose, St. Margarets, Westrop, Swindon, SN6 7HJ
GM4 TRS A. Pierce, Mains Of Auchreddie, New Deer, Aberdeenshire, AB53 6SL
G4 TRU A. Thompson, 47 The Signals, Feniton, Honiton, EX14 3UP
G4 TRV R. Pears, 24 Westbourne Drive, St Austell, PL25 5EA
G4 TRW K. Prior, 14 Bincombe Rise, Weymouth, DT3 6AS
G4 TRY A. Moriarty, 7 Meadow Drive, Bolton Le Sands, Carnforth, LA5 8HA
GM4 TRZ T. Mcleod, 1 Lochside Cottages, Otterston, Burntisland Fife, KY3 0RZ
G4 TSA R. Turley, Sunnybank, Matlock Road Kelstedge, Ashover, Chesterfield, S45 0DX
G4 TSB Dr R. Cooper, 8 Hollyfield Drive, Barnt Green, Birmingham, B45 8HP
G4 TSD R. Edwards, 8 Boney Hay Road, Burntwood, WS7 9AB
G4 TSF P. Scholefield, 10 Gainsborough Avenue, Leeds, LS16 7PG
GW4 TSG J. Williams, Cartref, Capel Garmon, Llanrwst, LL26 0RG
GI4 TSK J. Skillen, 3 Copeland Drive, Comber, Newtownards, BT23 5JJ
G4 TSN J. Lee, 46 Little Lane, Huthwaite, Sutton-in-Ashfield, NG17 2RA
G4 TSQ M. Levett, 5 Park Road, Yapton, Arundel, BN18 0JE
G4 TST D. Richardson, L'Ancresse, Uplands Road, Waterlooville, PO7 6HE
G4 TSV J. Robinson, 2 Bridge Mill Road, Nelson, BB9 7BD
G4 TSW Tiverton South West ARC, c/o A. Burnett, 2 Courtney Road, Tiverton, EX16 6EE
GW4 TTA Dragon ARC, c/o J. Parry, 'Ar Allt' Lon Hedydd Llanfairpwll Anglesey, North Wales, LL61 5JY
G4 TTB A. Gordon, Flat 3, Woodlands, Acer Grove, Ipswich, IP8 3RR
GM4 TTC P. Howes, 43 Tanzieknowe Road, Cambuslang, Glasgow, G72 8RD
GM4 TTD N. Loughrey, 47 Obsdale Road, Alness, IV17 0TU
G4 TTF Bishop Auckland AR Club, c/o T. Bevan, 6 Buttermere Grove West Auckland, Bishop Auckland, DL14 9LG
G4 TTG H. Bryant, 141 Shakespeare Road, Fleetwood, FY7 7HH
G4 TTJ J. Lee, 2 Rudgard Way, Liphook, GU30 7GW
GI4 TTL M. Corcoran, 20 Ringbuoy Cove, Cloughey, Newtownards, BT22 1LL
G4 TTM J. Betts, The Cottage, Meaford, Stone, ST15 0PX
G4 TTN G. Redgewell, 121 Gubbins Lane, Harold Wood, Romford, RM3 0DL
G4 TTO D. Sandland, 54 Bishopdale Drive, Rainhill, Prescot, L35 4QH
G4 TTQ R. Philpot, 38 Mountview Road, Clacton on Sea, CO15 6LN
G4 TTS C. Harrison, The Mobile Home, Langar Airfield, Nottingham, NG13 9HY
G4 TTX R. Smith, 405 Windmill Avenue, Kettering, NN15 6PS
G4 TTY E. Macdonald, 7 Alder Close, Crawley Down, Crawley, RH10 4UL
G4 TTZ R. Margolis, 12A Wyndham Close, Yateley, GU46 7TT
G4 TUA T. Higgs, 1 Merchants Row, Faraday Road, Kirkby Stephen, CA17 4AU
GW4 TUD I. Williams, 52 Bridge Street, Aberystwyth, SY23 1QB
G4 TUF A. Wilday, 12 Duke Street, Bamber Bridge, Preston, PR5 6FT
G4 TUH S. Elsdon, Church View, Chapel Lane Sharnford, Hinckley, LE10 3PE
GI4 TUJ W. Konos, 27 Hillhead Road, Ballynahinch, BT24 8LB
G4 TUK R. Scarfe, Owlswood, Dereham Road, Shipdham, Thetford, IP25 7NJ
G4 TUM J. Speakman, 33 Leyburn Avenue, Bispham, Blackpool, FY2 9AQ
G4 TUO E. Whitworth, 129A Broomhill, Downham Market, PE38 9QU
G4 TUP D. Norris, 26 Freckleton Road, Southport, PR9 9XE
GI4 TUV R. Bailie, 26 Moatview Park, Dundonald, Belfast, BT16 2BE
G4 TUX J. Baines, 12B Tall Trees Park, Old Mill Lane, Mansfield, NG19 0JP
GM4 TVB C. Burnet, 138 Hilton Drive, Aberdeen, AB24 4NH
G4 TVD G. Hector, 6 Benford Close, Bristol, BS16 2UD
GW4 TVE S. Edwards, 1 Maes Yr Efail, Penparc, Cardigan, SA43 1FB
G4 TVJ A. Johns, 5 Oakfields, Loddon, Norwich, NR14 6UT
G4 TVN B. Yates, 39 Moss Lane, Garstang, Preston, PR3 1PD
GW4 TVQ R. Thomas, 3 Tor Y Mynydd, Baglan, Port Talbot, SA12 8LE
G4 TVT G. Spencer, 322 Colchester Road, Ipswich, IP4 4QN
GW4 TVU V. Sedgebeer, 40 Pen Y Bryn, Croeserw Cymmer, Port Talbot, SA13 3SD
G4 TVW R. Stone, 51 Elaine Avenue, Rochester, ME2 2YW
G4 TVX R. Lamb, 27 The Ridgeway, Braintree, CM7 1EB
G4 TWC D. Powell, 8 Cranbrook Drive, Sittingbourne, ME10 1RF
G4 TWG S. Greenwood, Carter Place Farm, Hall Park, Rossendale, BB4 5BQ
G4 TWH G. Wood-Hill, 26 Bramerton Road, Hockley, SS5 4PJ
G4 TWK H. Hart, 20 Cowdray Drive, Goring-By-Sea, Worthing, BN12 4LH
G4 TWL T. Lee, 19A Imperial Avenue, Mayland, Chelmsford, CM3 6AQ
G4 TWP L. Miles, 21 Chaucer Walk, Eastbourne, BN23 7QT
G4 TWS S. Holmes, 7 Parkland Crescent, Old Catton, Norwich, NR6 7RQ
G4 TWT H. Holmes, 7 Parkland Crescent, Old Catton, Norwich, NR6 7RQ
G4 TWW T. Bevan, 30 Hawthorne Close, Stanton Hill, NG17 3NQ
G4 TXA D. Mccartney, 19 Friswell Road, Banbury, OX16 9NW
G4 TXD M. Robbins, 2 Tolview Terrace, Hayle, TR27 4AG
G4 TXE A. Goode, Tudor House, Chenhalls Road, Cornwall, TR27 6HJ
G4 TXF C. White, 7 Woodward Road, Pershore, WR10 1LW
G4 TXG N. Hamilton, North View, Chawston Lane, Bedford, MK44 3BH
G4 TXK T. Stanley, 35 Moorgate Road, Kippax, Leeds, LS25 7ET
G4 TXL A. Stevenson, Szabadság U. 32, VeresegyhµZ, Hungary, 2112
G4 TXM G. Porter, 20 Fitzwilliam Drive, Barton Seagrave, Kettering, NN15 6RG
GM4 TXN A. Newlands, 21 Castle Crescent, Inverbervie, Montrose, DD10 0SB
G4 TXO J. Middleton, 8 Cullen Close, Newark, NG24 1DF
G4 TXT D. Wales, 105 Olney Road, Lavendon, Olney, MK46 4ER
G4 TXV A. Turner, 11 Holmcroft Road, Kidderminster, DY10 3AQ
G4 TYA C. Carter, 12 Grove Street, Leamington Spa, CV32 5AJ
G4 TYD A. Kelly, Brook House, Tremar Coombe, Liskeard, PL14 5EN
GW4 TYH R. Roberts, Bryn Gwyfan, Hiraddug Road, Rhyl, LL18 6HS
G4 TYL M. Tyldesley, 24 Trinity Court Broughton Brigg N Lincolnshire Dn200Sj, Broughton, DN200SJ
G4 TYN D. Bell, 12 Parker Gardens, Stapleford, Nottingham, NG9 8QG
G4 TYO G. Lilley, 100 Trentham Drive, Nottingham, NG8 3NE
G4 TYP K. Ward, 9 Porlock Close, Long Eaton, Nottingham, NG10 4NZ
G4 TYR C. Miles, 23 Redacre Road, Sutton Coldfield, B73 5EA
G4 TYT A. Hunt, 141 Pickhurst Lane, Bromley, BR2 7HU
G4 TYW R. Wilson, 95 Longfield Road, Todmorden, OL14 6ND
G4 TYY J. Worley, 37 Fall Road, Heanor, DE75 7PQ
G4 TZA C. Read, 58 Somerset Road, Chiswick, London, W4 5DN
G4 TZF C. Toby, 32 Swallow Road, Langley Green, Crawley, RH11 7RF
G4 TZG S. Mellor, 52 Tetbury Drive, Bolton, BL2 5NS
G4 TZK G. Prater, 297 Highfield Road North, Chorley, PR7 1PH
G4 TZL K. Rogers, 7 Buckleigh Road, Wath-Upon-Dearne, Rotherham, S63 7JB
G4 TZM I. Paterson, 21 Beech Grove, Little Oakley, CO12 5NN
G4 TZO P. Pledger, Mas Tracbuch, Brunyola, Spain, 17441
G4 TZQ D. Rouse, 14 Kestrel Close, Downley, High Wycombe, HP13 5JN
G4 TZR R. Stringfellow, 18 Cline Court, Crownhill, Milton Keynes, MK8 0DB
G4 TZV E. Cronin, 10 Wellington Grove, Pudsey, LS28 8DG
G4 TZX G. Everest, 20 Seaway Road, St. Marys Bay, Romney Marsh, TN29 0RU
G4 UAA J. Gaffney, 77 South Street, Pennington, Lymington, SO41 8DY
G4 UAF J. Higgins, 124 Cromwell Road, South Kensington, London, SW7 4ET
G4 UAI P. Cockman, 29 Kensington Road, Southend-on-Sea, SS1 2SX
GW4 UAJ E. Allwood, 10 Fairfield Close, Aberdare, CF44 0PF

G4 UAL J. Guffogg, 8 Lincoln Road, Washingborough, Lincoln, LN4 1EQ
G4 UAM A. Gould, 3 Clarkson Road, Lingwood, Norwich, NR13 4BA
G4 UAQ I. Weston, 53 Dickens Road, Maidstone, ME14 2QR
G4 UAT A. Thomas, 10 Brisco Avenue, Loughborough, LE11 5HB
G4 UAU J. Parish, 50 Far Hey Close, Radcliffe, Manchester, M26 3GL
G4 UAV A. Waltham, 100 Middleton Road, London, E8 4LN
G4 UAY D. Grant, 115 Clayton Road, Newcastle, ST5 3EW
G4 UBB J. Brown, 21 Coulsdon Road, Sidmouth, EX10 9JJ
G4 UBC K. Durrant, 26 Dozule Close, Leonard Stanley, Stonehouse, GL10 3NL
GM4 UBF A. Pontiero, 1 Dalmeny Road, Hamilton, ML3 6PP
G4 UBI A. Priddy, 44 Frys Hill, Kingswood, Bristol, BS15 4QJ
GM4 UBJ W. Tracey, 65 Kirkland Street, Motherwell, ML1 3JW
G4 UBK K. Martin, 19 Rosevale Gardens, Luxulyan, Bodmin, PL30 5EP
G4 UBM R. Bryant, Plover, Hareby Road, Spilsby, PE23 4JB
GW4 UBQ C. Powles, 14 Willow Close, Four Crosses, Llanymynech, SY22 6NF
G4 UBR P. Richardson, 18 New Road, Heage, Belper, DE56 2BA
G4 UBT K. Stone, 63 Banks Road, Pound Hill, Crawley, RH10 7BS
G4 UCC W. Trinder, 354 Livesey Branch, Road, Blackburn, BB2 4QJ
G4 UCE B. Davies, 9 Paisley Avenue, Eastham, Wirral, CH62 8DL
G4 UCJ S. Gilbert, 3 Shenley Road, Whaddon, Milton Keynes, MK17 0LW
GW4 UCK G. Jones, Frondeg, 18 Chapel Rd. Three Crosses, Swansea, Sa4 3Pu., Swansea, SA4 3PU
G4 UCL A. Fallows, 72 Soutergate, Ulverston, LA12 7ES
G4 UCT A. Cooke, 7 School Lane, Warmingham, Sandbach, CW11 3QL
G4 UCU S. Hebel, 18 Castle Road, Colne, BB8 7AR
GW4 UCV I. Hynes, Bryn, Llanddona, Beaumaris, LL58 8UE
G4 UCX P. Johnson, 38 Bristol Road, Ipswich, IP4 4LP
G4 UCY C. Laird, 31 Foxlease, Bedford, MK41 8AP
G4 UCZ M. Kirk, 2 Denton Gardens, East Cowes, PO32 6EJ
G4 UDB C. Fay, 36 Shooters Hill Close, Southampton, SO19 1FW
G4 UDD S. Chapman, 2 Birds Croft, Great Livermere, Bury St Edmunds, IP31 1JJ
GW4 UDE M. Ellis, Seren Arian, Maesbury Hall Mill, Oswestry, SY10 8BB
G4 UDF I. Fox, 8 Priestfields, Leigh, WN7 2RG
G4 UDG C. Fawkes, 24 Moreton Close, Kidsgrove, Stoke-on-Trent, ST7 4HP
G4 UDH P. Harley, 6 Huntsbank Drive, Newcastle, ST5 7TB
GI4 UDI J. Mccullagh, 53 Fernagh Road, Omagh, BT79 0PL
G4 UDK R. Wood, 102 Ombersley Close, Redditch, B98 7UT
G4 UDN C. Peake, 279 Mansfield Road, Skegby, Sutton-in-Ashfield, NG17 3AP
G4 UDU P. Godbold, 13 Dawn Crescent, Upper Beeding, Steyning, BN44 3WH
G4 UDW P. Hersey, Cranbrook, Waghorns Lane, Uckfield, TN22 4JA
G4 UDY B. Moorecroft, 4 St. Davids Road, Locks Heath, Southampton, SO31 6EP
G4 UDZ S. Tyler, 28 Rushen Drive, Hertford Heath, Hertford, SG13 7RB
G4 UEA P. Robinson, 46 Waidshouse Road, Nelson, BB9 0SB
G4 UED G. Henstridge, 21 John Gay Road, Amesbury, Salisbury, SP4 7NN
G4 UEF K. Dalton, 17 Shute Avenue Watchfield, Swindon, SN6 8SX
GM4 UEH Rev. A. Ford, 14 Corsankell Wynd, Saltcoats, KA21 6HY
GW4 UEJ M. Charman, Noddfa, Dwrbach, Fishguard, SA65 9RL
G4 UEL G. Hollebon, Flat 24, Providence Place, Abbey Street, Farnham, GU9 7RQ
G4 UEN K. Foskett, 2 Ambleside Gardens, Southampton, SO19 8EY
G4 UEO D. Stewart, The Paddock, Allendale, Hexham, NE47 9EL
GW4 UEP J. Morgan, Arosfa, Upper Bridge Street, Newport, SA42 0PL
G4 UET J. Rolfe, 56 Elmhurst Road, Thatcham, RG18 3DH
G4 UFC P. Fretwell, 5 Main St, Brinsley, NG16 5BG
GM4 UFD R. Gall, 49A Ugie Street, Peterhead, AB42 1NX
G4 UFG A. Johnson, 27 Walden Avenue, Oldham, OL4 2PW
G4 UFJ N. Taylor, The Olde Barn, 369A Leymoor Road, Huddersfield, HD7 4QQ
G4 UFK A. Watts, 23 St. Marys Close, Torrington, EX38 8AS
G4 UFL N. Wood, 244 Leymoor Road, Golcar, Huddersfield, HD7 4QP
GM4 UFP C. Ross, The Old Cottages, Middlestead, Selkirk, TD7 5EY
GW4 UFQ B. Jackson, Bryn Tirion, Maes Y Waen, Bala, LL23 7SF
G4 UFR E. Horsfield, 13 St. Leonards Way, Ardsley, Barnsley, S71 5BS
G4 UFS D. Pearson, 48 Nuneham Grove, Westcroft, Milton Keynes, MK4 4DH
G4 UFU B. Steen, 30 Shady Grove, Alsager, Stoke on Trent, ST7 2NH
G4 UFX D. Blackwell, Rosegarth, 31 Main Street, Ilkeston, DE7 6AU
G4 UFZ R. Greenwood, 128 Towngate, Netherthong, Holmfirth, HD9 3XZ
G4 UGB R. Bracegirdle, 3 Westover Grove Warton, Carnforth, LA5 9QR
G4 UGD I. Clover, 80 Old Chester Road, Helsby, Helsby, wa6 9pq
GM4 UGF D. Duff, Felcanty, Monikie, Broughty Ferry, Dundee, DD5 3QN
GW4 UGI R. Crowley, 15 Rudry Street, Penarth, CF64 2TZ
G4 UGK C. Cattrall, 57 Stonebridge, Orton Malborne, Peterborough, PE2 5NT
G4 UGM D. Wade, 28 Hazel Road, Altrincham, WA14 1JL
GM4 UGN D. Duckworth, 16 Kennedy Court, Caol, Fort William, PH33 7PF
G4 UGO T. Farmer, York House, Old Gloucester Road, Bristol, BS35 3LQ
G4 UGQ Dr J. Davies, 42 Boxworth Road, Elsworth, Cambridge, CB23 4JQ
G4 UGR T. Burke, 10 Goodwood Road, Lancaster, LA1 4LZ
G4 UGT D. Hockin, 18 Lower Down Road, Portishead, Bristol, BS20 6PF
G4 UGU R. Hall, 19 Buckingham Place, Downend, Bristol, BS16 5TN
G4 UGV M. Hurrell, 74 Southcote Road, Bournemouth, BH1 3SS
G4 UGW P. Jones, 46 Sergeants Lane, Whitefield, Manchester, M45 7TS
GI4 UHA J. Maguire, 4 Lawnakilla Park, Enniskillen, BT74 7JN
GD4 UHB J. Parslow, Traie Vane, Lhergy Dhoo, Peel, IM5 2AE
G4 UHI D. Westby, 55 Tarn Road, Thornton-Cleveleys, FY5 5AY
G4 UHJ D. Lee, 1 West View Cottage The Level, Pillowell, Lydney, GL15 4QD
GW4 UHK J. Newell, 8 Belgrave Road, Abergavenny, NP7 7AL
G4 UHM S. Parsons, 248 Filey Road, Scarborough, YO11 3AQ
G4 UHQ B. Carr, 23 Belford Drive, Bramley, Rotherham, S66 3YW
G4 UHR D. Reekie, 37 Harvey Way, Saffron Walden, CB10 2AP
G4 UHS R. Rowlands, 18 Green Crescent, Rowner, Gosport, PO13 0DP
G4 UHT W. O'Reilly, 12 Singledge Avenue, Whitfield, Dover, CT16 3LQ
G4 UHU M. Rumens, 18 De Legh Grove, West Allington, Bridport, DT6 5QY
G4 UHZ D. Goulsbra, Delfour, 3 Chapel Street, Market Rasen, LN8 3AG
GW4 UIE S. Williams, 17 Brettenham St., Llanelli, SA15 3ED
G4 UIF A. Marston, Stormsfield, Station Road, Limerick, Ireland,
G4 UIH S. Woolgar, 8 Bowerland Avenue, Torquay, TQ2 8QH
G4 UII D. Woolgar, 8 Bowerland Avenue, Torquay, TQ2 8QH
GW4 UIL R. Lewis, Tal Engan, Tudweiliog, Pwllheli, LL53 8ND
G4 UIO A. Cochrane, 136 Osward, Courtwood Lane, Croydon, CR0 9HE
G4 UIQ J. Greenhough, 58 Gorsey Bank, Wirksworth, Matlock, DE4 4AD
GW4 UIR J. Patterson, Fairhaven, Tai Terfyn, Caerwys Road, Cwm Dyserth, Rhyl, LL18 6HT
G4 UIT D. Geraghty, Roselidden House Trevenen Bal, Helston, TR13 0PT
G4 UIW P. Wood, 61 Stoke Road, Bromsgrove, B60 3EP
G4 UIY S. Hamilton, 89 The Paddocks, Old Catton, Norwich, NR6 7HE
G4 UJA J. Adshead, 2 Gainsborough Avenue, Lostock Hall, Preston, PR5 5JG
GW4 UJF M. Finnigan, 3 Frances Avenue, Rhyl, LL18 2LW
G4 UJI E. Cowperthwaite, Woodlands, Garstang Road, Lancaster, LA2 0EG
G4 UJJ D. Bodman, 56A Martins Road, Keevil, Trowbridge, BA14 6NA
G4 UJL B. Poole, 1 Hungerford Piece Studley, Calne, SN11 9LR
G4 UJO M. Greer, The Pines, 5A Leek Road, Congleton, CW12 3HS
G4 UJP B. Smith, 17 Thornley Road, Wirral, CH46 6HB
G4 UJS R. Harrison, Green Lane House, Whixall, Whitchurch, SY13 2PT
G4 UJV J. Fenn, 40 Mildenhall Road, Fordham, Ely, CB7 5NR
G4 UJW C. Elliott, 52 Wellfield Road, Alrewas, Burton-on-Trent, DE13 7EZ
G4 UKA C. Hawkridge, 57 Wilkes Wood, Creswell, Stafford, ST18 9QR
G4 UKD B. Gibson, 161 Torbay Road, Harrow, HA2 9QF
GM4 UKG M. Manekshaw, 32 Inchcolm Drive, North Queensferry, Inverkeithing, KY11 1LD
G4 UKO N. Hill, 40A Hampden Road, Ashford, TN23 6JL
G4 UKR S. Blayer, 6 Lord Close, Poole, BH17 8QW
GW4 UKU P. Jones, 8 Tyn Y Pwll Estate, Llanbedrog, Pwllheli, LL53 7PG
G4 UKV I. Leonard, 11 St. Leonards Drive, Timperley, Altrincham, WA15 7RS
G4 UKW K. Wevill, 6 Henacre Wood Court, Queensbury, Bradford, BD13 2LJ
G4 UKX R. Miller, 31 Gladstone Road, Corton, Lowestoft, NR32 5HJ
G4 UKZ R. Rounce, Field House Farm, Blakeney Road, Fakenham, NR21 0BU
G4 ULD R. Todd, 616 Cowdin Drive, Glenwood Springs, Colorado, USA, 81601
G4 ULG J. Rawlings, Welwyn, Church Walk, Lydney, GL15 4NY
G4 ULI B. Long, 96 Lumbertubs Lane, Northampton, NN3 6AH
G4 ULM J. Martin, 25 Mcnish Court, Grenville Way, St Neots, PE19 8PE
G4 ULN I. Purdy, 76 Lea Road, Dronfield, S18 1SD
G4 ULP D. Pritchard, 1 Sandstone Cottages, Walton-In-Gordano, Clevedon, BS21 7AJ
G4 ULQ G. Judd, 1 Mayfield Way, Ferndown, BH22 9HP
G4 ULT L. Walker, Oaklea, 13 Manor Road, Sandown, PO36 9JA
G4 ULV D. Woodman, 30 Sheridan Way, Longwell Green, Bristol, BS30 9UE
G4 ULZ R. Ottway, 9 Grove Road, Burgess Hill, RH15 8LE
GM4 UMA S. Maclennan, Tree Tops, Ruilick, Beauly, IV4 7EY
G4 UMB P. Howard, 63 West Bradford Road, Waddington, Clitheroe, BB7 3JD
G4 UME H. Park, 11A Morecambe Road, Morecambe, LA3 3AA
G4 UMG M. Brassington, 42 Dartmouth Avenue, Newcastle, ST5 3NY
G4 UMJ R. Carslake, 38 Loppets Road, Tilgate, Crawley, RH10 5DW
G4 UMM A. Curran, 9 Forton Road, Newport, TF10 7JP
G4 UMP G. Daisley, 10 Arundel Road, Benfleet, SS7 4EF
G4 UMS M. Kinger, 5 Fore Street, Gunnislake, PL18 9BN
G4 UMT G. Elliott, 10 Farningham Close, Spondon, Derby, DE21 7DZ
G4 UMV P. Johnson, 52 Evesham Road, Cookhill, Alcester, B49 5LJ
G4 UMW R. Browning, 28 Mowbray Close Bromham, Bedford, MK43 8LF
G4 UMY M. Strong, 92 Cobham Road, Halesowen, B63 3JX
G4 UNB D. Williams, 13 Hendon Road, Nelson, BB9 9JL
G4 UNE S. Sharples, 1 Garners End, Chalfont St. Peter, Gerrards Cross, SL9 0HE
G4 UNF K. East, 39 Chapel Lane, Navenby, Lincoln, LN5 0ER
G4 UNH A. Pyne, 414 Beacon Road, Bradford, BD6 3DJ
G4 UNI T. Hepple, 18 King Charles Walk, London, SW19 6JA
G4 UNJ B. Walters, 17 Oakway, Birkenshaw, Bradford, BD11 2PG
G4 UNL R. Charlesworth, Po Box 841, 9506 Koronadal City, South Cotobato, Philippines
G4 UNM R. Bushell, 12 Sandham Close, Sandown, PO36 9DS
G4 UNO J. Dobson, 27 Darkfield Way, Woolavington, Bridgwater, TA7 8JB
G4 UNS D. Brown, 8 Gaynes Court, Upminster, RM14 2JH
G4 UNW P. Everard, The Bungalow, Toynton Fenside Road, Spilsby, PE23 5DB
G4 UNX J. Fry, 4 South Road, Brighton, BN1 6SB
GM4 UOD L. Drake-Brockman, 59 Sunnyside, Culloden Moor, Inverness, IV2 5ES
G4 UOI R. Butterfield, 33 Orchard Square, Wormley, Broxbourne, EN10 6JA
G4 UON P. Prowse, 9 Fairway, Carlyon Bay, St Austell, PL25 3QE
G4 UOO J. Bleaney, 58 Jeans Way, Dunstable, LU5 4PW
G4 UOR C. Bourke, 36 The Drive, Fareham, PO16 7NL
G4 UOS G. Newton, 5 Southend Gardens, Highbridge, TA9 3LD
G4 UOW J. Cosgrove, 8 Wandsworth Road, Newcastle upon Tyne, NE6 5AD
G4 UOZ E. Ball, 50 Keldgate, Beverley, HU17 8HY
G4 UPA J. Poxon, 22 Sandhills Road, Bolsover, Chesterfield, S44 6EY
GM4 UPB G. Read, 58 Limepark Crescent, Kelty, KY4 0FH
GI4 UPC W. Millar, 121 Ballypollard Road, Magheramorne, Larne, BT40 3JG
G4 UPD M. Parks, 240 Stainbeck Road, Leeds, LS7 2NN
GW4 UPG V. Jackson, 24 Bishop Road, Ammanford, SA18 3HA
G4 UPI J. Green, St. Annes, Poundfield Road, Crowborough, TN6 2BG
G4 UPK D. Thompson, 112 Lexton Drive, Churchtown, Southport, PR9 8QW
GM4 UPN P. Ingram, Flat 3 5 Munro Street, Alexandria, G83 0PU
G4 UPR J. Dickson, 33 Ringwood Grove, Weston-Super-Mare, BS23 2UA
G4 UPU R. Ainsworth, 14 Edge Fold Crescent, Worsley, Manchester, M28 7EX
GM4 UPX I. Wilson, 18 High Street, Jedburgh, TD8 6AG
G4 UPY A. Hodge, 116A Broad Road, Eastbourne, BN20 9RD
G4 UQA M. Goodman, Randoms, Holt Road, Holt, NR25 7UA
GM4 UQD A. Boyd, 86 Ravenswood Rise, Livingston, EH54 6PG
G4 UQE T. Fitzgerald, 11 Hillcrest Road, Camberley, GU15 1LF
G4 UQF M. Sole, 17 Hyholmes, Bretton, Peterborough, PE3 8LG
GM4 UQG R. Aitkenhead, 11 Elm Court, Quarter, Hamilton, ML3 7FB
G4 UQI A. Lether, 16 The Dingle, Fulwood, Preston, PR2 3EX
GM4 UQK J. Roberts, 1 East Mains Lodge, Kirkinner, Newton Stewart, DG8 9AQ
G4 UQM D. Grainger, 25 Westwood Heath Road, Leek, ST13 8LN
G4 UQN K. Stockley, 19 The Lawns, Wisbech, PE13 1SW
GD4 UQO P. Parker, 46 Ballaquane Park, Peel, Isle Of Man, IM5 1PX
G4 UQR J. Gibbs, 3 Holts Green, Great Brickhill, Milton Keynes, MK17 9AJ
G4 UQU D. Smith, 2 Niton Road, Weddington, Nuneaton, CV10 0BX
G4 UQW D. Beckett, 433 New St., Biddulph Moor, Stoke on Trent, ST8 7NG
G4 UQY M. Regan, 36 Moor Park Gardens, Leigh-on-Sea, SS9 4PY
G4 URA A. Haynes, Thorrington Nurseries, Tenpenny Hill, Colchester, CO7 8JB
GW4 URB R. Teesdale, 22 Cwmgelli Drive, Treboeth, Swansea, SA5 9BS

G4 URD R. Caira, 12 West Hill Road, Herne Bay, CT6 8HG
GW4 URG S. Richardson, Awel Y Ddol, Glanrafon, Corwen, LL21 0HA
G4 URM P. Butler, Tanglewood, Elms Lane, Wolverhampton, WV10 7JS
G4 URN M. Turvey, 106 Foxwell St., Worcxester, Worcester, WR5 2ET
G4 URP R. Powell, 57 Bartons Drive, Yateley, GU46 6DW
G4 URS J. Osborne, 64 Old Warren, Taverham, Norwich, NR8 6GA
G4 URT P. Hutchison, 1 Barkers Green, Wem, Shrewsbury, SY4 5JN
G4 URV Dr W. Peel, 34 Carlyn Avenue, Sale, M33 2EA
G4 URW J. Allison, 17 Gordon Terrace, Stakeford, Choppington, NE62 5UE
G4 URX T. Robinson, 26 Keeble Drive, Washingborough, Lincoln, LN4 1DZ
G4 USC K. Appleton, 5 Hart Hill Crescent, Full Sutton, York, YO41 1LX
G4 USD Bomchil Castro Goodrich Claro Arosmena, c/o D. Brill, 25 Boulevard Barbes, Paris, France, 75018
G4 USK B. Finlay, 4 Henden Mews, Maidenhead, SL6 4GY
G4 USN M. Havard, 61 Northwood End Road, Haynes, Bedford, MK45 3QB
G4 USP S. Hall, 22 Leam Road, Lighthorne Heath, Leamington Spa, CV33 9TE
G4 USQ T. Hodgetts, 14 St. Peters Road, Portishead, Bristol, BS20 6QY
G4 UST A. Forbes, Field Cottages, 44 Walkmills, Church Stretton, SY6 6NJ
G4 USW W. Jenkins, 5 Seatoller Place, Barrow-in-Furness, LA14 4NH
G4 USX E. Pritchard, 18 New Ridd Rise, Hyde, SK14 5DD
G4 UTE M. Chaudhry, 613 Service Road, G 10/4, Islamabad, Pakistan, ZZ7 9PO
G4 UTF A. Cockman, 31 Kensington Road, Southend-on-Sea, SS1 2SX
G4 UTG F. Collins, 31 Mount Pleasant Road, Poole, BH15 1TU
G4 UTJ J. Gorton, 17 Oxford Road, Colchester, CO3 3HW
G4 UTK D. James, 3 El Mirador Del Embalse, Los Romanes, Malaga, Spain, 29713
G4 UTM B. Dennis, Thistledown Yallands Hill, Monkton Heathfield, Taunton, TA2 8NA
G4 UTN G. Bromley, 46 Independent Hill, Alfreton, DE55 7DG
G4 UTQ M. Adamson, 13 Towers Close, Bedlington, NE22 5ER
G4 UTR M. Alder, 342 Church St., Edmonton, London, N9 9HP
GW4 UTS E. Bracey, 3 Dyffryn Road, Waunlwyd, Ebbw Vale, NP23 6UA
G4 UTV A. Cockerill, 90 Stockton Road, Middlesbrough, TS5 4AJ
G4 UUA M. Robinson, 2 Bridge Mill Road, Nelson, BB9 7BD
G4 UUB M. Lemin, Mill House, Lingwood Road, Blofield, Norwich, NR13 4AH
G4 UUE Dr L. Mcgrogan, 239 Haslingden Road, Rossendale, BB4 6RX
G4 UUF N. Kelly, 3 The Terrace, Gawcott, Buckingham, MK18 4HL
G4 UUG D. Payne, 23 Laburnum Avenue, Newbold Verdon, Leicester, LE9 9LQ
G4 UUH S. Rogers, 31 Coleridge Road, Ottery St Mary, EX11 1TD
G4 UUI S. Rooker, 67 Hawks Way, Ashford, TN23 5UW
G4 UUJ E. Jeffery, 11 Furze Hill Road, Shanklin, PO37 7PA
G4 UUQ T. Tallis, 1 High Peak Cottages, High Peak Junction, Whatstandwell, Matlock, DE4 5HN
G4 UUT A. Turner, 30 Wheatlands Road, Paignton, TQ4 5HU
G4 UUU C. Clayton, 17 Meadow Dene, East Ayton, Scarborough, YO13 9EL
G4 UUW D. Williams, 12 Springfield Road, Exmouth, EX8 3JX
G4 UVA P. Money, Meadow View, Podmore Lane, Dereham, NR19 2NS
G4 UVB P. Gibson, 9 Mallard Close, Aughton, Ormskirk, L39 5QJ
G4 UVD D. Asquith, 516 Old Bedford Road, Luton, LU2 7BY
G4 UVF J. Taylor, 6 The Stray, South Cave, Brough, HU15 2AL
G4 UVG D. Stewart, 4 Towles Pastures, Castle Donington, Derby, DE74 2RX
GW4 UVN J. Travers, 40 Birchgrove Road, Birchgrove, Swansea, SA7 9JR
G4 UVV D. Pike, 22 Stable Court, Gatchell Oaks, Taunton, TA3 7EG
G4 UVW E. Underhill, 5 Lyndhurst Croft, Eastern Green, Coventry, CV5 7QE
G4 UVX P. Lee, 51 Ashford Road, Faversham, ME13 8XN
G4 UVZ A. Whatmore, Hollybank, Sellicks Green, Taunton, TA3 7SD
G4 UWA M. Styne, 11 Paisley Walk, Church Gresley, Swadlincote, DE11 9FF
G4 UWF M. Kebbell, 56 King Edward Avenue, Hastings, TN34 2NQ
G4 UWG N. Dunn, 4 Swyneghyll, Temple Sowerby, Penrith, CA10 2AW
G4 UWM J. Mckenna, 18 Frobisher Close, Goring-By-Sea, Worthing, BN12 6EY
GM4 UWN R. Kane, 39 Tollohill Drive, Aberdeen, AB12 5DQ
G4 UWP L. Flynn, 20 Heather Lea Place, Sheffield, S17 3DN
GW4 UWR V. Thomas, 6 Hillside Court, Pontnewydd, Cwmbran, NP44 1LS
G4 UWS A. Walker, 53 Parkstone Avenue, Parkstone, Poole, BH14 9LW
G4 UWW A. Prior, 17 Millfield, Castleton Way, Eye, IP23 7DE
GM4 UWX J. Rennie, 19 Harbour Place, Portknockie, Buckie, AB56 4NR
G4 UXB R. Ball, 144 Broad Lane, Hampton, TW12 3BW
G4 UXC M. Butler, Field Farm Bungalow, Evesham, Wr11 7rp g
G4 UXD D. Brandon, 1 Woodlands Road, Saltney, Chester, CH4 8LB
G4 UXG J. Dew, 8 Silverbeck Way, Stanwell Moor, Staines-upon-Thames, TW19 6BT
G4 UXH C. Wilkinson, 14 Ryleyfield Road, Milnthorpe, LA7 7PT
G4 UXJ T. Ager, 5 Matthews Close, Bedhampton, Havant, PO9 3NJ
G4 UXL K. Jones, 10 Whetstone Hey, Great Sutton, Ellesmere Port, CH66 3PH
G4 UXO N. Emson, 9 Sands Close, Pattishall, Towcester, NN12 8LU
G4 UXP M. Huxham, 34 The Close, Brixham, TQ5 8RF
G4 UXV C. Osborn, 19 Maple Drive, Huntingdon, PE29 7JE
GM4 UXX A. Hood, 4 Murray Road, Law, Carluke, ML8 5HR
G4 UXY C. Boulter, 17 Forelands Way, Chesham, HP5 1QP
GM4 UYE H. Martin, 11 Ewing Court, Broomridge, Stirling, FK7 0QP
G4 UYF L. Aldhous, 5 Banks Lane, Heckington, Sleaford, NG34 9QY
G4 UYJ B. Crow, 690 Walmersley Road, Bury, BL9 6RN
GM4 UYK J. Caddis, 30 Newlands Drive, Kilmarnock, KA3 2DW
G4 UYM Dr F. Roberts, 5 Manor Farm Close, Broughton, Kettering, NN14 1SL
GM4 UYP J. Smith, 10 Witchknowe Avenue, Caprington, Kilmarnock, KA1 4LQ
G4 UYR R. Noble, Fallowfield, Chandler Road Stoke Holy Cross, Norwich, NR14 8RG
GW4 UYT R. Jenkins, 1 Lon Y Bryn, Glynneath, Neath, SA11 5BG
GM4 UYZ R. Glasgow, 7 Castle Terrace, Port Seton, Prestonpans, East Lothian, EH32 0EE
GW4 UZC D. Ralph, Tal-Y-Maes, Llanbedr, Powys, NP8 1SY
G4 UZE C. Mason, 145 Park Avenue, Ruislip, HA4 7UN
G4 UZF B. Matthews, 12 School Road, Thurston, Bury St Edmunds, IP31 3SP
G4 UZG G. Price, 28 Leewood Close, Brampton Bierlow, Rotherham, S63 6ET
G4 UZN A. Quest, 86 Buckstone Avenue, Leeds, LS17 5ET
G4 UZO K. Richards, Trigg Court, Trewetha, Port Isaac, PL29 3RU
GM4 UZR J. Low, 4 Smith Avenue, Inverness, IV3 5ES
GM4 VAC S. Murray, Taigh Nam Moireach, Croy, Inverness, IV2 5PN
G4 VAF N. Dorrington, Im Buergel 4, Woertham, Germany, 63939
GW4 VAG H. Green, 2 Whitchurch Road, Bangor On Dee, Wrexham, LL13 0AY
G4 VAH P. Hudson, 15 Fellows Road, Farnborough, GU14 6NU
G4 VAL V. Pellowe, 191 Preston New Road, Blackpool, FY3 9TN
G4 VAM P. Harrison, 2 Allington Close, Bainton, Stamford, PE9 3AG
G4 VAO M. Jordan, Petalouda, Low Street, Ketteringham, Wymondham, NR18 9RY
G4 VAP I. Kenyon, 4 Well Lane, Warton, Carnforth, LA5 9QZ
G4 VAS E. Cooper, 39 Violet Road, Southampton, SO16 3GZ
G4 VAV A. Brooks, 17 Grosvenor Avenue, Carshalton, SM5 3EJ
G4 VAX S. Cope, 24 Metcalf Road, Newthorpe, Nottingham, NG16 3NL
GM4 VAY A. Newlands, 7 Muir Close, Stewarton, Kilmarnock, KA3 3HG
GD4 VBA R. Harrison, 108 Ballacriy Park, Colby, Isle of Man, IM9 4NB
G4 VBD S. Mcadam, 92 Armstrong Close, Birchwood, Warrington, WA3 6DJ
GM4 VBE R. Fairholm, 28 Queensberry Avenue, Clarkston, Glasgow, G76 7DU
G4 VBI R. Harte, 32 Kingsgate Avenue, Kingsgate, Broadstairs, CT10 3QP
G4 VBJ B. Kay, 19 Langham Grove, Timperley, Altrincham, WA15 6DY
G4 VBK R. Deeprose, 70 Hollington Old Lane, St Leonards on Sea, TN38 9DP
GW4 VBM C. Leighton, 12 Morley Avenue, Connah'S Quay, Deeside, CH5 4RE
G4 VBO J. Mattock, 1633 Dufferin Cres, Nanaimo, Canada, V9S 5T4
G4 VBQ S. Largent, 31 Penzance Road, Kesgrave, Ipswich, IP5 1LU
G4 VBS P. Chapman, 10 School Cottages, Hargrave, Bury St Edmunds, IP29 5HR
GW4 VBV R. Beckers, 13 Taplow Terrace, Pentrechwyth, Swansea, SA1 7AD
G4 VBX A. Currell, 33 The Oval, Saham Toney, Thetford, IP25 7HW
G4 VCA G. Mason, Lilac Cottage, Tremar Coombe, Liskeard, PL14 5EL
G4 VCB C. Melvin, 5808 Sterling Trl, Mckinney, USA, 75071
G4 VCE S. Sewell, Medway, The Rosery, Norwich, NR14 8AL
G4 VCJ R. Percival, 6 Bulmer Place, Hartlepool, TS24 9BQ
GW4 VCL R. Parry, 2 Campanula Drive, Rogerstone, Newport, NP10 9JG
G4 VCN A. Soars, 118 Braddon Road, Loughborough, LE11 5YZ
G4 VCO D. Seddon, Zante, 31 Pembridge Road, Hemel Hempstead, HP3 0QN
G4 VCP C. Smith, 24 Watling Way, Whiston, Prescot, L35 7NG
G4 VCQ R. Hogan, 10 Lisle Place, Wotton-under-Edge, GL12 7BJ
G4 VCX M. Beaumont, 71 Lime Tree Avenue, Coventry, CV4 9EZ
GI4 VCZ P. Donnelly, 64 Aghnagar Road, Garvaghy, Dungannon, BT70 2EL
G4 VDB D. Brocklehurst, 73 Ridgeway, Clowne, Chesterfield, S43 4BD
G4 VDF A. Palmer, 1 Rosary Gardens, Yateley, GU46 6JT
GM4 VDG J. Rankin, 64 Forrest Walk, Uphall, Broxburn, EH52 5PN
G4 VDH W. Peak, 10 The Oval, Scarborough, YO11 3AP
G4 VDJ B. Lee, 61 Pendleway, Pendlebury, Manchester, M27 8QS
GW4 VDP D. James, Brig Y Gwynt, Penrhyn Geiriol, Trearddur Bay, Holyhead, LL65 2YW
G4 VDX J. Menguy, 6 Laurel Grove Lowton, Warrington, WA3 2EE
GW4 VEB D. Lintern, 108 Pontygwindy Road, Caerphilly, CF83 3HF
G4 VEC M. Elliott, 20 Haysel, Sittingbourne, ME10 4QE
G4 VEH L. Skinner, 15 Ridge Close, Portishead, Bristol, BS20 8RQ
GW4 VEI C. Barrett, 30 Brecon Road, Hirwaun, Aberdare, CF44 9ND
G4 VEL J. Smith, 43 Ash Close, Thetford, IP24 3HQ
G4 VEO G. Barratt, Charnwood, Great North Road, Retford, DN22 8NL
GW4 VEQ T. Jones, Penrhiw Bach, Bryngwran, Holyhead, LL65 3RD
G4 VER Verulam ARC, c/o P. King, 96 Mancroft Road, Caddington, Luton, LU1 4EN
G4 VET N. Greaves, Forty Shilling Cottage, Oghill, Co Kildare, Ireland
G4 VEW G. Davies, 40 Derby Road, Talke, Stoke-on-Trent, ST7 1SG
G4 VEY F. Havard, Whitehalgh Farm, Whitehalgh Lane, Blackburn, BB6 8ET
G4 VFC D. Monnery, 8 Reeds Lane, Southwater, Horsham, RH13 9DQ
GW4 VFE C. Davies, 3 Bryn Onnen, Flint, CH6 5QB
G4 VFG P. Lewis, 18 Bittaford Wood Bittaford, Ivybridge, PL21 0ET
G4 VFH G. Fuller, 61 The Underwood, London, SE9 3EP
G4 VFJ S. Shenton, 36 Walleys Drive, Newcastle, ST5 0NG
G4 VFK C. Archer, 118 Cator Lane, Beeston, Nottingham, NG9 4BB
G4 VFL A. Holland, 12 Riverside Drive, Egremont, CA22 2EH
G4 VFR K. Hackwell, 15 Standish Avenue, Billinge, Wigan, WN5 7TF
G4 VFU C. White, 4 Trenton Drive, Bradford, BD8 7SZ
G4 VFX C. Perkins, 32 Empshott Road, Southsea, PO4 8AU
GW4 VGB R. Harper, 114 Pantbach Road, Cardiff, CF14 1UE
G4 VGL S. Luckhaus, Wingertstrasse 5, Kleinwallstadt, Germany, 63839
G4 VGM E. Grindel, Bischofsheimer, Platz 24, Frankfurt, Germany, D 60326
G4 VGN V. Havran, Kurt-Schumacher, Ring 31, Dreieich, Germany, D-63303
GM4 VGR J. Buchanan, 114 Glasgow Road, Whins Of Milton, Stirling, FK7 0LJ
GM4 VGU A. Lyttle, 23 Heathfield Drive, Kirkmuirhill, Lanark, ML11 9SR
G4 VGY D. Cline, 68 Frenchgate, Richmond, DL10 7AG
G4 VHB C. Averill-Elias, 12 Bubwith Close, Chard, TA20 2BL
G4 VHE R. Haase, 674 Valley View Lane, Strafford, USA, 19087
G4 VHG J. Fowler, 6 Cridlake, Axminster, EX13 5BS
G4 VHI M. Sawyers, 20 Fairways, Ferndown, BH22 8BA
G4 VHJ J. Taylor, 29 Meadow Walk, Ewell, Epsom, KT17 2EF
G4 VHK Prof. R. Leslie, Tranquil, Rectory Lane Kingston, Cambridge, CB23 2NL
G4 VHL T. Langford, 11 The Grove Blackawton, Totnes, TQ9 7BA
G4 VHM M. Hindley, 12 Tremayne Avenue, Brough, HU15 1BL
GI4 VHO D. Calderwood, 43 Rathview Park Mullybritt, Lisbellaw, Enniskillen, BT94 5EW
GW4 VHP S. Murdoch, 55 Pendre Avenue, Prestatyn, LL19 9SH
GW4 VHS E. Williams, Newhaven, Kinmel Way, Abergele, LL22 9NE
G4 VHV A. Sims, 38 Giffard Drive, Welland, Malvern, WR13 6SE
G4 VHW C. Thompson, 2 Essen Lane, Kilsby, Rugby, CV23 8XQ
G4 VHX J. Allen, 18 Horsley Close, Chesterfield, S40 4XD
GM4 VHZ N. Brown, 7 Mid Road, Beith, KA15 2AJ
G4 VIA J. Mcsherry, 1 Station Houses, Corkickle, Whitehaven, CA28 7XG
G4 VIF R. Watts, 116 Hassall Road, Sandbach, CW11 4HL
G4 VII J. Lawrence, 25 Sylvia Crescent, Totton, Southampton, SO40 3LP
GM4 VIK T. Irwin, 6 Inverarnie Park, Inverarnie, Inverness, IV2 6AX
G4 VIL J. Fielding, 35 Amos Avenue, Litherland, Liverpool, L21 7QH
G4 VIM B. Pulfrey, 21 Emfield Road, Grimsby, DN33 3BW
G4 VIO A. Greenbank, 3 Cooperative Terrace, Stanley, Crook, DL15 9SE
GI4 VIP Sth Belfast, c/o P. Murphy, 11 Danesfort Apartments, Belfast, BT9 5QL
G4 VIQ P. Brushwood, 2 High Trees, Waterlooville, PO7 7XP
GM4 VIS H. Cameron, 14 Queen St., Castle Douglas, DG7 1HX
G4 VIT M. Wood, 42 Buckingham Drive, Willenhall, WV12 5TD
G4 VIX East Coast VHF Group, c/o D. Bartlett, Apartamentos Las Adelfas 8A, C/ San Borondon 111, Playa Honda, Spain, 35509
GI4 VIZ M. Jamieson, 59 Curragh Road, Coleraine, BT51 3RZ
G4 VJB V. Bloor, 22 Regency Close, Talke Pits, Stoke-on-Trent, ST7 1RH
GM4 VJH A. Douglas, 12 Leaburn Drive, Hawick, TD9 9NZ

G4 VJL J. Oldfield, 4 The Elms, Taunton, TA4 4AE
G4 VJN S. Matthews, 30 Broadgate Lane, Deeping St. James, Peterborough, PE6 8NW
G4 VJT K. Farmer, 61 Queens, Beckenham, Kent, BR3 4JJ
GI4 VJZ T. Wilson, Wilden, 18 Ahoghill Road, Antrim, BT41 3BJ
G4 VKC D. Lawrence, 6 Hollycombe Close, Liphook, GU30 7HR
G4 VKE R. Pearce, 1A Green Lane, Dalton-in-Furness, LA15 8LZ
GW4 VKG W. Weston, 3 Factory Terrace, Aberkenfig, Bridgend, CF32 9AF
GM4 VKI M. Kavanagh, 4 Old Auchans View, Dundonald, Kilmarnock, KA2 9EX
G4 VKJ T. Grant, 81 Hillworth Road, Devizes, SN10 5HD
G4 VKO J. Whittock, 18 Westons Brake, Emersons Green, Bristol, BS16 7BP
GI4 VKS A. Mccallion, 3 Lisky Road, Strabane, BT82 8NW
G4 VKV T. Linacre, 69 Elizabeth Road, Fazakerley, Liverpool, L10 4XL
G4 VKX I. Wade, 59 St. Annes Road, Kettering, NN15 5EQ
G4 VLA R. Trudgill, The Retreat, Kiln Lane, Bedford, MK45 4DA
G4 VLF J. Wingfield, 56 Manor Fields, Liphook, GU30 7BS
G4 VLH B. Pash, Dales Stores, Station Road, Cheltenham, GL54 4HP
G4 VLI R. Allen, 39 Deerpark, Co Meath, Ireland,
G4 VLK D. Haslehurst, 23 Yew Tree Drive, Shirebrook, Mansfield, NG20 8QH
G4 VLL C. Denham, 3 Glenmore Close, Flackwell Heath, High Wycombe, HP10 9DF
G4 VLN M. Evans, 2A Moreton Road, Worcester Park, KT4 8EZ
G4 VLP H. Knatchbull, 19 Riverside Road, West Moors, Ferndown, BH22 0LG
G4 VLS P. Turnham, 71 Theobald Road, Norwich, NR1 2NX
G4 VLT C. Tunna, 52 Shaftoe Road, Springwell, Sunderland, SR3 4EZ
GW4 VLU M. Hatwood, Calgary, Denbigh Circle, Rhyl, LL18 5HW
G4 VLV A. Flint, 4 Churchill Way, Painswick, Stroud, GL6 6RQ
G4 VLW R. Davey, 35 The Pines, Faringdon, SN7 8AT
GM4 VLX J. Brown, 33 Gartmore Road, Paisley, PA1 3NG
G4 VLZ M. Nettleship, 141 Hollybank Drive, Sheffield, S12 2BU
G4 VMA M. Anderson, 1 Ridgeway Square, Knottingley, WF11 0JY
G4 VMB D. Stoddart, 16 Market Place, Long Buckby, Northampton, NN6 7RR
G4 VMC P. Coates, 20 The Flashes, Gnosall, Stafford, ST20 0HL
G4 VMD C. Hackney, Mar Azul 9, Apt.20 2 Fase, Alicante, Spain, 03710 CALPE
G4 VME R. Youell, 1 Greenside Waterbeach, Cambridge, CB25 9HW
G4 VMF S. Schofield, 28 Rush Meadow Road, Cranbrook, Exeter, EX5 7GB
G4 VMG J. Holmes, 10 Chapel Road, Morley St. Botolph, Wymondham, NR18 9TF
G4 VMI M. Pickworth, 46 St. Andrews, Grantham, NG31 9PE
G4 VMM S. Tidmarsh, 4 The Grange, Earl Shilton, Leicester, LE9 7GT
G4 VMO J. Harris, 23 Brookvale Grove, Solihull, B92 7JH
G4 VMQ A. Hyde, 45 Camm Street, Sheffield, S6 3TR
G4 VMR J. Watkins, One Ash, Frogshall Lane, Haultwick, Ware, SG11 1JH
GW4 VMT G. Williams, 12 Heol Johnson, Talbot Green, Pontyclun, CF72 8HR
G4 VMW J. Curtis, 18 Carlidnack Close, Mawnan Smith, Falmouth, TR11 5HF
G4 VMX A. Ritchie, 24 Swift Close, Newport Pagnell, MK16 8PP
G4 VMY A. Cooper, 3 Marina Way, Ripon, HG4 2LJ
G4 VMZ A. Jones, 40 Alexandria Drive, Herne Bay, CT6 8HX
G4 VNA R. Bell, Long Meadows, Haddockstones, Harrogate, HG3 3LA
G4 VNC M. Leak, 32 Springdale Road, Market Weighton, York, YO43 3JT
G4 VNE D. Hunt, 233 Kingsley Road, Kingswinford, DY6 9RP
G4 VNG R. Mccallum, 9 Hardwick Close, Blackwell, Alfreton, DE55 5LL
GW4 VNK L. Smith, Blaenlluest, Cilcennin, Lampeter, SA48 8RP
G4 VNM S. Frost, 32 Hunters Lodge, Fareham, PO15 5NE
G4 VNR R. Sharp, 22 Sunderton Lane, Clanfield, Waterlooville, PO8 0NU
GW4 VNS R. Sims, 61 Constable Drive, Newport, NP19 7QB
G4 VNX W. Wood, 11 Walbert Avenue, Thurnscoe, Rotherham, S63 0TN
G4 VOG A. Hepworth, 9 Linden Grove, Kirkby-In-Ashfield, Nottingham, NG17 8JJ
G4 VOJ A. Tennant, 2 Chapel Hill Farm Cottage, Lower Lane, Preston, PR3 3SL
G4 VOK P. Dresser, 6 Acacia Avenue, Fencehouses, Houghton le Spring, DH4 6JG
G4 VOU A. Pinkney, 1 Hester Gardens. New Hartley. Whitley Bay., Whitley Bay., ne25 0sh
G4 VOV D. Hallsworth, Eastholme, Luddington Road, Scunthorpe, DN17 4PP
G4 VOW B. Pluckrose, 104 Edward Road, West Bridgford, Nottingham, NG2 5GB
G4 VOY R. Powell, Old School House, Broxwood, Leominster, HR6 9JQ
G4 VOZ J. Jennings, Mill Side, Mill Road, Lutterworth, LE17 5DE
G4 VPA J. Martindale, The Old School House, Ipswich Way, Stowmarket, IP14 6DJ
G4 VPC E. Ikin, 30 Kelsborrow Way, Kelsall, Tarporley, CW6 0NL
G4 VPD M. Pugh, 44 Simms Lane, Hollywood, Birmingham, B47 5HY
G4 VPE Dr R. Derricott, Birches, The Paddock, Stourbridge, DY9 0RE
G4 VPF O. Davies, 16 Central Way, Horninglow, Burton-on-Trent, DE13 0UU
G4 VPI R. Riley, 161 Botany Road, Kingsgate, Broadstairs, CT10 3SD
G4 VPJ D. Bridgnell, Penvale, 1 Tretherras Road, Newquay, TR7 2RB
G4 VPL J. Villena Bota, Santa Ana 74, Estartit, Gerona, Spain, 17258
G4 VPM A. Stafford, 233 Sparrow Branch Circle, Jacksonville, USA, 32259
G4 VPS B. Lewis, 6 Malt Dubs Close, Ingleton, Carnforth, LA6 3DZ
G4 VPW P. Wilcock, 12 Napier Road, Eccles, Manchester, m30 8ag
GW4 VPX A. Jones, Maes Y Llyn, Maesycrugiau, Pencader, SA39 9DH
G4 VPZ A. Hill, 36 Narrow Lane, Halesowen, B62 9NQ
G4 VQE R. Spencer, 6 Belland Drive, Charlton Kings, Cheltenham, GL53 9HU
G4 VQF P. White, 4 Barnett Lane, Wonersh, Guildford, GU5 0SA
G4 VQH M. Clutton, Cumberwell, Cumberland Lane, Whitchurch, SY13 2NJ
G4 VQI P. Hughes, 92 Freshwater Drive, Paignton, TQ4 7SD
G4 VQJ D. Northwood, 5 Beech Grange, Landford, Salisbury, SP5 2AL
GI4 VQK A. Ward, 50 Derry Road, Strabane, BT82 8LD
G4 VQL G. Dymond, 28 The Green, Exmouth, EX8 2QR
G4 VQP C. Smith, Foxgloves, Northlew, Okehampton, EX20 3PP
G4 VQR S. Reed, 139 Potovens Lane, Outwood, Wakefield, WF1 2LF
G4 VQS H. Jones, 47 Penkett Road, Wallasey, CH45 7QG
G4 VQT M. Pinnell, 3 Inmans Lane, Petersfield, GU32 2AN
GM4 VQY G. Leiper, 76 Martin Drive, Stonehaven, AB39 2LU
G4 VQZ J. Oakley, 152 Little Breach, Chichester, PO19 5UA
G4 VRB K. Raine, 30 St. Andrews Gardens, Shepherdswell, Dover, CT15 7LP
G4 VRC R. Doran, 28 Buckingham Road, Petersfield, GU32 3AZ
GM4 VRE P. Henderson, 134 Gray Street, Aberdeen, AB10 6JU
G4 VRG F. Margrave, Templars, Peerley Road, Chichester, PO20 8DW
G4 VRJ R. Clifft, 11 Hambleton St., Wakefield, WF1 3NW
G4 VRM A. Berry, 148 Maple Way, Gillingham, SP8 4RR
G4 VRN M. Blewett, 32 Miltons Crescent, Godalming, GU7 2NT
GW4 VRO P. Parsons, 9 Military Road, Pennar, Pembroke Dock, SA72 6SH
G4 VRP R. Porter, 47 Milford Avenue, Wick, Bristol, BS30 5PP
G4 VRS Aylesbury Vale Radio Society, c/o V. Gerhardi, 24 Putnams Drive, Aston Clinton, Aylesbury, HP22 5HH
G4 VRT B. Barker, The Nook, Park Lane, Roughbirchworth, Oxspring, Sheffield, S36 8WW
G4 VRU B. Stephenson, 12 Claremont Terrace, York, YO31 7EJ
G4 VRW K. Newbould, 188 Brooklands Avenue, Leeds, LS14 6RH
G4 VRX Dr G. Brown, 1 Dog Kennel Lane, Oldbury, B68 9LU
G4 VSB A. Brown, 6 The Firs, Rushbrooke Lane, Bury St Edmunds, IP33 2SY
G4 VSD P. Samuels, 6 Miriam Close Caister-On-Sea, Great Yarmouth, NR30 5PH
G4 VSI A. Stone, 29 Nottingham Road, Belper, DE56 1JG
G4 VSJ K. Drakeford, Sunnyside, Frolesworth Lane, Lutterworth, LE17 5AS
G4 VSK B. Skelton, 8 Dunelm Drive, West Boldon, East Boldon, NE36 0HJ
G4 VSL T. Watkins, One Ash, Frogshall Lane, Haultwick, Ware, SG11 1JH
G4 VSO R. Carter, Wendover, Three Ahes, HR2 8LU
G4 VSQ A. Bolton, 17 Lomond Avenue, Caversham Reading, Rg4 6pl
G4 VSR S. Alston, 21 Hilltop Road, Wingerworth, Chesterfield, S42 6RX
G4 VSS M. Isherwood, 32 Franklin Close Old Hall, Warrington, WA5 8QL
G4 VSV G. Ingham, Courthaven, South Duffield Road, Selby, YO8 5HP
G4 VSW M. Taylor, 2 Bickerton Drive, Hazel Grove, Stockport, SK7 5QY
G4 VSX P. Reilly, 40 Bollin Drive, Lymm, WA13 9QA
G4 VSY K. Hutton, 7 Roseveare Drive, Roseveare Park, Gothers, St Austell, PL26 8GY
G4 VTA J. Taylor, 219 Mandarin Way, Cheltenham, GL50 4SB
GM4 VTB M. Budas, 20 Oak Avenue, Bearsden, Glasgow, G61 3HD
G4 VTC A. Croydon, Harvesters, Newdigate Road, Dorking, RH5 4QB
G4 VTD I. Daniels, 24 Ockley Lane, Keymer, Hassocks, BN6 8BB
GW4 VTG E. Smith, 21 St. Davids Road, Pembroke, SA71 5JH
G4 VTM J. Hicks, Cory House, Kilworth Road, Lutterworth, LE17 6JW
G4 VTN J. Rodda, 22 Balmoral Drive, Felling, Gateshead, NE10 9TZ
G4 VTO P. Tanner, Beechcroft, Station Hill, Newton Abbot, TQ13 0EE
G4 VTQ D. Rainer, Twin Oaks Knightstone Lane, Ottery St Mary, EX11 1PR
G4 VTU R. George, 19 Apthorpe Street, Fulbourn, Cambridge, CB21 5EY
G4 VUA A. Burton, 26 Woffindin Close, Great Gonerby, Grantham, NG31 8LP
G4 VUD J. Head, 21 Reynell Avenue, Newton Abbot, TQ12 4HE
G4 VUF C. Tugman, 41 Chatsworth Road, Hunstanton, PE36 5DJ
GM4 VUG C. Green, 4 Gallowhill Gardens, Kinross, KY13 8RT
GW4 VUH J. Washington, Flat 9, Llys Canol, Holywell, CH8 7XG
G4 VUI P. Sweeney, 15 Alford Road, West Bridgford, Nottingham, NG2 6GJ
G4 VUK L. Wolfson, 7 Gilmore Drive, Prestwich, Manchester, M25 1NB
G4 VUM D. Stocks, 2 Newport Crescent, Mansfield, NG19 6BY
G4 VUN P. Norris, Thirn Farm Thirn, Ripon, HG4 4AU
G4 VUP G. Newton, 6 Yardley Way, Grimsby, DN34 5UQ
G4 VUR I. Daniels, 20A Stalham Road, Hoveton, Norwich, NR12 8DG
G4 VUV S. Valori, 7 Upton Close, Norwich, NR4 7PD
G4 VUW R. Kemp, 35 Rushett Drive, Dorking, RH4 2NR
G4 VVD P. Taylor, 8 High St., Clive, Shrewsbury, SY4 3JL
G4 VVE E. Macmanus, 41 Oldfield Crescent, Stainforth, Doncaster, DN7 5PE
GW4 VVF N. Allen, Lilac Cottage, Roddhurst, Presteigne, LD8 2LH
G4 VVK B. Murray, La Casa, 30 Middlegate Green, Rossendale, BB4 8PY
G4 VVM A. Bennett, 28 Kinglake Drive, Taunton, TA1 3RR
G4 VVP B. Gillard, Charmaine, Broadway, Chilcompton, Radstock, BA3 4GT
G4 VVQ F. Shead, 7 White Cottages, Fuller Street, Fairstead, Chelmsford, CM3 2AY
G4 VVS J. Blanchard, 41 Deane Drive, Galmington, Taunton, TA1 5PQ
G4 VVT A. Moss, 9 Summerfield Drive, Middleton, Manchester, M24 2TQ
GM4 VVX C. O'Hennessy, Savalbeg, Challenger Estate, Lairg, IV27 4ED
GM4 VVY D. Davis, 3 High Shore, Banff, AB45 1DB
G4 VVZ C. Wilson, 2 Bainton Close, Bradford-on-Avon, BA15 1SE
G4 VWA S. Ward, 88 Little Barn Lane, Mansfield, NG18 3JJ
GI4 VWC G. Christie, The Brambles, 9 Burnet Park, Newtownabbey, BT37 0XY
G4 VWD A. Davies, 85 Ridgewood Drive, St Helens, WA93XU
G4 VWE M. Courteney, 36 Nursery Close, Hellesdon, Norwich, NR6 5SJ
G4 VWF R. Hawkins, 43 The Courtyard, Taylor Avenue, Northampton, NN3 2DD
G4 VWG S. Vankassel, 9 Tarragon Close, Swindon, SN2 2SG
G4 VWI D. Hatton, 9 Gregory Close, Thurmaston, Leicester, LE4 8BP
G4 VWL J. Owen, 7 Linear Park, Wirral, CH46 6FL
GW4 VWO J. Bloodworth, Sibrwd Yr Awel, Penrhyndeudraeth, LL48 6AY
G4 VWS C. Davies, Essex House, 42 Boxworth Road, Cambridge, CB23 4JQ
G4 VWT B. Padgett, 11 Crow Croft Road, Pilsley, Chesterfield, S458HY
GM4 VWV R. Mcewan, 12 Valleyfield Drive, Cumbernauld, Glasgow, G68 9NW
G4 VWX A. Shone, 5 Mayflower Close, Malvern, WR14 2RH
GW4 VWY G. Whiteway, 4 Nicholas Court, Gorseinon, Swansea, SA4 4PR
GM4 VXA P. Williams, 21 St. Clair Way, Ardrishaig, Lochgilphead, PA30 8FB
G4 VXB M. Ellis, 16 Fielding Street, Faversham, ME13 7JZ
G4 VXD R. King, 1 Emmas Crescent, Stanstead Abbotts, Ware, SG12 8AZ
GW4 VXE T. Kirby, Parsonage House, St. Nicholas, Goodwick, SA64 0LG
G4 VXG G. Buxton, 12 Jute Road, York, YO26 5EN
G4 VXH A. Rean, 17 Mount Pleasant Road, Dawlish Warren, Dawlish, EX7 0NA
GM4 VXM I. Munro1, 12 Greenstone Place, Dundee, DD2 4XB
G4 VXN C. Bennett, 49 Keats Avenue, Redhill, RH1 1AF
G4 VXP R. Want, 19 Canterbury Road, Leyton, London, E10 6EE
G4 VXU J. Haig, 3 Hartland Court Gaping Lane, Hitchin, SG5 2JH
G4 VXV R. Boulton, 23 Stamford Bridge West, Stamford Bridge, York, YO41 1AQ
G4 VXW R. Seddon, 255 Westleigh Lane, Leigh, WN7 5PN
G4 VXX S. Oakes, 6 Wychwood Park, Weston, Crewe, CW2 5GP
G4 VYA J. Jacobs, 17 Cotswold Drive, Albrighton, Wolverhampton, WV7 3DQ
G4 VYC V. Packman, 241 Gurnard Pines Cockleton Lane, Cowes, PO31 8RL
G4 VYE J. Harris, 23 Balmoral Drive, Hednesford, Cannock, WS12 4LT
G4 VYF K. Thomas, 19 South End Hogsthorpe, Skegness, PE24 5NE
G4 VYG B. Roberts, 52 School Lane, Toft, Cambridge, CB23 2RE
G4 VYH N. Baker, 16 Boulderside Close, Norwich, NR7 0JJ
G4 VYI M. Dalley, 195 Marlcliffe Road, Sheffield, S6 4AH
G4 VYJ J. O'Sullivan, 30 Highbank Road, Kingsley, Frodsham, WA6 8AE
G4 VYK R. Vaughan, 73B Westward Drive, Pill, Bristol, BS20 0JR
G4 VYL R. Reilly, 4 Moreton Drive, Poulton-le-Fylde, FY6 8ED
G4 VYN Dr J. Lawrence, 1 Naples Close, Hopton, Great Yarmouth, NR31 9SB
G4 VYP D. Rimmer, 8A Mallee Avenue, Southport, PR9 8NL
GM4 VYQ W. Harvey, 32 Upper Glenfyne Park, Ardrishaig, Lochgilphead, PA30 8HH

G4 VYR G. Mccartney, 12 Timway Drive, West Derby, Liverpool, L12 4YR
GM4 VYU O. Jackson, Cossarshill Farm, Selkirk, TD7 5JB
G4 VZB I. Ray, Palmyra F2B-H8, Vila Sol, Portugal, 8125-307 Quarteira
G4 VZC P. Stokes-Herbst, 4 Thackeray Grove, Middlesbrough, TS5 7QX
G4 VZH A. Hobkirk, 216 Northwick Road, Worcester, WR3 7EH
GM4 VZI W. Lawrie, Sandana Old Woodhouselea, 2, Roslin, EH25 9QJ
G4 VZK D. Perkins, 10 The Foxes, Sutton Hill, Telford, TF7 4NH
G4 VZL J. Caddick, 5 Great Hay Drive Sutton Hill, Telford, TF7 4DT
G4 VZR D. Cormack, Lukes Orchard, Far Green, Dursley, GL11 5EL
G4 VZS L. Andrew, 10 Front Street, Rookhope, Bishop Auckland, DL13 2AY
G4 VZT P. Green, 61 Gravel Hill, Wimborne, BH21 3BJ
G4 VZW K. Rankine, Alfredo L. Jones 34, Building Perez Roche 3Rd Floor, Apartment 306, Las Palmas, Spain, 35008
GM4 VZY D. Deans, 17 Montrose Way, Dunblane, FK15 9JL
G4 WAB Worked All Britain Awards Group, c/o K. Hale, 58 St. Stephens Road, Saltash, PL12 4BJ
G4 WAC Wythall Radio Club, c/o M. Pugh, 44 Simms Lane, Hollywood, Birmingham, B47 5HY
G4 WAF A. Fewkes, 21 Tong Road Bishops Wood, Stafford, ST19 9AB
G4 WAG T. Morley, 2 Hawthorn Place, Woodbridge, IP12 4JZ
GI4 WAH J. Keenan, 24 Leode Road, Hilltown, Newry, BT34 5TJ
G4 WAK N. Rumbol, 66 The Avenue, Hadleigh, Benfleet, SS7 2DL
G4 WAL P. Walton, 6 Gorse Grove, Longton, Preston, PR4 5NP
G4 WAM M. Lockley, 37 Farmside Lane, Biddulph Moor, Stoke-on-Trent, ST8 7LY
G4 WAO J. Kimpton, 721/30 Lantana Avenue, Narrabeen, Australia, 2101
G4 WAP R. Southern, 31 Burnsall Road, Brighouse, HD6 3JS
G4 WAS K. Atack, 29 High Hill, Essington, Wolverhampton, WV11 2DW
G4 WAV A. Medland, Trevilla, 93 Pengelly Road, Delabole, PL33 9AT
G4 WAW South Bristol ARC, c/o A. Jenner, 24 The Willows, Nailsea, Bristol, BS48 1JQ
G4 WAX J. Moon, 25 Shotley Gardens, Gateshead, NE9 5DP
G4 WAZ N. Mackinnon, 49 Balmoral Way, Worle, Weston-Super-Mare, BS22 9AL
G4 WBA B. Westbrook, 14 Pickering Street, Maidstone, ME15 9RS
G4 WBF P. Finney, New Rivernook Farm, 10 Kinnersley Manor, Reigate, RH2 8QJ
G4 WBG R. Dunn, 13 Horton Gate, Giffard Park, Milton Keynes, MK14 5JQ
G4 WBH P. Jackson, 15 Bankside, Retford, DN22 7UW
G4 WBI S. Haydon, 58 Deanfield Road, Henley-on-Thames, RG9 1UU
G4 WBO K. Johnson, 23 Rotherhead Close, Horwich, Bolton, BL6 5UG
G4 WBP D. Hatfield, 29 Awbridge Road, Netherton, Dudley, DY2 0HZ
GW4 WBT S. Clifton, 15 Cae Clyd, Craig-Y-Don, Llandudno, LL30 1BL
GM4 WBU W. Swinburne, 29 Murray Place, Dollar, FK14 7HP
G4 WBV A. Fry, 128 Sylvan Way, Sea Mills, Bristol, BS9 2LU
G4 WBW K. Odlum, 17 Glebe Street, Talke, Stoke-on-Trent, ST7 1NP
G4 WCD D. Longstaff, 83 Spring Gardens, Anlaby Common, Hull, HU4 7QG
GM4 WCE Dr P. Kirsop, 79 Louis Braille Way, Gorebridge, EH23 4LD
G4 WCH W. Houghton, 28 Regent Street, Newton-le-Willows, WA12 9LS
G4 WCK C. Baverstock, 43 Tatnam Road, Poole, BH15 2DW
G4 WCO D. Foy, 37 Gorsey Croft, Eccleston Park, Prescot, L34 2RS
G4 WCP S. Richardson, 25 Kenmure Avenue, Patcham, Brighton, BN1 8SH
G4 WCY H. Bottomley, 8 Leyburn Place, Filey, YO14 0DQ
G4 WDA J. Curtis, 8 King Street, Wilton, Salisbury, SP2 0AX
G4 WDC G. Cooke, 106 Wirral Drive, Winstanley, Wigan, WN3 6LD
GM4 WDO A. Douglas, 3 North Lodge Cottages, Ladykirk, Berwick-upon-Tweed, TD15 1SU
G4 WDP D. Preston, 77 Wensley Road, Woodthorpe, Nottingham, NG5 4JX
G4 WDR West Devon Raynet, c/o I. Harley, 37 Crelake Close, Tavistock, PL19 9AX
G4 WDS R. Silvera, 10 White Hill, Kinver, Stourbridge, DY7 6AD
G4 WDZ K. Bennett, Lilac Cottage, St. Neots Road, Bedford, MK44 2ER
G4 WEC G. Leesley, Marsh View, Main Road, Alford, LN13 0JP
GM4 WED N. Tipping, Millburn Cott, Weisdale, ZE2 9LN
G4 WEE S. Leech, 9 Parkside Drive, Old Catton, Norwich, NR6 7DP
G4 WEH M. Pepper, 56 Meadow Lane, Burgess Hill, RH15 9JA
G4 WEL J. Bolton, 110 Vale Road, Ash Vale, Aldershot, GU12 5HS
G4 WEM A. Penney, 110 Vale Road, Ash Vale, Aldershot, GU12 5HS
G4 WEN K. Porter, Thornfield, Mount Carmel Road, Andover, SP11 7ES
G4 WEP W. Hewitt, 101 Sunnyside Avenue, Ball Green, Stoke-on-Trent, ST6 6DZ
G4 WET Triple B.C.G, c/o M. Butler, Field Farm Bungalow, Evesham, Wr11 7rp g
G4 WEV Dr A. Russell, 6 Bartlemy Road, Newbury, RG14 6JX
GM4 WEW C. Brown, Glencraig, Ballantrae, Girvan, KA26 0PA
G4 WEY B. Bush, 45 Mimosa Avenue, Wimborne, BH21 1TU
G4 WEZ K. Westley, 29 The Limes, Sawston, Cambridge, CB22 3DH
G4 WFC M. Morris, Fieldhead Farm, Denholme, Bradford, BD13 4LZ
GM4 WFE J. Longton, 5 Allanshaw Grove, Hamilton, ML3 8NZ
G4 WFF C. Mcguire, 17 Victoria Road, Emsworth, PO10 7NH
G4 WFK F. Seddon, 20 Pinfold Lane, Bottesford, Nottingham, NG13 0AR
G4 WFL P. Ford, 24 Tonstall Road, Epsom, KT19 9DP
GW4 WFM G. Moller, 31 Wyngarth, Winch Wen, Swansea, SA1 7EF
G4 WFR R. Cooper, 53 Saturn Close, Southampton, SO16 8BE
G4 WFT T. Kearsley, 142 Avenue Road, Rushden, NN10 0SW
GM4 WFV S. Duguid, 22 Cairn Crescent, Ayr, KA7 4PW
G4 WFW P. Edwards, 1 Radley Avenue, Wickersley, Rotherham, S66 2HZ
G4 WFZ P. Marsh, Columbia, 28 Orcheston Road, Bournemouth, BH8 8SR
G4 WGA R. Hall, Hillside, Potten End Hill, Hemel Hempstead, HP1 3BN
G4 WGB R. Vaughan, 6 Dellside Grove, St Helens, WA9 5AR
GM4 WGC P. Naughton, 16 Holton Crescent, Sauchie, Alloa, FK10 3DZ
G4 WGD C. Pearse, 77A Nutfield Road, Merstham, Redhill, RH1 3ER
G4 WGE A. Cross, 31 Mountcombe Close, Surbiton, KT6 6LJ
G4 WGF G. Fairhurst, 42 Chorley Road, Standish, Wigan, WN1 2SS
G4 WGJ M. Collins, 185 Church Road, Haydock, St Helens, WA11 0NB
G4 WGK G. Kemp, 38 Merlin Way, Leckhampton, Cheltenham, GL53 0LU
G4 WGN K. Wilson, 102 Waddicar Lane, Melling, Liverpool, L31 1DY
G4 WGR R. Gibson, 52 Broomfields, Denton, Manchester, M34 3TH
G4 WGT W. Taylor, 27 Netherley Road Coppull, Chorley, PR7 5EH
G4 WGU G. Tarry, 8 Wareham Road, Blaby, Leicester, LE8 4BE
G4 WGX B. Rivers, Maybank, Athelney Bridge, Bridgwater, TA7 0SB
G4 WGZ A. Brooker, 18 Honeybourne Way, Petts Wood, Orpington, BR5 1EZ
GM4 WHA G. Harper, 15 Seaforth Park, Annan, DG12 6HX
G4 WHF K. Wilson, 111 Marple Road, Stockport, SK2 5EP
G4 WHK R. Cann, 39 Grafton Road, Harwich, CO12 3BD
G4 WHL P. Callaghan, 8 Abbey Road, Edwinstowe, Mansfield, NG21 9LQ
G4 WHM P. Callaghan, 15 John Barrow Close, Rainworth, Mansfield, NG21 0GD
G4 WHT W. Tattersall, 45 Russell Avenue, Alsager, Stoke-on-Trent, ST7 2BN
G4 WHV M. Langdon, 58 Upper Marsh Road, Warminster, BA12 9PN
G4 WHY M. Foot, Oakfield Farm, Horton Way, Verwood, BH31 6JJ
G4 WHZ D. Cater, 104 St. Johns Road, Clacton-on-Sea, CO16 8DB
G4 WIA I. Whitmore, Sunny Bank, Commercial Road, St. Keverne, Helston, TR12 6LY
G4 WIG P. Lees, 107 Balmoral Road, Wordsley, Stourbridge, DY8 5JJ
G4 WIL J. Wilkinson, 147 Alder Lane, Hindley Green, Wigan, WN2 4ET
G4 WIM T. Forrester, Dow Brook House, Brades Lane, Freckleton, Preston, PR4 1HG
G4 WIN C. Mabbutt, 17 Lakeside, Peterborough, PE4 6QZ
G4 WIP A. Crickett, 40 Ousden Close Cheshunt, Waltham Cross, EN8 9RQ
G4 WIR I. Page, 127 Whyke Lane, Chichester, PO19 8AU
G4 WIS V. Gleek, Fieldgate, The Warren, Radlett, WD7 7DU
G4 WIY A. Clark, Applecroft Care Home, Sanctuary Close, Chilton Way, Dover, CT17 0ER
G4 WIZ D. Burleigh, 39 Neville Close, Basingstoke, RG21 3HG
GM4 WJA J. Fraser, Cherrybrae Croft, Aultmore, Keith, AB55 6QU
G4 WJB R. Barratt, 37 Cemetery Road, Whittlesey, Peterborough, PE7 1RT
G4 WJE M. Fenelon, 72 Fieldside, Epworth, Doncaster, DN9 1DP
G4 WJG D. Nicolson, 142 Shireburn Caravan Park, Edisford Road, Clitheroe, BB7 3LB
G4 WJH P. Mathews, 1 Erith Road, Belvedere, DA17 6HB
G4 WJJ P. Short, Flat 29, Orchard Court, Searle Street, Crediton, EX17 2HA
G4 WJM W. Cooper, 32 High St., Thurlby, Bourne, PE10 0EE
G4 WJQ B. Blain, 31 Crest Road, Bromley, BR2 7JA
G4 WJR J. Singleton, 11 Lound Place, Lound Street, Kendal, LA9 7FE
G4 WJS W. Somerville, Glendella, Wycombe Road Stokenchurch, High Wycombe, HP14 3RP
G4 WJV J. Forrest, 3 Martindale Park, Houghton le Spring, DH5 8EX
G4 WJW T. Murphy, 7 The Knapp, Templecombe, BA8 0JP
G4 WJX M. Kessel, 4 Harington Drive, Stoke-on-Trent, ST3 5ST
G4 WJZ A. Kerr, Braemoray, Dalditch Lane, Budleigh Salterton, EX9 7AS
G4 WKB H. Poulton, 1 Marnhull Close, Coventry, CV2 2JS
G4 WKD K. Dunstan, 41 Gravel Lane, Wilmslow, SK9 6LS
G4 WKG G. Rowley, Glassonby Lodge, Glassonby, Penrith, CA10 1DT
GW4 WKQ L. Jones, 8 Tyn Y Pwll Estate, Llanbedrog, Pwllheli, LL53 7PG
G4 WKT N. Bleek, 49 Lowhills Road, Peterlee, SR8 2DJ
G4 WKW R. Benton, 15 Polventon Parc, St Keverne, TR12 6PB
GW4 WKZ G. Fitch, Tides Reach, 39 Hen Gei Llechi, Y Felinheli, LL56 4PB
G4 WLA D. Dell, Bushmead, 12 Penfield Gardens, Dawlish, EX7 9NQ
G4 WLE N. Slater, 17 Hall Park Drive, Lytham St Annes, FY8 4QR
G4 WLG K. Dunwell, 8 Violet Grove, Thatcham, RG18 4DQ
G4 WLI P. Nutt, 1 Lime Gardens, Middleton, Manchester, M24 4AE
G4 WLJ N. Bell, 16 Amersham Close, Urmston, Manchester, M41 7WH
G4 WLK M. Morgan, 6 Blakeley Heath Drive, Wombourne, Wolverhampton, WV5 0HW
G4 WLP S. Mccombe, 14 Broadlands Avenue, Bournemouth, BH6 4HQ
G4 WLS C. Smith, 15 Bearsdown Close, Plymouth, PL6 5TX
GW4 WLT J. Williams, 7 Tynewydd, Nantybwch, Tredegar, NP22 3SG
G4 WLV D. Gladwin, Dorset House, St. Annes Road, Eastbourne, BN21 2HR
G4 WMA Dr P. Haslam, 55 Jesmond Park West, Newcastle upon Tyne, NE7 7BX
G4 WMB W. Bell, Rhydd Gardens, Worcester Road, Hanley Castle, Worcester, WR8 0AB
GW4 WMD W. David, Sirmione, Lawrenny Road, Kilgetty, SA68 0SY
GI4 WME F. Hull, 44 Killynether Walk, Belfast, BT8 7DB
G4 WMF G. Blake, Flat 5, 46 Marlborough Road, Ipswich, IP4 5AX
GU4 WMG J. Gallienne, Westward, Rue Des Marettes, St. Martin, Guernsey, GY4 6JW
G4 WMH W. Hall, 45 Dorchester Road, Solihull, B91 1LN
GM4 WMM W. Mcmillan, Rennabreck, Rendall, Orkney, KW17 2EZ
G4 WMN J. Robb, 3 Silver Dell, Watford, WD24 5LT
G4 WMO P. Stainton, Fairview, Mareham On The Hill, Horncastle, LN9 6PQ
G4 WMP M. Bangle, 21 Oakhill Road, Addlestone, KT15 1DH
G4 WMQ A. Richardson, 1 Silverton Terrace, Rothbury, Morpeth, NE65 7QS
G4 WMV R. Bridge, 3 Upmeadows Drive, Stafford, ST16 1PA
G4 WMY G. Kay, High Trees, Stockland Bristol, Bridgwater, TA5 2PZ
G4 WMZ K. Law, 50 Main Street, Little Downham, CB6 2ST
G4 WNA H. Williams, 37 Mickledales Drive, Marske-By-The-Sea, Redcar, TS11 6DF
G4 WND R. Banks, Crows Nest, Churchstoke, Montgomery, SY15 6TP
G4 WNF F. Rhodes, 248 Woolwich Road, London, SE2 0DW
G4 WNG T. Furness, 129 North Ridge, Bedlington, NE22 6DF
GI4 WNH E. Loughran, 6 Oaklea Road, Magherafelt, BT45 6NH
G4 WNI J. Howarth, 80 John F Kennedy Estate, Washington, NE38 7AL
G4 WNP R. Tant, 34 Manor Road, Wheathampstead, St Albans, AL4 8JD
GM4 WNQ P. Ramsey, 1 Skye Place, Stevenston, KA20 3DG
G4 WNU J. Smith, George Bungalow, The Street, Axminster, EX13 7RW
G4 WNV Dr S. Robinson, 18 Headley Grove, Tadworth, KT20 5JF
G4 WNW T. Almond, Maranatha, Lumber Lane, Warrington, WA5 4AX
G4 WNZ M. Watson, 5A Clarendon Road, Shanklin, PO37 7AG
G4 WOB J. Bazyk, Heyford Cedars, Watling Street, Northampton, NN7 4SB
G4 WOD J. Sheppard, 37 Oakfield Road. Kingswood., Bristol, BS15 8NT
G4 WOE D. Hudson, 2 Muirfield Rise, St Leonards-on-Sea, TN38 0XL
G4 WOH P. Thwaytes, 1 Sunningdale, Waltham, Grimsby, DN37 0UA
G4 WOI R. Allen, 115 Trerice Drive, Newquay, TR7 2TE
G4 WOL R. Tenwolde, 376 Buxton Road, Macclesfield, SK11 7ES
G4 WOS D. Flello, 1 St. Andrews Way, Tilmanstone, Deal, CT14 0JH
GD4 WOW J. Jones, Ballagarrow, Glen Auldyn, Ramsey, Isle Of Man, IM7 2AF
GW4 WPA T. Leary, 21 Gelli Glas Road, Morriston, Swansea, SA6 7PS
G4 WPB P. Bruce, Seascape, Surf Crescent, Sheerness, ME12 4JU
G4 WPE M. Bland, 18 Hill Street, Newhall, Swadlincote, DE11 0JR
G4 WPG L. Hatton, 51 Castner Avenue, Weston Point, Runcorn, WA7 4EH
GW4 WPH S. Valentine, Unit 21, Industrial Estate, Bala, LL23 7NL
G4 WPI J. Fuller, 22 Simmance Way, Amesbury, Salisbury, SP4 7TD
G4 WPO D. Bevan, 32 Thorley Park Road, Bishop's Stortford, CM23 3NQ
G4 WPR D. Trotman, 71 Bexley Street, Windsor, SL4 5BX
G4 WPT D. Jackman, 27 Vicarage Road, Birmingham, B14 7QA
G4 WQB K. Hamlyn, Penlan, 1 Elm Tree Cottages, The Common, Winchmore Hill, Amersham, HP7 0PN
GW4 WQC D. Williams, 149 Rhiwr Ddar, Taffs Well, Cardiff, CF4 7PD
G4 WQD J. Jocys, 28 Vaudrey Drive, Timperley, Altrincham, WA15 6HQ

GM4 WQH J. Naughton, 124 Churchill Street, Alloa, FK10 2JU
G4 WQL M. Bender, Ivy Chimney Villa, Skinners Bottom, Redruth, TR16 5DT
G4 WQO P. Truitt, 2A Queens Gate Place, London, SW7 5NS
G4 WQS N. Reading, 30 Clifton Rise, Windsor, SL4 5TD
G4 WQT R. Watson, 158 Kingsfold Drive, Penwortham, Preston, PR1 9EQ
G4 WQU P. Barrett, 9 Mabena Close, St. Mabyn, Bodmin, PL30 3BS
G4 WQZ J. Wiles, 12A Ashling Gardens Denmead, Waterlooville, PO7 6PR
G4 WRA Wordsley ARC, c/o S. Sands, 33 High Street, Kinver, Stourbridge, DY7 6HF
G4 WRB K. Beech, 40 Star Street, Wolverhampton, WV3 9BL
G4 WRC Wanlip Rad Club, c/o A. Wheeler, 11 Barley Way, Rothley, Leicester, LE7 7RL
G4 WRD N. Underwood, 44 East View, Barnet, EN5 5TN
GI4 WRJ R. Jennings, 12 Garnerville Gardens, Belfast, BT4 2PA
G4 WRK I. Edwards, 9 Long Lane, Nr Wellington, Salop, TF6 6HH
GU4 WRP D. Fletcher, Celicia, 5 La Neuve Rue Estate, La Neuve Rue, St. Peter Port, St Peter Port, Guernsey, GY1 1SF
G4 WRQ D. Wring, 8A Rectory Road, Easton-In-Gordano, Bristol, BS20 0QB
GJ4 WRR F. Leighton, 4 Victoria Village, Estate Trinity, Jersey, JE4 9VI
G4 WRX D. Cherrington, 4 Bloomfield Close, Wombourne, Wolverhampton, WV5 8HQ
G4 WSB A. Bowditch, 28 Selby Crescent, Freshbrook, Swindon, SN5 8PE
G4 WSE T. Saggerson, 18 Ploughmans Way, Great Sutton, Ellesmere Port, CH66 2YJ
G4 WSF R. Smith, 37 Lyngford Road, Taunton, TA2 7EF
G4 WSH F. Blaxland, Seton Villa, Scorrier Road, Scorrier, TR16 5AA
G4 WSI A. Brown, 81 Ipswich Crescent, Great Barr, Birmingham, B42 1LY
G4 WSL R. Cable, 4A Ermine Close, St Albans, AL3 4JZ
G4 WSV I. Swancott, 4 Ripple Close, Shrewsbury, SY2 6LS
G4 WSW M. Heaver, 24 Artillery Terrace, Guildford, Surrey, GU1 4NL
G4 WTA M. Jones, 57 Mountway Road, Bishops Hull, Taunton, TA1 5DS
G4 WTD C. Flatman, 36 Skoner Road, Bowthorpe Industrial Estate, Norwich, NR5 9AX
G4 WTE M. Rye, 33A Darnley Street, Gravesend, DA11 0PH
GM4 WTK R. Fortune, Stewarton Lodge, Eddleston, Peebles, EH45 8PP
GU4 WTN A. Hamon, 22 Mount Row, St Peter Port, Guernsey, GY1 1NT
G4 WTQ N. Harvey, 5 Harvey Gardens, Loughton, IG10 2AD
G4 WTR A. Morris, 18 Saxton Avenue, Doncaster, DN4 7AX
GM4 WTS W. Stevenson, 11 West Drive, Airdrie, ML6 8BL
GI4 WTT T. Mcdonnell, 52 Moira Road, Glenavy, Crumlin, BT29 4JL
G4 WTU D. Pay, Longmeadow House, Dunsford, Exeter, EX6 7AD
G4 WTX V. Hansford, Whitehouse Farm, Hardway, Bruton, BA10 0RJ
G4 WTZ J. Hall, 23 St. James Avenue, Congleton, CW12 4DY
G4 WUA G. Brown, 13 Francis Avenue, Moreton, Wirral, CH46 6DH
G4 WUB D. Farr, 10 Yeomanside Close, Whitchurch, Bristol, BS14 0PZ
G4 WUG K. Medley, 3 Beck Lane, Horsham St. Faith, Norwich, NR10 3LD
G4 WUH I. Hopkins, Kempley Green, Dymock, Newent, GL18 2BW
G4 WUI J. Marr, 11 Morley Crescent, Kelloe, Durham, DH6 4NN
G4 WUJ N. Plant, 73 Robert Burns Avenue, Cheltenham, GL51 6NX
G4 WUK D. Dyer, 64 Churchill Close, Sturminster Marshall, Wimborne, BH21 4BH
G4 WUM F. Amos, 53 Valley View, Jarrow, NE32 5QT
G4 WUO P. Bullock, 4 Yarmouth Road, Blofield, Norwich, NR13 4JS
G4 WUQ P. Harding, Flat D, 106 Bushey Hill Road, London, SE5 8QQ
GM4 WUR C. Phillips, Lonnie Sanday, Orkney, KW17 2BA
G4 WUS W. Bingham, 16 Carlin Park, Carlin How, Saltburn-by-the-Sea, TS13 4DF
G4 WUU P. Williamson, The Laurels, Norwich Road, Cawston, NR10 4HA
G4 WUV C. Baker, 48 Hazell Road, Farnham, GU9 7BP
G4 WUW M. Baker, 48 Hazell Road, Farnham, GU9 7BP
G4 WUX P. Bourne, 6 Blythe Mount Park, Blythe Bridge, Stoke-on-Trent, ST11 9PP
GW4 WVB J. Williams, Arfryn, Windsor Road, Wrexham, LL14 1ST
G4 WVC M. Jones, 24 Whitford Road, Birkenhead, CH42 7JA
G4 WVD M. Bundy, 5 Dawe Crescent, Bodmin, PL31 1PY
G4 WVF C. Farley, 8 Church Road, Mellor, Stockport, SK6 5PR
G4 WVH T. Hathaway, 24 Oxford Meadow, Sible Hedingham, Halstead, CO9 3QN
GW4 WVK D. Davies, 20 Broadlands Way, Oswestry, SY11 2YD
G4 WVM M. Keating, 29 Pimm Road, Paignton, TQ3 3XA
GI4 WVN Prof. H. Gilbody, 5 The Plateau, Piney Hills, Belfast, BT9 5QP
GW4 WVO Wenvoe ARC, c/o P. Owen, 13 Highland Close, Sarn, Bridgend, CF32 9SB
G4 WVP F. Russell, 56 Gatesgarth Road, Middleton, Manchester, M24 4JJ
G4 WVQ T. Friesner, 8 Dolphin Close Fishbourne, Chichester, PO19 3QP
G4 WVR Welland Valley ARS, c/o S. Day, 14 The Crescent, Market Harborough, LE16 7JJ
G4 WVT J. Stageman, Sunray, Kennford, Exeter, EX6 7XS
G4 WVW J. Lane, 41 Ravenswood Crescent, Harrow, HA2 9JL
G4 WVY J. Joynt, Gurtymadden, Loughrea, Ireland, County Galway
G4 WWB W. Bennett, 61L Mansion Drive, Croxteth, Liverpool, L11 9DP
GI4 WWF V. Fails, 38 Fortsandel Avenue, Coleraine, BT52 1TL
G4 WWG Dr A. Brown, 44 Earlswood, Skelmersdale, WN8 6AT
G4 WWH Dr P. Pavelin, 7A Castletown, Portland, DT5 1BD
G4 WWL I. Rowe, 19 Poplar Avenue, Wetherby, LS22 7RA
GW4 WWN M. Rowles, 7 Gelli Deg, Bryncoch, Neath, SA10 7PL
G4 WWP D. Barry, Plough End, 8 Dell Lane, Bishop's Stortford, CM22 7SJ
G4 WWR Three Counties ARC, c/o D. Kamm, Delabole Head, Week St. Mary, Holsworthy, EX22 6UU
GM4 WWU R. Steel, 4 Howford Road, Nairn, IV12 5QP
G4 WWY P. Brown, White Cottage, Woodside, Epping, CM16 6LF
G4 WWZ A. Artt, 5 Barnard Field, Amesbury, Salisbury, SP4 7FF
G4 WXC S. Vaughan, 8 Letcombe Avenue, Abingdon, OX14 1EQ
G4 WXF R. Logan, 52 Holbeach Drive Kingsway, Quedgeley, Gloucester, GL2 2BF
G4 WXI F. Mckeown, 1 Thirlmere Road, Preston, PR1 5TR
G4 WXJ R. Harnett, 41 Stepney Road, Scarborough, YO12 5BT
G4 WXK G. Sears, 36 Cedars Road, Exhall, Coventry, CV7 9NJ
G4 WXO J. Pemberton, Dunkirk Cottage, Dunkirk Lane, Chester, CH1 6LU
GM4 WXQ W. Goudie, 5 North Lochside, Lerwick, Shetland, ZE1 0PA
G4 WXR B. Hayes, 363 Watnall Road, Hucknall, Nottingham, NG15 6EP
G4 WXT G. Shead, 37 Shalford Road, Rayne, Braintree, CM77 6BY
G4 WXX J. Charnock, 20 Clifton Road, Ashton-In-Makerfield, Wigan, WN4 0AZ
G4 WYC S. Powell, 10 Foresters Square, Bracknell, RG12 9ES
GI4 WYE P. Doran, 143 Gransha Road, Bangor, BT19 7RB
G4 WYF C. Ellison, 31 Dudley Avenue, Blackpool, FY2 0TU
G4 WYH A. Mcphail, 300 Fletcher Road, Preston, PR1 5HJ
G4 WYI A. Huff, 4 Greding Walk, Brentwood Essex, CM13 2UF
G4 WYL C. Levett, 5 Park Road, Yapton, Arundel, BN18 0JE
G4 WYN D. Harries, 1 St. Michaels Close, Ashby-de-la-Zouch, LE65 1ES
G4 WYO K. Brewer, 14 Poplar Road, Kensworth, Dunstable, LU6 3RS
G4 WYW M. Fisher, 147 Outer Circle, Southampton, SO16 5HB
GW4 WYX W. Thomas, 24 Pontneathvaughan Road, Glynneath, Neath, SA11 5NT
G4 WYZ M. Prescott, Rathgael, 44 Glamis Drive, Chorley, PR7 1LX
G4 WZA A. Nokes, 24 Braces Lane, Marlbrook, Bromsgrove, B60 1DY
G4 WZB H. Worley, 22 Cross Road, Wellingborough, NN8 4AT
GM4 WZD J. Nicholl, Trees, 7 Holmisdale, Glendale, Isle of Skye, IV55 8WS
GM4 WZG B. Mcintosh, 14 River View, Dalgety Bay, Dunfermline, KY11 9YE
G4 WZH A. Le Couteur Bisson, 36 Gibson Way, Porthleven, Helston, TR13 9AW
G4 WZJ M. Wiblin, 60 Shepherds Lane, Bracknell, RG42 2BT
G4 WZM S. Johnston, Burn Moor End Farm, Wheathead Lane, Nelson, BB9 6LD
GM4 WZP J. Gentles, 8 Leadervale Terrace, Edinburgh, EH16 6NX
G4 WZQ I. Smith, 24 Sea View Road, Herne Bay, CT6 6JA
GW4 WZS A. Glynn, Cartref, Llanfachraeth, Holyhead, LL65 4UY
G4 WZT N. Thomas, Flat 18, Livability, Hereford, HR4 9HP
G4 WZU L. Thompson, 12 Long St., Great Gonerby, Grantham, NG31 8LN
GM4 WZY B. Pennycook, 3 Pirnie Mill, Forfar, DD8 3ES
G4 WZZ B. Hunt, 5 Osprey Close, Whitstable, CT5 4DT
GI4 XAA C. Mccann, Drumbally Hue House, Rock, Tyrone, BT70 3JY
G4 XAB R. Hunt, 3 Osprey Close, Whitstable, CT5 4DT
G4 XAE A. Cole, 13 Centre Close, Beccles, NR34 9JJ
G4 XAG B. Mahany, 3 Portland Road, Frome, BA11 4JA
G4 XAH D. Mahany, 3 Portland Road, Frome, BA11 4JA
G4 XAL P. Lawrence, 4 Monkshood Close, Wokingham, RG40 5YE
G4 XAN C. Goddard, 75 Downs Road, Slough, SL3 7DA
G4 XAR M. Kearns, 16 Fieldton Road, Liverpool, L11 9AG
G4 XAT R. Evans, 7 Westland Drive, Bromley, BR2 7HE
GW4 XAU J. Rutkowski, 7 Beach Road, Holyhead, LL65 1ES
GM4 XAV J. Stevens, 10 Tiel Path, Glenrothes, KY7 5AX
GM4 XAW P. Nelson, Croc Ard, Botany Street, Newton Stewart, DG8 9JG
GW4 XAZ I. Mitchell, 18-19 Hendre-Wen Road, Blaencwm, Treorchy, CF42 5DR
G4 XBC A. Turner, 19 Trelawney Road, St Austell, PL25 4JA
G4 XBD G. Nash, 36 Lynton Avenue, Arlesey, SG15 6TS
G4 XBF M. Ray, Willow Mead House, Willow Mead, Godalming, GU8 5NR
G4 XBG K. Murphy, 34 Hawkenbury Way, Lewes, BN7 1LT
G4 XBI D. Parslow, 1 Willington Close, Harlescott, Shrewsbury, SY1 3RH
G4 XBJ P. Kemp, 9 Moorfield Way, Wilberfoss, York, YO41 5PL
G4 XBS C. Smith, 1 Langley Court, St Ives, PE27 5WX
G4 XBU J. Atkinson, 8 Woodcock Road, Flamborough, Bridlington, YO15 1LJ
G4 XBW R. Head, The White House, School Lane, St Austell, PL25 3TJ
G4 XBX A. Wallman, 8 Oakbank Avenue, Manchester, M9 4EX
G4 XBZ Dr A. Roberts, Apartment 17, Fleur De Lis Duttons Road, Romsey, SO51 8LH
G4 XCB K. Rook, 232 Wick Road, Brislington, Bristol, BS4 4HN
G4 XCE A. Tamplin, Browtop, Old Lane, Crowborough, TN6 2AD
G4 XCK S. Boden, 14 Potters Way, Ilkeston, DE7 5EX
G4 XCM J. Tavener, The Cube, North Drive, Wirral, L60 0BD
G4 XCQ L. Prescott, 58 Blenheim Road, Ashton-In-Makerfield, Wigan, WN4 9JN
G4 XCR G. Wood, The Old Corner Smithey, New Road, Dereham, NR20 5TA
G4 XCV R. Barnett, 7 Chapel Terrace, Pendeen, Penzance, TR19 7SY
G4 XCX C. Clarke, 33 James Road, Kidderminster, DY10 2TP
G4 XCY F. Mills, 14 Seagram Close, Aintree, Liverpool, L9 0NA
G4 XDB A. Parry, 189 Kimbolton Road, Bedford, MK41 8DR
G4 XDC M. Taylor, 26 Summerfield Place, Wenlock Road, Shrewsbury, SY2 6JX
G4 XDE S. Crosskey, 25 Meadow Gardens, Baddesley Ensor, Atherstone, CV9 2DA
G4 XDG D. Humphreys, 129A Chester Road, Northwich, CW8 4AA
G4 XDH D. Humphreys, 15 Burton Road, Warrington, WA2 9AJ
G4 XDJ B. Fields, 64 Collins Street, Waikouaiti, New Zealand, 9510
G4 XDK N. Heasman, 5 Hackneys Corner Great Blakenham, Ipswich, IP6 0JQ
G4 XDL M. Norman, 52 Turkdean Road, Cheltenham, GL51 6AL
G4 XDM S. Comis, 178 Lordswood Road, Birmingham, B17 8QH
G4 XDP P. Knowles, 18 Brookside, Pill, Bristol, BS20 0JX
GW4 XDR A. Walker, Stanley Cottage, Station Road, Wrexham, LL13 0LJ
G4 XDT L. Booth, 40 St. Georges Road, New Mills, High Peak, SK22 4JT
G4 XDU D. Chislett, Hilltops, 2A St. Marks Road, Maidenhead, SL6 6DA
G4 XDV R. Hall, 9 Stone Court, South Hiendley, Barnsley, S72 9DL
G4 XDW A. Chidwick, 4 Burgess Close, Whitfield, Dover, CT16 3NP
G4 XDX S. Garbett, 2 Redruth Court, Launceston Road, Wigston, LE18 2FU
GU4 XEA P. Carre, La Petite Miellette, La Miellette Lane, Vale, Guernsey, GY3 5EN
G4 XED S. Cowdell, 6 Pearl St., Bedminster, Bristol, BS3 3EA
G4 XEE D. Bate, 15 Martins Drive, Ferndown, BH22 9SG
GW4 XEF B. Passmore, 16 Epworth Road, Rhyl, LL18 2NU
G4 XEI D. Mcloughlin, 23 St. Marys Court, Clayton Le Moors, Accrington, BB5 5LA
G4 XEJ A. Allen, 86 Grayswood Park Road, Quinton, Birmingham, B32 1HE
G4 XEL S. Evans, 72 Sandown Road, Toton, Nottingham, NG9 6JW
G4 XEO D. Holmes, 32 Eastcheap, Rayleigh, SS6 9JZ
GW4 XES D. Johns, 16 Maes Yr Haf, Llansamlet, Swansea, SA7 9ST
G4 XET J. Rawlinson, Hollydene, Newbiggin, Penrith, CA10 1TA
G4 XEW I. Rosenberg, 11 Parkside Drive, Edgware, HA8 8JU
G4 XEX P. Rivers, 34 Coales Gardens, Leicestershire, LE16 7NY
G4 XEZ A. Smith, 3 The Fold, Penn, Wolverhampton, WV4 5QY
G4 XFA R. Ball, 4 The Brow Hesketh Bank, Preston, PR4 6SJ
G4 XFC J. Fulton, 8 Park View Legsby, Market Rasen, LN8 3QP
GI4 XFE A. Calvin, 20 Orangefield Crescent, Armagh, BT60 1DS
G4 XFF J. Holdsworth, 37 Harewood Crescent, Old Tupton, Chesterfield, S42 6HS
G4 XFG N. Goodman, 3 Fury Avenue, Grimoldby, Louth, LN11 8UN
G4 XFM D. Steer, 24 Manor Drive, Ivybridge, PL21 9BD
GI4 XFN G. Smith, 4 Conway Court, Belfast, BT13 2DR
GI4 XFS C. Gilbody, 5 The Plateau, Piney Hills, Belfast, BT9 5QP
G4 XFT J. Tranter, 275 Bosty Lane, Aldridge, Walsall, WS9 0QE
GM4 XFU W. Davidson, 31 Glenmuir Crescent, Logan, Cumnock, KA18 3EY
G4 XFV A. Mckechnie, 2 Batts Pond Lane, Dropping Holms, Henfield, BN5 9YU
GI4 XFX R. Reid, 6 Sperrin Heights, Townhill Road, Ballymena, BT44 8AD
GI4 XFY E. Townley, 27 Windmill Road, Kilkeel, Newry, BT34 4LP
G4 XFZ R. Griffin, 53 St. Johns Avenue, Warley, Brentwood, CM14 5DG
G4 XGD P. Harman, 25 Pitts Road, Slough, SL1 3XG

G4 XGI R. Hales, 239 Charlton Road, Shepperton, TW17 0SH
G4 XGN P. Riggott, 1 Mill Lane Queensbury, Bradford, BD13 1LP
GI4 XGO G. Armstrong, 45 Rathmena Drive, Ballyclare, BT39 9HZ
G4 XGP M. Kelly, Birkby Lodge, Brickley Park Road, Kent, BR1 2AT
GI4 XGQ T. Devine, 141 Longland Road, Dunamanagh, Strabane, BT82 0PP
G4 XGR S. Clark, 1 Holcroft, Orton Malborne, Peterborough, PE2 5SL
G4 XGT J. Dawson, 21 Church Street, Needingworth, St Ives, PE27 4TB
GM4 XGY G. Smith, 1/2 1 Seres Court, Clarkston, Glasgow, G76 7PL
G4 XHC F. Jackson, 34 High Street, Blyton, Gainsborough, DN21 3JY
G4 XHE R. Cook, 7 New Road, Worthing, BN13 3JG
GM4 XHH R. Franklin, 8 Hawdene, Broughton, Biggar, ML12 6FW
G4 XHK L. Soutter, 2 Hyde Barton, Churchill Way, Northam, Bideford, EX39 1NX
GI4 XHO F. Orr, 29A Mccraes Brae, Whitehead, Carrickfergus, BT38 9NZ
G4 XHP D. Daniels, 40 Rennie Street, Dean Bank, Ferryhill, DL17 8NG
GM4 XHQ G. Mcinnes, 14 East Croft, Ratho, Newbridge, EH28 8PD
G4 XHT E. Wilkinson, 30 Old Bridge End, Roach Street, Blackford Bridge, Bury, BL9 9TA
GM4 XHV G. Horsburgh, 3 Dumyat Road, Alva, FK12 5NN
G4 XHX M. Powers, 12 Roman Avenue North, Stamford Bridge, York, YO41 1DP
G4 XHZ F. Jolley, 30 Oban Drive, Shadsworth, Blackburn, BB1 2HY
G4 XIE R. Shard, 76 Clipsley Lane, Haydock, St Helens, WA11 0UB
G4 XIL B. Hurst, 25 Hoadly Road, Cambridge, CB3 0HX
G4 XIM R. Bradfield, 118 East Road, Langford, Biggleswade, SG18 9QP
G4 XIN A. Henstock, 16 The Coppice, Enfield, EN2 7BY
G4 XIP J. Baker, 11 London Road, Old Basing, Basingstoke, RG24 7JE
GI4 XIR W. Bird, 198 Ashmount Gardens, Lisburn, BT27 5DB
GU4 XIT R. Bird, Redroof, La Mare De Carteret, Guernsey, GY5 7XD
G4 XIU M. Kelleway, Greenhills, Newport Road, Ventnor, PO38 2QW
G4 XIW S. Mackenzie, 6 Bridge Farm Close, Grove, Wantage, OX12 7QF
G4 XIX M. Purnell, The Olde Cottage, Lewdown, Okehampton, EX20 4DQ
G4 XIZ R. Heath, 9 Woodside Lane, Leek, ST13 7AN
GI4 XJD J. Doherty, 75 Drumflugh Road, Benburb, Dungannon, BT71 7QF
G4 XJE D. Brawn, 16 Mansel Close Cosgrove, Milton Keynes, MK19 7JQ
GM4 XJF J. Park, The Stables, Whiting Bay, Isle of Arran, KA27 8QH
G4 XJG R. Hirst, 47A Rowley Lane, Fenay Bridge, Huddersfield, HD8 0JG
GW4 XJK R. King, 30 Railway Terrace, Llanelli, SA15 2RH
G4 XJL D. Rogers, Gabwell House, Stokeinteignhead, Newton Abbot, TQ12 4QS
G4 XJN J. Williamson, 12 Honeysuckle Road Widmer End, High Wycombe, HP15 6BW
G4 XJS J. Smith, 84 Oakwood Drive, St Albans, AL4 0XA
GM4 XJY D. Mcminn, Crestholme, East Bay, Mallaig, PH41 4QF
G4 XKA P. Adams, 19 Thistledown, Tilehurst, Reading, RG31 5WE
G4 XKC A. Sieroslawski, 8 Poot Hall, Dewhirst Road, Rochdale, OL12 0AS
G4 XKD K. Dixon, 23 Dorking Walk, Corby, NN18 9JL
G4 XKF D. Browne, 67 Benfield Way, Portslade, Brighton, BN41 2DN
GI4 XKI J. O'Neill, 225 Dungannon Road, Killeshill, Dungannon, BT70 1TH
G4 XKK K. Burston, Can Singala, Hope Corner Lane, Taunton, TA2 7PB
G4 XKL R. Whetton, 117 Tutbury Road, Burton-on-Trent, DE13 0NU
G4 XKM P. Andrews, 88 Connegar Leys, Blisworth, Northampton, NN7 3DF
GM4 XKP K. Macgillivray, 87 Castle St., Forfar, DD8 3AG
G4 XKR R. Coward, 10 Market Street, Hambleton, Poulton-le-Fylde, FY6 9AP
G4 XKV G. Pigott, 67 Mayplace Road West Bexleyheath, Kent. Da7 4Jl, Bexleyheath, DA7 4JL
G4 XKZ M. Collins, 39 Denver Road, Dartford, DA1 3JU
G4 XLA T. Carruthers, Flat 43, House 119, St Petersberg, Russian Federation, 193024
GI4 XLB Sunspots RAC, c/o G. Curry, 87 Burren Road, Ballynahinch, BT24 8LF
G4 XLC E. Metcalfe, 18 Kirkstone Drive, Morecambe, LA4 5XP
G4 XLG M. Rollings, 39 Summerleys, Edlesborough, Dunstable, LU6 2HR
G4 XLM J. Todd, Dorothy House, 127 Dorothy Avenue North, Peacehaven, BN10 8DS
GM4 XLN J. Durrand, 9 Breadalbane Crescent, Wick, KW1 5AS
G4 XLO K. Tatlow, 24 Princes Road West, Torquay, TQ1 1PB
GM4 XLU E. Wallace, 10 Gean Court, Cumbernauld, Glasgow, G67 3LU
G4 XLY E. Grint, 15 Ivythorn Road, Street, BA16 0TE
G4 XMA B. Easton, 8 Church View Road, Camborne, TR14 8RQ
GM4 XMD W. Mcdicken, 4 Baillie Drive, Logan, Cumnock, KA18 3HS
G4 XME L. Shone, 3 Ascot Drive, Dudley, DY1 2SN
G4 XMJ G. Wiggins, Cherry Trees Thorney Road, Emsworth, PO10 8BN
G4 XML P. Glydon, 24 Imperial Road, Knowle, Bristol, BS14 9ED
G4 XMO S. Ashfield, 28 Long Grove, Baughurst, Tadley, RG26 5NY
G4 XMP C. Balderston, 22A Maplecresent, Basingstoke, RG215SS
G4 XMQ T. Cooling, 17 Hawthorn Avenue Cherry Willingham, Lincoln, LN3 4JS
GW4 XMR M. Richardson, 4 Waterloo Road, Ammanford, SA18 3SF
G4 XMS C. Munton, 86 Amsbury Road, Hunton, Maidstone, ME15 0QH
GW4 XMU D. Jones, 30 Sorrell Drive, Penpedairheol, Hengoed, CF82 8LA
G4 XMX L. Johnson, 6 Hurst Court, Bunbury, Tarporley, CW6 9QX
G4 XMY J. Colson, 718 East Buckingham Drive, Lecanto, USA, 34461
G4 XMZ P. Toms, Longfleet, Shipton Lane, Bridport, DT6 4NQ
G4 XNA A. Willis, 16 Redwood Close, Lymington, SO41 9LT
GM4 XND W. Clark, 173 Dunnikier Road, Kirkcaldy, KY2 5AD
G4 XNE F. Handy, 429 Penn Road, Penn, Wolverhampton, WV4 5LN
G4 XNF J. Cameron, 23 Farley Crescent, Oakworth, Keighley, BD22 7SH
G4 XNK H. Johnson, 7, Deer Close, Walsall, WS3 3EA
G4 XNO M. Goodearl, Glenhurst, Wood Lane, Dartmouth, TQ6 0DP
G4 XNP D. Rayner, 69 Saracen Road, Hellesdon, Norwich, NR6 6PB
GM4 XNQ D. Muir, 28 Grange Crescent, Edinburgh, EH9 2EH
G4 XNR P. Morris, The Overlands, Church Minshull, Nantwich, CW5 6DX
G4 XNS N. Speak, Le Bois Trainard, Lizant, France, 86400
G4 XNV D. Owen, 39 Smithford Walk Tarbock Green, Prescot, L35 1SF
G4 XNW J. Simmonds, 19 Red Admiral Apartments, Worcester Street, Stourbridge, DY8 1AJ
GD4 XOD W. Jones, Ballanard Road, Onchan, Douglas, IM4 5EA
G4 XOE M. Swaby, 16 Daimler Avenue, Herne Bay, CT6 8AE
G4 XOG C. Wood, 9 Tamar Close, Walsall, WS8 7LH
G4 XOH D. Blackwell, 10 High Oaks Gardens, Bournemouth, BH11 9LJ
G4 XOJ N. Wade, 6 Aisthorpe, Capel St. Mary, Ipswich, IP9 2HT
G4 XOL M. Osborne, 27 Silverdale Road, Newton-le-Willows, WA12 0JT
G4 XOM R. Egan, 56 Walker Avenue, Stourbridge, DY9 9EL
G4 XOP T. Cooper, 55 Meadway, St Austell, PL25 4HT
G4 XOW D. Lomas, Galmpton, Cannon Lane, Maidenhead, SL6 3NR
G4 XPD K. Nicholls, 2 Sherrier Way, Lutterworth, LE17 4NW
G4 XPI P. O'Dea, 5 Matthews Court, Blackpool, FY4 2BT
G4 XPJ A. Gridley, 13 Brockwell, Oakley, Bedford, MK43 7TD
G4 XPP J. Davies-Bolton, 39 Newholme Estate, Station Town, Wingate, TS28 5EJ
G4 XPT A. Fernandez, 2 Silverston Ave, Bognor Regis, PO21 2RB
G4 XPU M. Bennett, 7 Woburn Avenue, Firwood Ind Est, Bolton, BL2 3AY
G4 XPV P. Maisey, 155 Parkfield Drive, Birmingham, B36 9TY
G4 XPY D. Fuller, 51 Evenlode, Banbury, OX16 1PQ
G4 XQA K. James, 6 Holly Grove, Paddington, Warrington, WA1 3HB
G4 XQB T. Rowe, 1 Mere Road, Marston, Northwich, CW9 6DR
G4 XQD D. Cast, 100 Priory Court, Priory Park, Ipswich, IP10 0JX
G4 XQE A. Turner, 4 Taylor Road, Ashtead, KT21 2HY
G4 XQF N. Clacher, 3 Annan Crescent, Marton, Blackpool, FY4 4RQ
G4 XQG K. Icke, 85 Evendene Road, Evesham, WR11 2QA
GM4 XQJ B. Waddell, 3A Polmont Road, Laurieston, Falkirk, FK2 9QQ
G4 XQQ J. Freeman, 5A Beech Avenue, Briar Bank Park, Bedford, MK45 3WE
G4 XQV T. Sismey, Southland House, 1B West End Lane, Leeds, LS18 5JP
G4 XQX D. Oliver, 6 Kensington Road, Gosport, PO12 1QY
G4 XQY A. Clack, 1 Meadow View, Southwell, NG25 0EQ
G4 XQZ J. Fisher, 6 Castle Way, Havant, PO9 2RZ
G4 XRA R. Avery, 64 Burnmill Road, Market Harborough, LE16 7JF
G4 XRB J. Gagg, 20 Stanstead Avenue, Tollerton, Nottingham, NG12 4EA
G4 XRD G. Pope, 16 Catchpole Close, Corby, NN18 8DE
G4 XRG E. Godlieb, 4 Tytherington Park Road, Macclesfield, SK10 2EL
G4 XRJ J. Mills, Aquila, 4 Westhill Road South, Winchester, SO21 3HP
G4 XRK J. Lord, 16 Lark Valley Drive, Fornham St. Martin, Bury St Edmunds, IP28 6UG
G4 XRM B. Foster, 38 Oakfield Road, Street, BA16 0RE
G4 XRO S. Hill, 32 Hunters Croft, Haxey, Doncaster, DN9 2NX
GM4 XRP J. Porter, 1 Loney Crescent, Denny, FK6 5EG
G4 XRR M. Willgoss, 28 Southdown Avenue, Weymouth, DT3 6HS
GM4 XRT M. Taylor, The Old Croft, Forse, Lybster, KW3 6BX
G4 XRV R. Bullock, Putnams, Hawridge, Chesham, HP5 2UQ
GW4 XRW R. Wood, Bwlcyn, Eifl Road, Caernarfon, LL54 5HG
G4 XRX R. Headland, 18 Blucher Street, Liverpool, L22 8QB
GM4 XRY A. Rimmer, 16 Johnston Drive, Barassie, Troon, KA10 6SD
G4 XSA A. Boniface, 33 Caraway Place, Wallington, SM6 7AG
G4 XSB W. Bridgen, 11 Turnesc Grove, Thurnscoe, Rotherham, S63 0TY
G4 XSC G. Trim, 731 Dorchester Road, Weymouth, DT3 5LF
GI4 XSF M. Stevenson, 69 Portaferry Road, Cloughey, Newtownards, BT22 1HP
G4 XSG S. Stuart, 102 Mitton Road, Whalley, Clitheroe, BB7 9JN
G4 XSI J. Saunders, Top Hill Farm, Woodside Green, Maidstone, ME17 2ET
G4 XSM G. Davey, 49 Maltward Avenue, Bury St Edmunds, IP33 3XQ
G4 XSR Burton on Trent Wireless Club, c/o M. Ohara, 36 Balance Hill, Uttoxeter, ST14 8BT
G4 XST P. Cheeseman, 10 Limden Close, Stonegate, Wadhurst, TN5 7EG
GW4 XSX M. Tovey, Frondewi, Aberaeron, SA46 0JS
G4 XTA P. Godolphin, Shepherds Cottage, Flakebridge, Appleby-in-Westmorland, CA16 6JZ
GI4 XTC W. Armstrong, 8 Killowen Crescent, Lisburn, BT28 3DS
G4 XTE J. Johnson, Winterwood, West Lodge Crescent, Huddersfield, HD2 2EH
G4 XTF N. Hancocks, Wesley House Allensmore, Hereford, HR2 9BE
G4 XTG I. Brown, 453 Blackburn Road, Edgworth Turton, Bolton, BL7 0PW
G4 XTK A. Kurnatowski, 24 Eversleigh Rise, Darley Bridge, Matlock, DE4 2JW
G4 XTO W. Reade, 106 Wellington Road, Bollington, Macclesfield, SK10 5HT
G4 XTR N. Hearn, Horsebrook Farm, South Brent, TQ10 9EU
G4 XTS J. Strutt, Woodland, Gardiners Lane North, Billericay, CM11 2XE
GD4 XTT W. Brown, Cleckheaton, Ballaragh, Laxey, Isle Of Man, IM4 7PW
G4 XTU J. Jones, 3 Blackstope Lane, Retford, DN22 6NW
G4 XTW A. Bowes, 3 Cameron Mews, Mill St., Bury St Edmunds, IP28 7DP
G4 XTX C. Cooper, 31 Beacon Park Drive, Skegness, PE25 1HE
G4 XTZ A. Taylor, 36 Bodmin Avenue, Slough, SL2 1SL
G4 XUA R. Walton, 275 Ridgacre Road, Quinton, Birmingham, B32 1EG
GW4 XUE D. Thomas, 5 Parcydelyn, Carmarthen, SA31 1TS
GW4 XUG G. Patterson, 10 Clettwr Terrace Pontsian, Llandysul, SA44 4TU
G4 XUI J. Gordon, 54 Guibal Road, London, SE12 9LX
GM4 XUJ K. Traill, 57 Ashfield Drive, Dumfries, DG2 9BP
GD4 XUM M. Platt, Moaney Moar House Ronague Road, Ballasalla, Isle Of Man, IM9 3BA
G4 XUQ S. Winters, 16 Rushton Grove, Harlow, CM17 9PR
GM4 XUS G. Smith, 80 Deanburn Park, Linlithgow, EH49 6HA
G4 XUV D. Bevan, 46 Park Lane, Hartford, Northwich, CW8 1PY
G4 XUW D. Hudson, 54 Montfitchet Walk, Stevenage, SG2 7DT
G4 XUZ R. Chandler, 43 The Drive, Shoreham-by-Sea, BN43 5GD
G4 XVE J. Francis, Pintail Cottage, St. Helena Westleton, Saxmundham, IP17 3ED
GM4 XVF A. Henk, 3 Well Road, Tweedmouth, Berwick-upon-Tweed, TD15 2BB
G4 XVH D. Van Haaren, 7 Middle Boy, Abridge, Romford, RM4 1DT
G4 XVI J. Ames, 58 The Street, Ringland, Norwich, NR8 6AB
G4 XVM M. Brett, 25A First Avenue, Galley Hill, Waltham Abbey, EN9 2AL
G4 XVO Dr S. Bate, High Trees, Station Road, Much Hadham, SG10 6AX
G4 XVP P. Hart, 4 Kings Ride, Penn, High Wycombe, HP10 8BL
G4 XVS K. Hughes, Annies Cottage, Gravel Walk, Malpas, SY14 8JQ
G4 XVV E. Davies, 11 Herons Close, Fareham, PO14 2HA
G4 XVW I. Dobson, Pine View, Forest Dale Road, Marlborough, SN8 2AS
G4 XVY D. Bastin, 94 Clyfton Close, Broxbourne, EN10 6NY
G4 XWA M. Cohen, 7 Northdale Park, Swanland, North Ferriby, HU14 3RH
GW4 XWC W. Crooks, 52 St. Catherines Road, Baglan, Port Talbot, SA12 8AS
G4 XWD J. Cookson, Arcadia. 11 St Winefrides Road, Littlehampton, BN17 5HA
G4 XWE L. Perrett, 1 Churchill Close, Wells, BA5 3HY
GD4 XWF J. Harrison, 33 Tynwald Close, St. Johns, Douglas, Isle Of Man, IM4 3LZ
GM4 XWL S. Gaw, 10 Scotstoun Park, South Queensferry, EH30 9PQ
G4 XWM F. Walton, 36 Cranfield Road, Milton Keynes, MK17 8AW
GW4 XWN H. Walker, 46 Golden Grove, Rhyl, LL18 2RS
G4 XWP C. Boyce, 41 Furlong Close, Buckfast, Buckfastleigh, TQ11 0ER
G4 XWQ D. Cottle, The Brambles, Landkey Road, Barnstaple, EX32 9BW
G4 XWR P. Grainger, 26 Beattie St., South Shields, NE34 0NJ

GM4 XWS D. Munro, Eriskay, 4 Boswell Crescent, Inverness, IV2 3ET
G4 XWT F. Donachie, 57 Avon Road West, Christchurch, BH23 2DF
G4 XWW G. Winyard, 76 West Elloe Avenue, Spalding, PE11 2BJ
G4 XWZ D. Lerner, 6 Willow Road, Kings Stanley, Stonehouse, GL10 3HS
G4 XXA F. Mills, 66 Beeches Road, Charlton Kings, Cheltenham, GL53 8NQ
G4 XXB E. Mills, 66 Beeches Road, Charlton Kings, Cheltenham, GL53 8NQ
G4 XXD S. White, 15 Spurway Road, Canal Hill, Tiverton, EX16 4ER
GW4 XXF B. Morris, 62 Gerllan, Tywyn, LL36 9DE
G4 XXG Stockton & District ARG, c/o S. Bourne, 1 Humewood Grove, Stockton-on-Tees, TS20 1JU
G4 XXH R. Miles, Lone Oak, Clappers Farm Road, Silchester, Reading, RG7 2LH
G4 XXI G. Lee, 5 Morton, Tadworth, KT20 5UA
GW4 XXJ N. Jones, Hillesley, Montpellier Park, Llandrindod Wells, LD1 5LW
G4 XXK D. Hart, 52 Scalwell Lane, Seaton, EX12 2DJ
G4 XXM D. Frederick, 16 Phoenix Drive, Eastbourne, BN23 5PG
GM4 XXO I. Carbry, 24 Craigenhill Road, Kilncadzow, Carluke, ML8 4QT
GW4 XXP J. Bancroft, 101 Meliden Road, Prestatyn, LL19 8LU
G4 XXS G. Cooper, 44 Nursery Close, Hucknall, Nottingham, NG15 6DQ
G4 XXT J. Cassidy, 137 Heath Park Road, Gidea Park, Romford, RM2 5XJ
G4 XXW J. Groeger, Waldweg 11, Schneverdingen, Germany, D-29640
G4 XXX F. Pullen, 35 Berrycroft, Willingham, Cambridge, CB24 5JX
G4 XXZ D. Palfreman, 43 Southfield Close, Scraptoft, Leicester, LE7 9UR
G4 XYB M. Kingdon, Watersmeet Cottage, Brewers Lane, Calne, SN11 8EZ
G4 XYC A. Reynolds, 90 Windfield, Leatherhead, KT22 8UJ
G4 XYD R. Young, 5 Edge Hill, Chellaston, Derby, DE73 6RP
G4 XYG S. Randall, 129 Ryeland Way, Augusta Park, Andover, SP11 6RH
G4 XYH W. Stock, The Cottage, Hollow Road, Winscombe, BS25 1TG
GW4 XYI P. Coombs, 28 Cae Braenar, Holyhead, LL65 2PN
G4 XYK P. Mitchell, 19 Ashbourne Avenue, Whetstone, London, N20 0AL
G4 XYM J. Hoskins, 37 Green Close, Didcot, OX11 8TE
G4 XYN R. Savin, 7 Bannard Road, Maidenhead, SL6 4NG
G4 XYP R. Jobes, 11 Eglinton Street North, Sunderland, SR5 1DY
G4 XYR W. Clarkson, 5A Sutton Avenue, Bradford, BD2 1JP
G4 XYS J. Mundy, 19 Brickfield Grove, Halifax, HX2 9AZ
G4 XYU P. Davis, 5 Well Close, New Milton, BH25 6TA
G4 XYW A. Pevy, 2 Oaktree Way, Sandhurst, GU47 8QS
G4 XYY J. England, 2 Clifford Road Bramham, Wetherby, LS23 6RN
G4 XZA E. Wardle, 57 Brook View Drive, Keyworth, Nottingham, NG12 5RA
G4 XZC P. Gass, 7 Chipperfield Close, Upminster, RM14 3EA
G4 XZF B. Irwin, Rockside, Frog Lane, Braunton, EX33 1BB
G4 XZG P. Lawton, 5 Belvedere Gardens, Leeds, LS17 8BS
G4 XZI G. Hall, 22 Templenewsam View, Leeds, LS15 0LW
GW4 XZJ H. Jones, Hafan, 7 Tan Y Bryn Street, Tywyn, LL36 9UY
G4 XZK S. Bond, 12 Richmond Road, Farsley, Pudsey, LS28 5DY
G4 XZM K. Pickles, 79 Mill Lane, Hanging Heaton, Batley, WF17 6DZ
GM4 XZN J. Macdonald, 15 Muir Wood Drive, Currie, EH14 5EZ
GW4 XZP E. Wood, Bwlcyn, Eifl Road, Caernarfon, LL54 5HG
G4 XZS R. Weston, 2 Gill Park, Efford, Plymouth, PL3 6LX
G4 YAA S. Barnwell, 4 Railway Cottages, Reforne, Portland, DT5 2AR
GM4 YAA J. Sinclair, 3 Ben More Drive, Paisley, PA2 7NU
G4 YAB J. Livesley, 79 Mellor Road, New Mills, High Peak, SK22 4DP
GM4 YAC N. Howarth, Kilbeg House Teangue, Isle of Skye, IV44 8RQ
G4 YAF A. Trudgen, 14 Park An Pyth, Pendeen, Penzance, TR19 7ET
G4 YAH G. Warnes, 20 Clivedon Way, Halesowen, B62 8TB
G4 YAJ S. Woodhead, 804 Huddersfield Road, Dewsbury, WF13 3LZ
G4 YAK C. Dobinson, 37 Ladram Road, Thorpe Bay, Southend-on-Sea, SS1 3PX
G4 YAL S. Tuffin, 21 Garraways, Royal Wootton Bassett, Swindon, SN4 8NQ
G4 YAM C. Atkin, 23 Brewster Avenue, Immingham, DN40 1DW
G4 YAN R. Page, 26 Colne Road, High Wycombe, HP13 7XN
G4 YAP G. South, 9 Common Lane Royston, Barnsley, S71 4JA
G4 YAQ B. Setter, Briarwood Alexandra Road, Crediton, EX17 2DH
G4 YAS E. Lucas-Davis, 27 Cadbury Road, Sunbury-on-Thames, TW16 7NA
GM4 YAT T. Turner, 5 Braemore Place, Fort William, PH33 6HX
GM4 YAU W. Scott, Garden House, Fetternear, Inverurie, AB51 5LY
G4 YAV R. Dixon, 91 Bondicar Terrace, Blyth, NE24 2JR
G4 YAX D. Diss, 130 Beridge Road, Halstead, CO9 1JU
G4 YAZ H. Sheer, Sea Echo, 53 Leonard Road, New Romney, TN28 8RX
G4 YBA G. Collis, 13 Westbrook Close, Horsforth, Leeds, LS18 5RQ
G4 YBD G. Reed, 6 Tentergate Close, Knaresborough, HG5 9BJ
G4 YBG A. White, Northdale, Goughs Lane, Bracknell, RG12 2RA
G4 YBH B. Hawkins, Andorra, Haw Lane, High Wycombe, HP14 4JG
G4 YBI P. Aust, 28 The Green, Wennington, Rainham, RM13 9DX
G4 YBJ J. Davies, 8 Randle Meadow Court, Great Sutton, Ellesmere Port, CH66 2BL
GJ4 YBM A. Alexandre, Merryvale Cottage, La Vallee De St. Pierre, St Lawrence, Jersey, JE3 1EZ
G4 YBN I. Ansell, 9 Sewell Harris Close, Harlow, CM20 3HB
G4 YBP P. Darcy, Cherry Blossom Cottage, Hunts Lane, Netherseal, DE12 8BJ
G4 YBS Morecambe Bay ARS, c/o D. Jackson, 6 Bolton Lane, Carnforth, LA5 8BL
G4 YBT E. Tracey, 100 Booth Close, Kingswinford, DY6 8SP
GU4 YBW P. Wadley, Gironde, Lorier Lane, Vale, Guernsey, GY3 5JG
G4 YBX E. Scleparis, 8 Devonshire Park, Reading, RG2 7DX
G4 YCD M. Lowe, Crossley Farm, Bristol, BS17 1RH
G4 YCE L. Ball, 16 Kelston View, Whiteway, Bath, BA2 1NW
G4 YCG C. Beeston, 14 Valley Close, Waterlooville, PO7 5DX
GW4 YCJ A. Clift-Jones, Coed Tew Mill, Nant Glas, Llandrindod Wells, LD1 6PD
GW4 YCO D. Gill, 19 Rowling St., Williamstown, Tonypandy, CF40 1QY
G4 YCP G. Newman, 2 Grange Road, Eldwick, Bingley, BD16 3DH
G4 YCS I. Carby, Sunny Bank, North End Road Yapton, Ford, Arundel, BN18 0DH
G4 YCV C. Vickery, 7 Higher Redgate, Tiverton, EX16 6RJ
G4 YCW C. Croxford, Bodley Cottage, Parracombe, Barnstaple, EX31 4PR
GI4 YCZ J. Rainey, 40 Cranny Lane, Portadown, Craigavon, BT63 5SW
G4 YDB F. Heald, Brightling House, Alexandra Road, Heathfield, TN21 8ED
GM4 YDC S. Hunt, 5 Highland Road, Crieff, PH7 4LE
G4 YDD W. Davidson, 171 Ramsey Road, St Ives, PE27 3TZ
G4 YDE E. Metcalf, Beech Lee, Vicarage Lane, Alresford, SO24 0DU
G4 YDH G. Innes, Stonehaven, Holwell Road, Hitchin, SG5 3SL
G4 YDI R. Benbow, 54 Park Lea, Bradley Grange, Huddersfield, HD2 1QH
G4 YDL P. Martin, 3 Switchback Road North, Maidenhead, SL6 7UF
G4 YDM J. Allsopp, 30 Manor Park, Concord, Washington, NE37 2BT
G4 YDO P. Boaler, 21 Harts Close, Birmingham, B17 9LE
GI4 YDP G. Moore, 12 Irish Green Street, Limavady, BT49 9AD
G4 YDQ D. Hannant, 36 Coslany St., Norwich, NR3 3DT
G4 YDR D. Reed, 14 Glenholt Road, Plymouth, PL6 7JA
G4 YDT J. Craddock, 1 Brookes Road, Broseley, TF12 5SB
G4 YDW P. Grant, 3 Craggwood Close, Horsforth, Leeds, LS18 4RL
GW4 YDX K. Gill, 16 Hafodarthen Road, Llanhilleth, Abertillery, NP13 2RY
G4 YDZ M. Massen, The Old School, Horning Rd, Norwich, NR12 8JH
G4 YEB D. Whitton, 61 Greenacre Park, Gilberdyke, Brough, HU15 2TY
G4 YEE P. Hall, Barnlea, Knapp Lane, Romsey, SO51 9BT
G4 YEF B. Renner, 95 Reids Piece, Purton, Swindon, SN5 4BA
G4 YEG R. Paganuzzi, 3 St. Johns Close, Hook, RG27 9HW
G4 YEI S. Masterman, 5 Leggs Lane, Heyshott, Midhurst, GU29 0DJ
G4 YEJ A. Ayton, 3 Links Avenue, Norwich, NR6 5PE
G4 YEK S. Clack, 23 Cameron Grove, York, YO23 1LE
G4 YEO L. Gillain, 1 Willow Avenue, Denham, Uxbridge, UB9 4AG
GM4 YEQ Gala & Dist ARS, c/o J. Campbell, 50 Glebe Place, Galashiels, TD1 3JW
G4 YER D. Davies, 248 West Street, Hoyland Nr, Barnsley, S749EE
G4 YES J. Thompson, 3 Newport Mount, Headingley, Leeds, LS6 3DB
G4 YET R. Littlewood, 2A High Street Scotton, Gainsborough, DN21 3QZ
G4 YEX J. Kennedy, 60 Burnway, Albany, Washington, NE37 1QQ
G4 YFC T. Hill, 11 Paget Cottages, Munden Road, Ware, SG12 0NL
G4 YFF E. Reynolds, 4 Underwood Close, Stafford, ST16 1TB
G4 YFI R. Larter, 12 Ashby Road, Hinckley, LE10 1SL
G4 YFJ N. Von Fircks, 4 Park St., Salisbury, SP1 3AU
G4 YFK M. Aitchison, 21 St. Pauls Road West, Dorking, RH4 2HT
G4 YFO G. Wardy, 7 Yew Tree Road, Wistaston, Crewe, CW2 8BN
G4 YFS S. Van Praag, 12 Derby St., Darlington, DL3 0NW
G4 YFT B. Page, 7 Muirfield Crescent Tividale, Oldbury, B69 1PW
G4 YFU M. Parker, 85 Elston Road, Aldershot, GU12 4HZ
G4 YFV N. Banham, Lodge Bungalow, Norwich Road, Diss, IP21 4EE
G4 YFX P. Johnson, 119 Riverstone Way, Northampton, NN4 9QW
G4 YFZ B. Jones, Jetza, Rose Avenue, Burton-on-Trent, DE13 0DQ
G4 YGA S. Howarth, 44 Church Street, Old Catton, Norwich, NR6 7DR
G4 YGB A. Graph, 39 Poets Corner, Margate, CT9 1TR
G4 YGD G. Aldred, 212 Reepham Road, Norwich, NR6 5SW
G4 YGE C. Oakley, 9 Fitzroy Road, Landport, Lewes, BN7 2UB
G4 YGH D. Hart, 30 Dartford Avenue, Edmonton, London, N9 8HD
G4 YGJ L. Rozentals, 67 Linden Close, Eastbourne, BN22 0TT
G4 YGL G. Reece, 34 Priestley Gardens, Romford, RM6 4SL
G4 YGM P. Reynolds, 321 Sopwith Crescent Merley, Wimborne, BH211XQ
GM4 YGN F. Albers, 71 Old Evanton Road, Dingwall, IV15 9RB
G4 YGP J. Vinson, 11 Ripon Way, Carlton Miniott, Thirsk, YO7 4LR
G4 YGQ M. Tate, Stone Cottage, The Street, Norwich, NR13 3PL
GM4 YGS G. Wells, Squaredoch, Deskford, Buckie, AB56 5YD
G4 YGT W. Waldron, 16 Barke St., Highley, Bridgnorth, WV16 6LQ
G4 YGU J. Hetherington, 1 Downs Wood, Vigo, Gravesend, DA13 0SQ
G4 YGV FL R. Barnes, 7 Maycroft Close, Ipswich, IP1 6RG
G4 YGW Washington ARC, c/o F. Waters, 96 Stockley Road, Barmston Village, Washington, NE38 8DR
G4 YGY R. Fawke, 36 Lavender Meadows, Corelli Close, Stratford-upon-Avon, CV37 9FZ
G4 YGZ K. Ghillyer, 54 Longmeadow Road, Saltash, PL12 6DR
G4 YHG J. Hubner, 30 Orchard Rise, Olveston, Bristol, BS35 4DZ
G4 YHK H. Jackman, 8 The Buchan, Camberley, GU15 3XB
G4 YHN A. Gehammar, The Lodge, 119 Ashdon Road, Saffron Walden, CB10 2AJ
G4 YHP C. Jobling, Joycliff, 20A Poplar Road, Grimsby, DN41 7RD
GM4 YHS S. Grant, 16 Netherton Place, Westmuir, Kirriemuir, DD8 5LD
G4 YIA B. Robinson, 196 Bristol Avenue, Farington, Leyland, PR25 4QZ
G4 YIC A. Pitt, The Long Barn, Millmans Farm, Southend, Wotton under Edge, GL12 7PD
GW4 YID M. James, 14 Carmel Road, Winch Wen, Swansea, SA1 7JY
G4 YIE C. Kelley, Dunkeld, Bridge Street, Southam, CV47 2XY
G4 YIF S. Lumbard, 26 Waverleigh Road, Cranleigh, GU6 8BZ
G4 YIG J. Cluley, 24 Avon Green, Wyre Piddle, Pershore, WR10 2JE
G4 YIH W. Maycey, 21 Brook Drive, Wickford, SS12 9EQ
G4 YIM J. Cameron, 20 Fellmead, East Peckham, Tonbridge, TN12 5EQ
G4 YIP V. Ulfik, 31 Tarragon Way, East Hunsbury, Northampton, NN4 0SF
G4 YIS J. O'Farrell, Shannon, Gunville Road, Salisbury, SP5 1PP
G4 YIT I. Toon, 56 Cockbank, Turves, Peterborough, PE7 2HN
G4 YIV B. Whyle, 4 Britannia Gardens, Rowley Regis, B65 8DT
G4 YIZ A. Mansfield, 90 Vicarage Road, Mickleover, Derby, DE3 0EE
G4 YJA B. Lambert, Firbank, East Street Rusper, Horsham, RH12 4RE
G4 YJB R. Briggs, 32 Waterside, Evesham, WR11 1BU
G4 YJC G. Thornton, 115 High Street, Studley, B80 7HN
G4 YJD J. Donin, 2 Crablands, Selsey, Chichester, PO20 9AX
G4 YJH J. Hockey, 11 Amulet Way, Shepton Mallet, BA4 4TL
GW4 YJI T. Tilley, 47 Stratton Way, Neath Abbey, Neath, SA10 7BU
G4 YJK P. Duke, 4 Doggets Lane, Fulbourn, Cambridge, CB21 5BT
G4 YJM M. Leonard, 7 Moorside Parade, Drighlington, Bradford, BD11 1HR
G4 YJN M. Griggs, Tudor Rose Cottage, Malting Green Layer-De-La-Haye, Colchester, CO2 0JE
G4 YJP S. Stanton, 10 Jenner Crescent, Northampton, NN2 8NB
G4 YJQ D. Cutts, 36 Lodge Road, Little Oakley, Harwich, CO12 5EE
G4 YJS B. Parsons, 65 Foster Street, Widnes, WA8 6ET
G4 YJT C. Roberts, 59A Wharncliffe Road, Loughborough, LE11 1SL
G4 YJU C. Mount, 6 Almond Close, Countesthorpe, Leicester, LE8 5TG
G4 YJW G. Grundy, 47 Northiam Road, Eastbourne, BN20 8LP
G4 YJX J. Boyes, 1 Fuller Close, Shepton Mallet, BA4 5PX
G4 YJY W. Marsden, 33 Kilton Crescent, Worksop, S81 0AX
G4 YK B. Morrissey, 50 Fingringhoe Road, Langenhoe, Colchester, CO5 7LB
G4 YKB H. Ramsden, 23 Nandywell, Little Lever, Bolton, BL3 1JU
G4 YKE K. Howell, 25 Shelleycotes Road, Brixworth, Northampton, NN6 9NE
G4 YKG J. Pether, Stead Farm, Quarry Land Lane, Nr Axbridge, BS26 2QW
G4 YKH C. Littler, 11 Richards Road, Stoke D'Abernon, Cobham, KT11 2SX
G4 YKK A. Babbage, 247-248 Molesey Avenue, West Molesey, KT8 2ET
GW4 YKM K. Martich, 25 Pentwyn Isaf, Energlyn, Caerphilly, CF83 2NR

G4 YKQ P. Welford, 11 Ridgeside, Bledlow Ridge, High Wycombe, HP14 4JN
G4 YKR K. Ramsdale, 770 Warrington Road, Risley, Warrington, WA3 6AQ
G4 YKV K. Wood, 257 Church Road, Haydock, St Helens, WA11 0LY
GW4 YKW A. Hopkins, 30 Wavell Drive, Newport, NP20 6QN
G4 YKX M. Blakeley, Stowe House, Preston Gubbals Road, Shropshire, SY4 3LY
G4 YKZ R. Harvey, Richlyn House Cedar Road, Norwich, NR9 3JY
G4 YLG E. Jacquemai, 26 The Crescent, Brighton, BN2 4TD
G4 YLI R. Barnes, The Cottage, Hutton Row, Skelton, Penrith, CA11 9TR
G4 YLK D. Adams, 16 Centurion Close, Birchwood, Warrington, WA3 6NE
GM4 YLN C. Grierson, 6 Baberton Mains Court, Edinburgh, EH14 3ER
G4 YLO H. Timbrell, Crossing Cottage, Lamyatt, Shepton Mallet, BA4 6NG
GJ4 YLP C. Landor, L'Auge, La Rue Du Puchot, Jersey, JE3 6AS
G4 YLQ R. May, Flat 4, Hyde Bank Court, Hyde Bank Road, New Mills, High Peak, SK22 4NE
G4 YLT J. Smith, 43 Pagitt Street, Chatham, ME4 6RE
G4 YLW P. Yaxley, 10 Leybourne Close, Brighton, BN2 4LU
G4 YMB M. Brass, 11 Lealholm Way, Guisborough, TS14 8LN
GM4 YMC J. Mccallum, 86A Meadowfoot Road, Gladstone House, West Kilbride, KA23 9BY
GM4 YMD W. Macdiarmid, 167 Glasgow Road, Whins Of Milton, Stirling, FK7 0LH
G4 YME M. Elsey, Trekeek Farm, Camelford, PL32 9UB
G4 YMF J. Cracklow, 4 The Lawns, Sidcup, DA14 4ET
G4 YMG P. Chorley, 6 Conference Close, Warminster, BA12 8TF
G4 YMH M. Harney, 14 Druce Way, Thatcham, RG19 3PF
GM4 YMI A. Dodds, 9 Wellheads, Dunfermline, KY11 3JG
G4 YMJ W. Fitzgerald, 4 Sungold Place, Carterton, OX18 1DN
GW4 YML E. Jones, 15 St. Joseph Place, Llantarnam, Cwmbran, NP44 3HH
GM4 YMM C. Dons, 37 Ashley Drive, Edinburgh, EH11 1RP
G4 YMQ A. Kimm, 9 Tennis St., Burnley, BB10 3AG
G4 YMT M. Taylor, 64 Elmdene Road, Kenilworth, CV8 2BX
GJ4 YMX D. Warncken, Flat 53, La Tour Vert Le Coie Springfield Road, St Saviour, Jersey, JE2 7DN
G4 YMY R. Nicol, 32 Mayfair Drive, Newbury, RG14 6EE
G4 YMZ J. May, Glanrhyd Uchaf, Stags Head, Llangeitho, Tregaron, SY25 6QU
G4 YNG M. Garlick, Church View, School Lane, Islip, Kettering, NN14 3LQ
G4 YNH S. Payas, 36 Tintern Close, Popley, Basingstoke, RG24 9HE
G4 YNI H. Beckman, 16 Wilton Road, Crumpsall, Manchester, M8 4WQ
G4 YNK N. Fletcher, 11 Parkgate Drive, Bolton, BL1 8SD
G4 YNL A Team Contest Group, c/o R. Banks, Crows Nest, Churchstoke, Montgomery, SY15 6TP
G4 YNM B. Spencer, 33 New King Street, Bath, BA1 2BL
G4 YNO P. Barnes, 69 Southborne, Overcliff Drive, Bournemouth, BH6 3NN
G4 YNS S. Shakeshaft, Belvedere Cottage, Wrexham Road, Pulford, Chester, CH4 9DG
G4 YNT M. Yallop, 22, Warrenne Keep, Stamford, PE9 2NX
G4 YNU J. Scriven, 1 Holgate Road, Pontefract, WF8 4ND
G4 YNV H. Snaden, 92 Avon Way, Portishead, Bristol, BS20 6LU
G4 YNX B. Bassford, 12 Little Brum, Grendon, Atherstone, CV9 2ET
G4 YOA F. Harvey, 137 Epping New Road, Buckhurst Hill, IG9 5TZ
G4 YOC D. Gully, 9 Shellards Road Longwell Green, Bristol, BS30 9DU
G4 YOF G. Coomber, 6 Birch Way, Birch, Colchester, CO2 0NQ
G4 YOR R. Greenwood, 26 Littlefield Walk, Bradford, BD6 1UU
G4 YOS D. Corallini, 8 Britannia, Puckeridge, Ware, SG11 1TG
G4 YOT L. Zalicks, 3 Retford Path, Harold Hill, Romford, RM3 9NL
G4 YOV J. Metcalfe, 3 Castle Close, Stockton-on-Tees, TS19 0SL
GU4 YOX R. Beebe, San Grato, Les Houguette, Castel, Guernsey, Guernsey, GY5 7DZ
G4 YOZ B. Starkey, 52 Bermuda Road, Nuneaton, CV10 7HP
G4 YPA J. Aisher, 44 Cranleigh Road, Portchester, Fareham, PO16 9DN
G4 YPC P. Croucher, 66 Loop Road, Kingfield, Woking, GU22 9BQ
G4 YPD F. Thorne, 42 Roseworth Avenue, Liverpool, L9 8HF
G4 YPE N. Hanney, 62 Avonfield Avenue, Bradford-on-Avon, BA15 1JF
G4 YPF W. Taylor, 3 Westcroft, Leominster, HR6 8HE
G4 YPG R. Mason, Flat 4, Lysander House, Washington Road, Pulborough, RH20 4RF
G4 YPH D. Rothwell, 37 Eamont Avenue, Crossens, Southport, PR9 9YX
G4 YPI A. Maires, 26 Dunmow Road, Thelwall, Warrington, WA4 2HQ
G4 YPK P. Knowles, 6 Dorchester Close, Basingstoke, RG23 8EX
GM4 YPL Dr R. Thompson, Lochview West, 2 St. Ninians Avenue, Linlithgow, EH49 7BP
G4 YPQ K. Simpson, 5 Plover Fields, Madeley, Crewe, CW3 9EG
GI4 YPR W. Swail, 30 The Gables, Ballyphilip Road, Newtownards, BT22 1RB
G4 YPS A. Bradley, 22 Alexandra Crescent, Wigan, WN5 9JP
G4 YPV D. Ramsden, 76 Brigg Lane, Camblesforth, Selby, YO8 8HD
G4 YQA M. Lawson, 2 Low Lane, Embsay, Skipton, BD23 6SD
G4 YQC P. Whiting, 77 Melford Way, Felixstowe, IP11 2UH
G4 YQD T. Mayfield, 14 Wheatley Grange, Coleshill, Birmingham, B46 3LZ
G4 YQG B. Hodgetts, 3 Garsdale Road, Weston-Super-Mare, BS22 8PT
G4 YQH J. Frampton, 161 Longmead Avenue, Bristol, BS7 8QG
G4 YQJ F. Collie, 58 Waarem Avenue, Canvey Island, SS8 9DZ
G4 YQK K. Taylor, 22 Anderson, Dunholme, Lincoln, LN2 3SR
G4 YQL R. Silvey, 9 Kempe Road, Finchingfield, Braintree, CM7 4LE
G4 YQP M. Simmens, 1 Meaver Cottages, Meaver Road, Mullion, Helston, TR12 7DN
G4 YQQ A. Booth, 656 Southmead Road, Filton, Bristol, BS34 7RD
G4 YQS T. White, Rosewall Bungalow, Towednack Road, St Ives, TR26 3AL
G4 YQW K. Lawton, 52 Gamble Lane, Leeds, LS12 5LP
G4 YRA J. Hills, 27 Wellington Road, Denton, Newhaven, BN9 0RD
G4 YRC York Radio Club, c/o A. Palfrey, 5 Ings View, York, YO30 5XE
GM4 YRE E. Marcus, Woodmyre House, Edzell, Brechin, DD9 7UX
G4 YRF K. Amos, 1 Byron Close, Upper Caldecote, Biggleswade, SG18 9DF
G4 YRM R. Maynard, Clarnard, 7 Phillipps Avenue, Exmouth, EX8 3HY
GM4 YRO W. Patterson, 11 Almond Place, Comrie, Crieff, PH6 2BB
GI4 YRP T. Hutchinson, 47 Ballyhough Road, Donaghcloney, Craigavon, BT66 7PQ
G4 YRT G. Cogger, 40 The Crescent, Southwick, Brighton, BN42 4LA
G4 YRV M. White, 34 Pain'S Way, Amesbury, Salisbury, SP4 7RG
G4 YRX J. Hilliard, 44 Lerwick Way, Corby, NN17 2DZ
G4 YRY M. Holloway, 70 Baring Road, Southbourne, Bournemouth, BH6 4DT
G4 YRZ R. Denton, 48 Shireoaks Common, Shireoaks, Worksop, S81 8PE
G4 YSB J. Leary, 18A Chestnut Avenue, Andover, SP10 2HE
G4 YSE G. Ring, 31 Studland Park, Westbury, BA13 3HQ
G4 YSF J. Stimpson, 12 Fairhaven Court, Pittville Circus Road Roa, Cheltenham, GL52 2QR
G4 YSG A. Cooper, 85 Mansfield Road, Aston, Sheffield, S26 2BR
G4 YSH C. Bowers, Cornbury, Seymour Plain, Marlow, SL7 3BZ
G4 YSJ P. Rowland, 17 Hemel Hempstead Road, Redbourn, St Albans, AL3 7NL
GM4 YSN I. Brown, Redland House, Westruther, Gordon, TD3 6NF
G4 YSO M. Nixon, 32 Gilbert Sutcliffe Court, Cleethorpes, DN35 0SF
G4 YSP K. Metcalf, 34 Framland Drive, Melton Mowbray, LE13 1HY
G4 YSQ T. Rogers, Lodge 12 Benson Waterfront, Benson, Wallingford, Oxford, OX10 6SJ
G4 YSS J. Earnshaw, Dunelm, Ayton Road, Irton, Scarborough, YO12 4RQ
G4 YSZ R. Painton, 17 Brookside, Pill, Bristol, BS20 0JX
G4 YTB D. Slade, B2 2A Brisas Del Mar, Formentera Del Segura, Formentera Del Segura, Spain, 3179
G4 YTC M. Tyson, 1 Warwick Road, Bude, EX23 8EU
G4 YTD T. Booth, 37 Barco Avenue, Penrith, CA11 8LX
G4 YTF P. Godber, 3 Chalvington Close, Evington, Leicester, LE5 6XT
G4 YTG A. Gilbey, 83 Chignal Road, Chelmsford, CM1 2JA
G4 YTH T. Handford, 20 Minehead Road, Knowle, Bristol, BS4 1BN
G4 YTI N. Terry, 2 Crosley House, Crosley Wood Road, Bingley, BD16 4QD
G4 YTJ J. Pagett, 26 Rednal Hill Lane, Rednal, Birmingham, B45 9LR
G4 YTK S. Hopley, 35 Norton Grange, Norton Canes, Cannock, WS11 9QZ
G4 YTL D. Hilton-Jones, Home Farm, Lillingstone Lovell, Buckingham, MK18 5BJ
G4 YTM I. Pettinger, 266 West Street, Hoyland, Barnsley, S74 9EQ
G4 YTN L. Thomas, 31 Claude Avenue, Oldfield Park, Bath, BA2 1AE
G4 YTO M. Yeomans, 6 Badsey Close, Northfield, Birmingham, B31 2EJ
G4 YTQ D. Richardson, Holmlea, Town Street, Immingham, DN40 3DA
G4 YTR A. Hough, 27 Doncaster Road, Askern, Doncaster, DN6 0AL
G4 YTT M. Curran, 40 Barnpark Road, Teignmouth, TQ14 8PN
G4 YTU K. Maskell, 2 Birkhall Close, Chatham, ME5 7QD
G4 YTV R. Guttridge, Ivy House, Rise Road, Skirlaugh Hull, HU11 5BH
G4 YTY A. Dodd, 109 Perinville Road, Babbacombe, Torquay, TQ1 3PD
G4 YUA M. Rowland, 27 Wilmot Close, Witney, OX28 5NL
G4 YUF C. Sharon, 7 Waverley Gardens, Barkingside, Ilford, IG6 1PJ
G4 YUG C. Rogers, 221 Dales Road, Ipswich, IP1 4JY
G4 YUI R. Smith, 27 Laburnum Road, Bournville, Birmingham, B30 2BA
G4 YUK A. Ince, 3. Craycroft Road. Westwoodside., Doncaster., DN9 2DG
G4 YUL R. Hudson, 5 Common Lane, Hemingford Abbots, Huntingdon, PE28 9AN
G4 YUN M. Fox, 10 Alderhay Lane, Rookery, Stoke-on-Trent, ST7 4RQ
G4 YUO E. Hodges, 2 Joeys Field, Bishops Nympton, South Molton, EX36 4PX
G4 YUR J. Saunders, 40 Walkley Street, Sheffield, S6 3RG
G4 YUV H. Boehner, Not, Applicable, Resides, Germany, OUTSIDE OF THE UK IN
G4 YUZ I. Parker, 7 Cherry Tree Road, Hoddesdon, EN11 9JS
G4 YVA J. Guest, 57 Park Road, Quarry Bank, Brierley Hill, DY5 2HT
G4 YVB J. Finch, Hillside House, Chapel Street, Camelford, PL32 9UP
G4 YVD P. Challinor, Las Ciguenas, 33 Hill Road, Telford, TF2 8NA
G4 YVE C. Herwig, 29 New Road, Cupernham, Romsey, SO51 7LL
G4 YVF F. James, 70 Broadway West, Walsall, WS1 4DZ
G4 YVI Dr P. Rimmer, 1 Pear Tree Close, Weaverham, Northwich, CW8 3HD
G4 YVJ L. Beal, 17 Park Street, Cleethorpes, DN35 7NG
G4 YVK J. Felgate, 31 Melbourne Road, Ipswich, IP4 5PP
G4 YVM D. Perry, 11 St. Lawrence Close, Stratford Sub Castle, Salisbury, SP1 3LW
GW4 YVN G. Bertos, Farallon, Beach Road, Pembroke Dock, SA72 6TP
G4 YVO G. Howarth, 79 Eden Avenue, Edenfield, Bury, BL0 0LD
G4 YVQ R. Hargreaves, Lawnswood, Lee Road, Blackpool, FY4 4QS
G4 YVV P. Leetham, 26 Petersham Drive, Alvaston, Derby, DE24 0JU
G4 YVW P. Speed, 52 Hunter Avenue Shenfield, Brentwood, CM15 8PF
GW4 YVX J. Rafferty, Pen Y Bont, Llanfachraeth, Holyhead, LL65 4UY
G4 YVY T. Williams, 31 Manor Road, Holbury, Southampton, SO45 2NQ
G4 YVZ I. Cooling, 1 Grosvenor Gardens, Southampton, SO17 1RS
G4 YWA P. Cartwright, Danae, Glen Road, Kingsdown, Deal, CT14 8DD
G4 YWD W. Davies, Davies Electrical Services (N W) Ltd, 104 Bromborough Village Road, Bromborough, Wirral, CH62 7EX
G4 YWG D. Fowler, 22 Larchwood Crescent, Leyland, PR25 1RJ
GM4 YWI T. Ross, 36 Appin Drive, Prestonpans, EH32 9FB
G4 YWJ P. Ganley, 60 Bole Hill Road, Sheffield, S6 5DD
GW4 YWM W. Matthews, Cornerways, William Street, Swansea, SA9 1AT
G4 YWN G. Morris, Rivendell, The Street, Braintree, CM7 5HN
GM4 YWS G. Mckay, Reay House, St. Vigeans, Arbroath, DD11 4RA
GI4 YWT J. Crichton, 10 Bann Drive, Londonderry, BT47 2HW
GM4 YWU J. Bledowski, 24 Arbirlot, Arbroath, DD11 2NX
GM4 YWV W. Watson, 21 Cameron Way, Bridge Of Don, Aberdeen, AB23 8QD
G4 YWX A. Bell, 43 Bigsby Road, Retford, DN22 6SF
G4 YWZ C. Winning, 14 Fairmead Way, Totton, Southampton, SO40 7JH
G4 YXB A. Utley, 384 Skipton Road, Utley, Keighley, BD20 6HP
GM4 YXI Prof. K. Kerr, East Loanhead, Auchnagatt, Ellon, AB41 8YH
G4 YXJ T. Trethewey, 8 Sunningdale Road, Saltash, PL12 4BN
G4 YXR P. Waygood, 89 George St., Wellington, TA21 8HZ
G4 YXS F. Lake, 77A Wood Lane, Chapmanslade, Westbury, BA13 4AT
G4 YXU S. Rawcliffe, 4 Rue Des Contamines, Gex, France, 1170
G4 YXX N. Varnes, Kelneath, West Hill, Wincanton, BA9 9BZ
G4 YYC D. Craig, 48 Fairholme, Bedford, MK41 9DD
G4 YYD A. Birtwistle, 6 Solness Street, Bury, BL9 6PP
G4 YYE K. Smith, 27 Fairleas, Branston, Lincoln, LN4 1NW
G4 YYF S. Collings, 44 Church Road Wootton Bridge, Ryde, PO33 4PY
G4 YYG G. Plant, 74 Elmwood Drive, Blythe Bridge, Stoke-on-Trent, ST11 9NX
G4 YYH R. Blemings, 1 Trethern Close, Troon, Camborne, TR14 9ER
G4 YYI S. Tear, 18 The Chase, Sinfin, Derby, DE24 9PD
G4 YYL T. Anderton, 12 Oaklands Close, Halvergate, Norwich, NR13 3PP
G4 YYM M. Remnant, Redwood Court, Tolcarne Road, Camborne, TR14 9AA
G4 YYO P. Sutcliffe, Rosemead, Cheadle Road, Stoke-on-Trent, ST10 4BH
G4 YYP P. Plant, 74 Elmwood Drive, Blythe Bridge, Stoke-on-Trent, ST11 9NX
G4 YYR S. Gibbs, 14 Castle Mead, Kings Stanley, Stonehouse, GL10 3LD
G4 YYS S. Jones, 2 Rutland Way, Southampton, SO18 5PG
G4 YZA D. Brown, 104 Kineton Green Road, Solihull, Solihull
G4 YZC E. Smith, Willow Cottage, Mill Road, Louth, LN11 9TF
G4 YZD F. Benstead, 19 Davis Court, Eastland Road, Bristol, BS35 1DP

G4 YZF B. Alston-Pottinger, 16 Vincent Close, Great Yarmouth, NR31 0HR
G4 YZH B. Calvert-Toulmin, Brandesby House, 31 West End, Scunthorpe, DN15 9NR
G4 YZK R. Marsh, Hazatree, 71 Station Road, Worcester, WR3 7UP
G4 YZL G. Woollams, Pebbles, Pebsham Lane, Bexhill-on-Sea, TN40 2NT
G4 YZM S. Green, 4 Countess Drive, Walsall, WS4 1HT
G4 YZN K. Chapman, 10 Beck Lane, Collingham, Wetherby, LS22 5BW
G4 YZP J. Mcmahon, 15 Chatteris Park, Runcorn, WA7 1XE
G4 YZR M. Baker, 62 Court Farm Road, Whitchurch, Bristol, BS14 0EG
GM4 YZT J. Simpson, Flat 1/B, Isla Court, Perth, PH2 7HJ
GD4 ZAB L. Taylor, Burnt Mill House, Mount William, Summer Hill, Douglas, Isle of Man, IM2 4PE
G4 ZAC M. Lewis, Oak Fruit Farm, Devils Highway, Reading, RG7 1XS
GW4 ZAG G. Woodworth, 136 Wepre Park, Connah'S Quay, Deeside, CH5 4HW
GI4 ZAH F. Anderson, Flat 6, 25 Main Street, Coleraine, BT51 4RA
G4 ZAI A. Stevenson, 11 Alexandra Road, Malvern, WR14 1HA
G4 ZAL N. Head, 11 Crowden Crescent, Tiverton, EX16 4ET
G4 ZAM R. Lowe, The Chimes, 4 Broadway, Swindon, SN25 3BT
G4 ZAO D. Holmes, 17 Green Lane, Scarborough, YO12 6HL
G4 ZAP A1 Contest Group, c/o C. Wilson, 2 Bainton Close, Bradford-on-Avon, BA15 1SE
GW4 ZAR D. Flanagan, 13 Bryn Awelon, Flint, CH6 5QA
G4 ZAS J. Searle, 21 Chetwynd Drive, Southampton, SO16 3HY
GW4 ZAW J. Aspinall, 66 Lake Road East, Cardiff, CF23 5NN
G4 ZAX S. Jones, 71 Milford Road, Pennington, Lymington, SO41 8DN
G4 ZAY J. Tench, 20 Waterfield Meadows, North Walsham, NR28 9LD
G4 ZBC L. Brazier, 65 Cadman Crescent, Fallings Park, Wolverhampton, WV10 0SH
G4 ZBE T. Anderson, 38 Redwood Drive, Chase Terrace, Burntwood, WS7 2AS
G4 ZBG T. Green Two, 1 Conduit Road, Stamford, PE9 1QQ
G4 ZBH A. Holder, 47 Church Road, Gurnard, Cowes, PO31 8JP
G4 ZBK J. Olsen, Klintevej 218, Hjertebjerg, Stege, Denmark, DK 4780
G4 ZBL S. Bateman, 29 Nags Head Hill, St. George, Bristol, BS5 8LN
GW4 ZBN L. Connery, 37 Thomas St., Abertridwr, Caerphilly, CF8 4AU
G4 ZBO R. Parker, 34 Sandgate, Kendal, LA9 6HT
G4 ZBQ J. Thomas, 113 Southwood Drive, Coombe Dingle, Bristol, BS9 2QR
G4 ZBS A. Appleyard, 5 Rowan Close, Puriton, Bridgwater, TA7 8AL
GW4 ZBU J. Johns, 16 Maes Yr Haf, Llansamlet, Swansea, SA7 9ST
G4 ZBW J. Sceal, South Low, Lyth, Kendal, LA8 8DJ
G4 ZBZ A. Cooper, 26 Burlington Road, Skegness, PE25 2EW
G4 ZCA R. Mccormick, 22 Eric Road, Wallasey, CH44 5RQ
G4 ZCG A. Ashworth, 210 Liverpool Road, Hutton, Preston, PR4 5HB
GI4 ZCH C. Harkin, 39 Philiip Street, Derry City, BT48 7PN
G4 ZCJ H. Cresswell, 34 Kingsgate Avenue, Birstall, Leicester, LE4 3HB
GW4 ZCL P. Jones, 14 Fonmon Road Rhoose, Barry, CF62 3DZ
GW4 ZCM K. Frowd, 9 Heol Bryn Fab, Nelson, CF46 6JF
G4 ZCN B. Grylls, 22 Aldeburgh Close, Hartlepool, TS25 2RG
G4 ZCP B. Roberts, 70 Coombe Park Road, Coventry, CV3 2PE
G4 ZCR C. Glenn, 61 Ansell Road, Erdington, Birmingham, B24 8LX
G4 ZCS C. Saunders, Garlands, Malthouse Lane, Burgess Hill, RH15 9XA
G4 ZCT C. Thomas, 3 Poldice Terrace Poldice, St. Day, Redruth, TR16 5QA
GM4 ZCV E. Prietzel, 4 Cherry Lane, Cupar, KY15 5DA
G4 ZCW D. Reed, 4 Allwood Drive, Carlton, Nottingham, NG4 3EH
GW4 ZCY F. Barwell, Galahad, Penisarwaun, Caernarfon, LL55 3BN
G4 ZDD B. Brookfield, 17 St. Stephens Drive, Aston, Sheffield, S26 2EP
G4 ZDE R. Boss, 11 Penkridge Road, Church Gresley, Swadlincote, DE11 9FH
G4 ZDF T. Langham, 1 Chatsworth Avenue, Radcliffe-On-Trent, Nottingham, NG12 1DG
G4 ZDG R. Mather, 27 Bridgeacre Gardens, Coventry, CV3 2NQ
G4 ZDH D. Hepworth, Gander Green, Ings Lane Lastingham, York, YO62 6TD
G4 ZDN G. Titterington, 5 Stratford Crescent, Retford, DN22 7NX
G4 ZDP S. Williams, 18 Croft Road, Newbury, RG14 7AL
G4 ZDQ A. Siddons, 18 Earlswood Road, Evington, Leicester, LE5 6JB
G4 ZDR A. Perrett, 99 Welsford Avenue, Wells, BA5 2HZ
G4 ZDT D. Brodie, Oakdales, Chapel Road, Dilham, Dilham, NR28 9PZ
GW4 ZDU S. Kirkwood, Glanrafon, Llanddeusant, Holyhead, LL65 4AH
G4 ZDX A. Staniforth, 2 Park View, Mapperley, Nottingham, NG3 5FD
G4 ZDY R. Haining, 2 Keswick Close, Kirby Cross, Frinton-on-Sea, CO13 0TG
GW4 ZEA E. Hawkins, 12 Marine Drive, Ogmore-By-Sea, Bridgend, CF32 0PJ
G4 ZEB A. Richardson, 117 Polgrean Place, St. Blazey, Par, PL24 2LH
G4 ZEG E. Cross, 15 Carisbrooke Crescent, Barrow-in-Furness, LA13 0HU
G4 ZEJ R. Coombes, Aaru - 15 Semington Strand, Swindon, SN1 7DJ
G4 ZEL D. Rampton, Chalemar, Eddeys Lane, Bordon, GU35 8HU
G4 ZEN G. Gardner, 10 Chestnut Close, Ventnor, PO38 1DQ
G4 ZES R. Mills, 5 Summerlands Road, Marshalswick, St Albans, AL4 9XB
G4 ZEU W. Dix, 2 Churchdown Close, Boldon Colliery, NE35 9HA
G4 ZEW D. Adams, 4 St. Georges Close, Brampton, Huntingdon, PE28 4US
GM4 ZEX G. Duncan, Cromletvilla, South Road, Oldmeldrum, Inverurie, AB51 0AB
G4 ZEY E. Gough, 41 Matlock Green, Matlock, DE4 3BT
G4 ZEZ J. Curwen, 1 Oak Drive, Halton, Lancaster, LA2 6QJ
G4 ZFC R. Marks, 14 Carnation Road, Rochester, ME2 2YE
G4 ZFD D. Roper, Lunesdale, Halifax Road, Nelson, BB9 0EG
G4 ZFE R. Everitt, 6 Ormathwaites Corner, Warfield, Bracknell, RG42 3XX
G4 ZFJ C. Roberts, 122 Lower Road, Hullbridge, Hockley, SS5 6BH
G4 ZFP P. Lewis, 12 St. James Park, Tunbridge Wells, TN1 2LH
G4 ZFQ A. Reeves, 41 Nodes Road, Cowes, PO31 8AD
G4 ZFR Felixstowe Dars, c/o P. Whiting, 77 Melford Way, Felixstowe, IP11 2UH
GM4 ZFS S. Graham, 8 Kirkton Crescent, Dundee, DD3 0BN
G4 ZFT M. Nurse, 67 Grasleigh Way, Allerton, Bradford, BD15 9BD
G4 ZFV D. Green, 56 Southfields Road, Littlehampton, BN17 6PA
G4 ZFX J. Blades, 3 Briery Croft, Stainburn, Workington, CA14 1XJ
G4 ZFY J. Bowes, 8 Coxford Drove, Southampton, SO16 5FD
G4 ZGE T. Stocks, 1 Church Street Messingham, Scunthorpe, DN17 3SB
G4 ZGG B. Storey, 8-9 Chadley Lane, Godmanchester, Huntingdon, PE29 2AL
G4 ZGM C. Macdonald, 3 Shaftesbury Avenue, Doncaster, DN2 6DT
G4 ZGP G. Pritchard, 26 Anglesey Drive, Poynton, Stockport, SK12 1BU
G4 ZGQ D. Richardson, 55 Barton Road, Central Treviscoe, St Austell, PL26 7PT
G4 ZGZ S. Harris, Pentillie Cottage Quethiock, Liskeard, PL14 3SQ
G4 ZHA D. Raine, 160 Aragon Road, Morden, SM4 4QN
G4 ZHD P. Crofts, 8 Sandown Avenue, Mickleover, Derby, DE3 0QQ
G4 ZHE D. Cannon, 69 Hayfield Road, Oxford, OX2 6TX
G4 ZHG J. Nevin, 26 Beech Avenue, Newark, NG24 4DY
G4 ZHI J. Howell-Pryce, Bwlch Teulu, Tynygraig, Ystrad Meurig, SY25 6AJ
G4 ZHK D. Lennard, 24 Southdown Road, Shoreham-by-Sea, BN43 5AN
GM4 ZHL G. Graham, 18 Hamarsgarth, Mossbank, Shetland, ZE2 9TH
G4 ZHN D. Young, The Gables, Aerodrome Road, Canterbury, CT4 5EX
G4 ZHT Dr C. Strevens, 15 Mere House, Frodsham, WA6 0FN
G4 ZHX J. Burton, 23 Dorchester Close, Dartford, DA1 1ND
G4 ZHZ A. Nash, 10 Broome Close, Yateley, GU46 7SY
G4 ZIB A. Roberts, 25 Sebright Road, Wolverley, Kidderminster, DY11 5TZ
G4 ZID L. Chapman, 6 Barholm Avenue, Lutton, Spalding, PE12 9HS
G4 ZIF M. Taylor, Holly House, Faussett Hill, Canterbury, CT4 7AH
G4 ZIH R. West, 51 Glen Avenue, Herne Bay, CT6 6HU
G4 ZII A. Taylor, 50 Long Hill Rise, Hucknall, Nottingham, NG15 6GN
GM4 ZIL A. Brown, Skellies Knowes East, Leswalt, Stranraer, DG9 0RY
G4 ZIS R. Beech, 131 Bounces Road, Lower Edmonton, London, N9 8LJ
GM4 ZIT J. Brown, 51 Braeside Park, Balloch, Inverness, IV2 7HN
G4 ZIU M. O'Connell, 5 Beckwith Close, Harrogate, HG2 0BJ
G4 ZIW M. Hutchings, 31 Newtown Road, Little Irchester, Wellingborough, NN8 2DX
G4 ZIY M. Haddon, 1 Victoria Place, Weston, Portland, DT5 2AA
G4 ZIZ R. Barrett, Langton House Gullet Lane, Kirby Muxloe, LE9 2BL
G4 ZJC P. Berry, 3 Village Farm Road, Preston, Hull, HU12 8QH
G4 ZJD Dr L. Taylor, 14 Spring Grove, Chiswick, London, W4 3NH
G4 ZJE K. Faichney, 57 Moorside Road, Brookhouse, Lancaster, LA2 9PJ
G4 ZJH I. Tickle, 7 Ashfords Close, Saxmundham, IP17 1WB
GM4 ZJI C. Claydon, 33 Craigievar Drive, Glenrothes, KY7 4PH
G4 ZJK R. White, 29 Princes Road, Clacton-on-Sea, CO15 5LA
G4 ZJL D. Wood, 29 Oakville Road, Higher Heysham, Near Morecambe, LA3 2TB
G4 ZJO H. Docherty, 40 Norwood Drive, Morecambe, LA4 6LU
G4 ZJP M. Ward, Laurels, Eastergate Lane, Chichester, PO20 3SJ
G4 ZJR E. Knibb, The Cottage, Cold Newton Road, Leicester, LE7 9DA
G4 ZKA J. Watson, 158 Kingsfold Drive, Penwortham, Preston, PR1 9EQ
G4 ZKD J. Moule, Silver Dale, Callow Hill Rock, Kidderminster, DY14 9DB
G4 ZKE M. Chapman, 18 The Winter Knoll, Littlehampton, BN17 6ND
G4 ZKG J. Corfield, 5 Beasley Close, Great Sutton, Ellesmere Port, CH66 2SX
G4 ZKH M. Curtis, 11 Pentreath Terrace, Lanner, Redruth, TR16 6HP
G4 ZKI M. Day, 76 Freeman Road, Didcot, OX11 7DB
G4 ZKJ W. Applebee, 9 The Glade, Bucks Horn Oak, Farnham, GU10 4LU
G4 ZKM W. Ingram, 39 Ainsdale Drive, Peterborough, PE4 6RL
G4 ZKN P. Robinson, 1 Stennack, Troon, Camborne, TR14 9JT
G4 ZKQ J. Garner, Craythorne, Amberstone, Hailsham, BN27 1PJ
G4 ZKR J. Olbrien, 14 Ryecroft Close, Middlewich, CW10 0PJ
G4 ZKS A. Howland, Hollydene, Station Road, Colchester, CO7 8LJ
G4 ZKT A. Hale, 27 Danesbury Meadows, New Milton, BH25 5GX
G4 ZKW T. Davies, 7 Medway, Sturton By Stow, Lincoln, LN1 2DY
GI4 ZLD G. Breslin, 85 Whitehouse Park, Londonderry, BT48 0QA
G4 ZLF L. Forde, 3 Heather Way, Rosudgeon, Penzance, TR20 9PT
G4 ZLI T. Schofield, 25 Kingsfield, Ringwood, BH24 1PH
G4 ZLJ P. Aspinall, 20 Carr Lane, New Hall Hey, Rossendale, BB4 6BE
G4 ZLK S. Pike, 32 Rosewood Road, Dudley, DY1 4DZ
G4 ZLN B. Phillips, 2 Oriole Grove, Kidderminster, DY10 4HG
G4 ZLP N. Crook, 10 Shuttle Close, Rossington, Doncaster, DN11 0FR
G4 ZLT R. Winkup, 92 Barnes Crescent, Bournemouth, BH10 5AW
G4 ZLU T. Clark, Thaw House, Brunswick Street, Nelson, BB9 0HZ
G4 ZLX A. Whillock, 74 Chettell Way, Blandford St. Mary, Blandford Forum, DT11 9PH
G4 ZMA J. Smith, 7A The Green, East Leake, Loughborough, LE12 6LD
G4 ZMB D. Sharples, 11 Lina St., Accrington, BB5 1SL
G4 ZMH G. Robinson, 8 Fenlands Crescent, Lowestoft, NR33 9AW
GM4 ZMK R. Coyle, 216 Faifley Road, Clydebank, G81 5EG
G4 ZML D. Coupe, 14 Maltby Road, Thornton, Middlesbrough, TS8 9BU
G4 ZMM J. Roberts, 1 Grange Court, Hixon, ST18 0GQ
G4 ZMN P. Shepherd, 315 Daws Heath Road, Benfleet, SS7 2TY
G4 ZMP D. Butler, 42 Coombe Farm Avenue, Fareham, PO16 0TR
G4 ZMR M. Reynolds, Flat 2, Abbeyfield Court, 63 Abbey Foregate, Shrewsbury, SY2 6BG
G4 ZMU Dr A. Vernon, ? 29 Pinewood Avenue New Haw, Addlestone, KT15 3AA
G4 ZMY A. Wakely, 177 St.Hermans Estate, Hayling Island, PO11 9NE
GM4 ZNC W. Findlay, 46 Rowallan Drive, Kilmarnock, KA3 1TU
G4 ZNI A. Wragg, 11A Fall Road, Heanor, DE75 7PQ
GM4 ZNS J. Callaghan, 31 Hillview Road, Darvel, KA17 0DQ
GM4 ZNX D. Stockton, 13 Dunvegan Court, Crossford, Dunfermline, KY12 8YL
G4 ZNY K. Brown, 28 Blenheim Avenue, Stony Stratford, Milton Keynes, MK11 1EX
G4 ZNZ D. Neilson, 11 Craigs Way, Thirsk, YO7 1UD
GM4 ZOA S. Mcgregor, 35 Pentland Gardens, Edinburgh, EH10 6NN
G4 ZOB P. Harris, 47 North Park Grove, Roundhay, Leeds, LS8 1EW
G4 ZOC J. Lawton, Grenehurst, Pinewood Road, High Wycombe, HP12 4DD
G4 ZOD J. Greenberg, 12 Broadhurst Avenue, Edgware, HA8 8TR
G4 ZOF A. Hughes, Kemble Motors, Unit 9, Coundon, BISHOP AUCKLAND
G4 ZOG D. Andrews, 3 St. Davids Road, Thornbury, Bristol, BS35 2JE
G4 ZOH C. Hetherington, 10 Westway, Cowes, PO31 8QP
G4 ZOI D. Hunsdale, 1 Lesh Lane, Barrow-in-Furness, LA13 9EA
G4 ZON A. Edwards, 50 Wyatts Green Lane, Wyatts Green, Brentwood, CM15 0PY
G4 ZOQ J. Dennis, 44 The Drive, Uckfield, TN22 1BZ
G4 ZOR E. Wand, 13 Weetmans Drive, Colchester, CO4 9EA
GI4 ZOS W. Boyd, 51 South Sperrin, Knock, Belfast, BT5 7HW
G4 ZOT P. Varey, 6 Ludlow Avenue Garforth, Leeds, LS25 2LY
G4 ZOU D. Nuthall, 29 Bloxham Crescent, Hampton, TW12 2QG
G4 ZOX C. Moore, Spion Cop, Blacksmiths Lane, Lincoln, LN5 9SW
G4 ZOY D. Elliott, 6 Linden Close, Stakeford, Choppington, NE62 5LD
G4 ZPA B. Watling, 18 Deverell Place, Waterlooville, PO7 5ED
G4 ZPB R. Alexander, 1 Locarno Road, Swanage, BH19 1HY
G4 ZPC P. Collier, 7 Cavendish Close, Bicton Heath, Shrewsbury, SY3 5PG
G4 ZPH F. Machniak, 18 Wyatt Road, Kempston, Bedford, MK42 7EN
G4 ZPI D. Madden, 17 Canberra Gardens, Birmingham, B34 7LP
G4 ZPJ C. Marks, 85 Madrona, Tamworth, B77 4EJ
GW4 ZPL C. Barwell, Galaghad, Penisarwaun, Caernarfon, LL55 3BN
GW4 ZPM D. Thompson, 1 West Kinmel Street, Rhyl, LL18 1DA
G4 ZPN M. Brown, 47 Threlfall Road, Blackpool, FY1 6NW

G4 ZPO G. Belt, 45.Prospect Road, Dorchester, DT1 2PF
G4 ZPP B. Hope, 19 Seaview Court, Hillfield Road, Chichester, PO20 0JS
G4 ZPQ S. Drury, 24 Mollison Road, Hull, HU4 7HB
G4 ZPR R. Wilson, 1 Larkfield Avenue, Harrow, HA3 8NQ
G4 ZPW D. Mckie, 16 Guys Close, Addison Square, Ringwood, BH24 1PQ
G4 ZPZ I. Macpherson, 18 Mountbatten Avenue, Dukinfield, SK16 5BU
G4 ZQC R. Alderson, Old School House, Tattersett, King's Lynn, PE31 8RS
G4 ZQF R. Worth, 2 Orchard Drive, Otterton, Budleigh Salterton, EX9 7JL
GM4 ZQH J. Howell1, 26 Bonaly Crescent, Colinton, Edinburgh, EH130EW
G4 ZQJ A. Mayes, 103 Lionel Road, Canvey Island, SS8 9DJ
G4 ZQL N. Higgins, 7 Staveley Close, Middleton, Manchester, M24 4RU
G4 ZQM J. Neary, 29 Willow Avenue, Torquay, TQ2 8DH
G4 ZQS L. Brown, 17 Chaucer Walk, Langney, Eastbourne, BN23 7QT
G4 ZQT J. Wright, 85 Kingfisher Drive, Beacon Park Home Village, Skegness, PE25 1TQ
GW4 ZQV I. Bradford, The Meadows, Penyrheol, Pontypool, NP4 5XS
GW4 ZQY M. Couch, 37 Heol Rhosyn, Morriston, Swansea, SA6 6ER
G4 ZR5 R. Shaw, 1B Battle Close, Lindholme, Doncaster, DN7 6DA
G4 ZRA G. Moffatt, 30 Rose Walk, St Albans, AL4 9AF
G4 ZRB Dr W. Gerrard, 9 St. Marys Gardens, Bagshot, GU19 5JX
G4 ZRC M. Cole, 25 Holly Gardens, West Drayton, UB7 9PE
G4 ZRD B. Rosewarn, 16 Stoke Park Close Bishops Cleeve, Cheltenham, GL52 8UL
G4 ZRF T. Emery, 23 Richmondfield Way, Barwick In Elmet, Leeds, LS15 4HJ
GM4 ZRH A. Hutton, 4 Linn Road, Stanley, Perth, PH1 4QS
G4 ZRM D. Line, 28 Wykeham Road, Higham Ferrers, Rushden, NN10 8HU
GM4 ZRR I. Watt, 21 Clerwood Way, Edinburgh, EH12 8QA
G4 ZRT M. Johnson, 5 Donigers Dell, Swanmore, Southampton, SO32 2TL
G4 ZRV T. Russell, 56 Curzon Avenue, Cleethorpes, DN35 9HF
GW4 ZRW T. George, 80 Yew Street, Troedyrhiw, Merthyr Tydfil, CF48 4EE
GM4 ZRX J. Lindsay, 6 Netherhouse Avenue, Lenzie, Glasgow, G66 5NG
G4 ZRY A. Clemons, 2 Cherry Tree Road, Rainham, Gillingham, ME8 8JU
G4 ZRZ G. Baker, 12 Patrick Road, Caversham, Reading, RG4 8DD
G4 ZSA A. Smith, 16 Burley Close, South Milford, Leeds, LS25 5BT
G4 ZSC A. Kent, 9 Tolmers Gardens, Cuffley, Potters Bar, EN6 4JE
G4 ZSD W. Guy, 102 Bonington Road, Mansfield, NG19 6QQ
G4 ZSG J. Savegar, 39 Little Lane, Roundfield, Reading, RG7 6RA
G4 ZSO N. Dakin, 3 Tavistock Road, West Bridgford, Nottingham, NG2 6FH
G4 ZSP G. Marriott, 6 The Pastures, Barrow Upon Soar, Loughborough, LE12 8LA
G4 ZSR D. Woollams, Pebbles, Pebsham Lane, Bexhill-on-Sea, TN40 2NT
G4 ZSS S. Simpson, 8 Hallfield Avenue, Micklefield, Leeds, LS25 4AU
G4 ZST D. Nuttall, 92 Long Road, Lowestoft, NR33 9DH
G4 ZSV J. Broughton, 55 Webbs Close, Wolvercote, Oxford, OX2 8PX
G4 ZSW P. Withall, 19 Highfield Drive, Ewell, Epsom, KT19 0AU
G4 ZSX P. Johnson, 56 Sycamore Avenue, Lowestoft, NR33 9PJ
G4 ZSY C. Ingram, Woodwind, Kingstone, Hereford, HR2 9HD
G4 ZSZ B. Watts, 74 Westfield Road, Caversham, Reading, RG4 8HJ
G4 ZTA J. Reed, Easton Villa, Grangemoor Road, Morpeth, NE61 5PU
G4 ZTC T. Cleghorn, 12 Rennington Close, Stobhill Gate, Morpeth, NE61 2TQ
G4 ZTD K. Wright, 63 Dudley Avenue, Leicester, LE5 2EF
G4 ZTF J. Scott, Kemsley Street Cottage, Kemsley Street, Gillingham, ME7 3LS
GW4 ZTG K. Ford, Tan Y Bryn, Llanbedr, LL45 2ND
G4 ZTM N. Rohsler, 107 Quinton Lane, Quinton, Birmingham, B32 2TT
GM4 ZTO J. Mcilwraith, 54 Foreland, Ballantrae, Girvan, KA26 0NQ
G4 ZTQ S. Chappell, 67 Swanfield Drive, Chichester, PO19 6GL
G4 ZTR J. Lemay, Carlton House, White Hart Lane, Colchester, CO6 3DB
G4 ZTS C. Thorne, High Trees, Bradford On Tone, Taunton, TA4 1EX
GI4 ZTU H. Morgan, 42 Ardmore Road, Holywood, BT18 0PJ
G4 ZTW C. Gallagher, 9 St. Marys Road, New Romney, TN28 8JB
G4 ZTX C. Hall, 2 Brambling Lane, Wath-Upon-Dearne, Rotherham, S63 7GT
G4 ZTY D. Dalton, 22 Fernleigh Avenue, Mapperley, Nottingham, NG3 6FL
G4 ZTZ K. Taylor, 107 Trelowarren Street, Camborne, TR14 8AW
GW4 ZUA W. Webb, 62 Llewelyn Street Trecynon, Aberdare, CF44 8LA
G4 ZUC G. Allin, 1 Brookhill Court, Sutton-in-Ashfield, NG17 1EP
G4 ZUD B. Perry, Preswylfa, Carno, Caersws, SY17 5JP
G4 ZUE R. Hopkins, 259 Croft Road, Nuneaton, CV10 7EE
G4 ZUH J. Rowles, The Haven, 5 Honey Lane, Chatteris, PE16 6LG
G4 ZUI P. Bevington, 40 Carnarthen Street, Camborne, TR14 8UP
GW4 ZUJ B. Willis, 5 Park Street, Penrhiwceiber, Mountain Ash, CF45 3YW
GM4 ZUK A. Duncan, Barrhill House, Peterculter, AB14 0LN
G4 ZUL S. Cocks, 1 Church Ponds, Castle Hedingham, Halstead, CO9 3BZ
G4 ZUN C. Gee, 100 Plantation Hill, Worksop, S81 0QN
GW4 ZUS G. Smith, 8 Springfield Terrace, Llanhilleth, Abertillery, NP13 2RQ
GD4 ZUU P. Chambers, 15, Ramsey, Isle Of Man, IM7 1HE
GW4 ZUW A. Hockley, 44 Brookfields, Crickhowell, NP8 1DJ
G4 ZUX N. Sparks, 36 Tormynton Road, Worle, Weston-Super-Mare, BS22 9HT
G4 ZVA T. Webster, 42 The Meadow, Mount Pleasant Residential Park, Crewe, CW4 8JU
G4 ZVB G. Mantovani, 74 Barnsley Road, South Kirkby, Pontefract, WF9 3QE
G4 ZVD J. Birse, 3 Main Road, Rathmell, Settle, BD24 0LH
GM4 ZVF M. Sheriff, Schoolhouse Dunmore, Kilberry Road, Tarbert, PA29 6XY
G4 ZVK J. Taylor, 123 Lancaster Road, Hindley, Wigan, WN2 4JA
GW4 ZVL M. Higgins, 8 Clos Rheidol, Caldicot, NP26 4JD
G4 ZVN P. Baxter, 20 Thorpe Street, Thorpe Hesley, Rotherham, S61 2RP
GW4 ZVO B. Froley, 20 Sandymeers, Porthcawl, CF36 5LP
G4 ZVP B. Rhodes, 13 Amanda Road, Harworth, Doncaster, DN11 8HP
GW4 ZVQ S. Evans, 6 Eastfield Way, Caerleon, Newport, NP18 3EU
G4 ZVS C. Ford, 19 Listowel Road, Kings Heath, Birmingham, B14 6HH
G4 ZVU T. Chadwick, 102 Feltham Road, Ashford, TW15 1DP
GW4 ZVV B. Raby, 4 Tyfica Road, Pontypridd, CF37 2DA
G4 ZVX M. Russell, 67 Dugard Road, Cleethorpes, DN35 7SD
G4 ZVZ S. Josko, 69 Newborough Road, Shirley, Solihull, B90 2HB
G4 ZWA G. Johnson, The Cottage, Mareham On The Hill, Horncastle, LN9 6PQ
G4 ZWB R. Ayers, 5 Cornflower Close, Leiston, IP16 4UQ
G4 ZWD C. Spicer, 27 Carden Crescent, Brighton, BN1 8TQ
G4 ZWE P. Burtenshaw, 9 Winchester Way, Eastbourne, BN22 0JP
G4 ZWI F. Cooper, 29 Mayfair Avenue, Mansfield, NG18 4EQ
GM4 ZWJ S. Macfarlane, 6 Edward Drive, Helensburgh, G84 9QP
G4 ZWM H. Hoy, The Meadows Cottage, Stow Heath Road, North Walsham, NR28 0LR
GW4 ZWN T. Anziani, 42 Tyn Rhos Estate, Penysarn, LL69 9BZ
GW4 ZWO A. Green, Bryn Y Coed, Llanfair Road, Abergele, France, LL22 8DH
G4 ZWQ P. Smith, 16 Church St., Owston Ferry, Doncaster, DN9 1RG
G4 ZWR D. Edwards, 2 Mason Close, Headless Cross, Redditch, B97 5DF
G4 ZWX Dr P. Harrison, 17 Kings Road, Ascot, SL5 9AD
G4 ZWY S. Icke, 11 Church Lane, Bromyard, HR7 4DZ
G4 ZXA R. Smith, 1 Hall Lane, Wolvey, Hinckley, LE10 3LF
G4 ZXB A. Tomlins, 44 Newlands, Balcombe, Haywards Heath, RH17 6JA
G4 ZXF K. Kimber, Greenacres, Garrigill, Alston, CA9 3DY
GW4 ZXG L. Thomas, Roughton, Corntown Road, Bridgend, CF35 5BH
G4 ZXI N. Parnell, 4 Forge Lane, Headcorn, Ashford, TN27 9QQ
GM4 ZXJ J. Burns, 7 Johns Road, Eyemouth, TD14 5DX
G4 ZXN M. Ward, 25 Margeson Close, Coventry, CV2 5NU
G4 ZXP V. Legge, 26 Goldcroft Avenue, Weymouth, DT4 0ET
G4 ZXQ A. Mainwaring, The Old Smoke House, Lodge Farm Barns, Hereford, HR4 8NN
G4 ZXS E. Loach, 99 Gorse Lane, Clacton-on-Sea, CO15 4RJ
G4 ZXT M. Twigg, 30 Valley Drive, Yarm, TS15 9JQ
G4 ZXV W. Bailey, 225 Holburne Road, London, SE3 8HF
G4 ZXZ W. Johnson, 12A Kings Road, Spalding, PE11 1QB
G4 ZYH M. Timms, 5 Lytchett Way, Nythe, Swindon, SN3 3PJ
G4 ZYL J. Anderson, 44 Overhill Road, Burntwood, WS7 4SU
GW4 ZYM W. Williams, Plum Tree Farm, Commonwood Road, Wrexham, LL13 9TA
G4 ZYN M. Sherlock, Flat 6, 34 Duke Street, Southport, PR8 1JA
G4 ZYO B. Lawrance, 14 Warren Close, Porthleven, Helston, TR13 9BL
G4 ZYR H. Webber, 6 Barn Ground, Highnam, Gloucester, GL2 8LJ
GW4 ZYV J. Raymond, 23 Castle Pill Crescent, Steynton, Milford Haven, SA73 1HD
G4 ZYY G. Fildes, 62 Higher Days Road, Swanage, BH19 2LB
G4 ZYZ T. Seymour, 21 Chainhouse Road, Needham Market, Ipswich, IP6 8ER
G4 ZZD A. Hellier, 64 Penlee Park, Torpoint, PL11 2PZ
GM4 ZZH H. Firth, Edan, Berstane Road, Kirkwall, KW15 1NA
G4 ZZK B. Tutt, 15 Alexandria Drive, Herne Bay, CT6 8HX
G4 ZZL R. Lyons, 15 Winston Avenue, Tiptree, Colchester, CO5 0JU
GD4 ZZN A. Rickward, 14 Ballakneale Avenue, Port Erin, IM9 6ND
G4 ZZP K. Lock, 5 Copthorne Crest, Shrewsbury, SY3 8RU
G4 ZZS D. Bamber, 3 Abbotts View, Sompting, Lancing, BN15 0NG
G4 ZZV M. Singh-Gill, 30 King Edwards Gardens, London, W3 9RQ
GM4 ZZW R. Watts, 1C Kirklands, 100 Greenock Road, Largs, KA30 8PG
G4 ZZY T. Watts, Carne Grey Cottage, Trethurgy, St Austell, PL26 8YE
G4 ZZZ T. Smith, Lower Carniggey Farm, Greenbottom, Truro, TR4 8QL

G5

G5 ADY A. Jones, 67 Mandara Grove, Abbeydale, Gloucester, GL4 5XT
GI5 ALP Benbradagh Us Navcommsta, c/o Dr E. Harper, 404 Foreglen Road Dungiven, Londonderry, BT47 4PN
GW5 AMS Amplitude Modulation ARS, c/o S. Taylor, 43 Toronnen, Bangor, LL57 4TG
G5 AOZ Burnham Beeches, c/o U. Grunewald, Nuptown Orchard, Nuptown, Warfield, Bracknell, RG42 6HU
G5 ART A. Tait, The Reddings, Dirty Lane, Beausale, Warwick, CV35 7AQ
G5 AT RSGB Transatlantic Test Century Club No2, c/o N. Totterdell, Moscar Cross House, Hollow Meadows, Sheffield, S6 6GL
G5 ATT A. Horne, 38 Bradley Road, Waltham, Grimsby, DN37 0UZ
G5 AU A. Albinson, 24 Kasanka Ave, Brabham, Australia, WA 6055
GM5 AUG M. Topple, Emmaus, 30 Robshill Court, Newton Mearns, G77 6UG
G5 BBC London BBC Radio Group, c/o S. Richards, 6 Heathfield, Royston, SG8 5BW
G5 BBL J. Verduyn, 14 Ragleth Grove, Trowbridge, BA14 7LE
G5 BCO P. Gautier-Lynham, 95 Oxford Road, Marlow, SL7 2PL
GM5 BDW R. Watson, 37A Muirfield Crescent, Dundee, DD3 8PY
GM5 BDX B. Corkindale, 8 Rockland Park, Largs, KA30 8HB
G5 BH M. Coleman, Flat A, 53 De Parys Avenue, Bedford, MK40 2TR
G5 BK Cheltenham A.R.A, c/o A. Woolford, 39 Apple Orchard, Prestbury, Cheltenham, GL52 3EH
G5 BW W. Waugh, 67 Cragside, Whitley Bay, NE26 3EF
G5 CDC J. Kornreich, 35 Charlotte Drive, Spring Valley, USA,
GI5 CEO G. Hill, 10 Carntall Rise, Mossley, Newtownabbey, BT36 5UF
GM5 CGA L. Landers, 33 Newmanswalls Avenue, Montrose, DD10 9DD
G5 CH Chester and District Radio Society, c/o P. Hughes, 27 Hemsworth Avenue Little Sutton, Ellesmere Port, CH66 4SG
G5 CTX C. Thorne, 67 Devon Road, Cadishead, Manchester, M44 5HB
G5 CWP C. Peters, 15 Jasmine Crescent Trimdon Village, Trimdon Station, TS29 6QE
GM5 CX R. Ferguson, 19 Leighton Avenue, Dunblane, FK15 0EB
GM5 DAV D. Strachan, 30 Belhaven Park, Muirhead, G69 9FB
G5 DJW A. Osmond, 10 Deerhurst Park, Forest Row, RH18 5GD
GW5 DRE D. Eade, 16 Glynllifon Street, Blaenau Ffestiniog, LL41 3AF
GM5 DVS Dv Scotland, c/o I. Suart, 37 Meldrum Mains, Glenmavis, Airdrie, ML6 0QQ
G5 EDQ S. Hahn, 26 Watling St., Gillingham, ME7 2YH
G5 EDW E. Wright, 2 Wulfrath Way, Ware, SG12 0DN
G5 FM M. Wheeler, 114 Boundary Way, Glastonbury, BA6 9PH
G5 FRG Farringdon Radio Group, c/o D. Talaber, 54 Southfield Park, North Harrow, Harrow, HA2 6HE
G5 FZ Lincoln Short Wave Club, c/o P. Rose, Pinchbeck Farmhouse, Mill Lane, Sturton By Stow, Lincoln, LN1 2AS
G5 GIH G. Horobin, 21 Welwyn Avenue, Allestree, Derby, DE22 2JR
G5 GRG Genesis Radio Group, c/o A. Royds, 3A Fairfield Avenue, Rossendale, BB4 9TG
G5 GWH G. Hunt, 62 Pilkington Avenue, Sutton Coldfield, B72 1LG
G5 GX J. Smith, 4 Townend Villas, Humbleton, Hull, HU11 4NR
G5 HAM Ham Radio Network, c/o A. Whybrow, 64 Church Road Sevington, Ashford, TN24 0LF
G5 HBM H. Meyer, 10 Cairnside, Ilfracombe, EX34 8EW
G5 HI R. Birch, 15 Chester Street, Cirencester, GL7 1HF
G5 HOW Dr G. Howling, Sysonby Knoll Hotel, Asfordby Road, Melton Mowbray, LE13 0HP
G5 HY D. Wilkins, 8 Gainsborough Road, Ashley Heath, BH24 2HY
G5 JBQ E. Lorenzoni, 63 River Mill One, Station Road, London, SE13 5FL
G5 JCB Last Post Museum, c/o K. Hudson, 20 Claude Street, Crumpsall, Manchester,

M8 5AW
GM5 JDG M. Evans, Inverveddie Cottage, Peterhead, AB42 0LX
G5 JHE L. Batten, Y Felin Barn, Llawr-Y-Glyn, Caersws, SY17 5RH
G5 JIM J. Lang, 7 Marion Grove, Liverpool, L18 7HY
G5 JJ Taunton & District ARC, c/o A. Walrond, Leigh Hill Cottage, Lowton, Taunton, TA3 7SU
GM5 JNJ Ayrshire DX Group, c/o D. Anderson, 34 Culzean Crescent, Kilmarnock, KA3 7DT
G5 JPR J. Prince, Wellfield Avenue, Luton, LU33AT
G5 KC C. Quarton, Flaxton Gatehouse, Flaxton, York, YO60 7QT
GM5 KCC Straight Key Century Club, c/o M. Topple, Emmaus, 30 Robshill Court, Newton Mearns, G77 6UG
G5 KCP K. Pugh, 4 Salt Boxes, Pinvin, Pershore, WR10 2LB
G5 KN Kettering & District ARS, c/o K. Doswell, 15A Queen Street, Desborough, Kettering, NN14 2RE
G5 KW UK Six Metre Group, c/o D. Toombs, 1 Chalgrove, Welwyn Garden City, AL7 2QJ
G5 LK Reigate Amateur Transmitting Society, c/o P. Tribe, Island View, 108 Portsmouth Road, Lee-on-the-Solent, PO13 9AF
G5 LP L. Parker, 128 Northampton Road, Wellingborough, NN8 3PJ
G5 LSI Dr L. Soares Indrusiak, 6 Trafalgar House, Piccadilly, York, YO1 9QP
G5 LUX J. Ratty, 12 Challoners Close, East Molesey, KT8 0DW
GM5 LWD G. Ledgerwood, 28 Cannerton Park, Milton Of Campsie, Glasgow, G66 8HR
GM5 MAJ J. Devlin, 200 Second Avenue, Clydebank, Glasgow, G81 3LE
G5 MHZ 5 MHz Pioneers Club, c/o P. Gaskell, 131 Greenfield Road, Dentons Green, St Helens, WA10 6SH
G5 MS Manchester & District ARS, c/o K. Hudson, 20 Claude Street, Crumpsall, Manchester, M8 5AW
G5 MUN H. Van Driel, 20 Links Avenue Little Sutton, Ellesmere Port, CH66 1QT
G5 MW Medway Amateur Receiving & Transmitting Society, c/o J. Burton, 22 Pear Tree Lane, Hempstead, Gillingham, ME7 3PT
G5 MY H. Mee, 268 Victoria Rd East, Leicester, LE5 0LF
G5 NB N. Brown, 3 Mulberry Tree Close, Filby, Great Yarmouth, NR29 3HD
GW5 NF R. Ward, Lower Ton-Y-Felin Farm, Croespenmaen, Crumlin, Newport, NP11 3BE
GW5 NIC J. Nicholas, Reservoir House, St. Lythan'S, Wenvoe, CF5 6BQ
G5 NKW Key + Wire ARG, c/o K. Walker, 20 Thornhill Close, Bramcote, Nottingham, NG9 3FS
G5 NLD D. Van Dijk, 76 High Street, Tetsworth, Thame, OX9 7AE
G5 OD Wey Valley ARG, c/o A. Vine, Hilden, Woodland Avenue, Cranleigh, GU6 7HZ
G5 OW W. Wigg, 7 Brendon Way, Long Eaton, Nottingham, NG10 4JS
G5 OYY N. Weiner, 3 Bluebell Court Lower Mardyke Avenue, Rainham, RM13 8GF
G5 PAT P. Smyth, 61 Bridge Street, Kington, HR5 3DJ
GM5 PEB P. Bingham, 129 Livingstone Terrace, Irvine, KA12 9ER
G5 PI Philips Tele Lt, c/o J. Wilson, 20B High Green, Great Shelford, Cambridge, CB22 5EG
G5 PJK P. Kiernan, 4 Bradley Street, Southport, PR9 9HW
G5 PMJ P. Stoneham, The Croft, Blind Lane, Billericay, CM12 9SN
G5 QK Southend & Dis AR, c/o A. Radley, 16 Kingsley Lane, Thundersley, Benfleet, SS7 3TU
GM5 RAB R. Drummond, 11 Firwood Drive, Bo'ness, EH51 0NX
G5 RAH R. Hardy, 4 Tideswell Close, Staveley, Chesterfield, S43 3TE
G5 RC Skyline, c/o A. Crespo, 266 Trinity Road, London, SW18 3RQ
G5 RET G. Menendez, 3 Maybrick Road, Bath, BA2 3PT
G5 REV Rev. B. Topham, 2 Highgrove Gardens, Stamford, PE9 2GR
G5 RFL Radio Fraternity Lodge (8040), c/o A. Boyd, 5 Walmer Close, Southwater, Horsham, RH13 9XY
G5 RGS G. Jordan, 15 Ley Hill Road, Sutton Coldfield, B75 6TF
G5 RJH R. Harlow, 28 Dovecliff Crescent, Stretton, Burton-on-Trent, DE13 0JH
G5 ROB R. Hawkins, 1 Northbank Close, The Reddings, Cheltenham, GL51 6UA
GM5 RP Vowhars, c/o Dr I. White, 2 Appleby Cottages, Whithorn, Newton Stewart, DG8 8DQ
G5 RR Hucknall Rolls Royce A.R.C, c/o S. Sorockyj, 8 Bowden Avenue, Bestwood Village, Nottingham, NG6 8XN
G5 RS Guildford Contest Group, c/o P. Croucher, 66 Loop Road, Kingfield, Woking, GU22 9BQ
G5 RV Mid Sussex ARS, c/o J. Gibson, Four Oaks, Tylers Green, Haywards Heath, RH17 5DZ
GI5 SBZ Prof. K. Zepf, 33 Rannoch Road, Holywood, BT18 0NB
G5 SHO S. Hodgkiss, 14 Dales Close, Biddulph Moor, Stoke-on-Trent, ST8 7LZ
G5 SIX North London Six Metre Group, c/o K. Rochester, 22 Langford Road, Cockfosters, Barnet, EN4 9DS
G5 SRC Swinton (Amateur) Radio Club, c/o T. Ward, 173-175 Station Road, Pendlebury, Swinton, Manchester, M27 6BU
G5 STU S. Green, Flat 12, Canute House, Strand Street, Poole, BH15 1EJ
G5 TAM University of The Third Age, c/o M. Meadows, 8 Beeches Park Hampton Fields, Minchinhampton, Stroud, GL6 9BA
GM5 TDX E. Blakeway, 25 Allanton Grove, Wishaw, ML2 7LL
G5 TEO T. Orzechowski, 15 Rothesay Terrace, Northampton, NN2 7ER
GM5 TIM S. Doonan, West Clanfin Farm Waterside, Kilmarnock, KA3 6JQ
GI5 TKA Radio Security Service Memorial ARS, c/o W. Bradley, 14 Ardmore Grange, Ballygowan, Newtownards, BT23 5TZ
G5 TO Sheffield & District Wireless Society, c/o P. Day, 38 Broomhill Road, Old Whittington, Chesterfield, S41 9DA
G5 TV B. Watson, 173 Churchfield Lane, Darton, Barnsley, S75 5EA
G5 UI R. Perkis, 8 Mill Road, Cottingham, Market Harborough, LE16 8XP
GM5 VG OBO Windy Yett Contest Group, c/o W. Miller, 15 Glenalla Crescent, Ayr, KA7 4DA
G5 VH P. Chapman, 112 Sharpland, Leicester, LE2 8UP
G5 VNX A. Wright, 32 Temple Grove, Leeds, LS15 0HT
G5 VO N. Clarke, Brimham Lodge Farm, Brimham Rocks Road Burnt Yates, Harrogate, HG3 3HE
G5 VRA K. Cope, 9 Amber Heights, Ripley, DE5 3SP
G5 VZ C. Pearson, 4 Brentwood Close, Thorpe Audlin, Pontefract, WF8 3ES
G5 WQ I. Williams, Alma Cottage, Old Vicarage Lane, South Marston, Swindon, SN3 4SN
G5 WS RSGB Transatlantic Test Centenary Club, c/o N. Totterdell, Moscar Cross House, Hollow Meadows, Sheffield, S6 6GL
G5 XDX C. Macleod, 21 Halsetown, St Ives, TR26 3LY
G5 XX Ariel Radio Group, c/o P. Richmond, 57 The Fairway, Daventry, NN11 4NW
G5 YAX D. Wilkins, Malt Hill Cottage, Malt Hill, Warfield, Bracknell, RG42 6JG
G5 YC Icars, c/o Dr S. Bunting, 17 Sunnydene Avenue, Highams Park, London, E4 9RE
G5 YL K. Haywood, 126 Derby Street, Sheffield, S2 3NF
G5 YSS D. Turner, 21 Arnot Way, Higher Bebington, Wirral, CH63 8LP
G5 YZI B. Gray, 5 Oldfield Grove, London, SE16 2NA
G5 ZG Bishops Stortford AR Society, c/o A. Judge, 44 Thorley Lane, Bishop's Stortford, CM23 4AD

G6

G6 AAB T. Sloane, 42 Ashbury Drive, Blackwater, Camberley, GU17 9HH
G6 AAC P. Mcgoldrick, 23 Coleman Drive, Kemsley, Sittingbourne, ME10 2EA
GW6 AAG F. Steadman, 10 Oaktree Avenue, Sketty, Swansea, SA2 8LL
GM6 AAJ G. Scattergood, 14 Market Street, Forfar, DD8 3EY
G6 AAK J. Smith, 16 Cross Keys, Ossett, WF5 9SJ
G6 AAR D. Bolingford, Cobb Gate, School Lane, Pulborough, RH20 4LL
G6 AAZ K. Woodward, 19 Hazel Grove, Winchester, SO22 4PQ
G6 ABA P. Dobson, 16 Glenair Avenue, Parkstone, Poole, BH14 8AD
G6 ABG D. Coldbeck, 101 Westlands Road, Hull, HU5 5NX
G6 ABJ M. Claydon, 4 Sandringham Gardens, London, N12 0NX
G6 ABM A. Chick, The Rowans, Bourne View, Allington, Salisbury, SP4 0AA
G6 ABO R. Campbell, 207 Seabank Road, Wallasey, CH45 1HD
G6 ABP C. Cave, 31 Mill Road, Rearsby, Leicester, LE7 4YN
G6 ACJ D. Frampton, 28 Horsham Road, Owlsmoor, Sandhurst, GU47 0YY
G6 AD A. Fairclough, Forders House, Forders Lane, Marston Jabbett, Bedworth, CV12 9SG
G6 ADD T. Hallam, 98 Keppel Road, Sheffield, S5 0TY
G6 ADG M. Kennedy, 96 Kingsway, Boston, PE21 0AU
G6 ADO S. Nicholas, Greenbank, Chester High Road, Neston, CH64 7TR
G6 AEB S. Neil, 55 Colne Road, Brightlingsea, Colchester, CO7 0DU
G6 AEC D. Nicholls, 22 Yeo Way, Clevedon, BS21 7UP
G6 AEK D. Molyneux, 8 Ullswater Close, Hambleton, Poulton-le-Fylde, FY6 9EE
GM6 AES M. Clark, 12 Achaphubil, Fort William, PH33 7AL
G6 AFA P. Paskin, 36 Lewarne Road, Newquay, TR7 3JT
GD6 AFB N. Bazley, 77 Royal Park, Ramsey, Isle Of Man, IM8 3UH
G6 AFE R. Plested, 33 Hartbury Close, Cheltenham, GL51 0NZ
G6 AFG A. Afford, 2 Holly Court, Sandiway, Northwich, CW8 2PP
G6 AFK J. Adams, 6 Austen Road, Guildford, GU1 3NP
G6 AFL P. Blay, Treetops, Mount Pleasant, Crewkerne, TA18 7AH
GW6 AFQ R. Bambrey, Glangwili, Felinfach, Lampeter, SA48 7PG
G6 AFS D. Bell, 7 Chichester Drive, Cotgrave, Nottingham, NG12 3JJ
G6 AFT A. Carr, 4 Tansor Close, Corby, NN17 2QP
G6 AFX A. Crickett, 40 Ousden Close Cheshunt, Waltham Cross, EN8 9RQ
G6 AGA G. Clark, 2 Whitton Manor Road, Isleworth, TW7 7NL
G6 AGN D. Darby, 12 Laburnum Close, Clacton-on-Sea, CO15 2DD
G6 AGO B. Bean, 19 Coleshill Road, Sutton Coldfield, B75 7AA
G6 AGP A. Patterson, 10 Pear Tree Close, Wirral, CH60 1YD
G6 AGR M. Taylor, 8 Clifford Close, Long Eaton, Nottingham, NG10 3BT
GW6 AGS R. Thomas, 4 Duffryn Avenue, Cardiff, CF23 6LF
G6 AGY A. Smith, 103 Station Road, Seaham, SR7 0BD
G6 AGZ G. Smith, 103 Station Road, Seaham, SR7 0BD
G6 AHC T. Snook, 116 Rosemary Road, Poole, BH12 3HE
G6 AHD M. Sumner, Jaggen, Maldon Road, Chelmsford, CM3 6LF
G6 AHE P. Young, 8 The Slype, Wheathampstead, St Albans, AL4 8RY
G6 AHF C. Waterworth, 16 Fountains Walk, Lowton, Warrington, WA3 1EU
G6 AHH C. Walden, The Briers, Scures Hill, Nately Scures, Hook, RG27 9JS
G6 AHK C. Wallwork, 40-44 Henwood Green Road, Pembury, Tunbridge Wells, TN2 4LF
G6 AHN S. Reynolds, 12 Lowlands Crescent, Great Kingshill, High Wycombe, HP15 6EG
G6 AHO A. Oakes, 12 Bridge Mill Court, Chorley, PR6 9DU
G6 AHR R. Redpath, 11 View Terrace, The Platt, Dormansland, Lingfield, RH7 6QX
G6 AHV J. Spriggs, Kreuzstr. 18, Tuerkenfeld, Germany, D-82299
G6 AHX S. Evans, 18 Hillview Lane, Twyning, Tewkesbury, GL20 6JW
G6 AIB G. Farline, Willow Cottage, Manor View Road, Scarborough, YO11 3PB
G6 AIC R. Field, 266 Rochester Road, Burham, Rochester, ME1 3RJ
G6 AIG H. Gibson, 10 Trafalgar Street, Cambridge, CB4 1ET
G6 AII R. George, Timbers, Lake Lane, Bognor Regis, PO22 0AD
G6 AIK J. Gill, Millside, Mill Road, Steyning, BN44 3LN
G6 AIO P. Hillier, 20 Firtree Road, Norwich, NR7 9LG
G6 AIQ M. Homer, 29 Holmefield Avenue, Fareham, PO14 1EF
G6 AIU L. Harland, 16 Burford Close, Dagenham, RM8 3ST
G6 AIZ M. Holmes, 15 Anderton Way, Garstang, Preston, PR3 1RF
G6 AJ Barnsley & District ARC, c/o D. Gillott, 132 Racecommon Road, Barnsley, S70 6JY
GM6 AJA M. Hunt, Gaoith The Saorsa, 30 Kanachrine Place, Ullapool, IV26 2TX
G6 AJC I. Hodgkins, 2 Seagrave Road, Coventry, CV1 2AA
G6 AJG T. Jenkins, 134 Frankland Road, Croxley Green, Rickmansworth, WD3 3AU
GW6 AJK P. Jones, Ty?N Y Coed, Penley, LL13 0LN
G6 AJS A. Sharp, 17 Beechwood Avenue, Flanshaw, Wakefield, WF2 9JZ
G6 AJT B. Kenneally, 5 Havengore, Pitsea, Basildon, SS13 1JU
G6 AJV G. Larcombe, 55 Fairways International Caravan & Camping Park, Bath Road Bawdrip, Bridgwater, TA7 8PP
G6 AJW D. Lucas, 42 Falcon Way, Ashford, TN23 5UR
G6 AJX S. Lampard, 111 Whitworth Way, Wilstead, Bedford, MK45 3EF
G6 AK J. Brister, 49 Tiverton Road, Loughborough, LE11 2RU
G6 AKG R. Ayley, 1 Ballam Close, Upton, Poole, BH16 5QT
G6 AKK P. Archer, 26 Freshfield Drive, Macclesfield, SK10 2TU
G6 AKN M. Bentley, 9 Tinkers Castle Road, Seisdon, Wolverhampton, WV5 7HF
GW6 AKS F. Barwell, Galahad, Penisarwaun, Caernarfon, LL55 3BN
G6 AKX T. Blackburn, 42 Thames Drive, Biddulph, Stoke-on-Trent, ST8 7HL
G6 ALB A. Burge, 32 High Street, Swaffham Prior, Cambridge, CB25 0LD
G6 ALG N. Cutmore, 3 Linden Close, Tadworth, KT20 5UT
G6 ALJ T. Collins, 11 Sutton Road, Maidstone, ME15 9AE
G6 ALN G. Colclough, 20 Pembroke Drive, Whitby, Ellesmere Port, CH65 6TD
G6 ALR R. Delamare, 28 Blandford Road, Plymouth, PL3 5DU

G6 ALU S. Drury, 25 Crosslands, Stantonbury, Milton Keynes, MK14 6AY
G6 ALW B. Darby, 96 Bassnage Road, Halesowen, B63 4HG
G6 ALZ A. Davis, 2 Wolverhampton Road, Essington, Wolverhampton, WV11 2DB
G6 AMF B. Elliott, 41 Henwick Lane, Thatcham, RG18 3BN
GW6 AMK W. Needham, Cnwc Y Rhedyn, Aberporth, Cardigan, SA43 2DA
G6 AML J. Newcombe, Elmcroft, Upper Basildon, RG8 8LS
G6 AMV L. Mcconnell, 12 Marlborough Drive, Weston-Super-Mare, BS22 6DQ
G6 AMW C. Williams, 133 Devon Drive, Chandler'S Ford, Eastleigh, SO53 3GJ
G6 AMX P. Helm, 74 Neston Road Walshaw, Bury, BL8 3DB
G6 ANA P. Miller, Flat 907, De Montfort House, Leicester, LE1 5XR
GI6 ANC A. Murphy, 53 Whitehouse Park, Newtownabbey, BT37 9SH
G6 ANI J. Baverstock, Meadow View, Newbridge, Cadnam, Southampton, SO40 2NW
G6 ANJ C. Perrott, 15 Chestnut Drive, Claverham, Bristol, BS49 4LN
G6 ANO L. Goodwin, Gallifrey, The Village, Chelmsford, CM3 1AS
G6 ANR S. Garfirth, 19 Ingleside Drive, Stevenage, SG1 4RN
G6 ANV B. Gulliford, 12 Hawthorn Road, Eynsham, Witney, OX29 4NT
GM6 ANZ J. Howat, 61 Carolside Avenue, Clarkston, G76 7AD
G6 AOB A. O'Brien, 25 Sands Road, Paignton, TQ4 6EG
G6 AOF J. Henshaw, 5 Charlton Beeches Charlton Marshall, Blandford Forum, DT11 9NP
G6 AOH R. Hoblin, 4 Portiswood Close, Pamber Heath, Tadley, RG26 3UQ
GM6 AOJ W. Hay, 11 Lovat Road, Glenrothes, KY7 4RU
GM6 AOR G. Robertson, 32 The Square, Ellon, AB41 9JB
G6 AOS S. Pilbeam, 74 Southbank Avenue, Blackpool, FY4 5BX
G6 AOV C. White, 20A The Beacon, Ilminster, TA19 9AH
G6 APB G. Taylor, 1 Haigh Street, Greetland, Halifax, HX4 8JF
G6 APD L. Sawford, 16 Queens Close, Lee-on-the-Solent, PO13 9NA
G6 APE A. Schofield, 38 Edinburgh Road, Broseley, TF12 5PE
G6 APH C. Shiradski, 69 Masefield Avenue, Borehamwood, WD6 2HG
G6 APJ Rev. G. Smith, 6 Birtley Rise, Bramley, Guildford, GU5 0HZ
GW6 APK G. Sinclair, 4 Nant Y Mynydd, Seven Sisters, Neath, SA10 9BU
G6 APQ F. Hill, 12 Woodbine Walk, Chelmsley Wood, Birmingham, B37 6SB
G6 APW T. Harvey, 2 Trefor Jones Court, Brookfield Avenue, Dover, CT16 2QP
G6 APX S. Handley, 8 Nabb Close, St. Georges, Telford, TF2 9PT
GM6 AQB A. Riddell, 16 Lewis Drive, Old Kilpatrick, Glasgow, G60 5LE
G6 AQI T. Smith, 1 St. Jude Gardens, Colchester, CO4 0QJ
GM6 AQL A. Ryan, 6 Cumloden Court, Newton Stewart, DG8 6AB
GM6 AQR Dr H. Wynne, 103 New City Road, Glasgow, G4 9JX
G6 AQW N. Wiltshire, 66 Neville Road, Shirley, Solihull, B90 2QW
G6 ARC Andover Radio Amateurs Club, c/o C. Keens, Toad Hall, 69 Lillywhite Crescent, Andover, SP10 5NA
GD6 ARJ C. Jennings, 13 Barrule Drive, Ballasalla, Ballasalla, Isle Of Man, IM9 2HA
G6 ARM N. Kett, 10 Carrel Road, Great Yarmouth, NR31 7RF
G6 ARO I. Kendall, 65 Olive Grove, Swindon, SN25 3DB
G6 ARR S. Kimber, 3 Gloucester Way, Glossop, SK13 8RZ
G6 ART S. Langton, Corner Cottage, Harlow Road Sheering, Bishop's Stortford, CM22 7NB
G6 ASA Prof. N. Lipman, Meadowcroft, Cotswold Road, Oxford, OX2 9JG
G6 ASH N. Ash, 16 St. Marys Road, Sawston, Cambridge, CB22 3SP
G6 ASJ M. Bradley, Flat 20, Crown Court, Crown Street, Portsmouth, PO1 1QN
G6 ASK J. Matthews, Moor View, Oldways End, East Anstey, Tiverton, EX16 9JQ
GI6 ATD G. Rodgers, 23 Rathmore Park, Bangor, BT19 1DQ
G6 ATK K. Austin, 13 North End Grove, Portsmouth, PO2 8NF
G6 ATS D. Bowen, Flat 9, Moorfields Court, Silver Street, Bristol, BS48 2AG
GW6 ATT M. Bryan, 10 Woodlands Road, Barry, CF63 4EF
G6 ATW R. Czajkowski, 37 The Great Court, Royal Naval Hospital, Great Yarmouth, NR30 3JU
GI6 ATZ G. Curry, 87 Burren Road, Ballynahinch, BT24 8LF
G6 AUC Prof. H. Whitfield, Apartment 88, Rishworth Palace, Rishworth Mill Lane, Rishworth, Sowerby Bridge, HX6 4RZ
G6 AUD S. Challis, 38 Blacksmiths Lane, Wickham Bishops, CM8 3NR
G6 AUE G. Cosham, 85 Capsey Road, Ifield, Crawley, RH11 0UF
GI6 AUI D. Doherty, 42 Silverbrook Park, Newbuildings, Londonderry, BT47 2RD
G6 AUO M. Graffham, 106 Barford Road, Edgbaston, Birmingham, B16 0EF
G6 AUP B. Goodyear, 13 Moorland Avenue, Barnsley, S70 6PQ
G6 AUR B. Golding, 67 Milford Avenue, Wick, Bristol, BS30 5PP
GW6 AUS P. Humby, 126 Middleton Road, Oswestry, SY11 2XA
G6 AUW R. Howes, 8 Kitchener Road, Weymouth, DT4 0LN
G6 AUY D. Hawley, The Old Dairy, Edgefield Hall Barns, Edgefield, NR24 2RD
G6 AVI R. Tucker, Foxhall Cottage, Dukes Lane, Attleborough, NR17 1BL
G6 AVK C. Thomson, 160 Down Hall Road, Rayleigh, SS6 9PD
G6 AVL H. Thompson, 6 Alexandra Chase, Casa Rosa, Cramlington, NE23 6AA
G6 AVN D. Shaw, 33 The Fairway, Halifax, HX2 9PZ
G6 AVP A. Rowe, 2 Broom Mead, Bexleyheath, Da6 7ny
G6 AVS J. Russell11, 13 Stonebridge Lea, Orton Malborne, Peterborough, PE2 5LY
G6 AVT G. Stanhope, 39 Denham Close, Stubbington, Fareham, PO14 2BQ
G6 AVY D. Lane, 230 Raeburn Avenue, Eastham, Wirral, CH62 8BB
G6 AWF D. Miller, 33 Springfield Park, Twyford, Reading, RG10 9JG
G6 AWM C. Montgomery, 70 Campbell Road, Twickenham, TW2 5BY
G6 AWO R. Mansel, Ashcroft House, Ashfield Road, Bury St Edmunds, IP30 9HJ
G6 AWP A. Mchardy, The Haven, Hull Road, Hull, HU12 0TE
G6 AWY N. Armstrong, 27 High Street, Billinghay, LN4 4AU
G6 AWZ P. Ashdown, 1 Wheelers Patch, Emersons Green, Bristol, BS16 7JL
G6 AXC R. Beaumont, The New Hall, Fletchergate, Hull, HU12 8ET
G6 AXE G. Broad, 14 Albion Road, Westcliff-on-Sea, SS0 7DR
G6 AXH P. Brothers, 101 Bridgewater Drive, Northampton, NN3 3AF
G6 AXK P. Butler, 25 Orrishmere Road, Cheadle Hulme, Cheadle, SK8 5HP
G6 AXV R. Baldock, 12 Chippendale Close, Baughurst, Tadley, RG26 5HF
G6 AXY P. Coombes, Harleyford, Lower Wokingham Road, Crowthorne, RG45 6BT
GM6 AXZ K. Cocks, 60A Palmerston Place, Edinburgh, EH12 5AY
G6 AY G. Kellaway, 55 Ladbrooke Drive, Potters Bar, EN6 1QW
G6 AYD D. Chorley, Sunnylands, Sandpitts Hill, Langport, TA10 0NG
G6 AYE S. Cotterill, Arcadia, Leicester Lane, Leicester, LE9 9JJ
G6 AYH K. Cooke, 28 Curland Place, Longton, Stoke-on-Trent, ST3 5JL
G6 AYI B. Collett, 264A Lichfield Road, Sutton Coldfield, B74 2UH
GW6 AYR R. Shearing, Woodstock Fairview Old Winchfawr Road Clwydyfagwr, Merthyr Tydfil, CF48 1HW
G6 AYS T. Ramsden, 1A Fox Grove, Walton-on-Thames, KT12 2AT
G6 AYU P. Rice, 4 Council St., Walton, Peterborough, PE4 6AQ
G6 AYX D. Robinson, 3 King Edward Avenue, Wickham Market, Woodbridge, IP13 0SL
G6 AYY T. Rumbold, 23 Montague Road, Saltford, Bristol, BS31 3LA
G6 AZE A. Roberts, 9 Littlemoor Lane, Newton, Alfreton, DE55 5TY
GW6 AZG S. Pearless, Glanrafon, Dolgellau, LL40 2AH
G6 AZL P. Tarmey, 48 Merlin Crescent, Burton-on-Trent, DE14 3Jf
G6 AZP D. Glover, 16 Cardigan Grove, Trentham, Stoke-on-Trent, ST4 8XY
G6 AZR A. Granshaw, 38 Tudor Gardens, Stony Stratford, Milton Keynes, MK11 1HX
GW6 AZX R. Hughes, 4 Brittania Terrace, Porthmadog, LL49 9NB
G6 BAD C. Dovener, The Old Barn, 39 Sun Street, Haworth, Keighley, BD22 8BY
GW6 BAH G. Davis, 2 New House, Ponthir Road, Gwent, NP6 1PE
G6 BAL D. De La Haye, 14 Palace Meadow, Chudleigh, Newton Abbot, TQ13 0PJ
G6 BAM J. Draper, 42 Pitt Street, Broadwaters, Kidderminster, DY10 2UN
GM6 BAO A. Devine, 12 Auchengate, Barassie, Troon, KA10 6UG
G6 BAT D. Falstein, 3 Gracefields, 121 The Avenue, Fareham, PO14 3AA
G6 BAY C. Howes, 1 Wharrage Road, Alcester, B49 6QY
G6 BBD R. Hancock, 16 Buttermere, Wellingborough, NN8 3ZA
G6 BBG A. Harland, 23 Shelley Drive, Stratford Sub Castle, Salisbury, SP1 3JZ
G6 BBH N. Burton, 5/190 Little Marine Parade, Cottesloe, Australia, 6011
G6 BBI P. Ward, 63 Salcombe Drive, Glenfield, Leicester, LE3 8AG
G6 BBK S. Nelson, 10 Wragg Drive, Newmarket, CB8 7SD
G6 BBN J. Temple-Heald, Shires, 28 West End, Cambridge, CB22 4LX
G6 BBR M. Thomas, 17 Rectory Park Avenue, Sutton Coldfield, B75 7BL
G6 BBW J. Witts, 35 Warton Road, Basingstoke, RG21 5HL
G6 BCG R. Whitehouse, 5 Parkland Drive, Darlington, DL3 9DT
G6 BCL N. Miller, Bc House, East Hanningfield Road, Chelmsford, CM3 8EW
G6 BCM S. Ward, 33 All Saints Way, Aston, Sheffield, S26 2FJ
G6 BD M. Farmer, 16 Beckside, Nettleham, Lincoln, LN2 2PH
G6 BDH J. Kennard, 52 Lavender Lane, Stourbridge, DY8 3EF
GI6 BDI A. King, 43 Orby Gardens, Belfast, BT5 5HS
GW6 BDM C. Parker, Dolifor Llanwrthwl, Llandrindod Wells, LD1 6NU
GI6 BDN R. Larke, 11 Ballymaconnell Road South, Bangor, BT19 6DG
G6 BDW A. Sibley, 25 Vesta Avenue, St Albans, AL1 2PG
G6 BDY R. Southern, 208 Puxton Drive, Kidderminster, DY11 5HJ
G6 BEB J. Lines, Karen House, 11 Hill St., Brierley Hill, DY5 2AY
G6 BEH K. Penaluna, 5 Holkham Close, Rushmere St. Andrew, Ipswich, IP4 5DW
G6 BEL S. Fairweather, 65 Ambleside Avenue, Hornchurch, RM12 5EU
G6 BEN A. Burke, 24 Wentworth Close, Farnham, GU9 9HJ
G6 BER S. Boote, The Shippen, Downgate, Callington, PL17 8JX
GM6 BEY M. Craig, 7 Hallyards Cottages, Kirkliston, EH29 9DZ
G6 BFM A. Green, 117 Acanthus Road, Liverpool, L13 3DY
G6 BFP L. Humphrey, Four Gables, Gilletts Lane, High Wycombe, HP12 4BB
G6 BGA K. Turvey, St. Vincents Cottage, St. Vincents Lane, West Malling, ME19 5BW
G6 BGH I. Macdiarmid, 73 Stadium Avenue, Blackpool, FY4 3QA
GM6 BGJ I. Maclennan, 70 Kenneth St., Stornoway, HS1 2DS
GM6 BGL K. Maclean, 177 Lamond Drive, St Andrews, KY16 8JP
GM6 BGQ D. Small, 50 Toll Court, Lundin Links, Leven, KY8 6HH
G6 BGY J. Meek, Flat 26, Wickham Court, Clevedon, BS21 7TN
G6 BHA R. Smart, 67 Corkland Road, Chorlton Cum Hardy, Manchester, M21 8XT
G6 BHB J. Seager, 9 Lodge Close, Brighstone, Newport, PO30 4BX
G6 BHE V. Rogers, 66 East Beach Park, 66 East Beach Park, Shoeburyness, SS3 9SG
G6 BHH D. Palmer, Firdene, Abbey Road, Alton, GU34 5PB
G6 BHI A. Palmer, Firdene, Abbey Road, Alton, GU34 5PB
GW6 BHQ K. Williams, 19 Narberth Crescent, Llanyravon, Cwmbran, NP44 8RJ
GM6 BHR R. Warbrick, 8 Bathurst Drive, Alloway, Ayr, KA7 4QN
G6 BHS J. Watson, 58 St. Georges Drive, Cheltenham, GL51 8NX
G6 BHX C. Walker, 19 Springfield Grove, Corby, NN17 1EN
G6 BHY R. Vicarage, 10 Fleming Avenue, Sidford, Sidmouth, EX10 9NY
G6 BIA R. Thompson, 39 Grotto Road, South Shields, NE34 7AQ
GM6 BIG D. Anderson, 20 Greenrig Road, Hawksland, Lesmahagow, Lanark, ML11 9QA
G6 BIM J. Bowers, 6 Fairview Park, Hetton-Le-Hole, Houghton le Spring, DH5 0SE
G6 BIT Dr D. Crossley, 25 Newhaven Close, Bury, BL8 1XX
G6 BIU D. Carter, 23 First Street, Low Moor, Bradford, BD12 0JQ
G6 BIX E. Donbavand, 6 Springmeadow, Charlesworth, Glossop, SK13 5HP
G6 BJB Dr A. Forsyth, 14 Highgrove Road, Lancaster, LA1 5FS
G6 BJG I. Hancock, 64 Swanswell Road, Solihull, B92 7EY
G6 BJJ I. Harley, 37 Crelake Close, Tavistock, PL19 9AX
G6 BJL R. Harding, 12 Aller Vale Close, Exeter, EX2 5NH
G6 BJO P. Mctaggart, 33 Manor Farm Close, Barton-Le-Clay, Bedford, MK45 4TB
G6 BJR K. Hulbert, 15 St. Germans Road, Forest Hill, London, SE23 1RH
G6 BJY D. Vivash, 16 Whitchurch Close, Maidenhead, SL6 7TZ
GW6 BK Blackwood Contest Group, c/o R. Jones, 2 Pen-Y-Cwarel Road, Wyllie, Blackwood, NP12 2HP
G6 BKD J. Scotney, 30 Trinity Road, Rothwell, Kettering, NN14 6HY
G6 BKL P. Metcalfe, 65 Saville Road, Whiston, Rotherham, S60 4DZ
G6 BKY N. Arkwright, Bay Hill, Woodhouse Lane, Heversham, Milnthorpe, LA7 7EW
G6 BLA S. Woodford, The Lord Nelson, 1 Hale Road, Thetford, IP25 7RA
G6 BLC B. Conway, 29 Mandeville Road, Southgate, London, N14 7NJ
G6 BLK A. Johnson, Edelweiss, Boxley Road, Chatham, ME5 9JG
G6 BLU B. Nicholls, 29 Wittmead Road, Mytchett, Camberley, GU16 6ER
G6 BME D. Gibb, 46 School Road, Charing, Ashford, TN27 0JN
G6 BMG J. Hind, 80 Forge Fields, Sandbach, CW11 3RD
GM6 BML A. Ramsay, 15 Dunalistair Gardens, Broughty Ferry, Dundee, DD5 2RJ
GW6 BMP A. Roberts, 16A High Street, Llangefni, LL77 7NA
G6 BMY R. Satterthwaite, 47 Aberford Road, Baguley, Manchester, M23 1JY
G6 BMZ M. Williams, 3 Teesdale Road, Nottingham, NG5 1DA
GI6 BNI D. Mawhinney, 14 Cayman Avenue, Bangor, BT19 6XG
G6 BNO D. Dallaway, 17 Bantams Close, Birmingham, B33 0YL
GM6 BNS S. Lewis, Eyin Helga, Evie, Orkney, KW17 2PJ
G6 BNT M. Dronfield, 1 Kestrel Rise Swallownest, Sheffield, S26 4SD
G6 BNU Dr P. Entwistle, Waverley Cottage, Sherfield Road, Bramley, Tadley, RG26 5AG
G6 BNW Dr J. Garcia-Rodriguez, St. Albans, Mill Lane, Dover, CT15 4HR
G6 BOF G. Hollidge, 10 Newfoundland Close, Worth Matravers, Swanage, BH19 3LX

G6 BOK P. King, 10 Heath Hey, Woolton, Liverpool, L25 4TJ
G6 BOP A. Reid, 115 Robingoodfellows Lane, March, PE15 8JH
G6 BOQ E. Parker, Jasmine Cottage, Apperley, Gloucester, GL19 4DE
G6 BOX S. Wilson, 21 Plumian Way, Balsham, CB21 4EG
G6 BPH F. Bennewitz, 1 Millfield Avenue, Saxilby, Lincoln, LN1 2QN
G6 BPK Dr S. Cook, 50 Bath Road, Swindon, SN1 4AY
G6 BPN R. Edmondson, 91 Lewin Road, London, SW16 6JX
G6 BPY W. Roe, 39 Marlborough Road, Southwold, IP18 6LR
G6 BQC M. Stuart, 207 Saunders Lane, Mayford, Woking, GU22 0NT
G6 BQE P. Tilley, 22 Meadowsweet, Waterlooville, PO7 8RS
G6 BQM P. Bentley, Sandy Ridge, Church Street, Stoke-on-Trent, ST7 4RS
G6 BQQ M. Barnes, Drovers, Crampshaw Lane, Ashtead, KT21 2UF
G6 BRA Bracknell ARC, c/o I. Pawson, 3 Orion, Bracknell, RG12 7YX
GW6 BRC Barry ARS, c/o S. Trahearn, 148 Gladstone Road, Barry, CF62 8ND
G6 BRD W. Hammond, 245 Broadoak Road, Ashton-under-Lyne, OL6 8RP
G6 BRP P. Walter, 2 Hallams Lane, Beeston, Nottingham, NG9 5FH
GM6 BRU J. Steele, 54 Myrtle Crescent, Bilston, Roslin, EH25 9SB
G6 BRV R. Shelford, 1 All Lands Cottages, Highcross, Rotherfield, Crowborough, TN6 3QA
G6 BRW S. Sumner, 27 High Street, Flore, Northampton, NN7 4LL
G6 BRY C. Thomas, 52 Derwent Road, Burton-on-Trent, DE15 9FR
G6 BSP S. Guest, Flat 2 7 Corscombe Close, Weymouth Dorset Uk, DT4 0UE
G6 BSS G. Higgs, 68 Otterfield Road, Yiewsley, West Drayton, UB7 8PF
G6 BTB C. Pringle, 38 Priory Road, Littlemore, Oxford, OX4 4NE
G6 BTC A. Layton, 17 Maplehurst, Leatherhead, KT22 9NB
G6 BTM N. Rice, 31 Bold Street, Heysham, Morecambe, LA3 1TS
G6 BTP E. Beswarick, 2 Hurst, Beaminster, DT8 3ES
G6 BTR M. Challis, 18 Castlefield Close, Eastleaze, Swindon, SN5 7EG
G6 BTX K. Holmes, 313 Havering Road, Romford, RM1 4BZ
G6 BUH J. Walsh, 13 Byam Street, London, SW6 2RB
G6 BUP C. Chan, 11 The Paddocks, Welwyn Garden City, AL7 2BW
G6 BUT Harlow & District ARS, c/o M. Simkins, 37 St. Andrews Meadow, Harlow, CM18 6BL
G6 BUU R. Costello, 6 Qua Fen Common, Soham, Ely, CB7 5DH
G6 BUV A. Cutts, Highthorns Cottage, North Frodingham, Driffield, YO25 8LS
GW6 BUW I. Davies, Garthewyn, Caernarfon, LL55 2RL
G6 BUY R. Gingell, 23 Woodfarm Road, Malvern Wells, Malvern, WR14 4PL
G6 BV South Birmingham Radio Society, c/o J. Storey, 34 Austin Rise, Longbridge, Birmingham, B31 4QN
G6 BVF Prof. N. Linge, 21 Pennant Drive, Prestwich, Manchester, M25 3BT
G6 BVO S. Green, 3 Mulberry Close, March, PE15 9FH
GI6 BVQ T. Finlay, 4 Station Road, Eglinton, Londonderry, BT47 3PR
G6 BVR R. Gammage, 12 The Butts, Warwick, CV34 4SS
GW6 BVS J. Hayman, 22 Princess St., Abertillery, NP3 1AR
G6 BWA C. Clarke, 11 Eastmoor Villas, Epworth Road, Doncaster, DN9 2LH
G6 BWE K. Edwards, 289 Monks Walk, Buntingford, SG9 9DZ
G6 BWJ J. Richards, 44 Swain Street, Watchet, TA23 0AG
G6 BWK T. Wallis, 17 Alderbank, Wardle, Rochdale, OL12 9NH
G6 BWM R. Smith, 49 Aubourn Avenue, Lincoln, LN2 2JW
G6 BWN J. Stewart, 101 West Way, Lancing, BN15 8LZ
G6 BWO J. Taberner, 20 Stevenson Drive, Wirral, CH63 9AH
G6 BWP D. Weaver, 8 Strathmore Close, Worthing, BN13 1PQ
G6 BWT A. Bajjon, 35Ablackford Rd, Shirely, B90 4BU
G6 BXO C. Blackwell, 20 Southworth Avenue, Blackpool, FY4 3LH
G6 BXR R. Calvert, 3A Panxworth Road, South Walsham, Norwich, NR13 6DY
G6 BXS D. Ellison, Riverside, Old Mill Drive, Colne, BB8 0TX
G6 BXT M. Fry, 61 Swift Road, Abbeydale, Gloucester, GL4 4XH
GW6 BXU E. Hatherall, 101 Park Crescent, Abergavenny, NP7 5TL
G6 BXV D. Willis, Rivendell, Shirnall Hill, Alton, GU34 3EJ
G6 BYF C. Gomez, The Gazebo, Military Road, Rye, TN31 7NY
G6 BYK J. Parkes, 65 Ferrier Road, Stevenage, SG2 0NZ
G6 BYL D. Lycett, 1 Saredon Close, Pelsall, Walsall, WS3 4DH
G6 BYR T. Frangopulo, Flat 11, Jack Edwards Court 5 Lapwing Lane, Manchester, M20 2NT
G6 BZE M. James, 9 Wyke Mark, Winchester, SO22 5DJ
G6 BZG L. Green, 37 Park Road, Northville, Bristol, BS7 0RH
G6 BZL M. Adams, The Vicarage, Intake Lane, Ormskirk, L39 0HW
G6 BZQ G. Doubleday, 1 St. Johns Avenue, Chelmsford, CM2 0UA
G6 BZW E. Butt, 97 Hawthorn Crescent, Yatton, Bristol, BS49 4RG
G6 CAC J. Hallett, 16 Streche Road, Swanage, BH19 1NF
GI6 CAG W. Millar, 9 Lynnehurst Drive, Comber, Newtownards, BT23 5LN
G6 CAR A. Baldwin, Rathlin, Dromnea, Kilcrohane, Bantry, Ireland, P75 Y300
G6 CBB D. Beddow, 34 Loweswater Road, Stourport-on-Severn, DY13 8LP
G6 CBL D. Leslie, 8 The Avenue, Swarland, Morpeth, NE65 9JL
G6 CBP A. Pidgeon, 106 Winchester Avenue, St Johns, Worcester, WR2 4JQ
G6 CBY M. Jeeves, 52 Castlefields, Istead Rise, Gravesend, DA13 9EJ
G6 CCB A. Stonehouse, 105 Humberston Avenue, Humberston, Grimsby, DN36 4ST
G6 CCN L. Armstrong, Deuteros House 1 Bank Top, Earsdon, NE25 9JS
G6 CCQ R. Powell, Manuela 25 Jack Haye Lane Light Oaks, Stoke-on-Trent, ST2 7NG
G6 CDT G. Henshaw, 18 Queens Avenue, Ilkeston, DE7 4DL
G6 CDU G. Keeble, 4 Bardfield Way, Frinton-on-Sea, CO13 0AN
G6 CDV A. Morling, 33 Russell Court, Chesham, HP5 3JH
G6 CDW N. Miller, 3 Upwood Gorse, Tupwood Lane, Caterham, CR3 6DQ
G6 CEM E. Weir, 10 St. Georges Crescent, Whitley Bay, NE25 8BJ
G6 CEP A. Kneebone, 34 Henver Road, Newquay, TR7 3BN
G6 CEZ R. Brand, 17 Park Road, Fordingbridge, SP6 1EQ
G6 CFA J. Carrick Smith, 15 The Vale, Oakley, Basingstoke, RG23 7LB
G6 CFC G. Purchon, 33 Lancaster Avenue, Hitchin, SG5 1PA
G6 CFU N. Shaw, The Gables, Camp Lane, Banbury, OX17 1DH
G6 CGC R. Sheppard, 51 Marks Road, Wokingham, RG41 1NR
G6 CGI M. Rowat, 154 Hollingwood Lane, Bradford, BD7 4DB
G6 CGO E. Parr, 18 Arundel Close, Macclesfield, SK10 2NS
G6 CGQ R. Hatch, 4 Springfield Crescent, Parkstone, Poole, BH14 0LL
G6 CGY R. Percival, 6 Bulmer Place, Hartlepool, TS24 9BQ
G6 CHA E. Povey, Hillcroft, Schoolfields, Henley-on-Thames, RG9 4DH
G6 CHC V. Appleton, 15 Pinewood Crescent, Ramsbottom, Bury, BL0 9XE
G6 CHD P. Bridle, Flat 25, Riverview Gardens, 289 Old Chester Road, Birkenhead, CH42 3XQ
G6 CHI A. Bowley, Plum Tree House, Walk Close, Derby, DE72 3PN
G6 CHJ M. Carter, 5 Orchard Brook, Long Melford, Sudbury, CO10 9LF
G6 CHT M. Hall, 31 Meendhurst Road, Cinderford, GL14 2EF
G6 CHX P. Holland, High Lea Cottage, Witchampton Lane, Wimborne, BH21 5AF
G6 CIA T. Kenyon, 31 Marble Hill Gardens, Twickenham, TW1 3AU
G6 CID J. Andrews, Top Flat, 211 Brighton Road, South Croydon, CR2 6EJ
G6 CIE R. Townsend, 3 Cranfield View, Darwen, BB3 2HP
G6 CIF D. Taylor, 8 Russell Drive, Wollaton, Nottingham, NG8 2BH
G6 CII K. Sutton, 9 Babbacombe Drive, Ferryhill, DL17 8DA
G6 CIO J. Robinson, 31 Church Road, Banks, Southport, PR9 8ET
G6 CIP P. Ralston, Laund House, 9 College Avenue, Liverpool, L37 3JL
G6 CIT R. Young, 143 Rodmell Avenue, Saltdean, Brighton, BN2 8PH
G6 CJB P. White, 8 Kingswood Court, Maidenhead, SL6 1DD
GW6 CJJ Dr J. Alexander, 2 Meadow Park, Burton, Milford Haven, SA73 1NZ
G6 CJR S. Barber, Homedale, St. Monicas Road, Tadworth, KT20 6ET
G6 CJT B. Bradshaw, 19 Wren Garth, Beeford, Driffield, YO25 8FQ
G6 CKD L. Newbury, Skeyton Corner Chapel, Skeyton Corner, Norwich, NR10 5AP
G6 CKE C. Evans, 21 Snowdrop Close, Crawley, RH11 9EG
G6 CKH J. Muir, 150 Thorntree Road, Thornaby, Stockton-on-Tees, TS17 8LX
G6 CKJ D. Morris, 255 Lichfield Road, Wolverhampton, WV113EW
G6 CKK R. Martin, 1 Rosemount Court, Rochester, ME2 3NF
G6 CKL I. Martin, 24 Heddington Close, Trowbridge, BA14 0LH
G6 CKM D. Langdon, 17 Forest Grove, Eccleston Park, Prescot, L34 2RY
GM6 CKN R. Morrison, 38 Burnfoot Road, Hawick, TD9 8EN
G6 CKW R. Beattie, 11 Pine Grove, Bricket Wood, St Albans, AL2 3ST
G6 CKY M. Gray, 20 Ravenstone Street, London, SW12 9SS
G6 CKZ P. Berwick, 4 Brewer Road, Crawley, RH10 6BP
G6 CLA G. Blacksell, 152 Hawthorn Avenue, Colchester, CO4 3YA
G6 CLC N. Carter, 58 Meadow Road, Barlestone, Nuneaton, CV13 0HQ
G6 CLD G. Coker, 46 Clarendon Road, Ipplepen, Newton Abbot, TQ12 5QS
G6 CLK P. Carter, 19 Felix Road, Walton-on-Thames, KT12 2LB
G6 CLP J. Miller, 7 Malvern Crescent, Ashby-de-la-Zouch, LE65 2JZ
G6 CLU D. Lawes, 8 High Beech Chalet Park, Battle Road, St Leonards-on-Sea, TN37 7BS
G6 CLW B. Lloyd, 243 Stand Lane, Radcliffe, Manchester, M26 1JA
G6 CLX D. Lloyd, 31 Lever Park Avenue, Horwich, Bolton, BL6 7LF
GI6 CMA R. Dawson, 31 Clonmore Manor, Lisburn, BT27 4EW
G6 CMB I. Dalton, 10 St. Vincents Villas, Temple Hill, Dartford, DA1 5HT
G6 CMD C. Driver, 23 Mercers Row, St Albans, AL1 2QS
G6 CMF A. Daborn, 49 Crescent Road, Locks Heath, Southampton, SO31 6PE
G6 CML J. Sykes, 20 Woodend Road, Bournemouth, BH9 2JQ
G6 CMN A. Shaw, 14 Delph Crescent, Clayton, Bradford, BD14 6RY
GM6 CMQ D. Robson, 35 Lady Nairne Road, Dunfermline, KY12 9YD
G6 CMS M. Robertson, 13 Orchard Cottages, Main Road, Boreham, Chelmsford, CM3 3AD
G6 CMV D. Palmer, Spidrift, Landsdown Road, Malvern, WR14 1HX
G6 CMX J. Pell, 33 Low Street, Winterton, Scunthorpe, DN15 9RT
G6 CND J. Oliver, 3 Savile Walk, Brierley, Barnsley, S72 9HJ
G6 CNF J. Payne, 71 Waarden Road, Canvey Island, SS8 9AB
G6 CNK R. Freshwater, 82 Sandford Road, Chelmsford, CM2 6DH
G6 CNL P. Farnell, 40 Thorney Lane, Luddendenfoot, Halifax, HX2 6UX
G6 CNQ T. Genes, 28 Hillside Road, Burnham on Crouch, CM0 8EY
GW6 CNS J. Graham, 23 Somerset Road, Barry, CF62 8BL
G6 CNW J. Gibson, Penrose Cottage, Carne, St Austell, PL26 8DB
G6 CNX J. Goodwin, 10 Abingdon View, Worksop, S81 7RT
G6 COB J. Hodkinson, 3 Cypress Close, Market Drayton, TF9 3HJ
G6 COE C. Hill, Manor Farm Cottage, Portington, Goole, DN14 7LZ
G6 COG D. Holdsworth, Middle Pasture, Heath Lane, Halifax, HX3 0AG
G6 COL Lincoln Shortwave, c/o P. Rose, Pinchbeck Farmhouse, Mill Lane, Sturton By Stow, Lincoln, LN1 2AS
G6 COZ R. Turner, 73 Digby Court, Nottingham, NG7 1RG
G6 CP D. Cutter, David Cutter Engineering, 34 Greengate Lane, Knaresborough, HG5 9EL
G6 CPE K. Stanley, 35 St. Blaize Road, Romsey, SO51 7JY
G6 CPF J. Stephenson, 16 Greenways, Driffield, YO25 5HX
G6 CPO N. Wysocki, 6 Rose Dene, Stourport-on-Severn, DY13 8SU
G6 CPS A. Yates, 12 Graham Drive, Middleton, King's Lynn, PE32 1RL
G6 CPX M. Waples, 24 Constable Drive, Wellingborough, NN84UX
G6 CPY E. Whitham, 72 Bole Hill, Treeton, Rotherham, S60 5RE
G6 CQB M. Wilson, 23 Claydown Way, Slip End, Luton, LU1 4DU
G6 CQC A. Varty, Wisteria, Hillcrest, Burnhope, Durham, DH7 0BQ
G6 CQH J. Abbishaw, Hastings House Farm, Littletown, Durham, DH6 1QB
G6 CQR C. Bailey, 32 Ryland Road, Moulton, Northampton, NN3 7RE
G6 CRC Cheshunt & District A.R.S, c/o R. Gray, 51 Wyatt Close, Ickleford, Hitchin, sg53xy
G6 CRD S. Brown, 5 Keepside Close, Ludlow, SY8 1BQ
G6 CRF T. Bailey, 65 Edge Lane, Chorlton Cum Hardy, Manchester, M21 9JU
G6 CRG B. Bowes, 1 Rockall Close, Southampton, SO16 8EH
G6 CRR R. Solomons, 32 Church Road, Pembury, Tunbridge Wells, TN2 4BT
GM6 CRX F. Mcleod-Stangroom, 6 Leonach, Strathlachlan, Cairndow, PA27 8DB
G6 CSC W. Skidmore, 29 The Meadows, Grisedale Road, Bakewell, DE45 1TP
G6 CSK A. Beal, 115 Maldon Road, Witham, CM8 1HR
G6 CSL C. Redding, 20 Bromley Street, Workington, CA14 2TP
G6 CSN G. Chadwick, 25 Passmonds Crescent, Rochdale, OL11 5AW
G6 CSR H. Calloway, 6 Franchise Gardens, Wednesbury, WS10 9RQ
G6 CTA J. Davidson, 12 Hanbury Close, Dronfield, S18 1RF
G6 CTC Covtry Tech ARC, c/o J. Witt, 67 Dillotford Avenue, Coventry, CV3 5DS
G6 CTE L. Duffill, 14 Leonard Street, Hull, HU3 1SA
G6 CTH E. Dunne, 16 Ulleswater Close, Little Lever, Bolton, BL3 1UD
G6 CTP H. Wakefield, 32 Mandene Gardens, Great Gransden, Sandy, SG19 3AP
G6 CTV E. Eggs, 62 Laurel Manor, 18 Devonshire Road, Sutton, SM2 5EJ
G6 CTX M. Brown, 31 Kingscote Road, Cowplain, Waterlooville, PO8 8QD
G6 CTY C. Edwards, 54 Thoroughgood Road, Clacton-on-Sea, CO15 6DP
G6 CUA H. Erridge, 15 Maurice Road, Southsea, PO4 8HH
G6 CUE J. Frampton, 54 Hudson Road, Bexleyheath, DA7 4PG
G6 CUK A. Fisher, 1 Elm Walk, Catterick, Richmond, DL10 7PB
G6 CUQ N. Wedgbury, 12 The Ridgeway, Astwood Bank, Redditch, B96 6LT
GW6 CUR S. Williams, 371 Coed-Y-Gores Llanedeyrn, Cardiff, CF23 9NR

G6 CUT J. Whitehurst, Serendipity, 97 Noke Common, Newport, PO30 5TY
G6 CUV K. Wyeth, 3 West Palace Gardens, Weybridge, KT13 8PU
G6 CUY J. Wildsmith, Lingmoor, 7 Lambert Road, Uttoxeter, ST14 7QG
G6 CVB J. Taylor, 12 Fairview Drive, Westcliff-on-Sea, SS0 0NY
G6 CVD C. Thornley, Sylvastone House, Herne Street, Herne, CT6 7HG
G6 CVE R. Tanfield, 8 Rede Close, Bedford, MK41 7UH
G6 CVP D. Wilkins, The Workshop, Rear Of 59 Jasmine Grove, London, SE20 8JY
G6 CVR J. Geary, 2 Eastgate Lane, Terrington St. Clement, King's Lynn, PE34 4NU
G6 CVV M. Gumbrell, 2A The Avenue, Carlby, Stamford, PE9 4NA
G6 CVW W. Griffiths, 6 Stanway Close, Middleton, Manchester, M24 1HE
GW6 CVX R. Griffith, The Hollies, Morton, Oswestry, SY10 8AJ
G6 CVY H. Gibbons, 27 Bentham Road Chesterfield, Chesterfield, S40 4EZ
G6 CWF C. Hazell, 18 Cleeve Hill, Downend, Bristol, BS16 6HN
G6 CWH S. Harwood, 24 Firle Crescent, Lewes, BN7 1QG
G6 CWP D. Hartley, 4 Park Gate, Euston Road Fakenham Magna, Thetford, IP24 2QS
G6 CWW V. Holbrook, 84 Haddon Street, Derby, DE23 6NQ
GW6 CWZ D. Mccallum, Glan Alaw Llanddeusant, Holyhead, LL65 4AG
GI6 CXD R. Mcwhirter, 200 Townhill Road, Portglenone, Ballymena, BT44 8AR
G6 CXI A. Long, 35 Heath Court, Grampian Way, Derby, DE24 9NG
G6 CXN K. Lankshear, 28 Monmouth Place, Newcastle-under-Lyme, ST5 3DF
G6 CXO D. Lloyd, 16 Kingsley Road, Brighton, BN1 5NH
G6 CXV K. Phillips, Stockings Barn, Whitbourne, Worcester, WR6 5SR
G6 CXY R. Revan, 50 Woodland Rise, Welwyn Garden City, AL8 7LF
G6 CYA R. King, 55 Coppins Road, Clacton-on-Sea, CO15 3HS
G6 CYE A. Read, 36 West St., Tollesbury, Maldon, CM9 8RJ
G6 CYF D. Richards, 433-435 Cronton Road, Widnes, WA8 5QG
G6 CYH I. Roberts, 32 Priory Drive, Plymouth, PL7 1PU
G6 CYO I. Jarvis, The Garden House, Walkley Wood, Stroud, GL6 0RT
G6 CYT R. Kempton, 14 Bloxam Gardens, Rugby, CV22 7AP
G6 CYU M. Kendrick, 157 Pinar De Gariata, La Nucia, Alicante, Spain, 3530
G6 CYV P. Kirkham, 9 Bluebell Close, Biddulph, Stoke-on-Trent, ST8 6TJ
G6 CZB R. Poffley, 3 Bowerhill Road, Salisbury, SP1 3DN
G6 CZD M. Swanwick, 45, Coach Way, Willington, Derby, DE65 6ES
GW6 CZE C. Peacock, 8 Heol Ewenny, Pencoed, Bridgend, CF35 5QA
GM6 CZM I. Mcaulay, 9 Randolph Cliff, Edinburgh, EH3 7TZ
G6 CZO D. Mcghie, 54 School Road, Newborough, Peterborough, PE6 7RG
G6 CZX W. Aitchison, 18 Kerensa Green, Falmouth, TR11 2HE
G6 CZZ J. Abram, 3 Frenchies View Denmead, Waterlooville, PO7 6SH
G6 DAC M. Bounds, 5 Ingleby Close, Heacham, King's Lynn, PE31 7SA
G6 DAD D. Blagburn, 10 Tottington Avenue, Springhead, Oldham, OL4 4RY
G6 DAH D. Budd, 81 Bohemia Chase, Leigh-on-Sea, SS9 4PW
G6 DAI N. Brickwood, 4 Vale Cottages, Shillingstone, Blandford Forum, DT11 0SS
G6 DAN B. Daniels, 113 Orchard Way, Wymondham, NR18 0NZ
G6 DAO G. Bradbury, 3 Westfield Bank, Barlborough, Chesterfield, S43 4EG
G6 DAP J. Balmford, Upper Brook Farm House, The Avenue, Aylesbury, HP18 9LD
G6 DAQ A. Boonham, 1 Oakleigh Drive, Sedgley, Dudley, DY3 3LH
G6 DAU P. Bidwell, 156 Elstree Park, Barnet Lane, Borehamwood, WD6 2RP
G6 DAY M. Pemberton, 37 Bardsley Close, Croydon, CR0 5PS
G6 DBC A. Norfolk, 18 Middle Lane, Amcotts, Scunthorpe, DN17 4AT
GM6 DBJ J. Fairhurst, 1. Ackmore Court, Kyleakin, Isle of Skye, IV418PT
GW6 DBP J. Firmstone, 1 Holly Grange, Rhoswiel, Weston Rhyn, Oswestry, SY10 7TU
G6 DBQ D. Fryer, Norwood, 105 Chester Road, Stockport, SK7 6HG
G6 DBU R. Gambles, 5 College Way, Horspath, Oxford, OX33 1SQ
G6 DBX A. Grover, 44 Stirling Court Road, Burgess Hill, RH15 0PT
G6 DBY P. Gould, Derna, Surrey Lane, Colchester, CO5 0QT
G6 DBZ S. Griffin, 50 Cherrybrook Drive, Broseley, TF12 5SH
G6 DCH J. Molyneux, 18 Bay Close, Horley, RH6 8LF
G6 DCS G. Norris, 1 Pear Tree Avenue, Newhall, Swadlincote, DE11 0LZ
G6 DCT D. Littlewood, 50 Industry Road, Sheffield, S9 5FQ
GI6 DCX D. Lyons, 64 Drumgavlin Road, Ballynahinch, bt24 8qy
G6 DDA A. Moss, 20 Black-A-Tree Court, Black-A-Tree Road, Nuneaton, CV10 8BD
G6 DDC J. Leatherbarrow, 10 Henley Drive, Southport, PR9 7JU
GW6 DDF J. Morris, 45 Ffordd Pentre Mynach, Barmouth, LL42 1EN
G6 DDJ S. Pillinger, Calle Sileno 10, Mailbox (Buzon) 470, Fortuna, Spain, 30620
G6 DDO R. Owen, 36 Foley Road, Stourbridge, DY9 0RT
G6 DDP R. Oakden, 38 Brookfield Avenue, Hucknall, Nottingham, NG15 6FF
G6 DDR L. Horne, 8 Kingsway Avenue, Broughton, Preston, PR3 5JN
G6 DDU G. Goddard, 30 Western Avenue, Holbeach, Spalding, PE12 7QD
G6 DDX M. Wild, 46 All Saints Drive, North Wootton, King's Lynn, PE30 3RY
G6 DEG T. Hampson, 6 Rushmere Drive, Bury, BL8 1DW
GW6 DEP M. Harris, 11 Lower Rawlinson Terrace, Tredegar, NP2 4JD
G6 DER K. Hewitt, 6 Church Grove, Monk Bretton, Barnsley, S71 2EY
G6 DET M. Heighton, 3 Warner Road, Codsall, Wolverhampton, WV8 1SA
G6 DEV D. Harris, 15 Millwood Road, Orpington, BR5 3LG
GI6 DEY F. Hunter, Flat 9 50 Edenvale Crescent, Belfast, BT4 2BH
G6 DFA C. Willies, 17 Campion Way, Sheringham, NR26 8UN
G6 DFB C. Smith, 9 Barratts Close, Bewdley, DY12 2ED
G6 DFC P. Johnson, 3 Lance Drive, Chase Terrace, Burntwood, WS7 1FA
G6 DFH J. Roberts, 155 Langley Hall Road, Olton, Solihull, B92 7HB
G6 DFM J. Phelps, Windy Dene, Green Lane, Chessington, KT9 2DT
G6 DFR T. Parfitt, 4 Back Street, Lakenheath, Brandon, IP27 9HF
G6 DFV A. Parker, 13 Hartley Street, Colne, BB8 9DF
GW6 DFX D. James, 5 Lon Y Parc, Cardiff, CF14 6DF
G6 DFY Dr G. Joly, 116 Hind Grove, London, E14 6HP
G6 DFZ M. Jones, 28 Winston Avenue, Colchester, CO3 4NQ
G6 DGK G. Keegan, 12 Allington Road, Newick, Lewes, BN8 4NA
G6 DGQ J. Baines, 2 Moor Close, Radcliffe, Manchester, M26 4QF
G6 DGR N. Bean, 19 Coleshill Road, Sutton Coldfield, B75 7AA
GW6 DGU R. Britton, Llwynon, 95 North Road, Cardigan, SA43 1LT
G6 DGV C. Brock, 37 Ashington Drive, Bury, BL8 2TS
G6 DGW E. Ball, Ashleigh 111 Boslowick Road, Falmouth, TR11 2ER
G6 DGX J. Raby, Cedar House, Coppenhall, Stafford, ST18 9DA
G6 DHD A. Rollason, Fern Lodge, The Parade, Newton Abbot, TQ13 0JH
G6 DHI D. Kennedy, 1 Lynton Road, Hindley, Wigan, WN2 4EH
G6 DHT P. Chace, Flat 14, Tudor Rose Court, South Parade, Southsea, PO4 0DE
G6 DHU M. Chace, 26 Stillwater Drive, Unit 9, Westbrook, USA, 4092
G6 DHW I. Clayton, 15 Ashbourne Drive, Desborough, Kettering, NN14 2XG
G6 DIC P. Dickinson, 7 Church Croft, Coton In The Elms, Swadlincote, DE128HG
G6 DID J. Davis, 38 Dover Close, Southwater, Horsham, RH13 9XX
G6 DIE G. Drohan, 23 Lindholme Drive, Rossington, Doncaster, DN11 0UP
G6 DIF Rev. V. Van Den Bergh, St. Francis C Of E Church, Masefield Drive, Tamworth, B79 8JB
G6 DIM T. Eves, Banks Farm, Manor Road, Romford, RM4 1NH
G6 DIO R. Everson, Eversons Farm Bardfield Road Shalford, Braintree, CM7 5HU
G6 DIQ J. Wilkins, 14 Prospect Road, Shanklin, PO37 6AE
G6 DIR M. Wray, 18 Cleveland Street, Loftus, TS13 4JB
G6 DIZ D. Feeley, 177 Rock Street, Sheffield, S3 9JF
G6 DJH D. Harvey, 23 Sprules Road, Brockley, London, SE4 2NL
G6 DJQ G. Tomlinson, 10 Ashbourne Road, Underwood, Nottingham, NG16 5EH
G6 DJS D. Sojkowski, 7 Spenlow Drive, Chelmsford, CM1 4UQ
G6 DJX M. Turner, 24A Cedar Road, Balby, Doncaster, DN4 9DT
G6 DJY W. Telford, The Walnuts, Main Road, Boston, PE20 2LQ
G6 DKE E. Reynolds, 11 New St., Sudbury, CO10 1JB
G6 DKF L. Marsh, 18 Northgate, Hornsea, HU18 1ES
G6 DKI R. Tew, 11 Huson Road, Warfield, Bracknell, RG42 2QX
G6 DKK S. Simes, 53 Waterford Lane, Cherry Willingham, Lincoln, LN3 4AN
G6 DKM L. Sandford, 150 Tipton Road, Woodsetton, Dudley, DY3 1AL
G6 DKS R. Saverton, Flat 8 2 Christ Church Road, Surbiton, KT5 8JJ
G6 DLJ P. Bridges, Nutwood, Coldridge, Crediton, EX17 6AY
G6 DLM Q. Borthwick, 106 Westpole Avenue, Cockfosters, Barnet, EN4 0BB
G6 DLT J. Bartlett, 5 Park View Gypsyville, Hull, HU4 6NG
G6 DLZ P. Bosanquet-Bryant, Flat 21, Westcliff Court, Clacton-on-Sea, CO15 1LA
G6 DMC D. Crabtree, 145 Mendip Road, Yatton, Bristol, BS49 4ER
G6 DMF D. Wilkins, 124 Fullers Mead, Harlow, CM17 9AU
G6 DMG S. Wellon, 71 Toftdale Green, Lyppard Bourne, Worcester, WR4 0PE
G6 DMM K. Webster, 27 Glendale Close, Horsham, RH12 4GR
G6 DMQ P. Singleton, 37 Victoria Road, Harborne, Birmingham, B17 0AQ
G6 DNA T. Cattermole, 24 Cromwell Road, Colchester, CO2 7EN
G6 DNH M. Carvell, 12 Liskeard Drive, Allestree, Derby, DE22 2GW
GI6 DNI D. Chapman, 3 Brustin Lee, Ballygally, Larne, BT40 2QA
G6 DNL K. Snellin, 5 Kenley Close, Chislehurst, BR7 6QT
G6 DNV G. Taylor, 10 Scott Close, Hexham, NE46 2QB
GW6 DOC R. Yarnold, 47 Small Meadow Court, Caerphilly, CF83 3RT
G6 DOD M. Wheeler, 105 High Street, Wootton Bridge, Ryde, PO33 4LU
G6 DOI C. Wigginton, 4 Copes Haven, Shenley Brook End, Milton Keynes, MK5 7HA
GW6 DOK C. Williams, Caermai, Stad Pen Y Berth, Llanfairpwllgwyngyll, LL61 5YT
G6 DON J. Walsh, 7 Unicorn Place, Ball Green, Stoke-on-Trent, ST6 6LX
G6 DOQ H. Davies, 76 Brook Lane, Timperley, Altrincham, WA15 6RS
G6 DOR D. Durrant, 22 St. Martinsfield, Martinstown, Dorchester, DT2 9JU
G6 DOV L. Dunn, 24 Mynchen Road, Beaconsfield, HP9 2BA
G6 DOW A. Deacon, 1 Connaught Gardens, Crawley, RH10 8NB
G6 DOZ L. Dell, 205 Thelwall Lane, Warrington, WA4 1NF
G6 DPE D. Evans, 631 Chatsworth Road, Chesterfield, S40 3NT
G6 DPH B. Flinn, 65 Marina Avenue, Great Sankey, Warrington, WA5 1JH
G6 DPL L. Green, 76 Dibleys, Blewbury, Didcot, OX11 9PU
G6 DPS M. Harrison, 33 Campion Park, Up Hatherley, Cheltenham, GL51 3WA
G6 DQA M. Morgan, 125 Holymoor Road, Holymoorside, Chesterfield, S42 7DR
GW6 DQB Dr J. Mitchell, Y Graigwen, Cadnant Road, Menai Bridge, LL59 5NG
GW6 DQH D. Moore, 71 Woodlands Avenue, Talgarth, Brecon, LD3 0AT
G6 DQK P. Mcbride, 14 Hob Hill Meadows, Glossop, SK13 8LW
G6 DQO I. Martin, 6 Hollow Oak Lane, Cuddington, Northwich, CW8 2XN
G6 DQT W. Lasbury, Sonserra Flats, Flat 4 Fekruna St, St Pauls Bay, Malta,
G6 DQU L. Goodison, 16 Springfield, Sowerby Bridge, HX6 1AD
G6 DQY J. Orrells, Perry Willows, Yeaton, Shrewsbury, SY4 2HY
G6 DQZ N. Perry, 10 Carlyle Avenue, Kidderminster, DY10 3QZ
G6 DRC D. Cooper, 20 Simon De Montfort Drive, Evesham, WR11 4NR
G6 DRG T. Place, 73 Williams Street, Langold, Worksop, S81 9NX
G6 DRH D. Hickton, 27 Vanguard Road, Long Eaton, Nottingham, NG10 1DX
GI6 DRK I. Humes, 160 North Road, Belfast, BT4 3DJ
G6 DRN P. Haylor, 76 Beauchamp Road, Billesley Common, Birmingham, B13 0NR
G6 DRP D. Hemmins, 18 Burn Walk, Burnham, Slough, SL1 7EW
G6 DSA R. Jeffery, 7 Corfe Way, Winsford, CW7 1LU
G6 DSD R. Jones, 20 Bibsworth Avenue, Moseley, Birmingham, B13 0BA
G6 DSG N. Austin, 184 Tunstall Road, Knypersley, Stoke-on-Trent, ST8 7AH
G6 DSP C. Addis, 1 Newchurch Lane, Culcheth, Warrington, WA3 5RW
G6 DTH A. Allnutt, The Squirrels, Nutcombe Lane, Dorking, RH4 3DZ
G6 DTN D. Crake, Kentolop, Holyhead Road, Montford Bridge, Shrewsbury, SY4 1EE
G6 DTT A. Campbell, Eden Park, Den Cross, Edenbridge, TN8 5PW
G6 DTW Dr P. Lee, Links Corner Cottage Liks Road, Ashtead, Surrey, Uk, KT212EG
G6 DU G. Dunstan, 29 Simon Street, Victoria Point Qld, Australia, 4165
G6 DUC A. Rowlands, Hill House, Bridge Road Leigh Woods, Bristol, BS8 3PE
G6 DUH D. Crewe, 71 Ladybalk Lane, Pontefract, WF8 1LA
G6 DUI I. Castle, 26 Lonsdale Drive, Sittingbourne, ME10 1TS
G6 DUN R. Burrows, 32 Frenchs Farm Road, Poole, BH16 5RT
G6 DUT M. Bluck, 26 Mayfield Avenue, Scarborough, YO12 6DF
G6 DVE A. Redshaw, 417 Marston Road, Marston, Oxford, OX3 0JG
G6 DVO H. Warehand, 79 Woodlands Road, Hertford, SG13 7JF
G6 DVP R. Vickers, 51 Charlecote Close, Redditch, B98 0TQ
G6 DWM G. Sohal, 15 Icknield Road, Luton, LU3 2NY
G6 DWO R. Smart, 52 Devonshire Avenue, Southsea, PO4 9EF
G6 DWS N. Shearer, 64 Balsall Heath Road, Edgbaston, Birmingham, B5 7NE
GI6 DWZ T. Duffin, 16 Park Lane, Newcastle, BT33 0AR
G6 DXC C. Ellis, 43 Epsom Walk, Hereford, HR4 9NJ
G6 DXD A. Edwards, 35 Eldon Road, Cheltenham, GL52 6TX
G6 DXP M. Gentry, Maeldune, Orsett Road, Stanford-le-Hope, SS17 8NS
G6 DXY T. Dix, Willow Cottage 31 London Road, Woolmer Green, SG3 6JE
G6 DYK S. Hicks, 15 Chalice Close, Hampton Gardens, Peterborough, PE7 8RL
G6 DYM G. Hudgell, 18 Fellowes Lane, Colney Heath, St Albans, AL4 0QA
G6 DYR D. Bettie, 54 Grendon Road, Polesworth, Tamworth, B78 1NU
G6 DYU L. Horn, 9 Musson Close, Irthlingborough, Wellingborough, NN9 5XW
GW6 DYW C. Hughes, Cym Lane, Rogerstone, Newport, NP10 9EN
G6 DZI C. Kuss, 20 Windermere Road, Haydock, St Helens, WA11 0ES
G6 DZJ S. Kitchener, 101 Highfield Road, Tring, HP23 4DS
G6 DZT D. Anstock, 12 Raymoor Avenue, St. Marys Bay, Romney Marsh, TN29 0RD
G6 DZX J. Beardmore, 6 Essex Close, Congleton, CW12 1SH
G6 EAH R. Carrington, 45 Crompton Road, Pleasley, Mansfield, NG19 7RG

G6 EAM J. Calder, Wyrley Lodge, Hill Farm, Northwood Lane, Bewdley, DY12 1AT
G6 EAR P. Dowler, 21A Wash Lane, Clacton-on-Sea, CO15 1UW
G6 EAX S. Hufschmied, 99 Leverstock Green Road, Hemel Hempstead, HP3 8PR
G6 EAY M. Hughes, 15 Feckenham Road Headless Cross Redditch B97 5As, Redditch, B97 5AS
G6 EAZ R. Hildebrand, Meadow View, Cunningham Place, Bakewell, DE45 1DD
G6 EBL M. Brundle, 36 Campion Street, Derby, DE22 3EF
G6 EBO B. Beckers, 6 Patmore Way, Collier Row, Romford, RM5 2HF
GI6 EBX S. Bird, 70 Greencastle Road, Kilkeel, Newry, BT34 4JJ
G6 ECG G. Bradley, Greentree Cottage, Town End, Broadclyst, Exeter, EX5 3HW
G6 ECN B. Clay, 3 Sandy Close, Bollington, Macclesfield, SK10 5DT
G6 ECS P. Buckingham, Thrimley House, Thrimley Lane, Bishop's Stortford, CM23 1HX
G6 ECT R. Close, 208 Northampton Road, Wellingborough, NN8 3PW
G6 EDC F. Davis, 28 Western Drive, Claybrooke Parva, Lutterworth, LE17 5AG
G6 EDD S. Donald, 5 Windsor Road, Royston, SG8 9JF
G6 EDF D. Evans, 107 Bradbury Road, Solihull, B92 8AL
G6 EDM D. Evans, Caithness, Greenlands Road, Sevenoaks, TN15 6PG
G6 EDR R. Fletcher, 31 Snowdrop Close, Broadfield, Crawley, RH11 9EG
G6 EDT M. Fletcher, 9 The Causeway Carlton, Bedford, MK43 7LT
G6 EDU M. Firth, Kasamily, 73 Lions Lane, Ashley Heath, Ringwood, BH24 2HH
G6 EEB W. Moodie, 141 Wood Lane, Handsworth, Birmingham, B20 2AQ
G6 EED N. Mockridge, 2 Palm View, Waterloo Cross Caravan Park, Uffculme, Cullompton, EX15 3ES
G6 EEE A. Mead, 17 Beadle Way, Great Leighs, Chelmsford, CM3 1RT
G6 EEF D. Malekout, 59 Glebelands Avenue, Ilford, IG2 7DL
GI6 EEH S. Mccullagh, 18 Village Walk, Portadown, Craigavon, BT63 5TL
G6 EER G. Middleton, 37 Hamdon Close, Stoke-Sub-Hamdon, TA14 6QN
G6 EES P. Morris, 8 Millfield, Lambourn, Hungerford, RG17 8YQ
G6 EET D. Monk, 311 Birmingham Road, Lickey End, Bromsgrove, B61 0ER
G6 EEU M. Meredith, 55 New Barn Lane, Cheltenham, GL52 3LB
GU6 EFB K. Le Boutillier, Tiverton, Bailiffs Cross Road, St Andrew, Guernsey, GY6 8RT
G6 EFE S. Weiss, 7 Tennyson Avenue, Grays, RM17 5RG
GJ6 EFW E. Walscharts, Le Creux Country Park, La Route Orange, St. Brelade, Jersey, JE3 8GQ
GI6 EGE R. Hadden, 28 Belfast Road, Comber, Newtownards, BT23 5EW
GI6 EGJ J. Potts, 217 Donaghanie Road, Beragh, Omagh, BT79 0RZ
G6 EGO D. Pink, 31 The Fairway, Daventry, NN11 4NW
G6 EGU B. Nixon, 87 Field Avenue, Canterbury, CT1 1TS
G6 EGY I. Niven, Keepers Cottage, Sulby, Northampton, NN6 6EZ
G6 EHE W. Ward, 88 Central Road, Cromer, NR279BW
G6 EHG R. Quiney, 59 Malham Road, Stourport-on-Severn, DY13 8NT
G6 EHJ J. Parker, 16 Southland Road, Leicester, LE2 3RJ
G6 EHL D. Partington, 6 Celandine Avenue, Cowplain, Waterlooville, PO8 9BE
G6 EIH R. Mccracken, 16 Station Road, Rolleston-On-Dove, Burton-on-Trent, DE13 9AA
G6 EIO A. Mitchell, 85 Farriers Green, Monkton Heathfield, Taunton, TA2 8PP
GI6 EIR D. Mullan, 5 Mountfield Drive, Coleraine, BT52 1TW
G6 EIU M. Parkins, 6 Quantock Avenue Caversham, Reading, RG4 6PY
G6 EIZ J. Austin, 17 New Road, Ascot, SL5 8QB
G6 EJD D. Bird, 59 Speedwell Close, Melksham, SN12 7TE
G6 EJF R. Amos, 89 Stanstrete Field, Great Notley, Braintree, CM77 7JW
G6 EJH J. Bradley, 66 Belmont Road, Parkstone, Poole, BH14 0DB
G6 EJI P. Barrett, 4 Hazel Road, Middleton, Manchester, M24 2WB
G6 EJM T. Burrows, 11 Louis Close, Old Catton, Norwich, NR6 7BG
G6 EJT J. Bibby, 19 Richmond Crescent, Mossley, Ashton-under-Lyne, OL5 9LQ
G6 EJU C. Biddles, 129 Hallam Crescent East, Leicester, LE3 1FG
GI6 EJW W. Mccormick, 46 Gortlane Drive, Greenisland, Carrickfergus, BT38 8SY
G6 EKM R. Perks, 1 Tothill Court Shaldon, Teignmouth, TQ14 0EJ
G6 EKS A. Stelfox, 6 Surrey Street, Glossop, SK13 7AH
G6 EKT Hornsea ARC, c/o R. Guttridge, Ivy House, Rise Road, Hull, HU11 5BH
G6 ELG M. Wright, 6 Tregalister Gardens, St. Germans, Saltash, PL12 5NQ
G6 EMB G. Collins, 33 West Hay Grove, Kemble, Cirencester, GL7 6BE
G6 EML C. Exelby, The Old Farm House Lower Denford, Hungerford, RG17 0UN
G6 ENA M. Pyrah, 53 St. Georges Road, Ramsgate, CT11 7EF
G6 ENN D. Gordon, 38 Deer Park Road, Langtoft, Peterborough, PE6 9RB
G6 ENO B. Garrett, 226 Rydal Drive, Bexleyheath, DA7 5DG
GJ6 ENP J. Gready, Avon Cottage, La Rue D'Elysee, St Peter, Jersey, JE3 7DT
G6 ENQ G. Greenwood, 27 Delph Mount, Great Harwood, Blackburn, BB6 7QF
G6 ENR S. Grant, The Vicarage, Downe Street, Driffield, YO25 6DX
G6 ENS S. Gordon, 20 Hawkins Crescent, Shoreham-by-Sea, BN43 6TP
G6 ENT D. Gordon, 20 Swinburne Court, 143 Brighton Road, Lancing, BN15 8HX
G6 ENU I. Gordon, 9 Park Road, Camberley, GU15 2SP
G6 ENY N. Graham, Millers Croft, Queens Road, Freshwater, PO40 9ES
G6 ENZ G. Holmes, 10 Birch Road, Stamford, PE9 2FB
G6 EOK R. Lewis, 12 Station Road, Wimborne, BH21 1RG
G6 EON P. Martin, 40 Carnarthen Street, Camborne, TR14 8UP
G6 EOO R. Machin, 236 Tamworth Road, Kettlebrook, Tamworth, B77 1BY
G6 EOR W. Power, 31 Darbys Hill Road, Tividale, Oldbury, B69 1SE
G6 EPD J. Howe, 103 York Road, Bury St Edmunds, IP33 3EG
G6 EPL R. Jonas, 49 Clarendon Road, Aylesham, Canterbury, CT3 3AQ
G6 EPN P. Knight, Hawkwind, Elcot Lane, Marlborough, SN8 2AZ
G6 EPQ C. Kingston, 3 Hill Street, Cheslyn Hay, Walsall, WS6 7HR
GM6 EPU B. Sherman, 240 Annan Road, Dumfries, DG1 3HE
G6 EPX P. Shuttleworth, 12 Oak Avenue, Penwortham, Preston, PR1 0XQ
G6 EQB J. Singleton, 48 Pennine Way, Ashby-de-la-Zouch, LE65 1EW
G6 EQF R. Skinner, 23 Woodstock Road, Worcester, WR2 5ND
G6 EQI R. Smith, 3 Mendip Edge, Weston-Super-Mare, BS24 9JF
G6 EQL S. Thomas, 64 Victoria Road, Aigburth, Liverpool, L17 0DP
G6 EQP T. Thompson, 7 West Bank, Dorking, RH4 3BZ
G6 EQS J. Hastings, 2 Coltsfoot Road, Rushden, NN10 0GE
G6 EQT J. Aston, 3 Valley Road, Darley, Harrogate, HG3 2QE
G6 EQZ R. Bracken, 72 Brampton Way, Portishead, Bristol, BS20 6YT
G6 ERI R. Couch, 54 Hill Park Road, Gosport, PO12 3EB
G6 ERJ A. Croucher, 73 Loxley Close, Church Hill North, Redditch, B98 9JH
G6 ERK A. Cunliffe, 28 Rosebank Close, Ainsworth, Bolton, BL2 5QU
G6 ESJ P. Wookey, 16 Danvers Way, Westbury, BA13 3UE
G6 ESK D. Whittle, 3 Mackley Way, Harbury, Leamington Spa, CV33 9NP
G6 ESM D. Tankaria, 23 Oakwood Avenue, Southall, UB1 3QD
G6 ESQ P. Baker, 12 College Close, Coltishall, Norwich, NR12 7DT
G6 ETC J. Brown, 44 Perowne Way, Sandown, PO36 9BX
G6 ETL P. Cooke, 55 Priory Road, Portbury, Bristol, BS20 7TQ
G6 ETP J. Cookson, Barker Fold Farm, Tockholes Road, Tockholes, Darwen, BB3 0LU
GI6 ETQ A. Campbell, 16 Parkwood, Lisburn, BT27 4EF
G6 ETX M. Carter, 22 John Morgan Close, Hook, RG27 9RP
G6 ETZ C. Chalmers, Flat 30, Luna Apartments, 272 Field End Road, Ruislip, HA4 9DL
GM6 EUC D. Cruickshank, 61 Woodside Road, Banchory, AB31 4EN
G6 EUF P. Raynor, 29 Kilvin Drive, Beverley, HU17 9PG
G6 EUG C. Slater, 70 Windsor Avenue, Ashton-On-Ribble, Preston, PR2 1JD
G6 EUI C. Shaw, 19 Church Road, Teversham, Cambridge, CB1 9AZ
G6 EUO J. Slater, 47 Broom Road, Lakenheath, Brandon, IP27 9EZ
GW6 EUR P. Williams, Llwyn, Manafon, Welshpool, SY21 8BJ
GW6 EUS R. Williams, Llwyn, Manafon, Welshpool, SY21 8BJ
GW6 EUT A. Williams, Brynfield, Kingswood Forden, Welshpool Powys, SY21 8TS
G6 EUU A. Wilson, Flat 5, Shelley House, London, E2 0HE
G6 EUW A. Sheridan, 6 Mill Road, Burnham-on-Crouch, CM0 8PZ
G6 EUY W. Shadwell, 2 Poppy Close, Yaxley, Peterborough, PE7 3FA
G6 EVC C. Sleight, Orchard House, School Hill, Napton, Southam, CV47 8NN
G6 EVW S. Wem, 10 Astley Crescent, Halesowen, B62 9SX
G6 EVX A. Wood, 54 Wilton Park Road, Shanklin, PO37 7BU
G6 EVY H. Woolrych, 20 Meadow Drive, Devizes, SN10 3BJ
G6 EWH E. Turton, 27 Langdale Avenue, Hesketh Bank, Preston, PR4 6TD
G6 EWK D. Mason, 15 Windmill Gardens, Prenton, CH43 7YQ
GI6 EWO B. Davis, 49 The Roddens, Larne, BT40 1QL
G6 EWP D. Davy, 22 Scott Gardens, Lincoln, LN2 4LX
GW6 EWQ C. Dormer, 39 Eastmoor Road, Newport, NP19 4NX
GW6 EWX N. Evans, Abbey Dingle Care Home Abbey Road, Llangollen Denbighshire, LL20 8DD
G6 EXC P. Gibson, 9 Mallard Close, Aughton, Ormskirk, L39 5QJ
G6 EXE M. Graham, 11 Robert Moffat, High Legh, Knutsford, WA16 6PS
G6 EXG M. Gee, 100 Plantation Hill, Worksop, S81 0QN
G6 EXN E. Hall, 9 Valance Avenue, Chingford, London, E4 6DR
G6 EXU A. Jobber, Church Hill, Kings North, Ashford, TN23 3EG
G6 EXX B. Kent, 4 Bedmond Road, Pimlico, Hemel Hempstead, HP3 8SH
G6 EXZ A. Kent, 166 Louth Road, Scartho, Grimsby, DN33 2LG
G6 EYA P. Kershaw, 8 Stoppers Hill, Brinkworth, Chippenham, SN15 5AW
G6 EYI C. Moore, Glen View, Fosseway, Radstock, BA3 4BB
G6 EYJ D. Morton, 27 Beechfield Way, Hazlemere, High Wycombe, HP15 7TP
G6 EYS A. Morne, 16 Warmden Avenue, Baxenden, Accrington, BB5 2PR
G6 EZG I. Prince, 21 Aberfield Drive, Crigglestone, Wakefield, WF4 3PT
G6 EZH S. Pepper, 149 The Hill, Glapwell, Chesterfield, S44 5LU
G6 EZI J. Oldroyd, 357 Commercial St #704, Boston, USA, 2109
G6 EZM D. Winters, 13A St. Catherines Road, Bournemouth, BH6 4AE
G6 EZR F. Thompson, 28 Lordsmead, Cranfield, MK43 0HP
G6 EZY D. Powell, 30 Fernley Road, Southport, PR8 5AU
G6 FAF C. Narroway, 26 Fern Way, Watford, WD25 0HG
G6 FAH K. Lawrence, 54 Sheldrake Road, Christchurch, BH23 4BP
G6 FAL R. Stoneman, 9 Winchester Road, Northampton, NN4 8AZ
G6 FAX P. Brooks, Flat 4, 8 Glencathara Road, Bognor Regis, PO21 2SF
G6 FBA J. Butters, 21 Erleigh Road, Reading, RG1 5LR
G6 FBB R. Chidgey, 14 Drury Road, Colchester, CO2 7UX
G6 FBH G. Davis, Westbury House, 3 Windermere, Tamworth, B77 5TD
G6 FBJ J. Endicott, 1 Elm Tree Park, Yealmpton, Plymouth, PL8 2ED
GW6 FBV T. Howell, 19 Uwchgwendraeth, Drefach, Llanelli, SA14 7AR
G6 FCI C. Mcmahon, 6 Layton Road, Blackpool, FY3 8HS
G6 FCJ P. Magnus-Watson, 95 Sutton Lane, Slough, SL3 8AU
G6 FCL J. Mahoney, Winton Dene, The Street, Sudbury, CO10 8JP
G6 FCS D. Lane, 10 Whylands Close, Worthing, BN13 3HB
G6 FDD R. Pinchin, 10 Epping Drive, Melton Mowbray, LE13 1UH
G6 FDG I. Rivers, 35 Cloverville Approach, Bradford, BD6 1ET
G6 FDI B. Raymer, 19 Caithness Drive, Crosby, Liverpool, L23 0RG
G6 FDK S. Maskrey, The Hayloft, Stamford Lane, Chester, CH3 7QD
G6 FDO I. Moody, 54 Lansdowne Road, Studley, B80 7RD
G6 FDP S. Litobarski, 7 Exeter Road, Southsea, PO4 9PZ
GM6 FDQ G. Allan, Corse Farm, Kininmonth, Peterhead, AB42 4JU
G6 FDU R. Butterworth, 49 Swandene, Pagham, Bognor Regis, PO21 4UR
G6 FDX C. Bicknell, Flat 25, Madderfields Court, London, N11 2JL
GW6 FED D. Corsi, 4 Horsley Drive, Wrexham, LL12 8BE
G6 FEI D. Harris, 53 Welwyn Drive, Salford, M6 7PQ
G6 FEJ R. Hawkes, 1 The Fairway, Wellingborough, NN9 5YS
G6 FEM A. Harris, April Cottage, Sheepcote Green, Clavering, Saffron Walden, CB11 4SJ
GI6 FEN P. Irwin, 6 Cairnburn Avenue, Belfast, BT4 2HT
G6 FEQ P. Jolly, 22 Wellhouse Road, Barnoldswick, BB18 6DD
GW6 FES S. Jones, 12 Meadow Croft, Cross Lanes, Wrexham, LL13 0UJ
G6 FEX J. Sandford, 23 South Lawn, Locking, Weston-Super-Mare, BS24 8AD
G6 FFB D. Meaker, 181 Dovecote, Yate, Bristol, BS37 4PF
G6 FFH T. Sallis, 54 West Way, Hove, BN3 8LQ
G6 FFL T. Short, Freemans Farm, Itchington, Alveston, Bristol, BS35 3TL
G6 FFQ F. Bilton, 50 Coldwell Road, Crossgates, Leeds, LS15 7HA
G6 FFR B. Berry, 7 Barlow Close, Telford, TF3 2NQ
G6 FFU P. Coogan, 39 Sycamore Crescent, Macclesfield, SK11 8LW
G6 FGA R. Holyhead, 42 Dockham Road, Cinderford, GL14 2BH
G6 FGC C. Hawkins, 80 Duston Wildes, Northampton, NN5 6NR
G6 FGJ C. Tandy, 7 The Swallows, Patrons Way West, Uxbridge, UB9 5PB
G6 FGL T. Toulson, 30 Old Park Avenue, Sheffield, S8 7DR
G6 FGV J. Webber, 5 Leda Mews, Achilles Close, Hemel Hempstead, HP2 5WR
G6 FGW R. Weekes, 84 Vera Road, Yardley, Birmingham, B26 1TT
G6 FGY E. Westbrook, 66 Nelson Close, Croydon, CR0 3SW
G6 FHB M. Williams, 20 Wignall Road, Stoke-on-Trent, ST6 5LE
GI6 FHD A. Mcpartland, 4 Clanbrassil Gardens, Portadown, Craigavon, BT63 5YD
G6 FHK C. Leonard, 138 Sundridge Drive, Chatham, ME5 8JD
G6 FHM D. Sunderland, 1 Allfield Cottages, Condover, Shrewsbury, SY5 7AP
G6 FHR R. Plant, 32 Buckland Road, Pen Mill Trading Estate, Yeovil, BA21 5HA

G6 FIB T. Wicks, 123 The Crescent, Andover, SP10 3BN
GM6 FIK D. Stevenson, Flat C, 2 Melbourne Court Braidpark Drive, Giffnock, Glasgow, G46 6LA
G6 FIL D. Smith, 323 Colchester Road, Ipswich, IP4 4SF
G6 FIN A. Stevens, 16 Tremlett Grove, Ipplepen, Newton Abbot, TQ12 5BZ
G6 FIO J. Slater, 154 Ralph Road, Shirley, Solihull, B90 3JZ
G6 FIT A. Lewis, 81 Ashton Avenue, Rainhill, Prescot, L35 0QR
G6 FJA F. Aunger, 2, Lowick Woodthorpe, York, YO24 2RF
G6 FJE L. Plewa, 174 Dorset Avenue, Chelmsford, CM2 8YY
G6 FJG N. Pinkney, 4 St. Hughs Road, Buckden, St Neots, PE19 5UB
G6 FJI M. Richards, 1 Ashenden Close, Abingdon, OX14 1QE
G6 FJL P. Standen, 17 Canberra Road, Worthing, BN13 3HH
G6 FJO S. Turner, 14 The Poplars, Launton, Bicester, OX26 5DW
G6 FJP R. Wild, 15 Cartridge Street, Heywood, OL10 3AF
G6 FKB E. Taylor, 19 Chester Road, Saltney Ferry, Chester, CH4 0AQ
G6 FKE K. Redmond, 8 George Street, Morecambe, LA4 5SU
G6 FKL L. Taylor, 127 Dundee Close, Fearnhead, Warrington, WA2 0UJ
G6 FKN Dr M. Lee, 55 Wodeland Avenue, Guildford, GU2 4LA
GW6 FKP S. Moore, 25 Overdale Avenue, Mynydd Isa, Mold, CH7 6US
G6 FKR D. Roberts, 10 Woodville Terrace, Darwen, BB3 2JH
G6 FKS S. Robinson, 4 Grayling Close, Cambridge, CB4 1NP
G6 FKY T. Norris, The Old Post Office, Arundel Road, Arundel, BN18 0SD
G6 FLE N. Scott, 33 Leaze Close, Thornbury, Bristol, BS35 2FH
G6 FLH J. Smith, 25 Seafield Close, Seaford, BN25 3JR
G6 FLK J. Walton, 168 Park Road, Stanley, DH9 7AJ
GM6 FLL A. Simpson-Fraser, 430 Millcroft Road, Cumbernauld, Glasgow, G67 2QW
G6 FLQ G. Smith, Top Of The Hill, Hockerton Road, Kirklington, Newark, NG22 8PB
G6 FLR J. Smith, 33 Hop Pole Green, Leigh Sinton, Malvern, WR13 5DP
GW6 FLU C. Mock, Homelea, Royal Oak Hill, Newport, NP18 1JF
G6 FLW C. Thompson, 27 Queensland Drive, Colchester, CO2 8UD
G6 FLY H. Lee, 26 Ratcliffe Avenue, Branston, Burton-on-Trent, DE14 3DA
G6 FMF D. Timson, 40 Rockwood Road, Calverley, Pudsey, LS28 5AA
G6 FMN P. Rogers, 12 St. Peters Rise, Headley Park, Bristol, BS13 7LY
G6 FMS M. Peers, Viewpoint, Wicker Lane Guilden Sutton, Chester, CH3 7EL
G6 FMU I. Muir, 7 Bovarde Avenue Kings Hill, West Malling, ME19 4BS
GW6 FNB D. Morris, 11 Ffordd-Y-Mynach, Pyle, Bridgend, CF33 6HT
G6 FNJ R. Oglesby, 1 High Street, Littleton Panell, Devizes, SN10 4EL
G6 FNQ A. Smith, 16 Hazel Way, Barwell, Leicester, LE9 8GP
G6 FNY T. Strand, 24 Speedwell Close, Melksham, SN12 7TE
G6 FOF C. Morris, Flat Iv, Brummel Court, Worcester Road, Droitwich, WR9 0DF
G6 FOI Dr A. Regan, 153 Acre Lane, Cheadle Hulme, Cheadle, SK8 7PB
GI6 FOR N. Lane, 117A Hillhall Road, Lisburn, BT27 5BT
GM6 FOT T. Armour, 69 Hillend Road, Clarkston, Glasgow, G76 7XT
G6 FOV D. Aldridge, 17 Priory Close, Tavistock, PL19 9DJ
G6 FOW P. Atkinson, 30 Spital Terrace, Gainsborough, DN21 2HQ
G6 FOX West Midland Rdf, c/o T. Ray, 1 Providence Lane, Leamore, Walsall, WS3 2AQ
G6 FPC J. Body, The Durhams, Church Street, Bowerchalke, Salisbury, SP5 5BH
G6 FPF Dr G. Barnes, Rockleigh, 17 Savile Park, Halifax, HX1 3EA
G6 FPH M. Cole, 45 Gainsborough Road, Tilgate, Crawley, RH10 5LD
G6 FPK N. Cooper, 53 Stanway Road, Benhall, Cheltenham, GL51 6BU
G6 FPN Y. Dunn, 117 All Saints Way, West Bromwich, B71 1RU
G6 FPO G. Faulkner, Flat 2, 1265 Melton Road, Syston, Leicester, LE7 2EN
G6 FPP A. Ford, 8 Merganser Drive, Bicester, OX26 6UQ
G6 FPQ C. Groves, 7 Mill House, Mill Lane, Wick, Littlehampton, BN17 7PJ
G6 FPX S. Hill, 7 Meadowcroft Court, Runcorn, WA7 2NS
G6 FQL R. Heath, Flat 2, Portland View, 62 High Street, Great Yarmouth, NR31 6RQ
G6 FQP B. Kneebone, 1 Chapel Terrace, Carnkie, Helston, TR13 0DT
GI6 FQT D. Mcconville, 28 Derrycor Lane, Derryadd, Craigavon, BT66 6QW
G6 FRB J. Pearce, 25 Boughton Street, St. Johns, Worcester, WR2 4HE
G6 FRS Farnborough & District, c/o M. Hearsey, Halycon, Lawday Link, Farnham, GU9 0BS
G6 FS Leiston ARC, c/o M. Danfer, The Nook, Mill Common, Blaxhall, Woodbridge, IP12 2ED
GM6 FSG W. Chamberlain, Cwmmelyn, Kings Road, Whithorn, Newton Stewart, DG8 8PP
G6 FSK D. Fisher, 6 Small Holdings Road Clenchwarton, King's Lynn, PE34 4DY
G6 FSP D. Helliwell, 1 Beechfield Avenue, Barton, Torquay, TQ2 8HU
G6 FSU M. Apperly, Chaundlers, Church Lane, Kingston, Cambridge, CB23 2NG
G6 FTA M. Everall, 17 Golden Park Avenue, Torquay, TQ2 8LR
GM6 FTE P. Bondar, 4 The Steadings Swinside Townhead Farm, Jedburgh, TD8 6ND
G6 FTH D. Clark, 43 Glenfield Crescent, Chesterfield, S41 8SF
G6 FTJ P. Carter, 145 Wakefield Road, Dewsbury, WF12 8AJ
G6 FTL P. Dixon, 37 Carlton Close, Parkgate, Neston, CH64 6RB
GI6 FTM J. Dynes, 30 Breagh Road, Portadown, Craigavon, BT63 5LT
G6 FTR T. Fabbri, 53 Langdale Road, Cheltenham, GL51 3LX
G6 FTY R. Miles, 60 Aylesham Way, Yateley, GU46 6NT
G6 FUT I. Donn, Charsleys, Weedon Hill Hyde Heath, Amersham, HP6 5RN
GW6 FUY R. Fishwick, The Elms, New Road, Brynteg, Wrexham, LL11 6PD
G6 FVB R. Baker, 23 Disraeli Road, London, W5 5HS
G6 FVD J. Henville, 5 Station Road, Hemyock, Cullompton, EX15 3SE
G6 FVF P. Fenn, 38 Harwood Close, Welwyn Garden City, AL8 7SN
G6 FVJ A. James, 82 Sandringham Drive, Spondon, Derby, DE21 7QA
G6 FVL R. Young, 134 Harport Road, Redditch, B98 7PD
G6 FVM A. Williamson, 31 Poulton Road, Southport, PR9 7BE
G6 FVZ R. Munns, 2 The Tyleshades, Romsey, SO51 5RJ
G6 FWK D. Booth, 54 Shaw Drive, Knutsford, WA16 8JR
G6 FWO P. Dover, 92 The Roundway, Claygate, Esher, KT10 0DW
G6 FWU T. Pillar, 14 Thoresby Mews, Bridlington, YO16 7GZ
G6 FXE C. Walton, 49 Blandford Drive, Walsgrave, Coventry, CV2 2JD
G6 FXR D. Ainslie, Brackendene, 17 Sandhurst Road, Crowthorne, RG45 7HR
GI6 FXY E. Connolly, 21 Clanrye Avenue, Newry, BT35 6EH
GM6 FXZ A. Carnall, 3 Main St., Glenluce, Newton Stewart, DG8 0PN
G6 FYA C. Collins, 29, Seaford, BN25 1SP
G6 FYC J. Cowee, 26 Arundel Road, Heatherside, Camberley, GU15 1DL
G6 FYD D. Dale, 81 Burton Manor Road, Stafford, ST17 9PR
G6 FYE C. Das Neves Pedro, The Leas, Bearwood, Leominster, HR6 9EE
G6 FYL N. Harris, 104 Blandford Drive, Walsgrave, Coventry, CV2 2NE
G6 FYR A. Johnson, 14 Norman Road, Newhaven, BN9 9LJ
G6 FYT G. Whiting, Glyn South View, Liskeard, PL14 3EX
G6 FYU J. Walker, 33 Erica Way, Horsham, RH12 5XL
G6 FYX N. Holden, 1 Brooklands Way, Lincoln, LN6 0RH
G6 FZC D. Hall, 1 Westfall, Wearhead, Bishop Auckland, DL13 1JD
GI6 FZI G. Mcbriar, 15 Ambleside Drive, Bangor, BT20 4QB
G6 FZV W. Day, 4 Queenswood Drive, Worcester, WR5 3SZ
G6 FZW A. Eaves, 3 Station Cottages, Station Road, Leighton Buzzard, LU7 0SQ
G6 GA R. Rushton, 53 Crossfield Avenue, Blythe Bridge, Stoke-on-Trent, ST11 9PL
G6 GAB W. Honey, 20 Pennor Drive, St Austell, PL25 4UW
G6 GAC G. Jenkins, 75 Rectory Road, Coltishall, Norwich, NR12 7HW
G6 GAF A. Franklin, C/O 4 Princes Close, Seaford, BN25 2EW
GI6 GAG N. Orr, 405 Enniskeen, Drumgor, Craigavon, BT65 4AB
G6 GAK M. Tyrrell, 189 Runcorn Road, Barnton, Northwich, CW8 4HR
GI6 GAQ H. Warke, 5 Meadow View, Ballymoney, BT53 7AH
G6 GAW D. Peters, 46 Sheridan Road, Worthing, BN14 8ET
GI6 GBK J. Anderson, 1 Claragh Hill Drive, Kilrea, Coleraine, BT51 5YR
G6 GBL A. Abbott, 164 Bath Road, Reading, RG30 2HA
G6 GBT I. Cole, 21 Quincey Drive, Erdington, Birmingham, B24 9LX
G6 GBU A. Dixon, 33 Colgrove Road, Loughborough, LE11 3NL
G6 GBY P. Pearce, Criggion Mw Radio Station, Back Lane, Criggion, Shrewsbury, SY5 9BE
G6 GCI C. Burnett, 36 Mill Lane, Romsey, SO51 8EQ
G6 GCJ J. Burnett, 44 Bourne Vale, Hungerford, RG17 0LL
GW6 GCK J. Cook, St. Davids, Chepstow Road Langstone, Newport, NP18 2JR
G6 GCM T. Collings, 55 Flora Thompson Drive, Newport Pagnell, MK16 8SR
G6 GCO D. Davies, 79A Spenser Road, Bedford, MK40 2BE
G6 GCW L. Wiltshire, 4 Nether Close, Eastwood, Nottingham, NG16 3DL
G6 GCY J. Robinson, 84 Hereford Way, Middleton, Manchester, M24 2NN
G6 GDI V. Gerhardi, 24 Putnams Drive, Aston Clinton, Aylesbury, HP22 5HH
G6 GDR C. Price-Gore, 22 Oakham Close, Desborough, Kettering, NN14 2FH
G6 GEK A. Elliott, Knowle House, Hooke Road, Leatherhead, KT24 5DY
G6 GEL K. Inman, 15 Waterbridge Court, Appleton, Warrington, WA4 3BJ
G6 GEN R. Ainsworth, 18 Washington Drive, Slough, SL1 5RE
G6 GEP A. Holdup, Tunnel Farm, Tunnel Rd, Imbil (PO 155), Australia, 4570
G6 GES R. Haywood, 16 The St., Kingston, Canterbury, CT4 6JB
G6 GEV D. Ashton, 12 Little Lees, Charlbury, Chipping Norton, OX7 3HB
G6 GEX C. Farley, 1 Wesley Cottages, Mutley, Plymouth, PL3 4RB
G6 GFA P. Arscott, 122 Woodlands Road, Ashurst, Southampton, SO40 7AL
G6 GFC J. Burrows, 4 Cavendish Crescent, Alsager, Stoke-on-Trent, ST7 2EF
G6 GFG P. Cook, 109 Crosthwaite Avenue, Wirral, CH62 9DF
G6 GFJ H. Goozee, 45 Brighton Road, Purley, CR8 2LR
GM6 GFL Dr D. Begg, 12 Broomhill Road, Penicuik, EH26 9EE
G6 GFO N. Atrill, 22 Lester Close, Hr Compton, Plymouth, PL3 6PX
GM6 GFQ C. Barnard, 122 Union Grove, Aberdeen, AB10 6SB
G6 GFR T. Crook, 21 Cleveland Close, Maidenhead, SL6 1XE
G6 GGN M. Hoskin, 7 Worrall Mews, Clifton, Bristol, BS8 2HF
G6 GGT M. Huntley, 81-82 The Avenue, Sunderland, SR2 7EZ
G6 GGV B. Hollngworth, 62 Illingworth Avenue, Halifax, HX2 9JD
G6 GGW N. Gautrey, 11 Bracken Close, Crawley, RH10 8JR
G6 GGY L. Fitzwater, The Old Cottage, Babylon Lane, Tadworth, KT20 6XE
G6 GGZ M. Ferne, 24 Essex Gardens, Leigh-on-Sea, SS9 4HG
G6 GHE A. Rawdon, 44 Southgate, Hornsea, HU18 1AL
G6 GHU R. Wood, 12 Roundhead Drive, Thame, OX9 3DG
GI6 GIE J. Pinkerton, 40 Seacon Park, Seacon, Ballymoney, BT53 6QB
G6 GIF M. Oram, 25 Jerome Close, Marlow, SL7 1TX
G6 GIH P. Sewell, 6 Hawthorne Avenue, Bedford, MK40 4HJ
G6 GIU A. Stephens, 4 Falcon House, Gurnell Grove, London, W13 0AE
G6 GIV D. Newington, 6 Walkdale Brow, Glossop, SK13 6PX
G6 GJD C. Harper, Flat 2, Dove Tree Court, Blackpool, FY4 4NA
G6 GJN T. Biggs, 3 Pentathlon Way, Cheltenham, GL50 4SE
G6 GJV W. Willis, 24 Old Hall Lane, Walton on the Naze, CO14 8LE
GM6 GJW J. Leith, Appiehouse, Stenness, Stromness, KW16 3LB
G6 GJY S. Smith, 36 Greenfields, Earith, Huntingdon, PE28 3QH
G6 GKG W. Hodson, 27 Belvedere Grove, London, SW19 7RQ
G6 GKK A. Barton, Orchard Bungalow, Westfield Road, Retford, DN22 7BT
G6 GKL M. Borrow, 189 Crofton Road, Orpington, BR6 8JB
GW6 GKP J. Coyne, 44 Brompton Avenue, Rhos On Sea, Colwyn Bay, LL28 4TF
G6 GKT I. Houldridge, 57 Heads Lane, Hessle, HU13 0JH
G6 GLB G. Walker, 141 Deerleap, Bretton, Peterborough, PE3 9YD
G6 GLH P. Burfield, 33 St. Ediths Road, Kemsing, Sevenoaks, TN15 6PT
G6 GLO Gloucestershire County Raynet, c/o A. Webb, 47 Granville Street, Gloucester, GL1 5HL
G6 GLR Gt Lumley ARS, c/o B. Corker, 46 Danelaw, Great Lumley, Chester le Street, DH3 4LU
G6 GLT Dr R. Bennett, 11 Powys Close, Haslingden, Rossendale, BB4 6TH
G6 GLW J. Fisher, 4 Chancery Close, Lincoln, LN6 8SD
G6 GLZ D. Clews, 4 Chancellors Close, Coventry, CV4 7ED
GW6 GMF M. Inness, 6 Denning Road, Wrexham, LL12 7UG
G6 GMH G. Russell, 57 Oak Tree Drive, Cutnall Green, Droitwich, WR9 0QY
G6 GMR Greater Manchester Raynet, c/o N. Czernuszka, 12 Durham Drive, Ashton-under-Lyne, OL6 8BP
G6 GMU J. Oldfield, 30 Village Farm Caravan Site, Bilton Lane, Harrogate, HG1 4DL
G6 GMW Thornton Cleveleys ARS (Tcars), c/o J. Rodway, 9 York Avenue, Thornton-Cleveleys, FY5 2UG
GM6 GMZ N. Saunders, 6 Haughs Of Clinterty, Kinellar, Aberdeen, AB21 0TZ
GI6 GNA H. Wright, 2 Duncans Road, Lisburn, BT28 3LP
G6 GNC J. Thornber, 7 Buckland Close, Peterborough, PE3 9UH
G6 GND R. Lambert, 10 Ambleside, Rugby, CV21 1JB
G6 GNE J. Sugden, 3 Castle Keep, Hibaldstow, Brigg, DN20 9JG
G6 GNO B. Cooper, 8 Stanley Road, Doncaster, DN5 8RR
G6 GOG A. Kerr, 10 Hillcrest Road, Crosby, Liverpool, L23 9XS
G6 GOS M. Jones, 58 Newton Road, Lewes, BN7 2SH
G6 GOV B. Wood, 8 Chichester Drive, Chelmsford, CM1 7RY
G6 GOW R. Wheeler, 2 Heather Close, Brereton, Rugeley, WS15 1BB
G6 GOX L. Timbrell, 3 Rushden Road, Wymington, Rushden, NN10 9LN
G6 GPF J. Woodhouse, 102 Newland Road, Worthing, BN11 1LB
GM6 GPH J. Robertson, 2 Cults Bungalow Cults, Cupar, KY15 7TF

G6 GPR Dr D. Jefferies, 96 Broad Street, Wood Street Village, Guildford, GU3 3BE
G6 GPV K. Patching, 1 Peak House 84 Trinity Street, Fareham, PO16 7SJ
G6 GQF Dr G. Martin, 9 Clarkes Avenue, Kenilworth, CV8 1HX
G6 GQG I. Moston, 19 Wegnalls Way, Leominster, HR6 8TQ
G6 GQI R. Swann, 3 Elizabeth Avenue, Newmarket, CB8 0DJ
G6 GQJ J. Davies, 1 Woodland Road, Halesowen, B62 8JS
GI6 GRV J. Barnett, 2 Donegall Park, Whitehead, Carrickfergus, BT38 9ND
G6 GS Guildford & D Rd, c/o A. Pevy, 2 Oaktree Way, Sandhurst, GU47 8QS
G6 GSF K. Edwards, Whitehaven, High Street, Uckfield, TN22 4JU
G6 GSG R. Grimley, 11 Sewell Wontner Close, Kesgrave, Ipswich, IP5 2GB
G6 GSI D. Millington, 9 Roxburgh Croft, Leamington Spa, CV32 7HT
GW6 GSR South Glamorgan Raynet Group, c/o P. Williams, 5 Whitewell Drive, Llantwit Major, CF61 1TA
G6 GSV J. Williams, 54 Fleminghouse Lane, Huddersfield, HD5 8QG
G6 GTB J. Tracey, 100 Booth Close, Kingswinford, DY6 8SP
G6 GTC P. Willson, 37 The Grove, Sidcup, DA14 5NG
G6 GTH S. Leonard, 231 Hale Road, Hale, Altrincham, WA15 8DN
G6 GTJ L. Baldwin, The Fields, Pinewood Road, Market Drayton, TF9 4QE
GW6 GTS P. Dudman, Chapel House, Pen Y Bryn, Wrexham, LL14 1UA
G6 GTZ P. Wilson, 162 Bowerdean Road, High Wycombe, HP13 6XW
G6 GUC D. Ellis, Field End, Northwood Green, Westbury-on-Severn, GL14 1NB
G6 GUD M. Everley, 5 Firs Close Hazlemere, High Wycombe, HP15 7TF
G6 GUH P. Bonds, The Gables, Crosslane Head, Bridgnorth, WV16 4SJ
G6 GUT B. Turley, 12 Legh Drive, Woodley, Stockport, SK6 1PT
G6 GVF K. Waters, 25 Edwin Road, Twickenham, TW2 6SP
G6 GVH J. Marks, 124 Stowey Road, Yatton, Bristol, BS49 4EB
G6 GVI R. Wilkinson, 84 Park Road, Bolton, BL1 4RQ
G6 GVL M. Longley, 78 Priory Road, Eastbourne, BN23 7BE
G6 GVM P. Martin, 21 Baldwin Avenue, Eastbourne, BN21 1UJ
G6 GVO M. Pearce, 64 Goongarrie Drive, WA, Australia, 6169
G6 GVR G. Whittle, 5 Chantry Close, Westhoughton, Bolton, BL5 2LY
G6 GVS R. Wood, 6 Timberlaine Road, Pevensey Bay, Pevensey, BN24 6DE
G6 GVU S. Wood, 30 Ramsay Way, Eastbourne, BN23 6AL
G6 GVZ E. Rigby, 12 Sorrel Avenue, Tean, Stoke-on-Trent, ST10 4LY
GW6 GW Blackwood & District Amateur Radio Soc., c/o D. Lewis, 23 Gelligroes Road, Pontllanfraith, Blackwood, NP12 2JU
G6 GWE M. Ranger, 13 Springfield Close, Crowborough, TN6 2BN
G6 GWP J. Briggs, Wood Lea, Bawtry Road, Doncaster, DN10 5BS
G6 GWU P. Hopkinson, 59 Mulberry Drive, Upton-Upon-Severn, Worcester, WR8 0ET
G6 GWX C. Hore, 45 Medrose Street, Delabole, PL33 9BN
G6 GWY K. Dodd, 1 Nansen St., Bulwell, Nottingham, NG6 9JE
G6 GXE L. Jordan, 20 Coniston Road, Folkestone, CT19 5JF
G6 GXG D. Ridden, 6 Maple Drive, Witham, CM8 2LH
G6 GXK D. Wrigley, 45 Norford Way, Rochdale, OL11 5QS
G6 GXO C. Parks, 29 Heighams, Harlow, CM19 5NU
G6 GXS D. Mclean, Quartier Les Tourres, Pourcieux, France, 83470
G6 GXY M. White, 8 Browning Avenue, Droylsden, Manchester, M43 6QG
G6 GXZ B. Vaslet, Heatherlea, Adbury Holt, Newbury, RG20 9BN
G6 GYC D. Oultram, 61 Bolton Road, Westhoughton, Bolton, BL5 3DN
G6 GYF M. Marshman, 12 Neelands Grove, Cosham, Portsmouth, PO6 4QL
G6 GYG D. Langridge, 4 The Puddledocks Puddledock Lane, Sutton Poyntz, Weymouth, DT3 6LZ
G6 GYM D. Popely, 24 Lawson Avenue, Stanground, Peterborough, PE2 8PL
G6 GYN P. Price, 67 Bennetts Road, Keresley End, Coventry, CV7 8HY
G6 GYV B. Thompson, 17 Avenue Road, Askern, Doncaster, DN6 0AR
G6 GYX J. Walton, Copper Beech Lutton Gowts, Spalding, pe12 9lq
G6 GZC A. Jeffries, 29 Amberleigh Close Appleton Thorn, Warrington, WA4 4TD
G6 GZZ A. Hammond, 23 St. Andrews Road, Fremington, Barnstaple, EX31 3BS
G6 HAA A. Johns, Glan Y Nant, Murcot, Broadway, WR12 7HS
G6 HAT P. Calpin, 36 Chatsworth Grove, Harrogate, HG1 2AS
G6 HBF A. Bischtschuk, 30 Livingstone Road, Wirral, CH46 2QR
G6 HBJ T. Charman, 1 Bowler Lea, Downley, High Wycombe, HP13 5UD
G6 HBQ A. Ford, 1 Hem Heath Cottage, Longton Road, Stoke on Trent, ST4 8HP
G6 HBZ S. Jenkinson, Field End, Castleton, Hope Valley, S33 8WB
GD6 HCB A. Kennaugh, 36 Seafield Close, Onchan, Douglas, Isle Of Man, IM3 3BU
G6 HCF L. Carter, Hattersbrick Farm, Lancaster Road, Preston, PR3 6BN
G6 HCH C. Back, The Gift Shop, 3 Albion Villas, Main Road, Wareham, BH20 5RQ
G6 HCQ S. Crawford, 71 Harewood Road, Bedford, MK42 9TH
G6 HCT Home Counties ATV Group, c/o T. Grady, 63 Bridport Close, Lower Earley, Reading, RG6 3DG
G6 HCW D. Fieldsend, 3 Rosehall Close, Redditch, B98 7YD
G6 HDD P. Ingham, 1411 Helderberg Avenue, Rotterdam, USA, 12306
G6 HDF S. Kelly, 4 Franklyn Close, Wolverhampton, WV6 7SB
G6 HEB P. Ballance, 6 Coronation Terrace, Knaresborough, HG5 8JN
G6 HEF D. Bailey, Ardmore, Chapel Lane, Bootle, Millom, LA19 5UE
G6 HEJ G. Stewart, 3 Harvest Crescent, Carterton, OX18 1FF
G6 HFB A. Wedgwood, Ardmore, Chapel Lane, Bootle, Millom, LA19 5UE
G6 HFF G. Bates, The Anvil, 4 Eastgate, North Newbald, YO43 4SD
GM6 HFH I. Baker, 31 Strathaven Road, Stonehouse, Larkhall, ML9 3EN
G6 HFK L. Dutton, 5 Beaver Close, Stoke-on-Trent, ST4 6PR
G6 HFS B. Shaw, 43 Egremont Road, Hardwick, Cambridge, CB23 7XR
G6 HFW J. Graham, 142 Shakerley Lane, Atherton, Manchester, M46 9TZ
G6 HFZ S. Homer, 31 Shaftmoor Lane, Acocks Green, Birmingham, B27 7RU
G6 HGE D. Heale, 3 Evans Wharf, Hemel Hempstead, HP3 9WU
G6 HGG R. Ireson, 6 Walker Square, Wellingborough, NN8 5PQ
G6 HGI P. Johnston, 566 Woodchurch Road, Prenton, CH43 0TT
G6 HGK E. Kesterton, 24 Alexandra Road, Illogan, Redruth, TR16 4DY
G6 HGM R. Buckle, Bissom House, Parrotts Lane, Tring, HP23 6NE
G6 HGR K. Potts, 31 Sparnon Close, Redruth, TR15 2RJ
GM6 HGW C. Topping, 26 Crathes Close, Glenrothes, KY7 4SS
G6 HGX B. Waterloo, 55 Solent Road, Hill Head, Fareham, PO14 3LB
G6 HHE J. Avern, 8 Napier Crescent, Fareham, PO15 5BL
G6 HHH G. Dowse, 60 Lower Mortimer Road, Southampton, SO19 2HF
G6 HHK D. Birkbeck, Low Farm Brotton, Saltburn-by-the-Sea, TS12 2QX
G6 HIA A. Cook, Woodlands House, Hempstead Road, Hemel Hempstead, HP3 0DS
G6 HIB C. Craven, 24 Links Drive, Bexhill-on-Sea, TN40 1TE
G6 HIE B. Edwards, 28 Poppy Drive, Horam, Heathfield, TN21 9BL
G6 HIG G. Edmonds, Wellwood End, Waterworks Lane, Dover, CT15 5JW
G6 HIO D. Ollerton, 91 Church Road, Bickerstaffe, Ormskirk, L39 0EB
G6 HIQ C. Lavis, 88 Boundary Way, Glastonbury, BA6 9PH
G6 HIU N. Lasher, 1C Clarendon Road, London, E18 2AW
G6 HIV J. Martin, 1 Marsh Street, Strood, Rochester, ME2 4BB
G6 HIX J. O'Hagan, Brubell, 13 Chapel Road, Faringdon, SN7 8LE
G6 HJU J. Binns, 2 Gawsworth Close, Poynton, Stockport, SK12 1XB
G6 HJV J. Evill, 54 Copsey Grove, Farlington, Portsmouth, PO6 1NB
G6 HKA J. Moss, 801 Manchester Road, Linthwaite, HD7 5NF
GI6 HKE W. Leitch, 7 Burghley Mews, Belfast, BT5 7GX
G6 HKF R. Mew, Tehig, 16 La Mustais, Sion Les Mines, France, 44590
G6 HKH P. Randell, 38 Hanover Drive, Brackley, NN13 6JS
G6 HKL D. Martin, 9 Twinberrow Lane, Woodmancote, Dursley, GL11 4AP
G6 HKN W. Mccue, 2 Downham Avenue, Culcheth, Warrington, WA3 5RU
G6 HKP D. Merrington, 31 North Road, Wellington, Telford, TF1 3ED
G6 HKS R. Mason, 11 Dyers Mews, Neath Hill, Milton Keynes, MK14 6ER
G6 HKY W. Metcalfe, 81 Westminster Drive, Bromborough, Wirral, CH62 6AN
G6 HKZ R. Moses, 80 Edgeworth, Yate, Bristol, BS37 8YW
G6 HL Club Hut, c/o W. Ward, 88 Central Road, Cromer, NR279BW
G6 HLL B. Allman, 38 Whinchat Drive, Birchwood, Warrington, WA3 6PB
G6 HLR G. Marshall, 118 Heather Road, Small Heath, Birmingham, B10 9TB
GM6 HLT J. Melville, 6 Dixon Avenue Kirn, Dunoon, PA23 8NA
G6 HLU T. Miller, 23 Manchester Road, Altrincham, WA14 4RQ
G6 HMA P. Matthews, 47 Slyne Road, Morecambe, LA4 6PD
G6 HMF R. Venison, Brooklands, Sharnbrook Road, Souldrop, Bedford, MK44 1EX
G6 HMG R. Trowsdale, 422 Leatherhead Road, Chessington, KT9 2NN
GW6 HMJ A. Upcott, 67 Hunters Ridge, Brackla, Bridgend, CF31 2LJ
G6 HMN R. Sunter, 15 Wellhead, Winewall, Colne, BB8 8BW
G6 HMS E. Veall, 24 Meadow Drive, Tickhill, Doncaster, DN11 9ET
G6 HMV R. Tilley, 41 Rookery Road, Knowle, Bristol, BS4 2DX
G6 HMX D. Tucker, 2 Chardonnay Crescent, Thornton-Cleveleys, FY5 3UH
G6 HNE E. Shirt, 213 Carlton Road, Barnsley, S71 2BL
G6 HNI D. Baker, 16 Warners Bridge Chase, Rochford, SS4 1JE
G6 HNN M. Bugg, 39 Glencoe Road, Ipswich, IP4 3PP
G6 HNP P. Beever, 33 Masterton Road, Stamford, PE9 1SN
G6 HNQ K. Blackburn, 57 Hope Street, Leigh, WN7 1NB
G6 HNR J. Ball, 94 Marshall Lake Road, Shirley, Solihull, B90 4PN
G6 HNS R. Ball, 139 Bedford Road, Sutton Coldfield, B75 6DB
G6 HOB D. Brebner, 2 Oldborough Drive, Loxley, Warwick, CV35 9HQ
G6 HOC A. Bird, 95 Hundred Acre Road, Streetly, Sutton Coldfield, B74 2BS
G6 HOR D. Padfield, 9 Dunster Close, Minehead, TA24 6BY
G6 HOS C. Playford, 6 Nutberry Close, Teynham, Sittingbourne, ME9 9SP
G6 HPE P. Simms, 42 Bridgeside Close, Walsall, WS87BN
G6 HPK D. Scott, 8 Lynton Road, Chesham, HP5 2BU
G6 HPL R. Seddon, 255 Westleigh Lane, Leigh, WN7 5PN
G6 HPR A. Swinburne, 69 Northfield Avenue, Wigston, LE18 1FX
G6 HPT D. Sumner, 34 Japonica Close, Bicester, OX26 3YB
G6 HQ P. Bushell, The Fairway, Well Lane, Neston, CH64 4AN
G6 HQX A. Cook, 90 Ramsbury Walk, Trowbridge, BA14 0UX
G6 HRA J. Chesterman, 69 Heath Lane, Bladon, Woodstock, OX20 1RZ
GW6 HRU R. Walker, 54 Woolpitch Wood, Chepstow, NP16 6DW
G6 HRX T. Whitehead, 14 Somerset Road, Willenhall, WV13 2RY
G6 HSC V. Williamson, 28 Mill Park Drive, Braintree, CM7 1XF
G6 HSD R. Willmott, 85 Malthouse Lane, Earlswood, Solihull, B94 5RZ
G6 HSG A. Walsh, 14 The Rydings, Langho, Blackburn, BB6 8BQ
G6 HSI S. Wallace, 26 Parsons Drive, Glen Parva, Leicester, LE2 9NS
G6 HSR J. Hague, 33 East Rise, Royal Sutton Coldfield, B75 7TH
G6 HSS P. Hardiman, 12 Brempsons, Basildon, SS14 2AZ
G6 HSW L. Hagger, 48 Little Meadow Bar Hill, Cambridge, CB23 8TD
G6 HTA P. Hartas, 6 Newton St., Whitby, YO21 1QX
G6 HTB E. Brodie, 116 Pagham Road, Pagham, Bognor Regis, PO21 4NN
G6 HTH C. Hall, Oakenhill, North Pole Road, Maidstone, ME16 9HH
G6 HTS R. Hooper, 11 Joy Wood, Boughton Monchelsea, Maidstone, ME17 4JY
G6 HTT G. Reece, 9 Lambert Close, Framlingham, Woodbridge, IP13 9TE
G6 HTY A. Rollitt, St. Peters, 29 High Street, Lincoln, LN5 0EE
G6 HTZ A. Rogers, 11 Avebury Close, Curzon Park, Calne, SN11 0EP
GW6 HUD R. Rees, 5 Rhydyffynnon, Pontyates, Llanelli, SA15 5UG
G6 HUI B. Tanner, 2 Doveys Cottage, Kington Langley, Chippenham, SN15 5NT
G6 HUN A. Thompson, Hillier Garden Centre, Priors Court Road, Thatcham, RG18 9TG
G6 HUO J. Thompson, Belmoor Lodge, Pilton Lane, Exeter, EX1 3RA
G6 HUP M. Thompson, 2 Cotman Road, Lincoln, LN6 7PA
GW6 HUR N. Thursfield, Highmoor, Llanymynech, SY22 6HB
GW6 HVA M. Vernon, 33 Ffordd Morfa, Llandudno, LL30 1ES
G6 HVD N. Dunford, 8 Fair Mead, Mountsorrel, Loughborough, LE12 7BN
G6 HVE N. Martin, 12 Coppice Side, Hull, HU4 6XJ
G6 HVJ J. Durrant, 35 Britford Avenue, Wigston, LE18 2RF
G6 HVQ L. Dodson, 24 Ashcombe Terrace, Tadworth, KT20 5EW
G6 HVX D. Gladwish, 36 All Saints Street, Hastings, TN34 3BJ
GM6 HVY R. Goodwins, 3 Westmost Close, Edinburgh, EH6 4TE
G6 HWA F. Glover, 11 Esk Valley, Grosmont, Whitby, YO22 5BG
G6 HWI B. Wilson, 102 Woodlands Road, Woodlands, Doncaster, DN6 7JZ
G6 HWR M. Fern, 8 Hackney Road Hackney, Matlock, DE4 2PW
G6 HWT C. Freeman, Mill House Great Bricett, Ipswich, IP7 7DE
G6 HXB M. Aston, Flat 51 Acton House 253 Horn Lane, Acton, W3 9EJ
G6 HXI T. Ardern, 7 North Drive, Harwell, Didcot, OX11 0PE
G6 HXL D. Latham, 89 Kestrel Park, Skelmersdale, WN8 6TA
G6 HXU Dr E. Loader, 13 Vale Road, Hartford, Northwich, CW8 1PL
G6 HXW L. Leighton, 177 Terringes Avenue, Worthing, BN13 1JS
G6 HXZ P. Lovett, Betula, High Halden, Ashford, TN26 3LY
G6 HYD R. Ison, 48 Staples Hill Partridge Green, Horsham, RH13 8LF
G6 HYF C. Ironmonger, 77 Boston Road, Spilsby, PE23 5HH
G6 HYI P. Ingle, 8 Slayleigh Delph, Sheffield, S10 3RZ
G6 HYP N. Jones, 7 Church Terrace, Church Road, Norwich, NR16 2NA
G6 HZG K. Purser, 6 Parkway, Ryde, PO33 3UX
G6 HZH S. Prosser, 53 Broadlands Rise, Lichfield, WS14 9SF
GW6 HZJ D. Pemberton, Yr Efail, Capel Coch, Llangefni, LL77 7UR
G6 HZK G. Partridge, 53 Acres Road, Brierley Hill, DY5 2XY
G6 HZX R. Purdy, 49 Mansfield Road, Eastwood, Nottingham, NG16 3DY

G6 IAN I. Brooks, 10 Windermere Close, Dunstable, LU6 3DD
G6 IAO J. Sothcott, Flat 4, 19 Auckland Road East, Southsea, PO5 2HA
G6 IAT T. Bruce, 17 Blaydon Road, Luton, LU2 0RP
G6 IBD D. Bowles, 23 Broughton Way, Rickmansworth, WD3 8GW
GI6 IBL M. Barr, 4 Sandelwood Avenue, Coleraine, BT52 1JW
G6 IBN M. Bodill, 24 Dalbeattie Close, Arnold, Nottingham, NG5 8QX
G6 IBO S. Blay, 68 Springvale, Gayton, King's Lynn, PE32 1QZ
GM6 IBP G. Burnett, 314 Highcliffe, Spittal, Berwick-upon-Tweed, TD15 2JN
G6 IBU M. Squance, Flat 19, Brock House 2 Batter Street, Plymouth, PL4 0EF
G6 IBW R. Savigar, 95 Hillier Road, Devizes, Wiltshire, SN10 2FB
G6 ICC I. Campbell, 273 Crystal Palace Road, London, SE22 9JH
G6 ICH R. Brothwood, Amberley, Coombe Cross Bovey Tracey, Newton Abbot, TQ13 9EP
GD6 ICR M. Webb, Coastguard House, 1 Mount Morrison, Peel, Isle Of Man, IM5 1PN
G6 ICV I. Whittaker, 20 Manor Drive, Leicester, LE4 1BL
G6 ICZ R. Waller, 48 Lambfield Way, Ingleby Barwick, Stockton-on-Tees, TS17 5BG
GM6 IDF Dr H. Stinton, Lower Inchlumpie, Strathrusdale, Alness, IV17 0YQ
G6 IDG C. Stringer, Meadowbank, Station Road, Devon, EX10 0ER
G6 IDL M. Waud, 7 Chalkpit Lane, Candlesby, Spilsby, PE23 5SE
G6 IDO A. Davies, 8 Alexandra Road, Wednesbury, WS10 9LH
G6 IDU I. Rose, 144 Overton Road, Benfleet, SS7 4DT
G6 IDW T. Roberts, 9 Dixons Road, Market Deeping, Peterborough, PE6 8AG
G6 IEE M. Elsley, 25 Elmsdale Road, Wootton, Bedford, MK43 9JW
G6 IEI P. Williams, Peanjays, 4 Cutbush Close, Reading, RG6 4XA
G6 IEQ P. Hawkridge, 211 Goring Way, Worthing, BN12 5BU
GI6 IES C. Hagan, 15D Ballygalget Road, Portaferry, Newtownards, BT22 1NE
G6 IET S. Hargreaves, Dakyn Cottage, Kirby Hill, Richmond, DL11 7JH
G6 IFE P. Holland, 3 Manor Villas, Chilton Road, Chearsley, Aylesbury, HP18 0DN
G6 IFH J. Rimington, 8 Harvesters, Tolleshunt D'Arcy, Maldon, CM9 8UF
G6 IFN L. Rouse, 69 Shackerdale Road, Wigston, LE18 1BR
G6 IFQ S. Howcroft, Warwick Cottage, 5 Ecclesgate Road, Blackpool, FY4 5DW
G6 IFR G. Horwood, 25 Briar Road, Shepperton, TW17 0JB
G6 IFS N. Hollinshead, 35 Parkside Drive, May Bank, Newcastle-under-Lyme, ST5 0NL
G6 IFV J. Hunt, 77 Scott Street, Burnley, BB12 6NJ
G6 IGK L. Glasscock, 37 Huntingfield Road, Bury St Edmunds, IP33 2JA
G6 IGO D. Gospel, Sunnydene Farm, The Common, Fakenham, NR21 9JB
G6 IGU A. Greenleaf, The Lindens, Frating Road, Ardleigh, Colchester, CO7 7SY
G6 IGV D. Gregson, 8 Lennox Gate, Blackpool, FY4 3JQ
G6 IGW N. Gutten, 8 Chalfield Close, Crewe, CW2 6TJ
GW6 IGY J. Mead, 1 Tudor Court, Fagl Lane, Wrexham, LL12 9PJ
G6 IHB F. Norton, 62 Moorlands Drive, Shirley, Solihull, B90 3RE
G6 IHD S. Maxwell, 64 St. Georges Park, Wallasey, CH45 9LW
G6 IHG H. Mitchell, 17 Burners Close, Burgess Hill, RH15 0QA
GI6 IHM R. Mcdowell, 3 Lord Warden'S Park, Bangor, BT19 1YG
G6 IHO V. Marks, 176 Middlemarch Road, Coventry, CV6 3GL
G6 IHU J. Measom, 41 Glebe Road, Thringstone, Coalville, LE67 8NU
G6 IHW A. Mackinlay, 26 Anderson Road, Erdington, Birmingham, B23 6NN
G6 IIA A. Stansfield, 22 Low Stobhill, Morpeth, NE61 2SG
G6 IIF R. Sharpe, 14 Dansie Court, Compton Road, Colchester, CO4 0EA
G6 IIK D. Gill, 79 Heather Walk, Bolton-Upon-Dearne, Rotherham, S63 8BZ
G6 IIM P. Jones, 2 Farmers Heath Great Sutton, Ellesmere Port, CH66 2GX
G6 IIN P. Currigan, 5 Gayton Avenue, Wallasey, CH45 9LJ
G6 IIP J. Clarke, 19 Kensington Road Gaywood, King's Lynn, PE30 4AT
G6 IIU R. Cooper, 69 Vicarage Lane, Elworth, Sandbach, CW11 3BU
G6 IIZ Dr J. Clark, Brooklyn Cottage, Milton Combe, Yelverton, PL20 6HP
G6 IJE P. Chapman, 29 Ashwell Grove, Rotherham, S65 1NF
G6 IJK A. Clayphon, Downshire Lodge, Park Lane, Finchampstead, Wokingham, RG40 4PT
G6 IJQ W. Cartwright, 3 Masefield Rise, Halesowen, B62 8SH
G6 IJW G. Stoelwinder, Hampt Cottage, Middle Hampt Luckett, Callington, PL17 8NR
G6 IKC S. Saunders, 16 Hill Close, Pennsylvania, Exeter, EX4 6HG
G6 IKE S. Lynch, 21 Mill Lane, Bolton Le Sands, Carnforth, LA5 8HR
G6 IKH A. Simpson, 7 Hartington Street, Newcastle, ST5 8DR
G6 IKM D. Tarbuck, 23 Kingsway, Newton-le-Willows, WA12 8LZ
GM6 IKN K. Towns, Pluscarden Abbey, Pluscarden, Elgin, IV30 8UA
G6 IKO M. Smith Mbe, 34 Diamond Road, Watford, WD24 5EW
G6 IKU J. Stockton, 183 Eskdale, Tanhouse, Skelmersdale, WN8 6ED
G6 ILC G. Sword, Garden Cottage, Kiln Road, Salisbury, SP5 2HT
G6 ILD P. Southgate, Flat 1, The Old Yard, Mill Road, Holsworthy, EX22 7RT
G6 ILH A. Davies, 27 Whitecroft Lane, Mellor, Blackburn, BB2 7HA
G6 ILN J. Dodge, 5 Moat Way, Queenborough, ME11 5BU
G6 ILT E. Elliston, 117 Willbye Avenue, Diss, IP22 4NP
G6 ILU E. Edmonds, 1 Chalkhole Cottages, Flete Road, Margate, CT9 4LL
G6 ILX B. Edward, 27 Barford Close, Ainsdale, Southport, PR8 2RS
GW6 ILY W. Evans, Treetops, Whitchurch Road, Wrexham, LL13 0BL
G6 ILZ I. Fullerton, 54 Brashland Drive, Northampton, NN4 0SS
G6 IMH R. Firth, Kasamily, 73 Lions Lane, Ashley Heath, Ringwood, BH24 2HH
G6 IMJ K. Walker, 37 Willingdon Road, Liverpool, L16 3NE
G6 IML I. Walsh, 7 Winchester Avenue, Ashton-under-Lyne, OL6 8BU
G6 IMN K. Wetherell, 12 Parc Stephney, Budock Water, Falmouth, TR11 5EJ
G6 IMQ J. Wild, 20 Sandy Lane, Cholsey, Wallingford, OX10 9PY
GW6 IMS T. Vernalls, 5 Min Y Traeth, Minffordd, Penrhyndeudraeth, LL48 6EG
G6 IMW T. Rogers, Lodge 12 Benson Waterfront, Benson, Wallingford, Oxford, OX10 6SJ
G6 INA P. Reidy, Famagusta Avenue 45, House 2, Sotira, Cyprus, 5390
GW6 INF J. Markham, 4 Ty Arfon, Tywyn, LL36 0TA
G6 ING S. Meigh, Flat 94, Reginald Mitchell Court, Stubbs Lane, Stoke on Trent, ST1 3SP
G6 INI C. Mahony, 20 Kenchester, Bancroft, Milton Keynes, MK13 0QP
G6 INK E. Mcglen M.B.E., 22 Stratford Avenue, City of Sunderland, SR2 8RX
G6 INM K. O'Reilly, 1 Parkfield, Crewe, CW1 4TT
G6 INO R. Ottolini, 154 Barwick Road, Leeds, LS15 8SW
G6 INU D. Port, 8 Betterton Drive, Sidcup, DA14 4PS
G6 INV D. Pratley, 2 Haseldine Meadows, Hatfield, AL10 8HE
G6 INW M. Purnell, 14 Cranleigh Close, Bournemouth, BH6 5LD
G6 INX G. Pryke, 38 Colne Drive, Walton-on-Thames, KT12 3SQ
GW6 IOA C. Crow, Lindisfarne, Pen Y Waun, Cardiff, CF15 9SJ
G6 IOB P. Coghlan, 96 Cambridge Road, Ely, CB7 4HU
G6 IOE G. Crawford, 4 Beverley Gardens, Gedling, Nottingham, NG4 3LF
G6 IOM M. Cunningham, 16 Cherry Waye, Eythorne, Dover, CT15 4BT
G6 ION R. Civil, 7 Sunnybanks, Hatt, Saltash, PL12 6SA
G6 IOT P. Langley, 40 Kingshayes Road, Aldridge, Walsall, WS9 8RU
GI6 IOU M. Pollock, 33 Seahill, Donaghadee, BT21 0SH
G6 IOV P. Phelps, 57 Southbrook Road, Havant, PO9 1RL
G6 IOW D. Peachey, 4 Windermere Drive, Great Notley, Braintree, CM77 7UA
G6 IOX A. Pearce, 49 Bishopswood Road, Tadley, RG26 4HF
G6 IPB B. Doyle, Ashwell Croft, Brunthwaite Lane, Keighley, BD20 0ND
G6 IPC I. Downes, 21 Caldbeck Court, Beeston, Nottingham, NG9 5NH
G6 IPH R. Distin, The Martletts, Broad Oak, Rye, TN31 6DN
G6 IPN M. Davies, 54 Helmside Road, Oxenholme, Kendal, LA9 7HA
G6 IPQ R. Deakin, 55 Pendeen Crescent, Plymouth, PL6 6RE
GW6 IPR P. Drew, 6 Clos Cae'R Wern, Caerphilly, CF83 1SQ
G6 IPT S. Edwards, The Kas, Chapel Lane Souldrop, Bedford, MK44 1HB
G6 IPW S. Featherstone, 36 Denton Avenue, Grantham, NG31 7JL
G6 IQC S. Leak, 97 Lees Road, Ashton-under-Lyne, OL6 8BQ
G6 IQF R. Harris, Greve De Lec, 14A All Saints Lane, Clevedon, BS21 6AY
GM6 IQH R. Wickenden, 2 Buail-Bhan, Ballinluig, Pitlochry, PH9 0NH
G6 IQI P. White, 16 Charnwood Close, Chandler'S Ford, Eastleigh, SO53 5QP
G6 IQM M. Wooding, 5 Ware Orchard, Barby, Rugby, CV23 8UF
G6 IQP D. Marriott, 80 Andrew Avenue, Ilkeston, DE7 5DW
G6 IQY J. Price, 67 Broadway, Oldbury, B68 9DP
G6 IRE R. Aynge, 9 Sedgebrook Road, Blackheath, London, SE3 8LR
G6 IRF S. Atwell, 10 Belding Avenue, Manchester, M40 3SE
G6 IRG M. Andrews, 1 Garrick Close, Dudley, DY1 3DF
G6 IRJ G. Andronov, 90 Overbury Close, Northfield, Birmingham, B31 2HD
GI6 IRL J. Agnew, 23 Berwick Heights Moira, Craigavon, BT67 0SZ
G6 IRP C. Gardner, 7 Lesley Close, Bexley, DA5 1LX
G6 IRU B. Green, 23 Freemantle Avenue, Blackpool, FY4 1SX
G6 IRW K. Holmes, Gable Cottage, Low Hill, Helsby Warrington, WA6 0NW
G6 IRX C. Holt, 1 Vale View, Common Road, Wincanton, BA9 9RB
G6 IRY R. Hobbs, 120 Misbourne Road, Hillingdon, Uxbridge, UB10 0HP
G6 IRZ W. Hughes, 7 Richards Court, Bednall, Stafford, ST17 0SP
G6 ISA J. Hutchins, 18 Derby Road, Sale, M33 5PR
G6 ISB A. Hunt, 10 Sturton Street, Forest Fields, Nottingham, NG7 6HU
G6 ISG P. Hancock, 2 Gulistan Road, Leamington Spa, CV32 5LU
G6 ISM J. Hancock, 7 Hollies Close, Houghton-On-The-Hill, Leicester, LE7 9GW
G6 ISY L. Hill, 21 Liddiards Way, Purbrook, Waterlooville, PO7 5QW
GW6 ITB J. Imperato, 118 Heol Uchaf, Rhiwbina, Cardiff, CF14 6SS
GW6 ITJ M. Jones, 4 Plastirion Avenue, Prestatyn, LL19 9DU
G6 ITM D. Jupp, 26 Audley Avenue, Gillingham, ME7 3AY
G6 ITU M. Bunn, 45 Red Cat Lane, Burscough, Ormskirk, L40 0RA
G6 ITV L. Parker, 15 Savile Place, Mirfield, WF14 0AJ
G6 ITW M. Blundell, 68 Alton Road, Leicester, LE2 8QA
G6 IUD R. Bracey, 50 Harrow Way, Watford, WD19 5ET
G6 IUF R. Brittain, Sarenchel, St Cross, Harleston Norfolk, IP20 0NY
GW6 IUK R. Bastable, Gwynfryn, Ffordd Caergybi, Llanfairpwllgwyngyll, LL61 5SZ
G6 IUQ M. Gaylard, 66 Runnymede Road, Yeovil, BA21 5SU
G6 IUS B. Gilbert, 22 Oaklands Way, Hildenborough, Tonbridge, TN11 9DA
G6 IVB A. Gornall, 28 Woodward Close, Winnersh, Wokingham, RG41 5NW
G6 IVC M. Griffiths, 25 Lethbridge Road, Southport, PR8 6JA
G6 IVD P. Guy, 11 Ludlow Crescent, Redcar, TS10 2LQ
GI6 IVJ R. Brown, 157 Newtownards Road, Bangor, BT20 4HS
G6 IVP J. Burton, 22 Pear Tree Lane, Hempstead, Gillingham, ME7 3PT
G6 IVR Itchen Valley ARC, c/o K. Hastie, 3 The Woodlands, Kings Worthy, Winchester, SO23 7QQ
G6 IVW R. Balderson, 15 Woodrush Way, Moulton, Northampton, NN3 7HU
GW6 IVY C. Somerville, 1 Glyn Isaf, Llandudno Junction, LL31 9HF
GW6 IWC M. Saunders, Rose Cottage, Pleasant Lane, Wrexham, LL11 5DH
G6 IWD L. Sherratt, Anlyn, Norbury Drive, Brierley Hill, DY5 3DP
G6 IWK R. Skingley, Kynance, Church Cove The Lizard, Helston, TR12 7PQ
G6 IWT M. Rea, Osmary, Station Road, Bishop's Stortford, CM22 6LG
G6 IWU J. Rodwell, 20 Nelson Street, King's Lynn, PE30 5DY
G6 IWZ D. Jefferys, 22 Cleveland Gardens, Cricklewood, London, NW2 1DY
G6 IXE M. James, 58 Spitfire Way, Hamble, Southampton, SO31 4RT
G6 IXH D. Hodges, 5 Greenlands, Leighton Buzzard, LU7 3UJ
G6 IXM G. Hares, 30 Copped Hall Drive, Camberley, GU15 1NP
G6 IXN G. Hewitt, 66 Portland Drive, Forsbrook, Stoke-on-Trent, ST11 9AU
G6 IXP L. Holden, 16 George Street, Clayton Le Moors, Accrington, BB5 5QJ
GW6 IYA H. Woodnutt, Latitude, Gannock Park West, Conwy, LL31 9HQ
G6 IYD D. Noakes, 117 Kingsmead Park, Allhallows, Rochester, ME3 9TA
GM6 IYJ M. Plested, Cregneash, Platcock Wynd, Fortrose, IV10 8SQ
G6 IYM N. Perkins, Heathcote Kents Road, Torquay, TQ1 2NL
GW6 IYP R. Parry, Glengarriff, Rhyl Road, Rhyl, LL18 2TP
G6 IYS I. Porter, 25 Wolds Retreat, Brigg Road, Market Rasen, LN7 6RU
G6 IYY R. Adamek, 11 Top Road, Belaugh, Norwich, NR12 8XB
G6 IZA I. Alderton, 6 Hurford Drive, Thatcham, RG19 4WA
G6 IZK A. Collier, 2 The Hollies, Trerise Road, Camborne, TR14 7HB
GM6 IZU K. Frame, 102 High St., Galashiels, TD1 1SQ
GW6 IZZ C. Evans, 1 Ashdale Lane, Llangwm, Haverfordwest, SA62 4NU
G6 JAC P. Dalley, 32 Albert Road, Erdington, Birmingham, B23 7LT
G6 JAF I. Evans, 19 Grange Road, Stone, ST15 8PR
G6 JAK S. Deacon, 3 Blenheim Grove, Offord D'Arcy, St Neots, PE19 5RD
G6 JAL A. Dixon, 23 Appleby Drive, Barrowford, Nelson, BB9 6EX
G6 JAM M. Dainty, 5 Woodman Close, Wednesbury, WS10 9UA
G6 JAP G. Denison, Greengage Cottage 30 Long Street, Thirsk, YO7 1AP
G6 JAR M. Drake, 7 Orient Road, Paignton, TQ3 2PB
G6 JAS A. Balding, 8 Winston Way, Farcet, Peterborough, PE7 3BU
G6 JAY R. Luckett, 20 Leicester Villas, Hove, BN3 5SQ
GM6 JBF D. Mardlin, 35 Uist Road, Aberdeen, AB16 6FN
G6 JBL G. Moore, 2 Meadow Lea, Worksop, S80 3QJ
GW6 JBN R. Thomas, Post Office, Llanbedr, LL45 2HH
G6 JBQ G. Taylor, 54 Bowershott, Letchworth Garden City, SG6 2EU
G6 JBY J. Bibby, 24 Assarts Lane, Malvern, WR14 4JR
G6 JCI W. Henson, 1 Bonser Close, Carlton, Nottingham, NG4 1DP
GW6 JCK A. Harvey, Cornerways, Mount Pleasant, Holyhead, LL65 1SN

G6 JCM J. Hatfield, Tenter Close, Husthwaite, York, YO61 4PF
G6 JCV A. Haslehurst, Westlands, Stinting Lane, Mansfield, NG20 8EQ
G6 JCX F. Hewitt, 12 Woodside, North Walsham, NR28 9XA
G6 JCY B. Hedge, Birchwood Lodge, Barnards Road, Norwich, NR28 9RG
G6 JDC T. Kemp, 85 Rosehill Road, Rawmarsh, Rotherham, S62 7BX
GW6 JDF G. Walker, Lluesty, Bryn Hyfryd Road, Tywyn, LL36 9HG
G6 JDH G. Webster, 153 Frogmore Lane, Waterlooville, PO8 9RD
G6 JDO N. Wright, 99 School Road, Saxon Street, Newmarket, CB8 9RY
G6 JDP J. Mott-Gotobed, 14 Copse Road, New Milton, BH25 6ES
G6 JDW T. Scarfe, 26 Norman Keep, Warfield, Bracknell, RG42 7UY
G6 JEB J. Bailey, 22 Hainfield Drive, Solihull, B91 2PL
G6 JEF S. Wardley, 5 Swindon Street, Bridlington, YO16 4JD
G6 JEM P. Stoneman, 111 Fletemoor Road, Plymouth, PL5 1UL
GM6 JEP F. Cassidy, 9 Spey Road, Troon, KA10 7DY
G6 JEU P. Chrysostomou, 45 Leyborne Avenue, Ealing, London, W13 9RA
G6 JEY Lord M. Cooper, Flat 3, 32 Lansdowne Road, Worthing, BN11 5HB
G6 JFE S. Fantom, Cedar Ridge, The Dimple Fritchley, Belper, DE56 2HP
G6 JFL D. Barsby, 42 New Terrace, Pleasley, Mansfield, NG19 7PY
G6 JFN P. Brazier, Stud House, Mentmore, Leighton Buzzard, LU7 0QE
GM6 JFP D. Brown, 15 Eliots Park, Peebles, EH45 8HB
G6 JFU A. Mayman, Lingmell, Cedar Grove, Aldbrough, HU11 4QH
GW6 JFV T. Morris, 29 Heol Croes Faen, Nottage, Porthcawl, CF36 3SW
GI6 JGB M. Mcninch, 5 Bangor Road, Groomsport, Groomsport, BT19 6JF
G6 JGF A. Morgan, 8 Shaftesbury Road, Watford, WD17 2RQ
GM6 JGH Dr J. Massheder, 2B Pentland Park, Loanhead, EH20 9PA
G6 JGP A. Lawrence, Columbine Cottage, Ford, Salisbury, SP4 6DJ
G6 JGR J. Richardson, 30 Shaftesbury Avenue, Chandler'S Ford, Eastleigh, SO53 3BS
G6 JGT A. Davis, 7 Kennedy Crescent, Gosport, PO12 2NL
G6 JHG J. Grieve, 65 Royal Lane, Uxbridge, UB8 3QU
GM6 JHH W. Gunn, Tullochard, Scouriemore, Lairg, IV27 4TG
GD6 JHP J. Pauley, 32 King Edward Bay Apartments, Sea Cliff Road, Onchan, Isle of Man, IM3 2JF
G6 JHS A. Winrow, 14 Green Lane, Bayston Hill, Shrewsbury, SY3 0NS
G6 JHT M. Stannard, 16 Rose Court Primrose Road, Dover, CT17 0FP
GM6 JIC L. Paget, 40 Davaar Drive, Kilmarnock, KA3 2JG
G6 JIF J. Purdy, 99 Ashford Road, Hastings, TN34 2HY
GM6 JIL R. Mcmillan, 17B Kingston Road, Neilston, Glasgow, G78 3JA
G6 JIM J. King, 4 Glenhurst Avenue, Ruislip, HA4 7LZ
G6 JIR N. Brown, 11 Tudor Green, Jaywick, Clacton-on-Sea, CO15 2PA
G6 JJA N. Billingham, 35 Scalwell Park, Seaton, EX12 2DB
G6 JJB B. Banks, 16 Park Road, Burntwood, WS7 0EE
G6 JJF C. Byrne, 104 Ripon Hall Avenue, Ramsbottom, Bury, BL0 9RE
G6 JJG K. Breakwell, 91 Lynton Avenue, Claregate, Wolverhampton, WV6 9NQ
G6 JJI A. Bromfield, 11 Blackthorn Croft, Clayton-Le-Woods, Chorley, PR6 7TZ
G6 JJK J. Bourne, 91 Burwell Road, Exning, Newmarket, CB8 7DU
GM6 JJN R. Berry, Sylvan House, Glenmoriston, Inverness, IV63 7YJ
G6 JJP J. Pinson, 10 Kenelm Close, Clifton-On-Teme, Worcester, WR6 6EB
GI6 JJR N. Loughrey, 12 Billys Road, Newry, BT34 2NA
G6 JJT E. Ferguson, Willowmead, Church End, Ravensden, Bedford, MK44 2RP
GW6 JJX R. Price, Arfryn, Trecastle, Brecon, LD3 8UP
G6 JKF M. Lowe, 4 Virginia Avenue, Stafford, ST17 4YA
G6 JKK G. Orchard, 189 Sopwith Crescent, Wimborne, BH21 1SR
GM6 JKU G. Henderson, 34 Soutar Crescent, Perth, PH1 1QB
G6 JKV R. Henneman, Lydbury, Wistanswick, Market Drayton, TF9 2BB
G6 JKY B. Hadley, 60 Chapel St., Pensnett, Brierley Hill, DY5 4EF
G6 JLI P. Dixey, Rose Cottage, 197 Raikes Lane, Batley, WF17 9QF
G6 JLL S. Douglas, 1030 Shields Road, Walkerville, Newcastle upon Tyne, NE6 4SR
G6 JLU A. Millar, 8 Eisenhower Road, Shefford, SG17 5UP
G6 JMB J. Mountain, Thurlow House, Aldenham Avenue, Radlett, WD7 8HJ
GW6 JMC D. Miller, Maes Hyfryd, Llanfynydd, Wrexham, LL11 5HH
GI6 JMD J. Moller, 9 River Hill Lane, Newtownards, BT23 7GQ
G6 JMG P. Parton, 12 Duchess Drive, Bridgnorth, WV16 4JD
G6 JMJ K. Renton, 87 Shirley Gardens, Barking, IG11 9XB
G6 JMO J. Page, 9 Ascot Close, Elstree, Borehamwood, WD6 3JH
G6 JMX B. Wendon, 89 Palewell Park, London, SW14 8JJ
GM6 JNJ D. Anderson, 34 Culzean Crescent, Kilmarnock, KA3 7DT
GM6 JNQ I. Cox, 8 Traill St., Castletown, Thurso, KW14 8UG
G6 JNS P. Crosland, Sprackets Orchard, Curry Rivel, Langport, TA10 0PP
G6 JNV M. Carter, 17 Mcwilliam Road, Woodingdean, Brighton, BN2 6BE
G6 JNW S. Carter, 84 Barnett Road, Brighton, BN1 7GH
G6 JNY I. Eliade, 44 Brookside Road, Stratford-upon-Avon, CV37 9PH
G6 JNZ W. Caine, 53 Cromford Road, Crich, Matlock, DE4 5DJ
GM6 JOA A. White, Brodiescroft, Banff, AB45 3BR
GM6 JOD T. Lawless, 2 Lawers Place, Bourtreehill North, Irvine, KA11 1LR
G6 JOL R. Young, 53-55 Kirby Road, Leicester, LE3 6BD
GI6 JOP A. Wallace, 61 Locksley Park, Belfast, BT10 0AS
G6 JOR D. Webb, 1 Corelli Road, Basingstoke, RG22 4NB
GM6 JOS K. Arrowsmith, 2, Dornoch Road, Ness Castle, Inverness, IV2 6EQ
GW6 JPC C. Jenkins, 10 Marsh Court, Abergavenny, NP7 5HQ
G6 JPE A. Jephcott, 12 Clos Du Beauvoir, Rue Cohu, Castel, Guernsey, X X
G6 JPG J. Gilliver, 5 Yew Tree Park Homes Charing, Ashford, TN27 0DD
G6 JPM S. Green, 7 Brook View, Totnes, TQ9 5FH
GI6 JPO H. Graham, 104 Tattygare Road, Lisbellaw, Enniskillen, BT94 5FB
G6 JPQ J. Gould, 108 Newton Road, Burton-on-Trent, DE15 0TT
G6 JPR Dr M. Glover, 3 Sammons Way, Coventry, CV4 9TD
G6 JPS J. Skertchly, 132 Derby Road, Spondon, Derby, DE21 7LX
G6 JPT D. Gleave, 1 Fearnley Way, Newton-le-Willows, WA12 8SQ
G6 JQD R. Skinner, 23 Hardy Road, Greatstone, New Romney, TN28 8SF
G6 JQE G. Tannahill, 15 Bowfell Avenue, Newcastle upon Tyne, NE5 3XB
GU6 JQF M. Trenchard, Mont Gibel, 3 Clifton Stairs, St Peter Port, Guernsey, GY1 2PL
G6 JQH J. Williams, 18 The Leas, Barkston, Grantham, NG32 2PD
G6 JQX J. Wingfield, 56 Manor Fields, Liphook, GU30 7BS
G6 JRE S. Stanton, 6 Trevor Road Beeston, Nottingham, NG9 1GR
G6 JRI I. Wright1, 3 Sykes Court, Wheldrake, YO19 6GE
G6 JRL M. Bernard, 1 Foxglove Way, Hambleton, Selby, YO8 9UB
G6 JRM H. Bottomley, Nerefield, Aylesbury Road, Aylesbury, HP18 0BL
G6 JRS A. Cuthbertson, 72 Bulford Road, Durrington, Salisbury, SP4 8DJ
GM6 JRX D. Fraser, Kylerhea, Harbour Road, Castletown, Thurso, KW14 8TG
GI6 JRY R. Getty, 6 Rocheville, Cookstown, BT80 8QE
G6 JRZ S. Gunn, 55 Station Road, West Byfleet, KT14 6DT
G6 JSF M. Hayward, 1 Station Road, Grateley, Andover, SP11 8LG
G6 JSI A. Haswell, 66 White Hart Lane, Fareham, PO16 9BQ
GW6 JSJ D. James, 6 Chave Terrace, Maesycwmmer, Hengoed, CF82 7RZ
G6 JSN A. Sym, 1 Beech Close, Spetisbury, Blandford Forum, DT11 9HG
G6 JSR A. Mason, 5 Birch Road, Kippax, Leeds, LS25 7DY
G6 JTC M. Whiteley, 9 Fernie Close, Barton Seagrave, Kettering, NN15 6RE
G6 JTD D. Walker, 100 Clifton Road, Kingston upon Thames, KT2 6PN
G6 JTI H. Martin, 80 Topcliffe Road, Sowerby, Thirsk, YO7 1RT
G6 JTK R. Nokes, 99 Harmers Hay Road, Hailsham, BN27 1TW
G6 JTO J. Parfrey, 97 Gordon Road, Camberley, GU15 2JQ
G6 JTT J. Trett, 1 Moorland Way, Bridgwater, TA6 4JL
G6 JTV R. Allen, 65 Atherstone Road, Measham, Swadlincote, DE12 7EG
G6 JTW H. Marshall, Mardachrob, 4 Howell Road, Sleaford, NG34 9RX
GW6 JTX E. Bielawski, Rowanlea, Quarry Brow, Wrexham, LL12 8SJ
GM6 JUA D. Brown, 10 Culmore Place, Falkirk, FK1 2RP
G6 JUE N. Cockayne, 46 Canterbury Way, Stevenage, SG1 4DQ
G6 JUI K. Dare, One Bee, 1 Gloucester Road, Reading, RG30 2TH
G6 JUP J. Sutton, 252 Rawling Road, Gateshead, NE8 4UH
G6 JUQ G. Williams, 21 Arden Close, Southport, PR8 2RR
G6 JUT J. Whiting, Old Hall, Sykes Lane, Pickworth, Sleaford, NG34 0TZ
G6 JVA J. Greevy, 11A Norman Road, Walsall, WS5 3QJ
GW6 JVB R. Griffiths, 26 Brynglas, Gilwern, Abergavenny, NP7 0BP
G6 JVI C. Hedges, Canel, Thezac, France, 47370
G6 JVK M. Jeffery, 14 Rosemary Avenue, Earley, Reading, RG6 5YQ
G6 JVO M. Kidd, 99 Ferry Road West, Scunthorpe, DN15 8UG
G6 JVT C. Santer, 51 Limbrick Lane, Goring-By-Sea, Worthing, BN12 6AB
G6 JVX H. Schofield, 15 Deerfield Road, March, PE15 9AH
G6 JWD J. Davies, Welfare, High Street, Borth, SY24 5JD
GM6 JWF A. Paul, 20 Upper Bridge Street, Alexandria, G83 0LL
GM6 JWH D. Taylor, 3 Abbotsgrange Road, Grangemouth, FK3 9JD
GW6 JWL H. Roberts, Pen Yr Erw, Graigfechan, Ruthin, LL15 2EY
G6 JWM D. Le Grove, Apartment 3, Beechwood, Ilkley, LS29 8AH
G6 JWO A. Legg, 2 Alkington Farm Lane Cottage, Heathfield, Berkeley, GL13 9PL
G6 JXA K. Brown, 165 Canterbury Road, Morden, SM4 6QG
G6 JXC P. Chadbund, 20 Northlands Road, Adstock, Buckingham, MK18 2JH
GI6 JXG W. Collins, 33 New Row, Kilrea, Coleraine, BT51 5TA
G6 JXS W. Hughes, 27 Winchester Close, Ashington, NE63 9QJ
G6 JYB M. Niman, 55 Harrow Way, Chelmsford, CM2 7AU
G6 JYN D. Watkinson, 1 Melor View, Amesbury, Salisbury, SP4 7RL
G6 JYO C. Allen F B S, 20 Hollywood Lane, Hollywood, Birmingham, B47 5PX
G6 JYR D. Benton, 231 Prestwood Road, Wolverhampton, WV11 1RF
G6 JYX R. Drew, Derwent House, Landing Lane, Selby, YO8 6RA
G6 JZE P. Graham, 14 Carlaw Road, Birkenhead, CH42 8QA
G6 JZN A. Ogden, 30 Buccaneer Street, Penzance, TR18 2GD
G6 JZV K. Lummis, The Bothy, Upper Town Wetherden, Wetherden Upper Town, Stowmarket, IP14 3NF
G6 JZW C. Muller, 5 Ash Close, Flitwick, Bedford, MK45 1JY
G6 KAI M. Brighton, 11 West Close, Norwich, NR5 0NH
GM6 KAM A. Drummond, Flat 4F, Crossfolds Crescent, Peterhead, AB42 1RD
GW6 KAV Dr H. Hughes, Hendre Bach, Cerrigydrudion, Corwen, LL21 9TB
G6 KAW G. Instone, 19 Dickenson Road, Swindon, SN25 1WG
GM6 KAY C. Bates, 10 Swanston Avenue, Edinburgh, EH10 7BU
G6 KBC C. Philpot, 17 Jervis Court, Ilkeston, DE7 8PX
GW6 KBD D. Potts, 11 Walmer Road, Newport, NP19 8NU
G6 KBN R. Morby, The Rectory, High Street, Edgmond, Newport, TF10 8JR
G6 KBQ T. Williams, 2 Hazelwood, Greasby, Wirral, CH49 2RQ
G6 KBS J. Musgrave, 57 Chiltern Road, Baldock, SG7 6LT
GI6 KBX Rev. J. Turner, 45 Gloonan Hill, Ahoghill, Ballymena, BT42 1PU
G6 KBZ N. Wilkins, Norjen, Thorpe Market Road, Norfolk, NR11 8NG
G6 KCG P. Sharpe, 46 Beaumont Road, New Costessey, Norwich, NR5 0HG
G6 KCJ D. Wynters, 11 Heritage Lane, Ascott-Under-Wychwood, Chipping Norton, OX7 6AD
G6 KCV P. Willmott, C/O Oil Management Serv. Ltd., P.O. Box Hm 1751, Hamilton Hm Gx, Bermuda, X X
GM6 KDB K. Lee, West Skares, Glens Of Foudland, Huntly, AB54 6AT
GM6 KDD D. Scobbie, 17 Roselea Drive, Brightons, Falkirk, FK2 0TJ
GI6 KDN K. Mcinnes, 39 St. Johns Place, Belfast, BT7 3HA
G6 KDU M. Thorley, 5 Burland Road, Newcastle, ST5 7ST
G6 KDW G. Morton, 42 Fairycroft Road, Saffron Walden, CB10 1LZ
G6 KDY A. Perkins, 3 Greenway Close, Radcliffe-On-Trent, Nottingham, NG12 2BU
G6 KEH J. Golightly, 1 Pannier Mews, Castle Street, Torrington, EX38 8EE
G6 KEN K. Dasilva-Hill, 5 Station Road, Charing, Ashford, TN27 0JA
GM6 KEV D. Smith, Mandala, Belhaven Road, Dunbar, EH42 1NW
G6 KEX S. Smyth, 18 Weston Road, Birmingham, B19 1EH
G6 KEZ P. Pattison, 18 Broadgate Lane, Deeping St. James, Peterborough, PE6 8NW
G6 KFD P. Stockwell, 62 Golden Cross Road, Ashingdon, Rochford, SS4 3DQ
GM6 KFO G. Gordon, 31 Stoneyhill Avenue, Musselburgh, EH21 6SB
G6 KFR D. Jones, 2 The Orchard Mill Lane, Kings Sutton, Banbury, OX17 3RG
G6 KGA L. Coleman, Lilac Cottage, Coley Lane, Stafford, ST18 0XB
G6 KGK G. Gudgin, 7 Merchant Place, Middleton, Milton Keynes, MK10 9JL
G6 KGL G. Gudgin, 890 W Iowa Ave, Sunnyvale, USA, 94086
GW6 KGR M. Buck, Upper Glaisfer, Llangynidr, Crickhowell, NP8 1LN
G6 KGU Dr D. Craig, Pear Tree Cottage, Cripps Corner, Staplecross, TN32 5QS
G6 KHA T. Hyde, 14 Wyley Road, Coventry, CV6 1NW
G6 KHD K. Bierton, 44 Stalmine Hall Park, Hall Gate Lane, Poulton-le-Fylde, FY6 0LD
G6 KHG R. Champion, 25 Congreve Road, Worthing, BN14 8EL
G6 KHM L. Edwards, 71 Gleneagles Road, Yardley, Birmingham, B26 2HT
G6 KHN S. Harvey, 53 Winleigh Road, Handsworth Wood, Birmingham, B20 2HN
G6 KHW I. Bultitude, 48 Forty Acres Road, Devizes, SN10 3DG
G6 KIA C. Duckles, 8 Railway Cottages, Skillings Lane, Brough, HU15 1EN
G6 KIB P. Duesbury, The Bungalow, Robins Lane, Cambridge, CB23 8HH
G6 KIE D. Banks, 145 Compton Crescent, Chessington, KT9 2HG
G6 KIH P. Ball, 38 The Ridgeway, Knottingley, WF11 0JQ
G6 KIV S. Blythe, 17 Ashlea Road, Wirral, CH61 5UG
GW6 KIW M. Dennis, 16 Gwel Y Llan, Llandegfan, LL59 5YH

G6 KIZ M. Griffiths, 70 Towcester Road, Far Cotton, Northampton, NN4 8LQ
G6 KJA P. Hoyle, Sunset Cottage, Water End Lane, Ayot St. Peter, Welwyn, AL6 9BB
GI6 KJC Dr W. Abram, 1 The Briggs, Groomsport, Bangor, BT19 6HY
G6 KJE A. Dolby, 27 Tucker Road, Ottershaw, Chertsey, KT16 0HD
G6 KJH P. Horobin, 12 Laurel Road, Blaby, Leicester, LE8 4DL
G6 KJK J. Chappell, 15 Edmund Avenue, Stafford, ST17 9FT
G6 KJM J. Mirams, 29 Martello Court, Jevington Gardens, Eastbourne, BN21 4HR
G6 KJT S. Brabbins, 8 Park Drive, Bingley, BD16 3DF
G6 KJY L. Cartwright, 18 High Causeway, Much Wenlock, TF13 6BZ
G6 KKA J. Edmondson, 6 Park Lea, Bradley, Huddersfield, HD2 1QH
G6 KKN P. Clowes, 14 Derek Drive, Sneyd Green, Stoke-on-Trent, ST1 6BY
G6 KKW R. Rogers, 31 Westgate Bay Avenue, Westgate on Sea, CT88AH
GW6 KLC A. Morris, Bodvel Hall, Llannor, Pwllheli, LL53 6DW
G6 KLF A. Lythaby, 25 Greenhill Road, Otford, Sevenoaks, TN14 5RR
G6 KLH R. Taylor, 57 Walnut Tree Road, Shepperton, TW17 0RP
GW6 KLQ J. Laing, Penyboncyn, Pen-Y-Garnedd, Oswestry, SY10 0AN
G6 KMG I. Turnbull, 45 Elton Road, Darlington, DL3 8HU
GM6 KMK S. Windsor, Hillside Overbrae, Fisherie, Turriff, AB53 5QP
G6 KMQ C. Meadows, 47 Widney Lane, Solihull, B91 3LL
G6 KNE J. Wright, Rhumbles, Station Approach, Dorking, RH5 5HT
G6 KNK J. Solomon, 11 Angle Close, Hillingdon, Uxbridge, UB10 0BS
G6 KNM R. Suttenwood, 51 High St., Rowhedge, Colchester, CO5 7ET
G6 KNU H. Man, 115 Northdown Park Road, Margate, CT9 3PX
G6 KOB J. Sobanski, 10 Robert Avenue, Barnsley, S71 5RB
G6 KOE A. Reilly, 14 Carleton Gardens, Carleton, Poulton-le-Fylde, FY6 7PB
GM6 KON T. Wilkins, Aelart, Lyth, Wick, KW1 4UD
GM6 KOR K. Osborne, 42 India St., Edinburgh, EH3 6HB
G6 KPA G. Osborne, 19 Orchard Close, Biggleswade, SG18 0NE
G6 KPD J. Perrett, 12 Horne Close, Stratton-On-The-Fosse, Radstock, BA3 4SS
G6 KPJ P. Vaughan, The Views, Bedford Road West, Northampton, NN7 1HB
GM6 KPL A. Wilson, 193 Irvine Road, Kilmarnock, KA1 2LA
G6 KPT N. Woodley, 20 St. Edwards Road, Cheddleton, Leek, ST13 7JP
G6 KPW S. Taylor, 17 Crays Hill, Leabrooks, Alfreton, DE55 1LN
G6 KPX A. Thorne, 31 Oak Farm Close, Sutton Coldfield, B76 1PJ
G6 KQ K. Spicer, Grove Cottage, Dallinghoo, Woodbridge, IP13 0LR
G6 KQD G. Morris, 20 Victoria Way, Stafford, ST17 0NU
G6 KQJ H. Moon, 14 Elmwood Road, Eaglescliffe, Stockton-on-Tees, TS16 0AQ
G6 KQN G. Robertson, 24 Begonia Avenue, Farnworth, Bolton, BL4 0DS
G6 KQS J. Newton, Shestnadeseta Street 6, Mindya, Bulgaria, 5044
G6 KQZ B. Wiseman, 307 Kempshott Lane, Basingstoke, RG22 5LY
G6 KRG J. Freeman, 18A Five Bells, Watchet, TA23 0HZ
G6 KRJ J. Jones, 52 Woodleigh Drive, Sutton-On-Hull, Hull, HU7 4YZ
GW6 KRK E. Karklins, Lonlas House, Lonlas, Neath, SA10 6SD
G6 KRN M. Everitt, Mallory, St. Johns Road, Ventnor, PO38 3EF
G6 KRS N. Ashall, 21 Buxton Lane, Droylsden, Manchester, M43 6HL
G6 KRY C. Pieters, 32 Olde Farm Drive, Blackwater, Camberley, GU17 0DU
G6 KSK A. Hodgson, 33 Higham Road, Wainscott, Rochester, ME3 8BE
G6 KSO P. Lash, 7 Park Road, Stockport, SK4 4PY
G6 KSR F. Patman, Northcote, 31 Church Road, Lincoln, LN6 5UW
G6 KSV A. Sayers, 145 Campkin Road, Cambridge, CB4 2NP
G6 KTB W. Curtis, Innsbruck, Trevingey Crescent, Redruth, TR15 3DF
G6 KTC W. Bramwell, 15 Chadswell Heights, Lichfield, WS13 6BH
G6 KTE D. Brunt, 31 The Green, Kingsley, Stoke-on-Trent, ST10 2AG
G6 KTG D. Clements, 3 Tilefields, Hollingbourne, Maidstone, ME17 1TZ
G6 KTK C. Handley, Torcroft, 40 New Village Road, Little Weighton, HU20 3XH
G6 KTN M. Lamerick, 12 Woodhouse Close, Birchwood, Warrington, WA3 6QP
G6 KTO J. Martyn Clark, 127 Blackpool Road North, Lytham St Annes, FY8 3DB
GM6 KTP K. Morrison, 8 St. Helena Crescent, Hardgate, Clydebank, G81 5PD
G6 KTR G. Birch, 18 Hill Farm Way Boxted, Colchester, CO4 5RD
G6 KTX A. King, 2 Longstaff Gardens, Fareham, PO16 7RR
G6 KUI P. Walker1, 23 Denstone Drive, Alvaston, Derby, DE24 0HZ
G6 KUJ F. Moulding, 28 Woodbine Road, Bolton, BL3 3JH
G6 KVA R. Dresser, 6 Acacia Avenue, Fencehouses, Houghton le Spring, DH4 6JG
G6 KVE C. Payne, 4 George Street, Helpringham, Sleaford, NG34 0RS
G6 KVG S. Reid, 223 The Greenway, Epsom, KT18 7JE
G6 KVI B. Gosling, 15 Cherry Chase, Tiptree, Colchester, CO5 0AE
G6 KVK G. Howell, 19 Constable Avenue, Eaton Ford, St Neots, PE19 7RH
G6 KVR P. Roberts, 30 Baldwins Lane, Hall Green, Birmingham, B28 0QX
GI6 KVS H. Porter, 30 Twinburn Road, Newtownabbey, BT37 0EL
G6 KVY S. Trotter, 61 Trinity Road, Billericay, CM11 2RY
G6 KWA D. King, 20 Trinity Close, Haslingfield, Cambridge, CB23 1LS
G6 KWH J. Dixon, 16 Forest Lane, Martlesham Heath, Ipswich, IP5 3ST
GW6 KWM P. Holt, 5 Tudor Road, Treboeth, Swansea, SA5 9HF
GM6 KWU R. Adamson, 29 South Hermitage Street, Newcastleton, TD9 0QE
G6 KWY D. Tipping, 30 Bickerton Point, South Woodham Ferrers, Chelmsford, CM3 5YG
G6 KWZ J. Manning, 280 Ledbury Road, Hereford, HR1 1QL
G6 KXB R. Linzey, 29 Arkle Court, Alnwick, NE66 1BS
G6 KXD M. Parker, Hazel House, Talkin, Brampton, CA8 1LE
G6 KXJ K. Turner, 16 Orford Street, Liverpool, L15 8HX
G6 KXN R. Perry, 6 Morgan Close, Arley, Coventry, CV7 8PR
GM6 KXP D. Flanagan, Ryan Mar, Stair Drive, Stranraer, DG9 8EY
G6 KXW A. Blair, 35 South Court Avenue, Dorchester, DT1 2BY
G6 KYE M. Davis, 86 Upper Shaftesbury Avenue, Southampton, SO17 3RT
G6 KZI R. Gregory, 75 Station Road South, Belton, Great Yarmouth, NR31 9LZ
G6 LAE J. Clifton, 6 Chester Close, Newbury, RG14 7RR
G6 LAO S. Parkinson, Meadow Barn, Watery Lane, Astbury, Congleton, CW12 4RR
G6 LAU D. Tanswell, Highstead Farmhouse, Bradford, Holsworthy, EX22 7AA
G6 LAW C. Rudge, 1 Mill Lane, Alfington, Ottery St Mary, EX11 1PF
G6 LBE J. Massey, 10 Rapley Avenue, Storrington, Pulborough, RH20 4QL
G6 LBG N. Orgill, 32 Upland Avenue, Chesham, HP5 2EB
G6 LBJ P. Shadbolt, 39 Ringstead Crescent, Weymouth, DT3 6PT
G6 LBO K. Batty, 19 Breckland Close, Stalybridge, SK15 2QQ
G6 LBQ A. Hunter, 22 Lynthorpe Road Cadishead, Manchester, M44 5JQ
G6 LBR A. Ledger, 9 Fox Wood, Westlea, Swindon, SN5 7AW
G6 LCL T. Mallett, 11 Caragh Road, Chester le Street, DH2 3EA
G6 LCP P. Muzyka, 2 Engine Fold Off Lindale Lane Kirkhamgate, Wakefield, WF2 0PP
G6 LCS J. Mcneill, 2 Greenwood Close, Weaverham, Northwich, CW8 3RH
G6 LCU J. Retter, 12 Palmerston Road, Grays, RM20 4YR
G6 LCX M. Pomfret, 17 Lovers Lane, Atherton, Manchester, M46 0PG
G6 LD Denby Dale ARS, c/o J. Stocks, 96 North Street, Lockwood, Huddersfield, HD1 3SL
G6 LDA J. Round, 53 Furlong Lane, Halesowen, B63 2TB
GM6 LDG P. Clements, Dhualton Cottage, Kirtomy, Thurso, KW14 7TB
G6 LDJ R. Wilkinson, 2 Conway Avenue, Billingham, TS23 2HX
G6 LDM D. Shippen, 11A Pear Tree Drive, Wincham, Northwich, CW9 6EZ
G6 LDO C. Seeney, 91 Dovehouse Close, Eynsham, Witney, OX29 4EW
G6 LDP D. Scott, 7 Greenfield Mount, Wrenthorpe, Wakefield, WF2 0TJ
G6 LDW J. Tottle, 327A Edificio, Calle Miguel Machado, Son Caliv, Calvia, Spain, 07181,
G6 LDY J. Seddon, 11 Hilda St., Leigh, WN7 5DG
G6 LEB T. Leader-Chew, 10 Hawmead, Crawley Down, Crawley, RH10 4XY
G6 LEI S. Meadwell, 25 Redland Road, Oakham, LE15 6PH
G6 LEK R. Mason, 23 Fulmodeston Road, Stibbard, Fakenham, NR21 0LT
G6 LEU D. Last, Hillview, New Road, Bridport, DT6 4NY
G6 LEY D. Miller, 44 Long Lane, Ickenham, Uxbridge, UB10 8TA
GM6 LEZ J. Mcdermott, 12A Margaret Street, Greenock, PA16 8AS
G6 LFA A. Machin, 10 Buttlehide, Maple Cross, Rickmansworth, WD3 9TZ
G6 LFC J. Mchale, Glen Elg, The Green, Skipton, BD23 4LB
G6 LFD J. Corderoy, 1 Alandale Drive, Pinner, HA5 3UP
G6 LFG J. Bradbury, 281 Peter Street, Macclesfield, SK11 8EX
G6 LFJ J. Aslan, 16 Guildford Street, Brighton, BN1 3LS
G6 LFQ R. Cross, 84A Cranborne Avenue, Surbiton, KT6 7JT
G6 LFR L. Carr, 29 Hill Drive, Whaley Bridge, High Peak, SK23 7BH
G6 LFT Dr G. Cooke, 37 Hertford Close, Woolston, Warrington, WA1 4EZ
G6 LFW J. Ford, 24 Tonstall Road, Epsom, KT19 9DP
G6 LGM I. Rogers, Tremayne Cottage, Calloose Lane, Hayle, TR27 5ET
G6 LGR A. Picot, 14 Ringshall Road, St. Pauls Cray, Orpington, BR5 2LZ
G6 LHA F. Priestnall, 56 Badger Gate, Threshfield, Skipton, BD23 5EN
GW6 LHF D. Owen, Grasmere, Pontycleifion, Cardigan, SA43 1DW
G6 LHG B. O'Shea, 37 Gardeners Road, Halstead, CO9 2TA
G6 LHQ R. Harber, 7 Hamilton Avenue, Cobham, KT11 1AU
G6 LI Lincolnshire Poachers Contest Group, c/o D. Johnson, Dean Cottage, Dowsby Fen, Bourne, PE10 0TU
G6 LIB J. Baker, 5 Larkspur Close, Bishop's Stortford, CM23 4LL
G6 LIJ D. Chilton, 6 Deneside Close, Yarm, TS15 9NT
G6 LIK L. Clark, 56 Rembrandt Avenue, South Shields, NE34 8RU
GM6 LIN J. Quinn, 8 Cluny Drive, Newton Mearns, Glasgow, G77 6YG
G6 LJC G. Weston, Meadowside Wardle Lane St2 7Lp, Stoke on Trent, ST2 7LP
GM6 LJE R. Waitt, Orchard Cottage, Claygate, Canonbie, DG14 0RZ
G6 LJF J. White, 8 Well Side, Marks Tey, Colchester, CO6 1XG
G6 LJH M. Wilson, Hillside, Chapel Lane, Kettering, NN14 4EA
G6 LJR D. Twyman, 77 Essex Road, Maldon, CM9 6JH
G6 LJU J. Whitehouse, The Paddock, Westmancote, Tewkesbury, GL20 7EP
G6 LJX S. Williams, 187 London Road, Northwich, CW9 8AR
G6 LKA B. Woolnough, 99 Abbey Road, Leiston, IP16 4TA
G6 LKB D. Warburton, 36 Bigland Drive, Ulverston, LA12 9PD
G6 LKG R. Milne, 9 Brunstath Close, Wirral, CH60 1UH
G6 LKH C. Dunlop, 32 Court Way, Twickenham, TW2 7SN
G6 LKJ J. Depledge, 37 Higher Bents Lane, Bredbury, Stockport, SK6 1EE
G6 LKV G. Ashbee, 6 The Green, Wimbledon, London, SW19 5AZ
G6 LKW T. Ashbee, Plough Heights, Main Road, Itchen Abbas, Winchester, SO21 1BQ
G6 LKZ D. Bentley, 302 Bordesley Green East Stechford, Birmingham, B33 8ST
G6 LLD G. Bell, 4 Dallymore Drive, Bowburn, Durham, DH6 5ES
G6 LLF P. Bennett, 1 The Briars, Newcastle, ST5 9PU
G6 LLG M. Broadway, 69 The Brambles, Crowthorne, RG45 6EF
G6 LLL D. Burrows, 32 Whitfield Cross, Glossop, SK13 8NW
G6 LLP R. Farey, 38 Trent Close, Yeovil, BA21 5XQ
G6 LLU D. Setterfield, 10 Birch Walk, Bride Street, Todmorden, OL14 5ET
G6 LMB P. Steadman, 41 The Linkway, Brighton, BN1 7EJ
G6 LMC P. Webb, 63 Trinity Road, Halstead, CO9 1ED
GW6 LMI J. Evans, 91 Queens Avenue, Flint, CH6 5JP
G6 LMJ Dr G. Eardley, 45 Little Moss Lane, Scholar Green, Stoke-on-Trent, ST7 3BL
G6 LMR K. Fisher, 26 Manila Street, Sunderland, SR2 8RS
G6 LNA P. Cooper, 2 Meadowsweet Way, Wimblebury, Cannock, WS12 2GS
G6 LNF N. Clare, 4 Arlington, Weymouth, DT4 9SG
G6 LNL I. Dobson, 73 Stanley Street, Seaham, SR7 0AU
G6 LNS J. Duxbury, Woodlands, Wallace Lane, Preston, PR3 0BB
G6 LNU J. Durban, 62 Westfield Way, Charlton, Wantage, OX12 7EP
G6 LNV J. Cunliffe, 142 Hall Road, Hull, HU6 8SB
G6 LOC T. Stirrup, 23 Round Wood, Penwortham, Preston, PR1 0BN
G6 LOJ N. Pettit, 10 Broom Road, Lakenheath, Brandon, IP27 9ES
G6 LOR E. Wood, 57D Halesowen St., Rowley Regis, B65 0HF
G6 LPB R. Steele, 27 Beasley Grove, Birmingham, B43 7HG
G6 LPC A. Samways, 61 Cooper Road, Rye, TN31 7BG
G6 LPD A. Tucker, 63 Oakes Road, Bury St Edmunds, IP32 6PU
G6 LPG S. Taylor, 76 Queensdown Gardens, Bristol, BS4 3JF
G6 LPS T. Biddle, 55 Barley Mow Lane, Bromsgrove, B61 0LU
G6 LPT R. Bruckner, Flat 2 Byron Court Fairfax Road, London, NW6 4HB
G6 LPV D. Blackmore, 2 Witten Gardens Northam, Bideford, EX39 3RE
G6 LPX P. Brown, 5 Fairview Close, Amington, Tamworth, B77 3LA
G6 LQE D. Byrom, 206 Didsbury Road, Stockport, SK4 2AA
G6 LQG D. Bate, 45 Arlington Way, Stoke-on-Trent, ST3 7WH
G6 LQI N. Bird, 2 Manor Valley, Weston-Super-Mare, BS23 2SY
G6 LQM G. Barker, 99 Sheffield Road, Wymondham, NR18 0HS
G6 LQP D. Brown, Hillingswood, Acton, Newcastle, ST5 4FD
G6 LQR B. Walker, 5-6 Cochrane Terrace, Willington, Crook, DL15 0HN
G6 LRT C. Johnson, 52 Evesham Road, Cookhill, Alcester, B49 5LJ
G6 LRU R. Jones, 53 Wavertree Road, Blacon, Chester, CH1 5AF
G6 LRY C. Kelland, 11 The Meads, West Hanney, Wantage, OX12 0LJ
G6 LSB N. Key, Three Corner Cross, Rosecare, St. Gennys, Bude, EX23 0BE
G6 LSC M. Kitchener, 5 Whinbush Grove, Hitchin, SG5 1PT
G6 LSD P. Kerry, 35 Victoria Drive, Blackwell, Alfreton, DE55 5JL
GW6 LSL S. Wood, 2 Radyr Road, Llandaff North, Cardiff, CF14 2FU

G6 LSO C. Wolf, 35A Moorhouse Road, Carlisle, CA2 7LU
G6 LST D. Rhodes, 1 Tanpit Cottages, Winstanley, Wigan, WN3 6JY
G6 LSW A. Stevenson, 37 Hillside Road East, Bungay, NR35 1JU
G6 LTB P. Townrow, 64 Millham Road, Bishops Cleeve, Cheltenham, GL52 8BG
G6 LTD P. Sutton-Atkins, 91 Brook Road, Tolleshunt Knights, Tiptree, Colchester, CO5 0RH
G6 LTK K. Wilson, 44 Campbell Road, Caterham, CR3 5JN
G6 LTN A. Wanford, 4 Willows Close, Tydd St. Mary, Wisbech, PE13 5QR
G6 LTR J. Warner, 32 Rolleston Road, Wigston, LE18 2EP
G6 LUD A. Ryan, 153 Sutton Road, Maidstone, ME15 9AB
G6 LUE T. Yates, 5 Manor Garth, Kellington, Goole, DN14 0NW
G6 LUF A. Yates, 59 Worden Lane, Leyland, PR25 3BD
G6 LUJ R. Perry, Thornaby, Queen Street, Colyton, EX24 6JU
G6 LUK J. Russell, 9 Batten Close, Christchurch, BH23 3BJ
G6 LUM J. Papworth, 339 Gayfield Avenue, Brierley Hill, DY5 3JE
GW6 LUT A. Mills, 1 Derwen Fawr, Llandybie, Ammanford, SA18 2UY
G6 LUU A. Marshall, 8 Barn Owl Close, Northampton, NN40RQ
G6 LUY N. Mattey, 27 Middleton Road, Daventry, NN11 8BH
G6 LVB H. Long, 40 Ashburn Place, London, SW7 4JR
G6 LVC J. Shergold, 35 Orchard Grove, New Milton, BH25 6NZ
G6 LVG S. Normandale, 5 The Beacon, Ilminster, TA19 9AH
G6 LVI P. Hickey, 36 Station Road, Alderholt, Fordingbridge, SP6 3RB
G6 LVJ R. Hickey, Mallorin, Blackfield Road, Southampton, SO45 1EG
G6 LVM C. Holderness, 7 Oakfield Avenue, Clayton Le Moors, Accrington, BB5 5XG
G6 LVN R. Hope, 35 Pinewood Gardens, North Cove, Beccles, NR34 7PQ
G6 LVS D. Mallalieu Howard, Flat 17, The London Well Street, Ryde, PO33 2SS
G6 LVT C. Harvey, 619 West Street, Crewe, CW2 8SH
G6 LWA C. Hall, 147 Gordon Avenue, Camberley, GU15 2NR
G6 LWC R. Hardman, 4 Alverstone Road, Wallasey, CH44 9AA
G6 LWD P. Hurley, 18 Pear Tree Lane, Wolverhampton, WV11 1BD
G6 LWK M. Horsfield, 13 St. Leonards Way, Ardsley, Barnsley, S71 5BS
G6 LWT J. Mills, Smiths Hill Petrockstow, Okehampton, EX20 3EZ
G6 LWZ L. Miles, 1 Wyndham Wood Close, Fradley, Lichfield, WS13 8UZ
G6 LXE T. Doyle, Leal House, Sparrow Pit, Buxton, SK17 8ET
G6 LXF P. Duley, 4 Brean Road, Stafford, ST17 0PA
G6 LXL J. Ellis, Goosters Green, Hope Bagot, Ludlow, SY8 3AE
G6 LXP D. English, 14 Elm Close, Ryde, PO33 1ED
G6 LXU S. Westall, 4 South View Great Harwood, Blackburn, BB6 7NL
G6 LXV D. Woods, 110 Sandy Lane, Warrington, WA2 9JA
G6 LXW J. Weigh, 167 Farm View Road, Rotherham, S61 2BL
G6 LYA Dr P. Whysall, 1 Greenlees Close, Fareham, PO17 5GS
G6 LYD C. Weaver, Linton Lodge, Pluckley Road, Ashford, TN27 0AQ
G6 LYE G. Whiles, 7 Thorndale Street, Hellifield, Skipton, BD23 4JE
GM6 LYJ J. Young, 1 Stevenson Place, Annan, DG12 6BU
G6 LYM D. Miller, 9 High Mead, Hockley, SS5 4QG
G6 LZB P. Adams, 4 Cherry Coft, Croxley Green, Rickmansworth, WD3 3AL
G6 LZM G. Beddington, Konrei, Tower Hill, Norwich, NR8 5AX
G6 LZX B. Broad, 1 Sussex Close, Laindon, Basildon, SS15 6PR
G6 LZZ J. Bolton, Huds House, Cowgill, Sedbergh, LA10 5TQ
G6 MAA G. Bishop, Oyston Lodge, Lynstone Road, Bude, EX23 8LR
GW6 MAB J. Barwick, 13 Greenfields Avenue, Bridgend, CF31 4SR
G6 MAC B. Mcdonnell, 68 Chaigley Road, Longridge, Preston, PR3 3TQ
G6 MAD T. Morley, 2 Hawthorn Place, Woodbridge, IP12 4JZ
G6 MAJ A. Mulvaney, 38 Ramwells Brow, Bromley Cross, Bolton, BL7 9LL
G6 MAM K. Wright, 63 Dudley Avenue, Leicester, LE5 2EF
G6 MAR G. Wratten, 6 Buckle Drive, Seaford, BN25 2QN
G6 MAT A. Valentine, 21 Naseby Road, Congleton, CW12 4QX
G6 MAW R. Underwood, 35 Greenfields, Langley Mill, Nottingham, NG16 4GJ
G6 MAY D. Pool, 6 Rivett Close, Clothall Common, Baldock, SG7 6TW
G6 MBF D. Surgey, 4 Down Lane Bathampton, Bath, BA2 6UE
G6 MBH W. Stiling, 11 Carrol Grove, Cheltenham, GL51 0PP
G6 MBI D. Stainton, Tilton House, 39 Redland Grove, Nottingham, NG4 3ET
G6 MBL M. Snow, 32 Orchard Avenue, Worthing, BN14 7PY
G6 MBR Mid Beds R/Net, c/o I. Mciver, 3 Asgard Drive, Bedford, MK41 0UP
G6 MBV C. Sutcliffe, 1 St. James Street, Waterfoot, Rossendale, BB4 7HN
G6 MC Brimham Contest Group, c/o S. Clarke, Brimham Lodge Fm, Harrogate, HG3 3HE
G6 MCB M. Baldry, 10 Kingfisher Court, Lowestoft, NR33 8PJ
G6 MCC L. Crompton, 6 Moss Avenue, Ashton-On-Ribble, Preston, pr21sh
G6 MCE P. Garde, 21 Leicester Avenue, Timperley, Altrincham, WA15 6HR
G6 MCG C. Garnham, 1 Ennerdale Close, Felixstowe, IP11 9SS
G6 MCN A. Gillespie, Elm Tree Cottage, Chilbolton, Stockbridge, SO20 6BA
G6 MCQ J. Grane, 15 Pinelands Way, Osbaldwick, York, YO10 3QJ
GM6 MCV J. Mcvicar, 2 Lilliardsedge Par, Mr Ancrum, Roxburghshire, TD8 6TZ
G6 MCX P. Garland, 6 Barn Piece, Chandler'S Ford, Eastleigh, SO53 4HP
G6 MCY M. Goddard, 65 Langley Hall Road, Solihull, B92 7HE
GM6 MD Clyde Coast Contest Club, c/o A. Dunn, 50 Pemberton Valley, Ayr, KA7 4UB
G6 MDC M. Green, 9 Greencroft Avenue, Northowram, Halifax, HX3 7EP
G6 MDG J. Tyson, 1102 Rochdale Road, Blackley, Manchester, M9 7EQ
G6 MDM W. Smyth, 4 Dereham Road, Pudding Norton, Fakenham, NR21 7NA
G6 MDN G. Tillett, 43 Chippenham Road, Harold Hill, Romford, RM3 8HJ
G6 MDR I. Stanley, 6 Kennedy Avenue, Long Eaton, Nottingham, NG10 3GF
G6 MDS A. Scott, 3 Majestic Road, Hatch Warren, Basingstoke, RG22 4XD
G6 MDX D. Davies, 9 Lawnsfield Walk, Stafford, ST16 1TS
G6 MED P. Cartmell, 16 Churchfield Drive, York, yo322fl
G6 MEH J. Turner, 155 Bure Lane, Christchurch, BH23 4HB
G6 MEI C. Thacker, 1 Pine Grove, Chorley, PR6 7BW
G6 MER M. Roberts, 32 Orion, Bracknell, RG12 7YX
G6 MEW A. Hunt, 7 Wood Lodge, Calmore, Southampton, SO40 2UP
G6 MFB R. Hodges, 21A Preston Lane, Lyneham, Chippenham, SN15 4AR
G6 MFU N. Cowley, 126 Racecourse Road, Swinton, Mexborough, S64 8DS
G6 MGA S. Cooper, 27 Huntsmans Gate, Bretton, Peterborough, PE3 9AU
G6 MGH M. Cooper, 3 Marina Way, Ripon, HG4 2LJ
G6 MGN D. Richardson, 14 Wingfield Avenue, Lakenheath, Brandon, IP27 9HS
G6 MGQ A. Reddish, Wheelwrights House, Luckeys Corner, Hitcham, Ipswich, IP7 7LR
G6 MGZ J. Middleton, 187 Balcombe Road, Horley, RH6 9EA
GM6 MHC R. Mcnaught, 5 Bonnymuir Crescent, Bonnybridge, FK4 1GD
G6 MHF A. Marshall, Thistledome, First Avenue, Watford, WD25 9PS
G6 MHO I. Pomfret, 20 Sandown Road, Bury, BL9 8HN
G6 MHR J. Castelow, 7 Langford Close, Burley In Wharfedale, Ilkley, LS29 7NP
GW6 MHV B. Cooke, 51 Celyn Avenue, Cardiff, CF23 6EJ
G6 MHY Grimsby & Cleethorpes District Sas Radio Scouting Team, c/o A. Carlile, Top Flat 13B Mill Road, Cleethorpes, DN35 8HZ
G6 MIC M. Clayden, 121 North Lane, East Preston, Littlehampton, BN16 1HB
G6 MID P. Croft, Exchange Buildings, Exchange Street, Normanton, WF6 2AA
GW6 MIH M. Cleverley, 33 Tylchawen Crescent, Tonyrefail, Porth, CF39 8AL
G6 MIS S. Ransom, 1 Bilberry Road, Clifton, Shefford, SG17 5HB
G6 MIU W. Livesey, 20 West Way, Little Hulton, Manchester, M38 9GL
G6 MJA M. Addison, Berrymead, Oxford Street, Great Missenden, HP16 9JH
G6 MJB D. Lloyd, Rangelands, Old Guildford Road, Camberley, GU16 6PH
G6 MJM S. Parker, 19 Sundour Crescent, Wolverhampton, WV11 1AP
G6 MJT G. Painting, 35 Selsdon Road, New Haw, Addlestone, KT15 3HP
G6 MJW G. Davis, 30 Bonny Wood Road, Hassocks, BN6 8HR
GM6 MJY C. Donald, 126 Newburgh Circle, Bridge Of Don, Aberdeen, AB22 8XB
G6 MKD M. Douglass, 20 Cadshaw Close Birchwood, Warrington, WA3 7LR
G6 MKJ N. Ellis, 140 Wollaston Road, Irchester, Wellingborough, NN29 7DH
G6 MKL R. Ellis, 4A Elmdale Road, Earl Shilton, Leicester, LE9 7HQ
G6 MKO S. Everett, 11 Chepstow Road, Felixstowe, IP11 9BU
G6 MKQ G. Evans, 31 Queen Elizabeth Crescent, Accrington, BB5 2AS
GW6 MKR C. Foster, Pentwyn House, Delfryn, Aberdare, CF44 0TU
G6 MKZ P. Fisher, 2 Leas Drive, Iver, SL0 9RD
G6 MLH G. Marshall, Fern House, Church Road, Newark, NG23 7ED
GW6 MLI Dr D. Morgan, Northwood Hotel, 47 Rhos Road, Colwyn Bay, LL28 4RS
G6 MLJ T. Maker, 25 Walthams Place, Pitsea, Basildon, SS13 3PR
GW6 MLL B. Murphy, 22 Deepglade Close, St. Thomas, Swansea, SA1 8EJ
G6 MLS T. Abson, 177 Meadowhall Road, Kimberworth, Rotherham, S61 2JW
G6 MLV K. Barker, 8 Shelley Gardens, Wembley, HA0 3QG
G6 MMA M. Barlow, 56 Pasturegreen Way, Irlam, Manchester, M44 6TE
G6 MMB J. Bulbrook, 33 Stonecross Way, March, PE15 9DH
G6 MMD S. Burrows, 2 Luscombe Farm Cottages, Heath End, Stratford-upon-Avon, CV37 0PP
G6 MMG D. Brown, 28 Bishop Drive, Whiston, Prescot, L35 3JL
G6 MMJ P. Bromley, 3 Georgia Avenue, Broadwater, Worthing, BN14 8AZ
G6 MML V. Bates, The Anvil, 4 Eastgate, North Newbald, YO43 4SD
GW6 MMM J. Bowen, 18 Admirals Walk, Sketty, Swansea, SA2 8LQ
G6 MMR A. Salter, 143 Eastwood Road North, Leigh-on-Sea, SS9 4NB
G6 MMS J. Young, 45 Eaves Lane, Chadderton, Oldham, OL9 8RG
G6 MMT J. Ward, 64 Gladstone Road, Ipswich, IP3 8AT
G6 MNB M. Bulmer, Highfield, 7 Fountain Avenue, Altrincham, WA15 8LY
GW6 MNC W. Turner, 37 Dan-Y-Bryn Avenue, Radyr, Cardiff, CF15 8DD
G6 MNI R. Andrews, 10 Summerfield Close, Mevagissey, St Austell, PL26 6TZ
G6 MNJ P. Andrews, 10 Summerfield Close, Mevagissey, St Austell, PL26 6TZ
G6 MNL A. Butler, 45 Roewood Close, Holbury, Southampton, SO45 2JT
G6 MOD P. Boden, 54 Avill, Hockley, Tamworth, B77 5QF
G6 MOI A. Bates, 44 Chalfont Drive, Sileby, Loughborough, LE12 7RQ
G6 MOT S. Kilmister, 9 Wheal Jane Meadows, Threemilestone, Truro, TR3 6EN
G6 MOZ P. Sealey, 45 Haydon Way, Coughton, Alcester, B49 5HY
G6 MPE J. Simmons, 282 Bishopton Road West, Stockton-on-Tees, TS19 7LY
G6 MPJ W. Smith, Boleyn Service Station, 77 River Road, Barking, IG11 0DS
G6 MPK T. Smith, 87 Swanland Road, Hessle, HU13 0NS
G6 MPN A. Shalders, 29 Princess Drive, Sandbach, CW11 1BS
G6 MPT P. Pritchard, 5 Charlemont Road, Stone Cross, West Bromwich, B71 3HX
GW6 MPX D. Prince, 40 Ffordd Gryffydd, Llay, Wrexham, LL12 0RT
G6 MQD P. Mullins, The Birches, Clifton, Ashbourne, DE6 2GL
G6 MQG C. Northrop, Mayfield, 47B Hardhorn Road, Poulton-le-Fylde, FY6 7SR
G6 MQH M. Pickard, 9 Beatrice Way, Trowbridge, BA14 7TX
G6 MQI G. Pointon, 448 Stockport Road, Thelwall, Warrington, WA4 2TR
G6 MQJ P. Racher, 2 Heron Way, Horsham, RH13 6DG
G6 MQK I. Stinton, 57 Wildfields Road, Clenchwarton, King's Lynn, PE34 4DE
G6 MQN R. Wyatt, 1 Ivy Villa, Kings Hill, Haverhill, CB9 7NA
G6 MQP N. Roberts, 81 Broad Lane, Coventry, CV5 7AH
G6 MQU B. Plumtree, Sunnyside, Station Road, Skegness, PE24 5ES
G6 MQY P. Wilson, Laurel Cottage, 43 Newnham Road, Ryde, PO33 3TE
G6 MQZ T. Wilson, Orchard House, Whitmoor Lane, Guildford, GU4 7QB
G6 MRN N. Parr, 24 Park Avenue, Awsworth, Nottingham, NG16 2RA
GW6 MRO J. Reddaway, Voltaire House, Ffordd Uchaf, Gwynfryn, Wrexham, LL11 5UN
G6 MRP K. Playford, 1 Cherwell Close, Abingdon, OX14 3TD
G6 MRW P. Grant, 117 Hazel Avenue, Farnborough, GU14 0DW
G6 MRY C. Guy, 78 Park Road, Bolton, BL1 4RQ
G6 MSC T. Glover, 70 Sandown Road, Toton, Nottingham, NG9 6JW
G6 MSY Merseyside ARS, c/o A. Birch, 6 Crescent Road, Wallasey, CH44 0BQ
G6 MTB D. Hesketh, Flat 3, Redwood Court Plantation Terrace, Dawlish, EX7 9FD
G6 MTE S. Heath, 12 The Medway, Daventry, NN11 4QU
G6 MTF T. Horn, 9 Gipton Wood Avenue, Leeds, LS8 2TA
G6 MTG D. Knight, 119 Bracebridge Street, Nuneaton, CV11 5PD
GI6 MTL M. Mccutcheon, 10 Chestnut Brae, Gilford, Craigavon, BT63 6FA
G6 MTV D. Lovell, 5 Were Close, Warminster, BA12 8TB
G6 MTY M. Matthews, 30 Broadgate Lane, Deeping St. James, Peterborough, PE6 8NW
G6 MUJ C. James, 5 Arbroath Road, Luton, LU3 3LA
GW6 MUP W. Jones, Pen-Y-Berth, Pen-Y-Garth, Gwynedd, LL55 1EY
G6 MUQ T. Jones, 4 Anne Close, Christchurch, BH23 2NW
G6 MUW B. Kent, 6 Church Walk Mancetter, Atherstone, CV91NX
G6 MUX A. Kelly, 9 Cotswold Close, Dibden Purlieu, Southampton, SO45 5QW
GM6 MUZ Dr C. Duncan, 12 Juniper Park Road, Juniper Green, EH14 5DX
G6 MVD P. Sweeting, 35 The Ridings, Market Rasen, LN8 3EE
G6 MVF C. Stokes, 7 St. Nicholas Close, Arnold, Nottingham, NG5 6GU
G6 MVN R. Suckling, 21 Warren Court, Meadowside, Dartford, DA1 2RZ
G6 MVQ J. Simmonds, Overbeck South, Stokesley Road, Guisborough, TS14 8DL
G6 MVR D. Scott, 20 Belmont View, Harwood, Bolton, BL2 3QN
G6 MVS M. Sandler, 21 Kenilworth Gardens, Hornchurch, RM12 4SE
G6 MVW E. Sayer, 27 Glenmere Park Avenue, Benfleet, SS7 1SS
G6 MWB T. Gordon, 101 Dorset Road, Bexhill-on-Sea, TN40 2HU
G6 MWD Dr C. Goodhand, 22 Somin Court, Doncaster, DN4 8TN
G6 MWL D. Henderson, 7 Glenhaven Avenue, Borehamwood, WD6 1AY

G6 MWM B. Hall, 32 Danby Road, Newton, Hyde, SK14 4DL
G6 MWQ D. Horne, Flat 3, 9 Carlton Avenue, Ramsgate, CT11 9BP
G6 MWS A. Hueck, 9 Corden Avenue, Mickleover, Derby, DE3 9AQ
G6 MXE R. Peeling, 13 Greenview Crescent, Hildenborough, Tonbridge, TN11 9DR
G6 MXL C. Redwood, 53 Woodpecker Drive, Poole, BH17 7SB
G6 MXV A. Poupard, Woodlea, Crouch House Road, Edenbridge, TN8 5EN
G6 MYH C. Mallory, 11 Baymead Meadow, North Petherton, North Petherton, ta6 6qw
G6 MYL C. Jones, 63 Hockenhull Avenue, Tarvin, Chester, CH3 8LR
G6 MYO B. Johnson, 2 Plumtree Cottages, Hill Street, Swadlincote, DE12 7PW
G6 MYT C. King, 3 Huntingdon Gardens, Christchurch, BH23 2TW
GW6 MYY D. Davies, Penrallt, Abercaseg Road, Gerlan, Bangor, LL57 3SP
G6 MYZ S. Doorey, 11 Langley Gardens, Petts Wood, Orpington, BR5 1AB
G6 MZF L. Cromar, 17 Ipley Way, Hythe, Southampton, SO45 3LG
G6 MZN M. Esser, 10 Van Diemans Road, Wombourne, Wolverhampton, WV5 0BQ
G6 MZT B. Fereday, 16 Glentworth Gardens, Wolverhampton, WV6 0SF
G6 MZV R. Titmuss, 70 Mallards Rise, Church Langley, Harlow, CM17 9PL
G6 MZW D. Speak, 42 Penn Lea Road, Twerton, Bath, BA1 3RB
G6 NAD M. Mcdermott, 91 Hargwyne Street, London, SW9 9RH
G6 NAG D. Lang, 8 Church Hill, Cheddington, Leighton Buzzard, LU7 0SY
G6 NAH P. Proudlove, 14 Heath Avenue, Rode Heath, Stoke-on-Trent, ST7 3RY
G6 NAJ Rev. T. Leyland, 4 Ellsmore Meadow, Lichfield, WS13 6NJ
G6 NAL R. Pain, 200 Butchers Lane Mereworth, Maidstone, ME18 5QF
G6 NAP E. Lester, 178 Newtown Road, Malvern, WR14 1PJ
GI6 NAQ S. Mccullagh, 7 Clanbrassil Gardens, Portadown, Craigavon, BT63 5YD
G6 NAV R. Martin, 110 Binscombe, Godalming, GU7 3QJ
G6 NAX S. Moring, 1 Burrows Cottages, Toot Hill Road, Ongar, CM5 9QN
G6 NBF J. Baron, 9 Milton Avenue, Doncaster, DN5 8ER
G6 NBI J. Brett, 127 Cranbrook Drive, Maidenhead, SL6 6RY
G6 NBL R. Barnett, 5 Overbrook, Evesham, WR11 1DE
G6 NBM G. Bryant, 37 Broad Leas Court, Broad Leas, St Ives, PE27 5XG
G6 NBP P. Blease, 15 Shadybrook Lane, Weaverham, Northwich, CW8 3PN
G6 NCE M. Craig, 194 Elm Grove, Brighton, BN2 3DA
G6 NCL R. Pickstone, 33 Shore Mount, Littleborough, OL15 8EN
GU6 NCZ P. Wild, Honfleur, La Rue Du Le Hurel, Vale, Guernsey, GY3 5AF
G6 NDA C. Venn, Stantor, High Road, Templecombe, BA8 0DN
G6 NDH P. Walker, 37 Cromwell Road, Grimsby, DN31 2DN
G6 NDJ A. Wilson, 23 Claydown Way, Slip End, Luton, LU1 4DU
GI6 NDM G. O`Boyle, 27A Drapersfield Road, Cookstown, BT808RS
G6 NDS Northampton Sct, c/o I. Rivett, 30 Millside Close, Kingsthorpe, Northampton, NN2 7TR
G6 NEA J. Dean, 15 Park Close, Sonning Common, Reading, RG4 9RY
G6 NEK P. Diss, 130 Beridge Road, Halstead, CO9 1JU
G6 NEZ R. Emerson, 4 Freeford Gardens, Lichfield, WS14 9RJ
G6 NFB T. Wright, 182 Lansdowne Road, Oxton, Prenton, CH43 7SQ
G6 NFC A. Young, 1 Rapley Green, Bracknell, RG12 7PS
G6 NFE R. White, 10 Melbourne Rise Bicton Heath, Shrewsbury, SY3 5DA
G6 NFJ J. Eden, 23 Elm Green Close, Worcester, WR5 3HD
GI6 NFK S. Ferguson, 20 Old Road, Loughgall, Armagh, BT61 8JD
G6 NFR R. Foden, 1A Garden Cottages, Eaton Road, Liverpool, L12 3HQ
G6 NGA J. Lamble, 17 Willowvale, Lowestoft, NR324UB
G6 NGF M. Tatlow, 20 Windmill Way, Tysoe, Warwick, CV35 0SB
G6 NGM S. Cross, 7 April Place, Buckhurst Road, Bexhill-on-Sea, TN40 1UE
G6 NGN D. Simpson, The Hawthorns, Slacken Lane, Stoke-on-Trent, ST7 1NQ
GW6 NGR P. Thornton, 11 Fron Hyfryd, Llithfaen, LL536NT
G6 NGV W. Taylor, 29 Holmdale Road, Syston, Leicester, LE7 2JN
G6 NHA M. Malyon, 16 Tintern Road, Gosport, PO12 3QN
GW6 NHB G. Mahoney, 684 Beechley Drive, Pentrebane, Cardiff, CF5 3SS
G6 NHG S. Marshall, 25 Carlcroft, Wilnecote, Tamworth, B77 4DL
G6 NHK N. Martin, Stonea House, Middle Road, March, PE15 0AJ
GW6 NHL A. Mccallum, Glan Alaw, Llanddeusant, Holyhead, LL65 4AG
G6 NHO R. Smith, 20 Dryden Way, Higham Ferrers, Rushden, NN10 8DH
G6 NHU K. Maton, 41 Bemerton Gardens, Kirby Cross, CO13 0LQ
G6 NHV D. Meakins, 19 Booth Lane North, Northampton, NN3 6JQ
G6 NHW P. Minchin, 122 Mildenhall Road, Great Barr, Birmingham, B42 2PQ
G6 NHY K. Marriott, 1 Holbeck Road, Hucknall, Nottingham, NG15 7SR
GM6 NIA D. Mccall, 11 Craiglockhart Dell Road, Edinburgh, EH14 1JW
GM6 NIC J. Mcaulay, 9 Randolph Cliff, Edinburgh, EH3 7TZ
G6 NID J. Matthew, 24 Southgate, Rochdale, OL128UQ
G6 NIO M. Smith, 39 Seliot Close, Poole, BH15 2HQ
G6 NIW F. Shaw, 43 Egremont Road, Hardwick, Cambridge, CB23 7XR
G6 NIX E. Samuels, 19 Glynleigh Drive, Polegate, BN26 6LU
G6 NIZ A. Scott, The Conifers, Back Lane, York, YO30 2DF
G6 NJE J. Smith, 414 Sparrowhawk Drive, Willow Grove Park, Poulton-le-Fylde, FY6 0RS
G6 NJJ M. Swift, Spa Cottage, Spa Lane, Ormskirk, L40 6JQ
GM6 NJL M. Spittle, Stablecleugh, Ewes, Langholm, DG13 0HJ
G6 NJO P. Askham, 1 Park House Cottage, Carr Lane, Thirsk, YO7 3PF
G6 NJR P. Nikolic, 29 Tern View, Market Drayton, TF91DU
G6 NJT F. Neill, 7 Bellevue Terrace, Southampton, SO14 0LB
G6 NKI K. Brown, 73 Church Avenue, Preston, PR1 4UD
G6 NKL D. Baldock, Deeside, Platts Lane, Bucknall, Woodhall Spa, LN10 5DY
G6 NKS G. Pearn, 178 Lloyds Avenue Kessingland, Lowestoft, NR33 7TU
G6 NLC C. Rabe, 79 Rectory Avenue, Corfe Mullen, Wimborne, BH21 3EZ
G6 NLD A. Reed, 6 Brancaster Close, Nottingham, NG6 8SL
G6 NLE H. Roberts, 3 Short Avenue, Allestree, Derby, DE22 2EH
G6 NLG M. Ritchie, Bruern Abbey, Bruern, Chipping Norton, OX7 6QA
G6 NLN J. Burdass, Cedarwood, High Road, Brightwell-Cum-Sotwell, Wallingford, OX10 0PT
GW6 NLO T. Bott, 13 Cheriton Road, Pennar, Pembroke Dock, SA72 6RN
GW6 NLP M. Bryant, The Nook, Llanarmon Road, Wrexham, LL11 5YP
G6 NLQ T. Fradley, 40 Higher Green, Poulton-le-Fylde, FY6 7BL
G6 NLS D. Budd, Valhalla, 81 Bohemia Chase, Leigh-on-Sea, SS9 4PW
G6 NLU S. Buxton, 111 Digby Avenue, Nottingham, NG3 6DT
G6 NLX G. Richardson, 21 Broadlands Road, Hickling, Norwich, NR12 0YG
G6 NLZ M. Reynolds, 24 Mill Road, Lydd, Romney Marsh, TN29 9EJ
G6 NMA P. Ayers-Hunt, 15 Kelvin Road, Leamington Spa, CV32 7TF
G6 NME I. Griffiths, 34 Dobbins Oak Road, Pedmore, Stourbridge, DY9 9HX
G6 NMK M. Grimes, 73 Ryston Road, Denver, Downham Market, PE38 0DP
G6 NMQ G. Goodyer, Flat, 54 Wyndham Road, Petworth, GU28 0EQ
G6 NMU J. Greenhough, 58 Gorsey Bank, Wirksworth, Matlock, DE4 4AD
G6 NNA Dr A. Liggins, 97 Vessel Crescent, Scarborough, Ontario, Canada, M1C 5K5
GW6 NNB D. Owen, Pen Y Bont Maerdy, Corwen, LL21 0PE
G6 NNK P. Frampton, 118 Ramnoth Road, Wisbech, PE13 2JD
G6 NNO J. Evans, 74 Trejon Road, Cradley Heath, B64 7HJ
GI6 NNP V. Hagan, 7 Emania Terrace, Armagh, BT60 4AS
G6 NNS J. Hunt, Honeytiles Culford, Bury St Edmunds, IP28 6DT
G6 NNU J. Archer, 117 Heath Way, Erith, DA8 3LZ
G6 NOI J. Whelan, 8 Welland Road, Wirral, CH63 2JU
G6 NOL J. Weir, 17 Pasteur Drive, Apley, Telford, TF1 6PQ
GM6 NOO C. Wood, 5 Damhead Steading, Kinloss, Morayshire, IV36 3UA
G6 NOW R. Beck, 26 Cheshire Court, Ravenall Close, Birmingham, B34 6PZ
G6 NPC R. Carlson, 45 Firs Road, Milnthorpe, LA7 7QF
G6 NPE A. Coates, 2 Kipling Avenue, Burntwood, WS7 2HS
G6 NPJ J. Copeland, Little Cophall, Dowlands Lane, Crawley, RH10 3HX
G6 NPW S. Caine, 19 Turner Drive, Tingley, Wakefield, WF3 1UD
G6 NPZ P. Chard, 29 Nettle Gap Close, Wootton, Northampton, NN4 6AH
G6 NQB S. Clements, 39 Redland Close, Marlbrook, Bromsgrove, B60 1DZ
G6 NQL J. Wilkinson, The Old Joinery, Garsdale, Sedbergh, LA10 5PJ
G6 NQM K. Whitchurch, 65 Honey Hill Road, Kingswood, Bristol, BS15 4HN
G6 NQQ A. Wilson, 17 Rook Way, Horsham, RH12 5FR
GW6 NQU J. Hoy, 39 Blackbird Road, Caldicot, NP26 5RE
G6 NQY R. Heath, 222 Congleton Road, Talke, Stoke-on-Trent, ST7 1LW
G6 NRH D. Hallifax, 22 Wendover Way, Welling, DA16 2BN
G6 NRK A. Hunt, 39 Circular Road West, Liverpool, L11 1AY
G6 NRL C. Hargreaves, Viridis, Retford Road, Retford, DN22 0BY
G6 NRM S. Hanscombe, 24 St. Marks Drive, Wellington, Telford, TF1 3GA
GW6 NSG J. Jones, 26 Spring Road, Wrexham, LL11 2LU
GW6 NSK A. Jones, 4 Park View, Llanddew, Brecon, LD3 9RL
G6 NSQ P. James, 9 Smallholding, Tutbury Road, Burton-on-Trent, DE13 0AL
G6 NSU P. Lewis, 18 Bittaford Wood Bittaford, Ivybridge, PL21 0ET
G6 NSZ D. Lawton, 48 Woodlands Road, Woodlands, Doncaster, DN6 7JZ
G6 NTE J. Lyons, 8 Anstie Close, Devizes, SN10 2EN
G6 NTM E. Murphy, 25 Warrington Road, Ashton-In-Makerfield, Wigan, WN4 9PJ
GI6 NTP D. Mcalpine, 35 Carnamena Avenue, Belfast, BT6 9PJ
G6 NTQ R. Morgan, 1 Hillmeads Drive, Dudley, DY2 7TS
G6 NTW K. Gosbee, 64 Connaught Gardens, Palmers Green, London, N13 5BS
G6 NTY B. Griffiths, 18 Julius Drive, Coleshill, Birmingham, B46 1HL
G6 NUI D. Chamberlain, 44 Parsonage Chase, Minster On Sea, Sheerness, ME12 3JX
GM6 NUL R. Crawford, Glengarry, East Terrace, Kingussie, PH21 1JS
G6 NUQ S. Coward, 100 Lytham Road, Freckleton, Preston, PR4 1XB
G6 NUS A. Croft, 15 St. Marys Road, Bozeat, Wellingborough, NN29 7JU
G6 NUX S. Clack, Flat 9, The Grange, Emsworth, PO10 7QP
G6 NUZ A. Charlton, 26 Saundergate Lane, Wyberton, Boston, PE21 7BZ
G6 NVC K. Castley, Zone 2, (Opposite High School), Cagayan de Oro City, Philippines, 9000
G6 NVD C. Webb, 2 Rykhill, Chadwell St. Mary, Grays, RM16 4RR
G6 NVF D. Mcglasson, 19 Kennedy Street, Ulverston, LA12 9EA
G6 NVH M. Morris, 19 Gowy Close, Alsager, Stoke-on-Trent, ST7 2HX
G6 NVI S. Mindel, Longwood House, Arkley Lane, Barnet, EN5 3JR
GW6 NVJ C. Marlow, 27 Sandy Way, Connah'S Quay, Deeside, CH5 4SH
G6 NVS P. Harrison, 41 Chestnut Close, Handsacre, Rugeley, WS15 4TH
G6 NVU M. Haynes, 10 Cypress Grove, Denton, Manchester, M34 6EA
G6 NVW P. Higgins Jp, 49 Milton Road, Hoyland, Barnsley, S74 9AX
G6 NVY A. Hilbourne, 24 Tamarisk Avenue, Reading, RG2 8JB
G6 NWC P. Holley, Wotton Farm, Buckfastleigh, TQ11 0HB
G6 NWK M. Parker, 31 Sandholme Drive, Burley In Wharfedale, Ilkley, LS29 7RG
G6 NWN I. Poyser, 24 Overstone Close, Sutton-in-Ashfield, NG17 4NL
G6 NWS J. Hildreth, 69 Mason Street, Sutton-in-Ashfield, NG17 4HQ
G6 NWT W. Taylor, 33 Lancaster Avenue, Dawley, Telford, TF4 2HS
GM6 NX Stirling and District ARS, c/o H. Martin, 11 Ewing Court, Broomridge, Stirling, FK7 0QP
GW6 NXH W. Rees, 1 St. Marys Close, Briton Ferry, Neath, SA11 2JU
GW6 NXL T. Rees, Ty Goleu, Llwyngwril, LL37 2UZ
G6 NXM R. Rixon, 11 The Ridings, Waltham Chase, Southampton, SO32 2TR
G6 NXP S. Rafferty, 57 Essex Crescent, Billingham, TS23 4AW
G6 NXV M. Shannon, 129 Hampton Lane, Blackfield, Southampton, SO45 1WF
G6 NXW K. Sykes, 68 Newtown Avenue, Cudworth, Barnsley, S72 8DY
G6 NYF J. Aylward, 7 Manygates Lane, Wakefield, WF1 5NT
G6 NYG R. Adams, 6 Worcester Road Wychbold, Droitwich, WR9 7PE
G6 NYH G. Austin, 21 St. Georges Place, Northampton, NN2 6EP
G6 NYL P. Baylis, 118 Eastgate, Deeping St. James, Peterborough, PE6 8RD
GW6 NYR A. Davis, 9 Taliesin Street, Llandudno, LL30 2YE
GM6 NYT J. Danton, 12 Laburnum Road, Methil, Leven, KY8 2HA
G6 NZA M. Davenport, 8 Cedar Avenue, Chesterfield, S40 4ES
G6 NZG S. Edson, Inglewood, Camelot Gardens, Mablethorpe, LN12 2HP
G6 NZH J. Edwards, 69 Eastford Road, Warrington, WA4 6EY
G6 NZL P. Fletcher, 43 Merlin Way, Woodville, Swadlincote, DE11 7QU
G6 NZN G. Fowler, 10 Ullswater Road, Wimborne, BH21 1QT
G6 NZO M. Finney, 49 Ashcroft Drive, Old Whittington, Chesterfield, S41 9PA
G6 NZW D. Smith, 90 Endhill Road, Kingstanding, Birmingham, B44 9RP
G6 NZY B. Sparke, 1 Norwich Street, Mundesley, Norwich, NR11 8DN
G6 OAI S. Baverstock, 43 Tatnam Road, Longfleet, Poole, BH15 2DW
G6 OAN C. Bryan, 113 Hoe View Road, Cropwell Bishop, Nottingham, NG12 3DJ
G6 OAS A. Inglis, 59 Chapel Street, Forsbrook, Stoke-on-Trent, ST11 9DA
G6 OAU M. Jones, 17 Puddingmoor, Beccles, NR34 9PL
G6 OAV C. Jones, 1 Stonehill Close, Leigh-on-Sea, SS9 4AZ
GW6 OAW T. Jones, Marvor, Madyn Road, Amlwch, LL68 9DL
G6 OBA M. Kaznowski, 85 St. Albans Road, Kingston upon Thames, KT2 5HH
G6 OBB P. Kerr, Burrow Farm, Burrowbridge, Bridgwater, TA7 0RH
G6 OBD T. Keeling, 1 New Lane, Brown Edge, Stoke-on-Trent, ST6 8TQ
G6 OBE S. Kimblin, 4 Horsham Close, Westhoughton, Bolton, BL5 2GR
G6 OBG J. Kay, 12 Williams Avenue, Newton-le-Willows, WA12 0NN
G6 OBJ G. Webster, Flat 3 Charlotte Broadwood, Vicarage Lane, Dorking, RH5 5LL
G6 OBO L. Weiss, 7 Tennyson Avenue, Grays, RM17 5RG
G6 OBT H. Wilson, 11 Palmerston Close, Haslington, Crewe, CW1 5QE

G6 OBU S. Wright, 21 Poplars Close, Watford, WD25 7EW
G6 OCB D. Byers, 11 Heath Road, Ashton-In-Makerfield, Wigan, WN4 9DY
GI6 OCC K. Brennan, 1 Ballyscullion Lane, Bellaghy, Magherafelt, BT45 8NQ
G6 OCF R. Wallis, 187 Langtons Meadow, Farnham Common, Slough, SL2 3NT
G6 OCM E. Bunyan, 1 Talman Close, Ifield, Crawley, RH11 0RB
G6 OCO A. Bruce, 48 Durham Road, Gillingham, ME8 0JN
G6 ODA A. Bardy, 67 Chase Side, Enfield, EN2 6NQ
G6 ODF D. Adamson, 22 Longacres, Cannock, WS12 1LD
G6 ODT K. Lamford, 41 Drayton Road, Irthlingborough, Wellingborough, NN9 5TA
G6 ODU R. Leong, 55 Liverpool Road, Aughton, Ormskirk, L39 5AP
G6 ODW L. Liffchak, 6 Ashmore Grove, Welling, DA16 2RU
G6 OEJ A. Barnard, 36 St. Pauls Road, Walton Highway, Wisbech, PE14 7DN
G6 OEM J. Bolland, 18 Ward Avenue, Formby, Liverpool, L37 2JD
G6 OER J. Topping, 3 Dean Road, Handforth, Wilmslow, SK9 3AF
G6 OES M. Smith, Almond House, Waste Lane, Balsall Common, Coventry, CV7 7GG
G6 OET M. Telford, 11 Twyford Close, Swinton, Mexborough, S64 8UH
G6 OEW C. Thorn, 20 Kiln Road, Shaw, Newbury, RG14 2HA
GM6 OFB J. Mcardle, 40 Rodney Drive, Girvan, KA26 9DZ
G6 OFD L. Morrison, 29 Mead Hatchgate, Hook, RG27 9PU
G6 OFM J. Cordial, Flat 30, Homeview House, Seldown Road, Poole, BH15 1TT
GM6 OFO M. Clark, 38 Dunsinane Drive, Perth, PH1 2DU
G6 OFV S. Crossland, 16 Holland Road, High Green, Sheffield, S35 4HF
G6 OFZ J. Roberts, 8 Woodgate Close, Market Harborough, LE16 8EX
GW6 OGD S. Dawber, 7 Heol Y Pentir, Rhoose, CF62 3LQ
GM6 OGN A. Sives, 4 Fir Grove, Craigshill, Livingston, EH54 5JP
G6 OGT A. Scholes, 45 Howden Road, Blackley, Manchester, M9 0RQ
G6 OGZ J. Mitchell, 17 Spring Close, Lutterworth, LE17 4DD
GM6 OHF D. Macliver, 20 Lancaster Avenue, Beith, KA15 1AR
G6 OHK P. Butler, 219 Ridge Avenue, Burnley, BB10 3JF
G6 OHM A. Dunham, 28 Kingfisher Close, Chatteris, PE16 6TP
G6 OHQ G. Finch, 77 Furnivall Crescent, Lichfield, WS13 6DB
G6 OHR R. Edwards, 11 Litlington Court, Surrey Road, Seaford, BN25 2NZ
G6 OIA P. Rattenbury, 2 Main Road Upper Heyford, Northampton, NN7 3LZ
G6 OIB J. Riley, 11 Sutton Road, Mepal, Ely, CB6 2AQ
G6 OIF A. Postans, 62 Elm Grove, Bromsgrove, B61 0DX
G6 OIH R. Phillips, 1 Forge Close, Ashendon, Aylesbury, HP18 0HJ
G6 OIN A. Talbot, 159 Edgeside Lane, Rossendale, BB4 9TR
GW6 OIO M. Thomas, 13 Chestnut Grove, The Bryn, Blackwood, NP12 2PU
G6 OIX J. Roberts, 9 Tower Close, North Weald, Epping, CM16 6HA
G6 OIY J. Roberts, 6 Weavers Close, Braintree, CM7 2WB
GI6 OJC O. Okane, 39 Harberton Park, Ballymena, BT43 6NF
G6 OJH S. Page, 4 Trenley Close, Holbury, Southampton, SO45 2HN
GW6 OJK A. Mayall, 7 Cortay Park, Llanyre, Llandrindod Wells, LD1 6DT
G6 OJN K. Michael, 59 Mattock Lane, Ealing, London, W13 9LA
G6 OJV J. Greenley, 22 Langley Drive, Norton, Malton, YO17 9AR
G6 OJX E. Grayson, Manor Gate, 2 Polsue Way, Truro, TR2 4BE
G6 OJZ P. Anstock, 12 Raymoor Avenue, St. Marys Bay, Romney Marsh, TN29 0RD
G6 OKA C. Glover, 16 Woodfield Road, Radlett, WD7 8JD
G6 OKB R. Gilham, Wren Cottage, Wayborough Hill, Ramsgate, CT12 4HR
G6 OKC M. Gerrard, 29 Forest Drive, Broughton, Chester, CH4 0QT
G6 OKH P. Vanner, 23 Logwell Court, Northampton, NN3 9DJ
G6 OLJ Dr D. Hill, 33 Cleveland Close, Thornbury, Bristol, BS35 2YD
GM6 OLM R. Hendry, 43 Barone Road, Rothesay, Isle of Bute, PA20 0DY
G6 OLU P. Hobson, 220 Station Road, Burton Latimer, Kettering, NN15 5NT
G6 OLV A. White, 20 Wyles St., Gillingham, ME7 1ND
G6 OLY A. Williams, 1 Leslie Drive, Leigh-on-Sea, SS9 5NW
G6 OMH B. Staddon, 311 Cheney Manor Road, Swindon, SN2 2PE
G6 OMN G. Rogers, 7 Flordon, Birch Green, Skelmersdale, WN8 6PA
GW6 OMV P. Rea, 159 Mill View Estate, Maesteg, CF34 0DP
G6 ONE D. Williams, 16 Church St., Owston Ferry, Doncaster, DN9 1RG
G6 ONI B. Steponitis, Flat 7, 6 Second Avenue, Hove, BN3 2LH
G6 ONV M. Johnson, 16 Gardner Close, Raunds, Wellingborough, NN9 6HN
G6 ONW S. Jackson, 256 Perry Road, Sherwood Rise, Nottingham, NG5 1GP
GW6 ONZ P. Kinsey, Glyn Elwy Allt Goch, St Asaph, LL17 0BP
GM6 OOA D. King, Marionville, Donibristle, Cowdenbeath, KY4 8EU
G6 OOH J. Stone, 13 Winchester Close, Newport, PO30 1DR
G6 OOK M. Stewart, Fieldhead, New Barns Road, Arnside, Carnforth, LA5 0BH
G6 OOT C. Atkins, 11 Brambledown, West Mersea, Colchester, CO5 8RY
G6 OPD L. Middleton, 24 Townshend Road, Worle, Weston-Super-Mare, BS22 7FW
G6 OPK M. Lee, 23 Camford Close, Beggarwood, Basingstoke, RG22 4UJ
G6 OPV E. Starkey, 71 Elwick Drive, Liverpool, L11 4UW
G6 OPY R. Van Cleak, 19 Hanbury Road, Stoke Heath, Bromsgrove, B60 4LS
G6 OQJ W. Castle, 2 Wellington Close, Mundesley, Norwich, NR11 8JF
GI6 OQL J. Craig, 8 Muckamore View, Muckamore, Antrim, BT41 2EU
GM6 OQN R. Campbell, 32 Harvie Avenue Newton Mearns, Glasgow, G77 6LQ
G6 OQO J. Douthwaite, 38 Burnside Road, Newcastle upon Tyne, NE3 2DU
G6 OQV B. Everitt, The Hermitage, The Rookery, Galley Common, Nuneaton, CV10 9PB
GW6 ORE R. Trangmar, Ffynnon Bach Isaf, Tregarth, Bangor, LL57 4PA
G6 ORH D. Wright, 23 Oakenhall Avenue, Hucknall, Nottingham, NG15 7TF
G6 ORJ A. Weller, 104 Medina Avenue, Newport, PO30 1HG
G6 ORL D. Woodhouse, 5 Swallow Wood, Fareham, PO168UF
G6 ORM S. Whiley, 58 Audley Drive, Kidderminster, DY11 5NE
G6 ORO G. Walsh, 36 Westminster Street, Newtown, Wigan, WN5 9BH
G6 ORS A. Bennett, 39 West View, Parbold, Wigan, WN8 7NT
G6 ORT L. Bailey, 27 Birch Road, Congleton, CW12 4NN
G6 OSH D. Ridley, 37 Harewood Close, Whickham, Newcastle upon Tyne, NE16 5SZ
G6 OSJ J. Roberts, 1 Ollerdale Close, Allerton, Bradford, BD15 9BT
G6 OSK E. Robinson, 8 Barrier Mews, Stainforth, Doncaster, DN7 5PT
G6 OSO D. Parr, Lordings, Station Road, Pulborough, RH20 1AH
G6 OSR P. Price, 20 Froglands Way, Cheddar, BS27 3NY
G6 OSV I. Woodward, 20 Boyle Avenue, Warrington, WA2 0EZ
GM6 OSZ B. Williams, 29 St. Ternans Road, Newtonhill, Stonehaven, AB39 3PF
GW6 OTD P. Sizer, Gambos End, Reynoldston, Swansea, SA3 1BR
G6 OTE D. Shaw, 19 Upper Moors, Great Waltham, Chelmsford, CM3 1RB
G6 OTL G. Blades, 11 Willard Grove, Stanhope, Bishop Auckland, DL13 2XY
G6 OTP M. Rainbow, 38 Moselle Drive, Churchdown, Gloucester, GL3 2RY
G6 OTQ T. Roddy, 26 Chapeltown Road, Radcliffe, Manchester, M26 1YF
G6 OTS W. Peck, 5 Stirling Crescent, Horsforth, Leeds, LS18 5SJ
G6 OTV A. Ricalton, 33 Tintagel Close, Cramlington, NE23 1NZ
G6 OTW A. Ricalton, 84 Wansdyke, Morpeth, NE61 3RA
G6 OTZ A. Shaw, 4 Jones Lane, Burntwood, WS7 9DS
G6 OUA R. Saunders, 8 Norfolk Road, Luton, LU2 0RE
G6 OUI M. Burnell, 49 Ashfield Road, Carterton, OX18 3QZ
G6 OUJ B. Bozman, 33 Maple Road, Loughborough, LE11 2JL
GM6 OUL N. Bowry, 18 Mortonhall, Park Gardens, Edinburgh, EH17 8SR
G6 OUM B. Breaden, 6 Breydon Road, Sprowston, Norwich, NR7 8EE
G6 OUO P. Burgess, 232 Hightown Road, Luton, LU2 0DN
G6 OUT D. Andrew, 14 Westfield Grove, Morecambe, LA4 4LQ
G6 OUX S. Smith, 71 Rockford Close, Redditch, B98 7SZ
G6 OVA N. Styne, 2 Greenway, Burton-on-Trent, DE15 0AR
G6 OVC B. Thurlow, 1 Sheffield Way, Earls Barton, Northampton, NN6 0PF
GW6 OVD M. Clee, 29 Heol Uchaf Nelson, Treharris, CF46 6NT
G6 OVL J. Hopkins, 15 Wallace Drive, Wickford, SS12 9NA
G6 OVX R. Hadfield, 2 Bridge Street, Shaw, Oldham, OL2 8BG
G6 OWB E. Hill, 213A Leicester Road, Markfield, LE67 9RF
G6 OWI P. Haworth, 19 Arnside Crescent, Blackburn, BB2 5DU
G6 OWS K. Jones, 1 Chadlow Road, Liverpool, L32 7QR
G6 OWT G. Kelly, Brook House, Liskeard, PL14 5EY
GD6 OXG J. Williams, Brookfield, Douglas Road, Ballabeg, Isle of Man, IM9 4EF
G6 OXI C. Webb, 50 Ridgeway, Eynesbury, St Neots, PE192QY
G6 OXJ D. Webb, 10 Nuns Meadow, Gosfield, Halstead, CO9 1UB
GM6 OXL I. Wilkins, 133 Gavin Street, Motherwell, ML1 2RL
G6 OXN I. Walker, 66 Wood Street, Kettering, NN16 9SB
G6 OXQ A. Day, 7 Seagers, Great Totham, Maldon, CM9 8PB
G6 OXZ M. Charlton, 20 Bailey Crescent, South Elmsall, Pontefract, WF9 2TL
G6 OYF M. Matthews, 22 Elm Drive Bradley Stafford St18 9Ds, Stafford Uk, ST18 9DS
G6 OYU R. Spinner, 9 Lindholme Road, Lincoln, LN6 3RQ
G6 OYV T. Silvers, 15 Stanford Way, Walton, Chesterfield, S42 7NH
G6 OZH D. Martin, 2 Farm View Road, Kirkby-In-Ashfield, Nottingham, NG17 7HF
G6 OZT P. White, 3 South View, Whitwell, Worksop, S80 4NP
G6 OZU A. Wilson, 67 Sandpits, Leominster, HR6 8HT
G6 OZZ G. Llewellyn, 3 Ilex Close, Pamber Heath, Tadley, RG26 3DW
G6 PAA J. Brownsett, 10 Great Aldens, Bedford, MK41 8JS
G6 PAE R. Hillum, 48 Lydiard Way, Trowbridge, BA14 0UJ
G6 PAJ R. Green, 1 Knightsbridge Road, Messingham, Scunthorpe, DN17 3RA
G6 PAO M. Hale, 32 Oakfield Grove, Biddulph, Stoke-on-Trent, ST8 6UH
G6 PAP S. Hale, 19 Nailers Drive, Burntwood, WS7 0ES
G6 PAR P. Rhodes, 1 Killinghall Avenue, Bradford, BD2 4SA
GI6 PAZ W. Mcconnell, 17 Beech Green, Doagh, Ballyclare, BT39 0QB
G6 PBG N. Munnery, 3 Monnington Lane, Poundbury, Dorchester, DT1 3RJ
G6 PBI K. Partington, 38 Queensgate Drive, Royton, Oldham, OL2 5SD
G6 PBO J. Tobin, 5 Ashley Close, Ringwood, BH24 1QX
G6 PBW J. Wainwright, 8 Common Lane, Cutthorpe, Chesterfield, S42 7AN
G6 PBZ P. Wright, 75 Preston Road, Abingdon, OX14 5NG
G6 PCC R. Slade, 2 South Lodge Drive, Fornham St. Genevieve, Bury St Edmunds, IP28 6TQ
G6 PCE R. Stamford, 30 Craft Way, Steeple Morden, Royston, SG8 0PF
G6 PCP J. Brown, 1 Whitehouse Cottages, Woodham Walter, Maldon, CM9 6LR
GM6 PCW P. Boyd, 144 Brown Street, Paisley, PA1 2JE
G6 PCX J. Beresford, 1 Russell Place, Maltby, Rotherham, S66 7HB
G6 PDA A. Beacher, 9 Gleneagles Court, Normanton, WF6 1WW
G6 PDE C. Irish, 128 Rushmere Road, Ipswich, IP4 4JX
G6 PDJ S. Jubb, 4 Manor End, Worsbrough, Barnsley, S70 5JB
G6 PDM S. Procter, 1B York Villas, York Street, Colne, BB8 0ND
GW6 PDR S. Riggs, 3 Lawrence Terrace, Llanelli, SA15 1SW
G6 PEG C. Price, 42 Kipling Road, Kettering, NN16 9JZ
G6 PEH A. Rands, 20 Riby Road, Keelby, Grimsby, DN41 8ER
G6 PEP Dr J. Morris, 22 St. Amand Drive, Abingdon, OX14 5RQ
G6 PFC P. Cooper, 27 Ashgrove Road, Ashford, TW15 1NS
G6 PFF A. Willis, Kilncroft, Broadlayings, Newbury, RG20 9TS
GM6 PFJ G. Gott, 21 Hamilton Avenue, Dumfries, DG2 7LW
GW6 PFK L. Griffiths, 36 Lonydd Glas, Llanharan, Pontyclun, CF72 9FZ
G6 PFN A. Hewitt, 29 Brabazon Road, Oadby, Leicester, LE2 5HF
G6 PFP S. Hill, 10 Honeycrft Drive, St Albans, AL4 0GE
G6 PFX I. Harris, 1 Greenways, Mill Lane, Credenhill, Hereford, HR4 7EH
G6 PFZ A. Holroyd, 59 Southern Parade, Preston, PR1 4NJ
G6 PGG Dr D. Jones, Bramble Cottage, Newtown, Nantwich, CW5 8BG
G6 PGJ J. Kyle, 1A Lynmouth Gardens, Greenford, UB6 7HR
G6 PGM D. Kaye, The Pantiles, Bildeston Road, Stowmarket, IP14 2JT
G6 PGN C. King, 18812 Thornwood Circle, Huntington Beach, California, USA,
G6 PGO B. Key, 65 Ravenhurst Road, Birmingham, B17 9TB
G6 PGP S. Kinton, 7 Ferndale Drive, Ratby, Leicester, LE6 0LH
G6 PGQ M. Karaszy-Kulin, 30 Thatchers Close, Horsham, RH12 5TL
G6 PGT J. Chapman, 43 Balas Drive, Sittingbourne, ME10 5AS
G6 PGV N. Phillips, 5A Nutter Road, Thornton-Cleveleys, FY5 1BG
G6 PHC P. Dewick, Corner House, High Street, Gainsborough, DN21 4SW
G6 PHF M. Dent, 23 Spruce Avenue, Lancaster, LA1 5LB
G6 PHH P. Dickens, 2 Millfield Avenue, Marsh Gibbon, Bicester, OX27 0HP
G6 PHM P. Durbin, 2 Keswick Gardens, Pill, Bristol, BS20 0DR
G6 PHT S. Fitzhugh, 25 Bridge Meadow, Denton, Northampton, NN7 1DA
G6 PHU D. Ford, 6 Bluebell Grove Needham Market, Ipswich, IP6 8JH
G6 PHX C. Johnson, 37 Oakfield Drive, Mirfield, WF14 8PX
G6 PHZ P. Maddox, 7 Keats Road, Flitwick, Bedford, MK45 1QD
G6 PIB M. Livingston, 22 Oak Avenue, Elloughton, Brough, HU15 1LA
G6 PII D. Simpson, 20 Belvoir Place, Balderton, Newark, NG24 3HH
G6 PIM P. Lawford, 44 Clarendon Road, Broadstone, BH18 9HY
G6 PJC P. Brown, 13 Hillside Close, Biddulph Moor, Stoke-on-Trent, ST8 7PF
G6 PJD M. Belshaw, Tara Cottage, 11 Hectors Way, Blandford Forum, DT11 9QP
G6 PJE D. Bull, 2 School Road, St. Johns Fen End, Wisbech, PE14 8JR
G6 PJL J. Blacker, 74 Benomley Crescent, Almondbury, Huddersfield, HD5 8LU
G6 PJP L. Bealing, 18 Avon Road, Oakley, Basingstoke, RG23 7DJ
G6 PKG R. Davis, 72 Windyridge, Bisley, Stroud, GL6 7DA
G6 PKM Dr J. Allen, 27 Grafton Road, Whitley Bay, NE26 2NR
GM6 PKP J. Allardyce, 17 Hallglen Terrace, Glen Village, Falkirk, FK1 2AP

G6 PKS B. Bean, 46 Grand Drive, Herne Bay, CT6 8JS
G6 PKV G. Branagan, 434 Manchester Road West, Little Hulton, Manchester, M38 9XU
G6 PKX P. Bishop, 38 Parkside Gardens East Barnet, Barnet, EN4 8JS
G6 PKY R. Bush, 3 Charnwood Avenue Keyworth, Nottingham, NG12 5JX
G6 PLF J. Smoker, 9 Anson Way, Bicester, OX26 4UH
GM6 PLG P. Sloan, 24 Hythe View, Lossiemouth, IV31 6TP
G6 PLL C. Leach, 109 Congreve Road, Worthing, BN14 8EN
GI6 PLO I. Bell, 3 Stratford Drive, Bangor, BT19 6ZW
G6 PLR L. Chandless, 16 Crest Gardens, Ruislip, HA4 9HD
G6 PLT E. Cheetham, 172A Hesketh Lane, Tarleton, Preston, PR4 6UD
G6 PLU A. Chenery, 43 Wessex Estate, Ringwood, BH24 1XD
GW6 PMC R. Evans, 16 Monmouth Grove, Prestatyn, LL19 8TS
G6 PMD C. Eagling, 96 Regent Road Brightlingsea, Colchester, CO7 0NZ
G6 PMF H. Langsley, 39 Lavender Road, Basingstoke, RG22 5NN
G6 PMJ S. Murphy, 1 Orchard Cottage, Golden Valley, Newent, GL18 1HN
G6 PMO I. Parker, 27 St. Audries Road, Worcester, WR5 2AL
G6 PMR P. Shaw, 52 Belvedere Parade, Bramley, Rotherham, S66 3WA
G6 PMW G. Goodier, 3 The Paddock, Beckingham, Lincoln, LN5 0FD
G6 PNG P. Hill, 28 Somerton Grove, Thatcham, RG19 3XE
GM6 PNJ S. Hammond, Graywalls, Denny, FK6 5JF
G6 PNO P. Hill, 33 The Pastures, South Beach, Blyth, NE24 3HA
G6 POC R. Kinrade, 23 Crofthill Road, Slough, SL2 1HG
G6 POE J. Knott, 3 Lords Wood, Welwyn Garden City, AL7 2HF
G6 POG R. Williams, 29 Bridle Close, Banbury, OX16 9SZ
G6 POI J. Wright, Chez Mon, Burton Road, Carnforth, LA6 1QN
G6 POJ I. Worthy, 7 The Paddocks, Pilsley, Chesterfield, S45 8ET
GW6 POO R. Smallwood, 12 Oak Close, Connah'S Quay, Deeside, CH5 4GG
G6 POV M. Walker, 232 Bideford Green, Leighton Buzzard, LU7 2TS
G6 POW D. Pow, 16 Ancaster Close, Trowbridge, BA14 9DA
G6 POZ R. Farrall, 7 The Meadows, South Cave, Brough, HU15 2HR
G6 PPA T. Farmer, 35 Ascot Drive, Dudley, DY1 2SN
G6 PPD A. Morgan, High Bow Cottage, Bow, Carlisle, CA5 6EN
G6 PPU H. Chappell, Oanley, East Lyng, Taunton, TA3 5AU
G6 PPV P. Caswell, 94 Dewsbury Road, Luton, LU3 2HJ
G6 PPY S. Carter, 3 Mary Street, Burnley, BB10 4AJ
G6 PQI J. Finch, The Croft, Dalby Road, Melton Mowbray, LE14 3EX
G6 PQP P. Brooks, Cherating, Hanning Road Horton, Ilminster, TA19 9QH
G6 PRA J. Whittaker, 6 Bradley Gardens, Burnley, BB12 6JT
G6 PRE E. Snell, 156 Brookdale Avenue South, Greasby, Wirral, CH49 1SS
G6 PRL D. Brown, 63A Great Northern Street, Huntingdon, PE29 7HJ
G6 PRP W. Barker, 297 Williamthorpe Road North Wingfield, Chesterfield, S42 5NT
G6 PSA N. Turnham, 153 Canterbury Road, Urmston, Manchester, M41 0PY
G6 PSC M. Horn, 3 Church Cottages, Pound Lane, Beccles, NR34 0EX
G6 PSO I. Russell, 24 Standard Avenue, Tile Hill, Coventry, CV4 9BW
G6 PSQ R. Langton, Dawn Cliffe, Goodwin Road, Dover, CT15 6ED
G6 PSZ T. Shackleton, 27 Court Crescent, Kingswinford, DY6 9RJ
G6 PTF J. Wilson, Belle Vue House, Common Side, Workington, CA14 4PU
G6 PTT P. Hedley, 7 Midhill Close, Langley Park, Durham, DH7 9YA
GM6 PTX G. Gane, 28 Queens Croft, Kelso, TD5 7NN
G6 PUE M. Mackmin, 15-3 Koumasion, Peyia, Cyprus, 8560
G6 PUO A. Quantrill, 105 High Street, Sawston, Cambridge, CB22 3HJ
G6 PUR J. Howells, 66 Rochester Avenue, Burntwood, WS7 2DL
G6 PUV W. Holding, 20 Lingfield Crescent, Wigan, WN6 8QA
GM6 PVA M. Green, Fulwood House, Coronation Road, Bellshill, ML4 2RT
G6 PVC P. Coates, Jacaranda, Cotswold Close, Staines, TW18 2DD
GW6 PVK G. Jones, 12 The Nurseries, Cymau, Nr Wrexham, LL11 5LE
G6 PVT L. Adamson, 6 Castle View Estate, Derrington, Stafford, ST18 9NF
G6 PVV C. Burt, 1 Chapter Court, Vicarage Road, Egham, TW20 8NL
G6 PVW G. Bayliffe, 179 Breedon Street, Long Eaton, Nottingham, NG10 4EW
G6 PWF S. Choules, 43 Ashbrook Road, Old Windsor, Windsor, SL4 2LT
G6 PWJ J. Chiddick, Unioninkatu 45A10, Helsinki, Finland, 170
G6 PWL R. Cloutman, 35 Camlet Way, St Albans, AL3 4TL
G6 PWQ D. Dick, 140 Chatham Street, Stockport, SK3 9JU
G6 PWS R. Fuller, 18 St. Leonards Crescent, Sandridge, St Albans, AL4 9EH
G6 PXJ A. Harrison, Nirvana Cottage, 42 Bell Lane, Goole, DN14 8RP
G6 PXN R. Bee, 80 Hospital Road, Burntwood, WS7 0EQ
G6 PXQ R. Boyce, 3 Castleton Cottages, Westhide, Hereford, HR1 3RF
GW6 PXW E. Davies, 62 Heol Y Coedcae, Cwmllynfell, Swansea, SA9 2FY
G6 PXX L. Dempsey, 24 James Street, Great Harwood, Blackburn, BB6 7JE
GM6 PYD A. Dunnett, 11 Silverknowes View, Edinburgh, EH4 5PY
G6 PYE Cambridge Reapeater Group, c/o P. Nice, 31 Elizabeth Drive, Chapel St. Leonards, Skegness, PE24 5RS
G6 PYF D. Hills, 9 Brook Gardens, Devizes, SN10 2FX
G6 PYI M. Jones, 15 Rowan Rise, Kingswinford, DY6 8EE
G6 PYL P. Hatter, 14 Morland Avenue, Bromborough, Wirral, CH62 6BE
G6 PYM A. Hedges, 25 The Lanes, Cheltenham, GL53 0PU
GI6 PYP A. Gault, 134 Leighan Road, Randalshough, Monea, Enniskillen, BT93 7DN
G6 PYR H. Adams, Hill Sixty, Happisburgh, Norwich, NR12 0RB
G6 PZ P. Beecham, The Haybarn, Church Street, Bridgwater, TA7 9AT
G6 PZE D. Jefferson, 48 Neston Road, Walshaw, Bury, BL8 3DB
G6 PZF P. James, 18 Brackens Drive, Warley, Brentwood, CM14 5UE
G6 PZH B. Hickman, 7 Nina Close, Stourport-on-Severn, DY13 9RZ
G6 PZN M. Mcloughlin, 13 Old Manor Gardens, Wymondham, Melton Mowbray, LE14 2AN
G6 PZS D. Carr, 5 Church Meadow, Hyde, SK14 4RT
GM6 PZY Dr Z. Yang, 48 Foxglove Road, Newton Mearns, Glasgow, G77 6FP
G6 QA L. Jopson, 68 Greenmount Park Kearsley, Bolton, BL4 8NT
G6 QM Southgate ARC, c/o D. Berry, 4 Holly Hill, Winchmore Hill, London, N21 1NP
G6 QN T. Blakeman, 31 Walsingham Gardens, Epsom, KT19 0LS
G6 RAF Royal Air Force ARS, c/o M. Garrett, 489 Dorchester Road, Weymouth, DT3 5BP
G6 RAH R. Hammond, 126 Otley Drive, Ilford, IG2 6QY
GM6 RAK D. Brown, 14 Newton Crescent, Carnoustie, DD7 6HW
GW6 RAO D. Griffiths, 35 Greystones Crescent, Mardy, Abergavenny, NP7 6JY
G6 RAQ S. Hayter, Rookery Farm, Mill Road, Battisford, Stowmarket, IP14 2LT
GW6 RAV W. Keeley, 93 Park Crescent, Abergavenny, NP7 5TL
G6 RAZ J. Paton, 23 Courville Close, Alveston, Bristol, BS35 3RR
GI6 RBD K. Brady, 26 Kilbroney Road, Rostrevor, Newry, BT34 3BJ
G6 RBM R. Jeffery, 15 Greenway, Hulland Ward, Ashbourne, DE6 3FE
G6 RBO W. Bennett, 44 Wood Lane, Streetly, Sutton Coldfield, B74 3LR
G6 RBP R. Pearsey, 21 Ashwood Drive, Newbury, RG14 2PN
G6 RBR M. Allen, 1 Allens Yard, Chatteris, PE16 6QE
GW6 RBZ R. Coombes, 10 Goodrich Court, Llanyrafon, Cwmbran, NP44 8RY
G6 RC Crawley ARC, c/o R. Hadfield, 45 Erica Way, Copthorne, Crawley, RH10 3XG
G6 RCD P. Clark, 166 Attenborough Lane, Attenborough, Nottingham, NG9 6AB
GW6 RCK H. Fray, 17 Homelands Road, Cardiff, CF14 1UH
G6 RCT T. Stellar, 27 Blackmore Chase, Wincanton, BA9 9SB
GW6 RCX D. Smithies, 26 Coed Mor, Penyffordd, Holywell, CH8 9HY
G6 RCY D. Reed, 11 Grenville Close, Corby, NN17 2RP
G6 RDD I. Senter, 33 King Coel Road, Colchester, CO3 9AQ
G6 RDO A. Shaw, 1 Chapel Lane, Clifford, Wetherby, LS23 6HU
GW6 RDV B. Clarke, Flat 3, Carling Court, Haig Place, Cardiff, CF5 4PH
G6 REA J. Gilpin, River Bank House 28A Ivel Road, Sandy, SG19 1AX
G6 REC M. Hart, 7 Ullswater Avenue, South Wootton, King's Lynn, PE30 3NJ
GW6 REF D. Jones, 16 New Road, Llandovery, SA20 0ED
G6 REG A. Joyce, Ashdene, Highworth Road, Swindon, SN3 4SE
G6 REH J. Staplehurst, 12 Trotter Way, Epsom, KT19 7EW
G6 REM Maj. I. Atkinson, 537753, 7 Headquarter Squadron, Bfpo 36, AA1 1AA
GW6 REQ W. Vize, Cefn Rhos, Bethel, Caernarfon, LL55 1YB
G6 REV J. Yam, 4537 Mossburg Court, Marietta, USA, 30066
G6 REW G. Seymour-Smith, Glencarne, Bridgerule, Holsworthy, EX22 7ED
G6 REY R. Mcminn, 1C Bickley Avenue, Sutton Coldfield, B74 4DY
G6 RFH D. Ruck, 50 Kiel Walk, Corby, NN18 9DE
G6 RFJ L. Waite, The Towers, Castle St., Nottingham, NG2 4AE
G6 RFL R. Rothery, 12 Reevy Crescent, Bradford, BD6 2BT
G6 RFM A. Warren, 20 Wolverhampton Road, Stafford, ST17 4BP
G6 RFR A. Brammer, 5 The Green, Reepham, Lincoln, LN3 4DH
G6 RFS A. Brown, 22 Mount Wise Crescent, Plymouth, PL1 4GQ
G6 RFU P. Csapo, 87 Latchmere Road, Kingston upon Thames, KT2 5TU
G6 RGA J. Ewen, 26 Court Road, Eastbourne, BN22 9EZ
GM6 RGD T. Murray, 2 The Glebe, Edzell, Brechin, DD9 7SZ
G6 RGI A. Shead, 95 Sea Front, Hayling Island, PO11 0AW
G6 RGN W. Stockley, 10 Swan Road, Timperley, Altrincham, WA15 6BX
GW6 RGT M. Morgan, 17 Dunstable Road, Newport, NP19 9NE
GM6 RGY W. Hardie, 96 Carmuirs Avenue, Camelon, Falkirk, FK1 4PB
G6 RHA S. Howes, 46 High Street, Upwood, Huntingdon, PE26 2QE
G6 RHB R. Howlett, 4 Station Road, Willoughby, Alford, LN13 9NG
G6 RHJ M. Swain, 17 Sponnes Road, Towcester, NN12 6ED
G6 RHK C. Spencer, 21 Playford Road, Ipswich, IP4 5QZ
G6 RHL P. West, 6 Iveldale Drive, Shefford, SG17 5AD
G6 RHN Dr D. Morris, Rowley Farm, Rowley Lane, Borehamwood, WD6 5PE
G6 RHT D. Owens, 17 Priory Place, Greenham, Thatcham, RG19 8XT
G6 RHV I. Smith, 126A High Street, Teddington, TW11 8JB
G6 RIC M. Ellis, 28 High Meadows, Romiley, Stockport, SK6 4PT
G6 RIG N. Golding, Coppice View, 16 Littlewood Road, Walsall, WS6 7EU
G6 RII M. Dodson, Tree Tops, Badgeworth Lane, Cheltenham, GL51 4UW
G6 RIJ R. Fletcher, 33 Littlewood Lane, Cheslyn Hay, Walsall, WS6 7EJ
G6 RIM C. Berry, 258 Lowerhouse Lane, Burnley, BB12 6NG
G6 RIQ G. Dunn, 11 Ellesmere Rise, Grimsby, DN34 5PE
G6 RIY A. Wilkinson, 34 Coppice Lane, Hellifield, Skipton, BD23 4JW
G6 RIZ B. Jones, 96 Somerset Road, Farnborough, GU14 6DS
G6 RJH J. Proffitt, 38 Hockley Road, Poynton, Stockport, SK12 1RW
G6 RJU M. Scaife, 4 Arden Close, Wallsend, NE28 9YB
G6 RJW R. Woodgate, Roger Woodgate, C/O Ratana Paa, P.D.C., Turakina, New Zealand, 4548
G6 RKF R. Taylor, 20 Scraley Road, Heybridge, Maldon, CM9 4BL
G6 RKG J. Walters, The Gables, Lavenham Road, Sudbury, CO10 0RN
G6 RKJ R. Butland, 4 Park Close, Sonning Common, Reading, RG4 9RY
G6 RKQ D. Hudson, 19 Worcester Close, Lichfield, WS13 7SP
G6 RKS R. Domville, 12 Craig Terrace, Peterlee, SR8 3AJ
G6 RLG J. Kerr, 3 Lime Kiln Way, Salisbury, SP2 8RN
G6 RLM R. Maddison, Tom Butt, Hope Street, Elstead, Godalming, GU8 6DE
G6 RMA D. Glover, 14 Fitzgerald Avenue, Herne Bay, CT6 8LN
G6 RMJ W. Clements, 3 May Street, Durham, DH1 4EN
GI6 RMO D. Johnston, Olanda, Lisreagh, Co Fermanagh, BT94 5BX
G6 RMV R. Sharp, The Old School, Bridge Street, Bridport, DT6 5LS
GW6 RNA T. Lovell, 4 Maes Rathbone, Waen, St Asaph, LL17 0AD
G6 RNF R. Smith, 12 East View, West Bridgford, Nottingham, NG2 7QN
G6 RNR S. Allison, Rudd Hall Cottage, East Appleton, Richmond, DL10 7QD
G6 RNT G. Kingdon, Wymering, Copley Drive, Barnstaple, EX31 2BH
GW6 RNV C. Brewster, 35 Ffordd Las, Sychdyn, Mold, CH7 6DU
G6 RNZ D. Brook, 50 Ashley Down Road, Bristol, BS7 9JW
GI6 ROI J. Polson, 6 Castlemara Drive, Carrickfergus, BT38 7RB
G6 ROS J. Warwick, 12 Oak St., Sutton in Ashfield, NG17 3FF
G6 RPD R. Montford, 394 Selbourne Road, Luton, LU4 8NU
G6 RPH G. Tibbert, 5 Upper Havelock Street, Wellingborough, NN8 4PN
G6 RPK N. Waterton, 1270 Killaby Drive, Mississauga, Ontario, Canada, L5V 1B1
G6 RPW Lt. Col. S. Andrews, 1 Swynford Close, Kempsford, Fairford, GL7 4HN
G6 RQA D. Nicholls, 15 Poplar Avenue, Heacham, King's Lynn, PE31 7EB
G6 RQJ R. Sherlock, 34 St. Cecilias Road, Belle Vue, Doncaster, DN4 5EG
GM6 RQU B. Hamilton, 51 Grange Road, Grange, Edinburgh, EH9 1UF
GM6 RQW T. Christie, 4 Glebe Park, Bressay, Shetland, ZE2 9ER
G6 RQZ B. Cripps, 3 Sabre Court, Aldershot, GU11 1YY
G6 RRJ R. Urwin, 21 Winchester Avenue, Leicester, LE3 1AX
G6 RRS D. Rotgans, 18 Minter Avenue, Densole, Folkestone, CT18 7DS
G6 RRV R. Weston, The Old Dairy, Slate Cross, Bridgwater, TA7 8QR
G6 RRY M. Dunbar, 42 Wickham Way, Shepton Mallet, BA4 5YG
G6 RSI L. Hart, 25 Murcroft Road, Stourbridge, DY9 9HT
G6 RSU R. Anstee, 12 Ashmore Avenue, Stockport, SK3 0QY
G6 RTD G. Small, 6 Mary St., Longridge, Preston, PR3 3WN
G6 RTE Capt. J. Menhinick, Coburg, Barton Estate, East Cowes, PO32 6NT
G6 RTG J. Sutton, 29 Victory Avenue, Darlaston, Wednesbury, WS10 7RR
G6 RTM R. Ashberry, 30 Factory Lane Roydon, Diss, IP22 4EG
GM6 RTN K. Bone, School House Makerstoun, Kelso, TD5 7PB
G6 RTY D. Johnson, Penvern, Nacton, Ipswich, IP10 0EW

GW6 RUE J. Newey, Springwood Cottage, Tyntaldwyn Road Troedyrhiw, Merthyr Tydfil, CF48 4NG
G6 RUM I. Nice, 6 Malden Road, Sidmouth, EX10 9LS
GW6 RUO J. Griffiths, 4 Lon Elan, Meliden, Prestatyn, LL19 8LP
G6 RUP J. Horner, 43 Birch Close, Patchway, Bristol, BS34 5SA
G6 RUU S. Burns, 289 Wallasey Village, Wallasey, CH45 3HA
G6 RUY J. Heaney, 15 Perth Street, Nelson, BB9 8EE
G6 RVH R. Jamieson, 3 Waterpark Road, Prenton Park, Birkenhead, CH42 9NZ
G6 RVP C. Pung, 73 John Mace Road, Colchester, CO2 8WW
G6 RVS R. Sohst, 2 Shaftesbury Drive, Maidstone, ME16 0JS
G6 RVZ D. Carruthers, 168A Wanstead Park Avenue, London, E12 5EF
GU6 RWD S. Hancock, L'Hirondel, Hubits De Bas, St. Martin, Guernsey, GY4 6NB
GW6 RWJ D. Silcox, Troedyrhiw, Penparc, Cardigan, SA43 2AE
G6 RXD M. Yirrell, 66 Park Lane, Sandbach, CW11 1EP
G6 RXF G. Priestley, 7 Affleck Avenue, Radcliffe, Manchester, M26 1HN
G6 RXK A. Orchard, Kilimani, Cuilfail, Lewes, BN7 2BE
G6 RXP S. Dwyer, 10 Swan Street, Darwen, BB3 2LW
GM6 RXQ A. Gordon, 2, Merse Avenue, Kirkcudbright, DG6 4RN
G6 RXV K. Keen, 26 Brogden Close, Botley, Oxford, OX2 9DS
G6 RXY D. Ferguson, Aneataprint Four Ltd 3-5 Lord Street, Watford, WD17 2LN
G6 RYM H. Wagg, 43 Highfield Road, Birkenhead, CH42 2BU
G6 RYW V. Smith, 9 Pinewood Drive, Mansfield, NG18 4PG
G6 RZR N. Stevens, 151 Ferme Park Road, London, N8 9BP
G6 RZS R. Wood, 115 Anchorway Road, Green Lane, Coventry, CV3 6JH
G6 RZY N. Harper, 15 Epsom Close, Dosthill, Tamworth, B77 1QT
G6 SAQ G. Hines, 11 Montagu Gardens, Wallington, SM6 8EP
GW6 SBD G. Davies, 2 Ffordd Aled, Wrexham, LL12 7PP
G6 SBG A. Lubrani, Ranscombe Manor, Sherford, Kingsbridge, TQ7 2DP
G6 SBI D. Smith, 8 Corunna Drive, Horsham, RH13 5HG
G6 SBN M. Searl, Sinodun Road, Didcot, OX118HW
GI6 SBW A. Alcock, 22 Chippendale Avenue, Bangor, BT20 4PT
G6 SCB A. Bean, 66 Wickham Lane, London, SE2 0XN
G6 SCG M. Lockwood, 33 Elmtree Road, Calverton, Nottingham, NG14 6QA
G6 SCM J. Webb, Oakdene, 22 Meeting House Lane, Coventry, CV7 7FX
G6 SDC B. Simmons, Wootton Leas, 35 Benenden Green, Alresford, SO24 9PE
G6 SDE A. Curley, 21 Trinity Rise, Penton Mewsey, Andover, SP11 0RE
G6 SDI P. Hall, 28 Grangeway, Rushden, NN10 9EZ
GM6 SDV J. More, 51 Hilton Drive, Aberdeen, AB24 4NJ
G6 SDW M. Rowan, 1 Yanleigh Close, Bristol, BS13 8AQ
G6 SDY R. Beecroft, 28 Hall Garth Lane, West Ayton, Scarborough, YO13 9JA
G6 SEE Z. Feast, 2 Dyrham Close, Burnham-on-Sea, TA8 2TT
G6 SEF J. Feast, 2 Dyrham Close, Burnham-on-Sea, TA8 2TT
G6 SEK P. Ovey, 35 Lower Fairfield, St. Germans, Saltash, PL12 5NH
GM6 SEL M. Munro, 5 Corran Cismaol Horve, Isle of Barra, HS9 5ZE
GM6 SEV I. Carr, 36A Broomieknowe, Lasswade, EH18 1LN
G6 SFC T. Foulds, Deer Park, Detling Avenue, Broadstairs, CT10 1SR
G6 SFE S. Al-Kattan, 8 Little John Drive, Rainworth, Mansfield, NG21 0JJ
G6 SFF C. Hewes, 1 Broad Valley Drive Bestwood Village, Nottingham, NG6 8XA
G6 SFH P. Barber, 17 Wheelwright Avenue, Leeds, LS12 4UW
GI6 SFO G. Miskimmin, 332 Rathfriland Road, Dromara, Dromore, BT25 2HN
G6 SFR Flt Refuelng ARS, c/o A. Baker, Highleaze, Deans Drove, Poole, BH16 6EQ
G6 SFW P. Dodd, 9 Rudge Croft, Kitts Green, Birmingham, B33 9NZ
G6 SFY G. Barker, 18 Penryn Close, Nuneaton, CV11 6FF
G6 SGA S. Thornber, 18 Lichfield Road Talke, Stoke on Trent, ST7 1SQ
G6 SGD D. Carding, Mill House, Walcot, Telford, TF6 5ER
G6 SGE P. Bayliss, 36 Slingates Road, Stratford-upon-Avon, CV37 6ST
G6 SGM C. Macey, 29 Burleigh Road, Sutton, SM3 9NE
G6 SGU A. Aristotelous, 8 Sudbury Drive, Huthwaite, Sutton-in-Ashfield, NG17 2SB
G6 SGV M. Durkin, Selsdon House, 23 Jameson Road, Bexhill-on-Sea, TN40 1EG
G6 SGW J. Miller, 6 Saunders Mews, Southsea, PO4 9XZ
G6 SGY B. Sales, 2 Highview, Hurley, Atherstone, CV9 2RP
G6 SGZ J. Smith, 1 Markby Close, Moorside, Sunderland, SR3 2RG
G6 SHD G. Mcbrien, 26 Lumb Carr Avenue, Ramsbottom, Bury, BL0 9QG
G6 SHF M. Trolan, Hilbre, East Taphouse, Liskeard, PL14 4NJ
G6 SHQ M. Brown, 6 Snell Hatch, West Green, Crawley, RH11 7JB
G6 SHS K. Eldridge, 44 Merley Gardens, Merley, Wimborne, BH21 1TB
G6 SHZ S. Congrave, 28 London Road, St. Ippolyts, Hitchin, SG4 7NG
G6 SIG G. Timbrell, Crossing Cottage, Lamyatt, Shepton Mallet, BA4 6NG
G6 SIM J. Simarpi, 6 Berryman Court Lethbridge Road, Wells, BA5 2FF
G6 SIQ W. Whitcombe, 11 The Elms, Deerton Street, Teynham, Sittingbourne, ME9 9LH
GW6 SIX P. Macmillen, Spinney Cottage, Sychnant Pass Road, Conwy, LL32 8NS
G6 SJA W. Barnes, 17 Greenhill Road, Long Buckby, Northampton, NN6 7PU
G6 SJD A. Evans, Spring Bank, Coventry Road Kingsbury, Tamworth, B78 2LW
G6 SJG T. Hurton, 4 Athlone Close, Enham Alamein, Andover, SP11 6JY
G6 SJV P. Cliffe, 63 Mill Lane, Upton, Chester, CH2 1BS
G6 SKF L. Hopson, 39A Fenside Road, Boston, PE21 8HY
G6 SKK T. Parkin, 8 Horsley Crescent, Holbrook, Belper, DE56 0UB
G6 SKM W. Taylor, 5 Gadbury Avenue, Atherton, Manchester, M46 0LQ
G6 SKP A. Whitgreave, 2 Oaklea Avenue, Hoole, Chester, CH2 3RE
G6 SKR G. Walker, 81 Normanshire Drive, Chingford, London, E4 9HE
G6 SKS C. Learoyd, Leofric House, 31 Leofric Avenue, Bourne, PE10 9QT
G6 SKT W. Learoyd, Leofric House, 31 Leofric Avenue, Bourne, PE10 9QT
G6 SL C. Pettitt, 12 Hennals Avenue, Redditch, B97 5RX
G6 SLG A. Berry, Emberley Leys, Ratley, Banbury, OX15 6DS
G6 SLN C. Gleave, 10 Henley Road, Neston, CH64 0SG
GW6 SLO P. Charnley, 9 Bryn Crescent, Rhuddlan, Rhyl, LL18 5RF
G6 SLY D. Lewis, 30 Printers Park, Hollingworth, Hyde, SK14 8QH
G6 SLZ J. Mackenzie, 11 Upper Heyshott, Petersfield, GU31 4QA
G6 SMI G. Langstaff, Flat 22 Benwell Close, Benwell Grange, Newcastle-upon-Tyne, NE15 6RZ
G6 SMJ M. Pitts, 30 Sandhurst Avenue, Surbiton, KT5 9BS
GM6 SMW G. Harper, 15 Seaforth Park, Annan, DG12 6HX
G6 SNA D. Heather, 65 Fairview Road, Headley Down, Bordon, GU35 8HQ
G6 SND W. Jarvis, Owlpen, 20 Park Road, Wiltshire, SN10 4ED
G6 SNI G. Platt, 15 Mount Close, Nantwich, CW5 6JJ
G6 SNN D. Ramsey, 11 Pendle Close, Basildon, SS14 3NA
GJ6 SNQ A. Leighton, 2 Le Petit Menage, Fountain Lane, St Saviour, Jersey, JE2 7RL
G6 SNV R. Smith, 3 Payne Road, Wootton, Bedford, MK43 9JL
G6 SNZ I. Stones, 1 Valdene Close, Farnworth, Bolton, BL4 9NE
G6 SOA R. Wilday, 4 Kenelm Road, Clifton-On-Teme, Worcester, WR6 6DW
G6 SOO P. Williams, 2 St. Christophers Drive, Romiley, Stockport, SK6 3BE
G6 SOX J. Bradley, 68 Rosedale Avenue, Alvaston, Derby, DE24 0FJ
G6 SOY P. Berry, Bundys Cottage, Colwood Lane, Haywards Heath, RH17 5QQ
G6 SOZ A. Byrne, Holly Cottage, Deacons Lane, Thatcham, RG18 9RJ
G6 SPB D. Corder, 140 Edward Road, Somerford, Christchurch, BH23 3EW
G6 SPG P. Cesnavicius, 52 Boundary Road, Irlam, Manchester, M44 6HD
G6 SPH J. Crookbain, 11 Champlain Avenue, Canvey Island, SS8 9QL
G6 SPI S. Carwood, 34 Flemming Avenue, Ruislip, HA4 9LF
G6 SPN R. Barton, 82 Buckingham Road, South Woodford, London, E18 2NJ
G6 SPQ W. Holmes, 37 Barmpton Lane, Darlington, DL1 3HH
G6 SQL A. Lowthian, 38 Arthur Street, Ryde, PO33 3BU
G6 SQS F. Sivyer, 22 Boxley Road, Walderslade, Chatham, ME5 9LF
G6 SQT C. Wall, 151 Bisley Road, Stroud, GL5 1HS
G6 SRE B. Stone, Reindene, Faversham Road, Ashford, TN25 4PQ
G6 SRJ A. Waring, 2 Wroxton Close, Thornton-Cleveleys, FY5 3EY
G6 SRS Stourbridge & District ARS, c/o D. Scott, Hyde Bungalow, The Hyde, Stourbridge, DY7 6LS
G6 SRT D. Armstrong, 103 Victoria Road, Oxford, OX2 7QG
G6 SRU C. Alford, 43 Cleeve, Glascote, Tamworth, B77 2QD
G6 SRV R. Andrews, 11 Holly Grove, Verwood, BH31 6XA
G6 SRX I. Aram, 4 Severn Road, Chilton, Didcot, OX11 0PW
G6 SRY S. Aram, The Barn, Charleston Place, Eastbury, Hungerford, RG17 7JN
G6 SRZ W. Baxter, 19 Westbury Road, Nottingham, NG5 1EP
G6 SSH V. Bates, 382 Hindley Road, Westhoughton, Bolton, BL5 2DT
G6 SSN K. Burton, 2 Council House, Stainfield Road, Kirkby Underwood, Bourne, PE10 0SG
G6 SSQ D. Bolton, Huds House, Cowgill, Sedbergh, LA10 5TQ
G6 STB G. Munn, 1 Sovereign Close, Hastings, TN34 2UB
G6 STD D. Macey, Affric House, New Street, Banbury, OX15 0SR
G6 STE B. Stevens, 16 Fowey Close, Wellingborough, NN8 5WW
G6 STF G. Smith, Sunny Patch, Western Backway, Kingsbridge, TQ7 1QB
G6 STI H. Staddon, 45 Saxony Parade, Hayes, UB3 2TQ
G6 STJ S. Smith, 9 Beadon Drive, Salcombe, TQ8 8NU
GW6 STK R. Sweet, 9 Seafield Road, Colwyn Bay, LL29 7HB
GW6 STS G. Jones, The Bungalow, Castle Street, Pontypool, NP4 9QL
G6 SUC J. Hudson, 6 Brooks Close, Ringwood, BH24 1NE
G6 SUK B. Trevor, 39 Clayton Crescent, Brentford, TW8 9PT
G6 SUR P. Thornsby, 25 Kipling Way, Stowmarket, IP14 1TS
G6 SUV J. Barlow, 3 Shaw Brook Close, Rishton, Blackburn, BB1 4ES
G6 SVH K. Henderson, 42 Chartwell Avenue, Wingerworth, Chesterfield, S42 6SP
G6 SVJ S. Harvey, 148 Smithfield Road, Uttoxeter, ST14 7LB
G6 SVL S. Harris, 34 Butterfly Crescent, Nash Mills Wharf, Hemel Hempstead, HP3 9GS
G6 SVV R. Gray, Willett, 28 Hoe Lane, Romford, RM4 1AX
G6 SW Cannock Chase ARS, c/o M. Starkey, 23 Arthur Street, Cannock, WS11 5HD
G6 SWD A. Gibbings, 3 Bonville Crescent, Tiverton, EX16 4BN
G6 SWJ J. Askey, The Maltings, Brewery Yard, Kettering, NN14 3BT
G6 SWO M. Fincher, Brickyard Farm, Lincoln Road, Horncastle, LN9 5NW
G6 SWT S. French, 47 Horn Lane, Woodford Green, IG8 9AA
G6 SWW S. Ellin, 7 Crawshaw Avenue, Beauchief, Sheffield, S8 7DZ
G6 SWZ M. Davy, 22 Scott Gardens, Lincoln, LN2 4LX
G6 SXB L. Dunham, 5 King Street, Wimblington, March, PE15 0QF
G6 SXC J. Mallichan, 17 Napier Road, Gillingham, ME7 4HB
G6 SXD E. Drinkwater, 57 Ludlow Road, Bridgnorth, WV16 5AH
G6 SXN G. Dixon, 4 Yarborough Road, Keelby, Grimsby, DN41 8HG
G6 SYA M. Mills, 6 Bower Road, Hextable, Swanley, BR8 7SE
G6 SYB J. Malcom, 62 Linden Avenue, Ruislip, HA4 8UA
G6 SYI P. Somerfield, 27 Ormerod Street, Worsthorne, Burnley, BB10 3NU
G6 SYW B. Bauly, Poplar Farm Mendlesham, Stowmarket, IP14 5SN
G6 SYX G. Brookes, 47 Lucas Avenue, York, YO30 6HL
G6 SZB J. Barton, 76 Elvaston Road, North Wingfield, Chesterfield, S42 5HH
GM6 SZJ M. Burke, 3 Tarvit Avenue, Cupar, KY15 5BW
G6 SZS R. Crook, 26 Chapel Street, Rishton, Blackburn, BB1 4NP
G6 TAF P. Penny, 79 Grove Avenue, New Costessey, Norwich, NR5 0JA
G6 TAH D. Palmer, 17 Atyeo Close, Burnham-on-Sea, TA8 2EJ
G6 TAI J. Peel, 9 Hillspring Road, Springhead, Oldham, Ol44sj
G6 TAK P. Reay, 26 Clifton Court, Workington, CA14 3HR
G6 TAN M. Shread, 21 The Strand, Mablethorpe, LN12 1BQ
G6 TAP D. Squire, Green Valley, Raleigh Road, Barnstaple, EX31 4HY
G6 TAS R. Wroe, 11 Malvern Close, Banbury, OX16 9EL
GI6 TBC V. Loughran, 10 Oakwood, Armagh, BT60 1QR
GM6 TBE P. Lowrie, 11 Berrymoss Court, Kelso, TD5 7NP
G6 TBJ J. Larssen, 228A Barnsole Road, Gillingham, ME7 4JB
G6 TBV K. Cheers, 112 Rickerscote Road, Stafford, ST17 4HB
G6 TCV J. Halliday1, Calle Barrio 18 Ribera Baja, Alcala la Real, Spain, 23691
G6 TDG K. Hodges, 18 Leycester Close, Birmingham, B31 4SS
G6 TDJ G. Ward, 83 Buxton Road, Congleton, CW12 2DX
G6 TDW J. Mead, 12 Coltsfoot Close, Ixworth, Bury St Edmunds, IP31 2NJ
G6 TDX D. Yarrow, 193 Ladygate Lane, Ruislip, HA4 7RD
G6 TEB A. Varga, 2 Yew Tree Lane, Malvern, WR14 4LJ
G6 TEL S. Mayer, 453 Wimborne Road, Poole, BH15 3EE
GW6 TEO G. Smith, 11 Sandy Leys, Castlemartin, Pembroke, SA71 5HJ
G6 TEQ I. Stuckey, Rancliffe Higher Downs Road, Torquay, TQ1 3LD
G6 TER R. Monksummers, 29 Cloverfields, Peacemarsh, Gillingham, SP8 4UP
G6 TET B. Smith, 8 Devon Street, Leigh, WN7 2NG
G6 TEX T. Speak, 40 Orchard Close, Bolsover, Chesterfield, S44 6DY
G6 TFE D. Standen, Sunnybank, 54 Park Way, Hastings, TN34 2PJ
GI6 TFF R. Symington, 8 Thorndene Park, Carrickfergus, BT38 9EA
G6 TFJ R. Smith, Flat 1 2 St. Philips Place, Eastbourne, BN22 8LW
G6 TFL C. Sugars, 6 Church Street, Denby Village, DE5 8PQ
G6 TFP C. Mann, Woodford, Listowel Co Kerry, Ireland
G6 TFV D. Owen, 18 Prescott Avenue, Atherton, Manchester, M46 9LN
G6 TGB A. Pennells, 56 Wilverley Place, Blackfield, Southampton, SO45 1XW
G6 TGE G. Holman, 5 Ingleton Road, Newsome, Huddersfield, HD4 6QX
G6 TGJ J. Hirons, Furlong House, Racecourse Lane Bicton Heath, Shrewsbury, SY3

5BJ
G6 TGM W. Howard, 2 Heather Drive, Rise Park, Romford, RM1 4SP
G6 TGQ S. Houghton, 259B St. Faiths Road, Norwich, NR6 7BB
GW6 TGR G. Jones, 3 Tan Y Buarth Estate, Bethel, Caernarfon, LL55 1UP
G6 TGW R. Jones, 49 Sycamore Drive, Huntingdon, PE29 7JA
G6 THC M. Johnson, Happy Valley, Highfield Road, Westerham, TN16 3UX
G6 THM C. Smith, 8 Terry Close, Stoke-on-Trent, ST3 6NS
G6 THP L. Curwen, 12 Garden Close, St Ives, PE27 3XZ
GM6 TIB I. Campbell, 35 Thornwood Avenue, Lenzie, Glasgow, G66 4EL
G6 TID M. Coleman, 1 Burdon Drive, Bartestree, Hereford, HR1 4DL
G6 TIQ A. Price, Flat 2, South Elms, 69 Silverdale Road, Eastbourne, BN20 7EU
G6 TIU M. Rainer, 101 Gwydir Street, Cambridge, CB1 2LG
G6 TIW D. Reeve, 12 Lambourne Road, Birstall, Leicester, LE4 4FU
G6 TJC S. Deville, 39 Acre Close, Maltby, Rotherham, S66 8BL
GM6 TJD J. Doull, 52 Howburn Road, Thurso, KW14 7ND
G6 TJE R. Dawson, 10 St. Julien Close, New Duston, Northampton, NN5 6QX
G6 TJJ R. Downham, 11 Churchills Rise, High Street, Cullompton, EX15 3AU
G6 TJK A. Dowell, 54 Station St., Castle Gresley, Burton on Trent, DE14 1BS
G6 TJY I. Randle, 12 Cuckoo Avenue Hanwell, London, W7 1BT
G6 TJZ P. Rendell, 6 The Park, Bradley Stoke, Bristol, BS32 0AP
G6 TKB K. Slaughter, 652 Newchurch Road, Newchurch, Rossendale, BB4 9HG
GU6 TKE C. Wild, Honfleur, La Rue Du Le Hurel, Vale, Guernsey, GY3 5AF
G6 TKH J. Torring, Ivy Cottage, Royal Oak, Filey, YO14 9QE
G6 TKR E. Tratt, 162 Stoddens Road, Burnham-on-Sea, TA8 2EL
G6 TKV T. Tyrer, 85 Swann Lane, Cheadle Hulme, Cheadle, SK8 7HU
G6 TKW P. Tomlinson, 158 Seamore Avenue, Benfleet, SS7 4LA
G6 TKY E. Caligari, 209 Ormskirk Road, Upholland, Skelmersdale, WN8 0AA
G6 TLA P. Curran-Bilbie, 198 Birchwood Lane, Somercotes, Alfreton, DE55 4NF
G6 TLB P. Curran, 422 Carlton Road, Worksop, S81 7QW
G6 TLN D. Allen, 35 Fortescue Chase, Thorpe Bay, Southend-on-Sea, SS1 3SS
G6 TLP D. Attree, 36 Furze Road, Norwich, NR7 0AS
G6 TLX C. Bull, 35 Manor Road, Wokingham, RG41 4AR
GM6 TMH D. Bell, 11 Shebster Court, Thurso, KW14 7ES
G6 TMN H. Bryan, 2 Ashbrook Close, Hesketh Bank, Preston, PR4 6LY
G6 TMQ L. Saagi, 17 Broughton Close, Anstey, Leicester, LE7 7EU
G6 TNA C. Walton, 6 Gorse Grove, Longton, Preston, PR4 5NP
G6 TNE G. Skulski, 30 Eastfield Road, Laindon, Basildon, SS15 4JE
G6 TNI B. Telford, 18 Kirkstall Close, South Anston, Sheffield, S25 5BA
G6 TNJ D. Thomas, 97 The Gardens, Doddinghurst, Brentwood, CM15 0LX
G6 TNK J. Turnbull, 34 Bridge Avenue, Hanwell, London, W7 3DJ
G6 TNP R. Brooks, 8 Chesle Way, Portishead, Bristol, BS20 8JB
G6 TNQ N. Bosanquet-Bryant, 2 Dupont Close, Clacton-on-Sea, CO16 8YD
G6 TNR D. Blackman, 115 Ringwood, Bracknell, RG12 8XU
G6 TNW I. Webb, Cornerways Orchard Road, Eaton Ford, St Neots, PE19 7AN
G6 TOI A. Edgcombe, 2 Providence Place, Fore Street, Loddiswell, Kingsbridge, TQ7 4QP
G6 TOT A. White, Flat A, 106A Palmerston Road, Chatham, ME4 5SJ
GW6 TOX B. Taylor, Swn-Y-Don, Beaumaris, LL58 8RW
G6 TOY D. Williams, Hollybank, Royston Road, Taunton, TA3 7RE
GJ6 TPD E. Langlois, L'Amarrage, La Route Orange, St Brelade, Jersey, JE3 8GP
G6 TPE M. Long, 28 Wentworth Drive, Wirral, CH63 0JA
G6 TPG J. Littler, 39 Wigan Road, Golborne, Warrington, WA3 3TZ
G6 TPI C. Bryan, 3 Hales Place, Longton, Stoke-on-Trent, ST3 4NF
G6 TPO D. Bathe, Moel Tryfan 37 Mallows Green, Harlow, CM19 5SA
G6 TQC W. George, 28 Melbourn Close, Duffield, Belper, DE56 4FX
G6 TQF P. Game, 15 Nightingale Close, Gosport, PO12 3EU
GW6 TQH G. Giudice, 31 Woodfield Cross, Tredegar, NP22 4JG
G6 TQL T. Law, 41 Lime Trees, Tonbridge, TN12 0SS
G6 TQZ R. Andrews, 10 Fourth Avenue, Havant, PO9 2QX
G6 TRA D. Andrew, The Willows, Nordelph, Downham Market, PE38 0BY
G6 TRG Todmorden Ryt G, c/o P. Rigg, 1 Stones Hey Gate, Widdop Road, Hebden Bridge, HX7 7HD
G6 TRM M. Bryant, 104 Manor Road, Dover, CT17 9JZ
G6 TRN D. Best, 64 New Hey Road, Cheadle, SK8 2AQ
G6 TRO B. Wilcox, 1 Parklands, Stanwick, Wellingborough, NN9 6QX
G6 TRQ J. Wright, 9 Willow Close, Broadmeadows, Alfreton, DE55 3AP
G6 TRW A. Toas, 116 Rownhams Road, North Baddesley, Southampton, SO52 9EU
G6 TRX J. Sugrue, 124 Hall Lane, Upminster, RM14 1AL
G6 TRY G. Smillie, 5 Fleckers Drive, Up Hatherley, Cheltenham, GL51 3BB
G6 TSC V. Simmons, 88 Wellcome Avenue, Dartford, DA1 5JW
G6 TSE A. Sullivan, 20 Crockerne Drive, Pill, Bristol, BS20 0LF
G6 TSF P. Shayler, 38 Maryside, Slough, SL3 7ET
G6 TSJ P. Hannam, 7 Bodenham Close, Buckingham, MK18 7HR
G6 TSL R. Hill, Marclecote, Ledbury Road, Ross-on-Wye, HR9 7BE
G6 TSM W. Hirst, 8 Moss Road, Alderley Edge, SK9 7HZ
G6 TSP K. Hendry, 23 Briscoe Way, Lakenheath, Brandon, IP27 9SA
G6 TSX C. Heater, 8 Whitethorn Close, Ash, Aldershot, GU12 6NZ
G6 TSZ D. Hall, 282 Dereham Road, Norwich, NR2 3TL
GW6 TTA M. Harris, 7 Washington Street Landore, Swansea, SA1 2QE
G6 TTD R. Hewitt, 11614 Waesche Drive, Mitchellville, USA, 20271
G6 TTX W. Kenyon, 13 Baskerfield Grove, Woughton On The Green, Milton Keynes, MK6 3ES
GW6 TUD M. Prosser, 18 Thornhill Way, Rogerstone, Newport, NP10 9FT
GM6 TUE P. Mclaren, Dalriada, Fogwatt, Moray, IV30 8SY
G6 TUG I. Metcalfe, 12 Clarence Way, Horley, RH6 9GT
G6 TUS R. Page, 53 The Brambles, Bar Hill, Cambridge, CB23 8SZ
G6 TVA D. Biram, 124 Keresforth Hill Road, Red Gables, Barnsley, S70 6RG
G6 TVB R. Steele, 98 Obelisk Rise, Boughton Green, Northampton, NN2 8QU
G6 TVC D. Spooner, Thorny How, Canon Pyon, Hereford, HR4 8NT
GW6 TVD E. Sims, Hafan, Engedi, Holyhead, LL65 3RR
G6 TVE J. Tyreman, 12 Richmond Close, Rochdale, OL16 4RJ
G6 TVI G. Sculthorpe, 50 Station Road, Dersingham, King's Lynn, PE31 6PR
G6 TVJ I. Bennett, 47 Bakers Ground, Stoke Gifford, Bristol, BS34 8GD
G6 TVK A. Baker, 10 Kingscroft Court, Northampton, NN3 9BH
G6 TVP S. Burke, 17 The Crescent, Wragby, Market Rasen, LN8 5RF
GM6 TVR J. Black, Solway View, Carlisle Road, Annan, DG12 6QX
G6 TVX M. Collins, Coburg Cottage Mount Road, East Cowes, PO32 6NT
G6 TW South Cheshire ARS, c/o P. Walker, 11 Flixton Drive, Crewe, CW2 8AP
G6 TWA A. Woollard, 30 John Grinter Way, Wellington, TA21 9AR
G6 TWB South Cheshire ARS, c/o B. Rigby, 76 Woodland Road, Rode Heath, Stoke-on-Trent, ST7 3TL
G6 TWD W. West, Lectric, Alexandra Road, Crediton, EX17 2DH
GD6 TWF C. Wood, Deep Water, Glen Rushen Road, Peel, IM5 3BA
G6 TWR K. Simmonds, 1 Fallow Close, South Molton, EX36 3FL
G6 TWX A. Tatterton, 28 Kinloch Drive, Bolton, BL1 4LZ
G6 TXB B. Thompson, 12 Albion Road, Chatham, ME5 8SR
G6 TXH R. Webb, 90 Queens Road, Tunbridge Wells, TN4 9JU
G6 TXP A. Cronk, 65 Russell Court, Chatham, ME4 5LE
G6 TXQ V. Covell-London, 15 St. Nicholas Way, Potter Heigham, Great Yarmouth, NR29 5LG
G6 TXV S. Calver, 144A Smugglers Club Ground, Bridgemarsh Lane, Chelmsford, CM3 6DQ
G6 TXY B. Coulstock, 32 Climping Park, Bognor Road, Littlehampton, BN17 5DW
G6 TYB J. Cooke, 106 Wirral Drive, Winstanley, Wigan, WN3 6LD
G6 TYF R. Duley, Denova, Hornsby Lane, Grays, RM16 3AU
GW6 TYO B. Young, 3 Bryn Road, Pontlliw, Swansea, SA4 9ED
G6 TYT S. White, 8 Bilford Avenue, Worcester, WR3 8PJ
GM6 TYX Dr C. Macleod, Morven, Marybank, Isle of Lewis, HS2 0DD
G6 TZE J. Richards, 17 Chaffers Mead, Ashtead, KT21 1NA
G6 TZO N. Roberts, 11 Mallard Road, Barrow Upon Soar, Loughborough, LE12 8BF
G6 TZT T. Rogers, 36 Goodacre Road, Ullesthorpe, Lutterworth, LE17 5DL
G6 UAJ P. Longstaff, 27 Feather Wood, Westlea, Swindon, SN5 7AG
G6 UAN M. Moseley, Flat 40, Oceana Boulevard, Briton Street, Southampton, SO14 3HU
G6 UAP E. Magnuszewski, 49 Elvaston Road, Nottingham, NG8 1JU
GW6 UAS J. Mcmurray, 30 St. Martins Crescent, Llanishen, Cardiff, CF14 5QA
G6 UAW M. Egerton, 3 Boundary Close, Salisbury, SP2 9FZ
G6 UBH M. Faithfull, 99 Bramble Road, Hatfield, AL10 9SB
G6 UCI G. Miller, 11 Friars Avenue, Great Sankey, Warrington, WA5 2AR
GM6 UCN W. Mckenzie, 14 Bridgend, Dunblane, FK15 9ES
G6 UCO G. Bee, 80 Hospital Road, Burntwood, WS7 0EQ
G6 UCQ M. Burgess, 20 Norfolk Road, Luton, LU2 0RE
G6 UCT P. Bowron, 52 Eastcotes, Tile Hill, Coventry, CV4 9AU
G6 UCW A. Brookes, 8 Cedar Close, Ruskington, Sleaford, NG34 9FH
GM6 UCX M. Bellerby, Hamnavoe, Eabost West, Struan, Isle of Skye, IV56 8FL
G6 UCY K. Porter, 60 Spitfire Road, Wallington, SM6 9GL
G6 UDA R. Pyrah, Whispering Waves, The Shore, Poulton-le-Fylde, FY6 9EA
G6 UDB F. Scott, 15 Sunningdale Close, Kirkby-In-Ashfield, Nottingham, NG17 8NW
G6 UDF P. Phelps, 14 The Warren, Hazlemere, High Wycombe, HP15 7ED
G6 UDG A. Price, Brook House, Drury Lane, Shrewsbury, SY4 1DT
G6 UDI S. Phillips, 79 Selwyn St., Stoke, Stoke on Trent, ST4 1ED
G6 UDX B. Oldford, 16 Ludlow Drive, Stirchley, Telford, TF3 1EG
G6 UED T. Lloyd, 18 Coleville Road, Minworth, Sutton Coldfield, B76 1XR
G6 UEG I. Norman, Bridge Farm Snaffers Lane, Whaplode, Spalding, PE12 6RX
G6 UEH E. Naylor, 18 Mackenzie Crescent, Cheadle, Stoke-on-Trent, ST10 1LU
G6 UEI P. Norman, 20 Meadow Close, Budleigh Salterton, EX9 6JN
G6 UEQ R. Rix, Patterdale, Roe Downs Road, Alton, GU34 5LG
G6 UER V. Rice, 24 Harewood Close, Tuffley, Gloucester, GL4 0SR
G6 UEV M. Piper, 26 Hare Law Gardens, Stanley, DH9 8DG
GW6 UFH S. Fry, 10 Heaseland Place, Killay, Swansea, SA2 7EQ
G6 UFI Rev. P. Fanning, 1 Naburn Grove, Moreton, Wirral, CH46 0SN
G6 UFL S. Duckett, 35 Fowlmere Road, Foxton, Cambridge, CB22 6RT
GI6 UFO J. Fitzgerald, 15 Bunnahesco Road, Bunnahesco, Enniskillen, BT94 5HJ
GI6 UFU J. Campbell, 22 Sheridan Drive, Helens Bay, Bangor, BT19 1LB
G6 UFV C. Spencer, 18 Coatsby Road Kimberley, Nottingham, NG16 2TH
G6 UFZ K. Chamba, 63 Patricia Avenue, Wolverhampton, WV4 5AQ
G6 UGA M. Pinkney, 169 Sandringham Road, Perry Barr, Birmingham, B42 1PZ
GW6 UGC G. Phillips, 83 Heol Y Llwynau, Trebanos, Swansea, SA8 4DB
G6 UGE C. Power, 296 Alderley, Digmoor, Skelmersdale, WN8 9NB
G6 UGG A. Panton, 35 Long Water Drive, Gosport, PO12 2UP
G6 UGS M. Allison, 6 Eden Road, Beverley, HU17 7HD
G6 UGT T. Aherne, 21 Burbage Place, Alvaston, Derby, DE24 8NP
G6 UGW M. Bell, 61 Oldbury Orchard, Churchdown, Gloucester, GL3 2PU
G6 UGZ A. Scott, 23 Wingfield Road, Great Barr, Birmingham, B42 2QB
GM6 UHC A. Stewart, Skerry Alvha, Torphins, Banchory, AB31 4NB
G6 UHD B. Scott, Linda Cottage, St. Giles-On-The-Heath, Launceston, PL15 9RT
GM6 UHE A. Wilson, Lochend, Ayrshire, KA15 2LN
G6 UHF CQ Radio Group, c/o A. Page, 3 Reading Avenue, Nuneaton, CV11 6HE
G6 UHL B. Ritchie, 65 Ransome Avenue, Worcester, WR5 3AL
G6 UHS A. Coe, 22 St. Annes Way, Spalding, PE11 3PN
G6 UIF I. Clark, 41 Brook Close, Jarvis Brook, Crowborough, TN6 2ET
G6 UIM S. Daniels, 46 Freshwater Drive, Paignton, TQ4 7SD
G6 UIP G. Dodd, 6 Highfield Avenue Kirkby-In-Ashfield, Nottingham, NG17 8GF
G6 UIT W. Dillon, 49 Goring Way, Greenford, UB6 9NN
G6 UJC C. Dukes, 32 Greenacres Woolton Hill, Nr. Newbury, RG20 9TA
GM6 UJG V. Simpson, 43 Fortingall Place, Perth, PH1 2NF
G6 UJI B. Staton, 99 Linden Avenue, Prestbury, Cheltenham, GL52 3DT
G6 UJJ N. Stoker, 6 Beech Grove, Gateshead, NE9 7RE
G6 UJR M. Severs, 125 Hawthorne Way, Shelley, Huddersfield, HD8 8QF
G6 UKC M. Brown, 33 Stonegate, Cowbit, Spalding, PE12 6AH
G6 UKM S. Brown, 22 Asquith Close, Biddulph, Stoke-on-Trent, ST8 7LN
G6 UKN W. Bailey, 35 Elton Lane, Winterley, Sandbach, CW11 4TN
GW6 UKO C. Barwell, Galaghad, Penisarwaun, Caernarfon, LL55 3BN
G6 UKQ J. Riley, 56 Church St., Bignall End, Stoke on Trent, ST7 8PE
G6 ULD R. Humphrys, 10 St. Andrews Road, Bexhill-on-Sea, TN40 2BQ
G6 ULJ R. Green, Branford House, Valley Road, Tasburgh, Norwich, NR15 1NG
G6 ULS P. Kent-Woolsey, 32 Yaxham Road, Dereham, NR19 1AJ
G6 UMH E. Harding, 49 Compass Close, Murdishaw, Runcorn, WA7 6DL
G6 UML T. Reader, 76 West View Road, Dartford, DA1 1TR
G6 UMN L. Gibson, 27 Farm Street, Barrow-in-Furness, LA14 2RX
G6 UMS C. Hadler, Dowles Brook Lodge, Rock Cross Rock, Kidderminster, DY14 9SF
G6 UMT A. Handcocks, Woodpeckers, Chapel Lane, Southampton, SO45 1YX
GW6 UMU A. Haigh, Nant Fawr, Corwen, LL21 9AA
G6 UMX J. Hibbert, 125 Chase Hill Road, Arlesey, SG15 6UF
G6 UNA W. Stoneman, 5 Creaton Road, Hollowell, Northampton, NN6 8RP
GM6 UNQ E. Leask, 2/7 Barnton Avenue West, Edinburgh, EH4 6EB

G6 UNR R. Rogers, 97 Sutherland Avenue, Biggin Hill, Westerham, TN16 3HH
G6 UNU A. Lunn, 45 St. Anthonys Avenue, Eastbourne, BN23 6LN
G6 UOH I. Thacker, 3 Webster Way, Gonerby Hill Foot, Grantham, NG31 8GH
G6 UOO D. Wilde, 3 Canal Cottages, Buxworth, High Peak, SK23 7NF
G6 UOX M. Walker, 94 Lambert Road, Uttoxeter, ST14 7QY
G6 UPA D. Wiseman, 22 Queens Crescent, Clapham, Bedford, MK41 6DA
G6 UPH M. Hackney, Mar Azul 9, Apt.20 2 Fase, Alicante, Spain, 03710 CALPE
G6 UPI B. Hurrell, 33 Meadow Way, Hellesdon, Norwich, NR6 5NN
G6 UPL T. Hayhurst, The Paddock, Crooklands, Milnthorpe, LA7 7NL
G6 UPM A. Hayhurst, The Paddock, Crooklands, Milnthorpe, LA7 7NL
G6 UPQ D. Holloway, 48 Wenrisc Drive, Minster Lovell, Witney, OX29 0RQ
G6 UPR B. Hingston, Hazelwood Farm Marldon, Paignton, TQ3 1SQ
G6 UQ Stockport Rs, c/o B. Naylor, 47 Chester Road Poynton, Stockport, SK12 1HA
G6 UQA S. Duckles, 8 Railway Cottages, Skillings Lane, Brough, HU15 1EN
G6 UQI E. Payne, 4 Richmond Crescent, Barons Cross, Leominster, HR6 8RX
G6 UQO D. Oliver, 37 Milford Avenue, Elsecar, Barnsley, S74 8DT
G6 UQZ A. Parkhurst, 14 Church Street, Clare, Sudbury, CO10 8PD
G6 URD J. Ratcliffe, 63 Dickens Lane, Poynton, Stockport, SK12 1NN
G6 URF A. Hartley, 18 Smithy Close, Cronton, Widnes, WA8 5BT
G6 URK J. Jennings, 354 Williamthorpe Road, North Wingfield, Chesterfield, S42 5NS
G6 URM B. Johnson, 6 Winston Avenue, Plymouth, PL4 6AZ
GM6 URP R. Gray-Jones, Flat C 7 Nelson Street, Aberdeen, AB24 5EP
G6 URR I. Kirk, 12 Edinbane Close, Rise Park, Nottingham, NG5 5DU
G6 URT C. Kapoutsis, 7A East Lane, Morton, Bourne, PE10 0NW
G6 URU J. Keats, 50 Ringway Road, Park Street, St Albans, AL2 2RD
G6 USA P. Love, 2 Meadway, Dover, CT17 0PS
G6 USD M. Matthews, 213 Hucclecote Road, Gloucester, GL3 3TZ
G6 USG G. Comer, 27 Peckforton View, Kidsgrove, Stoke-on-Trent, ST7 4TA
G6 USL B. Cowell, 46 Gattison Lane, New Rossington, Doncaster, DN11 0NQ
G6 USO P. Chamings, 52 Crown Street, Redbourn, St Albans, AL3 7PF
G6 USR M. Davis, Sunny Bank, Headcorn Road, Maidstone, ME17 2AN
G6 UST P. Drury, 5 Bede Place, Peterborough, PE1 4EE
G6 USU T. Derbyshire, 32 Hardie Avenue, Wirral, CH46 6BJ
G6 USX W. Dennison, 41 Tarbert Walk, Stepney, London, E1 0EE
G6 USZ D. Deverell, 23 Frankmarsh Park, Barnstaple, EX32 7HN
G6 UTB H. Campbell, 175 Thrupp Lane, Brimscombe, Stroud, GL5 2RG
GW6 UTF D. Foster, 11 Dingle Road, Leeswood, Mold, CH7 4SN
G6 UTK G. Fisher, 16 Somerset Lane, Lansdown, Bath, BA1 5SW
G6 UTL S. Foulser, 32 Langhorn Road, Southampton, SO16 3TN
G6 UTO N. Simpkin, 2 Redland Close Beeston, Nottingham, NG9 5LA
G6 UTT P. Sheppard, Round Corners, 7 First Avenue, Bognor Regis, PO22 6ED
GI6 UUC J. Thompson, 21 Watch Hill Road, Ballyclare, BT39 9QW
G6 UUQ L. Sheward, 7 Harlington Avenue, Grove, Wantage, OX12 7NQ
G6 UUR S. Whitehead, 94 Cranmore Boulevard Shirley, Solihull, B90 4RU
GI6 UUT Dr W. Page, 4 Glebe Manor, Hillsborough, BT26 6NS
G6 UVB C. Wilson, Rustleigh, 7 Stagbury Close, Coulsdon, CR5 3PH
G6 UVN M. Henman, 4 Lyne Walk, Hackleton, Northampton, NN7 2BW
G6 UVO C. Heritage, 29 Hill Head, Glastonbury, BA6 8AW
G6 UVS P. Hannington, 21 Little Gate, Westhoughton, Bolton, BL5 2SD
G6 UVU J. Handy, 77 Abbeyfield Road, Wolverhampton, WV10 8TH
G6 UW Cambridge UWS, c/o Dr J. Keeler, 67 Perne Avenue, Cambridge, CB1 3RY
GM6 UWF J. Allan, 87 Needless Road, Perth, PH2 0LD
G6 UWI N. Bradshaw, 26 Suffolk Gardens, South Shields, NE34 7JF
G6 UWK J. Barden, 2 Pondhall Cotts. Bradfield Rd, Manningtree, co112sp
G6 UWO D. Bullock, 1 Selby Close, Beeston, Nottingham, NG9 6HS
G6 UWS M. Byles, 108 Kingsway, Wellingborough, NN8 2EN
GW6 UWW M. Williams-Davies, Plas Penrhos, Llwyngwril, LL37 2QB
G6 UWX I. Whiting, 15 Highfield Court, Grace Way, Stevenage, SG1 5EH
G6 UWY D. Williams, 16 Blaydon Walk, Wellingborough, NN8 5YU
G6 UX I. Truslove, 36 Leicester Road, Hinckley, LE10 1LS
G6 UXE R. Wheeler, 36 Kimbolton Crescent, Stevenage, SG2 8RJ
G6 UXF K. Young, 8 Magnolia Close, Worcester, WR5 3SJ
G6 UXG A. Webb, 35 Hill House Drive, Minster, Ramsgate, CT12 4BE
G6 UXK D. Wookey, 3 Westland Close, Boscombe Down, Amesbury, Salisbury, SP4 7QS
G6 UXM S. Vinnicombe, 8A Cross Road, Cholsey, Wallingford, OX10 9PE
G6 UXU C. Stanley, 494 Blackburn Road, Darwen, BB3 0AJ
G6 UXW P. Mundy, 25 Lonsdale Avenue, Cosham, Portsmouth, PO6 2PU
G6 UXX P. Leese, 4 Harefield, Harlow, CM20 3EF
G6 UXY A. Lightly, 8 Smithville Close, St. Briavels, Lydney, GL15 6TN
G6 UYJ A. Page, 35 Acorn Close, Christchurch, New Zealand, 8023
G6 UYK P. Russell, 1 Larch Grove, Kendal, LA9 6AU
G6 UYM D. Richards, 25-27 Burnivale, Malmesbury, SN16 0BL
G6 UYN A. Rumney, Church House Farm Cottage, Cheltenham, GL51 0TW
G6 UZA A. Kotowicz, 47 Portree Drive, Rise Park, Nottingham, NG5 5DT
G6 UZG P. Ashby, 26 Van Diemens Lane, Bath, BA1 5TW
G6 UZJ J. Austen, 13 Coverdale, Whitwick, Coalville, LE67 5BP
G6 UZL P. Bunn, Yew Trees Main Road, Little Haywood, Stafford, ST18 0TS
G6 UZM S. Byford, 21 Clarke Drive, Shaw, Swindon, SN5 5SH
G6 UZO M. Brunsdon, 7 Oldberg Gardens, Brighton Hill, Basingstoke, RG22 4NP
G6 UZR A. Brown, Badgers Way, Holton, Wincanton, BA9 8AL
G6 UZY M. Owen, 49 Southdale Drive Carlton, Nottingham, NG4 1DA
G6 VAA G. Perks, 55 Andrew Road, Tipton, DY4 0AJ
G6 VAD P. Purdy, 4 Hethersett Road, East Carleton, Norwich, NR14 8HX
G6 VAE T. White, 117A Western Road, Southall, UB2 5HN
G6 VAL A. Oughton, 326 Ware Road, Hailey, Hertford, SG13 7PG
G6 VAR J. Smith, 6 Hollams Road, Tewkesbury, GL20 5DG
G6 VAW C. Soars, 118 Braddon Road, Loughborough, LE11 5YZ
G6 VAX R. Saunders, 93 Oaks Avenue, Worcester Park, KT4 8XG
G6 VAZ D. Thomas, 25 Lime Close, Mildenhall, Bury St Edmunds, IP28 7PR
G6 VBA D. Townend, 25 De Trafford St., Huddersfield, HD4 5DR
G6 VBD J. Savage, 2 Alvecote Cottages, Alvecote Lane, Tamworth, B79 0DJ
G6 VBE R. Ransom, 1 Bilberry Road, Clifton, Shefford, SG17 5HB
G6 VBJ P. Tasker, Oaktree Cottage, Bunkers Hill, Ridley, Sevenoaks, TN15 7EY
G6 VBK D. Hatton, 24 Langdale Road, Leyland, PR25 3AR
GW6 VBN M. Hunt, 23 Swansea Road, Pontardawe, Swansea, SA8 4AL
G6 VBQ A. Haddock, 1 Heron Way, St Ives, PE27 6SS
GW6 VBR L. Cowley, 3 Pleasant Villas, Pontarddulais, Swansea, SA4 8QF
G6 VCF P. Brackstone, 3 Wentworth Close, Beverley, HU17 8XB
GI6 VCG J. Brownlees, 8 Cairnbeg Park, Larne, BT40 1UB
GI6 VCL K. Cunningham, 4 Garvaghy Road, Portglenone, Ballymena, BT44 8EF
G6 VCR H. Eden, 142 Ringway, Thornton-Cleveleys, FY5 2NW
G6 VDA A. Sutton, 3 Cornflower Close, Willand, Cullompton, EX15 2TT
G6 VDD P. Waddington, 20 Littlington Court Surrey Road, Seaford, BN25 2NZ
G6 VDK P. Lutas, 616 Queens Drive, Swindon, SN3 1AZ
G6 VDW R. Olliver, 39 Nutshalling Avenue, Rownhams, Southampton, SO16 8AY
G6 VDX I. Ogilvie, 8 Devonshire Road, Prenton, CH43 4UL
G6 VDY H. Jeffery-Wright, 55 Burland Avenue, Wolverhampton, WV6 9JJ
GW6 VED R. Straughan, 1 Crossroads, Gilwern, Abergavenny, NP7 0DX
G6 VEG T. Gray, 8 Holystone Grange, Holystone, Newcastle upon Tyne, NE27 0UX
GW6 VEI D. Pierce, Coed Duon, Tremeirchion, St Asaph, LL17 0UH
G6 VEJ F. Stone, 51 The Glen, Yate, Bristol, BS37 5PJ
GW6 VEN A. Rose, 4 Llys Clwyd, Kinmel Bay, Rhyl, LL18 5EW
GW6 VET J. Goodson, 22 Pant Gwyn, Bridgend, CF31 5BA
G6 VEY I. Haver, 4 Campion Way, Bourne, PE10 0QE
G6 VEZ G. Helm, 31 Faringdon Avenue, South Shore, Helmsman Electronics Ltd, Blackpool, FY4 3QQ
G6 VF S. Illman, 66 Frieth Road, Marlow, SL7 2QU
G6 VFA M. Hine, Tall Trees, Lime Lane, Derby, DE21 4RF
G6 VFB W. Hogan, 279 Halliwell Road, Bolton, BL1 3PE
G6 VFC D. Hooton, 80 Portland Road, Rushden, NN10 0DJ
G6 VFF S. Jackson, 47 Gurnard Pines, Cockleton Lane, Cowes, PO31 8RF
GW6 VFH R. Jenkins, 29 Pemberton St., Llanelli, SA15 2RB
G6 VFI A. Jones, The Studio, Fullers Vale Headley Down, Bordon, GU35 8NR
G6 VFO J. Stokes, 109 Hollyhedge Road, West Bromwich, B71 3BT
G6 VGA C. Mccall, Flat 6, Kent House, Park Cottages, Hawkhurst, Cranbrook, TN18 4JH
G6 VGC R. Woolley, 82 Pennycroft Road, Uttoxeter, ST14 7ET
G6 VGG Bromsgrove + District ARC, c/o A. Kelly, 8 Green Slade Crescent, Marlbrook, Bromsgrove, B60 1DS
G6 VGH I. Allen, 14 Bettridge Place, Wellesbourne, Warwick, CV35 9LY
GM6 VGO M. Barrett, 1 Walterstead Cottage, Ladykirk, Berwick upon Tweed, TD15 1XW
G6 VGS M. Bradbury, 55 Crowthorp Road, Northampton, NN3 5EY
G6 VGT D. Bowlas, 38 Senneleys Park Road, Northfield, Birmingham, B31 1AL
G6 VGV I. Craig, 1 Whitton Drive, Chester, CH2 1HF
G6 VGZ D. Cheriton, 5 Cornwall Close, Warwick, CV34 5HX
GM6 VHA M. Deverill, Flannan House, Aird Uig Timsgarry, Isle of Lewis, HS2 9JA
G6 VHE Dr M. Entwistle, 34 Webbs Court, Lyneham, Chippenham, SN15 4TR
G6 VHG A. Foster, 35 Gloucester Place, Peterlee, SR8 2HB
GW6 VIC V. Jones, Gwel Y Mor, Porth Y Felin Road, Holyhead, LL65 1BG
G6 VIF B. Morris, 21 Loxley Gardens, Southdown, Bath, BA2 1HS
G6 VIK I. King, 11 Cockhall Close, Litlington, Royston, SG8 0RB
G6 VIN J. Walker, 44 Albany Road, Kilnhurst, Mexborough, S64 5UG
G6 VIQ M. Watson, Salt Pie Farm, Birdsedge, Huddersfield, HD8 8XP
GM6 VIU A. Wilson, 1 High Street, Dysart, KY12TS
G6 VIY A. Wood, 12 Bishops Meadow, Sutton Coldfield, B75 5PQ
G6 VJA I. Taylor, 97 George St., Cleethorpes, DN35 8PL
G6 VJC J. Taylor, 17 Aintree Way, Castle Security Systems And Networking, Dudley, DY1 2SL
G6 VJK A. Maslin, 2 Clarks Cottages, White Horse Road, Colchester, CO7 6TX
G6 VJM D. Lynch, 30 Whitecroft View, Baxenden, Accrington, BB5 2QP
G6 VJP D. Pilkington, 45 High Meadows, Midsomer Norton, Radstock, BA3 2RZ
G6 VJR D. Reading, 23 Elwy Circle, Ash Green, Coventry, CV7 9AU
G6 VKA C. Thompson, Fourwinds, Walton Hill, Gloucester, GL19 4BT
GW6 VKI C. Richardson, Moorcroft, Kinnerley, Oswestry, SY10 8DW
G6 VKL D. Mayers, 15 Oakfield Road, Poynton, Stockport, SK12 1AR
G6 VKP J. Littlewood, 5 Laburnum Grove, Harrogate, HG1 4EH
G6 VKS Dr I. Morgan, Leigh House, 64 Widney Road, Solihull, B93 9AW
G6 VKX M. Webber, 23 Ramsey Close, Horley, RH6 8RE
GW6 VKY A. White, 86 Derlwyn, Dunvant, Swansea, SA2 7QE
G6 VLC S. Paxton, 11 Synderford Close, Didcot, OX11 7UT
G6 VLT J. Higgins, 190 Little Glen Road, Glen Parva, Leicester, LE2 9TT
G6 VLV D. Colman, 22 Peerley Close, East Wittering, Chichester, PO20 8PB
GI6 VLY Dr J. Earle, 25 Carnesure Park, Comber, Newtownards, BT23 5LT
G6 VMB C. Gibson, 103 Lydalls Road, Didcot, OX11 7DT
G6 VMF R. Hope, 30 Greendale Gardens, Hetton-Le-Hole, Houghton le Spring, DH5 0EF
G6 VMI D. Milne, 22 Eastnor Road, Reigate, RH2 8NE
G6 VMR M. Adams, 122 Green Lane Castle Bromwich, Birmingham, B36 0BX
G6 VMV S. Brocklehurst, Bank View, Reades Lane, Congleton, CW12 3LL
G6 VNC R. Davies, 27 Smiths Way, Water Orton, Birmingham, B46 1TW
G6 VNI G. Duggan, 28 Higher Rads End, Eversholt, Milton Keynes, MK17 9ED
G6 VNO N. Hanson, 100 Bassett Green Road, Southampton, SO16 3EF
G6 VNW B. Major, 3 Tithebarn Grove Wavertree, Liverpool, L15 6TG
G6 VOE D. Simpkins, 34 Rose Avenue, Weldon, Corby, NN17 3HB
G6 VOV R. Leavold, 8 Wilkinson Way, North Walsham, NR28 9BB
G6 VPH R. Gorton, 3 Pickford Avenue, Little Lever, Bolton, BL3 1PN
G6 VPJ G. Hall, 54 Townfields, Sandbach, CW11 4PQ
G6 VPL J. Hopkinson, 4 Marwood Croft, Streetly, Sutton Coldfield, B74 3JU
G6 VPN B. Jameson, 42 Eastgate, Fleet, Spalding, PE12 8NA
G6 VPU S. Mulligan, 406 St. Helens Road, Leigh, WN7 3PQ
G6 VPV J. Wake, 15 Deepdale Way, Darlington, DL1 2TA
G6 VPW R. Stoate, 19 Jean Road, Brislington Bs4 4Jt, Bristol, BS4 4JT
G6 VQC A. Read, Huenibachstrasse 75, Huenibach, Switzerland, CH-3626
G6 VQN A. Morris, 67 Broad Oak Way, Cheltenham, GL51 3LL
G6 VQV S. Shenfield, 3 Blackberry Grove, Bradwell-On-Sea, Southminster, CM0 7QE
G6 VQW R. Seaton, Wisteria Cottage, Welsh Road, Leamington Spa, CV33 9AQ
GM6 VRC J. Brown, Inchbeag Cottage, Inchcoonans, Perth, PH2 7RB
G6 VRF B. Crowther, 10 Askrigg Close, Marton Moss, Blackpool, FY4 5RE
G6 VRI G. Eccleshare, 22 Barley Close, Herne Bay, CT6 7XG
GW6 VRN M. Jones, 66 Brondeg, Heolgerrig, Merthyr Tydfil, CF48 1TP
GM6 VRU G. Giles, Monachan, Culrain, IV24 3DW
G6 VSE A. Bansal, Fernley, 2 Seaview Cotts, Chideock, DT6 6JE
G6 VSG D. Harris, 9 Garden City, Tern Hill, Market Drayton, TF9 3QB
G6 VSM J. Morris, 10 Batt Hall Kitchen Hill, Bulmer, Sudbury, CO10 7EZ
G6 VSQ F. Whitehurst, Roselands, Clarke Lane, Macclesfield, SK10 5AH

G6 VSY C. Sheppard, 42 Freeman Road, Didcot, OX11 7DD
G6 VTA M. Fisher, 46 Hedgerow Walk, Andover, SP11 6FD
G6 VTE S. Chambers, 1 Tatling Grove, Walnut Tree, Milton Keynes, MK7 7EG
G6 VTH R. Carney, 29 Hayton Close, Sunderland, SR5 2BU
G6 VTN P. Green, 79 The Spinney, Bar Hill, Cambridge, CB23 8SU
G6 VTX M. Brindley, 53B Seabridge Road, Newcastle, ST5 2HU
GW6 VTZ D. Campbell, 61 Maes Y Crofft, Cardiff, CF15 8FE
G6 VUE S. Butler, 231 Newman Road, Wincobank, Sheffield, S9 1LU
G6 VUF F. Caulfield Kerney, 47 Freemans Close, Stoke Poges, Stoke Poges, SL2 4ER
G6 VUG R. Collins, 37 Warwick Road, Twickenham, TW2 6SW
G6 VUJ J. Davis, 69 Bryanston Road, Solihull, B91 1BS
G6 VUN M. Hodson, 17 Marshfield Close, Redditch, B98 8RW
G6 VUX S. Challoner, Grosvenor Farm, Holme Street, Chester, CH3 8EQ
G6 VVE S. Banks, 29 Froxmere Close, Crowle, Worcester, WR7 4AP
GM6 VVG G. Caldwell, 10 Craigmath, Dalbeattie, DG5 4EB
G6 VVL K. Hotchen, 6 Nourse Close, Leckhampton, Cheltenham, GL53 0NQ
G6 VVS R. Jackson, 46 Ashford Road, Maidstone, ME14 5BH
G6 VVU Dr S. Adkins, 8 Kensey Valley Meadow, Launceston, PL15 9NB
G6 VVZ T. Butler, 103 Spring Gardens, Anlaby Common, Hull, HU4 7QH
G6 VWF J. Holbrook, 1 Segrave Grove, Hull, HU5 5DJ
G6 VWI C. Kowcun, 27 Mill Crescent, Kingsbury, Tamworth, B78 2LX
GI6 VWS J. Quigg, 9 Springhill Terrace, Limavady, BT49 9BS
G6 VWV S. Cresswell, 7 Japonica Drive, Nottingham, NG6 8PU
G6 VXC R. Callaghan, 5A Chapel Garth, West Ayton, Scarborough, YO13 9HH
G6 VXL T. Buck, 178 Rover Drive, Castle Bromwich, Birmingham, B36 9LL
G6 VXN A. Hart, 10 Walsgrave Close, Solihull, B92 9PQ
G6 VXR H. Metcalf, Beech Lee, Vicarage Lane, Alresford, SO24 0DU
G6 VXZ A. Sorab, Woodgaston Cottage, Woodgaston Lane, Hayling Island, PO11 0RL
G6 VYK E. Williams, 2 Ennerdale Court, Bridlington, YO16 6HL
GM6 VYY A. Mcminn, Siarardh, Mallaig, PH41 4QY
GM6 VYZ W. Mcminn, Glengyle, East Bay, Mallaig, PH41 4QF
GW6 VZB M. Lennox, 17 Coed Y Fron, Holywell, CH8 7UJ
G6 VZF A. Dawes, 1A Lower Olland Street, Bungay, NR35 1BY
G6 VZG M. Frosdick, 48 Woodfield, Briston, Melton Constable, NR24 2JY
G6 VZM J. Johnson, 62 Julien Road, Ealing, London, W5 4XA
G6 VZS D. Goodall, 94 Camp Mount, Pontefract, WF8 4BX
G6 VZU C. Hunt, 23 Beccles Road, Gorleston, Great Yarmouth, NR31 0PW
G6 VZZ J. Hackett, 18 Brow Edge, Rossendale, BB4 7TT
GW6 WAG D. Jones, Bradford House, The Square, Corwen, LL21 0DL
G6 WAM P. Woodyard, Sunny Nook Chapel Lane Scaleby Hill, Carlisle, CA6 4LY
G6 WAN R. Rayner, 12 Weedon Way, King's Lynn, PE30 4YY
G6 WAO N. Austin, 8 Chandler'S Court, Norwich, NR4 6EY
G6 WAR Mid Warwickshire ARS, c/o Q. Wright, 9 Browning Avenue, Warwick, CV34 6JQ
G6 WAS D. Carpenter, 2 Milton Road, Little Irchester, Wellingborough, NN8 2DY
G6 WAU Dr S. Connor, 1 Tallis Walk, Grange Park, Swindon, SN5 6BQ
G6 WAY J. Randall, 3 Steins Lane, Humberstone, Leicester, LE5 1ED
GM6 WAZ A. Ronnie, 7 Beechwood Avenue, Stranraer, DG9 0AU
G6 WBG P. Smith, Juniper Cottage, Palestine, SP11 7ER
G6 WBS S. Siggins, 5 Arrow Lane Halton, Lancaster, LA26QW
G6 WBT I. Thorp, Pinelodge, Carleton Green, Pontefract, WF8 3NJ
G6 WBX P. Yorke, 27 Luard Court, Havant, PO9 2TN
G6 WCI M. Richards, 72 Carlton Avenue, Westcliff-on-Sea, SS0 0QL
G6 WCW M. Smith, 234 Big Meadow Road, Wirral, CH49 9AW
G6 WCX D. Mardle, 22 Wayfield Link, Avery Hill, London, SE9 2LP
G6 WDC L. Baldwin, 26A Cheney Hill, Heacham, King's Lynn, PE31 7BS
G6 WDH A. Hutchison, 28 Allee De Quiberon, Colomiers, France, 31770
GJ6 WDK M. Monteil, Kalimera, Six Rues Villas 1 La Rue, St. Lawrence, Jersey, JE3 1GL
G6 WDM J. Million, 11 Derwent Mews, Blackhill, Consett, DH8 8TU
G6 WDR A. Tett, 19 Park Road, Shoreham-by-Sea, BN43 6PF
G6 WDS Dr F. Deravi, 22 Clifton Gardens, Canterbury, CT2 8DR
G6 WEH R. Burrows, 6 Frensham Drive, Hitchin, SG4 0QP
GM6 WEI G. Cockcroft, The Old Schoolhouse, Scatwell, Strathconon, Muir of Ord, IV6 7QG
G6 WEL J. Reynolds, 4 Rosewood Drive, Winsford, CW7 2UW
GW6 WEU K. Turner, 115 Newton Road, Newton, Swansea, SA3 4SW
G6 WEW J. Fitzsimons, 63 School Lane, Chapel House, Skelmersdale, WN8 8EN
G6 WFF G. Solkow, 12A Manor Court, Penkhull, Staffs, ST4 5DW
G6 WFM K. Farr, 3 Sheppard Drive, Chelmsford, CM2 6QE
GM6 WFP S. Pollok, School House Watten, Wick, KW1 5YJ
G6 WFS G. Quantrill, 47 Lambeth Road, Leigh-on-Sea, SS9 5XR
GW6 WFW A. Humphreys, 45 Cwm Place, Llandudno, LL30 1LP
GI6 WFX R. Johnston, 51 Kennedy Drive, Lisburn, BT27 4JA
G6 WGA A. Swift, 56 Birch Hall Avenue, Darwen, BB3 0JB
G6 WGE N. Riding, 15 Church Lane, Dewsbury Moor, Dewsbury, WF13 4EN
G6 WGM M. Reilly, Flat 5, 57 Cheriton Road, Folkestone, CT20 1DF
G6 WGY R. Clague, 11 Trebor Avenue, Bryntirion Park, Flintshire, CH66DP
G6 WGZ D. Collier, 133 Woodstock Road, Moston, Manchester, M40 0DG
G6 WHH S. Martin, 19 Old Manor Road, Rustington, Littlehampton, BN16 3QU
G6 WHS N. Read, 296 Westdale Lane, Mapperley, Nottingham, NG3 6EU
G6 WHT K. Willard, 5 Waltham Way, Frinton-on-Sea, CO13 9JE
G6 WHY K. Daniels, Greenacre, 71 Little Yeldham Road, Halstead, CO9 4LN
GI6 WHZ R. Freeburn, 6 Killycurragh Road, Cookstown, BT80 9LB
G6 WIG G. Crowton, 64 Atlantic Road, Birmingham, B44 8LQ
G6 WIL G. Wilden, 39 The Laurels, Morris Avenue, Jaywick, CO15 2JN
G6 WIO B. Mchugh, 63 Three Butt Lane, Liverpool, L12 7HE
G6 WIT G. Anderton, 12 Oaklands Close, Halvergate, Norwich, NR13 3PP
G6 WJD J. Dobson, 13 Elgin Close, Bedlington, NE22 5HJ
G6 WJJ A. Kendal, 3 Benbeck Grove, Tipton, DY4 8AJ
G6 WJW H. Hutton, Cassiobury, The Street, Diss, IP22 2PS
G6 WJX E. Jackson, Melford, 36 Ickleton Road, Cambridge, CB22 4RT
G6 WKI R. Lewis, 42 Launceston Close, Romford, RM3 8HQ
G6 WKN C. Reed, 14 Fletcher Drive, Wickford, SS12 9FA
G6 WKO J. Richards, 8 Westminster Crescent, Burn Bridge, Harrogate, HG3 1LY
G6 WKQ P. Rowe, 131 Cambridge Road, Great Shelford, Cambridge, CB22 5JJ
GW6 WKU W. Walker, 18 Parc Sychnant, Conwy, LL32 8SB
G6 WKZ M. Jaques, 3 The Rowans, Baldock, SG7 6HJ
G6 WLE R. Bailey, The Malt House, Great Shefford, Hungerford, RG17 7ED
GM6 WLJ D. Milne, 30 Bruceland Road, Elgin, IV30 1SF
G6 WLM S. Simmonds, 3 Robert Cramb Avenue, Tile Hill, Coventry, CV4 9LA
G6 WLP G. Smith, High Croft, 91 Rannerdale Drive, Whitehaven, CA28 6JZ
G6 WLQ M. Smith, 10 Riffams Court, Riffams Drive, Basildon, SS13 1BQ
G6 WLX A. Davey, Highdale, 82 Silver Street, Nailsea, BS48 2DS
GM6 WMA D. Elam, Achnacree, 38 Hunter Avenue, Loanhead, EH20 9SN
G6 WME B. Gray, 21 Litester Close, North Walsham, NR28 9JA
G6 WMG D. Hastings, Westering, Norwich, NR13 6RQ
G6 WML J. Barrasford, 34 Barnard Avenue, Ludworth, Durham, DH6 1LS
G6 WMR West Midlands County Raynet, c/o J. Barnett, 11 Ridge Street, Stourbridge, DY8 4QF
G6 WMT B. Roper, 3 Whites Close, St Agnes, TR5 0TU
G6 WMU D. Pearce, 247 Wigston Lane, Aylestone, Leicester, LE2 8DJ
GJ6 WMZ M. L'Amy, Tamarind, Le Mont De St. Anastase, St Peter, Jersey, JE3 7ES
G6 WNB G. Bennett, 6 Danescroft, Bridlington, YO16 7PZ
G6 WNG J. Haines, The Westlands, Wilcott, Shrewsbury, SY4 1BJ
GM6 WNX D. Mitchell, 65 Robb Place, Castle Douglas, DG7 1LW
GW6 WOB H. Stevens, Parc Y Dilfa, Talley, Llandeilo, SA19 7YT
GM6 WOF A. Firth, Edan, Berstane Road, Kirkwall, KW15 1NA
G6 WOI G. Flint, 782 College Road, Birmingham, B44 0AL
G6 WOT D. Fishlock, 93 Shackstead Lane, Godalming, GU7 1RL
G6 WPD V. Demicoli, 37 Elm Rd, Birmingham, B30 2AX
G6 WPE S. Mason, 46 Frankton Close, Redditch, B98 0HJ
G6 WPJ M. Phillips, Woodside, Bures, CO8 5BN
G6 WPK J. Puttock, 8 Millfield, St. Margarets-At-Cliffe, Dover, CT15 6JL
G6 WPL S. Lawson, 33 Country Meadows, Market Drayton, TF9 3LP
G6 WPO A. Brislin, Greengage, Plough Road, Droitwich, WR9 7NL
G6 WPR D. Fleetwood, The Watch House, Cadgwith, Ruan Minor, Helston, TR12 7JX
G6 WQH J. Wilson, 127 James Reckitt Avenue, Hull, HU8 7TJ
GW6 WQJ A. Tidswell, 9 Dewi Avenue, Holywell, CH8 7UG
G6 WQN W. Convery, 20 Grove Road, Hethersett, Norwich, NR9 3JP
G6 WRC Warrington ARC, c/o J. Lang, 7 Marion Grove, Liverpool, L18 7HY
GM6 WRY G. Smith, 41 Glebe Place, Galashiels, TD1 3JW
G6 WSF M. Strickland, 25 Coniston Drive, Aylesham, Canterbury, CT3 3HZ
G6 WSN D. Westgate, 72 Bosworth Street, Leicester, LE3 5RA
G6 WSX W. Carter, 49 The Oval, Holmfirth, HD9 3ET
G6 WSZ J. O'Hara, 4 Lower Mill Close, Goldthorpe, Rotherham, S63 9BY
G6 WTD R. Kenward, The Bungalow, 20 Church Road, Coventry, CV8 3ET
GW6 WTK B. Wiegold, 8 Nant Ddu, Caerphilly, CF83 3BU
G6 WTM M. Higlett, 3 Clover Way, Killinghall, Harrogate, HG3 2WE
GM6 WTT D. Anderson, 11 Longside Road, Mintlaw, Peterhead, AB42 5EJ
G6 WUD R. Green, 10 Torwood Court, Cramlington, NE23 2BZ
G6 WUR T. Price, 54 Medeway, Lake, Sandown, PO36 9HQ
GW6 WVD J. Williams, 48 Belvedere Drive, Plas Coch, Wrexham, LL11 2BG
G6 WVL J. Parr, 114 Ashton Road, Golborne, Warrington, WA3 3UX
G6 WVM D. Harrison, 22 Oswin Grove, Coventry, CV2 5GJ
G6 WVO Dr C. Hunt, Greenfield House, Heapham, Gainsborough, DN21 5PT
G6 WVR S. Worner, 1 Tynedale, Hull, HU7 6EL
G6 WVS P. Child, 36 Crosslands, Caddington, Luton, LU1 4ER
G6 WWA T. Banham, 28 Norwood Avenue, High Lane, Stockport, SK6 8BJ
G6 WWR Three Counties ARC, c/o D. Kamm, Delabole Head, Week St. Mary, Holsworthy, EX22 6UU
G6 WWS B. Smith, 17 Thornley Road, Wirral, CH46 6HB
G6 WWV P. Mann, 9 Holcombe Road, Blackpool, FY2 0SR
G6 WWY G. Miller, Silvermine, Cooks Lane, Axminster, EX13 5SQ
G6 WXI G. Ball, Ciss Green Farm, Watery Lane, Congleton, CW12 4RS
G6 WXJ L. Bagnall, 15 Ypres Road, Allestree, Derby, DE22 2NA
G6 WXK I. Buckie, 156 Greenfield Crescent, Horndean, Waterlooville, PO8 9EW
G6 WXM A. Burt, 17 Western Road, Wolverton, Milton Keynes, MK12 5AY
G6 WXN C. Bennett, 12 Sherwood Road, Winnersh, Wokingham, RG41 5NJ
G6 WXS R. Archer, 40 Caroline Street Preston, Preston, PR1 5UY
G6 WXZ A. Collier, 2 Viceroy Court Gordon Road, Horndon-On-The-Hill, Stanford-le-Hope, SS17 8NL
G6 WYD S. Chambers, 52 Chapel Lane, Spondon, Derby, DE21 7JW
G6 WYE A. Clack, 1 Meadow View, Southwell, NG25 0EQ
G6 WYF R. Cook, Arnel Ltd, Arnel House, 1 Peerglow Centre, Ware, SG12 9QL
G6 WYH N. Daniels, 2 Homelye Lane, Dunmow, CM6 3AW
G6 WYL J. Williamson, 5 Frensham Close, Stanway, Colchester, CO3 0HP
G6 WYQ S. Quade, 60 Carlton Mews, Birmingham, B36 0AD
G6 WYS W. Patching, 7 Bursledon Road, Hedge End, Southampton, SO30 0BP
G6 WZA D. Wickens, Auchensail, 3 Bews Lane, Chard, TA20 1JU
G6 WZC D. Slatter, 5 Opendale Road, Burnham, Slough, SL1 7LY
G6 WZD C. Sillence, 104 Coleford Bridge Road, Mytchett, Camberley, GU16 6DT
G6 WZE P. Robinson, 108 Station Road, Mickleover, Derby, DE3 9FP
G6 WZL B. Walker, 81 Stacey Avenue, Wolverton, Milton Keynes, MK12 5DN
G6 WZM H. Collinson, 28 Tadcaster Avenue, Leicester, LE2 9GA
G6 WZN M. Hodges, 2 Coral Avenue, Westward Ho, Bideford, EX39 1UW
G6 WZP D. Rogers, No. 10 The Saltings, Seaton, EX12 2XW
G6 WZY A. Gerrard, 51 Sheringham Drive, Crewe, CW1 3XJ
G6 WZZ B. Gibson, 55 Ledward Street, Winsford, CW7 3EN
G6 XAG A. Higgs, Neatsfold, Hilton, Blandford Forum, DT11 0DQ
G6 XAK C. Harding, 15 The Stampers, Tovil, Maidstone, ME15 6FF
G6 XAN S. Harding, 29 Wey Barton, Byfleet, West Byfleet, KT14 7EF
G6 XAR L. Hall, 170 Macers Lane, Wormley, Broxbourne, EN10 6EE
G6 XAT D. Armstrong, 69 Station Crescent, Rayleigh, SS6 8AR
G6 XAV G. Lawrence, 5 Longwood View, Furnace Green, Crawley, RH10 6PB
G6 XAW D. Lawrence, 2205 Southwest 44Th Terrace Cape Coral, Florida, 33904, 33904
G6 XBD S. Meakin, 25 Derby Road, London, E18 2PZ
G6 XBG J. Lines, 6 Hawthorn Road, Denmead, Waterlooville, PO7 6LJ
G6 XBS J. Newman, 21 Stains Close, Cheshunt, Waltham Cross, EN8 9JJ
GW6 XBV K. Simpson, 6 New Market Street, Usk, NP15 1AT
G6 XCC J. Sayer, 19 Arras Boulevard, Hampton Magna, Warwick, CV35 8TY
G6 XCD E. Ashworth, 232 Clifton Road, Darlington, DL1 5EA
G6 XCK S. Bishop, 1 Walsh Close, Hitchin, SG5 2HP

G6 XCM M. Paul, Hunters Cottage, Church Road, Wreningham, Norwich, NR16 1BA
G6 XCO R. Piper, 3 The Haven, Langley Park, Durham, DH7 9UW
G6 XCU R. Willis, 10 Nayling Road, Braintree, CM7 2RZ
G6 XD J. Taylor, 14 Woodway Close, Teignmouth, TQ14 8QG
G6 XDB J. Woodnutt, 17 Hill Farm Road, Chalfont St. Peter, Gerrards Cross, SL9 0DD
G6 XDI C. Packman, 4 Angel Lane, Hayes, UB3 2QX
G6 XDK V. Oag, Parkside, Stratton Park, Biggleswade, SG18 8QS
G6 XDY K. Gibson-Ford, 123 Hawthorn Crescent, Cosham, Portsmouth, PO6 2TJ
G6 XDZ N. Glover, 21A Jason Close, Bridlington, YO16 6JA
G6 XEB D. Green, 47 Siston Common, Bristol, BS15 4PA
G6 XEF P. Hammond, 31 Honey Way, Royston, SG8 7ES
G6 XEL D. Hawkins, Travellers Lodge, Bere Road, Wareham, BH20 7PA
G6 XEN R. Hill, 114 Moorside Crescent, Sinfin, Derby, DE24 9PT
G6 XEX A. Croft, Exchange Buildings, Exchange Street, Normanton, WF6 2AA
G6 XFB B. Roe, 11 Abbotts Way, Louth, LN11 8BS
G6 XFR F. Fielder, 103 Acworth Court, Acworth Crescent, Luton, LU4 9JE
G6 XFU A. Edge, 1 Newquay Drive, Macclesfield, SK10 3NQ
GW6 XGA D. Collins, 12 Penybedd, Pembrey, Burry Port, SA16 0HJ
G6 XGF D. Cadman, 32 Breedon Hill Road, Derby, DE23 6TG
G6 XGJ J. Davis, 446 Upper Wortley Road, Scholes, Rotherham, S61 2SS
G6 XGK M. Drinkall, 11 Rossefield Gardens, Bramley, Leeds, LS13 3RQ
G6 XGT M. Thornsby, 2 Shelley Way, Bacton, Stowmarket, IP14 4TP
G6 XGV M. Valenti, 545 Gander Green Lane, North Cheam, Sutton, SM3 9RF
G6 XHF S. Richards, 58 Holm Lane, Oxton, Prenton, CH43 2HS
GD6 XHG E. Rixon, 65 Friary Park Road Ballabeg, Castletown, Isle of Man, IM9 4EP
G6 XHI K. Ridgwell, 23 Peter Bruff Avenue, Clacton-on-Sea, CO16 8UA
G6 XHJ P. Raxworthy, 32 St. Marys Avenue, Alverstoke, Gosport, PO12 2HX
G6 XHK I. Roper, 109 Birstall Pk Ct., Birstall, WF17 9DL
G6 XID S. Mann, 3 Home Close, Wrington, BS40 5PX
G6 XIF C. Milton, 31 Morley Road, Tiptree, Colchester, CO5 0AA
G6 XII J. Miller, The Oast House, Houghton Green Lane, Rye, TN31 7PJ
G6 XIR M. Bennett, Ravenswood, The Shires, Southampton, SO3 4BA
G6 XIW D. Bye, 2 Valentine Mansions, London, N21 1BA
G6 XJB D. Wratten, 42 North Road, Petersfield, GU32 2AX
G6 XJC L. Whitehead, Flat 2, Masons Court, Clacton-on-Sea, CO15 3SE
G6 XJD D. Whysall, Christ Church Vicarage, 587 Nuthall Road, Nottingham, NG8 6AD
G6 XJE Rev. J. Whysall, Christ Church Vicarage, 587 Nuthall Road, Nottingham, NG8 6AD
G6 XJF A. Webb, 255 Bambury Street, Stoke-on-Trent, ST3 5QY
G6 XJI D. Wiblin, 98 Pemberton Road, Slough, SL2 2JY
G6 XJJ S. Mckay, 11 Brough Meadows, Catterick, Richmond, DL10 7LQ
G6 XJN G. Valenti, 31 Stratton Court, Bognor Regis, PO22 8DP
G6 XJT D. Ramsden, 76 Brigg Lane, Camblesforth, Selby, YO8 8HD
G6 XJZ D. Rowe, 5 Kelburn Close, Chandler'S Ford, Chandlers's Ford, SO53 2PU
G6 XKE H. Papworth, 339 Gayfield Avenue, Brierley Hill, DY5 3JE
G6 XKF A. Parfitt, 242 Hook Road, Chessington, KT9 1PL
G6 XKJ I. Pinkard, 10 Westminster Green, Handbridge, Chester, CH4 7LE
G6 XKK H. Parrott, 3 Fox Gardens, Lymm, WA13 9EY
G6 XKO R. Mclellan, 74 Mount Ambrose, Redruth, TR15 1QR
G6 XKV D. Bodenham, High Dale, Besbury, Minchinhampton, Stroud, GL6 9EP
G6 XKX R. Newell, 57 Evendene Road, Evesham, WR11 2QA
G6 XKY G. Ogden, 10 Hartington Drive, Standish, Wigan, WN6 0UA
G6 XLB R. Morris, 10 Danetre Drive, Daventry, NN11 4GY
G6 XLC J. Mills, 6 Borrowdale Road, Halfway, Sheffield, S20 4HL
G6 XLG P. Pulley, 7 St. Peters Close, Pirton, Worcester, WR8 9EH
G6 XLR S. Nightingale, Cefn Ydfa, Bartwood Lane, Ross-on-Wye, HR9 5TA
G6 XMA S. Butler, 45 Roewood Close, Holbury, Southampton, SO45 2JT
G6 XMB T. Betts, 3 Burns Avenue, Mansfield Woodhouse, Mansfield, NG19 9JR
G6 XML W. Barnes, 49 Sunningdale Road, Haydon Wick, Swindon, SN25 3AZ
G6 XMM T. Bugg, Gravel Hill, Nayland, Colchester, CO6 4BJ
G6 XMT M. Samson, 115A Far Gosford Street, Coventry, CV1 5EA
G6 XMU G. Smith, 71 Mount Pleasant Road, Wisbech, PE13 3NQ
G6 XN Wey Valley ARG, c/o A. Vine, Hilden, Woodland Avenue, Cranleigh, GU6 7HZ
G6 XND P. Smith, 6 Nuthatch, Longfield, DA3 7NS
G6 XNI A. Taylor, 20 Mythop Road, Marton, Blackpool, FY4 4UZ
G6 XNJ J. Taylor, 21 Crestfield Crescent, Elland, HX5 0LS
G6 XNK J. Theedom, 5 Rodbridge Drive, Southend-on-Sea, SS1 3DF
G6 XNN E. Townsend, 10 Little Oak Avenue, Kirkby-In-Ashfield, Nottingham, NG17 9BG
G6 XNP A. Trett, 236 Avondale, Ash Vale, Aldershot, GU12 5NQ
G6 XNQ R. Taylor, 53 Hutton Park, Hutton Moor Lane, Weston-Super-Mare, BS24 8RZ
G6 XNU V. Williams, 24 Sunny Bank Avenue, Blackpool, FY2 9EQ
G6 XOD E. Whitby, 37 Regeneration Way, Beeston, Nottingham, NG9 1NJ
G6 XOE F. Whitby, 37 Regeneration Way, Beeston, Nottingham, NG9 1NJ
G6 XOG C. Wells, Troutbeck, Arthington Lane, Pool In Wharfedale, Otley, LS21 1JZ
G6 XOR D. Winfield, 1 Underhill Close, Derby, DE23 1RH
G6 XOU H. Yeldham, 19 Wade Reach, Walton on the Naze, CO14 8RG
G6 XOX A. Patrick, 22 Falcon Way, Dinnington, Sheffield, S25 2NY
G6 XPB R. Partner, 22 Moordale Avenue, Priestwood, Bracknell, RG42 1RT
G6 XPF G. Love, 8 Scotts Way, Tunbridge Wells, TN2 5RG
G6 XPY R. Chappell, 17 Redcar Avenue, Hereford, HR4 9TJ
G6 XPZ A. Carter, 28A Smithwell Lane, Heptonstall, Hebden Bridge, HX7 7NX
G6 XQB R. Carter, 56 Main Road, Naphill, High Wycombe, HP14 4QB
G6 XQO P. Gait, 6 Martindale Road, Churchdown, Gloucester, GL3 2DW
G6 XQP G. Garner, Tredore, Haugh Road Banham, Norwich, NR16 2DE
G6 XQR F. Gizzi, 19 Kings Field, Bursledon, Southampton, SO31 8EN
G6 XQT N. Godwin, 9 Broadway, Barnsley, S70 6QQ
G6 XQY J. Griffin, 6 Heathfield, Royston, SG8 5BW
GW6 XRE B. Helsdon, Bryn Hedd, Cyffylliog, LL15 2DW
G6 XRF J. Hicks, Flat 213, Enterprise House, 112 Kings Head Hill, London, E4 7ND
G6 XRH J. Hoare, 8 Sunnyheath, Havant, PO9 3BW
G6 XRI S. Hobbs, 19 Ashfield Road, Kenilworth, CV8 2BE
G6 XRK M. Huggins, Black Firs, Pinewood Road, Iver, SL0 0NJ
G6 XRS Leicester Radio Society, c/o P. Taylor, 104 Winstanley Drive, Leicester, LE3 1PA
G6 XRY G. Kobiela, 61 Earith Road, Willingham, Cambridge, CB24 5LS
G6 XSB M. Dower, 19 Fullwell Court, Fullwell Avenue, Ilford, IG5 0RZ
G6 XSC I. Denison, 5 Hazelwood Close, Cheltenham, GL51 5RX
G6 XSK E. Firth, 2 Gladstone Close, Littlemoor, Weymouth, DT3 6RH
G6 XSL C. Franklin, Troy Cottage, Hyde Heath, Amersham, HP6 5RW
G6 XSS B. Gell, 27 Park Road, Barnstone, Nottingham, NG13 9JF
G6 XSY J. Goodey, 62 Rose Hill, Binfield, Bracknell, RG42 5LG
G6 XSZ D. Graham, 127 Shephall View, Stevenage, SG1 1RP
G6 XTC A. Tripp, 3 Ash Close, Oathills, Malpas, SY14 8JB
G6 XTD R. Hallsworth, 27 Westfield Avenue, Heanor, DE75 7BN
G6 XTG B. Haynes, 6 Epping Walk, Furnace Green, Crawley, RH10 6LX
G6 XTJ R. Harris, 88 Earles Meadow, Horsham, RH12 4HR
G6 XTK D. Harris, Claws Cottage, Crablands, Chichester, PO20 9AY
G6 XTT R. Holgate, 5 Exley Gardens, Halifax, HX3 9EE
G6 XTZ T. Jarvis, 1 Whitehall Avenue, Mirfield, WF14 0AQ
G6 XUD S. Justin, Garth, Park View Road, Pinner, HA5 3YF
G6 XUV D. Lee, 188 Manstone Avenue, Sidmouth, EX10 9TJ
G6 XUX R. Mettam, 12 School Lane, Marsh Lane, Sheffield, S21 5RS
G6 XVH G. Allen, 68 Hawthorn Avenue, Armthorpe, Doncaster, DN3 2ET
G6 XVQ P. Braybrooke, 6 Tubbenden Lane, Orpington, BR6 9PN
G6 XVY T. Barker, 2 The Beeches Chapel Lane, Overton, Morecambe, LA3 3HU
G6 XVZ A. Barker, 33 Willoughby Avenue, Kenilworth, CV8 1DG
GM6 XW A. Winton, 2 Castlehill Cottages, Brisbane Glen Road, Largs, KA30 8SN
G6 XWD C. Breckons, Low Wood Farm, Lamonby, Penrith, CA11 9SS
G6 XWM R. King, 52 Ford Road, Tiverton, EX16 4BE
G6 XWY D. Clarke, 10 Dorchester End, Colchester, CO2 8AR
G6 XX RSGB Contest Club, c/o N. Totterdell, Moscar Cross House, Hollow Meadows, Sheffield, S6 6GL
G6 XXB D. Cook, Stepping Stones, 31 Vicarage Hill, Paignton, TQ3 1NH
G6 XXE S. Crowther, 17 Carr Gate Crescent, Carr Gate, Wakefield, WF2 0QR
G6 XXJ D. Clubley, 37 Appleton Road, Beeston, Nottingham, NG9 1NE
G6 XXL G. Carter, Rivendell, North Reston, Louth, LN11 8JD
G6 XXN A. Clarke, 138 High Street, Barwell, Leicester, LE9 8DR
GW6 XXY K. Dobson, 152 Foryd Road, Kinmel Bay, Rhyl, LL18 5LS
G6 XYD K. Elsworth, 88 Mungo Park Way, Orpington, BR5 4EQ
G6 XYF R. Ediss, 5 Stirling Crescent, Totton, Southampton, SO40 3BN
G6 XYL J. Luxton, 2 Trinity Court, Westward Ho, Bideford, EX39 1LT
G6 XYO J. Fazey, 90 Beecher Road, Halesowen, B63 2DW
G6 XYR J. Scothern, 24 Cavendish Crescent, Kirkby-In-Ashfield, Nottingham, NG17 9BN
G6 XYS A. Searle, 22 Crowther Close, Southampton, SO19 1BX
G6 XYU J. Stanton, Waters & Stanton Plc, 22 Main Road, Hockley, SS5 4QS
G6 XYV E. Strode, 26 Churchill Close, Congleton, CW12 4QU
G6 XYX B. Slater, 47 Broom Road, Lakenheath, Brandon, IP27 9EZ
G6 XZA M. Scott, 28 Penwarden Way, Bosham, Chichester, PO18 8LF
G6 XZC C. Shaw, 17 South Street, Pilsley, Chesterfield, S45 8BQ
G6 XZM C. Smith, 104 Warren Road, Banstead, SM7 1LB
G6 XZP R. Sammons, 42 Woodcote Avenue, Wallington, SM6 0QY
G6 XZS J. Thorn, 20 Kiln Road, Shaw, Newbury, RG14 2HA
G6 YAH C. Wheeler, 11 Brooklands Way, Redhill, RH1 2BN
G6 YAI I. Wilson, 2 Kingswood Close, Owlthorpe, Sheffield, S20 6SD
G6 YAK P. Willetts, 49 Summervale Road, Hagley, Stourbridge, DY9 0LX
G6 YAQ F. Barker, 13 Ashbourne Road, Eccles, Manchester, M30 0HW
G6 YAR R. Porteus, 22 North View, Meadowfield, Durham, DH7 8SQ
G6 YAS M. Brashill, 42 Bannister Street, Withernsea, HU19 2DT
G6 YB City Bristol Gr, c/o D. Bailey, 70A Park Road, Staple Hill, Bristol, BS16 5LG
G6 YBC D. Anderson, 142 Tyldesley Road, Atherton, Manchester, M46 9AB
G6 YBH A. White, 85 Goddard Way, Saffron Walden, CB10 2EB
G6 YBV S. Hunt, 33 Rutland Street, Ashton-under-Lyne, OL6 6TX
G6 YCA J. Cooper, 17C Suttons Lane, Deeping Gate, Peterborough, PE6 9AA
G6 YCE A. Brooke, 14 Counting House Road, Disley, Stockport, SK12 2DB
G6 YCF M. Bartlett, 206 Victoria Road, Romford, RM1 2NP
G6 YCG A. Bennett, 10 Burlease, Chippenham, SN15 2AY
G6 YCI M. Buck, 178 Rover Drive, Castle Bromwich, Birmingham, B36 9LL
G6 YCL M. Banner, 7 Lowdham Road, Gedling, Nottingham, NG4 4JP
G6 YCM R. Brookes, 52 Larch Grove, Kendal, LA9 6AU
G6 YCN R. Brassington, Above Park Farm, Leek Road, Stoke on Trent, ST10 2PT
G6 YCO J. Baddeley, 52 Stephens Way Bignall End, Stoke-on-Trent, ST7 8PL
GW6 YCT M. Le Ves Conte, 74 Glan Road, Aberdare, CF44 8BW
G6 YCW B. Lancaster, 1 Belgrave Close, Dodleston, Chester, CH4 9NU
G6 YCZ J. Massey, 10 Rapley Avenue, Pulborough, RH20 4QL
G6 YDN J. Mountain, 15 Eldon Close, Chapel En le Frith, SK23 0PX
G6 YDO F. Mirams, 5 Shaftesbury Avenue, Cheadle Hulme, Cheadle, SK8 7DB
G6 YDP A. Gallagher, 1A Wynsome Street, Southwick, Trowbridge, BA14 9RB
GW6 YDT E. Gittins, 40 Melyd Avenue, Prestatyn, LL19 8RN
G6 YEA N. Guy, 43 Hereford Road, Bolton, BL1 4NJ
G6 YEK D. Heard, 103 Moorland Road, Weston-Super-Mare, BS23 4HU
G6 YEY R. Hope, 26 Chaucer Avenue, Andover, SP10 3DS
G6 YFF G. Hunter, 57 The Cedars, Hailsham, BN27 1TU
G6 YFG S. Lles, 3 Petersway Gardens St. George, Bristol, BS5 8TA
G6 YFH R. Ingle, 48 Barlborough Road Clowne, Chesterfield, S43 4RF
G6 YFL H. Jones, 15 Bonchurch Walk, Manchester, M18 8BP
G6 YFY I. Pitfield, 27 Winchester Crescent, Fulwood, Sheffield, S10 4ED
G6 YFZ D. Paul, Enfield, Gunton Road, Wymondham, NR18 0QP
G6 YGB R. Preston, 188 Dumers Lane, Radcliffe, Manchester, M26 2GF
G6 YGC A. Pilkington, 1 Woodshaw Grove Worsley, Manchester, M28 7XX
G6 YGH A. Richardson, 9 Webbers Way, Puriton, Bridgwater, TA7 8AS
GW6 YGI B. Rogers, Fronucha, Rhewl, Oswestry, SY10 7AS
G6 YGJ R. Robinson, 2 Badminton Close, Martongate, Bridlington, YO16 6GD
G6 YGP D. Lee, 4 Blythe Cottages, Blythe Lane, Ormskirk, L40 5UA
G6 YGV M. Lane, Harewood Villa, Harewood Place, Halifax, HX2 7PN
GM6 YGW B. Finch, Anchor Cottage, Lybster, KW3 6AS
G6 YHE G. Mander, 70 Copthall Way, New Haw, Addlestone, KT15 3TU
G6 YHF R. Marchant, 12 Poplar Close, Huntingdon, PE29 7BP
G6 YHK T. Miller, 6 Captains Walk, Falmouth, TR11 4HR
G6 YHL C. Miller, 5 Lodge Lane, Bewsey, Warrington, WA5 0AG
G6 YHP C. Molyneux, 23 Kemp Close, Chatham, ME5 9SP
G6 YHW G. Murly, 1128 Route Du Trieux, Vigne Redonde, France, 24360
G6 YIE S. Forbes, 8 Nutmeg Close, Earley, Reading, RG6 5GX
G6 YII K. Everington, 1 Norfolk Road, Wigston, LE18 4WH

G6 YIJ J. Elford, 7 Cunliffe Road, Stoneleigh, Epsom, KT19 0RJ
G6 YIK N. Drury, 444 Upper Shoreham Road, Shoreham-by-Sea, BN43 5NE
G6 YIO T. Chapman, 17 Trevor Road, Swinton, Manchester, M27 0YH
G6 YIP A. Cohen, 9 Terrace Rd, 9 Terrace Rd, Plymouth Meeting, USA, 19462
G6 YIQ J. Dixon, 8 East View, St. Ippolyts, Hitchin, SG4 7PD
G6 YIS R. Chell, 3 Elderberry Close, Stourport-on-Severn, DY13 8TF
G6 YIU P. Dawson, Ivy Dene, Middle Lane, Wolverhampton, WV8 2BE
G6 YIW W. Gilroy, Little Harewood Farm, Clamgoose Lane, Kingsley, Stoke-on-Trent, ST10 2EG
G6 YJD J. Govier, 111 Pearson Crescent, Wombwell, Barnsley, S73 8SF
G6 YJH A. Haills, The Cherries, Main Road, Chelmsford, CM3 1NR
G6 YJJ P. Hambly, 22C Windsor Road, London, W5 5PD
G6 YJO D. Arscott, 20 Orchid Vale, Kingsteignton, Newton Abbot, TQ12 3YS
G6 YJR J. Angel, 33 Grovewood Close, Chorleywood, Rickmansworth, WD3 5PX
G6 YLA J. Howard, 11 Lightwood, Crown Wood, Bracknell, RG12 0TR
G6 YLB G. Howse, 1 Sutherland Close, Woodloes Park, Warwick, CV34 5UJ
G6 YLD G. Hope, 17 Church Road, Sutton At Hone, Dartford, DA4 9EX
G6 YLN M. Hobbs, 22 Swan Place, Reading, RG1 6QD
G6 YLO P. Hizzey, Borde Neuve, Maurens, France, 31540
G6 YLQ D. Harrop, C/Mariano Aguilo 2A, Edificio Formentera 1, Mallorca, Spain, 7181
G6 YLR K. Harris, 20 Rose Walk, Wicken Green Village, Fakenham, NR21 7QE
G6 YLV J. Cromack, 45 Chelsea Road, Aylesbury, HP19 7BG
G6 YLW T. Cannon, 36 St. Margarets Drive, Wigmore, Gillingham, ME8 0NR
G6 YLX A. Crabtree, 15 Richmond Gardens, Redhill, Nottingham, NG5 8JS
G6 YLZ P. Cornes, 46 Newland Avenue, Stafford, ST16 1NL
GI6 YM City of Belfast YMCA Radio Club, c/o W. Mcaleer, 90 Gortin Park, Belfast, BT5 7EQ
G6 YMA N. Clark, 2 Barleycroft, Stevenage, SG2 9NP
G6 YMD M. Cooke, 22 Durham Close, Grantham, NG31 8RL
G6 YMH C. Hughes, 85 Benson Gardens, Wortley, Leeds, LS12 4LA
G6 YMI A. Harris, 10 Egroms Lane, Withernsea, HU19 2LZ
GW6 YML D. James, 68 Orchard Park, St. Mellons, Cardiff, CF3 0AQ
GW6 YMS P. Humphreys, Tyn Llan, Bodffordd, Llangefni, LL77 7DZ
G6 YMU D. Hutchings, 2 Burghley Avenue, Bishop's Stortford, CM23 4PD
G6 YMY P. Jacques, Caprius, The Parks, Evesham, WR11 8JP
G6 YNL R. Perry, Straight Mile Cottage, Gloucester Road, Rudgeway, Bristol, BS35 3SB
G6 YNT S. Pentecost, 3 Delamare Road Cheshunt, Waltham Cross, EN8 9AP
G6 YNV S. Raddy, 32 Berry Park, Saltash, PL12 6EN
G6 YNW M. Reeves, 17 Newark Avenue, Putnoe, Bedford, MK41 8NX
G6 YOG M. Rutt, 15 Salmons Road, Chessington, KT9 2JE
G6 YOP P. Harding, 54 Manor Road, Stretford, Manchester, M32 9JB
G6 YOR J. Gillott, 132 Racecommon Road, Barnsley, S70 6JY
G6 YOZ J. Addison, 20 Wychwood Rise, Great Missenden, HP16 0HB
GW6 YPA M. Attfield, 16 Rhodfar Eos, Cwmrhydyceirw, Swansea, SA6 6TF
G6 YPD S. Aldridge, Flat 7, Rosie Court, Newnham Street, Ely, CB7 4PQ
G6 YPF J. Armstrong, 14 Rickwood Park, Horsham Road, Dorking, RH5 4PP
G6 YPJ J. Brown, 30 The Avenue Brookville, Thetford, ip26 4rf
G6 YPK A. Bradbury, 20 Adrden Close, Warwick, CV34 5SN
G6 YPM J. Willats, 17 Purcell Road, Crawley, RH11 8XJ
G6 YPY S. Davis, 30 Bonny Wood Road, Hassocks, BN6 8HR
GM6 YQA C. Davies, 2 Sweyn Road, Thurso, KW14 7NW
G6 YQI E. Fletcher-Cowen, 18 Buckingham Avenue Horwich, Horwich, Bolton, BL6 6NR
G6 YQJ D. Fisher, 86 Parsons Lane, Littleport, Ely, CB6 1JS
G6 YQN N. Fox, 32 Westmorland Avenue, Kidsgrove, Stoke-on-Trent, ST7 1AT
G6 YQT S. Forbes, 11 Henfield View, Warborough, Wallingford, OX10 7DB
G6 YQU R. Fuller, The New House, Main Street, Lutterworth, LE17 6NT
G6 YQW J. Taylor, 7 Caddick Road, Birmingham, B42 2RL
G6 YRB J. Stewart, 107 Turnberry, Skelmersdale, WN8 8EG
G6 YRC A. Smith, 4 Wesley Grove, Burnley, BB12 0JJ
GM6 YRH A. Smith, Robsland, Strathaven Road, Lanark, ML11 0HY
G6 YRI S. Sizmur, 38 Longbourne Way, Chertsey, KT16 9ED
G6 YRJ T. Simmons, 3 West Hill Place, Brighton, BN1 3RU
GM6 YRN A. Stewart, 6 Lawers Place, Aberfeldy, PH15 2BE
G6 YRV D. Bedford, 28 Durfold Drive, Reigate, RH2 0QA
G6 YRY R. Bearchell, 81 Leaves Green Road, Keston, BR2 6DG
G6 YSB J. Bates, 16 Harewood Avenue, Great Barr, Birmingham, B43 6QE
G6 YSL S. Watts, 15 Churchill Way, Northam, Bideford, EX39 1DF
G6 YSN K. Ward, 8 Hinckley Road, St Helens, WA11 9HU
G6 YSO P. Wayer, 4 Chatburn Avenue, Waterlooville, PO8 8UB
G6 YSQ S. Tricker, 1 Drewitt Court, 75 Godstow Road, Oxford, OX2 8PE
G6 YSZ P. Tonge, 1 View Hill, Stalybridge, SK15 2TH
G6 YTB R. Watts, 41 Watford Road, Crick, Northampton, NN6 7TT
G6 YTO R. Cassidy, 9 Langham Way, Ely, CB6 1DZ
G6 YTR R. Broughton, Brookside, Blagdon Terrace, Newcastle upon Tyne, NE13 6EY
G6 YTV A. Black, Redholme, The Street, Thetford, IP25 6NL
G6 YTW A. Bennett, 29 Kennington Road, Kennington, Oxford, OX1 5NZ
G6 YTX R. Burnett, 46 Dorset Waye, Heston, Hounslow, TW5 0ND
G6 YTY A. Bournes, 115 Abbotts Ann Down, Andover, SP11 7BX
GW6 YUC E. Brooksbank, 22 King St., Carmarthen, SA31 1BS
G6 YUX B. Clough, Ashby Powerboat School, 31 Countess Road, Salisbury, SP4 7AS
G6 YUY F. Crockford, 41 Coram Green, Hutton, Brentwood, CM13 1LW
G6 YVD G. Wood, Tethers End, Angarrack Lane, Hayle, TR27 5JF
G6 YVJ C. Ward, 416A Portsmouth Road, Southampton, SO19 9AT
G6 YVS J. Wilson, 36 North Warren Road, Gainsborough, DN21 2TU
G6 YWL A. Griffiths, 45 Clarence Road, Bilston, WV14 6NZ
G6 YWN P. Groom, 2 Alms Road, Doveridge, Ashbourne, DE6 5JZ
G6 YWU D. Harding, 20 D'Arcy Road, Tiptree, Colchester, CO5 0RP
G6 YWV M. Harrison, Barn Cottage, Parwich, Ashbourne, DE6 1QB
G6 YWZ E. Heath-Coleman, 1 Longmead Oakford, Tiverton, EX16 9DW
G6 YXB D. Hewson, Woodwells, 52 Elmham Road, Beetley, Dereham, NR20 4BW
G6 YXO S. Fisher, 37 Elmlands Grove, York, YO31 1ED
G6 YXT B. Evans, C/O Angel 18, Xerta Tarragona, Spain, 43592
G6 YXV K. Faulkner, 5 Tregarrick, West Looe, Looe, PL13 2SD
G6 YXW P. Foulkes, 23 Callowbrook Lane, Rubery, Birmingham, B45 9HW
G6 YXX N. Frederick, 72 Cheltenham Street, Barrow-in-Furness, LA14 5HW
G6 YXY C. Edwards, Seymore, Greenhill Park Road, Evesham, WR11 4NL
G6 YYN K. Mccann, Treverven, Back Lane, Hemingbrough, Selby, YO8 6QP
G6 YYU A. Mutimer, 52 Sycamore Avenue, Wymondham, NR18 0HX
G6 YZB L. Nunn, 103 Bladindon Drive, Bexley, DA5 3BT
G6 YZF S. Alston-Pottinger, 86 Main Street, Walton, Street, BA16 9QN
G6 YZR M. Smith, 5 Derwent Close, North Anston, Sheffield, S25 4GD
G6 YZU L. Nixon, 87 Field Avenue, Canterbury, CT1 1TS
G6 ZAA J. Wellard, 19 South Motto, Kingsnorth, Ashford, TN23 3NJ
G6 ZAC A. Wilson, 21 Lakes Close, Chilworth, Guildford, GU4 8LL
G6 ZAF D. Walker, 27 Daltons Close, Langley Mill, Nottingham, NG16 4GP
GM6 ZAK A. Sutton, 22 St. Michaels Drive, Cupar, KY15 5BS
G6 ZAL S. Ward, 125 Heys Lane, Blackburn, BB2 4NG
G6 ZAM P. Waldron, 12 Lady Drive, Pengegon, Camborne, TR14 7UF
G6 ZAX R. Hollick, 7 Grenfell Road, Bournemouth, BH9 2UD
G6 ZAY R. Hope, 129 Lunedale Road, Dartford, DA2 6JX
G6 ZBL K. Jones, 18 Goldhurst Drive, Tean, Stoke-on-Trent, ST10 4LS
G6 ZBO M. Julians, 29 Trentdale Road, Carlton, Nottingham, NG4 1BU
G6 ZBT D. Green, 6 Garth Villas, Rimswell, Withernsea, HU19 2DB
G6 ZBV A. Higham, 12 Lakenheath Drive, Sharples, Bolton, BL1 7RJ
GW6 ZCR J. Phillips, 39 Bryn Glas, Rhosllanerchrugog, Wrexham, LL14 2EA
GM6 ZCX M. Rochester, Eadar Da' Sloc, Achmelvich, Lairg, IV27 4JB
GM6 ZCY M. Rochester, Eadar Da' Sloc, Achmelvich, Lairg, IV27 4JB
G6 ZDB G. Reddington, 2 South St., Newton, Alfreton, DE55 5TT
G6 ZDE D. Ellingworth, 3 Leighton Park West, Westbury, BA13 3RW
GW6 ZDH R. Roberts, All-Y-Coed, Sychnant Pass Road, Conwy, LL32 8EU
G6 ZDP K. Baum, 25 Lakers Meadow, Billinghurst, RH14 9NP
G6 ZDS R. Baldwin, 24 Keys Court, Banbury, OX16 2AZ
G6 ZDV A. Beales, Broomhill Bungalow, Mappleton Road, Hull, HU11 4UW
G6 ZDY R. Bethell, 5 Dorset Way, Maidstone, ME15 7EL
G6 ZEM J. Hollerbach, 119 Mead End, Biggleswade, SG18 8JU
G6 ZEN J. Homan, 55 Ark Royal, Bilton, Hull, HU11 4BN
G6 ZEQ R. Hubert, 62 Bluebell Woods, Shalloak Road, Broad Oak, Canterbury, CT2 0QB
G6 ZET D. Jackson, 6 Bolton Lane, Carnforth, LA5 8BL
G6 ZEW M. Jennings, 6 Broomroyd Worsbrough, Barnsley, S70 5DU
G6 ZEY K. Johnson, 66 Godwin Way, Cambridge, CB1 8QR
G6 ZEZ C. Jones, 709 Bath Road, Taplow, Maidenhead, SL6 0PB
G6 ZFA J. Justice, 6 Stanley Terrace, Devizes, SN10 5AJ
G6 ZFG P. Wood, 31 Larches Lane, Tettenhall, Wolverhampton, WV3 9PX
GM6 ZFI D. Smith, 39 High Croft, Kelso, TD5 7NB
G6 ZFK E. Toohey, No6 Block E, Peabody Avenue, London, SW1V 4AS
G6 ZFO D. Tate, 73 Sparth Avenue, Clayton Le Moors, Accrington, BB5 5QH
G6 ZFT H. Smithey, 35 Regent Road, Blackpool, FY1 4NB
G6 ZFU C. Stephen, 12 Beaufort Close, Leegomery, Telford, TF1 6XU
G6 ZFV T. Wynne-Jones, 62 Moor Lane, Rickmansworth, WD3 1LQ
G6 ZFX P. Senior, 13 St. Michaels Avenue, Swinton, Mexborough, S64 8NX
G6 ZFZ M. Turner, 1 Bakers Gardens Codsall, Wolverhampton, WV8 1HA
G6 ZG Gorleston ARS, c/o B. Alston-Pottinger, 16 Vincent Close, Great Yarmouth, NR31 0HR
G6 ZGA M. Smith, 6 Norton Crescent, Towcester, NN12 6DN
G6 ZGB C. Salmon, 20 Lime Close, Sandbach, CW11 1BZ
G6 ZGC M. Tann, 11 St. Margarets Grove, Redcar, TS10 2HW
G6 ZGF D. Simpson, 18 Croft House View, Morley, Leeds, LS27 8NS
G6 ZGH R. York, 8 The Rookery Brogden Street, Ulverston, LA12 0DB
G6 ZGI C. Butler, 8 Douglas Walk, Chelmsford, CM2 9XQ
G6 ZGK G. Weston, 2 Gill Park, Efford, Plymouth, PL3 6LX
G6 ZGU J. Brown, 10 Cherry Trail, Coldwater, Ontario, Canada, L0K 1E0
GW6 ZGY R. Bennett, 88 Coychurch Road, Pencoed, Bridgend, CF35 5NA
G6 ZHB J. Booth, 9 New Street, Abingdon, OX14 3PE
G6 ZHF S. Bailey, Silverthorne House, North Piddle, Worcester, WR7 4PR
G6 ZHJ G. Butler, 23 Roman Meadow, Downton, Salisbury, SP5 3LB
G6 ZHL M. Leack, 68 Dale Street, Lancaster, LA1 3AW
GW6 ZHM R. Lannon, 16 Heol Mabon, Rhiwbina, Cardiff, CF14 6RL
G6 ZHO G. Lattin, 5 Seymour Road, Broadfield, Crawley, RH11 9ES
G6 ZHS K. Lupton, Oak Tree Cottage, Post Office Lane, Frodsham, WA6 8JJ
G6 ZHU P. Lightfoot, 7 Fearns Avenue, Newcastle, ST5 8ND
GW6 ZHY A. Mayers, 2 Wyndham Gardens, Wrexham, LL13 9LY
G6 ZIC J. Mccomb, Bridge End Cottage, Bridge End, Hexham, NE48 2RY
G6 ZIO N. Dessau, 20 Coventry Circle, Mahopac, USA, 10541
GI6 ZIR A. Duffy, 81A Arney Road, Bellanaleck, Enniskillen, BT92 2DL
G6 ZIY A. Fairhurst, 16 Waverley Road, Hindley, Wigan, WN2 3BN
G6 ZJD E. Fensome, 77 Church Green Road, Bletchley, Milton Keynes, MK3 6BY
G6 ZJI A. Washby, 57 Cromwell Road, Hedon, Hull, HU12 8GF
G6 ZJK F. Webster, 1 Fir Tree Cottages, Lower Ansford, Castle Cary, BA7 7JY
G6 ZJM D. Leese, 22 Elm Road, Abram, Wigan, WN2 5XG
G6 ZJN R. Williams, 220 Euston Grove, Morecambe, LA4 5LJ
G6 ZJS L. Wright, 17 Drayton St., Alumwell Estate, Walsall, WS2 9QB
G6 ZJV A. Winterbottom, 38 Heaton Avenue Earlsheaton, Dewsbury, WF12 8AQ
G6 ZKC D. Usher, 26 Meneth, Gweek, Helston, TR12 6UW
G6 ZKM J. Cornell, 10 Craneswater Park, Southsea, PO4 0NT
G6 ZKS M. Staniland, 2 Epsom Road, Cantley, Doncaster, DN4 6HX
G6 ZKU B. Sawyers, 36 Frome Road, Bath, BA2 2QB
G6 ZKX R. Smith, Smith Farms, Herne Lane, Dereham, NR19 1QE
G6 ZKY E. Stebbings, 1 Coupland Road, Wootton, Abingdon, OX13 6DU
G6 ZKZ V. Smith, 40 Princess Gardens, Blackburn, BB2 5EJ
G6 ZLD P. Bent, 7 Bandon Rise, Wallington, SM6 8PT
G6 ZLJ M. Adams, 62 Woodlands Road, Holmcroft, Stafford, ST16 1QP
G6 ZLS G. Ashbee, 34 Manorgate Road, Kingston upon Thames, KT2 7AL
GM6 ZLY D. Brasenell, 18 Whitelaw Avenue, Castle Douglas, DG7 1GB
G6 ZMD S. Roberts, 36 Hill Crescent, Dudleston Heath, Ellesmere, SY12 9NA
G6 ZME Telford and District ARS, c/o J. Humphreys, 15 Colemere Drive, Wellington, Telford, TF1 3HH
G6 ZMG G. Mills, 57 Holborough Road, Snodland, ME6 5PA
GW6 ZMN W. Mcdowall, 36 Adenfield Way, Rhoose, Barry, CF62 3EA
G6 ZMO J. Mooney, 2 Madford Lane, Launceston, PL15 9EB
G6 ZMU G. Randall, 15 Bincombe Rise, Weymouth, DT3 6AS
G6 ZMX A. O'Shaughnessy, Southby, Buckland, Faringdon, SN7 8QR
G6 ZNJ A. Reeve, 188 Dorset Avenue, Great Baddow, Chelmsford, CM2 8YY
G6 ZNT I. Cross, 5 Upper Crescent, Minster Lovell, Witney, OX29 0RT

G6 ZNW S. Cascino, 3 Connaught Road, Folkestone, CT20 1DA
G6 ZOB A. Crowther, 16 Linden Avenue, Tuxford, Newark, NG22 0JR
G6 ZOE C. English, 124 Hillside Road, Portishead, Bristol, BS20 8LG
G6 ZOJ A. Buchan, 5 Copythorne Close, Brixham, TQ5 8QG
G6 ZOL P. Lancaster, 134 Wigan Road, Euxton, Chorley, PR7 6JW
G6 ZOT J. Leary, 24 Howard Drive, Old Whittington, Chesterfield, S41 9JU
G6 ZPL P. Manning, 21 Whitethorn Way, Oxford, OX4 6ER
G6 ZPO K. Miller, 47 Hermitage Green Hermitage, Thatcham, RG18 9SL
G6 ZPR A. Morris, 32 New Road, Wonersh, Guildford, GU5 0SE
G6 ZPV N. Mansfield, 2 Little Halt Portishead, Bristol, BS20 8JQ
G6 ZQA G. Nolan, 94 St. Andrews Road, Burgess Hill, RH15 0PH
G6 ZQJ A. Doughty, 42 Thornton Road, Ilford, IG1 2ER
G6 ZQS M. Charlton, 104 Foundry Street, Horncastle, LN9 6AF
G6 ZQU D. Crook, Bedford Road, Sherington, Newport Pagnell, MK16 9NQ
G6 ZRL K. Sellens, 7 The Knapps, Semington, Trowbridge, BA14 6JG
G6 ZRO B. Stoner, Montrose, Wesley Road, Robin Hoods Bay, Whitby, YO22 4RW
G6 ZRS B. Starr, 121 Pretoria Road, Patchway, Bristol, BS34 5PY
G6 ZRV P. Stainton, 9 Park Lane Reepham, Norwich, NR10 4JZ
G6 ZSF Dr D. Neely, 3 Sidestrand Road, Newbury, RG14 6HP
G6 ZSG L. Onions, 8 Prince Charles Close, Rubery, Birmingham, B45 0NB
G6 ZSH P. Owen, 288 Chervil Rise, Wolverhampton, WV10 0HR
G6 ZSQ J. Pepper, Flat 55, Seaward Court, West Street, Bognor Regis, PO21 1XJ
G6 ZSU P. Payton, 11 Hexham Way, Dudley, DY1 2UN
G6 ZTD B. Robinson, 23 Croft Drive, Millhouse Green, Sheffield, S36 9NE
G6 ZTF A. Bowler, 12 Wrenbury Drive, Coventry, CV6 6JZ
G6 ZTH M. Richardson, Better View, Back Lane, Sutton-in-Ashfield, NG17 2LL
G6 ZTL B. Rogers, 24 Marmion Road, Coningsby, Lincoln, LN4 4RG
G6 ZTP G. Down, 8A Abbeville Close, Exeter, EX2 4SJ
G6 ZTR S. Davies, 111 Southend, Garsington, Oxford, OX44 9DL
G6 ZTT OBO Mid-Cheshire Contest Group, c/o M. Baguley, 2 Kensington Way, Northwich, CW9 8GG
G6 ZTZ S. French, 22 Amity Street, Reading, RG13LP
G6 ZUE W. Edwards, 31 Cumberland Avenue, Benfleet, SS7 5NU
G6 ZUO D. Gibson, 14 Lowfield Road, Dewsbury Moor, Dewsbury, WF13 3SR
GW6 ZUS J. Gray, 36 Heol Pentre Felen, Llangyfelach, Swansea, SA6 6BY
G6 ZUV J. Griffin, 35 Cottage Street, Kingswinford, DY6 7QE
G6 ZUZ J. Hampshire, 14 Fellows Road, Cowes, PO31 7JN
G6 ZVB K. Harris123, 14 Dunstall Close, St. Marys Bay, Romney Marsh, TN29 0QX
G6 ZVD M. Hicken, 8, Calle Francisco Villaespesa, Oria, Spain, 4810
G6 ZVL M. Hoskins, 22 Rosedale Gardens, Thatcham, RG19 3LE
G6 ZVO K. Howarth, 79 Eden Avenue, Edenfield, Bury, BL0 0LD
G6 ZVR D. Inskip, 40 Burway Meadow Alrewas, Burton-on-Trent, DE13 7EB
G6 ZVU S. Hughes, 50 Albany Road, Dalton, Huddersfield, HD5 9UW
G6 ZVV N. Hull, 60 Portreath Place, Chelmsford, CM1 4DN
G6 ZWC C. Brown, 16 Old Croft Close, Good Easter, Chelmsford, CM1 4SJ
G6 ZWL C. Wright, 19 Redwood Glen, Chapeltown, Sheffield, S35 1EA
G6 ZWM R. Wade, 104 Brookehowse Road, London, SE6 3TW
G6 ZXN I. Carter, 17 Collingham Close, Templecombe, BA8 0LR
G6 ZXO J. Crowe, 15 Lambert Road, Kendray, Barnsley, S70 3AA
GW6 ZYI B. Jones, 10 Hughes Street, Penygraig, Tonypandy, CF40 1LX
G6 ZYL P. Jones, 274 Cannock Road, Westcroft, Wolverhampton, WV10 8QG
G6 ZYM K. Keeble, Hall Cottage, Hardwick Road, Harleston, IP20 9PU
G6 ZYX G. Spruce, 158 Wolverhampton Street, Wednesbury, WS10 8UB
G6 ZYZ P. Skerritt, Room 2, 157 Hagley Road, Edgbaston, Birmingham, B16 8UQ
G6 ZZ RSGB Transatlantic Test Centenary Club, c/o N. Totterdell, Moscar Cross House, Hollow Meadows, Sheffield, S6 6GL
G6 ZZE P. Read, 11 Fairview Avenue Whetstone, Leicester, LE8 6JQ
GW6 ZZF P. Thomas, 42 Wyndham Road, Abergavenny, NP7 6AF
G6 ZZR D. Wiltshire, 19 Heron Way, Basingstoke, RG22 5QF
G6 ZZS D. Watts, 176 Blatchcombe Road, Paignton, TQ3 2JP

G7

G7 AAI Dr A. Hickey, 11 Barker Road, Wirral, CH61 3XH
GM7 AAJ P. Mcmanus, 59 Mauchline Road, Hurlford, Kilmarnock, KA1 5AB
G7 AAR P. Comben, Undine Cottage, Sedrup, Hartwell, Aylesbury, HP17 8QN
G7 AAS D. Hawkins, 8 Braybrook Street, East Acton, London, W12 0AP
GW7 AAU H. Studdart, 33 Linden Avenue, Connah'S Quay, Deeside, CH5 4SN
GW7 AAV S. Studdart, 33 Linden Avenue, Connah'S Quay, Deeside, CH5 4SN
G7 AAY K. Richardson, 514 Obelisk Rise, Northampton, NN2 8SX
G7 ABE C. Duberley, 2 The Grove, Greenford, UB6 9BY
G7 ABF K. Austin, 6 Boothey Close, Biggleswade, SG18 0DG
G7 ABO Dr D. Bescoby, 40 High Street, Wickham Market, Woodbridge, IP13 0QS
G7 ABQ D. Ferns, 18 Sandelswood End, Beaconsfield, HP9 2AE
G7 ABR A. Clark, Brookside, Milford, Bakewell, DE45 1DX
G7 ABT D. Hepworth, 1 Greengate Crescent Epworth, Doncaster, DN9 1HA
G7 ABZ M. Bromage, 14 Rhuddlan Way, Kidderminster, DY10 1YH
G7 ACA B. Pearce, 39 Fairholme Park, Ollerton, Newark, NG22 9AS
G7 ACD R. Cariss, 6 Granville Avenue, Newport, TF10 7DX
G7 ACG J. Baker, 19 Green Lane, Rugeley, WS15 2AR
G7 ACJ G. Mantle, 6 North Green, Wolverhampton, WV4 4RQ
G7 ACK G. Bromfield, 63 Herondale Road, Mossley Hill, Liverpool, L18 1JZ
G7 ACM S. Pinkney, 39 Butterley Drive, Loughborough, LE11 4PX
G7 ACN C. Hayes, 9 Grenville Way, Thetford, IP24 2JH
G7 ACO R. Horton, 1 Stonehill Rise, Doncaster, DN5 9HD
G7 ACR P. Blakemore, 50 Longley Farm View, Sheffield, S5 7JX
G7 ADF I. Bradbury, 11 St. Stephens Avenue, Wigan, WN1 3UQ
G7 ADH G. Williams, 18 Luther Road, Bournemouth, BH9 1LH
G7 ADP C. Baker, 17 Coronation Road, Illogan, Redruth, TR16 4SG
G7 ADS A. Cresswell, 31 New Street, Doddington, March, PE15 0SP
GM7 ADU Dr L. Morrison, 22 Lodge Park, Kilmacolm, PA13 4PY
G7 ADW G. Laycock, 18 Montague Crescent, Garforth, Leeds, LS25 2EP
GM7 ADY Dr M. Morrison, 22 Lodge Park, Kilmacolm, PA13 4PY
G7 AEA Gloucestershire County Raynet, c/o R. Large, 5 Jasmine Close, Abbeydale, Gloucester, GL4 5FJ
G7 AEC Cheltenham Rynt, c/o P. Kent, 82 Despenser Road, Tewkesbury, GL20 5TW
G7 AEE Tewkesbury Rynt, c/o C. Davis, 38 Courtney Close, Tewkesbury, GL20 5FB
G7 AEF FRST of Dean Ry, c/o G. Harden, 13 Greenfield Road, Coleford, GL16 8BY
G7 AEG Gloucestershire County, c/o A. Ayres, Bryn Hyffryd, Phocle Green, Ross-on-Wye, HR9 7TW
G7 AEH Cotswold Raynet, c/o G. Hayter, 19 Austin Road, Cirencester, GL7 1BT
G7 AEQ R. Murphy, 17 Valley View Road, Paulton, Bristol, BS39 7QB
G7 AES P. Crook, 40 St. Aubins Avenue Brislington, Bristol, BS4 4NX
G7 AEY D. Martin, 12 The Willows, Kemsley, Sittingbourne, me102te
GW7 AFC M. Grant, 11 Golwg Yr Eglwys, Pontarddulais, Swansea, SA4 8EE
GM7 AFE A. Erwood, Lunna House, Lunna, Vidlin, ZE2 9QF
G7 AFL A. Fountaine, 19 Metcalfe Grove, Blakelands, Milton Keynes, MK14 5JY
G7 AFO K. Hollingsworth, 34 Marconi Drive, Yaxley, Peterborough, PE7 3ZR
G7 AFQ K. Marlow, Computer Science, Po Box 363, Edgbaston Birmingha, B15 2TT
G7 AFS M. Sheldon, 5 Runnymede Mews, Faversham, ME13 8RU
G7 AFT K. Brazier, 6 Sadlers Lane, Dibden Purlieu, Southampton, SO45 4LZ
G7 AFV M. Weston, 10 Flete Avenue, Newton Abbot, TQ12 4EH
G7 AFW M. Towers, 44 Ravenscroft Drive, Chaddesden, Derby, DE21 6NX
G7 AFZ D. Payea, 10 Royal Drive, Seaford, BN25 2XW
G7 AGA K. Askew, 3A Craven Drive, Broadheath, Altrincham, WA14 5JF
G7 AGB P. Rothwell, 20 Henbury Close, Corfe Mullen, Wimborne, BH21 3TF
G7 AGC M. Collis, 2 Westwood Avenue, Urmston, Manchester, M41 9NG
GW7 AGG R. Ricketts, 2 Brynystwyth, Penparcau, Aberystwyth, SY23 1SS
G7 AGI D. De Silva, 22 Bishop Road, Bristol, BS7 8LT
G7 AGO J. Lee, 188 Manstone Avenue, Sidmouth, EX10 9TJ
G7 AGR Aylesbury Ryn Group, c/o R. Clark, 9 Conigre, Chinnor, OX39 4JY
GM7 AHA V. Turnbull, 18 Easterfield Court, Livingston Village, Livingston, EH54 7BZ
G7 AHB T. Green, 12 Springfields, Ambrosden, Bicester, OX25 2AH
G7 AHO P. Russell, 27 Main Street, Haconby, Bourne, PE10 0UR
G7 AHP S. Crask, 14 Southfield Road, Paignton, TQ3 2SW
GW7 AHR M. Brett, 106 Trinity Avenue, Llandudno, LL30 2YQ
G7 AIB N. Denton, 41 Monks Dale, Yeovil, BA21 3JB
G7 AIC V. Newman, 35 Netherton Road, Yeovil, BA21 5NY
G7 AIF D. Grevatt, 17 Foxdale Drive, Angmering, Littlehampton, BN16 4HF
G7 AIH R. Whitenstall, 4 Monksmead, Borehamwood, WD6 2LQ
GW7 AIY V. Lamb, 19 Pemba Drive, Buckley, CH7 2HQ
G7 AIZ P. Pugh, 9 Red Kite Close, Gateford, Worksop, S81 8WA
G7 AJE F. Salt, 6 Bodycoats Road, Chandler'S Ford, Eastleigh, SO53 2GX
G7 AJG T. Ellis, 29 St. Annes Road, Clacton-on-Sea, CO15 3NF
G7 AJJ A. Hammond, 5 Durness Close, Kettering, NN15 5BN
G7 AJK A. Knowler, 385 Capstone Road, Gillingham, ME7 3JE
G7 AJN H. Lister, 68 Spring Avenue, Gildersome, Leeds, LS27 7BT
G7 AJP B. Staniforth, 1 Pylon Cottages, Donington-On-Bain, Louth, LN11 9RQ
G7 AJR S. Godfrey, 7 Laburnum Close, North Baddesley, Southampton, SO52 9JT
G7 AJS T. Green, 34 Thorn Close, Kettering, NN16 9BU
G7 AJT M. Carter, 17 Ash Crescent, Higham, Rochester, ME3 7BA
G7 AJX R. King, 31 Lambert Road, Sprowston, Norwich, NR7 8AA
G7 AKI P. Shambrook, 7 The Close, Cheltenham, GL53 0PQ
G7 AKJ W. Wrench, 2 Maunders Place, Otterton, Budleigh Salterton, EX9 7JE
G7 AKM D. Pearson, 8 Walnut Way, Swanley Kent, BR8 7TW
G7 AKP Y. Branch, 38 Kynaston Road, Didcot, OX11 8HD
G7 AKV E. Grantham, 18 Fen End Lane, Spalding, PE12 6AD
G7 ALC L. Civita, 53 Rockhurst Drive, Eastbourne, BN20 8XD
GI7 ALH J. Bailie, 42D John Street Lane, Newtownards, BT23 4LY
GM7 ALI A. Crighton, Tighnacreag, Pacemuir Road, Kilmacolm, PA13 4JJ
G7 ALR P. Goodman, 85 Rantree Fold, Basildon, SS16 5TW
G7 AMD G. Blakemore, 6 Pine Tree Close Hednesford, Cannock, WS12 4JT
G7 AMQ J. Morstatt, 32 Elwy Circle, Ash Green, Coventry, CV7 9AU
GW7 AMS A. Steel, 74 Caradoc Road, Prestatyn, LL19 7PF
G7 AMW P. Jackson, 4 Abbotsbury, Orton Malborne, Peterborough, PE2 5PS
G7 ANA R. Jackson, 5 Home Farm Court, Hooton Pagnell, Doncaster, DN5 7BL
G7 ANB D. Ball, 16 Kelston View, Whiteway, Bath, BA2 1NW
GM7 ANE W. Jamieson, 20 Cairn Road, Cumnock, KA18 1HN
G7 ANG A. Santagata, 25 Swyncombe Avenue, London, W5 4DR
G7 ANH F. Pattinson, 4 Carlisle Road, Brampton, CA8 1SR
G7 ANK C. Bryan, Beau Rivage, Seaholme Road, Mablethorpe, LN12 2DF
G7 ANO T. Hyde, 10 Castleton Avenue, Riddings, Alfreton, DE55 4AG
G7 ANQ J. Hedges, 31 Meadow Road, Hartshill, Nuneaton, CV10 0NL
G7 ANV S. O'Malley, 140 Allerburn Lea, Alnwick, NE66 2QP
G7 ANY A. King, 31 Springhill, Pennycross, Plymouth, PL2 3QZ
G7 AOA P. Gash, 2 Betjeman Walk, Yateley, GU46 6YP
GW7 AOE G. Williams, 10 Strawberry Place Morriston, Morriston, SA6 7AG
GJ7 AOG C. Eve, Flat 17, Le Petit Hurel, Queen'S Road, St. Helier, Jersey, JE2 3SY
G7 AOK R. Swynford-Lain, 19 Kenton Road, Earley, Reading, RG6 7LQ
GM7 AOM J. Curr, 56 Drygate Street, Larkhall, ML9 2DA
G7 AOQ J. Johnson, 32 Bradlea Rise, Rotherham, S625QJ
G7 AOU S. Wilkins, 15 Roundway, Egham, TW20 8BS
GU7 APA P. Ash, Trigale House, Alderney, Guernsey, GY9 3TZ
G7 APD Rugby Am Tra So, c/o S. Tompsett, 9 Ashlawn Road, Rugby, CV22 5ET
G7 API S. Garlick, 37 Edith Road, Kettering, NN16 0QB
G7 APL S. Bonham, 4 St. Martins Avenue, Studley, B80 7JJ
GW7 APM C. Mockford, Denamby Cottage, King Street, Leeswood, Coed-Llai, CH7 4SB
G7 APO C. Monckton, 219 Chaldon Road, Chaldon, Caterham, CR3 5XN
GW7 APP I. Capon, Pentre Garreg Bach, Marianglas, LL73 8PP
G7 APQ A. Jones, 6 Heatherbreea Gardens, Rushden, NN10 6EH
G7 APS S. Ellison, 16 Beechtree Road, Walsall, WS9 9LS
G7 APU L. Young, 7 Tudor Rose, 28 Northgate, Hunstanton, PE36 6AP
G7 AQA M. Hawkshaw, 5 Carr Hill Road, Calverley, Pudsey, LS28 5PZ
G7 AQD W. Williams, 2 Lightfoot Lane Fulwood, Preston, PR2 3LP
G7 AQF A. Gregory, 13 Combe Avenue, Portishead, Bristol, BS20 6JR
G7 AQK N. Mcgrath, 48 Willersley Avenue, Kent, BR6 9RS
G7 AQL B. Burbage, 9 Westerdale Drive, Frimley, Camberley, GU16 9RB
G7 AQN I. Cooper, Ceylon, 70 Carshalton Park Road, Carshalton, SM5 3SW
GI7 AQO R. Todd, 14 Glencroft Road, Newtownabbey, BT36 5GD
G7 AQV A. Russell, 73 Seymour Road, Newton Abbot, TQ12 2PX
G7 ARF S. Wright, 63 Cambridge Road, St Albans, AL1 5LF
G7 ARJ J. Baber, 130 Lumsden Road, Southsea, PO4 9LR
G7 ARK A. Wright, 3 Wyborn Close, Hayling Island, PO11 9HY
G7 ARP R. Orchard, 31 Chiswick House, Bell Barn Road, Birmingham, B15 2AA
GD7 ARS W. Wrigley, 20 Fairy Hill Close, Ballafesson, Port Erin, Isle Of Man, IM9 6TJ
G7 ART M. Blake, 20 Triumphal Crescent, Plymouth, PL7 4RW

G7 ASF Coventry ARS, c/o J. Beech, 124 Belgrave Road, Coventry, CV2 5BH
GW7 ASL P. Jones, 27 Hawthorn Road East, Llandaff North, Cardiff, CF14 2LR
G7 ASY P. Matkin, 2 Raven Close, Huntington, Cannock, WS12 4TQ
G7 ASZ Dr N. Blair, 19 Church Street, Bourn, Cambridge, CB23 2SJ
G7 ATJ R. Williams, 10 Barton Close, Whippingham, East Cowes, PO32 6LS
G7 ATW G. Johnson, Wood Nook Blennerhasset, Wigton, CA7 3RJ
G7 AUE B. Oubridge, 54 Cantle Avenue, Downs Barn, Milton Keynes, MK14 7QS
G7 AUF K. Gebhardt, 15 Jubilee Road, Corfe Mullen, Wimborne, BH21 3NH
G7 AUP A. White, Tioram, Garthends Lane, Selby, YO8 6QW
GW7 AUQ N. Smith, 7 Lili Mai, Barry, CF63 1DW
G7 AUR S. Davis, 7 Kennedy Crescent, Gosport, PO12 2NL
G7 AUU R. Selwood, 33 Chandlers, Sherborne, DT9 3RT
GM7 AUW J. Milne, 24 Lorne Street, Edinburgh, EH6 8QP
GM7 AUX E. Ramsay, Tighnduin, 2 Queen Street, Dundee, DD5 4HG
GI7 AUY I. Potts, 46 Richmond Park, Omagh, BT79 7SJ
GW7 AVB D. Daymond, 15 Constance Street, Newport, NP19 7DB
G7 AVF P. Honeybone, 30 The Hordens, Barns Green, Horsham, RH13 0PJ
G7 AVU R. Fisk, 25 Cromwell Street, Gainsborough, DN21 1DH
G7 AVZ L. Collings, 37 Armstrong Road, Mansfield, NG19 6HZ
G7 AWG V. Watson, 3 Anderton Rise, Millbrook, Torpoint, PL10 1DA
GM7 AWK D. Easton, 86 Dryburn Road, Kelloholm, Sanquhar, DG4 6SN
G7 AWP T. Noyes, 59 Abbots Leys Road, Winchcombe, Cheltenham, GL54 5QX
G7 AWS D. Henderson, 19 Stuart Road, York, YO24 3AX
G7 AWW M. Gynane, 164 Stockbridge Lane, Huyton, Liverpool, L36 8EH
G7 AXL A. Marchington, 30 Warwick Avenue, Golcar, Huddersfield, HD7 4BX
G7 AXM J. Smith, 7 Ainsworth Court, Cameron Close, Freshwater, PO40 9JH
G7 AXN R. Moxham, 8 Dunroyal Close, Helperby, York, YO61 2NH
G7 AXW Dr A. Parkes, 19 Malt Mill Lane, Halesowen, B62 8JA
G7 AYA G. Jessup, 25 Harrier Green, Holbury, Southampton, SO45 2EY
G7 AYB R. Upton, 35 Weston Street, Swadlincote, DE11 9AT
G7 AYE R. Phipps, 39 Perrinsfield, Lechlade, GL7 3SD
G7 AYI L. Faragher, 4 Kirloe Avenue, Leicester Forest East, Leicester, LE3 3LA
G7 AYL P. Thompson, Flat 11, Old Gaol, 16 Grove Street, Bath, BA2 6PJ
G7 AYO L. Hutt, Fenwick Crossing House, Fenwick Lane, Doncaster, DN6 0EZ
G7 AYP A. Gregory, 9 Fordbridge Road, Ashford, TW15 2TD
G7 AYQ A. Tregay, 53 Haverscroft Close, Taverham, Norwich, NR8 6LT
G7 AYS D. Webster11, 5 Eastfield Road, Princes Risborough, HP27 0JA
GM7 AYW J. Hunter, 1 Mitchell Drive, Rutherglen, Glasgow, G73 3QP
G7 AZA J. Cash, 89 Peacocks, Harlow, CM19 5NZ
G7 AZC A. Rogers, Yoke Farm, Upper Hill, Leominster, HR6 0JZ
G7 AZH V. Barber, 8 Hollyberry Close, Redditch, B98 0QT
G7 AZJ B. Sayers, 13 Mulberry Close, Cambridge, CB4 2AS
G7 AZT V. Mills, 18 Muir Road, Maidstone, ME15 6PX
G7 AZV R. Quick, Halfway House, Upton Scudamore, Warminster, BA12 0AE
G7 AZW M. Ney, 4 Rathen Road, Withington, Manchester, M20 4GH
G7 BAB W. Foden, 209 Lord Lane, Failsworth, Manchester, M35 0PX
G7 BAC M. Gohl, 4 Yewtree Drive, Hull, HU5 5YH
G7 BAE T. Searle, Haven Orchard, Exwick Lane, Exeter, EX4 2AP
GM7 BAS W. Hunter, 2 Wallace Cottages, Southend, Campbeltown, PA28 6RX
G7 BAV A. Roberts, 4 Rocky Park Road, Plymouth, PL9 7DQ
G7 BBC London BBC Radio Group, c/o J. Lee, 44 Howard Road, Nuneaton, CV10 7ES
G7 BBD C. Hartley, 102A Bedford Road, Cranfield, Bedford, MK43 0HA
G7 BBJ B. Jenkinson, 14 Sandhill Way, Harrogate, HG1 4JN
G7 BBN E. Crookall, 17 Dundee St., Moorlands, Lancaster, LA1 3DS
G7 BBT R. Pitman, 21 Old Church Road, St Leonards on Sea, TN38 9HB
G7 BBU Dr C. Sharp, Dept. Of Astronomy, University Of Arizona, Tuscon, USA, 85721
GW7 BBY M. Jones, Awelfa, Llangeler, Llandysul, SA44 5EP
GM7 BCC R. Sutherland, Tigh - Na - Coille, Mill Road, Nairn, IV12 5EW
G7 BCI V. White, 6 Laburnum Close, South Anston, Sheffield, S25 5GL
G7 BCK N. Gillies, 5 Pickmere Terrace, Dukinfield, SK16 4JJ
G7 BCO A. Adey, 8 Spinners Court, Telford, TF5 0PG
GM7 BDD A. Mankin, 10 Kerloch Crescent, Banchory, AB31 5ZF
G7 BDK P. Blackett, 32 Woodstock Road, Carshalton, SM5 3DZ
G7 BDR A. Davis, 5 Ludlow Close, Loughborough, LE11 3TB
G7 BDS J. Mills, 42 Maple Grove, Welwyn Garden City, AL7 1NL
G7 BEJ W. Young, 24 Peebles Road, Newark, NG24 4RW
G7 BEP M. Van Der Steeg, Horsebrook Farm, South Brent, TQ10 9EU
GI7 BET R. Griffin, 19 Jubilee Park, Cookstown, BT80 8LJ
G7 BFE C. Broom, 74A Newton Road, Torquay, TQ2 7BN
G7 BFH N. Lambert, 23 Knightlands Road, Irthlingborough, Wellingborough, NN9 5SU
G7 BGM D. Allison, 57 Algarth Road, Pocklington, York, YO42 2HJ
G7 BGO P. Toll, 83 Pepper St., Lymm, WA13 0JT
G7 BGT R. Pykett, The Lilacs, Station Road, Thurgarton, Nottingham, NG14 7HD
G7 BGV C. Barker, 5 Crossgates Wadworth, Doncaster, DN11 9TE
G7 BGY M. Bellaby, 21 Sprydon Walk, Nottingham, NG11 9ET
G7 BGZ M. Hickman, 8 Caxton Road, Woodlands, Doncaster, DN6 7SL
G7 BHE P. Gledhill, 36 Tylers Ride, South Woodham Ferrers, Chelmsford, CM3 5ZT
G7 BHG R. Gill, 24 Larkfield Crescent, Rawdon, Leeds, LS19 6EH
G7 BHR D. Coombes, 2 Ormesby Drive, Potters Bar, EN6 3DZ
G7 BHU T. Mayfield, 184 Wharf Road, Pinxton, Nottingham, NG16 6LQ
G7 BHW J. Wilson, 61 New Lane, Hilcote, Alfreton, DE55 5HT
G7 BHY P. Bailey, 21 Westhall Road, Mickleover, Derby, DE3 0PA
G7 BIK D. Clark, 33 Landers Reach, Lytchett Matravers, Poole, BH16 6NB
GW7 BIL W. Knox, 9 Harrow Close, Caerleon, Newport, NP18 3EF
G7 BIM S. Beazley, 10 Barnecut Close, St. Cleer, Liskeard, PL14 5RU
G7 BIP G. Roffey, 31 Saxville Road, Orpington, BR5 3AN
G7 BIQ K. Lloyd, 9 Hornbeam Walk, Witham, CM8 2SZ
G7 BIV R. Hudson, 12 Magnus Drive, Colchester, CO4 9WQ
G7 BIX T. Houlihane, Blackgate House, Scotland Road, Dry Drayton, Cambridge, CB23 8BX
G7 BIY B. Ansell, 26 Stubby Lane, Wolverhampton, WV11 3NL
G7 BJB C. Foster, 10 Handel Street, Derby, DE24 8AZ
G7 BJC M. Wiggins, 158 Prince Charles, Avenue, Derby, DE3 4LQ
G7 BJD R. Ward, 12 Meadow Lea Knighton Fields, Worksop, S80 3QJ
G7 BJE Rev. I. Godlington, 46 Bren Way Hilton, Derby, DE65 5HP
G7 BJG I. Harvey, 367 Stone Road, Stafford, ST16 1LD
G7 BJN J. Barlow, 38 St. Pauls Road, Newcastle, ST5 2PQ
G7 BJR G. Mitchell, 8 Addison Road, Mexborough, S64 0DJ
G7 BKJ C. Watney, 23 The Wad, West Wittering, Chichester, PO20 8AH
G7 BKL B. Moulton, 70 St. Georges Avenue, Westhoughton, Bolton, BL5 2EU
G7 BKN R. Hatcher, 61 Holland Road, Oxted, RH8 9AU
G7 BLD C. Holden, 26 Valebridge Drive, Burgess Hill, RH15 0RW
G7 BLJ R. Maytum, 62 Coronation Close, Great Wakering, Southend-on-Sea, SS3 0JG
G7 BLK T. Dicks, 4 Nicholas Drive, Reydon, Southwold, IP18 6RE
G7 BLL B. Weaver, 13 Atlay Street, Hereford, HR4 9PF
G7 BLT W. Manthorp, 49 Cassell Road, Bristol, BS16 5DE
G7 BLX M. Rowe, 31 Thornhill Avenue, Thornhill, Southampton, SO19 6PS
G7 BMC P. Hollis, 2 Falcon Drive, Whittington, Lichfield, WS14 9PF
G7 BME D. Watts, 9 Filwood Drive, Kingswood, Bristol, BS15 4HT
G7 BMM S. Hallam, 125 Charnwood Road, Shepshed, Loughborough, LE12 9NL
G7 BMP N. Gough, Oak Den, Park View, Swynnerton, Stone, ST15 0QG
G7 BMT S. Hodges, 15 Middlewich Street, Crewe, CW1 4BS
G7 BMY J. Moore, Fairview, 28 Bulmer Lane, Great Yarmouth, NR29 4AF
G7 BNB S. Reynolds, Dowgill Head House, North Stainmore, CA17 4EX
GW7 BNC J. Beadle, The Coach House, Cwmdauddwr, Rhayader, LD6 5HA
G7 BND B. Welthy, 8 Du Cane Place, Witham, CM8 2UQ
GM7 BNF M. Harrington, Mount Pleasant House, North Road, Wick, KW1 4DN
G7 BNI N. Pope, 21 Moultrie Road, Rugby, CV21 3BD
G7 BNK D. Wood, 16 Church Road, Pelsall, Walsall, WS3 4QN
G7 BNL A. Creek, Westmoor House, Wisbech Road, Ely, CB6 1RQ
G7 BNM A. Ison, 32 Station Road Lode, Cambridge, CB25 9HB
G7 BNN K. Sheppard, Woodlands Bungalow, Gunby Road, Skegness, PE24 5HT
G7 BNO C. Sheppard, The Old Mineral Water Works, Pinfold Lane, Irby in the Marsh, Pe245dh
G7 BNS T. Healey, 5 St. Johns Crescent, Huddersfield, HD1 5DY
G7 BNW G. Ingmire, 93 Havelock Road, Luton, LU2 7PP
G7 BNZ H. Williams, 20 Barham Mews, Teston, ME185BL
G7 BOH P. Craig, 3 Rothsbury Drive, Eastleigh, SO53 4QQ
GD7 BOJ M. Rodgers, 1 Kings Court, Ramsey, Isle of Man, IM8 1LJ
GM7 BOW R. King, 118 Boswell Road, Inverness, IV2 3EW
GW7 BOY B. Hodgkinson, 16 Swain Avenue, Buckley, CH7 3BR
GM7 BOZ A. Bowie, 374A High Street, Leslie, Glenrothes, KY6 3AX
G7 BPF D. Rose, 8 Ambrose Avenue, Hatfield, Doncaster, DN7 6QQ
G7 BPG D. Dixon, 5 Denbigh Close, Newcastle under Lyme, ST5 3DL
G7 BPI C. Thompson, 13 Wentworth Avenue, Luton, LU4 9EN
G7 BPM N. Hemingway, 24 Ealees Road, Littleborough, OL15 0HQ
G7 BPN K. Pentecost, 46 Austen Way, Crook, DL15 9UT
G7 BPQ M. Meadows, 4 The Grove, Wharncliffe Side, Sheffield, S35 0EA
G7 BPR S. Winlove-Smith, 7 Maughan Street, Shildon, DL4 1AP
G7 BPX A. Gifford, 56 Seymour Road, Gloucester, GL1 5QD
G7 BPZ J. May-Golding, 9 St. Barts Road, Sandwich, CT13 0BG
G7 BQA R. Golding, 9 St. Barts Road, Sandwich, CT13 0BG
G7 BQM Dr E. Turk, Sunny Meadow, Three Bridges Road, Long Buckby Wharf, Long Buckby, Northampton, NN6 7PP
G7 BQO J. Hirons, Furlong House, Racecourse Lane, Bicton Heath, Shrewsbury, SY3 5BJ
G7 BQS M. Rodgers, 14 North Street, Rawmarsh, Rotherham, S62 5NH
G7 BQT B. Miller, 22B Avondale Road, Fleet, GU51 3BS
G7 BQU G. Daynes, 25 Redwood Close, Keighley, BD21 4YG
G7 BQY A. Brighton, 22 Langport Drive Vicars Cross, Chester, CH3 5LY
G7 BRA Dr J. Riley, 132 Barrs Road, Cradley Heath, B64 7EZ
G7 BRB J. Marsh, 39 Palace Gate, Odiham, Hook, RG29 1JZ
G7 BRF D. Oliver, 10. Stepping Stones, Bidford on Avon, B50 4PH
G7 BRJ J. Mills, 70 Crescent Road, Rochdale, OL11 3LG
GM7 BRL D. O'Donnell, 188 Warriston Street, Glasgow, G33 2LD
G7 BRM G. Jeffery, 48 Minnis Lane, River, Dover, CT17 0PR
G7 BRP J. Biggs, 5 Churchgate St., Soham, Ely, CB7 5DS
G7 BRS J. Rushton, 391 Rossendale Road, Burnley, BB11 5HP
G7 BRU J. Rhodes, Little Meadow, Roke, Wareham, BH20 7JF
G7 BRX R. Bell, 10 Old Mill Way, Weston-Super-Mare, BS24 7AS
G7 BRZ C. Gaunt, 39 Sonja Crest, Immingham, DN40 2EQ
GW7 BSC R. Snelling, 91 Oakfield Road, Newport, NP20 4LP
G7 BSF A. Lewis, 20 Annes Walk, Caterham, CR3 5EL
G7 BSG Bemerton Sc Grp, c/o A. Carter, 28 Springfield Road, Wellington, TA21 8LG
G7 BSK J. Ingram, 201 Ocean Drive, #06-11, the Azure Singapore, Singapore, 98584
G7 BSL P. Bedford, 48 Trentham Drive, Bridlington, YO16 6EZ
G7 BSO A. Noble, 42 Upper Street, Salisbury, SP2 8LY
G7 BSP S. Farmer, Horton Brook Cottage, Horton, Wem, Shrewsbury, SY4 5NB
GW7 BTC Cardiff and District ARC, c/o R. Magwood, 13 Inverness Place, Cardiff, CF24 4RU
G7 BTI Madley ARG, c/o N. Prosser, 35 Holmfirth Close, Belmont, Hereford, HR2 7UG
G7 BTP P. Jensen, 16 Hawthorn Avenue, Immingham, DN40 1AR
G7 BUF P. Parkin, 2 The Knoll, Dronfield, S18 2EH
G7 BUK D. Brinnen, 134 Victoria Road, Mablethorpe, LN12 2AJ
G7 BUL E. Williamson-Brown, Meadowside, Green Lane, Stowmarket, IP14 5DS
G7 BUN H. Ellis, Home Cottage, Drury Square, King's Lynn, PE32 2NA
G7 BUR G. Robinson, 228 Bradford Road, Riddlesden, Keighley, BD20 5JT
G7 BUS G. Payne, 155 Camping Hill, Stiffkey, Wells-Next-the-Sea, NR23 1QL
G7 BVH M. Gurr, Elan, Sandown Road, Sandwich, CT13 9NY
G7 BVL K. Lambert, 1 Langton Road, Chichester, PO19 3LY
G7 BVS N. Hill123, 16 Bittern Avenue, Abbeydale, Gloucester, GL4 4WA
G7 BVZ B. Allen, 84 Holland Road, Little Clacton, Clacton-on-Sea, CO16 9RS
G7 BWE E. Gibbons, 19 Queens Park Road, Caterham, CR3 5RB
G7 BWF A. Gray, 147 Kirby Road, Walton on the Naze, CO14 8RL
G7 BWI J. White, 11 Rowdown, Upper Lambourn, Hungerford, RG17 8RF
G7 BWO D. Morgan, 26 Lyndhurst Road, Exmouth, EX8 3DT
G7 BWV R. Fosbraey, 122 East St., Sittingbourne, ME10 4RX
G7 BWW M. Widdows, 27 Market Close, Barnham, Bognor Regis, PO22 0LH
G7 BXA P. Austin, 24 Fairfield Terrace, Bramley, Leeds, LS13 3DH
G7 BXG F. Clarke, 37 Cambridge Road, Rainworth, Mansfield, NG21 0ax
G7 BXJ P. Turner, 260 New Lane, Huntington, York, YO32 9LY
G7 BXL M. Bickerton, 54 Swinnel Brook Park, Grane Road, Rossendale, BB4 4FN

G7 BXS R. Wake, 55 Bearsdown Road, Eggbuckland, Plymouth, PL6 5TR
G7 BXT R. Cross, 9 Chesham House Leyburn Crescent, Romford, RM3 8RU
G7 BXU S. Welton, 18 Coningham Road, Reading, RG2 8QP
GM7 BYB A. Mcintyre, 18 Seal Craig Gardens, Altens, Aberdeen, AB12 3SH
G7 BYE J. Hersom, 10 Young St., Gilesgate, Durham, DH1 2JU
G7 BYG A. Found, 18 Mead Fields, Bridport, DT6 5RF
G7 BYI M. Baldry, 10 Kingfisher Court, Lowestoft, NR33 8PJ
G7 BYK R. Barry, 14 Home Mead Creswicke Road, Bristol, BS4 1UQ
G7 BYN D. Bendrey, 73 Kestrel Close, Chipping Sodbury, Bristol, BS37 6XB
G7 BYS J. Pollard, 25 Heath Avenue, Ramsbottom, Bury, BL0 9UN
G7 BYU W. Mcguffie, 1 Norbury Drive, Marple, Stockport, SK6 6LL
G7 BYV T. Connolly, 7 Springfield Crescent, Sherborne, DT9 6DN
G7 BYW C. Stone, 60 Staddon Park Road, Plymouth, PL9 9HJ
G7 BZC R. Pickett, 1 The Cottage, Hospital Road, Wingland, Spalding, PE12 9YR
G7 BZD P. Yates, Kingsomborne The Broadway, Totland Bay, PO39 0BL
G7 BZE W. Gillott, 14 Oakham Place, Barnsley, S75 2ND
G7 BZM R. Brown, Apartment 313 2 Austin Way, Birmingham, B31 3GG
G7 BZQ G. Reynolds, 187 Steelhouse Lane, Wolverhampton, WV2 2AU
GW7 BZR J. Shurmer, 126 Gaerwen Uchaf Estate, Gaerwen, LL60 6JW
G7 BZU A. Mccoll, 105 North East Road, Southampton, SO19 8AF
GW7 BZY P. Mcfarland, 20 Caer Sarn, Caernarfon, LL54 7TW
G7 CAA N. Royle, 62 Cynthia Close, Poole, BH12 3JW
G7 CAF S. Bond, 38 Hampsfell Drive, Morecambe, LA4 4TU
G7 CAG A. Malpass, 48 Geoffrey Barbour Road, Abingdon, OX14 2ES
GW7 CAH M. Astley, 30 Lon Ceirios, Newtown, SY16 1PR
G7 CAS W. Welburn, 31 West Bank, Scarborough, YO12 4DX
G7 CAT G. Cutter, 48 Hyndley Road, Bolsover, Chesterfield, S44 6RX
G7 CBI M. Higgins, 2 Walden Road, Keynsham, Bristol, BS31 1QW
GW7 CBU J. Griffiths, 143 Brynglas, Hollybush, Cwmbran, NP44 7LL
G7 CBW S. Duffield, 21 Calley Close, Tipton, DY4 8XY
G7 CBY L. Cotton, 6 Blacksmith Row, Lytham St Annes, FY8 4UE
G7 CBZ S. Cotton, 6 Blacksmith Row, Lytham St Annes, FY8 4UE
G7 CCL S. Cullingworth, 19 Springbank Garforth, Leeds, LS25 1DD
GW7 CCR Flintshire Raynet, c/o M. Ellett, 14 Canon Drive, Bagillt, CH6 6LS
G7 CCS G. Oliver, 17 Jack Stephens Estate, Penzance, TR18 2QE
G7 CCV P. Upton, 73 Allington Close, Taunton, TA1 2NA
G7 CDI P. Matthews, 128 Thealby Gardens, Doncaster, DN4 7EG
G7 CDO A. Corps, 6A Salisbury Road, Leigh-on-Sea, SS9 2JX
GW7 CEA D. Smith, 62 Cheshire View. Brymbo., Wrexham, LL11 5AW
G7 CEB J. Redpath, 41 Sandford Green, Banbury, OX16 0SB
G7 CEC J. Evans, 28, Kenneth Gamble Court, West Avenue, Wigston, LE18 2FP
G7 CED K. Barnett, 126 Oldham St., Latchford, Warrington, WA4 1EX
G7 CEN T. Martin, 10 Hardy Close, Galley Common, Nuneaton, CV10 9SG
GW7 CEQ W. Jones, 26 Cwm Silyn, The Park, Caernarfon, LL55 2AG
G7 CER C. Riley-Moxon, 51 Nuttall Lane, Ramsbottom, Bury, BL0 9JX
G7 CEW D. Everard, 6 Leith Hill Green, St. Pauls Cray, Orpington, BR5 2SB
G7 CEY P. Copeland, 7 Stuart Avenue, Draycott, Stoke-on-Trent, ST11 9AA
G7 CFC A. Chamberlain, 11 Woden Crescent, Wolverhampton, WV11 1PR
G7 CFS D. Halliday, 33 Brentnall Close, Great Sankey, Warrington, WA5 1XN
G7 CFT G. Taylor, 48 Westwood Heath Road, Leek, ST13 8LL
G7 CFW K. Morrison, 7 Turnstone End, Yateley, GU46 6PE
G7 CFX A. Newell, 4 Dexter Square, Cricketers Way, Andover, SP10 5DB
G7 CGC P. Oliver, 32 Pearmain Way, Stanway, Colchester, CO3 0NP
G7 CGI C. Honey, 1 Brimley Court, Lower Brimley Road, Teignmouth, TQ14 8LW
G7 CGN S. Hodkinson, 17 Thorn Well, Westhoughton, Bolton, BL5 2PJ
G7 CGT A. Day, 8 The Garth, Ash, Aldershot, GU12 6QN
G7 CHB Lord R. Montague, 71 Middlethorpe Road, Cleethorpes, DN35 9PP
G7 CHC J. Anderson111, Tre Noce Cottage, 75 St. Michaels Road, Paignton, TQ4 5NA
G7 CIA Dr J. Adams, 14 Kedleston Close, Northampton, NN4 0WF
G7 CIH C. Hayes, 2 Castleford House, Castle Road, Okehampton, EX20 1HZ
G7 CIK M. Kielthy, 35 Alexandra Road, Sheringham, NR26 8HU
G7 CIQ A. Wiseman, 61 Hilton Avenue, Horwich, Bolton, BL6 5RH
G7 CIT T. Mcguigan, 18 Whalton Close, Gateshead, NE10 8SW
G7 CIU P. Burbury, 43 Locomotive Street, Darlington, DL1 2QF
G7 CIV B. Perrin, 8 Station Road, Gretton, Corby, NN17 3BU
G7 CIY K. Gaunt, 21 Abbey Close, Rendlesham, Woodbridge, IP12 2UD
G7 CJC J. Hughes, Hollies Bungalow, Valeswood, Shrewsbury, SY4 2LH
G7 CJD C. Dyer, 6 Witcombe, Yate, Bristol, BS37 8SA
G7 CJG G. Stringer, 13 Garfield St, Kettering, NN15 7HX
G7 CJO J. Graves, 7 Matthews Chase, Binfield, Bracknell, RG42 4UR
G7 CJS D. Evans, 16 Cruden Road, Gravesend, DA12 4HD
G7 CJW J. Wardle, 9 Leefield Road, Chapel-En-Le-Frith, High Peak, SK23 0LF
G7 CKE D. Scriven, 59 Breck Lane Dinnington, Sheffield, S25 2LJ
G7 CKG R. Varley, 23 Manor Court, Bingley, BD16 1QD
G7 CKL B. Taylor, 32 Marples Avenue Mansfield Woodhouse, Mansfield, NG19 9HA
G7 CKP J. Surman, 122 Burwell Meadow, Witney, OX28 5JQ
G7 CKQ K. Hartley, 39 Raleigh Road, Sunderland, SR5 5RD
G7 CKS D. Davies, 2 London Road, Battle, TN33 0EU
G7 CLG P. Ord, 52 Hillside Road, Stockton on Tees, TS20 1JQ
G7 CLH D. Smith, 7 Dunlin Close, Norton, Stockton-on-Tees, TS20 1SJ
GM7 CLM J. Harrington, Mount Pleasant House, North Road, Wick, KW1 4DN
G7 CLO C. Gaukroger, Meadow View, Lansallos, Looe, PL13 2PU
G7 CLX K. Marsden, 3 Lane Head, Heptonstall, Hebden Bridge, HX7 7PB
G7 CLY J. Hill, 55 The Oval, Welton, Brough, HU15 1DA
G7 CMB D. Kennett, 6 The Holdings, Hatfield, AL95HQ
GI7 CMC N. Moore, 164 Ardenlee Avenue, Belfast, BT6 0AE
GU7 CMH G. Simon, 3 Mahaut Villas, Collings Road, St Peter Port, Guernsey, GY1 1FP
G7 CMI Prof. B. Birch, 4 Kynnesworth Gardens, Higham Ferrers, Rushden, NN10 8NH
GW7 CMM G. Jones, 13 Palace Close, Flint, CH6 5YE
G7 CMN Cheshire County Raynet, c/o B. Williams, 3 Welton Close, Wilmslow, SK9 6HD
G7 CMP B. Fielding, The Copse Charmouth Road, Axminster, EX13 5SZ
G7 CNC D. Gray, Flat 57, Wesley Court, 1 Millbay Road, Plymouth, PL1 3LB
G7 CND J. Bell, 2 Rake Lane, Milford, Godalming, GU8 5AB
GU7 CNI M. Elliston, La Guillard Lane, St Andrew, Guernsey, GY6 8YJ
G7 CNM D. Clark, 33 Antrim Road, Lincoln, LN5 8TF
G7 CNP H. Weatherhead, 39 Meadow Park, Dawlish, EX7 9BU
GM7 CNW G. Dryburgh, 86 Normand Road, Dysart, Kirkcaldy, KY1 2XP
G7 CNX P. Hammond, 23 Peppers Close, Brandon, IP270PU
G7 CNZ K. Ford, 8 Blakedon Road, Wednesbury, WS10 7HY
G7 COA R. Johnson, 7 West Parade, Warminster, BA12 8LY
GW7 COB S. Coburn, 54 Queensway, Hope, Wrexham, LL12 9PE
G7 COC N. Whelan, 54 Boroughbridge Road, Northallerton, DL7 8BN
G7 COD A. Kitchen, 4 Dairy Cottages, Bank Newton, Skipton, BD23 3NT
G7 COG J. Lister, 6 Fordlands Crescent, Fulford, York, YO19 4QQ
G7 COP P. Payton, 3 Astor Close, Winnersh, Wokingham, RG41 5JZ
G7 COQ K. Raxworthy, 9 Harrow Drive, Edmonton, London, N9 9EQ
G7 COY C. Keevil, 82 Chatham Grove, Chatham, ME4 6LY
GM7 CPJ G. Currie, The Old Post Office, Cuminestown, AB53 5TQ
GM7 CPL C. Scott, 115 Tarvit Terrace, Springfield, Cupar, KY15 5SE
G7 CPN S. Burgess, 59 Back Lane, Congleton, CW12 4PY
G7 CPQ C. Ambrose, 47 Whitton Close, Swavesey, Cambridge, CB24 4RT
GM7 CPR J. Wright, 9 Meadowhead Road, Plains, Airdrie, ML6 7JF
GM7 CPY S. Leggat, Ailach, St. Aethans Road, Elgin, IV30 2YR
G7 CQA R. Ginn, 91 High Street, Shoeburyness, Southend-on-Sea, SS3 9AR
GW7 CQB D. Locock, Bank House, Selattyn, Oswestry, SY10 7DX
G7 CQG R. Bradshaw, 11 Glebe Avenue, Orton Waterville, Peterborough, PE2 5EN
G7 CQH M. Smith, 14B Witham Bank West, Boston, PE21 8PU
G7 CQK Capt. I. Phillips, Goldsworthy Farm, Stony Lane, Gunnislake, PL18 9BL
GU7 CQN J. Gardner, The Ferns, Rue De La Girouette, St. Saviour, Guernsey, GY7 9NN
GM7 CQQ A. Donaldson, 36 Rothes Park, Leslie, Glenrothes, KY6 3LH
G7 CQW A. Riley, 4 Birtle Drive Astley, Tyldesley, Manchester, M29 7RE
G7 CQX D. Read, 31 Grace Gardens, Bishop's Stortford, CM23 3EU
G7 CQZ E. Courtnell, 38 Woodchurch Lane, Birkenhead, CH42 9PH
G7 CRA I. Brelsford, 78 Borough Road, Redcar, TS10 2EQ
G7 CRK S. Halbertsma, 65 Gareth Grove, Bromley, BR1 5EG
G7 CRM D. Stump, 9 Shipton Grove, Swindon, SN3 1BZ
G7 CRN R. Phin, 35 Parkland Close, Newquay, TR7 3EB
G7 CRQ A. Haywood, Flat 21, Atholl House, 178 Woodcote Road, Wallington, SM6 0PB
G7 CRR J. Wheeler, 8 Slimbridge Close, Worcester, WR5 3SH
G7 CRS Maxpak, c/o M. Hall, 35 Bunns Lane, Dudley, DY2 7RA
G7 CRU P. Wallis, 6 Lancelott Court, Pershore, WR10 1RE
G7 CRV B. Burnside, The Rectory, Normanby, Sinnington, York, YO62 6RH
G7 CRY J. Bain, 7 Wrights Lane, Sutton Bridge, Spalding, PE12 9RH
G7 CSD N. Whittle, 31 St. Georges Avenue, Southall, UB1 1PZ
G7 CSE R. Rowthorn, 25 Branscombe Drive, Wootton Bassett, Swindon, SN4 8HS
G7 CSF J. Plowright, 11 Coddington Street, Newport, USA,
G7 CSI M. Morris, 20 Bracken Way, Chobham, Woking, GU24 8PR
G7 CSJ K. Chapman, 19 St. Johns Rise, Woking, GU21 7PN
GW7 CSK D. Winter, 25 Pembroke St., Thomastown, Porth, CF39 8DU
G7 CSL R. Dent, 9 Dovey Close, St. Ives, Huntingdon, PE27 6HW
G7 CSM R. Knight, 32 Linnet Close Abbeydale, Gloucester, GL4 4UA
G7 CSS M. Budd, 103 Old Charlton Road, Shepperton, TW17 8BT
G7 CST A. Clarke, 86 Roebuck Road, Walsall, WS3 1AL
G7 CSV Spen Valley ARS, c/o T. Clough, 37 Park Avenue, Mirfield, WF14 9PB
G7 CSX A. Keen, 20 Horam Park Close, Horam, Heathfield, TN21 0HW
G7 CTE N. Farmer, 4 Frank Hughes Avenue, Sandbach, CW11 3TA
G7 CTG E. Ives, 15 Northlands, Adwick-Le-Street, Doncaster, DN6 7AX
G7 CTN M. Puddephatt, Church Barn, Kirklington Road Eakring, Newark, NG22 0DA
G7 CTT S. Lock, 17 Elgar Crescent, Droitwich, WR9 7SP
GM7 CTV A. Smith, 13 Park Terrace, Markinch, Glenrothes, KY7 6BN
GI7 CTW E. Regan, 4 Lecumpher Road, Desertmartin, Magherafelt, BT45 5LY
G7 CUA R. Cookson, Briarswood, Snow Hill Lane, Preston, PR3 1BA
G7 CUB D. Price, Summer Fields, Mayfield Road, Oxford, OX2 7EN
G7 CUD J. Richardson, 68 Place Farm Way, Monks Risborough, Princes Risborough, HP27 9JY
G7 CUF G. Adrian, 140 Shawfield Road Ash, Aldershot, GU12 6SG
G7 CUL R. Bourn, 7 Clitheroes Lane, Freckleton, Preston, PR4 1SD
G7 CUO J. Folland, 41 Rydal Avenue, Billingham, TS23 1HX
G7 CUP P. Ingle, 8 Burnstone Gardens, Moulton, Spalding, PE12 6PS
G7 CUU D. Stainforth -Small, 10 Balland Park, Ashburton, Newton Abbot, TQ13 7BS
G7 CUW B. Minish, Raheens, Castlebar, Ireland,
G7 CUY Dr K. Martin, 8 Taylors Close, Meppershall, Shefford, SG17 5NH
G7 CVA E. Curnow, 9 Moreton Bay, Bilton, Hull, HU11 4ER
G7 CVC M. Porter, 19 Thistle Downs, Northway, Tewkesbury, GL20 8RE
G7 CVF S. Evans, 34 Kent Road, Southport, PR8 4BJ
G7 CVM D. Illman, 27 Blackborough Road, Reigate, RH2 7BS
G7 CVY K. Helgesen, 13 Mara Court, White Road, Chatham, ME4 5TW
G7 CVZ A. Bevins, 12 Wheatstone Road, Formby, Liverpool, L37 6BF
G7 CWE D. Ishmael, 38 Greenford Close, Orrell, Wigan, WN5 8RH
G7 CWI I. Green, 25 Riley Avenue, Lytham St Annes, FY8 1HZ
G7 CWM J. Denton, 48 Seas End Road, Surfleet, Spalding, PE11 4DQ
G7 CWN F. Merchant, 3 Main Road, Shortwood, Mangotsfield, BS16 9NH
G7 CWO D. Ward, 107 Oundle Road, Birmingham, B44 8ER
G7 CWT Prof. M. Joy, Cheddon Corner, Cheddon Fitzpaine, Taunton, TA2 8LB
G7 CWX C. Tarling, 36 Lancaster Road, North Weald, Epping, CM16 6JA
G7 CXB T. Mcinnes, 7 Hilary Drive, Merry Hill, Wolverhampton, WV3 7NJ
G7 CXO V. Cassar, 51 Aylesford Avenue, Beckenham, BR3 3SB
G7 CXT S. Haywood, 7 Stamford Gardens, Dagenham, RM9 4ET
G7 CXU S. Power, 8 Green Lane, Chislehurst, BR7 6AG
G7 CYD A. Jenkins, 15 Tilstone Avenue, Eton Wick, Windsor, SL4 6NF
G7 CYF C. Wardill, 85 Station Road, Chellaston, Derby, DE73 5SU
G7 CYN C. Casper, 22 Eshton Road, Gargrave, Fell View, Skipton, BD23 3SE
G7 CYQ E. Hornby, 14 Essex Road, Stevenage, SG1 3EZ
GW7 CYT D. Phillips, 15 Herbert Street, Treorchy, CF42 6AW
GM7 CZC R. Johnson, 3 Hamilton Gardens, Edinburgh, EH15 1NH
G7 CZF J. Jenkins, 18 High Beeches, Gerrards Cross, SL9 7HX
G7 CZL G. Miles, 7 Dobbin Close Rawtenstall, Rossendale, BB4 7TH
GM7 CZU J. Mclaughlan, 2 Donaldson Drive, Irvine, KA12 0QG
G7 DAB I. Weeks, 19 St. Michaels Road, Tunbridge Wells, TN4 9JG
G7 DAH C. Moore, 168 Church End Lane, Runwell, Wickford, SS11 7DN
GU7 DAI J. Smith, Ravello, 35 Le Villocq Estate, Le Villocq, Castel, Guernsey,

Guernsey, GY5 7SQ
GM7 DAJ R. Hepburn, 44 Macindoe Crescent, Kirkcaldy, KY1 2JG
G7 DAL J. Ratigan, 81 Cunningham Drive, Unsworth, Bury, BL9 8PD
GM7 DAP A. Lord, 5 Windsor Terrace, Brechin, DD9 6SD
G7 DAR R. Charlton, 13 Hollywood Avenue, Walkerville, Newcastle upon Tyne, NE6 4TN
G7 DAZ J. Watson, Flat 1, 53 Castle Street, Bolton, BL2 1AD
G7 DBN E. Turner, Rectory Cottage, Little Marsh, Marsh Gibbon, Bicester, OX27 0AP
G7 DBO I. Leaver, 2 Marnhull Close, Coventry, CV2 2JS
G7 DBT R. Claridge, 3 Wentworth Avenue, Leagrave, Luton, LU4 9EN
G7 DBV D. Ritson, C/O 12 Tudor Grange, Easington Village, Peterlee, SR8 3DF
GI7 DBZ W. Hollinger, 51 Collin Road, Ballyclare, BT39 9JS
G7 DCF A. Mclennan, 2 Phillip St, Beachmere, Australia, 4510
G7 DCJ S. Warner, 96 Walter Nash Road East, Kidderminster, DY11 7BY
G7 DCM L. Bant, 58 Severn Way, Cressage, Shrewsbury, SY5 6DS
G7 DCT A. Horsfall, 2 Temple Walk, Halton, Leeds, LS15 7SQ
G7 DDF P. Johnson, 6 Rugby Road, Lilbourne, CV23 0SP
G7 DDN C. Rolinson, 534 Haslucks Green Road, Shirley, Solihull, B90 1DS
G7 DDQ W. Roberts, 12 Camberley Drive, Penn, Wolverhampton, WV4 5RP
G7 DDR R. Murray, 8 Church Lane, Kirk Langley, Ashbourne, DE6 4NG
G7 DDV C. James, Flat 69-70, Silk House, Kilmington, Warminster, BA12 6QZ
G7 DEC A. Grundy, 647 Preston Old Road, Feniscowles, Blackburn, BB2 5ER
G7 DEE K. Denniss, 4 Aysgarth Road, Sheffield, S6 1HU
G7 DEF J. Kingsley, 3 The Orchard, Swarland, Morpeth, NE65 9NB
G7 DEG R. Williams, 6 Ralfland View, Shap, Penrith, CA10 3PF
G7 DEH H. Carpenter, 44 Bowbridge Road, Newark, NG24 4BZ
G7 DEI S. Courtney-Crowe, 28 Brymore Close, Prestbury, Cheltenham, GL52 3DY
G7 DEU G. Renton, 58 St. Christophers Road, Humberston, Grimsby, DN36 4EA
G7 DEY P. Knowles, 35 Raby Park Road, Neston, CH64 9SW
G7 DFC J. Worsnop, 1217 Thornton Road, Thornton, Bradford, BD13 3BE
GM7 DFI K. Trinder, 29 Woodside Road, Brookfield, Johnstone, PA5 8UB
G7 DFP J. Fitzpatrick, 22 Ferry Road Surlingham, Norwich, NR14 7AR
G7 DFV G. Jelley, 28 Blanches Road, Partridge Green, Horsham, RH13 8HZ
G7 DFW A. Crisp, 17 Gaitskell House, Howard Drive, Borehamwood, WD6 2PB
G7 DFX G. Allan, Kent House, 106 Kent Road, Sheffield, S8 9RL
G7 DGC M. Lewis, 21 Woodlands Road, Ashton-under-Lyne, OL6 9DU
G7 DGD B. Walton, 28 Durham Terrace, Durham, DH1 5EH
G7 DGE A. Biggin, 14 Coultas Avenue, Deepcar, Sheffield, S36 2PT
G7 DGF D. Coupe, 22 West Street, South Normanton, Alfreton, DE55 2AJ
G7 DGP B. Barrass, 7 The Crescent, Easton On The Hill, Stamford, PE9 3LZ
GM7 DHA K. Pugh, 1 Barrington Gardens, Beith, KA15 2BA
G7 DHD D. Dyson, 5 Warwick Street, Church, Accrington, BB5 4AL
GW7 DHG A. Gardner, 28 Usk Court, Thornhill, Cwmbran, NP44 5UN
G7 DHJ M. Harding, 79 Beaumont Walk, Leicester, LE4 0PP
G7 DHQ G. Davies, 17 Remington Road, Walsall, WS2 7EJ
G7 DHW D. Neale, Greyhills Farm, Diptford, Totnes, TQ9 7NQ
G7 DIB A. Finon, Radford House, Hall Lane, Wymondham, NR18 9TB
G7 DIE S. Salmon, 35 Westgate Road, Lytham St Annes, FY8 2SG
G7 DIG R. Dee, 10 Sanderson Street, Coxhoe, Durham, DH6 4DG
GW7 DIL P. Davis, The Willows, Park View, Cwmbran, NP44 1RB
G7 DIO C. Harper, 132 Park Avenue East, Dallas, Ga, USA, 30157
G7 DIR A. Brinton, 136 Efford Road, Plymouth, PL3 6NQ
G7 DIS C. Baker, Moffat House, Church Road, Broughton Moor, Maryport, CA15 7SS
GI7 DIT D. Roberts, 6 Plantation Road, Bangor, BT19 6AF
G7 DIU D. Leech, 4 Rydal Close, Huntingdon, PE29 6UF
G7 DIW A. Saul, 18 Elm Bank Close, Cubbington, Leamington Spa, CV32 6LR
G7 DIZ M. Beatrup, 34 Springfield Drive, Halesowen, B62 8EU
G7 DJA M. Rayner, 38A Fen Road Milton, Cambridge, CB24 6AD
GW7 DJL T. Davis, 127 Lon Glanyrafon, Newtown, SY16 1QT
G7 DJN A. Coates, 19 Bretton Avenue, Bolsover, Chesterfield, S44 6XN
G7 DJT D. Tinley, 2 Rosemount Close, Loose, Maidstone, ME15 0AJ
G7 DJY N. Monkman, 59 Fairway Approach, Normanton, WF6 2LX
G7 DKB D. Simons, 65 Dolphin Court Road, Paignton, TQ3 1AB
G7 DKY G. Rowntree, 20 Lancaster Lane, Leyland, PR25 5SN
G7 DKZ J. Stearn, Half Acre, Hatch Green, Taunton, TA3 6TN
GW7 DLD R. Hilton, 9 Waterloo Fields, Kingswood, Welshpool, SY21 8LF
G7 DLE J. Hodges, 70 Chestnut Drive, Brixham, TQ5 0DD
GM7 DLY J. Whitcomb, 1/1 30 Highburgh Road, Glasgow, G12 9DZ
G7 DME R. Gornall, 58 Homelawn House Brookfield Road, Bexhill on Sea, tn401pn
G7 DMG W. Hetherington, 32 Almond Way, Lutterworth, LE17 4XJ
G7 DMH S. Hetherington, 8 Kings Lane Yelvertoft, Northampton, NN6 6LX
G7 DMK P. Drage, Cranford House, 167 Rockingham Road, Kettering, NN16 9JA
GM7 DMN Y. Benting, Suthainn, Askernish, Isle of South Uist, HS8 5SY
G7 DMP J. Barnes, 26 Fairthorn Road, Sheffield, S5 6LX
G7 DMQ S. Rafferty, 22 Hengist Close, Horsham, RH12 1SB
G7 DMS M. Horsfall, 8 Greenbrook Road, Burnley, BB12 6NZ
G7 DMX N. Taylor, 5 Miranda Road, Preston, Paignton, TQ3 1LE
G7 DMZ B. Knight, Conchardrin, Tibberton, Gloucester, GL2 8EB
G7 DNF T. Haye, Woodside, Sutton Wood Lane, Alresford, SO24 9SG
G7 DNG I. Casey, 38 Wordsworth Road, Salisbury, SP1 3BH
GJ7 DNI S. Mcadams, Gemaur, St. Clements Road, St Helier, Jersey, JE2 4PX
GJ7 DNJ I. Meade, Etape De Base, La Rue Des Platons, Trinity, Jersey, JE3 5AA
G7 DNM E. Ashworth, 10 Wisteria Drive, Lower Darwen, Darwen, BB3 0QY
G7 DNP A. Patel, 11716 Vista Meadow Ln, Frisco, USA, 75035
G7 DNQ S. Howard, 5 Grummock Avenue, Ramsgate, CT11 0RR
G7 DNR K. Crookes, 64 Heron Drive, Audenshaw, Manchester, M34 5QX
G7 DNT K. Handscombe, 8 Fletcher Road, Ipswich, IP3 0LF
G7 DNV L. Carter, 3 Cleviscroft, Stevenage, SG1 1UJ
G7 DNX M. Morris, 4 Meadow Brook Road, Northfield, Birmingham, B31 1NE
G7 DOA D. Morris, Flat 3, 748 Melton Road, Leicester, LE4 8BD
G7 DOD S. Mann, 55 Taunton Street, Wavertree, Liverpool, L15 4ND
G7 DOF K. Mount, 4 Hermitage Road, Abingdon, OX14 5RN
G7 DOL Royal Naval ARS, c/o J. Kirk, 111 Stockbridge Road, Chichester, PO19 8QR
G7 DOR Dorking & District Radio Society, c/o G. Brind, 9 Becket Wood, Newdigate, Dorking, RH5 5AQ
G7 DOS B. Smith, 7 School Walk, Chase Terrace, Burntwood, WS7 1NQ
G7 DOW G. Smith, 59 Radipole Lane, Weymouth, DT4 9RR
G7 DOY J. Baddeley, 22 Scott Road, Denton, Manchester, M34 6FT
G7 DPE J. Glass, 8 Hazlemere View, Hazlemere, High Wycombe, HP157BY
G7 DPF G. Brightman, 5 Meadow Rise, Lacey Green, Princes Risborough, HP27 0QY
GD7 DPG J. Wrigley, 20 Fairy Hill Close, Ballafesson, Port Erin, Isle Of Man, IM9 6TJ
GM7 DPI P. Blacklaw, Flat 15, Servite House 21A High Street, Monifieth, Dundee, DD5 4AA
G7 DPR F. Overbury, 47 The Maltings, Dunmow, CM6 1BY
G7 DPV A. Marks, 57A Bulcote Drive Burton Joyce, Nottingham, NG14 5AZ
G7 DPW A. Butterworth, 3 Fir Tree Avenue, Worsley, Manchester, M28 1LP
G7 DPZ E. Curd, 11 Ashkirk Close, Waldridge, Chester le Street, DH2 3HY
G7 DQA J. Hallin, 12 Church Park Road, Plymouth, PL6 7SA
G7 DQE M. Meerman, University Of Surre, Dept Elec. Eng., Surrey, GU2 5XH
G7 DQL P. Perkins, 29 Parkhill, Middleton, King's Lynn, PE32 1RJ
G7 DQQ M. Lampett, 130 Clopton Road, Birmingham, B33 0RL
G7 DQZ D. Keen, 5 London Road, Uckfield, TN22 1HU
G7 DRD T. Platt, 30 Great Ellshams, Banstead, SM7 2BA
G7 DRG R. Moseley, 307 Archer Road, Stevenage, SG1 5HF
G7 DRO W. Webber, Springfield Lodge, Broadway, Winscombe, BS25 1UE
G7 DRR D. Bellinger, Holly Cottage, Deacons Lane, Thatcham, RG18 9RJ
G7 DRT M. Dickinson, 1 Tregaron Avenue, Cosham, Portsmouth, PO6 2JU
G7 DRU A. Tink, 13 The Wicketts, Filton, Bristol, BS7 0SR
G7 DRW V. Wynn, 12 Holly Road, Orpington, BR6 6BE
G7 DRX J. Stelmasiak, Golden Rocks, Coed Lane, Montgomery, SY15 6AB
GM7 DRY S. Graham, 15 Stone Crescent, Mayfield, Dalkeith, EH22 5DT
G7 DSA S. Jeffery, 7 Corfe Way, Winsford, CW7 1LU
GU7 DSB P. Blampied, 9 Rue Des Grons Estate, St Martin, Guernsey, GY4 6JT
G7 DSO P. Yarnold, 162 Harwill Crescent, Aspley, Nottingham, NG8 5LF
G7 DSQ R. Roberts, 16 Poplar Drive, Pucklechurch, Bristol, BS16 9QF
G7 DST Dr D. Thomson, 96 Rydal Avenue, Loughborough, LE11 3RX
G7 DSU C. Tong, 24 James Road Cuxton, Rochester, ME2 1DJ
G7 DSV D. Crinson, 16 Lomax Street, Geeat Harwood, Bb67dj
GJ7 DTA A. Lange, Les Bois, La Rue De La Pointe, St Peter, Jersey, JE3 7AQ
GW7 DTB R. Dixon, 19 The Burrows, Porthcawl, CF36 5AJ
GM7 DTC J. Arthur, 15 St. Andrews Place, Beith, KA15 1JE
G7 DTG S. Le Poer Trench Brown, Holy Mill, Longville, TF13 6ED
G7 DTK T. Timms, 16 Claverdon Road, Coventry, CV5 7HP
G7 DTR M. Penny, 79 Grove Avenue, New Costessey, Norwich, NR5 0JA
G7 DTS A. Lowe, 33 Dandies Chase, Eastwood, Leigh-on-Sea, SS9 5RF
G7 DTT A. Reeves, 13 Maple Way, Leavenheath, Colchester, CO6 4PQ
G7 DTV H. Partridge, 11 Elm Close, Stourbridge, DY8 3JH
G7 DUB P. Spencer, 11 Two Trees Estate, Wadebridge, PL27 7PG
G7 DUC B. Tonkin, 9 Penhallick Road, Carn Brea, Redruth, TR15 3YJ
G7 DUE W. Davis, 6 Bushy Mead, Waterlooville, PO7 5DY
GW7 DUI C. Teague, 27 Pond Mawr, Maesteg, CF34 0NG
G7 DUK M. Clarke, 12 Moat Court, Shaw Close, Chertsey, KT16 0PH
GJ7 DUX S. Raynes, Mon Plaisir, La Rue De Samares, St Clement, Jersey, JE2 6LZ
G7 DUY R. Jones, 14 Lunsford Road, Liverpool, L14 0NU
GD7 DUZ S. Kelly, 44 Westhill Avenue, Castletown, Isle of Man, IM9 1HY
GW7 DVJ D. Maxted, 33 Bryn Dryslwyn, Bridgend, CF31 5BT
G7 DVO T. Spearing, 139 Holt Road, Hellesdon, Norwich, NR6 6UA
GI7 DWF J. Murphy, 19 Kilburn Park, Armagh, BT61 9HA
G7 DWH E. Shaw, 5 Charlock Grove, Cannock, WS11 7FR
G7 DWI A. Davies, 23 Holly Place, Eastbourne, BN22 0UT
G7 DWM C. Hadjigeorgiou, 26 Priory Gardens, Hampton, TW12 2PZ
G7 DWN A. Keen, 29 Churchman Close Melton, Woodbridge, IP12 1RN
G7 DWO S. Prisk, 86 Wycliffe Grove, Werrington, Peterborough, PE4 5DF
G7 DWQ P. Donovan, 35 Ash Drive, Wardley, Swinton, M27 9QP
G7 DWU M. Stapleton, 15 Haviland Way, Cambridge, CB4 2RA
G7 DWV Dr K. Webster, 9 King John Avenue, Fareham, PO16 9AP
G7 DWX M. Twells, Camels, Annscroft, Shrewsbury, SY5 8AN
G7 DWY I. Barraclough, Maru, 25 Blaithroyd Lane, Halifax, HX3 9PS
G7 DXB L. Watson, 9 Croft Close Cumwhinton, Carlisle, CA4 8FG
G7 DXC J. Maxwell, Tysties, Tile Barn, Newbury, RG20 9UY
GM7 DXE A. Todd, Waterfurrows, Breakachy, Beauly, IV4 7AE
G7 DXH R. Ife, 43 Bracknell Crescent, Nottingham, NG8 5EU
G7 DXN P. Chapman, 1 Bader Close, Watton, Thetford, IP25 6FF
G7 DXQ K. Salt, 4 Burghwood Road, Ormesby, Great Yarmouth, NR29 3LT
GM7 DXT J. Macleod, 59 Fife Street, Keith, AB55 5EG
G7 DXV P. Shepherd, 25 Tomkins Close, Stanford-le-Hope, SS17 8QU
G7 DXX J. Walker, 121 Park Drive, Upminster, RM14 3AU
G7 DYB A. Rudling, 1 St. Anthonys Close, Ottery St Mary, EX11 1EN
G7 DYD M. Green, 59 Brand End Road Butterwick, Boston, PE22 0JD
G7 DZD Dr K. Quinlan, Polidoris Cottage, Polidoris Lane, High Wycombe, HP15 6XD
GI7 DZE S. Thompson, 19 Windsor Heights, Larne, BT40 1UL
GM7 DZK J. Malone, 8 St. Margarets Crescent, Polmont, Falkirk, FK2 0UP
G7 DZR G. Shand, 5 Bromyard Drive, Chellaston, Derby, DE73 6PF
G7 DZY B. Daniel, Tamar Bay Rd, Freshwater Bay, Isle of Wight, PO40 9QS
G7 EAA D. Mills, 3 Brookside Glasbury, Hereford, HR3 5NF
G7 EAH Dr P. Stewart, 34 North Park Road, Bramhall, Stockport, SK7 3JS
G7 EAQ D. Potten, 151 Sherborne Road, Yeovil, BA21 4HF
G7 EAR Echelford ARS, c/o P. Gray, 2 Bryan Close, Sunbury Oppn Thames, TW16 7UA
G7 EAT Dr J. Hatfield, 22 Blackhorse Crescent, Amersham, HP6 6HP
G7 EBF P. Langfield, 88 Counthill Road, Oldham, OL4 2PE
G7 EBI M. Evans, 87 Tintagel Close, Andover, SP10 4DB
G7 EBL G. Orlebar, 21 Field Lane, Willersey, Broadway, WR12 7QB
GI7 EBM A. Stewart, 122 Dublin Road, Antrim, BT41 4SA
G7 EBR M'hd & E.Berk Rd, c/o R. Mclachlan, Heathersett, Lightlands Lane Cookham, Cookham, SL6 9DH
G7 ECA C. Price, 18 Armley Park Road, Leeds, LS12 2PG
G7 ECE Dr S. Holmes, 31 Brightside Avenue, Staines, TW18 1NE
G7 ECG K. Thompson, 32 The Crescent Pattishall, Towcester, NN12 8NA
G7 ECQ N. Murray, East End House, Oak Lane, Sheerness, ME12 3QR
G7 ECU Eastbourne & Wealden Raynet, c/o J. Eade, High Shaw Sandrock Hill, Sedlescombe, TN33 0QR
G7 EDA R. Woodcock, 5 Walton Road, Sidcup, DA14 4LJ
G7 EDF D. Hall, 29 Airedale Avenue, Tickhill, Doncaster, DN11 9UH
G7 EDK I. Barkley, 39 Fulbeck Avenue, Wigan, WN3 5QN

G7 EDZ M. Hedges, 17 Fairview Road, Dudley, DY1 2RT
G7 EED N. Winter, 66 Priory Wharf, Birkenhead, Wirral, CH41 5LD
G7 EEE A. Mothew, 7 Ashfields, Loughton, IG10 1SB
G7 EEG T. Cole, Walden, The Common, Lydney, GL15 6NT
G7 EEJ D. Swift, 63 Guiness Trust Buildings, Fulham Palace Road, London, W6 8BD
G7 EEN S. Paget, 2 Willow Cottages, Huxley Lane, Chester, CH3 9BE
G7 EFA C. Horsfield, Flat 2, Rosemary Court 53 Chantrey Road, Sheffield, S8 8QU
G7 EFG M. Marston, Apt 12B La Palmeras De Benavista, Calle Virgo, Malaga, Spain, 29680
G7 EFL A. Crooks, 7 The Cleave, Harpenden, AL5 5SJ
G7 EFV W. Waterton, 23 Mill Drive, Leven, Beverley, HU17 5NR
GM7 EGQ I. Harrop, South Millburn Cottage, Benslie, Kilwinning, KA13 7QY
G7 EGU P. Adams, 12 The Birches, Benfleet, SS7 4NT
G7 EGX M. Miller, 12 Leighfields Avenue, Leigh-on-Sea, SS9 5NN
GW7 EHD D. Cotton, 135 Main Road, Bryncoch, Neath, SA10 7TW
GM7 EHN J. Baird, 26 Bearside Road, Stirling, FK7 9BY
G7 EHR R. Watson, 2 Stanley Cottages, Fair View Lane, Colyford, Colyton, EX24 6QZ
G7 EHS I. Tooley, L'Eree, Burnham Road, Chelmsford, CM3 6DP
G7 EHU B. Walley, 52 Main St., Rosliston, Swadlincote, DE12 8JW
G7 EHY D. Robinson, 130 Magnolia Drive, Colchester, CO4 3LX
G7 EIA S. Ralph, 70 Mickleburgh Hill, Herne Bay, CT6 6DX
G7 EIE T. Jacobs, 43 Winfields, Pitsea, Basildon, SS13 1HA
G7 EIK G. Edlin, 2 Ashby Road, Donisthorpe, Swadlincote, DE12 7QG
G7 EIS E. Shaddick, 6 Haylands, Portland, DT5 2JZ
G7 EJH J. Tyerman, 7 Veronica Close, Branston, Lincoln, LN4 1PU
G7 EJK J. Turner, Shuckstone Lodge, Shuckstone Lane, Tansley, Matlock, DE4 5GT
G7 EJN G. Prosser, 23 Guiting Road, Selly Oak, Birmingham, B29 4RD
G7 EJO I. Oxley, 29 The Gables, Newhall, Swadlincote, DE11 0TG
G7 EKC A. Taylor, 106 Raeburn Avenue, Surbiton, KT5 9EA
G7 EKG P. Ainscow, 10 Rectory Road, Felling, Gateshead, NE10 9DH
G7 EKH S. Rishton, 14A The Green, Settle, BD24 9HL
G7 EKJ S. Cox, 137 Perry Walk, Blackrock Road, Birmingham, B23 7XL
G7 EKL N. Denker, 103 Springhill Road, Burntwood, WS7 4UJ
G7 EKM K. Mcgeough, 57 Stonehouse Park, Thursby, Carlisle, CA5 6NS
G7 EKT G. Aucott, 21 Riverway, Wednesbury, WS10 0DN
G7 EKW S. Foxall, 25 Western Road, Sutton Coldfield, B73 5SP
G7 ELA M. Dunn, 45 Chaddock Lane, Worsley, Manchester, M28 1DE
G7 ELC N. Fountain, The Venture, Green Lane, Upton, Huntingdon, PE28 5YE
G7 ELE L. Dean, Apperley Bridge Marina, Waterfront Mews, Bradford, BD10 0UR
G7 ELG A. Scarisbrick, 106 Edward Street, Grantham, NG31 6JG
G7 ELH K. Graham, 44A Adelaide Drive, Colchester, CO2 8UB
G7 ELS A. Bartram, 47 Temple Gate Crescent, Leeds, LS15 0EZ
G7 ELV P. Lockwood, 61 Beverley Road, Whitley Bay, NE25 8JQ
G7 ELX J. Northfield, 48 Gleanings Drive, Halifax, HX2 0PA
G7 ELZ T. Leahy, Flat 15 Old Brewery House 294 London Road, Wallington, SM6 7DD
G7 EME Worcs Mnbounce, c/o D. Palmer, Spidrift, Landsdown Road, Malvern, WR14 1HX
G7 EMH A. Adem, 38 Cantley Gardens, Ilford, IG2 6QA
GW7 EMO D. Brough, 19 Cameron Street, Cardiff, CF24 2NW
GW7 EMV M. O'Reilly, 40 St. Anthony Road, Heath, Cardiff, CF14 4DJ
G7 EMZ I. Mellor, 124 Ryknield Road, Kilburn, Belper, DE56 0PF
G7 ENA D. Neal, 33 Swallow Drive, Louth, LN11 0DN
G7 ENC D. Flatters, 7 Cornwall Crescent, Diggle, Oldham, OL3 5PW
GM7 ENM S. Yates, 67 Closeburn, Thornhill, DG3 5HR
G7 ENQ W. Donald, 15 Kingsland Parade, Portobello, Dublin, Ireland,
G7 ENR C. Davies, 138 Cannons Gate, Clevedon, BS21 5HN
G7 ENS N. Swift, 19 Carlton Road, Caversham, Reading, RG4 7NT
G7 ENT S. Alexander, 18 Southbourne Grove, Hockley, SS5 5EE
G7 EOA G. Harrold, Birches, 2 Mill Hill Lane, Wymondham, NR18 9DD
G7 EOC C. Hopkins, 16 Maypole Road, Gravesend, DA12 2LP
G7 EOD A. Capone, North Lodge Farm Cottage Fixby, Huddersfield, HD2 2EW
G7 EOE E. David, 105 Kingsdown Park, Whitstable, CT5 2DH
G7 EOG C. Flynn, 2 Trafalgar Avenue, Grimsby, DN34 5RE
G7 EOH G. Newnham, 22 Warren Place Calmore, Southampton, SO40 2SD
G7 EOK P. Wilson, 18 Riversdale Road, Halton, Runcorn, WA7 2AP
G7 EPE D. Chamberlain, 33 Drake Close, New Milton, BH25 5JG
G7 EPL L. Wong, 320 Wilbraham Road, Chorlton Cum Hardy, Manchester, M21 0UX
G7 EPM J. Enever, 21 Waldegrave Way, Lawford, Manningtree, CO11 2DT
G7 EPN J. Webb, 36 Westfield Drive, Knutsford, WA16 0BN
G7 EPR R. Jeeves, 172 Aldwick Road, Bognor Regis, PO21 2YQ
G7 EPX B. Coffin, 3 Berkeley Close, Melksham, SN12 6AZ
G7 EPY W. Wright, 53 Forshaw Avenue, Grange Park Estate, Blackpool, FY3 7PW
G7 EQG J. Sturman, 41 Jay Close, Haverhill, CB9 0JR
G7 EQK A. Grime, 7 Egremont Road Milnrow, Rochdale Ol16 4Ep, Rochdale, OL16 4EP
G7 EQO M. Ryan, 13 Normandy Road, Heavitree, Exeter, EX1 2SR
G7 EQR D. Gammans, 35 Chute Avenue, High Salvington, Worthing, BN13 3DS
G7 EQX P. Wickers, 4 Little Foxburrows, Colchester, CO2 7UG
G7 ERC Hastings & Rother Raynet Grp, c/o G. Hodge, 8 Stainsby Street, St Leonards-on-Sea, TN37 6LA
GW7 ERI A. Brown, 3 Wyebank View, Tutshill, Chepstow, NP16 7DR
G7 ERS J. Hauton, 15 Bourne Close, Lincoln, LN6 7DR
G7 ESE C. Tully, 19 Glyn Place East Melbury, Shaftesbury, SP7 0DP
GW7 ESF T. Ford, 14 Hillsnook Road, Ely, Cardiff, CF5 5DD
G7 ESI S. Gregory, 73 Princess Way, The Walshes, Stourport-on-Severn, DY13 0EL
G7 ESL M. Kemble, 41 Princes Way Bletchley, Milton Keynes, MK2 2FB
GM7 ESM J. Grundey, 6 Ternemny Villas, Knock, Huntly, AB54 7LR
G7 ESO K. Reynolds, 20 Wentwood Gardens, Plymouth, PL6 8TD
GD7 ESU C. Ellis, Ballahams, 3 Glen Road, Laxey, Isle Of Man, IM4 7AP
G7 ESX S. Bowman, 39 Pearson Street, Spennymoor, DL166HP
G7 ESY I. Bowman, 22 Bryan Street, Spennymoor, DL16 6DW
G7 ESZ D. Foster, 5 Newman Road, Plymouth, PL5 2DX
G7 ETC S. Abel, 121 Angela Road, Horsford, Norwich, NR10 3HF
G7 ETK S. Yates, 14 Wright Street, Horwich, Bolton, BL6 7HZ
G7 ETM A. Sherratt, 22 Lane Green Avenue, Codsall, Wolverhampton, WV8 2JT
G7 ETS S. Hambleton, 6 Wylde Green Road, Sutton Coldfield, B72 1HB
G7 EUB R. Tabor, 9 Liskeard Road, Saltash, PL12 4HE
G7 EUF G. Rhodes, 54 Chell Green Avenue, Stoke-on-Trent, ST6 7JY
G7 EUT D. Richards, Orchard Cottage, Ashbourne Road, Ashbourne, DE6 4NJ
G7 EVC P. Stone, 2 The Russets, Lees Close, Ashford, TN25 6RW
G7 EVF G. Keene, 28 Little Hoddington, Upton Grey, Basingstoke, RG25 2RN
GW7 EVG G. Nicholas, 37 Lon Y Berllan, Abergele, LL22 7JF
G7 EVI J. Galbraith, 13 Simeons Walk, Quarry Bank, Brierley Hill, DY5 2EL
G7 EVK A. Wade, 3 Ashendene Grove, Stoke-on-Trent, ST4 8NW
G7 EVP M. Pritchard, 155 Elliott Road, March, PE15 8HF
G7 EVQ M. Jordan, 139 Camping Hill, Stiffkey, Wells-Next-the-Sea, NR23 1QL
G7 EVR S. Sprint, 14 Monterey Drive, Allerton, Bradford, BD15 9LP
G7 EVT C. Mills, 16 Broom Close, Wath-Upon-Dearne, Rotherham, S63 7JU
G7 EVY G. Lawton, 125 Glovers Way Burscough, Ormskirk, L40 5AA
G7 EWA A. Hampton, 5 Willow Crescent, Worthing, BN13 2SU
GW7 EWD D. Siviter, Cilgeraint Farm, St. Anns Bethesda, Bangor, LL57 4AX
G7 EWH D. Hughes, Balshult, Eriksmala, Sweden, 361 94
G7 EWK V. Thomas, 4 The Common, Whissonsett, Dereham, NR20 5SZ
G7 EWL S. Hogarth, 34 High Street, Irthlingborough, Wellingborough, NN9 5TN
G7 EWS P. Breck, 209 Eden Park Avenue, Beckenham, BR3 3JW
G7 EWV P. Ridge, 8 Hazel Coppice, Hook, RG27 9RH
G7 EWX N. Price, 245 Anchor Road, Longton, Stoke-on-Trent, ST3 5DX
G7 EWY P. Kember, 2 Sandhills Crescent, Wool, Wareham, BH20 6HB
G7 EXD R. Fletcher, 160 Barnsley Road, Denby Dale, Huddersfield, HD8 8QW
GW7 EXH B. Mee, Anncott, Hylas Lane, Rhuddlan, Rhyl, LL18 5AG
G7 EXO K. Brown, 15 Gloucester Road, Aldershot, GU11 3SL
GW7 EXQ R. Morris, 40 Maes Pedr, Carmarthen, SA31 3BR
G7 EXT H. Jarvis, Dovecote Farm, Patmans Lane, Boston, PE22 8QJ
G7 EXX E. Gwilliam, 15 Sheppard Way, Minchinhampton, Stroud, GL6 9BZ
G7 EXZ S. Hobbs, 15 The Valley, Salisbury, SP2 9EJ
G7 EYA R. Hatcher, 53 Lancaster Drive, Broadstone, BH18 9EH
G7 EYE S. Finnegan, 25 Westcliff Gardens, Margate, CT9 5DT
G7 EYL M. Dixon, 19 Stanford Way, Broadbridge Heath, Horsham, RH12 3LH
G7 EYM M. Parkyn, Brookfield, Clee St. Margaret, Craven Arms, SY7 9DX
GW7 EYP M. Jenkins, 23 Guenever Close, Thornhill, Cardiff, CF14 9AH
G7 EYR P. Wiles, 16 Churchill Road, Broadheath, Altrincham, WA14 5LT
G7 EYS C. Chance, 19 White Beam Rise, Clanfield, Waterlooville, PO8 0LQ
G7 EYV A. Little, 444 Dunsbury Way, Leigh Park, Havant, PO9 5BJ
G7 EZE A. Byrne, 23 The Deansway, Kidderminster, DY10 2RH
G7 EZH P. Higginson, 93 Oakfield Road, Wollescote, Stourbridge, DY9 9DE
G7 FAD V. Ritson, 24 Chapel Road, Pawlett, Bridgwater, TA6 4SH
G7 FAG P. Lunn, 1 Burman Road, Wath-Upon-Dearne, Rotherham, S63 7NE
G7 FAQ E. Cloude, 39 Wentworth Close, Weybourne, Farnham, GU9 9HJ
G7 FAR Raf Waddington ARC, c/o M. Farmer, 16 Beckside, Nettleham, Lincoln, LN2 2PH
GM7 FAS A. Warnock, 21B Melbost Point, Isle of Lewis, HS2 0BG
G7 FAZ A. Mould, 8, Foxwood, Brierley Hill, DY5 2PH
G7 FBE E. Dare, 17 Montgomery Drive, Spencers Wood, Reading, RG7 1BQ
G7 FBT E. Woodhouse, 7 Cow Heys, Dalton, Huddersfield, HD5 9RG
GW7 FBV S. Hathaway, 9 Mirehouse Place, Angle, Pembroke, SA71 5BD
G7 FBY R. Furniss, 7 Elizabeth Road, Sutton Coldfield, B73 5AR
G7 FCC D. Bent, 21 Loughborough Avenue, Nottingham, NG2 4LN
G7 FCJ P. Honeywell, 7 Hopton Gardens Hopton, Great Yarmouth, NR31 9DF
G7 FCL D. Hames, 10 Downs Close, East Studdal, Dover, CT15 5BY
GI7 FCM S. Fleming, 15 Castle Green, Ballynure, Ballyclare, BT39 9GN
G7 FCO G. Howat, 44 Castle Street Thornbury, Bristol, BS35 1HB
GI7 FCP J. Mccormick, 14 Ballyoran Park, Portadown, Craigavon, BT62 1JN
G7 FCR Fylde Coast Ray, c/o R. Knighton, 262 Victoria Road West, Thornton-Cleveleys, FY5 3QB
G7 FCU F. Smith, 3 Downside Close, Findon Valley, Worthing, BN14 0EZ
GI7 FCW P. Quinn, 53 Dernanaught Road, Dungannon, BT70 3BU
G7 FDD W. Cooper, 33 Elm Drive Cherry Burton, Beverley, HU17 7RJ
G7 FDS T. Whitehead, 23 Rossendale Road, Lytham St Annes, FY8 3HY
G7 FDW C. Milburn, 9 Woodhall Avenue, Bradford, BD3 7BY
G7 FEA M. Codling, 18 Ash Grove Pinehurst, Swindon, SN2 1RX
G7 FED E. Wells, 1A Brocklewood Avenue, Poulton-le-Fylde, FY6 8BZ
G7 FEE F. Harvey, 39 Simonside Terrace, Heaton, Newcastle upon Tyne, NE6 5JY
G7 FEF J. Pipkin, 46 Charles Avenue, Albrighton, Wolverhampton, WV7 3LF
G7 FEG I. Kilkenny, 23 Hazelhurst Road, Stalybridge, SK15 1HD
G7 FEL J. Slater, 313 Southend Road, Stanford-le-Hope, SS17 8HL
G7 FEP D. Birt, 3 South Brent Close, Brent Knoll, Highbridge, TA9 4BS
G7 FEQ B. Shipton, 4 School Close, Kilcott Road, Wotton-under-Edge, GL12 7RH
G7 FFB D. Egleton, 16A Frescade Crescent, Basingstoke, RG21 3NF
G7 FFC V. Prall, 20 Marlowe Close, Basingstoke, RG24 9DD
G7 FFI K. Harding, 21 Doulton Way, Ashingdon, Rochford, SS4 3BX
G7 FFK A. James, 66 Rydal Crescent, Worsley, Manchester, M28 7JD
G7 FFM P. Malpass, 30 Countisbury Road, Norton, Stockton-on-Tees, TS20 1PZ
G7 FFR J. Rutherford, 270 Milburn Road, Ashington, NE63 0PL
G7 FFS A. Pike, 63 Mill Lane, Bentley Heath, Solihull, B93 8NN
G7 FFV I. Miller, 5 Avenue Terrace, Sunderland, SR2 7HB
G7 FFW S. Lonsdale, 16 Hinkler St., Cleethorpes, DN35 8PR
G7 FFZ J. Humphries, 23 Sycamore Drive, Lutterworth, LE17 4TR
G7 FGA D. Moreland, 179 Carr Lane, York, YO26 5HQ
G7 FGD J. Brown, St. Winnolls House, St. Winnolls, Torpoint, PL11 3DX
GM7 FGH J. Bartolo, 84 Calderbraes Avenue, Uddingston, Glasgow, G71 6ED
GI7 FGQ P. Faulkner, 40 Glenariff Drive, Comber, Newtownards, BT23 5HA
G7 FGR G. Cluley, 1 Shepherds Lane, Greetham, Oakham, LE15 7NX
GJ7 FGS J. Bette- Bennett, 2 Aspley Villas, Bagatelle Road, Jersey, JE2 7TA
G7 FGZ D. Mitchell, 55 Halewick Lane, Sompting, Lancing, BN15 0ND
G7 FHA C. Brailsford, 65 Cherry Orchard, Codford, Warminster, BA12 0PW
G7 FHU I. Davies, 12 Kaye Avenue, Culcheth, Warrington, WA3 5SA
G7 FHV T. Beeching, 11 Kents Road, Haywards Heath, RH16 4HL
GI7 FHZ E. Mccrystal, 33 Richmond Park, Omagh, BT79 7SJ
G7 FIA P. Page, 42 Alamein Gardens, Stone, Dartford, DA2 6BN
GM7 FIE I. Stevenson, 68 Silverknowes Eastway, Edinburgh, EH4 5NE
G7 FIJ B. Parsons, 20 High Park Road, Halesowen, B63 2JA
G7 FIK S. Regan, 92-94 Lytham Road, Blackpool, FY1 6DZ
GM7 FIS J. Russell, 15 Glen View, Cumbernauld, Glasgow, G67 2DA
G7 FJC P. Fellingham, Flat 6, Duncan House, Collingwood Close, Peacehaven, BN10 8BA

G7 FJU C. Ovenden, 2 Firemans Cottage, Fortis Green, London, N10 3PB
GI7 FJY N. Gamble, 21 Anderson Crescent, Waterside, Londonderry, BT47 2BY
G7 FJZ P. Selley, 2 Coronation St., Barnstaple, EX32 7AY
G7 FKF R. Phillips, 24 Harris Lane, Wistow, Huntingdon, PE28 2QG
G7 FKJ C. Holloway, 23 Ryecroft Road, Stretford, Manchester, M32 9BS
G7 FKP D. Henderson, 2 Beverley Court, Beverley Road, York, YO43 3NB
G7 FKS H. Ellis, 10 Gardens Quay, Pitwines Close, Poole, BH15 1XL
G7 FKX C. Wood1, 2 Plain Cottages, Plain Road, Marden, Tonbridge, TN12 9LS
G7 FKZ R. Bowden, 35 Glebelands, Biddenden, Ashford, TN27 8EA
GM7 FLG D. Pegg, 11 Glenward Avenue, Lennoxtown, Glasgow, G66 7EP
G7 FLI J. Moyse, 2 Kestle Drive, Truro, TR1 3PT
G7 FLS K. Lawson, 60 Minster Avenue, Beverley, HU17 0ND
G7 FLX B. Timms, 74 Park Gwyn, St. Stephen, St Austell, PL26 7PN
GM7 FLZ E. Chesters, Blackhill, Blackhill Road, Kirkwall, KW15 1FP
G7 FMB R. Burns, 43 Gibson St., Bickershaw, Wigan, WN2 5TF
G7 FMI M. Kensall, 40 Eskdale Avenue, Ramsgate, CT11 0PB
G7 FMJ G. Mitchell, 7 Buxton Close, Whetstone, Leicester, LE8 6NT
G7 FML P. Clarke, 93 Commercial Road, Spalding, PE11 2YU
G7 FMQ J. Sutton, 10 Cathcart Road, Stourbridge, DY8 3UZ
G7 FMU D. Appleby, Unterrohr 194, at-8294, Austria, Rohr bei Hartberg
G7 FMV D. Sweet, 50 Mereside, Soham, Ely, CB7 5XE
G7 FMW P. Olson, 23 Dennett Close, Liverpool, L31 5PD
G7 FND E. Millership, 16 Bramble Way, Wirral, CH46 7UP
G7 FNM J. Walmsley, Bank Field, 5 Dimples Lane, Preston, PR3 1RD
G7 FNN A. Morley, 87 Epsom Drive, Ipswich, IP1 6SS
GI7 FNP H. Massey, 156 Killaughey Road, Donaghadee, BT21 0BQ
GW7 FNQ D. Jones, Noddfa, High Street, Bodorgan, LL62 5AS
G7 FNT S. Taylor, 39 Hookhills Road, Paignton, TQ4 7LR
G7 FNU R. Beaumont, 49 Vincent Close, Broadstairs, CT10 2ND
GI7 FOD P. Mccollam, 32 Robinson Way, Bangor, BT19 6NR
G7 FOT W. Sanger, Tregonning Lea, Laddenvean, Helston, TR12 6QD
G7 FOX Melton Mowbray ARS, c/o G. Mason, 120 Scalford Road, Melton Mowbray, LE13 1JZ
G7 FPJ J. Marks, Flat 112 Weavers Quay, 51 Old Mill Street, Manchester, M4 6GB
G7 FPM K. Pontin, Flat 41 Park Farm The Street, Moredon Swindon, SN25 3ES
GM7 FPN A. Mcpherson, 3 Fulmar Road, Elgin, IV30 4HL
G7 FPR G. Flanagan, Flat 9, West Cliff Court, 25 Portarlington Road, Bournemouth, BH4 8BX
G7 FPS S. Harrison, 3 Smallbridge Close, Worsley, Manchester, M28 7XS
G7 FPU C. Wissun, 111 Berkeley Vale Park, Berkeley, GL13 9TQ
G7 FPW P. Howard, 3 Hollies Close, Shepton Mallet, BA4 5LG
GM7 FPX A. Stewart, 8 Crinan Place, Ardrossan, KA22 7PT
G7 FPZ D. Foster, 29 Harrowfield Road, Stechford, Birmingham, B33 9BU
G7 FQE J. Whiffen, 5 Sharpthorpe Close, Lower Earley, Reading, RG6 4DB
GW7 FQP S. Earle, Clydfan Nanternis, New Quay, SA45 9RP
G7 FQY G. Spark, Park Lodge, Cheltenham Drive, Sale, M33 2DQ
GM7 FQZ J. Brooks, The Steading, West Mosstown, Lonmay, Fraserburgh, AB43 8RU
GM7 FRC Fife Raynet Group, c/o J. Burke, 25 Duncan Road, Auchmuty, Glenrothes, KY7 4HS
G7 FRH A. Russ, 21 Francis Road, St. Pauls Cray, Orpington, BR5 3LY
G7 FRR M. Samson, 277 Upton Lane, London, E7 9PR
G7 FRW K. Mccaffery, 34 Ringwood Road, Luton, LU2 7BG
G7 FSA R. Colclough, 8 Parker Jervis Road, Stoke-on-Trent, ST3 5RP
G7 FSC K. Davis, 26 Mendip Drive, Nuneaton, CV10 8PT
G7 FSD A. Holles, 20 Sapcote Road, Burbage, Hinckley, LE10 2AU
G7 FSH F. Pearson, 15 York Road, Driffield, YO25 5AT
G7 FSJ P. O'Connor, 149 Roxeth Green Avenue, Harrow, HA2 0QJ
G7 FSR A. Wyard, 85 Swaledale, Bracknell, RG12 7ET
G7 FTA M. Collins, 15 Trevethan Rise, Falmouth, TR11 2DX
G7 FTD K. Taber, 110 Uplands, Peterborough, PE4 5AF
GM7 FTK M. Long, 52 Stirling Road, Milnathort, Kinross, KY13 9XG
G7 FTM S. Clayton, 22 Orchard Avenue, North Anston, Sheffield, S25 4BW
G7 FTS C. Bessant, 15 Burgess Green Close St Annes Park, Bristol, BS4 4DG
G7 FTS O. Whiteside, 9 Beech Grove House, Beech Grove, Harrogate, HG2 0ES
G7 FUM N. Seath, 6 Harvester Way, Sibsey, Boston, PE22 0YD
G7 FUQ A. Vincent, 12 Spelman Road, Norwich, NR2 3NJ
G7 FUV I. Marsh, 56B Oliver Crescent, Farningham, Dartford, DA4 0BE
G7 FUW J. Birch, 15 Adstone Grove, Birmingham, B31 4AU
G7 FVH R. Barrick, Orchard Bungalow, Pasture Lane, Middlesbrough, TS6 8EH
G7 FVR C. Heywood, 29 Smallwood Mews, Wirral, CH60 6TE
GM7 FWA A. Pratt, 129 Brodie Court, Glenrothes, KY7 4UE
G7 FWD B. Nicholls, 15 Canal Way, Devizes, SN10 2UB
G7 FWE A. Smith, 12 Northgate, Beccles, NR34 9AS
G7 FXO P. Werba, 47 Ulwell Road, Swanage, BH19 1LG
G7 FXW P. Whitworth, 67 Staddiscombe Road, Staddiscombe, Plymouth, PL9 9LU
GW7 FXX B. Harries, 12 Panteg, Llanelli, SA15 3TF
G7 FXY P. Hallett, 30 Summerdown Walk, Trowbridge, BA14 0LJ
G7 FXZ G. Hodgetts, 2 Friars Gorse, Stourton, Stourbridge, DY7 6SP
GM7 FYB D. Wemyss, 24 Brucklay Court, Peterhead, AB42 2UF
GW7 FYG C. Wright, 12 Bryn Teg, Arddleen, Llanymynech, SY22 6PZ
G7 FZB Dr G. Ridgeway, 4 Russell Avenue, Alsager, ST7 2BL
G7 FZJ M. Whatley, Woodside West, Wood Lane, Halifax, HX3 8HB
G7 FZN P. Du Plessis, 42 La Providence, Rochester, ME1 1NB
GW7 FZW M. Heel, 27 Englefield Drive Oakenholt, Flint, CH6 5SB
G7 GAB R. Hagues, 40 Barton Road, Rugby, CV22 7PT
GM7 GAE I. Mackenzie, 52/5 Craighall Road, Edinburgh, EH6 4RU
G7 GAG J. O'Neill, 24 Lily Lane, Bamfurlong, Wigan, WN2 5JN
GW7 GAH R. Dore, Maespoeth, Corris, Machynlleth, SY20 9RD
G7 GAK D. Garbutt, 8 Yorkshire Road, Partington, Manchester, M31 4GW
G7 GAP J. Cartwright, 109 Kneller Road, Twickenham, TW2 7DT
G7 GAZ B. Kerrison, 45 Bramley Crescent, Bearsted, Maidstone, ME15 8JZ
GM7 GBD G. Macgregor, 6 Kincaidfield Milton Of Campsie, Glasgow, G66 8ER
G7 GBE S. Burgoine, 47 Squirrel Close, Hounslow, TW4 7NU
G7 GBJ J. Kaczmarek, 2 Westgate Terrace, London, SW10 9BJ
G7 GBN P. Baird, 168 Plumberow Avenue, Hockley, SS5 5AT
G7 GBZ P. Leach, 21 Abbess Close, Chelmsford, CM1 2SE
G7 GCD S. Lee, 51 Ireland Crescent, Red Deer, Canada, T4R 3K8
G7 GCF P. Kell, Flat25, Church Court, Church Rd., Haywards Heath, RH16 3UE
G7 GCI M. Collett, Flat 9, Connaught Court, 13 Connaught Avenue, London, E4 7AG
G7 GCU M. Edge, 2 Yew Tree Place, Walsall, WS3 3DG
G7 GCW Dr P. Andrew, 3 Grayway Close, Highfields Caldecote, Cambridge, CB23 7UZ
G7 GDA T. Wootton, 1 Lingfield Drive, Walsall, WS6 6LS
G7 GDC A. Gosden, 10 Radcliffe Way, Northolt, UB5 6HP
GM7 GDE A. Hood, 26 Annan Avenue, East Kilbride, Glasgow, G75 8XT
G7 GDV A. Porteous, 73 Dowgate Close, Tonbridge, TN9 2EJ
G7 GEA J. Broadfoot, 65A Swan Meadow, Pewsey, SN9 5HP
G7 GEE J. Gee, 51 Hattons Lane, Childwall, Liverpool, L16 7QR
G7 GEF G. Duthie, 15 Wagtail Close, Twyford, Reading, RG10 9ED
G7 GEI M. Arliss, 22 Rowena Avenue, Edenthorpe, Doncaster, DN3 2JF
G7 GEL J. Mansell, 8 Himley Gardens, The Straits, Dudley, DY3 3AS
G7 GEP C. Danks, Neuadd Las, Llanddewi Brefi, Tregaron, SY25 6NY
G7 GES B. Norcott, 5 The Shrubbery, Upminster, RM14 3AH
G7 GEU V. Bruntnell, 4 Cypress Avenue, Dudley, DY3 2JF
G7 GEX N. Potter, 4 Eastleigh Drive, Mickleover, Derby, DE3 9HZ
G7 GFC D. Mullock, 18 Tewkesbury Close, Upton, Chester, CH2 1NF
G7 GFH W. Baker, 41 Kenwood Park Road, Sheffield, S7 1NE
G7 GFK K. Percival, 1608 Scant Row, Chorley Old Road, Bolton, BL6 6PZ
G7 GFM J. Hunt, 43 Felton Close, Redditch, B98 0AG
G7 GFP I. Bishop, 115 Burman Road Shirley, Solihull, B90 2BQ
G7 GFQ M. Charlwood, 60 Alfred Road, Feltham, TW13 5DJ
G7 GFR B. Clifford, 8 Caldbeck Place, North Anston, Sheffield, S25 4JY
G7 GFX P. Everard, 56 Hawkins Crescent, Shoreham-by-Sea, BN43 6TP
G7 GGA N. Dawson, 4 Bathurst Close, Staplehurst, Tonbridge, TN12 0NA
G7 GGF C. Martin, 73299, Bfpo 5442, BF19DB
G7 GGG G. Richardson, 11 Queensway, Forest Town, Mansfield, NG19 0BX
G7 GGH G. Hurrell, 13 Hinton Road, Newport, PO30 5QZ
G7 GGJ A. Edwards, 45 Chilton Grove, Yeovil, BA21 4AW
G7 GGM D. Thomalla, 14 Walkers Lane, Penketh, Warrington, WA5 2PA
G7 GGN J. Williams, 41 Cote Green Lane, Marple Bridge, Stockport, SK6 5EB
G7 GGT S. Mullins, 549 Bromford Lane, Washwood Heath, Birmingham, B8 2EA
GI7 GHC T. Lyons, 3 Clanbrassil Gardens, Portadown, Craigavon, BT63 5YD
GW7 GHE D. Pearson, Warren Cottage Pontfadog, Llangollen, LL20 7AT
G7 GHH I. Wraith, 7 Bowman Close, Sheffield, S12 3LR
G7 GHP G. Fellows, 34 The Ridings, Bexhill-on-Sea, TN39 5HU
G7 GHT M. Alexander, 51 Park Lane, Bootle, L20 6DJ
GM7 GIF K. Juner, 56 Queens Gardens, East Calder, Livingston, EH53 0EG
G7 GIG R. Vincent, 14 Trevenson Street, Camborne, TR14 8JB
G7 GIJ J. Barnett, 20 Springford Gardens, Southampton, SO16 5SW
G7 GIK D. Keene, Firemark Cottage West Street Odiham, Hook, RG29 1NT
GM7 GIO W. Mackinnon, 31 Kirk Bauk, Symington, Biggar, ML12 6LB
GM7 GIS M. Glendinning, 148 Gala Park, Galashiels, TD1 1HD
G7 GJA P. Cockayne, 7A Wrekin Drive, Bradmore, Wolverhampton, WV3 7HZ
G7 GJI D. Sager, 29 Station Road, Mickleover, Derby, DE3 9GH
GM7 GJM C. Unsworth, 3 Thorfinn Place St. Margarets Hope, St Margaret's Hope, KW17 2TR
G7 GJN A. Khachaturian, 377 Watford Road, St Albans, AL2 3DD
G7 GJO P. Morris, 117 Lonsdale Avenue, Doncaster, DN2 6HF
G7 GJS N. Cheesewright, 5 Duberly Close, Perry, Huntingdon, PE28 0BP
G7 GJT W. Everton, Fencott, Fen Road, Lincoln, LN4 1AE
G7 GJU G. Darby, 5 Lumsden Terrace, Catchgate, Stanley, DH9 8EQ
G7 GJV A. Gordon, 1 Surrey Street, Hetton-Le-Hole, Houghton le Spring, DH5 9LX
GI7 GJX N. Simmons, 57 Moyra Road, Doagh, Ballyclare, BT39 0SQ
G7 GJY J. Chapman, 77A Carnforth Gardens, Elm Park, Hornchurch, RM12 5DR
G7 GJZ C. Brown, 73 Ringstone, West Huntspill, Highbridge, TA9 3RF
G7 GKC I. Boyd, 43 Lower Landedmans, Westhoughton, Bolton, BL5 2QL
G7 GKD L. Tryhorn, 46 Mill Green Road, Amesbury, Salisbury, SP4 7RE
GW7 GKN S. Gordon, 6 Oakridge Acres, Tenby, SA70 8DB
G7 GKQ L. Measures, 163 Huddersfield Road, Meltham, Holmfirth, HD9 4AJ
GM7 GKT R. Smith, 27 Elm Lane, Foresters Lodge, Glenrothes, KY7 5TD
GW7 GKX J. Rough, 10 Beaconsfield Road, Shotton, Deeside, CH5 1EZ
G7 GLA J. Mitchinson, 93 Hinckley Road Leicester Forest East, Leicester, LE3 3GN
GM7 GLJ A. Potter, 42 Pender Gardens, Rumford, Falkirk, FK2 0BJ
G7 GLL L. Carlile, 26 The Bungalows, Stonebroom, Alfreton, DE55 6LH
G7 GLP J. Walton, 14 Chapel Street, Stanhope, Bishop Auckland, DL13 2NB
G7 GLQ D. Cottrell, 2 Foss Court, Summerhill Road, Bristol, BS5 8HF
G7 GLR Grtr Lndn Rayne, c/o I. Jackson, 5 Vivien Close, Chessington, KT9 2DE
G7 GLS J. Pinna, 31 Bowness Road, Little Lever, Bolton, BL3 1UB
G7 GLW R. Cains, 58 Sunnydale Road, Lee, London, SE12 8JN
G7 GLZ R. Hourston, 12 The Warren, Chesham, HP5 2RY
G7 GMB J. Craig, 1 Eldon Road, Eastbourne, BN21 1UD
G7 GMD M. Ollerton, 1 Hammy Way, Shoreham-by-Sea, BN43 6GH
G7 GMQ D. Smith1, 65 St. Anthonys Road, Kettering, NN15 5JB
G7 GMR B. Golland, 15 Turpin Close, Gainsborough, DN21 1PA
G7 GMU G. Lamb, Parisfield, Headcorn Road, Tonbridge, TN12 0BT
G7 GMZ W. Newton, 7 Moss Close, Bridgwater, TA6 4NA
G7 GNA L. Smith, 13 Eagle Avenue, Waterlooville, PO8 9UB
GM7 GNO N. Goodall, 26 Greenbank Loan, Edinburgh, EH10 5SJ
G7 GNS J. Olive, 2 Wyke Cottage Wotton Road, Rangeworthy, Bristol, BS37 7NA
G7 GNU P. Brayshaw, 38 Chilfrome Close, Canford Heath, Poole, BH17 9WE
G7 GOA S. Constable, 18 Salvington Gardens, Worthing, BN13 2BH
GM7 GOE M. Doig, 18 Gotterstone Drive, Broughty Ferry, Dundee, DD5 1QW
G7 GOK N. Breckell, Barn Hill Lodge, Barn Hill Road, Broadwell, Coleford, GL16 7BL
G7 GOV M. Hill, 31 Brocklesby Avenue, Immingham, DN40 2AS
GM7 GPG A. Jakowuik, 167 Magdala Terrace, Galashiels, TD1 2HZ
G7 GPI A. Baily, 13 Longleigh Lane, Bexleyheath, DA7 5SL
G7 GPJ R. Banks, Highview, New Road, Sturminster Newton, DT10 2HF
G7 GPL N. Giles, 6 Bridgewater Mews, London Road, Warrington, WA4 6LF
G7 GPU A. Sharman, 9 Silver Close, Minety, Malmesbury, SN16 9QT
G7 GQA A. Doswell, 14 Carisbrooke Drive, Charlton Kings, Cheltenham, GL52 6YA
G7 GQB M. Woodhouse, 18 Soame Close, Aylsham, Norwich, NR11 6JF
G7 GQC G. Beckingham, 20 Baptist Close, Abbeymead, Gloucester, GL4 5GD
G7 GQD D. Pearce, 2 Mell Avenue, Hoyland, Barnsley, S74 9HF
G7 GQH R. Hannemann, 112 Northern Road, Aylesbury, HP19 9QY
G7 GQL J. Sutton, 15 Lowther Street, Penrith, CA11 7UW
G7 GQM E. Sutton, 15 Lowther Street, Penrith, CA11 7UW
G7 GQO R. Harman, The Briars, Brambleberry Lane, Skegness, PE24 5DQ

G7 GQW D. Williams, 28 Mill Lane, Great Sutton, Ellesmere Port, CH66 3PF
G7 GQX D. Howard, 6 Draycote Close, Solihull, B92 9PT
G7 GRC Grantham ARC, c/o K. Burton, 2 Council House, Stainfield Road, Kirkby Underwood, Bourne, PE10 0SG
GM7 GRH N. Hardie, 38 Sentry Knowe, Selkirk, TD7 4BG
G7 GRM J. Booth, 2 Fairfax Mews, London, E161TY
G7 GRO D. Stimpson, 19 Moss Bank, Winsford, CW7 2ED
G7 GRR S. Everett, 4 Ilkley Place, Newcastle, ST5 6QP
G7 GRU M. Lucas, 22 Ferny Brow Road, Wirral, CH49 8EE
GI7 GRY S. Gordon, 138 Mullalelish Road, Richhill, Armagh, BT61 9LT
GI7 GSB A. Wiese, 105 Milltown Avenue, Lisburn, BT28 3TR
G7 GSC N. Godden, 23 Rapsons Road, Willingdon, Eastbourne, BN20 9RJ
G7 GSD K. Osborn, 2A Sullington Gardens, Worthing, BN14 0HR
G7 GSF S. Blandford, Flat 18, Avro House, 5 Boulevard Drive, London, NW9 5HF
G7 GSR J. Shrubsall, 54 Park Avenue, Sittingbourne, ME10 1QY
G7 GSX C. Penfold, 149 Shuttlewood Road Bolsover, Chesterfield, S44 6NX
G7 GTG A. Hyndman, Norman House, Railway Terrace, Herts, WD4 8JE
G7 GTH A. Marriott, Norman House, Railway Terrace, Herts, WD4 8JE
GM7 GTS C. Richman, 18 Nigel Rise, Livingston, EH54 6LT
G7 GTU G. Sharples, 24 Kelboro Avenue, Audenshaw, Manchester, M34 5UH
GM7 GTX M. Kaye, 146 Newlands Road, Grangemouth, FK3 8NZ
G7 GUA A. Dennis, Chy An Ros, Riverside, Hayle, TR27 5JD
G7 GUB A. Alderton, 7 Bigland Drive, Ulverston, LA12 9NU
G7 GUG G. Wales, 1 Ryton Fold, North Anston, S25 4AG
GM7 GUL C. Jordan, 3 Birch Avenue, Rosemount, Blairgowrie, PH10 6XE
G7 GUO S. Falconer, 6 Ogilvie Road, High Wycombe, HP12 3DS
GI7 GUT D. Watt, 51 Rashee Road, Ballyclare, BT39 9HT
GM7 GVD D. Innes, 6 Mamore Terrace, Inverness, IV3 8PF
GI7 GVI T. Henderson, 7 Legaloy Road, Ballyclare, BT39 9PS
G7 GVJ S. Fletcher, Fernleigh, Ash Lane, Gloucester, GL2 9PS
G7 GVP C. Price, 16 Woodlands Drive, Warton, Preston, PR4 1UQ
G7 GWA A. Jakins, 29 Burchnall Close, Deeping St. James, Peterborough, PE68QJ
GW7 GWO S. Evans, 1 Brynheulwen Blaenannerch, Cardigan, SA43 2AH
G7 GWT G. Taylor, 33 Heol Aberwennol, Borth, SY24 5NP
GM7 GWW S. Gardiner, Kyendigaet, Whiteness, Shetland, ZE2 9GJ
G7 GXE P. Kitson, 15 Louvain Road, Derby, DE23 6DA
GM7 GXI G. Cowan, 15 Waterhaughs Grove, Glasgow, G33 1RS
G7 GXR B. Clewes, 19 Church Mews, Denton, Manchester, M34 3GL
GI7 GXZ S. Dornan, 3 Hampton Lane, Bangor, BT19 7GB
G7 GYN C. Barlow, 2/5 Hospital Steps, Gibraltar, Gibraltar, GX11 1AA
G7 GYR K. Wade, Eccleston Hall, Lydiate Lane, Chorley, PR7 6LY
G7 GZB C. Davies, 84 Hob Hey Lane, Culcheth, Warrington, WA3 4NW
G7 GZC D. Coles, 20 James Darby House 11 Mereway Road, Twickenham, TW2 6SA
G7 GZJ K. Oliver, 155 Old Road, East Cowes, PO32 6AX
G7 GZK I. Croft, 34 Laburnum Drive, Armthorpe, Doncaster, DN3 3HE
G7 GZU S. Selwyn, 65 Porterhouse Road, Ripley, DE5 3FL
G7 GZV H. Houldershaw, The First Bungalow, Fen Road, Boston, PE22 8EX
G7 GZZ E. Gaffney, 1 White Hart Lane, Wistaston Green, Crewe, CW2 8EX
GW7 HAE C. Davies, Afallon, 3 Penygraig, Aberystwyth, SY23 2JA
G7 HAF W. Hunton, 60A Bondgate, Helmsley, York, YO62 5EZ
G7 HAH Finningley ARS, c/o M. Hotchin, 122 Buckingham Avenue, Scunthorpe, DN15 8NS
G7 HAR B. Ferris, 5 Guildway, Todwick, Sheffield, S26 1JN
G7 HAS A. Newton, Rockburn, Victoria Road, Malvern, WR14 2TE
G7 HBN P. Osborne, 11 Galston Road, Luton, LU3 3JZ
G7 HBO R. Cornell, 161 Tuckers Road, Loughborough, LE11 2PH
G7 HBU T. Hickling, 6 Harrold Road, Bozeat, Wellingborough, NN29 7LP
G7 HBV C. Heard, 42 Hallowell Down South Woodham Ferrers, Chelmsford, CM3 5FS
G7 HCB B. Atterbury, 7 Ross Court, Stevenage, SG2 0HD
G7 HCC D. Jones, 120 Heathfield Road, Keston, BR2 6BF
G7 HCJ A. Parr, 52B Trent Boulevard, West Bridgford, Nottingham, NG2 5BD
G7 HCL P. Good, 80 Meredith Road, Stevenage, SG1 5QS
G7 HCN A. Jones, 179 Blandford Road, Efford, Plymouth, PL3 6JZ
G7 HCO N. Lambert, Bradfields Farm, Burntmills Road, Wickford, SS12 9JX
G7 HCQ D. Browne, 293 St. Albans Road, Hemel Hempstead, HP2 4RP
G7 HCR G. Richardson, The Homestead Washway Road, Holbeach, Spalding, PE12 7PP
G7 HCT K. Moore, 8 Lilac Close, Toftwood, Dereham, NR19 1JY
GW7 HDC A. Rowe, 5 Church Street, Knighton, LD7 1AG
G7 HDR D. Horder, 77 Grove Avenue, Harpenden, AL5 1EZ
GW7 HDS S. Barker, 11 Prosser Street, Treharris, CF46 5LN
G7 HDU J. Tombs, Mariedown, Bustards Lane, Wisbech, PE14 7PQ
G7 HDW J. Bigger, 128 Trueway Drive South, Shepshed, Loughborough, LE12 9DY
G7 HEJ G. Atkinson, 23 Fielding Road, Blackpool, FY1 2QL
G7 HEK A. Owen, 57 Melrose Avenue, Vicars Cross, Chester, CH3 5JB
G7 HEN M. Priestley, 29 Birchlands Avenue, Wilsden, Bradford, BD15 0HB
G7 HEP A. Ellis, Eikly Tregada, Launceston, PL15 9NA
G7 HEY L. Morrell-Cross, Delta Lodge, 14 Rushton Crescent, Bournemouth, BH3 7AF
G7 HEZ J. French, 25 Twickenham Court, Stourbridge, dy8 4qg
G7 HFE S. Hitches, 7 Church Close, Chedgrave, Norwich, NR14 6NH
G7 HFL C. Elphick, 2 Vine Way, Brentwood, CM14 4UU
G7 HFP C. Castle, 2 Wellington Close, Mundesley, Norwich, NR11 8JF
G7 HFS I. Harling, 71 De Cham Road, St Leonards on Sea, TN37 6HF
G7 HFW A. Wood, Flat 413, Manchester, M1 2FA
GW7 HFZ A. Strachan, 16 Clos Y Wiwer, Llantwit Major, CF61 2SG
G7 HGB J. Dunn, 10 Endsleigh Close, Upton, Chester, CH2 1LX
G7 HGD P. Allott, 1 Abbey Court, Abbey Road, Knaresborough, HG5 8HX
G7 HGF I. Simpson, Honeysuckle Cottage, 39 Chewton Street, Nottingham, NG16 3GY
G7 HGI R. Roberts, 13 Tudor Way, Wickford, SS12 0HS
G7 HGQ D. Horwood, 12 Curtis Close, Mill End, Rickmansworth, WD3 8QA
G7 HGT P. Stimpson, 93 Chaucer Road, Farnborough, GU14 8SR
GW7 HGU M. Howard, 64 Lawrenny St., Neyland, Milford Haven, SA73 1TB
GM7 HHB J. Brown, 133 Meadowbank Road, Kirknewton, EH27 8BH
G7 HHI S. Curry, Barnabus Communications, Barnabus Cottage, Egley Road, Mayford, Mayford, Woking, GU220NQ
G7 HHK R. Johnson, Honeysuckle Cottage, Front Street, Northallerton, DL6 2AA
G7 HHM L. Dring, 22 Castle Street, Eastwood, Nottingham, NG16 3GW
G7 HHN K. Glover, 7 Mill Lane, Cressing, Braintree, CM77 8HN
G7 HHQ R. Saunders, The Grange, High Road, Wisbech, PE13 4RG
G7 HHT M. Gotts, 23 Beechcroft Avenue, Croxley Green, Rickmansworth, WD3 3EG
G7 HHU A. Edwards, 3 Simonside Close, Morpeth, NE61 2XY
G7 HHW G. Phillips, 14 Orchard Close, Plymouth, PL7 2GT
G7 HHZ J. Whelan, 1 Chevin Road, Milford, Belper, DE56 0QH
G7 HIC K. Bow, 16 Brook Road, Ivybridge, PL21 0AX
G7 HID M. Burgess, 63 Chalvey Park, Slough, SL1 2HX
G7 HIH R. Pedro, 65 Glebe Crescent, Harrow, HA3 9LB
G7 HII D. Lloyd, No.5 The Close, Burton Gardens, Hereford, HR4 8RQ
G7 HIJ J. Gunia, 21 Campbell Avenue, Leek, ST13 5RR
G7 HIK J. Doherty, 101 Padacre Road, Torquay, TQ2 8QQ
G7 HIN P. Riddell, 4 Pear Tree Road, Addlestone, KT15 1SR
G7 HIO W. Austin, 53 Giantswood Lane, Congleton, CW12 2HQ
G7 HIQ J. Hickey, 53 Norwood Avenue, Hasland, Chesterfield, S41 0NN
GM7 HIR A. Pert, 56 Lochiel Drive, Milton Of Campsie, Glasgow, G66 8ET
G7 HIT P. Chambers, 7 Redland Close, Beeston, Nottingham, NG9 5LA
G7 HIU R. Hurst, 33 Northern Road, Aylesbury, HP19 9QT
G7 HIX R. Gray, 12 St. Francis Close, Deal, CT14 9LS
G7 HIY T. Jefford, 7 Bellevue Street, Folkestone, CT20 1HY
G7 HJD G. Holland, 11 Swanton Drive, Dereham, NR20 4DW
G7 HJG R. Blewitt, Fazeley Mill Marina, Coleshill Road, Fazeley, Tamworth, B78 3SE
G7 HJJ H. Holman, 62 The Ridge, Kennington, Ashford, TN24 9EU
G7 HJK R. Kearnes, 25 Epsom Close, Clacton-on-Sea, CO16 8FE
GW7 HJN S. Tweed, Y Tardis 257 Penybanc Road, Ammanford, SA18 3QW
G7 HJQ M. Erber, 75 St. Andrews Road North, Lytham St Annes, FY8 2JF
G7 HJR T. Rudderham, 24 Casswell House, Grimsby, DN32 7SB
G7 HJT T. Reynard, 12 Acorn Close, Selsey, Chichester, PO20 9HL
G7 HJX D. Raybould, 63 Rochester Avenue, Burntwood, WS7 2DL
G7 HKN P. Walsh, 2 Elm Road, Winwick, Warrington, WA2 9TW
G7 HKQ I. Tideswell, 2 Pangbourne Avenue, Urmston, Manchester, M41 0GF
G7 HKT C. Fowle, 70 The Parade, Greatstone, New Romney, TN28 8RE
G7 HKU J. Turner, 7 Highfield Crescent, Baildon, Shipley, BD17 5NR
G7 HKZ T. Allen, 15 Manning Road, Cotford St. Luke, Taunton, TA4 1NY
G7 HLB A. Coombs, Treveth, Boyton, Launceston, PL15 9RJ
G7 HLG B. Morrell-Tourle, 77 Mallard Road, Bournemouth, BH8 9PJ
G7 HLP K. Baldock, 66 Port Road, Northampton, NN5 6NL
G7 HLU V. Meads, May Tree Barn, Main Street Upper Benefield, Peterborough, PE8 5AN
G7 HLV J. Jordan, 18 Sunningdale Crescent, Cullingworth, Bradford, BD13 5BA
G7 HLW G. Burn, 4 Goston Gardens, Thornton Heath, CR7 7NQ
GW7 HLZ R. Davies, 8 Deri Road, Uskabergavenny, NP75SY
G7 HMA I. Smith, 4 Stour Road, Grays, RM16 4BS
G7 HMB G. Bull, 48 Spragg House Lane, Stoke-on-Trent, ST6 8DX
G7 HMF E. Last, 134 New Queens Road, Sudbury, CO10 1PJ
G7 HMI R. Shelford, 3 Browning Chase, Littleport, Ely, CB6 1FH
G7 HMK A. Baldwin, B M Box 6902, London, WC1N 3XX
G7 HMN C. Boutell, 6 Willow Way, Harwich, CO12 4HR
G7 HMQ B. Boult, 20 Perry Road, Long Ashton, Bristol, BS41 9FE
G7 HMS RNARS London (HMS Belfast) Group, c/o C. Read, 58 Somerset Road, Chiswick, London, W4 5DN
G7 HMU J. Stratton, 22 Tufton Gardens, West Molesey, KT8 1TE
G7 HMV M. Wood, 26 Parkfield Crescent Kimpton, Hitchin, SG4 8EQ
G7 HMW W. Knight, 30 Stretford Road, Urmston, Manchester, M41 9JZ
G7 HMZ A. Murfin, 31 Kings Road, St Neots, PE19 1LD
G7 HNF J. Baldwin, 71 Norfolk Road, Littlehampton, BN17 5HE
G7 HNG A. Lord, 59 Monks Orchard Road, Beckenham, BR3 3BJ
G7 HNL M. Carter, Meadowview Bransford Road Rushwick, Worcester, WR2 5SJ
G7 HNM G. Greatrix, West Cottage, Main Road, Boston, PE20 3PZ
G7 HNR A. Ball, 39 Deepdale Avenue, Birmingham, B26 3EL
G7 HOA Widnes & Runcorn ARC, c/o D. Wilson, 12 New Street, Elworth, Sandbach, CW11 3JF
GW7 HOC D. Warburton, 71 Richards Terrace, Cardiff, CF24 1RW
G7 HOE P. Goode, 23 Byworth Road, Farnham, GU9 7BT
G7 HOK P. Kellingley, 290 Calmore Road, Calmore, Southampton, SO40 2RF
G7 HOL D. Martin, Alken, The Covert, Orpington, BR6 0BT
GW7 HOM V. Cole, 77 Parc Castell Y Mynach, Creigiau, Cardiff, CF15 9NZ
G7 HON S. Martin, Broad Oak House, Pheasant Lane, Maidstone, ME15 9QR
G7 HOT J. Scott, 16 Hawton Road, Newark, NG24 4QB
G7 HOV J. Bertram, 21 Mayfair Avenue, Twickenham, TW2 7JG
G7 HPI C. Vance, 64 Caulfield Road, Swindon, SN2 8BT
G7 HQC I. Sorrell, 67 Northfield Drive, Pontefract, WF8 2DJ
G7 HQF P. Smith, 17 Beverley Avenue, Canvey Island, SS8 0DN
G7 HQJ A. Baker, 34 Clare Street, Stoke-on-Trent, ST4 6ED
GW7 HQL J. Caswell, 31 Pontalun Close, Barry, CF63 1QJ
G7 HQP J. Woods, 1 Dean Road, Cosham, Portsmouth, PO6 3DG
GM7 HQW B. Currie, Fawn House, Abriachan, Inverness, IV3 8LB
G7 HQY K. Walton, Springfield, Green Lane, Uckfield, TN22 5LA
G7 HRB P. Briggs, Mistleberry Cottage, Newtown, Sixpenny Handley, Salisbury, SP5 5PF
G7 HRF S. Crutchley, 8 Cloverland Drive, Hemsby, Great Yarmouth, NR29 4JY
G7 HRH R. Conway, 9 Whitworth Lane, Loughton, Milton Keynes, MK5 8EB
G7 HRJ T. Jackson, 18 Shepherds Close East Runton, Cromer, NR27 9PQ
G7 HRL T. Turner, 21 Spurgate, Hutton, Brentwood, CM13 2LA
G7 HRM S. Baker, 35 Cosmo Street, Westerly, USA, 2891
G7 HRP I. Booth, 16 Sandstone Drive, Leeds, LS12 5SU
G7 HRR Hucknall Rolls Royce ARC, c/o S. Sorockyj, 8 Bowden Avenue, Bestwood Village, Nottingham, NG6 8XN
G7 HRZ S. Haynes, 10 Cypress Grove, Denton, Manchester, M34 6EA
G7 HSA A. Cramp, 7 St Margarets Road, Ludlow, SY8 1XN
G7 HSB A. Green, Moss View, Southport Road, Ormskirk, L39 7JU
G7 HSL T. Reddish, 72 Edgmond Close, Redditch, B98 0JQ
G7 HSN J. Calder, Grassington, Station Road, Bedale, DL8 1SX
G7 HSO D. Hardinges, 4 The Close, Eastcote, Pinner, HA5 1PH
G7 HSS J. East, 102 Westfield Lane, Wyke, Bradford, BD12 9LS
GW7 HSW I. Edgington, 108 Ger Y Llan, Penrhyncoch, Aberystwyth, SY23 3TR
G7 HSY B. Stanton, 111 Beaconside, South Shields, NE34 7PT

GD7 HTG S. Hill, 54 Wybourn Drive Onchan, Isle of Man, IM3 4AT
G7 HTI Hilderstone Radio Society, c/o I. Lowe, 54 College Road, Margate, CT9 2SW
G7 HTN P. Seitz, 6 Meadow Rise Iwade, Sittingbourne, ME9 8SB
GW7 HTU P. Jenkins, 28 King Edward Road, Brynmawr, Ebbw Vale, NP23 4SD
GJ7 HTV A. Mourant, Little Mead, Claremont Road, St Saviour, Jersey, JE2 7RT
G7 HUC M. Fiorentini, Crompton, Draffin Lane, Camber, Rye, TN31 7RA
G7 HUG N. Markley, 79 Peake Close, Peterborough, PE2 9JE
G7 HUJ S. Telford, 44 Northcote Crescent, Leeds, LS11 6NN
G7 HUK P. Hart, 104 St. Austell Drive, Wilford, Nottingham, NG11 7BQ
G7 HUO C. Terry, Sandfields, Long Lane, Newbury, RG14 2TH
G7 HUP M. Terry, Sandfields, Long Lane, Newbury, RG14 2TH
GW7 HVA N. Callan, 24A Ynysmeurig Road, Abercynon, Mountain Ash, CF45 4SY
GI7 HVC T. Kennedy, 1 Inverleith Drive, Belfast, BT4 1RJ
G7 HVF S. Garlick, 20 Hill St, Upper Gornal, DY3 2DF
G7 HVL C. Spires, 15 Staple Hill Road, Bristol, BS16 5AA
G7 HVN M. Templeman, 28 Kewstoke Road, Kewstoke, Weston-Super-Mare, BS22 9YD
G7 HVO R. Gerrard, 12 Goldrill Gardens, Bolton, BL2 5NL
G7 HVU S. Lamb, 18 The Green, North Burlingham, Norwich, NR13 4SZ
G7 HWM A. Brookes, 8 Peppersgate, Lower Beeding, Horsham, RH13 6ND
G7 HXF J. Newton, 18 Siddeley Close, Broughton, Chester, CH4 0SG
G7 HXI I. Duffin, Mirabella, Bush Drive, Bush Estate, Norwich, NR12 0SF
G7 HYG R. Barber, 180 Beechfield, Hoddesdon, EN11 9QN
G7 HYM N. Singer, 11 Langley Road, Beckenham, BR3 4AE
G7 HYS D. Germaney, 22 Westbrook Road, Weston-Super-Mare, BS22 8JX
G7 HYZ M. Thompson, 23 Hare Park Lane, Crofton, Wakefield, WF4 1HS
G7 HZQ S. Breen, 20 Goodwood Close, Clophill, Bedford, MK45 4FE
G7 HZS G. West, West Cottage, Eaudyke Road, Boston, PE22 8RU
G7 HZU A. Bateman, 4 Fair Meadows, High Street, Rugeley, WS15 3LD
G7 HZZ A. Clayton, 6 Albert Road, Bunny, Nottingham, NG11 6QE
G7 IAE R. Lindley, 23 Quadrant Close, Murdishaw, Runcorn, WA7 6DW
GW7 IAK C. Hughes, 16 Morgraig Avenue, Newport, NP10 8UP
G7 IAM M. Chrzanowski, 53 Lamb Street, Kidsgrove, Stoke-on-Trent, ST7 4AL
G7 IAS R. Bell, 3 Haywards Heath Road, Balcombe, Haywards Heath, RH17 6NG
GW7 IAT Lt. Cmdr. M. Howells, 34 Cobden Street, Cross Keys, Newport, NP11 7PF
G7 IAU C. Penney, 9 Elm Lane, Minster On Sea, Sheerness, ME12 3SQ
G7 IAW C. Walsh, 4 Musbury Crescent, Rossendale, BB4 6AY
G7 IBD S. Beaumont, 29 Tiln Lane, Retford, DN22 6RT
G7 IBF R. Waller, 6 Pitchcombe, Yate, Bristol, BS37 4JX
G7 IBH K. Ashton, 13 Laceys Avenue, Leverton, Boston, PE22 0BG
G7 IBL M. Whatley, 2 Thompsons Hill, Sherston, Malmesbury, SN16 0PZ
GM7 IBM M. Robertson, Woodside House, Feabuie, Inverness, IV2 5EQ
G7 IBN F. Goodes, 17 Ashmead Close, Lords Wood, Chatham, ME5 8NY
GW7 IBT A. Earp, 42 Tudor Gardens, Neath, SA10 7RX
G7 IBU D. Nicholls, 19 Kimmeridge, Crown Wood, Bracknell, RG12 0UD
G7 IBX V. Finlayson, 92 Herlington, Orton Malborne, Peterborough, PE2 5PR
G7 ICD J. Mcdowall, 19 Plaistow Court, Hallwood Park, Runcorn, WA7 2GR
G7 ICE R. Catlow, 137 Haven Lane, Oldham, OL4 2QQ
G7 IDE I. Evans, 18 Plemstall Way, Mickle Trafford, Chester, CH2 4QJ
G7 IDH M. Newton, 5 Granville Avenue, Newcastle, ST5 1JH
G7 IEB R. Emberton, 10 Lodway Close, Pill, Bristol, BS20 0DE
G7 IED R. Martin, 45 Quail Holme Road, Knott End-On-Sea, Poulton-le-Fylde, FY6 0BT
G7 IEF D. Roadnight, 14 Newquay Crescent, Harrow, HA2 9LJ
GD7 IEH M. Blackburn, 63 Westbourne Drive, Douglas, Isle Of Man, IM1 4BB
G7 IEO A. Cook, 26 Worcester Road, Stourport-on-Severn, DY13 9PB
G7 IER A. Bent, Three Gables, Craggs Hill, Carnforth, LA6 1DJ
G7 IET Dr A. Dunlop, High View, Milton Avenue, Sevenoaks, TN14 7AU
GM7 IEU A. Steele, 20 Stewart Way, Alford, AB33 8UB
G7 IEY D. Chenery, 25 Aldreth Road, Haddenham, Ely, CB6 3PW
GI7 IEZ T. Mc Geown, 1 Drumcairn Road, Armagh, BT61 7SA
G7 IFB S. Thompson, 39 Greenbank Road, Ambleside, LA22 9BD
G7 IFI B. Jones, 25 Milton Drive, Wistaston Green, Crewe, CW2 8BT
G7 IFJ R. Page, 68 The Ridgeway, St Albans, AL4 9PS
G7 IFL P. King, 1 Rue Du Canelots, Saint Frajou, France, 31230
G7 IFM J. Hewitt, 9 Alford Fold, Fulwood, Preston, PR2 3UU
G7 IFO N. Rigby, 2 Mill Lane, Sutton Manor, St Helens, WA9 4HW
G7 IFR T. Chibnell - Smith, Nursery Cottage, Whitney-On-Wye, Hereford, HR3 6HT
G7 IFU M. Mccartney, 1 Tollemache Close, Manston, Ramsgate, CT12 5LX
GI7 IFW S. Boskett, 314 Shore Crescent, Belfast, BT15 4JU
GM7 IFX J. Barnett, 72 Cameron Toll Gardens, Edinburgh, EH16 4TG
G7 IGF I. Fields, Boarzell Cottage, London Road, Etchingham, TN19 7QY
G7 IGR C. Crowhurst, 143 Drayton High Road, Drayton, Norwich, NR8 6BD
G7 IGU L. Evans, 58 Westminster Drive, Bromborough, Wirral, CH62 6AW
G7 IHD J. Bolsover, 18 Millers Ford Low Bentham, Lancaster, LA2 7BF
G7 IHE H. Robinson, 16 Coniston Avenue, Ashton-In-Makerfield, Wigan, WN4 8AY
GM7 IHH A. Tolson, 4 Albert Place, Langholm, DG13 0AT
GM7 IHJ M. Alexander, 38 The Wynd, Dalgety Bay, Dunfermline, KY11 9SJ
G7 IHN S. Harvey, Gabled Cottage, Shipton Oliffe, Cheltenham, GL54 4HZ
G7 IHP K. Weston, 2 Beech Grove, Somerton, TA11 6LG
GM7 IHR R. Brodie, Midgeloch Cottage, Arbuthnott, Scotland, AB3 1NX
G7 IHV G. Havell, Flat 13, Waldron House, London, SW2 1PA
G7 IHX J. Allan, 60 Godfrey Road, Halifax, HX3 0SU
GM7 IHZ Dr G. Hayes, Flat 6, 87 London Road, Edinburgh, EH7 5TT
G7 IIB P. Shields, 3 Hawthorn Road, Tavistock, PL19 9DL
G7 IIC S. Oliphant, Homeside, Compton, Paignton, TQ3 1TD
G7 IID E. Caunt, 5 Littledale, Pickering, YO18 8PS
G7 IIF A. Barrett, 17 Wimborne Avenue, Wirral, CH61 7UL
G7 IIH J. Bond, 1 Old House Courtyard Southover High Street, Lewes, BN7 1HT
G7 IIN M. Hewitt, 2 Hill View, Worstead, North Walsham, NR28 9SD
G7 IIO B. Bellamy, 71 High Road, Benfleet, SS7 5LH
G7 IIS C. Beatrup, Bon Air, 34 Springfield Drive, Halesowen, B62 8EU
G7 IIZ G. Dooley, 93 Springfields, Walsall, WS4 1JX
G7 IJC D. Wells, 21 Kings Road, Barnetby, DN38 6HF
G7 IJI D. Gibbs, Flat 1 74 Priory Road Hall Green, Birmingham, B28 0TE
G7 IJL D. Mcclew, 135 Hermitage Street Rishton, Blackburn, BB1 4ND
G7 IJW Dr B. Rushton, Cherrydene, New Road, Windermere, LA23 2LA
G7 IJY B. Evans, 51 Katrina Grove, Featherstone, Pontefract, WF7 5LW
GM7 IKB G. Riddell, Lawhead Croft, Tarbrax, West Calder, EH55 8LW
G7 IKG A. Thynne, 1 Earlston Way, Birmingham, B43 5JR
G7 IKM W. Willan, 31 St. Oswalds Lane, Bootle, L30 5QD
G7 IKS A. Raistrick, 10 Orchard Way, Chinnor, OX39 4UD
G7 ILA T. Brennan, 9 Mill Lane, Felixstowe, IP11 7RL
G7 ILD P. Brown, 4 Solent Drive, Barton On Sea, New Milton, BH25 7AW
G7 ILG I. Glossop, 1 Harborough Hill Cottages, Birmingham Road, Kidderminster, DY10 3LH
G7 ILI A. Page, 29 Lambourne Close, Fareham, PO14 1SL
G7 ILJ Banbury Ray Grp, c/o B. Thornton, 21 Valley Road, Banbury, OX16 9BQ
G7 ILL P. Atherton, Findern Lane, Willington, Derby, DE65 6DW
G7 ILP K. Naylor, 3 Windrush Close, Bicester, OX26 2AR
G7 ILS I. Warrilow, 84 Marple Road, Stockport, SK2 5RN
G7 ILX R. Voges, 43 Eastgate, Fulwood, Preston, PR2 3HS
G7 ILY D. Barber, 2 St. Jamess Mews, Church, Accrington, BB5 4JR
G7 IMB S. Jeffcoate, 25A Northampton Road, Lavendon, Olney, MK46 4EY
G7 IMD A. Spittlehouse, 7 Fernbank, Battle Green, Doncaster, DN9 1LJ
G7 IMH M. Fortescue, 98 Campbell Road, Oxford, OX4 3NU
G7 IMO L. Handley, 11 Brook Close, Blythe Bridge, Stoke-on-Trent, ST11 9PX
G7 IMQ P. Bannister, 222 Haslucks Green Road, Shirley, Solihull, B90 2LN
G7 IMR M. Taft, 44 Langcomb Road, Shirley, Solihull, B90 2PR
G7 IMT D. Gerard, 15 Nyetimber Lane, Bognor Regis, PO21 3HQ
GI7 IMU A. Reid, 18 Orby Grove, Belfast, BT5 6AL
G7 IMV R. King, Old Orchard, South Milton, Kingsbridge, TQ7 3JZ
G7 IMY S. Kemp1, 16 Douglas Road, Aylesbury, HP20 1HW
G7 IMZ G. Smith, 19 Parker Road, Humberston, Grimsby, DN36 4TT
G7 INC G. Bacon, 36 Warnadene Road, Sutton-in-Ashfield, NG17 5BD
G7 ING M. Darbyshire, 50 Gaythorne Avenue, Preston, PR1 5TA
GI7 INR A. Greer, 6 Ashley Gardens, Banbridge, BT32 4BN
G7 INY J. Coady, Sunset, Station Road Wisbech St. Mary, Wisbech, PE13 4RT
G7 IOB C. Knowlson, 28 Hill Drive, Handforth, Wilmslow, SK9 3AR
G7 IOC R. Mitchell, 2 Corbar Road, Stockport, SK2 6EP
G7 IOF K. Roebuck, 20 Ryecroft Close, Outwood, Wakefield, WF1 2LW
G7 IOI N. Telford, 18 Kirkstall Close, South Anston, Sheffield, S25 5BA
G7 ION M. Tennant, 64 Aldenham Road, Kemplah Park, Guisborough, TS14 8LD
G7 IOO P. Horton, 408 Woodcrest Way, Forney, USA, 75126
G7 IPA M. Clements, 23 Pudding Lane, Gadebridge, Hemel Hempstead, HP1 3JU
G7 IPH P. Baker, 6 Firework Close, Kingswood, Bristol, BS15 4LT
G7 IPI P. Crane, 64 Bridge Avenue, Cheslyn Hay, Walsall, WS6 7EP
GI7 IPO H. Stokes, 32 Islay Street, Antrim, BT41 2TS
G7 IPR N. Daniels, 136 Womersley Road, Knottingley, WF11 0DQ
GW7 IPS S. Hamlyn, 6, New Road, Newcastle Emlyn, SA38 9BA
G7 IPX C. Bowden, 36 Aspin Drive, Knaresborough, HG5 8HQ
G7 IQD R. Cook, 17 Beech Gardens, Sandpits Road, Ludlow, SY8 1UT
G7 IQM P. Jaggs, 218 New Road, London, E4 9SJ
G7 IQO D. Flatters, 87 Albert Promenade, Loughborough, LE11 1RD
G7 IQZ R. Norman, 87 Edenfield Gardens, Worcester Park, KT4 7DX
GW7 IRD T. Jones, 3 Woodlands Close, St. Arvans, Chepstow, NP16 6EF
G7 IRF A. Saunders, 24 Hallam Close Midsomer Norton, Radstock, BA3 2FG
G7 IRG G. Wisbey, 4 Avenue Road, Streatham, London, SW16 4HL
G7 IRH W. Moth, 145 Carisbrooke Road, Newport, PO30 1DG
GM7 IRI K. Baxter, Flat 2, 29B Corbiehill Road, Edinburgh, EH4 5BQ
GI7 IRJ P. Mcateer, 36 Ballyquillan Road, Aldergrove, Crumlin, BT29 4RH
G7 IRK D. Deacon, 14 Dukes Road, Braintree, CM7 5UE
G7 IRN M. Scott, 3 Summerhill, Ticehurst, Wadhurst, TN5 7JA
G7 IRP D. Williams, 66 Gover Road, Hanham, Bristol, BS15 3JZ
G7 IRU C. Hosegood, 4 The Orchard, Sixpenny Handley, Salisbury, SP5 5QL
G7 ISD Dr C. Rizzo, Downside Downs Road Funtington, Chichester, PO18 9LS
G7 ISE G. Walters, 12 Portstone Close, Northampton, NN5 6QP
G7 ISR G. Lines, 11A Gloucester Road North, Bristol, BS7 0SG
GI7 ISX S. Butler, 25 Chippendale Avenue, Bangor, BT20 4PX
G7 ITB D. Davies, Greyroofs Albert Place, Washington, NE38 7BW
G7 ITM G. Clarkson, 40 Wharf Road, Ash Vale, Aldershot, GU12 5AY
G7 ITO M. De Banks, 56 Blackwater Drive, Aylesbury, HP21 9RX
G7 ITS M. Fasham, 29 Granville Avenue, Ramsgate, CT12 6DX
G7 ITT S. Import, The Old Rectory, Dufton, Appleby-in-Westmorland, CA16 6DA
G7 ITU S. Marlow, 14 Lightgate Road, South Petherton, TA13 5AJ
G7 ITW D. Fennelly, 23 Trent View Gardens, Radcliffe-On-Trent, Nottingham, NG12 1AY
G7 ITX T. Emblem-English, 4 Mark Avenue, London, E4 7NR
G7 ITZ B. Calvert, Wall To Wall Communications, Unilink House, 21 Lewis Road, Sutton, SM1 4BR
G7 IUB L. Porter, 324-326 Lillie Road, London, SW6 7PP
G7 IUE G. Clem, 25 Alexander Close, Waterlooville, PO7 5TB
GM7 IUF W. Howie, 24 Newfield Drive, Dundonald, Kilmarnock, KA2 9EW
G7 IUI Prof. W. Fagan, 32 Eastfield Road Thurmaston, Thurmaston, Leicester, LE48FP
G7 IVF C. Flux, 35 Oaklyn Gardens, Shanklin, PO37 7DF
G7 IVG K. Graham, 10 Summerfields, Dalston, Carlisle, CA5 7NW
G7 IVN P. Jagdev, 10 St. Johns Road, Southall, UB2 5AN
G7 IVU R. Walker, 16 Norman Drive, Stilton, Peterborough, PE7 3RS
GI7 IVX R. Connolly, 21 Eleastan Park, Kilkeel, Newry, BT34 4DA
G7 IWA A. Maunder, 2 Downhouse Road, Waterlooville, PO8 0TX
G7 IWE T. Haggie, 6 Rose Villas Middleburg Street, Hull, HU9 2QR
G7 IWK G. Blackburn, 10 Lodge Close, Redhill, Nottingham, NG5 8NZ
G7 IWM S. Gusterson, 19 Blackberry Close, Higham Ferrers, Rushden, NN10 8FJ
G7 IWU H. Judge, 8 Fontenoy Road, Balham, London, SW12 9LU
G7 IWV A. Godley, 177 Cheriton Road, Folkestone, CT19 5HG
G7 IWW R. Gibbs, 32 Beswick Avenue, Ensbury Park, Bournemouth, BH10 4EY
G7 IWZ R. Murray, 92 North Lane, East Preston, Littlehampton, BN16 1HE
G7 IXC R. Barkley, 39 Fulbeck Avenue, Wigan, WN3 5QN
G7 IXG D. Doermann, 19 Jackman Close, Fradley, Lichfield, WS13 8PW
G7 IXH P. Lawton, 207 Eachelhurst Road, Sutton Coldfield, B76 1EA
G7 IXK G. Smith, 129 Chiltern Way, Duston, Northampton, NN5 6BW
G7 IXP P. Hammersley, 30 Bonner Grove, Aldridge, Walsall, WS9 0DU
G7 IYA B. Whittock, 12 Hillside Crescent, Midsomer Norton, Radstock, BA3 2NB
G7 IYG N. Hobbs, 24 Falkland Road, Southport, PR8 6LG
G7 IYH Dr L. Hobbs, 24 Falkland Road, Southport, PR8 6LG

G7 IYK D. Trewren, 184 High Street Oldland Common, Bristol, BS30 9QQ
G7 IYM T. Mann, 34 Crows Grove, Bradley Stoke, BS32 0DA
G7 IYN A. Attack, 8 Lewis Road, Hornchurch, RM11 2AJ
G7 IYQ K. Fulcher, Derventio, High Street, Gainsborough, DN21 5LY
G7 IYX R. Dodds, 33 Westgate, Warley, Oldbury, B69 1BA
G7 IYY P. Hollyoake, 79 Borough Cres, Oldbury, B69 1AJ
GW7 IZA G. Griffiths, Newcastle Court, Evancoyd, Presteigne, LD8 2PA
G7 IZC R. Phillips, 25 Carlile Hill, Hemlington, Middlesbrough, TS8 9SL
G7 IZE K. Palmer, 6 Parklands Close, Arnold, Nottingham, NG5 9QU
G7 IZM F. Lucas, Ivella, Recreation Street, Dudley, DY2 9EU
G7 IZN M. Dunn, 39 Gainsbrook Crescent, Norton Canes, Cannock, WS11 9TN
G7 IZU A. Smith, 7 Hartley Avenue, Plymouth, PL3 5HW
G7 IZV C. Funnell, 61 Blackwatch Road, Coventry, CV6 3GS
G7 IZW F. Chilton, 127 Nicholls Field, Harlow, CM18 6EB
G7 JAE C. Pritchard, 11 Willow Green, Needingworth, St Ives, PE27 4SW
G7 JAF A. Lambert, 50 Clarendon Road, Sheffield, S10 3TR
GI7 JAM Dr K. Gibson, 4 Ilford Avenue, Belfast, BT6 9SF
G7 JAN J. Martyn, Aspiration, Queens Road, Crowborough, TN6 1QQ
G7 JAO C. King, 39 West Street, Huntingdon, PE29 1WT
G7 JAQ R. Adam, 8 Lexington Court, Purley, CR8 1JA
G7 JAS D. Harris, 68 Tomlinson Avenue, Luton, LU4 0QW
G7 JAX D. Hall, The Crow'S Nest 9A Cheveley Road, Newmarket, CB8 8AD
G7 JBD C. Storrie, 3 Stocken Hall Mews, Stretton, Oakham, LE15 7RL
G7 JBN V. Bertram, 21 Mayfair Avenue, Twickenham, TW2 7JG
G7 JBW P. Hoath, 1 Red Lodge Drive, Bilton, Rugby, CV22 7TT
G7 JBZ R. Cone, 6 Renault Drive, Bracebridge Heath, Lincoln, LN4 2QG
G7 JCD M. Jones, 4 Bell Street, Tipton, DY4 8HZ
G7 JCF S. Beamish, The Old Vicarage, Vicarage Road, Woodbridge, IP13 8DT
G7 JCQ G. Blunt, 36 Whitefield Way, Liverpool, L6 2NB
G7 JCX J. Price, 37 The Court, Anderby Creek, Skegness, PE24 5YQ
G7 JDA A. Roberts, 5 Colmar Close, Daventry, NN11 9BT
G7 JDB J. Blackburn, 2 Heath Drive, Sutton, SM2 5RP
G7 JDF J. Hope, 48 Holbeck, Bracknell, RG12 8XE
G7 JDH A. Nevill, 10 Grangefield Crescent, New Rossington, Doncaster, DN11 0LU
G7 JDI D. Carslake, 21 Kestrel Drive, Bingham, Nottingham, NG13 8QD
G7 JDK R. Rothwell, 11 St Marks Road, Stourbridge, DY9 7DT
G7 JDN M. Collins, 8 Newfield Road, Marlow, SL7 1JW
G7 JDR C. Bennett, The Old Cottage, Waterside Road, Southminster, CM0 7QT
GM7 JDS B. Reid, 10 Badenoch Road Kirkintilloch, Glasgow, G66 3NX
GW7 JDX Dr M. Ghassempoory, 102 Colchester Avenue, Penylan, Cardiff, CF23 9AZ
GI7 JEB M. Gibson, 1 Downshire Park, Bangor, BT20 3TP
GM7 JED I. Macdonald, 3 Anderson Road, Stornoway, HS1 2PG
G7 JEJ T. Hyder, 83 Beam Hill Road, Burton on Trent, DE13 0AD
GI7 JEM D. Branagh, 146 Craigs Road, Carrickfergus, BT38 9XA
G7 JFI T. Steeper, 16 High Street, Eagle, Lincoln, LN6 9DH
G7 JFM S. Smith, 26 Broadsands Avenue, Paignton, TQ4 6JN
GM7 JFN K. Maclean, 10B Knockaird, Port Of Ness, Isle of Lewis, HS2 0XF
G7 JFU R. Evison, Sandpiper Fibbards Road, Brockenhurst, SO42 7RD
G7 JGB A. Kemp, 7 Hartley Way, Bishopdown, Salisbury, SP1 3WS
G7 JGE C. Hobson, 28 Withering Road, Swindon, SN1 4GU
GW7 JGF T. Froggatt, Gwyndy, Moelfre, Anglesey, LL72 8LN
GM7 JGH A. Bruce, 20 Weir Crescent, Milton, Wick, KW1 5SS
G7 JGQ A. Greenland, 19 The Ridgeway, Potton, Sandy, SG19 2PS
GM7 JGR Dr J. Howie, 29, Coates Gardens, Edinburgh, EH12 5LG
G7 JGS G. Swindells, 52 Western Avenue, Blacon, Chester, CH1 5PP
GI7 JGT M. Mc Namee, 22 St. Patricks Park, Rosslea, Enniskillen, BT92 7QY
G7 JGW W. Holroyd, 8 Carr Dene Court, Preston Street, Preston, PR4 2XA
G7 JGY P. Smith, 174 Willerby Road, Hull, HU5 5JW
G7 JGZ R. Brooks, 8 Chichester Place, Tiverton, EX16 4BW
GW7 JHC T. Christie, 7 Hayes View, Oswestry, SY11 1TP
G7 JHE G. Beckett, 34 Bradwall Road, Sandbach, CW11 1GF
GW7 JHK P. Brettle, 27 Neath Road, Resolven, Neath, SA11 4AA
G7 JHM J. Mccollin, 17 Lamsey Road, Hemel Hempstead, HP3 9HB
G7 JHU S. Birchall, 83 Wilton Avenue, Chapel St. Leonards, Skegness, PE24 5YN
G7 JHV D. Gervais, Seven Gables Lodge, Buckingham Road, Buckingham, MK18 3NA
G7 JHW R. Johnson, 30 Thorpe Downs Road, Church Gresley, Swadlincote, DE11 9FB
G7 JHX Dr J. Williams, 40 Tythe Barn Lane, Shirley, Solihull, B90 1RW
G7 JHZ D. Randles, 20 Felix Road, London, W13 0NT
G7 JIB L. Evans, Polvellan, School Hill, St Austell, PL26 6TG
G7 JIF S. Ruffell, 2 Beulah Cottage, Church Street, West Stour, SP8 5RL
G7 JIM W. Barton, 4 Hawthorn Flats, Hawthorn Road, Dorchester, DT1 2PE
G7 JIN C. Willis, 9 Avington Close, Sedgley, Dudley, DY3 3LN
G7 JJC P. Gerrard, 6 Ellabank Road, Heanor, DE75 7HF
G7 JJG K. Watts, 68 Kentwood Hill, Tilehurst, Reading, RG31 6DE
G7 JJJ C. Marshall, Gladstan House, 70 Chester Road, Runcorn, WA7 3DY
G7 JJP L. Towler, 8 Stowehill Road, Peterborough, PE4 7PY
G7 JJW S. Coffin, 5 Colt Close, Streetly, Sutton Coldfield, B74 2EA
G7 JJX R. Wallace, 31 Salts Road, West Walton, Wisbech, PE14 7EJ
GI7 JKA J. Mccullagh, 2 Holestone Road, Doagh, Ballyclare, BT39 0SB
G7 JKD M. Coward, 7, Brackenrigg, Armathwaite, Carlisle, CA4 9PX
G7 JKH C. Hyde, 42 Fern Road, Whitby, Ellesmere Port, CH65 6PB
GW7 JKK J. Mossman, 13 Tynrhos Estate, Caergeiliog, Holyhead, LL65 3HS
GI7 JKM S. Glendinning, 2 Scotts Road, Moneymore, Magherafelt, BT45 7TW
G7 JKW S. Avery, Wilding Farm Cottage Cinder Hill, Lewes, BN8 4HP
G7 JKY Dr S. Smith, 73, Station Street, Rippingale, Bourne, PE10 0SX
G7 JLC A. Edwards, 34 Albion Road, Malvern Link, Malvern, WR14 1PU
GI7 JLD J. Hunter, 29 Mullaghacall Road, Portstewart, BT55 7EG
G7 JLF R. Pike, 6 Bridens Way Haddenham, Aylesbury, HP17 8DH
GW7 JLG A. Williams, 2 Nant Y Berllan, Llanfairfechan, LL33 0SN
G7 JLK R. Elliott, 39 Amanda Way, Pensilva, Liskeard, PL14 5RA
G7 JLO N. Townend, 124 Rylands Road, Southend-on-Sea, SS2 4LJ
G7 JLS D. Bryant, Knowle Barns, Broadhempston, Totnes, TQ9 6DA
G7 JLT K. Bryant, 18 Loundyes Close, Thatcham, RG18 3EB
G7 JMB J. Baker, Green Lane Farmhouse, Rugeley, WS15 2AR
G7 JME P. Good, 11 Moorland Road, Didsbury, Manchester, M20 6BB
G7 JMQ M. Tidmarsh, 16 Castleton Road, Mitcham, CR4 1NY
G7 JMU D. Butterworth, 27 Royds Avenue, Linthwaite, Huddersfield, HD7 5QU
G7 JMW A. Weaver, 116 Maldon Road, Tiptree, Colchester, CO5 0BN
G7 JMZ J. Bache, 62 Whittingham Road, Halesowen, B63 3TP
G7 JNM A. White, 6 Greenbank, Hadfield, Glossop, SK13 1PD
G7 JNS S. Mclennan, 179 King John Avenue, Bear Wood, Bournemouth, BH11 9SJ
G7 JOA RSC of Cheshire, c/o C. Rickerby, Brownsea, 113 Cliftonville Road, Woolston, Warrington, WA1 4BJ
G7 JOW J. Ashbee, 49 Sandwich Road, Whitfield, Dover, CT16 3LT
G7 JPN M. Bateman, 22 Bowling Green Lane, Albrighton, Wolverhampton, WV7 3HL
G7 JQF W. Booth, 8 Park Crescent, Bacup, OL13 9RL
GD7 JQI A. Kissack, 30 High View Road, Douglas, Isle Of Man, IM2 5BH
G7 JQT E. Barry, 8 Astley Crescent, Scotter, Gainsborough, DN21 3SL
G7 JQW H. Derrick, 28 Great Parks, Holt, Trowbridge, BA14 6QP
G7 JQZ D. Beadle, 4 Harlaxton Drive, Lincoln, LN6 3NR
G7 JRC D. Smith, 26 Mill Fields Todwick, Sheffield, S26 1JS
G7 JRD T. Alwyn-Clark, 1 Blackfriars Road, Lincoln, LN2 4WS
GI7 JRG A. Mcnerlin, 27 Roeview Park, Limavady, BT49 9BQ
G7 JRJ C. Wainwright, 31 Queens Road, Leytonstone, London, E11 1BA
G7 JRK P. Dixon, 7 Pincey Mead, Basildon, SS13 3EW
G7 JRM C. Hinton, 65 South Street, Tarring, Worthing, BN14 7NE
G7 JRP T. Pratley, 28 Charles Avenue, Watton, Thetford, IP25 6BZ
GW7 JRT J. Tonge, Bracken Brae, Gwalchmai, Holyhead, LL65 4SL
G7 JRU A. Martin, 36 Saxon Road, Lowestoft, NR33 7BT
G7 JSC R. Brotherton, 167 Pershore Road Hampton, Evesham, WR11 2NB
G7 JSE S. Almond, 2 King Street, Swinton, Mexborough, S64 8ND
G7 JSG R. Maynard, 8 Badgers Walk Pool Lane, Clows Top, Kidderminster, DY14 9NT
GW7 JSH J. Field, Dan-Y-Coed, North Beach Road, Aberystwyth, SY23 3DT
G7 JSQ P. Domachowski, 39 Wycliffe Road West, Coventry, CV2 3DX
G7 JSS C. Watson, 26 Jupiter Gate, Stevenage, SG2 7ST
G7 JST Jubilee Sailing Trust(ARS), c/o J. Wheatley, 8 Winchester Close, Feniton, Honiton, EX14 3EX
G7 JSV W. Mcareavey, 3 Hall Farm Cottage, East Heckington, Boston, PE20 3QG
G7 JSW R. Steward, 2 Glenister House, 238 Avondale Drive, Hayes, UB3 3PP
G7 JTB R. Pluck, The Garden House, St. Leonards Avenue, Blandford Forum, dt11 7pa
G7 JTD D. Lockett, 10 Cornwall Drive, Bayston Hill, Shrewsbury, SY3 0ER
G7 JTF A. Harvey, Rose House, Rose Grove, Doncaster, DN3 3AJ
G7 JTH J. Carter, 30 Swift Way, Sandal, Wakefield, WF2 6SR
G7 JTI G. Cuskin, 57 Aln Street, Hebburn, NE31 1XT
G7 JTK S. Bell, 6 Broom Wood Court, Prudhoe, NE42 6RB
G7 JTR D. Lock, Pelican House, Chilton Candover, Alresford, SO24 9TX
G7 JTV J. Caswell, 3 Birch Road, Finchampstead, Wokingham, RG40 3LB
G7 JTZ R. Smith, 17 Julian Road, Spixworth, Norwich, NR10 3QA
GW7 JUB T. Jones, 37 Bro'R Dderwen, Clynderwen, SA66 7NR
G7 JUC K. Marsh, 21 Edward Road, Eynesbury, St Neots, PE19 2QF
GI7 JUH T. Cox, 13 Shrewsbury Gardens, Belfast, BT9 6PJ
G7 JUJ P. Moss, 23 Lees Row, Padfield, Glossop, SK13 1EN
G7 JUL A. Whitcher, 12 Battersby Street, Bury, BL9 7SG
G7 JUN M. Steadman, 26 Walkers Green, Marden, Hereford, HR1 3DU
G7 JUP J. Beckingham, 20 Baptist Close, Abbeymead, Gloucester, GL4 5GD
G7 JUR P. Lock, 1 Carters Walk, Farnham, GU9 9AY
GW7 JUV C. Broadbent, 12 Aelybryn Ceinws, Machynlleth, SY20 9EZ
GM7 JUX W. Dyer, 24 Southfield Road, Cumbernauld, Glasgow, G68 9DZ
G7 JUZ R. Shams-Nia, 1090 Eastern Avenue, Ilford, IG2 7SF
G7 JVB P. Wade, 41 Prospect Avenue, Stanford-le-Hope, SS17 0NH
G7 JVC M. Hewitt, 1 Harpswell Hill Park, Hemswell, Gainsborough, DN21 5UT
G7 JVE N. Cook, 35 Glanville Road, Hadleigh, Ipswich, IP7 5SQ
G7 JVF S. Mobley, 2 Lingham Close, Solihull, B92 9NW
G7 JVG A. White, 19 Haswell Close, Wardley, Gateshead, NE10 8UE
G7 JVJ E. Peacock, Octon Lodge, Langtoft, Driffield, YO25 3BJ
G7 JVK R. Hardie, 12 Hopland Close, Longwell Green, Bristol, BS30 9XB
G7 JVN D. Greywolf, 3 Denham Close, St Leonards-on-Sea, TN38 9RS
G7 JVO K. Saxby, 184 Brodrick Road, Eastbourne, BN22 9RH
G7 JVQ F. Sparks, 36 High View Road, Guildford, GU2 7RT
G7 JWD T. Place, 34 Holcroft, Orton Malborne, Peterborough, PE2 5SL
G7 JWE A. Liddell, 4 Russet Court, Kingswood, Wotton-under-Edge, GL12 8SG
G7 JWH A. Butler, 22 Willow Park, Minsterley, Shrewsbury, SY5 0EH
G7 JWI G. Harrison, 58 Hollywall Lane, Stoke-on-Trent, ST6 5PP
G7 JWJ E. Hickman, Eriska, 33 Romany Way, Stourbridge, DY8 3JR
G7 JWL E. Oakes, 30 Linden Avenue, Stourport-on-Severn, DY13 0EQ
G7 JWO R. Allcock, 44 Newmount Road, Stoke-on-Trent, ST4 3HQ
G7 JWQ B. Priestley, Priorswood Cottage, Tyndale Road, Gloucester, GL2 7DJ
G7 JWV R. Ebbetts, Markway House, Blackbush Road, Lymington, SO41 0PB
G7 JWW S. Charters, Beechgrove, Haselor Lane Hinton-On-The-Green, Evesham, WR11 2QZ
G7 JWX B. Maley, 10 Wolsey Place, 49-51 London Road, Hailsham, BN27 3FU
G7 JWY C. Ainley, 23 Forresters Close, Norton, Doncaster, DN6 9HX
G7 JXB K. Cox, 16 Henty Close, Walberton, Arundel, BN18 0PW
G7 JXD J. Pritchard, 22 Osborne Way, Haslingden, Rossendale, BB4 4DZ
G7 JXF M. Forknell, 24 Sherbourne Avenue, Nuneaton, CV10 9JH
G7 JXJ C. Smith, 30 Rookery Close, St Ives, PE27 5FX
G7 JXL D. Kerridge, 7 Haslers Place, Haslers Lane, Dunmow, CM6 1AJ
G7 JXR G. Wiseman, 7 Barton Road, Woodbridge, IP12 1JQ
G7 JXT I. Ballantyne, 2 Dunvegan Close Manea, March, PE15 0LU
G7 JXU M. Barker, 103 Friarswood Road, Newcastle, ST5 2EF
G7 JXX I. Thaiss, 4A Union Street, Market Rasen, LN8 3AA
G7 JXY I. Guffick, 13 Alderwood Close, Hartlepool, TS27 3QR
G7 JYG H. Odd, Verona, Harrow Road, Sevenoaks, TN14 7JU
GW7 JYJ T. Gittoes, Oak Farm, Builth Wells, LD2 3EN
GI7 JYK P. Lowrie, 13 Carwood Park, Newtownabbey, BT36 5JU
G7 JYL J. Sage, 8 Foxwood Road, Bean, Dartford, DA2 8BH
G7 JYQ T. Dabbs, 4 Caverleigh, Cadogan Road, Surbiton, KT6 4DH
GM7 JYW P. Lawrence, Gateside Smithy, Munlochy, IV8 8PA
G7 JYY M. Penn, 5 Angus Close, Kenilworth, CV8 2XH
G7 JYZ S. Turley, 22 Powlers Close, Stourbridge, DY9 9HH
G7 JZC A. Upchurch, 68 Lindleys Lane, Kirkby-In-Ashfield, Nottingham, NG17 8AD
G7 JZI W. Hilton, 8 Ashfield Avenue, Hindley Green, Wigan, WN2 4RG

G7 JZJ M. Doyle, 133A Pope Lane, Penwortham, Preston, PR1 9DD
G7 JZK W. Hancox, Flat 34, Millbank Court, Barlows Lane, Liverpool, L9 9HQ
G7 JZM G. Priestley, 24 Saxton Avenue, Bradford, BD6 3SW
G7 JZS M. Budd, 37 Cheyne Walk, Hornsea, HU18 1BX
G7 JZY K. Long, Manor Farm, 27 Church Street, Hull, HU11 4RN
G7 KAK I. Clewley, 31 Kenilworth Road, Basingstoke, RG23 8JF
GD7 KAM A. Swearman, 56 Garth Avenue, Surby, Isle of Man, IM9 6QU
G7 KAO D. Clarke, 2 Wilmot Road, Dartford, DA1 3BA
G7 KAV N. Stemp, 3 Loxwood, East Preston, Littlehampton, BN16 1DT
G7 KBD A. Carlton, 32 Culver Road, Bradford-on-Avon, BA15 1HZ
G7 KBE B. Mcintyre, West Abbey Nursing Home, Stourton Way, Yeovil, BA21 3UA
G7 KBH K. Wainwright, 25 Tithebarn Road, Rugeley, WS15 2QW
GW7 KBI G. Dreiling, Picton Farm, Holywell, CH8 9JQ
GM7 KBK E. Pratt, 46 Sheddocksley Drive, Aberdeen, AB16 6NX
G7 KBR P. Phillips, 10 Byron Grove, East Grinstead, RH19 1SG
G7 KBZ S. Hutchinson, 32 Uppleby, Easingwold, York, YO61 3BB
G7 KCC J. Durdin, 16 Barnwood Close, Kingswood, Bristol, BS15 4JA
G7 KCE J. Hannaford, 22 Barn Park, Stoke Gabriel, Totnes, TQ9 6SR
G7 KCK P. Langley, 321 Maidstone Road, Rochester, ME1 3EF
G7 KCN B. Elcoate, 9 Parsonage Lane Laindon, Basildon, SS15 5YN
G7 KDG P. Edmondson, 20 Mill Road, Impington, Cambridge, CB24 9PE
G7 KDH D. Edmondson, 4 Elm View, Steeton, Keighley, BD20 6SZ
GW7 KDI P. Stevenson, Nant Fach Cerrigydrudion, Corwen, LL21 0SB
G7 KDJ A. Chadwick, 2 Auden Place, Longton, Stoke-on-Trent, ST3 1SJ
G7 KDM C. Campbell, 21 Sellwood Drive, Carterton, OX18 3AZ
G7 KDN A. Thomas, 49 Tristan Close, Calshot, Southampton, SO45 1BN
G7 KDQ K. Roan, 133 Woodhouse Lane, Beighton, Sheffield, S20 1AD
G7 KDR B. Hopkins, 14 Falkenham Rise, Basildon, SS14 2JQ
GW7 KDU M. Lewis, 111 Willowbrook Gardens, St. Mellons, Cardiff, CF3 0BY
G7 KDX R. Bell, 5 Byron Avenue, Blyth, NE24 5RN
G7 KEA R. Chapman, Flat 3, Goda Court, Littlehampton, BN17 6AS
GI7 KEC J. Stafford, 31 Shimna Close, Belfast, BT6 0DZ
G7 KEE B. Daw, 19 Rowan Close, Yarnfield, Stone, ST15 0EP
G7 KEI B. Edgley, 2 Queens Close, Hyde, SK14 5RE
G7 KEK R. Horsfall, 7 Lytham Close, Doncaster, DN4 6UT
G7 KEP A. Reeve, 97 Mendip Vale, Coleford, Radstock, BA3 5PP
G7 KFM I. Hasman, Fleetway, The Spinney, Newark, NG24 2NT
G7 KFN C. Hasman, Fleetway, The Spinney, Newark, NG24 2NT
G7 KFP G. Spicer, 19 Byfield Way, Bury St Edmunds, IP33 2SN
G7 KFQ N. Camp, 1 Higher Tresillian Cottages, Tresillian., Newquay, TR8 4PL
GM7 KFS A. Wood, Seaward, Toward, Dunoon, PA23 7UA
G7 KFZ R. May, 153 Station Road, Winsford, CW7 3DE
GW7 KGD H. Wrighton, 43 Bryn Celyn, Colwyn Bay, LL29 6DH
G7 KGH M. Forder, 157 Kennington Road, Kennington, Oxford, OX1 5PE
G7 KGI E. Gould, 53 Green Road, Kidlington, OX5 2EU
G7 KGP J. Chisholm, 162 Ardington Road, Northampton, NN1 5LT
G7 KGR C. Saunders, 148 Downs Barn Boulevard, Downs Barn, Milton Keynes, MK14 7RR
G7 KGV I. Lewis, Whitehill Lodge, Hextalls Lane, Redhill, RH1 4QT
GM7 KHA S. Grant, 2 Clayton Avenue, Irvine, KA12 0TR
G7 KHE M. Knowlson, 23 Hawthorne Avenue, Shipley, BD18 2JB
G7 KHF S. Bates, 29 Hill Veiw Kirby Hill, Boroughbridge, Yo51 9be
G7 KHL S. Smith, 287 Campkin Road, Cambridge, CB4 2LD
GI7 KHR W. Smyth, 35 Davarr Avenue, Dundonald, Belfast, BT16 2NT
G7 KHT A. Haw, 16 Sunnybank Crescent, Yeadon, Leeds, LS19 7TE
G7 KHV R. Irvine, 1 Nutana Avenue, Hornsea, HU18 1JU
G7 KHW D. Nock, 112 Helmsley Close, Bewsey, Warrington, WA5 0GB
G7 KHZ R. Hobbs, 3 Duncombe Close, Bridgwater, TA6 4UT
G7 KID Dr C. Baily, 25 Rocks Park Road, Uckfield, TN22 2AT
G7 KIE N. Kirkman, 4 Woodhall Crescent, Saxilby, Lincoln, LN1 2HZ
G7 KIF C. Davis, 91 Station Road, Barton Under Needwood, Burton-on-Trent, DE13 8DS
G7 KII M. Chilcott, 16 Mount Gould Avenue St. Judes, Plymouth, PL4 9EZ
G7 KIL C. Hunt, 39 Withdean Crescent, Brighton, BN1 6WG
G7 KIN B. Kinsella, 8 Sherwood Park Road, Sutton, SM1 2SQ
GW7 KIO G. Hawthorn-Slater, Ty Croes, Garndolbenmaen, LL51 9UJ
G7 KIQ P. Hyde, 10 Highfield Crescent, Taunton, TA1 5JH
GW7 KIS B. Latta, 17 Park Lane, Holywell, CH8 7UR
G7 KIT D. Hogg, 26 Grenville Drive, Church Crookham, Fleet, GU51 5NR
GW7 KIV R. Gadney, 6 Dan Yr Eppynt Tirabad, Llangammarch Wells, LD4 4DR
G7 KIW R. Henery, 117 Marlborough Road, Swindon, SN3 1NJ
GM7 KIY J. Webster, 31 Harperland Drive, Kilmarnock, KA1 1UH
G7 KJA R. Early, 11 Wenlock Drive, Newport, TF10 7HH
G7 KJD J. Smallwood, 6 Thatchers Croft, Copmanthorpe, York, YO23 3YD
G7 KJE A. Wilkes, 51 Shrewsbury Drive, Newcastle, ST5 7RQ
G7 KJI D. Lean, 6 Orchard Close Edmondsham, Wimborne, BH21 5RQ
GW7 KJO M. Wray, Dinas Bran, Ceidio, Pwllheli, LL53 8UG
G7 KJP R. Mcmahon, 8 Meadow Close, Holburn Estate, Ryton, NE40 3RU
G7 KJR C. Baxter, 6 Merrington Close, Kirk Merrington, Spennymoor, DL16 7HU
G7 KJT S. Mills, 49 Temple Gate Crescent, Leeds, LS15 0EZ
G7 KJV M. Litchman, 26 Oak Tree Close, Loughton, IG10 2RE
G7 KJW P. Haylock, 25 Whitehouse Road, Sawtry, Huntingdon, PE28 5UA
G7 KJX R. Tebbutt, 37 Christchurch Drive, Daventry, NN11 4RX
G7 KKW J. Marsden, 11 Firethorn Drive, Hyde, SK14 3SN
G7 KLJ S. Kerr, 2 Shaw Cross, Kennington, Ashford, TN24 9JY
G7 KLN J. Abbey, 4 Northway, Curzon Park, Chester, CH4 8BB
G7 KLP C. Hartigan, Doonagore, Doolin, Ireland,
G7 KLR L. Pooley, 51 Lincroft, Cranfield, Bedford, MK43 0HS
G7 KLS A. Macaulay, 14 Shipcote Lane, Gateshead, NE8 4JA
G7 KLT T. Hassall, 5 Ashworth Street, Bacup, OL13 9LS
G7 KLV G. Lovegrove, 64 Vicarage Lane, Great Baddow, Chelmsford, CM2 8HY
G7 KLZ J. Fowler, Quinnhaven, Banton Shard, Bridport, DT6 3EB
G7 KMA S. Balkham, 49 St. Georges Road, Hastings, TN34 3NH
GI7 KMC J. Magee, 2 Gilbourne Court, Belfast, BT5 7JB
GW7 KMD N. Hilton, 9 Waterloo Fields, Kingswood, Welshpool, SY21 8LF
G7 KME D. Silverton, 49 Brighton Road Holland-On-Sea, Clacton-on-Sea, CO15 5SR
G7 KMF R. Lythall, 6 Belmont Crescent, Little Houghton, Barnsley, S72 0HT
G7 KMH S. Smith, 12 Holgate Close, Malton, YO17 7YP
G7 KMK N. Deacon, 159 Waterworks Road, Coalville, LE67 4HZ
GM7 KMM S. Linksted, 1 Stevenson Avenue, Polmont, Falkirk, FK2 0GU
G7 KMO P. Butler, 15 Roxby Close, Bessacarr, Doncaster, DN4 7JH
G7 KMP J. Davies, 14 Cullen View, Probus, Truro, TR2 4NY
G7 KMT P. Jones, 24 Valley Lane, Lichfield, WS13 6SU
G7 KMW A. Brown, 4 Kimberley Close, Redditch, B98 8RL
G7 KNA A. Jenner, 24 The Willows, Nailsea, Bristol, BS48 1JQ
G7 KNK H. Arrowsmith, 15 Hermitage Close, Frimley, Camberley, GU16 8LP
G7 KNM M. Giles, 9 Bower Green, Lords Wood, Chatham, ME5 8TN
GW7 KNN B. Jones, Rivendell, Heol Llewelyn, Coedpoeth, Wrexham, LL11 3PB
G7 KNQ C. Martin, 27 Sheepfold Lane, Ruddington, NG11 6NS
G7 KNR P. Grimshaw, 12 Field Maple Drive, Ribbleton, Preston, PR2 6EU
G7 KNS G. Bubb, Clearways Hadlow Stair, Tonbridge, TN10 4HD
G7 KNU P. Davis, 29 Wiltshire Drive, Trowbridge, BA14 0RX
G7 KNW D. Hatfield, Ballyheane, Castlebar, Ireland, F23KT92
G7 KOF L. Barr, 7 Southwold Gardens, New Silksworth, Sunderland, SR3 1LG
G7 KOI G. Russ, 12 Marconi Road, Chelmsford, CM1 1QB
G7 KON W. Humphreys, 43 Arundel Street, Bolton, BL1 6RR
G7 KOS S. Mccormick, 22 Eric Road, Wallasey, CH44 5RQ
GM7 KPE J. Reid, 10 Fernhill Gardens, Windygates, Leven, KY8 5DZ
G7 KPF A. Gayne, 119 Lower Lickhill Road, Stourport-on-Severn, DY13 8UQ
G7 KPH M. Wood, 2 Ridings Lane, New Mill Road, Huddersfield, HD7 2SQ
G7 KPM J. Haywood, 5 Canada Lane Caistor, Caistor, LN7 6RN
G7 KQL D. Walker, 34 Kingsford Street, Salford, M5 5HX
GW7 KQN C. Parsons, Little Foxes, Craig Penllyn, CF71 7LE
G7 KQT S. Schrier, 163 West Lane, Hayling Island, PO11 0JW
G7 KRB S. Wells, 55 Staverton Road, Daventry, NN11 4EY
G7 KRC Keighley ARS, c/o K. Conlon, 4 Hill Crest Drive, Slack Head, Milnthorpe, LA7 7BB
G7 KRE T. Benjamin, 24 Moat Farm Drive Hillmorton, Rugby, CV21 4HG
G7 KRG Keighley Ray Gr, c/o T. Binns, Cross Farm, 1. Cross Lane Oxenhope, Keighley, BD22 9LE
G7 KRH T. Hurley, 18 Manewas Way, Newquay, TR7 3AJ
G7 KRI J. Tilley, 40 Marlborough Road, Stretford, Manchester, M32 0AN
G7 KRM B. Walker, 3 Moorlands Drive, Mayfield, Ashbourne, DE6 2LP
G7 KRO D. Willis, 5 St. Andrews Place, Brightlingsea, Colchester, CO7 0RH
GM7 KRQ H. Gordon, The Cedars, Methlick, Ellon, AB41 7DU
G7 KRS Kettering & District A.R.S, c/o C. Woodward, Flat 3, Burley House Rockingham Road, Cottingham, Market Harborough, LE16 8XS
G7 KRT H. Leong, 38 Woodland Road, Sawston, Cambridge, CB22 3DU
GW7 KRY A. Ryall, 1 Vine Tree, Rumble Street, Usk, NP15 1QG
G7 KRZ S. Pountain, 21 Hayfield Road, Chapel-En-Le-Frith, High Peak, SK23 0JF
GM7 KSA R. Vennard, 4 Braehead, Girdle Toll, Irvine, KA11 1BD
G7 KSE A. Hill, 53 Fairladies, St Bees, CA27 0AR
G7 KSH C. Coleman, 16 Greyhound Road, Glemsford, Sudbury, CO10 7SJ
G7 KSP G. Hampson, 11 Gladstone Grove, Stockport, SK4 4BX
G7 KSQ S. Little, 25 Thrift Wood, Bicknacre, Chelmsford, CM3 4HT
G7 KSS M. Watts, 70 Kentwood Hill Tilehurst, Reading, RG31 6DE
G7 KSV D. Pickering, 15 Primrose Close, Purley On Thames, Reading, RG8 8DG
G7 KTH G. Pargeter, 2 Mayfair Drive Kingsmead, Northwich, CW9 8GF
G7 KTL C. Lake, 16 Falkland House, Bromley Road, London, SE6 2RN
G7 KTP T. Daniels, Three Yew Trees, Newton St. Margarets, Hereford, HR2 0QG
G7 KTQ J. Klunder, 58 Windsor Drive, Brinscall, Chorley, PR6 8PX
G7 KTR A. Slinn, Santon, Pound Lane, Sevenoaks, TN14 7NA
GI7 KTU P. Donnelly, 18 Marcella Park, Newtownards, BT23 4SF
GM7 KTY P. May, 6 Hillpark Way, Edinburgh, EH4 7BJ
G7 KUB R. Warrell, Rose Cottage, Brookbottom, High Peak, SK22 3AY
G7 KUG Dr D. Rutherford, 9 College Drive, Ruislip, HA4 8SD
G7 KUM A. Yorke, 45 Ling Road, Chesterfield, S40 3HT
GM7 KUN C. Schofield, Airidh Ghrianach Knock, Carloway, Isle of Lewis, HS2 9AU
G7 KUR P. Rennison, 30 Millfield Road, Chorley, PR7 1RE
G7 KUU K. Bates, Newhaven Cottage, Star Green, Stroud, GL6 6AD
GM7 KVB A. Whyte, 3 Glenfield Road, Cowdenbeath, KY4 9EP
GI7 KVR P. Mcdonald, 13 Heathfield, Culmore, Londonderry, BT48 8JD
G7 KVT B. Moorey, 132 Queensway, Hereford, HR1 1HQ
GM7 KVU G. Kilgour, 2/1 6 Thornwood Place, Glasgow, G11 7PP
G7 KVZ J. Ashmore, 46 Mease Close, Measham, Swadlincote, DE12 7NA
G7 KWA J. Billam, 46 Rugby Road, Rainworth, Mansfield, NG21 0AU
G7 KWD M. Savin, Flat 3, 30 Thurso Close, Reading, RG30 4YJ
G7 KWF A. Richards, 18 Orchard Way, Lower Kingswood, Tadworth, KT20 7AD
G7 KWM R. Saunders, 3 Curtismill Close, Orpington, BR5 2JX
G7 KWN A. Dance, 8 Eversley Road, Arborfield Cross, Reading, RG2 9PU
G7 KWO J. Lewis, 6 Abbots Way, Beckenham, BR3 3RL
G7 KWP G. Lewis, 7 Hollam Drive, Dulverton, TA22 9EL
G7 KWQ T. Holliday, 131 Skinburness Road, Silloth, Wigton, CA7 4QH
G7 KWS S. Riches, 5 Norfolk St., Forest Gate, London, E7 0HN
G7 KWT J. Ruddock, 13A Murray Road, Northwood, HA6 2YP
G7 KXN M. Bonser, 24 Meend Garden Terrace, Cinderford, GL14 2EB
G7 KXS P. Adam, 50 Lower Edge Road, Rastrick, Brighouse, HD6 3LD
G7 KXT G. Belt, 3 Prospect Hill, Whitby, YO21 1QE
G7 KXV I. Eastham, 51 Chapman Road, Fulwood, Preston, PR2 8NY
G7 KXZ C. Holdford, 23 Willow Close, Newbury, RG14 7FX
G7 KYD S. Walker-Kier, 45 Anstey Road, Peckham, London, SE15 4JX
G7 KYG J. Hope, 29 Horner Road, Taunton, TA2 8DZ
G7 KYH J. Mann, Hyatts Mead, East End, Banbury, OX15 5LH
G7 KYI Rev. M. Wilcockson, Queen'S House 16 High Street, Linton, Cambridge, CB21 4HS
G7 KYJ A. Clark, 10 Garfield Close, Lincoln, LN1 3QP
G7 KYL M. Lack, 39 Riverview, Church Laneham, Retford, DN22 0FL
G7 KYW C. Mellings, 4 Kiln Lane, Horley, RH6 8JG
G7 KYX G. Stones, Ropercroft, Chapel Road, Boston, PE22 9PW
G7 KZG C. Cain, Rydal House, Audley Road, Newport, TF10 7DT
G7 KZJ M. Woodland, 8 Berkeley Crescent, Stourport-on-Severn, DY13 0HJ
GM7 KZL J. Mawson, 5 Forth View, Kirknewton, EH27 8AN
G7 KZV C. Dodson, 64 Stoneleigh Road, Solihull, B91 1DQ
G7 KZY L. Stirrup, 16 Berwyn Grove, St Helens, WA9 2AR
GM7 LAC P. Green, Clochcan School Cottage, Auchnagatt, Ellon, AB41 8UJ
G7 LAF M. Kidman, 465 Grove Green Road, London, E11 4AA

G7 LAK C. Wilkinson, 9 Cheddar Close, Rainworth, Mansfield, NG21 0HX
G7 LAL I. Mazura, 45 Bolingbroke Road, Scunthorpe, DN17 2NQ
G7 LAN D. Halsey, 67 Watling St., Rochester, ME2 3JH
G7 LAS R. Cridland, 47 Stanhope Road, Swadlincote, DE11 9BQ
GD7 LAV A. Gawne, Keristal House, Marine Drive, Douglas, Isle Of Man, IM4 1BJ
G7 LAW J. Danner, 16 Batemans Acre South, Coventry, CV6 1BE
G7 LAX W. Keeys, 9 Broomfield Avenue, Rayleigh, SS6 9EJ
G7 LBD L. Lewis, 29 Sefton Avenue, Hove Edge, Brighouse, HD6 2NA
G7 LBH Dr A. Champion, 5 Airedale Cliff, Leeds, LS13 1EA
GW7 LBI M. Durdin, 52 Norton Road, Penygroes, Llanelli, SA14 7RS
G7 LBL Dr A. Batey, 9 Rampton Drift, Longstanton, Cambridge, CB24 3EH
G7 LBM T. Howard, 21 Church Lane, Thornhill, Dewsbury, WF12 0JZ
G7 LBO D. Wilson, 109 Nightingale Drive, Taverham, Norwich, NR8 6TR
G7 LBP M. Akiki, 103 Main Street, Tupper Lake, USA, 12986
G7 LCD A. Sermons, 18 Crispin Way, Uxbridge, UB8 3WS
G7 LCK J. Berry, Roseneath, Walcote Road, Lutterworth, LE17 6EQ
GI7 LCQ C. Serplus, 14 Claggan Park, Aghadowey, Colraine, BT51 4BD
G7 LCS A. Daniels, Wiscombe, Cleveland Road, Worcester Park, KT4 7JQ
G7 LCV M. Sims, 4 Arran Close, Stapleford, Nottingham, NG9 8LT
G7 LCW C. Simpson, 124 Tattershall Road, Boston, PE21 9LR
G7 LDD R. Newton, 114 Kingston Road, Taunton, TA2 7SP
GW7 LDP P. Martin, 19 Clos Bevan, Gowerton, Swansea, SA4 3GY
G7 LDR E. Woolfenden, 20 Belvedere Avenue Atherton, Manchester, M46 9LQ
GM7 LDU W. Adie, 16 Gordon Crescent, Methlick, Ellon, AB41 7DH
G7 LEB F. Stevens, 4 Pennine Road, Bedford, MK41 9AS
G7 LED D. Miles, 2 Barrington Road, Solihull, B92 8DP
G7 LEL D. Hawkins, 93 Buxton Drive, Bexhill-on-Sea, TN39 4AS
G7 LEN West Lincs Rynt, c/o A. Clark, Emergency Planning Department, Fire Brigade Headquarters, Lincoln, LN5 8EL
G7 LET I. Maughan, 95 York Road, Swindon, SN1 2JR
G7 LEX S. Wilkes, The Coach House, Astley Abbotts, Bridgnorth, WV16 4SP
G7 LEY Essex Packet Gr, c/o G. Lloyd, 9 Hornbeam Walk, Witham, CM8 2SZ
G7 LFC D. Hughes, 86 Colinmander Gardens, Ormskirk, L39 4TF
G7 LFL S. Botterill, 7 Plumtree Road, Cotgrave, Nottingham, NG12 3HT
G7 LFM A. Cocker123, 30 Shaw Road, Rochdale, OL16 4SH
G7 LFQ J. White, 56A Clarendon Street, Herne Bay, CT6 8LZ
GM7 LFT A. Monk, 36 North Road, Saline, Dunfermline, KY12 9UQ
GM7 LFX A. Wilson, 8 Grahamsdyke Road, Bo'ness, EH51 9EG
G7 LFZ W. Mumford, Agden Green Farm, The Green, Great Staughton, St Neots, PE19 5DQ
G7 LGI G. Bryce, 135 Fairbridge Road, Upper Holloway, London, N19 3HF
G7 LGS N. Green, 2 Whittaker Mews, High Street, Rocester, Uttoxeter, ST14 5JU
G7 LGV M. Smith, 3 Tithe Barn, Merton, Bicester, OX25 2NF
G7 LGY H. Abbott, 1 St. Lawrence Close, Heanor, DE75 7AN
G7 LHK T. Reynolds, 11 Duncalfe Drive, Sutton Coldfield, B75 5EX
G7 LHS S. Gray, 26 Hatfield Gardens, Appleton, Warrington, WA4 5QJ
G7 LHT F. Wilson, 3A Vernon Road, Kirkby-In-Ashfield, Nottingham, NG17 8EJ
G7 LHV G. Beaumont, 16 Chelburn View, Littleborough, OL15 9QQ
G7 LIE B. Lovatt, 12 Nelson Street, Leek, ST13 6BB
G7 LIH S. Warren, 41 Barton Road, Rugby, CV22 7PT
G7 LII J. Eccles, 30 The Stour, Daventry, NN11 4PR
G7 LIK C. Tunbridge, 12 Burnham Road, Latchingdon, Chelmsford, CM3 6EU
G7 LIT G. Blaxall, 27 St. Davids Road, Hextable, Swanley, BR8 7RJ
G7 LIW D. Kent, 10 Goldgarth, Grimsby, DN32 8QS
G7 LJA P. Gibson, 62 Glen Park Pensilva, Liskeard, PL14 5PW
G7 LJB C. Mott-Gotobed, 5 Cotswold Close, Basingstoke, RG22 5BA
GM7 LJE J. Freer, 30 Kilmarnock Drive, Cruden Bay, Peterhead, AB42 0NG
GJ7 LJJ N. Utting, Oberon, Bagatelle Road, Jersey, JE2 7TX
G7 LJL S. Murton, 10 James Allchin Gardens, Kennington, TN24 9SD
G7 LJQ G. Roser, 26 Willow Road, Larkfield, Aylesford, ME20 6QZ
G7 LKC J. Radtke, 22 Spinney Drive, Banbury, OX16 9TA
G7 LKI D. Connor, Green Pastures, Stratton Road, Holcombe, Radstock, BA3 5ED
G7 LKL S. Titterington, 33 Victoria Road, Urmston, Manchester, M41 5BZ
G7 LKR A. Maciver, 55 Nordale Park, Rochdale, OL12 7RT
G7 LKV R. Spray, 132 Mansfield St., Sherwood, Nottingham, NG5 4BD
G7 LKY D. Parkinson, 36 Henley Road, Ipswich, IP1 3SA
G7 LKZ C. Bowden, 20 Parc Peneglos, Mylor Bridge, Falmouth, TR11 5SL
G7 LLD M. Bewley, 75 Sugden Road, Worthing, BN11 2JG
G7 LLY J. Wharton, 74 Brompton Park, Brompton On Swale, Richmond, DL10 7JP
G7 LMI J. Hollowood, 10 Rossendale Close, Shaw, Oldham, OL2 8JJ
G7 LMR K. Lavin, 35 Manor Bend, Galmpton, Brixham, TQ5 0PB
G7 LMT D. Pantrey, 10 Columbine Close, East Malling, West Malling, ME19 6ES
G7 LNB A. West, 142 The Street, Kingston, Canterbury, CT4 6JQ
G7 LND R. Williams, 45 Station Road, Westbury, BA13 3JW
G7 LNG J. Tucker, 2 Ivydene Road, Ivybridge, PL21 9BH
G7 LNI S. Czarnota, 11 Spring Park, Chapel Road, Ipswich, IP6 9NX
G7 LNJ R. Woolridge, 8 Alastair Drive, Yeovil, BA21 3BT
G7 LNK P. Knox, 24 Bannister Drive, Banbury, OX16 1GQ
G7 LNM D. Gilham, 53 The Close, Bradwell, Great Yarmouth, NR31 8DR
GM7 LNO G. Cash, 3 Hallydown Crescent, Eyemouth, TD14 5TB
G7 LNP A. Jones, 1 Abbey Way, Rushden, NN10 9HF
G7 LNT P. Cundall, 40 Union Court, Otley, LS21 3NW
G7 LNU N. Sparrow, 46 Thomas Bell Road, Earls Colne, Colchester, CO6 2PF
G7 LNV N. Turland, 2 Ludlow Close, Beeston, Nottingham, NG9 3BY
G7 LNY H. Mascall, 37 Carnival Close, Ilminster, TA19 9DG
G7 LOA L. Fisher, 195 Malvern Road, Billingham, TS23 2PJ
G7 LOE J. Bhogal, 36 Titford Road, Warley, Oldbury, B69 4QA
G7 LOG T. Smallwood, 51 Barlow Road, Barlow, Blaydon-on-Tyne, NE21 6JU
GM7 LOK D. Barr, 17 Ballantrae, East Kilbride, Glasgow, G74 4TZ
G7 LOV E. Farrar, 23 Grovehill Road, Filey, YO14 9NL
G7 LOW M. Bosberry, 31 St Lukes Road, Gosport, PO12 3JN
G7 LOY A. Powell, 76 Glendale Avenue, Washington, NE37 2JS
G7 LOZ K. Blackham, 86 Heather Road, Small Heath, Birmingham, B10 9TA
G7 LPB A. Sellick, 15 Thorpe Street, Raunds, Wellingborough, NN9 6LS
G7 LPD P. Wilkinson, 43 Polperro Drive, Freckleton, Preston, PR4 1YD
G7 LPF C. Hewitt, 9 Alford Fold, Fulwood, Preston, PR2 3UU
G7 LPG P. Garcia, 52 Pilot Road, Hastings, TN34 2AN
G7 LPK R. Hilliard, 8 Cromwell Crescent, Sleaford, NG34 7HW
GW7 LPM L. La Traille, 33 Festival Crescent, New Inn, Pontypool, NP4 0NB
G7 LPN T. Snape, 4 Back Street, Abbotsbury, Weymouth, DT3 4JP
G7 LPO A. Perry, 63A Brookland Road, Huish Episcopi, Langport, TA10 9TH
G7 LPP F. Rice, 42 Donegal Road, Knowle, Bristol, BS4 1PL
G7 LPT A. Page, The Farmhouse, Budges Shop, Trerulefoot, Saltash, PL12 5DA
G7 LPV G. Soden, 21 Bracknell Drive, Alvaston, Derby, DE24 0BP
G7 LPW K. Sharples, 11 West Drove North, Walpole St. Peter, Wisbech, PE14 7HU
G7 LPZ D. Williams, 7 Hampton Drive, Great Sankey, Warrington, WA5 1JF
G7 LQD M. Baguley, 2 Kensington Way, Northwich, CW9 8GG
G7 LQK R. Dunn, 12 Roseberry St., Beamish, Stanley, DH9 0QR
G7 LQN C. King, 33 Alexandra Road, Swallownest, Sheffield, S26 4TA
G7 LQO Dr L. Brown, 19 Stephen Drive, Sheffield, S10 5NX
G7 LRB P. Stevens, 62 Lansdowne Road, Bayston Hill, Shrewsbury, SY3 0JG
G7 LSB L. Brown, 4 Loraine Gardens, Ashtead, KT21 1PD
G7 LSD P. Wainwright, 3 Ashridge Close, Nuneaton, CV11 4XG
G7 LSF J. Blain, 91 Deanfield Road, Henley-on-Thames, RG9 1UU
G7 LSG Dr P. Campbell, 13 Springfield Close, Marden, Hereford, HR1 3EH
GM7 LSI J. Stuart, 3 Pringle Road, Elgin, IV30 4HN
G7 LSP P. Harness, 16 Norfolk Street, Boston, PE21 6PW
G7 LSZ Dr M. Foreman, Vallgatan 10, Vara, Sweden, 534 31
G7 LTG P. Savage, 60 Colonial Road, Bordesley Green, Birmingham, B9 5NG
G7 LTO M. Milns, 3 Merlin Court, Batley, WF17 0RG
G7 LTP P. Sawyer, 96 Violet Lane, Croydon, CR0 4HG
G7 LTR D. Ingham, 19 Recreation Avenue, Ashton-In-Makerfield, Wigan, WN4 8SU
G7 LTT M. Phillips, 2 Hemwood Road, Windsor, SL4 4YU
G7 LTU G. Smith, 36 Sandalwood Road, Loughborough, LE11 3PS
G7 LTW T. Metcalfe, 39 Chobham Road, Frimley, Camberley, GU16 8PS
GM7 LTX A. Warner, 41 Gaynor Avenue, Loanhead, EH20 9LU
G7 LUB J. Broome, Henbant Fach, Penuwch, Tregaron, SY25 6QZ
G7 LUF G. Whitehouse, 27 Kings End Road, Powick, Worcester, WR2 4RB
G7 LUK P. Preston, 45 Saxons Heath, Long Wittenham, Abingdon, OX14 4PU
G7 LUL F. Russell, 61A Fleet Street, Plymouth, PL2 2BU
GM7 LUN J. Keddie, Garrion, Bowland Road, Clovenfords, Galashiels, TD1 3ND
G7 LUO N. Head, 12 Heston Walk, Redhill, RH1 5JB
G7 LUR J. Nolan, 33 Cambridge Road, Langford, Biggleswade, SG18 9PS
G7 LVA S. Sorrell, 19 College Road, Hockwold, Thetford, IP26 4LD
G7 LVE D. Wright, Flat 1 The Annexe, Uxbridge, UB9 5HJ
G7 LVG J. Ashton-Jones, Kiddley Kopse, Mordiford, Hereford, HR1 4LR
G7 LVM J. Maule, 12 Edith Cavell Way, Steeple Bumpstead, Haverhill, CB9 7EE
G7 LVN M. Odam, 10 The Orchards, Meare, Glastonbury, BA6 9PU
G7 LVS M. Unsworth, 41 Aylesbury Crescent, Hindley Green, Wigan, WN2 4TY
GM7 LWA S. Leith, 3 County Houses, Roseisle, Elgin, IV30 5YE
G7 LWF J. Totten, 28 Newman Road, Devizes, SN10 5LE
G7 LWH E. Dalley, 5 Anstey Mill Close, Alton, GU34 2QT
G7 LWU S. Porter, 1 Belt Drove, Elm, Wisbech, PE14 0BA
G7 LWY D. Northeast, 11 Repton Road, Earley, Reading, RG6 7LJ
G7 LXA C. Staff, Uphill Road South, Weston Super Mare, BS23 4TU
G7 LXB W. Roberts, 36 Wray Court, Emerson Valley, Milton Keynes, MK4 2GF
G7 LXH D. Hayzen, 79 Swinburne Avenue, Hitchin, SG5 2QZ
GW7 LXI J. Baines, Pentre Clawdd Cottage, Gobowen, Oswestry, SY10 7AE
G7 LXV N. Hobbs, 224 Belchers Lane, Bordesley Green, Birmingham, B9 5RY
G7 LXY J. Hopkins, 7 Montgomery Close, Coventry, CV3 4FS
G7 LYB R. Brown, 61 Paddockhurst Road, Gossops Green, Crawley, RH11 8EU
G7 LYH J. Briggs, 16 Belmont Place, Colchester, CO1 2HU
G7 LYL C. Nixon, 52 Gloucester Drive, Basingstoke, RG22 4PH
G7 LYN S. Laugher, Jasmine Cottage, Healey, Ripon, HG4 4LH
G7 LYS C. Ameigh, 45 Manley Road, Ilkley, LS29 8QP
G7 LZB A. Howat, 6 Richmond Road, London, N2 8JT
G7 LZM L. Mountain, 45 Westway Gardens, Redhill, RH1 2JB
G7 LZY W. Eatwell, 45 Admirals Walk, Minster On Sea, Sheerness, ME12 3BB
G7 MAB M. Dodson, 64 Stoneleigh Road, Solihull, B91 1DQ
GM7 MAG P. Budgen, 12 Boggs Holdings, Pencaitland, Tranent, EH34 5BB
GD7 MAN Three Legs VHF Contest Group, c/o A. Kissack, 30 High View Road, Douglas, Isle Of Man, IM2 5BH
G7 MAR J. Rivers, Wind In The Willows 1 Hazelwood Close, Ryde, PO33 2UP
G7 MAT K. Hinton, 10 Hillview Road, Basingstoke, RG22 6BQ
G7 MAV A. Goodall, 21 Sladburys Lane, Clacton-on-Sea, CO15 6NX
GM7 MBB L. Millar, 34 Brora Drive, Renfrew, PA4 0XA
G7 MBH M. Davis, 3 Thornley Close Ushaw Moor, Durham, DH7 7NN
GI7 MBP W. Kane, 21 Mount Coole Gardens, Belfast, BT14 8JY
G7 MBU R. Marks, 23 Parfitt Way, Dover, CT16 2QW
G7 MBY D. Richards, 6 Kingley Close, Wickford, SS12 0EN
G7 MCE D. Wilkinson, 56 Cobden Street, Dalton-in-Furness, LA15 8SE
G7 MCK K. Singleton, Spring Cottage, Barcombe Lane, Paignton, TQ3 2QS
G7 MCS B. Mcshea, 5 Frensham Avenue, Fleet, GU51 3EL
G7 MCT C. Taylor, 36 Harewood Road, Shaw, Oldham, OL2 8EA
G7 MDI G. Hawkes, 17 Beacon Hill, Burnham Market, King's Lynn, PE31 8ET
GI7 MDJ S. Clarke, 86 Roddens Crescent, Castlereagh, Belfast, BT5 7JP
GI7 MDK D. Robinson, 4 Ballylesson Road, Magheramorne, Larne, BT40 3HL
G7 MDM S. Wilkins, 5, Blackthorn Close, Gainsborough, DN21 1WB
GI7 MDP S. Mcilvenna, 10 Sycamore Court, Drumaness, Ballynahinch, BT24 8QZ
G7 MDT D. Limb, 34 Elmwood Avenue, Boston, PE21 7RU
G7 MDV C. Prowse, 125 Hill Road, Portchester, Fareham, PO16 8JY
G7 MEA R. Thomas, 25 Heath Hill, Heathfield, Newton Abbot, TQ12 6SP
G7 MEE A. Wood, 14 Anatase Close, Sittingbourne, ME10 5AN
G7 MEG D. Cash, 3 Marsh Lane, Wolverhampton, WV10 6RU
G7 MER C. Hurst, 28 Hengistbury Road, Barton-on-Sea, BH25 7LU
G7 MES M. Stevens, Autumn Cottage, Silver Street, Horncastle, LN9 5NH
G7 MEU D. Hughes, 25 Highfield Road, Carnforth, LA5 9BE
G7 MEX Mexborough & District ARS, c/o J. Saiger, 10 Markham Avenue, Armthorpe, Doncaster., DN3 2AZ
G7 MEZ J. Arter, 18 Essex Road, Westgate-on-Sea, CT8 8AP
G7 MFA A. Sejwacz, 20 Wellington Gardens, Newton-le-Willows, WA12 9LT
G7 MFE S. Morris, The Grange, Downash Farm, Rosemary Lane, Wadhurst, TN5 7PS
G7 MFH B. Fifield, Clyro, Lower Coombses, Chard, TA20 2SX
G7 MFN I. Douglas, 13 Castlereagh Street, New Silksworth, Sunderland, SR3 1HJ
G7 MFO R. Parkes, 7 Main Street, Preston, Hull, HU12 8UB
G7 MFP A. Dresser, 7 Torcross Grove, Calcot, Reading, RG31 7AT

G7	MFR	C. Jenkins-Powell, 43 Cambridge Road, Lee-on-the-Solent, PO13 9DH
G7	MFW	D. Burdett, 17 Brambledown, Chatham, ME5 0DY
G7	MFX	P. March, 39 Rochford Garden Way, Rochford, SS4 1QH
G7	MFY	A. Wakeling, The Willows, Litcham Road, King's Lynn, PE32 2LJ
G7	MFZ	M. Sherratt, 21 Tweedale Close, Mursley, Milton Keynes, MK17 0SB
G7	MGA	R. Thorley, 9 Birchendale Close, Tean, Stoke-on-Trent, ST10 4LT
G7	MGC	T. Gerrard, 41 Auberson Road, Bolton, BL3 3AU
G7	MGG	R. Shirley, 1 St. Richards Court, Bellingham Crescent, Hove, BN3 7FW
G7	MGM	D. Barnes, 36 Westbrook Crescent, Cockfosters, Barnet, EN4 9AS
G7	MGQ	M. Ball, 11 Plantation Road, Thorne, Doncaster, DN8 5EA
G7	MGT	P. Cox, 17 Hyde Lane, Upper Beeding, Steyning, BN44 3WJ
G7	MGV	C. Chadburn, 31 Darwin Close, Top Valley, Nottingham, NG5 9LN
GW7	MGW	E. Palmer, 10 Maes Gwyn, Llanfair Caereinion, Welshpool, SY21 0BD
G7	MGX	P. Asbury, 67 Orchard Way, Measham, Swadlincote, DE12 7JZ
G7	MGY	S. Welger, 55 Burford Avenue, Swindon, SN3 1BX
G7	MGZ	S. Plant, 99 Pegwell Road, Ramsgate, CT11 0ND
GW7	MHB	M. Burt, 44 Overton Close, Buckley, CH7 2AX
G7	MHD	A. Thorp, 34 Third Avenue, Hightown, Liversedge, WF15 8JU
GW7	MHF	R. Johnston, Gledrid Cottage, Oaklands Road, Wrexham, LL14 5DW
G7	MHL	J. Britton, Salters Rest, Salters Mill, Shrewsbury, SY4 5NW
G7	MHO	S. Fell, 14 Rectory Avenue, Corfe Mullen, Wimborne, BH21 3EZ
G7	MHQ	G. Taylor, 21 New Road, Kirkheaton, Huddersfield, HD5 0JB
G7	MHV	S. Stillwell, 130 London Road, Chatteris, PE16 6SF
G7	MID	A. Haydon, 9 Ash Close, Newport, PO30 5UR
G7	MIE	S. Hudson, 20 Churchill Road, Gravesend, DA11 7AQ
G7	MIF	B. Dickenson, 22 Ford Close, Herne Bay, CT6 8AN
G7	MII	D. Burgin, 7 Bramble Close, Halliford, Shepperton, TW17 8RR
G7	MIM	T. Wheeler, 60 Bredhurst Road, Gillingham, ME8 0PE
G7	MIN	A. Jones, 17 Maybush Drive, Chidham, Chichester, PO18 8SR
G7	MIP	Sir H. Kneale, 57 Danforth Close, Framlingham, Woodbridge, IP13 9HP
G7	MIS	A. Trott, 8A Wyatt Road, Kempston, Bedford, MK42 7EH
G7	MIT	T. Good, 11 Moorland Road, Didsbury, Manchester, M20 6BB
G7	MIZ	J. Locker, Delamere, 8 Concordia Avenue, Wirral, CH49 6JD
G7	MJD	A. Bruring, 5 Church Lane, Hartford, Huntingdon, PE29 1XP
G7	MJI	P. Sayers, 23 Roseveare Road, Eastbourne, BN22 8RS
G7	MJJ	R. Delves, 66 Palmeira Road, Bexleyheath, DA7 4UX
G7	MJP	C. Edwards, 16 Martin Street, Normanton, WF6 1DA
G7	MJS	Dr G. Davies, 78 Chatsworth Road, Southport, PR8 2QF
G7	MJV	A. Watts, 32 Hedley Davis Court, Cherry Orchard Lane, Salisbury, SP2 7UE
G7	MJX	D. Hanson, 64 Laxfield Way, Lowestoft, NR33 7HH
G7	MKB	J. Humphries, 25 Wrekenton Row, Wrekenton, Gateshead, NE9 7JD
G7	MKF	B. Stracey, 31 Westfield Road, Margate, CT9 5PA
G7	MKG	P. Bradbury, 40 Titty Ho, Raunds, Wellingborough, NN9 6DF
G7	MKJ	N. Austin, 30 Cardinal Avenue, Borehamwood, WD6 1EP
G7	MKP	A. Brooks, 7 Lindford Drive, Norwich, NR4 6LT
G7	MKQ	A. Airey, 2 Rossmere, Greenways Estate, Spennymoor, DL16 6TZ
G7	MKV	P. Nicholls, 53 Fastolff Avenue Gorleston, Great Yarmouth, NR31 7ND
G7	MLC	G. Bunn, 1 Twelve Acre Road, Norwich, NR2 3PZ
G7	MLJ	P. Skinner, 84 Beresford Avenue, Tolworth, Surbiton, KT5 9LW
G7	MLK	D. Rose, 87 Second Avenue, Sudbury, CO10 1QX
G7	MLL	M. Birtles, 65 Hemsworth Road, Sheffield, S8 8LJ
GW7	MLN	J. Corcoran, 23 Tan Yr Allt, Abercrave, Swansea, SA9 1XF
G7	MLO	J. Large, 5 Raynsford Rise, Stanningfield Road, Great Whelnetham, Bury St Edmunds, IP30 0TS
G7	MLT	A. Armstrong-Bednall, 63 Wellington Street, Heanor, DE75 7FW
G7	MLU	S. Murray, 26 Alfreda Avenue, Hullbridge, Hockley, SS5 6LT
G7	MLW	G. Kilbey, 38 Midland Road, Stonehouse, GL10 2DH
G7	MLX	G. Crisp, Hoppers Farm, Great Kingshill, High Wycombe, HP15 6EY
G7	MMC	S. Reed, 32 Plantation Road, Amersham, HP6 6HL
G7	MME	E. Hughes-Lai, 18 Ramillies Avenue, Plymouth, PL5 2NU
GW7	MMG	P. Pike, 19 Hillrise Park, Clydach, Swansea, SA6 5DX
GW7	MMH	E. Cooke, 32 Chapel Road, Three Crosses, Swansea, SA4 3PU
GM7	MMI	J. Wilson, 32 Silverburn Road, Bridge Of Don, Aberdeen, AB22 8RW
G7	MMJ	S. Pratt, 57 Regency Court, Bradford, BD8 9EX
G7	MMK	D. Baggaley, 6 Bylands Place, Newcastle, ST5 3PQ
G7	MMV	K. Foster, 5 Newman Road, Plymouth, PL5 2DX
G7	MMW	P. Francis, 14 Fulmar Place, Meir Park, Stoke on Trent, ST3 7QF
G7	MND	W. South, Dufonis, Dorchester Road, Wareham, BH20 6EQ
G7	MNE	B. Altman, 5 Ridgemount Gardens, Enfield, EN2 8QL
G7	MNG	M. Whale, 499 Maidstone Road, Wigmore, Gillingham, ME8 0JX
G7	MNK	J. Lambe, 4 St. Georges Road, Enfield, EN1 4TX
G7	MNL	South Devon Raynet Group, c/o C. Coker, 46 Clarendon Road, Ipplepen, Newton Abbot, TQ12 5QS
G7	MNO	R. Nightingale, 58 Nutfield Grove, Filton, Bristol, BS34 7LJ
G7	MNP	G. Turner, 23 Withycombe Road, Penketh, Warrington, WA5 2QL
G7	MNQ	P. Bland, 19 Sookholme Drive, Warsop, Mansfield, NG20 0DN
G7	MNS	A. Cartwright, 118 High Road West, Felixstowe, IP11 9AL
G7	MNT	B. Woods, 64 Yarningale Road, Coventry, CV3 3EQ
G7	MNZ	C. Gaskin, 2 The Briars, West Kingsdown, Sevenoaks, TN15 6EZ
G7	MOB	P. Thain, 26 Haston Lee Avenue, Blackburn, BB1 9QT
G7	MOD	D. Rust, 26 Mill Road, Wiggenhall St. Germans, King's Lynn, PE34 3HL
G7	MOH	E. Middleton, Fairwinds, Southella Road, Yelverton, PL20 6AT
G7	MOK	N. Rieger-Ridd, 3 Rockland Close, Swaffham, PE37 7SP
G7	MOO	K. Parker, 11 Ringer Way, Clowne, Chesterfield, S43 4DW
G7	MOW	K. Starnes, 17 Boughey Place, Lewes, BN7 2EN
G7	MOX	W. Jones, 62 Mallings Drive, Bearsted, Maidstone, ME14 4HG
G7	MOY	B. Jenkins, 27 Glendale, South Woodham Ferrers, Chelmsford, CM3 5TS
G7	MPF	R. Ransome, High Winds, High Town Green, Bury St Edmunds, IP30 0SZ
G7	MPH	R. Cole, 18 Borrowdale Close, Benfleet, SS7 3HE
G7	MPJ	J. Tweedy, 59 St. Aloysius View, Hebburn, NE31 1RH
G7	MPV	J. Woods, Rozel, Bigbury Road, Canterbury, CT4 7ND
G7	MPZ	C. Atkins, 278 Walderslade Road, Chatham, ME5 9AA
G7	MQC	C. Thomas, 1 George Gent Close, Steeple Bumpstead, Haverhill, CB9 7EW
GW7	MQE	D. Smith, 11 Cymau Lane Caergwrle, Wrexham, LL12 9DH
G7	MQF	A. Kirkham, 49 Macclesfield Road, Leek, ST13 8LD
G7	MQP	S. Richardson, 73 Primrose Copse, Horsham, RH12 5PZ
G7	MQQ	H. Griffiths, 11 Gensing Road, St Leonards-on-Sea, TN38 0ER
G7	MQU	D. Sandever, 57 Hayes Lane, Wimborne, BH21 2JB
G7	MQW	R. Carroll, 71 Pelham St., Manton, Worksop, S80 2TT
G7	MRF	M. Farmer, 3 Brackenberry, Cross Heath, Newcastle, ST5 9PS
G7	MRH	E. Cole, 11 Ainsworth House, Wellington Road, Brighton, BN2 3BG
G7	MRJ	E. Stead, Cranford, Church Road, Walpole St. Peter, Wisbech, PE14 7NS
G7	MRL	N. Williams, 1 Dorset Close, Whitehaven, CA28 8JP
G7	MRO	B. Bowker, 205 Smallshaw Lane, Ashton-under-Lyne, OL6 8RJ
G7	MRZ	R. Thompson, 4 Hill Top Road, Birdwell, Barnsley, S70 5QZ
G7	MSC	B. Knight, 3 Burgess Cottages, Mongeham Road, Deal, CT14 8JW
G7	MSF	K. Sanderson, 45 Bygrove, New Addington, Croydon, CR0 9DG
G7	MSG	M. Olivant, 2 Vicarage Gardens, Flamstead, AL3 8EF
G7	MSH	H. Samwells, 2 Dudley Walk, Macclesfield, SK11 8SD
G7	MSK	T. Mann, 21 Glastonbury Court, Yeovil, BA21 3TW
G7	MSN	D. Stead, 15 Reeves Close, Porthleven, Helston, TR13 9PB
G7	MSQ	G. Shelley, 41 Thornley Road, Stoke-on-Trent, ST6 7AL
G7	MSS	D. Forward, 4B Cowper Road, Deal, CT14 9TW
G7	MST	T. Bennett, Rose Cottage, High Street, Rotherham, S62 6LN
G7	MTA	C. Parr, 13 Peartree Avenue, Southampton, SO19 7JN
G7	MTE	M. Coote, 22 Tennyson Close, Boston, PE21 8DL
G7	MTF	T. Foley, Flat 38, Windmill Court, Uxbridge Road, Swindon, SN5 8RT
G7	MTG	A. Blakeston, 8 Victor Street, Cutsyke, Castleford, WF105HB
G7	MTI	F. Paley, 68 Dennil Road, Leeds, LS15 8SD
G7	MTJ	C. Chase, Asholt, Ermine Street, Scunthorpe, DN15 0AD
G7	MTQ	S. Saunders, 3 The Terrace, High Street, Cavendish, Sudbury, CO10 8AS
G7	MTV	M. Bourne, 100 Dimsdale View West, Newcastle, ST5 8EL
G7	MTW	R. Powell, 4 Diana Close, Spencers Wood, Reading, RG7 1HP
G7	MUB	R. Harcourt, 7 Lightfoot Close, Newark, NG24 2HT
G7	MUD	Christchurch ARS, c/o D. Layne, 5 Howe Close, Christchurch, BH23 3JA
G7	MUE	S. Roper, 1 Holywell Road, Kilnhurst, Mexborough, S64 5UQ
GM7	MUN	J. Smith, 14 John Collins Crescent, Galashiels, TD1 2FA
G7	MUT	T. Cannon, 5 Barn Close, Upton, Poole, BH16 5RX
G7	MUY	A. Sadler, 19 Lyndhurst Park Home Estate Sea Lane, Ingoldmells, PE25 1PD
G7	MVE	M. Cotton, 2 Redhill View, Castleford, WF10 4QL
GW7	MVG	H. Millington, Arran Clayton Road, Mold, CH7 1SU
G7	MVU	N. Brown, 6 Hundon Place, Haverhill, CB9 0AP
G7	MVX	G. Trudgill, 61 Lansdowne Road, Coxhoe, Durham, DH6 4DN
G7	MVY	R. Stockley, 10 Swan Road, Timperley, Altrincham, WA15 6BX
GI7	MWA	S. Stewart, 3 Killyfaddy Road, Magherafelt, BT45 6EX
G7	MWB	W. Bone, 217 Bensham Road, Gateshead, NE8 1US
G7	MWC	R. Moss, 6 Adelaide Gardens, Stonehouse, GL10 2PZ
G7	MWH	P. Cross, Churchill House, Churchill Road, Louth, LN11 7QW
G7	MWI	L. Hansen, 19 Market St., Appledore, Bideford, EX39 1PW
G7	MWJ	A. Holloway, 31 Gays Road, Hanham, Bristol, BS15 3JR
G7	MWK	J. Arkle, 16 Sea View, Ashington, NE63 0XH
GM7	MWL	H. Murray, 23 Denmore Gardens, Bridge Of Don, Aberdeen, AB22 8LJ
G7	MWM	A. Scott, 22 Planters Grove, Lowestoft, NR33 9QL
G7	MWS	P. Parrish, 5 Kestrel Lane, Cheadle, Stoke-on-Trent, ST10 1RU
G7	MWU	G. Haswell, 16 Hither Green, Jarrow, NE32 4LP
G7	MWW	J. Smith, 19 The Crescent, Mitcheldean, GL17 0SB
GM7	MWX	R. Raynor, La Pergola, Kilmuir, Inverness, IV1 1XG
G7	MXL	T. Grange, 7 The Cherries, Canvey Island, SS8 0BB
G7	MXM	C. Turner, 55 Fordfield Road, Ford Estate, Sunderland, SR4 6XG
G7	MXN	L. Orchard, 678 Devonshire Road, Blackpool, FY2 0AW
G7	MXQ	B. Gilbraith, 19 Bullcote Green, Royton, Oldham, OL2 6NJ
G7	MXS	A. Harrison, Midhope Lodge Midhopestones, Sheffield, S36 4GW
G7	MXT	D. Harris, 12 Turner Avenue, Billingshurst, RH14 9PU
GU7	MXZ	B. Heath, 21 Clos De Bas, Green Lanes, St. Peter Port, Guernsey, GY1 1TS
GW7	MYD	P. Williams, 5 Bright St., Cross Keys, Newport, NP1 7PB
GM7	MYF	C. Dennett, Lyn-Ard, Smollett Street, Alexandria, G83 0DW
G7	MYI	A. Stride, 3 Barnfield Cottages, Edmondsham, Wimborne, BH21 5RD
G7	MYJ	R. Ball, 5 Miller Fold Avenue, Accrington, BB5 0NT
G7	MYM	D. Roberts, Chatterbox, 3A, Station Road, Pershore, WR10 1NQ
G7	MYN	C. George, 22 Elgar Drive, Shefford, SG17 5RZ
G7	MYO	C. Mciver, 2 Abbey Meadows, Chertsey, KT16 8RA
G7	MYT	P. Hilton, 40 Megstone Avenue, Whitelea Chase, Cramlington, NE23 6TU
G7	MYY	L. Fuller, 78C Seal Road, Sevenoaks, TN14 5AT
G7	MZA	R. Loukes, Fagus, The Street, Staple, Canterbury, CT3 1LL
G7	MZE	T. Ingle, 68 Wooldale Drive, Filey, YO14 9ER
G7	MZJ	I. Mitchell, 87 Bluebell Avenue, Penistone, Sheffield, S36 6AF
G7	MZK	D. Mitchell, 28 Southgate, Penistone, Sheffield, S36 6EA
G7	MZL	N. Baker, 43 Little Park, Wadhurst, TN5 6DL
G7	MZS	L. Terry, 8 Carters Close, Slyfield, Guildford, GU1 1FR
G7	MZW	A. Calvert, 122 Grampian Way, Thorne, Doncaster, DN8 5YW
G7	MZX	R. Barron, 15 Fernhill Close, Poole, BH17 8SQ
G7	MZY	I. Sharp, 6 Ullswater Drive, Bath, BA1 6NP
GM7	MZZ	G. Whiting, 21 Leckethill Court, Cumbernauld, Glasgow, G68 9EG
GM7	NAA	I. Skeoch, 1 Castleton Crescent, Grangemouth, FK3 0BH
G7	NAE	A. Westwood, 3 Walgrave Close, Belper, DE56 1UF
G7	NAI	J. Stock, 31 Grange Road, Wickham Bishops, Witham, CM8 3LT
G7	NAL	H. Wood, 31 Goring Avenue, Gorton, Manchester, M18 8WW
G7	NAO	I. Langmuir, 2 Nelson Road, Newport, PO30 1QT
G7	NAP	D. Gee, 28 Rein Road, Morley, Leeds, LS27 0JA
G7	NBE	M. Goodwin, 23 Saxon Way, Ashby-de-la-Zouch, LE65 2JR
G7	NBF	A. Sadler, 23 Wolsey Road, Moor Park, Northwood, HA6 2HN
G7	NBG	L. Mcguire, 200 Wellingborough Road, Rushden, NN10 9SX
G7	NBI	P. Webb, 42 Holland Road, Ampthill, Bedford, MK45 2RS
G7	NBJ	D. Corfield, 177 Hurst Rise, Matlock, DE4 3EU
G7	NBL	C. King, Broadlea, Honey Hill, Fen Drayton, Cambridge, CB24 4SF
G7	NBP	C. Williams, 28 Sundorne Crescent, Shrewsbury, SY1 4JE
G7	NBQ	S. Ambrose, 3 Three Mile Pond, Sawbridgeworth, CM21 9ED
G7	NBR	W. Hayward, 15 Whitehouse Road, South Woodham Ferrers, Chelmsford, CM3 5PF
G7	NBU	K. Bassett, Manor Farm, Marsh Green, Exeter, EX5 2EX
G7	NBV	F. Young, 6 Birchvale Court, Desborough, Kettering, NN14 2UY
G7	NBZ	P. Millerchip, 6 Washbrook View, Ottery St Mary, EX11 1EP
G7	NCD	T. Willis, 15 Cedar Court, Congleton, CW12 3JP
G7	NCE	K. Derbidge, 1 Batch View, Grange Avenue, Street, BA16 9PE
G7	NCG	E. Turner, 16 The Rowans, Doddington, March, PE15 0SE
G7	NCP	J. Merrington, Cartref, Ball Lane, Frodsham, WA6 8HP

G7 NCV K. Hobbs, 61 Fairway, Waltham, DN37 0NB
G7 NCW G. Hinton, 25 Linden Close, Prestbury, Cheltenham, GL52 3DX
GU7 NCZ Dr N. Turner, Camellia Lodge, L'Aumone, Castel, Guernsey, GY5 7RT
G7 NDB K. Marshall, Doveysmead, Chapel Street, Basingstoke, RG25 2BZ
G7 NDC P. Hirst, 47A Rowley Lane, Fenay Bridge, Huddersfield, HD8 0JG
G7 NDI D. Bunney, 3 Hilmanton, Lower Earley, Reading, RG6 4HN
G7 NDN S. Ward, Russet House, Beech Road, Haslemere, GU27 2BX
G7 NDO H. Blackburn, 4 Hawkridge, Furzton, Milton Keynes, MK4 1BQ
G7 NDQ G. Bettyes, 44 Springfield Road, Oundle, Peterborough, PE8 4LT
G7 NDS M. Fry, 14 The Lawns Collingham, Newark, NG23 7NT
G7 NDT R. Walker, 46 Lodge Road, Little Houghton, Northampton, NN7 1AE
GI7 NEB J. Conlon, 30 Drumglass Way, Dungannon, BT71 4AG
G7 NEC C. Watson, Westwood, Laneside, Queensbury, BD13 1NE
G7 NED G. Gini, Tenuta Buzzoletto Nuovo 6, Garbagna, Italy, 28070
G7 NEE E. Maloney, 56 Westonfields Drive, Longton, Stoke-on-Trent, ST3 5JA
G7 NEG R. Smith, 32 Water Lane, Wootton, Northampton, NN4 6HE
G7 NEH G. Pemberton, 2 Hockenhull Avenue, Tarvin, Chester, CH3 8LP
G7 NEM J. Barber, 2 Gresley Way, March, PE15 8QA
G7 NER T. Stokes, 33 Talbot Drive, Euxton, Chorley, PR7 6PD
GI7 NET K. Nolan, 34 Lisgoole Park, Drumgallan, Enniskillen, BT74 5ND
GI7 NFB M. Robinson, 92A Dromore Road, Hillsborough, BT26 6HU
GM7 NFF P. Salmon, Mill House, Monreith, Newton Stewart, DG8 9LJ
G7 NFG M. Holdsworth, 9 Beaumont Close, Bowburn, Durham, DH6 5QA
G7 NFK J. Watmough, 2 Burbage Heights, Buxton, SK17 6YU
GW7 NFM E. Jones, Maes Y Coed, Vownog Road, Mold, CH7 6ED
G7 NFN R. Bailey, 29 Priory Close, Bath, BA2 5AL
G7 NFO M. Hall, 30 Kingsley Avenue, Rugby, CV21 4JY
G7 NFR J. Thomas, 2 Alexandra Road, Uxbridge, UB8 2PQ
GW7 NFT J. Parry, Charlbury, Usk Road, Newport, NP18 1LP
GW7 NFY M. Mee, Anncott, Hylas Lane, Rhyl, LL18 5AG
G7 NGB J. Alger, Church Hill Cottage Church Hill, Caterham, CR36SA
G7 NGF D. Hatcher, 8 Churchfield, Monks Eleigh, Ipswich, IP7 7JH
G7 NGI J. Price, 32 Wiltshire Drive, Trowbridge, BA14 0RE
G7 NGN R. Williams, 73 Quedgeley Park, Greenhill Drive, Gloucester, GL2 5NZ
G7 NGQ A. Bauer, 5 Horse Fayre Fields, Spalding, PE11 3FA
GW7 NGU J. Vaughan, Montrose, 5 Trewarren Drive, Haverfordwest, SA62 3TR
G7 NGX A. Mckenna, 12 Sunnyside Road, Beeston, Nottingham, NG9 4FH
G7 NHB R. Griffiths, 4 Wolrige Way, Plympton, Plymouth, PL7 2RU
G7 NHC D. Scholes, 71 Pelham St, Ashton under Lyne, OL70DU
G7 NHD T. Carroll, 32 Marfords Avenue, Wirral, CH63 0JW
G7 NHE J. Turnbull, 32 Haydon, Washington, NE38 8PF
G7 NHF K. Williams, 1 St. Ives Way, Halewood, Liverpool, L26 7YW
G7 NHL K. Mitchell, 12 Lon Lafant, Llandudno Junction, SK17 9PL
G7 NHQ K. Cotterill, 1 Molineux Avenue, Broadgreen, Liverpool, L14 3LT
G7 NHR P. Dunlop, 4 Birket Avenue Moreton, Wirral, CH46 1QZ
GM7 NHS Dr R. Johnson, 3 Hopetoun Green, Bucksburn, Aberdeen, AB21 9QX
GM7 NHU G. Devereux, 44 Greenhead Road, Dumbarton, G82 2PN
G7 NHV K. Johnson, 43 Glencoe Road, Great Sutton, Ellesmere Port, CH66 4NA
G7 NHW K. Mckane, 60 Hazelwood Road, Callington, PL17 7EU
GU7 NHX A. Dorrian, Le Petit Jardin, Clos Des Emrais, Castel, Guernsey, GY5 7YB
G7 NHY V. Brooker, Flat 6, 46 Foxglove Way, Wallington, SM6 7JU
G7 NHZ A. Greatbatch, 46 Java Crescent, Trentham, Stoke-on-Trent, ST4 8RT
G7 NIA H. Seldon, 22 Downside Avenue, Plymouth, PL6 5SD
G7 NIB I. Clark, 1 Hayward Parade, Oakengates, Telford, TF2 6EZ
G7 NID T. Thorpe, 7A Rosalind Close, Colchester, CO4 3JH
G7 NIH A. Davies, 1 Fire Station Yard, Rochdale, OL11 1DT
G7 NII Prof. R. Kalawsky, 23 Brook Lane, Loughborough, LE11 3RA
G7 NIL J. Chin, 198 Bermondsey Wall East, London, SE16 4TT
G7 NIN D. Townsend, 40 Popes Lane, Sturry, Canterbury, CT2 0JZ
G7 NIR L. Jones, 53 Ennisdale Drive, Wirral, CH48 9UF
G7 NIU S. Tanner, 31 Four Acres, Portland, DT5 2JG
GW7 NIW G. Durno, Lothlorien, Upper Denbigh Road, St Asaph, LL17 0BH
G7 NIX C. Shurety, O Fran Villa, Camp Road, Norwich, NR8 6LD
G7 NIZ A. Bowers, Flat 2, Jevington House, Upperton Road, Eastbourne, BN21 1LW
G7 NJB G. Harris, 58 The Leas, Minster On Sea, Sheerness, ME12 2NL
G7 NJD J. Stewart, 22 Garden Road, Kendal, LA9 7ED
G7 NJE C. White, 13 Peel St., Heywood, OL10 4QD
G7 NJG D. Godwin, 2 Barncroft Drive, Hempstead, Gillingham, ME7 3TJ
G7 NJI M. Tribe, The Paddock, Wix Hill, Leatherhead, KT24 6ED
GW7 NJM P. Martin, 2.Gwarllyn.Tudweiliog.Pwllheli, Pwllheli.Gwynedd, Ll538ng
G7 NJP M. Neal, 3 Nursery Way, Grimston, King's Lynn, PE32 1DQ
GW7 NJQ S. Richardson, Holmleigh, Broughton, Cowbridge, CF71 7QR
GW7 NJT J. Jones, 8 Manor Court, Ewenny, Bridgend, CF35 5RH
G7 NJW G. Bullen, 24 Meadowside Road, Sutton Coldfield, B74 4SJ
G7 NJX S. Mullen, 18 Helens Road, Sandford, BS25 5PD
G7 NJZ J. O'Rourke, 39 Rutherglen Road, Corby, NN17 1ER
G7 NKH D. Smith, 7 Salisbury Road, Carshalton, SM5 3HA
G7 NKI A. Davin, 7 Paynes Park, Hitchin, SG5 1EH
G7 NKJ T. Westbrook, 5 Newlands, Northallerton, DL6 1SJ
G7 NKS Dr J. Cowburn, 26 Birch Close, Broom, Biggleswade, SG18 9NR
G7 NKU C. Prout, 1 Westbrook Lustrells Vale, Saltdean, Brighton, BN2 8EZ
G7 NKV A. Taylor, Moonlight Cottage, 4 Alderley Road, Macclesfield, SK11 9AP
G7 NKZ K. Bradley, 161 Mortimer Road, Southampton, SO19 2HJ
G7 NLA C. Milburn, Greenleas, Furlongs Lane, Horncastle, LN9 6LD
G7 NLF G. Bandara, 26 Undine Street, London, SW17 8PR
G7 NLJ R. Watts, 33 Rockside View, Matlock, DE4 3GP
G7 NLP J. Greenacre, 30 Ramsey Grove, Bury, BL8 2RE
G7 NLR North Lancashire Raynet Group, c/o D. Andrew, 14 Westfield Grove, Morecambe, LA4 4LQ
G7 NLY J. James, 14 Fairview Drive, Bayston Hill, Shrewsbury, SY3 0LE
G7 NLZ R. Bennett, 16 Graham Close, Portslade, Mile Oak, Brighton, BN41 2YE
G7 NMB Cumbria Emergency Planning Unit, c/o K. Bennett, 38 Northumberland St., Workington, CA14 3EY
G7 NME B. Caldicott, 1 Naish Road, Burnham-on-Sea, TA8 2LE
G7 NMI K. Osborne, 42 Barbrook Lane, Tiptree, Colchester, CO5 0EF
GI7 NMK L. Breadon, 32 Ashley Crescent, Millisle, Newtownards, BT22 2BG
G7 NMT M. Beach, 11 Lawday Link, Farnham, GU9 0BS
GW7 NNA A. Watkin Ba Hnd, Aarburg, Windsor Close, Oswestry, SY11 2UA
GW7 NNM J. Day, Tynwtra, Bwlch-Y-Ffridd, Newtown, SY16 3HX
G7 NNR B. Hughes, Dorfstrasse 71, Waldfeucht, Germany, 52525
GM7 NNS A. Strachan, Mormond View, New Leeds, Peterhead, AB42 4HX
G7 NNU J. Wood, 19 Arbour Crescent, Macclesfield, SK10 2JB
G7 NNZ D. Dukeson, Wadsley Nook, Far Lane, Sheffield, S6 4FD
G7 NOI G. Evans, 20 Bleasdale Court, Longridge, Preston, PR3 3TX
G7 NOQ J. Howarth, 61 Poplar Drive, Lamaleach Park, Lamaleach Drive, Preston, PR4 1EG
G7 NOR M. Bowers, Uprising, Shottendane Road, Margate, CT9 4NE
G7 NOS A. Roxburgh, 11 Briscoe Drive Moreton, Wirral, CH46 0TN
GI7 NOW R. Mcmaster, 40 Woodlands, Ballycarry, Carrickfergus, BT38 9JD
G7 NPL C. Daniel, 24 Canterbury Road, Dewsbury, WF12 7LA
GM7 NPR G. White, 2 Keill Cottage, Isle of Gigha, PA41 7AD
G7 NPT J. Boyd, 26 Pear Tree Place, Warrington, WA4 1AX
G7 NQJ B. Silcocks, 2 Derham Road, Bristol, BS13 7SA
GM7 NQP G. Kinnell, 61 Hallforest Avenue, Kintore, Inverurie, AB51 0TF
G7 NQR E. Purvis, 36 Birchington Avenue, Middlesbrough, TS6 7EZ
G7 NQU J. Varnham, 1 Burgin Road, Anstey, Leicester, LE7 7FA
G7 NQW A. Symon, Flat3, 66A Clyde Road, Croydon, CR06SW
G7 NQX J. Bailes, 48 Harlech Close, Eston, Middlesbrough, TS6 9SZ
G7 NQZ D. Harrison, 18 The Crescent, Eaglescliffe, Stockton-on-Tees, TS16 0JB
G7 NRB L. Kirk, 10 Dyers Mews, Neath Hill, Milton Keynes, MK14 6ER
G7 NRG P. Atkinson, 8 Thanet Terrace, Appleby-in-Westmorland, CA16 6TU
G7 NRO C. Flanagan, 2 Wynyerd House, Durham Road, Wolviston, Billingham, TS22 5LP
G7 NRP R. Crossley, 52 Richard Road, Darton, Barnsley, S75 5NP
G7 NRR P. Morris, Antler Cottage, High Street, Scaldwell, Northampton, NN6 9JS
G7 NRS A. Saunders, Christ Church Vicarage Schofield Street, Leigh, WN7 4HT
G7 NRV R. Wheeldon, 10 Mill View Court, School Lane, St Neots, PE19 8GJ
GW7 NSJ S. Shufflebotham, 9 Smale Rise Oswestry, Shropshire, SY11 2YL
G7 NSK P. Blunden, 20 Fiskerton Road, Reepham, Lincoln, LN3 4EB
G7 NSN J. Vinters, 106 Halifax Road, Ripponden, Sowerby Bridge, HX6 4AG
GW7 NTA T. Blunsdon, 3 Railway Terrace, Aberbeeg, Abertillery, NP13 2AD
G7 NTG J. Smith, 54 Greenfield Avenue, Kettering, NN15 7LL
G7 NTI A. Wood, 23 Cross Ryecroft Street, Ossett, WF5 9EW
G7 NTO D. Brown, 5 Ash Close, Watlington, OX49 5LW
GW7 NTP P. Banks, Ysgol Emrys Ap Iwan, Rhuddlan Road, Abergele, LL22 7HE
G7 NTQ C. Thomson, 17 Upperfield Grove, Corby, NN17 1HN
G7 NUC D. Poulet, 3 Barton Cottages, Newton St. Cyres, Exeter, EX5 5DA
G7 NUE A. Isted, 22 Tavy Road, Worthing, BN13 3PG
G7 NUG T. Brown, 125 Godinton Road, Ashford, TN23 1LN
GW7 NUL J. Page, 39 Rest Bay Close, Porthcawl, CF36 3UN
G7 NUM M. Exton, Thorn Cottage, 42 High Street, Bourne, PE10 0SR
G7 NUN S. Latham, 4 Shaston Road, Stourpaine, Blandford Forum, DT11 8TA
GM7 NUQ C. Mair, 23 Strathburn Gardens, Inverurie, AB51 4RY
G7 NUT I. Nutley, Czechers, Potten End Hill Water End, Hemel Hempstead, HP1 3BN
GW7 NUU R. Clegg, The Cottage At Fron Isaf, Pentrecelyn, LL15 2HR
G7 NVB K. Fowler, Flat 10, Westwood House Edinburgh Road, Norwich, NR2 3RL
GM7 NVG C. Park, Flat 4, Corrow, Cairndow, PA24 8AD
G7 NVI P. Lancaster, 2 North Farm Road, Lancing, BN15 9BS
GW7 NVM R. Skelton, 7 Whitethorn Place, Sketty, Swansea, SA2 8HR
G7 NVS D. Poulton, 93 Pretoria Road, Ibstock, LE67 6LP
G7 NVZ G. Moore, 1 Sibland Way, Thornbury, Bristol, BS35 2EJ
G7 NXV D. Ross, 37 Cartmell Drive, Leeds, LS15 0NQ
GM7 NYB N. Macfarlane, 3 Kilmore Terrace, Devaig Isle of Mull, PA75 6GN
G7 NYD B. Collinge, 4 Ash Grove, Preesall, Poulton-le-Fylde, FY6 0EW
G7 NYF P. Ridley, 11 Thorney Close, Fareham, PO14 3AF
GW7 NYP Gloucester Repeater Group, c/o Prof. N. Negus, Llain, Llanycefn, Clynderwen, SA66 7XT
GM7 NZI R. Simpson, 2/1 53 Jedworth Avenue, Glasgow, G15 7QE
G7 NZM G. Clifton, 21 Park Road, Featherstone, Wolverhampton, WV10 7HS
G7 NZO S. Bate, 5 Turnpike Way, Ashbourne, DE6 1UD
G7 NZR A. Haslam, The Goldings, Hayton, Brampton, CA8 9JA
G7 NZU L. Elliott, 62 Holmsley Lane Woodlesford, Leeds, LS26 8RY
G7 NZV R. Easting, 3 Ellistons Yard, Ballingdon Street, Sudbury, CO10 2BU
G7 NZY C. Wells, 6 Craister Court, Cambridge, CB4 2SH
G7 NZZ D. Poole, 239 Forest Road, Fishponds, Bristol, BS16 3QY
G7 OAA J. Davies, 78 Chatsworth Road, Southport, PR8 2QF
GM7 OAF E. Capstick, 24 Dalmore Crescent, Helensburgh, G84 8JP
G7 OAH K. Adams, Queena, Bicton, Liskeard, PL14 5RF
G7 OAI J. Cannell, 53 Thimble Close, Rochdale, OL12 9QP
G7 OAJ S. Walker, 3 Sefton Villas Spook Hill, North Holmwood, Dorking, RH5 4JW
G7 OAS A. White, 1 Little Orchard, Stanway Road, Broadway, WR12 7NQ
G7 OAV Dr A. Holohan, 8 School House Terrace, Kirk Deighton, Wetherby, LS22 4EH
G7 OAX K. Morrison, 25 Holywell Close, Poole, BH17 9BG
G7 OBC R. Hudman, 27 Egerton Road, Streetly, Sutton Coldfield, B74 3PQ
G7 OBD M. Peach, 48 Melrose Avenue Portslade, Brighton, BN41 2LS
G7 OBF J. Aston, 9 Beaufort Close, Reigate, RH2 9DG
GM7 OBM D. Macpherson, 138 Broomhill Crescent, Alexandria, G83 9QL
G7 OBP G. Turner, 11 Royds Crescent, Rhodesia, Worksop, S80 3HF
G7 OBR M. Biddles, 338 High Road, Whaplode, Spalding, PE12 6TG
G7 OBS M. Simkins, 37 St. Andrews Meadow, Harlow, CM18 6BL
G7 OBX N. Finbow, 17 Bartholomew Close Bardney, Lincoln, LN35XT
G7 OCC A. Graham, 19 Talbot Road, Rushden, NN10 9NS
G7 OCH J. Doy, 11A Shrubland Avenue, Ipswich, IP1 5EA
G7 OCK B. Phillipson, 27 Victoria Avenue, Crook, DL15 9DB
G7 OCQ W. Horwood, 2A Bellrope Lane, Roydon, Diss, IP22 5RG
GM7 OCU G. Rule, 105/19 Causewayside, Edinburgh, EH9 1QG
G7 OCX J. Turton, 60 Shafton Lane, Leeds, LS11 9RE
G7 OCY D. Norton, 52 Letchworth Road, Leicester, LE3 6FG
G7 ODB R. Evans, 18 Lilac Close, Keyworth, Nottingham, NG12 5DN
G7 ODG R. Beadle, 8 Erica Gardens, Croydon, CR0 8LG
G7 ODM G. Wane, 1A Rickyard Close, Polesworth, Tamworth, B78 1DE
G7 ODN L. Nicoletti, 6 Laverock Close, Kimberley, Nottingham, NG16 2QX
GW7 ODP M. Jones, Flat 2 Block 15 Heol Eifion Gorseinon, Swansea, SA4 4PH
G7 ODR D. Cartwright, 6 Peveril Road, Castleton, Hope Valley, S33 8UA
G7 ODT C. Wright, Top Farm Bungalow, Ermine Street, Huntingdon, PE28 4EW
G7 ODV G. Bagley, Woodcroft, Gatton Bottom, Redhill, RH1 3BH

G7 ODZ B. Fowler, 4 Langley Street, Derby, DE22 3GL
G7 OEA P. Foulkes, 60 Hornby Boulevard, Litherland, Liverpool, L21 8HG
G7 OED R. Stanley, 58 Wells Gardens, Basildon, SS14 3QS
G7 OES W. Robinson, 5 North View, Newfield, Chester le Street, DH2 2SD
G7 OET L. Hitchen, 40 Methuen Avenue, Fulwood, Preston, PR2 9QX
G7 OEW S. Yohn, Little Cottage, Hale, Milnthorpe, LA7 7BL
G7 OEY G. Hodges, 12 Linwal Avenue, Houghton-On-The-Hill, Leicester, LE7 9HD
G7 OFI Capt. P. Smith, 11A Springwell Close, Maltby, Rotherham, S66 7HG
G7 OFM R. Squires, 2 Fishergreen, Ripon, HG4 1NW
G7 OFU N. Patterson, 63 Squires Wood, Fulwood, Preston, PR2 9QA
G7 OFV T. Wordsworth, 61 Crane Road, Kimberworth, Rotherham, S61 3HN
G7 OGL D. Parker, 9 Warwick Gardens, Thrapston, Kettering, NN14 4XB
G7 OGN D. Arnold, 18 Pheasant Way, Spring Park, Northampton, NN2 8BJ
G7 OGO K. Mann, 89 Wootton Village, Boars Hill, Oxford, OX1 5HW
G7 OGR D. Arthurs, 32 Lowfield Avenue, Rotherham, S61 4PD
GM7 OGS D. Rushmer, 1A Low St., New Pitsligo, Fraserburgh, AB43 6NQ
G7 OGT T. Mcdonald, 3 Widden Close, Sway, Lymington, SO41 6AX
G7 OHD P. Martindale, 4 The Crayke, Bridlington, YO16 6YP
G7 OHM R. Jarvis, 5 Caldecote Avenue, Cockermouth, CA13 9EQ
G7 OHO J. Hislop, 10 Park Wood Close, Broadstairs, CT10 2XN
G7 OHW L. Blanchard, 1 Dibden Lane, Alderton, Tewkesbury, GL20 8NT
G7 OIA M. Larcombe, 52 Orchard Road, Burgess Hill, RH15 9PL
G7 OIB S. Johnson, 85 Bradley Road, Trowbridge, BA14 0QS
G7 OIE S. Viney, 5 Hawthorne Grove, Dudley, DY3 2QQ
GW7 OIK D. Todd, Tyrcae, Gwernogle, Carmarthen, SA32 7SA
GM7 OIN J. Cowan, 1 Treebank Crescent, Ayr, KA7 3NF
G7 OIR A. Grundy, 21 Ribston Close, Shenley, Radlett, WD7 9JW
G7 OIT I. Guest, 46 Brasenose Drive, Kidlington, OX5 2EQ
G7 OJA P. Mann, 11 New Mills Road, Hayfield, SK22 2JG
GM7 OJJ J. Alexander, Newton Of Kinmundy Cottage, Kinmundy, Peterhead, AB42 5AY
G7 OJO R. Brown, 34 Fallowfield Road, Solihull, B92 9HH
GW7 OJT Dr E. Wolfenden, 18 Edison Crescent, Clydach, Swansea, SA6 5JF
G7 OJU F. Dixon, 9 Lincoln Road, Fenton, Lincoln, LN1 2EP
G7 OJX K. Trigg, 41 Veasey Road, Hartford, Huntingdon, PE29 1TA
G7 OJY V. Holyoake, 14 Maudlin Court, De Cham Road, St Leonards-on-Sea, TN37 6JY
G7 OJZ J. Neale, 20 Oakfield Road, Wollescote, Stourbridge, DY9 9DL
G7 OKF M. Hawkins, 294 Norton Lane, Earlswood, Solihull, B94 5LP
G7 OKI W. Cornish, 21 Centaur Street, Portsmouth, PO2 7HB
G7 OKO F. Webb, 50 Hassam Avenue, Newcastle, ST5 9ET
G7 OKR J. Campbell, 94 Liscard Road, Wallasey, CH44 8AB
G7 OKT S. Smith, 17 Thackers Way, Deeping St. James, Deeping St James, Nr Peterborough, PE68HP
G7 OKV K. Porter, 47 Pick Hill, Waltham Abbey, EN9 3LD
GM7 OKX G. Chesworth, Auchinway, Skares, Cumnock, KA18 2RE
G7 OKY D. Schofield, 26 The Chase, Coulsdon, CR5 2EG
G7 OLC P. Broadhead, 45 Priory Close, Dudley, DY1 3ED
G7 OLF W. Levick, 50 Wintern Court, Lea Road, Gainsborough, DN21 1NA
G7 OLG M. Hodge, 271 Marsh Lane, Bootle, L20 5BG
G7 OLH A. Barnett, 20 Mortlake Drive, Mitcham, CR4 3RQ
G7 OLT I. Edwards, 31 The Twistle, Byfield, Daventry, NN11 6UR
G7 OLU W. Wood, The Alley Off Of Gajdoru St, Xaghra, Gozo, Malta, XRA 104
G7 OLW P. Newton, 22 Barrow Rise, Weymouth, DT4 9HJ
G7 OLX J. Chambers, 46 Violet Road, West Bridgford, Nottingham, NG2 5HA
G7 OMA M. Heales, 11 Cardinals Walk, Hampton, TW12 2TR
G7 OMF K. Gill, 358 Moor End Road, Halifax, HX2 0RH
G7 OMI J. Patel, 1 The Glade, Furnace Green, Crawley, RH10 6JS
G7 OMM R. Stroud, 22 Marvell Close, Crawley, RH10 3AL
G7 OMN J. Eyes, 31 Langdale Road Wistaston, Crewe, CW2 8RS
G7 OMQ J. Gillman, 4 Yeosfield, Riseley, Reading, RG7 1SG
GM7 OMU S. Macmillan, 74 Canberra Avenue, Clydebank, G81 4LN
GI7 OMY D. O'Buitigh, 11 Rossnareen Avenue, Belfast, BT11 8LP
G7 ONB C. Robinson, Jordan, The Green, Stowmarket, IP14 3AB
G7 ONE R. Jacobs, Rose Mount, Grove Road, Ventnor, PO38 1TH
G7 ONF M. Whitley, St. Marys, Chapel Lane, Langtoft, Driffield, YO25 3TD
G7 ONI J. Churchill, 30 Brigade Place, Caterham, CR3 5ZU
GM7 ONJ A. Martin, The Cairn, Duntrune, by Dundee, DD4 0PP
G7 ONL R. Ramsey, 17 Derby Road, Guisborough, TS14 7DP
G7 ONR P. Green, 23 Singleton Court, Patrington, Hull, HU12 0SF
G7 ONV D. Das, 4 Farcliff, Sprotbrough, Doncaster, DN5 7RE
G7 OOB K. Barnes, The Old School House, 32 Church Street, Peterborough, PE6 8DA
G7 OOE J. Bone, Flat 11, Homebreeze House, Beach Street, Morecambe, LA4 6BT
G7 OOF E. Cottle, 3 Mainstone, Romsey, SO51 8HG
G7 OOH K. Mcallister, Willow Croft, 109A Kaye Lane, Huddersfield, HD5 8XT
G7 OOI R. Phillipson, 22 Bagmere Close, Brereton, Sandbach, CW11 1SG
GI7 OOM D. Magowan, 35 Princeton Avenue, Lurgan, Craigavon, BT66 8LW
G7 OOO Scarborough Seg, c/o R. Clayton, 9 Green Island, Irton, Scarborough, YO12 4RN
G7 OOP A. Constantine, Fairways, Birchington Close, Bexhill-on-Sea, TN39 3TF
G7 OOS J. Smalley, 88 Gledhow Wood Road, Leeds, LS8 4DH
G7 OOT S. Godrich, 11 The Ringway, Queniborough, Leicester, Le7 3dn
G7 OOU Dr D. Witts, Heelands, Heelands, Milton Keynes, MK13 7PZ
G7 OOV R. Nunn, 49 Lulworth Drive, Roborough, Plymouth, PL6 7DT
G7 OPB A. Adams, 44 Berkeley Vale Park, Berkeley, GL13 9TG
G7 OPD P. Hundy, 101 Goodway Road, Great Barr, Birmingham, B44 8RS
G7 OPG R. Palmer, 3 Oldfield Close, Maidstone, ME15 8DY
G7 OPI D. Pink, 87 Lillybrook Estate, Lyneham, Chippenham, SN15 4AS
G7 OPJ J. Buttery, 38 Wigmore Gardens, Worle, Weston-Super-Mare, BS22 9AQ
GM7 OPN W. Cairns, 74 Jean Armour Drive, Mauchline, KA5 6DT
G7 OPS P. Whiting, 77 Melford Way, Felixstowe, IP11 2UH
G7 OPY L. Kelly, 8 Solent Hill, Freshwater, PO40 9TG
G7 OQB E. Dockray, 2 The Gardens, Farsley, Pudsey, LS28 5HW
GM7 OQE G. Murray, 10 Mcgregor Court, Crossgates, Cowdenbeath, KY4 8ER
G7 OQG S. Williams, 7 Wilton Crescent, Macclesfield, SK11 8TH
G7 OQL C. Broad, 96 Kingsley Court, Fraddon, St Columb, TR9 6PD
G7 OQO A. Thomas, 8 Woodlands Fold, Birkenshaw, Bradford, BD11 2LG
G7 OQQ S. Ayers, 20 Wytham View, Eynsham, Witney, OX29 4LU
G7 OQT K. Peacock, 20A Pool View Caravan Park, Buildwas, Telford, TF8 7BS
G7 OQU Dr M. Jones, 30 Redruth Street, Manchester, M14 7PX
GW7 ORB D. Cullen, 16 Chapel Street, Upper Brynamman, Ammanford, SA18 1AD
G7 ORE Essex Raynet, c/o N. Smith, Clare Cottage, White Ash Green, Halstead, CO9 1PD
G7 ORG R. Gunner, White House, The Whiteway, Cirencester, GL7 7BA
GM7 ORJ A. Ross, 16 Croft Road, Kiltarlity, Beauly, IV4 7HZ
G7 ORK D. Love, Woodland View, Lower Street, Shepton Mallet, BA4 6BB
G7 ORN S. Tideswell, 35 Didcot Drive Marchington, Uttoxeter, St14 8LT
G7 ORS J. Lorenzen, 40 Boundary Road, Ramsgate, CT11 7NW
G7 ORT D. Buckley, 22B Anerley Grove, Kingstanding, Birmingham, B44 9QH
G7 ORV S. Edmonds, 170 Halton Road, Sutton Coldfield, B73 6NZ
G7 ORW R. Garnett-Frizelle, 17 Bridport Avenue, New Moston, Manchester, M40 3WP
GM7 ORX J. Lee, 2/5 Heriot Bridge, Edinburgh, EH1 2HR
G7 OSB C. Baker, 22 Court Park, Thurlestone, Kingsbridge, TQ7 3LX
G7 OSH Watcombe Radio Club, c/o P. Worlledge, 181 Roselands Drive, Paignton, TQ4 7RN
G7 OSJ A. Ramm, 17 Sharrington Road, Bale, Fakenham, NR21 0QX
G7 OSK D. Ramm, 24 Rowan Way, Holt, NR25 6TZ
G7 OSO A. Kinnersley, Weathertop, Barthomley Road, Stoke-on-Trent, ST7 8HU
GM7 OSQ D. Clark, Benmhor, Baluachrach, Tarbert, PA29 6TF
G7 OST S. Timms, 7 Portway Drive, High Wycombe, HP12 4AU
G7 OTE S. Barlow, 16 Arundel Avenue, Urmston, Manchester, M41 6NQ
GW7 OTQ A. Dibbins, 2 Edwards Close, Briggs Lane, Oswestry, SY10 8PS
GM7 OTT I. Alexander, Newton Of Kinmundy Cottage, Kinmundy, Peterhead, AB42 5AY
G7 OUG K. Gatfield, 76 Barnwood Avenue, Gloucester, GL4 3AJ
G7 OUT M. Reynolds, 15 Foxfield Drive, Stanford-le-Hope, SS17 8HH
G7 OUZ B. Donkin, 13 Saddlebow Road, King's Lynn, PE30 5BQ
G7 OVB G. Hutton, 17 Fonteyn Place, Stanley, DH9 6XE
G7 OVE D. Brown, 9 Lancaster Way, East Winch, King's Lynn, PE32 1NY
G7 OVK C. Carson, 47 Stratford Close, Cramlington, NE23 8HW
G7 OVM N. Ward, 79 Ulwell Road, Swanage, BH19 1QU
G7 OVS J. Dilks, Handley Farm Bungalow, Brant, Beckingham, LN5 0RN
G7 OWB K. Moorcroft, 58 Oakley Road, Dovercourt, Harwich, CO12 4QU
G7 OWP J. Oliphant, 16 Sylvias Close, Amble, Morpeth, NE65 0GB
G7 OWQ B. Clifton, 3 Kirton Road, Cosham, Portsmouth, PO6 2ES
GM7 OWU B. Brander, 3 Spartleton Place, Dundee, DD4 0UJ
G7 OWV M. Baines, 21 Acre Moss Lane, Kendal, LA9 5QE
G7 OWX D. Allen, 130 Seamer Road, Scarborough, YO12 4EY
G7 OWZ D. Payne, 147 Upper Marehay Road, Marehay, Ripley, DE5 8JG
G7 OXA D. Giles, 73 Barsby Drive, Loughborough, LE11 5UJ
G7 OXB S. Richardson, 6 Dane Ghyll, Barrow-in-Furness, LA14 4PZ
G7 OXH R. Wilkins, 85 St. Richards Road, Otley, LS21 2AL
G7 OXK J. Leach, 2 Andover Close, Feltham, TW14 9XG
G7 OXN B. Burdis, Toledillo 19, Malaga, Spain, 29570
G7 OXP R. Singleton, 91 Robins Lane, St Helens, WA9 3NF
G7 OXV R. Blott, Chateau Perigord Ii Bloc E Apt 5, 6 Lacets Saint Leon, Monaco, Monaco, MC98000
G7 OXY J. Mathew, 139 Dowthorpe Hill, Earls Barton, Northampton, NN6 0PX
G7 OYD P. Thompson, Stanville Cowick Road Snaith, Goole, DN14 9JG
G7 OYF B. Gilbert, 3 Williams Way, West Row, Bury St Edmunds, IP28 8QB
G7 OYP S. Hollis, 89 Longfield Lane, Cheshunt, Waltham Cross, EN7 6AN
GU7 OYU A. Stoaling, Carando, La Petite Mare De Lis Clos, La Rocquette, Castel, Guernsey, GY5 7BN
G7 OYX B. Dorey, 8 Richmond Road, Swanage, BH19 2PZ
G7 OZA G. Johnston, 94 Abercorn Crescent, Harrow, HA2 0PU
G7 OZE K. Baldry, 160 Rover Drive, Castle Bromwich, Birmingham, B36 9LL
G7 OZH D. Albury, 40 Mulberry Gardens, Fordingbridge, SP6 1BP
G7 OZI A. Morley, 6 Millway, Chudleigh, Newton Abbot, TQ13 0JN
G7 OZJ S. Morley, 7 Bank Avenue, Mitcham, CR4 3DW
GW7 OZP J. Barrett, Tree Tops, Comins Coch, Aberystwyth, SY23 3BL
G7 OZQ T. Pluck, 29 Templegate View, Leeds, LS150HQ
G7 OZU M. Knight, 30 Mountbatten Drive, Biggleswade, SG18 0JJ
G7 PAE C. Bailes, 2 Church Road, Catworth, PE28 0PA
G7 PAF R. Scaife, 50 Springbank Road, Gildersome, Leeds, LS27 7DJ
G7 PAG P. Gould, 10 Heron Park, Lychpit, Basingstoke, RG24 8UJ
G7 PAK A. Smith, 153 Seymour Way, Sunbury-on-Thames, TW16 7NL
G7 PAN I. Leather, 44 Newlands Road, Intake, Sheffield, S12 2FZ
G7 PAY C. Wilson, 107 Hamilton Avenue, Uttoxeter, ST14 7FE
GM7 PBB J. Gray, 5 North Dell, Isle of Lewis, HS2 0SW
G7 PBC P. Sherburn, 70 Briarwood Road, Stoneleigh Park, Epsom, KT17 2NG
G7 PBH H. Parrish, 5 Kestrel Lane, Cheadle, Stoke-on-Trent, ST10 1RU
G7 PBK J. Oliver, 27 Rosamund Avenue, Pickering, YO18 7HF
G7 PBO M. Sewell, 4 Cherfield, Minehead, TA24 5TD
GW7 PBP V. Roberts, 44 Mount Crescent, Morriston, Swansea, SA6 6AP
GI7 PBQ R. Young, 8 Glenside Avenue, Drumbo, Lisburn, BT27 5LQ
G7 PBT R. Spirrell, 32 Churchfield Drive, Castle Cary, BA7 7LA
G7 PBV J. Mason, 56 Skegby Road, Sutton-in-Ashfield, NG17 4EZ
G7 PCE K. Toop, 10 Hunt Road, Blandford Forum, DT11 7LZ
G7 PCF J. Banfield, Highbury, 81 Clophill Road, Bedford, MK45 2AD
G7 PCG M. Bartlett, 6 St Vincent Chase, Braintree, CM7 9UJ
G7 PCT P. Treadwell, 22 Meynell Close, Melton Mowbray, LE13 0RA
G7 PCV A. Newman, 115 Wolverhampton Road, Cannock, WS11 1AR
G7 PCW C. Naylor, 25 Tavistock Way, Wakefield, WF2 7QS
GW7 PCX G. Bellis, 70 Osborne St., Rhos, Wrexham, LL14 2HT
G7 PDH J. Swallow, 26 Balmoral Road, Abbots Langley, WD5 0ST
G7 PDO M. Galea, 17 Waterloo Road, Horsham St. Faith, Norwich, NR10 3HS
G7 PDR J. Martin, 45 Quail Holme Road, Knott End-On-Sea, Poulton-le-Fylde, FY6 0BT
G7 PDU S. Willis, 180 Thisselt Road, Canvey Island, SS8 9BL
G7 PEB J. Edwards1, 37 The Orchard, Swanley, BR8 7UR
G7 PEC Essex Raynet, c/o G. Tiller, 15 Woodlands Gardens, Romsey, SO51 7TE
G7 PEE T. Griffiths, 19 James Copse Road, Waterlooville, PO8 9RG
G7 PEH P. Gater, 166 Rolls Ave, Crewe, CW1 3QD
G7 PEN E. Penn, 53 Manfield Avenue, Walsgrave, Coventry, CV2 2QF
GW7 PEO P. Bennett, 14 Harlech Crescent, Prestatyn, LL19 8DG
G7 PER D. Boughton, 59 Redland Drive, Kirk Ella, Hull, HU10 7UX

G7 PEU R. Chapman, 12 Lynton Road, Chesham, HP5 2BU
GW7 PEX K. Barker, 12 Swn Yr Afon Kenfig Hill, Bridgend, CF33 6AJ
G7 PFD A. Thompson, Forge House, Newark, Newark, NG22 0PN
G7 PFG J. Smith, 6 Aspen Close Ss89Jj, Canvey Island Essex, SS89JJ
G7 PFI M. Green, Runnymede, Aston Common, Aston, Sheffield, S26 2AD
GW7 PFK E. Gillet, 4 Camrose Court, Caldy Close, Barry, CF62 9DR
G7 PFL M. Dormer, 5 Kipling Walk, Basingstoke, RG22 6BN
GM7 PFQ D. Chadwick, 19 Lochslin Place, Balintore, Tain, IV20 1UP
G7 PFT J. Richardson, Unit F, Tollgate Business Centre, Stafford, ST16 3HS
G7 PFY J. Ellis, 11 Moorland Crescent Guiseley, Leeds, LS20 9EF
G7 PGH M. Card, 11 Manifold Road, Eastbourne, BN22 8EH
G7 PGY G. Sleeman, 21 Millbank, Kintbury, Hungerford, RG17 9UW
G7 PHB S. Beesley, 15 Byron Close, Cheadle, Stoke-on-Trent, ST10 1XB
G7 PHC M. Porter, 16 The Oval, Scarborough, YO11 3AP
G7 PHD I. Connor, 34 Mace Road Mildenhall, Bury St Edmunds, IP28 7FP
G7 PHE A. Lassman, 69 St. Ladoc Road, Keynsham, Bristol, BS31 2EQ
G7 PHF Dr G. Lloyd, 256 Penns Lane, Sutton Coldfield, B76 1LQ
G7 PHG D. Harbron, 48 Sheridan Road, South Shields, NE34 9JJ
G7 PHI T. Larsen, 47 Lizard Lane Whitburn Village, Sunderland, SR6 7AL
G7 PHK J. Wilkes, 229 Merland Rise, Tadworth, KT20 5JQ
G7 PHL R. Marshall, 3 Lawrence Crescent, Sutton-in-Ashfield, NG17 4HX
G7 PHR W. Stewart, 1 Laing Close, Bardney, Lincoln, LN3 5XS
G7 PHT K. Dennis, 4 Ash Grove, Sheringham, NR26 8PT
G7 PHW S. Dobson, 166 Lynfield Drive, Bradford, BD9 6EZ
G7 PHY D. Dobson, 166 Lynfield Drive, Bradford, BD9 6EZ
GW7 PIB J. Challenger, 33 Blossom Close, Langstone, Newport, NP18 2LT
G7 PIG C. Wood, 2 Longfield Avenue, Heald Green, Cheadle, SK8 3NH
G7 PIJ M. Beeson, 1 Tamar Grove, Cheadle, Stoke-on-Trent, ST10 1QQ
G7 PIK B. Bashford, 51 Broadwater Road, Worthing, BN14 8AH
GW7 PIN W. Griffiths, 147 High Street, Tonyrefail, Porth, CF39 8PL
G7 PIP R. Oswald, 17 Dunclutha Road, Hastings, TN34 2JA
G7 PIR J. Briggs, 8 Premier Court, 100 Monyhull Hall Road, Birmingham, B30 3QJ
G7 PIX D. Atkins, 99 Mariners Way, Maldon, CM9 6YX
GI7 PIZ S. Hewitt, 11 Erindee Close, Donaghadee, BT21 0NS
G7 PJD R. Howes, 58 Mayfield Avenue, Orpington, BR6 0AQ
GI7 PJF R. Stewart, 1 Portmore Hall, Ballydonaghy Road, Crumlin, BT29 4WT
G7 PJG D. Gohill, Flat 71, Hatton Place, Luton, LU2 0FD
GI7 PJU C. Robinson, 19D Divis Tower, Belfast, BT12 4QB
G7 PKD R. Brown, 1 Octavian Close, Hatch Warren, Basingstoke, RG22 4TY
G7 PKG B. Jenkins, 3 Wisteria Avenue, Branston, Lincoln, LN4 1QB
G7 PKH P. Precious, 99 Sherwood Avenue, St Albans, AL4 9PW
G7 PKJ D. Alway, 79 Landseer Avenue, Bristol, BS7 9YW
G7 PKK A. Sharp, 170 Kingshill Road, Swindon, SN1 4LL
G7 PKP J. Constance, 1 Grayling Close, Grimsby, DN37 9HA
G7 PKQ P. Troll, 18 Bowness Road, Millom, LA18 4LS
GM7 PKT R. Morrison, Corran Gardens, Corran Gardens, Onich, Fort William, PH33 6SJ
G7 PKY E. Plant, 22 Bournville Road, London, SE6 4RN
G7 PLE J. Goodliffe, 25 Stansgate Avenue, Cambridge, CB2 0QZ
G7 PLP I. Brown, 9 Larford Walk, Stourport-on-Severn, DY13 0HE
G7 PLS N. Partridge, 13 Alderney Way, Immingham, DN40 1RB
G7 PLV A. Miller, 10 Limerick Close, Ipswich, IP1 5LR
GW7 PMA J. Beach, 11 St Annes, Western Lane, Swansea, SA3 4EW
G7 PMB R. Cooke, 2 Harvey Court, Warrington, WA2 9SD
G7 PMF V. Lennox, 64 Oak Avenue, Blidworth, Mansfield, NG21 0TL
G7 PMG S. Rose, 84 Woodhill Park, Pembury, TN2 4NP
G7 PMI D. Williams, 6 Raven Crescent, Billericay, CM12 0JF
G7 PMK W. Harrison, 67 Connaught Gardens, Shoeburyness, Southend-on-Sea, SS3 9LR
G7 PMQ P. Slight, 4 Field Close Welton, Lincoln, LN2 3TT
G7 PMU S. Lawrence, 85 West Avenue, Clacton-on-Sea, CO15 1HB
G7 PMV G. Gimber, 10 Harrowdene Gardens, Teddington, TW11 0DH
G7 PMW G. Wall, Westerland, Sandhill Road, Buckingham, MK18 2LZ
G7 PMX M. Firth, 5 Courtenays, Seacroft, Leeds, LS14 6JZ
G7 PMY B. Whittington, Flat 5, Anjou Court, 8 Hereward Road, Eastbourne, BN23 6TQ
G7 PNE D. Head, 76 Dryden Crescent, Stevenage, SG2 0JH
G7 PNF M. Cleverley, 43 Friesian Gardens, Newcastle, ST5 6BB
G7 PNG R. Owen, 53 Huntingdon Drive, Castle Donington, Derby, DE74 2SR
G7 PNM P. Smith, 41A Thornhill Road Middlestown, Wakefield, WF4 4RU
G7 PNP P. Collins, Summerley, Hollow Road, Widdington, Saffron Walden, CB11 3SL
GM7 PNX N. Armstrong, 4 Arboretum Road, Edinburgh, EH3 5PD
G7 POA Dr F. Greaves, Ratooragh, Schull, Ireland, WEST CORK
G7 POC N. Phillips, 9 Symonds Close, Chandler'S Ford, Eastleigh, SO53 3TP
G7 POI L. Selway, Rua Do Rochio No 7, Poco Redondo, Tomar, Portugal,
G7 POQ R. Erdinc, 811 York Road, Leeds, LS14 6AA
G7 POS D. Ford, 44 Cardinal Square, Beeston, Leeds, LS11 8HR
G7 POT S. Eastwood, Cross Trods Spellowgate, Driffield, YO25 5UP
G7 POV A. Lickley, 18 Byron Street, Macclesfield, SK11 7PL
G7 POW J. Sanderson, 40 Sheldon Close, Bransholme, Hull, HU7 4RU
G7 PPC B. Lowe, 19 Wolverhampton Road, Bloxwich, Walsall, WS3 2EZ
G7 PPL A. Pardivalla, 2 Mcdowell Way, Narborough, Leicester, LE19 2RA
GM7 PPN H. Waugh, 93 Denholm Road, Musselburgh, EH21 6TU
G7 PPS H. Lodge, 69 Helena Road, Rayleigh, SS6 8LQ
G7 PQB K. Hunter, 30 Loxley Road, Lowestoft, NR33 9PG
G7 PQD B. Harrison, 145 St. Leonard St., Hendon, Sunderland, SR2 8QB
G7 PQL M. Robinson, 1 Selby Close, Baxenden, Accrington, BB5 2TQ
G7 PQM R. Buchan, 16 Lomond Drive, Kettering, NN15 5DE
G7 PQP D. Little, 20 Vicarage Close, Shillington, Hitchin, SG5 3LS
GW7 PQS M. Waite, 4 Clos Bryngwyn, Garden Village, Swansea, SA4 4BJ
G7 PQW N. Wills, 18 Hollis Way, Southwick, Trowbridge, BA14 9PH
G7 PQX P. Smith, 4 Stone Lodge Lane, Ipswich, IP2 9PA
G7 PRB W. Ross, 8 Mayall Court, Waddington, Lincoln, LN5 9PY
G7 PRC S. Newstead, Rectory Cottage, Church Road, Norwich, NR12 8YL
G7 PRD R. Ware, 23 Harmsworth Drive, Stockport, SK4 4RP
G7 PRI P. Newton, 5 Sandford Road, Winscombe, BS25 1HD
GW7 PRK R. Zeal, 5 Llanthewy Road, Newport, NP20 4JR
G7 PRO P. Julian, 45 Rectory Avenue, Corfe Mullen, Wimborne, BH21 3EZ
GW7 PRW M. Price, Graig Hill Chapel, Skenfrith, Abergavenny, NP7 8UF
G7 PRZ B. Reed, Blackthorn, Wanborough Lane, Cranleigh, GU6 7DS
G7 PSC P. Rivers, 39 Ashton Road, Birmingham, B25 8NZ
G7 PSF J. Block, 40 Cromwell Road, Cambridge, CB1 3EF
GM7 PSH A. Stevens, 69 Polwarth Terrace, Prestonpans, EH32 9PX
G7 PSK N. Kingsley-Lewis, Forge Cottage, Rudham Road Helhoughton, Fakenham, NR21 7BY
G7 PSL M. Lamb, 52 Crookham Grove, Morpeth, NE61 2XF
G7 PSS P. Allnutt, 37 Moss Mead, Chippenham, SN14 0TN
G7 PST T. Ellis, 19 Cavendish Avenue, Colchester, CO2 8BP
G7 PSU A. Drummond, 10-12 Tottington Road, Turton, Bolton, BL7 0HS
G7 PSV M. Bennett, 83 Middlethorpe Road, Cleethorpes, DN35 9PP
G7 PSW D. Wieloch, 43 Northampton Grove, Langdon Hills, Basildon, SS16 6ED
G7 PSZ A. Laurence, Brookvale, Nooklands, Preston, PR2 8XN
G7 PTA M. Masterman, 7 Pond Bank, Blisworth, Northampton, NN7 3EL
G7 PTB R. Player, 49 St. Johns Road, Tilney St. Lawrence, King's Lynn, PE34 4QJ
G7 PTC M. Watson, 12 Milk Thistle Close, Stainton, Middlesbrough, TS8 9FQ
G7 PTD J. Midwood, 8 Thompson Avenue, Holt, NR25 6EN
G7 PTH R. Graham, 8 Pecche Place, Chineham, Basingstoke, RG24 8AA
G7 PTM J. Taylor, 8 Worsley Avenue, Blackpool, FY4 2DH
G7 PTT A. Myers, 24 Milburn Street, Crook, DL15 9DY
G7 PTV M. Howse, 28 Courtiers Drive, Bishops Cleeve, Cheltenham, GL52 8NU
G7 PTX P. Castle, 26 Chestnut Walk, Pulborough, RH20 1AW
G7 PTZ R. Mold, 134 Kipling Avenue, Brighton, BN2 6UE
G7 PUA J. Campbell, 48 Renforth Street, Gateshead, NE11 9BE
GI7 PUG G. Clegg, 45 Strandburn Drive, Belfast, BT4 1NA
G7 PUK D. Glass, 2 Thorne Square, Sunderland, SR3 4PA
G7 PUL R. Hoggard, 1 Whiphill Close, Bessacarr, Doncaster, DN4 6DX
G7 PUN D. Russon, 123 Queens Drive, Newton-le-Willows, WA12 0LN
G7 PUP A. Hurd, 27 Atlow Close, Chesterfield, S40 4LQ
G7 PUW S. Frizzell, 58 Crisp Road, Lewes, BN7 2TX
G7 PUZ L. Martyn, 1 Canewdon Hall Close, Canewdon, Rochford, SS4 3PY
G7 PVE G. Eddy, 102 Springfield Close, Andover, SP10 2QT
G7 PVF M. Folland, 14 High St., Shoreham, Sevenoaks, TN14 7TD
G7 PVG B. Fox, 10 Materman Road, Stockwood, Bristol, BS14 8SS
GU7 PVI M. Major, East Liberty, Gibauderie, St Peter Port, Guernsey, GY1 1XJ
G7 PVL C. Watts, 8 Clark Close, Wraxall, Bristol, BS48 1JL
G7 PVQ S. Denton, 9 High Meadow, Tollerton, Nottingham, NG12 4DZ
G7 PVU G. Rouse, 18 Westfield Avenue, Woking, GU22 9PH
G7 PVY D. Driver, 53 Brabant Way, Westbury, BA13 3UW
G7 PVZ W. Sefton, 87 Lillibrooke Crescent, Maidenhead, SL6 3XL
G7 PWA N. Padley, 12A Wey Close, Ash, Aldershot, GU12 6LY
G7 PWI C. Thornton, 1 Elizabeth Drive, Tring, HP23 5HL
G7 PWJ J. Churchill, 68 Anthony Road, London, SE25 5HB
G7 PWK M. Bibb, 9 Nelson Court, Old Nelson Street, Lowestoft, NR32 1EH
G7 PWL M. Turnbull, 11 Waverley Avenue, Whitley Bay, NE25 8AU
GI7 PWQ G. Mccormick, 24 Warren Park Drive, Lisburn, BT28 1HF
G7 PWS C. Collins, 32 St. Martins Road, New Romney, TN28 8JY
G7 PWU H. Tomlinson, 42 Gawsworth Avenue, Crewe, CW2 8PB
G7 PWV P. Woolhouse, 21 Coombe Wood Hill, Purley, CR8 1JQ
GM7 PXJ L. Michie, 5 Torridon Place, Rosyth, Dunfermline, KY11 2EZ
GM7 PXL Lt. D. Warner, Torran, Letterfinlay, Spean Bridge, PH34 4DZ
G7 PXM J. Cordell, 45 Queens Gardens, Dartford, DA26HZ
G7 PXR P. Gilbert, 4 Ruby Street, Bristol, BS3 3DY
G7 PXS G. Mape, 34 Amberwood Drive, Manchester, M23 9NZ
G7 PXX E. Spires, 8 Regent Terrace Barrow Road, New Holland, Barrow-upon-Humber, DN19 7QB
G7 PYB J. Bailey, 1 Greenwood, Ascot, SL5 8LL
G7 PYN A. Wentworth, 5 York Avenue, Prestwich, Manchester, M25 0FZ
G7 PYQ N. Gell, 1 Lawton Road, Rushden, NN10 0DX
G7 PYR V. Tuff, 8 Millcroft Court, Blyth, NE24 3JG
G7 PYT G. Coleman, 23 Graham Avenue, Patcham, Brighton, BN1 8HA
G7 PYV A. Turner, 20 Kipling Gardens, Upper Stratton, Swindon, SN2 7LJ
G7 PYW K. Houghton, 42 Pear Tree Avenue, Coppull, Chorley, PR7 4NL
G7 PZB R. Dewsbery, 8 Westfield Close, Market Harborough, LE16 9DX
G7 PZE F. Eastham, 4 Dunkirk Avenue, Fulwood, Preston, PR2 3RY
G7 PZF W. Bailey, 15 Norfolk Road, Congleton, CW12 1NY
GM7 PZH M. Drennan, 6 Hillpark Way, Edinburgh, EH4 7BJ
G7 PZL A. Morton, 54 Rose Farm Approach, Normanton, WF6 2RZ
G7 PZM S. Carter, 6 Bramalea Close, London, N6 4QD
G7 PZQ P. Breese, 6 Wheatley Close, Birmingham, b75 5ej
G7 PZT J. Keen, 30 Fielding Crescent, Blackburn, BB2 4TD
G7 PZU A. Haworth, 8 Boulsworth Crescent, Nelson, BB9 8DF
G7 RAB D. Evans, 31 Kinsbourne Way, Thornhill, Southampton, SO19 6HB
G7 RAE J. Kirkwood, 6 Trinity Court, Rothwell, Kettering, NN14 6YQ
G7 RAF Royal Air Force ARS (Rafars), c/o K. Sellens, 7 The Knapps, Semington, Trowbridge, BA14 6JG
G7 RAG J. Dennis, The Old Chapel House, Alford, LN13 9PH
GI7 RAH T. Tweedie, 17 Schomberg Park, Belfast, BT4 2HH
G7 RAI M. Moorhouse, 11 Hazel Grove, Huddersfield, HD2 2JP
G7 RAJ D. Eggett, 68 Forest Lane, Kirklevington, Yarm, TS15 9ND
GM7 RAK J. Boyd, 102 Provost Milne Grove, South Queensferry, EH30 9PL
G7 RAL Loughborough & District ARC, c/o I. Hewitt, 26 Outwoods Drive, Loughborough, LE11 3LT
GI7 RAM J. Christie, 3 Victoria Drive, Sydenham, Belfast, BT4 1QT
G7 RAT Reigate Amateur Transmitting Society, c/o P. Tribe, Island View, 108 Portsmouth Road, Lee-on-the-Solent, PO13 9AF
G7 RAU D. Edwards, Blue Stones, 9-10 Mile End. Lizard, Helston, TR12 7AS
G7 RAY B. Kagelmacher, Rauhe Häge 5, Wismar, Mecklenburg-Vorpommern, Germany, 23966
G7 RAZ M. Wager, 115 Queensway, Taunton, TA1 4NL
G7 RBA M. Sims, 23 Winding Way, Alwoodley, Leeds, LS17 7RB
G7 RBB A. Perkins, 9 St. Martins Close, Canterbury, CT1 1QG
G7 RBC I. Rodgers, 89 Braemar Road, Worcester Park, KT4 8SN
G7 RBL C. Johnson, 23 St Georges Avenue, Stoke-on-Trent, ST6 7JR
G7 RBQ P. Dodman, 15 Goscote Close, Redditch, B97 6UF
G7 RBR J. Pavia, 67 Browns Rock Road, Burnt Hill, Rd1, Oxford, New Zealand, 7495
G7 RBS A. Sercombe, 28 Strumpshaw Road, Brundall, Norwich, NR13 5PA

G7 RBY M. Flett, 8 Church Road, Darley Dale, Matlock, DE4 2GG
G7 RBZ J. Hudson, Atholgarth House, St. Johns Lane, Bewdley, DY12 2QZ
G7 RCC R. Wendes, 24 Whitehead Crescent Wootton Bridge, Ryde, PO33 4JF
GI7 RCH G. Walker, 16 Stormount Crescent, Belfast, BT5 4NT
G7 RCK S. Motala, 28 Fishwick View, Preston, PR1 4YB
G7 RCL R. Abbott, 2 Leybourne Drive, Springfield, Chelmsford, CM1 6TX
G7 RCP D. Baines, 157 Hall Green Road, West Bromwich, B71 2DY
G7 RCS J. Harris, 172 Shenstone Avenue, Stourbridge, DY8 3DZ
G7 RCU A. Walker, 37 East Road, Brinsford, Wolverhampton, WV10 7NP
G7 RCW J. Matthews, 126 Ingrave Road, Brentwood, CM13 2AG
G7 RDA P. Brownsett, 10 Great Aldens, Bedford, MK41 8JS
GM7 RDH R. Spence, Leyan Harray, Orkney, KW17 2LQ
G7 RDJ R. Middleton, 32 West Busk Lane, Otley, LS21 3LW
G7 RDP B. Gawthorpe, 19 Tower Hill, Clitheroe, BB7 1PD
G7 RDQ T. Rochford, 1C Oak Villa, Moors Avenue, Hartlebury, Kidderminster, DY11 7YL
G7 RDT Dorset Raynet, c/o S. Hawkins, Forest Edge, Deer Park, Milton Abbas, Blandford Forum, DT11 0AY
GM7 RDY J. Mowat, Nether Bigging, Shapinsay, Orkney, KW17 2EB
G7 REC D. Allison, 52 Boyn Valley Road, Maidenhead, SL6 4ED
GM7 REF Epping Forest Raynet Group, c/o M. Harrington, Mount Pleasant House, North Road, Wick, KW1 4DN
GM7 REG J. Robertson, 13 Swanston View, Edinburgh, EH10 7DG
G7 REH P. Evans, 45 Chiltern Drive, Charvil, Reading, RG10 9QF
G7 REJ S. Hutchinson, 42 Greenham Mill, Mill Lane, Newbury, RG14 5QW
G7 RES G. O'Neill, Wing Cottage, Bedbury Lane, Freshwater, PO40 9PD
G7 REV T. Jones, 7 Sedum Close, Huntington, Chester, CH3 6BL
GM7 REY J. Macdonald, 27 Melantee, Fort William, PH33 6PY
GW7 RFA A. Lord, The Mount, Trefecca, Brecon, LD3 0PW
G7 RFC Essex Raynet, c/o G. Farrell, 95 Washington Road, Maldon, CM9 6JF
G7 RFD P. Johnson, Sixpenny Cottage, Farthings Fold, Bourne, PE10 0RN
G7 RFE R. Johnson, 8 Merlin Close, Bourne, PE10 0BZ
G7 RFH J. Fearns, 23 Homestead St., Stoke on Trent, ST2 0RQ
G7 RFM G. Hunt, 7 Kevington Drive, St. Pauls Cray, Orpington, BR5 2NT
G7 RFO R. Thomson, 123 Oak Avenue, Todmorden, OL14 5PE
GW7 RFP P. Barry, 44 Heol Onen, Brynmawr, Ebbw Vale, NP23 4TS
G7 RFS K. Abeynayake, 25 Anderson Avenue, Earley, Reading, RG6 1HD
G7 RFT K. Whittle, 26 Beachs Drive, Chelmsford, CM1 2NJ
G7 RFX R. Oxlade, 3 Thyme Court, Northampton, NN3 8HY
G7 RFY G. Annett, 8 Wheatsheaf Way, Linton, Cambridge, CB21 4XB
G7 RFZ Dr J. Bilmen, 145 The Maples, Harlow, CM19 4RD
G7 RGA P. Cattanach, Grosse Neugasse 5/4, Wien, Austria, 1040
G7 RGG S. Emmett, 14 Ernle Road, Calne, SN11 9BT
G7 RGI H. Jones, Four Acres, Bungalow Farm Winwick Gated Road, West Haddon, NN6 7BH
G7 RGJ P. Musselwhite, 80 Craven Road, Orpington, BR6 7RT
G7 RGO E. Allan, 282 Bilton Road, Rugby, CV22 7EG
G7 RGR R. Jones, 18 Ash Grove, Burnham-on-Crouch, CM0 8DP
G7 RGU R. Oxlade, 3 Thyme Court, Lumbertubs, Northampton, NN3 8HY
G7 RGV H. Jump, 4 Bankwood, Shevington, Wigan, WN6 8EY
G7 RHD M. Clarke, 6 Oldbrook Fold, Timperley, Altrincham, WA15 7PA
G7 RHE S. Payne, 19 Weavers Lane, Sevenoaks, TN14 5BT
G7 RHF A. Richards, 3 Marsh Gate, Clee St. Margaret, Craven Arms, SY7 9DU
G7 RHI A. Mclocklin, 43 Forbes Avenue, Potters Bar, EN6 5NB
G7 RHM K. Pang, 30 Barnwood Avenue, Gloucester, GL4 3AH
G7 RHT P. Bennett, 47 Bakers Ground, Stoke Gifford, Bristol, BS34 8GD
G7 RIA M. Cook, 19 Reigate Avenue, Clacton-on-Sea, CO16 8FB
GW7 RIB P. Nicholls, 11 Ifor Hael Road, Rogerstone, Newport, NP10 9FB
G7 RIE J. Glenn, School House, 70 Norwich Road, Norwich, NR12 7EG
G7 RIJ E. Devine, 23 Radley Avenue, Wickersley, Rotherham, S66 2HZ
G7 RIO W. Care, 29 Wheal Gorland Road, St. Day, Redruth, TR16 5LT
G7 RIS I. Higton, 25 Shady Grove Hilton, Derby, DE65 5FX
G7 RIU D. Kirk, 19 The Meads, Hildersley, Ross-on-Wye, HR9 7NF
GM7 RJG A. Forbes, 28 Innes Street, Inverness, IV1 1NS
G7 RJO C. Compton, 55 Lulot Gardens, London, N19 5TR
G7 RJR D. Crossley, 33 Town Lane, Castle Acre, Kings Lynn, Pe322au
G7 RJW D. Lamden, 6 Ashbourne Way, Thatcham, RG19 3SH
GW7 RKC A. Hall, 29 Ely Street, Tonypandy, CF40 1BY
G7 RKE J. Bottomley, Grove House, 2 Woodlane, Falmouth, TR11 4RG
G7 RKJ C. Hindmarsh, 5 Jackman Drive, Horsforth, Leeds, LS18 4HS
GW7 RKQ S. Rudge, 1 Marl Mews, Marl View Terrace, Conwy, LL31 9BJ
G7 RKT P. Jones, 14 Westerleigh Road, Clevedon, BS21 7US
G7 RKU P. Dickinson, Haven, The Row, Bury St Edmunds, IP29 4DL
G7 RKV D. Wilson, 32 Laurel Bank, The Highlands, Whitehaven, CA28 6SW
G7 RKW J. Hart, 75 Falconers Road, Luton, LU2 9ET
G7 RKX C. Hill-Smith, Top Flat, The Warehouse, West St., Newton Abbot, TQ13 7DU
G7 RLK S. Drury, 5 Hawthorn Close, Healing, Grimsby, DN41 7SR
G7 RLO L. Van Beers, Cob Cottage, Tram Inn, Hereford, HR2 9AN
G7 RLQ T. Winton, 84 Pembroke Road, Clifton, Bristol, BS8 3EG
GW7 RLS City & County of Swansea, c/o J. Gray, City And County Of Swansea, Emergency Planning Unit, Swansea, SA1 3SN
G7 RLV C. Pitchford, 84 New Road, Rubery, Birmingham, B45 9HY
G7 RLX G. Coleman, 120 Kidderminster Road South, Hagley, Stourbridge, DY9 0JH
GW7 RLZ G. Roberts, 4 Fawnog Wen, Penrhyndeudraeth, ll48 6ps
G7 RMD D. Devlin, 5 Kelsall Avenue, Sutton Manor, St Helens, WA9 4DQ
G7 RME M. Buckley, 8 Highthorne Street, Armley, Leeds, LS12 3LB
GM7 RMF E. Walker, 38 Greenbank Gardens, Edinburgh, EH10 5SN
G7 RMG G. Chapman, Crockers Farm, Stoke Wake, Blandford Forum, DT11 0HF
G7 RMJ M. Amies, Home Farm, Hulme Walfield, Congleton, CW12 2JJ
G7 RMQ R. Scarce, 7 Mussidan Place, Theatre Street, Woodbridge, IP12 4NN
G7 RMW Mid Warks Raynet Group, c/o R. Medcalf, 19 All Saints Road, Warwick, CV34 5NL
G7 RMX N. Taylor, West Mede, Exeter Road, Honiton, EX14 1AX
G7 RMZ East Chehire Raynet Group, c/o B. Williams, 3 Welton Close, Wilmslow, SK9 6HD
G7 RNA OBO North Anglia Raynet, c/o K. Kent, 5 Jubilee Road, Heacham, King's Lynn, PE31 7AR
G7 RNB S. Bieber, Tonkins Quay, Mixtow, Fowey, PL23 1NB
GW7 RNC T. Heywood-Bell, 4 Aberthaw Close, Newport, NP19 9QA
G7 RNF T. Roberts, 6 Petworth Road, Southport, PR8 2QL
GM7 RNJ M. Dennis, 47 Viewfield Road, Aberdeen, AB15 7XP
G7 RNN North Norfolk Raynet, c/o A. Farrow, 18 The Green Trimingham, Norwich, NR11 8ED
G7 RNQ R. Young, 12 Elmwood Close, Stokesley, Middlesbrough, TS9 5HX
G7 RNX A. Linney, 5 Elliscales Avenue, Dalton-in-Furness, LA15 8BW
GI7 ROB R. Degossely, 82 Knightsbridge, Lisburn, BT28 3DG
GM7 ROC J. Armstrong, 15B Lamberton, Berwick-upon-Tweed, TD15 1XB
G7 ROI J. Naylor, 46 Loxley Drive, Mansfield, NG18 4FB
G7 ROM A. Boardman, 147 Musgrave Road, Bolton, BL1 4HW
G7 ROP R. Sykes, 46 Crescent Road, Netherton, Dudley, DY2 0NW
G7 ROY R. Clayton, 9 Green Island, Irton, Scarborough, YO12 4RN
G7 RPJ J. Barnard, 39 Ecclestone Close, Bradwell, Great Yarmouth, NR31 8RG
G7 RPK L. Goffin, The Hollies, Belaugh Green Lane, Norwich, NR12 7AJ
G7 RPP I. Gurney, 28 Barrington Drive, Basingstoke, RG24 9RS
GM7 RPT D. Hutchison, 55 Springfield Road, Tarbolton, Mauchline, KA5 5QU
G7 RPW S. Pike, 1 Barley Garth, Burton Pidsea, Hull, HU12 9AF
G7 RQD M. Folkes, 3 Colindale Road, Ferring, Worthing, BN12 5JF
GW7 RQI D. Pearson, 142 Heol Bryngwili, Cross Hands, Llanelli, SA14 6LY
GM7 RQK S. Skidmore, 6 Blairlinn View, Cumbernauld, Glasgow, G67 4AD
G7 RQO B. Wylie, 54 Cromwell Street, Lincoln, LN2 5LP
GW7 RQV N. Jenkins, 41 Park Street, Bridgend, CF31 4AX
G7 RRD G. Smith, 12 Oakwood Glade, Holbeach, Spalding, PE12 7JS
G7 RRJ A. Mcconnachie, 16 Poplar Avenue, Wyre Piddle, Pershore, WR10 2RJ
GW7 RRM Dr S. Whitehouse, 6 Dol Y Dderwen, Ammanford, SA18 2GA
G7 RRO G. Gardner, 47 Old Road, Stanningley, Pudsey, LS28 6BG
GW7 RRS A. Davis, 5 Jubilee Road, Bridgend, CF31 3BA
G7 RRY M. Saltmer, 12 Beechings Mews, Whitby, YO21 3DW
G7 RSA C. Hawkes, 103 Station Road, Roydon, King's Lynn, PE32 1AW
GW7 RSE L. Clarke, 83 Lancaster Street, Blaina, Abertillery, NP13 3EQ
G7 RSK A. Scott, 62 Berry Meade, Ashtead, KT21 1SG
G7 RSM Dr R. Bloor, 7 Highfield Court, Clayton Road, Newcastle, ST5 3LT
G7 RTA S. Harding, 39 Clayton Road, Lidget Green, Bradford, BD7 2LX
GI7 RTB P. Mccrory, 24 Drumcoo Green, Dungannon, BT71 4AJ
G7 RTC G. Darby, 69 Churchill Road, Earls Barton, Northampton, NN6 0PQ
G7 RTI K. Werner, 85 Brecon Way, Downley, High Wycombe, HP13 5NW
G7 RTJ D. Bransby, 7 West Cliff Avenue, Whitby, YO21 3JB
G7 RTL Radio-Tele Lincolnshire Group, c/o M. Pell, 7 Churchfleet Lane, Gosberton, Spalding, PE11 4NE
G7 RTN J. Burrows, 37 Braydeston Crescent Brundall, Norwich, NR135LD
G7 RTO B. Theaker, 25 Pinewood Drive, Plymouth, PL6 7SP
G7 RTQ M. Cowley, 72 Warley Road, Warley, Oldbury, B68 9TB
G7 RTR D. Freeman, 59 Grange Road, Somersham, Huntingdon, PE28 3JT
G7 RTX K. Brookes, Kohima, Spout Lane, Stoke-on-Trent, ST2 7LR
G7 RUC D. Millen, 75 Mill Hill Lane, Burton on Trent, DE15 0BA
G7 RUH R. Peggram, Starcroft Janes Close Blackfield, Southampton, SO45 1WJ
G7 RUJ D. Brain, 3 Mill Lane, Skipsea, Driffield, YO25 8SP
G7 RUN M. Graves, 20 Stace Way, Worth, Crawley, RH10 7YW
G7 RUP J. Shamash, 45 Elder Street, Lincoln, LN5 8QX
G7 RUQ L. Murphy, Flat 3, Evelyn Court, 187 South Coast Road, Peacehaven, BN10 8NS
G7 RUR S. Todorovic, 2 Greenslade Gardens, Nailsea, BS48 2BL
G7 RUS R. Parkin, 25 Kent House Lane, Beckenham, BR3 1LE
G7 RUX J. Gardner, 195 Upper Elmers End Road, Beckenham, BR3 3QU
G7 RUY D. Ager, 11 Tilbury Close, St. Pauls Cray, Orpington, BR5 2JR
G7 RVC P. Sutherland, 9 Lely Close, Bedford, MK41 7LS
G7 RVG P. Doble, High Street, Stoke-Sub-Hamdon, TA146PT
G7 RVH R. Bush, Church View, Overcross, Banham, Norwich, NR16 2BY
G7 RVI T. Hankins, Cawdor House, Cawdor, Ross-on-Wye, HR9 7DN
GD7 RVP S. Rand, 3 Yn Aittin Voar, Bretney Road, Jurby, Isle Of Man, IM7 3EU
GM7 RVR N. Moir, 34 Souter Drive, Inverness, IV2 4XJ
G7 RVT T. Smith, 9 Crofters Way, Westlands, Droitwich, WR9 9HU
G7 RVW R. Crofts, Little Isle, Woodgate Green, Tenbury Wells, WR15 8LX
G7 RVY H. Branch, 326 Springfield Road, Chelmsford, CM2 6BA
G7 RWC C. Halbert, 3 Third Row, Ellington, Morpeth, NE61 5HF
G7 RWF J. Buck, 14 Crosstree Walk, Colchester, CO2 8QF
G7 RWN D. Taylor, 48 Southcroft Road, Gosport, PO12 3LD
GJ7 RWT A. Cutland, Little Gables, La Route Orange, St Brelade, Jersey, JE3 8GQ
G7 RWW J. Kirkham, 100 Prince Charles Avenue, Derby, DE22 4FL
G7 RWY B. Sankey, 121 Green Lane, Coventry, CV3 6EB
G7 RXB N. Larson, 90 Lingfield Ash, Coulby Newham, Middlesbrough, TS8 0SU
G7 RXE V. Donald, 20 Parkhill Road, Barnby Dun, Doncaster, DN3 1DP
G7 RXI V. Ball, 30 Park Drive, Worlingham, Beccles, NR34 7DJ
G7 RXJ J. Dyson, 21 Highmoor, Morpeth, NE61 2AS
G7 RXK R. Thompson, 7 Rufford Close, Sutton-in-Ashfield, NG17 4BX
GM7 RXL D. Winton, 273 Hilton Drive, Aberdeen, AB24 4NT
G7 RXO N. Larsen, 6 Shrewsbury Close, Barwell, Leicester, LE9 8JX
G7 RXW M. Lockitt, 19 Roundway Down, Perton, Wolverhampton, WV6 7SX
G7 RXX S. Cooper, 30 Pinta Drive, Stourport-on-Severn, DY13 9RY
G7 RXZ D. Cooper, 1A Kent Street, Dudley, DY3 1UU
G7 RYA D. Tomlin, 154 Court Lane, Erdington, Birmingham, B23 5RG
GM7 RYK G. Pollard, 127 Braeside Park, Mid Calder, Livingston, EH53 0TE
G7 RYL D. Sheridan, 78 Oaklands Park, Buckfastleigh, TQ11 0BP
G7 RYM R. Pugh, 41 East Beach Park, Shoeburyness, Southend-on-Sea, SS3 9SG
G7 RYN D. Proctor, 11 Bedford Rise, Winsford, CW7 1NE
G7 RYO K. Turner, 34 Amherst Road, Kenilworth, CV8 1AH
GM7 RYT D. Weller, 66 Dolphin Road, Currie, EH14 5SA
G7 RYW F. Trainer, 23 Woodend Avenue, Hunts Cross, Liverpool, L25 0NY
GM7 RZE Dr D. Bremner, 'Braedine', Johnshill, Lochwinnoch, PA12 4EL
GW7 RZN E. Taylor, 8 First Avenue, Prestatyn, LL19 7LP
G7 RZQ N. Waterman, 1 Wood Lane Close, Sonning Common, Reading, RG4 9SP
G7 RZW A. Davies, 16 Sutton Road, Bolton, BL3 4QR
G7 SAC Sutton & Cheam Radio Society, c/o J. Puttock, Sutton & Cheam Rs, 53 Alexandra Avenue, Sutton, SM1 2PA
G7 SAI E. Birt, 10 Wilden Lane, Stourport on Severn, DY13 9LR
GM7 SAK A. Jardine, 17 Louisa Drive, Girvan, KA26 9AH
GW7 SAQ R. Turner, 60 Clos-Y-Deri, Porthcawl, CF36 3PR

G7 SAT D. Marritt, 96 Kynaston Road, Orpington, BR5 4JZ
G7 SAX R. Newman, 31 Oval Gardens, Alverstoke, Gosport, PO12 2RA
GI7 SBF Dr J. Henderson, 1 Brook Lodge Ballinderry Lower, Lisburn, BT28 2GZ
GW7 SBJ E. Birtwistle, 29 Church View, Pentre, Deeside, CH5 2DP
G7 SBK D. Hunt, 298 Cavendish Road, Carlton, Nottingham, NG4 3QH
G7 SBN M. Royal, 3 Bethany Place, St Just, Penzance, TR19 7HB
GW7 SBO R. Thomas, 25 Lon Lwyd Isaf, Pentraeth, LL75 8LN
G7 SBP N. Hancocks, 9A St. Philip Street, Penzance, TR18 2DN
G7 SBZ M. Newton, 24 Chestnut Avenue, York, YO31 1BR
G7 SCE P. Farman, 298 Laburnum Grove, Portsmouth, PO2 0EX
GM7 SCJ G. Deas, 81 Speirs Road, Bearsden, Glasgow, G61 2LT
G7 SCL J. Robinson, 4 Gardner Close, Loughborough, LE11 5YB
G7 SCN P. Brotherton, 73 Thorneywood Rise, Nottingham, NG3 2PE
G7 SCO D. Brooke, 34 Park Road, Burwell, Cambridge, CB25 0ES
G7 SCP D. Wain, 51 Foxstone Way, Eckington, Sheffield, S21 4JX
G7 SCR Suffolk Coastal Raynet, c/o R. Keen, 13 Mill View Close, Woodbridge, IP12 4HR
G7 SCT G. Rutherford, 24 Chestnut Avenue, Hedon, Hull, HU12 8NH
G7 SCU D. Ibrahim, 14 Dunvegan Road, London, SE9 1SA
G7 SCV J. Straughan, 16 Garner Close, Chapel Park, Newcastle upon Tyne, NE5 1SQ
G7 SCX P. O'Rourke, 186 Cottingham Road, Corby, NN17 1SY
G7 SCZ D. Kiteley, 13 Chiltern Close, Astley Cross, Stourport-on-Severn, DY13 0NU
G7 SDC D. Coe, 105 Raynham Road, Bury St Edmunds, IP32 6ED
G7 SDD M. Smith, 38 Vestry Road, Street, BA16 0HX
GW7 SDE I. Jones, 14 Clare Court, Loughor, Swansea, SA4 6UH
G7 SDG R. Martin, 8 Short Lane, Bricket Wood, St Albans, AL2 3SE
G7 SDM G. Davies, 11 Ninfield Close, Carlton Colville, Lowestoft, NR33 8SD
GM7 SDP D. Ryan, 1 Clashbenny Place, St. Madoes, Perth, PH2 7TS
G7 SDQ M. Smith, Flat 4, Stoneleigh Court Lansdown Road, Bath, BA1 5TL
G7 SEG A. Harrison, 44 Rosslyn Road, Whitwick, Coalville, LE67 5PT
G7 SEJ M. Baskeyfield, 3 Merlewood, Bracknell, RG12 9PA
G7 SEK R. Newham, 26A Kilwardby Street, Ashby-de-la-Zouch, LE65 2FQ
G7 SEO R. Plant, 22 The Woodlands, Wokingham, RG41 4UY
G7 SER Sutton Coldfield & Dist Raynet, c/o J. Trickey, 59 Shelley Drive, Sutton Coldfield, B74 4YD
G7 SEU Dr E. Kershaw, 83 Foxhunter Drive, Oadby, Leicester, LE2 5FH
G7 SEY P. Simpson, The Conifers, Woodhouse Lane, Telford, TF4 3BJ
G7 SFA M. Stevens, Flat 7, 1A Woodstock Road, Croydon, CR0 1JS
G7 SFD M. King, 4 Keith Avenue, Ramsgate, CT12 6JQ
GM7 SFE R. Lawrie, 84 Redlawood Road, Cambuslang, Glasgow, G72 7TP
G7 SFF D. Hartshorn, 21 Hucklow Avenue, Chesterfield, S40 2LT
G7 SFI S. Merrifield, 2 Larkspur Glade, Telford, TF3 2AQ
G7 SFJ S. Pratt, 15 Springwell Close, Cowling, Keighley, BD22 0AP
G7 SFL M. James, 8 Swan Quay Bath Lane, Fareham, PO16 0DX
G7 SFM R. Wiltshire, 8 Hilltop Lane, Heswall, Wirral, CH60 2TT
G7 SFS L. Banner, 7 Lowdham Road, Gedling, Nottingham, NG4 4JP
G7 SFY B. Purkiss, 99 Westland Road, Yeovil, BA20 2AZ
G7 SGH B. Williams, Hillside, Wigmore Lane Eythorne, Dover, CT15 4AW
G7 SGK R. Ward, 16 Southgate Crossgates, Scarborough, YO12 4NB
G7 SGM R. Gifford, 100 Gadebridge Road, Hemel Hempstead, HP1 3EW
G7 SGO R. Percival, 145 Queen Street, Whitehaven, CA28 7AW
G7 SGR South Gloucestershire Raynet, c/o T. Humphreys, 93 Cornwall Crescent Yate, Bristol, BS37 7RU
G7 SHI C. Conce, 35 Mortimer Drive, Sandbach, CW11 4HS
G7 SJD S. Fitzpatrick, 21 Corn Close, South Normanton, Alfreton, DE55 2JD
G7 SJK T. Masson, Apple Tree Cottage, Neath Gardens, Reading, RG3 4UL
G7 SJS P. Roberts, 5 Snelston Crescent, Littleover, Derby, DE23 6BL
G7 SJX B. Shields, 20 Gresley Court, Grantham, NG31 7RH
G7 SKA P. Burnett, 4 Lavendon Court, Barton Seagrave, Kettering, NN15 6QH
GM7 SKB Dr D. Fortune, 26 Newton Grove, Newton Mearns, Glasgow, G77 5QJ
GW7 SKC OBO West Glamorgan Cc, c/o J. Gray, City And County Of Swansea, Emergency Planning Unit, Swansea, SA1 3SN
G7 SKH G. Murray, Brookside, Thirlby, Thirsk, YO7 2DJ
G7 SKL J. Koops, 64 Winchester Avenue, Nuneaton, CV10 0DW
G7 SKR D. Tarbatt, 9 Dashwood Close, Warrington, WA4 3JA
G7 SKV D. Graham, 11 Hibernia St., Deane, Bolton, BL3 5PQ
G7 SKW B. Mcinnes, Thistledome, Tumby Road, Coningsby, Lincoln, LN4 4RQ
G7 SKX A. Wilkinson, 21 Solbys Road, Basingstoke, RG21 7TG
G7 SLB R. Machin, 11 Victoria Street, Irthlingborough, Wellingborough, NN9 5TR
G7 SLJ D. Lloyd-Jones, 2 Leyside, Rayne, Braintree, CM77 6DE
G7 SLL G. Peach, 120 Craven Road, Newbury, RG14 5NR
GI7 SLN G. Mcafee, 12 Skerryview, Craigahullier, Portrush, BT56 8NJ
G7 SLP P. Hardcastle, 45 Dean Park Drive, Drighlington, Bradford, BD11 1AL
GJ7 SLU C. Whittaker, Coeur Joyeux, La Rue Des Sapins, St Peter, Jersey, JE3 7AD
G7 SLV R. Walker, 210 London Road, Worcester, WR5 2JT
G7 SLY P. Taylor, 46 Ralph Road, Staveley, Chesterfield, S43 3PY
G7 SLZ T. Gill, 2 Church View Clatworthy, Taunton, TA4 2EQ
G7 SMC G. Jameson, 17 Lansbury Avenue, Mastin Moor, Chesterfield, S43 3AG
G7 SMD J. Cook, 40 Preston Avenue, Alfreton, DE55 7JY
G7 SME P. Helliwell, 1 Beechfield Avenue, Barton, Torquay, TQ2 8HU
G7 SMH G. Newton, 8 Lynch Mead, Winscombe, BS25 1AT
G7 SMN A. Holden, 1 Rose Cottage, Little Bramford Lane, Ipswich, IP1 2PH
G7 SMQ B. Cottee, 41 Colesbourne Road, Clifton, Nottingham, NG11 8JG
G7 SMT F. Claydon, 5 Mill Gardens, Ringmer, Lewes, BN8 5JD
GW7 SMV L. Ashford, 13 Cefn Court, Rogerstone, Newport, NP10 9AH
G7 SMZ R. Walker, 24 Colin St., Alfreton, DE55 7HT
G7 SNB O. Newland, 22A Cromwell Road, Basingstoke, RG21 5NR
G7 SNC I. Palmer, 182 Salhouse Road, Norwich, NR7 9AD
G7 SNJ R. Chaytor, 19 Granville Avenue, Hartlepool, TS26 8ND
G7 SNP K. Jordan, 7 Park Avenue, Bedlington, NE22 7EH
G7 SNQ S. Taylforth, 1 Clough Terrace, Barnoldswick, BB18 5PD
G7 SNR S. Brodie, Waterloo Cottage, Tanners Green, Norwich, NR9 4QS
G7 SNT B. Jordan, 40 High Street, Coltishall, Norwich, NR12 7HD
G7 SNW J. Ward, 40 Lancaster Drive, Long Sutton, Spalding, PE12 9BD
G7 SNX M. Pearce, 42 Pine Close, Rudloe, Corsham, SN13 0LB
GI7 SOB K. Elgin, 50 Ballinteer Road, Macosquin, Coleraine, BT51 4LZ
G7 SOE M. Howard, East Dean House, East End Langtoft, Peterborough, PE6 9LP
G7 SOH C. Brown, 9 Marjorie Street, Rhodesia, Worksop, S80 3HR
G7 SOV C. Howarth, 5 West Mount, Orrell, Wigan, WN5 8LX
G7 SOZ S. Jude, 9 Winchfield, Great Gransden, Sandy, SG19 3AN
GM7 SPA J. Brown, 11 Oak Gardens, Oak Drive, Lenzie, Glasgow, G66 4BF
GM7 SPB M. Garrington, South Orrock Bungalow, Balmedie, Aberdeen, AB23 8XY
G7 SPE R. Keep, 14 Foster Road, Kempston, Bedford, MK42 8BU
G7 SPL D. Pomfret, 52 Warwick Close, Bury, BL8 1RT
G7 SPM C. Jones, Nb Guanche Bradford On Avon Marina, Widbrook Bradford-on-Avon, BA15 1UD
G7 SPN S. Townsley, 222 Prince Consort Road, Gateshead, NE8 4DX
G7 SPP H. Conrad, 22 Low Stobhill, Morpeth, NE61 2SG
G7 SPZ R. Brown, 19 Comberton Road, Toft, Cambridge, CB23 2RY
G7 SQH G. Chew, 45 Brackley, Weybridge, KT13 0BL
G7 SQM N. Crawford, 20 Fearnley Crescent, Kempston, Bedford, MK42 8NL
G7 SQW A. Woods, 10 Radcliffe Road, Drayton, Norwich, NR8 6XZ
G7 SQY D. Colton, 9 Thornemead, Peterborough, PE4 7ZD
G7 SRA Sudbury and District Radio Amateurs, c/o A. Harman, 107 Kempson Drive, Great Cornard, Sudbury, CO10 0YF
G7 SRB D. Shorten, 32 Stoneleigh Drive, Carterton, OX18 1ED
G7 SRC Essex Raynet, c/o N. Hull, C/O 95 Washington Road, Maldon, CM9 6JF
G7 SRH M. Harper, 31 Lorland Road, Cheadle Heath, Stockport, SK3 0JJ
G7 SRI M. Lowe, 2 White Post Bungalows, North Leverton, Retford, DN22 0AS
GM7 SRJ S. Jones, Smiddy Cottage, Auchencrow, Eyemouth, TD14 5LS
G7 SRK R. Carder, 45 Chalklands, Linton, Cambridge, CB21 4JQ
G7 SRL A. Gallichan, 4 Wigston Road, Rugby, CV21 4LT
G7 SRV P. Everett, 26 Tennyson Close, Horsham, RH12 5PN
G7 SSA M. Addicott, Orchardleigh, The Street, Radstock, BA3 4HG
G7 SSB D. Jones, 429 Redmires Road, Sheffield, S10 4LF
G7 SSG J. Smye, 24 Eastfield Road, Wincanton, BA9 9LT
G7 SSJ D. Sutton, 32 Queensway, Euxton, Chorley, PR7 6PW
G7 SSK R. Walton, Easingmoor House, Thorncliffe Road, Leek, ST13 7LW
GW7 SSN N. Cole, 40 Primrose Court, Ty Canol, Cwmbran, NP44 6JJ
GW7 SSQ P. Cole, 18 Juniper Crescent, Henllys, Cwmbran, NP44 6EH
G7 SSU J. Stewart, 43A Independent Place, London, E8 2HE
G7 SSW J. Haywood, 7 Anna Walk, Stoke-on-Trent, ST6 3BX
G7 STC K. Gater, 110 Byrds Lane, Uttoxeter, ST14 7NB
G7 STD L. Goodridge, 110 Quarrendon Road, Amersham, HP7 9EP
G7 STF D. Taylor, 147A Callington Road, Saltash, PL12 6JA
G7 STG B. Spavins, 8 Berkeley Avenue, Briar Bank Park, Bedford, MK45 3WH
GM7 STI I. Pearce, 1 Mount Farm Cottage, Cupar, KY15 4NA
G7 STL M. Anderson, 94 Tolworth Road, Surbiton, KT6 7SZ
G7 STM M. Wyatt, 8 St Mary'S Drive, Sutterton, PE20 2LU
G7 STQ M. Oura, The Quoins, Gloucester Road, Bath, BA1 8AD
G7 SUA D. Wiseman, 12 Hamilton Way, Acomb, York, YO24 4LE
G7 SUM G. Fewings, 22 Watcombe Road, West Southbourne, Bournemouth, BH6 3LU
G7 SUQ A. Jobson, 7 Dunlin Close, Norton, Stockton-on-Tees, TS20 1SJ
G7 SUS R. Biss, 1 Fairey Crescent, Gillingham, SP8 4PE
G7 SUT C. James, (James), Lower Kenneggy Farm, Lower Kenneggy, Rosudgeon, Penzance, TR20 9AR
G7 SUU R. Wolk, Calle Zaragoza 48, Castalla, Spain, 3420
G7 SUV J. Patterson, 28 Woodridge Avenue, Thornton-Cleveleys, FY5 1PR
G7 SVE A. Jackson, 14 West Field Gardens, Sandy, SG19 1HF
G7 SVF K. Ingram, 15 Kent Avenue, East Cowes, po326qn
G7 SVI C. Lambert-Hutchinson, 63 Chalbury Close, Canford Heath, Poole, BH17 8BP
G7 SVM D. Bradley, 22 Grosvenor Road, Ettingshall Park, Wolverhampton, WV4 6QY
G7 SVQ R. Holmes, 18 Dresden Close, Mickleover, Derby, DE3 0RD
G7 SVT D. Bultitude, 1 Pembroke Gardens, Northampton, NN5 7ES
G7 SVU N. Hinchliffe, 19 Grange Road, Blidworth, Mansfield, NG21 0RN
G7 SWB Dr R. Bambrey, 76 Boundary Way, Glastonbury, BA6 9PH
G7 SWE F. Rowbotham, 56 Farnborough Road, Clifton, Nottingham, NG11 8GF
G7 SWH A. Howell, 35 Melton Road, Wakefield, WF2 7PR
G7 SWQ I. Wild, 153 Alexandra Road, Sheffield, S2 3EH
G7 SWR M. Prentice, 26 Meir View, Stoke-on-Trent, ST3 6AH
G7 SWS A. Bird, 18 Welbeck Drive, Spalding, PE11 1PD
G7 SWV Dr C. Smith, 11 Woods Close, Haskayne, Ormskirk, L39 7JL
G7 SWW R. Jones, Flat 32 Tottenhoe Court Colville Road Cherry Hinton, Cambridge, Cb5 8se
GM7 SWX D. Curran, 104 Mcpherson Crescent, Chapelhall, Airdrie, ML6 8XL
G7 SWZ J. Halliday, 14 Heath Gardens, Halifax, HX3 0BD
G7 SXB D. Phillips, 197 Downall Green Road, Ashton-In-Makerfield, Wigan, WN4 0DW
G7 SXG D. Dean, 17 Drayton Close, Runcorn, WA7 4TW
GM7 SXI A. Williams, Gardeners Cottage, Ardrossan, KA22 8PH
G7 SXJ J. Farrow, 74 The Droveway, St Margarets Bay, Nr. Dover, CT15 6DD
GW7 SXN D. Davies, 35 Ty Llwyd Parc Estate, Quakers Yard, Treharris, CF46 5LA
GW7 SXU I. Harries, Gwastad, Maenygroes, New Quay, SA45 9RJ
G7 SYC W. Jarvill, 66 Gloucester Road, Newbury, RG14 5JN
G7 SYD S. Applegate, 180 Logan Street, Nottingham, NG6 9FU
G7 SYE T. Laskey, 72 Windermere Avenue, Ramsgate, CT11 0PL
G7 SYI R. Hutchinson, 10 Clifton Avenue, Eaglescliffe, Stockton-on-Tees, TS16 9BA
G7 SYJ M. Hogg, 55 Ardenfield Drive Wythenshawe, Manchester, M22 5DJ
G7 SYL G. Thaxter, Leymoon, Gayton Road East Winch, King's Lynn, PE32 1NW
G7 SYQ A. Orchiston, 16 Windsor Close, Collingham, Newark, NG23 7PR
G7 SYS R. Baxter, 107 Kendale Road, Bridgwater, TA6 3QE
G7 SYT C. Denman, 12 Woodland Close, Northampton, NN5 6NH
G7 SYU Dr D. Bowers, 88 Stamford Avenue, Springfield, Milton Keynes, MK6 3LQ
G7 SYY S. Howarth, 14 Eaves Lane, Chorley, PR6 0PY
GM7 SZA S. Mussell, Dunelm, Thornhill Road, Cuminestown, Turriff, AB53 5WH
G7 SZB N. Kendal-Ward, 29 Denmark Street, Gateshead, NE8 1NQ
G7 SZF N. Hartley, 66 Broad Lane, Norris Green, Liverpool, L11 1AN
G7 SZG K. Gardner, 27 Lindon Drive Alvaston, Derby, DE24 0LP
G7 SZO R. Collinson, 56 Orchard Valley, Hythe, CT21 4EA
GI7 SZV A. Maclaine, 172 Moylagh Road, Seskanore, Omagh, BT78 2PN
G7 SZW D. Green, 43 James Street, Selsey, Chichester, PO20 0JG
G7 SZZ R. Roberts, Connemara, High Hesket, Carlisle, CA4 0JF
G7 TAE S. Wersby, Oak Barn, 6 Timothys Field Abbotts Ann, Andover, SP11 7AT

G7 TAF M. Hawes, 78 Martyns Way, Bexhill-on-Sea, TN40 2SH
G7 TAT J. Moye, 33 Prince Charles Road, Colchester, CO2 8NS
G7 TAV S. Houghton, 28 Heron Way, Mayland, Chelmsford, CM3 6TP
G7 TAX F. Roullier, 19 Terling Road, Dagenham, RM8 1DS
G7 TBC P. Stockdale, 77 Fort Hill Road, Sheffield, S9 1BA
G7 TBF N. Smith, 47 Kiveton Lane, Todwick, Sheffield, S26 1HJ
G7 TBJ J. Kewn, 31 Trescoe Road, Long Rock, Penzance, TR20 8JY
G7 TBO L. Shergold, 8 The Moors, Lydiard Millicent, Swindon, SN5 3LE
G7 TBP R. Gill, 17 Old Hall Close, Calverton, Nottingham, NG14 6PU
G7 TBU S. Fitzjohn, 10 Samsons Close, Brightlingsea, Colchester, CO7 0RP
G7 TBW T. Polain, 22 Hilltop Avenue, Hullbridge, Hockley, SS5 6BN
G7 TBX A. Siddle, 5 Neneside, Benwick, March, PE15 0YF
G7 TCB P. Hubberstey, 10 Dove Avenue, Penwortham, Preston, PR1 9RP
G7 TCD G. Ward, 162 Greenbank Road, Darlington, DL3 6ES
G7 TCH Hastings College Radio Club, c/o D. Grandfield, Hastings College, Arts & Technology, St Leonards on Sea, TN38 0HX
G7 TCQ S. Preston, 18 Station Road, Great Wyrley, Walsall, WS6 6LQ
G7 TCW C. Haslewood, 66 Hunter Road, Cannock, WS11 0AF
GI7 TDA J. Mckeever, 19 Corrycroar Road, Pomeroy, Dungannon, BT70 3DY
G7 TDN A. Baines, 60 Norton Drive, Halifax, HX2 7RB
GW7 TDQ D. Banister, 41 Tynycoed Road, Great Orme, Llandudno, LL30 2QA
G7 TDR R. Smith, 47 Kiveton Lane, Todwick, Sheffield, S26 1HJ
G7 TEA A. Goddard, 50 Ardmore Walk, Manchester, M22 5QG
GI7 TEB J. Mathers, 14 Castlewood Avenue, Coleraine, BT52 1JR
G7 TEG G. Fletcher, 171 Obelisk Rise, Northampton, NN2 8TX
GW7 TEO P. Taylor, 8 First Avenue, Prestatyn, LL19 7LP
G7 TEP K. Blain, 27 Prospect Road, Ash Vale, Aldershot, GU12 5ED
G7 TET I. Mowbray, 23 Rhodes Avenue, Bishop's Stortford, CM23 3JN
G7 TEZ G. Masters, 85 Petersham Road, Creekmoor, Poole, BH17 7DW
G7 TFA B. Wrampling, 18D May Avenue, Canvey Island, SS8 7EE
G7 TFG H. Orchel, Gildertofts, Ingleby Greenhow, Middlesborough, TS9 6JF
GI7 TFK S. Mccormick, 74 Belsize Road, Lisburn, BT27 4BH
G7 TFL S. Dodds, 4 Claremont Road, Wisbech, PE13 2JR
GM7 TFN C. Paton, 4 Abbeyhill, Dhailling Road, Dunoon, PA23 8FG
G7 TFU B. George, 43 Claverton Road West, Saltford, Bristol, BS31 3DU
G7 TFX J. Patterson, 11 Elmway, Chester le Street, DH2 2LD
G7 TFZ D. Thomas, 27 Kingsbury Road, Coventry, CV6 1PW
GW7 TGB C. Bristow, 18 Clarendon Close, Chepstow, NP16 5TL
G7 TGF A. Gibson, 100 Top Row, Darton, Barnsley, S75 5JQ
G7 TGG C. Preston, 34 Forrester Street Precinct, Walsall, WS2 8RE
GI7 TGJ G. Heaney, 38 Derryvore Lane, Portadown, Craigavon, BT63 5RS
G7 TGK C. Coombe, 123 Farleigh Road, Pershore, WR10 1JY
G7 TGN P. Dawson, 1 Eastfield Road, Bridlington, YO16 7DZ
G7 THF M. Thompson, 4 Saxony Way, Donington, Spalding, PE11 4YA
GI7 THH T. White, Shallamar, 3A Park Road, Strabane, BT82 8EL
G7 THI F. Gillespie, Low Fold, Hoff, Appleby-in-Westmorland, CA16 6TA
G7 THJ B. Mills, 37 Ashley Road, Hildenborough, Tonbridge, TN11 9ED
G7 THK K. Grover, 6 Wren Court, Battle, TN33 0DU
GI7 THY R. Larimer, 131 Carnalea Road, Seskanore, Omagh, BT78 2PP
G7 THZ J. Reid, 12 Marlay Grove, Crownhill, Milton Keynes, MK8 0AT
G7 TIB D. Cross, 91 Ilges Lane, Cholsey, Wallingford, OX10 9PA
G7 TID S. Lindsay, 15 Westmorland Gardens, Gateshead, NE9 6HP
G7 TIE I. Chamberlain, 14 High House Avenue, Wymondham, NR18 0HY
G7 TIK C. Mcqueen, 2 Stamford Road, Weldon, NN17 3JL
G7 TIM G. Jones, 42 Everard Road, Southport, PR8 6NA
G7 TIN R. Martin, 2 East View, North Walsham Road, North Walsham, NR28 0PJ
G7 TIR D. Thomas, 5 Minster Drive, Urmston, Manchester, M41 5HA
G7 TIV J. Askew, 22 Cowslip Grove, Calne, SN11 9QQ
GW7 TIX D. Price, Sabrina, Pool Road, Newtown, SY16 1DW
G7 TIY D. Miller, 92 Eton Drive, Wirral, CH63 1JS
G7 TJD M. Crosfill, Polmennor Farmhouse, Heamoor, Penzance, TR20 8UL
GW7 TJM M. Roberts, 2 Donnen Street, Port Talbot, SA13 1NE
G7 TJQ C. Shaw, 30 Southern Way, Stoke-on-Trent, ST6 1PX
G7 TJV C. Ho, Po Box 900, Fanling Post Office, Hong Kong, Hong Kong
G7 TJW G. Willoughby, 31 Trenoweth Road, Penzance, TR18 4RS
G7 TJZ I. Smith, 39 Hollingsworth Road, Lowestoft, NR32 4AU
G7 TKB F. Coles, Le Bouillo, Estampes, France, 32170
G7 TKG B. Mersi, 4 Westdown Road, Bournemouth, BH11 9EQ
G7 TKI R. Pettett, 2 Windmill Close, Great Dunmow, Dunmow, CM6 3AX
G7 TKM M. Hewitt, 17 Farquhar Road, Maltby, Rotherham, S66 7PD
G7 TKO M. Smith, 11 Martigny Road, Melksham, SN12 7PG
G7 TKP M. Hewitt, 3 Orchard Rise, Bourne Lane, Reading, RG7 5NS
G7 TKT Dr P. Ashford, 3 Valley Road, Cheadle, SK8 1HY
G7 TKW M. Peppiatt, 31E LIverton St, Kentish Town, London, NW5 2PE
G7 TLC D. Benton, Hawthorn Cottage, Penrose, Wadebridge, PL27 7TB
G7 TLD M. Clare, 43 Birchfield Close Blackbird Leys Cowley, Oxford, OX4 6DL
G7 TLK K. Hemsil, Silkcot, Burngullow Lane High Street, St Austell, PL26 7TQ
G7 TLL H. Hodson, 1 Chevin Avenue, Borrowash, Derby, DE72 3HR
G7 TLR K. Marshall, 28 Deerness Grove, Esh Winning, Durham, DH7 9LY
G7 TMC M. Conlon, 3 Selside, Brownsover, Rugby, CV21 1PG
G7 TMF T. Foster, 98 Station Road, Carlton, Nottingham, NG4 3DA
G7 TMH A. Hunt, 63A Toms Lane, Kings Langley, WD4 8NJ
G7 TMM A. Kirkham, Flat 6, The Laurels, 14 Marlborough Road, Buxton, SK17 6RD
G7 TMO P. Foster, 218 Stoops Lane, Bessacarr, Doncaster, DN4 7JQ
GI7 TMQ J. Bell, 72 Coleraine Road, Portrush, BT56 8HN
G7 TMR R. Nelson, 15 Poplars Close, Burgess Hill, RH15 9SZ
G7 TMU V. Swanwick, 43 Hormare Crescent, Storrington, Pulborough, RH20 4QX
G7 TNO D. Lunn, 23 Moynton Close, Crossways, Dorchester, DT2 8TX
G7 TNQ M. Mrzyglod, 8 Beech Road, Shillingford Hill, Wallingford, OX10 8LU
GW7 TNS M. Davies, 33 Hazel Mead, Brynmenyn, Bridgend, CF32 9AQ
G7 TNT B. Scarsbrook, Salix, 96 Moss Lane, Alderley Edge, SK9 7HW
G7 TNU P. Sparke, 18 Gordon Road, Haywards Heath, RH16 1EJ
G7 TNZ M. Wells, 37 Water Meadows, Worksop, S80 3DF
G7 TOA S. Haigh, 2 Locker Avenue, Warrington, WA2 9PS
G7 TOB R. Wardell, 1 Enfield Close, Norden, Rochdale, OL11 5RT
G7 TOF I. Pardington, 36 Rivermeads Avenue, Twickenham, TW2 5JJ
G7 TOI P. Goodayle, 2 Downs Road, Seaford, BN25 4QL
G7 TOO P. Crabtree, 106 Sagecroft Road, Thatcham, RG18 3BF
G7 TOU M. Mussard, 35 Oakfield Gardens, Beckenham, BR3 3AY
G7 TOY A. Peet, 95 Recreation St., Mansfield, NG18 2HP
G7 TOZ J. Whytock, 48 Lythe Fell Avenue, Halton, Lancaster, LA2 6NL
G7 TPB J. Kilminster, 499 Hagley Road West, Quinton, Birmingham, B32 2AA
G7 TPD T. Morton, 28 Turnfields, Ickford, Aylesbury, HP18 9HP
G7 TPG B. Barber, 114 Scrogg Road, Newcastle upon Tyne, NE6 4HA
G7 TPH R. Hand, 70 Flansham Lane, Bognor Regis, PO22 6AH
GI7 TPO G. Hodgkinson, 675 Crumlin Road, Belfast, BT14 7GD
G7 TPS B. Seed, 10 South Place, Calne, SN11 0JA
G7 TPW A. Grigor, 48 Valebridge Drive, Burgess Hill, RH15 0RW
G7 TQA D. Legge, 28 Dresser Road, Prestwood, Great Missenden, HP16 0NA
G7 TQC C. Banister, York Avenue, East Cowes, PO32 6JT
G7 TQE T. Brown, 138 Holmesdale Road, South Norwood, London, SE25 6HY
G7 TQT R. Denton, 37 Tenby Road, Cheadle Heath, Stockport, SK3 0UN
GU7 TQX E. Grisley, Les Clercs, Contree Des Clercs, St Pierre Du Bois, Guernsey, GY7 9DA
G7 TRB P. Stevenson, 50 Field Lane, Beeston, Nottingham, NG9 5FJ
G7 TRG K. Liddle, 36 Vicarage Lane, Grasby, Barnetby, DN38 6AU
G7 TRL D. Perry-Wright, 1 Marlfield Close, Preston, PR2 7AL
G7 TRM K. White, 20 Agnes Close, Bude, EX23 8SB
G7 TRV D. Glover, 59 Shelley Close, Abingdon, OX14 1PP
G7 TSB E. Jones, 26 Wood End, Bluntisham, Huntingdon, PE28 3LE
GW7 TSO K. Jones, 40 Ffordd Garnedd, Y Felinheli, LL56 4QY
G7 TSP C. Leman, 92 Queens Crescent, Eastbourne, BN23 6JP
G7 TSQ J. Stafford, 89 Mossley Road, Ashton-under-Lyne, OL6 9RH
G7 TTH G. Quint, 4 Gibson Grove, Malvern, WR14 1NX
GI7 TTO D. Dunlop, 63 Cloyfin Road, Coleraine, BT52 2NY
G7 TTP R. Hughes, 5 Cornwallis Close, Bromham, Bedford, MK43 8LG
GM7 TTU R. Emmott, 5 Thorter Loan, Dundee, DD1 3AW
GW7 TTX M. Tahla, Penrhiw, Ffestiniog, Blaenau Ffestiniog, LL41 4PN
G7 TTY A. Hubbard, 40 Field Drive, Shirebrook, NG20 0BP
GM7 TUD J. Pedley, 4 Tinwald View Back Road, Locharbriggs, Dumfries, DG1 1RT
G7 TUG N. Mitchell, 49 Kersey Road, Felixstowe, IP11 2UL
G7 TUH P. Ferguson, 152 Chestnut Drive, Sale, M33 4HR
G7 TUK J. Steel, 10 Green Courts, Winterton-On-Sea, Great Yarmouth, NR29 4AQ
G7 TUM J. Moore, Waveney, Abbotts Way, Bush Estate, Norwich, NR12 0TA
G7 TUP R. Irwin, 7 Hameau Des Peupliers, Rue Du Vert Pre, Lys Lez Lannoy, France, 59390
GW7 TUQ B. Forhead, Autumn Villa, Chapel Street Newbridge, Wrexham, LL14 3JH
G7 TUS R. Munden, 2 Hain Villa, Forest Road, Ruardean, GL17 9XR
G7 TUV J. Hewitt, 6 Crawley Walk, Warley, Cradley Heath, B64 5EX
G7 TVL E. Roberts, 800 Walsall Road, Great Barr, Birmingham, B42 1EU
G7 TVQ J. Gilbert, Mills Caravan, Garland Cross Kings Nympton, Umberleigh, EX37 9TT
G7 TVT L. Whiteside, 8 The Orchards, Eaton Bray, Dunstable, LU6 2DD
GI7 TVV A. Mccready, 25 Glendun Park, Bangor, BT20 4UX
G7 TWA D. Bullard, 20 Chaney Road, Wivenhoe, Colchester, CO7 9QZ
G7 TWC M. Ruttenberg, 90 Heath View, London, N2 0QB
G7 TWJ P. Edwards, Cleveland, Blackberry Road, Lingfield, RH7 6NQ
GM7 TWM I. Hipkin, 1 Maclennan Place, Dufftown, Keith, AB55 4EF
G7 TWU F. Clarkson, 313 Normanby Road, Middlesbrough, TS6 0BQ
G7 TWW C. Papaioannou, 2 Temple Lane, Temple, Marlow, SL7 1SA
G7 TXF A. Scott, 60 Lowndes Park, Driffield, YO25 5BG
G7 TXR S. Loyd, Maple House, Pangbourne Road, Reading, RG8 8LN
G7 TXV C. Wood, 20 Tedworth Close, Guisborough, TS14 7PR
G7 TXW M. Oliver, 14 Harwood Road, Gosport, PO13 0TT
G7 TXX D. Williams, 57 Hillside Avenue, Kidsgrove, Stoke-on-Trent, ST7 4LW
G7 TYB J. Hawley, 89 Mansfield Avenue, Denton, Manchester, M34 3NS
G7 TYH S. Furminger, 9 Amberley Gardens, Wokingham, RG41 1LN
G7 TYJ J. Pennington, 6 Penny Hapenny Crt, Atherstone, CV9 2AA
G7 TYO S. Birtwhistle, 14 Woodley Street, Bury, BL9 9HZ
G7 TYP B. Cook, 40 Preston Avenue, Alfreton, DE55 7JY
G7 TYR OBO West Kent Raynet, c/o D. Collins, 71 Trench Road, Tonbridge, TN10 3HG
G7 TYT M. Claxton, 9 Thompson Avenue, Beverley, HU17 0BG
G7 TZB D. Vincent, 6 Nathan Gardens, Poole, BH15 4JZ
G7 TZD P. Jones, 361 Wellingborough Road, Rushden, NN10 6BA
GW7 TZG P. Kelly, 33 Yeo St., Resolven, Neath, SA11 4HS
GW7 TZI M. Tonkin, 185 Pentregethin Road, Cwmbwrla, Swansea, SA5 8AU
G7 TZN S. Buckingham, 8 Tedder Avenue, Buxton, SK17 9JU
G7 TZO C. Turner, 308 North Road, Yate, Bristol, BS37 7LL
G7 TZQ R. Darby, 25 Bramley Road, Marsh Lane, Sheffield, S21 5RD
G7 TZU T. Stalker, 172 Kirkby Road, Barwell, Leicester, LE9 8FS
G7 TZV G. Broughton, 111 Broadway, Manchester, M40 3NL
G7 TZW R. Wheatley, 288 Bennett Street, Long Eaton, Nottingham, NG10 4JA
G7 TZX D. Johnson, 12 Heron Close, Broughton, Chester, CH4 0RL
G7 TZZ J. Eyre, 41 Wood Street, Kettering, NN14 1BG
GM7 UAC E. Edwards, 12 Highfield Place, Girdle Toll, Irvine, KA11 1BW
G7 UAI UK DX, c/o P. Blizzard, 12 Hilton Road, Malvern, WR14 3NP
G7 UAK S. Hunter, 30 Adelaide Street, Barrow-in-Furness, LA14 5TX
G7 UAL D. Witherall, 221 Poynters Road, Dunstable, LU5 4SH
G7 UAT J. Johnson, 10 Croft Avenue, Newcastle, ST5 8EY
G7 UAV I. Morris, 60 Moorland Avenue, Lincoln, LN6 7RD
G7 UAY D. Pickering, 28 Keepers Wood Way, Chorley, PR7 2FU
G7 UBB E. Knight, 26 Skylark Avenue, Chalkers Rise, Peacehaven, BN10 8GF
G7 UBD T. Thomas, 166 Bluebell Road, Southampton, SO16 3LP
G7 UBK G. Reddecliffe, 5 Stanley Close, Dymchurch, Romney Marsh, TN29 0TY
G7 UBO J. Pearson, 22 Ashburnham Close, Norton, Doncaster, DN6 9HJ
G7 UBP C. Howard, 144 Fairfield Road, Heysham, Morecambe, LA3 1LR
G7 UBQ T. Bray, The Dell, Burnlee Road, Holmfirth, HD9 2LF
G7 UBX P. Pleydell, 6 The Croft, Meriden, Coventry, CV7 7NQ
GI7 UBY C. Lunnon, 3 Parkfield, Crumlin, BT29 4SG
G7 UCB P. Hudson, 47 Hall Farm Road, Duffield, Belper, DE56 4FJ
G7 UCG J. Woodward, 108 Tamworth Road, Sutton Coldfield, B75 6DH
G7 UCL S. Dixon, 5 Swanmore Road, Havant, PO9 4LG
G7 UCN A. Allport, 55 Byrds Lane, Uttoxeter, ST14 7NF
G7 UCO D. Reed1, 8 Wolverstone Drive Hollingdean, Brighton, BN17FB
G7 UCP D. Hornby, 7 Shawfield Grove, Rochdale, OL12 7SU

G7 UCR K. Yeo, 48 Great Goodwin Drive, Guildford, GU1 2TY
GI7 UCS M. Grainger, 1 Knocknamoe Bungalows, Omagh, BT79 7LA
G7 UCT B. Lord, 13 Park Ave, Norden, Timperley, WA14 5AQ
G7 UCZ D. Evans, 3 Dalkeith Close, Bransholme, Hull, HU7 5AS
G7 UDE D. Clark, 32 Laburnum Grove, Burstead Close, Brighton, BN1 7HX
G7 UDJ C. Edwards, 6 Blacksmiths Close, Nether Broughton, Melton Mowbray, LE14 3EW
G7 UDM D. Bonfield, 49 Linden Grove, Chandler'S Ford, Eastleigh, SO53 1LE
G7 UDU J. Selwyn, 5 Main Road, Billockby, Great Yarmouth, NR29 3BG
G7 UDX C. Harris, 8 Trelawney Rise, Callington, PL17 7PT
G7 UEC D. Denyer, 85 Highlands Road, Horsham, RH13 5ND
G7 UEI D. Longhurst, Burston, Wood Road, Hindhead, GU26 6PZ
G7 UEJ S. Kitchen, 344 Windward Way, Castle Bromwich, Birmingham, B36 0UH
G7 UEK A. Jones, 60 Heywood Drive, Starcross, Exeter, EX6 8SD
G7 UEL R. Dean, Sandshadow, Stow Road, King's Lynn, PE34 3PF
G7 UET A. Levy, 29 Ferndale Avenue, Reading, RG30 3NQ
G7 UEV C. Fox, 3 Manor Drive, Wragby, Market Rasen, LN8 5SL
G7 UEX P. Cardwell, 2 Hayfield Place, Sheffield, S12 4XH
G7 UFF L. Brackstone, 276 Ladyshot, Harlow, CM20 3EY
G7 UFI B. Courtenay, 251 Smeeth Road Marshland St. James, Wisbech, PE14 8ES
GM7 UFN T. Graham, 265 Gilmartin Road, Linwood, Paisley, PA3 3SU
G7 UFT R. Elliott, 16 Prince Philip Road, Colchester, CO2 8PA
G7 UFV D. Riseborough, 2 The Barn, Grigsons Wood, Norwich, NR16 2LW
G7 UFW N. Brickwood, 4 Patterson Close, Northampton, NN3 3PE
G7 UGA M. Turner, 14 The Rookery, Barrow Upon Soar, Loughborough, LE12 8JZ
G7 UGB M. Howard, Netherwood, Shortthorn Road, Stratton Strawless, Norwich, NR10 5NU
G7 UGC A. Fellows, 343 Wake Green Road, Moseley, Birmingham, B13 0BH
G7 UGR D. Barnett, 81 Bankside West Lynn, King's Lynn, PE34 3JH
G7 UGW J. Smith, 32 Aberdeen Street, Hull, HU9 3JU
G7 UGY N. Thornley, Purt Ny Shee, Marton Road Willingham By Stow, Gainsborough, DN21 5JU
G7 UHE G. Tiller, 15 Woodlands Gardens, Romsey, SO51 7TE
G7 UHG D. Tropman, 91 Reindeer Road, Fazeley, Tamworth, B78 3SW
G7 UHL S. Yuill, 20 Buttercup Court Deeping St. James, Peterborough, PE6 8TF
G7 UHS C. Jewell, 43 Rannoch Road, Bristol, BS7 0SA
G7 UHT C. Griffiths, 33 Westwood Road, Newport, PO30 1TD
G7 UHW M. Mcdermott, 4 Tolcairn Court, 28 Lessness Park, Belvedere, DA17 5BT
G7 UHX B. Anderson, 22 The Drive, Clacton-on-Sea, CO15 4NN
G7 UHY R. Blewitt, 62 Vicarage Road West, Dudley, DY1 4NP
G7 UID S. Clarke, 75 Beaumont Street, Netherton, Huddersfield, HD4 7HE
G7 UII C. Savage, 24 Park Hill, Awsworth, Nottingham, NG16 2RD
G7 UIO N. Johnson, 64 Rotten Row, Pinchbeck, Spalding, PE11 3RH
GI7 UIP K. O'Reilly, 1 Commons, Irvinestown, BT941je
GJ7 UIT C. Totty, Flat 4, Beech Court, Woodlands Apartments, La Rue Des Cotils, Grouville, Jersey, JE3 9AY
G7 UIU S. Palmer, 54 Hawthorn Road, Exeter, EX2 6EA
GW7 UIZ J. Hughes, Maes Y Ffynnon, 7 Meadow Gardens, Llandudno, LL30 1UW
G7 UJC G. Taylor, 34 Hockley Road, Poynton, Stockport, SK12 1RW
GM7 UJJ J. Scott, 1 Carrick Knowe Drive, Edinburgh, EH12 7EB
GM7 UJO S. Maxwell, 24 Castle Drive, Airth, Falkirk, FK2 8GD
G7 UJS C. Sheffield, 1 Benson Close Perton, Wolverhampton, WV6 7LU
G7 UJT S. Dransfield, Gardener Ground House, West End, Goole, DN14 8RW
G7 UJY M. Poole, 184 Woodgates Lane, Swanland, North Ferriby, HU14 3PR
G7 UKA T. Collier, 23 The Riggs, Brandon, Durham, DH7 8PQ
G7 UKF M. Ellis, 64 Coppice Drive, Dordon, Tamworth, B78 1QZ
G7 UKK A. Firth, 59 Station Road, Shepley, Huddersfield, HD8 8DS
G7 UKN K. Riley, 27 Limewood Close, Blythe Bridge, Stoke-on-Trent, ST11 9NZ
G7 UKR M. Blackburn, 36 Mardale Grove, Barrow-in-Furness, LA13 9QG
G7 ULC C. Probert, 25 Elizabethan Way, Rugeley, WS15 2EE
GI7 ULG S. Murdoch, 78 Landgarve Manor, Crumlin, BT29 4SF
G7 ULJ P. White, 1 Hazel Hill Place, Nottingham, NG5 5FA
G7 ULL P. Craig, 6 Marsham Close, Chislehurst, BR7 6JD
G7 ULM P. Howarth, 4 Ringwood Avenue, London, N2 9NS
G7 ULN J. Grundy, 47 Northiam Road, Eastbourne, BN20 8LP
G7 ULS K. Hurst, 94 East Park, Harlow, CM17 0SB
G7 ULW S. Widdowson, 45 Limes Avenue, Staincross, Barnsley, S75 6JP
G7 UMA N. Tindall, 87 The Grove, Marton-In-Cleveland, Middlesbrough, TS7 8AN
G7 UMF D. Griffiths, Home Farm House Cottage, Leebotwood, Church Stretton, SY6 6LX
GW7 UMS K. Keepin, 45 Heol Y Groes, Cwmbran, NP44 7lt
GW7 UMW A. Banner, 14 Carlson Drive, Wrexham, LL11 2YF
G7 UMY D. Rockliffe, 3 Hewell Lane, Barnt Green, Birmingham, B45 8NZ
G7 UNB A. Bevington, 54 Pheasant Road, Smethwick, B67 5PD
G7 UNU N. Davies, 16 St. Leonards Close, Scole, Diss, IP21 4DW
GW7 UNV E. Jones, Crungoed Farm Llanbister Road, Llandrindod Wells, LD1 5UR
G7 UNW N. Othen, 234A Regents Park Road, London, N3 3HP
G7 UNZ W. Scott, Rose Brae, Lazonby, Penrith, CA10 1AJ
G7 UOD H. Golding, 238 Greenkeepers Road, Great Denham, Bedford, MK40 4GW
GW7 UOH S. Lupton, Egryn, Ffordd Dewi Sant Nefyn, Pwllheli, LL53 6EA
G7 UOL R. Bennion, 3 Dorrington Close, Ruskington, Sleaford, NG34 9EQ
G7 UOQ N. Birt, 60 Church Road Woodley, Reading, RG5 4QB
G7 UOS B. Yates, 2 Larchwood, Countesthorpe, Leicester, LE8 5RH
G7 UOU A. Colville, 34 Great North Road, Welwyn, AL6 0PS
GM7 UPD C. Edwards, The Bennachie Craft Centre Chapel Of Garioch, Inverurie, Ab51 5HE
G7 UPL S. Northeast, 143 Henderson Road, Southsea, PO4 9JE
G7 UPN C. Jackson, 2 Northway, Guildford, GU2 9SB
G7 UPP R. Hall, Greenviews, Lower Kingsbury, Sherborne, DT9 5ED
GI7 UPQ M. Cunningham, 4 Garvaghy Road, Portglenone, Ballymena, BT44 8EF
GI7 UPU F. Gillespie, 33 Clonliffe Park, Londonderry, BT48 8NT
G7 UPZ I. Sansom, 26 Finedon Road, Wellingborough, NN8 4EB
G7 UQA D. Haigh, 29 Victoria Grove, Wakefield, WF2 8UP
G7 UQG Newcastle District Scouts Radio Club, c/o Dr R. Bloor, 7 Highfield Court, Clayton Road, Newcastle, ST5 3LT
GW7 UQJ M. Mee, Cerrig Gwynion, Penisarwaun, Caernarfon, LL55 3PW
GM7 UQM M. Horne, 10 Blair Place, Kirkcaldy, KY2 5SQ
G7 UQQ T. Wakeling, 42 Albany Road, Chislehurst, BR7 6BQ
G7 UQV M. Willoughby, 30 Kipling Road, Ipswich, IP1 6EW
GI7 UQW B. Neill, 81 Orangefield Road, Belfast, BT5 6DD
GI7 URC A. Brown, 3 Clara Road, Belfast, BT5 6FN
G7 URJ J. O'Brien, 45 Rossall Promenade, Thornton-Cleveleys, FY5 1LP
G7 URL D. Foster, Pentlow, Crowle Bank Road, Scunthorpe, DN17 3HZ
G7 URM T. Heartfield, 69 Great Thrift, Petts Wood, Orpington, BR5 1NF
G7 URP D. Palmer, Edison House, Bow Street, Great Ellingham, Attleborough, NR17 1JB
G7 URR S. Easter, Flat 11, Saxon Court, Hitchin, SG4 9TB
G7 URS R. Bird, 9 Orchard Lane, Wembdon, Bridgwater, TA6 7QY
G7 URT C. Langham, 9 Laurence Close, Shurdington, Cheltenham, GL51 4SZ
G7 URW N. Tucker, 15 Mount Pleasant Road, Dawlish Warren, Dawlish, EX7 0NA
GI7 USA A. Niblock, Apartment 1, 6 Glenburn Court, Glynn, Larne, BT40 3FF
G7 USB J. Ainsworth, 42 Buttfield Road, Hessle, HU13 0AS
GM7 USC G. Mckelvie, 37 Carskeoch Drive, Patna, KA67LR
G7 USG J. Sutherland, 4 Cherbury Close, Bracknell, RG12 9HT
G7 USI R. Hayselden, 400 Heath End Road, Nuneaton, CV10 7HG
G7 USJ T. Cogan, 11 Highgrove Walk, Weston-Super-Mare, BS24 7EF
G7 USM K. Reavill, 11 Clarence Road, Beeston, Nottingham, NG9 5HY
G7 USO N. Brodt-Savage, 5 Granville Road, Westfield, Woking, GU22 9ND
G7 USQ B. Siddall, 6 Delside Avenue, Manchester, M40 9LF
G7 USV Dr D. Atkins, 14 Ryde Place, Lee-on-the-Solent, PO13 9AU
G7 USX M. Woollard, Barnside, Colchester Road, Colchester, CO7 7EG
G7 UTB H. Scott-Telford, 9 Squires Close, Rochester, ME2 2TZ
G7 UTC M. Bean, Ashmore, Belle Vue Road, Sudbury, CO10 2PP
GM7 UTD D. Forrest, 15 Invergarry Avenue, Thornliebank, Glasgow, G46 8UR
G7 UTE B. Spencer, 80 Horncastle Road, Boston, PE21 9HY
G7 UTG J. Dodds, 84 Borrowdale Avenue, Walkerdene, Newcastle upon Tyne, NE6 4HL
G7 UTH R. Banks, 50 Vale Road, Portslade, Brighton, BN41 1GG
G7 UTI G. Ducros, 21 Wardlow Gardens, Plymouth, PL6 5PU
G7 UTR G. Kelsall, 3 Raven Street, Bingley, BD16 4LB
G7 UTS M. James, 7 Greenfield Park, Portishead, Bristol, BS20 6RG
G7 UTT R. Pearce, 15 St. Andrews Road, Backwell, Bristol, BS48 3NR
G7 UTY C. Lewis, 3 Jacobs Close, Stantonbury, Milton Keynes, MK14 6EJ
G7 UUA P. Matthews, 6 West Road, Halstead, CO9 1EH
G7 UUB F. Gibbs, 62 Wenvoe Avenue, Bexleyheath, DA7 5BT
G7 UUC M. West, 69 Frampton Crescent, Bristol, BS16 4JD
G7 UUD K. Matthews, 6 West Road, Halstead, CO9 1EH
G7 UUG N. Griffiths, 125 Coleridge Way, Crewe, CW1 5LF
GW7 UUH A. Hughes, 30 Liddell Drive, Llandudno, LL30 1UH
G7 UUK M. Hooper, 12 Meare, Dunster Crescent, Weston-Super-Mare, BS24 9DY
G7 UUL A. Riggs, Lower House, Stockland Bristol, Bridgwater, TA5 2PY
G7 UUN R. Wood, 7 Lilac Grove, Luston, Leominster, HR6 0EF
G7 UUO J. Harbidge, Low Balk Farm, Finkle Street, Bishop Burton, Beverley, HU17 8QP
G7 UUP J. Chapman, 8 Oakfield Court, Stanley Common, Ilkeston, DE7 6XB
G7 UUQ R. Moorhouse, 66 Wicor Mill Lane, Fareham, PO16 9EG
G7 UUT A. Wilson, 36 Davey Crescent, Great Shelford, Cambridge, CB225JF
G7 UUW S. Hearn, 28 Neithrop Avenue, Banbury, OX16 2NF
G7 UVB D. Anstie, 20 Keyes Road, Norwich, NR1 2JX
G7 UVF G. Cheetham, 35 South Park Grove, New Malden, KT3 5BZ
G7 UVL D. Croot, 58 Dixie Street, Jacksdale, Nottingham, NG16 5JZ
G7 UVN I. Cross, 25 Yatesbury Avenue, Blakelaw, Newcastle upon Tyne, NE5 3SZ
GW7 UVO R. Moss, 10 Llys Eleanor Shotton Lane, Shotton, Deeside, CH5 1EH
G7 UVP J. Swanwick, Ramblers, Clarks Farm Road, Chelmsford, CM3 4PH
GM7 UVS J. Graham, 265 Gilmartin Road, Linwood, Paisley, PA3 3SU
G7 UVV I. Perry, Meadow Cottage, Mill Lane, Halstead, CO9 2NW
G7 UVW D. Mills, 11 Northfield Road, Dagenham, RM9 5XH
G7 UVY C. Carr, 10 Bonds Road, Hemblington, Norwich, NR13 4QF
G7 UWB B. Wright, 2 Butterfly Gardens, Rushmere St. Andrew, Ipswich, IP4 5TF
G7 UWC C. Wright, 32 The Pastures, Rushmere St. Andrew, Ipswich, IP4 5UQ
G7 UWE P. Smith, 12A Sandicroft Place, Preesall, Poulton-le-Fylde, FY6 0PB
G7 UWG R. Hancox, 13 Regnum Close, Eastbourne, BN22 0XH
G7 UWI M. Jones, 41 Milton Brow, Weston-Super-Mare, BS22 8DD
GM7 UWL D. Cottage, Greystones, Wick, KW1 4XP
G7 UWO G. Holland, 15 Rollis Park Road, Oreston, Plymouth, PL9 7LU
G7 UWP P. Groves, Flat 3, County Chambers Station Road, Gloucester, GL1 1DH
G7 UWS D. Brunt, 91 Shaftesbury Avenue, Feltham, TW14 9LW
G7 UWV I. Brown, 35 Lees Terrace, Bilston, WV14 8EL
G7 UWW C. Harding, 1 Saddleton Grove, Saddleton Road, Whitstable, CT5 4LY
G7 UWZ D. Pooley, 25 Wharncliffe Road, Highcliffe, Christchurch, BH23 5DB
G7 UXD R. Wade, The Limes, Hunston, Bury St Edmunds, IP31 3EL
G7 UXH E. Gaunt, 71 Hebron Road, Stokesley, TS9 5DF
G7 UXK S. Hedges, 25 Rudland Close, Thatcham, RG19 3XW
G7 UXO P. Stokes, Pl17 8Fd, Callington, PL17 8FD
G7 UXQ J. Manwaring, 17 Pleasant View, Burnhope, Durham, DH7 0BA
G7 UXR M. Taylor, 27 Lincoln Road, Newark, NG24 2BU
G7 UXU H. Andrews, 24 Belvoir Road, Widnes, WA8 6HR
GW7 UXY R. Williams, 16 Tir Dafydd, Pontyates, Llanelli, SA15 5TP
G7 UYB C. Major, 17 Jubilee Cottages, Station Road, Bedford, MK43 0PN
G7 UYI B. Williamson, 12 Middleton Close, Southampton, SO18 2FP
G7 UYJ J. Jardine, 41 Charles Drive, Anstey, Leicester, LE7 7BH
G7 UYT J. O'Toole, 4 Lindisfarne Road, Dagenham, RM8 2RA
G7 UYW J. Kitchener, 101 Highfield Road, Tring, HP23 4DS
G7 UZA J. Rodinson, 21 Graylands Road, Liverpool, L4 9UG
G7 UZG A. Mcwilliam, 43 Hylder Close, Swindon, SN2 2SL
G7 UZI P. Pullen, 12 Kimpton Road, Sutton, SM3 9QJ
G7 UZN D. Dawson, 12 Thurlow Terrace, Kentish Town, London, NW5 4JB
G7 UZO D. Lock, 1 Heaton Avenue, Huddersfield, HD5 0LJ
G7 UZS N. Thompson, 30 Dene View, Ashington, NE63 8JF
G7 UZY A. Musther, 2 Fakenham Close, Lower Earley, Reading, RG6 4AB
G7 VAB M. Richards, Sunnymead, Ardley End, Bishop's Stortford, CM22 7AJ
G7 VAD M. Beeney, Oakville Farm, Lewes Road, Uckfield, TN22 5JH
G7 VAE J. Beeney, 37 Coppice Avenue, Eastbourne, BN20 9PP
G7 VAG G. Podmore, 9 Pendlebury Street, Warrington, WA4 1TU
G7 VAS M. Kay, 12 The Crescent, Ashton-On-Ribble, Preston, PR2 1JP
G7 VAY R. Dingle, 87 Eighth Avenue, Bridlington, YO15 2NA

G7	VBD	M. Ennis, 1 Nairn Road, Cramlington, NE23 1RQ
GW7	VBE	D. Harris, 29 Queen Street, Blaengarw, Bridgend, CF32 8AH
G7	VBF	J. Barwell, Flat 9, Corinth House 33 Barley Lane, Ilford, IG3 8XE
G7	VBJ	D. Wager, 162 Harvest Fields Way, Sutton Coldfield, B75 5TJ
G7	VBL	J. Munday, 20 Highcroft, Wood Road, Hindhead, GU26 6PW
G7	VBN	B. Richards, 52 Ripon Way Carlton Miniott, Thirsk, YO7 4LR
G7	VBU	D. Firth, 5, Birchfield Grove, Skelmanthorpe, Huddersfield, HD8 9BS
GW7	VBY	Dr D. Morrison-Smith, 1 Neptune House, Upper Corris, Machynlleth, SY20 9BQ
G7	VBZ	P. Bunce, 66 Berry Park, Saltash, PL12 6EN
G7	VCB	L. Tooze, Flat 1, 91 Harbour Road, Seaton, EX12 2NJ
G7	VCE	M. Flack, 31 Harebell Close, Cambridge, CB1 9YL
G7	VCF	J. O'Donnell, 3 Linden Avenue, Altrincham, WA15 8HA
G7	VCG	I. Firby, 19 St. Georges Drive, Manchester, M40 5HL
G7	VCJ	C. Hansford, 14 Parsonage Crescent, Castle Cary, BA7 7LT
G7	VCK	M. Robertson, 67 Oatland Gardens, Leeds, LS7 1SL
G7	VCN	R. Bawley, 52 Pitville Avenue, Liverpool, L18 7JG
G7	VCP	P. Stubbs, 2 Cynthia Road, Runcorn, WA7 4TX
GI7	VCR	S. Robertson, 32 Castle Meadows, Carrowdore, Newtownards, BT22 2TZ
GM7	VCV	A. Brown, 96 Barony Terrace, Kilbirnie, KA25 6DB
G7	VCY	D. Seymour, 24 Farley Dell, Coleford, Radstock, BA3 5PJ
G7	VCZ	S. Ldr. P. Weaver, 42 Erithway Road, Coventry, CV3 6JT
G7	VDA	I. Singer, 197 Rosalind Street, Ashington, NE63 9BB
G7	VDD	P. Mcgowan, Rua Oliva De La Frontera, 29, 1Esq, Caldas Da Rainha, Portugal, 2500-886
G7	VDH	J. Brook, 2 New Laithe Bank, Holmfirth, HD9 1HL
G7	VDI	D. Norris, Flat 11, 10 Cromartie Road, London, N19 3SJ
G7	VDJ	Dr S. Henry, Hertford College, Catte Street, Oxford, OX1 3BW
G7	VDK	G. Taylor, 4 Brown Crescent, Eighton Banks, Gateshead, NE9 7EX
GM7	VDL	W. Steele, 35 Devlin Court, Whins Of Milton, Stirling, FK7 0NP
G7	VDN	H. May, 18 Pennant Hills, Bedhampton, Havant, PO9 3JZ
G7	VDQ	D. Butterworth, 6 Fir Grove, Weaverham, Northwich, CW8 3JD
G7	VDS	M. Hudson, 5 Berkeley Court High Street, Cheltenham, Gl526da
G7	VDT	G. Baines, 20 Whitehall Rise, Wakefield, WF1 2AL
G7	VDU	A. Marston, 111 Averil Road, Leicester, LE5 2DE
G7	VDV	N. Keech, 14 Simpson Court, Ashington, NE63 9SD
G7	VDX	S. Taverner, 8 The Rye Lea, Droitwich, WR9 8SS
G7	VEB	D. Wilson, 210 Stanks Lane South, Swarcliffe, Leeds, LS14 5PD
GM7	VEC	S. Fearn, 2 Carabhat, Carinish, HS6 5HR
G7	VEE	A. Saunders, 25 Southern Drive, South Woodham Ferrers, Chelmsford, CM3 5NY
G7	VEF	R. Parkin, 17 Roberts Road, Watford, WD18 0AY
GW7	VEH	J. Humphrey, Bryn Ebbw, Beaufort Hill, Ebbw Vale, NP23 5QR
G7	VEI	A. Stripp, 87 Elthorne Park Road, London, W7 2JH
GW7	VEL	C. Lee, 23 Forest Close, Coed Eva, Cwmbran, NP44 4TE
G7	VEX	N. Hindle, 19 Barkway Road, Royston, SG8 9EA
G7	VEY	J. Martin, 3 The Rise, Calne, SN11 0LQ
G7	VFA	J. Juggins, 5 Charter Close, Helston, TR13 8SR
G7	VFC	O. Dewberry, The Stables, Barrack Street, Manningtree, CO11 2RB
G7	VFE	R. Wills, 14 Penwood Heights, Penwood, Highclere, Newbury, RG20 9EY
GW7	VFJ	S. Magee, Lle Da, Cefn Bychan Road, Mold, CH7 5EL
G7	VFL	K. Sherman, 12 Portland Drive, Stourbridge, DY9 0SD
G7	VFQ	A. Latham, 22 Moorland Road, Leek, ST13 5BW
GM7	VFR	J. Smith, 28 Tollerton Drive, Irvine, KA12 0QE
G7	VFU	A. Mullord, 296 City Way, Rochester, ME1 2BL
G7	VFV	G. Somers, 16 Button Drive, Newquay, TR7 3FB
G7	VFX	R. Watts-Read, 43 Whyteleafe Hill, Whyteleafe, CR3 0AJ
G7	VFY	S. Walters, 16 North Lodge, 46 Somerset Road, New Barnet, Barnet, EN5 1RJ
G7	VGA	S. Bonney, 22 Gordon Drive, Abingdon, OX14 3SW
GW7	VGB	D. Beynon, 129 Eureka Place, Ebbw Vale, NP23 6LN
G7	VGC	B. Goody, Flat 31, Homeweave House, Robinsbridge Road, Colchester, CO6 1UL
G7	VGE	R. Teague, 18 Aspen Close, Great Blakenham, Ipswich, IP6 0HQ
G7	VGH	M. Smith, 46 Bentham Way, Ely, CB6 1BS
G7	VGJ	A. Cole, 58 Stradbroke Drive, Chigwell, IG7 5QZ
G7	VGK	W. Parrett, 5 Coniston Close, Walton, Liverpool, L9 0NG
G7	VGL	D. Pemberton, 12 Victor Road, Thatcham, RG19 4LX
G7	VGM	S. Taylor, 22 Raby Square, Hartlepool, TS24 8HH
G7	VGN	A. Chamberlain, 7 Mccalmont Way, Newmarket, CB8 8HU
G7	VGO	A. Hurst, 12 Spilsby Close, Hartlepool, TS25 2RD
GI7	VGR	C. Dorrian, 47 Albany Drive, Carrickfergus, BT38 8BF
G7	VGT	P. Schranz, 42 South Townside Road, North Frodingham, Driffield, YO25 8LE
G7	VGX	H. Dolman, 28 The Downs, Middleton, Manchester, M24 1TJ
G7	VGY	D. Childs, 7 Grange Road, East Cowes, PO32 6EA
G7	VHC	C. Spires, 5 Springhead, Sutton Veny, Warminster, BA12 7AG
GW7	VHD	A. Jones, 57 Dinerth Road, Rhos On Sea, Colwyn Bay, LL28 4YG
G7	VHF	East Anglian Six Meter Group, c/o A. Garry-Durrant, Casa Santosa Llano Del Espino, Albox, Spain, 4800
G7	VHG	S. Bryan, 18 Whalley Crescent, Wroughton, Swindon, SN4 9EP
G7	VHJ	P. Gow, 11 Rodley Square, Lydney, GL15 5AZ
G7	VHN	J. Hart, 35 Aintree Close, Uxbridge, UB8 3HS
G7	VHO	B. Hart, 35 Aintree Close, Uxbridge, UB8 3HS
GM7	VHQ	I. Helie, 22 Mcclue Road, Renfrew, PA4 9BL
G7	VHS	J. Mclaughlin, 16 East Street, Batley, WF17 5QY
G7	VHU	D. Beastall, 11 Hopwood Bank, Horsforth, Leeds, LS18 5AW
G7	VHX	J. Golding, 65 Longworth Avenue, Tilehurst, Reading, RG31 5JU
G7	VHZ	E. Gilowski, 126 Owlsmoor Road, Owlsmoor, Sandhurst, GU47 0ST
G7	VIB	D. Airs, Cornerways Cottage, Poffley End, Witney, OX29 9UW
G7	VIE	J. Cooper, 9 Highfield Crescent, Halesowen, B632BD
G7	VIG	A. Smith, 19 Gibsons Gardens, North Somercotes, Louth, LN11 7QH
G7	VIH	P. Wilson, 117 Naseby Road, Kettering, NN16 0LL
G7	VIK	N. Higgins, 6 Larksfield Avenue, Bournemouth, BH9 3LP
G7	VIL	G. Mason, 6 Willowtree Avenue, Gilesgate Moor, Durham, DH1 1EB
G7	VIP	F. Marston, 1 Weaver Road, Leicester, LE5 2RL
G7	VIR	A. James, 19 Coach Lane, Redruth, TR15 2TP
G7	VIV	A. Nussey, 9 Brent Street, Brent Knoll, Highbridge, TA9 4DU
GI7	VIW	A. Harvey, 5 Kilmaine Road, Bangor, BT19 6DT
G7	VIX	I. Rogers, 29 Longlands Road, Emsworth, PO10 8HL
G7	VIY	A. Harper, 3 Eskdale Crescent, Blackburn, BB2 5DT
G7	VJA	K. Sharman, 1 The Greenwoods, Hartland, Bideford, EX39 6JA
G7	VJD	D. Skidmore, Weavers, Kingsdale Road, Berkhamsted, HP4 3BS
G7	VJE	C. Rohrer, Alpenrose, Bedlars Green Great Hallingbury, Bishop's Stortford, CM22 7TP
G7	VJG	P. Veitch, 58 Beaulieu Close, Toothill, Swindon, SN5 8AQ
G7	VJH	T. Scanlon, 11 Caterhouse Road Framwellgate Moor, Durham, DH1 5HP
G7	VJI	G. Dawes, 19 Benbridge Avenue, Bournemouth, BH11 9HN
G7	VJJ	T. Wood, 39 Baker Road, Bournemouth, BH11 9JD
GW7	VJK	N. Cole, Tycoch, Llandovery, SA20 0UP
G7	VJM	C. Margetts, 16 Lahn Drive, Droitwich, WR9 8TQ
G7	VJQ	C. Radford, 12 Homewood Drive, Kirkby-In-Ashfield, Nottingham, NG17 8QB
G7	VJT	S. Tibbetts, 113 Highfield Crescent, Halesowen, B63 2AY
G7	VJU	R. Jones, 20 Carnoustie Close, Southport, PR8 2FB
G7	VJY	T. Hill, 15 Catkin Walk, Rugeley, WS15 2NS
G7	VKA	K. Wandless, 1 Lee Moor Cottages, Rennington, Alnwick, NE66 3RL
G7	VKB	A. Cunnington, 131 Colson Road, Loughton, IG10 3QY
G7	VKG	M. Gibson, 6 Harrison Road, Mansfield, NG18 5RG
G7	VKJ	C. Buckley, 14 Sunny Drive, Prestwich, Manchester, M25 3JJ
GM7	VKN	R. Beharie, Isengard, Norseman Village, Firth, KW17 2NY
G7	VKY	B. Shrimpling, 45 Fairmont Road, Grimsby, DN32 8DZ
G7	VLA	M. Sandham, 7 Mill Close, Caverswall, Stoke-on-Trent, ST11 9HA
G7	VLB	S. Kirkbright, 48 Plant Crescent, Stafford, ST17 4EH
GM7	VLC	A. Haines, 164 North High St., Musselburgh, EH21 6AR
G7	VLD	K. Howard, 43 Hazeldell, Watton At Stone, Hertford, SG14 3SN
G7	VLF	N. Faiz, 48 Cox House, Field Road, London, W6 8HN
G7	VLH	C. Hill, 14 Blenheim Close Chandler'S Ford, Eastleigh, SO53 4LD
G7	VLJ	E. Baker, 29 Ashcroft Road, Ipswich, IP1 6AB
G7	VLL	J. Woodhouse, 5 Dolphin Villas, Hazlerigg, Newcastle upon Tyne, NE13 7NG
G7	VLR	Leicester Raynet Group, c/o A. Holmes, 5 Launde Park, Market Harborough, LE16 8BH
GM7	VLZ	A. Pearce, 105 Gyle Park Gardens, Edinburgh, EH12 8NQ
G7	VME	P. Schofield, 22 Atherton Court, Meadow Lane, Windsor, SL4 6BN
G7	VML	L. Alden, 20 Kings Walk, Shoreham-by-Sea, BN43 5LG
G7	VMO	S. Fawcett, 45 Forresters Close, Norton, Doncaster, DN6 9HX
G7	VMQ	T. Jones, Ockton House, 24 Station Road, Okehampton, EX20 1EA
GW7	VMT	E. Wetherall, 38 Argyle Street, Pembroke Dock, SA72 6HL
G7	VNC	C. Cave, Little Meadow, Brewham Road, Bruton, BA10 0JD
G7	VND	Dr G. Morris, 17 Bradshaw Road, Inkersall, Chesterfield, S43 3HJ
G7	VNG	A. Elmes, Pookeezows, 10 Farnham Avenue, Hassocks, BN6 8NS
G7	VNJ	M. Swain, Cottage Farm, Langley Marsh, Wiveliscombe, Taunton, TA4 2UL
G7	VNK	R. Cronshaw, Flat 2, 28 Adelaide Terrace, Blackburn, BB2 6ET
G7	VNL	C. Lambert, 43 Church Road, Guildford, GU1 4NQ
G7	VNM	A. Melham, 4 Constantine Road, North Bitchburn, Crook, DL15 8AG
G7	VNN	C. Backhouse, Elm House, The Green Saxlingham Nethergate, Norwich, NR15 1TH
G7	VNO	C. Brown, 6 Ellesmere Avenue, Derby, DE24 8WD
G7	VNP	K. Knights, 13 Millfield Place, Wilburton, Ely, CB6 3SA
G7	VNQ	K. Staddon, 1 Aller Grove, Whimple, Exeter, EX5 2TJ
G7	VOA	D. Hughes, 60 Martingale Place, Downs Barn, Milton Keynes, MK14 7QN
G7	VOH	S. Holland, 49 Oxland Road, Illogan, Redruth, TR16 4SH
G7	VOI	T. Nicholas, Talmont, Chester Road, Tarporley, CW6 0SD
G7	VOK	A. Bracey, 42 Lampton Grove, Bristol, BS13 0QA
G7	VOM	J. Snelgrove, 22 Plains Avenue, Maidstone, ME15 7AU
G7	VON	A. Snelgrove, 22 Plains Avenue, Maidstone, ME15 7AU
GW7	VOO	P. Hockey, 98 Meadow Rise, Brynna, Pontyclun, CF72 9TF
G7	VOQ	S. Casey, 5 Willow Road, Leyland, PR26 8NP
G7	VOT	A. Moseley, 15 Hawthorn Avenue, Billingham, TS23 1EE
G7	VOX	M. Ingram, Foxhill, Lower Daggons, Fordingbridge, SP6 3EE
G7	VPA	N. Muncey, 2 Ladysmith Avenue, Whittlesey, Peterborough, PE7 1XX
G7	VPD	R. Friend, High Hedges, Church Road, Norwich, NR12 8YL
G7	VPN	A. Berry, 13 Collimer Close, Chelmondiston, Ipswich, IP9 1HX
G7	VPQ	J. Bishop, 27 Southway, Blacon, Chester, CH1 5NW
G7	VPS	D. Barwood, 6 Hemington Close, King's Lynn, PE30 3YB
GM7	VPT	D. Leask, Avonmuir, The Loan, Muiravonside, Eh496Lw, Linlithgow, EH49 6LW
G7	VPU	P. Ansell, White Hatch, Uvedale Road, Oxted, RH8 0EW
GW7	VQA	N. Parker, 47 Rosehill Road, Rhyl, LL18 4TN
GM7	VQB	T. Roy, 1 Rose Terrace, Leven, KY8 4DF
G7	VQC	D. Driver, 27 Cricketers Way, Chatteris, PE16 6UR
G7	VQE	D. Frost, 24 Woodland Close, Northampton, NN5 6NH
G7	VQI	G. Burch, 6 The Barracks, Parkend, Lydney, GL15 4HR
G7	VQJ	W. Willmott, Walton House Sandwich Road, Eastry, CT13 0DP
G7	VQL	M. Endean, 17 Dryden Place, Tilbury, RM18 8HQ
G7	VQM	C. Davis, 8 Mulberry Grove, Wallasey, CH44 6PZ
G7	VQO	W. Good, 53 Harrow Lane, St Leonards-on-Sea, TN37 7JY
G7	VQR	P. Mawdsley, 7 Aldebert Terrace, London, SW8 1BH
G7	VQW	P. Jessup, 2 Tile Lodge Cottages, Hoath Road, Canterbury, CT3 4JN
G7	VQX	J. Hunter, 7 Berry Hill, Nunney, Frome, BA11 4NR
G7	VRJ	M. Holland, 31 The Avenue, Andover, SP10 3EP
G7	VRK	S. Balding, 13 Church Close, Colby Road Banningham, Norwich, NR11 7DY
G7	VRX	R. Croft, Wallbury Lodge, Dell Lane, Bishop's Stortford, CM22 7SQ
G7	VRY	P. Bambridge, 8 Temple Lane, Tonwell, Ware, SG12 0HP
GM7	VSB	J. O'Neill, 39 Ardneil Court, Ardrossan, KA22 7NQ
G7	VSE	D. Briggs, 15 Orkney Close, Manchester, M23 2AT
GW7	VSF	W. Thomas, 2 Ffordd Trecastell, Llanharry, Pontyclun, CF72 9ND
G7	VSJ	D. Jones, 6 Eastville, Bath, BA1 6QN
G7	VSL	R. Taylor, 11 Ranscombe Close, Brixham, TQ5 9UR
G7	VSM	J. Skinner, 85 Main St., Barton Under Needwood, Burton on Trent, DE13 8AB
G7	VSN	L. Franklin, 4 Rossington Close, Metheringham, Lincoln, LN4 3DS
GW7	VSO	D. Lewis, 45 Llewellyn St., Pontygwaith, Ferndale, CF43 3LF
G7	VSP	S. Wooster, 44 King Johns Road, North Warnborough, Hook, RG29 1EJ
GW7	VST	G. Davies, 41 Woodlands Road, Barry, CF63 4EF
G7	VSW	S. Weston, 11 Friars Road, Abbey Hulton, Stoke-on-Trent, ST2 8DQ
G7	VTC	L. Dodd, Mulberry Lodge, Hallaze Road, Hallaze, St Austell, PL26 8YW
G7	VTE	G. Forster, 33 Deer Valley Road, Holsworthy, EX22 6DA
G7	VTH	D. Reacher, 33 Cator Crescent, New Addington, Croydon, CR0 0BL

G7 VTJ T. Scott, 50 Davison Avenue, Whitley Bay, NE26 1SH
G7 VTL C. Davis, 38 Courtney Close, Tewkesbury, GL20 5FB
G7 VTN P. Mccaulay, 33 Millmoor Way, North Hykeham, Lincoln, LN6 9PJ
G7 VTQ Prof. D. Parsons, 1 Kent Drive, Congleton, CW12 1SD
G7 VTR C. Taylor, Tookeys House, Tookeys Drive, Astwood Bank, B96 6BB
G7 VTS P. Green, 81 Victoria Road, Farnborough, GU14 7PP
G7 VTT J. King, Portland Manor Care Home, Thornhill Road, Newcastle upon Tyne, NE20 9PZ
G7 VTW D. Jacques, 4 Cedar Close, Thorpe Willoughby, Selby, YO8 9QL
G7 VUH C. Squire, 19 Southfield Road, Burley In Wharfedale, Ilkley, LS29 7PA
G7 VUL J. Cook, 32 Ash Bank Road, Stoke-on-Trent, ST2 9DR
G7 VUM D. Riches, 118 Drayton Road, Norwich, NR3 2DL
G7 VUP J. Milner, Sweetcroft Brentor, Tavistock, PL19 0NJ
G7 VUU K. Coe, 5 George St., Enderby, Leicester, LE19 4NQ
G7 VVF D. Rossiter, 37 Meadway, Enfield, EN3 6NT
G7 VVK P. Bradley, 22 Cavalier Close, Romford, RM6 5EJ
G7 VVL N. Quest, 21 Neave Crescent, Romford, RM3 8HN
G7 VVO I. Anderson, 18 St. Anthonys Drive, Wick, Bristol, BS30 5PW
G7 VVX A. Renton, 18 Stoneworks Garth, Crosby Ravensworth, Penrith, CA10 3JE
G7 VWA D. Lever, 35 Carteret Road, Luton, LU2 9JZ
G7 VWG R. Evans, 113 Highbridge Road, Burnham-on-Sea, TA8 1LW
G7 VWM J. Hazell, 7 Higher Road, Woolavington, Bridgwater, TA7 8EA
G7 VWN J. Blackwell, 7 Church Road, Darley Dale, Matlock, DE4 2GG
G7 VWO D. Lisle, Kent Ii, Broadmoor Hospital, Crowthorne, RG45 7EG
G7 VWW J. Brook, 45 Colonial Court, Senoia, USA, 30276
GI7 VXC A. Crozier, 38 Hawthorn Hill, Dromara, Dromore, BT25 2HY
G7 VXK B. Amare, Po Box 30464, Addis Ababa, Ethiopia,
G7 VXQ C. Elcombe, 10 Northport Drive, Wareham, BH20 4DR
GM7 VXR P. Crankshaw, 3 North Neuk, Troon, KA10 6TT
G7 VXS G. Burchell, 23B Luff Meadow, Stowmarket Road, Ipswich, IP6 8DP
G7 VYB R. Patel, 30 Buckingham Drive, Luton, LU2 9RA
G7 VYF K. Tadesse, Po Box 60229, Addis Ababa, Ethiopia,
G7 VYI R. Smith, 82 Long Row, Shrewsbury, SY1 4DD
G7 VYN M. Jarman, 143 Rotherham Road, Barnsley, S71 2LL
G7 VYQ I. Holman, 7 The Silent Woman Park, Tavistock, PL19 9LQ
GM7 VYR I. Findlay, 2 Bothwell Road, Uddingston, Glasgow, G71 7ET
G7 VYT J. Graver, 15 Cartwright Road, Charlton, Banbury, OX17 3DG
G7 VYW W. Moreton, 17 Hadley Road, Bilston, WV14 6RX
G7 VYY S. Middleton, 22 Hall Villa Lane, Toll Bar, Doncaster, DN5 0LH
G7 VYZ M. Lancastle, 31 Ridgeside, Kirk Merrington, Spennymoor, DL16 7HF
G7 VZD J. Payne, 15 Belmont Road, Tiverton, Ex166ar
G7 VZI H. Charles, 6 Bridewell Street, Wymondham, NR18 0AR
G7 VZL D. Forster, 15 Bracondale, Norwich, NR1 2AL
G7 VZM M. Blacklock, 39 Birtwistle Avenue, Colne, BB8 9RS
G7 VZQ D. Polley, 33 Wye Close, Crawley, RH11 9QZ
G7 VZR C. Gain, 14 Battens Avenue, Overton, Basingstoke, RG25 3NL
G7 VZS A. Walsh, 21 Rydal Avenue, Darwen, BB3 2SA
GM7 VZV D. Henry, 106 Whinhill Gate, Aberdeen, AB11 7WF
G7 VZY Dr M. Page-Jones, 2 Chestnut Close, Romsey, SO51 5SP
G7 WAA E. Donaghy, Mendips, 36 Attwood Road, Salisbury, SP1 3PR
G7 WAB Worked All Britain Awards Group, c/o K. Hale, 58 St. Stephens Road, Saltash, PL12 4BJ
G7 WAC Wythall Contest Group, c/o L. Volante, Richmond House, Icknield Street, Birmingham, B38 0EP
G7 WAE T. Ward, 173-175 Station Road, Pendlebury, Swinton, Manchester, M27 6BU
G7 WAF D. Keeble, 71 St. Lawrence Avenue, Bolsover, Chesterfield, S44 6HS
G7 WAQ A. Moss, 21 Shrubbery Lane, Weymouth, Dt4 9ly
G7 WAS S. Staines, 6 The Quantocks, Flitwick, Bedford, MK45 1TQ
G7 WAW D. Thompson, 12 Dam Head Road, Barnoldswick, BB18 5NH
G7 WAY S. Foster, 137 Cheltenham Road, Longlevens, Gloucester, GL2 0JH
G7 WBA R. Grandshaw, Treehaven, South Lane, Salisbury, SP5 2BZ
G7 WBE D. Welch, 38 Little Sammons Chilthorne Domer, Yeovil, BA22 8RB
G7 WBH D. Small, 17 Claygate Road, Wimblebury, Cannock, WS12 2RN
G7 WBJ W. Naylor, 5 Burman Close, Shirley, Solihull, B90 2DR
G7 WBL J. Wheeler, 92 Holford Road, Bridgwater, TA6 7NZ
G7 WBM Dr P. Longhurst, Burston, Wood Road, Hindhead, GU26 6PZ
G7 WBO M. Stenning, 56 Hampshire Court, Upper St. James'S Street, Brighton, BN2 1JZ
GM7 WBP F. Kelly, 50 Farm Road, Blantyre, Glasgow, G72 9DT
G7 WBR E. Davis, 33 Truggers, Handcross, Haywards Heath, RH17 6DQ
G7 WBU A. Hopley, 124 Lonnen Road, Wimborne, BH21 7AZ
G7 WBW A. Millward, 26 Osprey Close, Scotton, Catterick Garrison, DL9 3RA
G7 WBY B. Mulder, 8 Chapel Close, Little Gaddesden, Berkhamsted, HP4 1QG
G7 WBZ M. Malone, 11 Pine Close, Rishton, Blackburn, BB1 4JX
G7 WCB A. Bennett, 12 Barns Close, Walsall, WS9 9BD
G7 WCF C. Rose, The Barn, Lower Killigorrick, St Keyne, Liskeard, PL14 4QP
G7 WCG D. Seabrook, 44 Village Centre, Richmond Letcombe Centre, Letcombe Regis, OX12 9RG
G7 WCN K. Packer, 47 Sheppard Road, Basingstoke, RG21 3JH
G7 WCP C. Sharpe, 30 Mardale Way, Loughborough, LE11 3SS
GW7 WCR J. Pitkin, 29 Dolwerdd Estate, Pen Y Parc, Cardigan, SA43 1RF
GI7 WCS J. Stitt, 199 Gobbins Road, Islandmagee, Larne, BT40 3TX
G7 WDC M. Kiteley, 13 Chiltern Close, Astley Cross, Stourport-on-Severn, DY13 0NU
G7 WDD B. Bird, 4 Berkeley Crescent, Frimley, Camberley, GU16 8YN
G7 WDG P. Wyatt, 6 Bridge Road, Coalville, LE67 3PW
G7 WDM N. Feetham, 154 Magdalen Lane, Hedon, Hull, HU12 8LB
G7 WDN A. Goodridge, 9 Radcliffe Way, Oundle, Peterborough, PE8 4QE
G7 WDO C. Barker, 15 Epping Green, Hemel Hempstead, HP2 7JP
G7 WDS A. James, 36 Lemon Hill, Mylor Bridge, Falmouth, TR11 5NA
G7 WEB D. Raxter, 2 Lower Croft, Cropthorne, Pershore, WR10 3NA
GM7 WED R. Feilen, 131 Croftend Avenue, Glasgow, G44 5PF
G7 WEK C. Taylforth, 1 Clough Terrace, Barnoldswick, BB18 5PD
G7 WEM T. Hewitt, 6 Mayfield, Catforth Road, Preston, PR4 0HH
G7 WEN J. Marron, 30 York Road, Nunthorpe, Middlesbrough, TS7 0EZ
G7 WEP M. Williams, 59 Thistledene, Thames Ditton, KT7 0YH
G7 WER P. Fisher, 21 Charlotte Close, Mount Hawke, Truro, TR4 8TS
G7 WEW A. Cossey, 17 Hazel Close, Norwich, NR8 6YE
G7 WFC R. Smith, 35 Montsale, Pitsea, Basildon, SS13 1JL
G7 WFD M. Dockerty, 16 Valley Way, Stalybridge, SK15 2QZ
G7 WFH T. Ford, 3 Greenmore Road, Bristol, BS4 2LA
G7 WFK G. Tew, 5 Hill Top Avenue, Tamworth, B79 8QB
G7 WFQ J. Stevens, Springfield Cottage, 57 Brindley Street, Stourport-on-Severn, DY13 8JG
GM7 WFT G. Edwards, 19 Howe Park, Edinburgh, EH10 7HF
G7 WFV J. Kendall, 3 High Street, Cottenham, Cambridge, CB24 8SA
G7 WFZ R. Stone, 222 Dedworth Road, Windsor, SL4 4JP
G7 WGA J. Potter, 198 Battle Road, St Leonards-on-Sea, TN37 7AL
G7 WGD D. Price, 199 Central Drive, Bilston, WV14 8JE
G7 WGE T. Forster, 20 Bryant Avenue, Slough, SL2 1LG
G7 WGI J. Gordon, 19 Heywood Gardens, Havant, PO9 4HR
G7 WGL K. Firth, Chimneys, 30 Kingscroft, King's Lynn, PE31 6QN
GM7 WGM S. Andrew, 16 Colthill Road, Milltimber, AB13 0EF
G7 WGN J. Hutton, 42 Priory Road, Cottingham, HU16 4SA
G7 WGO G. Bradshaw, 3 Falmouth Avenue, Haslingden, Rossendale, BB4 6QN
G7 WGP A. Brook, 163 Station Road, Mickleover, Derby, DE3 9FL
G7 WGX D. Sayles, 82 Molineaux Road, Shiregreen, Sheffield, S5 0JY
G7 WGY D. Wilson, 75 Gainsborough Road, Scotter, Gainsborough, DN21 3RU
G7 WGZ E. Morley, 91 Allerton Road, Stoke-on-Trent, ST4 8PQ
G7 WHA O. Morley, 91 Allerton Road, Stoke-on-Trent, ST4 8PQ
G7 WHI D. Page, 55 Hinckley Road, Stoney Stanton, Leicester, LE9 4LL
G7 WHM A. Howgate, 7 Caledonian Way, Belton, Great Yarmouth, NR31 9PQ
G7 WHP W. Jones, 7 Hampstead Gardens, Hockley, SS5 5HN
GM7 WHQ S. Gray-Thompson, Vagastie, Lairg, IV27 4AD
G7 WHU M. Nock, Mesquida, Lorraine Road, Newhaven, BN9 9QB
G7 WHX J. Bodle, 48 Bolsover Road, Hove, BN3 5HP
G7 WHZ T. Crane, 15 Belchamps Way, Hawkwell, Hockley, SS5 4NT
G7 WIC G. Probyn, 24 Woollaton Close, Grange Park, Swindon, SN5 6BB
G7 WID G. White, 21 Tollfield Road, Boston, PE21 9PN
G7 WIG R. Bilsland, 56 Cowleigh Bank, Malvern, WR14 1PH
G7 WIQ S. Pack, 245A Beacon Road, Loughborough, LE11 2QZ
G7 WIY M. Downing, 12 Martindale Road, Woking, GU21 3PJ
G7 WJC B. Webster, 50 Blackburn Road, Rishton, Blackburn, BB1 4BH
G7 WJE H. Coots, 40 Essex Close, Romford, RM7 8BD
G7 WJJ R. Morton, 29 Lanmoor Estate, Lanner, Redruth, TR16 6HN
G7 WJK J. Stephens, 19 Aspen Fold, Oswaldtwistle, Accrington, BB5 4PH
GM7 WJP A. Anderson, 232 Annan Road, Dumfries, DG1 3HE
G7 WJV R. Stroud, 55 Haymeads Lane, Bishop's Stortford, CM23 5JJ
G7 WJW G. Cripps, 52 Cleveland, Tunbridge Wells, TN2 3NQ
G7 WJZ P. Clarke, 21 Long Furlong Road, Sunningwell, Abingdon, OX13 6BL
G7 WKC T. Hasted, Springfield House, Birds End, Bury St Edmunds, IP29 5HE
G7 WKG A. Roche, Flat 22, Trident Court, Birmingham, B20 2NX
G7 WKH P. Clark, 21 Sandfield Road, Arnold, Nottingham, NG5 6QA
G7 WKP A. Andrew, Thrift, Madles Lane, Ingatestone, CM4 9QA
G7 WKV B. Jewell, 49 Spinney Road, Burton Latimer, Kettering, NN15 5ND
G7 WKW M. Davis, Cherry Pie Bay Road, Freshwater, PO40 9QS
GI7 WLA D. Calvin, 65 Tannaghmore Road, Markethill, Armagh, BT60 1TW
G7 WLC D. Evans, 1 Brigadier House Captain Gardens, Colchester, CO2 7LD
G7 WLL I. Irlam, 31 Wyatt Road, Dartford, DA1 4SN
G7 WLM J. Tamlyn, Hedge Rise, Sidmouth Road, Exeter, EX2 5QJ
GM7 WLO J. Burt, Olivet, Lanton Road, Jedburgh, TD8 6SD
G7 WLV G. Southall, 6 Dudley Wood Avenue, Dudley, DY2 0DG
G7 WLY R. Bagwell, 20 School Approach, South Shields, NE34 6DP
G7 WRG Walsall Raynet Group, c/o S. Glazzard, 109 Highfields Road, Chasetown, Burntwood, WS7 4QS
G7 WRS T. Misselbrook, City Fields, Wakefield, WF3 4GD
G7 WSH R. Munt, Box 166, Se-177 23 Jarfalla, Sweden, SWEDEN
G7 WWW B. Cole, 6 Parkstone Parade, Hastings, TN34 2PS
G7 XPC P. Chorley, Boone Hill House, Mount Boone Hill, Dartmouth, TQ6 9NZ
G7 YAB A. Brewerton, The Vicarage, Highthorn Road, Kilnhurst, Mexborough, S64 5TX
G7 ZMS M. Larcombe, 65 Western Road, Burgess Hill, RH15 8QW
G7 ZRT R. Thayne, 213 Carlton Road, Boston, PE21 8NG
G7 ZZY P. Pile, Apartment 836, Lagos, Portugal, 8600

G8

G8 AA M. Keilty, 25 Lathom Avenue Wallasey, Wirral, Ch44 5uh
G8 AAC J. Billingham, 14 St. Matthews Court, Sutherland Road, Brighton, BN2 2EX
G8 AAD B. Blight, 43 North Street, Oxon, OX9 3BJ
G8 AAE D. Phillips, 2 Walkers Close, Chelmsford, CM1 6UW
GW8 AAF F. Blake, 3 Morfa Gaseg, Llanfrothen, Penrhyndeudraeth, LL48 6BH
G8 AAI M. Bues, 7A Alice Parkins Close, Hadleigh, Ipswich, IP7 6FE
G8 AAR F. May, Quatre Vents, Church Road, Sudbury, CO10 0QP
G8 AAT R. Pye, 7 Meadow View, Potterspury, Towcester, NN12 7PH
G8 AAU N. Stanners, 22 Brands Hill Avenue, High Wycombe, HP13 5QA
G8 ABB G. Rogers, 10 The Laurels, Bletchley, Milton Keynes, MK1 1BL
G8 ABX G. Catling, 3 The Tene, Baldock, SG7 6DG
G8 ACL H. Cosford, 3 Applewood, Park Gate, Southampton, SO31 7HQ
G8 ACQ R. Whattam, The Aviary No1, Arkwright Rd, Beds, MK44 1SE
G8 ADA J. Robinson, 7 Rhyl St., Liverpool, L8 6QL
G8 ADC J. Haile, 145 Dunstable Road, Caddington, Luton, LU1 4AN
G8 ADD B. Carter, 51 Smirrells Road, Birmingham, B28 0LA
GM8 ADK M. Ritchie, 11 Cromwell Road, Aberdeen, AB15 4UH
G8 ADQ J. Taylor, 21 Launcestone Close, Earley, Reading, RG6 5RY
G8 ADX E. Lawley, 3 Barnicott Close, Newton Ferrers, Plymouth, PL8 1BP
G8 ADY P. Harrison, 2 The Barns, Bridge End, Carlton, Bedfordshire, MK43 7LP
G8 ADZ N. Shepherd, 7 High St., Kelvedon, Colchester, CO5 9AG
G8 AEN P. Helm, 74 Neston Road, Walshaw, Bury, BL8 3DB
G8 AER J. Tanner, Merlins Mill, Toadsmoor Road, Stroud, GL5 2UG
G8 AEU J. Nightingale, 6 Aubrey Close, Chelmsford, CM1 4EJ
G8 AFA C. Atkins, 2 Eastlands, Yetminster, Sherborne, DT9 6NQ
G8 AFI P. Funnell, 25 Broadyates Road, Yardley, Birmingham, B25 8JF
G8 AFN P. Cleall, 139 Preston Grove, Yeovil, BA20 2DB
G8 AFQ T. Hall, 24 Church Street, Kelvedon, Colchester, CO5 9AH
GI8 AFS M. Granville, 33 Dunfield Terrace, Londonderry, BT47 2ES
G8 AFU P. Gilby, 191 Send Road, Send, Woking, GU23 7ET

G8 AGJ J. Evans, 1 Grosvenor Close, Hatch Warren, Basingstoke, RG22 4RQ
GM8 AGM M. Collar, Shoemakers Croft, Hatton, Peterhead, AB42 0TB
G8 AGN Dr B. Chambers, 5 The Ridge, Sheffield, S10 4LL
G8 AGQ A. Strong, 3 Ellorslie Drive, Stocksbridge, Sheffield, S36 2BB
GW8 AHB P. Swinbank, 13 Mundy Place, Cardiff, CF24 4BZ
G8 AHE L. Arnold, 402 Bournville Gardens, 49 Bristol Road South, Birmingham, B31 2FT
G8 AHN J. Barnes, 2 Mappins Road, Catcliffe, Rotherham, S60 5TH
G8 AHR P. Rushworth, 2 Aberdeen Close, Coventry, CV5 7NE
G8 AIE P. Willcocks, 27 Manor Road, Barnet, EN5 2LE
G8 AIM F. Tarver, 14 Southview Road, Leamington Spa, Cv32 7JD
G8 AIP M. Osment, Flat 2, Weavers Court, Shoreham-by-Sea, BN43 5ES
GI8 AIR W. Parkes, 15 Bushfoot Park, Portballintrae, Bushmills, BT57 8YX
GW8 AJA D. Hardy, 7 Coed Y Go Cottages, Coed Y Go, Oswestry, SY10 9AU
G8 AJM C. Payne, 14 Watts Lane, Louth, LN11 9DG
G8 AJP J. Eade, High Shaw Sandrock Hill, Sedlescombe, TN33 0QR
G8 AJZ R. Boardall, 9 Oxford Street, Bury, BL9 7EL
G8 AKA T. Wiltshire, Bramblings, Pelican Road, Tadley, RG26 3EL
G8 AKC C. Bell, Croftner, Mary Tavy, Tavistock, PL19 9QD
G8 AKE J. Warrington, 26 Lynton Road, Melton Mowbray, LE13 0NN
G8 AKF J. Ballantyne, Brookeside, Ashwellthorpe Road, Wreningham, Norwich, NR16 1AW
G8 AKL G. Ashcroft, Wood End House, Wood End, Bluntisham, Huntingdon, PE28 3LE
G8 AKM G. Roper, 19 Normay Rise, Newbury, RG14 6RY
G8 AKP P. Mcquade, The Old Swan, Holt Road Sharrington, Melton Constable, NR24 2PH
G8 AKQ S. Birkill, St. Anns, Ecclesall Road South, Sheffield, S11 9PX
G8 AKU B. Willson, Hilltop, Cryers Hill Road, Cryers Hill, High Wycombe, HP15 6LJ
G8 AKX M. Perry, 216 Marlpool Lane, Kidderminster, DY11 5DL
G8 ALD M. Lunt, 18 Longhurst Road, Hindley Green, Wigan, WN2 4PL
G8 ALE M. Brereton, Gleaston Water Mill, Gleaston, Ulverston, LA12 0QH
G8 ALQ A. Whitlock, 23 Daly Way, Aylesbury, HP20 1JW
G8 ALR J. Cull, 2 Drybrook Cottages, Amesbury Road, Cholderton, Salisbury, SP4 0ER
G8 ALS M. Stevenson, 15 Wall Hill Road, Allesley, Coventry, CV5 9EN
G8 AMC Amc Radio Club, c/o D. Millward, 77A Meadowcroft, St Albans, AL1 1UG
G8 AMD H. Bate, 88 Darnick Road, Sutton Coldfield, B73 6PG
G8 AMG M. Foster, 9 Norman Way, Irchester, Wellingborough, NN29 7AT
G8 AMJ D. Woolley, Tweddell'S Garth, West End, Leyburn, DL8 3HN
G8 AMK L. Parry, 13 Cannon Hill, Bracknell, RG12 7QA
G8 AMU C. Saveker, 23 Southlands Avenue, Horley, RH6 8BS
G8 ANN G. Townsend, 61 Richmond Park Road, London, SW14 8JU
G8 ANO D. Lawton, Grenehurst, Pinewood Road, High Wycombe, HP12 4DD
G8 ANT S. Holland, 14 The Vineries, Eastbourne, BN23 7TP
GD8 ANU C. Howard, 5 Ballure Grove, Ramsey, Isle Of Man, IM8 1NF
GM8 AOB J. Briscoe, 2 Peebles Place, Fort William, PH33 6UG
G8 AOE B. Duffell, 7 Potto Close, Yarm, TS15 9RZ
G8 AOG M. Browne, 143 Thatch Leach Lane, Whitefield, Manchester, M45 6EP
G8 AOI T. Knight, 3 Eaton Close, Rainworth, Mansfield, NG21 0AR
G8 AOJ G. Smith, Forest View Cottage, Gorsty Knoll, Coleford, GL16 7LR
G8 AOK A. Porch, 17 Purcell Close, Brighton Hill, Basingstoke, RG22 4EL
G8 AOO B. Hills, 3 Frithmead Close, Basingstoke, RG21 3JW
G8 AOZ P. Hughes, 247 High Greave, Sheffield, S5 9GS
G8 APB C. Plummer, Barley House Farm, Newtown, Stoke-on-Trent, ST8 7SW
G8 APF D. Chaplin, 4 Blenheim Close, Loughborough, LE11 4SA
G8 APL G. Parsons, 21 Wild Ridings, Fareham, PO14 3BS
G8 APM G. White, 1 Drakes Close, Hythe, Southampton, SO45 5BP
G8 APW D. Taylor, 87 Grasmere Road, Chester le Street, DH2 3EU
G8 APY J. Bond, Folly House, The Reddings, Cheltenham, GL51 6RL
G8 APZ S. Lucas, 84 Woodman Road, Warley, Brentwood, CM14 5AZ
G8 AQA P. Nickalls, Holy Mill, Longville, Much Wenlock, TF13 6ED
G8 AQB M. Ballance, 24 Western Road, Wolverton, Milton Keynes, MK12 5BE
G8 AQH R. Hine, 147/149 Bolton Hall Road Bolton Woods, Bradford, BD2 1BQ
G8 AQN A. Hibberd, 20 Barby Lane, Rugby, CV22 5QJ
G8 AQO A. Copperwaite, 71 Gladbeck Way, Enfield, EN2 7EL
G8 AQP S. Warner, 14 Andrews Way, Aylesbury, HP19 8WA
G8 ARA B. King, 15 Newstead Road, West Southbourne, Bournemouth, BH6 3HJ
GW8 ARC Dr A. Craggs, 15 Pen-Y-Groes Avenue, Cardiff, CF14 4SP
G8 ARF L. Thompson, 44 Tillmouth Avenue, Holywell, Whitley Bay, NE25 0NP
G8 ARH N. Blackmore, 35 Weyhill Gardens, Weyhill, Andover, SP11 0QT
G8 ARM B. Pickrell, Perrans, Ludgvan, Penzance, TR20 8AJ
GW8 ARR P. Edwards, Trevland, Felindre, Knighton, LD7 1YL
GW8 ASA G. Wyatt, 3 Creidiol Road, Mayhill, Swansea, SA1 6TZ
G8 ASC P. Richards, 134 Downhills Park Road, Tottenham, London, N17 6BP
GW8 ASD A. Pugh, Willcroft, Mold Road, Gwersyllt, LL11 4AF
G8 ASG M. Farrell, Hobberley House, Hobberley Lane, Leeds, LS17 8LX
G8 ASJ G. Swan, Morogar, Post Office Lane Kempsey, Worcester, WR5 3NX
G8 ASP I. Gurton, 28 Bloomfield Road, Harpenden, AL5 4DB
G8 ASV D. Skinner, Latch Cottage, Nursery Lane, Blackboys, Uckfield, TN22 4EU
G8 ASW R. Warrender, 102 Turnberry Road, Great Barr, Birmingham, B42 2HT
G8 ASX A. Hoggan, 25 Clingan Road, Bournemouth, BH6 5PY
GM8 AT W. Beattie, Alastrean House, Tarland, Aboyne, AB34 4TA
G8 ATC Dr R. Gayton, 20 Barton Close, Exton, Exeter, EX3 0PE
G8 ATD A. Barter, 503 Northdown Road, Margate, CT9 3HD
G8 ATE R. Turlington, 2 Laithwaite Close, Leicester, LE4 1BX
G8 ATG M. Williamson, 120 Warbreck Hill Road, Blackpool, FY2 0TR
G8 ATK M. Hearsey, Halycon, Lawday Link, Farnham, GU9 0BS
G8 ATL M. Lankester, 154 Gorse Lane, Clacton-on-Sea, CO15 4RJ
G8 ATP K. Mintern, 71 Crafts End, Chilton, Didcot, OX11 0SB
G8 AUL P. Buck, 41 Marion St., Brighouse, HD6 2BJ
G8 AUN R. Chiddick, 87 Aylsham Road, Norwich, NR3 2HW
G8 AUU C. Partridge, 6 Blagdon Walk, Teddington, TW11 9LN
G8 AVB C. Dickson, 5 Arrow View, Ledbury, HR8 2FR
G8 AVC R. Evans, Mansfield, 1 Horsehead Lane, Chesterfield, S44 6HU
G8 AVK R. Kimberley, 8 Nutwell Road, Weston-Super-Mare, BS22 6EN
GM8 AVM I. Macdonald, Benvoir, Lightlands Avenue, Wigtown, Newton Stewart, DG8 9EE
G8 AVO J. Wainwright, 33 Station View, Nantwich, CW5 7BJ
G8 AVQ J. Florentin, 17 Campden Hill Gardens, London, W8 7AX
G8 AVZ M. Keeping, 8 Calderdale Close, Southgate West, Crawley, RH11 8SQ
G8 AWB R. Lawrence, 16 Westover Road, Callington, PL17 7HD
G8 AWE M. Wellspring, 21 Rue De La Gendarmerie, Aigre, France, 16140
G8 AWI C. Smith, 129 Earls Road, Nuneaton, CV11 5HP
GW8 AWM F. Evans, Ty Cryr, Chepstow Road, Usk, NP15 1HN
G8 AWN B. Procter, 28 Holme Grove, Burley In Wharfedale, Ilkley, LS29 7QB
G8 AWY J. Ward, 44 Rugby Road, Barby, CV238UB
G8 AXN C. Amery, 9 View Close, Biggin Hill, Westerham, TN16 3XE
G8 AXO A. Nunn, 9 Elmhurst Court, Hamblin Road, Woodbridge, IP12 1HB
G8 AXR R. Moore, 22 Cardan Drive, Ilkley, LS29 8PH
G8 AXV K. Shail, Veeda Glenta, Blackmore Park Road, Malvern, WR13 6NN
G8 AYC N. Walker, 36 Meyrick Drive, Wash Common, Newbury, RG14 6SX
G8 AYJ J. Hanson, 22 Church Way, Falmouth, TR11 4SG
G8 AYM N. Pritchard, 108 Kynaston Avenue, Aylesbury, HP21 9DS
G8 AYV J. Lewis, Newnham House, Shurton, Bridgwater, TA5 1QG
G8 AZB S. Smith, 11 Grayshott Laurels, Lindford, Bordon, GU35 0QB
G8 AZM D. Johnson, 195 Staplers Road, Newport, PO30 2DP
G8 AZN R. Barnes, 18 Battle Road, Tewkesbury, GL20 5TZ
G8 AZR J. Dimmock, 93 Barton Road, Harlington, Dunstable, LU5 6LG
G8 AZT J. Jones, 9 Queens Walk, Thornbury, Bristol, BS35 1SR
G8 AZZ G. Craddock, Westlands, Bowden Hill, Yealmpton, Plymouth, PL8 2JX
G8 BAD D. Donati, 53 Smithbarn, Horsham, RH13 6DT
G8 BAG T. Rowley, 7 Hall Farm Close, Castle Donington, Derby, DE74 2NG
G8 BAJ P. Southby, 51 Teddington Park, Teddington, TW11 8DE
G8 BAK P. Knight, 4 Dimmock Road, Wootton, Bedford, MK43 9DW
G8 BAL C. Robinson, 17 Fairview Close Hythe, Southampton, SO45 5EX
G8 BAQ B. Kneller, Mystic Flight, Brackenhill Road, Eastlound, Doncaster, DN9 2LR
G8 BAS D. Gardiner, 31 Alexander Drive, Cirencester, GL7 1UG
G8 BAZ P. Talbot, 19 Bladen Valley, Briantspuddle, Dorchester, DT2 7HP
G8 BBK R. Nelson, 10 Wragg Drive, Newmarket, CB8 7SD
G8 BBV J. Goulty, 1 Larksway, Felixstowe, IP11 2PN
G8 BBZ P. Barker, 3 Hudson Fold, Heptonstall, Hebden Bridge, HX7 7PH
G8 BCA R. Chambers, 11 Thetford Road, Mildenhall, Bury St Edmunds, IP28 7HX
G8 BCF G. Podmore, Crownfield, Kings Lane, Faringdon, SN7 7SS
G8 BCG P. Taylor, The Byre Coombe Farm, St. Keyne, Liskeard, PL14 4RS
G8 BCI E. Rowlands, Wychanger Cottage, Luccombe, Minehead, TA24 8TA
G8 BCJ A. Unsworth, Meadow View, Clockhouse Lane, North Stifford Rm165Ur., Grays, RM16 5UR
GW8 BCL H. Bottomley, Llwyn Y Berllan, Battle, Nr Brecon, LD3 9RN
G8 BCO C. Boys, 34 Firacre Road, Ash Vale, Aldershot, GU125JT
G8 BDF J. Hanney, 16 Parsonage Barn Lane, Ringwood, BH24 1PX
G8 BDM J. Adams, 1 Powell Close, Creech St. Michael, Taunton, TA3 5TE
G8 BDQ G. Hedley, 260E 100 Sts, Raymond, Alberta, Canada, TOK 250
GM8 BDX A. Scott, 20 Treaty Park, Birgham, Coldstream, TD12 4NG
G8 BDZ K. Cowdell, 6 Pearl Street, Bristol, BS3 3EA
G8 BEH D. Hill, Care Uk, Prince George House, 102 Mansbrook Boulevard, Ipswich, IP3 9GY
G8 BEK C. Dunn, 75 Waddington Avenue, Burnley, BB10 4LA
G8 BEQ K. Greenough, 2 Bexley Close, Glossop, SK13 7BG
G8 BFA S. Davis, 21 Cordville Close, Chaddesden, Derby, DE21 6WX
G8 BFC P. Johnson, 15 Elvaston Lane, Alvaston, Derby, DE24 0PX
G8 BFH J. Marriott, 104 Whinbush Road, Hitchin, SG5 1PN
G8 BFK S. Ballard, 26 Crafts End, Chilton, Didcot, OX11 0SA
G8 BFL B. Jayne, 38 Townfields, Lichfield, WS13 8AA
G8 BFM A. Whittaker, 6 Kingsbridge Way, Bramcote, Nottingham, NG9 3LW
GW8 BFO R. Hayter, Glanyrafon, Talywern, Machynlleth, SY20 8NY
G8 BFV D. Edwards, 34 Campkin Road, Wells, BA52DG
G8 BGI B. Hepburn, 52 Hibiscus Grove, Bordon, GU35 0XA
G8 BGL R. Gilliatt, 21 Main St., Thorpe On The Hill, Lincoln, LN6 9BG
G8 BGM M. Lee, 32 Fernham Road, Faringdon, SN7 7LB
G8 BGT A. Dermont, 7 Pool Close, Little Comberton, Pershore, WR10 3EL
G8 BGV P. Selwood, 43 Keene Way, Galleywood, Chelmsford, CM2 8NT
G8 BHC J. Richmond-Hardy, 45 Burnt House Lane, Kirton, Ipswich, IP10 0PZ
G8 BHE N. Gutteridge, 68 Max Road, Quinton, Birmingham, B32 1LB
G8 BHK J. Vickers, 242B High Road, Trimley St. Martin, Felixstowe, IP11 0RG
GM8 BHR G. Pearson, 2 Hamilton Terrace, Edinburgh, EH15 1NB
G8 BHX M. Berry111, 27 Greenway Road, Heald Green, Cheadle, SK8 3NR
G8 BHY A. Heath, 7 Coral Close, Coventry, CV5 7AD
G8 BIG M. Stebbings, 15 St. Helena Way, Horsford, Norwich, NR10 3EA
G8 BIH J. Akam, 10 Apple Tree Road, Alderholt, Fordingbridge, SP6 3EW
G8 BII B. Hunt, 53 The Sands, Milton-Under-Wychwood, Chipping Norton, OX7 6ER
G8 BIR R. Harris, 35 Freemantle Road, Eastville, Bristol, BS5 6SY
G8 BIS P. Lyon, Frogs Hall, Cannon Street, New Romney, TN28 8BJ
G8 BIW R. Booth, 16 Darwynn Avenue, Swinton, Mexborough, S64 8DU
G8 BIX A. Parcell, Birdies Barn, Minions, Liskeard, PL14 5LE
G8 BJA D. Couchy, 8 Chapel St., Wincham, Northwich, CW9 6DA
G8 BJB G. King, 62 Heathfield Road, Sholing, Southampton, SO19 1DP
GM8 BJF Dr B. Flynn, 15 Riselaw Crescent, Edinburgh, EH10 6HN
GM8 BJJ A. Morton, 4 Mountstuart St., Millport, KA28 0DP
G8 BJO J. Barfoot, 21 Richard Crampton Road, Beccles, NR34 9HN
G8 BJQ L. Case, 58 Brookdale, Widnes, WA8 4TB
G8 BKD P. Scotney, 30 Trinity Road, Rothwell, NN14 6HY
G8 BKE C. Towns, 21 Seafield Close, Barton On Sea, New Milton, BH25 7HR
G8 BKG D. Wright, 61 Potton Road, St Neots, PE19 2NN
G8 BKH G. Shepherd, 64 Dawley Road, Arleston, Telford, TF1 2JF
G8 BKL E. Danks, 18 Lichfield Street, Stourport-on-Severn, DY13 9EU
G8 BKQ C. Clark, 21A Headland Park Road, Paignton, TQ3 2EN
G8 BLB P. Blakeney, 45 Hampden Avenue, Chesham, HP5 2HL
G8 BLD J. Draper, 11 Parva Close, Little Barningham, Norwich, NR11 7NJ
G8 BLK M. Keightley, 20 Longrood Road, Rugby, CV22 7RG
G8 BLP C. Bond, 5 Rushley Close, Sheffield, S17 3EG
G8 BME F. Burrow, 51 Stanhope Avenue, Morecambe, LA3 3AJ
G8 BMH J. Parry, 29 Heath Road Upton, Chester, CH2 1HT
G8 BMI G. Theasby, 115 Bevercotes Road, Sheffield, S5 6HB
G8 BMP M. Taylor, 96 Woodhouses Road, Burntwood, WS7 9EJ
G8 BMQ B. Cedar, 29 Velsheda Court, Hythe Marina Village, Southampton, SO45

6DW
G8 BMZ P. Cowling, 94 Welholme Road, Grimsby, DN32 0NG
G8 BNB R. Gibbs, 15 Gosford Hill Court, Bicester Road, Kidlington, OX5 2XP
GI8 BNC J. Mccann, 61 Glengawna Road, Glengawna, Omagh, BT79 7WJ
G8 BNE R. Kendall, Random Stones, Arkendale Road, Knaresborough, HG5 0QA
G8 BNG A. Green, 37 Bramcote Lane, Nottingham, NG8 2NA
GM8 BNH I. Gall, Cluaran, Bridge Of Don, Aberdeen, AB23 8BD
G8 BNK P. Banbury, 16 Gloucester Road, Whitstable, CT5 2DS
G8 BNR R. Wells, 279 Hatfield Road, St Albans, AL4 0DH
G8 BOB A. Robinson, 29 Thomas Manning Road, Diss, IP22 4HL
G8 BOI M. Simpson, 9 Brock House, 2 Batter Street, Plymouth, PL4 0EF
G8 BOJ K. Agombar, 54 Julien Road, London, W5 4XA
G8 BOP M. Palmer, 109 Longfellow Road, Dudley, DY3 3EF
G8 BOQ K. Phillips, 1140 Riverberry Drive, Reno Nv, USA, 89509
G8 BOS B. Saunders, 88 Bramwoods Road, Chelmsford, CM2 7LT
G8 BPH J. Rome, 1 Bridge Cottages, Downhall Road, Hatfield Heath, CM22 7AS
G8 BPN G. Wilkerson, Hill House, Newton, Leominster, HR6 0PF
G8 BPQ J. Wiseman, 147 Hilton Road, Nottingham, NG3 6AR
G8 BPS C. Booth, 11 High St., Haxey, Doncaster, DN9 2HX
G8 BPU H. Skelhorn, 9 Moss Lane, Bollington, Macclesfield, SK10 5HJ
G8 BPW A. Stoker, 35A Church End Lane, Runwell, Wickford, SS11 7JE
G8 BPY P. Hollis, 5 Salisbury Road, New Malden, KT3 3HZ
G8 BQF A. Dixon, 2 Yorkdale Drive, Hambleton, Selby, YO8 9YB
G8 BQH M. Marsden, Hunters Moon, Buckingham Road Hardwick, Aylesbury, HP22 4EF
GW8 BQK G. Oatway, 21 Victoria Park, Colwyn Bay, LL29 7AX
G8 BQT I. Hudson, Flat 32, Three Crowns House, King's Lynn, PE30 5DT
G8 BQZ P. Plunkett, 30 Broadlands Avenue, Shepperton, TW17 9DQ
G8 BRD Dr C. Dawson, 33 Rough Common Road, Rough Common, Canterbury, CT2 9DL
G8 BRG P. Mitchell, 3 Goodwin Court, Farnsfield, Newark, NG22 8LU
G8 BRL B. Ward, 10 Upper Moorfield Road, Woodbridge, IP12 4JW
G8 BRU G. Gallamore, 30 Orchard Avenue, Partington, Manchester, M31 4DL
G8 BSD J. Ceresole, 7 Stokes Bay Home Park, Stokes Bay Road, Gosport, PO12 2QU
G8 BSP A. Wicks, 1 Castle Hill Close, Shaftesbury, SP7 8LQ
GM8 BSQ A. Shepherd, 2 Westwood Place, Skene, Westhill, AB32 6WS
GM8 BSU A. Weller, 18 Froghall Road, Aberdeen, AB24 3JL
G8 BTC B. Fenwick, 16 Pine Walk, Uckfield, TN22 1TU
G8 BTD P. Sladen, 2 Burlea Close, Crewe, CW2 8SZ
G8 BTL H. Futcher, Sarum, 12 Thursby Road, Woking, GU21 3NZ
G8 BTU J. Dowson, The Granary, St. Peters Road, Arnesby, Leicester, LE8 5WJ
G8 BTV P. Marlow, 1 Vineries Close, Leckhampton, Cheltenham, GL53 0NU
GW8 BTX T. Storeton-West, Tan Y Banc, Blaenpennal, Aberystwyth, SY23 4TT
G8 BTY M. Dennis, Thistledown, Yallands Hill, Taunton, TA2 8NA
G8 BUB B. Goodall, 10 Westoby Close, Shepshed, Loughborough, LE12 9SS
GD8 BUE I. Rae, 65 Lezayre Park, Ramsey, IM8 2PT
G8 BUF M. Higgins, 59 Clinton Crescent, Ilford, IG6 3AH
G8 BUI Dr C. Nowikow, 10 Windmill Road, Whitstable, CT5 4NL
G8 BUV C. Chapman, 6 Pickhurst Green, Hayes, Bromley, BR2 7QT
G8 BUX Buxton Radio Amateurs, c/o D. Carson, 21 Harris Road, Harpur Hill, Buxton, SK17 9JS
G8 BUZ J. Paine, 1 Elm Close, London, SW20 9HX
G8 BVB P. Power, 8 The Fairway, Camberley, GU15 1EF
G8 BVF J. Wearing, 122 Dixon Drive, Chelford, Macclesfield, SK11 9BX
G8 BVL M. Porter, Birklands, 16 The Oval, Scarborough, YO11 3AP
G8 BVQ R. Straker, 26 Constance Crescent, Hayes, Bromley, BR2 7QJ
G8 BVR G. Oddy, 2 Manor Farm, Chard, TA20 2EB
G8 BVU P. Reilly, 19 Maunders Court, Liverpool, L23 9YU
G8 BVY G. Spinks, 40 Ferndale Avenue Walthamstow, London, E17 9EH
G8 BWA M. Pollard, 3 Highfield Road, Chertsey, KT16 8BU
G8 BWH R. Robinson, 1 John Dixon Lane, Darlington, DL1 1HG
G8 BWK J. Harper, 2 Wolves Mere Woolmer Green, Knebworth, SG3 6JW
G8 BWP C. Jones, 2 Windmill Crescent, Wolverhampton, WV3 8HY
GW8 BWX A. Hancock, 38 High Street, Pontycymer, Bridgend, CF32 8HY
G8 BXA A. Nicol, 18 Lower End, Swaffham Prior, Cambridge, CB25 0HT
G8 BXC R. Clark, 41 Avenue Road, Bexleyheath, DA7 4
G8 BXD Dr R. Edgecombe, 48 Birchwood Road, Woolaston, Lydney, GL15 6PE
G8 BXH J. Pryke, 52 Oaklands Avenue, Watford, WD19 4LW
G8 BXJ A. Pullen, 22700 Gault Street, West Hills, Ca, USA, 91307-2306
G8 BXO J. Stacey, 3 West Park, South Molton, EX36 4HJ
G8 BXQ T. Hordley, 9 Newtown, Charlton Marshall, Blandford Forum, DT11 9NN
G8 BYB A. Hebden, 1 Ringwood Avenue, Newbold, Chesterfield, S41 8RA
G8 BYC C. Keen, Brighton Road, Radio Relay, Lewes, BN7 3JL
G8 BYI R. Burrows, 76 Southfield, Southwick, Trowbridge, BA14 9PW
G8 BZJ A. Matheson, 1 St. Edmunds Close, Bromeswell, Woodbridge, IP12 2PL
G8 BZL G. Lindsay, 4 Downs View, Hove, BN3 8EN
GW8 BZN D. Goadby, Ty Mawr, Bryncroes, Pwllheli, LL53 8EH
GM8 BZP D. Joiner, 8 Damask Crescent, Newmachar, Aberdeen, AB21 0NG
G8 BZR P. Clark, 10 Chez Gueunie, St Leger Magnazeix, France, 87190
G8 BZT D. Allen, 156 Middlecotes, Tile Hill, Coventry, CV4 9AZ
G8 CA Axe Vale ARC, c/o P. Cross, Balls Farm Cottage, Musbury Road, Axminster, EX13 8TT
G8 CAA C. Broomfield, 8 Woodview Crescent, Hildenborough, Tonbridge, TN11 9HD
G8 CAB J. Sawford, 68 Harlyn Drive, Pinner, HA5 2DA
G8 CAF R. Price, Flat 3, Dippons House Dippons Drive, Wolverhampton, WV6 8HJ
G8 CAH A. Parsons, 153 Denman Drive, Ashford, TW15 2AP
GW8 CAK P. Kenyon, The Elvins, Norton, Presteigne, LD8 2EP
G8 CAM I. Foster, 22 Margetts Place, Lower Upnor, Rochester, ME2 4XF
G8 CAU J. Borradaile, 25 Inglewood Crescent, Carlisle, CA2 6JJ
G8 CAV C. Isenman, Bracklinn House, Broadlands Road, Brockenhurst, SO42 7PB
G8 CBA G. Tipler, Scotts House, Chorley, Bridgnorth, WV16 6PR
G8 CBE K. Quarman, 127 Highfield Lane, Hemel Hempstead, HP2 5JG
G8 CBO K. Smith, 6 Hermitage Close, North Mundham, Chichester, PO20 1JZ
G8 CBU R. Aldous, 23 Aldhous Close, Luton, LU3 2LZ
G8 CCD J. Hodge, 71 Rawcliffe Road Walton, Liverpool, L9 1AN
G8 CCF S. Hall, Knackershole Barn, Dulverton, TA22 9RU
G8 CCJ D. Petri, 42 Lucas Road, Snodland, ME6 5PY
G8 CCL J. White, 22 Millfields Station Road, Burnham-on-Crouch, CM0 8HS
G8 CCN R. Read, 76 School Road, Downham, Billericay, CM11 1QN
G8 CCO J. Hess, 3 Havana Court, Eastbourne, BN23 5UH
G8 CCV M. O'Donnell, 40 Mercers Drive Bradville, Milton Keynes, MK13 7AY
G8 CDA M. Richards, Copperknobs, High Street, Stockbridge, SO20 6HE
G8 CDB P. Strudwick, 20 New Road, Broomfield, CM1 7AN
G8 CDC Capt. P. Jones, March House, Burnthurst Lane, Princethorpe, Rugby, CV23 9QA
G8 CDD R. Leman, Crundalls Farmhouse Gedges Hill Matfield, Tonbridge, TN12 7EA
G8 CDG N. Broadbent, 2 Market Hill, Clare, Nr Sudbury, CO10 8NN
G8 CDV T. Jeacock, 9 Parkwood Rise, Barnby Dun, Doncaster, DN3 1LY
GM8 CEA R. Spencer, Pitagown House, Cluny, Newtonmore, PH20 1BS
G8 CEE C. Carr, 6 Jervaulx Road, Morton On Swale, Northallerton, DL7 9RA
G8 CEP D. Clough, 165 Pilgrims Way, Andover, SP10 5HT
G8 CET W. Marsden, 163 Buxton Old Road, Disley, Stockport, SK12 2AY
G8 CEX B. Turner, 50 Bosworth Road, Leigh-on-Sea, SS9 5AB
GJ8 CEY A. Hearne, Hearnes Hastle, 2 Teighmore Park, La Chevre Rue, Grouville, Jersey, JE3 9EF
G8 CEZ R. Fuller, 35 Chichester Walk, Merley, Wimborne, BH21 1SL
G8 CFD R. Rimmer, 6 The Dene, Blackburn, BB2 7QS
G8 CGM P. Raybould, 115 Curlew Crescent, Bedford, MK41 7HY
G8 CGW J. Elliott, 92 Hinckley Road, Barwell, Leicester, LE9 8DN
G8 CHA N. Blackburn, 158 Dyas Road, Great Barr, Birmingham, B44 8SW
G8 CHC B. King, 32 Mayfield, Buckden, St Neots, PE19 5SZ
G8 CHI A. Tidder, 3 Fernway Close, Wimborne, BH21 2ST
G8 CHK R. King, 28 Jenkinson Road, Towcester, NN12 6AW
G8 CHN G. Barber, 666 Bradford Road, Birkenshaw, Bradford, BD11 2EE
G8 CHO S. Humm, 235 Felmongers, Harlow, CM20 3DP
G8 CHY K. Twort, 39 Mile End Lane, Stockport, SK2 6BN
GM8 CIF D. Macdonald, 22 Drummie Road, Devonside, Tillicoultry, FK13 6HT
G8 CIG P. Tester, Gable Crest, Longburton, Sherborne, DT9 5PD
G8 CIJ F. Fyfe, 28 Whitton Close, Greatworth, Banbury, OX17 2EH
G8 CIT W. Mckillop, 2 Moores Green, Wokingham, RG40 1QG
G8 CIX M. Maynard, 41 Liverpool Avenue, The Pyramid, Southport, PR8 3NP
G8 CJA Dr M. Dowson, The Granary, St. Peters Road, Leicester, LE8 5WJ
G8 CJD C. Hutton, 25 Fiddlers Lane, East Bergholt, Colchester, CO7 6SJ
GM8 CJG R. Kirsch, Milntack House Laurieston, Castle Douglas, DG72PW
G8 CJH D. Fletcher, 17 Durley Chine Road South, Bournemouth, BH2 5JT
G8 CJL A. Dorling, 4 The Pastures, Rushmere St. Andrew, Ipswich, IP4 5UQ
G8 CJM A. Croft, 15 Blenheim Avenue, Chatham, ME4 6UU
G8 CJQ R. Barnes, 3 Ivy Cottages, Church Lane, Knutsford, WA16 7RD
G8 CJT C. Coles, 15 Somerdale Avenue, Bath, BA2 2PG
GM8 CJW J. West Of Stow, Stow Mill, Stow, Galashiels, TD1 2RB
G8 CKB P. Ebsworth, Olamyra 20, Forland, Steinsland, Norway, 5379
GW8 CKJ A. Williams, 54 St. Augustine Road, Griffithstown, Pontypool, NP4 5EZ
G8 CKK A. Zerafa, 2 Furnwood, St. George, Bristol, BS5 8ST
G8 CKN R. Powers, The Dell, Hussell Lane, Alton, GU34 5PF
G8 CKS J. Sargent, The Coach House, Speltham Hill, Waterlooville, PO7 4RU
G8 CKV S. Dale, 30 Almond Road, Peterborough, PE1 4LT
G8 CLI D. Hall, 9 Fairfax Close Barford, Warwick, CV35 8ER
G8 CLJ I. Richmond, 58 Cauldron Barn Road, Swanage, BH19 1QF
G8 CLK K. Woollven, 7 Heatherstone Avenue, Dibden Purlieu, Southampton, SO45 4LR
G8 CLW J. Griffin, 185 Eastcote Avenue, West Molesey, KT8 2EX
G8 CLY J. Lythgoe, 18 Ranleigh Walk, Harpenden, AL5 1SR
G8 CLZ QRZ ARG of Sussex, c/o J. Eade, High Shaw Sandrock Hill, Sedlescombe, TN33 0QR
G8 CMD A. Ashford, 56 Guarlford Road, Malvern, WR14 3QP
G8 CMG R. Williams, 18 Woodford Crescent, Plymouth, PL7 4QY
G8 CMK W. Blankley, 16 Charles Road, St Leonards-on-Sea, TN38 0QA
G8 CMO R. Grounds, 101 Honeysuckle Way, Witham, CM8 2XQ
G8 CMP C. Heymans, 10 Rushmore Drive, Widnes, WA8 9QB
G8 CMU M. Adcock, Rudhall Farm, Phocle Green, Ross-on-Wye, HR9 7TL
GW8 CNF S. Biddiscombe, 20 Arlington Close, Malpas, Newport, NP20 6QF
GW8 CNS W. Mathias, Grenan Bungalow, Highland Avenue, Bridgend, CF32 9YH
G8 CON J. Beith, 18 Avenue Road, New Milton, BH25 5JP
G8 COR G. Peters, 156 Preston Road, Whittle-Le-Woods, Chorley, PR6 7HE
G8 CPA J. Vizor, 31 Somerset Road, Swindon, SN2 1NE
G8 CPF M. Edwards, 14 Cheyney Walk, Westbury, BA13 3UH
G8 CPJ I. Lever, 23 Anton Road, Andover, SP10 2EN
G8 CPK FL D. Hibbin, 95A Thorpe Acre Road, Loughborough, LE11 4LF
G8 CPM C. Mortlock, 27 Baldwin Road, Greatstone, New Romney, TN28 8SY
G8 CPN J. Hawkins, Westhay Farm, Higher Clovelly, Bideford, EX39 5SH
G8 CPQ V. Humphrey, 5 Wistow Road, Luton, LU3 2UR
G8 CQG P. Cornell, 1 Orient Drive, Winchester, SO22 6NZ
G8 CQH Dr P. Best, 21 Greening Drive, Edgbaston, Birmingham, B15 2XA
G8 CQQ A. Paterson, 36 Bracadale Road, Nottingham, NG5 5EE
G8 CQV W. Hunter, 2 Green Acre, Goosnargh, Preston, PR3 2BQ
G8 CQX J. Hawes, 193 Leckhampton Road, Cheltenham, GL530AD
G8 CQZ C. Powlesland, The Ferns, Broad Street, Gloucester, GL19 3BN
G8 CRB S. Blunt, 53 Butt Lane Milton, Cambridge, CB24 6DG
G8 CRC C. Callegari, 16 Rustington Court, St. Johns Road, Eastbourne, BN20 7HS
GW8 CRH I. Troughton, Rhiwbina, Pentre Lane, Cwmbran, NP44 3AP
G8 CRM P. Watson, Tall Oak, 6 New Road, Bury St Edmunds, IP29 5QL
G8 CRV J. Christian, 5 Towers Way Corfe Mullen, Wimborne, BH21 3UA
G8 CRX S. Winford, Mayflower, South Hanningfield Road, Chelmsford, CM3 8HJ
G8 CRZ P. Hunt, 17 Selfridge Avenue, Southbourne, Bournemouth, BH6 4NB
G8 CSA Silverthorn Radio Club, c/o L. Butterfields, 22 Horsley Road, London, E4 7HX
GM8 CSE H. Hogarth, 32 Broomhall Park, Edinburgh, EH12 7PU
G8 CSK S. Browning, 12 Sunderland Close, Woodley, Reading, RG5 4XR
G8 CSQ P. Benson, Ashbank Bungalow, Bentham, Lancaster, LA2 7HX
G8 CSR J. Credland, Lieu-Dit Cornier, Prayssas, France, 47360
G8 CTB K. Chambers, 24 Primrose Close, Flitwick, Bedford, MK45 1PJ
G8 CTD A. Tait, Birch Glen, 71 Twemlows Avenue, Whitchurch, SY13 2HD
G8 CTJ M. Maxey, 28 Herald Way, Burbage, Hinckley, LE10 2NX
G8 CTR Dr D. Upton, Polwin, Budock Water, Falmouth, TR11 5DT
G8 CTX C. Havercroft, 28 Anglers Way, Cambridge, CB4 1TZ
G8 CUA R. Boittier, 5 The Crescent, Harlow, CM17 0HN

G8 CUB R. Ray, Little Mallards, Mallard Way, Brentwood, CM13 2NF
G8 CUG P. Cockram, 14 Langshott Close, Woodham, Addlestone, KT15 3SE
G8 CUL M. Stevens, 67 New Road, East Hagbourne, Didcot, OX11 9JX
G8 CUN G. Rawlings, 109 The Upway, Basildon, SS14 2JD
G8 CUW S. Thackery, 19 Pyne Point, Clevedon, BS21 7RL
G8 CUX D. Stanton, 106 Scrapsgate Road, Minster On Sea, Sheerness, ME12 2DJ
G8 CVF P. Dobson, 3 Wallingford Road, Wirral, CH49 6PW
GM8 CVN J. Struthers, 79 Woodfield Park, Colinton, Edinburgh, EH13 0RA
G8 CVP R. Perry, 49 Harwich Road, Little Clacton, Clacton-on-Sea, CO16 9NE
G8 CVQ A. Parr, 8 Kingston Avenue, North Cheam, Sutton, SM3 9TZ
G8 CVS J. Jenkinson, 26 Blenheim Drive, Oxford, OX2 8DG
G8 CW Essex CW Contest Club, c/o P. Tittensor, 47 St. Johns Road, Chelmsford, CM2 0TY
G8 CWE T. Cook, 141 Station Road, Watlington, King's Lynn, PE33 0JG
G8 CWJ J. Abbott, 20 Highbury Avenue, Salisbury, SP2 7EX
G8 CWQ G. Horsfall, Lancaster New Road, Garstang, Preston, PR3 1AD
G8 CXA D. Froggatt, 2 Cobden Avenue, Mexborough, S64 0AD
G8 CXF J. Lucas, 48 Sycamore Drive, Ash Vale, Aldershot, GU12 5PR
G8 CXI D. Phillips, 13 Bowford Avenue, Bexleyheath, DA7 4ST
G8 CXK G. Peck, 45 Bentley Close, Northampton, NN3 5JS
G8 CXT D. Coxhill, 82 Williams Close, Hanslope, Milton Keynes, MK19 7BT
G8 CXV R. Brown, 19C Arlington Drive, Mapperley Park, Nottingham, NG3 5EN
G8 CXW P. Appleby, 23 Oban Drive, Ashton-In-Makerfield, Wigan, WN4 0SJ
G8 CXZ M. Mills, 145 Park St., Haydock, St Helens, WA11 0BL
G8 CYA N. Parker, 10 Lockhart Close, Kenilworth, CV8 1RB
G8 CYE S. Cook, 24 Beaufort Court, Beaufort Road, Richmond, TW10 7YG
G8 CYF M. Bucknall, Driftaway, White Oak Green, Hailey, Witney, OX29 9XP
G8 CYG W. Steer, Downside, Membury, Axminster, EX13 7AF
G8 CYK W. Poel, Hockham Hill, Spring Elms Lane, Little Baddow, Chelmsford, CM3 4SD
G8 CYL P. Smith, Andelain, Drift Road Whitehill, Bordon, GU35 9DZ
G8 CYT F. White, 12 Burcombe Road, Bournemouth, BH10 5JT
G8 CYU P. York-Jones, 18 Solway Road, Cheltenham, GL51 0LZ
G8 CYW S. Wisher, 17 Kenmore Crescent Greenside, Ryton, NE40 4QY
G8 CYX D. Storey, 43 Harwood Close, Welwyn Garden City, AL8 7ST
G8 CZE F. Beesley, 9 Northway, Droylsden, Manchester, M43 6EF
G8 CZG Lord D. Bell, 25A Mill Lane, Great Harwood, Blackburn, BB6 7UQ
G8 CZI D. Paterson, 3 Shawcroft Close, Shaw, Oldham, OL2 7DA
G8 CZJ J. Meredith, 25 Frankel Avenue, Redhouse, Swindon, SN25 2NJ
G8 CZM K. Jones, 3 Webb Avenue, Perton, Wolverhampton, WV6 7YH
G8 CZP V. Maund, 24 Elliott Crescent, Bedford, MK41 0HL
G8 CZQ I. Bayliss, West Common Lodge, West Common Close, Gerrards Cross, SL9 7QR
GM8 CZU I. Davidson, 3 Hillcrest Avenue, Kirkcaldy, KY2 5TU
G8 DAI A. Justin, Garth, Park View Road, Pinner, HA5 3YF
G8 DAM D. Goodway, 35 South Avenue, Buxton, SK17 6NQ
G8 DBD R. Taylor, 54 Portsmouth Road, Lee-on-the-Solent, PO13 9AG
G8 DBH C. Wallwork, Honeywicke Cottage, Honeywick Lane, Dunstable, LU6 2BJ
G8 DBK P. Barker, 24 Main Street, South Croxton, Leicester, LE7 3RJ
G8 DBO K. Smith, Wilson Hall Farm, Slade Lane, Wilson, Derby, DE73 8AG
G8 DBP J. Mills, 93 Gays Road, Hanham, Bristol, BS15 3JX
G8 DBU N. Greensted, High View Oust Care Home, Poulton Lane, Canterbury, CT3 2NH
G8 DCD J. Durrant, 27 Trafford Road, Willerby, Hull, HU10 6AJ
G8 DCJ P. Mcquail, 3 Post Office Lane, Draycott, Moreton-in-Marsh, GL56 9JZ
G8 DCX R. Sangster, 10 Addison Road, Banbury, OX16 9DH
G8 DD South Notts ARC, c/o D. Hill, 86 The Downs, Nottingham, NG11 7EB
G8 DDC Dunstable Dwn Rd, c/o C. Asquith, 36 Sunningdale, Luton, LU2 7TE
G8 DDH M. Lelliott, Well Lane Corner, Lower Froyle, Alton, GU34 4LJ
G8 DDN P. Bennett, Whitelands, Common Mead Lane, Gillingham, SP8 4RB
G8 DDY P. Thompson, 1A Downside Avenue, Niton, Ventnor, PO38 2DE
G8 DEC A. Malcolm, 68 Old Birmingham Road, Lickey End, Bromsgrove, B60 1DG
G8 DEJ T. Ray, 1 Providence Lane, Leamore, Walsall, WS3 2AQ
G8 DEL D. Coppen, 100 Atbara Road, Teddington, TW11 9PD
G8 DEM B. Willetts, 11 Albert Road, Warley, Oldbury, B68 0NA
G8 DER R. Richardson, Hazeldene, Sutton Road Fovant, Salisbury, SP3 5LF
G8 DET J. Bowen, 6 Bishops Court Gardens, Chelmsford, CM2 6AZ
G8 DEX J. Hosking, 21 Yeo Valley Way, Wraxall, Bristol, BS48 1PS
G8 DEY D. Parr, 58 Ritson St., Toxteth, Liverpool, L8 0UF
GM8 DFC Dr R. Cliff, 32 Lochardil Road, Inverness, IV2 4LD
G8 DFI B. Oliver, 6 Catherton Road, Cleobury Mortimer, Kidderminster, DY14 8EB
GM8 DFX Rev. J. Lincoln, 59 Obsdale Park, Alness, IV17 0TR
GI8 DGB B. Moore, 34A Feumore Road, Ballinderry Upper, Lisburn, BT28 2LH
G8 DGC S. Hall, 3 Sleepers Delle Gardens, Winchester, SO22 4NU
G8 DGH E. Townsend, The Manor House, Leicester, LE8 0AP
G8 DGR R. Smallwood, The Island, Hyde End Lane, Brimpton, RG74TH
G8 DGW M. Wickham, 8 Verlands Close Niton, Ventnor, PO38 2BG
G8 DHA D. Bishop, Oyston Lodge, Lynstone Road, Bude, EX23 8LR
G8 DHE G. Mather, 72 Cranleigh Road, Worthing, BN14 7QW
G8 DHF S. Matthews, 213 Hucclecote Road, Gloucester, GL3 3TZ
G8 DHI G. Roberts, 56 Horse Shoes Lane, Birmingham, B26 3HY
G8 DHJ C. Pickering, 28 George V Avenue, Margate, CT9 5QA
G8 DHS D. Digby, 73 Bedford Street, Crewe, CW2 6JB
GW8 DHT J. Clifford, Dippers Barn Coppice Lane Pool Quay, Welshpool, SY21 9JY
G8 DHU M. Baxter, 11B The Leys, Roade, Northampton, NN7 2NR
G8 DHV N. Eaton, 3 Thirslet Drive, Heybridge, Maldon, CM9 4YN
GI8 DHW J. Hendron, 9 Drumahiskey Road, Bendooragh, Ballymoney, BT53 7QL
G8 DIQ T. Hall, 7 Sweetlake Cottage, Nobold, Shrewsbury, SY5 8NH
G8 DIR K. Walker, 12 Willow Park, Minsterley, Shrewsbury, SY5 0EH
G8 DIU B. Cannon, 52 Goodhew Close, Yapton, Arundel, BN18 0JA
G8 DIY P. Geeson, 109 Folly Road, Mildenhall, Bury St Edmunds, IP28 7BT
G8 DJF A. Dickson, 7 Sandford Gardens, High Wycombe, HP11 1QT
G8 DJL J. Renaut, 4 Brune Way, West Parley, Ferndown, BH22 8QG
G8 DJO M. Adcock, 37 Ashpole Road, Braintree, CM75LW
G8 DJT G. Platts, 1 Blacksmiths Court, Kingham, Chipping Norton, OX7 6GE
G8 DJU J. Frisby, 66 Clear Crescent, Melbourn, Royston, SG8 6JD
G8 DJW G. Membury, 21 Webbers Piece, Maiden Newton, Dorchester, DT2 0AQ
GM8 DKB E. Taynton, 42 Craigmount Park, Edinburgh, EH12 8EE
G8 DKD C. Weale, Fair View, Abbots Lench, Evesham, WR11 4UP
GM8 DKG Dr C. Pegrum, 4 Northampton Drive, Glasgow, G12 0LE
G8 DKI D. Lucas, The Old Barn, The Street, Malmesbury, SN16 9DL
G8 DKK B. Harber, 45 Brandles Road, Letchworth Garden City, SG6 2JA
G8 DKV M. Coldicott, The Old Cottage, Church Lane, Morley, Ilkeston, DE7 6DE
G8 DKW M. Solomons, 389 B Alexandra Avenue, Harrow, HA2 9EF
G8 DLH A. Hall, 19 Crewkerne Road, Chard, TA20 1EZ
G8 DLL M. Monro, 6 Yew Tree Road, Hayling Island, PO11 0QE
G8 DLP R. Baker, Royal Oak House, Crich, Derbyshire, DE4 5BH
G8 DLX M. Crampton, 55 Gilbert Ave, Bilton, Rugby, CV22 7BZ
G8 DLZ P. Lea, 7 Cressex Road, High Wycombe, HP12 4PG
G8 DML J. Hughes, 12 Plough Garth, Kellington, Goole, DN14 0PD
G8 DMN S. Rundle, 4 Bridge Close, Evercreech, Shepton Mallet, BA4 6LZ
G8 DMT M. Caley, 40 Spenser Way, Jaywick, Clacton-on-Sea, CO15 2QT
G8 DMU A. Frazer, 11A, Leadhall Way Keld House, Harrogate, HG29PG
G8 DNH J. Webber, 21 Highfield Court, Wigton, CA7 9DR
G8 DNL K. Smith, 19 Westfield Avenue, South Croydon, CR2 9JY
G8 DNP P. Donoghue, Hillcrest, The Green, Harlow, CM17 0QR
GW8 DOA G. Pollard, 3 Carey Walk, Neath, SA10 7DD
G8 DOB I. Stuart, 87 Redgrove Park Hatherley Lane, Cheltenham, GL51 6QZ
G8 DOF P. White, 6 Curzon Court, Curzon Street, Chester, CH4 8PA
G8 DOH Dr A. Seeds, 114 Beaufort Street, London, SW3 6BU
GM8 DOR A. Barrett, Mains Of Glasclune Farm, Blairgowrie, PH10 6SF
G8 DOW B. Lee, 19 Lizard Head, Littlehampton, BN17 6RY
G8 DOY R. Elliott, Flat 27, Queen Mother Court, 151 Sellywood Road, Birmingham, B30 1TH
G8 DPE V. Brooks, 19 Malham Avenue, Wigan, WN3 5PR
G8 DPH T. Booth, 155 Oxford Road, Windsor, SL4 5DX
G8 DPQ D. Hendon, 2 Ellis Avenue, Onslow Village, Guildford, GU2 7SR
GM8 DPV J. Hunting, 77 Califer Road, Forres, IV36 1JB
G8 DPW D. Holden, 63 High St., Queenborough, ME11 5AG
G8 DQD T. Taylor, 15 Kennard Road, Bristol, BS15 8AA
G8 DQE R. Lees, 6 Library Road, Ferndown, BH22 9JP
G8 DQF L. Johnston, 9 Tunbridge Close, Burwell, Cambridge, CB25 0EL
G8 DQK A. Symonds, 19 Danby Terrace, Exmouth, EX8 1QS
G8 DQN N. Hunter, 33 Chapel Court, Billericay, CM12 9LX
G8 DQP J. Peden, 51A Bewdley Road, Kidderminster, DY11 6RL
GM8 DRA R. Macleod, 9 Croftcroighn Gate, Glasgow, G33 5JJ
G8 DRB K. Slee, 4 Dibbinview Grove, Wirral, CH63 9FW
G8 DRE D. Atkinson, Colne House, Robinson Road, Brightlingsea, CO7 0ST
G8 DRK R. Vince, 5 Bay Tree Road, Bath, BA1 6NA
G8 DRQ Dr R. Cochrane, 134 Moor Lane South, Ravenfield, Rotherham, S65 4QR
G8 DSG W. Jones, Elm Hurst, Station Road, Shrewsbury, SY4 2BB
G8 DSM J. Witherspoon, 109 Bromsgrove Road, Redditch, B97 4RL
GW8 DSO C. Warwick, 33 Ceri Road, Townhill, Swansea, SA1 6LS
G8 DST G. Smith, 23 Whaggs Lane, Whickham, Newcastle upon Tyne, NE16 4PF
G8 DSU R. Gill, 61 Cross Deep Gardens, Twickenham, TW1 4QZ
G8 DTA A. Parsons, 20 Paddocks Lane, Prestbury, Cheltenham, GL50 4NX
G8 DTE M. Pusey, 6 Blagdon Close, Martinstown, Dorchester, DT2 9JT
G8 DTF R. Price, 29 Birchfield Drive Worsley, Manchester, M28 1ND
G8 DTM F. Partington, 21 East Road, Wymeswold, Loughborough, LE12 6ST
G8 DTQ B. Petifer, 14 Wood Lane, Caterham, CR3 5RT
G8 DTS B. Norcliffe, 2 Alexander Drive, Heswall, Wirral, CH61 6XT
G8 DTT W. Moore, 26 Richard Moon St., Crewe, CW1 3AX
G8 DTX I. Sanderson, 15 Gorse Road, Huddersfield, HD3 4BN
G8 DUF R. Bird, 129 Park Road, Formby, Liverpool, L37 6AD
G8 DUI D. Cox, 52 Avill Crescent, Taunton, TA1 2PL
G8 DUO I. Casewell, 7 Pine Drive, Finchampstead, Wokingham, RG40 3LD
GW8 DUP R. Harris, 64 Frederick Place, Llansamlet, Swansea, SA7 9SX
G8 DUT H. Orgel, 1 Taunton Grove, Whitefield, Manchester, M45 6TJ
G8 DUV Dr C. Zammit, 9 Sandbanks Drive, Hatch Warren, Basingstoke, RG22 4UL
G8 DUW I. Redfern, 8 Lilac Grove, Stourport-on-Severn, DY13 8SR
GW8 DUY C. Davies, 14 Twynpandy, Pontrhydyfen, Port Talbot, SA12 9TW
G8 DVB J. Sandon, 461 Archer Road, Pin Green, Stevenage, SG1 5QP
G8 DVF T. Jones, 5 Bush Hatch, Frodsham, WA6 7QJ
G8 DVJ G. Wilks, 8 Chestnut Grove, East Barnet, Barnet, EN4 8PU
G8 DVK D. Aram, 4 Severn Road, Chilton, Didcot, OX11 0PW
G8 DVN D. Smith, 2 Burnor Pool, Calverton, Nottingham, NG14 6FL
G8 DVS A. Sterry, 9 Finch Avenue, Wakefield, WF2 6SE
G8 DVU R. West, 55 Burney Bit, Pamber Heath, Tadley, RG26 3TL
G8 DVW R. Leadbeater, The Birches, Torpenhow, Wigton, CA7 1JF
G8 DWF N. Earl, 162 Winchmore Hill Road, London, N21 1QP
G8 DWP P. Lee, 223 Chelmsford Road, Shenfield, Brentwood, CM15 8SA
G8 DWW C. Garcia, 8 Lyme Road, Bath, BA1 3LN
G8 DWX G. Haslip, 1 Sea Cottages, 28 Steyne Road, Seaford, BN25 1QF
GW8 DX J. White, Keepers Lodge Pumpsaint, Llanwrda, SA19 8DX
G8 DXF C. Tarran, Woodlands, School Road, Romsey, SO51 6AR
G8 DXI W. O'Connor, 3 Sterndale Close, Desborough, Kettering, NN14 2XL
G8 DXM C. Taylor, 45 Greenfield St., Shrewsbury, SY1 2PY
G8 DXO R. Humble, 3 Plover Close, Milborne Court, Sherborne, DT9 5DD
G8 DXP A. Cheasley, 25 Normanhurst Road, Walton-on-Thames, KT12 3EQ
G8 DXU B. Pollard-Wilkins, Seacall Limited, 16 Seabeach Lane, Eastbourne, BN22 7JG
G8 DXV H. King, 11 Priory Mead, Doddinghurst, Brentwood, CM15 0NB
G8 DXZ J. Sandys, 28 The Maultway, Camberley, GU15 1PS
G8 DYA C. West, 14 Ashleigh Gardens, Wymondham, NR18 0EX
G8 DYG M. Marshallsay, 2 Prospect Cottages, Lime Street, Gloucester, GL19 4NX
G8 DYI K. Holdway, 18 Pennymore Close, Stoke-on-Trent, ST4 8YQ
GM8 DYT J. Hotchin, 2 Moorfield Place, Gatehead, Kilmarnock, KA2 0AX
G8 DZC P. Martin, 58 Hearn Road, Woodley, Reading, RG5 3QG
G8 DZH J. Ray, 7 Barnmead, Theydon Bois, Epping, CM16 7ET
G8 DZJ G. Booth, 68 Tarragon Drive, Meir Heath, Stoke-on-Trent, ST3 7YE
G8 DZN B. Bird, 7 Old Kingsdown Close, Broadstairs, CT10 2HG
G8 DZW R. Brookes, 29 Ripley Road, Liversedge, WF156QE
G8 EAD M. Hutchings, 109 Longlands Way, Heatherside, Camberley, GU15 1RU
G8 EAH I. Carress, 1 Riplingham Road, Skidby, Cottingham, HU16 5TR
G8 EAJ Prof. P. Cannon, Field Cottage, Mathon Road, Malvern, WR13 6ER
G8 EAM John Newton Memorial Radio Club, c/o R. Newton, 8 Old Farm Road,

Minehead, TA24 8AS
G8 EAN J. Cunningham, 62 Kings Hill, Beech, Alton, GU34 4AN
G8 EAX S. Herod, 8 Deben Way, Felixstowe, IP11 2NS
G8 EBD G. Welch, 18 Alderdale, Wolverhampton, WV3 9JF
G8 EBM S. Haseldine, Newton House, Bretby Lane, Newton Solney, Burton-on-Trent, DE15 0RY
G8 EBQ R. Martin, 10 Westways, Stoneleigh, Epsom, KT19 0PQ
G8 EBT R. Lees, Hurlands, Hurlands Lane, Godalming, GU8 4NT
G8 EBX P. Starling, 14 Merton Place, Littlebury, Saffron Walden, CB11 4TH
G8 ECG K. Montgomery, The Old Village Post Office, High St., Oxford, OX44 9HP
G8 ECI D. Brown, 8 Waddingham Place, New Waltham, DN36 4QY
G8 ECR P. Jago, 39 Royal Avenue, Flat 2, London, SW3 4QE
G8 ECZ P. Barker, 14 Elsworth Green, Newcastle upon Tyne, NE5 3YB
G8 EDN T. Gallagher, 35 Wilhelmina Avenue, Coulsdon, CR5 1NL
G8 EDQ C. Soundy, 16 Crane Cottages, West Cranmore, Shepton Mallet, BA4 4QN
G8 EDS W. Hind, 3 Birds Hill, Letchworth Garden City, SG6 1PH
G8 EDX C. Vitiello, 1 North Street, Rothersthorpe, Northampton, NN7 3JB
G8 EEA D. Hill, 872 Oldham Road, Rochdale, OL11 2BN
G8 EEK B. Bruce, Three Ways, Wisbech Road, Wisbech, PE14 9RF
G8 EEM C. Gill, 77 Main Road Hambleton, Selby, YO8 9HW
G8 EEY A. Mobbs, 149 The Paddocks, Old Catton, Norwich, NR6 7HR
G8 EFG M. Vaughan, 69 Seamore Avenue, Benfleet, SS7 4EZ
G8 EFK E. Carter, 44 Plattes Close, Shaw, Swindon, SN5 5SA
G8 EFU C. Bloxidge, 33 Rosemary Hill Road, Sutton Coldfield, B74 4HL
G8 EGE J. Denton, 32 Highfields Mead East Hanningfield, Chelmsford, CM3 8XA
G8 EGG D. Hemingway, Conygore Farm, Howell Hill, Yeovil, BA22 7QZ
G8 EGL C. Burton, 13 Newells Terrace, Misterton, Doncaster, DN10 4DP
G8 EGM M. Booth, 16 Falcon Drive, Birdwell, Barnsley, S70 5SN
G8 EGU M. Smith, 35 Queen Street, Balderton, Newark, NG24 3NS
G8 EHD P. Brenton, 40 Furneaux Road Milehouse, Plymouth, PL2 3ET
G8 EHE K. Emerson, Stone Gables, Upper Minety, Malmesbury, SN16 9PR
G8 EHF J. Healen, 12 Primrose Lane, Standish, Wigan, WN6 0NR
G8 EHM Sir E. Vavasour, 15 Mill Lane, Earl Shilton, Leicester, LE9 7AW
GW8 EHO N. Holms, 22 Heol Isaf Radyr, Cardiff, CF15 8AL
GW8 EHQ J. Brown, 106 Marlborough Road, Penylan, Cardiff, CF23 5BY
G8 EHS Dr A. Fletcher, 40, Dereham Avenue, Ipswich, IP3 0QB
G8 EHX M. Melbourne, 42 Pasture Road, Stapleford, Nottingham, NG9 8GL
G8 EIE R. Forster, 7 Western Way, Alverstoke, Gosport, PO12 2NE
G8 EII M. Smith, 17 Girton Close, Owlsmoor, Sandhurst, GU47 0UP
G8 EIN N. Shepherd, 166 Chaldon Way, Coulsdon, CR5 1DF
G8 EJC R. Drew, 9 Sona Merg Close Heamoor, Penzance, TR18 3QL
G8 EJQ P. Vaughan, 15 Humber Gardens, Wellingborough, NN8 5WE
G8 EKD M. Nilson, 9 Middlemead, Folkestone, CT19 5UB
GM8 EKF F. Benson, 53 Warriston Drive, Edinburgh, EH3 5NA
G8 EKG G. Newstead, 97 Hawthorn Crescent, Burton-on-Trent, DE15 9QN
G8 EKN M. Biltcliffe, 4 Fleming Close, Bicester, OX26 2YA
G8 EKW G. Thornton, 4 Fir Tree Close, Exmouth, EX8 4EU
G8 EKZ A. Jones, 97A Bakers Ground, Stoke Gifford, Bristol, BS34 8GD
G8 ELG E. Joyce, 34 Milton Avenue, Eaton Ford, St Neots, PE19 7LE
G8 ELH D. Fisher, 17 Thrushel Close, Swindon, SN25 3PP
G8 ELP A. Stockley, Blacksole House, The Boulevard, Herne Bay, CT6 6GZ
G8 ELW R. Straker, 15 Rue Robert Garnier, le Mans, France, 72000
G8 EMA D. Pedley, Heronsway, West Street, Barford St. Martin, Salisbury, SP3 4AH
G8 EMB W. Tickell, 26 Shear Brow, Blackburn, BB1 7EX
G8 EMH D. Roebuck, 7 Elm Tree Close, North Anston, Sheffield, S25 4FG
G8 EMU J. Wheeler, 9 Elmer Close, Malmesbury, SN16 9UE
G8 EMX G. Hankins, 92 Sunningdale Road, Birmingham, B11 3QJ
G8 EMY K. Britain, Blenheim Cottage, Falkenham, Ipswich, IP10 0QU
G8 ENA E. Fellows, 343 Wake Green Road, Birmingham, B13 0BH
G8 ENB R. Whitby, 138 Browns Lane, Stanton-On-The-Wolds, Nottingham, NG12 5BN
G8 END I. Bodie, Thye Linney, Higher Poldown Farm, Truro, TR13 0FB
G8 ENM A. Hall, Rose Cottage, Burstall, Ipswich, IP8 3DX
G8 ENS J. Morris, 6 Barrowby Gate, Grantham, NG31 7LT
G8 ENW P. Baker, Top Of The Hill, Post Office Lane, Cheltenham, GL52 3PS
G8 ENY R. Hersey, 7 Tower Close, Brandon, IP27 0LJ
G8 EOH G. Simpkins, 12 Eastwood End, Wimblington, PE15 0QJ
G8 EOJ E. March, 23 Pebworth Close, Redditch, B98 9JX
G8 EOM D. Garrard, 48 Shorefields, Benfleet, SS7 5BQ
G8 EOV B. Cross, 3 The Meads, Haslemere, GU27 1LA
G8 EOZ K. Waight, 13 Kilda Road, Highworth, Swindon, SN6 7HS
G8 EPC M. Dyke, Cortijo Las Marrojas, Buzon 48 Palancar, 1820 Granada, Spain,
G8 EPH C. Kilvington, 53 Hall St.Skegby, Sutton-in-Ashfield, NG17 3EJ
G8 EPK D. Skye, 16 Lulworth Avenue, Poole, BH15 4DQ
G8 EPQ R. Prew, 16 Stokenchurch Place, Bradwell Common, Milton Keynes, MK13 8AT
G8 EPS G. Phelan, 113 Albert Road, Epsom, KT17 4EN
G8 EPZ C. Ward, 4 The Hawthorns, Charvil, Reading, RG10 9TS
G8 EQC D. Cliffe, Common Farm, Riley Hill, Lichfield, WS13 8JE
G8 EQD D. Wright, 22 West Hill, Rotherham, S61 2HB
GW8 EQI J. Fellows, 8 The Links Gwernaffield, Mold, CH7 5DZ
G8 EQO B. Tyler, 842 Handsworth Road, North Vancouver, Canada, V7R ZA2
G8 EQY F. Butler, 511 Fulbridge Road, Peterborough, PE4 6SB
G8 EQZ C. Reynolds, 49 Westborough Way, Anlaby Common, Hull, HU4 7SW
GW8 ERA M. Voss, 9 Chapel Close, Garndiffaith, Pontypool, NP4 7QS
G8 ERN R. Walker, 12 Foldyard Close, Sutton Coldfield, B76 1QZ
G8 ERV K. Blackman, 7 Deanery Close, Ripley, DE5 3TR
G8 ESK B. Kermode, 7 Midgeham Grove, Harden, Bingley, BD16 1DA
G8 ESL W. Miller, 5 Givendale Close, Bridlington, YO16 6GQ
G8 ESW W. Brade, 51 Coventry Gardens, Herne Bay, CT6 6SB
G8 ETD T. Rumble, 5 Mulberry Way, Spalding, PE11 2QJ
G8 ETI N. Foggin, 12 Linnetsdene, Covingham, Swindon, SN3 5AG
GM8 ETJ K. Mccartney, Greystones, Eskdaill Street, Langholm, DG13 0BG
G8 ETN S. Last, 72 Humber Road, Chelmsford, CM1 7PG
G8 ETP M. Furnival, The Ballroom, Stokeley Manor, Stokenham, Kingsbridge, TQ7 2SE
G8 ETR R. Cooke, 4 New Forest Close, Far Forest, Kidderminster, DY14 9TJ
G8 ETS D. Swale, 369 Scalby Road, Scarborough, YO12 6TG
G8 ETU A. Metcalf, 10 Manor Bend, Galmpton, Brixham, TQ5 0PB
G8 ETV P. Richardson, 3 Butlers Close, Amersham, HP6 5PY
G8 EUE M. Gasper, The Barn, Back Road, Halesworth, IP19 9DZ
G8 EUF C. Hall, The Orchard, Arkholme, Carnforth, LA6 1AX
GM8 EUG N. Robertson, 10 Warrenpark Road, Largs, KA30 8EF
GD8 EUH D. Pickard, Mont Y Mer. St. Georges Crescent., Port Erin, IM9 6HR
G8 EUV C. Fenton-Coopland, 14 Chevril Court, Wickersley, Rotherham, S66 2BN
G8 EVD T. Cartwright, 132 Mere Road, Wigston, LE18 3RL
G8 EVI Dr A. Clark, 3 North Street, Owston Ferry, Doncaster, DN9 1RT
G8 EVR K. Taylor, 39 Bowerfield Crescent, Hazel Grove, Stockport, SK7 6JB
G8 EWC A. Rouse, Clinton, Church Road, Colchester, CO7 8HS
G8 EWD M. Smith, 47 Salisbury Road, Market Drayton, TF9 1AR
G8 EWF B. Gilbert, 1 Wilmington Drive, Sutton-On-Sea, Mablethorpe, LN12 2JU
G8 EWL C. Burgess, Jalna, 12 Foley Close, Ashford, TN24 0XA
G8 EWN D. Edmonds, Great House Cottage, Great House Lane, Ripponden, Sowerby Bridge, HX6 4LQ
G8 EWT G. Diacon, 45 Woodpecker Way, Witney, OX28 6NN
G8 EXF R. Slatter, Ashwell House, Stratford Road, Oversley Green, Alcester, B49 6PG
GD8 EXI Dr S. Baker, Ballanarran House, Surby Road, Ballafesson, Port Erin, Isle Of Man, IM9 6TE
G8 EXJ B. Jones, 38 Wyresdale Road, Lancaster, LA1 3DU
G8 EXK M. Hatch, 6 Portland Street, Blyth, NE24 1NP
G8 EXN C. Briggs, 22 Woodlesford Crescent, Halifax, HX2 0RB
G8 EXQ T. Connell, 28 Tasman Close, Corringham, Stanford-le-Hope, SS17 7LD
G8 EXS P. Atherton, 10 Cheriton Drive, Ravenshead, Nottingham, NG15 9DG
GM8 EXU J. Steven, Andor, Skitten, Wick, KW1 4RX
G8 EXZ S. Warren, 269 Upper Weston Lane, Southampton, SO19 9HY
G8 EYA W. Rimmer, 79 Brookhurst Avenue, Wirral, CH63 0LA
G8 EYM N. Kearey, 73 Wellesley Drive, Crowthorne, RG45 6AL
G8 EYP Dr A. Faulkner, 79A West Drive, Highfields Caldecote, Cambridge, CB23 7RY
G8 EYQ J. Clee, 34 Knebworth Road, Bexhill-on-Sea, TN39 4JJ
G8 EYY M. Hancock, 12 Mellor Road, Hillmorton, Rugby, CV21 4BP
G8 EZB M. Whitlock, 85 Antrobus Road, Sutton Coldfield, B73 5EL
G8 EZD A. Gifford, Broncroft, Rock Green Bank, Ludlow, SY8 2DT
G8 EZE P. Swallow, 1 Auden Crescent, Ledbury, HR8 2UU
G8 EZG A. Pybus, Elm Bank Care Home 81, Northampton Road, Kettering. Northant?S, Kettering, NN16 7JZ
G8 EZL T. Lambert, 40 Deepdale Road, North Shields, NE30 3AN
G8 EZR K. James, 67 Drakes Way, Portishead, Bristol, BS20 6LD
G8 EZT R. Elgy, 130 Stebbing House, Queensdale Crescent, London, W11 4TG
G8 EZU K. Darbyshire, 24 Neston Road, Walshaw, Bury, BL8 3DB
G8 EZV G. White, 94 Wingate Road, Luton, LU4 8PY
G8 EZZ R. Chambers, 15 Barnfield Close, Braunton, EX33 2HL
G8 FAB Southampton ARC, c/o M. Troy, 22 Jackie Wigg Gardens Totton, Totton, Southampton, so409lz
G8 FAD W. Chown, 7840 Sw 136Th Avenue, Beaverton, USA, 97008
G8 FAK S. Sherratt, 21 Tweedale Close, Mursley, Milton Keynes, MK17 0SB
G8 FAR R. Elms, Fernside, Great Burches Road, Benfleet, SS7 3NA
G8 FAS S. Hotham, 54 Devon Drive, Westbury, BA133XQ
G8 FAT B. Haines, 20 Westfield Gardens, Harrow, HA3 9EJ
G8 FAX E. Bye, 117 Bull Lane, Rayleigh, SS6 8LZ
G8 FBA M. Hearne, Thicket Cottage, Crawley Down Road Felbridge, East Grinstead, RH19 2PS
G8 FBF D. Fellows, 10 Benning Way, Wokingham, RG40 1XX
G8 FBK L. West-Knights, 4 Paper Buildings, Temple, London, EC4Y 7EX
G8 FBM M. Bates, 11 The Rise Partridge Green, Horsham, RH13 8JB
G8 FBQ B. Corker, 46 Danelaw, Great Lumley, Chester le Street, DH3 4LU
G8 FBW A. Williams, 16 Hillside Road, Penn, High Wycombe, HP10 8JJ
G8 FC Royal Air Force ARS, c/o R. Hyde, 25 The Pastures, Cottesmore, Oakham, LE15 7DZ
G8 FCO G. Onions, 3 Tower Rise, Tividale, Oldbury, B69 1NP
G8 FCQ M. Lister, 246 Wigston Lane, Aylestone, Leicester, LE2 8DH
G8 FCT R. Chadwick, Ithaca, Heck Lane, Goole, DN14 0RD
G8 FDE B. Mcmanus, 6 Rowley Road, St Neots, PE19 1UF
G8 FDF J. Bastable, 94 Baymead Lane, North Petherton, Bridgwater, TA6 6RN
GW8 FDI G. Felton, 10 Penbodeistedd, Llanfechell, Gwynedd, LL68 0RE
G8 FDR M. Bingham, 18 Ladywell Gate, Welton, Brough, HU15 1NL
G8 FDZ D. Targett, 10 Thames Mews, Poole, BH15 1JY
G8 FEJ M. Woudstra, Flat 1, 2 Upper Park Road, St Leonards-on-Sea, TN37 6SJ
G8 FEK E. Gawthorpe, 35 Highfield Way, North Ferriby, HU14 3BG
G8 FET J. Guppy, 16 Barnfield Close, Hastings, TN34 1TS
G8 FEZ F. Stuart, 70 Peartree Road, Herne Bay, CT6 7EQ
G8 FFA E. Davis, 24 Redcar Avenue, Hereford, HR4 9TJ
G8 FFC C. Mcmanus, 6 Rowley Road, St Neots, PE19 1UF
G8 FFF C. Player, 28 Darwin Walk Withersfield, Haverhill, CB9 7ST
GM8 FFH Dr D. Brown, 14 Barloan Place, Dumbarton, G82 3QW
GM8 FFK G. George, 13 Balmoral Terrace, Elgin, IV30 4JH
G8 FFM B. Jackson, 23 Rylands Heath, Luton, LU2 8TZ
G8 FFU C. Burrows, 6 Brook Way, Lower Somersham, Ipswich, IP8 4PE
G8 FFW P. Rycroft, Shore View House, 100 Pilling Lane, Poulton-le-Fylde, FY6 0HG
GM8 FFX G. Knight, 6 Findon Road, Findon, Aberdeen, AB12 3RN
G8 FFZ P. Ewington, 26 Dickens Road, Rugby, CV22 5RW
G8 FGB S. Whitehead, 74 Manchester Road, Haslingden, Rossendale, BB4 5TE
G8 FGN H. Elstob, 25 Lindfield Avenue, Seaford, BN25 4DU
G8 FGQ H. Brittan, Meadowhurst Cottage, Woodcock Heath, Uttoxeter, ST14 8QS
G8 FGR K. Balch, 10 Heming Place, Stoke-on-Trent, ST2 9DF
G8 FGY P. Griffiths, 5 Chestnut Crescent, Carlton Colville, Lowestoft, NR33 8BQ
G8 FGZ C. Boon, Corner Cottage, Brackenhill, Nottingham, NG14 7EF
G8 FHC M. Passam, Birchenbower, Birchendale, Stoke-on-Trent, ST10 4HL
G8 FHI M. Clarke, 2 The Grove Penton Grafton, Andover, SP11 0RS
GM8 FHK J. Gallacher, 23 East Avenue, Carluke, ML8 5TS
G8 FHL Dr D. Green, 83 Wigan Lower Road, Standish Lower Ground, Wigan, WN6 8LJ
G8 FIE Dr N. Mcfetridge, 16 Blagrove Lane, Wokingham, RG41 4BA
G8 FIF Dr D. Howlett, 11 Barfleur Rise, Lyme Regis, DT7 3QY
G8 FIG C. Cole, 157 Cherry Tree Road, Beaconsfield, HP9 1BD
G8 FJA P. Webster, 3 Eden Avenue, Bare, Morecambe, LA4 6QL

G8 FJG R. Shoulder, 264 Wennington Road, Rainham, RM13 9UU
G8 FJR D. Jowett, 59 Old Road, Thornton, Bradford, BD13 3DQ
G8 FKF C. Sargeant, Northview, 20 South Marsh Road, Grimsby, DN41 8AN
G8 FKH D. Balharrie, 27 Norfolk Road, Uxbridge, UB8 1BL
GM8 FKL G. Twibell, Mambeg, Dervaig, Tobermory, Isle of Mull, PA75 6QN
G8 FLL D. Roseaman, 101 Westbrook, Bromham, Chippenham, SN15 2EE
G8 FLS I. Maciver, 160 Marsden Road, Burnley, BB10 2QP
G8 FLV A. Nicholson, 29 Quaker Lane, Northallerton, DL6 1EE
G8 FMA E. Sillars, 34 Sandown Road, Stevenage, SG1 5SF
G8 FMC D. Keston, 8 Copse Gate Winslow, Buckingham, MK18 3HX
G8 FMD C. Wells, 5 Hepplewhite Close, Baughurst, Tadley, RG26 5HD
G8 FME A. Hilton, 28 Eastern Esplanade, Broadstairs, CT10 1DR
G8 FMI F. Steed, 19 Chancery Lane, Debenham, Stowmarket, IP14 6RN
GM8 FMR D. Taylor, 14 Fenton Street, Alloa, FK10 2DT
G8 FMT P. March, Devonholme, Bedford Road, Hitchin, SG5 3RX
G8 FMW R. Whitehouse, 92 Willenhall Road, Bilston, WV14 6NP
G8 FMX D. Beard, 9 Bowgate, Gosberton, Spalding, PE11 4ND
G8 FMZ P. Mcnamara, Sunnybank Cottage, Lower Swell, Cheltenham, GL54 1LG
G8 FNG P. Robinson, 52 Lea Court, New Road, Crewe, CW3 9DN
G8 FNH M. Nash, 12 Ruston Park, Rustington, Littlehampton, BN16 2AB
GW8 FNO R. Gregory, 5 Bryn Castell, Radyr, Cardiff, CF15 8RA
G8 FNR D. Stone, 165 Wellington Hill West, Bristol, BS9 4QW
GW8 FOL G. Spencer, Tyn Cae, Llanfwrog, Anglesey, LL65 4YL
G8 FOT B. Butterworth, 21 Higher Drive, Purley, CR8 2HQ
GW8 FOY L. Oakes, Flat 2 Brython, 54-56 Lloyd St., Llandudno, LL30 2YP
G8 FPA D. Hoult, 1 East Mill Gate, Cherry Willingham, LN3 4BZ
G8 FPG S. Banner, Oedhofstrasse 18, Amstetten, Austria, 3300
G8 FPU R. Hutton, 5 Tollemache Road, Prenton, CH43 8SU
G8 FPW F. Brown, The Bungalow, Oxcroft Bank, Shepeau Stow, Spalding, PE12 0TY
G8 FQN R. Schneider, 15 Hope Lane, Upper Hale, Farnham, GU9 0HY
G8 FQS Dr P. Simpson, 17 Reynard Close, Horsham, RH12 4GX
G8 FQZ C. Stocker, 8 Brook Drive, Astley, Manchester, M29 7HS
G8 FRH P. Lyall, 20 Horn Lane, Woodford Green, IG8 9AA
G8 FRI J. Lucas, 42 Westerleigh Road, Bath, BA2 5JE
G8 FRJ P. Hayes, 28 Rochester Road, Barnsley, S71 2NJ
G8 FRS K. Gurr, 119 Vaisey Road, Stratton, Cirencester, GL7 2JW
G8 FRY N. Friday, 140 Chapel Point Village, Skegness, PE245UZ
G8 FSJ R. Page, 39 Carlton Street, Kettering, NN16 8EB
G8 FSL A. Benham, 141 The Close, Salisbury, SP1 2EY
GW8 FSN B. Steadman, 8 Machno Place, Denbigh, LL16 3YA
GU8 FSU V. Rees, Le Chene Lodge, Le Chene Hill, Forest, Guernsey, GY8 0AJ
G8 FSV A. Mason, Ulitsa Svoboda 20, Banevo, Bulgaria, 8125
G8 FTE R. Cowley, 20 Mill Road, Willingham, Cambridge, CB24 5JU
G8 FTP P. Jarrett, 15 Groveside, East Rudham, King's Lynn, PE31 8RL
G8 FTW R. Goodchild, 48 Coral Drive, Ipswich, IP1 5HS
G8 FTX D. Gotch, The Bungalow, West Lane, Shipley, BD17 5DW
G8 FTY C. Gray, 135 Long Drive, Ruislip, HA4 0HL
G8 FUB L. Jones, 52 New Lane, Aughton, Ormskirk, L39 4UD
G8 FUH S. Melling, 15 Woodbridge Hill Gardens, Guildford, GU2 8AR
G8 FUI W. Raybould, 33 Roberts Green Road, Dudley, DY3 2BB
G8 FUL J. Masterton, 15 Maylins Drive, Sawbridgeworth, CM21 9HG
G8 FUO R. Britton, 12 Bulkeley Avenue, Windsor, SL4 3LP
G8 FVC D. Mclay, 6 Burton Road, Castle Gresley, Swadlincote, DE11 9HD
G8 FVE K. Lake, 79 Sherrards Way, Barnet, EN5 2BP
GW8 FVI C. Reeves, 37 Arnold Gardens, Kinmel Bay, Rhyl, LL18 5NH
G8 FVJ D. Still, 133A Feltham Road, Ashford, TW15 1AB
G8 FVK Prof. P. Marks, Ovinswell House, Low Street, Lastingham, York, YO62 6TJ
G8 FVM P. Peake, 34 Blackhalve Lane, Wolverhampton, WV11 1BH
GM8 FVN G. Adams, Heath Court, Morven Way, Ballater, AB35 5SF
G8 FVT D. Bainton, 86 Holywell Avenue, Whitley Bay, NE26 3AD
G8 FWA J. Errington, The Woodlands, Station Road, Leicester, LE8 9FP
G8 FWC D. Sharp, 2 Millbank Street, Goole, DN14 5XF
G8 FWD T. Mckee, 19 Wall Lane Terrace, Cheddleton, Leek, ST13 7ED
G8 FWE J. Maidment, Down Along, Cherry Lane, Barrow-upon-Humber, DN19 7AX
G8 FWH J. Hill, 21 Somersby Road, Mapperley, Nottingham, NG3 5QB
G8 FWK J. Cranfield, 65 Broome Manor Lane, Swindon, SN3 1NB
G8 FWY P. Russell, 57 Norburn Park, Witton Gilbert, Durham, DH7 6SG
G8 FXA G. Griffiths, 51 Bempton Road, Liverpool, L17 5DB
G8 FXC M. Bradford, 3 Veysey Close, Hemel Hempstead, HP1 1XQ
G8 FXG N. Lay, Orchard Cottage, Parkham, Bideford, EX39 5PL
G8 FXL A. Patterson, 139 Lowther Road, Bournemouth, BH8 8NP
G8 FXM D. Toombs, 1 Chalgrove, Welwyn Garden City, AL7 2QJ
G8 FXN R. Thackeray, 104 Stag Leys, Ashtead, KT21 2TL
G8 FXU D. Percival, Trebakken, 11 Lamborne Close, Sandhurst, GU47 8JL
G8 FXV Dr M. White, 2 Mill Close, Denmead, Waterlooville, PO7 6PE
G8 FXX R. Limb, Charnwood House, Station Road, Henley-on-Thames, RG9 3JS
G8 FYK K. Payne, Flat 4, Monton Bridge Court, Eccles, M30 8UW
G8 FYX N. Fensch, Glen Cottage, Bowl Road, Ashford, TN27 0HB
G8 FZI M. Logsdon, Pilgrims Cottage, Langford Budville, Wellington, TA21 0RH
G8 FZT T. Unsworth, Heathview, 15 Fenton Road, Huntingdon, PE28 2SD
G8 FZV D. Ryan, Turners Oak, Barrs Lane, Woking, GU21 2JN
G8 FZW J. Brown, 16 Greenwood Close, Moulton, Northampton, NN3 7RD
G8 GAR H. Taylor, 21 Windermere Road, Coulsdon, CR5 2JF
G8 GAT Dr M. Smith, 241 Sandbanks Road, Poole, BH14 8EY
GM8 GAX P. Howson, 1 Howetown Fishcross, Alloa, FK10 3AW
G8 GBE P. Richardson, 50 Amberley Road, Gosport, PO12 4EW
G8 GBM R. Head, 29 Kingslea Road, Solihull, B91 1TQ
G8 GBP C. Fawdon, 21 Bevan Close, Southampton, SO19 9PE
G8 GBU D. Barker, 311 Uttoxeter Road, Mickleover, Derby, DE3 9AH
G8 GBY Hull and District ARS, c/o B. Atkinson, 165 Alliance Avenue, Hull, HU3 6QY
G8 GCC Cannock Chase ARS, c/o B. Gallear, 5 Oak Avenue, Cannock, WS12 4QA
G8 GCK G. Croome, 10 Axford Close Gedling, Nottingham, NG4 4BB
G8 GCM J. Price, 6 Kernick Road, Penryn, TR10 8NX
G8 GCO N. Wall, 9 North Close, Ipswich, IP4 2TL
G8 GCS C. Coker, 46 Clarendon Road, Ipplepen, Newton Abbot, TQ12 5QS
G8 GDC R. Laver, 40 Middleton Close, Tysoe, Warwick, CV35 0SS
G8 GDH D. Brown, 56 Paddock Road, Staincross, Barnsley, S75 6LE
GM8 GDN M. Brunton, 2 Easter Place, Portlethen, Aberdeen, AB12 4XL
G8 GDZ R. Thompson, 23 Fox Hill, Selly Oak, Birmingham, B29 4AG
G8 GEA K. Warriner, Windover, 16 The Ridgeway, Eastbourne, BN20 0EU
G8 GEB S. Rowsby, 10 Echells Close, Bromsgrove, B61 7EB
G8 GEE R. Sherwood, 19 Norton Drive, Warwick, CV34 5FE
G8 GET J. Shepherd, 6 The Jordans, Coventry, CV5 9JT
G8 GEV Dr J. Moore, 17 Kings Grove, Barton, Cambridge, CB23 7AZ
G8 GEZ L. Wooller, 4 Old Court Close, Brighton, BN1 8HF
G8 GFA R. Marshall, The Village School, Upleatham, Redcar, TS11 8AG
G8 GFB C. Jones, 1 Primrose Hill Road Euxton, Chorley, PR7 6BA
G8 GFF N. Sanderson, 54 Kelvedon Close, Chelmsford, CM1 4DG
G8 GFQ E. Hill, 22 Botham Grove, Stoke-on-Trent, ST6 5NX
G8 GFS M. Winiberg, Summerhill, Smallhythe Road, Tenterden, TN30 7NB
G8 GFW J. Douglas, 1030 Shields Road, Newcastle upon Tyne, NE6 4SR
G8 GFY D. King, 108 Huddersfield Road, Meltham, Holmfirth, HD9 4AG
G8 GFZ T. Cockram, The Bungalow, Dyke Hill, Chard, TA20 2PY
G8 GGI R. Geddes, 107 Dukes Avenue, New Malden, KT3 4HR
G8 GGM P. Burfoot, 18 Ember Road, Langley, Slough, SL3 8ED
G8 GGO P. Carson, 16 Gaynes Park Road, Upminster, RM14 2HJ
G8 GGP T. Hague, 2 Winns Row, Godolphin Road, Helston, TR13 8QH
G8 GGR C. Coleman, 19 Megdale, Matlock, DE4 3JW
G8 GGS M. Clarke, Ruskin, Ashurst Drive, Tadworth, KT20 7LS
GW8 GGW N. Dudman, Chapel House, Pen Y Bryn, Wrexham, LL14 1UA
G8 GHB A. Hunt, 21 Plumpton Gardens, Cantley, Doncaster, DN4 6SN
G8 GHH C. Gibbs, 32 Gresham Avenue, Margate, CT9 5EH
G8 GHK W. White, 60 Parklands, Rochford, SS4 1SH
G8 GHL S. Garland, 53 The Crescent, Horsham, RH12 1NA
G8 GHO J. Wood, 17 Yew Tree Park Road Cheadle Hulme, Cheadle, SK8 7EP
G8 GHP R. Whyte, 64 Ash Tree Drive, West Kingsdown, Sevenoaks, TN15 6LF
GM8 GHQ P. Laverock, 3/6 Hopetoun Crescent, Edinburgh, EH7 4AY
G8 GHR B. Farey, 5 Ivel View, Sandy, SG19 1AU
GM8 GHV W. Sherriffs, Hillcrest, Disblair, Aberdeen, AB21 0RJ
G8 GIF K. Turner, 20 Rawdon Way, Faringdon, SN7 7YT
G8 GIG A. Patterson, Rose Cottage, Ingatestone Road, Ingatestone, CM4 0RS
G8 GIH K. Foster, 52 Bottesford Avenue, Scunthorpe, DN16 3EN
G8 GIK J. Hart, 28A Dunton Road Stewkley, Leighton Buzzard, LU7 0HZ
G8 GIL M. Dimmock, 44 Diddington Close, Bletchley, Milton Keynes, MK2 3EB
GM8 GIQ C. Wearing, 16 Campbell Drive, Troon, KA10 6XE
G8 GIU J. Harman, 13 Linthorpe Court, South Shields, NE34 9BU
G8 GIZ D. Ollerhead, 15 Kingsley Road, Chester, CH3 5RR
G8 GJA P. Reeves, 77 Cale Way, Wincanton, BA9 9BS
G8 GJC J. Tillin, 76 Holbrook Road, Belper, DE56 1PB
G8 GJG N. Giltrow, 7 The Square, Milton-Under-Wychwood, Chipping Norton, OX7 6JN
GM8 GJI Dr E. Smith, The Steading, Craigmyle, Banchory, AB31 4LS
G8 GJM R. Harwood, 9 Cornwall Close, Woosehill, Wokingham, RG41 3AG
G8 GJO A. Heasman, 170 Plum Lane, London, SE18 3HF
G8 GJU M. Bernard, 33 Station Road, Over, Cambridge, CB24 5NJ
G8 GJV T. England, 30 Sparrow Way, Burgess Hill, RH15 9UL
G8 GJW C. Drouet, Ash Trees, Hallfield, Dalston, Carlisle, CA5 7QH
G8 GKC C. Ridley, 39 Lancelot Road, Welling, DA16 2HX
G8 GKH R. Hadley, 36 Folly Lane, Cheltenham, GL50 4BY
G8 GKL C. Rauch, 40 Russett Close, King's Lynn, PE30 3HB
G8 GKR M. Fellows, 343 Wake Green Road, Birmingham, B13 0BH
G8 GKX D. Nicholson, Avenida Espaã±A, Edificio Sorrento, Malaga, Spain, 29793
G8 GLB P. Brown, 4 King Edgar Close, Ely, CB6 1DP
G8 GLC I. Cooper, 77A Benhill Wood Road, Sutton, SM1 3SL
G8 GLD M. Bounford, 62 High Street, Wood Lane, Stoke-on-Trent, ST7 8PB
G8 GLI J. Husk, Brandhu, Common Moor, Liskeard, PL14 6EP
G8 GLL J. Woolford, Daedalian, Trewoon Road Mullion, Helston, TR12 7DS
G8 GLP B. Barnett, 21 Primrose Walk, Maldon, CM9 5JJ
G8 GLS G. Wimlett, Yew Tree Lodge White Horse Lane Barton, Preston, PR3 5AH
G8 GLV A. Brown, 23 Vincents Way, Naphill, High Wycombe, HP14 4RA
G8 GLY A. Higgins, 86B Cranleigh Road, Bournemouth, BH6 5JL
G8 GLZ A. Findlay, 7 Market Square, Winslow, MK18 3AB
G8 GMA D. Elliott, 56 Lincoln Avenue, Willenhall, WV13 1JQ
G8 GMB S. Bradshaw, 82 Arden Way, Market Harborough, LE16 7DD
G8 GML P. Melbourne, 2 Jubilee Cottages, Jubilee Lane, Colchester, CO7 7RY
G8 GMU B. Leathley-Andrew, 4 Robinson Road, Bedworth, CV12 0EL
G8 GNI Dr A. Thomas, The Stone Barn, 1 Home Farm Close, Chesterton, OX26 1TZ
G8 GNO S. Ferdenzi, 4 Ashworth Road, Rossendale, BB4 9JE
G8 GNX J. Bartholomew, 33 Manor Way, Woodmansterne, Banstead, SM7 3PN
G8 GNZ G. Blake, 22 Cannon Leys, Galleywood, Chelmsford, CM2 8PD
GW8 GOC M. Black, Mediascene Ltd, Unit A-D, Bowen Industrial Estate, Bargoed, CF81 9AB
G8 GOM A. Ireson, 32 The Avenue, Wellingborough, NN8 4ET
G8 GON A. Jefford, 37 Marions Way, Exmouth, EX8 4LF
GW8 GOO P. Nelson, 15 Hill Street Gerlan, Bethesda, Bangor, LL57 3TD
G8 GOR A. Pearce, 153 Henver Road, Newquay, TR7 3EJ
G8 GOS K. Roche, 96 Porter Road, Basingstoke, RG22 4JR
G8 GOT D. Parkin, 252 Standbridge Lane, Crigglestone, Wakefield, WF4 3JA
G8 GP J. Homersham, 35 Meeting Street, Ramsgate, CT11 9RT
G8 GPF D. Clark, 6 Bradley Park Road, Torquay, TQ1 4RD
G8 GPO OFRAC Baldock, c/o D. Thorpe, 70 Willow Way, Ampthill, MK45 2SP
GW8 GQE J. Moore, Oak House, Falcondale Drive, Lampeter, SA48 7SB
G8 GQF I. Mcenteggart, 46 Bissley Drive, Maidenhead, SL6 3UZ
G8 GQG J. Crow, 58 Cooden Drive, Bexhill-on-Sea, TN39 3AX
G8 GQJ R. Clark, 9 Conigre, Chinnor, OX39 4JY
G8 GQS B. Summers, 9 Prior Croft Close, Camberley, GU15 1DE
G8 GRB R. Day, 20 Linacre Road, Torquay, TQ2 8LF
G8 GRC J. Drakeley, Rowan Cottage, Four Crosses Lane, Cannock, WS11 1RU
G8 GRD L. Hetherington, 5 Withey Close West, Bristol, BS9 3SX
GD8 GRE C. Wilkinson, The Grange 14 Montreux Court, Douglas, Isle Of Man, IM2 6AF
G8 GRL K. Edwards, 11 Foxes Road, Ashen, Sudbury, CO10 8JS
G8 GRO R. Nicholls, 10 Polmeere Road, Penzance, TR18 3PD
G8 GRP E. Poole, Ramillies Hall School, Ramillies Avenue, Cheadle, SK8 7AJ
G8 GRQ A. Plail, 46 Hayling Rise, Worthing, BN13 3AG
G8 GRS R. Woodward, 158 Highridge Road, Bishopsworth, Bristol, BS13 8HU
G8 GRT R. Oakley, 17 Windmill Close, Ellington, Huntingdon, PE28 0AJ

G8 GSL I. Liston-Brown, 20 Chatterton Avenue, Lichfield, WS13 8EF
G8 GSU R. Wade, The Hermitage 15 Heath Hill Road North, Crowthorne, RG45 7PD
G8 GSY G. Young, 6 The Maypole Thaxted, Dunmow, CM6 2QZ
GW8 GT F. Clare, Glen View, Newport Road, Caldicot, NP26 3BZ
G8 GTD S. Porter, 20 Newbridge Road, Ambergate, Belper, DE56 2GR
G8 GTI K. Barnes, 75 Southmeade, Liverpool, L31 8EG
G8 GTR Dr D. Murray, 27 Station Avenue, Walton-on-Thames, KT12 1NF
G8 GTU P. Stephens, 3 Inett Way, Droitwich, WR9 0DN
G8 GTV B. Raby, 10 Bulverton Park, Sidmouth, EX10 9EW
G8 GTZ N. Matthews, 12 Petrel Croft, Basingstoke, RG22 5JY
G8 GUA R. Wood, 339 Horse Road, Hilperton Marsh, Trowbridge, BA14 7PE
G8 GUH G. Ohara, 107 Castlesteads Drive, Carlisle, CA2 7XD
GW8 GUJ J. Stubbs, 14 The Glen, Langstone, Newport, NP18 2NR
G8 GUN H. Parker, 7 The Hollies, Clee Hill, Ludlow, SY8 3NZ
G8 GUS M. Board, 48 Skipper Way, Lee-on-the-Solent, PO13 9EY
GM8 GUX J. Thomson, 2 Wilton Hill, Hawick, TD9 8BA
GW8 GVI M. Ritchie, Bryngolman Farm, Llangolman, Clynderwen, SA66 7QL
G8 GVL K. Woods, 7 Ives Close, West Bridgford, Nottingham, NG2 7LU
G8 GVN E. Shield, 14 Wellwood Street, Amble, Morpeth, NE65 0EL
G8 GVO C. Byard, 32 Castlefields, Leominster, HR6 8BJ
G8 GVV P. Richmond, 57 The Fairway, Daventry, NN11 4NW
G8 GVW P. Shillito, Little Orchard, Thorney Road Kingsbury Episcopi, Martock, TA12 6BG
G8 GVZ L. Sullivan, 89 Richmond Crescent, Mossley, Ashton-under-Lyne, OL5 9LQ
G8 GWB N. Awcock, 16A Ongar Road, Writtle, Chelmsford, CM1 3NU
G8 GWJ J. Vincent, 12 Spelman Road, Norwich, NR2 3NJ
G8 GWM N. Hay, 20 The Ridgeway, Fetcham, Leatherhead, KT22 9AZ
G8 GWP G. Atkinson, Parkland Stables, Landmere Lane, Nottingham, NG11 6ND
G8 GWR Dr S. Linney, 164 Kineton Green Road, Solihull, B92 7ES
G8 GWX R. Howells, 52 Upton Park, Upton, Chester, CH2 1DG
G8 GXF J. Ashmore, 3 The Cedars, Stockwell Road, Wolverhampton, WV6 9AZ
G8 GXN K. Raynor, 17 Kirkstone Walk, Nuneaton, CV11 6EZ
G8 GXO P. Rogers, 26 Hall Lane, Sutton, Macclesfield, SK11 0EP
G8 GXS Rev. J. Hadjioannou, The Vicarage, Wakefield Road, Pontefract, WF9 5BX
G8 GYB V. Vesma, Durvale, 5 Jonas Drive, Wadhurst, TN5 6RJ
G8 GYI G. Stanley, 133 Park Lane, Kidderminster, DY11 6TE
G8 GYK G. Carter, 19 Wych Elms, Park Street, St Albans, AL2 2AR
G8 GYL I. Bishop, 5 Trent Close, Tolpuddle, Dorchester, DT2 7HA
G8 GYM R. Claridge, 124 Pemdevon Road, Croydon, CR0 3QP
G8 GYP V. Holmes, 104 York Avenue, Hayes, UB3 2TP
G8 GYS P. Wright, 1 Lambourne Way, Thruxton, Andover, SP11 8NE
G8 GYV B. Simms, 17 Peregrine Close, Weston-Super-Mare, BS22 8UY
G8 GYX T. Ellinor, 53 Hillside, Banstead, SM7 1HG
G8 GYY C. Gregory, 5 Fox Close, Wigginton, Tring, HP23 6ED
G8 GZC G. Tew, 73 King Cerdic Close, Chard, TA20 2JB
GI8 GZM J. Mawhinney, 12 Shane Park, Lurgan, Craigavon, BT66 7HD
G8 GZN A. Hill, 1 Greenways, Highcliffe, Christchurch, BH23 5BA
G8 GZV R. Duke, 5 Pembroke Close, Billericay, CM12 0PF
G8 GZW Rev. A. Davis, 8 Roberts Road, Greatstone, New Romney, TN28 8RL
G8 GZX J. Eadie, 5 Silver Street Cublington, Leighton Buzzard, LU7 0LJ
GW8 HAG T. Regan, 8 Park View Gardens, Bassaleg, Newport, NP10 8JZ
G8 HAM S. Collins, 2 St Teresa Drive, Chippenham, SN152BD
G8 HAU R. Lambarth, 38 Kirkley Park Road, Lowestoft, NR33 0LG
G8 HBQ P. Davies, 24 Upland Grove, Leeds, LS8 2SX
GM8 HBY C. Ross, 16 Glebe Crescent, Airdrie, ML6 7DH
G8 HBZ S. Stephenson, 6 Livingstone Close Rothwell, Kettering, NN14 6HT
G8 HCK A. Rutter, The Uplands, Castle Howard Road, Malton, YO17 6NJ
G8 HCL V. Menday, Huf House, Horseshoe Ridge, Weybridge, KT13 0NR
G8 HCS H. Stratton, 26 Marjorie Road, Chaddesden, Derby, DE21 4HQ
G8 HCW C. Morgan, 24 High Mead Royal Wootton Bassett, Sn4 8Lw, Swindon, SN4 8LW
G8 HCZ Dr P. Iredale, Mayfield, Woodlands Road, Raydon, Ipswich, IP7 5LJ
GW8 HDH J. Dowdall, 56 Goetre Bellaf Road, Dunvant, Swansea, SA2 7RP
G8 HDJ P. Muxlow, 17 Station Road, Grasby, Barnetby, DN38 6AP
G8 HDL M. Connell, 38 White Close, High Wycombe, HP13 5NG
G8 HDM I. Arnold, 44 Elwick Avenue, Acklam, Middlesbrough, TS5 8NT
G8 HDP R. Jenkins, 10 Ulstan Close, Woldingham, Caterham, CR3 7EH
GW8 HEB T. Brady, 8 Cefn Hawys, Red Bank, Welshpool, SY21 7RH
G8 HER A. Lambert, 2 Huxley Close, Locks Heath, Southampton, SO31 6RR
G8 HEU P. Whitehead, 7 Vulcan Road, Freckleton, Preston, PR4 1JN
GW8 HF D. Phillips, 34 Graig Terrace, Graig, Pontypridd, CF37 1NH
G8 HFL L. Caine, 25 Smallbrook Road, Broadway, WR12 7EP
G8 HFW T. Hall, 23 Burcott Gardens, Addlestone, KT15 2DE
G8 HGG P. Abernethy, Enfield House, Halford, Shipston-on-Stour, CV36 5DA
G8 HGI M. Warriner, Rapps Lodge, Rapps, Ilminster, TA19 9LG
G8 HGL D. Lambert, 4 Tamworth Road, Bedford, MK41 8QY
G8 HGM K. Ellis, 11 Ringwood Close, Eastbourne, BN22 8UH
G8 HGN R. Harrison, 59 Grange Road, Billericay, CM11 2RQ
GM8 HHC P. Dick, Napier House, 8 Colinton Road, Edinburgh, EH10 5DS
G8 HHO M. Strange, 60A Manor Road, Dersingham, King's Lynn, PE31 6LH
G8 HHR J. Bardell, 239 Meadow Road, Droitwich, WR9 9BZ
G8 HHZ P. Woods, 14 Cromwell Road, Muswell Hill, London, N10 2PD
G8 HI K. Burnitt, 15 St. Bedes, East Boldon, NE36 0LE
G8 HIG L. Cole, 151 Carshalton Park Road, Carshalton, SM5 3SF
G8 HIO T. Ellis, Hollybush House, Hawley Green, Blackwater, Camberley, GU17 9BP
G8 HIQ S. Whitehouse, 2 Lindholme Drive, Rossington, Doncaster, DN11 0UR
GW8 HJC C. Harper, Deunant, Capel Curig, Betws-Y-Coed, LL24 0DS
G8 HJD C. Tubis, Rockleaze Mews, Rockleaze Avenue, Bristol, BS9 1NG
G8 HJF C. Williams, Kingsclere House, Fox'S Lane, Newbury, RG20 5SL
G8 HJG C. Williams, Kingsclere House, Fox'S Lane, Newbury, RG20 5SL
G8 HJH M. Norton, 179B Kimbolton Road, Bedford, MK41 8DR
G8 HJK P. Hunt, 40 Leighton Road, Toddington, Dunstable, LU5 6AL
G8 HKK M. North, 194 North Road, Combe Down, Bath, BA25DN
G8 HKN R. Meakins, 335 Court Road, Orpington, BR6 9BZ
G8 HKP E. Jakins, 2 South View Place, Midsomer Norton, Bath, BA3 2AX
G8 HKS M. Booth, 30 Manor Green, Harwell, Didcot, OX11 0DQ
G8 HLE R. Marshall, 54 Tudor Avenue, Maidstone, ME14 5HJ
G8 HLH R. Wheeler, 14 Robins Lane, St Helens, WA9 3NF
G8 HLJ E. Edwards, Flat 27, Barncroft, Wirral, CH61 6YH
G8 HLM S. Catlin, 3 Manor Lane, Langham, Oakham, LE15 7JL
G8 HLQ E. Birch, 17 Canalside Cottages Chester Road, Preston Brook, Runcorn, WA7 3AQ
G8 HMA R. Smith, 5 Newton Close, Loughborough, LE11 5UU
G8 HMG P. Walker, 12 Brownlow Road, Redhill, RH1 6AW
G8 HMJ M. Kellett, Wistow Gate, Glen Road, Leicester, LE8 9FH
G8 HMV J. Nicholas, 4 Lion Lane, Clee Hill, Ludlow, SY8 3NJ
G8 HMZ P. Cheseldine, 6 Lissett Close, Lincoln, LN6 0SY
G8 HNA S. Clark, 1 Roman Road, Broadstone, BH18 9DF
G8 HNM R. Parker, 1 Whitmore Orchard, Whitmore Lane, Taunton, TA2 6SR
G8 HNS R. Stanleigh, Shallowpool Bungalow , Shallowpool, PL13 2ND
G8 HNT T. Thompson, 25 Meadow Avenue, Codnor, Ripley, DE5 9QN
G8 HOI R. Warner, Barley Hill Farm, Combe St. Nicholas, Chard, TA20 3HJ
G8 HOR E. York, Combe Brune, Prayssac, France, 46220
GW8 HOS V. Mander, Meadowlands, Severn Lane, Welshpool, SY21 7BB
G8 HOU H. Cox, 21 North Avenue, Hayes, UB3 2JE
G8 HPF Dr I. Mclenaghan, 82 Cheam Road, Epsom, KT17 1QP
G8 HPJ P. Beaumont, 1 Byron Road, Mexborough, S64 0DG
GW8 HPL W. Taylor, Bywell, Chester Road, Rossett, Wrexham, LL12 0HN
G8 HPN G. Staniewicz, Flat 1, Jubilee Farm, Gillingham, SP8 5SJ
G8 HPS A. Hancock, 9 Elmside, Willand, Cullompton, EX15 2RN
G8 HPV D. Green, 67 Coombe Park Road, Binley, Coventry, CV3 2NW
G8 HPW M. Hanaghan, 9 Goole Road, Grindon Broadway, Sunderland, SR4 8HT
G8 HPY A. Mander, 18 Bridge Avenue, Otley, LS21 2AA
GW8 HQM S. Bastow, Bryn Goleu, Rhosgadfan, Caernarfon, LL54 7LB
G8 HQO N. Johnson, 56 Clarkson Avenue, Wisbech, PE13 2EG
G8 HQW P. Kirby, 2 Kneeton Park, Middleton Tyas, Richmond, DL10 6SB
G8 HRA C. Ryalls, 15 Belmont Way, South Elmsall, Pontefract, WF9 2BT
G8 HRC Havering & District ARC, c/o D. Nuttall, 92 Long Road, Lowestoft, NR33 9DH
G8 HRF K. Dodman, 10 Newark Road, Lowestoft, NR33 0LY
G8 HRW S. Watkin, 9 Longden Close, Haynes, Bedford, MK45 3PJ
G8 HSI J. Carey, 7 Church Road, Walton on the Naze, CO14 8DF
G8 HSR E. Warren, 37 Kingston Drive, Mangotsfield, Bristol, BS16 9BQ
G8 HSS M. Saxon, 4 The Coppice, Impington, Cambridge, CB24 9PP
G8 HST M. Sanders, 19 Brunswick Gardens, Hainault, Ilford, IG6 2QU
G8 HSV S. Otter, Redlands Back Lane Bilsby, Alford, LN13 9PT
GM8 HSY Dr H. Reekie, 5 Golf Course Road, Bonnyrigg, EH19 2EU
G8 HTA K. Parker, 20 River Avenue, Hoddesdon, EN11 0JS
G8 HTB A. Barker, Bank Royd Barn, Bank Royd Lane, Halifax, HX4 0EW
G8 HTF D. Fletcher, 40 Bentham Drive, Liverpool, L16 5EU
G8 HTM J. Taylor, Perry House, 188 Walstead Road, Walsall, WS5 4DN
G8 HTN A. Kettlety, 106 Denton Road, Audenshaw, Manchester, M34 5BD
G8 HTO A. Farrell, 206 London Road, Northampton, NN4 8AU
G8 HTW P. Allen, 46 Slade Avenue, Burntwood, WS7 2EL
G8 HTZ S. Druitt, 25 Holcroft, Orton Malborne, Peterborough, PE2 5SL
GI8 HUD T. Huddleston, 29 North Parade, Belfast, BT7 2GF
G8 HUF S. Carpenter, Fernlea, Fernhill Lane, Camberley, GU17 9HA
G8 HUG I. Coulson, 56 Potterdale Drive Little Weighton, Cottingham, HU20 3UX
G8 HUH T. Rabbitts, Laurel Cottage, Wick Lane, Highbridge, TA9 4BU
G8 HUO D. Sharpe, 37 Oulton Avenue, Bramley, Rotherham, S66 2SS
G8 HUR N. Mills, 3 Whitfield Close, Wilford, Nottingham, NG11 7AU
GW8 HUS A. Mead, 12 Wyelands View, Mathern, Chepstow, NP16 6HN
G8 HUT N. Onions, Windy Ridge, Dunmow Road, Dunmow, CM6 3PJ
G8 HUV M. Rowlands, 3 Littledown View, Great Durnford, Salisbury, SP4 6AU
G8 HUY J. Hill, 24 Hunters Hill Close, Guisborough, TS14 7FH
G8 HVF C. Billson, Knotts End, Bateman Road, Loughborough, LE12 6NN
G8 HVT M. Evans, 25 Walnut Close, Nailsea, Bristol, BS48 4YH
G8 HVV C. Goadby, Heligan, 12 School Road, Newmarket, CB8 9RX
G8 HVX A. Staniforth, 25 Brown Ct, East Brunswick, USA, 8816
G8 HVZ R. Anderson, 23 Callington Road, Saltash, PL12 6DU
G8 HWI J. Simons, 7A Walton Way, Stone, ST15 0JF
G8 HWJ E. Smith, Brickyard House, Wainfleet Road, Irby-In-The-Marsh, Skegness, PE24 5AT
GW8 HWL P. Jenkins, 20 Dimbath Avenue, Blackmill, Bridgend, CF35 6ED
G8 HWQ D. Mappin, 13 Willow Close, Filey, YO14 9NY
GW8 HWS J. Mills, 13 Egerton Street, Cardiff, CF5 1RF
G8 HXD M. Ledger, 58 Mount Pleasant Close, Lightwater, GU18 5TR
G8 HXE K. Haywood, 6 Lydney Road, Urmston, Manchester, M41 8RN
G8 HXR M. Brooke, 70 Wootton Avenue, Peterborough, PE2 9EG
G8 HXW A. Sargent, 22 Duckmill Crescent, Duckmill Lane, Bedford, MK42 0AF
GW8 HYI E. Whitfield, Erw Mor, Nebo, Amlwch, LL68 9NE
G8 HYK J. Brockwell, 10 Tregony Rise, Lichfield, WS14 9SN
G8 HYL H. Tuff, 4 Battery Terrace, Mevagissey, St Austell, PL26 6QS
G8 HYM S. Bradley, 247 Filey Road, Scarborough, YO11 3AE
G8 HYP M. Peers, 45 Carlton Crescent, East Leake, Loughborough, LE12 6JF
GW8 HYT P. Madden, Ty Gwyn, Llandovery, SA20 0NT
G8 HZI M. Holdsworth, 2 Newman Drive, Branston, Burton-on-Trent, DE14 3DZ
G8 HZJ R. Ingamells, Moor View, Small Banks, Moorside Ilkley, LS29 0QQ
GW8 HZL D. Wildman, 7 Alder Way, West Cross, Swansea, SA3 5PD
G8 HZN R. Orchard, Tredinneck Moor, Newmill, Penzance, TR20 8XT
G8 HZQ P. Healy, 93 Tile Kiln Lane, Leverstock Green, Hemel Hempstead, HP3 8NW
G8 HZS T. Storey, 50 Longfield Road, Darlington, DL3 0HX
G8 IAJ J. Richardson, 43 Front St., Leadgate, Consett, DH8 7SB
G8 IAK R. Thomas, 88 Parkway, London, SW20 9HG
GW8 IAM S. Lloyd, 4 Cwmdu Court, Cwmdu, Crickhowell, NP8 1RU
G8 IAN M. Lees, 175 Overdale Road, Romiley, Stockport, SK6 3EN
G8 IAR P. Smith, 60 Foxdown, Overton, Basingstoke, RG25 3JQ
G8 IBC D. Herke, 24 The Lawns, Farnborough, GU14 0RF
G8 IBE R. Bailey, 6 Kestrel Close, Horsham, RH12 5WD
G8 IBK M. Murray, Heads Nook Hall, Heads Nook, Brampton, CA8 9AA
G8 IBL H. Hallybone, Birch Bassett, 52 Busbridge Lane, Godalming, GU7 1QQ
G8 IBO T. Gill, 21 Winn Road, London, SE12 9EX
G8 IBP R. May, 10 Lime Close, Wokingham, RG41 4AW
G8 IBR N. Davies, 1 Helens Close, The Street, Redgrave, Diss, IP22 1RW
G8 IC M. Dawson, 60 Ashenhurst Road, Todmorden, OL14 8DS
GM8 ICC A. Campbell, 2 Cairndhu Cottage, Cairnbaan, Lochgilphead, PA31 8SQ
GW8 ICT C. Hopley, Clayton Cottage, Alltami Road, Mold, CH7 6RW

G8 IDE J. Pimlott, 40 Queens Road, Higher St. Budeaux, Plymouth, PL5 2NW
G8 IDJ I. Judd, 33 Coles Mede, Otterbourne, Winchester, SO21 2EG
G8 IDK R. Voisey, 2 Chester Place, Malvern, WR14 1RQ
G8 IDL D. Smith, The Old Forge, High Street, Brinkley, Newmarket, CB8 0SE
G8 IEA S. Parham, 132 Wrotham Road, Gravesend, DA11 7LB
G8 IEI J. Mckillop, 2 Moores Green, Wokingham, RG40 1QG
G8 IEL R. Tust, 28 Osprey Close, Beechwood, Runcorn, WA7 3JH
GM8 IEM M. Hall, 199 Clashmore, Lochinver, Lairg, IV27 4JQ
G8 IER P. Nice, 31 Elizabeth Drive, Chapel St. Leonards, Skegness, PE24 5RS
G8 IEV B. Guy, Hawthorn Folly, Cul De Sac, Boston, PE22 8EY
G8 IEW C. Davies, Applecroft, St. Johns Road, Gloucester, GL2 7DF
G8 IEZ C. Moss, 11 Sheepfold Crescent Barrow, Clitheroe, BB7 9XR
G8 IFF N. Gunn, 1865 El Camino Drive, Xenia, USA, 45385
G8 IFH K. Thomas, 1 Byways, Yateley, GU46 6NE
G8 IFN N. Hinderwell, 1 Bower Grove, West Mersea, CO5 8GJ
G8 IFT I. Gordon, 40 Grange Crescent, Rubery, Birmingham, B45 9XB
G8 IHA J. Gregory, 2 Abbey Dale Close, Kilburn, Belper, DE56 0PY
G8 IHC S. Styler, 85 Fairoaks Drive, Great Wyrley, Walsall, WS6 6HA
G8 IHF D. Cochrane, 18 Russell Avenue, Dunchurch, Rugby, CV22 6PX
G8 IHT S. Chambers, 7 Mowbray Road, Northallerton, DL6 1QT
GM8 IID N. Paterson, 4 Cambridge Road, Renfrew, PA4 0SL
G8 IIG G. Punter, 18 Lodge Road, Sharnbrook, Bedford, MK44 1JP
GM8 IIH W. Jarvie, Berryhill Farm, Tak-Ma-Doon Road, Kilsyth, Glasgow, G65 0RY
G8 III L. Roberts, 32 Orion, Bracknell, RG12 7YX
G8 IIK D. Hooker, 2 Fernlea Court, Lydd Road, Camber, Rye, TN31 7RS
GM8 IIO W. Robson, 18 Colinton Mains Green, Edinburgh, EH13 9AG
G8 IIS B. Heaney, 19 Ormonde Drive, Liverpool, L31 7AN
G8 IIZ W. Rush, 17 Hagden Lane, Watford, WD18 0HQ
G8 IJC Dr C. Phillipson, 24 Wyatt Close, Martin, Lincoln, LN4 3RN
G8 IJE B. Laxton, 1 Stoney Lane, Walsall, WS3 3RF
G8 IJG D. Adams, 77 Chestnut Crescent, Shinfield, Reading, RG2 9HA
G8 IJI K. Williamson, 4 Lynwood Drive, Wakefield, WF2 7EF
G8 IJM H. Wallington, 5 Glebe Road, Royal Wootton Bassett, Swindon, SN4 7DU
G8 IJS R. Sayer, Vignouse, Paimpont, France, 35380
GW8 IJT M. Cawood, 51 Mayflower Drive, Marford, Wrexham, LL12 8LD
G8 IK V. Morse, 42 Kingscote Road, Dorridge, Solihull, B93 8RA
G8 IKA D. Poll, 66 Southlands Avenue, Orpington, BR6 9NF
G8 IKG K. Raper, 26 Lancaster Way, Scalby, Scarborough, YO13 0QH
GW8 IKH R. Rolley, Glas Cwm, Dyffryn Crawnon, Crickhowell, NP8 1NU
G8 IKK J. Channon, 49 Fallow Road, Helston, TR13 8WH
G8 IKS D. Warwick, Orchard Cottage, Colber Lane, Bishop Thornton, Harrogate, HG3 3JR
G8 IKW P. Nutt, 1 Lime Gardens, Middleton, Manchester, M24 4AE
G8 ILB N. Allinson, 42 Rook Lane, Stockton-on-Tees, TS20 1SB
G8 ILD R. Barrow, 50 Redhill Drive, Bredbury, Stockport, SK6 2HQ
G8 ILG J. Law, 29 Brackenwood, Orton Wistow., Peterborough, PE2 6YP
G8 ILJ S. Nutt, 23A Hesketh Drive, Southport, PR9 7JX
G8 ILN M. Grindrod, 20 Castle Mead, Kings Stanley, Stonehouse, GL10 3LD
G8 ILP T. Voller, 179 High Street, Harriseahead, Stoke-on-Trent, ST7 4JU
G8 ILU J. Parker, 15Burton Road, Heckington, NG34 9QR
G8 ILW D. Couse, 6 Reading Drive, Sale, M33 5DL
G8 ILZ I. Walker, 113 Whitlock Drive, Wimbledon, London, SW19 6SH
G8 IMB M. Stubbs, Crofters, Harry Stoke Road, Bristol, BS34 8QH
G8 IMH M. Fereday, 35 Manor House Park, Codsall, Wolverhampton, WV8 1ES
G8 IMI C. Kitchener, Hamrest 5, Mill Road, Cromer, NR27 0BG
G8 IMJ R. Head, 21 Church Street, Fleetwood, FY7 6JR
G8 IMM R. Keeley, 6 Standings Rise, Whitehaven, CA28 6SX
G8 IMS M. Stroud, 39 Brocks Drive, Guildford, GU3 3NE
G8 IMZ A. Palfrey, 5 Ings View, York, YO30 5XE
G8 INA D. Harris, 102 Greatmeadow, Northampton, NN3 8DF
G8 INC K. Davenport, 10 Woodend Lane, Hyde, SK14 1DT
G8 INL B. Miller, 1 The Meadows, Monk Fryston, Leeds, LS25 5PJ
G8 INO A. Brown, 25 Birch Lane, Haxby, York, YO32 3RP
G8 INS P. Williams, 2 Rosamund Road, Crawley, RH10 6QF
G8 INZ T. Prentice, 36 Ives Close, Yateley, GU46 7RD
G8 IOA P. Crockford, 24 High Street, Easton On The Hill, Stamford, PE9 3LN
G8 IOJ D. Martin, 54 The Crossway, Portchester, Fareham, PO16 8PB
G8 IOK J. Noden, 1 Ashley Court, Providence Hill, Southampton, SO31 8AT
GM8 IOL R. Thomson, Middlerig Farm, Bathgate, EH482HH
G8 ION Dr J. Hollis, 19 Burlington Grove, Sheffield, S17 3PH
G8 IOS K. Evans, 12 Moxhull Drive, Sutton Coldfield, B76 1LZ
G8 IOW P. Wright, 70 Hardy Barn, Shipley, Heanor, DE75 7LY
G8 IPA A. Powell, 8 Penzoy Avenue, Bridgwater, TA6 5BT
G8 IPF H. Billingham, Tanglewood, Brookside Orchard, Pulborough, RH20 3BD
G8 IPG A. Shaw, 92 Freemantle Road, Romsey, so510ax
G8 IPK C. Knight, The Lodge, 16A Cromptons Lane, Liverpool, L18 3EX
G8 IPN C. Foote, 3 Mere Road, Weybridge, KT13 9NU
G8 IPQ A. Badcock, 7 Heathfield Road, Chandler'S Ford, Eastleigh, SO53 5RP
G8 IPT P. Hughes, 27 Hemsworth Avenue Little Sutton, Ellesmere Port, CH66 4SG
G8 IPY B. Hewitt, 177 Avery Hill Road, London, SE9 2EX
G8 IQA F. Hall, 34 Dronfield Road, Eckington, Sheffield, S21 4BR
GW8 IQC M. White, 5 Marlowe Close, Rogerstone, Newport, NP100bt
G8 IQF C. Newell, 16A Pembroke Road, Framlingham, Woodbridge, IP13 9HA
G8 IQT T. Spicer, 3 Parkers Fields, Quorn, Loughborough, LE12 8EJ
G8 IQX M. Dixon, 57 Northease Drive, Hove, BN3 8PP
G8 IRC D. De Fraine, Block 8 Lot 3, Rosalina Village 1, Upper Libby Road, Davao City, Philippines, 8023
G8 IRL Dr K. Brown, 56 Haydock Close, Alton, GU34 2TL
G8 IRM M. Emery, 45 Old Pasture Road, Frimley, Camberley, GU16 8RT
G8 IRN A. Telford, 9 Fellside, Tower Wood, Windermere, LA23 3PW
G8 IRS J. Wiles, 12A Ashling Gardens Denmead, Waterlooville, PO7 6PR
G8 ISE G. Sharp, 46 Coronation St., Monk Bretton, Barnsley, S71 2ES
G8 ISI F. Breame, 68 Church Road, Bramshott, Liphook, GU30 7SH
G8 ISJ J. Witt, 67 Dillotford Avenue, Coventry, CV3 5DS
G8 ISM P. Goldsmith, 5 Old School Court, Rock Hill Road, Ashford, TN27 9DW
G8 ITB R. Perzyna, 29 Lakeside Drive, Bromley, BR2 8QQ
GI8 ITD T. Davidson, 26 Lower Parklands, Dungannon, BT71 7JN
GU8 ITE D. Eaton, Glenfield, Le Foulon, St Andrew, Guernsey, GY6 8UF
G8 ITG P. Levitt, 21 Station Road, Firsby, Spilsby, PE23 5PX
GW8 ITI J. Evans, Rosegarth, Woodbine Road, Blackwood, NP12 1QH
G8 ITJ M. Adnams, 8 Webb Street, Nuneaton, CV10 8JQ
G8 ITU P. Wragg, 7 St. Johns Mount, Thirsk Road, Easingwold, York, YO61 3HG
G8 ITX O. Williams, Meadow View, Irthington, Carlisle, CA6 4NN
G8 IUB Birmingham ARS, c/o D. Cottam, 14 Barnard Close, Rednal, Birmingham, B45 9SZ
G8 IUC R. Glover, 8 Woodberry Way, Chingford, London, E4 7DX
G8 IUD B. Sermons, 17 Well Side, Marks Tey, Colchester, CO6 1XG
G8 IUG P. Tewkesbury, 267 York Road, Stevenage, SG1 4HD
GW8 IUM M. Richardson, 15 Brynfa Avenue, Welshpool, SY21 7TS
G8 IUN S. Tolputt, Walnut Lodge, Annings Lane, Bridport, DT6 4QN
G8 IUP I. Walukiewicz, Louise Cottage, Branksome Avenue, Stockbridge, SO20 6AH
G8 IUQ M. Wareing, 20 Middlesex Avenue, Burnley, BB12 6AA
G8 IVB P. Samson, 49 Crest View Drive, Petts Wood, Orpington, BR5 1BZ
G8 IVO R. Hartland, Three Gables, Crozens Lane, Hereford, HR1 1XY
GM8 IVR P. Ager, 50 Pitbauchlie Bank, Dunfermline, KY11 8DP
G8 IWB A. Parker, 33 Colerne Drive, Hucclecote, Gloucester, GL3 3SX
G8 IWE R. Thomas, 6 Copeland Drive, Poole, BH14 8NW
G8 IWF T. Bierney, 5318 N 106 Avenue, Glendale, USA, 85307
G8 IWI J. Pearce, 34 Fleetwood Avenue, Westcliff-on-Sea, SS0 9RA
G8 IWJ G. Strange, 12 Bronington Avenue, Bromborough, Wirral, CH62 6DT
GM8 IWL Dr C. Sutherland, 15 Stanley Drive, Brookfield, Johnstone, PA5 8UF
G8 IWO N. Jones, 14 Salcombe Grove, Swindon, SN3 1ER
G8 IWQ A. Jacques, 17 Pyrethrum Way, Willingham, Cambridge, CB24 5UX
GM8 IWR T. Campbell, 5 Coastguard Station, Heugh Road, Portpatrick, Portpatrick, DG9 8TF
G8 IWT R. Shears, 15 Hale Pit Road, Great Bookham, Leatherhead, KT23 4BS
G8 IWX B. Homer, 116 Shorncliffe Road, Folkestone, CT20 2PQ
G8 IXC L. Prior, 64 Montfort Road, Walderslade, Chatham, ME5 9HA
G8 IXK B. Owen, 21 Marlborough Road, Luton, LU3 1EF
G8 IXL P. Baker, Doules Mead, Heath Lane, Farnham, GU10 5PA
G8 IXM R. Pickett, 126-127 Cuckoofield Lane, Mulbarton, Norwich, NR14 8BA
G8 IXN K. Watkins, 23 Mount Ambrose, Redruth, TR15 1NX
G8 IXP R. Lister, 8 Carlton Avenue, Wilmslow, SK9 4EP
G8 IXX J. Brister, 49 Tiverton Road, Loughborough, LE11 2RU
GM8 IXZ A. Legood, 25 Frankfield Place, Dalgety Bay, Dunfermline, KY11 9LR
G8 IYD D. Hancock, 4 Elmside, Willand, Cullompton, EX15 2RN
G8 IYE P. Shore, 1 Whatsill, Hopton Wafers, Kidderminster, DY14 0QB
G8 IYH A. Bevington, Malthouse, Hoggs Lane, Purton, Swindon, SN5 4HQ
G8 IYJ C. Buckland, 7 The Maltings, Royal Wootton Bassett, Swindon, SN4 7EZ
G8 IYK R. Sayers, 3 Riversdale Cottages, The Staithe, Stalham, Norwich, NR12 9BY
G8 IYN C. Marsh, 6 De Burgh Hill, Dover, CT17 0BS
G8 IYS J. Simkins, 18 Riding Hill, South Croydon, CR2 9LN
G8 IYZ A. Barker, 8 Manor Avenue, Attenborough, Nottingham, NG9 6BP
GI8 IZB Beaneaters DX Group, c/o J. Crawford-Baker, 131 Gobbins Road, Islandmagee, Larne, BT40 3TX
G8 IZR P. Higginson, 18 Park Meadow, Westhoughton, Bolton, BL5 3UZ
G8 IZW P. Cain, 22 Ditton Green, Luton, LU2 8RU
G8 IZY S. Eldridge, 6 Cobbles Crescent, Crawley, RH10 8HA
G8 JAB A. Berriman, Meadowside, Little-In-Sight, St Ives, TR26 1AX
G8 JAC A. Jackson, 59 Leas Road, Warlingham, CR6 9LP
G8 JAD J. Townsend, 56 Seymour Road, Northfleet, Gravesend, DA11 7BN
G8 JAG D. Williams, Flat 7, Yewbarrow Lodge, Main Street, Grange-over-Sands, LA11 6EB
G8 JAI A. Livesley, Gates Garth, Barbon, Carnforth, LA6 2LJ
G8 JAN P. Biggadike, 49 Willow Road, Downham Market, PE38 9PG
G8 JAQ J. Walker, 2 Morris Drive, Stafford, ST16 3YE
G8 JAW B. Heed, 3 Woodcote Green Downley, High Wycombe, HP13 5UN
G8 JAY A. Jay, Jasper, The Reddings, Cheltenham, GL51 6RT
G8 JBC C. Jervis, 8 Portobello Close, Willenhall, WV13 3QA
G8 JBD P. Godfrey, 3 Lowry Way, Lowestoft, NR32 4LW
GM8 JBJ J. Berry, Willowburn, Kirkton, Hawick, TD9 8QJ
G8 JBM Dr S. Wood, 18 Grange Road, Shanklin, PO37 6NN
G8 JBP G. Head, 34 Balds Lane, Stourbridge, DY9 8SG
G8 JBQ R. Hughes, Court Church View, South Perrott, Beaminster, DT8 3HU
G8 JBT D. Bellingham, 22 Princes Drive, Codsall, Wolverhampton, WV8 2DJ
G8 JBV D. Dawe, 7 Princes Road, Romford, RM1 2SR
G8 JCB P. Pullinger, 1 Sycamore Cottages, Upper Wield, Alresford, SO24 9RP
G8 JCC C. Purchase, 35 Pasture Way, Bridport, DT6 4DW
G8 JCD M. Northey, Achill Mist House, Kilmeaney, Listowel, Ireland, V31 VX95
GM8 JCF P. Carnegie, 29 Castle Terrace, Cullen, Buckie, AB56 4SD
G8 JCL J. Essex, 40 Lincoln Walk, Heywood, OL10 3JB
G8 JCN W. Allen, 143 Cherry Crescent, Rawtenstall, Rossendale, BB4 6DS
G8 JCS A. Bunting, Manor House, Market Place, Binbrook, LN8 6DE
G8 JCV P. Hewitt, 28 Amersham Avenue, Langdon Hills, Basildon, SS16 6SJ
GW8 JDB V. Grayson, Willow Lodge, Croeslan, Llandysul, SA44 4SJ
G8 JDC T. Robinson, 17 Balliol Road, Brackley, NN13 6LY
G8 JDD R. Kelsall, The Cottage, Denford Road, Stoke-on-Trent, ST9 9QG
G8 JDN B. Deefholts, Yew Tree House, 28 St Marys Road, Meare, Glastonbury, BA6 9SP
G8 JDQ K. Few, 35 Whitton Close, Swavesey, Cambridge, CB24 4RT
GW8 JEI N. Cross, Glan Alaw, Llanddeusant, Holyhead, LL65 4AG
G8 JEM E. Cheer, 15 Stibbs Way, Bransgore, Christchurch, BH23 8HG
G8 JET D. Higginson, 43 North Street West Butterwick, Scunthorpe, DN17 3JR
G8 JFC F. Wilmott, 2 Manor Close, Misson, Doncaster, DN10 6HE
G8 JFL D. Crough, 32 Roundaway Road, Ilford, IG5 0NP
G8 JFT N. Hewitt, 36 Princes Terrace, Kemp Town, Brighton, BN2 5JS
G8 JFX T. Simmons11, Cedar House 5, Blueberry Close, Maidwell, NN6 9XL
GM8 JGB W. Fleming, 65 Dundonald Park, Cardenden, Lochgelly, KY5 0DG
G8 JGE C. Newbury, 37 Johns Avenue, Hendon, London, NW4 4EN
G8 JGF P. Walters, 3 Inkerman Street, Selston, Nottingham, NG16 6BQ
G8 JGL N. Owen, 59 Fernwood Drive, Leek, ST13 8JA
G8 JGM J. Martin, 19C Willow Tree Road, Altrincham, WA14 2EQ
G8 JGU R. Hallam, 37 Dingle Avenue Appley Bridge, Wigan, WN6-9LF
G8 JHA T. White, 24 Chapel Street, Tingley, Wakefield, WF3 1RE
G8 JHC I. Whitworth, 104 The Dormers, Highworth, Swindon, SN6 7PD
G8 JHE M. Brogan, 31 Rempstone Road, East Leake, Loughborough, LE12 6PW

G8 JHG Dr J. Collins, Hill Crest, The Hill, Millom, LA18 5HB
G8 JHH M. Baugh, 71 Hatch Lane, Old Basing, Basingstoke, RG24 7EF
G8 JHL J. Lovell, 2 Moran Close, Wilmslow, SK9 3UF
G8 JHM I. Carney, 39 Blenheim Crescent, Luton, LU3 1HB
G8 JHO P. Evans, 5 Hunters Close, Bilston, WV14 7BN
G8 JIE C. Riding, 14 The Coppice, Clayton Le Moors, Accrington, BB5 5RU
G8 JIP G. Miller, 39 Scrivens Mead, Thatcham, RG19 4FQ
G8 JIS T. Macey, Whitegates, Histons Hill, Wolverhampton, WV8 2HA
G8 JIT J. Mckinnon, 142 Hughes Street, Bolton, BL1 3EZ
G8 JIU P. Dunham, 19 The Lunds, Kirk Ella, Hull, HU10 7JJ
G8 JJF J. Puddifoot, 23 Alwyn Street, Liverpool, L17 7DT
G8 JJK T. Barrett, Flat 38, Queens Court, Cheltenham, GL50 2LU
GM8 JJN J. Pryde, 7 The Engine Green, Fishcross, Alloa, FK10 3JN
GW8 JJP P. Tabberer, 8 Wynnstay Road, Old Colwyn, Colwyn Bay, LL29 9DS
G8 JJR K. Mcmahon, 27 Marlborough Avenue, Doncaster, DN5 8EH
GW8 JJZ R. Merrick-Jenkins, 17 Marlais Park, Carmel, SA14 7UF
G8 JKB C. Hemmings, 11 Brookside, Desborough, Kettering, NN14 2UD
GW8 JKC J. Kendall, 26 Bryn Seiri Road, Conwy, LL32 8NR
G8 JKD C. Littman, 70 Orbel Street, London, SW11 3NY
G8 JKV D. Leary, Blackers Hill Farm, Lowndes Drove, Needingworth, St Ives, PE27 4NE
G8 JLA K. Turner, 13 Stanhope Street, Saltburn-by-the-Sea, TS12 1AL
G8 JLB B. Silver, 280 Britten Road, Brighton Hill, Basingstoke, RG22 4HR
G8 JLD J. Garters, Sun Patch, Garfield Road, Hailsham, BN27 2BT
G8 JLM P. Higham, 56 Coopers Avenue, Heybridge, Maldon, CM9 4YX
G8 JLY L. Leach, 2 Nightingale Place, Droitwich, WR9 7HG
G8 JMB J. Button, 16 Meadow Rise, Broadstone, BH18 9ED
G8 JMG J. Gartland, 175 Talbot Street, Whitwick, Coalville, LE67 5AY
G8 JMK D. Butler, 144 Longridge Way, Weston-Super-Mare, BS24 7HS
G8 JMO B. Justin, 1704 Cottontown, Forest Virginia, USA, 24551
G8 JMP D. Beech, 8 Copthorne Drive, Lightwater, GU18 5TE
G8 JMS S. Miller, 44 Greenway Road, Galmpton, Brixham, TQ5 0LZ
G8 JMU J. Potter, 15 Alterton Close, Goldsworth Park, Woking, GU21 3DD
G8 JMY D. Hugman, 6 Barley Close, Henley-in-Arden, B95 5HU
G8 JNI D. Bookham, 1 Monks Rise, Fleet, GU51 4HB
G8 JNO S. Munday, 25 Southend Road, Weston-Super-Mare, BS23 4JY
G8 JNR R. Hedderley, 17 Linford Close, Handsacre, Rugeley, WS15 4EF
G8 JNZ K. Crowder, 15 Fleetwood Close, Minster On Sea, Sheerness, ME12 3LN
GI8 JOA D. Thompson, 16 Lynden Gate Park, Portadown, Craigavon, BT63 5YJ
G8 JOC Dr E. Powell, 49 Normanby Road, Worsley, Manchester, M28 7TS
G8 JOX J. Dobson, Home Farm, Mansmore Lane, Charlton On Otmoor, Kidlington, OX5 2US
GW8 JOY T. Bowen, 7 Bedford Close, Greenmeadow, Cwmbran, NP44 5HN
G8 JPA J. Hunt, Woodstone Farm, High Common Road, Diss, IP22 2HS
GI8 JPF T. Phillips, 52 Belfast Road, Bangor, BT20 3PU
G8 JPJ D. Jones, 17 Stanmore Close, Clacton-on-Sea, CO16 7HQ
G8 JPU D. Potts, 25 Southlands Road, Congleton, CW12 3JY
G8 JPV M. Parnell, 101 Ridgeway, Wellingborough, NN8 4RZ
G8 JPW J. Abbott, 11 Red House Road, Bodicote, Banbury, OX15 4BB
G8 JQG J. Hough, 77 Pennine Court, Macclesfield, SK10 2RN
G8 JQH P. Wright, 33B Slack Lane, Crofton, Wakefield, WF4 1HX
G8 JQS G. Greensmith, Japonica, Hawthorne Avenue, Biggin Hill, Westerham, TN16 3SG
G8 JQV D. Marchant, 11 Derehams Lane, Loudwater, High Wycombe, HP10 9RH
G8 JQW R. Thomas, 8 Sherwood Court, Sherwood Street, Bolsover, Chesterfield, S44 6GF
GI8 JRE J. Donnelly, 9 Lomond Heights, Cookstown, BT80 8XW
G8 JRF M. Willson, 19 The Willows, Highworth, Swindon, SN6 7PG
G8 JRN R. Stockdale, 53 Brightwalton, Newbury, RG20 7BT
G8 JRW M. Austin, The House, Four Seasons Village, Winkleigh, EX19 8DP
G8 JRZ A. Mills, 42 Mora Avenue, Chadderton, Oldham, OL9 0EJ
G8 JSC K. Austin, 139 Sewall Highway, Coventry, CV2 3NG
G8 JSE F. Cowlin, 9 Zealand Close, Hinckley, LE10 1TJ
G8 JSF R. Williams, 35 Broadhurst Grove, Lychpit, Basingstoke, RG24 8SB
G8 JSL P. Smith, 13 Manor Garth, Pakenham, Bury St Edmunds, IP31 2LB
G8 JSM C. Wood, 57 Holly Crescent, Rainford, St Helens, WA11 8ER
G8 JSN P. Bailey, 50 Amis Avenue, New Haw, Addlestone, KT15 3ET
G8 JSR V. Hinksman, 1 Shaw Lane, East Woodburn, Hexham, NE48 2SL
G8 JTD Otley ARS, c/o J. Castelow, 7 Langford Close, Burley In Wharfedale, Ilkley, LS29 7NP
G8 JTG E. Spanton, 14 Days Lane, Sidcup, DA15 8JN
G8 JTL M. Davies, 25 Walker Avenue, Quarry Bank, Brierley Hill, DY5 2LY
G8 JUC J. Wheatley, 44 Kingswood Close, Boldon Colliery, NE35 9LG
G8 JUG N. Spenceley, 18 Rectory Road, Broadmayne, Dorchester, DT2 8EG
G8 JUK B. Storeton-West, Nazdar, Camps Heath Oulton, Lowestoft, NR32 5DW
G8 JUS T. Gale, 58 Westwood Road, Newbury, RG14 7TL
G8 JUT S. Linton, 41 Long Close, Bristol, BS162UF
G8 JUV S. York, 10 Beechwood Avenue, Wallasey, CH45 8NX
GM8 JUY R. Mcmillan, 12 Parkthorn View, Dundonald, Kilmarnock, KA2 9EZ
G8 JVE M. Rowe, 97 Old Worthing Road, East Preston, Littlehampton, BN16 1DU
G8 JVI A. Hicks, 10 Evans Close, Eynsham, Witney, OX29 4QY
G8 JVM R. Bown, Park View, Chapel Street, Telford, TF4 3DD
G8 JVS Rev. M. Fairey, 10 Mallard Close, York, YO10 3BS
G8 JVU A. Johnson, Clematis Cottage, Wheatlow Brooks, Stafford, ST18 0EW
G8 JVV J. Burchell, Gooseleys Farm, Harrow Hill, Halstead, CO9 4LX
G8 JVW P. Boswell, 10 The Grange, Wombourne, Wolverhampton, WV5 9HX
GM8 JVZ Dr M. Nimmo, The Court, 6 Farington Street, Dundee, DD2 1PJ
G8 JWC D. Luscombe, 31 Tewkesbury Drive, Prestwich, Manchester, M25 0HR
G8 JWD I. Rees, Knowle Cottage, Whittonditch Road, Marlborough, SN8 2PX
G8 JWE J. Hickman, 41 Field Road, Ramsey, Huntingdon, PE26 1JP
G8 JWK R. Staveley, 52 New Road, Wootton Bassett, Swindon, SN4 7DG
GW8 JWL G. Smith, 13 Lapwing Close, Penarth, CF64 5GA
GW8 JWP J. Griffiths, Llygad Yr Haul, Ferwig, Cardigan, SA43 1PX
GM8 JWQ K. Faloon, Moss-Side Croft, 6 Rothiemay, Huntly, AB54 5NY
G8 JWT R. Trett, Low Barn, Norwich Road, Woodton, Bungay, NR35 2LP
G8 JWX J. Wright, 7 Basket Gardens, London, SE9 6QP
G8 JXG J. Dean, 6 Greenleas, Pembury, Tunbridge Wells, TN2 4NS
G8 JXK S. Blew, 24 Batts Park, Taunton, TA1 4RE
G8 JXP D. Mccabe, 78 Oakleigh Road, Stratford-upon-Avon, CV37 0DN
G8 JXS M. Stephenson, 6 Cedar Road, Tewkesbury, GL20 8PX
G8 JXU C. West-Bulford, 25 Sunnyside Close, Heacham, King's Lynn, PE31 7DX
G8 JXV T. Trew, Stockers Lodge, Bere Farm Lane, Fareham, PO17 6JJ
G8 JYN Basingstoke ARC, c/o P. Cresswell, 108 Hawthorn Way, Basingstoke, RG23 8NH
G8 JYS M. Fletcher, 20 Leahurst Close, Norton, Malton, YO17 9DF
G8 JYV K. Dumbill, 30 Caithness Drive, Crosby, Liverpool, L23 0RQ
G8 JYX P. Johnson, 42 College Gardens, London, E4 7LG
G8 JZI P. Smith, 4 Fellstone Vale, Withnell, Chorley, PR6 8UE
G8 JZO J. Gibbs, 6 Southampton Close, Blackwater, Camberley, GU17 0HB
G8 JZT S. Osborn, 67 Chessington Avenue, Bexleyheath, DA7 5NP
G8 JZX C. Stephenson, Armanby, Main Street, Selby, YO8 8QT
G8 JZZ R. Taylor, Higher Priestacott, Belstone, Okehampton, EX20 1QX
G8 KAE R. Bushell, 102 Winchester Gardens, Northfield, Birmingham, B31 2QB
G8 KAM J. Hurnandies, 70 Orchard Rise West, Sidcup, DA15 8SZ
G8 KAP D. Patrick, Quarryside, Stockdalewath, Dalston, Carlisle, CA5 7DP
G8 KAS B. Buschl, 27 The Drive, Court Farm Road, Newhaven, BN9 9DJ
G8 KB P. Johnson, 55 Rodney Hill, Loxley, Sheffield, S6 6SG
G8 KBB D. Roberts, 32 Woodbridge Close, Appleton, Warrington, WA4 5RD
G8 KBG A. Price, 21 Packwood Close, Bentley Heath, Solihull, B93 8AN
G8 KBH D. Ward, 3 Sherbourne Close, Poulton-le-Fylde, FY6 7UB
GW8 KBK A. Short, 70 Tremains Court, Brackla, Bridgend, CF31 2SS
G8 KBM I. Whiting, 11 Woodsend Close, Burton Joyce, Nottingham, NG14 5DY
G8 KCB J. Nally, 313 Wyndhurst Road, Stechford, Birmingham, B33 9DL
GW8 KCH K. Houston, 6 Ashgrove, Llanellen, Abergavenny, NP7 9HP
GW8 KCY M. Bover, Glynfach Bungalow, Pontyates, Llanelli, SA15 5TF
G8 KDD D. Coton, 17 Flambards Close, Meldreth, Royston, SG8 6JX
G8 KDF Dr M. Sach, Old School, Cambridge Road, St Neots, PE19 6ST
G8 KDM A. Smith, 2A Chesterfield Road, Barlborough, Chesterfield, S43 4TR
G8 KDO Dr P. Topham, 5 Kings Road, Cambridge, CB3 9DY
G8 KDU R. Eager, 45 Fleetwood Avenue, Herne Bay, CT6 8QW
G8 KEA M. Sutton, 178 Cole Lane, Borrowash, Derby, DE72 3GN
G8 KED C. Mullineaux, 27 Ashfield Avenue, Lancaster, LA1 5EB
G8 KEJ M. Johnson, 23 The Crest, Surbiton, KT5 8JZ
G8 KEK P. Wilson, 5 Mons Close, Harpenden, AL5 1TD
G8 KEO P. Dickinson, Halshanger Farm, Ashburton, Newton Abbot, TQ13 7HY
GI8 KEP K. Bones, 54 Derryvolgie Park, Lisburn, BT27 4DA
GW8 KEV K. Shafto, 3 Harding Close, Boverton, Llantwit Major, CF61 1GX
G8 KFD R. Gwynn, 36 Woodstock Close, Burbage, Hinckley, LE10 2EG
G8 KFF R. Parker, 17 Valley Road, Streetly, Sutton Coldfield, B74 2JE
GI8 KFG P. Douglas, 21 Hillhead Road, Ballycarry, Carrickfergus, BT38 9HE
G8 KFJ D. Greig, 23 Parsons Walk, Walberton, Arundel, BN18 0PA
G8 KFK P. Loten, 15 Hornsea Burton Road, Hornsea, HU18 1TP
G8 KFN R. Heron, 46 Bradvue Crescent, Bradville, Milton Keynes, MK13 7AJ
G8 KFS B. Russell, 56 Kingsmead Avenue, Surbiton, KT6 7PP
G8 KGE S. Bailey, 50 Quantock Close, Warmley, Bristol, BS30 8UT
G8 KGG A. Ward, 49 Spielplatz, Lye Lane, St Albans, AL2 3TD
G8 KGR R. Tidswell, Helloplane, Clubhurn Lane, Spalding, PE11 4BQ
G8 KGS C. Suslowicz, 2366 Coventry Road Sheldon, Birmingham, B26 3LS
G8 KGV P. Jessop, 84 Common Road Kensworth, Dunstable, LU6 3RG
G8 KHF J. Dove, 33 The Haystack, Daventry, NN11 0NZ
G8 KHH C. Young, 26 Horsham Avenue, Peacehaven, BN10 8HX
G8 KHI R. Partridge, 5 Beck Close, Mundesley, NR11 8QL
G8 KHU D. Fielding, 216 Andover Road, Newbury, RG14 6PY
G8 KHV R. Evans, 6 Park End, Lichfield, WS14 9US
G8 KIG P. Winwood, 2 The Warren, Abingdon, OX14 3XB
G8 KIH J. Sargent, 9 Lee Woottens Lane, Basildon, SS16 5HD
G8 KIK D. Bland, 17 Knowles Close, Kirklevington, Yarm, TS15 9NL
GM8 KIQ J. Harper, 11 Cathburn Holding, Cathburn Road, Wishaw, ML2 9QL
G8 KIW H. Muller, 118 Park Lane, Northampton, NN5 6PZ
G8 KIZ S. Morris, 23 Ellesmere Way, Carlisle, CA2 6LZ
G8 KJI J. Richardson, 4 Torrington Lane, East Barkwith, Market Rasen, LN8 5RY
G8 KJJ L. Haywood, 9 Canberra Crescent, West Bridgford, Nottingham, NG2 7FL
GW8 KJK G. Park, 34 Delafield Road, Abergavenny, NP7 7AW
GM8 KJO D. Moodie, 1 Lageonan Road, Grandtully, Aberfeldy, PH15 2QY
G8 KJP P. King, 25 Lamellyn Drive, Truro, TR1 3JR
G8 KJT R. Burgess, 3 Deeside Avenue, Chichester, PO19 3QF
G8 KKA B. Stevens, 2 Hawthorn Crescent, Shepton Mallet, BA4 5XR
G8 KKD D. Jones, Little Haynes Barn, Kemble, Cirencester, GL7 6BS
G8 KKG M. Elliott, 9 Moyleen Rise, Marlow, SL7 2DP
G8 KKH C. Hills, 8 Blackdale, Cheshunt, Waltham Cross, EN7 6DF
G8 KKN D. Mcfarlane, 10 Green Lane, Vicars Cross, Chester, CH3 5LA
G8 KKU J. Walker, 21 Garden Hedge, Leighton Buzzard, LU7 1DJ
G8 KLA C. Bonner, 6 Camley Park Drive, Maidenhead, SL6 6QF
G8 KLC P. Webber, 37 Rasen Road, Market Rasen, LN8 3XL
G8 KLT P. Valteris, 25 Copthill Way, Houghton le Spring, DH4 4FB
G8 KMK Kirkless Raynet, c/o G. Edinburgh, 77 Westerley Lane, Shelley, Huddersfield, HD8 8HP
G8 KMM J. Bryant, 12 Dale Tree Road, Barrow, Bury St Edmunds, IP29 5AD
G8 KMP M. Pollock, 25 Meadow Lane, Burgess Hill, RH15 9HZ
G8 KMR M. Davis, 8 Mead Close, Leckhampton, Cheltenham, GL53 7DX
G8 KNC S. Wheatley, The Gables, Green Road, Wivelsfield Green, Haywards Heath, RH17 7QA
G8 KNF D. Hawkins, 109 Elphinstone Road Walthamstow, London, E17 5EY
GW8 KNJ J. Blinco, 2 Agnes Hunt Drive, Park Hall, Oswestry, SY11 4FE
G8 KNN J. Bigwood, 133 Gilbert Road, Cambridge, CB4 3PA
G8 KNS M. Jelfs, Adams Acre, Chapel Lane, Wimborne, BH21 3SL
G8 KNU R. Jacobs, 1 Coverdale, Northampton, NN2 8UU
G8 KOC R. Backham, 15 Rushmead Close, South Wootton, King's Lynn, PE30 3LY
G8 KOD R. Adams, Swinneys, Station Road, Carterton, OX18 3PR
GM8 KOF D. Mcnaughton, 6 Wilderhaugh Court, Galashiels, TD1 1QL
G8 KOL D. Slocombe, 7 Talbot Avenue, Herne Bay, CT6 8AD
G8 KOM D. Hanson, 42 Choseley Road, Knowl Hill, Reading, RG10 9YT
G8 KOQ N. Morris, 88 Tynesbank, Worsley, Manchester, M28 0SL
G8 KOS S. Head, 3 Ripon Gardens, Waterlooville, PO7 8ND
G8 KOZ A. Clarke, 42 Chamberlain Way, Biddulph, Stoke-on-Trent, ST8 7BB
G8 KPD B. Fothergill, 53 Meadow Court Ponteland, Newcastle upon Tyne, NE20 9RA

G8 KPE E. Howard, 15 Amherst Road, Bexhill on Sea, TN40 1QH
G8 KPG G. Wright, 58 Lifton Croft, Kingswinford, DY6 8RZ
GM8 KPH M. Hobson, 17 Well Brae, Pitlochry, PH16 5HH
G8 KPL S. Williams, Flat 7, Yewbarrow Lodge, Main Street, Grange-over-Sands, LA11 6EB
G8 KPV G. Hickman, Pine Tree Cottage, Calverton Road, Blidworth, Mansfield, NG21 0NW
G8 KPY D. Pratt, 77 Hayfield Road, St. Mary Cray, Orpington, BR5 2DL
G8 KQA R. Laslett, Dinnages, Street End Lane, Heathfield, TN21 8SA
G8 KQB S. Prior, East Brantwood, Manor Road, Barnstaple, EX32 0JN
G8 KQV Dr S. Evans, 4 Holcot Lane, Anchorage Park, Portsmouth, PO3 5TR
G8 KQZ G. Dawkins, 8 Chancery Lane, Eye, Peterborough, PE6 7YF
G8 KRB K. Barnes, Fairseat Close, Totnes, TQ9 5AN
G8 KRG Prof. C. Harrison, 11 Ringley Park, Whitefield, Manchester, M45 7NT
G8 KRV J. Cottier, 83 Elizabeth Drive, Tamworth, B79 8DE
G8 KSA W. Hall, 67 Selwyn Drive, Stockton-on-Tees, TS19 8XF
G8 KSC D. Goodwin, 41 Newpool Road, Knypersley, Stoke-on-Trent, ST8 6NT
G8 KSD A. Hewett, 1 Mountside, Westfield Lane, Folkestone, CT18 8BY
GW8 KSE W. Salisbury, 28 Dyke Street, Brymbo, Wrexham, LL11 5AH
GW8 KSF A. Salisbury, 28 Dyke St., Brymbo, Wrexham, LL11 5AH
G8 KSH A. Wilkins, 2 Beechfield Crescent, Banbury, OX16 9AR
GM8 KSJ D. Cowie, 8 Centre Street, Kelty, KY4 0EQ
GW8 KSL R. Cleaver, 61 Llewellyn Park Drive Morriston, Swansea, SA6 8PF
G8 KSM R. Beament, Midlands Farm, Horndon, Mary Tavy, Tavistock, PL19 9NQ
G8 KST T. Mayer, 61 Rawley Crescent, New Duston, Northampton, NN5 6PU
G8 KSW J. Wood, 38 Beech Lane, West Hallam, Ilkeston, DE7 6GU
G8 KSX A. Thompson, Carloway, Turner Lane, Ilkley, LS29 0LE
G8 KSZ I. Newbold, 40 Heath Close, Stonnall, Staffordshire, WS9 9HU
G8 KTA P. Thomas, 76 Church Road, Braunston, Daventry, NN11 7HQ
G8 KTC M. Rhys, 2 Sun Lane, Teignmouth, TQ14 8EF
G8 KTG D. Smith, 76 Reigate Road, Brighton, BN1 5AG
G8 KTX M. Butler, 7 Bassett Road, Coventry, CV6 1LF
G8 KUA C. Bridgland, 10 Eastlands Grove, Stafford, ST17 9BE
G8 KUV A. Simonds, 3 Links Close, Seaford, BN25 4NU
G8 KUZ J. Wiggins, 35 Downing Avenue, Newcastle, ST5 0LB
G8 KVN A. Nelson, 29 Coxford Road, Southampton, SO16 5FG
G8 KVO C. Miller, Broomwood, South Park, Sevenoaks, TN13 1EL
G8 KVU Ham Radio Builders Club, c/o Rev. G. Smith, 6 Birtley Rise, Bramley, Guildford, GU5 0HZ
G8 KW R. Shears, 15 Hale Pit Road, Great Bookham, Leatherhead, KT23 4BS
G8 KWD G. Bettley, 1 Dovetrees, Covingham, Swindon, SN3 5AX
G8 KWH N. Liddle, 67 Algarth Road, Pocklington, York, YO42 2HJ
G8 KWJ D. Barnwell, Bernagh, Duncombe Street, Kingsbridge, TQ7 1LR
G8 KWN R. Bryant, 81 Dukes Drive, Halesworth, IP19 8TJ
G8 KWP A. Darragh, The Gables, Belle Vue Lane, Chester, CH3 7EJ
G8 KWV J. Bailey, 20 Smitham Downs Road, Purley, CR8 4NB
GM8 KXF G. Robb, 3 Doonholm Park, Ayr, KA6 6BH
G8 KXO B. Gamble, 79 Humphries House, Lindon Drive, Walsall, WS8 6DL
GW8 KXW J. Watts, 6 Castle View, Haverfordwest, SA61 2JA
GI8 KYI T. Carlisle, 46 Middle Road, Carrickfergus, BT38 9DN
G8 KYK C. Keens, 3 Kirk Gardens, Totton, Southampton, SO40 9UZ
G8 KYP J. Buckley, 11 Salisbury Grove Giffard Park, Milton Keynes, MK14 5QA
GW8 KZA J. Wells, 30 St. Andrews Road, Barry, CF62 8BR
G8 KZG P. Delaney, 6 East View Close, Wargrave, Reading, RG10 8BJ
G8 KZJ E. Lockyear, 140 Andover Road, Orpington, BR6 8BL
G8 KZN W. Clinton, 5 Moorland Crescent, Castleside, Consett, DH8 9RF
G8 KZO R. Edgeley, 6 Hearne Gardens, Shirrell Heath, Southampton, SO32 2NR
G8 KZY C. Denison, 40 Leysholme Drive, Leeds, LS12 4HQ
G8 LAB R. Harste, 2 Park Drive, Ingatestone, CM4 9DT
G8 LAM R. Lambley, 31 Ridgeway Road, Redhill, RH1 6PQ
G8 LAU D. Peck, 3 Dearnford Avenue, Wirral, CH62 6DX
G8 LAY E. Hibbett, Trumps Lodge, Broad Street, Ottery St Mary, EX11 1BY
GM8 LBC C. Dalziel, 2 Alder Avenue, Hamilton, ML3 7LL
G8 LBG J. Cook, Highlands, Littledown, Shaftesbury, SP7 9HD
G8 LBS C. Ranson, 281 Hawthorn Drive, Ipswich, IP2 0QG
G8 LBT M. Rigby, 16 Juniper Way, Penrith, CA11 8UF
G8 LCA J. Scott, 123 Cotswold Way, Tilehurst, Reading, RG31 6SR
G8 LCE M. Perrett, 21 Tredova Crescent, Falmouth, TR11 4EQ
G8 LCI A. Goode, 42 Fourth Street, Black Rock, Australia, 3193
GI8 LCJ D. Craig, 40 Chilton Road, Carrickfergus, BT38 7JT
G8 LCK L. Reynolds, 1845 Van Buren Road, Caswell, USA, 4750
G8 LCL S. Tames, 21 Lind Close, Earley, Reading, RG6 5QX
G8 LCM K. Day, Powys Lodge, 6 Court Road, Worcester, WR8 9LP
G8 LCP N. Jamieson, 1 Langdale Place, Newton Aycliffe, DL5 7DX
G8 LCS J. Monte, 11 Woodfield Avenue, Hyde, SK14 5BB
G8 LCU A. Frankling, 5 Chadborn Avenue Gotham, Nottingham, NG11 0HT
G8 LCZ J. Sellick, 24 Windsor Road, Lytham St Annes, FY8 1ET
G8 LDB K. Oldham, 165 Mountsorrel Lane, Rothley, Leicester, LE7 7PU
G8 LDC Dr J. Salthouse, 10 Ramillies Avenue, Cheadle Hulme, Cheadle, SK8 7AL
G8 LDJ C. Douglas, 22 Connaught Road, Sittingbourne, ME10 1EH
G8 LDU Dr G. Noble, 9 Sunrise Way Kings Hill, West Malling, ME19 4DL
G8 LDV B. Harrad, 32 Woodfield Avenue, Northfleet, Gravesend, DA11 7QG
G8 LDW P. Harness, 7 Castlegate, Gipsey Bridge, Boston, PE22 7BS
G8 LDY R. Tompkins, 16 Garden Close, Watford, WD17 3DP
GM8 LEA N. Adam, Bridaig Villa, Gladstone Avenue, Dingwall, IV15 9PG
G8 LEB R. Hill, Rose Lodge, 35 Colne Fields, Huntingdon, PE28 3DL
G8 LED Northampton Radio Club, c/o J. Cockrill, 28 Northampton Road, Harpole, Northampton, NN7 4DD
G8 LEG K. Hardy, 2 Forest Hill, Maidstone, ME15 6UU
G8 LEM R. Griffith, 9 Devonshire Road, West Kirby, Wirral, CH48 7HR
G8 LES M. Sanders, 39 Telegraph Lane, Four Marks, Alton, GU34 5AX
G8 LF E. Byrne, 40 Wentworth Avenue, Ascot, SL5 8HQ
GM8 LFB J. Rabbitts, 38 Murchison Street, Wick, KW1 5HW
GM8 LFI Prof. M. Cartmell, 33 Orrok Park, Edinburgh, EH16 5UW
GI8 LFY A. Penn, 9 Milltown Road, Donaghcloney, Craigavon, BT66 7NE
G8 LGA R. Ward, 1 Horton, Downswood, Maidstone, ME15 8TN
G8 LGE P. Devine, 3 The Hawthorns, Outwood, Wakefield, WF1 3TL
G8 LGM R. Field, 20 Hill Road, Watlington, OX49 5AD
G8 LGP K. Harris, 20 Westminster Close, Devizes, SN10 1BF
G8 LGS P. Chitty, 109 Bannings Vale, Saltdean, Brighton, BN2 8DH
G8 LGT D. Blakemore, 20 Derwent Road, Coventry, CV6 2HB
G8 LGU R. Milliken, 15 Lee Grove, Chigwell, IG7 6AD
G8 LGW R. Liddiard, 26 Dowgate Close, Tonbridge, TN9 2EL
G8 LGY R. Tyson, 18 Blackthorn Close, Gedling, Nottingham, NG4 4AU
G8 LHD D. Allen, 21 Goldings Close, Haverhill, CB9 0EQ
G8 LHF P. Earl, Holly Cottage, Popes Lane, Colchester, CO6 2DZ
G8 LHI M. Levy, 22 Boyne Road, Lewisham, London, SE13 5AW
G8 LHP Dr A. Milne, 49 Cleevemount Road, Cheltenham, GL52 3HF
G8 LHQ M. Tuffrey, 50 Lynette Avenue, London, SW4 9HD
G8 LHS D. Waters, 84 Littlehaven Lane, Horsham, RH12 4JB
G8 LHT I. Harwood, 38 Spring Crescent, Sprotbrough, Doncaster, DN5 7QF
G8 LHZ P. Avon, 81 Parsonage Barn Lane, Ringwood, BH24 1PU
G8 LID N. Dowler, 1 Cottage Walk, Clacton-on-Sea, CO16 8DG
G8 LIE N. Borrell, Chapel Cottage Cotherstone, Barnard Castle, DL12 9PQ
G8 LIH G. Storey, 27 Dyche Road, Sheffield, S8 8DQ
G8 LII J. Lee, 225 Avenue Road, Rushden, NN10 0SN
G8 LIK S. Hurst, Fareview, Woodhead Road, Holmfirth, HD9 2PX
G8 LIP B. Greenbeck, 10 Campbell Avenue, Bottesford, Scunthorpe, DN16 3SA
G8 LIU N. Clyne, 78 Halford Road, Ickenham, Uxbridge, UB10 8QA
G8 LIX R. Keates, 35 Walsh Grove, Birmingham, B23 5XE
GW8 LJJ E. Edwards, 11 Old Village Road, Barry, CF62 6RA
G8 LJO W. Ricketts, 22 Westfield Close Durrington, Salisbury, SP4 8BY
G8 LJQ C. Asquith, 142B Newbegin, Hornsea, HU18 1PB
G8 LJU J. Spicer, 6 Avenue Road, Worcester, WR2 4ES
G8 LJY A. Griffiths, 17 Ferenberge Close, Farmborough, Bath, BA2 0DH
G8 LKA S. Whitehead, 4 Colleton Crescent, Exeter, EX2 4DG
G8 LKB I. Rabson, 50 Burwell Meadow, Witney, OX28 5JQ
G8 LKK R. Horsford, 2 Old Mill, Mill Lane, Chard, TA20 2ND
GM8 LKL A. Hogg, 43 Muir Wood Road, Currie, EH14 5JN
G8 LKP J. Duchscherer, 36 Hamdon Close, Stoke-Sub-Hamdon, TA14 6QN
G8 LKQ D. Falkner, 45 Westwood Carleton, Skipton, BD23 3DW
G8 LKS D. Burton, 48 West Beeches Road, Crowborough, TN6 2AG
G8 LKW H. Colville, Flat 33, Hamilton Court 165 Northfield Road, Birmingham, B30 1DU
GW8 LKX M. Corrigan, 3 Heathway, Heath, Cardiff, CF14 4JQ
G8 LLD P. Pritchard, 15 Hilldene Close Flitwick, Bedford, MK45 1AQ
G8 LLJ M. Tutt, 9 Russell Drive, Dunbridge, Romsey, SO51 0RA
G8 LLS S. Perkins, 17 Lime Tree Avenue, Malvern, WR14 4XE
G8 LLV G. Christie, 9 / 49 Democrat Drive, The Basin, Victoria 3154, Australia, Melbourne, Australia, 3154
G8 LLZ S. Hickling, Shamba Slade Lane Galmpton, Brixham, TQ50PE
G8 LM J. Jennings, Mill Side, Mill Road, Lutterworth, LE17 5DE
G8 LMC R. Lovell, 16 North View, Staple Hill, Bristol, BS16 5RU
G8 LMF P. Rigby, 92 Albany Road, Ansdell, Lytham St Annes, FY8 4AR
G8 LMI D. Morgan, 23 Banstead Road, Caterham, CR3 5QH
G8 LMW C. Smith, 73 Desford Road, Newbold Verdon, Leicester, LE9 9LG
G8 LMY D. Sweetland, 15 Wasdale Close, Owlsmoor, Sandhurst, GU47 0YQ
G8 LNC D. Golding, 27 Wesermarsch Road, Cowplain, Waterlooville, PO8 8JJ
G8 LNG D. Severn, 20 Somerton Avenue, Wilford, Nottingham, NG11 7FD
GM8 LNH R. Pascal, 19 Cloch Na Strom, Whiteness, Shetland, ZE2 9LG
G8 LNQ C. Tindill, The Old School, Bellerby, Leyburn, DL8 5QN
G8 LNU L. Tucker, 10 The Meadow, Waterlooville, PO7 6YJ
G8 LOF S. Champion, 4 Oldcastle Croft, Tattenhoe, Milton Keynes, MK4 3EN
G8 LOJ S. Dorrington-Ward, Higher Dairy, Stoke Abbott, Beaminster, DT8 3JT
GM8 LON R. Bruce, 10 John Huband Drive, Birkhill, DD2 5RY
G8 LOP P. Coomber, 10 Streeton Way, Earls Barton, Northampton, NN6 0HX
G8 LOU Dr P. Mattos, Olive House, Rock Road Rock, Wadebridge, PL27 6NW
G8 LOZ J. Ramsay, Strathmore, 5 Parkhurst Road, Guildford, GU2 8AP
G8 LPA N. Hilbery, 16 Albert Road, Ashford, TW15 2LU
G8 LPC R. Cawley, 59 The Horseshoe, Hemel Hempstead, HP3 8QS
G8 LPI R. Bray, 2 Hill Park, Walsall Wood, Walsall, WS9 9RD
G8 LPN K. Edwards, 22 Claverton Estate, Stoulton, Worcester, WR7 4RH
G8 LPX C. Morgan, 43 Ferndown Road, Manchester, M23 9AW
G8 LQB W. Morrison, 14 Browns Grove, Kesgrave, Ipswich, IP5 2GP
G8 LQF J. Pettifor, 12 Windmill Road, Atherstone, CV9 1HP
GW8 LQH M. White, Penrhos, Llanddewi, Llandrindod Wells, LD1 6SL
GM8 LQL W. Cowell, High Clachaig, Kilmory, Isle of Arran, KA27 8PG
G8 LQM P. Green, Nut House, 2 Warren Barns, Warren Lane, Bedford, MK45 4AS
G8 LQN G. Bryce, 6A Kingfisher Drive, Whitby, YO22 4DY
G8 LQO C. Mckenzie, 6 Pasturefield Close, Sale, M33 2LD
G8 LQP R. Lines11, 5 Dowling Drive, Pershore, WR10 3EF
G8 LQZ R. Banfield, 2 Laleham Close, Eastbourne, East Sussex, BN21 2LQ
G8 LRD P. Hutchings, 59 Braemor Road, Calne, SN11 9DU
GW8 LRO A. Williams, 1 Glyncoch Terrace, Pontypridd, CF37 3BW
G8 LRS D. Massey, 28 Rufus Close, Rownhams, Southampton, SO16 8LR
G8 LSA H. Potter, Burwood House, Salisbury Road, Woking, GU22 7UR
G8 LSC P. Wheeler, 3 Oatfield Road, Orpington, BR6 0ER
G8 LSD A. Wyatt, 75 Millbrook Road, Crowborough, TN6 2SB
G8 LSH D. Oakley, 136 Chanctonbury Way, London, N12 7AD
G8 LSI R. Dungan, 2 Lamorna Close, Orpington, BR6 0TD
G8 LSS A. Tompson, 15 Plumbley Meadows, Winterborne Kingston, Blandford Forum, DT11 9BY
GI8 LTB R. Mcwilliams, 4 Wheatfield Drive, Coleraine, BT51 3RD
G8 LTC R. Hore, 19 Warrenside Close, Ramsgreave, Blackburn, BB1 9PE
G8 LTD S. Vaslet, 4 Coniston Crescent, Redmarshall, Stockton-on-Tees, TS21 1HT
G8 LTN A. Brown, Casita, The Ridge, Cold Ash, Thatcham, RG18 9HT
GW8 LTV G. Snellgrove, 142 Arail St., Six Bells, Abertillery, NP13 2NQ
G8 LTY A. Harman, 107 Kempson Drive, Great Cornard, Sudbury, CO10 0YF
GM8 LUK J. Macassey, 8 Balgavies Avenue, Dundee, DD4 7 NR
G8 LUL R. Myers, 33 Withenfield Road, Manchester, M23 9BT
G8 LUP A. Semark, 11 Fir Tree Close, Thorpe Willoughby, Selby, YO8 9PF
GI8 LUR A. Hewitt, 18 Knockview Avenue, Newtownabbey, BT36 6TZ
G8 LUV G. Fairbrass, 230 Kirkby Road, Barwell, Leicester, LE9 8FS
G8 LVC P. Johnson, 54 Beechwood Close, Chandler'S Ford, Eastleigh, SO53 5PB
G8 LVF A. Sierota, 20 Marder Road, London, W13 9EN
G8 LVL D. Holmes, 30 Roydale Close, Loughborough, LE11 5UW

G8 LVM A. Holmes, 5 Launde Park, Market Harborough, LE16 8BH
G8 LVQ White Rose ARS, c/o E. Hannaby, 170A Weston Drive Otley, Leeds, LS21 2DT
G8 LVU E. Brown, 37 Lynton Way Sawston, Cambridge, CB22 3EA
G8 LVW C. Snell, 138 Main Road, Great Leighs, Chelmsford, CM3 1NP
G8 LWA D. Tyler, Wayside View, Orsett Road, Stanford-le-Hope, SS17 8PN
G8 LWC Dr J. Stuart, 13 Gloucester Close, Petersfield, GU32 3AX
G8 LWO F. Merritt, 17 Blakes Way Eaton Socon, St Neots, PE19 8PU
G8 LWQ S. Wood, Lucerne, Berrycroft, Soham, Ely, CB7 5BL
G8 LWS Ariel Ra Gp Lws, c/o G. Rowlands, C/O Gareth Rowlands, Engineering Pigeon Holes, Acton, W3 0RP
G8 LXS G. Pascoe, Newhaye, Broadhempston, Totnes, TQ9 6DB
G8 LXY S. Clarke, 128 Putteridge Road, Luton, LU2 8HQ
G8 LYB S. Tompsett, 9 Ashlawn Road, Rugby, CV22 5ET
G8 LYG W. Leach, 15 Beech Lea, Blunsdon, Swindon, SN26 7DE
GM8 LYO P. Mahood, 4 Irvine Court, Glasgow, G40 3LE
GM8 LYQ I. Lindsay, 10/4 Mertoun Place, Edinburgh, EH11 1JZ
G8 LYV K. Kearns, 79 Church Road, Hatfield Peverel, Chelmsford, CM3 2LB
G8 LYW B. Theedom, 5 Rodbridge Drive, Southend-on-Sea, SS1 3DF
G8 LZG G. Allen, 21 Dale Road, Welton, Brough, HU15 1PE
G8 LZK M. Ball, 46A Daniels Crescent, Long Sutton, Spalding, PE12 9DS
G8 LZO J. Hibbert, 80 High Street, Newchapel, Stoke-on-Trent, ST7 4PT
G8 LZS P. Martin, 35 Martineau Lane, Hurst, Reading, RG10 0SF
G8 LZV A. Fulcher, Keepers, The Street, Ashford, TN27 0QF
GW8 LZY S. Brown, Maes Yr Haidd, 8 Glanceulan, Aberystwyth, SY23 3HF
G8 MAA G. Chaplin, 8 Manor House Drive, Northwood, HA6 2UJ
G8 MAD P. Tostevin, 20 Wallace Avenue, Worthing, BN11 5QY
G8 MAF T. Beckham, 2 Sandbanks Place, Ersham Road, Hailsham, BN27 3LJ
G8 MAR M. Sibley, 10 Ainley Close, Huddersfield, HD3 3RJ
G8 MAV P. Lewis, Westbank, 46 Weyside Road, Guildford, GU1 1HX
G8 MAY A. Lake, 9 Grafton Close, King's Lynn, PE30 3EZ
G8 MBE S. Fouracres, 4A Newton Square Bampton, Tiverton, EX16 9NE
G8 MBJ J. Parsons, 34 Mill Hill, Brancaster, King's Lynn, PE31 8AQ
G8 MBK P. Bland, 17 Knowles Close, Kirklevington, Yarm, TS15 9NL
G8 MBM C. Proctor, 15 Chiltern Street, Aylesbury, HP21 8BN
G8 MBQ R. Jones, 46 Wilmington Close, Woodley, Reading, RG5 4LR
G8 MBS R. Vitiello, 6 Meeting Oak Lane, Winslow, MK18 3JU
G8 MBU R. Williams, 8 Gurnard Heights, Gurnard, PO31 8EF
G8 MBV I. Wood, Tessian Lodge, Lydden Road, Dover, CT15 7HF
G8 MCA G. Bryan, 34 Shelbury Close, Sidcup, DA14 4BE
G8 MCC C. Divall, 12 Latchmount Gardens, Axminster, EX13 5JT
G8 MCJ B. Pritchard, 14 Rugby Way, Croxley Green, Rickmansworth, WD3 3PH
G8 MCR V. Eagles, 3 Church Road, Buckhurst Hill, IG9 5RU
G8 MCT C. Bate, 28 Argyle Avenue, Tamworth, B77 3PH
G8 MCW P. Elkins, 615 Blandford Road, Upton, Poole, BH16 5ED
G8 MCY M. Dannatt, 46 Laburnham Road, Biggleswade, SG18 0NX
G8 MDG D. Shaw, 35 Tinshill Lane, Leeds, LS16 6BU
G8 MEA C. Wilson, 2 Blackthorn Garth, Beverley, HU17 8FZ
G8 MEC D. Uttley, 1 Edgeside, Great Harwood, Blackburn, BB6 7JS
G8 MED P. Shirtliff, 2 Birch Avenue, Newton, Preston, PR4 3TX
G8 MEE K. Patman, 5 Lime Grove, Holbeach, Spalding, PE12 7NG
G8 MEH L. Steele, Caprice, Woodville Road, Bude, EX23 9JA
G8 MEI R. Whitby, 24 Macaulay Avenue, Great Shelford, Cambridge, CB22 5AE
G8 MEM A. Lillywhite, 1 Roblin Close, Aylesbury, HP21 9DT
GW8 MER M. Busson, 14 Squires Gate, Rogerstone, Newport, NP10 0BP
G8 MEX I. Glenn, 257 Wimpole Road, Barton, Cambridge, CB23 7AE
G8 MFF R. Hedley, 20 Spencer Drive, Tiverton, EX16 4PY
G8 MFH R. Lake, 2A Hyde Lane, Bovingdon, Hemel Hempstead, HP3 0EG
G8 MFI S. Mcguigan, 16 Queen Street, Middleton Cheney, Banbury, OX17 2NP
G8 MFM R. Wood, 36 New England Road, Haywards Heath, RH16 3JS
G8 MFO T. Sorensen, 22 The Cottrells, Angmering, Littlehampton, BN16 4AF
GW8 MFQ A. John, 79 Harding Close, Boverton, Llantwit Major, CF61 1GX
G8 MFR R. Irwin, Copperfield, 97 Offerton Lane, Stockport, SK2 5BS
G8 MFU D. Parry, 19 Norton Lane, Great Wyrley, Walsall, WS6 6PE
G8 MFV R. Hickmott, Brisley Cottage, Canterbury Road, Ashford, TN25 4DW
GM8 MFZ Dr N. Kennedy, Deveron, North Deeside Road, Pitfodels, Aberdeen, AB15 9PL
G8 MGD D. Marshall, 7 Aesops Orchard, Woodmancote, Cheltenham, GL52 9TZ
G8 MGE G. Young, 30 Degenhardt Streett, South Australia, Australia, 5545
GW8 MGF J. Tait, 2 Bron-Y-Coed, Coed-Y-Glyn, Wrexham, LL13 7QJ
G8 MGG W. Whiteside, 5 Church Close, Levens, Kendal, LA8 8QE
G8 MGK J. Dosher, 40 Bromfield Road, Redditch, B97 4PN
G8 MGO J. Marshall, 34 Derwent Drive, Swindon, SN2 7NJ
G8 MGP A. Hill, 5 Lilac Walk, Kempston, Bedford, MK42 7PE
G8 MGQ D. Garwood, 13 Market Street, Bradford-on-Avon, BA15 1LL
G8 MGZ P. Haynes, 2 The Chase, Furnace Green, Crawley, RH10 6HW
G8 MHA L. Humphrey, 1 Falkenham Road, Kirton, ip100np
G8 MHD C. Cooper, 16 Paulton Drive, Bishopston, Bristol, BS7 8JJ
G8 MHE G. Cross, 117 Broadway, Eccleston, St Helens, WA10 5PB
G8 MHI K. Russell, 12 Evans Close, Greenhithe, DA9 9PG
G8 MHN S. Scrase, 5 Clinton Road, Self Employed Person, KT22 8NU
G8 MHO A. Fraser, 184 Old Road, Harlow, CM17 0HQ
G8 MHT G. Dallaway, Flat 2, School Court, Meyer Street, Stockport, SK3 8JE
GM8 MHU I. Fraser, 12 Auchlea Place, Aberdeen, AB16 6PD
G8 MIA A. Malbon, The Lodge, Blithbury Road, Rugeley, WS15 3HJ
G8 MIC M. Williams, Flat 2, High Point, Highgate, N6 4BA
G8 MIE K. Croucher, 140 Dane Road, Coventry, CV2 4JW
G8 MIF F. Golding, 16 Lessness Park, Belvedere, DA17 5BG
G8 MIH R. Green, 33 Bulkington Avenue, Worthing, BN14 7HH
G8 MII T. Ashton, 30 Highfields Road, Chasetown, Burntwood, WS7 4QU
G8 MIN R. Welsh, 14 Drayton Close, High Halstow, Rochester, ME3 8DW
G8 MIT C. Wyatt, 273 Nuthurst Road, Birmingham, B31 4TQ
GI8 MIV G. Hutchinson, 40 Oldstone Hill, Muckamore, Antrim, BT41 4SB
G8 MIW J. West, 21 Gardenia Crescent, Mapperley, Nottingham, NG3 6JA
G8 MJF K. Bottomley, Whispering Winds, 15 Marvell Rise, Harrogate, HG1 3LT
G8 MJH P. Harrison, 154 Cherrydown Avenue, Chingford, London, E4 8DZ
GM8 MJV T. Melvin, Blue House, Remote, Pathhead, EH37 5UP
G8 MJX D. Coomber, 1 Brympton Road, Coventry, CV3 1GW
G8 MKC Milton Keynes ARS, c/o D. White, 1 Whaddon Road, Shenley Brook End, Milton Keynes, MK5 7AF
G8 MKE C. Rose, 45 Clent Road, Warley, Oldbury, B68 9ES
G8 MKG G. Barraclough, 1 Meadowcroft Road, Middlesbrough, TS6 0JD
G8 MKN I. Wager, 106 Turner Road, Colchester, CO4 5JT
G8 MKO R. Pocock, 3 Brewery Cottages, Netherley Road, Prescot, L35 1QG
G8 MKQ A. Bullock, 35 Parkstone Avenue, Thornton-Cleveleys, FY5 5AE
G8 MKS P. Moore, 3A High Street, Mow Cop, Stoke on Trent, ST7 3ND
G8 MKW J. Green, Huntley, Chesham Road, Tring, HP23 6HH
G8 MKX J. Donnithorne, 6 Bulbourne Court, Tring, HP23 4TP
G8 MLA P. Richardson, 11 Overstone Road Coldham, Wisbech, PE14 0ND
G8 MLB N. Bourner, 11 Richborough Road, Sandwich, CT13 9JE
G8 MLD M. Warren, 17 Bolehill Park, Hove Edge, Brighouse, HD6 2RS
G8 MLK J. Owen, The Old Coach House, Callow Hill, Virginia Water, GU25 4LD
G8 MLW D. Carr, 39 Fallowfield Road, Walsall, WS5 3DH
GM8 MMA W. Williamson, Leeskol, Camb, Shetland, ZE2 9DA
G8 MMF P. Dorrington, 57 Ferring Lane, Ferring, Worthing, BN12 6QS
G8 MMG D. Bentley, 55 Saddlers Road, Quedgeley, Gloucester, GL2 4SY
G8 MMM G. Nicholas, Greenbank, Chester High Road, Neston, CH64 7TR
G8 MMN Dr M. Holmes, 8 High Street, Norley, Frodsham, WA6 8JS
G8 MMP M. Swain, 38 Longdale Lane, Ravenshead, Nottingham, NG15 9AD
GM8 MMW W. Dick, 58 Kirkland Road, Glengarnock, Beith, KA14 3AJ
G8 MNC M. Bilkey, Pebble Flek, The Green, St Austell, PL25 5TA
GM8 MNG C. Raine, Broomhill Edgehead, Pathhead, EH37 5RS
G8 MNL P. Carruthers, 16 Wivenhoe Close, Rainham, Gillingham, ME8 7QB
GM8 MNM R. Hood, Milton Of Auchindoir House, Rhynie, Huntly, AB54 4JB
G8 MNO W. Stewart, 9 Ashley Road, Marnhull, Sturminster Newton, DT10 1LQ
GM8 MNR D. Jenkins, 16 Bentinck Street, Galston, KA4 8HT
G8 MNY J. Stockley, 27 Campden Road, South Croydon, CR2 7ER
G8 MOF F. Bellamy, 3 Manor Road, Crowle, Scunthorpe, DN17 4ET
G8 MOG D. Dale, Blackwood Hall, Felton, Morpeth, NE65 9QW
GM8 MOI C. Stirling, 20 Craigford Drive, Bannockburn, Stirling, FK7 8NQ
G8 MOL P. Marshall, 134 Gladbeck Way, Enfield, EN2 7EN
G8 MOS A. Reale, 20 Wickham Close, Alton, GU34 1RR
GI8 MOV F. Warwick, 34 Shaws Wood Cullybackey, Ballymena, BT42 1SB
GW8 MOZ G. Elliott, 9 Hove Avenue, St. Julians, Newport, NP19 7QP
G8 MPG G. Rigby, 1 Route Danton, Petit Caudos, Mios, France, 33380
G8 MPM W. Brock, 15 Picketleaze, Chippenham, SN14 0DN
G8 MQF M. Cooper, Woodstock, Snow Hill, Crawley, RH10 3EG
G8 MQK J. Lindley, 17 Leyfield Bank, Holmfirth, HD9 1XU
G8 MQT T. Smith, 416 Charminster Road, Bournemouth, BH8 9SG
G8 MQX R. Eccles, 6 Queens Drive, Barnsley, S75 2QJ
G8 MQY B. Densham, 47 High Street, Paulerspury, Towcester, NN12 7NA
G8 MRI R. Davey, 23 Campbell Close, Hunstanton, PE36 5PJ
G8 MRN M. Watch, 9 High Drive, Rowner, Gosport, PO13 0QS
GM8 MST Dr G. Kelly, 36 Craigleith Drive, Edinburgh, EH4 3JU
G8 MSY J. Wilkinson, 11 Wigmore Road, Tadley, RG26 4HH
G8 MTB M. Greenfield, 8 The Spinney, Clayton, Newcastle, ST5 4DA
G8 MTI M. Dibsdall, 28 Court Farm Avenue, Epsom, KT19 0HF
G8 MTV J. Wood, Coach House, Croft On Tees, Darlington, DL2 2SL
G8 MUF J. Ames, 16 Vere Gardens, Henley Road, Ipswich, IP1 4NZ
G8 MUV B. Clarke, Flat 26, Compton Grange, Whitehall Road, Cradley Heath, B64 5BG
G8 MUX J. Mottram, Church View, New Road, High Peak, SK23 7NH
G8 MVC R. Westlake, Flat 9, Grosvenor Court, 135-139 The Grove, London, W5 3SL
G8 MVD K. Wilks, 72 Grasmere Road, Bradford, BD2 4HX
G8 MVH J. Armstrong, 30A Abbey Fields, Faversham, ME13 8JA
G8 MVJ C. Chambers, Hollybank, Back Street, Driffield, YO25 3TD
G8 MVS N. Fuller, 48 White Street, Easterton, Devizes, SN10 4PA
G8 MVY E. Phillips, 2 Primrose Cottage, The Street, Reading, RG7 1QY
G8 MWA Medway Amateur Receiving & Transmitting Society, c/o J. Burton, 22 Pear Tree Lane, Hempstead, Gillingham, ME7 3PT
G8 MWD D. Lewing, 94 Carville Crescent, Brentford, TW8 9RD
G8 MWE K. Knight, 54 Vicarage Lane, Water Orton, Birmingham, B46 1RU
G8 MWN K. Harris, Bella Vista, Station Road, Yelverton, PL20 7JS
G8 MWU P. Stafford, 5 Westmead Drive, Newbury, RG14 7DJ
G8 MWW W. Westlake, West Park, Clawton, Holsworthy, EX22 6QN
G8 MWX A. Priestley, 55 Derwent Avenue, Garforth, Leeds, LS25 1HN
G8 MXD G. York, 13 Cherwell Close, Thornbury, Bristol, BS35 2DN
G8 MXQ A. Taylor, 311 Smeeth Road, Marshland St. James, Wisbech, PE14 8ES
G8 MXR W. Pitt, 1 Windy Ridge, James Street, Stourbridge, DY7 6ED
G8 MXT L. Mansfield, 25 Carlton Road, Derby, DE23 6HB
G8 MXV K. Ayriss, 6 Langstons, Trimley St. Mary, Felixstowe, IP11 0XL
G8 MXW C. Down, 100 Lynwood Drive, Merley, Wimborne, BH21 1UQ
G8 MYF M. Johnson, 42 Marlborough Road, Ryde, PO33 1AB
G8 MYG C. Hunt, Rowan Bank, 2 Cranston Rise, Bexhill-on-Sea, TN39 3NJ
G8 MYJ C. Drewe, 37 Baker Street, Chelmsford, CM2 0SA
G8 MYK A. Rowley, Holly Cottage, 368 Highters Heath Lane, Birmingham, B14 4TE
GM8 MYO C. Tyler, 26 Sinclair Way, Knightsridge, Livingston, EH54 8HW
G8 MYV D. Webster, 35 Raymond Road, Maidenhead, SL6 6DF
G8 MZA D. Garrett, Brookside Farm, Tonge, Derby, DE73 8BD
G8 MZD P. Diggins, 8 Gloucester Gardens, Bagshot, GU19 5NU
G8 MZQ Dr W. Katz, The Beacon, Goathland, Whitby, YO22 5AN
GW8 MZR R. Harris, 15 Quarry Rise, Undy, Caldicot, NP26 3JU
G8 MZW S. Adams, 20 Queens Drive, Skegness, PE25 1RE
G8 MZY Prof. D. Cushman, 50 St. Peters St., Syston, Leicester, LE7 1HJ
G8 MZZ P. Boam, 36 Copeland Drive, Stone, ST15 8YP
GW8 NAC K. Davies, 45 Castle View, Simpson Cross, Haverfordwest, SA62 6EN
G8 NAG M. Smith, 18 Manor Lane, Verwood, BH31 6HX
G8 NAI J. Lazzari, 3 Terson Way, Weston Coyney, Stoke-on-Trent, ST3 5RQ
GM8 NAL P. Corbishley, Tweedbank House, Cardrona, Peebles, EH45 9HX
G8 NAM P. Buttress, 18 Taffrail Gardens, South Woodham Ferrers, Chelmsford, CM3 5WH
G8 NAP P. Beacon, 67 St. Helena Road, Polesworth, Tamworth, B78 1NJ
G8 NAU C. Searle, 55 Chenies Close, Tunbridge Wells, TN2 5LN
G8 NAV N. Vernon, 11 Alexandra Road, Birchington, CT7 0DX
GW8 NBF C. Morgan, 84 Treowen Road Newbridge, Newport, NP11 3DP
G8 NBI A. Buxton, 95 Cantelupe Road, Ilkeston, DE7 5HT

G8 NBO L. Phillips, 14 Heal Park Crescent, Fremington, Barnstaple, EX31 3AP
GM8 NBV C. Davies, 35 Laverock Avenue, Hamilton, ML3 7DD
G8 NCK N. Brown, 9 Redhill Close, Tamworth, B79 8EJ
GW8 NCN D. Sanford, 35 Summerfield Avenue, Cardiff, CF14 3QA
G8 NCS M. Green, 21 Hill View Rise, Northwich, CW8 4XA
G8 NCU J. Watkins, Llwynteg, Glanwern, Borth, SY24 5LT
G8 NDB Dr G. Jarrett, 1 Church Street, Twycross, Atherstone, CV9 3PJ
G8 NDE J. Turner, 9 Clifton Avenue Culcheth, Warrington, WA3 4PD
G8 NDF D. Simpson, 10 Buckingham Way, Byram, Knottingley, WF11 9NN
G8 NDK K. Lindley, 25 Lindsey Court, Epworth, Doncaster, DN9 1SD
G8 NDN C. Keens, Toad Hall, 69 Lillywhite Crescent, Andover, SP10 5NA
G8 NDR N. Burridge, 8 Cedar Close, Ware, SG12 9PG
G8 NDT N. Thompson, 125 Summer Road, Erdington, Birmingham, B23 6DX
G8 NDV P. Fay, 42 Roberts Road, Salisbury, SP2 9BY
G8 NED Wisbech AR & Electronics Club, c/o A. Bridgeland, 17 Oldfield Lane, Wisbech, PE132RJ
G8 NEF Dr R. Peel, 76 Cypress Grove, Ash Vale, Aldershot, GU12 5QW
G8 NEI K. Marsh, 1 Parr Close, Exeter, EX1 2BG
G8 NEL S. Nightingale, 75 Gordon Road, Herne Bay, CT6 5QX
G8 NEO D. Edwards, 3 Murton Close, Burwell, Cambridge, CB25 0DT
GM8 NET A. Fraser, 50 Tannin Crescent, East Kilbride, G75 9FS
G8 NEY D. Millard, Weavern House, Hartham Lane, Biddestone, Chippenham, SN14 7EA
G8 NFD K. Gardiner, 8 Foxlands Drive, Sutton Coldfield, B72 1YZ
GM8 NFG J. Aitken, Hamabo, 10 Lynnpark, Kirkwall, KW15 1SL
G8 NFM F. Turner, 46 Main Street, Kings Newton, Derby, DE73 8BX
G8 NFP A. Crockett, 57 Upland Road, Sutton, SM2 5HW
GM8 NFT N. Leitch, 4 Dullatur Road, Dullatur, Glasgow, G68 0AF
G8 NFZ S. Sims, 35 Garfield Road, Hailsham, BN27 2BU
G8 NGE K. Ebborn, 18 St. Marys Park, Ottery St Mary, EX11 1JA
G8 NGF D. Stone, Aston Hill Cottage, 5 Aston Hill, Westbury, SY5 9JS
G8 NGJ P. Richardson, Domaine De Calcat, 47150 La Sauvetat Sur Lede, Lot Et Garonne, France, la Sauvetat Sur Lede, France, 47150
G8 NGM N. King, 42 Constance Close, Witham, CM8 1XY
G8 NGR M. Roberts, 62 Oakwood Road, Bricket Wood, St Albans, AL2 3QA
G8 NGZ V. Edwards, 33 Eyrescroft, Bretton, Peterborough, PE3 8ES
G8 NHD P. Mart, 6 Rose Creek Gardens, Great Sankey, Warrington, WA5 3TT
G8 NHG R. Wilkins, New Copse House, Fishbourne Lane, Ryde, PO33 4EZ
G8 NHO J. Austin, 5 Mercia Road, Baldock, SG7 6RZ
G8 NIE D. Sharpe, 5 Drydales, Kirk Ella, Hull, HU10 7JU
G8 NIL D. Bales, 30 Railway Road, Wisbech, PE13 2QA
GU8 NIS Guernsey ARS, c/o D. Eaton, Glenfield, Le Foulon, St Andrew, Guernsey, GY6 8UF
G8 NIU R. Whiting, Heather Bank Glebelands, Minehead, TA24 8DH
G8 NJA Torbay ARS, c/o D. Webber, 43 Lime Tree Walk, Milber, Newton Abbot, TQ12 4LF
G8 NJI P. Woodhead, 24 North Bar Without, Beverley, HU17 7AB
G8 NKJ L. Reid, 26 Mansion Avenue, Whitefield, Manchester, M45 7SS
G8 NKM A. O'Donovan, 2 Mackenzie Road, Beckenham, BR3 4RU
G8 NKN S. Gorwits, 29 Howitt Drive, Bradville, Milton Keynes, MK13 7DY
G8 NLF I. Munro, 30 Willow Close, Bordon, GU35 0TH
G8 NLK M. Bennett, 68 Meadow Hill Road, Birmingham, B38 8DA
G8 NLS S. O'Brien, Flat 2, 99A Howard Street, North Shields, NE30 1NA
G8 NMH B. O'Regan, 10 School Hill, Little Sandhurst, Sandhurst, GU47 8LD
G8 NMK C. Eccles, 18 Infinity View, Stockton-on-Tees, TS18 2FN
G8 NMM C. Reid, 138/17, Bang Saray, Pattaya, Thailand, 20230
G8 NMO D. Pechey, Jays Lodge, Crays Pond, Reading, RG8 7QG
G8 NMT J. Hicks, 7 North Croft, High Wycombe, HP10 0BP
G8 NNA B. Crellin, 60 College Fields, Woodhead Drive, Cambridge, CB4 1YZ
GW8 NNF R. Galpin, 23 Heol Y Delyn, Lisvane, Cardiff, CF4 5SR
G8 NNL J. Hurley, 64 Carleton Road, Chorley, PR6 8UB
G8 NNP B. Gower, 132 Goldsworthy Way, Slough, SL1 6AY
G8 NNS G. Stamp, 41 Willoughby Road, Wallasey, CH44 3DZ
G8 NNU T. Rowe, 68 Cobourg Road, Montpelier, Bristol, BS6 5HX
G8 NNX M. Cohen, 41 South Station Road, Liverpool, L25 3QE
G8 NNZ N. Carrington, 22 Parrs Wood Avenue, Manchester, M20 5ND
G8 NOB N. Bean, 33 Badger Close, Guildford, GU2 9PJ
G8 NOD M. Stamford, The Old Wheelwrights, East Street, Leominster, HR6 9HB
G8 NOF R. Holt, Tile House, Vicarage Hill, Solihull, B94 5EB
G8 NOP P. Price, Calwich View, Dove Street, Ellastone, Ashbourne, DE6 2GY
G8 NOS A. Swallow, 67A Strines Road, Marple, Stockport, SK6 7DT
GW8 NP Highfields ARC, c/o S. Williams, 371 Coed-Y-Gores Llanedeyrn, Cardiff, CF23 9NR
G8 NPD J. Hodnett, 126 Northwood Lane, Newcastle, ST5 4BN
G8 NPH A. Arnold, 2 Duck Lane, Haddenham, Ely, CB6 3UE
G8 NPP A. Brown, Dunlop Hiflex Powerbend, Pennywell Industrial Estate, Sunderland, SR4 9EN
G8 NPR J. Blackshaw, 23 Cherry Orchard, Oakington, Cambridge, CB24 3AY
G8 NPT A. Work, Vosselaan 52, Hillegom, Netherlands, 2181CD
G8 NPZ P. Whiteman, 22 Hartsbourne Road, Earley, Reading, RG6 5PY
G8 NQC P. Manser, 61 Galsworthy Drive, Caversham, Reading, RG4 6QB
G8 NQI J. Gartside, 12 Starfield Avenue, Hollingworth Lake, Littleborough, OL15 0NG
G8 NQK G. English, 25 Powell Gardens, Newhaven, BN9 0PS
G8 NQN M. Bancroft, Westfield, Towngate Southowram, Halifax, HX3 9QZ
G8 NQO A. Whyatt, 11 The Perrings, Nailsea, Bristol, BS48 4YD
G8 NQY W. Lea, 20 Gloucester Road, Walsall, WS5 3PN
G8 NRC S. Deighton, 132 Horsehead Lane Bolsover, Chesterfield, S44 6XH
G8 NRF G. Wood, 3 Cleveleys Road, Great Sankey, Warrington, WA5 2SR
G8 NRP M. Andrew, 80 Hamble Drive, Abingdon, OX14 3TE
G8 NRR R. Bambrook, 26 Croft Road, Thame, OX9 3JF
G8 NRU D. Carr, 78 Kingsleigh Road, Heaton Mersey, Stockport, SK4 3PG
G8 NSD F. Taylor, 96 Elvaston Road, North Wingfield, Chesterfield, S42 5HH
G8 NSE F. Wood, 96 Manchester Road, Astley, Manchester, M29 7EJ
G8 NSK J. Barnes, 23 Spenser Road, King's Lynn, PE30 3DP
G8 NSO S. Fleetham, Hill House Leckhampton Hill, Cheltenham, GL53 9QG
G8 NST J. Leek, 30 Casuarina Road, Bucklands Beach, Auckland, New Zealand, 1706
G8 NSX R. Miller, 89 Moorside, Spennymoor, DL16 7DZ
G8 NSZ M. Stanway, 72 Sheldons Court, Winchcombe Street, Cheltenham, GL52 2NR
G8 NTD K. Johnson, 24 Capers Close, Enderby, Leicester, LE19 4QD
G8 NTG W. Howell, 6 Unity Avenue, Sneyd Green, Stoke-on-Trent, ST1 6DE
G8 NTH A. Hewat, 41 Summersbury Drive, Shalford, Guildford, GU4 8JG
G8 NTJ K. Hand, 75 Hill Street, Hednesford, Cannock, WS12 2DW
G8 NTQ M. Roper, 6 Ilmington Close, Hatton Park, Warwick, CV35 7TL
G8 NTR Dr J. Williams, 133A Wiltshire Lane, Pinner, HA5 2NB
G8 NTS J. Smart, Greystone, High Street, Swindon, SN26 7AR
G8 NTY C. Mallows, 9 Chestnut Drive, Shenstone, Lichfield, WS14 0JH
G8 NTZ D. Kowalczyk, 5 Priestthorpe Lane, Bingley, BD16 4ED
G8 NVB N. Brown, 9A Decoy Drive, Eastbourne, BN22 0AB
G8 NVC B. Ellis, 7 Highmoor Close, Corfe Mullen, Wimborne, BH21 3PU
GM8 NVE D. Watters, 28 Bruce Road, Crossgates, Cowdenbeath, KY4 8AZ
GM8 NVG A. Wilson, Lochend, Beith, KA15 2LN
G8 NVH S. Reynolds, 242 Butchers Lane, Mereworth, Maidstone, ME18 5QH
G8 NVI A. Stevens, 67 New Road, East Hagbourne, Didcot, OX11 9JX
G8 NVS S. Hindle, Innisfree, Barton Terrace, Dawlish, EX7 9QH
G8 NVT R. Hatfield, 1 Slade Close, Ottery St Mary, EX11 1SY
G8 NVX M. Moss, 24 Magna Lane, Dalton, Rotherham, S65 4HH
G8 NVZ Dr G. Evans, 4 Holcot Lane, Anchorage Park, Portsmouth, PO3 5TR
G8 NWC G. Boor, 27 Welbeck Drive, Spalding, PE11 1PD
G8 NWI J. Vine, 117 Betterton Road, Rainham, RM13 8ND
G8 NWL J. Mason, 46 Bradford Street, Chelmsford, CM2 0FJ
G8 NWM V. Maxfield, 50 Hanthorpe Road, Morton, Bourne, PE10 0NT
G8 NWS J. Caddick, 58 Beachcroft Road, Kingswinford, DY6 0HX
G8 NWU Maj. M. Wright, 69 Wroxham Drive, Nottingham, NG8 2QR
G8 NWZ M. Percy, 73 Ridgeway, Wellingborough, NN8 4RY
G8 NXA T. Ehlen, 58B Warriner Gardens, London, SW11 4DU
G8 NXB N. Borrett, 44 Oakhill Road, Ashtead, KT21 2JG
G8 NXD M. Waterfall, 12A Boskenna Road, Four Lanes, Redruth, TR16 6LS
G8 NXE S. Eyles, 2 Salisbury Close, Lichfield, WS13 7SN
G8 NXG R. Burn, 3 Barnsite Gardens, Rustington, Littlehampton, BN16 3QG
G8 NXJ I. Livesey, 26 Hilltop Road, Twyford, Reading, RG10 9BN
GW8 NXK G. Garner, 31 Clare St., Manselton, Swansea, SA5 9PG
G8 NXQ W. Povey, 31 Baddlesmere Road, Whitstable, CT5 2LB
G8 NXS D. Stevenson, 86 Kingston Road, Luton, LU2 7SA
G8 NXY W. Godwin, Heathwood House, Burton Road, Duddon Heath, Duddon, Tarporley, CW6 0GJ
G8 NYB D. Reed, 59 Cowley Avenue, Chertsey, KT16 9JJ
G8 NYC J. Primmer, 45 Roman Way, Felixstowe, IP11 9NP
G8 NYD M. Perry, 23 Victors Crescent Hutton, Brentwood, CM13 2HZ
G8 NYH R. Adams, 2 Longwill Avenue, Melton Mowbray, LE13 1UR
G8 NYJ I. Gibbs, 3 Badger Drive, Lightwater, GU18 5TS
G8 NYK M. Nicholson, Flat 14, Whyke Court, Chichester, PO19 8TP
G8 NYM M. Lister, 97 Hightown Road, Liversedge, WF15 8DG
G8 NYR B. Rabey, 36 Park Way, St Austell, PL25 4HR
GM8 NYV S. Richardson, Rowan Bank, Melvich, Thurso, KW14 7YJ
G8 NYZ N. Cooper, 4 Mossfield Crescent, Kidsgrove, Stoke-on-Trent, ST7 4YA
G8 NZA P. Baker, 12 Hockeredge Gardens, Westgate-on-Sea, CT8 8AN
G8 NZB B. Durrant, 16 Merrymeet, Whitestone, Exeter, EX4 2JP
G8 NZC K. Edmunds, 44 Antonia Circuit, Hallett Cove, Australia, SA 5158
G8 NZD C. Atkinson, 8 Southwood Road, Dunstable, LU5 4EA
G8 NZK N. O'Hagan, 5 Bankside, Finchampstead, Wokingham, RG40 3QB
GM8 NZL E. Hogg, 43 Muir Wood Road, Currie, EH14 5JN
GW8 NZN D. Roberts, 12 Erw'R Llan, Nannerch, Mold, CH7 5RF
G8 NZO J. Crozier, 43 Shepherds Way, Birmingham, B23 5XR
G8 NZR K. Pullan, 18 Heathfield, Mirfield, WF14 9BJ
G8 OAD G. Baxter, 4 Deeping Road, Baston, Peterborough, PE6 9NP
GM8 OAH W. Easton, 21 Cameron Avenue, Bishopton, PA7 5ES
G8 OBB T. Hooker, Inglewood, Woodside Road, Luton, LU1 4DJ
G8 OBP D. Payne, 11 Welbeck Close, Blaby, Leicester, LE8 4HF
G8 OBR J. Readle, 17 Rossendale View, Todmorden, OL14 6HN
G8 OBT T. Graham, 22 Locker Park, Wirral, CH49 2RZ
G8 OCA J. Astle, River View, Brough, Kirkby Stephen, CA17 4BZ
G8 OCE J. Hardy, 42 Fir Tree Drive, Wales, Sheffield, S26 5LZ
G8 OCF R. Harris, 25 Pear Tree Drive, Chard, TA20 2FP
G8 OCM E. Dubbins, 2 Elizabeth Avenue, Rose Green, Bognor Regis, PO21 3EL
G8 OCO M. Hughes, 49 Reedings Road, Barrowby, Grantham, NG321AU
GI8 OCR J. Mcilveen, 31 Edenaveys Crescent, Armagh, BT60 1NT
G8 OCS D. Simpson, 6 St. Martins Close, Stratford-upon-Avon, CV37 9QW
G8 OCT S. Terry, 207 Lakeside Drive, Walhalla Sc, USA, 29691
G8 OCV C. Smart, Old Queens Head, Ipswich Road, Diss, IP21 4XP
G8 ODK R. Varley, 41 Lang Lane, West Kirby, Wirral, CH48 5HQ
GM8 OEG A. Swiffin, Glebe House, Kellas, Dundee, DD5 3PD
G8 OEJ E. Bray, Rothesay, 6 Empshott Road, Southsea, PO4 8AU
G8 OEK P. Brown, Estate Yard House, Beverley, HU17 7PN
G8 OEO J. Thompson, 4 The Grove, Ponteland, Newcastle upon Tyne, NE20 9HQ
G8 OEU T. Hipwood, 3 Camview, Paulton, Bristol, BS39 7XA
G8 OFA Dr M. Cranage, Corris House, West Gomeldon, Salisbury, SP4 6LS
G8 OFI G. Radivan, 15 Agecroft Road West, Prestwich, Manchester, M25 9RE
G8 OFN R. Pashley, 24 Essendine Crescent, Sheffield, S8 8PB
G8 OFO R. Short, Langtree House, Castle Hill, Fordingbridge, SP6 2AX
GM8 OFQ G. Dobson, Ocean View, Stromness, KW16 3PQ
G8 OFR R. Coole, Courtyard Cottage, Horse Fair Lane, Swindon, SN6 6BN
G8 OFX A. Nelson, 37 Brook Way, Romsey, SO51 7JZ
G8 OFZ I. Mcgowan, Meld House, Hawthorn Road, Shrewsbury, SY3 7NB
G8 OGP S. Martin, Aldon, The Hayes, Cheddar, BS27 3HS
G8 OGR J. Holton, 24 Great Austins, Farnham, GU9 8JQ
G8 OHC G. Scholes, 14 Braemar Road, Bulwell, Nottingham, NG6 9HN
G8 OHG J. Myall, 52 Princethorpe Way, Binley, Coventry, CV3 2HF
G8 OHH J. Morgan, 41 Lingen Avenue, Hereford, HR1 1BY
G8 OHM South Birmingham Radio Society, c/o N. Gutteridge, 68 Max Road, Quinton, Birmingham, B32 1LB
G8 OHP C. Cainsford-Betty, 19 Reads Street, Stretham, Ely, CB6 3JT
G8 OHS M. Emery, 25 Bradgate Drive, Sutton Coldfield, B74 4XG
GW8 OIJ A. Stark, 5 Ladyhill Road, Newport, NP19 9RY
G8 OIU Dr J. Cave, 4 Grove Road, Newbury, RG14 1UH

G8 OIY Dr S. Robertson, 249 Ware Road, Hertford, SG13 7EJ
G8 OJK V. Willett, 20 The Green, Sharlston Common, Wakefield, WF4 1EF
G8 OJR W. Pearce, 160 Philip Lane, Tottenham, London, N15 4JN
G8 OKB R. Mccann, Goss House, Clark Street, Stourbridge, DY8 3UF
G8 OKD M. Bailey, 28 St. Pauls Hill Road, Hyde, SK14 2SW
G8 OKE R. Brown, 8 Grassmere Way, Waterlooville, PO7 8QD
G8 OKI L. Mather, 8 Carnoustie Avenue, Chesterfield, S40 3NN
G8 OKN M. Gallagher, 50 Warwick Road, Southam, CV47 0HW
GW8 OKR B. Kirkpatrick, 88 Gaer Park Drive, Newport, NP20 3NR
G8 OKS B. Dawson, 9 Redmayne Close, Billingham, TS23 3HG
G8 OKZ D. Shillington, 6 Moss Close, Willaston, Neston, CH64 2XQ
G8 OLA R. Elwood, 28 Manor Lane, Selsey, Chichester, PO20 0NX
GI8 OLH T. Lavery, 21 Mussenden Grange, Articlave, Coleraine, BT51 4US
G8 OLK P. Smith, 21A Meadow Way, Bracknell, RG42 1UE
G8 OLL R. Porter, 9 School Avenue Brownhills, Walsall, WS8 6AG
G8 OLP M. Matthews, 19 Perrylands, Charlwood, Horley, RH6 0BL
G8 OLY D. Curwell, 9 St. Georges Road, Aldershot, GU12 4LD
G8 OMB D. Parker, 146 Merlin Avenue, Nuneaton, CV10 9QJ
G8 OMC D. Smith, 71 Ashbourne Avenue, Aspull, Wigan, WN2 1HW
G8 OMK C. Robins, 49 Brockenhurst Gardens, London, NW7 2JY
G8 OMQ D. Bliss, 7 Silver Lane, West Wickham, BR4 0SG
G8 OMW F. Rowan, 91 St. Nicholas Road, Littlemore, Oxford, OX4 4PW
G8 ONH J. Sager, Well Cottage, Fenn Lane, Woodbridge, IP12 4NZ
GW8 ONP Dr J. Eastwood, 30 Gerddi Rheidol Trefechan, Aberystwyth, SY23 1DB
G8 ONR M. Loader, 20 Edgcumbe Drive, Tavistock, PL19 0ET
G8 ONS K. Creighton, 10 Oram Close, Allery Banks, Morpeth, NE61 1XF
G8 ONY B. Goodhew, 101 Brier Road, Sittingbourne, ME10 1YL
G8 OO A. Holdsworth, Millfield, The Green, Dereham, NR20 5LL
G8 OOC J. Morecroft, 217A Longhurst Lane, Mellor, Stockport, SK6 5PN
G8 OOF Dr G. Ellison, 36 Park Hill, Clapham, London, SW4 9PB
G8 OOQ Dr M. Barton, 23 Caledonia Place, Bristol, BS8 4DL
G8 OOS M. Reeson, 18 Hazelnut Way, Louth, LN11 7BZ
G8 OPA P. Barry, 32 Rutland Avenue, Sidcup, DA15 9DZ
G8 OPC D. Crawley, 9 Gwynns Walk, Hertford, SG13 8AD
G8 OPE M. De Rouffignac, 2 Westgate, Old Malton, Malton, YO17 7HE
G8 OPI J. Spooner, 59 Woodlands Park Drive, Dunmow, CM6 1WT
G8 OPO G. Bartels, 37 Faircross Avenue, Romford, RM5 3SX
G8 OPP J. Birkett, 13 The Strait, Lincoln, LN2 1JD
G8 OPX T. Willford, 15 Foxglove Close, Broughton Astley, Leicester, LE9 6YU
G8 OPY G. Winston, 8 Linnet Close, Shoeburyness, Southend-on-Sea, SS3 9YE
G8 OQC J. Kent, 106 Victoria Road, Barnet, EN4 9PA
G8 OQG P. Jobbins, 35 Keys Avenue Horfield, Bristol, BS7 0HQ
G8 OQK A. Taylor, 77 Bedford Road, Marston Moretaine, Bedford, MK43 0LA
G8 OQP T. Spacagna, 3A Station Road, Romsey, SO51 8DP
G8 OQR M. Crossman, 127 The Grove, Southend-on-Sea, SS2 4DA
G8 OQT J. Lambert, 125 Tudor Way, Mill End, Rickmansworth, WD3 8HT
G8 ORM M. Baguley, 42 Kendall Avenue, Shipley, BD18 4DY
G8 ORO D. Coulter, 9 Taylors Way, Whitehaven, CA28 9PD
G8 ORR C. Brown, 179 Bournville Lane, Birmingham, B30 1LY
G8 ORX K. Rashleigh, 43 Oxshott Way, Cobham, KT11 2RU
G8 OSG N. Mcalpine, 15 Sparrows Herne, Basildon, SS16 5JH
G8 OSH N. Hubbard, 31 Bridlington Crescent, Monkston, Milton Keynes, MK10 9HG
G8 OSJ D. Halliwell, 9 Berkeley Avenue, Alsager, Stoke-on-Trent, ST7 2BW
G8 OSX K. Dawson, 3 The Green, Bonehill, Tamworth, B78 3HW
G8 OSZ S. Ashley, 12 Dene Close, Wellingborough, NN8 5QP
G8 OTA H. O'Tani, 51A Horley Road, Bristol, BS2 9TL
G8 OTC C. Anderson, 42 Elizabethan Way, Rugeley, WS15 2EE
G8 OTD S. Ballard, 11 Laburnum Gardens Quedgeley, Gloucester, GL2 4WF
G8 OTG R. Cannon, 111 Brangbourne Road, Bromley, BR1 4LP
G8 OTH C. Churchill, 87 Bradley Crescent, Shirehampton, Bristol, BS11 9SR
GM8 OTI Dr J. Cooke, 6/2 Greenbank Terrace, Edinburgh, EH10 6ER
G8 OTZ D. Logan, 33 Foxwood Drive, Kirkham, Preston, PR4 2DS
G8 OUG L. Gray, 20 Turner Way, Clevedon, BS21 7YN
G8 OUH I. Harfield, White Gates, Crofton Avenue, Lee on the Solent, PO13 9NJ
G8 OUI D. Baines, 1 Carole Close, Sutton Leach, St Helens, WA9 4PW
GW8 OUM D. Briggs, 43 Monmouth Walk, Markham, Blackwood, NP12 0QR
G8 OUS S. Greendale, 15 Bosworth Road, Cambridge, CB1 8RG
G8 OUY D. Smith, 41 Mitcham Road, Camberley, GU15 4AR
G8 OVO N. Lihou, 47 Lichfield Drive, Brixham, TQ5 8DG
G8 OVZ S. Gosby, 20 Woodland Mount, Hertford, SG13 7JD
G8 OWA Dr R. Lewin, 24 Gilliver Close, Burton-on-Trent, DE14 2FL
G8 OWG Dr K. Rowe, 43 Knighton Close, Duston, Northampton, NN5 6NE
G8 OWS J. Greenall, 6 Neasham Drive, Darlington, DL1 4LG
G8 OWV Prof. J. Everard, 4 Lloyd Close Heslington, York, YO10 5EU
G8 OWZ O. Cockram, 446 Holdenhurst Road, Bournemouth, BH8 9AE
G8 OXD P. Brown, 41 School St., Castleford, WF10 2SB
G8 OXE M. Brooks, 3 Wood Side, Wood Street, March, PE15 0SB
G8 OXG N. Powell, 42 Sheraton Drive, Kidderminster, DY10 3QR
G8 OXI E. Mannix, La Vieille Scierie, Les Allues, France, 73550
G8 OXS A. Lambert, Electronic Media Services Ltd, Building 51, Whitehill & Bordon Enterprise Park, Budds Lane, Bordon, GU35 0FJ
G8 OXU C. Madge, 89 Heron Gardens, Rayleigh, SS6 9TU
G8 OXX P. Bailey, 236 Sandy Lane, Droylsden, Manchester, M43 7JX
G8 OYA S. Mckinty, Sands House 5 Nightingale Gardens, Moreton-in-Marsh, GL56 0FX
G8 OYB K. Armstrong, 2 Blackthorn Grove, Shawbirch, Telford, TF5 0LL
G8 OYF J. Popplewell, 6 Roseleigh Avenue, Manchester, M19 2NP
G8 OYL W. Shave, 26 Hessle Avenue, Boston, PE21 8DA
G8 OYM P. Taylor, 15 De Montfort Road, Lewes, BN7 1SP
G8 OYQ M. Everitt, 48 Rant Meadow, Hemel Hempstead, HP3 8EQ
GW8 OYT B. Jones, 6 Rhodfa Maes Hir, Rhyl, LL18 4JF
G8 OYX C. Blanchard, 19 Cuckfield Avenue, Ipswich, IP3 8RZ
G8 OYY J. Bishop, No1 Billhurst Cottage, Plaistow Street, Lingfield, RH7 6EY
G8 OZD A. Batty, 23 Sandyshot Walk, Wythenshawe, Manchester, M22 5AQ
G8 OZH J. Burrell, 6 Blenheim Croft, Brackley, NN13 7ET
G8 OZP R. Platts, 43 Iron Walls Lane, Tutbury, Burton-on-Trent, DE13 9NH
G8 OZQ S. Pallett, 6 Lancaster Close, Coalville, LE67 4TG
G8 OZT N. Morley, Mazongill, Orton, Penrith, CA10 3RZ
G8 OZY P. Harrison, 91 Obelisk Rise, Northampton, NN2 8QU
G8 PAB M. Humphries, 20 Taunton Street, Swindon, SN1 5EE
G8 PAD J. Field, Flat 24, Dibden Court, Honeywall, Stoke-on-Trent, ST4 1PB
G8 PAE H. Jackson, 22 Harkness Drive, Waterlooville, PO7 8SH
G8 PAG M. Rose, 20 Broad Piece, Soham, Ely, CB7 5EL
GM8 PAH D. Schofield, 166 Harbour Place, Dalgety Bay, Dunfermline, KY11 9AA
G8 PAI D. Rout, Two Akers, Wrabness Road, Harwich, CO12 5NE
G8 PAK C. Hollier, The Sheiling, Ryall Road, Ryall, Upton-Upon-Severn, Worcester, WR8 0RH
G8 PAL P. Hankinson, 37 Victoria Avenue, Whitefield, Manchester, M45 6DP
G8 PAN S. Day, 14 The Crescent, Market Harborough, LE16 7JJ
G8 PAT P. Mcguinness, 9 Farmdale Road, Carshalton Beeches, SM5 3NG
G8 PBH Dr A. Kent, 46 Russley Road, Bramcote, Nottingham, NG9 3JE
G8 PBI I. Murphy, The Nurseries, Carnon Crease, Truro, TR3 6LJ
G8 PBM A. Royston, 4 Cedars Walk, Dunstable Street, Ampthill, Bedford, MK45 2JY
G8 PBY G. Coles, Dairy Farmhouse, West Winterslow, Salisbury, SP5 1RE
GJ8 PCY P. Falle, 2 Greystones, Gorey Village Main Road, Grouville, Jersey, JE3 9EP
G8 PDE W. Burin, 35 Beaconsfield Road Low Fell, Gateshead, NE9 5EU
GI8 PDK Dr D. Courtney, 79 Fort Road, Belfast, BT8 8LX
G8 PDM C. Commander, 8 Cannon Place, Hampstead, London, NW3 1EJ
G8 PDP R. Hinchliffe, 34 Oaklea, Ash Vale, Aldershot, GU12 5HP
G8 PDY S. Procter, 8 Pond End Road, Sonning Common, Reading, RG4 9SA
G8 PEA K. Wibberley, 5A Marston Road, Croft, Leicester, LE9 3GX
GM8 PEB Dr P. Fineron, The Courtyard, Crauchie, East Linton, EH40 3EB
G8 PEN C. Vernon, 48 Long Beach, Hemsby, Great Yarmouth, NR29 4JD
G8 PEW A. Pewsey, Pineholm, High Close, Bovey Tracey, Newton Abbot, TQ13 9EX
G8 PFL J. Turner, 32 Petunia Crescent, Chelmsford, CM1 6YP
G8 PFR M. Gibson, Eccles Wall Farm, Bromsash, Ross-on-Wye, HR9 7PW
GW8 PFT P. Hinson, 7 Awel Tywi, Llangunnor, Carmarthen, SA31 2NL
G8 PFZ R. Harrison, Badgers Oak, Redbrook Street, Ashford, TN26 3QU
G8 PGA R. Martinus, 62 High Oaks Close, Locks Heath, Southampton, SO31 6SX
G8 PGE D. Sinclair, 12A Sunnydown Road, Winchester, SO22 4LD
G8 PGF A. Price, 11 Gatcombe Gardens, Titchfield, Fareham, PO14 3DR
G8 PGH K. James, 44 The Oakfield, Littledean Hill Road, Cinderford, GL14 2DE
G8 PGI P. Lord, Beechwood House, Main Street, Lutterworth, LE17 5QA
GI8 PGJ D. Campbell, 18 The Counties, Mark Street, Portrush, BT56 8QA
G8 PGO D. Carter, 49 Hinckley Road, Sapcote, LE9 4LG
G8 PHB P. Mckenzie, 21 Arnside Walk, Chapel House, Newcastle upon Tyne, NE5 1BT
G8 PHG Dr B. Cook, Hinsley Mill House, Hinsley Mill Lane, Market Drayton, TF9 1HP
G8 PHJ M. Palmer, 38 Windermere Close, Dartford, DA1 2TX
G8 PHM M. Kent, Meadow Bank, Rye Lane, Sevenoaks, TN14 5JF
G8 PHQ C. Challender, 9 Blick Close, West Winch, King's Lynn, PE33 0UA
G8 PHS E. Campbell, 2 Russell Avenue, March, PE15 8EL
G8 PHV R. Wetton, 8 St. Moritz Close, Northwick, Worcester, WR3 7ND
G8 PIC C. Pomphrett, 47 North Leas Avenue, Scarborough, YO12 6LJ
G8 PIN R. Bannister, 14 Amery Close, Worcester, WR5 2HL
G8 PIO O. Futter, 25 Amhurst Gardens Belton, Great Yarmouth, NR31 9PH
G8 PIP P. Elwell, 4 Richmond Grove, Wollaston, Stourbridge, DY8 4SF
GM8 PIV E. Souter, 3/2 10 James Gray Street, Glasgow, G41 3BS
G8 PIY D. Clifton, 10 Scotney Road, Basingstoke, RG21 5SR
G8 PJC J. Mcdonald, 17 Highfield Close, Wokingham, RG40 1DG
G8 PJD P. Deffee, 18 Poplar Road, Kensworth, Dunstable, LU6 3RS
GW8 PJE P. Evans, 10 Cae Perllan Brackla, Bridgend, CF31 2HL
G8 PJF E. Summers, 161 North Cray Road, Sidcup, DA14 5LT
G8 PJQ C. Cole, 70 Throgmorton Road, Yateley, GU46 6FA
G8 PK B. Wilson, 58 High Street, Saxilby, Lincoln, LN1 2HA
GW8 PKB L. Rudge, 8 Penrallt Estate, Llanystumdwy, Criccieth, LL52 0SR
G8 PKG I. Bosworth, 7 Sandbourne Court, 54-56 West Overcliff Drive, Bournemouth, BH4 8AB
G8 PKJ G. Rowland, 18 Heights Way, Leeds, LS12 3SN
G8 PKM C. Mitchell, 25 Belper Road, Ashbourne, DE6 1BB
GW8 PKV M. James, 28 Bloomfield Gardens, Narberth, SA67 7EZ
G8 PL A. Garry-Durrant, Casa Santosa Llano Del Espino, Albox, Spain, 4800
G8 PLI N. Vranic, 30 Mitchell Street, Sheffield, S3 7NL
G8 PLJ J. Bailey, 8 Hild Avenue Cudworth, Barnsley, S72 8RN
G8 PLO R. Clubley, Church Hill House, High Street, Braintree, CM7 4BY
G8 PMA L. Pennell, 182 Northampton Road, Wellingborough, NN8 3PJ
G8 PMJ D. Hughes, 18 Bailey Close, Pewsey, SN9 5HU
G8 PMR L. Morris, 17 Kestrel Close, Hornchurch, RM12 5LS
GW8 PNE P. Griffiths, 13 Wesley Court Warren Street, Tenby, SA70 7JT
G8 PNM N. Cocking, 15 Spring View Road, Sheffield, S10 1LS
G8 PNN G. Emmerson, 72 The Gables, Widdrington, Morpeth, NE61 5RB
G8 POE J. Phillips, 235 Barn Mead, Harlow, CM18 6ST
G8 POG P. Wood, 23 Shipley Avenue, Newcastle upon Tyne, NE4 9QY
G8 POI D. Evans, 14 Lower Wharf, Wallingford, OX10 9AA
G8 POK G. West, 6 Willerton Close, Chidswell, Dewsbury, WF12 7SQ
G8 POL M. Williams, 22 Charlecote Drive, Nottingham, NG8 2SB
G8 POO S. Robinson, 23 Jameson Drive, Corbridge, NE45 5EX
G8 POP R. Mundy, 12 Cantors Way, Minety, Malmesbury, SN16 9QZ
G8 POQ B. Salter, 19 Garnier Park, Wickham, Fareham, PO17 5LD
G8 POS A. Axon, 7 Tudor Grove, Groby, Leicester, LE6 0YL
G8 PPA S. Ornstein, Sandy Lodge, Trusthorpe Road, Sutton-On-Sea, Mablethorpe, LN12 2LN
G8 PPD A. Saunders, Suffolk House, Main Road, Ipswich, IP9 1DX
G8 PPN A. Osmond, The Moorings, 10 North Road, Shanklin, PO37 6DB
G8 PPQ G. Boakes, 16 Terminus Drive, Herne Bay, CT6 6PP
GD8 PPU G. Brookes, 44 Magherchirrym, Port Erin, Isle of Man, IM9 6DB
G8 PQA A. Gapper, 9 St. Annes Drive Wick, Bristol, BS30 5PN
G8 PQB G. Grantham, 18 Fen End Lane, Spalding, PE12 6AD
G8 PQH F. Rowsell, 4 Coxswain Way, Selsey, Chichester, PO20 0UA
G8 PQJ J. Robinson, Rose Cottage, Wavering Lane West, Gillingham, SP8 4NR
G8 PQN M. Henshaw, 23A Bedford Road, Northill, Biggleswade, SG18 9AH
G8 PQZ G. Collier, 6 Copse Close, Tilehurst, Reading, RG31 6RH
G8 PRC Plymouth Radio Community, c/o P. Connor, 20 Longfield, Lutton, Ivybridge, PL21 9SN
G8 PRH A. Hartley, 16 Old Thorne Road, Hatfield, Doncaster, DN7 6ER
G8 PRJ S. Sanders, 19 Brunswick Gardens, Hainault, Ilford, IG6 2QU

G8 PRK R. Holmwood, 3 Stanstead Road, Caterham, CR3 6AD
G8 PRN R. Morley, 21 Meadow View Skelmanthorpe, Huddersfield, HD8 9ET
G8 PRP B. Youster, 24 Sunningdale Road, Weston-Super-Mare, BS22 6XP
G8 PSC J. Benoy, 46 Bickham Road, Plymouth, PL5 1SB
G8 PSF A. Ball, 20 Inverness Avenue, Enfield, EN1 3NT
GW8 PSJ L. Finch, Hafan Deg, Waungilwen, Felindre, Llandysul, SA44 5YG
G8 PSO R. Gould, 20 Southwood Drive, Coombe Dingle, Bristol, BS9 2QU
G8 PSS J. Meldrum, 18 Oatlands Way, Pity Me, Durham, DH1 5GL
GM8 PSV B. Thomson, 51 Main Street, Newmill, Keith, AB55 6UR
G8 PSZ C. Wood, 5 Farm Close, Market Drayton, TF9 3UH
G8 PTF R. Duddin, 16 Gateley Road, Warley, Oldbury, B68 0NU
G8 PTH A. Emmerson, 71 Falcutt Way, Northampton, NN2 8PH
G8 PTL M. Fleming, 16 Church Street, Chasetown, Burntwood, WS7 3QL
G8 PTN D. Stoney, 7 Sandwell Close, Long Eaton, Nottingham, NG10 3RG
GW8 PTS W. Leddington, 4 Cherry Walk, Monmouth, NP25 5DE
G8 PTW C. Wallace, Windy Ridge, Langley Priory, Derby, DE74 2QQ
G8 PTY P. Thornton-Evison, Greyfriars, Townsend, Wantage, OX12 0AT
G8 PUB S. Lucas, 84 Woodman Road, Warley, Brentwood, CM14 5AZ
G8 PUE J. Taylor, The Jays, 5 Watling Close, Bourne, PE10 9XL
G8 PUH B. Merrell, 16 Box Close, Broadfield, Crawley, RH11 9QT
G8 PUK S. Mann, 1 Blackthorn Avenue Bramley, Rotherham, S66 2LU
G8 PUN J. Keleher, 9 Broadwell Drive, Leigh, WN7 3NE
G8 PUR T. Rose, 41 Keats Way, Hitchin, SG4 0DP
G8 PUT C. Townsend, 2 Netherfield Drive, Netherthong, Holmfirth, HD9 3ES
G8 PUY N. Dowsett, 4 Pankhurst Avenue, London, E16 1UT
G8 PVG D. Hobbs, 46 Gloucester Road, Bridgwater, TA6 6DZ
G8 PVK R. Still, The Manor House, 5 Beechwood Avenue, Bournemouth, BH5 1LY
GJ8 PVL P. Bertram, Roz-Den, La Rue De La Guilleaumerie, St Saviour, Jersey, JE2 7HQ
G8 PVR J. Riggs, 21 Janeva Court, Liskeard Road, Saltash, PL12 4FD
G8 PVV M. Walton, 12 Davison Street, Newcastle upon Tyne, NE15 8NB
G8 PWE I. Ashford, 28 Gorsey Lane, Great Wyrley, Walsall, WS6 6JA
G8 PWK M. Forsey, 84 Garner Road, Walthamstow, London, E17 4HH
G8 PWO J. Thwaites, 15 Spring Head Road, Kemsing, Sevenoaks, TN15 6QL
G8 PWQ N. Wilson, 114 Northgate Way, Terrington St. Clement, King's Lynn, PE34 4LH
G8 PWT L. Cook, 16 Florence Road, Maidstone, ME16 8EN
G8 PWU J. Crossland, 1 Carter Lane, Flamborough, Bridlington, YO15 1LW
G8 PX Oxford and District ARS, c/o G. Diacon, 45 Woodpecker Way, Witney, OX28 6NN
G8 PXI M. Parker, 65 Shoreham Drive, Penketh, Warrington, WA5 2HY
G8 PXM K. Townsend, 4 Respryn Close, Liskeard, PL14 3TE
G8 PXO G. Murray, Old Court, Westleigh, Tiverton, EX16 7HT
G8 PXU B. Gascoigne, 108 Blandford Avenue, Castle Bromwich, Birmingham, B36 9JD
G8 PXW A. Stevens, 23 Millers Green Drive, Kingswinford, DY6 0DA
G8 PYD G. Farrell, 95 Washington Road, Maldon, CM9 6JF
G8 PYE P. Barrett, 30 Rosslyn Park Road, Plymouth, PL3 4LN
G8 PYU I. Walton, Tall Trees, Tredington, Shipston-on-Stour, CV36 4NG
G8 PZD T. Wills, 66 Kipling Road, St. Marks, Cheltenham, GL51 7DQ
G8 PZF B. Simpson, 14 Priestthorpe Lane, Bingley, BD16 4EE
G8 PZI W. Nolan, South Lawn, 77 Reigate Road, Reigate, RH2 0RE
GW8 PZS J. Coady, 21 Garth Wen, Llanfaes, Beaumaris, LL58 8PT
G8 PZX F. Gunn, 8 College Gardens, Hornsea, HU18 1EF
G8 QM V. Flowers, Eothen Homes Ltd, 45 Elmfield Road, Newcastle upon Tyne, NE3 4BB
G8 QZ D. Sager, 29 Station Road, Mickleover, Derby, DE3 9GH
G8 RAC J. Maines, Brick House Farm, Marden, Hereford, HR1 3ET
G8 RAF Royal Air Force ARS, c/o S. Mckinnon, 145 Enville Road, Kinver, Stourbridge, DY7 6BN
G8 RAJ R. Shore, 42 King George Avenue, Bournemouth, BH9 1TX
G8 RAN K. Reeman, 4 Alfreda Avenue, Hullbridge, Hockley, SS5 6LT
G8 RAO A. Yates, 87 Princess Road, Warley, Oldbury, B68 9PW
GW8 RAS R. Neville, 11 Heol Urban, Llandaff, Cardiff, CF5 2QP
G8 RAU R. Lewis, 66A Derek Gdns, Southend on Sea, SS26QY
G8 RAV K. Jackson, 17 Copperfield Close, Kettering, NN16 9EW
G8 RAX C. Hill, 183 Manchester Road, Swinton, Manchester, M27 4FA
G8 RBI C. Allen, 8 Shoulbard, Fleckney, Leicester, LE8 8TX
G8 RBK I. Brown, 56 Church Lane, Darley Abbey, Derby, DE22 1EY
GM8 RBR W. Egerton, Croft House, Upper Breakish, Isle of Skye, IV42 8PY
G8 RBS P. Bickersteth, Tregarth, Fernsplatt, Truro, TR4 8RJ
G8 RBU T. Dewey, 93 Calverton Road, Arnold, Nottingham, NG5 8FQ
G8 RBV D. Deighton, 3 Bluebell Way, Huncoat, Accrington, BB5 6TD
G8 RBW C. Ellison, 29 Ashton Road, Clay Cross, Chesterfield, S45 9FA
G8 RBX L. Fitzpatrick-Browne, 24 Beechmount Avenue, Hanwell, London, W7 3AG
G8 RBY P. Hodson, 43 Thorpe Road, Melton Mowbray, LE13 1SE
G8 RCE K. Shergold, 47 Moorcroft Gardens, Redditch, B97 5WG
G8 RCK R. Woollard, 68 Trunk Furlong, Aspley Guise, Milton Keynes, MK17 8HX
G8 RCL G. Whiston, 86 Elmsett Close, Great Sankey, Warrington, WA5 3RX
G8 RCO D. Russell, 53 The Campions, Borehamwood, WD6 5QE
G8 RCZ G. Fermor, 26 Byron Road, Exeter, EX2 5QN
G8 RDA K. Forster, 10 Springfield Oval, Witney, OX28 6EG
G8 RDB R. George, Juniper Cottage, Hillesden, Buckingham, MK18 4BX
G8 RDG R. Maltby, Meadow Croft, Bishop Lane, Henfield, BN5 9DG
GW8 RDI R. Colclough, Login Fach, Waunarlwydd, Swansea, SA5 4NJ
G8 RDJ J. Davies, 11 Bromley Road, Macclesfield, SK10 3LN
G8 RDK L. Mayhew, 47 Beeches Avenue, Worthing, BN14 9JE
G8 RDN T. Sale, 20 Redwood Drive, Chase Terrace, Burntwood, WS7 2AS
G8 RDP J. Webb, 6 Chatsworth Avenue, Fleetwood, FY7 8EG
G8 RDQ P. Williams, 30 Duchess Drive, Bridgnorth, WV16 4JD
G8 RDT K. Williams, Jasmine Cottage, Evesham Road, Dodwell, Stratford-upon-Avon, CV37 9ST
G8 REF P. Ellis, 15 Alexander Close, Bognor Regis, PO21 4PS
GM8 REG R. Bell, Fairview, Main Street, Huntly, AB54 7SY
G8 REO R. Mitchell, 4 Friendly Fold Road, Halifax, HX3 5QF
G8 REQ F. Robinson, 13 Dorset Drive, Wirral, CH61 8SX
G8 RER J. Fothergill, 53 Meadow Court, Ponteland, Newcastle upon Tyne, NE20 9RA
G8 RES M. Howard, Lodge Cottage, Stody Estate, Stody, Melton Constable, NR24 2EW
GW8 REV C. Marsh, 50 Timothy Rees Close, Cardiff, CF5 2AU
G8 REX G. Stephens, 71 Quicks Road, London, SW19 1EX
G8 RF F. Raby, 20 Lime Tree Road, Codsall, Wolverhampton, WV8 1NT
G8 RFC R. Cassell, 1 St. Saviour Close, Colchester, CO4 0PW
GW8 RFD P. Short, 6 Broadmead Pontllanfraith, Blackwood, NP12 2NL
G8 RFE M. Wallace, 26 Parsons Drive, Glen Parva, Leicester, LE2 9NS
G8 RFF K. Richardson, 12 Badger Hill, Brighouse, HD6 3QX
G8 RFL D. Robinson, 25 Angelica, Amington, Tamworth, B77 3JZ
G8 RFP D. Clarke, 57 Stubley Drive, Dronfield, S18 8QY
G8 RFV R. Bulmer, 4 Valerian Drive, Stafford, ST16 1FJ
G8 RFW G. Sharpe, Holborn House 28 Sheffield Road, Dronfield, S18 2GG
G8 RFY C. Garner, Flat 10, Chapel Court, Leicester Road, Narborough, Leicester, LE19 2FX
G8 RFZ D. Cooke, 27 Goosehill Court, Balby, Doncaster, DN4 8SX
G8 RGN J. Harling, 6 Fontwell Road, Little Lever, Bolton, BL3 1TE
GM8 RGO M. Robson, Whistlefield Cottage, Loch Eck, Dunoon, PA23 8SG
G8 RGU M. Burt, Hartcliff Farm, Okeford Fitzpaine, Blandford Forum, DT11 0EF
G8 RHC J. Cranage, Corris House, West Gomeldon, Salisbury, SP4 6LS
G8 RHM K. Hoggett, 14 Wyld Court, Allesley, Coventry, CV5 9LQ
G8 RHN M. Kirkham, 60 Westminster Drive, Grimsby, DN34 4TY
GW8 RHP Rev. J. Williams, Monte Vista, Llandyrnog, Denbigh, LL16 4HH
G8 RHQ S. Ormondroyd, 15 Meadowlands, Blundeston, Lowestoft, NR32 5AS
G8 RHU E. Carvill, 61 Midhurst Drive, Ferring, Worthing, BN12 5BQ
G8 RHZ A. Robertson, 22 Court Way, Twickenham, TW2 7SN
G8 RIB P. Fallon, 17 Blundell Road, Widnes, WA8 8SS
G8 RIC K. Murphy, 79 Torkington Road, Hazel Grove, Stockport, SK7 6NR
G8 RIK R. Milner, Lyndene, Holyhead Road, Shrewsbury, SY4 1EE
G8 RIM J. Boardman, 81 Shipston Road, Stratford-upon-Avon, CV37 7LW
G8 RIP M. Walmsley, 27 Russell Avenue, Preston, PR1 5TP
G8 RIR S. Newbury, 19 Kings Meadow Drive, Winkleigh, EX19 8HD
G8 RIS R. Merriman, 8 Abbots Meadow, Chittlehampton, Umberleigh, EX37 9QE
G8 RIW B. Harvey, 56 Oakwood Drive, Grimsby, DN37 9RN
G8 RJB R. Bridgwater, 31 Pembroke Avenue, Worthing, BN11 5QS
G8 RJF K. Freer, 54A High Lane East, West Hallam, Ilkeston, DE7 6HW
G8 RJM S. Reap, The Staddles, Romsey Road, Stockbridge, SO20 8DB
G8 RJQ M. Corke, 6 Dhow Street, Sun Valley, Cape Town, South Africa, 7975
G8 RJZ M. Wills, 9 Allerdale Close, Thirsk, YO7 1FW
G8 RKG D. Peck, Flat 1, Shrewsbury Court, 21-23 Manor Road, Worthing, BN11 3RU
G8 RKH L. Hunt, 15 Oxford Street, Cowes, PO31 8PT
G8 RKO J. Butler, 36 Park Road, Bracknell, RG12 2LU
G8 RKX A. Titley, 6 Spring View, Luddendenfoot, Halifax, HX2 6EX
G8 RLD R. Dowdell, 2 Coles Courtyard, 5 Black Horse Way, Horsham, RH12 1NU
GI8 RLE J. Ashe, 49 Deans Walk, Richhill, Armagh, BT61 9LD
G8 RLF R. Dickerson, 7 Sixpenny Close, Titchfield Common, Fareham, PO14 4SY
GI8 RLG H. Emerson, Little Castle Dillo, Co Armagh, BT61 7DF
GW8 RLI V. Banfield, New Haven, Main Road, Pontypridd, CF38 1RY
G8 RLN J. Barnett, 11 Ridge Street, Stourbridge, DY8 4QF
G8 RLW G. Woodward, 6 Lang Road, Huntington, York, YO32 9SD
G8 RMI S. Blake, 10 Dimore Close, Hardwicke, Gloucester, GL2 4QQ
G8 RML M. Juby, Silver Birch, High Road, Diss, IP22 5RU
G8 RMP J. Bond, 19 Compton Avenue, Mannamead, Plymouth, PL3 5DA
GM8 RMR E. Scott, 81 Rosehaugh Road, Inverness, IV3 8SR
GI8 RNG W. Smyth, 11 Alexander Park, Armagh, BT61 7JB
G8 RNM I. Lucking, 32 Nolton Place, Edgware, HA8 6DL
G8 RNT P. Walking, Flat 36, Highlands House, Wharncliffe Road, Southampton, SO19 7GG
G8 RNU I. Strange, Holly Lane, Tansley, Matlock, DE4 5FF
G8 RNV N. Jefferies, 8 Cambridge Green, Fareham, PO14 4QX
G8 ROC Radio Operators Cornwall, c/o C. Macleod, 21 Halsetown, St Ives, TR26 3LY
G8 ROF C. Taylor, 26 Bernard Road, Brighton, BN2 3EQ
G8 ROG Dr A. Johnston, 12 Whitby Court Caversham, Reading, RG4 6SF
G8 RON R. Eyes, 6 Bakers Lane, Southport, PR9 9RN
G8 ROS R. Platt, 40 Solent Drive, Darcy Lever, Bolton, BL3 1RN
G8 ROU D. Hardy, 394 Slade Road, Birmingham, B23 7LG
G8 RPA K. Mendum, 47 The Crescent, Letchworth Garden City, SG6 1SW
G8 RPD J. Fennell, Broad Park Cottage, Stanbury Copse, Ilfracombe, EX34 8DW
GM8 RPE J. Robinson, 18 Craigshannoch Road, Wormit, Newport-on-Tay, DD6 8ND
G8 RPI G. Atkinson, 25 Appletrees, Bar Hill, Cambridge, CB23 8SJ
GI8 RPP M. Elder, 44 Learmount Road, Claudy, Londonderry, BT47 4AQ
GI8 RPT N. Copeland, 34 Glenkyle Park, Newtownabbey, BT36 6SP
G8 RQF J. Duffy, 5 Birch Court, Prudhoe, NE42 6PZ
G8 RQH I. Sherer, 1A Appleyard Drive, Barton-upon-Humber, DN18 5TD
GI8 RQI D. Allen, 40 Brambledown Drive, Banbridge, BT32 4RA
G8 RQN P. Needham, 2 Woodridge Close, Bracknell, RG12 9QX
GM8 RQW J. Kennedy, 6 Ryedale Terrace, Dumfries, DG2 7DL
G8 RRC P. Sharpe, 27 Record Road, Emsworth, PO10 7NS
G8 RRN M. Jones, 26 Cliff Road, Felixstowe, IP11 9PJ
GJ8 RRP J. Parry, 2 Thornley, Bagatelle Road, Jersey, JE2 7TZ
G8 RRR H. Potter, 134 Ifield Road, Crawley, RH11 7BW
G8 RRS M. Ellison, 22 Cotebrook Drive, Upton, Chester, CH2 1RD
G8 RSA Dr S. Hasko, 105 High Street Brampton, Huntingdon, PE28 4TQ
GM8 RSC J. Chinnock, 3B Dundee Street, Letham, Forfar, DD8 2PQ
G8 RSE J. Murphy, 15 Loders Close, Poole, BH17 9BF
G8 RSI G. Whitney, 4 Sefton Crescent, Sale, M33 7EN
G8 RSK P. Tyrell, 14 Park Farm Road, Horsham, RH12 5EW
G8 RSQ L. Williamson, 17 Ray Bond Way, Aylsham, Norwich, NR11 6UT
G8 RSV S. Staniforth, 4 Moses View, Shireoaks, Worksop, S81 8NH
G8 RSX T. Beck, 10 Rookery Close, Hatfield Peverel, Chelmsford, CM3 2DF
G8 RTB R. Breeze, 119 Sundorne Road, Shrewsbury, SY1 4RP
GM8 RTI Dr J. Grieve, 1 Orchard Way, Inchture, Perth, PH14 9QB
G8 RTK L. Man, 4 Back Lane, Yeadon, Leeds, LS19 7SQ
G8 RTN G. Smith, 1 Abbey Road Goldington, Bedford, MK41 9LG
G8 RUX R. Gough, 3 Meadowlands, Havant, PO9 2RP
G8 RVB A. Augustus, 35 Cedar Grove, Wolverhampton, WV3 7EB
G8 RVO Dr A. Parsons, 8 Queen Annes Gardens, Ealing, London, W5 5QD
GJ8 RVT Jersey ARS, c/o M. Turner, 4 Le Clos Sara, St Lawrence, Jersey, JE3 1GT
G8 RVY P. Lee, 8 Sandringham Gardens, Ellesmere Port, CH65 9EY

G8 RVZ I. Martin, Maycroft, Hatchett Lane, Edingale, B79 9JG
G8 RW P. Standley, 9 Capelands, New Ash Green, Longfield, DA3 8LG
G8 RWG N. Montanana, 91 Coulsdon Road, Coulsdon, CR5 2LD
G8 RWH I. Jackson, 5 Vivien Close, Chessington, KT9 2DE
G8 RWJ R. Adams, Ground Floor Flat, 91 Mount Pleasant Road, Hastings, TN34 3SL
G8 RWM F. Box, 11 Cook Avenue, Newport, PO30 2LL
G8 RWN M. Mckenzie, Flat 40, Broomfield House, Huddersfield, HD3 4RS
G8 RWU A. Capron, 28 Windmill Road, Hemel Hempstead, HP2 4BN
G8 RWZ K. Hodson, 117 High Lane, Brown Edge, Stoke-on-Trent, ST6 8RT
G8 RXY G. Alcock, 61 Henshall Hall Drive, Congleton, CW12 3TY
G8 RXZ P. Allgood, 11 Dover Drive, Leegomery, Telford, TF1 6TD
G8 RYE D. Cope, 29 Desford Road, Newbold Verdon, Leicester, LE9 9LG
G8 RYJ P. Tegg, Glendale, 36 Wrecclesham Hill, Farnham, GU10 4JW
G8 RYK R. Taylor, 18 Spruce Avenue, Selston, Nottingham, NG16 6DX
G8 RYL I. Smith, 12 Windmill Lane, Fulbourn, Cambridge, CB21 5DT
G8 RYO P. Sargent, Hill View, North Road, Tetford, Horncastle, LN9 6QH
G8 RYX S. Jones, The Old Granary, Broomsmead, Lapford, EX17 6NA
GM8 RYZ I. Mclaren, 2 Cleeve Drive, Perth, PH1 1HH
G8 RZ P. Webster, 15 Napier Street, Workington, CA14 2PT
G8 RZL T. Claydon, 20 Ivy Lane, Royston, SG8 9DQ
G8 RZN D. Dunn, Valarms, The Pigeons, Wisbech, PE13 4JU
G8 RZS E. Humpston, 2 The Glebe, Hildersley, Ross-on-Wye, HR9 5BL
G8 RZZ K. Colman, 10 South Rise, North Walsham, NR28 0EE
G8 SAA I. Adamson, 5 Rose Meadows Somersham, Huntingdon, PE28 3YU
G8 SAL Saltash Dist AR, c/o K. Hale, 58 St. Stephens Road, Saltash, PL12 4BJ
G8 SAN Dr R. Charlton, 31 Meriden Road, Hampton-In-Arden, Solihull, B92 0BS
GM8 SAP D. Cooper, 4 Ruskie Avenue, Callander, FK17 8LA
G8 SAR M. Elliott, 54 Bankhouse Road, Trentham, Stoke-on-Trent, ST4 8EL
GM8 SAU B. Titmarsh, Caberfeidh, Clachan Na Luib, HS6 5HD
G8 SAX P. Wilkinson, 60 Whalley Drive, Aughton, Ormskirk, L39 6RF
G8 SBH H. Cromack, 6 West Park, South Molton, EX36 4HJ
G8 SBJ T. Blankley, 16 Charles Road, St Leonards-on-Sea, TN38 0QA
GW8 SBK L. Cleak, Danetre, Newport Road, Cwmbran, NP44 3AE
GW8 SBN J. Kemp, Poldhu 259 Delfford, Rhos, Swansea, SA8 3EP
GW8 SBO P. Sibert, Chibembe Lodge Lamborough Lane, Clarbeston Road, SA63 4XD
G8 SBQ D. Wooller, 95 Havant Road Drayton, Portsmouth, PO6 2JE
G8 SBS J. Westlake, 41D Shirley Road, Southampton, SO15 3EW
G8 SCG J. Downes, 6 Lagonda, Glascote, Tamworth, B77 2RY
G8 SCI M. Bynorth, 374 Bloxwich Road, Walsall, WS2 7BG
G8 SCO Street Club Operators, c/o D. Barber, 3 Vestry Road, Street, BA16 0HY
G8 SCT B. Cater, 50 Woodside Darras Hall, Ponteland, NE20 9JB
G8 SCY C. Rosewall, 12 Treloggan Lane, Newquay, TR7 2JN
G8 SDE R. Pitts, 84 Prospect Avenue, Pye Nest, Halifax, HX2 7HP
G8 SDN E. Mciver, 31 Hartshill, Bedford, MK41 9AL
G8 SDS South Dorset Rs, c/o W. Barton, 4 Hawthorn Flats, Hawthorn Road, Dorchester, DT1 2PE
G8 SDU R. Clayton, 1 Raymond Road, Hellesdon, Norwich, NR6 6PL
G8 SDX S. Dale, 25 Copeland Avenue, Tittensor, Stoke-on-Trent, ST12 9JA
G8 SEA Capt. J. Pearce, Clematis Cottage, 52 Cheselbourne, Dorchester, DT2 7NP
G8 SED P. Starling, 5 Ash Close, Bacton, Stowmarket, IP14 4NR
G8 SEE R. Stone, 10 Rosemullion Gardens, Tolvaddon, Camborne, TR14 0EY
G8 SEI J. Snell, 18 Green Lane Grendon, Atherstone, CV9 2PL
G8 SEK C. Watts, 5 Pennycress Way, Bridgwater, TA5 2FQ
G8 SEQ J. Beech, 124 Belgrave Road, Coventry, CV2 5BH
G8 SEV P. Matthews, 10 Norway Close, Corby, NN18 9EG
G8 SEW N. Humphreys, 13 Gordon Drive, East Boldon, NE36 0TD
G8 SEY A. Graver, 8 Avenue Road, Bishop's Stortford, CM23 5NU
G8 SFA S. Milsom, 30 Beechwood Drive, Prudhoe, NE42 5PN
G8 SFD C. Williams, 14 Milton Place, Bideford, EX39 3BN
G8 SFF M. Watson, 7 Grange Lane, Willingham By Stow, Gainsborough, DN21 5LB
G8 SFI S. Firth, 8 Lyndale Avenue, Osbaldwick, York, YO10 3QB
G8 SFM K. Saunders, 4 Rue Des Camelias, Villalier, France, 11600
G8 SFQ T. Mcnamara, 12 Scarsdale Road, Great Barr, Birmingham, B42 2JW
GW8 SFT D. Mansell, 57 Ger Y Llan Penrhyncoch, Aberystwyth, SY23 3HQ
G8 SGB P. Houseago, 11 Arnstones Close, Colchester, CO4 3AS
G8 SGF P. Gilliland11, 34 Cavan Drive, St Albans, AL3 6HP
G8 SGH P. Marshall, 123 Rochford Garden Way, Rochford, SS4 1QJ
G8 SGI S. Pascoe, 34 Ravens View, Witham St Hughs, Lincoln, LN6 9JE
G8 SGM A. Boyce, 22 Peggotty Close, Chelmsford, CM1 4XU
G8 SGP G. Wheeler, 14 Sparkmill Terrace, Beverley, HU17 0PA
G8 SGV M. Williams, The Hideaway, 39 Terrys Avenue, Victoria, Australia, 3160
G8 SGX S. Saltmer, 12 Beechings Mews, Whitby, YO21 3DW
G8 SH J. Storey, 34 Austin Rise, Longbridge, Birmingham, B31 4QN
G8 SHC P. Hammond, 35 West Green, Barrington, Cambridge, CB22 7RZ
G8 SHE R. Shears, 20 Regency Gardens, Grantham, NG31 9JW
G8 SHF C. Scrase, 2 Park View, Seaton Junction, Axminster, EX13 7PS
G8 SHR P. Goodfellow, 10 St. Agnes Walk, Knowle, Bristol, BS4 2DL
GW8 SIE R. Stark, Roneragh, Llanrhaeadr, Denbigh, LL16 4NN
G8 SIG A. Jeffery, 14 Holly Mount Shavington, Crewe, CW2 5AZ
G8 SIK A. Sturt, 6 Kenley Road, Kingston upon Thames, KT1 3RW
G8 SIM J. Green, 7 Russell Road, Runcorn, WA7 4BG
G8 SIN D. Dobson, 93 Old Road, Headington, Oxford, OX3 8SX
GW8 SIT M. Shewring, 2 Glan Hafan, Trefechan, Aberystwyth, SY23 1AT
G8 SIU D. Stillwell, 2B Lesley Owen Way, Shrewsbury, SY1 4RB
G8 SJA P. Farrar, 17 Clough Lane, Halifax, HX2 8SG
G8 SJK D. Monk, Manor Farmhouse, Willen Road, Milton Keynes Village, Milton Keynes, MK10 9AF
G8 SJO S. Ootam, 9 Harewood Road, Isleworth, TW7 5HB
GI8 SJS R. Hoey, 28 Hanwood Heights, Dundonald, Belfast, BT16 1XU
G8 SKA R. Holden, 5 Lawrence Grove, Kidderminster, DY11 7DR
G8 SKG L. Challis, 30 London Road, Kirton, Boston, PE20 1JA
GI8 SKN D. Reid, 2 New Line, Carrickfergus, BT38 9DL
GI8 SKR G. Bannister, 65 Osborne Drive, Belfast, BT9 6LJ
G8 SLB P. Lockwood, 36 Davington Road, Dagenham, RM8 2LR
G8 SLE G. Leake, Flat 3, Edward May Court, Bournemouth, BH11 8AW
G8 SLP J. Barry, 18 Hough Green, Chester, CH4 8JG
G8 SLU M. Hack, Anmee The Ride, Iford, Billinghurst, RH14 0TF
G8 SMA C. Ward, 25 Blewbury Drive Tilehurst, Reading, RG31 5HJ
G8 SMH K. Hempsall, 69 Wantage Road, Didcot, OX11 0AE
G8 SMR Sth Mnchester Rd, c/o J. Heath, 19 Anson Road Swinton, Manchester, M27 5GZ
G8 SMZ C. Shaw, 1 Guilford Cottages, East Langdon, Dover, CT15 5JD
GM8 SNB G. Allan, 13 Mitchell Drive, Rutherglen, Glasgow, G73 3QP
G8 SND W. Hoskins, 89 Boston Road, Lytham St Annes, FY8 3PS
GM8 SNE P. Walter, 4 Rose Gardens Cairneyhill, Dunfermline, KY12 8QS
G8 SNF I. Hewitt, 26 Outwoods Drive, Loughborough, LE11 3LT
G8 SNJ P. Townsend, 7E Back Lane, Holmfirth, HD9 1HQ
G8 SNQ R. Knock, 18 The Hawthorns, Eccleston, Chorley, PR7 5QW
G8 SNR D. Mangnall, 23 Blundell Road, St Annes, FY83AG
G8 SNV Dr M. Dey, Windy Lodge, 18 Ripley Road, Hampton, TW12 2JH
G8 SOI D. Carter, 35 Upland Road, West Mersea, Colchester, CO5 8DR
GM8 SOK J. Sturrock, 4 Ann Street, Edinburgh, EH4 1PJ
G8 SOU R. Topping, 47 Celtic Road, Deal, CT14 9EF
G8 SPC R. White, Flat 1 Oaklands, Elton Road, Clevedon, BS21 7QZ
G8 SPD M. Beevers, Pool House Cottage, Astley, Stourport-on-Severn, DY13 0RH
G8 SPE R. Armstrong, 21 Gunnersbury Gardens, London, W3 9AE
G8 SPG G. Wood, 40 Larch Grove, Peterborough, PE1 4JY
G8 SPM Dr P. Armitage, 5 Park Court, Heath Road, Brixham, TQ5 9AX
G8 SPP C. Parkinson, 77 Lime Grove, Doddinghurst, Brentwood, CM15 0QX
G8 SPU R. Doughty, 47 Red Lion Close, Tividale, Oldbury, B69 1TP
G8 SPW A. Rouse, 49 Ghyllside Avenue, Hastings, TN34 2QB
G8 SQA T. Povey, 24 Townley Way, Earls Barton, Northampton, NN6 0HR
G8 SQH D. Hutchinson, Ryton Villa, Horsecroft Lane, Dymock, GL18 2EJ
G8 SQK L. Mcmahon, 25 Belmont Avenue, Warrington, WA4 1LY
G8 SQP R. Marquiss, 66 Oakwood Rise, Tunbridge Wells, TN2 3HF
G8 SQY S. Cade, 2 Grosvenor Crescent, Louth, LN11 0BD
G8 SQZ S. Westlake, 11 Mount Road, Evesham, WR11 6BE
G8 SRC Swindon ARC, c/o D. Forrest, 166 Meadowcroft, Swindon, SN2 7LE
G8 SRN E. Day, 10 Carlrayne Lane, Menston, Ilkley, LS29 6HH
G8 SRS Stockport Rs, c/o B. Naylor, 47 Chester Road Poynton, Stockport, SK12 1HA
G8 SRV A. Ashe, 34 College Avenue, Maidenhead, SL6 6AX
G8 SRZ B. Atkinson, 3 Sandy Close, Whitwell, Worksop, S80 4PY
G8 SSE K. Lawrence, 7 Canada Way, Pak House, Worcester, WR2 4DJ
G8 SSL A. Marwood, 65 Castleton Avenue, Arnold, Nottingham, NG5 6NH
G8 SSP C. Horswell, Hungereckstrasse 60/3, Vienna, Austria, A-1230
G8 SSS Exmoor Radio Club, c/o J. Stacey, 3 West Park, South Molton, EX36 4HJ
G8 SSX D. Baker, 99 Repton Road, Wigston, LE18 1GD
G8 SSY E. Davies, 5 Cheapside, Horsell, Woking, GU21 4JG
G8 STD St Dunstans ARS, c/o E. John, Obo St. Dunstans Ars, 52 Broadway Avenue, Wallasey, CH45 6TD
G8 STF T. Woods, 12 Primrose Court, Egerton Street, Wallasey, CH45 2PE
G8 STI J. Maiden, 42 Timberdine Avenue, Worcester, WR5 2BD
G8 STJ D. Carter, 4 Sandale Close, Gamston, Nottingham, NG2 6QG
G8 STM A. Bilton, 34 Wimberley Way, South Witham, Grantham, NG33 5PU
G8 STR B. Beestin, 45 Pinehill Road, Crowthorne, RG45 7JP
G8 STW J. Ferguson, Meridian, The Street, Hepworth, Diss, IP22 2PX
G8 STY J. Holmes, 45 College Avenue, Gillingham, ME7 5HY
G8 SUG G. Peterson, 15 Hindhead Green, Watford, WD19 6TR
G8 SUJ W. Shambrook, 49 Beaufort Road, Church Crookham, Fleet, GU526AY
G8 SUM K. Smith, 11 Church Street, Earl Shilton, Leicester, LE9 7DA
G8 SUN S. Williams, 11 Cotman Drive, Hinckley, LE10 0GB
G8 SUQ J. Corbidge, 11 Berkeley Close, Folkestone, CT19 5NA
G8 SUV B. Pont, 56 Ravenhill Road, Bristol, BS3 5BT
G8 SUW N. Pont, Maisemoor, 17 Vicarage Lane, Bridgwater, TA7 9LR
G8 SVR J. Allart, 54 Urban Gardens, Concord, Washington, NE37 3DE
G8 SVT T. Ellis, 58 Greenland Drive, Sheffield, S9 5GJ
G8 SWC H. Moyle, 9 Park Approach, Welling, DA16 2AW
G8 SWK D. Taylor, 19 Armley Grange Oval, Leeds, LS12 3QJ
G8 SWL E. Theodorson, 7 Kingfisher Court, Overstone Lakes, Ecton Lane, Northampton, NN6 0BD
G8 SWO C. Watts, 42 Truscott Avenue, Bournemouth, BH9 1DB
G8 SWW P. Copeman, 1 Chestnut Avenue, Welney, Wisbech, PE14 9RG
G8 SXA J. Davies, Ballards Piece, Forest Hill, Marlborough, SN8 3HN
G8 SXB D. Mullenger, 6 Churchfields, Kingsley, Bordon, GU35 9PJ
G8 SXD B. Davies, Ballards Piece, Forest Hill, Marlborough, SN8 3HN
GW8 SXI Rev. G. Howells, 53 Abbey Road, Rhos On Sea, Colwyn Bay, LL28 4NR
G8 SXJ F. Hutchings, 21 School Lane, St. Ives, Ringwood, BH24 2PF
GM8 SXQ A. Leigh, Duaig, Lochavich, Taynuilt, PA35 1HJ
G8 SXU J. Simmons, 167 Bourne Vale, Hayes, Bromley, BR2 7LX
G8 SYA K. Parker, 3 Cross Roads, East Stour, Gillingham, SP8 5LW
G8 SYD M. Thomson, 11 Uranus Road, Hemel Hempstead, HP2 5QF
G8 SYE N. Trotman1, 38 Oldbury Road, Nuneaton, CV10 0TD
G8 SYM D. Whittle, 16 Garner Drive, Astley, Manchester, M29 7RT
G8 SYV J. Morgan, Linden Lea, Fakes Road, Great Yarmouth, NR29 4JL
GW8 SZC P. Henry, 1 Afan Valley Road, Neath, SA11 3SS
G8 SZG C. Just, 2 The Old Rectory, The High Road Felmersham, Bedford, MK43 7HN
GW8 SZL D. Phillips, 9 Baldwin Street, Newport, NP20 2LT
G8 SZP D. Price, 196 Monument Court, Stevenage, SG1 3BT
G8 SZR M. Matthews, 11 Church End, Ashdon, Saffron Walden, CB10 2HG
GM8 SZS B. Mccaffrey, 5 The Old Orchard, Limekilns, Dunfermline, KY11 3HS
G8 SZZ S. Jackson, 18 Blakesware Gardens, Edmonton, London, N9 9HU
G8 TAE G. Woolterton, 155 El Alamein Way, Bradwell, Great Yarmouth, NR31 8SX
G8 TAQ A. Dyce, 26 Forest Road, Winford, Sandown, PO36 0JY
G8 TAU A. Fisher, 2 Hillside Mansions, Barnet Hill, Barnet, EN5 5RH
GI8 TAX R. Mcloughlin, 27 The Manor, Portadown, Craigavon, BT62 3QU
G8 TAY R. Manning, North Waterhouse Farm, Runnon Moor Lane, Hatherleigh, Okehampton, EX20 3PL
G8 TBB J. O'Meara, 117 Little Sutton Lane, Sutton Coldfield, B75 6SN
G8 TBF R. Jenkins, 11 Westfield Drive, Worksop, S81 0JS
GW8 TBG M. Terry, 265 Delffordd, Rhos, Swansea, SA8 3EP
G8 TBL N. Mosedale, Flat 113, Altitude Apartments 9 Altyre Road, Croydon, CR0 5BP
G8 TBU N. Doe, Longclose, Langtree, Torrington, EX38 8NR
G8 TBV A. Kyle, 6 Mill Hill Drive, Halesworth, IP19 8DB
G8 TBW R. Crathorne, 340 Farnborough Road, Castle Vale, Birmingham, B35 7PD
G8 TBX S. Pybus, 26 White Laithe Court, Leeds, LS14 2EQ

GW8 TBY D. Padley, 31 South Drive, Rhyl, LL18 4SU
GM8 TCG J. Blackie, Drumcharry, Montrose Road, Auchterarder, PH3 1BZ
GM8 TCH M. Bell, 1 Dow Brae, Town Yetholm, Kelso, TD5 8SA
G8 TCP C. Dawe, 21 Portmellon Park, Mevagissey, St Austell, PL26 6XD
G8 TCQ A. Dening, 42 Grove Avenue, Yeovil, BA20 2BD
G8 TDP D. Cooke, 19 St. Aldwyn Road, Seaham, SR7 0AN
G8 TEB D. Clarke, 59 Baden Powell Crescent, Pontefract, WF8 3QT
G8 TEC G. Cook, Flat 3, Southdown Court, Southdown Road, Winchester, SO21 2BX
G8 TEF A. Crute, Shangri La, Winsor Estate, Looe, PL13 2JY
G8 TEK K. Worley, 33 Lynbrook Close, Netherton, Dudley, DY2 9HE
G8 TEL P. Hynes, 3 Holt Park Gardens, Leeds, LS16 7RB
G8 TEO B. Jay, 13 Oakhurst Road, West Moors, Ferndown, BH22 0DW
G8 TEQ Dr D. Linsdall, 2B Linkswood Road, Burnham, Slough, SL1 8AT
G8 TFB S. Haywood, 12 Elm Terrace, Tividale, Oldbury, B69 1UD
G8 TFR S. Cottis, 61 Oaken Grove, Maidenhead, SL6 6HN
G8 TFU P. Simpson, 1 Emmit Field Close, Chesterfield, S40 2UH
G8 TFW L. Stamp, 41 Willoughby Road, Wallasey, CH44 3DZ
G8 TFY N. Richards, 38 Parsons Road, Irchester, Wellingborough, NN29 7EA
G8 TGB M. Verrall, 1 Speedwell Avenue, Weedswood, Chatham, ME5 0SB
G8 TGD D. Troop, 10 Mellowdew Road, Coventry, CV2 5GL
G8 TGH B. Wilmott, 27 Apple Grove, Bognor Regis, PO21 4NB
GW8 TGS Dr W. Williams, 17 Llys Yr Onnen, Coity, Bridgend, CF35 6FA
G8 THE Dr R. Hill, 12 Winchelsea Lane, Hastings, TN35 4LG
G8 THH D. Baker, 5 Larkspur Close, Bishop's Stortford, CM23 4LL
GW8 THL Dr F. Morgan, Glantowy Lodge, Capel Dewi Road, Carmarthen, SA32 8AA
GW8 THM M. Griffin, 3 Pritchard Close, Danescourt, Cardiff, CF5 2QS
G8 THR P. Crossley, Firpark Farm, Fir Park, Market Rasen, LN8 3YL
G8 THZ A. Tipper, 24 Waverley Road Hoylake, Wirral, CH47 3DD
G8 TIA D. Trickett, 25 Spring Street, Halesowen, B63 2SY
G8 TIO P. Dear, Oakdale, Gadbridge Lane Ewhurst, Cranleigh, GU6 7RW
G8 TIU A. Brierley, Thorndale House, Brockhurst Farm, Watersfield, Pulborough, RH20 1NX
GW8 TIX G. May, 19 Mclaren Cottages, Abertysswg, Rhymney, Tredegar, NP22 5BH
G8 TJG F. Starkey, 13 Thorncliffe Drive, Darwen, BB3 3QA
G8 TJI M. Oldfield, Willows, Stablebridge Road, Buckinghamshire, HP22 5ND
G8 TJR C. Colebrook, 14 Yiewsley Drive, Darlington, DL3 9XS
G8 TKD D. Hensby, 28 Moorland Crescent, Whitworth, Rochdale, OL12 8SU
G8 TKQ J. Ackerley, 24 Macaulay Road, Lutterworth, LE17 4XB
G8 TKY T. Bootyman, 14 Vale View, Ackworth, Pontefract, WF7 7HQ
G8 TLC S. Parker, 22 Lincoln Drive, Syston, Leicester, LE7 2JW
G8 TLH R. Rogers, 14 Coningsby Drive, Franche, Kidderminster, DY11 5LU
G8 TLL L. Stewart, The Spinney, Holmes Lane, Scunthorpe, DN15 9QY
G8 TLP R. Pearce, Doral, The Grove, Sevenoaks, TN15 6JJ
G8 TLT J. Beveridge, 18 Sir Christopher Court, Hythe, SO45 6JR
G8 TMA H. Colborn, Orchard Cottage, Blakeney Hill, Blakeney, GL15 4BS
G8 TMD T. Clint, 11 Home Lea, Rothwell, Leeds, LS26 0PP
GI8 TME J. Campbell, 115 Dromore Road, Ballynahinch, BT24 8HU
G8 TMH H. Caseley, 7 Lacy Way, Leominster, HR6 9AY
G8 TMJ P. Faulkner, 8 Parkfield Road, Cheadle Hulme, Cheadle, SK8 6EX
G8 TML J. Foster, 14 Braemar Grove, Heywood, OL10 3RR
G8 TMM E. Gilbert, 34 School Lane, Harpole, Northampton, NN7 4DR
G8 TMQ P. Stevens, 17 Weaver Close, Brierley Hill, DY5 4QN
G8 TMR P. Taylor, 22 Windermere Drive, Rainford, St Helens, WA11 7LD
G8 TMV C. Tuckley, 98 Woodland Road, Sawston, Cambridge, CB22 3DU
G8 TNA S. Thompson, Hollies, Chapel Hill, Sticker, St Austell, PL26 7HG
G8 TNB P. Thompson, Lyndhurst Cottage, Main Street, Newark, NG23 6ST
G8 TND C. Schiffman, 14 Caspian Way, Purfleet, RM19 1LE
G8 TNE D. Pickford, 80 Hollowood Avenue, Littleover, Derby, DE23 6JD
G8 TNH P. Jeffries, 22 Ingrams Way, Hailsham, BN27 3NP
G8 TNK North Kent Radio Society, c/o S. Osborn, 67 Chessington Avenue, Bexleyheath, DA7 5NP
G8 TNS S. Ward, 2 Nursery Road, Rugeley, WS15 1EZ
G8 TNU A. Lambert, 69 Anvil Crescent, Broadstone, BH18 9DZ
G8 TOI R. Hempstead, 21 Lymington Avenue, Clacton-on-Sea, CO15 4PJ
G8 TOP N. Huggins, 4 Kelmscott Close, Goldings, Northampton, NN3 8XN
G8 TOQ J. Jackson, 1 Dolly Garth, Arkengarthdale Road, Richmond, DL11 6QX
G8 TOT D. Lodge, 134 Deyne Road, Huddersfield, HD4 7EP
GW8 TOX K. Taylor, Swn Y Don, Llanfaes, Beaumaris, LL58 8RG
G8 TPC B. Taylor, 161 Sidegate Lane, Ipswich, IP4 4JN
G8 TPF N. Sanvoisin, 11 Ter Rue Lecoq, St Nom la Breteche, France, 78860
G8 TPM N. Wellsbury, 15 Woodlands Road, Cookley, Kidderminster, DY10 3TL
G8 TPP M. Strudwick, 65 Neave Crescent, Harold Hill, Romford, RM3 8HN
G8 TQH A. Mcmullin, Fair View, Rickham East Portlemouth, Near Salcombe, TQ8 8PJ
G8 TQI P. Herod, 4 St. James Road, Little Paxton, St Neots, PE19 6QW
G8 TQJ R. Markfort, 105 Woodlands Way, Southwater, Horsham, RH13 9TF
G8 TQK A. Mayhew, 51 Upland Road, Sutton, SM2 5HW
G8 TQO G. Harrison, 1 Wellis Gardens, St Leonards-on-Sea, TN38 0UX
G8 TQP R. Healey, 14 Jardine Drive Bishops Cleeve, Cheltenham, GL52 8XQ
G8 TQV R. Tuckett, 89 Hillbrook Road, Tooting, London, SW17 8SF
G8 TQZ B. Woods, 84 Beauly Way, Rise Park, Romford, RM1 4XR
G8 TRF Invicta Contest Group, c/o I. Hope, 5 The Crescent, Northfleet, Gravesend, DA11 7EB
G8 TRG Tower Radio Group, c/o R. Green, 2 Ragley Walk, Rowley Regis, B65 9NT
GW8 TRO K. Prosser, 12 Sycamore Court, Woodfieldside, Blackwood, NP12 0DA
G8 TRQ M. Walker, 20 Littlewood Lane, Cheslyn Hay, Walsall, WS6 7EJ
G8 TRR W. Pickard, 19 Canham Close, Kimpton, Hitchin, SG4 8SD
G8 TRS Tamworth ARS, c/o T. Robertson, 10 Athelstan Way, Tamworth, B79 8LB
G8 TRU S. Lynch, 4 Tanglewood, Welwyn, AL6 0RU
G8 TRY G. Scott, 19 Penkett Road, Wallasey, CH45 7QF
G8 TSC J. Collins, 12 Malham Road, Thatcham, RG19 3XB
G8 TSG D. Johansen, 45 Marfords Avenue, Bromborough, Wirral, CH63 0JJ
GI8 TSI I. Raine, 48 Gardners Road, Lisburn, BT27 5PD
G8 TSV G. Rowlands, 39 Nelthorpe Street, Lincoln, LN5 7SJ
G8 TSZ A. Twyford, 70 Ellesfield Drive, West Parley, Ferndown, BH22 8QW
G8 TTD Dr P. Palin, Flat 2, Springfield Heversham, Heversham, Milnthorpe, LA7 7EJ
G8 TTE Dr E. Thomas, Fairfield, St. Marys Road, Oakham, LE15 8SU
G8 TTI D. Kearns, 14 Draycot Cerne, Chippenham, SN15 5LD
G8 TTJ J. Holland, 2 Hadfield Close, Staunton, Gloucester, GL19 3QY
G8 TTP R. Nicholls, 57 Mandalay Court, London Road, Brighton, BN1 8QW
G8 TTU C. Smithson, 4 Calder Avenue, Littleborough, OL15 9JE
G8 TTX K. Sugg, 28 Well Close, Winscombe, BS25 1HQ
G8 TUH D. George, East Winch Road, Blackborough End, Kings Lynn, PE32 1SF
G8 TUN C. Denton, 34 Brook Lane, Ormskirk, L39 4RE
G8 TUU R. Boyce, 9 Kestrel Close, Bexhill-on-Sea, TN40 1UG
G8 TVC T. Webb, 25 Wheatfield Drive, Ramsey, Huntingdon, PE26 1SH
G8 TVM K. Biggs, 33 Blanford Gardens, West Bridgford, Nottingham, NG2 7UQ
G8 TVU A. Cole, 14 Ellesmere Grove, Stainforth, Doncaster, DN7 5BS
GM8 TVV N. Coote, 130 Castle Gardens, Paisley, PA2 9RD
G8 TVW D. Young, 58 Furzefield Road, Welwyn Garden City, AL7 3RJ
GW8 TVX R. Hope, 75 Priors Way, Dunvant, Swansea, SA2 7UH
G8 TVZ R. Midgeley, 2 Digswell Park Cottages, Digswell Park Road, Welwyn Garden City, AL8 7NN
G8 TWA G. Jasper, Clann Farm, Clann Lane, Bodmin, PL30 5HD
GI8 TWB B. Mitchell, 7 Crea Road, Randalstown, Antrim, BT41 3DX
G8 TWR J. Evans, 49 Inverness Avenue, Enfield., EN1 3NU
G8 TWS M. Corbett, 32 Bibury Road, Cheltenham, GL51 6BA
G8 TWT B. Fisher, 24 Vessey Road, Worksop, S81 7PG
G8 TWZ I. Goodman, 271 Alcester Road, Hollywood, Birmingham, B47 5HJ
G8 TXA R. Heeley, 4 Cherry Tree Lane, Halesowen, B63 1DU
GM8 TXC J. Hedley, 3/2 48 White Street, Glasgow, G11 5EA
G8 TXJ D. Shaw, Cotehouse, Bleatarn, Appleby-in-Westmorland, CA16 6PX
G8 TXK P. Shulver, Skanes House Redmoor, Bodmin, PL30 5AT
G8 TXL J. Sillitoe, 42 Marsham Road, Kings Heath, Birmingham, B14 5HD
G8 TXT B. Linkins, Curlew Cottage, Higher Wringworthy Farm, Looe, PL13 1PR
G8 TXW G. Sutcliffe, 41 Rose Avenue, Irlam, Manchester, M44 6AQ
G8 TXX J. Taylor, The Jays, 5 Watling Close, Bourne, PE10 9XL
G8 TYB R. Uccellini, 49 Slaithwaite Road, Meltham, Holmfirth, HD9 5PG
G8 TYD D. Prouse, 2 Grove View, Bristol Road, Hambrook, Bristol, BS16 1RD
G8 TYF H. Sasse, Flat 8 The Beeches, 43 Queens Road, Leicester, LE2 1WQ
G8 TYH R. Marsh, 58 Statham Avenue, Lymm, WA13 9NL
G8 TYX J. Hodgson, 21 Portcullis Drive, Wallingford, OX10 9LY
G8 TYY Dr T. Hopkins, 5 Rochester Close, Bacup, OL13 8RN
G8 TZE D. Pritt, 387 London Road, Clanfield, Waterlooville, PO8 0PJ
G8 TZJ A. Sellers, 2 Dunkenshaw Crescent, Lancaster, LA1 4LQ
G8 TZN R. North, 5 George Road, Guildford, GU1 4NP
G8 TZU R. Collis, 17 Belvedere Close, Guildford, GU2 9NP
G8 TZW G. Dunn, 29 Sundridge Road, Kingstanding, Birmingham, B44 9NY
G8 UAD R. White, 72 Green Lane, Bournemouth, BH10 5LF
G8 UAE D. Scott, Hyde Bungalow, The Hyde, Stourbridge, DY7 6LS
G8 UAF G. Scargill, 98 Southleigh Road, Leeds, LS11 5SG
G8 UAI D. Lockwood, 61 Green Lane, Tickton, Beverley, HU17 9RH
GW8 UAM L. Wright, 19 Lon Y Fran, Caerphilly, CF83 2RX
GW8 UAP T. Williamson, Fforch Farm House, Cemetery Road, Treorchy, CF42 6TF
G8 UAY J. Earley, 36 Camelot, Cornish Avenue, Weltevredenpark, South Africa, 46
G8 UBD A. Baker, 19 Rockington Way, Crowborough, TN6 2NJ
G8 UBF G. Beadle, 8 Woodland View, Southwell, NG25 0AG
G8 UBJ R. Lester, 71 Ronelean Road, Surbiton, KT6 7LL
G8 UBN G. Hodgson, East Cottage, Chineham Lane, Basingstoke, RG24 9LR
G8 UBP A. Jacketts, Flat 7, Jenneth Court, 44 Mauldeth Road, Stockport, SK4 3NB
G8 UBU R. Jarvis, 39 Moy Road, Colchester, CO2 8NZ
G8 UBX R. Manning, 2 Reydon Close, Haverhill, CB9 7WG
G8 UCC I. Bradley, 8 Hunt Avenue, Heanor, DE75 7QB
G8 UCK T. Colligan, 53 Datchworth Turn, Hemel Hempstead, HP2 4PB
G8 UCL N. Cook, 61 Redland Road, Malvern, WR14 1LY
G8 UCN A. Crookes, 23 Helliwell Lane, Deepcar, Sheffield, S36 2NH
G8 UCP M. Culling, 101 Orchard Drive, Park Street, St Albans, AL2 2QL
G8 UCR D. Davis, 2 Bowen Road, Rotherham, S65 1LH
GI8 UCS A. Edwards, 6 Carwood Park, Newtownabbey, BT36 5JU
G8 UCY D. Walker, 43 Wimborne Road, Corfe Mullen, Wimborne, BH21 3DS
G8 UCZ C. Wren, 24 Willow Way, Martham, Great Yarmouth, NR29 4SH
G8 UDA B. Watson, 3 Anderton Rise, Millbrook, Torpoint, PL10 1DA
G8 UDD S. James, 42 Wilshire Avenue Springfield, Chelmsford, CM2 6QW
G8 UDG D. Roberts, 25 Metcalfe Road, Cambridge, CB4 2DB
G8 UDI P. Newman, 4 Old Barn Court, Ludford, Market Rasen, LN8 6AZ
GW8 UDJ M. Loach, Rhiwgwraidd, Llanilar, sy23 4sq
G8 UDS J. Farrant, Partida Barrancs 16, Orba, Alicante, Spain, 3790
G8 UDV A. Frost, 10 Ramsden Square, Cambridge, CB4 2BJ
G8 UDZ A. Gilbertson, The Corn Store, Manor Farm, Old Alresford, Alresford, SO24 9DH
G8 UED W. Harris, 1 Marsh Cottages, Middle Street, Yeovil, BA22 7AP
G8 UEE S. Melvin, 2 Salters Court, Newcastle upon Tyne, NE3 5BH
G8 UEF P. Mobberley, The Willows, Hobro, Kidderminster, DY11 5ST
G8 UEI A. Howells, 27 Swallow Lane, Aylesbury, HP19 7HW
GW8 UEK N. Hosker, Laburnum Cottage, Church Lane, Old Aston Hill, Ewloe, CH5 3BF
G8 UEU J. Thornburn, 19 Pingate Lane South, Cheadle Hulme, Cheadle, SK8 7NP
G8 UEY J. Rice, 2 Medalls Path, Stevenage, SG2 9DX
G8 UEZ C. Southall, 40 Tathall End, Hanslope, Milton Keynes, MK19 7NF
G8 UFO E. Charlton, 2 Bullock Road, Washingley, Peterborough, PE7 3SH
G8 UFX I. Fowler, 1 Mayfields, Shefford, SG17 5AU
G8 UGK P. Warburton, 4384 Henneberyy Road, Manilus, USA, 13104
GM8 UGO W. Kay, 22 Linton Terrace, Perth, PH1 1LE
G8 UGS P. Marks, 106 Darlton Drive, Arnold, Nottingham, NG5 7LW
G8 UHJ P. Couch, 6 Quantock Gardens, Ramsgate, CT12 6SW
G8 UHK A. Rolls, 9 Mareschal Road, Guildford, GU2 4JF
G8 UHL J. Rangeley, Tanglewood Buckden, Skipton, BD23 5JA
G8 UHM W. Rosser, Fanshawgate House, Fanshaw Gate Lane, Holmesfield, Dronfield, S18 7WA
G8 UHO D. Reay, 78 Wyresdale Road, Lancaster, LA1 3DY
G8 UHT P. Shaw, Poole Bank Cottage, Poole, Nantwich, CW5 6AL
G8 UHW C. Mobbs, 5 Garth Avenue, Leeds, LS17 5BH
G8 UID R. Nock, 83 Coles Lane, West Bromwich, B71 2QW
G8 UIG M. Chedzoy, Helena, Picts Hill, Langport, TA10 9EZ
G8 UIL J. Gale, Barn End, Highampton, Beaworthy, EX21 5LT
G8 UIO D. Salter, 9 Old Milverton Road, Leamington Spa, CV32 6BA
GI8 UIU Dr P. Moore, 13 Ballygallum Road, Downpatrick, BT30 7DA
G8 UIV Dr P. Morton-Thurtle, 23 Wife Of Bath Hill, Canterbury, CT2 8PQ

G8	UIW	S. Threlfall-Rogers, 43 Nanpantan Road, Loughborough, LE11 3ST
G8	UJF	D. Headland, Hazelwood, Haywards Lane, Cheltenham, GL52 6RF
G8	UJO	B. Leveton, Orman House, 17A Grove Avenue, Norwich, NR5 0JD
G8	UJS	A. Mcdermott-Roe, 100 Dorset Ave, West Parley, Ferndown, Ferndown, BH22 8DZ
G8	UJV	M. Johnson, 9 South Road, Brampton, Huntingdon, PE28 4PX
G8	UKH	L. Luck, 1313 Rodney Lane, Winchester, Canada, K0C 2K0
G8	UKI	T. Gawn, 15 Barradon Close, Torquay, TQ2 8QE
G8	UKO	P. Coupe, 10A West End Avenue, Brundall, Norwich, NR13 5RF
G8	UKV	M. Vincent, 9 Sleapford, Long Lane, Telford, TF6 6HQ
G8	UKY	R. Mills, 3 Hallam Moor, Liden, Swindon, SN3 6LS
GW8	UKZ	Dr A. Walker, Glandenys, Cross Inn, Llanon, SY23 5NA
G8	ULH	J. Wilson, 2 Reston Court, Cleethorpes, DN35 0JQ
G8	ULJ	G. Sowter, 21 Seawell Road, Bude, EX23 8PD
G8	ULL	M. Williams, 9 Fir Tree Drive, West Winch, King's Lynn, PE33 0PR
G8	ULM	D. Petty, 16 Audley End, Saffron Walden, CB11 4JB
G8	ULQ	C. Copsey, Flat 17, Cherrywood Court, Moordown Avenue, Solihull, B92 8QS
G8	UMA	J. Spencer, 78 Copse Avenue, Farnham, GU9 9EA
G8	UMB	R. Kennedy, Little Thatch, Canterbury Road, Dover, CT15 7HJ
G8	UML	M. Spencer, 79 Salisbury Close, Alton, GU34 2TP
GM8	UMN	J. Norrie, 13 Pentland Crescent, Dundee, DD2 2BU
G8	UMO	M. Walker, Winterfell, Fen Road, Boston, PE22 8HA
G8	UMY	P. Mahoney, 2 Elm Lodge, Elm Avenue, Ruislip, HA4 8PH
G8	UND	J. Warburton, 164 High Street, Lewes, BN7 1XU
G8	UNO	R. Clarke, 58 Orpin Road, Merstham, Redhill, RH1 3EY
G8	UNP	R. Burningham, 26 St. Margarets Drive, Sibsey, Boston, PE22 0ST
G8	UOJ	A. Abrahams, 69 Culverhouse Road, Luton, LU3 1PY
G8	UOL	B. Cooper, 2 Vernon Way, Plot 58 - Poltair Vale, Penryn, TR10 8SJ
G8	UOZ	M. Freestone, 12 St. Martins Approach, Ruislip, HA4 7QD
GM8	UPC	A. Wells, 64 Denwell Road, Insch, AB52 6LH
G8	UPD	R. Payne, 1 Mill Hill, Horning, Norwich, NR12 8LQ
G8	UPF	K. Hutchinson, 8 Innage Crescent, Bridgnorth, WV16 4HU
GM8	UPI	D. Mcalpin, Birchwood, Belzies, Lockerbie, DG11 1SA
GW8	UPJ	Rev. P. Nunn, 1 Talwrn Court, Coedpoeth, Wrexham, LL11 3NN
G8	UPK	D. Akester, 19 Bracken Park, Bingley, BD16 3LG
G8	UPO	D. Steele, 43 Lancastria Mews, Boyndon Road, Maidenhead, SL6 4SA
G8	UPX	M. Westwood, 59 Woodland Drive, St Albans, AL4 0EN
GW8	UQC	Dr D. Bolton, Mill Farm, Manorowen, Fishguard, SA65 9PT
G8	UQR	T. Riley, 3 Hoefield Crescent, Nottingham, NG6 8AY
G8	UQV	F. Hall, 478 Darwen Road, Bromley Cross, Bolton, BL7 9DX
GM8	UQX	S. Roberts, Blaruaine, Dunoon, PA23 8TH
G8	UQY	G. Waywell, 14 Causeway, Great Harwood, Blackburn, BB6 7HU
G8	URB	F. Mcgilp, High Beech Ashley, Tiverton, EX16 5PA
G8	URG	P. Ridgeon, 1 Fiennes Road Herstmonceux, Hailsham, BN27 4LN
G8	URI	G. Cross, 11 Highfield App/Ch, Billericay, Essex, CM11 2PD
G8	URZ	N. Bird, 15 Bramley Close, Powick, Worcester, WR2 4SR
G8	USA	M. Dawes, 20 Alden Street, Danvers, USA, 1923
G8	UST	M. Freeman, Sunnyside, Long Lane, Mansfield, NG20 8AZ
G8	UTH	G. Sunderland, 28 Tillotson Avenue, Sowerby Bridge, HX6 1BX
GW8	UTK	B. Davies, Rhosyr, Llanfair Pg, LL61 5JB
G8	UTQ	W. Vander Byl, 45 Scotby Road, Scotby, Carlisle, CA4 8BD
G8	UTW	H. Mirams, 58 Ing Head Terrace, Shelf, Halifax, HX3 7LB
G8	UTY	J. Wardle, Spring Cottage, Chapel Road, Hayle, TR27 6BA
G8	UUC	G. York, 10 Beechwood Avenue Wallasey Village, Wallasey, CH45 8NX
GI8	UUN	G. Curtis-Smith, 19A Glenavy Road, Lisburn, BT28 3UT
G8	UUO	D. Drizen, 3 Crittall Close, Silver End, Witham, CM8 3SY
G8	UUR	P. Dicken, 99 Stanton Road, Burton-on-Trent, DE15 9SE
G8	UUS	P. Owen, 2 Plantation Road, Wollaton, Nottingham, NG8 2ER
G8	UUV	M. Copsey, 1 Cooper Terrace, Dereham Road, Dereham, NR19 2BJ
GM8	UUW	C. Fyfe, 20 Larch Grove, Galashiels, TD1 2LB
G8	UVF	T. Cassell, 3 Rose Hill, Waterlooville, PO8 9QU
G8	UVG	C. Parker-Larkin, 40 Head Street Goldhanger, Maldon, CM9 8AZ
G8	UVN	B. Rigby, 76 Woodland Road, Rode Heath, Stoke-on-Trent, ST7 3TL
G8	UVU	M. Soble, Whitethorn Farm, Carey, Hereford, HR2 6NG
G8	UVY	J. Haste, 11 Corporation Road, Chelmsford, CM1 2AR
G8	UVZ	B. Hart, 63 Newcastle Road, Congleton, CW12 4HL
G8	UWD	R. Hornby, 63 Shearwater Drive, Bicester, OX26 6YR
G8	UWE	M. Jefford, 37 Marions Way, Exmouth, EX8 4LF
G8	UWG	A. Kay, 19 Chesnut Grove, Higher Tranmere, Birkenhead, CH42 0LB
G8	UWI	A. Downing, 21 Firfield Road, Thundersley, Benfleet, SS7 3UU
G8	UWL	D. Cooper, 18 Stockfield Road, Stoke on Trent, St37ap
G8	UWM	M. Crossley, 3 Derby Street, Stockport, SK3 9HF
GW8	UWP	N. Elgood, Rhyd Y Paith Rhydyfelin, Aberystwyth, SY23 4PU
G8	UXB	B. Payne, 45 Kellaway Avenue, Westbury Park, Bristol, BS6 7XS
G8	UXD	A. Gill, 4 Cornubia Close, Hayle, TR27 4RL
G8	UXL	P. Nicholson, 3 Eastleigh, Skelmersdale, WN8 6AX
G8	UXW	J. Benton, 2 Appleton Road, Fareham, PO15 5QH
G8	UXX	K. Brazington, 38 Tamworth Road, Amington, Tamworth, B77 3BT
G8	UYB	K. Reece, Winsford Grange Nursing Home, Station Road By Pass, Winsford, CW7 3NG
G8	UYF	R. Ritchie, Flat 2, Sundon Park Parade, Luton, LU3 3BH
G8	UYK	F. Rowntree, 5 Cornel House, Osborne Road, Windsor, SL4 3SQ
G8	UYL	R. Rumbelow, The Spinney, The Chase, Leatherhead, KT22 0HR
G8	UYM	J. Sanderson, 5 Babbacombe Drive Rudd Hill Estate, Ferryhill, DL17 8DA
G8	UYR	B. Smith, 3 Harwin Close, Wolverhampton, WV6 9LF
G8	UYW	T. Carrig, 12 Longmoor Drive, Liphook, GU30 7XA
G8	UYY	M. Daish, 27 Westbourne Road, Portsmouth, PO2 7LB
G8	UZM	Dr R. Jefferson, 23 Valerian Avenue, Heddon-On-The-Wall, Newcastle upon Tyne, NE15 0EA
G8	UZQ	S. Haywood, 19 Crich Way, Newhall, Swadlincote, DE11 0UU
G8	UZW	J. Durrant, 114 Rosebank Avenue, Hornchurch, RM12 5QS
G8	UZY	D. Fisher, 8 Beech Road, Stibb Cross, Torrington, EX38 8HZ
G8	VAD	J. Goodings, 133 Lache Lane, Chester, CH4 7LU
G8	VAE	A. Griffiths, Tranby Croft, Matt Pits Lane, Wainfleet, Skegness, PE24 4LY
G8	VAF	K. Chittenden, Ponders, Hall Road, Colchester, CO6 3DX
GM8	VAM	G. Brazier, 117A East Clyde Street, Helensburgh, G84 7PL
G8	VAN	M. Goodwin, 15 Meadow Close, Repton, Derby, DE65 6GT
G8	VAR	B. Harper, 51 Cross Lane, Scarborough, YO12 6DQ
G8	VAT	G. Denton, 90 Willow Lane, Knottingley, WF11 8AJ
G8	VBA	R. Webb, 78 Station Road, Rolleston-On-Dove, Burton-on-Trent, DE13 9AB
G8	VBC	R. Timms, 20 Driftside, Blackfordby, Swadlincote, DE11 8BD
G8	VBE	C. Thomas, Apartment 231, Bournville Gardens Village, Birmingham, B31 2FS
G8	VBI	D. Sproson, 1 The Old Orchard, Whitehall, South Petherton, TA13 5AQ
G8	VBK	R. Penver, 56 Cottesmore Avenue, Ilford, IG5 0TG
G8	VBW	M. Corbett, 86 Jordan Avenue, Stretton, Burton-on-Trent, DE13 0JD
GM8	VBX	D. Coulthart, 23 Larchfield Road, Dumfries, DG1 4HU
GW8	VCA	D. Dyer, Ty Newydd, 24D Fforest Hill, Neath, SA10 8HD
G8	VCH	J. Grevatt, 17 Foxdale Drive, Angmering, Littlehampton, BN16 4HF
G8	VCI	G. Gwynne, 5 Stanstead Avenue, Nottingham, NG5 5BL
G8	VCJ	C. Gunn, Oakridge, Chapel Road Indian Queens, St Columb, TR9 6JZ
G8	VCL	R. German, 29 Glenthorne Gardens, Sutton, SM3 9NL
G8	VCN	M. Newport, 53 Southfiels Drive, Yeovil, BA213FJ
G8	VCO	B. Nicholls, 35 Lynn Road, Downham Market, PE38 9NJ
G8	VCQ	W. Norman, 27 Newport Road, Barnstaple, EX32 9BG
G8	VCU	S. Morgan, 5 Parklands, Ufford, Woodbridge, IP13 6ES
G8	VDJ	R. Pauley, Lamplight, Casterton Lane, Tinwell, Stamford, PE9 3UQ
G8	VDP	K. Roberts, 35A Rockley Avenue, Birdwell, Barnsley, S70 5QY
G8	VDQ	C. Parnell, 213A Northfield Avenue, London, W13 9QU
GW8	VEE	G. Blore, Ty Newydd, Cymau, Wrexham, LL11 5EU
G8	VEM	A. Challis, 12 Moorland Crescent, Boultham Moor, Lincoln, LN6 7NL
G8	VEN	H. Chapman, 24 Croft Road, Cosby, Leicester, LE9 1SE
G8	VEQ	A. Stone, 47 Oakford Villas, North Molton, South Molton, EX36 3HJ
G8	VER	Verulam ARC, c/o P. King, 96 Mancroft Road, Caddington, Luton, LU1 4EN
G8	VEZ	T. Wagg, 15 Barncroft Way, Havant, PO9 3AA
GW8	VFF	A. Wilkins, 11 Redhouse Road, Ely, Cardiff, CF5 4FG
G8	VFI	D. Franklin, 49 Hope Road, Benfleet, SS7 5JQ
G8	VFM	J. Callaghan, 271 Belvoir Road, Coalville, LE67 3PL
G8	VFP	D. Evans, 2 Cottage Lane, Marlbrook, Bromsgrove, B60 1DW
GW8	VFQ	R. Elliott, 19 Pencoed, Dunvant, Swansea, SA2 7PQ
G8	VG	A. Windle, Flat 11, Parham House 15 King George'S Drive, Liphook, GU30 7GB
GW8	VGB	R. Morgan, 4 Underhill Lane, Horton, Swansea, SA3 1LB
G8	VGI	A. Lilly, 47 Horton Street, Frome, BA11 3DP
G8	VGQ	P. Andrews, 1 Waite Meads Close, Purton, Swindon, SN5 4ET
G8	VGY	E. Davis, 850 Lantana Rd, Lantana, USA, 33462
G8	VHB	M. Fitzgibbons, 8 Lundhill Close, Wombwell, Barnsley, S73 0RW
G8	VHF	Bewlay Bros ARC, c/o J. Goodier, 20 Poleacre Lane, Woodley, Stockport, SK6 1PG
G8	VHG	I. Gower, 10 Homethorpe, Hull, HU6 9EU
G8	VHI	R. Woolley, 103 Mancetter Road, Nuneaton, CV10 0HP
G8	VHK	M. Stanford, Fircroft, Mutton Hall Lane, Heathfield, TN21 8NR
G8	VHL	S. Price, 35 Western Road, Goole, DN14 6QW
G8	VHO	S. Pratt, 4 Bussex Square, Westonzoyland, Bridgwater, TA7 0HD
G8	VHX	G. Shering, 66 Oliver Street, Ampthill, Bedford, MK45 2QL
G8	VIB	J. Sim, 22 Dene View, Ashington, NE63 8JT
G8	VIC	M. Rump, 24 Stoneleigh Avenue, Brighton, BN1 8NP
G8	VIV	K. Bilke, 8 Ibworth Lane, Fleet, GU51 1AU
G8	VJG	K. Halls, Little Rema, Cray Road, Crockenhill, Swanley, BR8 8LP
G8	VJO	C. Green, 12 Spenser Grove, Great Harwood, Blackburn, BB6 7JU
G8	VJP	L. French, 14 Manor Farm Court, Thrybergh, Rotherham, S65 4NZ
G8	VJR	D. Fowles, 52 Lucy Lane South, Stanway, Colchester, CO3 0HY
G8	VJU	K. Earl, 210 Churchill Avenue, Chatham, ME5 0JS
G8	VJW	G. Davey, 147 Deeds Grove, High Wycombe, HP12 3PA
G8	VJY	A. Crafer, 155 Upham Road, Swindon, SN3 1DR
GI8	VKA	R. Coulter11, 34 Toberdowney Valley, Ballynure, Ballyclare, BT39 9TS
GM8	VKN	M. Tarr, 1 Methven Drive, Dunfermline, KY12 0AH
G8	VKO	G. Tandy, 24 Coast Road, Hopton Great Yarmouth, NR31 9BT
G8	VKQ	C. Sparke, 47 Jobes, Balcombe, Haywards Heath, RH17 6AF
G8	VKR	A. Williams, 10 Catalina Close, Woodley, Reading, RG5 4UG
GW8	VKS	Dr J. Williams, Glenview, Penrhos, Usk, NP15 2LF
GM8	VL	GMDX Group, c/o R. Ferguson, 19 Leighton Avenue, Dunblane, FK15 0EB
G8	VLL	A. Kett, 476 Earlham _Road, Norwich, NR4 7HP
G8	VLP	M. Clark, 6 Shalcross Drive, Cheshunt, Waltham Cross, EN8 8UX
G8	VLR	R. Law, 19 Central Drive, Bramhall, Stockport, SK73JU
G8	VLS	D. Leeder, 2 Robin Court East Rainton, Houghton le Spring, DH5 9RW
G8	VLY	R. Macbeth, 9 Woodside, Stroud, GL5 1PL
G8	VMF	M. Clayton, 54 Banks Road Golcar, Huddersfield, HD7 4RE
G8	VML	L. Gibson, 57A Heritage Park, Hatch Warren, Basingstoke, RG22 4XT
G8	VMP	K. Webster, 5 Ridgmont Rd, St Albans, AL1 3AG
G8	VMQ	Dr A. Parker, 78 Whitbarrow Road, Lymm, WA13 9BA
G8	VMY	D. Whitfield, Framingham, Manor Road, Hayling Island, PO11 0QR
G8	VMZ	M. Walpole, 9 The Paddocks, Brandon, IP27 0DX
G8	VNF	B. Benson, 31 Helston Close Brookvale, Runcorn, WA7 6AA
G8	VNL	J. Darlington, 111 Maas Road, Northfield, Birmingham, B31 2PP
G8	VNN	A. Dowsett, 70 Warren Drive, Broughton, Chester, CH4 0PT
G8	VNO	G. Edmonds, 29A Manor Park, Woolsery, Bideford, EX39 5RH
G8	VNP	R. Elden, 124 Larchcroft Road, Ipswich, IP1 6PQ
G8	VNX	J. Dixon, 19 Pheasant Drive, Wincham, Northwich, CW9 6PX
G8	VOB	K. Fisher, 50 Queen Street, Henley-on-Thames, RG9 1AP
G8	VOH	Dr P. Renshaw, Hayes Pond Cottage, Hayes Plat, Rye, TN31 6HQ
G8	VOI	R. Reeves, 4 Elmwood Avenue, Waterlooville, PO7 7LG
G8	VOJ	J. Read, 99 Blackheath Road, Lowestoft, NR33 7JF
G8	VOQ	R. Rogalewski, 47 Partridge Crescent, Dewsbury, WF12 0HT
G8	VOY	J. Mcparlane, 47 Aragon Road, Kingston upon Thames, KT2 5QB
G8	VPD	I. Morley, 2 Livingstone Walk, Park Wood, Maidstone, ME15 9JB
G8	VPE	J. Noy, 14 Poplar Drive, Filby, Great Yarmouth, NR29 3HU
G8	VPG	S. O'Sullivan, 15 Witney Close, Saltford, Bristol, BS31 3DX
G8	VPH	B. Aveling, 6 Brambling, Wilnecote, Tamworth, B77 5PQ
G8	VPO	J. Chalmers, 10 Hornbeam Close, Wokingham, RG41 4UR
GW8	VPP	S. Croft, 43 York Road, Colwyn Bay, LL29 7EY
G8	VPR	B. Gallear, 5 Oak Avenue, Cannock, WS12 4QA
G8	VPX	A. Davies, 14 Primrose Grove, Keighley, BD21 4NP
G8	VPY	R. Davies, 1 Seaham Close, Norton, Stockton-on-Tees, TS20 1RT
G8	VQA	S. Foulser, 9 Oak Coppice Close, Eastleigh, SO50 8PH
G8	VQE	B. Haylett, 160 Hookfield, Harlow, CM18 6QN

G8 VQH M. Russ, 71 Farriers Close, Martlesham Heath, Ipswich, IP5 3SN
G8 VQJ M. Otterson, 161 Tollgate Lane, Bury St Edmunds, IP32 6DF
G8 VQK M. Nightingale, 31 Cradge Bank, Spalding, PE11 3AB
G8 VQN R. Martin, Suite 248, 548-550 Elder House Elder Gate, Milton Keynes, MK9 1LR
G8 VQQ J. Locke, 2 Norton Close, Daventry, NN11 4GW
G8 VQX A. Hyde, 5-7 St. Nicholas Drive, Caister-On-Sea, Great Yarmouth, NR30 5QT
G8 VR K. Rochester, 22 Langford Road, Cockfosters, Barnet, EN4 9DS
G8 VRN D. Sutton, 14 Brocklesby Close, Gainsborough, DN21 1TT
GW8 VRS Dr D. Fone, 29 South Rise, Cardiff, CF14 0RF
G8 VRV G. Dyer, Chez Nous, 11 Fore Street, St Austell, PL26 7NN
G8 VRW M. Davis, 4 Joel Close Earley, Reading, RG6 5SN
G8 VSF J. Williams, Staithe Marsh House, The Staithe, Norwich, NR12 9DA
G8 VSH P. Taylor, 62 Westfield Avenue, Ashchurch, Tewkesbury, GL20 8QP
G8 VSI M. Sutcliffe, 3 Sunningdale Way, Neston, CH64 0UY
G8 VSN G. Tennant, 85 Coronation Drive, South Normanton, Alfreton, DE55 2HS
G8 VSO C. Tompkinson, 24 Rodney Close, Exmouth, EX8 2RP
G8 VSR J. Rowley, 10 Friars Close, Cheadle, Stoke-on-Trent, ST10 1AT
G8 VSV D. Petty, 7 Luscombe Close, Ipplepen, Newton Abbot, TQ12 5QJ
G8 VSX R. Hammond, 52A Mill Road, Cleethorpes, DN35 8JA
GI8 VTK A. Boston, 14 Galloway Point, Donaghadee, BT21 0ES
G8 VTV A. Hart, 78 Shepherds Way, Rickmansworth, WD3 7NR
G8 VTX I. King, Mill Ghyll, Low Lane, Kendal, LA8 8AT
G8 VTY N. Knight, 36A Abbey Street, Rugby, CV21 3LH
G8 VTZ M. Keay, 492 Falmer Road, Brighton, BN2 6LH
G8 VU D. Blair, 121 Longstomps Avenue, Chelmsford, CM2 9BZ
GW8 VUG I. Wilkinson, 6 Cwm Teg, Old Colwyn, Colwyn Bay, LL29 8ZA
G8 VUK A. Palmer, 24 Vellum Drive, Carshalton, SM5 2TL
G8 VUM P. Reade, 30 Hayleigh House, Silcox Road, Bristol, BS13 0JG
G8 VUN R. Roberts, 5480 Laburnum Ave., British Columbia, Canada, V8A 4MB
G8 VUO D. Roberts, 32 Central Road, Gloucester, GL1 5BY
G8 VUS A. Branton, 20 Sling Lane, Malvern, WR14 2TU
G8 VUU R. Clifford, 1 Darell Croft, Sutton Coldfield, B76 1HU
GW8 VUV A. Gravell, 49 Rehoboth Road, Five Roads, Llanelli, SA15 5DJ
G8 VVB C. Heath, 3 Trebellan Drive, Hemel Hempstead, HP2 5EL
G8 VVC C. Haver, 31 Edenham Road, Hanthorpe, Bourne, PE10 0RB
G8 VVG S. Hackett, 11 Fairfield Avenue, Upminster, RM14 3AZ
G8 VVM D. Merrick, 259 Wigmore Road, Gillingham, ME8 0LZ
G8 VVP P. North, 84A Park Road, Great Sankey, Warrington, WA5 3ET
G8 VVR C. Key, 23 Oxford Road, Kesgrave, Ipswich, IP5 1EL
G8 VVU M. Strange, 15 Waterside, Willesborough, Ashford, TN24 0AX
GW8 VVX R. Williams, The Basement, 9 Charlton St., Llandudno, LL30 2AA
G8 VVY R. Shelley, 69 Avenue De Gien, Malmesbury, SN16 9GX
G8 VVZ A. Stephens, 12 Sutherland Walk, Aylesbury, HP21 7NS
G8 VWH S. Hall, 31 Somerton Gardens, Earley, Reading, RG6 5XG
G8 VWJ M. Hoare, 45 Tilehurst Road, Reading, RG1 7TT
G8 VWU J. Marriott, 9 Albany Walk, Peterborough, PE2 9JN
G8 VWV A. Ball, White Cottage, Barracks Lane, Reading, RG7 1BB
G8 VXB D. Young, 66 Porchester Road, Kingston upon Thames, KT1 3PS
G8 VXR C. Hunt, 41 Maylands Way, Harold Wood, Romford, RM3 0BQ
G8 VXU D. Llewelyn, 56 Marlpit Lane, Seaton, EX12 2HN
G8 VXY P. Nicol, 38 Mitten Avenue, Rubery, Birmingham, B45 0JB
G8 VYK Selex Galileo Sports & Leisure Club, c/o M. Purser, 17 Firecrest Road, Chelmsford, CM2 9SN
G8 VYO A. Swain, 6 Abbots Grove, Belper, DE56 1BX
G8 VYP C. Syms, 24 Warmdene Road, Patcham, Brighton, BN1 8NL
G8 VYQ G. Todd, 100 Avebury Drive, Washington, NE38 7DB
G8 VYT D. Tombs, 24 Ferndale Road, Northville, Bristol, BS7 0RP
GM8 VYZ A. Raine, 10 Castle View, Airth, Falkirk, FK2 8GE
G8 VZB A. Poole, 6 Rutland Avenue, Willsbridge, Bristol, BS30 6EZ
G8 VZD C. Ramsey, 10 Lindley Road, London, E10 6QT
G8 VZI M. Warren, 39 St. Marks Road, Weston-Super-Mare, BS22 7PF
G8 VZJ M. Webb, 24 College Avenue, Grays, RM17 5UW
G8 VZR N. Giddings, Colliford Lake Park St. Neot, Liskeard, PL14 6PZ
G8 VZS I. Chapman, 188 Goodhart Way, West Wickham, BR4 0HA
G8 VZT D. Hall, 4 Steventon Road, Wellington, Telford, TF1 2AS
G8 VZY B. Levie, 51 Budges Road, Wokingham, RG40 1PL
G8 VZZ P. Marks, Flat 3, 47 The Thoroughfare, Woodbridge, IP12 1AH
G8 WAE P. Vella, Chez Tillet, Rougnac, France, 16320
G8 WAJ N. Treanor, 23 Norton Avenue, Penketh, Warrington, WA5 2RB
G8 WAL P. Taylor, 14 Cedern Avenue, Elborough, Weston-Super-Mare, BS24 8PA
G8 WAM G. Weeks, Forge Cottage, The Bury, Hook, RG29 1ND
G8 WAP R. Warren, 22 Tyndale, North Wootton, King's Lynn, PE30 3XD
G8 WAV C. Jacobs, 133 Fordham Road, Isleham, Ely, CB7 5QX
G8 WAW I. Howard, 85 Mollington Avenue, Liverpool, L11 3BQ
G8 WBG S. Netherton, 33 Bethel Road, St Austell, PL25 3HB
G8 WBK A. Maufe, 28 Dale View, Ilkley, LS29 9BP
G8 WBL T. Mole, 20 Horns Park, Bishopsteignton, Teignmouth, TQ14 9RP
G8 WBN D. Neale, 24 Addison Road, Reading, RG1 8EN
G8 WBO S. Holley, 4 St. Andrews Close, Wilton, Salisbury, SP2 0LJ
G8 WBP P. Humphreys, 910 High Lane, Stoke-on-Trent, ST6 6HE
G8 WBT A. Farnborough, 9 Mitchelmore Road, Yeovil, BA21 4BA
G8 WBU A. Greenall, Flat 11, Sutherland House Royal Herbert Pavilions, Gilbert Close, London, SE18 4PS
G8 WBY J. Osborne, Little Martins Langham, Colchester, CO4 5PY
GI8 WBZ A. Smith, 12 Sandringham Heights, Carrickfergus, BT38 9EG
GW8 WCA K. Winter, Derwen, Hillside, Monmouth, NP25 4LY
G8 WCD A. Thomas Smith, Lyndhurst House, 19 Fleet End Close, Havant, PO9 5ED
G8 WCH R. Shepherd, 299 West Wycombe Road, High Wycombe, HP12 4AA
G8 WCQ V. Mcclure, 43 Roman Way, Seaton, EX12 2NT
G8 WCT A. Grindrod, 54 Priestley Drive, Pudsey, LS28 9NQ
G8 WCX A. Essex, 32 Crossfield Drive, Skellow, Doncaster, DN6 8RJ
G8 WDC Wirral& Dist ARC, c/o G. Scott, 19 Penkett Road, Wallasey, CH45 7QF
G8 WDX R. Lamkin, 3 Homestead Close, Upton, Aylesbury, HP17 8XQ
G8 WEM M. O'Neill, Coolrake, Moone, Co Kildare, Ireland
GW8 WEY T. Jones, 80 Taff Embankment, Cardiff, CF11 7BG
GI8 WFA W. Harvey, 25 Shanes Hill Road, Kilwaughter, Larne, BT40 2PA
G8 WFP C. Kershaw, 50 Wellgarth, Halifax, HX1 2BJ
GW8 WFS J. Lawson-Reay, The Nook, Conway Road, Llandudno, LL30 1PY
G8 WGD P. Randall-Cook, 3 Wellmeadow, Staunton, Coleford, GL16 8PQ
G8 WGE I. Robinson, 26 Wick Road, Teddington, TW11 9DW
G8 WGN J. Marks, Stam 69, Huizen, Netherlands, 1275 CG
G8 WGP Gilwell Park Scout Radio Club, c/o S. Barber, Homedale, St. Monicas Road, Tadworth, KT20 6ET
G8 WGQ D. Onione, 19 Chapman Close, Kempston, Bedford, MK42 8RU
GM8 WGU A. Irving, 23 Woodlea Park, Sauchie, Alloa, FK10 3BG
G8 WHB H. Couchman, Pond Cottage, Woodside Green, Maidstone, ME17 2EU
G8 WHD P. Whittington, 7 Bowden Rise, Seaford, BN25 2HZ
GI8 WHP S. Craig, 8 Andrew Avenue, Larne, BT40 1EB
G8 WHR S. Wood, 90 Plymyard Avenue, Bromborough, Wirral, CH62 6BR
G8 WIJ B. Magrath, 14 Strines Road, Marple, SK67BT
G8 WIM Wimbledon&Dis Rd, c/o G. Cripps, 115 Bushey Road, Raynes Park, London, SW20 0JN
G8 WIR J. Vousden, 44 Castle Road, Tankerton, Whitstable, CT5 2DY
GI8 WIU S. Douthart, 75 Market St., Ballycastle, BT54 6DS
G8 WIW T. Elsey, Flat 4, 32 Pembridge Square, London, W2 4DT
G8 WJB J. Geer, 31 The Beeches, Salisbury, SP1 2JH
GM8 WJK J. Nicolson, Clickhimin, Serrigar, Orkney, KW17 2RL
GI8 WJN A. Humphreys, 20 Ballyreagh Road, Tempo., Enniskillen, BT94 3EH
G8 WJY M. Garton, 13 Damaskfield, Worcester, WR4 0HY
G8 WKA R. Reich, Cob Barn, Northlew, Okehampton, EX20 3NR
G8 WKE J. Bloxham, 15 Windmill Road, Breachwood Green, Hitchin, SG4 8PG
G8 WKH I. Jones, 2 Castle Keep Mews, Newcastle, ST5 2SD
G8 WKK M. Daniels, 6 Middlemead Stratton-On-The-Fosse, Radstock, BA3 4QH
G8 WKL Downside Scl AR, c/o M. Daniels, 6 Middlemead Stratton-On-The-Fosse, Radstock, BA3 4QH
G8 WKX R. Denton, 18 Sealand Court, Esplanade, Rochester, ME1 1QH
G8 WKZ K. Spragg, 88 Low Lane, Middlesbrough, TS5 8EB
G8 WLB S. Austen, Shiralee, The Plain Road, Ashford, TN25 6RA
G8 WLD W. Parrott, 11 St. Georges, Chester, CH1 3HG
G8 WLL S. Lown, 50 Fall Birch Road, Lostock, Bolton, BL6 4LG
G8 WLV R. Barber, 10 St. Leonards Close, Upper Minety, Malmesbury, SN16 9QB
G8 WLY J. List, 41 Westbury Crescent, Dover, CT17 9QQ
G8 WMC E. Holman, Weavers Cottage, The Shoe, Chippenham, SN14 8SA
G8 WMF A. Dawe, Highcroft, Upper House Lane, Guildford, GU5 0SX
G8 WMG J. Bassnett, 105 Edgemoor Drive, Crosby, Liverpool, L23 9UF
G8 WMK G. Bessant, 4 Sleigh Road, Sturry, Canterbury, CT2 0HR
G8 WMW A. Rowell, 25 Headcorn Gardens, Cliftonville, Margate, CT9 3ES
GW8 WNB K. Phillips, Lluest Y Coed, 39 Llwyn Ynn, Talybont, LL43 2AG
GW8 WNK J. Davies, Fronallt Llanbedrog, Pwllheli, LL53 7PB
G8 WNQ R. Harrison, Margaty, Pencoys, Redruth, TR16 6LR
G8 WOX A. Hartland, 16 Hillgrove Crescent, Kidderminster, DY10 3AP
G8 WOZ D. Hopkins, 14 Abraham Drive, Silver End, Witham, CM8 3SP
G8 WPA Dr B. Lloyd, 238 Brecknock Road, London, N19 5BQ
G8 WPF A. Middleton, 13 Ragleth Road, Church Stretton, SY6 7BN
G8 WPU I. Rivett, 30 Millside Close, Kingsthorpe, Northampton, NN2 7TR
G8 WPV A. Reason, 71 Cavendish Road, Hazel Grove, Stockport, SK7 6HU
G8 WPX G. Ratcliffe, 68 Priory Close, Tavistock, PL19 9DG
G8 WQ Weymouth and District Short Wave Club, c/o G. Watts, 3 Maple Grove Knightsdale Road, Weymouth, DT4 0FE
G8 WQC J. Maxworthy, 3 Hoylake Close, Slough, SL1 5UR
G8 WQE A. Vaughan, 12 Kingsley Road, Frodsham, WA6 6SG
G8 WQT T. Rickard, 137 Hugin Avenue, Broadstairs, CT10 3HN
G8 WQW J. Shergold, 35 Orchard Grove, New Milton, BH25 6NZ
G8 WQZ D. Mead, 9 Abraham Drive, Silver End, Witham, CM8 3SP
G8 WRB Dr D. Kirkby, Stokes Hall Lodge, Burnham Road, Chelmsford, CM3 6DT
GW8 WRC T. Harston, Ogilvie House, St. Ishmaels, Haverfordwest, SA62 3TD
G8 WRG Warks Raynet Gr, c/o D. Salter, 9 Old Milverton Road, Leamington Spa, CV32 6BA
G8 WRI W. Lawrence, 15 Rissington Road, Tuffley, Gloucester, GL4 0HP
G8 WRL R. Williamson, 35 Villiers Avenue, Twickenham, TW2 6BL
G8 WRV R. Bygrave, 35 East St., St. Neots, Huntingdon, PE19 1JU
G8 WRY G. Brock, 54 Lord Haddon Road, Ilkeston, DE7 8AW
G8 WSB M. Beardsley, 121 Wood Road, Lower Gornal, Dudley, DY3 2LR
G8 WSC R. Burg, 20 Rowan Way, Witham, CM8 2LJ
G8 WSF F. Price, 26 Teviot Gardens, Pensnett, Brierley Hill, DY5 4QL
G8 WSH M. French, 32 St. Michaels Road, Long Stratton, Norwich, NR15 2PH
G8 WSM Weston Super Mare Radio Society, c/o D. Dyer, 26 Locking Road, Weston-Super-Mare, BS23 3DF
G8 WSP P. Arup, Alma House, Broadway Road, Windlesham, GU20 6BU
G8 WSQ A. Beeston, 8 Meadow Close, Repton, Derby, DE65 6GT
G8 WSR Wirral Schools Radio Club, c/o S. Wood, 90 Plymyard Avenue, Bromborough, Wirral, CH62 6BR
G8 WSS M. Blair, 12 Medoc Close, Pitsea, Basildon, SS13 1NR
G8 WSU J. Hoggarth, Cotherstone, Rockingham Paddocks, Kettering, NN16 9JR
G8 WSV M. Cartwright, 9 Montgomery Close, Kettering, NN15 5BY
G8 WSW R. Carter, 46 Arterial Road, Leigh-on-Sea, SS9 4DA
G8 WSX Chichester ARC, c/o G. Goodyer, Flat, 54 Wyndham Road, Petworth, GU28 0EQ
G8 WSY P. Bloor, 216 Waterloo St., Burton on Trent, DE14 2NB
G8 WSZ J. Foster, 1 Thorn Court, Four Marks, Alton, GU34 5BY
G8 WTB D. Crowe, 27 Hartfield Court, Collett Road, Ware, SG12 7LT
G8 WTM R. Britt, 2 Lindisfarne Court, Maldon, CM9 6UQ
G8 WTN J. Capon, 24 Furness Close, Chadwell St. Mary, Grays, RM16 4JB
G8 WTZ D. Holland, 29 Lily Crescent, Sunderland, SR6 7HN
G8 WUF D. Legg, 2 Birkbeck Road, Wimbledon, London, SW19 8NZ
G8 WUG R. Spence, 8 Stoneleigh, Sawbridgeworth, CM21 0BT
GW8 WUM H. Matthews, 24 Clos Y Berllan, Rhuddlan, Rhyl, LL18 2UL
G8 WUO K. Baker, 57 Hedingham Road, Hornchurch, RM11 3QH
G8 WUR S. Browning, 360 Aureole Walk, Newmarket, CB8 7AZ
G8 WUS P. Besley, Higher Minzies Down Farm, Bolventor, Launceston, PL15 7TT
G8 WUU J. Cooper, 156 Church Road, Benfleet, SS7 4EN
G8 WVB S. Ayer, 335 Ings Road, Kingston Upon Hull, Hull, HU7 4UY
G8 WVH J. Bull, 12 Eastfield Crescent, Laughton, Sheffield, S25 1YT
G8 WVO P. Dawson, 35 Crofton Road, Ipswich, IP4 4QP
G8 WVZ C. Edwards, 9 Bradworth Close, Osgodby, Scarborough, YO11 3PZ

G8 WWC G. Ludlow, 48 Clifford Avenue, Walton Cardiff, Tewkesbury, GL20 7RW
G8 WWD G. Hunter, 151 Norwich Drive, Wirral, CH49 4GD
G8 WWF P. O'Ryan, 12 Minton Close, Congleton, CW12 3TD
G8 WWI P. Leverington, 28 Burymead, Stevenage, SG1 4AY
G8 WWJ J. Kirton, 13 Saltersford Road, Grantham, NG31 7HH
G8 WWM A. Morgan, 316 Middle Road, Southampton, SO19 8NT
G8 WWO J. Jackson, 12 Lower Laith Avenue, Todmorden, OL14 5RU
G8 WWW M. Harrington, 17 Church Road, Penponds, Camborne, TR14 0QE
GM8 WWY W. Kemp, 35 Quarry Drive, Kirkintilloch, Glasgow, G66 3RY
GW8 WXP R. Hadland, 41 Colby Road, Burry Port, SA16 0RH
G8 WXV A. Faulkner, 8 Wayside Trull, Taunton, TA3 7HS
G8 WYI P. Herring, 52 Mellowship Road, Eastern Green, Coventry, CV5 7BY
G8 WYR Leeds & Dis ARS, c/o M. Howes, Yarnbury Rufc, Brownberrie Lane, Leeds, LS18 5HB
GW8 WYW C. Burn, Ynysfallen, Church Street, Wrexham, LL14 2RL
G8 WZJ A. Collier, 44 Cockington Close, Leigham, Plymouth, PL6 8RQ
G8 WZK D. Collins, 71 Trench Road, Tonbridge, TN10 3HG
G8 WZO P. Evans, 63 Broadfield Road, London, SE6 1NQ
GW8 WZR D. Gale, 50 Pickle Line Road, Newport, NP19 4DL
G8 WZW K. Aspden, Langriggs, Goose House Lane, Darwen, BB3 0EH
G8 XAA Bristol Raynet, c/o A. Williams, 38 Seneca Street, Bristol, BS5 8DX
G8 XAJ T. Sherman, 44 Cleveland Avenue, Weymouth, DT3 5AG
G8 XAK C. Porteous, 1 Dawson Cottage Lords Hill Common, Shamley Green, Guildford, GU5 0TJ
G8 XAO G. Woodman, 51 Beech Hall Crescent, London, E4 9NW
GW8 XAS G. Evans, Wynona, Esplanade, Penmaenmawr, LL346LY
G8 XAX K. Tully, 225 Main Road, Harwich, CO12 3PL
G8 XBY P. Allwood, 4 Wightmans Orchard, Piddletrenthide, Dorchester, DT2 7QQ
G8 XCE D. Baker, 48 Elmwood Street, Burnley, BB11 4BP
G8 XCJ I. Coton, 77 Lockesfield Place, London, E14 3AJ
G8 XCL I. Davis, 28 Sycamore Close, Lydd, Romney Marsh, TN29 9LE
G8 XCW R. Thomson, Shire Jee Neevas, Cold Ash Hill, Cold Ash, Thatcham, RG18 9PH
G8 XCY A. Worsfold, 5 Turner Close, Langney, Eastbourne, BN23 7PF
G8 XDD D. Lucas, 43 Larcombe Road, Petersfield, GU32 3LS
G8 XDL R. Medcalf, 19 All Saints Road, Warwick, CV34 5NL
G8 XDM P. Mutter, 129 Demesne Road, Wallington, SM6 8EW
G8 XDR C. Johnstone, 24 Elibank Road, Eltham, London, SE9 1QH
G8 XDV S. Huyton, 33 Hide Gardens Rustington, Littlehampton, BN16 3NP
G8 XDY K. Lloyd, 11 Ingham Close, Broughton, Chester, CH3 5UH
G8 XEF M. Mciver, 31 Hartshill, Bedford, MK41 9AL
G8 XEI V. Nolan, 127 Martins Lane, Blakehall, Skelmersdale, WN8 9BQ
G8 XEN H. Hughes, 101 Mousehold Avenue, Norwich, NR3 4RX
G8 XER J. Smith, 32 Station Crescent, Lidlington, Bedford, MK43 0SD
G8 XET C. Street, Russets, Isle Brewers, Taunton, TA3 6QN
G8 XEU R. Stephens, 21 St. James Avenue, Lancing, BN15 0NN
G8 XEZ M. Ward, 11 Rogate Gardens, Portchester, Fareham, PO16 8DS
G8 XFK R. Young, 34 Wharfedale Drive, Bridlington, YO16 6FB
G8 XFY I. Downie, 17 Clyfton Crescent, Immingham, DN40 2AZ
G8 XGB K. Dickson, 29 Sunnyfields Drive, Minster On Sea, Sheerness, ME12 3DH
G8 XGG S. Gwilliam, 40 Falcon Close, Droitwich, WR9 7HF
G8 XGK P. Manford, Smithy Hay, Hay Lane, Rugeley, WS15 4QG
G8 XGO P. Mckellow, 155 Pittmans Field, Harlow, CM20 3LE
G8 XGS J. Hindmarsh, Roseworth Cottage, Roseworth Cottage West, Hexham Road, Newcastle upon Tyne, NE15 9EB
G8 XGT M. Saul, 23 Rockingham Road, Bury St Edmunds, IP33 2SA
G8 XGV J. Schofield, Wildwinds, St. Johns Road, Wroxall, PO38 3EH
G8 XGW N. Shearing, 51 Mill Lane, Huthwaite, Sutton-in-Ashfield, NG17 2SJ
G8 XHD P. Riebold, 7 Clitsome View, Roadwater, Watchet, TA23 0RH
G8 XHK K. Prior, 9 Tangmere Road, Crawley, RH11 0JJ
G8 XHN R. Harman, Dirleton Cottage, Church Hill, Godshill, Godshill Ventnor, PO38 3HY
G8 XHU G. Arrowsmith, 2 Orchard Drive, Bishops Hull, Taunton, TA1 5ES
G8 XIM I. Churchill, 12 Wyedale Avenue, Coombe Dingle, Bristol, BS9 2QQ
G8 XIN M. Chapman, 4 Amberley Court, Sidcup, DA14 6JT
G8 XIR K. Church, 11 Cambria Crescent, Gravesend, DA12 4NJ
G8 XIY A. Tee, 136 Burstellars, St Ives, PE27 3TJ
G8 XIZ H. Tillotson, 30 St. Laurence Road, Northfield, Birmingham, B31 2AX
G8 XJB B. Simmons, 88 Wellcome Avenue, Dartford, DA1 5JW
GW8 XJC R. Smith, 6 Lavender Court, Brackla, Bridgend, CF31 2ND
G8 XJE J. Williams, 32 Fair St., Broadstairs, CT10 2JL
G8 XJJ G. Hayes, 58 Church Road, Woodley, Reading, RG5 4QB
G8 XJL M. Halford, 35 The Limes, Stony Stratford, Milton Keynes, MK11 1ET
G8 XJN W. Hefferman, 74 Balmoral Drive, Borehamwood, WD6 2RB
G8 XJO S. Hedicker, 1 Hares Close Cottages, Selborne Road, Liss, GU33 6HG
G8 XKD R. Dance, 402 Wimborne Road East, Ferndown, BH22 9NB
G8 XKH W. Flood, 3 March Meadow, Wavendon Gate, Milton Keynes, MK7 7TB
G8 XKI G. Fowler, 26 Laburnum Drive, Armthorpe, Doncaster, DN3 3HE
G8 XKT D. Last, 77 Brunswick Road, Ipswich, IP4 4BS
GM8 XKW J. Ness, Fenway, Dalbeattie Road, Dumfries, DG2 8LN
G8 XLA L. Mayes, Stone House, Goathland, Whitby, YO22 5AN
G8 XLB J. Martin, Thatched Cottage, Thaxted Road, Saffron Walden, CB11 3BJ
G8 XLE W. Metcalf, 30 Rosemary Road, Waterbeach, Cambridge, CB25 9NB
G8 XLG C. Proctor, 69 Goodrington Road, Paignton, TQ4 7HZ
G8 XLH A. Ralph, 15 Portchester Close, Stanground, Peterborough, PE2 8UP
G8 XLI J. Rigby, 93 Birch Grove, Ashton-In-Makerfield, Wigan, WN4 0QX
GW8 XLL R. Stubbs, 35 Laburnum Drive, Rhyl, LL18 4JH
G8 XLZ K. Riley, 122 Dryden Road, Gateshead, NE9 5TX
G8 XMH D. Higgins, 80 Hill Morton Road, Sutton Coldfield, B74 4SG
G8 XML J. Hopper, 21 Knowles Avenue, Crowthorne, RG45 6DU
G8 XMO H. Houghton, 21 John Gwynn House, Newport St., Worcester, WR1 3NY
G8 XMS L. Sellar, 'Kiawah', Ringfield Drive, Hereford, HR1 4PR
G8 XMU E. Jones, 3 Byland Close, Boston Spa, Wetherby, LS23 6PU
GW8 XMW D. Jones, 7 Llys Y Godian, Trimsaran, Kidwelly, SA17 4BQ
G8 XMZ R. Linton, 11 Keats Lane Wincham, Northwich, CW9 6PP
G8 XNA J. Lane, 12 Penarwyn Woods, St. Blazey Gate, Par, PL24 2DG
G8 XNB R. Elliott, Smugglers Cottage, Oreham Common, Henfield, BN5 9SB
G8 XNC R. Lacey, 12 Melville Avenue, Frimley, Camberley, GU16 8NA
G8 XND D. Lucas, 6 Holborns Site, Main Road, Spalding, PE12 9PF
G8 XNL J. Rigby, 43A Corser Street, Stourbridge, DY8 2DE
G8 XNN H. Vadgama, 20 Hollies Walk, Wootton, Bedford, MK43 9LB
G8 XNO P. Lambert, 92 Winterslow Drive, Leigh Park, Havant, PO9 5DZ
G8 XOB P. Ashcroft, Fendley Corner, Common Lane, Harpenden, AL5 5DW
G8 XOC D. Bird, 119 Brandon Road, Watton, Thetford, IP25 6LL
G8 XOE B. Baker, Linden Lea, Fivehead, Taunton, TA3 6PU
G8 XOM P. Cook, Orchard Cottage New Road, Elmswell, IP30 9BS
G8 XOR D. Sparrow, 23 Tranmere Grove, Ipswich, IP1 6DU
G8 XOU I. Spinks, 30 Lime Tree Walk, Watton, Thetford, IP25 6EU
G8 XOV L. Sedgwick, 28 Fairhaven Road, Redhill, RH1 2LA
G8 XOX R. Sneath, 16 Wavish Park, Torpoint, PL11 2HJ
G8 XPB K. Chadwick, 5 Mason Close, Great Sutton, Ellesmere Port, CH66 2GU
G8 XPD M. Dawkins, 24 Beaufort Drive, Barton Seagrave, Kettering, NN15 6SF
G8 XPQ B. Whitehead, 17A Home Close, Histon, Cambridge, CB24 9JL
G8 XPZ S. Lovell, 98B Baker Road, Newthorpe, Nottingham, NG16 2DP
G8 XQA P. Lineham, 10 Streetsbrook Road, Shirley, Solihull, B90 3PL
G8 XQB M. Lees, 127 Mayfield Gardens, London, W7 3RA
G8 XQD T. Miller, 35 Caudle Avenue, Lakenheath, Brandon, IP27 9AU
G8 XQH E. Massey, 21 Arlington Drive, Macclesfield, SK11 8QL
G8 XQL J. Alcock, Shirley Cottage, Welland Road, Worcester, WR8 0SJ
G8 XQN A. Cleave, 37 Alledge Drive, Woodford, Kettering, NN14 4JQ
G8 XQS M. Chapple, 10 Alderley Heights, Lancaster, LA1 2HR
G8 XQT C. Dodds, Cornhill, Magpie Close, Flackwell Heath, High Wycombe, HP10 9DZ
G8 XQZ Dr G. Farmer, 39 Plough Rise, Upminster, RM14 1XR
G8 XRG R. Margetts, Mowbray, Arbor Road, Leicester, LE9 3GE
G8 XRL R. Mills, 131 High Road East, Felixstowe, IP11 9PS
G8 XRP R. Pryor, 27 Hollickwood Avenue, London, N12 0LS
G8 XRR J. Nicholson, 3 Eastleigh, Skelmersdale, WN8 6AX
G8 XRS G. Nuttall, 120 Cleevelands Avenue, Cheltenham, GL50 4PX
G8 XRW D. Owen, 18 Bushey Close, Capel St. Mary, Ipswich, IP9 2HW
G8 XRX A. O'Kavanagh, 24 Greentrees Crescent, Sompting, BN159SY
G8 XSA W. Ash, 53 Waxland Road, Halesowen, B63 3DN
GI8 XSB F. Aughey, 239 Bridge Street, Portadown, Craigavon, BT63 5AR
G8 XSD J. Atkinson, 8 Grove End, Luton, LU1 5PF
G8 XSF M. Ainley, 152 Bourne View Road, Huddersfield, HD4 7JS
G8 XST W. Butchers, 12 Church Road, St. Marychurch, Torquay, TQ1 4QY
G8 XSU M. Bond, 58 St. Pauls Street, Clitheroe, BB7 2LS
GI8 XSY K. Steenson, 108 Morgans Hill Road, Cookstown, BT80 8BW
G8 XTD R. Cavendish, 66 Coachmans Drive, Liverpool, L12 0HX
G8 XTE P. Connor, 20 Longfield, Lutton, Ivybridge, PL21 9SN
G8 XTJ J. Fitzgerald, 21 Honor Road, Prestwood, Great Missenden, HP16 0NJ
G8 XTO R. Evans, 53 Dolphin Court Road, Paignton, TQ3 1AG
G8 XTR P. Emmans, 16 Foresters Close, Rags Lane, Waltham Cross, EN7 6TF
G8 XTU M. Fowler, 28 St. Hildas Road, Doncaster, DN4 5EE
G8 XTW P. Seaford, 14 Nevis Close, Leighton Buzzard, LU7 2XD
G8 XUB N. Reddish, 15 Drakes Close, Redditch, B97 5NG
G8 XUE L. Radcliffe, 25 Oakleigh Drive, Codsall, Wolverhampton, WV8 1JP
G8 XUH J. Pearson, 14 Gorse Close, Brampton Bierlow, Rotherham, S63 6HW
GM8 XUK D. King, 59 South Knowe, Crossgates, Cowdenbeath, KY4 8AW
G8 XUL D. James, 19 Estuary Drive, Felixstowe, IP11 9TL
GW8 XUM P. Jeavons, Manora, Penisarwaun, Caernarfon, LL55 3PW
G8 XUN M. Hickman, 24 Calverley Road, Kings Norton, Birmingham, B38 8PW
G8 XUU E. White, 97 Fillongley Road, Meriden, Coventry, CV7 7LW
G8 XUW D. Shields, 54 Wildmoor Lane, Catshill, Bromsgrove, B61 0PA
G8 XVJ E. Gedvilas, 23 Pennington Drive Newton Le Willows, Newton-le-Willows, WA12 8BA
G8 XVO C. Hetherington, 23 Falkland Court, Braintree, CM7 9LL
G8 XWH C. Langham, 27 Fyfield Avenue, Swindon, SN2 5ED
G8 XWR M. Izzard, 17 Greenfields Avenue, Alton, GU34 2ED
G8 XXA J. Harrison, 10 Gaia Lane, Lichfield, WS13 7LW
G8 XXC P. Prince, 21 Ash Close, Appley Bridge, Wigan, WN6 9HU
G8 XXG S. Richardson, 52 Nailsea Park, Nailsea, Bristol, BS48 1BB
G8 XXI J. Akines, 105 Sutcliffe Avenue, Grimsby, DN33 1EZ
G8 XXJ J. Allchin, 40 Vale Road, Seaford, BN25 3EZ
G8 XXM C. Beecher, 6 Brices Meadow Shenley Brook End, Milton Keynes, MK5 7HB
G8 XXU M. Caulton, 115 Delves Green Road, Walsall, WS5 4NH
G8 XXV G. Clarke, 28 Little Potters, Bushey, WD23 4QT
G8 XXZ P. Grace, 6 Davis Grove, Yardley, Birmingham, B25 8LQ
G8 XYA N. Southorn, 20 Bratton Avenue, Devizes, SN10 5BA
G8 XYJ M. Porter, 20, Southfield Road, Much Wenlock, TF13 6AX
G8 XYQ D. Stanford, Laurel House, Top Road, Woodbridge, IP13 6JF
G8 XYR R. Tiller, Wayside, Ockley Lane, Hassocks, BN6 8NU
G8 XYS R. Travett, 39 Amwell Road, Cambridge, CB4 2UH
G8 XZB J. Payne, 25 Ringwood Road, Bath, BA2 3JL
G8 XZC A. Pinder, 2 Eleanor Road, Woodlands, Harrogate, HG2 7AJ
G8 XZQ M. Fowler, 1 Mayfields, Shefford, SG17 5AU
G8 XZX J. Tyler, 81 Mill Farm Drive, Tibshelf, DE555QL
G8 XZZ J. Tabberer, 22 Ogden View Close, Halifax, HX2 9LY
G8 YAE C. Wenn, 11 Bysouth Close, Ilford, IG5 0XN
GM8 YAQ R. Wroblewski, 1 Normandy Place, Rosyth, Dunfermline, KY11 2HJ
G8 YAS A. Miller, 113 West Front Road, Bognor Regis, PO21 4TB
G8 YAT I. Naylor, 8 Churchill Close, Uttoxeter, ST14 8BB
G8 YAU R. Newton, Cascades, Top Road, Brigg, DN20 0NN
G8 YAZ G. Oates, 21 Churchill Mansions, Cooper Street, Runcorn, WA7 1DH
G8 YBC S. Ford, 14 Merryfield, Fareham, PO14 4SF
G8 YBH A. Bristow, 2 Nursery Cottages, Staplehurst Road, Tonbridge, TN12 9BS
G8 YBO R. Colebrook, 21 Hillclose Avenue, Darlington, DL3 8BH
G8 YBR I. Davidson, 1 Mooracre Lane, Bolsover, Chesterfield, S44 6ER
G8 YBT N. Dilley, 26 Linhey Close, Kingsbridge, TQ7 1LL
GI8 YBU M. Dunne, 26 Duncreggan Road, Londonderry, BT48 0AD
G8 YBZ M. Hampson, 7 Merryfield Close, Bransgore, Christchurch, BH23 8BS
G8 YCI A. Lewis, 8 Arundel Road, Hartford, Huntingdon, PE29 1YW
G8 YCJ G. Vickery, 56 Hilden Park Road Hildenborough, Tonbridge, TN11 9BL
G8 YCK K. Tomlinson, 27 Brackens Lane, Alvaston, Derby, DE24 0AQ
G8 YCL S. Turner-Smith, 26 Ash Church Road Ash, Aldershot, GU12 6LX
G8 YCP J. Sergeant, 5 Jedburgh Close, North Shields, NE29 9NU

G8 YCQ N. Storey, 15 Tower Avenue, Upton, Pontefract, WF9 1ED
G8 YDB R. Merry, Havenwood, Oak Farm Lane, Sevenoaks, TN15 7JU
G8 YDC J. Jebb, 30 Runnymede, Nunthorpe, Middlesbrough, TS7 0QL
G8 YDE S. Inns, 11 Hodds Wood Road, Chesham, HP5 1SQ
G8 YDJ M. Alexander, 101 Richmond Street, Stoke-on-Trent, ST4 7DZ
GW8 YDR D. Catleugh, 48 Tyn Y Celyn, Glan Conwy, Colwyn Bay, LL28 5NN
GM8 YEC P. Eunson, Sandwick Cottage, Bridge End, Burra Isle, ZE2 9LD
G8 YEF A. Eaton, 16 Wood Road, Godalming, GU7 3NN
G8 YEJ J. Glover1, 5 Meadows Rise, Wymondham, LE14 2AP
G8 YEN M. Stevens, 29 Luscombe Close, Ipplepen, Newton Abbot, TQ12 5QJ
G8 YEO Yeovil ARC, c/o R. Spirrell, 32 Churchfield Drive, Castle Cary, BA7 7LA
G8 YEP B. Meyer, 6 Barrington Road, Sutton, SM3 9PP
G8 YEQ N. Littleboy, 22 Sylvaner Court, Vyne Road, Basingstoke, RG21 5NZ
G8 YFA A. Regnart, 3 Preston Avenue, North Shields, NE30 2BW
G8 YFH D. Oliver, 30 Lipscombe Rise, Alton, GU34 2HP
G8 YFK J. Mason, 80 Swallow Drive, Milford On Sea, Lymington, SO41 0XG
G8 YFP J. Wells, 24 Tomlinson Way Ruskington, Sleaford, NG34 9TW
GI8 YGG P. Foley, 5 Woodland Drive, Cookstown, BT80 8PL
GM8 YGI P. Sime, 29 Huntingtower Road, Baillieston, Glasgow, G69 7BH
G8 YGK W. Standing, 72 Ivydore Avenue, Durrington, Worthing, BN13 3JD
G8 YGM D. Southward, 3A Carnoustie Close, West Derby, Liverpool, L12 9NE
G8 YGO G. Tarr, 40 The Garth, Coniston, LA21 8EQ
G8 YGT B. Senior, 1 Bedale Close, Coalville, LE67 3BE
G8 YHF S. Kenyon, 8 Dunedin Gardens, Ferndown, BH22 9EQ
G8 YHO R. Hodge, 4 Hidcote Road, Kenilworth, CV8 2PP
G8 YIG C. Fawcett, 11 Hurst Close, Glossop, SK13 8UF
GM8 YIK Dr A. Robson, Flat 12, 57 Hesperus Broadway, Edinburgh, EH5 1FT
G8 YIN S. Wood, 246 Rush Green Road, Romford, RM7 0LA
GI8 YJD R. Perver, 6 Gransha Road, Bangor, BT20 4TG
GI8 YJF D. Roxburgh, 5 Forestbrook Park, Rostrevor, Newry, BT34 3DX
G8 YJL C. Pogmore, Sunnybanks, West End Barlborough, Chesterfield, S43 4HE
GW8 YJN A. Price, 45 Baring Gould Way, Haverfordwest, SA61 2SB
G8 YJQ P. Holt, Flat 13, Norbiton Hall, Kingston upon Thames, KT2 6RA
G8 YJS G. Hammond, 21 Cawston Road, Reepham, Norwich, NR10 4LU
G8 YJT C. Jarvis, 516 Kingsbury Road, Erdington, Birmingham, B24 9NF
GI8 YJV P. Lloyd, 18 Demesne Road, Holywood, BT18 9NB
G8 YJZ P. Rayson, 38 Ashburnham Way, Lowestoft, NR33 8SJ
G8 YKE C. Andrew, 17 St. James Close, Kettering, NN15 5HB
G8 YKG M. Armour, 22 Langcliffe Close, Culcheth, Warrington, WA3 4LR
G8 YKM A. Browne, 140 Tongham Road, Aldershot, GU12 4AT
G8 YKO S. Bardsley, 73 Highlands, Royton, Oldham, OL2 5HL
GW8 YKS D. Barton, Salisbury Hill Barn, St. Brides Netherwent, Caldicot, NP26 3AT
GM8 YKT E. Brumby, 141 Morriston Road, Elgin, IV30 4NB
G8 YKV A. Cragg, 28 Damian Way, Hassocks, BN6 8BJ
G8 YKY D. Canham, 82 Rugby Road, Binley Woods, Coventry, CV3 2AX
G8 YLA R. Cato, Orrell House, Winterpit Lane, Horsham, RH13 6LZ
G8 YLC R. Cura, 105A Bexley Road, Erith, DA8 3SN
GW8 YLK B. Evans, Mynyddmelin, Pontfaen, Fishguard, SA65 9SL
G8 YLM M. Farnworth, 16 Lees Court, Ribble Avenue, Darwen, BB3 0HW
G8 YLR R. Foss, 4 Sandy Close, Wimborne, BH21 2NG
G8 YLS D. Fox, 4 Lacey Grove, Annesley, Nottingham, NG15 0EG
G8 YMD J. Carins, 26 Roman Way, St. Margarets-At-Cliffe, Dover, CT15 6AH
G8 YMM P. Stevenson, 6 Dighton Gate, Stoke Gifford, Bristol, BS34 8XA
G8 YMN M. Shorter, 10 Lodgefield Road, Chestfield, Whitstable, CT5 3RF
G8 YMR A. Snow, 28 The Maltings Station Street, Tewkesbury, GL20 5NN
G8 YMS P. Swarbrook, 14 The Willows, Leek, ST13 8XF
G8 YMT D. Smith, 7 Peterdale Road, Brimington, Chesterfield, S43 1JA
G8 YMU L. Shaw, 108 Brookvale Road, Solihull, B92 7JA
G8 YMW A. Sneath, 21 Garrick Close, Lincoln, LN5 8TG
G8 YMZ J. Trent, The Hollies Bourne Road, West Bergholt, Colchester, CO6 3EP
G8 YNB A. Taylor, 4 Tarrant Drive, Harpenden, AL5 1RP
G8 YNC P. Tuck, 30 Brownlow Road, New Southgate, London, N11 2DE
G8 YNE S. Horner, 15 Newhouse Road, Huddersfield, HD2 1ED
G8 YNF G. Holman, 62 The Ridge, Kennington, Ashford, TN24 9EU
G8 YNG A. Hall, 33 Deanwood Road, Dover, CT17 0NT
G8 YNH M. Hall, 20 Cubitt House, Black Bull Road, Folkestone, CT19 5SH
G8 YNK M. Higton, 12 Chestnut Avenue, Mickleover, Derby, DE3 9FT
G8 YOC M. Witchard, 110 Bradley Road, Huddersfield, HD2 1QY
G8 YOE C. Victory, Pennros, Treworgans, Cuberts, TR8 5HH
G8 YOG J. Woodard, 213 Leicester Road, Ibstock, LE67 6HP
G8 YOK J. Ward, 3 Sherbourne Close, Poulton-le-Fylde, FY6 7UB
G8 YOX A. Munday, 77 Postland Road, Crowland, Peterborough, PE6 0JB
G8 YOY M. Maxwell, 962 Bury Road, Bolton, BL2 6NX
G8 YPH T. Mcknight, 31 Cavendish Road, Eccles, Manchester, M30 9EE
G8 YPK V. Maddex, 140A Kents Hill Road, Benfleet, SS7 5PH
G8 YPL P. Martin, 3 Grange Avenue, Southport, PR9 9AH
G8 YPN P. Lutman, 47 Conan Drive, Richmond, DL10 4PQ
G8 YPQ M. Waring, Woodside Cottage, Mansfield Road, Ollerton, Newark, NG22 9DX
GW8 YPR R. Williams, 54 Woodlands Avenue, Talgarth, Brecon, LD3 0AT
G8 YPV G. Williams, 54 Greenacre, Wembdon, Bridgwater, TA6 7RF
G8 YPY D. Wilson, 120 Poulton Road, Fleetwood, FY7 7ar
G8 YQA D. Arnold, 10 Shaw Place, Leek, ST13 6ES
G8 YQC M. Beetlestone, 19 Tenbury Road, Birmingham, B14 6AD
G8 YQH V. Carter, 69 Angela Crescent, Horsford, Norwich, NR10 3HE
G8 YQN P. Gebbie, 333 Scalby Road, Scarborough, YO21 6TG
G8 YQO D. Henderson, Reverie Pennys Lane, Margaretting, Essex, CM4 0HA
G8 YQU G. Lenihan, 28 Paddock Crescent, Sheffield, S2 2AR
GM8 YRE J. Firth, 6 Upper Burnside Drive, Thurso, KW14 7XB
G8 YRF R. Foxley, 20B Alder Copse, Horsham, RH12 1LD
G8 YRG G. Williams, 2 Windsor Court Watton, Thetford, IP25 6XB
G8 YRL B. Trim, Endon Cottage, 63B Rose Street, Wokingham, RG40 1XS
GM8 YRT W. Stewart, 20 Corrie Place, Scone, Perth, PH2 6QE
G8 YRW R. Williams, 29 Woodfield Road, Bude, EX23 8JB
GM8 YRX E. Saxon, 73 Upper Burnside Drive, Thurso, KW14 7XB
G8 YSA P. Powers, 5 Bracken Close, Hugglescote, Coalville, LE67 2GP
G8 YSH L. Jannetta, 1 Lake Road, Hadston, Morpeth, NE65 9TF
G8 YSJ W. Bannerman, 3 The Cornfield, Langham, Buttercups, Holt, NR25 7DQ
G8 YTF G. Mcgowan, 281 Ashgate Road, Chesterfield, S40 4DB
GI8 YTH S. Moore, 7 Cyprus Avenue, Belfast, BT5 5NT
GW8 YTO A. Ham, 46 Celtic Way, Rhoose, Barry, CF62 3FT
G8 YTP S. Holgate, 91 Valley Road, Stockport, SK4 2DB
G8 YTR S. Higgs, 5 Lawnswood Close, Cowplain, Waterlooville, PO8 8RU
G8 YTU F. Adams, 27 Challenger Close, Malvern, WR14 2NN
G8 YTX K. Bagshaw, 36 St. Peters Road, Buxton, SK17 7DX
G8 YTZ J. Cockett, 2A Priory Avenue, Petts Wood, Orpington, BR5 1JF
GM8 YUI G. Mcclintock, 13 St. Andrews Drive, Gourock, PA19 1HY
GW8 YUJ J. Milburn, Orme View, Anglesey, LL73 8PE
G8 YUK A. White, 10 Stott Drive, Urmston, Manchester, M41 6WA
GM8 YUM G. Walker, 24 George Street, Cellardyke, Anstruther, KY10 3AU
G8 YUO M. Taylor, 4 Yew Tree Court, Botley Road, Swanwick, Southampton, SO31 1EA
G8 YUP B. Stevens, 77 Dean Lane, Hazel Grove, Stockport, SK7 6EJ
G8 YUR M. Robelou, 12 Cooks Drove, Earith, Huntingdon, PE28 3QG
G8 YVC M. Smith, 31 Burringham Road, Scunthorpe, DN17 2BD
G8 YVM N. Matthes, 24 Albany Road, Fleet, GU51 3LY
G8 YVP M. Nicholson, 33 Painshawfield Road, Stocksfield, NE43 7PX
G8 YVQ Dr C. Harper, Chusan, Farley Court, Church Road, Reading, RG7 1TT
G8 YVS R. Hillan, 128A Bridge Street, Deeping St. James, Peterborough, PE6 8EH
G8 YVW C. Stacey, 157 Ormond Road, Sheffield, S8 8FT
GI8 YWE M. Anderson, 17 Leydene Court, Lisburn, BT28 3LL
G8 YWJ A. Frost, 76 Tregrea Estate, Beacon, Camborne, TR14 7SU
G8 YWK W. Gleave, 6 Sidlaw Avenue, Chester le Street, DH2 3DD
G8 YWL G. Pitt, 17 Penfound Gardens, Bude, EX23 8FF
G8 YXI D. Shemeld, 13 Arran Road, Sheffield, S10 1WQ
G8 YXJ R. Skells, 31 Perry Road, Leverington, Wisbech, PE13 5AE
G8 YXQ D. Chatterton, 27 Victoria Road, Folkestone, CT19 5AT
G8 YXR E. Ferris, Karravas, Osborne Road, Deal, CT14 8BT
G8 YXZ R. Dominy, 8 Meadow Road, Claygate, Esher, KT10 0RZ
G8 YYA H. Duesbury, 4 Harbour View Close, Poole, BH14 0PF
G8 YYC G. Miller, 93 Shepherds Grove Park Stanton, Bury St Edmunds, IP31 2BN
GW8 YYF K. Jones, 3 Penffordd, Pentyrch, Cardiff, CF15 9TJ
G8 YYL Lady G. Johnson, Kilmurry House, Kilmurry Fermoy, Co Cork, Ireland, blank
GI8 YYM I. Ferris, 48 Abbey Gardens, Belfast, BT5 7HL
G8 YYW M. Freeman, 2 Poolthorne Farm Cottage, Cadney, Brigg, DN20 9HU
G8 YYX A. Layton, 7 Higher Saxifield, Harle Syke, Burnley, BB10 2HB
G8 YZA K. Sawday, 15 Moorland View, Buckfastleigh, TQ11 0AF
G8 YZC R. Smith, 86 Manor Road, Borrowash, Derby, DE72 3LN
G8 YZF M. Bishop, 6 Tiverton Close, Kingswinford, DY6 8PD
G8 YZL P. Thackeray, 19 Moneyfly Road, Verwood, BH31 6BL
G8 YZT P. Wing, 14 Huntingdon Road, Kempston, MK42 7EX
G8 YZY D. Spencer, 28 Watery Lane, Minehead, TA24 5NZ
G8 ZAD R. Mantle, 37 Willis Road, Stockport, SK3 8HQ
G8 ZAJ C. French, 26 Wood Street, Ash Vale, Aldershot, GU12 5JG
GM8 ZAK H. Gemmell, 53 Southesk Avenue, Bishopbriggs, Glasgow, G64 3AD
G8 ZAT J. Haslip, 18 Downsview Drive, Wivelsfield Green, Haywards Heath, RH17 7RW
G8 ZAU D. Hoodless, 21 Meadow Close, Eastwood, Nottingham, NG16 3DQ
G8 ZAX R. Rees, 69 Pewley Way, Guildford, GU1 3PZ
G8 ZBC C. Lucas, 8 Hawker Close, Broughton, Chester, CH4 0SQ
G8 ZBJ W. Sheldon, 15 Hawthorn Place, Walsall, WS2 0HZ
G8 ZBN T. Nye, 18 Kingsway, Chandler'S Ford, Eastleigh, SO53 2FE
G8 ZCJ J. Skidmore, 55 Elmsleigh Road, Heald Green, Cheadle, SK8 3UD
G8 ZCK C. Wilson, 19 Chace Avenue, Potters Bar, EN6 5LX
GM8 ZCS A. Westerman, 5 Eldon Gardens, Bishopbriggs, G64 2EU
GI8 ZDB R. Logue, 46 Brunswick Road, Londonderry, BT47 5SZ
G8 ZDS P. Hocking, 10 South Terrace, Camborne, TR14 8ST
G8 ZDT P. Langford, 1 Kingaby Gardens, Rainham, RM13 7PH
G8 ZEE A. Hudson, 1 Laburnum Court, Cheltenham, GL51 0XE
GW8 ZEI E. Whitham, 44 Tyddyn Isaf, Menai Bridge, LL59 5DA
G8 ZEK P. Jacobi, Highbury, Furzehill, Wimborne, BH21 4HD
GM8 ZEQ M. Smith, Haremuir Bungalow, Benholm, Montrose, DD10 0HX
G8 ZES P. Street, 50 Dickson House, Ridgway Road, Stoke on Trent, ST1 3BA
G8 ZEV C. Hartt, 9 Laura Grove, Paignton, TQ3 2LR
G8 ZEW A. Joy, 15 Wymersley Close, Great Houghton, Northampton, NN4 7PT
G8 ZEX S. Lauritson, 42 Woodstock Road, Kingswood, Bristol, BS15 9UE
G8 ZFD C. Askin, 54 York Road, Hull, HU6 9RA
G8 ZFI P. Bryant, 21 Devonshire Close, Stevenage, SG2 8RY
G8 ZFL A. Butcher, 4 Maple Close, Bristol, BS30 9PX
G8 ZFQ M. Kanelis, 57 Ringwood Avenue, Redhill, RH1 2DY
G8 ZFS P. Wiley, 36 Hungate Lane, Hunmanby, Filey, YO14 0NN
G8 ZFT R. Thompson, 329 Prestbury Road, Prestbury, Cheltenham, GL52 3DF
G8 ZFU G. Taylor, 8 Ullathorne Road, Streatham, London, SW16 1SN
GM8 ZFW J. Morris, 1 Wealthyton Cottages Keig, Alford, AB33 8BH
G8 ZFX P. Blake, 35 Kings Court 71-76 Wright Street, Hull, HU2 8JR
GI8 ZFZ D. Alexander, 33 Greenan Road, Newry, BT34 2PJ
GM8 ZGC C. Dowers, 71 Mosshead Road Bearsden, Glasgow, G61 3EZ
G8 ZGF R. Mackrell, 17 Townfield Avenue, Worsthorne, Burnley, BB10 3JG
G8 ZGK A. Mockford, 58 Wendover Heights, Wendover, HP22 6PH
G8 ZGM S. Berks, 14 Austen Way, Hastings, TN35 4JH
G8 ZGQ A. Longuet, 10 Severnmead, Grovehill, Hemel Hempstead, HP2 6DX
G8 ZGS J. Holden, 128 Greenways, Norwich, NR4 6HA
G8 ZGY R. Bareham, 49 Wharf Road, Crowle, Scunthorpe, DN17 4HU
G8 ZHA R. Morrall, 32 Broadstone Avenue, Walsall, WS3 1EW
G8 ZHN P. Gibbons, 13 Canon Park, Berkeley, GL13 9DF
G8 ZHR N. Lawes, 87 Glebelands, Crayford, Dartford, DA1 4RY
G8 ZHS P. Lester, 1C Eastwood Road, London, E18 1BN
GI8 ZHW J. Mcdonnell, 6 Sandhurst Park, Bangor, BT20 5NU
G8 ZIA A. Bowman, Evergreen, Durham Road, Stockton-on-Tees, TS21 3LT
G8 ZIC C. Harrison, 2 Bridgemere Close, Radcliffe, Manchester, M26 4FS
G8 ZIH J. Eady, Pytchley Lodge, Pytchley, Kettering, NN14 1EE
G8 ZIK E. Serwa, 102 Cornwall Road, Wolverhampton, WV6 8UZ
GW8 ZIL I. Bell, 102 Ewenny Road, Bridgend, CF31 3LN
G8 ZIP K. Lake, 22 Chapmans Close Stirchley, Telford, TF3 1ED
G8 ZIW G. Ludar-Smith, 2 Springmead, Queenborough Lane, Braintree, CM77 7PX
G8 ZIY P. Eyre, 27 Holborn View, Codnor, Ripley, DE5 9RB
G8 ZJH B. Mccourt, 3 Littlemore Close, Upton, Wirral, CH49 4GS

G8 ZJK R. Cole, Flat 6, Barton Court, 19 Southwood Road, Hayling Island, PO11 9PS
G8 ZJO S. Tomschey, 36, Lord Lytton Ave, Coventry, CV2 5jw
G8 ZK ZK Contest Group, c/o C. Archer, 118 Cator Lane, Beeston, Nottingham, NG9 4BB
GM8 ZKF D. Robson, 6 Ladywood Estate, Milngavie, Glasgow, G62 8BE
GM8 ZKN I. Diment, 22 Academy Place, Bathgate, EH48 1AS
GM8 ZKU S. Hawley, Hill Of Ardiffery, Hatton, Peterhead, AB42 7TB
G8 ZLF T. Gilleard, 3 Paul Crescent, Humberston, Grimsby, DN36 4DF
G8 ZLL I. Thomas, 30 Alcot Close, Crowthorne, RG45 7NE
G8 ZLN P. Thompson, 81 Ashmead Road, Banbury, OX16 1AA
GW8 ZLT M. Chambers, 3 Manod Road, Blaenau Ffestiniog, LL41 4DD
G8 ZLU M. Wright, 17 Colwyn Crescent, Stockport, SK5 7LL
G8 ZMC A. Mccalden, 127 Kings Road, Godalming, GU7 3EU
G8 ZME M. O'Toole, Daffodil Cottage, Dunsmore, Aylesbury, HP22 6QH
GM8 ZMF M. Osborn, 3 Lovers Lane, South Queensferry, EH30 9UP
G8 ZMG S. Watson, 61 Glenview Road, Shipley, BD18 4AR
G8 ZMH K. Robinson, 33 Cranford Road, Northampton, NN2 7QU
G8 ZML B. Ewart, 36 Sycamore Rise, Holmfirth, HD9 7TJ
G8 ZMM R. Bunney, 35 Grayling Mead, Romsey, SO51 7RU
G8 ZMQ P. Burnley, 45 Ashwell Road, Heaton, Bradford, BD9 4AX
G8 ZNB A. Harris, 55 Frenchgate, Richmond, DL10 7AE
G8 ZNK G. Barnes, 18, Wellesley Avenue, Goring-By-Sea, Worthing, West Sussex, BN12 4PN
GW8 ZOE S. Trott, 6 Mounton Drive, Chepstow, NP16 5EH
G8 ZOJ G. Barrett, The Old Chapel, 5 Tappers Lane, Bridgwater, TA6 6SJ
G8 ZOO J. Molinghen, 16 Dumpers Lane, Chew Magna, Bristol, BS40 8SS
G8 ZOV Dr R. Nicholson, 24 Barnmead, Haywards Heath, RH16 1UZ
GM8 ZOW P. Oram, 24 John Smith Place, Kelty, KY4 0NL
G8 ZOY G. Page, 1A Montagu Gardens, Wallington, SM6 8EP
G8 ZPD P. Davies, 46 Spring Street, Colley Gate, Halesowen, B63 2SZ
G8 ZPE P. Cooper, The Bungalow, Clopton, Kettering, NN14 3DZ
G8 ZPH D. Bucknell, 46 Heath Row, Bishop's Stortford, CM23 5DE
G8 ZPO R. Blackwell, 46 Wyatts Drive, Thorpe Bay, Southend-on-Sea, SS1 3DG
G8 ZPW A. Martin, 23 Portfield Road, Christchurch, BH23 2AF
G8 ZQA P. Stonebridge, 207 Henley Road, Ipswich, IP1 6RL
G8 ZQB J. Smith, 7 Mill Hill Close, Whetstone, Leicester, LE8 6NF
G8 ZQG S. Wood, 8A Glendale Avenue, Glenfield, Leicester, LE3 8GF
G8 ZQJ D. Young, 9 Larchfield House, Highbury Estate, London, N5 2DE
G8 ZQM K. Pascoe, 21 Cotswold Avenue, Sticker, St Austell, PL26 7ER
GM8 ZQY S. Frey, 2 Balgeddie Gardens, Glenrothes, KY6 3QR
G8 ZRD I. Gilzean, 35 Pieces Terrace, Waterbeach, Cambridge, CB25 9NE
G8 ZRE D. Hewitt, 31 Broadmead, Vicars Cross, Chester, CH3 5PT
G8 ZRG B. Hawes, 129 Wycombe Lane, Wooburn Green, High Wycombe, HP10 0HJ
G8 ZRM R. Myers, 9 Romney Road, Rottingdean, Brighton, BN2 7GG
G8 ZRN G. John, 29 Park Road, Northville, Bristol, BS7 0RH
G8 ZRQ R. Knight, 2 Stag Place, Pocklington, York, YO42 2FN
G8 ZRU D. Moger, 47 Powys Grove, Banbury, OX16 0UG
G8 ZRV G. Sargant, 9 Orchard Way, Reigate, RH2 8DS
G8 ZSD I. Worthington, 7 Bowness Close, Gamston, Nottingham, NG2 6PE
G8 ZSK A. Allcock, 30 Clyde Grove, Crewe, CW2 8NA
G8 ZSM L. Barlow, 4 Bucknell Place, Thornton-Cleveleys, FY5 3HZ
G8 ZSP A. Blanchard, 41 Deane Drive, Galmington, Taunton, TA1 5PQ
G8 ZSZ I. Dickinson, 16 Heathfield Grove, Beeston, Nottingham, NG9 5EB
G8 ZTB S. Fenn, 21 Waarem Avenue, Canvey Island, SS8 9DS
G8 ZTD J. Francis, 9 Holland Close, Bognor Regis, PO21 5TW
G8 ZTF J. Hargraves, 321 Northway, Maghull, Liverpool, L31 0BW
G8 ZTG J. Harman, 20 Sunview Avenue, Peacehaven, BN10 8PJ
G8 ZTM N. Ledeux, 14 Jubilee Close, Cam, Dursley, GL11 5JQ
G8 ZTN F. Lock, Monks Rest, The Street, Charmouth, Bridport, DT6 6PE
G8 ZTR J. Macdonald, 74 Bradford Road, Boston, PE21 8BJ
GM8 ZTV F. Millar, 13 Edzell Park, Kirkcaldy, KY2 6YB
G8 ZUF K. Rogers, 36 Goodacre Road, Ullesthorpe, Lutterworth, LE17 5DL
G8 ZUL R. Yates, 16 Arnold Grove, Shirley, Solihull, B90 3JR
G8 ZUU M. Smith, 2 Newbury Close, Mapperley, Nottingham, NG3 5QW
G8 ZUZ D. Unwin, 1 Bentinck Close, Nuncargate, Nottingham, NG17 9ET
G8 ZVI L. Hart, 28A Dunton Road, Stewkley, Leighton Buzzard, LU7 0HZ
G8 ZVK B. Ackroyd, 91 Bulford, Wellington, TA21 8DH
G8 ZVM M. Atkinson, Menamber Farm, Trenear, Helston, TR13 0HE
G8 ZVS R. Bird, 80 Clearmount Road, Weymouth, DT4 9LE
G8 ZVX A. Breeds, 26 Heighton Road, Newhaven, BN9 0JU
G8 ZVZ I. Collins, Knapp Cottage, Pixley, Ledbury, HR8 2QB
G8 ZWA P. Collins, 40 Shacklegate Lane, Teddington, TW11 8SH
G8 ZWC L. Curtis, 34 Gaisford Road, Worthing, BN14 7HW
G8 ZWF R. Cowling, 20 Claremont Hill, Shrewsbury, SY1 1RD
G8 ZWN M. Davies, Sunningdale, Sulhamstead Hill, Reading, RG7 4DE
G8 ZWU K. Graham, 670 Stafford Road, Ford Houses, Wolverhampton, WV10 6NW
G8 ZXL Lord R. Pretty, 52 Queens Road, Hersham, Walton-on-Thames, KT12 5LW
GM8 ZXQ J. Mcdermott, Milking Green Gate, Eliock, Sanquhar, DG4 6LD
G8 ZXT J. Marshall, 58 Sandbed Court, Leeds, LS15 8JJ
G8 ZXU P. Mcguinness, 83 Beaconsfield, Telford, TF3 1NH
G8 ZXY W. Mason, 365 Heath Road South, Birmingham, B31 2BJ
G8 ZXZ D. Holmes, 17 Spring Hall Close, Shelf, Halifax, HX3 7NE
G8 ZYC Zycomm Elect Lt, c/o M. Sneap, Ivy Farm Bungalow, Farm Close, Pentrich, Ripley, DE5 3RR
G8 ZYH E. Hitch, 35 Hawthorndene Road, Hayes, Bromley, BR2 7DY
G8 ZYI N. Hitch, 1B Greenlands, Platt, Sevenoaks, TN15 8LL
G8 ZYM I. Hammond, 1 Old Rectory Close, Barham, Ipswich, IP6 0PY
G8 ZYR P. Hodgkinson, 25 Polisken Way, St. Erme, Truro, TR4 9RB
G8 ZYT S. Higlett, 28 Oak Crescent, Potton, Sandy, SG19 2PY
G8 ZZB D. Kellet, April Cottage, 10 Yorkdale Drive, Selby, YO8 9YB
G8 ZZF A. Leatherbarrow, 33 Chester Avenue, Sale, M33 4NS
G8 ZZG T. Lock, 40 Chertsey Road, Ashford Common, Ashford, TW15 1SQ
G8 ZZK D. Lee, 14 Woodview Close, West Kingsdown, Sevenoaks, TN15 6HP
G8 ZZL P. Lake, 166 Burrs Road, Clacton on Sea, CO15 4LH
G8 ZZR P. Vince, 19, Links Road, Ashtead, KT21 2HB
G8 ZZS D. Vaughan, Orchard Farm House, Framsden, Stowmarket, IP14 6HD
G8 ZZT J. Tonks, Flat, 3 Greystone Passage, Dudley, DY1 1SL
G8 ZZV A. Tye, 3 Parkwood Court, Forest Park, Nottingham, NG6 9FB
G8 ZZW I. Shepherd, 12 Grains Road, Delph, Oldham, OL3 5DS
G8 ZZY A. Smart, 101 Bardon Road, Coalville, LE67 4BF

M0

GBM0SSB Glenrothes & District ARC, c/o T. Brown, 11 Approach Row, East Wemyss, Kirkcaldy, KY1 4LB
M0 AAA Reading and District ARC, c/o V. Robinson, 4 Hilltop Road, Caversham, Reading, RG4 7HR
M0 AAC P. Bergin, 15 Monks Way, Harmondsworth, West Drayton, UB7 0LE
M0 AAD M. Stockton, 37 Ney Street, Ashton-under-Lyne, OL7 9NL
M0 AAF D. Hodgson, 1B Court Farm Avenue, Epsom, KT19 0HD
M0 AAG A. Salt, 1 Chantry Close, Harrow, HA3 9QZ
M0 AAK M. Pearson, 56 Parkwood Green, Parkwood, Gillingham, ME8 9PP
M0 AAM R. Armstrong, 71 Bradshaw View, Queensbury, 71 Bradshaw View, Queensbury, Bradford, BD13 2FF
M0 AAN W. Glover, 21 West End Way, Lancing, BN15 8RL
M0 AAP I. Parker, 23 Southdown Road, Benham Hill, Thatcham, RG19 3BF
M0 AAR J. Kemp, 394 Great Thornton Street, Hull, HU3 2LT
M0 AAS J. Whittaker, 10 Pownall Court, Wilmslow, SK9 5QE
M0 AAV S. Bates, 6 Foxdell, Northwood, HA6 2BU
MI0 AAW S. Blakley, 123 Mount Merrion Avenue, Belfast, BT6 0FN
MI0 AAZ J. Anderson, 1 Claragh Hill Drive, Kilrea, Coleraine, BT51 5YR
M0 ABA T. Hackett, 16 Eagle Way, Shoeburyness, SS3 9RJ
MM0 ABB C. Kane, 46 Hillmoss, Kilmaurs, Kilmarnock, KA3 2RS
M0 ABC D. Cracknell, 120 Woodhill, London, SE18 5JL
MI0 ABD J. Mccarrison, 11 Boretree Island Park, Newtownards, BT23 7BW
M0 ABF K. Molyneux, 220 Woodlands Holiday Homes Pk, Dowles Road, Bewdley, DY12 3AE
M0 ABG A. Powell, Crosstrees, Main Road, Theberton, Leiston, IP16 4RX
M0 ABH R. Tyler, 343 Broadwater Crescent, Stevenage, SG2 8EZ
M0 ABI M. Lennon, 12 Byron Road, Barton On Sea, New Milton, BH25 7NX
MM0 ABJ C. Ewart, 13 Princes St., Innerleithen, EH44 6JT
M0 ABK M. Gray, 19 Marsh View Newton, Preston, PR4 3SX
MI0 ABN N. Crawford, 10 White Mountain Road, Lisburn, BT28 3QY
M0 ABO J. Valle Espin, 203 Broadway, Horsforth, Leeds, LS18 4HL
M0 ABP J. Barker, Karma, 6 Acredykes, Bridlington, YO15 1LY
M0 ABQ Sir W. Couse, 68/29 Moo 3 Rattanapron Village, Tambon Khungkong, Chiang Mai, Thailand, 50230
M0 ABT S. Little, 46 Marine Drive, Seaford, BN25 2RU
M0 ABU K. Simkin, 19 Summercourt, Hailsham, BN273AW
MW0 ABV P. Plummer, Hill Road, Neath Abbey, sa108nd
M0 ABY A. Soane, 24 Nurseries Road, Wheathampstead, St Albans, AL4 8TP
M0 ABZ A. Allbright, Greenacre, Carne Road, Newlyn, Penzance, TR18 5QA
M0 ACA E. Morley, 91 Allerton Road, Stoke-on-Trent, ST4 8PQ
M0 ACB E. Mcdonald, 32 Butterwick Road, Messingham, Scunthorpe, DN17 3PB
M0 ACC A. Dixon, 17 Coppice Court, Weymouth, DT3 5SA
M0 ACI S. Stacey, Trehill, Trekenner, Launceston, PL15 9PH
M0 ACK M. Jackson, 121 Kiln Lane, Eccleston, St Helens, WA10 4RH
M0 ACM D. Forrest, 166 Meadowcroft, Swindon, SN2 7LE
MM0 ACN J. Green, Tigh Callum, Culkein Drumbeg, Lairg, IV27 4NL
MM0 ACR L. Skinner, Manse Hall, Drumoak, Banchory, AB31 5HA
MM0 ACT R. Skinner, Manse Hall, Drumoak, Banchory, AB31 5HA
M0 ACU M. Eddyvean, 41 Liddell Road, Cowley, Oxford, OX4 3QU
M0 ACV T. Bevan, 6 Buttermere Grove West Auckland, Bishop Auckland, DL14 9LG
M0 ACW Over The Hill DX Group, c/o R. Williams, Dyffryn Coed, Union Road, Coleford, GL16 7QB
M0 ADA C. Rule, Marconi House, Meaver Road, Helston, TR12 7AH
M0 ADB N. Pringle, 21 Petersmiths Drive, New Ollerton, Newark, NG22 9RZ
M0 ADD A. Saville, 1 Bellingham Close, Shaw, Oldham, OL2 7UU
M0 ADG D. Morris, 86 Richardson St., Carlisle, CA2 6AG
M0 ADH M. Mettam, 190 Scotter Road, Scunthorpe, DN15 7EQ
M0 ADL A. Jermaks, 52 Laburnum Crescent, Allestree, Derby, DE22 2GR
M0 ADO D. Kennard, 16 Cantilupe Crescent, Aston, Sheffield, S26 2AT
M0 ADR G. Galbraith, 24 Airedale, Hadrian Lodge West, Wallsend, NE28 8TL
M0 ADW R. Latham, 47 Oldfield Park, Westbury, BA13 3LQ
M0 ADY A. Grundy, 21 Ribston Close, Shenley, Radlett, WD7 9JW
M0 ADZ A. Gall, 26 West Acre Drive, Norwich, NR6 7HX
M0 AEC Dr S. Roper, 15 St. Gerards Road, Solihull, B91 1TZ
M0 AEE M. Clarke, 19 Hodroyd Cottages, Brierley, Barnsley, S72 9JA
M0 AEJ V. Trend, 64 Shutlock Lane Moseley, Birmingham, B13 8NZ
M0 AEK J. Sloan, Flat 39, Colonel Stevens Court, 10A Granville Road, Eastbourne, BN20 7HD
MW0 AEL S. Townsend, 42 Burns Crescent, Bridgend, CF31 4PY
M0 AEN M. Austen, 11 Corn Avill Close, Abingdon, OX14 2ND
M0 AEP G. Dawes, 11 Ferriby Road, Barton-upon-Humber, DN18 5LE
M0 AEQ M. Bardell, 47 Calverleigh Crescent, Furzton, Milton Keynes, MK4 1HY
M0 AEU F. Heritage, 50 Laurel Close, North Warnborough, Hook, RG29 1BH
MW0 AEV E. Jones, 18 Madryn Terrace, Llanbedrog, Pwllheli, LL53 7PF
MI0 AEX J. Smith, 54A Blackstaff Road, Kircubbin, Newtownards, BT22 1AF
M0 AEZ M. Herpe, 25 Gordon Street Sutton In Craven, Keighley, BD20 7EU
M0 AFC T. Boon, 27 Meadowside Avenue, Clayton Le Moors, Accrington, BB5 5XF
MW0 AFD S. Edwards, 59 St. Andrews Road, Colwyn Bay, LL29 6DL
M0 AFF F. Hallsworth, Flat 2 Derwent Court Salt Ayre Lane, Lancaster, LA1 5JP
M0 AFJ T. Hague, 2 Winns Row, Godolphin Road, Helston, TR13 8QH
M0 AFQ B. Eagleton, 12 Park Court, Hadley, Telford, TF1 6AD
M0 AFR P. Walker, 1 Vicarage Lane, Fordington, Dorchester, DT1 1LH
M0 AFS P. Whiteley, 53 Sharp Lane, Almondbury, Huddersfield, HD4 6SS
MI0 AFT J. Stewart, 23 Swifts Quay, Carrickfergus, BT38 8BQ
M0 AFV R. Rippin, Lyndhurst, Myatts Field Harvington, Evesham, WR11 8NG
M0 AFW C. Parkinson, 4 Campion Drive, Killamarsh, Sheffield, S21 1TG
M0 AFX D. Waters, Station House, Station Road, Manningtree, CO11 2LH
M0 AFY R. Ford, 70 Jubilee Road, Darnall, Sheffield, S9 5EH
M0 AFZ P. Nairne, 137 Barden Road, Tonbridge, TN9 1UX
M0 AGA K. Gunstone, 67 Woodside, Sutton in Ashfield, NG17 3EB
MW0 AGE J. Chinnock, 22 Mill Road, Pyle, Bridgend, CF33 6AP
M0 AGJ A. Bowker, 120 Broomhouse Lane, Doncaster, DN4 9DB

M0 AGP M. Weber, Tall Chimneys, Malacca Farm, West Clandon, West Clandon, GU4 7UG
M0 AGR M. Bray, 2 Camborne Drive, Fixby, Huddersfield, HD2 2NF
M0 AGS E. Smeaton, 27 Sandringham Avenue, Burton-on-Trent, DE15 9BJ
M0 AGT R. Markham, 8 Railway Cuttings, Ilminster, TA19 9FG
M0 AGU J. Shorthouse, 84 Mount Pleasant, Ackworth, Pontefract, WF7 7HU
MM0 AGV T. Mcguigan, 42 Fidra Avenue, Burntisland, KY3 0AZ
M0 AGW W. Mason, 104 Chester Road, Poynton, Stockport, SK12 1HG
M0 AGY M. Griffin, 15 Victoria Road, St Austell, PL25 4QF
MM0 AHC M. Collins, Redwoods, Barcaldine, Oban, PA37 1SG
M0 AHD M. Carter, 7 Dawlish Green, Middlesbrough, TS4 3NW
M0 AHF G. May, 14 Tennyson Avenue, Dukinfield, SK16 5DP
MI0 AHH C. Doris, 9 Gortalowry Park, Cookstown, BT80 8JH
MI0 AHI J. Doris, 92 Coolnafranky Park, Cookstown, BT80 8PW
M0 AHJ C. John, 5 Highfield Gardens, Aldershot, GU11 3DB
M0 AHS M. Nicholas, 4 Chesterfield Mews, Chesterfield Road, Ashford, TW15 3PF
M0 AHT W. Burt, 5 Aged Miners Homes, Springwell Terrace, Hetton-Le-Hole, Houghton le Spring, DH5 0BA
M0 AHV H. Banks, 104 Viking Road, Bridlington, YO16 6TB
M0 AHY G. Parsons, Gull Cottage, Briar Close, Hastings, TN35 4DP
M0 AHZ R. Brown, 17 Ridgeway, North Seaton, Ashington, NE63 9TJ
M0 AIB S. Budd, 19 Queen Street, Worthing, BN14 7BL
M0 AIC S. Deary, 43 Old Road, Tintwistle, Glossop, SK13 1LH
M0 AID K. Marsh, Highgrove, Creech Heathfield, Taunton, TA3 5EW
MW0 AIE R. Duncombe, 1 Pennar Court, Pembroke Dock, SA72 6NW
MI0 AIH D. Martin, 34 Lower Kildress Road, Cookstown, BT80 9RN
M0 AIJ C. Blake, 30 Pine Tree Walk, Poole, BH17 7EH
MM0 AIK Scottish DX Contest Club, c/o B. Devlin, 112 Benview, Bannockburn, Stirling, FK7 0HJ
M0 AIQ B. Fryett, 9 Trethiggey Crescent, Quintrell Downs, Newquay, TR8 4LF
M0 AIS A. Benns, 7 Brooklands Road, Burnley, BB11 3PR
M0 AIT R. Holt, 41 Garden Avenue, Ilkeston, DE7 4DF
M0 AIY R. Carter, 16 Holts Lane, Clayton, Bradford, BD14 6BL
MW0 AIZ R. Ramm, Mor Welir, Sarnau, Llandysul, SA44 6QY
M0 AJB North West 320 DX Club, c/o A. Birch, 6 Crescent Road, Wallasey, CH44 0BQ
M0 AJC M. Mcinally, Flat 7, 32-33 Edgar Road, Margate, CT9 2EJ
M0 AJD M. Saxton, Heathercrest, 7 Boston Road, Horncastle, LN9 6EY
MW0 AJH J. Donnell, 42 Wentworth Crescent, Mayals, Swansea, SA3 5HT
M0 AJI S. Nursey, 9 Tydd Low Road, Long Sutton, Spalding, PE12 9AR
M0 AJJ P. Olson, 23 Dennett Close, Liverpool, L31 5PD
MM0 AJQ J. Stone, 1 Seafield Crescent, Bilston, Roslin, EH25 9TD
M0 AJT C. Towle, 116 Stainton Drive, Grimsby, DN33 1JB
M0 AJX G. Jones, 7 Hardwick View, Skegby, Sutton-in-Ashfield, NG17 3BW
M0 AKD Dr G. Dublon, 25 Carr Lane, Sandal, Wakefield, WF2 6HJ
M0 AKE R. Johnson, 24 Balmoral Avenue, Stanford-le-Hope, SS17 7BD
M0 AKF M. Temblett, 42 Westward Road, Bristol, BS13 8DB
M0 AKI E. Woollen, 6 Back Lane, Kington Magna, Gillingham, SP8 5EL
M0 AKJ R. Hunt, 8 Spicer Close, Cullompton, EX15 1QD
M0 AKK S. Elden, 124 Larchcroft Road, Ipswich, IP1 6PQ
MM0 AKM J. Hood, 88/2 Craighouse Gardens, Edinburgh, EH10 5LW
M0 AKQ R. Gawan, 39 The Filberts, Fulwood, Preston, PR2 3YS
M0 AKR K. Daniels, 122 Furzehatt Road, Plymstock, Plymouth, PL9 9JT
M0 AKS R. Lusty, 483 Bacup Road, Rossendale, BB4 7JA
MI0 AKU Foyle and District ARC, c/o T. Campbell, 27 Silverbrook Park, Newbuildings, Londonderry, BT47 2RD
MM0 AKX ATC Scotland & N.Ireland Region ARC, c/o J. Ramsay, 150 City Road, Dundee, DD2 2PW
M0 AKY T. Money, 119 Twyford Way, Canford Heath, Poole, BH17 8SR
M0 AKZ R. Taylor, 46 Crescent Road, Netherton, Dudley, DY2 0NW
M0 ALB N. Hixson, Flat 35, Milward Court, Reading, RG2 7BG
M0 ALD J. Britten, 10 Broadgate Avenue, Horsforth, Leeds, LS18 5DT
M0 ALE P. Johnson, 91 Highlands Road, Andover, SP10 2PZ
M0 ALF R. Faithfull, 5 Hadleigh Road, Portsmouth, PO6 3RD
MW0 ALG D. Burge, Ucheldir, Maenygroes, New Quay, SA45 9TH
M0 ALH S. Case, 5 Haldon Grove, Birmingham, B31 4LN
M0 ALK R. Cook, 3 Mill Close, Hartford, Huntingdon, PE29 1YL
MM0 ALM D. Wood, West Raedykes, Rickarton, Stonehaven, AB39 3SY
MW0 ALN A. Lane, 14 Hertford Place, Newport, NP19 7SN
M0 ALO D. Hooper, 21 High Street, Great Linford, Milton Keynes, MK14 5AX
M0 ALQ G. Denby, 14 Talman Grove, Stanmore, HA7 4UQ
M0 ALR T. Knight, 117 Ennerdale Road, Cleator Moor, CA25 5LR
MI0 ALS E. Stanford, 33 Glenview Gardens, Belfast, BT5 7LY
M0 ALT I. Halliwell, 61 Cliffe Road, Shepley, Huddersfield, HD8 8AG
M0 ALX M. Feasey, 4 Abbeydale, Carlton Colville, Lowestoft, NR33 8WJ
MM0 ALY A. Brown, 4 Averon Park, Blackburn, Aberdeen, AB21 0LH
M0 ALZ T. Thompson, 23 Oaklands, Paulton, Bristol, BS39 7RP
M0 AMB B. Metcalfe, 5 Oakdale Avenue, Bradford, BD6 1RP
M0 AME D. Draper, 36 Highfield Gardens, Combe Martin, Ilfracombe, EX34 0HQ
M0 AMF R. Jefferies, 38 Towbury Close, Redditch, B98 7YZ
MW0 AMI R. Hall, 33 Heol Y Garreg Las, Llandeilo, SA19 6EB
MW0 AMJ L. Carter, 13 Maes Dolau, Idole, Carmarthen, SA32 8DQ
M0 AMM G. Smith, East Lodge, Woodlands Drive, Bradford, BD10 0NX
MW0 AMN G. Thomas, Stonehall Mill Farm, Wolfscastle, Haverfordwest, SA62 5NT
M0 AMP A. Davies, 27 Foxley Grove, Bicton Heath, Shrewsbury, SY3 5DF
MW0 AMQ G. Thomas, Stonehall Mill Farm, Wolfscastle, Haverfordwest, SA62 5NT
M0 AMS M. Burke, 8 Childwall Gardens, Ellesmere Port, CH66 1RL
MM0 AMV R. Moodie, 18 Bennecourt Drive, Coldstream, TD12 4BY
MM0 AMW D. Gillies, 10 Killeonan, Campbeltown, PA28 6PL
M0 AMX J. Howell, Orchard House, Blennerhasset, Wigton, CA7 3QX
M0 AMZ J. Williams, 61 Longfield Road, South Woodham Ferrers, Chelmsford, CM3 5JJ
M0 ANC R. Jones, 31 Main Street, Awsworth, Nottingham, NG16 2RH
M0 ANH J. Waller, 56 Daventry Road, Dunchurch, Rugby, CV22 6NS
M0 ANK S. Cotterill, 320 Hamstead Road, Great Barr, Birmingham, B43 5EH
M0 ANN G. Wardale, 25 The Crescent, Huyton, Liverpool, L36 6ER
M0 ANO R. Spencer, 19 Trafalgar Road, Cirencester, GL7 2EJ
M0 ANP N. Crooks, 3 Grove Court, Settle, BD24 9QR
M0 ANQ J. Chadwick, 6 Harper Fold Road, Radcliffe, M26 3RU
M0 ANS A. Rawlings, 57 High Street, Nash, Milton Keynes, MK17 0EP
M0 ANU G. Coolledge, 49A Enfield Avenue, New Waltham, Grimsby, DN36 4RB
MW0 ANV R. Davies, 4 Maes Derlwyn, Llanberis, Caernarfon, LL55 4TW
MW0 ANX J. Jensen, Pistyll Canol Farm, Llandeilo Road, Ammanford, SA18 2LQ
M0 AOA D. Young, 47 Horseshoe Crescent Pocklington, York, YO42 1UN
M0 AOB J. Allen, 149 Penistone Road, Waterloo, Huddersfield, HD5 8RP
M0 AOD D. Kay-Newman, Kay-Spray, Pottery Road, Ilminster, TA19 9QN
MM0 AOF D. Henry, 106 Whinhill Gate, Aberdeen, AB11 7WF
M0 AOG G. Dyson, 32 Farleigh Fields, Orton Wistow, Peterborough, PE2 6YB
M0 AOH J. Barber, 7 Thomas Street, Carlisle, CA2 5DZ
M0 AOI J. Russell, 46 Eastleigh Drive, Tingley, Wakefield, WF3 1PF
M0 AOJ A. Elliott, 26 Watery Lane, Minehead, TA24 5NZ
M0 AOK S. Millar, 4 Broomfield, Benfleet, SS7 2ST
MM0 AOL R. Bloomfield, 35 Shaw Street, Dunfermline, KY11 4AX
M0 AOM M. Goodrich, Urb Les Basetes B3, Adsubia, Alicante, Spain, 3786
MM0 AOQ C. Greig, 5 Mitchell Place, Stuartfield, Peterhead, AB42 5WE
M0 AOT M. Stanley, 16 Fenton, Keswick, CA12 4AZ
MM0 AOY D. Stephen, 16 The Square, Portlethen, Aberdeen, AB12 4QA
M0 AOZ M. Boothman, Ballysax, Curragh, Ireland, R56 PX47
M0 APC M. Brown, 6 Rose Court, Garforth, Leeds, LS25 1NS
M0 APD J. Udall, 4 Church Lane Chilcote, Swadlincote, DE12 8DL
MM0 APF Inverclyde Contest Group, c/o J. Fisher, High Birches, Culbokie, Dingwall, IV7 8JS
M0 APH A. Gilbert, 79A Station Road, Brimington, Chesterfield, S43 1LJ
M0 APK D. Allen, 162 Wood Lane, Newhall, Swadlincote, DE11 0LY
M0 APL B. Tucker, 2 Hundall Court, Grasscroft Close, Chesterfield, S40 4HN
M0 APN A. Nelson, 29 Coxford Road, Southampton, SO16 5FG
M0 APY R. Arey, 4 Iveson Lawn, Leeds, LS16 6NA
M0 APZ F. Piper, 6 Russell Street, Little Hulton, Manchester, M38 0LW
M0 AQA G. Shaw, 6 Bromstone Road, Broadstairs, CT10 2HA
M0 AQE E. Entwistle, 43 Brock Road, Chorley, PR6 0DB
M0 AQF T. Davies, 20 The Coppice, Impington, Cambridge, CB24 9PP
M0 AQH E. Blackburn, 2 Stockwell Drive, Knaresborough, HG5 0LW
MJ0 AQJ N. Jones, 1 Cornucopia Court, Le Mont Pinel, St Helier, Jersey, JE2 4RS
M0 AQK K. Hesketh, 13 Elm Road, St Helens, WA10 3NE
M0 AQO G. Willson, 40 Grace Gardens, Bishop's Stortford, CM23 3EX
M0 AQP A. Bellamy, 8 Dorothy Road, Kettering, NN16 0PH
M0 AQQ K. Evans, 5 Garswood Avenue, Rainford, St Helens, WA11 8JW
MW0 AQT E. Lucocq, 96 Carisbrooke Way, Cardiff, CF23 9HX
M0 AQW Processed Audio Group, c/o M. Storkey, 9 Waterman Court, Acomb, York, YO24 3FB
MI0 AQX J. May, 8 Oak Vale Avenue, Newry, BT34 2BQ
MW0 AQZ L. Griffiths, Tros Y Garreg Plas Road, Holyhead, LL65 2LU
M0 ARA J. Layton, 6 Granby Road, Cheadle Hulme, Cheadle, SK8 6LS
M0 ARC East Yorkshire Contest, c/o V. Lindgren, 143 Hull Road, Anlaby., Hull, HU106ST
MW0 ARD A. Davies, 19 Maes-Y-Dderwen, Dinas Cross, Newport, SA42 0XF
M0 ARH N. Ravilious, 17 Halls Green, Weston, Hitchin, SG4 7DR
M0 ARK B. Shepherd, 17 Huntock Place, Brighouse, HD6 2NW
MW0 ARL H. Davies, Garth Wen, 1, Bryn Siriol, Coedpoeth, Wrexham, LL11 3PZ
M0 ARO J. Parsons, 36 Gainsborough, Milborne Port, Sherborne, DT9 5BD
M0 ARQ J. Churchill, 59 Highfield, Letchworth Garden City, SG6 3PY
MW0 ARV M. Thomas, 12 School Road, Rhosllanerchrugog, Wrexham, LL14 1BB
M0 ARX M. Richardson, 39 Wilson Avenue, Deal, CT14 9NL
M0 ARY M. O'Rourke, Brookside Farm, Walpole, Halesworth, IP19 9BH
M0 ARZ S. Hurst, 25 Florence Road Abington, Northampton, NN1 4NA
MM0 ASB R. Barbour, 40 Mannerston Holdings, Linlithgow, EH49 7ND
M0 ASC A. Clayton, 7 Salisbury Avenue, Broadstairs, CT10 2DT
M0 ASD A. Gallichan, 4 Wigston Road, Rugby, CV21 4LT
M0 ASG C. Nehmzow, 26 Woodlands, Colchester, CO4 3JA
M0 ASI N. Johns, 85 South Hill, Hooe, Plymouth, PL9 9PT
M0 ASJ S. Griggs, 16 Sharleston Drive Stainforth, Doncaster, DN7 5PU
MW0 ASL J. Phillips, 57 Ffordd Llanerch, Penycae, Wrexham, LL14 2ND
M0 ASN J. Marron, 190 Cotswold Crescent, Billingham, TS23 2QH
M0 ASO P. Bluthner, 21 Gambier Perry Gardens, Gloucester, GL2 9RD
M0 ASR D. Campanario, 3 Foxearth Hall, Leek Road, Stoke on Trent, ST9 0DG
M0 ASU K. Hamer, Flat 7, Red Court, 66 Upper Park Road, Salford, M7 4JA
MI0 ASV G. Best, 1 Bensons Road, Lisburn, BT28 3QX
M0 ASY S. Werner, 4225 Place Sainte-Helene, Laval, Canada, H7W 1P3
M0 ATA A. Rundle, 9 Windsor Terrace, East Herrington, Sunderland, SR3 3SF
M0 ATB R. Hutton, 57 Sandy Lane, Upton, Poole, BH16 5EJ
M0 ATC Dorset & Wilts Wing ARC, c/o R. Ley, 23 Heronbridge Close, Westlea, Swindon, SN5 7DR
M0 ATD A. Holzapfel, Flat 1-4, 4 Tanner Street, London, SE1 3LD
MW0 ATG H. Thomas, 15 Coronation Terrace, Pontypridd, CF37 4DP
MW0 ATI G. Roberts, Glanafon, Drefach, Llanybydder, SA40 9YB
MW0 ATK S. Brewer, 16 Oxwich Close, Cefn Hengoed, Hengoed, CF82 7JB
M0 ATL P. Nash, 110 Cranborne Road, Potters Bar, EN6 3AJ
M0 ATQ J. Torry, 41 Nevill Road, Rottingdean, Brighton, BN2 7HH
MW0 ATR D. Williams, 17 Brynawelon, Llanelli, SA14 8PU
M0 ATS J. Ammundsen, 62 Linden Avenue, Broadstairs, CT10 1HR
MW0 ATT E. Cooke, Anelog, Rhewl Fawr Road, Holywell, CH8 9HJ
M0 ATV A. Reilly, 19 The Ridgway, Romiley, Stockport, SK6 3EE
M0 ATX E. Williams, Dyffryn Coed, Union Road, Coleford, GL16 7QB
M0 ATY C. Kirkland, 9 Holland Way, Newport Pagnell, MK16 0LL
M0 ATZ C. Hardy, 56 Vyner Road, Wallasey, CH45 6TF
M0 AUA D. Hazel, 7 Arley Close, Upton, Chester, CH2 1NW
M0 AUF N. Handforth, 26 Appleby Crescent, Mobberley, WA16 7GB
M0 AUG G. Ashton, 95 Willow Drive, Lamaleach Park, Lamaleach Drive, Freckleton, Preston, PR4 1DF
M0 AUK J. Sporton, 199 Glaisdale Drive West, Nottingham, NG8 4GY
M0 AUR A. Taylor, 18 Chestnut Road, Glemsford, Sudbury, CO10 7PS
M0 AUS A. Dowie, 12 Malvern Drive, Gonerby Hill Foot, Grantham, NG31 8GA
M0 AUT D. Randles, 12 Wain Court. Rakeway, Saughall. Chester, CH16BF
M0 AUW R. Hull, 1 Northfield Cottage, Withington Road, Cheltenham, GL54 4LL
M0 AUY L. Jeffries, 37 Woodfield Road, Bournemouth, BH11 9EU
M0 AVA D. Salsbury, 1 Somerset Avenue, Tyldesley, Manchester, M29 8LQ

M0 AVF A. De Araujo, 13 Fifth Avenue Shaws Trailer Park, Knaresborough Road, Harrogate, HG2 7NJ
M0 AVH D. Eaton-Watts, 129 Blake Road, West Bridgford, Nottingham, NG2 5LA
MI0 AVI Newry High School Radio Club, c/o G. Millar, 1 Mullybrannon Road, Dungannon, BT71 7ER
M0 AVK D. Swift, 8 Grove Lane, Buxton, SK17 9HG
M0 AVL D. Meakin, 47 Sheridan Street, Walsall, WS2 9QX
M0 AVN A. Oatey, Robin Hill, Blackpost Lane, Totnes, TQ9 5RF
M0 AVP A. Baughan, Camino Tigalate No. 31-33, Villa De Mazo, St Cruz de Tenerife, Spain, 38730
M0 AVQ J. Worthington, 23 Sefton Avenue, Congleton, CW12 3DB
M0 AVS V. Saundercock, 14 Rashleigh Avenue, Plymouth, PL7 4DA
M0 AVU M. Scott, 36 Glebe Crescent, Newcastle upon Tyne, NE12 7JR
M0 AVW C. Spence, 32 Woodford Walk, Thornaby, Stockton-on-Tees, TS17 0LT
M0 AVY A. Johnson, 131 Rylands Road, Kennington, Ashford, TN24 9LU
M0 AVZ D. Clutterbuck, 2 Spring Valley Drive, Leeds, LS13 4RN
M0 AWB A. Boom, Oakthorpe House, 8A Peterborough Road, Peterborough, PE6 0BA
M0 AWD M. Mansfield, Piso 4 (Izq), Avda Jaime I - 14, Altea (Alicante), Spain, 3590
M0 AWE A. Ellis, 50 Taylors Crescent, Cranleigh, GU6 7EN
M0 AWH P. Bush, 144 Stoke Lane, Westbury-On-Trym, Bristol, BS9 3RN
M0 AWI J. Ross, 13 Wensleydale Crescent, Oakridge Park, Milton Keynes, MK14 6GX
MM0 AWJ K. Gray, 18 Greenmantle Place, Glenrothes, KY6 3QQ
MI0 AWL A. Smith, 12 Sandringham Heights, Carrickfergus, BT38 9EG
M0 AWN C. Gladman, 24 Priory Road, Chessington, KT9 1EF
MW0 AWO S. Jones, 6 Heol Will Hopkin, Llangynwyd, Maesteg, CF34 9ST
M0 AWP P. Oliver, 21 Charlotte Close, Mount Hawke, Truro, TR4 8TS
MM0 AWU G. Moffat, 16/1 Laichpark Loan, Edinburgh, EH14 1UH
M0 AWX G. Schoof, 4-5 Canal Row, Haigh, Wigan, WN2 1NA
M0 AWY D. Ashdown, Cartwheels, 4 Honeysuckle Close, Hailsham, BN27 3TP
MW0 AXA W. Townsend, 133 Hazeldene Avenue, Brackla, Bridgend, CF31 2JR
M0 AXC D. Russell, 8 Norburton, Burton Bradstock, Bridport, DT6 4QL
M0 AXE D. Marshall, 34 Brentwood Road, Sheffield, S11 9BU
M0 AXG K. Wheeler, 26 Melverton Avenue, Wolverhampton, WV10 9HN
M0 AXJ A. Clay, 22 Park Street, Wallasey, CH44 1AT
M0 AXL J. Cook, 36 Kotuku Street, Coffs Harbour, Australia, NSW 2450
M0 AXN J. Davis, 60 West Bar Street, Banbury, OX16 9RZ
M0 AXO G. Morris, 7 Rowley View, Bilston, WV14 8DE
MM0 AXR T. Rees, 23 Doune Road, Dunblane, FK15 9AT
M0 AXV M. Amplett, 44A Darby Road, Coalbrookdale, Telford, TF8 7EW
M0 AXW J. Davies, 243A Bradford Road, Winsley, Bradford-on-Avon, BA15 2HL
M0 AXX E. Moody, The Apiaries, Rufford Lane, Newark, NG22 9DG
M0 AXZ P. Morgan, 20 Bishops Way, Buckden, St Neots, PE19 5TZ
M0 AYA S. Sellman, Inselhof, Banbury Road Gaydon, Warwick, CV35 0HH
M0 AYB T. Davies, 95 High Brigham, Brigham, Cockermouth, CA13 0TJ
M0 AYC J. Soakell, 162 Manor Road, New Milton, BH25 5ED
MM0 AYE J. Welsh, 7 South Cathkin Cottage, Rutherglen, Glasgow, G73 5RG
M0 AYF D. Kostryca, 12 High Street, Upton, Gainsborough, DN21 5NL
M0 AYG M. Wood, Weavers, Kingsdale Road, Berkhamsted, HP4 3BS
M0 AYI G. Waring, 7 Tynedale Terrace, Stanley, DH9 7TZ
M0 AYO H. Parker, 21 Mayfield Street, Hull, HU3 1NS
M0 AYS C. Pocock, 4 Broadfields, Harpenden, AL5 2HJ
M0 AYU I. Gibson, 7 Peverells Wood Close, Chandler'S Ford, Eastleigh, SO53 2FY
M0 AYV A. Eyles, 25 Chatham Road, Winchester, SO22 4EE
M0 AYX A. King, 28 King Street, Cirencester, GL7 1JT
M0 AYY R. Corfield, 35 Taplings Road, Winchester, SO22 6HE
M0 AZB R. Goddard, 6 Upper Ley Dell, Chapeltown, Sheffield, S35 1AL
M0 AZC P. Niel, 19 Fountains Close, Whitby, YO21 1JS
M0 AZE M. Surplice, 43A Cremorne Road, Sutton Coldfield, B75 5AQ
M0 AZG J. Kisiel, Wayside Cottage, South Stoke Road, Reading, RG8 0PL
M0 AZJ G. Gould, 32 Archer Road, Kenilworth, CV8 1DJ
M0 AZK D. Hill, 35 Bridle Lane, Sutton Coldfield, B74 3QE
MW0 AZN R. Cullis, 20 Larch Close, New Inn, Pontypool, NP4 0RT
M0 AZP P. Lemasonry, 7 Eastwood Road, Sittingbourne, ME10 2LZ
M0 AZR S. Gale, 39 Thorley Park Road, Bishop's Stortford, CM23 3NG
M0 AZS R. Buckle, 25 Portsmouth Close, Rochester, ME2 2QY
M0 AZT M. Thomas, 36 Seaview Avenue, Peacehaven, BN10 8SA
M0 AZV N. Devine, 46 Tytton Lane West, Wyberton, Boston, PE21 7HL
M0 AZW K. Coe, 5 George St., Enderby, Leicester, LE19 4NQ
M0 AZY F. Willis, 99 Kenilworth Court, Coventry, CV3 6JB
M0 AZZ A. Bond, 6 Meadway, Knebworth, SG3 6DN
MW0 BAA Blacksheep Contest + DX Group, c/o S. Purser, Penbrey, Llanfair Caereinion, Welshpool, SY21 0DG
MM0 BAC C. Mackay, 27 Barleyknowe Terrace, Gorebridge, EH23 4EQ
M0 BAD T. Brusel De La Torre, Flat 5, Douglas Hall, 3 Victoria Street, Preston, PR1 7QR
M0 BAE A. Radford, 25 Priory Avenue, Kirkby-In-Ashfield, Nottingham, NG17 9BU
MM0 BAG G. Craig, 9 Green Drive, Inverness, IV2 4EX
M0 BAH A. Tyler, 15 Chanctonbury Close, Washington, Pulborough, RH20 4AR
M0 BAI P. Byrne, 9 Glenluce Road, Liverpool, L19 9BX
M0 BAJ D. Nelson, 46 The Cunnery, Kirk Langley, Ashbourne, DE6 4LP
M0 BAK K. Williams, Cranesbie, 6 Dore Road, Sheffield, S17 3NB
M0 BAL F. Johnson, 7 Pharos Court, Pharos St., Fleetwood, FY7 6BG
M0 BAM A. Tobin, 17 Brockhampton, Cheltenham, GL54 5XH
M0 BAO A. Edwards, 75 Combe Park, Yeovil, BA21 3BE
M0 BAP W. Stewart, Hillfield Bungalow, Grunt Lane, Stroud, GL6 6PH
M0 BAR B. Bartley, 5 Cookes Wood Broompark, Durham, DH7 7RL
MI0 BAT S. Gilmore, 8 Fortfield, Dromore, BT25 1DD
M0 BAU G. Hoyle, 39 Randle Meadow, Great Sutton, Ellesmere Port, CH66 2BG
M0 BAV L. Evans, 16 Kynaston Drive, Wem, Wem, SY4 5DE
M0 BAW D. Rose, 31 Mount Crescent, Warley, Brentwood, CM14 5DB
M0 BAY G. Bilson, Fieldgate, 55 Littlemoor Lane, Alfreton, DE55 5TY
M0 BAZ J. Waterfield, 287 Turves Green, Birmingham, B31 4BS
M0 BB Burnham Beeches Radio Club, c/o U. Grunewald, Nuptown Orchard, Nuptown, Warfield, Bracknell, RG42 6HU
M0 BBE J. Hayward, 76 Lincoln Road, Skegness, PE25 2EE
MI0 BBF D. Doherty, 175 Bridge Road, Glarryford, Ballymena, BT44 9QA
M0 BBH M. Redman, 19 Richmond Road, Rugby, CV21 3AB
M0 BBK J. Meakin, White House Farm, Osmotherley, Northallerton, DL6 3QA
MW0 BBL A. Hadden, 164 Derwen Fawr Road, Sketty, Swansea, SA2 8DP
MW0 BBM B. Meredith, 27 Hyde Place, Llanhilleth, Abertillery, NP13 2RT
M0 BBO S. Woodford, 31 Seaborough View, Crewkerne, TA18 8JB
M0 BBQ K. Taylor, 241 Daventry Road, Cheylesmore, Coventry, CV3 5HH
M0 BBT T. Pirrie, Walnut Thatch, Tysoe Road, Warwick, CV35 0UE
MW0 BBU S. Lloyd, 41 Coombs Drive, Milford Haven, SA73 2NU
M0 BBV R. Lyford, 4 Wrentham Estate, Old Tiverton Road, Exeter, EX4 6ND
M0 BBW S. Gleadall, 59 Old Chapel Road, Warley, Smethwick, B67 6HU
M0 BCC R. Clapp, 11 Kensington Gardens, Ilkeston, DE7 5NZ
M0 BCE Dr W. Johnstone, 67 Station Lane, Birkenshaw, Bradford, BD11 2JE
M0 BCF R. Cranwell, 7 Central Drive, Elston, Newark, NG23 5NT
M0 BCG I. Williams, Alma Cottage, Old Vicarage Lane, South Marston, Swindon, SN3 4SN
M0 BCH C. Chadburn, 31 Darwin Close, Top Valley, Nottingham, NG5 9LN
M0 BCI N. Armstrong, 112 Chandos Street, Netherfield, Nottingham, NG4 2LW
M0 BCJ G. Lewis, 42 Ladywood Road, Ilkeston, DE7 4NE
M0 BCK K. Bell, 71 Wheatfield Road, Stanway, Colchester, CO3 0YA
M0 BCL P. Williams, 37 Winyards View, Crewkerne, TA18 8JA
M0 BCN Maj. D. James, 54 Woolacombe Lodge Road, Birmingham, B29 6PX
M0 BCQ Craven Radio Amateur Group (Aka Crag), c/o G. Hanson, 11 Churchill Way, Cross Hills, Keighley, BD20 7DN
MM0 BCR L. Haynes, 29 Invercauld Road, Aberdeen, AB16 5RP
M0 BCT M. Danfer, The Nook, Mill Common, Blaxhall, Woodbridge, IP12 2ED
M0 BCV S. Graham, 4 Oakland Avenue, Ellenborough, Maryport, CA15 7BU
M0 BCW P. Mason, 15 Granton Avenue, Clifton, Nottingham, NG11 9AL
M0 BCZ R. Burton, 1 Avenue Court, Mount Avenue, London, W5 1PY
MM0 BDA Dr R. August, Smiddyhill House, Stracathro, Brechin, DD9 7QE
M0 BDB R. Taylor, 86-88 Hillside Crescent, Leigh-on-Sea, SS9 1HQ
M0 BDD D. Webster, 210 Walesby Lane, New Ollerton, Newark, NG22 9UU
M0 BDE B. Thorburn, 3 Victoria Road, Bexhill-on-Sea, TN39 3PD
M0 BDF J. Reid, Rosebury, Soldridge Road, Alton, GU34 5JF
M0 BDH P. Fisher, 21 Charlotte Close, Mount Hawke, Truro, TR4 8TS
MM0 BDJ R. Hawkins, 7 Ola Drive, Scrabster, Thurso, KW14 7UE
M0 BDL D. Ferris, 167 Lonsdale Avenue, Doncaster, DN2 6HF
M0 BDQ K. Kisselev, 37 Stanley Av., Barking, Essex, IG11 0LD
M0 BDS Dr G. Butler, 11 Keppel Drive, Bridlington, YO16 6ZD
M0 BDU D. Goodwin, 15 Tennyson Road, Bentley, Doncaster, DN5 0EG
M0 BDW P. Hayes, 1 Stile Plantation, Royston, SG8 9HP
MI0 BDX A. Patterson, 33 Marlborough Park, Carryduff, Belfast, BT8 8NL
MI0 BDZ M. Chancellor, 55 Brae Hill Park, Belfast, BT14 8FP
MW0 BEA C. Lewis, 60 Caeau Gleision, Rhiwlas, Bangor, LL57 4UA
M0 BEC R. Millerchip, 16 Kennedy Crescent, Gosport, PO12 2NN
MM0 BED J. Macdonald, 22 New Parliament Place, Campbeltown, PA28 6GY
M0 BEE Whitehaven ARC (T.S.Bee), c/o N. Williams, 1 Dorset Close, Whitehaven, CA28 8JP
M0 BEH P. Mutter, 129 Demesne Road, Wallington, SM6 8EW
M0 BEJ G. Moody, 25 Norbiton Common Road, Kingston upon Thames, KT1 3QB
M0 BEK C. Dunn, 75 Waddington Avenue, Burnley, BB10 4LA
MW0 BEL A. Owen, Rafael Fawr House, The Fraich, Fishguard, SA65 9QJ
M0 BEM M. Taperell, 16 Parkhall Croft, Birmingham, B34 7BU
M0 BEO K. Anderson, 53 Priory Grove, Hull, HU4 6LU
M0 BEQ H. Walsh, 38 Potter Hill, Greasbrough, Rotherham, S61 4PA
MW0 BER D. Jones, Hafan Mynydd Bodafon, Llanerchymedd, LL71 8BG
MI0 BES J. May, 8 Oak Vale Avenue, Newry, BT34 2BQ
MW0 BET V. Hughes, Manley, 1 Garden Drive, Llandudno, LL30 3LL
M0 BEV R. Knapp, 85 Eastern Avenue, Liskeard, PL14 3TD
M0 BEX J. Hrycan, 40 Marina Drive, Marple, Stockport, SK6 6JL
M0 BFA D. Wilson, 30 Little Avenue, Swindon, SN2 1NL
M0 BFB K. Francks, 63 Parc Godrevy, Newquay, TR7 1TY
M0 BFM S. Jones, 69 Colville Street, Liverpool, L15 4JX
M0 BFT S. Smith, 225 South Drove, Lutton Marsh, Spalding, PE12 9NT
M0 BFV L. Papazoglou, 37A Eaton Road, West Kirby, Wirral, CH48 3HE
M0 BGE T. Parker, 24 Burrows Close, Lawford, Manningtree, CO11 2HE
MM0 BGO R. Herd, 4 Smithy Lane, Balmullo, St Andrews, KY16 0FG
M0 BGR H. Howard, 10 Lawnside, London, SE3 9HL
M0 BGS G. Steedman, 5 Allerton Grange Gardens, Leeds, LS17 6LL
M0 BGT G. Dyson-Bawley, Grahil, 33 Ridgetor Road, Liverpool, L25 6DG
M0 BGU J. Moran, 8 Doffcocker Lane, Bolton., BL1 5RG
MM0 BGW A. Munro, 2 Woodlands View, Inshes Wood, Inverness, IV2 5AQ
M0 BHA C. Baker, 1 Astley Green Darleyhall, Luton, LU2 8TS
M0 BHE M. Sadler, Hill View, Horton, Ilminster, TA19 9QU
M0 BHH D. Newing, 13 Maxwell Road, Broadstone, BH18 9JG
M0 BHJ P. Worlledge, 181 Roselands Drive, Paignton, TQ4 7RN
M0 BHK G. Robertson, 22 Carlton Villas, Hatt, Saltash, PL12 6PS
M0 BHM A. Whitehouse, 19 Cleeve Road, Marlcliff, Alcester, B50 4NX
M0 BHN P. Jarvis, 26 Nally Drive, Woodcross, Bilston, WV14 9UT
M0 BHO S. Coe, 113 Highfield Road, Yeovil, BA21 4RJ
M0 BHP G. King, 12 Bracken Close, Blackburn, BB2 5AH
M0 BHQ F. Lugg, 4 Newbury Close, Walsall, WS6 6DF
M0 BHR G. King, 8 Oak Lane, Burghill, Hereford, HR4 7QP
M0 BHV J. Cocks, 12 Birch Pond Road, Plymouth, PL9 7PG
M0 BHW G. Jeckells, Hogals End, Mill St., Thetford, IP25 7QN
MI0 BHX T. Costford, 19 Cornacully Road, Meenarainy, Belcoo, Enniskillen, BT93 5BR
M0 BIC J. Brocklebank, Springfield House, Sixhills Lane, Market Rasen, LN8 6AN
M0 BIH R. Deakin, 40 Brussels Road, Stockport, SK3 9QG
M0 BII S. Keightley, 11 Sandringham Avenue, Wisbech, PE13 3ED
M0 BIJ C. Harden, 19 Nutshalling Avenue, Rownhams, Southampton, SO16 8AY
M0 BIT P. Smith, Karinya, Rectory Road Haddiscoe, Norwich, NR14 6PG
MM0 BIX E. Cameron, 5 King Street Ferryden, Montrose, DD10 9RR
M0 BIZ M. Thomas, 39 Treworder Road, Truro, TR1 2JZ
M0 BJD B. Duffy, 27 Kinloch Close Halewood, Liverpool, L26 9XZ
M0 BJE A. Cockram, 70 Arlington Drive, Marston, Oxford, OX3 0SJ
M0 BJJ S. Miyake, Hakata Radio, P.O.Box 232, Hakata-North, Japan, 812-8799
M0 BJK G. Scothorn, School House, Kirk Balk, Barnsley, S74 9HU
M0 BJL S. Jarvis, Kellow, Old Lyndhurst Road, Southampton, SO40 2NL
MD0 BJM M. Rodgers, 1 Kings Court, Ramsey, Isle of Man, IM8 1LJ
M0 BJN F. Humphris, 169 Bloxham Road, Banbury, OX16 9JU

M0 BJO D. Greenway, 37 Primrose Hill Park Homes, Primrose Hill, Somerton, TA11 7AP
M0 BJP R. Pearce, Kolner, 86 Thrupp Lane, Stroud, GL5 2DG
M0 BJR M. Brown, 3 The Vines, Kelsale, Saxmundham, IP17 2PU
M0 BJS P. Hall, 25 Ham Green, Pill, Bristol, BS20 0EY
M0 BJT K. Davison, 2 Sitwell Close, Spondon, Derby, DE21 7GT
MJ0 BJU A. Mourant, Little Mead, Claremont Road, St Saviour, Jersey, JE2 7RT
M0 BJX R. Glover, 89 Cambridge Road, Linthorpe, Middlesbrough, TS55LD
M0 BKA J. Slough, 2554 Hamilton Rd., Lebanon Oh, USA, 45036
M0 BKD P. Mccormack, 3 Greenway Close, Torquay, TQ2 8EF
M0 BKF C. Brodrick, 6 Carlton St., Hartlepool, TS26 9ES
M0 BKG G. Rundle, 15 Sandown Road, Paignton, TQ4 7RL
M0 BKJ P. Mulliner, 23 Churchill Way, Burton Latimer, Kettering, NN15 5RX
M0 BKK Dr J. Rowe, The Old Rectory, Wickenby, Lincoln, LN3 5AB
M0 BKL S. Passmore, 35 David Road, Paignton, TQ3 2QF
M0 BKN S. Sherwin, 1 Nursery Close, Wroughton, Swindon, SN4 9DR
M0 BKS K. Sim, 3 Thorngate Close Penwortham, Preston, PR1 0XN
M0 BKV D. Kamm, Delabole Head, Week St. Mary, Holsworthy, EX22 6UU
M0 BLD S. Tsuzuki, Flat 4, Lancing House, Watford, WD24 4RL
M0 BLF D. Smith, 67 Lambs Lane, Cottenham, Cambridge, CB24 8TB
M0 BLH S. Laddiman, 27 Morgans Way Hevingham, Norwich, NR10 5PD
M0 BLI J. Crangle, 43 Scarfell Close, Peterlee, SR8 5PF
M0 BLN K. Hopps, 100 Etherley Lane, Bishop Auckland, DL14 6TU
M0 BLO R. Jackson, 4 Hornbrook Gardens, Plymouth, PL6 6LS
M0 BLR A. Cresswell, 31 New Street, Doddington, March, PE15 0SP
M0 BLS G. Hickford, Sanclare, 94 Manor Road, Fleetwood, FY7 7HY
M0 BLT M. Waldron, 32 Windmill Street, Upper Gornal, Dudley, DY3 2DQ
MW0 BLU W. Jepson, 3 Marchog, Holyhead, LL65 2HD
M0 BLV G. Peacock, 7 Pensclose, Witney, OX28 2EG
M0 BLY S. Young, 126 Stevens Road, Dagenham, RM8 2QL
MM0 BLZ A. Blackburn, 6 Dunadd View, Kilmichael Glassary, Lochgilphead, PA31 8QA
MM0 BMA W. Irwin, 15 Inchcolm Place, East Kilbride, Glasgow, G74 1DR
M0 BMB B. Bentham, 89 Westborough Way, Hull, HU4 7SW
M0 BMD J. Green, 788 The Ridge, St Leonards-on-Sea, TN37 7PS
MI0 BME P. Maile, 3 Cairnmore Avenue, Lisburn, BT28 2DW
M0 BMF D. Anger, 17 Dell Road, Andover, SP10 3JT
MM0 BMG N. Stewart, 160 Carrick Knowe Drive, Edinburgh, EH12 7EW
M0 BMJ I. Singer, 197 Rosalind Street, Ashington, NE63 9BB
MI0 BML P. Doris, 92 Coolnafranky Park, Cookstown, BT80 8PW
MI0 BMM B. Moore, 8 Orken Lane, Aghalee, Craigavon, BT67 0ED
M0 BMN P. Webb, 41 Lancaster Gardens, Wolverhampton, WV4 4DN
M0 BMR P. Pirrazzo, 30 Coronation Road, Middlewich, CW10 0DL
M0 BMT D. Bunting, 6 Mill Gardens, Worksop, S80 3QG
M0 BMU J. Moritz, Carillon, 6 Bell Lane, Hatfield, AL9 7AY
M0 BMW K. Wrack, 18 Carrs Road, Cheadle, SK8 2EE
M0 BMX M. Fitchett, Barkenroy, Ludgvan, Penzance, TR20 8AJ
M0 BMY L. Bilson, Fieldgate, 55 Littlemoor Lane, Alfreton, DE55 5TY
M0 BMZ A. Martin, 11 The Mount, Worcester Park, KT4 8UD
MW0 BNB T. Rogers, The Willows, 48 Hillock Lane, Wrexham, LL12 8YL
M0 BNC N. Clark, 28 Thickthorn Close, Kenilworth, CV8 2AF
M0 BND A. Sherman, 45 Norfolk Road, Weymouth, DT4 0PW
M0 BNF I. Copping, 54 Hartley Road, Kirkby-In-Ashfield, Nottingham, NG17 8DP
M0 BNO G. Ryder, 11 Claremont Gardens, Farsley, Pudsey, LS28 5BF
M0 BNP J. Taylor, 3 Inhams Close, Murrow, Wisbech, PE13 4HS
M0 BNR N. Rodley, 268 Grovehill Road, Beverley, HU17 0HP
M0 BNS British Naturist ARS, c/o C. Beesley-Reynolds, Kaos Roams, Palmerston Close, Leicester, LE8 0JJ
M0 BNZ D. Brooks, The Elms, Trewoon Road, Helston, TR12 7DS
M0 BOB R. Adlington, 7 Kynance Close, Romford, RM3 7LB
M0 BOC G. Tomlinson, 96 Highfield Drive Farnworth, Bolton, BL4 0RN
M0 BOH S. Saiger, 10 Markham Avenue, Armthorpe, Doncaster, DN3 2AZ
M0 BOI B. Johnson, 10 Saffron Road, Tickhill, Doncaster, DN11 9PW
MI0 BOK P. Connolly, 94 North Parade, Belfast, BT7 2GJ
M0 BOL R. Rose-Round, 2 Lee Road, Blackpool, FY4 4QS
M0 BOM R. Wilson, 22 Leadhills Way, Bransholme, Hull, HU7 4ZA
M0 BOQ R. Slater, 79A Ainsworth Road, Radcliffe, Manchester, M26 4FA
MI0 BOU J. Orr, 17 Argyll View, Larne, BT40 2JR
M0 BOX S. Wilson, 21 Plumian Way, Balsham, CB21 4EG
M0 BOY L. Sanduly, 1 Archford Croft, Emerson Valley, Milton Keynes, MK4 2EZ
MI0 BPB A. Mulholland, 83 Tullyrain Road, Donaghcloney, Craigavon, BT66 7PP
M0 BPC K. Hunt, 4 Oak Avenue, Willington, Crook, DL15 0BJ
MM0 BPF R. Armstrong, 98 Burnbank Road, Ayr, KA7 3QJ
M0 BPM D. Barclay, 13 Avondale Terrace, Chester le Street, DH3 3ED
M0 BPN N. Taplin, 149 Frindsbury Road, Strood, Rochester, ME2 4JD
MM0 BPP K. Lindsay, 34 Park Crescent, Newtown St. Boswells, Melrose, TD6 0QS
M0 BPQ Dr S. Bunting, 17 Sunnydene Avenue, Highams Park, London, E4 9RE
M0 BPS D. Lawrence, 42 Upper Packington Road, Ashby-de-la-Zouch, LE65 1UL
M0 BPT R. Walker, Fists The International Morse Preservation Society, Po Box 6743, Tipton, DY4 4AU
M0 BPU T. Lyne, The Ark, 10 Blackfields Avenue, Bexhill-on-Sea, TN39 4JL
MM0 BPV A. Finlayson, 10B Flesherin, Isle of Lewis, HS2 0HE
M0 BPW A. Shelswell, Waterway, Dock Lane, Melton, Woodbridge, IP12 1PE
MM0 BPX Dr K. Scott, Kirklands, Craigend Road, Galashiels, TD1 2RJ
M0 BPY P. Henson, 35 Westbrook Drive, Rainworth, Mansfield, NG21 0FB
M0 BQB T. Lee, 91 Old Vicarage Park, Narborough, King's Lynn, PE32 1TG
M0 BQC D. Bradley, 1 Traddles Court, Chelmsford, CM1 4XZ
M0 BQD E. Lee, 16 Phoenix Chase, North Shields, NE29 8SS
M0 BQE C. Margetts, 16 Lahn Drive, Droitwich, WR9 8TQ
M0 BQF M. Preece, 51 Drancy Avenue, Willenhall, WV12 5RD
M0 BQH K. Goodacre, 22 New Ferry Road, Wirral, CH62 1BJ
MM0 BQI J. Martin, 3 Lismore Avenue, Edinburgh, EH8 7DW
MM0 BQJ T. Cassidy, 44 Wellpark Road, Saltcoats, ka21 5lj
MM0 BQL A. Cromack, 10 Manse Road, Ardersier, Inverness, IV2 7SR
M0 BQO U. Rose, 45 Ringstead Crescent, Weymouth, DT3 6PT
M0 BQT S. Smith, 55 Market Street, Ilkeston, DE7 5RB
M0 BQZ J. Romanis, 23 Old Farm Lane, Stubbington, Fareham, PO14 2BZ
M0 BRA Brackwell Radio Amateur Contest Group, c/o G. Leonard, 65 Qualitas, Bracknell, RG12 7QG
M0 BRE P. Mann, 38 Green Lane, Halesowen, B629LP
MM0 BRG R. Harman, 2 Dornoch Court, Kilwinning, KA13 6QN
M0 BRH T. Linham, 44 Vestry Road, Street, BA16 0HX
M0 BRI P. Walker, 20 Arbury Banks, Chipping Warden, Banbury, OX17 1LU
M0 BRL T. Wells, Flat Above Londis, 150 High Street, Boston Spa, LS23 6BW
M0 BRM E. Turner, 16 The Rowans, Doddington, March, PE15 0SE
MW0 BRO M. Lewis, 8 Rose Close, Pembroke, SA71 4TR
M0 BRP M. Wastie, 5 Pensclose, Witney, OX28 2EG
M0 BRU W. Griffin, 48 Wardle Way, Kidderminster, DY11 5UJ
M0 BSB W. Carr, 37 Keats Road, Greenmount, Bury, BL8 4EP
M0 BSC P. Bonsey, Wood View, Lower Kelly, Calstock, PL18 9RY
M0 BSD B. Daly, 10 Parfitts Close, Farnham, GU9 7DH
M0 BSF N. Rigazzi-Tarling, 3 The Planes, Bridge Road, Chertsey, KT16 8LE
M0 BSH W. Kerslake, 42 Silverdale Road, Newcastle, ST5 2TB
M0 BSI L. Preston, Hillside, Trewennack, Helston, TR13 0PQ
M0 BSJ P. Poore, 42 Kibblewhite Crescent, Twyford, Reading, RG10 9AX
M0 BSL M. Pell, 7 Churchfleet Lane, Gosberton, Spalding, PE11 4NE
MM0 BSM S. Mcquillian, 68 Strathmore Drive, Cornton, Stirling, FK9 5BE
M0 BSP E. Doyle, 33 Bodenham Road, Birmingham, B31 5DP
M0 BSQ J. Doyle, Orchard Corner, Piddington Road, Ludgershall, Aylesbury, Hp189pj
MI0 BSU D. Stanley, 59 Gransha Road, Kircubbin, Newtownards, BT22 1AJ
M0 BSV A. Ilett, 34 Westbrook Park Road, Woodston, Peterborough, PE2 9JG
M0 BSW P. Collins, 10 Robinswood, Wansford, Peterborough, PE8 6JQ
MM0 BSX G. Scattergood, 14 Market Street, Forfar, DD8 3EY
M0 BSZ A. Wanford, 4 Willows Close, Tydd St. Mary, Wisbech, PE13 5QR
MM0 BTD J. Ganson, 1 Beechwood Terrace West, St. Fort, Newport-on-Tay, DD6 8JH
M0 BTG G. Barrett, 114 William Street, Long Eaton, Nottingham, NG10 4GD
MW0 BTI E. Thomas, Craig-Y-Don, Conwy, LL27 0JJ
M0 BTJ J. Ingram, 58 Belben Road, Poole, BH12 4PJ
MI0 BTK A. Cobb, 62 Katesbridge Road, Dromara, Dromore, BT25 2PN
M0 BTL A. Withers, 5 Tintern Road, Skelton-In-Cleveland, Saltburn-by-the-Sea, TS12 2YN
MI0 BTM G. Duffy, Carrageen Cottage, Enniskillen, BT74 6ET
M0 BTN J. Escreet, Colfield, Carlton Lane, Hull, HU11 4RA
M0 BTO S. Collins, 69 Walkers Heath Road, Kings Norton, Birmingham, B38 0AL
M0 BTP E. Edmunds, 5 Nelsons Quay, St. Helens, Ryde, PO33 1TA
M0 BTR J. Springett, 31 Mountbatten Court, Andover Road, Winchester, SO22 6BA
MW0 BTU G. Rowlands, Dalar Wen, Rhosmeirch, Llangefni, LL77 7SJ
M0 BTX R. Elliott, 8 Bridge Place, Amersham, HP6 6JF
M0 BTZ R. Harrison, 55 Stapleford Close, Romsey, SO51 7HU
M0 BUA R. Miller, 1 Mews Cottages, Penview Crescent, Helston, TR13 8RX
M0 BUC P. Buck, 6 Bedford Road 2 Bracebridge House, Sutton Coldfield, B756AA
M0 BUE J. Page, Cherry View, 28 Ellerslie Lane, Bexhill-on-Sea, TN39 4LJ
M0 BUF E. Smith, Lamorna, Broadmead Road, Woking, GU23 7AD
M0 BUG N. Jamieson, 1 Langdale Place, Newton Aycliffe, DL5 7DX
MM0 BUH Capt. N. Smith, Nether Dallachy Farmhouse, Boyndie, Banff, AB45 2JT
M0 BUI R. Styles, 4 Cameron Close, Bude, EX23 8SP
M0 BUR N. Armer, 17 Keswick Road, Lancaster, LA1 3HJ
M0 BUT T. Gale, 33 Watson Close Upavon, Pewsey, SN9 6AF
M0 BUV A. Zerafa, 2 Furnwood, St. George, Bristol, BS5 8ST
M0 BUY J. Swann, 1 Sunnindale Drive, Tollerton, Nottingham, NG12 4ES
M0 BVD C. Turner, North Holme, Drain Lane, York, YO43 4DQ
M0 BVF G. Kapranos, 89 Spohr Terrace, South Shields, NE33 3LQ
MI0 BVG T. Wedlock, 13 Drumawhey Road, Newtownards, BT23 8RS
M0 BVI Dr P. Rogers, Flat 4 Holmdale, 2 Osborne Road, Poole, BH14 8SD
M0 BVM D. Hancock, 4 Elmside, Willand, Cullompton, EX15 2RN
M0 BVN A. Tavender, 38 Hesley Grove, Chapeltown, Sheffield, S35 1TX
M0 BVO J. Cull, 2 Drybrook Cottages, Amesbury Road, Cholderton, Salisbury, SP4 0ER
M0 BVQ G. Stokes, 9 The Haven, Harwich, CO12 4LA
M0 BVT J. Colles, The Orangery, Ufford Place Ufford, Woodbridge, IP13 6DP
M0 BVU S. Noble, 30 Flude Road, Coventry, CV7 9AQ
M0 BVV F. Hibberd, 58 Stoke Green, Coventry, CV3 1AN
M0 BVW B. Cox, 13 The Graylands, Coventry, CV3 6EW
M0 BVX D. Poulton, 14 George Street, Gun Hill, Coventry, CV7 8HL
M0 BVY N. Marshall, Key Cottage, 1 Haynall, Little Hereford, Nr. Ludlow, SY8 4AY
M0 BVZ P. Poulton, 14 George Street, Gun Hill, Coventry, CV7 8HL
M0 BWB J. Ridout, 136 Church Hill Road, Cheam, Sutton, SM3 8NA
M0 BWC J. Barton, 27 Francis Way, Salisbury, SP2 8EF
M0 BWF D. Surman, 27 Stanley Road, Hinckley, LE10 0HP
M0 BWH S. Jones, 34 Bury Green, Little Downham, Ely, CB6 2UH
M0 BWI W. Wilson, 8 Nora Street, South Shields, NE34 0RA
M0 BWL S. Vinnicombe, 8A Cross Road, Cholsey, Wallingford, OX10 9PE
MW0 BWM P. Lane, 51 Maesgwyn, Cwmdare, Aberdare, CF44 8TH
M0 BWN H. Seldon, 22 Downside Avenue, Plymouth, PL6 5SD
M0 BWO B. Watts, 24 Carisbrooke Way, Redcar, TS10 2LJ
M0 BWP R. Bebbington, 64 Stafford Road, Toll Bar, St Helens, WA10 3JH
M0 BWQ H. Robinson, 5 Coppice Close, Haxby, York, YO32 3RR
M0 BWS D. Pacheco Iii, 35 Baker Close, Caversfield, Bicester, OX27 8FQ
M0 BWU A. Hall, 16 Brushfield Avenue, Sileby, Loughborough, LE12 7NX
M0 BWW K. Allen, 55 Brandish Crescent, Clifton, Nottingham, NG11 9JZ
M0 BWY D. Hill, 86 The Downs, Nottingham, NG11 7EB
M0 BWZ A. Foad, 42 Catkin Way, New Balderton, Newark, NG24 3DT
M0 BXA L. Jensen, 17 Middleton Close, Tysoe, Warwick, CV35 0SS
M0 BXB E. Fuller, 36 North Road, Hull, HU4 6LJ
M0 BXC S. Coulston, 15 West View, Clitheroe, BB7 1DG
M0 BXD C. Robinson, Flat 1, St. Michaels Court, Worcester, WR2 5QR
M0 BXF P. Pickup, 13 Siddows Avenue, Clitheroe, BB7 2NX
M0 BXG J. King, 155 Barcombe Avenue, London, SW2 3BQ
M0 BXM D. Grimshaw, Apartment 52 Albion Mill, Blackburn, BB2 4lx
M0 BXP P. Bramberg, 18 North Bailey, Durham, DH1 3RH
M0 BXT A. Burge, 32 High Street, Swaffham Prior, Cambridge, CB25 0LD
M0 BXU P. White, 61 North St., Pewsey, SN9 5ES
MM0 BYB A. Lee, Burrowgate, Stronsay, Kw17 2AN
M0 BYI D. Moorey, 35 Broadway Manor, The Broadway, Hull, HU9 3PN
M0 BYJ R. Stoddart, 4 Belmont Road, Rednal, Birmingham, B45 9LW
M0 BYK B. Kurtcebe, Apartment 22, Rose Court, Baltic Avenue, Brentford, TW8 0FU
M0 BYL British Young Ladies Club, c/o J. Jones, 21 The Maltings, Warminster, BA12

8JR
M0 BYM J. Robson, Flat 6, 57 Lewisham Park, London, SE13 6QP
MI0 BYR D. Christie, 5 Moneydig Park Garvagh, Coleraine, BT51 5JP
MW0 BYS W. Reed, 2 St. Marys Park, Jordanston, Milford Haven, SA73 1HR
MW0 BYT G. Roscoe, 45 Bro Infryn, Glasinfryn, Bangor, LL57 4UR
M0 BYU C. Dodshon, 62 Moor Road, Melsonby, Richmond, DL10 5PE
M0 BYV J. Goldsbrough, 63 Aske Road, Redcar, TS10 2BP
M0 BYY J. Ellner, 21 Cranmer Road, Hampton Hill, Hampton, TW12 1DW
M0 BYZ K. Crossman, 24 Coxs Drive, Baltonsborough, Glastonbury, BA6 8RG
M0 BZA R. Stoddart, 163 Flatts Lane, Middlesbrough, TS6 0PP
M0 BZB S. Thirlaway, 10A Sea View, Blackhall Colliery, Hartlepool, TS27 4AX
M0 BZC A. Phillips, 39 Stonechat Road, Billericay, CM11 2NZ
M0 BZE Dr B. Allen, 1 St. Marys Close, Mursley, Milton Keynes, MK17 0HP
M0 BZH M. Wilkinson, 124 Doncaster Road Darfield, Barnsley, S73 9JA
M0 BZI F. Western, 12 St. Oswalds Crescent, Brereton, Sandbach, CW11 1RW
M0 BZK D. Mapeley, 6 Green Lane, Wolverton, Milton Keynes, MK12 5HB
M0 BZN L. Evans, 184 West St., Dunstable, LU6 1NX
M0 BZO G. Major, 17 Jubilee Cottages, Station Road, Bedford, MK43 0PN
M0 BZQ J. Doxey, 34 Lime Tree Crescent, New Rossington, Doncaster, DN11 0BT
M0 BZR M. Taylor, 38 Manor Road, Slyne, Lancaster, LA2 6LB
M0 BZS R. Cornthwaite, 18 Slaidburn Drive, Accrington, BB5 0JJ
M0 BZV P. Wade, 11 Hillside Crescent, Puriton, Bridgwater, TA7 8AP
M0 BZX S. Turner, 75 Keir Hardie Avenue, Stanley, DH9 6JU
M0 BZY M. Haywood, 4 Wentworth Gate, Birmingham, B17 9EB
M0 BZZ R. Bunker, 12 Bedford Avenue, Birkenhead, CH42 4QX
MW0 CAB Dr I. Jones, 4 Crowhill, Haverfordwest, SA61 2HL
MI0 CAC H. Mcgoldrick, 2 Carsdale, Mullanahoe Road, Dungannon, BT71 5GA
M0 CAD B. Lovatt, 12 Nelson Street, Leek, ST13 6BB
MM0 CAE J. Gauson, 112A High Street, New Pitsligo, Fraserburgh, AB43 6NN
M0 CAG I. Solly, 16 The Maples, Broadstairs, CT10 2PE
M0 CAJ D. Tysoe, 21 Burnt Close, Luton, LU3 3SU
M0 CAM Granta Contest Group, c/o M. Marsden, Mill Cottage, Shrowle, East Harptree, Bristol, BS40 6BJ
M0 CAN J. Wallis, 8 Kennedy House Hainworth Lane, Keighley, BD21 5BD
M0 CAR S. Robertson, 5 Sear Hills Close, Balsall Common, Coventry, CV7 7QL
M0 CAS M. Cassidy, 6 Mosley Street, Blackburn, BB2 3ST
M0 CAT J. Isaksson, 4 Horse Chestnut Gardens, Chellaston, Derby, DE73 6SB
M0 CAV J. Pearson, 82 Devoke Avenue, Worsley, Manchester, M28 7EN
M0 CAX D. Deakin, Restholme Cottage, Mosham Road, Doncaster, DN9 3BA
M0 CAZ A. Dahalay, The Manor, 80 Beach Road, Weston Super Mare, BS22 9UU
M0 CBA G. Gunn, The Old Rectory, Coreley, Ludlow, SY8 3AW
M0 CBD W. Eldridge, Minafon, Llangeitho, Tregaron, SY25 6TT
M0 CBF N. Lock, 57 Western Way, Basingstoke, RG22 6DF
M0 CBG A. Godney, 38 Botley Drive, Havant, PO9 4QY
MM0 CBI J. Isaacs, 5 Main Road, Glencraig, KY5 8AL
M0 CBK H. Purves, 2 Bourtree Close, Wallsend, NE28 9AA
MM0 CBL G. Black, 9 Mcculloch Road, Girvan, KA26 0EF
M0 CBM R. Newman, 11 Pine View Close, Woodfalls, Salisbury, SP5 2LR
M0 CBN R. Liversidge, 17 Millbank Close, High Green, Sheffield, S35 4NS
M0 CBP B. Muizelaar, Birtley Cb Services, Phoenix Communications, 33 Penshaw View, Portobello Road, Birtley, DH3 2JL
M0 CBT L. Betts, 33 Four Wells Drive, Sheffield, S12 4JB
MI0 CBX R. Graham, 21 Meadowvale Crescent, Bangor, BT19 1HQ
M0 CCA S. Ramsden, 24 Western College Road, Plymouth, PL4 7AG
MM0 CCC J. Maclean, 15 Muirpark Terrace, Tranent, EH33 2AS
M0 CCD J. Newsome, 241 Skellow Road, Skellow, Doncaster, DN6 8JL
M0 CCF Firepower Museum, c/o M. Buckley, Springfield, 12 Ranmore Avenue, Croydon, CR0 5QA
M0 CCG N. Griffiths, 85 Foljambe Road, Chesterfield, S40 1NJ
MW0 CCK K. Dancer, 63 Romilly Crescent, Cardiff, CF11 9NQ
MW0 CCL K. Dancer, 118 Fairwater Grove West, Cardiff, CF5 2JR
MW0 CCN J. Jones, 4 Westfield Road, Rhyl, LL18 4PN
M0 CCQ P. Burgess, 27 Watergate Street, Ellesmere, SY12 0EX
MW0 CCS A. Patterson, 3 Trem Y Foryd, Kinmel Bay, Rhyl, LL18 5JE
M0 CCU J. Buckley, 14 Eastfield, Foxholes, Driffield, YO25 3QW
M0 CCV L. Parsons, Gull Cottage, Briar Close, Hastings, TN35 4DP
M0 CCW M. Thornton, Bramble Lodge, Youngers Lane, Skegness, PE24 5JQ
M0 CCZ A. Cubitt, 132 Cauldwell Hall Road, Ipswich, IP4 5BP
M0 CDB T. Faris, 7 Mount Close, Fetcham, Leatherhead, KT22 9EF
M0 CDF M. Brooks, 9 The Green, Blaby, Leicester, LE8 4FQ
MW0 CDG C. Gozzard, Craig Dulas, Rhydyfoel Road, Llanddulas, Abergele, LL22 8EG
M0 CDJ S. Spevack, Pips Hill, Garden Close, Leatherhead, KT22 8LR
MM0 CDK D. Dickson, 115 Hatton Gardens, Glasgow, G52 3PU
M0 CDL J. Griffin, 35 Cottage Street, Kingswinford, DY6 7QE
M0 CDN N. Bullough, 29 Redfern Road, Stone, ST15 0LF
MW0 CDO P. Tomlinson, 17 Heol Yr Orsedd, Margam, Port Talbot, SA13 2HL
M0 CDQ J. Grant, 15A Elmdon Road, Marston Green, Birmingham, B37 7BU
M0 CDS S. Sanders, 52 Hazelwood Road, Callington, PL17 7EU
M0 CDU E. Linley, 40 Belvoir Road, Cleethorpes, DN35 0SE
MM0 CDW A. Bryce, 23 Primrose Avenue, Inverkip, Greenock, PA16 0DS
M0 CDY J. Elsworth, 15 Elm Avenue, Christchurch, BH23 2HJ
M0 CDZ K. Mountford, 22 Hollington Drive, Oxford, Stoke-on-Trent, ST6 6TZ
M0 CEB M. Bridge, 100 Pennine Road, Bacup, OL13 9PH
M0 CEC T. Ditchfield, 13 Alexandra Road, Waterloo, Liverpool, L22 1RJ
M0 CEG A. Shipp, 12 Rainbow Court Paston Ridings, Peterborough, PE4 7UP
M0 CEO B. Lewin, 68 Brackley Square, Woodford Green, IG8 7LS
M0 CEQ N. Taylor, 18 Chestnut Road, Glemsford, Sudbury, CO10 7PS
M0 CER T. Oka, 65 Curzon Street, London, W1J 8PE
M0 CES D. Hadden, 52 Brant Road, Lincoln, LN5 8SH
M0 CEW J. Lehane, 174 Stamfordham Drive, Liverpool, L19 6PZ
M0 CEX P. Newell, Thwaites Bank, Spring Avenue, Keighley, BD21 4TD
MM0 CEZ P. Moran, 3 Dunottar Avenue, Coatbridge, ML5 4LL
M0 CFB A. Bennett, 11A Ratcliffe Road Haydon Bridge, Hexham, NE47 6ER
M0 CFD N. Hould, 53 Laurel Avenue, Forest Town, Mansfield, NG19 0DW
MM0 CFE S. Campbell, 9 Arbuthnott Place, Stonehaven, AB39 2JA
M0 CFF Dr R. Fukuda, 2-6-17-218 Mita, Meguro-Ku, Tokyo, Japan, 153-0062
M0 CFH A. Hewett, 1 Mountside, Westfield Lane, Folkestone, CT18 8BY
MW0 CFQ P. Brennan, 1 Gerddi Mair, St. Clears, Carmarthen, SA33 4ET
M0 CFR R. Gould, 22 Vereker Drive, Sunbury-on-Thames, TW16 6HF
M0 CFT K. Pye, 5 Teme Avenue, Wellington, Telford, TF1 3HU
M0 CFZ B. Farrington, Flat 5 4 Southfield Rise, Paignton, TQ3 2NE
M0 CGA A. Desoer, 53 Highfield Road South, Chorley, PR7 1RH
M0 CGB A. Wiseman, 28 Inchfield, Worsthorne, Burnley, BB10 3PS
M0 CGE R. Corneloues, 21 Brook Street, Manningtree, CO11 1DL
M0 CGF I. Schofield, 100 Purley Downs Road, Sanderstead, CR2 0RB
M0 CGO R. Pratt, 11 Park Road, Ryde, PO33 2BG
MW0 CGP T. Peters, 37 Lon Coed Bran, Cockett, Swansea, SA2 0YD
M0 CGR J. Clarey, 2 Sebastopol Cottages, Redmere, Ely, CB7 4SS
M0 CGS S. Hancock, Monrad, Back Street, Gainsborough, DN21 3DL
M0 CGT G. Thompson, 160 Rempstone Road, Wimborne, BH21 1SX
MI0 CGV T. Keery, 51 School Road, Ballyroney, Banbridge, BT32 5JF
M0 CGW C. Selwyn-Smith, Miranda, East Molesey, KT8 9AN
MM0 CGZ C. Cregan, 4 Fowlers Court, Prestonpans, EH32 9AT
M0 CHD J. Neale, 20 Oakfield Road, Wollescote, Stourbridge, DY9 9DL
M0 CHE D. Letton, 21 Westfield, Bradninch, Exeter, EX5 4QU
MW0 CHI M. Broxton, 4 Owen St, Orange Gardens, Pembroke, SA71 4EP
M0 CHJ S. Birbeck, 3 Accrington Road Hapton, Burnley, BB11 5QL
M0 CHL D. Fitzpatrick, 21 Stambridge Road, Clacton-on-Sea, CO15 3JR
MU0 CHN J. Gardner, The Ferns, Rue De La Girouette, St. Saviour, Guernsey, GY7 9NN
M0 CHO G. Burton, 41 Etchingham Road, Langney, Eastbourne, BN23 7DS
M0 CHP S. Nicholls, 50 Gorleston Road, Lowestoft, NR32 3AQ
M0 CHR D. Whitelock-Wainwright, 21 Whelan Gardens, St Helens, WA9 5TD
M0 CHS M. Stanley, 9 Clare Way, Clacton on Sea, CO168BX
M0 CHU T. Hodby, 1 Hawksworth Close, Rotherham, S65 3JX
MM0 CHV W. Adamson, 47 Clarinda Gardens, Dalkeith, EH22 2LW
MM0 CHX G. Mckenna, 19 Gordon Terrace, Ayr, KA8 0EF
MW0 CIA N. Lemon, 75 Tai Llwyd Road, Neath, SA10 7DY
MI0 CIB P. Bell, 1Knockbracken Drive, Coleraine, BT521WN
M0 CIC R. Partington, 47 Sandmoor Road, New Marske, Redcar, TS11 8DJ
M0 CIE M. Coles, 1 Church Street, Taunton, TA1 3JE
M0 CIF D. Rosewarn, 16 Charles Crescent, Taunton, TA1 2XN
MW0 CIH J. Jones, 14 Heol Tywysog, Pentre Halkyn, Holywell, CH8 8HA
MM0 CIK D. Cox, Seye Hethen, Sanquhar, DG4 6JZ
M0 CIO R. Wyatt, 8 Millbrook Road, Bushey, WD23 2BU
M0 CIP L. Thompson, 9 Elmwood Drive, Ponteland, Newcastle upon Tyne, NE20 9QQ
M0 CIR C. Ryalls, 15 Belmont Way, South Elmsall, Pontefract, WF9 2BT
MW0 CIS A. Bray, Coedcelyn, Talley, Llandeilo, SA19 7YR
MW0 CIT V. Bray, Coedcelyn, Talley, Llandeilo, SA19 7YR
M0 CIW H. Delafield, 205 South Avenue, Abingdon, OX14 1QU
MW0 CJB D. Newton-Goverd, 2 Blaen Y Morfa, Morfa, Llanelli, SA15 2BG
M0 CJC G. Fowle, 12 Lytham Road, Broadstone, BH18 8JS
M0 CJD C. Rhodes, Wayside, Hardstoft, Pilsley, Chesterfield, S45 8AH
M0 CJE N. Murphy, 9 Eliot Gardens, Newquay, TR7 2QE
MM0 CJF J. Smith, 41 Dickie Drive, Peterhead, AB42 1HB
M0 CJG D. Jesinger, C/O B Jesinger, 29 Breakspeare Close, Watford, WD24 6DA
MM0 CJH N. Fowler, 59 Milnefield Avenue, Elgin, IV30 6EJ
M0 CJI R. Turner, 53 Queens Drive, Sandbach, CW11 1BN
M0 CJJ R. Brown, 1 Octavian Close, Hatch Warren, Basingstoke, RG22 4TY
M0 CJK S. Coulthard, 90 Rochester Crescent, Crewe, CW1 5YQ
M0 CJM N. Toombes, 46 The Vale, Oakley, Basingstoke, RG23 7LD
M0 CJN J. Gilmore, 30 Fairhaven Road, Caversfield, Bicester, OX27 8TU
M0 CJO A. Kay, Pear Tree Cottage, Hale House Lane, Farnham, GU10 2JG
M0 CJR N. Porter, 27 Severn Road, Aveley, South Ockendon, RM15 4NR
M0 CJS M. Elliott, 42 Ryelands Crescent, Stoke Golding, CV13 6EP
MM0 CJT A. Mctaggart, 1/4 Lady Nairne Place, Edinburgh, EH8 7LZ
M0 CJY I. James, 56A Ridgeway, East Herringthorpe, Rotherham, S65 3NN
M0 CJZ G. Dobson, 35 Parkhill Road, Doncaster, DN3 1DP
M0 CKA P. Webb, 42 Holland Road, Ampthill, Bedford, MK45 2RS
M0 CKB K. Prior, The Annex, 26 St David?S Road, Norwich, NR93DH
M0 CKC P. Wakelam, 6 Frankland Road, Durham, DH1 5HZ
M0 CKE J. Balls, 70 Risegate Road, Gosberton, Spalding, PE11 4EY
MM0 CKF J. Lewis, 9 Cessnock Road, Troon, KA10 6NJ
M0 CKG D. Kilburn, Shepherds Cottage, Burradon, Morpeth, NE65 7HF
M0 CKI S. Edwards, 5 Chalk Lane, Sutton Bridge, Spalding, PE12 9YF
MM0 CKK A. Mcluckie, 25 Churchill Avenue, Kilwinning, KA13 7JN
M0 CKL G. Sell, 135 Northfields, Norwich, NR4 7ET
M0 CKM K. More, 18 Douglas Close, Ford, Arundel, BN18 0TG
M0 CKN W. Dodge, 78 Littleworth Road, Dover, USA, 03820-4331
M0 CKO S. Westall, 4 South View Great Harwood, Blackburn, BB6 7NL
M0 CKP D. Warner, 6 Desborough Road, Rushton, Kettering, NN14 1RG
M0 CKS R. Cockings, 4 Freeman Gardens, High Green, Sheffield, S35 4NT
MW0 CKT C. Campbell-Moore, Maes Celyn Henllan Street, Denbigh, LL16 3PE
M0 CKU W. Meisenbach, 5 Byland Court, Whitby, YO21 1JJ
M0 CKV R. Wiltshire, 8 Hilltop Lane, Heswall, Wirral, CH60 2TT
M0 CKX F. Graseley, 11 Wilberforce Road, South Anston, Sheffield, S25 5EG
MW0 CLB M. Youlden, 33 Treseifion, Porthdafarch Road, Holyhead, LL65 2NN
M0 CLD I. Arnold, 4 Larkspur Close, Tanfield Lea, Stanley, DH9 9UH
M0 CLE A. Pascoe, 53 Priory, Bovey Tracey, Newton Abbot, TQ13 9HP
M0 CLG G. Gundry, 181 Stoneleigh Avenue, Worcester Park, KT4 8YA
M0 CLI E. Donaghy, Mendips, 36 Attwood Road, Salisbury, SP1 3PR
M0 CLJ J. Townsend, 28 West Street, Darfield, Barnsley, S73 9NF
M0 CLK P. Fitzpatrick, 16 Lichens Crescent, Oldham, OL8 2NS
M0 CLL J. Nixon, 25 Grove Court, Alsager, Stoke-on-Trent, ST7 2DS
M0 CLM M. Bromley, Chartley, Norton Lea, Warwick, CV35 8JX
M0 CLN P. Tose, Flat 4, 33 Rohilla Close, Whitby, YO22 4BU
M0 CLO K. Farthing, 86 Coldnailhurst Avenue, Braintree, CM7 5PY
MI0 CLP M. Hunter, 11A Maydown Road, Armagh, BT61 8BU
M0 CLR W. Huddleston, 17 Moorside Road, Brookhouse, Lancaster, LA2 9PJ
M0 CLW S. Pearson, 8 The Pastures, Edlesborough, Dunstable, LU6 2HL
M0 CMC M. Veary, 4 Millennium Way, Stone, ST15 8ZQ
M0 CME D. Viney, 5 Waters Edge, Bognor Regis, PO21 4AW
M0 CMF T. Beaumont, Po Box 109, Camberley, GU15 4ZF
M0 CMH M. Hemmings, 117 Kingston Hill Avenue, Romford, RM6 5QP
M0 CMI D. Lane, 230 Raeburn Avenue, Eastham, Wirral, CH62 8BB

M0 CMK L. Taylor, 15 Montgomery Court, 50 Walcourt Road, Kempston, Bedford, MK42 8SY
M0 CMN B. Pearce, 12 Bean Avenue, Bracebridge, Worksop, S80 2EW
MM0 CMO E. Skea, Craigard, Craigton, Inverness, IV1 3YG
M0 CMP J. Miller, 23 Wetherby Gardens, Farnborough, GU14 6BW
M0 CMQ E. Brendish, Flat 1, 19 Blake Hall Road, London, E11 2QQ
M0 CMS J. Lewis, 3 Jacobs Close, Stantonbury, Milton Keynes, MK14 6EJ
M0 CMT J. Donald, 67 Cradley, Widnes, WA8 7PL
M0 CMW J. Hudson, 3 Vogan Avenue, Crosby, Liverpool, L23 0SG
M0 CMZ I. Okanoue, 142-1 Yoshida, Okoh-Cho, Kochi, Japan, 783-0045
MW0 CNA M. Evans, 322 Heol Gwyrosydd, Penlan, Swansea, SA5 7BR
MW0 CNB B. Polgreen, 344 Heol Gwyrosydd, Penlan, Swansea, SA5 7BP
MW0 CNC T. Symons, Brynchwyth Farm, Fairyland Road, Neath, SA11 3QE
MW0 CND M. Davies, 10 Torrington Road, Gendros, Swansea, SA5 8DU
M0 CNE P. Black, 43 Malvern Avenue, Rugby, CV22 5JN
MM0 CNF J. Clark, 16 Bonnyton Avenue, Drongan, Ayr, KA6 7DG
M0 CNG D. Pearce, 2 Mell Avenue, Hoyland, Barnsley, S74 9HF
M0 CNH B. Cockerill, 4 Foxglove Close, Rugby, CV23 0TS
MI0 CNI W. Hamilton-Sturdy, 243 The Woods, Larne, BT40 1BD
M0 CNK J. Serrano, 42 Fleetwood Walk, Murdishaw, Runcorn, WA7 6DZ
M0 CNL P. Glover, 14 Holbrook Close, Clacton-on-Sea, CO16 8TH
M0 CNM A. Williams, 18 Sturdee Close, Thetford, ip24 2lf
M0 CNN S. Mcnally, 14 Cornelius Drive, Pensby, Wirral, CH61 9PR
M0 CNP D. Edwards, 3 Murton Close, Burwell, Cambridge, CB25 0DT
M0 CNS Essex DX Group, c/o T. Hackett, 16 Eagle Way, Shoeburyness, SS3 9RJ
M0 CNU T. Norman, 113 Dracaena Avenue, Falmouth, TR11 2ER
MM0 CNV W. Aitken, 63 Newlands Road, Grangemouth, FK3 8NT
M0 CNW H. Martin, 23 St. Marks Road, Gorefield, Wisbech, PE13 4QQ
M0 CNX P. Frampton, 118 Ramnoth Road, Wisbech, PE13 2JD
M0 CNY S. Hayes, 36 Mayfield Road, Chaddesden, Derby, DE21 6FW
M0 CNZ P. Enrico, Trengrouse House, Polhorman Lane, Helston, TR12 7JD
M0 COA C. Egginton, Flat 117, Seaward Tower, Gosport, PO12 1HH
MW0 COB R. Cobb, 107 High Street, Neyland, Milford Haven, SA73 1TR
M0 COC R. Dean, 10 Livingstone Road, Ellesmere Port, CH65 2BE
MW0 COD C. Queeley, 63 Tonna Road, Caerau, Maesteg, CF34 0RU
MW0 COE R. Thomas, 11 Heol-Y-Parc, North Cornelly, Bridgend, CF33 4LT
MW0 COF L. Thomas, 11 Heol-Y-Parc, North Cornelly, Bridgend, CF33 4LT
M0 COI N. Burgess, 12 Glenfield Road, Grimsby, DN37 9EE
M0 COJ D. Clenshaw, 4 Spring Meadow, Glemsford, Sudbury, CO10 7PN
M0 COM D. Fryer, 16 Elston Place, Aldershot, GU124HY
M0 CON J. Bell, 5 Burntwood Close, London, SW18 3JU
M0 COO M. Goulbourne, 9 Hawksworth Close, Liverpool, L37 7EX
M0 COP P. Wesley, Stalden, Ludlow Road, Church Stretton, SY6 6RB
M0 COQ Dr C. Cane, 69 Stoughton Drive North, Leicester, LE5 5UD
M0 COT J. Pears, 8 The Hawthorns, Ellesmere, SY12 9ER
M0 COV V. Fairhurst, 44 Harold Road, Coventry, CV2 5LG
MW0 COZ J. Gearey, 46 Priory St., Carmarthen, SA31 1NN
M0 CPB D. Tucker, 137 Seaford Road, London, W13 9HS
M0 CPC C. Cook, Hilbry Cottage, Douglas Road, Crowborough, TN6 3QT
MW0 CPD P. Jennings, 5 Pantydwr, Nantybwch, Tredegar, NP22 3RZ
M0 CPE H. Broyles, 16 Clifton Road, Shefford, SG17 5AE
M0 CPF K. Payne, 20 Laburnum Road, Exeter, EX2 6EG
M0 CPG Capel Battery Preservation Group, c/o J. Button, 4 Capstone Ridge, Hempstead, Gillingham, ME7 3AQ
M0 CPK A. Childs, 5 Barnes Wallis Drive, Leegomery, Telford, TF1 6XT
M0 CPL W. Harrison, 2 Mount House Road, Formby, Liverpool, L37 3LB
MW0 CPN M. Watkins, Highwinds, Bryn Pydew, Llandudno Junction, LL31 9QF
MM0 CPS Cockenzie & Port Seton ARC, c/o R. Glasgow, 7 Castle Terrace, Port Seton, Prestonpans, East Lothian, EH32 0EE
M0 CPT P. Smith, 3 Glenfield Square, Farnworth, Bolton, BL4 7TG
M0 CPU N. Bartlett, 61 Uplands Road, West Moors, BH220BU
M0 CPW P. Thompson, 73 Erlstoke Close, Plymouth, PL6 5QN
MM0 CPZ R. Roberts, 16 Swanston Avenue, Edinburgh, EH10 7BX
M0 CQF C. Warhurst, 26 Cherry Tree Close, Ilkeston, DE7 4HQ
M0 CQH I. Hirst, 30 Lincoln Road, Skellingthorpe, Lincoln, LN6 5UU
M0 CQO P. Hunter, 31 Itchen Grove Perton, Wolverhampton, WV6 7QY
M0 CQQ D. Newman, 89 Sea Place, Goring-By-Sea, Worthing, BN12 4BH
MW0 CQR D. Conde, 13 Broxton Road, Wrexham, LL13 9BA
MM0 CQT S. Forsyth, West Park, Innes Road, Fochabers, IV32 7NL
M0 CQV J. Hubbard, 99 Tuckers Road, Loughborough, LE11 2PH
M0 CQW J. Drummond, 60 Park Lane, Exeter, EX4 9HP
MW0 CRA M. Clarke, Freshwinds Halkyn, Holywell, CH88ES
M0 CRD P. Mayne, 17 School Street, Cottingley, Bingley, BD16 1QB
M0 CRG Calderdale Raynet ARG, c/o M. Cox, Haugh Shaw Hall Haugh Shaw Road, Halifax, HX1 3LE
M0 CRH A. Roberts, 23 Church St., Great Harwood, Blackburn, BB6 7NF
MW0 CRI D. Marston, Rosario, High Street, Cardigan, SA43 3EF
M0 CRJ R. Jones, 8 Downing Avenue, Newcastle, ST5 0JY
M0 CRM N. Williams, 1 Dorset Close, Whitehaven, CA28 8JP
M0 CRN S. Corbett, 80 Helmsdale Lane, Great Sankey, Warrington, WA5 1SY
M0 CRO G. Mansell, 87 Bifield Road, Stockwood, Bristol, BS14 8TT
M0 CRP A. Harradine, 4 Hesketh Drive, Lostock Gralam, Northwich, CW9 7QJ
MI0 CRQ K. Mcauley, Layde View, 19 Rathlin Avenue, Ballycastle, BT54 6DQ
MI0 CRR P. Quinn, 11 Blackpark Road, Ballyvoy, Ballycastle, BT54 6QZ
M0 CRT N. Hall, 1 Lynton Drive, Edenthorpe, Doncaster, DN3 2QF
M0 CRU J. Bradley, 47 Brunswick Park Road, Wednesbury, WS10 9HH
M0 CRW J. Roebuck, 2 Royston Close, Walton, Chesterfield, S42 7NE
M0 CRZ S. Crabtree, 107 Rochdale Road, Shaw, Oldham, OL2 7JT
M0 CSB M. Tait, 2 Wilsford Close, Walsall, WS4 1QP
MW0 CSC C. Osborn, Lowfield, Station Road, Usk, NP15 2EP
M0 CSD T. Quinn, 11 Meadowfield Stokesley, Middlesbrough, TS9 5EL
M0 CSE J. Burdett, 27 Hayfield Road, Woolston, Warrington, WA1 4PE
M0 CSF A. Prandoczky, 18 Kestrel Court Birtley, Chester le Street, DH3 2PT
M0 CSG City and Sticks ARG, c/o A. Hosking, 30 Edrick Road, Edgware, HA8 9JD
M0 CSO C. Mckenzie, 15 Belmont Drive, Saltney Ferry, Chester, CH4 0AL
M0 CSP E. Watson, 49 Beechfield Road, Bolton, BL1 6HZ
M0 CSQ R. Bates, 51 Boyton Road, Ipswich, IP3 9PD
M0 CSR J. Gardiner, 18 Granville Terrace, Guiseley, Leeds, LS20 9DY
M0 CST A. Holbrook, 24 Victoria Street, Sheerness, ME121YA
M0 CSU M. Deacon, 26 Brecon Chase, Minster On Sea, Sheerness, ME12 2HX
M0 CSV R. Tavener, 3 Mill Close Roxwell Chelmsford Essex Cm1 4Pg 3 Mill Close, Roxwell, Chelmsford, CM1 4PG
M0 CSZ S. Clarey, 36 Birchin Bank, Elsecar, Barnsley, S74 8DP
M0 CTC P. Ridgeon, 1 Fiennes Road Herstmonceux, Hailsham, BN27 4LN
M0 CTF D. Kaye, 20 Tryfan Close, Redbridge, Ilford, IG4 5JX
M0 CTI B. Johnson, 268 Badsley Moor Lane, Rotherham, S65 2QP
M0 CTJ G. Tepper, 5 Herbert Street, Mexborough, S64 0JZ
M0 CTK M. Stocking, 125 Bassnage Road, Halesowen, B63 4HD
M0 CTL S. Ball, 64 Stallington Road, Blythe Bridge, Stoke-on-Trent, ST11 9PD
M0 CTM R. Pearson, 21 West Parade, Spalding, PE11 1HD
M0 CTN D. Haughton, 31 Holmes Carr Road, New Rossington, Doncaster, DN11 0QF
M0 CTP G. Hyde, 4 Albright Close, Pocklington, York, York, YO42 2PE
M0 CTQ C. Greaves, 2 Attlee Avenue, New Rossington, Doncaster, DN11 0QX
M0 CTR A. Smith, 25 Hill Corner Road, Chippenham, SN15 1DW
MM0 CTT D. Stewart, 45 Kilwinning Road, Irvine, KA12 8RZ
MM0 CTU C. Stewart, 45 Kilwinning Road, Irvine, KA12 8RZ
MW0 CTX B. Jones, 87 Heol Llanelli, Pontyates, Llanelli, SA15 5UB
MW0 CUA D. Herbert, 50 Clos Cilsaig, Dafen, Llanelli, SA14 8QU
M0 CUD S. Marsh, 6 Mayland Drive, Streetly, Sutton Coldfield, B74 2DG
M0 CUF S. Hall, 28 Pugneys Road, Wakefield, WF2 7JT
MM0 CUG G. Grant, 11 Auchriny Circle, Bucksburn, Aberdeen, AB21 9JJ
M0 CUH D. Coisson, 21 Medalls Path, Stevenage, SG2 9DX
M0 CUI K. Lucas, 7 Wales Road, Kiveton Park, Sheffield, S26 6RA
M0 CUK M. Slade, 7 Glebe Field, Chaddleworth, Newbury, RG20 7EZ
M0 CUL M. Stevens, 67 New Road, East Hagbourne, Didcot, OX11 9JX
MI0 CUN P. Alexander, 59A Lismurn Park, Ahoghill, Ballymena, BT42 1JW
M0 CUP M. Phillips, Roseneath, 4C Valley Road, Kenley, CR8 5DG
M0 CUQ Dr G. Cooke, 37 Hertford Close, Woolston, Warrington, WA1 4EZ
M0 CUS G. Mack, 1085 Evesham Road, Astwood Bank, Redditch, B96 6EB
M0 CUT S. Cruise, 18 Morris Court Close, Bapchild, Sittingbourne, ME9 9PL
M0 CUU C. Wardell, 8 Lower End, Bricklehampton, Pershore, WR10 3HL
M0 CUY S. Odell, 41 Pevensey Park Road, Westham, Pevensey, BN24 5HW
M0 CVA V. Dewey, 100D Bromley High Street, London, E3 3EG
M0 CVB J. Sherbourne, 4 Chelston Terrace, Chelston, Wellington, TA21 9HT
M0 CVC C. Blackburne, 1 Watson Road, Leeds, LS14 6AE
M0 CVG B. Watmough, 28 Aspin Oval, Knaresborough, HG5 8EL
M0 CVH I. Cowie, 25 Penrice Close, Colchester, CO4 3XN
M0 CVJ C. Dale, The Common, Alsager, Stoke on Trent, ST7 2TQ
M0 CVK R. Henshall, 19 Townson Road, Ashmore Park, Wolverhampton, WV11 2PP
M0 CVN A. Fulcher, Keepers, The Street, Ashford, TN27 0QF
M0 CVO N. Booth, 68 Vernon Road Kirkby-In-Ashfield, Nottingham, NG17 8ED
M0 CVP B. Sutherland, 9 Park Drive South Hoole, Chester, CH2 3JT
M0 CVR P. Gurney, Perhams Green, Plymtree, Cullompton, EX15 2LW
M0 CVS F. Sadler, 12 Yokecliffe Drive, Wirksworth, Matlock, DE4 4EX
MW0 CVT R. Evans, The Brae, Coed-Cae-Ddu Road, Pontllanfraith Blackwood, NP12 2DA
M0 CVU A. Forrest, 2 Otterburn Grove, Blyth, NE24 4QP
MW0 CVW P. Wright, 145 Park Avenue, Bryn-Y-Baal, Mold, CH7 6TR
M0 CVZ D. Haynes, 113 The Glade, Croydon, CR0 7QP
MM0 CWB J. Benson, 15 Hawkhill Place, Stevenston, KA20 4HN
MW0 CWF S. Beer, 4 Churchfields, Barry, CF63 1FP
MM0 CWI C. Mcgowan, 82 Normand Road, Dysart, Kirkcaldy, KY1 2XP
MM0 CWJ J. Cameron, 407 Smerclate, Isle of South Uist, HS8 5TU
M0 CWN J. Thorne, Willow Lodge, The Street, Bury St Edmunds, IP29 5AP
MW0 CWS B. Chapman, Gwyddfan, Mountain Road, Kidwelly, SA17 4EY
M0 CWT M. Kozakowski, 18 Town Barn Road, Crawley, RH11 7EB
M0 CWX Prof. A. Morris, Shenlea, Leasowes Lane, Halesowen, B62 8QE
M0 CWY G. Andrews, 18 Vyne Road, Sherborne St. John, Basingstoke, RG24 9HX
M0 CWZ K. Summers, 30 David Street, Kirkby-In-Ashfield, Nottingham, NG17 7JW
MM0 CXA A. Burns, Flat D, 22 Elliot Street, Arbroath, DD11 3BZ
MM0 CXB K. Gillen, Flat 1/R, 12 Fergus Drive, Glasgow, G20 6AG
MI0 CXE G. Rea, 50 Culrevog Road, Dungannon, BT71 7PY
MW0 CXH P. Evans, 35 Trinity Road, Llanelli, SA15 2AB
M0 CXL D. Findlay, 78 South View Road, Bradford, BD4 6PJ
M0 CXO A. Preece, 1 Springfield Close, Thirsk, YO7 1FH
M0 CXQ C. Zdziech, 4 Pine Tree Mews, Barlby, YO8 5GZ
MW0 CXW I. Gray, 4 Llundain Fach, Felinfoel, Llanelli, SA15 4PF
M0 CXY E. Threadingham, 2 Cullum Close, Chichester, PO19 6GG
MM0 CXZ V. Madden, 38 Hunters Avenue, Dumbarton, G82 2RZ
M0 CYB Reading University ARC, c/o T. Cannon, 35 Loddon Bridge Road, Woodley, Reading, RG5 4AP
M0 CYD S. Bourne, 1 Humewood Grove, Stockton-on-Tees, TS20 1JU
M0 CYE J. Taylor, 46 Lever House Lane, Leyland, PR25 4XL
M0 CYG P. Davies, Devildchies, West Bay Road, Bridport, DT6 4EH
M0 CYI T. Witherspoon, Po Box 2, Swannanoa, USA, 28778
M0 CYJ Dr D. Nicole, 53 Cobbett Road, Southampton, SO18 1HJ
M0 CYM E. Tewsley, 20 Crookhorn Lane, Waterlooville, PO7 5QF
M0 CYR Dr P. Palin, Flat 2, Springfield Heversham, Heversham, Milnthorpe, LA7 7EJ
M0 CYT Prof. S. Warrillow, Po Box 452, Canterbury, Victoria, Australia, 3126
M0 CYU D. Holdroyd, 2 Vicarage Lane, Naburn, York, YO19 4RS
M0 CYX G. Yoxall, 33 Glasgow Road, Southsea, PO4 8HR
M0 CZA A. Moss, Lime Kiln Basin, Whitebridge Estate, Stone, ST15 8LQ
M0 CZB J. Webber, 5 Leda Mews, Achilles Close, Hemel Hempstead, HP2 5WR
M0 CZC M. Wade, 42 New Road, Burnham-on-Crouch, CM0 8EH
M0 CZE M. Garton, 13 Damaskfield, Worcester, WR4 0HY
MI0 CZF W. Belshaw, 9 Ashmount Gardens, Lisburn, BT27 5BZ
MM0 CZH D. Mitchell, 3 Lade Crescent, Bucksburn, Aberdeen, AB21 9HJ
MM0 CZK C. Rogers, 18 Primrose Lane Rosyth, Dunfermline, KY11 2SL
MM0 CZM A. Stewart, G/1 290 Dumbarton Road, Old Kilpatrick, Glasgow, G60 5LJ
M0 CZN W. Martin, 17 Dickens Road, Maidstone, ME14 2QW
M0 CZO J. Dixon, 83 Portmeads Rise, Chester le Street, DH3 2NW
M0 CZP T. Ostley, 30 Ashley Way Brighstone, Newport, PO30 4HH
M0 CZR B. Richtering, 20 Shaftesbury Avenue, Hornsea, HU18 1LX
M0 CZT D. Ward, 42 Fern Grove, Cherry Willingham, Lincoln, LN3 4BG
M0 CZU D. Gregson, 15 Alice Street, Oswaldtwistle, BB5 3BL
M0 CZX F. Boele, 301 Cell Barnes Lane, St Albans, AL1 5QB

MM0 DAA S. Mackie, Kierycraigs Lodge, Blairadam, Kelty, KY4 0JF
M0 DAB D. Bowles, 23 Broughton Way, Rickmansworth, WD3 8GW
M0 DAC D. Tanner, 55 Arundel Drive, Bramcote, Nottingham, NG9 3FN
M0 DAD D. Roper, 84 Tynedale Drive, Cowpen, Blyth, NE24 4DS
M0 DAE M. Haladij, 50 Liberty Drive, Duston, Northampton, NN5 6TU
M0 DAG D. Godden, 94 The Common, South Normanton, Alfreton, DE55 2EP
M0 DAH S. Brown, 4 Foundry Lane, Manchester, M4 5LB
M0 DAL A. Pounder, 6 Barbondale Grove, Knaresborough, HG5 0DX
M0 DAN D. Black, 8 Cornwood Close, Finchley, London, N2 0HP
MW0 DAR P. Drew, 14 Longfield Court, Hirwaun, Aberdare, CF44 9NG
M0 DAS A. Cockburn, Grangewood, Blenheim Road, Littlestone, New Romney, TN28 8RD
MM0 DAT D. Thain, 20 Spey Street, Fochabers, IV32 7EH
M0 DAW D. Wilcox, Medlar Cottage, Faringdon Road, Swindon, SN6 8AJ
MW0 DAX C. Evans, Y Rhosfa, High Street, Llanfyllin, SY22 5AF
M0 DAZ D. Drake, 38 Frinton Road Broxtowe, Nottingham, NG8 6GQ
M0 DBA Dr M. Brown, 2 Beacon Road, Marazion, TR17 0HF
M0 DBB J. Hayhurst, 18 Ducket Close, Richmond, DL10 5QD
MM0 DBC D. Brown, 10 Culmore Place, Falkirk, FK1 2RP
M0 DBD L. Dean, 24 Spennithorne Avenue, Leeds, LS16 6JA
MM0 DBF W. Callanan, 3 Eden Place, Aberdeen, AB25 2YF
M0 DBG G. Hodges, 12 Linwal Avenue, Houghton-On-The-Hill, Leicester, LE7 9HD
M0 DBH S. Palik, 10 Clare Road, Northborough, Peterborough, PE6 9DN
M0 DBI Dr A. Wylie, 15 Worcester Gardens, Greenford, UB6 0BH
M0 DBJ O. Peters, 3 Churchill Close, Sutton, Ely, CB6 2QF
MI0 DBK A. Quinn, 11 Derrynaught Road, Collone, Armagh, BT60 1LZ
M0 DBM C. Bloxidge, 33 Rosemary Hill Road, Sutton Coldfield, B74 4HL
M0 DBO N. Head, 12 Heston Walk, Redhill, RH1 5JB
MM0 DBR A. Butler, 68 Laws Road, Aberdeen, AB12 5LJ
M0 DBT C. Gregson, 11 Coupe Green, Hoghton, Preston, PR5 0JR
M0 DBX R. Batten, 118 Marryat Road, New Milton, BH25 5JF
M0 DBY A. Burton, Normanby View, Otby Lane, Market Rasen, LN8 3UT
M0 DCB D. Brown, 7 Limber Hill, Cheltenham, GL50 4RJ
MM0 DCC R. Rutherford, 3 Stevenson Street, Oban, PA34 5NA
M0 DCD A. Ripley, 3 Linden Way Wetherby, Leeds, LS22 7QU
M0 DCG Dr T. Spence, 38 Burtonwood Road, Great Sankey, Warrington, WA5 3AJ
MW0 DCM D. Maydew, 40 Penrhys Road, Tylorstown, Tylorstown, CF43 3BD
M0 DCO B. Tutty, 14 Nursery Walk, Canterbury, CT2 7TF
M0 DCP Y. Ochiai, 1-8-7 Tamagawa Den-En-Chofu, Setagaya-Ku, Tokyo, Japan, 1580085
MW0 DCQ W. Ashton, 44 Bryn Awel, Bettws, Bridgend, CF32 8SA
M0 DCS D. Taylor, Garth Farm, Hull Road, Selby, YO8 6NH
MW0 DCT P. Grace, 14 Langcliff, Swansea, SA3 4JF
M0 DCU S. Quantrill, 7 Clare Road, Kessingland, Lowestoft, NR33 7PS
M0 DCV P. Howell, 18 High Street Foxton, Cambridge, CB22 6SP
M0 DCW D. Eggleton, 79 Hazel Close, Twickenham, TW2 7NP
M0 DCY G. Tarr, 40 The Garth, Coniston, LA21 8EQ
M0 DCZ J. Farrer, 3 Pitt Garth, Haggs Lane, Grange-over-Sands, LA11 6PH
M0 DDA B. Waterloo, 55 Solent Road, Hill Head, Fareham, PO14 3LB
M0 DDB C. Parr, 13 Peartree Avenue, Southampton, SO19 7JN
M0 DDC A. Copperwaite, 71 Gladbeck Way, Enfield, EN2 7EL
MW0 DDE A. Bullock, 3 Cae Celyn, Berriew, Welshpool, SY21 8BT
M0 DDI P. Cassidy, 34 Belgrave Avenue, Penwortham, Preston, PR1 0BH
M0 DDK R. Milne, 8 Connaught Walk, Rayleigh, SS6 8UY
M0 DDT Dr C. Potter, 12 Beech Road, Headington, Oxford, OX3 7RR
M0 DDU L. Shepherd, 334 Copnor Road, Portsmouth, PO3 5EL
M0 DDV M. Watts, 51 Chester Road, Sidcup, DA15 8RX
MI0 DDW S. Cooper, 31 Kilbroney Valley, Rostrevor, Newry, BT34 3SR
M0 DDY J. Todd, 108 Clee Road, Grimsby, DN32 8NX
M0 DEA N. Jacovides, 11 Stravonos Street, Nicosia, Cyprus, 2335
M0 DEB 759 Sqn(Beccles)ATC, c/o E. Lugmayer, 17 Borough End, Beccles, NR34 9YW
MM0 DEC I. Street, 5 Calder Road, Bellsquarry, Livingston, EH54 9AA
M0 DEF G. Thacker, 4 Duffy Place, Rugby, CV21 4EF
M0 DEI Aeroventure ARS, c/o R. Liversidge, 17 Millbank Close, High Green, Sheffield, S35 4NS
M0 DEJ A. Neumann, 5 Denfield Avenue, Halifax, HX3 5NL
M0 DEK D. Dyde, 10 Essex Close, Worcester, WR2 5RW
M0 DEL A. Lindley, 2 Mickle Hill Farm Cottage, Mickle Hill Road, Blackhall Colliery, Hartlepool, TS27 4DF
M0 DEN D. Miller, Brookhill Gardens Pinxton, Nottingham, NG16 6JX
M0 DEO C. Wheldon, 39 Felton Avenue, South Shields, NE34 6RY
M0 DEP R. Speed, 48 Stony Lane, Burton, Christchurch, BH23 7LE
M0 DEQ M. Clarke, 21 Sycamore Road, Greenstead Estate, Colchester, CO4 3NF
M0 DER R. Hannigan, 4 Westlands Avenue, Tetney, Grimsby, DN36 5LP
M0 DES R. Beck, 26 Corcoran Street, Duncraig, Perth, Australia, 6023
M0 DEV M. Twells, Camels, Annscroft, Shrewsbury, SY5 8AN
MW0 DEW D. Walters, 17 Heol Islwyn, Llanrhystud, SY23 5BW
M0 DEX J. Constable, 438 Old Road, Clacton-on-Sea, CO15 3SB
M0 DEY H. Watt, 40 Long Wood Road, Bristol, BS16 1FD
M0 DFA D. Crake, Kentolop, Holyhead Road, Montford Bridge, Shrewsbury, SY4 1EE
M0 DFD S. Sparkes, Flat 1, 14 Hall Road, Wilmslow, SK9 5BN
M0 DFF M. Bywater, 16 Grove Road, Cromer, NR27 0BY
M0 DFH C. Miles, 26 Meadowside, Grindleton, Grindleton, Clitheroe, Lancs, BB7 4RR
M0 DFL C. Houghton, 20 St. Peters Way, Thurston, Bury St Edmunds, IP31 3RZ
MW0 DFN D. Thomas, 48 Gilbert Road, Llanelli, SA15 3RA
MI0 DFO R. Kennedy, 83 Craigstown Road, Randalstown, Antrim, BT41 2PN
M0 DFQ P. Dickman, 1 Old Hall Close, Henley, Ipswich, IP6 0RJ
M0 DFW D. Wright, 61 Potton Road, St Neots, PE19 2NN
M0 DFX D. Riley, 9 Century Avenue, Mansfield, NG18 5EE
M0 DFY M. Watts, 14 Beverley Close, Lowestoft, NR33 8QQ
M0 DGA A. Bevins, 12 Wheatstone Road, Formby, Liverpool, L37 6BF
M0 DGB D. Balharrie, 27 Norfolk Road, Uxbridge, UB8 1BL
MM0 DGI S. Spence, Halley, Deerness, Orkney, KW17 2QL
M0 DGJ G. Hannam, 26 Cornflower Way, Melksham, SN12 7SW
M0 DGK M. Luby, 19 Robin Lane Bentham, Lancaster, LA2 7AB
M0 DGQ B. Zarucki, 26 Heathfield Road, Kings Heath, Birmingham, B14 7DB
MM0 DGR Scottish-Russian ARS, c/o J. Phunkner, 7 Plenshin Court, Glasgow, G53 6QW
M0 DGT P. Kelly, 4 The Crescent, Guildford, GU2 8AL
M0 DGU C. Thompson, 80 Aston Road, Willerby, Hull, HU10 6SG
MI0 DGX N. Mccully, 12 Cargygray Road, Hillsborough, BT26 6BL
M0 DHE J. Coxon, Links View Farm, Fairy Lane, Sale, M33 2JT
MW0 DHF P. King, 11 Lord Street, Penarth, CF64 1DD
M0 DHI B. Phillips, 79 Leen Valley Drive, Shirebrook, Mansfield, NG20 8BJ
M0 DHM E. Dean, 26 Silverdale, Maidstone, ME16 9JG
M0 DHN G. Kirkpatrick, 23 Hornhatch, Chilworth, Guildford, GU4 8AY
M0 DHO D. Honey, Bluebell Cottage, Crondall Road, Fleet, GU51 5SU
M0 DHP R. Benitez, 4 Raphael Drive, Thames Ditton, KT7 0BL
MM0 DHQ A. Clark, 20 Church Street, Kilwinning, KA13 6BE
M0 DHU M. Tachibana, 81334076949, Tokyo, Japan, 1070062
M0 DHX H. Wood, 402 Chemin De Peyrebelle, Valbonne, France, 6560
MM0 DHY A. Hart, Tigh Na Coille, Daviot, Inverness, IV2 5EP
M0 DID R. Macgregor, 125 Spring Lane, Birmingham, B24 9BY
M0 DIG J. O'Mahoney, 36 Scotton Gardens, Catterick Garrison, DL9 4HX
M0 DIJ J. Allen, 2 Chichester Walk, Chichester Road, Ramsgate, CT12 6NX
M0 DIL S. Lowe, 31 Court Farm Road, Bristol, BS14 0EH
M0 DIN Dr A. Bottrill, 4 St Luke'S Mews Gilesgate, Durham, DH11JA
M0 DIQ R. Mullen, 18 Chandlers Ridge, Nunthorpe, Middlesbrough, TS7 0JL
MM0 DIS I. Elder, 24 Birniehill Avenue, Bathgate, EH48 2RR
M0 DIT J. Tildesley, 16 Calver Grove, Keighley, BD21 2RX
M0 DIW D. Walls, 15 Brant Road, Lincoln, LN5 8RL
M0 DJA A. Garner, The Chestnuts, Surfleet, Spalding, PE11 4BA
M0 DJB A. White, 368 Hall Lane, Whitwick, Coalville, LE67 5PF
M0 DJD D. Gould, 11 Secret Garden, Hunslet, Leeds, LS9 8FB
M0 DJF D. Fradley, 33 Churchill Way Riseing Brook Stafford, Stafford, St17 9nz
M0 DJH D. Hale, 51 Lynhurst Avenue, Sticklepath, Barnstaple, EX31 2HY
M0 DJI J. Ford, 1 Cherry Road, Enfield, EN3 5SE
M0 DJQ B. Cushing, 51 Anderida Road, Eastbourne, BN22 0PZ
MW0 DJT D. Turner, Minyrafon, Whitemill, Carmarthen, SA32 7EN
M0 DJW Dr D. Westland, 8 Faris Barn Drive, Woodham, Addlestone, KT15 3DZ
M0 DKD Dr D. Bains, 2 Arundel Road, Brighton, BN2 5TD
M0 DKJ I. Rose, 26 Beckford Road, Cowes, PO31 7SG
M0 DKL E. Whittle, 15 Weavers Court, Scorton, Preston, PR3 1NQ
M0 DKN D. Turney, 2 Beult Meadow, Cage Lane, Ashford, TN27 8PZ
M0 DKP J. Jones, 3 St. Judes Walk, Cheltenham, GL53 7RU
M0 DKR D. Austin, 11 Seabourne Avenue, Blackpool, FY4 1EH
M0 DKS D. Kees, 23 Walsingham Way, Billericay, CM12 0YE
M0 DKT D. Knott, 80 Melville Court, Chatham, ME4 4XJ
M0 DKU P. Giles, Fenimora Cottage, Paines Hill, Bicester, OX25 4SQ
M0 DKV D. Bowers, Oak House, Church Stile Lane, Exeter, EX5 1HP
M0 DKX J. Horsfield, 91 Harlington Road, Mexborough, S64 0DT
M0 DLB S. Orange, 20 Borrowdale Avenue, Fleetwood, FY7 7LF
M0 DLC P. Bartlett, 43 Chamberlain Way, Pinner, HA5 2AU
M0 DLE R. Tingay, 1 Ullswater Road, Sompting, Lancing, BN15 9UF
M0 DLG J. Conway, 27 Victoria Road, Gorleston, Great Yarmouth, NR31 6EF
MM0 DLH A. Dunsmore, 21 East Croft, Ratho, Newbridge, EH28 8PD
M0 DLI Wellesley House Radio Club, c/o J. Hislop, 10 Park Wood Close, Broadstairs, CT10 2XN
M0 DLL D. Gray, 68 Sixth Cross Road, Twickenham, TW2 5PD
M0 DLM D. Mann, Hunters Lodge, Grange Road, Bedford, MK44 3NT
M0 DLP D. Birch, 4 Avon Bank Cottages, Avon Bank, Pershore, WR10 3JP
M0 DLR K. Jones, 21 Wharton Crescent, Beeston, Nottingham, NG9 1RJ
M0 DLS M. Bell, 56 Boyd Road, Wallsend, NE28 7SQ
M0 DLX D. Cleal, 9 Ladymere Place, Ockford Road, Godalming, GU7 1AH
M0 DLY C. Kerrison, 18 Parks Road, Dunscroft, Doncaster, DN7 4AH
M0 DLZ P. Parry, 31 Swinburne Way, Daybrook, Nottingham, NG5 6BX
M0 DMA D. Marshall, 75 North Hill Road, Sheffield, S5 8DT
M0 DMB C. Simpson, 6 Dalton Close, Driffield, YO25 6YE
M0 DMD M. Coe, 2 Burnslack Road, Ribbleton, Preston, PR2 6EX
M0 DME A. Pell, 5 Gallery Close, Northampton, NN3 5NT
M0 DMF C. Price, 27 Naunton Way, Cheltenham, GL53 7BQ
M0 DMI J. Enderby, 42 Claremont Avenue, Chorley, PR7 2HL
M0 DMJ C. Peters, 9 Evelyn Close, Twickenham, TW2 7BL
MM0 DMK D. Mckenzie, 17 Alexander Drive, Livingston, EH54 6DB
M0 DMR D. Rolf, 40 Gunton Drive, Lowestoft, NR32 4QB
M0 DMS D. Schofield, 9 Manor View, Shafton, Barnsley, S72 8NQ
MI0 DMT J. Hyndman, 4 Larchfield Gardens Kilrea, Coleraine, BT51 5SB
MM0 DMU J. Stewart, 76 Caroline Terrace, Edinburgh, EH12 8QU
M0 DMX B. Donnachie, 3 Downs View Close, Scaynes Hill, Haywards Heath, RH17 7EQ
M0 DMY C. Hicken, 76 Peaksfield Avenue, Grimsby, DN32 9QG
M0 DMZ M. Hughes, 9 North Street, Owston Ferry, DN9 1RT
MI0 DNB E. Hyndman, 4 Larchfield Gardens Kilrea, Coleraine, BT51 5SB
M0 DND D. Brown, 97 Hewett Road, Portsmouth, PO2 0QS
MW0 DNF C. Owen, 58 Bedwellty Road, Cefn Fforest, Blackwood, NP12 3HB
MM0 DNH Dr B. Shippey, Binn Ericht, 438 Perth Road, Dundee, DD2 1JT
M0 DNJ D. Cook, Flat 6, 7 Heath Court, Felixstowe, IP11 0YQ
MW0 DNK R. Law, 12 Gwel Y Llan Llandegfan, Menai Bridge, LL59 5YH
MI0 DNM W. Wilson, 68 Ballygowan Park, Banbridge, BT32 3AW
M0 DNN A. Wharton, 184 Surbiton Road, Stockton-on-Tees, TS19 7SH
M0 DNO C. Anderson, Flora, 61 Merrilees Crescent, Clacton-on-Sea, CO15 5XY
M0 DNR R. Dickson, 102 Blakemere Crescent, Portsmouth, PO6 3SH
M0 DNU A. Bateman, 3 Church Croft, Bramshall, Uttoxeter, ST14 5DE
M0 DNV A. Truman, 92 Spring Meadow, Sutton Hill, Telford, TF7 4AQ
M0 DNW S. Dyson, 18 Coniston Road, Chorley, PR7 2JA
MM0 DNX D. Barrett, 88 Camp Rd Baillieston, Glasgow, G69 6qp
M0 DNY P. Crump, 41 Vernon Way, Guildford, GU2 8DE
M0 DNZ D. Pauley, Charente, Westerfield Road, Ipswich, IP6 9AJ
M0 DOA D. Brace, 10 Sawles Road, St Austell, PL25 4UD
M0 DOB S. Dobbs, Firbeck House Farm Cottage, Steetley, Worksop, S80 3EB
M0 DOC J. Jones, 16 Laurel Avenue, Darwen, BB3 3AG
M0 DOD T. Bodily, Flat 5, Denbeigh House, Rushden, NN10 0AT
M0 DOH R. Golsby, 19 Glascote Close, Shirley, Solihull, B902TA
M0 DOK A. Abrams, The Cottage, Victoria Road, Rushden, NN10 0AS
M0 DOM S. Cassidy, 71 Kensington Avenue, Penwortham, Preston, PR1 0EE

M0 DON Leicester Amateur Radio Show, c/o J. Theodorson, 7 Kingfisher Court, Overstone Lakes, Ecton Lane, Northampton, NN6 0BD
M0 DOP A. Hoskins, 38 Tasmania Close, Basingstoke, RG24 9PQ
MW0 DOR E. Roberts, 6 Trem Y Moelwyn, Tanygrisiau, Blaenau Ffestiniog, LL41 3SS
M0 DOS A. Yearp, 29 Humber Road, Ferndown, BH22 8XN
MM0 DOT D. Macnaughton, 24 Kepplehills Drive, Bucksburn, Aberdeen, AB21 9PQ
M0 DOW A. Holden, 21 East View Meadowfield, Durham, DH7 8RY
M0 DOY R. Kendrick, Forest View, 126 Ameysford Road, Ferndown, BH22 9QE
MW0 DOZ Heads of The Valleys ARC, c/o A. Sneddon, 3 Marigold Close, Gurnos, Merthyr Tydfil, CF47 9DA
M0 DPF S. Thompson, Flat 3, 225 Westmacott Drive, Feltham, TW14 9XB
MD0 DPG P. Taylor, 14 Royal Park, Ramsey, Isle Of Man, IM8 3UF
M0 DPH P. Tullock, 17 Owthorne Walk, Bridlington, YO16 7GB
M0 DPJ B. Crawshaw, 112 Waller Road, Sheffield, S6 5DQ
M0 DPK P. Tucker, 2 Kemps Field, Cranbrook, Exeter, EX5 7AZ
M0 DPQ R. Meadley, 2 Lower Hollacombe Cottages, Torquay Road, Paignton, TQ3 2DP
M0 DPS C. Seager, 77 Stonewood, Bean, Dartford, DA2 8BZ
M0 DPV A. Clark, 195 Ivyhouse Road, Dagenham, RM9 5RS
M0 DPW D. Wakefield, Rosebank, 109 Spring Road, Stoke-on-Trent, ST3 7JA
M0 DPY K. Wilks, 22 Uppercroft, Haxby, York, YO32 3GD
M0 DQB J. Brown, Ladythorn, Cleeve Hill, Cheltenham, GL52 3QB
M0 DQD P. Mcgonigal, 4 West View, Evenwood, Bishop Auckland, DL14 9QH
M0 DQH G. Waugh, 5008 Spartanburg Cove, Austin, Texas, USA, 78730
M0 DQK H. Beckett, 5 Chatsworth, Benfleet, SS7 3BB
M0 DQL D. Porteus, 2 Piers Close, Malvern, WR14 3JH
M0 DQN R. Richards, Broad Oak Bascombe Road, Churston, Brixham, TQ5 0JZ
M0 DQO C. Bloy, 6 Cornfield, Fareham, PO16 8UE
MM0 DQP J. Mckay, 28 Lumsden Crescent, St Andrews, KY16 9NQ
M0 DQS P. Mccleaft, Gas House Farm, Shavington Park, Market Drayton, TF9 3SY
MM0 DRA A. Cattanach, 8 Auchencairn Place, Monifieth, Dundee, DD5 4TS
M0 DRB F. Dawson, Silverthorne, Lower Sea Lane, Bridport, DT6 6LR
M0 DRD Dr D. Homer, 118 Brampton Way, Portishead, Bristol, BS20 6YT
M0 DRE J. Barnett, 5 Manknell Road, Chesterfield, S41 8LZ
M0 DRF D. Falkner, 45 Westwood Carleton, Skipton, BD23 3DW
M0 DRG D. Green, 39 Heritage Park, Hatch Warren, Basingstoke, RG22 4XT
M0 DRI A. Lyons, 20A Russell Street Devonport, Auckland, New Zealand, 624
M0 DRK D. Carman, 30 Parsonage Close, Burwell, Cambridge, CB25 0ER
M0 DRL D. Lane, The Coltmoor, Peterchurch, Hereford, HR2 0SW
M0 DRM Rose & Crown Radio Club, c/o J. Mahoney, 27 Linby Drive, Bircotes, Harworth, Doncaster, DN11 8FP
M0 DRN A. Small, 28 Capel Street Capel-Le-Ferne, Folkestone, CT18 7LZ
M0 DRO Capt. R. Bell, Sandgate House, Gough Road, Sandgate, Folkestone, CT20 3BE
M0 DRQ M. Munn, 54 Longbeech Park, Canterbury Road, Ashford, TN27 0HA
M0 DRS Dr S. Sampathkumar, 52 Crowstone Road, Westcliff-on-Sea, SS0 8BD
MM0 DRT D. Taylor, Hillcrest, Woodside Place, Banchory, AB31 5XW
MW0 DRU D. Underwood, 891 Heol Y Ffynon, Penrhys, Ferndale, CF43 3RN
M0 DSC D. Westwood, 60 Selwyn Drive, Stockton-on-Tees, TS19 8XF
M0 DSF Dr D. Fenna, 84 High Park Road, Ryde, PO33 1BX
M0 DSI D. Shaw, 35 Tinshill Lane, Leeds, LS16 6BU
M0 DSL P. Zatylny, 100 Larkspur Way, West Ewell, Epsom, KT19 9LU
MM0 DSM E. Mcneill, 3 Sunnybrae Terrace, Maddiston, Falkirk, FK2 0LP
M0 DSN D. Tomlinson, 3 Holgate Road, Nottingham, NG2 2EB
M0 DSO R. De Ieso, 57 Kingfisher Road, Thrapston, NN14 4GN
M0 DSR N. Passam, 177 Uttoxeter Road, Blythe Bridge, Stoke-on-Trent, ST11 9HQ
M0 DSS D. Smith, Church House Farm, Church Terrace, Alnwick, NE66 2YD
M0 DSW J. Wright, 31 Cherry Orchard, Marlborough, SN8 4AS
M0 DSX D. Mir, 2 Chadhurst Cottages, Coldharbour Lane, Dorking, RH4 3JH
M0 DSY D. Bailey, 28C Cliff Road Dovercourt, Harwich, CO12 3PP
MW0 DSZ D. Church, Elm Cottage, Pant, Oswestry, SY10 9RB
M0 DTA R. Cheverall, 1 Clarkwood Cottages, Twitty Fee, Chelmsford, CM3 4PG
M0 DTB T. Belton, Flat 6, 28 First Avenue, Hove, BN3 2FF
MW0 DTD R. Edwards, 59 Allt-Yr-Yn Close, Newport, NP20 5EE
MW0 DTH R. Howes, 8 Birbeck Road, Caldicot, NP26 4DX
M0 DTJ Dr J. Holloway, 76 Sir Thomas Whites Road, Coventry, CV5 8DR
M0 DTK D. Cox, 45 Victoria Road, Walderslade, Chatham, ME5 9HB
MM0 DTL Aberdeenshire Contest Group, c/o I. Ross, Idlewilde, Kintore, AB51 0XA
M0 DTR D. Anderson, 76 South Road, Northfield, Birmingham, B31 2QY
M0 DTS R. Swinbank, Oxhill Farm, Hilton, Yarm, TS15 9LB
MM0 DTW R. Clark, 52 Loons Road, Dundee, DD3 6AQ
M0 DUB D. Wood, 3 Ripley Close, Wakefield, WF3 2FG
M0 DUP S. Cox, 63 Netherfield Avenue, Eastbourne, BN23 7BT
M0 DUQ A. Bridgeland, 17 Oldfield Lane, Wisbech, PE132RJ
MM0 DUR M. Mckay, Tryggo, Sarclet, Wick, KW1 5TU
M0 DUT J. Barley, 2 Little Hobbyvines, Duckend, Stebbing, CM6 3BP
M0 DUU D. Allen, 20 Evenley Road, Northampton, NN2 8JR
M0 DUV D. Tootill, Apartment 18, Ley Gardens, Lawley Close, Church Stretton, SY6 6GA
M0 DUY D. Bates, Bold Gate Lodge, Praze, Camborne, TR14 0NQ
MM0 DVB A. Paton, 17 Union Terrace, Keith, AB55 5EQ
M0 DVD P. Barnes, 20 Benbow Drive, South Woodham Ferrers, Chelmsford, CM3 5FP
M0 DVF I. Dummer, 29 Chisholm Close, Southampton, SO16 8GU
M0 DVG W. Bosworth, 87 Tregorrick Road, Exhall, Coventry, CV7 9FH
MI0 DVH M. Mclaughlin, 22 Duncreggan Road, Londonderry, BT48 0AD
M0 DVK S. Christodoulou, 27 Manor Place, Cambridge, CB1 1LE
MW0 DVM D. Morris, 14 Church Terrace, Porth, CF39 0ET
M0 DVQ A. Yates, 12 Rowsley Avenue, Derby, DE23 6JY
M0 DVR P. King, 130 Langer Lane, Chesterfield, S40 2JJ
M0 DVT J. Adlington, 23 Newstead Road, Abbey Hulton, Stoke-on-Trent, ST2 8HU
M0 DVU T. Wilf, Flat 10, Church View Church Walk, Bourne, PE10 9UQ
M0 DVW G. Mallinson, 6 Deerplay Court, Bacup, OL13 8GE
MM0 DVZ J. Craig, Borrowhill Farmhouse, Strichen, Fraserburgh, AB43 6TJ
M0 DWB M. Tidman, 3 Stannet Way, Wallington, SM6 8BE
M0 DWC D. Wraight, 2 Silver Mead, Congresbury, Bristol, BS49 5EX
MI0 DWD D. Hamilton, 7 Bolea Park, Limavady, BT49 0SH
MI0 DWE D. Eames, 108 Rabbitburrow Road, Farnamullan, Lisbellaw, Enniskillen, BT94 5FL
MM0 DWF Dr L. Boehme, 26 Sandylands Road, Cupar, KY15 5JS
M0 DWG Dr R. Gooch, 14 Cotterill Road, Surbiton, KT6 7UN
M0 DWK Dr C. Jewell, 4 Springfield, Bentham, Lancaster, LA2 7BA
M0 DWM D. Cartlidge, 2 Walton Road, Walsall, WS9 8HN
M0 DWP D. Peters, 5 Riverside, Buntingford, SG9 9HJ
M0 DWS D. Summerwill, 52 Lanmoor Estate, Lanner, Redruth, TR16 6HN
M0 DWT D. Todd, 20 Wasdale Close, Plymouth, PL6 8TL
MW0 DWU A. Fleming, 30 Corbett Close, Tywyn, LL36 0BL
M0 DWW M. Giffin, 15 Chalk Lane, Sutton Bridge, PE12 9YF
M0 DWX D. Mcintosh, 14 Anne Crescent, Barnstaple, EX31 3AF
M0 DWZ R. Webster, 74 Bescar Brow Lane, Scarisbrick, Ormskirk, L40 9QG
MW0 DX Burnham Beeches Radio Club, c/o T. Clapp, Crunns Farm. Coxhill, Narberth, SA67 8EH
M0 DXA T. Vasdaris, 86 Hall Cross Road, Huddersfield, HD5 8LD
MM0 DXC C. Stevenson, 8 Carlaverock Grove, Tranent, EH33 2EB
MM0 DXD J. Wilson, 20 Ballumbie Gardens, Dundee, DD4 0NR
M0 DXF P. Binswanger, 115 Hawthorn Bank, Spalding, PE11 1JQ
MM0 DXH J. Hume, 8/11 Leslie Place, Edinburgh, EH4 1NH
M0 DXJ C. Humphris, Glebe House, School Lane, Spilsby, PE23 4AU
M0 DXM G. Pesch, Nikolaus-Jansen-St. 10, Simmerath, Germany, D52152
M0 DXN T. Green, 9 Craiglands Park, Ilkley, LS29 8SX
M0 DXP D. Poole, 17 London St., Chertsey, KT16 8AP
M0 DXQ P. Dougherty, 79 Beverly Road, West Caldwell, New Jersey, USA, 07006-6532
M0 DXR M. Haynes, 32 Bentley Drive, Harlow, CM17 9PA
M0 DXS D. Sheppard, 75 St. Nicholas Road, Littlestone, New Romney, TN28 8QA
M0 DXT W. Tinnion, 3 Brayton Road, Aspatria, Wigton, CA7 3DJ
M0 DXV M. Whitehead, 29 Coulsons Road, Bristol, BS14 0NN
MW0 DXX S. Pettipher, Min Yr Afon, Llandysul, SA44 5AT
M0 DXZ J. Arnell, Jersey Farm, Little London, Bishop's Stortford, CM23 1BD
M0 DYA O. Haselden, 15 Broadmeadow Close, Totton, Southampton, SO40 8WB
M0 DYB Dr S. Kemp, 5 Oakdale Avenue, Frodsham, WA6 6PY
M0 DYG P. Swan, 44 Gillingham Road, Gillingham, ME7 4RR
M0 DYH P. Lockley, 2 Valley View, Bowling Green, Constantine, TR11 5AP
M0 DYO M. Blair, 20 Hazel Drive, Burn Bridge, Harrogate, HG3 1NY
M0 DYQ P. Davis, 9 Chartwell Road, Kirkby-In-Ashfield, Nottingham, NG17 7HB
M0 DYR S. Durham, 24 Morgan Close, Yaxley, Peterborough, PE7 3GE
MW0 DYS A. Davies, 17 Queens Road, Merthyr Tydfil, CF47 0NB
M0 DYU Rev. W. Walker, 9 Malthouse Lane, Dorchester-On-Thames, Wallingford, OX10 7LF
M0 DYV R. Hitchins, 1 The Villas, Romansleigh, South Molton, EX36 4JW
M0 DYW D. Donnelly, 11 Anton Street, London, E8 2AD
MM0 DYX D. Francis, 2 Morlich Crescent, Dalgety Bay, Dunfermline, KY11 9UW
MW0 DYZ R. Jeffery, 66 Redhouse Road, Cardiff, CF5 4FH
M0 DZA M. Brinnen, 82 Victoria Road, Mablethorpe, LN12 2AJ
M0 DZB K. Johnson, 16 Lamberts Close, Weasenham, King's Lynn, PE32 2TE
M0 DZC R. Barton, 57 Croxteth Drive, Rainford, St Helens, WA11 8LA
M0 DZD M. Smith, 16 Thornton Drive, Brierley Hill, DY5 2BS
M0 DZG B. Banks, 1 Worthington Road, Dunstable, LU6 1PN
M0 DZH P. Holloway, 43 Little Sammons, Chilthorne Domer, BA228RB
M0 DZL M. Fitzpatrick, 3 Orchard Close, Yealmpton, Plymouth, PL8 2JQ
M0 DZM J. Enright, Flat 11/A, Cold Springs Farm, Manchester Road, Buxton, SK17 6ST
M0 DZO A. Hambidge, 25 Lock Drive, Stechford, Birmingham, B33 8AB
M0 DZT R. Clark, 10 Clarendon Road, Bournemouth, BH4 8AL
M0 DZV J. Maynard, 5 Farm Close, Crowthorne, RG45 6SE
M0 DZW I. Massey, 129 Church Street, Milnthorpe, LA7 7DZ
M0 DZX A. Nolan, 41 Taylor Street, Rochdale, OL12 0HX
M0 EAB S. Pochojka, 38 West Cotton Close, Northampton, NN4 8BY
M0 EAD J. Hart, 1 Cuckoo Nest, Harden, Bingley, BD16 1BD
M0 EAE K. Kotarba, 14 North Park, Bristol, BS15 1UW
M0 EAF R. Astbury, 14 Thornberry Drive Dy12Pl, Dudley, DY1 2PL
MM0 EAI D. Jamieson, Drumrae, Barbour Road, Helensburgh, G84 0JN
M0 EAK F. Cappleman, 8 The Woodlands, Lilleshall, Newport, TF10 9EN
M0 EAL G. Gauld, 251 Fleetwood Road South, Thornton-Cleveleys, FY5 5EA
M0 EAM G. Bourne, 72 Cornish Way, Royton, Oldham, OL2 6JY
MW0 EAN R. Westcott, 8 Pen Y Bigyn, Llanelli, SA15 1PB
M0 EAO C. Mcdonnell, Sunholme, Witham Bank West, Boston, PE21 8PU
M0 EAQ Dr T. Gale, 38 Lantree Crescent, Trumpington, Cambridge, CB2 9NJ
MM0 EAR Dalridia ARC, c/o K. Carroll, 32E Meadowburn Place, Campbeltown, PA28 6ST
M0 EAS P. Humphrey, 19 Donnay Close, Gerrards Cross, SL9 7PZ
MW0 EAT A. Phillips, 85 Gorseinon Road, Penllergaer, Swansea, SA4 9AB
M0 EAU D. Harrison, 21 Springfield Estate, Scopwick, Lincoln, LN4 3NP
MM0 EAX D. Thomson, Boat House, Finstown, Isle of Orkney, KW17 2EH
M0 EAY C. Knight, Shamrock, Stow Lane, Wisbech, PE13 2JU
M0 EAZ R. Bennett, Hawthorn Cottage, Mill Lane, Norwich, NR12 8HP
M0 EBD R. Chew, 1 Exeter Close, Aintree, Liverpool, L10 8LU
M0 EBG P. Good, 80 Meredith Road, Stevenage, SG1 5QS
M0 EBI Prof. N. Billingham, 19 Tumulus Road, Saltdean, Brighton, BN2 8FR
M0 EBJ S. Rope, 12 Edrich Close, Norwich, NR134JD
M0 EBN S. Bywater, Birch Wood, Norwich Road, Cromer, NR27 0HG
M0 EBO I. Bryant, 17 Kent Road, Southampton, SO17 2LJ
M0 EBP G. Joyce, 8 Christ Church Street, Preston, PR1 8PJ
M0 EBQ A. Medhurst, 44 Battle Road, Hailsham, BN27 1DS
M0 EBR M. Shuttleworth, 19 Edgerton Drive, Tadcaster, LS24 9QW
MW0 EBT J. Larden, Pavilla, 2 Craignant, Nantmel, Llandrindod Wells, LD1 6EW
M0 EBU K. Naylor, 35 Whiston Road, Northampton, NN2 7RR
M0 EBV H. Hepworth, The Leas, Main Street Bishop Wilton, York, YO42 1RX
M0 EBX M. Tween, 79 West Close, Fernhurst, Haslemere, GU27 3JS
M0 ECC E. Cassidy, 3 The Elms, Great Chesterford, Saffron Walden, CB10 1QD
MW0 ECF D. Edwards, 25 Bryn Coed, Gwersyllt, Wrexham, LL11 4UE
M0 ECK HMS Cavaller Radio Club, c/o B. Lucas, 8 Gilbert Close, Hempstead, Gillingham, ME7 3QQ
M0 ECL E. Lerpiniere, The Windmill, Millwrights, Colchester, CO5 0LQ
M0 ECM C. Martin, 14 Freeston Terrace, St. Georges, Telford, TF2 9HD
M0 ECP K. Fujita, Flat A, 33/F, Block 5, 10 Sheung Ning Road, Tseung Kwan O, Kowloon, Hong Kong, Kowloon, Hong Kong

M0 ECQ I. Nutt, Green Acres, Chapple Road, Newton Abbot, TQ13 9JY

M0 ECR East Cheshire Radio Group, c/o G. Hannan, 20 Arlington Drive, Stockport, SK2 7EB

M0 ECT M. Mrozinski, 6 Grange Road, Northampton, NN3 2AZ

MM0 ECV J. Sleet, Keppel Gate, Forglen, Turriff, AB53 4LX

M0 ECW East Cheam Wireless Society, c/o T. Watts, 26 Woodger Close, Guildford, GU4 7XR

M0 ECX Dr N. Rothwell Hughes, Cefn Glas, Clyro, Hereford, HR3 5JT

MW0 ECY H. Fingerhut-Holland, Osnok, 1 South Cliff Street, Tenby, SA70 7EB

M0 ECZ E. Williams, 22 Sherwood Avenue, Melksham, SN12 7HL

M0 EDA S. Hemmings, Leylands, Leigh Road, Frome, BA11 3LR

M0 EDE A. Holmes, 10 Doe Park, York, YO30 4UQ

MI0 EDF P. Hawthorne, 77 Pollock Drive, Lurgan, Craigavon, BT66 8JP

MU0 EDN B. Gray, 21 Auderville, Alderney, Guernsey, GY9 3XE

M0 EDO S. Williams, 6 Grasmere Road, Dewsbury, WF12 7PU

M0 EDP J. Barr, 41 Salisbury Road, London, E12 6AA

MW0 EDQ N. Heyne, New House, Hope, Welshpool, SY21 8JD

M0 EDR S. Pritchard, 18 Cumberland Avenue, Basingstoke, RG22 4BG

MW0 EDS G. Edwards, Ogwen Terrace, High Street Bethesda, Bangor, LL57 3AY

M0 EDU M. Wade, Watts Palace Cottage, Chitcombe Road, Rye, TN31 6EX

MW0 EDX A. Koval, 287 Heol Y Coleg, Newtown, SY16 1RA

M0 EEB M. Brady, 3 Bransdale, Worksop, S81 0XY

M0 EEG Dr C. Pomfrett, 17 Manifold Close, Sandbach, CW11 1XP

M0 EEH P. Halpin, 50 Celtic Road, Deal, CT14 9EF

M0 EEK D. Edgar, 31 Albany Villas, Hove, BN3 2RT

M0 EEL S. Connelly, 79 Pettycot Crescent, Gosport, PO13 0SJ

M0 EEP J. Nethercott, 30 Goldcrest Road, Chipping Sodbury, Bristol, BS37 6XF

MM0 EFI F. Wenseth, 2 Sunnybank Cottage, Logie Coldstone, Aboyne, AB34 5PQ

MM0 EFJ M. Donnachie, Roselea, Cluny Road, Dingwall, IV15 9NJ

MI0 EFM E. Mulligan, 27 Hillside Park, Belfast, BT9 5EL

MU0 EFR D. Robert, Nos Treis 7 Liberation Drive, Route Des Clos Landais, St. Saviour, Guernsey, GY7 9PH

M0 EGA R. Price, 2 Wordsworth Avenue, Easington Lane, Houghton le Spring, DH5 0NR

M0 EGC East of Greenwich Radio Amateur Club, c/o D. Green, 6 Garth Villas, Rimswell, Withernsea, HU19 2DB

M0 EGL C. Dunne, 48 Drury Road, Harrow, HA1 4BW

M0 EGN C. Lewis, 3 Sovereign Way, Calcot, Reading, RG31 4US

M0 EGV I. Robinson, 5 The Meadows, Bempton, Bridlington, YO15 1LU

M0 EHA G. Beam, 35A Moor Lane, York, YO24 2QX

M0 EHF Essex Raynet, c/o G. Tiller, 15 Woodlands Gardens, Romsey, SO51 7TE

M0 EHL M. Longbottom, 32 Ann'S Hill Road, Gosport, PO12 3JY

M0 EHS P. Shaw, 16 Sutherland Road, Cradley Heath, B64 6EA

M0 EIJ P. Micuda, 9 Manor Court, Manorgate Road, Kingston upon Thames, KT2 7AN

M0 EIW J. Walsh, 35 Carisbrooke Road, Bushbury, Wolverhampton, WV10 8AB

M0 EJB D. Banks, 9 Woodbank, Egremont, CA22 2RL

M0 EJF C. Briggs, 21 Peak View Road, Chesterfield, S40 4NW

M0 EJG J. Freeman, High Meadow, Martens Lane, Colchester, CO6 5AG

M0 EJL P. Kendall, 3 Hurstwood Close, Lincoln, LN2 4TX

M0 EJW Dr M. Bishop, 22 Herrington Road, Dorchester, DT1 2BS

M0 EKB G. Patrick, Athena, 121 Ringmer Road, Worthing, BN13 1DX

M0 ELA A. Mckenzie, Little Wishmore, Whitbourne, Worcester, WR6 5SR

M0 ELC L. Chesters, 5 Lingley Fields, Frizington, CA26 3RU

MM0 ELF W. Mccue, 188 Redburn, Alexandria, G83 9BU

M0 ELO C. Bootz, 10 Davenport Avenue, Nantwich, CW5 5QJ

MM0 ELP C. Maxwell, 29 Ambleside Rise, Hamilton, ML3 7HJ

M0 ELS J. Randall, 32 Holm Oak Close, Canterbury, CT1 3JL

MM0 EMC E. Mcpherson, 12 Lambourn, Wolfhill, Perth, PH2 6TQ

M0 EMD E. Deeley, 26 Eversley Crescent, Isleworth, TW7 4LS

M0 EME P. Tomlinson, 217 Old Hall Road, Tapton, Chesterfield, S40 1HQ

M0 EMM D. Martin, 14 Freeston Terrace, St. Georges, Telford, TF2 9HD

M0 EMR C. Wright, 8 Kendal Road, Sheffield, S6 4QG

M0 EMW E. Wheeler, 3 Praze Road, Porthleven, Helston, TR13 9LR

MI0 ENR R. Mcfadden, 36 Trinity Drive, Ballymoney, BT53 6EQ

M0 EOT B. Podmore, 78 Ridge Road, Stoke-on-Trent, ST6 5LP

M0 EOU J. Hinds, 83 Feckenham Road, Astwood Bank, Redditch, B96 6DE

MM0 EPC European Psk Club, c/o J. Phunkner, 7 Plenshin Court, Glasgow, G53 6QW

M0 EPR E. Rippon, 319 Beechdale Road, Nottingham, NG8 3FF

M0 EPX L. Stone, 32 Watermeadow Lane, Storrington, Pulborough, RH20 3GU

M0 EQD D. Wright, 22 West Hill, Rotherham, S61 2HB

MM0 EQE A. Thompson, 22 Lochend Road, Carnoustie, DD7 7QF

MW0 EQL J. Sneddon, 3 Marigold Close, Gurnos, Merthyr Tydfil, CF47 9DA

M0 EQM R. Agacy, 23 Highgate Lane, Bolton-Upon-Dearne, Rotherham, S63 8HR

M0 EQY M. Campbell, 377 Bushbury Lane, Wolverhampton, WV10 8JZ

M0 ERG Eagle Radio Group, c/o T. Stow, 38 The Strand, Mablethorpe, LN12 1BQ

M0 ERJ E. Jones, 37 Sluice Road, Denver, Downham Market, PE38 0DY

MM0 ERK B. Murray, Sherwood Cottage, Farnell, Brechin, DD9 6UH

M0 ERN E. Coleby, 13 Farm Close, Sunniside, Newcastle upon Tyne, NE16 5PP

M0 ERS J. Hauton, 15 Bourne Close, Lincoln, LN6 7DR

M0 ERY M. Samborskyy, St Johns College, St Johns Street, Cambridge, CB2 1TP

M0 ESB D. Baldwin, 356 Uppingham Road, Leicester, LE5 2BF

M0 ESP Ilera, c/o L. Marobin, Flat 60, Tudor Court, King Henrys Walk, London, N1 4NU

M0 ESR Alderley Explorer Scout A R U, c/o P. Phillips, 2 Millstream Close, Goostrey, Crewe, CW4 8JG

M0 ESU M. Bown, 47 Ullswater Crescent, Weymouth, DT3 5HF

M0 ESW E. Swanepoel, Bridge Cottage, Tinhay, Lifton, PL16 0AH

M0 ESZ M. Owens, 66 Woodlands Road, Bishop Auckland, DL14 7LZ

M0 ETA G. Andrews, 158 Latchmere Road, Kingston upon Thames, KT2 5TU

M0 ETE R. Home, Beech Cottage, The Green, Huntingdon, PE28 9NA

M0 ETP J. Bolton, 2 Patterdale Street, Hetton-Le-Hole, Houghton le Spring, DH5 0BH

M0 ETQ D. Bolton, 2 Patterdale Street, Hetton-Le-Hole, Houghton le Spring, DH5 0BH

M0 ETS C. Lyon, 3 Doodstone Avenue, Lostock Hall, Preston, PR5 5TY

M0 ETY S. Hindle, 35 Heyhead Street, Brierfield, Nelson, BB9 5BN

M0 EUI G. Plant, 11 Shinwell Grove, Stoke-on-Trent, ST3 7UG

M0 EUK G. Stoker, 32 Willow Way, Ponteland, Newcastle upon Tyne, NE20 9RF

M0 EUS A. Jones, 3 Warren Houses, Tile Lodge Road, Ashford, TN27 0BX

M0 EUY A. Swan, 47 Warren Close, Whitehill, Bordon, GU35 9EX

M0 EVE P. Beier, 20 Markham Avenue, Armthorpe, Doncaster, DN3 2AZ

M0 EVI A. Butler, 88 Manor Road, New Milton, BH25 5EJ

M0 EVK R. Kidd, 4 Oakfields Close, Norwich, NR4 6XH

M0 EVT E. Haralampiev, 13 South Hill Grove, Harrow, HA1 3PR

M0 EWG B. Read, 111 Fitzpain Road, West Parley, Ferndown, BH22 8SF

M0 EWW R. Moreton, 25 Holyoake Place, Rugeley, WS15 2NP

M0 EXM B. Wheeler, 2 Rose Street, Houghton le Spring, DH4 5BB

MW0 EYE Dr C. Varghese, 15 Crestacre Close Newton, Swansea, SA3 4UR

M0 EYT P. Marsh, 10 Pardys Hill, Corfe Mullen, Wimborne, BH21 3HW

M0 EZO R. Griffin, 15 Donside Close, Boldon Colliery, NE35 9BS

M0 EZP D. Brewerton, 10 Porter Avenue, York, YO19 4AG

M0 FAK R. Chick, 15 Bonfire Close, Chard, TA20 2EG

MU0 FAL C. Fallaize, Lorbert, Pleinheaume Road, Vale, Guernsey, GY6 8NR

M0 FAR A. Rawson, 64 Dukes Mead, Fleet, GU51 4HE

M0 FAT A. Moffatt, 10 Fenn Road, Barnsley, S75 3DE

M0 FAZ D. Fower, 31 Hillswood Avenue, Leek, ST13 8EQ

M0 FBB S. Smale, 230 Wareham Road, Corfe Mullen, Wimborne, BH21 3LW

M0 FBM A. Lock, 2 Knutscroft Lane, Thurloxton, Taunton, TA2 8RL

MU0 FBO R. Stockwell, Fleurs Des Champs, La Colline Des Bas Courtills, Saint Saviours, Guernsey, GY7 9YQ

M0 FCA W. Mannerfelt, 16 Suffolk Road, London, SW13 9NB

M0 FCB F. Brunt, 74 Bardley Crescent, Tarbock Green, Prescot, L35 1RJ

M0 FCD M. Christieson, September Cottage, Rushlake Green, Heathfield, TN21 9PP

M0 FCG M. Hardy, 4 Kirk Balk Hoyland, Barnsley, S74 9HU

M0 FCI D. Houghton, 106 Lawn Avenue Woodlands, Doncaster, DN6 7TT

MM0 FCM C. Sheridan, 8 South Park Grove, Biggar, ML12 6GJ

M0 FCP F. Parsons, 17 Hannah More Close, Wrington, Bristol, BS40 5QG

M0 FCR A. Crespo, 266 Trinity Road, London, SW18 3RQ

M0 FCT P. Ryder, 6 Lodge Drive, Moulton, Northwich, CW9 8RQ

M0 FCW M. Ballard, 41 Middlefield Avenue, Halesowen, B62 9QJ

M0 FCY D. Thornton, 63 Houghtonside, Houghton le Spring, DH4 4BW

MW0 FDG J. Myszka, 65 Heol-Y-Frenhines, Bridgend, CF31 4RN

M0 FDX G. Marsden, 71 Sedgley Avenue, Rochdale, OL16 4TY

M0 FEU M. Pullan, Hauptstrasse 13/2, St Radegund Bei Graz, Austria, 8061

M0 FEY E. Fey, 56 Bathurst Close, Burnham on Sea, TA8 2SZ

MM0 FFC I. Douglas1, 15 Henderson Crescent, Broxburn, EH52 6HA

M0 FFX T. Wootten, Trinity Hall, Cambridge, CB2 1TJ

M0 FGA D. Mccarty, Po Box :4910, the Woodlands Tx, USA, 77387

M0 FGB F. Buck, 89 Marsh St., Barrow in Furness, LA14 2AD

M0 FGC T. Seed, 1A, Purok, Brngy Arenas. Arayat, Pampanga, Philippines

M0 FHM D. Sunderland, 1 Allfield Cottages, Condover, Shrewsbury, SY5 7AP

M0 FIL P. Graham, 29 Lancaster Street, Colne, BB8 9AZ

M0 FIS P. Fisher, 12 St. Anns Avenue, Grimsby, DN34 4PW

MD0 FIX N. Wallace, 85 Erin Vale, Lezayre Park, Ramsey, IM8 3PU

M0 FJM J. Hudson, 62 Old Road, Churwell, Morley, Leeds, LS27 7RT

M0 FJS F. Stevenson, 33 Highfield Close, Amersham, HP6 6HG

M0 FLC I. Pollard, Ilderton Glebe Cottage, Ilderton, Alnwick, NE66 4YD

M0 FLF C. Wilson, 12 Desmond Avenue, Hornsea, HU18 1AF

M0 FMT P. March, Devonholme, Bedford Road, Hitchin, SG5 3RX

M0 FMY J. Mallichan, 17 Napier Road, Gillingham, ME7 4HB

M0 FOG N. Brereton, 10 Coverdale Close, Stoke-on-Trent, ST3 7RZ

M0 FOR G. Mcinnes, 32 Bis Rue D'Ezy, 32 Bis Rue D'Ezy, Ivry la Bataille, France, 27540

M0 FOX P. Leicester, 30 Knighton Street, North Wingfield, Chesterfield, S42 5JA

M0 FPA R. Etchells, 6 Woodbank Court, Canterbury Road, Manchester, M41 7DY

M0 FPQ C. Groves, 7 Mill House, Mill Lane, Wick, Littlehampton, BN17 7PJ

M0 FRA T. Fray, 20 St. Catherines Road, Blackwell, Bromsgrove, B60 1BN

M0 FRC Franklin Radio Group, c/o R. Topliss, 12 Dorothy Avenue, Skegness, PE25 2BP

M0 FRD M. Taylor, 10 Piers Road, Glenfield, Leicester, LE3 8BN

M0 FRG A. Howard, 4 Woodgarth Avenue, Manchester, M40 1QE

M0 FRH I. Fraser, Flat 2, 77 Bayford Road, Littlehampton, BN17 5HN

M0 FRS C. Boys, 34 Firacre Road, Ash Vale, Aldershot, GU125JT

MW0 FRY R. Fry, Old Police Station, Parkmill, Swansea, SA3 2EQ

M0 FSH N. Harris, 23 Winchester Avenue, Chatham, ME5 9AR

M0 FSK F. Kennedy, 3 Brookfield View, Bolton Le Sands, Carnforth, LA5 8DJ

M0 FSN P. Wells, 7 Kings Meadow, Overton, Basingstoke, RG25 3HP

M0 FTL R. Metcalfe, 33 Midland Terrace, Hellifield, Skipton, BD23 4HJ

M0 FTR Five Towns ARC, c/o J. Adlington, 23 Newstead Road, Abbey Hulton, Stoke-on-Trent, ST2 8HU

MW0 FUN B. Fisher, 93 Heol Llanelli, Pontyates, Llanelli, SA15 5UH

M0 FVD M. Kittika, 51 Overlea Drive, Burnage, Manchester, M19 1QY

M0 FVV A. Hanner, 8 Countryside Farm Park, Church Lane, Steyning, BN44 3HF

MM0 FWG R. Gaisford, 9 Rattray St., Boness, EH51 9PE

M0 FWM F. Mifflin, Windsor House, Harras Road, Whitehaven, CA28 6SG

M0 FWO J. Thomson, 51 Birch Avenue, Cuerden Residential Park, Leyland, PR25 5PD

M0 FXB A. Macrides, 16 Newtons Road, Kewstoke, Weston-Super-Mare, BS22 9LG

M0 FXX R. Limb, Charnwood House, Station Road, Henley-on-Thames, RG9 3JS

M0 FYA A. Young, 39 Thornton Drive, Hoghton, Preston, PR5 0LX

M0 FZR R. Wickenden, Selwood Cottage, Moor Lane, Wincanton, BA9 9EJ

M0 FZU C. Walcott, 28 Balfour Road, London, W13 9TN

MM0 FZV G. Bourhill, 30C Salters Road, Wallyford, Musselburgh, EH21 8AA

M0 FZW N. Crampton, 7 Barneveld Avenue, Canvey Island, SS8 8NZ

M0 FZX S. Norman, 27 Ashburton Road, Ickburgh, Thetford, IP26 5JA

M0 GAC G. A'Court, Three Oaks, Greenhill Lane, Winscombe, BS25 5PE

M0 GAE G. Errington, 22 Willoughby Drive, Whitley Bay, NE26 3DY

M0 GAG R. Burrell, 38 Standhill Crescent, Barnsley, S71 1SU

M0 GAH A. Cunningham, 23 Heathgate Close, Birstall, Leicester, LE4 3GW

MM0 GAI T. Spencer, 3 Tramore Crescent, Prestwick, KA9 1LT

MM0 GAL J. Vilar Ameijeiras, 76 Willowbank Road, Aberdeen, AB11 6XL

M0 GAN P. Street, 50 Dickson House, Ridgway Road, Stoke on Trent, ST1 3BA

M0 GAQ K. Ingram, 15 Kent Avenue, East Cowes, po326qn

M0 GAV A. Burton, 51, Wilcox Road Foxhill, Sheffield, S6 1BQ

M0 GAX C. Ponder, 12 Wood Street, Doddington, March, PE15 0SA

M0 GBA G. Allison, 16 Copse Road, Plymouth, PL7 1PZ

M0 GBB Winteringham Wireless Society, c/o D. Ogg, 36 Cliff Road, Winteringham, Scunthorpe, DN15 9NQ

M0 GBC W. Jefferies, 26 Norcutt Road, Twickenham, TW2 6SR
M0 GBF B. Findler, 1 Gordon Avenue, Stoke-on-Trent, ST6 2LY
M0 GBH C. Johnston-Stuart, 35 Robbins Close, Bradley Stoke, Bristol, BS32 8AS
M0 GBK S. Nash, 6 Berry Road, Meltham, Holmfirth, HD9 5PL
M0 GBO J. Zemlicka, 37 Ascot Gardens, Southall, UB1 2SA
MI0 GBU Causeway Radio Club, c/o N. Bolt, 32 Bush Gardens, Bushmills, BT57 8AE
MW0 GBW B. Collins, 58 Brockhill Way, Penarth, CF64 5QD
M0 GBZ E. Mcpherson, 138 Shephall View, Stevenage, SG1 1RR
M0 GCA T. Sheridan, 4 Stane Close, Bishop's Stortford, CM23 2HU
M0 GCB T. Kim, 9 Temeraire Heights, Folkestone, CT20 3TL
M0 GCC G. Cook, 61 Mortomley Lane, High Green, Sheffield, S35 3HS
MM0 GCF J. Brown, 78 Egilsay St., Glasgow, G22 7RG
M0 GCH G. Holmes, 11 Claudeen Close, Southampton, SO18 2HQ
M0 GCI M. Rowley, Flat 23, Minstrel Court, 170 High Street, Harrow, HA3 7AX
M0 GCN D. Bourne, 47 Bondend Road, Upton St. Leonards, Gloucester, GL4 8DZ
M0 GCQ G. Soteriou, Whiteleys Cottages, 32 Mornington Avenue, London, W14 8UW
M0 GCR G. Rumsey, 13 Greenhills Road, Northampton, NN2 8EL
MW0 GCS L. Powell, 2 Gelliderw Pontardawe, Swansea, SA8 4NB
MW0 GCT S. Tweddle, 3 Bron Ffinan, Pentraeth, LL75 8UT
M0 GCU J. Jordan, The Cottage, Papcastle, Cockermouth, CA13 0LA
MI0 GCV T. Conlon, 7 Waringfield Gardens, Moira, Craigavon, BT67 0FQ
M0 GCX V. Narinian, 15 Headley Gardens, Great Shelford, Cambridge, CB22 5JZ
MM0 GCY C. Bewley, Fodderlee Dell, Hawick, TD9 8JE
M0 GDC R. Bakken, Flat 6, Marlow House, Abbey Street, London, SE1 3DW
MM0 GDG Dr A. Curlis, 94 Kirkhill Road, Aberdeen, AB11 8FX
M0 GDH G. Hogg, 7 Elbra Farm Close, Ellenborough, Maryport, CA15 7RG
MM0 GDI B. Massie, Flat 4/R, 74 Commercial Street, Dundee, DD1 2AP
M0 GDJ B. Holt, 36 Tenter Hill Lane, Sheepridge, Huddersfield, HD2 1EJ
MM0 GDL D. Lindsay, 114 Strathblane Road, Milngavie, Glasgow, G62 8HD
MW0 GDM P. De Mengel, Fern Cottage, 1 The Gail, Haverfordwest, SA62 4HJ
M0 GDP R. Parkinson, 18 Lime Tree Gardens Lowdham, Nottingham, NG14 7DJ
M0 GDS G. De Sousa, 29 Slippers Hill, Hemel Hempstead, HP2 5XT
M0 GDT D. Holland, 13 Linley Drive, Boston, PE21 7EJ
M0 GDU R. North, 24 Gadesden Road, Epsom, KT19 9LB
M0 GDV D. Harbron, 6 West View, Penshaw, Houghton le Spring, DH4 7HP
M0 GDX D. Hayes, 43 Linden Avenue, Sheffield, S8 0GA
M0 GEB G. Beesley, Stone Barn, Black Dog, Crediton, EX17 4QX
M0 GEC G. Clennell, 69 Seventh Row, Ashington, NE63 8HX
M0 GED F. Holt, 8 Pleasant View, Coppull, Chorley, PR7 4PH
M0 GEF G. Freeman, 11 Westward Road, Malvern, WR14 1JX
MW0 GEI S. Walmsley, 29 Shelley Court Machen, Caerphilly, CF83 8TT
M0 GEK L. Gange, 15 Ham Close, Worthing, BN11 2QE
M0 GEL S. Attwood, 60 Underwood Avenue, Ash, Aldershot, GU12 6PL
M0 GEN G. Barusevicus, 6 Middlebrook Crescent, Bradford, BD8 0EN
MM0 GEO G. Muir, 51 Lindsay Way, Livingston, EH54 8LQ
M0 GEP A. Holdup, Tunnel Farm, Tunnel Rd, Imbil (PO 155), Australia, 4570
M0 GEU A. Nicholson, 14 Rossinyol, Los Arcos, Alicante, Spain, 3530
M0 GEX C. Farley, 1 Wesley Cottages, Mutley, Plymouth, PL3 4RB
M0 GEY M. Spinks, 26 Church Hill, Royston, Barnsley, S71 4NH
MM0 GFA C Roy Hill Club, c/o P. Mcbride, 1 Hillside, Croy, Glasgow, G65 9HJ
M0 GFD J. Smith, 3 Glamis Road, London, E1W 3EE
MI0 GFE Antrim & District ARS, c/o R. Robinson, 31 Brantwood Gardens, Antrim, BT41 1HP
M0 GFF B. Courtney, 44 Uxbridge Road, Rickmansworth, WD3 7AR
M0 GFJ D. Russell, Halfway House, Holbrook Road, Ipswich, IP9 1BP
M0 GFK A. Ochot, 6 Ufford Close, Harrow, HA3 6PP
M0 GFM G. Dawson, Bramwell, Winchester Road, Southampton, SO32 2LG
M0 GFN J. Bell, 50 Colchester Terrace, Sunderland, SR4 7RY
M0 GFO R. Mcdermott, 2 Monument Close, Wellington, TA21 9AL
MM0 GFP A. Stewart, 21 Mansfield Avenue, Newtongrange, Dalkeith, EH22 4SJ
MM0 GFR G. Forster, 4 Kirk Brae, Morvern, Oban, PA80 5XW
M0 GFX P. Hull, 1 Sawpits Close, Stogumber, Taunton, TA4 3TX
MI0 GGB S. Quigg, 100 Whispering Pines, Limavady, BT49 0UF
MM0 GGD G. Duncan, 5 Jarvis Place, Carnoustie, DD7 7BR
MM0 GGG D. Banks, 60 Leander Crescent, Bellshill, ML4 1JB
M0 GGH J. Gubric, 279 Nottingham Road, Eastwood, Nottingham, NG16 2AP
M0 GGK D. Lawson, 30 Meadowcroft, St Helens, WA9 3XQ
M0 GGL C. Lester, 21 Barwell Way, Witham, CM8 2TY
M0 GGM G. Markey, Trebrown Farm, Horningtops, Liskeard, PL14 3PU
M0 GGO C. Loughran, 8 Douglas Road, Dover, CT17 0BD
M0 GGP Angel of The North ARC, c/o S. Townsley, 222 Prince Consort Road, Gateshead, NE8 4DX
M0 GGQ A. Shaw, 2 Dowber Court, Thirsk, YO7 1SP
M0 GGT B. Ashton, 31 Home Close Renhold, Bedford, MK41 0LB
M0 GGU G. Medlicott, Lower Medlicott Farm, Wentnor, Bishops Castle, SY9 5EL
M0 GGW G. Milsom, 31 Chichester Close, Bowerdean Road, High Wycombe, HP13 6AU
M0 GGX J. Patient, 4 Bucklebury Heath, South Woodham Ferrers, Chelmsford, CM3 5ZU
M0 GGZ A. Chaplin, 33 The Crofts, Little Wakering, Southend-on-Sea, SS3 0JS
M0 GHA M. Dudley, 8 Woolpack Meadows, North Somercotes, Louth, LN11 7QG
M0 GHC M. Wojcik, 43 Connaught Road, London, W13 0TF
M0 GHE L. Nordgren, 41 Forest Road, London, E7 0DN
MI0 GHI A. Murphy, 13 Torrens Park, Lislagan Upper, Ballymoney, BT53 7DE
MM0 GHK T. Lee, 54 Shielfield Terrace, Tweedmouth, Berwick-upon-Tweed, TD15 2EE
MM0 GHM G. Cochrane, 33 Portland Road, Galston, KA4 8EA
MM0 GHN N. Inglis, Orchard House 35 Portland Park, Hamilton, ML3 7JY
M0 GHO G. Hopkins, 27 The Templars, Worthing, BN14 9JT
M0 GHR I. Millar, 3A South Street, Wiveliscombe, Wiveliscombe, TA4 2LZ
MM0 GHT K. Brown, 21 Strain Crescent, Airdrie, ML6 9ND
M0 GHV S. Young, 27A Norton Road, London, E10 7LQ
M0 GHX C. Painter, 45 Meadow Lane, Beeston, Nottingham, NG9 5AE
M0 GHY P. Hollas, 46 Askham Fields Lane, Askham Bryan, York, YO23 3PS
M0 GHZ D. Millard, Weavern House, Hartham Lane, Biddestone, Chippenham, SN14 7EA
M0 GIA S. Amesbury, 222 Bond Street, Macclesfield, SK11 6RG
M0 GIB D. Gibbons, 4 Ivychurch Mews, Runcorn, WA7 5AR
M0 GID G. Dunne, Swan Hotel, Market Place, Sturminster Newton, DT10 1AR
M0 GIE P. Ellis, 40 Grasmere Road Royton, Oldham, OL2 6SR
M0 GIF R. Manser, Flat 38, St. Johns Court, Portsmouth, PO2 8NA
M0 GIG D. Wharlley, 15 Crampton Court, Grosvenor Road, Broadstairs, CT10 2XU
MI0 GIJ J. Thompson, 119 Rathkyle, Antrim, BT41 1LN
M0 GIL G. Wildman, 55 Hill Street, Bradley, Wolverhampton, WV14 8SB
M0 GIM Dr G. Mitchener, Cabins, Wenham Road, Ipswich, IP8 3EY
MW0 GIN S. Peel, 28 Dan Yr Allt, Llanelli, SA14 8AT
M0 GIP W. James, 3 Midfield Close, Gillow Heath, Stoke-on-Trent, ST8 6RD
M0 GIQ G. Griffiths, 16 Back Lane, Winteringham, Scunthorpe, DN15 9NW
M0 GIU P. Tier, 16A Burcombe Road, Bournemouth, BH10 5JT
M0 GIW D. Ryan, Goosehill House, West Street, Thorne, Doncaster, DN8 5QU
M0 GIY Dr P. Swansbury, 119A Trelowarren Street, Camborne, TR14 8AW
M0 GIZ C. Melia, 45 Sheriff Street, Hartlepool, TS26 8EZ
M0 GJA K. Nyquist, 25 Marsh View, Newton, Preston, PR4 3SX
MM0 GJC G. Costa, 54 High Street, Dollar, FK14 7BA
M0 GJD J. Farrant, Orchard Cottage, Claycastle, Crewkerne, TA18 7PB
M0 GJH A. Vine, Hilden, Woodland Avenue, Cranleigh, GU6 7HZ
MM0 GJJ G. Johnson, Speur Mor, Gifford Road, Longformacus, Duns, TD11 3NZ
M0 GJK G. Knight, 20 Crossway, Welwyn Garden City, AL8 7EE
M0 GJL R. Brodie, 8 West View Terrace, Main Road, Salcombe, TQ8 8AB
MI0 GJN M. Edwards, 15 Highgrove Road, Carrickfergus, BT38 9AG
M0 GJS G. Suter, 1 Laburnham Lodge, Worthing, BN13 3dn
M0 GJU J. Freeman, 38 City Road, Cambridge, CB1 1DP
M0 GJV E. Jones, 43 Wesley Road, Wimborne, BH21 2QB
M0 GJX A. Jordan, 21 Madison Avenue, Exeter, EX1 3AH
M0 GKA S. Len, 13 Griffiths Gardens, Caversfield, Bicester, OX27 8FL
MM0 GKB K. Mackintosh, Chez Nous, Scatwell, Strathconon, Muir of Ord, IV6 7QG
M0 GKD A. Mockford, 29 Kingston Close, Blandford Forum, DT11 7UQ
M0 GKG R. Ley, 23 Heronbridge Close, Westlea, Swindon, SN5 7DR
M0 GKJ R. Frencham, 1 Eggerslack Cottages, Windermere Road, Grange-over-Sands, LA11 6EX
M0 GKK Surrey Space Centre, c/o D. Fishlock, 93 Shackstead Lane, Godalming, GU7 1RL
MI0 GKL Bushvalley ARC, c/o J. Traynor, 8 Roeville Terrace, Limavady, BT49 0BH
MM0 GKN J. Hogg, 31 Woodlea Court, Crosshouse, Kilmarnock, KA2 0ES
M0 GKO G. Thorpe, 81 Knoll Drive, Coventry, CV3 5PJ
M0 GKP Dr P. Haydn Smith, 3 King Henrys Road, Lewes, BN7 1BT
M0 GKR A. Steel, 78 Water Meadows, Worksop, S80 3DB
MM0 GKT Dr D. Bushby, Coach House, Dalginross, Crieff, PH6 2HB
MM0 GKU T. Mccall, 119 Claremont, Alloa, FK10 2ER
MW0 GKV M. Williams, 30 Elm Drive, Risca, Newport, NP11 6HJ
MW0 GKW M. Sullivan, 14 Clerks Court, Mill Lane, Welshpool, SY21 7JA
M0 GLE G. Tomkins, 12 Jarrah Court, Bathurst, Australia, 2795
M0 GLF Dr J. Stanford, 1941 Ute Creek Drive, Longmont, Co, USA, 80504
MI0 GLG T. Currie, 26 High Street, Portaferry, Newtownards, BT22 1QT
M0 GLI M. Shasby, 19 Crawshaw Grange, Crawshawbooth, Rossendale, BB4 8LY
MD0 GLK A. Dorman, 1 Sprucewood Rise, Foxdale, Douglas, Isle Of Man, IM4 3JP
M0 GLL C. Smith, 2 Blankney Close, Fareham, PO14 3RX
M0 GLP G. Parker, 420 Meadow Lane, Nottingham, NG2 3GD
M0 GLQ C. Senior, 13 Oak Crescent, Woolaston, Lydney, GL15 6PF
MW0 GLS S. Day, 2 Cae Job, Piercefield Lane, Aberystwyth, SY23 1RJ
M0 GLT M. Rosenbrand, 12 Greville Road, Cambridge, CB1 3QL
M0 GLU A. Vincz, 9 St. Brelades Road, Crawley, RH11 9RQ
M0 GLV M. Jusko, 13 Ellerby Grove, Hull, HU9 3PR
MM0 GLX B. Burt, 182 Old Inverkip Road, Greenock, PA16 9JG
M0 GMA G. May, 95 Moorfield Avenue, Denton, Manchester, M34 7TX
M0 GMC G. Collier, 6 Copse Close, Tilehurst, Reading, RG31 6RH
M0 GMD D. Gray, 68 Endeavour Way, Hythe Marina Village, Southampton, SO45 6LA
M0 GME G. Ellis, 46 The Uplands, Scarborough, YO12 5HX
M0 GMG R. Bell, 92 Dean Drive, Wilmslow, SK9 2EY
MW0 GMH O. Williams, 39 Camden Road, Maes-Y-Coed, Brecon, LD3 7RT
M0 GMI C. Woodbridge, 53 Baffins Road, Portsmouth, PO3 6BE
M0 GMK C. Dawson, 9 Mulberry Close, Poringland, Norwich, NR14 7WF
M0 GMN W. Owen, 8 Sandhurst Avenue, Lytham St Annes, FY8 2DA
M0 GMO P. Cheshire, 29 Madison Avenue, Exeter, EX1 3AH
M0 GMQ P. Hall, 13 Sheard Avenue, Ashton-under-Lyne, OL6 8DS
M0 GMS Rev. S. Smith, 5 Melhuish Close, Witheridge, Tiverton, EX16 8AZ
M0 GMT D. Clapp, 150 Brougham Road, Worthing, BN11 2PH
M0 GMU P. Sweatman, 62 Barncroft Way, Waterlooville, PO93AQ
M0 GMW C. Watts, 10 Kemble Gardens, Bristol, BS11 9RY
MW0 GMZ H. Hughes, 21 Maes Geraint, Pentraeth, LL75 8UR
M0 GNA A. Shaw, 21 Laburnum Road, Prenton, CH43 5RP
M0 GNB M. Dreszer, 47 Lydgate Court, Nuneaton, CV11 5RR
M0 GNC A. Ellison, 24 The Grove, Brentwood, CM14 5NS
MW0 GNF C. Hughes, 26 Tan Y Bryn, Valley, Holyhead, LL65 3ES
M0 GNG S. Gee, 20 Nesham Place, Houghton le Spring, DH5 8AG
MM0 GNH K. Foreman, 16 Beveridge Place, Kinross, KY13 8QY
M0 GNJ D. Coventry, 1 Seacrest Avenue, Fleetwood, FY7 6FG
M0 GNK R. Jennings, 8769 Greengrass Way, Parker, USA, 80134
M0 GNL Dr C. Price, Byways, Taylors Lane, Chichester, PO18 8QQ
M0 GNM Dr H. Donnelly, 6 Famet Walk, Purley, CR8 2DY
M0 GNO E. Whiten, 17 Scott Close, Ashby-de-la-Zouch, LE65 1HT
M0 GNP D. Salter, 142 Brays Road, Birmingham, B26 2PP
MM0 GNS C. Stewart, 9 Rousay Wynd, Kilmarnock, KA3 2GP
M0 GNU D. White, 3 West Street, South Normanton, Alfreton, DE55 2AJ
MM0 GNX A. Messner, 6 Elistoun Drive, Tillicoultry, FK13 6NT
M0 GNY M. Zlobinski, 16 Birkdale Avenue Atherton, Manchester, M46 9PY
M0 GOA J. Goacher, 41 Clay Hill, Two Mile Ash, Milton Keynes, MK8 8AY
M0 GOB B. Holland, 11 Silverlands Park, Buxton, SK17 6QX
M0 GOC T. Ward, 1 Darrismere Villas, Edinburgh Street, Hull, HU3 5AS
MM0 GOF J. Mcculloch, 2 Riverbank Wynd, Gatehouse Of Fleet, Castle Douglas, DG7 2EA
MM0 GOG D. Baillie, 126 Main St., Fauldhouse, Bathgate, EH47 9BW
M0 GOH P. Preston, 49 Cowpasture Lane, Sutton-in-Ashfield, NG17 5AR
M0 GOI K. Hornby, 39 Parkland View, Barnsley, S71 5LG
M0 GOK D. Richards, 73 Greenfields Avenue, Alton, GU34 2EW
M0 GOL T. Goldsmith, 37 Cowdray Road, Sunderland, SR5 3PG
MW0 GOM J. Roissetter, 2 The Willows Usk Road, Caerleon, Newport, NP18 1JB

MM0 GON G. Craig, 1 Butt Avenue, Helensburgh, G84 9DA
M0 GOO J. Brook, The Clock Tower, Rectory Lane, Chichester, PO20 9DT
M0 GOP G. Oliver, 17 Jack Stephens Estate, Penzance, TR18 2QE
M0 GOQ J. Barbieri, 20 Gilbard Court, Chineham, Basingstoke, RG24 8RG
M0 GOT S. Martin, 3 Houndsmill, Horsington, Templecombe, BA8 0ED
MW0 GOV C. Davis, 132 Steynton Road, Steynton, Milford Haven, SA73 1AN
M0 GOW C. Gowing, 5 Curson Road, Tasburgh, Norwich, NR15 1NH
MW0 GOX J. Davies, Penralt, Abercaseg Road, Bangor, LL57 3SP
M0 GOY R. Klima, 87 Campion Avenue, Hull, HU4 7AR
MI0 GOZ V. Maksimavicius, 14 Leckagh Walk, Magherafelt, BT45 6JU
MI0 GPB G. Bunting, 8 Moor Park Avenue, Belfast, BT10 0QE
M0 GPC S. Withnall, 2 Lansdown Close, Cheltenham, GL51 6QP
M0 GPD J. Fuller, Bramble Cottage, Leggatt Hill, Lodsworth, Petworth, GU28 9DP
M0 GPE T. Loker, 24 St. Albans Hill, Hemel Hempstead, HP3 9NG
M0 GPG G. Dyson, 111 Chester Road, Whitby, Ellesmere Port, CH65 6SB
M0 GPH T. Hall, 18 Common Lane New Haw 18 Comman Lane New Haw, Addlestone, kt153lh
M0 GPJ D. Waite, 1 Naseby Court, Bradville, Milton Keynes, MK13 7EP
M0 GPK Dr W. Gasser, 76 Empress Road, Derby, DE23 6TE
MM0 GPL C. Jones, Croy Lodge, Shandon, Helensburgh, G84 8NN
M0 GPN B. Taylor, 11 Halyard Croft, Hull, HU1 2EP
M0 GPO G. Otter, 3 Glen Park Avenue, Glenfield, Leicester, LE3 8GH
MW0 GPP B. Doyle, 3 Bryn Road, Flint, CH6 5HU
M0 GPQ A. Wieckowski, 110 Scrubs Lane, London, NW10 6QY
M0 GPU A. Norrie, 45 Eastern Way, Ponteland, Newcastle upon Tyne, NE20 9RD
M0 GPV W. Tommasini, 49 Taverner Close, Poole, BH15 1UP
M0 GPW P. Andrew, 13 Luke Road, Aldershot, GU11 3BW
M0 GPX B. Wagiel, 116 Hurworth Avenue, Slough, SL3 7FQ
M0 GPY Prof. H. Schmidt, 88 Candlemas Lane, Beaconsfield, HP9 1AE
MM0 GPZ G. Paterson, 33 Parkneuk Road, Blantyre, Glasgow, G72 0TR
M0 GQB M. Cox, Haugh Shaw Hall Haugh Shaw Road, Halifax, HX1 3LE
M0 GQD J. Parrett, 6 Shelley Road, East Grinstead, RH19 1TA
M0 GQE G. Moss, 10 Thistlegreen Road, Dudley, DY2 9JT
MM0 GQF Z. Biorka, 150 St. Michaels Road, Newtonhill, Stonehaven, AB39 3XW
MI0 GQG B. Crozier, 33 Cullentragh Road, Poyntzpass, Newry, BT35 6SD
MI0 GQI M. Crozier, 33 Cullentragh Road, Poyntzpass, Newry, BT35 6SD
M0 GQJ D. Downer, 19 Watergate Road, Newport, PO30 1XN
M0 GQM T. Kidwell, 2 Batts Farmyard, Wilton, Marlborough, SN8 3SS
MW0 GQO R. Burton, 6 Troed Y Garn Llangybi, Pwllheli, LL53 6DQ
M0 GQP B. Sims, 4 New Cottages, Cranwich Road, Thetford, IP26 5EQ
M0 GQR A. Matheson, 21 Warren Hill Road, Woodbridge, IP12 4DU
M0 GQS D. Roguszczak, 4 Home Farm Close, Reading, RG2 7TD
M0 GQT S. Hubin, 10 Coneygeare Court, Eynesbury, St Neots, PE19 2UL
M0 GQU M. Burzynski, 2 Somerset Avenue, Luton, LU2 0PJ
M0 GQV P. Langabeer, 1 Newfield Crescent, Middlesbrough, TS5 8RE
M0 GQW E. Tart, Sunnybank Farm, Wattlesborough Heath, Shrewsbury, SY5 9EG
M0 GRA G. Hickford, 56 Alexander Close, Abingdon, OX14 1XB
M0 GRB N. Harris, 45 Sleigh Road, Sturry, Canterbury, ct2 0ht
M0 GRE W. Greenall, 356 Warrington Road, Abram, Wigan, WN2 5XA
M0 GRF D. Blyth, 45 Clarence Road, Bilston, WV14 6NZ
MI0 GRG M. Mcgrory, 10 Bramley Court, Red Lion Road, Kilmore, Armagh, BT61 8ND
M0 GRH G. Hart, 55 Runswick Drive, Nottingham, NG8 1JE
M0 GRI R. Ingham, 84 Marian Court, Gateshead, NE8 2JB
MW0 GRJ G. Jones, 12 Field Close, Flint, CH6 5RQ
MI0 GRN A. Cartin, 64 Ashgrove Park, Magherafelt, BT45 6DN
M0 GRP G. Priestley, 53 Millfield Gardens, Crowland, Peterborough, PE6 0HA
M0 GRR S. Turner, 12 Park Street, Morecambe, LA4 6BN
M0 GRT N. Rotari, 81 School Road, Dagenham, RM10 9QD
M0 GRU A. Webb, 17 Dickins Way, Horsham, RH13 6BQ
M0 GRW C. Gibbs, 112 Barkham Ride, Finchampstead, Wokingham, RG40 4EN
M0 GRX Aldridge & Barr Beacon ARC, c/o E. Roberts, 117 Walstead Road, Walsall, WS5 4LU
M0 GRY G. Collis, 16 Hill Grove, Barrow Hill, Chesterfield, S43 2NW
MW0 GRZ G. Woloszun, 44 Cowbridge Road, Bridgend, CF31 3DA
M0 GSC Rev. M. Bracci, 12 Bowling Green Close, Bognor Regis, PO21 4HB
M0 GSI C. Nelmes, 119 Exeter Road, Dawlish, EX7 0AN
M0 GSK M. Silver, 52 Park Crescent, Elstree, Borehamwood, WD6 3PU
M0 GSL G. Walker, 6 Tenbury Drive, Shrewsbury, SY2 5YB
M0 GSN P. Newton, 61 Ashbourne Crescent, Taunton, TA1 2RA
M0 GSO R. Harris, 1 Hollinhey Close, Bootle, L30 7RN
M0 GSP S. Palmer, 58 Highlands Way, Whiteparish, Salisbury, SP5 2SZ
MM0 GSQ A. Young, 4/4 Prestonfield Terrace, Edinburgh, eh165ee
MW0 GSR S. Poyser, Glandwr Snowdon Street, Porthmadog, LL49 9DF
MM0 GSS G. Smith, 40 Pirleyhill Drive, Shieldhill, Falkirk, FK1 2EA
MM0 GSW I. Wishart, 7 Cairngorm Crescent, Kirkcaldy, KY2 5RF
M0 GSX P. Stocking, 6 Royal Oak Road, Rowley Regis, B65 8NX
MU0 GSY L. Roithmeir, Flat 2, Maison Ramee, Route De La Ramee, St. Peter Port, Guernsey, GY1 2ES
M0 GSZ G. Starling, 4 Three Corner Drive, Norwich, NR6 7HA
M0 GTE P. Allen, 6 Helston Close, Wigston, LE18 2JH
M0 GTH R. Killen, 3 Great Charles Close, St. Stephen, St Austell, PL26 7PW
MI0 GTI A. Jamison, 11 Richmond Gardens, Newtownabbey, BT36 5LA
M0 GTJ R. Henderson, 14 Oxford Avenue, St Albans, AL1 5NS
M0 GTL K. Cossey, 34 Pinewood Road, Hordle, Lymington, SO41 0GP
MI0 GTM J. Sills, 145 Ballycolman Estate, Strabane, BT82 9AJ
MW0 GTN Prof. D. Embrey, 21 Rockfield Glade, Parc Seymour, Penhow, Caldicot, NP26 3JF
M0 GTO J. Bateman, 26 Thackeray Road, East Ham, London, E6 3BW
M0 GTP G. Perry, 173 Litchfield Road, Wednesfield, WV11 3HX
M0 GTQ N. Bennett, 16 Dickens Road, Worksop, S81 0DP
M0 GTR P. Henderson, 24 Farrow Road Whaplode Drove, Spalding, PE12 0TS
M0 GTT R. Wilcox, 107 Somerset Avenue Yate Bristol Bs377Sj, Bristol, BS377SJ
MM0 GTU A. Cumming, 15 Cockburn Crescent Whitecross, Linlithgow, EH49 6JT
M0 GTV D. Suplatowicz, 93 Huntingtower Road, Grantham, NG31 7AZ
MM0 GTX B. Thomson, 21 Sound Of Kintyre, Machrihanish, Campbeltown, PA28 6NZ
MW0 GTY G. Jones, 182 Pontardulais Road, Tycroes, Ammanford, SA18 3RD
M0 GTZ Derwent Valley Radio Group, c/o K. Sanderson, 39 Kirkland Street Pocklington, York, YO42 2BX
M0 GUC M. Elkington, 32 The Knoll, Kingswinford, DY68JT
M0 GUD G. Gash, 61 Beaconsfield Road, Rotherham, S60 3HB
MM0 GUE J. Mcmorland, 382 Maryhill Road, Glasgow, G20 7YQ
M0 GUF G. Jones, 57 Oxford Road, Banbury, OX16 9AJ
M0 GUG E. Govan, 9 Willowbank, Sandwich, CT13 9QA
M0 GUH E. Cobb, 8 Manor Avenue, Poole, BH12 4LD
M0 GUJ D. Tarrant, 17 Orchard Close, Corfe Mullen, Wimborne, BH21 3TW
M0 GUL C. Repton, Hp6 6Qu, Amersham, HP6 6QU
M0 GUM D. Adshead, 16 Moat Way, Swavesey, Cambridge, CB24 4TR
M0 GUO P. Fry, 4 Stretham Road, Wicken, Ely, CB7 5XH
M0 GUR J. Harris, Flat 3 49 Enys Road, Eastbourne, BN21 2DN
M0 GUU G. Moore, 40 Main Street, South Rauceby, Sleaford, NG34 8QG
MW0 GUV A. Hubbard, Pant-Y-Meillion, Velindre Penboyr, Llandysul, SA44 5JA
MM0 GUW M. Mccabe, 15 Laggan Road, Glasgow, G43 2SY
MM0 GUX M. Potts, 14 Constarry Road, Croy, Kilsyth, Glasgow, G65 9HF
M0 GVB W. Brennan, Flat 34, Compton Court Shopfield Close, Rustington, Littlehampton, BN16 3JQ
MI0 GVC Castlerock ARS, c/o J. Mcpeake, 47 Dunclug Gardens, Ballymena, BT43 6NN
M0 GVE M. Kentell, 24 Kendal Court, Congleton, CW12 4JN
M0 GVI D. Capon, 144 Stow Road, Magdalen, King's Lynn, PE34 3BD
M0 GVK D. Lyon-Mckeil, 1372 Turnstone Way, Sunnyvale, Ca, USA, 94087-3736
M0 GVL P. Smith, 24 Bede Crescent, Benington, Boston, PE22 0DZ
M0 GVN P. Keane, Pembroke Lodge, Byes Lane, Reading, RG7 2QB
M0 GVP Lv21 Lightship Museum, c/o C. Turner, 84 Gravel Hill Way Dovercourt, Harwich, CO12 4XN
M0 GVQ A. Sibley, 27 Sherwood Road, Tetbury, GL8 8BU
M0 GVT C. Lee, Grove House, Harrowbarrow, Callington, PL17 8JN
M0 GVW M. Wibberley, 5 The Row, Broadwell, Rugby, CV23 8HF
M0 GVX A. Farrar, 8 Wensley Street, Thurnscoe, Rotherham, S63 0PX
M0 GVY M. Hall, 29 The Spinney, Finchampstead, Wokingham, RG40 4UN
M0 GVZ C. Turton, 32 Northfield Crescent, Driffield, YO25 5ES
M0 GWA G. Rodmell, 2 Meadow Way, Walkington, Beverley, HU17 8SD
M0 GWB M. Baker, 39 Exham Close, Warwick, CV34 5UL
M0 GWC G. Chaloner, 9 Fairthorne Rise, Old Basing, Basingstoke, RG24 7EH
M0 GWD M. Ponsford, 83 Grant Road, Farlington, Portsmouth, PO6 1DU
M0 GWE K. Graffham, 15 Hayes Road, Clacton-on-Sea, CO15 1TX
M0 GWF A. Randles, 62 Brookside Avenue, Poynton, Stockport, SK12 1PW
M0 GWH D. Iveson, 11 Newport Road, North Cave, Brough, HU15 2NU
M0 GWK N. Green, 11 Wythburn Way, Rugby, CV21 1PZ
MW0 GWL J. Gwilliam, 39 Wyndham Street, Glynfach, Porth, CF39 9HT
M0 GWM R. Poole, 57 Loxley Avenue, Shirley, Solihull, B90 2QF
MM0 GWO H. Storie, 33 Harbour Street, Plockton, IV52 8TN
M0 GWQ J. Baker, Birdsong, Princes Close Redlynch, Salisbury, SP5 2HQ
M0 GWR J. Akinin, 70 Valley Road, West Bridgford, Nottingham, NG2 6HQ
MW0 GWT G. Thomas, 20 Ael Y Bryn, Caerau, Maesteg, CF34 0YG
M0 GWU W. Probst, Flat 19, Lindsay Park, 16 Lindsay Road, Poole, BH13 6AU
MW0 GWV A. Warner, 35 Lon Y Berllan, Abergele, LL22 7JF
MW0 GWY Dr I. Williams, 19 Stryd Y Brython, Ruthin, LL15 1JA
M0 GXB G. Bichard, 9 Kelburne Close, Winnersh, Wokingham, RG41 5JG
MW0 GXC G. Zaza, 42 Borras Road, Wrexham, LL12 7EP
MW0 GXE T. Banks, 18 Leicester Road, Newport, NP19 7ER
M0 GXH J. Hayward, 91 Hoyland Road, Hoyland, S74 0AP
M0 GXK J. Rodriguez Cemillan, Flat 51, Waxham, London, NW3 2JJ
M0 GXM Dr M. Roe, 68 Argyle Street, Cambridge, CB1 3LR
M0 GXN S. Woodmore, 66 Imperial Way, Chislehurst, BR7 6JR
M0 GXO I. Sheppard, 13 Meynell Close, Chesterfield, S40 3BL
MM0 GXQ G. Milne, 139 Rannoch Drive, Cumbernauld, Glasgow, G67 4ES
MM0 GXU G. Sutherland, 7 Abbotsgrange Road, Grangemouth, FK3 9JD
M0 GXV C. Berry, 60 Copthorne Road, Leatherhead, KT22 7EE
M0 GXW R. Lee, 39 Pickering Close Cramlington, Whitley Bay, NE23 6QB
MM0 GXY R. Mannifield, 2 Plewlands Avenue, Edinburgh, EH10 5JY
M0 GYA R. Moody, 372 Walsall Road, Perry Barr, Birmingham, B42 2LX
M0 GYB M. Peterson, 29 Warwick Close, Saxilby, LN1 2FT
M0 GYC D. Fletcher, 18 Bruce Avenue, Warrington, WA2 9BB
MM0 GYD A. Young, 21 Corrour Road, Glasgow, G43 2DY
MW0 GYF 610 Sqn City of Chester ATC ARC, c/o M. Buxton, 25 Pen Y Bryn, Sychdyn, Mold, CH7 6EE
MM0 GYG A. Fletcher, 164 Mayfield Road, Edinburgh, EH9 3AR
M0 GYH M. Pearce, 38 Salisbury Road, Beaconsfield Upper, Victoria, Australia, 3808
M0 GYI D. Leigh, 39 Hill Chase, Chatham, ME5 9HE
M0 GYK A. Roberts, 14 Cowper Close, Newport Pagnell, MK16 8PG
M0 GYL Dr M. Redman, 4 Deauville Avenue, Cowes, PO31 7GA
M0 GYM P. Mcewen, 7 Springfield, Longhoughton, Alnwick, NE66 3NT
M0 GYN K. Hulme, Sutherland Road, Longsdon, Stoke-on-Trent, ST9 9QD
M0 GYO D. Parker, 53 Brisbane Way, Cannock, WS12 2GR
M0 GYP S. Gillard, 1 Chevening Close, Stoke Gifford, Bristol, BS34 8NJ
M0 GYR East Yorkshire Emergency Communications Group, c/o A. Russell, 3 St. Nicholas Close, North Newbald, York, YO43 4TT
M0 GYS D. Garner, Flat 4, Joseph Nye Court, Portsmouth, PO1 3RD
M0 GYU L. Karchev, 6 Croombs Road, London, E16 3RY
MW0 GYV P. Oseland, 6 Oaklands Close, Bridgend, CF31 4SJ
MM0 GYX I. Watson, 10 Christie Place, Elgin, IV30 4HX
MW0 GYY G. Lewis, Bryn Cottage, Clydach, Abergavenny, NP7 0LL
MM0 GZA S. Hargreaves, 4 Oxenfoord Avenue, Pathhead, EH37 5QD
M0 GZB A. Armitage, 6 Rosebery Avenue Hythe, Southampton, SO45 3HJ
M0 GZC I. Coulson, 2 Marl Hurst, Edenbridge, TN8 6LN
M0 GZD Sherwood ARC, c/o E. Rippon, 319 Beechdale Road, Nottingham, NG8 3FF
M0 GZE P. Slup, 1 The Meadow, Copthorne, Crawley, RH10 3RG
M0 GZF M. Lamport, Bartwood, Dancing Green, Ross-on-Wye, HR9 5TE
M0 GZH N. Smith, 2 Norton Villas, Vicarage Road, Maidstone, ME18 6DX
M0 GZI Medway Radio Society, c/o M. Sharp, Pentober, Firmingers Road, Orpington, BR6 7QG
M0 GZK C. Yung, Flat 8, Bridge Court, London, E10 7JS
MW0 GZL A. Burleton, 9 Waghausel Close, Caldecott, Np26 4qr
M0 GZM J. Oldman, High Waters, Bentfield Green, Stansted, CM24 8HX
M0 GZW T. Lorn, 152 Brougham Court, Peterlee, SR8 1PZ
MM0 GZZ D. Taylor, 1 Mayfield Farm Cottages, Reston, Eyemouth, TD14 5LG

MW0 HAB M. Mainwaring, 36 Oak Street, Gilfach Goch, Porth, CF39 8UG
MW0 HAC C. Thomas, 2 Ffordd Donaldson, Copper Quarter, Swansea, SA1 7FJ
M0 HAF R. Hambly, 144 Station Road, Irchester, Wellingborough, NN29 7EW
M0 HAG G. Hill-Adams, 6 Broadleaze Way, Winscombe, BS25 1JX
M0 HAH M. Summers, 21 Quantock Avenue, Caversham, Reading, RG4 6PY
M0 HAL P. Musselwhite, 80 Craven Road, Orpington, BR6 7RT
M0 HAM C. Bays, 116 Rochester Road, Durham, DH1 5PN
M0 HAN P. Maennel, 4 Central Buildngs, Market Place, York, YO61 3AB
M0 HAO M. Martin De La Fuente, 35 Silchester Road, Reading, RG30 3EJ
MW0 HAP A. Phillips, 3 Pen Y Llys, Rhyl, LL18 4EH
MM0 HAR H. Stuart, 31 Robertson Road Lhanbryde, Elgin, IV30 8PE
MW0 HAT R. Hatfield, 35 Victoria Road, Penarth, CF64 3HY
M0 HAU J. Goodale, 82 Farnborough Road, Farnborough, GU14 6TH
M0 HAW Harpenden and Wheathampstead District Scout Group, c/o M. Wood, 26 Parkfield Crescent Kimpton, Hitchin, SG4 8EQ
M0 HAZ A. Freeman, 34 Marmion Road, Coningsby, Lincoln, LN4 4RG
M0 HBC B. Broad, 22 Minchin Acres Hedge End, Southampton, SO30 2BJ
M0 HBE M. Robins, 17 Old Turnpike, Fareham, PO16 7HB
M0 HBH E. Mathieson, 30 Lynfield Road, Frome, BA11 4JB
M0 HBJ S. Blaikie, 22 Juno Close, Goring-By-Sea, Worthing, BN12 4UB
M0 HBL Prof. P. Richmond, 7 Softley Drive, Norwich, NR4 7SE
M0 HBM Dr B. Denyer-Green, Dunsley South, Park Road, Forest Row, RH18 5BX
M0 HBN J. Bell, 255 Willington Street, Maidstone, ME15 8EP
M0 HBO K. Such, 38 Hornby Grove, Hull, HU9 4PG
M0 HBT D. Nelson, 110 Chandag Road, Keynsham, Bristol, BS31 1QF
M0 HBU I. Duffie, Trebeighan Farm, Saltash, PL12 5AE
M0 HBV D. Ingrey, 1 Ponders Road, Fordham, Colchester, CO6 3LX
M0 HBX Dr J. Pelham, 5 The Crescent, Shortstown, MK42 0UJ
M0 HBY P. Watkins, 135 Lodge Road, Writtle, Chelmsford, CM1 3JB
MW0 HCA F. Price, 2 Bryniau Duon Estate, Llandegfan, Menai Bridge, LL59 5PP
MW0 HCC C. Dumitrescu, Aelfryn, Pen Y Cefn Road, Caerwys, CH75BE
M0 HCE M. De Jong, Grachtstraat 64, Oirsbeek, Netherlands, 6938 HP
M0 HCI G. Burton, 26 Church View, Egremont, CA22 2DT
M0 HCM L. Michalowski, 136 St. Bernards Road, Newcastle, ST5 6HL
M0 HCN D. Mills, 261 West Wycombe Road, High Wycombe, HP12 3AS
MM0 HCO S. Mckenzie, 0/2 69 Glenkirk Drive, Glasgow, G15 6AU
M0 HCP J. Hunt, 14 Nevill Close, Hanslope, Milton Keynes, MK19 7NY
M0 HCR North Anglia Raynet, c/o K. Kent, 5 Jubilee Road, Heacham, King's Lynn, PE31 7AR
M0 HCT M. Fitzjohn, 96 Nightingale Gardens, Nailsea, Bristol, BS48 2BN
M0 HCV C. Wallace, 16 Morley Square, Bristol, BS7 9DW
MW0 HCW J. Morgan, 3 Maes Yr Hebog, Penrhyn Bay, Llandudno, LL30 3EY
M0 HCY Blackwater Radio Contest Group, c/o A. Copperwaite, 71 Gladbeck Way, Enfield, EN2 7EL
M0 HCZ C. Lycett, 2 Royce Avenue, Hucknall, Hucknall, Nottingham, NG15 6FU
MM0 HDA R. Fairfull, Blackford Farm Gartocharn, Alexandria, G83 8SD
M0 HDC Norfolk County Raynet, c/o S. Lucas, 31 Lilian Close, Norwich, NR6 6RZ
M0 HDE A. Morris, 71 Lurdin Lane, Standish, Wigan, WN6 0AQ
M0 HDG Hallam DX Group, c/o N. Totterdell, Moscar Cross House, Hollow Meadows, Sheffield, S6 6GL
M0 HDJ D. Hall, 8 Colston Close, Bristol, BS16 4PQ
M0 HDK E. Erbes, 488 Birkfield Drive, Ipswich, IP2 9JE
M0 HDN B. Richter, C/O Dr Steffen Grant, Wolfson College, Oxford, OX2 6UD
M0 HDP P. Bolton, 2 Alexander Court Chute Lane, Gorran Haven, St Austell, PL26 6NU
M0 HDQ G. Van Breemen, 58 Horseshoe Lane, Bromley Cross, Bolton, BL7 9RR
M0 HDR R. Scholey, Barleycroft, Lower Road, Ipswich, IP6 9AR
M0 HDS Hinckley District Scouts, c/o M. Smith, 14B Witham Bank West, Boston, PE21 8PU
M0 HDT D. Simpson, 50 Castle Hill, Berkhamsted, HP4 1HF
M0 HDU J. Legrain, 17 Route De La Cote, St Laurent Sur Gorre, France, 87310
M0 HDV D. Cowling, 11 Shakespeare Avenue, Scunthorpe, DN17 1SA
MM0 HDW J. Duncan, 36 Bank Row, Wick, KW1 5EY
M0 HEJ G. Hatt, 4H Colman House, Earlham Road, Norwich, NR4 7TJ
M0 HEM J. O'Toole, 4 Lindisfarne Road, Dagenham, RM8 2RA
M0 HEN T. Kindts, Northwood, Marhamchurch, Bude, EX23 0HH
M0 HEP G. Zorzi, 3B Ambleside Avenue, Telscombe Cliffs, Peacehaven, BN10 7LS
M0 HET S. Cordner, 29 Buxton Road, Aylsham, Norwich, NR11 6JD
M0 HEW T. Johnson, 15 Tennyson Road, Creswell, Worksop, S80 4DW
M0 HEX J. Ash, 47 Stein Road, Emsworth, PO10 8LB
M0 HEY M. Hickford, 56 Alexander Close, Abingdon, OX14 1XB
M0 HFA A. Birkett, 67A Branston Road, Burton-on-Trent, DE14 3BY
M0 HFB P. Szewczyk, 514 Whaddon Way, Bletchley, Milton Keynes, MK3 7LD
M0 HFC Humber Fortress DX ARC, c/o J. Cunliffe, 142 Hall Road, Hull, HU6 8SB
M0 HFE Barnsley and District ARC, c/o J. Sobanski, 10 Robert Avenue, Barnsley, S71 5RB
M0 HFF E. Bray, 28, Henshall Avenue, Latchford, Warrington, WA4 1PY
M0 HFH J. Rowden, 7 Regents Close, Thornbury, Bristol, BS35 1HX
M0 HFI Clan Maclean ARS, c/o J. Mclean, 24 Durham Drive, Oswaldtwistle, BB5 3AT
M0 HFO M. Jessop, Department Of Electronic Engineering, Claverton Down, BA2 7AY
M0 HFQ C. Lai, Storeys Way, Cambridge, CB3 0DG
M0 HFR G. Hall, 93 Berkeley Avenue, Bexleyheath, DA7 4TZ
MM0 HFU E. Horn, 3 Mckay Place, Newton Mearns, Glasgow, G77 6UZ
M0 HFW De Havilland Heritage Radio Group, c/o J. Newton, 6413 Hillegass Avenue, Oakland, USA, 94618
M0 HFX A. Walker, 17 Carr House Road, Halifax, HX3 7QY
M0 HFY B. Eames, 22 Ashgrove Close, Hardwicke, Gloucester, GL2 4RT
M0 HFZ B. Cox, 7 Wolsey Avenue, London, E6 6HG
M0 HGA D. Corless, 3 Barn Close, Clifford Chambers, CV37 8HJ
M0 HGD D. Molloy, 187 Babylon Lane, Heath Charnock, Chorley, PR6 9ET
M0 HGG Dr C. Regan, 1 Fairways, Birkenhead, CH42 8JZ
MW0 HGK W. Tse, Marino Room, Fulton Houose, Swansea, SA2 8PP
MW0 HGM A. Pritchard, 12 Llys Le Breos, Mayals, Swansea, SA3 5DL
MM0 HGN D. Higgins, 1 Meggatland Farm Cottage, Inchture, PH14 9QL
M0 HGO S. Westwood, 118 Abbey Lane, Leigh, WN7 5NU
M0 HGS M. Shepherd, North Waver Cottage, Bells Road Belchamp Walter, Sudbury, CO10 7AR
M0 HGV R. Dodds, West Villa, The Green, Wallsend, NE28 7PG
M0 HGY J. Read, 31 Merebrook Road, Macclesfield, SK11 8RH
M0 HHA M. Meehan, 14 Grosvenor Road, Walton, Liverpool, L4 5RB
M0 HHB G. Willard, 4 Varrier Jones Place, Papworth Everard, Cambridge, CB23 3XP
M0 HHC K. Jackson, 4 Milfoil Close, Marton-In-Cleveland, Middlesbrough, TS7 8SE
M0 HHD P. Rogers, 16 Begonia Close, Basingstoke, RG22 5RA
M0 HHE Dr G. Panico, 7 Hollybush Lane, Orpington, BR6 7QN
M0 HHF C. Greenwood, 1 Bentinck Close, Boughton, Newark, NG22 9HP
M0 HHG G. Aldridge, Greenridge, Fore Street Bishopsteignton, Teignmouth, TQ14 9QR
MD0 HHH H. Dorman, 1 Sprucewood Rise, Foxdale, Douglas, Isle Of Man, IM4 3JP
M0 HHI J. Hughes, Milestone House, Easole Street, Nonington Dover, CT15 4HE
M0 HHM W. Roberts, 113 Somerset Avenue, Luton, LU2 0PL
M0 HHP M. Kasprzyk, 80 Ederline Avenue, London, SW16 4sa
M0 HHR M. Lee, 34 Astley Road, Liverpool, L36 8DA
MI0 HHU R. Benko, 23 Six Mile Water Mill Drive, Antrim, BT41 4FG
MI0 HHV B. Craney, 8A Drumhoy Drive, Carrickfergus, BT38 8NN
M0 HHX A. Currie, 31 Launceston Road, Bodmin, SO50 6AY
M0 HIC Hounslow A.R. Instruction Centre, c/o M. De Silva, 31 Rosemary Avenue, Hounslow, TW4 7JQ
MW0 HID S. Leo, 37 The Coldra, Newport, NP18 2LS
M0 HIG B. Hultquist, 37 New Road, Tiptree, Colchester, CO5 0HN
M0 HIH K. Manos, 102 Goodwood Avenue, Sale, M33 4QL
M0 HIJ Norfolk & Suffolk 4x4 Response, c/o J. Whiteside, The Old Antique Shop, Bank Street Pulham Market, Diss, IP21 4TG
M0 HIL D. Hill, 11 Paddock Lane, Metheringham, Lincoln, LN4 3YG
M0 HIM P. Newsome, 47 Bramhall Drive, Washington, NE38 9DE
M0 HIN Hinckley Sea Cadets, c/o V. Hopkins, 6 Daimler Road, Coventry, CV6 3GD
M0 HIO D. Wood, 10 Tadgedale Avenue, Market Drayton, TF94DD
M0 HIP Hippings Methodist Primary School Amateur Radio Cl, c/o J. Mclean, 24 Durham Drive, Oswaldtwistle, BB5 3AT
M0 HIQ D. Cotton, 1 Fieldfare Close, Penwortham, Preston, PR1 9NG
M0 HIW P. Jones, Tallonhouse, Mill Lane, Pulham St. Mary, Diss, IP21 4QY
M0 HIX A. Holmes, 49 Elm Grove South, Barnham, Bognor Regis, PO22 0EJ
M0 HIY A. Thomas, 1 Millers Close, Ruardean Hill, Drybrook, GL17 9AU
M0 HIZ W. Easdown, 11 Mulcaster Avenue, Kidlington, OX5 2HG
M0 HJB M. Stillman, 58 Highfield Road, Bognor Regis, PO22 8PH
MM0 HJC Clydebank Cadet Centre, c/o K. Brown, 10 Richmond Street, Clydebank, G81 1RF
M0 HJD D. Harbron, 6 West View, Penshaw, Houghton le Spring, DH4 7HP
M0 HJE P. Frost, 11 Church Road, Swainsthorpe, Norwich, NR14 8PH
M0 HJF H. Felstead, Rosmede, Windmill Drive, Rustington, Littlehampton, BN16 3HW
MW0 HJG N. Williams, 27 Meadway Rogiet, Caldicot, NP26 3SA
M0 HJI R. Havart, 2 Holly Farm Road, Reedham, Norwich, NR13 3TH
M0 HJJ A. Wierdis, 39 Milton Road, London, SW19 8SF
M0 HJL R. Taylor, 27 The Holt, Hailsham, BN27 3ND
M0 HJN W. Tomczyk, 3L, 76 Berry Street, New York, USA, 11249
M0 HJO J. Brooks, Treven House, Treven, Tintagel, PL34 0DT
M0 HJQ P. Garrett, 21 Wychbury Road, Wolverhampton, WV3 8DN
M0 HJR D. Vale, 21 Chelston Road, Ruislip, HA4 9SA
M0 HJW I. Ftaiha, 8 Parkside, London, NW7 2LH
M0 HJY R. Green, 44 Aldwyn Place, Larchwood Drive, Egham, TW20 0RZ
MW0 HKA M. Day, 11 Troedrhiw-Trwyn, Pontypridd, CF37 2SE
M0 HKB K. Brunning, 45 Dover Road, Ipswich, IP3 8JQ
M0 HKC K. Cullum, 7 Gate Farm Road, Shotley Gate, Shotley, Ipswich, IP9 1QH
M0 HKE A. Mullin, 111 Arps Road, Codsall, Wolverhampton, WV8 1SG
M0 HKG M. Clarke, 40 Fingringhoe Road, Langenhoe, CO5 5AD
M0 HKH Dr A. Fronters, Flat 54, Central Quay North, Bristol, BS1 4AU
M0 HKI L. Todman, 17 Hall Road, St. Dennis, St Austell, PL26 8BE
M0 HKK Dr A. Doe, 26 Beachfield Road, Bembridge, PO35 5TN
M0 HKL G. Alberti, Josef-Retzer-Strasse 48, MšNchen, Germany, 81241
M0 HKM R. Richardson, 11 Packman Green, Countesthorpe, Leicester, LE8 5WS
M0 HKP Dr D. Potts, 3A Westfield, Blean, Canterbury, CT2 9ER
M0 HKS M. Booth, 30 Manor Green, Harwell, Didcot, OX11 0DQ
M0 HKT Mexborough ARS, c/o A. Farrar, 8 Wensley Street, Thurnscoe, Rotherham, S63 0PX
MM0 HKU E. Duncan, 3 George Street, Banff, AB45 1HS
M0 HKV P. Bull, 87 Braemor Road, Calne, SN11 9DU
M0 HKW A. Brand, 6 Walnut Close, Milton, Cambridge, CB24 6ET
M0 HLB D. Slater, 13 Longford Close, Rainham, Gillingham, ME8 8EW
M0 HLC C. Taylor, 1 Jasmine Gardens, Warrington, WA5 1GU
M0 HLD D. Hicks, 36 Middlesex Road, Maidstone, ME15 7PL
M0 HLF A. Higton, 4 Paddocks Close, Pinxton, Nottingham, NG16 6JR
M0 HLI D. Hyde, 136 Station Road, Woodmancote, Cheltenham, GL52 9HN
M0 HLM K. Schmidt, Church, Corner, Mareham-le-Fen, PE22 7RA
MM0 HLN H. Mason, 20 David'S Crescent, Kilwinning, KA13 6JJ
M0 HLP G. Bunting, 31 Hardwick Avenue, Allestree, Derby, DE22 2LN
MM0 HLQ G. Gilmour, 3 Campsie Drive Milngavie, Glasgow, G62 8HX
M0 HLR A. Brown, 3 Alston Road, New Hartley, Whitley Bay, NE25 0ST
M0 HLS C. Murphy, 17 Shepherd Street, Littleover, Derby, DE23 6GA
MM0 HLU I. Konstas, 25/4 Milton Street, Edinburgh, EH8 8HA
M0 HLV C. Hicks, 11 Patch Street, Bath, BA25BN
MW0 HLW W. Maxwell, 25 Laurel Drive, Buckley, CH7 2QP
M0 HLX D. Bailey, 2B Queens Road, Enfield, EN1 1NE
M0 HLY A. Nisbet, 6 Tomline Road, Felixstowe, IP11 7QW
M0 HLZ M. Meachen, 20 Wilkinson Road, Rackheath, Norwich, NR13 6SG
M0 HMB R. Stratford, 32A Priory Avenue, High Wycombe, HP13 6SW
M0 HME R. Campbell, 17 Elgar Road, Southampton, SO19 0JG
M0 HMF M. Smith Ii, The White House, Old Avenue, West Byfleet, KT14 6AE
M0 HMI G. Tamir, Service Now, Strata, 1 Bridge Street, Staines-upon-Thames, TW18 4TW
M0 HMJ D. Sadauskas, Flat 18, Newport House Newport Street, Tiverton, EX16 6FJ
M0 HMO Dr H. Nickalls, Holy Mill, Longville, Much Wenlock, TF13 6ED
M0 HMR C. Harmer, Spring Corner, Rockness Hill, Nailsworth, Stroud, GL6 0PJ
M0 HMS E. Purvis, 36 Birchington Avenue, Middlesbrough, TS6 7EZ
M0 HMU Fleetwood Radio Enthusiasts Group, c/o J. Earnshaw, 128 Shakespeare Road, Fleetwood, FY7 7HJ
MW0 HMV C. Josey, 726 Llangyfelach Road, Treboeth, Swansea, SA5 9EL

M0 HMX R. Sykes, 11 Lodwells Orchard, North Curry, Taunton, TA3 6DX
MI0 HMY A. Hamill, 15 Maythorn Avenue, Coleraine, BT52 2EU
M0 HMZ P. Iljin, 15 The Green, Newton Burgoland, LE67 2SS
M0 HNA Southern Microwave Group, c/o D. Austen, Tudorlands, Silchester Road, Bramley, Tadley, RG26 5DG
M0 HNC A. Ribeiro, 38A Galpins Road, Thornton Heath, CR7 6EB
M0 HND Technical Experimenters Group, c/o B. Smith, 73 Devon Street, Hull, HU4 6PL
M0 HNE R. Ashley, 15 Wimbourne Drive, Gillingham, ME8 9EN
M0 HNF P. Dickson, 49 Signal Road, Grantham, NG31 9BL
M0 HNG A. Douglas, Gobbins Cottage, Sandy Lane, Lathom, Ormskirk, L40 5TU
M0 HNH A. Reason, 1 Iles Cottages, St. Marys, Stroud, GL6 8NX
M0 HNI R. Weaver, 15 Sharps Field, Headcorn, Ashford, TN27 9UF
M0 HNJ P. Evans, 67 Grenville Street, Stokport, SK3 9ER
M0 HNK Dr R. Tofts, Elmcroft, Redhill Road, Ross-on-Wye, HR9 5AU
M0 HNL Dr R. Campbell, 2 Hesketh Bank, York, YO10 5HH
MW0 HNM L. Coleman, Felin Newydd, Ciliau Aeron, Lampeter, SA48 7PX
M0 HNN T. Walsh, 6 Brass Thill, Durham, DH1 4DS
M0 HNO H. Nishio, 28 New Lane, Havant, PO9 2NQ
MI0 HNQ Hilltop ARC Co.Down, c/o A. Mcgarvey, 66A Scaddy Road, Downpatrick, BT30 9BS
M0 HNT A. Angus, 51 Osprey Drive, Blyth, NE24 3QS
M0 HNX R. Raeburn, 145 Paddock Road, Basingstoke, RG22 6QQ
M0 HOB G. Brotherhood, 17 Baldwin Close, Forest Town, Mansfield, NG19 0LR
M0 HOF B. Ajeti, 88 Bushfield Crescent, Edgware, HA8 8XJ
M0 HOI Dr S. Musgrave, Orchard Cottage, Stalmine, Poulton le Fylde, FY6 0LZ
M0 HOJ F. Costa, 7 Aylesborough Close, Cambridge, CB4 2HH
M0 HOK L. Carberry, 23 Greens Beck Road, Stockton-on-Tees, TS18 5AR
MM0 HOL C. King, 19 Gleneagles Way, Deans, Livingston, EH54 8EW
M0 HOM M. Hotchin, 122 Buckingham Avenue, Scunthorpe, DN15 8NS
M0 HOO A. Hodgeon, 30 Rock Bank, Buxton, SK17 9JF
M0 HOP Sir A. Hopson, 1 Hall Lane, Leicester, LE2 8SF
M0 HOQ J. Johnson, 226 Preston New Road, Southport, PR9 8NY
M0 HOT H. Rose, 10 St. Vincents Close, Girton, Cambridge, CB3 0PE
M0 HOU H. Atifeh, 57 Lincoln Drive, Rugby, CV23 1BS
M0 HOV A. Hristov, Flat A, 71 Beckenham Lane, Bromley, BR2 0DN
M0 HOY S. Curtis, 354 St. Helens Road, Leigh, WN7 3PQ
MI0 HOZ M. Na BpÝob, 94 Curlyhill Road, Strabane, BT82 8LS
M0 HPB D. Bisbey, 17 Benson Close, Lichfield, WS13 6DA
MI0 HPE P. Dorris, 29 Eia Street, Belfast, BT14 6BT
M0 HPF G. Cheeran, 37 Farnol Road, Dartford, DA1 5NG
M0 HPG Cumbria Raynet Group, c/o P. Woodburn, 21 The Row, Silverdale, Carnforth, LA5 0UG
MW0 HPH Ystrad Mynach College, c/o P. Jones, 23 Pinecroft Avenue, Aberdare, CF44 0HY
M0 HPJ J. Whiteside, The Old Antique Shop, Bank Street Pulham Market, Diss, IP21 4TG
M0 HPL R. Masshedar, 6 Hutton Avenue, Hartlepool, TS26 9PN
M0 HPP G. Fleming, 17 Greenfield Close, Kippax, Leeds, LS25 7PX
M0 HPR H. Richardson, 7 Regent Road, Leyland, PR25 2LJ
M0 HPS Dr H. Powell, 6 Sowbury Park, Chieveley, Newbury, RG20 8TZ
M0 HPT D. Prior, 10 Birley Close, Appley Bridge, Wigan, WN6 9JL
M0 HPU D. Rudling, Rose Cottage, Ludwells Lane, Southampton, SO32 2NP
M0 HPV D. Green, 67 Coombe Park Road, Binley, Coventry, CV3 2NW
M0 HPW M. Phillips, 59 Bradeley Road Haslington Crewe, Crewe, CW1 5PX
MI0 HPX A. Sweeney, 117 Lisnablagh Road, Coleraine, BT52 2HD
M0 HPZ L. Edmonds, 3 Waterlow Road, London, N19 5NJ
M0 HQA P. Massolt, 26 Redgate Heights, Hunstanton, PE36 5EA
M0 HQB K. Kulpinski, 85 Severn Drive, Taunton, TA1 2PW
M0 HQC M. Copse, 3 The Limes, Market Overton, Oakham, LE15 7PX
MM0 HQD D. Searle, 32 Linwood Terrace, Hamilton, ML3 9AF
M0 HQG South Normanton and District ARC, c/o J. Mason, 56 Skegby Road, Sutton-in-Ashfield, NG17 4EZ
MM0 HQI L. Richings, 2 St. Margarets Place, Edinburgh, EH9 1AY
M0 HQJ H. Quigg, 60 Oak Road, Ripon, HG4 2NB
M0 HQL S. Mitchell, 42 Fairfield Avenue, Felixstowe, IP11 9JJ
M0 HQM R. Givens, 13 Wakehurst Drive, Crawley, RH10 6DL
M0 HQO P. Freeman, 57 Ruffa Lane, Pickering, YO18 7HN
M0 HQP N. Marley, Penstemons, Chapel Lane Pen Selwood, Wincanton, BA9 8LY
M0 HQQ P. Taylor, 54 Church Road, Stanley, Liverpool, L13 2BA
M0 HQR P. Naik, 82 Misbourne Road, Uxbridge, UB10 0HW
M0 HQU C. Lombao, 82 Cirrus Drive, Shinfield, Reading, RG2 9FL
M0 HQZ J. Widdowson, 26 Woodville Gardens West, Boston, PE21 8BW
M0 HRA H. Alderson, 66 Houghtonside Estate, Houghton le Spring, DH4 4BW
M0 HRC W. Nicholas, 16 Withymoor Road, Netherton, Dudley, DY2 9LA
MW0 HRD C. Hughes, 88 Derlwyn Street, Phillipstown, New Tredegar, NP24 6BA
MI0 HRG Hill Top Radio Group, c/o B. Vaughan, 32 Claremore Road, Castlederg, BT81 7RF
M0 HRH C. Morley, 191 Purbrook Way, Havant, PO9 3RS
MM0 HRI I. Candy, 6 Provost Milne Gardens, Arbroath, DD11 5FG
MM0 HRL I. Gourlay, 76 Largo Road, St Andrews, KY16 8NJ
M0 HRM C. Greenwood, 21 Valley Drive, Thornhill Dewsbury, wf120he
MI0 HRO C. Stockdale, 3 Hightown Drive, Newtownabbey, BT36 7TG
M0 HRP R. Huelin, 15 Hill Chase, Walderslade, Chatham, ME5 9HE
M0 HRT R. Bryan, 1 White Cottage, Old Warwick Road, Lapworth, Solihull, B94 6LN
MI0 HRV T. Darrah, 42 Pinewood Avenue, Carrickfergus, BT38 8EW
M0 HRW A. Parker, 9 Milecastle Court, Newcastle upon Tyne, NE5 2PA
M0 HRY S. Wheeler, 98 Charterhouse Road, Orpington, BR6 9EW
M0 HRZ D. Irving, 8 Durlston Road, Swanage, BH19 2DL
MM0 HSB W. Forrester, 149 Whyterose Terrace, Methil, Leven, KY8 3AR
M0 HSC Northeast ARS, c/o G. Cockburn, 20 Hexham Avenue, Hebburn, NE31 2HN
M0 HSG P. Scrimshaw, 126 Nelson Road, Leighton Buzzard, LU7 3EG
M0 HSH Dr J. Brooks, 44 Rowan Drive, Seaton, EX12 2UH
MW0 HSI Clwud Portable Operating Group, c/o M. Allington, Merton Place Nursing Home, 8 Pwllycrochan Avenue, Colwyn Bay, LL29 7BU
M0 HSJ H. Jones, 116 Dark Lane, Bedworth, CV12 0JH
M0 HSQ I. Tsimperidis, Flat 410, Birch Court, Howlands, Welwyn Garden City, AL7 4LR
MM0 HSR Brannock High Radio Group, c/o P. Bainbridge, 49 Hare Moss View, Bathgate, EH47 0DN
M0 HSS A. Perrow, 16 Bannister Walk, Cowling, Keighley, BD22 0NU
M0 HSU S. Challis, 73 Rivenhall Way, Hoo, Rochester, ME3 9GF
MM0 HSV K. Baird, 24 Main Street, Sorn, KA5 6HU
M0 HSW H. Scott Whittle, 7 Skyline House, Dickens Yard, Longfield Avenue, London, W5 2BJ
M0 HSX M. Josi, 10 Robert Close, Billericay, CM12 9DS
M0 HSZ J. Merritt, 41 Great Grove, Bushey, WD23 3BQ
M0 HTA I. Cooke, 11 Farriers Gate, Chatteris, PE16 6AY
M0 HTB H. Banasiak, 16 St Christophers Close, Bath, BA2 6RG
M0 HTE J. Taylor, 90 Village Road, Gosport, PO12 2LG
M0 HTF C. Rose, 132 Golf Green Road, Jaywick, Clacton-on-Sea, CO15 2RW
MW0 HTG G. Edwards, 17 Glan Y Mor Road, Penrhyn Bay, Llandudno, LL30 3NL
M0 HTI S. Storey, 10 Amble Way, Trimdon Station, TS29 6DZ
M0 HTJ Hamtests.Co.UK, c/o P. Gibson, 60 Raglan Road, Bromley, BR2 9NW
M0 HTK H. Kassier, 26 Higher Port View, Saltash, PL12 4BX
MM0 HTL D. Hegarty, 13 Gaitschaw Lane, Selkirk, TD7 4HS
MW0 HTO Brecon and Radnor ARS, c/o D. Bowen, 25 Maendu Terrace, Brecon, LD3 9HH
M0 HTQ K. Numata, 28 Guildhouse Street, London, SW1V 1JJ
M0 HTR Ashton In Makerfield ARC, c/o P. Williams, 35 Cansfield Grove, Ashton In Makerfield, Wigan, wn4 9se
M0 HTS C. Mellor, 104 Rocky Lane, Eccles, Manchester, M30 9LY
M0 HTU J. Stokoe, 77 High Street Market Deeping, Peterborough, PE6 8ED
M0 HTV M. Gregson, 10 Eden Avenue, Consett, DH8 6EZ
M0 HTW H. Chan, Flat 20A, Pine Manor, 61 Waterloo Road, Mongkok, Hong Kong, 852
M0 HTX D. Anderson, 53 Collywell Bay Road Seaton Sluice, Whitley Bay, NE264RG
M0 HTY M. Tointon, 13 Ridgeway, Broadstone, BH18 8DY
M0 HUA A. Brown, 63 Pound Green Lane, Shipdham, Thetford, IP25 7LH
MM0 HUF S. Harvey, 63 Darley Road, Cumbernauld, Glasgow, G68 0JR
M0 HUG S. Eyre, St. Michael Mead, The Common Barton Turf, Norwich, NR12 8BA
M0 HUH P. Tan, Cambridge, Cb3 0bn, UNITED KINGDOM
M0 HUI I. Magness, Orchard House Ravenswood Drive, Camberley, GU15 2BU
M0 HUL A. Molloy, Rue Apesenia 12, Urrugne, France, 64122
M0 HUN J. Hunt, Flat 1, Strand House, 16 Wells View Drive, Bromley, BR2 9UL
MM0 HUQ M. Flaws, West Voe, Sumburgh, Shetland, ZE3 9JN
M0 HUS H. Steers, 39 Upwood Road, London, SE12 8AE
MW0 HUU M. Pope, 4 Croft Villas, Narberth, SA67 7DY
M0 HUV E. Valdez, 65 Broken Cross, Charminster, Dorchester, DT2 9QB
M0 HUW G. Gentile, Via B, Vecchia, Preganziol, Italy, 31022
MW0 HUY K. Saltmarsh, 15 Colbourne Road, Beddau, Pontypridd, CF38 2LN
M0 HUZ C. Warwick, 104 Church Road, Formby, Liverpool, L37 3NH
MM0 HVA J. Hawkins, 113 Meadowpark Avenue, Bathgate, EH48 2ST
MW0 HVB H. Bancroft, Stop And Call, Goodwick, SA64 0EX
M0 HVC R. Barnard, 3 Heaths Close, Enfield, EN1 3UP
M0 HVD D. King, 25 Church Road, Worthing, BN13 1ET
M0 HVE L. Sargent, 18 Lyndhurst, Maghull, L316DY
MM0 HVF Z. Bak, 62/6 North Gyle Loan, Edinburgh, EH12 8LD
M0 HVI M. Kurczab, 159 Huddersfield Road, Halifax, HX3 0AH
M0 HVK D. Ackrill, 62 Lapsley Drive, Banbury, OX16 1EW
MW0 HVL E. England, 2 Luton Street, Blaenllechau, Ferndale, CF43 4PB
M0 HVM J. Vollbrecht, Reinsdorf, Steingasse 3, Nebra (Unstrut), Germany, 6642
M0 HVN D. Connolly, 2 Layton Close, Birchwood, Warrington, WA3 6PT
M0 HVO I. Bailey, 8 Willow Drive, Ringwood, BH24 3BE
M0 HVP 1466 Holmfirth, c/o N. Tindall, Royds Mount, Linthwaite, Huddersfield, HD7 5QX
M0 HVQ D. Holland, 7 Hayward Close, Walkington, Beverley, HU17 8YB
M0 HVR J. Brawn, 5 Downs Cote View, Westbury On Trym, Bristol, BS9 3TU
M0 HVS N. Bethell, 4 Magazine Road, Wirral, CH62 3LH
MM0 HVU D. Smith, 25 High Academy Street, Armadale, Bathgate, EH48 3HG
M0 HVV M. Hickman, 13 Millfields Avenue, Rugby, CV21 4HJ
MM0 HVW D. Plummer, 39 St. Nicholas Drive, Banchory, AB31 5YG
M0 HVX C. Pattison-Hart, 133 Park Road, Bingley, BD16 4EJ
M0 HWC Hadley Wood Contest Group, c/o M. Ruttenberg, 90 Heath View, London, N2 0QB
M0 HWD D. Levy, Flat 36, Claydon House, London, NW4 1LS
MI0 HWG P. Moore, 32 Kinnegar Rocks, Donaghadee, BT21 0EZ
M0 HWH K. Quigley, 12 Silver Lane, Billingshurst, RH14 9RJ
M0 HWI A. Trzepietowski, 8 Parsons Nook, Coventry, CV2 4QY
M0 HWJ Dr A. Boldireff Strzeminski, 47 Waters Edge, Canterbury, CT1 1WX
M0 HWL R. Riches, Flat 21, Hawthornden, 84 Bradford Road, Otley, LS21 3LE
M0 HWM S. Baddeley, 50 Western Esplanade, Herne Bay, CT6 8JA
M0 HWN E. Wilcockson, Flat 17, Dale Court, Seymour Road, Slough, SL1 2NU
M0 HWO G. Phillips, 73 Gotham Road, Wirral, CH63 9NG
M0 HWP J. Tarrant, 70 Sunnymead, Midsomer Norton, Radstock, BA3 2SD
M0 HWQ P. Browne, 151 North Road, St. Andrews, Bristol, BS6 5AH
M0 HWS G. Hewis, 10 Albert Road, New Malden, KT3 6BS
M0 HWT G. Mutch, 94 Abbotswood Road, Brockworth, Gloucester, GL3 4PF
MW0 HWU Raynet Pembrokeshire, c/o I. Baker, 28 Kensington Road, Neyland, Milford Haven, SA73 1TL
M0 HWV P. Heiney, 12 Church Lane, Walberswick, IP186UZ
M0 HWW B. Vemic, 7A Selby Road, London, SE20 8SF
M0 HWZ D. Mccann, 41 Weighton Road, Harrow, HA3 6HY
M0 HXA A. Crosland, 18 Duncan Crescent, Bovington, Wareham, BH20 6NN
MI0 HXB T. Browne, 7 Hawthorn Park Greysteel, Londonderry, BT47 3YE
MW0 HXC F. Kroon, Flat 10, Court Rise, Hoggan Park, Brecon, LD3 9SZ
M0 HXE R. Hill, 108 Hitchin Close, Romford, RM3 7EQ
M0 HXF Dr R. Thorpe, 6 Millthorpe, Sleaford, NG34 0LD
M0 HXH J. Matthewson, 20 Yew Tree Road, New Ollerton, Newark, NG22 9UL
M0 HXI C. Baker, 29 Green Lane, Bristol, BS11 9JD
M0 HXK H. Carrythers, 31 Baden Street, Hartlepool, TS26 9BJ
M0 HXM D. Estevez, Oceano Atlantico, 38, Tres Cantos, Spain, 28760
M0 HXN J. Orme, 42 Dovecote, Newport Pagnell, MK16 8BB
M0 HXO J. Neal, 48 Mansfield Road, South Normanton, Alfreton, DE55 2ER
M0 HXS E. Haener, 110 Great Stone Road, Manchester, M16 0HD
M0 HXV A. Yeomans, 17 Hollyfield Close, Tring, HP23 5PL

MW0 HXX D. Machon, 22 Albert Street, Caerau, Maesteg, CF34 0UF
M0 HXZ M. Mallette, Not, Applicable, Resides, Germany, OUTSIDE OF THE UK IN
MW0 HYA J. Starbuck, 28 Plas Panteidal, Aberdyfi, LL35 0RF
M0 HYC T. Rutt, Granthorpe, Hull Road, Hull, HU11 5RN
M0 HYD F. Hyde, 10 Devonshire Drive, Barnsley, S75 1EE
M0 HYE T. Byers, 1 Hazelwood Avenue, Sunderland, SR5 5AH
M0 HYG H. Hope, 51 Margravine Gardens, London, W6 8RN
M0 HYH C. Glass, The Old Homestead, Havikil Lane, Knaresborough, HG5 9HN
M0 HYJ A. Rand, 17 Fairways Drive, Harrogate, HG2 7ES
MW0 HYK K. Dutfield-Cooke, Tan Yr Efail, Segurinside, Llandudno Junction, LL31 9QE
M0 HYL A. Robnett, 38B Woodmere Avenue, Watford, WD24 7LN
MM0 HYM W. Jackson, 3 Annick Road, Dreghorn, Irvine, KA11 4EY
M0 HYN B. Davies, 12 Scalebor Gardens, Burley In Wharfedale, Ilkley, LS29 7BX
MW0 HYP D. Thomas, 67 Crynallt Road, Neath, SA11 3RN
MI0 HYQ A. Zakrzewski, 11 Millbrook Gardens, Kilrea, Coleraine, BT51 5RZ
M0 HYX Magnetic Fields Contest Group, c/o P. Marchant, 16 Melrose Drive, Peterborough, PE2 9DN
M0 HZ Horizontal Net, c/o A. Gravell, 21 Wickridge Close, Stroud, GL5 1ST
M0 HZA G. Charlesworth, 6 Eastfield Close, Sutterton, Boston, PE20 2JF
M0 HZB Z. Yao, Room 4C, Unit 12-2, Zhonghaibanshanxigu Garden, No.15 Zhongqing Rd., Yantian District, Shenzhen, China, 518000
M0 HZC V. Perovic, Staudenbühlstrasse 126, Zurich, Switzerland, 8052
MI0 HZD K. Mikicki, 17 Glenhoy Drive, Belfast, BT5 5LB
M0 HZE P. Preston, 12 Backney View, Greytree, Ross-on-Wye, HR9 7JP
M0 HZF G. Craioveanu, Flat 8, Apex House, Burch Road, Northfleet, Gravesend, DA11 9FF
MM0 HZI T. Johnston, The Old Schoolhouse, Luggate Burn, Haddington, EH41 4QA
MM0 HZJ S. Thomas, Macsherry, Glenachulish, Ballachulish, PH49 4JZ
M0 HZK D. Pearson, 37 Elmridge, Leigh, WN7 1HN
MM0 HZL H. Mckay, 3 Flemington Gardens, Whitburn, Bathgate, EH47 0NS
M0 HZM D. Greenland, 1 Hilltop, Tuesley Lane, Godalming, GU7 1SB
MM0 HZO N. Clark, 27 Deansloch Crescent, Aberdeen, AB16 5UY
M0 HZP D. Morrow, 57 Bispham Road, Poulton-le-Fylde, FY6 7PE
M0 HZR N. Barker, 17 Pippin Walk, Hardwick, Cambridge, CB23 7QD
M0 HZT J. Jones, 21 The Maltings, Warminster, BA12 8JR
M0 HZU D. Eate, 69 Dunyeats Road, Broadstone, BH18 8AE
M0 HZV M. Mirchev, 82 Collingwood Road, Uxbridge, UB8 3EL
M0 HZW W. Dawkins, 2 Nativity Close, Sittingbourne, ME10 1ET
M0 HZX M. Stephens, 52 Waterslea Drive, Bolton, BL1 5FJ
M0 HZY J. Strandberg, Apartment 201, Satin House, 15 Piazza Walk, London, E1 8PW
M0 IAA I. Astley, 6 Shay Court, Crofton, Wakefield, WF4 1SL
M0 IAD I. Macdonald, Broomhill Mill Lane, Worthing, BN13 3DH
M0 IAE Anglo-European School Radio Club, c/o M. Adcock, 37 Ashpole Road, Braintree, CM75LW
M0 IAF I. Fletcher, 19 Church Street, St. Day, Redruth, TR16 5JY
M0 IAG S. Donath, 12A Comerford Road, London, SE4 2AX
M0 IAH I. Pryke, 9 Charles Avenue, Grundisburgh, Woodbridge, IP13 6TH
M0 IAJ I. Jones, 21 Kennet Green, Worcester, WR5 1JQ
M0 IAK I. Karbhari, Flat B, 226 Westbourne Park Road, London, W11 1EP
MM0 IAL I. Lindsay, 265 Stirling Street, Denny, FK6 6QJ
M0 IAM C. Collins, 31 Warren Road, Godalming, GU7 3SH
M0 IAS G. Reywer, 1 Tiverton Close, Houghton le Spring, DH4 4XR
M0 IAT I. Chick, 7 Furzegood Marldon, Paignton, TQ3 1PH
M0 IAX M. Bumstead, Middle Leys, Leys Farm, Withypool, Minehead, TA24 7RU
M0 IAZ R. Dykes, Gorse Cliff, West Hill, Heybrook Bay, Plymouth, PL9 0BB
M0 IBD W. Thiele, 50B, The Highway, London, E1W 2BG
MM0 IBE P. Woods, 92 Preston Crescent, Prestonpans, EH32 9RD
MW0 IBH K. Davies, 16 Lon Ffawydd, Abergele, LL22 7DU
MW0 IBI Taff Vale ARC, c/o A. Burns, 34 Lakeside Gardens, Merthyr Tydfil, CF48 1EN
M0 IBK H. Schr÷der, Hamptstrabe 8, Dieblich, Germany, 56332
MM0 IBL C. Niven, The Coachmans Cottage, Balmullo Farm, Balmullo, St Andrews, KY16 0AQ
M0 IBN W. Parish, 104 Blackamoor Lane, Maidenhead, SL6 8RH
MM0 IBO J. Moreno, 1-19 Albion Street, Glasgow, G1 1LH
M0 IBQ M. Savage, 3 Marlborough Close, Cheltenham, GL53 7RY
M0 IBR B. Clayton, 26 Wood Walk, Mexborough, S64 9SG
MW0 IBT D. Jones, 44 Stryd Y Wennol, Ruthin, LL15 1QN
M0 IBW Dr S. Harrison, 8 St. Michaels Close, Buckland Dinham, Frome, BA11 2QD
M0 IBX A. Beacham, Sushael Cottage, Denton Lane Denton Lane, Wootton, CT4 6RN
M0 IBY Utc Sheffield ARC, c/o M. Rigby, 75 Manchester Road, Deepcar, Sheffield, S36 2QX
MW0 IBZ I. Baker, 28 Kensington Road, Neyland, Milford Haven, SA73 1TL
M0 ICA P. Rushby, 16 Foxhill Lane, Selby, YO8 9AR
MM0 ICB I. Buchner, 33 Blakewell Gardens Tweedmouth, Berwick-upon-Tweed, TD15 2HJ
MW0 ICE C. Evans, 25 Beech Drive, Hengoed, CF82 7JP
M0 ICG G. Tagg, Tinkers Cottage, Nevendon Road, Wickford, SS12 0QB
M0 ICI J. Le Roux, 20 Varsity Drive, Twickenham, TW1 1AG
M0 ICJ L. Zywicki, 18 Springbank, Brigg, DN20 8PW
M0 ICK M. Heywood, 16 Edinburgh Drive, Hindley Green, Wigan, WN2 4HL
M0 ICL J. Salek, 19 Eskmont Ridge, London, SE19 3PZ
MW0 ICO P. Jones, 76 Pengwern, Llangollen, LL20 8AS
M0 ICP I. Pass, 69 Cotswold Road, Bath, BA2 2DL
MD0 ICS C. Schofield, Rockside, Dreemskerry Road, Maughold, Isle Of Man, IM7 1BL
M0 ICT M. Gascoyne, 31 Dale View, Hemsworth, Pontefract, WF9 4TA
M0 ICZ D. Jones, Drove Farm, Sheepdrove, Lambourn, Hungerford, RG17 7UN
M0 IDC J. Clark, 27 The Gabriels, Newbury, RG14 6PZ
M0 IDE C. Laycock, 35A High Street, Henlow, SG16 6AA
M0 IDG I. Garrard, 33 Uplands Road, Hockley, SS5 4DL
M0 IDI P. Askew, 6 Claremont Avenue, Newcastle upon Tyne, NE15 7LB
M0 IDJ S. Lo, Fieldgate, The Avenue, Claverton Down, Bath, BA2 7AX
M0 IDK I. King, 7, Greenacres Avenue, Blythe Bridge, Stoke on Trent, ST11 9HU
M0 IDL G. Stockley, Flat 1, The Pentagon, 94 Stanley Green Road, Poole, BH15 3AG
M0 IDM A. Ferriroli, 142 Hillbury Road, Warlingham, CR6 9TD
M0 IDR I. Reeve, 36 Stone Pippin Orchard, Badsey, Evesham, WR11 7AA
MW0 IDT I. Booth, 18 Clos Y Wiwer, Pentre Cwrt, Llantwit Major, CF61 2SG
M0 IDW D. Byrne, Suite 214, 1 Hanley Street, Nottingham, NG1 5BL
M0 IDY S. Gorski, 2 Samphire Close, Didcot, OX11 6HP
M0 IEA C. Bowler, 42A Honor Road Prestwood, Great Missenden, HP16 0NL
M0 IEB Dr C. Bridges, 23 Bramley Vale, Cranleigh, GU6 7FY
M0 IED A. Wedge, 30 Primrose Way Locks Heath, Southampton, SO31 6WX
MW0 IEH D. Lockyer, 19B Drury Lane, Buckley, CH7 3DU
MM0 IEJ J. Macdonald, 24 St. Pauls Drive, Armadale, Bathgate, EH48 2LT
M0 IEK P. Phipps, Meakers Cottage, Long Load, Langport, TA10 9JX
MM0 IEL I. Lee, Roehill, Crossroads, Keith, AB55 6LQ
M0 IEM M. Feast, 30 Peveril Drive, Riddings, Alfreton, DE55 4AP
M0 IEO M. Sanderson, 2 East Crescent, Canvey Island, SS8 9HL
M0 IEP V. Williams, 11 Priory Green Highworth, Swindon, SN6 7NU
M0 IEQ M. Priest, 35 Albert Road, Chaddesden, Derby, DE21 6SJ
M0 IER A. Coote, 148 Clarendon Street, Dover, CT17 9RB
M0 IES M. Reaney, Odessa Marine, Little London, Newport, PO30 5BS
M0 IET C. Blount, 55 Silverthorne Drive, Caversham, Reading, RG4 7NR
MJ0 IEW Dr C. Poole, Kayalami, La Ruelle Du Clos Du Parcq, St Brelade, Jersey, JE3 8AQ
M0 IEY C. Ryder, Sunnymead, Well Head Road Newchurch-In-Pendle, Burnley, BB12 9LW
M0 IEZ N. Cohen, 8 Henry Gepp Close, Adderbury, Banbury, OX17 3FE
M0 IFB D. Endean, 11 Forrester Drive, Brackley, NN13 6NE
M0 IFC A. Ward, 77 Malvern Road, St Helens, WA92EZ
MI0 IFG Blacks Hillbillies ARC, c/o P. Hosey, 13 Glenelly Gardens, Omagh, BT79 7XG
M0 IFH I. Harley, 1 Portland Crescent, Meden Vale, Mansfield, NG20 9PJ
MW0 IFK R. Richards, 77 Church Road, Llanstadwell, Milford Haven, SA73 1EA
M0 IFP M. Byzdra, 3 Brownlow Street, Whitchurch, SY13 1QW
M0 IFT R. Hodson, 99 Alcester Road, Hollywood, Birmingham, B47 5NR
M0 IGB I. Bennett, 44 Haig Avenue, Whitley Bay, NE25 8JG
M0 IGF M. Lane, Greenacres, Bickington Road, Barnstaple, EX31 2JG
M0 IGG S. Wright, 23 Kitchener Street, Walney, Barrow-in-Furness, LA14 3QW
M0 IGJ Tintagel and District Radio Amateur Club, c/o J. Brooks, Treven House, Treven, Tintagel, PL34 0DT
MI0 IGL D. Neill, 8 Castle Meadows Carrowdore, Newtownards, BT22 2TZ
M0 IGM R. Powley, 8 Treadgold Avenue Great Gonerby, Grantham, NG31 8PD
MM0 IGO W. Mcbain, 56 Scotstoun Park, South Queensferry, EH30 9PQ
M0 IGP E. Coles, 41 Venn Court Brixton, Plymouth, PL82AX
M0 IGW A. Griffiths, 67 Griffiths Rd, Upwey, Australia, 3158
M0 IGX A. Griffiths, 67 Griffiths Rd, Upwey, Australia, 3158
M0 IGY J. Hardman, 45 Doncaster Avenue, Manchester, M20 1DH
M0 IHB R. Monaghan, Reservoir Cottage, Tavistock Road Roborough, Plymouth, PL6 7BD
M0 IHC D. Michalczyk, 155 Ewart Road, Nottingham, NG7 6HG
MM0 IHE I. Hepworth, Bronte Cottage, Inverugie, Peterhead, AB42 3DN
M0 IHJ M. Rose, 149 Claremont Road, Blackpool, FY1 2QJ
M0 IHM I. Millman, 3 Oyster Mews, 1-3 Forest Road, Poole, BH13 6EN
M0 IHN K. Missenden, 47 Roseacre Drive, Elswick, Preston, PR4 3UQ
M0 IHR E. Squires, 2, Northgate Path, Borehamwood, WD6 4EX
M0 IHT H. Lennertz, 18 Church Terrace, Exeter, EX2 5DU
M0 IHU P. Farley, 1 Holders Road Amesbury, Salisbury, SP4 7PW
MW0 IHW G. Reason, 454 Cowbridge Road West, Cardiff, CF5 5BZ
M0 IHX M. Allinson, 60 King Street Aspatria, Wigton, CA7 3AH
M0 IHY B. Page, 12 Hitchens Close, Hemel Hempstead, HP1 2PP
M0 IIA A. Bullard, 15 Rowan Drive, Lutterworth, LE17 4SP
M0 IIE J. Bennet, 53 Haven Road, Barton-upon-Humber, DN18 5BS
MU0 IIF S. Bougourd, La Petite Folie, Folie Lane, Guernsey, GY3 5SE
MI0 IIG I. Gibb, 1 Shankill Road, Garvary, Enniskillen, BT94 3DB
M0 IIM C. Gordon, 90 Sunholme Drive, Wallsend, NE28 9YW
M0 IIQ A. Birkett, 21 Cedar Drive Wyke, Bradford, BD12 9HL
M0 IIU J. Fogg, 17 Lyppiatt Road, Bristol, BS5 9HW
M0 IIZ D. Saunders, 6 Cherry Tree Drive Sedgefield, Sedgefield, TS21 3DN
M0 IJA C. Heal, 2 Thorn Gardens, Ramsgate, CT11 7AS
M0 IJC P. Drobny, Flat 5, Busby Court Ceylon Place, Eastbourne, BN21 3JG
MW0 IJE Dr J. Woore, 10 Haverfordwest Road, Letterston, Haverfordwest, SA62 5UA
M0 IJG I. Mcmullan, Glenayr, Glen Cuileann, Kilquade, Greystones, Ireland, A63 XH90
M0 IJI M. Harper, 15 Furrows End, Drayton, Abingdon, OX14 4GN
M0 IJJ M. Cox, 17A Church End, Weston Colville, Cambridge, CB21 5PE
M0 IJL T. Dix, Ebb House Downhouse Lane Higher Eype, Bridport, DT6 6AH
M0 IJO Dr E. Melikyan, 39 St. Evox Close Rownhams, Southampton, SO16 8FS
M0 IJP A. Jamieson, 48 Shackleton Road, Crawley, RH10 5BX
M0 IJR P. Jones, 446 Winchester Road, Southampton, SO16 7DG
M0 IJT J. Rushby, 68 Trowbridge Road, Bradford-on-Avon, BA15 1EN
M0 IJU A. Wakefield, Kuleana, 4 Barnrigg Barbon, Carnforth, LA6 2LJ
M0 IJY K. Scotney, Flat 39, Richmond Hill Gate, 1 Richmond Hill Drive, Bournemouth, BH2 6LT
M0 IKA G. Mills, Bowness, Moortown Lane Brighstone, Newport, PO30 4AN
M0 IKB A. Young, 15 Shelton Avenue, East Ayton, Scarborough, YO13 9HB
M0 IKC G. Bryant, 11 Cadwallon Road, London, SE9 3PX
M0 IKD M. Draper, 160 Chanctonbury Road, Burgess Hill, RH15 9HA
M0 IKE D. Bilson, 31 Middleton Drive, Inkersall, Chesterfield, S43 3HS
MM0 IKG J. Dick, 85 Mellerstain Road, Kirkcaldy, KY2 6UD
MW0 IKH C. Eyre, Croes Yr Onen, Chapel Street Newbridge, Wrexham, LL14 3JH
M0 IKI J. Swann, 5 Lanark Close, Hazel Grove, Stockport, SK7 4RU
M0 IKM T. Palmer, 29 Field End, Maresfield, Uckfield, TN22 2DJ
M0 IKN Dr F. Kuttikkate, 34 Shetland Crescent, Rochford, SS4 3FJ
M0 IKT D. Capstick, 3 Andrew Close Dibden Purlieu, Southampton, SO45 4LS
M0 IKV P. Day, 1 Pine Close, Lutterworth, LE17 4UT
M0 IKW G. Cannon, 30 Main Street, Flixton, Scarborough, YO11 3UB
M0 ILA D. Mielczarek, 7 Bryant Close, Camberley, GU16 8AD
MM0 ILC S. Jordaan, 16 Hillside Terrace, Selkirk, TD7 4LT
M0 ILF B. Devon, 8 Bruckner Place, Claremont Meadows, Australia, 2747
MW0 ILG J. Lewis, 26 Penlan Crescent, Swansea, SA2 0RL
MI0 ILH A. Storey, 31 Willowbrook Park, Bangor, BT19 7GY
M0 ILI B. Neville, 14 Alabury House Birch Close, Huntington, York, YO31 9PP
MI0 ILJ D. Fisher, 11 Summer Street, Belfast, BT14 6ES
M0 ILK D. Critoph, 26 Pevensey Close, Aylesbury, HP21 9UB
M0 ILM M. Miller, Barn Cottage, Wingfield Hall, Manor Road, Alfreton, DE55 7NH

M0 ILN D. Bee, 25 Blatcher Close, Minster On Sea, Sheerness, ME12 3PG
M0 ILO M. Noblet, 1 Lingdale Road, Wirral, CH48 5DG
M0 ILT P. Evans, 12 Cottage Corner, Ilton, Ilminster, TA19 9ER
M0 ILU J. Taylor, 41 Waters View, Yarwell Mill, Yarwell, Peterborough, PE8 6EU
M0 ILX D. Vaughan, 1 Boulnois Avenue, Poole, BH14 9NX
M0 ILY A. Trueman, 10 Mountbatten Ave, Dukinfield, SK165BU
M0 ILZ P. Moore, 22 Audit Hall Road, Empingham, Oakham, LE15 8PH
M0 IMD I. Douglas, 13 Castlereagh Street, New Silksworth, Sunderland, SR3 1HJ
M0 IME M. Scobie, Suncourt, Meadfoot Sea Road, Torquay, TQ1 2LQ
M0 IMF G. Brown, 29 Park Close, Little Eaton, Derby, DE21 5DY
M0 IML B. Vile, 24 Hudson Close, Dover, CT16 2SG
M0 IMM S. Imms, 26 Greenwood Avenue, Rowley Regis, B65 9NJ
M0 IMP I. Pollard, 24 Terminus Road, Littlehampton, BN17 5BX
M0 IMQ A. Tomaszewski, Tw34Ad, London, TW34AD
M0 IMS M. Sims, 5 Sandy Leaze, Bradford-on-Avon, BA15 1LX
M0 IMT I. Turner, 1 Elmwood Rise, Sedgley, DY3 3QJ
MM0 IMU K. Swierczynski, 103 Hendry Road, Kirkcaldy, ky2 5db
M0 IMV Kimbolton School ARC, c/o D. Brattle, 34 York Street, Bedford, MK40 3RJ
M0 IMW I. Walker, 24 Hawthorn Road, Norwich, NR5 0LP
M0 IMY O. Fienko, 42 Fenside Road, Boston, PE21 8JH
M0 IMZ E. Preda, 4 Playsteds Lane, Great Cambourne, Cambridge, CB23 6GA
M0 INB I. Barraclough, Maru, 25 Blaithroyd Lane, Halifax, HX3 9PS
MW0 INC I. Curnock, 62 Heol Y Banc, Bancffosfelen, Llanelli, SA15 5DL
M0 IND P. Ind, 30 Thompson Road, Stroud, GL5 1SY
MM0 INE P. Thompson, Windward Honeyfield Road, Jedburgh, TD8 6JW
M0 INF Dr A. Fugard, Flat 136, Willowbrook House, Coster Avenue, London, N4 2ZT
MM0 INH W. Goodfellow, 1 Yester Place., Haddington, EH41 3BE
M0 INI M. Smith, Church Farm, Mucklestone, Market Drayton, TF9 4DN
MI0 INL O. Hart, 18 Sandringham Court Portadown, Craigavon, BT63 5BF
M0 INM L. Dettman, 55 Canada Drive, Cottingham, Hull, HU16 5EH
M0 INP I. Popgueorguiev, 239 Westborough Road, Westcliff-on-Sea, SS0 9PR
MM0 INS C. Ralph, 37 Seaview Terrace, Edinburgh, EH15 2HE
MM0 INT Outer Hebrides Iota Group, c/o C. Mcgowan, 21 Franchi Drive, Stenhousemuir, Larbert, FK5 4DX
M0 INU Chelsea Pensioners Radio Club, c/o R. Petrie, Royal Hospital Chelsea, Royal Hospital Road, London, SW3 4SR
MW0 INW J. Evans, 9 Cynfi Terrace Deiniolen, Caernarfon, LL55 3LG
M0 INX S. Griffiths, 14 Vicarage Lane, Oxford, OX1 4RQ
M0 INY I. Davis, Top Pub Brown Edge, Hill Top, Stoke-on-Trent, ST6 8TX
M0 IOA Isle of Avalon ARC, c/o M. Wheeler, 114 Boundary Way, Glastonbury, BA6 9PH
MM0 IOB A. Macleod, 2 Eoligarry, Isle of Barra, HS9 5YD
M0 IOC I. O'Connor, 7 Grove Court, Shotton Colliery, Durham, DH6 2QD
M0 IOE K. Malinowski, 46, Bennett Court, 2 Pitcher Lane, Ashford, TW15 2BN
M0 IOI Dr S. Leask, 1 Collington Street, Beeston, Nottingham, NG9 1FJ
M0 IOK D. Proctor, 4 The Green, Sproatley, Hull, HU11 4XF
MM0 IOL A. Morrison, 6A Upper Barvas, Isle of Lewis, HS2 0QX
MD0 IOM M. Perry, 48 Anagh Coar Road, Douglas, Isle of Man, IM2 2AR
M0 IOO M. Kowalczyk, 10 Hillberry Close, Narborough, LE193EW
M0 IOS B. Stone, 8 Gissons Lane, Kennford, Exeter, EX6 7UB
M0 IOT C. Norris, 115 Sutton Road, Walpole Cross Keys, PE344HE
MI0 IOU T. Herbison, 22 Dernaveagh Road, Ballymena, BT43 6SX
M0 IOV Dr K. Singh, Flat 401 Wolsey House, Princess Street, Ipswich, IP1 1RS
M0 IOW B. Cant, 15 Mountbatten Drive, Newport, IOW., PO30 5SG
M0 IOZ Dr K. Singh, Flat 401 Wolsey House, Princess Street, Ipswich, IP1 1RS
M0 IPB S. Buckley, 1 Columbia Street Fairview, Cheltenham, GL52 2JR
MM0 IPD L. Flis, 30 Bruce Avenue, Inverness, IV3 5HE
M0 IPE R. Fulcher, 1 Edwards Close Hutton, Brentwood, CM13 1BU
M0 IPH A. Grzegorek, Basement Flat 1, 17 Fitzroy Street, Bristol, BS4 3BY
M0 IPI R. Hall, 1 The Willows Brent Knoll, Highbridge, TA9 4EJ
M0 IPJ B. Mcnamara, Flat 6, Nightingales 5 Milner Road, Bournemouth, BH4 8AD
MW0 IPL Men's Shed Radio Club, c/o C. Beech, 9 Wesley Court, Pembroke Dock, SA72 6NE
M0 IPO B. Davies, 16 Pearmains Close, Orwell, Royston, SG8 5QY
M0 IPQ R. Zielinski, 33 Ellis Road, Cambridge, CB2 9BG
M0 IPR I. Ridings, 25 Mond Road, Irlam, Manchester, M44 6QA
M0 IPS A. Hollings, 39 Rendham Road, Saxmundham, IP17 1EA
M0 IPV D. Steel, 18 Sandal Hall Mews, Wakefield, WF2 6ED
M0 IPZ A. Griffin, 4 Bramley Close Kingswood, Wotton-under-Edge, GL12 8SF
M0 IQC R. Hall, 47 Woodlands Drive Skelmanthorpe, Huddersfield, HD8 9DB
M0 IQF M. Blazejewski, 73 Byron Road, Luton, LU4 0HX
M0 IQG G. Robinson, 91 Tilstock Crescent, Shrewsbury, SY2 6HH
M0 IQH D. Pemberton, 142 Norfolk Road, Huntingdon, PE29 1RH
MM0 IQK I. Andrews, 29 The Riggs Auchtermuchty, Cupar, KY14 7DX
M0 IQL H. Taylor, 25 Northolme Avenue, Nottingham, NG6 9AP
M0 IQM A. Bailey, 52 Berkeley Road, Shirley, Solihull, B90 2HT
MM0 IQN D. Petrovic, 206 Mayfield Drive Armadale, Bathgate, EH48 2JL
M0 IQQ T. Bell, 28 Inwood Drive, Coleford, GL168EZ
M0 IQR C. Wrobel, 33 Harsnett Road, Colchester, CO1 2HS
M0 IQX A. Emmerson, 8 Weston Close, Heath Hayes, Cannock, WS11 7YX
M0 IQY C. Cromie, 140 Whalley Road Wilpshire, Blackburn, BB1 9LJ
MW0 IQZ G. Cattle, 12 Claerwen, Gelligaer, Hengoed, CF82 8EW
M0 IRB S. Gandy, 54 Saxty Way, Sowerby, Thirsk, YO7 1SB
MM0 IRC C. Fraser, Rockside, Locheport, Isle of North Uist, HS6 5EU
M0 IRD I. Day, 137 Tuffley Lane, Tuffley, Gloucester, GL4 0NZ
M0 IRF Radio Group O, c/o S. Constable, 8 Hurst Lane, Herstmonceux, Hailsham, BN27 4TS
M0 IRG W. Linton, 29 Lancaster Park, Richmond, TW10 6AB
M0 IRH P. Geary, 51 Heathermount Drive, Crowthorne, RG456HJ
M0 IRI R. Trelease, 23 Torridon Close, Woking, GU21 3DB
MM0 IRJ I. Johnstone, 14 Carledubs Crescent, Uphall, Broxburn, EH52 6TH
M0 IRK P. Holmes, 1 Leonards Place, Bingley., BD16 1AD
M0 IRL G. Maddden, 192 Ardilaun, Portmarnock, Ireland, D13 FA03
M0 IRO J. Ramirez Gonzalez, 107 Maidstone Road, London, N11 2JS
M0 IRP I. Pipe, 9 Sherlock Hoy Close, Broseley, Telford, Tf125jb
M0 IRS D. Spinks, 15 Brunlees Drive, Telford, TF3 2NH
M0 IRT I. Thomas, 47 Salisbury Avenue, Coventry, CV3 5DA
M0 IRU A. Whybrow, 64 Church Road Sevington, Ashford, TN24 0LF
M0 IRV L. Jones, 116 Captain Fold Road Little Hulton, Manchester, M38 9UB
MI0 IRX A. Mcguigan, 30 Cogry Hill, Ballyclare, BT39 0RY
MI0 IRZ D. Gregg, 9 Willowfield, Tandragee, Craigavon, BT62 2EJ
MW0 ISC S. Charles, Tower Side, Pantymwyn, Mold, CH7 5HY
M0 ISE W. Su, No.27, Ln. 800, Yingde St. Qianzhen Dist., Kaohsiung, Taiwan, Province Of China, 80651
MW0 ISF C. Astbury, Old Police House, Llanegryn, Tywyn, LL36 9SL
M0 ISG F. Rundle, 3 Winchester Court, Jarrow, NE32 4TN
M0 ISI A. Cucchiara, 172 Gonville Crescent, Stevenage, SG2 9LZ
M0 ISJ Windmill Amateur Radio DX Group, c/o J. Connor, 28 Church Street, Hungerford, RG17 0JE
M0 ISK G. Howard, 8 Paddock Road, Woodford, Kettering, NN14 4FL
M0 ISL A. Paulick, Wormbacher Weg 27, Berlin, Germany, 12207
M0 ISN Chorley & District A.R.S, c/o E. Entwistle, 43 Brock Road, Chorley, PR6 0DB
M0 ISQ R. Lilley, 3 Coultshead Avenue, Billinge, Wigan, WN5 7HS
M0 IST N. Lasseter, 7 Huntington Mews, York, YO31 8JB
M0 ISU H. Baker, 300 Lowerhouse Lane, Burnley, BB12 6LZ
M0 ISW I. Singlehurst-Ward, 39 Nadder Close, Tisbury, Salisbury, SP3 6JL
MI0 ISY C. Maguire, 1 Churchill Street Antrim Road, Belfast, BT15 2BP
M0 ISZ B. Hardy, 10 Spring Farm Road, Burton-on-Trent, DE15 9BN
M0 ITA R. Ritossa, 182, Rue Chateau Des Rentiers, Paris, France, 75013
M0 ITB M. Brown, 99 Apprentice Drive, Colchester, CO45SE
M0 ITC R. Fearn, 79 Maudlin Drive, Teignmouth, TQ14 8SB
M0 ITG N. Crudgington, Appledore Blackness Lane, Keston, BR2 6HL
M0 ITI M. Bruce, 28 Pheasants Way, Rickmansworth, WD3 7ES
MI0 ITS P. Ford, 25 Carnhill, Londonderry, BT48 8BA
M0 ITT J. Griffiths, 83 Golborne Road, Ashton-In-Makerfield, Wigan, WN4 8XA
M0 ITX N. White, 99 The Common, Mellis, Eye, IP23 8EF
M0 ITY J. Culak, 128 Dunmow Road, Bishop's Stortford, CM23 5HN
M0 ITZ S. Proverbs, 14 Spring Lane, Shepshed, Loughborough, LE12 9JE
M0 IUA Dr C. Oxley, Appletree House Halam Road, Southwell, NG25 0AH
M0 IUB J. Fisher, 4 Orchard Close, East Leake, Loughborough, LE12 6PL
M0 IUC M. Curtis, 9 Oaklands Crescent, Holt, NR25 6UD
M0 IUE D. Cassidy, 186 Kingsman Drive, Boorley Green, Botley, SO32 2TG
M0 IUG S. Campbell, 50 Northwood Road, Whitstable, CT5 2ES
M0 IUI J. Pinto, 21 Steed Crescent, Colchester, CO2 7SJ
M0 IUK D. Grayson, 79 Errington Avenue, Sheffield, S2 2EA
M0 IUM G. Clark, 65 Chyvelah Vale, Gloweth, Truro, TR1 3YJ
MW0 IUN I. Jones, 21 Albert Street, Maesteg, CF34 0UF
M0 IUQ G. Hinson, The Leys, Brierley Hill, DY5 3UJ
M0 IUR E. Vrentzos, 12 Floyer Close, Queens Road, Richmond, TW10 6HS
M0 IUT R. Williams, 10 Bramley Close, Twickenham, TW2 7EU
M0 IUU I. Stephens, 14 Vardon Close Kingston Hill, Stafford, ST16 3YW
M0 IUV S. Helm, 10 St. Annes Avenue, Middlewich, CW10 0AE
M0 IUZ D. Waterhouse, 6 Sands Lane, South Ferriby, Barton-upon-Humber, DN18 6JS
M0 IVC I. Marques, 7A Kings Park Road, Southampton, SO15 2AS
M0 IVE I. Valkov, Flat 10, Warlingham House, London, SE16 3DQ
M0 IVJ J. Raybould, 33 Lincoln Road, Dorrington, Lincoln, LN4 3PT
M0 IVK A. Parkin, 22 Glorney Mead, Badshot Lea, Farnham, GU9 9NL
M0 IVL P. O'Shea, 37 Barclay Court, Ilkeston, DE7 9HJ
M0 IVM A. Nel, 12 Wain Court, Rakeway, Saughall, Chester, CH1 6BF
M0 IVN P. Mocker, Flat 24, The Gardens, Clapton Common, London, E5 9AZ
M0 IVO W. Gissing, 2 Yeo Moor, Clevedon, BS21 6UQ
M0 IVS M. Mann, Woodlands, Coston, Norwich, NR9 4DT
M0 IVU M. Holland, 32 Saltersgate Drive, Birstall, Leicester, LE4 3FF
M0 IVW D. Barrett, 183 Wilson Avenue, Brighton, BN2 5PD
M0 IVZ C. Cumming, 2 Alexander Close, New Milton, BH25 5NS
M0 IWA D. Brooks, 61 Carisbrooke High St., Newport, PO30 1NR
M0 IWB I. Bunting, 14 Mill Pightle, Aylsham, Norwich, NR11 6LX
M0 IWG M. Tyrrell, 47 Woodley Hill, Chesham, HP5 1SL
M0 IWI D. Gilbertson, 6 Lewens Lane, Wimborne, BH21 1LE
M0 IWJ J. Siviter, 96 Byne Road, London, SE26 5JD
M0 IWN P. Taylor, 138 Paulhan Street, Bolton, BL33DT
M0 IWO S. Noller, 3 Thor Road, Norwich, NR7 0JS
M0 IWQ T. Warner, 6 Cotswold Court, Skelmersdale Road, Clacton on Sea, CO15 6EN
MM0 IWS J. Greene, Ingleneuk, Beltie Road, Torphins, AB31 4JU
M0 IWT A. Gale, 22 Graham Street, Swindon, SN1 2EY
MW0 IWU Eryri DX, c/o J. Pritchard, 1 Tan Y Coed, Maesgeirchen, Bangor, LL57 1LU
M0 IWV R. Molyneux, 23 Birks Drive, Bury, BL8 1JA
M0 IWW R. Molyneux, 23 Birks Drive, Bury, BL8 1JA
MW0 IWX Blaenau Gwent Radio Group, c/o A. Lewis, 33 Heol Helig, Brynmawr, Ebbw Vale, NP23 4TY
M0 IWZ A. Hanna, 35 Orchard Drive, Mayland, Chelmsford, CM3 6EP
M0 IXC Dr A. Mazur, 1 Oak Trees New Road, Stoborough, BH20 5BB
M0 IXF A. Smythson, 64 Blinco Rd, Rushden, NN10 0EA
M0 IXG I. Omenaca-Gavin, 2 Mayflower Court Highbridge Wharf, Reading, RG1 3AJ
MI0 IXH M. Mcquillan, 21 Glen River Park Glenavy, Crumlin, BT29 4FX
MM0 IXJ K. Brown, 10 Richmond Street, Clydebank, G81 1RF
M0 IXL M. Norfield, 122 Huntingdon Road, Upwood, Ramsey, Huntingdon, PE26 2QQ
M0 IXM R. Gaskell, Cottonwood California, Baldock, SG7 6NU
M0 IXN R. Hughes, 22 Roman Close, Blue Bell Hill, Chatham, ME5 9DJ
M0 IXP D. Hall, 31 Edinburgh Drive, Didcot, OX11 7HS
M0 IXQ O. Reyes Salazar, P.O Box 117, Heidelberg, Australia, 3084
MM0 IXT S. Kelso, 98 Highmains Avenue, Dumbarton, G82 2QB
M0 IXU A. Foote, Flat One, Kimber'S Close Kennet Road, Newbury, RG14 5JF
MW0 IXV W. Groves, 3 Tetbury Close, Newport, NP20 5HX
M0 IXW A. Royds, 3A Fairfield Avenue, Rossendale, BB4 9TG
M0 IXX S. Colquhoun-Lynn, 76 Chestnut Drive, Sale, M33 4HL
M0 IXY A. Atkinson, 21 Dennington Crescent, Basildon, SS14 2FF
M0 IXZ P. Rollason, 68 Heathfield Lane West, Wednesbury, WS10 8QP
M0 IYC Basingstoke Makerspace, c/o K. Roche, 96 Porter Road, Basingstoke, RG22 4JR
MM0 IYD G. Mann, Halcyon, Strachan, Banchory, AB31 6NL
M0 IYE A. Gibbons, 305 Monks Road, Lincoln, LN2 5LB
M0 IYF M. Christensen, 40 Aireview Terrace, Broughton Road, Skipton, BD23 1RX

M0 IYG D. Stevenson, 37 Larkhill, Skelmersdale, WN8 6TE
M0 IYH T. Machii, 3-6-17-311 Nakacho, Musashino City, Tokyo, Japan, 180-0006
M0 IYJ P. Reeves, 11 Quines Hill Road Forest Town, Mansfield Notts, NG190NW
M0 IYK C. Walson, 30 West Crescent, Duckmanton, Chesterfield, S44 5HE
M0 IYM B. Clark, 8 Langdale Close, Farnborough, GU14 0LQ
M0 IYN G. Moss, 125 Lavender Avenue, Mitcham, CR4 3RS
M0 IYP G. Wood, 28 Marford Crescent, Sale, M33 4DH
M0 IYQ D. Buchan, 40 Bradstone Road, Winterbourne, Bristol, BS36 1HQ
MM0 IYT M. Mercanti, 45 Brueacre Drive, Wemyss Bay, PA18 6HA
M0 IYV D. Tennant, 9 Beaumaris Road, Hindley Green, Wigan, WN2 4NB
M0 IYZ Dr P. Duvoisin, 6A Newland Street, Witham, CM8 2AQ
M0 IZA A. Nikitits, 270A The Ridgeway, St Albans, AL4 9XQ
M0 IZF A. Robeson, 19 Cameron Drive Woodlands, Ivybridge, PL21 9TS
M0 IZG A. Evans, 8 Plowden Close, Bolton, BL3 3NU
M0 IZM P. Stuart, 5 Welbeck Gardens, Woodthorpe, Nottingham, NG5 4NX
MI0 IZN Orchard County DX Club, c/o E. Simpson, 10 Woodview Park, Tandragee, Craigavon, BT62 2DD
MI0 IZP J. Henshaw, 80 Ballystrudder Road, Islandmagee, Larne, BT40 3SJ
M0 IZR D. Butler, Westbrook, Lower Farm Road, Ringshall, Stowmarket, IP14 2JE
M0 IZS D. Sexton, 16 Rufus Isaacs Road, Caversham, Reading, RG4 6DD
M0 IZV P. Gardner, 7 Park Avenue, Shipley, BD18 3LW
M0 IZW P. Raftery, Ballybride, Roscommon, Ireland, F42TK49
MW0 IZX K. Earnshaw, 5 Castle Mews George Street, Pontypool, NP4 6BU
M0 JAD P. Holland, 30 Knighton Park Road, London, SE26 5RJ
M0 JAE J. Allen, 20 Spa Hill, Kirton Lindsey, Gainsborough, DN21 4BA
M0 JAF J. King, 22 Latchmere Gardens, Leeds, LS16 5DN
M0 JAG A. Pegg, 18 Blythe Way, Shanklin, PO37 7NJ
M0 JAI P. Gaur, 34 Queensberry Avenue, Copford, Colchester, CO6 1YN
M0 JAJ J. Stedman, 60 Sandown Road, Ipswich, IP1 6RE
M0 JAK J. Swain, 84 Sunnymead Drive, Waterlooville, PO7 6BX
M0 JAM J. Mortimer, 4 Nethercliffe Crescent, Guiseley, Leeds, LS20 9HN
MW0 JAN J. Day, 20 St. Johns Drive Pencoed, Bridgend, CF35 5NF
M0 JAO J. Sansom, 4 Vicarage Road, Eastbourne, BN20 8AU
M0 JAP Dr D. James, Bramble Cottage, Tray Lane, Atherington, Umberleigh, EX37 9HY
M0 JAQ J. Malia, 55 West Farm Avenue Longbenton, Newcastle upon Tyne, NE12 8LS
MI0 JAR J. Rice, 42 The Crescent, Ballymoney, BT53 6ES
MI0 JAT J. Mcgoldrick, 23 Lettercarn Road, Clare, Castlederg, BT81 7QY
M0 JAV Dr J. Rogers, 9 Cherry Tree Avenue, Shireoaks, Worksop, S81 8PH
MW0 JAW J. Stevens, 8 St. Josephs Court, Llanelli, SA15 1NR
M0 JAX J. Edwards, 45 Bramshaw Gardens, Bournemouth, BH8 0BT
MI0 JAY W. Graham, 19 Margaret Square, Ballymoney, BT53 6BZ
M0 JAZ J. Sadler, 10 Spindle Warren, Havant, PO9 2PU
M0 JBA J. Baines, 7 Willerby Low Road, Cottingham, HU16 5JD
M0 JBC J. Crank, 38 Harley Avenue, Harwood, Bolton, BL2 4NU
M0 JBD J. Day, 124 Radstock Road, Southampton, SO19 2HU
M0 JBF J. Cobb, 32 Dellmont Road, Houghton Regis, Dunstable, LU5 5HU
MW0 JBH A. Jozwik, 160 Broad Mead Park, Newport, NP19 4PF
MI0 JBK J. Mackenzie, 30 Dalriada Gardens, Ballycastle, BT54 6DZ
MM0 JBS J. Summers, 1 Main Road, Fairlie, Largs, KA29 0DP
MI0 JBT J. Traynor, 8 Roeville Terrace, Limavady, BT49 0BH
M0 JBV M. Harvey, May Tree Cottage, Kelvedon Road, Tiptree, Colchester, CO5 0LJ
M0 JBW J. Mclaughlin, 34 Cambridge Road, Birstall, Batley, WF17 9JF
M0 JBY E. Riddle, 37B Stubbs Lane, Braintree, CM7 3NR
M0 JBZ J. Chalmers, 19 Brettenham Crescent, Ipswich, IP4 2UB
M0 JCC I. Jefferson, 19 Orchard Way, Flitwick, Bedford, MK45 1LF
M0 JCD J. Dalgliesh, 61 Clonners Field, Stapeley, Nantwich, CW5 7GU
M0 JCE J. Crewe, 22 Myrtle Tree Crescent Sandbay, Weston-Super-Mare, BS22 9UL
M0 JCG C. Grant, 27 Bulrush Close, Chatham, ME5 9BN
M0 JCH J. Paul, East Park, Church Road, Cowes, PO31 8HA
M0 JCK E. Beechill, Belleroyd Farm Blackshaw Head, Hebden Bridge, HX7 7JP
M0 JCL J. Plant, 67 Kenley Road, London, SW19 3JJ
M0 JCM J. Murray, 2 The Cuttings Hampstead Norreys, Thatcham, RG18 0RR
M0 JCP J. Haynes, 16 Mountsfield, Frome, BA11 5AR
M0 JCQ J. Stevens, 51 Cheddington Road, Pitstone, Leighton Buzzard, LU7 9AQ
M0 JCR J. Reynolds, 15 Chestnut Mead, Oxford Road, Redhill, RH1 1DR
M0 JCS J. Stevenson, 18 Drakehouse Lane, Sheffield, S20 1FW
M0 JCT J. Townsend, 47 Main St., Wolston, Coventry, CV8 3HH
M0 JCV GNRS, c/o R. Moseley, 20 Bleakley Avenue, Notton, Wakefield, WF4 2NT
M0 JCY R. Manciu, 3 Delta Court, Standard Road, Hounslow, TW4 7AN
M0 JCZ M. Ciechan, 109 Rose Vale, Liverpool, L5 3PD
M0 JDA J. Dale, Corydon, Church Street, Sevenoaks, TN14 7SW
M0 JDB J. Pollard, 72 Windy Arbour, Kenilworth, CV8 2BB
M0 JDD J. Dedier, 10 Lowry Close, Haverhill, CB9 7GH
M0 JDE D. Foxall, 1 Doe Hey Grove, Farnworth, Bolton, BL4 7HS
M0 JDF NCC Group Radio Society, c/o P. Hodges, 191 Broadstone Road, Stockport, SK4 5HP
M0 JDL J. Dowdeswell, 18 Lechlade Gardens, Fareham, PO15 6HF
M0 JDP J. Page, 5 Riddimore Avenue, Hereford, HR2 7LJ
M0 JDS J. Sweatman, 14 Clover Court, Jasmine Grove, Waterlooville, PO7 8BP
M0 JDT J. Tyrrell, 28 Park Avenue, Princes Avenue, Hull, HU5 3ER
M0 JDU J. Dowson, 4 Thorntree Close, Goole, DN14 6HJ
MW0 JDW J. Williams, 1 Nant Terrace, Pentraeth, LL75 8YE
M0 JDY Z. Zhao, Christs College, Cambridge, CB2 3BU
M0 JDZ R. Griffin, 15 Donside Close, Boldon Colliery, NE35 9BS
M0 JEA M. Augustus, 3 Heathend Cottages Heathend, Wotton-under-Edge, GL12 8AS
M0 JEB J. Braithwaite, 129 Cherry Tree Gardens, Blackpool, FY4 4PY
M0 JEC N. Bland, 63 Swindon Road, Wroughton, Swindon, SN4 9AG
M0 JEG B. Mcdonald-Watson, 39 Heald Way, Nantwich, CW5 6SQ
M0 JEH D. Mackie, 41 Fairfield Avenue, Sandbach, CW11 4BP
M0 JEI O. Phillips, 54 Marshall Road, Cambridge, CB1 7TY
M0 JEJ S. Hunt, 7 Cokefield Avenue, Nuthall, Nottingham, NG16 1AU
M0 JEK A. Skarzynski, 1 River View Moorings, Bridge Road, Stoke Ferry, King's Lynn, PE33 9TS
M0 JEL J. Loughrey, 54 Marshall Drive, Bramcote, Nottingham, NG9 3LD
M0 JEM J. Mayo, 134 Bromsgrove Road, Redditch, B97 4SP
M0 JEO R. Bishop, 12A Goseley Avenue, Hartshorne, Swadlincote, DE11 7EZ
M0 JEP J. Price, Oxted Place East, Broadham Green Road, Oxted, RH8 9PF
M0 JEQ D. Tynan, 2 Horsham Grove Whelley, Wigan/Greater Manchester, WN2 1AP
MJ0 JER R. Taylor, 21 Samares Avenue La Grande Route De St. Clement, St Clement, Jersey, JE2 6NY
M0 JES J. Saunders, 105 Wembdon Hill, Wembdon, Bridgwater, TA6 7QB
MM0 JET 49f Son Air Cadet Radio Club, c/o B. Burt, 182 Old Inverkip Road, Greenock, PA16 9JG
M0 JEU R. Tadhunter, 44 Queensway, Melksham, SN12 7LD
M0 JEZ J. Powell, 46 Woodmancote, Yate, Bristol, BS37 4LL
M0 JFB J. Button, 1 Amber Close, Rainworth, Mansfield, NG21 0FU
M0 JFD J. Dixon, 23 Dee Way, Winsford, CW7 3JB
M0 JFE J. Earnshaw, 128 Shakespeare Road, Fleetwood, FY7 7HJ
M0 JFH D. Flood, 506 Preston Old Road, Blackburn, BB2 5LY
M0 JFI S. Kennett, 321 Green Lane, Chertsey, KT16 9QR
M0 JFJ Y. Alghurair, 12 Second Avenue, Bath, BA2 3NN
M0 JFK P. Waine, 14 Grace Street, Sutton, St Helens, wa93ng
M0 JFM J. Marsh, 14 Eyam Road, Hazel Grove, Stockport, SK7 6HP
M0 JFO J. Owen, 8 Highridge Crescent, Bristol, BS13 8HN
M0 JFR L. Vencel, 12 Alston Road, Ipswich, IP3 8EU
M0 JFW J. Wheeler, 428 Bromsgrove Road, Hunnington, Halesowen, B62 0JL
MW0 JFX K. Boxley, 9 Llys-Y-Pentre, Afonwen, CH7 5UY
M0 JGB J. Greenway-Brown, 207 Lowe Avenue, Wednesbury, WS10 8NS
MW0 JGE N. Lewis, 1 Clyne Drive, Blackpill, Swansea, SA3 5BU
M0 JGF P. King, 14 Glenhurst Drive, Whickham, Newcastle upon Tyne, NE16 5SJ
M0 JGH J. Hunt, 15 Greenway, London, SW20 9BQ
M0 JGK A. Day, 3 Park Close, Sudbury, CO10 2XZ
M0 JGM J. Marlett, 6 Delamere Avenue Sutton Manor, St Helens, WA9 4AP
MM0 JGP J. Pirie, Millhouse, Watermill, Fraserburgh, AB43 7ED
M0 JGR J. Glover, 12 Willow Street, London, E4 7EG
M0 JGS J. Seaton, 88 Stanley Road, Cambridge, CB5 8LB
M0 JGU P. Shaw, 32 Hardwick Road East, Worksop, S80 2NT
M0 JHB J. Butcher, 3 Basket Gardens, London, SE9 6QP
MW0 JHC J. Clarke, 29 Sisial Y Mor, Rhosneigr, LL64 5XB
M0 JHD J. Hardy, Lambda House, Seanor Lane, Chesterfield, S45 8DH
M0 JHF J. Foster, Westfield House, 23 High Street, Cumnor, OX2 9PE
M0 JHG J. Ginever, 66 London Road, Maidstone, ME16 8QU
MM0 JHL J. Hutchinson, Hawthorn Cottage, Muirhall, West Calder, EH55 8NL
M0 JHM J. Rymer, 23 Chetwode Road, Tadworth, KT20 5PS
M0 JHN M. Rees, Hillcrest, Cumeragh Lane, Whittingham, Preston, PR3 2AL
M0 JHO B. Hodgson, 28 Grove Drive, Woodhall Spa, LN10 6RT
M0 JHP H. Alava Moreira, 185 Riverdale Road, Erith, DA8 1PZ
M0 JHV G. Dulcu, 49 Pearson Road, Crawley, RH10 7AJ
M0 JHW J. Wheeldon, 11 Stathern Walk, Grantham, NG31 7XG
M0 JIB J. Bickers, 3 The Old Brickyard, West Haddon, Northampton, NN6 7GP
M0 JIC M. Matusiewicz, 6 Westminster Crescent, Doncaster, DN2 6JQ
M0 JID T. Gore, 22 Stoppard Road, Burnham-on-Sea, TA8 1QB
M0 JIE K. Reynolds, 20 Wentwood Gardens, Plymouth, PL6 8TD
M0 JIJ J. Matthews, 2 Farm Close, Bungay, NR35 1JG
M0 JIL G. Heyes, 5 Ashgarth Way, Harrogate, HG2 9LD
M0 JIM J. Heinen, The Coach House, St. Mary Bourne, Andover, SP11 6EW
M0 JIO A. Edwards, 23 Brittany Avenue, Ashby-de-la-Zouch, LE65 2QY
M0 JIP J. Pownall, 75 Park Barn Drive, Guildford, GU2 8ER
M0 JIQ D. Samarin, Long Gables, Gorse Avenue, Kingston Gorse, East Preston, BN16 1SQ
MJ0 JIS Jersey Scouts ARC, c/o C. Totty, Flat 4, Beech Court, Woodlands Apartments, La Rue Des Cotils, Grouville, Jersey, JE3 9AY
M0 JIU J. Uren, 4 Killivose Road, Camborne, TR14 7RN
M0 JIX Dr C. Fletcher, 7 Highfield Crescent, Baildon, Shipley, BD17 5NR
M0 JJA R. James, 51 The Lampreys, Gloucester, GL4 6QU
M0 JJB J. Barton, 93 Cardigan Road, Bridlington, YO15 3JU
M0 JJC A. Coghlan, 1 Bell View Cross Houses, Shrewsbury, SY5 6JR
M0 JJD J. Dignan, Plum Tree Cottage, Station Road, Marsh Gibbon, Bicester, OX27 0HN
M0 JJE M. Chisholm, 19 Tudor Close, Newtoft, Market Rasen, LN8 3NQ
M0 JJH G. Cavie, Dawn, Maypole Road, Colchester, CO5 0EN
M0 JJI A. Shakesby, 14 Dawnay Road, Bilton, Hull, HU11 4HB
M0 JJK J. King, 18 Ross Road, Wallington, SM6 8QB
M0 JJM D. Oliver, 20 Five Oaks Close, Malvern, WR14 2SW
M0 JJN J. Nicholls, 6 Marasca End, Holt Drive, Colchester, CO2 0DL
M0 JJP J. Walker, Flat A 14 Elswick Road, London, SE13 7SR
MM0 JJQ M. Butterworth, 97A Dunearn Drive, Kirkcaldy, KY2 6AL
M0 JJR J. Reilly, 22 Charlecote Gardens, Sydenham, Leamington Spa, CV31 1GE
MM0 JJU M. Love, 17 Lindsay Road, East Kilbride, Glasgow, G74 4HZ
MM0 JJV J. Vennard, 4 Braehead, Girdle Toll, Irvine, KA11 1BD
M0 JJX D. Jarman, 62 Combe Drive, Dunstable, LU6 2AE
MM0 JJZ K. Vielhaber, Nether Littlefold, Crieff, PH7 3NY
M0 JKB S. Lucas, 31 Lilian Close, Norwich, NR6 6RZ
M0 JKE J. Howarth, 19 Farnham Croft, Leeds, LS14 2HR
M0 JKF J. Ferrol, 29 Westlands, Haltwhistle, NE49 9BS
M0 JKG J. Gaskin, Badgers Barn, Canterbury Road, Folkestone, CT18 8DF
MM0 JKH D. Smith, 9 Ceum A Bhealaich, Stornoway, HS1 2UB
M0 JKI A. Munns, Peverels, Stoke Farm Drive, Battisford, Battisford Tye, Stowmarket, IP14 2NA
M0 JKM J. Lugsden, 21 Overhill Way, Beckenham, BR3 6SN
M0 JKN J. Wilson, Flat 5, Blake House, London, SE1 7DX
M0 JKP M. Howden, 11 Marsh Lane Gardens, Kellington, Goole, DN14 0PG
M0 JKQ C. Poulson, 9 Scattergate Green, Appleby-in-Westmorland, CA16 6SP
M0 JKS Dr D. Pegler, September Cottage, East Bank, Winster, DE42DT
MW0 JKU C. Taylor, 23 Heol Derw, Brynmawr, Ebbw Vale, NP23 4TT
M0 JKV A. Allan, The Oxford Health Co Ltd, Unit 4, Longlands Road, Bicester, OX26 5AH
M0 JKW J. Whitehead, 20 Seamill Park Crescent, Worthing, Bn11 2pn
M0 JKX Invicta Contest Group, c/o I. Hope, 5 The Crescent, Northfleet, Gravesend, DA11 7EB
MM0 JKY A. Chambers, 35 Echline Grove, South Queensferry, EH30 9RU
MI0 JLC S. Mulligan, 70 Lisnoe Walk, Lisburn, BT28 1QD
M0 JLE J. Cafe, Flat 6, Smiths Court, 73 East Borough, Wimborne, BH21 1PJ

M0 JLI I. Bertea, 100 Felbrigge Road Ilford Ig3 9Xj United Kingdom, Ilford, IG3 9XJ
MW0 JLN N. Howells1, 21 Coed Bach, Pencoed, Bridgend, CF35 6TF
M0 JLP J. Powell, 23 Park Road, Norton, Malton, YO17 9DZ
M0 JLR J. Redhead, 28 Sandfields, Frodsham, WA6 6PT
M0 JLS A. Stevens, 16 Hunters Chase, March, PE15 9EL
M0 JLT L. Taylor, 33 Priestley Avenue, Darton, Barnsley, S75 5LG
M0 JLW J. Welford, 26 Templewood, Welwyn Garden City, AL8 7HX
M0 JLY A. Page, 207 Brooklyn Road, Cheltenham, GL51 8DZ
MM0 JMB J. Brown, 11 Fairway Avenue, Elgin, IV30 6XF
M0 JMC J. Mccutcheon, 15 Maytrees, Hitchin, SG4 9LT
M0 JME J. Blundell, 21 Walmsley Street, Fleetwood, FY7 6LJ
M0 JMF J. Foster, 41 Coinagehall Street, Helston, TR13 8ER
MM0 JMI Prof. J. Davies, 5 County Council Houses, Kingston, North Berwick, EH39 5JE
M0 JMJ J. Jukes, 22 Hazelmere Road, Creswell, Worksop, S80 4HS
MM0 JMK J. Mckechnie, 8 Waulker Avenue, Stirling, FK8 1SA
MI0 JML J. Mccaw, 62 High Street, Ballymena, BT43 6DT
M0 JMN R. Andrews, Owls Rest, Park Lane, Worcester, WR2 6PQ
M0 JMP J. Poole, 18 Grosvenor Avenue, Kidderminster, DY10 1SS
M0 JMQ M. Ireland, 45 Pheasant Rise, Bar Hill, Cambridge, CB23 8SA
M0 JMS J. Stimpson, 149 Hungate Street, Aylsham, Norwich, NR11 6JZ
M0 JMT J. Tanner, 4 Pool Meadow Close, Birmingham, B13 9YP
M0 JMU Dr J. Hunt, 10 Couzens Close, Chippenham, SN15 1US
M0 JMV D. Vanstone, 99 Tudor Road, Sudbury, CO10 1NT
M0 JMY J. Mcmullan, 17 Banbury Close, Accrington, BB5 4BZ
M0 JNF B. Coley, 17 Livingstone Road Handsworth, Birmingham, B20 3LS
M0 JNJ J. Butcher, 14 Park Road, Lowestoft, NR32 1SW
MM0 JNL G. Crawford, 4 Windram Road Chirnside, Duns, TD11 3UT
M0 JNP J. Perry, 7 Lancaster House Belle Vue Road, Paignton, TQ4 6HD
M0 JNQ S. Rhenius, Baythorne Cottage Baythorne End, Halstead, CO9 4AB
M0 JNS J. Shatford, 31 Pinner Park Avenue, Harrow, HA2 6LG
M0 JNU J. Nuttall, 73 Severn Drive, Walton-Le-Dale, Preston, PR5 4TD
M0 JNV S. Pomeroy, 55 Woodmancote Vale, Woodmancote, Cheltenham, GL52 9RJ
M0 JNX J. Hall, 1 Nash Close, Earley, Reading, RG6 5SL
MM0 JNY Dr E. Newman, Yewbank, Bamff Road, Alyth, Blairgowrie, PH11 8DR
M0 JOB J. O'Brien, 76 Berkeleys Mead, Bradley Stoke, Bristol, BS32 8AU
M0 JOC J. O'Connell, 23 Halstead Road, Gosfield, CO9 1PG
M0 JOD J. Preece, 51 Drancy Avenue, Willenhall, WV12 5RD
M0 JOH J. Godfrey, 17 Lichfield Road, Sneinton, Nottingham, NG2 4GF
M0 JOI P. Edwards, 4 Stodart Road, London, SE20 8ET
M0 JOJ J. Green, 26 Foxhill Road Burton Joyce, Nottingham, NG14 5DB
MM0 JOK J. Burgoyne, 5 Shankston Crescent, Cumnock, KA18 1HA
M0 JOL M. O'Leary, 4 Park Farm Close, Martinstown, Dorchester, DT2 9TW
MM0 JOM J. Mann, 10 Brimmond Walk, Westhill, AB32 6XH
M0 JOO A. Williams, 12 St. Wilfrids Crescent, Brayton, Selby, YO8 9EU
MW0 JOP R. Bowen, 16 Lon Hywel, Whitland, SA34 0BE
M0 JOR J. Orr, 13 Haldane Close, Brierley, Barnsley, S72 9LL
M0 JOW D. Hamilton, 16 Conifer Crest, Newbury, RG14 6RT
MM0 JOX P. O'Hara, 13/3 135 Kirkton Avenue, Glasgow, G13 3EP
M0 JOY J. Bilson, 31 Middleton Drive, Inkersall, Chesterfield, S43 3HS
M0 JPA J. Wake, 60 Cloverville Approach, Bradford, BD6 1ET
M0 JPB J. Bull, 91 Lime Road, Wednesbury, WS10 9NF
MI0 JPC J. Cosgrove, 91 Church Street, Newtownards, BT23 4AN
MI0 JPD J. Doyle, 25 Parkmore Road, Magherafelt, BT45 6PF
M0 JPE J. Poole, 61 Lower Vickers Street, Miles Platting, Manchester, M40 7LX
M0 JPG J. Gleeson, 124 Rushes Mead, Harlow, CM18 6QE
M0 JPK G. Anestopoulos, 29 Constable Close, Woodley, Reading, RG5 4US
MI0 JPL J. Jones, 107 Belfast Road, Whitehead, Carrickfergus, BT38 9SU
M0 JPM J. Meijer, Birchwood East End, Gooderstone, King's Lynn, PE33 9DB
M0 JPN T. Nakagawa, 7 Milton Street, Barrowford, Nelson, BB9 6HE
MI0 JPO J. Alexander, 24 Alexandra Avenue, Ballymoney, BT53 6EX
M0 JPP S. Clark, 20 Herbert Street, Loughborough, LE11 1NX
M0 JPR C. Kwok, 16 Overlinks Drive, Salford, M6 7PF
M0 JPS J. Styles, 42 Brook Street, Woodbridge, IP12 1BE
M0 JPT J. Tan, 22, Jalan Geikie, Miri, Malaysia, 98000
M0 JPW J. Woolvin, 62 Whitewood Park, Liverpool, L9 7LG
M0 JPX J. Parkes, 24 Kenilworth Road, Lichfield, WS14 9DP
M0 JQB J. Bovey, 17 North Bank Road, Bingley, BD16 1UH
M0 JQD D. Start, The Rise, Valley Lane Swaby, Alford, LN13 0BH
M0 JQJ N. Brierley, Whitewalls Farm, Warleggan, Mount, Bodmin, PL30 4HF
M0 JQK T. Firth, 126 Tombridge Crescent, Kinsley, Pontefract, WF9 5HE
M0 JQL M. Dhur, 3 Sharpes Corner, Lakenheath, Brandon, IP27 9LA
M0 JQO Daventry Amateur Radio Repeater Group, c/o A. Baker, 10 Kingscroft Court, Northampton, NN3 9BH
M0 JQP P. Kilby, 1 Home Farm Barns, South End, Milton Bryan, Milton Keynes, MK17 9HS
MI0 JQS Tri County ARC, c/o V. Mcfarland, 108 Fernagh Road, Whiteabbey, Newtownabbey, BT37 0BE
M0 JQV G. Huff, 135 Westminster Road Toothill, Swindon, SN5 8JE
M0 JQW J. Wang, Flat 31, 74 Arlington Avenue, London, N1 7AY
MI0 JQY Rev. M. Donald, Station Road, Garvagh, BT51 5LA
M0 JRA A. Jessop, 4 Katherine Street, Thurcroft, Rotherham, S66 9LG
M0 JRE J. Eaton1, 38 Litchford Road, New Milton, BH25 5BQ
MM0 JRF J. Fyfe, 53A Ware Road, Glasgow, G34 9AR
M0 JRI J. Dyson, 77 Grantham Road, Southport, PR8 4LT
M0 JRJ J. Jenkins, 31 Pendrell Street, London, SE18 2PH
M0 JRL J. Lynn, 72 Badger Close, Guildford, GU2 9WA
M0 JRN D. Higton, 1 Allendale Road, Caister-on-Sea, NR30 5ES
M0 JRO J. Roberts, 'Meadowside', Keswick Lane, Bardsey, Leeds, LS17 9AD
M0 JRP R. Pearson, 14 Highfields, Lakenheath, Brandon, IP27 9DZ
M0 JRQ C. Pearson, 4 Brentwood Close, Thorpe Audlin, Pontefract, WF8 3ES
MM0 JRR J. Rayne, 8 Bankton Grove, Livingston, EH54 9DW
M0 JRU C. Denton, 100 Lincoln Road Deeping Gate, Peterborough, PE6 9BA
M0 JRW J. Wilson, 1 Locarno Avenue, Runwell, Wickford, SS11 7HX
MW0 JRX O. Bross, 8 Queens Drive, Buckley, CH7 2LJ
M0 JRZ J. Robb, 37 Wroxham Road, Woodley, Reading, RG5 3AX
M0 JSA A. Jones, 5 Meadowlands, Kirton, Ipswich, IP10 0PP
M0 JSD J. Delaney, 33 Deepdale Close, Ibstock, LE67 6LW
M0 JSE M. Egan, 32 Hawksworth Avenue Guiseley, Leeds, LS20 8EJ
MM0 JSG J. Galloway, 144 Strathkinnes Road, Kirkcaldy, KY2 5PZ
M0 JSH J. Yan, St Edmund'S College, Mount Pleasant, Cambridgeshire, CB3 0BN
MI0 JSJ J. Smith, 54A Blackstaff Road, Kircubbin, Newtownards, BT22 1AF
M0 JSK G. Booton, 69 The Street, Deal, CT14 0AJ
M0 JSL J. Swales, 90 Earlswood Road, Dorridge, Solihull, B93 8RN
M0 JSN J. Sowman, 53 Newton Wood Road, Ashtead, KT21 1NN
M0 JSP J. Swiffen, Dragonelle, Marina Drive, Shireoaks, S81 8NQ
M0 JSR J. Street, 22 Roman Acre, Wick, Littlehampton, BN17 7HN
MM0 JSU J. Smit, Annabells Cottage, Deirdre, Connel, Oban, PA37 1PL
M0 JSW J. Woodland, 14 Kelham Green, Nottingham, NG3 2LP
M0 JSX J. Sawyer, 27 Croft Road, Wallingford, OX10 0HN
M0 JSY B. Josyfon, 25 Norrice Lea, London, N2 0RD
M0 JSZ F. Jackson, 5 Chalmers Avenue, Haversham, Milton Keynes, MK19 7AG
M0 JTB J. Brown, 55 Barrington Road, Rubery, Birmingham, B45 9EU
M0 JTD D. Baker, 65 Madison Street, Tunstall, St6 5hs
MI0 JTE J. Elliott, 183 Kilraughts Road, Ballymoney, BT53 8NL
MW0 JTG B. Williams, Bryn Celyn, Hendre Road, Conwy, LL32 8RJ
M0 JTH J. Thomas, 77 Hawthorn Avenue, Lowestoft, NR33 9BB
M0 JTJ J. Talbot-Jones, 21 Downsview Drive, Wivelsfield Green, Haywards Heath, RH17 7RN
M0 JTL I. Bardell, 17 Stanton Avenue, Bradville, Milton Keynes, MK13 7AR
M0 JTM G. Ridley, 12 Garforth Avenue, Steeton, Keighley, BD20 6SP
M0 JTN M. Chivers, 4 Hunters Lodge, Fareham, PO15 5NF
M0 JTQ J. Cook, 28 Wenny Estate, Chatteris, PE16 6UX
MM0 JTX Dr J. Wills, 2 Old Dalmore Gardens Auchendinny, Penicuik, EH26 0RR
M0 JUA J. Thirlwell, 6A Brook Street, Warminster, BA12 8DN
MM0 JUC R. Cook, Flat 6, Howards Court, Caledonian Road, Perth, PH1 5NJ
M0 JUI P. Stavropoulos, 60 Trent Valley Road, Stoke-on-Trent, ST4 5JA
M0 JUJ M. Smith, 24 Fifth Avenue, Portsmouth, PO6 3PE
M0 JUK J. Allison, 72 Chantry Croft, Kinsley, Pontefract, WF9 5JL
MM0 JUL R. Mitchell, Broadhills, Isle of Coll, PA78 6TB
M0 JUN J. Unwin, 39 Whinfell Drive, Normanby, Middlesbrough, TS6 0BG
M0 JUP Rev. S. Perry, 8 Collingwood Avenue, March, pe15 9ef
M0 JUQ Capel Battery Radio Group, c/o I. Hope, 5 The Crescent, Northfleet, Gravesend, DA11 7EB
M0 JUT D. Heron, 36A Forest Road, Hartwell, Northampton, NN7 2HE
M0 JUU R. Bazant, 1807 128Th St, College Point, USA, 11356
M0 JUV D. Howard, 31 White Mullein Drive, Redlodge, IP288XP
M0 JUW Gilmore Radio Club, c/o K. Sale, 19 Mayfield Grove, Stockport, SK5 7JB
M0 JUX A. Seedhouse, 29 Great Northern Road, Dunstable, LU5 4BN
M0 JVC K. Francis, 203 Colchester Road Lawford, Manningtree, CO11 2BU
M0 JVE K. Hedges, 28 Hill Park, Congresbury, Bristol, BS49 5BT
M0 JVG J. Geisau, Huelchratherstr 37, Koeln, Germany, D-50670
M0 JVL R. Scott, 34 Moorfield Road, Birmingham, B34 6QY
M0 JVM D. Turton, 8 Lightwoods Road, Warley, Smethwick, B67 5AY
M0 JVR D. Galloway, Riverview Barn, Ferry Lane, Twyford, Barrow-On-Trent, Derby, DE73 7AA
M0 JVT J. Turner, White Haven 7 Lawrence Hill Gardens, Dartford, DA1 3AP
M0 JVU J. Connolly, 2 Waring Avenue, St Helens, WA9 2QG
M0 JVV J. Waddy, 70 Linden Avenue, Prestbury, Cheltenham, GL52 3DS
M0 JVW J. Wild1, 11 Whitefield Avenue, Newton-le-Willows, WA12 8BY
M0 JVX D. Hudson, 30 Sarmatian Fold, Ribchester, Preston, PR3 3YG
M0 JVY P. Smythe, 15 Falcon Drive, Trowbridge, BA14 7GE
M0 JVZ L. Bailey, Flat 14 Walter Hull Court King Edward Road Loughborough, Loughborough, LE11 1SU
M0 JWA J. Arrow, 29 Billington Gardens, Hedge End, Southampton, SO30 2AX
M0 JWC H. Griffiths, The Old Church Hall, Christmas Common, Watlington, OX49 5HL
M0 JWD J. Wright, 32 Carlton Road, Nottingham, NG10 3LF
M0 JWE J. Williamson, 64 Sandy Lane, Irlam, Manchester, M44 6WJ
MM0 JWH J. Hosea, 61 John Street, Helensburgh, G84 9JZ
M0 JWJ J. Jordan, 31 Rotherham Road, Dinnington, Sheffield, S25 3RG
M0 JWL M. Lee, Up To Date House, Shore Road, Boston, PE22 0NA
M0 JWM J. Mccown, 2105 Viking Drive, Colorado Springs, USA, 80910
MW0 JWP J. Pritchard, 1 Tan Y Coed, Maesgeirchen, Bangor, LL57 1LU
M0 JWQ M. Krol, 74 Boakes Drive, Stonehouse, GL10 3QW
M0 JWR J. Richardson, 6 Clarence Road Scorton, Richmond, DL10 6EE
M0 JWT J. Wieczorek, 135 Pinkerton Road, Basingstoke, RG22 6RX
M0 JWU R. Jones, St. Fillans, The Warren, East Horsley, Leatherhead, KT24 5RH
M0 JWV P. Preston, 4 Barn Close, Torriano Avenue, London, NW5 2SY
MW0 JWW J. Williams, 12 Pantycelyn, Fishguard, SA65 9EH
M0 JWX P. Preston, 4 Barn Close, Torriano Avenue, London, NW5 2SY
MI0 JWY Newry and District ARC, c/o M. Keenan, 30 Ballynabee Road Camlough, Newry, BT35 7HD
M0 JWZ J. Blakemore, 16 Brentwood Avenue, Newbiggin-by-the-Sea, NE64 6JH
M0 JXD J. Drinkell, 11 Valley Walk, Kettering, NN16 0LY
M0 JXE D. Ennion, 347 Parkgate Road, Chester, CH1 4BE
M0 JXF B. Power, 20 Marlpool Place, Kidderminster, DY11 5BB
M0 JXG J. Guess, Flat 257, Helen Gladstone House Nelson Square, London, SE1 0QB
MM0 JXI J. Innes, 33 Monktonhall Place, Musselburgh, EH21 6RR
M0 JXM D. Easterling, 14 Brunswick Close, Biggleswade, SG18 0DA
M0 JXR B. Glen, 37 Martingale Way, Portishead, Bristol, BS20 7AW
M0 JXS C. Ember, 35 Mattock Lane, Ealing, London, W5 5BH
M0 JXY Red Brick Stables ARC, c/o J. Hall, 4 Dorking Crescent, Clacton-on-Sea, CO16 8FQ
MW0 JYC M. Coleman, Felin Newydd Ciliau Aeron Lampeter Sa48 7Px, Lampeter, SA48 7PX
M0 JYF D. Gaskell, 26 Stoneyland Drive, New Mills, High Peak, SK22 3DL
M0 JYG E. Mccormick, 526 Watling Street Road, Ribbleton, Preston, PR2 6TU
M0 JYJ J. Lundrigan, 9 Richards Close, March, PE15 8UH
M0 JYM J. Sangster, 139 Higher Road, Liverpool, L26 1UN
M0 JYN G. Kermeen, 31 Coppice Close, Sedgley, Dudley, DY3 3NP
M0 JYP K. Sale, 19 Mayfield Grove, Stockport, SK5 7JB
M0 JYQ B. Sadler, 56 Hayden Lane, Hucknall, Nottingham, NG15 8BS
MM0 JYR J. Rae, 4 Hillside Crescent, Langholm, DG13 0EE
M0 JYS B. Sadler, 56 Hayden Lane, Hucknall, Nottingham, NG15 8BS
MM0 JYV S. Spoor, The Old School, Main Street, Whitsome, Duns, TD11 3NB

MW0 JYW G. Uvner, The Stratford House B&B 8 Craig Y Don Parade, Llandudno Ll30 1Bg, Llandudno, LL30 1BG
M0 JYZ N. Honey, 29 Mill House Lane, Winterton, Scunthorpe, DN15 9QP
MM0 JZB J. Brown, 60 Laburnum Lea, Hamilton, ML3 7LZ
MW0 JZE A. David, 45 Amanwy, Llanelli, SA14 9AH
M0 JZF I. Johnston, 4 Fieldway Crescent, Cowes, PO31 8AJ
M0 JZG A. Loukes, 14 Batchwood View, St Albans, AL3 5TD
MM0 JZI A. Mclean, 10 Seaforth Gardens, Annan, Dumfries And Galloway, DG12 6UH
M0 JZK J. Howlett, 29 Little London, Heytesbury, Warminster, BA12 0ES
MW0 JZM S. Kedward, 9 Lawrence Avenue, Aberdare, CF44 9EW
M0 JZN J. Nichols, 4 Elmlee Close, Chislehurst, BR7 5DU
M0 JZO B. Rusu, 47 Elmgrove Crescent, Harrow, HA1 2QT
M0 JZS K. Mueller, 33 Rylands Road, Southend-on-Sea, SS2 4LW
M0 JZT M. Hunt, 189 Gibbins Road, Birmingham, B29 6NH
M0 JZU K. Mueller, 33 Rylands Road, Southend-on-Sea, SS2 4LW
M0 JZV Z. Nikolic, 6 Warbank Lane, Kingston upon Thames, KT2 7ES
MM0 JZW J. Burton, 6 Kilfinnan Lodges, Spean Bridge, PH34 4EB
MI0 JZZ C. Mclelland, 14 Clifton Park, Coleraine, BT52 2HW
M0 KAB K. Bull, 12 Hinksford Mobile Home Park, Kingswinford, DY6 0BG
M0 KAC 1127 (Kendal) Sqn ATC, c/o R. Walker, 35 Romany Close, Letchworth Garden City, SG6 4LA
M0 KAD K. Allen, 48 Beaumont Rise, Worksop, S80 1YG
M0 KAE K. Ely, 6 Baxter Square, Town End Farm Estate, Sunderland, SR5 4ND
MI0 KAG E. Rantin, 8A Buchanans Road, Newry, BT35 6NS
M0 KAI K. Boothman, 5 Millwood Road, Balby, Doncaster, DN4 9DA
M0 KAJ D. Ramsell, 36 West Street, Burton-on-Trent, DE15 0BW
M0 KAK D. Knight, Flat 9, Loddon House, London Road, Ruscombe, Reading, RG10 9BW
MI0 KAM K. Mullan, 19 Parklea, Portstewart, BT55 7HA
M0 KAN K. Nicholson, 11 Lancaster Way, Skellingthorpe, Lincoln, LN6 5UF
M0 KAO K. Jeffery, 9 Gordon Road, Tunbridge Wells, TN4 9BL
M0 KAP P. Smith, 3 Watts Road, Colchester, CO2 9DZ
M0 KAR K. Lott, 6 Centurion Close, Sandhurst, GU470HH
M0 KAU K. Sharpe, 18 Dudhill Road, Rowley Regis, B65 8HT
M0 KAW K. Wells, 2 Holmefield, Farndon, Newark, NG24 3TZ
MW0 KAX M. Reynolds, 2 Scales Row, Aberdare, CF44 0PW
M0 KBA A. Bhakoo, 4 Bryden Cottages, High Street, Uxbridge, UB8 2NY
M0 KBB K. Bright, 20 Radley Road, Bristol, BS16 3TL
M0 KBC K. Aird, 11 Minter Avenue, Densole, Folkestone, CT18 7DS
M0 KBD P. Smiths, 12 Lambton Road, Stockton-on-Tees, TS19 0ER
M0 KBG V. Hayes, 68 Billingsley Road, Birmingham, B26 2EA
M0 KBH N. Kimber, 127 Beaconsfield, Withernsea, HU19 2EW
M0 KBJ S. Harvey, 110 Vicarage Road, Wednesfield, Wolverhampton, WV11 1SF
M0 KBK Prof. R. Houlston, 51 Adelaide Road, Surbiton, KT6 4SR
MI0 KBL P. Brennan, 23 Ardchrois, Donaghmore, Dungannon, BT70 3LB
MW0 KBN Holy Island ARC, c/o C. Thorley, Helston, The Mountain, Holyhead, LL65 1YR
M0 KBP B. Al-Rawi, Flat 17, Harrow Lodge, London, NW8 8HR
M0 KBR W. Townsend, 8 Beech Avenue Kirkham, Preston, PR4 2UE
MM0 KBT D. Bisset, Jordieland Cottage, Kirkcudbright, DG6 4XT
M0 KBU C. Norris, Heather View, Forest Front, Hythe, Southampton, SO45 3RJ
M0 KBW M. Cerveny, Apartment 214, Piccadilly Heights, Wain Avenue, Chesterfield, S41 0GF
M0 KBX K. Hewson, 48 Ruskin Road, Belvedere, DA17 5BB
M0 KBY M. Criscione, 46 South Molton Street, London, W1K 5RX
M0 KBZ K. Tietz, 112 Over Lane, Belper, DE56 0HN
M0 KCA J. Cater, 5 Shady Grove, Hilton, Derby, DE65 5FX
M0 KCC K. Cartwright, 53 Sedgley Road, Dudley, DY1 4NE
MM0 KCD K. Davies, 1 Myreton Way, Falkirk, FK1 5NZ
M0 KCE T. Peterson, 59 Fleet Street, Holbeach, Slane Lodge, Spalding, PE12 7AU
M0 KCF G. Mccaffery, 7 Cliffe Court, Sunderland, SR6 9NT
M0 KCI Ashtead Radio Club, c/o Dr P. Lee, Links Corner Cottage Liks Road, Ashtead, Surrey, Uk, KT212EG
M0 KCJ C. Kennedy, 81 Corporation Road, Audenshaw, Manchester, M34 5LZ
M0 KCO K. Cornwell, 19 Forest Road, Chandler'S Ford, Eastleigh, SO53 1NA
M0 KCP R. Hutton, 22A Victoria Road, Maldon, CM9 5HF
MM0 KCS M. Brunsdon, 25 Buckstone Lea, Edinburgh, EH10 6XE
M0 KCV K. Conlon, 136 Chart Downs, Dorking, RH5 4DG
M0 KCW C. Wade, 31 Melton Green, Wath-Upon-Dearne, Rotherham, S63 6AA
M0 KCY J. Crabtree, 5408 Oaklawn Avenue, Edina, USA, 55424
M0 KCZ R. Robinson, 22 Riddings Court, Timperley, Altrincham, WA15 6BG
M0 KDA S. Bolton, 201 Lime Tree Avenue, Crewe, CW1 4HZ
M0 KDE A. Lee, 14 Bernice Avenue, Chadderton, Oldham, OL9 8QJ
M0 KDH D. Hughes, 75 Suncote Avenue, Dunstable, LU6 1BN
MW0 KDL K. Lewis, 21 Wheatley Place, Merthyr Tydfil, CF47 0TA
M0 KDM P. Sephton, 11 Moss Avenue, Leigh, WN7 2HH
M0 KDO J. Saxon, 134 Sherwood Drive, Wigan, Wn5 9rs
M0 KDR K. Roberts, 4 Quartz Close, Tolvaddon, Camborne, TR14 0FT
M0 KDT J. Hobbs, 2 Eccles Road, Wittering, Peterborough, PE8 6AU
M0 KDU M. Shingler, 4 Church Lane, Checkley, Stoke-on-Trent, ST10 4NJ
M0 KDV D. Craven, 69 Markham Avenue, Rawdon, Leeds, LS19 6NE
M0 KDX P. Martin, 26 Chart Lane, Reigate, RH2 7DY
MM0 KDY D. Latto, 8 Aspen Avenue, Glenrothes, KY7 5TA
M0 KEB K. Legg, Bennetts, High Street, Thorpe-Le-Soken, Clacton-on-Sea, CO16 0EG
M0 KED A. Keddie, 6 Vulcan Crescent, North Hykeham, Lincoln, LN6 9SB
M0 KEE J. Charlton, Hillside House, Ham Lane, Bristol, BS41 8JA
M0 KEF P. Munson, 8 Longley Lane, Spondon, Derby, DE21 7AT
M0 KEG I. Bain, 45 Larpool Crescent, Whitby, YO22 4JD
M0 KEJ K. Borszlak, 37 Bank Lane, Little Hulton, Manchester, M38 9UH
M0 KEL K. Gale, 4 Field Court, Sea Road, Littlehampton, BN16 1JS
M0 KEO Prof. I. Neal, 8 Rushey Gill, Brandon, Durham, DH7 8BL
M0 KEP T. Keep, Coombe Cottage, Coombe Lane, Cradley, Malvern, WR13 5JF
MW0 KEQ K. Mogford, 49 Cefn Road, Rogerstone, Newport, NP10 9AQ
MM0 KES E. Wigram, Inglenook, North Road, Lerwick, Shetland, ZE1 0PR
M0 KEU T. Barris, Vogelgartenstr. 41/1, Eislingen/Fils, Germany, 73054
MW0 KEV K. Dawson, 57 Fair View, Blackwood, NP12 3NR
M0 KEW D. Askew, 19 Oliver Road Staplehurst, Tonbridge, TN12 0TE
M0 KEY B. Wagstaff, 4 Fleets Road, Sturton By Stow, Lincoln, LN1 2BU
M0 KFA A. Ferenc, 2A Rosedene Avenue, London, SW16 2LT
M0 KFB K. Bell, 11 Mill Lane, Hogsthorpe, Skegness, PE24 5NF
M0 KFF M. Harrington, Tanglewood House, Station Road, Tilbrook, PE28 0JY
M0 KFH I. Hemmens, 26 Grove Gate, Staplegrove Taunton, TA26DF
M0 KFI S. Gilbert, 34 Ramsey Road, St Ives, PE27 5RD
M0 KFJ J. Moyler, 38 Golden Hill, Whitstable, CT5 3AR
M0 KFK D. Westwood, 28 Weybridge Mead, Yateley, GU46 7UY
MW0 KFL Dr E. Flikkema, 7 St. James Mews, Great Darkgate Street, Aberystwyth, SY23 1DW
MM0 KFP Dr M. Sutcliffe, 11 Low Borland Way, Eaglesham, Glasgow, G76 0BP
MM0 KFR K. Fisher, 9 Haymarket Crescent, Livingston, EH54 8AP
M0 KFU D. Sweeney, 3 Robin Hood Close, Woking, GU21 8SS
M0 KFV K. O'Connell, 63 Hazelton Road, Colchester, co4 3ds
M0 KFW K. Whittaker, 32 Ashleigh Mount Road, Exeter, EX4 1SW
M0 KFY J. Gilvarry, 16 Fox Walk, Sheffield, S6 3QZ
M0 KGA S. Pollak, 22 Gloucester Road, Avonmouth, Bristol, BS11 9AD
M0 KGE Dr D. Rodriguez, 23 Thorpe Way, Cambridge, cb5 8uj
MW0 KGG D. Owen, Tanrallt, Blaenpennal, Aberystwyth, SY23 4TP
M0 KGI S. Byatt, High Trees, High Street, Pavenham, Bedford, MK43 7NJ
MM0 KGK D. Munro, 10A Steinish, Isle of Lewis, HS2 0AA
M0 KGM K. Graham, 22 Repton Avenue, Oldham, OL8 4JB
MW0 KGP Dr P. Kelly, Arosfa, Westminster Road, Wrexham, LL11 6DN
MM0 KGS G. Sinclair, 33 Keptie Road, Arbroath, DD11 3EF
MW0 KGU P. Bennett, 2 Tan Yr Wylfa, Abergele, LL22 7DX
M0 KGV C. Moulding, 28 Queens Avenue, Highworth, Swindon, SN6 7BA
M0 KGW Dr B. Minnis, 16 Dene Tye, Crawley, RH10 7TS
MW0 KGY K. Yearsley, Garth Lea, Lon St. Ffraid, Holyhead, LL65 2YH
M0 KHA C. Wale, 6 West Howe Close, Bournemouth, BH11 8AE
M0 KHB K. Bridge, 43 Highbank, Blackburn, BB1 9SX
M0 KHD S. Gaspar, 54 Herberts Park Road, Wednesbury, WS10 8QH
MM0 KHG Craighalbert Radio Club, c/o J. Brown, 78 Egilsay St., Glasgow, G22 7RG
M0 KHK C. Cheung, 23 Mosley Road, Timperley, Altrincham, WA15 7TF
M0 KHM K. Maddy, 56 Coachwell Close, Telford, TF3 2JB
M0 KHN J. Riches, 4 Kings Road, Chalfont St Giles, HP8 4HU
M0 KHO D. Pluright, 21 Buscot Drive, Abingdon, OX14 2BJ
M0 KHQ Swinton Radio Club, c/o T. Ward, 173-175 Station Road, Pendlebury, Swinton, Manchester, M27 6BU
M0 KHS K. Shuttleworth, 27 Queens Drive, Fulwood, Preston, PR2 9YJ
M0 KHW K. Wright, 12 Bushmead Road, Luton, LU2 7EU
M0 KHX M. Cooper, 49 Bolton Lane, Ipswich, IP4 2BX
M0 KHZ K. Wheatley, 1 Braithwaite Court, Egremont, CA22 2DN
M0 KIB C. Mckinney, Swallow Cottage, Chickney Road, Henham, Bishop's Stortford, CM22 6BG
M0 KIC N. Jaggs, 15 St. Anthonys Road, Kettering, NN15 5HT
M0 KID N. Fairbairn, 15 Hewitt Road, Dover, CT16 1TH
M0 KIF M. Chalkley, 36 Cowper Road, Bournemouth, BH9 2UJ
M0 KIG K. Gamble, 67 Queen Street, Burntwood, WS7 4QQ
MW0 KIJ N. Sugg, 12 Caerleon Grove Castle Park Cf48 1Jh, Merthyr Tydfil, CF48 1JH
M0 KIL K. Lockstone, Oceana, The Parade Pevensey Bay, Pevensey, BN24 6LX
M0 KIN A. Clarke, 57 Welland Avenue, Grimsby, DN34 5JP
M0 KIO Kingston Immortals Contest Group, c/o N. Newby, 167 Watersplash Road, Shepperton, TW17 0EN
M0 KIQ S. Dowling, 34 Kit Hill Avenue Walderslade, Chatham, ME5 9EX
M0 KIR M. Kirkman, 8 Ashington Drive, Arnold, Nottingham, NG5 8GH
M0 KIW A. Thorn, 12 Hardy Drive Hardingstone, Northampton, NN4 6UX
M0 KJC K. Cole, 4 Marsham Road Hazel Grove, Stockport, SK7 5JB
MM0 KJG K. Glacken, 14 Hailes Avenue, Edinburgh, EH13 0NA
M0 KJK K. Kolesnik, 15 Steer Road, Swanage, BH19 2RU
MM0 KJM K. Martin, 4 Hunter Crescent, Troon, KA10 7AH
MW0 KJN K. Nedin, 28 Llys-Y-Coed Birchgrove, Swansea, SA7 9PR
M0 KJT K. Todman, 12 Winscombe, Bracknell, RG12 8UD
M0 KKA C. Wheatley-Hince, Jasmine Cottage, Main Street, Chaddleworth, Newbury, RG20 7EH
M0 KKB S. King, 23 Queens Avenue, Canterbury, CT2 8BA
MI0 KKD K. Dorman, 25 Blackthorn Road, Newtownabbey, BT37 0GH
MM0 KKF W. Mackenzie, 7 Urquhart Grove, Elgin, IV30 8TB
M0 KKH N. Haigh, 10 Moor Park Gardens, Dewsbury, WF12 7AS
M0 KKM K. Minett, Rosedene, Honey Hill, Wimbotsham, King's Lynn, PE34 3QD
M0 KKO T. Stack, 31A Chester Road South, Kidderminster, DY10 1XJ
M0 KLA S. Collins, 3 Sedgemoor Road, Camps Bay, South Africa,
M0 KLB E. Gomez Lozano, 2 Annesley Road, Oxford, OX4 4JQ
M0 KLG R. Searby, 72 Polstain Road Threemilestone, Truro, TR3 6DH
M0 KLH K. Hawes, 837 Garratt Lane, London, SW17 0PG
M0 KLJ J. Athersmith, 20 Fell Close, Ulverston, LA12 0AE
M0 KLK W. Foster, 55 Drake Avenue Minster On Sea, Sheerness, ME12 3SA
M0 KLL K. Lloyd, 1 Fordenbridge Square, Sunderland, SR4 0BA
M0 KLM B. Whiteley, 2A Beechfield Close, Thorpe Willoughby, Selby, YO8 9QJ
MM0 KLN K. Nevins, 4 Ubbanford, Norham, Berwick-upon-Tweed, TD15 2LA
MM0 KLR Kilmarnock and Loudoun ARC, c/o A. Clark, 20 Church Street, Kilwinning, KA13 6BE
M0 KLT G. Clark, 28 Manor Road, Woolton, Liverpool, L25 8QG
MW0 KLW A. Jones, 2 Erw Terrace, Bethel, Caernarfon, LL55 1YT
M0 KLX R. Harper, 19 Tennyson Avenue, King's Lynn, PE30 2QG
M0 KMB A. Bailey, 58 Billy Buns Lane, Wombourne, Wolverhampton, WV5 9BP
M0 KMI K. Mills, 6 West Coombe, Bristol, BS9 2BA
MI0 KMJ A. Mcguinness, 42 Downshire Road, Carrickfergus, BT38 7LD
MM0 KMK C. Murphy, 3 Bolestyle Crescent, Kirkmichael - Maybole, KA19 7PW
M0 KML K. Leesmith, 31 Dogger Lane, Wells-Next-the-Sea, NR23 1BE
M0 KMR Medway Raynet, c/o R. Sohst, 2 Shaftesbury Drive, Maidstone, ME16 0JS
MW0 KMS K. Smith, 62 Waterloo Road, Talywain, Pontypool, NP4 7HJ
M0 KMT M. Sadler, 12 John Corbett Drive, Amblecote, Stourbridge, DY8 4BW
M0 KMW M. Ballard, 41 Middlefield Ave, Halesowen, B62 9QJ
M0 KMX K. Murphy, 120 Elmway, Chester le Street, DH2 2LQ
M0 KNB M. Ronan, 49 Dorset Street, Nottingham, NG8 1PU
MM0 KNE T. Kane, 40B Brisbane Street, Greenock, PA16 8NP
M0 KNH K. Holman, 39 Trellech Court, Yeovil, BA21 3TE
M0 KNK K. Kariraman, 26 Tythe Barn Lane, Shirley, Solihull, B90 1RW

M0 KNM M. Hemmings, 62 Spencer Way, Stevenage, SG2 8GD
MM0 KNN C. Kennedy, 16 New Garrabost, Isle of Lewis, HS2 0PL
M0 KNV K. Taylor, 44 Main Street, Willoughby, Rugby, CV23 8BH
M0 KNX M. Mattiello, Via Luigi Gaudio, 21, Faedis, Udine, Italy, 33040
M0 KOB A. Prestwich, Highfield, Exminster, Exeter, EX6 8AT
M0 KOG C. Haw, 72 Mayflower Road, Boston, PE21 0EZ
M0 KOH K. Hough, 15 Moorside Road Endmoor, Kendal, LA8 0EN
M0 KOI P. Burrows, 37 Dorrington Close, Murdishaw, Runcorn, WA7 6JR
M0 KOM L. Wilson, 6 Marrick Road, Middlesbrough, TS3 7RX
M0 KOO K. Milone, 381 Mail Boxes Etc, Viking House, 13 Micklegate, York, YO1 6RA
M0 KOT N. Baulf, 1 Lower Chart Cottages, Brasted Chart, Westerham, TN16 1LS
M0 KOV F. Limbert, Knowlecroft Little Ribston, Wetherby, LS22 4ET
MM0 KOZ R. Thomson, 30 Sealstrand, Dalgety Bay, Dunfermline, KY11 9NG
MI0 KPA S. Frazer, 2 Cavanballaghy Road, Killylea, Armagh, BT60 4NZ
M0 KPB K. Blanshard, 30 Torquay Crescent, Symonds Green, Stevenage, SG1 2RS
M0 KPC P. Gagliardi, 7 Saxon Way, Jarrow, NE32 3QA
M0 KPD R. Simpson, 48 Weatherhill Road, Horley, RH6 9LY
M0 KPE F. Keane, 14 Pollard Court, Beverley Road, Hull, HU5 2TP
M0 KPI R. Mackay, 7 Darwin Close, Lee-on-the-Solent, PO13 8LS
M0 KPJ M. Barrett, 57 Marlborough Avenue, Hornsea, HU18 1UA
M0 KPK D. Smith, 333A, Forton Road, Gosport, PO12 3HF
M0 KPO S. Warren, 7 Crich Way, Newhall, Swadlincote, DE11 0UU
M0 KPP P. Baker, 69 Cavendish Road, Walsall, WS27HH
M0 KPT K. Towler, 3 Elm Grove, Barnham, Bognor Regis, PO22 0HF
M0 KPU C. Weight, 166 Prince Henry Road, London, SE7 8PJ
M0 KPW C. Leviston, 13 Pryors Walk, Askam-in-Furness, LA16 7JG
MI0 KQU S. Homer, 10 Bann Drive, Londonderry, BT47 2HW
M0 KRA S. King, Box 14, West Union South Carolina, USA, 296960014
MM0 KRC K. Carroll, 32E Meadowburn Place, Campbeltown, PA28 6ST
M0 KRD D. Niggemann, 35 Holm Court, Twycross Road, Godalming, GU7 2QT
M0 KRE K. Knight, 11 Sweetbriar Lane Holcombe, Dawlish, EX7 0JZ
M0 KRK Dr D. De-Cogan, 52 Gurney Road, New Costessey, Norwich, NR5 0HL
M0 KRL K. Fell, 5 Henry Road, Wath-Upon-Dearne, Rotherham, S63 7NF
M0 KRM Medway Raynet, c/o R. Sohst, 2 Shaftesbury Drive, Maidstone, ME16 0JS
M0 KRO Kent Active Radio Amateurs, c/o K. Richardson, 35 Vidgeon Avenue, Hoo, Rochester, ME3 9DE
M0 KRP R. Bullen, 2 Redlands Cottages, East Coker, Yeovil, BA22 9HF
M0 KRR A. Kerr, 23 Manor Park, Duloe, Liskeard, PL14 4PT
MW0 KRS C. Young, 34 Penlan Crescent, Uplands, Swansea, SA2 0RL
M0 KRU P. Crewe, 31 Seas End Road Moulton Seas End, Spalding, PE12 6LD
M0 KRW I. Bricknell, 171 Springthorpe Road, Birmingham, B24 0SN
M0 KRX K. Rosema, Apartment 801, 25 Goswell Road, London, EC1M 7AJ
M0 KSA R. Love, 48 Langland Drive, Dudley, DY3 3TH
M0 KSB Kingsmead School, c/o J. Matthews, Moor View, Oldways End, East Anstey, Tiverton, EX16 9JQ
M0 KSE S. Shields, 28 Bank Field Westhoughton, Bolton, BL5 2QG
M0 KSG M. Wood, 26 Parkfield Crescent Kimpton, Hitchin, SG4 8EQ
M0 KSO Kings School Radio Society, c/o D. Lee, 188 Manstone Avenue, Sidmouth, EX10 9TJ
M0 KSR K. Mcinnes, 5A Northbrook Road, London, N22 8YQ
MM0 KSS K. Scott, 76 Fowler Avenue, Aberdeen, AB16 7YR
MW0 KST S. Davies, 5 Maldwyn Street, Cardiff, CF11 9JR
M0 KSX F. Stepien, 11A Wirral Gardens, Wirral, CH63 3BD
MM0 KTE K. Taylor, 77 Queen Margaret Fauld, Dunfermline, KY12 0RL
MM0 KTL K. Lane, 23 Mayfield Avenue, Tillicoultry, FK13 6HB
M0 KTN A. Snell-Pym, 50 Newton Avenue, Gloucester, GL4 4NU
M0 KTR A. Rowbottom, 8 Hedge Drive, Colchester, CO2 9DT
M0 KTT C. Jacobs, Flat 33, The Lodge, Lavender Road, Waterlooville, PO7 8BX
M0 KTV B. Chauhan, 45 Burnham Drive, Whetstone, Leicester, LE8 6HY
M0 KUH Hull & District ARS, c/o R. Pike, 44 Falkirk Close Bransholme, Hull, HU7 5BX
MI0 KUJ T. Calka, 71 Willowfield Street, Belfast, BT6 9AW
M0 KUK J. Kowalski, 126 Harland Avenue, Sidcup, DA15 7PA
M0 KUL K. Richards, 207 Beaulieu Gardens, Blackwater, Camberley, GU17 0LG
M0 KUP A. Anderson, 89A Malmesbury Park Road, Bournemouth, BH8 8PS
M0 KUR S. Campion, Flat 6, Carousel Steps, 10 Hawtree Close, Southend-on-Sea, SS1 2TZ
M0 KUV I. Barnes, 12 Sunderton Lane, Waterlooville, PO8 0NU
M0 KUY G. Michilin, 46 Via N.Sauro, Preganziol, Italy, 31022
M0 KVA A. Dokic, 28 Tudor Gardens, Shoeburyness, Southend-on-Sea, SS3 9JG
M0 KVB A. Taylor, 60 Wood Ride, Petts Wood, Orpington, BR5 1PY
M0 KVF K. Emery-Ford, 48 Welham Grove, Retford, DN22 6TS
M0 KVK K. Sim, 49 St. Julians Wells, Kirk Ella, Hull, HU10 7AF
M0 KVM K. Mills, 6A Deacons Walk, Schofield Street, Mexborough, S64 9NH
M0 KVN K. Finn, 132 Lansdowne Grove, Wigston, LE18 4LY
M0 KVR A. Burfield, 4 Eastern Crescent, Chelmsford, CM1 4JQ
M0 KWA Headcorn Aerodrome, c/o P. Blunt, 17 Offens Drive, Staplehurst, Tonbridge, TN12 0LR
M0 KWB K. Bradd, Flat 2, Montrose Court, London, NW9 5BS
M0 KWK D. Hall, 1 Pendreth Place, Cleethorpes, DN35 7UR
M0 KWM D. Wright, 203 Winn Street, Lincoln, LN2 5EY
M0 KWP J. Simon, 14 Le Peyrefus, Daignac, France, 33420
M0 KWR K. Royce, 11 Church Lane, Stibbington, Peterborough, PE8 6LP
M0 KWS S. Kendrick, 29 Waterside Silsden, Keighley, BD20 0LQ
M0 KWV M. Evans, 16 Colville Grove, Sale, M33 4FW
M0 KWW K. Willson, Ludpit Cottage, Ludpit Lane, Etchingham, TN19 7DB
M0 KWY D. Wells, 27 Victoria Avenue, Camberley, GU15 3HT
M0 KXD D. Rivron, Spring Cottage, Mill Lane, Bellerby, Leyburn, DL8 5QN
M0 KXK R. Britt, Thoroughfare House, South Burlingham Road, Norwich, NR13 4FA
M0 KXL K. Laing, 16 Cherrywood Drive Gonerby Hill Foot, Grantham, NG31 8QL
MW0 KXN K. Nicholls, 35 Partridge Road, Cardiff, CF24 3QW
M0 KXQ P. Hardacre, 13 St. Johns Street, Bridlington, YO16 7NL
M0 KYI K. Armstrong, 29 Thorntree Avenue, Crofton, Wakefield, WF4 1NU
M0 KYL A. Kyle, Greengates, Ainsworth Street, Ulverston, LA12 7EU
M0 KYR K. Orfanidis, Flat 36, Cumberland Court, London, W1H 7DP
M0 KYX I. Morgan, 30 Farm Road, Hutton, Weston-Super-Mare, BS24 9RH
MM0 KZA A. Marques Gomes, 34 Campbell Close, Hamilton, ML3 6BF
M0 KZB E. Arkinstall, 79 Sundorne Road, Shrewsbury, SY1 4RU
M0 KZC M. Clack, 42A Provost Street, Fordingbridge, SP6 1AY
M0 KZH D. Taylor, 49 Boggart Hill Gardens, Seacroft, Leeds, LS14 1LJ
MM0 KZJ A. Thomson, 5 Gib Grove, Dunfermline, KY11 8DH
M0 KZM M. Osborne, 9 Sunningdale Court, Jupps Lane, Goring-By-Sea, Worthing, BN12 4TU
M0 KZP N. Simmonds, 3 Noneley Hall Barns, Noneley, Wem, Shrewsbury, SY4 5SL
M0 KZT L. Kinzett, 52 Thistle Drive, Peterborough, PE2 8HX
M0 LAA S. Jefferson, 145 Duke St Fenton, Stoke-on-Trent, ST2 43NR
M0 LAB M. Labourn, 6 Healey Drive, Ossett, WF5 8NA
M0 LAE Rev. L. Clark, 226 Philip Lane Tottenham, London, N15 4HH
M0 LAF N. Smith, 40 Fairdale Drive, Newthorpe, Nottingham, NG16 2FG
M0 LAG D. Whelan, 431 Leeds Road, Huddersfield, HD2 1XT
M0 LAI P. Tolcher, 15 Langstone Close, Torquay, TQ1 3TX
M0 LAL C. Mole, 6 Clements Road, Chorleywood, Rickmansworth, WD3 5JT
MW0 LAO A. Powell, 31 Highmead, Pontllanfraith, Blackwood, NP12 2PF
M0 LAR G. Jackson, 14 Southdown Crescent, Cheadle Hulme, Cheadle, SK8 6EQ
M0 LAS L. Milford, 82B Oakley Lane, Oakley, Basingstoke, RG23 7JX
M0 LAT A. Laity, 9 Haverhill Road, Stapleford, Cambridge, CB22 5BX
M0 LAW M. Martin, 2821 Bissonnet St, Houston, USA, 77005-4014
M0 LAY A. Brooks, 19 Hawthorn Close, Knodishall, Saxmundham, IP17 1XW
M0 LAZ A. Burton, 1 Turnstile Walk, Trowbridge, BA14 0NA
MI0 LBA Church Island ARG, c/o T. Flanagan, 18 Hunters Park, Bellaghy, Magherafelt, BT45 8JE
MW0 LBB B. Bush, 2 Penrhiw Cottages, Brynithel, Abertillery, NP13 2AU
M0 LBD D. Lagan, 24 Milvain Close, Gateshead, NE8 3RS
MM0 LBF R. Bertram, 46 Main Street Main Street, Pathhead, EH37 5QB
M0 LBJ B. Jenson, 10 Tintern Close, Paulsgrove, Portsmouth, PO6 4LS
M0 LBK L. Karthauser, 17 Manor Close Abbotts Ann, Andover, SP11 7BJ
M0 LBL Marine Radio Museum Society, c/o W. Cross, 31 Joshua Close, Liverpool, L5 0TD
M0 LBM P. Matthews, 15 Tennyson Way, Melton Mowbray, LE13 1LJ
MW0 LBR Dr F. Labrosse, 72 Ger Y Llan Penrhyncoch, Aberystwyth, SY23 3HQ
MI0 LBS L. O'Sullivan, 24 Swifts Quay, Carrickfergus, BT38 8BQ
MM0 LBX J. Cattigan, Lunan Home Farm Cottage, Lunan Bay, Arbroath, DD11 5ST
M0 LBY L. Lay, 17 Herbert Road, Hornchurch, RM11 3LD
M0 LCA E. Taylor, 32 Knoll Drive, Warwick, CV34 5YQ
M0 LCC Lymington Community Association Radio Club, c/o K. Cromar, 74 Shakespeare Drive, Totton, Southampton, SO40 3NS
MW0 LCH L. Holder, 133 Maple Drive, Brackla, Bridgend, CF31 2PR
MW0 LCK M. Heathcote, Ingledown, Trelogan, Holywell, CH8 9BZ
M0 LCM L. Micallef, 1 Brockholme Mews, Great Cambourne, Cambourne, CB23 6GU
M0 LCN Lnc Activities and Training Community Interest Company, c/o D. Poulton, 93 Pretoria Road, Ibstock, LE67 6LP
M0 LCR V. Lynch, 16 Okehampton Crescent, Sale, M33 5HR
M0 LCS A. Selwood, 3 Warborough Cottages Warborough Road, Letcombe Regis, Wantage, OX12 9LE
M0 LCW Lids CW & Data Club, c/o A. Hill, 53 Fairladies, St Bees, CA27 0AR
M0 LCX Dr L. Clift, 150 Greenstead Road, Colchester, CO1 2SN
M0 LCY M. Taylor, 18 Mallow Walk, Westgate, Morecambe, La3 3Qa
M0 LDC L. Spriggs, 19 Mackenzie Square, Shephall, Stevenage, SG2 9TT
MW0 LDD M. Ladd, 50 Brynmelyn Avenue, Llanelli, SA15 3RT
M0 LDE G. Matts, 34 Barry Road, Leicester, LE5 1FA
M0 LDG Lundy DX Group, c/o J. Edmunds, Caroline Cottage, New Passage Road, Bristol, BS35 4LZ
M0 LDH L. Hawkins, 121 Selmeston Road, Eastbourne, BN21 2TL
MW0 LDJ L. Jessup, 23 Penallt Estate Llanelly Hill, Abergavenny, NP7 0RA
M0 LDK S. Alexander, 13 Padgate, Thorpe End, Norwich, NR13 5DG
M0 LDP D. Parker, 51 Parsonage Road, Henfield, BN5 9HZ
M0 LDQ D. Arnold, The Chase, Rectory Road, Penzance, TR19 6BB
M0 LDR L. Reynolds, 12 Providence Crescent, Boundary Way, Hull, HU4 6EF
M0 LDV D. Curtis, 7 Neale Close, Aylsham, Norwich, NR11 6DJ
M0 LDX S. Cape, 92 Davison Avenue, Whitley Bay, NE26 3SY
M0 LDY J. Davies, 5 Beauchamp Road, Kenilworth, CV8 1GH
M0 LDZ L. Cook, 43 Midge Hall Drive, Rochdale, OL11 4AX
MW0 LEA P. Price, Stable Cottage, Pendre, Cardigan, SA43 1JU
M0 LEB Prof. A. Sadka, 8 Wilson Close, Upper Heyford, Bicester, OX25 5BE
M0 LED L. Dixon, 23 Gipsy Lane, Buckfastleigh, TQ11 0DL
M0 LEE L. Jones, 30 Creola Court, Louisiana Drive, Warrington, Wa53Tl, England, Warrington, WA53TL
MW0 LEF M. Verardi, 21 Snowdon Street, Y Felinheli, LL56 4HQ
M0 LEH M. Reynolds, 24 Burton Close, Corringham, Stanford-le-Hope, SS17 7SB
M0 LEK Leek and District ARC, c/o S. Jefferson, 145 Duke St Fenton, Stoke-on-Trent, ST2 43NR
MJ0 LEL L. Langlois, Brookfield, La Rue D'Empierre, Trinity, Jersey, JE3 5QF
MM0 LEN L. Cochrane, 2 Muir Terrace, Paisley, PA3 4LT
M0 LEO L. Boberschmidt, 3928 Denfeld Court, Maryland, USA,
M0 LEP Sir R. Hewett, 17 Westfield, Dursley, GL11 4EP
MM0 LER M. Dickeson, 44 Mossmill Park, Mosstodloch, Fochabers, IV32 7JY
M0 LET R. Collis, 20 Lille Meadows, Haxby, York, YO32 3YY
M0 LEV M. Leveridge, 17 Gladstone Court, Dewsbury, WF13 4DQ
MW0 LEW L. Thomas, 2 Goytre Crescent, Goytre, Port Talbot, SA13 2YD
M0 LEX R. Styles, Padcroft, Weir, OL13 8QL
M0 LEY C. Kirby, 8 Reynolds Way, Swindon, Sn25 4GF
M0 LEZ L. Robinson, 15 Seldown Lane, Poole, BH15 1UA
M0 LFC Friskney + East Lincolnshire Communications Club, c/o B. Derbin-Sykes, 1, Lentons Lane, Friskney, Boston, PE22 8RR
M0 LFF Leicestershire Foxes Contest Group, c/o A. Fairclough, Forders House, Forders Lane, Marston Jabbett, Bedworth, CV12 9SG
M0 LFS Dr L. Spacek, 33 Wellesley Road, Colchester, CO3 3HE
M0 LGA O. Parker, 37 Springbank Crescent, Gildersome, Leeds, LS27 7DN
M0 LGB G. Benson, 2 Guisborough Road, Nunthorpe, Middlesbrough, TS7 0LB
M0 LGC Letchworth Garden City ARC, c/o M. Russell, 107 Cambridge Road, Hitchin, SG4 0JH
MI0 LGD Lord G. Drummond, 44 Camphill Park, Ballymena, BT42 2DJ
MW0 LGE R. Samphire, Courtlands, Newport Road, Magor, Caldicot, NP26 3BZ
M0 LGL L. Layland, 3 Thirlmere Road, Golborne, Warrington, WA3 3HH
M0 LGN B. Fitzgerald-O'Connor, 24 Routh Street, London, E6 5XX
M0 LGP L. Porter, 134 Grimshaw Lane, Middleton, Manchester, M24 2AH
MM0 LGR D. Boden, 42, Kirkwynd, Maybole, KA197AE

MM0 LGS J. Gorczynski, 54 Lindsay Gardens, Bathgate, EH48 1DU
MM0 LGT M. Mckay, 75 Cardross Road, Dumbarton, G82 4JL
M0 LGW W. Grocott, 12 Marigold Road, Stratford-upon-Avon, CV37 7DW
M0 LHA M. Burton, 139 Avenue Road, Erith, DA8 3BA
M0 LHB L. Bramley, 140 Nevill Road, Hove, BN3 7QB
M0 LHC A. Walrond, Leigh Hill Cottage, Lowton, Taunton, TA3 7SU
M0 LHF Fenland Portable Group, c/o C. Day, 14 Windsor Drive, Ramsey Forty Foot, Ramsey, Huntingdon, PE26 2XX
M0 LHK L. King, The Old School, Coombe Cross Bovey Tracey, Newton Abbot, TQ13 9EP
M0 LHR P. Lonsdale, 77 Burtons Road, Hampton Hill, Hampton, TW12 1DE
M0 LHS R. Silcox, 103 Oakdale Road, Downend, Bristol, BS16 6EG
M0 LIE E. St Quinton, Mill Cottage, The Thorofare, Woodbridge, IP13 8BB
M0 LIJ D. Smout, Sunrays, Warbage Lane, Bromsgrove, B61 9BH
M0 LIO Licolnshire Scout Radio Club, c/o A. Hull, 1 Occupation Lane, New Bolingbroke, Boston, PE22 7LW
M0 LIS E. Buckland, 11 Veronica Close, Basingstoke, RG22 5NW
MM0 LIV Livingstone District ARS, c/o N. Morris, 23 Sedgebank, Sedgebank, Livingston, EH54 6HE
MM0 LJA Dr L. Auchterlonie, 2 Lawhead Road East, St Andrews, KY16 9ND
M0 LJC L. Carpenter, Shardlow Marina, London Road, Derby, DE72 2GL
M0 LJD L. Goldsmith, Hunters Cottage, 61 Fengate Drove, Weeting, Brandon, IP27 0PW
M0 LJH L. Hopgood, 62 Briarwood Drive, Blackpool, FY2 0EB
M0 LJK L. Marriott, 94 Lyndhurst Road, Worthing, BN11 2DW
M0 LJL L. Lewis, 28 Brow Hey, Bamber Bridge, Preston, PR5 8DS
M0 LJT A. Tranter, 122 Summerhill Road, Bristol, BS5 8JU
M0 LKD L. Kahlbau, 7 Hamilton Court, De La Warr Road, Lymington, SO41 0PR
M0 LKE L. Kett, 52 Northgate, Hornsea, HU18 1EU
M0 LKJ K. Mccarthy, 34 Shawley Way, Epsom, KT18 5PB
M0 LKL L. Brigham, 42 Cayley Close, Clifton, York, YO30 5PT
M0 LKN C. Emery, 50 Hobkirk Drive, Sinfin, Derby, DE24 3DT
M0 LKR R. King, 12 South Road, Marden, TN12 9EN
M0 LKS R. Boruch, 4 Hafton Road, Salford, M7 3TF
M0 LKT L. Taylor, Marralomeda 7 Gatekeeper Close, Great Park, Newcastle upon Tyne, NE13 9EH
MW0 LKX W. Dabrowski, 2 Underhill Crescent, Knighton, LD7 1DG
M0 LKY A. Lee, 27 Victoria Avenue, Camberley, GU15 3HT
MI0 LLG S. Horner, 10 Meadow Court, Bushmills, BT57 8SD
MW0 LLK C. Tanner, Pen Y Gogarth Llaneilian, Amlwch, LL68 9NH
MI0 LLM R. Hetherington, 112 Screeby Road, Fivemiletown, BT75 0LG
MW0 LLO C. Hill, 9 Oliver Road, Newport, NP19 0HU
M0 LLS I. Powell, 9 Cardinal Crescent, Bromsgrove, B61 7PR
M0 LLW A. Bostock, 26 Ingham Road, Bawtry, Doncaster, DN10 6NW
M0 LLX L. Andrews, 72 Grange Road, Alresford, SO24 9HF
MW0 LLY R. Martin, Llan Owen, Rhulen, Builth Wells, LD2 3UY
M0 LMB B. Savage, Rufford, Barnes Lane, Milford On Sea, Lymington, SO41 0RR
MM0 LMC L. Chan, 5 Lansdowne Drive, Cumbernauld, Glasgow, G68 0JB
M0 LMH L. Hudson, 68 Eleanor Road, Harrogate, HG2 7AJ
M0 LMI L. Mikolka, Flat, 1 Scotney Court, Romney Marsh, TN29 9JP
M0 LMN D. Robinson, Height End Farm, Kirk Hill Road, Haslingden, Rossendale, BB4 8TZ
M0 LMO M. Moody, 16 Tyersal Court, Bradford, BD4 8EN
M0 LMQ R. Lewis, 20 Hillary Road, Rugby, CV22 6EU
M0 LMR D. Stanley, 58 Wells Gardens, Basildon, SS14 3QS
M0 LMS D. Partridge, 44 Trumpet Terrace, Cleator, CA23 3DY
MW0 LMW D. Jenkins, 36 Brynawel Road, Gorseinon, SA4 4UX
M0 LNE D. Bruce, Nunfield House, Bull Lane, Sittingbourne, ME9 7SL
MI0 LNL M. Robinson, 84 Windsor Crescent, Cookstown, Bt80 8ez
M0 LNY M. Mccoy, The Oaks, Liverton, Newton Abbot, TQ12 6EZ
M0 LOB M. Garry, 34 Conway Road, Paignton, TQ4 5LH
M0 LOC D. Lock, 20 Jasmine Close Trimley St. Martin, Felixstowe, IP11 0UY
M0 LOF D. Loftus, 5 Knowle Mount, Burley, Leeds, LS4 2PP
MM0 LOG W. Stuart, Meikleton, Aith, Bixter, Shetland Isles, ZE2 9NE
M0 LOH P. Rasmussen, 7 Portal Drive North, Upper Heyford, Bicester, OX25 5TH
M0 LOU D. Cave, 22 Longsight Road, Mapplewell, Barnsley, S75 6HB
M0 LOW D. Barnes, 11 Yewside, Gosport, PO13 0ZD
M0 LOX S. Cilliers, 61 Heathwood Gardens, London, SE7 8ET
MM0 LOZ D. Leech, The Croft House, 9, Ruilick, Beauly, IV4 7AB
M0 LPA N. Hilbery, 16 Albert Road, Ashford, TW15 2LU
M0 LPB L. Brown, 28 Farley Way, Stockport, SK5 6JD
M0 LPF B. Hall, 30 Worcester Road, Dudley Westmidlands, dy29ln
MW0 LPG L. Parsons, Ty Crwn, Rhosgadfan, Caernarfon, LL54 7HU
M0 LPK M. Napieralski, Flat 2, Reeves House, Crawley, RH10 7SW
M0 LPL G. Roberts, 25 Chalfont Way, Liverpool, L28 3QB
M0 LPM A. Monaghan, Briar Patch 117 Elm High Rd, Wisbech, pe14 0dn
MI0 LPO J. Mcerlean, 24 Mccorley Road, Toomebridge, Antrim, BT41 3NH
M0 LPR Dulverton Junior School, c/o J. Matthews, Moor View, Oldways End, East Anstey, Tiverton, EX16 9JQ
M0 LPT P. Perreas, 108 Lakonikis St., Kalamata, Greece, 24133
M0 LPW L. Walker, 7 Stroudley Close, Ashford, TN240TY
M0 LPZ G. Harmath, 101 Whitton Avenue East, Greenford, UB6 0QE
MM0 LQF J. Bennet, 24/11 Greenpark, Edinburgh, EH177TA
M0 LQR T. Longmore, 3 Dairy Farm Cottages, Northlands Road, Gainsborough, DN21 5DN
M0 LQY I. Gilmore, 19 Green Sward Lane, Redditch, B98 0EN
M0 LRA Leeds Radio Amateurs, c/o S. Priestley, 49 Victoria Crescent, Leeds, LS28 7SS
M0 LRB R. Baechle, Dr.-Schuhwerk-Strasse 32A, St Blasien, Germany, 79837
MI0 LRC S. Davison, 60 Cornation Place, Craigavon, BT66 7AN
M0 LRD R. Taylor, 89 St. Johns Road, Pelsall, Walsall, WS3 4EZ
M0 LRG Leicester Radio Group, c/o F. Barkhouse, 312 Humberstone Lane, Leicester, LE4 9JP
M0 LRH Dr R. Lecybyl, 49 Thorndon Close, Orpington, BR5 2SH
MD0 LRK C. Larkham, Monte Rosa, 7 Ballaughton Close, Douglas, IM2 1JE
M0 LRO M. Street, Flat 6, Derwent Court, Solihull, B92 7BU
M0 LRS L. Smith, 20 St. Loyes St., Bedford, MK40 1ZL
M0 LRY L. Bown, 4 Oak Leys, Brewood, Stafford, ST19 9EH
M0 LRZ S. Carr, 24 Park Road, Blyth, NE24 3DH
M0 LSA L. Mossop, 4 Brookdale Way, Waverton, Chester, CH3 7NT
M0 LSG G. Holland, 6 Moorfield Road, Widnes, WA8 3JE
M0 LSH M. Humphreys, 12 Pixiefields, Cradley, Malvern, WR13 5ND
M0 LSI N. Highfield, 298 Mersea Road, Colchester, CO2 8QY
M0 LSL A. Wood, 4 St. Andrews, The Common, Cranleigh, GU6 8NX
MM0 LSM A. Halcrow, Da Cro, Branchiclate, Burra Isle, ZE2 9LA
M0 LSN Rev. D. Harding, 9, Gilbert Street, Blenheim, New Zealand, 7201
M0 LSO G. Coltman, The Oaks Rushden, Buntingford, SG9 0SN
M0 LSS L. Storry, The Chimes, Madeira Drive, Bude, EX23 0AJ
M0 LSV R. Derham, Netherwood, Copse Lane, Hook, RG29 1SX
M0 LSW South Lancashire ARC, c/o D. Horner, 21 Ainsworth Road, Little Lever, Bolton, BL3 1RG
M0 LSX A. Dale, 37 Bussey Road, Norwich, NR6 6JF
M0 LSY Y. Li Song, 8 Birkin Court, Welwyn Garden City, AL7 3FA
M0 LSZ S. Deakin, 20 Riccat Lane, Stevenage, SG1 3XY
M0 LTA A. Tokely, 17 Sycamore Avenue, Horsham, RH12 4TP
M0 LTD C. Brink, 138 Brookside, Burbage, Hinckley, LE10 2TN
M0 LTE T. Fanning, 26 Mandeville Close, Tilehurst, Reading, RG30 4JT
M0 LTG M. Champion, 155 Walton Road, Walton on the Naze, CO14 8NF
M0 LTK J. Forsyth, 2 Littleworth, Oxford, OX33 1TR
M0 LTL K. Leung, Apartment 915, Metropolitan House, 1 Hagley Road, Birmingham, B16 8HU
M0 LTN A. Rademaker, 26 Elm Park Close, Houghton Regis, LU5 5PN
M0 LTP Z. Lukasz, 28 Heather Lane, West Drayton, UB7 8AW
M0 LTS L. Stant, Ears, University Of Surrey Student'S Union, Guildford, GU2 7XH
M0 LTT M. Lovatt, 3 Withington Close, Atherton, Manchester, M46 0EZ
M0 LTW L. Farrant, 21 Lime Tree Walk, Newton Abbot, TQ12 4LF
MI0 LTX I. Kondrashenkov, 44 Forge Manor, Magheralin, Craigavon, BT67 0XP
M0 LUC A. Gee, 18 St. James Avenue West, Stanford-le-Hope, SS17 7BB
M0 LUD G. Stanley, 11 Marlborough Street, Ossett, WF5 8JW
M0 LUF A. Clack, 3 Darwin Close, Swindon, SN3 3NF
M0 LUH N. Porter, 114 Kingston Avenue, Worcester, WR3 8PP
MW0 LUK C. Moreton, 20 Millbrook Court, Little Mill, Pontypool, NP4 0HT
MM0 LUP A. Young, Broomloan, Staffin, Portree, IV51 9JX
M0 LUS C. Campbell, 5 Ryebank, Holmfirth, HD9 1EU
M0 LUT A. Lutley, Springfield, Rookery Hill, Ashtead, KT21 1HY
M0 LUY L. Isaac Sneath, 21 Garrick Close, Lincoln, LN5 8TG
M0 LVL S. Kiel, 32 Weavers Avenue, Frizington, CA26 3AT
M0 LVR Dr O. De Peyer, Flat 5, Molasses House, Clove Hitch Quay, London, SW11 3TN
M0 LVW L. Van Wezel, 1 Waveney Road, Felixstowe, IP11 2NT
M0 LWM B. Blackstone, 2 Sneating Hall Cottages, Sneating Hall Lane, Frinton-on-Sea, CO13 0EW
MM0 LWS M. Strachan, 62 Charleston Drive, Dundee, DD2 2EZ
M0 LWT C. Jenkins, 25 Longmeadow Grove West Heath, Birmingham, B31 4SU
M0 LWZ Royal Signals Museum, c/o G. Budden, 7 Ashburton Gardens, Bournemouth, BH10 4HP
M0 LXA C. Staff, Uphill Road South, Weston Super Mare, BS23 4TU
M0 LXD D. Sawyer, 46 Brendon Gardens, Fair Oak, Eastleigh, SO50 7GG
M0 LXG C. Whatmough, 11 Blackchapel Drive, Rochdale, OL16 4QU
M0 LXI E. Pestano, 38 Third Avenue, Bexhill-on-Sea, TN40 2PA
M0 LXS C. Claxton, Forest Edge, Deer Park, Milton Abbas, Blandford Forum, DT11 0AY
M0 LXV D. Edwards, 212, Eastern Avenue North Kingsthorpe, Northampton, NN2 7AT
M0 LXY R. Thompson, Croft Michael Farm, Croft Mitchell, Troon, Camborne, TR14 9JJ
M0 LYB P. Lyba, 6 Ambassadors Way, North Shields, NE29 8ST
M0 LYD R. Beck, Moorings, Pleasance Road Central, Romney Marsh, TN29 9NP
M0 LYF K. Fletcher, Beverley Hotel, 55 Old Brumby Street, Scunthorpe, DN16 2AJ
M0 LYI M. Lynn, 47 Hamsterley Drive, Crook, DL15 9PT
M0 LYN G. Molyneaux, 3 Wilson Close, Thelwall, Warrington, WA4 2ET
M0 LYQ E. Gudziunas, 66 Packwood Close, Bentley Heath, Solihull, B93 8AW
M0 LYR Lowestoft and Gt Yarmouth Repeater Group, c/o J. Crawford, 23 Meadow Road, Bungay, NR35 1LE
M0 LZH L. Herman, 28 Millwood Court, Bury, BL9 9SD
M0 LZI I. Duxbury, 334 Linnet Drive, Chelmsford, CM2 8AL
M0 LZM P. Radford, 43 Bells Lane, Nottingham, NG8 6EX
M0 LZN J. Marshall, 6 Foster Walk, Sherburn in Elmet, LS25 6EU
M0 LZQ D. Bate, 14 Bromley Drive, Leigh, WN7 5NA
M0 LZS J. Kelly, Whiteley Bank Lodge, Canteen Road, Whiteley Bank, Ventnor, PO38 3AF
M0 LZU K. Kostov, 47 Wapshott Road, Staines-upon-Thames, TW18 3HB
M0 LZX J. Mutter, 27 Snowdonia Way, Huntingdon, PE29 6XP
MW0 LZZ C. Stubbs, 50 Laburnum Close, Rogerstone, Newport, NP10 9JQ
M0 MAC J. Mcgowan, 72B Adelphi Crescent, Hornchurch, RM12 4JZ
M0 MAF M. Milne, Flambards, Manor Road, Dunmow, CM6 2JR
M0 MAG M. Tinnion, 3 Hillhead Road, Newcastle upon Tyne, NE5 5AP
MW0 MAH M. Arnett, 5 Ffordd Cerrig Mawr, Caergeiliog, Holyhead, LL65 3LU
M0 MAI M. Mahoney, Hurdle Cottage, Mannington, Wimborne, BH21 7JZ
M0 MAJ M. Jones, 20 Chelsea Drive, Sutton Coldfield, B74 4UG
M0 MAL M. Elliott, 4 Maple Close, Keelby, Grimsby, DN41 8EL
MD0 MAN R. Cunningham, 3 Kellets Cottage, Lhergy Cripperty, Union Mills, Isle Of Man, IM4 4NF
M0 MAO M. Boland, 24 Hallam Close, Moulton, Northampton, NN3 7LB
MI0 MAP J. Phillips, 23 Edenderry Gardens, Banbridge, BT32 3BQ
M0 MAQ E. Macgurk, 10 Elmore Road, Lee on Solent, PO139DU
M0 MAR A. El Khalidi, 66 Trevor Crescent, Ruislip, HA4 6ND
M0 MAT M. Jeffery, 25A Stockwell Drive, Mangotsfield, Bristol, BS16 9DW
MW0 MAU M. Uphill, 1 Brynview Avenue, Ystrad Mynach, Hengoed, CF82 7DB
M0 MAW N. Cheesewright, 5 Duberly Close, Perry, Huntingdon, PE28 0BP
M0 MAX M. Skinley, 69 George Street, Wellington, TA21 8HZ
MW0 MAY M. Stokes, 23 Goetre Fawr Road Killay, Swansea, SA2 7QS
M0 MAZ M. Stevenson, 127 Walton Road, Chesterfield, S40 3BX
M0 MBA Dr Z. Derzsi, 217 Bensham Road, Bensham, Gateshead, NE8 1US
M0 MBB M. Bowell, 28 Jubilee Close, Byfield, Daventry, NN11 6UZ
MM0 MBC J. Curtis, 11 Haston Crescent, Perth, PH2 7XD

M0 MBD D. De La Haye, 4 Nicola Mews, Ilford, IG6 2QE
M0 MBE S. Holt, 14 Fir Street, Cadishead, Manchester, M44 5AU
M0 MBG M. Cooper, 9 Conway Close, Crewe, CW1 3XN
M0 MBH Dr M. Holbrook, 140 High Street, Steyning, BN44 3LH
M0 MBI A. Fox, 53 Shakespeare Way, Taverham, Norwich, NR8 6SL
M0 MBM M. Bridgehouse, 43 Age Croft, Oldham, OL8 2HG
M0 MBO M. Bay, 10 Mitchinson Street Steeple Claydon, Buckingham, MK18 2GS
M0 MBR M. Mutkin, 13 The Grove, Radlett, WD7 7NF
M0 MBS M. Hyman, 6 Belvedere Court, St. Anns Road, Manchester, M25 9LB
M0 MBT M. Mckenna, 54 Whickham Road, Hebburn, NE31 1QU
M0 MBV M. Hennessey, 3 Northgate Cottage, Falmer Road, Rottingdean, Brighton, BN2 7DT
M0 MBW F. Smith, 32 Amesbury Drive, London, E4 7PZ
M0 MBZ M. Bray, 13 Rosebay Close, Hartlepool, TS26 0ZL
M0 MCA A. Howden, 7 West Vale, Filey, YO14 9AY
MI0 MCB J. Mcbride, 21 Mosside Gardens, Mosside, Ballymoney, BT53 8QQ
MI0 MCC C. Mcclelland, 2 Stuart Park, Ballymoney, BT53 7BE
M0 MCE D. Mcewan, 29 St. Andrews Avenue, Weymouth, DT3 5JS
M0 MCG Moors Contest Group, c/o R. Edgar, 45 Exeter Road, Dawlish, EX7 0AB
M0 MCH M. Chapman, 3 Whitton Close, Doncaster, DN4 7RB
M0 MCI P. Illidge, 55 East Park Road, Spofforth, Harrogate, HG3 1BH
M0 MCL K. Winton, 130 George V Avenue, Worthing, BN11 5RX
M0 MCO M. Tinsell-Stanton, 38 Comberton Road, Kidderminster, DY10 3DT
M0 MCP R. Van-Der-Wijst, 6 Willow Street, Romford, RM7 7LJ
M0 MCT M. Bradbury, 104 Lilly Hall Road Maltby, Rotherham, S66 8AT
M0 MCV R. Treacher, 93 Elibank Road, London, SE9 1QJ
M0 MCW K. Phillips, 31B Waterloo Close, Blackburn, BB2 4RQ
M0 MCY M. Rolph, 17 Moorlands Park, Sinope, LE67 3BD
M0 MDC M. Clements, 21 Mallard Place, Twickenham, TW1 4SW
M0 MDE D. Edmondson, 21 Hawthorne Close, Heathfield, TN21 8HP
M0 MDG Middlesex DX Group, c/o S. Smith, 12 Stoneleigh Avenue, Hordle, SO41 0GS
MM0 MDH M. Herridge, The Hollies, Petticoat Lane, Orkney, KW17 2RP
MW0 MDJ M. Johns, 151 Somerset Street, Abertillery, NP13 1DR
M0 MDO D. Mcauslan, Casa Arco Iris, Via Variante Nascente, 8005-491, Portugal, SANTA BARBARA DE NEX
M0 MDP P. Murphy, 41 Tower Street, Sunderland, SR2 8NF
MW0 MDT M. Griffiths, Mandalay, Bromfield Street, Wrexham, LL14 1NF
MW0 MDV D. Morgan, 28 Harbour Village, Goodwick, SA64 0DY
M0 MDZ M. Day, 14 Windsor Drive, Ramsey Forty Foot, Ramsey, Huntingdon, PE26 2XX
M0 MEA M. Attlesey, 1 The Landway, Borough Green, Sevenoaks, TN15 8RG
M0 MEB E. Bias, 10 Riverdale Road, Shrewsbury, SY2 5TA
M0 MED D. Creighton, 8 Stockton Road West, Hawthorn, Seaham, SR7 8RS
M0 MEF M. Jasiorkowski, 275 Hartland Ave, Sheffield, S20 2PZ
M0 MEG A. Stephenson, 1 Northrop Close, Sunnybrow, Crook, DL15 0NS
M0 MEH M. Horton, 12 Kelburn Close, Chandler'S Ford, Eastleigh, SO53 2PU
M0 MEI M. Eilec, 5 Balata Way, Basingstoke, RG24 9YP
M0 MEL M. Kirk, 128 Perry Hill Road, Oldbury, B68 0BJ
M0 MEN M. Norris, 35 Sudbrooke Road, London, SW12 8TQ
M0 MEO Mexborough & District ARS, c/o S. Saiger, 10 Markham Avenue, Armthorpe, Doncaster, DN3 2AZ
MI0 MEV D. Lynas, 7 Regency Avenue, Dollingstown, Craigavon, BT66 7TY
M0 MEW T. Garcia-Quismondo, 11 Half Moon Lane, Worthing, BN13 2EN
MW0 MEX G. Johnson, 47 Heol Fawr, Penyrheol, Caerphilly, CF83 2JU
M0 MEY T. Hart, 17A Meyrick Park Crescent, Bournemouth, BH3 7AG
M0 MFA F. Alfrey, 16 Walls Road, Bembridge, PO35 5RA
MW0 MFB Dr M. Brown, Bryn Siriol Mount Road, St Asaph, LL170DB
M0 MFC M. Chester, 84 Edinburgh Drive, Spalding, PE11 2RT
M0 MFH M. Huson, 21 Fairfield Avenue, Stoke-on-Trent, ST3 4NU
MI0 MFI E. Taylor, 17 Rutherglen Street, Belfast, BT13 3LR
M0 MFL K. Bantock, 22 Deepdale Drive, Consett, DH8 7EH
M0 MFP C. Reed, 2 Drapers Lane, Hedon, Hull, HU12 8BG
M0 MFS A. Brook, Inkerman House 113 Clovelly Road, Bideford, EX39 3BY
M0 MFT M. Tweedie, 14 St. Cuthbert Drive Romanby, Northallerton, DL7 8JF
M0 MFX Dr M. Fox, 11 Park End, London, NW3 2SE
M0 MGA M. Smyth, 111 Forest Road, Whitehill, Whitehill, GU359BA
MM0 MGB A. Britton, 15 Glenbrook, Balerno, EH14 7JE
M0 MGF J. Gower, 10 Dann Court, Hedon, Hull, HU12 8GT
M0 MGI M. Isbell, 20 Woodland Crescent, Wolverhampton, WV3 8AS
MI0 MGJ M. James, 6 Portaferry Road, Newtownards, BT23 8NN
M0 MGK G. Marley, 144 Moreland Road, South Shields, NE34 8NJ
MM0 MGM M. Macfarlane, 9 Dreghorn Park, Colinton, Edinburgh, EH13 9PH
M0 MGP G. Champion, 34 Greenfields, Edenside, Kirby Cross., Frinton-on-Sea, CO13 0SW
M0 MGS M. Smith, 313 Stourbridge Road, Dudley, DY1 2EF
M0 MGW S. Dean, 154 Broad Lane, Walsall, WS3 2TQ
M0 MHG G. Hamilton, 5 Felixstowe Road, Sunderland, SR4 0BF
M0 MHO B. Silveira, 6 Amesbury, Waltham Abbey, EN9 3LQ
MM0 MHP H. Percival, 3/2 21 Prince Albert Road, Glasgow, G12 9JU
M0 MHQ Braunstone Troop Military Radio Group, c/o E. Scott, 3 School Drive, Coalville, LE67 4AN
M0 MHT M. Thompson, 4 Jubilee Court Ravenscroft, Holmes Chapel, Crewe, CW4 7HA
M0 MHU M. Hubbard, 14 Parkfield Crescent, Kimpton, Hitchin, SG4 8EQ
M0 MHW G. Hardman, 12 Fernleigh Chorley New Road, Horwich, Bolton, BL6 6HD
M0 MHY J. Mahoney, 27 Linby Drive, Bircotes, Harworth, Doncaster, DN11 8FP
MM0 MHZ Backpackers Radio Activity Group, c/o P. Thompson, 31 St. Marys Drive, Perth, PH2 7BY
M0 MIB V. Ball, 24 Carr Lane, Warsop, Mansfield, NG20 0BN
M0 MID P. Staite, Chestnut Farm, Eastville, Boston, PE22 8LX
MW0 MIE M. Ireland, Pen Y Gadlas, Ffordd Bryniau, Prestatyn, LL19 8RD
M0 MIG M. Navarro, 22 Lark Crescenent Harford, Huntingdon, PE29 1YN
MM0 MIJ J. Smith, 15E Afton Road, Cumbernauld, Glasgow, G67 2DW
M0 MIM M. Pearce, 1 Briars Wood, Horley, RH6 9UE
M0 MIQ M. Iqbal, 22 Rupert Avenue, High Wycombe, HP12 3NG
M0 MIR M. Lesniowski, 3 Woodland Avenue, Worksop, S80 2RB
M0 MIT B. Mitchell, 34 St. Marys Avenue, Gosport, PO12 2HX
M0 MIZ S. Stebbings, 4 Coltsfoot Lane, Bull'S Green, Knebworth, SG3 6SB
MW0 MJB M. Lee, 1 Maes Y Frenni, Crymych, SA41 3QJ
M0 MJD M. Davis, The Innings, Cricketts Lane, Chippenham, SN15 3EG
M0 MJF M. Firth, 209 High Street, Wickham Market, Woodbridge, IP13 0RQ
M0 MJG M. Garrett, 489 Dorchester Road, Weymouth, DT3 5BP
M0 MJH M. Hickford, 3 Ashen Road, Clare, Sudbury, CO10 8LQ
M0 MJK M. Keyte, 3 Lower High St., Mow Cop, Stoke on Trent, ST7 3PB
M0 MJS M. Sykes, Hope Cottage, 8 Brookside Road, Wimborne, BH21 2BL
M0 MJT N. Tyerman, 44 Hawkstone Close, Guisborough, TS14 7PE
M0 MJW M. Whitfield, 10 Bede Haven Close, Bude, EX23 8QF
M0 MJX M. Johnson, 54 Birchwood Drive, Ulverston, LA12 9PN
MM0 MJY M. Yarrow, Lomond Villa, Downies Village, Aberdeen, AB12 4QX
M0 MKE King Edward Vii School, c/o P. Treadwell, 22 Meynell Close, Melton Mowbray, LE13 0RA
MW0 MKG M. Gray, 15 The Circle, Cwmbran, NP44 7JP
M0 MKH M. Hadfield, 22 Mansfield Road, Clowne, Chesterfield, S43 4DH
M0 MKO M. O'Connor, 28 Cardigan Road, Southport, PR8 4SF
M0 MKR Milton Keynes Raynet, c/o J. Breen, 68 Honeysuckle Way, Bedford, MK41 0TF
M0 MKV A. Crawford, 4 Trimpley Drive, Kidderminster, DY11 5LB
M0 MKY J. Wilkinson, 8 Hunters Point, Chinnor, OX39 4TG
MM0 MLB M. Burgess, 11 Cromar Drive, Dunfermline, KY11 8GE
MM0 MLD W. Lawson, 60 Inglis Avenue, Port Seton, EH32 0AQ
M0 MLE J. Statham, Oakwoods, School Lane Upper Basildon, Reading, RG8 8LT
M0 MLG M. Goff, 27 Harley Road, Oxford, OX2 0HS
M0 MLH F. Peters, 56 Moresdale Lane, Leeds, LS14 6SY
M0 MLJ R. Tattersall, 70 Selwyn Street, Hillstown, Bolsover, Chesterfield, S44 6LR
M0 MLK M. Kipling, 12 Jolly Brows, Harwood, Bolton, BL2 4LZ
M0 MLM M. Millen, Flat 9, Sussex Court, Park Road, Bognor Regis, PO21 2PY
MM0 MLO M. Olesen, 19 Holmston Crescent, Ayr, KA7 3JJ
M0 MLS A. Stevenson, 252 Perry Wood Road, Great Barr, Birmingham, B42 2BH
M0 MLT M. Titcombe, 1A Langdale Avenue, Harpenden, AL5 5QU
M0 MLV T. Jones, Flat 92, Berkeley Court, London, NW1 5ND
M0 MLW M. Wren, Church Farm, Pointon Fen, Sleaford, NG34 0LF
M0 MLY J. Malley, 18 Park View, Seaton Delaval, Whitley Bay, NE25 0AL
M0 MLZ M. Mills, 17 Hornby Street, Plymouth, PL2 1JD
M0 MMC K. Sansom, 23 Victoria Crescent, Poole, BH12 2JQ
M0 MMD M. Moore, 52 Limefield Street, Accrington, BB5 2AF
MM0 MMG M. Gourlay, 14 Holmes Holdings, Broxburn, EH52 5NS
M0 MMJ M. Majhail, 3 Poynders Hill, Hemel Hempstead, HP2 4PQ
MM0 MMN I. Campbell, Fossoway Lodge, Kinross, KY13 0PD
M0 MMO J. Summerhill, 43 Rangers Walk, Bristol, BS15 3PW
M0 MMP Y. Weng, 20 Kendal Grove, Leeds, LS3 1NS
M0 MMR S. Quinn, 7 Poppleton Court, Tingley, Wakefield, WF3 1UY
M0 MMS Dr M. Mattingley-Scott, Bergheimer Strasse 28, Heidelberg, Germany, 69115
M0 MMT M. Johnson, 3 Bersted Mews Bersted Street, Bognor Regis, PO22 9RR
M0 MMU A. Moss, Winstons, Mayfield Lane Durgates, Wadhurst, TN5 6DG
M0 MMX D. Watt, 4 Spring Gardens Terrace, Padiham, Burnley, BB12 8JB
M0 MNB J. Davidson, 78 Old Heath, Shrewsbury, SY1 4SE
M0 MNG E. Spicer, 3 Golden Avenue Close East Preston, Littlehampton, BN16 1QS
M0 MNO D. Edge, 122 Aldbanks, Dunstable, LU6 1AJ
MM0 MNS M. Stuart, 3 Arran View, Largs, KA30 9ER
M0 MNU P. Richardson, 14 Portland Street, Worksop, S80 1RZ
M0 MNV P. Bienko, 94 Foster Street, Lincoln, LN5 7QF
MW0 MNX Artie Moore ARS, c/o K. Dawson, 57 Fair View, Blackwood, NP12 3NR
M0 MNZ L. Bennett, 19 Campion Crescent, Cranbrook, TN17 3QJ
MM0 MOB M. Overthrow, 63 Primrose Avenue, Larkhall, ML9 1JX
MM0 MOC Museum of Communication A.R.S., c/o K. Horne, 10 Blair Place, Kirkcaldy, KY2 5SQ
MI0 MOD T. Thompson, 8 Knockburn Avenue, Lisburn, BT28 2QF
M0 MOI S. Pettitt, 11 Derling Drive, Raunds, Wellingborough, NN9 6LF
MM0 MOK L. Matchett, 9 Colcoon Park, Gorebridge, EH23 4RS
M0 MOL G. Mollard, 1 Barnard Street, Barrow-in-Furness, LA13 9TD
MW0 MON D. Williams, 10 Bronllys, Gaerwen, LL60 6JN
M0 MOR A. Turner, 29 Welling Road, Orsett, Grays, RM16 3DW
M0 MOS M. Beckett, 59 Broadacre, Caton, Lancaster, LA2 9NH
MM0 MOT A. Smith, 56 Ayr Road Douglas, Lanark, ML11 0QA
M0 MOW J. Fautley, 71 Pullman Lane, Godalming, GU7 1YB
MW0 MOX Aberkenfig & District ARC, c/o B. Price, 156 Parc Bryn Derwen, Llanharan, Pontyclun, CF72 9TX
MM0 MPA D. Panton, 64 Dochart Crescent, Polmont, Falkirk, FK2 0RE
M0 MPB M. Budd, 28 Ladymeadow Court Middleton, Milton Keynes, MK10 9HZ
MW0 MPD D. Jenkins, 16 Celtic Road, Whitchurch, Cardiff, CF14 1EG
M0 MPF M. Finn, Atzenbach 38, Idar-Oberstein, Germany, 54473
M0 MPI M. Ibbett, 76 Chalkdown, Stevenage, SG2 7BN
M0 MPM M. Meerman, 24 Horseshoe Crescent, Burghfield Common, Reading, RG7 3XW
M0 MPQ J. Dales, 6 Woodfield Drive, Sawtry, Huntingdon, PE28 5TZ
M0 MPS B. Hopkins, 28 Dean Lodge Grange Road, Southbourne, Bournemouth, BH6 3ND
M0 MPT M. Travis, Cherrydayle, 6 Lingwood Close, Southampton, SO16 7GJ
MM0 MPW F. Dove, 5 Cairnorchies, Mintlaw, Peterhead, AB42 4LH
M0 MPY S. Gray, 2 Gloucester Road, Pilgrims Hatch, Brentwood, CM15 9ND
M0 MRB Dr M. Brickley, The Shearings, Milborne Port Road, Charlton Horethorne, Sherborne, DT9 4NH
M0 MRC C. Robinson, 9 Chatsworth Avenue, Culcheth, Warrington, WA3 4LD
MI0 MRG Marconi Radio Group, c/o P. Quinn, 11 Blackpark Road, Ballyvoy, Ballycastle, BT54 6QZ
M0 MRH A. Hawksworth, 87 Bradeley Road, Haslington, Crewe, CW1 5PX
M0 MRI A. Titmus, The Old Police House, Arundel Road, Fontwell, Arundel, BN18 0SX
M0 MRJ M. Jebbett, 16 Hastings Meadow Close, Kirby Muxloe, Leicester, LE9 2DR
M0 MRK M. Newton, Hall Farm Bungalow, Holbeck, Worksop, S80 3NF
M0 MRL D. Weight, 19 Shakespeare Drive, Upper Caldecote, Biggleswade, SG18 9DD
MM0 MRM A. Moe, 43 Huntlyburn Terrace, Melrose, TD6 9BH
MM0 MRO C. Munro, 11 Craigleith Hill Green, Edinburgh, EH4 2ND
M0 MRP M. Phillips, 71 Stour View Gardens, Corfe Mullen, Wimborne, BH21 3TL
M0 MRQ Dr P. Lawrence, 20 Pond Lane, Drayton, Norwich, NR8 6PP

MW0 MRS Marches ARS, c/o M. Bobby, Hafan, Church Street, Penycae, LL14 2RL
M0 MRT K. Turner, 3 Park Close, Pinxton, Nottingham, NG16 6QQ
MI0 MRV M. Kashkoush, 41 Dunamallaght Road, Ballycastle, BT54 6PF
M0 MRW M. Williams, 124 Ringwood Road, Christchurch, BH23 5RF
M0 MRX M. Roper, 26 Malpas Close, Bransholme, Hull, HU7 4HH
M0 MRY J. Mullins, 61 St. Johns Road, Slough, SL2 5EZ
M0 MSA Mid Somerset ARC, c/o C. Lavis, 88 Boundary Way, Glastonbury, BA6 9PH
MI0 MSB B. Campbell, 3, B Crewe Road, Ballinderry Upper, Lisburn, BT28 2PL
M0 MSE M. Edmonds, 60 Shenstone Road, Maypole, Birmingham, B14 4TJ
M0 MSF T. Reed, Seafield, Charing Hill, Charing, Ashford, TN27 0NG
M0 MSG M. Gibbons, 117 Ettingshall Road, Bilston, WV14 9XF
MM0 MSH B. Mccosh, 5 Muckhart Road, Dollar, FK14 7AE
MI0 MSM D. Dellett, 5 Larchfield Gardens, Kilrea, Coleraine, BT51 5SB
MI0 MSO P. Hosey, 13 Glenelly Gardens, Omagh, BT79 7XG
M0 MSS M. Simpson, 7 New Hall Farm, Cowling, Keighley, BD22 0JQ
M0 MSV S. Morozov, Haroldene, Towpath, Shepperton, TW17 9LL
M0 MSX M. Smith, 6 Neeps Terrace, Middle Drove, Wisbech, PE14 8JT
M0 MSY Merseyside ARS, c/o J. Woolvin, 62 Whitewood Park, Liverpool, L9 7LG
M0 MSZ M. Strange, 101 Southbroom Road, Devizes, SN10 1LY
M0 MTA M. Atfield, 42 Pauls Croft Cricklade, Swindon, SN6 6AJ
M0 MTC Wirral & DARC, c/o G. Brown, 13 Francis Avenue, Moreton, Wirral, CH46 6DH
M0 MTD T. Davidson, 2 Ridgeway, Rotherham, S65 3PQ
M0 MTF M. Falkowski, 23 Casson Street, Crewe, CW1 3EG
M0 MTG G. Clarke, 2 Gort Lomie, Clonlara, Ireland, V94 H96H
M0 MTI M. Juvonen, 23 Harwood Road, East Hagbourne, Didcot, OX11 9LX
M0 MTJ M. Smith, 6 Peverill Road, Perton, WV6 7PH
M0 MTN C. Martin, 14 Campbell Road, Eastleigh, SO50 5AD
MM0 MTO R. Foulds, 83 Croftfoot Road, Glasgow, G44 5JU
M0 MTQ D. Hartley, 102 Moss Lane, Sale, M33 5BE
MW0 MTR M. Roynon, 16 Greenwood Avenue, Pontnewydd, Cwmbran, NP44 5JE
M0 MTS C. Small, Riddings Barn, Hope Bagot, Ludlow, SY8 3AE
M0 MTW J. Bailey, 22 Wilford Drive, Ely, CB6 1TL
M0 MTX A. Price, 67 Mansfield Road, Glapwell, Chesterfield, S44 5QA
M0 MUC M. Wolfson, 4 Crabmill Lane, Easingwold, York, YO61 3DE
M0 MUI M. Tsun, Dyson'S Farm, Long Row, Tibenham, NR16 1PD
MM0 MUL A. Jackson, Union Farm, Craigrothie, Cupar, KY15 5PJ
MW0 MUM A. Sneddon, 3 Marigold Close, Gurnos, Merthyr Tydfil, CF47 9DA
MM0 MUN E. Munro, 55 Abergeldie Road, Aberdeen, AB10 6ED
MM0 MUR G. Murray, The Barn House, Springfield Farm, Carluke, ML8 4QZ
M0 MUU S. Bunting, Highfield, Three Ashes, HR2 8LU
M0 MUZ M. Hickman, 40 Tredington Grove, Caldecotte, Milton Keynes, MK7 8LR
M0 MVB S. Norman, 38 The Croft, Christchurch, Wisbech, PE14 9PU
M0 MVE M. Lambev, 2 Keats Way, Hitchin, SG4 0DR
M0 MVL M. Lloyd, 17 Williams Mead, Bartestree, Hereford, HR1 4BT
MW0 MVM A. Holt, Tyshoni, New Street, Llandrindod Wells, LD1 6BU
M0 MVO R. Zakrzewski, 31 Kingston Crescent, Chelmsford, CM2 6DN
M0 MVS Maritime Volunteer Service Radio Club, c/o L. Miller, 28 Arthur Road Cliftonville, Margate, CT9 2EN
MM0 MVX D. Wilson, Rivendell Lodge, Glenkindie, Alford, AB33 8RN
MW0 MWA A. Nisbet, 8 Pen Dinas, Tonypandy, CF40 1JD
M0 MWB S. Brown, 18 Goring Ave. Gorton., Manchester, M18 8WW
MW0 MWJ J. Jillings, Cane Garden, Dolau, Llandrindod Wells, LD1 5TE
MW0 MWL D. Mead, 35 Holly Street, Rhydyfelin, Pontypridd, CF37 5DA
M0 MWN M. Singer, 1 Bentley Road, Slough, SL1 5BB
M0 MWR M. Redstall, 56 Westmorland Road, Felixstowe, IP11 9TJ
M0 MWS M. Smith, Ashcroft, Black Horse Lane, Winterbourne Earls, Salisbury, SP4 6HW
M0 MWT R. Wilkes, 157 Saltwells Road, Dudley, DY2 0BN
MM0 MWW Orkney ARC, c/o E. Holt, Ashwell, St. Ola, Kirkwall, KW15 1SX
MW0 MWX Mike-Whiskey DX Group, c/o C. Morris, Hideaway Bettws Cedewain, Newtown, Powys, SY16 3DS
M0 MXC M. Craven, 78 Connaught Road, Brookwood, Woking, GU24 0HF
MU0 MXF O. Borisov, The Palms La Couture St Peter Port, Guernsey, GY1 2DZ
M0 MXN R. Lovell, Formby, Formby, Livepool, L37 4BP
M0 MXX M. Day, 33 Ryndle Walk, Scarborough, YO12 6JT
M0 MYA D. Passey, 5 The Croftings, Felton Close, Ludlow, SY8 1DS
M0 MYB H. Ibbitson, Tor View, Whitstone, Holsworthy, EX22 6TB
M0 MYC R. Browne, 2 Martham Close, London, SE28 8NF
M0 MYE D. Myers, 39 Rowan Avenue, Shildon, DL4 2AS
M0 MYG M. Young, 72 Goddard Way, Saffron Walden, CB10 2EB
M0 MYJ A. Frost, 23 Judeland, Chorley, PR7 1XJ
M0 MYK M. Knowles, 86 West Shore Road, Walney, Barrow-in-Furness, LA14 3UD
MM0 MYL C. Williamson, 31 Medrox Gardens, Cumbernauld, Glasgow, G67 4AJ
M0 MYM D. Crane, 3 Middlemead Close West Hanningfield, Chelmsford, CM2 8UR
M0 MYN C. George, 22 Elgar Drive, Shefford, SG17 5RZ
M0 MZC M. Carpenter, The Retreat, High Lane Manaccan, Helston, TR12 6HT
M0 MZD B. Greenwood, 13 Mayflower Street, Blackburn, BB2 2RX
M0 MZE D. Pilkington, 197 Saltings Road, Snodland, ME6 5HP
M0 MZX A. Watts, 12 Duchy Close, Dorchester, DT1 2EL
M0 NAA G. Porter, Higher Bramble, Trusham, Newton Abbot, TQ13 0NW
MW0 NAB N. Hockenhull, 6 Bryn Gannock, Deganwy, Conwy, LL31 9UG
M0 NAE B. Smith, Maple Lodge, Burtoft Lane South, Boston, PE20 2PF
M0 NAG N. Parry, 125 Lawsons Road, Thornton-Cleveleys, FY5 4PL
M0 NAI R. Neufeld, 19 Douai Grove, Hampton, TW12 2SR
MI0 NAJ P. Thompson, 57 Ballyduff Road, Newtownabbey, BT36 6PA
M0 NAK C. Marshall, 51 Hedgerow Close, Redditch, B98 7QF
M0 NAL Dr P. Shaw, 25 Headcorn Road, Platts Heath, Maidstone, ME17 2NH
M0 NAM N. Matthes, 24 Albany Road, Fleet, GU51 3LY
M0 NAO N. Tateishi, 4-24-14-402 Taihei, Sumida-Ku, Tokyo, Japan, 1300012
M0 NAP A. Newell, 17 Southlands Grove, Thornton, Bradford, BD13 3BG
M0 NAQ R. Smith, 15 Hollybush Road, North Walsham, NR28 9XT
M0 NAR Northwest ARC, c/o R. Seddon, 255 Westleigh Lane, Leigh, WN7 5PN
M0 NAS N. Smith, Clare Cottage, White Ash Green, Halstead, CO9 1PD
M0 NAU N. Coventry, 1 Seacrest Avenue, Fleetwood, FY7 6FG
M0 NAW N. Carey, 28 Tremayne Road, St Austell, PL25 4NE
M0 NAX G. Mosner, Zum Roehrbrunnen 16, Dreieich, Germany, 63303
M0 NAY C. Pegrum, 3 Bretland Road, Tunbridge Wells, TN4 8PS
M0 NAZ A. Davies, 4 Capella Path, Hailsham, BN27 2JY
M0 NBA B. Chalmers, 19 Brettenham Crescent, Ipswich, IP4 2UB
M0 NBC North Bristol Amateurradio Club, c/o P. Stevenson, 6 Dighton Gate, Stoke Gifford, Bristol, BS34 8XA
M0 NBD Norfolk County Raynet Group, c/o A. Mobbs, 149 The Paddocks, Old Catton, Norwich, NR6 7HR
M0 NBJ N. Jones, 26B Wellington Road, Wallasey, CH45 2NG
M0 NBK G. Hicks, 8 Mill Lane, Stockton-on-Tees, TS20 1LG
M0 NBL F. Noble, 1045, 45Th Street Apartment A, California, USA, 94608
M0 NCA Norfolk Coast ARS, c/o R. Leeds, 1A Clare Road, Cromer, NR27 0DD
M0 NCC Northampton DX, c/o G. Bansil, 15 Abington Close, Crewe, Cw13tl
M0 NCE N. Irvine, 100 Cavendish Road, Sunbury-on-Thames, TW16 7PL
M0 NCG M. Dumpleton, 3 Gooch Close Bacton, Norwich, NR12 0FA
M0 NCI Lifeboat ARS, c/o D. Hughes, 86 Colinmander Gardens, Ormskirk, L39 4TF
M0 NCJ C. Nicholson, 97 Station Road, Burgess Hill, RH15 9ED
M0 NCK N. Jewitt, 10 Gorse Lane, Oadby, Leicester, LE2 4RQ
M0 NCL M. Seabrook, 1 The Covers, Morpeth, NE61 2RU
M0 NCN Rev. M. Gillingham, 14 Nethergreen Gardens, Killamarsh, S21 1FX
M0 NCR Norfolk County Raynet Group, c/o A. Mobbs, 149 The Paddocks, Old Catton, Norwich, NR6 7HR
M0 NCV G. Killpack, 20 Fisher Close, Banbury, OX16 3ZW
M0 NCZ N. Czernuszka, 12 Durham Drive, Ashton-under-Lyne, OL6 8BP
M0 NDA Nuneaton & District ARC, c/o D. Parker, 146 Merlin Avenue, Nuneaton, CV10 9QJ
M0 NDC N. Caulfield, Smallbrook Road, Whitchurch, SY13 1BS
M0 NDE N. Evans, 234 Cauldon Road, Stoke-on-Trent, ST4 2BS
M0 NDF D. Finlay, 23 Glen Way, Oadby, Leicester, LE2 5YF
M0 NDJ D. Noe, 21 Gale Crescent, Banstead, SM7 2HZ
MI0 NDK N. Jameson, 15A Ednagee Road, Castlederg, BT81 7QF
M0 NDL S. Emary, Mallards, Fishers Lane, Mark, Highbridge, TA9 4LZ
MM0 NDM N. Maclucas, Lochnell Lodge, Benderloch, Oban, PA37 1QS
M0 NDO A. Nall, 22 Park Grove, Swillington, Leeds, LS26 8UN
M0 NDP N. Plunkett, 11 Stoneleigh Gardens, Grappenhall, Warrington, WA4 3LE
M0 NDT D. Farrant, 13 Bramham Down, Guisborough, TS14 7BY
M0 NDU J. Knight, 30 Ash Meadow, Lea, Preston, PR2 1RX
MM0 NDX C. Mcgowan, 21 Franchi Drive, Stenhousemuir, Larbert, FK5 4DX
M0 NDY R. Potter, 8 Hansard Way, Kirton, Boston, PE20 1QN
M0 NDZ T. Daskalov, 40 Elm Road, Norton Canes, Cannock, WS11 9QN
M0 NEC N. Eccles, 55 New Street, Lymington, SO41 9BP
M0 NEG K. Metcalfe, 33 Corsican Drive, Hednesford, Hednesford, WS12 4SS
M0 NEH Dr N. Hoare, 5 Kelsey Head, Port Solent, Portsmouth, PO6 4TA
M0 NEM P. Sanders, 6 Primrose Hill, Warwick, CV34 5HW
MM0 NEO N. Thomson, Four Winds, Holland Bush Hightae, Lockerbie, DG11 1JL
M0 NER A. Fraser, 18 Donside Close, Boldon Colliery, NE35 9BS
M0 NEU D. Newgas, 4207 3Rd Ave Nw, Seattle, USA, 98107
M0 NEV L. Creek, 382 Ripon Road, Stevenage, SG1 4NQ
MM0 NEW B. Newcombe, 9 Calder Road, Bellsquarry, Livingston, EH54 9AA
M0 NEX L. Jepson, 143 Walnut Avenue Weaverham, Northwich, CW8 3DX
M0 NFB N. Bisiker, 31 Lansdowne Avenue, Waterlooville, PO7 5BL
M0 NFD Northern Fells Contest Group, c/o C. Davies, 28 Neville Road, Darlington, DL3 8HY
M0 NFI N. Mooney, 60 Rhyddings Street, Oswaldtwistle, Accrington, BB5 3EY
M0 NFN P. Dernikos, P.O Box 599, Ashburton, Australia, 3147
M0 NFR New Forest ARS, c/o R. Ferguson, 31 Barton Court Road, New Milton, BH25 6NW
M0 NFY N. Young, 139 Northumberland Street, Norwich, NR2 4EH
M0 NGB N. O'Brien-Bird, 18 Milsted Close, Sunderland, SR3 2RF
M0 NGC C. Austin, 12 Laburnum Grove, Newbury, RG14 1LF
M0 NGH C. Ng, 9A Romsey Road, Southampton, SO16 4BY
M0 NGI P. Strachan-Buckley, 9 Short Street, Aldershot, GU11 1HA
M0 NGK K. Charazinski, 114 Marmet Avenue, Letchworth Garden City, SG6 4QF
M0 NGL N. Nash, Roann, Bedmond Road, Pimlico, Hemel Hempstead, HP3 8SH
M0 NGN N. Green, 44 Rushyford Drive, Chilton, Ferryhill, DL17 0EQ
M0 NGS Northamptonshire Grammar School ARC, c/o R. Tickle, 5 Bramley Court, Harrold, Bedford, MK43 7BG
M0 NGY J. Fautley, 71 Pullman Lane, Godalming, GU7 1YB
M0 NGZ S. Lawrance, 69 Athelstan Gardens, Wickford, SS11 7EF
M0 NHK Newbury and District Hackspace, c/o N. Bland, 63 Swindon Road, Wroughton, Swindon, SN4 9AG
MM0 NHM N. Morris, 23 Sedgebank, Sedgebank, Livingston, EH54 6HE
M0 NHY M. Pittas, 89 Seddon Road, Morden, SM4 6ED
M0 NIB N. Bown, 14 Parsons Mead, Abingdon, OX14 1LS
M0 NIC N. Bellamy, 102 Bye Mead Emerson?S Green, Bristol, BS16 7DQ
MI0 NID N.I Dxer's Group, c/o S. Barnes, 191 Marlacoo Road, Portadown, Craigavon, BT62 3TD
M0 NIE B. Niewiadomski, 41 The Crescent, Keresley End, Coventry, CV7 8LB
M0 NIF G. Calder, 41 Wood End Way, Chandler'S Ford, Eastleigh, SO53 4LN
M0 NIG N. Howe, 45 Kettering Road, Islip, Kettering, NN14 3JT
M0 NIL Dr R. Blackwell, Vikings Hall, Baylham, Ipswich, IP6 8JS
M0 NIW N. Wiseman, 23 Kingsway, Langley Park, Durham, DH7 9TB
M0 NJD J. Davies, 35 Semley Road, Hassocks, BN6 8PD
M0 NJE N. Eustice, 22 Lower Wear Road, Exeter, EX2 7BQ
M0 NJI N. Isherwood, 41 Livingstone Road, Blackburn, BB2 6NE
M0 NJJ N. Pipkin, 46 Charles Avenue, Albrighton, Wolverhampton, WV7 3LF
M0 NJP N. Pettefar, 44 Duck Lane, Laverstock, Salisbury, SP1 1PU
MM0 NJS N. Sheridan, Cemetery Lodge, Lochmaben, Lockerbie, DG11 1RL
M0 NJW N. Wears, 25 Topcliffe Mews, Morley, Morley, LS27 8UL
M0 NJX Dr M. Nassau, 4A London Road, Liphook, GU30 7AN
M0 NKE N. Yorke, 30 Bramdene Avenue, Nuneaton, CV10 0DH
M0 NKR A. Goldsmith, 61 Fengate Drove, Weeting, Brandon, IP27 0PW
M0 NKS B. Maggs, 44 Coldharbour Road, Hungerford, RG17 0AZ
MD0 NKX D. Wilson, 23 Snugborough Avenue, Union Mills, Braddan, Isle of Man, IM4 4LT
M0 NKY J. Atkinson, 17 Agricola Gardens Hadrian Park, Wallsend, NE28 9RX
M0 NLI Sos Radio Group, c/o D. Hughes, 86 Colinmander Gardens, Ormskirk, L39 4TF
M0 NLP C. Bowman, 26 Albany Hill, Tunbridge Wells, TN2 3RX
M0 NLR Dr A. Clark, 3 North Street, Owston Ferry, Doncaster, DN9 1RT

M0 NLT P. Austin, Bush Farmhouse Clee St. Margaret, Craven Arms, SY7 9DT
M0 NLW Newton-Le-Willows Raynet, c/o P. Williams, 2 Sycamore Avenue, Newton-le-Willows, WA12 8LT
MI0 NLY S. Carlin, 9 Mullandra Park, Kilcoo, Newry, BT34 5LS
M0 NMA S. Marsh, 31A Broad Street, Stamford, PE9 1PJ
M0 NMC N. Mcintyre, 27 Chapel Close, St Ann'S Chapel, Gunnislake, PL18 9JB
M0 NMD N. Davison, 1 Retford Close Breadsall Estate, Derby, DE21 4DX
M0 NMH N. Hilton, 20 Darbyshire Close, Deeping St. James, Peterborough, PE6 8SF
M0 NMI D. Blake, Pound Farm, Swan Lane, Leigh, Swindon, SN6 6RD
M0 NMO Monitoring Monthly, c/o K. Nice, 19 Southill Road, Poole, BH12 3AW
MM0 NNA Dr V. Vyshemirsky, 2103 Great Western Road, Glasgow, G13 2XX
M0 NNB A. Hopper, 7 Holmesdale Villas, Swallow Lane, Dorking, RH5 4EY
M0 NNH G. Bansil, 15 Abington Close, Crewe, Cw13tl
M0 NNL Dr N. Lutte, Oak House, Brandy Hole Lane, Chichester, PO19 5RX
MW0 NNX R. Cotterell, 49 Graham Court, Caerphilly, CF83 1RF
M0 NOA A. De Broise, 113 Conisbrough Grove, Leeds, LS25 2QB
M0 NOC P. Bolton, 1 Acorn Rise, Hollesley, Woodbridge, IP12 3JT
M0 NOE M. Hodgson, 10A Myrtle Grove, Enfield, EN2 0DZ
M0 NOI C. Birkin, 16 Marystow Close, Allesley, Coventry, CV5 9EA
M0 NOK B. Handley, 68 Northfield Avenue, Rothwell, Leeds, Ls260sw
M0 NOM M. Wickens, Haven Lea, Queens Drive, Windermere, LA23 2EL
MI0 NOR N. Mckee, 54 Castlemore Park, Belfast, BT6 9RP
M0 NOS A. Hayward, 54 Eastern Avenue, Mitcheldean, GL17 0DF
M0 NOV E. Lane, 50 Oakhurst Close, Belper, DE56 2TR
M0 NOW N. Walker, 79 Arklow Drive, Hale Village, Liverpool, L24 5RR
M0 NOY D. Noyek, 34 The Spinney, Sidcup, DA14 5NF
M0 NOZ J. Norrington, 32 Fulfen Way, Saffron Walden, CB11 4DW
M0 NPA N. Aleksander, 3 Elm Walk, London, NW3 7UP
M0 NPB N. Prater, 100 Pitfold Road, London, SE12 9HY
MW0 NPC Letterston ARC, c/o H. Bancroft, Stop And Call, Goodwick, SA64 0EX
M0 NPD N. Du Pre, Honeysuckle, Donald Way, Winchelsea Beach, Winchelsea, TN36 4HF
M0 NPG North Pennines Radio Group, c/o B. Lenton, 32 Forstersteads, Allendale., NE47 9AS
M0 NPH N. Holden, Plum Cottage, Avon Dassett, Southam, CV47 2AP
M0 NPL N. Livingstone, 2 Mickleton, Wilnecote, Tamworth, B77 4QY
M0 NPQ N. Ubonis, 8 Burleigh Close, Great Yarmouth, NR30 2RU
MI0 NPR N. Prentice, 26 Claranagh Road, Claranagh, Enniskillen, BT94 3FJ
M0 NQB N. Berrie, 12 Packenham Road, Basingstoke, RG21 8XT
M0 NQU E. Wagner, 3 Sarre Road, London, NW2 3SN
MM0 NQY P. Davis, 4 Daisy Park, Baltasound, Unst, Shetland, ZE2 9EA
M0 NRC Newton Le Willows ARC, c/o K. Horsfield, 59 Queens Drive, Newton-le-Willows, WA12 0LY
M0 NRG N. Grigsby, 67 Abshot Road, Titchfield Common, Fareham, PO14 4NB
M0 NRH N. Hickson, 27 Cressing Road, Witham, CM8 2NP
M0 NRJ N. Johnson, Belair, Western Road, Crediton, EX17 3NB
M0 NRK E. Musselle, 2 Rectory Crescent, Middle Barton, Chipping Norton, OX7 7BP
M0 NRP A. Andrew, 80 Hamble Drive, Abingdon, OX14 3TE
M0 NRS N. Stoker, 11 Hewley Crescent, Throckley, Newcastle upon Tyne, NE15 9AT
M0 NRW M. Reeve, 9 Kingfisher Walk, Loddon, Norwich, NR14 6FB
M0 NRY L. Brackstone, 276 Ladyshot, Harlow, CM20 3EY
MW0 NSC Neath & District Sea Cadet Unit, c/o J. Mason, 2 Golwg-Y-Bryn, Off Woodland Road, Skewen, Neath, SA10 6SP
M0 NSI B. Taylor, 15 Gledhall Street, Stalybridge, SK15 1LE
M0 NSP G. Jurgaitis, 70B Ingleby Road, Ilford, IG1 4RY
M0 NSR Norfolk Scout Radio, c/o C. Rolph, 21 South End Hogsthorpe, Chapel St Leonard, PE245NE
MD0 NSS N. Smith, 4 Cooil Farrane, Douglas, IM2 1NX
M0 NTA T. Brookes, 94 Newton Road, Lowton, Warrington, WA3 1DG
M0 NTC G. Bull, 9 Kilburn Place, Dudley, DY2 8HP
M0 NTG 93 Contest Group, c/o J. Spurgeon, Whitgift House, Whitgift, Goole, DN14 8HL
M0 NTH D. Higgs, 4 Rowsley Road, Stretford, Manchester, M32 9QA
M0 NTI T. Ward, 173-175 Station Road, Pendlebury, Swinton, Manchester, M27 6BU
M0 NTK J. Carrington, 15 Astley Court, Newcastle upon Tyne, NE12 6YR
M0 NTN N. Norris, 6 Tell Grove, London, SE22 8RH
M0 NTT J. Naylor, 15 Cawder Road, Skipton, BD23 2QE
M0 NTY C. Shane, 21 Avon Walk, Leighton Buzzard, LU7 3DE
M0 NTZ M. Montgomery, 4 Merryhill Country Park, Telegraph Hill, Norwich, NR9 5AT
M0 NUC Brede Steam ARS, c/o D. Adkin, 31 Fieldway, Broad Oak, Rye, TN31 6DL
M0 NUD N. Cull, 8 Eaton Road, Norwich, NR4 6PY
M0 NUG N. Lewis, 81 Long Lane, Upton, Chester, CH2 1JG
MM0 NUO Dr C. Brown, 4 Damselfly View, Edinburgh, EH17 8XH
M0 NUT D. Mainwaring, 1 Buckingham Close, Didcot, OX11 8TX
M0 NUX J. Horn, 8 Princess Close, Watton, Thetford, IP25 6XA
M0 NUZ A. Charlton, 26 Saundergate Lane, Wyberton, Boston, PE21 7BZ
M0 NVJ C. Drury, 129 Greenhill Road, Mossley Hill, Liverpool, L18 7HQ
M0 NVK E. Ridoutt, The Bungalow, Main Road, Porchfield, Newport, PO30 4LP
M0 NVQ R. Lynch, 2 Launceston Close, Oldham, OL8 2XE
M0 NVT N. Tennant, 22 The Lizard, Wymondham, NR18 9BH
MW0 NVY W. Oliver, Pwllmeyric, Chepstow, NP16 6LE
MI0 NWA J. Baker, 324 Clonmeen, Drumgor, Craigavon, BT65 4AT
M0 NWC North West ARC, c/o J. Mcinnes-Boylan, 17 Hindsford Bridge Mews, Atherton, Manchester, M46 9QZ
M0 NWE A. Wood, 85 Love Lane, Rayleigh, SS6 7DX
MI0 NWG North West ARC, c/o D. Keys, 71 Madison Avenue, Eglinton, Londonderry, BT47 3PW
M0 NWI P. Martin, 5 Shropshire Drive, Wilpshire, Blackburn., BB1 9NF
MW0 NWJ Dr N. Jones, 54 Glanrhyd, Coed Eva, Cwmbran, NP44 6TY
M0 NWK A. Leggett, 5 Syerston Way, Newark, NG24 2SU
MW0 NWM S. Taylor, 43 Toronnen, Bangor, LL57 4TG
MI0 NWO D. Adams, 65 Rose Park, Limavady, BT49 0BF
M0 NWT J. Turner, 2 South Drive, Padiham, Burnley, BB12 8SH
M0 NWW N. Warner, 12 Bay Road, Harwich, CO12 3JZ
M0 NWY S. Newhouse, 28 Hillmorton Lane, Lilbourne, Rugby, CV23 0SS
M0 NXA D. Pounder, 15 Eldon Grove, Hartlepool, TS26 9LY
M0 NXB N. Beck, 2 Killerton View, Wyndham Road, Silverton, Exeter, EX5 4JZ
M0 NXD N. Downes, 17 Knightswood Close, Rosliston, Swadlincote, DE12 8JJ
M0 NXF W. Rowland, 5 Gwinear Downs, Leedstown, Hayle, TR27 6DJ
M0 NXP M. Whitaker, 5 Horns Drove, Rownhams, Southampton, SO16 8AH
MI0 NYC J. Murdie, 9 Henderson Park, Bangor, BT19 1NS
M0 NYG N. Cox, 182 North Tenth Street, Milton Keynes, MK9 3AY
M0 NYM Guisborough & District ARC, c/o R. Dutton, 3 Kilkenny Road, Guisborough, TS14 7LE
M0 NYP S. Vzor, 40 Henlow Road, Birmingham, B14 5DS
M0 NYX Dr J. Hyde, The Grove, 7 Mill Lane, Kidderminster, DY10 3ND
M0 NYY N. Woodruffe, 139 North Home Road, Cirencester, GL7 1DY
M0 NZA V. Vesma, Durvale, 5 Jonas Drive, Wadhurst, TN5 6RJ
M0 NZL D. Foster, 42 Cranborne Avenue, Eastbourne, BN20 7TT
M0 NZR D. Bond, 4 Alfred Road, Haydock, St Helens, WA11 0QD
M0 OAA J. Chatterton, 6 Bayliss Road, Wargrave, Reading, RG10 8DR
M0 OAB B. Hodson, 176 Carters Mead, Harlow, CM17 9EU
M0 OAC D. Ion, 78 Blackmore Street, Derby, DE23 8AX
M0 OAD S. Davis, 74 The Driveway, Canvey Island, SS8 0AD
M0 OAE Q. Wright, 9 Browning Avenue, Warwick, CV34 6JQ
M0 OAJ A. Johnson, 19 Meadow Vale, Bristol, BS5 7RG
M0 OAL A. Weller, 169 Ashgate Road, Chesterfield, S40 4AN
M0 OAR P. Wallace, 7 Trinity View, Ketley Bank, Telford, TF2 0DX
M0 OAT G. Walker, 2 Cliffe Bank Cottages, Piercebridge, DL2 3SX
M0 OAU L. Boylan, 30 Pembrey Way, Liverpool, L25 9SN
MI0 OBC D. Best, 13 Cranley Green, Bangor, BT19 7FE
M0 OBD M. Hopkins, 27 Girtford Crescent, Sandy, SG19 1HR
MI0 OBE J. Watt, 23 Riverview Park, Ballymoney, BT53 7QS
M0 OBK D. Cull, 6 Compass Way, Bromsgrove, B60 3GP
M0 OBL M. Orbell, 21 Reedings Road, Barrowby, Grantham, NG32 1AU
M0 OBM D. Peacock, 41 Oxford Meadow, Sible Hedingham, Halstead, CO9 3QW
MI0 OBR A. Savage, 469 Old Belfast Road, Bangor, BT19 1RQ
MM0 OBT R. Hutcheon, 12 Denbecan, Alloa, FK10 1QZ
M0 OBU J. Haddleton, 3, Sawley Avenue, Whitefield, M45 8PP
M0 OBW D. Wilson, 12 New Street, Elworth, Sandbach, CW11 3JF
M0 OBY D. Clavey, 32 Apollo Close, Dunstable, LU5 4AQ
M0 OBZ J. Mcinnes-Boylan, 17 Hindsford Bridge Mews, Atherton, Manchester, M46 9QZ
M0 OCC Oxo Contest Club, c/o C. Wilmott, 60 Church Hill, Royston, Barnsley, S71 4NG
M0 OCE P. Clarke, 14 Wellfield Road, Culcheth, Warrington, WA3 4JP
M0 OCJ C. Overson, Studio Flat, 6 Grenville Street, Bideford, EX39 2EA
M0 OCK A. Mock, 60 Effra Road Wimbledon, London, SW19 8PP
M0 OCL L. Hendry, 109 Grove Avenue, New Costessey, Norwich, NR5 0HZ
M0 OCM I. Johnson, 9 Brook Road, Pontesbury, Shrewsbury, SY5 0QZ
M0 OCP D. Drynski, Flat 22, Nicholas Court, Corney Reach Way, Chiswick, London, W4 2TS
M0 OCV N. Powis, 24 Rosemullion Close, Exhall, Coventry, CV7 9NQ
M0 ODB G. Rodriguez, 32 Mount Pleasant, Prestwich, Manchester, M25 2SD
M0 ODD I. Rotheram, 60 Whitewood Park, Liverpool, L9 7LG
M0 ODE S. Hawkins, Forest Edge, Deer Park, Milton Abbas, Blandford Forum, DT11 0AY
MM0 ODI R. Kelly, 11 Kelvin Drive, Chryston, Glasgow, G69 0LZ
MM0 ODL F. Gordon, Crofts Of Torrancroy, Strathdon, AB36 8UJ
M0 ODM D. Merridale, The Granary, Falledge Lane, Upper Denby, HD8 8YH
M0 ODS D. Eccles, 123 New Inn Lane, Stoke-on-Trent, ST4 8HA
MM0 ODX R. Johnstone, 15 Barassiebank Lane, Troon, KA10 6SH
M0 ODZ G. Fenton, 40 High Street, Easington Lane, Houghton le Spring, DH5 0JN
M0 OEB C. Brennan, 19 The Furrow, Littleport, Ely, CB6 1GL
M0 OED G. Marfell, Springfields Bungalow, Drybrrok, GL17 9BW
M0 OEG S. Finch, 25 Bluebell Avenue, Wigan, WN6 8NS
M0 OEK C. Taylor, 51 Barnes Close, Sturminster Newton, DT10 1BN
M0 OER Dr M. Cianni, 121 Springfield Park Avenue, Chelmsford, CM2 6EW
M0 OES I. Jones, 33 Cobham Avenue, Liverpool, L9 3BP
M0 OET B. Mead, 8 Wordsworth Road, Kettering, NN16 9LB
M0 OFE H. Bond, Flat 4, Athrington Court, First Avenue, Felpham, Bognor Regis, PO22 7LB
M0 OFF R. Withers, 50 Coneygear Road, Hartford, Huntingdon, PE29 1QL
M0 OFL D. Gunn, 40 The Pastures, Oadby, Leicester, LE2 4QD
M0 OFM P. Joyce, 2 Harold Road, Cuxton, Rochester, ME2 1EE
M0 OGD G. Davies, 6 Bayleys Close, Empingham, Oakham, LE15 8PJ
M0 OGI M. Shopland, 128 Whitewood Park, Liverpool, L9 7LG
M0 OGS M. Jordan, 10 Wilmot Green, Great Warley, Brentwood, CM13 3DD
M0 OGX K. Fujita, 3-21 Denenchofu-Honcho, Ota-Ku, Tokyo, Japan, 1450072
M0 OGZ G. Singleton, 2 Rome Avenue, Burnley, bb115lq
M0 OHA Ormiston Horizon Academy Radio Club, c/o L. Preece, 29 Elliott Street, Newcastle, ST5 1JL
M0 OHI G. Chaffey, 63 Underwood Road, Eastleigh, SO50 6FX
M0 OIA P. Rutkowski, 17 Beaminster Gardens, Ilford, IG6 2BN
M0 OIC B. Downes, 6 Greenland Crescent, Beeston, Nottingham, NG9 5LB
MI0 OIM M. Edwards, 58 Rosemount Park, Jordanstown, BT37 0N
M0 OIO R. Hubbard, Southbroom School House, Estcourt Street, Devizes, SN10 1LW
M0 OJC C. Curry, 41 Bargate, Richmond, DL10 4QY
M0 OJG J. Canning, The Mount, Birmingham Road, Alcester, B49 5EG
MM0 OJJ J. Jarvie, Berryhill Farm, Tak-Ma-Doon Road, Kilsyth, Glasgow, G65 0RY
M0 OJO N. Hudson, Woodpecker Cottage Red Lane, Aldermaston, Reading, RG7 4PA
M0 OJS O. Spurway, 6 Alder Grove, Crewkerne, TA18 7DJ
M0 OJX T. Yamamoto, 1141-5, Kozukue, Kouhoku, Kanagawa, Japan, 222-0036
M0 OKB B. Bailur, 27 Russell Road, Felixstowe, IP11 2BG
M0 OKD D. Shelsher, Gables, Colchester Road, Ardleigh, Colchester, CO7 7PQ
MM0 OKG Dr J. Bowes, 1 Greendyke Cottage, Falkirk, FK2 8PP
M0 OKK G. Cooper, Holmfield, Chelmorton, Buxton, SK17 9SG
M0 OKL A. Gates, 46 Gloucester Place, Littlehampton, BN17 7AL
M0 OKQ M. Broum, 137 Culvers Avenue, Carshalton, SM5 2BA
M0 OKS S. Mookerjee, 6 Dipper Drive, Altrincham, WA14 5YF
M0 OKT C. Law, 23 Yeldersley Close, Chesterfield, S40 4LG
MM0 OKY J. Williamson, Clunie Cottage, Tullibardine Road, Auchterarder, PH3 1LX
M0 OLD S. Old, Firtrees, Main Street, Scarborough, YO11 3UD
MW0 OLE O. Thomas, Garth Celyn, St. Davids Road, Aberystwyth, SY23 1EU
M0 OLG M. Hirst, 69 Potton Road, St Neots, PE19 2NN

M0 OLM G. Dewey, West Riding Five Ash Down, Uckfield, TN22 3AP
M0 OLS A. Lee, 1 Lower Stoke Limpley Stoke, Bath, BA2 7FU
M0 OLW Rev. N. Cooper, Frindsbury Vicarage, 4 Parsonage Lane, Rochester, ME2 4UR
M0 OMA Dr M. Hocking, 52 Ashbourne Drive, Newcastle, ST5 6RL
MW0 OMB R. Rimmer, Dwyfor, Heol Las, Llantrisant, Pontyclun, CF72 8EG
M0 OMC Holsworthy ARC, c/o D. Roomes, View Field, Milton Damerel, Holsworthy, EX22 7NY
M0 OMD D. Haigh, 80 Saddlers Road, Quedgeley, Gloucester, GL2 4SY
MM0 OMG G. Robinson, Asgard, 12 Upper Waston Road, Burray, KW172TT
M0 OMI J. Jones, 1 Knebworth Road, Bexhill-on-Sea, TN39 4JH
MW0 OMK C. Morris, Hideaway Bettws Cedewain, Newtown, Powys, SY16 3DS
MM0 OMS M. Scullion, 24 Langmuir Road, Kirkintilloch, Glasgow, G66 2QE
M0 OMT S. Thompson, 30 Southport Parade, Hebburn, NE31 2AQ
M0 OMV J. Barton, 37 Lytton Road, Sheffield, S5 8AX
MW0 OMZ N. Shepherd, Prospect, Newchapel, Llanidloes, SY18 6JY
M0 ONB Eaaro ARC, c/o J. Williams, 66 Oakfield Avenue, Hitchin, SG4 9JD
M0 OND J. Isherwood, 11 Manor Crescent, Chesterfield, S40 1HU
M0 ONH A. Brown, 51 Towncroft, Chelmsford, CM1 4JX
M0 ONI G. Swift, 43 Storth Lane, Kiveton Park, Sheffield, S26 5QS
M0 ONL Online Radio Club, c/o A. Amos, Willow Tree House, Deers Green, Clavering, Saffron Walden, CB11 4PX
M0 ONO R. Vermeulen, 7 Butterfly Crescent, Evesham, WR11 1BP
M0 ONQ A. Richardson, The Chalet, Lincoln Road, Lincoln, LN4 2EX
M0 ONS C. Stone, 26 Chesham Road North, Weston-Super-Mare, BS22 8AD
MM0 ONX R. Walker, 10 Westpark Gate, Saline, Dunfermline, KY12 9US
M0 ONY P. Bown, 19 Victory Villas, Hatherop Road, Fairford, GL7 4JU
M0 ONZ A. Cross, 12 Appleby Drive, Langdon Hills, Basildon, SS16 6NU
M0 OOD Capt. J. Newman, Reeds, The Street, Cranbrook, TN17 4DB
MM0 OOF A. Bennett, 60 Rosemount Buildings, Edinburgh, EH3 8DD
MD0 OOH A. Elliott, Round Table House, Ronague, Castletown, Isle of Man, IM9 4HJ
M0 OOO A. White, 1 The Red House, Old Gallamore Lane, Market Rasen, LN8 3US
M0 OOR D. Thomas, 8 Cedar Avenue, Weston-Super-Mare, BS22 8HL
M0 OOS A. Norton, 9 The Common, West Tytherley, Salisbury, SP5 1NS
M0 OOT J. Logan, Whetstead, Grange Road, Gillingham, ME7 2UN
M0 OOZ Sir G. Cowne, 8 Arthur Road, New Malden, KT36LX
MI0 OPC Oville Amateur Radio Portable Club, c/o H. Evans, Oville House 404 Foreglen Road, Dungiven, Londonderry, BT47 4PN
M0 OPG O. Griffiths, 272 Worcester Road, Malvern, WR14 1BD
M0 OPK P. Kirby, 11 Bembridge Court, Crowthorne, RG45 6BN
M0 OPL P. Lounton, 107 Browning Hill Coxhoe, Durham, DH6 4SA
MI0 OPM D. Kirkwood, 1 Rural Cottages, Front Road, Lisburn, BT27 5LF
MW0 OPS H. Willott, 14 Warwick Close, Chepstow, NP16 5BU
MW0 OPY D. Pingel, 7 Duffryn Close Bassaleg, Newport, NP10 8PD
M0 OQO G. Brindle, 25 Chedworth Drive, Witney, OX28 5FS
MM0 OQR P. Taylor, 2 Laurel Grove, Aberdeen, AB22 8YJ
M0 ORC J. King, Plum Tree Cottage, Royston Place, Barton On Sea, New Milton, BH25 7AJ
M0 ORE G. Moore, 2 Spinacre, Barton On Sea, New Milton, BH25 7DF
M0 ORI D. Dart, Ticklebelly Cottage Lower Charlton Trading Estate, Shepton Mallet, BA4 5QE
MM0 ORK 59 Degrees North ARG, c/o E. Holt, Ashwell, St. Ola, Kirkwall, KW15 1SX
M0 ORM Quantum Amateur Radio & Technology Society, c/o D. Hughes, 86 Colinmander Gardens, Ormskirk, L39 4TF
M0 ORN S. Pafrey, 75 New Queen St., Bristol, BS15 1DE
M0 ORR C. Hale, 24 Wolverhampton Road, Kidderminster, DY10 2UT
M0 ORS D. Holman, 38 Polyear Close, Polgooth, St Austell, PL26 7BH
M0 ORY B. Shephard, 13 Forest Street, Annesley Woodhouse, Kirkby In Ashfield, Nottingham, NG17 9HE
M0 OSB G. Webster, 15 Bridge Road, Chichester, PO19 7NW
M0 OSE R. Allan, 44 Elderdene, Chinnor, OX39 4EJ
M0 OSH W. Rogalski, 22 Sadler Road, Walsall, WS8 6BG
M0 OSI S. Haigh, 17 Glebe Street, Swadlincote, DE11 9BW
M0 OSL S. Latimer, 40 Petersham Road, Long Eaton, Nottingham, NG10 4DD
M0 OSM C. Penfold, 14 Romney Road, Tetbury, GL8 8JU
M0 OSO C. Dinu, Apartment 8, 16 Abbey Road, London, NW8 9GS
M0 OSX A. Logan, 23 Cherry Tree Rise, Walkern, Stevenage, SG2 7JL
M0 OSY S. Houssart, Flat 3, Virginia Court, London, SE16 6PU
M0 OTA G. Hutchinson, 128 Crescent Drive North, Woodingdean, Brighton, BN2 6SF
M0 OTE D. Barlow, 21 Yellow Brook Close, Aspull, Wigan, WN2 1ZH
MW0 OTG D. James, Coombe House, Coombe, Presteigne, LD8 2HL
MW0 OTH A. Hodgson, Browns Holiday Park Towyn Road, Conwy, LL22 9HD
M0 OTJ Dr M. Ibison, 40 Regent Drive, Fulwood, Preston, PR2 3JB
M0 OTL G. Humphrey, 324 Snarlton Lane, Melksham, SN12 7QW
M0 OTO M. Pesendorfer, 13 Blake Road, London, N11 2AD
M0 OTS 126 (City of Derby) Sqn ATC, c/o R. Bateman, 81 Stanton St., Derby, DE23 6NF
M0 OTT Dr C. Darby, Brookfield, Forest Green, Dorking, RH5 5SG
MW0 OUC J. Bidwell, 26 Lone Road, Clydach, Swansea, SA6 5HR
M0 OUS J. Holley, Lookers, Blenheim Road Littlestone, New Romney, TN28 8PR
MI0 OUT P. Gibson, 118 Coleraine Rd, Portstewart, bt55 7hs
MM0 OUU C. Mcconnochie, 72 Duddingston Avenue, Kilwinning, KA13 6RS
M0 OVB R. Gowers, 43 Tungstone Way, Market Harborough, LE16 9GA
MM0 OVD D. Adamson, 5 Central Quadrant, Ardrossan, KA22 7DY
M0 OVI O. Popa, 5 Lanark Close, Horsham, RH13 5RY
MM0 OVK M. Mcallister, 36 Girvan Crescent, Newmilns, KA16 9HZ
MM0 OVV A. Bernard, 200 Carden Avenue, Cardenden, Lochgelly, KY5 0EN
M0 OVW G. Sandell, 1 St. Margaret Road, Ludlow, SY8 1XN
M0 OWG D. Burgfeld, 20 Wilson Row, Crowthorne, RG45 6WE
MM0 OWL C. Barclay, 3 Kildrummy Drive, Gartcosh, Glasgow, G69 8LE
MW0 OWM R. Cooke, 6 The Forestry, Trecastle, Brecon, LD3 8YA
M0 OWO A. Roworth, 17 Davidson Avenue, Congleton, CW12 2EQ
M0 OWS P. Stocks, 12 Bredbury Drive Farnworth, Bolton, BL4 7QD
M0 OXD C. Romocea, 21 Hurst Lane, Cumnor, Oxford, OX2 9PR
M0 OXO C. Wilmott, 60 Church Hill, Royston, Barnsley, S71 4NG
M0 OXR S. Wyatt, 55 Ridgefield Road, Oxford, OX4 3BX
M0 OXW D. Ellis, 4 Vane Terrace, Darlington, DL3 7AT
MM0 OXX A. Berry, 8 Hill Street Striling Fk7 0Dh, Striling, FK7 0DH
M0 OXZ P. Rodley, 27 Tollgate Close, Northampton, NN2 6RP
M0 OYH C. Staples, 32 Browns Lane, Netherton, Bootle, L30 5RW
M0 OYQ S. Long, 14 Petley Close, Flitwick, Bedford, MK45 1XP
M0 OYR S. Roberts, 77 Lambwath Road, Hull, HU8 0HB
M0 OYZ G. Brierley, 35 Ochrewell Avenue, Deighton, Huddersfield, HD2 1LL
M0 OZD R. Pounder, 65 Stubsmead, Swindon, SN3 3TB
MW0 OZI C. Osborne, Gwinwydden, Tremont Road, Llandrindod Wells, LD1 5BH
M0 OZJ B. Aicheler, Rose Cottage, Chilsworthy, Holsworthy, EX22 7BQ
M0 OZO I. Taylor-Hayward, Roslyn House, 57 Sheriff Highway, Hedon, Hull, HU12 8HA
MM0 OZY D. Leiper, 29 Thornton Avenue, Bonnybridge, FK4 1AR
M0 PAA Dr P. Thompson, 3 Floyers Field, West Stafford, Dorchester, DT2 8FJ
M0 PAC P. De Camps, 22 Osier Road, Spalding, PE11 1UU
M0 PAF G. Batty, 85 Cobcar Lane, Elsecar, Barnsley, S74 8BW
M0 PAG T. Pagden, 199 Woad Farm Road, Boston, PE21 0EN
M0 PAI A. Dodd, 14 Davies Street, Macclesfield, Sk10 1GE
M0 PAJ P. Alley, 58 Osprey Close, Watford, WD25 9AR
M0 PAL B. Beck, 21 Winston Grove, Retford, DN22 6SQ
M0 PAM A. Martins, 32 Godwin Road, Canterbury, CT1 3UF
M0 PAO P. Mcfadden, Maple Cottage, Leighton Buzzard, LU7 9DZ
M0 PAQ P. Meerman, 24 Horseshoe Crescent, Burghfield Common, Reading, RG7 3XW
M0 PAR A. Holland, 18 Mason Close, Malvern, WR14 2NF
M0 PAV S. Richards, 18 Lowfields Staxton, Scarborough, YO12 4SR
M0 PAW K. Pawley, 12 Barchington Avenue, Torquay, TQ2 8LB
M0 PAX G. Levine, 65 Clitheroe Road, Romford, RM5 2SL
M0 PAY J. Houghton, 7 West View, Skeffling, Hull, HU12 0US
MM0 PAZ S. Mckinnon, 8 Rowanlea Avenue, Paisley, PA2 0RP
M0 PBC P. Burt, 56 Winslade Road, Sidmouth, EX10 9EX
M0 PBD C. Smith, 21 Earl Spencer Court, Peterborough, PE2 9PQ
M0 PBF P. Bone, 11 Fox Hill Drive, Stalybridge, SK15 2RP
M0 PBN P. Biggin, Galadean, Farriers Way, Newport, PO30 3JP
M0 PBO M. Waistell, 23 Halton Court, Sheffield, S12 4ND
M0 PBR P. Rawlinson, 15 Elmbourne Drive, Belvedere, DA17 6JE
M0 PBT P. Burgess, 61 Grosvenor Avenue, Torquay, TQ2 7JX
M0 PBX P. Batson, 71 North Parade, Falmouth, TR11 2TE
M0 PBZ P. Bond, 16 Little Avenue, Swindon, SN2 1NL
M0 PCA P. Asbury, 67 Orchard Way, Measham, Swadlincote, DE12 7JZ
M0 PCB I. Kelly, 261 Bodiam Avenue, Tuffley, Gloucester, GL4 0XW
M0 PCC P. Bishop, 18 Holmwood Avenue, South Croydon, CR2 9HY
M0 PCE P. Crema, Not, Applicable, Resides, Italy, OUTSIDE THE UK IN
M0 PCH C. Morgan, 28 Tewther Road, Bristol, BS13 0NL
MI0 PCJ P. Hume, 2 Seabourne Parade, Belfast, BT15 3NP
M0 PCK P. Clay, Sylvesterweg 35, Viktring, Austria, 9073
M0 PCN Pelican Radio Group, c/o D. Perry, 11 St. Lawrence Close, Stratford Sub Castle, Salisbury, SP1 3LW
M0 PCO P. James, 44 Narbonne Avenue Ellesmere Park Eccles, Manchester, M30 9DL
M0 PCR P. Rudd, 27 Loxwood Avenue, Worthing, BN14 7QY
M0 PCS P. Sefton, 27 Donovan Avenue, London, N10 2JU
MW0 PCT S. Gau, Disgwylfa, The Downs, Cardiff, CF5 6SB
MI0 PCW Dr R. Bishop, 2 Alexander Park, Carrickfergus, BT38 7LL
M0 PCX P. Chronopoulos, Flat 11, Eton Hall, London, NW3 2DW
M0 PCZ P. Colyer, 23 Florida Road, Torquay, TQ1 1JY
M0 PDA P. Stallibrass, 12 Sheerwater Close, Bury St Edmunds, IP32 7HR
M0 PDB B. Beed, 72 Looseleigh Lane, Plymouth, PL6 5HH
M0 PDC P. Collins, 59 Portman Road, Scunthorpe, DN15 8PE
MM0 PDD S. Burnside, Woodend Farm, Buchlyvie, Stirling, FK8 3PD
M0 PDE Dr C. Clark, 2 Orchard Close, Elmstead, Colchester, CO7 7AS
M0 PDF Dr A. Clark, 2 Orchard Close, Elmstead, Colchester, CO7 7AS
M0 PDG J. Nicholls, 2 Karen Rise, Arnold, Nottingham, NG5 8GE
M0 PDH P. Hardwick, 2 Cliffe Cottages, Sandy Lane, Liss, GU33 7JE
M0 PDL S. Symonds, 301 North Fairlee Farm, Fairlee Road, Newport, PO30 2JU
M0 PDN N. Dinsdale, 3 Pond Cottages, Faulkland, Radstock, BA3 5XB
M0 PDP S. Martin, 77 Chatford Drive, Shrewsbury, SY3 9PH
M0 PDQ C. Almey, 152 Queensgate, Bridlington, YO16 6RW
MW0 PDR P. Randall, 24 Ffordd-Y-Goedwig, Pyle, Bridgend, CF33 6HY
M0 PDS P. Schoenmaker, 24 Greenheys Drive, London, E18 2HB
M0 PDU L. Fuller, Rosemar Lodge Westford, Wellington, TA21 0DX
MW0 PDV P. Devlin, Brynteg, Fron Bache, Llangollen, LL20 7BP
M0 PDW P. Whiteley, Grantham House, Grantham Road, Halifax, HX3 6PL
M0 PDX N. Pagdin, 74 Thelwall New Road, Thelwall, Warrington, WA4 2HY
M0 PDY P. Dyer, 19 Church Road, Evesham, WR11 2NE
M0 PDZ P. Harper, 36 Barrow Close, Marlborough, SN8 2BD
M0 PEA G. Pearson, 41 Myrica Grove, Hoole, Chester, CH2 3EW
M0 PEB P. Burke, 38 Bosworth Square, Rochdale, OL11 3QG
M0 PEG P. Grainger, 36 Orchard Road, Wigton, CA7 9JL
MW0 PEH B. Sellers, 86 St. John Street, Ogmore Vale, Bridgend, CF32 7BB
M0 PEM C. Rawlin, 5 Japonica Hill, Immingham, DN40 1LT
M0 PEQ S. Grammenos, Flat 2B, Chelverton Road, Putney, London, SW15 1RH
M0 PER A. Perkins, 3 Intake Close Willaston, Neston, CH64 2XG
M0 PES R. Blacker, 1 Ashwindham Court, Woking, GU21 8AW
M0 PET P. Gough, 7 The Crossings, Cannock, WS110EZ
M0 PEW P. Woolley, 84 Bowthorpe Road, Norwich, NR2 3TP
M0 PEX D. Munn, 36 Moor Lea, Braunton, EX33 2PE
M0 PFC Coalvile Radio Club, c/o D. Poulton, 93 Pretoria Road, Ibstock, LE67 6LP
M0 PFF G. Peters, Curlew Court, Guys Head Road, Sutton Bridge, Spalding, PE12 9QQ
MM0 PFH Pentland Firth Radio Hams, c/o D. Morrison, 4 West Murkle, Murkle, Thurso, KW14 8YT
M0 PFJ Splitters, c/o K. Baker, 64 Pendle Drive, Basildon, SS14 3LZ
MM0 PFN P. Darmady, Breckster Upper Camster, Lybster, KW3 6BD
M0 PFO P. Noble, 14 Park Street, Swallownest, Sheffield, S26 4UP
M0 PFT A. Perfect, 3 Chelmarsh Close, Chellaston, Derby, DE73 6PB
M0 PFW P. Borer, 88 Beechings Way, Gillingham, ME8 6LX
M0 PGC P. Corley, 90 Hill Road, Benfleet, SS7 1AL
M0 PGD P. Dann, 73 Ingoldsby Road, Northfield, Birmingham, B31 2HW
M0 PGH G. Hart, 11 Sadlers Ride, West Molesey, KT8 1SU

M0 PGI G. Hartless, 32 Long Acre, Mablethorpe, LN12 1JF
M0 PGL A. Md Ali, 67 Thorley Lane Timperley, Altrincham, WA15 7BA
M0 PGM P. Meadows, 6A College Road, Maidenhead, SL6 6BE
M0 PGN P. Niewiadomski, Flat1 79A Dartmouth Road, London, SE23 3HT
M0 PGO S. Mansfield, The Old Piggery, Ham Lane, Compton Dundon, Somerton, TA11 6PQ
M0 PGS P. Smith, The Dingle, 27 Habberley Road, Bewdley, DY12 1JH
M0 PGW P. Whiffing, 38 Green Close, Stannington, Morpeth, NE61 6PE
M0 PGX P. Graham, 28 Newburgh House, Highworth, Swindon, SN6 7DW
M0 PHB P. Bartlett, 58 Ashdown Road, Chandler'S Ford, Eastleigh, SO53 5QJ
MM0 PHD P. Dutton, 20/9 Craighall Crescent, Edinburgh, EH6 4RZ.
M0 PHL P. Stephens, 100A Foxglove Road, Eastbourne, BN23 8BX
M0 PHM P. Matthews, Sky Reach, Appletree Farm, StAustell, PL268RT
M0 PHN Dr P. Mason, 29 Grantham Road, Bristol, BS15 1JR
M0 PHO P. Honey, 3 Peterswood, Harlow, CM18 7RJ
M0 PHP C. Rodway, 26 Redesdale Avenue, Newcastle upon Tyne, NE3 3PP
M0 PHV P. Velzeboer, The Highnings, Coneygree Fold, Chipping Campden, GL55 6JL
M0 PHX Phoenix Radio Group, c/o A. Clayton, 6 Albert Road, Bunny, Nottingham, NG11 6QE
M0 PIA C. Kolderman, 6 Flanders Close, Kemsley, Sittingbourne, ME10 2PX
M0 PIB P. Badley, 37 Martins Lane, Dorchester-On-Thames, Wallingford, OX10 7JE
MW0 PIC R. Miles, 63 Phillip Street, Mountain Ash, CF45 4BG
MM0 PID T. Hamilton, 57/6 North Street, Bo'ness, EH51 0AE
M0 PIE R. Cockroft, 8 Lumb Lane, Huddersfield, HD4 6SZ
M0 PIK B. Pike, 19 Cardigan Gardens, Reading, RG1 5QP
M0 PIP C. Sidey, Inish Mor 19 Greenwood Court, Bideford, EX39 3SF
M0 PIR J. Clark, 4 Exeter Street, North Tawton, EX20 2HB
M0 PIT P. Hayes, 4 London Road, Roade, Northampton, NN7 2NL
M0 PIX R. Bibby, 40 Morval Crescent, Runcorn, WA7 2QS
M0 PJA P. Archer, 31 Stoney Bank Drive, Kiveton Park, Sheffield, S26 6SJ
M0 PJC P. Crabtree, 106 Sagecroft Road, Thatcham, RG18 3BF
M0 PJD P. Davies, 53 Lammas Road, Cheddington, Leighton Buzzard, LU7 0RY
M0 PJF P. Franklin, 1 Aberdeen Court, Newcastle upon Tyne, NE3 2XU
M0 PJG P. Galer, 62 Court Mount, Canterbury Road, Birchington, CT7 0BT
MW0 PJJ P. Jones, 23 Pinecroft Avenue, Aberdare, CF44 0HY
M0 PJK P. Knappett, Hope Cottage, The Green, Clacton-on-Sea, CO16 0BU
MI0 PJL P. Letters, 24 Old Grange Avenue, Carrickfergus, BT38 7UE
M0 PJM P. Mcmillan, 82 Front Street, Tudhoe Colliery, Spennymoor, DL16 6TJ
M0 PJP P. Pearson, 22 Norris Street, Darwen, BB3 3DR
MM0 PJQ P. Quinn, 24 Highfield Avenue, Paisley, PA2 8LG
MW0 PJR P. Rees, 8 Pencae Terrace, Llanelli, SA15 1NZ
MI0 PJS P. Smiley, 100 Lislaban Road, Cloughmills, Ballymena, BT44 9HZ
M0 PJT P. Tomlinson, 11 Haynes Close, Clifton, Nottingham, NG11 8JN
M0 PJX D. Dickson, 6 St. Johns View, Old Hutton, Kendal, LA8 0NG
M0 PJY P. Yarwood, 3 Old Blundells Court, Station Road, Tiverton, EX16 4LF
M0 PKE P. Walsh, 181 Hermes Close, Hull, HU9 4DR
M0 PKH P. Halloway, 82 Northwall Road, Deal, CT14 6PP
M0 PKL C. Wilson-Shah, 42 Glenthorne Road, London, N11 3HJ
M0 PKV P. Slade, End Cottage, Monyash Road, Bakewell, DE45 1FG
M0 PKW P. Watson, 10 Whitelands Crescent, Baildon, Shipley, BD17 6NN
MM0 PLB P. Bromley, Broadwood Treovis Upton Cross, Liskeard, G68 9JY
MI0 PLC R. Thomson, 1 Litchfield Park, Coleraine, BT51 3TN
M0 PLE C. Pryke, 50 Raglan Gardens, Watford, WD19 4LL
M0 PLG P. Gyngell, 54 Association Walk, Rochester, ME1 2XD
M0 PLH P. Hamnett, 13 Breakwater Court West, Berry Head Road, Brixham, TQ5 9AG
M0 PLN J. Kendrick, 29 Waterside, Silsden, Keighley, BD20 0LQ
M0 PLO P. Lesiecki, 165 New Road, Stoke Gifford, Bristol, BS34 8TG
M0 PLP P. Hallson, 4 Cranbrook Drive, Esher, KT10 8DL
M0 PLR D. White, 10 Meaux Road, Wawne, Hull, HU7 5XD
M0 PLS J. Walczak, 18 Heathfield, Chippenham, SN15 1BQ
M0 PLT G. Myers, 6 Ullswater Close, Biggleswade, SG18 8LX
M0 PLV P. Le Vallois, 14 London Row, Arlesey, SG15 6RX
M0 PLW P. Latham, 135 Ashgate Road, Chesterfield, S40 4AN
M0 PLX J. Telecki, 5 Stonegate, Cowbit, Spalding, PE12 6AH
M0 PLY R. Austen, Holly Tree Cottage, Station Road, Immingham, DN40 3AX
MJ0 PMA P. Ahier, Les Trois Carres, La Rue D'Aval, Jersey, JE3 6ER
M0 PMC P. Curnow, 13A Warden Close, Maidstone, ME16 0JL
M0 PMH P. Holmquest, 6 Rhyme Hall Mews, Fawley, Southampton, SO45 1FX
M0 PMI P. Mccormick, Fieldview, Crown East Lane, Lower Broadheath, Worcester, WR2 6RH
M0 PMJ P. Mullen, 14 Anderson Road, Hemswell Cliff, Gainsborough, DN21 5XP
M0 PMM P. Levetsky, 9 Moat Walk, Pound Hill, Crawley, RH10 7ED
MD0 PMN P. Best, 5 The Willows, Ballasalla, Isle Of Man, IM9 2EW
M0 PMO C. Breen, 64 Linkstor Road Woolton, Liverpool, L25 6DH
M0 PMR S. Macauley, 1 Moricambe Crescent, Anthorn, Wigton, CA7 5AS
M0 PMV P. Mansfield, 27 Popplechurch Drive, Swindon, SN3 5DE
MM0 PMW M. Mclauchlan, 8 Craigie St., Ballingry, Lochgelly, KY5 8NS
MI0 PMX G. Todd, 27 Ardreagh Road, Aghadowey, Coleraine, BT51 4DN
M0 PMY P. May, 33 Minsterley Avenue, Shepperton, TW17 8QS
M0 PNA P. Fulbrook, 167 Droitwich Road, Fernhill Heath, WR3 7TZ
M0 PNB P. Bozikis, 336 Higham Hill Road, London, E17 5RG
MW0 PNC M. Bloore, Halfway House, Hyfrydle Road, Talysarn, Caernarfon, LL54 6HG
M0 PNN P. Bowen, 12 Powell Place, Newport, TF10 7BS
M0 PNZ R. Maddock, 48 Collygree Parc, Goldsithney, Penzance, TR20 9LY
M0 POA A. Polesel, 33 The Maltings, Leighton Buzzard, LU7 4BS
MW0 POB J. Lewis, 2 Tymaen Crescent, Cwmavon, Port Talbot, SA12 9EA
M0 POC Network Radios Group, c/o N. Knapton, 4 Crabmill Lane, Easingwold, York, YO61 3DE
MM0 POD A. Conlon, Kilrae, Barrpath, Glasgow, G65 0EX
M0 POE A. Chlebikova, St. Catharine'S College, Cambridge, CB2 1RL
M0 POG Stafford Portable Operating Group, c/o P. Wilkes, 8 Cloverdale, Stafford, ST17 4QJ
M0 POI Points of Historical Interest, c/o M. Boland, 24 Hallam Close, Moulton, Northampton, NN3 7LB
M0 POQ R. Finch, 19B Kiln Road, Newbury, RG14 2LS
M0 POS G. Postlethwaite, 20 Birkett Drive, Ulverston, LA12 9LS
M0 PPC Central Radio Amateur Circle, c/o M. Hallard, 77 Banklands Road, Dudley, DY2 8BT
M0 PPG G. Beacher, 22 Trowbridge Gardens, Luton, LU2 7JY
MW0 PPM F. Miers, Pentre Isaf, Bryneglwys, Corwen, LL21 9NA
MW0 PPO V. Frostick, Clawdd Llwyd, Ceunant, Caernarfon, LL55 4RR
M0 PPP G. Batty, 85 Cobcar Lane, Elsecar, Barnsley, S74 8BW
M0 PPR P. Rozenek, 10 South Road, Portsmouth, PO1 5QT
M0 PPS I. Underwood, 12 Forge Lane, Gillingham, ME7 1UG
MI0 PPW J. Macfarlane, 1 Main Street, Uttony, Magheraveely, Enniskillen, BT92 6NB
M0 PPX A. James, 36 Westcote Close, Solihull, B92 8PL
M0 PPY J. Martin, 20 Hall Green Road, West Bromwich, B71 3LA
M0 PPZ P. Zanek, Mulberry Hill, Violet Lane, Tadley, RG26 5JX
M0 PQA P. Casado Arias, 24 Oldridge Road, London, SW12 8PJ
M0 PQI T. Pearsall, 16 Langdale Road, Leyland, PR25 3AR
MM0 PQN R. Tripney, 7 Sunnyside St., Camelon, Falkirk, FK1 4BJ
MI0 PQR Greenisland Electronics ARS, c/o B. Mckeen, 27 Old Grange Drive, Carrickfergus, BT38 7HG
M0 PRA D. Runyard, Tilecroft, Shortheath Crest, Farnham, GU9 8SA
MM0 PRB P. Bacon, 12 The Greens, Maddiston, Falkirk, FK2 0FN
MW0 PRC P. Randall, 24 Ffordd-Y-Goedwig, Pyle, Bridgend, CF33 6HY
M0 PRD P. Denham, Flat 1, 10 Prince Alfred Avenue, Skegness, PE25 2UH
M0 PRF J. Petch-Harrison, 13 Church Lane, Shepton Mallet, BA4 5LE
MW0 PRI D. Price, 11 Cefn Melindwr, Capel Bangor, Aberystwyth, SY23 3LS
MI0 PRM E. Simpson, 10 Woodview Park, Tandragee, Craigavon, BT62 2DD
M0 PRN P. Norman, Mellon, Hincaster, Milnthorpe, LA7 7ND
MW0 PRO J. White, Keepers Lodge Pumpsaint, Llanwrda, SA19 8DX
MW0 PRP P. Pugh, 27 Bank Street, Tonypandy, CF40 1PJ
M0 PRT B. Dey, 15 Bradenham Road, Grange Park, Swindon, SN5 6EB
M0 PRV D. Parvin, 11 Stanhope Way, Sevenoaks, TN13 2DZ
MM0 PSA P. Smith, 13 Newmills Grove, Balerno, EH14 5SY
M0 PSB J. Bell, 8 Firsleigh Park, Roche, St Austell, PL26 8JN
M0 PSC P. Croxford, 1 Meteor Close, Bicester, OX26 4YA
M0 PSD P. Davies, 2 Lynfords Drive, Runwell, Wickford, SS11 7PP
M0 PSE P. Glandfield, Flat 5, 7 Cargate Avenue, Aldershot, GU11 3EP
MW0 PSG 1st Pencoed Scout Group, c/o L. Ward, 11 Verlands Way, Pencoed, Bridgend, CF35 6TY
M0 PSI Dr A. Al-Azzawi, 33 Clare Mead, Rowledge, Farnham, GU10 4BJ
M0 PSK Dr C. Gibson, 1 Ryelands Orchard, Leominster, HR6 8QQ
MM0 PSM S. Milne, 5 Moriston Court, Grangemouth, FK3 0JJ
M0 PSR R. Tickle, 5 Bramley Court, Harrold, Bedford, MK43 7BG
M0 PSS P. Shaw, 10 Godbold Road, London, E15 3AL
M0 PST P. Stevens, 19 Elmfield Place, Newton Aycliffe, DL5 7BD
M0 PSW J. Godfrey, 4 Cherry Close, Houghton Conquest, Bedford, MK45 3LQ
M0 PSY D. Shuttleworth, 27 Union St., Egerton, Bolton, BL7 9SP
M0 PSZ S. Macdonald, Woodside Cottage, Horton Way, Verwood, BH31 6JJ
M0 PTA LSWC Portable Group, c/o S. Mcbain, 13A St. Lukes Close, Cherry Willingham, Lincoln, LN3 4LY
M0 PTB P. Singleton, 9 Sherbourne Road, Middleton, Manchester, M24 6FF
MM0 PTE P. Bainbridge, 49 Hare Moss View, Bathgate, EH47 0DN
M0 PTG P. Threakall, 83 Gregory Avenue, Birmingham, B29 5DG
M0 PTO M. Reeve, 9 Kingfisher Walk, Loddon, Norwich, NR14 6FB
M0 PTR P. Clifford, 90 Sherwood Avenue, Poole, BH14 8DL
M0 PTS P. Boultwood, 32 Makepiece Road, Bracknell, RG42 2HJ
M0 PTT R. Winthrop, 120 Stanhope Road, Carlisle, CA27BS
MM0 PTX A. Morgan, Broomrig House, Harviestoun Road, Dollar, FK14 7PT
MW0 PTY P. Matthews, The Chateau, Wynnstay Hall Estate, Ruabon, Wrexham, LL14 6LA
M0 PTZ Prof. P. Curtis, Cotswold, Salisbury Road, Abbotts Ann, Andover, SP11 7NX
M0 PUC Dr J. Woods, Haycocks Farm, Haycocks Lane, Colchester, CO5 8SS
M0 PUD A. Eyre, St. Michael Mead, The Common, Norwich, NR12 8BA
M0 PUH Dr M. Foster, 58 New Terrace, Staverton, Trowbridge, BA14 6NY
M0 PUR R. Janezko, 10A Lens Road, Allestree, Derby, DE22 2NB
M0 PUS B. Beard, 8 Monks Close, Newcastle-under-Lyme, ST5 3QU
M0 PUT K. Puttock, 12 Beechfields, School Lane, Petworth, GU28 9DH
M0 PVI P. Handley, 97 Applegarth Avenue, Guildford, GU2 8LX
M0 PVN P. Nicholls, 23 Bishops Gate, Birmingham, B31 4AJ
M0 PVO K. Chippindall-Higgin, 173 St. Augustine Road, Southsea, PO4 9AB
M0 PVP Baron B. Mendham, 252 Gregson Lane, Hoghton, Preston, PR5 0LA
M0 PVS V. Sterea, 8 Hamilton Street, Stalybridge, SK15 1LL
M0 PVU R. Coomer, 30 Torquay Road, Kingskerswell, Newton Abbot, TQ12 5EZ
MW0 PVW P. Witts, 82 Park View, Llanharan, Pontyclun, CF72 9SB
M0 PWB P. Booth, 12 Heathgate, Wickham Bishops, Witham, CM8 3NZ
M0 PWC P. Clark, 4 Chestnut Crescent, Blythe Bridge, Stoke-on-Trent, ST11 9NH
M0 PWD P. Woodburn, 21 The Row, Silverdale, Carnforth, LA5 0UG
M0 PWF W. Feather, Long House Farm, Ellers Road, Keighley, BD20 7BH
MD0 PWI C. Ingles, 1 Hillberry View, Onchan, Isle Of Man, IM3 3GB
M0 PWL M. Mynn, The Town House, Parsons Field, St. Mary'S, Hugh Town, TR21 0JJ
MM0 PWM P. Mackie, 8 Letham Avenue, Pumpherston, Livingston, EH53 0NG
M0 PWS P. Snelson, 103 Queens Road, Vicars Cross, Chester, CH3 5HF
M0 PWT Dr D. Garnett, Hill View, Snailbeach, Shrewsbury, SY5 0NS
M0 PWV P. Wells, 6 Westmead Road, Chichester, PO19 3JD
MW0 PWY L. Jones, Ty'R Ysgol, Holland Street, Ebbw Vale, NP23 6HT
M0 PXD P. Donaghy, 67 Brockenhurst Way, Bicknacre, Chelmsford, CM3 4XN
M0 PXI L. Evans, 8 Bishops Drive, Copplestone, Crediton, EX17 5HR
M0 PXL J. Dyson, 5 Welton Park, Daventry, NN11 2JW
M0 PXM P. Matthew, 24 Jubilee Close, Pamber Heath, Tadley, RG26 3HP
M0 PXO P. Offord, 1 Adare Close, Dunmow, CM6 2GR
M0 PXP J. Maudsley, Knight Stainforth Hall, Little Stainforth, Settle, BD24 0DP
M0 PXR D. Eggett, 5 Winifred Road, Poole, BH15 3PU
M0 PXY A. Schofield, 4 St. Guthlacs Close, Crowland, Crowland, PE6 0ES
M0 PXZ Dr C. Fox, 45 Park Road, Wivenhoe, Colchester, CO7 9LS
M0 PYA W. Allen, 109 Barston Road, Oldbury, B68 0PU
M0 PYB P. Bunting, 29 Marion Avenue, Alverthorpe, Wakefield, WF20BJ
M0 PYC D. Polley, 6 Coneygear Road, Hartford, Huntingdon, PE29 1QL
M0 PYE N. Atkins, The Old Rectory, Church Lane, Chippenham, SN14 6DE
M0 PYG G. Fielding, Chapel Court, Chapel Lane, Malvern, WR13 5HX
MI0 PYN S. Pynappels, 38 Dora Avenue, Newry, BT34 1JW
MM0 PYR Elderslie ARS, c/o P. Temple, 23 Ramsay Place, Johnstone, PA5 0EX
MM0 PYS Elderslie ARS, c/o K. Gillen, Flat 1/R, 12 Fergus Drive, Glasgow, G20 6AG

M0 PYT P. Kimberlee, 24 Jacey Road, Shirley, Solihull, B90 3LJ
M0 PZC Dr C. Hulbert, Nutmeg Cottage, 7 St. Leonard'S Street, Stamford, PE9 2HU
M0 PZD A. Jakstas, 36 Boxridge Avenue, Purley, CR8 3AQ
M0 PZR P. Hanman, 7 Tremenheere Road, Penzance, TR18 2AH
M0 RAB Dr C. Finnegan, 249 Winchester Road, Basingstoke, RG22 6EP
M0 RAC R. Cochrane, 7 Lawn Terrace, Blackheath, London, SE3 9LJ
M0 RAD Avon Valley Ara, c/o P. Badham, 201 York Avenue, East Cowes, PO32 6BH
MM0 RAG J. Hutson, 13 Greenan Road, Ayr, KA7 4ET
MM0 RAI T. Vanderydt, 33 Portland Road, Galston, KA4 8EA
MM0 RAM D. Stevenson, 51 Shannon Drive, Falkirk, FK1 5HU
M0 RAN M. Moran, 27 Burnet Close, Padgate, Warrington, WA2 0UH
M0 RAP R. Peech, 17 Chestnut Avenue, Crossgates, Leeds, LS15 8ED
M0 RAR N. Booth, Greenfield, Westmancote, Tewkesbury, GL20 7EP
M0 RAT N. Mcwilliam, 13 Rawlins Street, Liverpool, L7 0JE
M0 RAU R. Johnson, 15 Magazine Close, Wisbech, PE13 1LH
M0 RAW R. Wyeth, 112 Main Road, Crockenhill, Swanley, BR8 8JL
M0 RAX M. Pike, 5 Rowan Drive, Heybridge, Maldon, CM9 4BW
M0 RAZ Dr R. Bowman, 48 Eliot Drive St. Germans, Saltash, PL12 5NL
MW0 RBA R. Arnould, 13 Laurel Place, Sketty, Swansea, SA2 8JL
M0 RBB R. Brier, 48 Burton Rise, Kirkby-In-Ashfield, Nottingham, NG17 9BR
M0 RBC R. Colman, 197 Coppins Road, Clacton-on-Sea, CO15 3LA
M0 RBD M. Czerski, 3 Ladymead Close, Whaddon, Milton Keynes, MK17 0LL
M0 RBE R. Smith, 4 London Road, Lindal La12 0Ll, Ulverston, LA12 0LL
M0 RBF R. Ferguson, 31 Barton Court Road, New Milton, BH25 6NW
M0 RBG R. Blandford, 60 Benomley Road, Almondbury, Huddersfield, HD5 8LS
M0 RBH R. Hookham, 20 White Lodge Park Portishead, Bristol, BS20 7HH
M0 RBI C. Bennett, The Old Cottage, Waterside Road, Southminster, CM0 7QT
M0 RBJ N. Roberts, 40 Armour Road, Tilehurst, Reading, RG31 6HN
M0 RBK Dr R. Bleaney, 40 Broadstone Road, Harpenden, AL5 1RF
MW0 RBL R. Lovesey, 33 Ty Isaf Park Avenue, Risca, Newport, NP11 6NB
M0 RBM R. Medland, 5 Bay Tree Cottages, Hospital Road, Bude, EX23 9BP
MM0 RBN R. Corkey, 11 Golf View Cardenden, Lochgelly, KY5 0NW
M0 RBQ R. Simpson, 22 Kenworthy Road, Stocksbridge, Sheffield, S36 1BZ
MM0 RBR R. Hutton, 2 Watson Place, Dunfermline, KY12 0DR
M0 RBT Dr R. Hunt, 7 Knotley Hall Cottages, Chiddingstone Causeway, Tonbridge, TN11 8JH
M0 RBU D. Perrin, 54 High Street, West Wratting, Cambridge, CB21 5LU
MW0 RBV R. Briant, Talarvor, Llanon, SY23 5HG
M0 RBX R. Buckland, 34 Beechwood Drive, Meopham, Gravesend, DA13 0TX
M0 RBY R. Hall, Concorde Cottage, Ellingstring, Ripon, HG4 4PW
M0 RCC R. Chadwick, 4 Gleneagles Drive, Haydock, St Helens, WA11 0YS
M0 RCD E. Capstick, 130 Ship Lane, Farnborough, GU14 8BJ
M0 RCE G. Morse, 34 Headford Road, Bristol, BS4 1QE
MI0 RCF Lough Erne ARC, c/o H. Graham, 104 Tattygare Road, Lisbellaw, Enniskillen, BT94 5FB
MW0 RCH Christleton High School A.R.C, c/o S. Smith, 102 Gresford Road, Llay, Wrexham, LL12 0NW
M0 RCI R. Chappell, 11 Highfields Road, Darton, Barnsley, S75 5ER
M0 RCK R. Wells, 27 Victoria Avenue, Camberley, GU15 3HT
M0 RCL Dr A. Jackson, 40 Richardby Crescent, Durham, DH1 3TY
M0 RCM R. Moxham, 8 Dunroyal Close, Helperby, York, YO61 2NH
M0 RCN R. Hart, 4 Glade Mews, Guildford, GU1 2FB
M0 RCP Dr R. Peterson, 9 Moseley Wood View, Leeds, LS16 7ES
M0 RCR R. Room, 197 Newbridge Road, Bath, BA1 3HH
M0 RCT R. Tomkinson, 24 Beech Drive, Wistaston Green, Crewe, CW2 8RE
M0 RCV South East Hampshire Raynet, c/o P. Raxworthy, 32 St. Marys Avenue, Alverstoke, Gosport, PO12 2HX
M0 RCX R. Rawson, 30 Harty Road, Haydock, St Helens, WA11 0YY
M0 RCY D. Osborne, 12 Sandringham Close, Brackley, NN13 6JQ
MW0 RCZ R. Shipman, 1 Lledfair Place, Heol Pentrerhedyn, Machynlleth, SY20 8DL
M0 RDA A. Rivers, 34 Brookfield, Mawdesley, Ormskirk, L40 2QJ
M0 RDB Capt. R. De Savigny-Bower, 55 Fenwick Close, Woking, GU21 3BZ
M0 RDC R. Cooke, 1 Fiona Walk, Fazakerley, Liverpool, L10 4YW
MM0 RDD R. Duncan, 12 Douglas Loan, Kirkwall, KW15 1FU
M0 RDI A. Riddick, 30 Britannia Road, Banbury, OX16 5DW
M0 RDK Phase Array DX Group, c/o S. Gadsby, 30 Woodside Close, Knaphill, Woking, GU21 2DD
M0 RDP R. Parker, 53 Tunstall Road, Canterbury, CT2 7BX
M0 RDR R. Rawlinson, 17 Walmer Place, Winsford, CW7 1HA
M0 RDS R. Staniland, The Cottage, Martin Moor, Lincoln, LN4 3BQ
MM0 RDT R. Tripney, 7 Sunnyside St., Camelon, Falkirk, FK1 4BJ
M0 RDV R. Vincent, 141 Timberleys, Littlehampton, BN17 6QD
M0 RDX J. Scott, 443 Ford Green Road, Stoke-on-Trent, ST6 8LX
M0 RDY R. Hawkins, 3 Fairways Drive, Harrogate, HG2 7ES
M0 RDZ R. Dell, 18 Greenacres, Fulwood, Preston, PR2 7DA
M0 REA A. Duffield, 4 Crabmill Lane, Easingwold, York, YO61 3DE
M0 REB D. Hook, 28 Rifford Road, Exeter, EX2 5JT
M0 REC R. Clare, 26 Hall Park, Swanland, North Ferriby, HU14 3NL
M0 RED G. Sharp, 18 Whitbred Road, Salisbury, SP2 9PE
M0 REG Worthing Radio Events Group, c/o M. Folkes, 3 Colindale Road, Ferring, Worthing, BN12 5JF
MW0 REH R. Harlow, Swyn-Y-Mor, Penrallt Road, Holyhead, LL65 2UG
MI0 REI R. Bestek, 111 Cloughwater Road, Ballymena, BT43 6SZ
M0 REJ R. Jones, 39 Dalton Lane, Barrow-in-Furness, LA14 4LE
M0 REM M. King, 6 Aylesbury Close, Hockley Heath, Solihull, B94 6PA
M0 REP Dr A. Wallman, Oakwood, 30 Elmsway, Bramhall, Stockport, SK7 2AE
M0 REQ E. Stammers, 40 Tillingbourne Road, Shalford, Guildford, GU4 8EY
M0 REV Rev. J. Drake, 37 Weston Lane, Southampton, SO19 9GN
M0 REX R. Duffy, 45 Chatham Road, Winchester, SO22 4EE
M0 REZ S. Brown, 4 Dorado Gardens, Orpington, BR6 7TD
MM0 RFA R. Aird, 42 Kelvin Walk, Largs, KA30 8SJ
M0 RFK D. Gardner, 122 All Saints Avenue, Maidenhead, SL6 6LT
M0 RFM R. Mannock, Craybourne, Higher Brill, Falmouth, TR11 5QG
M0 RFU J. Byrne, 316 Turncroft Lane, Stockport, SK1 4BP
M0 RFW Dr R. White, 2 Uplands Cottages, Rattle Road, Pevensey, BN24 5DT
M0 RFY D. Murray, 9 Woodend, Sutton, SM1 3LW
M0 RGB G. Baker, Ling Cottage, Crag Foot, Carnforth, LA5 9SA
M0 RGC G. Henshall, 43 Cumberworth Road, Skelmanthorpe, Huddersfield, HD8 9AB
M0 RGD R. Dale, 17 Spencer Gardens, Brackley, NN13 6AQ
M0 RGE R. Edwards, 46 Lavers Oak, Martock, TA12 6HG
M0 RGF A. Mobbs, 149 The Paddocks, Old Catton, Norwich, NR6 7HR
M0 RGH R. Smith, Five Elms, Lullington Road Edingale, Tamworth, B79 9JA
M0 RGI I. Reichenfeld, 7 Hazelbank Close, Liphook, GU30 7BY
M0 RGL Gloucestershire County Raynet, c/o M. White, 7 Overthwart Crescent, Worcester, WR4 0JW
M0 RGN P. Williams, 35 Cansfield Grove, Ashton In Makerfield, Wigan, wn4 9se
M0 RGO D. Robertson, 53 Moor Lane, Weston-Super-Mare, BS22 6RA
M0 RGP Redditch Amateur Radio & Contest Group, c/o R. Jones, 6 Wychwood Drive Hunt End, Redditch, B97 5NW
MJ0 RGR R. Bisson, A122, Le Capelain House, Castle Quay, La Rue De L'Etau, St Helier, Jersey, JE2 3EA
M0 RGS LRGS ARC, c/o D. Saul, 78 Ingleton Drive, Lancaster, LA1 4QZ
M0 RGV G. Green, 90 Princes Way, Fleetwood, FY78DX
MI0 RGX R. Gilmore, 11 Abbots Gardens, Newtownabbey, Belfast, BT379QZ
M0 RGY A. Mobbs, 149 The Paddocks, Old Catton, Norwich, NR6 7HR
M0 RHA P. Van Staveren, 14 Fortune Green Road Flat 3, London, NW6 1UE
M0 RHB R. Burton, 23 Freston, Paston, Peterborough, PE4 7EN
M0 RHC Dr R. Hopper, 255 Wellbrook Way, Girton, Cambridge, CB3 0GL
MW0 RHD R. Burton, 6 Troed Y Garn Llangybi, Pwllheli, LL53 6DQ
M0 RHE R. Head, 24 Beaufort Road, Church Crookham, Fleet, GU52 6AZ
M0 RHG D. Bines, 39 School Close, Bretton, Peterborough, PE3 9FS
M0 RHI R. Hunt, 4 Rue Josy Printz, Hesperange, Luxembourg, L-5841
MM0 RHL Edinburgh Hacklab, c/o T. Hawes, 83 Dinmont Drive, Edinburgh, EH16 5RY
M0 RHO R. Morgan, 14 Ash Road, Ashurst, Southampton, SO40 7AT
M0 RHQ J. Middleton, 8 Cullen Close, Newark, NG24 1DF
M0 RHR A. Thomas, 1 Millers Close, Ruardean Hill, Drybrook, GL17 9AU
M0 RHS R. Hawkins, Forest Edge, Deer Park, Milton Abbas, Blandford Forum, DT11 0AY
MW0 RHT R. Titcombe, 82 Liverpool Road, Buckley, CH7 3NB
MM0 RHU R. Humphrey, 11 Pearce Grove, Edinburgh, EH12 8SP
M0 RHW W. Westlake, 2 Chegwin Court, Newquay, TR7 2DE
M0 RHX R. Haynes, 28 Ridgeway View, Montgomery, SY15 6BF
M0 RIA T. Lea, 1 Roseland Close, Keyworth, Nottingham, NG12 5LQ
MI0 RIB W. Nicholl, 58 Dunnalong Road, Bready, Strabane, BT82 0DW
M0 RIC D. Moore, 379 Main Road, Harwich, CO12 4DW
M0 RIG A. Rigg, 12 Bydales Drive, Marske-By-The-Sea, Redcar, TS11 7HJ
M0 RIK F. Woodhams, Greenways, Mill Lane, Fareham, PO15 5DU
M0 RIS M. Baines, 21 Acre Moss Lane, Kendal, LA9 5QE
M0 RIU D. Simmons, 8 Lower Grange, Huddersfield, HD2 1RU
M0 RIV Riviera ARC, c/o S. Crask, 14 Southfield Road, Paignton, TQ3 2SW
M0 RJB R. Blaney, 16 Pages Close, Wymondham, NR18 0TU
M0 RJE R. Edwards, 23 Queens Walk, Ruislip, HA4 0LX
M0 RJH Dr J. Reynolds, 38 Spring Lane, Hockley Heath, Solihull, B94 6QY
MM0 RJJ R. James, The Garret, Alyth, Blairgowrie, PH11 8HQ
M0 RJK R. Kavanagh, Chatslea, 57 Mill Lane, Littlehampton, BN16 3JP
M0 RJM R. Millington, Quaintways, The Avenue, Tarporley, CW6 0BA
MI0 RJN R. Neill, 15 Hawthorn Place, Coleraine, BT52 2ES
M0 RJO R. Odell, 41 Pevensey Park Road, Westham, Pevensey, BN24 5HW
MM0 RJR R. Renshaw, Smithy House, Scotscalder, Halkirk, KW12 6XJ
M0 RJS R. Somerville Roberts, Bank House, 7 Mill Lane, Stoke-on-Trent, ST7 3LD
M0 RJT R. Gilbert, 61 Coltstead, New Ash Green, Longfield, DA3 8LN
MI0 RJW R. Wylie, 9A Kinnegar Drive, Holywood, BT18 9JQ
M0 RJX R. Harrison, 18-20 Hall Lane, Kirkburton, Huddersfield, HD8 0QW
MI0 RJY J. Young, 7 Dunmore Close, Cookstown, BT80 8AS
M0 RJZ N. Noda, 14 Widmer Court, Vicarage Farm Road, Hounslow, TW3 4NL
M0 RKA M. Hawker, Popes Corner Marina, Holt Fen, Little Thetford, Ely, CB6 3HR
MW0 RKB M. Brady, Ty Mawr Uchaf, Dulas, LL70 9DQ
MW0 RKD R. Hark, 5 Victoria Park, Bagillt, CH6 6JS
M0 RKE D. Roake, 9 Falcondale Walk, Westbury On Trym, Bristol, BS93JG
M0 RKF C. Whitelaw, 18 Marine Drive, Bishopstone, Seaford, BN25 2RT
M0 RKH R. Hayes, 59 Elm Road, Folksworth, Peterborough, PE7 3SX
MD0 RKI R. Kijak, 13 Falcon Cliff Court, Douglas, Isle of Man, IM2 4AH
M0 RKK R. Pike, 44 Falkirk Close Bransholme, Hull, HU7 5BX
MM0 RKN C. Welsh, 28 Peacock Wynd, Motherwell, ML1 4ZL
M0 RKR G. Hirst, 94 Upper Brighton Road Sompting, Lancing, BN15 0LB
MM0 RKT R. Towers, 34 South Park Road, Hamilton, ML3 6PN
M0 RKW R. Watson, 5 Angrove Gardens, Sunderland, SR4 7TB
M0 RKX M. Hemming, 11 Blackberry Way, Evesham, WR11 2AH
M0 RKY R. Brown, 24 Malthouse Court, Wellington, Telford, TF1 1QJ
M0 RLC Dr R. Coley, 165 Westerfield Road, Ipswich, IP4 3AB
MW0 RLD B. Shelley, Sunray, Pendine, Carmarthen, SA33 4PD
M0 RLF R. Beardmore, 28 Broadway, Ilkeston, DE7 8TD
M0 RLI D. Thomas, 130 Norwich Avenue, Southend-on-Sea, SS2 4DH
MW0 RLJ R. Johns, 1 Llanferran St. Nicholas, Goodwick, SA64 0LL
M0 RLM R. Lima Matos, 12C Crouch End Hill, London, N8 8AA
MM0 RLN D. Nixon, 17 Semple Place, Linwood, Paisley, PA3 3RT
M0 RLP R. Le Piez, 279 Oakley Road, Southampton, SO16 4NR
MD0 RLS R. Smith, 3 Rheast Barrule, Castletown, IM9 1HW
M0 RLV R. Oliver, 3 Histon Road Cottenham, Cambridge, CB24 8UF
M0 RLW R. Williams, 16 Irving Road, Norwich, NR4 6RA
MM0 RMD R. Nelson, 8 South Street Cambus Fk102Pa, Stirling, FK102PA
M0 RMF G. Coupe, 112 Greenwood Drive, Kirkby-In-Ashfield, Nottingham, NG17 8GH
M0 RMG G. Chapman, Crockers Farm, Stoke Wake, Blandford Forum, DT11 0HF
M0 RMH R. Halliwell99, 38 Larch Grove, Kendal, LA9 6AU
M0 RMI R. Miller, 23 Clarendon Road, Sevenoaks, TN13 1EU
M0 RMJ R. Jeffs, 45 Forest Road, Bingham, Nottingham, NG13 8RL
MI0 RMK A. Mackenzie, 30 Dalriada Gardens, Ballycastle, BT54 6DZ
M0 RML Radio Millenium Lodge, c/o P. Greenhalgh, 13 Primrose Avenue, Urmston, M41 0TY
M0 RMM R. May, 18 The Glebe, Camborne, TR14 7EW
M0 RMN M. Watmough, 6 Blair Park, Knaresborough, HG5 0TH
M0 RMO C. Larner, 98 Allandale, Hemel Hempstead, HP2 5AT
M0 RMP R. Perkin, 26 Hall Avenue, Leek, ST13 6BU
M0 RMT J. Gorlinski, Flat 2, 5 Motcombe Lane, Eastbourne, BN21 1PS
M0 RMU A. Murayama, 248-4 Nakano-Cho, Ise-City, Japan, 516-0034
M0 RMW R. Williams, 11 Solent Way, Selsey, PO20 0JR

M0 RMY T. Rowlands, 7 Northfield Crescent, Beeston, Nottingham, NG9 5GR
M0 RMZ R. Mansfield, 8 Haysoms Drive, Greenham, Thatcham, RG19 8EY
M0 RNC M. Brigham, 21 Overdale Close, York, YO24 2RT
M0 RND A. Greig, 3 Fir Grange Avenue, Weybridge, KT13 9AR
M0 RNG Radio Nutters Group, c/o D. Wressell, 6 St. Josephs Close, Bishopdown, Salisbury, SP1 3FX
M0 RNH I. Justice, 4 Saxon Street, Droylsden, Manchester, M43 7FR
M0 RNI D. Harris, 13 Horsecroft, Ewyas Harold, Hereford, HR2 0EQ
M0 RNP P. Rainer, 6 Highland Close, Folkestone, CT20 3SA
M0 RNR B. Pickup, 3 Mews Court, Houghton le Spring, DH5 8GB
M0 RNU M. Nutt, 110 Birkinstyle Lane, Shirland, Alfreton, DE55 6BT
M0 RNW R. Wellsted, 127 Goldthorn Hill, Wolverhampton, WV2 4PS
M0 RNX J. Faulks, 11 Fishguard Spur, Slough, SL1 1TS
M0 RNZ N. Cooper, Romer Cottage, Long Reach, Ockham, Woking, GU23 6PF
M0 ROC M. Russell, 107 Cambridge Road, Hitchin, SG4 0JH
M0 ROJ R. Reeves, Goldford House, Goldford Lane, Malpas, SY14 8LL
M0 ROL A. De Mora, 36 West Park, Minehead, TA24 8AN
M0 ROM R. Pasika, 192 Longfield Lane, Cheshunt, Waltham Cross, EN7 6AQ
M0 RON A. Eustace, 3 Linworth Road, Bishops Cleeve, Cheltenham, GL52 8PF
M0 ROO R. Smith, South Cottage, Radley Green Road, Chelmsford, CM1 4NW
MM0 ROR I. Learmonth, 14 Deansloch Terrace, Aberdeen, AB16 5SN
MM0 ROV M. Gerrard, 10 Whinhill Gardens, Aberdeen, AB11 7WD
M0 ROW TS Gambia / Thorne Sea Cadets, c/o P. Ormsby, Wood Cottage, Little Heck, Goole, DN14 0BU
M0 ROX Dr R. Pediani, Old School House Great Coxwell, Faringdon, SN7 7NB
M0 ROY R. Henson, 2 Byron Street, Shirebrook, Mansfield, NG20 8PJ
MW0 RPB R. Bennett, 28 Neyland Path Fairwater, Cwmbran, NP44 4PX
M0 RPD I. Handley, Rosedale, Chapman Street, Market Rasen, LN8 3DS
MW0 RPE R. Evans, Brackenwood, 30 Northop Country Park, Mold, CH7 6WD
M0 RPF R. Fullagar, 6 Locke Way, Stafford, ST16 3RE
M0 RPI D. Akerman, The Brick Barn, Coppice Farm, Ross-on-Wye, HR9 7QW
M0 RPJ Dr R. Jepsen, 20 The Mount, Aspley Guise, Milton Keynes, MK17 8EA
M0 RPK R. King, 188 Providence Road, Sheffield, S6 5BE
M0 RPO R. Powell, 26 Fenwick Avenue, South Shields, NE34 9AJ
M0 RPR M. Roper, 13 St. Cuthbert Street, Worksop, S80 2HN
MI0 RPT R. Tomalin, 22 Drumfad Road, Millisle, Newtownards, BT22 2JQ
M0 RPZ R. Whiteway, 53 Tavistock Road, Weston-Super-Mare, BS22 6NX
M0 RQD S. Cownley, 5 Pumphouse Lane, East Cowes, PO32 6FJ
M0 RQK B. Dare, 1 St. Johns Villas, Sivell Place, Exeter, EX2 5ES
M0 RQN J. Foster, 23 High Street, Cumnor, Oxford, OX2 9PE
M0 RQQ D. Cockram, 4 Greenleaze, Marston Meysey, Swindon, SN6 6LJ
M0 RQX L. Coneley, 4 Primrose Close, Gosport, PO13 0WP
M0 RRC Rustyradios A.R.C.G., c/o S. Williams, 32 Waterdell Lane, St. Ippolyts, Hitchin, SG4 7QZ
MI0 RRE R. Rantin, 8A Buchanans Road, Newry, BT35 6NS
M0 RRF I. Sharpe, 4 Low Dowfold, Crook, DL15 9AE
M0 RRG Richmond Raynet Group, c/o J. Kirby, 2 Kneeton Park, Middleton Tyas, Richmond, DL10 6SB
M0 RRL K. Woodhams, 83 Langdale Place, Newton Aycliffe, DL5 7DY
MM0 RRM R. Murray, The Barn House, Springfield Farm, Carluke, ML8 4QZ
M0 RRN D. Barker, 12 The Weavers, Denstone, Uttoxeter, ST14 5DP
M0 RRR R. Reeves, 15 Higher Albert Street, Chesterfield, S41 7QE
M0 RRX R. Ridge, Roskellan House, Maenlay, Helston, TR12 7QR
MW0 RRY G. Sherry, 22 York Street, Oswestry, SY11 1LX
M0 RSA D. Silburn, 34 Northfields, Strensall, York, YO32 5XW
M0 RSB R. Brown, 22 Blaire Park, Yateley, GU46 7QP
M0 RSC Chesterfield & District Scouts A R C, c/o K. Greatorex, 54 Lilac Grove, Glapwell, Chesterfield, S44 5NG
M0 RSD K. Winwood, 146 Chapel Street, Pensnett, Brierley Hill, DY5 4EQ
M0 RSE Radio Society of Great Britain, c/o S. Thomas, 2 Myrtle Cottages, Sandy Lane, Saxmundham, IP17 1HR
M0 RSF C. Darlow, 418 Broad Lane, Bramley, Bramley, Leeds, LS13 3DF
M0 RSG E. Flint, The Bell House, Kingston Deverill, Warminster, BA12 7HE
M0 RSH R. Hansford, 17 Dolver Close, Corby, NN18 8NB
MM0 RSI R. Inglis, 13 Princes Street, California, Falkirk, FK1 2BX
M0 RSJ R. Dunstan, 53 Church View Road, Camborne, TR14 8RQ
MI0 RSN R. Robinson, 30 Trasnagh Drive, Newtownards, BT23 4PD
MI0 RSO S. Mcclean, 28A Ashfield Court, Donaghadee, BT20 0BF
M0 RSP Dr R. Paden, 21 The Rookery, Balsham, Cambridge, CB21 4EU
M0 RST C. Taylor, 48 Northdown Park Road, Cliftonville, Margate, CT9 3PT
M0 RSU R. Suchocki, 8 High Street, Bluntisham, Huntingdon, PE28 3LD
MW0 RSV G. Jones, 31 Liverpool Road, Buckley, CH7 3LH
M0 RSW R. Weatherup, Sidney Sussex College, Cambridge, CB2 3HU
M0 RSX R. Stone, 63 Sands Lane, Lowestoft, NR32 3ER
M0 RSY A. Davies, Penthouse Caravan, Shutt Green Lane, Stafford, ST19 9LX
MM0 RTD D. Robertson, 17 Keswick Drive, Hamilton, ML3 7HN
M0 RTE A. Green, 40 Claines Road, Northfield, Birmingham, B31 2EE
M0 RTH R. Duffield, 4 Crabmill Lane, Easingwold, York, YO61 3DE
M0 RTK Dr R. Tempo, 35 Warminster Road, Bath, BA2 6XG
M0 RTL A. Garthwaite, 278 Carlton Road, Barnsley, S71 2BA
M0 RTM C. Bell, 4 Main Street, Newbold, Rugby, CV21 1HW
M0 RTP R. Paster, 8 Rachaels Lake View, Warfield, Bracknell, RG42 3XU
M0 RTQ K. Jones, 15 Wheatfield Close, Epworth, Doncaster, DN9 1SY
MM0 RTT R. Turpie, 11 Ashkirk Place, Dundee, DD4 0TN
M0 RTV R. Coombs, 55 Highfield Road, Hemsworth, Pontefract, WF9 4EA
M0 RTW C. Colless, 128 Ditton Lane, Fen Ditton, Cambridge, CB5 8SS
MI0 RTY M. Strawbridge, 9 Wheatfield Crescent, Coleraine, BT51 3RA
M0 RUB D. Golding, Windrush Cottage 84-85 Bradenstoke, Chippenham, SN15 4EL
MI0 RUC N. Bolt, 32 Bush Gardens, Bushmills, BT57 8AE
M0 RUG J. Fry, Deal Cottage, Ipswich Road, Long Stratton, Norwich, NR15 2TF
MW0 RUH D. Thomas, 23 Merthyr Dyfan Road, Barry, CF62 9TG
M0 RUK C. Lote, 8 Warren Place, Walsall, WS8 6BY
M0 RUM M. Symmonds, 24 Woodville Grove, Stockport, SK5 7HU
MI0 RUR D. Boyd, 11 Abbey Gardens, Belfast, BT5 7HL
M0 RUT J. Wright, 4 Sweeters Field Road, Alfold, GU6 8UD
M0 RUX J. Dunn, 3 Hobbs Way, Rustington, Littlehampton, BN16 2QU
M0 RUZ R. Brierley, 39 Hatfield Road, Alvaston, Derby, DE24 0BU
M0 RVB Dr J. Harmer, 1 Wynford Rise, Leeds, LS16 6HX
MW0 RVC R. Gripp, 23 Edmond Locard Court, Chepstow, NP16 6FA
M0 RVD R. Dewes, 31 Woodlea Avenue, Lutterworth, LE17 4TU
M0 RVE Rother Valley Emergency Communications Group, c/o G. Swift, 43 Storth Lane, Kiveton Park, Sheffield, S26 5QS
MI0 RVH T. Nelson, 25 Monaghan Road Annashanco Rosslea, Belfast, bt927pt
M0 RVJ Rev. J. Goodman, The Vicarage, 4 Austenway, Chalfont St Peter, SL9 8NW
M0 RVK V. Rotaru, 15 Leicester Drive, Glossop, SK13 8SH
M0 RVN R. Vern, 104 High Street, Netheravon, Salisbury, SP4 9PJ
M0 RVT D. Rajanayagam, 87 Riffel Road, London, NW2 4PG
M0 RVV R. Harvey, 1 Bickley Moss, Whitchurch, SY13 4JF
M0 RWA R. Anderson, 56A Cheriton Avenue, Adwick-Le-Street, Doncaster, DN6 7BT
M0 RWB R. Broadbridge, 41 Benmoor Road, Poole, BH17 7DS
M0 RWD D. Eastwood, 13 Riverwood Drive, Halifax, HX3 0TH
M0 RWG R. Grout, 5 Branton Close, Great Ouseburn, York, YO26 9SF
M0 RWH R. Hornby, 61 Fulwood Heights, Fulwood, Preston, PR2 9AW
MM0 RWJ R. Welsh, 28 Peacock Wynd, Motherwell, ML1 4ZL
M0 RWK West Kent Raynet, c/o D. Parvin, 11 Stanhope Way, Sevenoaks, TN13 2DZ
M0 RWL R. Lane, 9 Hartoft Road, Hull, HU5 4JZ
M0 RWM R. Mayfield, 75 Cartwright Street, Loughborough, LE11 1JW
M0 RWN R. Nock, 83 Coles Lane, West Bromwich, B71 2QW
M0 RWR Riverway ARS, c/o E. Reynolds, 4 Underwood Close, Stafford, ST16 1TB
M0 RWS R. Stokes, 20 Elizabeth Road, Waterlooville, PO7 7LY
M0 RWW R. Wells, 57 The Ridings, Paddock Wood, Tonbridge, TN12 6YA
M0 RXA P. Costall, 3 Gaynsford Place, Little Canfield, Dunmow, CM6 1WB
M0 RXB R. Badami, Flat F 373 Camden Road, London, N7 0SH
MW0 RXD R. Dutton, Burn Naze, Old Mill Road, Penmaenmawr, LL34 6TE
M0 RXG M. Robson, 270 Calder Road, Lincoln, LN5 9TL
M0 RXM A. Matynka, 43 Fairfoot Close, Chippenham, SN14 0PL
M0 RXS J. Brewer, 1 Bentley Road, Forncett St. Peter, Norwich, NR16 1LH
M0 RXV R. Mansell, 1412 Warwick Road, Knowle, Solihull, B93 9LG
M0 RXX S. Southern, 37 Conway Road, Calcot, Reading, RG31 4XP
M0 RXZ B. Lupton, 124 Wolsey Crescent, New Addington, Croydon, CR0 0PF
M0 RYA J. Kay, Uplands Farm, Dallington, Heathfield, TN21 9NG
M0 RYB Dr P. Lock, The Firs, The Butts, Norwich, NR16 2EQ
MM0 RYE R. Rothon, 112 Ravenswood Rise, Livingston, EH54 6PG
M0 RYG Ramsbury Amateur Radio DX Group, c/o J. Connor, 28 Church Street, Hungerford, RG17 0JE
M0 RYK M. Granatt, 16 Culverden Avenue, Tunbridge Wells, TN4 9RF
MI0 RYM R. Murphy, 40 Stoneypath, Londonderry, BT47 2AF
M0 RYO R. Makioka, 77-38 Kawai-Honcho, Asahi-Ku, Yokohama-City, Japan, 2400052
MM0 RYR R. Clow, 25 Scott Street, Newcastleton, TD9 0QQ
M0 RYS R. Sayre, 8 Lorne Road, Richmond, TW10 6DS
MW0 RZC G. Bodley, 34 Claremont Road Newbridge, Newport, NP11 5DL
MJ0 RZD R. Luscombe, Flat 3, 1 Rouge Bouillon, St. Helier, Jersey, JE2 3ZA
M0 RZE I. Laidler, 5 South Street, West Rainton, Houghton le Spring, DH4 6PA
M0 RZO A. Brzosko, 32 Lower Park Street, Cambridge, CB5 8AR
MW0 RZS M. Beasley, Ffynnon Wen, Bontnewydd, Aberystwyth, SY23 4JJ
M0 RZX B. Forrest, 32 Idonia Road, Wolverhampton, WV6 7NQ
M0 RZY E. Moore, 33 Avon Drive, Congleton, CW12 3RQ
M0 SAA B. Matthews, 30 Oaklands Drive, Brandon, IP27 0NR
M0 SAB A. Brackstone, 3 Petunia Close, Basingstoke, RG22 5NX
M0 SAC M. Couchman, 140 The Tideway, Rochester, ME1 3QE
M0 SAD D. Platt, 50 Poplars Road, Stalybridge, SK15 3EN
MM0 SAH S. Henderson, 13 Dunnottar Place, Kirkcaldy, KY2 5YX
MI0 SAI S. Barnes, 191 Marlacoo Road, Portadown, Craigavon, BT62 3TD
MM0 SAJ S. Smith, 10 Munro Street, Stenhousemuir, Larbert, FK5 4QF
MM0 SAK A. Jardine, 17 Louisa Drive, Girvan, KA26 9AH
M0 SAL D. Salter, 142 Brays Road, Birmingham, B26 2PP
MI0 SAM S. Christie, 17 Kilburn, Belfast, BT12 6JS
M0 SAO A. Dossa, 24 Warwick Drive, Cheshunt, Waltham Cross, EN8 0BW
MI0 SAP S. Murray, 117 Knockview Drive, Tandragee, Craigavon, BT62 2BL
M0 SAQ D. Astley, 34 Church Terrace, Glossop, SK13 7RL
M0 SAR S. Roy, 28 Kingston Rise, New Haw, Addlestone, KT15 3EY
M0 SAV A. Smithies, 35 Dialstone Lane, Stockport, SK2 6AA
MM0 SAX G. Sproul, 25 Mulben Place, Glasgow, G53 7UP
M0 SAY D. Sayles, 82 Molineaux Road, Shiregreen, Sheffield, S5 0JY
M0 SAZ M. Parker, Ridgeways, Mill Common, Westhall, Halesworth, IP19 8RQ
M0 SBA S. Walker, 33 Parkside Somercotes, Alfreton, DE55 4LA
M0 SBB A. Southwell, 56 Lambrook Road, Taunton, TA1 2AF
M0 SBC K. Smith, 7 Rosebery Avenue, Morecambe, LA4 5RU
M0 SBD M. Denut, 17 Quillet Road, Newlyn, Penzance, TR18 5QR
M0 SBF S. Larkins, 4 Water Lane, Greenham, Thatcham, RG19 8SS
M0 SBH Dr S. Nandalan, 22 The Common, Parbold, Wigan, WN8 7DA
MW0 SBJ D. John, 29 Eleanor Street, Tonypandy, CF40 1DW
M0 SBK S. Johnson, 2 North Square, Edlington, Doncaster, DN12 1ED
M0 SBL P. Trembath, 48 Treveneth Crescent, Newlyn, Penzance, TR18 5NG
MM0 SBO S. Boyd, 1 St. Marks Lane, Edinburgh, EH15 2PX
M0 SBR H. Hamilton, Flat B, 9 Cambridge Drive, London, SE12 8AG
M0 SBT S. Burtsal, 69A Pewley Way, Guildford, GU1 3PZ
MW0 SBX M. Price, 24, Highland Gardens Neath Wales, Neath Port Talbot, SA10 6PJ
M0 SBY S. Bassett, 3 Lower Merryfield, Anchor Road, Radstock, BA3 5PG
M0 SBZ D. Smith, 105 Princes Street, Dunstable, LU6 3AS
M0 SCA S. Light, 16 Cabot Road, Yeovil, BA21 5FQ
M0 SCB T. Bacon, Norreum, Church Road, Reading, RG7 1TJ
M0 SCE D. Hagan, 8 Charles Close, Westcliff-on-Sea, SS0 0EU
M0 SCG Sands Amateur Radio Communications Group, c/o B. Watson, 7 Branksome Drive, Morecambe, LA4 5UJ
M0 SCN S. Nuttall, 17 Redgate, Northwich, CW8 4TQ
M0 SCO S. Court, 16 Worcester Road, Woodthorpe, Nottingham, NG5 4HY
M0 SCP D. Purbrick, 94A Polperro Way, Hucknall, Nottingham, NG15 6JW
M0 SCR Cornwall Raynet Group, c/o K. Harris, 8 Trelawney Rise, Callington, PL17 7PT
M0 SCS S. Smith, 113 Deaconsfield Road, Hemel Hempstead, HP3 9JA
M0 SCT G. Rutherford, 24 Chestnut Avenue, Hedon, Hull, HU12 8NH
M0 SCU S. Culshaw, 37 Netherby Road, Wigan, WN6 7PU
M0 SCW S. Warren, 7 Crich Way, Swadlincote, DE11 0UU
M0 SCX Dr S. Wing, 107 Highlands Boulevard, Leigh-on-Sea, SS9 3TH
M0 SCY Sandringham School ARC, c/o A. Gray, 5 Meadow Close, Marshalswick, St

Albans, AL4 9TG
M0 SDA E. Gedvilas, 23 Pennington Drive Newton Le Willows, Newton-le-Willows, WA12 8BA
M0 SDB D. Bower, 4 Winsford Road, Sheffield, S6 1HT
M0 SDC 7m Contest Group, c/o C. Wilson, 2 Frankham Close, Dinnington, Sheffield, S25 3QG
MW0 SDD Swansea & District ARC, c/o J. Bidwell, 26 Lone Road, Clydach, Swansea, SA6 5HR
M0 SDE Sugar Delta ARC, c/o S. Preston, The Chapel, Robson St., Shildon, DL4 1EB
M0 SDF P. Ashton, 32 Sycamore Road New Ollerton, Nr Newark, NG22 9PS
M0 SDG M. Torrington, 4 Aylesby Gardens, Grimsby, DN33 1SB
MM0 SDK M. Bartlett, 93 Lumsden Crescent, Almondbank, PH1 3UA
M0 SDM S. Mason, 8 Barrowby Gate., Grantham, NG31 7LT
M0 SDP S. Plows, York House, York Close, Measham, Swadlincote, DE12 7JH
MI0 SDR D. Reid, 13 Gelvin Grange, Londonderry, BT47 2LD
M0 SDS S. Stocker, 2 Peveril Avenue, Borrowash, Derby, DE72 3JJ
M0 SDT S. Theaker, 10 Grange Fields Mount, Leeds, LS10 4QN
M0 SDU L. Soldan, 35 Lingfield Gate, Leeds, LS17 6DB
M0 SDW Dr S. Willoughby, 33 Viscount Close, Diss, IP22 4GL
M0 SDY P. Cattermole, Blaxhall Hall Crossing, Little Glemham, Woodbridge, IP13 0BP
M0 SEA A. Newns, 3 Fox'S Yard, Harbour Village, Penryn, TR10 8GF
M0 SEB S. Banach, 13 Hilton Crescent Worsley, Manchester, M28 1FY
MW0 SEC L. Hayward, Cefn Gribyn Carmel, Llannerch-Y-Medd, LL71 7BU
M0 SED D. Cockburn, 4 Tranmere Avenue, Heysham, Morecambe, LA3 2BB
M0 SEJ S. Adams, 1 Byford Way, Winslow, Buckingham, MK18 3RJ
MM0 SEK J. Mcphillips, 86 Glenburn Avenue, Motherwell, ML1 5EF
M0 SEL S. Elliott, 50 West End Road, Mortimer Common, Reading, RG7 3TH
M0 SEM M. Skinner, 5 Sycamore Avenue, Upminster, RM14 2HR
M0 SEO Dr R. Seo, Flat 130, Oslo Court, London, NW8 7EP
M0 SER C. Lewis, 9 Chatsworth Gardens, Sydenham, Leamington Spa, CV31 1WA
M0 SET P. Harvey, 35 Isaac Street, Liverpool, L8 4TH
M0 SEV P. Holmes, 53 Bishops Hull Road, Bishops Hull, Taunton, TA1 5EP
M0 SEW J. Sewell, 8 Anglesmede Crescent, Pinner, HA5 5SP
MM0 SEY N. Rogers, 108 Beechwood Road, Cumbernauld, Glasgow, G67 2NP
MW0 SEZ S. Ezard, 59 Station Farm, Croesyceiliog, Cwmbran, NP44 2JW
M0 SFA S. Astbury, 131 Denton Avenue, Grantham, NG31 7JG
M0 SFD Dr F. Derry, 13 Fraucup Close, Ford, Aylesbury, HP17 8XU
M0 SFI F. Sidzhimov, 51 Barnwood Road, Guildford, GU2 8JD
MM0 SFM S. Forrest-Mcneill, 34 Maitland Hog Lane, Kirkliston, EH29 9DX
M0 SFR Dr A. Shafarenko, Church House, Kimbolton Road, Bolnhurst, Bedford, MK44 2ES
M0 SFT D. Swift, 31 Meadow Lane, Westbury, BA13 3AE
M0 SGA A. Suttle, 61 Albert Street, Shildon, DL4 2DN
MW0 SGD S. Doherty, 104 Cromwell Road, Milford Haven, SA73 2EN
M0 SGF S. Francis, 17 Garden Close, Rough Common, Canterbury, CT2 9BP
M0 SGG A. Rowan, 14 Craven Lea, Liverpool, L12 0NF
M0 SGH S. Hall, Orchard End, Asenby, Thirsk, YO7 3QR
M0 SGJ S. James, 94 North Road, Withernsea, HU19 2AY
M0 SGK S. Knott, 24 John Street, Leek, ST13 8BL
MM0 SGQ S. Gill, 5 Ramornie Place, Kingskettle, Cupar, KY15 7PT
M0 SGS S. Priestley, 49 Victoria Crescent, Leeds, LS28 7SS
M0 SGV S. Vanstone, 2 Walker Crescent, Weymouth, DT4 9AU
M0 SGW S. Whalley, 1 Cambridge Road, Gatley, Cheadle, Stockport, SK8 4AE
MW0 SGX J. Bidwell, 26 Lone Road, Clydach, Swansea, SA6 5HR
M0 SGZ J. Alincastre, 90 York Crescent, Durham, DH1 5PT
M0 SHA Surbiton Heritage Amateur Radio, c/o T. Fell, 24 Ardmay Gardens, Surbiton, KT6 4SW
MM0 SHB S. Boyd, 6 Schaw Road, Prestonpans, EH32 9HA
M0 SHD S. Hyde-Dryden, 90 Broadoaks Grange, Carlisle, CA1 2TA
M0 SHF N. Newman, 1 Hadham Park Cottages, Cradle End, Ware, SG11 2EH
M0 SHH P. Val, 72 Devonport Road, London, W12 8NU
M0 SHI M. Joshi, 14 Doyle Close, Erith, DA8 3QT
M0 SHK K. Holloway, 6 Britons Lane Close, Beeston Regis, Sheringham, NR26 8SH
M0 SHM S. Marriott, 4 Stone Cross Gardens, Catterall, Preston, PR3 1YQ
M0 SHN A. Al-Shakarchi, 17 Fairfax Place, London, NW6 4EJ
M0 SHO S. Howarth, 19 Farnham Croft, Leeds, LS14 2HR
M0 SHP S. Shepherd, 24 Brayton Road, Whitehaven, CA28 6EF
M0 SHQ S. Hedgecock, 37 Tennyson Road, Maldon, CM9 6BE
M0 SHR St. Helens Raynet Group, c/o P. Gaskell, 131 Greenfield Road, Dentons Green, St Helens, WA10 6SH
M0 SHV A. Sidhu-Brar, White Gates, Main Road, Northampton, NN7 3NA
M0 SHY S. Wildman, 55 Hill Street, Bradley, Bilston, WV14 8SB
M0 SHZ P. Shires, 30 Philip Garth, Wakefield, WF1 2LS
MM0 SIA Scottish Isles DX Group, c/o B. Titmarsh, Caberfeidh, Clachan Na Luib, HS6 5HD
M0 SIH S. Hammond, Ellsworth, Thrigby Road, Filby, Great Yarmouth, NR29 3HJ
M0 SII Salford University ARS, c/o V. Lynch, 16 Okehampton Crescent, Sale, M33 5HR
MM0 SIL J. Connelly, 60 Frankfield Street, Glasgow, G33 1BU
M0 SIN T. Brundrett, 45 Talbot Crescent, Whitchurch, SY13 1PH
MW0 SIP A. Ferguson, The Mount Stables, Salem, Llandeilo, SA19 7HD
MJ0 SIT S. Whitfield, Ceylon Cottage, Journeaux Street, St Helier, Jersey, JE2 3XQ
M0 SIY S. Shaul, 31 Chatterton Avenue Ermine West, Lincoln, LN1 3SZ
M0 SJD S. Davies, 2 Greenways, Hyde Lea, Stafford, ST18 9BD
M0 SJG S. Goodwin, 9 Downsview, Warminster, BA12 9DU
M0 SJJ S. Jones, 39 Dalton Lane, Barrow-in-Furness, LA14 4LE
M0 SJK S. Kearley, 36 Priory Road, Wirral, CH48 7EU
M0 SJL S. Low, 11 Bitterley Close, Ludlow, SY8 1XP
M0 SJR S. Roberts, 7 Alberta Grove, Prescot, L34 1PX
M0 SJS S. Stanhope, 61 Heathfield St., Manchester, M40 1LF
M0 SJV S. Viney, 5 Hawthorne Grove, Dudley, DY3 2QQ
M0 SJW S. Whitehead, 55 Crombie Road, Sidcup, DA15 8AT
M0 SKA D. Bennett, 2 Broadway, Blackburn, BB1 8QZ
M0 SKC S. Clay, Akers Lodge, 6 Penn Way, Rickmansworth, WD3 5HQ
MW0 SKD E. Edwards, 6 Kerslake Terrace, Tonypandy, CF40 1EQ
M0 SKF S. Keating-Fry, 18 Hewitt Avenue, London, N22 6QD
M0 SKG Strood Kent Contest Group, c/o B. Howard, 8 St. Margarets Close, Seasalter, Whitstable, CT5 4ST
M0 SKI Dr J. Skittrall, 14 Tamarin Gardens, Cambridge, CB1 9GH
M0 SKM S. Marshall, 96 Bidwell Hill, Houghton Regis, Dunstable, LU5 5EP
M0 SKN Dr M. Romenskyy, 11A Kensington Park Road, London, W11 3BY
M0 SKO A. Skolik, 13 Locks Meadow, Owls, Dormansland, RH7 6AW
M0 SKP R. Ibbotson, 33 St. Peters Avenue, Caversham, Reading, RG4 7DH
M0 SKR South Kesteven ARS, c/o S. Mason, 8 Barrowby Gate., Grantham, NG31 7LT
M0 SKT D. Webb, 52 Simpkin Close, Eaton Socon, St Neots, PE19 8PD
M0 SKV M. Sherrey, 25 Aspen House, Stratton-On-The-Fosse, Radstock, BA3 4RW
MM0 SKX S. Kirkbride, 18 North Roundall, Limekilns, Dunfermline, KY11 3JY
M0 SKY L. Sparks, 9 Hawk Place, Moresby Parks, Whitehaven, CA28 8YG
MM0 SLB Stromness Academy ARC, c/o D. Smith, 23 Torness, Kirkwall, KW15 1UU
M0 SLC K. Molnar, 201 Fold Croft, Harlow, CM20 1SW
M0 SLD S. Lovell, 20 Courtenay Walk, Weston-Super-Mare, BS22 7TQ
MI0 SLE D. Tarnowski, 11 Millbrook Gardens, Kilrea, Coleraine, BT51 5RZ
M0 SLF S. Farnell, 16 Lily Way, Lowestoft, NR33 8NN
M0 SLG D. Logan, Cedar House Reading Road North, Fleet, GU51 4AQ
MW0 SLH J. Hewitt, 1 Highfield, Gloucester Road, Chepstow, NP16 7DF
M0 SLL D. Firth, 5 Mowhay Gardens, Hatherleigh, EX20 3FE
M0 SLO Dr A. Saje, 72 Bedworth Road, Bulkington, Bedworth, CV12 9LL
M0 SLP S. Richardson, 89 Mead End, Biggleswade, SG18 8JR
M0 SLR South Lancashire ARC, c/o J. Bridson, 10 Clegg Street, Astley, Tyldesley, Manchester, M29 7DB
M0 SLY S. Lyon, 10 Sycamore Close, Preston, Hull, HU12 8TZ
M0 SMA B. Beckett, 38A Whinney Banks Road, Middlesbrough, TS5 4HG
MM0 SMB B. Mcsherry, 3 Taylor Road, Whitburn, Bathgate, EH47 0NL
M0 SMC S. Mcgregor, 16 Dibbins Green, Wirral, CH63 0QF
MM0 SMD J. Nicol, 18 Tininver Street, Dufftown, Keith, AB55 4AZ
M0 SME G. Bystryakov, 20 Elmhurst Gardens, Leeds, LS17 8BG
M0 SMF C. Smith, 21 Mill House Drive, Cheltenham, GL50 4RG
M0 SMG A. Booth, 16 Coronation Street, Wessington, Alfreton, DE55 6DX
M0 SMH S. Hassan, 69 Waltham Close, West Bridgford, Nottingham, NG2 6LD
M0 SMJ M. Seaward, 7 St. Olafs Road, Stratton, Bude, EX23 9AF
MW0 SML Sparks At The Shed, c/o A. Adams, 1 Nant Y Ffynnon, Letterston, Haverfordwest, SA62 5SX
M0 SMN Dr D. Richardson, 89A Bean Oak Road, Wokingham, RG40 1RJ
M0 SMP S. Peel, 21 Fairfield Avenue, Ormesby, Middlesbrough, TS7 9BB
M0 SMQ S. Mansfield, 545 Portswood Road, Southampton, SO17 3SA
M0 SMS W. Pinkhardt, 43 Cambrian Way, Calcot, Reading, RG31 7DD
M0 SMT S. Tasker, 94 Davenport Drive, Cleethorpes, DN35 9JR
MI0 SMV S. Mcveigh, 28 Waringfield Avenue, Moira, Craigavon, BT67 0FA
MI0 SMY S. Dallas, 101 Coagh Road Stewartstown, Dungannon, BT71 5JL
M0 SMZ O. Carp, Rothera Research Station, Stanley, Antarctica, FIQQ 1ZZ
MM0 SNA P. Gacek, 66C Church Street, Berwick upon Tweed, TD15 1DU
M0 SNB Secret Nuclear Bunker Contest Group, c/o Dr G. Smart, 30 Cornmills Road, Soham, CB7 5AT
M0 SND J. Popple, 11, Chapel Close, Waterbeach, Cambridge, CB25 9JW
MI0 SNG S. Gilmour, 14G Malcolm Road, Lurgan, Craigavon, BT66 8DF
M0 SNJ N. Jones, 8 Regent Court Belvedere Close, Guildford, GU2 9GA
MM0 SNK J. Dow, 58 Beatty Crescent, Kirkcaldy, KY1 2HS
MI0 SNY G. English, 15 Murrays Hollows, Ballyroney, Banbridge, BT32 5ES
M0 SNZ S. Dodd, Bradden Lane, Gaddesden Row, Hemel Hempstead, HP2 6JB
M0 SOA G. Mccourty, The Orchard, Eaton, Tarporley, CW6 9AJ
M0 SOC Second Class Operators Club (UK), c/o R. Pike, 6 Bridens Way Haddenham, Aylesbury, HP17 8DH
M0 SOE B. Macmillan, 27 Fiveways Rise, Deal, CT14 9QN
M0 SOL Solway DX Group, c/o C. Wolf, 35A Moorhouse Road, Carlisle, CA2 7LU
M0 SOO J. Barker, Pearl Bungalow, Killerby Cliff, Cayton Bay, Scarborough, YO11 3NR
M0 SOR S. Orchard, 30 Wilkes Court, Ipswich, IP5 2EQ
M0 SOT A. Cowan, 217 South Park Road, Wimbledon, London, SW19 8RY
M0 SOU J. Lovelock, Sea Spray, Lighthouse Road The Lizard, Helston, TR12 7NU
M0 SOV S. Riesenberg, 25 Westbere Road, London, NW2 3SP
M0 SOX G. Galliver, 29 Archery Fields, Odiham, Hook, RG29 1AE
M0 SPA Staffordshire ARC, c/o N. Briggs, 20 Broad Lane, Pelsall, Walsall, WS4 1AP
M0 SPB R. Evans, 21 Quilter Close, Bilston, WV14 9AX
M0 SPC C. Smith, 59 Moneyfly Road, Verwood, BH31 6BL
M0 SPD S. Davies, 10 Knutsford Green, Wirral, CH46 8TT
M0 SPH S. Hodkinson, 17 Thorn Well, Westhoughton, Bolton, BL5 2PJ
M0 SPJ P. Snook, 7 Sandhurst Avenue, Kwazulu Natal, South Africa, 3610
M0 SPK P. Susa, 3 Ainsdale Drive, Whitworth, Rochdale, OL12 8QB
MM0 SPL S. Ling, Leadburnlea Leadburn, West Linton, EH46 7BE
M0 SPN S. Netting, 39 Poulton Street, Swindon, SN2 1BH
M0 SPS A. Hutley, 90 Main Road, Crick, Northampton, NN6 7TX
M0 SPV A. Vitiello, 8 Pegasus Road, Leighton Buzzard, LU7 3NJ
M0 SPX Spixworth Scout Radio Group, c/o P. Burgess, 26 William Peck Road, Spixworth, Norwich, NR10 3QB
M0 SPZ Dr S. Pearce, 15 Hillfield Court Road, Gloucester, GL1 3QS
M0 SQC Polish ARC, c/o B. Niewiadomski, 41 The Crescent, Keresley End, Coventry, CV7 8LB
M0 SRA Simpson ARS, c/o I. Ridings, 25 Mond Road, Irlam, Manchester, M44 6QA
M0 SRB S. Britten, 10 Second Avenue, Wolverhampton, WV10 9PP
M0 SRJ R. Shenton, 14, Leighton Close, Uttoxeter, st148sj
MI0 SRM S. Mccormick, 8 New Close, Portavogie, Newtownards, BT22 1DZ
M0 SRN P. Holland, 2 Blythorpe, Hull, HU6 9HG
M0 SRO D. Coupe, 6 Berry Avenue, Kirkby-In-Ashfield, Nottingham, NG17 8GE
M0 SRP P. Skidmore, 36 Princes Drive, Harrow, HA1 1XH
MI0 SRR D. Poots, 18 Upper Quilly Road, Dromore, BT25 1NP
MM0 SRX Strathclyde 4x4 Response, c/o T. Kane, 40B Brisbane Street, Greenock, PA16 8NP
M0 SRZ S. Robottom-Scott, 73 St. Bernards Road, Solihull, B92 7DF
MW0 SSB I. Rowlands, 22 Maes William Williams Vc, Amlwch, LL68 9DS
M0 SSD G. Birkby, 8 Kestrel Drive, Dalton-in-Furness, LA15 8QA
M0 SSE J. Mossman, 12 Cheviot Crescent, Hadston, Morpeth, NE65 9SP
M0 SSF Canon P. Midwood, 4 Larch Crescent, Holt, NR25 6TU
MM0 SSG C. Haldane, 6A Earls Gate, Bothwell, Glasgow, G71 8BP
M0 SSH Dr S. Herman, Barbary House, California Lane, Bushey, WD23 1EX
M0 SSJ P. Dekkers, 21 Nodens Way, Lydney, GL15 5NP
M0 SSK K. Baker, 64 Pendle Drive, Basildon, SS14 3LZ
M0 SSM S. Mcmurtrie, 5 Hill Road, Carshalton, SM5 3RA

M0 SSN B. Woods, 28 Delph Drive, Burscough, Ormskirk, L40 5BE
M0 SSO M. Slater, 8 Flaxman Rise, Oldham, OL1 4QB
M0 SSP Waterlooville ARC, c/o R. Shillabeer, 29 Newlease Road, Waterlooville, PO7 7BX
M0 SSR S. Stewart, 2 Broadlands, Clinton Way, Fairlight, TN35 4DL
M0 SST South Staffordshire AR Tutors Grp, c/o R. Finch, 12 Simcox Street, Hednesford, Cannock, WS12 1BG
M0 SSV S. Vickers, 22 Thistle Green, Birmingham, B38 9TT
M0 SSW Silcoates School AR & Elec.Club, c/o N. Wears, 25 Topcliffe Mews, Morley, Morley, LS27 8UL
M0 SSX Sussex 4x4 Response, c/o D. Green, St. Annes Poundfield Road, Crowborough, TN6 2BG
M0 SSY R. Moss, 27 Ravens Close, Bignall End, Stoke-on-Trent, ST7 8QE
M0 STA R. Stafford, 133 Essex Road, Stamford, PE9 1LA
M0 STC S. Cheal, 12 Heather Lane, Worthing, BN13 3BU
M0 STF S. Binns, 174 Enfield Chase, Guisborough, TS14 7LQ
M0 STI D. Smith, Heath Farm, Heath Road, Woolpit, Bury St Edmunds, IP30 9RL
M0 STJ C. Cherry, 12 Scarisbrick New Road, Southport, PR8 6PY
M0 STK P. Roberts, 121 Hartshill Road, Stoke-on-Trent, ST4 7LU
M0 STL A. Palmer, 18 Windsor Avenue, Great Yarmouth, NR30 4EA
M0 STN S. Neale, 48 Five Acres Fold, Northampton, NN4 8TQ
M0 STS G. Sowden, The Grange Lodge, Rodley Lane, Calverley, Pudsey, LS28 5QH
M0 STT S. Gordon, 6 Aspinall Grove, Hailsham, BN27 3GP
MM0 STU S. Macpherson, 41 Mull Avenue, Port Glasgow, PA14 6DP
M0 STV S. Ridgeon, 7 Southlands, Haxby, York, YO32 2PB
M0 SUD S. Griffiths, 37 Stourton Close, Knowle, Solihull, B93 9NP
M0 SUF S. Batley, 2 Boulge Road, Hasketon, Woodbridge, IP13 6LA
M0 SUG D. Eastlake, 148 Pursey Drive, Bradley Stoke, Bristol, BS32 8DP
M0 SUI R. Isnenghi, Flat 24, Weller Court, Melvill Road, Falmouth, TR11 4ES
M0 SUN C. Tate, 11A Nether Lea, Cranage, Crewe, CW4 8HX
M0 SUR St George's Academy ARC (Sga-ARC), c/o P. Dickson, 49 Signal Road, Grantham, NG31 9BL
MM0 SUS I. Macdonald, The Cottage, High Craigton, Glasgow, G62 7HA
M0 SUU W. Malcolm-Brown, Flat 11, Chiltern Court, Harpenden, AL5 5LY
M0 SUY D. Valaris, 18 Charnwood Gardens, Gateshead, NE9 5SB
M0 SUZ S. Coombes, 33 Clarence Park Road, Bournemouth, BH7 6LF
M0 SVA T. Papadopoulos, 77 Cottrell Road, Bristol, BS5 6TN
M0 SVB S. Bell, 1 Cherwell Road, Aylesbury, HP21 8TW
MM0 SVE S. Shaw, 2 Highfield, Dalry, KA24 4HP
M0 SVH Dr S. Hill, 36 The Woodlands, Market Harborough, LE16 7BW
M0 SVN S. Farnsworth, 26 Burrows Close, Headington, Oxford, OX3 8AN
M0 SVP G. Mylonas, 68 Lancaster Gate, Cambourne, CB236AT
M0 SVR S. Ring, 35 Sturmer Close Yate, Bristol, BS37 5UR
M0 SVT Dr A. Dickson, 1 Roebuck Drive Baldwins Gate, Newcastle-under-Lyme, ST5 5FE
M0 SVV S. Wade, 42 Beauclerk Green, Winchfield, Hook, RG27 8BF
M0 SVX A. Thorpe, 8 Syke Avenue, Tingley, Wakefield, WF3 1LU
MW0 SWB A. Jones, 69 Hendre Gwilym, Tonypandy, CF40 1HF
M0 SWC D. Brough, 38 Tynedale Avenue, Crewe, CW2 7NY
M0 SWE M. Sweeney, 3 Orchard Cottages, Asenby, Thirsk, YO7 3QW
M0 SWF S. Fry, Mount Pleasant Farm, Thorpe Fendykes, Skegness, PE24 4QR
M0 SWH S. Heard, 29 Grange Farm Road, Yatton, Bristol, BS49 4RB
M0 SWI R. Dixon, Coach House, Chapel Lane, Ellel, Lancaster, LA2 0PN
M0 SWL B. Bosson, 80 White Horse Road, Marlborough, SN8 2FE
MW0 SWN P. Barnes, 95 Marcroft Road, Port Tennant, Swansea, SA1 8PN
M0 SWO A. Sword, Plevna 100 Eaton Road, Norwich, NR4 6PS
MW0 SWR G. Waters, 47 Close Yr Eryr, Bridgend, CF35 6HE
M0 SWT M. Colpman, 20 Rochford Road, Basingstoke, RG21 7TQ
M0 SWV B. Elms-Lester, Ferndale House Kerry'S Gate, Hereford, HR2 0AH
M0 SWZ I. Swindells, 69 Danby Close, Newton Moor, Hyde, SK14 4AF
M0 SXA Essex Ham, c/o P. Sipple, 52 Fillebrook Avenue, Leigh-on-Sea, SS9 3NT
M0 SXE Downland Radio Group, c/o G. Keegan, 12 Allington Road, Newick, Lewes, BN8 4NA
M0 SXH S. Hunter, 9 Gelt Burn, Didcot, OX11 7TZ
M0 SXM S. Morris, 23 De Courtenai Close, Bournemouth, BH11 9PG
M0 SXN S. Neale, 43 Crompton Road Pleasley, Mansfield, NG19 7RG
M0 SYG A. Sygerycz, 75 O'Brien Road, Cheltenham, GL51 0UP
M0 SYJ P. Krzeminski, Flat 46, Polden House, Bristol, BS3 4LG
M0 SYL S. Krolak, 179 Beatrice Street, Swindon, SN2 1BD
M0 SYM S. Ludlam, Lower Wolves, Bosham Hoe, Bosham, Chichester, PO18 8EU
MI0 SYN J. Bradshaw, 51 Albany Drive, Carrickfergus, BT38 8BF
M0 SYS S. Strange, 94 Digby Avenue, Nottingham, NG3 6DY
M0 SYW S. Ward, Wellbeck, Wheel Road, Alpington., Norwich, NR14 7NH
M0 SYY C. Gibson, 3 Conway Drive, Billinge, Wigan, WN5 7LH
MM0 SZC S. Gohl, 40 Tantallon Gardens, Bellsquarry, Livingston, EH54 9AT
M0 SZD S. Denman, 12 Dyke Vale Road, Sheffield, S12 4ER
M0 SZL Z. Szot, 45 Ealing Park Gardens, London, W5 4EX
M0 SZQ S. Jones, Marvin House, Ryhill Pits Lane, Cold Hiendley, Wakefield, WF4 2DU
MW0 SZR A. Lamb, 30 Dale Road, Queensferry, Deeside, CH5 1XE
M0 TAA E. Slevin, Woodcock Hall, Cobbs Brow Lane, Newburgh, WN8 7NB
M0 TAB A. Brotherhood, 5 Longcliffe Road Shepshed, Loughborough, LE12 9LW
M0 TAD B. Catchpoole, 8 Buckland Avenue, Basingstoke, RG22 6JL
MW0 TAF E. Brookes, 40 Llancayo Street, Bargod, Bargoed, CF81 8TG
M0 TAJ T. Kemp, 30 Tawny Sedge, King's Lynn, PE30 3PW
M0 TAK T. Cooper, Flat 6, Smiths Court, 73 East Borough, Wimborne, BH21 1PJ
M0 TAL C. Travis, 4 Kingsdale, Worksop, S81 0XJ
M0 TAM M. Cox, Havenview Stables, Main Road, Havenstreet, Ryde, PO33 4DR
M0 TAN T. Nichols, 12 Ivy Grove, Shipley, BD18 4JZ
M0 TAO Dr O. Bock, Okenstr. 34, Jena, Germany, D-07745
M0 TAP W. Cooper, 20 Staple Close, Waterlooville, PO7 6AH
M0 TAQ F. Clements, 40 Ellison Fold Terrace, Darwen, BB3 3EB
M0 TAR T. Balls, Rostan 99 Front Road, Murrow, Wisbech, PE13 4JQ
M0 TAT F. Felix, Flat 41, King Edward Court, Wembley, HA9 7DQ
M0 TAV V. Hopkins, 6 Daimler Road, Coventry, CV6 3GD
M0 TAW T. Woodhouse, The Old Granary, 12 Limekiln Lane, Newport, TF10 9EZ
M0 TAX E. Underhill, 61 Goldthorne Avenue, Sheldon, Birmingham, B26 3LA
M0 TAY A. Ayres, Bryn Hyffryd, Phocle Green, Ross-on-Wye, HR9 7TW
M0 TAZ D. Cutts, 38 Berkeley Drive, Hornchurch, RM11 3PY
M0 TBA A. Baker, 6 Bayliss Avenue, Wolverhampton, WV4 6NW
MW0 TBB C. Morris, 17 Percy Road, Wrexham, LL13 7EA
MI0 TBD A. Kelly, 19 Union Street Mews, Coleraine, BT52 1EN
MI0 TBE E. Hill, 24 Whitehouse Park, Newtownabbey, BT37 9SQ
M0 TBG Team Thunderbox, c/o C. Moulding, 28 Queens Avenue, Highworth, Swindon, SN6 7BA
MM0 TBH J. Kelly, 41 Glenshee Street, Glasgow, G31 4RT
M0 TBI S. Smith, 12 Stoneleigh Avenue, Hordle, SO41 0GS
M0 TBJ T. Buck, 6 Lynn Road, Terrington St. Clement, King's Lynn, PE34 4JX
M0 TBK E. Cree, 24 Old Lincoln Road, Caythorpe, Grantham, NG32 3EJ
MI0 TBN S. Donnelly, 14 Derryloste Road, Derrytrasna, Craigavon, BT66 6PS
M0 TBQ D. Nicholls, 62 Queen Elizabeth Way, Telford, TF3 2JW
M0 TBR R. Thorogood, 4 Deerhurst Close, Calcot, Reading, RG31 7RX
M0 TBS T. Tiesdell-Smith, 4 Godwin Close, Epsom, KT19 9LD
MI0 TBV T. Mckee, 4 Earlford Heights, Newtownabbey, BT36 5WZ
M0 TBW R. East, 6 Ashley Road, Worcester, WR5 3AY
MM0 TBY S. Turnbull, 15 Woodruff Gait, Dunfermline, KY12 0NL
M0 TBZ Dr C. Cowen, Rosita, White Street Green, Sudbury, CO10 5JN
M0 TC T. Palmer, Edison House, Bow Street, Great Ellingham, NR17 1JB
M0 TCA T. Codner-Armstrong, 22 Thoresby Road, Rainworth, Mansfield, NG21 0DS
M0 TCB D. Howarth, 32 Cotswold Drive Rothwell, Leeds, LS26 0QZ
M0 TCC Dr T. Choy, 39 Netherton Road, Manchester, M14 7FN
M0 TCD A. Allen, Milverton, Mill Road West Chiltington, Pulborough, RH20 2PZ
M0 TCE Dr C. Eaglen, 46 Sark Close, Hounslow, TW5 0PZ
M0 TCF L. Allen, 481 Topsham Road, Exeter, EX2 7AQ
MW0 TCJ T. Jones, 4 Llys Y Nant Glais, Swansea, SA7 9Jb
M0 TCL D. Mort, 7 Sheldon Avenue, Congleton, CW12 3LD
M0 TCM Thorpe Camp Museum Radio Group, c/o A. Nightingale, 42 Spilsby Road, Horncastle, LN9 6AW
M0 TCN C. Lyne, 4 Bridge Close, Catterick Garrison, DL9 4PG
M0 TCO T. Corcoran, 191 Queensway, Rochdale, OL11 2NA
MM0 TCP K. Brown, 33 Windsor Gardens, Largs, KA30 9DN
MM0 TCQ T. Campbell, 10 Barra Gardens, Old Kilpatrick, Glasgow, G60 5HR
M0 TCR T. Rozier, 124 Deansfield Road, Wolverhampton, WV1 2LD
M0 TCT T. Collins, 11 Joseph Gardens, Silver End, Witham, CM8 3SN
MM0 TDB D. Gartshore, 85 Springhill Street, Douglas, Lanark, ML11 0NZ
M0 TDC R. Stevenson, 97 Queen Street, Crewe, CW1 4AL
M0 TDD M. Xian, 39 Belson Road, London, SE18 5PU
M0 TDE Triode ARG, c/o N. Knapton, 4 Crabmill Lane, Easingwold, York, YO61 3DE
MW0 TDF W. Welch, Kenilworth, School Lane, Gobowen, Oswestry, SY11 3LD
M0 TDG T. Grant, 51A Hursley Road, Eastleigh, SO53 2FS
M0 TDK A. Tyrwhitt-Drake, Holly Cottage, Church Lane, Beccles, NR34 0AU
M0 TDM R. Hydes, 60 Handsworth Grange Road, Sheffield, S13 9HH
M0 TDP A. Pickett, 4 Trembel Road, Mullion, Helston, TR12 7DY
MW0 TDQ J. Grzywaczewski, 16 Dyffryn Road, Port Talbot, Sa132ug
MM0 TDS D. Scott, Farewell, Arnage, Auchnagatt, Ellon, AB41 8UW
MW0 TDV R. Smith, Hendafarn, Sarnau, Llanymynech, SY22 6QJ
M0 TDW B. Woodroffe, 2 Little Mead, Shalbourne, Marlborough, SN8 3QB
M0 TDY A. Askam, 8 The Pastures, Weston-On-Trent, Derby, DE72 2DQ
MW0 TDZ T. Clapp, Crunns Farm. Coxhill, Narberth, SA67 8EH
M0 TEA A. Goddard, 50 Ardmore Walk, Manchester, M22 5QG
M0 TEB M. Bell, 36 Schneider Road, Barrow-in-Furness, LA14 5DW
M0 TEF A. Smith, 101 Chaucer Drive, Lincoln, LN2 4LT
M0 TEG D. Horner, 21 Ainsworth Road, Little Lever, Bolton, BL3 1RG
MM0 TEI A. Wright, 17A James Court, 493 Lawnmarket, Edinburgh, eh1 2pb
M0 TEK E. Moore, 44 Bridge Street, Oxford, OX2 0BB
M0 TEN E. Williams, 15 Tenth St., Peterlee, SR8 4NE
M0 TER B. Ashcroft, 16 Edge Lane, Crosby, Liverpool, L23 9XE
M0 TES C. Brown, Town End House, Ulverston Road, Gleaston, Ulverston, LA12 0PZ
M0 TET A. Ford, 5 Prenede, Roches, France, 23270
M0 TEX R. Rushlow, 94 Tennyson Street, Guiseley, Leeds, LS20 9LW
M0 TEZ T. Mullaney, 8 Westerham Close, Macclesfield, SK10 3BG
M0 TFB B. Titmus, 68 Hobart Road, Cambridge, CB1 3PT
M0 TFC Thanet Radio and Electronics Club, c/o P. Kirkden, 22 Leas Green, Broadstairs, CT10 2PL
M0 TFF R. Cichocki, 12 Crossland Crescent, Wolverhampton, WV6 9JY
M0 TFH T. Hull, 12 Durley Road, Gosport, PO12 4RT
MI0 TFK R. Vage, 80 Chinauley Park, Banbridge, BT32 4JL
M0 TFN T. Nolan, 2 Shore Road, Cowes, PO31 8LB
M0 TFO R. Styles, 52 Vernham Grove, Bath, BA2 2TB
M0 TFS T. Smith, 43 Hall Street, St Helens, WA94XN
MM0 TFU I. Macalister, 33 King Street Crosshill, Maybole, KA19 7RE
M0 TFX T. Fisk, 2 Hall Farm Cottage, Caston Road, Caston, Attleborough, NR17 1BW
M0 TFY D. Butler, Church Cottage, Church Road, Badminton, GL9 1HT
MM0 TGB T. Brown, 11 Approach Row, East Wemyss, Kirkcaldy, KY1 4LB
M0 TGC J. Hay, Hudson House, Fremington Road, Seaton, EX12 2HX
M0 TGE T. Gentry, 175 Huddersfield Road, Halifax, HX3 0AS
M0 TGF M. Leach, 64 Grove Street, Wantage, OX12 7BG
MM0 TGG G. Jamieson, 6 Maryville Park, Aberdeen, AB15 6DU
MI0 TGL R. Greer, 11, Mullaghcarton Road, Lisburn, BT28 2TE
M0 TGM D. Pask, Apartment 403, 1314 Tower Road, Halifax, Canada, NS B3H 4S7
M0 TGN D. Trudgian, 18 Hart Close, Wootton Bassett, Swindon, SN4 7FN
M0 TGS Altrincham Grammar School For Boys, c/o G. Binns, 21, Rydal Close, Winsford, CW7 2SE
M0 TGT S. Faulkner, Mount Pleasant, Elkstones, Buxton, SK17 0LU
M0 TGV G. Cooke, 12 Marcus Road, Dartford, DA1 3JX
M0 TGW M. Rigby, 75 Manchester Road, Deepcar, Sheffield, S36 2QX
M0 TGX T. Green, 35 Park Road, Allington, Grantham, NG32 2EB
M0 TGY T. Guy, 16 Cogdeane Road, Poole, BH17 9AS
M0 THA T. Hurren, 4 Grove Bungalows, Upper Street, Horning, Norwich, NR12 8NF
M0 THB T. Barratt, 17 Main Road Collyweston, Stamford, PE9 3PF
MM0 THE A. Lang, 202 Devonside Road, Carmichael, Biggar, ML12 6PQ
M0 THJ A. Howell-Jones, 11 Staffick Close, Kenton, Exeter, EX6 8NS
M0 THM T. Mcconnell, 51 Langney Road, Eastbourne, BN21 3QD
M0 THN R. Blane, Redfield, Buckingham Road, Buckingham, MK18 3LZ
M0 THO A. Boato, Via A Diaz 20, Marcon, Venezia, Italy, 30020

MJ0 THP M. Thorpe, Dolphin Cottage, Union Road, Grouville, Jersey, JE3 9ER
M0 THT Laser ATC Rac (South), c/o T. Toon, 9 Boundstone Lane, Sompting, Lancing, BN15 9QL
M0 THY H. Tang, Flat 31, 74 Arlington Avenue, London, N1 7AY
MM0 TIE R. Adamson, 6 Camdean Crescent, Rosyth, Dunfermline, KY11 2TJ
M0 TIF J. Housego, 16 Ligo Avenue, Stoke Mandeville, HP225TX
M0 TIK C. Pascoe, Treleven, Primrose Hill, Goldsithney, Penzance, TR20 9JR
M0 TIL Coalhouse Fort, c/o J. Parker, 76 Elm Road, Grays, RM17 6LD
M0 TIN D. Le Grove, Apartment 3, Beechwood, Ilkley, LS29 8AH
MI0 TIP W. Thompson, 25 Darby Road, Carrickfergus, BT38 7XU
MM0 TIR M. Apaza Machaca, 12 Donald Street, Dunfermline, KY12 0BY
M0 TIU A. Beaumont, 112 Whittington Road, Hutton, Brentwood, CM13 1JZ
M0 TIW A. Thornton, 78 Wellington Road, Ryde, PO33 3QJ
M0 TIX R. Cowles, Bonnie Rock, 76 Fordham Road, Ely, CB7 5AL
M0 TIZ A. Tyrrell, Flat 1, 61 Vicarage Road, Eastbourne, BN20 8AH
M0 TJB T. Barnes, Flat 38, Mill Court, Edinburgh Gate, Harlow, CM20 2JG
M0 TJC J. Hill, 10 Elizabeth Avenue, Downham Market, PE38 9EQ
MW0 TJD T. Davies, 58A Ynyswen Road, Treorchy, CF42 6ED
M0 TJJ T. Jinkerson, 104 Foxcote, Finchampstead, Wokingham, RG40 3PE
M0 TJL T. Leavold, 129 Aylsham Road, Norwich, NR3 2AD
MI0 TJM T. Mulholland, 215 Finaghy Road North, Belfast, BT11 9ED
M0 TJN T. Newton, 4 Manor Close, Bradford Abbas, Sherborne, DT9 6RN
MM0 TJR T. Thorne, Top Flat, 26 Mary Elmslie Court, Aberdeen, AB24 5BE
MW0 TJS T. Scott, 44 Croft Road Broad Haven, Haverfordwest, SA62 3HY
MM0 TJT Jaggy Thistles, c/o W. Findlay, 46 Rowallan Drive, Kilmarnock, KA3 1TU
M0 TJU E. Duffield, 92 Crosby Street, Stockport, SK2 6SP
M0 TJV C. Vernon, 29 Alice St., Deane, Bolton, BL3 5PJ
M0 TJW T. Beardwood, Flat 9, Alma House, Ripon, HG4 1NG
M0 TJX T. Humphreys, 93 Cornwall Crescent Yate, Bristol, BS37 7RU
M0 TKA T. Kay, 64 Cowcliffe Hill Road, Huddersfield, HD2 2PE
M0 TKC T. Canning, Whitegates, Mayfield Avenue New Haw, Addlestone, KT15 3AG
M0 TKD K. Raistrick, 2 Greenacres Grove, Shelf, Halifax, HX3 7RN
MM0 TKE Prof. T. Kerby, 1 St. Mark'S Lane, Edinburgh, EH15 2PX
M0 TKF R. Nicholson, 27 Bishopton Way, Hexham, NE46 2LR
M0 TKG Dr T. Bishop, 2 Jennings Street, Stockport, SK3 9BX
M0 TKM M. Gosi, 49 Elms Drive, Marston, Oxford, OX3 0NW
M0 TKR C. Barker, 11 Long Meadows, Chorley, PR7 2YA
M0 TKS Dr T. Kyriacou, 54 Sutton Avenue, Silverdale, Newcastle-under-Lyme, ST5 6TB
M0 TKT R. Bradshaw, 272 Councillor Lane, Cheadle Hulme, Cheadle, SK8 5PN
M0 TKU D. Carter, 3 Queens Road, Brentwood, CM14 4HE
M0 TKW T. Kelly, 50 Ivanhoe Road, Herne Bay, CT6 6EQ
M0 TKX A. Sood, Parima, Sewardstone Road, London, E4 7RA
M0 TLC R. Ainsworth, 181 Carlton Road, Boston, PE21 8NG
MI0 TLF T. Flanagan, 18 Hunters Park, Bellaghy, Magherafelt, BT45 8JE
MI0 TLG R. Greer, 11, Mullaghcarton Road, Lisburn, BT28 2TE
M0 TLJ R. Mcwilliam, 3 Fountains Close, Riccall, York, YO19 6QN
M0 TLL B. Smith, 24 Oakwood Park, Leeds, LS8 2PJ
M0 TLM I. Williams, 36 Telford Road, Tamworth, B79 8EY
M0 TLN S. Moissejev, 41 Queens Road, Caversham, Reading, RG4 8DN
M0 TLO R. Hunter, 3 Sandy Way, Croyde, Braunton, EX33 1PP
M0 TLR G. Taylor, Southlands, Church Lane, Great Holland, Frinton-on-Sea, CO13 0JS
M0 TLX D. Burdsall, 37 Fulmar Walk, Whitburn, Sunderland, SR6 7BW
M0 TLY M. Casey, 7 Cobham Avenue, Manchester, M40 5QW
M0 TMA T. Moncaster, 24 Minster Yard, Lincoln, LN2 1PY
M0 TMB R. Cook, 6 Aster Road, Ipswich, IP2 0NQ
M0 TMC E. Newby, 22 Acton Road, Liverpool, L32 0TT
M0 TMD H. Melhuish, 22 Mayflower Close, Glossop, SK13 8UD
M0 TMF A. Fullwood, 16 Hollands Place, Walsall, WS3 3AU
MM0 TMG K. Cussick, 15A Finlow Terrace, Dundee, DD4 9ND
MW0 TMH T. Mitchell, 9 Rhiw Grange, Colwyn Bay, LL29 7TT
MW0 TMI D. Willis, 51 Fforchaman Road, Cwmaman, Aberdare, CF44 6NG
MW0 TMJ T. James, Penrallt, Mountain, Holyhead, LL65 1YR
M0 TMM M. Elliott, 60A Forest Street, Shepshed, Loughborough, LE12 9DA
M0 TMN T. Nguyen, 9 Green Street, Cambridge, CB2 3JU
M0 TMO K. Chadwick, 17 Nettlebed Nursery, New Road, Shaftesbury, SP7 8QS
MI0 TMP T. Mcelwee, Orchard Bank, 1A Dunover Road, Ballywalter, Newtownards, BT22 2LE
M0 TMS T. Schwabe, 112 Clarkson Court, Hatfield, AL10 9GW
M0 TMT R. Aldridge, 37 Vincent Road, Luton, LU4 9AN
M0 TMX D. Mcglone, 32 Shipley Mill Close, Kingsnorth, Ashford, TN23 3NR
MM0 TMZ A. Miles, 9 Buchanan Drive, Lenzie, Kirkintilloch, Glasgow, G66 5HS
M0 TNB M. Jakubowski, 75 Ashcombe Road, London, SW19 8JP
M0 TNC A. Burton, 12 Munden Grove, Watford, WD24 7EE
M0 TNE T. Newman, 10 Dereham Road, Garvestone, Norwich, NR9 4AD
M0 TNF G. Barnard, Red Ridges Ghyll Road, Crowborough, TN6 1SU
M0 TNG S. Adaway, 20 Foundry Street, Barnsley, S70 1PL
M0 TNL V. Behal, 21 Bromley Road, Walthamstow, London, E17 4PR
M0 TNT A. Roberts, Chy Kerenza, Parc Morrep, Penzance, TR20 9TE
M0 TNV M. Clough, 8 Skeldyke Road, Kirton, Boston, PE20 1LR
M0 TNX K. Haworth, 11 Petersfield Close, Bootle, L30 1SG
MM0 TOB T. Burnett, Inner Lodge Achnacloich, Oban, PA37 1PR
M0 TOF J. Morgan, Glas Y Dorlan, Pontrhydfendigaid, Ystrad Meurig, SY25 6EJ
M0 TOG D. White, Woodpeckers, Top Green, Romsey, SO51 0JP
M0 TOL Tolmers Scout Campsite, c/o A. Rixon, 17 Brimmers Way, Aylesbury, HP19 7HR
M0 TOM T. Wilcox, 17 Westminster Avenue, Royton, Oldham, OL2 5XY
M0 TOP A. Topsfield, Wild Willow Cottage, Hancock Lane, Truro, TR2 5DD
M0 TOR J. Dearden, 7 Wadworth Street, Denaby Main, Doncaster, DN12 4EN
M0 TOZ G. Agostinelli, 50 The Furrows, Southam, CV471TA
M0 TPA A. Patrick, 2 Beacon Grange, Malvern, WR14 3EU
MM0 TPD J. Watson, 64 Anstruther Street Law, Carluke, ML8 5JG
M0 TPE T. Pearce, 101 Primrose Hill, Widmer End, High Wycombe, HP156NT
M0 TPG A. Gravell, 21 Wickridge Close, Stroud, GL5 1ST
M0 TPH G. Emsden, Flat 47, Cedar Court, London, N10 1EG
M0 TPJ T. Mallaband, 29 Ferndale Road, Burgess Hill, RH15 0HB
M0 TPL A. Capon, 1 Windermere Way, North Common, BS305XN
M0 TPT D. Millward, 77A Meadowcroft, St Albans, AL1 1UG
M0 TPW T. Winyard, 48 Windsor Drive, Yate, Bristol, BS37 5DY
M0 TPY T. Pollett, 10 Bridport Road, Poole, BH12 4BS
MM0 TQH R. Hay, Roddach Cottage East, Cummingston, Burghead, Elgin, IV30 5XY
M0 TQV R. Tuckett, 89 Hillbrook Road, Tooting, London, SW17 8SF
M0 TRB T. Bowden, Carmel, Swallowcliffe, Salisbury, SP3 5PW
MI0 TRC P. Clarke, 26 Derryhale Lane, Portadown, Craigavon, BT62 4HL
M0 TRE Dr A. Kicman, 15 Leaden Close, Leaden Roding, Dunmow, CM6 1SD
M0 TRF T. Knox, 21 St. Annes Avenue, Bournemouth, BH6 3JR
M0 TRK R. Tarling, Moonrakers, Ashley, Box, Corsham, SN13 8AN
M0 TRN T. Horsten, Kastelsvej 4, 2.Tv, Copenhagen E, Denmark, 2100
MW0 TRO A. Roberts, 14 Beech Road, Monmouth, NP25 5DA
M0 TRP A. Pursglove, 78 Alfreton Road, Westhouses, Alfreton, DE55 5AJ
MM0 TRS S. Troscheit, 20 James Street, St Andrews, KY16 8YA
M0 TRU A. Hartwell, 41 Orchard Grove, Brixham, TQ5 9RH
M0 TRV T. Hammett, 9 Coral Close, Aughton, Sheffield, S26 3RB
M0 TRW T. Wormald, 12 Church Lane, Owermoigne, Dorchester, DT2 8HS
M0 TRX Dr M. Cooper, Apartment 33, St. Peters Court, 2 St. Peters Street, Worcester, WR1 2PJ
M0 TRY R. Barnes, 1 Moira Close, Chaddesden, Derby, DE21 4RL
M0 TRZ S. Jordan, 8 Averham Close, Swadlincote, DE11 9SG
M0 TSA I. Macfarlane, 70 Ashby Drive, Rushden, NN10 9HH
MM0 TSB J. Morris, 42B Church Street, Borve, Isle of Lewis, HS2 0RT
M0 TSD S. Smith, 103 Comberford Road, Tamworth, B79 8PE
M0 TSM M. Bull, Sunrise, Ram Lane, Norwich, NR15 2DG
M0 TSN M. Lee, 46 Little Lane, Huthwaite, Sutton-in-Ashfield, NG17 2RA
M0 TSW T. Walker, 11 Banburies Close, Bletchley, Milton Keynes, MK3 6JP
M0 TSZ T. Skorzewski, 40 Hedges Way, Luton, LU4 9FD
M0 TTB A. Bright, 86 Fourth Avenue, Watford, WD25 9QQ
M0 TTE S. Fairbourn, 17 Perry'S Lane, Wroughton, Swindon, SN4 9AX
MW0 TTF D. D'Mellow, 2 Twynpandy Pontrhydyfen, Port Talbot, SA12 9TW
M0 TTG Tall Trees Contest Group, c/o B. Gale, Tall Trees Farm, Noah'S Ark Lane, Great Warford, WA16 7AX
M0 TTH T. Haley, 3 Akeman Rise, Ramsden, Chipping Norton, OX7 3BJ
M0 TTI Dr S. White, Upton Farm, Upper Strode, Bristol, BS40 8BG
MW0 TTK M. Buxton, 25 Pen Y Bryn, Sychdyn, Mold, CH7 6EE
M0 TTL A. Dickinson, 18 Cullingworth Avenue, Hull, HU6 7DD
M0 TTN R. Colman-Whaley, 37 Suters Drive, Taverham, Norwich, NR8 6UU
M0 TTO G. Grant, 15 Watson Close, Rugeley, WS15 2PE
MW0 TTR Aberdare ARS, c/o B. Werrell, 26 Glynhafod Street, Cwmaman, Aberdare, CF44 6LD
MW0 TTU M. Evans, The Brae, Coed-Cae-Ddu Road, Blackwood, NP12 2DA
M0 TTX G. Watkins, 21 Comberton Avenue, Kidderminster, DY10 3EG
M0 TTY D. Spence, 30 Chestnut Drive, Shirebrook, Mansfield, NG20 8NH
MI0 TUB D. Given, 15 Middle Road, Lisburn, BT27 6UU
M0 TUI H. Drysdale, Waunycaerau Ffynonn Gynnydd, Hereford, HR3 5ND
M0 TUK P. Ray, 136 Haselbury Road, London, N18 1QD
M0 TUM R. Robinson, 4 Limetree Court, Taverham, Norwich, NR8 6QY
M0 TUN G. Bergeret, 20 Rue Labrouste, Paris, France, 75015
M0 TUR D. Biyikli, Basement, 300 Portobello Road, London, W10 5TA
M0 TUT S. Prescott, 210 Inver Road, Blackpool, FY2 0LW
M0 TUV A. Dingwall, 48 Village Farm Caravan Site, Bilton Lane, Harrogate, HG1 4DL
M0 TUW R. Harris, 7 Fosse Lane, Shepton Mallet, BA4 4PS
M0 TUX B. Sutton, 25 Mead Road, Folkestone, CT19 5QY
M0 TVA C. Beresford, 13 Chaseside Avenue, Twyford, Reading, RG10 9BT
M0 TVG M. Shurley, 43 Charles Close, Wroxham, Norwich, NR12 8TU
M0 TVL C. Mackay, 665A Edenfield Road, Rochdale, OL11 5XE
M0 TVR T. Parker, 100 Horsebridge Hill, Newport, PO30 5TL
M0 TVS A. Travis, 2 Home Farm Cottage, Ossington, Newark, NG23 6LH
M0 TVT K. Wilson, 26 Mill Field, Sutton, Ely, CB6 2QB
M0 TVU P. Swingewood, 9 Goodall Grove, Great Barr, Birmingham, B43 7PQ
M0 TVV M. Hillman, Flat 5, 32 South Terrace, Littlehampton, BN17 5NU
M0 TVX R. Taylor, 6 Appleton Drive, Belper, DE56 1FQ
M0 TWC Travelling Wave Contest Group, c/o K. Haywood, 6 Lydney Road, Urmston, Manchester, M41 8RN
M0 TWG P. Hallewell, 32 Shaldon Grove, Aston, Sheffield, S26 2DH
M0 TWJ W. Twemlow, Flat 6, 27 Marmion Road, Liverpool, L17 8TT
MM0 TWK C. Hall, 3 Academy Street, Tain, IV19 1ED
M0 TWM J. Nethercott, 15 French Gardens, Cobham, KT11 2AJ
M0 TWO P. Dunn, 13 Stanton Avenue, Newsham Farm Estate, Blyth, NE24 4PL
M0 TWS T. Wood, 44 Wincobank Lane, Sheffield, S4 8AA
M0 TWW T. Larman, 861B London Road, Westcliff on Sea, SS0 9SZ
MM0 TWX Dr P. Calvi-Parisetti, 1 Aytoun Road, Glasgow, G41 5RL
M0 TXA D. Harvey, 5 Tithe Barn Drive Bray, Maidenhead, SL6 2DF
M0 TXB C. Howell, 47 Birch Park Coalway, Coleford, GL16 7RU
M0 TXD G. Rowberry, 32 Tiree Avenue, Worcester, WR5 3UA
M0 TXH H. Gruen, Allfrey House, Herstmonceux, Hailsham, BN27 4RS
M0 TXK M. Fletcher, 7 Richard Street, Bacup, OL113 8QJ
M0 TXL P. Dunnicliffe, 19 Woodland Road, Chelmsford, CM1 2AT
MI0 TXM A. Mcgarvey, 66A Scaddy Road, Downpatrick, BT30 9BS
MM0 TXN S. Moore, 35 Niddrie House Park, Edinburgh, EH16 4UH
MM0 TXO A. Reid, Johnston Farm, Leslie, Insch, AB52 6PD
M0 TXP P. Cassells, 49 Dodds Lane, Maghull, Liverpool, L31 0BD
M0 TXS M. Townsend, 25 Barton Road, Bedford, MK42 0NA
M0 TXX G. Acton, 39 Craig Road, Macclesfield, SK11 7YH
M0 TYG D. Moore, Camara, 379 Main Road, Harwich, CO12 4DW
M0 TYH T. Hill, 14 Hunters Mead, Motcombe, Shaftesbury, SP7 9QG
M0 TYK J. Shephard, 19 Duffryn, Telford, TF3 2BU
M0 TYN G. Cockburn, 20 Hexham Avenue, Hebburn, NE31 2HN
MM0 TYR C. Taylor, 47 Seaview, Knock, Isle of Lewis, HS2 0PD
M0 TYW W. Hibberd, 169 Highbury Grove, Cosham, Portsmouth, PO6 2RL
M0 TZD A. Maiden, 79 Green End Road, Manchester, M19 1LE
M0 TZM T. Menzies, 4 Meadow Road, Muxton, Telford, TF2 8JH
M0 TZO P. Gibson, 60 Raglan Road, Bromley, BR2 9NW
M0 TZR P. Haygarth, 5 Forth Close, Peterlee, SR8 1DG
M0 TZT S. Emmett, Middle Farm, East Side, North Littleton, Evesham, WR11 8QW
M0 TZY S. Crabb, 1 Council Houses, Hall Lane, Crostwick, Norwich, NR12 7BB
M0 TZZ P. Moore, 24 Plough Road, Dormansland, Lingfield, RH7 6PS

MW0 UAA D. Bowen, 25 Maendu Terrace, Brecon, LD3 9HH
M0 UAC D. Carter, Bungalow 7, Higher Ingsdon Quarry, Liverton, Newton Abbot, TQ12 6JA
M0 UAR UK Amateur Radio Discord, c/o D. Keene, Firemark Cottage West Street Odiham, Hook, RG29 1NT
M0 UAS D. Williams, 2 Tyning Road, Peasedown St. John, Bath, BA2 8HT
M0 UAT I. Marsh, 56B Oliver Crescent, Farningham, Dartford, DA4 0BE
M0 UAV T. Ward, Flat 26, Bassett Court, Bassett Avenue, Southampton, SO16 7DR
MI0 UBE L. Calderwood, 43 Rathview Park, Mullybritt, Enniskillen, BT94 5EW
M0 UCD J. Turner, 17 Beechwood Road, Dronfield, S18 1PW
M0 UCH C. Howard, 1 Beale Road, Cheltenham, GL51 0JN
M0 UCI R. Cooke, 56 Park Avenue, Mitcham, CR4 2EN
M0 UCK A. Manning, 10 The Quarry, Kidderminster, DY10 2QD
MW0 UCL K. Ucele, 30 Britannia Apartments, Phoebe Road, Copper Quarter, Pentrechwyth, Swansea, SA1 7FG
M0 UDA A. Cattell, 2 St. James Close, Ruscombe, Reading, RG10 9LJ
M0 UDB D. Beard, 1 Bond Close, Leonard Stanley, Stonehouse, GL10 3GQ
MM0 UDI R. Duncan, South Backieley, Turriff, AB53 4GS
M0 UDL D. Woodhams, 83 Langdale Place, Newton Aycliffe, DL5 7DY
M0 UEE R. Goldsack, 5 Parc Dellen, Croft Farm Park, Luxulyan, Bodmin, PL30 5EW
M0 UEH Dr S. Smith, 557, Riverside Island Marina, Isleham, Ely, CB7 5SL.
M0 UEI I. Dukes, 36 Red Barn Road Brightlingsea, Colchester, CO7 0SJ
M0 UET R. Taylor, 68 Charter Road, Chippenham, SN15 2RA
M0 UEZ R. Scholefield, 4 Minnie Street, Haworth, Keighley, BD22 8PR
M0 UFA M. Atherton, 4 Bakers Park Saltney, Chester, CH4 8FB
M0 UFC M. Bryant, 284 Brantingham Road Chorlton Cum Hardy, Manchester, M21 0QU
MI0 UFL A. Davidson, 8 Rashee Court, Ballyclare, BT39 9SE
MI0 UFT J. Martin, 23 Winters Gardens, Omagh, BT79 0DZ
M0 UGD D. Underwood, 24 Wheatcroft Road, Rawmarsh, Rotherham, S62 5ED
M0 UGE R. Evans, 11 Swane Road Stockwood, Bristol, bs14 8nq
M0 UGG Dr R. Weller, 67 Uplands Way, Diss, IP22 4DF
M0 UGH A. Stevenson, Antonienstrasse 40, Berlin, Germany, 13403
M0 UGM G. Mountain, 34 Albert Road, Warlingham, CR6 9EP
M0 UGR C. Luckett, 257 Folkestone Road, Dover, CT17 9LL
M0 UGX M. Street, 7 Salisbury Street, Sowerby Bridge, HX6 1EE
M0 UHH E. Hanna, 35 Orchard Drive Mayland, Chelmsford, CM3 6EP
MM0 UIG M. Mackinnon, 17 Valtos, Miavaig, Isle of Lewis, HS2 9HR
M0 UJD K. Colman, 10 South Rise, North Walsham, NR28 0EE
M0 UKA R. Wood, 7 Wishart Green, Old Farm Park, Milton Keynes, MK7 8QB
M0 UKC UK Young Contesters Group, c/o S. Pearson, 8 The Pastures, Edlesborough, Dunstable, LU6 2HL
M0 UKF F. Hennigan, 50 Fairford Crescent, Downhead Park, Milton Keynes, MK15 9AE
M0 UKI UK Islands Group, c/o C. Wilmott, 60 Church Hill, Royston, Barnsley, S71 4NG
M0 UKM M. Busch, Dammstrabe 4, Neuwied, Germany, 56564
M0 UKN A. Mcdonald, 11 Micklegate Murdishaw, Runcorn, WA76HT
M0 UKO A. Shaw, 20 Hillcrest Close, Thrapston, Kettering, NN14 4TB
M0 UKS J. Banham, Timandra, Mill Road, Hardwick, Norwich, NR15 2ST
MM0 UKW C. Houston, 1 Langmuir Avenue, Perceton, Irvine, KA11 2DR
MW0 UKX G. Ralls, 37 Gresham Place, Treharris, CF46 5AF
M0 ULC W. Biernacki, 2A Tewkesbury Terrace, London, N11 2LT
M0 ULD M. Elford, 10 Meadowlands, Lymington, SO41 9LB
MI0 ULK S. Morrow, 769 Farranseer Park, Macosquin, Coleraine, BT51 4NB
MM0 ULL K. Mackenzie, Alderwood, Braes, Ullapool, IV26 2TB
M0 ULR J. Gromadzki, 13 Merrill Heights, Maidenhall Approach, Ipswich, IP2 8GA
M0 UMG M. Ribbands, Dyson'S Farm, Long Row, Tibenham, NR16 1PD
MM0 UMH L. Mitchell Hynd, Smithy House, Bruichladdich, Isle of Islay, PA49 7UN
M0 UMM A. Insarov, Broadway Chambers, 20 Hammersmith Broadway, London, W6 7AF
M0 UMS M. Salt, 1 Chantry Close, Harrow, HA3 9QZ
MI0 UNA U. Murray, 80 Canterbury Park, Londonderry, BT47 6DU
M0 UNI G. Rigby, Gas House Farm, Shavington Park, Market Drayton, TF9 3SY
M0 UNJ A. Perek, 28 Sefton Avenue, Plymouth, PL4 7HB
M0 UNN S. Jukna, 85 St. Davids Crescent, Aspull, Wigan, WN2 1SZ
M0 UNU H. Ilie, Sos. Iancului Nr. 33 Bloc 105A Scara B Apt 63, Bucharest, Romania, 21717
M0 UOE University of Essex ARS (Ears)., c/o U. Nehmzow, University Of Essex, Department Of Biological Sciences, Colchester, CO4 3SQ
M0 UOG University of Greenwich, c/o P. Smith, 1 Lambourne Place, Blackheath, SE3 7BH
M0 UOK B. Eddy, 345 Moo 2, Phon Thong., Chaiyaphum, Thailand, 36000
M0 UOO R. Bone, 6 Danehurst Place, Locks Heath, Southampton, SO31 6PP
M0 UOY University of York ARC, c/o M. Walker, 19 Highbury Place, Headingley, Leeds, LS6 4HD
M0 UPA J. Van Der Elsen, 6 Kent Close, Churchdown, Gloucester, GL3 2HQ
MW0 UPH A. Williams, 8 Old Tanymanod Terrace, Blaenau Ffestiniog, LL41 4BU
M0 UPL C. Panaitescu, 131 Stafford Road, Croydon, CR0 4NN
M0 UPS V. Vaznais, 35 Lynwood Drive, London, KT4 7AA
M0 UPU A. Stirk, 59 West Avenue, Lightcliffe, Halifax, HX3 8TJ
M0 UQB T. Chan, 29 Marthall Drive, Sale, M33 2XP
M0 URB G. Urban, 33 High Meadow, Hathern, Loughborough, LE12 5HW
M0 URF A. Vincent, Post Office House, Station Road, Andoversford, Cheltenham, GL54 4HP
M0 URJ S. Spencer, 18 Goseley Avenue, Hartshorne, Swadlincote, DE11 7EZ
M0 URL P. Gavin, 11 Campbell Close, Yateley, GU46 6GZ
MM0 URN I. Quinnell, Acarsaid, Kinlochbervie, Lairg, IV27 4RP
M0 URX T. Beaumont, 83 Limbrick Avenue, Tile Hill, Coventry, CV4 9EX
M0 USB R. Davies, 24 Evesham Avenue, Whitley Bay, NE26 1QR
M0 USC S. Chaney, 54 Clementine Avenue, Seaford, BN25 2XG
MW0 USK C. Burke, 84 Elgam Avenue, Blaenavon, Pontypool, NP4 9QU
M0 USM 5th Reigate Scout Group, c/o G. Mountain, 34 Albert Road, Warlingham, CR6 9EP
MI0 UST S. Dockery, 70 Main Street, Greyabbey, BT22 2NG
M0 USV D. Soames, 40 Woodland Drive, North Anston, Sheffield, S25 4EP
M0 USY P. Shields, 34 Dryden Close, Grantham, NG31 9QS
M0 UTA A. Emmerson, 31 Culver Road, Stockport, SK3 8PG
M0 UTD M. Jones, 110 Becconsall Drive Leighton, Crewe, CW1 4RP
M0 UTG J. Dodds, 84 Borrowdale Avenue, Walkerdene, Newcastle upon Tyne, NE6 4HL
M0 UTH G. Guinan, 5A Temple Lane, Silver End, Witham, CM8 3QY
M0 UTK S. Kembrey, 101 Yew Tree Drive, Bristol, BS15 4UF
MW0 UTT B. Bull, Swan Cottage, Swan Road, Welshpool, SY21 0RH
M0 UTX J. Swift, Raf Holmpton Rysome Lane, Holmpton, HU19 2RG
MI0 UTY D. Cartin, 6 Grange Avenue, Magherafelt, BT45 5RP
M0 UUU A. Sheard, 15 Bent Lanes, Urmston, Manchester, M41 8PB
M0 UVZ Glanford Electronics, c/o G. Cowling, Laissez Faire, Reedness, Goole, DN14 8ET
M0 UWD G. Deacon, 32 Gloucester Road, Exwick, Exeter, EX4 2EF
M0 UWS I. Lindsay, 17 Middleforth Green, Penwortham, Preston, PR1 9TB
M0 UXB D. Coomber, 14 Francis Green Lane, Penkridge, Stafford, ST19 5HF
M0 UXO M. Gritton, 53 Brinkburn Grove, Banbury, OX16 3WX
M0 UXS C. Dutton, Twillingate Farm, Tiptoe, Lymington, SO41 6EJ
MI0 UYD P. Page, 259 Bridge Street, Portadown, Craigavon, BT63 5AR
M0 UYR R. Brooker, 18 Honeybourne Way, Petts Wood, Orpington, BR5 1EZ
MW0 UZO D. White, 222 St. Fagans Road, Cardiff, CF5 3EW
M0 VAA G. Mcgowan, 281 Ashgate Road, Chesterfield, S40 4DB
MI0 VAC V. Crothers, 5 Thornleigh Park, Ballymoney, BT53 7BX
M0 VAD D. Cook, 44 Statfold Lane, Fradley, Lichfield, WS13 8NY
M0 VAG A. Grant, 26 Fountains Avenue, Boston Spa, Wetherby, LS23 6PX
M0 VAH E. Whitehouse, 16 Rue Gaston De Caillavet, Paris, France, 75015
M0 VAI D. Vainas, 51 Magister Road, Bowerhill, Melksham, SN12 6FD
M0 VAM M. Medcalf, 47 Paddock Drive, Chelmsford, CM1 6UX
M0 VAP M. Alcaino Pizani, Flat 45, Brian Redhead Court, 123 Jackson Crescent, Manchester, M15 5RR
M0 VAR Belvoir Vale AR, c/o B. Hiley, 9 Pinfold Lane, Harby, Melton Mowbray, LE14 4BU
M0 VAS V. Papanikolaou, 104 West Drive Gardens, Soham, Ely, CB7 5EX
M0 VAT A. Rodgers, 123 Mill Lane, Northfield, Birmingham, B31 2RP
M0 VAU Dr M. Vaughan, Hillside Tregorrick, St Austell, PL26 7AG
M0 VAW V. Werrett, 3 Hardingham Drive, Sheringham, NR26 8YE
M0 VAY R. Bewick, 357 Franklands Village, Haywards Heath, RH16 3RP
MM0 VBC B. Casling, 18 Inglis Avenue, Port Seton, Prestonpans, EH32 0AD
M0 VBE B. Barwise, 10 Morland Road, London, E17 7JB
M0 VBR J. Baughan, Chestnut Farm, Eastville, Boston, PE22 8LX
MW0 VBT M. Kveksas, 25 Barnard Way, Church Village, Pontypridd, CF38 1DQ
M0 VBV V. Be`Dard, 53 Cottingley Crescent, Leeds, LS11 0HZ
M0 VBW B. Whall, 3 Farrow Close, Great Moulton, Norwich, NR15 2HR
M0 VBY D. Potter, 30 Mersham Gardens, Goring-By-Sea, Worthing, BN12 4TQ
M0 VCA J. Davis, 29 Willow Tree Rise, Bournemouth, BH11 8EE
M0 VCB C. Fox, Millstone Cottage, Prior Wath Road, Scarborough, YO13 0AZ
MW0 VCC H. Morris, The Heights Hotel, 74 High Street, Llanberis, Caernarfon, LL55 4HB
M0 VCE N. Baker, 56 Chalklands, Bourne End, SL8 5TJ
M0 VCP S. Pryke, 12 Seaward Avenue, Leiston, IP16 4BB
M0 VCR P. Woodhouse, 8 Greenhill Road, Halesowen, B62 8EZ
M0 VCS V. Stocker, 25 Davies Drive, Uttoxeter, ST14 7EQ
M0 VCX J. Woods, Invicta Cottage, Carbrooke Road, Thetford, IP25 6SD
M0 VDO A. Collier, 65 Sandpits, Leominster, HR6 8HT
M0 VDQ A. Bolster, 45 Headlands Drive, Hessle, HU13 0JP
MW0 VDX Group Two, c/o J. White, Keepers Lodge Pumpsaint, Llanwrda, SA19 8DX
M0 VEC R. Trevan, 35 Oaktree Drive, Hook, RG27 9RA
M0 VEL Dr S. Favell, Lodge Farm, Moor Lane, Reepham, Lincoln, LN3 4EE
M0 VES M. Thompson, 2 Old Coastguard Cottages, Holmpton, Withernsea, HU19 2QU
MW0 VET M. Williams, 10 Clettwr Terrace Pontsian, Llandysul, SA44 4TU
M0 VEX T. Cope, 47 Kings Avenue, Atherstone, CV9 1JY
M0 VEY P. Sidwell, 7 Spring Field Close Sigglesthorne, Hull, HU11 5QP
M0 VEZ S. Debenham, 10 Elizabeth Close, Wellingborough, NN8 2JA
M0 VFC R. Chipperfield, 13 Harlestones Road, Cottenham, Cambridge, CB24 8TR
M0 VFG P. Hawkins, Broadaford Farm, Bittaford, Ivybridge, PL21 0LD
M0 VFR S. Tomlinson, 7 Springwell Close, Crewe, CW2 6TX
MI0 VFW Mid Ulster ARC, c/o J. Lappin, 46 Grange Road, Kilmore, Armagh, BT61 8NX
M0 VGA D. Silkstone, 169 Otley Road, Harrogate, HG2 0DA
M0 VGC R. West, 557 East Bank Road, Sheffield, S2 2AG
M0 VGG J. Tricklebank, 1 Hewell Road, Barnt Green, Birmingham, B45 8NG
M0 VGH R. Ashworth, 10 Mulberry Close, Wigan, WN5 9QL
M0 VGK A. Green, 28 Queensway, Old Dalby, Melton Mowbray, LE14 3QH
M0 VGL A. Lunn, 57 Greets Green Road, West Bromwich, B70 9ES
M0 VGT R. Newell, 44 New Street, Chagford, Newton Abbot, TQ13 8BB
M0 VGV G. Venugopalan, 3 Southwater Close, London, E14 7TE
M0 VHC T. Oliver, 17 East Lea, Newbiggin-by-the-Sea, NE64 6BQ
M0 VHG V. Greatwood, 11 The Green, Long Preston, Long Preston, Skipton, BD23 4PQ
M0 VIG A. Smith, 19 Gibsons Gardens, North Somercotes, Louth, LN11 7QH
M0 VII A. Price, Barn Owl Roost, Astwith, Chesterfield, S45 8AN
MM0 VIK E. Crawford, The Cottage, Cumliewick Sandwick, Shetland Islands, ZE2 9HH
M0 VIN C. Vincent, 64 Park End Road, Romford, RM1 4AU
M0 VIR D. Smith, 48 Shirley Gardens, Tunbridge Wells, TN4 8TH
M0 VIS R. Morris, 45 St. Kildas Road, Bath, BA2 3QL
M0 VIT J. Franks, 14 The Hamlet, Slades Hill, Templecombe, BA8 0HJ
M0 VJR Cambridge District ARC, c/o S. Rhenius, Baythorne Cottage Baythorne End, Halstead, CO9 4AB
M0 VJX B. Walker, 255 Packington Avenue, Birmingham, B34 7RU
M0 VKB R. Bearcroft, 45 Broad Marston Road, Pebworth, Stratford-upon-Avon, CV37 8XT
M0 VKC N. Williams, 17 Sunnyside, Malpas, SY14 7AA
M0 VKG A. Smith, 305, 1414 5Th St Sw, Calgary, Canada, T2R 0Y8
M0 VKJ A. Yiangou, 153 Hoppers Road, London, N21 3LP
M0 VKK R. Cresswell, Meadow View, Hulver Road, Beccles, NR34 7UW
M0 VKP C. Schroth, Flat 3, 15B Cavendish Road, Bournemouth, BH1 1QX
M0 VKR L. Bullen11, 5 West View, Long Sutton, TA10 9LT
M0 VKS D. Vickers, 178 Bakewell Road, Matlock, DE4 3BA
M0 VKX R. Routledge, 42 Eighth Row, Ashington, NE63 8JX
M0 VKY S. Billingham, Kewell House, Wombourne Road, Swindon, Dudley, DY3 4NF
M0 VLA A. Howsen, Oakland Villa, Seaton Road, Maryport, CA15 8ST

M0 VLC F. Harwood, 1, South Highall Cottage, Lincolnshire, LN10 6UR
M0 VLI W. Toher, The Chapel, Station Road, Darlington, DL2 1JG
M0 VLK V. Lucock, 34 Wentworth Drive, Ipswich, IP8 3RX
M0 VLL V. Leppard, 39 Queensland Drive, Colchester, CO2 8UD
MW0 VLO B. Henley, Rhewin Glas, Glynarthen, Llandysul, SA44 6PR
M0 VLP G Qrp Club, c/o P. Barville, Felucca, Pinesfield Lane, West Malling, ME19 5EN
M0 VLT A. Macdonald, Woodside Cottage, Horton Way, Verwood, BH31 6JJ
M0 VLX D. Cassidy, 172 Lyde Road, Yeovil, BA21 5PN
M0 VMC D. Burkin, 26 Rampton Road, Cottenham, Cambridge, CB24 8UL
MD0 VMD P. Birchall, 7 Richmond Close, Douglas, Isle Of Man, IM2 6HR
M0 VMH V. Hocking, 80 Barton Tors, Bideford, EX39 4HA
M0 VMV R. Vale, 611 College Road, Birmingham, B44 0AY
M0 VMW Vintage & Military ARS, c/o S. Mckinnon, 145 Enville Road, Kinver, Stourbridge, DY7 6BN
M0 VMX M. Seaney, 56 Winnham Drive, Fareham, PO16 8QG
M0 VNG M. White, 7 Overthwart Crescent, Worcester, WR4 0JW
M0 VNO D. Harwood, 36 Seaview Drive, Great Wakering, Southend-on-Sea, SS3 0BE
M0 VNR N. Ramsey, Dalestones, Lansdown Road, Bath, BA1 5TB
M0 VNY V. Brindle, 185 Brunshaw Road, Burnley, BB10 4DL
MU0 VOE Dr H. Voehrs, 50 High Street, Alderney, Guernsey, GY9 3TG
M0 VOG Vintage Operating Group, c/o M. Buckley, Springfield, 12 Ranmore Avenue, Croydon, CR0 5QA
M0 VOK R. Remnant, 172 Burnham Road, Highbridge, TA9 3EH
M0 VOL C. Brayshaw, Wayside Cottage, Main Street, Harpham, Driffield, YO25 4QY
M0 VOM N. Curran, 8 Daneswood Close, Whitworth, Rochdale, OL12 8UX
M0 VON G. Smith, 92 Brighton Road, Banstead, SM7 1BU
M0 VOS S. Devos, Applecross Cottage, Main Road, Newark, NG23 7HR
M0 VOZ M. Crockford, Centre Cottage Kelk, Kelk, YO258HL
M0 VPC J. Elstone, 54 Oakfield, Woking, GU21 3QS
M0 VPE I. Stirzaker, Avenida De Huelva 7, Conjunto Latinos 44, Rojales, Spain, 3170
MM0 VPF H. Phillips, Maplebank, Leithen Road, Innerleithen, EH446NJ
M0 VPG R. Killington, 5 Ladymead Close, Maidenbower, Crawley, RH10 7JH
M0 VPK M. Smith, 18 Hawthorn Road. Old Leake, Boston, PE22 9NY
M0 VPL M. Wells, 23 Eastmead, Bognor Regis, PO21 4QT
MM0 VPM A. Cowan, 32 Esk Valley Terrace, Dalkeith, EH22 3FT
MM0 VPR P. Rice, 255 Eskhill, Penicuik, EH26 8DF
M0 VQJ Raf Holmpton Ara, c/o J. Swift, Raf Holmpton Rysome Lane, Holmpton, HU19 2RG
M0 VQP A. Majoch, 66 Boughton Green Road, Northampton, NN2 7SP
M0 VRD R. Drage, 4 Bruce'S Close Conington, Peterborough, PE73QW
M0 VRG Vintage Radio Group, c/o A. Clayton, 6 Albert Road, Bunny, Nottingham, NG11 6QE
M0 VRI J. Groves, 350 Middle Deal Road, Deal, CT14 9SN
M0 VRP I. Bell, 53 Shurland Avenue, Sittingbourne, ME10 4QT
MW0 VRQ S. Trahearn, 148 Gladstone Road, Barry, CF62 8ND
M0 VRS J. Strange, Culloden, Ulting Road Hatfield Peverel, Chelmsford, CM3 2LU
M0 VRT L. Hummerstone, 70 Salisbury Road, Plymouth, PL4 8TA
M0 VRW P. Wilson1, 45 Newquay Close, Hartlepool, TS26 0XG
M0 VRZ M. Lee, 496A Kam Sheung Road, Sheung Tsuen, Hong Kong
M0 VSD Dr L. Kirkcaldy, 17 Central Avenue, Exeter, EX4 8NG
M0 VSE P. Taylor, 104 Winstanley Drive, Leicester, LE3 1PA
MM0 VSG Vital Sparks Group, c/o A. Cushley, 56 Riverside Park, Lochyside, Fort William, PH33 7RB
M0 VSP N. Briggs, 20 Broad Lane, Pelsall, Walsall, WS4 1AP
M0 VSQ Vulture Squadron Contest Group, c/o I. Kelly, 261 Bodiam Avenue, Tuffley, Gloucester, GL4 0XW
M0 VSR T. Wai Ming, 313 Devizes Road, Salisbury, SP2 9LU
MM0 VSU L. Bradley, Amon Sul, Kiltarlity, Beauly, IV4 7HT
M0 VSW S. Whall, 17 Vicarage Road, Deopham, Deopham, NR18 9DR
MM0 VTA Dr D. Mcnicholl, 42 Dean Street, Edinburgh, EH4 1LW
M0 VTC P. Robins, 20 Saffron Close, Chineham, Basingstoke, RG24 8XQ
M0 VTD S. Iles, Bigbury Bay Holiday Park, Challaborough, Kingsbridge, TQ7 4HS
M0 VTE A. Gallop, 75 Shearmans, Fullers Slade, Milton Keynes, MK11 2BQ
M0 VTG D. Howlett, 21 Chandlers, Orton Brimbles, Peterborough, PE2 5YW
M0 VTJ T. Scott, 50 Davison Avenue, Whitley Bay, NE26 1SH
MW0 VTK J. Martin, 78 Llwyn Ynn, Talybont, LL43 2AG
M0 VTR M. Newell, 55 Station Road Brimington, Chesterfield, S43 1JU
M0 VTS P. Wilkes, 8 Cloverdale, Stafford, ST17 4QJ
MM0 VTV R. Farrer, 23 Upper Craigour, Edinburgh, EH17 7SE
M0 VUB S. Daley, 1 North Green, Calverton, Nottingham, NG14 6NT
M0 VUE C. Suddell, Lynhurst, Littleworth Lane Partridge Green, Horsham, RH13 8JX
MM0 VUV R. Fraser, 72 Ferguson Drive, Denny, FK6 5AG
M0 VUW C. Smith, 2 Burley Gardens, Street, BA16 0SN
M0 VVA A. Amos, Willow Tree House, Deers Green, Clavering, Saffron Walden, CB11 4PX
M0 VVC M. Walker, 6 Broadoak, Tadley, RG26 3UZ
M0 VVG Elkstones ARS, c/o R. King, 8 Rydal Court, Congleton, CW12 4JL
M0 VVM T. Aldred, 31 Cock Road, Bristol, BS15 9SH
MW0 VVO S. Barry, 1 Pearson Cottages, St. Brides, Haverfordwest, SA62 3BN
M0 VVQ N. Ham, 4 Heighes Drive, Alton, GU34 2fj
M0 VVR C. Chak, P.O. Box 691 Tuen Mun Central Post Office, Hong Kong, Hong Kong, 999077
M0 VVT M. Mcgregor, 141 Herne Road, Ramsey St. Marys, Huntingdon, PE26 2SY
M0 VVV J. Worthington, The Old Hundred, Farm Lane, Farnham, GU10 5QE
M0 VVX J. Platt, 12 Tawny Grove Four Marks, Alton, GU34 5DU
MW0 VWC W. Wiggans, Bronysgawen, Llanboidy, Whitland, SA34 0EX
M0 VWD V. Downes, 55 Ashfield Road Bromborough, Wirral, CH62 7EE
M0 VWK M. Poole, 15 Roberts Place, Dorchester, DT1 2JJ
M0 VWP R. Russell, 7 Holt Place, Coach House Mews, Ferndown, BH22 9UX
MM0 VWR D. Green, 35 Douglas Avenue, Brightons, Falkirk, FK2 0HB
M0 VWS M. Smith, The Lawns Tylers Green, High Wycombe, HP108BH
M0 VWT C. Poole, 15 Devon Close, Macclesfield, SK10 3HB
M0 VWW J. Bielen, 5 Long Hale, Pitstone, LU7 9GF
M0 VXC V. Cepraga, 9 Kynon Close, Gosport, PO12 4LW
M0 VXD S. Finlayson, 41 Low Catton Road, Stamford Bridge, York, YO41 1DZ
M0 VXX T. Quiney, 20 Britannia Gardens, Stourport-on-Severn, DY13 9NZ
M0 VXY K. Poulton, 21 East View, London, E4 9JA
M0 VYB Vision Youth Centre (Bovington), c/o S. Hawkins, Forest Edge, Deer Park, Milton Abbas, Blandford Forum, DT11 0AY
M0 VYC A. Handley, 4 Southwood Drive, Thorne, Doncaster, DN8 5QS
M0 VYW A. Willsher, 1 Tolputt Court Gladstone Road, Folkestone, CT19 5NE
M0 VZA Stisted Contest Group, c/o G. Nurse, Orchard Bungalow, Priors Green, Stisted, Braintree, CM77 8BP
M0 VZR C. Gain, 14 Battens Avenue, Overton, Basingstoke, RG25 3NL
M0 VZS D. Clewer, 45 Ashfield Road, Andover, SP10 3PE
M0 VZT R. Clay, 75 Trinity View, Ketley Bank, Telford, TF2 0DY
M0 VZV M. O'Donovan, 10 Rockfield Close, Teignmouth, TQ14 8TS
M0 WAB W. Baxter, 19 Westbury Road, Nottingham, NG5 1EP
M0 WAC M. Friesch, Stag Hill Court 39-E, University Campus, Guildford, GU2 7JG
M0 WAD A. Waddington, 8 Redbrook Close, Bromborough, Wirral, CH62 6EA
M0 WAE L. Severe, 19324 Paddock View Drive, California, USA, 33647
M0 WAF P. Marchant, 16 Melrose Drive, Peterborough, PE2 9DN
M0 WAG O. Prin, 19 The Colliers, Heybridge Basin, Maldon, CM9 4SE
M0 WAH W. Horsewell, 15 Highcroft Lane, Waterlooville, PO8 9NX
M0 WAI C. Lam, 58 Sparrow Hill, Loughborough, LE11 1BU
M0 WAJ A. Hagland, 11 Coppice View, Heathfield, TN21 8YS
MM0 WAK W. Laurie, 306 Lanark Road West, Currie, EH14 5RR
M0 WAM D. Beet, 1 Shottesford Avenue, Blandford Forum, DT11 7XU
M0 WAO B. Walstra, 138 Tanhouse Farm Road, Solihull, B92 9EY
MM0 WAP F. Pudsey, 21/2 Bathfield, Edinburgh, EH6 4DU
M0 WAQ M. Welland, 76 Lovel Road, Chalfont St. Peter, Gerrards Cross, SL9 9NX
M0 WAR D. Warwick, 3 The Pines, Stoke Ferry, King's Lynn, PE33 9XW
M0 WAS O. Staines, 52 James Place, Flitwick, MK45 1GW
M0 WAU J. Lynch, Beechway, Raddel Lane, Warrington, WA4 4EE
M0 WAV A. Snelson, 6 Rayleigh Close, Braintree, CM7 9TX
MM0 WAX B. Hendry, 9 Glen Aray View, Inveraray, PA32 8TW
M0 WAY W. Thomas, 5 Thornley Road, Wolverhampton, WV11 2HR
M0 WAZ W. Payne, 1 Niton Road, Rookley, PO38 3NP
M0 WBB W. Brown, 126 Alexandra Road, Ashington, NE63 9LU
M0 WBC J. Phillips, 56 Rosemary Avenue, Hounslow, TW4 7JG
M0 WBD D. Blake, The Bramleys, Gaysfield Road, Fishtoft, Boston, PE21 0SF
M0 WBF W. Millington, 93 Feiashill Road, Trysull, Wolverhampton, WV5 7HT
M0 WBG N. Challis, 48 Brunsfield Close, Wirral, CH46 6HE
M0 WBJ B. Webb, 88 Stanley Road, Cambridge, CB5 8LB
M0 WBK W. Knapp, 32 Turner Close, Shoeburyness, Southend-on-Sea, SS3 9TL
M0 WBM A. Birch, 22 Ullswater Road, Burnley, BB10 4HX
M0 WBR R. Walker, 1A Winifred Way, Caister-On-Sea, Great Yarmouth, NR30 5AB
M0 WBS W. Bennison, 21 Ashdene Close, Chadderton, Oldham, OL1 2QG
M0 WBY J. Willby, 10 Sunbury Road, Birmingham, B31 4LJ
M0 WBZ K. Hunt, 11 De Marnham Close, West Bromwich, B70 6RJ
M0 WCA M. Bostock, 86 Beauvale Drive, Ilkest on, DE7 8SJ
M0 WCB Wessex Contest Group ARS, c/o D. Trudgian, 18 Hart Close, Wootton Bassett, Swindon, SN4 7FN
MM0 WCD C. Docherty, 23 The Maltings, Haddington, EH41 4EF
MM0 WCG Woodpecker Contest Group, c/o R. Fraser, Hopefield Cottage, Gladsmuir, Tranent, EH33 2AL
M0 WCH C. Haynes, 4 Thorn Close, Rugby, CV21 1JN
M0 WCK C. Kakoutas, Trinity College, Trinity Street, Cambridge, CB2 1TQ
M0 WCL C. Lonie Jr, 41 De La Hay Avenue, Plymouth, PL3 4HS
M0 WCM W. Maddox, 28A Redcar Avenue, Ingol, Preston, PR2 3YY
MW0 WCP Dr E. Harries, Ty Traeth, Caerwedros, Llandysul, SA44 6BS
M0 WCR M. Mcsherry, 5 Briery Croft, Stainburn, Workington, CA14 1XJ
M0 WCS D. Sewell, 19 St. Leonards Way, Ashley Heath, Ringwood, BH24 2HS
MM0 WCT T. Woods, Marsden, Lochard Road, Aberfoyle, Stirling, FK8 3SZ
M0 WCW C. Wise, 32 Commercial Street, Willington, DL15 0AD
M0 WCZ D. Wells, 96 Tennyson Avenue, Rugby, CV22 6JF
M0 WDD D. Mcarthur, 7 Gore Avenue, Salford, M5 5LF
M0 WDE F. Windridge, 84 Queens Road, Skegness, PE25 2JE
M0 WDG D. Wressell, 6 St. Josephs Close, Bishopdown, Salisbury, SP1 3FX
M0 WDJ D. Watson, 56 Lambton Avenue, Delves Lane Industrial Estate, Consett, DH8 7JE
M0 WDK P. Barrows, 5A Magdalen Road, Willoughby, Rugby, CV23 8BJ
M0 WDL D. Lee, The Old Barn, Hatch Beauchamp, Taunton, TA3 6AE
M0 WDO T. Smith, Chy Crowshensy, Clifton Road, Redruth, TR15 3UD
M0 WDP W. Phillips, 55 Kilton Crescent, Worksop, S81 0AX
M0 WDU D. Walsh, 8 Prestwold Way, Aylesbury, HP19 8GZ
M0 WDZ S. Horne, 29 Shaftesbury Street, Fordingbridge, Hampshire, SP6 1JF
M0 WEB B. Munro-Smith, 8 Billings Way, Cheltenham, GL50 2RD
M0 WEC P. Wagstaff, 49 The Paddock, Earlsheaton, Dewsbury, WF12 8BY
MW0 WEE A. Brown, Oakridge, 6 Bro Hafan, Llandysul, SA44 6NQ
MM0 WEI E. Ireland, The Steading, Blairmains, Shotts, ML7 5TJ
M0 WEL D. Wells, 40 Barnham Broom Road, Wymondham, NR18 0DF
M0 WEN C. Owen, Garden Cottage, Holbeck Woodhouse, Worksop, S80 3NQ
M0 WEO D. Hart, 7 Penrose Road, Ferndown, BH22 9JF
M0 WES H. Partridge, 19 Dickens Drive, Melton Mowbray, LE13 1HZ
M0 WET T. Clarke, 80 Bendall Road, Birmingham, B44 0SN
M0 WEV J. Wedge, 9 Claremont Mews, Wolverhampton, WV3 0EB
M0 WEW A. Mcewen, 4 The Pantyles, Nightingale Lane, Sevenoaks, TN14 6BX
M0 WFA A. Walker, 14 Maritime Avenue, Hartlepool, TS24 0XF
MW0 WFB L. Bowman, Chanrick, Penderyn Road, Aberdare, CF44 9RU
M0 WFC C. Cooper, 25 Waterside Close, Loughborough, LE11 1LP
M0 WFI C. Tam, Room B2.18, Uninn Infinity, 10 Parkside, Coventry, CV1 2PQ
M0 WFK P. Ashton, 14 Poppy Close, Boston, PE21 7TJ
M0 WFM M. Deeley, Unit 4 Beechwood Business Park, Cannock, WS11 7GB
M0 WFN W. Newton, 7 Moss Close, Bridgwater, TA6 4NA
M0 WFO S. Harris, 1 Eastbank Drive, Worcester, WR3 7BH
M0 WFR F. Waller, 19 Wortley Avenue S738Sb, Wombwell, S738SB
M0 WFX C. Bolton, 201 Lime Tree Avenue, Crewe, CW1 4HZ
M0 WGA R. Mahorney, Walnut Cottage, Church Lane, Wallingford, OX10 0SD
M0 WGB G. Beale, 34 Teville Road, Worthing, BN11 1UG
M0 WGC C. Watkins, 25 Citadilla Close, Gatherley Road, Richmond, DL10 7JE
M0 WGF E. Lewis, 105 Wards Hill Road, Minster On Sea, Sheerness, ME12 2LH
M0 WGI S. Sugihara, Southfield, Park Lane, Wokingham, RG40 4PY
MI0 WGL W. Leonard, 57 Old Coach Road Mullanavehy, Enniskillen, BT92 2EW
MI0 WGM G. Mccusker, 10 Birchdale, Lurgan, Craigavon, BT66 7TR

M0 WGO I. Paterson, 11 Ocho Rios Mews, Eastbourne, BN23 5UB
MW0 WGR West Glamorgan Raynet, c/o W. Britten-Jones, 101 Mill View Estate, Maesteg, CF34 0DE
M0 WGS Wings Museum, c/o B. Bloomfield, 2 Walstead Manor Cottages, Scaynes Hill Road, Haywards Heath, RH16 2QG
MI0 WGW E. Kyle, 2 Wattstown Crescent, Coleraine, BT52 1SP
MM0 WHA W. Anderton, 15 Queens Crescent, Lockerbie, DG11 2BA
M0 WHB W. Bray, 46 Alexandra Road, Lostock, BL6 4BB
M0 WHC W. Clayton, 403 Queens Drive Walton, Liverpool, L4 8TY
MI0 WHG Windy Hill Contest Group, c/o S. Frazer, 2 Cavanballaghy Road, Killylea, Armagh, BT60 4NZ
M0 WHJ W. Hoar, 46 Pendean Avenue, Liskeard, PL14 6DA
M0 WHK K. White, 4 Top Birches, St Neots, PE19 6BD
M0 WHL A. Lancefield, 19 Tawny Sedge, King's Lynn, PE30 3PW
MM0 WHM J. Scott, Mid Henshilwood Farm, Braehead Forth, Lanark, ML11 8HB
M0 WHO M. Sims, 133 Canterbury Road, Hawkinge, Folkestone, CT18 7BS
M0 WHP R. Hoppe, 1 Grove Road Houghton Regis, Dunstable, LU5 5PD
M0 WHQ Norfolk County Raynet, c/o A. Mobbs, 149 The Paddocks, Old Catton, Norwich, NR6 7HR
M0 WHR D. Williams, Flat 3 The Old Council House, Market Street, Atherstone, CV9 1ET
M0 WHT K. Snipe, 5 Draycott Road, Chiseldon, Swindon, SN4 0LT
MI0 WHX S. Gibson, 22 Station Road, Bangor, BT19 1HD
M0 WIA W. Armes, 11 Rutland Road, Broadheath, Altrincham, WA14 4HW
MM0 WIC C. Aitken, Windybraes, Upper Gills, Canisbay, Caithness, Canisbay, KW1 4YB
M0 WIG S. Biggs, 81 St Abbs Drive, Bradford, BD6 1EJ
M0 WIK Dr K. Morris, 44 Leamington Road, Weymouth, DT4 0EZ
MW0 WIL W. Howe, 78 Coychurch Road, Pencoed, Bridgend, CF35 5NA
M0 WIN O. Prosser, 2 Caroline Close, Ventonleague, Hayle, TR27 4EX
M0 WIO D. Owen, 8 Crag Bank Crescent, Carnforth, LA5 9EQ
M0 WIT D. Whitley, 10 Kenmore Drive, Cleckheaton, BD19 3EJ
M0 WIV R. Laidler, Swallow Cottage, Wiverton, Plympton, Plymouth, PL7 5AA
MW0 WIW Wireless In Wales Operating Group, c/o D. Pierce, Coed Duon, Tremeirchion, St Asaph, LL17 0UH
M0 WIZ I. Moore, Sun House, 33 Church Lane, Trowbridge, BA14 0TE
M0 WJA D. Redmayne, 10 The Square, Kington, HR5 3BA
MI0 WJC W. Campbell, 9 Rochester Court, Coleraine, BT52 2JL
M0 WJG W. Garvey, 254 Bury Road, Tottington, Bury, BL8 3DT
M0 WJH W. Haddock, 5 Bradley Close, Middlewich, CW10 0PF
M0 WJL G. Hayers, 87 Bradleigh Avenue, Grays, RM17 5RH
MI0 WJM W. Murray, 80 Canterbury Park, Londonderry, BT47 6DU
M0 WJT W. Taylor, 99 St. Marys Close, Littlehampton, BN17 5QQ
M0 WJW W. Wellington, 57 Hillcrest, Whitley Bay, NE25 9AF
M0 WKG K. Latham, 45 Hollybank Close, Northwich, CW8 4GS
MM0 WKJ W. Jenkins, 3A Manse Grove, Stoneyburn, Bathgate, EH47 8EW
M0 WKL W. Lam, 29 Marthall Drive, Sale, M33 2XP
M0 WKO P. Holton, 66 Mill Road, Gillingham, ME7 1JB
M0 WKR N. Clarke, Brimham Lodge Farm, Brimham Rocks Road Burnt Yates, Harrogate, HG3 3HE
M0 WKT G. Williams, 18 Elmsleigh Road, Farnborough, GU14 0ET
MM0 WKY R. Mackay, 12 Robertson Square, Wick, KW1 5NF
M0 WLA West Coast Rollers (Science and Engineering Club), c/o R. Wynne, South Graceholme, High Lorton, Cockermouth, CA13 9UQ
M0 WLB L. Barlow, 51 Hare Hill Road, Littleborough, OL15 9HE
M0 WLD B. Wild, 1 Sunnymount, Midsomer Norton, Radstock, BA3 2AS
M0 WLF I. Prater, 470 Bishport Avenue, Bristol, BS13 0HS
M0 WLH Prof. W. Lionheart, Marsham, Start Lane, Whaley Bridge, High Peak, SK23 7BP
M0 WLK R. Readman, 1 Millside Close, Kilham, Driffield, YO25 4SF
MM0 WLL W. Fleming, 65 Dundonald Park, Cardenden, Lochgelly, KY5 0DG
MW0 WLP P. Williams, 127 Penchwintan Road, Bangor, LL57 2YG
M0 WLS W. Le Serve, 120 Cheam Road, Cheam, Sutton, SM1 2EB
MU0 WLV A. Prosser, Woodlands, La Vassalerie, St Andrew, Guernsey, GY6 8XL
M0 WLY A. Omar, 84 Beaumont Hill, Darlington, DL1 3ND
M0 WMB M. Chanter, 7 Woodford Crescent, Plymouth, PL7 4QY
M0 WMG R. Shepherd, 19 Elford Avenue, Newcastle upon Tyne, NE13 9AP
MW0 WML R. Davison, 2 Marlow Terrace, Mold, CH7 1HH
M0 WMO D. Tordoff, 49 Dale Edge, Eastfield, Scarborough, YO11 3EP
M0 WMR W. Ross, 62 Derwent Drive, Tewkesbury, GL20 8BB
M0 WMS M. Smith, 12 Lincoln Drive Waddington, Lincoln, LN5 9NH
M0 WMT M. Lawrence, 83 Gresham Drive, Northampton, NN4 9SB
MW0 WMW M. Suddaby, Bryn Eiddion, Rhydymain, Dolgellau, LL40 2AS
M0 WMX D. Colver, 85 Whitemoor Lane, Belper, DE56 0HD
M0 WNA A. Toal, 29 Highland Drive, Oakley, Basingstoke, RG23 7LF
M0 WNF N. Fellingham, 23 Brooklands, Colchester, CO1 2WA
M0 WNI R. Karpinski, 55 Cambridge Avenue, New Malden, KT3 4LD
MW0 WNL R. Bartrum, 1A Bakers Way, Bryncethin, Bridgend, CF32 9RJ
M0 WNT A. Taylor, 90 Coppice Avenue, Eastbourne, BN20 9QJ
M0 WNV T. Higginson, 109 London Road, Biggleswade, SG18 8EE
MM0 WNW N. White, 2 Appleby Cottages, Whithorn, Newton Stewart, DG8 8DQ
MM0 WOA Prof. G. Woan, 6 Sandpiper Road, Lochwinnoch, PA12 4NB
M0 WOB D. Bowden, 58 Southville, Yeovil, BA21 4JF
M0 WOD G. Norgrove, The Rockings Alcester Road Burcot, Bromsgrove, B60 1pj
M0 WOF Woofferton Transmitting Station, c/o M. Porter, 20, Southfield Road, Much Wenlock, TF13 6AX
M0 WOJ A. Landless, 2 Aspen Way, Banstead, SM7 1LE
M0 WOM D. Shingleton, 6 Newsham Walk, Manchester, M12 5QB
M0 WON H. Moore, 52 Limefield Street, Accrington, BB52AF
M0 WOS W. Barnes, Cushendall, Lyngate Road, North Walsham, NR28 0DH
M0 WOT D. Watts, 3 Witney Road, Crawley, RH10 6GJ
M0 WOW D. Dunne, 1 Burton Gardens, Brierfield, Nelson, BB9 5DR
M0 WPA S. Robbins, Sugar Mouse Luxury Confectionary Ltd, 2 Central Buildings, Market Place, Easingwold, York, YO61 3AB
M0 WPI T. Hobson-Smith, 15 Henconner Lane, Chapel Allerton, Leeds, LS7 3NX
M0 WPJ P. Joyner, 3 Barton Road, Canterbury, CT1 1YG
M0 WPL P. Loda, St. Albans Court, Sandwich Road Nonington, Dover, CT15 4HH
M0 WPN W. Nichols, Newcourt Farmhouse, Silverton, Exeter, EX5 4HT
M0 WPR W. Rees, 67 Chine Walk, West Parley, Ferndown, BH22 8PS
M0 WPS W. Phillips, 36 Beeches Road, Great Barr, Birmingham, B42 2HF
M0 WPT J. Rosa, Flat 7, Wick Hall, Abingdon, OX14 3NF
M0 WPX Data - Dxers, c/o K. Holloway, 6 Britons Lane Close, Beeston Regis, Sheringham, NR26 8SH
M0 WPY R. Hampson, 10 Lindisfarne Close, Sandy, SG19 1TT
M0 WQK D. Blackie, 8 Kingswood Road, Manchester, M14 6SB
M0 WQR V. Keeley, Hawthorns, Cowbit Drove, West Pinchbeck, Spalding, PE11 3TG
M0 WRA B. Wray, 10 Winstanley Road, Sale, M33 2AR
M0 WRC Workington & District ARC, c/o K. Matthews, St. Helens Cottage, Flimby, Maryport, CA15 8RX
M0 WRD D. Whitehouse, 6 Larch Close, Heathfield, TN21 8YW
M0 WRE G. Norris, 12 Westfield Placescholes, Bradford, BD19 6DU
M0 WRI P. Hodge, 141 Linden Place, Newton Aycliffe, DL5 7BQ
M0 WRJ Dr Y. Suzuki, Eng. 5-203, Shizuoka University, 3-5-1 Johoku, Hamamatsu-Naka, Japan, 432-8561
MD0 WRK R. Kissack, 6 Falcon Cliff Court, Douglas, Isle of Man, IM2 4AQ
MM0 WRO M. Lorenowicz, Oban Bay Hotel Corran Esplanade, Oban, PA34 5AE
MW0 WRP W. Powell, 40 Heol Ty Newydd, Cilgerran, Cardigan, SA43 2RT
MW0 WRQ Tonypandy Scout Group, c/o B. Jones, 10 Hughes Street, Penygraig, Tonypandy, CF40 1LX
M0 WRR Waveney Wireless, c/o L. Cropley, San Ferryann, Brundish Road, Wilby, Eye, IP21 5LS
M0 WRS W. Smith, Flat 48, Winehala Court, 50A Sandbeds Road, Willenhall, WV12 4GA
M0 WRU W. Rudge, 33 Wyrley Rd, Wolverhampton, WV11 3NY
MM0 WRX K. Mccormick, 51 Millford Drive, Paisley, PA3 3EJ
MW0 WRY J. Richardson, 15 Calland Street, Plasmarl, Swansea, SA6 8LE
M0 WSA S. Williams, Flat 35, Winterton House, London, E1 2QR
M0 WSB A. Craddock, 58 Vicarage Road, Mickleover, Derby, DE3 0ED
M0 WSC T. Kwok, 313 Devizes Road, Salisbury, SP2 9LU
MW0 WSD N. Orchard, 152 Garden Suburbs, Trimsaran, sa174af
M0 WSE A. Champion, 4 Oldcastle Croft, Tattenhoe, Milton Keynes, MK4 3EN
MM0 WSG Glasgow University Wireless Society, c/o T. Storkey, Flat E, 29 Herbert Street, Glasgow, G20 6NB
MW0 WSH Dr J. Blaxland, 7 Maes Y Sarn, Pentyrch, Cardiff, CF15 9QQ
MM0 WSK J. Muchowski, 71 The Braes, Tullibody, Alloa, FK10 2TT
M0 WSN R. Swinburne, 32 Hollywell Road, Birmingham, B26 3BX
M0 WSP S. Peare, 15 Clydesdale Gardens, Bognor Regis, PO22 9BE
M0 WSR B. Harrison, 43A Rumbridge Street, Totton, Southampton, SO40 9DR
MI0 WST D. Caiden, 14 Church Street, Greyabbey, Newtownards, BT22 2NQ
M0 WSW S. Whittaker, 25 Cleveleys Road, Blackburn, BB2 3JS
M0 WSX Wessex Ham, c/o R. Thomas, 28 Clarks Meadow, Shepton Mallet, BA4 4FD
M0 WSZ J. Hocking, 26 Musket Road, Heathfield, Newton Abbot, TQ12 6SB
M0 WTC J. Paradas, 1A Brocks Ghyll, Eastbourne, BN20 9RQ
M0 WTG D. Cooper, Little Heath, Bradfield Common, North Walsham, NR28 0QR
M0 WTJ T. Weston, 4 The Pightle, Peasemore, Newbury, RG20 7JS
MW0 WTK P. Iles, 150 Pen-Y-Bryn, Caerphilly, CF83 2LA
M0 WTL O. Fallon, 26 Central Avenue, Corfe Mullen, Wimborne, BH21 3JD
M0 WTN J. Withington, 20 Bond Way, Hednesford, Cannock, WS12 4SN
M0 WTO M. Roberts, Flat 6, 463 Brighton Road, Lancing, BN15 8LF
M0 WTW W. Walker, 247 Forest Road, Tunbridge Wells, TN2 5HT
M0 WTX S. Jackson, 64 Main Road, Moulton, Northwich, CW9 8PB
M0 WTY R. Clare, Kimberley, Boston Road, Bicker, Boston, PE20 3AP
MM0 WTZ K. Waitz, C/O Waitz-Rainey, 16 Inverkeithing Road, Aberdour, Burntisland, KY3 0RS
M0 WUL W. Stewart, 5 St. Catherines Close, Uttoxeter, ST14 8EF
M0 WUS S. Burns, 22 Pendle Close, Peterlee, SR8 2JS
MW0 WVA Wales Digital Radio Group, c/o D. D'Mellow, 2 Twynpandy Pontrhydyfen, Port Talbot, SA12 9TW
M0 WVE M. Huggett, 12 West View Cottages Lewes Road, Lindfield, Haywards Heath, RH16 2LJ
M0 WVL C. Lacey, 2 Purbeck Cottages Acton, Langton Matravers, Swanage, BH19 3LU
MI0 WWB W. Bradley, 14 Ardmore Grange, Ballygowan, Newtownards, BT23 5TZ
M0 WWD J. Godfrey, 4 Glenfield Road, Bideford, EX39 2LU
MI0 WWF S. Nash, 5 Drumard Cottages, Dans Road, Ballymena, BT42 2PX
MM0 WWH N. Robertson, Ladyburn, Port William, Newton Stewart, DG8 9QN
MM0 WWM R. Jowett, Fearnoch Ardentallen, Oban, PA34 4SF
MW0 WWR West Wales Radio Group, c/o I. Gray, 4 Llundain Fach, Felinfoel, Llanelli, SA15 4PF
M0 WWV A. Norden, 10 School Lane, Watton At Stone, Hertford, SG14 3SF
MM0 WXD Dr D. Fisher, 1 Inverleith Row, Edinburgh, EH3 5LP
MM0 WXE A. Barclay, 21 Netherlea, Scone, Perth, PH2 6QA
M0 WXF A. England, 12 Bronte Court Swinburne Road, Wellingborough, NN8 3BF
M0 WXO M. Shibata, 18-41 Moegino Aobaku, Yokohama, Japan, 156-0045
M0 WXP S. Martin, 6 Cherry Tree Drive, Sedgefield, Stockton-on-Tees, TS21 3DN
MM0 WXS A. Mccall, 1 Finlayson Drive, Airdrie, ML6 8LU
M0 WXU P. Elsey, 62B Coleraine Road, London, SE3 7PE
M0 WXX D. Sharpen, 52 Woodsend Road, Urmston, Manchester, M41 8QT
M0 WXY S. Baldwin, 143 Oxford Road, Swindon, SN3 4JA
M0 WYB J. Scully, 10 Eckweek Road, Peasedown St. John, Bath, BA2 8EQ
M0 WYC The Radio Club, c/o D. Jones, Marvin House, Ryhill Pits Lane Cold Hiendley, Wakefield, WF4 2DU
M0 WYE H. Burnham, 13 The Close, Wye, Ashford, TN25 5BD
M0 WYH D. Lester, 171 Glenavon Road, Birmingham, B14 5BT
M0 WYM C. Ivermee, 42 Daniell Street, Truro, TR1 2DN
MW0 WYN D. Davies, 2 Hendre Ddu, Manod, Blaenau Ffestiniog, LL41 4BH
M0 WYP J. Marvel, Dean House Farm, Nordan, Leominster, HR6 0AW
M0 WYR Wyre ARG, c/o K. Haworth, 11 Petersfield Close, Bootle, L30 1SG
M0 WYT T. Webster, 1 Fen Close, Newton, Alfreton, DE55 5TD
M0 WYW Y. Wong, 15B Block 27, City One Shatin N.T., Hong Kong, .
M0 WYZ K. Winwood, 146 Chapel Street, Pensnett, Brierley Hill, DY5 4EQ
M0 WZM M. Kitt, 18 Brickmakers Road, Colden Common, Winchester, SO21 1TT
M0 WZT M. Fulbrook, 2 Cob Place, Westbury, BA13 3GS
MW0 WZX C. Davies, 23 Nottingham Street, Cardiff, CF5 1JP
MM0 WZZ W. Ramsay, 1 Northburn Road, Eyemouth, TD14 5AU
MM0 XAB A. Lark, 20 Lawfield, Coldingham, Eyemouth, TD14 5PB

M0 XAC G. Dean, 62 Baptist Close, Abbeymead, Gloucester, GL4 5GD
MW0 XAD S. Gordon, 8 Maesteg, Cymau, Wrexham, LL11 5EP
M0 XAE B. Smith, 8 Mill Field Close, South Kilworth, Lutterworth, LE17 6FE
M0 XAJ A. Calvert, 5 Pond Cottages, Butts Pond, Sturminster Newton, DT10 1BE
M0 XAK A. Kent, 4 Sellerdale Drive, Wyke, Bradford, BD12 9DA
M0 XAM A. Morgan, 18 Keysworth Drive, Wareham, BH20 7BD
MM0 XAO Dr A. Onken, 19/7 Damside, Edinburgh, EH4 3BB
M0 XAR S. Halliday, 8 Newby Farm Road, Scarborough, YO12 6UN
M0 XAS A. Shepherd, 33 Meadow Way, Turton, Bolton, BL7 0DE
M0 XAT M. Harwood, 36 Coronation Avenue, Seaton, Workington, CA14 1DW
MM0 XAU H. Stoeteknuel, C/O Marriott Parkview, Dunrossness, Shetland, ZE2 9JG
M0 XAV J. Hazeltine, 21 Hassock Way, Wimblington, March, PE15 0PJ
M0 XAW R. Watson, 8 Bourne Close, Warminster, BA12 9PT
M0 XBA G. Harvey, 55 Trelawney Road Hainault, Ilford, IG6 2NJ
M0 XBB R. Boland, 10 Kenilworth Drive, Kidderminster, DY10 1YD
MM0 XBD B. Donnelly, 19 Douglas Drive, Dunfermline, KY12 9YG
M0 XBI A. Romanov, 10 Gloucester Walk, Westbury, BA13 3XG
M0 XBM C. Atkinson, 7 Hamilton Road, Grantham, NG31 9QG
M0 XBN B. Johnson, 6 Trevor Road, Swinton, Manchester, M27 0YH
M0 XBQ Dr F. Sedgemore, 12 The Brambles, Royston, SG8 9NQ
M0 XBR A. Brade, Sand Gap, Bursea Lane, York, YO43 4DF
M0 XBW B. Woollett, 23 Kinglake House, Southall, UB24FZ
M0 XBY Bromley & District ARS, c/o R. Perzyna, 29 Lakeside Drive, Bromley, BR2 8QQ
M0 XCA C. Ashley, 22 Pasture Close, Lower Earley, Reading, RG6 4UY
M0 XCF I. Titchener, 18 King Edgar Close, Ely, CB6 1DP
M0 XCH C. Harding, 27 Eston Avenue, Malvern, WR14 2SR
M0 XCJ C. Jackson, 84 Ogley Road, Walsall, WS8 6BB
M0 XCL C. Loud, 24 Harrington Avenue, Lowestoft, NR32 4JU
M0 XCO M. Macdonell, 54 Cinque Foil, Peacehaven, BN10 8DZ
MM0 XCP J. Reid, 69 Limekiln Wynd, Mossblown, KA6 5BE
M0 XCR C. Raphael, 86 Main Street, South Rauceby, Sleaford, NG34 8QQ
MM0 XCS C. Sharp, 12 Manse Place, Inverkeithing, KY111AZ
M0 XCT D. Aldred, 14 The Meadows, Radcliffe, Manchester, M26 4NS
M0 XCX P. Houghton, 6 Olivers Court, Calne, SN11 0FL
M0 XCZ M. Hartley, 24 Burnham Avenue, Bognor Regis, PO21 2JU
M0 XDA D. Sullivan, 16 Wadley Close Tiptree, Colchester, CO5 0SL
M0 XDC D. Copsey, Fairview, Mill Lane, Hook End, Brentwood, CM15 0PP
M0 XDF D. Ferrington, 20 Innings Lane, Warfield, Bracknell, RG42 3TR
M0 XDJ K. Gribben, 33 Strawberry Close, Birchwood, Warrington, WA3 7NT
M0 XDK C. Griffiths, 12 Bank View, Northampton, NN4 0RS
M0 XDL R. Coles, 10 Littlemoor Road, Weymouth, DT3 6AA
MW0 XDN Dr A. Newsome, Dros-Dro, Station Road, Letterston, SA62 5RY
M0 XDS D. Sharp, 8 Beechfield, Hoddesdon, EN11 9QH
MW0 XDT R. Snape, Bodlondeb, North Road, Whitland, SA34 0AX
M0 XDV A. Dalla-Volta, 88 Claygate Lane, Esher, KT10 0BJ
M0 XDX P. Dumpleton, 20 Cambridge Road North, Mablethorpe, LN12 1QR
MM0 XEA J. Dunlop, 5 Loudon Road, Glasgow, G33 6NJ
M0 XED C. Butcher, 60 Barton Road, Canterbury, CT1 1YH
M0 XEE M. Toher, The Chapel, Station Road, Darlington, DL2 1JG
M0 XEF C. Tacon, 2 Pavillion Court, 74-76 Northlands Road, Southampton, SO15 2NN
M0 XEK E. Kemp, 4 Foundry Flats, Foundry Square, Hayle, TR27 4AE
M0 XEM S. Christie, 124 Bickershaw Lane, Abram, Wigan, WN2 5PP
M0 XER L. Bodnar, 47 Alchester Court, Towcester, NN12 6RL
M0 XEY C. Blain, 8 Fern Way, Weaverham, Northwich, CW8 3EZ
M0 XFG F. Grande, Restrup Engvej 26, Aalborg, Denmark, 9000
M0 XFL C. Coles, 2 Fern Square, Chickerell, Weymouth, DT3 4NZ
MW0 XFU C. Jenkins, Flat 5, The Lawns, Usk, NP15 1BA
M0 XFX J. Hawkes, 53 Mill Hill, Derby, DE24 5AF
M0 XG Braintree & District ARS, c/o G. Nurse, Orchard Bungalow, Priors Green, Stisted, Braintree, CM77 8BP
M0 XGB K. Dickson, 29 Sunnyfields Drive, Minster On Sea, Sheerness, ME12 3DH
M0 XGG M. Bayliff, 80 The Meadows, Leominster, HR6 8RE
M0 XGK J. Haig, 1 Vallibus Close, Lowestoft, NR32 3DS
M0 XGL G. Lund, 1 Thrush Close, Gloucester, GL4 4WZ
M0 XGN G. Norbury, 3 Sherard Croft, Birmingham, B36 0LS
M0 XGR W. Smith, 81 Hazelgrove Residential Park Milton Street, Saltburn-by-the-Sea, TS12 1FE
M0 XGS G. Stanley, 95 Old Vicarage, Westhoughton, Bolton, BL5 2EG
M0 XGT Dr T. Thomas, 55 Bath Street, Southampton, SO14 6GR
M0 XGW G. Whall, 10 Hillcrest Court, Ipswich Road, Diss, IP21 4YJ
M0 XGX G. Round, 128 Leicester Road, Shepshed, Loughborough, LE12 9DH
M0 XHD I. Hammond, 88 Great Innings North, Watton at Stone, SG14 3TD
M0 XHN Hubnet Amateur Radio & Digital Online Network, c/o S. Curtis, 354 St. Helens Road, Leigh, WN7 3PQ
M0 XHS Hammersmith ARS, c/o S. Alpuvan, 51 Limes Avenue, Chigwell, IG7 5NX
M0 XIA I. Alderman, 107 Manton Drive, Luton, LU2 7DL
M0 XIC M. Savage, 5 Mere Hall Barns, Mere Lane, Enville, Stourbridge, DY7 5JL
M0 XID G. Hurst, 173 Halling Hill, Harlow, Cm20 3jp
M0 XIG J. Wakefield, Oakhurst, Lower Common Road, Romsey, SO51 6BT
M0 XIK J. Lambert Hurley, 64 Henry Road, West Bridgford, Nottingham, NG2 7ND
M0 XJG J. Goodman, 70 Bradford Road Eccles, Manchester, M30 9FT
M0 XJL W. Pickles, 31 Longfield Road, South Woodham Ferrers, CM3 5JL
M0 XJM J. Meek, Rose Tree Cottage, Charlbury, OX7 3RX
M0 XJN J. Neal, 5 Shelley Close, Huntingdon, PE29 1NF
M0 XJT P. Hodson, 21 Green Hill, London Road, Worcester, WR5 2AA
MM0 XKA C. Robertson, 5 Broomlands Place, Irvine, KA12 0DU
M0 XKD L. Booth, 8 Rowthorne Close, Northampton, NN5 4WB
MW0 XKL R. Jones, Flat 2, Tan Y Geraint, 33 Princess Street, Llangollen, LL20 8RD
M0 XKM M. Koster, Stalworthy Manor Farm, Suton Lane, Suton, Wymondham, NR18 9JG
M0 XKO P. Goodridge, 22 Horefield, Porton, Salisbury, SP4 0LE
M0 XKW K. Williams, 35 Lord Street, Coventry, CV5 8DA
M0 XKX Kent County Raynet, c/o M. Granatt, 16 Culverden Avenue, Tunbridge Wells, TN4 9RF
M0 XLA J. Tompkins, 3 Hartwell Road, Portsmouth, PO3 5TN
M0 XLB S. Borrell, Rose Cottage, Colchester Main Road, Colchester, CO7 8DD
M0 XLH L. Hollingworth, 43 Wingfield Road, Hull, HU9 4PR
MI0 XLK S. Baird, 11 Laral Park, Newtownabbey, BT37 0LH
M0 XLR D. Roberts, 19 East Avenue, Warrington, WA2 8ad
M0 XLT K. Jackson, 7 River Place, Gargrave, Skipton, BD23 3RY
M0 XLX H. Knight, 10 Welford Road, Barton, Alcester, B50 4NP
M0 XLY M. Oxley, 49 Dalton Crescent, Shildon, DL4 2LE
M0 XMB P. Foster, 100 Howe Road, Norton, Malton, YO17 9BL
M0 XMC M. Coad, 227D Woodham Lane, New Haw, Addlestone, KT15 3NR
M0 XMD M. Davidson, 19 Mason Street, Workington, CA14 3EH
MW0 XMG P. Provis, Dingle Gardens, Croesbychan, Aberdare, CF44 0EJ
M0 XMH M. Hopewell, 4 Cotes Crescent Bicton Heath, Shrewsbury, SY3 5AS
MW0 XMI R. Lacey, 7 Oak Tree Drive, Cefn Hengoed, Hengoed, CF82 8FN
M0 XMK M. Rose, 19 Hawthorn Street, Peterlee, SR8 3LY
M0 XML Ex-Military Land Rover Assoc., c/o J. Butcher, Mount Pleasant, Trampers Lane, North Boarhunt, Fareham, PO17 6DG
M0 XMP M. Pope, Drove Lodge 39 The Drove Barroway Drove, Downham Market, PE38 0AJ
MM0 XMQ L. Anderson, 6 St. Keiran Crescent, Stonehaven, AB39 2GQ
M0 XMS M. Smith, Higher Alterhay, Combe St. Nicholas, Chard, TA20 3LT
M0 XMT P. Setter, 199 Southbourne Grove, Westcliff-on-Sea, SS0 0AN
M0 XMX M. Lewis, 1 Kingsmead Stretton, Burton on Trent, DE13 0FQ
M0 XMZ Dr D. Boocock, 47 Long Close, Anstey, Leicester, LE7 7QG
MM0 XNC R. Biggart, Lodgebush Cottage, Craigie, Kilmarnock, KA1 5NA
M0 XNG J. Blackwell, 164 York Road Haxby, York, YO32 3EL
M0 XNR N. Reeve, 16 Hospital Drove, Little Sutton, Long Sutton, Spalding, PE12 9EL
M0 XNW North West Radio Group, c/o D. Roberts, 19 East Avenue, Warrington, WA2 8ad
M0 XOB J. Dominy, 19 Church Hill Avenue, Warton, Carnforth, LA5 9NU
M0 XOC M. Cox, 55 Malvern Crescent, Ince, Wigan, WN3 4QA
M0 XOD G. Smith, 47 Percy Road, Carlisle, CA2 6ER
M0 XOK D. Mitchell, 3 Ivy Cottage, Main Road, Theberton, Leiston, IP16 4RX
M0 XOL T. Brownen, 43 Great Rea Road., Brixham., TQ5 9SW
M0 XOM R. Smith, 21 Canal Road Crossflatts, Bingley, BD16 2SR
M0 XON K. Handscombe, 8 Fletcher Road, Ipswich, IP3 0LF
M0 XOR M. Hauser, 23 Prince William Way Sawston, Cambridge, CB22 3SZ
M0 XOS O. Snowdon, Churchill College, Cambridge, CB3 0DS
MW0 XOT J. Messenger, 34 Goylands Close, Llandrindod Wells, LD1 5RB
M0 XOU I. Spinks, 30 Lime Tree Walk, Watton, Thetford, IP25 6EU
M0 XOX J. Hill, 45 Venus Street, Congresbury, Bristol, BS49 5HA
M0 XPA P. Hekman, 28 Beechcroft Avenue, Crewe, CW2 6SQ
M0 XPB P. Bannon, 73 London Road, Worcester, WR5 2DU
M0 XPD Dr P. Darlington, 8 Uplands Road, Urmston, Manchester, M41 6PU
M0 XPJ J. Parfitt, 12 Jodrell Place, Selsey, Chichester, PO20 0FQ
M0 XPK P. Davies, 2 Lynfords Drive, Runwell, Wickford, SS11 7PP
M0 XPL C. Lawrence, Croft House, Station Road, Lancaster, LA2 8ER
M0 XPM P. Mullen, 12 Poplar Grove, Conisbrough, Doncaster, DN12 2JG
MM0 XPP L. Johnstone, 67 Belfast Quay, Irvine, KA12 8PR
M0 XPS P. Standley, 9 Capelands, New Ash Green, Longfield, DA3 8LG
MM0 XPT C. Watkinson, The Hillock Farmhouse, Lumphanan, Banchory, AB31 4QL
MM0 XPZ S. Groves, 1/1 99 Belville Street, Greenock, PA15 4SX
M0 XQB C. Louie, 29 Marthall Drive, Sale, M33 2XP
M0 XQS P. Jarvis, 50 Northcott, Bracknell, RG12 7WR
M0 XQX T. Tzvetkov, 52 Clearwell Gardens, Cheltenham, GL52 5GH
M0 XRA J. Batsman, 141 Bury Street, Ruislip, HA4 7TQ
M0 XRC Axholme Radio Club, c/o J. Fennell, Bajamar House, Belton Road, Doncaster, DN9 1JL
M0 XRF J. Margarson, 26 David Street, Grimsby, DN32 9NL
M0 XRH Eccentricity Radio Hams Club, c/o J. Rolfe, 56 Elmhurst Road, Thatcham, RG18 3DH
MM0 XRI R. Irvine, 9 Pearce Grove, Edinburgh, EH12 8SP
M0 XRL R. Langford, 103 Huddersfield Road, Elland, HX5 0EE
M0 XRM D. Bingham, 33 Sheffield Road, Creswell, Worksop, S80 4HN
MW0 XRT D. Burt, 2 Cae Masarn, Pentre Halkyn, Holywell, CH8 8JY
MW0 XRU T. Nutbeem, 6 Morris Rise, Blaenavon, Pontypool, NP4 9PA
M0 XRV J. Coates, 2 Holstein Drive, Scunthorpe, DN16 3TT
M0 XRW A. Diaz, 29, Parkside Gardens, Widdrington, Morpeth, NE61 5RP
MM0 XRZ M. Nicholls, Grahams Onsett Farm, Newcastleton, TD9 0TT
M0 XSD C. Catlin, 27 Main Street, Frizington, CA26 3SA
M0 XSG C. Braisby, 4 Langmans Way, Woking, GU21 3QY
M0 XSH D. Shaw, 81 Chesterfield Road Tibshelf, Alfreton, DE55 5NJ
M0 XSJ Northwich Repeater Group, c/o S. Jackson, 64 Main Road, Moulton, Northwich, CW9 8PB
M0 XTD C. Morgan, 5 Montgomery Avenue Hampton-On-The-Hill, Warwick, CV35 8QP
M0 XTG C. Day, 14 Windsor Drive, Ramsey Forty Foot, Ramsey, Huntingdon, PE26 2XX
MW0 XTK P. Hoath, 8 Liverpool Terrace, Llithfaen, Pwllheli, LL53 6NN
MM0 XTW A. Wallis, Pine Cottage, South Smallburn, Peterhead, AB42 5BL
M0 XTX A. Cristofoletti, 39 Jessop Street, Codnor Ripley, DE5 9RN
MW0 XTZ M. Digby, 40 Waterloo Road, Ammanford, SA18 3SF
M0 XUA Dr M. Sinclair, 40 Grotto Road, South Shields, NE34 7AH
M0 XUB B. Xu, 138 King'S College, Cambridge, CB2 1ST
M0 XUH Dr G. Thomas, 3 The Croft, Wilton, Egremont, CA22 2PW
M0 XUI J. Ribbands, Dyson'S Farm, Long Row, Tibenham, NR16 1PD
M0 XUM W. Webb, 84 Bruce Street, Swindon, SN2 2EN
M0 XUU R. Gopan, 84 Hilmanton, Lower Earley, Reading, RG6 4HN
M0 XVF J. Smith, 8 Mayfields, Spennymoor, DL16 6RN
M0 XVI M. Collins, Chilterns, Little Frieth, Henley-on-Thames, RG9 6NR
M0 XVK R. Redmond, 28 Common Lane, Polesworth, Tamworth, B78 1LS
M0 XVL G. Elsigan, Traunuferstr 143 A, Haid, Austria, A-4053
M0 XVN M. Tarrant, Holly Cottage Nanstallon, Bodmin, PL30 5JZ
M0 XVX Dr A. Smith, Borrowdale House, 10 Borrowdale Road, Malvern, WR14 2DS
M0 XWD J. Brown, 8 Chatsworth Street, Sutton-in-Ashfield, NG17 4GG
M0 XWP M. Meyer, 22 Orchard Grove, Newton Abbot, TQ12 1FZ
M0 XWS M. Swann, 56 Mansel Drive Old Catton, Norwich, NR6 7NB
M0 XXC Thames ARG, c/o A. Atkinson, 21 Dennington Crescent, Basildon, SS14 2FF
M0 XXG M. Heenan, 15 Woodacre Green, Bardsey, Leeds, LS17 9AB
M0 XXI S. Lindberg, 6 Meadow View, Belper, DE56 1UT

M0 XXJ J. Creaser, 8 Millwood Road, Hounslow, TW3 2HH
M0 XXK K. Mitchell, 1 Denstroude Cottages, Denstroude Lane, Canterbury, CT2 9JX
M0 XXL D. Tinn, 28 South Road, Kirkby Stephen, CA174sn
M0 XXM M. Jennings, Springfield Farm, The Causeway, Stow Bridge, King's Lynn, PE34 3PP
MM0 XXP A. Pitkethley, 99 Margaretvale Drive, Larkhall, ML9 1EH
M0 XXT Lapworth Radio Club, c/o C. Mccormick, 65 Glendon Way, Dorridge, Solihull, B93 8SY
M0 XXV Yorkshire DX Club, c/o C. Smith, 199A Richardshaw Lane, Stanningley, Pudsey, LS28 6AA
MM0 XXW M. Whyte, 147/2 Lower Granton Road, Edinburgh, EH5 1EX
M0 XXX H. Taylor, Sunnyside Well, Chaingate Lane, Bristol, BS37 9XN
M0 XYA P. Hodges, 191 Broadstone Road, Stockport, SK4 5HP
M0 XYL M. Turner, 2 Oakleigh Road, Droitwich, WR9 0RP
M0 XYM K. Polston, 10 Marsh Farm Road, South Woodham Ferrers, Chelmsford, CM3 5WP
M0 XYN Yorkshire Net UK, c/o L. Riley, 51 Brier Lane, Havercroft, Wakefield, WF4 2AT
M0 XYT West Wight Radio Society, c/o C. Oliver, 3 Windsor Drive, Freshwater, PO40 9GB
M0 XYX A. Loyd, Maple House, Pangbourne Road, Reading, RG8 8LN
M0 XYZ A. Clark, 10 Garfield Close, Lincoln, LN1 3QP
M0 XZG Dr G. Welch, Amazonas, Sandy Lane, Hightown, Liverpool, L38 3RP
M0 XZS A. Coetzee, 6 Covent Gardens, Colwall, Malvern, WR13 6FA
M0 XZT P. Driver, 68 Ripon Road, Dewsbury, WF12 7LG
M0 XZW C. Set, Hughes Hall, Wollaston Road, Cambridge, CB1 2EW
M0 XZX S. Lowe, 14 Windmill Rise, York, YO26 4TX
MM0 YAB C. Phillips, 8 The Square, Newtongrange, Dalkeith, EH22 4QD
MW0 YAC K. Smith, 3 Pendoylan Walk, Cwmbran, NP44 7JX
MW0 YAD Cwmbran ARS, c/o K. Smith, 3 Pendoylan Walk, Cwmbran, NP44 7JX
MW0 YAE G. Thatcher-Sharp, 20 Dilys Street, Blaencwm, Treorchy, CF42 5DT
MW0 YAG A. Graham, 2 Heol Undeb, Beddau, Pontypridd, CF38 2LB
M0 YAH W. Coburn, 42 Hinton Wood Avenue, Christchurch, BH23 5AH
M0 YAL W. Dowkes, Woodlea, Gillamoor Road, York, YO62 6EL
MI0 YAM D. Foley, 14 Chestnut Hall Court, Moira, Maghaberry, BT67 0GJ
MD0 YAU D. Smith, Alwyn, Four Roads Port St. Mary, Isle of Man, IM9 5LH
M0 YAV W. Jones, 8 Oakbrook Close, Ewyas Harold, Hereford, HR2 0NX
M0 YAW D. Cooper, Seinna Bryne Lane, Padbury, MK18 2AL
M0 YAY D. Young, Thring Lodge, Granley Gardens, Cheltenham, GL51 6LQ
M0 YBC D. Croft, 33 Roughaw Road, Skipton, BD23 2PY
MI0 YBH J. Mccourt, 70A Sessiagh Scott Road Rock, Dungannon, BT70 3JU
M0 YBT M. Harrison, 2 Rosemount Court, Holly Bank Road, York, YO24 4EG
MW0 YBZ P. Smith, 29 Heol Cwarrel Clark, Caerphilly, CF83 2NE
M0 YCB C. Button, 59 Lindsey Road, Harworth, Doncaster, DN11 8QJ
M0 YCG Yorkshire Dales Contest Group, c/o D. Falkner, 45 Westwood Carleton, Skipton, BD23 3DW
MM0 YCH C. Haws, 1/1 4 Fenella Street, Glasgow, G32 7JT
MM0 YCJ Prof. C. Jones, 11B Ettrick Road, Edinburgh, EH10 5BJ
MI0 YCK D. Mayock, 27 Woodvale, Bessbrook, BT35 7FD
M0 YCQ S. Burgess, 10B Scotland Street, Ellesmere, SY12 0EG
M0 YCS C. Steuwe, Moor Barn Farm, Madingley Road, Cambridge, CB23 7PG
M0 YCX G. Bevan, 2 Geddes Street, Blackburn, BB2 5LQ
M0 YCY A. Capitan, 41 Cunningham Drive, Runcorn, WA7 4DL
M0 YDB D. Breed, 8 Tudor Street, New Rossington, Doncaster, DN11 0JG
M0 YDF D. Fowler, 5 Highfield Cottages, Everingham, York, YO42 4JG
M0 YDH D. Holman, 20 Green Drive, Wolverhampton, WV10 6DW
M0 YDJ D. Jackson, 3 Laburnum Road, Cadishead, Manchester, M44 5AS
M0 YDK A. Morrell, 19 Nairn Road, Stamford, PE9 2YR
M0 YDM Dr D. Maldoom, Hyde Manor, The Street, Kingston, Lewes, BN7 3PB
M0 YDP S. Allington, 57 Lightfoot Drive, Carlisle, CA1 3BP
M0 YDW D. White, 9 Wyatts Lane, Tavistock, PL19 0EU
MW0 YDX B. Dallimore, 4 Llys Dyffryn, St Asaph, LL17 0SX
M0 YEE A. Chapman, 6 Chaffer Lane, Birdham, Chichester, PO20 7EZ
MM0 YEK East Kilbride ARS, c/o A. Hood, 26 Annan Avenue, East Kilbride, Glasgow, G75 8XT
M0 YEP D. Stinson, 1 The Croft, Earls Colne, Colchester, CO6 2NH
MM0 YEQ G. Pearce, 2 Kirkriggs, Forfar, DD8 2AT
M0 YES P. Shaw, 32 Hardwick Road East, Worksop, S80 2NT
MM0 YET G. Burnett, 1B Craig Road, Troon, KA10 6DA
M0 YFT A. Murdoch, 128 Whalley Road, Langho, Blackburn, BB6 8DD
M0 YGB A. Birch, 3 Partridge Way, High Wycombe, HP13 5JX
M0 YGG A. Mansfield, 20 High Street, Broughton, Kettering, NN14 1NG
MW0 YGJ G. Owen, 14 Bideford Road, Newport, NP20 3BJ
M0 YGW J. Newton, 6413 Hillegass Avenue, Oakland, USA, 94618
M0 YHA Aries ARC, c/o A. Clayton, 6 Albert Road, Bunny, Nottingham, NG11 6QE
MM0 YIA A. Maclennan, 5 Baluachrach, Culbokie, Dingwall, IV7 8FP
M0 YIG G. Coleman, 15 Redwood Drive, Ormskirk, L39 3NS
M0 YIJ G. Chesters11, 50 Primrose Chase, Goostrey, Crewe, CW4 8LJ
M0 YIM M. Ellwood, 27 Bath Meadow, Halesowen, B63 2XH
M0 YIT D. Gresty, 81 Old Hall Road, Sale, M33 2HU
M0 YJD J. Davidson, 4 Willows Avenue, Alfreton, DE55 7ER
M0 YJL C. Liu, Flat 603, Nexus Point, 10 Edwards Road, Birmingham, B24 9EQ
M0 YJO J. Hatton, 49 Buxton Street, Morecambe, LA4 5SR
M0 YJT C. Jarvis, 516 Kingsbury Road, Erdington, Birmingham, B24 9NF
M0 YJW J. Williams, 66 Oakfield Avenue, Hitchin, SG4 9JD
M0 YKB D. Yakub, 42 Swift Close, Blackburn, BB1 6LF
M0 YKC D. Forshaw, 14 Hope Carr Road, Leigh, WN7 3ET
M0 YKR Yorkshire Resistors, c/o G. Bystryakov, 20 Elmhurst Gardens, Leeds, LS17 8BG
M0 YKS S. Davison, 5 Denby Drive, Baildon, Shipley, BD17 7PQ
M0 YLA R. Cato, Orrell House, Winterpit Lane, Horsham, RH13 6LZ
M0 YLG G. Roddis, 61 Everton Road, Potton, Sandy, SG19 2PB
MW0 YLS S. Smith, 102 Gresford Road, Llay, Wrexham, LL12 0NW
MI0 YLT S. Mccormick, 46 Lany Road, Moira, Craigavon, BT67 0NZ
M0 YLY S. Myland, 2 Willhays Close, Kingsteignton, Newton Abbot, TQ12 3YT
M0 YMA A. Banks, 2 Holt Close, Farnborough, GU14 8DG
MW0 YMB Dr W. Dickson, The Rowans, Pwllmeyric, Chepstow, NP16 6LA
MI0 YMF M. Foley, 44 Gallows Street, Dromore, BT25 1BD
MM0 YMG M. Gibson, 18 Pentland View, Edinburgh, EH10 6PS
M0 YMJ P. Coppin, 3 Firtree Close, Rough Common, Canterbury, CT2 9DB
M0 YMM N. Trangmar, 8 Maxstoke Close, Meriden, Coventry, CV7 7NB
M0 YNK Y. Watkins, 1 St. Saviour Close, Colchester, CO4 0PW
M0 YNY N. Oldrid, 4 Bar Lane Mapplewell, Barnsley, S75 6DQ
M0 YOB D. Ferguson, 39 Fouracres, Maghull, Liverpool, L317BP
M0 YOJ J. Boone, Amberley, Pinewood Road, High Wycombe, HP12 4DA
M0 YOL D. Parker, 16 Aldborough Road, Dagenham, RM10 8AS
M0 YOM J. Thresher, Quarry Grange, Nuneaton Road Over Whitacre, Coleshill, Birmingham, B46 2NH
MM0 YOS O. Sturm, Lochinvar House, Dalry, Castle Douglas, DG7 3XJ
M0 YOT J. Partington, 56 Rutherford Drive, Bolton, BL5 1DL
M0 YPJ P. Kirby, 30 New Street, Eccleston, Chorley, PR7 5TW
M0 YPS Hunmanby Primary ARC, c/o C. Fox, Millstone Cottage, Prior Wath Road, Scarborough, YO13 0AZ
M0 YPW Dr P. Woodfin, Laurel Cottage, Barrow Street, Much Wenlock, TF13 6EN
M0 YQB R. Suda, 3-11-206, Hiyoshi-Honcho, Kohoku-Ku, Yokohama-City, Kanagawa-Pref, Japan, 223-0062
M0 YQC D. Berry, 27 Harcourt Terrace, Headington, Oxford, OX3 7QF
M0 YQQ S. Evans, 30 Dalglish Creasent Radbrook, Shrewsbury, SY3 9FW
M0 YRC Mango, c/o M. Hodgkinson, 34 Pennine Way, Brierfield, Nelson, BB9 5DT
M0 YRF Yorkshire Radio Friends, c/o R. Potter, 30 High Street, Doncaster, DN7 6RY
M0 YRG A. Tring, 12 Ainsdale Close, Orpington, BR6 8DJ
M0 YRM M. Radulov, 14 Grove Road, Chatham, ME4 5HS
M0 YSE K. Moysey, 109 Langbrook Cottages, Langbrook, Ivybridge, PL21 9JX
M0 YSG St George's School, c/o M. Hubbard, 14 Parkfield Crescent, Kimpton, Hitchin, SG4 8EQ
MM0 YSK S. Ram, 28 Craigievar Gardens, Kirkcaldy, KY2 5SD
M0 YSR R. Moys, 12A Palmerston Avenue, Fareham, PO16 7DP
MM0 YTA J. Anderson, The Grange, Leslie Road, Scotlandwell, Kinross, KY13 9JE
M0 YUG G. Bates, 230 Brook Street, Erith, DA8 1DZ
M0 YUX S. Fabris, Via Marchesan 43, Treviso, Italy, 31100
M0 YVG D. Cordes, 22 Holywell Avenue, Newcastle upon Tyne, NE6 3RY
MW0 YVK E. Howells, 72 Holywell Crescent, Abergavenny, NP7 5LG
MW0 YVT C. Williams, 1 South View, Freeholdland Road, Pontypool, NP4 8LL
M0 YVX B. Tomlinson, 7 Springwell Close, Crewe, CW2 6TX
M0 YWA M. Pack, 59 West End Falls, Nafferton, YO25 4QA
M0 YWO M. Bruyneel, 14 Riversmead, St Neots, PE19 1HA
M0 YXR C. Richardson, Heathercroft, Kirkby Mills, Kirkbymoorside, York, YO62 6NN
M0 YYA G. Bridge, 16 Victoria Street, Ramsbottom, Bury, BL0 9ED
M0 YYT S. Bennett, 7 Holme Park, High Bentham, LA2 7ND
M0 YYV M. Lomax, 7 Planetree Road, West Derby, Liverpool, L12 6RE
M0 YYY Yate Contest Group, c/o H. Taylor, Sunnyside Well, Chaingate Lane, Bristol, BS37 9XN
MM0 YZE West Scotland Air Cadet Radio Club, c/o K. Brown, 10 Richmond Street, Clydebank, G81 1RF
M0 YZF C. Preece, 14 Dock Street, Widnes, WA8 0QX
M0 YZG G. Castledine, 11 Cam Road, Cheltenham, GL52 5QS
M0 YZT Dr H. Orridge, Stonecross Cottage, Wadworth Hall Lane, Wadworth, Doncaster, DN11 9BH
M0 YZV L. Sadler, 2 Birkdale Avenue, Dinnington, Sheffield, S25 2SX
M0 YZY S. Collins, 5 Fernleigh Gardens, Stafford, ST16 1HA
M0 ZAA J. Wellard, 19 South Motto, Kingsnorth, Ashford, TN23 3NJ
MW0 ZAB A. Bubb, Ground Floor Flat 5 Temple Street, Newport, NP202GJ
M0 ZAE H. Ehm, 17 Stuart Road, Kempston, Bedford, MK42 8HS
M0 ZAF R. Barter, 8 Orchard Close, Newton Abbot, TQ12 3DF
M0 ZAI M. Grice, 48 St. Ives Road, Coventry, CV2 5FZ
M0 ZAK J. Steel, 6 Central Avenue, Shepshed, Loughborough, LE12 9HP
MM0 ZAL B. Keiller, Da Cro, Branchiclate, Burra Isle, ZE2 9LA
M0 ZAM G. Campbell, 10 Welbeck Road, Rochdale, OL16 4XP
MW0 ZAP J. Davies, Rose Villa, Creigiau, Cardiff, CF15 9NN
MW0 ZAQ C. Rayment, Brambles, Alltami Road, Mold, CH7 6RT
M0 ZAR Dr S. Smith, Po Box 446, Clanwilliam, Clanwilliam, South Africa, 8135
M0 ZAV R. Amos, 6 Eccles Road, Wittering, Peterborough, PE8 6AU
MM0 ZAW A. Woodford, Nordkette, Evnabrek, Levenwick, Shetland, ZE2 9GY
M0 ZAY H. Jones, 17 Doves Yard, London, N1 0HQ
MM0 ZBD D. Brown, 181/1 (Gf) Gorgie Road, Edinburgh, EH11 1TT
MM0 ZBH P. Mclaren, 1 Morayvale, Aberdour, Burntisland, KY3 0XE
M0 ZBT S. Green, 6 Garth Villas, Rimswell, Withernsea, HU19 2DB
M0 ZBZ M. Carvell, 10 Burns Close, Stevenage, SG2 0JN
MW0 ZCE M. Douglas, 486 Malpas Road, Newport, NP20 6NB
MM0 ZCG Shetland Contest Group, c/o H. Stoeteknuel, C/O Marriott Parkview, Dunrossness, Shetland, ZE2 9JG
M0 ZCJ C. Jonas, 198 High Street, Chesterton, Cambridge, CB4 1NX
M0 ZCM A. Adams, 45 Four Oaks Road Tedburn St. Mary, Exeter, EX6 6AP
M0 ZCO C. O Broin, Dewhurst, Orchard End, Weybridge, KT13 9LS
M0 ZCP C. Parker, The Grange, Watercombe Cornwood, Nr. Ivybridge, PL21 9RB
MM0 ZCT C. Thompson, Lochview West, 2 St. Ninians Avenue, Linlithgow, EH49 7BP
M0 ZCW P. Smith, 101 Brunel Avenue, Newthorpe, Nottingham, NG16 3RE
M0 ZDB D. Brownsea, 47 Southill Road, Bournemouth, BH9 1SH
M0 ZDC Dr D. Cooke, Apartment 9, 27 Sheldon Square, London, W2 6DW
M0 ZDD A. Henderson, Clocktower Lodge, Cragside, Rothbury, Morpeth, NE65 7PU
M0 ZDE D. Kirkden, 57 Crow Hill Road, Margate, CT9 5PF
M0 ZDG D. Griffin, 101 Kingsway, Duxford, Cambridge, CB22 4QN
M0 ZDH D. Hardwick, 30 Halfcot Avenue, Stourbridge, DY9 0YB
M0 ZDJ S. Scott, 13 Silver Close, Harrow, HA3 6JT
M0 ZDM D. Mason, 94 Guessburn, Stocksfield, NE43 7QR
M0 ZDO A. Hipkiss, 2 Brooklands, Walsall, WS5 4DJ
M0 ZDU A. Hawkes, Coachmans Emsworth Road, Lymington, SO41 9BL
MW0 ZDX G. Western, 5 Meredith Close Acrefair, Wrexham, LL14 3GB
M0 ZEB D. Featherby, 14 Station Road, Sutton, Ely, CB6 2RL
M0 ZED P. Phillips, 2 Millstream Close, Goostrey, Crewe, CW4 8JG
M0 ZEE J. Rufes, Flat 1 & 3-8 12 Smyrna Road, London, NW6 4LY
M0 ZEH S. Hendy, Flat 2, 33 Kingston Road, Leatherhead, KT22 7SL
M0 ZEL S. James, 35 Prospect Road, Dronfield, S18 2EA
M0 ZEM R. Donaldson, 10 Berry Avenue, Trimdon Grange, Trimdon Station, TS29 6EE
MW0 ZEN A. Bolton, Valley Lodge, Upper Redbrook, Monmouth, NP25 4LU
M0 ZEQ F. Trigg, Pendle, 2 Langley Common Road, Wokingham, RG40 4TS
MM0 ZET Eshaness Radio Club, c/o H. Hassel, Sumra, Eshaness, Shetland, ZE2 9RS

M0 ZEY N. Horton, 51 Walsingham Gardens, Epsom, KT19 0LS
M0 ZFF D. Almond, 55 Forde Park, Yeovil, BA21 3QP
MM0 ZFG S. Street, Tangaroa, Fairfield Gardens, Kilcreggan, Kilcreggan, G84 0HS
M0 ZGB J. Hobbs, 82 Perry'S Lane, Wroughton, Swindon, SN4 9AP
M0 ZGT E. Ayre, 1 Spring Gardens, Broadmayne, Dorchester, DT2 8PP
M0 ZID S. Frampton, 20 Winslow Close, Boldon Colliery, NE35 9LR
MM0 ZIF M. Hazel-Mcgown, 9 Barra Wynd, Irvine, KA11 1DB
M0 ZIG J. Stoppard, 15 South Lodge Court, Old Road, Chesterfield, S40 3QG
M0 ZIM M. Raynor, 68 Cambridge Street, South Elmsall, Pontefract, WF9 2AR
M0 ZIP Cross Border Contest Group, c/o K. Pritchard, 9 Golf Close, Pyrford, Woking, GU22 8PE
M0 ZJB M. Collier, 32 London Road, Dereham, NR19 1AW
M0 ZJO J. Rawlinson, Westfield Farm, Risden Lane, Cranbrook, TN18 5DU
M0 ZJQ A. Rawlinson, Westfield Farm, Risden Lane, Hawkhurst, Sandhurst, Cranbrook, TN18 5DU
M0 ZJV S. De Vries, 46 Chaulden House Gardens, Hemel Hempstead, HP1 2BP
M0 ZKA F. Hruszka, 20 Winchester Avenue, Leicester, LE3 1AU
M0 ZKK M. Bayman, Marsworth, Tring, HP23 4LX
MI0 ZKX S. Mcnulty, 13 Syenite Place, Rostrevor, Newry, BT34 3EP
M0 ZLE M. Holmes, 6 Wells Court, Saxilby, Lincoln, LN1 2GY
M0 ZLF T. Lovell, 7 Victoria Wharf Victoria Street North, Grimsby, DN31 1PQ
M0 ZLH A. Stabler, 11 Lincolns Avenue, Gedney Hill, Spalding, PE12 0PQ
M0 ZLI D. Challis, 3 West Leaze Place, Bradley Stoke, Bristol, BS32 8AF
M0 ZLK C. Forber, 32 Larch Avenue, Newton-le-Willows, WA12 8JF
M0 ZLP R. Edmondson, Kamway, Stanhoe Road, King's Lynn, PE31 8NJ
M0 ZMB P. Smart, 142 Finch Road, Chipping Sodbury, Bristol, BS37 6JB
M0 ZMM Dr R. Hodgkinson, 39 Oxford Road, Carlton-In-Lindrick, Worksop, S81 9BD
M0 ZMO L. Whitfield, 49 Northview Road, Dunstable, LU55HB
M0 ZMS M. Strickland, Ancoats, Piercy End, York, YO62 6DQ
M0 ZMT M. Thompson, 133 Redford Avenue, Horsham, RH12 2HH
M0 ZMX M. Hardingham, Prospect House, High Street Isle Of Grain, Rochester, ME3 0BS
M0 ZNP C. Gray, 2 Gloucester Road, Pilgrims Hatch, Brentwood, CM15 9ND
M0 ZNZ Dr G. Richardson, Berwick Cottage, Bailes Lane, Guildford, GU3 2AX
M0 ZOE Z. Dunne, 1 Burton Gardens, Brierfield, Nelson, BB9 5DR
MM0 ZOG P. Riddle, Carngeal, Pitlochry, PH16 5JL
M0 ZOM C. Langdon, 6 Glebe Close, Stockton, Southam, CV47 8LG
M0 ZOO Worcester Radio Amateurs Association, c/o P. Badham, 201 York Avenue, East Cowes, PO32 6BH
M0 ZOR B. Trayhurn, 15 Wight Drive, Caister-On-Sea, Great Yarmouth, NR30 5UN
M0 ZOV J. Renmans, 17 Cartmel Crescent, Chadderton, Oldham, OL9 8DA
M0 ZPA P. Davies, 68 Sidmouth Avenue, Stafford, ST17 0HF
M0 ZPC E. Krebser, P.O.Box 621 Port Edward, Kwazulu, Natal, South Africa, 4295, Port Edward, South Africa, 4295
M0 ZPD P. Davies, 46 Spring Street, Colley Gate, Halesowen, B63 2SZ
M0 ZPG P. Morris, 10 Haslam Avenue, Sutton, SM3 9ND
M0 ZPL D. Janowicz, 20 Salisbury Road, St Leonards-on-Sea, TN37 6RX
M0 ZPM P. Mcgillewie, 64 Caradoc View, Hanwood, Shrewsbury, SY5 8ND
M0 ZPU R. Compton, 18 Drove Road, Gamlingay, SG193NY
M0 ZPZ C. Parry, 27 Tynedale Close, Stockport, SK5 7NA
MM0 ZRC R. Chroston, North Stonganess, Cullivoe, Shetland, ZE2 9DD
M0 ZRD D. Hind, 19 Ellington Road, Arnold, Nottingham, NG5 8SJ
M0 ZRF R. Fidler, 44 Windermere Avenue, Ramsgate, CT11 0PF
M0 ZRG G. Reynolds, 43 Orchard Drive, Watford, WD17 3DX
M0 ZRR T. Cooper, 9 Websters Close, Shepshed, Loughborough, LE12 9AT
M0 ZRS R. Styles, 4 Coningsby Close, Gainsborough, DN21 1SS
M0 ZRX K. Bianchini, 10 St. Leonards Road, Headington, Oxford, OX3 8AA
MI0 ZSC J. Sinclair, 29 Tern Crescent, Carrickfergus, BT38 7RU
M0 ZSJ J. Gibson, 22 Woodburn Drive Chapeltown, Sheffield, S351YS
M0 ZSM S. Sissens, 20 Fallow Drive, Eaton Socon, St Neots, PE19 8QL
M0 ZSS M. Chamberlain, 30 Roxton Rd, Great Barford, MK44 3 LR
M0 ZTD Dr T. Digman, 74 Baddlesmere Road, Whitstable, CT5 2LA
M0 ZTE S. Broom, 128 Springhill Road, Wolverhampton, WV11 3AQ
M0 ZTG A. Hill, 5 Park Road, Thurnscoe, Rotherham, S63 0TG
M0 ZUB D. Zubrzycki, 11 Bishopdale Holme, Bradford, BD62AB
M0 ZUI D. Bill, 5 Kennington Road, Wolverhampton, WV10 9RJ
MW0 ZUS T. Lewis, 34 Erw Goch, Ruthin, LL15 1RR
M0 ZVB P. Carpenter, 11 Lakeside, Beckenham, BR3 6LX
M0 ZVF R. Baxter, 30 Croft Gate, Bolton, BL2 3JJ
M0 ZVR B. Bateman, 27 Imperial Avenue, Kidderminster, DY10 2RA
MU0 ZVV J. Bligh, The Bounty, Salines Lane, St Sampson, Guernsey, GY2 4FL
MW0 ZWR G. Spicer, 6 Cromwell Road, Neath, SA10 8DR
M0 ZWT I. Lonsdale, 23 Hunts Field Clayton-Le-Woods, Chorley, PR6 7TT
M0 ZWW Dr W. Warwicker, 28 Porters Wood, Petteridge Lane, Matfield, Tonbridge, TN12 7LR
M0 ZXG G. Carless, Silver Cottage, Silver Street, South Petherton, TA13 5BY
MM0 ZXI J. Stewart, 6 Sutherland Way, East Kilbride, Glasgow, G74 3DL
M0 ZXJ J. Wildsmith, Flat 34, Romsley Hill Grange, Farley Lane, Romsley, Halesowen, B62 0LN
M0 ZXQ I. Talbot, 41 Elmwood Close, Cannock, WS11 6LX
M0 ZXW D. Bambrough, 7 Barnwell View, Herrington Burn, Houghton le Spring, DH4 7FB
MW0 ZXY A. Pugh, 27 Cae'R Pandy, Penrhyn-Coch, Aberystwyth, SY23 3FT
M0 ZYD D. Moger, 23 Elmsleigh Road, Paignton, TQ4 5AX
M0 ZYF Wessex DX Group, c/o Dr T. Langdon, 58 Upper Marsh Road, Warminster, BA12 9PN
MM0 ZYT J. Mccoll, 25 Nelson Ave, Coatbridge, Ml5 5Lr, Coatbridge, ML5 5LR
M0 ZYZ F. Wagstaff, 19 Grange Park Avenue, Bedlington, NE22 7EF
M0 ZZA A. De Maillet, Brock Cottage, The Park, Lower Brailes, Banbury, OX15 5JB
M0 ZZE S. Sims, 2 Kintbury, Duxford, Cambridge, CB22 4RR
M0 ZZI A. Downing, Chrysostomou Nicolaou 2 8577 Tala, Paphos, Cyprus, 8577
MM0 ZZO J. Mcginty, 1/1 119 Neilston Road, Paisley, PA2 6ER
M0 ZZT S. Webber, 59 Mincinglake Road, Stoke Hill, Exeter, EX4 7DY

M1

M1 AAC L. Healy, 31 Roach Road, Sheffield, S11 8UA
MW1 AAH D. Creber, 8 George Manning Way, Gowerton, Swansea, SA4 3HB
M1 AAS A. Jackson, 29 Bramble Avenue, Birkenhead, CH41 0AX
M1 AAY Dr N. Thomson, 11 School Close Stamford Bridge, York, YO41 1PT
MM1 ABA I. Hopley, 53 Redmoss Road, Aberdeen, AB12 3JJ
M1 ABC B. Johnson, 20 Valleyside, Hemel Hempstead, HP1 2LN
M1 ABF K. Perkin, 25 Rownall View, Leek, ST13 8JN
M1 ABG H. Mallin, Riverside House, Rope Walk, Southampton, SO31 4HD
M1 ABM M. Chapman, Woodcroft, Windmill Green, Pevensey, BN24 5DY
MW1 ABT R. Macleod, 24 Heol Powis, Gungrog Hill, Welshpool, SY21 7TP
M1 ABU S. Norman, 53 Saddlers Park, Eynsford, Dartford, DA4 0HA
M1 ABV B. Breet, 23 Mitchell Street, Eccles, Manchester, M30 8AJ
M1 ABX S. Painting, Claytons, Inkpen, Newbury, RG17 9QE
M1 ABY A. Highams, Rua Das Americas, 17, Fim Da...Rua America Do Norte, Portal Das Americas, Morretes, Brazil, 83350-000
MW1 ABZ I. Jones, 11 Glamorgan Street Canton, Cardiff, CF5 1QS
M1 ACA G. Barnett, 63 Sandcroft, Sutton Hill, Telford, TF7 4AB
M1 ACB S. Thomas, 2 Myrtle Cottages, Sandy Lane, Saxmundham, IP17 1HR
M1 ACC J. Chambers, 9 Farnborough Road, Swindon, SN3 2DR
M1 ACE P. Musk, 38 Glenwood, Welwyn Garden City, AL7 2JS
M1 ACF ACF/CCF Radio Club, c/o M. Buckley, Springfield, 12 Ranmore Avenue, Croydon, CR0 5QA
M1 ACJ S. Shearing, 42 Meadow Park, Wesham, Preston, PR4 3DN
M1 ACK R. Mackay, Conifers, The Street, Guildford, GU4 7TJ
M1 ACL K. Green, 42 Dartmouth St., Stoke on Trent, ST6 1HB
M1 ACN M. Goom, 47 Sandringham Court, Slough, SL1 6JU
M1 ACO R. Hankin, Solway View, Whitrigg, Torpenhow, Wigton, CA7 1JG
M1 ACQ S. Thomas, 111 Jersey Avenue, Bristol, BS4 4QX
M1 ACT C. Leeds, 12 Northfields, Norwich, NR4 7EU
M1 ADK G. Pratt, 2 Houghton Road, Newbottle, Houghton le Spring, DH4 4EF
M1 ADN T. Whiting, 28 Legarde Avenue, Hull, HU4 6AP
M1 ADP K. Eastwood, 35 Finisterre Rise, Great Yarmouth, NR30 5TT
M1 ADT R. Vickerstaff, 16 Sewell Wontner Close, Kesgrave, Ipswich, IP5 2GB
M1 ADV I. Owen, 104 Hall Farm Road, Melton, Woodbridge, IP12 1RW
M1 ADX J. Towns, 72 Longfields Road, Norwich, NR7 0NA
M1 ADZ N. Davis, 71 Brettenham Road, London, E17 5AZ
M1 AEA M. Waldron, 32 Windmill Street, Upper Gornal, Dudley, DY3 2DQ
M1 AEB G. Haughie, 100 Henton Road, Edwinstowe, Mansfield, NG21 9LE
M1 AED M. Cattell, 12 Fairway Road, Warley, Oldbury, B68 8BE
M1 AEG A. Green, Michigan, North Road, Whitemoor, Nanpean, St Austell, PL26 7XN
M1 AEH D. Cossey, 11 Halden Avenue, Norwich, NR6 6UX
M1 AEI D. Bowyer, East Foldhay, Zeal Monachorum, Crediton, EX17 6DH
M1 AEJ A. Benjamin, Fieldview, Field Common Lane, Walton-on-Thames, KT12 3QH
M1 AEK D. Mulliner, 23 Nostell Way, Bridlington, YO16 6FY
MM1 AEL C. Haswell, 6 Lochlann Road, Culloden, Inverness, IV2 7HB
M1 AEO R. East, 27 Caddywell Meadow, Torrington, EX38 7NZ
M1 AEP M. Griffin, 76 Waylands, Swanley, BR8 8TN
M1 AEQ F. Allenby, 9 Church View, Holme-On-Spalding-Moor, York, YO43 4BG
M1 AEV A. Aiello, Llamedos, Walnut Road, Wisbech, PE14 7NP
M1 AEX R. Pyman, 16 Bramshott Close, Allington, ME16 0RX
MM1 AEZ R. Barrett, 59 Dunrobin Road, Kirkcaldy, KY2 5YT
M1 AFF R. Dyer, 4 Downleaze, Durrington, Salisbury, SP4 8AB
M1 AFP P. Jefford, 61 Willow Way, Flitwick, Bedford, MK45 1LN
M1 AFQ A. Brooks, 86 Violet Road, Norwich, NR3 4TS
M1 AFU S. Derwin, 5 Hawthorne Grove, Yarm, TS15 9EZ
M1 AFV S. Wells, 1 Neath Gardens, Leeds, LS9 6RG
MW1 AFW C. Jones, Arosfa House, 7 Wood Street, Bargoed, CF81 8NW
M1 AFX D. Hill, 3 Morcar Road, Stamford Bridge, York, YO41 1PR
M1 AFY S. Jones, Dairy Cottage, Menith Wood, Worcester, WR6 6UB
M1 AFZ C. Grizzell, 16 Lower Park, Minehead, TA24 8AX
M1 AGA R. Taylor, 5 Thirlmere Drive, Bury, BL9 9QE
M1 AGE D. Thorley, 53.The Spires, Moreton on Lugg, HR4 8FJ
M1 AGH D. Hirst, 2 The Birches, Marlborough Road, Swindon, SN3 1PT
M1 AGK R. Large, 5 Jasmine Close, Abbeydale, Gloucester, GL4 5FJ
M1 AGP J. Davies, 10 Gorselands, Hollesley, Woodbridge, IP12 3QL
M1 AGR M. Skinner, 44 Westminster Crescent, Sheffield, S10 4EX
M1 AGW S. Whitehouse, 47 Mulberry Road, Bloxwich, Walsall, WS3 2NG
M1 AGY J. Cuddy, 4 Thames Gardens, Plymouth, PL3 6HD
M1 AHA S. Lefevre, 9 Old Barn Crescent, Hambledon, Waterlooville, PO7 4SW
M1 AHF M. Byatt, 15 Lower Farm Road, Plympton, Plymouth, PL7 1JJ
M1 AHJ P. Clarke, 36 Eldred Drive, Orpington, BR5 4PF
MM1 AHL D. Mcarthur, 12 Laburnum Grove, Lenzie, Kirkintilloch, Glasgow, G66 4DF
M1 AHN G. Boyce, 10 Quarry Close, Ross-on-Wye, HR9 7DR
M1 AHR K. Peters, 82 Blackmoor Road, Moortown, Leeds, LS17 5JP
M1 AHT L. Russell, 106 Stambridge Road, Rochford, SS4 1DP
MW1 AHU Dr G. Armstrong, 61 Victoria Drive, Llandudno Junction, LL31 9PF
M1 AHY M. Rhodes, 1 Chetwode, Overthorpe, Banbury, OX17 2AB
M1 AIB P. Lewis, 58 Ocean Close, Fareham, PO15 6QP
M1 AIK T. Bardgett, 49 St. James St., South Petherton, TA13 5BN
M1 AIM A. Moore, Silver Trees, Woodlands Lane, Pulborough, RH20 3HG
M1 AIN D. White, 19A Gravenhurst Road, Campton, Shefford, SG17 5NY
M1 AIX W. Greenall, 356 Warrington Road, Abram, Wigan, WN2 5XA
M1 AIY M. Bastin, 14 Golvers Hill Road, Kingsteignton, Newton Abbot, TQ12 3BP
M1 AJA Dr L. Mason, 45 Willow Rise, Thorpe Willoughby, Selby, YO8 9PP
M1 AJG M. Ramskill, 7 Hobart Road, Dewsbury, WF12 7LS
M1 AJM S. Hoskins, 21 Wicken House, London Road, Maidstone, ME16 8QP
M1 AJQ W. Clarke, 10 Athol Close, Sinfin, Derby, DE24 9LZ
M1 AJT P. Amos, 16 Eastry Road, Erith, DA8 1NN
M1 AJU S. Bradford, 28 Downs Road, Walmer, Deal, CT14 7SY
M1 AKF P. Wilson, 9 The Brooklands, Wrea Green, Preston, PR4 2NQ
M1 AKH D. Mulvana, 17 Wildlake Orton Malborne, Peterborough, PE2 5PG
M1 AKL V. Parsons, 20 Old Top Road, Hastings, TN35 5DJ
M1 AKN R. Day, 20 Gardiner Close, Bury St Edmunds, IP33 2UB
M1 AKQ S. Hutton, Astrid, Main Street, Reedness, DN14 8ER
M1 AKT D. Thomas, 10 The Willows Oxspring, Sheffield, S36 8ZZ
M1 AKV R. Kowalski, 2 Newcross Park, Kingsteignton, Newton Abbot, TQ12 3TJ
M1 AKZ M. Dean, 164 Ambleside Road, Lightwater, GU18 5UW
M1 ALA D. Mawson, 84 Walnut Avenue, Weaverham, Northwich, CW8 3DX
M1 ALE A. Ratcliff, 71 Spring Gardens, Leek, ST13 8DD
M1 ALF R. Coston, 272 Warley Road, Blackpool, FY2 0UG

M1 ALG J. Sindall, 9 James Street, Rotherham, S60 1JU
M1 ALH K. Whinney, 6 Eve Balfour Way, Haughley, Stowmarket, IP14 3NW
M1 ALM R. Hodgkins, 38 Byron Road, Gillingham, ME7 5QH
M1 ALO P. Rogers, Flat 11, Green Court, Lewes, BN7 1HY
M1 ALR C. Moore, Colchester House, Farrington Road, Bristol, BS39 7LW
M1 ALT M. Oram, 43 Peverell Avenue West, Poundbury, Dorchester, DT1 3SU
M1 ALU T. Summers, Caravan 26, Waitegates Caravan Site, Doncaster, DN7 4EJ
M1 ALX D. Philip, 10 Rosedale Gardens, Bodmin, PL31 2HE
M1 AMA A. Day, 9 Arundel Road, Tewkesbury, GL20 8AS
M1 AMB D. Snow, 7 Aynsley Close, Cheadle, Stoke-on-Trent, ST10 1DP
M1 AMI D. Chambers, 94 Hawthorn Avenue, Colchester, CO4 3JR
M1 AMJ D. Bonnett, 254 Norwich Road, Wisbech, PE13 3UT
M1 AMP H. Withers, 23 Fernie Road, Guisborough, TS14 7LZ
M1 AMW C. Whitehead, 18 Victoria Quay, Ashton-On-Ribble, Preston, PR2 2YW
M1 AMZ K. Birch, 16 Brentwood Avenue, Thornton-Cleveleys, FY5 3QR
M1 ANC C. Mclean, 18 Chatfield Road, Gosport, PO13 0TN
MW1 AND M. Griffiths, 32 Hill Street, Aberdare, CF44 6YG
M1 ANK R. Taylor, 86-88 Hillside Crescent, Leigh-on-Sea, SS9 1HQ
M1 ANL D. Clapp, Wenlock Edge, Park Hill, Shepton Mallet, BA4 4AZ
M1 ANN A. Webb, 20 The Fleet, Stoney Stanton, LE9 4DY
M1 ANO C. Worlledge, 181 Roselands Drive, Paignton, TQ4 7RN
MM1 ANP Dr J. Tobias, Gowanpark House, Gowanpark, Cumnock, KA18 2NZ
M1 ANQ D. Hirst, 66 Turncroft Lane, Stockport, SK1 4AB
M1 ANR R. Hall, 12 West Lane, Edwinstowe, Mansfield, NG21 9QT
M1 ANT D. Simcock, 51 Broadway, Stockport, SK2 5SF
M1 AOB R. Pentney, 11 Beech Park Holsworthy Beacon, Holsworthy, EX22 7NB
M1 AOD V. Bolger, Little Annaside, Bootle, Millom, LA19 5XL
M1 AOF A. Wainwright111, 8 Mount Drive, Purdis Farm, Ipswich, IP3 8UU
M1 AOG S. Moriarty, 31 Guernsey Way, Banbury, OX16 1UE
M1 AOL J. Pepper, 16 Chartwood, Loggerheads, Market Drayton, TF9 4RJ
M1 AOR S. Darrigan, 70 Somerset West Kirby, Wirral, CH486EJ
M1 AOU R. Pinchen, 9 Orwell Close, Swindon, SN25 3LZ
M1 AOX J. O'Neill, 4 Heathlands Road, Little Sutton, Ellesmere Port, CH66 5PB
M1 APB S. Sanders, 52 Hazelwood Road, Callington, PL17 7EU
M1 APC J. Hulme, 28 Chapel Close, Gunnislake, PL18 9JB
M1 APF J. Thompson, 97 King Lane, Leeds, LS17 5AX
M1 APH P. Hildebrand, 82 Reed Drive, Redhill, RH1 6TB
M1 APL A. Speakman, 28 Eden Gardens, Leeds, LS4 2TQ
M1 APQ M. Hodson, 93 Deer Park Road, Fazeley, Tamworth, B78 3SZ
MM1 APS C. Stuart, 5 Deloraine Court, Hawick, TD9 7QE
M1 APT I. Waterhouse, Little Bracken, 9 Willingdon Drove, Eastbourne, BN23 8AL
M1 APV G. Andrews, 18 Elms Close, Whitefield, Manchester, M45 8XR
M1 AQI M. Sanderson, 14 Hazelwood Avenue, York, YO10 3PD
M1 AQJ P. Jackson, 39 Chapel Street, Hazel Grove, Stockport, SK7 4HW
M1 AQN P. Wilson, 14 Jerome Way, Shipton-On-Cherwell, Kidlington, OX5 1JT
M1 AQP C. Chapman, 9 Edinburgh Avenue, Sawston, Cambridge, CB22 3DW
M1 AQR A. Cooper, 61 Greenhaze Lane, Great Cambourne, Cambridge, CB23 5EF
M1 AQX A. Pell, 4 Mill Close, Braunston, Daventry, NN11 7HY
M1 AQY G. Cottam, 26 Ayton Court, Bedlington, NE22 6NS
M1 ARF D. Riley, 9 Century Avenue, Mansfield, NG18 5EE
M1 ARH W. Mountford, 3 Spurstow Close, Prenton, CH43 2NQ
M1 ARL D. Edwards, 28 Solingen Estate, Blyth, NE24 3EP
MW1 ARM D. Rees, Y Coed, Tan Lan Hill, Holywell, CH8 9JB
M1 ARS H. Cawley, 11 Cleveland Way, Winsford, CW7 1QL
M1 ART F. Melhuish, Alverdean, Mile End Road, Coleford, GL16 7QD
M1 ARU R. Stanley, 113 Upper Brents, Faversham, ME13 7DL
M1 ARX S. Russell, 13 Burdett Road, Crowborough, TN6 2EN
MI1 ASN P. Mcmahon, 26 Ballycraigy Road, Newtownabbey, BT36 5ST
M1 ASR G. Jefferies, Fenland Lodge, Landing Lane, Broomfleet, Brough, HU15 1RE
M1 ASV A. Evans, 14A New Road, Tiptree, Colchester, CO5 0HJ
M1 ATA A. Betts, 31 Porters Wood, Petteridge Lane, Matfield, Tonbridge, TN12 7LR
M1 ATB G. Gale, 2 Manston Crescent, Crossgates, Leeds, LS15 8QZ
M1 ATC Air Training Corps, c/o R. Courtney, 5 Bute Close, Highworth, Swindon, SN6 7HN
M1 ATI R. Proctor, 11 Bedford Rise, Winsford, CW7 1NE
M1 ATJ P. Whitby, 90 Manor Road, Martlesham Heath, Ipswich, IP5 3SY
M1 ATP A. Plitsch, 64 Oxford Road, Lowestoft, NR32 1TP
MM1 ATR L. Robinson, 17 Burn Brae Avenue, Westhill, Inverness, IV2 5RG
M1 ATU M. Bloss, 20 Barry Walk, Brighton, BN2 0HP
MM1 ATY D. Shirley, 17 Carlaverock Terrace, Tranent, EH33 2PL
MM1 AUF C. Mcclintock, 13 St. Andrews Drive, Gourock, PA19 1HY
MM1 AUG M. Mcclintock, 30 Findhorn Road, Inverkip, Greenock, PA16 0HX
M1 AUH A. Easton, 1 Wood Street, Warrington, WA1 3AY
MI1 AUI V. Hughes, 7 Craiglands Manor, Newtownabbey, BT36 5FG
M1 AUK B. Pittaway, 66 Montrose Avenue, Leamington Spa, CV32 7DY
M1 AUN J. Yarnall, 85 Wombourne Park, Wombourne, Wolverhampton, WV5 0LX
M1 AUO Rev. D. Eady, The Rectory, Rectory Lane, Cheltenham, GL51 9RD
M1 AUP A. Bailey, 47 Whiteridge Road, Kidsgrove, Stoke-on-Trent, ST7 4TH
M1 AUR M. Garlick, 59 Foundry Avenue, Leeds, LS9 6BY
MW1 AUV P. Davies, 4 Caradog Place, Townhill, Swansea, SA1 6NH
M1 AUW T. Roberts, 109 Gordon Avenue, Norwich, NR7 0DS
M1 AUZ G. Wright, School House Farm, The Gravel, Mere Brow, Preston, PR4 6JX
M1 AVB J. Pidgeon, 6 Marlborough Road, Musbury, Axminster, EX13 8AH
M1 AVM S. Cooper, 93 Langton Road, Norton, Malton, YO17 9AE
MM1 AVR S. Mciver, 9 Balvicar Road, Oban, PA34 4RP
M1 AVU G. Purrier, Archways, Forge Hill, Lydbrook, GL17 9QS
M1 AVV S. Linney, 3 Severn Road, Walney, Barrow-in-Furness, LA14 3TS
M1 AVW R. Hedges, 10 The Green, Goldenbank, Falmouth, TR11 5PR
M1 AWC M. Worrall, 15 Whitegate Drive, Bolton, BL1 8SF
MI1 AWM D. Reid, 179 Melmount Road, Sion Mills, Strabane, BT82 9LA
M1 AWN J. Sanders, 135 Windmill Avenue, Kettering, NN15 7DZ
M1 AWS A. Jones, 35 St. Marys Close, Aspull, Wigan, WN2 1RL
MW1 AWT P. Mccarthy, 43 Bodnant Road, Llandudno, LL30 1LT
MM1 AWV R. Lynch, 21 Carnoustie Avenue, Gourock, PA19 1HF
M1 AWX S. Yendell, 35 Chester Road, Newquay, TR7 2RH
M1 AXD P. Gartell, 1 Springfields, Richards Castle, Ludlow, SY8 4EP
M1 AXE G. Ecclestone, 11 Longrood Road, Rugby, CV22 7RG
M1 AXG D. Tucker, 2 New Cottages, Bill Hill, Wokingham, RG40 5QU
M1 AXM D. Russell, 39 Mowries Court, Somerton, TA11 6NF
M1 AXP B. Cornall, Fernholm, Taylors Lane, Preston, PR3 6AB
M1 AYA P. Booth, 61 Coalpit Lane, Rugeley, WS15 1EW
M1 AYC A. Booth, 35 Gillamore Drive, Whitwick, Coalville, LE67 5PA
M1 AYG R. Sanders, 17 Shelley Lane, Kirkburton, Huddersfield, HD8 0SW
MI1 AYL B. Byrne, 6 Holymount Road, Gilford, Craigavon, BT63 6AT
M1 AYN J. Hewlett, 28 Coombs Road, Coleford, GL16 8AY
M1 AYR K. Miller, 15A Holly Close, Cherry Willingham, Lincoln, LN3 4BH
M1 AYU R. Freeman, 6 Sutton Road, Leverington, Wisbech, PE13 5DW
M1 AZA M. Gould, 36 Wistaria Road, Wisbech, PE13 3RH
M1 AZB J. Titterton, Longfield Farm, Ifield Road, Horley, RH6 0DR
M1 AZF S. Arbuckle, 34 Swallowcliffe Gardens, Yeovil, BA20 1DQ
M1 AZG A. Ruston, 42 The Straits, Dudley, DY3 3BH
MW1 AZI S. Dunlop, 34 Dan-Y-Coed Clydach, Abergavenny, NP7 0LS
M1 AZJ A. Bottrell, 36 Tremodrett Road, Roche, St Austell, PL26 8JA
M1 AZM P. Jefferson, 20 Buckstone Grove, Leeds, LS17 5HW
M1 AZO K. Austen, 12B Downs Road, Folkestone, CT19 5PW
M1 AZQ A. Beale, Flat 2, Brabstone House, Medway Drive, Greenford, UB6 8LN
MW1 AZR Gwent Raynet Group, c/o R. Snelling, 91 Oakfield Road, Newport, NP20 4LP
M1 AZV R. Aston, 2 Brockwood Crescent, Keyworth, Nottingham, NG12 5HQ
M1 AZY J. Drummond, 60 Park Lane, Exeter, EX4 9HP
M1 BAA M. Crow, 180 Bois Moor Road, Chesham, HP5 1SS
M1 BAC J. Booth, 35 Gillamore Drive, Whitwick, Coalville, LE67 5PA
M1 BAD D. Redding, 11 Camley Gardens, Maidenhead, SL6 5JW
M1 BAI A. Saunders, 128 Foxcroft Drive, Wimborne, BH21 2LA
MW1 BAJ J. Alexander, 19 Rhes Brickyard Row, Llanelli, SA152DZ
M1 BAN Dr T. Baldwin, 17 Warn Crescent, Oakham, LE15 6LZ
M1 BAR Bar-Packers Contest Group, c/o N. Roscoe, 35 Kenilworth Road, Cheadle Heath, Stockport, SK3 0QL
M1 BAS C. Bastin, 42 Peterborough Road, Exwick, Exeter, EX4 2EG
M1 BAV G. Noble, 96 Foxroyd Lane Estate, Dewsbury, WF12 0BD
M1 BBB P. Marshall, 1 Prospect Cottages, St. Anns Chapel, Gunnislake, PL18 9HH
M1 BBH C. Tan, 62A Jalan Sepah Puteri 5/6, Kota Damansara, Selangor, Malaysia,
M1 BBR M. Deglos, 405 Armidale Place, Bristol, BS6 5BQ
M1 BBS Dr D. Hawkes, The Bank East Northdown Close, Margate, CT9 3YA
M1 BBU G. Price, 118 Broadstone Road, Heaton Chapel, Stockport, SK4 5HS
M1 BCB D. Ball, 14 Ayot Path, Borehamwood, WD6 5BJ
M1 BCM J. Worthing, 27 Mayfield Close, Shrewsbury, SY1 4BF
M1 BCR A. Richards, 3 Beeston Close, Watford, WD19 6LF
M1 BCU A. Howe, 6 Crabapple Close, Wymondham, NR18 0XT
M1 BCY T. Keeler, 72 Grafton Road, Selsey, Chichester, PO20 0JB
M1 BCZ R. Craggs, 1 Hylton Court, Bowmonts Road, Tadley, RG26 3SH
M1 BDD E. Reynolds, Cloonagh, Ballinagore, Ireland,
M1 BDH T. Woods, Swan House Tarrington, Hereford, HR1 4EU
M1 BDJ G. Hamlin, Down Farm Bungalow, Stockbridge, SO20 8EA
M1 BDL A. Statham, 24 Fulton Close, High Wycombe, HP13 5SP
M1 BDO W. Lodeweegs, 68 Totterdown Lane, Weston-Super-Mare, BS24 9NJ
M1 BDR Essex Raynet (Braintree District), c/o G. Farrell, 95 Washington Road, Maldon, CM9 6JF
M1 BDS P. Colwell, 56 Hamelin Road, Gillingham, ME7 3EX
MW1 BDV W. Davis, Cartref, Blaenannerch, Cardigan, SA43 1SN
M1 BDY V. Beard, 13 Mayesford Road, Romford, RM6 4NU
M1 BEC Eton College ARC, c/o Rev. M. Wilcockson, Queen'S House 16 High Street, Linton, Cambridge, CB21 4HS
M1 BED Bedford and District ARC, c/o H. Ehm, 17 Stuart Road, Kempston, Bedford, MK42 8HS
M1 BEO D. Tatlow, Mulberry House, Bettys Grave, Cirencester, GL7 5ST
M1 BEP A. Amos, 118 Mount Hill Road, Bristol, BS15 8QR
MW1 BEQ A. Strange, 8 Tregarn Close, Langstone, Newport, NP18 2JL
MW1 BEW C. Sampson, Stable Cottage, Lower House Farm, Welshpool, SY21 8LA
M1 BEX G. Olsen, 36 Bluebell Way, South Shields, NE34 0BZ
MM1 BFE J. Murray, 27 Wellpark Road, Banknock, Bonnybridge, FK4 1TP
M1 BFF A. Buckley, 191 Broadway, Stoke-on-Trent, ST3 5PW
M1 BFG H. Tribe, 20 Penrith Road, Bournemouth, BH5 1LT
M1 BFI Z. Billington, 27 Saxon Gardens, Caister-On-Sea, Great Yarmouth, NR30 5AH
M1 BFO P. Aplin, 30 Cheviot Drive, Charvil, Reading, RG10 9QD
M1 BFR A. Day, 21 Coronation Road, Prestbury, Cheltenham, GL52 3DA
M1 BFV F. Richardson, 3 Carl Moult House, Regent St., Swadlincote, DE11 9PH
M1 BFX J. Belcher, 101 Colne Drive, Romford, RM3 9LA
M1 BFY A. Davey, 34 Monkswood, Littleport, Ely, CB6 1JD
M1 BGF M. Shearman, 4 Mcdonough Close, Fitton Hill, Oldham, OL8 2PD
M1 BGK J. Lewis, 12 Eastleigh Road, Staple Hill, Bristol, BS16 4SQ
M1 BGL G. Langford, 22 Kensington Way, Oakengates, Telford, TF2 6NA
M1 BGS M. Elvers, 3 St. Michaels Road, Maidstone, ME16 8BS
M1 BGT R. Williams, 19 Venice Close, Chellaston, Derby, DE73 5BX
M1 BGY I. Townson, 53 Brompton Road, Bradford, BD4 7JD
M1 BHC M. Lee, 11 Sturrocks, Vange, Basildon, SS16 4PQ
M1 BHE B. Vickers, 17 Linden Close, Dewsbury, WF12 8PL
M1 BHN J. Chambers, 16 Wood Way, Huntington, York, YO32 9QG
MM1 BHO R. Hopkins, 15 Station Drive, Dalbeattie, DG5 4FA
M1 BHP D. Hartley, 10 Tamworth Grove, Clifton, Nottingham, NG11 8JA
M1 BHZ T. Shepherd, 40 Pheasant Way, Cirencester, GL7 1BL
M1 BIB P. Brooke, 34 Park Road, Burwell, Cambridge, CB25 0ES
M1 BIG T. Read, 57 Ollard Avenue, Wisbech, PE13 3HF
M1 BIK C. Blackmur, 9 Cameron Close, Chatham, ME5 0DD
M1 BIL C. Pugh, 16 Park Gate, Somerhill Road, Hove, BN3 1RL
M1 BIX G. Perrins, 1 Cornhill Gardens, Leek, ST13 5PZ
M1 BIY J. Laker, 7 Station Court, West Lane, Hayling Island, PO11 0FP
MW1 BJB S. Mandal, 1 High Ridge Drive, Bersham Road, Wrexham, LL14 4JD
M1 BJC P. Marshall, 75 Drewstead Road, London, SW16 1AA
M1 BJE S. Robinson, 140 The Street, Kirtling, Newmarket, CB8 9PD
MM1 BJG O. Ferula, 4 Castledyke Road, Carstairs, Lanark, ML11 8SU
MM1 BJO R. Lowe, 9 South Quarry Gardens, Gorebridge, EH23 4GX
M1 BJS G. Ausher, 94 New Road, Ditton, Aylesford, ME20 6AE
MM1 BJT D. Smith, 1076 Aikenhead Road, Glasgow, G44 4TJ
MM1 BJZ P. Fraser, 27 Rigghead Avenue Cumbernauld Village, Glasgow, G67 2AY
M1 BKE P. Hewlett, 28 Coombs Road, Coleford, GL16 8AY

M1 BKF W. Hill, 492 Earlham Road, Norwich, NR4 7HP
M1 BKI J. Carins, 26 Roman Way, St. Margarets-At-Cliffe, Dover, CT15 6AH
M1 BKL P. Coddington, 2 Canal View Chemistry, Whitchurch, SY13 1BZ
M1 BKQ T. Beadman, 6 Gaiafields Road, Lichfield, WS13 7LT
M1 BKS M. Kevern, Wheal Bal, Trewellard, Penzance, TR19 7SP
M1 BKW S. Plant, 17 New Road, Driffield, YO25 5DJ
MW1 BLE C. Beech, 9 Wesley Court, Pembroke Dock, SA72 6NE
M1 BLJ S. Brion, 165 Kings Head Hill, London, E4 7JG
M1 BLO P. Hoggard, 41 Malpas Close, Bransholme, Hull, HU7 4HH
M1 BLW E. Banks, 165 Burstall Hill, Bridlington, YO16 7NH
M1 BLX P. Buxton, Bower House, Thornham Road, Eye, IP23 8HP
MI1 BLZ D. Kyle, Sea Breezes, Rathlin Island, Ballycastle, BT54 6RT
M1 BMC K. Termie, 14 Hollins Lane, Marple Bridge, Stockport, SK6 5BB
M1 BMD W. Curtis, 5 Cambridge Road, Kesgrave, Ipswich, IP5 1EN
MM1 BMK E. Mitchell, Carradale Yondertonhill, Hatton, Peterhead, AB42 0RE
M1 BML M. Carroll, 712 Hoover Street, Nelson, Canada, VIL 4X4
M1 BMQ J. Waters, 71 Sixth Avenue, Blyth, NE24 2SU
M1 BMR C. Robinson, 76 Ingrams Way, Hailsham, BN27 3NX
M1 BMU E. Woodward, 6 Lang Road, Huntington, York, YO32 9SD
M1 BMV J. Fox, Stonecroft, Horley, Banbury, OX15 6BJ
M1 BMW J. Burrill, 3 Town Farm Close, Pinchbeck, Spalding, PE11 3SG
M1 BNG R. Smith, 45 The Avenue, Mortimer Common, Reading, RG7 3QU
M1 BNH P. Walton, 135 Starling Road, Bury, Greater Manchester, BL8 2HF
M1 BNI M. Clarke, 1 Burbury Close, Bedworth, CV12 8DU
M1 BNK A. Wood, 1 Minch Road, Hartlepool, TS25 3QY
MI1 BNO B. Bonnar, 49 Cullycapple Road, Aghadowey, Coleraine, BT51 4AR
M1 BNR Holsworthy Community College, c/o G. Forster, 33 Deer Valley Road, Holsworthy, EX22 6DA
MW1 BNY B. Fitzpatrick, 1 Pistyll Newydd Mynyddygarreg, Kidwelly, SA17 4NW
M1 BOA G. Heard, The Saddlers, The Street, Thurgarton, Norwich, NR11 7PD
M1 BOB R. Allen, 43 Vowell Close, Bristol, BS13 9HS
M1 BOD P. Hanfrey, 49 Allotment Road, Niton, Ventnor, PO38 2DZ
MI1 BOE A. Prenter, 5 Knockview Gardens, Newtownabbey, BT36 6UA
M1 BOL D. Harding, Po Box 11755 Apo, Grand Cayman, Cayman Islands,
M1 BOP M. Riley, 3 Foxley Close, Ipswich, IP3 8BW
M1 BOZ A. Thompson, 26 Balmoral Avenue, Clitheroe, BB7 2QH
M1 BPD A. Collins, Old Orchards, Romsey Road, Stockbridge, SO20 6PR
M1 BPK J. Bloor, Hillview House, Whitegates, Bromyard, HR7 4ES
M1 BPN A. Burchell, 25 Cherbury Close, London, SE28 8PG
M1 BPS A. Wellman, Lyndale, Northleach, Cheltenham, GL54 3JJ
M1 BPU B. Lake, 1 Eunice Grove, Chesham, HP5 1RL
M1 BPW P. Williams, 26 Downs View Road, Westbury, BA13 3AQ
M1 BPY D. Dixey, 102 Blackberry Road, Stanway, Colchester, CO3 0RZ
M1 BQC S. Sparks, 36 Tormynton Road, Worle, Weston-Super-Mare, BS22 9HT
M1 BQD K. Sparks, 29 Pennycress, Weston-Super-Mare, BS22 8QH
M1 BQE C. Sparks, 36 Tormynton Road, Worle, Weston-Super-Mare, BS22 9HT
M1 BQF J. Stewart, 45 The Cross, Wivenhoe, Colchester, CO7 9QH
M1 BQM G. Preedy, 12A Grange Court, Prescot Road, Stourbridge, DY9 7LA
MW1 BQO E. Buckley, 3 Cae Eithin, Minffordd, Penrhyndeudraeth, LL48 6EF
M1 BQS F. Gibson, 125 Chelveston Drive, Corby, NN17 2QJ
M1 BQT F. Lloyd, 8 Balfour Cottages Burcot, Abingdon, OX14 3DR
M1 BQU S. Daniels, 9 Beechwoods, Burgess Hill, RH15 0DE
M1 BQW Dr V. Edwards, Elder Cottage, Skeyton Common, Skeyton, NR10 5AU
M1 BQY Trowbridge and District ARC, c/o D. Birch, 32 Union Street, Trowbridge, BA14 8RY
M1 BQZ W. Rowley, 6 Sea King Crescent, Colchester, CO4 9RJ
MI1 BRA Belfast Royal Academy Amateur, c/o N. Moore, 164 Ardenlee Avenue, Belfast, BT6 0AE
MI1 BRS R. Dickey, 8 Coachmans Way, Hillsborough, BT26 6HQ
M1 BRU D. Pope, 32 Barn Crescent, Newbury, RG14 6HD
M1 BRX D. Seddon, 22 Newton Road, Lowton, Warrington, WA3 1EB
M1 BRY J. Fisher, 18 The Smooting, Tealby, Market Rasen, LN8 3XZ
M1 BRZ D. Lee, 53 Portmellon Park, Mevagissey, St Austell, PL26 6XD
M1 BSB W. Charman, 30 Beaufort Road, Bournemouth, BH6 5AL
M1 BSE J. Wharton, Flat 12, Brent Court, Leicester, LE3 2XQ
M1 BSF S. Ogden, 39 Levens Drive, Heysham, LA3 1jn
M1 BSI C. Bowen, 45 Morris Road, Nottingham, NG8 6NE
M1 BSM G. Meyer, 447-449 Manchester Road, Stockport, SK4 5DJ
M1 BSN J. Riley, Peacehaven, Great Street, Stoke-Sub-Hamdon, TA14 6SH
M1 BSO B. Ford, 7 Courtwick Road, Wick, Littlehampton, BN17 7NE
M1 BSP M. Ford, 27 The Street, Rustington, Littlehampton, BN16 3PA
M1 BSU T. Osborne, 134 Merridale Road, Wolverhampton, WV3 9RJ
M1 BSV M. Black, 37 Castle Street, Lancashire, BB9 0TW
M1 BSX Dr A. May, 22 Hillside, Abbotts Ann, Andover, SP11 7DF
M1 BSY W. Ginger, Wenick House, 152 Hawks Road, Hailsham, BN27 1NA
MW1 BTA G. Haines, 12 Tyn Rhos Estate, Gaerwen, LL60 6HL
M1 BTD D. Wilkinson, 89 The Northern Road, Liverpool, L23 2RD
M1 BTI D. Piggin, Blacksmiths Cottage Horsehouse, Leyburn, DL8 4TS
MW1 BTM C. Jones, 17 Grove House Court, Pontygwaith, Ferndale, CF43 3LJ
M1 BTO N. Martin, 28 Churchmead Close, Lavant, Chichester, PO18 0AY
M1 BTR J. Charles, Ash Tree, Priory Lane Grimoldby, Louth, LN11 8SP
M1 BTU P. Mason, 23 Glade Close, Little Billing, Northampton, NN3 9SN
M1 BUC A. Benson, 12 Longfellow Road, Caister-On-Sea, Great Yarmouth, NR30 5RH
M1 BUG M. Dugdale, 57 Macauley Avenue, Blackpool, FY4 4YF
M1 BUJ I. Lewis, 15 Margaret Close, Thurmaston, Leicester, LE4 8GL
MW1 BUN D. Luke, 56 Maerdy Park, Pencoed, Bridgend, CF35 5HX
M1 BUP N. Foyen, Sherborne Valley Kennels, Sherborne, DT9 4SZ
M1 BUQ I. Houghton, 39 Fiskerton Way, Oakwood, Derby, DE21 2HY
M1 BUR B. Murray, 7 Herdwick Place, Middlewich, CW10 9QY
M1 BUU C. Evans, 10 Holme Park Bentham, Lancaster, LA2 7ND
M1 BUX S. Leaker, 166 Beckett Road, Doncaster, DN2 4BB
M1 BVI K. Worrall, 66 Elm St., Hollingwood, Chesterfield, S43 2LH
M1 BVP M. Taylor, 56 Newgate Lane, Mansfield, NG18 2LQ
M1 BVT M. Beardsley, 121 Wood Road, Lower Gornal, Dudley, DY3 2LR
MM1 BVW A. Gordon, The Paddock, Greenhead Farm, West Saltoun, EH34 5EH
M1 BVX S. Simpson, 20 Staveley Grove, Keighley, BD22 7DH
M1 BWH F. Laycock, 8 Melling Way, Liverpool, L32 1TP
M1 BWJ J. Bottle, 15B Elizabeth House, Alexandra Street, Maidstone, ME14 2BX
M1 BWN S. Jarrett, 17 Wolmers Hey, Great Waltham, Chelmsford, CM3 1DA
M1 BWR Dr E. Oakley, Brooklands Lodge, Park View Close, Ventnor, PO38 3EQ
M1 BWS Dr A. Kerr, 14 Glamorgan Close, St Helens, WA10 3XT
M1 BWZ P. Quick, 70 Trent Avenue, Maghull, Liverpool, L31 9DE
M1 BXC A. Blakeney, 7 Gayton Road, Eastcote, Towcester, NN12 8NG
M1 BXD M. Cross, 51 Edgehill Drive Lang Farm, Daventry, NN11 0GR
MM1 BXF G. Nesbitt, 12 Wester Boghead Crosshill Road, Lenzie, G66 4SR
M1 BXJ M. Ellis, 62 Peterborough Road, Crowland, Peterborough, PE6 0BA
M1 BXM M. Forster, 10 Weaver Valley Road, Winsford, CW7 3JU
M1 BXO D. Ellis, 67 Ambersham Crescent, East Preston, Littlehampton, BN16 1AJ
M1 BXQ J. Squire, 57 The Avenue, Chinnor, OX39 4PE
M1 BXU D. Napper, 47 Mallard Walk, Sidcup, DA14 6SG
MW1 BXX M. Broxton, 4 Owen St, Orange Gardens, Pembroke, SA71 4EP
M1 BYG A. Chatel, 17 Star Holme Court, Star Street, Ware, SG12 7EA
M1 BYH A. Moss, Festina Lente Macclesfield Canal Centre Brook Street, Macclesfield, SK11 7AW
M1 BYI P. Stockley, 41 Fairway Court, Cleethorpes, DN35 0NN
M1 BYQ R. Josephs, 113 Patrick Street, Grimsby, DN32 9PQ
M1 BYT H. Bloomfield, 49 Oak Crescent, Garforth, Leeds, LS25 1PW
M1 BZD D. Atkins, 63 Elmwood Road, Keighley, BD22 7DW
M1 BZF S. Gore, 10 Cambridge Street, Guiseley, Leeds, LS20 9AU
M1 BZG L. Raybould, 7 Tenbury House, Highfield Lane, Halesowen, B63 4RN
M1 BZH J. Rose, 200 Burntwood Road Norton Canes, Cannock, WS11 9UR
M1 BZI F. Lee, 26 Lache Hall Crescent, Chester, CH4 7NF
M1 BZJ P. Buer, 71 Belvedere Road, Ashton-In-Makerfield, Wigan, WN4 8RX
M1 BZK D. Riches, 15 Ashton Way, Saltash, PL12 6JE
M1 BZR D. Wright, 18 Allensway, Stanford-le-Hope, SS17 7HE
M1 BZZ P. Lonsdale, 14 Donne Close, Wirral, CH63 9YJ
MM1 CAC G. Mathers, 46 Castle Street, Fraserburgh, AB43 9DH
M1 CAE R. Naylor, 93 Woodland Road Halton, Leeds, LS15 7DN
M1 CAH C. Bennet, 32 Angelica Avenue, Stotfold, Hitchin, SG5 4HH
M1 CAO R. Cameron, 23 Ravenscroft, Hook, RG27 9NP
M1 CAR R. Griffiths, 22 Quarry Road, Hereford, HR1 1SS
M1 CAV J. Morrison, West View, Sunk Island Road, Ottringham, Hull, HU12 0DX
M1 CAX K. Biggs, 22 Wallingford Close, Bracknell, RG12 9JE
M1 CAY M. Harris, Weathercock Cottage, East Mersea Road, Colchester, CO5 8SL
M1 CBC P. Wainwright, 5 Pulcroft Road, Hessle, HU13 0ND
M1 CBH J. Tomlins, 3 Turnstone Crescent, Mansfield, NG18 3SP
M1 CBK R. Scott, 198 Slade Green Road, Erith, DA8 2JG
M1 CBO R. Appleby, 3 St. Judes Way, Burton-on-Trent, DE13 0LR
M1 CBT C. Taylor, 48 Northdown Park Road, Cliftonville, Margate, CT9 3PT
M1 CBU R. Stansfield, Sundene, 157 Hollin Lane, Wakefield, WF4 3EG
M1 CBV G. Leeder, 89 Chesterton Avenue, Harpenden, AL5 5ST
M1 CBY D. Howse, 24 Sandown Road, Bishops Cleeve, Cheltenham, GL52 8BY
M1 CBZ A. Howse, 24 Sandown Road, Bishops Cleeve, Cheltenham, GL52 8BY
M1 CCA B. Thomas, Hazel Mount, Lockhams Road, Southampton, SO32 2BD
M1 CCF M. Buckley, Springfield, 12 Ranmore Avenue, Croydon, CR0 5QA
M1 CCG G. Smales, 6 Chestercourt Cottages, Camblesforth, Selby, YO8 8HZ
M1 CCL R. Chantler, 68 Chilton Lane, Ramsgate, CT11 0LQ
M1 CCN P. Terry, 13 Lamsey Lane, Heacham, King's Lynn, PE31 7LA
M1 CCQ A. Bantoft, 110 St. Peters Road, Wiggenhall St. Peter, King's Lynn, PE34 3HF
MM1 CCR A. Annan, Easter Cottage Blairlogie, Stirling, FK9 5PX
MI1 CCT B. Vaughan, 32 Claremore Road, Castlederg, BT81 7RF
MI1 CCU I. Morrow, 90 Bracky Road Sixmilecross, Omagh., BT79 9PH
M1 CCX Lord P. Tully, 3 Doe Crag Houses, Otterburn Camp, Otterburn, Newcastle upon Tyne, NE19 1NX
M1 CCY R. Coxon, 7 Elworthy Road, Longhoughton, Alnwick, NE66 3LS
M1 CDJ F. Waite, 91 Priors Hill, Wroughton, Swindon, SN4 0RL
M1 CDL D. Hall, 4 Burns Close, Peterborough, PE1 3JJ
M1 CDP A. Mcdade, 20 Westbury Walk, Corby, NN18 0AE
M1 CDQ R. Damm, 18 Mayfair Crescent, Waltham, Grimsby, DN37 0EE
M1 CDT C. Behan, St Chads Close, Hornjnglow, Burton upon Trent, DE13 0ND
M1 CDV A. Mcewen, 23 Cobholm Road, Great Yarmouth, NR31 0BU
M1 CDX K. Leach, 6 Tewkesbury Avenue, Blackpool, FY4 2NF
M1 CEA T. Gladman, 14 Chalford Close, West Molesey, KT8 2QL
M1 CEM B. Harper, 36 Percy Street, Oswaldtwistle, Accrington, BB5 4LY
M1 CEW M. Trueblood, 44 Wallgate Road, Liverpool, L25 1PR
M1 CEY J. Page, 5 Grassington Place, Thatcham, RG19 3XD
MW1 CFA K. Thorley, Helston, Mountain, Holyhead, LL65 1YR
MM1 CFC D. Goodfellow, 4 West Grange Street, Monifieth, Dundee, DD5 4LD
MW1 CFE A. Evans, Flat 5, Nanthir Lodge, Nanthir Road, Bridgend, CF32 8BL
M1 CFG S. Appleby, 2 Stella Farm, Narborough, King's Lynn, PE32 1HY
M1 CFS J. Mac, Lyndale, Torbryan, Newton Abbot, TQ12 5UR
M1 CFW R. Powell, 151 Bury Hill Close, Anna Valley, Andover, SP11 7LL
M1 CFZ A. Moore, 23 Ashgrove Court, Oakwood, Derby, DE21 2LH
M1 CGB M. Jones, 64 St Kenelms Road, Halesowen, B62 0PG
M1 CGF L. Griffiths, 91 Worrall Road Wadsley, Sheffield, s6 4ba
M1 CGI A. Fishwick, Causeway House Farm, Coppice Lane, Heapey, Chorley, PR6 9DA
M1 CGJ S. Fishwick, Causeway House Farm, Coppice Lane, Chorley, PR6 9DA
M1 CGM K. Foster, 48 The Street, Newbourne, IP12 4NY
M1 CGO N. Hewgill, 40 Lime Tree Place, Stowmarket, IP14 1BT
M1 CGP D. Doyle, 100 New Road, South Darenth, Dartford, DA4 9AR
M1 CGQ N. Mulryan, Flat 31, Chatsworth Lodge, Buxton, SK17 6XX
M1 CGR P. Bowles, 25 North Down, Staplehurst, Tonbridge, TN12 0PG
M1 CHF B. Crossley, 19 Westwick Close, Walsall, WS9 9EA
M1 CHM M. Bailey, 10 Argyll Avenue, Doncaster, DN2 6LG
MM1 CHQ D. Wildridge, 1 Glamis Gardens, Dalgety Bay, Dunfermline, KY11 9TD
M1 CHS J. Arundale, 29 Deepdale Avenue, Scarborough, YO11 2UQ
M1 CHU W. Llewellyn, 105 Sandford Avenue, Church Stretton, SY6 7AB
M1 CIE J. Morgan, 15 Town Head, Dearham, Maryport, CA15 7JW
M1 CIG S. Spurr, 12 Rushmoor Close, Rickmansworth, WD3 1NA
M1 CIJ J. Turk, 25 Berkeley Road, Newbury, RG14 5JE
M1 CIM B. Kidane, Po Box 10130, Addis Ababa, Ethiopia,
MM1 CIR P. Merckel, 1 Mortimer Court, Dalgety Bay, Dunfermline, KY11 9UQ
M1 CIS P. Jameson, 1 White Acres Road, Mytchett, Camberley, GU16 6EY
M1 CIZ S. Jones, 34 Marlborough Road, Rugby, CV22 6DD

M1 CJB L. Holyer, 45 Crabble Hill, Dover, CT17 0RX
M1 CJE A. Eastland, 4 Bergamot Close, Manton, Marlborough, SN8 4HT
M1 CJF R. Barrett, Upland, Tidings Hill, Halstead, CO9 1BJ
M1 CJM R. Wallis, 26 Heather Bank, Osbaldwick, York, YO10 3QH
M1 CJN S. Ogiela, 107 Osbaldwick Lane, York, YO10 3AY
M1 CJT A. Brotherhood, 5 Longcliffe Road Shepshed, Loughborough, LE12 9LW
M1 CJX T. Cotterell, 52 The Crofts, Hatch Warren, Basingstoke, RG22 4RF
M1 CJZ A. Roberts, 3 Jaynes Close, Banbury, OX16 9ES
M1 CKJ G. Reeds, 26 Holme Leaze, Steeple Ashton, Trowbridge, BA14 6EH
MW1 CKK S. Lowe, 2 Bryn Eglwys, Llanfachreth, Dolgellau, LL40 2EF
M1 CKO S. Chapman, 9 Edinburgh Avenue, Sawston, Cambridge, CB22 3DW
M1 CKQ B. Roth, 46 Newport Road, Saffron Walden, CB11 4BS
M1 CKU M. Brooks, 13 Weatherside, Blaydon-on-Tyne, NE21 5QL
MM1 CKW A. Johnson, 20 Falkirk Road Glen Village, Falkirk, FK1 2AG
M1 CLI M. Poulter, 26 West Crescent, Duckmanton, Chesterfield, S44 5HE
M1 CLO G. Capon, 24 Beech Drive, Brackley, NN13 6JH
MM1 CLR R. Vause, 100 Carmuirs Avenue, Camelon, Falkirk, FK1 4PB
M1 CLW E. Brown, 76 West Park Drive, Swallownest, Sheffield, S26 4UY
M1 CLX A. Robinson, 11 The Crescent, Whalley, Clitheroe, BB7 9JW
M1 CLZ J. Taylor, 307 Birmingham Road, Lickey End, Bromsgrove, B61 0ER
M1 CML K. Grout, 36 Churchill Road, Exmouth, EX8 4DN
M1 CMM J. Timmis, Upper House, Abdon, Craven Arms, SY7 9HX
M1 CMN M. Curtis, 20 Alder Road, Folkstone, CT19 5BZ
M1 CMR B. Jarvis, 26 Longhouse Road, Halifax, HX2 8RE
MM1 CMU J. Freeland, 12 Mccathie Drive, Newtongrange, Dalkeith, EH22 4BW
M1 CMW R. Cottington, 3 Dickens Drive, East Malling, West Malling, ME19 6SJ
M1 CMX M. Hurley, 5 Borough Crescent, Stourbridge, DY8 3UT
MJ1 CNB N. Fryer, 25 Walter Benest Court, La Route Des Quennavais, St Brelade, Jersey, JE3 8NS
M1 CND T. Wooldridge, 12 Redwood Avenue, Leyland, PR25 1RN
M1 CNE M. Howard, 18 Hydehurst Close, Crowborough, TN6 1EN
M1 CNH J. Gibb, 25 Ferndown Gardens, Cobham, KT11 2BH
M1 CNI G. Lambley, Jasmic, Main Road, Spilsby, PE23 4BE
M1 CNJ F. Manley, 37 Goodrington Close, Banbury, OX16 0DB
M1 CNK P. Wilton, 217 Chamberlayne Road, Eastleigh, SO50 5HZ
M1 CNL R. Tew, 66 St. Nicholas Estate, Baddesley Ensor, Atherstone, CV9 2EZ
MW1 CNN D. Hayward, 1 Elidyr Road, Newbridge, Newport, NP11 3EE
M1 CNP C. Robinson, 9 Chatsworth Avenue, Culcheth, Warrington, WA3 4LD
M1 CNS E. Mathias, 17 St. Johns Terrace, Lewes, BN7 2DL
M1 CNX S. Russell, 11 Lowgate Avenue, Bicker, Boston, PE20 3DF
M1 CNY K. Wilson1, 12 New Street, Elworth, Sandbach, CW11 3JF
MW1 COB K. Sands, 5 Ynysgau Street, Ystrad, Pentre, CF41 7UE
M1 COE R. Alford, 225 N Main Box 174, Gas, USA, 66742
MW1 COJ A. Jones, 22 Wendover Avenue, Towyn, Abergele, LL22 9LP
M1 COL Colchester Radio Amateurs Club, c/o H. Yeldham, 19 Wade Reach, Walton on the Naze, CO14 8RG
MJ1 COO D. Gallichan, 20 Cranham Court La Rue Des Chenes, St. Helier, Jersey, JE2 4RY
M1 COQ M. Panton, 116 Kings Road Glemsford, Sudbury, CO10 7QZ
MM1 COS S. Laepong, 29D Hill St., Montrose, DD10 8AZ
M1 COV S. Rollinson, College Farm Cottage, Humber Lane, Hull, HU12 0UX
MW1 COY S. Beer, 4 Churchfields, Barry, CF63 1FP
M1 CPB K. Ellison, 33 Priory Grove, Sunderland, SR4 7SU
M1 CPC F. Moy, 86 Manningford Road, Birmingham, B14 5LX
M1 CPD F. Marrai, 19 Hind Close, Chigwell, IG7 4EA
M1 CPL R. Ramsay, Fairview, Briar Close, Hastings, TN35 4DP
M1 CPP I. Mcleary, 11 Malcolm Court, Whitley Bay, NE25 8NN
M1 CQC S. Eglinton, 2 Victoria Road, Saltash, PL12 4DL
M1 CQF M. Barnes, 58 Prince St., Dalton in Furness, LA15 8EU
M1 CQI A. Oxlade-Gotobed, 22 St. Peters Road, Basingstoke, RG22 6TD
M1 CQK N. Ore, Willowdene, Rode Lane, Norwich, NR16 1NW
M1 CQL A. Ore, Willowdene, Rode Lane, Norwich, NR16 1NW
M1 CQM B. Suyat, 24 Lynmouth Road, London, E17 8AF
M1 CQN B. Adkins, 4 Orion Close, Ward End, Birmingham, B8 2AU
MW1 CQP K. Blackwell-Chambers, 69 Burford Gardens, Cardiff, CF11 0AP
M1 CQQ M. Sheanon, 3 Bridge Road, Little Sutton, Long Sutton, Spalding, PE12 9EG
M1 CQR R. Field, 3 Waveney Drive Belton, Great Yarmouth, NR31 9JU
M1 CQS G. Browne, 30 Dereham Road Easton, Norwich, NR9 5EJ
M1 CQT A. Wheeler, 60 Bredhurst Road, Gillingham, ME8 0PE
M1 CQU N. Anderson, 19 Berrylands, Liss, GU33 7DB
M1 CQX M. Jones, 44 Purley Road, Sunderland, SR3 1QS
M1 CRA WACRL(World Association of Christian Ras), c/o P. Jackson, 8 Buttree Court, South Kirkby, Pontefract, WF9 3NB
MW1 CRE S. Woolley, 139 Prince Of Wales Avenue, Flint, CH6 5JU
M1 CRF N. Ovenden, 4 Montrose Way, Thame, OX9 3XH
M1 CRL J. Edwards, 44 Hunter Road, Norwich, NR3 3PY
M1 CRO Colchester Contest Group, c/o J. Lemay, Carlton House, White Hart Lane, Colchester, CO6 3DB
M1 CRP P. Booth, 39 New Close, Eyam, Hope Valley, S32 5QX
M1 CRQ J. Nuttall, 114 Plumstead Road, Norwich, NR1 4JX
M1 CRZ A. Tudge, 7 Moreton Avenue, Whitefield, Manchester, M45 8GG
MI1 CSA J. Higgins, 43 Temple Road, Garvagh, Coleraine, BT51 5BJ
M1 CSC S. Swancutt, 100 Oundle Road, Birmingham, B44 8EN
M1 CSE C. Childs, 43 Eastdale Road, Burgess Hill, RH15 0NJ
M1 CSG G. Millsott, 11 Kingsthorn Road, Poundbury, Dorchester, DT1 3RR
M1 CSI D. Avery, 38 Junction Road, Burgess Hill, RH15 0JN
M1 CSL D. Cooke, 125 Glenhills Boulevard, Leicester, LE2 8UH
M1 CSU I. Bliss, 3 Ford Road, Ashford, TW15 2RF
M1 CSZ S. Eggleton, 5 Ladywood Grange, Lady Margaret Road, Ascot, SL5 9QH
M1 CTB C. Dale, 3 Ivatt Close, Bawtry, Doncaster, DN10 6QF
M1 CTG M. Hopkins, The Black Swan, Burn Bridge Road, Harrogate, HG3 1PB
MW1 CTJ T. Joyes, 4 The Smithy, Devauden, Chepstow, NP16 6QA
M1 CTK D. Hunt, 4 Warmdene Road, Brighton, BN1 8NL
M1 CTM A. Mcmullen, 70 Sylvan Avenue, Timperley, Altrincham, WA15 6AB
M1 CTO L. Chung, 104 Penland Road, Haywards Heath, RH16 1PH
MI1 CTQ J. Murphy, 19 Fernagh Road, Omagh, BT79 0HX
M1 CTY M. Gray, 8 Middleton Terrace Cross Street, Cowes, PO31 7TE
M1 CUB K. Hope, 37 Hollyhurst Road, Darlington, DL3 6HT
M1 CUC J. Mcgowan, 72B Adelphi Crescent, Hornchurch, RM12 4JZ
MI1 CUS J. Woods, 18 Mullaghdrin Road, Dromara, Dromore, BT25 2AF
M1 CUX R. Matthews, 18 Hawkins Close, Daventry, NN11 4JQ
M1 CUY J. Wood, 3 Harold Collins Place, Colchester, CO1 2GQ
M1 CVB P. Bull, 8 Mayfield Lane, Martlesham Heath, Ipswich, IP5 3TZ
M1 CVF C. Block, 13 Beatrice Road, Capel-Le-Ferne, Folkestone, CT18 7LH
M1 CVG D. Wood, 17 St. Peters Close, Henley, Ipswich, IP6 0RH
M1 CVH S. Blewitt, 9 Durlston Close, Amington, Tamworth, B77 3QG
M1 CVK K. Bennett, 34 Shrubbery Close, Barnstaple, EX32 9DG
M1 CVL M. Crossley, 196 Middleton Road Hopwood, Heywood, OL10 2LH
MW1 CVM D. Giles, 9 Ty Newydd Court, Pontnewydd, Cwmbran, NP44 1LJ
M1 CVT A. Mallett, 8 Shaws Close, Prestwood, Great Missenden, HP16 0SL
M1 CVU K. Smolkovic, 26 Keeling Way, Attleborough, NR17 1YF
M1 CVX S. Taylor, 17 York Close, Clayton Le Moors, Accrington, BB5 5RB
M1 CWA Prof. J. Gough, 2 Brunswick Cottages, Cambridge, CB58DL
M1 CWB I. Bryant, 17 Kent Road, Southampton, SO17 2LJ
M1 CWD D. Taberer, 4 Hillfields Road, Brierley Hill, DY5 2NG
M1 CWG T. Foreman, 39 Railway Street, North Fleet, DA11 9DU
M1 CWN B. Bailey, 8 Alderton Road, Grittleton, Chippenham, SN14 6AN
M1 CWO J. Collinson, 50 Willoughby Park, Alnwick, NE66 1ET
M1 CWV A. Dykes, 149 Mayfield Road, Chaddesden, Derby, DE21 6FZ
M1 CWY O. White, 35 Drage Street, Derby, DE1 3RW
M1 CXA J. Marcus, 115 Kimberley Road, Solihull, B92 8QA
M1 CXK A. Cordier, 49 Laburnum Avenue, Dartford, DA1 2QN
M1 CXN H. Neal, 141 Manor Road, Erith, DA8 2AQ
MM1 CXO J. Donnelly, 21 Mcdonald Drive, Irvine, KA12 0QS
M1 CXP R. Gill, 45 Biggin Lane, Ramsey, Huntingdon, PE26 1NB
M1 CXV G. Allison, 16 Copse Road, Plymouth, PL7 1PZ
M1 CXW D. Hines, 31 Clegge Street, Warrington, WA2 7AT
MJ1 CYD C. Paland, 19 Maison St. Louis, St Saviour, Jersey, JE2 7LX
M1 CYJ G. Jenkinson, 4 Brundish House, Braithwell Road, Rotherham, S66 8JT
M1 CYK E. Doran, 96 Lovat Road, Preston, PR1 6DQ
M1 CYL K. Langhamer, 26 Maple Drive, Burgess Hill, RH15 8AW
M1 CYM M. Husband, 31 Crescent Road, Colwall, Malvern, WR13 6QW
M1 CYN W. Hamilton, 22 Rayford Close, Dartford, DA1 3AJ
M1 CYP B. Millard, 11A Fourways, Tetney, Grimsby, DN36 5NF
M1 CYR K. Scott, 362 Cannock Road, Heath Hayes, Cannock, WS12 3HA
M1 CYT S. Davey, 2 Staveley Road, Dunstable, LU6 3QQ
M1 CYX D. Bryan, 14 Fairfield Way, Totland Bay, PO39 0EF
M1 CZA K. Churchill, 76 Preston Drive, Bexleyheath, DA7 4UE
M1 CZF M. Lewis, The Manor House, The Green, Banbury, OX17 1BU
M1 CZI I. Johnson, 2 Latton Close, Didcot, OX11 0SU
M1 CZL S. Bond, Powell Cottage, 1 Powell Close, Leamington Spa, CV33 9PX
M1 CZM J. Stocks, 10 Hollycroft Road, Emneth, Wisbech, PE14 8AY
M1 CZO D. Charles, 29 Acacia Gardens, Upminster, RM14 1HT
M1 CZX T. Ruane, Ventnor, High Lane, Haslemere, GU27 1AZ
M1 CZY D. Mageehan, 37 Gosbecks Road, Colchester, CO2 9JR
M1 CZZ D. Gallier, 86 Pine Tree Road, Oldham, OL8 3LQ
M1 DAB M. Mcphail, 126 Welbeck St., Creswell, Worksop, S80 4AN
M1 DAH J. Saiger, 10 Markham Avenue, Armthorpe, Doncaster., DN3 2AZ
MM1 DAK I. Mcdonald, 5 Well Street, Rosehearty, Fraserburgh, AB43 7NW
MW1 DAM A. Cartwright, 7 Pen Parc, Malltraeth, Bodorgan, LL62 5BG
M1 DAN D. Black, 8 Cornwood Close, Finchley, London, N2 0HP
M1 DAP M. Purcell, 14 Adelaide Road, Blacon, Chester, CH1 5SY
M1 DAS D. Nicolson, Woodbridge House, Wembworthy, Chulmleigh, EX18 7SN
MW1 DAU C. Cater, 40 Frances Avenue, Wrexham, LL12 8BN
MI1 DAW R. Bamber, 15 Ladybrook Parade, Belfast, BT119ER
MW1 DBA P. Kinley, 18 Larchwood Road, Wrexham, LL12 7SG
M1 DBC R. Carroll, 27 Sheraton Grange, Stourbridge, DY8 2BE
M1 DBF G. Jones, 12 Birks Holt Drive, Maltby, Rotherham, S66 7JZ
M1 DBK M. Lawrance, 18 The Green Road, Sawston, Cambridge, CB22 3LP
M1 DBM B. Barrett, Kite Hill Camping Park, Firestone Copse Road, Wootton Bridge, PO33 4LQ
M1 DBW P. Roberts, 7 Boscombe Road, Swindon, SN25 3EZ
M1 DCE P. Rollinson, 4 Turmarr Villas, Easington, Hull, HU12 0TJ
MW1 DCF G. Hopkins, 132 Laurel Road, Bassaleg, Newport, NP10 8PT
M1 DCH P. Wakefield, Flat 3, 21 Priests Road, Swanage, BH19 2RG
MW1 DCI P. Bevan, 61 Dinas St., Plasmarl, Swansea, SA6 8LQ
M1 DCK W. Curry-Peace, 14 Springfield Road, Stoke-on-Trent, ST4 6RU
M1 DCV M. Mcbride, 127 Leicester Causeway, Coventry, CV1 4HL
M1 DCX T. Dore, 53 Queens Avenue, Shipston on Stour, CV36 4DJ
M1 DCY A. Dore, 7 Cavendish Drive, Kidderminster, DY10 2SX
M1 DDB T. Smith, 69 Sunningdale, Grantham, NG31 9PF
M1 DDF O. Spevack, 16 Ranmore Road, Dorking, RH4 1HD
M1 DDI K. Skidmore, 239 Alfreton Road Blackwell, Alfreton, DE55 5JN
M1 DDR D. Carter, 30 Swift Way, Sandal, Wakefield, WF2 6SR
M1 DDW J. Reed, 9 Mercer Drive, Harrietsham, Maidstone, ME17 1AY
M1 DDY T. Reed, Seafield, Charing Hill, Charing, Ashford, TN27 0NG
MM1 DEA G. Leadbetter, 8 Tomtain Brae, Cumbernauld, Glasgow, G68 9ER
MM1 DEE G. Hall, 0/2 1 Thistle St Kirkintilloch, Glasgow, G661nu
M1 DEG R. Smith, 41 Middle Deal Road, Deal, CT14 9RG
M1 DEJ M. Hibbert, 5 Cliff View Road, Cliffsend, Ramsgate, CT12 5ED
M1 DER A. Cooper, 21 Hayton Way, Skipton, BD23 1DQ
M1 DEY K. Armstrong, 8 Caxton Garth, Threshfield, Skipton, BD23 5EZ
MI1 DEZ T. Reid, 15 Gillistown Road, Ahoghill, Ballymena, BT42 2RJ
M1 DFB A. Dunster, 113 Canterbury Road, Folkestone, CT19 5NR
M1 DFC S. Bell, 14 Charlotte Avenue, Wickford, SS12 0DX
M1 DFK C. Ansell, 51 East Road, Brinsford, Wolverhampton, WV10 7NP
M1 DFM K. Davies, 58 Popes Lane Sturry, Canterbury, CT2 0LA
M1 DFO A. Bruce, 4 Drayton Manor 507 Parrswood Road, Manchester, M20 5GJ
MW1 DFQ B. Howard, 64 Lawrenny St., Neyland, Milford Haven, SA73 1TB
M1 DFW Dr K. Prakash, 14 Masham Road, Harrogate, HG2 8QF
M1 DGE D. Cockayne, 32 Shaw Close, Garforth, Leeds, LS25 2HA
M1 DGK J. Kirkham, Flat 31, 123 St. Anns Road, London, W11 4BT
M1 DGL R. Walsh, 10 Standen Road Bungalows, Clitheroe, BB7 1LA
M1 DGP C. Anderson, 15 John Gunn Close, Chard, TA20 1DG
M1 DGQ P. Talbot, 5 Stones Walk, Burghfield Common, Reading, RG7 3JA
MW1 DGR C. Smith, 22 Berth Y Glyd Road, Old Colwyn, Colwyn Bay, LL29 9HT

M1 DGS D. Snell, 154 Oaks Cross, Stevenage, SG2 8NA
M1 DGW M. Wharton, 9 Orchard View, Linton Colliery, Morpeth, NE61 5SP
M1 DGX J. Griffiths, 10 Cote Road, Telford, TF5 0NQ
M1 DGY H. Shemming, 6 Smiths Place, Kesgrave, Ipswich, IP5 2YR
M1 DHA A. Davis, 19 Grange Street, Barnoldswick, BB18 5LB
M1 DHC A. Killing, 102 Coquet Grove, Newcastle upon Tyne, NE15 9LH
M1 DHG M. Hilton, 40 Megstone Avenue, Whitelea Chase, Cramlington, NE23 6TU
M1 DHI J. Edwards, Willows, Sunray Avenue, Whitstable, CT5 4EQ
M1 DHJ I. Maltas, 20 Suddaby Close, Hull, HU9 3RG
M1 DHM R. Fraser, 12 Birchen Road, Halewood, Liverpool, L26 9TL
M1 DHO R. Bloxam, 39 Claremont Drive, Ravenstone, Coalville, LE67 2ND
M1 DHT D. Shackleton, 29 Windmill Green, Ditchingham, Bungay, NR35 2QP
MM1 DHU P. Mcbride, 1 Hillside, Croy, Glasgow, G65 9HJ
M1 DHV G. Turner, 35 Horncastle Road, Wragby, Market Rasen, LN8 5RB
M1 DHW J. Simlat, 7 Coventry Close, Wroughton, Swindon, SN4 9BB
M1 DHY D. Sandell, 29 Manor Road, Herne Bay, CT6 6RF
M1 DIB D. Beck, 25C Lickless Gardens, Horsforth, Leeds, LS18 5QU
M1 DIE J. Halsall, 83 Poole Road, Leeds, LS15 7HD
M1 DIL B. Jones, 39 Rosewood Avenue, Burnham-on-Sea, TA8 1HE
M1 DIM K. Ingram, 1 Hazel Close, Brackley, NN13 6PE
M1 DIN D. Blythe, 51 Lea Side, Halton-Lea-Gate, Brampton, CA8 7LA
M1 DIR R. Duncan, Lake View, 2 Upper New Road, Cheddar, BS27 3DH
M1 DJA A. Garner, The Chestnuts, Surfleet, Spalding, PE11 4BA
M1 DJB C. Meakin, 56 Coronation Walk, Gedling, Nottingham, NG4 4AQ
M1 DJC G. Griffiths, 8 Grays Lane, Paulerspury, Towcester, NN12 7NW
M1 DJG K. Miller, 52 Stanway Road, Shirley, Solihull, B90 3JE
M1 DJI A. Waddington, 5 Glenview Avenue, Bradford, BD9 5PA
MM1 DJJ G. Waddington, Wester Lathallan, Leven, KY8 5QP
M1 DJN R. Francis, 50 Parsonage Chase, Minster On Sea, Sheerness, ME12 3JX
M1 DJO B. Kynaston, 76 Thorncliffe Avenue, Dukinfield, SK16 4UD
M1 DJP A. Saltmarsh, 55 Wentworth Grove, Winsford, CW7 2LJ
M1 DJS I. Turk, 11 Medway Crescent, North Hykeham, Lincoln, LN6 8UB
MI1 DJW J. Campbell, 2 Lakeview, Crumlin, BT29 4YA
M1 DJX A. Baker, 27 Liney Road, Westonzoyland, TA70EU
M1 DKA A. Lewis, 111A Cheltenham Road, Longlevens, Gloucester, GL2 0JG
M1 DKF R. Saward, Vigeland, 61A Old Main Road, Boston, PE20 2BU
M1 DKK N. Hall, 98 Newbiggin Road, Ashington, NE63 0TH
M1 DKL D. Liddard, 81 Tattersall Gardens, Leigh-on-Sea Essex, SS9 2QS
MW1 DKM A. Williams, 25 Tre Rhosyr, Newborough, Llanfairpwllgwyngyll, LL61 6TG
M1 DKP A. Maylin, 221 Branksome Avenue, Stanford-le-Hope, SS17 8DD
M1 DKW R. Illman, 35 Courtenay Park, South Brent, TQ10 9BT
M1 DKY H. Sanders, Copper Coins, Deans Drove, Poole, BH16 6EQ
M1 DKZ D. Forster, 3 West View, Middleton, Ludlow, SY8 3ED
M1 DLE M. Gifford, 22 St. Agnes Way, Kesgrave, Ipswich, IP5 1JZ
M1 DLG D. Smith, 19 Victoria Grove, Newbury, RG14 7RA
M1 DLM D. Russell, 6 Deansgate Lane North, Liverpool, L37 7ER
M1 DLR R. Wood, Bank View, Bilham Road, Huddersfield, HD8 9PA
M1 DLS J. Wharton, 18 Queen Street, Newbiggin-by-the-Sea, NE64 6DE
M1 DLX Castle House School ARC, c/o J. Griffiths, 10 Cote Road, Telford, TF5 0NQ
M1 DMB S. Atton, A1 Manor Park, Happisburgh, Norfolk, NR12 0PW
MM1 DME R. Fuggle, 14 Quebec Drive, East Kilbride, Glasgow, G75 8SA
M1 DMH J. Hardman, 2 Well Orchard, Bamber Bridge, Preston, PR5 8HJ
M1 DMN R. Gilbert, 8 Church Road, West Kingsdown, Sevenoaks, TN15 6LL
M1 DMR B. Pilcher, 283 London Road, Portsmouth, PO2 9HE
M1 DMT J. Hull, 68 Meadow Avenue, West Bromwich, B71 3EE
MM1 DMU D. Mccann, 69 Davies Drive, Lomond Industrial Estate, Alexandria, G83 0UF
M1 DNA D. Bruce, 15 St. Richards Road, Deal, CT14 9JR
M1 DNC L. Hall, 15 Fullwood Avenue, Newhaven, BN9 9SP
M1 DNE S. Moppett, 59 Piccadilly, Tamworth, B78 2ER
M1 DNG R. Steward, Long Meadow, Seven Acres Lane, Southwold, IP18 6UL
M1 DNJ D. Houbart, 10 Lancelot Close, Rochester, ME2 2YT
M1 DNQ J. Golding, Flat 6, Daver Court, London, SW3 3TS
MD1 DNT D. Hughes, 13 Julian Road, Douglas, Isle Of Man, IM2 6HW
MW1 DNY D. Reid, 11 Caer Delyn, Bodffordd, Llangefni, LL77 7EJ
M1 DNZ D. Herridge, 93 Freshbrook Road, Lancing, BN15 8DE
M1 DOA S. Jefferson, 145 Duke St Fenton, Stoke-on-Trent, ST2 43NR
MI1 DOG S. Mcauley, Layde View, 19 Rathlin Avenue, Ballycastle, BT54 6DQ
MW1 DOO V. Lee, 40 Willins Coed Eva, Cwmbran, NP44 4TJ
M1 DOR D. Stuart, 58 Woodplace Lane, Coulsdon, CR5 1NF
M1 DOS C. Smith, 13 West Winds Road, Winterton, Scunthorpe, DN15 9RU
M1 DOT S. Myall, 71 Barnes Avenue, Fearnhead, Warrington, WA2 0BL
MW1 DOU M. Jones, 3 St. Catherines Close, Llanfaes, Beaumaris, LL58 8LH
M1 DOX N. Price, 162 Stamshaw Road, Portsmouth, PO2 8LX
M1 DOZ D. Sampson, Spirits Hall Cottage, Mountains Road, Great Totham, CM9 8BY
MM1 DPC M. Mclauchlan, 8 Craigie St., Ballingry, Lochgelly, KY5 8NS
M1 DPE L. Stockwell, 167 Hathaway Road, Grays, RM17 5LW
MM1 DPH J. Crichton, 51 Obree Avenue, Prestwick, KA9 2NN
M1 DPI E. Thompson, 9 Elmwood Drive, Ponteland, Newcastle upon Tyne, NE20 9QQ
M1 DPJ A. Bonner, Flat 15, Enderleigh House, Havant, PO9 1LQ
MI1 DPL J. Stewart, 45 Mull Road, Antrim, BT41 2TR
M1 DPO J. Gould, 14 Homestead Road, Orpington, BR6 6HW
M1 DPQ Dr P. Orr, 74 Amalfi Tower, Lakeside Village, Sunderland, SR3 3AL
M1 DPU P. Cain, 108 Spencer Road, Norwich, NR6 6DG
M1 DPW P. Whiffing, 38 Green Close, Stannington, Morpeth, NE61 6PE
M1 DPX D. Collins, 1 New Street Close, Stradbroke, Eye, IP21 5JH
M1 DPY J. Bowes, 40 Nursery Road, Angmering, Littlehampton, BN16 4FH
MI1 DQB C. Gardner, 50 Kirkliston Park, Belfast, BT5 6ED
M1 DQE I. Fletcher, Priory House, 56 Fairfield Road, Saxmundham, IP17 1BA
M1 DQG R. Kennedy-Bright, Sandiacre, Orchard Lane, Hanwood Sy58Le, Shewsbury, SY58LE
M1 DQH A. Duffield-Dyche, 1A The Hawthorns, Brockton, Shrewsbury, SY5 9JY
M1 DQI M. Jones, 35 Pendle Way, Meole Brace, Shrewsbury, SY3 9QS
M1 DQQ S. Stanley, 35 Statham Close, Lymm, WA13 9NN
M1 DQU A. Bedford, 44 Kirtling Place, Haverhill, CB9 0AU
MM1 DQV A. Gibbs, Cathlawhill Farm, Torphichen, Bathgate, EH48 4NW
MM1 DQW G. Steven, 36 Springhill Terrace, Springside, Irvine, KA11 3AL
M1 DQX P. Ormerod, 14 Fanny Moor Crescent, Huddersfield, HD4 6PL
M1 DRB K. Glover, 14 Crawley Crescent, Eastbourne, BN22 9RN
M1 DRK A. Thomson, 33 Finchley Close, Hull, HU8 0AN
M1 DRL D. Luff, 12 Swan Lane, Sellindge, Ashford, TN25 6EP
M1 DRM M. Bird, Driftwood, 37 Beachwood Avenue, Kingswinford, DY6 0HL
MI1 DRP P. Mcdaid, 66 Laurel Drive, Strabane, BT82 9PN
M1 DRZ J. Stevens, Springfield Cottages, 57 Brindley Street, Stourport-on-Severn, DY13 8JG
MM1 DSD G. Duncan, 5 Jarvis Place, Carnoustie, DD7 7BR
M1 DSE P. Gibson, 77 Sunlea Avenue, North Shields, NE30 3DT
M1 DSI D. Palmer, 2 Merrill Road, Thurnscoe, Rotherham, S63 0NN
M1 DSQ N. Taylor, 36 Bodmin Avenue, Slough, SL2 1SL
M1 DSU J. Henderson, 12 Chathill Terrace, Newcastle upon Tyne, NE6 3BB
M1 DSV A. Newell, Thwaites Bank, Spring Avenue, Keighley, BD21 4TD
MM1 DSX J. Spiers, 29B Carlyle Gardens, Haddington, EH41 3LS
M1 DSZ S. Turner, 27 Huntspill Road, Highbridge, TA9 3DQ
M1 DTG M. Whitchurch, 94 Hundred Acres Lane, Amersham, HP7 9BN
MM1 DTN W. Gray, 1 Regent Court, Regent Street, Keith, AB55 5ED
M1 DTO T. Jones, 40 Chester Road South, Kidderminster, DY10 1XJ
M1 DTS E. Cawte, Woodpeckers, Rectory Gardens, Church Stretton, SY6 6DP
MW1 DTT S. Walters, 31 John Street, Pentre, CF41 7JT
MM1 DTU J. Weddell, 10 High Street, Eyemouth, TD14 5EU
M1 DUA D. Riley, Flat 2, Crown Crest Court, Sevenoaks, TN14 5AS
M1 DUB D. Stalley, 42 Gadby Road, Sittingbourne, ME10 1TJ
M1 DUC J. Parker, 76 Elm Road, Grays, RM17 6LD
M1 DUD R. Burrows-Ellis, 49 Highland Drive, Worlingham, Beccles, NR34 7AR
MW1 DUJ D. Jones, 13 Pontardulais Road, Cross Hands, Llanelli, SA14 6NT
M1 DUO R. Easthope, 8 Gilbert House, 6 Mill Park, Cambridge, CB1 2FJ
M1 DUV S. Crook, 24 Gainsborough Drive, Northfleet, Gravesend, DA11 8NH
MM1 DVC I. Hendry, 47 Fraser Place, Keith, AB55 5EB
M1 DVJ C. Wood, 4 Waterdale Farm Park, Kingskerswell, TQ12 5EX
M1 DVO R. Waters, Romosco, Mill Lane, Bradfield, Manningtree, CO11 2QP
M1 DVV P. Wise, 13 Waltham Road, Newton Abbot, TQ12 1LH
M1 DWQ I. Lowcock, Sunflower Cottage Loddiswell, Kingsbridge, TQ7 4QJ
M1 DWT D. Hamilton, 120 Hall Road, Hull, HU6 8SB
MM1 DWU G. Mcvittie, 46 Mote Hill Road, Girvan, KA26 0EB
M1 DWV Dr M. Stitson, 91 St. Judes Road Englefield Green, Egham, TW20 0DF
M1 DWW W. Bennett, Hillrise, Kings Road, Malvern, WR14 4HL
M1 DXA P. Beech, Grange Farm North Road, Atwick, YO25 8DW
M1 DXB B. Smith, 39B Palace Avenue, Paignton, TQ3 3EQ
M1 DXG R. Williamson, Beverley, Swineshead Road, Boston, PE20 1SG
M1 DXL C. Churchward, Unit 52 Bizspace, Atlantic Street, Altrincham, WA14 5NQ
M1 DXO N. Onions, 34 Redwing Court, Southsea, PO4 8PB
M1 DXQ M. Rhead, 11 Shelley Road, Stoke-on-Trent, ST2 8JN
MM1 DXU M. Richards, Rowan Lea, Weyland Gait, Kirkwall, KW15 1QR
MD1 DXW W. Griffiths, 7 Cooyrt Shellagh, Ballasalla, Isle of Man, IM9 2EU
M1 DYC J. Guilford, 2 Lacey Close, Ilkeston, DE7 9LF
M1 DYD F. Frost, Flat 34, Kingsley Court, 21 Pincott Road, Bexleyheath, DA6 7LA
M1 DYE D. Ejugue, Po Box 62449, Addis Ababa, Ethiopia,
M1 DYF A. Teffera, Po Box 819, Addis Abba, Ethiopia,
M1 DYG N. Teklehaimanot, Po Box 21866, Addis Abba, Ethiopia,
M1 DYH M. Belete, Po Box 181922, Addis Ababa, Ethiopia,
M1 DYI E. Melaku, 77 Chaucer Drive, Lincoln, LN2 4LT
M1 DYJ A. Williams, Alwent Farm, Staindrop, Darlington, DL2 3NS
M1 DYK H. Davison, 15 High St., Rippingale, Bourne, PE10 0SR
M1 DYL D. Davison, 15 High St., Rippingale, Bourne, PE10 0SR
M1 DYO A. Cruise, Badgers Holt, Park Grove, Chalfont St Giles, HP8 4BG
M1 DYS R. Broadbridge, 41 Benmoor Road, Poole, BH17 7DS
M1 DYU G. Rogers, 55 Upholland Road, Billinge, Wigan, WN5 7JA
M1 DYW O. Barnes, 14 Caroline Close, Wivenhoe, Colchester, CO7 9SD
M1 DZM J. Nichols, 51 Ashbourne Road, Barnsley, S71 3QD
M1 DZR R. Daglish, Beck Lea, Pasture Road, Frizington, CA26 3XN
MM1 DZW R. Heath, Roadside Cottage, Glenkindie, Alford, AB33 8SH
MW1 EAA G. Tucker, 18 Plymouth Road, Penarth, CF64 3DH
M1 EAB A. Thornton, 11 St. Nicholas Close, Richmond, DL10 7SP
M1 EAI A. Beale, 6 Meadow View, Belper, DE56 1UT
M1 EAJ Y. Bessell-Baldwin, 10 Tudor Close, Barton-Le-Clay, Bedford, MK45 4NE
M1 EAK C. Day, 35 Rochford Road St. Osyth, Clacton-on-Sea, CO16 8PH
M1 EAN B. Authers, 91 Hay Green Lane, Bournville, Birmingham, B30 1RF
M1 EAS J. Schofield, 19 Shottery Walks, Bredbury, Stockport, SK6 2HR
M1 EAW K. Gallacher, 63 Holst Avenue, Basildon, SS15 5RH
M1 EAZ J. Barker, 53 Derby Street, Colne, BB8 9AA
M1 EBC J. Best, Longview, Central Road, Maryport, CA15 7ER
M1 EBD D. Best, Longview, Central Road, Maryport, CA15 7ER
M1 EBH L. Emmerson, Flat 18, 12 Merley Lane, Wimborne, BH21 1RX
M1 EBI P. Bird, 10A Shackleton Road, Bloxwich, Walsall, WS3 3BZ
M1 EBK M. Rowley, 20 Long Leasow, Selly Oak, Birmingham, B29 4LT
M1 EBL C. Venables, Deepdene, Rickford, Guildford, GU3 3PQ
M1 EBN R. Bunce, 4 Newlands, Gainsborough, DN21 1QZ
M1 EBS P. Beckwith, 4 Hunters Yard, Riseley, MK44 1EN
M1 EBU W. Mitchell, 8 Woodland Crescent, Burgess Hill, RH15 0LJ
M1 EBW I. Merrill, 26 Catkin Drive, Giltbrook, Nottingham, NG16 2UB
M1 EBY P. Clark, Flat Six Haynes House Booker Place, High Wycombe, HP12 4QD
M1 ECB K. Cronin, East Cottage Westviile Rd, Thornton le Fen, LN4 4YJ
M1 ECC D. Wright, 74 Witchards, Basildon, SS16 5BN
M1 ECD North-Northants Raynet Group, c/o M. Wright, 27 Willow Road, Kettering, NN15 7BA
M1 ECH S. Kirby, 2 Kneeton Park, Middleton Tyas, Richmond, DL10 6SB
M1 ECI A. Funnell, 15 Hendham Road, London, SW17 7DH
M1 ECM M. White, 100 Burnham Road, Coventry, CV3 4BQ
M1 ECQ C. Hamilton, 101 Gipsy Lane, Swindon, SN2 8DL
M1 ECT M. Procter, 141 Ruddington Lane, Nottingham, NG11 7BY
M1 ECV D. Bould, 38 Curlew Grove, Bridlington, YO15 3NX
M1 ECW N. Mcmahon, 23 St. James Close, Bramley, Tadley, RG26 5XH
M1 EDA S. Fabian, 3 Manor Cottages Manor Orchard Horley, Banbury, OX15 6DZ
M1 EDF G. Powell, Sycamore Cottage, Church Lane, Tamworth, B79 0LD
M1 EDL A. Wolverson, 28 Thorness Close Alvaston, Derby, DE24 0UY
M1 EDO J. Hurst, 13 Peregrine Road, Hainault, Ilford, IG6 3SR
M1 EDW P. Tinkler, 27 Cavendish Drive, Carlton, Nottingham, NG4 3DX

MM1 EDY J. Goldstraw, 26 Craigmill Gardens, Carnoustie, DD7 6HT
M1 EEN G. Butterfield, Pasadena, St. Helens Road, Walcott, Norwich, NR12 0LU
M1 EEP J. Nethercott, 30 Goldcrest Road, Chipping Sodbury, Bristol, BS37 6XF
M1 EEQ A. Waite, 221 Hasler Road, Poole, BH17 9AH
M1 EER G. Ball, 11 Jersey Close, Congleton, CW12 3TW
M1 EEW A. Beckwith, 19 Westmorland Avenue, Dukinfield, SK16 5JA
M1 EEY N. Beckley, 76 Keir Hardie Way, Barking, IG11 9NY
M1 EEZ A. Kypriadis, 119 Whitfield Villas, South Shields, NE33 5NH
M1 EFP J. Carter, 5 Hastings Avenue, Seaford, BN25 3LB
M1 EFT P. Swanton, 54 South Avenue, Warrington, WA2 8BQ
M1 EGC G. Hancock, 24 Kings Avenue, Corsham, SN13 0EG
M1 EGD S. Sykes, 12 Banksville, Holmfirth, HD9 1XP
M1 EGG P. Bird, 37 Beachwood Avenue, Kingswinford, DY6 0HL
M1 EGL P. Ritchley, 34 Chesildene Avenue, Throop, Bournemouth, BH8 0DS
M1 EGM B. Ritchley, 25 Branwell Close, Christchurch, BH23 2NP
M1 EGN J. Eyres, 13 Newburn Crescent, Swindon, SN1 5ES
M1 EGP R. Argent, 48 Church Green, Staplehurst, Tonbridge, TN12 0BE
MM1 EGS T. May, 34 Dee Place, East Kilbride, Glasgow, G75 8RZ
M1 EGV C. Mills, 6 Levisham Gardens, Bewsey, Warrington, WA5 0GD
M1 EGW D. Green, 5 St. Benedicts Close, Cranwell Village, Sleaford, NG34 8DB
M1 EGX M. Beddard, 18 Hyacinth Way, Burbage, Le10 2uh
M1 EGZ W. Gravestock, 23A Murrell Road, Ash, Aldershot, GU12 6ST
M1 EHB J. Nixon, 1 Appletree Close Long Riston, Hull, Hu115fb
M1 EHD P. Leadill, 7 Keldale, Haxby, York, YO32 3GG
M1 EHF G. Gore-Thorne, 19 Poplar Way, Ringwood, BH24 1UY
M1 EHI I. Croasdale, 16 Buttermere Avenue, Chorley, PR7 2JG
M1 EHJ R. Roychoudhuri, 62A Parkway, Eastbourne, BN20 9DY
MM1 EHO S. Mcneil, 35 Sutors Avenue, Nairn, IV12 5AZ
M1 EHV C. Aram, 1 Snuggs Lane, East Hanney, Wantage, OX12 0HU
M1 EHZ J. Brown, 3 Malton Close, Blyth, NE24 5AS
M1 EIE S. Stephenson, 31 Sherbrooke Avenue, Hull, HU5 4AG
MI1 EIH G. Brennan, 15 Kinnegar Rocks, Donaghadee, BT21 0EZ
M1 EIJ S. Sweetlove, 36 Park Avenue, Corsham, SN13 0JT
M1 EIO A. Rixon, 17 Brimmers Way, Aylesbury, HP19 7HR
M1 EIR E. Fishbourne, 8 Somers Walk, Tupsley, Hereford, HR1 1QX
M1 EIU C. Martin, Flat 3, Dodds House, Vicarage Lane, Tarporley, CW6 9BP
M1 EIW G. Kinney, 1 Eden Park, Brixham, TQ5 9LS
M1 EIZ L. Rutherford, 197 Rosalind Street, Ashington, NE63 9BB
M1 EJD D. Pickering, 2 Priory Green, Highworth, Swindon, SN6 7NU
M1 EJE D. Clarke, 1 Combe Bank, Lindthorpe Way, Brixham, TQ5 8PB
M1 EJG J. Clarke, 21 Mill Lane, Blakedown, Kidderminster, DY10 3ND
M1 EJI B. Hunt, 2A Golf Road, Radcliffe-On-Trent, Nottingham, NG12 2GA
M1 EJJ M. Laurie, 2 The Steps, Phocle Green, Ross-on-Wye, HR9 7TW
M1 EJL P. Langfield, 21 Amy Johnson Court, Great Passage Street, Hull, HU1 2AJ
M1 EJO P. Matthews, The Stables, Alkham Road, Dover, CT16 3EE
M1 EJQ M. Cross, 7 Hallside Road, Enfield, EN1 4AD
M1 EJS S. Birchall, 11 Rosebery Road, Felixstowe, IP11 7JR
M1 EJX M. Heley, 22 St. Lawrence Road Dunscroft, Doncaster, DN7 4AS
M1 EKA B. Parker, 38 Cross St., Thurcroft, Rotherham, S66 9NJ
M1 EKB K. Baldwin, 5 Rosedale Close, Belmont, Hereford, HR2 7ZD
M1 EKD S. Pounder, 4 Otley Mount, East Morton, Keighley, BD20 5TD
M1 EKH D. Bowker, 54 Edward Street, Middleton, Manchester, M24 6BN
M1 EKK K. George, 12 The Maltings, Clayton, Bradford, BD14 6DF
M1 EKM G. Harden, 13 Greenfield Road, Coleford, GL16 8BY
M1 EKP D. Meeds, 5 Lenton Way Frampton, Boston, PE20 1AU
M1 EKU P. Lancaster, 16 Wiltshire Close, Bury, BL9 9EY
M1 ELB C. Mitchell, A 3, Pakkalanrinne 14, Vantaa, Finland, 1510
MM1 ELE D. Marshall, 10 Spencer Crescent, Carnoustie, DD7 6DQ
M1 ELI D. Welch, 41 Mersey Way, Bletchley, Milton Keynes, MK3 7PS
M1 ELM A. Dent, 5 Loader Close, Bournemouth, BH9 1LR
M1 ELN M. Pike, 1 Sevelm, Up Hatherley, Cheltenham, GL51 3RZ
M1 ELQ P. Houghton, 19 Gilthwaites Lane, Denby Dale, Huddersfield, HD8 8SG
M1 ELR C. Davies, 94 Alnwick Drive, Moreton, Wirral, CH46 6ET
M1 ELS J. Matias, 23 The Approach, London, W3 7PA
M1 ELW H. Watson, 5 Arbroath, Ouston, Chester le Street, DH2 1QY
M1 EMB S. Buckley, 1 Chouler Gardens, Stevenage, SG1 4TB
M1 EMC K. Heselton, 3 Winterslow Road, Penhill, Swindon, SN2 5JJ
M1 EMG D. Woodcroft, 23 Wilkin Walk, Cottenham, Cambridge, CB24 8TS
M1 EMO P. Warden, 12A Landscape View, Saffron Walden, CB11 4AU
M1 EMP N. Douglas, 2 Huntingdon Close, Fareham, PO14 4JP
M1 EMR J. Dixon, 5 Laburnum Avenue, Moorends, Doncaster, DN8 4SF
M1 EMU P. Gilmore, 2 Ridgeway, Billericay, CM12 9NT
M1 EMX H. Harvey, 153 Stradbroke Grove, Clayhall, Ilford, IG5 0DL
M1 ENA J. Long, 1 Tangway, Chineham, Basingstoke, RG24 8SU
M1 ENE S. Marcot, 5 The Crescent, West Wickham, BR4 0HB
M1 ENJ C. Berry, 29 Marlborough Crescent, Long Hanborough, Witney, OX29 8JP
M1 ENK J. Berry, 29 Marlborough Crescent, Long Hanborough, Witney, OX29 8JP
M1 ENQ R. Townsend, 56 Seymour Road, Northfleet, Gravesend, DA11 7BN
M1 ENX S. Caine, Magnolia House, The Larches, East Grinstead, RH19 3QL
M1 ENZ R. Rouse, 15 Mulimbah Street, Eleebana, Australia, 2282
MW1 EOK P. Davies, 9 Cramer Court, Rhyl, LL18 2BX
MW1 EOO H. Matthews, 24 Clos Y Berllan, Rhuddlan, Rhyl, LL18 2UL
M1 EOP M. Jurkiewicz, 21 Porlock Avenue, Stafford, ST170HS
MW1 EOR M. Edwards, 2 The Twyn, Fleur De Lis, Blackwood, NP12 3UL
M1 EOU T. Nadin, 4 Firtree Rise, Chapeltown, Sheffield, S35 1QG
M1 EOV P. Spurgeon, 15 Ketts Close, Wymondham, NR18 0NB
M1 EOZ C. Cartwright, 8 Hudson Road, Blackpool, FY1 6LY
MJ1 EPG B. Allchin, Pont Marquet Cottage, La Rue Des Mans, St Brelade, Jersey, JE3 8BL
MW1 EPI D. Williams, Rhilin, Rhydwyn, Holyhead, LL65 4EA
M1 EPK K. Hichisson, 2 Tithe Farm Close, Langford, Biggleswade, SG18 9NE
M1 EPN R. Shaddick, 5 Shrewsbury Bow, Weston-Super-Mare, BS24 7SB
M1 EPR A. Wilson, 22 Ormesby Road, Raf Coltishall, Norwich, NR10 5JY
M1 EPU South Devon Raynet, c/o G. Coker, 46 Clarendon Road, Ipplepen, Newton Abbot, TQ12 5QS
M1 EPX M. Clark, 8 Willow Close, Clevedon, BS21 6HR
M1 EQA N. Trewin, 70 Trelowen Drive, Penryn, Cornwall, TR10 9WS
M1 EQB G. Trouse, 1 Amanda Close, Bexhill-on-Sea, TN40 2TB
M1 EQD P. Burton, 99 Western Avenue, Blacon, Chester, CH1 5QX
MM1 EQE A. Thompson, 22 Lochend Road, Carnoustie, DD7 7QF
MI1 EQI C. Blake, 9 Mullaghbrack Road, Hamiltonsbawn, Armagh, BT60 1JU
M1 EQO C. Hayward, 12 Trouvere Park, Hemel Hempstead, HP1 3HY
M1 EQV M. Edwards, 36 Mount Street, Coventry, CV5 8DE
M1 EQW S. Harrison, 20 Wisewood Avenue, Wisewood, Sheffield, S6 4WG
M1 ERA S. Trimble, Pentreath, Cury Cross Lanes, Helston, TR12 7BJ
M1 ERD A. Trimble, Pentreath, Cury Cross Lanes, Helston, TR12 7BJ
M1 ERF F. Tatlow, 45 Pasture Road, Stapleford, Nottingham, NG9 8HR
M1 ERH S. Bird, 12 Commercial Road, Shepton Mallet, BA4 5DH
M1 ERJ P. Chandler, 94 Shrubland St., Leamington Spa, CV31 3BD
MI1 ERL P. Cranston, 135 Saintfield Road, Lisburn, BT27 6YW
M1 ERN J. Baugh, 172 Pontefract Road, Cudworth, Barnsley, S72 8BE
M1 ERO D. Eastope, 9 St. Davids House, Willow Way, Redditch, B97 6PG
M1 ERP S. Carruthers, 13 Belah Road, Carlisle, CA3 9RE
M1 ERS S. Webster, 402 Windmill Lane, Sheffield, S5 6FZ
M1 ERU S. Carrington, 137 Richmond Park Road, Bournemouth, BH8 8UA
M1 ERV S. Chambers, 1 Northleaze, Corsham, SN13 0QW
M1 ERY K. Sylvester, 8 Beacon Park Close, Skegness, PE25 1HQ
M1 ESD J. Smith, 50 Chadcote Way, Catshill, Bromsgrove, B61 0JT
M1 ESH W. Inch, 17 Grantley Close Copford, Colchester, CO6 1YP
M1 ESI D. Crane, 132 Windermere Drive, Warndon, Worcester, WR4 9JD
M1 ESM L. Wallace, 20 Radworthy, Furzton, Milton Keynes, MK4 1JH
M1 ESV R. Scotland, 11 Edwards Court, Slough, SL1 2HY
M1 ETC M. Ribton, 80 Trafalgar St., Gillingham, ME7 4RN
M1 ETM A. Hayward, 67 Pinecroft, Carlisle, CA3 0DB
M1 ETN D. Allen, Ash Tree Farm, Small End, Friskney, Boston, PE22 8PF
M1 ETS E. Coleby, 13 Farm Close, Sunniside, Newcastle upon Tyne, NE16 5PP
M1 ETT T. Hindson, 73D Leigh Road, Wimborne, BH21 2AA
M1 ETW A. Talabi, 1 Crealock Grove, Woodford Green, IG8 9QZ
M1 EUB A. Wearne, 58 Mainstone Avenue, Princerock, Plymouth, PL4 9NB
M1 EUE J. Underwood, 27 Woodville Road, London, E17 7ER
M1 EUF J. Cunningham, 16 Welbeck Road, Doncaster, DN4 5EY
M1 EUL B. Fielding, 16 The Horseshoe, York, YO24 1LX
M1 EUM P. Thorne, 6 Wordsworth Terrace, Penrith, CA11 7QT
M1 EUN J. Fletcher, 66 Deightonby Street, Thurnscoe, Rotherham, S63 0JA
M1 EUR P. Allen, 32 Milward Road, Loscoe, Heanor, DE75 7JX
M1 EUX C. Bartlett, 25 Westfields, Buckingham, MK18 1DZ
MI1 EVD T. Carlisle, 12 Drumawhey Gardens, Bangor, BT19 1SR
M1 EVF S. Larden1, Flat 42, Worcester House, Hill Street, Halesowen, B63 4TJ
M1 EVH K. Wright, 60 Ashley Road, Walsall, WS3 2QF
MM1 EVJ J. Conway, 26 Kerse Avenue, Dalry, KA24 4DJ
M1 EVN J. Haughey, 31 Woodfield Drive, West Mersea, Colchester, CO5 8PX
M1 EVZ S. Punch, 39 Wilson Road Stalham, Norfolk, NR12 9FL
M1 EWB P. Dabell, 21 Winchcombe Drive, Burton-on-Trent, DE15 9EN
M1 EWD T. Beevers, 4 Cloud Avenue, Stapleford, Nottingham, NG9 8BN
M1 EWF R. Read, 111 Fitzpain Road, West Parley, Ferndown, BH22 8SF
MW1 EWJ E. Williams, 82 Maes Llwyn, Amlwch, LL68 9BG
M1 EWM A. O'Hea, 12 De Vitre Place, Grove, Wantage, OX12 0DA
M1 EWP B. Lester, 37 Cormorant Drive, St Austell, PL25 3BB
M1 EWT C. Berry, Berriscot, 7 Gloweth Villas, Truro, TR1 3LU
M1 EWV K. Trench, 10 Victoria Road, Morley, Leeds, LS27 9DS
M1 EXJ M. Wohlgemuth, 39 Great Mead, Denmead, Waterlooville, PO7 6HH
M1 EXO S. Andrews, 64 Bradgate Road, Markfield, LE67 9SN
M1 EXQ M. Peters, Yare House, Thuxton, Norwich, NR9 4QJ
M1 EXS G. Burton, 2 Derwent Street, Darwen, BB3 1EF
M1 EXW M. Gardner, 21 Tiptree, Castlehaven Road, London, NW1 8TL
M1 EYA R. Neale-Gardner, 72 Queensway Barwell, Leicester, LE9 8AP
M1 EYG R. Barrow, Fern House, Ripponden Old Lane, Sowerby Bridge, HX6 4PA
MW1 EYH F. Bailey, 28 Coopers Field, St. Martins, Oswestry, SY11 3BU
MM1 EYI N. Kingon-Rouse, 148 Oldwood Place, Livingston, EH54 6UX
M1 EYL A. Banks, 12 Taylor Court, Weston-Super-Mare, BS22 7LU
M1 EYO A. Poxon, 34 Conduit St., Tintwistle, Glossop, SK13 1LR
M1 EYP T. Read, 31 Merebrook Road, Macclesfield, SK11 8RH
M1 EYQ B. Davis, 104 Lever House Lane, Leyland, PR25 4XP
M1 EYS E. Shears, 22 Richborough Drive, Charlton, Andover, SP10 4EZ
M1 EYT A. Kok, 14 Throop Road, Templecombe, BA8 0HR
MW1 EYU D. Kok, 14 Castle View, Fron Goch, Caernarvon, LL55 4LE
MM1 EYZ D. Bruce111, Carnichal, Maud, Peterhead, AB42 4QG
M1 EZB C. Coonick, 36 Magdalen Way, Weston-Super-Mare, BS22 7PG
M1 EZC G. Sawyer, 101 Southern Drive, Loughton, IG10 3BY
M1 EZD G. Wesson, 12 Lea Close, Alcester, B49 6AP
M1 EZE S. Davies, 35 Queensland Crescent, Chelmsford, CM1 2DZ
M1 EZG M. Lane, 21 Winterbourne Road, Poole, BH15 2ES
M1 EZH L. Copley, Elmford, Mount Pleasant South, Whitby, YO22 4RQ
M1 EZJ M. Skelton, Annexe, 1 Muxton Lane, Muxton, Telford, TF2 8PB
M1 EZK K. Sheehan, 43 Central Avenue, Beverley, HU17 8LL
M1 EZL J. Anderson, 16 Hanham Road, Corfe Mullen, Wimborne, BH21 3PZ
M1 EZP M. White, 1 Nursery Close, Wroughton, Swindon, SN4 9DR
M1 EZR D. Birkenshaw, 309 New Parks Boulevard, Leicester, LE3 6nr
M1 EZT M. Marston, The Oaks, Nether Compton, Sherborne, DT9 4PZ
M1 EZU D. Bedwell, 12 Towcester Close, Chippenham, SN14 0XX
M1 EZX D. Bridle, 8 Cowleaze Road, Broadmayne, Dorchester, DT2 8EW
MI1 EZZ R. Bradley, 41 Graymount Road, Newtownabbey, BT36 7DR
M1 FAA T. Atkinson, 10 Invicta Close, Chislehurst, BR7 6SJ
M1 FAF T. Jones, 354 Bridgeman St., Bolton, BL3 6SJ
M1 FAI S. Taylor, Cherry Tree Cottage, Heslington, York, YO10 5DX
M1 FAJ D. Green, 7 Greenside Court, Mickleover, Derby, DE3 0RG
MI1 FAR A. Bryce, 123 Newtownards Road, Comber, Newtownards, BT23 5LD
MM1 FAS R. Krawczyk, 7 Anderson Crescent, Bishopmill, Elgin, IV30 4HJ
MW1 FAT C. Jayne, 65A Park Crescent, Abergavenny, NP7 5TL
M1 FAX D. Jones, 87 Forster St., Warrington, WA2 7AX
M1 FAY K. Rowsell, Elmtree Cottage, Chilworthy, Chard, TA20 3BH
M1 FBF A. Wilson, 2 Briar Close, Newhall, Swadlincote, DE11 0RX
M1 FBI S. Plews, 70 Baulkham Hills, Penshaw, Houghton le Spring, DH4 7RZ
M1 FBL J. Rowe, Hunters Brook, Fine Lane, Newport, PO30 3JY
M1 FBN I. Dalton, 10 Durkar Fields, Durkar, Wakefield, WF4 3BY
M1 FBS D. Faul, Ph 1113, 1200 The Esplanade North, Ontario, Canada, L1V6V3

M1 FBW J. Machalski, 28 Longdales Road, Lincoln, LN2 2JU
MI1 FCB T. Kilgore. Mbe, 8 Slieve Shannagh Park, Newcastle, BT33 0HW
M1 FCC C. Chuter, 35 Longford Road, Bognor Regis, PO21 1AB
M1 FCE D. Sampson, 32 The Gannets, Stubbington, Fareham, PO14 3SY
M1 FCF D. Patrick, 2 The Orchard, Old Totnes Road, Buckfastleigh, TQ11 0FG
M1 FCG R. Boyns, 65 Alma Road, Plymouth, PL3 4HE
M1 FCH T. Johnson, Orchard House, Tollerton, York, YO61 1PS
MI1 FCQ P. Mccauley, 5 Brookmount Rise, Omagh, BT78 5AL
M1 FCV C. Roberts, 57 Chandos Road, Lightpill, Stroud, GL5 3QT
M1 FCW M. Yeomans11, 7 Oak Avenue, Elloughton, Brough, HU15 1LA
M1 FCX A. Rudnicki, 1 Willoughby Court London Colney, St Albans, AL2 1HL
M1 FCZ E. Snowdon, 22 Twizziegill View, Easington, Saltburn-by-the-Sea, TS13 4NX
MM1 FDF S. Taylor, Sunnybrae, Kinellar, Aberdeen, AB21 0TY
M1 FDH F. Howker, 65 Boxley Drive, West Bridgford, Nottingham, NG2 7GN
M1 FDK G. Charnock, Oldhouse, Newton St Margarets, Hereford, HR2 0QR
MW1 FDN D. Morgan, Castle Cottage, Newbridge, Newport, NP11 3NT
M1 FDO C. Heading, 27 Broadlands Avenue, Eastleigh, SO50 4PP
M1 FDY L. Browning, 5 Redstart Avenue, Kidderminster, DY10 4JR
M1 FEK D. Stewart, 1 Twelve Acres, Welwyn Garden City, AL7 4TG
M1 FEM F. Priborsky, Fohrenstrasse 49, Tuttlingen, Germany, 78532
MM1 FEO P. Gaskin, Tft Electronics, Unit 1, Skeld Industrial Estate, Skeld, ZE2 9NL
M1 FEQ K. Watson, Bolehall Manor Club, Amington Road, Tamworth, B77 3LH
M1 FER G. Watson, 85 Thomas Street, Tamworth, B77 3PP
M1 FET E. Dodd, 2 Chichester Crescent, Chadderton, Oldham, OL9 0RW
MW1 FEU M. Williams, 68 Hengoed Road, Penpedairheol, Hengoed, CF82 8BR
M1 FEW A. Franklin, 11 Harden Hills, Shaw, Oldham, OL2 8NE
M1 FEX C. Wells, 37 Water Meadows, Worksop, S80 3DF
M1 FEY I. Trail, 10 Hillary Drive, Crowthorne, RG45 6QE
M1 FEZ S. West, 142 Burrowmoor Road, Cambridgeshire, PE15 9SS
M1 FFA Dr D. Byfield, 23 New Cross, Longburton, Sherborne, DT9 6EJ
M1 FFC P. Francis, Unit 15 Wellington Road, Bridgwater, TA6 5HA
MW1 FFE D. Evans, 5 Brunel Road, Fairwater, Cwmbran, NP44 4QT
M1 FFF D. Leech, 9 Little Gransden Lane, Great Gransden, Sandy, SG19 3BA
M1 FFG G. Siviter, Flat 106, Lancaster House, Rowley Regis, B65 0QE
M1 FFM M. Dawson, 140A Healey Road, Scunthorpe, DN16 1HT
M1 FFN J. Willingham, 3 Cherry Road, Nailsea, Bristol, BS48 2EE
M1 FFO E. Sanderson, Flat, Old Post Office, High Street, Yeovil, BA22 7NQ
M1 FFP B. White, 7 Park Street, Castle Cary, BA7 7EH
M1 FFR K. Singam, 44 Forty Lane, Wembley, HA9 9HA
M1 FFS J. Bates, Flat 7, Colyer House, London, SE2 0AJ
M1 FFV J. Cobb, 39 Bank House Road, Sheffield, S6 3TL
M1 FFX C. Burgess, 13 Glyne Drive, Pebsham, Bexhill-on-Sea, TN40 2PW
MW1 FFY R. Davies, 100 Lewis Road, Neath, SA11 1DQ
MW1 FGB R. Tremelling, 45 Bryntawe Road, Ynystawe, Swansea, SA6 5AD
M1 FGH Dr M. James, Oak Tree House St. Matthews Terrace, Leyburn, DL8 5EL
M1 FGM T. Beswick, 37 Grovewood Road, Misterton, Doncaster, DN10 4EF
M1 FGO Dr A. Kerr-Munslow, 28 Swallow Court, St Neots, PE19 1NP
MW1 FGV J. Rowe, Flat 27 Rhodfa Frank, Ammanford, SA18 2QE
M1 FHA T. Earp, 14 Drakes Avenue, Devizes, SN10 5AZ
M1 FHB R. Earp, 1 Baynton Way, Edington, Westbury, BA13 4PT
MI1 FHE K. Lunney, 20 Inniskeen Close, Enniskillen, BT74 6HD
M1 FHJ J. Bolton, 11 Forest Drive, Lytham St Annes, FY8 4PF
MM1 FHL J. Crockett, 70 Inchview Terrace, Edinburgh, EH7 6TH
MM1 FHO L. Norman, 16 Cotton Street, Balfron, Glasgow, G63 0PF
M1 FHP E. Parr, 36 Ridley Drive, Great Sankey, Warrington, WA5 1HP
M1 FHQ I. Anderson, 11 Rays Drive, Lancaster, LA1 4NT
MM1 FHR G. Hunt, 1 Love St., Kilwinning, KA13 7LQ
MM1 FHS N. Sampson, 47 Muirend Road, Perth, PH1 1JD
M1 FHT A. Di Domenico, 11 Mowbreck Court, Wesham, Preston, PR4 3AG
M1 FHX D. Jordan, 38 Weston Lane, Otley, LS21 2DB
MM1 FHZ Dr D. Bickle, Lon Mhor, 10 Grean, Isle of Barra, HS9 5XU
M1 FIB P. Heathcote, Flat 6, Balmoral House, 12 Balmoral Road, Westcliff-on-Sea, SS0 7AZ
M1 FIE Dr A. Phelan, Calle Misericordia 14, Jimena de la Frontera, Spain, 11330
M1 FIG S. Hunt, 1 Trefusis Cottages, Flushing, Falmouth, TR11 5TE
M1 FII C. Parker, 3 Lyon Close, Abingdon, OX14 1PT
M1 FIL P. Smith, 18 Cheltenham Grove, Newcastle, ST5 6QS
M1 FIP F. O'Sullivan, 10 Hampton Close, London, NW6 5LR
M1 FIR J. Coles, 45 Common Lane, Titchfield, Fareham, PO14 4BX
MI1 FIS D. Stewart, 16 Weavers Lodge, Donaghcloney, Craigavon, BT66 7LE
M1 FIV W. Wailes, 8 Staindrop Terrace, Stanley, DH9 8JW
M1 FJA R. Clifford-Smith, 60 Petworth Gardens, Southampton, SO16 8EF
M1 FJB S. Hunt, Maxxwave Ltd, Maxxwave House Unit 32, Hill Lane Close, Markfield, LE67 9PY
M1 FJC J. Soltysik, 24 Cottage Close, Hednesford, Cannock, WS12 1BS
M1 FJD J. Coleman, 52 Brook Street, Colchester, CO1 2UT
M1 FJF P. Griffiths, Probation Hostel Haworth House, Blackburn, Bb2 2hl
M1 FJG A. Roberts, 39 Colsterdale Carlton Colville, Lowestoft, NR33 8TN
M1 FJH F. Bate, 10Alder Close, Worcester, WR3 8QH
M1 FJJ A. Parkinson, 21 Lambton Street, Bolton, BL3 3LG
MW1 FJK K. Hughes, 33 Brynglas, Penygroes, Llanelli, SA14 7PY
M1 FJL R. Collins, Perch Hill Cottage, Perch Hill, Wells, BA5 1JA
MM1 FJM R. Moore, Ard Na Mara, Gulberwick, Shetland, ZE2 9TX
M1 FJP P. Blackman, 73 St. Marks Road, Chester, CH4 8DE
M1 FJQ D. Tinker, 116 Longley Avenue West, Sheffield, S5 8WF
MW1 FLY R. Eyre, 9 Poplar Close, Rogiet, Caldicot, NP26 3TL
M1 FMC A. Bailey, 2 Kingswood Road, Shrewsbury, SY3 8UX
M1 FNE G. Scott, 19 Witton Gardens, Jarrow, ne32 5yj
M1 FRB F. Barnes, 4 Pound Close, Ducklington, Witney, OX29 7TH
MI1 FRM Dr F. Montgomery, 4 Thornbrook, Lisburn, BT27 5LW
M1 FUR Coulsdon Amateur Transmitting Society, c/o A. Briers, 33 Deans Walk, Coulsdon, CR5 1HR
M1 FWD S. Davies, 17 Taw View Fremington, Barnstaple, EX31 2NJ
MM1 FZR S. Gillies, 49 Meadowside Road, Queenzieburn, Glasgow, G65 9EJ
M1 GAP A. Prince, 29 St. Stephens Road, West Bromwich, B71 4LR
M1 GAR A. Gardner, 39 Court Close, Twickenham, TW2 5JH
M1 GAS P. Cheeseman, 24 Meadowcroft Close Glenfield, Leicester, LE3 8QX
MM1 GBS D. James, 9 Dunbar Lane, Duffus, Elgin, IV30 5QN
M1 GCS G. Steedman, 61 Granville Street, Barnsley, S75 2TQ
M1 GDB G. Bell, 4 Fairley Way, Cheshunt, Waltham Cross, EN7 6LG
M1 GDE G. Edgar, 61 Winchester Avenue, Lancaster, LA1 4HX
M1 GDH D. Hayward, 16 Heathway, Dagenham, RM10 9PP
M1 GEO Dr G. Smart, 30 Cornmills Road, Soham, CB7 5AT
M1 GFE F. Erridge, 17 Head Street, Goldhanger, Maldon, CM9 8AY
M1 GGG G. Ma, 26 Church Lane, Chalgrove, Oxford, OX44 7TA
M1 GHT D. Buckerfield, 62 Springfield Crescent, Somercotes, Alfreton, DE55 4LH
M1 GIZ S. Bridger, 80 Springhill Crescent, Madeley, Telford, TF7 4DP
MW1 GLD G. Davies, 77 Tydraw St., Port Talbot, SA13 1BR
M1 GMO M. Hastry, 56 Kilsyth Close, Fearnhead, Warrington, WA2 0SQ
M1 GOH R. Horry, 1 Council House Nidds Lane, Kirton, Boston, PE20 1LZ
M1 GPC G. Carpenter, 67 Angela Crescent, Horsford, Norwich, NR10 3HE
M1 GPE G. Emmerson, 29 Dulsie Road, Talbot Woods, Bournemouth, BH3 7DY
M1 GRA G. Stephens, 25 Fore Street Langtree, Torrington, EX38 8NG
M1 GRP G. Pound, 19 Victory Green, Portsmouth, PO2 8RH
M1 GSM S. Watson, 6 Mount Pleasant, Stanley, Crook, DL15 9SF
M1 GSX I. Greenfield, 1 Dale Road, Shrewsbury, SY2 5TE
M1 GTI D. Burgin, 15 Birch Grove, Chippenham, SN15 1DD
M1 GUR P. Gurney, 12 Church Street, Wymeswold, Loughborough, LE12 6TX
M1 GUS J. Batchelor, 5 Gladden Fields, South Woodham Ferrers, Chelmsford, CM3 7AH
M1 GWA G. Warburton, 50 Clarendon Road, Sheffield, S10 3TR
M1 GWZ Dr P. Miller Tate, 19 Esher Avenue, Walton-on-Thames, KT12 2SZ
M1 GXL T. Higginson, 109 London Road, Biggleswade, SG18 8EE
M1 HFM M. Poole, 18 Lockway, Drayton, Abingdon, OX14 4LG
M1 HFX R. Ayers, 1 Handley Park, Sixpenny Handley, sp5-5pl
M1 HGV M. Rule, 23 Rue Du Puits Doux, Sacy, St Christophe A Berry, France, 2290
M1 HHL B. Miller, Oakhurst, 1 Southern Oaks Barton On Sea, New Milton, BH25 7JT
M1 HJE S. Elliott11, 7 Manor Close Harston, Cambridge, CB22 7QF
M1 HLG H. Glover, 14 Crawley Crescent, Eastbourne, BN22 9RN
M1 HLL S. Batchelor, 2 Belmont Avenue, Atherton, Manchester, M46 9RR
M1 HMP P. Grech, 108 Hind Grove, London, E14 6HU
MM1 HMV B. Shearer, Latheron, 113 Auchamore Road, Dunoon, PA23 7JJ
M1 HMZ B. Allison, 5 Mayfield Drive, Howwood, Johnstone, PA9 1BJ
M1 HOP Sir A. Hopson, 1 Hall Lane, Leicester, LE2 8SF
M1 HPR I. Hooper, 25 Honey Lane, Buntingford, SG9 9BQ
M1 HQX W. Hammond, 28 Fengate Mobile Home Park, Peterborough, PE1 5XD
M1 HVJ A. Jefferiss, 27 Sherbourne Drive, Maidenhead, SL6 3EP
MM1 HWB P. Oldham, 2 Old Bar Road, Nairn, IV12 5BX
M1 HZZ A. Barbour, 36 Roseacre Drive, Elswick, Preston, PR4 3UQ
M1 IAN I. Tennent, Flat 2, 97 Rydens Road, Walton-on-Thames, KT12 3AW
MM1 ICE A. Somerville, 8 Craiglockhart Park, Edinburgh, EH14 1HE
M1 ICL I. Leather, White Cottage, Aston Lane, Aston, Runcorn, WA7 3BU
M1 IFT Dr A. Bartle, 10 Holme Dene, Haxey, Doncaster, DN9 2JX
M1 IHM D. Jones, 19 Queens Road, Donnington, Telford, TF2 8DB
M1 IKE M. Collins, 3 Beacon View, Grayrigg, Kendal, LA8 9BT
M1 IMC I. Mcleod, 12 Seathwaite Avenue, Marton, Blackpool, FY4 4RL
M1 IOS J. Goody, 9 Garrison Lane, St. Mary'S, Isles of Scilly, TR21 0JD
M1 IOW P. Legg, 20 Arthur Moody Drive, Newport, PO30 5JR
M1 IRB I. Bush, 17 Queens Place, Shoreham-by-Sea, BN43 5AA
M1 IRM P. Rowley, 6 Duesbury Green, Stoke-on-Trent, ST3 2RZ
MM1 JAC J. Campbell, 119 Campbell Avenue, Dumbarton, G82 3PB
M1 JAK A. Hanson-Brown, 35 York Avenue, Bedworth, CV12 9EL
MW1 JAN J. Carfoot, 24 Marble Church Grove, Bodelwyddan, LL18 5UP
MM1 JAS J. Shankland, 2 Strathdoon Place, Ayr, KA7 4PB
M1 JCB T. Wightman, Laithbutts Farm, Cowan Bridge, Carnforth, LA6 2JL
M1 JCL J. Plant, 67 Kenley Road, London, SW19 3JJ
M1 JCS C. Starr, 64 Green Lane, Lambley, Nottingham, NG4 4QE
M1 JDH J. Hammond, 8 Rowntree Way, Saffron Walden, CB11 4DG
M1 JDW J. Mitchell, Sheet Hill Farmhouse, Winfield Lane, Sevenoaks, TN15 0LZ
M1 JEC J. Cook, 35 Holdbrook Way, Romford, RM3 0JD
M1 JES J. Gilbert, Thr Oaktree, Ellenbrook Lane, Hatfield, AL10 9NT
M1 JHG J. Green, 33 Edenvale Crescent, Lancaster, LA1 2NW
M1 JHL S. Jouhal, 35 Cherrywood Gardens, Nottingham, NG3 6LR
M1 JIM J. O'Hea, 12 De Vitre Place, Grove, Wantage, OX12 0DA
M1 JJN J. Nicholson, 6 Mill Gardens, West End, Southampton, SO18 3AG
M1 JJS P. Springate, 10 Pipers Close, Burnham, Slough, SL1 8AW
M1 JLM B. Murfitt, 21 Priors Drive, Norwich, NR6 7LJ
M1 JMB J. Bevan, 1 Condor Close, Weston-Super-Mare, BS22 8SE
MI1 JOE W. Murray, 80 Canterbury Park, Londonderry, BT47 6DU
M1 JON J. Godding, 58 Dukeswood Road, Longtown, Carlisle, CA6 5UJ
M1 JPS J. Patterson, 39 Coquet Drive, Ellington, Morpeth, NE61 5LN
M1 JSS J. Smout, 34 Kidderminster Road, Bromsgrove, B61 7JT
M1 JTA J. Tyers, Hillcrest, Papermill Lane, Evedon, Sleaford, NG34 9PD
MM1 JWF J. Frame, 24 Douglas Crescent, Erskine, PA8 6BJ
M1 JWM J. Machin, 42 Woodstock Road, Loxley, Sheffield, S6 6TG
M1 JWR J. Rutherford, Nook On Lyne, Longtown, Carlisle, CA6 5TS
M1 JWS J. Smith, 134 Blaker Court, Fairlawn, London, SE7 7EU
M1 KAZ Dr A. Forrest, 261 East End Road, London, N2 8AY
M1 KCB K. Crank, 319 Manchester Road, Clifton, Manchester, M27 6PT
M1 KDH K. Harvey, 29 The Hobbins, Bridgnorth, WV15 5HH
M1 KDJ K. Jowett, The Gables, Brinkworth Road, Royal Wootton Bassett, Swindon, SN4 8DT
MW1 KDP M. Heron, Heron House, Park Road, Gwynedd, LL42 1PL
M1 KEJ R. Jeffs, 45 Forest Road, Bingham, Nottingham, NG13 8RL
M1 KES M. Oconnor, 13 Ashburnham Road, Southend-on-Sea, SS1 1QB
M1 KEV K. Mahoney, 11 Leyland Walk, Bristol, BS13 8PY
M1 KEY M. Hardy, 4 Kirk Balk Hoyland, Barnsley, S74 9HU
M1 KGL K. Large, 6 Sylvden Drive, Wisbech, PE13 3UD
M1 KIP L. Kipling, 16 Northampton Close, Bracknell, RG12 9EF
M1 KIQ D. Rosser, P.O. Box 268, Enmore, Australia, 2042
M1 KMC A. Coathup, 54 Rydal Road, Kendal, LA9 6LB
M1 KOS Dr K. Tsioumparakis, 10 Lavender Close, Leatherhead, KT22 8LZ
M1 KPW K. Whitmarsh, 7 Foxs Furlong, Chineham, Basingstoke, RG24 8WN
M1 KSB K. Best, 42 Falmer Avenue, Goring-By-Sea, Worthing, BN12 4TD
M1 KTA R. Baines, 34 Bury Road, Stapleford, Cambridge, CB22 5BP
M1 KTY K. Mallows, 57 Top Road, Kingsley, Frodsham, WA6 8DA

M1 KVN K. Finn, 132 Lansdowne Grove, Wigston, LE18 4LY
M1 KWH K. Hargreaves, Langton Lodge, Fordcombe Road, Tunbridge Wells, TN3 0RB
M1 LAN A. Worsley, 10 Millfield View, Worksop, S80 3QB
M1 LAP L. Pollard, 45 Nanny Marr Road, Darfield, Barnsley, S73 9AB
MM1 LBA A. Bulloch, 4 Cartleburn Gardens, Kilwinning, KA13 7ND
M1 LCL A. Kvilums, 78 Wagon Lane, Solihull, B92 7PN
MM1 LDR A. Sloan, 36 Paterson Avenue, Irvine, KA12 9JJ
M1 LEO A. Bennett, 16 Manor Avenue, Crewe, CW2 8BD
M1 LES L. Rodger, 19 South Walk, West Wickham, BR4 9JA
M1 LIP A. Lippett, 2 Ralph Court, Stafford, ST17 9FR
MM1 LJB C. Newman, Upper Flat, 3 Lindsay Gardens, Alexandria, G83 0US
MW1 LLL M. Greatorex, Cwm Pennant, Moel View Road, Prestatyn, LL19 9SU
M1 LMJ L. Jones, 47 Pine Crescent, Chandler'S Ford, Eastleigh, SO53 1LN
M1 LMO N. Waring, 3 Sampson St., Eastoft, Scunthorpe, DN17 4PQ
M1 LOL A. King, 8 Rydal Court, Congleton, CW12 4JL
M1 LOU R. Cave, 26 Longsight Road, Mapplewell, Barnsley, S75 6HB
M1 LRX D. Horwood, 42 Southlands Drive, Timsbury, Bath, BA2 0HB
M1 LSD L. Dawes, 52 Ridley Road, Carlisle, CA2 4LD
MW1 LSG A. Jenkins, 25 Maes Hyfryd, Flint, CH6 5LN
M1 LTS L. Stone, 32 Watermeadow Lane, Storrington, Pulborough, RH20 3GU
M1 LXM A. May, 7 Stanton Close, Blandford Forum, DT11 7RT
M1 LYE F. Lye, 5 New Road, Hextable, Swanley, BR8 7LS
M1 LYN L. Asbury, 67 Orchard Way, Measham, Swadlincote, DE12 7JZ
M1 MAB I. Patrick, 17 Stamford Way, Fair Oak, Eastleigh, SO50 7JJ
M1 MAD M. Cottrell, 9 Woodland Terrace, Kingswood, Bristol, BS15 9PU
M1 MAJ Dr M. Johnson, 2A St. Margarets Road, Girton, Cambridge, CB3 0LT
M1 MAL M. Cadman, Flat 17, Harmon House, London, SE8 3AS
M1 MCL Dr M. Lawson, 9 Headingley Mews, Wakefield, WF1 3AB
M1 MCW S. Lace, 19 Methuen Street, Walney, Barrow-in-Furness, LA14 3PS
M1 MDE D. Elwood, High Farm Cottage 7, Newport Road, Market Drayton, TF9 2TH
M1 MDP M. Palmer, 57 Bemersley Road, Stoke-on-Trent, ST6 8JF
MW1 MFY D. Lee, 13 Yr Efail, Treoes, Bridgend, CF35 5EG
MM1 MHD P. Overton, Cluanie, Cairnballoch, Alford, AB33 8HQ
M1 MIC M. Hodgson, French Farm Bungalow, French Drove, Thorney, Peterborough, PE6 0PQ
M1 MIJ W. Waddington, 117 Dominion St., Walney, Barrow in Furness, LA14 3BP
M1 MIT T. Haynes, Jeetex Marine, Riverside Estate, Brundall, Norwich, NR13 5PL
M1 MKL M. Livesey, 33 Carrington Close, Birchwood, Warrington, WA3 7QA
M1 MLM A. Lomas, 32 Crestway Road, Baddeley Green, Stoke-on-Trent, ST2 7LD
M1 MNR Mid Norton Raynet Group, c/o L. Knighton, 5 Quidenham Road, East Harling, Norwich, NR16 2JD
M1 MOB M. Dimambro, 26 Fetcham Court, Bank Top, Newcastle upon Tyne, NE3 2UL
M1 MOD A. Straw, Flat 17, Poseidon Court, Homer Drive, London, E14 3UG
M1 MOG M. Taylor, 136 Lenthall Avenue, Grays, RM17 5AB
MM1 MOY J. Dye, Allt Na Slanaichd, Moy, Inverness, IV13 7YE
M1 MPA M. Allgar, 13 Deacon Avenue, Kempston, Bedford, MK42 7DU
M1 MPB M. Burfield, 11 Myrtle Grove, Wallasey, CH44 6QA
M1 MPK M. Kassai, 6 Cranhill Close, Littleover, Derby, DE23 3XU
M1 MPW M. Wooldridge, 26 Little Meadow Bar Hill, Cambridge, CB23 8TD
M1 MRB M. Butler, 210 Green Wrythe Lane, Carshalton, SM5 2SP
M1 MRS R. Shepperley, Flat F, London, N3 1QL
M1 MSF M. Forster, 4 West Street, Top Flat, Horncastle, LN9 5JF
M1 MST C. Walker, Fairfield House, Brimfield Cross, Ludlow, SY8 4ND
M1 MTV A. Mason, 76 Burlington Way Mickleover, Derby, DE3 9BD
M1 MUM P. Worlledge, 181 Roselands Drive, Paignton, TQ4 7RN
M1 MUS M. Suleyman, 26 Old Park Road South, Enfield, EN2 7DB
M1 MVX A. Webb, 5 Highfield Avenue, Bristol, BS15 3RA
M1 NAD D. Barratt, Chapel Cottage, Briantspuddle, Dorchester, DT2 7HX
M1 NAS N. Swann, 11 Erdyngton Road, Leicester, LE3 1JF
M1 NCC N. Cordell, 392 Laceby Road, Grimsby, DN34 5LX
M1 NER A. Taylor, 16 Bellmans Road Whittlesey, Peterborough, PE7 1TY
M1 NEW K. Mason, 12 Vicarage Close, Weston-Super-Mare, BS22 7PA
M1 NHR North Hertfordshire Raynet Assoc., c/o K. Edwards, 289 Monks Walk, Buntingford, SG9 9DZ
M1 NIS B. Ashcroft, 16 Edge Lane, Crosby, Liverpool, L23 9XE
M1 NIZ M. Austin, 11A Beverley Road, Ipswich, IP4 4BU
M1 NMG N. Gunnell, 8 Upper Dingle, Madeley, Telford, TF7 5RX
M1 NNN J. Earnshaw, Dunelm, Ayton Road, Irton, Scarborough, YO12 4RQ
M1 NPH N. Holland, 40 Marlborough Road, Castle Bromwich, Birmingham, B36 0EH
M1 NSP S. Naredla, 15 Sweet Briar Drive, Laindon, Basildon, SS15 4HA
M1 NTV N. Tartt, 47 Leatham Park Road, Featherstone, Pontefract, WF7 5DP
M1 NTY D. Basson, Penrhyn, Stonehouse Road, Sevenoaks, TN14 7HW
M1 NUS R. Cartwright, 14 Bromsgrove Avenue, Eccles, Manchester, M30 8WB
M1 NXX J. Lynch, 14 The Pastures, Cayton, Scarborough, YO11 3UU
M1 OBR D. O'Brien, 14 Lower Bettesworth Road, Ryde, PO33 3EL
M1 OCN C. Wilson, Houseboat Wolf Chandlers Quay, Maldon, CM9 4LF
M1 OJS O. Smith, 106 Middle Street, Blackhall Colliery, Hartlepool, TS27 4EB
M1 ONE C. Chambers, 75A Main St., Sedbergh, LA10 5AB
MI1 OPM D. Kirkwood, 1 Rural Cottages, Front Road, Lisburn, BT27 5LF
M1 OXR A. Garthwaite, 278 Carlton Road, Barnsley, S71 2BA
M1 PAB P. Bush, 1A Sherwood Close, Kennington, Ashford, TN24 9PT
M1 PAC P. Cole, 48 Emily Street, Keighley, BD21 3HY
M1 PAF P. Fletcher, 7 Gatesyde Place, Eskdale, Holmrook, CA19 1UD
M1 PAH D. Hardy, 1 Upper Steeping, Desborough, Kettering, NN14 2SQ
M1 PAM P. Rodger, 19 South Walk, West Wickham, BR4 9JA
M1 PAS W. James, 3 Midfield Close, Gillow Heath, Stoke-on-Trent, ST8 6RD
M1 PFS P. Symonds, 2 Gorling Close, Ifield, RH11 0TJ
M1 PGH P. Howell, 16 Everard Road, Bedford, MK41 9LD
M1 PGT P. Tomlin, 10 The Martins, Thatcham, RG19 4FD
M1 PJB P. Backx, 43 Tindale Avenue, Cramlington, NE23 2BP
M1 PJH P. Hill, Flat 12, Danecourt 14 St. Peters Road, Poole, BH14 0PA
M1 PKB P. Booth, 19 Gunville Crescent, Bournemouth, BH9 3PZ
M1 PKW P. Wilton, Downsview Cottage, Wappingthorn Farm Lane, Steyning, BN44 3AG
M1 PLC M. Woods, 76 Kings Road, Evesham, WR11 3BS
M1 PMR A. Marshall, 53 Birch Close, Corfe Mullen, Wimborne, BH21 3TB
M1 PRC Peterborough & District ARC, c/o T. Ralph, 15 Portchester Close, Stanground, Peterborough, PE2 8UP
M1 PTE P. Merrick, 12 Wilkinson Road, Wednesbury, WS10 8SH
M1 PTR P. Ridley, 6 Elm Close, Poynton, Stockport, SK12 1QH
MM1 PTT C. Morrison, 28 Solway Road Bishopbriggs, Glasgow, G64 1QW
M1 PUW Dr P. Rogers, 12 Cook'S Folly Road, Sneyd Park, Bristol, BS9 1PL
M1 PVC P. Craven, Hithe, Chuck Hatch Lane, Colemans Hatch, Hartfield, TN7 4EN
M1 PVF P. Flavell, 26 Hulles Way, North Baddesley, Southampton, SO52 9NS
M1 PWT P. Telco, 7 Brockswood Lane, Welwyn Garden City, AL8 7BA
M1 PXB J. Hopkins, Millinder House, Westerdale, Whitby, YO21 2DE
M1 PYE J. Pye, 19 Nellan Crescent, Stoke-on-Trent, ST6 1PS
M1 RAD D. Clapp, 150 Brougham Road, Worthing, BN11 2PH
MM1 RAH R. Hemesley, Overton Farm House, Kirknewton, EH27 8DD
M1 RAL R. Leach, 4 Honey Pot Drive, Baildon, Shipley, BD17 5TJ
MI1 RAY R. Maguire, 139 Carrowshee Park, Drumhaw, Enniskillen, BT92 0FS
MI1 RDR R. Ross, 13 Eureka Drive, Belfast, BT12 5NR
M1 RDX A. Casson, 5 Hazeltree Road, Ulverston, LA12 9JP
M1 REC G. Doughty, 1A The Crescent, Ketton, Stamford, PE9 3SY
M1 REJ R. Jacklin, 63 Ventnor Rise, Heathfield, Nottingham, NG5 1NW
M1 REK R. King, 8 Rydal Court, Congleton, CW12 4JL
MW1 RES F. Garrett, 11 Third Avenue, Flint, CH6 5LT
MI1 RGL R. Leigh, 42 Comber Road Killinchy, Newtownards, BT23 6PB
M1 RGW R. Warren, 23 Bramshaw Close, Winchester, SO22 6LT
M1 RIC R. Booth, 108 Newlands Gardens, Workington, CA14 3PE
M1 RIG D. Gleadell, 10 Swires Terrace, Halifax, HX1 2EP
MM1 RIK R. Irvine, 83 Glenacre Road, Cumbernauld, Glasgow, G67 2NT
M1 RJG R. Gooderham, 3 School Meadow, Barnby, Beccles, NR34 7QL
M1 RJJ J. Fowler, 48 Grangefields Road, Jacob'S Well, Guildford, GU4 7NP
M1 RJL Rev. R. Lapwood, Rose Cottage, Challow Road, Wantage, OX12 9DN
M1 RJS R. Speak, Flat 1, Queens Court, East Road, Colchester, CO5 8EB
M1 RKB R. Browning, 575 Rayleigh Road, Leigh-on-Sea, SS9 5HR
M1 RKY M. Harriott, 20 Old Road, Tean, Stoke-on-Trent, ST10 4EG
M1 RMO R. Oakley, 41 Cyrano Way, Grimsby, DN37 9SQ
MM1 RMS R. Scott, Morven, Isle of Lewis, HS2 0QX
M1 RMW R. Warry, 51 South Street, Crewkerne, TA18 8DB
M1 ROD R. Jones, 7 Warner Avenue, North Cheam, Sutton, SM3 9RH
M1 ROE A. Roebuck, Holford Farm, Chester Road, Knutsford, WA16 0TZ
MD1 RPC R. Woolley, Moaralyn, King Williams Road, Castletown, Isle Of Man, IM9 1BL
M1 RSB R. Bain, Oak Tree Cottage, Long Barn Road, Sevenoaks, TN14 6NH
M1 RST R. Dixon, 20B Ratcliffe Road, Hedge End, Southampton, SO30 4HA
M1 RTT R. Thong, 10 Lowry Close, Haverhill, CB9 7GH
M1 RWB R. Blears, The Pump House, Warrington Road, Chester, CH2 4DQ
M1 SAB S. Armatage, Hayleazes, Lincoln Hill, Hexham, NE46 4BE
M1 SAC S. Clark, Flat 11, Saxon House Aylward Drive, Stevenage, SG2 8UY
M1 SAM S. Evans, 44 Edwards Drive, Plymouth, PL7 2SU
M1 SAN R. Sanders, Magnolia Cottage, Lanreath, Looe, PL13 2NX
MW1 SAS A. Jones, 4 Crowhill, Haverfordwest, SA61 2HL
M1 SAZ S. Haynes, 4 Monkend Terrace, Croft On Tees, Darlington, DL2 2SQ
M1 SCS S. Smith, 113 Deaconsfield Road, Hemel Hempstead, HP3 9JA
M1 SCW S. Whiteman, 109 Gordon Road, West Bridgford, Nottingham, NG2 5LX
M1 SDE A. Raine, Stable Cottage, St. Martins, Richmond, DL10 4SJ
M1 SEM S. Rear, 18 Brook Lane Cottages Sellindge, Ashford, TN25 6HG
M1 SFS Summerfields ARC, c/o Rev. R. Lapwood, Rose Cottage, Challow Road, Wantage, OX12 9DN
M1 SGA S. Ash, 1 Twyford Mill, Pig Lane, Bishops Stortford, CM22 7PA
M1 SHA S. Creber, 4 Joyce Close, Swindon, SN25 4GX
M1 SHE J. Little, 7 Deerfern Close, Great Linford, Milton Keynes, MK14 5BZ
M1 SIM C. Simcock, 51 Broadway, Stockport, SK2 5SF
M1 SIN T. Brundrett, 45 Talbot Crescent, Whitchurch, SY13 1PH
M1 SJA A. Stockwell, 28 Scholars Walk, Chalfont St. Peter, Gerrards Cross, SL9 0EJ
M1 SJH S. Harrison, 64 Douglas Road, Leigh, WN7 5HG
M1 SKA D. Spencer, 4 Douglas Road, Portsmouth, PO3 6AU.
M1 SKI A. Grabianski, 29 Lismore Road, Highworth, Swindon, SN6 7HU
M1 SKY A. Johnston, 3 Troutbeck Gardens, Barrow-in-Furness, LA14 4LR
M1 SLH K. Taylor, 23 The Chestnuts, Abingdon, OX14 3YN
M1 SMF S. Flanagan, 33 Ullswater Road, Chorley, PR7 2JB
M1 SNM M. Wilson, 3 Brookhurst Close, Chelmsford, CM2 6DX
M1 SPW N. Walker, 79 Arklow Drive, Hale Village, Liverpool, L24 5RR
M1 SPY S. Pybus, Grewgrass Farm, Grewgrass Lane, Redcar, TS118EB
M1 SRC Surrey Raynet, c/o T. Dabbs, 4 Caverleigh, Cadogan Road, Surbiton, KT6 4DH
M1 SRH S. Howlett, 20 Long Perry, Capel St. Mary, Ipswich, IP9 2XD
M1 SRP S. Pickin, 138 Boreham Field, Warminster, BA12 9EF
M1 SSB S. Bygrave, 10 Spinney Close, North Cove, Beccles, NR34 7PT
M1 STI B. Hall, 5 Perche Court, Midhurst, GU29 9TE
MM1 STK T. Storkey, Flat E, 29 Herbert Street, Glasgow, G20 6NB
M1 SUE S. Gunn, 5 School Villas, Broxted, Dunmow, CM6 2BS
M1 SUM D. Sumner, 30E Malvern Avenue, Ellesmere Port, CH65 5AD
M1 SWB S. Bainbridge, 6 Sandyville Grove, Liverpool, L4 8UL
M1 SWL International Shortwave League, c/o A. Kinson, 6 Uplands Park, Broad Oak, Heathfield, TN21 8SJ
M1 SWR S. Rigby, 2 Tong Forge Lizard Lane Shifnal Shropshire Tf11 8Qd, Shifnal, TF11 8QD
M1 SWS S. Southworth, 157 Birkwood Avenue, Cudworth, Barnsley, S72 8JB
MM1 SYD S. Mccance, 34 Calside, Paisley, PA2 6DB
M1 SYG A. Sygerycz, 75 O'Brien Road, Cheltenham, GL51 0UP
M1 TAD T. Denby, 22 Okehampton Crescent, Welling, DA16 1DE
MW1 TAF M. Williams, 1 Gomer Court, Abergele, LL22 7UU
M1 TAP A. Prince, 1 Woodside, Inglesbatch, Bath, BA2 9DZ
M1 TAT J. Pymm, Larkfield, Goxhill Road, Barrow-upon-Humber, DN19 7EE
M1 TAZ C. Curtis, 69 Lerwick Croft, Bicester, OX26 4XX
M1 TCI T. Cheeseman, 1 Queens Avenue, Birchington, CT7 9QN
M1 TCP P. Braidwood, 77 Pheasant Way, Cirencester, GL7 1BJ
M1 TCQ J. Cheese, 43 Moorside Road, Brookhouse, Lancaster, LA2 9PJ
M1 TCR T. Rozier, 124 Deansfield Road, Wolverhampton, WV1 2LD
M1 TDD Lord T. Denham, 20 Kirby Road, Basildon, SS14 1RX
M1 TES J. Crawford, 23 Meadow Road, Bungay, NR35 1LE
M1 TET C. Causby, 5 Blenheim Drive Higher Folds, Leigh, WN7 2YR
M1 TMF A. Fullwood, 16 Hollands Place, Walsall, WS3 3AU
M1 TOD R. Todd, 26 Dene Road, Guildford, GU1 4DD

M1 TOM T. Boardman, 9 Elm Grove, Farnworth, Bolton, BL4 0AY
M1 TRC C. Marren, 7 Mill Lane, East Ardsley, Wakefield, WF32BL
M1 TSU I. James, 21 Evelyn Way, Irchester, Wellingborough, NN29 7AP
M1 TUG J. Wilson, 36A Havelock Road, Maidenhead, SL6 5BJ
M1 TVR M. Bryan, 16 Walesmoor Avenue, Kiveton Park, Sheffield, S26 5RG
M1 TXT P. Dimambro, 26 Fetcham Court, Bank Top, Newcastle upon Tyne, NE3 2UL
M1 UKC B. Welland, 171 Hillcrest Road, Newhaven, BN9 9EZ
M1 ULD G. Auld, Victoria Villa, Sheepwash, Choppington, NE62 5NG
MI1 UNA U. Murray, 80 Canterbury Park, Londonderry, BT47 6DU
MW1 USK Usk Side Radio Club, c/o D. Collins, 34 Tone Road, Bettws, Newport, NP20 7AW
MW1 VCD J. Roberts, 13 Maes-Y-Coed, Gwersyllt, Wrexham, LL11 4PF
M1 VGH A. Hutchinson, 112 York Road, Haxby, York, YO32 3EG
M1 VHF B. Burden, 18 Challenor Close, Finchampstead, Wokingham, RG40 4UJ
M1 VHT K. Morrison, 31 Simonside Crescent, Hadston, Morpeth, NE65 9YA
M1 VIP A. Evans, Apartment 28, Stocks Court, 2 Harriet Street, Manchester, M28 3JW
M1 VLS B. Wilson, 131 Denmark Road, Beccles, NR34 9DW
M1 VPL A. Westland, 8 Faris Barn Drive, Woodham, Addlestone, KT15 3DZ
M1 VPN D. Coombes, The Old School, Combe Raleigh, Honiton, EX14 4UL
M1 VRC R. Broadley, The Smithy, Elstronwick, Hull, HU12 9BP
M1 VSR C. Worthington, 25 Bradshaw Road Marple, Stockport, SK6 6PH
MM1 VTB C. Budas, 20 Oak Avenue, Bearsden, Glasgow, G61 3HD
M1 WAW W. Waller, 4 New Road, Sheerness, ME12 1BW
M1 WAZ S. Eggleton, 12 Cedar Grove, Trowbridge, BA14 0HS
M1 WDK S. Papworth1, Spring Cottage, Gringley Road, Walkeringham, Doncaster, DN10 4HT
M1 WDX F. Buck, 89 Marsh St., Barrow in Furness, LA14 2AD
M1 WEH M. Harrold, 29 Barnford Crescent, Warley, Oldbury, B68 8PP
MW1 WEJ W. Jones, 53 Bro Enddwyn, Dyffryn Ardudwy, LL44 2BG
M1 WHO S. Calver, 82 Kristiansand Way, Letchworth Garden City, SG6 1UE
M1 WIN C. Higgins, 52 Pittsfield, Cricklade, Swindon, SN6 6AW
MM1 WKD A. Cartledge, Ard Shonas, Rogart, IV28 3XE
M1 WRX J. Mallows, 57 Top Road, Kingsley, Frodsham, WA6 8DA
M1 WTL W. Leverett, 514 Arleston Lane, Stenson Fields, Derby, DE24 3AG
M1 WVS E. Brown, Rose Cottage, Grindlow, Buxton, SK17 8RJ
MI1 WWG R. Mccann, 96 Carncullagh Road, Stranocum, Ballymoney, BT53 8PS
M1 WWW A. Varley, Bank Farm, Matlock Road, Spitewinter, Chesterfield, S45 0LL
MW1 WYN W. Britten-Jones, 101 Mill View Estate, Maesteg, CF34 0DE
M1 XCG C. Gaskell, Folly Cottage, Watermillock, Penrith, CA11 0LS
MM1 XJS K. Brown, 21 Strain Crescent, Airdrie, ML6 9ND
M1 XPS W. Chu, Flat 153, 16 Sutton Plaza, Sutton, SM1 4FX
M1 XRC T. Crellin, 2 Senlac Green, Uckfield, TN22 1NN
M1 XTN S. Morris, Undercliffe Villas, 4 Carnarvon Road, Reading, RG1 5SD
M1 XXT C. Willetts, The Retreat, Wood Lane, Colchester, CO3 9TR
M1 XZG R. Mckenzie, 48 Fuller Close, Swindon, SN2 7TN
MM1 YAM C. Allanson, Five Acres, Lairg, IV27 4DG
M1 YOW D. Bland, 16 Tennyson Avenue, Grays, RM17 5RG
M1 ZAR J. Wilkinson, 26 Hazelwood Grove, Sanderstead, South Croydon, CR2 9DU
M1 ZEM J. Ogden, 14 Bishops Close, Little Downham, Ely, CB6 2TQ
M1 ZXG N. Moffat, 22 Churchill Way, Acklington, Morpeth, NE65 9DB
M1 ZXZ P. Clarke, 31 Northern Rise, Great Sutton, Ellesmere Port, CH66 4QY
M1 ZZA C. Thomas, 55 High Street, Aylburton, Lydney, GL15 6BZ
M1 ZZY T. Quinn, 43 Stirtingale Road, Bath, BA2 2NG

M3

MD3 AAI M. Rish, 3 Lime St., Port St Mary, IM9 5ED
M3 AAQ L. Handley, 11 Brook Close, Blythe Bridge, Stoke-on-Trent, ST11 9PX
M3 AAS P. Rixon, 52 New Road, Hatfield Peverel, Chelmsford, CM3 2JA
M3 AAY N. Sanderson, 54 Kelvedon Close, Chelmsford, CM1 4DG
M3 ABQ J. Sharp, Hunrosa, Crowlas, Penzance, TR20 8DS
M3 ABY J. Turnbull, 32 Haydon, Washington, NE38 8PF
M3 ACA G. Reeds, 26 Holme Leaze, Steeple Ashton, Trowbridge, BA14 6EH
M3 ACF M. Buckley, Springfield, 12 Ranmore Avenue, Croydon, CR0 5QA
M3 ACY J. Bailey, 13 Newark Road, Mexborough, S64 9EZ
M3 ADB Dr A. Bottrill, 4 St Luke'S Mews Gilesgate, Durham, DH11JA
M3 ADJ B. Waterloo, 55 Solent Road, Hill Head, Fareham, PO14 3LB
M3 ADL P. Jefford, 61 Willow Way, Flitwick, Bedford, MK45 1LN
MM3 ADM A. Mair, 8 Cockburn Crescent Whitecross, Linlithgow, EH49 6JT
M3 ADT R. Vickerstaff, 16 Sewell Wontner Close, Kesgrave, Ipswich, IP5 2GB
M3 AEA P. Ewington, 26 Dickens Road, Rugby, CV22 5RW
M3 AEE A. Buckman, 116 Ashling Park Road, Waterlooville, PO7 6EG
M3 AEJ A. Pearce, 8 Carworgie Way, St. Columb Road, St Columb, TR9 6PT
M3 AEZ S. Hall, 122 Norwich Road, New Costessey, Norwich, NR5 0EH
M3 AFF A. Barnes, 3 Sparks Villas, Black Torrington, Beaworthy, EX21 5PX
MW3 AFR K. Barry, Flat 1, 24 Vale Street, Denbigh, LL16 3BE
M3 AFS I. Dwyer, 39 Berry Road, Newquay, TR7 1AS
MW3 AFX L. Cook, 8 Swn Yr Afon Flats, Cefn Coed, Merthyr Tydfil, CF48 2SA
M3 AGA P. Bore, 29 Edgerton Road, Lowestoft, NR33 9BG
M3 AGB A. Bennett, 16 Manor Avenue, Crewe, CW2 8BD
M3 AGE D. Doroba, Flat 3, 305A London Road South, Lowestoft, NR33 0DX
M3 AGH M. Arnold, 27 Poplar Avenue, Bentley, Walsall, WS2 0NT
M3 AGI M. Taylor, 29 Ferndale Avenue, Reading, RG30 3NQ
MI3 AGR P. Mcdaid, 66 Laurel Drive, Strabane, BT82 9PN
M3 AHJ R. Selwood, 33 Chandlers, Sherborne, DT9 3RT
M3 AHL D. Furness, 9 Ouzel Drive, Bradford, BD6 3YN
M3 AHO M. Coulson, 64 Craddock Street, Spennymoor, DL16 7TA
M3 AHQ J. Maloney, 196 Finchale Road, Hebburn, NE31 2BW
M3 AHR T. Hamilton, 16 Weardale Street, Spennymoor, DL16 6ER
M3 AHS A. Stevenson, 97 Queen Street, Crewe, CW1 4AL
M3 AHU G. Taylor, 63 Millbrook Towers, Stockport, SK1 3NL
M3 AHZ A. Hardman, 47 Oatlands Road, Manchester, M22 1AH
M3 AIE C. Gibson, 11 Parkside Avenue, Queensbury, Bradford, BD13 2HQ
M3 AIG K. Rendell, 18 Bonfire Close, Chard, TA20 2EG
M3 AIL G. Langdon, 43 Daniel St., Ryde, PO33 2BH
MI3 AIN R. Martin, 23 Scaddy Road, Downpatrick, BT30 9BW
M3 AIR P. Robinson, 16 Bartlett Close, Liverpool, L31 8BZ
M3 AIS S. Martin, 25 Wellswood Road, Ellesmere Port, CH66 1JX
M3 AIZ J. Smith, 44 Knapp Way, Malvern, WR14 1SG
M3 AJA J. Crowhurst, 5 Hampshire Road, Canterbury, CT1 1SJ
M3 AJC A. Charbit, 65 Bourne Street, London, SW1W 8JW
M3 AJD I. Humberstone, 20 Kingswood Road, Colchester, CO4 5JX
MI3 AJK D. Poots, 18 Upper Quilly Road, Dromore, BT25 1NP
M3 AJN S. Panczel, 11 Chauncy Road, Manchester, M40 3GG
M3 AJU R. Smith, 41 Middle Deal Road, Deal, CT14 9RG
M3 AJV A. Brown, 28 Parkland Drive Wingerworth, Chesterfield, S42 6UU
M3 AJW A. Watmough, 37 Heath Park Road, Buxton, SK17 6NY
M3 AKE R. Hoey, 225 King Avenue, Bootle, L20 0BY
M3 AKH A. Hanson, Pilgrim Cottage, South Road, Truro, TR3 7AD
M3 AKQ R. Vickerstaff, 16 Sewell Wontner Close, Kesgrave, Ipswich, IP5 2GB
M3 AKR A. Beers, 5 Wayside Estate, Christchurch, Wisbech, PE14 9NY
M3 AKW K. Pentney, 17 Hilly Park, Norton Fitzwarren, Taunton, TA2 6RH
M3 ALB A. Borda, 4 Bow Arrow Lane, Dartford, DA1 1YY
MM3 ALG J. Freeland, 12 Mccathie Drive, Newtongrange, Dalkeith, EH22 4BW
M3 ALX B. Harrison, 15 Helmington Terrace, Hunwick, Crook, DL15 0LQ
M3 ALZ S. Matley, 67 Alexandra Road, Chandlers Ford, SO53 2BP
M3 AMA A. Ayling, 58 Lower Derby Road, Portsmouth, PO2 8EX
M3 AMB S. Hegarty, 6 Wymbush Crescent, Bristol, BS13 0BB
M3 ANE R. Odle, 24 Longfellow Road, Gillingham, ME7 5QG
M3 ANH G. Gore-Thorne, 19 Poplar Way, Ringwood, BH24 1UY
M3 ANW A. Dennis, Menlo Park, Salisbury Road, Marlborough, SN8 3RP
M3 AOC D. Fawcett, 6 Wand Hill, Boosbeck, Saltburn-by-the-Sea, TS12 3AW
M3 AOM R. Adkins, 91 Fernbank Road, Birmingham, B8 3LL
M3 AOP B. Smith, 45 Branson Avenue, Stoke-on-Trent, ST3 5LA
M3 AOQ I. Sephton, 131 Smeaton Road, Upton, Pontefract, WF9 1LG
M3 APA A. Sims, 127 Cooks Spinney, Harlow, CM20 3BW
M3 APE D. Baines, 157 Hall Green Road, West Bromwich, B71 2DY
M3 APO J. Watson, 38 Ryelands Road, Lancaster, LA1 2QW
M3 APQ D. Hawken, The Crest, Cuttinglye Road, Crawley Down, RH10 4LR
M3 AQF J. Paul, 67 Fleet Avenue, Upminster, RM14 1PZ
M3 AQG P. Harrison, 55 Hudson Close, Worcester, WR2 4DP
M3 AQJ L. Jesson, 17 Omaha Drive, Hinckley, LE10 0WU
M3 AQK P. O'Shea, 37 Barclay Court, Ilkeston, DE7 9HJ
MM3 AQM W. Fullerton, 28 Kerr Avenue, Saltcoats, KA21 5PS
M3 AQN G. Macauley, 1 Moricambe Crescent, Anthorn, Wigton, CA7 5AS
M3 AQP J. Burns, 34 Fleswick Avenue, Whitehaven, CA28 9PB
MM3 AQW C. Maclean, 16 Glamis Avenue, Elderslie, Johnstone, PA5 9NR
M3 ARB A. Bronze, 215 Ivyhouse Road, Dagenham, RM9 5RS
MW3 ARM D. Rees, Y Coed, Tan Lan Hill, Holywell, CH8 9JB
M3 ARS A. Simpson, 36 Little Sammons, Chilthorne Domer, Yeovil, BA22 8RB
M3 ARU A. Smith, 130 Watkin Street, Warrington, WA2 7DN
M3 ASC D. Goldsbrough, 45 Tithe Barn Road, Stockton-on-Tees, TS19 8SZ
MW3 ASG J. James, 77 Cypress Crescent St. Mellons, Cardiff, CF3 2WL
MI3 ASH A. Nicholl, 58 Dunnalong Road, Bready, Strabane, BT82 0DW
M3 ASN D. Hartless, 2 Brendon, Wilnecote, Tamworth, B77 4JW
M3 ASZ B. Gilligan, 4 Orion Close, Ward End, Birmingham, B8 2AU
M3 ATB T. Brierley, 6 Bridle Avenue, Wallasey, CH44 7BJ
M3 ATC I. Kilkenny, 23 Hazelhurst Road, Stalybridge, SK15 1HD
MM3 ATI B. Rodger, 95A Main Street, Coaltown, Glenrothes, KY7 6HX
MI3 ATT D. Cooke, 7 Killyclooney Road, Dunamanagh, Strabane, BT82 0LZ
M3 AUB D. Albison, 6 Rossendale Way, Shaw, Oldham, OL2 7TX
M3 AUC D. Edney, 49 Burns Road, Loughborough, LE11 4ND
M3 AUF R. Cassell, 10 Palmer Close, Branston, Burton-on-Trent, DE14 3DY
M3 AUK M. Blake, 125 Ludlow Road, Portsmouth, PO6 4AF
M3 AUL K. Rickard, 7 Thorngate Close, Penwortham, Preston, PR1 0XN
M3 AUN L. Rowley, 10 Derby Place, Newcastle, ST5 3DX
M3 AUP S. Burns, 22 Pendle Close, Peterlee, SR8 2JS
MM3 AUX J. Milne, 24 Lorne Street, Edinburgh, EH6 8QP
M3 AVF J. Hancox, 3 Hillfoot Road, Liverpool, L25 7UJ
MI3 AVJ D. Johnson, 6 Sand Pits, Fenaghy Road, Ballymena, BT42 1JL
MW3 AVU R. Taylor, 48 Clark Avenue, Pontnewydd, Cwmbran, NP44 1RZ
M3 AVZ R. Henshall, 14 Greenway, Congleton, CW12 4PS
MM3 AWC Dr J. Harrington, 1 Kynoch Terrace, Keith, AB55 5FX
M3 AWD S. Mcleman, 55 Anxey Way, Haddenham, Aylesbury, HP17 8DJ
MW3 AWI D. Davies, Penrallt, Abercaseg Road, Gerlan, Bangor, LL57 3SP
M3 AWN B. Procter, 28 Holme Grove, Burley In Wharfedale, Ilkley, LS29 7QB
M3 AWQ M. Stables, 58 Cedar Street, Derby, DE22 1GE
M3 AWS A. Smith, Woodlands, Old School Lane, Biggleswade, SG18 9JL
M3 AXB S. Balkham, 49 St. Georges Road, Hastings, TN34 3NH
M3 AXT S. Plant, Columbell, 43 Fairfield Road, Bromsgrove, B61 9JW
MW3 AXW D. Evans, 7 Bryn Piod, Llanfachreth, Dolgellau, LL40 2EE
M3 AXX D. Monnington, 77 Stanley Gardens, Paignton, TQ3 3NX
M3 AXZ K. Wood, Willow Brook, Stapley, Taunton, TA3 7QB
M3 AYC T. Symonds, 68 Manor Crescent, Pan, Newport, PO30 2BH
M3 AYJ A. Parkman, 27 Arctic Road, Cowes, PO31 7PE
M3 AYP A. Allen, 43 Marriott Road, Dudley, DY2 0JY
M3 AYQ D. Payne, 31 Cockering Road, Canterbury, CT1 3UP
MM3 AYS G. Mcgann, Strathview, 43 Main Street, Fintry, Glasgow, G63 0XE
M3 AYT W. Wilson, 35 Darbishire Road, Fleetwood, FY7 6QA
M3 AZE P. Harris, 18 The Broadway, Wombourne, Wolverhampton, WV5 0HY
M3 AZF D. Hastings, 43 Delmar Avenue, Leverstock Green, Hemel Hempstead, HP2 4LZ
M3 AZH P. Beresford, 58 Plumptre Way, Eastwood, Nottingham, NG16 3LR
M3 AZK A. Collier, 65 Sandpits, Leominster, HR6 8HT
M3 AZP R. Butland, 4 Park Close, Sonning Common, Reading, RG4 9RY
M3 AZR D. Jones, 31 Summerhill Drive, Liverpool, L31 3DN
M3 BAA M. Eggleton, 78 Toronto Avenue, Blackpool, FY2 0PD
M3 BAH B. Holmes, 11 Deerness Road, Bishop Auckland, DL14 6UB
M3 BAL A. Mccarthy, Lockinuar House, Treswell Road, Retford, DN22 0HU
M3 BAN J. Sergeant, 15 Tennyson Place, Walton-Le-Dale, Preston, PR5 4TT
M3 BAO L. Walker, Little Chapple, Skilgate, Taunton, TA4 2DP
M3 BAS B. Squance, 4 Glenholt Road, Plymouth, PL6 7JA
M3 BAT P. Batty, 134 Plymouth Road, Scunthorpe, DN17 1TS
M3 BAW B. Welthy, 8 Du Cane Place, Witham, CM8 2UQ
M3 BBA R. Banfield, 2 Laleham Close, Eastbourne, East Sussex, BN21 2LQ

M3	BBB	M. Goodwin, 23 Saxon Way, Ashby-de-la-Zouch, LE65 2JR
M3	BBC	J. Holme, 17 Oxlea Grove, Westhoughton, Bolton, BL5 2AF
M3	BBF	J. Mackett, 49 Tennyson Road, Cowes, PO31 7PY
M3	BBL	B. Blackham, 5 Reedham Drive, Bramley, Rotherham, S66 2SW
MW3	BBQ	G. Richards, 3 Pen Y Mynydd, Bettws, Bridgend, CF32 8SE
M3	BBS	R. Pilgrim, 36 Wessex Gardens, Twyford, Reading, RG10 0AY
M3	BBY	R. Taylor, 93 Blue Dolphin Park, Reculver Lane, Herne Bay, CT6 6SS
MM3	BCA	A. Macinnes, 377 South Boisdale, Isle of South Uist, HS8 5TE
MM3	BCC	R. Sutherland, Tigh - Na - Coille, Mill Road, Nairn, IV12 5EW
M3	BCH	B. Heirene, 9 Ryecroft Crescent, Barnet, EN5 3BP
MI3	BCM	A. Birkhead, 21 Carson Villas, Upperlands Maghera, BT46 5SH
M3	BCQ	M. Shepherd, 47 Ripley Grove, Barnsley, S75 2RX
MI3	BCR	T. Washbourne, 15 Apsley Street, Belfast, BT7 1BL
M3	BCS	B. Smith, 98 Orange Hill Road, Burnt Oak, Edgware, HA8 0TW
M3	BCW	S. Peake, 3 Marigold Walk, Bermuda Park, Nuneaton, CV10 7SW
M3	BDA	A. Yates, Kingsomborne, Broadway, Totland Bay, PO39 0BL
M3	BDC	C. Ciotti, 6 Bascott Road, Bournemouth, BH11 8RH
M3	BDH	T. Woods, Swan House Tarrington, Hereford, HR1 4EU
M3	BDQ	J. Harvey, Flat 15, Gloucester House, The Walk., Felixstowe, IP11 9DE
M3	BEE	P. Sykes, 2 Thornton Villas, Barrow Road, Barrow-upon-Humber, DN19 7QG
MI3	BEG	A. Nicholl, 58 Dunnalong Road, Bready, Strabane, BT82 0DW
M3	BEK	R. Cowlishaw, 23 Aldrich Drive, Willen, Milton Keynes, MK15 9HP
M3	BER	S. Berry, 4 Newlands Park Way, Newick, Lewes, BN8 4PG
M3	BET	A. Waters, 12 Anvil Court, Whittonstall, Consett, DH8 9JU
M3	BFB	M. Bellamy, 23 Hazelwood, Benfleet, SS7 4NW
M3	BFG	R. Hogwood, 22 Queen Elizabeth Drive Easington Lane, Houghton le Spring, DH5 0NW
M3	BFJ	E. Harvey, 62 Archibald Road, Romford, RM3 0RH
M3	BFU	L. Buttriss, 5 Church Close, Upper Sheringham, Sheringham, NR26 8UB
M3	BFX	M. Jones, 138 Brompton Farm Road, Rochester, ME2 3RE
M3	BFY	D. Bowler, 43 Stirtingale Road, Bath, BA2 2NG
M3	BGE	R. Allen, 77 Highfield Road, Stroud, GL5 1ES
M3	BGT	J. Ahmed, 59 Ramsgate, Lofthouse, Wakefield, WF3 3PX
MM3	BHD	R. Freeland, 12 Mccathie Drive, Newtongrange, Dalkeith, EH22 4BW
MM3	BHG	E. Hay, 11 Lovat Road, Glenrothes, KY7 4RU
M3	BHI	D. Wilkinson, 139 Church Road, Jackfield, Telford, TF8 7ND
M3	BHK	S. Prinnett, 29 Ford Park Road, Plymouth, PL4 6RD
M3	BHP	D. Rugen, 19 Jacksons Close, Haskayne, Ormskirk, L39 7LD
M3	BIB	A. Mcgoff, 55 Knights End Road, March, PE15 9QA
M3	BIC	T. Humphries, 10 Cropthorne Avenue, Leicester, LE5 4QL
MI3	BIE	D. Hamilton, 7 Bolea Park, Limavady, BT49 0SH
M3	BIK	D. Newell, 24 Melchester Grove, Stoke-on-Trent, ST3 7FW
M3	BIL	W. Harvey, 76 Barlaston Road, Stoke-on-Trent, ST3 3LF
M3	BIO	A. Ness, 15 Scotforth Road, Lancaster, LA1 4TS
M3	BIR	D. Cupit, Hazelnut Cottage, Weston Road, Newark, NG22 0HB
M3	BIZ	M. Turner, 2 Higher Farm Cottages, Swallowcliffe, Salisbury, SP3 5PE
M3	BJB	A. Berry, 4 Newlands Park Way, Newick, Lewes, BN8 4PG
M3	BJE	R. Thorpe, 129 Fairham Road, Stretton, Burton-on-Trent, DE13 0BT
M3	BJH	B. Hill, 26 Providence Close, Leamore, Walsall, WS3 2AL
M3	BJJ	B. Jenkinson, 7 Chestnut Avenue, Thorngumbald, Hull, HU12 9LD
M3	BJL	W. Lewis, 3 St. Martins Road, Folkestone, CT20 3LA
M3	BJV	B. Adkins, 4 Orion Close, Ward End, Birmingham, B8 2AU
M3	BJW	B. Moseley, 232 West Bromwich Road, Walsall, WS1 3HL
M3	BJZ	D. Koch, 83 Springfield Park, Maidenhead, SL6 2YU
MI3	BKA	J. Calvert, 29 Recreation Road, Larne, BT40 1EW
M3	BKB	A. Smith, Harrowstones, Harrowbeer Lane, Yelverton, PL20 6EA
M3	BKI	M. Davenport, 44 Vicarage Road, Hastings, TN34 3LY
M3	BKJ	T. Davenport, 44 Vicarage Road, Hastings, TN34 3LY
M3	BKU	P. Bridger, 2 Marline Avenue, St Leonards-on-Sea, TN38 9HP
M3	BKV	K. Blanch, 2C Brunswick Terrace, North St., Sandown, PO36 8BG
M3	BLF	D. Saunders, 42 Peascroft, Long Crendon, Aylesbury, HP18 9AU
M3	BLG	L. Grainger, 81 Windmill Rise, Tadcaster, LS24 9HR
MI3	BLN	W. Coates, 3 Thornleigh Park, Bangor, BT20 4NN
M3	BLO	M. Herdman, 23 Marshall Avenue, Huncoat, Accrington, BB5 6NB
M3	BLR	D. Griffiths, 48 Conygar View, Dunster, Minehead, TA24 6PW
M3	BLX	M. Mina, 15 Manor Drive, Manchester, M21 7QG
MW3	BMF	B. Francis, 105 Cilmaengwyn Road, Pontardawe, Swansea, SA8 4QN
M3	BMH	F. Patrovits, 30 St. Ronans Drive, Seaton Sluice, Whitley Bay, NE26 4JQ
M3	BMI	D. Richmond, 105 Lord Street, Crewe, CW2 7DP
M3	BMN	M. Smith, 8 Devon Street, Leigh, WN7 2NG
M3	BMQ	J. Nicholls, 55 Moat Avenue, Coventry, CV3 6BT
M3	BMU	R. Scott, 157 Fairview Road, Stevenage, SG1 2NE
M3	BMV	G. Brown, 3 Willow Lane, Goostrey, Crewe, CW4 8PP
M3	BMW	B. Ashcroft, 16 Edge Lane, Crosby, Liverpool, L23 9XE
M3	BNU	M. Hall, 20 Cubitt House, Black Bull Road, Folkestone, CT19 5SH
M3	BOB	F. Kennedy, 19 High Street, East Hoathly, Lewes, BN8 6DR
MW3	BOC	F. Hart, 60 Heol Bryncwils, Sarn, Bridgend, CF32 9UE
MM3	BOJ	D. Boden, 42 Kirkwynd, Maybole, KA19 7AE
M3	BON	N. Peters, Cemetery Lodge, 12 Mount Pleasant, Crewkerne, TA18 7AH
MW3	BOO	O. Williams, 39 Camden Road, Maes-Y-Coed, Brecon, LD3 7RT
MW3	BOP	K. Stimson, 10 Heol Twyn Du, Merthyr Tydfil, CF48 1LU
M3	BOR	R. Parfitt, 7 Water Lane Close, Barnstaple, EX32 9JX
M3	BOU	M. Mullis, 18 Springfield Grove, Southam, CV47 0ES
M3	BOV	D. Waters, 117 Lower Herne Road, Herne Bay, CT6 7NA
M3	BPF	L. Gaspar, 18 North Hill Close, Burton Bradstock, Bridport, DT6 4RY
M3	BPG	P. Robertson, 64 Castle St., Frome, BA11 3DY
M3	BPL	A. Browell, 67 Stadium Avenue, Blackpool, FY4 3QA
M3	BPN	M. Newell, 55 Station Road Brimington, Chesterfield, S43 1JU
MM3	BPR	S. Mclaughlin, 21 Shirrel Road, Motherwell, ML1 4RD
MI3	BPS	G. Brennan, 15 Kinnegar Rocks, Donaghadee, BT21 0EZ
M3	BPY	P. Hollis, 5 Salisbury Road, New Malden, KT3 3HZ
MW3	BPZ	N. Heyne, New House, Hope, Welshpool, SY21 8JD
MM3	BQK	M. Cleland, 85 Carfin St., Motherwell, ML1 4JL
M3	BQN	R. Coleman, 77 Millstrood Road, Whitstable, CT5 1QF
M3	BQT	J. Rabbitt, 66 Parkfield Avenue, Delapre, Northampton, NN4 8QB
MI3	BRJ	S. Molloy, 6 Glenloch Park, Coleraine, BT52 1TY
M3	BRL	P. Long, 17 Wellesley Way, Newport, PO30 2GA
M3	BRQ	T. Leaver, 132 Old London Road, Hastings, TN35 5LZ
MM3	BRR	R. Hall, 13 Cleat, Castlebay, Isle of Barra, HS9 5XX
M3	BRT	B. Ingham, 19 Recreation Avenue, Ashton-In-Makerfield, Wigan, WN4 8SU
M3	BRU	M. Brunsdon, 7 Oldberg Gardens, Brighton Hill, Basingstoke, RG22 4NP
M3	BRV	D. Prout, 2 Pine Crest Way, Bream, Lydney, GL15 6HG
M3	BRW	P. Buttery, 103 Elsie Street, Goole, DN14 6DY
MI3	BRX	H. Mairs, 9 Eureka Drive, Belfast, BT12 5NR
MM3	BSC	B. Clark, 6 The Links, Cumbernauld, Glasgow, G68 0EP
M3	BSF	G. Mate, 200 Chesterholm, Carlisle, CA2 7XY
M3	BSH	B. Hunt, 15 High Street North, West Mersea, Colchester, CO5 8JU
M3	BSI	A. Storer, 16 Eastfields, Braunston, Daventry, NN11 7JN
MW3	BSJ	B. Jones, 4 Old Tanymanod, Blaenau Ffestiniog, LL41 4BU
M3	BSM	L. Cookman, The Flat Above Cobblers Corner, Ewhurst Road, Cranleigh, GU6 7AA
MI3	BSN	J. Murphy, 19 Kilburn Park, Armagh, BT61 9HA
M3	BTG	G. Simpson, 17 Temple Crescent, Leeds, LS11 8BG
M3	BTI	J. Roddam, Birches Farm, Button Street, Preston, PR3 2LH
M3	BTJ	J. Meredith, 42 Hollins Crescent, Talke, Stoke-on-Trent, ST7 1JY
MW3	BTN	M. Luxton, 3 The Paddocks, Newgate Street, Brecon, LD3 8DJ
M3	BTZ	P. Bull, 8 Mayfield Lane, Martlesham Heath, Ipswich, IP5 3TZ
M3	BUA	J. Bull, 8 Mayfield Lane, Martlesham Heath, Ipswich, IP5 3TZ
M3	BUH	A. Brownley, 5 Skipton Road, Sheffield, S4 7DD
MI3	BUT	K. Martin, 19 North St., Ballycastle, BT54 6BW
M3	BUU	J. Olive, 2 Wyke Cottage Wotton Road, Rangeworthy, Bristol, BS37 7NA
MM3	BUZ	B. Cameron, 6/4 Parkgrove Green, Edinburgh, EH4 7RQ
M3	BVA	P. Booth, 19 Gunville Crescent, Bournemouth, BH9 3PZ
M3	BVK	E. Shirley, Ham House, Ham Lane, Shepton Mallet, BA4 5JW
M3	BVL	G. Sowden, The Grange Lodge, Rodley Lane, Calverley, Pudsey, LS28 5QH
M3	BVM	C. Timm, 14 Little Copse Chase, Chineham, Basingstoke, RG24 8GL
M3	BVP	R. Hirst, 105 Haregate Road, Leek, ST13 6PX
M3	BVQ	A. Waller, 2 Barkis Mead, Owlsmoor, Sandhurst, GU47 0GT
M3	BVX	J. Pickard, 39 Eafield Avenue, Milnrow, Rochdale, OL16 3UN
M3	BVY	A. Spinks, 10 Foxley Close, Norwich, Nr58dq
M3	BWF	P. Cook, 5 Home Farm, Highworth, Swindon, SN6 7EG
M3	BWT	B. Williams, Dunsmoir Cottage, Chapel Hill Road Wreay, Carlisle, CA4 0RP
MM3	BWV	A. Smith, 4 The Terrace, Lhanbryde, Elgin, IV30 8NY
M3	BWZ	K. Scully, 2 St. Michaels & All Angels Church, Canada Road, Deal, CT14 7BL
M3	BXC	W. Bray, 46 Alexandra Road, Lostock, BL6 4BB
M3	BXE	D. Morris, 17 Mallory Way, Daventry, NN11 0UN
M3	BXG	H. Chick, 15 Bonfire Close, Chard, TA20 2EG
M3	BXH	C. Wheeler, 17 Orchard Way, Timberscombe, Minehead, TA24 7UL
M3	BXN	B. Holland, 23 White Avenue, Langold, Worksop, S81 9PT
MI3	BXQ	B. Hannigan, 4 Silverhill Road, Strabane, BT82 0AE
M3	BXS	R. Wake, 55 Bearsdown Road, Eggbuckland, Plymouth, PL6 5TR
M3	BXW	B. Akiens, 444 Groby Road, Leicester, LE3 9QB
M3	BXX	J. Smith, 8 Albert Gardens, Halifax, HX2 0HT
M3	BXY	K. Walker, 77 Blackwood Grove, Halifax, HX1 4QG
M3	BYA	D. Passey, 5 The Croftings, Felton Close, Ludlow, SY8 1DS
M3	BYF	J. Milne, 9 Roman Road, Colchester, CO1 1UR
MI3	BYJ	D. Bell, 1 Knockbracken Drive, Coleraine, BT52 1WN
M3	BYL	B. Lewis, 20 Annes Walk, Caterham, CR3 5EL
MI3	BYQ	J. Martin, 19 North St., Ballycastle, BT54 6BW
M3	BYS	T. Scott, 157 Fairview Road, Stevenage, SG1 2NE
M3	BYX	N. Clayton, 280A Loughborough Road, Leicester, LE4 5LH
MD3	BZA	A. Radcliffe, Cronk-Vue, Ballayockey, Andreas, IM7 3HP
M3	BZC	S. Galloway, 1 Mount Pleasant, Leeds, LS10 3TB
M3	BZQ	R. Bostock, 5 Hethersett Way, New Rossington, Doncaster, DN11 0RZ
M3	CAA	T. Groves, 43 Grasmere Road Kennington, Ashford, TN249BQ
MI3	CAB	P. Gibson, 118 Coleraine Rd, Portstewart, bt55 7hs
M3	CAD	D. Coubrough, 16 Celandine Close, Billericay, CM12 0SU
M3	CAE	J. Chantler, 1 Glencoe Road, Margate, CT9 2SL
M3	CAP	P. Barusevicus, 6 Middlebrook Crescent, Bradford, BD8 0EN
M3	CAQ	C. Hosegood, 4 The Orchard, Sixpenny Handley, Salisbury, SP5 5QL
M3	CAW	S. Court, 16 Worcester Road, Woodthorpe, Nottingham, NG5 4HY
M3	CAZ	C. Mitchell-Watson, 144 Shakespeare Crescent, Dronfield, S18 1ND
M3	CBC	P. Eden, 22 Greenside, Stoke Prior, Bromsgrove, B60 4EB
M3	CBH	B. Kenny, 37 Coningswath Road, Carlton, Nottingham, NG4 3SF
MI3	CBJ	N. Mckittrick, The Coach House, 74 Lyle Road, Bangor, BT20 5LT
MI3	CBL	S. Wylie, 38 Elmfield Park, Donaghadee, BT21 0AX
M3	CBN	K. Hughes, High Lane Cottage, Congleton Road, Macclesfield, SK11 9RR
MM3	CBO	J. Mccash, 11 Tintagel Gardens, Chryston, Glasgow, G69 0PH
MW3	CBS	K. Hulme, 13 Lime Street, Gorseinon, Swansea, SA4 4AD
M3	CBV	B. Robertshaw, 14 Lawrence Avenue, Mansfield Woodhouse, Mansfield, NG19 8DJ
M3	CBW	A. Edwards, 6 St. Pauls Road, Nuneaton, CV10 8HL
MW3	CBX	K. Roberts, 8 Bronant, Lixwm, Holywell, CH8 8NG
M3	CBY	D. Smith, 58 Marriott Road, Leicester, LE2 6NT
MI3	CCA	G. Wright, 38 Fern Grove, Bangor, BT19 1FG
M3	CCB	A. Gunn, 6 St. Pauls Road, Nuneaton, CV10 8HL
MW3	CCE	C. Evans, 25 Beech Drive, Hengoed, CF82 7JP
M3	CCF	T. Toon, 9 Boundstone Lane, Sompting, Lancing, BN15 9QL
M3	CCJ	S. Grainger, 62 Meadowleaze, Longlevens, Gloucester, GL2 0PS
MI3	CCN	R. Reilly, 220 Ardmore Road, Londonderry, BT47 3TE
M3	CCP	P. Smith, 56 Darnhall Crescent, Nottingham, NG8 4PZ
M3	CCS	A. Sutton, 11 Arnhill Road, Gretton, Corby, NN17 3DN
MI3	CCT	T. Vaughan, 20 Killen Park, Killen, Castlederg, BT81 7TJ
M3	CCY	J. White, 22 Millfields Station Road, Burnham-on-Crouch, CM0 8HS
MI3	CDA	C. Adjey, 1 Foyle Park, Portstewart, BT55 7DL
M3	CDE	S. Martins, 46 Ruskin Road, Mansfield, NG19 7LX
M3	CDI	C. Pearson, Greystones Farm Cottage, Richmond, DL11 7AJ
MW3	CDJ	C. Josey, 726 Llangyfelach Road, Treboeth, Swansea, SA5 9EL
MW3	CDL	J. Roberts, 1 Seymour Drive, Rhuddlan, Rhyl, LL18 5PP
MJ3	CDP	C. Paland, 19 Maison St. Louis, St Saviour, Jersey, JE2 7LX
M3	CDV	A. Willoughby, 25 Maple Close, Louth, LN11 0DW
M3	CDY	J. Bennett, 39 West View, Parbold, Wigan, WN8 7NT
M3	CEB	B. Cooke, 2 Harvey Place, Andover, SP10 2BU
MI3	CEM	C. Murray, 80 Canterbury Park, Londonderry, BT47 6DU
M3	CEN	C. Nowell, 39 Brockfield Park Drive, York, YO31 9EL
M3	CER	M. Clews, 16 Chestnut Street, Worcester, WR1 1PA

M3 CET M. Tibbits, 8 Holly Road, Northampton, NN1 4QR
M3 CEZ B. Stratford, 5 The Sycamores, Peacehaven, BN10 8AB
M3 CFI C. Isherwood, 32 Franklin Close, Old Hall, Warrington, WA5 8QL
M3 CFJ S. Taylor, 12 Minehead Avenue, Burnley, BB10 2NP
M3 CFM K. Sawyers, 27 Dukeswood Road, Longtown, Carlisle, CA6 5UJ
M3 CFP C. Penfold, 149 Shuttlewood Road Bolsover, Chesterfield, S44 6NX
M3 CFU P. Flook, 17 Valentine Close, Bristol, BS14 9ND
MI3 CGA R. Doherty, 48 Drumard Park, Londonderry, BT48 0RL
M3 CGC J. Stanway, Flat 17, Charlton Court London Road, Gloucester, GL1 3QH
M3 CGH C. Horne, 56 Pilkington Road Braunstone, Leicester, LE3 1RA
M3 CGI D. Hunter, 39 Nicholas Avenue, Rudheath, Northwich, CW9 7LD
M3 CGJ J. Nenova, 18 Longtown Road, Romford, RM3 7QL
M3 CGM S. Allen, 9 Tiled House Lane, Brierley Hill, DY5 4LG
M3 CGO C. Oliver, 25 Mary Peters Drive, Greenford, UB6 0SS
M3 CGP B. Townsend, 21 Royds Drive, New Mill, Holmfirth, HD9 1LH
M3 CGS C. Shirley, Ham Houes, Ham Lane, Shepton Mallet, BA4 5JW
MI3 CGT C. Brown, 33 Broadlands Drive, Carrickfergus, BT38 7DJ
MI3 CGU A. Brown, 33 Broadlands Drive, Carrickfergus, BT38 7DJ
MI3 CGZ S. Buchanan, 162 Victoria Road, Bready, Strabane, BT82 0DZ
M3 CHE D. Letton, 21 Westfield, Bradninch, Exeter, EX5 4QU
M3 CHU Dr J. Waterhouse, 31 Donnington Place, Wantage, OX12 9YE
MW3 CHZ M. Bowen, 7 Ael Y Bryn, Beddau, Pontypridd, CF38 2AL
MI3 CID C. Morrow, 46 Tullyard Way, Belfast, BT6 9NU
M3 CIE C. Barker, 14 Hall Road, Wilmslow, SK9 5BN
M3 CIG J. Smith, 9 Trafalgar Road, Newport, PO30 1QD
M3 CIJ C. Martin, 92 Thrupp Lane, Thrupp, Stroud, GL5 2DG
M3 CIO T. Brookes, 370 Broxtowe Lane, Nottingham, NG8 5ND
M3 CIP C. Powis, 28 Kington Gardens, Birmingham, B37 5HS
M3 CIS C. Smith, 6 Birtley Rise, Bramley, Guildford, GU5 0HZ
MI3 CIV M. Semple, 58 Green Drive, Larne, BT40 2ER
MI3 CIW D. Ritchie, 58 Green Drive, Larne, BT40 2ER
MI3 CIZ C. Mccord, 23 Blackthorn Green, Larne, BT40 2JE
M3 CJA C. Alexander, 25 Diamedes Avenue, Stanwell, Staines-upon-Thames, TW19 7JE
M3 CJE R. Wilberforce, 106 Marlborough Road, Slough, SL3 7JY
M3 CJH C. Houghton, 23 Carpenters Way, Badshot Lea, Farnham, GU9 9FT
M3 CJI D. Sharp, 11 Dovedale Gardens, Leeds, LS15 8UP
MM3 CJP C. Paton, 4 Abbeyhill, Dhailling Road, Dunoon, PA23 8FG
M3 CJX C. Cross, 131 Arnold Lane, Gedling, Nottingham, NG4 4HF
M3 CKD P. Loomes, 107 Main Street, Sedgeberrow, Evesham, WR11 7UE
MI3 CKF D. Hamilton, 1 Meadow Bank, Ballysally Road, Coleraine, BT52 2QA
M3 CKH C. Hammett, 63 Treffry Road, Truro, TR1 1WL
M3 CKM S. Plumb, 94 Wellington Street, New Whittington, Chesterfield, S43 2BG
M3 CKO C. Plumb, 94 Wellington Street, New Whittington, Chesterfield, S43 2BG
MM3 CKP S. Aron, 2/3, 14 Woodend Road, Glasgow, G73 4DX
M3 CKU G. Moorhouse, 29 Dee Road, Walsall, WS3 1NW
MM3 CLA C. Henderson, 22 Bowmont Place, East Kilbride, Glasgow, G75 8YG
M3 CLO K. Williams, 36 Castle St., Tiverton, EX16 6RG
M3 CLP C. Palmer, 21 Ibbett Close, Kempston, Bedford, MK43 9BT
M3 CLT C. Turner, 1B Amberbanks Grove, Blackpool, FY1 6DW
MJ3 CMB C. Boudier, 213 Le Marais, St Clement, Jersey, JE2 6GH
MW3 CMG C. Griffiths, 43 Cysgod Y Graig, Denbigh, LL16 3TD
M3 CMI H. Dickinson, 1 Larch Close, Kirkheaton, Huddersfield, HD5 0NJ
M3 CMK A. Saville, 4 Shannon Court, Downs Barn, Milton Keynes, MK14 7PP
M3 CMM B. Perryman, 55 Alderfield, Penwortham, Preston, PR1 9HD
M3 CMP C. Pidd, 36 Hunster Close, Doncaster, DN4 6RE
M3 CMW C. Willimot, 5 Green Lane, Upton, Huntingdon, PE28 5YE
M3 CMX J. Roberts, 52 School Lane, Toft, Cambridge, CB23 2RE
M3 CNC A. Wood, 29 Hornbeam Drive, Wingerworth, Chesterfield, S42 6FY
M3 CND S. Bradley, 6 Downing Street, South Normanton, Alfreton, DE55 2HE
MW3 CNL D. Emanuel, 27 Gwysfa Road, Ynystawe, Swansea, SA6 5AE
MM3 CNS G. Kennedy, 1 Martins Buildings, High Street, Perth, PH2 7QP
M3 CNV C. Parry, 43 Castle Mount, Dewsbury, WF12 0DW
M3 COB J. Hodkinson, 3 Cypress Close, Market Drayton, TF9 3HJ
M3 COE R. Alford, 225 N Main Box 174, Gas, USA, 66742
MI3 CON A. Mclernon, 520 Carneety Terrace, Castlerock, Coleraine, BT51 4SZ
M3 CPH C. Harrap, 10 Newhaven Place, Portishead, Bristol, BS20 8EG
MD3 CPK C. Kelly, 16 Viking Road, Douglas, Isle Of Man, IM2 6PB
M3 CPN D. Dunford, 151 Doncaster Road, Rotherham, S65 2BY
M3 CPX N. Thorne, Barford Stream, Churt Road, Farnham, GU10 2QU
M3 CPY L. Goodby, 21 Kiniths Way, Hurst Green, Halesowen, B62 9HJ
MI3 CQB T. Elliott, 183 Kilraughts Road, Ballymoney, BT53 8NL
MI3 CQO R. Brennan, 15 Kinnegar Rocks, Donaghadee, BT21 0EZ
M3 CQP J. Gilbert, Sweden End, Ambleside, LA22 9EX
MI3 CQR R. Hendy, 10 Captains Road Forkhill, Newry, BT35 9RR
M3 CQT A. Loasby, 89 Jubilee Crescent, Wellingborough, NN8 2PQ
M3 CQW D. Wooding, 3 Chichele Court, North St., Rushden, NN10 6BU
MI3 CQX M. Finnegan, 18 Springfarm Heights, Newry, BT35 8XA
M3 CRD C. Matthews, 18 Tennyson Gardens, Darlington, DL1 5BJ
M3 CRL C. Finnis, 44 Disraeli Road, Christchurch, BH23 3NB
M3 CRV D. Priestner, 4 Oak Street, Northwich, CW9 5LJ
M3 CSF C. Finnis, 44 Disraeli Road, Christchurch, BH23 3NB
M3 CSI C. Massimo, 2 Beattie Close, Great Bookham, Leatherhead, KT23 3JF
M3 CSK C. Newton, 7 Moss Close, Bridgwater, TA6 4NA
M3 CSM M. Lawson, 233 Southwell Road West, Mansfield, NG18 4HF
M3 CSN J. Prichard, 1 Polton Dale, Swindon, SN3 5BN
M3 CSR D. Hunter, 9 Sleigh Road, Sturry, Canterbury, CT2 0HR
MI3 CSS R. Ennis, 91 Main Street Carrowdore, Newtownards, BT22 2HW
MI3 CST D. Browne, 26 Brooklands Gardens, Dundonald, Belfast, BT16 2PQ
M3 CSV P. Southern, 32 Mayville Avenue, Scarborough, YO12 7NP
MM3 CSX A. Thomson, 5 Gib Grove, Dunfermline, KY11 8DH
M3 CSZ D. Croft, 33 Roughaw Road, Skipton, BD23 2PY
M3 CTB M. Burnett, 16 Church Lane, Reepham, Lincoln, LN3 4DQ
M3 CTN M. Mcewen, 31 Holmes Carr Road, New Rossington, Doncaster, DN11 0QF
M3 CTO A. Kvilums, 78 Wagon Lane, Solihull, B92 7PN
M3 CTT M. Holroyd, 9 Coniston Green, Aylesbury, HP20 2AJ
M3 CUH S. Twynam, 129 Poplar Drive, Herne Bay, CT6 7QA
M3 CUU D. Taylor, 10 Church Rd, Northwich, CW9 5NT
MM3 CVB C. Budas, 20 Oak Avenue, Bearsden, Glasgow, G61 3HD
M3 CVD C. Dewberry, 2 Cleatham Villas, Cleatham, Gainsborough, DN21 3HY
M3 CVH C. Holley, 28 White Horse, Uffington, Faringdon, SN7 7SE
M3 CVL C. Egan, 20 Woodhatch Road, Brookvale, Runcorn, WA7 6BJ
M3 CVM D. Healey, 28 Witley Drive, Sale, M33 5NQ
M3 CVO A. Beal, 29 Bennetts Road North, Keresley End, Coventry, CV7 8JX
M3 CVW R. Twose, 10 Galingale Close, Bicester, OX26 3FD
M3 CWA T. Watkins, 6 Linnet Close, Waterlooville, PO8 9UY
M3 CWC J. Harrison, 12 Hillside Crescent, Skipton, BD23 2LE
M3 CWH C. Harlow, 12 Penhurst Court, Grove Road, Worthing, BN14 9DG
MM3 CWO J. Mills, 65 Strathkinnes Road, Kirkcaldy, KY2 5PX
M3 CWV C. Harding, 9 Westbourne Road, Middlesbrough, TS5 5BN
M3 CWZ A. Robinson, 30 Cope Street, Walsall, WS3 2AT
MI3 CXD W. Duffy, 4 Deramore Drive, Strathfoyle, Londonderry, BT47 6XL
MI3 CXM P. Boyd, 784 Shore Road, Newtownabbey, BT36 7DG
M3 CXX A. Boyle, 15 Slatter, Satchell Mead, London, NW9 5UQ
M3 CYJ G. Jenkinson, 4 Brundish House, Braithwell Road, Rotherham, S66 8JT
MW3 CYM D. Taylor, 7 Trellewelyn Close, Rhyl, LL18 4NF
MW3 CYQ D. Hughes-Burton, 6 Troed Y Garn, Llangybi, Pwllheli, LL53 6DQ
M3 CYS K. Limbert, 7 Acacia Avenue, Liverpool, L36 5TL
MW3 CYU S. Hughes-Burton, 6 Troed Y Garn, Llangybi, Pwllheli, LL53 6DQ
M3 CYW N. Duke, 5 Hannay Close, Barrow-in-Furness, LA14 1SZ
M3 CZB M. Hillary, 55 Lumsden Road, Portsmouth, PO4 9LN
M3 CZE M. Gray, 14 Florence Crescent, Gedling, Nottingham, NG4 2QJ
M3 CZJ G. Whitear, 19 Camelia Close, Littlehampton, BN17 6UT
M3 CZL R. Morley, 6 Ford Drive, Yarnfield, Stone, ST15 0RP
M3 CZM J. Stocks, 10 Hollycroft Road, Emneth, Wisbech, PE14 8AY
M3 CZW C. Walton, Old Drive, Sulby, Northampton, NN6 6EZ
M3 CZX P. Godfrey, Bramley End, Old Lane, Chester, CH3 6QX
M3 CZY D. Mageehan, 37 Gosbecks Road, Colchester, CO2 9JR
M3 DAB D. Bambrook, 18 Vervain Close, Bicester, OX26 3SR
M3 DAE D. Edge, 20 Parkers Court, Hallwood Park, Runcorn, WA7 2FP
M3 DAF D. Fearn, 15 Broome Acre, Broadmeadows, Alfreton, DE55 3AW
M3 DAM A. Petts, 19 Sandwell Avenue, Darlaston, Wednesbury, WS10 7RH
M3 DAV R. Halsall, 50 Leominster Drive, Manchester, M22 5DH
M3 DBA K. Miller, 8 Cooks Close, Ashburton, Newton Abbot, TQ13 7AN
MI3 DBB D. Brown, 17 Parkmore Drive, Strathfoyle, Londonderry, BT47 6XA
MW3 DBF J. Mcconnell, Ty Cerrig, Tremeirchion, St Asaph, LL17 0UP
M3 DBG B. Miller, The Piglet, Saddleback Barn, Staverton, TQ9 6AN
M3 DBS D. Baines, 21 Vera Road, Norwich, NR6 5HU
M3 DBU D. Burrows, Pen Parc, 9 Clough Hall Road, Stoke-on-Trent, ST7 1AR
M3 DBX D. Billinge, 29 Stanley Avenue, Wallasey, CH45 8JN
M3 DBY D. Pratt, 4 Bussex Square, Westonzoyland, Bridgwater, TA7 0HD
M3 DBZ S. Issatt, 7 Birch Road, Doncaster, DN4 6PD
M3 DCI T. Billington, 31 June Avenue, Blackpool, FY4 4LQ
M3 DCJ J. Whiffin, 335 High Street, Eastleigh, SO50 5NE
M3 DCL M. Phillips, Orchards, Brains Green, Blakeney, GL15 4AJ
MI3 DCM D. Christie, 1 Marino Park, Ballymoney, BT53 7BB
MM3 DCN A. Mclellan, 57 Hunter Street, Kirn, Dunoon, PA23 8JR
M3 DCP D. Palmer, Edison House, Bow Street, Great Ellingham, Attleborough, NR17 1JB
M3 DCS R. Cowles, Bonnie Rock, 76 Fordham Road, Ely, CB7 5AL
MI3 DDK E. Kashkoush, 41 Dunamallaght Road, Ballycastle, BT54 6PF
MM3 DDQ P. Lucas, 69A Broomhill Crescent, Alexandria, G83 9QT
MM3 DDS D. Scott, Farewell, Arnage, Auchnagatt, Ellon, AB41 8UW
M3 DDY M. Rote, 71B Headland Crescent, Exeter, EX1 3NP
M3 DDZ E. Perez-Mendez, 56 Gaunt Close, Sheffield, S14 1GD
M3 DEA D. Adams, 9 Brancaster Avenue, Charlton, Andover, SP10 4EN
M3 DEB D. Stanley, Harby, Brightlingsea Road, Colchester, CO7 8JH
MM3 DEC E. Mcneill, 3 Sunnybrae Terrace, Maddiston, Falkirk, FK2 0LP
M3 DEE D. Lewis, 10 Addington Road, Bolton, BL3 4QZ
M3 DEH M. Dollin, 102 Rossall Road, Thornton-Cleveleys, FY5 1HQ
MW3 DEI D. Jones, Maes Y Llwyn, Tan Y Foel, Borth-Y-Gest, Porthmadog, LL49 9UH
M3 DEJ D. Balls, 7 Rowan Close, Holbeach, Spalding, PE12 7BT
MW3 DEL A. Siddle-Ward, 40 Wynn Avenue North, Old Colwyn, Colwyn Bay, LL29 9RH
MW3 DEM M. Dancer, 281 Fishguard Road, Llanishen, Cardiff, CF14 5PW
M3 DFB A. Dunster, 113 Canterbury Road, Folkestone, CT19 5NR
M3 DFC D. Chambers, 94 Hawthorn Avenue, Colchester, CO4 3JR
MM3 DFG A. Graham, 72 India St., Montrose, DD10 8PW
M3 DFL D. Little, 3 Swallow Dale, Thringstone, Coalville, LE67 8LY
M3 DFM K. Davies, 58 Popes Lane Sturry, Canterbury, CT2 0LA
M3 DFP D. Wicks, Friars Piece, Parham, Woodbridge, IP13 9LY
MI3 DFR P. Clarke, 26 Derryhale Lane, Portadown, Craigavon, BT62 4HL
M3 DFS R. Mordaunt, 330 Harborough Avenue, Sheffield, S2 1UU
M3 DFU J. Mole, 11 Branfill Road, Upminster, RM14 2YX
M3 DFV G. Jelley, 28 Blanches Road, Partridge Green, Horsham, RH13 8HZ
M3 DFW D. Ferrow, 1 Temple Avenue, Blyth, NE24 5ET
MM3 DFZ A. Zimnowlocki, 44 Dunbar Place, Kirkcaldy, KY2 5SE
M3 DGD C. Densham, 27 Lloyds Crescent, Exeter, EX1 3JQ
M3 DGJ D. Gardiner, 28 Winfield, Newent, GL18 1QB
M3 DGN A. Lister, 15 Elmwood Drive, Breadsall, Derby, DE21 4GB
M3 DGR D. Riley, 7 High St., Bolsover, Chesterfield, S44 6HF
M3 DHA S. Webber, 124A Exeter Road, Kingsteignton, Newton Abbot, TQ12 3LY
MM3 DHE R. Murray, 11 Castleview, Dundonald, Kilmarnock, KA2 9HZ
MM3 DHG J. Blades, 58 Hunter Road, Crosshouse, Kilmarnock, KA2 0LD
MM3 DHN P. Traill, 38 Burnside Road, Gorebridge, EH23 4EU
MI3 DHR D. Richards, 70 Cherryhill Avenue, Dundonald, Belfast, BT16 1JD
M3 DHS D. Sherwin, 5 North Road, Buxton, SK17 7EA
M3 DHV S. Mccron, 72 Yeoman Way, Trowbridge, BA14 0QP
M3 DHW J. Mccron, 72 Yeoman Way, Trowbridge, BA14 0QP
M3 DIB K. Dibben, 82 Lenthay Road, Sherborne, DT9 6AF
M3 DIG K. Dignall, 11 Mottershead Road, Widnes, WA8 7LD
MW3 DIL D. Jenkins, 10 Ty Fry Close, Brynmenyn, Bridgend, CF32 8YB
M3 DIM H. Brace, 56A Patching Hall Lane, Chelmsford, CM1 4DA
M3 DIS R. Hensman, 24 Belchmire Lane, Gosberton, Spalding, PE11 4HG
M3 DIT C. Jacklin, 69 Prince William Drive, Butterwick, Boston, PE22 0JG
M3 DIU J. Neenan, 11 Shaftesbury Square, West Bromwich, B71 1DX
M3 DIW D. Brooks, 61 Carisbrooke High St., Newport, PO30 1NR

M3 DIY C. Ashworth, 79 Stonehouse Road, Rugeley, WS15 2LL
MM3 DIZ D. Mcarthur, 12 Laburnum Grove, Lenzie, Kirkintilloch, Glasgow, G66 4DF
M3 DJG K. Miller, 52 Stanway Road, Shirley, Solihull, B90 3JE
MI3 DJM J. Mcbride, 22 Birchwood, Omagh, BT79 7RA
MW3 DJV T. Lewis, 22 Munro Place, Barry, CF62 8BU
M3 DJW D. Wilson, 158 Higher Lane, Rainford, St Helens, WA11 8BH
MW3 DJZ K. Lewis, 22 Munro Place, Barry, CF62 8BU
MM3 DKA J. Blair, 47 Chapelhill Mount, Ardrossan, KA22 7LU
M3 DKC L. Johnson, 222 Norwich Road, Norwich, NR5 0EZ
M3 DKG D. Morris, 4 Carnarvon Road, Reading, RG1 5SD
M3 DKK A. Milne, 4 Bearsden Way, Broadbridge Heath, RH12 3AQ
MI3 DKN H. Gillespie, 30 Groarty Road, Rosemount, Londonderry, BT48 0JX
M3 DKO L. Best, 97 Pine Tree Avenue, Canterbury, CT2 7TA
M3 DKT D. Knott, 80 Melville Court, Chatham, ME4 4XJ
M3 DKW C. Hemingway, 78 Hildyard Street, Grimsby, DN32 7NJ
M3 DKZ G. Tetley, 21 Lowther Crescent, Leyland, PR26 6QA
M3 DLB D. Beach, 2 Millward Close, Telford, TF2 8AR
M3 DLC D. Cole, Amber Lights, Market Lane, Walpole St. Andrew, Wisbech, PE14 7LT
M3 DLE D. Edwards, 5 Chalk Lane, Sutton Bridge, Spalding, PE12 9YF
M3 DLJ D. Mcdougall, 15 Caldew Drive, Dalston, Carlisle, CA5 7NS
MI3 DLO M. Harte, 17 Main St., Carrickmore, Omagh, BT79 9AY
M3 DLP C. Thulborn11, 7 Damsire Close, Liverpool, L9 9EJ
M3 DLT J. Hankin, 271 Windrows, Skelmersdale, WN8 8NP
M3 DLU M. Salt, 84 Hayfield, Stevenage, SG2 7JR
M3 DLZ E. Harvey, 125 North Road, Clowne, Chesterfield, S43 4PQ
M3 DMG D. Glazebrook, 12 Kestrel Road, Haverhill, CB9 0PH
M3 DMJ D. Jodrell, 2 Charlesworth Street, Crewe, CW1 4DE
MI3 DMM D. Mcauley, Layde View, 19 Rathlin Avenue, Ballycastle, BT54 6DQ
M3 DMW R. Weir, 130 Alexander Square, Eastleigh, SO50 4BX
M3 DMY A. Hoskins, 38 Tasmania Close, Basingstoke, RG24 9PQ
MM3 DMZ A. Perks, The Lodge, Cemetery Drive, Dumbarton, G82 5HD
M3 DNA D. Roberts, 19 East Avenue, Warrington, WA2 8AD
M3 DNB N. Brown, 241 Bury Road, Tottington, Bury, BL8 3DY
M3 DNC B. Brown, 241 Bury Road, Tottington, Bury, BL8 3DY
MI3 DNN W. Crozier, 3 Carnhill Avenue, Newtownabbey, BT36 6LE
M3 DNX J. Cox, 5 Golden Avenue Close, East Preston, Littlehampton, BN16 1QS
M3 DOA K. Hodder, 12 Garden Crescent, Barnham, Bognor Regis, PO22 0AR
MI3 DOD K. Campbell, 11 Caldwell Drive, Portrush, BT56 8ST
M3 DOM D. Pritchard, 8 The Paddock, Dawlish, EX7 0EJ
M3 DOO J. Dooley, 29 The Drive, Alsagers Bank, Stoke-on-Trent, ST7 8BB
MM3 DOP J. Shields, 4 Rhindmuir Drive, Baillieston, Glasgow, G69 6ND
M3 DOR G. Membury, 21 Webbers Piece, Maiden Newton, Dorchester, DT2 0AQ
M3 DOX A. Bajjon, 35A Blackford Road, Shirley, Solihull, B90 4BU
M3 DPB M. Oram, Lancroft, West End Road, Boston, PE21 7NQ
MW3 DPF D. Fitzgerald, 60 The Links, Trevethin, Pontypool, NP4 8DQ
M3 DPG D. Ganner, 58 Kilngate, Lostock Hall, Preston, PR5 5UW
M3 DPI J. Spencer, 19 Meadowside Avenue, Bolton, BL2 2SS
M3 DPP D. Pitchfork, 57 Calvary Crescent, Bentilee, Stoke-on-Trent, ST2 0AQ
M3 DPQ R. Palmer, 34 Ditmas Avenue, Kempston, Bedford, MK42 7DP
M3 DPS D. Sager, 29 Station Road, Mickleover, Derby, DE3 9GH
MW3 DPV D. Vaughan, 40 Meadow Terrace, Phillipstown, New Tredegar, NP24 6BW
M3 DPX A. Holden, 21 East View Meadowfield, Durham, DH7 8RY
M3 DPY W. Ellis, 16 Furlong Drive, Tean, Stoke-on-Trent, ST10 4LD
MW3 DQB R. Cotterell, 49 Graham Court, Caerphilly, CF83 1RF
M3 DQJ A. Barker, 21 Raysmith Close, Southwell, NG25 0BG
M3 DQQ D. Horsley1, 1 Mead Close, Swanley, BR8 8DQ
MM3 DQV B. Mcrae, 29 Woodneuk Road, Gartcosh, Glasgow, G69 8AG
MM3 DQX K. Mcrae, 29 Woodneuk Road, Gartcosh, Glasgow, G69 8AG
M3 DRA D. Abbott, 29 Malvern Road, Peterborough, PE4 7TT
M3 DRH D. Horner, 38 Lyndhurst Drive, Bicknacre, Chelmsford, CM3 4XL
M3 DRM D. Mostyn, 3 Woodlands Avenue, Cheadle Hulme, Cheadle, SK8 5DD
M3 DSA D. Bartlett, 9 Macdonald Avenue, Hornchurch, RM11 2NF
M3 DSE D. Essery, Hilgay, First Avenue, Watford, WD25 9PS
M3 DSI A. Clarke, 3 Epps Road, Sittingbourne, ME10 1JD
MI3 DSM D. Mccord, 24 Craigstown Road, Moorfields, Ballymena, BT42 3DF
M3 DSU S. Jenner, 4 Christie Close, Chatham, ME5 7NG
M3 DSW D. Wadey, 15 Canberra Place, Tangmere, Chichester, PO20 2WB
MW3 DTC R. Charge, Llifon, Waunfawr, Caernarfon, LL55 4YY
M3 DTD G. Asher, Hillside, 23 Mill Road, Saxmundham, IP17 1DP
M3 DTH R. Hart, 11 Ivy Hall Road, Sheffield, S5 0GX
MW3 DTO D. Owens, 16 Blanche Street, Dowlais, Merthyr Tydfil, CF48 3PE
M3 DUL R. Clark, 10 Clarendon Road, Bournemouth, BH4 8AL
M3 DUO D. Flaherty, 24 Ansdell Drive, Eccleston, St Helens, WA10 5DW
M3 DUR M. Durrant, 40 Wood Street, Mow Cop, ST7 3PE
MD3 DUZ S. Kelly, 44 Westhill Avenue, Castletown, Isle of Man, IM9 1HY
M3 DVA S. Maskrey, The Hayloft, Stamford Lane, Chester, CH3 7QD
M3 DVB J. Moreton, 71 Bevendean Avenue, Saltdean, Brighton, BN2 8PF
MM3 DVD P. Gazinski, 14 Corstorphine Road, Edinburgh, EH12 6HN
M3 DVG D. Green, 89 Upper Ratton Drive, Eastbourne, BN20 9DJ
M3 DVH D. Swaby, 34 Lambourne Road, Barking, IG11 9PS
M3 DVM J. Diston, 15 Colletts Gardens, Broadway, WR12 7AX
M3 DVN D. Noel, 58 Easenhall Lane, Redditch, B98 0BJ
M3 DVP A. Whyman, 8 Staplers Close, Great Totham, Maldon, CM9 8UN
M3 DVQ A. Cree, 24 Old Lincoln Road, Caythorpe, Grantham, NG32 3EJ
M3 DVU J. Binnell, 146 Hales Crescent, Warley, Smethwick, B67 6QX
M3 DWA B. Hollis, 212 Rye Lane, Halifax, HX2 0QP
M3 DWD D. Wraight, 2 Silver Mead, Congresbury, Bristol, BS49 5EX
MI3 DWQ D. Mcglone, 10 O'Neill Terrace, Dromore, Omagh, BT78 3AW
M3 DWV D. Simmons, 39 Crosier Court, Upchurch, Sittingbourne, ME9 7AS
MW3 DWZ B. Chapman, Gwyddfan, Mountain Road, Kidwelly, SA17 4EY
M3 DXD A. Clark, Brookside, Milford, Bakewell, DE45 1DX
M3 DXI T. Townson, 4 Crawford Street, Bradford, BD4 7JJ
M3 DXL L. Sucharyna Thomas, Dame School House, 103 High Street, Milton Keynes, MK11 1AT
M3 DXN B. Worthington, 9 Greenway, Penwortham, Preston, PR1 0TD
M3 DXR R. Bunce, 4 Newlands, Gainsborough, DN21 1QZ
MD3 DXW W. Griffiths, 7 Cooyrt Shellagh, Ballasalla, Isle of Man, IM9 2EU
M3 DYF P. Beier, 20 Markham Avenue, Armthorpe, Doncaster, DN3 2AZ
M3 DYL W. Malin, 729 Wellingborough Road, Northampton, NN3 3JE
M3 DYR P. Burns, 395 Hastilar Road South, Sheffield, S13 8EH
MM3 DYT P. Mckenzie, 76 East Bankton Place, Livingston, EH54 9BZ
M3 DYU L. Horn, 9 Musson Close, Irthlingborough, Wellingborough, NN9 5XW
M3 DYY D. Ellison, 48 Keenan Drive, Bootle, L20 0AL
M3 DZC M. Blakely, 2957 Wilderness Blvd, Florida, USA, 34219
MI3 DZD W. Brown, 16 Wallace Park, Rasharkin, Ballymena, BT44 8QH
M3 DZK H. Cartwright, 142 Ferness Road, Hinckley, LE10 0SE
M3 DZN D. Newman, 11 Pine View Close, Woodfalls, Salisbury, SP5 2LR
M3 DZQ R. Sharpe, 31 Carolyn House, Littlehampton Road, Worthing, BN13 1RE
M3 DZT A. Dominy, 58C Church St, Harwich, CO12 3DS
MM3 DZW C. Graham, 4 Dodridge Cottages, Pathhead, EH37 5UJ
M3 EAE H. Carter, 23 Brookdale Avenue, Marple, Stockport, SK6 7HP
MW3 EAI A. Whitburn, 14 Westgil Pen Ffordd, Blackwood, NP12 3QS
MI3 EAQ S. Flanagan, 18 Hunters Park, Bellaghy, Magherafelt, BT45 8JE
M3 EAS A. Stacey, 21 Edward Terrace, Stanley, DH9 7JW
M3 EAT T. Bowskill, 522 New St., Hilcote, Alfreton, DE55 5HU
M3 EBA A. Barnett, 53A Walkford Road, Walkford, Christchurch, BH23 5QD
M3 EBF E. Fury, 1 Wigley Drive, Wigley, Ludlow, SY8 3DR
M3 EBG H. Crossley, 196 Middleton Road, Heywood, OL10 2LH
M3 EBK M. Hall, 22 Leam Road, Lighthorne Heath, Leamington Spa, CV33 9TE
M3 EBU E. Burrows, 9 Clough Hall Road, Kidsgrove, Stoke-on-Trent, ST7 1AR
M3 EBZ A. Cattermole, Blaxhall Hall Crossing, Little Glemham, Woodbridge, IP13 0BP
M3 ECD D. Snowdon, Holly Cottage, Romsey Road, Romsey, SO51 0HG
M3 ECF R. Maltby, 15 Haglane Copse, Pennington, Lymington, SO41 8DT
M3 ECJ P. Stringfellow, 5 Cowslip Way, Romsey, SO51 7RR
MM3 ECO W. Davidson, 44 Abercromby Crescent, Helensburgh, G84 9DX
M3 ECQ J. Ranger, 9 Mitchells Close, Romsey, SO51 8DY
M3 ECS R. Parkhouse, 3 Thornfield Close, Seaton, EX12 2SS
M3 ECU A. Caws, 61 Kinver Close, Romsey, SO51 7JW
M3 ECV D. Bould, 38 Curlew Grove, Bridlington, YO15 3NX
M3 ECW R. Parsons, 145 Middlemarch Road, Coventry, CV6 3GJ
MW3 ECZ M. Douglas, 486 Malpas Road, Newport, NP20 6NB
M3 EDC E. Cook, 13 High Street, Cawston, Norwich, NR10 4AE
M3 EDS S. Elliott, 25 Staunton Heights, 27 Dunsbury Way, Havant, PO9 5AR
M3 EDU M. Islam, 158 Somerville Road, Chadwell Heath, Romford, RM6 5AT
MM3 EDW C. Martin, 82 Biggart Road, Prestwick, KA9 2EQ
M3 EEJ J. Maguire, 15 Brown Lane, Heald Green, Cheadle, SK8 3RR
MD3 EEW E. Wood, The Hawthorns, Droghadfayle Road, Port Erin, Isle of Man, IM9 6EL
MU3 EFB K. Le Boutillier, Tiverton, Bailiffs Cross Road, St Andrew, Guernsey, GY6 8RT
M3 EFC P. France, 238 Strand Road, Bootle, L203HN
M3 EFL M. Arnold, 334 Stourbridge Road, Halesowen, B63 3QR
M3 EFQ D. Hillman, 132 Vicarage Road, Oldbury, B68 8HY
M3 EFV S. Leathes, Harrogate Ladies' College, Clarence Drive, Harrogate, HG1 2QG
M3 EFW C. Broadbent, 9 Orchard Road, Bromley, BR1 2PR
M3 EFX R. Lockyear, 114 Wishaw Close, Redditch, B98 7RF
MW3 EGB E. Bateman, 32 Park Avenue, Bodelwyddan, Rhyl, LL18 5TB
M3 EGC E. Colley, 14 Hawthorne Close, Tyldesley, Manchester, M29 8PH
M3 EGF R. Mahoney, 1 Warner Avenue, Barnsley, S75 2EQ
MI3 EGJ C. Hazlett, 13 Faughanview Park, Claudy, Londonderry, BT47 4HQ
M3 EGM T. Brain, 47 Bankwood Crescent, New Rossington, Doncaster, DN11 0PU
M3 EGU R. Knight, 35 Bayswater Road, Headington, Oxford, OX3 9PB
M3 EGV M. Johnson, 5 Blackbird Close, Thurston, Bury St Edmunds, IP31 3PF
M3 EGY J. Johnson, 5 Blackbird Close, Thurston, Bury St Edmunds, IP31 3PF
MW3 EGZ C. Cutcliffe, 12 Heol Fargoed, Bargoed, CF81 8PP
M3 EHF D. Austen, Tudorlands, Silchester Road, Bramley, Tadley, RG26 5DG
M3 EHH C. Mason, Apartment 23, Burnside House, Carleton Road, Skipton, BD23 2BE
M3 EHJ Sir S. Harris, 22 Westmoore Court, Nottingham, NG3 6EE
M3 EHK M. Ellaway, 3 Lamb Close, Thatcham, RG18 3UE
MM3 EHM R. Allan, 30 Woodside Way, Glenrothes, KY7 5DF
M3 EHP D. Moses, 121 Badger Avenue, Crewe, CW1 3JN
M3 EHY R. Cheesley, 1 Lechlade Road, Inglesham, Swindon, SN6 7RB
M3 EIA R. Southworth, 37 Pound Close, Lyneham, Chippenham, SN15 4PJ
M3 EIJ C. Bailey, 13 Newark Road, Mexborough, S64 9EZ
MM3 EJB J. Burgoyne, 5 Shankston Crescent, Cumnock, KA18 1HA
M3 EJH I. Haughton, 17 Rivermead, Byfleet, West Byfleet, KT14 7BZ
M3 EJL E. Lawrence, 4 Malvern Road, Gillingham, ME7 4BA
M3 EJM E. Mcgee, 107 Fakes Road Hemsby, Great Yarmouth, NR29 4JL
M3 EJR E. Trueman, 2 Nursery Close, Saxilby, Lincoln, LN1 2JD
MM3 EJV S. Waldron, 9 Fullarton Avenue, Dundonald, Kilmarnock, KA2 9DX
M3 EJX J. Bennett, 21 Scott Avenue, Sutton Manor, St Helens, WA9 4AN
M3 EKA E. Denman, 12 Woodland Close, Northampton, NN5 6NH
M3 EKC A. Taylor, 106 Raeburn Avenue, Surbiton, KT5 9EA
MM3 EKL R. Harrigan, 7 Almond Crescent, Paisley, PA2 0NG
M3 EKP J. Virdee, 29 Larkfield Crescent, Houghton le Spring, DH4 4PE
M3 EKR N. Harris, 45 Sleigh Road, Sturry, Canterbury, CT2 0HT
M3 EKU P. Lancaster, 16 Wiltshire Close, Bury, BL9 9EY
M3 EKY D. Osbourne, 34 Lambourne Road, Barking, IG11 9PS
M3 EKZ D. Smedley, 27 Stirling Avenue, Loughborough, LE11 4LJ
M3 ELD E. Dalton, 120 Goodway Road, Birmingham, B44 8RG
M3 ELN H. Van Schie, 135 Mellish Court, Bletchley, Milton Keynes, MK3 6PE
M3 ELP A. Ogburn, 88 Castle Rise, Runcorn, WA7 5XW
M3 ELS M. Skinner, 5 Sycamore Avenue, Upminster, RM14 2HR
MM3 ELT A. Kerr, 14 Ellisland Drive, Dumfries, DG2 9DZ
M3 ELV L. Tremble, 7 Allerton Grove, Birkenhead, CH42 5LR
M3 EMA E. Holmes, 11 Deerness Road, Bishop Auckland, DL14 6UB
M3 EMN K. Alexander, 25 Diamedes Avenue, Stanwell, Staines, TW19 7JE
M3 EMO E. O'Neal, 22 Hill Lane, Birmingham, B43 6NA
M3 EMS S. Calver, 82 Kristiansand Way, Letchworth Garden City, SG6 1UE
M3 EMU R. Searle, Hollies, 27 Cuckmere Rise, Heathfield, TN21 8PG
M3 EMX P. Hardwick, 2 Cliffe Cottages, Sandy Lane, Liss, GU33 7JE
M3 ENE R. Evans, 53 Faygate Road, Eastbourne, BN22 9RR
M3 ENF T. Evans, 53 Faygate Road, Eastbourne, BN22 9RR
M3 ENJ C. Berry, 29 Marlborough Crescent, Long Hanborough, Witney, OX29 8JP
M3 ENO E. Cross, 17 Nicholson Court, Tideswell, Buxton, SK17 8PX

MM3 ENP W. Wilson, Laurieston Farm, Hollybush, Ayr, KA6 6HB
M3 ENS R. Nelson, 10 Westmorland Avenue, Willington Quay, Wallsend, NE28 6SN
M3 ENY N. Wootton, 54 York Road, Harlescott, Shrewsbury, SY1 3RA
MI3 EOD N. Crawford, 10 White Mountain Road, Lisburn, BT28 3QY
MI3 EOH B. Mccalmont, 19 Drumsesk Place, Warrenpoint, Newry, BT34 3NL
M3 EOL B. Jenson, 10 Tintern Close, Paulsgrove, Portsmouth, PO6 4LS
M3 EOQ D. Horton, Glen View, New Road, Bude, EX23 9LE
M3 EOT G. Gidman, 8 Minerva Close, Knypersley, Stoke-on-Trent, ST8 6SZ
M3 EOX J. Allen, 2 Chichester Walk, Chichester Road, Ramsgate, CT12 6NX
MW3 EOY S. Mclaughlin, 7 Marine Terrace, Criccieth, LL52 0EF
M3 EOZ A. Hammond, 52 Esther Avenue, Wakefield, WF2 8BX
M3 EPC N. Ball, 12 Dixons Farm Mews, Clifton, Preston, PR4 0PA
MW3 EPJ S. Mcdonald, 56 Scotchwell View, Haverfordwest, SA61 2RE
MW3 EPK G. Llewellyn, Hazeldene, Abercrave, Swansea, SA9 1SP
M3 EPQ D. Caldwell, 44 Maxwell Road, Littlehampton, BN17 7BW
M3 EPR A. Wilson, 22 Ormesby Road, Raf Coltishall, Norwich, NR10 5JY
MW3 EQE O. Richards, 57 Maesgwyn, Aberdare, CF44 8TL
M3 EQL S. Suresh, 2 Amberley Walk, Kingsmead, Milton Keynes, MK4 4AX
M3 EQP T. Thompson, 7 West Bank, Dorking, RH4 3BZ
M3 EQQ J. Laney, 18 Dyrham Close, Thornbury, Bristol, BS35 1SX
MI3 EQS T. Mcdonnell, 52 Moira Road, Glenavy, Crumlin, BT29 4JL
M3 EQW M. Savage, 23 Queen Mary Road, Salisbury, SP2 9LD
M3 EQY S. Heard, 42 Hallowell Down South Woodham Ferrers, Chelmsford, CM3 5FS
MM3 ERD E. Davidson, 44 Abercromby Crescent, Helensburgh, G84 9DX
MM3 ERP P. Smith, 1 Hillside Cottages, Tillymorgan, Insch, AB52 6UN
M3 ERR D. Barnes, 11 Yewside, Gosport, PO13 0ZD
MW3 ESE S. Reed, 2 St. Marys Park, Jordanston, Milford Haven, SA73 1HR
MW3 ESF S. Reed, 2 St. Marys Park, Jordanston, Milford Haven, SA73 1HR
M3 ESG A. Pickersgill, 9 The Malthouse, Ashbury, SN6 8NB
MW3 ESH E. Hoy, 39 Blackbird Road, Caldicot, NP26 5RE
M3 ESK K. Crane, 15 Leighton Road, Ipswich, IP3 0LJ
M3 ESN J. Carragher, High Gorses, Henley Down, Battle, TN33 9BP
M3 ESQ M. Peters, 9 Evelyn Close, Twickenham, TW2 7BL
M3 ESS M. Stirling, 3 Rother Croft, New Tupton, Chesterfield, S42 6BE
MW3 ETB F. Llewellyn, 47 St. Teilos Road, Abergavenny, NP7 6HB
M3 ETH J. Goodyear, 30 Ashburton Road, Alresford, SO24 9HH
M3 ETI D. Mcspadden, 37 Halliday Crescent, Southsea, PO4 9JU
M3 ETQ N. Greene, 308 Cedar Road, Nuneaton, CV10 9DY
M3 EUE J. Underwood, 27 Woodville Road, London, E17 7ER
M3 EUF W. Atherton, 64 Dam Lane, Rixton, Warrington, WA3 6LB
M3 EUM N. Miller, 1 Alanbrooke Road, Colchester, CO2 8EG
M3 EUP C. Cutler, 18 Berkeley Road, Peterborough, PE3 9PA
M3 EUR F. Watt, 5 Brambling Road, Horsham, RH13 6AX
M3 EUU D. Broad, 34 Arderne Avenue, Crewe, CW2 8NS
M3 EUW A. Lomas, Scanderlands Farm, Gloves Lane, Alfreton, DE55 5JJ
M3 EUY S. Swan, 47 Warren Close, Whitehill, Bordon, GU35 9EX
M3 EVB E. Munn, 32A Brunswick Street, Wakefield, WF1 4PW
M3 EVC M. Lorimer, 1A Lingford Street, Hucknall, Nottingham, NG15 7SJ
M3 EVE D. Watson, 14 Gawber Road, Barnsley, S75 2AF
M3 EVF D. Redmayne, 10 The Square, Kington, HR5 3BA
M3 EVI A. Bowron, 11 Lealholme Grove, Fairfield, Stockton-on-Tees, TS19 7AP
M3 EVJ B. Green, 12 The Ridgeway, Coal Aston, Dronfield, S18 3BY
M3 EVM S. Burnand, 53 Sidley Road, Eastbourne, BN22 7JL
MW3 EVN D. Evans, 35 Caroline Road, Llandudno, LL30 2TY
M3 EVP A. Munton, 56 Jacklin Drive, Leicester, LE4 7SU
M3 EVR P. Smith, 77 Holymoor Road, Holymoorside, Chesterfield, S42 7EA
M3 EVV M. Harrison, 43 Erskine Road, South Shields, NE33 2TH
MD3 EVY J. Phillips, 1 Cronk Elfin, Ramsey, Isle Of Man, IM8 2EX
MM3 EWI J. Woods, 12 Westbank Terrace, Macmerry, Tranent, EH33 1QE
M3 EWN E. Nevard, Millinder House, Westerdale, Whitby, YO21 2DE
M3 EWQ E. Quinn, 20 Greenfield Road, Rotherham, S65 3NX
MW3 EWR E. Roberts, 10 Ael Y Bryn, Waunfawr, Caernarfon, LL55 4AZ
M3 EWV J. Kooner, 44 Headingley Road, Birmingham, B21 9QD
M3 EWW S. Cooper, 53 Queensway, Warton, Preston, PR4 1XU
M3 EWY N. Kellow, 17 Queensway, Warton, Preston, PR4 1XT
M3 EWZ R. Dobson, 1 Auster Crescent, Freckleton, Preston, PR4 1JL
M3 EXJ L. Taylor, 17 Lacy Street, Hemsworth, Pontefract, WF9 4NW
M3 EXK J. Wilkes, 47 Greenwood Park, Hednesford, Cannock, WS12 4DQ
MM3 EXW G. Rotherham, 43 Colliston Road, Dunfermline, KY12 0XW
M3 EXY R. Shuttleworth, 5 Eastwood Drive, Marple, Stockport, SK6 7PW
MI3 EYB P. Mckeown, 7 Knockoneill Road, Maghera, BT46 5NX
M3 EYH A. Carter, 37 Seathorne, Withernsea, HU19 2BB
M3 EYK J. Dodsworth, 12 Fowlmere Road, Birmingham, B42 2EA
MM3 EYM C. Somerville, 39 Edgehead Village, Pathhead, EH37 5RL
MM3 EYN R. Somerville, 39 Edgehead Village, Pathhead, EH37 5RL
M3 EYP J. Read, 31 Merebrook Road, Macclesfield, SK11 8RH
M3 EYR C. Greene, 308 Cedar Road, Nuneaton, CV10 9DY
M3 EYS C. Lewis, 41 Hazelholt Drive, Havant, PO9 3DL
M3 EYW S. Thornton, 2 Sceptre Grove, New Rossington, Doncaster, DN11 0RW
M3 EYX W. Thornton, 2 Sceptre Grove, New Rossington, Doncaster, DN11 0RW
M3 EYY E. Driver, 99 Queens Road, North Weald, Epping, CM16 6JQ
M3 EYZ C. Board, Pinmoor, Moretonhampstead, Newton Abbot, TQ13 8QA
M3 EZB G. Leake, 154 Wareham Road, Lytchett Matravers, Poole, BH16 6DT
MI3 EZF P. Rice, 11 Kirkwood Park, Saintfield, Ballynahinch, BT24 7DP
M3 EZH L. Copley, Elmford, Mount Pleasant South, Whitby, YO22 4RQ
M3 EZJ M. Pires, 7 Felstead Close Earley, Reading, RG6 5TP
MI3 EZK B. Flanagan, 50 Towncastle Road, Strabane, BT82 0AJ
M3 EZX W. Taylor, 5 Council Bungalows, Churchtown, Belton, Doncaster, DN9 1PD
M3 EZY M. James, 82 Hill Crescent, Sutton-in-Ashfield, NG17 4JA
M3 FAA D. Mcglone, 32 Shipley Mill Close, Kingsnorth, Ashford, TN23 3NR
M3 FAC C. Perkins, Havasu, Treragin, Callington, PL17 8BL
M3 FAE J. Modha, 95 Stanway Road, Shirley, Solihull, B90 3JF
M3 FAK M. O'Brien, 4 Teal Close, Hawkinge, Folkestone, CT18 7TG
M3 FAL F. Van Den Langenberg, Flat 26, Yew Tree Court, Shifnal, TF11 9BF
M3 FAY F. Eavis, 61 Hitchmead Road, Biggleswade, SG18 0NL
M3 FBG B. Jennings, 6 The Bungalow, St. Johns Road, Wroxall, Ventnor, PO38 3EL
M3 FBJ T. Mccann, 21 Ladyseat, Longtown, Carlisle, CA6 5XX
M3 FBN D. Whitehead, 89, Cowpes Close, Sutton-in-Ashfield, NG17 2BU
M3 FBP C. Leal, 41 The Orchard, Croston, Leyland, PR26 9HS
M3 FBR M. Morris, 12 Redver Gardens, Newport, PO30 5JJ
MI3 FBW P. Coulter, 6 Skelton Close, Carrickfergus, BT38 8GP
MI3 FBX C. Gardner, 10 Abbington Manor, Bangor, BT19 1ZQ
MI3 FCA S. Churchill, 20 Killen Park, Killen, Castlederg, BT81 7TJ
M3 FCB J. Mcdonald, 43 Barley Close, St Ives, PE27 3AJ
MM3 FCG W. Mccue, 188 Redburn, Alexandria, G83 9BU
MI3 FCK J. Morgan, 6 Gannet Way, Carrickfergus, BT38 7RT
M3 FCN P. Norman, 25 Hillswood Avenue, Leek, ST13 8EQ
M3 FCO B. Dawson, 29 Hillswood Avenue, Leek, ST13 8EQ
M3 FCR L. Moreland, 25 St. Georges Avenue, Bridlington, YO15 2ED
M3 FCS R. Horne, 1 Ireland Road, Ipswich, IP3 0EJ
M3 FDB J. Johnson, 5 Oakey Ley, Bradfield St. George, Bury St Edmunds, IP30 0AU
M3 FDM M. Goss, 80 Merryhills Drive, Enfield, EN2 7PD
MW3 FDO W. Corbett, 27 Waunfawr Gardens. Crosskeys, Newport, NP11 7AJ
M3 FDQ A. Proctor, 3 The Courtyard, Tattingstone Park, Ipswich, IP9 2NF
M3 FDV A. Goodwin, 36 Cambridge Street, Bridlington, YO16 4JZ
M3 FEA R. Morley, 191 Purbrook Way, Havant, PO9 3RS
M3 FEC F. Curry, 22 Caernarvon Close, Towcester, NN12 6UP
M3 FED F. Dunn, 71 Redfield Road, Midsomer Norton, BA3 2JH
M3 FEG J. Restall, 1 Johndory, Dosthill, Tamworth, B77 1NY
M3 FEL Dr E. Fellows, 95 Arnold Road, Eastleigh, SO50 5AS
MI3 FEO R. Robinson, 92 Groomsport Road, Bangor, BT20 5NT
M3 FES F. Shirley, Ham House, Ham Lane, Shepton Mallet, BA4 5JW
MM3 FET A. Galbraith, 22 Jeffrey Street, Kilmarnock, KA1 4EB
MI3 FEX D. Rantin, 8 Buchanans Road, Newry, BT35 6NS
M3 FEY J. Brinnen, 134 Victoria Road, Mablethorpe, LN12 2AJ
M3 FFA T. Johnson, 43 Cherry Orchard Avenue, Halesowen, B63 3RZ
M3 FFE W. Johnson, 43 Cherry Orchard Avenue, Halesowen, B63 3RZ
M3 FFI D. Ross, 27 The Meadows, Skegness, PE25 2JA
M3 FFK D. Lythall, 71 Bennett Street, Kimberworth, Rotherham, S61 2JZ
MW3 FFL B. Kendrick, 77 Heolddu Crescent, Bargoed, CF81 8US
M3 FFO P. Hoe, 12 Ashbridge Rise, Chandler'S Ford, Eastleigh, SO53 1SA
M3 FFU D. Deakin, 75 Dairyground Road, Bramhall, Stockport, SK7 2QW
M3 FFV P. Lloyd, 71 Grove Road, Stourbridge, DY9 9AE
M3 FGG S. James, 94 North Road, Withernsea, HU19 2AY
MM3 FGH N. Macaulay, 68 Lorn Road, Dunbeg, Oban, PA37 1QQ
MM3 FGI C. Gillespie, 18 Roslin Crescent, Rothesay, Isle of Bute, PA20 9HT
MI3 FGK N. Craig, 29 Oughtagh Road, Killaloo, Londonderry, BT47 3TR
MM3 FGL A. Macdonald, Manderley, Benvoullin Road, Oban, PA34 5EF
M3 FGO J. Taylor, 8 Orchard Grove, Dudley, DY3 2UU
M3 FGQ I. Prior, 81 Ladymeade, Ilminster, TA19 0EA
M3 FGR D. Rootes, 1 Shelfinch, Toothill, Swindon, SN5 8AR
M3 FGU S. Fellows, 24 Habberley Road, Rowley Regis, B65 9QN
MW3 FGV J. Rowe, Flat 27 Rhodfa Frank, Ammanford, SA18 2QE
M3 FGX J. Wells, 54 Queens Road, Everton, Liverpool, L6 2NG
M3 FHI R. Norwood, Flat 27, Cranfield Court, Galadriel Spring, South Woodham Ferrers Essex, Cm3 7Bd., Chelmsford, CM3 7BD
M3 FHK M. Tompkins, 4 Prospect View, Rawtenstall, Rossendale, BB4 8JG
MI3 FHM C. Patton, 13 Oldpark Avenue, Ballymena, BT42 1AX
M3 FHO G. Flack, 20 The Pastures, Hardwick, Cambridge, CB23 7XA
M3 FHP D. Haines, 29 Parks Road, Mitcheldean, GL17 0DQ
M3 FHQ C. Price, 10 St. James Park, Lower Milkwall, Coleford, GL16 7LG
M3 FHV B. Cahill, 56 Dene Road, Headington, Oxford, OX3 7EE
M3 FIA A. Smeed, Flatt 8 9 Peir Terrace, Lowestoft, Nr330ab
M3 FIB S. Watling, 1 Chediston Green, Chediston, Halesworth, IP19 0BB
M3 FIH G. Street, Flat 9, Weavers Cottages, Congleton, CW12 1AG
M3 FIM K. Meredith, 3 Abbots Road, Abbey Hulton, Stoke-on-Trent, ST2 8DU
M3 FIP J. Shingler, 19 Cherry Tree Avenue, Runcorn, WA7 5JJ
M3 FIW J. Watson, 32 Franklin Close, Old Hall, Warrington, WA5 8QL
M3 FIX A. Thompson, 51 Kempe Way, Weston-Super-Mare, BS24 7DZ
M3 FIY E. Watson, 32 Franklin Close, Old Hall, Warrington, WA5 8QL
M3 FIZ A. Finn, 105 Lynmouth Close, Biddulph, Stoke-on-Trent, ST8 6LS
MM3 FJA G. Cull, 2 Pitairlie Farm Cottages, Pitgaveny, Elgin, IV30 5PQ
M3 FJB J. Bell, 2 Rake Lane, Milford, Godalming, GU8 5AB
M3 FJC B. Hinchliffe, 272 South St., Rotherham, S61 2NP
M3 FJD A. Lythall, 6 Belmont Crescent, Little Houghton, Barnsley, S72 0HT
M3 FJE M. Jones, 6D Terrace Road, Walton-on-Thames, KT12 2SU
M3 FJN A. Siebert, Po Box 127, Nantwich, CW5 8AQ
M3 FJP J. Park, 18 Ladgate Grange, Middlesbrough, TS3 7SL
M3 FJQ M. Rafique, 21 Syddall Avenue, Heald Green, Cheadle, SK8 3AA
M3 FJR L. Siebert, Po Box 127, Nantwich, CW5 8AQ
MW3 FJW F. Finch, Porthgwyh, Lamb Road, Aberdare, CF44 9JU
MM3 FJX J. O'Connor, 23 Osborne Terrace, Cockenzie, Prestonpans, EH32 0BY
M3 FKK J. Waddington, 2 Heron Court, Daventry, NN11 0XT
M3 FKL M. Rose, 128 Boultham Park Road, Lincoln, LN6 7TG
M3 FKM G. Rose, 128 Boultham Park Road, Lincoln, LN6 7TG
M3 FKN M. Mellish, 302 Belvedere Road, Burton-on-Trent, DE13 0RD
MM3 FKO C. Lorimer, 70A Morningside Drive, Edinburgh, EH10 5NU
M3 FKS B. Hoare, 2 St. Peters Close, South Newington, Banbury, OX15 4JL
M3 FKV R. Johnson, 90 Regent Street Church Gresley, Swadlincote, DE11 9PJ
M3 FKW S. Papworth, 103 Station Road, Quainton, Aylesbury, HP22 4BX
MW3 FLA G. Backhouse, De10 Isaf, Bryneglwys, Corwen, LL21 9NP
M3 FLB A. Bean, 25 Riverfield Grove, Bolehall, Tamworth, B77 3NB
M3 FLC F. Hanmore, 7 Tarbert Walk, London, E1 0EE
M3 FLE D. Wallstone, 128 Maltby Road, Mansfield, NG18 3BL
MW3 FLI C. Bainbridge, The Brindles, Primrose Hill, Connahs Quay, CH5 4QA
M3 FLJ G. Jackson, 5 Woodside Close, Siddington, Macclesfield, SK11 9LQ
MW3 FLK G. Badham, 103 Stanfield Street, Cwm, Ebbw Vale, NP23 7TG
M3 FLL I. Hamilton, 36 North Parade, Hoylake, Wirral, CH47 3AJ
M3 FLP S. Dec, 101 Cranford Road, Northampton, NN2 7QY
MW3 FLU A. Yates, 4 High St., Abergele, LL22 7AR
M3 FLV J. Showell, 14A Station Approach, Hayes, Bromley, BR2 7EH
M3 FLZ M. Mccormick, Sarnia, 73 Abelia, Tamworth, B77 4EZ
MM3 FMB S. Markey, 232 Main Street, Renton, Dumbarton, G82 4QA
M3 FME D. Bennett, 29 Margraten Avenue, Canvey Island, SS8 7JD
M3 FMI N. Ashley, 3 Nightingale Road, Trowbridge, Wiltshire, BA14 9TP
M3 FMK A. Jones, 36 Sutherland Drive, Newcastle, ST5 3NZ

MD3 FMN T. Hardwick, 3 Poplar Terrace, Douglas, Isle of Man, IM2 4AR
M3 FMP D. Gilhooly, 50 Hillborough Crescent Houghton Regis, Dunstable, LU5 5NX
M3 FMQ A. Ritchie, 50 Hillborough Crescent Houghton Regis, Dunstable, LU5 5NX
M3 FMV C. Hatter, 14 Morland Avenue, Bromborough, Wirral, CH62 6BE
MM3 FMY L. Dickenson, 9 Naver Road, Thurso, KW14 7QA
M3 FNA A. Brooks, 93 Durham Road, Stockton-on-Tees, TS19 0DE
M3 FNC F. Chance, 128 Chapel St., Pensnett, Brierley Hill, DY5 4EQ
M3 FNH W. Stopforth, 52 Cypress Road, Southport, PR8 6HF
M3 FNM P. Hewitt, 166 Sheringham Avenue, London, E12 5PQ
M3 FNO J. Scott-Brown, 2 Haddon Close, Fareham, PO14 1PH
M3 FNR S. Pitchford, 419 Chell Heath Road, Stoke-on-Trent, ST6 6PB
M3 FNT S. Williams, 11 Hilda Street, Leigh, WN7 5DG
M3 FNY J. Eagle, 1B Kingsley Avenue, Daventry, NN11 4AN
M3 FOD L. Newby, 22 Acton Road, Liverpool, L32 0TT
MM3 FOE S. Espie, 70 Everard Rise, Livingston, EH54 6JD
MI3 FOJ D. Kane, 22 Rowan Road, Ballymoney, BT53 7AQ
M3 FOK J. Old, 33 Rookhill Road, Pontefract, WF8 2BY
M3 FOQ J. Woods, 3 Ingle Avenue Morley, Leeds, LS27 9NP
M3 FOR M. Baldwin, 52 Salisbury Road Me45Nn, Chatham, ME4 5NN
M3 FOS I. Woods, 3 Ingle Avenue, Morley, Leeds, LS27 9NP
M3 FOV D. Read, L'Eglise, Durley St., Southampton, SO32 2AA
M3 FPA R. Etchells, 6 Woodbank Court, Canterbury Road, Manchester, M41 7DY
MI3 FPB I. Buchanan, 162 Victoria Road, Bready, Strabane, BT82 0DZ
MI3 FPE J. Rice, 42 The Crescent, Ballymoney, BT53 6ES
MW3 FPF J. Briers, 117 Heath Mead, Cardiff, CF14 3PL
M3 FPG A. Jennings, 6 The Bungalow, St. Johns Road, Wroxall, Ventnor, PO38 3EL
M3 FPH P. Harris, Flat 33, Buckingham Court Shrubbs Drive, Bognor Regis, PO22 7SE
MM3 FPI C. Jones, Croy Lodge, Shandon, Helensburgh, G84 8NN
M3 FPM L. Noel, 58 Easenhall Lane, Redditch, B98 0BJ
MI3 FPN C. Mcintyre, 18 Glanroy Crescent, Newtownabbey, BT37 9JZ
M3 FPS J. Reid, 5 Hamlet Road, Fleetwood, FY7 7HW
M3 FPT P. Turner, 92 Lancashire Street, Leicester, LE4 7AE
M3 FPU S. Cash, 6 The Mariners, Valetta Way, Rochester, ME1 1FB
M3 FPZ D. Turner, 11 Weetwood Road, Congresbury, Bristol, BS49 5BN
M3 FQA S. Wadsworth, 47 Kilnhurst Road, Todmorden, OL14 6AX
M3 FQG J. Heagren, 14 Pepperbox Rise, Whaddon, Salisbury, SP5 3BF
MM3 FQI G. Robinson, 12 Hannahston Avenue, Drongan, Ayr, KA6 7AU
M3 FQM J. Dunning, 16 Shaggs Meadow, Lyndhurst, SO43 7BN
M3 FQN S. Dunning, 16 Shaggs Meadow, Lyndhurst, SO43 7BN
M3 FQT C. Inwood, 7 The Poplars, George Street, Mablethorpe, LN12 2BP
M3 FQX J. Wadeson, 75 Bedford Drive, Sutton Coldfield, B75 6AX
M3 FRB D. Munday, 29 Coombe Park, Wroxall, Ventnor, PO38 3PH
M3 FRD F. Mcloughlin, 128 Windrows, Church Farm, Skelmersdale, WN8 8NW
M3 FRE J. French, Eypes Mouth Country Hotel, Eypes, Bridport, DT6 6AL
M3 FRJ R. Fearnley, 31 Radburn Court, Dunstable, LU6 1HW
M3 FRQ S. Smith, 7 Rosebery Avenue, Morecambe, LA4 5RU
M3 FRS B. Page, 65 Meadow View, Charminster, Dorchester, DT2 9RE
M3 FRU G. Ison, 2 Hayes Road, Nuneaton, CV10 0NH
M3 FRX M. Breffit, 10 Garrard Place, Ixworth, Bury St Edmunds, IP31 2EP
M3 FSB S. Babic, 17 Ashwood Drive, Broadstone, BH18 8LN
M3 FSC F. Creese, 8A Brockman Road, Folkestone, CT20 1DL
M3 FSD D. Babic, 17 Ashwood Drive, Broadstone, BH18 8LN
M3 FSE D. Edge, Lymn Bank Cottage, Lymn Bank, Skegness, PE24 4PJ
M3 FSQ D. Gornall, 40 Welbrow Drive, Longridge, Preston, PR3 3TB
MI3 FSR J. Brown, 4 Stratford Gardens, Bangor, BT19 6ZH
M3 FSS A. Goodchild, Gravel Lane, Ringwood, Bh24 1ll
M3 FSU G. Lewis, 57 Oakwood Road, Sutton Coldfield, B73 5EH
M3 FSV P. Murray, 2 Thurlow Gardens, Bishop Auckland, DL14 7GH
MI3 FSW F. White, 28 Lord Warden'S Parade, Bangor, BT19 1YU
MI3 FSX T. Mulholland, 215 Finaghy Road North, Belfast, BT11 9ED
M3 FSY K. Doorbar, 23 Oaktree Road, Rugeley, WS15 1AD
M3 FTA M. Everall, 17 Golden Park Avenue, Torquay, TQ2 8LR
MW3 FTB S. Robson, 16 Dunraven Road, Sketty, Swansea, SA2 9LG
MW3 FTC D. Robson, 16 Dunraven Road, Sketty, Swansea, SA2 9LG
M3 FTE S. Manley, Hill Cottage, Exminster Hill Exminster, Exeter, EX6 8DW
M3 FTI F. Gibbs, 62 Wenvoe Avenue, Bexleyheath, DA7 5BT
M3 FTJ J. Lightly, 8 Smithville Close, St. Briavels, Lydney, GL15 6TN
M3 FTK C. Gale, 51 Heron Way, Horsham, RH13 6DW
MW3 FTP J. Mckenna, 33 Low Islwyw, Prestatyn, LL19 8HQ
M3 FTU P. Panayiotou, 7 Aireville Rise, Bradford, BD9 4ES
M3 FTV A. Dunham, 28 Kingfisher Close, Chatteris, PE16 6TP
M3 FTW C. Hughes, 80 Ayrshire Close, Buckshaw Village, Chorley, PR7 7DB
MW3 FTY R. Edwards, Hillside, 3 Treforis, Ammanford, SA18 2RA
M3 FTZ A. De Vries, 21 Brean Down Road, Plymouth, PL3 5PU
M3 FUB N. Phillips, First Floor Flat, 116 Lodge Road, Croydon, CR0 2PF
M3 FUD S. Merison, Flat 2, 5 Church Lane, Banbury, OX16 5LR
MM3 FUG W. Gillespie, 33 Lochnell Road, Dunbeg, Oban, PA37 1QJ
M3 FUH Dr C. Pomfrett, 17 Manifold Close, Sandbach, CW11 1XP
M3 FUQ S. Day, The Lodge, Attleborough Fish Farm Norwich Road, Attleborough, NR17 2LA
M3 FUR P. Henderson, 214 Marsh Street, Barrow-in-Furness, LA14 1BQ
M3 FUV P. Spowart, Ruggs Hall, Clatterway Hill, Matlock, DE4 2AH
M3 FVA S. Grimbleby, 96 Waldeck Street, Reading, RG1 2RE
MW3 FVC Rev. J. Huntington, 87 Dinerth Road, Rhos On Sea, Colwyn Bay, LL28 4YH
M3 FVE D. Millard, 114 Ainsdale Drive, Werrington, Peterborough, PE4 6RP
MW3 FVH M. Hallett, 24 Brynhyfrydst, Clydach Vale, Tonypandy, CF40 2DZ
M3 FVJ C. Brown, 44 Stanley Avenue, Inkersall, Chesterfield, S43 3SY
MI3 FVW D. Shaw, 9 The Ten Cottages, Newtownards Road, Donaghadee, BT21 0PU
M3 FVX P. Sarratt, Unit 20F, Brooke Business Park, Lowestoft, NR33 9LZ
M3 FWA P. Matthew-Brown, 57 The Limes Avenue, London, N11 1RD
M3 FWJ C. Green, 160 Ashbrook Road, London, N19 3DJ
M3 FWO K. Beckett, 95 Warrens Hall Road, Dudley, DY2 8DH
M3 FWR D. Marsh, 1 Caunts Crescent, Sutton-in-Ashfield, NG17 2FH
M3 FWS J. Steele, 70 The Crescent, Andover, SP10 3BU
M3 FWT J. Cooper, 1 Dearing Close, Lyndhurst, SO43 7JP
M3 FWU R. Wilson, 7 Cornwall Close, Kirton Lindsey, Gainsborough, DN21 4DF
MI3 FXE J. Higginson, 47 Ballycorr Road, Ballyclare, BT39 9DD
M3 FXM D. Toombs, 1 Chalgrove, Welwyn Garden City, AL7 2QJ
M3 FXQ H. Gregory, 178 Over Lane, Belper, DE56 0HL
M3 FXU F. Siviter, Flat 76, Lancaster House, Rowley Regis, B65 0QE
M3 FXX V. Hocking, 80 Barton Tors, Bideford, EX39 4HA
MW3 FYA C. Alloway, 9 Millands Park, Llanmaes, Llanwitmajor, CF61 3XR
MM3 FYF J. Fyfe, 5 Beaufort Avenue, Newlands, Glasgow, G43 2YL
M3 FYM A. Reynolds, 5 Broadlands, Broadmeadows, South Normanton, Alfreton, DE55 3NW
MM3 FYN D. Innes, 39 Mormond Place, Strichen, Fraserburgh, AB43 6SY
MW3 FYR C. Johnson, 50 Treberth Avenue, Newport, NP19 9TA
M3 FYV B. Cairns, 4 Spence Court, Great Ayton, Middlesbrough, TS9 6DW
M3 FYZ A. Shellam, 1 Trafalgar House, Nelson Drive, Cannock, WS12 2GH
M3 FZB P. Paduch, 291 Rochfords Gardens, Slough, SL2 5XH
M3 FZC J. Charter, 36 Northumberland Avenue, London, E12 5HD
M3 FZE R. Humphreys, 19 Monks Green, Fetcham, Leatherhead, KT22 9TL
MM3 FZI R. Mcdonald, 12 Queen Street, Tayport, DD6 9NE
M3 FZJ L. Larkins, 34 Guycroft, Otley, LS21 3DS
M3 FZM C. Ramsdale, 87 Mill Lane, Kirk Ella, Hull, HU10 7JN
M3 FZO B. Shepherd, 19 Washfield Lane, Treeton, Rotherham, S60 5PU
M3 FZS A. Green, 10 Howard Close, Teignmouth, TQ14 9NW
M3 FZT R. Carpenter, 218 Mansel Road West, Southampton, SO16 9LR
M3 FZV M. Reed, Channel Pool, Armathwaite, Carlisle, CA4 9QY
MD3 GAB J. Espey, 9A Hilltop View, Douglas, Isle Of Man, IM2 2LA
M3 GAE J. Law, 5 Sudbury Close, Chesterfield, S40 4RS
M3 GAF G. Allen, 39 Hallam Road, Newton Heath, Manchester, M40 2SY
M3 GAG P. Gagliardi, 7 Saxon Way, Jarrow, NE32 3QA
MI3 GAM G. Moucka, 5 Glebe Gardens, Moira, Craigavon, BT67 0TU
M3 GAP G. Porter, 65 Bartlett St., Wavertree, Liverpool, L15 0HN
M3 GAV C. Tomlinson, 9 Wells Close, Astley, Manchester, M29 7WF
M3 GBA G. Barlow, Ingleneuk, Hammersley Hayes Road, Stoke-on-Trent, ST10 2DW
M3 GBB G. Bartley, 19 South Avenue, Shadforth, Durham, DH61LB
M3 GBC M. Casey, 7 Cobham Avenue, Manchester, M40 5QW
MW3 GBD B. Dallimore, 4 Llys Dyffryn, St Asaph, LL17 0SX
MJ3 GBJ S. Boudier, 253 Le Marais, St Clement, Jersey, JE2 6GH
M3 GCD E. Elsworth-Wilson, 31 Douglas Avenue, Brixham, TQ5 9EL
M3 GCH S. Shreeves, 20 Selly Oak Road, Jordanthorpe, Sheffield, S8 8DU
M3 GCJ G. Johnson, 30 Trinidad Close, Basingstoke, RG24 9PY
M3 GCM G. Masters, 85 Petersham Road, Creekmoor, Poole, BH17 7DW
M3 GCN G. Newton, 18 Parks Road, Dunscroft, Doncaster, DN7 4AH
M3 GCP G. Papworth, 70 Edward Road, West Bridgford, Nottingham, NG2 5GB
M3 GCR G. Watt, 5 Brambling Road, Horsham, RH13 6AX
M3 GCS G. Sadler, 43 Laurel Grove, Stafford, ST17 9EF
M3 GCT P. Wylie, 40 Sheepwash Avenue, Choppington, NE62 5NN
MM3 GDC G. Cochrane, 33 Portland Road, Galston, KA4 8EA
M3 GDE G. Edgar, 61 Winchester Avenue, Lancaster, LA1 4HX
M3 GDI T. Whittam, 27 Dimples Lane, Garstang, Preston, PR3 1RD
M3 GDK P. Weaver, 1 Madeley Street, Newcastle, ST5 6LS
MW3 GDL G. Jones, 31 Parcy Mynach, Pontyberem, Llanelli, SA15 5EN
M3 GDV D. Bearne, 59 Foxhole Road, Foxhole Estate, Paignton, TQ3 3TD
M3 GDX J. Ormerod, 14 Fanny Moor Lane, Hall Bower, Huddersfield, HD4 6PJ
M3 GDY V. Dealey, 11 Ashcombe Close, Witney, OX28 6NL
MI3 GEI G. Convery, 12 Linen Grove, Belfast, BT14 8PP
M3 GEK C. Poolman, 5 Slessor Street, Waddington, Lincoln, LN5 9NE
M3 GEN B. Jones, 39 Dalton Lane, Barrow-in-Furness, LA14 4LE
MI3 GER G. Reilly, 220 Ardmore Road, Londonderry, BT47 3TE
MM3 GEW R. Brooks, 6 Amochrie Drive, Paisley, PA2 0BE
M3 GEX E. Rogers, Maes Gwersyll, Garthmyl, Montgomery, SY15 6RS
M3 GEZ D. Fowles, 2 The Red House, Gallamore Lane, Market Rasen, LN8 3UB
MI3 GFA G. Freeman, 817 Windyhall Park, Coleraine, BT52 1TU
M3 GFE T. Dunn, Rakers Rest, 31 Orleigh Avenue, Newton Abbot, TQ12 2TP
MW3 GFG R. Lightly Carver, 129 Carbonne Close, Monmouth, NP25 5EH
M3 GFH M. Dunn, 6 Hamilton Drive, Newton Abbot, TQ12 2TL
M3 GFO N. Fox, 15 Hawthorne Grove, Bentley, Doncaster, DN5 0PQ
M3 GFW P. Seaman, 18 Earlsford Road, Mellis, Eye, IP23 8DY
M3 GFZ M. Hewson, 27 Grange Crescent, Lincoln, LN6 8BT
M3 GGA K. Mccarthy, 260 Whalley Drive, Bletchley, Milton Keynes, MK3 6PJ
M3 GGE G. Townsend, 23 Lodgefield Park, Stafford, ST17 0YE
M3 GGV T. White, 25 Vicarage Close, Shillington, Hitchin, SG5 3LS
M3 GHA G. Halls, 16 Lovent Drive, Leighton Buzzard, LU7 3LR
M3 GHD D. Bell, 27 Kings Coombe Drive, Kingsteignton, Newton Abbot, TQ12 3YU
M3 GHE M. Barnes, 49 Harrowden Road, Bedford, MK42 0RS
MW3 GHF G. Creed, Great House Farm, Croesypant, Pontypool, NP4 0JD
M3 GHG P. Walton, 15 Arbour Close, Northwich, CW9 7BF
M3 GHH G. Hazlewood, 102 Throne Road, Rowley Regis, B65 9JX
M3 GHI J. Haslam, 25 Lulworth Road, Eccles, Manchester, M30 8WP
M3 GHL G. Law, 14 Sandpit Lane, Hilton, Bridgnorth, WV15 5PH
M3 GHO G. Cliffe, 5 Laurel Cottages, Ongar Hill Road, King's Lynn, PE34 4JB
M3 GHR R. Gill, 84 Leypark Road, Exeter, EX1 3NT
M3 GHS S. Stanhope, 61 Heathfield St., Manchester, M40 1LF
MI3 GHW G. Mclernan, Drumcor Hill, Enniskillen, BT74 6BQ
MI3 GHY I. Gibb, 1 Shankill Road, Garvary, Enniskillen, BT94 3DB
M3 GIE R. Harper, 19 Tennyson Avenue, King's Lynn, PE30 2QG
M3 GIF E. Roberts, 8 Skamacre Crescent, Lowestoft, NR32 2QG
M3 GIH E. Peck, 11 Blake Road, Stapleford, Nottingham, NG9 7HN
M3 GIK M. Haughey, 10 Sharp St., Hull, HU5 2AB
MM3 GIR K. Gibson, 136 Henrietta Street, Girvan, KA26 0AE
M3 GIX A. Scrutton, 35 Gainsborough Road, Warrington, WA4 6DA
M3 GIY J. Eaton, 10 Motcombe Farm Road, Heald Green, Cheadle, SK8 3RW
MI3 GJG G. Mcgill, 4 Grainan Park, Londonderry, BT48 7UA
MI3 GJI P. Bingham, 45 Gowanvale Drive, Banbridge, Bt323gd
M3 GJN P. Marriott, 38 Westfields, Tilney St. Lawrence, King's Lynn, PE34 4QS
M3 GJW G. Watson, 2 Bow St., Mansfield Woodhouse, Mansfield, NG19 9PJ
MW3 GKB D. Khan, 20 Cae Penrallt, Trearddur Bay, Holyhead, LL65 2WA
M3 GKE B. Luetchford, 89 Lime Grove, Gayton, King's Lynn, PE32 1QU
M3 GKG A. Boag, 53 Castlewood Road, London, N16 6DJ
M3 GKH G. Buxton, 18 Savernake Close Rubery, Rednal, Birmingham, B45 0DD
MW3 GKI N. Axon, 30 Rating Row, Beaumaris, LL58 8AF
M3 GKJ S. Willis, 1 Whiffins Orchard, Coopersale Common, Epping, CM16 7HT

M3 GKK M. Ahmed, 75 Drove Road, Swindon, SN1 3AE
M3 GKX J. Boot, 110 Wallace Road, Bradley, Bilston, WV14 8AU
M3 GKY A. Reed, 32 Hollis Gardens, Cheltenham, GL51 6JQ
M3 GLA G. Astbury, 12 Southall Road, Ashmore Park, Wolverhampton, WV11 2PZ
M3 GLC G. Colclough, Little Hallands, Norton, Seaford, BN25 2UN
MM3 GLH G. Bruce, 60 Kingsmills, Elgin, IV30 4BU
M3 GLM G. Parkins, 73 Orwell View Road, Shotley, Ipswich, IP9 1NW
M3 GLT G. Talbot, 26 Chevalier Grove, Crownhill, Milton Keynes, MK8 0EJ
M3 GMB S. Bradshaw, 82 Arden Way, Market Harborough, LE16 7DD
M3 GMG G. Mcgeough, 57 Stonehouse Park, Thursby, Carlisle, CA5 6NS
MI3 GMI M. Crozier, 33 Cullentragh Road, Poyntzpass, Newry, BT35 6SD
M3 GML G. Linfield, 82 Claremont Road, Swanley, BR8 7QT
MM3 GMP J. Williams, Baptist Manse, Balemartine, Isle of Tiree, PA77 6UA
M3 GMY A. Pickering, 16 Chestnut Grove, Accrington, BB5 0ND
MW3 GNB J. Carfoot, 24 Marble Church Grove, Bodelwyddan, LL18 5UP
M3 GNM S. Mcloughlin, 7 Wilmots Way, Pill, Bristol, BS20 0JT
M3 GNN A. Glover, 103A Latimer Street, Liverpool, L5 2RF
M3 GNY A. Hunt, Chesnut Cottage, Hine Town Lane, Shillingstone, Blandford Forum, DT11 0SN
MM3 GOE T. Macdonald, Main Road Farm, Balephuil, Isle of Tiree, PA77 6UE
MM3 GOI S. Adam, 231/1 Gogarloch Syke, Edinburgh, EH12 9JF
MM3 GOT E. Griffiths, Achnamara, Heanish, Isle of Tiree, PA77 6UL
M3 GOV A. Ward, 81 Northbrooks, Harlow, CM19 4DB
MM3 GOX C. Williams, Ormer, Kirkapol, Isle of Tiree, PA77 6TW
MM3 GOY E. Williams, Ormer Cottage, Kirkapol, Scarinish, PA77 6TW
M3 GOZ N. Gostling, 49 Roundhouse Road, Dudley, DY3 2AX
MM3 GPB A. Williams, Ormer Cottage, Kirkapol, Scarinish, PA77 6TW
M3 GPC G. Cockram, 28 Tabley Road, Bolton, BL3 4BR
MW3 GPG S. Griffiths, 8 Heol Cynwyd, Llangynwyd, Maesteg, CF34 9TB
MW3 GPJ G. Jackson, 40 Ellis Avenue, Old Colwyn, Colwyn Bay, LL29 9LB
MM3 GPL G. Lawrie, 4 Revoan Court, Castle Road, Grantown-on-Spey, PH26 3AE
M3 GPM P. Mansfield, 106 Field Lane, Burton-on-Trent, DE13 0NN
M3 GPN R. Orton, 38 Whitehill Avenue, Barnsley, S70 6PP
M3 GPP A. Ault, 89 Southbourne Coast Road, Bournemouth, BH6 4DX
M3 GPR G. Richards, 1 Maple Close, Seaton, EX12 2TP
M3 GPX S. Russell, 37 St Nicholas Close, Boston, PE21 0AE
M3 GQB J. Gyton, 10 Longcroft, Southdown Road, Shoreham-by-Sea, BN43 5AY
M3 GQD J. Reynolds, 3 Ardleigh, Basildon, SS16 5RA
MW3 GQE J. Doyle, 18 The Paddocks, Tonna, Neath, SA11 3FD
M3 GQI K. Cramp, 20 Combeland Road, Minehead, TA24 6BT
M3 GQL P. Ridgers, 231 The Greenway, Epsom, KT18 7JE
M3 GQM T. Sheppard, 1 Waveney Walk, Crawley, RH10 6RL
M3 GQP A. Hall, 21 Eardulph Avenue, Chester le Street, DH3 3PR
MM3 GQR S. Mciver, 9 Balvicar Road, Oban, PA34 4RP
MW3 GQS C. Williams, Pen Y Cae, Bodeiliog Road, Denbigh, LL16 5PA
MM3 GQT G. Mc Gregor, 34 Cairn Road, Cumnock, KA18 1HN
M3 GQW E. Tart, Sunnybank Farm, Wattlesborough Heath, Shrewsbury, SY5 9EG
MM3 GQY H. Dineley, Banks, Burray, Orkney, KW17 2ST
MW3 GRC G. Coombes, 25 Afan Valley Road, Cimla, Neath, SA11 3SS
M3 GRF A. Grace, 2 St. Peters Crescent, Bicester, OX26 4XA
M3 GRI T. Griffiths, 56 The Avenue, Totland Bay, PO39 0DN
M3 GRY G. Crane, 35 Betjeman Avenue, Wootton Bassett, Swindon, SN4 8JY
M3 GSI C. Nelmes, 119 Exeter Road, Dawlish, EX7 0AN
MM3 GSL G. Shaw, 1 Fir Park, Sorn, Mauchline, KA5 6HY
M3 GSM P. Taylor, 77 Ladstone Towers, Sowerby Bridge, HX6 2QP
M3 GSQ W. Howarth, 68 Valentine Road, Sheffield, S5 0NZ
M3 GSR W. Chave, 91 Newman Road, Exeter, EX4 1PQ
MI3 GSW G. Heggan, 19 Lough Road, Lisburn, Bt276ts
M3 GTA D. Langmead, 38 Milton Grove, London, N11 1AX
M3 GTB G. Bland, 20 Brereton Close, Castlefields, Runcorn, WA7 2LR
MM3 GTF F. Davidson, 27 Gordon Way, Livingston, EH54 8JG
M3 GTG R. Chisholm, 162 Ardington Road, Northampton, NN1 5LT
M3 GTH E. Jones, 43 Wesley Road, Wimborne, BH21 2QB
M3 GTK J. Walker, 34 Vian Road, Waterlooville, PO7 5TW
MW3 GTM G. Mainwaring, 3 Elias St., Neath, SA11 1PP
MI3 GTO G. Shaw, 49 Cloughey Road, Portaferry, Newtownards, BT22 1NQ
M3 GTQ G. Thompson, 28 St. Georges Road, Atherstone, CV9 3BP
M3 GTT G. Hines, 126 Linacre Lane, Bootle, L20 6ES
M3 GTV A. Burfield, 4 Eastern Crescent, Chelmsford, CM1 4JQ
M3 GUF R. Glossop, 21 Elizabeth Avenue, Tattershall Bridge, Lincoln, LN4 4JJ
M3 GUG K. Bell, 12A Mill Lane, Carlton, Goole, DN14 9NG
MW3 GUH S. Tolhurst, Gwernrynydd Fach, Nantmel, Llandrindod Wells, LD1 6EW
M3 GUJ A. Brimble-Brice, 174-176 Withycombe Village Road, Exmouth, EX8 3ba
M3 GUM K. Waterhouse, 74 Clifford Road, West Bromwich, B70 8JY
M3 GUO S. Shaw, 4 Perry Hill, Chelmsford, CM1 7RD
M3 GUQ M. Reynolds, 15 Michelham Close, Eastbourne, BN23 8JD
M3 GUU S. Whiting, 56 Station Road, Branston, Lincoln, LN4 1LH
M3 GVC P. Wayer, 4 Chatburn Avenue, Waterlooville, PO8 8UB
MM3 GVE C. Brown, 9 Newton Crescent, Rosyth, Dunfermline, KY11 2QW
MW3 GVF A. Roberts, 9 Llys Hendre, Rhuddlan, Rhyl, LL18 5YF
M3 GVJ L. Ballinger, 9 Somerville Court, Cirencester, GL7 1TG
M3 GVN J. Bacheta, 32 Essex Road, London, E12 6RE
M3 GVT G. Finney, 78 Lockley St., Stoke on Trent, ST1 6PQ
MW3 GVU J. Brennan, 1 Gerddi Mair, St. Clears, Carmarthen, SA33 4ET
M3 GWC S. Clarkson, Carisbrooke, Poolhouse Road, Wolverhampton, WV5 8AZ
MW3 GWH G. Haines, 12 Tyn Rhos Estate, Gaerwen, LL60 6HL
MI3 GWQ T. Cosgrove, 301 Russell Court, Claremont Street, Belfast, BT9 6JX
M3 GWW G. Wheelhouse, 86 Severn Street, Hull, HU8 8TQ
MW3 GWZ P. French, 4 Acacia Avenue, Newport, NP19 9AT
M3 GXG M. Sartorius, Holmwood, Priory Road, Sunningdale, Ascot, SL5 9RH
M3 GXI B. Saunders, 4 Mendip Road, Worthing, BN13 2LP
M3 GXX I. Bardell, 17 Stanton Avenue, Bradville, Milton Keynes, MK13 7AR
M3 GYA H. Denmead, 47 Holland Road, Clevedon, BS21 7YJ
M3 GYB M. Peterson, 29 Warwick Close, Saxilby, LN1 2FT
M3 GYH R. Blake, Taita, Linnards Lane, Northwich, CW9 6ED
M3 GYI L. Horton, 36 Merevale Crescent, Morden, SM4 6HL
MM3 GYU T. Bloomfield, Midyard House, Carnwath, Lanark, ML11 8LH
M3 GYY G. Clarke, 233 Muirhead Avenue, Liverpool, L13 0AY
M3 GZD D. Blackmore, 197 Cotton Lane Wa75Jb, Runcorn, WA7 5JB
M3 GZE S. Lee, 154 Grangeway, Runcorn, WA7 5JA
MM3 GZG K. Cunningham, 11 Glendoune Street, Girvan, KA26 0AA
M3 GZI A. Farmar, Hawkes Place, Horslett Hill, Holsworthy, EX22 6RS
M3 GZJ M. Strowger, 88 Castle Rise, Runcorn, WA7 5XW
M3 GZP I. Plain, 18 Arundel Road, Bath, BA1 6EF
M3 GZQ M. Parris, 19B Milfoil Drive, Eastbourne, BN23 8BR
M3 GZT A. Cain, 55 Lytham Green, Muxton, Telford, TF2 8SQ
M3 GZU R. Lang, 89 Dodthorpe, Hull, HU6 9HA
M3 GZW R. Shadbolt, 58 Westfield Road, Manea, March, PE15 0LN
M3 HAC J. Hilton, 32 Dowry St., Fitton Hill, Oldham, OL8 2LP
M3 HAD H. Rhymes, 12 Reedling Drive, Southsea, PO4 8UF
MW3 HAE C. Davies, Afallon, 3 Penygraig, Aberystwyth, SY23 2JA
MM3 HAF M. Hoskin, 15 East Den Brae, Letham, DD9 2PJ
M3 HAI G. Young, 47 Birdhill Road, Woodhouse Eaves, Loughborough, LE12 8RP
M3 HAJ I. Griffiths, 147 Greenlawns, St. Marks Road, Tipton, DY4 0SU
M3 HAK R. Siebert, Po Box 127, Nantwich, CW5 8AQ
M3 HAL S. Hall, Orchard End, Asenby, Thirsk, YO7 3QR
M3 HAM G. Roberts, 25 Chalfont Way, Liverpool, L28 3QB
MW3 HAQ C. Olding, 10 Ty Nant, Caerphilly, CF83 2RA
MW3 HAS H. Mustafa, 17 Furness Close, Ely, Cardiff, CF5 4PG
M3 HAT H. Kennedy, 19 High Street, East Hoathly, Lewes, BN8 6DR
M3 HAU W. Kent, Long Spring Cottage Gracious Lane, Sevenoaks, TN13 1TJ
M3 HAW S. Avery-Hawkins, 41 Daniels Welch, Coffee Hall, Milton Keynes, MK6 5DA
M3 HAZ H. Flower, 17 Scott Grove, Morecambe, LA4 4LN
MW3 HBC M. Lewis, 96 Roundhouse Close, Nantyglo, Ebbw Vale, NP23 4QY
MW3 HBF D. Williams, 54 Howell Street, Pontypridd, CF37 4NR
M3 HBM J. Baxter, 10 Speedwell Close, Bedworth, CV12 0NS
M3 HBP S. Bethell, 35 Fulford Road, Bristol, BS13 9RL
M3 HBS J. Bodie, 144 Alexandra Road, Mutley, Plymouth, PL4 7EQ
M3 HBT T. Richley, 30 Chicheley Road, Harrow, HA3 6QL
M3 HBX D. Glenn, 84 Cambridge Street, Normanton, WF6 1ER
M3 HCA E. Foster, 12 Dunham Grove, Leigh, WN7 3DS
M3 HCB H. Benton, Emiviz, The Ridge, Salisbury, SP5 2LQ
M3 HCE A. Macnauton, 27A Lincoln Road, Poole, BH12 2HT
M3 HCG C. Hawkins, 118 Aldebury Road, Maidenhead, SL6 7HE
M3 HCL C. Lott, 55 Juniper Road, Farnborough, Gu14 9xu
M3 HCP D. Hounslow, 3 Hengrave Green, Ivington, Leominster, HR6 0JL
MW3 HCW P. Webb, 15 Aber Road, Prestatyn, LL19 7HL
M3 HDL R. Guess, 69 Rowan Drive Kirkby-In-Ashfield, Nottingham, NG17 8FP
M3 HDV B. Hampson, 38 Parley Road, Bournemouth, BH9 3BB
M3 HEE A. Totterdell, Moscar Cross House, Hollow Meadows, Sheffield, S6 6GL
M3 HEI S. Collett, 81 Wycombe Road, Prestwood, Great Missenden, HP16 0HW
M3 HEJ E. Haycock, 55 Ashbourne Road, Rocester, Uttoxeter, ST14 5LF
M3 HEO A. Fagan, 77 Watling Street West, Towcester, NN12 6AG
M3 HER F. Nation, 1 Claydon Path, Stoke Mandeville, Aylesbury, HP21 9EF
M3 HET H. Thomas, 15 Ashgrove Way, Bridgwater, TA6 4UB
M3 HEV A. Sturgess, Hawks Barn, Long Lane, Shaftesbury, SP7 0BJ
M3 HFA E. Gainford, 10 The Spinney, Ashford, TN23 3LF
M3 HFH S. Overall, Flat 74, Douglas Buildings Marshalsea Road, London, SE1 1EL
MW3 HFO H. Foster, 11 Rosedale Gardens, Rhyl, LL18 4TY
M3 HFT E. Rogers, Maes Gwersyll, Garthmyl, Montgomery, SY15 6RS
MM3 HFU G. Johnson, Speur Mor, Gifford Road, Longformacus, Duns, TD11 3NZ
M3 HFX B. Mccann, 21 Ladyseat, Longtown, Carlisle, CA6 5XX
M3 HGA T. Winwood, 2 The Warren, Abingdon, OX14 3XB
M3 HGE A. Hitchens, 16 Harrisons Place, Northwich, CW8 1HX
M3 HGH K. Stewart, 17 Delamere Street, Bury, BL9 6NE
M3 HGL B. Peck, 60 Richmond Road, Ipswich, IP1 4DP
M3 HGM N. Peters, 57 High Street, Collingtree, Northampton, NN4 0NE
MW3 HGO R. Ward, Beech Cottage, Saron Road, Goytre, NP4 0BN
M3 HGP P. Bland, 5 Pembroke Way, Winsford, CW7 1QZ
MW3 HGR H. Roberts, Hen Ddol, Northfield Road, Barmouth, LL42 1PT
M3 HGT D. Leverton, 21 Laburnum Grove, Killamarsh, Sheffield, S21 1GR
M3 HGW M. Bancroft, Oak Lodge 3 The Oaks Scothern, Lincoln, LN2 2WB
M3 HGX D. Clark, 12 Wilson Crescent, Lostock Gralam, Northwich, CW9 7QH
M3 HGZ J. Sejwacz, Flat 9, Mayrick Court, Newton le Willows, WA12 9GB
M3 HHB M. Clarke, 40 Fingringhoe Road, Langenhoe, CO5 5AD
M3 HHC S. Crossley, 29 Rycroft Avenue, Bingley, BD16 1PU
M3 HHN A. Highfield, 29 Blewitt Street, Brierley Hill, DY5 4AW
M3 HHQ D. Sejwalz, 4 Ash Avenue, Newton-le-Willows, WA12 8HJ
M3 HHX M. Arnott, 2 Hambleton Close, Elsecar, Barnsley, S74 8DS
M3 HIE B. Edwards, 28 Poppy Drive, Horam, Heathfield, TN21 9BL
M3 HIG T. Higgins, 15 Ellen Street, Warrington, WA5 0LY
M3 HIM F. Doyle, 1 Claydon Path, Stoke Mandeville, Aylesbury, HP21 9EF
M3 HIN A. Watkinson, 34 Marble House, Felspar Close, London, SE18 1LN
M3 HIO S. Jarvis, 18 Savernake Close Rubery, Rednal, Birmingham, B45 0DD
M3 HIP J. Dixon, 6 Howland Close, Eastbourne, BN23 5AJ
M3 HIT A. Bryan, 16 Walesmoor Avenue, Kiveton Park, Sheffield, S26 5RG
MW3 HIX P. Owen, 24 Sirhowy Court, Green Meadow, Tredegar, NP22 4PL
M3 HJB H. Beier, 20 Markham Avenue, Armthorpe, Doncaster, DN3 2AZ
MM3 HJC C. Hazle, 6 Dalneigh Road, Inverness, IV3 5AH
M3 HJD L. Dixon, Stirk House Hotel, Gisburn, BB7 4lj
M3 HJE H. Evans, Littlefield House, Bolney Road, Haywards Heath, RH17 5AW
M3 HJF S. Gilchrist, Kening, Ashleigh Crescent, Barnstaple, EX32 8LA
M3 HJG A. Howe, 18 Co-Operation Street, Crawshawbooth, Rossendale, BB4 8AG
M3 HJJ S. Norris, 15 East View, Choppington, NE62 5UF
M3 HJU P. Webster, Magoos Gweek, Helston, TR12 6TX
M3 HJV C. Lishman, 6 Clarence Road, Accrington, BB5 0NA
M3 HJW R. Smith, 21 Lambert Close, Waterlooville, PO7 5XA
MM3 HKE L. Higgins, 11 Strathyre Place, Broughty Ferry, Dundee, DD5 3WN
MW3 HKH S. Crighton, 12 Cwm Road, Waunlwyd, Ebbw Vale, NP23 6TR
M3 HKM K. Monaghan, The Bulstone Hotel, Branscombe, Seaton, EX12 3BL
M3 HKV L. Flawn, Autumns, Jacks Lane, Bishop's Stortford, CM22 6NT
M3 HLA J. Meeks, 64 Belford St., Burnley, BB12 0DF
M3 HLD R. Metcalfe, 33 Midland Terrace, Hellifield, Skipton, BD23 4HJ
MM3 HLG S. Paul, 10 Beechwood Gardens, Westhill, AB32 6YE
M3 HLN H. Noel, 58 Easenhall Lane, Redditch, B98 0BJ
M3 HLP P. Hallas, 37 Oakfield Road, Bromborough, Wirral, CH62 7BA

M3 HLV J. Ferguson, 41 Brunswick St., Burnley, BB11 3NX
M3 HLX B. Buskin, 6 Elgin Close, Bedlington, NE22 5HJ
MW3 HLZ R. Davies, 8 Deri Road, Uskabergavenny, NP75SY
MJ3 HMA M. Haddon, Balik Pulau, Bradford Ave, La Route Des Genets, St Brelade, Jersey, JE3 8DP
M3 HME E. Cotton, 98 Severn Street, Hull, HU8 8TQ
M3 HMK H. Knighton, 5 Quidenham Road, East Harling, Norwich, NR16 2JD
M3 HML B. Just, 2 The Old Rectory, The High Road, Bedford, MK43 7HN
MM3 HMM D. Macmillan, 2 Fladda Road, Oban, PA34 4HZ
M3 HMT H. Tate, 52 Marlborough Road, London, N22 8NN
M3 HNE M. Ellis, 58 Egghill Lane, Northfield, Birmingham, B31 5NT
M3 HNK T. Lee, 68 Wharton Drive, Springfield, Chelmsford, cm1 6bf
M3 HNL J. Stone, 27 The Heathlands, Warminster, BA12 8BU
M3 HNM N. Evans, 38 Cockster Road, Longton, Stoke-on-Trent, ST3 2EG
MW3 HNP J. Nelson, 31 Y Drim, Ponthenry, Llanelli, SA15 5NY
M3 HNQ I. Rowland, 45 Birks Road, Mansfield, NG19 6JU
M3 HNU G. Sandell, 20 Kirkby View, Sheffield, S12 2NB
M3 HNV P. Breckell, 45 Gordon Avenue, Mansfield, NG18 3AZ
M3 HOD A. Hodson, 22 Walmley Ash Road, Sutton Coldfield, B76 1HY
M3 HOE A. Hoe, 12 Ashbridge Rise, Chandler'S Ford, Eastleigh, SO53 1SA
M3 HOM J. Homsey, 105 Lynwood, Folkestone, CT19 5DD
M3 HOQ N. Lambert, 17 Starcross Road, Weston-Super-Mare, BS22 6NY
M3 HOU E. Edwards, 5 Brindley Road, Silsden, Keighley, BD20 0LD
M3 HOV B. Brown, 3 Swaledale, Worksop, S81 0UY
MW3 HOY M. Hoy, 39 Blackbird Road, Caldicot, NP26 5RE
M3 HPF C. Jamieson, 12 Round Oak Grove, Cheddar, BS27 3BW
M3 HPM D. Woodward, 140 Ewe Lamb Lane, Bramcote, Nottingham, NG9 3JW
M3 HPN W. Hodgson, 11 Tudor Court, Hitchin, SG5 2BE
M3 HPO A. Riches, 84 Elgar Drive, Shefford, SG17 5RA
M3 HPT P. Taylor, Fenway Farm, Ten Mile Bank, Downham Market, PE38 0EU
M3 HPY K. Tokley, 9 Peel Road, Springfield, Chelmsford, CM2 6AQ
M3 HPZ A. Bidwell, 134 Milton Road, Weston-Super-Mare, BS23 2US
M3 HQB C. Smith, 71 Connaught Road, Luton, LU4 8ER
MM3 HQC J. Henry, 7 Wenlock Road, Paisley, PA2 6UJ
MW3 HQD R. Wilkes, 9 Brynmawr Road, Ebbw Vale, NP23 5FF
MM3 HQL G. Fuller, 20 Drumellan Road, Ayr, KA7 4XA
M3 HQN M. Haworth, 26 Willowhey, Marshside, Southport, PR9 9TW
M3 HQP J. Parrott, 2 Boyd Close, Wirral, CH46 1RX
M3 HQQ H. Dixon, 45 Penkhull Terrace, Stoke-on-Trent, ST4 5DH
M3 HQS M. Williamson, 5 Fernbank Close, Crewe, CW1 6ES
M3 HQU M. Copeman, 1 Chestnut Avenue, Welney, Wisbech, PE14 9RG
MW3 HQV E. Carter, 34 Wrexham Road Brynteg, Wrexham, LL11 6HR
M3 HQW A. Hillbeck, 28 Darent Avenue, Walney, Barrow-in-Furness, LA14 3NU
MW3 HRE S. Esp, 34 Wrexham Road Brynteg, Wrexham, LL11 6HR
M3 HRM D. Morgan, 171 Town Road, London, N9 0HJ
M3 HRN C. Fower, 31 Hillswood Avenue, Leek, ST13 8EQ
M3 HRT S. Brashill, 42 Bannister Street, Withernsea, HU19 2DT
M3 HRV D. Porter, 39 Panama Road, Burton-on-Trent, DE13 0SQ
M3 HRY P. Odle, 24 Longfellow Road, Gillingham, ME7 5QG
M3 HSC J. Taggart, 250 Thomas Drive, Liverpool, L14 3LF
M3 HSE C. Hoyle, 43 Helme Drive, Kendal, LA9 7JB
MM3 HSG M. Douglas, 195 Dumbuck Road, Dumbarton, G82 3NU
M3 HSH K. White, 30 Nuneaton Road, Bedworth, CV12 8AL
M3 HSI A. Mcgann, 8 Hertford Close, Whitley Bay, NE25 9XH
M3 HSJ D. Teasdale, 43 Easington Road, Stockton-on-Tees, TS19 8ES
M3 HSM C. Hayes, 7 Harries Court, Waltham Abbey, EN9 3NS
M3 HSR P. Hilton, 14 Masefield Road, Thatcham, RG18 3AF
M3 HSS J. Little, 41 Sevenoaks Road, Portsmouth, PO6 3JP
M3 HSV C. Srinivasan, 2 Hall Drive, Burley In Wharfedale, Ilkley, LS29 7LL
M3 HSW H. White, 96 Clarendon Road, Luton, LU2 7PJ
MW3 HSZ P. Bishop, 76 Heol Homfray, Cardiff, CF5 5SB
M3 HTA S. Fuller, 29 Beckley Road, Wakefield, WF2 9QB
M3 HTE E. Townley, Beetham, Water Hill Lane, Halifax, HX2 7SG
M3 HTF D. O'Flanagan, 16 Corbett Road, London, E11 2LD
M3 HTG A. Murphy, 22 Shenley Fields Drive, Birmingham, B31 1XH
M3 HTO P. Hardy, 21 West Avenue, Boston Spa, Wetherby, LS23 6EJ
M3 HTR S. Adlam, 50 High St., Westtown, Dewsbury, WF13 2QF
MM3 HTY D. Cunningham, 105 Lower Bathville Armadale, Bathgate, EH48 2JS
M3 HUB A. Hubbard, 70 Little Morton Road North Wingfield, North Wingfield, S42 5HN
M3 HUG R. Hughes, 7 Willow Place, Darlington, DL1 5LX
M3 HUN P. Hunter, 7 Fairfield Avenue Hilcote, Alfreton, DE55 5HL
M3 HUS N. Payne, 19 Sid Park Road, Sidmouth, EX10 9BW
M3 HUW H. Weatherhead, 39 Meadow Park, Dawlish, EX7 9BU
M3 HUX N. Gibson, 19 Nene Side Close, Badby, Daventry, NN11 3AD
M3 HUY D. Dolan, 29 Byland Way, Monk Bretton, Barnsley, S71 2JY
M3 HVA P. Robinson, 8 Clewley Road, Branston, Burton-on-Trent, DE14 3JE
M3 HVE R. Dolman, 3 Cloonmore Avenue, Orpington, BR6 9LE
M3 HVH M. Bell, 25 Derwent Crescent, Barnsley, S71 3QU
M3 HVL K. Phizacklea, 23 High Duddon Close, Askam-in-Furness, LA16 7EW
M3 HVN P. Harris, 17 Seymour Avenue, Great Yarmouth, NR30 4BB
M3 HVO G. Omar, 140 Twickenham Road, Isleworth, TW7 7DJ
M3 HVP T. Moss, 3 Haddon Close, Macclesfield, SK11 7YG
M3 HVS A. Hunt, 14 Offranville Close, Leicester, LE4 8NR
M3 HVU K. Jessop, 61 Fountayne Street, Goole, DN14 5HQ
M3 HVV K. Ozwell, 109 Abbey Road, Grimsby, DN32 0HN
M3 HVW J. Naylor, 12 Princess Avenue, Wesham, Preston, PR4 3BA
M3 HVX D. Proctor, 58 Hornby Drive, Newton, Preston, PR4 3SU
M3 HVY C. Hacker, 49 Lamaleach Drive, Freckleton, Preston, PR4 1AJ
MJ3 HWC H. Carrel, 5 Belmont Road, St Helier, Jersey, JE2 4SA
M3 HWN D. Wardman, 45 The Grainger, North West Side, Gateshead, NE8 2BG
M3 HWP D. Potts, 19 Clay Street, Workington, CA14 2XZ
M3 HWS J. Cleverley, 4A Godfrey Street, Netherfield, Nottingham, NG4 2JG
M3 HWV A. Hicks, 30 Manna Drive, Elton, Chester, CH2 4RP
M3 HWW H. Wilson, 2 Railway Close, Burwell, Cambridge, CB25 0DW
M3 HWX D. Beer, 46 The Mailyns, Gillingham, ME8 0DZ
M3 HWY G. Elsworthy, 40 Moorfield Way, Wilberfoss, York, YO41 5PL
M3 HXB B. Wheat, 62 Havenwood Rise, Nottingham, NG11 9HE
M3 HXF J. Merchant, 186 Manor Hall Road, Southwick, Brighton, BN42 4NH
M3 HXG D. Haestier, 18 Midhurst Rise, Brighton, BN1 8LP
M3 HXH J. Clarke, 144 St. Johns Avenue, Kidderminster, DY11 6AU
M3 HXM W. Cole, The Spinney, Holmes Chapel Road, Congleton, CW12 4SN
M3 HXO M. Shaw, 10 Beechwood Avenue, Shevington, Wigan, WN6 8EH
M3 HXQ T. Johnson, 27 Fonthill Road, Bristol, BS10 5SR
M3 HXS M. Campbell, 71 Sages Lea Woodbury Salterton, Exeter, EX5 1RA
M3 HXW K. Bouris, 3 Suffolk Court, Vicarage Road, Maidenhead, SL6 7DT
M3 HXZ N. Mcintyre, 27 Chapel Close, St Ann'S Chapel, Gunnislake, PL18 9JB
M3 HYD M. Douglas, 13 Castlereagh Street, New Silksworth, Sunderland, SR3 1HJ
M3 HYE D. Bruce, 6 Princes Way, King's Lynn, PE30 2QL
M3 HYF A. Sargent, 15 Wilton Road, Balsall Common, Coventry, CV7 7QW
MM3 HYG P. Mcarthur, 22 Bridgeway Terrace, Kirkintilloch, Glasgow, G66 3HJ
M3 HYI A. Reynolds, Fairview, Coombe Way, Teignmouth, TQ14 9QA
M3 HYO E. Hearne, 6 Hillview Road, Basingstoke, RG22 6BQ
M3 HYQ W. Oakley, 1 Southern Avenue, Henlow, SG16 6EY
M3 HYV K. Jones, Court House Farm, Holmes Chapel Road, Crewe, CW4 8AS
M3 HZA D. Pryor, 10 Thornton Crescent, Church Langton, Market Harborough, LE16 7TA
MW3 HZB K. Clark, 56 Morris Avenue, Llanishen, Cardiff, CF14 5JW
M3 HZC C. Chen, Conville And Cains College, Trinity Street, Cambridge, CB2 1TA
M3 HZD D. Evans, 7 Bowerwood Road, Fordingbridge, SP6 1BJ
M3 HZE D. Mannion, 17 Balmoral Road, Haslingden, Rossendale, BB4 4EA
M3 HZH D. Hubbard, 99 Tuckers Road, Loughborough, LE11 2PH
M3 HZK Dr R. Gooch, 14 Cotterill Road, Surbiton, KT6 7UN
M3 HZM M. Holden, 26 Valebridge Drive, Burgess Hill, RH15 0RW
M3 HZN R. Rippin, 28 Ridgeway West, Market Harborough, LE16 7LG
M3 HZO P. Sarll, 81 Austendyke Road, Weston Hills, Spalding, PE12 6BX
M3 HZP B. Holden, 26 Valebridge Drive, Burgess Hill, RH15 0RW
M3 HZW J. Trevarrow, 17 Harland Road, Elloughton, Brough, HU15 1JT
M3 IAA D. Rose, Homeberry House, 13 Ashcroft Gardens, Cirencester, GL7 1RU
M3 IAC I. Crabb, 9A Lonsdale Road, Southend-on-Sea, SS2 4LZ
M3 IAE S. Wilkinson, 144 West End Road, Morecambe, LA4 4EF
M3 IAF I. Firby, 19 St. Georges Drive, Manchester, M40 5HL
MM3 IAG I. Gerrard, 10 Station Road, Ardersier, Inverness, IV2 7ST
MI3 IAI T. Scott, 25 Lisavon Drive, Belfast, BT4 1LJ
M3 IAO A. Hirst, 34 Woodhall Avenue, Bradford, BD3 7BU
M3 IAP G. Evans, 57 Lock Crescent, Kidlington, OX5 1HF
M3 IAQ P. King, Philinda, Carmen Street, Saffron Walden, CB10 1NR
M3 IBE M. Hagan, 186 Saltwell Road, Gateshead, NE8 4XH
M3 IBJ D. Dickinson, 8 East Grange Garth, Leeds, LS10 3EJ
MM3 IBM C. Mckillop, 7 Auchneagh Farm Lane, Greenock, PA16 7BJ
M3 IBS F. Carter, 26 Union Road, Shirley, Solihull, B90 3DQ
M3 IBT J. Ferrol, 29 Westlands, Haltwhistle, NE49 9BS
M3 IBY J. Leaman, 40 Higher Budleigh Meadow, Newton Abbot, TQ12 1UL
M3 IBZ D. Langridge, 12 Battles Lane, Kesgrave, Ipswich, IP5 2XF
M3 ICA A. Schmidt, Church Corner, Fieldside, Boston, PE22 7RA
MM3 ICD W. Ton, 24 Craigmount Hill, Edinburgh, EH4 8DL
M3 ICH M. Green, 4 Tudor Court, Grimethorpe, Barnsley, S72 7NA
M3 ICN A. Groat, 23 Knightwake Road, New Mills, High Peak, SK22 3DQ
M3 ICO L. Widdowson, 11 Belmont Drive, Staveley, Chesterfield, S43 3PQ
M3 IDA D. Cothey, Summerley House, Skircoat Moor Road, Halifax, HX3 0HA
M3 IDB N. Dallen, 77 Hazon Way, Epsom, KT19 8HG
M3 IDC F. Harris, 51 Hillmans Road, Newton Abbot, TQ12 1AA
M3 IDD D. Barker, 21 Boundary Crescent Lower Gornal, Dudley, DY3 2HJ
M3 IDF J. Kelly, 1 Bramble Close, New Ollerton, Newark, NG22 9TN
M3 IDH K. Reynolds, 3 Lilac Close, Chelmsford, CM2 9NY
M3 IDK D. Norman, 22 Stirling St., Hull, HU3 6SL
M3 IDO A. Thompson, 3 Rufford Road, Long Eaton, Nottingham, NG10 3FP
M3 IDQ A. Stevenson, 2 Diddington Close, Bletchley, Milton Keynes, MK2 3EB
MM3 IDR K. Tait, 33 Bankton Avenue, Livingston, EH54 9LD
M3 IDW J. Valle Espin, 203 Broadway, Horsforth, Leeds, LS18 4HL
M3 IDY J. Taylor, Roseneath, 4C Valley Road, Kenley, CR8 5DG
M3 IEA P. Jays, 138 Lower Wear Road, Exeter, EX2 7BD
MM3 IEC E. Cohen, 234 Allison Street, Glasgow, G42 8RT
M3 IEF S. Painting, 15 Surrey Walk, Walsall, WS9 8JG
M3 IEG M. Luxton, 6 Tumbling Field Lane, Tiverton, EX16 4LN
M3 IEL R. Tust, 28 Osprey Close, Beechwood, Runcorn, WA7 3JH
M3 IEM T. Rogers, 45 Church Road, Westoning, Bedford, MK45 5LP
M3 IEP N. Freeman, 27 Montpelier Drive, Caversham, Reading, RG4 6QA
M3 IEQ J. Wheeler, 41 Winnards Park, Sarisbury Green, Southampton, SO31 7BX
M3 IET C. Blount, 55 Silverthorne Drive, Caversham, Reading, RG4 7NR
M3 IEU T. Bougourd, 1 Poplar Close, Newton Abbot, TQ12 4PG
M3 IEV M. Blount, 55 Silverthorne Drive, Caversham, Reading, RG4 7NR
M3 IEW P. Mellish, 302 Belvedere Road, Burton-on-Trent, DE13 0RD
M3 IFA A. Harrison, 16 Ingshead Avenue, Rawmarsh, Rotherham, S62 5BH
M3 IFB D. Symonds, 2 Montgomery Cottages, Stonham Road, Stowmarket, IP14 5LS
M3 IFE M. Ferguson, 80 Chester Road, Holmes Chapel, Crewe, CW4 7DR
M3 IFF P. Webster, 15 Napier Street, Workington, CA14 2PT
M3 IFG F. Gatenby, 6 Telford Close, Audenshaw, Manchester, M34 5FB
MI3 IFI D. Sloan, 15 Deramore Drive, Portadown, Craigavon, BT62 3HH
M3 IFJ B. Royce, 82 Ridge Lane, Watford, WD17 4TA
M3 IFK K. Holloway, 26 Aldgate Drive, Brierley Hill, DY5 3NT
M3 IFM J. Lomako, 23 Stretton Close Ackton, Pontefract, WF7 6HT
MI3 IFO E. Mcclements, 5 Eastbank, Strathfoyle, Londonderry, BT47 6UW
MW3 IFZ N. Bruines, 24 Trenel, Burry Port, SA16 0UT
M3 IGA Dr L. Igali, 22 Mile End Road, Norwich, NR4 7QY
M3 IGN I. Nicholls, 8 Northcroft Road, Corsham, SN13 0LS
MI3 IGO A. Blythe, 159 Victoria Road, Bready, Strabane, BT82 0DZ
MW3 IGZ M. Matthias, 18 Brynmally Park, Pentre Broughton, Wrexham, LL11 6BP
M3 IHA S. Duffy, 38 Well Lane, Newton, Chester, CH2 2HL
MW3 IHB G. Jones, Wern Lodge, Gobowen, Oswestry, SY10 7JY
M3 IHC G. Ward, 5 Allerton Road, Shrewsbury, SY1 4QQ
MW3 IHD I. Davies, 221 Trowbridge Green, Rumney, Cardiff, CF3 1RE
M3 IHN R. Hargeaves, 58 Horsewell Lane, Wigston, LE18 2HQ
M3 IHO K. Brown, 143 Princes Road, Ellesmere Port, CH65 8EP
M3 IHQ T. Wall, 50 Higham Gobion Road, Barton-Le-Clay, Bedford, MK45 4LT

M3 IHR D. Tilley, 15 Dowhills Park, Liverpool, L23 8SS
M3 IHS C. Ivers, 11 Twelve Acre Crescent, Farnborough, GU14 9PW
M3 IHU J. Smith, Wilson Hall Farm, Slade Lane, Derby, DE73 1AG
M3 IHV Dr H. Donnelly, 6 Famet Walk, Purley, CR8 2DY
MW3 IHX D. Eacott-Palfrey, 165 High Street, Blaina, Abertillery, NP13 3AW
MI3 IHY S. Quigg, 100 Whispering Pines, Limavady, BT49 0UF
MI3 IHZ K. Mcdonald, 37 Ardgarvan Cottages, Limavady, BT49 0NF
M3 IIA R. Dale, 17 Spencer Gardens, Brackley, NN13 6AQ
MI3 IIH T. Mcconnell, 41 Moyra Road, Doagh, Ballyclare, BT39 0SQ
MW3 IIJ R. Parry, 5 Accar Y Forwyn, Denbigh, LL16 3PW
MI3 IIL C. Mcconnell, 41 Moyra Road, Doagh, Ballyclare, BT39 0SQ
M3 IIN B. Stokes, 19 Hall Park, Barrow-On-Trent, Derby, DE73 7HD
M3 IIP K. Hunt, 13 Beaumaris Court, Spondon, Derby, DE21 7RG
MM3 IIT G. Saunders, Tower Guest House, 32 James Street, Stornoway, HS1 2QN
M3 IIV J. Hurkett, 9 Fair Field Park, Five Lanes, Launceston, PL15 7RQ
M3 IJD J. Doyle, 10 Greenall Court, Prescot, L34 1NH
M3 IJE I. Ewen, 26 Court Road, Eastbourne, BN22 9EZ
M3 IJF E. Livesey, Reevsmoor, Hollington, Ashbourne, DE6 3AG
MM3 IJI M. Carmichael, 39 Longsdale Crescent, Oban, PA34 5JR
M3 IJO P. Chaney, 246 Agar Road, Illogan Highway, Redruth, TR15 3NJ
M3 IJS I. Sapstead, 7 Shrubbery Grove, Royston, SG8 9LJ
M3 IJT C. King, 1 Victoria Court, Hadley, Telford, TF1 5FL
M3 IJV J. Millichip, 33 Lincoln Road, Stevenage, SG1 4PJ
M3 IJZ C. Young, 15 Shelton Avenue, East Ayton, Scarborough, YO13 9HB
MW3 IKC J. Kenchington, 36 Lando Road, Pembrey, Burry Port, SA16 0UR
M3 IKD J. Hockedy, 22 Victoria Road, Frome, BA11 1RR
M3 IKE M. Murray, Po Box 55, Calle San Jaime, Benijofar, Spain, 3178
M3 IKI J. Batson, 19 Seaview Road, Canvey Island, SS8 7PB
M3 IKJ A. Knitter, 15 Thompson Drive, Hatfield, Doncaster, DN7 6JX
M3 IKM T. Palmer, 29 Field End, Maresfield, Uckfield, TN22 2DJ
M3 IKN C. Staite, 11 Lavender Lane, Great Denham, MK40 4SB
M3 IKR D. Powis, Fircroft, Pound Lane, Woodbridge, IP13 0LN
MM3 IKS G. Frew, 20 Achlonan, Taynuilt, PA35 1JJ
M3 IKV E. Smith, Grange Farm, Main Street, Newark, NG23 5PX
M3 ILB N. Silveston, Caravan C, East End House, Oak Lane, Minster On Sea, Sheerness, ME12 3QR
M3 ILG B. Evans, 12 The Mead Thaxted, Dunmow, CM6 2PU
M3 ILJ E. Copper, 238 Canterbury Road, Kennington, Ashford, TN24 9QL
M3 ILM N. Hickson, 27 Cressing Road, Witham, CM8 2NP
M3 ILR I. Roper, 1 Holywell Road, Kilnhurst, Mexborough, S64 5UQ
M3 ILV P. Hinchliffe, 21 Prospect Hill, Haslingden, Rossendale, BB4 5EF
M3 ILY M. Andrews, 27 Bramble Avenue, Norwich, NR6 6LN
M3 ILZ B. Mcandrew, 8 Springhill Walk, Morpeth, NE61 2JT
M3 IMB I. Berry, 4 Newlands Park Way, Newick, Lewes, BN8 4PG
MM3 IMC I. Mccuaig, 1 Ericht Bank Drive, Kirn, Dunoon, PA23 8HB
M3 IMJ P. Coppin, 3 Firtree Close, Rough Common, Canterbury, CT2 9DB
MM3 IMK I. Mackinnon, 6 Glencruitten Rise, Oban, PA34 4RX
M3 IMM I. Margetts, 48 Spetchley Road, Worcester, WR5 2NL
MI3 IMO T. Mcnaughter, 36 Elms Park, Coleraine, BT52 2QE
M3 IMP I. Phillpott, 14 Buttercup Close, Paddock Wood, Tonbridge, TN12 6BG
M3 IMR A. Ashworth, 22 Crow Lane, Ramsbottom, Bury, BL0 9BR
M3 INC A. Nicholls, 117 Hurligham Road, Kingstanding, b440ng
M3 IND V. Lowe, 35 Elm Place, Armthorpe, Doncaster, DN3 2DE
M3 INH Dr E. Hayes, 11 Ashleigh Wood, Monaleen, Ireland,
M3 INJ A. Hayes, 11 Ashleigh Wood, Monaleen, Ireland,
M3 INL I. Lockyer, 11 Lorina Road, Ramsgate, CT12 6DD
MM3 INO G. Patterson, 28 Highcliffe, Spittal, Berwick-upon-Tweed, TD15 2JH
M3 INQ Z. Ardern, 6 Chaucer Avenue, Mablethorpe, LN12 1DA
MI3 INS D. Quigg, 9 Springhill Terrace, Limavady, BT49 9BS
MM3 INY L. Affleck, 1 Fank Brae, Mallaig, PH41 4RQ
M3 IOC R. Mitchell, 2 Corbar Road, Stockport, SK2 6EP
MM3 IOF M. Mitchell, Easter Kilwhiss Farm, Ladybank, Cupar, KY15 7UR
MI3 IOH W. Bradley, 16 Mullaghanagh Road, Dungannon, BT71 7AY
MJ3 IOJ N. Taylor, 21 Samares Avenue, La Grande Route De St. Clement South, St Clement, Jersey, JE2 6NY
M3 IOK R. Peel, 34 Pagdin Drive, Styrrup, Doncaster, DN11 8LU
MM3 IOM J. Grundey, 6 Ternemny Villas, Knock, Huntly, AB54 7LR
M3 IOQ B. Reynard, 90 Barnsley Road, Darton, Barnsley, S75 5NS
M3 IOT L. Bewley, 21 Duloe Gardens, Pennycross, Plymouth, PL2 3RS
M3 IOX S. Bridges, 140 Highbridge Road, Burnham-on-Sea, TA8 1LW
M3 IPD K. Barron, 80 Primrose Crescent, Norwich, NR7 0SF
MW3 IPK T. Vincent, 88 Lake St., Ferndale, CF43 4HE
M3 IPM S. Jackson, 1 The Avenue, Burton-Upon-Stather, Scunthorpe, DN15 9EX
M3 IPQ A. Badcock, 7 Heathfield Road, Chandler'S Ford, Eastleigh, SO53 5RP
M3 IPT W. Bull, 117 Walton Road, Wednesbury, WS10 0EU
M3 IPY L. Earnshaw, 63 Manor Road, Fleetwood, FY7 7LJ
M3 IPZ P. Wyles, Casa De La Rosa, Torbay Road, Torquay, TQ2 6RG
MM3 IQA G. Robbins, 33 Moffat Court, Glenrothes, KY6 1JR
MM3 IQD M. Gourlay, 14 Holmes Holdings, Broxburn, EH52 5NS
MW3 IQE K. Lowther, Cynon Villa, Main Road, Mountain Ash, CF45 4BX
M3 IQF D. Green, 12 Nostell Road, Ashton-In-Makerfield, Wigan, WN4 9XD
M3 IQG S. Parris, 3 Manor Close, Ringmer, Lewes, BN8 5PA
M3 IQJ C. Evans, Bridge Farm, Shrawley, Worcester, WR6 6TQ
M3 IQM J. Fitzpatrick, 21 Corn Close South Normanton, Alfreton, DE55 2JD
M3 IQN M. Davis, 63 Glebe Road, Barrington, Cambridge, CB22 7RP
M3 IQP L. Bailey, 18 Dudley Place, St Helens, WA9 1BL
M3 IQQ A. Harris, 32 King Edward Road, Gillingham, ME7 2RE
M3 IQS M. Cox, 7 Wilson Close, Daventry, NN11 9WH
MM3 IQU W. Curry, 36 Banklands Newburgh, Cupar, KY146DN
MW3 IQY S. Beer, 49 Central Street, Pwllypant, Caerphilly, CF83 2NJ
M3 IRF S. France, 13 Cirencester Close, Little Hulton, Manchester, M38 9HB
M3 IRH I. Harris, 19 Holmrook Road, Carlisle, CA2 7TB
M3 IRJ G. Rogers, 25 Easton Road, Wirral, CH62 1DR
M3 IRM J. Richardson, 3 Aylesbury Avenue, Urmston, Manchester, M41 0SB
M3 IRP P. Cannam, 1 Field Close, Hinckley, LE10 1TH
M3 IRQ S. Cannam, 82 Barwell Lane, Hinckley, LE10 1SS
M3 IRR A. Gregory, 140 Alder Street, Newton-le-Willows, WA12 8HP
M3 IRS D. Mountford, 189 Lloyd St., Stockport, SK4 1NH
M3 IRU P. Dawson, 88 Urmson Road, Wallasey, CH45 7LQ
MI3 IRV E. Mercer, 8 Woodside Gardens, Portadown, Craigavon, BT62 1EW
M3 IRX A. Ryan, 60 Stanstead Road, Halstead, CO9 1YB
MI3 IRY W. Cooney, 30 Clanbrassil Park, Portadown, Craigavon, BT63 5XT
MM3 ISA I. Mcdermid, 52 Main Street, Pathhead, EH37 5QB
MI3 ISC P. England, 30 Kernan Grove, Portadown, Craigavon, BT63 5RX
M3 ISG A. Evans, 58 Lime Tree Avenue, Crewe, CW1 4HL
M3 ISH S. Brooks, 7 Mayfield Road, Northwich, CW9 7AS
M3 ISI S. Russon, 165 Billington Avenue, Newton-le-Willows, WA12 0AU
M3 ISJ M. Turner, 72 Neville Street, Newton-le-Willows, WA12 9DB
M3 ISN A. Dickinson, 77 Ullswater Avenue, Warrington, WA2 0NQ
M3 ISO I. Butler, 23 Owen Way, Basingstoke, RG24 9GH
M3 ISQ R. Lilley, 3 Coultshead Avenue, Billinge, Wigan, WN5 7HS
MI3 ISX S. Butler, 25 Chippendale Avenue, Bangor, BT20 4PX
M3 ISY M. Edwards, Rouse Farm, Normans Lane, Warrington, WA4 4PY
MM3 ITA J. Veal, 6 Morrison Avenue, Tranent, EH33 2AR
M3 ITH P. Cowin, 16 West Lane, Shap, Penrith, CA10 3LT
M3 ITI S. Roberts, 7 Alberta Grove, Prescot, L34 1PX
M3 ITK D. Willson, Flat 26, King Charles Place, Shoreham-by-Sea, BN43 5JH
M3 ITL I. Buckton, 67 Tennyson Avenue, Middlesbrough, TS6 7ND
M3 ITM S. Garthwaite, 278 Carlton Road, Barnsley, S71 2BA
M3 ITT D. Jones, 77 Brinkburn Grove, Banbury, OX16 3WX
M3 ITU M. Stephenson, 15 Springwood Road, Hoyland, Barnsley, S74 0AZ
M3 ITZ S. Heywood, 16 Edinburgh Drive, Hindley Green, Wigan, WN2 4HL
M3 IUC M. Sturt, 2 Golden Villas, Heathfield Road, Freshwater, PO40 9LQ
M3 IUH A. Morris, 22 Dixon Avenue, Newton-le-Willows, WA12 0NE
M3 IUK A. Mason, 16 Newstead View, Fitzwilliam, Pontefract, WF9 5DP
MW3 IUS I. Canterbury, Brynllethryd Bungalow, Senghenydd, Caerphilly, CF83 4HJ
M3 IUV P. Watson, 32 Shrewsbury Way, Saltney, Chester, CH4 8BY
M3 IUX R. Higham, 17 Walkmill Gardens, Wellington, Seascale, CA20 1EF
M3 IUZ B. Homer, 7 King Street, Quarry Bank, Brierley Hill, DY5 2DH
M3 IVA K. Yates, 103 Raleigh Crescent, Stevenage, SG2 0EB
M3 IVD A. Taplin, 2 Old London Road, Rawreth, Wickford, SS11 8TZ
M3 IVI W. Ivison, 11 Durham St., Fence Houses, Houghton le Spring, DH4 6LA
M3 IVN G. Newman, 5 Edward Crescent, Skegness, PE25 3SA
M3 IVO W. Gissing, 2 Yeo Moor, Clevedon, BS21 6UQ
M3 IVV G. Merrington, Cartref, Ball Lane, Frodsham, WA6 8HP
M3 IVX R. Balm, 250 Coppice Road, Arnold, Nottingham, NG5 7HF
M3 IVY S. Dixon, 5 Swanmore Road, Havant, PO9 4LG
MW3 IWC M. Phillips, 45 Lewis St., Aberbargoed, Bargoed, CF81 9DZ
M3 IWG J. Pusey, 29 Arthur Moody Drive, Carisbrooke, Newport, PO30 5JR
M3 IWJ B. Larman, Cornhill, Mount Bovers Lane, Hawkwell, SS5 4JE
M3 IWK D. Judge, 12 Heelas Road, Wokingham, RG41 2TL
M3 IWN S. Reynolds, 2 Lawson Court, Boldon Colliery, NE35 9NH
M3 IWO W. Hall, 16 Barrington Close, Chelmsford, CM2 7AX
M3 IWR J. Chapman, South View, Mill End Rushden, Buntingford, SG9 0SU
M3 IWT M. Champness, 10 Isaac Square, Great Baddow, Chelmsford, CM2 7PP
M3 IWX S. Evans, 7 Gloster Ropewalk, Dover, CT17 9ES
M3 IWZ A. Hanna, 35 Orchard Drive, Mayland, Chelmsford, CM3 6EP
M3 IXC D. Watson, 10 Gimson Close, Tuffley, Gloucester, GL4 0YQ
M3 IXD D. Watson, 45 Kennel Lane, Brockworth, Gloucester, GL3 4NP
M3 IXE H. Hector, 71 Edinburgh Drive, North Anston, Sheffield, S25 4HB
M3 IXF M. Lucas, 38 Hazel Avenue, Braunton, EX33 2EZ
M3 IXH R. Maas, 15 Pine Court, Attleborough, NR17 2HU
MW3 IXJ A. Littleford, 19 Llys Arthur, Towyn, LL22 9PH
M3 IXK A. Pickett, 4 Trembel Road, Mullion, Helston, TR12 7DY
M3 IXM P. Bell, 7 Greenwood Way, Norwich, NR7 9HW
M3 IXO D. Lowe, 5 Daisy St., Bury, BL8 2QG
M3 IXT S. Widdowson, 11 Belmont Drive, Staveley, Chesterfield, S43 3PQ
M3 IXU D. Skinner, 77 Rolleston Avenue, Petts Wood, Orpington, BR5 1AL
M3 IXX D. Cattermole, 39 Moor Lea, Braunton, EX33 2PF
M3 IXY A. Guest, 1 Green Meadows, Cannock, WS12 3YA
MW3 IXZ H. Alban, 15 Catherine Close, Abercanaid, Merthyr Tydfil, CF48 1YY
M3 IYE P. Ellwood, 13 Wensley Drive, Manchester, M20 3DD
M3 IYG A. Dixon, Flat 4, Farmer House, London, SE16 4BY
MI3 IYH A. Wilson, 108A Salia Avenue, Carrickfergus, BT38 8NE
M3 IYO K. Peabody, 23 Grange Mount, West Kirby, Wirral, CH48 6ET
MI3 IYP S. Kelly, 1 Ardnamoyle Park, Londonderry, BT48 8HN
M3 IYX P. Chapman, 4 The Street, Sutton, Pulborough, RH20 1PS
M3 IYY J. Side, Railway Crossing Cottage, Ash Road, Sandwich, CT13 9JB
M3 IZB P. Lewis, 37 Speedwell Close, Melksham, SN12 7TE
M3 IZD G. Curtis, 11 Bloomery Way, Maresfield, Uckfield, TN22 2DP
M3 IZH J. Winson, 41 Windsor Crescent, Ilkeston, DE7 4HD
M3 IZI D. Groom, 10 Sunnymead Road, Burntwood, WS7 2LL
M3 IZJ J. Winson, 12 Rydal Ave, Long Eaton, NG10 4EB
M3 IZM K. Winson, 41 Windsor Crescent, Ilkeston, DE7 4HD
M3 IZN A. Lodge, 45 Laneside Avenue, Sutton Coldfield, B74 2BU
MM3 IZO E. Stuart, 6D Dundee Street, Letham, DD8 2PQ
M3 IZP B. Johnson, 199 Lynwood, Folkestone, CT19 5TA
M3 IZQ K. Coad, 17 Dilly Lane, Barton On Sea, New Milton, BH25 7DQ
MW3 IZV C. Sadler, Brimaston Cottage, Hayscastle, Haverfordwest, SA62 5PW
M3 IZW J. Reynolds, Fairview, Coombe Way, Teignmouth, TQ14 9QA
M3 JAB J. Butler, 219 Ridge Avenue, Burnley, BB10 3JF
M3 JAC J. Sefton, 87 Lillibrooke Crescent, Maidenhead, SL6 3XL
M3 JAL A. Lloyd, 10 Makepeace Close, Vicars Cross, Chester, CH3 5LU
MW3 JAP J. Phillips, 39 Bryn Glas, Rhosllanerchrugog, Wrexham, LL14 2EA
M3 JAZ J. Hudspeth, 108 Fir Tree Lane, Burtonwood, Warrington, WA5 4NE
M3 JBB J. Benson, 51 Hollowood Avenue, Littleover, Derby, DE23 6JD
M3 JBE J. Brocklesby, 34 Sinnington End, Highwoods, Colchester, CO4 9RE
M3 JBF N. Morphew, 5 Canterbury Close, Canterbury Road, Folkestone, CT19 5EL
M3 JBK S. Glass, 16 Norman Way, Colchester, CO3 4PS
M3 JBM M. Butchers, 67 Keepers Coombe, Bracknell, RG12 0TW
M3 JBQ J. Daniells, Holly Villa, Foxholes, Wem, Shrewsbury, SY4 5UJ
M3 JBW J. Wolohan, 24 Granby Close, Corby, NN18 0AG
M3 JBZ A. Bell, 28 Haven Baulk Avenue, Littleover, Derby, DE23 4BJ
M3 JCA C. Ashman, 40 St. Matthews Road, Kettering, NN15 5HE
MI3 JCB M. Crawford-Baker, George'S Nest, 131 Gobbins Road, Larne, BT40 3TX
M3 JCE T. Higgins, 7 Nobles Close, Coates, Peterborough, PE7 2BT

MW3 JCG J. Scott, Winfield, Templeton, Narberth, SA67 8SP
M3 JCH J. Hill, Anglecroft, Somerford Booths, Congleton, CW12 2JU
M3 JCQ J. Bond, Oakley, 19 Poplar Road, Tenterden, TN30 7NT
M3 JCS J. Sanderson, 54 Kelvedon Close, Chelmsford, CM1 4DG
M3 JCT L. Jenner, 231 Greeswood Rord, Tunbridge Wells, TN2 3HU
M3 JCU J. Turton, 2 Elkstone Road, Chesterfield, S40 4UT
M3 JCY J. Connolly, 2 Waring Avenue, St Helens, WA9 2QG
MW3 JDA J. Ruck, 286 Barry Road, Barry, CF62 8HF
M3 JDF J. Feather, 21 Cedar Avenue, Wickersley, Rotherham, S66 2NT
M3 JDG J. Godding, 58 Dukeswood Road, Longtown, Carlisle, CA6 5UJ
M3 JDJ J. Handley, 4 Manor Close, Draycott, Stoke-on-Trent, ST11 9AZ
M3 JDN J. Dobson, 17 Cadley Causeway, Fulwood, Preston, PR2 3RU
MI3 JDQ J. Quigg, 9 Springhill Terrace, Limavady, BT49 9BS
M3 JDS J. Smith, Clare Cottage, White Ash Green, Halstead, CO9 1PD
M3 JDX J. Porteious, 62B Church Close, Stilton, Peterborough, PE7 3RG
M3 JEE J. Edmondson, 11 Cedar Terrace, Fencehouses, Houghton le Spring, DH4 5ND
M3 JEH J. Hutt, 48 Hill Crest, Swillington, Leeds, LS26 8DL
MW3 JEK C. Gilbert, 11 Parc Y Deri, Neath, SA10 6BQ
M3 JEM J. Carvill, The Lodge, Oldbury Road, Worcester, WR2 6AA
M3 JEP M. Clarke, 138 Colne Road, Halstead, CO9 2HJ
M3 JER J. Higgins, 6 Larksfield Avenue, Bournemouth, BH9 3LP
M3 JFA P. Cowan, 230 High St., Felixstowe, IP11 9DS
M3 JFB J. Beezer, 23 Milburn St., Sunderland, SR4 6AU
M3 JFF J. Neighbour, 112 Holyfields, West Allotment, Newcastle upon Tyne, NE27 0EX
M3 JFP J. Payne, Jomayne, Farm Lane, Evesham, WR11 8TL
M3 JFS J. Skinner, 12 Hanbury House, Cardy Close, Redditch, B97 6LP
MM3 JFW A. Wilson, 20 Ballumbie Gardens, Dundee, DD4 0NR
M3 JGH J. Hewitt, 166 Ormskirk Road, Rainford, St Helens, WA11 8SW
M3 JGI J. Ives, 87 Sheepwalk, Paston, Peterborough, PE4 7BJ
M3 JGJ A. Skinner, Chevington, Carlton Road, Godstone, RH9 8LD
M3 JGN R. Wild, 90 Broadway East, Redcar, TS10 5DP
M3 JGQ A. Greenland, 19 The Ridgeway, Potton, Sandy, SG19 2PS
MM3 JGR J. Gracie, 34 Ayr Road, Dalmellington, Ayr, KA6 7SJ
MD3 JGS Dr A. Foxon, 39 Droghadfayle Road, Port Erin, Isle Of Man, IM9 6EN
MM3 JGT J. Thomson, 26 South Dean Road, Kilmarnock, KA3 7RB
M3 JGU M. Connell, 17 The Crescent, Stockport, SK3 8SL
M3 JGW W. Holroyd, 8 Carr Dene Court, Preston Street, Preston, PR4 2XA
M3 JGX A. Hadfield, 50 Eastbourne Road, Southport, PR8 4DT
M3 JHC J. Clark, 27 The Gabriels, Newbury, RG14 6PZ
M3 JHJ J. Hickey, 11 Greenfield Avenue, Hodthorpe, Worksop, S80 4XT
M3 JHL J. Locke, 2 Fairnley Road, Nottingham, NG8 4AH
M3 JHR J. Richardson, 44 Cross Tree Road, Wicken, Milton Keynes, MK19 6BT
MM3 JHS J. Hume, 8/11 Leslie Place, Edinburgh, EH4 1NH
M3 JHT J. Tarver, 14 South View Road, Leamington Spa, CV32 7JD
M3 JHV C. Smith, 11 Chesterton Road, Thatcham, RG18 3UH
M3 JHW J. Warren, 1 Acre Close, Rochester, ME1 2RE
M3 JIA J. Allanson, Tresco, Hampton, Swindon, SN6 7RL
M3 JIC P. Baker, 25 Regency Court, Winsford, CW7 1FE
M3 JID J. Douglas, 26 Walker Drive, Bootle, L20 6NN
M3 JIG L. Higgs, 14 Lydgate Close Lawford, Manningtree, CO11 2SU
M3 JIH D. Marsland, 154 Moss Lane, Litherland, Liverpool, L21 7NN
M3 JII A. Riley-Marsland, 154 Moss Lane, Litherland, Liverpool, L21 7NN
MW3 JIJ K. Powell, 11 Terrig Street, Shotton, Deeside, CH5 1XU
M3 JIK S. Sanders, 3 Edmunds Square, Mickleover, Derby, DE3 0DU
M3 JIL G. Hinsley, 38 Swindon Lane, Prestbury, Cheltenham, GL50 4NY
MM3 JIN J. Nicol, 18 Tininver Street, Dufftown, Keith, AB55 4AZ
M3 JIR S. Buckley, 35 Marlborough Road, Irlam, Manchester, M44 6HH
M3 JIT W. Carty, 49 Princess Gardens, Blackburn, bb25ej
M3 JIU G. Rigby, 106 Broadway Crescent, Binstead, Ryde, PO33 3QS
M3 JIV R. French, 19 Melstone Avenue, Stoke-on-Trent, ST6 6EX
M3 JIW J. Biggin, Galadean, Farriers Way, Newport, PO30 3JP
M3 JJB J. Barton, 93 Cardigan Road, Bridlington, YO15 3JU
MM3 JJC Dr A. Curlis, 94 Kirkhill Road, Aberdeen, AB11 8FX
M3 JJH A. Williams, 78 Hales Crescent, Smethwick, B67 6QS
M3 JJM J. Martin, 1 Collins Lane, West Harting, Petersfield, GU31 5NZ
M3 JJN J. Nicholson, 6 Mill Gardens, West End, Southampton, SO18 3AG
M3 JJS J. Stanton, Waters & Stanton Plc, 22 Main Road, Hockley, SS5 4QS
M3 JJT D. Thompson, 5 Vashon Drive, Droitwich, Wr97jp
M3 JJU M. Slee, 88 West Avenue, Lightcliffe, Halifax, HX3 8TJ
M3 JKA D. Moffatt, 5 Florence Place, Decoy Road, Newton Abbot, TQ12 1DX
M3 JKB Dr J. Arkinstall, 5 Dunnock Way, Colchester, CO4 3UP
M3 JKE N. Knapton, 4 Crabmill Lane, Easingwold, York, YO61 3DE
M3 JKG T. Shaughnessy, 220 Ladybank Road, Mickleover, Derby, DE3 0RS
M3 JKI D. Evans, 69 Westbourne, Honeybourne, Evesham, WR11 7PT
M3 JKJ S. Asling, 18 Cecilia Grove, St. Peters, Broadstairs, CT10 3DE
M3 JKM K. Mclaughlin, 34 Cambridge Road, Birstall, Batley, WF17 9JF
M3 JKP S. Asling, 18 Cecilia Grove, St. Peters, Broadstairs, CT10 3DE
M3 JKT J. Phillips, The Manor, Blackwoods, York, YO61 3ER
MM3 JKX J. Kirkpatrick, Brims School House, Longhope, Stromness, KW16 3NZ
M3 JKZ J. Hawkins, 294 Norton Lane, Earlswood, Solihull, B94 5LP
M3 JLA J. Astbury, 12 Southall Road, Ashmore Park, Wolverhampton, WV11 2PZ
M3 JLB J. Bevan, 18 Martin Road, Diss, IP22 4HR
M3 JLD J. Denny, 9 Hawthorn Way, Macclesfield, SK10 2DA
M3 JLE J. Isard-Brown, 5 Grove Crescent, Croxley Green, Rickmansworth, WD3 3JT
M3 JLF K. Bailey, 55 Bridgend Park Brewery Road, Wooler, NE716QG
M3 JLH J. Hood, 17 Grays Close, Motcombe, Shaftesbury, SP7 9QB
M3 JLI J. Breckon, Greenbank, Chester High Road, Neston, CH64 7TR
MW3 JLK J. Knowles, 72 Uplands Avenue, Connah'S Quay, Deeside, CH5 4LG
M3 JLR J. Ramsay, Lane Cottage, Weymore Cottages, Bucknell, SY7 0EP
MM3 JLS S. Bence, 14 Stein Terrace, Ferniegair, Hamilton, ML3 7FR
M3 JLV S. Matthews, 60 Brook Street, Erith, DA8 1JQ
M3 JLX S. Crouch, 152 Thornhill Road, Brighouse, HD6 3AH
MI3 JMC J. Mccaw, 62 High Street, Ballymena, BT43 6DT
M3 JMI J. Ioannou, 22 Somerly Close, Binley, Coventry, CV3 2LA
M3 JMJ J. Jenkinson, 7 Chestnut Avenue, Thorngumbald, Hull, HU12 9LD
M3 JMK J. Keegan, The Cottage, 11 Condor Grove, Lytham St Annes, FY8 2HE
M3 JMQ D. Baines, 3 Dunkirk Avenue, Houghton le Spring, DH5 8HN
M3 JMU M. Crowley, 133 Jessop Road, Stevenage, SG1 5LH
M3 JMW J. Moxley-Wyles, 7 Gidley Way, Horspath, Oxford, OX33 1RQ
M3 JMX J. Moore, 110 Pooles Lane, Willenhall, WV12 5HW
M3 JMY J. Varty, 4 St. Cuthberts Close, Burnfoot, Wigton, CA7 9HQ
M3 JNB J. Kearney, 12 Forshaw Lane, Burtonwood, Warrington, WA5 4ES
M3 JND J. Dukes, 79 Jubilee Avenue, Boston, PE21 9LE
MM3 JNJ A. Campbell, 3 North Shawbost, Isle of Lewis, MS2 9BD
M3 JNQ J. Clark, 1 Brooklime Road, Liverpool, L11 2YH
M3 JNR K. Challoner, 23 Chapel Lane, Queensbury, Bradford, BD13 2QA
M3 JNT J. Foulds, 7 Bridge Road, Little Sutton, Spalding, PE12 9EG
M3 JNU S. Robinson, 29 Grange Lane, Mountsorrel, Loughborough, LE12 7HY
MW3 JNX M. January, Blaenau Ucha Farm, Treudoyn, Flintshire, CH7 4NS
M3 JNY J. Hutton, Cassiobury, The Street, Diss, IP22 2PS
M3 JOA N. Lambert, 3 Nightingale Walk, Stockton-on-Tees, TS20 1SZ
M3 JOF J. Miles, 11 Enborne Gate, Newbury, RG14 6AZ
M3 JOJ J. Smith, 5 Manifold Gardens, Plymouth, PL3 6HL
M3 JOM C. Loughran, 8 Douglas Road, Dover, CT17 0BD
M3 JOS D. Bown, 34 Kings Gardens, Bedworth, CV12 8JG
M3 JOW J. Fletcher, 32 Chapel Lane, Barwick In Elmet, Leeds, LS15 4EJ
MW3 JPF J. Freelove, 12 Honeyborough Road, Neyland, Milford Haven, SA73 1RE
M3 JPG Rev. G. Bowen, 93 Pelham Road, Bexleyheath, DA7 4LY
M3 JPI J. Pickering, Batemill, Batemill Lane, Macclesfield, SK11 9BW
M3 JPM G. Meyer, 37 Marina Village, Preston Brook, Runcorn, WA7 3BH
M3 JPP H. Tonge, 38 Colemeadow Road, Billesley Common, Birmingham, B13 0JL
MM3 JPS J. Greig, 28 Suttieslea Crescent, Newtongrange, Dalkeith, EH22 4AR
M3 JPU J. Patient, 4 Bucklebury Heath, South Woodham Ferrers, Chelmsford, CM3 5ZU
MW3 JQC M. Breakwell, 9 Llys Y Dderwen, New Quay, SA45 9SY
MI3 JQD B. Young, 63 Scarvagherin Road Spamount, Castlederg, BT81 7NW
M3 JQF S. Walrond, 17 Madam Lane, Weston-Super-Mare, BS22 6PW
M3 JQG D. Johnson, 11 Horseshoe Avenue, Dove Holes, Buxton, SK17 8DP
M3 JQJ L. Berry, 6 Warren Park Close, Brighouse, HD6 2RU
MW3 JQK E. Williams, Criafol, Upper Llandwrog, Caernarfon, LL54 7PU
M3 JQM L. Ross, 2 Bedford Street, Blackburn, BB2 4EU
M3 JQN V. Greco, 38 Horbling Lane, Stickney, Boston, PE22 8DQ
M3 JQS J. Hellowell, Upper Hole Head Farm, Ash Hall Lane, Soyland, HX6 4NU
M3 JQT S. Watson, 32 Bradley View, Holywell Green, Halifax, HX4 9DN
M3 JQV P. Handy, 30 Kingfisher Drive, Cheltenham, GL51 0WN
M3 JQW D. Roscoe, 28A Princess St., Chorley, PR7 3AP
MW3 JQX T. Leaworthy, 7 Maesderwen Rise, Stafford Road, Pontypool, NP4 5SS
M3 JQY J. Johnson, 30 Thorpe Downs Road, Church Gresley, Swadlincote, DE11 9FB
M3 JRA A. Jessop, 4 Katherine Street, Thurcroft, Rotherham, S66 9LG
M3 JRF J. Francis, 11 Middle Stream Close, Bridgwater, TA6 6LF
M3 JRI J. Rowe, 45 Durham Road, Wilpshire, Blackburn, BB1 9NH
MM3 JRK G. Smith, 40 Pirleyhill Drive, Shieldhill, Falkirk, FK1 2EA
M3 JRM J. Marter, 4 Meadow Way, Seaford, BN25 4QT
M3 JRN J. Newman, 25 Milebush Road, Southsea, PO4 8NF
M3 JRQ L. Pearson, 4 Brentwood Close, Thorpe Audlin, Pontefract, WF8 3ES
M3 JRR J. Read, 26 Chaucer Road, Walsall, WS3 1DF
MM3 JSB J. Bence, 5 Braeside Gardens, Hamilton, ML3 7PN
M3 JSF J. Flores-Watson, 10 Bramwell Gardens, Coventry, CV6 6NB
MI3 JSH S. Hutchinson, 21 Lord Warden'S Grange, Bangor, BT19 1YN
M3 JSK J. Killian, 7 Dankworth Road, Basingstoke, RG22 4LJ
M3 JSM A. Mclaughlin, 34 Cambridge Road, Birstall, Batley, WF17 9JF
M3 JSO J. O'Shea, 56 Crummock Gardens, London, NW9 0DJ
M3 JSQ M. Woodruff, 14 Primatt Crescent, Shenley Church End, Milton Keynes, MK5 6AS
M3 JST J. Taylor, 5 Pyman Close, Martham, Great Yarmouth, NR29 4UR
MI3 JTB J. Black, 17 Fairymount Terrace, Taylors Avenue, Carrickfergus, BT38 7HN
MW3 JTJ J. Jones, Bronydd, Blaenffos, Boncath, SA37 0HZ
MI3 JTM J. Monteith, 58 Bells Hill, Limavady, BT49 0DQ
M3 JTO J. Bosworth, 10 Aston Street, Leeds, LS13 2BJ
MD3 JTT J. Talbot, 43 Harcroft Meadow New Castletown Road, Douglas, Isle of Man, IM2 1JT
M3 JTU A. Bailey, 58 Billy Buns Lane, Wombourne, Wolverhampton, WV5 9BP
M3 JTZ R. Smith, 17 Julian Road, Spixworth, Norwich, NR10 3QA
M3 JUC K. Marsh, 21 Edward Road, Eynesbury, St Neots, PE19 2QF
M3 JUF Dr J. Schofield, 6 Robin Royd Avenue, Mirfield, WF14 0LF
M3 JUL J. Townsend, 56 Seymour Road, Northfleet, Gravesend, DA11 7BN
MW3 JUM R. Jones, 10 Erw Wen Road, Colwyn Bay, LL29 7SD
M3 JUO S. Jacklin, 32 Edison Drive, Rugby, Cv21 1fb
M3 JUX C. Quilter, 473 Sidcup Road, Mottingham, London, SE9 4ET
M3 JUY R. Thomas, 52 Victoria Road, Saltney, Chester, CH4 8SS
M3 JUZ R. Shams-Nia, 1090 Eastern Avenue, Ilford, IG2 7SF
M3 JVA P. Smith, 101 Brunel Avenue, Newthorpe, Nottingham, NG16 3RE
MW3 JVH E. Chell, Mesen Fach, Llanybydder, SA40 9TY
MI3 JVJ P. Hannigan, 4 Silverhill Road, Strabane, BT82 0AE
M3 JVK J. Kirkham, 35 Central Avenue, Woodlands, Doncaster, DN67NW
M3 JVP J. Papworth, Flat 1, 70 Edward Road, Nottingham, NG2 5GB
M3 JVR B. Gibbs, Flat 6, Castleton Court, Southsea, PO5 3AU
MI3 JVV K. Hannigan, 4 Silverhill Road, Strabane, BT82 0AE
M3 JVW J. Wheway, 20 Radnor Street, Derby, DE21 6DZ
M3 JVX E. Mcgowan, Plovers, Moor Road, Walesby, LN8 3UR
M3 JWJ D. Shields, 42 Studland Park, Westbury, BA13 3HL
M3 JWM J. Watts, 10 Lacy Road, Ludlow, SY8 2NS
M3 JWN J. Newell, 7 Talbot, Tamworth, B77 2RS
M3 JWQ W. Johnson, 10 Archdale Road, Nottingham, NG5 6EB
MW3 JWV L. Percival, Blue Cedars, Gresford, Wrexham, LL12 8RN
M3 JWW J. Wainwright, 8 Common Lane, Cutthorpe, Chesterfield, S42 7AN
M3 JWZ S. Sewell, The Old Vicarage, Church Bank, Crewe, CW4 8PG
MI3 JXA C. Matchett, 28 Glendale Avenue East, Belfast, BT8 6LF
M3 JXE D. Ennion, 347 Parkgate Road, Chester, CH1 4BE
MI3 JXG C. Birney, 45 Sallys Wood, Irvinestown, Enniskillen, BT94 1HQ
M3 JXI H. Southall, 12 Prescot Close, Mickleover, Derby, DE3 0TB
M3 JXN P. Jackson, Langsmead Barn, Eastbourne Road, Blindley Heath, Lingfield, RH7 6JX

MI3 JXO D. Burke, 7 Edinburgh Villas, Omagh, BT79 0DW
M3 JXV C. Trew, Ringstone Lodge, 66 Oakwood Road, Horley, RH6 7BX
M3 JXX J. Driver, 99 Queens Road, North Weald, Epping, CM16 6JQ
M3 JXY A. Gowans, 38 Beech Way, Twickenham, tw25jt
M3 JYA D. Kemp, 7 Hillhurst Grove, Birmingham, B36 9TS
M3 JYE B. Edwards, 16 Whitland Close, Rednal, Birmingham, B45 8SJ
M3 JYG D. Batty, 168 Rotherhithe New Road, London, SE16 2AP
M3 JYH G. Swain, 3 Flaxfield Drive, Crewkerne, TA18 8DF
M3 JYO D. Perks, 6 Old School Gardens, Yatton Keynell, Chippenham, SN14 7BB
M3 JYP C. Lester, 21 Barwell Way, Witham, CM8 2TY
M3 JYW X. Chen, Harrogate Ladies' College, Clarence Drive, Harrogate, HG1 2QG
M3 JYZ C. Williamson, 53A High St., Whitwell, Hitchin, SG4 8AJ
M3 JZA K. Armstrong, 29 Thorntree Avenue, Crofton, Wakefield, WF4 1NU
M3 JZD K. Hart, 2 Springfield Cottages, Acton, Newcastle, ST5 4EF
M3 JZE S. Peregrine, 18 Gisborne Close, Mickleover, Derby, DE3 9LU
M3 JZF D. Wall, 96 Albert Street, Wigan, WN5 9EF
M3 JZI M. Rolls, 3 Gleneagles Crescent, New Holland, Barrow-upon-Humber, DN19 7TL
M3 JZK M. Stinton, 57 Wildfields Road, Clenchwarton, King's Lynn, PE34 4DE
M3 JZL D. Atkins, 20 Nappsbury Road, Luton, LU4 9AL
M3 JZN M. Johnson, 25 Rowan Drive, Kirkby-In-Ashfield, Nottingham, NG17 8FU
M3 JZO S. Bacon, 34 Fishers St., Kirkby In Ashfield, Nottingham, NG17 9AH
M3 JZP A. Bacon, 34 Fishers St., Kirkby In Ashfield, Nottingham, NG17 9AH
M3 JZT W. Soffe, 96 Urban Road, Doncaster, DN4 0EP
MW3 JZV J. Jenkins, 13 Birch Hill, Newport, NP20 6JD
M3 JZX M. Cooper, 69 Leicester Road, Kibworth Harcourt, Leicester, LE8 0NP
M3 KAC D. Carr, 19 Kingsmead Walk, Speedwell, Bristol, BS5 7RL
M3 KAE B. Mcfarlane, 11 Hill St., Barnsley, S71 5AL
M3 KAK K. Harden, 59 Violet Avenue, Edlington, Doncaster, DN12 1NW
M3 KAL K. Lawton, Meadowbank, Sutton St. Nicholas, Hereford, HR1 3BJ
M3 KAN K. Hudson, 3 Worcester Ave, Mansfield, NG19 8QJ
MM3 KAQ P. Cotton, 5 Carronhall Avenue, Carronshore, Falkirk, FK2 8AN
M3 KAU H. Leaver, 1 Litcham Close, Litcham, King's Lynn, PE32 2QX
M3 KAX D. Larkin, 19 Elizabeth Court, Hemsworth, Pontefract, WF9 4TQ
M3 KAY K. Limbert, 82 Kipling Avenue, Liverpool, L36 0TZ
M3 KBB K. Barrow, Fenway Farm, Ten Mile Bank, Downham Market, PE38 0EU
M3 KBE J. Kelly, 1 Bramble Close, New Ollerton, Newark, NG22 9TN
M3 KBF C. Harley, 1 Portland Crescent, Meden Vale, Mansfield, NG20 9PJ
M3 KBG J. Crowther, 16 Linden Avenue, Tuxford, Newark, NG22 0JR
M3 KBL K. Lee, 10 Queens Way, Pontefract, WF8 2LX
MU3 KBP K. Pratt, Avalon Le Clos Des Sablon, Sandy Lane, St Sampson, Guernsey, GY2 4RN
MJ3 KBQ C. Daniells, 7 Arlington Road, York, Jersey, JE3 1GP
M3 KBY M. Kirby, 76 Burton Road, Overseal, Swadlincote, DE12 6JJ
M3 KBZ R. Griffiths, 7 Macnaghten Road, Tankersley, Barnsley, S75 3DD
M3 KCA C. Jameson, Flat 9, Britannia Court, Poole, BH12 3HF
M3 KCC C. Evans, 158 Delamore St., Liverpool, L4 3SX
M3 KCG R. Jones, 79 Turpins Rise, Stevenage, SG2 8QZ
MW3 KCL M. Brennan, 37 Marguerites Way, Cardiff, CF5 4QW
M3 KCO K. Cornmell, 19 Forest Road, Chandler'S Ford, Eastleigh, SO53 1NA
M3 KCP K. Preen, 12 Sandpit Lane, Hilton, Bridgnorth, WV15 5PH
M3 KCQ T. Stansfield, 40 Rushmore House, Rubery, Birmingham, B45 9RU
MD3 KCT J. Kennaugh, White Gables, 25 Kissack Road, Castletown, Isle Of Man, IM9 1NW
M3 KCU M. Johnson, 25 Rowan Drive, Kirkby-In-Ashfield, Nottingham, NG17 8FU
M3 KDK S. Allington, 137 Marshall Lane, Northwich, CW8 1LA
M3 KDL C. Lote, 8 Warren Place, Walsall, WS8 6BY
M3 KDM D. Langley, 2 Holly Bush Cottages Holmesdale Road, Sevenoaks, TN13 3XN
MM3 KDN A. Traynor, 2 Mains Of Carmyllie Farm Cottages Carmyllie, Arbroath, DD11 2RJ
M3 KDO R. Smith, 3 Vernon Road, Southport, PR9 7EZ
MI3 KDR K. Dickson, 66 Lisnabreeny Road, Belfast, BT6 9SR
M3 KDV D. Craven, 69 Markham Avenue, Rawdon, Leeds, LS19 6NE
M3 KDY C. Lindsay, 152 Dinmore Avenue, Blackpool, FY3 7QS
M3 KEC S. Conlon, 6 Cardigan St., Ashton On Ribble, Preston, PR2 2AS
M3 KEF K. Forster, Meadow View, Cracow Moss, Crewe, CW3 9BS
M3 KEJ K. Jefferson, 45 Rutherford Crescent, Leighton Buzzard, LU7 3GE
M3 KEL K. Watwood, 57 Cliveden Road, Stoke-on-Trent, ST2 8LP
M3 KER M. Springett, 31 Mountbatten Court, Andover Road, Winchester, SO22 6BA
M3 KEV K. Graffham, 15 Hayes Road, Clacton-on-Sea, CO15 1TX
M3 KEW J. Taylor, 90 Aldam Road, Doncaster, DN4 9EL
M3 KEY K. Calderbank, 6 Heathfield, Heath Charnock, Chorley, PR6 9LA
M3 KEZ D. Brough, 57 Francis Road, Ashford, TN23 7UP
M3 KFE S. Evans, 54 Stafford Crescent, Newcastle, ST5 3EA
M3 KFH R. Drake, 42 Sunningdale Close, Doncaster, DN4 6UR
MI3 KFI T. Warmington, 57 Carbet Road, Portadown, Craigavon, BT63 5RJ
M3 KFL P. Hooper, 4 Castlemead Close, Saltash, PL12 4LF
M3 KFO W. Cromack, 45 Southroyd Park, Pudsey, LS28 8AX
M3 KFP J. Rouse, Nettleden, Galane Close, Northampton, NN4 9YR
M3 KFQ N. Camp, 1 Higher Tresillian Cottages, Tresillian., Newquay, TR8 4PL
M3 KFR F. Coles, 8 Moore Close, Church Crookham, Fleet, GU52 6JD
M3 KFT J. Ferrol, 29 Westlands, Haltwhistle, NE49 9BS
M3 KGE S. Angove, 62 Trelissick Fields, Hayle, TR27 6HZ
M3 KGG K. Gordon, 308 Claremont Road, Swanley, BR8 7QZ
M3 KGJ K. Weston, 114 Morland Road, Ipswich, IP3 0LZ
M3 KGO R. Dunn, 15 Catkins Close, Catshill, Bromsgrove, B61 0TT
MW3 KGP Dr P. Kelly, Arosfa, Westminster Road, Wrexham, LL11 6DN
MW3 KGQ J. Kelly, Arosfa, Westminster Road, Wrexham, LL11 6DN
M3 KGV C. Moulding, 28 Queens Avenue, Highworth, Swindon, SN6 7BA
M3 KHA A. Hughes, 8 Bullens Green Lane Colney Heath, St Albans, AL4 0QS
MW3 KHC A. Vick, Flat 8, Kingshill Court, Newport, NP20 4DT
M3 KHE K. Bradley, 9 Spruce Grove, Kirkby-In-Ashfield, Nottingham, NG17 7QB
MW3 KHH A. Peake, 24 Rhigos Gardens, Cardiff, CF24 4LS
M3 KHI C. Webb, 5 Pound Lane, Preston Bissett, Buckingham, MK18 4LX
M3 KHJ C. Ashman, 56 Farriers Close, Swindon, SN1 2QT
MW3 KHK S. Rosser, 25 Clos Tir Ypwll, Pantside, Newport, NP11 5GE
M3 KHM L. Lebaldi, 11 Artle Place, Lancaster, LA1 2QP
M3 KHT H. Kwan, Harrigate Ladies College, Clarence Drive, Harrogate, HG1 2QG
M3 KHW K. Stocker, 50 Mount Drive, Harrow, HA2 7RP
MW3 KHY J. Robinson, 41 Ashbrook, Brackla, Bridgend, CF31 2AT
MI3 KIL P. Hill, Flat 12 Kilcreggan Homes, Elizabeth Avenue, Carrickfergus, BT38 7EP
M3 KIN D. Kinsey, 161 Heath Road South, Weston, Runcorn, WA7 4RP
M3 KIO E. Smith, The Cabin, Nothe Parade, Weymouth, DT4 8TX
M3 KIQ M. Lebaldi, 11 Artle Place, Lancaster, LA1 2QP
M3 KIR K. Wheeler, 27 Elley Green, Neston, Corsham, SN13 9TX
MW3 KIS C. Heaton, Flat, St. Asaph Conservative Club, High Street, St Asaph, LL17 0RG
M3 KIT K. Cattermole, Blaxhall Hall Crossing, Little Glemham, Woodbridge, IP13 0BP
M3 KIU S. Moore, 10 Strathcona Avenue, Hull, HU5 4AD
M3 KIZ P. Lewis, 16 Valley Road, St Albans, AL3 6LR
M3 KJB K. Brooks, 24 Morris Drive, Weaverham, Northwich, CW8 3LP
M3 KJC K. Cole, 4 Marsham Road Hazel Grove, Stockport, SK7 5JB
M3 KJD K. Davies, 20A Hart Road, Wolverhampton, WV11 3QJ
M3 KJE W. Taylor, 99 St. Marys Close, Littlehampton, BN17 5QQ
MM3 KJG K. Glacken, 14 Hailes Avenue, Edinburgh, EH13 0NA
M3 KJK J. Mason, 22 Eskdale Avenue, Halifax, HX3 7NH
M3 KJM K. Marsh, 11 Apollo Road, Stourbridge, DY9 8YG
M3 KJS K. Sealey, 8 Esplanade, Burnham-on-Sea, TA8 1BE
M3 KJV D. Baker, 65 Madison Street, Tunstall, St6 5hs
M3 KJY C. Atkins, 87 Wentworth Road, Doncaster, DN2 4DA
M3 KKA N. Holdridge, 15 Ballam Avenue, Doncaster, DN5 9DY
M3 KKB S. Silvers, 39 Hickinwood Crescent, Clowne, Chesterfield, S43 4AQ
M3 KKF R. Taylor, 38 Edleston Road, Crewe, CW2 7HD
M3 KKG K. Gledhill, 19 Palmers Terrace, Treknow, Tintagel, PL34 0EH
M3 KKI I. Todorovic, 3 Braemar Close, Stoke-on-Trent, ST2 8NL
M3 KKN G. Allen, 60 Danefield Road, Cheshire, CW95PX
M3 KKO M. O'Neill, 15 School Road, Hockley Heath, Solihull, B94 6QH
MI3 KKP A. Temple, 27 Oakland'S Glenshane Rd Claudy Co Londonderry Bt47 4Ff, Londonderry, BT47 4ff
M3 KKQ G. Thomas, 15 Buckley Avenue, Byley, Middlewich, CW10 9NW
M3 KKS B. Roaf, 8 Weare Close, Portland, DT5 1JP
M3 KKX A. Jones, 27 Fishpond Lane, Holbeach, Spalding, PE12 7DQ
M3 KKZ R. Kirby, 44 Wilby Avenue, Little Lever, Bolton, BL3 1QE
M3 KLB M. Crombie, 12 Sir James Reckitt Haven, Hull, HU8 8QR
MM3 KLO S. Dobie, Flat 20, 3 Arneil Place, Edinburgh, EH5 2GU
M3 KLS K. Symonds, 68 Manor Crescent, Pan, Newport, PO30 2BH
M3 KLT M. Rogers Jones, Glanva, 20 Birchwood, Leyland, PR26 7QJ
M3 KLU N. Andrews, Temple View House, Shopland Road, Rochford, SS4 1LH
M3 KLV P. Bailey, 71 Norfolk Avenue, Leigh-on-Sea, SS9 3HA
M3 KLY K. Lingham, 102 Chancery Lane, St Helens, WA9 1SQ
MI3 KMB K. Barr, 64 Owenreagh Drive, Strabane, BT82 9DT
M3 KMH K. Haywood, 6 Lydney Road, Urmston, Manchester, M41 8RN
MW3 KML K. Iball, Highcroft, 18 Tan Y Coed, Mold, CH7 6TU
M3 KMN K. Macnauton, 27A Lincoln Road, Poole, BH12 2HT
M3 KMO K. Owen, 10 Pitcher Lane, Leek, ST13 5DB
M3 KMS K. Stanley, 3 Hale Way, Colchester, CO4 5BD
M3 KMT T. Ramsden, 37 Hyde Abbey Road, Winchester, SO23 7DA
MW3 KMU G. Cattle, 12 Claerwen Gelligaer, Hengoed, CF828EW
M3 KMW K. Ward, 127 Lower Lime Road, Oldham, OL8 3NP
MM3 KMX S. Mclachlan, 531 Blair Avenue, Glenrothes, KY7 4RF
MW3 KNE A. Mctaggart, Brick Hall, Hundleton, Pembroke, SA71 5QX
M3 KNF R. Curno, 19 Beckwith Road, Yarm, TS15 9TG
M3 KNK H. Arrowsmith, 15 Hermitage Close, Frimley, Camberley, GU16 8LP
MW3 KNR R. Seal, 5 Millfield, Lisvane, Cardiff, CF14 0RW
M3 KNT K. Royce, 11 Church Lane, Stibbington, Peterborough, PE8 6LP
M3 KNV A. Lebaldi, 11 Artle Place, Lancaster, LA1 2QP
MM3 KNY K. Brown, 21 Strain Crescent, Airdrie, ML6 9ND
M3 KOA J. Beecroft, 48 Manor Road, Alton, GU34 2PB
M3 KOF R. Johnson, 30 Avenue Road, Coalville, LE67 3PB
M3 KOJ K. Lowe, 7 King Street, Creswell, Worksop, S80 4ER
M3 KOR H. Ross, 9 First Avenue, Edwinstowe, Mansfield, NG21 9NZ
M3 KOU A. Rose, 133 Petersmith Drive, New Ollerton, Newark, NG22 9SG
M3 KPB K. Bromley, 40 Winfrith Road, Fearnhead, Warrington, WA2 0QE
M3 KPF J. Simmons, 246 Ruskin Road, Crewe, CW2 7JY
M3 KPG K. Stretton, 6 Highfields, Hilltop Drive, Rye, TN31 7HT
M3 KPL J. Kelly, 1 Bramble Close, New Ollerton, Newark, NG22 9TN
M3 KPO S. Warren, 7 Crich Way, Newhall, Swadlincote, DE11 0UU
M3 KPQ A. Jermyn, 3 Tudor Walk, Carlton Colville, Lowestoft, NR33 8NE
M3 KPU A. Parkes, 59 Wellington Gardens, Battle, TN33 0HD
M3 KPZ T. Wood, 33 Somerdale Avenue, Bristol, BS4 2XN
M3 KQB P. Broughton, 23 Ivy Place, Tantobie, Stanley, DH9 9PT
M3 KQC S. Chandler, 4 Gladstone House, Horton Crescent, Epsom, KT19 8BW
M3 KQD B. Minks, 132 North Road, Clowne, Chesterfield, S43 4PF
M3 KQF P. Woodcock, 27, Knighton, Stafford, ST20 0QH
MM3 KQI R. Houston, 19 Deansloch Place, Aberdeen, AB16 5SB
M3 KQP A. Eastwell, 11 Middlebere Drive, Wareham, BH20 4SD
M3 KQR B. Hall, 126 Eton Road, Burton-on-Trent, DE14 2SN
M3 KQS N. Ralph, 24 Back Street, Laxton, Goole, DN14 7TP
M3 KQT D. Berry, Flat5, 106 Braybrooke Road, Hastings, TN34 1TG
M3 KQW H. Malpas, 148 Queen Street, Crewe, CW1 4AU
M3 KQY R. Brisley, 15 Elm Fields, Old Romney, Romney Marsh, TN29 9SN
M3 KRB L. Kirby, 76 Burton Road, Overseal, Swadlincote, DE12 6JJ
M3 KRD K. Dukes, 127 Carlton Road, Boston, PE21 8LL
M3 KRE R. Jacobs, 35 Edgar Road, Canterbury, CT1 1NR
MI3 KRL K. Mccrystal, 85 Tamlaght Road, Omagh, BT78 5BB
M3 KRM S. Whitlock, Railway Crossing Cottage, Ash Road, Sandwich, CT13 9JB
MW3 KRN C. Richards, 2 Castle Lodge Crescent, Caldicot, NP26 4JL
M3 KRO G. Ticehurst, 118 Old Roman Bank, Terrington St. Clement, King's Lynn, PE34 4JP
M3 KRP K. Taylor, 3 The Drive, Lichfield, WS14 9QT
M3 KRQ N. Gadalla, 26 South Parade, Boston, PE21 7PN
MM3 KRR D. Corbett, 1F1, 13 Learmonth Place, Edinburgh, EH4 1AX
M3 KRS C. Cheverall, 3 Egerton Road, Room 3, Egerton House, Bexhill-on-Sea, TN39 3HH

M3 KRX D. Shires, 1 West Close, High Coniscliffe, Darlington, DL2 2LN
M3 KRY R. Dyson, 4 Royston Lane, Royston, Barnsley, S71 4NL
M3 KRZ B. Renowden, Hilrowenick, Polwithen Drive, St Ives, TR26 2SP
MW3 KSE N. Lane, 13 Traston Road, Newport, NP19 4RQ
M3 KSG K. Gordon, 308 Claremont Road, Swanley, BR8 7QZ
M3 KSH A. Wilkins, 2 Beechfield Crescent, Banbury, OX16 9AR
MW3 KSI M. Giudice, 31 Woodfield Cross, Tredegar, NP22 4JG
M3 KSK M. Sheppard, 107 Queen St., Swinton, Mexborough, S64 8NF
MD3 KSN F. Kelly, 5 Maynrys Castletown, Isle of Man, IM9 1HP
M3 KSP S. Eldridge, 20 Edendale Road, Melton Mowbray, LE13 0EW
M3 KSS A. Eades, Violet Bank, 18 Hillside Road, Leigh-on-Sea, SS9 2DT
MM3 KSV P. Mccluskey, 119 Tower Drive, Gourock, PA19 1SG
M3 KTA R. Baines, 34 Bury Road, Stapleford, Cambridge, CB22 5BP
M3 KTD K. Davidson, 5 Hanover Parc, Indian Queens, St Columb, TR9 6ER
M3 KTH K. Howard, 5 St Nicholas Street, Dereham, NR19 2BS
M3 KTT B. Gardner, 38 Apley Rd, Stourbridge, Dy84pa
M3 KTV B. Bean, 46 Grand Drive, Herne Bay, CT6 8JS
MM3 KUB C. Barber11, Quinish House, Dervaig, Tobermory, Isle of Mull, PA75 6QL
M3 KUE B. Hall, 65 Cavendish Road, Worksop, S80 2ST
M3 KUG M. Amos, 233B Abington Avenue, Northampton, NN1 4PU
M3 KUH A. Grannon, The Chestnuts, Church Lane, Hull, HU11 4PR
M3 KUJ D. Russell, 72 Langholm Drive, Cannock, WS12 2EZ
M3 KUK J. Jones, 15 Kinnaird Road, Sheffield, S5 0NN
M3 KUM C. Sellors, 73 Wilkinson Drive Kesgrave, Ipswich, IP5 2DS
M3 KUN J. Wilson, 20 Elgitha Drive, Thurcroft, Rotherham, S66 9PD
M3 KUO D. Holden, 2 Beacon Road, Bickershaw, Wigan, WN2 4AF
M3 KUQ T. Bush, 19 Spring Vale, Waterlooville, PO8 9DA
M3 KUS P. Connelly, 3 Finch Close, Weston-Super-Mare, BS22 8XS
MM3 KUU G. White, 119 Waggon Road, Brightons, Falkirk, FK2 0EJ
M3 KUV P. Baines, 4 Oxford Crescent, Hetton-Le-Hole, Houghton le Spring, DH5 9LJ
M3 KUY S. Church, The Willows, Warboys Road, Huntingdon, PE28 3AH
M3 KUZ K. Mckeown, 27 Lusty Glaze Road, Newquay, TR7 3AE
M3 KVC A. Stacey, 311 Hyde End Road, Spencers Wood, Reading, RG7 1DD
M3 KVD D. Speed, 137 Church Road North, Skegness, PE25 2QQ
M3 KVG D. Hannon, 20 High Street, Whittlebury, Towcester, NN12 8XJ
M3 KVH K. Harrison, 55 Hudson Close, Worcester, WR2 4DP
M3 KVI D. Swann, 37 Burgh Road, Skegness, PE25 2RA
M3 KVJ K. King, 16 Clare Way, Bexleyheath, DA7 5JU
M3 KVK J. Mckeown, 27 Lusty Glaze Road, Newquay, TR7 3AE
M3 KVL K. Martin, 19 Comrie Crescent, Burnley, BB11 5HX
MM3 KVN K. Clark, 38 Dunsinane Drive, Perth, PH1 2DU
M3 KVR K. Robertson, 189 Harrowby St., Farnworth, Bolton, BL4 7DF
M3 KVU E. Smith, 98 Chapel Fields Charterhouse Road, Godalming, GU7 2AA
MM3 KVV W. Morrison, 7 Knowehead Crescent, Kirriemuir, DD8 5ab
M3 KVW M. Keelan, 16 North Drive, Harwell, Didcot, OX11 0PE
MM3 KVX G. Marshall, West Inch Farm, Kinnordy, Kirriemuir, DD8 5ET
MM3 KVY M. Mcconnell, 6 Langlaw Road, Mayfield, Dalkeith, EH22 5AX
M3 KWF S. Mainzer, Lillypool House, Waldersea, Wisbech, PE14 0NR
M3 KWR M. Sutty, 14 Sedgwick Street, Cambridge, CB1 3AJ
M3 KWS S. Dunn, 64 Stucley Road, Bideford, EX39 3EQ
M3 KWZ D. Leese, 41 Woolston Avenue, Congleton, CW12 3DZ
M3 KXB R. Walsh, 8 Eyton Place, Dawley, Telford, TF4 2DL
M3 KXD C. Collins, 2 Kew Crescent, Sheffield, S12 3LP
M3 KXE P. Beresford, 23 High Lowe Avenue, Congleton, CW12 2EP
M3 KXF J. Gregory, 9 Longfields Crescent, Hoyland, Barnsley, S74 9HZ
M3 KXG R. Robinson, 18 O'Connell Road, Liverpool, L3 6JF
M3 KXI C. Day, 4 Marlborough Way, Market Harborough, LE16 7LW
M3 KXS E. Booth, 18 Maple Road, Kiveton Park, Sheffield, S26 5PH
M3 KXV D. Hollinrake, 4 Sandwood Avenue, Broughton, Chester, CH4 0RJ
M3 KXY D. Preston, Home View, Paradise Lane, Reading, RG7 6NU
M3 KXZ P. Millis, 26 Chalkland Rise, Brighton, BN2 6RH
M3 KYG S. Humphreys, Flat 1, Winslow Court 100 Fordwych Road, London, NW2 3NN
M3 KYH J. List, 41 Westbury Crescent, Dover, CT17 9QQ
M3 KYK J. Hall, 1 Nash Close, Earley, Reading, RG6 5SL
MM3 KYO K. Rafferty, 13 Robin Crescent, Buckhaven, Leven, KY8 1EZ
M3 KYQ K. Rennison, 49 Syston Avenue, St Helens, WA11 9JJ
M3 KYV S. Littlewood, Townside Lodge, Townside, Immingham, DN40 3PS
M3 KYZ D. Whitelock, 22 Anne Crescent, Waterlooville, PO7 7NA
M3 KZB C. Cascarino, 87 Esther Grove, Wakefield, WF2 8EX
M3 KZC M. Clack, 42A Provost Street, Fordingbridge, SP6 1AY
MM3 KZD S. Corstorphine, 33 Springfield, West Barns, Dunbar, EH42 1UF
M3 KZI A. Pateman, 37 Hemans Road, Daventry, NN11 9AL
M3 KZJ P. Lamb, 13 Pool End, St Helens, WA9 3RE
M3 KZP K. Page, 62 Farndon Avenue, Sutton Manor, St Helens, WA9 4DN
M3 KZR A. Bailey, 9 Park View, Abram, Wigan, WN2 5QR
M3 KZS P. Allen, 4 The Links, Northam, Bideford, EX39 1LS
M3 KZT D. Baugh, 36 Well Hay Close, Plymouth, PL9 8DT
M3 KZV P. Camplin, 16 Green Street, Hoyland, Barnsley, S74 9RF
M3 KZW S. Grainger, 15 Carr House Lane, Wirral, CH46 6EN
M3 LAG L. Gething, 106 Westgate, Elland, HX5 0BB
M3 LAJ L. Jarman, 53 Enderby Crescent, Gainsborough, DN21 1XQ
M3 LAP A. Clarke, 57 Welland Avenue, Grimsby, DN34 5JP
M3 LAQ T. Mynors, 6 Walcott Avenue, Christchurch, BH23 2NG
M3 LBC R. Kenton, 65 Warren Drive, Broughton, Chester, CH4 0PU
M3 LBD T. Donegan, 19 Teesgate, Thornaby, Stockton-on-Tees, TS17 9AN
M3 LBG D. Davis, 36 Arbour Street, Southport, PR8 6SQ
M3 LBJ I. Blundell, 43 Ponsonby Place, London, SW1P 4PS
M3 LBK L. Karthauser, 17 Manor Close Abbotts Ann, Andover, SP11 7BJ
M3 LBM N. Waller, 16 Rother Croft, New Tupton, Chesterfield, S42 6BE
M3 LBN G. Lowe, 8 Markland Crescent, Clowne, Chesterfield, S43 4NG
M3 LBP D. Clarke, 30 Chelford Road, Bromley, BR1 5QT
M3 LBQ M. Bradshaw, 342 Manchester Road, Blackrod, Bolton, BL6 5BG
M3 LBR S. Cross, 31 Parkfields, Abram, Wigan, WN2 5XR
M3 LBT R. Carter, 43 Sheldon Avenue, Standish, Wigan, WN6 0LW
MW3 LBX D. Roberts, 118 Ffordd Ddyfrdwy, Mostyn, Holywell, CH8 9PQ
M3 LBY P. Wallstone, 3 Wilson Street, Mansfield, NG19 7JW
M3 LBZ J. Fletcher, Paradise Barn, Bounds Lane, Chard, TA20 2TJ
MM3 LCC M. Pentland, Castleview, Salterhill, Elgin, IV30 5PT
M3 LCE R. Armstrong, 26 Lancaster Road, Carnforth, LA5 9LD
M3 LCF D. Lovell, 109 Aylesbury Crescent, Plymouth, PL5 4HX
M3 LCI M. Davidson, 19 Mason Street, Workington, CA14 3EH
M3 LCL C. Lewis, 19 Elgar Close, Great Sutton, Ellesmere Port, CH65 7AZ
M3 LCP L. Pentney, 4 Caley Road, Tunbridge Wells, TN2 3BL
M3 LCS M. Croxford Simmons, 37 Queens Road, Askern, Doncaster, DN6 0LU
M3 LCU P. Loose, Tae Ping, Main Road, King's Lynn, PE31 8BP
M3 LCZ R. Humpage, 10 Whalley Road, Lancaster, LA1 2HA
M3 LDC L. Cattermole, Blaxhall Hall Crossing, Little Glemham, Woodbridge, IP13 0BP
M3 LDE L. Davis, Romano House, Gorefield Road, Wisbech, PE13 5AS
M3 LDF S. Brown, 21 Woborrow Road, Heysham, Morecambe, LA3 2PW
M3 LDH L. Holmes, 48 Woodpecker Close, Wirral, CH49 4QP
M3 LDI J. Bircumshaw, 39 Woodthorpe Lane, Sandal, Wakefield, WF2 6JG
M3 LDJ J. Harvey, 125 North Road, Clowne, Chesterfield, S43 4PQ
MM3 LDK T. Mcfarlane, 39 Beechwood Road, Tarbolton, Mauchline, KA5 5RP
M3 LDL J. Parker, 1 Schoose Caravan Park, Workington, CA14 4JA
M3 LDM T. Craddock, 12 Wold Road, Burton Latimer, Kettering, NN15 5PN
MI3 LDO P. Logan, 18 Castle Lane, Lisnaskea, Enniskillen, BT92 0FW
M3 LDQ L. Paddon, 21 Oak Park Drive, Havant, PO9 2XE
MM3 LDR A. Sloan, 36 Paterson Avenue, Irvine, KA12 9JJ
M3 LDS R. Walker, 17 Brookside Paulton, Bristol, BS39 7NL
M3 LDT R. Taylor, 7 Wall Street Blackpool, Blackpool, FY1 2EG
MW3 LDY N. Cole, Tycoch, Llandovery, SA20 0UP
M3 LEB S. Bell, 221 Horninglow Road, Sheffield, S5 6SG
M3 LEF J. Peace, 237A Mapperley Plains, Nottingham, NG3 5RG
MD3 LEG H. Leslie, 2 Close Lhergy, Union Mills Braddan, Douglas, IM4 4LU
M3 LEK E. Little, 7 Deerfern Close, Great Linford, Milton Keynes, MK14 5BZ
M3 LEL L. Sargeant, 99 Pot Kiln Road, Great Cornard, Sudbury, CO10 0DX
M3 LEN L. Brackstone, 276 Ladyshot, Harlow, CM20 3EY
M3 LEO L. Flynn, 2 Trafalgar Avenue, Grimsby, DN34 5RE
MD3 LEP A. Le Prevost, 58 Meadow Crescent, Douglas, Isle of Man, IM2 1QX
MW3 LEW L. Jenkins, 12 The Dell, Bryncethin, Bridgend, CF32 9BJ
M3 LEX A. Green, Croft House, Welbeck Road, Chesterfield, S44 6DH
M3 LFC D. Hughes, 86 Colinmander Gardens, Ormskirk, L39 4TF
MI3 LFE A. Boylan, 36 Callan Bridge Park, Armagh, BT60 4BU
M3 LFG L. Gill, 2 Loxton Court, Mickleover, Derby, DE3 0PH
M3 LFH S. Gregory, 1 St Martin Street, Atherton, Manchester, M29 9DN
MM3 LFI D. Pomphrey, Flat 2/9, 109 Bell Street, Glasgow, G4 0TQ
MW3 LFL K. Morgan, Gwel-Yr-Afon, Penrhyncoch Road, Aberystwyth, SY23 3EA
M3 LFO S. Yeldham, 19 Wade Reach, Walton on the Naze, CO14 8RG
M3 LFP R. Brown, 9 Bayleaf Crescent, Oakwood, Derby, DE21 2UG
M3 LFQ W. Mcgill, 49 Anthony Close, Colchester, CO4 0LD
M3 LFU A. Heyes, 528 Manchester Road, Paddington, Warrington, WA1 3TZ
M3 LFV D. Mear, 10 Peters Court, Hatton, Derby, DE65 5JG
M3 LFX A. Lamb, 9 Budworth Avenue Seaton Sluice, Whitley Bay, NE26 4DB
M3 LFZ M. Haseldine, 59 Brackley Road, Bedford, mk42 9sh
M3 LGF D. Sporton, 40 Eastwood Park Drive, Hasland, Chesterfield, S41 0BD
M3 LGH S. Hallam, 18A Market Street, Hoylake, Wirral, CH47 2AE
M3 LGI I. Scott, 13 Fairmount Road, Bexhill-on-Sea, TN40 2HN
M3 LGJ P. Bishop, 2 Hall Farm Bungalows, Illington, Thetford, IP24 1RR
MI3 LGL L. Logue, 21 Moyagh Road, Cullion, Londonderry, BT47 2SL
M3 LGM M. Steeples, 17 Windsor Avenue, Thurlstone, Sheffield, S36 9RX
MW3 LGS S. Lewis, 11 Treseder Way, Cardiff, CF5 5NW
MM3 LGU R. Pennykid, 50 Queen Street, Edinburgh, EH2 3NS
M3 LGX G. Cash, 28 Bramblewood Close, Prenton, CH43 9YT
M3 LGY M. Cash, 28 Bramblewood Close, Prenton, CH43 9YT
MW3 LHA L. Hailstone, 1 Hornbeam Close, Cimla, Neath, SA11 3XA
M3 LHE J. Gammer, 12 West Rise, Tonbridge, TN9 2PG
M3 LHF D. Dunne, 1 Burton Gardens, Brierfield, Nelson, BB9 5DR
M3 LHG G. Gammer, 12 West Rise, Tonbridge, TN9 2PG
M3 LHI Z. Dunne, 1 Burton Gardens, Brierfield, Nelson, BB9 5DR
M3 LHM L. Marshall, Thistledome, First Avenue, Watford, WD25 9PS
M3 LHQ K. Brice, 10A Nelson Drive, Exmouth, EX8 2PU
M3 LHU C. Jenkins, 5 Marlborough Close, Eastbourne, BN23 8AN
M3 LHW A. England, Beech House, Vicarage Gardens, Bradford, BD11 2EF
M3 LHX A. Woodsford, 5 Eliot Close, Wickford, SS12 9ED
M3 LHZ S. Yates, 14 Rushden Road, Sandon, Buntingford, SG9 0QR
M3 LIB J. Bradburn, 4 Lathkil Grove, Buxton, SK17 7PH
M3 LIN L. Chesters, 8 Conway Grove, Blacon, Chester, CH1 5RU
M3 LIU I. Cartmell, 10 Derwent Avenue, Burnley, BB10 1HZ
M3 LIV Z. Bayliss, 16 Oakmere Close, Sandbach, CW11 1WN
M3 LIW J. Brown, 8 Chatsworth Street, Sutton-in-Ashfield, NG17 4GG
M3 LIX M. Knell, 8 Wimborne Gardens, Kirby Cross, Frinton-on-Sea, CO13 0TH
M3 LIY S. Mason111, 12 Blue Cedar Drive, Streetly, Sutton Coldfield, B74 2AE
M3 LJA L. Abel, 121 Angela Road, Horsford, Norwich, NR10 3HF
M3 LJB L. Bazley, 18 Wellington Street United Kingdom, Radcliffe, M26 2RB
M3 LJF L. Gudgeon, Shillingsworth Cottage, Leckhampstead Road, Milton Keynes, MK19 6BY
M3 LJI G. Billington, 47 Smithy Leisure Park, Cabus Nook Lane, Preston, PR3 1AA
M3 LJJ J. Lightfoot, Apple Tree Cottage, Flowers Hill, Reading, RG8 7BD
M3 LJK A. Smith, 46 Mulberry Close, Goldthorpe, Rotherham, S63 9LB
MI3 LJQ W. Phair, 34 Sketrick Island Park, Newtownards, BT23 7BN
MD3 LJS J. Keig, 60 Garth Avenue, Surby, Port Erin, Isle Of Man, IM9 6QZ
M3 LJX G. Jones, 7 St. Ives Road, Weston-Super-Mare, BS23 3XX
M3 LJZ L. Jennings, 29 Mountbatten Drive, Newport, PO30 5SJ
M3 LKD J. Dixon, 23 Dee Way, Winsford, CW7 3JB
M3 LKE E. Smith, 30 Teignmouth Road, Torquay, TQ1 4EA
M3 LKJ P. Manning, 1 Waverley Gardens, Ash Vale, Aldershot, GU12 5JP
M3 LKO K. Cope, 64 Queen St., Pensnett, Brierley Hill, DY5 4HA
M3 LKU A. Head, 34 Balds Lane, Stourbridge, DY9 8SG
MM3 LKV C. Duncan, 131 Croftend Avenue, Glasgow, G44 5PF
M3 LKY S. Smith, 10 Parkway South, Doncaster, DN2 4JS
M3 LLB L. D'Aubay-Butler, Copse Edge 16 Francis Road, Frodsham, WA6 7JR
M3 LLC L. Steele, 14 Rowley View, West Bromwich, B70 8QR
M3 LLK A. Soper, 16 Queen Elizabeth Drive, Crediton, EX17 2EJ
M3 LLM J. Proudman, 61 Iffley Road, Oxford, OX4 1EB
M3 LLN A. Sibley, 27 Sherwood Road, Tetbury, GL8 8BU

M3 LLQ J. Beards, 175 Blackhalve Lane, Wolverhampton, WV11 1AH
M3 LLT C. Moseley, 15 Holden Crescent, Walsall, WS3 1PY
MM3 LLU B. Gaudie, Sunnyside, Harray, Orkney, KW17 2JS
MW3 LLV H. Leonard, 11 Newton Road Grangetown, Cardiff, CF11 8AJ
M3 LLX J. Wright, 2 Regent Road, Church, Accrington, BB5 4AR
M3 LLZ P. Davies, 53 Lammas Road, Cheddington, Leighton Buzzard, LU7 0RY
M3 LMA L. Adkins, 4 Orion Close, Ward End, Birmingham, B8 2AU
M3 LMB P. Breslin, 8A Mountfield, Tennyson Road, Yarmouth, PO41 0PS
M3 LMC L. Mclaughlin, 34 Cambridge Road, Birstall, Batley, WF17 9JF
M3 LMD B. Dake, 100 Lodge Road, West Bromwich, B70 8PL
M3 LME G. Beardmore, 9 Ashmore Drive, Gnosall, Stafford, ST20 0RP
M3 LMH L. Holme, 11 Oxlea Grove, Westhoughton, Bolton, BL5 2AF
M3 LML M. Carney, 2 Lilac Meadows, Lawley Village, Telford, TF4 2NX
M3 LMQ R. Hines, Flat 1, 37 Wellesley Road, Great Yarmouth, NR30 1EU
MI3 LMR R. Spence, 299 Moyarget Road, Mosside, Ballymoney, BT53 8DL
MW3 LMU E. Meek, 7 High Tree Rise, Oakdale, Blackwood, NP12 0DP
MW3 LMV A. Harris, Flat 11, Tyn-Y-Coed, Cwmbran, NP44 4PQ
M3 LMY F. Rowlands, 45 Union Street, Market Rasen, LN8 3AA
MW3 LMZ M. Williams, 30 Elm Drive, Risca, Newport, NP11 6HJ
MI3 LNC H. Davis, 29 Fir Park, Broughshane, Ballymena, BT42 4DH
MM3 LNF I. Mcgurk, 16 Smith Crescent, Alexandria, G83 8BL
M3 LNJ R. Woolridge, 8 Alastair Drive, Yeovil, BA21 3BT
M3 LNM S. Lawton, 4 Astland Gardens, Tarleton, Preston, PR4 6SX
M3 LNN J. Tomlinson, 27 Elm Road, Seaforth, Liverpool, L21 1BJ
M3 LNQ C. Woodruff, 14 Primatt Crescent, Shenley Church End, Milton Keynes, MK5 6AS
MM3 LNT M. Paterson, 20 Loch Street, Rosehearty, Fraserburgh, AB43 7JT
M3 LNU M. Durban, 62 Westfield Way, Charlton, Wantage, OX12 7EP
M3 LNY L. Goff, 27 Harley Road, Oxford, OX2 0HS
M3 LOA R. Loader, Sunnyside, Main Street, Leyburn, DL8 4LU
M3 LOE D. Gosling, 19 Alcester Close, Plymouth, PL2 1EA
MM3 LOF H. Paterson, 20 Loch Street, Rosehearty, Fraserburgh, AB43 7JT
MW3 LOI L. Pring, 42 The Links, Trevethin, Pontypool, NP4 8DQ
M3 LOT D. Filby, 14 Jeffcut Road, Chelmsford, CM2 6XN
M3 LOX D. Loxley, 33 Longwood Road, Tingley, Wakefield, WF3 1UG
M3 LOY B. Taylor, 11 Halyard Croft, Hull, HU1 2EP
M3 LPE J. Wright, 26 Walmsley Close Church, Accrington, BB5 4HL
M3 LPF S. Hemmings, Leylands, Leigh Road, Frome, BA11 3LR
MI3 LPH L. Hesketh, 61 Elmwood Cottages, Newtownabbey, BT36 5WQ
M3 LPI R. Brierley, 26 Jacobsen Avenue, Hyde, SK14 4DW
M3 LPJ L. Renmans, 70 Burman Road, Liverpool, L19 6PW
M3 LPK G. Brierley, 26 Jacobsen Avenue, Hyde, SK14 4DW
M3 LPN B. Kersey, 61 Crown Road, Portslade, Brighton, BN41 1SJ
M3 LPR J. Main, 15 Byron Road, Lydiate, Liverpool, L31 0DB
M3 LPT L. Towler, 8 Stowehill Road, Peterborough, PE4 7PY
M3 LPU P. Higgins, 9 Claremont Grove, Exmouth, EX8 2JW
MD3 LPW L. Wernham, Rogane Cottage, Church Lane, Santon, Isle Of Man, IM4 1EZ
M3 LQA P. Ellis, 40 Grasmere Road Royton, Oldham, OL2 6SR
M3 LQB A. Foulds, 4 Kropacz Court, South Street, Doncaster, DN6 7JL
M3 LQC J. Bhart, 47 Fitzroy Avenue, Broadstairs, CT10 3LS
M3 LQD S. Moakes, 46 Parsonage St., Stockport, SK4 1HZ
M3 LQE M. Clarke, 48 Delves Wood Road, Huddersfield, HD4 7AS
M3 LQI S. Briggs, 56 Broadfields Road, Exeter, EX2 5RG
M3 LQJ P. Langford, 24 Asotte Way, Southam, CV47 1GH
MM3 LQK W. Caithness, 36 Wards Drive, Muir of Ord, IV6 7PX
MI3 LQN H. Mcerlean, 47 Barrack Road, Magherafelt, BT45 6ly
M3 LQO R. Claydon, 3 Birch Trees, Ambleside Road, Windermere, LA23 1EU
M3 LQP D. Brown, 21 Woborrow Road, Heysham, Morecambe, LA3 2PW
M3 LQV P. Newton, 61 Ashbourne Crescent, Taunton, TA1 2RA
M3 LQW R. Edwards, 46 Lavers Oak, Martock, TA12 6HG
M3 LQX M. Hall, 2 Hallcroft Road, Haxey, Doncaster, DN9 2HP
M3 LQY R. Vigors, 12 Sandfield Park, Lichfield Road, Brownhills, WS8 6LN
M3 LRF P. Callaghan, 41 Higher Ash Road, Talke, Stoke-on-Trent, ST7 1JN
M3 LRJ A. Smith, 101 Chaucer Drive, Lincoln, LN2 4LT
M3 LRK M. Davey, 67 Rotherham Baulk, Carlton-In-Lindrick, Worksop, S81 9LE
M3 LRN A. Page, 148 Waleton Acres, Carew Road, Wallington, SM6 8PY
M3 LRP R. Plater, Garsides, Keeling Street, Louth, LN11 7QU
MI3 LRR W. Turtle, 35 Buckna Road, Broughshane, Ballymena, BT42 4NJ
M3 LRU J. Fitzpatrick, 29 Delmar Road, Knutsford, WA16 8BG
M3 LRW L. Wilson, The Rectory, Church Close, Thetford, IP25 7LX
M3 LRX D. Horwood, 42 Southlands Drive, Timsbury, Bath, BA2 0HB
M3 LRZ L. Reynolds, 12 Providence Crescent, Boundary Way, Hull, HU4 6EF
M3 LSE M. Newton, 20 Orchard Way, Timberscombe, TA24 7UL
M3 LSF L. Caslin, 30D Holmewood, Holme, Peterborough, PE7 3PG
MW3 LSG C. Goodridge, 94 High Street, Nantyffyllon, Maesteg, CF34 0BP
M3 LSK T. Newton, 43 Hayfield Road, Minehead, TA24 6AD
M3 LSL K. Summers, 11 Jefferson Court, Franklin Street, Hull, HU9 1JB
MM3 LSO S. Arnott, 2 Nisbet Avenue, Eyemouth, TD14 5BF
M3 LSS L. Stephenson, 91 Hoyland Road, Hoyland Common, Barnsley, S74 0AP
M3 LSU J. Evans, 7 Westland Drive, Hayes, Bromley, BR2 7HE
M3 LSX A. Dale, 37 Bussey Road, Norwich, NR6 6JF
M3 LTA L. Talbot, 26 Chevalier Grove, Crownhill, Milton Keynes, MK8 0EJ
M3 LTG L. Gear, 26 Woodlands Way, Denaby Main, Doncaster, DN12 4LR
M3 LTH M. Preston, 53 Links Road, Birmingham, B14 4TW
M3 LTP P. Whittall, 165A High St., Brierley Hill, DY5 3BU
M3 LTR D. Ingham, 19 Recreation Avenue, Ashton-In-Makerfield, Wigan, WN4 8SU
M3 LTT A. Old, 51 Villiers Close, Plymouth, PL9 7QP
MW3 LTU M. Brady, Ty Mawr Uchaf, Dulas, LL70 9DQ
M3 LTV A. Walker, 76 Greenway, Birmingham, B20 1EQ
M3 LTW J. Bennett, 2 Victoria St., Pensnett, Brierley Hill, DY5 4LB
M3 LUD P. Ludders, 280 Hopewell Road, Bilton Grange, Hull, HU9 4HH
M3 LUK L. Mandeville, 6 Oxford Road, Benson, Wallingford, OX10 6LX
M3 LUO D. Shoubridge, 51 Pipers Field, Ridgewood, TN225SD
M3 LUP S. Ingram, Flat 1, Philip Howard Court, Glynne Street, Farnworth, Bolton, BL4 7DQ
M3 LUU A. Gillett, 441 Radipole Lane, Weymouth, DT4 0QF
M3 LUW G. Stocks, 62 Ridge Park Avenue, Plymouth, PL4 6QA
M3 LUZ A. Bent, 14 Pleasant Road, Eccles, Manchester, M30 0FS
M3 LVA L. Adlington, 21 Newstead Road, Stoke-on-Trent, ST2 8HU
MW3 LVF R. Hawkins, Nook Cottage, Common-Y-Coed, Caldicot, NP26 3AX
M3 LVK C. Street, Russetts, Isle Brewers, Taunton, TA3 6QN
M3 LVM K. West, 36 Watlington Road, Cowley, Oxford, OX4 6SS
M3 LVP J. Gilbert, 148 Purcell Road, Coventry, CV6 7LB
M3 LVR E. Mills, 76 Main St., Burton Joyce, Nottingham, NG14 5EH
MM3 LVT J. Dock, 75 Ferguslie Park Avenue, Paisley, PA3 1BE
M3 LVX P. Hampton, 4 Moorland View, Plymstock, Plymouth, PL9 8NW
M3 LVY A. Kendrick, 13 Queens Drive, Middlewich, CW10 0DG
MI3 LVZ M. Mccloy, 4 Audleys Park, Newtownards, BT23 8UA
M3 LWG J. Arrowsmith, 45 Hilderic Crescent, Dudley, DY1 2EU
MM3 LWJ R. Bertram, 46 Main Street Main Street, Pathhead, EH37 5QB
MM3 LWO G. Mallolm, 32 Holly Crescent, Dunfermline, KY11 8BT
M3 LWP J. Clowes, 52 Pennine Drive, St Helens, WA9 2BU
MD3 LWQ M. Corlett, 18 Queens Drive, Peel, Isle Of Man, IM5 1BQ
MM3 LWT S. Conway, 26 Rennie St., Kilmarnock, KA1 3AR
MI3 LWU A. Geary, 114 Maddan Road, Armagh, BT60 3LJ
M3 LWV M. Powell, 37 Newnham Close, Mildenhall, Bury St Edmunds, IP28 7PD
M3 LWX W. Alder, 21 Manor Gardens, London, SW20 9AB
MM3 LWZ C. Coore, 14 Craigs Drive, Edinburgh, EH12 8UW
M3 LXA S. Jones, 10 Litchborough Grove, Whiston, Prescot, L35 7NE
M3 LXB J. Wynne, 43 Lansdown Road, Broughton, Chester, CH4 0NZ
MI3 LXE Dr I. Stevenson, 55 Churchtown Road, Downpatrick, BT30 7AZ
M3 LXF S. Mills, 27 Boscow Crescent, St Helens, WA9 3SX
M3 LXH S. Wright, Flat 2, 19 Cearns Road, Prenton, CH43 2JL
MI3 LXJ B. Crozier, 33 Cullentragh Road, Poyntzpass, Newry, BT35 6SD
M3 LXK C. Wynne, 43 Lansdown Road, Broughton, Chester, CH4 0NZ
MI3 LXN D. Bryans, 1 Meadowvale Avenue, Bangor, BT19 1HG
M3 LXP P. Enfield, 1 Horton Park, Blyth, NE24 4JD
M3 LXR L. Dexter, 27 Underwood Avenue, Worsbrough, Barnsley, S70 4AU
M3 LXS M. Woolley, 4 Robert Street, Warrington, WA5 1TQ
M3 LXU T. Moscrop, 64 Gresham Road, Norwich, NR3 2NG
M3 LXV J. Thompson, 51 Grafton Road, Oldbury, B68 8BP
MI3 LXW M. Lewis, 7 Liester Park, Ballyrobert, Ballyclare, BT39 9RZ
MI3 LXZ Dr A. Mcdowell, 10 Lord Wardens Vale, Bangor, BT19 1GH
M3 LYA C. Cooper, The Haven, Ipswich Road, Norwich, NR15 2TA
M3 LYC L. Bentley, 1 Cotswold Road, Lupset, Wakefield, WF2 8EL
M3 LYG R. Dunkley, 25 St. Andrews Crescent, Wellingborough, NN8 2ES
MM3 LYH C. Rodger, 23 Harrysmuir Road, Pumpherston, Livingston, EH53 0NT
M3 LYP I. Hallatt, 11 Cheshire St., Audlem, Crewe, CW3 0AH
MW3 LYQ R. Lovesey, 33 Ty Isaf Park Avenue, Risca, Newport, NP11 6NB
M3 LYR T. Martin, 46 Hayes Crescent, Frodsham, WA6 7PG
MM3 LYS E. Smith, 27 Elm Lane, Foresters Lodge, Glenrothes, KY7 5TD
M3 LYU C. Arner, 6 Welshampton Close, Great Sutton, Ellesmere Port, CH66 2WL
M3 LYV C. Walsh, 133 Belfield Road, Accrington, BB5 2JD
M3 LYX D. Shaw, 37 Smirthwaite View, Normanton, WF6 1AW
M3 LYZ F. Martin, 1 Marsh Street, Strood, Rochester, ME2 4BB
MI3 LZA T. Conway, 203 Garrymore, Moyraverty, Craigavon, BT65 5JF
MW3 LZC T. Rule, 35 Pill St., Penarth, CF64 2JS
MM3 LZD J. Dinning, 1 South Brae Aiket Road, Dunlop, Kilmarnock, KA3 4BP
MI3 LZF J. Steele, 46 Circular Road, Newtownards, BT23 4BN
M3 LZK J. Delves, 14 Stanthorne Avenue, Crewe, CW2 8NH
M3 LZL F. Shields, 4 Occupation Lane, Earlsheaton, Dewsbury, WF12 8PY
M3 LZR N. Finlay, 50 Melchett Crescent, Rudheath, Northwich, CW9 7EP
M3 LZT D. Young, 20 Summerhouse, Tickenham, Clevedon, BS21 6SN
MM3 LZU C. Tait, 39 Baleshrae Crescent, Kilmarnock, KA3 2GN
M3 MAA M. Ahmed, 59 Ramsgate Lofthouse, Wakefield, WF3 3PX
MD3 MAN A. Espey, 9A Hilltop View, Douglas, Isle of Man, IM2 2LA
MM3 MAO M. Overthrow, 63 Primrose Avenue, Larkhall, ML9 1JX
M3 MAR M. Jeffery, 14 Holly Mount Shavington, Crewe, CW2 5AZ
M3 MBC M. Bridgeland, 17 Oldfield Lane, Wisbech, PE13 2RJ
M3 MBF A. Fryer, 9 The Oval, Guildford, GU2 7TS
M3 MBH D. Hemmings, 1 Sunray Grove, Hucknall, Nottingham, NG15 6RF
M3 MBI S. Stuart, 46 Breach Road, Heanor, DE75 7NJ
MI3 MBM M. Buchanan, 49 Glengiven Avenue, Limavady, BT49 0RW
M3 MBQ M. Daniells, 6 Carlton Road, Manchester, M16 8BB
M3 MBR M. Roberts, 13 St. Michaels Court, Stevenage, SG1 5TB
M3 MBV H. Su, 135 Devana Road, Leicester, LE2 1PN
M3 MBZ M. Stephens, 45 Ham Farm Lane, Emersons Green, Bristol, BS16 7BW
M3 MCA A. Matthews, 70 Branksome Hall Drive, Darlington, DL3 9SR
MD3 MCB B. Perrin, 18 Bellevue Park, Peel, Isle Of Man, IM5 1UF
M3 MCF M. Frame, 23 Greenside Court, Sunderland, SR3 4HS
MM3 MCG M. Gaston, Ellena, Lochans Mill Avenue, Stranraer, DG9 9BZ
M3 MCU J. Mccallum, 48 Heber St., Bristol, BS5 9JT
MM3 MDB M. Brunsdon, 25 Buckstone Lea, Edinburgh, EH10 6XE
M3 MDI J. Gleeson, 19 Alston Avenue, Shaw, Oldham, OL2 7SX
M3 MDK K. Sawyers, 27 Dukeswood Road, Longtown, Carlisle, CA6 5UJ
M3 MDN M. Dobson, 17 Cadley Causeway, Fulwood, Preston, PR2 3RU
M3 MDS A. Flemming, 20 Chatham Hill, Chatham, ME5 7AA
MI3 MDV A. Jess, 25 Gransha Road, Dundonald, Belfast, BT16 2HB
M3 MDW M. Webster, 2 Brook Close, Nottingham, NG6 8NL
M3 MDY M. De Young, 97 Kingfisher Road, Larkfield, Aylesford, ME20 6RE
M3 MEB M. Collins, 73 Westholme Road, Bidford-On-Avon, Alcester, B50 4AN
M3 MEE S. Parker, 100 Horsebridge Hill, Newport, PO30 5TL
M3 MEF M. Fry, 46 Butt Parks, Crediton, EX17 3HE
M3 MEG D. Cash, 3 Marsh Lane, Wolverhampton, WV10 6RU
MM3 MEH M. Mumford, 13 Galloway Drive, Culloden, Inverness, IV2 7ND
M3 MEI G. Childs, 144 Sturdee Avenue, Gillingham, ME7 2HL
M3 MEO D. Blake, 9 Malling Avenue, Eastfield, Scarborough, YO11 3FA
M3 MEP M. Porteious, 17 Church Walk, Yaxley, Peterborough, PE7 3YD
M3 MER R. Fox, 3 Cherry Blossom Close, Harlow, CM17 0EX
M3 MES N. Messenger, 7 Skinners Close, Swordy Park, Alnwick, NE66 1EU
M3 MEU D. Cowman, 23 Kirk Flatt, Great Urswick, Ulverston, LA12 0TB
M3 MEW M. Wright, 9 Dinmore Avenue, Blackpool, FY3 7RR
MW3 MEY G. Alker, Bryn Y Mor, Lon Ganol, Menai Bridge, LL59 5YA
MI3 MFD F. Doherty, 52 Madison Avenue Eglinton, Londonderry, BT47 3PW
M3 MFE P. Penfold, 2 The Leas, Essenden Road, St Leonards-on-Sea, TN38 0PU
M3 MFF M. Frohnsdorff, 75 Alexander Drive, Faversham, ME13 7TA

M3 MFG C. Goulty, Seletar, 22 Western Avenue, Felixstowe, IP11 9TS
MM3 MFN A. Harkess, 7 Gardiner Road, Prestonpans, EH32 9HF
MM3 MFR M. Mcminn, 28 Dunlop Road, Dumfries, DG2 9NN
M3 MFS S. Spencer, 55 Witton Lane, West Bromwich, B71 2AA
M3 MFT A. Hill, 1 Rochester Close, Mountsorrel, Loughborough, LE12 7UH
M3 MFU D. Mutlow, Dunvegan, Wood Lane, Nuneaton, CV13 0AU
M3 MFX C. Richardson, 16 Church Road, Boreham, Chelmsford, CM3 3EF
M3 MFZ J. Gosling, 11 Pinfold Place, Harby, Melton Mowbray, LE144BX
M3 MGD D. Riggs, 37 Moot Gardens, Downton, Salisbury, SP5 3LG
M3 MGI T. Rogers, 18 Field Road, Bridlington, YO16 4AU
M3 MGJ M. Minshull, 12 Dunnett Close, Attleborough, NR17 2NG
MM3 MGK S. Brown, 21 Whiteford Avenue, Dumbarton, G82 3JU
M3 MGL M. Talbot, 26 Chevalier Grove, Crownhill, Milton Keynes, MK8 0EJ
M3 MGN M. Naylor, 96 New Meadows, Rawmarsh, Rotherham, S62 7FE
M3 MGO M. Butler, Wood Green, Astley, Stourport-on-Severn, DY13 0RU
M3 MGP M. Champion, 155 Walton Road, Walton on the Naze, CO14 8NF
M3 MGQ J. Browne, 29 Longbridge Close, Tring, HP23 5HG
M3 MGU R. Gregory, Town End, Kirkby Road, Askam-in-Furness, LA16 7EY
M3 MGZ M. Curtis, 10 Woodstock Gardens, Blackpool, FY4 1JP
M3 MHD M. Downes, 38 Queensway, Warton, Preston, PR4 1XU
MW3 MHG G. Rees, 47 Loftus Street, Cardiff, CF5 1HL
M3 MHL A. Parrish, 5 Kestrel Lane, Cheadle, Stoke-on-Trent, ST10 1RU
M3 MHN M. Hurren, 257 Norwich Road, Wroxham, Norwich, NR12 8SL
M3 MHP E. Skinner, 11 Finch Crescent, Leighton Buzzard, LU7 2PE
MM3 MHQ A. Mccurdy, 5 Kestrel Place, Greenock, PA16 7BL
M3 MHR M. Reavell, 85 Mccarthy Close, Birchwood, Warrington, WA3 6RS
MM3 MHS M. Shearer, 113 Auchamore Road, Dunoon, PA23 7JJ
M3 MHV M. Vaughan, 12 Kingsley Road, Frodsham, WA6 6SG
M3 MHZ D. Lawrence, 23 Parkmead Road, Wyke Regis, Weymouth, DT4 9AL
MM3 MID K. Middleton, 1 Campbell Court, Lochmaben, Lockerbie, DG11 1NF
MI3 MIE Y. Wilson, 59 Crew Road, Upperlands, Maghera, BT46 5TU
M3 MIF J. Tranter, 64 Geneva Drive, Newcastle, ST5 2QH
M3 MIG P. Cattermole, Blaxhall Hall Crossing, Little Glemham, Woodbridge, IP13 0BP
M3 MII G. Elsworth, 367 West Dyke Road, Redcar, TS10 4PS
M3 MIJ J. Dean, 2 Wellington Street, Chesterfield, S43 2BJ
M3 MIN A. Jones, 17 Maybush Drive, Chidham, Chichester, PO18 8SR
M3 MIO A. Cox, 11 Windmill Drive, Audlem, Crewe, CW3 0BE
M3 MIP R. Parrish, 5 Kestrel Lane, Cheadle, Stoke-on-Trent, ST10 1RU
M3 MIQ M. Fradley, 43 Grange Drive, Penketh, Warrington, WA5 2JN
M3 MIR S. Carruthers, 13 Belah Road, Carlisle, CA3 9RE
M3 MIS J. Greatrix, West Cottage, Main Road, Boston, PE20 3PZ
M3 MIU F. Lie, Harrogate Ladies' College, Clarence Drive, Harrogate, HG1 2QG
M3 MIV T. Skinner, Flat 4, 25-27 Bridge Street, Leighton Buzzard, LU7 1AH
M3 MIX M. Cole, 9 Troopers Drive, Romford, RM3 9DE
M3 MJD M. Dennison, 27 Chapel St., Cawston, Norwich, NR10 4BG
M3 MJE M. Edwards, 3 George Street, Bourne, PE10 9HE
M3 MJH M. Hickford, 3 Ashen Road, Clare, Sudbury, CO10 8LQ
MI3 MJI R. Wylie, 69 Rubane Road, Kircubbin, Newtownards, BT22 1AU
M3 MJJ J. Miller, Flat 1 Block 2, St. Phillips Place, Eastbourne, BN22 8LW
M3 MJL M. Lee, Up To Date House, Shore Road, Boston, PE22 0NA
M3 MJM M. Marter, 4 Meadow Way, Seaford, BN25 4QT
M3 MJN M. Noon, 97 Cherrycroft, Skelmersdale, WN8 9EF
M3 MJV M. Verrechia, 20 The Wyvern Grafham, Huntingdon, PE28 0GG
M3 MJY M. Kirby, 2 Morton Crescent, Bradwell, Great Yarmouth, NR31 8NT
M3 MKB M. Baxter, 5 Farnborough Street, Farnborough, GU14 8AG
M3 MKD M. Davis, 15 Farmcroft Road Mansfield Woodhouse, Mansfield, NG19 8QU
M3 MKH M. Hall, 10 Darwin Walk, Northampton, NN5 6LR
M3 MKJ M. Allen, 8 Green Close, South Wonston, Winchester, SO21 3EE
M3 MKK M. Kilkenny, 23 Hazelhurst Road, Stalybridge, SK15 1HD
M3 MKM D. Kelly, Clearance Crescent Whitley Bay, Newcastle, Ne262dz
MW3 MKN M. Joseph, 32 Charles Street, Trealaw, Tonypandy, CF40 2UN
M3 MKO J. Duffield, 4 Church Hill, Easingwold, York, YO61 3JS
M3 MKV J. Beech, 124 Belgrave Road, Coventry, CV2 5BH
M3 MKZ J. Swift, 52 North Street, Burwell, Cambridge, CB25 0BB
M3 MLA P. Richardson, 11 Overstone Road Coldham, Wisbech, PE14 0ND
MD3 MLB M. Bazley, 9B Cronk Y Berry View, Douglas, Douglas, Isle Of Man, IM2 6HH
M3 MLF M. Firth, 126 Tombridge Crescent, Kinsley, Pontefract, WF9 5HE
M3 MLG M. Goff, 27 Harley Road, Oxford, OX2 0HS
M3 MLI M. Litt-Wilson, 14 Wastwater Rise, Seascale, CA20 1LB
M3 MLK S. Ellison, 23 Murphy Grove, St Helens, WA9 1QY
MM3 MMB M. Baird, Creag Saval, Lairg, IV27 4ED
MI3 MMC M. Mcclure, 12 St. Patricks Park, Ballymoney, BT53 6JG
M3 MMG M. Glover, 96 Byron St., Macclesfield, SK11 7QA
MM3 MMI F. Millar, 13 Edzell Park, Kirkcaldy, KY2 6YB
MW3 MMJ M. Jones, 47 Maes Derw, Llandudno Junction, LL31 9AN
M3 MML M. Lemer, 1 Holmbush Court, Brent St., London, NW4 2NS
M3 MMN G. Larrigan, 9 Sandpiper Gardens, Chippenham, SN14 6YH
MM3 MMO D. Frost, Flat 6, 88 Albion Street, Glasgow, G1 1NY
M3 MMP P. Evans, 4 Havelock Court, Havelock Street, Aylesbury, HP20 2NU
M3 MMZ M. Morse, 5 Northload Terrace, Glastonbury, BA6 9JW
M3 MND A. Harrop, 35 Langdale Crescent, Dalton-in-Furness, LA15 8NR
MU3 MNG R. Bougourd, 3 Bartholemew, Victoria Road, St Peter Port, Guernsey, GY1 1JB
M3 MNQ B. Gould, 26 Trembel Road, Helston, TR12 7DY
M3 MNR J. Redrup, 58 Shaftesbury Road, Bournemouth, BH8 8ST
M3 MNT T. Williamson, 286 Glynswood, Chard, TA20 1BX
M3 MNU P. Richardson, 14 Portland Street, Worksop, S80 1RZ
MW3 MNV N. Hill, 53 Broadmead, Pontllanfraith, Blackwood, NP12 2NJ
M3 MNY K. King, Chad Lane Farm, Chad Lane, St Albans, AL3 8HW
M3 MOB J. Howarth, 5 Sydenham Building, Bath, BA2 3BS
M3 MOF J. Jones, The Studio, Ferney Hoolet, Hook, RG27 8SW
M3 MOH G. Worrall, 94 Scotia Road, Stoke-on-Trent, ST6 4ET
MW3 MOJ K. Pitt, 21 Maes-Yr-Onen, Nelson, Treharris, CF46 6LF
M3 MOP A. Barden, 38 Silver Close, Tonbridge, TN9 2UY
M3 MOQ S. Hutchinson, 28 Willow Road, New Balderton, Newark, NG24 3DA
MI3 MOT A. Thompson, 23 Causeway End Park, Lisburn, BT28 2HX
M3 MOV M. Greenhow, 39 Boston Avenue, Runcorn, WA7 5XE
M3 MPC M. Coles, 29 Sydney Road, Exeter, EX2 9AH
MM3 MPK M. Kilday, 3 Union Drive, Whitburn, Bathgate, EH47 0AJ
MI3 MPL H. Currie, 58 Duneden Park, Belfast, BT14 7NF
M3 MPM M. Murrell, 11 Ajax Close, Hull, HU9 4BE
M3 MPT P. Denehy, 17 Coverdale, Hull, HU7 4AL
M3 MQA L. Woolley, 4 Robert Street, Warrington, WA5 1TQ
M3 MQB J. Hing, 6 Peartree Walk, Billericay, CM12 0PY
M3 MQC J. Taylor, 6 Hawks Close, Walsall, WS6 7LE
M3 MQH N. Hilton, 20 Darbyshire Close, Deeping St. James, Peterborough, PE6 8SF
M3 MQI A. Paxton, Havencroft, Dalbury Lees, Ashbourne, DE6 5BE
M3 MQJ G. Jackson, Long Pools Farm, Marsh Lane, Market Drayton, TF9 2TG
M3 MQM K. Rooney, Red Cap Farm, Green Fairfield, Derbyshire, SK17 7JF
M3 MQP M. Richards, 57 Coronation Close, Broadstairs, CT10 3DL
M3 MQR D. Rogers, 9 Prospect Place, Stafford, ST17 4HZ
M3 MQX R. Rowe, 16 Orchard Road, Plymouth, PL2 2QY
M3 MQY G. Robinson, Crowstone Mews, Syke House Lane, Greetland, HX4 8PA
M3 MRA N. Rogers, 4 Lawson Court, Millfield Avenue, Market Harborough, LE16 8XR
MW3 MRC M. Clayton, 12 The Broadway, Abergele, LL22 7DF
MI3 MRF T. Mccullough, 23 Edenvale Park, Antrim, BT41 1AY
MI3 MRG A. Colligan, 8 Mourneview Crescent, Lisburn, BT28 3HD
M3 MRJ J. Johnson, 3 Rumbold Road, Hoddesdon, EN11 0LP
MW3 MRK M. Knowles, 72 Uplands Avenue, Connah'S Quay, Deeside, CH5 4LG
MW3 MRL M. Williams, 5A Derllwyn Close, Tondu, Bridgend, CF32 9DH
M3 MRM J. Milner, Stone Cottage, Wistanstow, Craven Arms, SY7 8DG
M3 MRN J. Orange, 20 Borrowdale Avenue, Fleetwood, FY7 7LF
M3 MRO M. Ross, 143 Rose Lane, Romford, RM6 5NR
M3 MRQ N. Thain, 24 Wilmington Road, Hastings, TN34 2BT
M3 MRS L. Henley, 5 Gosselin Street, Whitstable, CT5 4LA
M3 MRU D. Edwards, 10 Queens Avenue, Ilfracombe, EX34 9LN
MM3 MRX G. Robertson, 24 Tippet Knowes Court, Winchburgh, Broxburn, EH52 6UW
M3 MRZ T. Hall, 8 Vicarage Close, Billesdon, Leicester, LE7 9AN
M3 MSB S. Bridge, 59 Alder Hey Road, St Helens, WA10 4DN
M3 MSC L. Nicklin, 59 Laurel Road, Armthorpe, Doncaster, DN3 2ES
M3 MSH M. Hunt, 57 Colsterdale, Worksop, S81 0XH
M3 MSJ M. Stocker, 1 Rectory Close, Carlton, Bedford, MK43 7JT
M3 MSL M. Witter, 62 Old Road, Churwell, Leeds, LS27 7RT
M3 MSN M. Austin, 38 Garden Road, Folkestone, CT19 5RA
M3 MSP M. Skinner, West Cottage, Fen Street, Rattlesden, Bury St Edmunds, IP30 0RW
M3 MSQ E. Bartlett, 86 Usk Road, Tilehurst, Reading, RG30 4HU
M3 MST M. Wilson, 234 Aylsham Drive Ickenham, Uxbridge, UB10 8UF
M3 MSU L. Wardle, 35 Woodland Close, Barnstaple, EX32 0eg
M3 MSX M. Jenkins, 1 Green End Road, Sawtry, Huntingdon, PE28 5UX
M3 MSY M. Spicer, 13 Strawberry Path, Oxford, OX4 6RA
M3 MSZ M. Swift, 8 Grove Lane, Buxton, SK17 9HG
MW3 MTB M. Buxton, 25 Pen Y Bryn, Sychdyn, Mold, CH7 6EE
M3 MTC C. Mathewson, 33 Thornton Road, Bootle, L20 5AN
M3 MTL M. Price, 9 Herbarth Close, Liverpool, L9 1JZ
MM3 MTM M. Mcleary, 146 Captains Road, Edinburgh, EH17 8DX
M3 MTP M. Powell, 2A Park Avenue, Uttoxeter, ST14 7AX
M3 MTQ S. Saunders, 3 The Terrace, High Street, Cavendish, Sudbury, CO10 8AS
M3 MTR M. Bryan, 16 Walesmoor Avenue, Kiveton Park, Sheffield, S26 5RG
M3 MUA J. Anderson, 121 Barton Road, Stretford, Manchester, M32 9AF
M3 MUF N. Wilson, 3 New Road Bovington, Wareham, BH20 6JZ
M3 MUI M. Bromfield, The Cottage, Huttoft, LN13 9RF
M3 MUO P. Riddle, 8 Lower Dingle, Oldham, OL1 4PB
M3 MUP N. Speight, Flat 14, Cranbrook, London, NW1 0LJ
M3 MUQ W. Mitchell, Flat 12, 1 Benwell Road, London, N7 7AY
M3 MUU S. Langton, 51 Fairfield Avenue, Datchet, Slough, SL3 9NF
M3 MUX D. Baseden, 27 Bayfield, Painters Forstal, Faversham, ME13 0EF
M3 MVI M. Pick, 290 Horsley Road, Washington, NE38 8HS
M3 MVJ R. Jefferiss, 21 East Street, Fritwell, OX27 7PX
M3 MVK M. Egerton, 43 New Street, Crewe, CW15PN
M3 MVM J. Goulding, 79 Dalston Drive, Manchester, M20 5LQ
M3 MVN C. Harding, 27 Eston Avenue, Malvern, WR14 2SR
M3 MVO D. Campbell, Beeches, Hammersley Lane, High Wycombe, HP10 8HG
M3 MVR A. Hobbs, 2 The Mead, Beaconsfield, HP9 1AW
MW3 MVT R. Williams, Royston, 18 Grove Street, Maesteg, CF34 0HY
M3 MVV J. Mullarkey, 41 Foyle Avenue, Chaddesden, Derby, DE21 6TZ
MM3 MVY M. Gerrard, 10 Whinhill Gardens, Aberdeen, AB11 7WD
MI3 MWA B. Allen, 48 Kevlin Gardens, Omagh, BT78 1QS
M3 MWG M. Gosling, 1 Zion Street, Plymouth, PL1 2HX
M3 MWM M. Martin, 80 Waveney Road, Hull, HU8 9LY
MW3 MWO A. Rowlands, 8 Cleveland Avenue, Tywyn, LL36 9EG
M3 MWQ B. Hall, 64 Synehurst Crescent, Badsey, Evesham, WR11 7XX
M3 MWT R. Waters, School House, East Moors Road Helmsley, York, YO62 5HJ
M3 MWV M. Williams, 6 Richmond Terrace, Barrow-in-Furness, LA14 5LH
M3 MXA M. Anderson, 5 Saffron Court, Wakefield, WF2 0FQ
MW3 MXC J. Baker, 43 Clyde St., Risca, Newport, NP11 6BP
M3 MXF R. Mason, 27 Meadway, Malvern, WR14 1SB
M3 MXG S. Turner, 28 Fox Lea, Kesgrave, Ipswich, IP5 2YU
M3 MXH M. Price, 43 Heckington Drive, Nottingham, NG8 1LF
M3 MXJ P. Smith, 1 Grappenhall Hall School House, Church Lane, Warrington, WA4 3ES
M3 MXK M. Kendall, 67 Slade Valley Road, Ilfracombe, EX34 8LG
M3 MXM R. Hardy, 35 Chilton Road, Ipswich, IP3 8PD
MM3 MXN S. Reid, 14 St. Marys, Monymusk, Inverurie, AB51 7HH
M3 MXO R. Kemp, 4 Scoones Close, Bapchild, Sittingbourne, ME9 9SW
M3 MXP A. Macgregor, 53 Napier Place, Orton Wistow, Peterborough, PE2 6XN
M3 MXV K. Moulder, 51A Aston Cantlow Road, Wilmcote, Stratford-upon-Avon, CV37 9XN
M3 MXW D. Platt, 50 Poplars Road, Stalybridge, SK15 3EN
M3 MXX T. Chapman, 16 Andover Place, Cannock, WS11 6EH
M3 MXZ M. Rowe, 15 Atlantic Close, Treknow, Tintagel, PL34 0EL
M3 MYE D. Sykes, 2 The Street, Claxton, Norwich, NR14 7AS
M3 MYG K. Toner, 17 Crooked End Place, Ruardean, GL17 9YN
M3 MYI C. Toner, 17 Crooked End Place, Ruardean, GL17 9YN
M3 MYK M. Lees, 28 Pleasant Avenue, Bolsover, Chesterfield, S44 6LL
M3 MYM D. Cox, 3 Besley Court, Lethbridge Road, Wells, BA5 2FE

M3 MYQ R. Jenkins, Glascoed, Garthmyl, Montgomery, SY15 6RT
M3 MYT N. Groat, 138 Freedom Road, Sheffield, S6 2XE
M3 MYW M. Edmond, 12 Yeoman Close, Worksop, S80 2RR
M3 MYZ M. Jones, 65 Montgomery Avenue, Bournemouth, BH11 8BN
M3 MZA R. Nicholson, 24 Barnmead, Haywards Heath, RH16 1UZ
M3 MZC A. Nicholson, 24 Barnmead, Haywards Heath, RH16 1UZ
M3 MZG D. Butterfield, 57 Holmes Road, Retford, DN22 6QU
M3 MZN S. Hamilton, 25 Keefe Close, Chatham, ME5 9AG
MM3 MZO I. Graham, 17 Royal Avenue, Stranraer, DG9 8ET
M3 MZP T. Burcombe, 49 Huntingdon Close, Mitcham, CR4 1XJ
M3 MZR E. Moon, Moon Marine, Rock Channel, Rye, TN31 7HJ
M3 MZT L. Jones, 13 Bracewell Close, Sutton, St Helens, WA9 3SH
M3 MZV A. Linton, 10 Adelaide Square, Shoreham-by-Sea, BN43 6LN
M3 MZW M. Weller, 27 Rochester Avenue, Woodley, Reading, RG5 4NA
MM3 MZX A. Bowers, 4A Pine Street, Greenock, PA15 4HW
MW3 NAE N. Edwards, 17 Queensway, Garnlydan, Ebbw Vale, NP23 5EE
M3 NAF N. Foster, 18 Austen Ave, Sawley, Nottingham, NG103GG
M3 NAH N. Higham-Hook, 31 Ringwood, Bracknell, RG12 8YG
M3 NAL C. Corbishley, 15 High St., Hardingstone, Northampton, NN4 7BT
M3 NAO D. Bradshaw, 65 Lichfield Court, Sheen Road, Richmond, TW9 1AX
MW3 NAQ S. Marles, 4 Maes Y Llan, Conwy, LL32 8NB
M3 NAR B. Johnson, 15 Oak Avenue, Willington, Crook, DL15 0BJ
M3 NAT N. Poate, 15 Jackdaw Rise, Eastleigh, SO50 9JT
M3 NAW N. White, 10 Elm Crescent, Alderley Edge, SK9 7PQ
M3 NBB N. Beith, 18 Avenue Road, New Milton, BH25 5JP
M3 NBD N. Draper, 107 Arkwrights, Harlow, CM20 3LY
M3 NBG L. Mcguire, 200 Wellingborough Road, Rushden, NN10 9SX
M3 NBH T. Dunne, 23 Warstone Lane, Birmingham, B18 6JQ
M3 NBI K. Fell, 5 Henry Road, Wath-Upon-Dearne, Rotherham, S63 7NF
M3 NBK A. Rooney, 44 Heritage Drive, Gillingham, ME7 3EH
M3 NBL N. Bland, 63 Swindon Road, Wroughton, Swindon, SN4 9AG
M3 NBQ S. Cross, 138 Crow Lane West, Newton-le-Willows, WA12 9YL
M3 NBU J. Shufflebotham, 316 Stockport Road, Hyde, SK14 5RU
M3 NBX V. Pomfrett, 17 Manifold Close, Sandbach, CW11 1XP
M3 NBZ N. Birnie, 61 Pipers Croft, Dunstable, LU6 3JZ
M3 NCB D. Lawson, 30 Meadowcroft, St Helens, WA9 3XQ
MI3 NCC P. Haughey, 10 Captains Road, Forkhill, Newry, BT35 9RR
M3 NCD N. Welsh, 18 Linkway, Runcorn, WA7 5EJ
M3 NCE M. Palmer, 22 Nightingale Drive, Poulton-le-Fylde, FY6 7UQ
M3 NCG M. Dumpleton, 3 Gooch Close Bacton, Norwich, NR12 0FA
M3 NCH P. Blackie, 30 Queens Avenue, Ilfracombe, EX34 9LS
M3 NCL N. Lees, 31 Cosford Drive, Dudley, DY2 9JN
MM3 NCM N. Cunningham, 11 Glendoune Street, Girvan, KA26 0AA
M3 NCN Rev. M. Gillingham, 14 Nethergreen Gardens, Killamarsh, S21 1FX
M3 NCO M. Collingswood, School House, Norton Canes High School, Burntwood Road, Norton Canes, Cannock, WS11 9SP
M3 NCP L. Steer, 51 Kings Chase, East Molesey, KT8 9DG
M3 NCQ M. Williams, 136 Courtfield Road, Quedgeley, Gloucester, GL2 4UF
MW3 NCS N. Sedgebeer, 16 Metcafe Street Caerau, Maesteg, CF340TB
M3 NCT N. Croft, 22 King Edward Crescent, Leeds, LS18 4BE
M3 NDC N. Crowley, 133 Jessop Road, Stevenage, SG1 5LH
M3 NDF D. Fard, 187 Fleetwood Road South, Thornton-Cleveleys, FY5 5NS
M3 NDJ C. Belham, 1 Kenmare Bank, Northwich, CW9 8BN
MW3 NDO J. Hoskins, 18 Bryn Yr Onnen, Southsea, Wrexham, LL11 6RG
M3 NDR N. Nash, Roann, Bedmond Road, Pimlico, Hemel Hempstead, HP3 8SH
MW3 NDU R. Williams, Bryn Mawr, Gwalchmai, Holyhead, LL65 4PY
M3 NDZ A. Humphriss, 44 Bishops Close, Stratford-upon-Avon, CV37 9ED
M3 NEA J. Rice, 2 Medalls Path, Stevenage, SG2 9DX
M3 NEC N. Chisholm, 162 Ardington Road, Northampton, NN1 5LT
M3 NEE M. Jay, Web-Stile Farm, Coley Hill, Bristol, BS39 5ED
M3 NEG P. Mcgarry, 10 Douglas Avenue, Soothill, Batley, WF17 6HG
MW3 NEI N. Mcloughlin, 19 Byron Road, Newport, NP20 3HJ
M3 NEL N. Snape, 20 Marlow Court, Adlington, Chorley, PR7 4LE
MI3 NEN N. Nicholl, 34 Berryhill Road, Artigarvan, Strabane, BT82 0HN
M3 NEP N. Payne, 7 Lane House, Eastfield Close, Worcester, WR3 7TT
M3 NFB S. Billingham, Kewell House, Wombourne Road, Swindon, Dudley, DY3 4NF
M3 NFE I. Jenkin, 62 Burns Road, Wellingborough, NN8 3RS
MW3 NFF B. Armstong, 3 Walton Cres, Llandudno Junction, LL31 9RR
M3 NFG N. Gonzalez, 46 Whitton View, Rothbury, Morpeth, NE65 7QN
M3 NFH E. Ashley, The Chapel, Ashford, Ludlow, SY8 4BX
M3 NFJ A. Williams, 327 Locking Road, Weston-Super-Mare, BS23 3LY
M3 NFK M. Watmough, 39 Ripon Gardens, Buxton, SK17 9PL
M3 NFQ S. Grainger, 132 Honiton Way, Hartlepool, TS25 2PY
M3 NFU A. Scarlett, 87 Coronation Avenue, Shildon, DL4 2AZ
M3 NFW M. Millward, 48 Nightingale Close, Farnborough, GU14 9QH
MM3 NFX I. Marshall, 6 Broom Wynd, Shotts, ML7 4HP
MW3 NFZ D. Matthews, 42 College Road, Oswestry, SY11 2SG
M3 NGC G. Champion, 34 Greenfields, Edenside, Kirby Cross., Frinton-on-Sea, CO13 0SW
M3 NGE N. Taggart, 61 Well Lane, Curbridge, Witney, OX29 7PB
M3 NGF R. Dickerson, 68 Chestnut Avenue, Spixworth, Norwich, NR10 3QQ
M3 NGG N. Clare, 123 Cunningham Road, Tamerton Foliot, Plymouth, PL5 4PU
MM3 NGJ A. Bernard, 200 Carden Avenue, Cardenden, Lochgelly, KY5 0EN
M3 NGK N. Kaye, 33A Aggborough Crescent, Kidderminster, DY10 1LQ
M3 NGM J. Gregory, 17 Meadowgarth, Belford, NE70 7PA
MW3 NGN B. Cook, 218 Ffordd Pennant, Mostyn, Holywell, CH8 9NZ
M3 NGO M. Hirst, 34 Welldon Crescent, Harrow, ha11qr
MW3 NGP A. Cook, 218 Ffordd Pennant, Mostyn, Holywell, CH8 9NZ
M3 NGU R. Bunting, 9 Hammond Way, Attleborough, NR17 2RQ
MM3 NGV M. Baird, 28 Loch Road, Bridge of Weir, PA11 3NB
M3 NGY A. Gromen-Hayes, 95 Maypole Road, Ashurst Wood, East Grinstead, RH19 3RB
M3 NGZ D. Boot, 10 Madehurst Rise, Sheffield, S2 3BJ
M3 NHA B. Ratcliff, 27 Furlong Road, Manchester, M22 1UD
MW3 NHC I. Meek, 30 St. Peters Road, Penarth, CF64 3PP
M3 NHD N. Harley, 1 Portland Crescent, Meden Vale, Mansfield, NG20 9PJ
M3 NHE J. Kelly, 12 Park Road, Milford On Sea, Lymington, SO41 0QU
M3 NHI S. Whitehead, 55 Crombie Road, Sidcup, DA15 8AT
MW3 NHN R. Williams, 92 Bowleaze, Greenmeadow, Cwmbran, NP44 4LF
M3 NHP N. Powell, 4 Holme Court Avenue, Biggleswade, SG18 8PF
M3 NHS N. Smith, 248A South Street, Romford, RM1 2AD
M3 NHU L. Gale, 9 Ely Close, Worthing, BN13 1BH
M3 NHV R. Gale, 9 Ely Close, Worthing, BN13 1BH
M3 NHW A. Hill, 39 Lambs Row, Lychpit, Basingstoke, RG24 8SL
M3 NHZ N. Hubbard, 7 Creake Road, Syderstone, King's Lynn, PE31 8SF
MW3 NIA H. Samuels, 11 Bennions Road, Wrexham, LL13 7AW
M3 NIC N. Rowland, 11 Babylon Way, Eastbourne, BN20 9DL
MI3 NIE C. Morton, 29 Lackaboy View, Enniskillen, BT74 4DY
M3 NIF F. Radford, 3 Pierpoint Terrace, Brighton Road, Hassocks, BN6 9TR
M3 NII C. Sims, 7 Ainthorpe Lane, Ainthorpe, Whitby, YO21 2JN
M3 NIT M. Paris, 13 Butfield, Lavenham, Sudbury, CO10 9SD
M3 NIZ G. Myers, 25 Sherwood Road, North Bersted, Bognor Regis, PO22 9DR
M3 NJA N. Atrill, 22 Lester Close, Hr Compton, Plymouth, PL3 6PX
M3 NJC N. Cook, Idle Shores Springfield Road, Woolacombe, EX34 7BX
M3 NJD N. Darby, 60 Pine St., Grange Villa, Chester le Street, DH2 3LX
M3 NJJ D. Wharlley, 15 Crampton Court, Grosvenor Road, Broadstairs, CT10 2XU
M3 NJK R. Mcallister, 57 Wigan Road, Standish, Wigan, WN6 0BE
M3 NJM N. Marsh, 16 Daytona Quay, Eastbourne, BN23 5BN
M3 NJO N. O'Hara, 41 Exeter Street, Blackburn, BB2 4AU
M3 NJQ C. Johnston, 15 Queens Road, Haydock, St Helens, WA11 0RH
MI3 NJU P. Mcmahon, 9 Sycamore Court, Drumaness, Ballynahinch, BT24 8QZ
MM3 NJV P. Vernon, 54 Brothock Way, Arbroath, DD11 4BH
M3 NJY N. Young, 5 Winslow Road, Boston, PE21 0EJ
M3 NKB J. Banks, 195 The Cornfields, Weston-Super-Mare, BS22 9DZ
M3 NKC D. Ansell, 30 Curzon Avenue, Horsham, RH12 2LB
MW3 NKG R. Rooker, 37 High Close, Nelson, Treharris, CF46 6HJ
M3 NKL D. Toyne, 19 Poachers Rest, Welton, Lincoln, LN2 3TR
M3 NKN J. Hickman, Ardoch, Harlestone Road, Northampton, NN6 8AW
M3 NKO N. March, 25 Emlyn Road, London, W12 9TF
M3 NKP J. Keeble, 2 Astley Cooper Place, Brooke, Norwich, NR15 1JB
M3 NKU C. Prout, 1 Westbrook Lustrells Vale, Saltdean, Brighton, BN2 8EZ
M3 NKW J. Dale, 37 Bussey Road, Norwich, NR6 6JF
M3 NKX D. Doggett, 82 Hurst Road, Kennington, Ashford, TN24 9RS
M3 NKZ M. Ashton, 31 Home Close Renhold, Bedford, MK41 0LB
M3 NLA N. King, 222 Prince Consort Road, Gateshead, NE8 4DX
M3 NLF J. Grosvenor, 10 Neves Close, Lingwood, Norwich, NR13 4AW
MM3 NLH S. Anderson, Newbigging Toll House, Drumsturdy Road, Dundee, DD5 3RE
M3 NLI D. Bale, 22 Highgrove Court, Rushden, NN10 0DH
M3 NLJ N. Jeffery, 7 Corfe Way, Winsford, CW7 1LU
M3 NLK N. Lake, 64 Womersley Road, Norwich, NR1 4QB
M3 NLM P. Maybin, 16 Appleby Road, London, E16 1LQ
M3 NLN A. Paulizky, 18 Cherry Tree Drive, London, SW16 2PE
M3 NLP N. Pearson, 34 Downside Road, Sutton, SM2 5HP
M3 NLQ M. Thompson, 4 Oat Hill Road, Towcester, NN12 6EZ
M3 NLW L. Boull, 80 Ascot Road, Baswich, Stafford, ST17 0AQ
M3 NLX R. Mcdermott, 2 Monument Close, Wellington, TA21 9AL
M3 NMB N. Beech, 94 Victoria Road, Runcorn, WA7 5ST
MI3 NMG N. Mcgonigle, 5 Woodend Road, Strabane, BT82 8LF
MM3 NMI D. Small, 30 Caledonia Crescent, Ardrossan, KA22 8LW
M3 NMJ S. Martin, 1 Buddleia Close, Weymouth, DT3 6SG
M3 NMK A. Kemplay, 8 Rue Du Lavoir, le Chillou, France, 79600
M3 NMM J. Jones, 40 Trembear Road, St Austell, PL25 5NY
M3 NMP M. Pedley, 60 Ack Lane East, Bramhall, Stockport, SK7 2BY
M3 NMR C. Purkiss, Flat 6, 220 Greenheys Lane West, Manchester, M15 5AF
M3 NMV L. Yates, 108 Hawthorn Avenue, Lowestoft, NR33 9BB
M3 NMX B. Scrivens, 7 Normandy Way, Fordingbridge, SP6 1NW
MW3 NNA A. Lydford, 93 Hardwick Avenue, Chepstow, NP16 5EB
M3 NNG R. Simms, 3 The Byeway, London, SW14 7NL
M3 NNH K. Bansil, 65 Hervey Street, Northampton, NN1 3QL
M3 NNI A. Mason, 4 Quay Mill Walk, Great Yarmouth, NR30 1JG
M3 NNJ J. Moore, 2 Newsons Meadow, Lowestoft, NR32 2NW
M3 NNM P. Morling, 7 Hobill Close, Leicester Forest East, Leicester, LE3 3PS
MM3 NNO C. Lewis, 9 Cessnock Road, Troon, KA10 6NJ
M3 NNQ J. Blamey, 46 First Avenue, Canvey Island, SS8 9LP
M3 NNV J. Paul, East Park, Church Road, Cowes, PO31 8HA
M3 NNY J. Brough, 10 Linnet Close, Huntington, Cannock, WS12 4TP
M3 NNZ B. James, 19 Dukes Crescent, Sandbach, CW11 1BL
M3 NOD N. Lightfoot, 4 Prospect Close, Hatfield Peverel, Chelmsford, CM3 2JE
M3 NOE D. Noe, 21 Gale Crescent, Banstead, SM7 2HZ
M3 NOF D. Donnelly, 72 Bagots Oak, Stafford, ST17 9SB
MI3 NOH N. O'Hagan, 55 Meadowside, Antrim, BT41 4HD
M3 NOJ J. Reynolds, 4 Perriclose, Chelmsford, CM1 6UJ
M3 NOM R. Smallman, 128 Bevan Lee Road, Cannock, WS11 4PT
M3 NON N. O'Sullivan, 20 Pond Lane, Drayton, Norwich, NR8 6PP
M3 NOR A. Norman, 91 Church Road, Radley, Abingdon, OX14 3QF
M3 NOS D. Leggett, Fools Watering, London Road, Beccles, NR34 8AQ
M3 NOW S. Wilson, 55 Kent Road, Reading, RG30 2EJ
M3 NOY S. James, 124 Alcock Avenue, Mansfield, NG18 2NF
M3 NPA A. Nicholson, 7 Lingfoot Crescent, Sheffield, S8 8DA
M3 NPC N. Collins, Hill Farm, Broadheath, Tenbury Wells, WR15 8QN
M3 NPE J. Nicholson, 7 Lingfoot Crescent, Sheffield, S8 8DA
MM3 NPG M. Maltman, 30 Haldane Place, Dundee, DD3 0JR
M3 NPH A. Arnold, 2 Duck Lane, Haddenham, Ely, CB6 3UE
M3 NPI M. Harrell, 34 Nelson Drive, Cannock, WS12 2GF
M3 NPK N. Kerner, Headingley Cottage, Ryehurst Lane, Bracknell, RG42 5QZ
M3 NPO D. Shuttleworth, 27 Union St., Egerton, Bolton, BL7 9SP
M3 NPR N. Robinson, 15 The Lavers, Hockley Road, Rayleigh, SS6 8ED
M3 NPS N. Sharpe, 23 Cheney Road, Faversham, ME13 8DG
M3 NPW P. Webster, 39 Farndale Terrace, Leeds, LS14 5BQ
M3 NPX S. Taylor-Mccormick, North View Boarding Kennels, Skitham Lane, Preston, PR3 6BD
M3 NPZ A. Paterson, 35 Darlington Road, Richmond, DL10 7BG
M3 NQA R. Clark, 12 Ash Drive Haughton, Stafford, ST18 9EU
MW3 NQE S. Walmsley, 29 Shelley Court Machen, Caerphilly, CF83 8TT
MW3 NQH S. Pope, 11 Haman Place, Gelligaer, Hengoed, CF82 8EG
M3 NQI P. Denham, Flat 1, 10 Prince Alfred Avenue, Skegness, PE25 2UH

MW3 NQK J. Argent, 7 Lloyds Hill, Buckley, CH7 3ER
M3 NQL L. Kelsey, 111-113 George Street, Mablethorpe, LN12 2BS
M3 NQN K. Roberts, 40 Portland Drive, Skegness, PE25 1HF
M3 NQO A. Whyatt, 11 The Perrings, Nailsea, Bristol, BS48 4YD
M3 NQS N. Turner, 18 The Green Road, Sawston, Cambridge, CB22 3LP
MM3 NQT M. Simon, 100 Findhorn Place, Edinburgh, EH9 2NZ
M3 NQU E. Wagner, 3 Sarre Road, London, NW2 3SN
M3 NQY G. Eklund, 26-28 Zulu Road, Nottingham, NG7 7DR
MI3 NRB N. Bolt, 32 Bush Gardens, Bushmills, BT57 8AE
M3 NRI A. Miles, 5 Pershore Road, Basingstoke, RG24 9BE
M3 NRJ N. Johnson, 27 Redford Crescent, Bristol, BS13 8SA
M3 NRK N. Kind, 18 Cunningham Road, Bentley, Walsall, WS2 0AY
M3 NRQ A. Nicholson, 7 Lingfoot Crescent, Sheffield, S8 8DA
M3 NRV D. Giering, 39 High House Drive, Inkberrow, Worcester, WR7 4EG
M3 NRW E. Cromwell, 92 Hatch Road, Pilgrims Hatch, Brentwood, CM15 9QA
M3 NRX J. Williams, 26 Raglan Close, Castleford, WF10 1PL
M3 NSB A. Oatey, Robin Hill, Blackpost Lane, Totnes, TQ9 5RF
MI3 NSF R. Foley, 6 Lislane Drive, Saintfield, Ballynahinch, BT24 7HU
M3 NSG N. Garry, 4 Fairstead, Skelmersdale, WN8 6RD
M3 NSH S. Shelley, 8 Harewood Close, Eastleigh, SO50 4NZ
M3 NSJ K. Ingham, 231 Regent Street, Nelson, BB9 8SQ
M3 NSM G. Coyle, 19 Mounsey Road, Bamber Bridge, Preston, PR5 6LS
M3 NSO N. Walch, 52 Marsh House Road, Sheffield, S11 9SP
M3 NSQ S. Beedham, 27 Malpas Close Bransholme, Hull, HU7 4HH
MI3 NSR G. Mccullough, 32 Thistlemount Park, Lisburn, BT28 2UN
M3 NSS M. Price, 25 School Crescent, Lydney, GL15 5TA
M3 NST N. Stirling, 3 Rother Croft, New Tupton, Chesterfield, S42 6BE
M3 NSX I. Thompson, 33 Longsight Road, Mapplewell, Barnsley, S75 6HD
MW3 NTE H. Fingerhut-Holland, Osnok, 1 South Cliff Street, Tenby, SA70 7EB
MU3 NTH N. Thomas, 6 Tunstall Terrace, Gibauderie, St Peter Port, Guernsey, GY1 1XJ
M3 NTI M. Gough, 57 Ravenglass Road, Westlea, Swindon, SN5 7BN
M3 NTJ N. Thompson, 33 Longsight Road, Mapplewell, Barnsley, S75 6HD
M3 NTQ M. Gough, 58 Church Street, Brierley, Barnsley, S72 9JG
M3 NTR M. Pratchett, Adastra Cottage, Letcombe Regis, Wantage, OX12 9JP
M3 NTW C. Northwood, Apartment 50, 2 Munday Street, Manchester, M4 7BB
MM3 NTX G. Askew, 49 Kittlegairy Road, Peebles, EH45 9LX
M3 NTZ S. Woodward, 19 Beech Court, Spondon, Derby, DE21 7TP
M3 NUB N. Vichitcheep1, 6 Ormathwaites Corner, Warfield, Bracknell, RG42 3XX
M3 NUE A. Lunn, 57 Greets Green Road, West Bromwich, B70 9ES
M3 NUH M. Gladders, 2 Albion Mansions, Saltburn-by-the-Sea, TS12 1JP
M3 NUI Rev. S. Scotson, 16 Merryfield Road, Dudley, DY1 2PD
M3 NUL S. Heaton, 13A Roche Avenue Bilton, Harrogate, Hg1 4es
M3 NUM A. Ogle, 22 Warwick Street, Daventry, NN11 4AL
M3 NUO K. Holdt, 18 Garrard Road, Banstead, SM7 2ER
MW3 NUP M. Lewis, 4 Coldwell Terrace, Pembroke, SA71 4QL
M3 NUQ M. Dickenson, 6 The Pavilions, Blandford Forum, DT11 7GF
M3 NUU D. Copsey, Fairview, Mill Lane, Hook End, Brentwood, CM15 0PP
MI3 NUW A. Kincaid, 428 Cushendall Road, Ballymena, BT43 6QE
MW3 NUX D. James, 10 Hafan Deg, Pencoed, Bridgend, CF35 6YG
M3 NVA A. Caine, 53 Cromford Road Crich, Matlock, DE4 5DJ
M3 NVC P. Brown, 13 Rydal Road, Weston-Super-Mare, BS23 3RT
MM3 NVD D. Baillie, 126 Main St., Fauldhouse, Bathgate, EH47 9BW
M3 NVE K. Whiteley, 14 Milton Street, Goole, DN14 6EL
M3 NVF N. Fletcher, 2 Handforth Road, Crewe, CW2 8PL
M3 NVG J. Webster, 72 Grosvenor Street, Derby, de248at
M3 NVH J. Boull, 80 Ascot Road Baswich, Stafford, ST17 0AQ
M3 NVK L. Bolton, 9 Nab Crescent, Meltham, Holmfirth, HD9 5LT
M3 NVL J. Jones, 5 Cranleigh Road, Liverpool, L25 2RP
M3 NVO L. Clift, 8 Kendal Road, Gloucester, GL2 0NB
M3 NVP M. Weir, 153 Tyndale Crescent, Birmingham, B43 7HX
MW3 NVQ R. Wright, 12 Bryn Teg, Arddleen, Llanymynech, SY22 6PZ
M3 NVR P. Gutteridge, 75A Collingwood Drive, Birmingham, B43 7JW
M3 NVS J. Caddick, 135 Broadway, Dunscroft, Doncaster, DN7 4HB
M3 NVV C. Elliot, 106 Occupation Road, Corby, NN17 1EG
MI3 NVX M. Mccay, 2 Riverview, Spamount, Castlederg, BT81 7NA
M3 NWD A. Broll1, 17 Broadway, Farcet, Peterborough, PE7 3AY
MM3 NWF K. Whyte, 27 Queens Road, Inverbervie, Montrose, DD10 0RY
M3 NWH F. Gear, 251 Abington Avenue, Northampton, NN3 2BU
M3 NWK S. Lofthouse, 32 Westbrook Park Road, Peterborough, PE2 9JG
MI3 NWO D. Adams, 65 Rose Park, Limavady, BT49 0BF
M3 NWQ F. Stone, Rrt, Gosw, 2 Rivergate, Bristol, BS1 6EH
MI3 NWU G. Mckeever, 45 Blackthorn Court, Coleraine, BT52 2EX
M3 NWY T. Bown, 16 Sandringham Court, Queen Elizabeth Road, Nuneaton, CV10 9AR
MM3 NWZ A. Smith, 17 High Street, Stranraer, DG9 7LL
M3 NXA A. Davenport, 10 Woodend Lane, Hyde, SK14 1DT
M3 NXC P. Waring, 1 Fanshaw Road Eckington, Sheffield, S21 4bw
MW3 NXD A. Downing, 3A Pant Hirgoed, Pencoed, Bridgend, CF35 6YD
M3 NXE S. Burrows, 78A Coronation Road, Earl Shilton, Leicester, LE9 7HJ
M3 NXF N. Forbes, 55 The Henrys, Thatcham, RG18 4LS
M3 NXH I. Rimell, 51 Woodlands Avenue, Woodley, Reading, RG5 3HF
M3 NXJ I. Livesey, 26 Hilltop Road, Twyford, Reading, RG10 9BN
M3 NXK S. Dudley, 28 Walnut Lane, Wednesbury, WS10 0BH
M3 NXO B. Bradford, 3 Beverly Close, Thornton-Cleveleys, FY5 5DR
M3 NXQ V. Stokes, 52 Brantley Avenue, Wolverhampton, WV3 9AR
MM3 NXY R. Hay, 12 Mitchell Brae, Balmedie, Aberdeen, AB23 8PW
M3 NXZ S. Robinson, 1 Woodgate Road, Manchester, M16 8LX
M3 NYA M. Hoggan, 19 The Drive, Uckfield, TN22 1BY
MI3 NYB N. Throne, 12 Mason Road, Magheramason, Londonderry, BT47 2RY
M3 NYF A. Probst, 37 Devonshire Street, Skipton, BD23 2ET
M3 NYG N. Carson, Gov Office For The South West, 2 Rivergate, Bristol, BS1 6EH
M3 NYI A. Parradine, 76 Parsonage Road, Rainham, RM13 9LF
M3 NYK K. Nye, 32 Seafield Close, Chichester, PO20 8DP
M3 NYM G. Whitehurst, 28 Vicarage Fields, Warwick, CV34 5NJ
M3 NYQ M. Hives, 12 Springwood Avenue, Chadderton, Oldham, OL9 9RR
M3 NYX J. Mock, 29 Tavistock Place, Paignton, TQ4 7NZ
MM3 NYY G. Cleary, 2 Merlinford Avenue, Renfrew, PA4 8XS
M3 NZG A. Wheeler, 14 Sparkmill Terrace, Beverley, HU17 0PA
M3 NZK J. Bayliss, 39 Elms Avenue, Littleover, Derby, DE23 6FB
M3 NZN C. Jones, 20 Freeman Road, Wednesbury, WS10 0HQ
M3 NZR D. Oddie, 5 The Bridleway, Forest Town, Mansfield, NG19 0QJ
M3 NZV L. Williams, 2 The Tannery, Dol-Y-Bont, Borth, SY24 5LX
M3 NZW A. Woods, 42 St. Pauls Drive, Wellington, Telford, TF1 3GD
MM3 NZX M. Mcdonald, 17 Ramsay Mews, Strathaven, ML10 6GN
MW3 NZZ I. Powe, 7 Wellington Drive, Greenmeadow, Cwmbran, NP44 5HH
M3 OAB A. Barker, 43 Ploughmans Drive, Shepshed, Loughborough, LE12 9SG
M3 OAC G. Dray, 2 Mulberry Drive, Malvern, WR14 4AT
M3 OAG W. Scott, 8 Woodacre, Whalley Range, Manchester, M16 8QQ
M3 OAJ S. Clay, Akers Lodge, 6 Penn Way, Rickmansworth, WD3 5HQ
M3 OAK D. Smith, 98 Orange Hill Road, Burnt Oak, Edgware, HA8 0TW
M3 OAL M. Edwards, 30 Morrison Road, Tipton, DY4 7PU
M3 OAM M. Parkin, 51 Far Lane, Rotherham, S65 2HQ
M3 OAQ S. Tingay, 9 Cottage Homes, Wakefield Road, Huddersfield, HD5 9XT
M3 OAS I. Miller, 64 Queens Road, Vicars Cross, Chester, CH3 5HD
M3 OAT J. Judson, 559 Colne Road, Burnley, BB10 2LG
M3 OAX F. Smith, 32 Amesbury Drive, London, E4 7PZ
M3 OAZ R. Mccarthy, 11 Britain Street, Bury, BL9 9PD
M3 OBB O. Boar, 19 Blyford Road, Lowestoft, NR32 4PZ
M3 OBD A. Dow, 5 Verney Crescent, Liverpool, L19 4UR
M3 OBM G. Scarr, 15 Church Park, Overton, Morecambe, LA3 3RA
M3 OBN L. Stott, 70 Elizabeth St., Ashton under Lyne, OL6 8SX
M3 OBO R. Goody, 113 Kenneth Road, Basildon, SS13 2BH
M3 OBS M. Simkins, 37 St. Andrews Meadow, Harlow, CM18 6BL
M3 OBU C. Camsey, 1 Park Close, Milford On Sea, Lymington, SO41 0QT
M3 OBX D. Sanderson, 65 Holm Flatt Street, Parkgate, Rotherham, S62 6HJ
M3 OBZ D. Thomas, 51 Sandringham Avenue, Vicars Cross, Chester, CH3 5JF
M3 OCA P. Bainbridge, 6 Waterland Lane, St Helens, WA9 3AF
M3 OCC B. O'Connor, 58 St Johns Way, Thetford, IP24 3NP
M3 OCJ S. Rafter, 30 Monmouth Grove, St Helens, WA9 1QB
M3 OCL H. Lister, 68 Spring Avenue, Gildersome, Leeds, LS27 7BT
M3 OCP M. Cave, 33 Forest Road Athersley North, Barnsley, S71 3BG
M3 OCQ R. Hall, 5 Lea Hill Road, Birmingham, B20 2AS
M3 OCR O. Crump, 3 Vosper Road, Southampton, SO19 9SS
M3 OCS A. Owen, 5 Croft Close, Rowton, Chester, CH3 7QQ
MM3 OCY D. Mcclelland, 5 Cambusmoon Terrace, Gartocharn, Alexandria, G83 8RU
M3 ODC P. Hatter, 14 Morland Avenue, Bromborough, Wirral, CH62 6BE
M3 ODH D. Hughes, 75 Suncote Avenue, Dunstable, LU6 1BN
M3 ODK L. Bedford, 29 Kent Road, Brookenby, Market Rasen, LN8 6EW
M3 ODL K. Bedford, 29 Kent Road Brookenby, Binbrook, Market Rasen, LN8 6EW
M3 ODO E. White, 3 Davy Drive, Maltby, Rotherham, S66 7EN
MM3 ODV A. Cairns, 57 Miller St., Dumbarton, G82 2JA
M3 OEB D. Norris, 24 Northway, Fulwood, Preston, PR2 9TP
MW3 OEC L. Williams, 35 Heol Y Sheet, North Cornelly, Bridgend, CF33 4EU
MD3 OED R. Britton, 3 Meadowfield, Port Erin, Isle Of Man, IM9 6PH
M3 OEE B. Hood, 52 Kent St., Preston, PR1 1RY
M3 OEF S. Tattum, 20 Drew Close, Poole, BH12 5ET
M3 OEG J. Cook, 42 Pampas Close, Colchester, CO4 9ST
M3 OEH A. Mclean, 47 Tarn Drive, Bury, BL9 9QB
MW3 OEJ M. Kay, 3 Protheroe Avenue, Pen-Y-Fai, Bridgend, CF31 4LU
M3 OEM B. Horton, 44 Chamberlain St., St Helens, WA10 4NL
M3 OEN J. Edwards, 36 Westerley Lane, Shelley, Huddersfield, HD8 8HP
M3 OEO M. Hammersley, 47 Shaftesbury Avenue, Timperley, Altrincham, WA15 7NP
MI3 OEQ A. Jamison, 11 Richmond Gardens, Newtownabbey, BT36 5LA
M3 OER S. Warner, 28 Jameson Bridge Street, Market Rasen, LN8 3EW
M3 OEV M. Cuff, 1 Ivy Close, St. Leonards, Ringwood, BH24 2QZ
M3 OFA N. Shaw, Greenacres Poultry, Three Lowes, Stoke on Trent, ST10 3BW
M3 OFB J. Woodruff, 10 Bailey Close, Blackburn, BB2 4FT
M3 OFC S. Dunne, 1 Burton Gardens, Brierfield, Nelson, BB9 5DR
M3 OFD D. Dunne, 1 Burton Gardens, Brierfield, Nelson, BB9 5DR
MM3 OFE J. Jackson, 25 Lomond Crescent, Alexandria, G83 0RJ
M3 OFH J. Wright, Flat 4, Consort House, Albert Street, Shrewsbury, SY1 2HT
M3 OFJ M. Jones, 20 Chelsea Drive, Sutton Coldfield, B74 4UG
M3 OFN C. Bamford, 71 Trent Road, Shaw, Oldham, OL2 7YQ
M3 OFS E. Marsh, 16 Laurel Close, North Warnborough, Hook, RG29 1BH
M3 OFU T. Smith, 151 Halfords Lane, West Bromwich, B71 4LQ
M3 OFV M. Kempson, 67 Esther Avenue, Wakefield, WF2 8BY
MI3 OFX T. Conlon, 7 Waringfield Gardens, Moira, Craigavon, BT67 0FQ
M3 OGC L. Li, Harrogate Ladies' College, Clarence Drive, Harrogate, HG1 2QG
M3 OGD R. Whiteside, Hill Crest, Farley Hill, Farley, Matlock, DE4 5LT
M3 OGL T. Woodhouse, The Old Granary, 12 Limekiln Lane, Newport, TF10 9EZ
M3 OGM B. Debenham, 80 Stewart Road, Chelmsford, CM2 9BD
MM3 OGS R. Keay, 26 Cherrywood Drive, Beith, KA15 2DZ
MM3 OGU T. Mussell, Dunelm, Thornhill Road, Cuminestown, Turriff, AB53 5WH
M3 OGV R. Bicknell-Thompson, 4 Linden Court, Greenfrith Drive, Tonbridge, TN10 3LW
M3 OHC M. Holgate, 10 Brecon Crescent, Ashton-under-Lyne, OL6 8UA
MM3 OHD D. Hague, 13 North Dell, Ness, Isle of Lewis, HS2 0SW
MI3 OHE E. Paterson, 1 Sycamore Grove, Belfast, BT4 2RB
MI3 OHF M. Graham, 21 Meadowvale Crescent, Bangor, BT19 1HQ
MI3 OHG N. Wylie, 26 Lisnoe Walk, Lisburn, BT28 1QD
M3 OHI G. Chaffey, 63 Underwood Road, Eastleigh, SO50 6FX
M3 OHJ B. Stanfield, 24 Rowan Close, Kingsbury, Tamworth, B78 2JR
M3 OHL B. Widdowson, 28 Highfield Lane, Chesterfield, S41 8AU
M3 OHN P. Martin, Flat 9, Paddock Court, Graham Avenue, Portslade, Brighton, BN41 2WU
M3 OHO M. Murray, 2 Meadway, Penwortham, Preston, PR1 0JL
MI3 OHP J. Leetch, 30 Murob Park, Ballymena, BT43 6JG
M3 OHQ W. Kong, Clarence House, Harrogate Ladies College, North Yorkshire, HG1 2QG
M3 OHR L. Gray, 29 Longview Road, Liverpool, L36 1TA
M3 OHX D. Taylor, 58 Shenstone Road, Great Barr, Birmingham, B43 5LN
M3 OHY H. Chan, Harrogate Ladies' College, Clarence Drive, Harrogate, HG1 2QG
M3 OHZ H. Fleming, 29 Model Village, Creswell, Worksop, S80 4BN
MI3 OIB D. Couser, 5 Colinbrook Park, Dunmurry, Belfast, BT17 0NZ
M3 OIC L. Crane, 32 Clarke Avenue, Newark, NG24 4NY
M3 OII T. Hoggan, 9 Nursery Field, Buxted, Uckfield, TN22 4NG

MW3 OIK W. Jaggard, 2 Aled Drive, Rhos On Sea, Colwyn Bay, LL28 4UU
M3 OIL O. Hutley, 1 John Ray Street, Braintree, CM7 9DZ
M3 OIN H. List, 41 Westbury Crescent, Dover, CT17 9QQ
MD3 OIS M. Wallace, 61 Vernon Road, Ramsey, IM8 2EG
MM3 OIX N. White, 29 Forgie Crescent, Maddiston, Falkirk, FK2 0LY
M3 OIY P. Yuen, Harrogate Ladies' College, Clarence Drive, Harrogate, HG1 2QG
MM3 OIZ J. Mcmonigle, 19 Monach Gardens, Dreghorn, Irvine, KA11 4EB
MM3 OJE T. Mcfarlane, 110C East Main Street, Darvel, KA17 0JB
M3 OJJ A. Chance, 24 Doddsfield Road, Slough, SL2 2AD
M3 OJK N. Schall, 39 Emery Avenue, Chorltonville, Manchester, M21 7LE
M3 OJN P. Moule, 30 Hillview Road, Chelmsford, CM1 7RX
M3 OJP D. Moulding, 5 Chalk Lane, Sutton Bridge, Spalding, PE12 9YF
MM3 OJR J. Rae, Bowhouse Farm Cottage, Auchtermuchty, Cupar, KY14 7ES
M3 OJS P. Rowlands, 314 Stourbridge Road Catshill, Bromsgrove, B61 9LH
M3 OJU P. Watling, 1 Chediston Green, Chediston, Halesworth, IP19 0BB
MM3 OJV C. Mcgougan, 6 Calder Place, Kilmarnock, KA1 3QL
M3 OJW S. Earl, 9A Florida Street, Daws Hill Lane, High Wycombe, HP11 1QA
M3 OJX L. Parkman, 9 Malim Way, Gonerby Hill Foot, Grantham, NG31 8QF
M3 OJY B. Mason, 4108 Hingston Avenue, Montreal, Canada, H4A 2J7
M3 OJZ J. Neale, 20 Oakfield Road, Wollescote, Stourbridge, DY9 9DL
M3 OKC P. Barnesok, 39 Prospect Avenue, Rochester, ME2 3BZ
M3 OKE A. Ewence, 9 Mount Pleasant, Bradford-on-Avon, BA15 1SJ
MD3 OKG K. Glaister, 42 Barrule Drive, Onchan, Douglas, Isle Of Man, IM3 4NR
MD3 OKH B. Glaister, 42 Barrule Drive, Onchan, Douglas, Isle Of Man, IM3 4NR
M3 OKP R. Bancroft, Ashrigg, Lazonby, Penrith, CA10 1AT
M3 OKY R. Eglington, 33 Bradley Lane, Bilston, WV14 8EW
M3 OKZ A. Cammish, 6 West Vale, Filey, YO14 9AY
M3 OLD S. Old, 49 Penrith Cresent, Castleford, WF10 2RG
M3 OLE S. Jervis, 45 Wyndham Road, Stoke-on-Trent, ST3 3LX
M3 OLF I. Scott, Croft Cottage, Cumwhinton, Carlisle, CA4 8ER
M3 OLI O. Palmer, 21 Ibbett Close, Kempston, Bedford, MK43 9BT
MI3 OLM C. Doole, 110 Moyagall Road, Knockloughrim, Magherafelt, BT45 8PJ
M3 OLN E. Hall, 5 The Paddocks, Thursby, Carlisle, CA5 6PB
M3 OLQ P. Oliver, 12 Walkmill Crescent, Carlisle, CA1 2WF
MW3 OLT L. Thomas, 15 Blaenwern, Newcastle Emlyn, SA38 9BE
M3 OLU A. Shepherd, 39 Minehead Road, Dudley, DY1 2NZ
M3 OLW A. Mills, 107 Blyth Ave, Southend-on-Sea, SS3 9nj
MW3 OLX W. Hance, 27 Maesybont, Glanamman, Ammanford, SA18 2AY
M3 OLZ L. Bullen11, 5 West View, Long Sutton, TA10 9LT
M3 OME Dr A. Sadanandam, 23 Clare Avenue, Hoole, Chester, CH2 3HT
M3 OMF O. Fury, 1 Wigley Drive, Wigley, Ludlow, SY8 3DR
MM3 OMI B. Bowman, 15 Aboyne Road, Aberdeen, AB10 7BS
M3 OMT T. Baggley, 16 Seaton Road, Seaton, Workington, CA14 1DT
M3 OMU J. Fletcher, 217 Homefield Road, Sileby, Loughborough, LE12 7TG
M3 OMX K. Boulton, 17 Grange Avenue, Goodrington, Paignton.Tq4 7Jy, Paignton, TQ4 7jy
M3 OMZ D. Jarvis, 7 Leonard Road, Greatstone, New Romney, TN28 8UJ
M3 ONB I. Woollen, 33 The Oaks, Taunton, TA1 2QX
M3 ONE C. Chambers, 75A Main St., Sedbergh, LA10 5AB
MW3 ONG J. Brydges, 9 Twynygarreg, Treharris, CF46 5RL
M3 ONH S. Fooks, 11 Breck Road, Darlington, DL3 8NH
MM3 ONI O. Stein, Library House, Stafford Street, Tain, IV19 1AZ
M3 ONK R. Bray, 49 Montacute Way, Wimborne, BH21 1TZ
M3 ONM P. Connor, 244 Gregory Avenue, Birmingham, B29 5DR
MD3 ONP J. Taylor, Ballafayle Cottage, Ballafayle, Ramsey, Isle Of Man, IM7 1ED
M3 ONV J. Bonar, Flat, 5A Friday Street, Minehead, TA24 6EE
MM3 ONX L. Paget, 40 Davaar Drive, Kilmarnock, KA3 2JG
M3 OOA T. Durant, 39 Snydale Road, Normanton, WF6 1NY
M3 OOC B. Cooper, 71 High Street, Birstall, Batley, WF17 9RG
M3 OOE G. Kelly, 141 Narbeth Drive, Aylesbury, HP20 1pz
M3 OOH S. Howroyd, 7 Garendon Road, Loughborough, LE11 4QB
M3 OOL T. Peterson, 9 Moseley Wood View, Leeds, LS16 7ES
M3 OOP J. Dietsch, 21 Lake View Avenue, Chesterfield, S40 3DR
M3 OOQ A. Shaw, 2 Montrose Avenue, Montrose Street, Hull, HU8 7RY
MM3 OOT J. Ferrans, 77 Knockinlaw Road, Kilmarnock, KA3 2AS
M3 OOU A. Hoskins, 38 Tasmania Close, Basingstoke, RG24 9PQ
M3 OOX F. Spencer, 29 Kliffen Place, Halifax, HX3 0AL
M3 OOY C. Spencer, 29 Kliffen Place, Halifax, HX3 0AL
M3 OPA N. Van-Den-Langenberg, 2 Grove Crescent, Bridgnorth, WV15 5BS
M3 OPB O. Blackburn, 128 High St., Crigglestone, Wakefield, WF4 3EF
M3 OPC J. Spencer, 29 Kliffen Place, Halifax, HX3 0AL
M3 OPD K. Spencer, 29 Kliffen Place, Halifax, HX3 0AL
M3 OPG P. Reed, 32 London Road, Warmley, Bristol, BS30 5JH
M3 OPM J. Newby, 22 Acton Road, Liverpool, L32 0TT
M3 OPN J. Shemwell, 4 Darvel Close, Bolton, BL2 6UD
M3 OPS A. Hindle, 41 Seedfield, Staveley, Kendal, LA8 9NJ
M3 OPT D. Wicks, 148 Long Lane, Staines, TW19 7AJ
M3 OPU K. Morris, 80 Bridge Street, Chatteris, PE16 6RN
M3 OPV K. Ashcroft, 9 Aldermere Crescent, Urmston, Manchester, M41 8UE
M3 OPW H. Foot, 1 South View, Piddletrenthide, Dorchester, DT2 7QS
M3 OPX A. Robinson, 75 De La Pole Avenue, Hull, HU3 6RD
M3 OQD H. Nehmzow, 16 Goldington Avenue, Bedford, Mk403by
M3 OQG A. Chruscinski, 39 Sherwood Rise, Mansfield Woodhouse, Mansfield, NG19 7NP
M3 OQH D. Cooper, 52 Meadow Lane, Birkenhead, CH42 3YE
M3 OQI G. Woodward, 5 Barnard Road, Chelmsford, CM2 8RR
M3 OQJ B. Male, 44 Lakefields, West Coker, Yeovil, BA22 9BT
M3 OQK G. Williamson, 26 Portland Mews, Bridlington, YO16 4EH
M3 OQL S. Davin, 40 Theynes Croft, Long Ashton, Bristol, BS41 9NA
M3 OQQ T. Garvey, 21 Oak Park Drive, Havant, PO9 2XE
MM3 OQR P. Taylor, 2 Laurel Grove, Aberdeen, AB22 8YJ
M3 OQS B. Daley, 129A Kingsway South, Warrington, WA4 1RW
MM3 OQV J. Campbell, 6 Dunard Court, Carluke, ML8 5RX
M3 OQZ P. Hall, 13 Sheard Avenue, Ashton-under-Lyne, OL6 8DS
M3 ORB A. Goold, 6 The Elms, Kempston, Bedford, MK42 7JN
M3 ORE G. Beale, 34 Teville Road, Worthing, BN11 1UG
M3 ORL A. Price, 67 Mansfield Road, Glapwell, Chesterfield, S44 5QA
M3 ORN O. Newton, 84 Ameysford Road, Ferndown, BH22 9QB
MW3 ORP J. Marlow, West Bulthy, Bulthy, Welshpool, SY21 8ER
M3 ORQ M. Winch, 2 Cranleigh Gardens, Cowes, PO31 8AS
M3 ORT B. Williams, 3 Welton Close, Wilmslow, SK9 6HD
M3 ORU C. Baker, 10 Kirton Close, Coventry, CV6 2PG
MW3 ORY N. Lewis, 1 Clyne Drive, Blackpill, Swansea, SA3 5BU
M3 ORZ G. Marshall, 12 Arthur Avenue, Caister-On-Sea, Great Yarmouth, NR30 5PQ
M3 OSA J. Ross, 142 Bridle Road, Croydon, CR0 8HJ
M3 OSC E. Walker, 2 The Green, Blencogo, Wigton, CA7 0DF
M3 OSF J. Cobbold, 2 The Green, Blencogo, Wigton, CA7 0DF
MW3 OSI R. Johnson, 25 Lon Tyrhaul Llansamlet, Swansea, SA7 9SF
MM3 OSK A. Twort, 17 Balallan, Isle of Lewis, HS2 9PN
M3 OSP P. Moss, 26 Woodlands Avenue, Farnham, GU9 9EY
MW3 OSQ R. Phillips, 188 Charston, Greenmeadow, Cwmbran, NP44 4LD
M3 OSS C. Dell, 18 Greenacres, Fulwood, Preston, PR2 7DA
M3 OSU T. Pollard, 35 Weatherall St. North, Salford, M7 4TH
M3 OSW K. Sheldon, 35 Weatherall St. North, Salford, M7 4TH
M3 OSY W. Shiu, Harrogate Ladies' College, Clarence Drive, Harrogate, HG1 2QG
M3 OTG A. Hill, 5 Park Road, Thurnscoe, Rotherham, S63 0TG
M3 OTI D. Scotcher, 17 St. Dominics Square, Luton, LU4 0UN
M3 OTM O. Morris, 1 Crawford Avenue, Peterlee, SR8 5EG
M3 OTP R. Tailford, 28 Paddock Wood, Prudhoe, NE42 5BJ
M3 OTQ T. Chapman, 17 Trevor Road, Swinton, Manchester, M27 0YH
M3 OTR S. Taylor, 49 Chestnut Avenue, West Drayton, UB7 8BU
M3 OTS G. Rees, 32 Glencroft Close, Burton-on-Trent, DE14 3GJ
M3 OTU R. Woolgar, 2 Solent Way, Milford On Sea, Lymington, SO41 0TE
M3 OTZ M. Wright, 8 St. Wilfrids Road, Oundle, Peterborough, PE8 4NX
MW3 OUC J. Bidwell, 26 Lone Road, Clydach, Swansea, SA6 5HR
M3 OUF R. Crewe, 12 Dimple Gardens, Ossett, WF5 8LJ
M3 OUG C. Rickwood, 7 Bromley Mount, Wakefield, WF1 5LB
M3 OUH P. Loxton, 32 Parkhill Crescent, Wakefield, WF1 4EZ
M3 OUI P. Marsh, 16 Laurel Close, North Warnborough, Hook, RG29 1BH
M3 OUL G. Martin, 12 Poolside, Phase 1, St Joseph, Trinidad And Tobago,
MI3 OUN S. Fulton, 120 Dunnalong Road, Bready, Strabane, BT82 0DP
M3 OUQ W. Speak, 20 Pear Tree Drive, Wincham, Northwich, CW9 6EZ
M3 OUS E. Mcfalls, 119 Park Avenue, Shelley, Huddersfield, HD8 8JZ
M3 OUV S. Clarke, 49 Torr View Avenue, Plymouth, PL3 4QN
M3 OVA T. Durant, 2 Tamar Way, North Hykeham, Lincoln, LN6 8TZ
M3 OVC P. Goodburn, 2 Sunray Cottages, Holt Street, Dover, CT15 4HZ
M3 OVE M. Love, 9 Firswood Drive, Swinton, Manchester, M27 5QY
M3 OVF M. Clay-Burley, 90 Huntington Terrace Road, Cannock, WS11 5HB
M3 OVG T. Speak, 20 Pear Tree Drive, Wincham, Northwich, CW9 6EZ
M3 OVM K. Nelson, 40 Staunton Road, Newark, NG24 4EX
M3 OVO A. Rowe, Southern Point, Grange View, Houghton le Spring, DH4 4HU
MW3 OVT J. Jones, 40 Ffordd Coed Marion, Caernarfon, LL55 2EF
MM3 OVV S. Monaghan, 13 Ballyhennan Crescent, Tarbet, Arrochar, G83 7DB
M3 OVX J. Alexander, 14 Barlow Road, Stretford, Manchester, M32 0RG
M3 OWF A. Holter, 42 Collingwood Close, Eastbourne, BN23 6HW
M3 OWI A. Hodgeon, 30 Rock Bank, Buxton, SK17 9JF
M3 OWN O. Dixon, 28 Manchester Road, Audenshaw, Manchester, M34 5GB
M3 OWO M. Middleditch, 8 Royal Close, Yeovil, BA21 4NX
M3 OWQ P. Mcspirit, 26 Horridge Avenue, Newton-le-Willows, WA12 0AS
M3 OWU D. Adlam, 31 Coxons Close, Huntingdon, PE29 1TS
M3 OWZ S. Waller, 17 Vere Road, Peterborough, PE1 3DZ
MM3 OXB B. Rodriguez, Sprouston House, Newtown St. Boswells, Melrose, TD6 0RY
M3 OXD C. Romocea, 21 Hurst Lane, Cumnor, Oxford, OX2 9PR
M3 OXN P. Jodrell, 2 Greggs Avenue, Chapel-En-Le-Frith, High Peak, SK23 9TU
MM3 OXQ S. Mckinnon, 8 Rowanlea Avenue, Paisley, PA2 0RP
MW3 OXV J. Evans, 71 Bangor Road, Johnstown, Wrexham, LL14 2SR
M3 OXY O. Bazar, 1 Claremont Road, London, NW2 1BP
MM3 OYB S. Morgan, 23 Duncan Road, Glenrothes, KY7 4HS
MW3 OYC S. Poyser, Glandwr Snowdon Street, Porthmadog, LL49 9DF
M3 OYE Lady C. Windsor, 44 Paragon Place, Norwich, NR2 4BL
M3 OYJ C. Rose, 32 Hobart Place, Thornton-Cleveleys, FY5 3DQ
MM3 OYL A. Shearman, 4 Millbrae Crescent, Clydebank, G81 1EH
M3 OYN R. Watson, 60 Beresford Avenue, Surbiton, KT5 9LJ
MI3 OYP C. Brennan, 1 Ballyscullion Lane, Bellaghy, Magherafelt, BT45 8NQ
M3 OYQ N. Loughran, 22 Edulf Road, Borehamwood, WD6 5AD
M3 OYR A. Kirby, 36 Baron Street, Darwen, BB3 1NP
M3 OYS A. Parkes, Oliver Court Bath Hill Terrace, Great Yarmouth, NR30 2LF
M3 OYU N. Carr, 6 Baldwin Avenue, Eastbourne, BN21 1UJ
MM3 OYW J. Waugh, 9 Kedar Bank, Mouswald, Dumfries, DG1 4LU
M3 OYZ M. Ward, 25 Highbury Crescent, Doncaster, DN4 6AL
M3 OZB J. Robinson, 34 High St., Dragonby, Scunthorpe, DN15 0BE
M3 OZC H. Derbyshire, 12 Trinity Homes, St. Clare Road, Deal, CT14 7PX
M3 OZD R. Pounder, 65 Stubsmead, Swindon, SN3 3TB
M3 OZE J. Baldry, 160 Rover Drive, Castle Bromwich, Birmingham, B36 9LL
M3 OZI S. Bird, 9 Almery Drive, Carlisle, CA2 4EX
MI3 OZK J. Sills, 145 Ballycolman Estate, Strabane, BT82 9AJ
M3 OZN P. Davies, 92 Thirlmere Road, Hinckley, LE10 0PF
M3 OZP C. Rolfe, 45 St Clements Court Wear Bay Crescent, Folkestone, CT19 6BP
MI3 OZT R. Hepburn, 34 Pinewood Crescent, Claudy, Londonderry, BT47 4AD
MM3 OZU G. Craig, 1 Butt Avenue, Helensburgh, G84 9DA
MM3 OZW W. Pauley, 43 Pringle Avenue, Tarves, Ellon, AB41 7NZ
M3 OZY O. Morris, 44 Leamington Road, Weymouth, DT4 0EZ
MM3 PAE C. Hume, Sundhopeburn, Yarrow, Selkirk, TD7 5NF
M3 PAI M. Norton, Springfield, Back Lane, Kingston, Sturminster Newton, DT10 2DT
M3 PAP J. Parrott, 2 Boyd Close, Wirral, CH46 1RX
M3 PAU P. Laing, 5 Talisman Close, Barrow-in-Furness, LA14 2UT
M3 PAX N. Haigh, 10 Moor Park Gardens, Dewsbury, WF12 7AS
M3 PBA P. Alce, 1/2 Arawa Street, Christchurch, New Zealand, 8013
M3 PBB R. Bannon, 18 Clavell Road, Liverpool, L19 4TR
M3 PBE R. Rudd, 11 Woodlands Way, Lepton, Huddersfield, HD8 0JA
M3 PBK B. Kellner, 95 Shakespeare Road Ipswich Suffolk, Ipswich, ip16et
M3 PBP D. Parsons, Barn Owl Cottage, Stoke St. Mary, Taunton, TA3 5BY
MM3 PBQ M. Hopkins, 15 Station Drive, Dalbeattie, DG5 4FA
M3 PBR T. Cumming, 2 Ash Grove, Perth Street, Hull, HU5 3PF
M3 PBU W. Wilkinson, 35, Fitzgerald Court, Haughton Green, M34 7LB
MW3 PBV S. Williams, Flat 28, Llys Celyn Cedar Crescent, Tonteg, Pontypridd, CF38

1LF
M3 PBW P. Roberts, 7 Boscombe Road, Swindon, SN25 3EZ
M3 PCC P. Crossley, Firpark Farm, Fir Park, Market Rasen, LN8 3YL
MI3 PCF P. Ford, 25 Carnhill, Londonderry, BT48 8BA
M3 PCP P. Papper, 50 Lincoln Road, Stevenage, SG1 4PL
M3 PCQ J. Campbell, 22 Horsewhim Drive, Kelly Bray, Callington, PL17 8GL
M3 PCW A. Maxwell, Tysties, Tile Barn, Newbury, RG20 9UY
MW3 PCX G. Bellis, 70 Osborne St., Rhos, Wrexham, LL14 2HT
M3 PDC P. Cooper, 62 Fredericks Road, Beccles, NR34 9UG
M3 PDD P. Bennett, 94 Queensway, Taunton, TA1 5QT
MW3 PDE P. Eckersley, 14 Bronantffer, Gwaun Cae Gurwen, Ammanford, SA18 1EN
M3 PDG E. Aitken, 20 Plover Drive, Bury, BL9 6JH
M3 PDK J. Fuller, Rosemar Lodge, Westford, Wellington, TA21 0DX
MI3 PDL P. Burns, 25 Orchard Road, Strabane, BT82 9QS
MM3 PDM P. Mckay, 7 Buchanness Drive, Boddam, Peterhead, AB42 3AT
MI3 PDN R. Neill, 84 Carnreagh, Craigavon, BT64 3AN
M3 PDP J. Clarkson, 56 Edward Bailey Close, Binley, Coventry, CV3 2LZ
M3 PDU L. Fuller, Rosemar Lodge Westford, Wellington, TA21 0DX
M3 PDY P. Dyer, 19 Church Road, Evesham, WR11 2NE
MW3 PEH B. Sellers, 86 St. John Street, Ogmore Vale, Bridgend, CF32 7BB
M3 PEQ M. Vincent, 8 Waldemar Park, Norwich, NR6 6TD
MD3 PER S. Perry, 1 Cronk Grianagh Estate, Strang, Douglas, Isle Of Man, IM4 4QP
MM3 PEV R. Stevenson, 17 Springbank Gardens, Lawthorn, Irvine, KA11 2BY
MM3 PEY D. Oates, 14 Craighlaw Avenue, Eaglesham, Glasgow, G76 0EU
MM3 PFA R. Maddock, 24 Dalhousie Terrace, Montrose, DD10 9BX
M3 PFE R. Laverick, 55 Bondicar Terrace, Blyth, NE24 2JW
M3 PFF A. Fisher, 17 Spicers Way, Totton, Southampton, SO40 9AX
M3 PFK J. Lennon, 6 Piccadilly Square, Burnley, BB11 4QG
M3 PFL K. Gallagher, Flat 1, 59 Trinity Road, Bridlington, YO15 2HF
M3 PFM P. Morris, Canary Cottage, Eye Road, Eye, IP23 7JX
M3 PFN J. Fisk, The Cottage In The Croft, The Croft, Costessey, Norwich, NR8 5DT
M3 PFU S. Silver, 2 Brandon Close, Grange Park, Swindon, SN5 6AA
M3 PFY P. Murthwaite, 34 Cambridge St., Bridlington, YO16 4JZ
M3 PGB D. Brierley, 639 Borough Road, Birkenhead, CH42 9QA
M3 PGD P. Danvers, Heath Farm, Aylsham Road, North Walsham, NR28 0JP
M3 PGH P. Howell, 16 Everard Road, Bedford, MK41 9LD
M3 PGI G. Pollard, 24 Terminus Road, Littlehampton, BN17 5BX
M3 PGK A. Farrar, 8 Wensley Street, Thurnscoe, Rotherham, S63 0PX
M3 PGL P. Lockwood, 42 The Gables, Sedgefield, TS21 3EU
MW3 PGN A. Gazi, 51 Cyncoed Road, Cardiff, CF23 5SB
M3 PGO P. Goodchild, 577 Parrs Wood Road, East Didsbury, Manchester, M20 5QS
M3 PGS P. Stevenson, 6 Dighton Gate, Stoke Gifford, Bristol, BS34 8XA
M3 PGU G. Farrar, 174 Houghton Road, Thurnscoe, Rotherham, S63 0SA
M3 PGY C. Graham, 19 Pontop View, Rowlands Gill, NE39 2JP
MM3 PHC T. Given, 26 Campbell Court, Cumnock, KA18 1NP
M3 PHF J. Austin, 66 Homewood Avenue, Sittingbourne, ME10 1XJ
M3 PHG P. Greenway, 26 Coleridge Gardens, Burnham-on-Sea, TA8 2QA
M3 PHO J. Wilson, 448 Hythe Road, Willesborough, Ashford, TN24 0JH
MM3 PHP P. Goodhall, 12 Templand Road, Lhanbryde, Elgin, IV30 8PP
M3 PHQ N. Benes, 2 Leasowes Close, Watling Street North, Church Stretton, SY6 7BB
M3 PHR P. Norman, 22 Stirling St., Hull, HU3 6SL
M3 PHS P. Saben, Tredinneck Moor, Newmill, Penzance, TR20 8XT
M3 PHX C. Duffill, 181 Foden Road, Great Barr, Birmingham, B42 2EH
M3 PHZ A. Billings, 46 Thorley Drive, Cheadle, Stoke-on-Trent, ST10 1SA
M3 PIA G. Mears, 59 Hastoe Park, Aylesbury, HP20 2AB
M3 PIH D. Huckle, 1 Glebe Road, Biggleswade, SG18 0PE
M3 PIK M. Gould, 57 Fowler Close, Leicester, LE4 0SF
M3 PIL R. Harvey, 50 Warren Walk, Ferndown, BH22 9LY
M3 PIO T. Nakagawa, 7 Milton Street, Barrowford, Nelson, BB9 6HE
M3 PIQ T. Bourne, 100 Dimsdale View West, Newcastle, ST5 8EL
M3 PIW J. Witchell, 7 Watercombe Lane, Yeovil, BA20 2ED
M3 PIY D. Dewsbury, 62 Yew Tree Drive, Leicester, LE3 6PL
M3 PJG P. Goodayle, 2 Downs Road, Seaford, BN25 4QL
M3 PJI P. Jones, 86 Oaks Lane, Rotherham, S61 3ND
M3 PJJ P. Johnson, 7 Harrington Court, Hertford Heath, Hertford, SG13 7QT
MI3 PJM P. Mccausland, 31 Oakleigh Fold, North Street, Craigavon, BT67 9BS
M3 PJN P. Northover, 181 Mullway, Letchworth Garden City, SG6 4BD
M3 PJS P. Seabrook, 29 Gadby Road, Sittingbourne, ME10 1TJ
MW3 PJU M. Sawford, 62 Heol Briwnant, Cardiff, CF14 6QH
M3 PJV P. Vipond, The Old Forge, Nentsbury, Alston, CA9 3LH
MW3 PKC L. Hill, 12 Heol Coedcae, Bargoed, CF81 8QJ
M3 PKE R. North, 11 Tintagel Close, Keynsham, Bristol, BS31 2NL
M3 PKH C. Bell, 12A Mill Lane, Carlton, Goole, DN14 9NG
M3 PKL B. North, 11 Tintagel Close, Keynsham, Bristol, BS31 2NL
M3 PKM P. Bliss, 6 Jubilee Gardens, Biggleswade, SG18 0JW
M3 PKQ J. Blackburn, 64 Marsh Lane, Birmingham, B23 6PJ
M3 PKR B. Parker, 117 Corporation Road, Dudley, DY2 7QT
M3 PKZ M. Woolley, 84 Bowthorpe Road, Norwich, NR2 3TP
M3 PLB P. Beier, 20 Markham Avenue, Armthorpe, Doncaster, DN3 2AZ
M3 PLI P. Lister, 73 Seabrook Court, Seabrook, Hythe, CT21 5RY
M3 PLN P. Lewin, The Hawthorns, Hawthorne Drive, Stafford, ST19 9NQ
M3 PLP P. Price, 4 Priory Avenue, North Ferriby, HU14 3AE
M3 PLU T. Palmer, Edison House, Bow Street, Great Ellingham, NR17 1JB
M3 PLV C. Mccollum, 25 Byron Road, Locking, Weston-Super-Mare, BS24 8AG
M3 PMI J. Segrove, 87 Henry Road, West Bridgford, Nottingham, NG2 7ND
M3 PMK P. Kidd, 78 Studfield Road, Sheffield, S6 4SU
M3 PML P. Lines, 56 Old Hall Close, Amblecote, Stourbridge, DY8 4JQ
M3 PMN P. Nicholls, 30 Nailbourne Court, Palm Tree Way, Lyminge, CT18 8LX
M3 PMO P. Brindle, 5 Showfield Close, Sherburn In Elmet, Leeds, LS25 6LW
MI3 PMR R. Catney, 32 Cairndore Avenue, Newtownards, BT23 8RF
MW3 PMU D. Jones, 34 Pen Y Bryn, Rassau, Ebbw Vale, NP23 5AJ
MI3 PMW M. Pollock, 5 St. Marys Terrace, Stream Street, Newry, BT34 1HL
M3 PMX A. Marlow, Yeomans Barn, Kingsbridge, TQ7 3BH
M3 PMY T. Wright, 11 Ash Close, Daventry, NN11 0XH
M3 PNA N. Phillpott, Eescroft, Stombers Lane, Folkestone, CT18 7AP
M3 PNB B. Crosswell, 5 Harty Ferry View, Whitstable, CT5 4TE
M3 PNF G. Taylor, 31 Ashfurlong Crescent, Sutton Coldfield, B75 6EN
M3 PNH N. Hoyle, 34 The Drive, Halifax, HX3 8NJ
M3 PNI E. Bishop, 29 Windsor Court, Poulton-le-Fylde, FY6 7UX
M3 PNO C. Turner, 28 Fox Lea, Kesgrave, Ipswich, IP5 2YU
MW3 PNR R. Williams, Plaen Cottage, Bodfari, Denbigh, LL16 4BS
M3 PNV J. Nelmes, 118 Silcoates Lane, Wrenthorpe, Wakefield, WF2 0PE
M3 PNY P. Robinson, 19 St. Wilfrids Crescent, Brayton, Selby, YO8 9EU
M3 PNZ R. Maddock, 48 Collygree Parc, Goldsithney, Penzance, TR20 9LY
MI3 POB P. O'Brien, 71 Whitepark Road, Ballycastle, BT54 6LP
M3 POH D. Dean, 119 Queens Drive, Newton-le-Willows, WA12 0LN
MM3 POI T. Penna, North Windbreck Deerness, Orkney, KW17 2QL
M3 POP J. Morris, Lurdin Lodge, 71 Lurdin Lane, Wigan, WN6 0AQ
M3 POQ R. Finch, 19B Kiln Road, Newbury, RG14 2LS
M3 POV N. Trangmar, 8 Maxstoke Close, Meriden, Coventry, CV7 7NB
M3 POW M. Davies, 5 Twyford Avenue, Great Wakering, Southend-on-Sea, SS3 0EZ
MM3 PPA S. Parsons, Linksview, Barrock, Thurso, KW14 8SY
MI3 PPD R. Reilly, 220 Ardmore Road, Londonderry, BT47 3TE
M3 PPG J. Godfrey, 4 Cherry Close, Houghton Conquest, Bedford, MK45 3LQ
MI3 PPI P. Pearson, 1 Rock Cottages, Springwell Road, Bangor, BT19 6LZ
M3 PPK N. Green, 11 Wythburn Way, Rugby, CV21 1PZ
M3 PPO A. Wade, 40 Throxenby Lane, Scarborough, YO12 5HW
MW3 PPQ P. Sharrock, 23 Grosvenor Gardens, Wrexham, LL11 1EF
M3 PPR P. Saving, Room 1, Abbeyfield, The Glebe Field, Sevenoaks, TN13 3DR
M3 PPU M. Whitten, 75 Regent Street, Whitstable, CT5 1JQ
M3 PPY J. Evans, 21 Quilter Close, Bilston, WV14 9AX
M3 PPZ R. Bullen, 67 Abberley Road, Liverpool, L25 9QY
M3 PQB D. Fagg, 62 Hawkins Road, Folkestone, CT19 4JA
MW3 PQE A. Paffey, 1 St. Vincent Road, Newport, NP19 0AN
MW3 PQF E. Paffey, 1 St. Vincent Road, Newport, NP19 0AN
M3 PQG J. Goldfinch, 138 Palmerston Road, Chatham, ME4 5SJ
M3 PQI M. Bhatia, Swaynes House, Room 2 Flat 2, Wirenhoe Park, CO4 3SQ
M3 PQJ J. Pratt, 4 Bussex Square, Westonzoyland, Bridgwater, TA7 0HD
M3 PQL A. Mellor, 112 Allerton Road, Stoke-on-Trent, ST4 8PL
MI3 PQM P. Millar, 37 Thorncroft, Ahoghill, Ballymena, BT42 1RX
M3 PQN A. Phillips, 15 Hertford Close, Woolston, Warrington, WA1 4EZ
M3 PQQ R. Fern, 3 Park Road, Featherstone, Wolverhampton, WV10 7HS
M3 PQT S. Broadbent, 86 Inverness Road, Dukinfield, SK16 5AB
M3 PQU A. Milton-Eldridge, 2 Partridge Close, Didcot, OX11 6AB
M3 PQV M. Nolan, Bath Road Post Office, Post Restante. Bath Road, Devizes, SN10 1QG
M3 PRN S. Martin, 35 Hermitage Green, Hermitage, Thatcham, RG18 9SL
M3 PRS G. Manchester, 251 Osmaston Park Road, Allenton, Derby, DE24 8DA
M3 PRU P. Broadbere, Refail, Resthill Road, Wirral, CH63 6HN
M3 PRY S. Mariott, 4 Stone Cross Gardens, Catterall, Preston, PR3 1YQ
M3 PRZ P. Radmall, Appleford, Bowcombe Road, Kingsbridge, TQ7 2DJ
M3 PSB S. Birch, 6 Crescent Road, Wallasey, CH44 0BQ
M3 PSC P. Cattel, 21 School Hill, Chickerell, Weymouth, DT3 4BA
M3 PSD P. Sheppard, 107 Queen Street, Swinton, Mexborough, S64 8NF
M3 PSE P. Elliott, 11 Forgefields, Herne Bay, CT6 7TB
M3 PSF R. Baroch, 15 Salters Lane, Redditch, B97 6JH
M3 PSI K. Morton, 47 Trinity Court, Halstead, CO9 1PY
MM3 PSL P. Leech, The Croft House, 9 Ruilick, Beauly, IV4 7AB
M3 PSM S. Maughan, 17 Upper Dane, Desborough, Kettering, NN14 2LB
M3 PSO W. Weaver, Challacombe House, Perrinpit Road, Bristol, BS36 2AT
M3 PSR P. Roberts, 43 Ashbourne Crescent, Sale, M33 3LQ
M3 PSS P. Swanepoel, Bridge Cottage, Tinhay, Lifton, PL16 0AH
M3 PSU J. Davidson, 5 Hanover Parc, Indian Queens, St Columb, TR9 6ER
M3 PSZ A. Laurence, Brookvale, Nooklands, Preston, PR2 8XN
M3 PTA P. Chambers, 257 Kings Acre Road, Hereford, HR4 0SR
M3 PTB T. Bishop, 4 Walnut Grove, Worlington, Bury St Edmunds, IP28 8SF
M3 PTG T. Green, Huntley, Chesham Road, Tring, HP23 6HH
M3 PTI B. Parton, 51 Marston Grove, Stoke-on-Trent, ST1 6EF
M3 PTQ W. Hughes, 6A Park Road, Melton Mowbray, LE13 1TT
M3 PTR P. Dryden, 27 Delaval Crescent, Blyth, NE24 4AZ
M3 PTV B. Fitzakerley, 38 Hazel Grove, Armthorpe, Doncaster, DN3 3HG
M3 PTX C. Hunter, 30 Glebelands, Pulborough, RH20 2JJ
M3 PUB P. Swynford, 6 The Rise, Cold Ash, Thatcham, RG18 9PD
M3 PUE A. Hannon, 8 Circular Road West, Liverpool, L11 1AZ
MI3 PUH J. Dunlop, 118 Ardenlee Avenue, Belfast, BT6 0AD
M3 PUL P. Stead, 36 Reeds Avenue East, Wirral, CH46 1RQ
M3 PUN J. Rideout, 35 Colmead Court, Northampton, NN38QE
M3 PUQ J. Hunt, 104 Hamilton Avenue, Cheam, Sutton, SM3 9RL
M3 PUT C. Townsend, 2 Netherfield Drive, Netherthong, Holmfirth, HD9 3ES
MW3 PUU B. Hill, 97 Maesglas Grove, Newport, NP20 3DN
M3 PUZ S. Turford, 1 Portland Crescent Bolsover, Chesterfield, s446eg
M3 PVB T. Ireland, 114 Alder Lane, Warrington, WA2 8AW
MW3 PVC M. Cook, 9 Drenewydd, Park Hall, Oswestry, SY11 4AH
M3 PVI P. Handley, 97 Applegarth Avenue, Guildford, GU2 8LX
M3 PVP A. Botley, Flat 1/B, 46 Trull Road, Taunton, TA1 4QH
M3 PVQ E. Rhodes, The Old Forge, Stoke Gabriel, Totnes, TQ9 6RL
M3 PVU J. Watson, 20 St. Marys Gardens, Hilperton Marsh, Trowbridge, BA14 7PG
M3 PVV T. Crisp, 6 Tumlins, All Cannings, Devizes, SN10 3PQ
M3 PVX Dr D. Jones, 22 Taylor Street, Hollingworth, Hyde, SK14 8PA
M3 PWE P. Ward, 69 Woodlands Avenue, Tadcaster, LS24 9HP
M3 PWK S. Platts, 59 Sea View Road, Drayton, Portsmouth, PO6 1EW
M3 PWM P. Mitchell, 13 Ashorne Close, Matchborough, Redditch, B98 0EY
M3 PWO D. Robertson, 53 Moor Lane, Weston-Super-Mare, BS22 6RA
M3 PWS P. Sykes, 2 Thornton Villas, Barrow Road, Barrow-upon-Humber, DN19 7QG
M3 PWW P. Wright, 16 Hainault Avenue, Giffard Park, Milton Keynes, MK14 5PA
M3 PWZ B. Pearson, 119 Tolkien Road, Eastourne, BN23 7AQ
M3 PXE K. Peel, 123 Cunningham Road, Tamerton Foliot, Plymouth, PL5 4PU
M3 PXF T. Gabriel, 57 West Down Road, Delabole, PL33 9DT
MM3 PXG S. Simpson, 63F Ballindean Road, Dundee, DD4 8NX
M3 PXK R. Ellery, 12 Sentry Close, St Issey Wadebridge, pl27 7qd
M3 PXL P. Houghton, 37 Cedar Avenue, Cottingham, HU16 4AL
MM3 PXO E. Mccook, 6 Elms Place, Stevenston, KA20 4EF
M3 PXP M. Williams, 9 Clarence Place, Stonehouse, Plymouth, PL1 3JN
M3 PXQ N. Kendall, 19 Clowance Lane, Mount Wise, Plymouth, PL1 4HU
M3 PXT P. Mutavdzic, 1 Hawthorne Drive, Kingwood, Henley-on-Thames, RG9 5WE

M3 PXU V. Parton, 51 Marston Grove, Stoke-on-Trent, ST1 6EF
M3 PXW B. Smith, 19 Alexandra Square, Winsford, CW7 2YR
M3 PXY C. Fox, 45 Park Road, Wivenhoe, Colchester, CO7 9LS
M3 PXZ Dr C. Fox, 45 Park Road, Wivenhoe, Colchester, CO7 9LS
M3 PYB D. Furlong, 6A Glebe Avenue, Ruislip, HA4 6QZ
M3 PYD M. Smith, 6 Neeps Terrace, Middle Drove, Wisbech, PE14 8JT
M3 PYG A. Webb, 104 Birds Nest Avenue, Leicester, LE3 9ND
M3 PYI W. Mcbain, Willow Cottage, Gedney Broadgate, Spalding, PE12 0DE
M3 PYJ S. Randall, 23 Onslow Road, Plymouth, PL2 3QG
M3 PYO D. Horner, 21 Ainsworth Road, Little Lever, Bolton, BL3 1RG
M3 PYR P. Rushby, 16 Foxhill Lane, Selby, YO8 9AR
M3 PYS E. Ransom, 10 Gillercomb, Redcar, TS10 4SG
M3 PYT C. Harris, 16 Downfield Way, Plymouth, PL7 2DU
M3 PYW B. Simmonds, 55 Pepys Road, St Neots, PE19 2EN
MI3 PYX D. Caiden, 14 Church Street, Greyabbey, Newtownards, BT22 2NQ
MW3 PYY C. Thomas, 22 Sea Road, Abergele, LL22 7BU
MW3 PZC M. Marston, The Retreat, The Catch, Holywell, CH8 8DU
M3 PZF J. Bealey, 17 Chelston Road, Newton Abbot, TQ12 2NN
MM3 PZJ S. Ling, Leadburnlea Leadburn, West Linton, EH46 7BE
M3 PZL R. Mckenzie, 4 Simpkin Street, Abram, Wigan, WN2 5QD
M3 PZN L. Mckenzie, 4 Simpkin Street, Abram, Wigan, WN2 5QD
MW3 PZO S. Connor, 8 Bro Arfon, Upper Llandwrog, Caernarfon, LL54 7BH
MI3 PZV C. Stewart, 99 Hillmount Road, Cullybackey, Ballymena, BT42 1NZ
M3 PZX P. Seabrook, 29 Gadby Road, Sittingbourne, ME10 1TJ
MW3 RAA A. Gordon, 5 Parc Hendy, Mold, CH7 1TH
M3 RAE J. Webb, 6 Chatsworth Avenue, Fleetwood, FY7 8EG
M3 RAK R. King, 79 Holmside Avenue, Minster On Sea, Sheerness, ME12 3EZ
MW3 RAU T. Rowlands, 3, Pool, Llanfairfechan, LL330TN
MM3 RBF D. Gemmell, 36 Church St., Dumfries, DG2 7AS
M3 RBI R. Gilbert, 61 Coltstead, New Ash Green, Longfield, DA3 8LN
MM3 RBJ B. Johnston, 71 Upper Mastrick Way, Aberdeen, AB16 5QG
MI3 RBM R. Abraham, 9 Milfort Gardens, Waringstown, Craigavon, BT66 7PD
M3 RBP R. Peacock, 27 Greenside, Kendal, LA9 5DU
M3 RBQ C. Boarer, 37 The Martlets, Rustington, Littlehampton, BN16 2UB
M3 RBT R. Kerr, The Dower House, Church Square, Derby, DE73 8JH
M3 RBU B. Upton, 1 Sunningdale Close, Eastleigh, SO50 8PU
M3 RBX R. Swynford, 6 The Rise, Cold Ash, Thatcham, RG18 9PD
M3 RCC N. Prescott, 3 View Fields, Station Road, Doncaster, DN9 3AE
M3 RCD R. Cooke, 22 Shepperton Close, Great Billing, Northampton, NN3 9NT
M3 RCE R. Edwards, 15 Burghley Street, Bourne, PE10 9NS
M3 RCG R. Gifford, Mill House, Mill Road, Topcroft, Bungay, NR35 2BW
M3 RCI K. Steele, Flat 22, Bradgate Court, Staunton Avenue, Derby, DE23 1PR
M3 RCQ M. Snowden, Amber Lights, Market Lane, Walpole St. Andrew, Wisbech, PE14 7LT
MM3 RCR J. Mcmartin, 19 Bruce Street, Bannockburn, Stirling, FK7 8UF
M3 RCS R. Swietlik, 3 Tarvin Close, Sutton Manor, St Helens, WA9 4DL
M3 RCT M. Russell, 107 Cambridge Road, Hitchin, SG4 0JH
M3 RCV R. Treacher, 93 Elibank Road, London, SE9 1QJ
M3 RCW R. Wiggins, 68 Beaconsfield Road, Burton-on-Trent, DE13 0NT
MM3 RCX K. Carroll, 32E Meadowburn Place, Campbeltown, PA28 6ST
MM3 RCZ A. Conlon, Kilrae, Barrpath, Glasgow, G65 0EX
M3 RDA R. Astbury, 12 Southall Road, Ashmore Park, Wolverhampton, WV11 2PZ
M3 RDH R. Hastings, 43 Delmar Avenue, Leverstock Green, Hemel Hempstead, HP2 4LZ
MM3 RDP D. Moore, 47 Lockhart Street, Germiston, Glasgow, G21 2AP
MW3 RDS R. Smith, Hendafarn, Sarnau, Llanymynech, SY22 6QJ
M3 RDV R. Dewes, 31 Woodlea Avenue, Lutterworth, LE17 4TU
M3 RDW R. Wells, 37 Water Meadows, Worksop, S80 3DF
M3 RDY R. Young, 4 Hammond Court, Mablethorpe, LN12 2EL
M3 REL R. Lowis, 53 Harewood Crescent, Louth, LN11 0JD
M3 REM A. Kernick, 40 Leyster Street, Morecambe, LA4 5NF
M3 REP A. Wilkinson, 6 Humbledon View, Sunderland, SR2 7RX
M3 REQ A. Williamson, 25 Manor Road, Rugby, CV21 2SZ
M3 RET M. Parkes, 12 Penderel Street, Walsall, WS3 3DX
M3 REX S. Thompson, Rutland, Quaker Lane, Wirral, CH60 6RD
M3 REZ A. Adkins, 91 Fernbank Road, Birmingham, B8 3LL
M3 RFF P. Richardson, 31 Castlefields Drive, Brighouse, HD6 3XF
M3 RFG R. Gray, Upper Bisterne Farmhouse, Bisterne, Ringwood, BH24 3BP
M3 RFH R. Henderson, 9 Green Mead, South Woodham Ferrers, Chelmsford, CM3 5NL
M3 RFI K. Hilton, 199A Sale Lane, Tyldesley, Manchester, M29 8PG
M3 RFK Rev. R. Eardley, Bridge Cottage, Martin, Fordingbridge, SP6 3LD
M3 RFO Dr J. Mccue, 40 Bradbury Road, Stockton-on-Tees, TS20 1LE
M3 RFR S. Fallows, 23 Howard Street, Burnley, BB11 4BJ
M3 RFW R. Ford, 30 Cartmel Close, Worcester, WR4 9NT
M3 RFX R. Ashworth, 10 Mulberry Close, Wigan, WN5 9QL
M3 RGC R. Cummings, Juan Rodriguez El Cusques, No. 25 (Plot 20A), Alicante, Spain
MW3 RGD R. Hogben, The Steppes, Presteigne Road, Knighton, LD7 1HY
M3 RGE M. Tate, 52 Marlborough Road, London, N22 8NN
M3 RGG J. Brown, 339 Manor Road, Brimington, Chesterfield, S43 1NU
MM3 RGH R. Heath, Roadside Cottage, Glenkindie, Alford, AB33 8SH
M3 RGJ R. Jamieson, 3 Waterpark Road, Prenton Park, Birkenhead, CH42 9NZ
M3 RGK K. Harley, Care Of: Mr K Harley 9 Amyas Way, Northam, Bideford, EX39 1UT
M3 RGN P. Frampton, 118 Ramnoth Road, Wisbech, PE13 2JD
M3 RGP R. Prangnell, 124 St. Marys Road, Cowes, PO31 7SR
M3 RGU A. Oxlade, 27 Spenfield Court, Northampton, NN3 8LZ
MM3 RGZ J. Cairney, 5 James Street, Bannockburn, Stirling, FK7 0NQ
MM3 RHA W. Hawthorn, 8 Drummond Place, Stirling, FK8 2JE
M3 RHB J. Palmer, 2 Dagonet Road, Bromley, BR1 5LR
M3 RHD R. Duffield, 4 Crabmill Lane, Easingwold, York, YO61 3DE
M3 RHG R. Greatrix, 24 Berwick Drive, Cannock, WS11 1NS
MW3 RHI R. Chalk, 42 Erskine Road, Colwyn Bay, LL29 8EU
M3 RHJ M. Dennis, 10 Welland Court, Burton Latimer, Kettering, NN15 5ST
M3 RHK S. Bharrich, 8 Ferrers Ave, Tutbury, DE13 9JR
M3 RHL R. Looker, 165 Mollison Drive, Wallington, SM6 9GX
M3 RHO R. Frylinck, 46 Buckingham Road, Richmond, TW10 7EQ
M3 RHP Lord R. Montague, 71 Middlethorpe Road, Cleethorpes, DN35 9PP
M3 RHR G. Kensett, 12 Rustics Close, Calvert, Buckingham, MK18 2FG
MM3 RHT G. Fyfe, 7 Coralmount Gardens, Kirkintilloch, Glasgow, G66 3JW
M3 RIA R. Wilkinson, 18 Green Road, Kendal, LA9 4QR
M3 RIE M. Hatton, Elisha Cottage, St. Peters Walk, Hull, HU7 5FB
MI3 RIF J. Smyth, 37 Ardfreelin, Newry, BT34 1JG
MI3 RIL L. Scott, 98 Aghafad Road Dunamanagh, Strabane, BT82 0QQ
M3 RIP M. Pearce, 104 Sea Lane, Goring-By-Sea, Worthing, BN12 4PU
M3 RIU M. Manser, 17 Emperor Way, Kingsnorth, Ashford, TN23 3QY
MI3 RIV S. Datchanamourty, 22 Craigmore Road, Bessbrook, Newry, BT35 6LF
MM3 RIX R. Guthrie, 27 Meadowbank Road, Kirknewton, EH27 8BH
M3 RJB R. Bird, 78 Arden Road, Hockley, Tamworth, B77 5JE
M3 RJF R. Fitzgerald, 3 Sefton Lane Maghull, Liverpool, L31 8AE
M3 RJH R. Hicks, 31 Arundel Road, Great Yarmouth, NR30 4LD
M3 RJI P. Cummings, 24 Spindle Road, Malvern, WR14 2WB
M3 RJK R. Kelso, 55D Lewisham Hill, London, SE13 7PL
M3 RJP R. Page, 4 Nursery Drive, March, PE15 8EQ
M3 RKE K. Simmons, 26 Red Hill Close, Studley, B807BZ
MM3 RKF R. Mannifield, 2 Plewlands Avenue, Edinburgh, EH10 5JY
M3 RKJ J. Mccoll, 6 Grenville Close, Bodmin, PL31 2FB
M3 RKK E. Whiten, 17 Scott Close, Ashby-de-la-Zouch, LE65 1HT
M3 RKN R. Neville, 4 Danson Gardens, Blackpool, FY2 0XH
M3 RKR R. Rudd, 43 Greenlands Road, East Cowes, PO32 6HT
M3 RKV A. Gallop, 75 Shearmans, Fullers Slade, Milton Keynes, MK11 2BQ
M3 RKZ P. Lewin, 42 Eastland Road, Yeovil, BA21 4EX
MI3 RLA A. Holmes, 5 Cambrai Cottages, Belfast, BT13 3PS
M3 RLB A. Marlborough, Maximillian Cottage, Manswood Common, Wimborne, BH21 5BH
MM3 RLG M. Geldart, 13B Greystone Place, Newtonhill, Stonehaven, AB39 3UL
M3 RLH M. Thompson, 9 Stratfield Place, New Milton, BH25 5XE
M3 RLM M. Rose, 71 Old Street, Ludlow, SY8 1NS
M3 RLO C. Rodway, 11 Cleveland Avenue, Bishop Auckland, DL14 6AR
M3 RLT B. Tarpey, 54 Lowforce, Wilnecote, Tamworth, B77 4LU
M3 RLX G. Cox, 6 Bullfinch Close, Poole, BH17 7UP
M3 RMD M. Rout, 2 Woods End Cottages, Kirby Bedon, Norwich, NR14 7EB
M3 RMG G. Chapman, Crockers Farm, Stoke Wake, Blandford Forum, DT11 0HF
M3 RMH R. Hunt, 4 Rue Josy Printz, Hesperange, Luxembourg, L-5841
M3 RMI J. Salmon, 25 Helston Road, Chelmsford, CM1 6JF
M3 RMQ J. Weston, 25 Cambridge Road, Orrell, Wigan, WN5 8PL
M3 RMS R. Stevenson, 97 Queen Street, Crewe, CW1 4AL
M3 RMU A. Teed, 57 Lymington Road, Torquay, TQ1 4BG
M3 RMV R. Ley, 23 Heronbridge Close, Westlea, Swindon, SN5 7DR
M3 RMX R. Moore, 47 Darwin Road, Walsall, WS2 7EN
M3 RMZ P. Randall, 289 Wilson Avenue, Rochester, ME1 2SS
M3 RNG P. Davies, Bluebells, Station Road, Bere Alston, Yelverton, PL20 7EP
M3 RNK V. Penprase, 62 California Gardens, Plymouth, PL3 6SZ
M3 RNM M. James, 7 Pixey Place, Oxford, OX2 8BB
MI3 RNN R. Nicholl, 30 Foyle Crescent, Newbuildings, Londonderry, BT47 2QR
M3 RNO R. Baker, 12 Byland Road, Skelton-In-Cleveland, Saltburn-by-the-Sea, TS12 2NJ
M3 RNS P. Rainey, 27 School Road, Silver End, Witham, CM8 3RZ
M3 RNU R. Hargate, 79 Boundary Road, Beeston, Nottingham, NG9 2QZ
M3 RNW R. Whitehead, 1 Smithy Site, Farnborough, Wantage, OX12 8NS
M3 RNY R. Kirk, 5 Sidcup Court, Southgate Way, Chesterfield, S43 2NR
M3 ROF D. Johnson, 184 Howbeck Road, Arnold, Nottingham, NG5 8QE
M3 ROI W. Chorlton, 25 Ash Grove, Orrell, Wigan, WN5 8NG
M3 ROQ R. Bird, 12 Windsor Road, Loughborough, LE11 4LL
M3 ROU P. Curnow, 301 Hill Top Drive, Rochdale, OL11 2AG
MM3 ROV D. Brown, Courtyard Cottage, Letters Farm, Argyll, PA27 8BX
M3 ROW F. Webley, 2 Octavian Drive, Bancroft, Milton Keynes, MK13 0PN
MW3 ROX I. Davies, 2 Dinas Terrace, Aberystwyth, SY23 1BT
M3 RPA O. Akanyeti, Sq/H8/4/A University Quays, Lightship Way, Colchester, CO2 8GY
M3 RPD I. Handley, Rosedale, Chapman Street, Market Rasen, LN8 3DS
M3 RPE R. Evenden, 1 Castle Place, 2 Castle Street, Tonbridge, TN9 1BN
M3 RPF R. Fullagar, 6 Locke Way, Stafford, ST16 3RE
M3 RPH R. Hales, 8 Barton Close, Kingsbridge, TQ7 1JU
M3 RPK H. Friberg, 19 Holmcroft, Newbiggin-by-the-Sea, NE64 6DQ
M3 RPQ J. Faulkner, 3 Britannia Quay, 37 River Road, Littlehampton, BN17 5DB
M3 RPR R. Roebuck, 50 Henson Avenue, Blackpool, FY4 3LY
M3 RPS P. Cooper, 10 Headon Gardens, Exeter, EX2 6LE
MW3 RPX A. Davies, 24 Ash Lane, Mancot, Deeside, CH5 2BR
MW3 RPZ M. Stokes, 23 Goetre Fawr Road Killay, Swansea, SA2 7QS
M3 RQB R. Wood, 1 Kildare Garth, Kirkbymoorside, York, YO62 6LN
MM3 RQC J. Livingstone, 17 Livingstone Drive, Bo'ness, EH51 0BQ
MM3 RQG H. Smith, Ryefield, Windyknowe Road, Galashiels, TD1 1RG
M3 RQJ M. Powell, 2 Walton Avenue, Twyford, Banbury, OX17 3LB
M3 RQO D. Rouse, 46 Frensham Drive, Bradford, BD7 4AS
MM3 RQP F. Pudsey, 21/2 Bathfield, Edinburgh, EH6 4DU
M3 RQQ Dr R. Baldwin, Red Barn, Elder Lane, Grimston, King's Lynn, PE32 1BJ
M3 RQR L. Robinson, 19 Adur Avenue, Shoreham-by-Sea, BN43 5NN
MI3 RRE R. Rantin, 8A Buchanans Road, Newry, BT35 6NS
M3 RRJ J. Roughley, 42 Thistledown Close, Wigan, WN6 7PA
M3 RRN D. Dunstan, 2 Trevarren Avenue, Four Lanes, Redruth, TR16 6NH
MW3 RRU G. Clements, 9 Esgair Y Gog, Bronllys, Brecon, LD3 0HY
M3 RRV R. Taylor, 2 Chadwick Road, Moorends, Doncaster, DN8 4NG
MW3 RRW G. Tucker, 18 Plymouth Road, Penarth, CF64 3DH
M3 RRZ S. Garrett, 44 Wardle Crescent, Leek, ST13 5PW
M3 RSH Dr R. Hodgkinson, 39 Oxford Road, Carlton-In-Lindrick, Worksop, S81 9BD
M3 RSN C. Keeley, 3A St. Marks Road, Huyton, Liverpool, L36 0XA
MM3 RSR R. Rogerson, 93 Auchencrieff Road, Locharbriggs, Dumfries, DG1 1UZ
MI3 RST J. Donaldson, 12 Drumcrow Road, Glenanne, Armagh, BT60 2JQ
M3 RSX R. Shippey, 43 Westbury Street, Bradford, BD4 8PB
M3 RTE R. Turner, 2 Gate House Cottages, Hunton Road, Tonbridge, TN12 9SG
MM3 RTH W. Fitzsimons, 34 Caledonian Road, Stevenston, KA20 3LG
M3 RTI R. Brew, 45 Stephenson Road, Braintree, CM7 1DL
M3 RTP S. Tingay, 9 Cottage Homes, Wakefield Road, Huddersfield, HD5 9XT
M3 RTR S. Davis, 104 Cairo Avenue, Peacehaven, BN10 7LA
MW3 RTU M. Kidner, 4 Tonypistyll Road, Newbridge, Newport, NP11 4HJ

MW3 RUH D. Thomas, 23 Merthyr Dyfan Road, Barry, CF62 9TG
M3 RUI R. Wang, 86 Sunnyside Road, Beeston, Nottingham, NG9 4FG
M3 RUK K. Moody, 114 Acomb Road, York, YO24 4EY
M3 RUL C. Rule, 109 Carshalton Park Road, Carshalton, SM5 3SJ
M3 RUO S. Mcguinness, 64 Newshaw Lane, Hadfield, Glossop, SK13 2AT
M3 RUR J. Stimpson, 2 Church Avenue, Kings Sutton, Banbury, OX17 3RJ
MI3 RUV D. Mccloskey, 1 Dernaflaw Cottages Dernaflaw Road, Dungiven, Londonderry, BT47 4PP
M3 RUW D. Jaynes, 81 Bude Crescent, Stevenage, SG1 2QL
MM3 RUZ G. Ruzgar, 22 Ochil Terrace, Dunfermline, KY11 4BW
M3 RVE G. Thorpe, 81 Knoll Drive, Coventry, CV3 5PJ
M3 RVJ D. Purser, 11 Barnards Close, Malvern, WR14 3NJ
M3 RVK G. Myall, 418 Chester Road, Warrington, WA4 6ES
M3 RVM D. Morley, 5 Pelham Close, Westham, Pevensey, BN24 5NL
MW3 RVN S. Jones, 15 Corn Hill, Porthmadog, LL49 9AT
MW3 RVP R. Lasbury, 57 Westbourne Road, Whitchurch, Cardiff, CF14 2BR
M3 RVQ R. Lester, 17 Clarence Road, Capel-Le-Ferne, Folkestone, CT18 7LW
M3 RVS R. Sohst, 2 Shaftesbury Drive, Maidstone, ME16 0JS
M3 RVX M. Brandon, 9 Holly Drive, Winsford, CW7 1DZ
M3 RWC R. Cornwall, 9 Bishop Close, Dunholme, Lincoln, LN2 3US
M3 RWD R. Davidson, 3 Eastridge Drive Bishopsworth, Bristol, BS13 8HQ
M3 RWI J. Marshall, 18 Dunnett Road, Folkestone, CT19 4BX
M3 RWK J. Lecaille, Tezlan, Colton Road, Norwich, NR9 5BB
M3 RWN R. Nock, 83 Coles Lane, West Bromwich, B71 2QW
M3 RWR D. Griffiths, 1 Ballard Crescent, Dudley, DY2 9EZ
M3 RWV R. Chown, 7 Foden Walk, Wilmslow, SK9 2HQ
M3 RXD R. Dryburgh, 21 Glebe Close, Stow On The Wold, Cheltenham, GL54 1DJ
MI3 RXF I. Smyth, 42 Mullintill Road, Claudy, Londonderry, BT47 4JN
M3 RXG R. Jolly, 102 Swanstree Avenue, Sittingbourne, ME10 4LF
MW3 RXH R. Rimmer, Dwyfor, Heol Las, Llantrisant, Pontyclun, CF72 8EG
MW3 RXK S. Merrifield, 37 South View Drive Rumney, Cardiff, CF3 3LX
MM3 RXM R. May, 12 Clochbar Gardens, Milngavie, Glasgow, G62 7JP
M3 RXO L. Milburn, 17 Hammingden Court, Crawley, RH10 3FR
M3 RXP Dr R. Whittle, 20 Marlbrook Lane, Marlbrook, Bromsgrove, B60 1HN
M3 RXQ M. Milne, Flambards, Manor Road, Dunmow, CM6 2JR
M3 RXT J. Bridson, 10 Clegg Street, Astley, Tyldesley, Manchester, M29 7DB
MI3 RXU I. Ophert, 5 Cloghboy Road, Bready, Strabane, BT82 0DN
M3 RXW R. Webb, 17 St. Marys Close, Chudleigh, Newton Abbot, TQ13 0PL
M3 RYA R. Petts, 19 Sandwell Avenue, Darlaston, Wednesbury, WS10 7RH
MI3 RYD S. Davison, 60 Cornation Place, Craigavon, BT66 7AN
M3 RYG R. Hughes, 117 Liverpool Road, Irlam, Manchester, M44 6EH
M3 RYI S. Ashcroft, 9 Aldermere Crescent, Urmston, Manchester, M41 8UE
MI3 RYJ J. Hazlett, 25 Gorteen Crescent, Limavady, BT49 9EW
M3 RYN R. Fowler, Ryland, Back Lane, Doncaster, DN9 3AJ
M3 RYO M. Shasby, 19 Crawshaw Grange, Crawshawbooth, Rossendale, BB4 8LY
M3 RYR R. Farrar, 41 Newtown Avenue, Cudworth, Barnsley, S72 8DY
M3 RYT J. Abbott, 22 Brent Close, Witham, CM8 1TJ
M3 RYY R. Crowther, 6 Kaliton, Church Street, Callington, PL17 7GB
M3 RYZ A. Clunnie, 19 Griffin Road, Warwick, CV34 6QX
M3 RZB R. Brittain, 159 Caledonia Road, Wolverhampton, WV2 1JA
MJ3 RZD R. Luscombe, Flat 3, 1 Rouge Bouillon, St. Helier, Jersey, JE2 3ZA
M3 RZE C. Russell, 255 Leeds Road, Shipley, BD18 1EH
M3 RZF S. Harris, Cross House, Mill Lane, Preston, PR3 2JX
M3 RZG J. Plant, The Cottage, Back Springfield Road, Lytham St Annes, FY8 1TN
M3 RZI O. Rabbitt, 20 Lysander Drive, Padgate, Warrington, WA2 0GL
M3 RZJ B. Trayhurn, 15 Wight Drive, Caister-On-Sea, Great Yarmouth, NR30 5UN
M3 RZL A. Linden, 12 Godstone House, Pardoner St., London, SE1 4DT
M3 RZM G. Kingstone, 17 Ullswater Drive, Leighton Buzzard, LU7 2QR
M3 RZN I. Seaman, 6 Aylsham Road, Buxton, Norwich, NR10 5EX
M3 RZO N. Bardell, 1 Walshs Manor, Stantonbury, Milton Keynes, MK14 6BU
M3 RZP R. Powell, 53 St. Marys Road, Adderbury, Banbury, OX17 3HA
MI3 RZT J. Thompson, 119 Rathkyle, Antrim, BT41 1LN
M3 RZU U. Ekpe, Cathedral Court, University Campus, Guildford, GU2 7JH
M3 RZV R. Millington, Quaintways, The Avenue, Tarporley, CW6 0BA
MW3 RZW R. Lancaster, 10 Railway Terrace, Tirphil, New Tredegar, NP24 6EY
M3 RZX J. Shorey, 47 Stanham Road, Dartford, DA1 3AN
M3 RZY S. Trotter, 62 Regent Street, Whitstable, CT5 1JQ
M3 SAA S. Atkinson, 10 Pond Lane, New Tupton, Chesterfield, S42 6BG
M3 SAB S. Hughes, 9 Melverton Avenue, Wolverhampton, WV10 9HN
MW3 SAI R. Blore, Ty Nwydd, Cymau, Wrexham, LL11 5EU
MM3 SAK A. Mcneil, 21 Dumbreck Terrace, Queenzieburn, Glasgow, G65 9EA
M3 SAO A. Osmond, 36 Knowles Road, Leicester, LE3 6LU
M3 SAR S. Abraham, 12 Graham Road, Halesowen, B62 8LJ
M3 SAS G. Deakin, 145 Duke Street, Stoke-on-Trent, ST4 3NR
M3 SAY S. Yapp, Dickers Farm, Beechy Road, Uckfield, TN22 5JG
M3 SAZ S. Greatorex, 54 Lilac Grove, Glapwell, Chesterfield, S44 5NG
M3 SBA A. Savory, 51 Catterick Way, Towcester, NN12 6NX
M3 SBB B. Stoneley, 44 Ilthorpe, Hull, HU6 9ER
M3 SBE S. Edwards, 5 Gorse Hill Road, Brickfields, Worcester, WR4 9TU
M3 SBJ S. Inman, 9 Colbert Avenue, Ilkley, LS29 8LU
M3 SBP S. Palin, Rose Tree Cottage, 17 Rowland Lane, Thornton-Cleveleys, FY5 2QX
M3 SBQ K. Walsh, Interval, Liverpool Marina, Liverpool, L3 4BP
M3 SBS D. Green, 144 Dilloways Lane, Willenhall, WV13 3HJ
M3 SBT B. Lockley, 35 High Street, Blackpool, FY1 2BN
M3 SBY B. Young, 25 Rombalds Drive, Skipton, BD23 2SP
M3 SCA S. Ahmed, 59 Ramsgate, Lofthouse, Wakefield, WF3 3PX
M3 SCF H. Fish, New House Peaton, Peaton, Craven Arms, SY7 9DW
M3 SCH S. Holden, 7 Macbeth Close, Colchester, CO4 3SZ
M3 SCJ C. Short, 35 Whitley Willows Lepton, Huddersfield, HD8 0GD
MM3 SCO G. Macleod, 12A Loyal Terrace, Tongue, IV27 4XQ
MM3 SCQ D. Macgregor, The Tundra, Upper Lybster, Lybster, KW3 6AT
M3 SCX S. Williamson, 19 Alcester Close, Plymouth, PL2 1EA
M3 SDB S. Bennett, 17 Knox Close, Norwich, NR1 4LN
M3 SDH S. Hodgson, 31 Ullswater Avenue, Crewe, CW2 8QQ
M3 SDJ S. Hackwood, 8 Ronald Walk Dresden, Stoke-on-Trent, St34sn
M3 SDK J. Donald, 11 Row Brow Park, Dearham, Maryport, CA15 7JU
M3 SDN N. Hurst, 74 Holden Road, Salterbeck, Workington, CA14 5LZ
MM3 SDP I. Lipkowitz, Chuccaby, Longhope, Stromness, KW16 3PQ
M3 SDQ M. Behrooz-Kafshdooz, 20 Byron Road, London, W5 3LL
M3 SDV Dr J. Pelham, 5 The Crescent, Shortstown, MK42 0UJ
M3 SEE S. England, 4 Ouse Close, Chandler'S Ford, Eastleigh, SO53 4RW
M3 SEJ J. Shepherd, 9 Wrea Head Close Scalby, Scarborough, YO13 0RX
MI3 SEK R. Thomson, 1 Litchfield Park, Coleraine, BT51 3TN
MI3 SEO S. Murray, 117 Knockview Drive, Tandragee, Craigavon, BT62 2BL
MM3 SES S. Smart, Cherrytrees, Top Street, Conon Bridge, Dingwall, IV7 8BH
MW3 SET S. Taylor, 43 Toronnen, Bangor, LL57 4TG
MI3 SEV M. Severn, 99 Crawfordsburn Road, Bangor, BT19 1BJ
M3 SEY M. Howes, 1 The Meadows, Herne Bay, CT6 7XB
MW3 SEZ S. Ezard, 59 Station Farm, Croesyceiliog, Cwmbran, NP44 2JW
M3 SFC A. Woodward, 19 Hazel Grove, Winchester, SO22 4PQ
M3 SFJ F. Smith, 9 Bramwell St., St Helens, WA9 2DP
M3 SFL S. Leitch, 12 Cavendish Avenue, Churchdown, Gloucester, GL3 2HW
M3 SFN A. Passey, 3 The Yard, Bayton, Kidderminster, DY14 9LH
MW3 SFP S. Parry, Aukland Terrace, Crymych, SA41 3QG
M3 SFZ S. Free, Mill Farm, Hargham Road, Attleborough, NR17 1DT
M3 SGE C. Sargent, Bradley, Holcombe Village, Dawlish, EX7 0JT
M3 SGF S. Blount, 55 Silverthorne Drive, Caversham, Reading, RG4 7NR
M3 SGG N. Evans, Flat 3, 240 Berrow Road, Burnham-on-Sea, TA8 2JG
M3 SGI S. Mallinson, 11 Union Road, Liversedge, WF15 7HW
M3 SGJ Rev. J. Scott, The Parsonage, 102A, Nutley Lane, Reigate, RH2 9HA
MM3 SGQ S. Gill, 5 Ramornie Place, Kingskettle, Cupar, KY15 7PT
M3 SGS S. Salmon, 35 Westgate Road, Lytham St Annes, FY8 2SG
M3 SGV R. Greaves, 7 Eller Brook Close, Heath Charnock, Chorley, PR6 9NQ
MW3 SGX J. Bidwell, 26 Lone Road, Clydach, Swansea, SA6 5HR
M3 SGZ J. Bentham, 18 Cauldon Avenue, Swanage, BH19 1PQ
M3 SHB S. Brown, 6 Good Avenue, Trimdon Grange, Trimdon Station, TS29 6EF
M3 SHI A. Shillabeer, 29 Newlease Road, Waterlooville, PO7 7BX
M3 SHJ S. Hughes, 4 Cobden Court, Birkenhead, CH42 3YH
M3 SHK R. Silversides, 7 Earles Lane, Kelsall, Tarporley, CW6 0QR
M3 SHN S. Neale, 28 Needham Drive, Sutton St. James, Spalding, PE12 0EG
M3 SHQ K. Browne, 24 Oaktree Avenue, Cuerden Residential Park, Leyland, PR25 5PJ
MM3 SHT D. Mcclure, 10 Greystone Close, Strathaven, ML10 6FW
M3 SHW K. Shaw, 2 Montrose Avenue, Montrose Street, Hull, HU8 7RY
M3 SHX A. Davies, 13 The Close, Stalybridge, SK15 1HU
M3 SHZ P. Bennett, 1 Queens Road, Carterton, OX18 3YB
M3 SII K. Borthwick, 15 Thomas Close, Ixworth, Bury St Edmunds, IP31 2UQ
MI3 SIL S. Linton, 68 Old Frosses Road, Cloughmills, Ballymena, BT44 9NA
M3 SIM S. Lord, 34 Alsop Street, Leek, ST13 5NZ
M3 SIS L. Simmons, 2 Blakemere Way, Sandbach, CW11 1XU
M3 SIW J. Wellings, 133 Griffins Brook Lane, Birmingham, B30 1QN
M3 SIY S. Shaul, 31 Chatterton Avenue Ermine West, Lincoln, LN1 3SZ
M3 SIZ J. Easdown, 38 North Street, Barming, Maidstone, ME16 9HF
M3 SJD S. Darby, 4 Whately Mews, Whately Road, Lymington, SO41 0XS
M3 SJH S. Hewitt, 4 Carrow Road, Dagenham, RM9 4TJ
M3 SJK S. Kerrison, 18 Parks Road, Dunscroft, Doncaster, DN7 4AH
M3 SJL J. Lowe, 46 Runshaw Avenue, Appley Bridge, Wigan, WN6 9JN
M3 SJM S. Whitaker, 34 Alder Grove, Poulton-le-Fylde, FY6 8EH
M3 SJQ J. Cleaver, 27 Lawton Crescent, Biddulph, Stoke-on-Trent, ST8 6EH
M3 SJR S. Rigby, 36 Richmond Road, Stoke-on-Trent, ST4 8RH
M3 SJV P. Mccarthy, 38 Lyndhurst Drive, Leyton, London, E10 6JD
M3 SJW S. Wills, 8 Frobisher Road, Yeovil, BA21 5FP
M3 SJX B. Shields, 20 Gresley Court, Grantham, NG31 7RH
M3 SJY K. Young, 14 Beechwood Avenue, Chatham, ME5 7HH
M3 SKB S. Brown, Rushbrook, Holly Grange Road, Lowestoft, NR33 7RR
M3 SKC D. Bryan, 3 George Street, Bourne, PE10 9HE
M3 SKD S. Kidd, 27 Hillswood Avenue, Leek, ST13 8EQ
M3 SKN P. Probst, 37 Devonshire Street, Skipton, BD23 2ET
M3 SKQ P. Henderson, 24 Farrow Road Whaplode Drove, Spalding, PE12 0TS
M3 SKT S. Taylor-Toms, 34 Larkspur Drive, Chandler'S Ford, Eastleigh, SO53 4HU
M3 SKU V. Leddington, 20 Bewell Head, Bromsgrove, B61 8HY
M3 SKV J. Hawkes, 53 Mill Hill, Derby, DE24 5AF
MW3 SKW M. Barber, 1 Gwernant, Cwmllynfell, Swansea, SA9 2FT
M3 SKX B. Martin, 111 The Avenue, Wallsend, NE28 6SD
M3 SKY S. Keevil, Gamekeepers Cottage, Snarehill, Thetford, IP24 2QA
M3 SKZ J. Parfitt, 5 Sheridan Road, Frimley, Camberley, GU16 7DU
MM3 SLB S. Berry, Willowburn, Kirkton, Hawick, TD9 8QJ
MM3 SLD J. Bradley, Tongue Of Bombie, Kirkudbright, DG6 4QD
M3 SLF R. Cave, 26 Longsight Road, Mapplewell, Barnsley, S75 6HB
MW3 SLI J. Backhouse, De10 Isaf, Bryneglwys, Corwen, LL21 9NP
MW3 SLL S. Lawton, Bryngarw Lodge, Brynmenyn, Bridgend, CF32 8UU
MW3 SLO D. Dash, 36 Rockvilla Close, Varteg, Pontypool, NP4 7QF
M3 SLQ N. Jones, 10 Leamington Close, Cannock, WS11 1PW
MI3 SLT S. Whitten, 2 Springwell Manor, Castlederg, BT81 7DR
M3 SLZ T. Gill, 21 Trevor Smith Place, Taunton, TA1 3RW
M3 SMD A. Davies, 7 Windermere Grange, Edlington, Doncaster, DN12 1NQ
MM3 SMI J. Smith, 14 John Collins Crescent, Galashiels, TD1 2FA
M3 SMK S. Mackimm, 16 Stanneybrook Close, Rochdale, OL16 2YH
M3 SML S. Lowe, 59 Knight Avenue, Gillingham, ME7 1UE
M3 SMM S. Mole, 17A Marlborough, Seaham, SR7 7SA
M3 SMN S. Kent, 4 Arden Close, Chesterfield, S40 4NE
M3 SMR R. Shepperley, Flat F, London, N3 1QL
M3 SMY S. Harkness, 114 Morland Road, Ipswich, IP3 0LZ
M3 SMZ S. Rdwards, 59 Laburnum Road, Tipton, DY4 9QS
MM3 SNB G. Mcgeouch, 59 Torogay Street, Glasgow, G22 7RA
M3 SNF I. Hewitt, 26 Outwoods Drive, Loughborough, LE11 3LT
MW3 SNH B. Jones, Browerdd, Llangybi, Lampeter, SA48 8NH
MW3 SNJ S. Jones, 14 Lower Cross Road, Llanelli, SA15 1NQ
M3 SNL R. James, 50 Andrew Allan Road, Rockwell Green, Wellington, TA21 9DY
M3 SNN N. Chapman, 8 Pennine Drive, Edith Weston, Oakham, LE15 8HY
M3 SNO R. Snowden, 5 Eastfield Road, Wisbech, PE13 3EJ
M3 SNQ L. Thornton, 11 Polruan Road, Truro, TR1 1QR
M3 SNR S. Rennalls, 7 Hollybush Close, Sowerby Bridge, HX6 1AH
MW3 SNW S. Williams, Hillsboro Aberkenfig, Bridgend, CF32 0EW
M3 SNX G. Rigden, Corner House, Ashford Road, Kent, TN27 0EE

M3 SNY R. Beardshall, 41 Hill Crest, Hoyland, Barnsley, S74 0BU
M3 SNZ S. Seath, 9 Winchester Avenue, Morecambe, LA4 6DX
MW3 SOC N. Davies, 65 Walters Road, Neath, SA11 2DW
M3 SOF S. Vaux, 171 Foxon Lane, Caterham, CR3 5SH
M3 SOG R. Stearn, 18 Kings Avenue, Chippenham, SN14 0UJ
M3 SOQ P. Swann, 2 Little Walton, Eastry, Sandwich, CT13 0DW
M3 SOT S. Gregory, 11 Ribblesdale Avenue, Congleton, CW12 2BS
M3 SOV P. Fernie, 39 North Parade, Falmouth, TR112TE
M3 SOY J. Rolph, The Hollies, Back Lane, Norwich, NR10 4HL
M3 SPA I. Beresford, 16A Holbeck Hill, Scarborough, YO11 2XD
M3 SPG S. Garthwaite, 278 Carlton Road, Barnsley, S71 2BA
M3 SPJ S. Hinds, 69 Carshalton Grove, Wolverhampton, WV2 2QZ
M3 SPL A. Ladell, 25 Harwood Avenue, Thetford, IP24 2LY
M3 SPP R. Penrose, 41 Milton Road, Eastbourne, BN21 1SH
M3 SPQ S. Schonborn, 116 Hough Lane, Wombwell, Barnsley, S73 0EF
M3 SPR S. Mclaughlin, 34 Cambridge Road, Birstall, Batley, WF17 9JF
M3 SPU P. Saunders, 62 Parkfield Avenue, Eastbourne, BN22 9SF
M3 SPY R. Gardner, 6 Meade Road, Liverpool, L13 9AA
MW3 SQA C. Davis, 132 Steynton Road, Steynton, Milford Haven, SA73 1AN
M3 SQE D. Mcdonald, 3 Lindley Street, Mansfield, NG18 1QE
M3 SQG M. Breslin, 15 Acorn Gardens, East Cowes, PO32 6TD
M3 SQH A. Lawrence, 11 Pembroke Court St. Johns Road, Chesterfield, S41 8NX
M3 SQI N. Thake, Easton House, Water Street, Berwick St. John, Shaftesbury, SP7 0HS
MM3 SQJ J. Morris, 10 Middlemas Road, Dunbar, EH42 1GJ
MM3 SQM D. Mchardy, 486 Kilmarnock Road, Glasgow, G43 2BW
M3 SQO P. Burke, 38 Bosworth Square, Rochdale, OL11 3QG
M3 SQP S. Scotching, 26 Newton Way, Leighton Buzzard, LU7 4YU
M3 SQQ S. Oxenham, 10 Arnside Close, Plymouth, PL6 8UU
M3 SQS R. Garland, 113 The Drive, Feltham, TW14 0AH
M3 SQT C. Eyre, 23 Nelson Street, Congleton, CW12 4BS
M3 SQU M. Van Den Bergh, The Parsonage, Masefield Drive, Tamworth, B79 8JB
M3 SQV S. Sandford, 11 Browning Close, Tamworth, B79 8NB
M3 SQX I. Okorji, 37 Lydwell Park Road, Torquay, TQ1 3TQ
M3 SQZ N. Swift, 59 Milton Avenue, Malton, YO17 7LB
MM3 SRF R. Farrer, 23 Upper Craigour, Edinburgh, EH17 7SE
MI3 SRG E. Coates, 148 Springwell Road, Groomsport, Bangor, BT19 6LX
M3 SRH S. Hubball, 24 Newstead Road, Stoke-on-Trent, ST2 8HX
M3 SRI S. Issatt, 69 St. Lawrence Avenue, Snaith, Goole, DN14 9JH
MM3 SRK A. Ross, 16 Croft Road, Kiltarlity, Beauly, IV4 7HZ
MI3 SRL S. Rea, 70 Raloo Road, Larne, BT40 3DU
M3 SRQ J. Stoppard, 15 South Lodge Court, Old Road, Chesterfield, S40 3QG
M3 SRT S. Thompson, 30 Southport Parade, Hebburn, NE31 2AQ
M3 SRV J. Matthews, 23 Elmhurst, Bridgnorth, WV15 5DJ
M3 SRY P. Seymour, 3 Larch Avenue Allington Gardens, Allington, Grantham, NG32 2FG
M3 SSG A. Butler, 12 South Bank Cottages South Stoke, Reading, RG8 0HX
M3 SSI S. Bangalore, 2 Amberley Walk, Kingsmead, Milton Keynes, MK4 4AX
M3 SSL M. Belcher, 52 Kynaston Road, Didcot, OX11 8HD
M3 SSO B. Hawes, 3 Orchard Close, Cassington, Witney, OX29 4BU
M3 SSU E. Little, 41 Sevenoaks Road, Portsmouth, PO6 3JP
M3 STJ S. Jordan, 10 Kirkby Avenue Garforth, Leeds, LS252BN
M3 STQ S. Fox, Hawthorns, School Lane, Martlesham, Woodbridge, IP12 4RR
M3 STR N. Soltysik, 24 Cottage Close, Hednesford, Cannock, WS12 1BS
MI3 STW R. Bradley, 41 Graymount Road, Newtownabbey, BT36 7DR
MI3 STY S. Nicholl, 89 Glenshane Road, Londonderry, BT47 3SF
MW3 SUF N. Smith, 7 Hawthorne Avenue, Connah'S Quay, Deeside, CH5 4TF
M3 SUI S. Allen, 33 Rookhill Road, Pontefract, WF8 2BY
M3 SUJ J. Cook, 40 Preston Avenue, Alfreton, DE55 7JY
M3 SUK A. Holland, 40 Sunnyside Road, Poole, BH12 2LQ
MM3 SUS S. Holt, Ashwell, Cannigall, Kirkwall, KW15 1SX
MM3 SUV A. Mccaig, 46 Patterson Drive, Law, Carluke, ML8 5LT
M3 SUW P. Hopkins, 40, Grange Close, Condover, Shrewsbury, SY5 7AT
M3 SUY T. Van Den Bergh, 19 Perrycrofts Crescent, Tamworth, B79 8UA
M3 SVB S. Black, 7 Harwood Close, Gosport, PO13 0TY
M3 SVC S. Cox, 19 Exbury Way, Andover, SP10 3UH
M3 SVD M. Hewitt, Redwood House, Adbury Holt, Newbury, RG20 9BW
M3 SVF L. Leung, Harrogate Ladies' College, Clarence Drive, Harrogate, HG1 2QG
M3 SVH K. Henderson, 42 Chartwell Avenue, Wingerworth, Chesterfield, S42 6SP
M3 SVJ R. Gee, Flat 1D, Quarmby Road, Huddersfield, HD3 4HQ
M3 SVL S. Leech, 9 Brocklehurst Mews, Macclesfield, sk102gy
MI3 SVM S. Murray, 80 Canterbury Park, Londonderry, BT47 6DU
M3 SVN S. Adkins, 4 Orion Close, Ward End, Birmingham, B8 2AU
M3 SVO L. Birdsall, 8 North Cote, Ossett, WF5 9RE
M3 SVP A. Chapman, 24 Eaton Grange Drive, Long Eaton, Nottingham, NG10 3QE
M3 SVT S. Taylor, 17 Mendip Drive, Bolton, BL2 6LQ
M3 SVY L. Beardsley, 1 Amber Villas, Sutton St. Nicholas, Hereford, HR1 3DF
M3 SVZ G. Brierley, 35 Ochrewell Avenue, Deighton, Huddersfield, HD2 1LL
MM3 SWA S. Anderson, 33 Dryden Avenue, Loanhead, EH20 9JT
MI3 SWD S. Mcauley, Layde View, 19 Rathlin Avenue, Ballycastle, BT54 6DQ
M3 SWF R. Jenkinson, Esperance, West End Road, Doncaster, DN9 1LB
MM3 SWG S. Groves, 1/1 99 Belville Street, Greenock, PA15 4SX
M3 SWK S. Walker, 64 Belmont Road, Rugby, CV22 5NY
MW3 SWO S. Owen, 500 Cowbridge Road West, Cardiff, CF5 5DA
M3 SWQ A. Marks, Grosvenor Hotel, 51 Grosvenor Road, Scarborough, YO112LZ
M3 SWS S. Lowe, 31 Court Farm Road, Bristol, BS14 0EH
MM3 SWU S. Jenkins, 66 Spruce Avenue, Johnstone, PA5 9RG
M3 SWV J. Moppett, 59 Piccadilly, Tamworth, B78 2ER
MM3 SWW D. Elliot, Thisleycrook, Torphins, AB214NR
M3 SXF S. Forbes, 55 The Henrys, Thatcham, RG18 4LS
MI3 SXI I. Mckeown, 19 Castlehill Comber, Newtownards, BT23 5XA
MM3 SXJ P. Duckles, 45 Redhouse Place Blackburn, Bathgate, EH47 7QB
M3 SXK C. Little, 33 Wigmores, Telford, TF7 5NB
MI3 SXM G. Hutton, 13 Meadowbank, Sepatrick, Banbridge, BT32 4PZ
M3 SXP S. Perring, 16 Salford Road, Bolton, BL5 1BL
MI3 SXQ C. Cunningham, 1 Ballykeel Court, Ballymartin, Newry, BT34 4XW
MI3 SXR A. Robb, 10 Rosepark East, Belfast, BT5 7RL
MM3 SXT S. Thorogood, 38 Forres Drive, Glenrothes, KY6 2JU
M3 SXU J. Jackson, 90 Horne Street, Bury, BL9 9HS
M3 SXV S. Simms, 9 Horsa Place, Sherburn In Elmet, Leeds, LS25 6QA
M3 SXZ D. George, 9 Winscombe Court, Frome, BA11 2DZ
MM3 SYB D. Nicholson, 4 Upper Barvas, Isle of Lewis, HS2 0QX
M3 SYC A. Chaplin, 33 The Crofts, Little Wakering, Southend-on-Sea, SS3 0JS
MI3 SYF D. Bates, 31 Drumard Park, Lisburn, BT28 2HU
M3 SYH D. Horton, 2 Brampton Way, Bulkington, Bedworth, CV12 9PR
MI3 SYI B. Mcdonald, 20 Aughan Park, Poyntzpass, Newry, BT35 6TW
M3 SYL S. Stratford, 23 The Fairway, Banbury, OX16 0RR
M3 SYN S. Lanaway, 1 Clovers Cottages, Faygate Lane, Faygate, Horsham, RH12 4SH
MM3 SYO S. Angus, 20 Norlands, Errol, Perth, PH2 7QU
MM3 SYQ A. Clark, 17 Glentilt Terrace, Perth, PH2 0AE
MM3 SYU C. Higgins, 65A Forthill Road, Broughty Ferry, Dundee, DD5 3DQ
M3 SYV T. Symons, Southgate, The Commons, Mullion, TR12 7HZ
M3 SYW C. Voke, 16 Exton Road, Chichester, PO19 8BP
M3 SYY B. Sweeney, 14 Eaves Lane, Chorley, PR6 0PY
M3 SYZ S. Symonds, 68 Manor Crescent, Pan, Newport, PO30 2BH
M3 SZC S. Crabtree, 107 Rochdale Road, Shaw, Oldham, OL2 7JT
MW3 SZD M. Musgrave, Hillside Cottage, Hiraddug Road, Rhyl, LL18 6HS
MW3 SZF G. Williams, Ty Newydd, Rhyd, Penrhyndeudraeth, LL48 6ST
MJ3 SZI M. Brown, 77 Andium Court, Langtry Gardens, St. Saviours Hill, St. Saviour, Jersey, JE2 7AH
M3 SZK M. Gridley, The Granary, Yarde Farm, West Hatch, Taunton, TA3 5RP
MM3 SZM S. Macdonald, 110 High Street, Cuminestown, AB53 5YH
M3 SZO R. Savery, 75 Bramley Road, Tewkesbury, GL20 8AQ
M3 SZQ S. Snelson, 212 Dickson Road, Blackpool, FY1 2JS
M3 SZS D. Forster, 23 Field Street, Padiham, Burnley, BB12 7AU
M3 SZT J. Smith, 32 Youlgreave Drive, Sheffield, S12 4SE
M3 SZY S. Holt, 108 Blandford Avenue, Castle Bromwich, Birmingham, B36 9JD
M3 TAE T. Eadon, Chapel Cottage, Newcastle Road South, Sandbach, CW11 1RS
MW3 TAF C. Williams, 96 Bryn Road, Swansea, SA2 0AT
M3 TAG A. Aldred, 78 The Drive, Horley, RH6 7NH
MM3 TAM T. Aitken, 27 Beeches Avenue, Clydebank, G81 6HX
M3 TAN S. Greenfield, 4 Charlesworth Square, Gomersal, Cleckheaton, BD19 4NX
MM3 TAV A. Mcconochie, 15 Slains Crescent, Cruden Bay, Peterhead, AB42 0PZ
M3 TAW T. Whittam, 27 Dimples Lane, Garstang, Preston, PR3 1RD
M3 TBF T. Ferguson, 6 Binley Close, Birmingham, B25 8NE
M3 TBG A. Archer, 23 St. Ives Road, Somersham, Huntingdon, PE28 3ER
M3 TBH T. Hobbs, 2 The Lynch, West Stour, Gillingham, SP8 5RN
M3 TBK E. Cree, 24 Old Lincoln Road, Caythorpe, Grantham, NG32 3EJ
MI3 TBL Dr T. Littler, 15 Belmont Grove, Lisburn, BT28 3YB
M3 TBP C. Parker, Red Leas, 22 Bent Lane, Colne, BB8 7AA
M3 TBQ I. Hill, 74 Clarence Road, Torpoint, PL11 2LT
M3 TBU T. Burnham, Creedy Barn, Kennerleigh, Crediton, EX17 4RU
M3 TBV D. Parker, 16 Aldborough Road, Dagenham, RM10 8AS
M3 TBW J. Humphrey, Flat 1, Kingswood House, 10 Lewes Road, Eastbourne, BN21 2BX
M3 TBZ D. Heathcote, 154 High Street, Harriseahead, Stoke-on-Trent, ST7 4JX
M3 TCG T. Graham, First Floor Flat, 43 Belgrave Crescent, Bath, BA1 5JU
M3 TCR P. Ryall, Windsor Lodge, Pantile Hill, Southminster, CM0 7BA
M3 TCT S. Farrar, 20 Cleveland Grove, Lupset, Wakefield, WF2 8LD
M3 TCU G. Read, Flat 3, Parkmead Court, Ryde, PO33 2HD
M3 TCX T. Carroll, 14 Glenpark Drive, Southport, PR9 9FA
M3 TCY T. Earnshaw, 63 Manor Road, Fleetwood, FY7 7LJ
M3 TDB T. Berry, Roseneath, Walcote Road, Lutterworth, LE17 6EQ
M3 TDH T. Hewitt, 6 Mayfield, Catforth Road, Preston, PR4 0HH
M3 TDM T. Reddington, 174 Home Farm Road, Wirral, CH49 7LH
M3 TDP T. Packham, Bradstow Lodge, 19 Crow Hill, Broadstairs, CT10 1HN
M3 TDT A. Dockerill, 8 Bennett Road, Swanton Morley, Dereham, NR20 4LY
M3 TEE D. West, 42 Scholars Green, Wigton, CA7 9QW
M3 TEG T. Mason, 31 Manor Park Road, Hailsham, BN27 3AT
M3 TEI R. Mcknight, Ardralla, Church Cross, Skibbereen, Ireland, P81 RK12
M3 TEL T. Pink, 11 Harmony Meadow, Roche, St Austell, PL26 8EJ
MI3 TEM S. Kirkwood, 1 Rural Cottages, Front Road, Lisburn, BT27 5LF
M3 TEN T. Newman, Sometimes (The Workshop), South Pew, Dorchester, DT2 9HZ
M3 TEP T. Payne, 2 Greenleas, Waltham Abbey, EN9 1SZ
MM3 TEQ A. Leiper, 6 Inchyra Place, Grangemouth, FK39EQ
M3 TEV S. Ball, 13 Yew Tree Road, Hayling Island, PO11 0QE
M3 TEY J. Hartshorne, 8 Ashbee Street, Bolton, BL1 6NT
M3 TFA A. Tran, Flat 4, Room 9, Rayleigh Tower, Colchester, CO4 3SQ
M3 TFB D. Parkinson, 4 Meadow View, Sherburn In Elmet, Leeds, LS25 6BY
M3 TFE J. King, Plum Tree Cottage, Royston Place, Barton On Sea, New Milton, BH25 7AJ
MI3 TFF F. Finlay, 62 Slieveboy Road, Claudy, Claudy, BT47 4AS
M3 TFG B. Jones, 75 Mamble Road, Stourbridge, DY8 3SY
M3 TFI D. Woods, 3 Brook Street, Port Sunlight, Wirral, CH62 5DB
M3 TFK P. Hemsley, 140 Greenhill Lane, Riddings, Alfreton, DE55 4EX
M3 TFM S. Lytollis, 91 Townfoot Park, Brampton, CA8 1RZ
M3 TFO R. Styles, 52 Vernham Grove, Bath, BA2 2TB
M3 TFP T. Plummer, 33 East St., Sudbury, CO10 2TU
M3 TFS R. Shoubridge, 2 Copestake Drive, Burgess Hill, RH15 0LD
M3 TFW T. Harlow, 10 Fraser Road, Poole, BH12 5AY
M3 TFX T. Fisk, 2 Hall Farm Cottage, Caston Road, Caston, Attleborough, NR17 1BW
M3 TFZ E. Miller, 8 Arthur Avenue, Caister-On-Sea, Great Yarmouth, NR30 5PQ
M3 TGA G. Arnold, 1 Louthe Way, Sawtry, Huntingdon, PE28 5TR
M3 TGC G. Cooper, 21 Thistle Bridge Road, Chivenor, Barnstaple, EX31 4FL
M3 TGD J. Park, 3 Flaxfield Drive, Crewkerne, TA18 8DF
M3 TGE R. Lindon, 134 Station Road, Sutton Coldfield, B73 5LD
M3 TGJ J. Alexander, 335 Canterbury Road, Birchington, CT79TY
M3 TGK P. Wale, Munstead Oaks, Hascombe Road, Godalming, GU8 4AB
M3 TGL G. Thorne, 72 Devonshire Road, London, E16 3NJ
M3 TGO G. Hope, 27 Clearmount Drive Charing, Ashford, TN27 0LH
M3 TGP T. Porter, 208 Clapgate Lane, Ipswich, IP3 0RG
M3 TGT H. Doman, 13 Cumbria Close, Maidenhead, SL6 3DD
M3 TGW T. Willis, 32 Sandover, Northampton, NN4 0TS
M3 TGZ D. Lines, 68 Rugby Place, Brighton, BN2 5JA

M3 THE M. Peck, 60 Riverside Drive, Tern Hill, Market Drayton, TF9 3QH
MW3 THI M. Price, 18 Rhiw Tremaen, Brackla, Bridgend, CF31 2JA
M3 THJ B. Mills, 37 Ashley Road, Hildenborough, Tonbridge, TN11 9ED
M3 THN P. Cobley, 58 John Street, Newhall, Swadlincote, DE11 0SR
M3 THQ J. Peerless, 503 Honeypot Lane, Stanmore, HA7 1JH
M3 THY T. Lupton, 81 Home Farm Lane, Bury St Edmunds, IP33 2QL
M3 TIC M. Tinsell-Stanton, 38 Comberton Road, Kidderminster, DY10 3DT
M3 TID B. Maddox, 72 Church Road, Hartshill, Nuneaton, CV10 0LY
M3 TIE L. Morrell-Cross, Delta Lodge, 14 Rushton Crescent, Bournemouth, BH3 7AF
M3 TIF C. Robertson, 4 Pleasant Street, Walshaw, BL8 3AU
M3 TII B. Aylward, 53 Overdown Rise, Portslade, Brighton, BN41 2YF
M3 TIJ J. Aylward, 53 Overdown Rise, Portslade, Brighton, BN41 2YF
M3 TIK M. Richardson, 1 Cedar Drive, Lowestoft, NR33 9HA
M3 TIL J. Tillson, 23 The Fitches, Knodishall, Saxmundham, IP17 1UX
M3 TIQ D. Perkins, 56 Cliff Street, Rishton, Blackburn, BB1 4EE
M3 TIY A. Green, 65 Rosamond Road, Bedford, MK40 3UG
M3 TIZ D. Marsh, 16 Laurel Close, North Warnborough, Hook, RG29 1BH
MW3 TJG T. Gwyther, 15 Denbigh Court, Caerphilly, CF83 2UN
M3 TJI T. Adams, 11 St. Georges Crescent, Gravesend, DA12 4AR
M3 TJJ J. Jones, 19 Southbank Street, Leek, ST13 5LS
M3 TJK J. Mackenzie, 30 Dalriada Gardens, Ballycastle, BT54 6DZ
M3 TJL T. Lake, 85 Clarkson Road, Norwich, NR5 8ED
MI3 TJM T. Moore, 43 Woodburn Park, Londonderry, BT47 5PS
M3 TJO T. Jones, 4 Anne Close, Christchurch, BH23 2NW
M3 TJQ A. Newton, 8 Trent Meadow, Taunton, TA1 2NP
MI3 TJR T. Ruddell, 30 Ballynacor Meadows, Portadown, Craigavon, BT63 5UU
M3 TJT T. Ticehurst, 10 Rushlake Road, Coldean, Brighton, BN1 9AD
M3 TJU E. Duffield, 92 Crosby Street, Stockport, SK2 6SP
MM3 TKE R. Mckie, 16 Silver Street, Creetown, Newton Stewart, DG8 7HU
MW3 TKI I. Hoyle-Jackson, 21 Kimberley Close, Sketty, Swansea, SA2 9DZ
M3 TKN M. Roche, 1 Lancaster Close, London, NW9 5RE
M3 TKO K. Smart, 33 East Street, Littlehampton, BN17 6AU
M3 TKQ J. Macey, 62 Uttoxeter Road, Hill Ridware, Rugeley, WS15 3QU
M3 TKT M. Turowski, 57 Millwood Road, Orpington, BR5 3LQ
M3 TKU P. Loveden, 57 St. Marys Road, Rawmarsh, Rotherham, S62 5BD
M3 TKV T. King, 24 Royston Avenue, Basildon, SS15 4EW
M3 TKW K. Rowley, 10 Mount Close, Wombourne, Wolverhampton, WV5 9ER
M3 TLB D. Weller, 6 Aldervale, Fermor Road, Crowborough, TN6 3BY
M3 TLD D. Tattersall, 17 Badger Close, Durkar, Wakefield, WF4 3QD
M3 TLG A. Dann, 18 Salcombe Way, Ruislip, HA4 6BA
MM3 TLH T. Holt, Ashwell, Cannigall, Kirkwall, KW15 1SX
M3 TLJ B. Hawkins, 60 High Street Barwell, Leicester, LE9 8DR
M3 TLL L. Nilon, 5 Denby Drive Baildon, Shipley, Bd17 7pq
M3 TLM T. Lockett, 14 Tildsley Crescent, Weston, Runcorn, WA7 4RN
M3 TLN C. Dean, 119 Queens Drive, Newton-le-Willows, WA12 0LN
M3 TLO D. Livings, 30, Grenfell Avenue, Holland-On-Sea, Clacton-on-Sea, CO15 5XH
MW3 TLP A. Buckley, 53 Derlwyn St., Phillipstown, New Tredegar, NP24 6AZ
MM3 TLQ D. Field, 7 Admiralty Street, Portknockie, Buckie, AB56 4NB
M3 TLT A. Mcgregor, 41 Breedon Close, Corby, NN18 9PG
M3 TLU M. Bailey, 71 Somerfield Road, Walsall, WS3 2EG
MI3 TLV M. Nicholl, 34 Berryhill Road, Artigarvan, Strabane, BT82 0HN
M3 TLW R. Crook, 80 Kings Road, Biggin Hill, Westerham, TN16 3XY
M3 TLX D. Burdsall, 37 Fulmar Walk, Whitburn, Sunderland, SR6 7BW
M3 TLY I. Bain, 45 Larpool Crescent, Whitby, YO22 4JD
M3 TLZ D. Pointon, 1 Cross Cottages, Alsager Road, Audley, Stoke-on-Trent, ST7 8JQ
M3 TME M. Trick, 24 King St., Tiverton, EX16 5JE
M3 TMG G. Thorpe, Jasmine, Sutton Road, Mablethorpe, LN12 2PT
M3 TMM T. Mckain, 21 Oakhurst Grove, East Dulwich, London, SE22 9AH
MI3 TMN A. Mcnulty, 2 Devenish Crescent, Devenish, Enniskillen, BT74 4RB
M3 TMQ M. Mutton, 74 Alexandra Road, Sheerness, ME12 2AT
MW3 TMR R. Thomas, 35 Under Ffrydd Wood, Knighton, LD7 1EF
M3 TMX J. Harrop, Apartment 5, Forest Edge, Sneyd Street, Stoke-on-Trent, ST6 2PY
M3 TMY T. Mulraney, 3 Salvia Close Churchdown, Gloucester, GL3 1LL
M3 TMZ A. Laister, 23 Berry Street, Greenfield, Oldham, OL3 7EF
M3 TNB M. Anthony, Magpie Bungalow, Goongumpas, St. Day, Crofthandy, Redruth, TR16 5JL
M3 TND N. Tam, Room 102, 30 Evelyn Gardens, London, SW7 3BG
M3 TNE C. Pickford, 80 Hollowood Avenue, Littleover, Derby, DE23 6JD
MM3 TNF J. Odonnell, 33 Broomward Drive, Johnstone, PA5 8HR
MM3 TNG D. Stewart, 45 Kilwinning Road, Irvine, KA12 8RZ
M3 TNH K. Stockley, 357 Clements Road, Ramsgate, CT12 6UG
M3 TNJ J. Armstrong-Taylor, Driftwood, Station Road, Yelverton, PL20 7JS
M3 TNK N. Fong, Harrogate Ladies' College, Clarence Drive, Harrogate, HG1 2QG
M3 TNL P. Dawson, 400 Ropery Road, Gainsborough, DN21 2TH
M3 TNM T. Nguyen, 9 Green Street, Cambridge, CB2 3JU
M3 TNN T. Ellis, 84 Revelstoke Road, London, SW18 5PB
M3 TNO J. Dean, 25 Chantry Avenue, Bexhill, TN40 2EA
M3 TNV M. Clough, 8 Skeldyke Road, Kirton, Boston, PE20 1LR
M3 TNW P. Stanford, 4 Barkway Road, Royston, SG8 9EA
M3 TNY T. Limbert, 7 Acacia Avenue, Liverpool, L36 5TL
MW3 TOB H. Matthews, 24 Clos Y Berllan, Rhuddlan, Rhyl, LL18 2UL
M3 TOE A. Ward, 39 Linley Close, Bridgwater, TA6 4HL
M3 TOF D. Holyoake, 281 Causeway, Green Road, West Midlands, B68 8LT
MW3 TOI M. Francis, 72 Bro Ednyfed, Llangefni, LL77 7WD
M3 TOJ H. Clough, 8 Skeldyke Road, Kirton, Boston, PE20 1LR
M3 TOR C. Jones, Po Box 293, Ford, Plymouth, PL2 1WT
M3 TOT E. Brown, Rose Cottage, Grindlow, Buxton, SK17 8RJ
MM3 TOV E. Wilson, The Old Schoolhouse, Fife, KY15 4NB
M3 TOY A. Ashworth, 79 Stonehouse Road, Rugeley, WS15 2LL
M3 TPD T. Dooley, 32 Coult Avenue, North Hykeham, Lincoln, LN6 9RG
MM3 TPF N. Mann, Ramsburn Cottage, Knock, Huntly, AB54 7LQ
M3 TPG T. Greenall, Hall Lane Farm, Hall Lane, Warrington, WA4 4AF
M3 TPH T. Hazel, 84 Rodwell Avenue, Weymouth, DT4 8SQ
M3 TPI A. Paxton, 20F Green End, Granborough, Buckingham, MK18 3NT
MW3 TPJ T. Price, 5 Rhodfa'R Pant, Pant, Merthyr Tydfil, CF48 2DG
M3 TPN T. Norrington, 32 Fulfen Way, Saffron Walden, CB11 4DW
MI3 TPR R. Thompson, 14 Gelvin Grange, Londonderry, BT47 2LD
M3 TPU R. Morgan'S, 7 Tennyson Close, Braintree, Essex, CM7 1AB
M3 TPW T. Wooldridge, 12 Redwood Avenue, Leyland, PR25 1RN
M3 TPY T. Purcell, 18 Millberg Road, Seaford, BN25 3ST
M3 TQA A. Madge, Flat, 17 Newcomen Road, Dartmouth, TQ6 9BN
M3 TQB D. Holmes, 4 Council House, Nidds Lane, Boston, PE20 1LZ
M3 TQD J. Lear, 6 South View Green, Bentley, Ipswich, IP9 2DR
M3 TQF C. Bailey, 55 Bridgend Park Brewery Road, Wooler, NE71 6QG
M3 TQG G. Joy, Fair Oak, Higher Furzeham Road, Brixham, TQ5 8QP
MM3 TQH R. Hay, Roddach Cottage East, Cummingston, Burghead, Elgin, IV30 5XY
MM3 TQI D. George, D C George, 91 Regent Street, Keith, AB55 5ED
M3 TQJ T. Kemp, 30 Tawny Sedge, King's Lynn, PE30 3PW
MM3 TQK J. Traill, 31 Sherwood Place, Bonnyrigg, EH19 3JY
M3 TQN N. Davies, 156 Britannia Avenue, Dartmouth, TQ6 9LQ
M3 TQP R. Messingham, 2 The Lodge, Sotherington Lane, Blackmoor, Liss, GU33 6DA
M3 TQT F. Goodall, 1 Parkfield Grove, Leeds, LS11 7LS
M3 TQU J. Bingham, 31 Wyre Close, Paignton, TQ4 7RU
M3 TQX A. Basterfield, Red 3, Purn Holiday Park, Bridgwater Road, Weston-Super-Mare, BS24 0AN
M3 TQY G. Rogers, 26 Chaucer Close, Waterlooville, PO7 6AQ
M3 TRC P. Lee, 15 Talkin Drive, Middleton, Manchester, M24 5LS
M3 TRH T. Hill, 10 Parade View, Walsall, WS8 7JA
M3 TRJ T. Jones, 5 Broomfields Road, Appleton, Warrington, WA4 3AE
M3 TRO S. Phillips, 37 Wensley Road, Barnsley, S71 1SB
M3 TRP S. Walker, 26 The Warren, Hardingstone, Northampton, NN4 6EW
MI3 TRR R. Elliott, 183 Kilraughts Road, Ballymoney, BT53 8NL
MM3 TRZ T. Reilly, 21 North Street, Motherwell, ML1 1LQ
M3 TSA T. Hickson, 27 Cressing Road, Witham, CM8 2NP
M3 TSE E. Hughes, 74 Westmorland Road, Coventry, CV2 5BT
M3 TSF P. Shayler, 38 Maryside, Slough, SL3 7ET
M3 TSG A. Ryder, 4 Edgeway, Strelley, Nottingham, NG8 6LY
M3 TSI P. Hewson, 30 Princess Road, Kirton, Boston, PE20 1JW
MW3 TSJ S. Trott, 6 Mounton Drive, Chepstow, NP16 5EH
M3 TSN M. Lee, 46 Little Lane, Huthwaite, Sutton-in-Ashfield, NG17 2RA
M3 TSO T. Owens, 74 Tees Crescent, Stanley, DH9 6JD
M3 TSV K. Reason, 28 St. Marys Grove, Swindon, SN2 1RQ
M3 TTA E. Davies, 58 Popes Lane, Sturry, Canterbury, CT2 0LA
M3 TTH T. Haley, 3 Akeman Rise, Ramsden, Chipping Norton, OX7 3BJ
M3 TTK S. Ridley, 123 Lanercost Drive, Newcastle upon Tyne, NE5 2DL
M3 TTS D. Brook, 140 Dearne Hall Road, Barugh Green, Barnsley, S75 1LX
MW3 TUB D. Butler, 12, Bro Gwynfaen, Llandysul, SA44 4ST
M3 TUC T. Gerrard, 16 Haig Road, Carlisle, CA1 3AS
M3 TUD C. Copeman, 1 Chestnut Avenue, Welney, Wisbech, PE14 9RG
M3 TUF J. Caswell, 3 Pavilion Court, Roydon, Diss, IP22 5SP
M3 TUH B. Tucker, 12 Alpha Place, Appledore, Bideford, EX39 1QY
M3 TUL A. Dodd, 14 Davies Street, Macclesfield, Sk10 1GE
MW3 TUO A. Browning, 11 Heather Close, Sirhowy, Gwent, NP224PW
MM3 TUR P. Turner, 99 Maitland Hog Lane, Edinburgh, eh29 9du
MI3 TUS W. Donnell, 71 Niblock Oaks, Antrim, BT41 2DP
M3 TUU G. Milsom, 31 Chichester Close, Bowerdean Road, High Wycombe, HP13 6AU
M3 TUW S. Tucker, 11 Maple Drive, Killamarsh, Sheffield, S21 1GA
MI3 TUZ A. Gault, 7 Gardenmore Place, Larne, BT40 1SE
M3 TVC R. Evans, 7 Westland Drive, Hayes, Bromley, BR2 7HE
M3 TVD D. Steele, 22 Grindle Close, Thatcham, RG18 3PD
M3 TVJ J. Evans, 16 Longfield Place, Poulton-le-Fylde, FY6 7DB
M3 TVK J. Hodgson, 5 Clifton Place, Freckleton, Preston, PR4 1RQ
M3 TVN T. Bunce, 31 Kensington Avenue, Middlesbrough, TS6 0QQ
M3 TVO N. Allen, 52 Storth Lane, Kiveton Park, Sheffield, S26 5QT
MM3 TVQ C. Smith, 37 Glebe Road, Mosstodloch, Fochabers, IV32 7JH
M3 TVV D. Render, 4 Station Terrace, Allerton Bywater, Castleford, WF10 2BS
M3 TVZ D. Darby, 116 Middle Road, Southampton, SO19 8FS
MM3 TWA I. Whiteford, 54 Bilby Terrace, Irvine, KA12 9DT
M3 TWB S. Bourdon, 35 Main Road, Woolverstone, Ipswich, IP9 1BA
MM3 TWG T. Galbraith, 77 Netherwood Park, Deans, Livingston, EH54 8RW
M3 TWK W. Tam, Harrogate Ladies' College, Clarence Drive, Harrogate, HG1 2QG
MI3 TWM C. Smallwoods, 73 Knightsbridge, Londonderry, BT47 6FE
M3 TWP D. Jenner, 116 Trench Road, Tonbridge, TN10 3HQ
MW3 TWQ P. Jones, 36 Hopkin Street, Treherbert, Treorchy, CF42 5HL
M3 TWS T. Stanford, 27 Mill Gardens, Elmswell, Bury St Edmunds, IP30 9DQ
M3 TWV J. Benbow, 20 Clifton Close, Thornton-Cleveleys, FY5 4NG
MM3 TWW E. Wallace, 57 Henderson Park, Windygates, Leven, KY8 5DL
M3 TWY M. Sewell, Flat 2-4, 6 Augusta Road, Ramsgate, CT11 8JP
M3 TXA T. Ruddick, Hazel Gill, Croglin, Carlisle, CA4 9RR
M3 TXF H. Northcote, 58 Warren Avenue, Wakefield, WF2 7JN
M3 TXG D. Mardlin, 13 Churchill Crescent Sonning Common, Reading, RG4 9RU
M3 TXH A. Hensman, 20 St. Marys Road, Braintree, CM7 3JR
MI3 TXI R. Carlin, 10 Top Of The Hill, Londonderry, BT47 2HA
M3 TXJ R. Williams, 198 Leverington Common, Leverington, Wisbech, PE13 5BP
M3 TXL T. Graham, Woodtown, Sampford Spiney, Yelverton, PL20 6LJ
M3 TXM M. Bristow, 9 Chadwick Drive, Harold Wood, Romford, RM3 0ZA
M3 TXP A. Barker, 37 Newbarns Road, Barrow-in-Furness, LA13 9SF
M3 TXQ P. Holman, 107 Eathorpe Close, Redditch, B98 0HH
M3 TXR E. Smith, 13 Eagle Avenue, Waterlooville, PO8 9UB
M3 TXS H. Huckle, 43 The Baulk, Biggleswade, SG18 0PX
MI3 TXT M. Kashkoush, 41 Dunamallaght Road, Ballycastle, BT54 6PF
M3 TXU J. Fesel, 3 Brook Street, Port Sunlight, Wirral, CH62 5DB
M3 TXV R. Fuller, 183 Nottingham Road, Alfreton, DE55 7FL
MM3 TYA M. Anthoney, 10 Cedar Road, Kilmarnock, KA1 2HP
MW3 TYC G. Lewis, Bryn Cottage, Clydach, Abergavenny, NP7 0LL
M3 TYG M. Kyriacou, 54 Sutton Avenue, Silverdale, Newcastle, ST5 6TB
M3 TYI I. Morris, 6 St. Nicholas Court, Gloucester, GL1 2QZ
M3 TYL M. Tyler, 40 Bullards Lane, Woodbridge, IP12 4HE
M3 TYM T. Martin, 14 Campbell Road, Eastleigh, SO50 5AD
M3 TYO G. Aldridge, Greenridge, Fore Street Bishopsteignton, Teignmouth, TQ14 9QR
M3 TYQ D. Brown, 9 Chisholm Close, Standish, Wigan, WN6 0QP
M3 TYS K. Sanchez-Garci, 74 Gorthorpe, Hull, HU6 9EZ

M3 TYU M. Rawlings, 3 Greenlands, Woolton Hill, Newbury, RG20 9TB
M3 TYV D. Atkinson, 133 Lingmoor Rise, Kendal, LA9 7PL
M3 TYW D. Gray, Flat 2 413 Weelsby Street, Grimsby, DN328BJ
M3 TYX D. Mcgrath, 48 Willersley Avenue, Orpington, BR6 9RS
M3 TYZ D. Prince, 29 St. Stephens Road, West Bromwich, B71 4LR
M3 TZB D. Vincent, 6 Nathan Gardens, Poole, BH15 4JZ
M3 TZE R. Tse, 1 Oaklands, Gallows Lane, Westham, BN24 5AW
M3 TZF F. Wells, 12 Portelet Place, Hedge End, Southampton, SO30 0LZ
M3 TZI J. Cowell, Mount Rivers, Bootle, Millom, LA19 5XN
M3 TZN M. Robins, 13 Sarum Way, Hungerford, RG17 0LJ
MM3 TZP S. Wright, 22 Beechwood, Linlithgow, EH49 6SF
M3 TZQ G. Wright, 2 Hillcrest Drive, Castleford, WF103QN
M3 TZS Q. Mdlongwa, 27 Faifax Avenue, Bierley, Bradford, BD4 6JY
MW3 UAA H. Lee, 3 Summerhill Park, Simpson Cross, Haverfordwest, SA62 6EU
M3 UAE T. Baines, 10 Croydon Avenue, Leigh, WN7 1TP
M3 UAG D. Garland, 8 Ladywell Gate, Welton, Brough, HU15 1NL
M3 UAJ A. Neale, 5 Millside, Wombourne, Wolverhampton, WV5 8JJ
M3 UAK M. Brown, 10 Terry Cooney Place, Newcastle upon Tyne, NE5 2FA
M3 UAM C. Byrne, 31 Graham Drive, Castleford, WF10 3EY
M3 UAO M. Lewis, 6 Remembrance Road, Newbury, RG14 6BA
MW3 UAP G. Davies, 45 Greensway, Abertysswg, Tredegar, NP22 5AR
M3 UAQ A. Quinton, 111 Spenser Road, Ipswich, IP1 6HP
M3 UAR A. Riches, 20 Western Gardens, Crowborough, TN6 3EB
M3 UAV G. Preece, 22 Crofters Green, Bradford, BD10 8RZ
M3 UAW T. Arrow, Crystalwood, Stonemans Hill, Newton Abbot, TQ12 5PZ
M3 UAX D. Foyston, 22 Ronaldsway Drive, Newcastle under Lyme, ST5 9HE
M3 UAY M. Scarr, 15 Church Park, Overton, Morecambe, LA3 3RA
MM3 UBB G. Goddard, 17 Burnside, Flotta, KW16 3NP
MM3 UBD D. Branson, Derelochy, Kingsteps, Nairn, IV12 5LF
M3 UBE A. Jones, 6 Barnes Avenue, Dalton-in-Furness, LA15 8NE
M3 UBF J. Bonney, 37 Watery Lane, Brackley, NN13 7NJ
M3 UBG G. Chambers, 26 Parkin Close, Cropwell Bishop, Nottingham, NG12 3DG
M3 UBH C. Davis, 1 Ashland Court, North Street, Crewkerne, TA18 7AP
M3 UBK B. Oakley, 8A Crooked Mile, Waltham Abbey, EN9 1PS
M3 UBL D. Hutchinson, 91, Pentland Avenue, Billingham, TS23 2RF
MW3 UBQ J. Robinson, 25 Stad Castellor, Cemaes Bay, LL67 0NP
MM3 UBR J. Branson, East Bank, South Road, Fochabers, IV32 7LU
M3 UBS D. Foord, 28 Ferndale, Teversham, Cambridge, CB1 9AL
M3 UBT C. Bowes, 11 Burghwallis Lane, Sutton, Doncaster, DN69JU
M3 UBU K. Kettle, 19 St. Trinians Drive, Richmond, DL10 7SS
MW3 UBY M. Davies, 70 Heol Bryncwils, Sarn, Bridgend, CF32 9UE
M3 UCA F. Bano, 14 Norman Trollor Court, Cromer, NR27 9RR
M3 UCC K. Wilton, 71 Aston Clinton Road, Weston Tuville, Buckinghamshire, HP22 5AB
M3 UCF T. Austin, 51 Ashburnham Road, Ramsgate, CT11 0BH
M3 UCH D. Francis, 1905 London Road, Leigh-on-Sea, SS9 2SY
MM3 UCI M. Mccallum, 15 Quarry Road, Law, Carluke, ML8 5HB
M3 UCJ C. Hewett, 20 Cornwallis Avenue, Herne Bay, CT6 6UQ
M3 UCL X. Liu, Harrigate Ladies College, Clarence Drive, Harrogate, HG1 2QG
M3 UCO D. Lawson, 2 The Blossoms, Fulwood, Preston, PR2 9RF
MI3 UCS M. Edwards, 15 Highgrove Road, Carrickfergus, BT38 9AG
M3 UCU S. Finch, 25 Bluebell Avenue, Wigan, WN68NS
M3 UCV T. Kilroy, 55 Summerfield Crescent, Brimington, Chesterfield, S43 1HB
M3 UCY L. Long, 25 St. Matthias Road, Deepcar, Sheffield, S36 2SG
M3 UCZ B. Smallbone, 46 Lesters Road, Cookham, Maidenhead, SL6 9LS
MW3 UDA G. Lloyd, 2 Bryn Y Coed, Holywell, CH8 7AU
MM3 UDB D. Brown, 18 Louisa Drive, Girvan, KA26 9AH
M3 UDC D. Moran, 47 Radcliffe Park Road, Salford, M6 7WP
M3 UDD M. Fisher, 12 Abbey Mews, Pontefract, WF8 1TD
M3 UDF D. Fagan, 58 Main Street, Linton, Swadlincote, DE12 6PZ
MM3 UDI P. Pirie, Willowbank, Kirkton Of Tough, Alford, AB33 8ER
M3 UDJ M. Pharoah, 116 Chatsworth Street, Barrow-in-Furness, LA14 5TP
M3 UDK J. Cunningham, 56 Askam Avenue, Pontefract, WF8 2PN
MM3 UDL B. Nielsen, House Of Shannon Wester Templands, Fortrose, IV10 8RA
M3 UDN Dr P. Thompson, 3 Floyers Field, West Stafford, Dorchester, DT2 8FJ
MW3 UDO R. Oliver, 6 Clevedon Avenue, Sully, Penarth, CF64 5SX
MM3 UDQ J. Smith, 10 High Street, Portknockie, Buckie, AB56 4LD
M3 UDS W. Sun, 10 Scholfield Way, Eastbourne, BN23 6HQ
M3 UDU S. Hadfield, 56 Risborough Road, Bedford, MK41 9QW
MM3 UDV D. Herd, 4 West Fairbrae Drive, Edinburgh, EH11 3SY
MW3 UDW M. Evans, 9 St Teilos Close, Ebbw Vale, Blaenau Gwent, NP23 6NE
M3 UDZ D. Mellor, 28 Winster Road, Staveley, Chesterfield, S43 3NJ
MM3 UEA E. Ewing, Arisaig, Priestland, Darvel, KA17 0LP
M3 UEC R. Griffiths, 8 Mount Street, Kings Lynn, PE30 5NH
M3 UED G. Henstridge, 47 Churchfield Drive, Castle Cary, BA7 7LB
M3 UEE J. Price, 3 Perlethorpe Close, Gedling, Nottingham, NG4 4GF
M3 UEF P. Marchant, 16 Melrose Drive, Peterborough, PE2 9DN
MW3 UEG A. Evans, 9 St Teilos Close, Ebbw Vale, Blaenau Gwent, NP23 6NE
M3 UEJ C. Liversidge, Flat 3-8 Cromwell Terrace, Scarborough, YO11 2DT
MW3 UEK N. Hosker, Laburnum Cottage, Church Lane, Old Aston Hill, Ewloe, CH5 3BF
M3 UEL S. Long, 25 St. Matthias Road, Deepcar, Sheffield, S36 2SG
M3 UEN A. Sammut, The Quaker Cottage, Wainfleet Bank, Skegness, PE24 4JP
M3 UEP A. Hudders, 10 Waterton Grove, Wakefield, Wf28hr
M3 UEQ G. Lambert, 17 Starcross Road, Weston-Super-Mare, BS22 6NY
M3 UER P. Stone, 2 Endeavour Close, Lower Stondon, Bracknell, SG166JR
MM3 UET M. Henderson, 22 Bowmont Place, East Kilbride, Glasgow, G75 8YG
MW3 UEU K. Rowney, 14A Fagwr Road, Craig-Cefn-Parc, Swansea, SA6 5TB
M3 UEW A. Allanson, 5 Kingsley Avenue, Crofton, Wakefield, WF4 1RN
M3 UEX D. Carr, 36 Rookwood Mount, Leeds, LS9 0LL
M3 UEY A. Holder, 12A Evenlode Road, Gloucester, GL4 0JT
M3 UEZ W. Sargeant, 2 Church Mews, Judith Road, Kettering, NN16 0QR
M3 UFA S. Leggett, 246 Delaware Road, Shoeburyness, Southend-on-Sea, SS3 9NT
MI3 UFD R. Stirrup, 211 Leckagh Drive, Magherafelt, BT45 6ND
M3 UFE J. Chau, Harrogate Ladies' College, Clarence Drive, Harrogate, HG1 2QG
M3 UFF A. Mullen, 81 Worcester St., Stourbridge, DY8 1AX
M3 UFG D. Gunn, 40 The Pastures, Oadby, Leicester, LE2 4QD
MW3 UFH C. Livingstone-Lawn, 72 Ty Mawr Avenue, Rumney, Cardiff, CF3 3AG
M3 UFJ S. Jarvis, 10 Wood Lane, Wolverhampton, WV10 8HJ
M3 UFK N. Nash, Flat 21, Farringdon House, Green Lane, Walsall, WS2 8NP
M3 UFL W. Oakey, 41 Beaford Grove, London, SW20 9LB
M3 UFS A. Brand, 6 Walnut Close, Milton, Cambridge, CB24 6ET
M3 UFT D. Redman, 5 Dee Road, Lancaster, LA1 2QX
M3 UFU K. Hillbeck, 28 Darent Avenue, Walney, Barrow-in-Furness, LA14 3NU
M3 UFW N. Mclean, 21 Matlock Avenue, Wigston, LE18 4NA
M3 UFX S. Blackwell, 38 Peatburn Avenue, Heanor, DE75 7RL
M3 UFY J. Poland, 14 Malleson Road, Liverpool, L139DF
M3 UFZ A. Saddington, 7 Baker Court, Thrapston, Kettering, NN14 4XA
M3 UGA T. Barnard, 8 Argyle Road, Poulton-le-Fylde, FY6 7EW
M3 UGB G. Boast, 102 Abbeyfield Road, Birmingham, B23 5LL
M3 UGD D. Underwood, 33 Meadow Avenue, Rawmarsh, Rotherham, S62 7EE
M3 UGF M. Sims, 133 Canterbury Road, Hawkinge, Folkestone, CT18 7BS
M3 UGH M. Cole, 25 Freemans Road, Minster, Ramsgate, CT12 4EL
MI3 UGI M. Downey, 18 Castlevue Park, Moira, Craigavon, BT67 0LN
M3 UGJ G. Douch, 63 Greenaways Ebley, Stroud, GL5 4UN
M3 UGK M. Rimmer, 15 Brade Street, Southport, PR9 8LS
MM3 UGL N. Rogers, 108 Beechwood Road, Cumbernauld, Glasgow, G67 2NP
M3 UGO D. Hill, 19 Farren Road, Birmingham, B31 5HH
M3 UGQ P. Woodyard, 65 Raglan Street, Lowestoft, NR32 2JS
M3 UGR C. Luckett, 257 Folkestone Road, Dover, CT17 9LL
M3 UGT G. Taylor, Southlands, Church Lane, Great Holland, Frinton-on-Sea, CO13 0JS
M3 UGV S. Jawor, 5 Cotteswold Rise, Stroud, GL5 1HD
MW3 UGW G. Wilcock, 42 Erskine Road, Colwyn Bay, LL29 8EU
M3 UGX W. Owen, 8 Sandhurst Avenue, Lytham St Annes, FY8 2DA
MD3 UGY M. Wernham, Fair Isle, Lhoobs Road, Douglas, Isle Of Man, IM4 3JB
M3 UGZ R. Dearden, 218 South Street, Highfields, Doncaster, DN6 7JQ
M3 UHB A. Hartley, 47 Windways, Little Sutton, Ellesmere Port, CH66 1JG
M3 UHC J. Bramley, 17 Oakholme Rise, Worksop, S81 7LJ
M3 UHG R. Smith, 20A Waverley, Skelmersdale, WN8 8BD
M3 UHH H. Hopkins, 3 Colegrave Road, Bloxham, Banbury, OX15 4NT
MI3 UHI E. Kyle, 2 Wattstown Crescent, Coleraine, BT52 1SP
M3 UHJ P. Hopkinson, 28 Stockdove Way, Thornton-Cleveleys, FY5 2AR
MM3 UHK W. Spiers, 19/2 150 Charles Street, Glasgow, G21 2QF
MI3 UHL W. Mark, 82 Glenleslie Road, Clough, Ballymena, BT44 9RH
M3 UHN P. Sherratt, 39 Vimy Road, Leighton Buzzard, LU7 1FQ
MW3 UHO B. Cayford, 13 Harford Square, Newtown, Ebbw Vale, NP23 5FA
M3 UHP C. Rodway, 26 Redesdale Avenue, Newcastle upon Tyne, NE3 3PP
M3 UHQ L. Richardson, 127A Mount Gould Road, Plymouth, PL4 7PY
M3 UHS I. Stephens, 2 Boniface Walk, Burnham-on-Sea, TA8 1RE
M3 UHV D. Lisi, 60 Middlecroft Road, Staveley, Chesterfield, S43 3XH
M3 UHW M. Troth, 21 Willow Road, Bromsgrove, B61 8PN
M3 UHX A. Anderson, 89A Malmesbury Park Road, Bournemouth, BH8 8PS
M3 UHY R. Bond, 21 Coleridge Close, Bletchley, Milton Keynes, MK3 5AF
MI3 UIA P. Dallas, 12 Glendun Crescent, Coleraine, BT52 1UJ
M3 UIB B. Wilde, Seaways, Cliff Rise, Fowey, PL23 1QQ
M3 UIC D. Simpson, 140 Church Road, Redfield, Bristol, BS5 9HN
M3 UIF M. Wheeler, Homeleigh, Station Road, Rochester, ME3 7RN
M3 UII S. Odonoghue, 15 Chandlers Close, New Waltham, Grimsby, DN36 4WH
M3 UIJ M. Rogers, 41 Barton Hill Drive, Minster On Sea, Sheerness, ME12 3NF
M3 UIK D. Mccrae, 37 Burnside, Wigton, CA7 9RE
M3 UIL T. Brice, 10A Nelson Drive, Exmouth, EX8 2PU
MI3 UIM P. Mckeever, 7 St Canices Park, Eglinton, Ireland, BT47 3AQ
MW3 UIN L. Collins, 283 Graig Road, Godrergraig, Swansea, SA9 2NZ
M3 UIP J. Dunkin, 6 Kingsley Grove, Grimsby, DN33 1NL
MW3 UIQ A. Henderson, 45 Brynamman Road, Lower Brynamman, Ammanford, SA18 1TR
M3 UIS A. Marshall, 13 The Markhams, New Ollerton, Newark, NG22 9QX
M3 UIU U. Ukommi, 30 Oregano, Room 3, Hazel Farm, Surrey, GU2 9TY
MI3 UIV V. Madden, 2, Rathbeg, Limavady, BT49 OAT
MI3 UIW J. Smyth, 39 Whitehill Park, Limavady, BT49 0QF
MM3 UIX C. Dyer, 55 Duthie Road, Gourock, PA19 1XS
MW3 UIY P. Henderson, 45 Brynamman Road, Lower Brynamman, Ammanford, SA18 1TR
M3 UJD K. Colman, 10 South Rise, North Walsham, NR28 0EE
M3 UJE M. Corbett, 23 Heathfield Road, Fleetwood, FY7 7LY
M3 UJF J. Finney, 40 Tempest Avenue, Darfield, Barnsley, S73 9BJ
M3 UJH J. Harbron, 48 Sheridan Road, Biddick Hall, South Shields, NE34 9JJ
M3 UJL J. Bailey, 38 Barlow Drive, Sheffield, S6 5HQ
M3 UJN C. Hall, 10 First Street, Pont Bungalows, Consett, DH8 6JG
M3 UJO J. Phillips, Flat L 15, International House, Guildford Court, Guildford, GU2 7JL
M3 UJP J. Fletcher, 2 Sunflower Meadow, Irlam, M44 6TD
MM3 UJQ A. Hackman, The Herdsman Cottage, Brighouse Bay, Kirkcudbright, DG6 4TT
M3 UJR J. Roberts, 93 Earlshall Road, Eltham, London, SE9 1PP
M3 UJS J. Suter, 11 Summerdown Close, Durrington, Worthing, BN13 3QG
M3 UJT I. Dransfield, Gardener Ground House, West End, Goole, DN14 8RW
M3 UJV G. Roth, 121 St. Annes Road, Wolverhampton, WV10 6SL
M3 UJX L. Parish, 6 Courtwick Road, Wick, Littlehampton, BN17 7NE
M3 UJY C. Meakin, 102 Ryknield Road, Kilburn, Belper, DE56 0PF
M3 UJZ J. Cooke, 9 Nyetimber Crescent, Pagham, Bognor Regis, PO21 3NN
M3 UKB C. Price, 9 Arlington Avenue, Aston, Sheffield, S26 2AA
M3 UKD J. Parfrey, 47 Ford Lane, Rainham, RM13 7AS
M3 UKF P. Seaton, Sunnymead, Newland, Barnstaple, EX32 0ND
MM3 UKG D. Gilmour, 35 Hailes Gardens, Edinburgh, EH13 0JH
M3 UKH G. Nelson, 812 Hessle Road, Hull, HU4 6RD
M3 UKJ J. Boyd, 14 James Street, Seaham, SR7 7QW
MW3 UKK M. Davison, Maes Y Neuadd Hotel, Talsarnau, LL47 6YA
M3 UKN N. Hewitt, Redwood House, Adbury Holt, Newbury, RG20 9BW
M3 UKO N. Batchelor, 15 Sufton Rise. Mordiford., Hereford, HR1 4EN
M3 UKR A. Rogers, 260 Griffiths Drive, Wolverhampton, WV11 2JS
M3 UKU M. Qassim, Winchester Road, Kings Somborne, Stockbridge, SO20 6NY
M3 UKV T. Vincent, 12 Lancaster Place, Kenilworth, CV8 1GL
MI3 UKW M. Mcerlean, 38A Culbane Road Portglenone, Portglenone, BT448NZ
M3 UKX J. Cornish, 2 Micklehill Drive, Shirley, Solihull, B90 2PU
M3 UKY D. Gordon, 32 Claremont Road, Stockport, SK2 7AR
M3 ULB S. Owen, 21 Market Place, Hingham, Norwich, NR9 4AF

M3 ULE L. Lawrence, Rookery Rise, 8 Woodside, Brede, TN31 6DS
M3 ULI C. Dunn, 13 Springfield Road, Leyland, PR25 1AR
M3 ULL M. Bull, 5 Roach Place, Rochdale, OL16 2DD
M3 ULM I. Cottom, 8 Bridgewater Rise, Brackley, NN13 6DA
M3 ULN R. Fricker, 2 Buttermere Drive, Allestree, Derby, DE22 2SN
M3 ULO S. Clarke, 2 Dawn Crescent, Upper Beeding, Steyning, BN44 3WH
M3 ULQ J. Gromadzki, 13 Merrill Heights, Maidenhall Approach, Ipswich, IP2 8GA
M3 ULS J. Firth, 36 Howley Grange Road, Halesowen, B62 0HW
M3 ULT S. Dickinson, 28 Ingleton Walk, Barnsley, S70 2NE
M3 ULU A. Eizzard, 41 Sycamore Drive, Waddington, Lincoln, LN5 9DR
M3 ULW L. Weston, 114 Morland Road, Ipswich, IP3 0LZ
M3 ULX G. Quilter, 83 Jameson Street, Wolverhampton, WV6 0NT
M3 ULZ A. Nokes, 3 Eastview, Ditcheat, Shepton Mallet, BA4 6PN
M3 UMA M. Abel, 121 Angela Road, Horsford, Norwich, NR10 3HF
M3 UMB M. Bryans, The Lodge, The Warren Croydon Road, Bromley, BR2 7AL
MI3 UMC J. Mcauley, 29 Ilse Court, Larne, BT40 3NT
M3 UMD B. Walton, 40 Princess Street, Mapplewell, Barnsley, S75 6ET
M3 UMG P. Haygarth, 5 Forth Close, Peterlee, SR8 1DG
M3 UMJ M. Hockin, 7 Gourders Lane, Kingskerswell, Newton Abbot, TQ12 5DZ
M3 UML M. Layton, 16 Gwelmor, Camborne, TR14 7BP
M3 UMM M. Case, 6 Boldventure, Close, St Austell, PL25 3DY
MD3 UMN D. Kneale, 4 Glashen Terrace, Ballasalla, Isle Of Man, IM9 2ET
M3 UMR S. Rutt, Granthorpe, Hull Road, Hull, HU11 5RN
M3 UMV P. Shaves, 33 Derwent Drive, Bletchley, Milton Keynes, MK3 7BG
M3 UMW M. Wilkins, 31 Stratton Audley Road, Fringford, Bicester, OX27 8ED
M3 UMX R. Bates, 61 Park View, Crowmarsh Gifford, Wallingford, OX10 8BN
MM3 UMY C. Brogan, 13 Mitchell Avenue, Cambuslang, Glasgow, G72 7SQ
MM3 UMZ J. Bain, 13 Mitchell Avenue, Cambuslang, Glasgow, G72 7SQ
M3 UNB K. Puttock, 12 Beechfields, School Lane, Petworth, GU28 9DH
M3 UNF H. Coram, 20 Eton Walk, Exeter, EX4 1FD
M3 UNH A. Hitchcott, 121 Oakhurst Road, Acocks Green, Birmingham, B27 7PB
M3 UNI N. Bradshaw, 291B Hull Road, Hull, HU4 7rh
M3 UNK M. Wright, 2 Regent Road, Church, Accrington, BB5 4AR
M3 UNL P. Webb, 104 Birds Nest Avenue, Leicester, LE3 9ND
M3 UNP N. Patel, 16 Camellia Court, 18 Copers Cope Road, Beckenham, BR3 1NB
MM3 UNQ N. Stevenson, 12 Glamis Place, Greenock, PA16 7NB
M3 UNR A. Worlledge, 181 Roselands Drive, Paignton, TQ4 7RN
M3 UNT P. Morris, 689 Tonge Moor Road, Bolton, BL2 3BW
M3 UNY L. Bain, 45 Larpool Crescent, Whitby, YO22 4JD
MW3 UNZ S. Valentine, 12 Cwrt Y Glyn, Carmel, Llanelli, SA14 7SA
M3 UOB C. Bradley, 6 Copeland Row, Evenwood, Bishop Auckland, DL14 9PY
M3 UOC Dr J. Mohammed, 2 Moffat Avenue, Ipswich, IP4 3JH
M3 UOD J. Cooke, 2 Church Street Close, Thurnscoe, Rotherham, S63 0QT
MM3 UOE R. Wilson, 12 Queen Road, Irvine, KA12 0XA
M3 UOF K. Mills, 6A Deacons Walk, Schofield Street, Mexborough, S64 9NH
M3 UOJ K. Barry, 25 Delabole Road, Merstham, Redhill, RH1 3PB
M3 UOK A. Abraham, Flat 2, 41 Francis Road, Birmingham, B33 8SL
M3 UOL L. Haworth, 139 Manchester Road, Accrington, BB5 2NY
M3 UOM M. Catterall, 2 Cedar Road, Bishop Auckland, DL14 6ET
M3 UON S. Kamal, 184 Aycliffe Road, Borehamwood, WD6 4EG
M3 UOO R. Bone, 6 Danehurst Place, Locks Heath, Southampton, SO31 6PP
MM3 UOR R. Paterson, 1/R 162, Glasgow Street, Ardrossan, KA228HA
MM3 UOS D. Young, 81 Leaven Place, Irvine, KA12 9PA
M3 UOX A. Tate, 24 Brentingby Close, Melton Mowbray, LE13 1ES
M3 UOY G. Fripp, 35 Kiln Close, Bovey Tracey, Newton Abbot, TQ13 9YL
M3 UOZ A. Knight, 39 Thurnview Road, Evington, Leicester, LE5 6HL
M3 UPB L. Banahan, 18 Lynn Road, Ely, CB6 1DA
M3 UPF M. Elkington, 32 The Knoll, Kingswinford, DY68JT
M3 UPJ P. Harley, 16 Clover Drive, Rushden, NN10 0TZ
M3 UPK J. Addy, 12 Wortley Avenue, Swinton, Mexborough, S64 8PT
M3 UPL B. Banks, 44 Manor Road, Swinton, Mexborough, S64 8PY
M3 UPN P. Needham, 9 Westwood, Broughton, Brigg, DN20 0AU
M3 UPO D. Clarke, 2 Dawn Crescent, Upper Beeding, Steyning, BN44 3WH
M3 UPP L. Wilkes, 24 Ilminster, Dunster Crescent, Weston-Super-Mare, BS24 9EB
M3 UPQ M. Davis, 3 Pollards Court, Rochford, SS4 1GH
M3 UPT S. Thomas, 103 Liverpool Road, Upton, Chester, CH2 1BB
MW3 UPX M. Davies, 11 High Street, Malltraeth, Bodorgan, LL62 5AS
MM3 UPY S. Dunbar, 34 Windyknowe Crescent, Bathgate, EH48 2BU
M3 UQA P. Allison, Flat 11, Forest Court, 5-11 Salisbury Road, Fordingbridge, SP6 1EG
M3 UQB C. Kerridge, 2 Allerdale Close, Thirsk, YO7 1FW
M3 UQD E. Sabbatella Riccardi, 97 Harp Island Close, London, NW10 0DQ
M3 UQE M. Burstow, 40 Wyndham Road, Petworth, GU28 0EQ
M3 UQF G. Hillbeck, 28 Darent Avenue, Walney, Barrow-in-Furness, LA14 3NU
M3 UQG H. Roxbrough, 17 Stanwell Close, Sheffield, S9 1PZ
M3 UQH J. Parrett, 6 Shelley Road, East Grinstead, RH19 1TA
M3 UQI G. Jones, 57 Oxford Road, Banbury, OX16 9AJ
M3 UQJ G. Dyson, 111 Chester Road, Whitby, Ellesmere Port, CH65 6SB
M3 UQL D. Smith, 131 Canons Walk, Thetford, IP24 3PT
MM3 UQN D. Taylor, 1 Mayfield Farm Cottages, Reston, Eyemouth, TD14 5LG
M3 UQO G. Smith, 92 Brighton Road, Banstead, SM7 1BU
M3 UQP K. Eccleston, 22 Lowes House, Rodney Drive, Woodley, Stockport, SK6 1SL
MM3 UQT C. Page, 15 Jackton View, East Kilbride, Glasgow, G75 9NW
M3 UQY R. Cox, 17 Hyde Lane, Upper Beeding, Steyning, BN44 3WJ
M3 UQZ D. Hartley, Flat 4, The Old Mill, Station Road, Ellesmere Port, CH66 1NY
M3 URA M. Barber11, Homedale, St. Monicas Road, Kingswood, Tadworth, KT20 6ET
M3 URD R. Dell, 18 Greenacres, Fulwood, Preston, PR2 7DA
M3 URE R. Eddy, 15 Western Place, Penryn, TR10 8HQ
M3 URF R. Freeman, 9 Bramley Road, Wisbech, PE13 3PA
MW3 URG R. Grayson, Willow Lodge, Croeslan, Llandysul, SA44 4SJ
M3 URH L. Harman, 7 Loughborough Road, Walton On The Wolds, Loughborough, LE12 8HT
MW3 URO D. Fullick, 69 Shakespeare Road, St. Dials, Cwmbran, NP44 4LW
MM3 URQ S. Urquhart, 1 Hatchery Cottage, Station Road, Duns, TD11 3HS
M3 URS R. Singh, Stainburn House, Barrowby Lane, Harrogate, HG3 1HY
M3 URT S. Mather, 35 Neargates Charnock Richard, Chorley, PR7 5EY
M3 URV R. Earnshaw, 53 Blue Waters Drive, Paignton, TQ4 6JF
M3 URW R. Woodford, South Calvadnack, Carnmenellis, Redruth, TR166PN
M3 URX Dr D. Craig, Pear Tree Cottage, Cripps Corner, Staplecross, TN32 5QS
M3 URZ C. Evans, 1 Rialto Road, Mitcham, CR4 2LT
M3 USB M. Perocevic, 12 Ash Road, Crewe, CW1 4DU
M3 USC S. Coleman, 32 Southwell Road, Wisbech, PE13 3LQ
M3 USF A. Willis, 17 Ladypit Terrace, Whitehaven, CA28 6AQ
M3 USH S. Hall, 14 Nicholson Place, East Hanningfield, Chelmsford, CM3 8UT
M3 USJ S. Hall, 5 Ropery Lane, Barton-upon-Humber, DN18 5TW
MW3 USK C. Burke, 84 Elgam Avenue, Blaenavon, Pontypool, NP4 9QU
MM3 USN J. Challis, Bay Villa, Strachur, Cairndow, PA27 8DE
MW3 USP S. Barwell, 50 Mill Close, Caerphilly, CF83 2LL
MW3 USS D. Provis, Dingle Gardens, Croesbychan, Aberdare, CF44 0EJ
M3 UST G. Evans, 2 Tower Farm Cottages, Featherbed Lane, Hemel Hempstead, HP3 0BT
MM3 USV A. Dunn, 66 Glen Doll Road Neilston, Glasgow, G78 3QP
M3 USW P. Jenkins, 48 The Pantiles, Bexleyheath, DA7 5HG
MW3 USX S. Tozer, 110 Glanffornwg, Wildmill, Bridgend, CF31 1RL
M3 USZ R. Lane, 9 Hartoft Road, Hull, HU5 4JZ
MM3 UTH P. Dower, 1670 Maryhill Road, Glasgow, G20 0HJ
M3 UTJ S. Crichton, 27 Rosewood Ave, Stockport, SK4 2DQ
M3 UTM A. Williams, 1 Nimmings Close, Birmingham, B31 4TA
M3 UTN K. Bailey, The Firs, South Chard, Chard, TA20 2RX
M3 UTP T. Pentz, 30 Lindrick Way, Harrogate, HG3 2SU
M3 UTQ D. Bell, 18 Julius Hill, Warfield, Bracknell, RG42 3UN
MM3 UTU M. Magee, 30 Burnfield Drive, Mansewood, Glasgow, G43 1BW
M3 UTV M. Compagno, 18 Bromford Crescent, Birmingham, B24 9RJ
MI3 UTY M. Regan, 80 Killowen Drive, Magherafelt, BT45 6DS
M3 UTZ S. Abdullah, 24 The Grove, Walsall, WS5 4BX
M3 UUE H. Roberts, 71A Brook Street, Stourbridge, DY8 3UX
M3 UUF R. Dicker, 38 Inkerman Road, Southampton, SO19 9DA
M3 UUG S. Lenton, Badgers Rise, The Willows, Torquay, TQ2 7TB
M3 UUL G. Rowe, 10 Alexander Avenue, Selston, Nottingham, NG16 6FW
M3 UUN T. Harding, 7 Chiltern Avenue, Poulton-le-Fylde, FY6 7DY
M3 UUO D. Lynch, 12 Shipley Close, Blackpool, FY3 7UJ
M3 UUS S. Mallinson, 63 Celandine Avenue, Locks Heath, Southampton, SO31 6WZ
MM3 UUT L. Thomson-Best, 5 Gib Grove, Dunfermline, KY11 8DH
M3 UUW J. Douch, 63 Greenaways, Ebley, Stroud, GL5 4UN
MW3 UUY D. Williams, 10 Bronllys, Gaerwen, LL60 6JN
M3 UUZ R. Chandler, 26 Chalky Bank, Gravesend, DA11 7NY
M3 UVA J. Grainger, 32 Ellenfoot Drive, Maryport, CA15 7DB
M3 UVC R. Milton, 66 Hoo Marina Park, Vicarage Lane, Rochester, ME3 9TG
M3 UVD D. Tregear, 73 Alderley Lane, Leigh, WN7 3DW
MM3 UVF W. Mcblain, 50 Hamilton Crescent, Stevenston, KA20 4JE
M3 UVJ J. Binfield, 55 Gladstone Road, Broadstairs, CT10 2HY
M3 UVK W. Penny, 159 Coxford Road, Maybush, Southampton, SO16 5JX
M3 UVM R. Murphy, 23 Lowndes Close, Stockport, SK2 6DW
M3 UVO A. Ruocco, 16 Conyers Avenue, Grimsby, DN33 2BY
M3 UVQ A. Barton, 11 Grove Avenue, Beeston, Nottingham, NG9 4ED
M3 UVS S. Thorne, 72 Devonshire Road, London, E16 3NJ
M3 UVT M. Garry, 34 Conway Road, Paignton, TQ4 5LH
M3 UVV A. Mackay, 16 Vicarage Close, Chard, TA20 2HH
M3 UVX D. Housden, 12 Regent House, Cheltenham Gardens, Southampton, SO30 2UD
M3 UVY A. Marrison, 316 Booth Walk, Newton Aycliffe, DL5 7NT
M3 UWB M. Tromans, 10 Crofters View, Little Wenlock, Telford, TF6 5AU
M3 UWE D. Scott, 7 Teal Close, Brookside, Telford, TF3 1NY
M3 UWF S. Roberts, Errys, Stunts Green, Herstmonceux, Hailsham, BN27 4PP
M3 UWI K. Jones, 41 Milton Brow, Weston-Super-Mare, BS22 8DD
M3 UWJ R. Wilmot, 41 Milton Brow, Weston-Super-Mare, BS22 8DD
M3 UWM T. Kitto, 4 Pennard Council Houses, St. Breock, Wadebridge, PL27 7LL
M3 UWR S. Dawson, 51 St. Edwards Road, Gosport, PO12 1PW
M3 UWT A. Chapman, 10 Derwent Road, Seaton Sluice, Whitley Bay, NE26 4JH
M3 UWU M. Redfern, 62 Meadow Road, Dudley, DY1 3JU
M3 UWV C. King, 21 Lowdham, Wilnecote, Tamworth, B77 4LX
M3 UWW R. Massimino, 115 Trelowarren Street, Camborne, TR14 8AW
MU3 UWX M. Barker, Rosee Terres, Les Effards Road, St Sampson, Guernsey, GY2 4YW
M3 UWY T. Baddeley, 44 Lowry Close, Willenhall, WV13 3BD
MW3 UWZ J. Evans, 2 Collfryn Cottages, Bethesda Bach, Caernarfon, LL54 5SF
M3 UXC A. Spaxman, 70 Park View, Shafton, Barnsley, S72 8PY
M3 UXE E. Bruce, 26 Queens Road, Wilbarston, Market Harborough, LE16 8QJ
M3 UXF M. Bridgehouse, 43 Age Croft, Oldham, OL8 2HG
M3 UXG M. Leech, 11 Westlake Close, Torpoint, PL11 2BZ
M3 UXH K. Sowter, 55 Ward Street, New Tupton, Chesterfield, S42 6XR
M3 UXI L. Allanson, 5 Kingsley Avenue, Crofton, Wakefield, WF4 1RN
M3 UXK K. Jones, 41 Tibbs Hill Road Abbott Langley, Watford, Wd5 0ee
M3 UXL A. Bond, 21 Coleridge Close, Bletchley, Milton Keynes, MK3 5AF
M3 UXM M. Bruce, 26 Queens Road, Wilbarston, Market Harborough, LE16 8QJ
M3 UXN P. Overton, 39 Bridle Road, Madeley, Telford, TF7 5HB
MM3 UXO L. Aitken, 92B Belville Street, Greenock, PA15 4TA
M3 UXR M. Griffiths, 11 Frogwell Park, Chippenham, SN14 0RB
M3 UXS C. Dutton, Twillingate Farm, Tiptoe, Lymington, SO41 6EJ
M3 UXU D. Almond, 2 Farm Veiw, New Tupton, Derbyshire, S426BD
M3 UXX M. Selvey, 52 Coppice Close, Cheslyn Hay, Bella Casa, Walsall, WS6 7EZ
M3 UYC D. Griffin, 101 Kingsway, Duxford, Cambridge, CB22 4QN
M3 UYF M. Heritage, High View, Common Lane, Corley, Coventry, CV7 8AQ
M3 UYG A. Goldsmith, 61 Fengate Drove, Weeting, Brandon, IP27 0PW
MW3 UYH G. Broadbent, 7 James Close, Llanon, SY23 5HP
MW3 UYJ J. Davies, 19 Falcon Place, Blaenymaes, Swansea, SA5 5NX
M3 UYK Y. Chiu, Harrigate Ladies College, Clarence Drive, Harrogate, HG1 2QG
M3 UYL G. Richards, 1 Brisbane Road, Weymouth, DT3 6RB
M3 UYO R. Dewis, 6 St Nicolas Close, Pevensey, BN245LB
M3 UYQ T. Hull, 12 Durley Road, Gosport, PO12 4RT
M3 UYU B. Elderbrant, 20 Loxley Road, Southport, PR8 6NR
M3 UYV J. Rudd, 5 St Andrews Close, Blofield, NR13 4JX
M3 UYW P. Underhill, 35 Windermere Road, Reading, RG2 7HU
MW3 UYX A. Harvey, 15 Pen Y Lan, Penclawdd, Swansea, SA4 3LL
M3 UYY M. Dale, 37 Bussey Road, Norwich, NR6 6JF
M3 UZA P. Greenhalgh, 33 Shepherds Lane, Chester, CH2 2DH

M3 UZB S. Bailey, 23 Maple Avenue, Tolladine, Worcester, WR4 9RD
M3 UZE B. Monksummers, 10 Breach Close, Bourton, Gillingham, SP8 5BB
MW3 UZH J. Alfei, 9 Brookside, Gowerton, Swansea, SA4 3AY
M3 UZK A. Lancefield, 19 Tawny Sedge, King's Lynn, PE30 3PW
M3 UZL S. Clarke, Brimham Lodge Fm, Harrogate, HG3 3HE
M3 UZN R. Truelove, 104 Malines Avenue, Peacehaven, BN10 7RL
MW3 UZO D. White, 222 St. Fagans Road, Cardiff, CF5 3EW
MW3 UZP J. Young, 10 Heol Fion, Gorseinon, Gorseinon, SA4 4PN
M3 UZV J. Porter, 18 Cornerway, Birmingham, B38 9RH
M3 UZW J. Boag, 60 Harebell, Amington, Tamworth, B77 4NA
M3 UZZ G. Hamilton, 11 The Spinney, West Lavington, Devizes, SN10 4HP
M3 VAE C. Johns, 6 Cranham Close Kingswood, Bristol, BS15 4QB
M3 VAF K. Brown, 41 Church Street, Swinton, Mexborough, S64 8EF
M3 VAG B. Gittings, 29 Highdown Way, Swindon, SN25 4YD
MM3 VAH T. Hamilton, 57/6 North Street, Bo'ness, EH51 0AE
M3 VAM M. Medcalf, 47 Paddock Drive, Chelmsford, CM1 6UX
M3 VAQ J. Bowley, 2 Cottage, Middle Battenhall Farm, Worcester, WR5 2JL
M3 VAR A. Dokic, 28 Tudor Gardens, Shoeburyness, Southend-on-Sea, SS3 9JG
M3 VAS V. Papanikolaou, 104 West Drive Gardens, Soham, Ely, CB7 5EX
M3 VAT D. Fielding, 2 Christchurch Road, Bradford-on-Avon, BA15 1TB
MM3 VBF S. Campbell, 78 Liddel Road, Cumbernauld, Glasgow, G67 1JE
M3 VBG B. Garry, 34 Conway Road, Paignton, TQ4 5LH
M3 VBH A. Townsend, 42 Grove Avenue, Yeovil, BA20 2BD
M3 VBI J. Broadhurst, Flat 2, Pennant Court, Rowley Regis, B65 8DW
M3 VBL J. Williams, 3 Corner Field, Kingsnorth, Ashford, TN23 3NH
M3 VBM V. Maynard, 34 Heath Farm Park, Barford St. Martin, Salisbury, SP3 4BQ
M3 VBN K. Sloan, Woodland Halt, Old Station Road, Winchester, SO21 1BA
M3 VBP G. Bramham, 1 Watson Avenue, Dewsbury, WF12 8PZ
M3 VBT V. Be-Dard, 53 Cottingley Crescent, Leeds, LS11 0HZ
M3 VBV G. Foster, 18 Austen Avenue, Long Eaton, Nottingham, NG10 3GG
M3 VBW B. Whall, 3 Farrow Close, Great Moulton, Norwich, NR15 2HR
M3 VBY D. Potter, 30 Mersham Gardens, Goring-By-Sea, Worthing, BN12 4TQ
M3 VBZ A. Bolton, 26, St Margarets Avenue, Sutton, SM3 9TT
M3 VCB C. Burbridge, 9 Victoria Road, Stirchley, Birmingham, B30 2LS
MI3 VCI G. Lyttle, 37 Cloyfin Park, Coleraine, BT52 2BL
M3 VCK M. Brown, 9 Warsop Road, Barnsley, S71 3NR
M3 VCM C. Mallory, 11 Baymead Meadow, North Petherton, North Petherton, ta6 6qw
M3 VCO A. Knowles, 260 Haunchwood Road, Nuneaton, CV10 8DL
M3 VCP S. Pryke, 12 Seaward Avenue, Leiston, IP16 4BB
M3 VCQ S. Wilson, 2 Frankham Close, Dinnington, Sheffield, S25 3QG
M3 VCV L. Smith, 11 Appleby Avenue, Timperley, Altrincham, WA15 7HY
M3 VCW C. West, 1 Willetts Mews, Hoddesdon, EN11 9DX
M3 VCY G. Shakespeare, 6 Waterworks Cottages, Clough Road, Hull, HU6 7QB
M3 VCZ M. Rigby, 75 Manchester Road, Deepcar, Sheffield, S36 2QX
M3 VDA A. Crowther, 11 Goodman Court, Central Drive, Chesterfield, S44 5BA
M3 VDE R. Ferguson, 31 Barton Court Road, New Milton, BH25 6NW
M3 VDF G. Bertola, 17 Caraway Drive Branston, Burton-on-Trent, DE14 3FQ
M3 VDH D. Hind, 116 Gowthorpe, Selby, YO8 4HA
M3 VDL M. Gosling, 12 The Grove, Studley, B80 7QL
M3 VDN J. Owen, 90 Granville Drive, Kingswinford, DY6 8LW
M3 VDO D. Oliver, 11 Crooked Creek Road, Rendlesham, Woodbridge, IP12 2GL
M3 VDT D. Cattermole, Blaxhall Hall Crossing, Little Glemham, Woodbridge, IP13 0BP
M3 VDU B. Marston, 111 Averil Road, Leicester, LE5 2DE
M3 VDV G. Haggas, 8 Retford Close, Stockton-on-Tees, TS19 9EJ
M3 VDZ G. Yuill, 14 Gardyn Croft, Taverham, Norwich, NR8 6UZ
MW3 VEH D. Thomas, 57 Brynhyfryd Street, Treorchy, CF42 6DT
M3 VEJ R. Buckwell, 75 Brookside Avenue, Polegate, BN26 6DQ
MW3 VEL L. Wilson, 48 The Woodlands, Brackla, Bridgend, CF31 2JG
M3 VEM C. Vernon, 80 Shirley Drive, Worthing, BN14 9BB
MW3 VEN P. Hoy, 39 Blackbird Road, Caldicot, NP26 5RE
MI3 VEQ B. Mcconnell, 14 Ballymacruise Park, Millisle, Newtownards, BT22 2NW
M3 VEW A. Allanson, 5 Kingsley Avenue, Crofton, Wakefield, WF4 1RN
M3 VEX K. Fox, 39 Felton Avenue, South Shields, NE34 6RY
M3 VEY A. Selvey, Bella Casa, 52 Coppice Close, Cheslyn Hay, Walsall, WS6 7EZ
MW3 VFB B. Pugh, Plas Newydd, 12 Fair Meadow Close, Milford Haven, SA73 3TF
M3 VFC M. Derringer, 19 Skylark Road, Trumpington, Cambridge, CB2 9AQ
M3 VFE P. Goddard, 62 Woodlands Drive, Thetford, IP24 1JJ
MI3 VFF J. Dunlop, 34 Keel Park, Moneyrea, Newtownards, BT23 6DE
MI3 VFJ N. Orr, 5 Manor Court, Donaghadee, BT21 0NR
MM3 VFK J. Scally, 15 Ardoch Court, Stevenston, KA20 3PU
M3 VFL A. Senior, 38 Haslemere, Way, Crewe, CW1 4JZ
M3 VFM V. Millard, 20 Droveway Gardens, St. Margarets Bay, Dover, CT15 6BS
MW3 VFN E. Thomas, 29 Maes Y Wern, Carway, Kidwelly, SA17 4HF
M3 VFP N. Roberts, 40 Armour Road, Tilehurst, Reading, RG31 6HN
M3 VFS M. Toher, The Chapel, Station Road, Darlington, DL2 1JG
M3 VFU M. Whitehead, 19 Wrose Brow Road, Shipley, BD18 2NT
MI3 VFZ T. Currie, 26 High Street, Portaferry, Newtownards, BT22 1QT
M3 VGF A. Lewington, 6 Brookhill Road, Darton, Barnsley, S75 5EL
M3 VGH G. Hunter, 5 Charlton Grove, Silsden, Keighley, BD20 0QG
MW3 VGJ A. Lloyd, 4 Gladstone Terrace, Miskin, Mountain Ash, CF45 3BS
M3 VGK B. Lunn, 204A Main Street, Horsley Woodhouse, Ilkeston, DE7 6AX
M3 VGP P. Temple, 136 Roborough Close, Bransholme, North Yorkshire, HU7 4RP
M3 VGT Viscount A. Andover, Bishoper Farmhouse, Brokenborough, Malmesbury, SN16 9SR
M3 VGX M. Price, 52 Newmarket Street, Norwich, NR2 2DW
M3 VGZ D. Adshead, 16 Moat Way, Swavesey, Cambridge, CB24 4TR
M3 VHA R. Parker, 29 Hill Lea Gardens, Cheddar, BS27 3JH
M3 VHB S. Allen, 27 Cottons Meadow, Kingstone, Hereford, HR2 9EW
M3 VHC W. Ho, Harrogate Ladies' College, Clarence Drive, Harrogate, HG1 2QG
M3 VHE J. Heinonen, Riittiontie 155, Vampula, Finland, 32610
M3 VHH K. Foster, 10 Bleaswood Road Oxenholme, Kendal, LA9 7EY
M3 VHI R. Silcox, 103 Oakdale Road, Downend, Bristol, BS16 6EG
MM3 VHM V. Moran, 31 Hermitage Crescent, Coatbridge, ML5 4NE
M3 VHO F. Ho, Clarence House, Harrogate Ladies College, North Yorkshire, HG1 2QG
M3 VHQ J. Winson, 35 Newington Avenue, Southend-on-Sea, SS2 4RD
M3 VHU J. Redfearn, 3 Taylor Hill Road, Huddersfield, HD4 6HN
M3 VHV L. Kelly, 9 Ham Lane, Farrington Gurney, Bristol, BS39 6TW
MI3 VHW T. Boyd, 40 Walnut Park, Larne, BT40 2WF
M3 VHY A. Blake, 53 Valley Road, Middlesbrough, TS4 2RY
M3 VHZ S. Southern, 37 Conway Road, Calcot, Reading, RG31 4XP
M3 VIA N. Purkiss, 357 Fair Oak Road, Eastleigh, SO50 8AA
M3 VIB M. Michalak, 100 Nursery Lane, Northampton, NN2 7TJ
M3 VIG D. Potter, 1 Wentworth Road, Rugby, CV22 6BG
M3 VII A. Price, Barn Owl Roost, Astwith, Chesterfield, S45 8AN
M3 VIJ D. Williams, 18 Lower Greave Road, Meltham, Holmfirth, HD9 4DY
M3 VIO C. Skinner, Beeston Marina Ltd, 1A The Quay, Beeston Marina, Riverside Road, Nottingham, NG9 1NA
M3 VIR S. Kim, 9 Magness Road, Deal, CT14 9JF
MM3 VIS J. Dowson, 19 Tweed Crescent, Wishaw, ML2 8QR
M3 VIU J. Bettles, 2 Ellfield Close, Bristol, BS13 8EF
MM3 VIV V. Smith, 10 Quilco, Dounby, Orkney, KW17 2HW
M3 VIW C. Wall, 26 Wallace Lane, Whelley, Wigan, WN1 3XT
M3 VJC J. Cooper, Gulean. Cross Common, the Lizard, TR12 7PE
M3 VJE J. Edmunds, 17 Stephens Road, Liskeard, PL143SX
M3 VJI D. Mcevoy, 33 Heathcote Drive, Hasland, Chesterfield, S41 0BB
M3 VJJ V. Bowkett, 9 Gwealmayowe Park, Helston, TR13 0PE
MW3 VJL V. Lea, 30 Cardiff Road, Pwllheli, LL53 5NU
M3 VJM J. Ball, 26 Verona Court, Yeo Vale Road, Barnstaple, EX32 7EN
MW3 VJN T. Thomas, Emlyn House Cawdor Terrace, Newcastle Emlyn, SA38 9AS
M3 VJO J. Sawyer, 27 Croft Road, Wallingford, OX10 0HN
M3 VJS V. Sansom, 70 Valley Road, West Bridgford, Nottingham, NG2 6HQ
M3 VJW J. Whittington, Meadowbank, Bridgerule, Holsworthy, EX22 7EN
M3 VJX P. Buckley, 9 Carton Close, Rochester, ME1 2QF
MW3 VKA A. Vincent, 88 Lake Street, Ferndale, CF43 4HE
M3 VKB V. Britton, Badgers Hollow, Witt Road, Salisbury, SP5 1PL
M3 VKF R. Mottershead, 10 St. Mary Close, Blackpool, FY3 7UB
M3 VKJ K. Jones, Railway Crossing Cottage, Ash Road, Sandwich, CT13 9JB
MW3 VKM R. Morgan, 14 Woodland Road, Pontllanfraith, Blackwood, NP12 2LS
M3 VKN I. Astley, 6 Shay Court, Crofton, Wakefield, WF4 1SL
MM3 VKO S. Macdonald, 366 Millcroft Road Cumbernauld, Glasgow, G67 2QW
MM3 VKP C. Allan, Ardroy, Kinclaven Road, Perth, PH1 4EY
M3 VKS D. Vickers, 178 Bakewell Road, Matlock, DE4 3BA
M3 VKT E. Kottis, Guildford Court Reception, University Campus, Guildford, GU2 7JL
M3 VLG C. Periam, 5 Elliott Walk, Preston, PR1 7TP
M3 VLH P. Harlow, 92 Eton Road, Burton-on-Trent, DE14 2SW
M3 VLI W. Toher, The Chapel, Station Road, Darlington, DL2 1JG
MW3 VLJ L. Jones, Ty'R Ysgol, Holland Street, Ebbw Vale, NP23 6HT
M3 VLL L. Burbidge, 33 Burcote Fields, Towcester, NN12 6TH
M3 VLN V. Grimmer, 48 Bingham Avenue, Sutton-in-Ashfield, NG17 3AR
M3 VLO P. Todd, 93 Derwent Drive, Tibshelf, Alfreton, DE55 5LT
M3 VLT P. Honey, 3 Peterswood, Harlow, CM18 7RJ
M3 VLX S. Mintram, 25 Blunden Drive, Slough, SL3 8WG
M3 VMA M. Collins, 8 Pictor Grove, Buxton, SK17 7TQ
MD3 VMD P. Birchall, 7 Richmond Close, Douglas, Isle Of Man, IM2 6HR
M3 VME C. Frost, 6 Link Way, Arborfield Cross, Reading, RG2 9PD
MD3 VMN V. Matthewman, Monte Rosa, 7 Ballaughton Close, Isle of Man, IM2 1JE
M3 VMQ J. Farrer, 37 Priory Grove, Ditton, Aylesford, ME20 6BB
M3 VMU J. Easterbrook, 2 Warden Road, Eastchurch, Sheerness, ME12 4EJ
M3 VMV R. Vale, 611 College Road, Birmingham, B44 0AY
MW3 VMY T. Peters, 74 Maes Y Capel, Pembrey, Burry Port, SA16 0EG
M3 VNG J. Hardy, Lambda House, Seanor Lane, Chesterfield, S45 8DH
M3 VNH N. Halford, 20 Albany Avenue, Manchester, M11 1HQ
M3 VNI J. Preece, 25 Broadmead Catford, London, SE6 3TG
M3 VNL C. Lockyear, 26 Wentworth Gardens, Exeter, EX4 1NH
M3 VNM A. Crawford, 39 Fownhope Close, Redditch, B98 0LA
M3 VNN V. Nikolaidis, 35-46 Ernst Chain Road, Manor Park, Guildford, GU2 7YW
M3 VNO D. Harwood, 36 Seaview Drive, Great Wakering, Southend-on-Sea, SS3 0BE
M3 VNP G. Simcock, 11 Bannatyne Close, Manchester, M40 3TD
M3 VNQ T. Mcbride, 53 Blackdown Grove, St Helens, WA9 2BD
MW3 VNR C. Mukans, 2 Ffordd Cottages, Johnstown, Carmarthen, SA33 5BL
M3 VNS Dr S. Sampathkumar, 52 Crowstone Road, Westcliff-on-Sea, SS0 8BD
MM3 VNT M. Robertson, 1A Church Street, Lochgelly, KY5 9JS
MM3 VNU R. Robb, 8 Morven View, Tarland, Aboyne, AB34 4UH
MW3 VNV M. Williams, 25 Clos Sant Paul, Llanlli, sa15 1hr
MM3 VNW A. Sim, 44 Hillmoss, Kilmaurs, Kilmarnock, KA3 2RS
M3 VNX L. Wolfe, 90 Alderney Road, Erith, DA8 2JD
MW3 VNZ R. Blackmore, 96 Vachell Road, Cardiff, CF5 4HJ
M3 VOA A. Sartorius, Holmwood, Priory Road, Sunningdale, Ascot, SL5 9RH
M3 VOB P. Sheargold, 7 Mendip Close, Rough Hills, Wolverhampton, WV2 2HF
M3 VOI R. Reeves, 4 Elmwood Avenue, Waterlooville, PO7 7LG
M3 VOJ A. Kreissl, 382 Oldfield Road, Altrincham, WA14 4QT
M3 VOL J. Garner, 2 Coniston Grove, Haresfinch, St Helens, WA11 9NH
M3 VON A. Ellerington, 4 Wathcote Close, Richmond, DL10 7DX
M3 VOR I. Mitchell, 4 Walhouse Drive, Penkridge, Stafford, ST19 5SP
M3 VOU M. Carr, 25 Malvern Avenue, Fareham, PO14 1QF
M3 VOW N. Pettefar, 44 Duck Lane, Laverstock, Salisbury, SP1 1PU
M3 VOY P. Williams, 45 Blackgate Lane, Tarleton, Preston, PR4 6US
M3 VOZ J. Morris, 4 Pleasant Terrace, Lincoln, LN5 8DA
M3 VPA A. Penfold, 179 Byron Rd, Thornhill, Southampton, SO19 6FB
M3 VPB T. Horsoo, 5 Kelmarsh Court, Great Holm, Milton Keynes, MK8 9EN
M3 VPD J. Bridges, 176 Coombe Valley Road, Dover, CT17 0HE
M3 VPH P. Hanson, 34 20Th Avenue, Hull, HU6 9JH
M3 VPJ P. Allen, 21 Chase Vale, Burntwood, WS7 3GD
MM3 VPK C. Hebenton, 43 East Avenue, Uddingston, Glasgow, G71 6LG
M3 VPN M. Thornton, 72 Northfield Lane, Wickersley, Rotherham, S66 2JA
MI3 VPO D. Smith, 164 Ballygowan Road, Hillsborough, BT26 6EG
M3 VPP L. Reeves, 38 Eagle View, Aston, Sheffield, S26 2GL
M3 VPQ S. Woods, 30 Kenilworth Drive, Earby, Barnoldswick, BB18 6NA
MW3 VPT G. Taylorr, 17, Llygad Yr Haul, Glynneath, SA11 5RL
M3 VPX P. Dodds, 22 Wheatear Lane, Ingleby Barwick, Stockton-on-Tees, TS17 0TB
M3 VPY P. Mcdonough, 91 Lever Street, Little Lever, Bolton, BL3 1BA
M3 VQC A. Roberts, 23 Seaton Avenue Houghton-Le-Spring, Sunderland, Dh5 8eq
M3 VQD K. Morgan, 43 Kenilworth Drive, Earby, Barnoldswick, BB18 6NA
M3 VQF A. White, 38 Hillcrest Road, Yeovil, BA21 4RA
M3 VQG S. Walcot, Chapel House, Hemington, BA3 5XU

MI3 VQH J. Kane, 5 Woodlawn Court, Carrickfergus, BT38 8DP
M3 VQI J. Watts, 6 Elm Court, Newhaven, BN9 9NR
MW3 VQJ G. Ellis, 3 Cae Bach, Talybont, Bangor, LL57 3YJ
M3 VQL T. Morton, 43 Hob Moor Drive, York, YO24 4JU
M3 VQN B. Roberts, 102 Brougham Road, Marsden, Huddersfield, HD7 6BJ
M3 VQP A. Majoch, 66 Boughton Green Road, Northampton, NN2 7SP
M3 VQQ C. Unwin, Mansells Farm, Mansells Lane, Hitchin, SG4 8TJ
M3 VQS B. Jamieson, 14 Ridgeway, Ashington, NE63 9TJ
M3 VQV D. Page, 12 Blacksmiths Meadow, Coads Green, Launceston, PL15 7FF
M3 VQZ N. Long, 25 Blendworth Lane, Southampton, SO18 5GY
M3 VRA V. Tomlinson, 180 Kendal Drive, Castleford, WF10 3QZ
M3 VRB J. Banks, 73 Buckthorn Avenue, Stevenage, SG1 1TN
M3 VRD R. Elias, Roganann, Dol Y Bont, Borth, SY24 5LX
MM3 VRI M. Mccallum, 15 Quarry Road, Law, Carluke, ML8 5HB
M3 VRL Dr S. Sethuraman, 9 Bramcote Close, Aylesbury, HP20 1QE
M3 VRM M. Veal, 2 Bernards Close, Christchurch, BH23 2EH
M3 VRN R. Noake, 44 Loxton Square, Bristol, BS14 9SF
M3 VRP S. Broadhurst, Flat 8 Pennant Court, Ross Heights, Rowley Regis, B65 8DW
M3 VRU T. Baxendale, 3 Greylands Close, Sale, M33 6GS
M3 VRV M. Hemstock, 6 Hucknall Crescent, Gedling, Nottingham, NG4 4HZ
MM3 VRX N. Thomson, Four Winds, Holland Bush Hightae, Lockerbie, DG11 1JL
M3 VRY J. Docherty, 208 Thornton Close, Newton Aycliffe, DL5 7NP
M3 VSB K. Dunne, 1 Burton Gardens, Brierfield, Nelson, BB9 5DR
MM3 VSC S. Clark, 75 Treeswoodhead Road, Kilmarnock, KA1 4PB
M3 VSF B. Nicholls, 5 Golden Miller Close, Newmarket, CB8 7RT
MW3 VSG D. Riley-Kydd, 26 Talwrn Road, Wrexham, LL11 3PG
M3 VSH A. Freedman, 276 Newchurch Road, Stacksteads, Bacup, OL13 0TA
M3 VSL V. Cronin, 4 Carnarvon Road, Reading, RG1 5SD
M3 VSO R. Langmuir, 24 Briar Road, Bexley, DA5 2HN
M3 VSQ M. Wilde, 18 Ledston Luck Cottages Kippax Leeds Ls257Bx, Leeds, LS257BX
M3 VST F. Mcdermott, 6 Bruce St., Swindon, SN2 2EL
MM3 VSU S. Rodwell, Bourtree, Kennethmont, Huntly, AB54 4NN
M3 VSW S. Whall, 17 Vicarage Road, Deopham, Deopham, NR18 9DR
MM3 VSX L. Forbes, Woodside, The Muirs, Huntly, AB54 4GD
M3 VSZ L. Schofield, 23 The Mount, Wrenthorpe, Wakefield, WF2 0NZ
M3 VTA C. Duffield, 32 Mount Close, Honiton, EX14 1QZ
MM3 VTB D. Hamilton, 57/6 North Street, Bo'ness, EH51 0AE
M3 VTG D. Howlett, 21 Chandlers, Orton Brimbles, Peterborough, PE2 5YW
M3 VTH A. Ball, 18 Seaview Crescent, Ostend Road, Walcott, NR120NL
MI3 VTJ T. Dorrian, 29 Sperrin Road, Limavady, BT49 0AS
MW3 VTK J. Martin, 78 Llwyn Ynn, Talybont, LL43 2AG
M3 VTL P. Wilson, 2 Mary Rose Close, Cheslyn Hay, Walsall, WS6 7BE
M3 VTN N. Mayall, 10 South Rise, North Walsham, NR28 0EE
M3 VTO S. Tanner, 30 Christina Park, Totnes, TQ9 5UR
M3 VTP M. Tuffs, Izzyinn 87, Foxhills Road, Scunthorpe, DN158LL
M3 VTQ E. Macgurk, 10 Elmore Road, Lee on Solent, PO139DU
M3 VTS P. Wilkes, 8 Cloverdale, Stafford, ST17 4QJ
M3 VTV A. Bedford, 1 Carder Crescent, Bilston, WV14 0JT
M3 VTX P. Hind, 116 Gowthorpe, Selby, YO8 4HA
M3 VUA J. Hunt, 101 Kinoulton Court, Grantham, NG31 7XR
M3 VUC N. Lay, 32 School Road, Billericay, Essex, CM129LH
M3 VUH S. Colman, 22 Shearwater Way, Stowmarket, IP14 5UG
M3 VUI S. Hefford, 5 Fenners, Worle, BS22 7DR
MW3 VUJ R. Williams, 34 Maendu Terrace, Brecon, LD3 9HH
M3 VUK B. Lewis, 68 Irwin Avenue, Rednal, Birmingham, B45 8QU
M3 VUN G. Diggins, 7 Minterne Road, Bournemouth, BH9 3EH
M3 VUO G. Twigg, 4 Crossway, Widnes, WA8 8SQ
M3 VUP A. Lowe, 65 North Road, Clowne, Chesterfield, S43 4PG
M3 VUQ G. Grimshaw, 1 Hardy Close, Pinner, HA5 1NL
M3 VUS V. Ban, Christs College, Cambridge, CB2 3BU
M3 VUV P. Larner, Flat1A 53, West Street, Horncastle, LN9 5JE
M3 VUX T. Loker, 24 St. Albans Hill, Hemel Hempstead, HP3 9NG
M3 VUY S. Heard, The Annexe, 2 Church Walk, Bideford, EX39 2BP
M3 VUZ V. Ball, 24 Carr Lane, Warsop, Mansfield, NG20 0BN
M3 VVA K. Jones, 4 Hawthorne Road, Castle Bromwich, Birmingham, B36 0HH
M3 VVB L. Cunningham, 96 Kingsleigh Drive, Castle Bromwich, Birmingham, B36 9DY
M3 VVH G. Spencer, 7 Squadron Close, Castle Vale, Birmingham, B35 7PF
MM3 VVI J. Mason, 27 Niddrie Marischal Gardens, Edinburgh, EH16 4LX
MW3 VVJ W. Jones, Bryn Golau, Mynytho, Pwllheli, LL53 7RL
MW3 VVO S. Barry, 1 Pearson Cottages, St. Brides, Haverfordwest, SA62 3BN
M3 VVQ N. Mcdougall, 15 Answell Avenue, Manchester, M8 4GG
MM3 VVS I. Lindsay, 265 Stirling Street, Denny, FK6 6QJ
M3 VVW S. Smith, 12 Stoneleigh Avenue, Hordle, SO41 0GS
M3 VVY M. Soper, 16 Queen Elizabeth Drive, Crediton, EX17 2EJ
MM3 VVZ F. Coombes, 44 Lochfield Road, Paisley, PA2 7RL
M3 VWD G. Clarke, 11 Blackfordby Lane, Moira, Swadlincote, DE12 6EX
MW3 VWE E. Jones, Afon Lodge Caravan Park, Parciau Bach, Carmarthen, SA33 4LG
M3 VWF T. Fentiman, 64 St. Nicholas Road, Faversham, ME13 7PD
M3 VWG P. Brown, 4 Heather'S Edge, Heather Lane, Hathersage, S32 1Dt, Hathersage, S32 1DT
MI3 VWH R. Hunter, 5 Castle Rise, Tandragee, Craigavon, BT62 2NE
M3 VWJ P. Kavanagh, 83 Imperial Avenue, Southampton, SO15 8PT
M3 VWK M. Poole, 15 Roberts Place, Dorchester, DT1 2JJ
MW3 VWO D. Best, 8 Carno Street, Rhymney, Tredegar, NP22 5EA
M3 VWP C. Johnson, 42 Reap Lane, Portland, DT5 2JX
M3 VWR A. Bartlett, 62 Kewstoke Road, Bath, BA2 5PU
M3 VWW S. Hegarty, 31 Beaconsfield Road, Deal, CT14 7DA
M3 VWY N. Jewitt, 10 Gorse Lane, Oadby, Leicester, LE2 4RQ
M3 VXB B. Walker, 18 Seals Green, Kings Norton, Birmingham, B38 9UW
M3 VXC D. Stinson, 1 The Croft, Earls Colne, Colchester, CO6 2NH
M3 VXG G. Kavanagh, 83 Imperial Avenue, Southampton, SO15 8PT
M3 VXH S. Liles, 62 Southwood Drive, Surbiton, Surbiton, KT5 9PH
MI3 VXI S. Henry, 105 Ramsey Park, Macosquin, Coleraine, BT51 4NG
M3 VXK J. Diplock, 8 Lodge Road, Messing, Colchester, CO5 9TU
MM3 VXL J. May, 12 Clochbar Gardens, Milngavie, Glasgow, G62 7JP
M3 VXM J. Stephenson, 54 Elizabeth Road, Haydock, St Helens, WA11 0PP
M3 VXN M. Parkinson, 4 Meadow View, Sherburn In Elmet, Leeds, LS25 6BY
M3 VXO O. Carpenter-Beale, 6 Betherinden Cottages, Bodiam Road, Cranbrook, TN18 5LW
MM3 VXP P. Mckay, Bailur 1 Calligary, Ardvasar, IV45 8RU
M3 VXX T. Quiney, 20 Britannia Gardens, Stourport-on-Severn, DY13 9NZ
M3 VXY S. Hill, 1 Meadow Crescent, Wesham, Preston, PR4 3BB
MM3 VYA A. Rodgers, 13 Mill Street, Caldercruix, Airdrie, ML6 7QB
M3 VYB P. Anstis, 112 West Street, Hartland, Bideford, EX39 6BQ
M3 VYD K. Lowcock, 43 Larch Street, Nelson, BB9 9RH
M3 VYE H. Ewer, 89 Bridle Close, Enfield, EN3 6EB
M3 VYF M. Ward, 93 Sandsfield Lane, Gainsborough, DN21 1BQ
M3 VYK L. Shallcross, 6 Wimbrick Close, Wirral, CH46 9RY
M3 VYM L. Clark, 16 Kibblewhite Crescent, Twyford, Reading, RG10 9AX
M3 VYN V. Roberts, 17 Houldsworth Crescent, Coventry, CV6 4HL
MM3 VYR I. Findlay, 2 Bothwell Road, Uddingston, Glasgow, G71 7ET
M3 VYS M. Clifford, 32 Tiverton Way, Chessington, KT92QS
M3 VYT S. Ferguson, 12 Summerfields, Dalston, Carlisle, CA5 7NW
MM3 VYU W. Young, 26 Needle Green, Carluke, ML8 4AF
M3 VYV J. Arblaster, 22 Wood Lane, Carlton, Barnsley, S71 3JJ
MM3 VYY T. Mason, 11 St Serf Road, Glenrothes, KY74EA
M3 VZC C. Mcnulty, 91 Barn Hey Crescent, Meols, CH47 9RW
M3 VZH S. Prichard, 4 Morecambe Road, Scale Hall, Lancaster, LA1 5JA
MM3 VZI D. O'Kane, 0/1 51 Girvan Street, Glasgow, G33 2DP
M3 VZL S. Haycock, 51 South Crescent, Southend-on-Sea, SS2 6TB
M3 VZP S. Arpino, 24 Broad Lane, Rochdale, OL16 4PG
M3 VZQ D. Riddick, 36 Shadygrove Road, Carlisle, CA2 7LD
M3 VZS D. Clewer, 45 Ashfield Road, Andover, SP10 3PE
M3 VZU A. Murphy, 16 Spencer Avenue, Peterborough, PE2 8QH
M3 VZV S. Thompson, 64 Church Road, Fordham, Colchester, CO6 3NJ
M3 VZZ C. Halls, 2 Cock Fen Road, Lakesend, Wisbech, PE14 9QE
M3 WAC W. Clarke, 41 Upton Road, Atherton, Manchester, M46 9RQ
M3 WAF W. Morgan, Little Sandyhurst House, 186 Sandyhurst Lane, Ashford, TN25 4NX
MW3 WAL S. Patchett, Ty Ucha Farm, Nantyr, Llangollen, LL20 7DD
M3 WAP A. Cosic, 35 Betteridge Drive, Sutton Coldfield, B76 1FN
M3 WAV A. Swain, 1 George Street, Brimington, Chesterfield, S43 1HG
M3 WAY P. Jones, 22 Blair Road, Trowbridge, BA14 9JZ
M3 WBA P. Allin, 8 Kiln Close, Dove Holes, SK178FQ
MD3 WBC J. Wernham, Fair Isle, Lhoobs Road, Douglas, Isle Of Man, IM4 3JB
M3 WBI W. Whitcher, 17 Watermead, Stratton St. Margaret, Swindon, SN3 4WE
M3 WBJ B. Williams, 2 Pokas Cottages, Chelveston, Wellingborough, NN9 6AL
M3 WBK W. Knapp, 32 Turner Close, Shoeburyness, Southend-on-Sea, SS3 9TL
MI3 WBL B. Lockhart, 5 Lisnalee Park, Mountnorris, Armagh, BT60 2UP
M3 WBM M. Leake, 20 Witchampton Road, Broadstone, BH18 8HZ
M3 WBN R. Cranston, Hyrton House, Middle Street, Lincoln, LN1 2RG
M3 WBQ W. Argyle, 62 Yew Tree Drive, Leicester, LE3 6PL
M3 WBR H. Lowthian, West Brownrigg, Penrith, CA11 9PF
M3 WBT B. Weston, 10 Clement Drive, Peterborough, PE2 9RQ
MI3 WBU S. Rantin, 8A Buchanans Road, Newry, BT35 6NS
MM3 WBV S. Verth, 23 The Quilts, Leith, Edinburgh, EH65RY
M3 WCA W. Alexander, 53 Woodlands Drive, Stanmore, HA7 3PB
MW3 WCE J. Thomas-Jones, 10 Greenwood Avenue, Gwersyllt, Wrexham, LL11 4EB
M3 WCI A. Hand, 150 Curtin Drive, Moxley, Wednesbury, WS10 8RN
M3 WCM W. Cornish, 21 Centaur Street, Portsmouth, PO2 7HB
M3 WCO T. Purcell, 28 Millberg Road, Seaford, BN25 3ST
M3 WCQ J. Winter, Flat 23, Knightlow Lodge Knightlow Avenue, Coventry, CV3 3HH
M3 WCR C. Wilson, 234 Aylsham Drive Ickenham, Uxbridge, UB10 8UF
M3 WCX C. Whall, 52 Spitfire Road, Upper Cambourne, Cambridge, CB23 6FN
M3 WCZ M. Beckett, 59 Broadacre, Caton, Lancaster, LA2 9NH
M3 WDB C. Jones, 34 Cadle Road, Wolverhampton, WV10 9SJ
M3 WDC W. Carless, 39 Harrison Road, Cannock, WS11 0AQ
M3 WDH Dr W. Henderson, 14 Highfield Road, Newcastle upon Tyne, NE5 5HS
MI3 WDI D. Wiggins, 12 Vauxhall Park, Belfast, BT9 5GZ
M3 WDK P. Shaw, 45 Wood End Road, Wolverhampton, WV11 1NW
M3 WDN E. Temple, 32 Lower Barresdale, Alnwick, NE66 1DW
MI3 WDO D. Dallas, 28 Kemp Park, Ballycastle, BT54 6LE
MM3 WDT W. Thom, 1 Bennan, Mossdale, Castle Douglas, DG7 2NG
M3 WDU A. Stanmore, 198 Heathfield Road, Southport, PR8 3HE
M3 WDV D. Golding, Windrush Cottage 84-85 Bradenstoke, Chippenham, SN15 4EL
M3 WDY W. Duffield, 26 Mount Close, Honiton, EX14 1QZ
M3 WDZ W. Davies, 17 Oakdale Avenue, Harrogate, HG1 2JN
M3 WEA T. Bradley, 1 Park Close, North Weald, Epping, CM16 6BP
M3 WEF P. Brown, 4 Parsonage Drive, St Helens, WA9 4ZW
MM3 WEI E. Ireland, The Steading, Blairmains, Shotts, ML7 5TJ
M3 WEJ J. Varley, 54 Richmond Park Road, Kingston upon Thames, KT2 6AH
M3 WEQ A. Schuler, 6 Tatham Court, Taunton, TA1 5QZ
MI3 WES W. Spence, 8 Kilmahamogue Road, Moyarget, Ballycastle, BT54 6JH
MM3 WEV G. Weir, 95 White Street, Whitburn, Bathgate, EH47 0BH
M3 WEZ L. Wayman, Oak Tree Lodge, Redbridge Road, Dorchester, DT2 8BG
M3 WFB D. Read, 9 Meadow Road, Albrighton, Wolverhampton, WV7 3DZ
M3 WFC J. Paradas, 1A Brocks Ghyll, Eastbourne, BN20 9RQ
M3 WFE A. Kitney, 3 Wordsworth Close, Torquay, TQ2 6EA
MW3 WFF R. Jones, 5 Heol Llwyn Gollen, Merthyr Tydfil, CF48 1LR
M3 WFG W. Griffiths, 68 Altcar Lane, Formby, Liverpool, L37 6AY
MW3 WFH W. Harries, 18 Bro Teify, Alltyblacca, Llanybydder, SA40 9SR
MD3 WFJ J. Hill, 54 Wybourn Drive, Onchan, Isle Of Man, IM3 4AT
M3 WFK P. Ashton, 14 Poppy Close, Boston, PE21 7TJ
M3 WFL M. Rimmer, 7 Brookdale, Southport, PR8 3UA
M3 WFO A. Jamieson-Colville, 77 Salters Way, Dunstable, LU6 1UG
MI3 WFT C. Mooney, 12 Curragh Walk, Derry City, BT48 8HX
MM3 WFU D. Green, 1 The Square, Tomintoul, Ballindalloch, AB37 9ET
M3 WFY C. Nelson, 14 Windy Harbour Road, Southport, PR8 3DU
M3 WGB K. Moore, Flat 8, Lindis Court, Boston, PE21 8SX
M3 WGI S. Sugihara, Southfield, Park Lane, Wokingham, RG40 4PY
M3 WGK L. Mobley, 2 Boxhedge Road West, Banbury, OX16 0BS
M3 WGM M. Brown, 27 Greenfield Close, Dunstable, LU6 1TS
M3 WGO I. Paterson, 11 Ocho Rios Mews, Eastbourne, BN23 5UB
M3 WGV V. Wright, Beech Cottage, Baron Wood, Carlisle, CA4 9TP

MM3 WGW G. Wallace, 29 Dunlop Street, Stewarton, Kilmarnock, KA3 5AT
M3 WGY D. Wilson, 75 Gainsborough Road, Scotter, Gainsborough, DN21 3RU
M3 WGZ G. Wells, 22 Mill Road, Deal, CT14 9AA
M3 WHA A. Ballinger, 9 Somerville Court, Cirencester, GL7 1TG
M3 WHB B. Balchin, 301 New Hall Lane, Preston, PR1 5XE
M3 WHG T. Tunstell, 23 Swallow Crescent, Innsworth, Gloucester, GL3 1BL
M3 WHH J. Cook, 20 Huntingdon Close, Totton, Southampton, SO40 3NX
M3 WHL D. Dawson, 11 Aukland Grove, St Helens, WA9 5LR
M3 WHQ G. Woods, 8 Wareham Road, Lytchett Matravers, Poole, BH16 6DP
M3 WHR D. Williams, Flat 3 The Old Council House, Market Street, Atherstone, CV9 1ET
MM3 WHS D. Mclean, 72 Bowfield Crescent, Glasgow, G52 4HJ
M3 WHV S. Everson, 41 Westminster Lane, Newport, PO30 5ZF
M3 WHX H. Smith, 6 Tynemouth Place, North Shields, NE30 4BJ
M3 WHY G. Cahill, 81 Albemarle Road, Willesborough, Ashford, TN24 0HJ
M3 WIA W. Armes, 11 Rutland Road, Broadheath, Altrincham, WA14 4HW
M3 WIC R. Ashwick, 98 Woodbury Avenue, East Grinstead, RH19 3UX
M3 WID J. Free, Flat 6, 60 Wyncroft Road, Widnes, WA8 8QE
M3 WIJ H. Parkinson, 61 Cinnamon Lane, Fearnhead, Warrington, WA2 0AG
M3 WIT J. Withers, 16 Tamworth Close, Etherley Dene, Bishop Auckland, DL14 0RN
MW3 WIV L. Elston, 11 Woodland Walk, Blaina, Abertillery, NP13 3JS
M3 WIX M. Wilson, 12 Gorsey Lane, Great Wyrley, Walsall, WS6 6JA
M3 WJA A. Whitelam, 107 Welholme Road, Grimsby, DN32 0NQ
MM3 WJD D. Wishart, Curcum, Swannay, Orkney, KW17 2NS
M3 WJK J. Knowles, 10 Grove Hill, Hessle, HU13 0RT
M3 WJN J. Gorman, 1 Patterdale Road, Ashton-In-Makerfield, Wigan, WN4 0EF
MI3 WJO W. Jordan, 24 Spelga Place, Newtownards, BT23 4ND
MW3 WJP P. Watson, Eirianfa Cwmduad, Carmarthen, SA33 6XJ
M3 WJU J. Mowlam, 46 Walpole Street, Weymouth, DT4 7HQ
M3 WJV C. Cherry, 12 Scarisbrick New Road, Southport, PR8 6PY
M3 WJY A. Margaswamy, 59 Grants Yard, Station Road, Burton-on-Trent, DE14 1BW
MM3 WJZ I. Cogle, 56 Dryburgh Avenue, Rutherglen, Glasgow, G73 3EU
M3 WKC A. Campbell, Gate House, The Bog, Shrewsbury, SY5 0NG
MM3 WKF D. Goodwin, Little Dens, Stuartfield, Peterhead, AB42 5DG
M3 WKK R. Horgan, 74 Inglewhite Road, Longridge, Preston, PR3 2NA
M3 WKL C. Winch, 2 Cranleigh Gardens, Cowes, PO31 8AS
M3 WKM B. Howard, 5 St Nicholas Street, Dereham, NR19 2BS
M3 WKV K. Dale, 26 Warwick Place, Langdon Hills, Basildon, SS16 6DU
M3 WKZ B. Drury, 6 Ellen Grove, Harrogate, HG1 4RH
MM3 WLA M. Dickeson, 44 Mossmill Park, Mosstodloch, Fochabers, IV32 7JY
MW3 WLB L. Pearce, 31 High Street, Abertridwr, Caerphilly, CF83 4DD
M3 WLD W. Douglas, 1 Sleetbeck Road, Roadhead, Carlisle, CA6 6PA
M3 WLG D. Flynn, Alberi, Manor Road, Chichester, PO20 0SF
M3 WLL W. Garnett, Starfish Cottage, Higher Clovelly, Bideford, EX39 5ST
M3 WLO M. Anderson, 38 Shellard Road, Filton, Bristol, BS34 7LU
MW3 WLS D. Elias, 31 Banc Y Gors, Upper Tumble, Llanelli, SA14 6BR
M3 WLU J. Hepburn, 32 Green Croft, Ashington, NE63 8EF
M3 WLV D. Burden, 16 Milnthorpe Lane, Wakefield, WF2 7DE
MI3 WLW L. Elliott, 19 Gosford Road, Collone, Armagh, BT60 1LQ
M3 WLX M. Gooch, Flat 16, Townfield Court, 32 Horsham Road, Dorking, RH4 2JE
M3 WLY R. Readman, 1 Millside Close, Kilham, Driffield, YO25 4SF
M3 WMC W. Carr, 18 Whiteway Close, Bristol, BS5 7QZ
MW3 WMI N. Mitchell, 37 Brookside, Glan Y Mor, Fairbourne, LL38 2BX
MI3 WMK M. Mckeen, 27 Old Grange, Carrickfergus, BT38 7HQ
M3 WMO R. Laughlin, 7 Catherine Hunt Way, Colchester, CO2 9HN
MW3 WMP W. Phillips, Annfield, Penrhyndeudraeth, LL48 6LS
MM3 WMQ N. Davidson, 25 Hopetoun Court, Bucksburn, Aberdeen, AB21 9QS
M3 WMS W. Stone, 79 Woodlands Road, Allestree, Derby, DE22 2HH
M3 WMU S. Breese, 11 Balham Grove, Birmingham, B44 0NF
M3 WMV D. Messenger, 18 Glebelands, Harlow, CM20 2PA
M3 WNC S. Willmott, Emborough Grove, Radstock, BA34SF
M3 WNF N. Fellingham, 23 Brooklands, Colchester, CO1 2WA
MM3 WNH A. Maitland, 6 Thorn Avenue, Coylton, Ayr, KA6 6NL
M3 WNI R. Karpinski, 55 Cambridge Avenue, New Malden, KT3 4LD
M3 WNM N. Edwards, 15 Penderry Rise, Catford, London, SE6 1EZ
MM3 WNP S. Murray, 25 Braefoot, Girdle Toll, Irvine, KA11 1BY
M3 WNT T. Corker, North Side, Wingerworth Hall Estate, Chesterfield, S42 6PL
M3 WNV J. Giffard, 5 Hazelwood Road, Oxted, RH8 0JA
M3 WNX S. Moissejev, 41 Queens Road, Caversham, Reading, RG4 8DN
M3 WNZ J. Strawbridge, 36 St. Dunstans Road, Salcombe, TQ8 8AN
M3 WOC C. Bellis, Cliffe Bungalow, Barnsley Road, Barnsley, S72 9JX
M3 WOD D. Wood, 27 St. Mildreds Avenue, Ramsgate, CT11 0HT
M3 WOI R. Burlong, 20 Jockey Mead, Horsham, RH12 1LF
M3 WOK B. Burden, 18 Challenor Close, Finchampstead, Wokingham, RG40 4UJ
M3 WOL P. Bickley, Smithy Cottage, Old Post Office Road, Bury St Edmunds, IP29 5RD
M3 WOQ K. Meyer, 42 Sandcross Lane, Reigate, RH2 8EL
M3 WOS W. Barnes, Cushendall, Lyngate Road, North Walsham, NR28 0DH
MW3 WOV H. Lyall, 29 King Alfreds Road Sedbury, Chepstow, NP16 7AQ
M3 WOW C. Constable, 29 Sheppeys, Haywards Heath, RH16 4NP
M3 WOX J. Ward, 64 Laxey Road, Blackburn, BB2 3LQ
M3 WOY C. Dunstan, 67 Knights Way, Mount Ambrose, Redruth, TR15 1PA
M3 WPC R. Brett, 3 Rectory Close, Chingford, London, E4 8BG
MW3 WPH S. Harrison, 2 Hendre, Newtown, Ebbw Vale, NP23 5FE
M3 WPI O. Prin, 19 The Colliers, Heybridge Basin, Maldon, CM9 4SE
M3 WPK H. Burch, 46 School Lane, Horton Kirby, Dartford, DA4 9DQ
M3 WPM M. Almeida, 20 Gresley Court, Grantham, NG31 7RH
MW3 WPN R. Blackett, 10 Acton Gardens, Wrexham, LL12 8DD
M3 WPO P. Woolley, 84 Bowthorpe Road, Norwich, NR2 3TP
M3 WPP D. Rayner, 42 Chapelgate, Sutton St. James, Spalding, PE12 0EE
M3 WPS W. Snowden, 5 Eastfield Road, Wisbech, PE13 3EJ
M3 WPU M. Mchugh, 51 Rutland Street, Hyde, SK14 4SY
M3 WPV M. Hardy, 66 Exeter Road, Doncaster, DN2 4LF
M3 WPW W. Whyatt, 686 Whitchurch Lane, Whitchurch, Bristol, BS14 0EJ
M3 WQA M. Herbert, 31 Mayfield Avenue, New Haw, Addlestone, KT15 3AQ
MI3 WQC S. Dillon, 2 Otter Park, Strathfoyle, Londonderry, BT47 6YU
MW3 WQE P. Pritchard, 1A Pant Hirgoed, Pencoed, Bridgend, CF35 6YD
M3 WQF D. Robinson, 1 Common Piece, Swinefleet, Goole, DN14 8DE
M3 WQG D. Pearson, 49 Longford Road, Twickenham, TW2 6EB
M3 WQI K. Squires, 10 Markham Avenue, Armthorpe, Doncaster, DN3 2AZ
M3 WQL H. Short, 71 Lilac Crescent, Burnopfield, Newcastle upon Tyne, NE16 6QF
M3 WQN J. Gordon, 17 Cateran Way Collingwood Grange, Cramlington, NE23 6EX
MM3 WQO E. Hughes, 12 Cults Drive, Tomintoul, Ballindalloch, AB37 9HW
MI3 WQT A. Mcbride, 2 Glenbrook Cottage, Lugan, Craigavon, BT66 8QT
MW3 WQV E. Morgan, Holly Cottage, Old Racecourse, Oswestry, SY10 7PQ
M3 WQX J. Neal, 75 Park Lane, Castle Donington, Derby, DE74 2JG
M3 WRA J. Turner, 35 Horncastle Road, Wragby, Market Rasen, LN8 5RB
M3 WRB R. Wheatley, 46 Victory Road, Steeple Claydon, Buckingham, MK18 2NY
MW3 WRH W. Hucker, 14 Greenway Court, Barry, CF63 2FE
M3 WRJ R. Waghorne, 5 Freelands Drive, Church Crookham, Fleet, GU52 0TE
M3 WRK S. Sarwar, Redfern 57A, Uni Of Warwick, Coventry, CV47AL
M3 WRL C. Lemin, 44 Barton Road, Berrow, Burnham-on-Sea, TA8 2LT
M3 WRM R. Dadge, 14 North Roskear Village, Camborne, TR14 0AS
M3 WRN J. Hills, 67 Thornham Road, New Milton, BH25 5AE
M3 WRO O. Woods, 8 Fairway Close, Croydon, CR0 7SH
M3 WRQ C. Smith, 224 Hither Green Lane, London, SE13 6RT
M3 WRS S. Webber, 59 Mincinglake Road, Stoke Hill, Exeter, EX4 7DY
M3 WRZ L. Coyne, 75 Newtown Road, Worcester, WR5 1HH
MW3 WSC G. Owen, 8 Masshyfrmd, Garndolbenmarn, Gwymedd, LL51 9SX
M3 WSE J. Cullen, 22 Longlands Road, Beeston, Nottingham, NG9 1LR
M3 WSH S. Holmes, 11 Holford Rise, Bremilham Road, Malmesbury, SN16 0EA
M3 WSI S. Warren, 9 Warning Tongue Lane, Doncaster, DN4 6TB
M3 WSJ J. Woodroof, 37 Danefield Road, Northampton, NN3 2LT
M3 WSN J. Spillett, Mockbeggar Cottage, Mockbeggar, Ringwood, BH24 3NQ
M3 WSO R. Pitman, 10 Somerville Way, Bridgwater, TA6 5SA
M3 WSQ L. Simpson, 101 Toftwood Road, Sheffield, S10 1SL
M3 WSR B. Harrison, 43A Rumbridge Street, Totton, Southampton, SO40 9DR
M3 WSS L. Shand, 52 Ten Acre Way, Rainham, Gillingham, ME8 8TL
M3 WSV Dr T. Kyriacou, 54 Sutton Avenue, Silverdale, Newcastle-under-Lyme, ST5 6TB
M3 WSW S. Ngai, Harrogate Ladies' College, Clarence Drive, Harrogate, HG1 2QG
M3 WTA D. Whitton, Sea View, Baycliff, Ulverston, LA12 9RL
M3 WTB B. Walden, 59 Brook View Drive, Keyworth, Nottingham, NG12 5RA
M3 WTC W. Caine, 53 Cromford Road, Crich, Matlock, DE4 5DJ
M3 WTD R. Burrow, 1 Tinshill Crescent, Cookridge, Leeds, LS16 7AS
M3 WTG D. Cooper, Little Heath, Bradfield Common, North Walsham, NR28 0QR
M3 WTL W. Leverett, 514 Arleston Lane, Stenson Fields, Derby, DE24 3AG
M3 WTN B. Watkin, 48 Peel Park Crescent, Little Hulton, Manchester, M38 0BU
M3 WTO S. Liu, Harrogate Ladies' College, Clarence Drive, Harrogate, HG1 2QG
M3 WTP E. Lowe, 21 Sherwood Avenue, Creswell, Worksop, S80 4DL
MW3 WTR W. Randall, 3 Penygraig, Aberystwyth, SY23 2JA
MI3 WTT A. Mcdonnell, 52 Moira Road, Glenavy, Crumlin, BT29 4JL
M3 WTU J. Townsend, 124 Rough Common Road, Rough Common, Canterbury, CT2 9BU
M3 WTY R. Clare, Kimberley, Boston Road, Bicker, Boston, PE20 3AP
M3 WUA A. Smith, 20 South Terrace, Northampton, NN1 5JY
M3 WUB C. Garner, 30 Pendula Road, Wisbech, PE13 3RR
M3 WUE D. Tidswell, 1 Cherrytree Grove, Spalding, PE11 2NA
M3 WUG J. Stainton, 40-41 Dyke End Golcar, Huddersfield, HD7 4LA
M3 WUH J. Middleton, 16 Kyme Road, Boston, PE21 8NQ
MM3 WUI P. Bingham, 129 Livingstone Terrace, Irvine, KA12 9ER
M3 WUJ J. Garner, 30 Pendula Road, Wisbech, PE13 3RR
M3 WUK W. Norwood, Flat 5, 20 Upperton Gardens, Eastbourne, BN21 2AH
M3 WUM R. Miles, Haseley Lodge, Birmingham Road, Warwick, CV35 7HF
M3 WUN I. Mcmahon, 8 Thackeray Close, Liverpool, L8 8NE
M3 WUO I. Taylor, Flat 65, Kemsley, Lewisham Park, London, SE13 6QW
MM3 WUP W. Steele, 35 Devlin Court, Whins Of Milton, Stirling, FK7 0NP
M3 WUQ I. Petropouleas, 16 Amfissis Street, Holargos, Athens, Greece, 155 62
M3 WUS M. Blenkinsop, 23 Pilmoor Drive, Richmond, DL10 5BJ
M3 WUV F. Harwood, 1, South Highall Cottage, Lincolnshire, LN10 6UR
M3 WUW M. Noakes, 26 Box Lane, Pontefract, WF8 2JW
M3 WUX K. Bailey, 58 Billy Buns Lane, Wombourne, Wolverhampton, WV5 9BP
M3 WVB P. George, 4 Mandelbrote Drive, Littlemore, Oxford, OX4 4XG
M3 WVC S. Howard, 11 Jolly Gardeners Court, Norwich, NR3 3HD
M3 WVD M. Bradshaw, 118 Queens Road, Vicars Cross, Chester, CH3 5HE
M3 WVF G. Bradshaw, 118 Queens Road, Vicars Cross, Chester, CH3 5HE
M3 WVG K. Pain, 200 Manor Road, Mitcham, CR4 1JF
M3 WVI J. Hanley, 5 Timline Green, Bracknell, RG12 2QP
M3 WVJ R. Brown, 8 Eliot Walk, Kidderminster, DY10 3XP
MI3 WVL J. Mooney, 12 Curragh Walk, Londonderry, BT48 8HX
MM3 WVN S. Clark, 5B Ladykirk Road, Prestwick, KA9 1JW
M3 WVO D. Willey, 17 Bridge Place, Saxilby, Lincoln, LN1 2QA
MM3 WVP R. Hutton, 1 Grianairigh, Northton, Isle of Harris, HS3 3JA
MM3 WVQ R. Robinson, 12 Hannahston Avenue, Drongan, Ayr, KA6 7AU
M3 WVT A. Bradshaw, 118 Queens Road, Vicars Cross, Chester, CH3 5HE
M3 WVX L. Gaynor, 225 Watson Court, Stadium Way, Watford, WD18 0FA
M3 WVY E. Pinvisase, 225 Watson Court, Stadium Way, Watford, WD18 0FA
M3 WWD L. Hornby, 4 Shakespeare Road, Prestwich, Manchester, M25 9GW
M3 WWH R. Fraser, 7 Hawthorne Avenue, Fleetwood, FY7 7PY
MI3 WWJ E. Simpson, 10 Woodview Park, Tandragee, Craigavon, BT62 2DD
MM3 WWM R. Jowett, Fearnoch Ardentallen, Oban, PA34 4SF
M3 WWN W. Northover, 13 Dagenham Avenue, Dagenham, RM9 6LD
MW3 WWO H. Golaszewski, 16 Wingate Drive, Llanishen, Cardiff, CF14 5LR
MM3 WWP C. Mccosh, 5 Bridgend Gardens, Windygates, Leven, KY8 5BP
MM3 WWQ A. Patrick, 2/L 38Glasgow Street, Millport, KA28 0DL
M3 WWR W. Witham, 4 King George Road, Colchester, CO2 7PE
M3 WWU G. Armitage, Windmill Cottage, Greens Gardens, Nottingham, NG2 4QD
MM3 WWV R. Kennedy, 45 Rodney Road, Gourock, PA19 1XG
M3 WWY G. Miller, Silvermine, Cooks Lane, Axminster, EX13 5SQ
M3 WWZ R. Alexander, 14 Ashfield Terrace, Appley Bridge, Wigan, WN6 9AG
M3 WXB B. Williamson, 114 Radburn Road, New Rossington, Doncaster, DN11 0SH
M3 WXD R. Fenn, 78 Sapphire Road, Bishops Cleeve, Cheltenham, GL52 7YU
M3 WXG G. Lewis, 7 The Barns Church Aston, Newport, TF10 9JJ
M3 WXH C. Renouf, 27 Ashburton Road, Croydon, CR0 6AP
M3 WXI C. Bossons, 31 Hanbridge Avenue, Newcastle, ST5 8HH
MW3 WXN B. Chandler, 100 Shakespeare Avenue, Penarth, CF64 2RX

M3 WXP D. Brookes, 13 Princess Street, Woodlands, Doncaster, DN6 7LX
MD3 WXS C. Ashworth, Ravenscourt Lodge, Peel Road, Douglas, Isle Of Man, IM1 5EQ
M3 WXU D. Grundy, 25 Albert Street, Bignall End, Stoke-on-Trent, ST7 8QB
M3 WXW C. Glitsun, 152 St. Awdrys Road, Barking, IG11 7QE
M3 WXX J. Child, 12 Beachill Road, Havercroft, Wakefield, WF4 2EJ
M3 WYA K. Willoughby, 11 Hardistry Drive, Pontefract, WF8 4BU
M3 WYG P. Engledow, 62 Purland Road, Norwich, NR7 9DZ
MM3 WYI N. Stewart, 220 Grieve Road, Greenock, PA16 7AL
M3 WYJ B. Gale, Barn End Burdon Lane, Highampton, EX21 5LT
MM3 WYM M. Stewart, 15 Fancy Farm Road, Greenock, PA16 7LH
M3 WYQ P. Holton, 66 Mill Road, Gillingham, ME7 1JB
M3 WYR C. Smith, 199A Richardshaw Lane, Stanningley, Pudsey, LS28 6AA
M3 WYT R. White, 2 Chambers Manor Cottages, Epping Upland, Epping, CM16 6PJ
M3 WYV E. Scott, 31 South Croft, Upper Denby, Huddersfield, HD8 8UA
M3 WYZ C. Pearman, Wicken Cottage Mill Hill, Edenbridge, TN8 5DB
M3 WZF D. Walker, 8 Wescoe Avenue, Great Houghton, Barnsley, S72 0DW
M3 WZG W. Power, 111 Woodlands Road, Ditton, Aylesford, Maidstone, ME206EF
MM3 WZH S. Armstrong, 85 Blantyre Court, Erskine, PA8 6BP
M3 WZJ R. Davies, 43 Woodfield Road, Holt, NR25 6TX
MM3 WZL J. Scott, 5 Barrwood Gate, Galston, KA4 8NA
M3 WZN J. Mitchell, 27 Tanager Close, Norwich, NR3 3QD
M3 WZP M. Williams, 59 Guilford Avenue, Dover, Ct16 3ng
M3 WZR P. Fry, 76 Mount Pleasant Road, New Malden, KT3 3LB
M3 WZS S. Thorne, 2 Ellfield Close, Bristol, BS13 8EF
M3 WZT M. Fulbrook, 2 Cob Place, Westbury, BA13 3GS
M3 WZV S. Norman, 44 Martival, Leicester, LE5 0PH
M3 WZY J. Dunn, 9 Wakefield Road, Stoke-on-Trent, ST4 5PU
MW3 WZZ S. Bobby, 56 Ffordd Offa, Rhosllanerchrugog, Wrexham, LL14 2EY
M3 XAC A. Curry, 30 Hillside Road, Norton, Stockton-on-Tees, TS20 1JG
MM3 XAF A. Ferries, Cairnbeathie, Lumphanan, AB31 4QA
M3 XAG C. Tame, 28 Tyrrells Way, Sutton Courtenay, Abingdon, OX14 4DF
M3 XAH A. Holmes, 614 City Road, Manor, Sheffield, S2 1GH
M3 XAI H. Parfitt, 5 Sheridan Road, Frimley, Camberley, GU16 7DU
MW3 XAJ A. Morgan, 46 Greensway, Abertysswg, Tredegar, NP22 5AR
M3 XAK M. Gaunt, 12 Glastonbury Abbey, Bedford, MK41 0TX
M3 XAM A. Morgan, 18 Keysworth Drive, Wareham, BH20 7BD
M3 XAN A. Brooks, 52 Houldsworth Drive, Chesterfield, S41 0BS
M3 XAO A. Hughes, 80C Royle Green Road, Manchester, M22 4WB
M3 XAR M. Roberts, 15 Pineside Avenue, Cannock Wood, Rugeley, WS15 4RG
M3 XAS J. Saxon, 134 Sherwood Drive, Wigan, Wn5 9rs
M3 XAU E. Landon, 24 Larchwood Close, Sale, M33 5RP
M3 XAV R. Hartle, 7 Boggard Lane, Charlesworth, Glossop, SK13 5HL
M3 XAY A. Yorkston, 26 Hamilton Road, London, NW10 1PA
M3 XBC B. Chamberlain, 10 Scott Road, Bishop's Stortford, CM23 3QH
MW3 XBE S. Best, 38 Greensway Abertysswg, Rhymney, Tredegar, NP22 5AR
M3 XBF M. Fisher, 25 Tennyson Road, Diss, IP22 4PY
M3 XBH B. Harrison, 24 Alderton Road, Nottingham, NG5 6DX
M3 XBL L. Rabone, 6 Cranwell Grove, Kesgrave, Ipswich, IP5 2YN
M3 XBN O. Popa, 25 Wells Park Road, London, SE26 6JQ
M3 XBO J. Kaby, 3 Kexby Mill Close, North Hykeham, Lincoln, LN6 9TB
M3 XBS O. Rogers, 22 Robson Drive, Hoo, Rochester, ME3 9EA
M3 XBT B. Totterdell, Moscar Cross House, Hollow Meadows, Sheffield, S6 6GL
M3 XBZ M. Stead, 38 Park Road, Bracknell, RG12 2LU
M3 XCA A. Clarke, 14 Tower Court, Haverhill, CB9 9DD
M3 XCB J. Gardiner, 146 Durley Drive, Prenton, CH43 3BB
M3 XCF R. Cockayne, 20 The Shrubbery, Rugeley, WS15 1JJ
M3 XCH C. Howard, 62 Bourne Valley Road, Poole, BH12 1DU
M3 XCJ J. Fergusson, 200 Tang Hall Lane, York, YO10 3RA
M3 XCN C. Norton, 34 The Grove, Little Aston, Sutton Coldfield, B74 3UD
M3 XCP C. Parker, The Grange, Watercombe Cornwood, Nr. Ivybridge, PL21 9RB
M3 XCS B. Porter, 5 Ribble Road, Liverpool, L25 5PN
M3 XCU J. Callis, 51 Pipistrelle Way, Oadby, Leicester, LE2 4QA
M3 XCV M. Abberley, 10 Cranesbill Close, Featherstone, Wolverhampton, WV10 7TY
M3 XCX D. Michael, 33 Garfield Close, Lincoln, LN1 3QL
MW3 XDB D. Barnett, 49 Parcyrhun, Ammanford, SA18 3HD
M3 XDD D. Baseden Butt, 29 Shearwater Way, Stowmarket, IP14 5UG
M3 XDH D. Harris, 23 Shearer Road, Portsmouth, PO1 5LL
M3 XDI J. Peain, 29 Wild Flower Way, Ditchingham, Bungay, NR35 2SF
MM3 XDP D. Paterson, 42 Third Avenue, Alexandria, G83 9BJ
M3 XDQ A. Green, 37 Fisher Close, Worsley Mesnes, Wigan, WN3 5UT
M3 XDV J. Hall, 9 Stone Court, South Hiendley, Barnsley, S72 9DL
MM3 XDW D. Woods, 39 Northfield, Tranent, EH33 1HU
M3 XDZ P. Neal, 14 Hilltop Close, Desborough, Kettering, NN14 2LQ
M3 XEA K. Bindley, 56 Iona Close, Beaumont Leys, Leicester, LE4 0QY
MI3 XEB R. Throne, 12 Mason Road, Magheramason, Londonderry, BT47 2RY
M3 XEF M. Goodwin, Bramble Cottage, Well Hill Lane, Orpington, BR6 7QJ
M3 XEG B. Pearce, 4 Mary Chapman Close, Norwich, NR7 0UD
M3 XEI W. Walther, 139 East Street, Epsom, KT17 1EJ
M3 XEJ E. Ellison, 5 Darwin Terrace, Darwin Street, Shrewsbury, SY3 8QQ
M3 XEL M. Morris, 8 Millfield, Lambourn, Hungerford, RG17 8YQ
M3 XEN J. Fautley, 71 Pullman Lane, Godalming, GU7 1YB
M3 XEO B. Hay, Riverside Road, Great Yarmouth, NR31 6PZ
M3 XEQ P. Whalan, The Old Post Office, South Street, Faversham, ME13 9NR
M3 XEX N. Farrow, 30 Highdown, Southwick, Brighton, BN42 4QS
MI3 XEY C. Cartin, 64 Ashgrove Park, Magherafelt, BT45 6DN
M3 XFA F. Alexis, 44 Osborne Road, Enfield, EN3 7RW
M3 XFB J. Emery, 63 Warren Road, Orpington, BR6 6JF
M3 XFC M. Calvert, 8 Brixham Drive, Wigston, LE18 1BH
M3 XFD N. Field, 9 Shepherds Fold Drive, Winsford, CW7 2UE
M3 XFH M. Ashton, Lodge Farm Bungalow, Wattisham Road, Ipswich, IP7 7LU
M3 XFI M. Radford, 3 Cockshott Drive, Armley, Leeds, LS12 2RL
MM3 XFM B. Burrows, 27 Bughtknowes Drive, Bathgate, EH48 4DP
M3 XFN K. Cross, 31 Parkfields, Abram, Wigan, WN2 5XR
MM3 XFP C. Lee-Marr, 65 Barry Road, Carnoustie, DD7 7QQ
M3 XFS J. Sewell, 56 Victoria Court, Luddesdown Road, Swindon, SN5 8HL
M3 XFT B. Dixon, 21 Pankhurst Road, Hoo, Rochester, ME3 9DF
MM3 XFX D. Sandilands, Cuil Moss Cottage, Ardgour, Fort William, PH33 7AB
M3 XFZ A. Coop, 47 Amy Street, Rochdale, OL12 7NJ
M3 XGA G. White, 89 Kings Drive, Thingwall, Wirral, CH61 9QA
M3 XGB I. Bennett, 4 Frances Close, Wivenhoe, Colchester, CO7 9RP
M3 XGC K. Emlay, 195 Barrington St., Manchester, M11 4FB
M3 XGD J. Challinor, 69 Brimrod Lane, Rochdale, OL11 4QF
M3 XGI E. Brook, 30 Pitchstone Court, Farnley, Leeds, LS12 5SZ
M3 XGL G. Clarke, 28 Mayfield Way, Mendlesham, Stowmarket, IP14 5SH
MM3 XGP G. Kelly, 203 Meldrum Court, Glenrothes, KY7 6UP
MI3 XGR J. Doherty, 62 Coolessan Walk, Limavady, BT49 9EN
MM3 XGS G. Suttie, 9B Pentland Crescent, Dundee, DD2 2BU
M3 XGU C. Frizzell, 85 Gibbon Road, Newhaven, BN9 9ER
M3 XGV M. Wills, 23 Moat Avenue, Coventry, CV3 6BT
M3 XGW G. Whall, 10 Hillcrest Court, Ipswich Road, Diss, IP21 4YJ
M3 XGY S. Kay, 476 North Drive, Thornton-Cleveleys, FY5 2HX
M3 XGZ J. Murray, 2 The Cuttings Hampstead Norreys, Thatcham, RG18 0RR
M3 XHB J. Kelly, 66 Denison Road, Feltham, TW13 4QG
M3 XHC J. Akinin, 70 Valley Road, West Bridgford, Nottingham, NG2 6HQ
M3 XHG M. Bolton, 11 Silvia Way, Fleetwood, FY7 7JF
M3 XHH S. Kiley, 178 Kingfisher Drive, Woodley, Reading, RG5 3LQ
M3 XHK P. Goodall, 61 Turf Hill Road, Rochdale, OL16 4XG
MW3 XHL J. Edwards, 19 Bryntirion, Henllan, Denbigh, LL16 5YL
M3 XHM P. Aitken, 25 Clunbury Road, Northfield, Birmingham, B31 3SY
M3 XHN S. Hutchinson, 17 Monsom Lane, Repton, Derby, DE65 6FX
M3 XHQ D. Jewitt, 26 Sands Lane, Barmston, Driffield, YO25 8PG
M3 XHT M. Shipham, 1 The Farmhouse, Farmhouse Lane, Hemel Hempstead, HP2 7AR
M3 XHU C. Hall, 28 Tidebrook Place, Stoke-on-Trent, ST6 6XF
M3 XHW H. Wright, 168 Spinney Hill Road, Northampton, NN3 6DN
M3 XHY B. Smith, 7 Kestrel Avenue, Bransholme, Hull, HU7 4ST
M3 XHZ M. Tickner, 36 Holtye Walk, Crawley, RH10 6QW
MM3 XIA I. Anderson, Cantyhaugh, Ogscastle, Lanark, ML11 8NE
M3 XID J. Roberts, 51 Bradfield Road, Broxtowe, Nottingham, NG8 6GP
M3 XIE D. Robinson, 19 Meadow Lane, Newcastle, ST5 9AJ
M3 XIF J. Williams, 41 Overton Lane, Hammerwich, Burntwood, WS7 0LQ
M3 XIG S. Abberley, 10 Cranesbill Close, Featherstone, Wolverhampton, WV10 7TY
M3 XIH A. Williams, 41 Overton Lane, Hammerwich, Burntwood, WS7 0LQ
MW3 XIJ A. Pritchard, 20 St. Malo Road, Cardiff, CF14 4HN
M3 XIK J. Lambert Hurley, 64 Henry Road, West Bridgford, Nottingham, NG2 7ND
M3 XIL S. Hoy, 114 Sheppey Beach Villas, Manor Way, Sheerness, ME12 4QY
M3 XIM D. Hackling, 20 Millers Lane, Norwich, NR3 3LU
M3 XIO M. Stevenson, 127 Walton Road, Chesterfield, S40 3BX
M3 XIP A. Pomfrey-Jones, 46 Hampton Road, Erdington, Birmingham, B23 7JJ
MI3 XIU M. Sinton, 65 Henderson Drive, Bangor, BT19 1NP
M3 XIV R. Treherne, 58 Cherry Orchard, Tewkesbury, GL20 8PJ
MM3 XIW T. Hunter, 1A Glen Avenue, Largs, KA30 8RQ
M3 XIY B. Kerry, 1, Churchway, Diss, IP22 1RN
MM3 XIZ I. Hepworth, Bronte Cottage, Inverugie, Peterhead, AB42 3DN
MM3 XJA J. Arthur, West Lodge, Murdoustoun, North & South Road, Motherwell, ML1 5LB
M3 XJE D. Holdsworth, 3 Briardale Road, Bradford, BD9 6PU
M3 XJF J. Ferrington, The Redwoods, 20 Innings Lane, Bracknell, RG42 3TR
M3 XJG J. Gaskin, Badgers Barn, Canterbury Road, Folkestone, CT18 8DF
M3 XJH A. Hylton, 3 Jubilee Cottages, Tring Road, Dunstable, LU6 2JU
M3 XJK A. Weaver, 77 East Acres, Widdrington, Morpeth, NE61 5NT
M3 XJL J. Landless, 2 Aspen Way, Banstead, SM7 1LE
M3 XJM L. Faik, Saarstr. 50, Dienheim, Germany, 55276
M3 XJO J. Boids, 2 Crown Street, Hoyland, Barnsley, S74 9HS
M3 XJP J. Plows, 187 Whitebeam Road, Birmingham, B37 7PA
M3 XJQ R. Barter, 8 Orchard Close, Newton Abbot, TQ12 3DF
M3 XJW J. Wood, Pear Tree Cottage, Agden, Whitchurch, SY13 3UA
M3 XJX A. Reay, 12 Victoria Avenue, South Hylton, Sunderland, SR4 0QZ
M3 XJZ J. Gabriel, 60 Goodwin Road, Ramsgate, CT11 0JJ
M3 XKF N. Tideswell, 19 Wish Court, Ingram Crescent West, Hove, BN3 5NY
MM3 XKH K. Hail, 70 Nobleston Estate, Alexandria, G83 9DB
M3 XKI L. Taute, 4 Mendelssohn Grove, Browns Wood, Milton Keynes, MK7 8DH
M3 XKJ J. Lewis, Millbrook, Church Street, Market Drayton, TF9 2TF
MW3 XKL R. Jones, Flat 2, Tan Y Geraint, 33 Princess Street, Llangollen, LL20 8RD
M3 XKM K. Mountford, 7 Flaxman Close, Barlaston, Stoke-on-Trent, ST12 9BD
M3 XKN L. Griffiths, 90 Keats Road, Wolverhampton, WV10 8NB
M3 XKO P. Goodridge, 22 Horefield, Porton, Salisbury, SP4 0LE
M3 XKP I. Pass, 69 Cotswold Road, Bath, BA2 2DL
M3 XKY G. Leverton, 24 Saxton Avenue, Bradford, BD6 3SW
M3 XLB P. Bailey, 44 Shelley Road, Wellingborough, NN8 3DB
M3 XLC G. Auld, 6 Pheabens Field, Bramley, Tadley, RG26 5BX
M3 XLJ L. Justin, Garth, Park View Road, Pinner, HA5 3YF
M3 XLK R. Crerar, 60 Gloucester Drive, London, N4 2LN
M3 XLM L. Matthewman, 2 St. Margaret Road, Ludlow, SY8 1XN
MM3 XLO J. Nattress, 44 Broadlands, Carnoustie, DD7 6JY
MM3 XLQ S. Rennie, 27 Whiting Road, Wemyss Bay, PA18 6EB
MW3 XLR J. Percival, Blue Cedars, Gresford, Wrexham, LL12 8RN
M3 XLS L. Turner, 16 Woodland Place, Scarborough, YO12 6EP
M3 XLW K. Weston, 44 Shelley Road, Wellingborough, NN8 3DB
M3 XMA Dr M. Ali, 4 The Crescent, Great Horkesley, Colchester, CO6 4EH
M3 XMB M. Brittain, 159 Caledonia Road, Wolverhampton, WV2 1JA
MW3 XME M. Elmer, 3 Maes Dolfor, Llanfairfechan, LL33 0rp
MW3 XMG P. Provis, Dingle Gardens, Croesbychan, Aberdare, CF44 0EJ
M3 XMH M. Higham, 30 Broome Road, Southport, PR8 4EQ
M3 XMJ S. Spicer, 34 Hillcrest Avenue, Halesowen, B63 2PR
M3 XMK M. Rose, 19 Hawthorn Street, Peterlee, SR8 3LY
M3 XMO A. Reed, 94 Moor Lane, Loughborough, LE11 1BA
M3 XMP G. Prasad, 19 Highcroft Road Oadby, Leicester, LE2 4RS
M3 XMQ D. White, 10 Meaux Road, Wawne, Hull, HU7 5XD
M3 XMS Dr F. Gavins, 85 Dance Square, London, EC1V 3AJ
MM3 XMT A. Stevenson, Starwood Croft, Craigellachie, Aberlour, AB38 9SQ
M3 XMU G. Hunter, 58 Repps Road, Martham, Great Yarmouth, NR29 4QT
MW3 XMY D. Worthington, 22 Rhys Avenue, Kinmel Bay, Rhyl, LL18 5NS
M3 XMZ R. Honeybourne, Flat 40, Napier Court West, Southend-on-Sea, SS1 1NH
M3 XNA H. Ball, Manor House, Tolgus Hill, Redruth, TR15 1AX
M3 XNB N. Newby, 22 Acton Road, Liverpool, L32 0TT

M3 XNC G. Reywer, 1 Tiverton Close, Houghton le Spring, DH4 4XR
M3 XNE A. Warr, 2 Fairfield Road, Bournheath, Bromsgrove, B61 9JN
M3 XNK N. Price, 22 Hanover Road, Warley, Rowley Regis, B65 9DZ
M3 XNM I. Stevenson, 79 Lunedale Road, Darlington, DL3 9AT
M3 XNN D. Elliott, 54 Grisedale Gardens, Gateshead, NE9 6NP
M3 XNO J. Manning, 9 Belmont Road, Taunton, TA1 5NS
MM3 XNP N. Page, 61 Calderglen Avenue, Blantyre, Glasgow, G72 9UP
M3 XNR K. Batt, 14 Milne Park West, New Addington, Croydon, CR0 0DN
M3 XNT P. Johannessen, 72 Duncombe Road South, Garston, Liverpool, L19 1QJ
M3 XNU A. Hollis, 9 St. Georges Road, Donnington, Telford, TF2 7NP
M3 XNV D. Dunn, 69 Broadwaters Drive, Kidderminster, DY10 2RY
MW3 XNW N. Williams, 1 Picton Terrace, Pontlottyn, Bargoed, CF81 9PT
M3 XNX N. Stubbs, 5 Newland Street, Wakefield, WF1 5AH
M3 XOA W. Dunstan, 57 Orchard Vale Flushing, Falmouth, TR11 5TT
M3 XOD I. Donnelly, 17 Jessop Close, Horncastle, LN9 6RR
M3 XOE D. Coe, 199 Newark Road, North Hykeham, Lincoln, LN6 8QS
M3 XOH M. Finn, Atzenbach 38, Idar-Oberstein, Germany, 54473
MI3 XOI G. Gorman, 11 Cootehall Road Crawfordsburn, Bangor, BT19 1JA
M3 XOJ M. Benson, 11 Hield Grove, Aston By Budworth, Northwich, CW9 6LN
MM3 XOK B. Hughes, 49 Marmion Drive, Kirkintilloch, Glasgow, G66 2BH
M3 XOL P. Stone, 60 Acorn Avenue, Braintree, CM7 2LR
M3 XOQ D. Vale, 221 Thurstone Road, Birmingham, B31 4PA
M3 XOR M. Nelmes, 119 Exeter Road, Dawlish, EX7 0AN
M3 XOT E. Blake, Ty Capel, Tynygraig, Ystrad Meurig, SY25 6AE
M3 XOU J. Howell, 56 Prouds Lane, Bilston, WV14 6PU
M3 XOV T. Harris, 2 Elm Green, Dudley, DY1 3RE
M3 XOW R. Woodworth, 2 Harrington Court, Meltham, Holmfirth, HD9 4ED
M3 XOY K. Chu, Harrogate Ladies' College, Clarence Drive, Harrogate, HG1 2QG
M3 XOZ W. Yung, 17 York Road, Harrogate, HG1 2QL
M3 XPF P. Collins, 14 Roundel Way, Marden, Tonbridge, TN12 9TW
M3 XPH P. Hennessey, 11 Monmouth Drive, Eaglescliffe, Stockton-on-Tees, TS16 9HU
M3 XPI A. Ackroyd, Prospect House, Causeway, Weymouth, DT4 9RX
M3 XPJ J. Parfitt, 12 Jodrell Place, Selsey, Chichester, PO20 0FQ
M3 XPK C. Park, 197 Occupation Road, Albert Village, Swadlincote, DE11 8HD
M3 XPL P. Rabone, 6 Cranwell Grove, Kesgrave, Ipswich, IP5 2YN
M3 XPN J. Shannon, 16 Croft Drive, Tickhill, Doncaster, DN11 9UL
M3 XPP P. Jarvis, 24 St. Peters Gardens, Leeds, LS13 3EH
M3 XPR P. Ryan, 10 Inchwood, Birch Hill, Bracknell, RG12 7ZX
M3 XPS P. Scarratt, 339 Utting Avenue East, Norris Green, Liverpool, L11 1DF
M3 XPU R. Wellburn, 86 Granville Street, Grimsby, DN32 9NU
M3 XPW J. Oglesby, 22 Elm Drive, Finningley, Doncaster, DN9 3EG
M3 XPY D. Chatzikos, 53 Benbow Court, Shenley Church End, Milton Keynes, MK5 6JE
M3 XQB M. Harbron, 39 Raleigh Road, Sunderland, SR5 5RD
MW3 XQE V. Wallace, 10 Maes Llydan, Benllech, Tyn-Y-Gongl, LL74 8RD
M3 XQG V. Littlewood, 31 Herriot Drive, Chesterfield, S40 2UR
M3 XQH J. Wilson, 46 Redwood Drive, Maltby, Rotherham, S66 8DL
MW3 XQJ P. Jones, 76 Pengwern, Llangollen, LL20 8AS
M3 XQK D. Goodfellow, 60 Pickering Green, Gateshead, NE9 7DX
M3 XQL M. Watson, 5 Birchwood Avenue, Whickham, Newcastle upon Tyne, NE16 5QS
M3 XQM A. Macrae, Oak Lodge, Verwood Road, Three Legged Cross, Wimborne, BH21 6RR
M3 XQO S. Lewis, 40 Bridle Road, Burton Latimer, Kettering, NN15 5QP
M3 XQQ D. Burman, 6 Goodyers Avenue, Radlett, WD7 8BA
M3 XQT J. Thompson, 13 Wentworth Avenue, Luton, LU4 9EN
M3 XQV R. Dutton, 473 Manchester Road, Lostock Gralam, Northwich, CW9 7QD
M3 XQW E. Slevin, Woodcock Hall, Cobbs Brow Lane, Newburgh, WN8 7NB
M3 XQX B. Cameron-Laker, The School Room, Haughley Green, Stowmarket, IP14 3RQ
M3 XQY J. Dutton, 473 Manchester Road, Lostock Gralam, Northwich, CW9 7QD
M3 XQZ J. Freeman, 38 City Road, Cambridge, CB1 1DP
M3 XRD D. Rogers, 44 County Street, Oldham, OL8 3RN
M3 XRG G. Duffy, 34 Twentyfifth Avenue, Blyth, NE24 2QW
M3 XRH S. Lawford, 26 Venetian Crescent, Darfield, Barnsley, S73 9PL
MW3 XRI J. Jones, Greystones, Rhewl, Oswestry, SY10 7AS
M3 XRK D. Richards, 73 Greenfields Avenue, Alton, GU34 2EW
M3 XRO R. Doughty, 1 Woodland Road, Wakefield, WF2 9DR
M3 XRP R. Potter, 1 Wentworth Road, Rugby, CV22 6BG
M3 XRQ B. Benson Jnr, 12 South Drive, Rudheath, Northwich, CW9 7JQ
MI3 XRT L. Murray, 80 Canterbury Park, Londonderry, BT47 6DU
M3 XRV R. Peacock, 21 Breamish Drive, Washington, NE38 9HS
M3 XRW R. Wainwright, 69 George A Green Road, Wakefield, WF2 8HA
M3 XRY G. Jones, 31 Cranage Close, Halton Lodge, Runcorn, WA7 5YN
M3 XSA K. Sproates, 33 Frome Road, Radstock, BA3 3JZ
M3 XSD P. Davies, 15 Kingsley Road, Chester, CH3 5RR
MM3 XSF S. Turnbull, 15 Woodruff Gait, Dunfermline, KY12 0NL
M3 XSG D. Levey, Heriots Wood, The Common, Stanmore, HA7 3HT
M3 XSI S. I'Anson, 5 Heather Gardens, Leeds, LS14 3HU
M3 XSJ J. Wildsmith, Flat 34, Romsley Hill Grange, Farley Lane, Romsley, Halesowen, B62 0LN
M3 XSK Dr J. Skittrall, 14 Tamarin Gardens, Cambridge, CB1 9GH
M3 XSN A. Thompson, 7 Lammermoor Road, Liverpool, L18 4QP
M3 XSP S. Purkiss, 19 The Hurstings, Maidstone, ME15 6YN
M3 XSR T. Came, 15 Brookland Road, Langport, TA10 9TA
M3 XST K. Wood, 52 Trench Road, Tonbridge, TN10 3HB
M3 XSU J. Martin, 52 Maple Avenue, Farnborough, GU14 9UR
M3 XSV M. Ridpath, The Grange, Main Street, Hull, HU12 0JF
M3 XSY C. Throup, Willow House, O'Keys Lane, Worcester, WR3 8RL
M3 XSZ Dr M. Bekara, 9 Southwood Court Pine Grove, Weybridge, KT13 9AT
M3 XTA D. Smith, 28A Bagshot Green, Bagshot, GU19 5JR
M3 XTE J. Sansom, 4 Vicarage Road, Eastbourne, BN20 8AU
MW3 XTF W. Welch, Kenilworth, School Lane, Gobowen, Oswestry, SY11 3LD
M3 XTG T. Greenaway, 11 Gribben Close, Tregonissey, St Austell, PL254EA
M3 XTK P. Mortiboy, 72 Uplands, Stevenage, SG2 7DW
M3 XTL M. Porter, 20, Southfield Road, Much Wenlock, TF13 6AX
M3 XTM J. Maguire, 14 Botha Road, St. Eval, Wadebridge, PL27 7TS
M3 XTP K. Hemmings, 11 Collenswood Road, Stevenage, SG2 9ER
M3 XTR P. Britton, 71 Upper Forster St., Walsall, WS4 2AB
M3 XTT N. Pearson, 116 The Stour, Daventry, NN11 4PT
M3 XTV T. Benson, 83 Glovers Road, Birmingham, B10 0LE
MD3 XUA C. James, 75 Silverburn Crescent, Ballasalla, Isle Of Man, IM9 2DY
MI3 XUC J. Mccollum, 26 Corkey Road, Loughgiel, Ballymena, BT44 9JJ
M3 XUE S. Leadbetter, 11 Cogos Park, Mylor Bridge, Falmouth, TR11 5SF
M3 XUF F. Leung, Harrogate Ladies' College, Clarence Drive, Harrogate, HG1 2QG
M3 XUG R. Barnes, 1 Moira Close, Chaddesden, Derby, DE21 4RL
M3 XUH Dr G. Thomas, 3 The Croft, Wilton, Egremont, CA22 2PW
MM3 XUI G. Taylor, 15 Ronaldsvoe, Kirkwall, KW15 1XE
M3 XUJ P. Hodgkinson, 122 Oulton Road, Stone, ST15 8DY
M3 XUO J. Steven, 1 Tree Terrace, Tree Road, Brampton, CA8 1TY
M3 XUR P. Andrews, 15 Park Lane, Bath, BA1 2XH
MI3 XUS S. Barnes, 191 Marlacoo Road, Portadown, Craigavon, BT62 3TD
M3 XUT R. Udall, 139 Leicester Road, Measham, Swadlincote, DE12 7JG
M3 XUU R. Gopan, 84 Hilmanton, Lower Earley, Reading, RG6 4HN
M3 XUV E. Paddison, 3 Westacre Gardens, Ormesby, Great Yarmouth, NR29 3SP
M3 XUW G. Marsh, 35 Wolverton Avenue, Bispham, Blackpool, FY2 9NU
MM3 XUX J. Munro, 5 Wallace Gait, Perth, PH1 2NS
MM3 XUY G. Nicholson, 4 John Street, Oban, PA34 5NS
MW3 XVB D. Jones, 26 Ffos Y Cerridden, Nelson, Treharris, CF46 6HQ
M3 XVC Lord V. Couchman, Flat 3, Arbroath Court, Sedlescombe Gardens, St Leonards-on-Sea, TN38 0TF
MM3 XVD I. Woods, 12 Westbank Terrace, Macmerry, Tranent, EH33 1QE
M3 XVF J. Smith, 8 Mayfields, Spennymoor, DL16 6RN
M3 XVJ M. Emmott, 1 Swallow Close, Kendal, LA9 7SN
M3 XVK L. Ward, 20 North Street, Maryport, CA15 6HR
M3 XVQ J. Mitchell, 27 Watts Close, Southampton, SO16 9WA
MW3 XVR A. Jones, 11 Bigyn Road, Llanelli, SA15 1NT
M3 XVU B. Lovius, 10 Templemore Avenue, Liverpool, L18 8AH
M3 XVW V. Walker, Kirby Welch & Co, West View, Longlands Lane, Wetherby, LS22 4BB
M3 XVZ P. Read, 53 Hill Top Road, Oldbury, B68 9DU
MM3 XWB J. Mccoll, 25 Nelson Ave, Coatbridge, Ml5 5Lr, Coatbridge, ML5 5LR
M3 XWC J. Conway, 18 Headland Close, Welford On Avon, Stratford-upon-Avon, CV37 8EU
MW3 XWE T. Banks, 18 Leicester Road, Newport, NP19 7ER
M3 XWF J. Coogan, 1 Langsett Rise, Sheffield, S6 2TY
M3 XWK L. Thompson, 34 Broadway, Gateshead, NE9 5PY
M3 XWM T. Sayers, 12 Hutton Terrace, Willington, Crook, DL15 0DS
M3 XWN C. Cartwright, 8 Hawes Grove, Bradford, BD5 9AN
M3 XWP A. Wright, 149 Burton Road, Overseal, Swadlincote, DE12 6JL
M3 XWR J. Halsall, 8 Woodcock Street, Wakefield, WF1 5LG
MM3 XWS W. Anderson, 4 Brackendene, Houston, PA6 7DE
M3 XWV J. Barbieri, 20 Gilbard Court, Chineham, Basingstoke, RG24 8RG
MI3 XWW D. Bradley, 33 Lilac Avenue, Limavady, BT49 0HS
M3 XWX S. Kerslake, 4 Guipavas Road, Callington, PL17 7PL
M3 XWY E. Jennings, 17 Manor Way, Worcester Park, KT4 7PH
M3 XXA J. Smith, 3 Glamis Road, London, E1W 3EE
M3 XXB S. Bunce, 15 Downs View Road, Bembridge, PO35 5QS
M3 XXE N. Dwyer, 82 Staunton Road, Kingston upon Thames, KT2 5TL
M3 XXG G. Collins, 110 Hawthorn Crescent Stapenhill Burton-On-Trent Staffs, Burton-on-Trent, DE15 9QW
MM3 XXI J. Redmond, 14 Bankfaulds Avenue, Kilbirnie, KA25 6AB
M3 XXK K. Mitchell, 1 Denstroude Cottages, Denstroude Lane, Canterbury, CT2 9JX
M3 XXL S. Pearce, 20 Barcote Walk, Plymouth, PL6 5QE
M3 XXM M. Jennings, Springfield Farm, The Causeway, Stow Bridge, King's Lynn, PE34 3PP
MM3 XXO S. Rogan, 21 Montrose Place, Selkirk, TD7 5BH
MM3 XXP A. Pitkethley, 99 Margaretvale Drive, Larkhall, ML9 1EH
M3 XXS S. Mellor, 11 Bolton Meadow, Leyland, PR26 7AJ
M3 XXU A. Friswell, 142 Aldermans Green Road, Coventry, CV2 1PP
MW3 XXY K. Dobson, 152 Foryd Road, Kinmel Bay, Rhyl, LL18 5LS
M3 XYA R. Pasika, 192 Longfield Lane, Cheshunt, Waltham Cross, EN7 6AQ
MI3 XYB J. Throne, 12 Mason Road, Magheramason, Londonderry, BT47 2RY
M3 XYC M. Cowan, Oak Haven, Smugglers Lane, Chichester, PO18 8QW
M3 XYH C. Bell, 60 East Vines, Sunderland, SR1 2DP
M3 XYI W. Cooper, 20 Staple Close, Waterlooville, PO7 6AH
M3 XYJ S. Wright, Emergency Planning Unit, Nycc County Hall, Northallerton, DL7 8AD
M3 XYK A. Raby, 209 Duke Of York Avenue, Wakefield, WF2 7DH
M3 XYM M. Leonard, 75 Skillings Lane, Brough, HU15 1BA
M3 XYN L. Marsh, 14 Herrick Road, Barnby Dun, Doncaster, DN3 1AW
M3 XYO J. Douglas, 14 Mountfields Walk, South Kirkby, Pontefract, WF9 3SJ
M3 XYP G. Thompson, 24 Fairmead Way, Sunderland, SR4 0NA
M3 XYT M. Carter, 17A Goodramgate, York, YO1 7LW
M3 XYU C. Coverley, 25 Fore Street, Milton Abbot, Tavistock, PL19 0PA
MW3 XYW A. Rosser, 25 Clos Tir-Y-Pwll Newbridge, Newport, NP11 5GE
M3 XYX W. Slater, 47 Broom Road, Lakenheath, Brandon, IP27 9EZ
M3 XYZ Dr R. Hill, 12 Winchelsea Lane, Hastings, TN35 4LG
MW3 XZB S. Gardner, 19 Crosscombe Terrace, Cwm, Ebbw Vale, NP23 7SP
M3 XZD J. Kelly, 2 Tamar Close, Higham, Barnsley, S75 1PS
M3 XZE P. Webb, 159 High Street, Arlesey, SG15 6SZ
M3 XZF X. Fang, Harrogate Ladies' College, Clarence Drive, Harrogate, HG1 2QG
M3 XZG D. Mestel, 1B Hayfield Road, Oxford, OX2 6TX
M3 XZH P. Snook, 7 Sandhurst Avenue, Kwazulu Natal, South Africa, 3610
M3 XZJ L. Aldred, 1 Eaton Grange Cottages, Eaton, Grantham, NG32 1ET
M3 XZK A. Smith, 20 Linden Road, Coxheath, Maidstone, ME17 4QS
M3 XZN M. Robinson, 10 Bramley Gardens, Poulton-le-Fylde, FY6 7RD
MW3 XZP C. Maggs, 15 Stuart Street, Treorchy, CF42 6SN
M3 XZR D. Davis, 9 Park Gate Mews, Upper Norwich Road, Bournemouth, BH2 5RA
M3 XZS S. Greaves, 9 Park Gate Mews, Upper Norwich Road, Bournemouth, BH2 5RA
M3 XZT K. Lewinton-Smith, 4 Old School Road, Barnstaple, EX32 9DP
MW3 XZV W. Jones, 2 Derwen Close, Connah'S Quay, Deeside, CH5 4AU
M3 XZY M. Reilly, Flat 59, The Keep, Stafford, ST17 9TW
M3 YAD A. Rossant, 4 Hands Lane, Abbotsbury, Weymouth, DT3 4JW

MW3 YAE G. Thatcher-Sharp, 20 Dilys Street, Blaencwm, Treorchy, CF42 5DT
M3 YAI H. Litten, 55 Downton View, Ludlow, SY8 1JE
M3 YAJ A. Jay, Jasper, The Reddings, Cheltenham, GL51 6RT
M3 YAL B. Loughran, 26 Squirrels Field, Mile End, Colchester, CO4 5YA
M3 YAO E. Bicknell, 12 Victory Road, Southampton, SO15 8QZ
M3 YAS G. Cummings, 18 Castleton Boulevard, Skegness, PE25 2TX
M3 YAV P. Meredith, Bottom Flat, 74 Earl Street, Grimsby, DN31 2PW
M3 YAW S. Johnson, 43 Terry Gardens, Kesgrave, Ipswich, IP5 2EP
M3 YAX L. Mason, 9 Trenethick Avenue, Helston, TR13 8LU
M3 YBB P. Ryalls, 22 Carr Lane, Riddlesden, Keighley, BD20 5HN
MM3 YBD W. Doull, 9 Mcdowall Avenue, Ardrossan, KA22 7AJ
M3 YBF P. Teszner, 21 Sprinkwood Grove, Stoke-on-Trent, ST3 6EQ
MM3 YBG C. Gerrard, 10 Whinhill Gardens, Aberdeen, AB11 7WD
MI3 YBI T. Quin, 165 Marlacoo Road, Portadown, Craigavon, BT62 3TD
M3 YBJ R. Marsh, 56B Oliver Crescent, Farningham, Dartford, DA4 0BE
MW3 YBK T. Jones, 25 Pritchard Terrace, Phillipstown, New Tredegar, NP24 6BS
M3 YBL B. Lace, 19 Methuen Street, Walney, Barrow-in-Furness, LA14 3PS
M3 YBN K. Stokes, 10 Heal Park Crescent Fremington, Barnstaple, Ex313ap
MM3 YBQ K. Verrall, 7 Roshven View, Arisaig, PH39 4NX
M3 YBR A. Knight, Flat 6, Brickworks, 9 Halstow Way, Ashford, TN23 4EQ
M3 YBT M. Fearon, 70 George Street, Heywood, OL10 4PW
M3 YBU M. Shields, 88 Thompson Avenue, Richmond, TW9 4JN
M3 YBW B. Warner, 15 Grosvenor Gardens, Shifnal, TF11 8EB
MW3 YBX D. Smethurst, 9 Lower Quay Road, Hook, Haverfordwest, SA62 4LR
M3 YCB C. Boston, 53 Bullock Road, Terrington St. Clement, King's Lynn, PE34 4PR
M3 YCD D. Clarke, 57 Glanville Place, Kesgrave, Ipswich, IP5 1NQ
MM3 YCG C. Graham, 64 Forgewood Road, Motherwell, ML1 3TH
MM3 YCI S. Burt, 182 Old Inverkip Road, Greenock, PA16 9JG
M3 YCJ V. Powell, 35 Bramber Close, Banbury, OX16 0XF
M3 YCK C. Fung, Harrogate Ladies' College, Clarence Drive, Harrogate, HG1 2QG
MW3 YCL C. Steer, 1 Park Way, Park, Merthyr Tydfil, CF47 8RH
M3 YCM M. Hemmings, 62 Spencer Way, Stevenage, SG2 8GD
M3 YCN W. Fowler, 20 The Court, Anderby Creek, Skegness, PE24 5YQ
M3 YCO D. Cope, 138 Leycett Road, Scot Hay, Newcastle, ST5 6AU
MW3 YCR C. Rayment, Brambles, Alltami Road, Mold, CH7 6RT
M3 YCS C. Sutton, 56 Neatherd Road, Dereham, NR20 4AY
M3 YCT P. Howells, 20 Warwick Street, Stourport on Severn, DY13 8JB
M3 YCU A. Wang, Harrogate Ladies' College, Clarence Drive, Harrogate, HG1 2QG
M3 YCV C. Watts, 35 Coldharbour Lane, Salisbury, SP2 7BY
M3 YCZ C. Landless, 2 Aspen Way, Banstead, SM7 1LE
M3 YDA A. Collins, Robin Post House, Robin Post Lane, Hailsham, BN27 3RA
M3 YDB D. Bush, 19 Spring Vale, Waterlooville, PO8 9DA
MI3 YDF D. Foley, 14 Chestnut Hall Court, Moira, Maghaberry, BT67 0GJ
MM3 YDH D. Hume, Sundhopeburn, Yarrow, Selkirk, TD7 5NF
M3 YDI M. Siddle, 8 Coleridge Close, Oulton, Leeds, LS26 8ET
M3 YDJ D. Wilkinson, 6 Brambledown Road, South Croydon, CR2 0BL
MM3 YDK D. Kilgour, 56 Green Street, Rothes, Aberlour, AB38 7BD
M3 YDL S. Thornton, 29 Farrar Avenue, Mirfield, WF14 9ED
M3 YDM T. Yardley, 19 Elms Close, Shareshill, Wolverhampton, WV10 7JT
MW3 YDS S. Morgan, 38 Ffordd Cadfan, Bridgend, CF31 2DP
MW3 YDT E. Gittins, 40 Melyd Avenue, Prestatyn, LL19 8RN
M3 YDV D. Brame, 7 Roche Garden, Exeter, EX2 6LS
M3 YDW D. White, 9 Wyatts Lane, Tavistock, PL19 0EU
M3 YDY D. Murfitt, 10 Benefield Road, Moulton, Newmarket, CB8 8SW
M3 YEA L. Pollard, 11 Alfriston Road, Worthing, BN14 7QU
MM3 YEC J. Dupont, 11 Golf View Cardenden, Lochgelly, KY5 0NW
M3 YEE H. Ngi, Harrogate Ladies' College, Clarence Drive, Harrogate, HG1 2QG
MW3 YEG J. Thorne, 11 Dowland Road, Penarth, CF64 3QX
M3 YEJ J. Marsh, 31 Clay Street, Soham, Ely, CB7 5HJ
M3 YEK A. Taylor, 16 Bellmans Road Whittlesey, Peterborough, PE7 1TY
M3 YEM Dr R. Browne, 30 Cromwell Road, Southowram, Halifax, HX3 9SE
MM3 YEQ G. Pearce, 2 Kirkriggs, Forfar, DD8 2AT
M3 YET A. Reilly-Cooper, 40 Clough Lane, Northwich, CW8 1JR
M3 YEU J. Mobbs, 6 School View, Banbury, OX16 4SD
M3 YEZ J. Dearden, 218 South Street, Highfields, Doncaster, DN6 7JQ
M3 YFG G. Walton, 11 Redwood Close, Hoyland, Barnsley, S74 0EJ
M3 YFH A. Sherman, 31 Peartree Avenue, Kingsbury, Tamworth, B78 2LG
M3 YFI T. Hall, 18 Common Lane New Haw 18 Comman Lane New Haw, Addlestone, kt153lh
M3 YFJ A. King, 6 Dunsfold Close, Crawley, RH11 8EY
M3 YFL J. Solomon, 17 Chadwick Terrace, Macclesfield, SK10 2DQ
M3 YFM A. Reader, 7 Church Road, Swindon, SN1 3HF
M3 YFN M. Green, Fire Beacon Cottage, East Hill, Sidmouth, EX10 0LR
M3 YFR D. Cockburn, 30 Queensberry Road, Burnley, BB11 4LH
MM3 YFT P. Finnie, 20 St. Margarets Road, Ardrossan, KA22 7ER
M3 YFW R. Stevens, 53 Keeble Way, Braintree, CM7 3JX
M3 YFX J. Loveridge, 96 High Road, Islington, King's Lynn, PE34 3BN
M3 YFY D. Rolfe, 49 Hillrise Avenue, Sompting, Lancing, BN15 0LU
M3 YFZ J. Blower, 4 Lamorna Close, Luton, LU3 2TH
M3 YGB A. Birch, 3 Partridge Way, High Wycombe, HP13 5JX
M3 YGC G. Cowley, 2 Manor Close, Farcet, Peterborough, PE7 3AA
M3 YGD G. Hyland-Davis, 34 Melody Close, Warden, Sheerness, ME12 4PU
M3 YGF G. Ferguson, 31 Barton Court Road, New Milton, BH25 6NW
MM3 YGI R. Stratton, 18 Dunnock Park, Perth, PH1 5FN
M3 YGK Y. Gopikrishna, 29 Alandale Drive, Pinner, HA5 3UP
M3 YGL D. Gribben, 44 Fern Close, Birchwood, Warrington, WA3 7NU
M3 YGO C. Chew, 10 Bruce Drive, South Croydon, CR2 8SL
M3 YGQ M. Smith, 7A Adamstiles, Barrowby, Grantham, NG32 1TG
M3 YGR J. Katz, 8 Astor Drive, Birmingham, B13 9QR
M3 YGS S. Dingmar, 10 Kertland Street, Savile Town, Dewsbury, WF12 9PU
M3 YGT G. Giannakopoulos, 3 Vauxhall Quay, Plymouth, PL4 0EZ
M3 YGU B. Chamberlain, 2 Stocks Loke, Cawston, Norwich, NR10 4BS
M3 YGV J. Harris, Flat 3 49 Enys Road, Eastbourne, BN21 2DN
M3 YGY C. Lee-Koo, Flat 5, 211 Sussex Gardens, London, W2 2RJ
M3 YGZ I. Mccourt, 181 Fircroft Road, Ipswich, IP1 6PS
MM3 YHA D. Morrison, 4 West Murkle, Murkle, Thurso, KW14 8YT
MW3 YHC D. Jones, Bryn Hyfryd, Pandy Tudur, Abergele, LL22 8UL
M3 YHD K. Marshall, 38 Staunton Road, Newark, NG24 4EX
M3 YHF G. Hand, Hollinhurst Farm, Park Lane, Stoke-on-Trent, ST9 9JB
M3 YHG J. Stringer, 31 Pipit Lane, Birchwood, Warrington, WA3 6NY
M3 YHH B. Hand, Hollinhurst Farm, Park Lane, Stoke-on-Trent, ST9 9JB
M3 YHJ A. Hussain, 789 Scarborough Street, Dewsbury, WF12 9AY
M3 YHL H. Lee, Harrogate Ladies' College, Clarence Drive, Harrogate, HG1 2QG
M3 YHM W. Mok, 17 York Road, Harrogate, HG1 2QL
M3 YHN M. Roberts, 7 Maxwell Place, Stoke-on-Trent, ST4 6RE
M3 YHP H. Pentz, 30 Lindrick Way, Harrogate, HG3 2SU
M3 YHQ Dr R. Mason, 9 Farmfields Rise, Woore, Crewe, CW3 9SZ
M3 YHR A. Rolland, Flat 29 Renfrew Court, Allfrey Road, Eastbourne, bn227sz
M3 YHT G. Winterbottom, 35 Abingdon View, Worksop, S81 7RT
M3 YHU C. Cleverley, 4A Godfrey Street, Netherfield, Nottingham, NG4 2JG
M3 YHV M. Chidgey, 46 Station Road, Shirehampton, Bristol, BS11 9TX
MW3 YHW J. Loughlin, 453 Heol-Y-Waun, Penrhys, Ferndale, CF43 3NW
M3 YHY F. Noble, 1045, 45Th Street Apartment A, California, USA, 94608
M3 YHZ C. Amos, 33 Douglas Road, Newcastle, ST5 9BP
M3 YIC L. Crabtree, 23 Ava Crescent, Richmond Hill, Ontario, Canada, L4B 2X1
M3 YIE A. Fuller, Flat 5 Maple House, 3 Fairfield Road, Havant, PO9 1AG
M3 YIF N. Spooner, 14 Glebe Road, Ongar, CM5 9HW
MM3 YIG C. Doolan, 56 Forfar Road, Greenock, PA16 0YL
MM3 YIH C. Rodgers, 3 Merrylee Avenue, Port Glasgow, PA14 5UT
M3 YII A. Boag, 60 Harebell, Amington, Tamworth, B77 4NA
M3 YIL A. Bloor, Springfield, Piltdown, Uckfield, TN22 3XN
MM3 YIO B. Fullerton, 55 Alexander Avenue, Stevenston, KA20 4BG
MM3 YIQ R. Doolan, 56 Forfar Road, Greenock, PA16 0YL
M3 YIT T. Yip, 6 Fulwith Grove, Harrogate, HG2 8HN
M3 YIV A. Little, Wisteria Cottage, 7 Shawfield Road, Havant, PO9 2SY
M3 YIX M. Norris, 35 Sudbrooke Road, London, SW12 8TQ
M3 YIY R. Egan, 26 Baldwin Road, Birmingham, B30 3LG
M3 YJA J. Appleby, 79 Glenwoods, Newport Pagnell, MK16 0NG
M3 YJB J. Birch, 6 Crescent Road, Wallasey, CH44 0BQ
M3 YJD J. Dowdeswell, 18 Lechlade Gardens, Fareham, PO15 6HF
MI3 YJE J. Elliott, 30 Moyle Road, Ballycastle, BT54 6AN
M3 YJF D. Moran, 23 Abbotsfield Crescent, Tavistock, PL19 8EY
M3 YJG R. Lawton, 41 Almond Avenue, Armthorpe, Doncaster, DN3 2HE
MM3 YJH J. Hume, 2 Tinnis Farm Cottage, Yarrow, Selkirk, TD7 5JZ
MW3 YJJ L. Gleed, 9 Medlock Close, Bettws, Newport, NP20 7EJ
M3 YJL J. Cairns1, 17 Alfred Avenue, Worsley, Manchester, M28 2TX
M3 YJM J. Patrick-Gleed, 6 Julius Way, Lydney, GL15 5QS
M3 YJN N. Allen, 5 Limecroft View, Wingerworth, Chesterfield, S42 6NR
M3 YJP G. Patrick-Gleed, 6 Julius Way, Lydney, GL15 5QS
M3 YJT H. Taylor, 21 Charlecote Drive, Chandler'S Ford, Eastleigh, SO53 1SF
M3 YJU P. Boreham, 67 Brent Lane, Dartford, DA1 1QT
M3 YJW J. Williams, 66 Oakfield Avenue, Hitchin, SG4 9JD
M3 YJY L. Bourne, 9A Partridge Croft, Lichfield, WS13 6SD
MW3 YKA A. Francis, 8 Cefn Adda Court, Newport, NP203DS
M3 YKC K. Comben, 9 West Lane, North Baddesley, Southampton, SO52 9GB
M3 YKF K. Fung, Harrogate Ladies' College, Clarence Drive, Harrogate, HG1 2QG
M3 YKH A. Harding, Sunnydene, Wellmead, Axminster, EX13 7SQ
M3 YKI A. Sulieman, 22 Warren Court, 80 Charlton Church Lane, London, SE7 7AD
MW3 YKL C. Warburton, 71 Richards Terrace, Cardiff, CF24 1RW
M3 YKN D. Bright, 103B Langer Road, Felixstowe, IP11 2EA
M3 YKO T. Chan, Harrigate Ladies College, Clarence Drive, Harrogate, HG1 2QG
MM3 YKR R. Murray, 18 Braids Road, Kirkcaldy, KY2 6JE
M3 YKT L. Jones, 16 Oxland Road, Illogan, Redruth, TR16 4SH
M3 YKZ D. Warren, 36 Milner Road, Heswall, Wirral, CH60 5RZ
M3 YLB R. Beck, 73 Crowborough Road, Southend-on-Sea, SS2 6LW
M3 YLJ G. Whitehead, 29 Coulsons Road, Bristol, BS14 0NN
M3 YLK J. Swift, 56 Leymoor Road, Huddersfield, HD3 4SW
M3 YLL J. Swift, 56 Leymoor Road, Huddersfield, HD3 4SW
M3 YLM Y. Lam, Harrogate Ladies' College, Clarence Drive, Harrogate, HG1 2QG
M3 YLN C. Mills, 118 Saxon Gardens, Shoeburyness, Southend-on-Sea, SS3 9PX
MW3 YLO A. Powell, 31 Highmead, Pontllanfraith, Blackwood, NP12 2PF
MM3 YLP C. Edwards, 18 Gelshfield, Halkirk, KW12 6UZ
M3 YLQ J. Painter, 58 Glenfield Road, Plymouth, PL6 7LN
MW3 YLR D. Trevelyan, 35 St. Kingsmark Avenue, Chepstow, NP16 5LY
M3 YLT L. Timmins, 83 Loxdale Sidings, Bilston, WV14 0TN
MW3 YLV K. Bell, 2 Hill Street, Risca, Newport, NP11 6QH
M3 YLW R. Anders, 6 Priors Close Ditton Priors, Bridgnorth, WV16 6TL
MD3 YLX D. Cain, 7 Cronk Y Berry Mews, Douglas, Isle Of Man, IM2 6HQ
M3 YLZ Capt. M. Brewster, Blackthorn Farm, Common Road, Dickleburgh, IP21 4PH
M3 YMC M. Comben, 9 West Lane, North Baddesley, Southampton, SO52 9GB
M3 YMD M. Dudley, 418 Sandon Road, Stoke-on-Trent, ST3 7LH
MI3 YMF M. Foley, 44 Gallows Street, Dromore, BT25 1BD
M3 YMG M. Crawley, 16 The Meadows, Herne Bay, CT6 7XF
M3 YMH M. Hurst, 20 Albany Avenue, Manchester, M11 1HQ
MM3 YMM M. Holmes, 1 Lauren Way, Paisley, PA2 9JW
MM3 YMN Dr J. Henderson, 7 Rowanhill Close, Port Seton, Prestonpans, EH32 0SY
M3 YMS M. Statham, 17 Nicholas Meadow, Higher Metherell, Callington, PL17 8DE
MM3 YMU R. Morrison, 4 West Murkle, Murkle, Thurso, KW14 8YT
M3 YMX J. Lewis, 4 Moor Park, Clevedon, BS21 6EH
MW3 YMY M. Ireland, Pen Y Gadlas, Ffordd Bryniau, Prestatyn, LL19 8RD
MW3 YNA M. Seagrave, The Firs, Llannon, Llanelli, SA14 6AP
M3 YNB N. Buttery, 22 Mallard Road, Rowlands Castle, PO9 6HN
M3 YNC C. Mackintosh, 18 Park Avenue, Castleford, WF10 4JT
M3 YND B. Higgins, 41 Lower Meadow, Harlow, CM18 7RE
M3 YNE J. Godfrey, 36 Greenwich Road, Hailsham, BN27 2PE
M3 YNH C. Arundel, 54 Broadmead, Castleford, WF10 4SE
M3 YNI R. Pike, 66 Prowses Hemyock, Cullompton, EX15 3QG
M3 YNJ A. Richards, 5 Bloomfield Drive, Bracknell, RG12 2JW
MW3 YNK G. Lane, 32 Caellepa, Bangor, LL57 1HF
M3 YNM R. Johnson, 50 Barnaby Rudge, Chelmsford, CM1 4YG
M3 YNN S. Linton, 89 Ragpath Lane, Stockton-on-Tees, TS19 9JS
M3 YNO A. Williamson, 32 Beech Close Eastfield, Scarborough, YO11 3QZ
MM3 YNP C. Haldane, 6A Earls Gate, Bothwell, Glasgow, G71 8BP
M3 YNS M. Gibson, 58 Byron Street, Barrow-in-Furness, LA14 5RL
M3 YNX D. Brewer, 15 Morella Road, London, SW12 8UQ
M3 YNY N. Oldrid, 4 Bar Lane Mapplewell, Barnsley, S75 6DQ

M3 YOB P. Lake, 19 Orchard Vale Midsomer Norton, Radstock, BA3 2RA
MM3 YOC R. Munro, 20 County Cottages, Piperhill, Nairn, IV12 5SE
M3 YOE A. Yorke, 33 Avon Crescent, Stratford-upon-Avon, CV37 7EX
M3 YOG K. Cartledge, Oysterber Farm, Burton Road, Lancaster, LA2 7ET
M3 YOH L. Mason, 432 Lichfield Road, Sutton Coldfield, B74 4BL
MM3 YOL D. Davidson, 44 Abercromby Crescent, Helensburgh, G84 9DX
M3 YOM J. Thresher, Quarry Grange, Nuneaton Road Over Whitacre, Coleshill, Birmingham, B46 2NH
M3 YOO A. Williams, 54 Longbridge, Willesborough, Ashford, TN24 0TA
M3 YOP T. Court, Eastgate Cottage, Perrys Lane, Norwich, NR10 4HJ
M3 YOQ M. Harris, Rodmarton House, Broad Town, Swindon, SN4 7RG
M3 YOT J. O'Malley, 140 Allerburn Lea, Alnwick, NE66 2QP
M3 YOU M. Young, 19 Tallents Close, Sutton At Hone, Dartford, DA4 9HS
M3 YOW G. Eycott, 1 Ham Road, Wanborough, Swindon, SN4 0DF
M3 YOX W. Hopkins, 30 Charles Darwin Road, Plymouth, PL1 4GU
M3 YOZ S. Hughes, 117 Liverpool Road, Irlam, Manchester, M44 6EH
M3 YPA S. Watts, 29 Brook Drive, Corsham, SN13 9AU
M3 YPB C. Bond, Tryfan, Vicarage Lane, Neston, CH64 5TJ
M3 YPD J. Scholz, 1930 Grant Ave, Williamstown, USA, 8094
M3 YPG W. Gratton, Park House, Brimham Rocks Road, Harrogate, HG3 3HE
MM3 YPH E. Bertram, 20 Kyles View, Largs, KA30 9ET
M3 YPI B. Deakin, 8 Patey St., Manchester, M12 5RP
M3 YPJ P. Kirby, 30 New Street, Eccleston, Chorley, PR7 5TW
MM3 YPN S. Hargreaves, 4 Oxenfoord Avenue, Pathhead, EH37 5QD
M3 YPP C. Snow, 19 Salters Road, Haylands, Isle of Wight, PO33 3HU
M3 YPR P. Ruocco, 8 Chain Road, Manchester, M9 6QX
M3 YPS L. Brady, 9 Wordsworth Close, Wootton Bassett, Swindon, SN4 8HJ
M3 YPU T. Galloway, 1 Farley Close Shadoxhurst, Ashford, TN26 1NB
M3 YPW R. Coleman, 5 Meeting Lane, Burton Latimer, Kettering, NN15 5LS
M3 YPX D. Parkhouse, 5 Long Yard, Briston, Melton Constable, NR24 2LB
M3 YQC P. Rogers, 16 Begonia Close, Basingstoke, RG22 5RA
M3 YQG A. Bandtock, 28 Campion Road, Westoning, Bedford, MK45 5LB
M3 YQH G. Smith, 6 Grange Crescent, Childer Thornton, Ellesmere Port, CH66 5NB
MW3 YQL J. Edwards, 2 Maes Merddyn, Gaerwen, LL60 6DG
M3 YQM D. Morbey, 44 Browning Road, Plymouth, PL2 3AP
MM3 YQP C. Pate, 75 Castings Avenue, Falkirk, FK2 7BJ
M3 YQQ Dr N. Lutte, Oak House, Brandy Hole Lane, Chichester, PO19 5RX
M3 YQR C. Hay, Sea Cadets, Riverside Road, Great Yarmouth, NR31 6PX
M3 YQT J. Best, 24 Suggitts Lane, Cleethorpes, DN35 7JJ
MW3 YQU J. Jones, 55 London Road, Holyhead, LL65 2NS
MM3 YQX K. Mcbride, 1 Cowal Place, Gourock, PA19 1EJ
M3 YRB D. Grantham, 56 Knapp Avenue, Eastwood, Nottingham, NG16 3JW
M3 YRC S. Evans, 2 Firbeck Crescent, Langold, Worksop, S81 9SB
M3 YRH R. Horner, 21 Ainsworth Road Little Lever, Bolton, BL3 1RG
M3 YRJ T. Lamont, 30 Shackleton Close, Old Hall, Warrington, WA5 9QE
M3 YRM R. Moles, 14 Dorsett Road, Stourport-on-Severn, DY13 8EL
M3 YRO P. Harvey, 35 Isaac Street, Liverpool, L8 4TH
M3 YRR C. Reid, 128 Main Street, Hensingham, Whitehaven, CA28 8PX
M3 YRS T. Godfrey, 12 Beacon House, Chulsa Road, London, SE26 6BP
M3 YRV S. Honywood, 169 Primrose Lane, Croydon, CR0 8YQ
M3 YRW J. Woods, 21 Appleyard Crescent, Norwich, NR3 2QN
M3 YRX M. Mckone, 12 Hawkshead Road, Knott End-On-Sea, Poulton-le-Fylde, FY6 0QE
M3 YRZ K. Lovell, Wimbledon Court, 3, Miiddlesbrough, TS5 5JP
M3 YSA J. Hallam, 63B Poppleton Road, London, E11 1LP
M3 YSC S. Croucher, 17 Sundridge Road, Woking, GU22 9AU
M3 YSD S. Dudley, 365 Sandon Road, Stoke-on-Trent, ST3 7LJ
M3 YSF S. Forrest, 58 Victoria Road, Manchester, M14 6BZ
MM3 YSJ J. Sinclair, 21 Oxford Avenue, Gourock, PA19 1XU
M3 YSL C. Brooks, 61 Boxfield Green, Stevenage, SG2 7DR
M3 YSM S. Marshall, 43 Glenkerry House, 98 Burcham St., London, E14 0SL
M3 YSN M. Beardsley, 2 Wingrove Avenue, Newcastle upon Tyne, NE4 9AL
M3 YSQ J. Powell, 46 Woodmancote, Yate, Bristol, BS37 4LL
M3 YSS S. Stewart, 8 Craig Street, Peterborough, PE1 2EJ
M3 YSU G. Cummings, 10 Perth Close, Skegness, PE25 2HY
M3 YSV D. Benwell, 10 Harvester Way, Crowland, Peterborough, PE6 0DA
M3 YSW G. Thrower, 8 Upton Gardens, Worthing, BN13 1DA
M3 YSY A. Zabalujevs, 30 Miles Close, London, SE28 0NJ
M3 YSZ C. Gao, 17 York Road, Harrogate, HG1 2QL
M3 YTA J. Marland, 8 Dulverton Gardens, Edinburgh Road, Bolton, BL3 1TR
MM3 YTB M. Martin, Flat A, 11 Craigpark Street, Clydebank, G81 5BS
M3 YTE A. Duffield, 4 Crabmill Lane, Easingwold, York, YO61 3DE
M3 YTF G. Garman, 11 Rye Close, Norwich, NR3 2LF
M3 YTG A. Clark, 330 Stafford Road, Caterham, CR3 6NJ
M3 YTH A. Shilliday, 26 Iskymeadow Road, Armagh, BT60 3JS
MM3 YTI G. White, 52 Union Road, Whitburn, Bathgate, EH47 0AP
M3 YTL T. Lee, Harrogate Ladies' College, Clarence Drive, Harrogate, HG1 2QG
M3 YTQ J. Hughes, 84 Rodwell Avenue, Weymouth, DT4 8SQ
M3 YTV D. Telford, 37 Swillington Lane, Swillington, Leeds, LS26 8QF
M3 YTZ J. Chaplin, 101 St. Cuthberts Drive, Gateshead, NE10 9AB
M3 YUA M. Boyd, 4 Crowton Cottages, Winsford Road, Cholmondeston, Winsford, CW7 4DP
M3 YUB D. George, 13 Cheltenham Way, Mablethorpe, LN12 2AX
M3 YUC A. Waudby, 7 Forest Grove, York, YO31 1BL
M3 YUD S. Nutt, 23A Hesketh Drive, Southport, PR9 7JX
M3 YUH P. Lai, Harrogate Ladies' College, Clarence Drive, Harrogate, HG1 2QG
M3 YUK J. Burman, 6 Goodyers Avenue, Radlett, WD7 8BA
M3 YUN R. Spalding, 7 Kingfisher Close, Scawby Brook, Brigg, DN20 9FN
M3 YUP B. Stevens, 77 Dean Lane, Hazel Grove, Stockport, SK7 6EJ
MD3 YUQ D. Kelly, 41 High Street, Port St Mary, Isle Of Man, IM95DN
M3 YUR C. Potter, 4 Tomlinson Street, Stoke-on-Trent, ST6 4NW
MM3 YUS P. Mccann, 3 Exmouth Place, Gourock, PA19 1JE
MI3 YUT D. Elliott, 15 Derrychara Park, Enniskillen, BT74 6JP
MM3 YUU M. Morrison, 8 Garallan, Kilwinning, KA13 6LU
M3 YUV L. Hudson, 68 Eleanor Road, Harrogate, HG2 7AJ
MM3 YUW D. Williamson, 31 Medrox Gardens, Cumbernauld, Glasgow, G67 4AJ
MM3 YUX C. Williamson, 31 Medrox Gardens, Cumbernauld, Glasgow, G67 4AJ
MM3 YUY D. Bendoris, 35 St. Michael'S Wynd, Kilwinning, KA13 6WH
MI3 YVB B. Craney, 8A Drumhoy Drive, Carrickfergus, BT38 8NN
M3 YVD T. Munro, 71 Zig Zag Road, Liverpool, L12 9EQ
M3 YVE Y. Neary, 3 Wordsworth Close, Torquay, TQ2 6EA
M3 YVF M. Lowe, 22 Ryelands Close, Market Harborough, LE16 7XE
M3 YVG M. Lowe, 22 Ryelands Close, Market Harborough, LE16 7XE
M3 YVJ I. Priest, 11 Dunlin Close, Kingswinford, DY6 8XP
MW3 YVK D. Bowen, 25 Maendu Terrace, Brecon, LD3 9HH
M3 YVN Dr R. Wall, 30 Church Street, Skerries, Co Dublin, Ireland,
M3 YVR L. Palir, 116 Carville Crescent, Brentford, TW8 9RD
M3 YVT J. Storey, 3 Woodside Road, Poole, BH14 9JH
MM3 YVU R. Corrieri, 160 Telford Road, East Kilbride, Glasgow, G75 0BX
M3 YVW L. Brisco, 1 Bescot Way, Thornton-Cleveleys, FY5 3QA
M3 YVY J. Hudd, South Crofty Cottage North Pool Road, Redruth, TR15 3JQ
M3 YVZ J. Smith, 98 Dorset Road, Coventry, CV1 4EB
MW3 YWC A. Johnston, 44 Cradoc Road, Brecon, LD3 9LH
M3 YWD W. Disney, 98 Widney Lane, Solihull, B91 3LL
M3 YWE J. Watkins, 6 Glebelands, Biddenden, Ashford, TN27 8EA
M3 YWF S. Griffiths, 22 Manor Rise, Arleston, Telford, TF1 2ND
M3 YWG W. Gradwell, 9 Nottingham Drive, Bolton, BL1 3RH
M3 YWH K. Skerry, 18 Park Avenue, Cheadle, Stoke-on-Trent, ST10 1LZ
M3 YWI F. Ingram, 1 Hazel Close, Brackley, NN13 6PE
M3 YWJ E. Mcmahon, 120 New Ferry Road, Wirral, CH62 1DY
M3 YWM P. Higgins, 35 Whittington Street, Plymouth, PL3 4EG
M3 YWO K. Lovell, 2 Beckingham Hall Cottages, Tolleshunt Major, Maldon, CM9 8EH
M3 YWP D. Spooner, 30 Clover Road, Norwich, NR7 8TF
MI3 YWT V. Crichton, 10 Bann Drive, Londonderry, BT47 2HW
M3 YWU J. Morris, 96 Bradford Crescent, Durham, DH1 1HW
M3 YXB R. Barrett, 18 Bullstake Close, Oxford, OX2 0HN
M3 YXC D. Jones, 18 Maxwell Drive, Hazlemere, High Wycombe, HP15 7BX
M3 YXD D. Wilson, 24 Hallamshire Mews, Wakefield, WF2 8YB
M3 YXE J. Peters, 11 Clockhouse Lane, Ashford, TW15 2EP
M3 YXF J. Barber, 13 Dock Road, Sharpness, Berkeley, GL13 9UA
M3 YXH G. Sweet, 12 Old Harrow Road, St Leonards-on-Sea, TN37 7EG
M3 YXJ A. Mcgreish, 36 Eastmoor Road, Oxborough, King's Lynn, PE33 9PX
M3 YXK Maj. G. Mcgreish, 36 Eastmoor Road, Oxborough, King's Lynn, PE33 9PX
M3 YXL A. Needham, 49 Macclesfield Road, Buxton, SK17 9AG
M3 YXM J. Pick, 178 Alcester Road South, Kings Heath, Birmingham, B14 6DE
MM3 YXN D. Cowie, 69 Broomfield Park, Portlethen, Aberdeen, AB12 4XT
M3 YXP J. White, 15 Norham Drive, Newcastle upon Tyne, NE5 5PR
M3 YXQ S. Pye, 23 Dene Way, Donnington, Newbury, RG14 2JL
M3 YXR X. Ren, Lincoln House, Clarence Drive, Harrogate, HG1 2QD
M3 YXS L. Reddall, 68 Broadhurst Green, Hednesford, Cannock, WS12 4LF
M3 YXT N. Wall, 6 Ashton Lane, Braithwell, Rotherham, S66 7AJ
M3 YXU P. Dossett, 92 Dale Valley Road, Southampton, SO16 6QU
M3 YXV C. Reynolds, 9 Skeyton Road, North Walsham, NR28 0BS
M3 YXW D. Cox, 9 Northbrook Copse, Bracknell, RG12 0UA
MI3 YXX L. Bradley, 4 Rathbeg Drive, Limavady, BT49 0BB
M3 YYE A. Hinckley, 114 Lawn Lane, Hemel Hempstead, HP3 9HS
M3 YYG S. Senior, 4 Flowers Meadow, Liverton, Newton Abbot, TQ12 6UP
M3 YYJ J. Whitford-Robson, 13 Perryman Close, Plymouth, PL7 4BP
M3 YYK K. Yardley, 4 Park Road, Hillton, Wolverhampton, WV107HS
M3 YYM I. Mcpherson, 86 Fletemoor Road, St. Budeaux, Plymouth, PL5 1UH
M3 YYO M. Scott-Martin, 14 Coles Gardens, Poole, BH15 4DX
MW3 YYQ R. Smith, 20 Fron Uchaf, Colwyn Bay, LL29 6DS
M3 YYR N. Brown, 85 Larkham Lane, Plympton, Plymouth, PL7 4PL
M3 YYS S. Hurrell, 4 Woodland Drive, Plympton, Plymouth, PL7 1SN
MI3 YYT D. Bradley, 33 Lilac Avenue, Limavady, BT49 0HS
M3 YYU R. Woolley, 84 Bowthorpe Road, Norwich, NR2 3TP
M3 YYV S. Chiew, 11 Svenskaby, Orton Wistow, Peterborough, PE2 6YZ
M3 YYW J. Elsmore, 142 St. Marks Road, Chester, CH4 8DH
M3 YZA S. Russell, 11 Morville Road, Dudley, DY2 9HR
M3 YZC D. Matheson, 21 Warren Hill Road, Woodbridge, IP12 4DU
M3 YZE T. Knight, 25 Allcot Road, Portsmouth, PO3 5DE
M3 YZH J. Kelly, Martins, Fairwarp, Uckfield, TN22 3BE
M3 YZI B. Sims, 4 New Cottages, Cranwich Road, Thetford, IP26 5EQ
M3 YZJ J. King, 2 Perys Court, Cracknore Hard Lane, Southampton, SO40 4UT
M3 YZK D. Lisi, 56 Gipsy Lane, Old Whittington, Chesterfield, S41 9JB
M3 YZN J. Murray, 6 Sheridan Court, St. Edmunds Road, Dartford, DA1 5NF
M3 YZO M. Lewis, Hillside, Caldy, Craven Arms, SY7 8QR
M3 YZP M. Tydeman, 6 Colin Close, Corfe Mullen, Wimborne, BH21 3QG
M3 YZQ J. Thompson, 32 Church Street, Warnham, Horsham, RH12 3QR
M3 YZU I. Dennison, 18 Tredington Grove, Caldecotte, Milton Keynes, MK7 8LR
M3 YZV M. Hughes, 19 Pendine Crescent, North Hykeham, Lincoln, LN6 8UW
M3 ZAA M. Walker, 4 Lemonroyd Marina, Fleet Lane, Leeds, LS26 9AJ
M3 ZAE R. Giles, 75 Bradstocks Way, Sutton Courtenay, Abingdon, OX14 4DA
M3 ZAI J. Elderfield, 2 Westwood Close, Amersham, HP6 6RP
M3 ZAL J. Million, 5 Passfield Square, Thornley, Durham, DH6 3DB
M3 ZAM N. Tweed, 42 Ophir Road, Worthing, BN11 2SS
M3 ZAN S. Tokley, 9 Peel Road, Springfield, Chelmsford, CM2 6AQ
MW3 ZAQ Z. David, 45 Amanwy, Llanelli, SA14 9AH
M3 ZAR R. Whitehouse, 29 Greswolde Road, Solihull, B91 1DY
M3 ZAU A. Cunningham, 96 Kingsleigh Drive, Birmingham, B36 9DY
M3 ZAV A. Cooper, 23 Ash Street, Manchester, M9 5XY
M3 ZAW C. Beresford, 13 Chaseside Avenue, Twyford, Reading, RG10 9BT
M3 ZAY T. Hooper, 71 Collins Parc, Stithians, Truro, TR3 7RB
M3 ZAZ Z. Ripley, 11 Gallows Hill Drive, Ripon, HG4 1UP
M3 ZBF J. Evans, 112 Seagrave Crescent, Sheffield, S12 2JP
M3 ZBH B. Hudson, 34 Eastwood Road, Bexhill-on-Sea, TN39 3PS
M3 ZBI A. Yarnold, 1 Ford Close, Ivybridge, PL21 9TQ
M3 ZBJ Z. Jennings, 29 Mountbatten Drive, Newport, PO30 5SJ
M3 ZBL R. Leese, 22 Southlands Road, Congleton, CW12 3JY
M3 ZBQ I. Marsh, 56B Oliver Crescent, Farningham, Dartford, DA4 0BE
M3 ZBR R. Offord, 16 Clive Avenue, Ipswich, IP1 4LU
M3 ZBS A. Knight, 37 Crispe House, 72 Dovehouse Mead, Barking, IG11 7EB
M3 ZBV A. Higham, 12 Lakenheath Drive, Sharples, Bolton, BL1 7RJ
M3 ZBW R. Weaver, 116 Carville Crescent, Brentford, TW8 9RD
M3 ZBX J. Gleeson, 124 Rushes Mead, Harlow, CM18 6QE
M3 ZBZ M. Carvell, 10 Burns Close, Stevenage, SG2 0JN

M3 ZCA A. Bevan, 6 Fryth Close, Haverhill, CB9 0ED
M3 ZCB C. Blackmun, 2A St. Margarets Road, Girton, Cambridge, CB3 0LT
M3 ZCE R. Sutton, 56 Neatherd Road, Dereham, NR20 4AY
M3 ZCG C. Gregory, 81 Fiskerton Way, Oakwood, Derby, DE21 2HY
MW3 ZCI P. Probert, 54 New Hall Road, Ruabon, Wrexham, LL14 6AT
M3 ZCJ J. Ma, Harrogate Ladies' College, Clarence Drive, Harrogate, HG1 2QG
M3 ZCM D. Griffin, 16A Kent Road, Fleet, GU51 3AH
M3 ZCN M. Barnes, 51 Lower Way, Great Brickhill, Milton Keynes, MK17 9AG
MW3 ZCO C. Owen, 97 Maesglas Grove, Newport, NP20 3DN
M3 ZCR C. Zarucki, 26 Heathfield Road, Kings Heath, Birmingham, B14 7DB
MM3 ZCS S. Kirkbride, 18 North Roundall, Limekilns, Dunfermline, KY11 3JY
MW3 ZCU D. Cook, 19 Almond Avenue Risca, Newport, NP11 6PF
M3 ZCW C. Lewis, 9 Chatsworth Gardens, Sydenham, Leamington Spa, CV31 1WA
M3 ZCX D. Stringer, 18 Townfield Close, Ravenglass, CA18 1SL
MI3 ZCY C. Wang, 32 Broadlands, Carrickfergus, BT38 7BL
MM3 ZDG Z. Graham, 15 Stone Crescent, Mayfield, Dalkeith, EH22 5DT
MM3 ZDI R. Cook, 5 Monkstadt, Linicro, Portree, IV51 9YN
M3 ZDK D. Osmand, Flat 5, Long Barn Rosevidney, Penzance, TR20 9BX
MW3 ZDQ A. Davies, 25 Llanfair Road, Tonypandy, CF40 1TA
M3 ZDS D. Scott, 33 Manor Crescent, Honiton, EX14 2DF
M3 ZDT I. Arrow, Crystalwood, Stonemans Hill, Newton Abbot, TQ12 5PZ
M3 ZDV D. Vasey, 22 Rickleton Village Centre, Washington, NE38 9ET
M3 ZDW D. Whyatt, 11 The Perrings, Nailsea, Bristol, BS48 4YD
M3 ZED A. Henderson, Clocktower Lodge, Cragside, Rothbury, Morpeth, NE65 7PU
M3 ZEH S. Hendy, Flat 2, 33 Kingston Road, Leatherhead, KT22 7SL
M3 ZEI D. Mctaggart, 59 Gainsborough Road, Richmond, TW9 2DZ
M3 ZEJ R. Fagg, 62 Hawkins Road, Folkestone, CT19 4JA
M3 ZER D. Nazer, 20 College Road, Ringwood, BH24 1NX
MM3 ZET H. Dally, 3 Gremmasgaet, Lerwick, Shetland, ZE1 0NE
M3 ZEV B. Dexter, 237 Wordsworth Avenue, Sheffield, S5 8NE
M3 ZEW E. Melman, 177 Grantham Road, London, E12 5NB
M3 ZEY M. Blagg, 17 Flint Avenue, Forest Town, Mansfield, NG19 0DS
M3 ZFB J. Creed, 11 Athelstan Road, Faversham, ME13 8QL
M3 ZFH S. Recht, 1 Ireton Close, Chalgrove, Oxford, OX44 7RZ
M3 ZFI C. Godfrey, 97 Whalley Drive, Bletchley, Milton Keynes, MK3 6HX
M3 ZFJ J. Tomlinson, 14 Shorecliffe Rise, 27 Radcliffe New Road, Radcliffe, Manchester, M26 1LE
MM3 ZFK W. Taylor, Garth Wood, Fishers Brae, Eyemouth, TD14 5NJ
M3 ZFL C. Rycott, 7 Crescent Grove, London, SW4 7AF
M3 ZFN D. Hoare, 47 High Street, Chalgrove, Oxford, OX44 7SJ
M3 ZFO D. Davies, Marshlands, Low Road, Thurlton, Norwich, NR14 6RL
M3 ZFS S. Fisher, 4 Beaufont Gardens, Bawtry, Doncaster, DN10 6RT
M3 ZFV P. Whiteley, 1 Newton Close, Fareham, PO14 3LF
M3 ZFW D. Tinn, 28 South Road, Kirkby Stephen, CA174sn
M3 ZFY R. Fye, 201 North Wing The Residence, Kershaw Drive, Lancaster, LA1 3SY
M3 ZGA G. Campbell, 10 Welbeck Road, Rochdale, OL16 4XP
M3 ZGC V. Jolliffe, 54 Glendale Avenue, Wash Common, Newbury, RG14 6RU
M3 ZGD G. Donaldson, Moddershall House, Moddershall, Stone, ST15 8TG
M3 ZGE C. Walton, 6 Hilltop Road, Bearpark, Durham, DH7 7DP
M3 ZGF P. Daniel, 57 Jephson Road, Sutton-in-Ashfield, NG17 5EH
M3 ZGG T. Turner, 86 Bevan Close, Huntingdon, PE29 1TJ
M3 ZGH G. Houlton, 7 Bartletts Hillside Close, Chalfont St. Peter, Gerrards Cross, SL9 0HH
M3 ZGI M. Jurczyszyn, 115 The Twitchell, Sutton-in-Ashfield, NG17 5AX
MM3 ZGK G. Munn, 28 Broomfield Park, Portlethen, Aberdeen, AB12 4XT
M3 ZGM G. Machin, 7 Lansdown, Yate, Bristol, BS37 4LS
M3 ZGO C. Ikeda-Chew, 10 Bruce Drive, South Croydon, CR2 8SL
MW3 ZGR G. Renshaw, 46 Forge Close Caerleon, Newport, NP18 3PW
M3 ZGS G. Somerville, 22 Woolven Close, Burgess Hill, RH15 9RR
M3 ZGT J. Throup, Willow House, O'Keys Lane, Worcester, WR3 8RL
M3 ZGU L. Retford, 111 Lander Close, Old Hall, Warrington, WA5 9PL
M3 ZGX P. Beltrami, 15 Woodroffe Square, Calne, SN11 8PW
M3 ZHC X. Beltrami, 20 Brunel Way, Calne, SN11 9FN
MD3 ZHD W. Callister, 13 Fairfield Avenue, Onchan, Isle Of Man, IM3 4BG
MU3 ZHF H. Fletcher, Le Villocq House, Le Villocq, Castel, Guernsey, GY5 7SA
M3 ZHG C. Barnes, 23 South Street, Crewe, CW2 6HN
M3 ZHI J. Brookes, 85 St. Johns Road, Rotherham, S65 1LT
M3 ZHM L. Martin, Little Acre, Swan Lane, Edenbridge, TN8 6AJ
M3 ZHP D. Pastwik, 451 Chorley Old Road, Bolton, BL1 6AH
M3 ZHQ J. Martin, Little Acre, Swan Lane, Edenbridge, TN8 6AJ
MW3 ZHU M. Hilliar-Mills, 1-3 Queens Road, Criccieth, LL52 0EG
M3 ZHV T. Gorbutt, 26 Whitethorn Avenue, Withernsea, HU19 2LN
M3 ZHW T. Hulme, 157 Birkinstyle Lane, Stonebroom, Alfreton, DE55 6LD
M3 ZHX K. Baker, 3 Manor Park, Duloe, Liskeard, PL14 4PT
M3 ZHY J. Mason, 9 Farmfields Rise, Woore, Crewe, CW3 9SZ
M3 ZHZ L. Kay, 20 Mytton Lane, Shawbury, Shrewsbury, SY4 4JE
M3 ZIA Z. Ul Haq, 7 Alden Walk, Stockport, SK4 5NW
M3 ZID R. Farrington, Sunny Brook, Broadway Rd, Evesham, WR11 7RN
M3 ZIE H. Ehm, 17 Stuart Road, Kempston, Bedford, MK42 8HS
M3 ZIF J. Whittick, 91 Godfrey Way, Dunmow, CM6 2SQ
M3 ZIH C. Warwick, 104 Church Road, Formby, Liverpool, L37 3NH
M3 ZII S. Savastano, Old Coach House, Hawkchurch, Axminster, EX13 5TX
M3 ZIL E. Greatorex, 22 Marlborough Way, Uttoxeter, ST14 7HL
M3 ZIM T. Fuller, 49 Scotby Avenue, Chatham, ME5 8ER
M3 ZIN P. Bruce, Pilgrims, Broadmead, Lymington, SO41 6DH
M3 ZIO M. Denon, 3 Duke Street, Clowne, Chesterfield, S43 4RZ
M3 ZIV I. Vickers, 3 Nesbit Road, St. Marys Bay, Romney Marsh, TN29 0SF
M3 ZIX S. Norris, 95 Waterloo Road, Ashton-On-Ribble, Preston, PR2 1BH
M3 ZIZ K. Druce, 20 Manwood Avenue, Canterbury, CT2 7AH
M3 ZJB M. Collier, 32 London Road, Dereham, NR19 1AW
M3 ZJD F. Stevens, 191 Parkinson Drive, Chelmsford, CM1 3GW
M3 ZJE M. Lake, 48 Sedgemoor Road, Bath, BA2 5PL
M3 ZJF P. Hayward, 14 Micklewright Avenue, Crewe, CW1 4DF
M3 ZJG P. Gibbs, 19 Lupin Close, Etherley Dene, Bishop Auckland, DL14 0TP
M3 ZJH J. Herant, 75 Victoria St., Chesterton, Newcastle, ST5 7EP
M3 ZJJ G. Chiu, Harrogate Ladies' College, Clarence Drive, Harrogate, HG1 2QG
M3 ZJK S. Knight, 18 Ludlow Close, Chippenham, SN15 3UG
MM3 ZJL J. Logan, 13 Hornel Road, Kirkcudbright, DG6 4LH
M3 ZJM J. Leavesley, 37 Western Road, Stourbridge, DY8 3XU
MI3 ZJN G. Connolly, 54 Granemore Park, Keady, Armagh, BT60 2GP
M3 ZJO J. Rawlinson, Westfield Farm, Risden Lane, Cranbrook, TN18 5DU
M3 ZJQ A. Rawlinson, Westfield Farm, Risden Lane, Hawkhurst, Sandhurst, Cranbrook, TN18 5DU
M3 ZJS G. Stokes, 24 Dayslondon Road, Waterlooville, PO7 5NN
M3 ZJV G. Hewis, 10 Albert Road, New Malden, KT3 6BS
MM3 ZJY G. Boyter, 36 Main Street, Springfield, Cupar, KY15 5SQ
M3 ZKA K. Allgar, 13 Deacon Avenue, Kempston, Bedford, MK42 7DU
M3 ZKD K. Dungey, 247 Park Road, Sittingbourne, ME10 1ER
M3 ZKE K. Pemberton, 38 Milford Drive, Bournemouth, BH11 9HJ
M3 ZKF K. Franklin, 1 Aberdeen Court, Newcastle upon Tyne, NE3 2XU
M3 ZKI I. King, 7 Shardlow Close, Haverhill, CB9 7RF
M3 ZKL Dr I. Jutting, 68 The Ridgeway, Tonbridge, TN10 4NN
M3 ZKO L. Bodie, Harrigate Ladies College, Clarence Drive, Harrogate, HG1 2QG
MM3 ZKQ G. Wareham, 12 The Knowe, Leven Road, Leven, KY8 5JH
M3 ZKT M. Sargeant, 3 Old Springfields Padbury, Buckingham, MK18 2ar
MW3 ZKW K. Watt, 28 Llysgwyn, Morriston, Swansea, SA6 6BJ
MW3 ZKX A. Griffiths, 6 South View, Taffs Well, Cardiff, CF15 7SE
MW3 ZKY J. Jones, 3 Rhedyw Road, Llanllyfni, Caernarfon, LL54 6SN
M3 ZLA G. Burrows, Aubrietia, Malpas, SY14 8AY
M3 ZLE L. Ellerington, 120 Stagsden Road Bromham, Bedford, MK43 8QJ
M3 ZLI P. Dee, 3 Cherry Orchard, Upton-Upon-Severn, Worcester, WR8 0LR
M3 ZLJ J. Priestman, 198 Felmongers, Harlow, CM20 3DW
M3 ZLL J. Giovinazzo, 3 Eleanor Avenue, Epsom, KT19 9HD
M3 ZLO M. Dring, 1 Hannaford Close, St Columb Rd, TR9 6FH
M3 ZLP Z. Gribben, 44 Fern Close, Birchwood, Warrington, WA3 7NU
M3 ZLR L. Rodrigues, 61 Grayling Way, Boston, PE21 8FS
M3 ZLS L. Sammut, 16 Queen Marys Road, New Rossington, Doncaster, DN11 0TS
M3 ZLV M. Ferenc, 2A Rosedene Avenue, London, SW16 2LT
M3 ZLW D. Wakefield, 12 Goddard Court, Wokingham, RG41 5HR
MW3 ZLX R. Miles, 63 Phillip Street, Mountain Ash, CF45 4BG
M3 ZLY T. Farr, 125 Rochford Way, Walton on the Naze, CO14 8SP
M3 ZLZ L. Wilmott, 60 Church Hill, Royston, Barnsley, S71 4NG
M3 ZMB M. Bateup, 47 St Johns Ave, Burgess Hill, RH15 8HJ
M3 ZME M. Davison, 1 Chancel Way, Barnsley, S71 2HS
MI3 ZMJ M. Johnston, 52 Lansdowne Road, Newtownards, BT23 4NT
M3 ZMM D. Mainwaring, 1 Buckingham Close, Didcot, OX11 8TX
M3 ZMN M. Nash, 11 Frederick St., Warrington, WA4 1HX
M3 ZMO A. Maguire, 132 Wigan Road, Ormskirk, L39 2BA
MI3 ZMP M. Paterson, 121 Ballybunden Road Killinchy, Newtownards, BT23 6RZ
M3 ZMQ M. Tew, Willowell, Spring Valley Lane, Colchester, CO7 7SD
M3 ZMR I. Nicholson, 2 Broom Close, Leyland, PR25 5RQ
M3 ZMS M. Strickland, Ancoats, Piercy End, York, YO62 6DQ
M3 ZMT M. Thompson, 133 Redford Avenue, Horsham, RH12 2HH
MW3 ZMU J. Taylor, 60 Rockfield Way, Undy, Caldicot, NP26 3FD
M3 ZMV P. Hayward, 14 Micklewright Avenue, Crewe, CW1 4DF
M3 ZMX I. Botham, 12 Lairgill, Bentham, Lancaster, LA2 7JZ
M3 ZMZ M. Bryant, 26 Coronation Road, Melksham, SN12 7PF
M3 ZNC D. Crosland, Simmonds Green, Varley Road, Huddersfield, HD7 5TY
M3 ZNF G. Panton, Dale View, Thorpe Fendykes, Skegness, PE24 4QN
M3 ZNG N. Galbraith, 213 Queens Road, Portsmouth, PO2 7LX
M3 ZNJ A. Harvey, 28 Langdown Road, Hythe, Southampton, SO45 6EW
M3 ZNL J. Ho, 231 Rush Green Road, Romford, RM7 0JP
M3 ZNM A. Sivyer, 63 Sugden Road, Worthing, BN11 2JG
MM3 ZNN Dr P. Holmes, Maraval, Doune Road, Dunblane, FK15 9AT
M3 ZNO T. Cao, Harrogate Ladies' College, Clarence Drive, Harrogate, HG1 2QG
M3 ZNP R. Young, 26 Silent Woman Park Coldharbour, Wareham, BH20 7PE
MM3 ZNQ W. Beaton, 4 Moorfield Gardens, Springfield, Cupar, KY15 5SH
M3 ZNR D. Belton, 4 Sandown Road, Toton, Nottingham, NG9 6GN
M3 ZNT I. Govan, 9 Willowbank, Sandwich, CT13 9QA
M3 ZNV N. Edwards, 36 Joseph Luckman Road, Bedworth, CV12 8BQ
M3 ZNX T. Hayward, 11 Radnor Close, Bodmin, PL31 2BZ
M3 ZNY P. Robinson, Flat 2, 46 Cliff Road, Sheringham, NR26 8BJ
M3 ZNZ L. Burke, 24 Pinecliffe Avenue, Bournemouth, BH6 3PZ
M3 ZOG J. Hyde, The Grove, 7 Mill Lane, Kidderminster, DY10 3ND
M3 ZOH N. Kerry, 3 Edinburgh Cottages, West Newton, King's Lynn, PE31 6AX
M3 ZOI J. Ogden, 295 Church Road, St. Annes, Lytham St Annes, FY8 3NP
M3 ZON C. Holdt, 18 Garrard Road, Banstead, SM7 2ER
M3 ZOO M. Lucas, 20 Collin Road, Kendal, LA9 5HN
M3 ZOP C. Poulson, 9 Scattergate Green, Appleby-in-Westmorland, CA16 6SP
M3 ZOR Z. Rabbitt, 66 Parkfield Avenue, Delapre, Northampton, NN4 8QB
M3 ZOU E. Nicholson, 24 Barnmead, Haywards Heath, RH16 1UZ
M3 ZPB P. Burgess, Tally Ho Cottage, High Street, Swindon, SN4 0AE
M3 ZPE P. Evans, 22 Northumberland Road, Wigston, LE18 4WL
M3 ZPJ P. Smyth, 91Beechcroft Ave, Darcy Lever, Bolton, BL26HB
M3 ZPM P. May, 95 Moorfield Avenue Denton, Manchester, M34 7TX
M3 ZPO L. Gravel, 16 North Street, Crowle, Scunthorpe, DN17 4NB
M3 ZPR P. Rogers, 11 Beech Crescent, Mexborough, S64 9EH
M3 ZPT P. Tatham, 54 High Street, Cleckheaton, BD19 3PX
M3 ZPW B. Matthews, 30 Oaklands Drive, Brandon, IP27 0NR
M3 ZPY L. Dobson, 8 Coronation Street, Darfield, Barnsley, S73 9HA
M3 ZPZ P. Shurmer, 1 The Glebe, East Harling, Norwich, NR16 2SZ
M3 ZQA P. Stonebridge, 207 Henley Road, Ipswich, IP1 6RL
M3 ZQB M. Plummer, 27 Bowness Court, Congleton, CW12 4JR
M3 ZQC M. Hackett, 26 Wantage Road, Didcot, OX11 0BP
M3 ZQF M. Ashworth, 123 Forest Road, Liss, GU33 7BP
M3 ZQG D. Bates, 80 St. Leonards View, Polesworth, Tamworth, B78 1JY
M3 ZQJ B. Yarwood, 12 Charminster Close, Waterlooville, PO7 7RP
M3 ZQM K. Pascoe, 21 Cotswold Avenue, Sticker, St Austell, PL26 7ER
M3 ZQN A. Wheeler, 8 Elsworth Grove, Birmingham, B25 8EJ
MM3 ZQP J. Docherty, 23 Turret Drive, Polmont, Falkirk, FK2 0QW
M3 ZQV L. Marshall, 65 Bacons Lane, Chesterfield, S40 2SX
MM3 ZQW C. Bryson, 29 Roull Road, Edinburgh, EH12 7JW
MM3 ZQX A. Falconer, 61 Mountcastle Drive North, Edinburgh, EH8 7SP
M3 ZRA R. Alford, 1 School Lane, Winmarleigh, Preston, PR3 0JY
M3 ZRB R. Brown, 194 Wymersley Road, Hull, HU5 5LN
MM3 ZRF R. Fletcher, Balnacraig Cottage, Mid Balnacraig, Alness, IV17 0XL

M3 ZRG R. Gladman, 18 Willingdon, Ashford, TN23 5YF
M3 ZRH R. Harrison, 16 Curlew Rise, Morley, Leeds, LS27 8US
MW3 ZRK D. Rowlands, 1 Forge Lane, Bassaleg, Newport, NP10 8NF
M3 ZRM R. Mcgregor, 84 Churchill Way, Burton Latimer, Kettering, NN15 5RS
M3 ZRO B. Stoner, Montrose, Wesley Road, Robin Hoods Bay, Whitby, YO22 4RW
M3 ZRP E. Stooke, 11 Westgate, Grantham, NG31 6LT
M3 ZRQ G. Crane, 33A Carlisle Gardens, Horncastle, LN9 5LP
M3 ZRR C. Taylor, 1 Jasmine Gardens, Warrington, WA5 1GU
M3 ZRS R. Stokes, 44 Broxhead Road, Havant, PO9 5LA
M3 ZRV R. Allwood, 1, 46, Guildford, GU2 7JN
M3 ZRW D. Williams, 15 Charwood Road, Wokingham, RG40 1RY
M3 ZRX C. Waterworth, 4 Mossdale Road, Ashton-In-Makerfield, Wigan, WN4 0EQ
M3 ZRY L. Acton, 39 Craig Road, Macclesfield, SK11 7YH
MM3 ZRZ D. Hibberd, 18 Whitestripes Path, Bridge Of Don, Aberdeen, AB22 8WF
M3 ZSA S. Arthur, 17 Bromeswell Road, Ipswich, IP4 3AS
M3 ZSB S. Bannister, 162 Dobcroft Road, Sheffield, S11 9LH
M3 ZSC S. Colman, 197 Coppins Road, Clacton-on-Sea, CO15 3LA
M3 ZSD E. Worthington, 7 Bowness Close, Gamston, Nottingham, NG2 6PE
MM3 ZSF S. Faccenda, 27 Dollar Avenue, Falkirk, FK2 7LF
M3 ZSH S. Hampson, 12 Flying Fields Drive, Macclesfield, SK11 7GE
M3 ZSI S. Airs, 6 The Willows, Culham, Abingdon, OX14 4NN
M3 ZSJ J. Gibson, 22 Woodburn Drive Chapeltown, Sheffield, S351YS
MM3 ZSK C. Brash, 4 Union Street, Lossiemouth, IV31 6BA
M3 ZSL R. Hudson, 35 Pear Tree Mead, Harlow, CM18 7BY
MW3 ZSM P. Leyshon, 34 South Street, Porth, CF39 0EG
MW3 ZSO R. Madson, 44 Manor Court, Church Village, Pontypridd, CF38 1DW
M3 ZST C. Butters, 4 The Dovecote, Pitsford, Northampton, NN6 9SB
M3 ZSU Capt. P. Westwell, Roden House, Dobsons Bridge, Whitchurch, SY13 2QL
M3 ZSV T. Bendelow, 3 St. Giles Close, Thirsk, YO7 3BU
M3 ZSW S. White, 1 The Red House, Old Gallamore Lane, Market Rasen, LN8 3US
M3 ZSY D. Bendelow, 17 Poppy Drive Sowerby, Thirsk, YO7 3SJ
MW3 ZTB T. Beach, 97 Van Road, Caerphilly, CF83 1LA
M3 ZTD G. Berry, 5 Oakholme Rise, Worksop, S81 7LJ
MW3 ZTH J. Evans, 8 Hengoed Crescent, Cefn Hengoed, Hengoed, CF82 7HF
M3 ZTK T. King, 215 Hartland Road, Reading, RG2 8DN
M3 ZTL R. Hincks, 79 Forest Street, Shepshed, Loughborough, LE12 9BZ
M3 ZTN A. Romanov, 132 Latchmere Drive, Leeds, LS16 5DY
M3 ZTP J. Wilson, 14 Elms Drive, Morecambe, LA4 6DQ
M3 ZTR T. Reed, Channel Pool, Armathwaite, Carlisle, CA4 9QY
MM3 ZTS C. Owens, 4 Union Road, Camelon, Falkirk, FK1 4PG
M3 ZTT A. Jeffery, 14 Holly Mount Shavington, Crewe, CW2 5AZ
M3 ZTU D. Stewart, 79 Eastfield Road, Driffield, YO25 5EZ
M3 ZTW A. Mackenzie, 91 Glenwood Drive, Romford, RM2 5AR
M3 ZTX J. Thornhill, 47 Hopton Lane, Mirfield, WF14 8JP
M3 ZUA P. Coleman, 18 Carr Road, Fleetwood, FY7 6QJ
M3 ZUB D. Burrows, 19 Fleming Avenue, Bottesford, Nottingham, NG13 0ED
M3 ZUC A. O'Keeffe, 1 Elmfield Road, Liverpool, L9 3BL
M3 ZUD D. Marshall, 117 Ogley Hay Road, Chase Terrace, Burntwood, WS7 2HU
MW3 ZUF D. Harris, 15 Mill Road, Pontllanfraith, Blackwood, NP12 2GE
M3 ZUG D. Duncan, 12 Carrington Avenue, Poppleton Road, York, YO26 4SH
M3 ZUI C. Shaw, 9 Pheasant Close, Mulbarton, Norwich, NR14 8BL
M3 ZUJ A. Trudgett, 103 Shandon Road, Worthing, BN14 9EA
M3 ZUL D. Stinson, 234 Pelsall Lane, Rushall, Walsall, WS4 1NG
MM3 ZUP C. Webster, 13 Harbour Road, Tayport, DD6 9EX
M3 ZUS S. Skinner, Chevington, Carlton Road, Godstone, RH9 8LD
M3 ZUT P. Streatfield, 66 Ockendon Road, London, N1 3NW
M3 ZUU R. Duncan, 12 Carrington Avenue, Poppleton Road, York, YO26 4SH
M3 ZUW J. Caithness, 6 Station Road, Catworth, Huntingdon, PE28 0PE
M3 ZUY B. Pike, 19 Cardigan Gardens, Reading, RG1 5QP
M3 ZUZ A. Finn, 29 Argyle Road, Weymouth, DT4 7LX
M3 ZVA S. Toher, The Chapel, Station Road, Darlington, DL2 1JG
MW3 ZVB G. Spencer, Erw Uchaf, Gorad, Holyhead, LL65 3BT
MW3 ZVD D. Jones, 7 Bryn Rhedyw, Llanllyfni, Caernarfon, LL54 6SS
M3 ZVF B. Lane, 26 Deben Valley Drive, Kesgrave, Ipswich, IP5 2FB
MW3 ZVH C. Nicholls, 26 Maes Geraint, Pentraeth, LL75 8UR
M3 ZVI F. Lees, 5 St. Winifred Road, Rainhill, Prescot, L35 8PY
M3 ZVK V. King, 215 Hartland Road, Reading, RG2 8DN
M3 ZVS K. Brown, 142 Moor Lane, Woodford, Stockport, SK7 1PJ
M3 ZVT M. Joynson-Ellis, 20 Morland Court, Skaters Way, Peterborough, PE4 6GW
M3 ZVU M. Flanagan, 804 New Hey Road, Huddersfield, HD3 3YW
M3 ZVW J. Woods, Invicta Cottage, Carbrooke Road, Thetford, IP25 6SD
M3 ZVX J. Barker, 26 Ardley Road, Fewcott, Bicester, OX27 7PA
MM3 ZVY K. Bourhill, Cherry Tree Cottage, 30C Salters Road, Wallyford, EH21 8AA
M3 ZWB N. Watts, 404 Marsh Road, Turves, Peterborough, PE7 2DW
M3 ZWD Z. Dixon, 18 Norfolk Close, Plymouth, PL3 6DB
M3 ZWF E. Watson, 4 Moorland Avenue, Blackburn, BB2 5EQ
M3 ZWG I. Garratt, 2 Hayclose Crescent, Kendal, LA9 7NT
M3 ZWH R. Simmons, 10 Whitehall Road, Kingswinford, DY6 9DY
M3 ZWI C. Riley, 15 Keddington Crescent, Louth, LN11 0AP
M3 ZWJ G. Hill-Adams, 6 Broadleaze Way, Winscombe, BS25 1JX
M3 ZWK C. Powell, 37 Newnham Close, Mildenhall, Bury St Edmunds, IP28 7PD
M3 ZWL R. Amos, 6 Eccles Road, Wittering, Peterborough, PE8 6AU
M3 ZWM I. Coulson, 2 Marl Hurst, Edenbridge, TN8 6LN
M3 ZWN N. Woodstock, 8 Fernheath Close, Bournemouth, BH11 8SL
MW3 ZWO S. Gibbon, 39 Pen-Y-Groes, Penyrheol, Caerphilly, CF83 2JL
M3 ZWP K. Bainbridge, 29 Bluebell Grove, Calne, SN11 9QH
M3 ZWQ K. Toohey, 197 Broad Oak Road, St Helens, WA9 2AQ
MW3 ZWR G. Spicer, 6 Cromwell Road, Neath, SA10 8DR
MW3 ZWS S. Todd, 4 Uplands Road, Pontardawe, Swansea, SA8 4AH
M3 ZWW Dr W. Warwicker, 28 Porters Wood, Petteridge Lane, Matfield, Tonbridge, TN12 7LR
M3 ZWX J. Turley, 19 Ibstock Drive, Stourbridge, DY8 1NW
M3 ZXA N. Gleaden, 25 Ridgway Avenue, Darfield, Barnsley, S73 9DU
M3 ZXC S. Makins, 10 Lower Mill St., Ludlow, SY8 1BH
M3 ZXE N. Egginton, Emergency Planning Unit, Shropshire County Council, Shrewsbury, SY2 6ND
M3 ZXG G. Carless, Silver Cottage, Silver Street, South Petherton, TA13 5BY
MM3 ZXL S. Fradley, 30 Polmont Park, Polmont, Falkirk, FK2 0XT
M3 ZXN A. Walton, 65 Broadway East, Rotherham, S65 2XA
M3 ZXQ I. Talbot, 41 Elmwood Close, Cannock, WS11 6LX
M3 ZXX K. Willets, 32 Yates Way, Ketley Bank, Telford, TF2 0AZ
MW3 ZYE P. Mason, 20 Coronation Road, Six Bells, Abertillery, NP13 2PJ
M3 ZYF H. Armstrong, 83 Hillary Grove, Carlisle, CA1 3JQ
M3 ZYG D. Zygadllo, 56 Scarisbrick Crescent, Liverpool, L11 7DW
M3 ZYI I. Ball, 17 Homecroft Road, Goldthorpe, Rotherham, S63 9DX
M3 ZYK M. Edge, 19 Burton Av, Rushall, West Mids, WS41NH
M3 ZYM D. Lavender, 39 Albany Crescent, Bilston, WV14 0HT
M3 ZYO B. Hall, 6 Marshall Close, Parkgate, Rotherham, S62 6DB
MW3 ZYQ M. Daymond, 58 Holmesdale Street, Cardiff, CF11 7HF
M3 ZYR A. Caulfield, 1 Carleton Close, Amesbury, Salisbury, SP4 7TU
MM3 ZYS S. Robertson, 20 Knockard Place, Pitlochry, PH16 5JF
M3 ZYT N. Bond, 21 Coleridge Close, Bletchley, Milton Keynes, MK3 5AF
MI3 ZYU A. Kelly, 19 Union Street Mews, Coleraine, BT52 1EN
M3 ZYV A. Brown, 7 Whilton Crescent, West Hallam, Ilkeston, DE7 6PE
M3 ZYW A. Pemberton, 38 Pond Green Way, St Helens, WA9 3SD
M3 ZYX T. Parker, 2 Kipling Close, Grantham, NG31 9ND
M3 ZYY P. Day, 46 Beatrice Avenue, Saltash, PL12 4NG
M3 ZYZ C. Wilmott, 60 Church Hill, Royston, Barnsley, S71 4NG
M3 ZZA M. Daniels, 10 Downland Avenue, Peacehaven, BN10 8TH
M3 ZZD D. Smith, 7 Kestrel Avenue Bransholme, Hull, HU7 4ST
M3 ZZE E. Olver, 41 Mount Tamar Close, Plymouth, PL5 2AL
M3 ZZF D. Webb, 18 Lavender Avenue, Minster On Sea, Sheerness, ME12 3UA
M3 ZZH H. Metson, Higher Churchtown Barn, North Hill, Launceston, PL15 7PQ
M3 ZZI D. Brooker-Evans, 379 Crownhill Road, Plymouth, PL5 2LN
M3 ZZJ L. Haley, 17 Oak Drive, Crownhill, Plymouth, PL6 5TZ
M3 ZZL P. Stretton, 38 Queens Way Melbourne, Derby, DE73 8FG
M3 ZZN A. Hobson, 8 Sycamore Road, Colchester, CO4 3NF
M3 ZZQ J. Snape, 2 Orchard Close, Fort Avenue, Preston, PR3 3YS
M3 ZZS J. Skinner, 36 Milton Road, Waterloo, Liverpool, L22 4RF
MW3 ZZU E. Jones, 39 Ger-Y-Llan, Velindre, Llandysul, SA44 5YB
M3 ZZV G. Jones, 39 Thurlow Way, Barrow-in-Furness, LA14 5XP
M3 ZZW G. Smith, 7 Kestrel Avenue, Bransholme, Hull, HU7 4ST
M3 ZZX I. Curtress, 7 Gwinnett Court, Shurdington, Cheltenham, GL51 4GQ

M5

M5 AA Mf Propagation Research Group, c/o R. Parsons, Netherhall Barn, Hallmoor Road, Darley Dale, Matlock, DE4 2HF
M5 AAT R. Nikolova, 206 Eastcote Av West Molesey, Surrey, KT82EX
M5 AAW C. Taylor, 40 Chilton Drive, Watnall, Nottingham, NG16 1HL
M5 ABC D. Last, Hillview, New Road, Bridport, DT6 4NY
M5 ABH D. Drew, 14 Greensfields, Skegby, Nottingham, NG17 3DN
M5 ABJ J. Bodle, 48 Bolsover Road, Hove, BN3 5HP
M5 ABN P. Herbert, Flat 1, 7 Cockington Lane, Paignton, TQ3 1EE
MW5 ABR P. Lovelock, Garn Cottage, Aberbechan, Newtown, SY16 3AY
M5 ABT R. Clark, 36 Southfields, Stanley, DH9 7PH
M5 ACD G. Steabler, 1 Westhill Road, Grimsby, DN34 4SG
M5 ACF Dr E. Mclusky, 11 Ripon Road, Killinghall, Harrogate, HG3 2DG
M5 ACJ R. Payne, 58 Sheepcote Lane, Amington, Tamworth, B77 3JW
M5 ACR A. Jakusz-Gostomski, 15 Goodliffe Gardens, Tilehurst, Reading, RG31 6FZ
M5 ACS M. Arnfield, Cleabarrow, Plumley Moor Road, Knutsford, WA16 0TU
M5 ACT E. Bluer, 8 Cedar Close, Waterlooville, PO7 7LN
M5 ACX M. Anderson, 1 Thames Close, Ferndown, BH22 8XA
M5 ADA Dr C. Goodhand, 22 Somin Court, Doncaster, DN4 8TN
MW5 ADD A. Dimmock, Gwyndy, Llandegfan, Menai Bridge, LL59 5PW
M5 ADE A. Deane, Flat 1-6, 76 Church Street, Tewkesbury, GL20 5RX
M5 ADF D. Hook, 15 Wordsworth Avenue, Sutton-in-Ashfield, NG17 2GG
M5 ADI D. Williams, 4 Dunkirk Rise College Bank Way, Rochdale, OL12 6UH
M5 ADL A. Lambert, 69 Anvil Crescent, Broadstone, BH18 9DZ
M5 ADM K. Marriott, 99 Stapleford Lane, Toton, Nottingham, NG9 6FZ
M5 ADQ B. Woolnough, 99 Abbey Road, Leiston, ip16 4ta
MW5 ADW P. Booth, 68 Trem Eryri, Llanfairpwllgwyngyll, LL61 5JF
M5 AEC M. Drinkwater, Green Quarter, Ward Green Old Newton, Stowmarket, IP14 4EZ
M5 AEE A. Steadman, 4 Vineyard Way, Buckden, St Neots, PE19 5SR
M5 AEF R. Burrows-Ellis, 49 Highland Drive, Worlingham, Beccles, NR34 7AR
M5 AEH C. Shackleton, 54A Blueleighs Park, Chalk Hill Lane, Great Blakenham, Ipswich, IP6 0ND
M5 AEI E. Howe, 22 Freston, Peterborough, PE4 7EN
M5 AEO J. Kempster, 4 Church Cottages, Great Gaddesden, Hemel Hempstead, HP1 3BU
MM5 AES J. Robertson, 138 East Main St., Armadale, Bathgate, EH48 2PB
M5 AEX Dr A. Thomas, The Stone Barn, 1 Home Farm Close, Chesterton, OX26 1TZ
M5 AFE W. Higgs, 955 Oldham Road, Rochdale, OL16 4SE
M5 AFG D. Hall, 4 Steventon Road, Wellington, Telford, TF1 2AS
M5 AFH J. Denmead, 47 Holland Road, Clevedon, BS21 7YJ
MI5 AFL I. Mccrum, 12 Bishops Court Road, Kilclief, Downpatrick, BT30 7NU
M5 AFV J. Jones, 10 Huntington Close, Redditch, B98 0NF
M5 AFX G. Rhodes, 60A Somerset Road, Cinderford, GL14 2HJ
M5 AFY S. Hall, 58 Lower Meadow Court, Northampton, NN3 8AX
M5 AGB A. Koeller, 116 Parham Road, Gosport, PO12 4UE
M5 AGG C. Ellis, Broken Ridge, Fir Tree Close, Ringwood, BH24 2QW
M5 AGI Rev. J. Addison, 20 St. Davids Drive, Callands, Warrington, WA5 9SB
MM5 AGM C. Campbell, 18 Parkview Avenue, Falkirk, FK1 5JX
M5 AGR G. Macadam, Acacia Lodge, 10 The Green, Herts, SG8 7AD
M5 AGS Capt. J. Lightfoot, Flat 18, The Cloister, Wokingham, RG40 1AW
M5 AGV D. Adkins, 196 High Road, North Weald, Epping, CM16 6EF
M5 AGW W. Green, 2 Irkdale Avenue, Enfield, EN1 4BD
M5 AGY A. Dermont, 7 Pool Close, Little Comberton, Pershore, WR10 3EL
M5 AGZ M. Gill, The Cottage, Barrowell Green, London, N21 3AU
M5 AHF M. Latimer-Sufit, Flat 615, Jacqueline House, 52 Fitzroy Road, London, NW1 8UB
MI5 AHG J. Campbell, 7A Desert Road, Mayobridge, Newry, BT34 2JB
MM5 AHM G. Welch, Beechlea, Yieldshields Road, Carluke, ML8 4QY
MM5 AHO G. Crowley, 3 Park View, Westfield, Bathgate, EH48 3PP
M5 AIB H. Whiteoak, 18 Gregory Springs Mount, Mirfield, WF14 8LG

MM5 AII L. Mccay, 4 South Mound, Houston, Johnstone, PA6 7DX
M5 AIO J. Dixon, 8 East View, St. Ippolyts, Hitchin, SG4 7PD
M5 AIQ A. Bain, 5 Norgrove Park, Gerrards Cross, SL9 8QT
M5 AJB J. Button, 1 Ross Cottages Southey Green, Sible Hedingham, Halstead, CO9 3RN
MI5 AJH E. Holmes, 7 Bamford Park, Dundrod, Crumlin, BT29 4JW
M5 AJK T. Dawson, 54 Graeme Road, Enfield, EN1 3UT
MM5 AJN D. Gerrie, 13 Martin Terrace Auchnagatt, Ellon, AB41 8TF
M5 AJO Dr M. Woodhouse, 5 Redhill Close, Bristol, BS16 2AH
M5 AJP D. Hone, 16 Newton Avenue, Aylesbury, HP18 0BN
MM5 AJW D. Mckay, Tryggo, Sarclet, Wick, KW1 5TU
M5 AJZ Dr S. Coles, 54 Brasslands Drive, Portslade, Brighton, BN41 2PN
M5 AKT M. Ellis1, 4 Magna Crescent, Great Hale, Sleaford, NG34 9JX
M5 AKW L. Howell, 18 High St., Foxton, Cambridge, CB2 6SP
M5 AKY D. Vosper, 5 Franklyn Terrace, Farrington Gurney, Bristol, BS39 6UD
MM5 AKZ I. Mcclelland, Parkburn, Dumfries, DG1 1RB
M5 ALA A. Shone, 50 Whitefield Avenue, Norden, Rochdale, OL11 5YG
M5 ALC R. Hatcher, 61 Holland Road, Oxted, RH8 9AU
M5 ALG A. Levy, 29 Ferndale Avenue, Reading, RG30 3NQ
MI5 ALJ Strabane ARS, c/o C. Hannigan, 4 Silverhill Road, Strabane, BT82 0AE
MI5 ALO E. Clementson, 84 Portaferry Road, Newtownards, BT23 8SN
M5 ALS D. Munro, 16 Gullimans Way, Leamington Spa, CV31 1LA
M5 ALU A. Horsfield, 45 Burchnall Close, Deeping St James, PE6 8QJ
MM5 ALX D. Ross, 37 Shillinghill, Alness, IV17 0SZ
M5 AMM Dr P. Adams, Tiniusstr. 29B, Berlin, Germany, 13089
M5 AMN A. Waddington, 41 Willow Way, Farnham, GU9 0NU
MI5 AMO J. Houston, 9 Lismore Drive, Dundonald, Belfast, BT16 1SL
MM5 AMV H. Cardwell, 26 South Road, Cupar, KY15 5JF
M5 AOI N. Pollard, 5 The Stackfield, Wirral, CH48 9XS
MM5 AON R. Henry, Woodyard House, Woodyard Road, Dumbarton, G82 4BG
M5 ARC Wisbech AR & Electronics Club, c/o J. Balls, 70 Risegate Road, Gosberton, Spalding, PE11 4EY
M5 ASK W. Booker, 3 Hollybank Avenue, Sheffield, S12 2BL
M5 ASR J. Richardson, 9 Hilton Avenue, Aylesbury, HP20 2EX
M5 ATR T. Ralph, 15 Portchester Close, Stanground, Peterborough, PE2 8UP
M5 AXA I. Bassett, 47 Queensdown Gardens, Brislington, Bristol, BS4 3JD
M5 BAD Dr C. Leese, 15 Ewden Way, Barnsley, S75 2JW
M5 BAE B A E Systems Great Baddow ARC, c/o P. Tittensor, 47 St. Johns Road, Chelmsford, CM2 0TY
M5 BAZ B. Carter, 25 Chester Close, Chafford Hundred, Grays, RM16 6ET
M5 BBB B. Bot, 52 Kenilworth Avenue, Wimbledon, SW19 7LW
M5 BFL S. Shenstone, Primrose Cottage, Stubb Lane, St Michael South Elmham, Bungay, NR35 1ND
M5 BGR D. Wrigley, 32 Avon Road, Chadderton, Oldham, OL9 0PH
M5 BIL W. Cooper, 16 Beaumont Hill, Dunmow, CM6 2AP
M5 BIR B. Jakubowski, 120 Chandos Street, Coventry, CV2 4HT
M5 BJC B. Crighton, 2 The Lake House, Savage Cat Farm, Gillingham, SP8 5QR
M5 BMW R. Read, 111 Fitzpain Road, West Parley, Ferndown, BH22 8SF
M5 BOP M. Riley, 3 Foxley Close, Ipswich, IP3 8BW
M5 BTB P. Brown, 42 Foxglove Close, Burgess Hill, RH15 8UY
M5 BUF R. Hanney, 74 Avon Road, Bournemouth, BH8 8SF
MI5 BUG G. Moucka, 5 Glebe Gardens, Moira, Craigavon, BT67 0TU
M5 BXB S. Burrows, 33 Pettys Close, Cheshunt, Waltham Cross, EN8 0EW
M5 CAB W. Daly, 85 Lordens Road, Huyton, Liverpool, L14 9PA
MW5 CAD G. Cadwaladr, Madog Yacht Club, Pen Y Cei, Porthmadog, LL49 9AT
M5 CAH C. Hunt, 105 Cropston Road, Anstey, Leicester, LE7 7BQ
M5 CBR A. Hutton, 29 Manor Road, Ashford, TW15 2SL
M5 CBS M. Beale, 8 Blakeney Avenue, Swindon, SN3 3NW
MM5 CFA C. Allan, 35 Hamewith Court, Alford, AB33 8QW
MI5 CFM L. Baine, 50 Lawnbrook Drive, Newtownards, BT23 8XD
M5 CHH C. Hollins, 56 Lovell Road, Cambridge, CB4 2QR
M5 CJH C. Hindle, 10 Barrington Meadows, Bishop Auckland, DL14 6NT
MW5 CKN S. Swinden, 4 Uwch Y Maes, Dolgellau, LL40 1GA
M5 CLO F. Jones, 19 Moreton Street, Prees, Whitchurch, SY13 2EG
M5 CMO B. Armstrong, 35 Northfields Crescent, Settle, BD24 9JP
M5 COL H. Craven, 4 Amanda Drive, Louth, LN11 0AZ
MW5 CYM I. Taylor, 7 Trellewelyn Close, Rhyl, LL18 4NF
MW5 DAD P. Stevenson, Nant Fach Cerrigydrudion, Corwen, LL21 0SB
M5 DAP D. Parker, 12 Sedge Close, Ivybridge, PL21 0WD
MI5 DAW R. Bamber, 15 Ladybrook Parade, Belfast, BT119ER
M5 DB Fawkes Contest Group, c/o A. Sheppard, 81 Deer Way, Horsham, RH12 1PX
M5 DHB D. Brattle, 34 York Street, Bedford, MK40 3RJ
M5 DIK R. King, 10 Bucks Avenue, Watford, WD19 4AS
M5 DJC M. Cressey, 33 Parklands Drive, Harlaxton, Grantham, NG32 1HX
MW5 DJO E. Owen, Pant-Y-Fedwen, 39 Glanrafon Estate, Bontnewydd, Caernarfon, LL55 2UW
M5 DLA R. Felds, 93 Bancroft Lane, Mansfield, NG18 5LL
M5 DND N. Read, 29 Welsh Street, Bishops Castle, SY9 5BS
M5 DNK D. Kennedy, Holmcroft, Lewis Road, Selsey, PO20 0RG
MM5 DOG K. Macdonald, 5 The Stances, Kilmichael Glassary, Lochgilphead, PA31 8QA
MW5 DQQ M. Poreba, 7 Waterloo Place, Brynmill, Swansea, SA2 0DE
M5 DR D. Rugen, 19 Jacksons Close, Haskayne, Ormskirk, L39 7LD
M5 DRW J. Noel, 58 Easenhall Lane, Redditch, B98 0BJ
M5 DUO P. Richards, 16 Fruiterers Arms Caravan Park, Uphampton Lane, Ombersley, Droitwich, WR9 0JW
MM5 DWW D. Wishart, Curcum, Swannay, Orkney, KW17 2NS
M5 DX Hallam DX Group, c/o N. Totterdell, Moscar Cross House, Hollow Meadows, Sheffield, S6 6GL
M5 DXX R. Morton, 16 Penshaw Avenue, Wigan, WN3 5NG
M5 DZH J. Lynch, 21 Worthington Avenue Hopwood, Heywood, OL10 2LN
M5 EAY G. Hobson, 30 Leigh Road, Westbury, BA13 3QL
M5 ECX T. Watts, 6 Keynsham Walk, Swindon, SN3 2AL
MI5 EEM W. Hesketh, 49 Mount Michael Park, Belfast, BT8 6JX
M5 EHG R. Bateman, 81 Stanton St., Derby, DE23 6NF
M5 EI Milton Keynes ARS, c/o Dr P. Gould, 152 High Street North, Stewkley, Leighton Buzzard, LU7 0EP
M5 ENM R. Gurowich, 1 St. Cuthberts Villas, Haybridge, Wells, BA5 1AH
M5 ERN E. Coleby, 13 Farm Close, Sunniside, Newcastle upon Tyne, NE16 5PP
M5 ESA A. La Fauci, 50 Gore Road, London, SW20 8JL
M5 EXY M. Brashill, 42 Bannister Street, Withernsea, HU19 2DT
M5 FAB S. Sawyers, 72 Langrigg Road, Carlisle, CA2 6DH
M5 FOX D. Ross, 37 Cartmell Drive, Leeds, LS15 0NQ
MM5 FWD S. Robertson, Grove Cottage, 30 Commerce Street, Insch, AB52 6HX
M5 GAC G. Pendrick, 23 Hazel Drive, Spondon.De21 7Ds, Derby, DE21 7DS
M5 GDK M. Gadeke, 3 Ripon Close, Towcester, NN12 6PL
M5 GHT G. Thompson, 15 Caithness Road, Hylton Castle Estate, Sunderland, SR5 3RE
M5 GJO G. Orlebar, 21 Field Lane, Willersey, Broadway, WR12 7QB
M5 GUS R. Guscott, 19 Springfield Way, Threemilestone, Truro, TR3 6BJ
M5 GUY G. Austin, 23 Ngunguru Heights Rise, Ngunguru, Whangarei, New Zealand, 173
M5 GVY N. Drumm, 8 Harpur Place, Thornhill, Egremont, CA22 2SG
M5 GWH L. Preece, 29 Elliott Street, Newcastle, ST5 1JL
M5 HDF Midland Contest Group, c/o M. Waldron, 32 Windmill Street, Upper Gornal, Dudley, DY3 2DQ
M5 HFJ J. Glover, 22 Hampden Road, Birkenhead, CH42 5LH
MI5 HIL B. Hill, 5 Heathers Close, Magheralin, Craigavon, BT67 0RN
MI5 HNA C. Archibald, 37 Jellicoe Drive, Belfast, BT15 3LA
MW5 HOC D. Warburton, 71 Richards Terrace, Cardiff, CF24 1RW
M5 HOT M. Palmer, 21 Ibbett Close, Kempston, Bedford, MK43 9BT
M5 IC Invicta Contest Group, c/o I. Lowe, 54 College Road, Margate, CT9 2SW
M5 IGE D. Russell, 90 Halleys Way, Houghton Regis, Dunstable, LU5 5HZ
MI5 IMB D. Barnes, 14 Old Station Road, Belleek, BT93 3EZ
M5 IMI C. Wilson, The Rectory, Church Close, Thetford, IP25 7LX
MM5 ISS G. Milne, 19 Fairview Crescent, Danestone, Aberdeen, AB22 8ZB
M5 ITE P. Hayler, 27 Birch Way, Heathfield, TN21 8BB
M5 JAO J. Owen, 10 Pitcher Lane, Leek, ST13 5DB
M5 JON J. Edmunds, Caroline Cottage, New Passage Road, Bristol, BS35 4LZ
M5 JPG J. Griffiths, 9 Cauldale Close, Middleton, Manchester, M24 5SU
M5 JWR J. Richardson, 6 Clarence Road Scorton, Richmond, DL10 6EE
M5 JWS J. Summers, 3 Thatchers Close, Burgess Hill, RH15 0QU
MI5 JYK P. Lowrie, 13 Carwood Park, Newtownabbey, BT36 5JU
M5 KEN J. Sharples, 21 Alexandra Pavilions, Stanleyfield Close, Preston, PR1 1QW
M5 KHH C. Young, 26 Horsham Avenue, Peacehaven, BN10 8HX
M5 KJM K. Murphy, 79 Torkington Road, Hazel Grove, Stockport, SK7 6NR
M5 KVK G. Howell, 19 Constable Avenue, Eaton Ford, St Neots, PE19 7RH
M5 KZI M. Hillary, 45 Frances Road, Purbrook, Waterlooville, PO7 5HH
M5 LAR S. Pounder, 4 Otley Mount, East Morton, Keighley, BD20 5TD
M5 LDF L. Farrington, 130 Millway Road, Andover, SP10 3AY
M5 LMG L. Griffiths, 51 Linley Avenue, Pontesbury, Shrewsbury, SY5 0TL
M5 LMY D. Sweetland, 15 Wasdale Close, Owlsmoor, Sandhurst, GU47 0YQ
M5 LRO C. Pickett, 1 Woolwich Close, Chatham, ME5 0HU
MI5 LYN E. Lynn, 60 Lurgan Tarry, Lurgan, Craigavon, BT67 9HN
M5 MC FWE Contest Group, c/o A. Smith, 12 Northgate, Beccles, NR34 9AS
M5 MDH M. Hampton, 126 The Crescent, Eastleigh, SO50 9BH
M5 MDX Stockport Radio Society, c/o B. Naylor, 47 Chester Road Poynton, Stockport, SK12 1HA
M5 MK "Milton Keynes Amateur, Radio Society", c/o T. Cowell, 102 Stamford Avenue, Springfield, Milton Keynes, MK6 3LQ
MI5 MTC M. Clarke, 19 Ardlougher Road Irvinestown, Enniskillen, BT94 1RN
M5 MUF M. Johnson, 1 Ferndale Drive, Ratby, Leicester, LE6 0LH
MW5 MWR M. Randall, 15 Erw Wen, Pencoed, Bridgend, CF35 6YF
M5 NEV N. Bridle, 95 Kings Stone Avenue, Steyning, BN44 3FJ
M5 NEX T. Jones, 5 Broomfields Road, Appleton, Warrington, WA4 3AE
M5 OOO H. Clayton, 3 Waterden Court, Queensdale Place, London, W11 4SQ
M5 OTA A. Hickson, Randolph Cottage, France Lynch Chalford Hill, Stroud, GL6 8LH
M5 PDL S. Symonds, 32 Redhoave Road, Poole, BH17 9DU
M5 PGC P. Challans, Flat 4, Sandringham Court, 2 Chandos Square, Broadstairs, CT10 1QN
M5 PIP Dr P. Nicolson, 56 Serpentine Road, Harborne, Birmingham, B17 9RE
M5 PM P. Molloy, 4 Tilt Meadow, Cobham, KT11 3AJ
M5 POO S. Robinson, 23 Jameson Drive, Corbridge, NE45 5EX
MM5 PSL P. Leybourne, 13 Sanblister Place, Virkie, Shetland, ZE3 9JX
M5 PSW P. Walkling, Flat 36, Highlands House, Wharncliffe Road, Southampton, SO19 7GG
M5 PWR P. Collins, 50 Seacroft Esplanade, Skegness, PE25 3BE
M5 PYE E. Boyd, 7 Fritton Court, Haverhill, CB9 8LX
M5 RAG R. Mullen, 4 Bay View Grove, Barrow-in-Furness, LA13 0EQ
M5 REG R. Barber, 35 Lower Park Crescent, Bishop's Stortford, CM23 3PU
M5 REV Rev. R. Moll, Penn Cottage, Green End, Buckingham, MK18 3NT
M5 REW R. Walker, 159 Cuckfield Road, Hurstpierpoint, Hassocks, BN6 9RT
M5 RFD C. Wardale, 18 Wolsey Way, Lincoln, LN2 4QH
M5 RHG R. Gower, 21 Saltings Crescent, West Mersea, Colchester, CO5 8GG
M5 RHS M. Krupa, 15 Pasture Avenue, Goole, DN14 6LG
M5 RIC R. Brokenshaw, Flat 11, Stonehill Court, 27 Westhill Road, Weymouth, DT4 9NB
M5 RJC R. Crowther, 14 Alice Close, Hereford, HR1 1XQ
MI5 RJS R. Scott, 98 Aghafad Road Dunamanagh, Strabane, BT82 0QQ
M5 RMF R. Fisher, 65 Kylemore Avenue, Mossley Hill, Liverpool, L18 4PZ
M5 ROB R. Johnson, 4 Ton Lane, Lowdham, Nottingham, NG14 7AR
MI5 ROY R. Arthur, 3 Stevenson Park Tullyally, Londonderry, BT47 3QS
M5 RPT R. Tickle, 5 Bramley Court, Harrold, Bedford, MK43 7BG
M5 RST D. Warren, 10 Meadow Bank, Meadow Lane, Alfreton, DE55 2BR
M5 SB Stewart Bryant Radio Station, c/o Dr S. Bryant, 154 London Road North, Merstham, Redhill, RH1 3AA
M5 SE R. Johnston, 60 Green Lane, Seaforth, Liverpool, L21 3UB
MM5 SHA Dr S. Stephenson, 7 Charlotte Gardens, Aberdeen, AB25 1LW
M5 SJM S. Mcbain, 13A St. Lukes Close, Cherry Willingham, Lincoln, LN3 4LY
M5 SJS J. Covel, 111A Balmoral Drive, Southport, PR9 8QH
M5 SKY Believe In Better Cg, c/o A. Mace, 14 High Street, Melbourn, Royston, SG8 6EB
M5 SLC S. Collis, 11 Luxton Way, Wiveliscombe, Taunton, TA4 2BW
M5 SRE P. Scott, 15 Victoria Drive, Blackwell, Alfreton, DE55 5JL
M5 SSB D. Rogers, 12 Archdale, High Wycombe, HP11 2JP
M5 SUE S. Coombs, 10 Horseshoe Walk, Widcombe, Bath, BA2 6DE
M5 SW M. Williamson, 5 John F Kennedy Walk, Tipton, DY4 0SF

M5 TAM J. Bore, 14 Westwood Avenue, Lowestoft, NR33 9RH
MI5 TCC T. Campbell, 27 Silverbrook Park, Newbuildings, Londonderry, BT47 2RD
M5 TLA M. Robinson, 19 St. Wilfrids Crescent, Brayton, Selby, YO8 9EU
MW5 TLE T. Evans, Gwauntrebeddau, Tregynon, Newtown, SY16 3ER
M5 TMG T. Green, 1 Forest Road, Blidworth, Mansfield, NG21 0SJ
M5 TNT S. Purdy, Wetheral House, Great Salkeld, Penrith, CA11 9NA
M5 TRT Dr C. Hardmeier, Robert Jacobsens Vej 80, 2.1., Copenhagen, Denmark, 2300
M5 TT M. Knowles, 11 Thorneycroft Avenue, Birkenhead, CH41 8HJ
M5 TTT J. Chisholm, 162 Ardington Road, Northampton, NN1 5LT
M5 TUE N. Wadsworth, Haygarth, Docker, Kendal, LA8 0DF
MM5 TUW G. Collie, Newton Cottage, Newton Avenue Elderslie, Johnstone, PA5 9BE
M5 TWO C. Van Zuilen, Stiermarkenweg 12, Alkmaar, Netherlands, 1827 EK
M5 TXJ D. Shaw, Cotehouse, Bleatarn, Appleby-in-Westmorland, CA16 6PX
M5 WAH W. Hetherington, 32 Almond Way, Lutterworth, LE17 4XJ
M5 WGD W. Dalzell, 9 Pyms Lane, Crewe, CW1 3PJ
MM5 WIG I. Macdonald, Benvoir, Lightlands Avenue, Wigtown, Newton Stewart, DG8 9EE
M5 WIZ R. Wiseman, 12A Breadcroft Lane, Harpenden, AL5 4TE
MI5 WJB W. Blanchflower, 7 Casaeldona Park, Belfast, BT6 9RB
M5 WJF W. Faulkner, 49 Oakfield Road, Shrewsbury, SY3 8AD
M5 WNS W. Sampson, Rowena, Clifford Street, Chudleigh, Newton Abbot, TQ13 0LH
M5 WSS Wellington School Radio Society, c/o P. Norman, 3 The Gables, Waterloo Road, Wellington, TA21 8JB
M5 XYZ C. Edgar, 9 Winchester Avenue, Morecambe, LA4 6DX
M5 YEX R. Hooperok, 1 Yale Close, Washingborough, Lincoln, LN4 1DX
MM5 YLO N. Marriott, Parkview, Dunrossness, Shetland, ZE2 9JG
M5 ZAP A. Morgan, 153 Beanfield Avenue, Coventry, CV3 6NY
M5 ZZZ S. Burke, 17 The Crescent, Wragby, Market Rasen, LN8 5RF

M6

MM6 AAA A. Martin, Cathcart, Farr, Inverness, IV2 6XJ
M6 AAC R. Johns, 27 Stoneby Drive, Wallasey, CH45 0LG
M6 AAD A. Mullinder, 15 Withington Close, Oakengates, Telford, TF2 6JR
M6 AAE A. Ellsom, 21 High Street, Isle of Grain, ME3 0BJ
M6 AAG A. Taylor, 4 Oxford Street, Carnforth, LA5 9LG
M6 AAJ J. O'Brien, 16 Arnold Road, Darlington, DL1 1JG
M6 AAK A. Comber, 7 Quantock Close, Rushmere St. Andrew, Ipswich, IP5 1AS
M6 AAL M. Castro, 11 Monfa Avenue, Stockport, SK2 7BH
MM6 AAM K. Mcclung, 23 Maple Grove, Troon, KA10 6QW
M6 AAO J. Holmes, 15 Nash Close, Farnborough, GU14 0HL
M6 AAP K. Metcalfe, 33 Corsican Drive, Hednesford, Hednesford, WS12 4SS
MM6 AAR R. Donaghy, 66 Newton Road, Dundee, DD3 0LT
M6 AAU C. Todd, 35 Main Road, Uffington, Stamford, PE9 4SN
MW6 AAW A. West, Bryn Goleu, Gwalchmai, Holyhead, LL65 4SW
M6 AAX A. Garman, 50 Heys Road, Prestwich, Manchester, M25 1JY
M6 ABB A. Bocutt, 15 Peel Avenue, Frimley, Camberley, GU16 8YT
M6 ABE A. Egan, 4 Rutter Avenue, Warrington, WA5 0HP
M6 ABG J. Larcombe, 3 Archer Terrace, Plymouth, PL1 5HD
M6 ABN P. Fulbrook, 167 Droitwich Road, Fernhill Heath, WR3 7TZ
MM6 ABO K. Balfour, 55B Cockels Loan, Renfrew, PA4 0NE
M6 ABQ R. Fiddy, 19 Ingham Road Stalham, Norwich, NR12 9DR
M6 ABR A. Cost, 55 Whitcliffe Grange, Richmond, DL10 4ET
M6 ABS A. Skarzynski, 1 River View Moorings, Bridge Road, Stoke Ferry, King's Lynn, PE33 9TS
M6 ABT E. Johnston, 57 Bishop Ken Road, Harrow, HA3 7HU
M6 ABV M. Magnall, 18 Osprey Avenue, Westhoughton, Bolton, BL5 2SL
M6 ABX S. Townsend, 8 Heather Gardens, Belton, Great Yarmouth, NR31 9PP
M6 ABZ A. Jones, 41 Milton Brow, Weston-Super-Mare, BS22 8DD
M6 ACA W. Shelley, 91 Canterbury House, Stratfield Road, Borehamwood, WD6 1NT
M6 ACB P. Flood, 115 Court Farm Road, Newhaven, BN9 9DY
M6 ACC A. Cheng, Pastures Green, Firwood Road, Virginia Water, GU25 4NG
M6 ACD T. Barker, 93 Hughenden Road, St Albans, AL4 9QN
M6 ACE D. Murphy, 23 Lowndes Close, Stockport, SK2 6DW
M6 ACF A. Fulton11, 117 Rokeby Park, Hull, HU4 7QE
MM6 ACI D. Simpson, Bridgefoot Of Ironside Cottage, New Deer, Turriff, AB53 6UP
M6 ACJ A. Petrie, 17 Brecon Close, Ashington, NE63 0HT
MM6 ACM A. Crawford, Hillview, Kintore, Inverurie, AB51 0XX
M6 ACP G. Dale, 26 Kensington Close, Houghton Regis, Dunstable, LU5 5TJ
M6 ACQ M. Ward, 22 Old Road, Leighton Buzzard, LU7 2RE
MM6 ACV R. Moore, 96 Queen Street, Castle Douglas, DG7 1EG
MM6 ACW A. Woodford, Nordkette, Evnabrek, Levenwick, Shetland, ZE2 9GY
M6 ACX A. Clark, 386 Wold Road, Hull, HU5 5QG
MM6 ACY A. Connelly, 40 Queen Street, Castle Douglas, DG7 1HS
MW6 ACZ S. Holmes, 26 Station Road, Llanrwst, LL26 0EP
M6 ADB A. Banks, 2 Holt Close, Farnborough, GU14 8DG
M6 ADD A. Dade, 15 Barton Road, Berrow, Burnham-on-Sea, TA8 2LT
M6 ADE A. Smith, 93 Sheriffs Highway, Gateshead, NE9 6QN
MW6 ADF E. Pierce, 26 Station Road, Aberconwy Aerials, Llanrwst, LL26 0EP
M6 ADG A. Gladman, 19 Colchester Road, Wymering, Portsmouth, PO6 3RH
M6 ADI W. Knight, 1 Caravan Site, Drakes Drive, St Albans, AL1 5AE
M6 ADJ H. Murray, 39 Warneford Way, Leighton Buzzard, LU7 4JG
M6 ADM A. Mccallum, 15 Leabank, Newcastle upon Tyne, NE15 7LN
MJ6 ADQ J. Crowder, 90 Hue Court, Hue Street, Jersey, JE2 3RX
MW6 ADS A. Scott, 11 Clive Road, St. Athan, Barry, CF62 4JD
M6 ADT A. Carman, 26 Coronation Close, Happisburgh, Norwich, NR12 0RL
MW6 ADU D. Jones, 1 Brig Y Nant, Llangefni, LL77 7QD
M6 ADX D. Richards, Flat 40, Leander Court, Teignmouth, TQ14 8AQ
M6 ADY A. Slim, Troublesome Reach, Playford Road, Woodbridge, IP13 6ND
MW6 ADZ A. Lewis, 3 Aster View, Port Talbot, SA12 7ED
MM6 AEB K. Hay, 13 Braehead, Langholm, DG13 0PS
M6 AEE S. Hedgecock, 37 Tennyson Road, Maldon, CM9 6BE
M6 AEG C. Holmes, 8 Byron Way, Caister-On-Sea, Great Yarmouth, NR30 5RW
M6 AEI S. Woodmore, 66 Imperial Way, Chislehurst, BR7 6JR
M6 AEJ T. Roberts, 62 Ullswater Avenue, Jarrow, NE32 4EY
M6 AEK J. Clark-Mcintyre, 6 Belvedere Road, Blackburn, BB1 9NS
M6 AEL J. Hofman, Brookside, High Street, Stockbridge, SO20 6EY
M6 AEM D. Barraclough, Flat 18 89 Park Road, London, SW19 2BD
MW6 AEN D. Field, 5 Clos Crugiau, Rhydyfelin, Aberystwyth, SY23 4RN
M6 AEQ B. Livesey, 14 Sycamore Avenue, Tyldesley, Manchester, M29 8WQ
M6 AER C. Fielden, 44 Hylion Road, Leicester, LE2 6JE
M6 AES A. Southwell, 56 Lambrook Road, Taunton, TA1 2AF
M6 AET G. Hunt, 35 Outram Street, Sutton-in-Ashfield, NG17 4BA
M6 AEU A. Seymour, Home Farm, Lodge Lane, Northampton, NN6 7PQ
M6 AEW A. Cockett, 10 San Marcos Drive, Chafford Hundred, Grays, RM16 6LT
MW6 AEX A. Bayliss, 2 Hillside Terrace, Tonypandy, CF40 2HJ
M6 AEY J. Smith, 140 Lowmoor Road, Kirkby-In-Ashfield, Nottingham, NG17 7JE
M6 AEZ R. Gore, 4 Westaway Park, Yatton, Bristol, BS49 4JU
M6 AFA A. Lambert, 1 Glebelands, Lympstone, Exmouth, EX8 5JD
M6 AFB A. Brown, 45 Saffron Park, Kingsbridge, TQ7 1RW
M6 AFC L. Hillier, 34 Kittiwake Close, Herne Bay, CT6 6JS
M6 AFF J. Hone, 12 Marlborough Close, Exmouth, EX8 4NA
M6 AFG C. Stacey, 9 Nile Road, Southampton, SO17 1PF
M6 AFH M. Hillam, Common Farm, Swinefleet, Goole, DN14 8DW
MI6 AFI G. Todd, Dp1, Shanaghy, Enniskillen, BT92 0EQ
MW6 AFK A. Studdart, 24 Wepre Park, Connah'S Quay, Deeside, CH5 4HN
M6 AFL A. Lawrence, 39 Newbury Drive, Daventry, NN11 0WQ
MI6 AFM J. Rice, Shanaghy, Lisnaskea, Enniskillen, BT92 0EQ
M6 AFN J. Hitchens, 57 Batchelor Way, Downton, Salisbury, SP5 3FN
MM6 AFQ J. Watt, 10 Marshall Gardens, Luncarty, Perth, PH1 3YX
M6 AFS P. Carter, 18 Park Road, Allington, Grantham, NG32 2EB
M6 AFU T. Turner, 23 Pankhurst Drive, Bracknell, RG12 9PS
M6 AFW S. Coombes, Pantiles, South Crescent, Skegness, PE24 5RQ
M6 AFX S. O'Donnell, 26 Park Avenue, Skegness, PE25 2TF
M6 AFY W. Haddock, 5 Bradley Close, Middlewich, CW10 0PF
M6 AGA G. Wright, 21 Larkfield Road, Redditch, B98 7PL
M6 AGD J. Salter, 20 Burrow Road, Chigwell, IG7 4HQ
MD6 AGF J. Kaighin, 10 Kerrocruin, Kirk Michael, Isle of Man, IM6 1AF
M6 AGG A. Greig, 3 Fir Grange Avenue, Weybridge, KT13 9AR
M6 AGH Dr A. Holt, 36 The Maltings, Malmesbury, SN16 0RN
M6 AGI R. Coombes, Rose Cottage, Choice Hill Farm, Choice Hill Road, Over Norton, Chipping Norton, OX7 5PZ
M6 AGK R. Hambly, 144 Station Road, Irchester, Wellingborough, NN29 7EW
MM6 AGL R. Perkins, 33 Station Road Newmachar, Aberdeen, AB21 0NS
M6 AGO Dr B. Denyer-Green, Dunsley South, Park Road, Forest Row, RH18 5BX
M6 AGQ K. Jones, 40 Sandrock Hill Road, Wrecclesham, Farnham, GU10 4RJ
M6 AGR A. Rayner, 12 Newhaven Drive, Lincoln, LN5 9UF
MW6 AGS D. Evans, 1 Heol Glyndwr, Fishguard, SA65 9LN
M6 AGT R. Macdonald, 31 Addison Drive, Stratford upon Avon, CV377PL
MI6 AGV A. Hamilton, 9 Slievenamaddy Avenue, Newcastle, BT33 0DT
M6 AGY D. Pearson, 73 Stackwood Avenue, Barrow-in-Furness, LA13 9HJ
M6 AGZ J. Gore, 88 Rowan Drive, Kirkby-In-Ashfield, Nottingham, NG17 8FR
M6 AHA P. Whitmore, 3 Crown Bank, Talke, Stoke-on-Trent, ST7 1PT
MM6 AHB A. Hearty, 42 Grange Avenue, Wishaw, ML2 0AQ
M6 AHD A. Dossa, 24 Warwick Drive, Cheshunt, Waltham Cross, EN8 0BW
M6 AHE E. Yohn, Ortner House Farm, Abbeystead, Lancaster, LA2 9BD
M6 AHF W. Starkey, Flat 13, Ramsden Court, Barrow-in-Furness, LA14 2HH
M6 AHH A. Hall, 14 Stanelow Crescent, Standon, Ware, SG11 1QF
MM6 AHJ W. Noon, 0/1 445 Royston Road, Glasgow, G21 2DE
MM6 AHK R. Renshaw, Smithy House, Scotscalder, Halkirk, KW12 6XJ
M6 AHL R. Butterfield, 35 Bede Crescent, Benington, Boston, PE22 0DZ
MM6 AHN A. Macdougall, 8 Mount Stuart Drive, Wemyss Bay, PA18 6DX
MI6 AHO A. Ismay, 21 Hillsborough Drive, Belfast, BT6 9DS
M6 AHP A. Meekins, 12 Myrtle Road, Kettering, NN16 9TW
MW6 AHQ D. Williams, 23 Parc Y Ffynnon, Ferryside, SA17 5TQ
MI6 AHR T. Grzybek, 75 Kilburn Street, Belfast, BT12 6JT
M6 AHS A. Strong, 55 Coley View, Halifax, HX3 7EB
MW6 AHT O. Davis, 18 Ty Gwyn Drive, Brackla, Bridgend, CF31 2QF
M6 AHU K. Lambert, 38 Whittleford Road, Nuneaton, CV10 9HU
MW6 AHV D. Burton, 11 Cwrt-Ucha Terrace, Port Talbot, SA13 1LD
M6 AHW A. Whitehead, 82 High Park Avenue, Stourbridge, DY8 3NA
MM6 AHX M. Mason, 101 Avalon Gardens, Linlithgow Bridge, Linlithgow, EH49 7PL
MM6 AHY J. Liddell, 49 Inchbrae Road, Glasgow, G52 3HA
M6 AHZ P. Dawes, 49 Altofts Lodge Drive, Altofts, Normanton, WF6 2LB
M6 AIA A. Barton, 51 Fieldhead Gardens, Dewsbury, WF12 7SN
M6 AIB C. Durrant, 12 Witney Close, Ipswich, IP3 9QF
MW6 AIC Dr R. Morgan, Spinning Wheel, Derwydd, Llandybie, Ammanford, SA18 2LX
M6 AIE C. Ring, 29 Shelley Close, Newport Pagnell, MK16 8JB
M6 AIF K. Ball, 29 Heather Grove, Wigan, WN5 9PJ
M6 AIH A. Bailey, 49 Grange Crescent, Gosport, PO12 3DS
M6 AII B. Ansell, 7 Bramley Green Road, Bramley, Tadley, RG26 5UE
M6 AIJ R. Tarling, Moonrakers, Ashley, Box, Corsham, SN13 8AN
MM6 AIK D. Seivwright, 35 Invercauld Gardens, Aberdeen, AB16 5RR
M6 AIL M. Barba, 5 Vetch Way, Andover, Hampshire, SP11 6RR
M6 AIN C. Campbell, 5 Ryebank, Holmfirth, HD9 1EU
M6 AIO A. Hadfield, 12 Manor Road, Caister-On-Sea, Great Yarmouth, NR30 5HG
M6 AIP K. Mcfadden, 3A Talbot Rd, Northampton, NN1 4HZ
M6 AIQ L. Fletcher, 110 West Street, North Creake, Fakenham, NR21 9LH
M6 AIR M. Ward, Flat 1, The Old Chapel, Chapel Street, Holsworthy, EX22 6AY
M6 AIT J. Barker-Gunn, 5 De Montfort Road, Lewes, BN7 1SP
M6 AIU D. Crashley, 14 Arran Close, Nuneaton, CV10 7JX
M6 AIV A. Barrett, 4 Wood Cottages, Cummings Cross, Liverton, Newton Abbot, TQ12 6HJ
M6 AIW A. Nutt, 9 Hereford Road, Southport, PR9 7DX
M6 AIZ M. Burbeck, 5 Wouldham Terrace, Saxville Road, Orpington, BR5 3AT
MI6 AJA J. Bingham, 27 Carrickdale Gardens, Portadown, Craigavon, BT62 3BN
M6 AJB A. Brown, Ponsharden Cottage, High Offley Road, Stafford, ST20 0LG
M6 AJC A. Cosham, 1 The Orchard, Redisham, Beccles, NR34 8PA
MW6 AJD R. Ware, 22 Harrington Street, Swansea, SA1 2BU
M6 AJF A. Fysh, 74 Kingsway, King's Lynn, PE30 2EL
M6 AJH A. Higham, 30 Broome Road, Southport, PR8 4EQ
MM6 AJI M. Mcgrorty, 59 Craighall Street, Stirling, FK8 1TA
M6 AJJ M. Baker, 8 Higher Polsham Road, Paignton, TQ3 2SY
M6 AJL A. Lashbrook, 1 Fortescue Road, Exeter, EX2 8LA
M6 AJM A. Mitchell, 18 Holly Leys, Stevenage, SG2 8JA

MI6 AJN A. Ruddell, 16 Beechfield Manor, Aghalee, Craigavon, BT67 0GB
MI6 AJO I. Hoey, 58 Tullynamullan Road, Shankbridge, Ballymena, BT42 2LR
M6 AJP A. Pilkington, 26 Ryelands Close, Market Harborough, LE16 7XE
M6 AKA G. Adgie, Flat 1, 6 Grayfield Avenue, Birmingham, B13 9AD
M6 AKE I. Brayley, 9 Church View Close, Southampton, SO19 8SJ
M6 AKF P. Edwards, 8 Aughton Way, Broughton, Chester, CH4 0QE
M6 AKG H. Turner, 39 Court Crescent, Kingswinford, DY6 9RJ
M6 AKH S. Ward, 11 Blenheim Drive, Hawkinge, Folkestone, CT18 7FA
M6 AKI B. Smith, Maple Lodge, Burtoft Lane South, Boston, PE20 2PF
M6 AKJ L. Lee, 8 William Avenue, Margate, CT9 3XT
M6 AKK A. Forsythe, 24 The Welkin, Lindfield, Haywards Heath, RH16 2PH
M6 AKO S. Clarke, 27 Coronation Road, Callington, PL17 7BX
M6 AKP A. Plastow, Bradcar Cottage, Bradcar Road, Attleborough, NR17 1EQ
MM6 AKQ S. Burns, 3/R 170 Lochee Road, Dundee, DD2 2NH
M6 AKT A. Millin, 79 Court View, Stonehouse, GL10 3PJ
M6 AKV A. Vivian, 15 Kew Klavji, Rhind Street, Bodmin, PL31 2FE
M6 AKW I. Duffie, Trebeighan Farm, Saltash, PL12 5AE
M6 AKY Dr R. Bleaney, 40 Broadstone Road, Harpenden, AL5 1RF
MM6 AKZ D. Henderson, 10 Rye Crescent, Glasgow, G21 3JS
M6 ALB A. Burns, 76, 76, Morprth, NE65 0TF
MW6 ALC A. Cotter, 3 George Street, Aberdare, CF44 6RY
M6 ALE D. Trollope, 80 Azalea Drive, Trowbridge, BA14 9GG
M6 ALG M. Yates, 8 Johnsons Street, Ludham, Great Yarmouth, NR29 5NZ
M6 ALH A. Higham, 30 Broome Road, Southport, PR8 4EQ
M6 ALJ A. Jones, 23 Cranley Road, Hersham, Walton-on-Thames, KT12 5BT
M6 ALK A. Kent, 4 Sellerdale Drive, Wyke, Bradford, BD12 9DA
MI6 ALL A. Armstrong, 45 Rathmena Drive, Ballyclare, BT39 9HZ
M6 ALM T. Hamilton, Flat 14, Kingsmead Court, Dunstable, LU6 1NQ
M6 ALO A. Nugorski, 49 Buxton Street, Morecambe, LA4 5SR
M6 ALP R. Mcleod, 75 Davis Street, Stanley, Falkland Islands (Malvinas), FIQQ 1ZZ
M6 ALQ S. Hassall, 21 Bridgnorth Grove, Chesterton, Newcastle under Lyme, ST5 7QP
M6 ALT D. Ingrey, 1 Ponders Road, Fordham, Colchester, CO6 3LX
M6 ALU J. Harrington, 69 Lexden Road, Colchester, CO3 3QE
M6 ALW J. Anderson, 15 Dickson Road, London, SE9 6RA
M6 ALX A. Jones, 5 Meadowlands, Kirton, Ipswich, IP10 0PP
M6 ALY A. Hall, 254 Walton Road, Walton on the Naze, CO14 8LT
MM6 ALZ F. Gordon, Crofts Of Torrancroy, Strathdon, AB36 8UJ
MM6 AMA A. Campbell, 1B Craig Road, Troon, KA10 6DA
M6 AMC A. Cottrell, 4 Oak Grove, Armthorpe, Doncaster, DN3 2DJ
M6 AMD Dr A. Dingle, 29 Castle View, Witton Le Wear, Bishop Auckland, DL14 0DH
M6 AMG J. Smith, 2 Sunfields Close, Polesworth, Tamworth, B78 1LW
M6 AMI T. Gladman, 39 Fairview Avenue, Rainham, RM13 9RL
M6 AMJ B. Jones, 9 St. James Close, Hanslope, Milton Keynes, MK19 7LF
M6 AML A. Lennon, 12 Lockgate East, Windmill Hill, Runcorn, WA7 6LB
M6 AMN C. Johnson, 58 Cheviot Road, London, SE27 0LG
M6 AMO A. Fleming, 39 Urswick Green, Barrow-in-Furness, LA13 0BH
M6 AMR A. Riddick, 30 Britannia Road, Banbury, OX16 5DW
M6 AMT A. Taylor, 11 Fillingfir Drive, Leeds, LS16 5EG
M6 AMV G. Lyon, 1 Eckersley Street, Wigan, WN1 3PP
M6 AMW A. Webb, 3 Blackmore Close, Thame, OX9 3ZH
M6 AMX T. Brown, 6 Wolstanholme Close, Congleton, CW12 3RX
MM6 ANB P. Rice, 255 Eskhill, Penicuik, EH26 8DF
M6 AND A. Faulkner, Northwood, Cranham, Gloucester, GL4 8HB
M6 ANE D. Mansfield, 1 Arun Court, Lower Street, Pulborough, RH20 2DD
M6 ANI P. Hanman, 7 Tremenheere Road, Penzance, TR18 2AH
M6 ANM E. Seymour, Home Farm, Lodge Lane, Northampton, NN6 7PQ
M6 ANN E. Mann, 6 London Road St. Georges, Telford, TF2 9LQ
M6 ANO J. Seymour, Home Farm, Lodge Lane, Northampton, NN6 7PQ
M6 ANP A. Coulthard, 47 Keston Crescent, Stockport, SK5 8NQ
M6 ANV G. Parkinson, 11 Curtis Road, Poole, BH12 3AQ
M6 ANX P. Tillotson, 9 Holker Street, Barrow-in-Furness, LA14 5RQ
M6 ANY M. Tuffill, 1 Madden Close, Nottingham, NG5 5US
M6 AOB A. Back, 34 Willow Road, Redhill, RH1 6LW
MW6 AOE S. Parfitt, 14 Heol Y Twyn, Rhymney, Tredegar, NP22 5DW
M6 AOF S. Salisbury, 6 Ryan Close, Leyland, PR25 2XW
M6 AOG M. Whelan, 134 Fields Farm Rd, Hyde.Cheshire., SK14 3QW
MM6 AOH S. Mckenzie, 0/2 69 Glenkirk Drive, Glasgow, G15 6AU
M6 AOJ J. Burnham, 5 Shirley Court, Torwood Gardens Road, Torquay, TQ1 1TZ
M6 AOK C. Macleod, 21 Halsetown, St Ives, TR26 3LY
M6 AOL P. Butler, 38 Oak Hill, Hollesley, Woodbridge, IP12 3JY
MM6 AON A. Watson, 10 Christie Place, Elgin, IV30 4HX
M6 AOP O. Ernster, 19 Rose Gardens, Farnborough, GU14 0RW
M6 AOQ M. Keilty, 17 Cliff Road, Wallasey, CH44 3DJ
MI6 AOR A. O'Reilly, 15 Killowen Point, Rostrevor, Newry, BT34 3AN
M6 AOT F. Fang, 49 Calthrope Road, Norwich, NR5 8RN
M6 AOV R. Simpson, 22 Kenworthy Road, Stocksbridge, Sheffield, S36 1BZ
M6 AOW A. Edwards, 2 Keystone Gardens, Steventon New Road, Ludlow, SY8 1LE
MI6 AOX T. Darrah, 42 Pinewood Avenue, Carrickfergus, BT38 8EW
MW6 AOY J. Regan, 30 Tynybedw Terrace, Treorchy, CF42 6RL
MI6 AOZ D. Mudd, 14 Bloomfield Road, Belfast, BT5 5LT
M6 APA A. Armstrong, 30 Tennyson Avenue, Hull, HU5 3TW
M6 APB A. Bradshaw, 130 Low Lane, Morecambe, LA4 6PS
M6 APC A. Cook, 37 Railway Road, Stretford, Manchester, M32 0RY
M6 APG N. Briggs, 20 Broad Lane, Pelsall, Walsall, WS4 1AP
M6 API P. Bishop, 18 Holmwood Avenue, South Croydon, CR2 9HY
M6 APM A. Munford, 4 Chatburn Park Avenue, Brierfield, Nelson, BB9 5QB
M6 APO A. Bais, 9 Lemongrass Road, Bicester, OX27 8BQ
M6 APR A. Ralph, 3 The Leys, St Albans, AL4 9HD
M6 APS T. Bell, 28 Inwood Drive, Coleford, GL168EZ
M6 APU D. Bolton, 88 Goldsborough, Wilnecote, Tamworth, B77 4DF
M6 APW A. Woodhouse, 4 Grafton Close, St Albans, AL4 0EX
M6 AQA A. Wright, 69 Thomas Street, Tamworth, B77 3PR
M6 AQD P. Hateley, 44 Painters Croft, Coseley, Bilston, WV14 8AP
M6 AQE M. Mohammed Shafi, 575 Wood Lane, Dagenham, London, RM8 1DR
M6 AQG P. Frost, 86 Grantham Road, Sleaford, NG34 7NW
M6 AQI C. Croasdale, 76 Kingsway, Euxton, Chorley, PR7 6PP
MW6 AQJ A. Hanley, 11 Heol Pearetree, Rhoose, Barry, CF62 3LB
M6 AQK T. Pearsall, 16 Langdale Road, Leyland, PR25 3AR
M6 AQL S. Melton, 2A The Orchard, Bishopthorpe, York, YO23 2RX
MM6 AQM C. Mcintyre, 2/1, 6 Castle Street, Glasgow, G81 4HH
M6 AQN E. Aksamit, 14 Popplewell Gardens, Gateshead, NE9 6TU
M6 AQO G. Foster, 22 Bradley Cottages, Consett, DH8 6JZ
M6 AQT L. Matthews, 20 Harbridge Road, Broughton, Chester, CH4 0FT
MW6 AQU R. Davison, 2 Marlow Terrace, Mold, CH7 1HH
M6 AQV V. Yotov, Sanford House 81 Skipper Way, St Neots, PE19 6LT
M6 AQW T. Newman, 39 Claremont Road, West Byfleet, KT14 6DY
M6 AQY L. Gibbons, 18 Langdale Road, Ribbleton, Preston, PR2 6AN
M6 ARB A. Boucher, 14 King Edward Road, Woodhall Spa, LN10 6RL
M6 ARC S. Froggatt, 140 Greenlea Court, Huddersfield, HD5 8QB
M6 ARD W. Peel, 15 Brockhurst Close, Horsham, RH12 1UY
M6 ARF M. Dailey, 58 Waincliffe Mount, Leeds, LS11 8AH
M6 ARH A. Hargreaves, 27 Meadow Head Close, Blackburn, BB2 4TY
M6 ARI B. Ansdell, 8 Oakland Drive, Dudley, DY3 2SH
M6 ARJ H. Clough, 8 Skeldyke Road, Kirton, Boston, PE20 1LR
M6 ARK M. Harris, 29 Queen Street, Halesowen, B63 3TZ
M6 ARL S. Skirving, 1 Hallington Close, Bolton, BL3 6YH
M6 ARM A. Martyn, 54 North Side, Hepthorne Lane North Wingfield, Chesterfield, S42 5hy
MM6 ARN A. O'Neill, 39 Ardneil Court, Ardrossan, KA22 7NQ
M6 ARP A. Holmes, 49 Elm Grove South, Barnham, Bognor Regis, PO22 0EJ
M6 ARQ J. Whitehead, 20 Seamill Park Crescent, Worthing, BN11 2PN
M6 ARR I. Clark, 21 James Street, Epping, CM16 6RR
M6 ARS B. Holland, 11 Silverlands Park, Buxton, SK17 6QX
MM6 ART A. Young, 4/4 Prestonfield Terrace, Edinburgh, eh165ee
M6 ARU J. Cherry, 12 Scarisbrick New Road, Southport, PR8 6PY
M6 ARV L. Williams, Broomstreet Farm, Porlock, Minehead, TA24 8JR
M6 ARW A. Winkley, 77 Lechlade Road, Birmingham, B43 5ND
MW6 ARX K. Bevan, 76 Tennyson Close, Pontypridd, CF37 5ER
M6 ARY H. Watts, 696 Knowsley Lane, Knowsley, Prescot, L34 9EH
M6 ASC A. Davies, 68 Wood Street, Castleford, WF10 1LN
M6 ASD A. Dhillon, 13 Weston Road, Guildford, GU2 8AU
M6 ASE A. Everett, 16 Robyns Road, Beeston Regis, Sheringham, NR26 8YJ
M6 ASF A. Foote, Flat One, Kimber'S Close Kennet Road, Newbury, RG14 5JF
M6 ASH D. Amos, 86 Royal Military Avenue, Folkestone, CT20 3EJ
M6 ASI T. Williams, Broomstreet Farm, Porlock, Minehead, TA24 8JR
M6 ASJ R. Bawden, 43 Enys Road, Camborne, TR14 8TW
M6 ASK A. Keatley, 156 Earlswood Way, Colchester, CO2 9NE
M6 ASM A. Mclachlan, 4 Stratton Close, Bexleyheath, DA7 4AJ
M6 ASO T. Roper, 48 Rowthorne Lane, Glapwell, Chesterfield, S44 5QD
M6 ASP T. Chapman, 1 East Dean Road, Lockerley, Romsey, SO51 0JL
M6 ASQ A. Hefford, 9 Chamomile Gardens, Farnborough, GU14 9XY
M6 AST B. Herrick, Honing Road, Dilham, NR28 9PL
M6 ASV C. Redmond, 6 Apsley Road, Southsea, PO4 8RH
M6 ASZ G. Wall, Flat 2, Coniston House Holyoake Road, Worsley, Manchester, M28 3DH
M6 ATA S. Waldock, 102 Beaconsfield Road, London, N15 4SQ
M6 ATB D. Zubrzycki, 11 Bishopdale Holme, Bradford, BD62AB
MW6 ATC A. Hare, 243 Heritage Park, St. Mellons, Cardiff, CF3 0DU
M6 ATD E. Gomez Lozano, 2 Annesley Road, Oxford, OX4 4JQ
M6 ATF N. Bisiker, 31 Lansdowne Avenue, Waterlooville, PO7 5BL
MW6 ATG A. Jones, 3 Manor Way, Kinmel Bay, Rhyl, LL18 5BP
M6 ATH A. Hicks, The Granary, Vann Lake Road, Dorking, RH5 5JB
M6 ATI A. Paricsi, Vine House, Northwick Road Pilning, Bristol, BS35 4HA
M6 ATJ M. Brasher, 48 Eldertree Road Thorpe Hesley, Rotherham, S61 2TQ
M6 ATK R. Atkins, 2 Sandpiper Crescent, Malvern, WR14 1UY
M6 ATL G. Hatt, 4H Colman House, Earlham Road, Norwich, NR4 7TJ
M6 ATM A. Malin, 65A Coventry Road, Burbage, Hinckley, LE10 2HL
M6 ATP D. Gibson, 25 Middleham Close, Ouston, Chester le Street, DH2 1TA
M6 ATQ I. Cooper, 54 St Margarets Road, Lowestoft, NR 324 HT
MM6 ATR A. Robertson, 6C Fergusson Road, Cumbernauld, Glasgow, G67 1LR
M6 ATS J. Watts, 70 Castleway North, Leasowe, Wirral, CH46 1RW
M6 ATU T. Mcbride, 15 Penkvale Road, Stafford, ST17 9EY
M6 ATW A. Wilkinson, 17 Stewkins, Audnam, Stourbridge, DY8 4YW
M6 ATX G. Williams, 19 Holden Walk, Wigan, WN5 9JQ
M6 ATY F. Payne, 45 Foxhill, Shaw, Oldham, OL2 7NQ
M6 AUA P. Cooke, 44 Brooklands Park, Craven Arms, SY7 9RL
M6 AUB S. Drake-Brockman, 13 St. Johns Place, Bury St Edmunds, IP33 1SW
M6 AUC S. Hall, 3 Cedar Grove, Prestwich, Manchester, M25 3DY
MJ6 AUD P. Ahier, Les Trois Carres, La Rue D'Aval, Jersey, JE3 6ER
M6 AUE G. Carter, 19 Brathay Crescent, Barrow-in-Furness, LA14 2BG
M6 AUF E. Washbrook, 47 Westgate, Leominster, HR6 8SA
MM6 AUG B. Moerman, 19/3 Pirniefield Bank, Edinburgh, EH6 7QQ
M6 AUH H. Beaumont, 1 Ashley Walk, Orleton, Ludlow, SY8 4HD
MM6 AUI D. Cunningham, 21 Constable Acre, Cupar, KY15 4AE
MM6 AUJ F. Linn, 14 Elphinstone Road, Tranent, EH33 2HR
M6 AUL H. Hylton, 214 School Road, Hall Green, Birmingham, B28 8PF
M6 AUM Lord W. Saint, Flat 74, Ferrier Point, London, E16 1QW
M6 AUO J. Mcmahon, 39 Hobmoor Croft, Birmingham, B25 8TJ
M6 AUP T. Mundell, 98 Westley Road, Bury St Edmunds, IP33 3SD
M6 AUQ P. Thearle, 12 Grange Walk, Bury St Edmunds, IP33 2QB
MI6 AUU M. Rushbrooke, 22 Dublin Road, Omagh, BT78 1ES
MW6 AUX D. Jones, 17 Miners Row, Aberdare, CF44 0TP
M6 AUZ S. Michaelis, Fieldgate, Coltstaple Lane, Horsham, RH13 9BB
M6 AVA C. Hughes, 10 Langford Road, Stockport, SK4 5BR
M6 AVD M. Darwen, Church Wood, Church Road, Lytham St Annes, FY8 5PX
MM6 AVE J. Rankin, 17 Dippin Place, Saltcoats, KA21 6AB
M6 AVF D. Bailey, 92 Morrell Street, Maltby, Rotherham, S66 7LH
M6 AVG L. Scott, 28 Cavendish Place, New Silksworth, Sunderland, SR3 1JW
MM6 AVH C. Lee, Flat 1, Uist Travel Accommodation, Daliburgh, HS8 5SS
M6 AVJ A. Johnson, 3 Lindens Close, Thorney Toll, Wisbech, PE13 4AR
MW6 AVK R. Gripp, 23 Edmond Locard Court, Chepstow, NP16 6FA
M6 AVL P. Smart, 142 Finch Road, Chipping Sodbury, Bristol, BS37 6JB
M6 AVM A. Hunter, 9 Gelt Burn, Didcot, OX11 7TZ
M6 AVO A. Craven, 45 Benhams Drive, Horley, RH6 8QT
M6 AVQ C. Brown, 77 Grange Road, Ramsgate, CT11 9LP

M6 AVR A. Mackay, 11 Greatheed Road, Leamington Spa, CV32 6ES
M6 AVS R. Paxman, 11 Gibsons Gardens, North Somercotes, Louth, LN11 7QH
M6 AVT R. Whalley, 188 Astley Street, Astley, Tyldesley, Manchester, M29 7AX
M6 AVU A. Pawlak, 8 Healey Close, Crewe, CW1 4RS
M6 AVV J. Noon, 108 Cardinal Avenue, Morden, SM4 4SX
M6 AVW A. Wilson, 11 Headland Way, Alton, Stoke-on-Trent, ST10 4AN
M6 AVY I. Clark, 22 Rosemary Avenue, Grimsby, DN34 4NJ
M6 AVZ K. Nicholson, 11 Lancaster Way, Skellingthorpe, Lincoln, LN6 5UF
M6 AWA A. Warnes, 13 Warren Avenue, Fakenham, NR21 8NP
M6 AWB A. Baker, 2 Stileway, Meare, Glastonbury, BA6 9SH
M6 AWC A. Coombes, 3 Marshall Close, Purley On Thames, Reading, RG8 8DQ
M6 AWE D. Barrett, 208 Doncaster Road, Rotherham, S65 2UE
M6 AWG M. Feltham, Flat 5 Rebecca Court, 9, Beckenham, BR3 1NN
M6 AWI B. Sturman, 26 Howes Avenue, Thurston, Bury St Edmunds, IP31 3PY
M6 AWJ M. Harland, Challenger Quay, Falmouth, TR11 3YL
M6 AWL M. Shepherd, North Waver Cottage, Bells Road Belchamp Walter, Sudbury, CO10 7AR
M6 AWN C. Smith, 44 Brooksfield, Bildeston, Ipswich, IP7 7EJ
M6 AWO M. Nelson, 30 Unicorn Place, Bury St Edmunds, IP33 1YP
M6 AWP L. Amunyela, 1 Artillery Street, Colchester, CO1 2JJ
M6 AWQ P. Vickers, 37 Faraday Road, Ipswich, IP4 1PU
M6 AWR A. Wright, 52 Thorpe Way, Cambridge, CB5 8UB
M6 AWS A. Stabler, 11 Lincolns Avenue, Gedney Hill, Spalding, PE12 0PQ
M6 AWU J. Harris, 4 Burgh Old Road, Skegness, PE25 2LN
MW6 AWV A. Williams, 88 Heol Homfray, Cardiff, CF5 5SB
M6 AWY J. Van-Boques-Tal, 9 Stubbins Lane, Gazeley, Newmarket, CB8 8RL
M6 AWZ A. Wilson, 4 Oxford Street, Doe Lea, Chesterfield, S44 5PH
M6 AXA M. Marsh, 6 Carr Green, Lowton, WA3 1EQ
M6 AXB A. Blamire, 21 The Laurels, Banstead, SM7 2HG
M6 AXC D. Howarth, 32 Cotswold Drive Rothwell, Leeds, LS26 0QZ
M6 AXD S. Denman, 12 Dyke Vale Road, Sheffield, S12 4ER
M6 AXF M. Atherton, 53 Gillars Green Drive, Eccleston, St Helens, WA10 5AU
M6 AXG T. Winter, 8 Thorpe Street, Hartlepool, TS24 0DX
M6 AXH A. Hanson, 14 Braithwaite Avenue, Keighley, BD22 6EU
M6 AXI A. Chapman, 6 Chaffer Lane, Birdham, Chichester, PO20 7EZ
M6 AXJ S. Mccluskey, 29 Hotspur Avenue, Bedlington, NE22 5TD
M6 AXL A. Lowery, 102 Brompton Park, Brompton On Swale, Richmond, DL10 7JP
M6 AXM M. Rai, Coldham Hall, Stanningfield, Bury St Edmunds, IP29 4SD
M6 AXP J. Christoforou, 55 Wood Street, Taunton, TA1 1UW
M6 AXR R. Williamson, 2 Hobbs Road, Shepton Mallet, BA4 4LS
M6 AXS N. Sutherland, 34 Little Heath Road, Chobham, Woking, GU24 8RL
MM6 AXT S. Glen, 5/3 Renfrew Chambers, 136 Renfield Street, Glasgow, G2 3AU
M6 AXU G. Briggs, 54 Behind Berry, Somerton, TA11 6JY
M6 AXW P. Hearnshaw, Flat 10, 83 Swallows Meadow, Solihull, B90 4PH
M6 AXX A. Marsden, 38 Sandhill Road, Rawmarsh, Rotherham, S62 5NT
MW6 AYA A. Young, 14 Ramsons Way, Cardiff, CF5 4QY
M6 AYC T. Jagger, 50 North Street, Lower Hopton, Mirfield, WF14 8PN
M6 AYE A. Yap, Harrogate Ladies' College, Clarence Drive, Harrogate, HG1 2QG
M6 AYG B. Leckey, 76 Cardigan Lane, Leeds, LS4 2LN
M6 AYH C. Wilson, 87 Levensgarth Avenue, Fulwood, Preston, PR2 9FP
M6 AYI R. Unsworth, 22 Meadow House Park, Badcocks Lane, Tarporley, CW6 9RT
M6 AYJ M. West, Flat 1, 32 High Street, Dawlish, EX7 9HP
M6 AYK D. Moger, 23 Elmsleigh Road, Paignton, TQ4 5AX
M6 AYL A. Sharam, 30 Heywood Avenue, Maidenhead, SL6 3JA
M6 AYM P. Armitage, 250 Abbeydale Road South, Totley Rise, Sheffield, S17 3LL
M6 AYN A. Taylor, 14 The Lawns, Hinckley, LE10 1DY
M6 AYP R. Styles, 4 Coningsby Close, Gainsborough, DN21 1SS
M6 AYQ J. Fudge, 25 Virginia Orchard, Ruishton, Taunton, TA3 5LP
MM6 AYR J. Mcmorland, 382 Maryhill Road, Glasgow, G20 7YQ
M6 AYS W. Abbott, 12 Yew Tree Gardens, Birchington, CT7 9AJ
M6 AYU C. Green, 14 St. Andrews Road, Bletchley, Milton Keynes, MK3 5DR
M6 AYV A. Lidster, 23 William Street, Rotherham, S60 2NG
M6 AYW K. Reeves, 5 Westby Crescent, Whiston, Rotherham, S60 4EA
M6 AYY C. Leviston, 13 Pryors Walk, Askam-in-Furness, LA16 7JG
M6 AYZ M. Tarrant, Wayside Cottage, Gabber Lane, Plymouth, PL9 0AW
M6 AZA A. Adams, 45 Four Oaks Road Tedburn St. Mary, Exeter, EX6 6AP
M6 AZD A. Lines, 26 Mcintyre Walk, Bury St Edmunds, IP32 6PF
M6 AZE A. Seedig, 21 Ambleside Close, Mytchett, GU16 6DG
M6 AZF S. Case, 136 Old Basin, Bridgwater, TA6 6LJ
M6 AZG P. Culverwell, 29 Elland Road, Brierfield, Nelson, BB9 5RX
M6 AZH E. Cross, 12B Oakridge, Three Rivers Country Park, Clitheroe, BB7 3JW
M6 AZK S. Tyler, 11 Windmill Cottages, Dilmore Lane, Worcester, WR3 7RX
M6 AZM P. Comley-Ross, 48 High Street, Topsham, Exeter, EX3 0DY
M6 AZN G. Villiers, 88 Redwald Road, Rendlesham, Woodbridge, IP12 2TE
MW6 AZP A. Dighton, 84 Trefelin, Aberdare, CF44 8LF
M6 AZQ M. Osband, 22 Samian Crescent, Folkestone, CT19 4JW
M6 AZR K. Braisher, 7 Ormond Road, Thame, OX9 3XN
M6 AZS P. Mcfadden, Maple Cottage, Leighton Buzzard, LU7 9DZ
M6 AZT S. Mason, 8 Barrowby Gate., Grantham, NG31 7LT
M6 AZV D. Simpson, 50 Castle Hill, Berkhamsted, HP4 1HF
M6 AZX N. Robinson, G19, Grange Country Park, Straight Road, Colchester, CO7 6UX
M6 AZY W. Millington, 93 Feiashill Road, Trysull, Wolverhampton, WV5 7HT
M6 BAA B. Gutteridge, 121 Station Road South, Walpole St. Andrew, Wisbech, PE14 7LZ
M6 BAC B. Warner, 8 Court Lane, Preston, Canterbury, CT3 1DJ
M6 BAD B. Barnes-Martin, 145 Farm Road, Barnsley, S70 3DW
M6 BAG D. Roberts, 27 Nairn Street, Jarrow, NE32 4HX
M6 BAH S. Bassett, 5 The Terrace, The Green, Stratford-upon-Avon, CV37 0JD
MI6 BAI B. Baird, 12 Manse Park, Newtownards, BT23 4TN
MI6 BAJ Dr A. Bell, 4 Mount Pleasant View, Newtownabbey, BT37 0ZY
M6 BAK C. Baker, 60 Belvedere Road, Danbury, Chelmsford, CM3 4RB
M6 BAM B. Butler, 42 Station Road, Stanbridge, Leighton Buzzard, LU7 9JF
M6 BAN B. Banner, 99 Swingate, Kimberley, Nottingham, NG16 2PU
M6 BAQ B. Williamson, 23 Tower Hamlets Street, Dover, CT17 0DY
M6 BAR D. Barry, Coasters, Station Road, Colchester, CO7 8LH
MW6 BAS B. Sweet, 14 Bryn Celyn, Colwyn Bay, LL29 6DH
MW6 BAU K. Burgess, 18 Fairmeadows, Maesteg, CF34 9JL
M6 BAV D. Woodbine, 29 Compass Tower, Munnings Road, Norwich, NR7 9TW
M6 BAW J. Middleton, 17 Woods Loke West, Lowestoft, NR32 3DN
M6 BAX R. Baxter, Rose Dene, Hornsby, Brampton, CA8 9HF
M6 BAY K. Linklater, 32 King Edward Road, Gillingham, ME7 2RE
M6 BAZ B. Pike, 36 Larkfield Avenue, Sittingbourne, ME10 2DP
M6 BBA J. Alincastre, 90 York Crescent, Durham, DH1 5PT
M6 BBB W. Kenway, 40 Grove Avenue, Gosport, PO12 1JX
M6 BBC A. Bright, 86 Fourth Avenue, Watford, WD25 9QQ
M6 BBD P. Symonds, 10 Cowper Court, Willunga, Australia, 5172
M6 BBE E. Puttock, 12 Beechfields, School Lane, Petworth, GU28 9DH
M6 BBF G. Symonds, 10 Cowper Ct, Willunga, Australia, 5172
M6 BBG P. Holland, 30 Knighton Park Road, London, SE26 5RJ
M6 BBH M. Ramsey, 21 Goldsmith Road, Eastleigh, SO50 5EN
MM6 BBK P. Robertson, 41 Hawthorn Street, Leven, KY8 4QE
M6 BBL N. Stanley, 253 Brownley Road Wythensawe, Manchester, M22 9UX
M6 BBM M. Harrison, 91 Rye Road, Hastings, TN35 5DH
M6 BBO A. Applegate, 13 Deacons Close, Kings Stanley, Stonehouse, GL10 3GA
M6 BBP A. Page, 207 Brooklyn Road, Cheltenham, GL51 8DZ
M6 BBQ A. Smith, 311 Albion Street, Southwick, Brighton, BN42 4AT
M6 BBR C. John, 27 Berberis Walk, West Drayton, UB7 7TZ
M6 BBS Dr P. Hyde, Flat 4, 1 Longhorn Avenue, Gloucester, GL1 2AR
M6 BBT M. Boswell, 5 Woods Avenue, Marsden, Huddersfield, HD7 6JX
M6 BBU R. Boardman, 12 St. Margarets Road, Alderton, Tewkesbury, GL20 8NN
MM6 BBX A. Wright, 17A James Court, 493 Lawnmarket, Edinburgh, eh1 2pb
M6 BCB R. Balmforth, 33 Lees Hall Road, Dewsbury, WF12 0RH
M6 BCC A. Boots, 36A Church Street, Charlton Kings, Cheltenham, GL53 8AR
MM6 BCF C. Flynn, 6C White Street, Ayr, KA8 9BW
M6 BCG A. Pickles, 87A Laburnum Road, Waterlooville, PO7 7EW
M6 BCH B. Chauhan, 45 Burnham Drive, Whetstone, Leicester, LE8 6HY
M6 BCJ J. Bullock, 49 Gallimore Close, Stoke-on-Trent, ST6 4DZ
M6 BCK T. Humphries, 32 Bonds Meadow, Oulton Broad, NR32 3QL
M6 BCN P. Kingston, 2 Deepdale, Great Easton, Market Harborough, LE16 8SS
M6 BCQ P. Chester, 33 Salehurst Road, London, SE4 1AS
M6 BCU B. Miles, 35 Plantation Drive, Walkford, Christchurch, BH23 5SG
M6 BCV M. Cooper, 9 Conway Close, Crewe, CW1 3XN
M6 BCW G. Fearnhead, 27 Lukins Drive, Dunmow, CM6 1XQ
M6 BCY J. Edwards, 45 Bramshaw Gardens, Bournemouth, BH8 0BT
M6 BCZ K. Percy, 55 Buxton Avenue, Heanor, DE75 7UN
MM6 BDA B. Adams, 18 Bellfield Road, North Kessock, Inverness, IV1 3XU
M6 BDD J. O'Brian, 83 Bramdean Crescent, London, SE12 0UJ
M6 BDG B. Greenberg, 94 Rivermead Court, Ranelagh Gardens, London, SW6 3SA
M6 BDI A. Bucknell, 12 Cliveden Grove, Hereford, HR4 0NE
MJ6 BDJ L. Langlois, Brookfield, La Rue D'Empierre, Trinity, Jersey, JE3 5QF
M6 BDL B. Little, 25 Thrift Wood, Bicknacre, Chelmsford, CM3 4HT
M6 BDM D. Mead, 32 Sherborne Road, Farnborough, GU14 6JT
M6 BDO A. Butcher, 31 Wittonwood Road, Frinton-on-Sea, CO13 9JZ
M6 BDQ D. Nelson, 110 Chandag Road, Keynsham, Bristol, BS31 1QF
M6 BDR B. Roberts, 10 Morningside Way, Liverpool, L11 1BD
MW6 BDS D. Bancroft, Stop And Call, Goodwick, SA64 0EX
M6 BDV B. Vile, 24 Hudson Close, Dover, CT16 2SG
M6 BDX T. Scott, 39 Neil Avenue, Holt, NR25 6TG
M6 BEB S. Smedley, Spring Cottage, Frys Well, Radstock, BA3 4HA
M6 BEC B. Beckett, 21 Horseshoes Lane, Langley, Maidstone, ME17 1SR
M6 BED J. Bratchley, Marrowbone House, Calstock Road, Gunnislake, PL18 9BU
M6 BEE B. Azzaro, 5 Rye Hill Close, Bere Regis, Wareham, BH20 7LU
MI6 BEF W. Tosh, 38 Ballycastle Road, Coleraine, BT52 2DY
MW6 BEG P. Sherwood, High Croft Jeffreyston, Kilgetty, SA68 0RG
M6 BEH B. Humphrey, 45 Rose Avenue, Hazlemere, High Wycombe, HP15 7PH
M6 BEI I. Whitlock, 109 Sorrell Drive, Newport Pagnell, MK16 8TZ
M6 BEJ J. Preston, 25 Hamlet Road, Haverhill, CB9 8EH
M6 BEK R. Clare, Kimberley, Boston Road, Boston, PE20 3AP
M6 BEL K. Whitelaw, 18, Seaford, Bn252rt
M6 BEM M. Hooper, 5 Pillowell Close, Cheltenham, GL525GJ
M6 BEN B. Robinson, 13 Dene View, Ellington, Morpeth, NE61 5HQ
M6 BEQ C. Bradley, 47 Long Meadow, Skipton, BD23 1BP
M6 BEU R. Bartha, 6 Chappell Close, Aylesbury, HP19 9QA
M6 BEX R. Whitehead, 29 Coulsons Road, Bristol, BS14 0NN
M6 BEZ P. Bailey, 81A Kings Parade Holland-On-Sea, Clacton-on-Sea, CO15 5JF
M6 BFB V. Walsh, 11 Coronation Drive Crosby, Liverpool, L23 3BN
M6 BFC B. Spaxman, 12 Stanhope Gardens, Barnsley, S75 2QB
M6 BFF K. Rivett, 89 Maidstone Road, Felixstowe, IP11 9EE
M6 BFG C. Bowman, 26 Albany Hill, Tunbridge Wells, TN2 3RX
MM6 BFH B. Haynes, 18 Drummond Street, Greenock, PA16 9DN
M6 BFI A. Jobbling, 3 Orchard Close, Cranfield, Bedford, MK43 0HX
MI6 BFJ J. Crozier, 9 Colinbrook Park, Dunmurry, Belfast, BT17 0NZ
M6 BFK A. Eaton, 40A Summer Street, Leighton Buzzard, LU7 1HT
M6 BFL K. Goldsworthy, Flat 2, Jordan House, Biggleswade, SG18 8FS
M6 BFO C. Murphy, 17 Shepherd Street, Littleover, Derby, DE23 6GA
M6 BFP R. Roberts, 25 Carlisle Avenue, Bootle, L30 1PX
M6 BFQ T. Dodd, 6 Victoria Terrace Prinsted Lane, Emsworth, Po108hx
MM6 BFR P. Riddle, Carngeal, Pitlochry, PH16 5JL
M6 BFS N. Grudgings, North Lodge, Templewood Lane, Slough, SL2 3HW
M6 BFU A. Elwin, 30 Kingsland Road, Aylesbury, HP21 9SL
M6 BFV O. Gledhill, 20 Curtis Way, Kesgrave, Kesgrave, IP5 2FX
M6 BFX S. Cave, Weymess Farm, Park Lane, Banbury, OX17 2RX
M6 BFZ N. Edwards, 11 Sandringham Road, Eccleston, Chorley, PR7 5SN
M6 BGA K. Baker, 27 St. Matthews Close, Cherry Willingham, Lincoln, LN3 4LS
M6 BGB A. Ajmani, Apartment 605, 7 Anchor Street, Ipswich, IP3 0BW
M6 BGC J. Moody, 16 Lingcrest, Gateshead, NE9 6SN
MI6 BGD B. Gilliland, 28 Baird Avenue, Donaghcloney, Craigavon, BT66 7LP
M6 BGE S. Saunders, 5 Park Court, Woking, GU22 7NW
M6 BGF P. Martin, 56 Devonshire Gardens, Bursledon, Southampton, SO31 8HE
M6 BGG M. Chiesa, 56 Waddon Road, Croydon, CR0 4JD
M6 BGH B. Higgins, 2 Bishops Yard, High Street, Huntingdon, PE28 3JB
M6 BGJ A. Jones, 22 Standish Road, Sheffield, S5 8xu
M6 BGK A. Yates, 1 Roberts Court, Whitwell, Hitchin, SG4 8AF
M6 BGL K. Oliver, 28 King Richards Hill, Earl Shilton, Leicester, LE9 7EY
M6 BGM P. James, 44 Narbonne Avenue Ellesmere Park Eccles, Manchester, M30

9DL
MW6 BGO D. Lacaman, 6 Falcon Road, Haverfordwest, SA61 2UE
M6 BGP B. Peacock, 17 Herril Ings, Tickhill, Doncaster, DN11 9UE
M6 BGS A. Somerville, 106 Bush Hill, Northampton, NN3 2PG
M6 BGT J. Summerhill, 43 Rangers Walk, Bristol, BS15 3PW
MM6 BGV A. Timmins, 16 Queens Crescent, Garelochhead, Helensburgh, G84 0DW
M6 BGW M. Walters, 110 Slade Road, Portishead, Bristol, BS20 6BB
M6 BGZ T. Chapman, 74 Kidderminster Road, Bewdley, DY12 1BY
M6 BHA T. Harris, 12 Maple Close, Stourport-on-Severn, DY13 8TA
M6 BHB R. Smith, Five Elms, Lullington Road Edingale, Tamworth, B79 9JA
M6 BHC A. Tipler, 27 Clumber Street, Hucknall, Nottingham, NG15 7PJ
MW6 BHD D. Rees, 49 Fair View, Hirwaun, Aberdare, CF44 9SA
M6 BHE S. Verity, 29 Patterdale Avenue, Fleetwood, FY7 8NW
M6 BHF M. Atherton, 4 Bakers Park Saltney, Chester, CH4 8FB
M6 BHH G. Hartless, 32 Long Acre, Mablethorpe, LN12 1JF
M6 BHI Dr R. Smyth, 46 Eagle Way, Abbeydale, Gloucester, GL4 4WS
M6 BHJ B. Hall, 1 Lytham Road, Leicester, LE2 1YD
M6 BHK C. Lai, Storeys Way, Cambridge, CB3 0DG
M6 BHM M. Biadon, 57 Fern Hill Road, Oxford, OX4 2JW
M6 BHN J. Hunter, 35 Inglefield, Hartlepool, TS25 1RN
MW6 BHO A. Edwards, Tir Brwyn, Rhydargaeau, Carmarthen, SA33 6BL
M6 BHP R. Earland, 11 Roseberry Avenue, Great Ayton, Middlesbrough, TS9 6EN
M6 BHQ L. Copeland, 78 Penderyn Crescent Ingleby Barwick, Stockton-on-Tees, TS17 5HQ
MM6 BHS J. Read, Knockenny Farm, Glamis, Forfar, DD8 1UE
M6 BHT T. Scales, 77 Upper Eastern Green Lane, Coventry, CV5 7DA
M6 BHU P. Chamberlain, Mulberry, Main Road, Edenbridge, TN8 6HZ
M6 BHY M. Worley, 12 Hall Farm Close, Melton, Woodbridge, IP12 1RL
M6 BHZ P. Palmer, 2 Dagonet Road, Bromley, BR1 5LR
M6 BIA R. Kijewski, 14 East Street, Heanor, DE75 7NE
M6 BIB N. Foster, Four Beeches, Gribthorpe, Goole, DN14 7NT
M6 BID M. Knights, 3 East View Cottages, Church Road, Woodbridge, IP13 0AT
M6 BIE J. Lucas, 64 Fitzroy Road, Whitstable, CT5 2LE
M6 BIF D. Sobey, Flat 2 73 Park Road, Blackpool, FY1 4JQ
M6 BIG D. Robson, 5 Rutherglen Drive, Hull, HU9 3PF
MM6 BIH G. Fordyce, 2 Church Street, East End, Earlston, TD4 6HS
MI6 BII P. Floyd, 25 Glenside, Omagh, BT79 7GL
M6 BIJ I. Jarvis, 6 Mullett Road, Wednesfield, Wolverhampton, WV11 1DD
M6 BIM H. Salter, 34 Hillside Villas, Millendreath Holiday Village, Millendreath, Millendreath Looe, PL13 1PE
M6 BIN R. Granger, 7 Webster Road, Winchester, SO22 5NT
MM6 BIO N. Mackenzie, 59 Plasterfield, Stornoway, HS1 2UR
MM6 BIP A. Stanley, 40 Coll, Isle of Lewis, HS2 0LP
M6 BIR B. Jakubowski, 120 Chandos Street, Coventry, CV2 4HT
M6 BIS D. Bisbey, 17 Benson Close, Lichfield, WS13 6DA
M6 BIU C. Andrews, 34 Russell Street, Kettering, NN16 0EL
M6 BIV O. Hall, 2 Beverley Lodge, Paradise Road, Richmond, TW9 1LL
M6 BIX J. Coward, 31 Oxford Road, Hyde, SK145QZ
MM6 BIY J. Corrigan, 27 Stonecraig Road, Wishaw, ML2 8BZ
M6 BJA B. Ardrey, 13 Grebe Avenue, St Helens, WA10 3QL
M6 BJB K. Richards, 5 Ramsay Close, Birchwood, Warrington, WA3 6PS
M6 BJC M. Holding, 8 Dunsop Close, Blackpool, FY1 6NP
M6 BJE B. Emmerson, 1 Tivydale Drive, Darton, Barnsley, S75 5PG
M6 BJF B. Froggatt, 11 Goldsmith Road, Walsall, WS3 1DL
MI6 BJG R. Gilmore, 11 Abbots Gardens, Newtownabbey, Belfast, BT379QZ
M6 BJI I. Taylor, 111 Kings Road, Lancing, BN15 8EQ
M6 BJK K. Ashton, 23 Wood Close, Wells, BA5 2GA
M6 BJL M. Blackmore, Timbers, Wolvershill Road, Banwell, BS29 6DG
M6 BJM M. Bunting, 31 Hardwick Avenue, Allestree, Derby, DE22 2LN
M6 BJN B. Potter, 55 Lindsworth Road, Kings Norton, Birmingham, B30 3RP
M6 BJO K. Hawkins, 2 Couford Grove, Huddersfield, HD2 1TH
M6 BJP P. Blackmore, Timbers, Wolvershill Road, Banwell, BS29 6DG
M6 BJQ R. Stokes, 3 Parham Walk, Grange Park, Swindon, SN5 6EQ
M6 BJT B. Claydon, 31 Riverside.Horley.Surrey, Horley, RH67LN
M6 BJW FL B. White, 53 Dacre Road, Brampton, Ca81bn
M6 BJX C. Thorpe-Morgan, 31 Dingledery, Olney, MK46 5ES
M6 BJY M. Walker, 20 Fernhurst Road, Mirfield, WF14 9LJ
MI6 BJZ P. Donnelly, 43 Ashfield Gardens Fintona, Omagh, BT78 2DD
M6 BKA M. Byrnes, 61 Furze Close, Peatmoor, Swindon, SN5 5DB
M6 BKC P. Glasper, 2 Iris Close, Stockton, TS18 1ax
MI6 BKD J. Tierney, 48 Ashfield Gardens Fintona, Omagh, BT78 2DD
M6 BKE C. Opie, 354 Beaumont Road, Plymouth, PL4 9EN
M6 BKF Dr C. Lane, 47 Glenarm Road, London, E5 0LY
M6 BKH D. Storer, 13 The Square, Lower Burraton, Saltash, PL12 4SH
M6 BKI S. Shaw, 3 Wellington Terrace, Littleborough, OL15 9DA
M6 BKJ P. Crow, 31 Lakeside, Overstone Park, Northampton, NN6 0QS
M6 BKK B. Tompkins, 13 Magazine Close, Wisbech, PE13 1LH
M6 BKL K. Jackson, 4 Milfoil Close, Marton-In-Cleveland, Middlesbrough, TS7 8SE
M6 BKM D. Turnbull, 63 Brecklands, Mundford, Thetford, IP26 5EG
M6 BKN S. Walker, 73 Sunnybank Road, Halifax, HX2 8RL
M6 BKO J. Legrain, 17 Route De La Cote, St Laurent Sur Gorre, France, 87310
MM6 BKP R. Adamson, 6 Camdean Crescent, Rosyth, Dunfermline, KY11 2TJ
MW6 BKS N. Adam, Tan Ffordd, Mynydd Llandygai, Bangor, LL57 4LX
M6 BKT S. Keeble, 38 Sandford Rise, Sandy, SG19 1ED
M6 BKU B. Cox, 7 Wolsey Avenue, London, E6 6HG
M6 BKV M. Archer, 4 The Bungalows, Mill Lane, Grays, RM20 4YD
M6 BLA J. Forder, 157 Kennington Road, Kennington, Oxford, OX1 5PE
M6 BLB J. Sell, 3 Powter Close, Elsenham, Cm226ut
M6 BLC B. Cripps, 215 Bournemouth Road, Poole, BH14 9HU
MM6 BLE S. Robertson, 43 Kindar Drive, New Abbey, Dumfries, DG2 8DA
M6 BLF M. Featherstone, 62 Poles Hill, Chesham, HP5 2QR
M6 BLG T. Clayton, 4 Stable Mews, Lords Hill, Coleford, GL16 8BJ
M6 BLH R. Pringle, 14 Marjorie Street, Cramlington, NE23 6XQ
M6 BLI S. Algar, 17 Riseway Close, Norwich, NR1 4NJ
M6 BLK P. Blake, 70 Front Street, South Hetton, Durham, DH6 2RG
M6 BLM K. Gallop, Airedale House, Airedale Drive, Castleford, WF10 2QA
M6 BLN M. Halliday, 1 Everard Close, Bury St Edmunds, IP32 6RU
M6 BLO L. Monshall, 7 Sweden Close, Harwich, CO12 4JU
M6 BLP B. Pearn, 230 Lloyds Avenue, Kessingland, Lowestoft, NR33 7TU
M6 BLQ N. Fahey, 5 Hillside, Felmingham, North Walsham, NR28 0LE
M6 BLS T. Malik, 83A Cleveland Road, Manchester, M8 4GT
M6 BLT W. Thomson, 72 Hurstwood Avenue, Bexleyheath, DA7 6SG
M6 BLV J. Moore, 53 The Boulevard, Great Sutton, Ellesmere Port, CH65 7DX
MI6 BLW E. Macgra, 197 Beraghvale, Skeoge Road, Londonderry, BT48 8UJ
M6 BLX R. Bowles, 7 Vineside, Gosport, PO13 0ZU
MM6 BLY W. Robb, 18 Buckie, Erskine, PA8 6EE
M6 BLZ J. Blezard, 10 North Row, Barrow-in-Furness, LA13 0HE
M6 BMB Dr J. Bell, 6 Highfields, Fetcham, Leatherhead, KT22 9XA
MI6 BMC W. Mccormick, 6 Church Street, Rosslea, Enniskillen, BT92 7DD
MW6 BME P. Toal, 17 Gronant Street, Rhyl, LL18 1PN
M6 BMF A. Ledger, 23 Wentworth Park, Freshbrook, Swindon, SN5 8QX
M6 BMI P. Davies, 2 Lynfords Drive, Runwell, Wickford, SS11 7PP
M6 BMJ B. Adamson, 21 The Hatches, Frimley Green, Camberley, GU16 6HG
M6 BMK P. Hardy, 50 Harridge Road, Leigh-on-Sea, SS9 4HA
MW6 BMM W. Murphy, 148 Caergynydd Road, Waunarlwydd, Swansea, SA5 4RE
M6 BMN D. Hagan, 8 Charles Close, Westcliff-on-Sea, SS0 0EU
M6 BMO B. Morris, 131 Littlehampton Road, Worthing, BN13 1QX
MM6 BMP A. Moerman, 11 Cupar Road, Kettlebridge, Cupar, KY15 7QD
MM6 BMQ J. Bannerman, 77/2 Park Avenue, Edinburgh, EH15 1JP
M6 BMT B. Thistlethwaite, 29 Lamorna Drive, Callington, PL17 7QH
M6 BMV R. Gocher, 20 Mulberry Crescent, South Shields, NE34 8DD
M6 BMW B. Martin, 11 Alpha Street, Toll Bar, Doncaster, DN50RA
M6 BMY S. Maqbool, 69 Waltham Close, West Bridgford, Nottingham, NG2 6LD
M6 BMZ J. Kay, 36 Winnington Road. Marple, Stockport, SK6 6PT
M6 BNA A. Piper, 3 Bakers Court, Bakers Court Lane, Lynton, EX35 6EW
M6 BNB J. Duffy, 19 Lydford Gardens, Bolton, BL2 6TU
M6 BNC B. Chambers, 13 Cherry Tree Crescent, Walton, Wakefield, WF2 6LQ
MW6 BNF S. Morgan, Holly Cottage, Old Racecourse, Oswestry, SY10 7PQ
M6 BNG R. Watson, 8 Bourne Close, Warminster, BA12 9PT
MW6 BNH G. Thomas, 4 Oak Tree Close, Buckley, CH7 3JU
M6 BNJ G. Hartley, 1 Manor View, Shafton, Barnsley, S72 8NQ
M6 BNK M. Wilsher, 70 Norris Road, Blacon, Chester, CH1 5DZ
M6 BNL D. Beck, 94 Shaldon Crescent, Plymouth, PL5 3RB
M6 BNM J. Sanders, 76 Fullerton Road, Plymouth, PL2 3AX
MM6 BNN O. Mckenzie, 179 Gordons Mills Road, Aberdeen, AB24 2XS
MM6 BNO K. Mckenzie, 179 Gordons Mills Road, Aberdeen, AB24 2XS
M6 BNP D. Brookes, 177 Charnwood Close, Rubery, Birmingham, B45 0JY
M6 BNQ C. Marshall, 51 Hedgerow Close, Redditch, B98 7QF
MM6 BNS Prof. T. Donaldson, Brawview Cottage, Maybole, KA19 8EN
MM6 BNT L. Waller, 1 Anne Arundel Court, Heathhall, Dumfries, DG1 3SL
M6 BNU S. Ditchburn, 18 St. Hilda Avenue, Waterlooville, PO8 0JF
M6 BNV I. Balboa, 7 Edenhall Close, Tilehurst, Reading, RG31 6RR
MM6 BNX W. Evans, Ninewar Farm, Duns, TD11 3PP
M6 BNY J. Beeney, 17 Norton Avenue, Herne Bay, CT6 7TA
M6 BOA T. Cocks, 9 Mountfield Way, Westgate-on-Sea, CT8 8HR
M6 BOB R. Manning, 15 Gurston Rise, Northampton, NN3 5HY
MW6 BOC G. Williams, 36 Park Street, Taibach, Port Talbot, SA13 1TD
MM6 BOD D. Lightbody, Glenorchy, Brownrigg Road, Falkirk, FK1 3BA
M6 BOE N. Marley, Penstemons, Chapel Lane Pen Selwood, Wincanton, BA9 8LY
M6 BOF B. Savage, Rufford, Barnes Lane, Milford On Sea, Lymington, SO41 0RR
M6 BOI K. Mecca, 38 Abbots Road, Faversham, ME13 8DE
M6 BOK V. Casambros, 27 Oaks Road, Folkestone, CT20 3JY
M6 BOP D. Holland, 7 Hayward Close, Walkington, Beverley, HU17 8YB
MM6 BOQ L. Mitchell, Carradale, Yondertonhill Hatton, Aberdeenshire, AB42 0RE
M6 BOQ A. Hill, 2 Liddon Road, Chalgrove, Oxford, OX44 7YH
M6 BOR G. Wren, Ashleigh, Yew Tree Hill, Holloway, Matlock, DE4 5AR
MM6 BOS A. Manson, Clochcan Schoolhouse, Ellon, AB41 8UJ
M6 BOT T. Gedvygas, 9 The Mill, Kirton, Boston, PE20 1LB
MW6 BOW T. Bowen, 7A Heol Maes Y Cerrig, Loughor, Swansea, SA4 6SW
M6 BOX C. Pegrum, 3 Bretland Road, Tunbridge Wells, TN4 8PS
M6 BOY C. Butler, 210 Green Wrythe Lane, Carshalton, SM5 2SP
MW6 BOZ E. Bosley, Glandwr, Snowdon Street, Porthmadog, LL49 9DF
M6 BPA A. Parnell-Brookes, 14 Greens Close, Hullavington, Chippenham, SN14 6EG
M6 BPB K. Furlong, 39 Britannia Avenue, Dartmouth, TQ6 9JT
M6 BPC G. Colson, 3 Dartford Road, Dartford, DA1 3EE
M6 BPD D. Taylor, 143 Sandhurst Road, London, SE6 1UR
M6 BPE P. Skinner, 27 Westcots Drive, Winkleigh, EX19 8JW
M6 BPG B. Patrick-Gleed, 6 Julius Way, Lydney, GL15 5QS
MW6 BPH B. Parsons, 29 Hillside Crescent, Buckley, CH7 2JS
M6 BPI W. Colquhoun, 92 Byfield Road, Woodford Halse, Daventry, NN11 3QS
M6 BPK J. Johnson, 4 Wallace Close, Hullbridge, Hockley, SS5 6NE
M6 BPL M. Lefton, 35 Hawkstone Avenue, Whitefield, Manchester, M45 7PR
MM6 BPM I. Wightman, 13 Gillsland, Eyemouth, TD14 5JF
M6 BPO R. Extrance, 1 Morngate Caravan Park, Bridport Road, Dorchester, DT2 9DS
M6 BPQ J. Chatterton, 6 Bayliss Road, Wargrave, Reading, RG10 8DR
M6 BPR K. Butt, 15 Hamble Park, Fleet End Road, Southampton, SO31 9JU
M6 BPS P. Dent, 25 Clyde Avenue, Hebburn, NE31 2JN
M6 BPU C. Shepherd, 1 Holley Park, Okehampton, EX20 1PL
M6 BPV J. Mcrobie, 6 Southill Gardens, Bournemouth, BH9 1SJ
M6 BPW B. Wang, Harrogate Ladies' College, Clarence Drive, Harrogate, HG1 2QG
M6 BPX J. Revell, 37 Tennyson Street, Goole, DN14 6EB
MM6 BPY L. Joseph, 3/4 60 Wilson Street, Glasgow, G1 1HD
MW6 BQA R. Staples, 16 Wellington Street, Aberdare, CF44 8EW
M6 BQB R. Babb, 15 Sylvan Lane, Hamble, Southampton, SO31 4QG
M6 BQC M. Powis, 3 Greenacres, Ludlow, SY8 1LU
M6 BQD H. Brewster, 1 Pinewood Drive, Camblesforth, Selby, YO8 8JU
M6 BQE M. Simpson, 32 Underhill Lane, Wolverhampton, WV10 8NS
M6 BQF G. Evans, 13 Lydgate Road, Sale, M33 3LW
MM6 BQG B. Wetton, Tigh Air Achnoc, West Helmsdale, Helmsdale, KW8 6HH
M6 BQJ R. Thompson, 16 West Leys Court, Moulton, Northampton, NN3 7UB
M6 BQK C. Holmes, 10 Southampton Street, Farnborough, GU14 6AX
MW6 BQL Dr J. Griffiths, Llain Bach, Beach Road, Porthmadog, LL49 9YA
MW6 BQM K. Williamson, 1 Rhug Gardens, Corwen, LL21 0EH
M6 BQN T. Gilmour, 83 Billington Road, Leighton Buzzard, LU7 4TG
M6 BQO W. Martin, 54 Merritt Road, Greatstone, New Romney, TN28 8SZ
M6 BQP J. Griffiths, 7 Tynedale Road, Blackpool, FY3 7UE

M6 BQQ S. Murray, 79 Nightingale Road, Liverpool, L12 0QN
M6 BQR T. Walton, 72 Burleigh Road, Frimley, Camberley, GU16 7EB
M6 BQS R. Brabazon, 10 St. Dominic Close, St Leonards-on-Sea, TN38 0PH
M6 BQV A. Brackpool, 79 St Leonards Road, Horsham, RH13 6EH
M6 BQW K. Jeffery, 9 Gordon Road, Tunbridge Wells, TN4 9BL
M6 BQZ M. Iqbal, 22 Rupert Avenue, High Wycombe, HP12 3NG
M6 BRB L. Bethell, 30 Finger Road, Dawley, Telford, TF4 3LB
M6 BRH S. Plummer, 2 Langdale Avenue, Outwood, Wakefield, WF1 3TX
M6 BRJ R. Green, 44 Aldwyn Place, Larchwood Drive, Egham, TW20 0RZ
M6 BRK A. Freshwater, 90A New Road, Minster On Sea, Sheerness, ME12 3PT
M6 BRL M. Burrell, 16 Atholl, Ouston, Chester le Street, dh2 1rs
M6 BRN M. Brundrit, 144 Reginald Road, Southsea, PO4 9HP
M6 BRO J. Smith, 38 The Vineries, Burgess Hill, RH15 0NF
M6 BRP S. Fiske, 12 Meliden Crescent, Bolton, BL1 6AJ
M6 BRQ A. Clements, 28 Durham Way Wyton, Huntingdon, PE28 2EQ
M6 BRR S. Ray, 28 Stenbury View, Wroxall, Ventnor, PO38 3DB
M6 BRS D. Glover, 11 Collbrook Avenue, Odsal, Bradford, BD6 1HL
M6 BRT A. Williams, 101 Horsebridge Hill, Newport, PO30 5TL
M6 BRU A. Dean, 14 Harvest Close, Worsbrough, Barnsley, S70 5AY
MM6 BRV D. Drysder, 37 Farburn Drive, Stonehaven, AB39 2BZ
M6 BRY B. Allen, 13 Woodgrove Road, Rotherham, S65 3RW
M6 BRZ B. Ager, 6 South Dibberford Farm, Beaminster, DT8 3HD
M6 BSA C. Jacobs, Flat 33, The Lodge, Lavender Road, Waterlooville, PO7 8BX
M6 BSB S. Hamill, 3 Pear Tree Croft, Brede, Rye, TN31 6EJ
M6 BSC B. Sewell, 12 Haylands Square, South Shields, NE34 0JB
M6 BSF A. Rizzo, 12 North Street, Wicken, Ely, CB7 5XW
M6 BSG B. Garton, 13 Damaskfield Road, Lyppard, Kettleby, WR4 0HY
M6 BSI A. Butler, 50 Scafell Way, West Bromwich, B71 1DQ
MM6 BSK I. Bannerman, 54 Broadleys Avenue, Bishopbriggs, Glasgow, G64 3AQ
M6 BSL T. Bilsel, 16 Hide Close, Sawston, Cambridge, CB22 3UR
M6 BSO B. Scott, 6 Congleton Edge Road, Congleton, CW12 3JJ
M6 BSP B. Smith, 8 Mill Field Close, South Kilworth, Lutterworth, LE17 6FE
M6 BSQ P. Bull, 87 Braemor Road, Calne, SN11 9DU
MW6 BSR B. Sturgess, 22 Heol Pant Y Deri, Cardiff, CF5 5PL
M6 BSS J. Chambers, 10 Derwent Street, Astley, Manchester, M29 7AT
M6 BST T. Hughes, 1 Sunnybank Road, Astley, Manchester, M29 7BJ
MW6 BSU A. Buckley, Cliff House, Sudbrook, Caldicot, NP26 5TB
M6 BSV C. Hayes, 7 Hadstock Close, Sandiacre, Nottingham, NG10 5LQ
M6 BSW B. Stone, 27 Mountbatten Close, Stretton, Burton-on-Trent, DE13 0FD
M6 BSX J. Cummins, 55 Rowley Street, Walsall, WS1 2AZ
M6 BSY D. Carter, 36 Sanderling Drive, Leigh, WN7 1HU
M6 BSZ C. Batchelor, 8 Howarde Court, Stevenage, SG1 3DF
M6 BTA L. Brown, 28 Farley Way, Stockport, SK5 6JD
M6 BTB R. Dalley, Birchley, Pine Avenue, Camberley, GU15 2LY
MW6 BTC V. Kennedy, Gerynant, Felin Ban Farm Estate, Cardigan, SA43 1PG
M6 BTE B. Ellerton, 4 Darfield Avenue, Owlthorpe, Sheffield, S20 6SU
M6 BTF P. Likitplug, Harrogate Ladies' College, Clarence Drive, Harrogate, HG1 2QG
M6 BTG M. Mutkin, 13 The Grove, Radlett, WD7 7NF
M6 BTH D. D'Souza, 3B Friend Street, London, EC1V 7NS
M6 BTK J. Barton, 8 Melton Lane, Sutton Bonington, Loughborough, LE12 5RQ
M6 BTM B. Gillett, 65 Kingsway, Wallasey, CH45 4PN
M6 BTN R. Nicholson, 57 Barnbridge, Tamworth, B77 1DF
M6 BTP P. Wilmot, 12 Brierholme Close, Hatfield, Doncaster, DN7 6EH
M6 BTQ C. Manning, 12 Whitehill Close, Camberley, GU15 4JR
MM6 BTR J. Rayne, 8 Bankton Grove, Livingston, EH54 9DW
M6 BTS B. Seward, 21 Chapel Close, Gunnislake, PL18 9JB
M6 BTT P. Moye, 13 Post Mill Gardens, Grundisburgh, Woodbridge, IP13 6UP
M6 BTY P. Matthews, The Old Chapel, High Street, Huntingdon, PE28 0PF
M6 BUA H. Bennett, 59 Scott Road, Bishop's Stortford, CM23 3QN
M6 BUB R. Hughes, 86 Colinmander Gardens, Ormskirk, L39 4TF
M6 BUC J. Chambers, 2 Farm Cottages, Magdalen Laver, Ongar, CM5 0ES
M6 BUD J. Woodland, 14 Kelham Green, Nottingham, NG3 2LP
M6 BUE J. Shepherd, 169 Northbrooks, Harlow, CM19 4DQ
MW6 BUF J. Hurst, 1 Castle Cottage, Aberedw, Builth Wells, LD2 3UL
MM6 BUG J. Millar, The Kennels, Lanton, Jedburgh, TD8 6SU
MW6 BUH I. Baylis, 248 Trebanog Road, Porth, CF39 9EL
M6 BUI D. Sewell, 19 St. Leonards Way, Ashley Heath, Ringwood, BH24 2HS
M6 BUJ G. Sheppard, 32 Bramble Drive, Hailsham, BN27 3EG
M6 BUL J. Ball, Conifers, Main Road, Spilsby, PE23 4BY
M6 BUR T. Skinner, 13 Sawbrook, Fleckney, Leicester, LE8 8TR
M6 BUU Dr D. Barrow, 48 Main Street, Greysouthen, Cockermouth, CA13 0UL
M6 BUW S. Bishton, 18 Galloway Road, Poole, BH15 4JX
M6 BUX P. Kerr, 35 Coppice Gardens, Stone, ST15 8BL
M6 BUY C. Marlow, 59 Purford Green, Harlow, CM18 6HN
M6 BVC C. Norwood, 94 Waverley, Woodside, Telford, TF7 5LU
M6 BVD R. Owen, 4 Aldersleigh Drive, Stafford, ST17 4RY
M6 BVE A. Hamilton, 56 Wyvern, Telford, TF7 5QH
MW6 BVG J. Campbell, 1B Bush Road, Mountain Ash, CF45 3BY
M6 BVK G. Kennedy-Brown, 4 Seymour Gardens, Brockley, London, SE4 2DN
M6 BVM S. Baynton, 50 Briton Way, Wymondham, NR18 0TT
MI6 BVN P. Mccullagh, 6 Striff Lane, Omagh, BT79 0WA
M6 BVP B. Hull, 3 Cavalry Crescent, Eastbourne, BN20 8NT
MW6 BVQ R. Unsworth, 20 Arran Drive, Rhyl, LL18 2NS
MW6 BVR J. Cowles, 76 Pendine Close, Barry, CF62 9DE
M6 BVS B. Watkins, 41 Kingshill Avenue, St Albans, AL4 9QH
MW6 BVU M. Ryall, 27 Clos Afon Twyi, Blackwood, NP12 3FX
M6 BVV Dr J. Morgan, Cedars, Springhill, Longworth, Abingdon, OX13 5HL
M6 BVW D. Edwards, 29 Larch Road, Maltby, Rotherham, S66 8AZ
M6 BVX A. Booth, 32 Acacia Avenue, Maltby, Rotherham, S66 8DS
M6 BVZ R. Ede, 14 Elm Close, Kidsgrove, Stoke-on-Trent, ST7 4HR
M6 BWB W. Beston, 79 Priestlands, Romsey, SO51 8FJ
M6 BWC W. Tunstall, 89 Lever Street, Little Lever, Bolton, BL3 1BA
M6 BWE G. Brown, 51 Arncliffe Drive, Knottingley, WF11 8RH
M6 BWF R. Leverington, 130 Osborn Road, Barton-Le-Clay, Bedford, MK45 4NY
MW6 BWG A. Madden, 51 Valley View, Cwmtillery, Abertillery, NP13 1JE
M6 BWH Dr V. Venugopalan, 6 Kennington Oval, Stoke-on-Trent, ST4 8FX
M6 BWJ L. Wale, 4 Essex Gardens, Market Harborough, LE16 9JS
M6 BWK S. Casey, Flat 4, Belmont Court, Plymouth, PL3 4DN
M6 BWL Z. Burningham, 255 Welland Park Road, Market Harborough, LE16 9DP
M6 BWN N. Bown, 14 Parsons Mead, Abingdon, OX14 1LS
M6 BWO M. Walters, 65 Bannawell Street, Tavistock, PL19 0DP
M6 BWP P. Walsh, 181 Hermes Close, Hull, HU9 4DR
M6 BWQ A. Carden, Hazelgrove, South Allington, Kingsbridge, TQ7 2NB
MW6 BWR P. Smith, 19 Grandison Street, Neath, SA11 2PG
M6 BWV A. Forber, 32 Larch Avenue, Newton-le-Willows, WA12 8JF
M6 BWW S. Laurence, 23 West Street, Bridlington, YO15 3DX
M6 BWZ S. Thirlwall, 2 Crossfield Avenue Blythe Bridge, Stoke-on-Trent, ST11 9PL
M6 BXA J. Lambert, Birchwood Norwich Road, Cromer, NR27 0HG
M6 BXB M. Stopper, 29 Daisy Dale, Boston, PE21 6DS
M6 BXD P. Hargreaves, 9 Croston Road, Lostock Hall, Preston, PR5 5LA
MM6 BXF D. Hook, The Delvine, Amisfield, Dumfries, DG1 3LH
MM6 BXH J. Curran, 355D Charleston Drive, Dundee, DD2 4HP
MW6 BXI R. Briant, Talarvor, Llanon, SY23 5HG
M6 BXJ M. Stillman, 58 Highfield Road, Bognor Regis, PO22 8PH
M6 BXK J. Street, 22 Roman Acre, Wick, Littlehampton, BN17 7HN
M6 BXL J. Lugsden, 21 Overhill Way, Beckenham, BR3 6SN
M6 BXM H. Phillips, Flat 10, Fitch Court, 59-63 Effra Road, London, SW2 1DD
M6 BXN C. Barthel, 176 Lumb Lane, Droylsden, Manchester, M43 7LJ
M6 BXO K. Cartwright, 53 Sedgley Road, Dudley, DY1 4NE
M6 BXP B. Pearson, 9 Dunbar Close, Kidderminster, DY10 3XS
MM6 BXQ A. Main, 1 Waulkmill Loan, Currie, EH14 5SS
MM6 BXR D. Reid, 111 Oswald Road, Ayr, KA8 8NX
M6 BXS A. Parker, 9 Milecastle Court, Newcastle upon Tyne, NE5 2PA
MM6 BXT B. Templeton, 43 Elm Park, Ardrossan, KA22 7BZ
M6 BXU B. Xu, 138 King'S College, Cambridge, CB2 1ST
MW6 BXV C. Astbury, Old Police House, Llanegryn, Tywyn, LL36 9SL
M6 BXW K. Morris, 95 Murrayfield Drive, Wirral, CH46 3RR
M6 BXX J. Baker, 29 Ravensmoor Close, North Hykeham, Lincoln, LN6 9AZ
M6 BXY P. Arnold, 20 Upper Seagry, Chippenham, SN15 5EX
M6 BXZ J. Walczak, 18 Heathfield, Chippenham, SN15 1BQ
M6 BYA K. Stowe, 7 Glebe Way, Corsham, SN13 9UL
M6 BYE T. Byers, 1 Hazelwood Avenue, Sunderland, SR5 5AH
M6 BYF D. Rogers, 5 Semple Gardens, Chatham, ME4 6QD
M6 BYH K. Broscomb, Flat 34, White Willows, 70 Dyche Road, Sheffield, S8 8DS
MM6 BYJ M. Stypka, 28/9 Halmyre Street, Edinburgh, EH6 8QD
M6 BYL W. Charlton, 2 Mallory Street, Earl Shilton, Leicester, LE9 7PH
M6 BYM T. Flynn, 15 Dale Garth, Scarborough, YO12 5NB
M6 BYN R. Jones, 62 Nathaniel Road, Long Eaton, Long Eaton, NG10 1GB
M6 BYO S. Downe, 9 Danesway, Exeter, EX4 9ES
M6 BYP P. Hoyle, Flat 3, 84 Beverley Road, Hull, HU3 1YD
M6 BYQ M. Abid, 16 Cliff Gardens, Scunthorpe, DN15 7PJ
M6 BYR J. Byrne, 316 Turncroft Lane, Stockport, SK1 4BP
M6 BYS A. Marriott, 11 Downing Street, South Normanton, DE55 2HE
MW6 BYT G. Pierce, Ley Farm, Green Lane, Halton, Wrexham, LL14 5BG
M6 BYU G. Webster, 15 Bridge Road, Chichester, PO19 7NW
MW6 BYV R. Potts, 4 Maes Glan, Rhosllanerchrugog, Wrexham, LL14 2DT
MW6 BYX N. Williams, 60 Denbigh Close, Wrexham, LL12 7TW
M6 BYY S. Bradley, 9 Crofton Road Crofton Road, Southsea, PO4 8NX
M6 BYZ G. Flinn, 38 Fir Grove, Whitehill, Bordon, GU35 9ED
M6 BZA D. Rudling, Rose Cottage, Ludwells Lane, Southampton, SO32 2NP
M6 BZB Y. Zhan, 65A Mason Street, Edge Hill, Liverpool, L7 3EN
MI6 BZC R. Benko, 23 Six Mile Water Mill Drive, Antrim, BT41 4FG
M6 BZD P. Elsey, 62B Coleraine Road, London, SE3 7PE
M6 BZE B. Evelyn, 19 Laudsdale Road, Rotherham, S65 3LG
M6 BZF K. Cunningham, 18 Dovenby Fold, Ince, Wigan, WN2 2PS
M6 BZG R. Paddock, 1 Harris Road, Bexleyheath, DA7 4QD
M6 BZH R. Packman, 62 Amsterdam Road, London, E14 3JB
MI6 BZI C. Mccormick, Flat 4, Legacorry House, Main Street, Armagh, BT61 9RW
M6 BZK R. Boyes, 63 Larch Road, New Ollerton, Newark, NG22 9SX
M6 BZM K. Slater, 63 29Th Avenue, Hull, HU6 8DG
M6 BZN Capt. G. Slater, 63 29Th Avenue, Hull, HU6 8DG
MW6 BZP M. Clifford, 27 Primrose Street, Tonypandy, CF40 1BW
MM6 BZQ Rev. A. Catterall, Asta House, Scalloway, Shetland, ZE1 0UQ
M6 BZT A. Smith, 305, 1414 5Th St Sw, Calgary, Canada, T2R 0Y8
M6 BZU E. St Quinton, Mill Cottage, The Thorofare, Woodbridge, IP13 8BB
M6 BZV S. Barker, Strait Hey Farm, Stock Hey Lane, Todmorden, OL14 6HB
M6 BZW F. Phillips, 34 Park Drive, Maldon, CM9 5JQ
MW6 BZX J. Palmer, Clettwr Hall, Pontshaen, Llandysul, SA44 4TU
M6 BZY I. Kazlauskaite, 4 Spencer Way, Redhill, RH1 5LY
M6 CAA C. Sparrey, 166 Abberley Avenue, Stourport-on-Severn, DY13 0LT
MI6 CAD A. Cahalan, 131 The Meadows, Randalstown, Antrim, BT41 2JD
MW6 CAE C. Catley, 40 Rockvilla Close, Varteg, Pontypool, NP4 7QF
M6 CAF C. Furlong, 22 Swaisland Road, Dartford, DA1 3DE
M6 CAG M. Sinclaire, 21 Ford Court, Winsford, CW7 1NJ
M6 CAH B. Heath, 108 Cow Lane, Bramcote, Nottingham, NG9 3BB
M6 CAI C. Ingamells, 2 St. Mary'S Drive, Sutterton, Boston, PE20 2LU
M6 CAJ P. Lewickyj, 37 Maple Street, Lincoln, LN5 8QS
MI6 CAK C. Doyle, 13 Tobin Park, Cookstown, BT80 0JL
MW6 CAN C. Davies, 3 Penhydd Houses, Oakwood Avenue, Pontrhydyfen, Port Talbot, SA12 9SE
M6 CAP C. Preece, 14 Dock Street, Widnes, WA8 0QX
M6 CAQ C. Allen, 29 Cedar Road, Hythe, Southampton, SO45 3PH
M6 CAR C. Taylor, 212 Plantation Hill, Worksop, S81 0HD
M6 CAS S. Hill, 1 Beresford Road, Walsall, WS3 1JX
M6 CAT D. Welford, 24 Hawthorn Crescent Quarrington Hill, Durham, DH6 4QW
MI6 CAV C. Mccammick, 23 Atkinson Avenue, Portadown, Craigavon, BT62 3HY
M6 CAW C. Waldron, 2 The Bourne, Eastleach, Cirencester, GL7 3NN
MI6 CAY M. Mullaney, 21 Aghagay Meadows, Aghagay, Enniskillen, BT92 8AE
M6 CBA T. Ingleby, 27 Burley Wood Lane, Leeds, LS4 2SU
M6 CBC C. Kent, 19 Coppice Rise, Harrogate, HG1 2DP
M6 CBD C. Arblaster, 22 Wood Lane, Carlton, Barnsley, S71 3JJ
MM6 CBE C. Ellison, 1 Newton Road, St. Fergus, Peterhead, AB42 3DD
MI6 CBG C. Gorman, 5 Linden Gardens, Bangor, BT19 6EB
M6 CBH C. Hunt, 105 Worlds End Lane, Weston Turville, Aylesbury, HP22 5RX
MM6 CBH C. Smith, 3 Distillery Cottages, Aberfeldy, PH15 2EB
MM6 CBI C. Addison, 8D Thomson Street, Johnstone, PA5 8RZ

M6 CBJ C. Jerome, Kennel Cottage, Hobland Road, Great Yarmouth, NR31 9AR
MW6 CBL S. Morgan, 31 Church Street, Briton Ferry, Neath, SA11 2JG
M6 CBM L. Clark, 18 Cutters Close, Narborough, Leicester, LE19 2FY
M6 CBN C. Nunnen, 16 Meden Avenue, Warsop, Mansfield, NG20 0PS
M6 CBO A. Carter, 18 Maynard Terrace, Clutton, Bristol, BS39 5PL
M6 CBP S. Bennett, 154 Humberstone Road, Grimsby, DN32 8HR
M6 CBQ A. Cranston, 196 West End Costessey, Norwich, NR8 5AW
M6 CBU W. Porter, 92 Turner Road, Tonbridge, TN10 4AJ
M6 CBY B. Eames, 22 Ashgrove Close, Hardwicke, Gloucester, GL2 4RT
M6 CCA M. Landon, 29 Portland Road, Hucknall, Nottingham, NG15 7SL
M6 CCB G. Winchester, 36 Crofters Green, Preston, PR1 7UG
M6 CCE C. Etchells, 7 Woodlands Drive, Sandford, Wareham, BH20 7QA
M6 CCF C. Finbow, Medlars Cottage, The Street, Woodbridge, IP13 7JP
MW6 CCG S. Williams, 1 Lawrence Terrace, Llanelli, SA15 1SW
M6 CCH A. Fisher, 63 Soloman Drive, Bideford, EX39 5XY
M6 CCI F. Roberts, 17 Ansisters Road, Ferring, Worthing, BN12 5JG
M6 CCJ J. Davidge, 45 Ferring Street, Ferring, Worthing, BN12 5JW
M6 CCM D. James, 123 Bruce Road, Woodley, Reading, RG5 3DY
M6 CCN M. Tucker, 182 Salisbury Road, Amesbury, Salisbury, SP4 7HW
MM6 CCO W. Hannah, 71 Herriot Avenue, Kilbirnie, KA25 7JB
M6 CCR A. Bhakoo, 4 Bryden Cottages, High Street, Uxbridge, UB8 2NY
MM6 CCS C. Stewart, 7 Hawkhill Place, Stevenston, KA20 4HN
MI6 CCU N. Morrow, 90 Bracky Road, Sixmilecross, Omagh, BT79 9PH
M6 CCV P. Marshall, 21 Lymn Avenue, Gedling, Nottingham, NG4 4EA
MM6 CCW J. Ransome, Evanton, Novar, Dingwall, IV16 9XH
M6 CCX M. Perkins, 2 Buckingham Orchard, Chudleigh Knighton, Chudleigh, Newton Abbot, TQ13 0EW
MM6 CCY A. Douglas, 163 Glenlora Drive, Glasgow, G53 6BI
M6 CDB A. Welch, 18 Monk Close, Tipton, DY4 7TP
M6 CDC C. Chalmers, 2 Canterbury Road, Bracebridge Heath, Lincoln, LN4 2TD
M6 CDD M. Wollaston, 5 Clover Court, Shardlow, Derby, DE72 2GE
M6 CDE G. Leedham Hawkes, 103 Wood Lane, Hednesford, Cannock, WS12 1BW
M6 CDG D. Chang, Montefiore House, Wessex Lane, Southampton, SO18 2NU
M6 CDH C. Hailstone, 2 Thornfield Avenue, Thornton-Cleveleys, FY5 5BH
M6 CDK J. Rowe, 22 Treaty Road Glenfield, Leicester, LE3 8LU
M6 CDL R. Winstanley, 173 Wimborne Road, Poole, BH15 2EF
M6 CDN D. Watkiss, 54 Rolston Close, Plymouth, PL6 6TN
M6 CDQ A. Camp, 3 Acre Close, Witnesham, Ipswich, IP6 9EU
M6 CDR M. Mason, 7 Langland Close, Malvern, WR14 2UY
M6 CDT J. Schleswick, 9 Wick House Close Saltford, Bristol, BS31 3BZ
M6 CDU E. Mortimer, 6 Lanes End, Gastard, Corsham, SN13 9QS
MW6 CDV K. House, Liddington, Dehewydd Lane Llantwit Fardre, Pontypridd, CF38 2EN
M6 CDW C. Wade, 31 Melton Green, Wath-Upon-Dearne, Rotherham, S63 6AA
MW6 CDZ G. Saltmarsh, 15 Colbourne Road Beddau, Pontypridd, CF38 2LN
M6 CEA M. Anderson, 27 Laing Road, Colchester, CO4 3UT
M6 CEB M. Bamber, 10 Sedgeley Mews, Freckleton, Preston, PR4 1PT
M6 CEE T. Massey, 2 Cannon Heath Farm Cottages, Cannon Heath, Basingstoke, RG25 3EJ
M6 CEF C. Davies, 1 Meadow Close, Bagworth, Coalville, LE67 1BR
MW6 CEG J. Martin, 2 Pant Heulog, Dyffryn Ardudwy, LL44 2BU
M6 CEH C. Halloway, 82 Northwall Road, Deal, CT14 6PP
M6 CEI K. Knapp, 46 Robin Hood Road, St. Johns, Woking, GU21 8SY
MI6 CEJ C. Jeffery, 197 Finvola Park, Dungiven, Londonderry, BT47 4ST
MM6 CEL R. Wood, 42 Moorhouse Avenue, Paisley, PA2 9NY
MM6 CEM C. Macdonald, 2 Porterfield Road, Inverness, IV2 3HW
M6 CEN R. Bird, 431 Ombersley Road, Worcester, WR3 7DQ
M6 CEO A. Humphries, 3 Suffolk Drive, Worcester, WR3 8QT
M6 CEP C. Pritchard, 8 Hoon Avenue, Newcastle, ST5 9NY
M6 CEQ D. Brown, 2 Kibworth Grove, Stoke-on-Trent, ST1 5QP
MM6 CES C. Canning, 17 Blackadder Crescent Greenlaw, Duns, TD10 6XN
M6 CET M. Knowles, 86 West Shore Road, Walney, Barrow-in-Furness, LA14 3UD
M6 CEU A. Burkitt, 53 Westwood Drive, Bourne, PE10 9PY
MM6 CEW C. Watson, 20 Norlands, Errol, Perth, PH2 7QU
MW6 CEX M. Brittle, 11 Haig Place, Gendros, Swansea, SA5 8BT
M6 CEY M. Green, 26 Drake Crescent, Kidderminster, DY11 6EE
M6 CEZ I. Hill, 3 Beresford Rd, Walsall, ws31jx
MM6 CFA S. Mackenzie, 20 Meadowhouse Road, Edinburgh, EH12 7HP
M6 CFD J. Seaton, 3 Tunstall Street, Middlesbrough, TS3 6PE
M6 CFE D. Cowdrey, 264 Stamford Road, Brierley Hill, DY5 2QF
MM6 CFH C. Fraser-Hopewell, 2/1 70 Albert Road, Glasgow, G42 8DW
M6 CFI K. Davis, 26 Mendip Drive, Nuneaton, CV10 8PT
M6 CFL M. Shortall, 16 Darnaway Close, Birchwood, Warrington, WA3 6TR
M6 CFN N. Layland, 3 Thirlmere Road, Golborne, Warrington, WA3 3HH
M6 CFO M. Summers, 21 Quantock Avenue, Caversham, Reading, RG4 6PY
M6 CFQ R. Smith, 15 Hollybush Road, North Walsham, NR28 9XT
M6 CFR J. Mather, 33A Forest Road, Southport, PR8 6JD
MM6 CFS F. Sturrock, 16 Carlyle Crescent, Buckhaven, Leven, KY8 1DW
M6 CFT C. Mole, 6 Clements Road, Chorleywood, Rickmansworth, WD3 5JT
M6 CFU S. Watson, 8 Church Close, Overstrand, Cromer, NR27 0NY
M6 CFV A. Stride, 2 Bailey Close, Pewsey, SN9 5HU
M6 CFW A. Comerford, 21 New Cross Road, Headington, Oxford, OX3 8LP
M6 CFX M. Lawrence, 61 Jack Warren Green, Cambridge, CB5 8US
M6 CFY A. Collison, Bramble Lodge, Walkford Lane, New Milton, BH25 5NL
M6 CFZ J. Gilpin, 17 Roundstone Crescent, East Preston, Littlehampton, BN16 1DG
M6 CGB C. Bradley, 3 Trecarne Gardens, Delabole, PL33 9DP
M6 CGC C. Cheadle, 55 Ellison Street, Sheffield, S3 7JH
M6 CGF S. Gateson, Flat 6, Kenley House, Croydon, CR0 6AQ
MW6 CGH M. Jones, 102 Thomas Street, Tonypandy, CF40 2AH
M6 CGJ R. Wheeler, 12 Drake Avenue, Didcot, OX11 0AD
M6 CGK Dr D. Richardson, 89A Bean Oak Road, Wokingham, RG40 1RJ
M6 CGL G. Livesey, 48 Kingsway, Leyland, PR25 1BL
M6 CGM C. Martin, 63 Oversetts Road, Newhall, Swadlincote, DE11 0SL
MI6 CGQ C. Glenn, 40 Deanfield, Londonderry, BT47 6HY
M6 CGS C. Smith, 17 Sunningdale Avenue, Sale, M33 2PJ
MW6 CGV C. Briant, Talarvor, Llanon, SY23 5HG
M6 CGW C. Watkins, 25 Citadilla Close, Gatherley Road, Richmond, DL10 7JE
M6 CGX R. Coy, 32 Windmill Way, Kirton Lindsey, Gainsborough, DN21 4FE
M6 CGY A. Lycett, 2 Royce Avenue, Hucknall, Nottingham, NG15 6FU
M6 CGZ A. Dalgliesh, 62 Newbury Street, Wantage, OX12 8DF
M6 CHC M. Corrigan, 33 Westbourne Road, Knott End-On-Sea, Poulton-le-Fylde, FY6 0BS
M6 CHD R. Kergozou, Lilac Cottage, The Street, Cheddar, BS27 3TH
M6 CHF Dr C. Ferguson, Royd Moor, Royd Moor Lane, Badsworth, Pontefract, WF9 1AZ
M6 CHG L. Hartley, 206 Latimer Road, Eastbourne, BN22 7JF
M6 CHH J. Wishart, 19 Chepstow Close, Chippenham, SN14 0XP
M6 CHI M. Boon, 45 Carlton Road, Wickford, SS11 7ND
M6 CHJ J. Chaplin-Madden, 96 Salisbury Road, Great Yarmouth, NR30 4LS
M6 CHL D. Cobbold, 65 St. Olaves Road, Bury St Edmunds, IP32 6RR
MM6 CHM R. Russell, 4 Valley Court, Patna, Ayr, KA6 7LQ
MM6 CHN C. Denny, Darnaway, Castleton Place, Braemar, Ballater, AB35 5ZQ
M6 CHP M. Cooper, 86 Grove Road, Tiptree, Colchester, CO5 0JG
MW6 CHQ A. Bowlzer, 3 Moreia Terrace, Harlech, LL46 2YW
M6 CHS M. Drury, 19 Cuffley Avenue, Watford, WD25 9RB
M6 CHT A. Trowse, 110 New Road, Hethersett, Norwich, NR9 3HQ
M6 CHU R. Last, 30 Abbot Road, Bury St Edmunds, IP33 3UB
MM6 CHV R. Gilchrist, 5 Glenburn, Leven, KY8 5BD
M6 CHW C. Wright, 14 Orchard Close, Poughill, Bude, EX23 9ES
M6 CHX M. Harrison, 2A Arundel Road, Camberley, GU15 1DL
MM6 CHY C. Hay, 30 Kincardine Place, East Kilbride, Glasgow, G74 3DN
M6 CHZ H. Eager, 45 Fleetwood Avenue, Herne Bay, CT6 8QW
MM6 CIA C. Anderson, 11 Willowpark Place, Aberdeen, AB16 6XY
M6 CIE W. Canavan, 9 The Ridings, Deanshanger, Milton Keynes, MK19 6JD
M6 CIG E. Black, 34 White Bank Road, Oldham, OL8 3JH
MM6 CIJ C. Thomson, 54 Alderman Road, Glasgow, G13 3YE
M6 CIK P. Turner, 35 Gattington Park, Hawthorn Hill, Dogdyke, Coningsby, Lincoln, LN4 4XA
M6 CIP A. Rowson-Brown, The Fox, Station Road, Baldock, SG7 5RN
M6 CIQ M. Walls, 38 Poplar Drive, Royston, SG8 7ER
M6 CIS Dr B. Rush, Springside, Uploders, Bridport, DT6 4NU
M6 CIU G. Brotherhood, 17 Baldwin Close, Forest Town, Mansfield, NG19 0LR
M6 CIW C. Waite, 69 Ellesmere Street Swinton, Manchester, M27 0JT
M6 CIX A. Skinner, 17 Hawkins Close, Harrow, HA1 4DJ
MW6 CIY M. Bennett, 24 Bryn Street, Merthyr Tydfil, CF47 0TG
M6 CJA C. Apps, 17 Kent Road, Formby, L37 6BG
MM6 CJC J. Scanlan, Aitnoch Farmhouse, Grantown-on-Spey, PH26 3PX
M6 CJD C. Dawson, 9 Mulberry Close, Poringland, Norwich, NR14 7WF
M6 CJE D. Chidgey, 42 Half Acre, Williton, Taunton, TA4 4NZ
M6 CJF C. Foster, 8 William Iliffe Street, Hinckley, LE10 0LY
M6 CJG C. Groves, 3 Hudson Davies Close, Pilley, Lymington, SO41 5PA
MW6 CJH C. Hill, 9 Oliver Road, Newport, NP19 0HU
M6 CJI J. Power, 43 Valley Road, Melton Mowbray, LE13 0DU
M6 CJJ P. Spry, 8 Maun Green, Newark, NG24 2HA
M6 CJK R. Tuffin, 133 Shirley Drive, Hove, BN3 6UJ
M6 CJM C. Martin, 2 Whitethorn Cottages, Dark Lane, Cheltenham, GL51 9RW
M6 CJN C. Jenkins, 5 Mayors Buildings, Bristol, BS16 5AU
M6 CJP C. Petrie, 14 Rotherfield Avenue, Eastbourne, BN23 8JQ
M6 CJQ B. Hiley, 9 Pinfold Lane, Harby, Melton Mowbray, LE14 4BU
M6 CJR C. Rundle, 1 Trezaise Close, Roche, St Austell, PL26 8HW
MW6 CJS C. Sweeney, 67 Blandon Way, Cardiff, CF14 1EH
M6 CJT C. Wright, Home Farm, Bretford, CV230LB
M6 CJU J. Brown, 34 Salisbury Road, Worthing, BN11 1RD
M6 CJV S. Clarke, 20 Woodlands Way, Southwater, Horsham, RH13 9HZ
M6 CJW C. Moss, 19 Tozer Close, Wallisdown, Bournemouth, BH11 8RB
M6 CJX I. Ftaiha, 8 Parkside, London, NW7 2LH
MM6 CJY C. Yohn, 3F/C 567 George Street, Aberdeen, AB25 3XX
MM6 CJZ J. Wright, 43 Spey Court, Stirling, FK7 7QZ
MM6 CKC M. Thomson, 30 Pladda Road, Saltcoats, KA21 6AQ
M6 CKD R. Basford, 4 Renoir Close, St Ives, PE27 3HF
M6 CKE L. Hardy, 221 Rookery Lane, Lincoln, LN6 7PJ
M6 CKF C. Farmer, 3 Laxton Way, Banbury, OX17 1GJ
M6 CKI R. Baines, 9 Deer Close, Walsall, WS3 3EA
M6 CKJ E. Taylor, Lanthorn Close, Broxbourne, EN107NR
M6 CKL P. Ridley, 218 Lichfield Road, Rushall, Walsall, WS4 1SA
MW6 CKM J. Jones, Isfryn Bungalow, Glan-Y-Nant, Llanidloes, SY18 6PQ
M6 CKO E. Redmond, 28 Common Lane, Polesworth, Tamworth, B78 1LS
M6 CKP C. Prior, 38 Windmill Road, Wombwell, Barnsley, S73 8PP
M6 CKQ Dr A. Colman, 5 Burn Heads Road, Hebburn, NE31 2TB
M6 CKR M. Sadler, 14 Woodlands Avenue, Water Orton, Birmingham, B46 1SA
M6 CKT C. Jenkings, 8 Tolworth Hall Road, Birmingham, B24 9NE
M6 CKU P. Naik, 82 Misbourne Road, Uxbridge, UB10 0HW
M6 CKV P. Pain, 12 Maple Drive, Bamber Bridge, Preston, PR5 6RA
M6 CKY N. Gamulea Schwartz, Flat 5, 127 Princes Avenue, Hull, HU5 3HH
M6 CKZ D. Osborne, 12 Sandringham Close, Brackley, NN13 6JQ
M6 CLA S. Clark, 43 Age Croft, Oldham, OL8 2HG
M6 CLB E. Williams, 35 Cansfield Grove, Ashton-In-Makerfield, Wigan, WN4 9SE
MM6 CLC C. Collins, Redwoods, Barcaldine, Oban, PA37 1SG
M6 CLF C. Studdart, 41 Queensgate Street, Hull, HU3 2TT
M6 CLG J. Hurlbutt, 55 Prospect Avenue, Seaton Delaval, Whitley Bay, NE25 0EL
M6 CLI C. Herlingshaw, 48 Keats Road, Normanby, Middlesbrough, TS6 0RW
MW6 CLJ C. Jones, 19 Crud Y Castell, Denbigh, LL16 4PQ
M6 CLK M. Clark, 34 Magdalene Road, Owlsmoor, Sandhurst, GU47 0UT
M6 CLL A. Mccall, 95 Newton Drive, Blackpool, FY3 8LX
MM6 CLM C. Houston, 115 Burnbank Road, Ayr, KA7 3QH
M6 CLO P. Clough, 101 Cathedral View, Houghton le Spring, DH4 4HN
MW6 CLP R. Piper, 16 Elm Rise, Bryncethin, Bridgend, CF32 9SX
M6 CLQ M. Eccleston, 15 Wood Road North, Manchester, M16 9GQ
M6 CLV C. Vernau, 62 Princethorpe Road, Ipswich, IP3 8NX
M6 CLW A. Sockett, 37 Windsor Road, Thorpe Hesley, Rotherham, S61 2QS
M6 CLX P. Cooke, 3 St. Stephens Court, Congleton, CW12 1QW
M6 CLZ C. Sinclair, 43 Newton Way, Leighton Buzzard, LU7 4SU
M6 CMB L. Gorecki, 6 Robinhood Lane, Winnersh, Wokingham, RG41 5LX
M6 CMF B. Hill, 41 Heath Road, Leighton Buzzard, LU7 3AB
M6 CMG C. Goodhand, 37 Westwick Gardens, Lincoln, LN6 7RQ
MM6 CMI D. Lyons, 2 Goswick Farm Cottages, Berwick-upon-Tweed, TD15 2RW

M6 CMK C. Kennedy, 30 Tatton Close, Cheadle, SK8 2LZ
MM6 CMM C. Mckinlay, 19 Ash Grove, Blackburn, Bathgate, EH47 7QJ
M6 CMO I. Steggles, 4 Hawley Vale, Hawley Road, Dartford, DA2 7RL
MI6 CMQ R. Donnan, 71 Victoria Avenue, Newtownards, BT23 7ED
M6 CMT C. Travis, 4 Kingsdale, Worksop, S81 0XJ
M6 CMW C. Walsh, 8 Kemp Road, Leicester, LE3 9PS
MM6 CMY C. Matheson, 322 Millfield Hill, Erskine, PA8 6JN
M6 CMZ M. Jones, 89 Gloucester Road, Coleford, GL16 8BN
M6 CNA M. Bradley, 11 Pike Road, Coleford, GL16 8DE
MM6 CNC C. Comrie, 11 Glendoune Street, Girvan, KA26 0AA
M6 CND D. Ward, 65 Water Street, Accrington, BB5 6QU
M6 CNG G. Southall, 28 Manor Road, Woodford Halse, Daventry, NN11 3QP
M6 CNI R. Johnson, 24 Fairfields, Upper Denby, Huddersfield, HD8 8UB
M6 CNK D. Stockton, 78B Alderfield Penwortham, Preston, PR1 9HA
M6 CNL H. Hope, 51 Margravine Gardens, London, W6 8RN
M6 CNN D. Bedford, 163 Mortimer Road, South Shields, NE34 0RR
MM6 CNO S. Leighton, 4 Earn Court, Alloa, FK10 1PT
M6 CNP M. Pratt, 1 Ashvale Gardens, Romford, RM5 3QA
M6 CNQ P. Phillips, 49 Lower Northam Road, Hedge End, Southampton, SO30 4HE
M6 CNR J. Taylor, 90 Village Road, Gosport, PO12 2LG
M6 CNS M. Phillips, 66 Minden Way, Winchester, SO22 4DU
MM6 CNV M. Jamieson, 5 Straid Bheag, Barremman, Helensburgh, G84 0QX
M6 CNX J. Mooney, 107 Tedder Road, South Croydon, CR2 8AR
M6 CNY R. East, 6 Ashley Road, Worcester, WR5 3AY
M6 CNZ J. Haddow, 28 Church Marks Lane, East Hoathly, Lewes, BN8 6EQ
M6 COB R. Williams, 11 Trinity Buildings, Carlisle, CA2 7DA
MW6 COD D. Codd, Gwachal-Tagy Farm, Haverfordwest, SA62 6HF
M6 COE C. Bassett, 3 Downshire Terrace, Street Lane, Haywards Heath, RH17 6UL
MM6 COF G. Steele, 1/2 50 Motehill Road, Paisley, PA3 4ST
M6 COH S. Parris-Hughes, 23 Hobney Rise, Westham, Pevensey, BN24 5NN
M6 COI M. Bernasinski, 1 Elizabeth Place, Clyde Road, London, N15 4LA
M6 COL C. Spicer, 3 School View, Tunstall, Sittingbourne, ME9 8DX
MI6 COM S. Carroll, Drumguiff Lane, Enniskillen, BT92 7HP
M6 CON C. Morgan, 41 Torwood Mount, Old Torwood Road, Torquay, TQ1 1PX
M6 COP L. Elphick, 30 The Meadows, Heskin, Chorley, PR7 5NR
M6 COU P. Coulsey, 25 Weavers Close, Horsham St. Faith, Norwich, NR10 3HY
M6 COV A. Hughes, 86 Colinmander Gardens, Ormskirk, L39 4TF
M6 COX A. Cox, 20 Dunns Dale, Maltby, Rotherham, S66 7NR
M6 COY C. Burden, Flat 2 Tower Court Tower Place, Kings Lynn, PE30 5DF
M6 CPA C. Godding, 11 Oakleaze Road, Thornbury, Bristol, BS35 2LL
M6 CPC C. Cook, 95 Station Road, Eccles, Manchester, M30 0PZ
MM6 CPE C. English, Easter Backlands, Roseisle, Elgin, IV30 5YD
M6 CPH C. Holden, 49 Whalley Road, Lancaster, LA1 2HE
MW6 CPJ D. Martin111, 5 Dolafon, Foel, Welshpool, SY21 0NJ
MM6 CPK E. Dungavel, 9C Anderson Crescent, Ayr, KA7 3RL
M6 CPN C. Norris, 115 Sutton Road, Walpole Cross Keys, PE344HE
M6 CPO S. Parker, 10 Wheelwrights Close, Sixpenny Handley, Salisbury, SP5 5SA
M6 CPP N. Barnard, 10 Whites Lane Kessingland, Lowestoft, NR33 7TF
M6 CPQ T. Walsh, 6 Brass Thill, Durham, DH1 4DS
M6 CPR G. Doxey, 2 Nettlecroft, Barnsley, S71 5SD
M6 CPS C. Stratford, 15 Ferndale Road, Banbury, OX16 0RZ
M6 CPT D. Baker, 21 Dolman Road, Gosport, PO12 1RB
MU6 CPV P. Martin, 1 Hazeldene, Grandes Maisons Road, Guernsey, GY2 4JS
M6 CPW P. Winkley, 8 Thelusson Court, Woodfield Road, Radlett, WD7 8JF
M6 CPX R. Gleave, 52 Cranborne Avenue, Warrington, WA4 6DE
M6 CPZ M. Spiers, 99 Chapel Street, Tiverton, EX16 6BU
M6 CQB I. Mckean, 14 Maltings Close, Cranfield, MK43 0EL
M6 CQD A. Fitton, 72 Newnham Court, Ipswich, IP2 9UE
M6 CQE D. Bishop, 62 Brindley Crescent, Hednesford, Cannock, WS12 4DS
MJ6 CQF J. Crill, Keukenhof, La Route Des Cotils, Grouville, Jersey, JE3 9AP
M6 CQH M. Fletcher, 7 Richard Street, Bacup, OL113 8QJ
M6 CQJ A. Tranter, 122 Summerhill Road, Bristol, BS5 8JU
MW6 CQL C. Summerfield, 11 Woodland Park, Penderyn, Aberdare, CF44 9TX
MW6 CQO C. Osborne, Gwinwydden, Tremont Road, Llandrindod Wells, LD1 5BH
M6 CQR P. Dunn, 9 Parkside Gardens, Widdrington, Morpeth, NE61 5RP
MI6 CQS S. Allen, 21 Derrymore Meadows, Bessbrook, Newry, BT35 7GA
M6 CQV C. Norman, 15 Maple Close, Sedbergh, LA10 5JE
M6 CQX G. Davies, 28 Danum Close, Hailsham, BN27 1UX
M6 CRA C. Ancill, 8 Ipley Way, Hythe, Southampton, SO45 3LJ
M6 CRE C. Etches, The Nook, Low Road Friskney, Boston, PE22 8NQ
M6 CRF C. Faulkner, Mount Pleasant, Elkstones, Buxton, SK17 0LU
M6 CRG C. Gloess, 5 Elmers Lane, Kesgrave, Ipswich, IP5 2GW
M6 CRH C. Hogan, 82 Abbott Road, Didcot, OX11 8HY
M6 CRJ C. Jackson, 30 Coronation Avenue, Mile Oak, Tamworth, B78 3NW
M6 CRO C. Ordish, 37 Mill Lane, Earl Shilton, Leicester, LE9 7AY
M6 CRP C. Pulford, 17 Canada Road, Cromer, NR27 9AH
MM6 CRQ C. Welsh, 28 Peacock Wynd, Motherwell, ML1 4ZL
M6 CRR R. Flevill, 62 Grosvenor Way, Horwich, Bolton, BL6 6DJ
M6 CRT C. Thomas, 17 The Green, Nether Heyford, Northampton, NN7 3LE
MW6 CRU C. Uphill, 167 Nantgarw Road, Caerphilly, CF83 1AN
MM6 CRW C. Watkinson, The Hillock Farmhouse, Lumphanan, Banchory, AB31 4QL
M6 CRX A. Haynes, 16 Hills Crescent, Colchester, CO3 4NU
M6 CRZ C. Roberts, 25 Queens Lea, Willenhall, WV12 4JA
M6 CSA S. Clark, 1 Hanover Cottages, Lippen Lane, Southampton, SO32 3LE
M6 CSC R. Metcalfe, Little Shernden, Shernden Lane, Edenbridge, TN8 5PS
M6 CSE C. Cartwright, 8 Charles Close, Westcliff-on-Sea, SS0 0EU
M6 CSF R. Powell, 13E Rothsay Road, Bedford, MK40 3PP
M6 CSG C. Greenwood, 21 Valley Drive, Thornhill Dewsbury, wf120he
M6 CSI C. Morris, 38 Beltony Drive, Crewe, CW1 4TX
M6 CSJ B. Appleby, 64 Lundy Close, Southend-on-Sea, SS2 6HB
M6 CSK D. Polley, 6 Coneygear Road, Hartford, Huntingdon, PE29 1QL
M6 CSL C. Langmaid, Flat 4, Woodlawn High Street, Partridge Green West Sussex, RH13 8HR
M6 CSM C. Moppett, 13 Piccadilly, Tamworth, B78 2ER
M6 CSN C. Snelling-Nash, 33 Church Lane, Kimpton, Hitchin, SG4 8RR
M6 CSP C. Pritchard, 9 Furnace Drive, Crawley, RH10 6HZ
M6 CSR C. Robinson, Watermill Farm Cottage, The Moor, Middleton, Saxmundham, IP17 3LW
M6 CSS K. Latham, 45 Hollybank Close, Northwich, CW8 4GS
M6 CSU C. Sharif, 38 King George Road, Ware, SG12 7DT
M6 CSW J. Cheng, Clarence Drive, Harrogate, HG1 2QG
M6 CSX A. Askam, 8 The Pastures, Weston-On-Trent, Derby, DE72 2DQ
M6 CSZ A. Chaplin, 10 St. Leonards Road, Malinslee, Telford, TF4 2EB
M6 CTA P. Kerton, 15 Barncroft Way, Havant, PO9 3AA
MW6 CTB J. Pritchard, 1 Tan Y Coed, Maesgeirchen, Bangor, LL57 1LU
MW6 CTD T. Cooper, Dorset Cottage, Alltyblacca, Llanybydder, SA40 9SU
MW6 CTE M. Sunderland, 16 Heather Avenue, Cardiff, CF5 5AH
MW6 CTG M. Adam, Tan Ffordd, Mynydd Llandygai, Bangor, LL57 4LX
MM6 CTH S. Fenton, 19 Rowan Road, Girvan, KA26 0BY
M6 CTI J. Roberts, 4 Oaks Road, Staines-upon-Thames, TW19 7LG
M6 CTJ C. Johnson, 29 Linden Road Creswell, Worksop, S80 4JT
M6 CTK C. Taylor, 9 Pendeen Crescent, Threemilestone, Truro, TR3 6SP
MM6 CTL C. Livingstone, 391 Dyke Road, Glasgow, G13 4QE
M6 CTN S. Knott, 15 Meadowlands Drive, Haslemere, GU27 2FD
M6 CTO D. Mcgleenan, Greenfields, Ellonby, Penrith, CA11 9SJ
MM6 CTQ A. Montgomery, 13 Rathad A'Mhaoir, Plasterfield, Stornoway, HS1 2UP
MW6 CTS D. Hooper, Nant Y Dryslwyn Cottage, Ty Mawr, Llanybydder, SA40 9RD
M6 CTT C. Rogers, 65 Darwin Close, Taunton, TA2 6TR
M6 CTV C. Leek, 46 The Hollies, Holbeach, Spalding, PE12 7JQ
M6 CTW C. Winnan, 133 Deepcut Bridge Road, Deepcut, Camberley, GU16 6SD
M6 CTX T. Rowlands, 7 Northfield Crescent, Beeston, Nottingham, NG9 5GR
M6 CTY P. Keane, 18 Main Road, Cannington, Bridgwater, TA5 2JN
M6 CTZ J. Richards, 5 Dumfries Place, Weston-Super-Mare, BS23 4LQ
MW6 CUA G. Young, 104 Ystrad Road, Pentre, CF41 7PW
M6 CUC R. Keast, 7 The Finches, Newport, PO30 5GU
M6 CUD S. Kneale, 59 Mayflower Avenue, Saxmundham, IP17 1BU
M6 CUE N. Connor, 28 Church Street, Hungerford, RG17 0JE
M6 CUJ J. Parsons, 3150 S Garland Way, Lakewood, USA, 80227
M6 CUL D. Richards, 58 Holm Lane, Oxton, Prenton, CH43 2HS
M6 CUN J. Colledge-Wiggins, Raffles, Southcombe, Chipping Norton, OX7 5QH
M6 CUP T. Cavanagh, 186 Cole Valley Road, Birmingham, B28 0DQ
M6 CUQ G. Fantom, 6 Middle Street, Farcet, Peterborough, PE7 3AX
MW6 CUR J. Troughton, Rhiwbina, Pentre Lane, Cwmbran, NP44 3AP
M6 CUS S. Williams, 16 Pinewood Avenue, New Haw, Addlestone, KT15 3AA
M6 CUU K. Boaler, 12 Belmont, Slough, SL2 1SU
M6 CUV A. Douglas, Gobbins Cottage, Sandy Lane, Lathom, Ormskirk, L40 5TU
MM6 CUW A. Allan, Skibhoull, Aith Bixter, Shetland, ZE2 9NB
M6 CUX A. Stott, 2 Douglas Drive, Maghull, Liverpool, L31 9DG
MI6 CUZ G. Angus, 59 Millisle Road, Donaghadee, BT21 0HZ
M6 CVB J. Adler, 1 Searles Meadow, Dry Drayton, Cambridge, CB23 8BW
MW6 CVC C. Smith, 29 Heol Cwarrel Clark, Caerphilly, CF83 2NE
M6 CVE P. Herron, 102 Garden City Villas, Ashington, NE63 0EU
M6 CVF L. Bishop, Franklin House, Canford School, Wimborne, BH21 3AF
MW6 CVH C. Hodgetts, 16 Myrtle Drive, Rogerstone, Newport, NP10 9EA
M6 CVJ J. Smith, Flat 9, Mount Gardens, 19 Davenport Road, Coventry, CV5 6QH
M6 CVK C. Saxton, 56 Coniston Crescent, Grimsby, DN36 4BA
MW6 CVN M. Beecroft, Hafod Y Wennol, Llanddoged, Llanrwst, LL26 0TY
MI6 CVO R. Lambe, 39 Tildarg Avenue, Belfast, BT11 9LU
M6 CVP L. Stamper, 22 Douglas Road, Workington, CA14 2QY
M6 CVU Z. Szikszai, 135 Devon Road, Newark, NG24 4JL
M6 CVV J. Szikszai, 135 Devon Road, Newark, NG24 4JL
MI6 CVW A. Logan, 15 Park Lane, Saintfield, Ballynahinch, BT24 7PR
M6 CWA G. Moon, 1 Brankenwall, Muncaster, Ravenglass, CA18 1RG
MM6 CWB C. Beedie, 21 Marywell Village, Arbroath, DD11 5RH
MI6 CWC C. Campbell, 21 Elms Park, Coleraine, BT52 2QF
M6 CWD B. Ledson, 16 Caton Close, Southport, PR9 9XF
M6 CWF A. Mills, 6 Kildare Drive, Peterborough, PE3 9TS
M6 CWG D. Levy, Flat 36, Claydon House, London, NW4 1LS
M6 CWI S. Chapman, 5 Bracton Drive, Nottingham, NG3 2LN
M6 CWJ B. Harrison, 104 Jeavons Lane, Great Cambourne, Cambridge, CB23 5FN
M6 CWK C. Austin, 1A Arundel Road, Peacehaven, BN10 8TE
MW6 CWL K. Saltmarsh, 15 Colbourne Road, Beddau, Pontypridd, CF38 2LN
MM6 CWN C. Cowan, 23 Park Road, Hamilton, ML3 6PD
M6 CWO J. Cox, 87 Richmond Road, Leighton Buzzard, LU7 4RF
M6 CWP C. Hicks, 6 Westlands Avenue, Weston-On-The-Green, Bicester, OX25 3RD
M6 CWQ A. Toth, 14 Runcorn Road, Sunderland, SR5 5ET
M6 CWR C. Ralphson, 20 Monsal Grove, Buxton, SK17 7TF
M6 CWU J. Henningway, 64 South Cliff, Bexhill-on-Sea, TN39 3EE
M6 CWV N. Symes, 23 Elm Road, Portslade, Brighton, BN41 1SA
M6 CWW C. Warhurst, 30 Nether Royd View, Silkstone Common, Barnsley, S75 4QQ
M6 CWX Dr T. Digman, 74 Baddlesmere Road, Whitstable, CT5 2LA
M6 CWZ L. Starrett, 50 Danes Road, Bicester, OX26 2LP
M6 CXB C. Burnham, 93 Colliers Break, Emersons Green, Bristol, BS16 7EB
M6 CXC B. Burr, Flat 87, Horatia House, Southsea, PO5 4AL
MM6 CXD A. Donald, 10 Fraser Road, Burghead, Elgin, IV30 5YN
M6 CXH R. Pilkington, 64 School Lane, Higher Bebington, Wirral, CH63 2LW
M6 CXI G. Megson, 147 Duke Of York Avenue, Wakefield, WF2 7DA
MM6 CXJ R. Inglis, The Shop, Roadside Skirza, Freswick, Wick, KW1 4XX
M6 CXM J. Van De Vondel, 50 Holmsley Lane, South Kirkby, Pontefract, WF9 3JF
M6 CXN F. Strickland, Hinds Cottage, Beverley Road, Driffield, YO25 9PF
M6 CXO M. Hickey, 100 Chester Road, Poynton, Stockport, SK12 1HG
M6 CXP B. Fitzgerald-O'Connor, 24 Routh Street, London, E6 5XX
M6 CXV A. Todd, 16 Worcester Close, Bracebridge Heath, Lincoln, LN4 2TY
M6 CXW A. Little, 13 Belmont Grove, Burnley, BB10 4NR
M6 CXZ M. Hack, 28 Horsefield View, Melton Mowbray, LE13 0TF
M6 CYA C. Rowe, 1 Avondale Street, Stoke-on-Trent, ST6 4NN
M6 CYB K. Williams, 18 Bye Road, Lidlington, Bedford, MK43 0RU
M6 CYC C. Carter, 142 Hall Street, Briston, Melton Constable, NR24 2LQ
M6 CYD A. Mcinnes, 87 Sovereigns Quay, Bedford, MK40 1TF
M6 CYE A. Forrester, 16 East Street, Hebburn, NE31 1HL
M6 CYF R. Silcock, 18 Saxon Road, Southampton, SO15 1JJ
M6 CYG C. Ashworth, 40 Fairholme, Sedbergh, LA10 5AY
M6 CYL Y. Chong, Harrogate Ladies' College, Clarence Drive, Harrogate, HG1 2QG
MW6 CYM G. Lewis, Upper Cwm Farm, Llantilio Crossenny, Abergavenny, NP7 8TG
M6 CYN G. Couzens, Oak Tree Cottage Kite Hill Wootton Bridge, Ryde, PO33 4LE
M6 CYO J. Mason, 77 Albutts Road, Walsall, WS8 7ND

M6 CYP Dr S. Mouradian, 1 Sandfield Road, St Albans, AL1 4JZ
MM6 CYQ G. Lewin, Larch Cottage, Lein Road, Fochabers, IV32 7NW
M6 CYS A. Owen, 21 Commercial Road, Chorley, PR7 1EU
M6 CYT D. Elias, 12 Dagmar Terrace, London, N1 2BN
MW6 CYU K. Evans, Maesyronnen, Sarnau, Llanymynech, SY22 6QL
M6 CYV K. Nicholson, 68 Brunswick Street, Leigh, WN7 2PL
M6 CYW J. Wood, 17 Harrington Avenue, Lincoln, LN6 7UP
MW6 CYX J. Shanahan, 23 Pentre Gwyn, Trewern, Welshpool, SY21 8DY
M6 CYY A. Brown, 114 Tavistock Road, Birmingham, B27 7LA
M6 CZA M. Singfield, 8 Barnes Crescent, Sutton-in-Ashfield, NG17 5BL
M6 CZB N. Shajan, 19 Sturgess Avenue, London, NW4 3TR
M6 CZC B. Crowhurst, 33 Clarendon House, Clarendon Road, Hove, BN3 3WW
M6 CZD T. Ayland, 225 Stroud Road, Gloucester, GL1 5JU
M6 CZF S. Amos, 5 Curbar Close, Mansfield, NG18 4XS
MM6 CZH S. Ram, 28 Craigievar Gardens, Kirkcaldy, KY2 5SD
M6 CZI J. Prodger, Piper Cottage, Alde House Drive, Aldeburgh, IP15 5EE
M6 CZJ J. Chalmers, 19 Brettenham Crescent, Ipswich, IP4 2UB
M6 CZL Dr T. Watson, Montrose Farm, Bury Road, Diss, IP22 2PY
M6 CZR R. Houghton, 17 Coronet Drive, Ibstock, LE67 6QG
MW6 CZS P. Hampton, Caretakers Flat, T.A. Centre, Newport, NP20 5XE
MW6 CZU L. Taylor, 140 Cotswold Way, Risca, Newport, NP11 6RG
MW6 CZV C. Parry, 91 Grove Road, Risca, Newport, NP11 6GL
M6 CZW T. Grayson, 20 Harrison Street, Tow Law, Bishop Auckland, Co Durham, DL13 4EE
M6 DAC D. Clough, 44 Small Drove, Weston, Spalding, PE12 6HS
M6 DAD J. Armstrong, Sherrylea, Coventry Road, Coventry, CV7 8BY
MM6 DAF D. Hannah, 7 Katrine Place, Irvine, KA12 9LU
M6 DAF D. Flunder, 17 Hampden Crescent, Kettering, NN16 0LA
M6 DAG D. Walker, Flat 5, Seward Court, 380-396 Lymington Road, Christchurch, BH23 5HD
MW6 DAI D. Evans, 29 Mount Pleasant, Bedlinog, Treharris, CF46 6SD
M6 DAK C. Wood, 18 Tufa Close, Chatham, ME5 9LU
M6 DAL D. Cannon, Hook 2 Sisters, Mautby Site Mautby, Great Yarmouth, NR29 3JB
M6 DAM J. Malia, 55 West Farm Avenue Longbenton, Newcastle upon Tyne, NE12 8LS
M6 DAN D. Trudgian, 18 Hart Close, Wootton Bassett, Swindon, SN4 7FN
M6 DAP D. Pasika, 192 Longfield Lane, Cheshunt, Waltham Cross, EN7 6AQ
MM6 DAQ D. Rooney, 19 Aurs Drive, Barrhead, Glasgow, G78 2LR
MW6 DAR D. May, 12 Marl Crescent, Llandudno Junction, LL31 9HS
M6 DAS D. Starling, Long Field Barn, Clarkes Lane, Beccles, NR34 8HR
MM6 DAT I. Mclachlan, Railway Cottage, Nether Falla, Peebles, EH45 8QZ
M6 DAU D. Skates, Flat 23, Orford Court Marsh Lane, Stanmore, HA7 4TQ
M6 DAW D. Harris, 13 Horsecroft, Ewyas Harold, Hereford, HR2 0EQ
M6 DAX D. Blackhorse-Hull, 1 Occupation Lane, New Bolingbroke, Boston, PE22 7LW
MI6 DAY A. Galbraith, 62 Millbrook Gardens, Castlederg, BT81 7DF
M6 DAZ D. Coleman, 9 Hogan Close, Newport, PO30 5UF
M6 DBA D. Austrin, 50 Lowestoft Road, Gorleston, Great Yarmouth, NR31 6LZ
MW6 DBB S. Elias, 20 Attlee Way, Cefn Golau, Tredegar, NP22 3TA
MW6 DBD M. Williams, 31 Waundeg, Nantybwch, Tredegar, NP22 3SN
M6 DBE D. Bateson, 19 Rothesay Road, Heysham, Morecambe, LA3 2UR
M6 DBF P. Allen, Flat 32, Whitworth Court, 9 Whitworth Road, Southampton, SO18 1JR
M6 DBG D. Bentley, 36 Byron Road, Mexborough, S64 0DG
M6 DBH D. Coote, 4 Hunters Oak Watton, Thetford, IP25 6HL
M6 DBI R. Buckland, 34 Beechwood Drive, Meopham, Gravesend, DA13 0TX
MM6 DBJ A. Brown, 17 Glamis Drive, Dundee, DD2 1QN
M6 DBK D. Wells, 27 Victoria Avenue, Camberley, GU15 3HT
M6 DBL D. Smith, 186 Weekes Drive, Slough, SL1 2YR
MM6 DBN W. Brannan, 16 Cairngorm Gardens Cumbernauld, Glasgow, G68 9JD
M6 DBP D. Robinson, 10 Pendragon Park, Glastonbury, BA6 9PQ
M6 DBQ D. Baldwin, 17 Teddy Gray Avenue Elworth, Sandbach, CW11 3AR
M6 DBS S. Ward, 22 St. Margarets Close, Horstead, Norwich, NR12 7ER
MM6 DBT J. Ross, 159 Grahams Road, Falkirk, FK2 7BQ
MW6 DBU D. Humphreys, 48 High Street, Abergwynfi, Port Talbot, SA13 3YW
M6 DBV A. Higgins, 7 Waterloo Terrace, Bideford, EX39 3DJ
M6 DBW D. Woodridge, 8 Goold Close, Corston, Bath, BA2 9AF
M6 DBX M. Hall, 29 The Spinney, Finchampstead, Wokingham, RG40 4UN
M6 DCA D. Cotz, 159 Ecclesfield Road, Sheffield, S5 0DH
M6 DCB P. Jones, Stonehead Farm Over Wyresdale, Lancaster, LA2 9DL
M6 DCD G. Clarke, 2 Beare Green Cottages, Horsham Road, Dorking, RH5 4PE
MI6 DCH D. Hickey, 83 Cairnmartin Road, Belfast, BT13 3PQ
MW6 DCI D. Morgan, Castle Cottage, Newbridge, Newport, NP11 3NT
M6 DCJ C. Davies, 83 Freeston Avenue, St. Georges, Telford, TF2 9EN
M6 DCK D. Renshaw, 25 Ashley Road, Worksop, S81 7JS
M6 DCL D. Lane, 46 Berkeley Vale Park, Berkeley, GL13 9TQ
MM6 DCM D. Mifsud, 25 Priory Road, Linlithgow, EH49 6BP
MW6 DCN D. Davies, 49 Heol Y Wal, Bradley, Wrexham, LL11 4BY
M6 DCO D. Close, 22 Station Road, Dodworth, Barnsley, S75 3JE
M6 DCP A. Pearson, 29 Broadoak Road, Langford, Bristol, BS40 5HD
M6 DCQ D. Manning, 153 Pavilion Road, Worthing, BN14 7EG
M6 DCS D. Shephard, 17 Grimsby Road, Laceby, Grimsby, DN37 7DF
MM6 DCT D. Tourish, 1 Langside Drive, Kilbarchan, Johnstone, PA10 2EL
M6 DCU D. Matthews, 81 Kipling Avenue, Goring-By-Sea, Worthing, BN12 6LH
M6 DCW D. Wetton, 1 Monksway, Birmingham, B38 9LW
M6 DCZ D. Mooney, 107 Tedder Road, South Croydon, CR2 8AR
M6 DDB D. Beavis, 17 Kingsbere Crescent, Dorchester, DT1 2DY
M6 DDC D. Chebsey, 21 Shortlands Lane, Walsall, WS3 4AG
M6 DDD V. Vesma, Pound House, Market Square, Newent, GL18 1PS
M6 DDF D. Poulton, 115 Highview, Vigo, Gravesend, DA13 0TQ
M6 DDH D. Muirhead, 13 Berry Street, Skelmersdale, WN8 8QZ
M6 DDI D. Barnett, 49 Loundes Road, Unstone, Dronfield, S18 4DE
M6 DDL P. Smith, 156 Esther Grove, Wakefield, WF2 8ET
M6 DDM D. Maycroft, 100 Benwick Road, Doddington, March, PE15 0UH
MM6 DDN D. Hornal, 95 Traprain Crescent, Bathgate, EH48 2BD
M6 DDO D. Daniels, 128 Woodcock Road, Norwich, NR3 3TD
M6 DDP D. Stallibrass, 12 Sheerwater Close, Bury St Edmunds, IP32 7HR
M6 DDQ D. James, 13 Lincoln Road, Fenton, Lincoln, LN1 2EP
MW6 DDR D. Reeves, 27 Beaufort Road, Pembroke, SA71 4PX
M6 DDT D. Cross, 4 Burns Avenue, Gloucester, GL2 5BJ
M6 DDU D. Hind, 116 Gowthorpe, Selby, YO8 4HA
M6 DDV D. Borrett, 84 Kingsway, Mapplewell, Barnsley, S75 6EX
MM6 DDX M. Dougan, 41D Balmerino Road, Dundee, DD4 8RP
M6 DDY J. Betteridge, 57 Wood Road, Chaddesden, Derby, DE21 4LY
M6 DEC D. Hunter, 9 Gelt Burn, Didcot, OX11 7TZ
MI6 DED A. Mccann, 6 Bowens Meadow, Lurgan, Craigavon, BT66 7UT
M6 DEE D. Daniels, 16 Bainbridge Road, Warsop, Mansfield, NG20 0ND
M6 DEF A. Clarke, 1 Riversdale Cottages, The Staithe, Stalham, Norwich, NR12 9BY
M6 DEG D. Edge, 18 Sandringham Avenue, Whitehaven, CA28 6XL
M6 DEI D. Beresford, 40 Fern Crescent, Congleton, CW12 3HQ
M6 DEJ D. Thomas, Stud Farm Bungalow, Stud Farm Drive, Tamworth, B78 3HS
M6 DEK D. Knighton, Holme Lea, Church Lane, Chesterfield, S44 5AL
M6 DEL D. Millard, 112 Avenue Road, Sandown, PO36 8DZ
M6 DEM L. Demirkaya, 11 Beech Park, Holsworthy Beacon, Holsworthy, EX22 7NB
M6 DEO D. Sproston, 22 Oakland Avenue, Haslington, Crewe, CW1 5PB
M6 DER I. Dermondy, 8 Glenwood Avenue, Baildon, Shipley, BD17 5RL
M6 DES D. Kirkden, 57 Crow Hill Road, Margate, CT9 5PF
M6 DEV M. Raynor, 68 Cambridge Street, South Elmsall, Pontefract, WF9 2AR
MJ6 DEY J. Bryant, 5 Louiseberg Court Queen'S Road, St Helier, Jersey, JE2 3GQ
M6 DEZ J. Butters, 7A Green Oak Avenue, Sheffield, S17 4FT
MW6 DFC D. Gee, 20 Davies Avenue, Brymbo, Wrexham, LL11 5AS
M6 DFD D. Pennison, 69 Caneland Court, Waltham Abbey, EN9 3DS
M6 DFE J. Wood, 3 Lion Lane, Haslemere, GU27 1JF
M6 DFG D. Gardner, 106 Willclare Road, Birmingham, B26 2NY
M6 DFI E. Garry, 34 Conway Road, Paignton, TQ4 5LH
M6 DFJ T. Saunders, 40 Southdown Avenue, Brixham, TQ5 0AN
M6 DFK J. Bailey, 22 Wilford Drive, Ely, CB6 1TL
M6 DFL D. Loveys, 3 Prestwood Court, Stafford, ST17 4DY
M6 DFM D. Mcauslan, Casa Arco Iris, Via Variante Nascente, 8005-491, Portugal, SANTA BARBARA DE NEX
M6 DFP D. Parker, 53 Brisbane Way, Cannock, WS12 2GR
M6 DFQ A. Singh, 19 Severn Crescent, Slough, SL3 8UU
M6 DFS D. Slade, 22 Oaklands Road, Mangotsfield, Bristol, BS16 9EY
M6 DFT I. Lonsdale, 23 Hunts Field Clayton-Le-Woods, Chorley, PR6 7TT
M6 DFU K. Pownall, 17 Horsebridge Road, Blackpool, FY3 7BQ
M6 DFW D. Whitley, 10 Kenmore Drive, Cleckheaton, BD19 3EJ
MW6 DFX M. Lima Barbosa, 65 Precelly Place, Milford Haven, SA73 2BW
M6 DFY R. Medway, 54 Peasland Road, Torquay, TQ2 8PA
M6 DFZ D. Antcliffe, 3 Wiltshire Mews, Cottam, Preston, PR4 0NP
M6 DGA D. Ashford, 56 Finch Close, Shepton Mallet, BA4 5GL
MM6 DGC D. Baillie, 77 Main Street, Fauldhouse, Bathgate, EH47 9AZ
M6 DGD D. Bailey, 2B Queens Road, Enfield, EN1 1NE
M6 DGG M. Harrison, 43 Second Avenue, Woodlands, Doncaster, DN6 7QQ
M6 DGI P. Kirby, 11 Bembridge Court, Crowthorne, RG45 6BN
M6 DGJ D. Cole, 4 Rosedale Road, Margate, CT9 2TD
M6 DGM D. Mouland, 2 Chafy Cottages, Holnest, Sherborne, DT9 6HX
M6 DGN I. Coleman, 24 Westwood Avenue, Plymouth, PL6 7HS
M6 DGO A. Thacker, 47 Hamilton Street, Walsall, WS3 3EN
M6 DGP P. Dransfield, 8 Church Mount, Mappleton, Hornsea, HU18 1XU
MW6 DGQ A. Evans, 27 Oaklands Park Drive, Rhiwderin, Newport, NP10 8RB
M6 DGR A. Deaves, 46-48 Charlemont Drive, Manea, March, PE15 0GA
M6 DGS D. Sole, 10 Glyn Place, East Melbury, Shaftesbury, SP7 0DP
M6 DGU Dr R. Jones, Mole Corner, Red Shute Hill, Thatcham, RG18 9QW
MW6 DGW D. Whitcombe, 4 Fairview Road, Llangyfelach, Swansea, SA5 7JJ
MM6 DGY D. Coyle, 3 Claddyburn Terrace, Cairnryan, Stranraer, DG9 8RD
MM6 DGZ D. Baillie, 8 Cairnsmore Road, Palnure, Newton Stewart, DG8 7AZ
M6 DHB A. Curd, 44 Elbridge Avenue, Bognor Regis, PO21 5AD
MM6 DHF D. Forsyth, Solano, Stirling Road, Dumbarton, G82 2PF
M6 DHG D. Lappage, 37 Matlock Drive, Cannock, WS11 6EN
MM6 DHI R. Buchan, 5 Fairview Terrace, Danestone, Aberdeen, AB22 8ZH
MW6 DHK D. Drake, 56 Bryn Llidiard, Bridgend, CF31 1QN
MM6 DHS K. Paterson, 69 Middlepart Crescent, Saltcoats, KA21 6LN
M6 DHT A. Herd, 7 Water Lane, Greenham, Thatcham, RG19 8SH
M6 DHU D. Maton, 41 Bemerton Gardens, Kirby Cross, Frinton-on-Sea, CO13 0LQ
M6 DHV D. Vincent, 38 Methuen Street, Walney, Barrow-in-Furness, LA14 3PR
M6 DHW D. Collins, 50 Woodville, Barnstaple, EX31 2HL
MM6 DHZ I. Smith, 32 Kaimes Avenue, Kirknewton, EH27 8AU
M6 DIB D. Bloxsome, 74 Dunclair Park, Plymouth, PL3 6DE
M6 DIF S. Goldsbrough, 2 Stutte Close, Louth, LN11 8YN
MW6 DIG D. Gozzard, Craig Dulas, Rhydyfoel Road, Abergele, LL22 8EG
M6 DIH A. Arnold, 1 Fairford Gardens, Wordsley, Stourbridge, DY8 5RF
M6 DII R. Wilson, 16 Thorndon Hall, Thorndon Park, Ingrave, Brentwood, CM13 3RJ
M6 DIJ D. Yates, 16 Sunnyfield Road, Prestwich, Manchester, M25 2RD
M6 DIL D. Yakub, 42 Swift Close, Blackburn, BB1 6LF
M6 DIM D. Vainas, 51 Magister Road, Bowerhill, Melksham, SN12 6FD
M6 DIO J. Cant, 42 Copandale Road, Beverley, HU17 7BW
M6 DIQ A. Collins, 14 Double Corner, Mendlesham Road, Cotton, Stowmarket, IP14 4RF
M6 DIT N. Baulf, 1 Lower Chart Cottages, Brasted Chart, Westerham, TN16 1LS
M6 DIU P. Sewell, 17 Chatham Close, Coventry, CV3 1LY
M6 DIV J. Inwood, Flat 13, Kelvestone House, 47 Park Road, Cannock, WS11 1NZ
M6 DIZ D. Balsdon, 25 Hanover Road, Plymouth, PL3 6BY
M6 DJA D. Aspital, 47 St. Augustines Park, Ramsgate, CT11 0DF
MM6 DJC D. Caveney, Westerhill, Clashnamuiach, Tain, IV20 1XP
M6 DJG Z. Zmajkovic, 67 Oakington Avenue, Little Chalfont, Amersham, HP6 6SX
M6 DJI D. Oliver, 20 Five Oaks Close, Malvern, WR14 2SW
M6 DJJ B. Dodson, 3 Bradley Road, Patchway, Bristol, BS34 5LF
M6 DJK L. Oxberry, 7 Kays Cottages Windy Nook, Gateshead, NE10 9ST
M6 DJL D. Reader, 190 Barnby Dun Road, Doncaster, DN2 4RF
M6 DJN D. Neville, 2 Oak View, Luddendenfoot, Halifax, HX2 6HQ
M6 DJQ M. Hodges, 4 Pond Fields Close, Little Hallingbury, Bishop's Stortford, CM22 7FF
M6 DJS D. Smith, 173 Leicester Road, Shepshed, Loughborough, LE12 9DG
MW6 DJT D. Terrell, 82 Baglan Street, Treherbert, Treorchy, CF42 5AR
M6 DJW D. Wheeler, 20 Hillwood Road Northfield, Birmingham, B31 1DJ
M6 DJX D. Jones, 102 Bryce Road, Brierley Hill, DY5 4ND
M6 DKC D. Cowan, 30 Crucian Way, Liverpool, L12 0AW

M6 DKF B. Hodgson, 28 Grove Drive, Woodhall Spa, LN10 6RT
MM6 DKI S. Young, 6 Ramsey Cottages, Bonnyrigg, EH19 3JG
M6 DKK P. Weston, 9 Clarendon Road, Smethwick, B67 6DA
M6 DKL P. Shaw, 21 Urmson Street, Oldham, OL8 2AN
M6 DKM W. Molloy, 32 Millers Barn Road, Jaywick, Clacton-on-Sea, CO15 2QB
MM6 DKN N. Ford, 11 Kenmore Way, Coatbridge, ML5 4FN
M6 DKO J. Greenwood, 18 Rookery Lane, Lincoln, LN6 7PY
M6 DKR D. Reeves, 67 The Cliff, Bryanston, Blandford Forum, DT11 0PP
M6 DKS G. Hutton, 8 Popples Drive, Halifax, HX2 9SQ
M6 DKT D. Carter, 18 Silent Woman Park, Coldharbour, Wareham, BH20 7PE
M6 DKU B. Burgess, 26 Bakers Way, Morton, Bourne, PE10 0XW
M6 DKW C. Nicholl, 36 Eylewood Road, London, SE27 9NA
M6 DKY O. Silva, Flat 71, Long Acre House, Pettacre Close, London, SE28 0PB
M6 DLA D. Aldred, 19 Birch Avenue, Bacton, Stowmarket, IP14 4NT
M6 DLB P. Booth, 7 Handley Crescent, East Rainton, Houghton le Spring, DH5 9QX
M6 DLC D. Smith, 106 Middle Street, Blackhall Colliery, Hartlepool, TS27 4EB
M6 DLD H. Rasooli Nia, 14 Larch Close, London, N11 3NN
M6 DLH D. De La Haye, 4 Nicola Mews, Ilford, IG6 2QE
M6 DLI J. Walters, 12 Leighton Road, Bilston, WV14 8LN
M6 DLJ D. Caldicott, 564 Fulbridge Road, Peterborough, PE4 6SA
M6 DLL D. Hallsworth, 29A Stephenson Court, Station Road, Stockport, SK5 6LE
M6 DLM D. Le Mare, The Sycamore, Church Bank, Barnard Castle, DL12 0AH
M6 DLO R. Coles, 10 Littlemoor Road, Weymouth, DT3 6AA
M6 DLP M. Palmer, New Haven, Stoneraise, Durdar, Carlisle, CA5 7AX
M6 DLQ P. Leng, The Barn, Gildersleets, Settle, BD24 0AH
M6 DLR T. Knight, 12 Vale Drive, Horsham, RH12 2JX
M6 DLU S. Bland, 61A Terminus Terrace, Southampton, SO14 3FE
M6 DLV C. Tompkins, 9 Billbrook Road, Hucclecote, Gloucester, GL3 3QS
M6 DLW D. Endean, 11 Forrester Drive, Brackley, NN13 6NE
M6 DLX B. Tufnell, 1 Moorlands Court Wath-Upon-Dearne, Rotherham, S63 6DD
M6 DLY A. Omar, 84 Beaumont Hill, Darlington, DL1 3ND
M6 DMA D. Stockton, 19 Chadwick Road, Middlewich, CW10 0EA
M6 DMB D. Booth, 5 Denham Close, Skegness, PE25 3PH
MM6 DMC D. Mccrae, 4 Castramont Road, Gatehouse Of Fleet, Castle Douglas, DG7 2JE
M6 DMD D. Mathieson-Dodd, 1 Dag Lane, North Kilworth, Lutterworth, LE17 6HD
M6 DME D. Elmy, 2 Mill Road Drive, Purdis Farm, Ipswich, IP3 8UT
M6 DMF J. Wilkins, 124 Fullers Mead, Harlow, CM17 9AU
M6 DMG D. Gillingham, 5 Hillfield, St. Marks, Cheltenham, GL51 7BQ
M6 DMN J. Robbins, 5 South Close, Greatworth, Banbury, OX17 2DZ
M6 DMO L. Wilburn, 23 Sutton Road, Kirk Sandall, Doncaster, DN3 1NY
M6 DMP D. Allen, 9 The Crescent Cootham, Pulborough, RH20 4JU
M6 DMQ D. Mcquirk, 6 French Gardens Blackwater, Camberley, GU17 9DP
M6 DMW D. Carey, 78 Bentley Road, Bramley, Rotherham, S66 1UH
M6 DMY A. Hoskins, 38 Tasmania Close, Basingstoke, RG24 9PQ
M6 DNB D. Eltham, 47 Glenville Close, Royal Wootton Bassett, Swindon, SN4 7EU
M6 DNC D. Swift, 31 Meadow Lane, Westbury, BA13 3AE
MI6 DND D. Fisher, 11 Summer Street, Belfast, BT14 6ES
M6 DNF D. Featherby, 14 Station Road, Sutton, Ely, CB6 2RL
M6 DNJ P. Hendry, 19 Parsons Close, Portsmouth, PO3 5LN
M6 DNL P. Swainson, 11 Conway Drive, Banbury, OX16 0QW
M6 DNM D. Mitchell, 4 Millstream Close, Goostrey, Crewe, CW4 8JG
M6 DNN D. Baker, 117 Locks Road, Locks Heath, Southampton, SO31 6LJ
M6 DNO C. Sheath, 53 Manor Road, Trowbridge, BA14 9HS
MW6 DNP Dr D. Morgan, Ty Bettws, Kilgwrrwg, Chepstow, NP16 6PN
M6 DNQ B. Southern, 25 Chilgrove Avenue, Blackrod, Bolton, BL6 5TR
M6 DNR L. Childs, 354 Linnet Drive, Chelmsford, CM2 8AL
M6 DNS M. Carter, 61 Catherine Avenue, Swallownest, Sheffield, S26 4RQ
M6 DNV G. Cater, 7 Seymour Street, Chelmsford, CM2 0RX
M6 DNX D. Garnham, 16 Cherrytree Road, Great Cornard Sudbury, CO10 0LJ
M6 DNZ D. Hart, 163 Wakering Road, Shoeburyness, Southend-on-Sea, SS3 9TN
M6 DOA D. Morris-Jones, 121 Plas Dinas, Blacon, Chester, CH1 5SW
M6 DOB A. Cowan, 111 Oaks Drive, St. Leonards, Ringwood, BH24 2QS
MM6 DOC Dr A. Davis, 13 High Road, Auchtermuchty, Cupar, KY14 7BE
MI6 DOD A. Henry, 10 Drumbreda Crescent, Armagh, BT61 7PE
M6 DOF A. Barker, 49 Rockingham Avenue, York, YO31 0TD
M6 DOG H. Godzisz, 1 Jutland Place, Egham, TW20 8ET
M6 DOJ A. Deery, 25 Ribblesdale Place, Preston, PR1 3NA
M6 DOM D. Chrumka, 94 Clock House Road, Beckenham, BR3 4JT
M6 DON D. Ashcroft, Don Ashcroft, Iken House, 8Acer Rd, Rendlesham, Woodbridge, IP12 2GA
M6 DOW J. Hogg, 31 Roseberry Grove, York, YO30 4SU
M6 DOX J. Armand, 95 Derby Road, Golborne, Warrington, WA3 3JJ
M6 DOZ D. Logan, Cedar House Reading Road North, Fleet, GU51 4AQ
M6 DPA C. Cooke, 42 Biddle Road, Leicester, LE3 9HG
M6 DPF P. Fletcher, 14 Snape Hill Crescent, Dronfield, S18 2GQ
M6 DPG D. Glover, 21 Monastery Road, Paignton, TQ3 3BU
M6 DPH D. Hancock, 2 Trevine Meadows, Indian Queens, St Columb, TR9 6NB
M6 DPI S. Hammond, 89 Beeston Road, Sheringham, NR26 8EJ
M6 DPJ P. Jones, 50 Clay Lane, Doncaster, DN2 4RJ
M6 DPL G. Payne, Jay Close, Eastbourne, Bn23 7RW
M6 DPN R. Fripp, 41 Sweyns Lease, East Boldre, Brockenhurst, SO42 7WQ
MM6 DPO P. Beck, Castle Clanyard Farm, Drummore, Stranraer, DG9 9HF
M6 DPP D. Parkinson, 24 New Road, Chatteris, PE16 6BW
MM6 DPQ M. Mcmath, 12 Townhead Crescent, Dalry, Castle Douglas, DG7 3UR
M6 DPR D. Runyard, Tilecroft, Shortheath Crest, Farnham, GU9 8SA
M6 DPU C. Lombao, 82 Cirrus Drive, Shinfield, Reading, RG2 9FL
MM6 DPV D. Peacock, Wyncum, Bridge Road, Castle Douglas, DG7 1TN
M6 DPW D. Wilde, 9 Redstone Park, Redhill, RH1 4AS
M6 DPX A. Potts, 28 Thistlebarrow Road, Salisbury, SP1 3RT
M6 DPY D. Hansford, 62 Bays Road, Pennington, Lymington, SO41 8HN
M6 DPZ D. Stansfield, 19 Cotman Fields, Norwich, NR1 4EN
M6 DQB A. Deiana, 41 Ashlar Drive, Eastfield, Scarborough, YO11 3FP
M6 DQC Dr R. Blackwell, Vikings Hall, Baylham, Ipswich, IP6 8JS
M6 DQD J. Earye, 28 Halls Drift, Kesgrave, Ipswich, IP5 2DE
MM6 DQG A. Stewart, 94 Dick Crescent, Burntisland, KY3 0BT
M6 DQH B. Hartland, 2 Brookland Close, Pevensey Bay, Pevensey, BN24 6RT
M6 DQK A. Walker, 4 Pretymen Crescent, New Waltham, Grimsby, DN36 4NS
M6 DQN J. Davis, 23 Blueberry Gardens, Andover, SP10 3XD
M6 DQO D. Stocker, 113 St. Marys Road, Bodmin, PL31 1NH
M6 DQP Rev. A. Lewis, Four Winds Cottage, Main Street, Brough, HU15 1RJ
MW6 DQQ D. Jones, 6 Frondeg, Tredegar, NP22 3NT
M6 DQR G. Youll, 4 Shaftsbury Court, Barnstaple Road, Scunthorpe, DN17 1YB
M6 DQS Dr T. Meaker, 1 Chemin Des Brugues, Roquevidal, France, 81470
M6 DQT W. Martindale, 57 Limefield Street, Accrington, BB5 2AF
M6 DQV P. Patterson, 3 Barnes Close, Southampton, SO18 5FE
M6 DQW A. Hose, 9 Cothey Way, Ryde, PO33 1QY
MM6 DQY J. Dow, 58 Beatty Crescent, Kirkcaldy, KY1 2HS
M6 DQZ S. James, 35 Prospect Road, Dronfield, S18 2EA
M6 DRG G. Lythgoe, 137 Dore Avenue, Fareham, PO16 8DU
M6 DRH D. Hughes, Warrenwood, Granville Rise, Totland Bay, PO39 0DX
M6 DRI T. Mckinley, 7 May Close, Godshill, Ventnor, PO38 3HB
MM6 DRJ M. Cormack, 2 Coghill Street, Wick, KW1 4PN
M6 DRK D. Kurn, 10 Dymoke Road, Mablethorpe, LN12 2BF
MW6 DRM D. Machon, 22 Albert Street, Caerau, Maesteg, CF34 0UF
M6 DRO D. Drought, 14 Mcminnis Avenue, St Helens, WA9 2PL
M6 DRP D. Rogers, 24 Bramblewood Way, Halesworth, IP19 8JT
M6 DRR D. Roderick, 88 Broadway, Wakefield, WF2 8LY
MW6 DRV D. Jones, 59 Llewelyn Street, Aberdare, CF44 8LA
MW6 DRW D. Williams, Long Nursery Coppice Walton, Presteigne, LD82RE
M6 DRZ A. Potts, 103 Etherstone Street, Leigh, WN7 4HY
MM6 DSC K. Henderson, 66 Rashierigg Place, Longridge, Bathgate, EH47 8AT
MM6 DSD D. Mcmillan, 45 Eton Avenue, Dunoon, PA23 8DG
M6 DSE S. Clarke, 23 Sinclair Court, Scarborough, YO12 7SD
M6 DSH D. Sheehan, 46 Darwin Close, London, N11 1TA
M6 DSJ D. Johnson, 5 Blackbird Close, Thurston, Bury St Edmunds, IP31 3PF
M6 DSO D. Osborne, 1 Bramble Close, Wilnecote, Tamworth, B77 5GG
MW6 DSP D. Pitman, 5 Maesgwyn Street, Port Talbot, SA12 6PF
M6 DSQ M. Bayman, Marsworth, Tring, HP23 4LX
M6 DSR D. Rhodes, 66 Lindale Gardens, Blackpool, FY4 3PQ
M6 DSS W. Stewart, 43 Newlands Drive, Halesowen, B62 9DX
M6 DST D. Stubbs, 39 Torcross Way, Redcar, TS10 2RU
MW6 DSU J. Wynne, 8 Coed Artro, Llanbedr, LL45 2LA
M6 DSV S. Townsend, 13 Cornwall Drive, Bury, BL9 9ET
M6 DSW D. Weight, 19 Shakespeare Drive, Upper Caldecote, Biggleswade, SG18 9DD
M6 DSY S. Waters, 28C Cliff Road, Dovercourt, Harwich, CO12 3PP
M6 DSZ D. Seedhouse, 8 Levett Road, Tamworth, B77 4AB
M6 DTA D. Easden, 20 Brunel Way, Calne, SN11 9FN
M6 DTC D. Cane, 98 Lancaster Road, Northolt, UB5 4TL
M6 DTD K. Worton, 39 Staite Drive, Cookley, Kidderminster, DY10 3UA
MI6 DTE D. Best, 13 Cranley Green, Bangor, BT19 7FE
M6 DTF M. Bell, 36 Schneider Road, Barrow-in-Furness, LA14 5DW
M6 DTH D. Hodgson, Flat 21, John Norgate House, Two Rivers Way, Newbury, RG14 5TF
M6 DTJ D. Brown, 8 Tees Cresent, Spennymoor, DL16 6QU
M6 DTL B. Jarvis, 398 Aldermans Green Road, Coventry, CV2 1NN
M6 DTM D. Mcgarrigle, 40 The Glade, Waterlooville, PO7 7PE
M6 DTN D. Thurman-Newell, 95 Drake Avenue, Minster On Sea, Sheerness, ME12 3RZ
M6 DTO G. Howe, 3 Halton Close, Lincoln, LN6 0YZ
M6 DTP M. Topham, Scardale, Thorn Bank, Hebden Bridge, HX7 5HS
M6 DTR S. Rhenius, Baythorne Cottage, Baythorne End, Halstead, CO9 4AB
M6 DTS D. Mason, 15 Lake Street, Dudley, DY3 2AU
M6 DTT D. Timbrell, 15A Firgrove Crescent Yate, Bristol, BS37 7AH
M6 DTU M. Woodruff, 5 Regency Heights, Caversham, Reading, RG4 7RH
MM6 DTV I. Campbell, 1 Carbostbeg, Carbost, Isle of Skye, IV47 8SH
M6 DTX P. Rose, 10 Milton Street, Worthing, BN11 3NE
M6 DTY C. Bell, 28 New Forest Motel, 230 Hurn Road, Ringwood, BH24 2BT
M6 DUB T. Price, 40 East Street, Kidderminster, DY10 1SE
M6 DUD M. Pugh, 31 Cambridge Street, Reading, RG1 7PA
M6 DUE D. Duell, Manor House, 144 Stonhouse Street, London, SW4 6BE
M6 DUF L. Duffy, 46 Church Road, Worcester, WR3 8NU
M6 DUH B. Brown, 40 Stansfield Road, London, SW9 9RZ
M6 DUI G. Walton, 7 Burlington Street, Ulverston, LA12 7JA
M6 DUJ A. Arnold, 25 Church Lane, Saxilby, Lincoln, LN1 2PE
M6 DUK S. House, 12 Glebe Close, Abbotsbury, Weymouth, DT3 4LD
MW6 DUL D. Davies, 2 Hendre Ddu, Manod, Blaenau Ffestiniog, LL41 4BH
M6 DUM J. Rowe, 9 Corfield Close, Finchampstead, Wokingham, RG40 4PA
M6 DUN A. Rigler, 10 The Ball, Dunster, Minehead, TA24 6SD
M6 DUO A. Richards, 16 Fruiterers Arms Caravan Park, Uphampton Lane, Ombersley, Droitwich, WR9 0JW
MI6 DUP R. Mcauley, 37 Ladyhill Road, Antrim, BT41 2RF
MI6 DUR A. Savage, 469 Old Belfast Road, Bangor, BT19 1RQ
M6 DUT T. Collins, 56 Grasvenor Avenue, Barnet, EN5 2DB
MM6 DUV D. Loughren, 25 Cuiken Avenue, Penicuik, EH26 0DR
M6 DUW T. Brooks, 200 Kingsway, College Estate, Hereford, HR1 1HE
M6 DUX S. Jones, 3 Brockton, Lydbury North, SY7 8BA
M6 DUZ B. Davies, 12 Scalebor Gardens, Burley In Wharfedale, Ilkley, LS29 7BX
M6 DVA J. Bligh-Wall, 3 George Street, Elworth, Sandbach, CW11 3BL
M6 DVE D. Thomson, 20 Westlea, Clowne, Chesterfield, S43 4QJ
M6 DVF K. George, Wylye, Auberrow, Hereford, HR4 8AN
M6 DVH D. Harrison, 30 Horseshoe Drive, Cannock, WS12 0FR
MM6 DVM D. Simpson, 35 Westbourne Avenue Tillicoultry Clackmannanshire Fk136Pu, Tillicoultry, Fk136pu
MI6 DVN M. Devlin, 17 Moninna Park, Newry, BT35 8PP
MW6 DVP D. Price, 11 Cefn Melindwr, Capel Bangor, Aberystwyth, SY23 3LS
MW6 DVQ A. Jones, 31 Russell Terrace, Carmarthen, SA31 1SZ
MM6 DVR R. Dover, Brooklet, Dodside Road, Newton Mearns, Glasgow, G77 6PZ
M6 DVS E. Johnson, 3 Conifer Way, Dunmow, CM6 1WU
M6 DVT D. Truscott, 37 Langley, Chulmleigh, EX18 7BQ
M6 DVU M. Hughes, 183 Station Road, Hednesford, Cannock, WS12 4DP
M6 DVV S. Storey, 10 Amble Way, Trimdon Station, TS29 6DZ
M6 DVW S. Challis, 73 Rivenhall Way, Hoo, Rochester, ME3 9GF
M6 DVY J. Sawyer, 18 The Mead, Dunmow, CM6 2PD
M6 DVZ J. Haywood, 8 Cedar Close, Market Rasen, LN8 3BE

M6 DWA Dr A. Dickson, 1 Roebuck Drive Baldwins Gate, Newcastle-under-Lyme, ST5 5FE
MM6 DWC K. Mclachlan, 13 Dick Terrace, Penicuik, EH26 8BW
M6 DWE D. Webb, 52 Simpkin Close, Eaton Socon, St Neots, PE19 8PD
M6 DWF D. Waring, 12 Mary Street, Farnhill, Keighley, BD20 9AU
M6 DWG D. Gaskell, 10 Freshford, St Helens, WA9 3WT
M6 DWI C. Hopper, 31 Mary Road, Deal, CT14 9HW
M6 DWJ D. Johnston, 34 Coxley Crescent, Netherton, Wakefield, WF4 4LR
M6 DWM D. Mitchard, 49 Gladstone Road, Broadstairs, CT10 2HY
M6 DWO A. Martin, 97 The Maltings, Dunmow, CM6 1BY
MM6 DWP D. Park, 54 Coblecrook Gardens, Alva, FK12 5BL
M6 DWS D. Saunders, 17 Sandy Lane, Prestwich, Manchester, M25 9RU
M6 DWT D. Potts, 103 Etherstone Street, Leigh, WN7 4HY
M6 DWV M. Tointon, 13 Ridgeway, Broadstone, BH18 8DY
M6 DWW D. Ward, 11 Pearce Road, Sheffield, S9 4JG
M6 DWY R. Friar, 16 Upper Field Close, Redditch, B98 9LE
M6 DWZ D. Whitaker, 67 Oakington Avenue, Amersham, HP6 6SX
M6 DX 350 DX, c/o L. Marsland, 154, Moss Lane, Litherland., Liverpool, L21 7NN
M6 DXA A. Kendrick, 29 Waterside, Silsden, Keighley, BD20 0LQ
M6 DXC S. Crawford, 18 Albatross Close, Standish, Wigan, WN6 0WB
M6 DXG D. Gaffney, 9 Tudor Road, Newton Abbot, TQ12 1HT
MM6 DXH I. Craig, 1/2 Drumlanrig Square, Hawick, TD9 0AS
M6 DXI J. Mee, 42 Potters Mead, Wick, Littlehampton, BN17 7HY
MM6 DXJ S. Lewington, 2 Coulliehare Cottages, Udny, Ellon, AB41 7PH
M6 DXL M. Groves, 15 Plains Lane, Littleport, Ely, CB6 1RJ
M6 DXN D. Curtis, 7 Neale Close, Aylsham, Norwich, NR11 6DJ
M6 DXQ D. Baker, 26 Lighthouse Close, Happisburgh, Norwich, NR12 0QE
M6 DXS D. Martin, 25 Broadmead Road, Blaby, LE8 4AB
M6 DXV G. Frayne, 98 De Lacy Court, New Ollerton, Newark, NG22 9RW
M6 DXW D. Sawyer, Realm House Heathfield Road, Bembridge, PO35 5UQ
M6 DXZ P. Jones, 18 Meadowbrook Road, Wirral, CH46 0RS
M6 DYA D. Greaves, 2 Willow Walk, Keynsham, Bristol, BS31 2TR
M6 DYB D. Brough, 38 Tynedale Avenue, Crewe, CW2 7NY
M6 DYC P. Richards, 1 Dorrs Drive, Watton, Thetford, IP25 6HB
M6 DYD M. Jones, 1 Elizabeth Court, Elizabeth Avenue, Norwich, NR7 0GY
MW6 DYF J. Jones, 40 Maes Mona, Amlwch, LL68 9AT
M6 DYG S. Aspinall, 3 Grasswood Road, Wirral, CH49 7NT
MM6 DYH D. Houston, The Knowe, Hardgate, Castle Douglas, DG7 3LD
M6 DYI L. Edmonds, 3 Waterlow Road, London, N19 5NJ
MW6 DYL D. Witts, 82 Park View, Llanharan, Pontyclun, CF72 9SB
MD6 DYM J. Dunbar, 3 Glenview Terrace, Port Erin, Isle of Man, IM9 6HA
M6 DYN C. Hall, 14 Marsh Road, Cowes, PO31 8JQ
M6 DYO D. Osborne, 3 Low Farm Road, Tunstall, Norwich, NR13 3PU
M6 DYP R. Laver, 20 Hall Street, Church Gresley, Swadlincote, DE11 9QU
M6 DYR A. Hateley, 26 Keelinge Street Tipton West Midlands Dy4 8Uq, Tipton, DY4 8UQ
M6 DYU Dr H. Coghlan, 1Bell View Cross Houses, Shrewsbury, SY5 6JJ
M6 DYV R. Dunn, 29 Hawe Lane, Sturry, Canterbury, CT2 0LL
M6 DYW S. Jones, 28A Park Road, Fowey, PL23 1ED
M6 DYX M. Carter, 15 Lashbrooks Road, Uckfield, TN22 2AY
M6 DYY N. Christopher, 161 Manor Road, Verwood, BH31 6DX
M6 DZA D. Gilbert, 34 Sullivan Way, Elstree, Borehamwood, WD6 3DH
M6 DZB M. Bennetts, 2 Chywoone Terrace, Newlyn, Penzance, TR18 5NR
MM6 DZC T. Whyte, Easter Unthank Farmhouse, Duffus, IV30 5RN
M6 DZF A. Tiling, 9 Coombe Street, Coventry, CV3 1GG
M6 DZH H. Kassier, 26 Higher Port View, Saltash, PL12 4BX
M6 DZN A. Bertoneri, Flat 1, 1 Albert Road, Nottingham, NG9 2GU
M6 DZP M. Marsh, 25 Southdown Road, Seaford, BN25 4PD
M6 DZQ E. Zieba, 12B Chingford Avenue, Chingford, London, E4 6RP
M6 DZR D. Malins, 36 Church Croft, Madley, Hereford, HR2 9LT
M6 DZS Dr Z. Derzsi, 217 Bensham Road, Bensham, Gateshead, NE8 1US
M6 DZT J. Emery, Mulberry Cottage, Quarry Lane, Combe St Nicholas, Chard, TA20 3PH
M6 DZV D. Abbott, 3 Brewhouse Lane, Soham, Ely, CB7 5JD
M6 DZX A. Norton, Flat 9, Brownhill Court, Southampton, SO16 9LB
M6 DZY M. Parker, 35 Prescelly Close, Nuneaton, CV10 8QA
M6 DZZ D. Hunt, 6 Lewisham Terrace, Newtown, Berkeley, GL13 9NP
M6 EAA E. Johnston, 67 Eversfield Road, Horsham, RH13 5JS
MI6 EAC E. Luchi, 68 Grovehill Gardens, Bangor, BT20 4NS
MW6 EAD C. Jones, 33 Graig Ebbw, Rassau, Ebbw Vale, NP23 5SF
M6 EAE J. Collier, 133 Woodstock Road, Moston, Manchester, M40 0DG
MI6 EAF G. Monteith, 66 Allen Park Dunamanagh, Strabane, BT82 0PD
M6 EAH M. Bell, 2 Hawthorne Road, Blyth, NE24 3DT
MI6 EAI E. Taylor, 17 Rutherglen Street, Belfast, BT13 3LR
M6 EAJ J. Lavery, Flat 2, Mentmore House, Mentmore Road, Ramsgate, CT12 6RY
M6 EAL E. Westwood, 17 Ennerdale Drive, Congleton, CW12 4FR
M6 EAN I. Parsons, 44 Hungerford Crescent, Bristol, BS4 5HQ
MW6 EAO J. Johnson, 1, Golygfar Eglwys Ruabon, Wrexham, LL14 6TD
M6 EAP R. Taylor, 18A Barnhall Road, Tolleshunt Knights, Maldon, CM9 8HA
MI6 EAS Dr J. Henderson, 1 Brook Lodge Ballinderry Lower, Lisburn, BT28 2GZ
M6 EAT R. Carr, Rambler House, Hill, Berkeley, GL13 9EB
M6 EAW E. Whitehouse, 16 Rue Gaston De Caillavet, Paris, France, 75015
M6 EAX F. Davies, 64 Main Road, Northwich, CW9 8PB
M6 EBA A. Thompson, 30 Birchwood Avenue, Lincoln, LN6 0JB
M6 EBC M. Petchey, 74 Avondale, Ellesmere Port, CH65 6RW
MM6 EBD A. Connell, 39 Glebe Crescent, Maybole, KA19 7HZ
M6 EBE E. Fogarty, Noke Farm, Hogscross Lane, Coulsdon, CR5 3SJ
M6 EBG J. Thomas, 128 Nuns Way, Cambridge, CB4 2NS
M6 EBI C. Bakes, Container City Building, 48 Trinity Buoy Wharf, London, E14 0FN
MM6 EBJ K. Mccuish, 6 Lighthouse Buildings, Birch Drive, Isle of Islay, PA43 7HZ
M6 EBL R. Clarke, 37B Northdown Park Road, Margate, CT9 2NH
M6 EBO B. Clayton, 26 Wood Walk, Mexborough, S64 9SG
M6 EBP D. Morton, 3 Pritchett Road, Birmingham, B31 3NL
M6 EBQ D. Stanley, 58 Wells Gardens, Basildon, SS14 3QS
M6 EBR A. Buckland, 21 Malton Close, Monkston, Milton Keynes, MK10 9HR
MI6 EBS E. Mcknight, 14 Marlacoo Beg Road, Portadown, Craigavon, BT62 3TF
M6 EBU L. Wells, 54 Giffords Cross Avenue, Corringham, Stanford-le-Hope, SS17 7NH
M6 EBW C. Rowles, 15 Rockhampton Walk, Colchester, CO2 8UJ
M6 EBX S. Lucas, Sunnyside, Church Road, Woodbridge, IP13 7NU
M6 EBY C. Barrett, 5 Oakapple Drive, Dereham, NR19 2SR
MM6 EBZ E. Ball, 32 Corton Lea, Ayr, KA6 6GJ
MW6 ECB E. Brady, 24 Gregory Avenue, Colwyn Bay, LL29 7ND
MI6 ECC G. Haslem, 124 Castle Rise, Tandragee, Craigavon, BT62 2NF
M6 ECD C. Dennis, 1 West Villa, Crathorne, Yarm, TS15 0BA
M6 ECE G. Mutch, 94 Abbotswood Road, Brockworth, Gloucester, GL3 4PF
M6 ECH E. Hearn, 41 Romney Avenue, Newcastle, ST5 7JR
M6 ECJ A. Meszaros, 9 Dunster Gardens, Cheltenham, GL51 0QT
M6 ECM G. Stapylton, 5 Coxcomb Walk, Crawley, RH11 8BA
MM6 ECO V. Nelson, 63 Melbourne Avenue, Clydebank, G81 4QD
MW6 ECR K. Gulliford, 5 The Square Abertridwr, Caerphilly, Glamorgan, CF83 4DH
M6 ECT E. Singleton, 1 Ermin Park, Brockworth, Gloucester, GL3 4BD
MI6 ECV C. Rafferty-Floyd, 25 Glenside, Omagh, BT79 7GL
M6 ECW C. Wilson, 31 Violet Road, South Woodford, London, E18 1DG
M6 ECX J. Steventon, 11 Wyvern Way, Blandford Forum, DT11 7XQ
M6 ECZ J. Hart, Flat 4, 17 Trinity Gardens, Folkestone, CT20 2RP
M6 EDA A. Bache, 62 Whittingham Road, Halesowen, B63 3TP
MM6 EDB J. Beck, Brigadoon, Stair Street, Drummore, Stranraer, DG9 9QE
M6 EDD E. Young, 210 High Street, Gt Wakering, SS3 0LS
M6 EDG N. Chamberlain, 8 Southfields, Binbrook, Market Rasen, LN8 6DX
M6 EDH E. Hitchins, 50-52 Moorland Road, Weston-Super-Mare, BS23 4HR
M6 EDI J. Cherry, 12 Scarisbrick New Road, Southport, PR8 6PY
M6 EDJ D. Jeffery, 4 Sandhurst Drive, Beeston, Nottingham, NG9 6NH
MI6 EDK K. Pearson, 17, Knocknamoe Road, Omagh, BT79 7LB
M6 EDL M. Arblaster, 22 Wood Lane, Carlton, Barnsley, S71 3JJ
M6 EDM P. Short, 15 Longcroft Court, Birchwood Crescent, Chesterfield, S40 2HT
M6 EDN A. Strange, 38 Manor Road, Martlesham Heath, Ipswich, IP5 3SY
M6 EDO D. Rasbarry, 27 Royle Close, Romford, RM2 5PS
MI6 EDP A. Connolly, 68 Willowbank Gardens, Belfast, BT15 5AJ
MW6 EDQ Dr J. Blaxland, 7 Maes Y Sarn, Pentyrch, Cardiff, CF15 9QQ
MW6 EDR E. Rees, 23 Northlands Park, Bishopston, Swansea, SA3 3JW
M6 EDS E. Kaye, 119 St. Bernards Avenue, Louth, LN11 8AS
M6 EDV P. Tranter, Swn Y Gwynt, Trefeglwys, Caersws, SY17 5PU
M6 EDW J. Tranter, Swn Y Gwynt, Trefeglwys, Caersws, SY17 5PU
M6 EDZ E. Twagirayezu, 33 Morton Street, Stoke-on-Trent, ST6 3PN
M6 EEA M. Rabl, 14 Bramble Close, Durrington on Sea, Worthing, BN13 3HZ
M6 EEB J. Cameron, 29 Webster Road, Stanford-le-Hope, SS17 0BE
MI6 EEC S. Carlin, 9 Mullandra Park, Kilcoo, Newry, BT34 5LS
M6 EED C. Wright, 24 Charlemont Crescent, West Bromwich, B71 3DA
M6 EEF A. Al Busaidi, 18 Sirdar Road, Southampton, SO17 3SJ
M6 EEG A. Gow-Barber, Plum Tree Villa, Butchers Lane, Ormskirk, L39 6SY
M6 EEH C. Branch, 62 Turnpike Road Connor Downs, Hayle, TR27 5DT
MW6 EEJ J. Nicholas, Reservoir House, St. Lythan'S, Wenvoe, CF5 6BQ
M6 EEL L. Layland, 3 Thirlmere Road, Golborne, Warrington, WA3 3HH
M6 EEM D. Roberts, 3 Heather Avenue, Melksham, SN12 6FX
M6 EEN P. Farren, 89 Fosterd Road, Newbold, Rugby, CV21 1DE
M6 EEO S. Davey, 38 Wordsworth Street, Barrow-in-Furness, LA14 5SE
M6 EEP M. Bartlett, 7 Thoresby, Tamworth, B79 7SQ
MM6 EEQ G. Brown, 37 Shandon Crescent, Bellshill, ML4 1LE
MM6 EER E. Grant, 6 Lindsay Place, Wick, KW1 4PF
M6 EET P. Sladen, 25 Linden Grove, Beeston, Nottingham, NG9 2AD
M6 EEU A. Bullard, 15 Rowan Drive, Lutterworth, LE17 4SP
M6 EEW A. George, 15 Ely Road, Croydon, CR0 2LW
MM6 EEX L. Grant, 6 Lindsay Place, Wick, KW1 4PF
M6 EEZ P. Murphy, 17 Birley Moor Avenue, Sheffield, S12 3AQ
M6 EFC D. Shannon, 68 Thornhill Road, Claydon, Ipswich, IP6 0DZ
MI6 EFD E. Hudson, 10 Kiloanin Crescent, Banbridge, BT32 4NU
M6 EFE T. Barnden, 27 Milton Road, Wokingham, RG40 1DE
M6 EFF E. Phipps, 12 Salisbury Close, Wokingham, RG41 4AJ
M6 EFG J. Jones-Robinson, 32 Verity Close, London, W11 4HE
MM6 EFH P. Donachie, 53 Seaforth Avenue, Wick, KW1 5NE
M6 EFJ A. Davies, 20 Hope Street, Halesowen, B62 8LU
MW6 EFK A. Williams, 27 Hick Street, Llanelli, SA15 1AR
M6 EFL A. Miariti, 19 Camelot Avenue, Nottingham, NG5 1DW
M6 EFM E. Falomir Montanes, 31 Albany Road, Birmingham, B17 9JX
M6 EFP P. Williamson, 22 Earls Road, Shavington, Crewe, CW2 5EZ
M6 EFR D. Richman, 42 Hathaway Road, Fleetwood, FY7 7JH
M6 EFU S. Nyakabau, 4 Vale View, Charvil, Reading, RG10 9SJ
MW6 EFV J. Dyer, Abergwawr House, Belmont Terrace, Aberdare, CF44 6UW
M6 EFW C. Steele, 40 Landor Road, Whitnash, Leamington Spa, CV31 2JX
M6 EFY R. Bell, 2 Hawthorne Road, Blyth, NE24 3DT
M6 EGA V. Santhosh, 119 Vaughan Road, Harrow, HA1 4EF
M6 EGB I. Kendrick, 10 Bankhouse Drive, Congleton, CW12 2BH
MM6 EGC J. Flannigan, 21 Kirkbean Avenue Rutherglen, Glasgow, G73 4EA
MW6 EGD P. Jones, 2 Fourth Avenue, Gwersyllt, Wrexham, LL11 4EE
M6 EGE J. Powell, 23 Park Road, Norton, Malton, YO17 9DZ
M6 EGF H. Reeves, 15 Mill Rise, Kidsgrove, Stoke-on-Trent, ST7 4UR
M6 EGG M. Jackson, 1 Kindred Barns, Ludlow, SY8 4LF
MW6 EGH E. Barker, 7 Upper High Street, Bedlinog, CF46 6RY
M6 EGJ A. Chudasama, 19 Walton Court, Bolton, BL3 6QP
M6 EGK C. Suddell, Lynhurst, Littleworth Lane Partridge Green, Horsham, RH13 8JX
M6 EGL K. Hendricks, Stranton, Darsham Road, Westleton, Saxmundham, IP17 3AH
M6 EGM L. Karaalp, 45 Baird Grove, Kesgrave, Ipswich, IP5 2DQ
M6 EGN A. Blackburn, 81 Belvedere Road, Ipswich, IP4 4AD
M6 EGO P. Bailey, 103 Jarden, Letchworth Garden City, SG6 2NZ
M6 EGP J. Churchill, West Winds, Brandheath Lane, New End, Astwood Bank, Redditch, B96 6NG
M6 EGS M. Sparrow, 21 Langwell Crescent, Ashington, NE63 8AB
M6 EGT J. Drake, 38 Fawcett Road, Stevenage, SG2 0EJ
M6 EGU A. Sharma, 152 Ladybarn Lane, Manchester, M14 6RW
M6 EGV C. Cousins, 43 Avon Close, Little Dawley, Telford, TF4 3HP
M6 EGW M. Delaney, 24 Lonsdale Road, Manchester, M19 3FL
M6 EGX M. Cross, 11 Polyplatt Lane, Scampton, Lincoln, LN1 2TL
M6 EGZ M. Rozier, 5 Pond Piece, Brandeston, Woodbridge, IP13 7AW
M6 EHA M. Philpott, Garmisch, Hazel Road, Aldershot, GU12 6HP
M6 EHB N. Livingstone, 2 Mickleton, Wilnecote, Tamworth, B77 4QY

M6 EHC E. Harris, 147 Longdown Road, Congleton, CW12 4QR
M6 EHD E. Delasalle, 31 West Hill Road, Hoddesdon, EN11 9DL
M6 EHE L. Hayward, The Chimes, Farley Way, Hastings, TN35 4AS
M6 EHF S. Bateson, 2 Green Crescent, Coxhoe, Durham, DH6 4BE
M6 EHH N. Hunt, 21 Reams Close, Fishtoft, Boston, PE21 0LL
M6 EHI D. Humm, 15 Sherborne Road, Farnborough, GU14 6JS
M6 EHK S. Buruma, 40, 40, Groningen, Netherlands, 9746PL
MM6 EHL S. Nicol, 9 Cowley Street, Methil, Leven, KY8 3QG
M6 EHM J. Jackson, 49 Leafield Rise, Two Mile Ash, Milton Keynes, MK8 8BX
M6 EHO G. Gibbs, 11 Fieldway Avenue, Leeds, LS13 1ED
M6 EHP P. Garraway, The Poplars, Crowell Road, Chinnor, OX39 4HP
M6 EHQ P. Simmen, 7 Thorpe Road, Thornton, Bradford, bd133at
M6 EHR P. Marshall, 5 Thorntree Court Crofton, Wakefield, WF4 1SP
M6 EHU D. Hodgson, 11 Harmony Place, Mountain, Queensbury, Bradford, BD13 1LD
M6 EHV W. Bradley, 4 Forest View Avenue, London, E10 6DX
M6 EHW P. Fellows, Flat 29, 1 Oundle Road, Peterborough, PE2 9PB
MW6 EHX B. Bull, Swan Cottage, Swan Road, Welshpool, SY21 0RH
M6 EHY K. Brundle, 17 The Paddocks, Hailsham, BN27 3AQ
M6 EIA J. Rutter, 19 Shaftesbury Avenue, Great Harwood, Blackburn, BB6 7ST
MW6 EIB A. Timms, Fistral, Red Bank, Welshpool, SY21 7PN
M6 EIC S. Hammond, 1 Elmer Close, Bognor Regis, PO22 6JU
M6 EIE R. Rose, 30 Brudenell Close, Cawston, Rugby, CV22 7GN
M6 EIF K. Goss, 57 Nursery Road, Leicester, LE5 2HQ
M6 EIG R. Webb, Norbury, Terrills Lane, Tenbury Wells, WR15 8DD
M6 EIH K. Harris, 78 Dovedale Road, Thurmaston, Leicester, LE4 8NB
M6 EII J. Rodgers, 5 Richil House 7 Ayston Road, Oakham, LE15 9RL
M6 EIJ J. Iason, 23 Baydon Grove, Calne, SN11 9AT
MW6 EIK G. Edwards, 17 Glan Y Mor Road, Penrhyn Bay, Llandudno, LL30 3NL
M6 EIL S. Wilkes, 24 Shelley Grove, Droylsden, Manchester, M43 7YG
M6 EIM J. Darmont, 114 Bowers Avenue, Norwich, NR3 2PS
M6 EIR A. Mulcahy, 85 Grifon Road, Chafford Hundred, Grays, RM16 6NP
M6 EIS I. Mahoney, 74 Radegund Road, Cambridge, CB1 3RS
M6 EIU G. Farnbank, 2 Market Hill, Foulsham, Dereham, NR20 5RU
M6 EIW D. Blake, Pound Farm, Swan Lane, Leigh, Swindon, SN6 6RD
MM6 EIX G. Edgar, 98 Barnton Road, Dumfries, DG1 4HN
M6 EIY J. Lee, 30 Wentworth, Yate, Bristol, BS37 4DJ
M6 EIZ G. Denman, 123 Links Avenue, Norwich, NR6 5PQ
M6 EJA A. Ashton, 46 Kingsland, Harlow, CM18 6XL
M6 EJB P. State, 53 Long Lane, Shirebrook, Mansfield, NG20 8AZ
M6 EJC E. Crosby, 16 Tweed Avenue, Ellington, Morpeth, NE61 5ES
M6 EJF R. Glynn, 106 Fairway Avenue, West Drayton, UB7 7AP
MW6 EJG A. Zennadi, 54 Llys Gwyrdd, Henllys, Cwmbran, NP44 7LS
MW6 EJI R. Colson, 1 Wildon Cottages, Twyn Allwys Road, Abergavenny, NP7 9RS
M6 EJJ M. Corr, Flat 7, Sovereign Court, 3 Winn Road, Southampton, SO17 1EH
MI6 EJK J. Milligan, 12 Rose Park, Limavady, BT49 0BF
M6 EJL M. Leggett, 10 Manor Road, Long Stratton, Norwich, NR15 2XR
MM6 EJO B. Holderness, 13 Glencairn Street, Camelon, Falkirk, FK1 4LY
M6 EJP J. Siddle, 7 Farebrother Street, Grimsby, DN32 0NH
M6 EJR A. Graham, 34 Balmoral Road, Stockport, SK4 4EB
M6 EJS E. Scott, 26 Fyfe Road, Corby, Nn172rd
MI6 EJT C. Rafferty, 25 Glenside, Omagh, BT79 7GL
M6 EJV A. Bowman, The Glen, Vicarage Road, Bude, EX23 8LN
M6 EJW E. Williamson, 70 Douglas Drive, Stevenage, SG1 5PH
M6 EJX R. Dickerson, 61 Highfield Terrace, Queensbury, Bradford, BD13 2BE
M6 EJY M. Boddy, 4 Witton Close, Reedham, Norwich, NR13 3HJ
M6 EKA E. Armstrong, 30 Tennyson Avenue, Hull, HU5 3TW
M6 EKB E. Beech, 124 Evering Avenue, Poole, BH12 4JH
M6 EKC S. Klymenko, Flat 9, Dinerman Court, 38-42 Boundary Road, London, NW8 0HQ
M6 EKD M. D'Arcy, 1 Moorfield Avenue, Denton, Manchester, M34 7TF
M6 EKE J. Welch, 130 Sandy Lane, Upton, Poole, BH16 5LY
M6 EKF G. Haime, 5 Poets Way, Dorchester, DT12fe
M6 EKI G. Lloyd, 1 Holmside Terrace, Stanley, DH9 6ET
M6 EKK A. Powell, Crosstrees, Main Road, Theberton, Leiston, IP16 4RX
M6 EKL S. Ross, 3 Highlands, Lakenheath, Brandon, IP27 9EU
M6 EKM M. Collyer, 26 Beaumont Drive, Northampton, NN3 8PS
M6 EKO S. Johnson, Willow End Cottage, Willow Corner, Thetford, IP25 6SS
M6 EKP S. Milner, Pavilion House, School Lane, Ormskirk, L40 3TG
M6 EKQ M. Isbell, 20 Woodland Crescent, Wolverhampton, WV3 8AS
M6 EKS J. Goodson, 109 Millside, Stalham, Norwich, NR12 9PB
MW6 EKT A. Price, 141 Attlee Way, Cefn Golau, Tredegar, NP22 3TE
M6 EKV A. Rimmer, 9 Weymouth Avenue, St Helens, WA9 3QX
M6 EKW D. Rimmer, 41 Ashburton Road, Wallasey, CH44 5XB
M6 EKX C. Evenden, 36 Castle View Road, Fareham, PO16 9LA
M6 EKY A. Irwin, 145 Coppermill Lane, London, E17 7HD
M6 ELB W. Astill, 14 Barlow Road, Exton, Oakham, LE15 6BL
M6 ELC J. Parkes, 24 Kenilworth Road, Lichfield, WS14 9DP
M6 ELD A. Titmus, The Old Police House, Arundel Road, Fontwell, Arundel, BN18 0SX
M6 ELE E. Nichols, 61B Norwich Common, Wymondham, NR18 0SW
M6 ELF K. Windass, 58 Nicholas Gardens, High Wycombe, HP13 6JG
M6 ELH N. Henderson, 14 Herbert Street, Carlisle, CA1 2QE
M6 ELI R. Wells, 27 Victoria Avenue, Camberley, GU15 3HT
M6 ELJ S. Girdwood, 8 Blaydon Walk, Wellingborough, NN8 5YU
M6 ELL K. Ellison, 117 Lichfield Road, Sunderland, SR52NX
M6 ELN D. Shuttleworth, 11 Bladen Close, Countesthorpe, Leicester, LE8 5SB
M6 ELQ L. Cullip, 11 Spiers Close, Tadley, RG26 3SF
M6 ELR E. Reeve, 12 Sime Street, Worksop, S80 1TD
M6 ELS J. Fay, 16 Foxhill, Whissendine, Oakham, LE15 7HP
MM6 ELU G. Ross, 5 Main Street, Alford, AB33 8QA
MM6 ELW P. Thornley, 13 Keose Lochs, Isle of Lewis, hs2 9jt
MW6 ELX E. Davies, 26 Heol Sant Gattwg, Llanspyddid, Brecon, LD3 8PD
M6 ELY E. Cockett, 2A Priory Avenue, Petts Wood, Orpington, BR5 1JF
M6 ELZ E. Luckett, 180 Clarendon Place, Dover, CT17 9QF
M6 EMA E. Bansil, 65 Hervey Street, Northampton, NN1 3QL
M6 EMC J. Landless, 2 Aspen Way, Banstead, SM7 1LE
M6 EME E. Dunstan, 57 Orchard Vale Flushing, Falmouth, TR11 5TT
M6 EMF P. Cotton, 29 Peake Avenue, Nuneaton, CV11 6DW
M6 EMG D. Gray, 68 Endeavour Way, Hythe Marina Village, Southampton, SO45 6LA
M6 EMH N. Asling, 9 First Avenue, Halifax, HX3 0DL
MI6 EMI K. Mikicki, 17 Glenhoy Drive, Belfast, BT5 5LB
M6 EMJ E. Jubb, 59 Buckingham Road Conisbrough, Doncaster, DN12 3DG
MW6 EMK M. Klimaszewski, 1 Woodwards Cottages, New Brighton, Wrexham, LL11 3ED
M6 EML R. Copus, 58 Colchester Rd, Holland on Sea, CO15 5DG
M6 EMM E. Barton, 86 Forge Lane, Kingswinford, DY6 0LG
M6 EMO J. Carey, 68 Queen Elizabeth Way, Woking, GU22 9AJ
M6 EMP P. Holman, 20 Green Drive, Wolverhampton, WV10 6DW
M6 EMQ A. Johnson, Avonworth, Grange Road, Bedford, MK43 7HJ
M6 EMS C. Shaw, 52 Margaret Avenue, Halesowen, B63 4BX
M6 EMT K. Titmarsh, 8 Bainbridge Close, North Walsham, NR28 9UP
M6 EMV E. Varley, Flat 8, Brookside Court, Liverpool, L23 0TT
M6 EMW S. Hunter, 9 Gelt Burn, Didcot, OX11 7TZ
M6 EMX B. Di-Giulio, 4 Highlands, Lakenheath, Brandon, IP27 9EU
M6 EMY E. Porter, 16 The Oval, Scarborough, YO11 3AP
M6 EMZ E. Stubbs, 39 Torcross Way, Redcar, TS10 2RU
M6 ENB T. Gurney, 24 Langley Way, Kettering, NN15 6HL
M6 ENC F. Pauling, Kingswood Farm House, Dalehouse Lane, Kenilworth, CV8 2JZ
M6 END D. Kataria, 87 Oakhill Road, Horsham, RH13 5LH
M6 ENE V. Roets, 7 Thorne Close, Harworth, Doncaster, DN11 8SN
M6 ENF E. Briggs, 20 Broad Lane, Pelsall, Walsall, WS4 1AP
M6 ENH F. Baker, 275 Bye Pass Road, Beeston, Nottingham, NG9 5HS
M6 ENI C. Passey, 1 Forest House, Baxworthy, Bideford, EX39 5SF
M6 ENJ O. Karaalp, 45 Baird Grove, Kesgrave, Ipswich, IP5 2DQ
M6 ENK A. Seelig, 9 Warren House Court 17 St Peters Avenue Caversham, Reading, RG4 7RG
M6 ENM T. Smythe, 19 Lime Close, Witham, CM8 2PA
M6 ENN S. Burton, 20 Flowerdown Avenue, Cranwell, Sleaford, NG34 8HZ
M6 ENP K. Matthews, St. Helens Cottage, Flimby, Maryport, CA15 8RX
MW6 ENQ M. Mckenna, 25 Heaton Place Norton Road, Rhos On Sea, Colwyn Bay, LL284TL
MI6 ENR M. Carruthers, 4 Kernan Avenue, Portadown, Craigavon, BT63 5TB
M6 ENS D. Spencer, 38 Town House Road, Nelson, BB9 9LL
M6 ENU C. Gordon, 9 Park Road, Camberley, GU15 2SP
M6 ENV D. Lock, Flat 33, Harrison Court, Harrison Close, Hitchin, SG4 9SG
M6 ENW R. Ferreira, 22 Vereker Road, London, W14 9JS
M6 ENX M. Reid, 17 Douglas Close, Hartford, CW81SH
MW6 ENY K. Miles, 25 Park Street, Penrhiwceiber, Mountain Ash, CF45 3YW
M6 ENZ S. Margrave, 30 Newtown Road, Bedworth, CV12 8QU
M6 EOA C. Shaw, 26 Grant Road, Spixworth, Norwich, NR10 3NN
M6 EOB B. Bland, 12 Park Lane, Pickmere, Knutsford, WA16 0JX
M6 EOC P. Flanagan, 71 Fellway, Pelton Fell, Chester le Street, DH2 2BY
MM6 EOE A. Mather, 38 Shandon Crescent, Balloch, Haldane, G83 8EX
M6 EOK P. Wilson, 45 Newquay Close, Hartlepool, TS26 0XG
M6 EOM D. Kawayida, Flat 7, Centre Point, London, SE1 5NU
M6 EON I. Patrick, Beech Cottage, Church Road, Friskney, Boston, PE22 8RD
MW6 EOP L. Rowlands, 14 Claerwen, Gelligaer, Hengoed, CF82 8EW
M6 EOQ R. Edmonds, 65 Green Lane, Redhill, RH1 2DF
M6 EOR D. Kershaw, 2 Croftlands, Neddy Hill, Carnforth, LA6 1JE
M6 EOS A. Beresford, 22 Tennyson Road, Rotherham, S65 2LR
M6 EOT L. Seaton, 4 Highfield Close, Empingham, Oakham, LE15 8QB
MM6 EOU A. Gorman, 36 Harestanes Road, Armadale, Bathgate, EH48 3LA
M6 EOV L. Marsh, 15 Buttercup Way, Southminster, CM0 7RY
M6 EOW E. Taylor, 11 Carlton Street, Featherstone, Pontefract, WF7 6AA
M6 EOX I. Smith, 4 Lammas Leas Road, Market Rasen, LN8 3AP
M6 EOY I. Jones, 90 Preston, Cirencester, GL7 5PR
MM6 EOZ G. Headridge, 79 Sheriffs Park, Linlithgow, West Lothian, EH49 7SR
M6 EPA T. Bannister, 6 Tanners Road, North Baddesley, Southampton, SO52 9FD
M6 EPD D. O'Shea, 37 Barclay Court, Ilkeston, DE7 9HJ
MW6 EPE P. Plummer, 26 Hill Road, Neath Abbey, Neath, SA10 7NR
M6 EPF A. Cook, 84 Clent View Road, Birmingham, B32 4LW
M6 EPH D. Hollis, 192 Rodbourne Road, Swindon, SN2 2AF
M6 EPJ P. Jensen, 36 Douglas Street, Derby, DE23 8LH
M6 EPL N. Wachs, 59 Broad Oak Way, Cheltenham, GL51 3LL
M6 EPN D. Cocks, 30 Great Arler Road, Leicester, LE2 6FF
M6 EPO A. Smith, 5 Boundary Close, Holcombe, BA3 5FP
M6 EPQ D. Tomlinson, 74 Bradshaw Avenue, Riddings, Alfreton, DE55 4AA
M6 EPS S. Cook, 270 Heneage Road, Grimsby, DN32 9NP
M6 EPT J. Rosa, Flat 7, Wick Hall, Abingdon, OX14 3NF
MM6 EPV A. Fraser, Garden Cottage, Moy, Inverness, IV13 7YQ
M6 EPW E. Waller, 27 Main Street, Frizington, CA26 3SA
MW6 EPX Dr D. Bentley, 9 Pen-Y-Lan Place, Cardiff, CF23 5HE
M6 EPZ N. Highfield, 298 Mersea Road, Colchester, CO2 8QY
MI6 EQA J. Mcgoldrick, 45 Stewarts Road, Dromara, Dromore, BT25 2AN
M6 EQB M. Prince, Crantrock, Fourth Avenue, Greytree, Ross-on-Wye, HR9 7HR
MI6 EQC E. Hamill, 24 Beechvalley, Dungannon, BT71 7BN
MI6 EQD A. Clearn, 182A Clonmore Road, Dungannon, BT71 6HX
M6 EQE K. Missenden, 47 Roseacre Drive, Elswick, Preston, PR4 3UQ
MW6 EQF P. Mcguinness, The Coach House, Cwmdauddwr, Rhayader, LD6 5HA
M6 EQG A. Nicholson, 147 Watling Avenue, Seaham, SR7 8JG
M6 EQJ A. Sherwin, 10 Closes Side Lane, East Bridgford, Nottingham, NG13 8NA
M6 EQK J. Tarrant, 70 Sunnymead, Midsomer Norton, Radstock, BA3 2SD
M6 EQL R. Paul, Mayo, Raleigh Park, Barnstaple, EX31 4JD
MM6 EQM D. Frost, Lanfair, Kennethmont, Huntly, AB54 4NN
M6 EQN D. Akerman, The Brick Barn, Coppice Farm, Ross-on-Wye, HR9 7QW
M6 EQO A. Walton, 11 Parkfield Road, Northwich, CW9 7AR
M6 EQQ M. Stiles, 13 Westover Close, Ivybridge, PL21 9BA
M6 EQR D. Leonard, 19 Nancy Street, Manchester, M15 4FZ
M6 EQS J. Khan, 123 Highfield Road, Hall Green, Birmingham, B28 0HR
M6 EQT D. Taylor, Flat 27, Laurel Court, 24 Stanley Road, Folkestone, CT19 4RL
M6 EQV A. Gould, 57 Northfield Road, Onehouse, Stowmarket, IP14 3HE
MM6 EQW C. Hutton, 1 Hillend Cottage, Roberton, Biggar, ML12 6RR
MM6 EQY K. Archibald, 31 Westwood Park, Deans, Livingston, EH54 8QP
M6 EQZ S. Carlisle, 103 Crossway Plympton, Plymouth, PL7 4HZ
M6 ERA C. Jenkins, 25 Longmeadow Grove West Heath, Birmingham, B31 4SU
M6 ERC E. Thornes, 11 Ringtale Place, Baldock, SG7 6RX

M6 ERD R. Overy, 62 Dykelands Road, Sunderland, SR6 8ER
M6 ERF E. Felton, 29 Pavitt Meadow, Galleywood, Chelmsford, CM2 8RQ
M6 ERH E. Halford, 69 Thirlmere Avenue, Astley, Manchester, M29 7PZ
M6 ERI M. Konstantynowicz, 12 Long Wall, Haddenham, Aylesbury, HP17 8DL
M6 ERN R. Springall, 18 Westbourne Park, Scarborough, YO12 4AT
MM6 ERO Z. Bak, 62/6 North Gyle Loan, Edinburgh, EH12 8LD
M6 ERP S. Jones, 36 Woodford Crescent, Burntwood, WS7 9AE
M6 ERR N. Wood, 10 Perriclose, Chelmsford, CM1 6UJ
M6 ERS T. Glenday, 52 Hollow Road, Bury St Edmunds, IP32 7AZ
M6 ERU M. Cook, 194 Exeter Road, Kingsteignton, Newton Abbot, TQ12 3NJ
M6 ERW I. Westby, 5 Rosklyn Road, Chorley, PR6 0NJ
MW6 ERZ E. Gibson, 8 Llanthewy Close, Croesyceiliog, Cwmbran, NP44 2PF
MW6 ESA S. Beavis, 65 Old Road, Baglan, Port Talbot, SA12 8TU
M6 ESH M. Young, 4 Priestburn Close, Esh Winning, Durham, DH7 9NF
MM6 ESL G. Robinson, Asgard, 12 Upper Waston Road, Burray, KW172TT
M6 ESO P. Melling, 23 Appletree Close, Cottenham, Cambridge, CB24 8UJ
M6 ESP E. Phiri, 26 Sedge Road, Andover, SP11 6RL
M6 ESR A. Hunt, 4 Weedon Road, Swindon, SN3 4EE
M6 ESS M. Chamberlain, 30 Roxton Rd, Great Barford, MK44 3 LR
M6 EST R. Aldridge, 37 Vincent Road, Luton, LU4 9AN
M6 ESV P. Taylor, 104 Winstanley Drive, Leicester, LE3 1PA
MW6 ESW E. Sweeney, 67 Blandon Way, Cardiff, CF14 1EH
M6 ESY Dr S. Prior, 17 Queens Walk, London, NW9 8ES
M6 ESZ E. Palo, 13 Welwyn Close, St Helens, WA9 5HL
M6 ETA D. Wall, 227 Wayfield Road, Chatham, ME5 0HJ
M6 ETC P. Hayes, 4 London Road, Roade, Northampton, NN7 2NL
M6 ETD R. Sindall, 16 Chantrell Road, Wirral, CH48 9XP
MI6 ETE W. Curry, 7 Ballyversal Road, Coleraine, BT52 2ND
M6 ETG P. Loades, 126 Wealcroft, Gateshead, NE10 8QS
M6 ETH E. Heath, 63 Meadway, Dunstable, LU6 3JT
M6 ETJ T. Widdowson, 18 Newland Road, Banbury, OX16 5HQ
M6 ETL D. Arthur, 10 Lyndhurst Court, Lyndhurst Road, Hove, BN3 6FZ
M6 ETN R. Bradshaw, 272 Councillor Lane, Cheadle Hulme, Cheadle, SK8 5PN
M6 ETP E. Price, 47 Albany Road, Reading, RG30 2UL
M6 ETR S. Guest, 19 Ellesmere Avenue, Ashton-under-Lyne, OL6 8UT
M6 ETS N. Luckett, Flat 3, 25 Upton Park, Slough, SL1 2DA
M6 ETU P. Gaur, 34 Queensberry Avenue, Copford, Colchester, CO6 1YN
MW6 ETW R. Atkins, 88 Park Crescent, Abergavenny, NP7 5TN
MW6 ETY D. Pesticcio, Ty Ffynnon Farm, St. Mellons Road, Cardiff, CF3 2TX
M6 ETZ M. Harvey-Ross, Flat 4, Harley Court, Church Road, Southampton, SO31 9GD
MM6 EUA C. Lockerbie, 22 Charlotte Street, Aberdeen, AB25 1LR
M6 EUB O. Wilson, 99 Farnham Road, Durham, DH1 5LN
MW6 EUC T. Coghlan, 13 Michaels Road, Rhyl, LL18 4SH
M6 EUE N. Austerfield, 22B Princes Avenue, Withernsea, HU19 2JA
M6 EUH D. Levett, 11 Love Lane, London, SE25 4NG
M6 EUI D. Nolan, Flat 7, Fonthill Court, London, SE23 3SJ
MW6 EUK S. Taylor, 9 Crud Yr Awel, Prestatyn, LL19 8YQ
M6 EUL N. Ramsey, Dalestones, Lansdown Road, Bath, BA1 5TB
MW6 EUO A. Jones, 75 Hollybush Road, Cardiff, CF23 6SZ
M6 EUP J. Wilson, 125 Langroyd Road, Colne, BB8 9ED
M6 EUU R. Scholey, 2 Newfield Crescent, Wath-Upon-Dearne, Rotherham, S63 6JN
M6 EUV A. Shitov, Flat 4, 31 St. Leonards Road, Exeter, EX2 4LR
M6 EUW M. Augustus, 3 Heathend Cottages Heathend, Wotton-under-Edge, GL12 8AS
M6 EUX R. Bradley, 116B Old Hawne Lane, Halesowen, B63 3ST
M6 EUZ R. Smith, 2 Queen Street, Boston, PE21 8XB
M6 EVA M. Knowles, 12 Dalestorth Avenue, Mansfield, NG19 6NT
M6 EVC J. Chapelle, 7 Elizabeth Way, Stowmarket, IP14 5AX
M6 EVD C. Harper, 4 Bentley Avenue, Jaywick, Clacton-on-Sea, CO15 2JW
M6 EVE B. Evered, Ivy Cottage, Rookham Hill, Rookham, Wells, BA5 3AL
M6 EVH H. Woodfin, 8 Bank Hall Close, Bury, BL8 2UL
M6 EVI C. Charles, 83 Dibdale Road, Dudley, DY1 2RX
M6 EVK P. Hadley, Narrow Boat Oregon The Moreings, Kinver, DY7 6LG
M6 EVL A. Hoile, 4 Lindale Mount, Wakefield, WF2 0BH
M6 EVM Capt. J. Thompson, Rose Cottage, Mickleton, Barnard Castle, DL12 0JD
MM6 EVO J. Rickerby, 12/10 Hermand Street, Edinburgh, EH11 1LR
M6 EVR S. Rodgers, 32 Merrimans Road Shirehampton Bristol, Bristol, BS11 0AG
M6 EVU M. Marrs, 43 Ely Close, Toothill, Swindon, SN5 8DB
M6 EVW E. Woodward, 309 Hartfields Manor, Hartfields, Hartlepool, TS26 0NW
M6 EVX C. Haynes, 25 Barnards Hill Lane, Seaton, EX12 2EQ
M6 EVY R. Dalton, 1 Ballard Estate, Four Lanes, Redruth, TR16 6QL
M6 EVZ D. Lynch, 7 Dollant Avenue, Canvey Island, SS8 9EJ
M6 EWC P. Jenkins, 137 Hawkhurst Road, Brighton, BN1 9EB
M6 EWD G. Dobson, 4 Durley Gardens, Orpington, BR6 9LL
M6 EWH M. Lovering, 16 Portland Avenue, Sittingbourne, ME10 3QY
M6 EWI R. Farhall, 2 Banks Cottages, Mountfield, Robertsbridge, TN32 5JZ
M6 EWJ R. Scott, 9 Burrows Close, Lawford, Manningtree, CO11 2HE
M6 EWK S. Dempster, 18 Salisbury Street, Swindon, SN1 2AN
M6 EWL C. Donovan, 130 Long Lynderswood, Basildon, SS15 5BG
M6 EWN D. Carmichael, 22 California Close Great Sankey, Warrington, WA5 8WU
M6 EWO P. Chadwick, 112 Sandy Lane, Warrington, WA2 9JA
M6 EWQ A. Williams, 5 Burleys Road, Crawley, RH10 7DB
MM6 EWR A. Robson, 100 Dawson Avenue, East Kilbride, Glasgow, G75 8LH
M6 EWT Dr A. Grounds, 49 Fletcher Way, Henfield, BN5 9FR
M6 EWU L. Schofield, 21 Wyche Close, Rudheath, Northwich, CW9 7TY
M6 EWV S. Wagstaff, 20 Hawfinch Road, Cheadle, Stoke-on-Trent, ST10 1RX
M6 EWW E. Whitewood, 2 Finn Farm Cottage, Finn Farm Road, Ashford, TN23 3EX
MM6 EWX A. Talplacido, 15 Park Avenue, Thurso, KW14 8JP
MM6 EWY A. Kinnersley, 5A Regent Terrace, Dunshalt, Cupar, KY14 7HB
M6 EWZ E. Wali Zangana, 78 Hivings Hill, Chesham, HP5 2PG
M6 EXA S. Morris, 5 Wensleydale Close, Stoke-on-Trent, ST1 6XR
M6 EXB T. Thompson, 14 Queen Street, Northwich, CW9 5JL
M6 EXC A. Hawksworth, 87 Bradeley Road, Haslington, Crewe, CW1 5PX
M6 EXD M. Chapman, 7 Dragons Lane, Shipley, Horsham, RH13 8GD
M6 EXE C. Rose, 55 Boddington Gardens, Biggleswade, SG18 0PH
M6 EXF S. Brown, 53 Prestwich Avenue, Worcester, WR5 1QF
M6 EXG A. Holt, 27 Ramsey Road, Middlestown, Wakefield, WF4 4QF
M6 EXH D. Phelps, 728 Sewall Highway, Coventry, CV6 7JJ
M6 EXI J. Edgson, 59 Gilmour Crescent, Worcester, WR3 7PJ
M6 EXJ S. Baldwin, 143 Oxford Road, Swindon, SN3 4JA
MW6 EXL L. Mead, 5 Hill Top, Ebbw Vale, NP23 6PJ
M6 EXN A. Gaskin, 44 Vale Road, Sutton, SM1 1QH
M6 EXO T. Crocker, 32 Godmanston Close, Poole, BH17 8BU
MI6 EXP A. Boyd, 27 The Meadows, Dungannon, BT71 6PW
M6 EXQ D. Harrison, 104 Gracemere Crescent, Birmingham, B28 0TZ
M6 EXR M. Jackson, 35 Marshall Road, Willenhall, WV13 3PB
M6 EXS J. Harrison, 69 The Cliff, Wallasey, CH45 2NN
M6 EXT R. Murray, 41 East Street, Rochdale, OL16 2EG
MI6 EXU T. Reid Bamford, 2 New Line, Carrickfergus, BT38 9DL
MW6 EXV E. Mcmorrow, 164 Maes Glas, Caerphilly, CF83 1JW
M6 EXW T. Williams, 5 Burleys Road, Crawley, RH10 7DB
M6 EXX B. Jackson, 2A Scrooby Street, Rotherham, S61 4PL
M6 EXY S. Lake, 85 Clarkson Road, Norwich, NR5 8ED
M6 EXZ G. Cockerell, 21 Coningsby, Bracknell, RG12 7BE
M6 EYA M. Rasell, 21 Avenue Sucy, Camberley, GU15 3EB
M6 EYB S. Crouch, 24 Church Road, Stanfree S44 6Aq, Chesterfield, S44 6AQ
M6 EYD S. Houssart, Flat 3, Virginia Court, London, SE16 6PU
M6 EYF A. Screen, Greenglade, Frith End, Bordon, GU35 0RA
M6 EYG P. Cooper, 3 Heath View Mill Lane, Grassmoor, Chesterfield, S42 5AD
M6 EYH S. Brown, 3 Grazebrook Croft, Birmingham, B32 3NL
MW6 EYI B. Ravenhill-Lloyd, 3 Bryn Awel, Conwy, LL32 8WB
M6 EYJ R. Miller, 23 Clarendon Road, Sevenoaks, TN13 1EU
M6 EYK J. Peden, 9 De Lacy Avenue Featherstone, Pontefract, WF7 6AY
M6 EYL D. Humphrey, 11 Colborne Close, Poole, BH15 1UR
MW6 EYM D. Smith, 3 Charles Close, Abergavenny, NP7 6AP
M6 EYO R. Sansom, 72 Wannock Lane, Eastbourne, BN20 9SQ
M6 EYP S. Strange, 94 Digby Avenue, Nottingham, NG3 6DY
M6 EYR K. Sim, 49 St. Julians Wells, Kirk Ella, Hull, HU10 7AF
M6 EYS P. Smith, 29 Ford Hayes Lane, Stoke-on-Trent, ST2 0HB
M6 EYT Dr G. Thomas, 6 Bankside, Headington, Oxford, OX3 8LT
MW6 EYU E. Parry, 22 Park Road, Tanyfron, Wrexham, LL11 5SH
M6 EYV J. Allen, 17 Inglemere Gardens Arnside, Carnforth, LA5 0BX
M6 EYW G. Littlechild, 85 Long Road, Canvey Island, SS8 0JB
MW6 EYX C. Taylor, 23 Heol Derw, Brynmawr, Ebbw Vale, NP23 4TT
M6 EYY D. Wilkinson, 20 Manchester Road, Barnoldswick, BB18 5PR
M6 EYZ D. Mullard, 46 Green Lane, Clanfield, Waterlooville, PO8 0JX
M6 EZA G. Iredale, Ship Cottage, Main Street, Maryport, CA15 7DX
M6 EZC M. Bentley, 5 Stokewell Road, Wath-Upon-Dearne, Rotherham, S63 6EL
M6 EZE M. Wood, 22 Prickett Road, Bridlington, YO16 4AT
M6 EZF J. Holland, 1 Ravenwood, Swadlincote, DE11 9AQ
M6 EZG R. Zerafa, 2 Furnwood, Bristol, BS5 8ST
M6 EZK J. Power, 12 Campbell Gordon Way, London, NW2 6RS
M6 EZM A. Metselaar, 5A Canons Corner, Edgware, HA8 8AE
M6 EZN A. Barrett-Sprot, 1 Malting End, Wickhambrook, Newmarket, CB8 8YH
MM6 EZO G. Irving, 55 Gillbank Avenue, Carluke, ML8 5UW
M6 EZP J. Crabb, 23 Chaffinch Crescent, Billericay, CM11 2YX
M6 EZR P. Harper, 36 Barrow Close, Marlborough, SN8 2BD
M6 EZS J. Marks, Chantry End, Oak Hill, Epsom, KT18 7BU
M6 EZT A. Wareham-Kirk, Orchard Cottage, Church Street, Haverhill, CB9 7SG
M6 EZV K. Gosney, 76 Westgate Lane, Lofthouse, Wakefield, WF3 3NS
M6 EZX D. Buchan, 40 Bradstone Road, Winterbourne, Bristol, BS36 1HQ
M6 EZY C. Cowley, Rushbrook, Leafield Road, Chipping Norton, OX7 6EA
M6 EZZ C. Garratt, 19 Cherry Tree Place, St Helens, WA9 2AF
M6 FAB L. Rolinson, 534 Haslucks Green Road, Shirley, Solihull, B90 1DS
M6 FAC B. Whelan, 147 Lawsons Road, Thornton-Cleveleys, FY5 4PL
M6 FAE S. Moorcroft, 44 Spencers Lane, Skelmersdale, wn9 8jr
M6 FAF J. Taylor, 19 Juniper Walk, Kempston, Bedford, MK42 7SX
M6 FAH S. Wilson, 63 Halford St, Tamworth, B79 7RA
M6 FAJ T. Loveland, 32 Whateley Lane, Whateley, Tamworth, B78 2ET
M6 FAK A. Ladd, 42 Plovers Way, Bury St Edmunds, IP33 2NJ
MW6 FAM A. Williams, 211A New Road, Skewen, Neath, SA10 6ET
MW6 FAN A. Burgess, 18 Fairmeadows, Maesteg, CF34 9JL
M6 FAS F. Southgate, 52 Jeffrey Lane, Belton, Doncaster, DN9 1LT
MI6 FAU W. Montgomery, 56 Hazelbank Road, Drumahoe, Londonderry, BT47 3NY
MW6 FAW S. Fawley, 21 Broad View Pontnewydd, Cwmbran, NP44 5JA
M6 FAX R. Stewart, 20 Siddal Street, Halifax, HX3 9BH
M6 FAY M. Furnivall, 10 Wilwick Lane, Macclesfield, SK11 8RS
MW6 FBA M. Young, 104 Ystrad Road, Pentre, CF41 7PW
M6 FBB M. Carr, 51 Langton Road, Holton-Le-Clay, Grimsby, DN36 5BH
M6 FBD D. Smith, Heath Farm, Heath Road, Woolpit, Bury St Edmunds, IP30 9RL
M6 FBE M. Money, 2 Lodge Farm Cottage, Black Horse Road, Norwich, NR10 5DJ
M6 FBF A. Robson, 6 Wood View, Maltby, Rotherham, S66 7PA
M6 FBH F. Hatfull, 16B Church Street, Easton On The Hill, Stamford, PE9 3LL
MW6 FBJ B. Jordan, 605 Monnow Way, Bettws, Newport, NP20 7DJ
M6 FBM R. Proudman, 7A Fithern Close, Dudley, DY3 1YA
M6 FBO J. Borthwick, 62A Rea Valley Drive, Birmingham, B31 3XE
M6 FBP B. Preskey, 14 Birchen Close, Chesterfield, S40 4JT
M6 FBQ J. Hewitt, 105 Princess Avenue, Pontefract, WF9 2qx
M6 FBR C. Chew, 9 Raven Park, Haslingden, Rossendale, BB4 4HN
M6 FBT I. Long, Flat 6, Kingfisher Court, Woking, GU21 6DQ
MM6 FBU A. Miller, 100 Kerrylamont Avenue, Glasgow, G42 0DW
M6 FBV S. Cooney, 56 Manor House Lane, Preston, PR1 6HN
M6 FBW S. Richards, 28 Lincoln Way, Thetford, IP24 1DG
M6 FBY S. Chau, Foley House, Heath Lane, Stourbridge, DY8 1QX
M6 FBZ A. Raven, 14 Paddock Close, Belton, Great Yarmouth, NR31 9NT
MM6 FCA R. Mcdonald, 135 Pappert, Alexandria, G83 9LG
M6 FCC F. Cornes, 18 Barke Street, Highley, Bridgnorth, WV16 6LQ
M6 FCD N. Fisher, 23 Matlock Road, Ferndown, BH22 8QT
M6 FCE C. Smith, 50 Oakwood Avenue, Wakefield, WF2 9JS
M6 FCF B. Taylor, 1 Whinchat Close, Stockport, SK2 5UU
M6 FCG G. Miles, 14 Woodlands Road, Great Shelford, Cambridge, CB22 5LW
M6 FCI A. Renny, 216 Malvern Road, Bournemouth, BH9 3BX
M6 FCJ D. Vincelli, 90 Broadbottom Road, Mottram, SK14 6JA
M6 FCM C. Boyle, 8 Westlees Close, North Holmwood, Dorking, RH5 4TN
M6 FCN A. John, Flat 10, Melbourne Court, 46 Seabourne Road, Bournemouth, BH5

2HT
M6 FCQ D. Aissa, 28 Sirdar Road, London, N22 6RG
M6 FCR B. Scannell, 60 Burnside Road, Dagenham, RM8 1XD
M6 FCS F. Spickernell, Stockstreet Farm, Mile Elm, Calne, SN11 0NE
M6 FCT M. White, 130 Main Street Walton, Street, BA16 9QX
M6 FCV G. Allen, 38 Edenside Kirby Cross, Frinton-on-Sea, CO13 0TQ
M6 FCW L. Heaney, 2 The Spinney, Eastleigh, SO50 8PF
M6 FCX L. Palmer, 39 Baird Grove, Kesgrave, Ipswich, IP5 2DQ
M6 FCZ C. Cooper, 25 Waterside Close, Loughborough, LE11 1LP
M6 FDD B. Evans, 2 Hastings Road, Eccles, Manchester, M30 8JR
M6 FDF D. Sankey, 30 Ballad Gardens, Manadon, Plymouth, PL5 3FF
M6 FDG D. Chapman, 27 Cuff Crescent, London, SE9 5RF
M6 FDH K. Wells, 45 Laburnum Avenue, Yaxley, Peterborough, PE7 3YQ
M6 FDI W. Dawson, 7 Field Close, Warboys, Huntingdon, PE28 2UT
M6 FDK A. Patmore, 5 Milton Close, Ramsey, Huntingdon, PE26 1LU
M6 FDL P. Dixon, 87, Ranger Drive, Wolverhampton, WV10 6BZ
M6 FDN M. Martin, 7 Carlton Drive, Priorslee, Telford, TF2 9SH
M6 FDR A. Raine, 91 Lulworth Avenue, Jarrow, NE32 3SB
M6 FDS S. Jackson, 5 Duchess Park Close, Shaw, Oldham, OL2 7YN
M6 FDU K. Gilpin, 181 Newport Road, Cowes, PO317ER
M6 FDW Dr T. Krishnamurthy, 64 Ingleby Road, Ilford, IG1 4RY
M6 FDX D. Mcmorrin, 24 Katherine Close, Addlestone, KT151NX
M6 FDY C. Sparks, 160 Wetmore Road, Burton-on-Trent, DE14 1QS
M6 FEA Dr F. Agboma, Flat 6, Mayfair Court, Edgware, HA8 7UH
MI6 FEB F. Beattie, 19 Derrin Road, Enniskillen, BT74 6AZ
M6 FEC P. Mills, 212 Carsic Road, Sutton-in-Ashfield, NG17 2BS
M6 FED H. Clark, 8 Snowberry Avenue, Belper, DE56 1RE
M6 FEE F. Mather, 98 Heathgate, Norwich, NR3 1PN
MW6 FEF M. Lewis, 6 Criccieth Close, Buckley, CH7 3QF
MW6 FEH R. Harris, 11 Maes Morgan, Llanrhaeadr Ym Mochnant, Oswestry, SY10 0LH
M6 FEI W. Darvill, 35 Allard Close, Northampton, NN3 5LZ
M6 FEJ A. Croston, Haldon, Poltimore Road, South Molton, EX36 4DA
M6 FEK R. Fekete, 22 St. Andrews Road, Ellesmere Port, CH65 5DG
M6 FEL M. Bryant, 19 Brooklands Road, Havant, PO9 3NS
M6 FEM S. Simpson, 48 Weatherhill Road Smallfield, Horley, RH6 9LY
M6 FEN S. Morris, 108 Templeside, Temple Ewell, Dover, CT16 3BA
M6 FEO R. Hunter, 3 Sandy Way, Croyde, Braunton, EX33 1PP
M6 FEP R. Phillips, 162 Glebelands, Pulborough, RH20 2JL
M6 FEQ C. Ford, 61 Gould Road, Barnstaple, EX32 8ET
M6 FEU P. Gould, 18 Tanfield Gardens, South Shields, NE34 7DY
MM6 FEX T. Mccallum, 3 Hillington Gardens, Glasgow, G52 2TP
M6 FEX S. Church, Sapphire Ridge, Mill Lane, Westbury, Brackley, NN13 5JS
M6 FEZ T. Ryder, 9A London Road, Slough, SL3 7RL
M6 FFB M. Hall, 67 Darlinghurst Grove, Leigh-on-Sea, SS9 3LF
M6 FFD A. Ramsbottom, 1 Barn Gill Close, Blackburn, BB2 3HU
M6 FFE E. Durkin, 30 Douglas Road West, Stafford, ST16 3NX
M6 FFF A. Beardsley, 10 Moreton Close, Church Crookham, Fleet, GU52 8NS
M6 FFI C. Lowe, 45 Winster Road, Chaddesden, Derby, DE21 4JY
M6 FFJ J. Preston, 3 Essex Drive, Kidsgrove, ST71HE
M6 FFK M. Ross, 11 Queens Place, Otley, LS21 3HY
M6 FFM S. Saville, Little Cranebrook, St. Michaels Road, Verwood, BH31 6JA
MW6 FFN G. Wilkins, 7 Byron Road, Newport, NP20 3HJ
M6 FFO Dr B. Issac, 9B Poplar Grove, Stockport, SK2 7JD
M6 FFP T. Skerton, 60 Oakington Avenue, Harrow, HA2 7JJ
MM6 FFQ D. Harvie, 58 Esk Drive, Livingston, EH54 5LE
M6 FFR R. Moys, 12A Palmerston Avenue, Fareham, PO16 7DP
M6 FFS R. Young, 48 Sussex Street, Cleethorpes, DN35 7NP
M6 FFT S. Merridale, The Granary, Falledge Lane, Upper Denby, HD8 8YH
M6 FFU L. Smith, 7 Hereson Road, Ramsgate, CT11 7DP
M6 FFV N. Calderley, 48 Woodlands, Horbury, Wakefield, WF4 5HH
M6 FFW Dr F. Harvey, 14 Wilton Road, Hornsea, HU18 1QU
M6 FFY M. Clarke, 4 Mill Lane, Brant Broughton, Lincoln, LN5 0RP
M6 FFZ R. Radley, 20 Thorntondale Drive, Bridlington, YO16 6GW
M6 FGA P. Meanwell, 20 Crow Park Avenue, Sutton-On-Trent, Newark, NG23 6QG
M6 FGC D. Newsome, 1 Freemans Wharf, Plymouth, PL1 3RN
M6 FGE M. Swan, 35 Colston Close, Plymouth, PL6 6AY
M6 FGG K. Legg, Bennetts, High Street, Thorpe-Le-Soken, Clacton-on-Sea, CO16 0EG
M6 FGH N. Speller, 2 Hurst Rise Road, Oxford, OX2 9HQ
M6 FGI K. Phillips, 10 Canhaye Close, Plymouth, PL7 1PG
MW6 FGJ M. Roberts, 38 Gwilliam Court, Monkton, Pembroke, SA71 4JL
M6 FGM B. Wilcock, 32 Mallard Road Scotton, Catterick Garrison, DL9 3NP
MW6 FGN N. Thomas, 8 Western Terrace, Blaengwynfi, Port Talbot, SA13 3YE
MW6 FGQ P. Passmore, 127 High Street, Neyland, Milford Haven, SA73 1TR
MM6 FGR A. Morrison, 4 West Murkle, Murkle, Thurso, KW14 8YT
M6 FGW D. Crouch, 7 Tresco Road, Berkhamsted, HP4 3JZ
M6 FGY B. Bestwick, 185 Ashbourne Road, Turnditch, Belper, DE56 2LH
M6 FGZ T. Crawley, 41 Lynmouth Drive, Ruislip, HA4 9BY
M6 FHB D. Bell, 27 Parkfields Avenue, London, NW9 7PG
M6 FHD D. Hartropp, 185 Leighton Road, London, NW5 2RD
M6 FHE J. Hawbrook, 7 Birkdale, Norwich, NR4 6AF
M6 FHF A. Bailie, 2 Loch Lane, Watton, Thetford, IP25 6HE
MI6 FHG Dr P. Donaghy, 41 Greenvale Manor Antrim, Antrim, BT41 1SB
M6 FHH F. Stisted, 19 Danehurst Street, London, SW6 6SA
M6 FHI C. Colless, 128 Ditton Lane, Fen Ditton, Cambridge, CB5 8SS
MI6 FHJ N. Cully, 42 Omerbane Road, Cloughmills, Ballymena, BT44 9PE
MW6 FHK N. Williams, 20 Railway Terrace, Pontyberem, Llanelli, SA15 5HN
M6 FHM B. Hodgkinson, 26 Daywell Rise, Rugeley, WS15 2RE
M6 FHN I. Popovic, 13 Walden Avenue, Arborfield, Reading, RG2 9HR
M6 FHO K. Hunt, Pound Farm, Gallows Hill, Diss, IP22 1RZ
M6 FHP P. Driver, 68 Ripon Road, Dewsbury, WF12 7LG
M6 FHQ J. Mount, The Limes, The Green, Wickhambreaux, Canterbury, CT3 1RQ
M6 FHR G. Daymond, 30 Elizabeth Drive, Newcastle upon Tyne, NE12 9QP
M6 FHS R. Goldup, 57 Partridge Way, Old Sarum, Salisbury, SP4 6PX
M6 FHV H. Peberdy, 53 Far Lane, Normanton On Soar, Loughborough, LE12 5HA
M6 FHX S. Powell, 29 Coppice End Road, Derby, DE22 2TA
M6 FHY D. Pearson, 37 Elmridge, Leigh, WN7 1HN
M6 FIA X. Zhang, Harrogate Ladies' College, Clarence Drive, Harrogate, HG1 2QG
M6 FIB M. Sutton, 4 Centenary Close Broughton, Chester, CH4 0FY
M6 FID P. Hack, Catte Street, Oxford, OX1 3BW
M6 FIG M. Fisher, 96 Reepham Road, Norwich, NR6 5PD
M6 FII J. Thurman, Warbanks Farm, Cockfield, Bury St Edmunds, IP30 0JP
M6 FIJ Dr D. De-Cogan, 52 Gurney Road, New Costessey, Norwich, NR5 0HL
MI6 FIK D. Wilson, 81 Parknasilla Way, Aghagallon, Craigavon, BT67 0AU
M6 FIL P. Dunnicliffe, 19 Woodland Road, Chelmsford, CM1 2AT
MW6 FIN F. Toomey-Langford, 8 Heol Dinas Isaf, Williamstown, Tonypandy, CF40 1NG
M6 FIO D. Smith, 62 Fulwoods Drive, Leadenhall, Milton Keynes, MK6 5LB
M6 FIR M. Firth, 209 High Street, Wickham Market, Woodbridge, IP13 0RQ
M6 FIT M. Bryan, 13 Elmwood Avenue, Sunderland, SR5 5AW
MI6 FIU P. Reid, 1 Nettlehill Mews, Lisburn, BT28 3HN
M6 FIV A. Foley, 23 Church Lane, Wymington, Rushden, NN10 9LW
M6 FIW R. Nicholson, 1 Isis Close, Aylesbury, HP21 9LY
MW6 FIX R. Hemming, 21 New Road, Jersey Marine, Neath, SA10 6JR
MW6 FIY S. Binnion, 55 Dythel Park, Pen-Y-Mynydd, Llanelli, SA15 4RR
M6 FIZ R. Faulkner, 31 Faulkland View, Peasedown St. John, Bath, BA2 8TG
M6 FJA A. Ferriroli, 142 Hillbury Road, Warlingham, CR6 9TD
M6 FJC C. Featherstone, 3 Pogmoor Road, Barnsley, S75 2EW
M6 FJD K. Dobson, 1 Howarth Road, Ashton-On-Ribble, Preston, PR2 2HH
M6 FJE C. Hinds, 15 Logan Way, Hemyock, Cullompton, EX15 3RD
M6 FJF J. Slater, 1 Highfield Court, Swinton, Mexborough, S64 8RF
M6 FJI J. Tonkyn, 82 Westfield Road, Harpenden, AL5 4JU
M6 FJJ F. Johnson, 12 Hollins Road, Harrogate, HG1 2JF
M6 FJL J. Robb, 37 Wroxham Road, Woodley, Reading, RG5 3AX
M6 FJM D. Marden, 1 Stroma Gardens, Hailsham, BN27 3AZ
M6 FJN M. Singer, 1 Bentley Road, Slough, SL1 5BB
M6 FJO I. Waddingham, 102 Nethershire Lane, Sheffield, S5 0QE
MW6 FJQ A. Lawson, Hafryn, Dyserth Road, Rhyl, LL18 5RB
MM6 FJS O. Sharp, 18 Urquhart Court, Kirkcaldy, KY2 5TX
MW6 FJT T. Cook, 19 Cae Bach Aur Estate, Bodffordd, Llangefni, LL77 7JS
M6 FJU P. Appleton, The Old Rectory, Station Road, Grimsby, DN36 5SQ
MM6 FJV W. Little, Burnside, Main Street, Lochans, Stranraer, DG9 9AW
M6 FJW S. Michael, 191 Sutton House, Scunthorpe, DN15 6SN
MW6 FJX C. Williams, 18 Caer Delyn, Llannerch-Y-Medd, LL71 8EJ
M6 FJZ S. Stewart-Sandeman, 1 Diomed Drive, Great Barton, Bury St Edmunds, IP31 2TF
M6 FKA L. Hall, 51 Bridgeacre Gardens, Coventry, CV3 2NQ
MM6 FKB S. Broll, 23 St. Medans, Monreith, Newton Stewart, DG8 9LL
MW6 FKC G. Stevenson, 3 Llwyn Rhosyn, Cardiff, CF14 6NS
MM6 FKD E. Rattray, Blashieburn Cottage, Blairingone, Dollar, FK14 7NT
M6 FKE K. Collings, 97 Jackmans Place, Letchworth Garden City, SG6 1RF
M6 FKF A. Cucchiara, 172 Gonville Crescent, Stevenage, SG2 9LZ
M6 FKG J. Shears, 161 Park Road, Keynsham, Bristol, BS31 1AS
M6 FKH J. Fensom, Foxdown House, Blandford Camp, Blandford Forum, DT11 8BP
M6 FKI W. Tennison, 85 Primrose Field, Harlow, CM18 6QT
M6 FKJ F. Johnson, 13 Honeymead Lane, Sturminster Newton, DT10 1EW
M6 FKN A. Blackwell, 27 Prince Charles Crescent, Farnborough, GU14 8DJ
M6 FKO G. Carver, 4 Andrews Road, Farnborough, GU14 9RY
M6 FKP C. Robinson, 3 Folly Hall Road, Bradford, BD6 1UL
M6 FKQ A. Smith, 116 Pilling Lane, Preesall, Poulton-le-Fylde, FY6 0HG
M6 FKR G. Howard, 8 Paddock Road, Woodford, Kettering, NN14 4FL
M6 FKS T. Nash, 32 Collington Rise, Bexhill-on-Sea, TN39 3RS
M6 FKU L. Harris-Pugh, 13 Myddelton Park, London, N20 0HT
M6 FKW A. Metcalf, 71 Harper Road, Coventry, CV1 2AL
M6 FKZ C. Spencer, 18 Coatsby Road Kimberley, Nottingham, NG16 2TH
MW6 FLA Dr F. Labrosse, 72 Ger Y Llan Penrhyncoch, Aberystwyth, SY23 3HQ
M6 FLB F. Buss, 44 Courtenay Road, Maidstone, ME15 6UL
M6 FLC C. Horridge, 6 Back Sreet, East Stockwith, Gainsborough, DN21 3DL
M6 FLE P. Threakall, 83 Gregory Avenue, Birmingham, B29 5DG
M6 FLF C. Wilson, 12 Desmond Avenue, Hornsea, HU18 1AF
MM6 FLG M. Bradshaw, 32 Greycraigs, Cairneyhill, Dunfermline, KY12 8XL
M6 FLH D. Adkin, 31 Fieldway, Broad Oak, Rye, TN31 6DL
M6 FLJ S. Bumstead, Middle Leys, Leys Farm, Withypool, Minehead, TA24 7RU
M6 FLK J. Kelly, 74 Hatfield Crescent, Bedford, MK41 9RB
M6 FLL D. Dobson, 1 Howarth Road, Ashton-On-Ribble, Preston, PR2 2HH
M6 FLN J. Horn, 8 Princess Close, Watton, Thetford, IP25 6XA
M6 FLQ K. Hobbs, 75 Coronation Walk, Gedling, Nottingham, NG4 4AS
M6 FLR J. Bicknell, 10 Farnefold Road, Steyning, BN44 3SN
M6 FLS D. Greenland, 1 Hilltop, Tuesley Lane, Godalming, GU7 1SB
M6 FLU J. Salter, 103 Cootes Avenue, Horsham, RH12 2AF
M6 FLW S. Miller, 14 Queens Croft, Queens Street, Swinton, Mexborough, S64 8NA
M6 FLX T. Clarke, 20 Cliff Road, Felixstowe, IP11 9PJ
M6 FLY J. Hall, 43 Norwood Drive, Brierley, Barnsley, S72 9EG
M6 FLZ R. Sutton, 80 Fishbourne Lane, Ryde, PO33 4EU
M6 FMC M. Straughan, 71 Silcoates Lane, Wrenthorpe, Wakefield, WF2 0PA
M6 FME F. Moulsdale, 63 Windermere Avenue, St Helens, WA11 7AG
M6 FMF C. Bassi, Cherries, Hadham Road, Ware, SG11 1LH
M6 FMG P. Booker, 17 Colton Copse, Chandler'S Ford, Eastleigh, SO53 4HQ
M6 FMI O. Earl, 46 Rowan Way, Chelmsley Wood, Birmingham, B37 7QT
M6 FMJ K. Browne, 4 Darcey Drive, Brighton, BN1 8LF
M6 FML F. Lowndes, 46 Danes Road, Manchester, M14 5JS
M6 FMN G. Morris, 6 Eastcourt Road, Worthing, BN14 7DB
M6 FMP F. Morson-Pate, 3 Tudor Lawns, Carr Gate, Wakefield, WF2 0UU
M6 FMS F. Mace, 40 Quarry House Gardens, East Rainton, Houghton le Spring, DH5 9RD
M6 FMT A. Webber, 73 Greenwood Road, Yeovil, BA21 3LF
M6 FMU J. D'Aubray-Butler, Copse Edge 16 Francis Road, Frodsham, WA6 7JR
MW6 FMV M. Plowman, Ffynnonwen, Plwmp, Llandysul, SA44 6EY
MW6 FMW C. Jones, 32 Vale Street, Denbigh, LL16 3BE
M6 FMY N. Mountford, 78B Meeting House Lane, London, SE15 2TX
M6 FMZ K. Bianchini, 10 St. Leonards Road, Headington, Oxford, OX3 8AA
MW6 FNA Dr E. Flikkema, 7 St. James Mews, Great Darkgate Street, Aberystwyth, SY23 1DW
M6 FNB N. Berry, The Mount, Deerfold, Bucknell, SY7 0EF
M6 FNE R. Merrifield, 6 Miersfield, High Wycombe, HP11 1TX
M6 FNF M. Byrd, 17A Castle Gates, Shrewsbury, SY1 2AB
M6 FNG M. Grice, 48 St. Ives Road, Coventry, CV2 5FZ

MW6 FNK F. Williams, 5 Tyn Rhos Estate, Gaerwen, LL60 6HL
MM6 FNL G. Cunningham, 17 Robb Place, Castle Douglas, DG7 1LW
M6 FNM R. Bound, 24 Buttington Road, Shrewsbury, SY2 5TS
MI6 FNO C. Johnston, 32 Maghereagh Road, Randalstown, Antrim, BT41 4NS
MM6 FNQ K. Jones, 6 Traill Street, Castletown, Thurso, KW14 8UG
M6 FNR F. Riches, 4 Priory Close, Chelmsford, CM1 2SY
MW6 FNT P. Taylor, 11 Watford Close, Watford Park, Caerphilly, CF83 1NQ
M6 FNU K. Cole, 30 Wood Road, Rotherham, S61 3RQ
MW6 FNV D. Robins, 25 Clos Mancheldowne, Barry, CF62 5AB
M6 FNX N. Farrington-Smith, 3 Milton Road, Wokingham, RG40 1DE
M6 FNY D. Chatterton, 3 Hunt Close, South Wonston, Winchester, SO21 3HY
MI6 FNZ R. Mckay, 31 Squires Hill Crescent, Belfast, BT14 8RE
M6 FOD C. Dempster, 18 Concorde Way, Gloucester, GL4 4PU
M6 FOE L. Downs, 18 Burnopfield Gardens, Newcastle upon Tyne, NE15 7DN
M6 FOJ D. Sutherland, 78 Holmden Avenue, Wigston, LE18 2EF
M6 FOK C. Fletcher, 18 Kings Road, Barnetby, DN38 6HJ
M6 FOL W. Bartle, 5 Crosskeys Row Chapel Milton, Chapel En le Frith, SK230QQ
M6 FOR I. Forester, 35 Thackeray Street, Sinfin, Derby, DE24 9GY
M6 FOT S. Moore, 170 Walsall Road, Aldridge, Ws90jt
M6 FOX S. Spence, 8 Teasdale Street, Consett, DH8 6AF
M6 FOY F. Foy, 4 The Square, East Rounton, Northallerton, DL6 2LB
M6 FOZ A. Foster, 47 Westmorland Drive, Desborough, Kettering, NN14 2XB
M6 FPC J. Lord, 5 Langworthy Avenue Little Hulton, Manchester, M38 9GQ
M6 FPD L. Coveney, 72 Bonney Road, Leicester, LE3 9NH
M6 FPE R. Coveney, 72 Bonney Road, Leicester, LE3 9NH
M6 FPF F. Clifton, Battery Road, Paull, Hull, HU12 8FP
M6 FPG D. Worth, 33 Lime Grove Close, Leicester, LE4 0UG
MW6 FPH M. Clarke, 48 Brynglas Road Llanbadarn Fawr, Aberystwyth, SY23 3QR
MM6 FPI A. Mcclements, 24 Glen Brae, Bridge of Weir, PA11 3BH
M6 FPJ D. Lavell, 51 Kingfisher Close, Newport, PO30 5XS
MW6 FPL A. Muldoon, 29 Eider Close, St. Mellons, Cardiff, CF3 0DF
M6 FPM T. Smith*, 1 Bents Close, Clapham, Bedford, MK41 6DY
M6 FPN L. Varga, 14. Tegla Str., Miskolc, Hungary, 3516
M6 FPP J. Misztela, 15 Severus House, 155 Varcoe Gardens, Hayes, UB3 2FJ
M6 FPQ E. Turner, 2 Shepherds Close, Fen Ditton, Cambridge, CB5 8XJ
M6 FPU E. Duell, 28 Watergall, Bretton, Peterborough, PE3 8NA
M6 FPV T. Van Den Bosch, 8 Stopford Garth, Wakefield, WF2 6RT
M6 FPW S. Bale, 121 Washbrook Road, Rushden, NN10 6UR
MM6 FPX T. Mcconnell, 2 Stewartgill Place, Ashgill, Larkhall, ML9 3BB
MM6 FPY I. Blackstock, 149D Carlisle Road, Crawford, Biggar, ML12 6TP
MW6 FQA E. Parry, Blaen Gwenin, Llanrhystud, SY23 5BZ
M6 FQC M. Roberts, 41 Mcneill Avenue, Crewe, CW1 3NW
MW6 FQD M. Roberts, 20 Upper Robinson Street, Llanelli, SA15 1TR
M6 FQE K. De La Hunty, Bangers Whistle, Yarmouth Road, Newport, PO30 4LZ
M6 FQF A. Goldstein, 33 Hughenden Avenue, Harrow, HA3 8HA
M6 FQH C. Goldsmith, 15 Stephenson Road, Cowes, PO31 7PP
M6 FQJ B. Boxall, 10 Honor Avenue, Wolverhampton, WV4 5HH
M6 FQK D. Briggs, 5 Links Avenue, Cromer, NR27 0EQ
M6 FQL G. Milner, 17 Haragon Drive, Amesbury, Salisbury, SP4 7FS
M6 FQM A. James, 27A Lilliana Way, Bridgwater, TA5 2GG
M6 FQN R. Vickers, 3 Grenville Avenue, Goring-By-Sea, Worthing, BN12 6JE
M6 FQO Dr A. Lipson, Field House, The Haven, Cambridge, CB21 5BG
M6 FQP A. Phillips, 43 Jutland Road, Hartlepool, TS25 1LP
M6 FQR A. Coote, 148 Clarendon Street, Dover, CT17 9RB
M6 FQU R. Parker, 53 Tunstall Road, Canterbury, CT2 7BX
M6 FQW L. Jones, 206 Lowe Avenue, Wednesbury, WS10 8NS
M6 FQX M. Rowland, 30 Tunnmeade, Harlow, CM20 3HL
MI6 FQY B. Maguire, 39 Carland Road, Dungannon, BT71 4AA
M6 FQZ A. Williams, Corner Cottage, South Brent, TQ10 9JF
M6 FRC F. Clement, 40 Ellison Fold Terrace, Darwen, BB3 3EB
MM6 FRD F. Redden, 31 Lawfield, Coldingham, Eyemouth, TD14 5PB
M6 FRF R. Abel, 43 Church Street, Didcot, OX11 8DG
M6 FRG J. Marshall, 1 Farriers Reach, Bishops Cleeve, Cheltenham, GL52 7UZ
M6 FRI J. Hewart, 14 Kestrel Close, Marple, Stockport, SK6 7JS
M6 FRJ D. Janowicz, 20 Salisbury Road, St Leonards-on-Sea, TN37 6RX
M6 FRK F. Waller, 19 Wortley Avenue S738Sb, Wombwell, s738sb
M6 FRO N. Froggatt, 6 Beech Grove, New Malden, KT3 3HR
M6 FRP N. Williams, The Bothy, Norton Bavant, Warminster, BA12 7BB
MI6 FRQ J. Coleman, 25 Kinnaird Street, Belfast, BT146BE
M6 FRR J. Hingley-Hickson, The White House, School Lane, Grantham, NG32 2ES
M6 FRT J. Parsons, 4 Saunders Close, Uckfield, TN22 2BX
M6 FRV F. Raven-Vause, 624 Hillbutts, Wimborne, BH21 4DS
MW6 FRW A. Middleton, 56 Caradoc Road, Prestatyn, LL19 7PF
M6 FRY C. Fryer, 14 Perks Road, Wolverhampton, WV11 2ND
M6 FRZ J. Potter, 21 Ulverston Crescent, Lytham St Annes, FY8 3RZ
M6 FSA S. Taylor, 18 Cycle Street, York, YO10 3LJ
MM6 FSB D. Kelly, 21 Dhailling Road, Dunoon, PA23 8EA
M6 FSC R. Cooper, 86 Grove Road, Tiptree, Colchester, CO5 0JG
M6 FSD M. Tayler-Grint, 94 Dairymans Walk, Guildford, GU4 7FF
M6 FSE L. Beaney, Brookside, Walkers Lane, Shorwell, Newport, PO30 3JZ
M6 FSG C. Moore, 25 Stonybeck Close, Westlea, Swindon, SN5 7AQ
M6 FSI D. Woolger, 17 Wyvern Close, Tangmere, Chichester, PO20 2GQ
M6 FSJ S. Burgess, 101 Brendon Way, Nuneaton, CV10 8NW
M6 FSM H. Morgan, 18 Chrystel Close, Tipton St. John, Sidmouth, EX10 0AY
M6 FSN A. Lillis, 4 Sinclair Close, Gillingham, ME8 9JQ
M6 FSO J. Forbes, Weald Barkfold Farm, Plaistow, Billingshurst, RH14 0PJ
MM6 FSP S. Paterson, Springbank, 2 Mains Of Cuffurach, Clochan, Buckie, AB56 5HP
MM6 FSQ C. Mackenzie, 89C Needless Road, Perth, PH2 0LD
MW6 FST W. Morris, Fairview, Trefonen, Oswestry, SY10 9DP
M6 FSU M. Cain, 59 Bailey Road, Leigh-on-Sea, SS9 3PJ
MM6 FSV P. Hadley, 51 Victoria Road, Buckhaven, Leven, KY8 1BG
M6 FSW P. Bowen, Arbor Tree Bungalow, Mill Street, Craven Arms, SY7 8EN
M6 FSZ D. Walker, 22 Fir Terrace, Esh Winning, Durham, DH7 9AG
MM6 FTA L. Pinkowski, 69 Nelson Avenue, Livingston, EH54 6BZ
MW6 FTC H. Lloyd, Apartment 126, Woodlands Hayes Road, Sully, Penarth, CF64 5QE
M6 FTF M. Goodman, 80 Clay Street, Soham, Ely, CB7 5HL
MM6 FTG F. Gorman, 11 Maxwell Drive, Baillieston, Glasgow, G69 6JB
M6 FTI T. Wetherill, Upper House Cottage, Holme Marsh, Kington, HR5 3JS
MW6 FTK M. Botham, 17 Laburnum Drive, Oswestry, SY11 2QW
M6 FTN D. Nathan, 138 Anchor Lane, Hemel Hempstead, HP1 1NS
M6 FTO G. Lawlor, 14 Woodkirk Close, Seghill, Cramlington, NE23 7TZ
M6 FTP R. Denham, 179 Pandon Court, Shield Street, Newcastle upon Tyne, NE2 1XY
M6 FTQ A. Street, 110 Magdalen Street, Colchester, CO1 2LF
M6 FTT M. Knowles, 11 Thorneycroft Avenue, Birkenhead, CH41 8HJ
M6 FTU R. Pownall, 17 Horsebridge Road, Blackpool, FY3 7BQ
M6 FTV S. Barber, 31 Aysgarth Avenue, Crewe, CW1 4QE
M6 FTW M. Bowman, 26 Albany Hill, Tunbridge Wells, TN2 3RX
M6 FTX V. Reeve, 29 Benfield Way, Portslade, Brighton, BN41 2DN
M6 FTY N. Baker, 13 Elm Way, Melbourn, Royston, SG8 6UH
M6 FUA C. Fear, 13 Frome Road, Chipping Sodbury, Bristol, BS37 6LD
M6 FUD A. Bentley, 40 Wyndale Road, Leicester, LE2 3WR
M6 FUE A. Davies, 15 The Crescent, Woodside Park, Poulton-le-Fylde, FY6 0QW
M6 FUF N. Irvine, 100 Cavendish Road, Sunbury-on-Thames, TW16 7PL
M6 FUH J. Reich, 11 Ravenscroft Road London, London, W4 5EQ
M6 FUJ H. Braithwaite, 20 Richard Moon Street, Crewe, CW1 3AX
M6 FUL M. Fuller, Rose Cottage, Wickmoor, Stathe, Bridgwater, TA7 0JR
M6 FUM Lord I. Kent, 111 Sinclair Avenue, Banbury, OX16 1BQ
MW6 FUN P. Hawkes, Brodawel Cynghordy, Llandovery, SA20 0LS
M6 FUQ A. Howard, 24 Ladybower Lane, Poulton-le-Fylde, FY6 7FY
M6 FUR D. Esdale, The Bell Inn, Central Lydbrook, Lydbrook, GL17 9SB
M6 FUT R. Harris, 63 The Drive, High Barnet, Barnet, EN5 4JG
M6 FUU S. Bloomfield, 20 Farmers Way, Copmanthorpe, York, YO23 3XU
M6 FUW C. Wilkinson, Flat 5, Hometide House Beach Road, Lee-on-the-Solent, PO13 9BP
M6 FUY A. Robnett, 38B Woodmere Avenue, Watford, WD24 7LN
MW6 FVA S. Matthews, Aberbran Fawr, Aberbran, Brecon, LD3 9NG
M6 FVC G. Gee, 17 Portherras Villas, Pendeen, Penzance, TR19 7TJ
MM6 FVD A. Mcdonald, 8 North Rayne Cottages, Meikle Wartle, Inverurie, AB51 5BY
M6 FVE A. Day, 47 Rembrandt Way, Bury St Edmunds, IP33 2LT
M6 FVH J. Mcmullen, 281 The Broadway, Cullercoats, North Shields, NE30 3LH
M6 FVJ P. Randerson, 7 Roman Crescent, Swindon, SN1 4HH
MM6 FVK R. Rothon, 112 Ravenswood Rise, Livingston, EH54 6PG
M6 FVL E. Smethurst, 4 Lower New Row, Worsley, Manchester, M28 1BE
M6 FVM J. Poriyath, 48 Howard Close, Cambridge, CB5 8QU
M6 FVN A. Storey, 1 Mill Hill Road, Bingham, Nottingham, NG13 8YR
M6 FVO M. Oates, 63 Longfellow Drive, Rotherham, S65 2LH
M6 FVP B. Straker, 37 Beech Grove Avenue, Garforth, Leeds, LS25 1EF
M6 FVR J. Rawley, 33 Dorset Way, Billericay, CM12 0UD
M6 FVS R. Sage, 9 North Hall Farm Bungalows, Barley Road, Royston, SG8 7PZ
MW6 FVT K. Young, 29 High Street, Coedpoeth, Wrexham, LL11 3RY
M6 FVV M. Roberts, Flat 6, 463 Brighton Road, Lancing, BN15 8LF
M6 FVW C. Fox, Millstone Cottage, Prior Wath Road, Scarborough, YO13 0AZ
M6 FVZ E. Preda, 4 Playsteds Lane, Great Cambourne, Cambridge, CB23 6GA
M6 FWB C. Andrews, 10 Hartford Road Hartley Wintney, Hook, RG27 8QW
M6 FWC H. Wilcox, 3 Hutton Street Hutton Wandesley, York, YO26 7ND
MI6 FWD J. Tipping, 16 The Oaks, Portadown, Craigavon, BT62 4HX
M6 FWE M. Scambell, 8 South Bank Road, East Cowes, PO32 6JE
M6 FWG S. Barclay, 64 Deepdene Avenue, Dorking, RH5 4AE
M6 FWL C. Pugsley, 86 West Town Lane, Bristol, BS4 5DZ
M6 FWM M. Tonkin, 8 Artis Avenue, Wroughton, Swindon, SN4 9BP
M6 FWO J. Capovila, 30 Church Street, Barrow-in-Furness, LA14 2JG
M6 FWP L. Churchill, 1 Fowlswick Cottages, Allington, Chippenham, SN14 6LU
M6 FWQ M. Duque, 2 Spekehill, London, SE9 3BN
M6 FWS M. Smith, 151 Rowlands Road, Worthing, BN11 3LE
M6 FWT M. Goslin, 44 Teasdale Road, Walney, Barrow-in-Furness, LA14 3SF
M6 FWV R. Sage, 29 Rosewarne Park, Connor Downs, Hayle, TR27 5LJ
M6 FWW D. Longson, 1 Beckside, Plumpton, Penrith, CA11 9PD
M6 FWX D. Jones, Drove Farm, Sheepdrove, Lambourn, Hungerford, RG17 7UN
M6 FWY D. Smith, 7 Oakley Grove, Wolverhampton, WV4 4LN
M6 FWZ Dr F. Da Dalt, 13 Johnstone Street, Bath, BA2 4DH
MW6 FXA S. Hodge, 2 Parc Cemlyn, Prestatyn, LL19 9NX
MU6 FXB R. Batiste, Asile De Paix, Clos Des Sablons, Sandy Lane, Guernsey, GY2 4RN
M6 FXE R. Hall, 134 Crowmere Road, Shrewsbury, SY2 5JY
M6 FXG T. Chapman, 37 Pheasant Way, Cirencester, GL7 1BJ
M6 FXI D. Thomas, 51 Barrhill Avenue, Brighton, BN1 8UE
M6 FXJ D. Parkes, 4 Round Saw Croft, Rubery, Birmingham, B45 9TT
M6 FXL T. Goodenough, 12 Jerounds, Harlow, CM19 4HE
MM6 FXM T. Johnston, The Old Schoolhouse, Luggate Burn, Haddington, EH41 4QA
M6 FXO K. Burton, 16 Lupton Close, Glasshouses, Harrogate, HG3 5QX
MM6 FXQ J. Anderson, The Grange, Leslie Road, Scotlandwell, Kinross, KY13 9JE
M6 FXU R. Ball, 7 Cliff Closes Road, Scunthorpe, DN15 7HT
MM6 FXW R. Bisset, 8/1 North Bughtlin Brae, Edinburgh, EH12 8XH
M6 FXX C. Freckelton, 44 Hawton Road Newark, Newark, NG24 4QB
M6 FXY L. Gregory, 11 Roundhill Road, Castleford, WF10 5AF
MM6 FXZ W. Goodfellow, 1 Yester Place., Haddington, EH41 3BE
M6 FYB P. Amond, Treetops, Gedding, Bury St Edmunds, IP30 0QD
M6 FYD L. Clark, 43 Grange View Harworth, Doncaster, DN11 8QP
M6 FYE P. Escott, 84 Salisbury Avenue, Bootle, L30 1PZ
MM6 FYF T. Burnett, 45 The Murrays Brae, Edinburgh, EH17 8UF
M6 FYG T. Weston, 4 The Pightle, Peasemore, Newbury, RG20 7JS
M6 FYH C. Cosgrove, 5 Kingswell Avenue, Wakefield, WF1 3DY
M6 FYJ S. Lo, Fieldgate, The Avenue, Claverton Down, Bath, BA2 7AX
M6 FYK E. Dutson, 19 Maythorn Drive, Cheltenham, GL51 0QH
M6 FYL M. Kipling, 12 Jolly Brows, Harwood, Bolton, BL2 4LZ
M6 FYM A. Armstrong, View Firth, Brewery Brow, Parton, Whitehaven, CA28 6PE
M6 FYN R. Gunson, 3 Kenilworth Drive, Halifax, HX3 8XP
MM6 FYR D. Fraser, Applecross, Easterton Hatton, Peterhead, AB42 0TQ
M6 FYT P. Jones, 186 Crowmere Road, Shrewsbury, SY2 5LA
M6 FYV V. Stanley, 82 Sycamore Grove, Bracebridge Heath, Lincoln, LN4 2RD
MM6 FYW A. Booth, 29 Golf Terrace, Insch, AB52 6JY
MW6 FYY C. Marchant, 8 Morningside Walk, Barry, CF62 9TE
M6 FZA R. Carr, 14 Southwell Close, Kirkby-In-Ashfield, Nottingham, NG17 8GP
M6 FZB P. Cox, 25 Coronation Crescent, Margate, CT9 5PN
M6 FZC I. Skeggs, 24 Kendall Road, Beckenham, BR3 4PZ

MM6 FZD S. Boyd, 6 Schaw Road, Prestonpans, EH32 9HA
M6 FZE A. Korben, 219 Leyland Road, Penwortham, Preston, PR1 9SY
M6 FZF R. Waterson, 43 Highland Road, Twerton, Bath, BA2 1DY
M6 FZG J. Williamson, 7 Dale Close, Wrecclesham, Farnham, GU10 4PQ
M6 FZJ J. Foster, Westfield House, 23 High Street, Cumnor, OX2 9PE
M6 FZK C. Bowler, 42A Honor Road Prestwood, Great Missenden, HP16 0NL
M6 FZL D. Finlay, 23 Glen Way, Oadby, Leicester, LE2 5YF
M6 FZM I. Jones, 8 Hengrove Avenue, Bristol, BS14 9TB
M6 FZN C. Williams, 31 Dent Close, South Ockendon, RM15 5DS
M6 FZQ C. Tacon, 2 Pavillion Court, 74-76 Northlands Road, Southampton, SO15 2NN
M6 FZR C. Bowskill, 5 Trent Walk, Daventry, NN11 4QF
M6 FZS M. Holmes, Lower Farm, Stony Moor, Newton-On-Rawcliffe, Pickering, YO18 8QJ
MM6 FZT R. Clow, 25 Scott Street, Newcastleton, TD9 0QQ
M6 FZV N. Walker, Wayfield Farm, Calstock Road, Gunnislake, PL18 9BY
M6 FZW T. Crampton, 7 Barneveld Avenue, Canvey Island, SS8 8NZ
M6 FZX T. Cash, 27 Robert Wynd, Bilston, WV14 9SE
M6 FZY J. Langdon, 6 Glebe Close, Stockton, Southam, CV47 8LG
MW6 GAA G. Green, 175 Garden Suburbs, Trimsaran, Kidwelly, SA17 4AF
M6 GAB G. Barnes, 24 Gainsborough Court, Andover, SP10 3SS
M6 GAC G. Moore, 40 Main Street, South Rauceby, Sleaford, NG34 8QG
M6 GAD G. Briggs, 5 Links Avenue, Cromer, NR27 0EQ
M6 GAE Dr P. Shaw, 25 Headcorn Road, Platts Heath, Maidstone, ME17 2NH
M6 GAF G. Furlong, 22 Swaisland Road, Dartford, DA1 3DE
M6 GAG A. Court, 3 Forsythia Close, Hythe, Southampton, SO45 3DJ
M6 GAH G. Hardill, 107 Leicester Road Whitwick, Coalville, LE67 5GN
M6 GAI G. Robinson, 16 Stanley Road South, Rainham, RM13 8AA
M6 GAK G. Cockburn, 20 Hexham Avenue, Hebburn, NE31 2HN
M6 GAL T. Kimberlee, 24 Jacey Road, Shirley, Solihull, B90 3LJ
M6 GAM A. Mawson, 14 Pontop View, Delves Lane, Consett, DH8 7JB
M6 GAN G. Rudley, 37 Cherry Way, Shepperton, TW17 8QQ
M6 GAO G. Webster, 24 Martin Avenue, Barrow Upon Soar, Loughborough, LE12 8LG
MI6 GAQ G. O'Reilly, 20 Lower Clonard Street, Belfast, BT12 4NH
M6 GAR G. Veale, 8 Duchy Cottages, Stoke Climsland, Callington, PL17 8PA
M6 GAS B. Hammond, 35 Stratford Avenue, Newcastle, ST5 0JS
MI6 GAU G. Ferguson, 22 Cloneen Drive, Ballymoney, BT53 6PT
MW6 GAV J. Ali, 5 Darent Close, Bettws, Newport, NP20 7SQ
M6 GAW G. Watson1, 70 Garden Hey Road, Moreton, Wirral, CH46 5NE
M6 GAX G. Austin, 4A Garlinge Road, Tunbridge Wells, TN4 0NR
MW6 GBA A. Jones, 2 Erw Terrace, Bethel, Caernarfon, LL55 1YT
M6 GBB D. Nock, 33 Hale Road Bradenham, Thetford, IP25 7RA
MI6 GBC G. Boyes, 50 Halfpenny Gate Road, Moira, Craigavon, BT67 0HW
M6 GBD G. Davage, 17 Howlett Close, North Walsham, NR280bf
M6 GBE H. Vecenans, 155 Upper Dale Road, Derby, DE23 8BP
M6 GBF G. Bouchier, Flat 24, Hine House, St Albans, AL4 0EY
MU6 GBG G. Lanoe, 1 Pinetrees Estate, Route De L'Islet, Guernsey, GY2 4EX
M6 GBH I. Appleby, 96 Cranbrook Road, Poole, BH12 3BT
M6 GBI C. Gibson, 17 Clyde Court, Grantham, NG31 7RB
M6 GBJ G. Walker, 25 Bear Close, Woodstock, OX20 1JT
M6 GBL L. Ulvenmoe, Flat 6, Turnstone Court, Greenfinch Way, Heysham, Morecambe, LA3 2JF
M6 GBM N. Cunningham, 115 Trafalgar Road, Washington, NE37 3DJ
M6 GBN G. Barton, 9 Tees Crescent, Stanley, DH9 6HX
M6 GBO B. Goodman, 6 Victoria Court, Penzance, TR18 2EX
MM6 GBP G. Holford, 5 Deerpark Cottages, Evanton, Dingwall, IV16 9XH
M6 GBQ A. Self, 10 Cecil Road, Hertford, SG13 8HR
M6 GBR P. Chamberlain, 22 Stanedge Grove, Wigan, WN3 5PL
MM6 GBS G. Somerville, 4 Kirkhill Way, Penicuik, EH26 8HH
M6 GBU D. Carter, Bungalow 7, Higher Ingsdon Quarry, Liverton, Newton Abbot, TQ12 6JA
M6 GBV A. Finn, 202 Northgate Road, Stockport, SK3 9NJ
M6 GBW G. Wright, Flat 27, Maudland House, Preston, PR1 2YJ
MM6 GBX G. Mcfarlane, 59 Millburn Road, Bathgate, EH48 2AF
M6 GBY R. Hughes, 96 Retallick Meadows, St Austell, PL25 3BZ
M6 GBZ D. Smith, 48 Shirley Gardens, Tunbridge Wells, TN4 8TH
M6 GCB G. Beynon-Fisher, 21 Scopsley Green, Whitley, Dewsbury, WF12 0NF
M6 GCC G. Clements, 4 Hobart Square, Norwich, NR1 3JB
M6 GCD G. Lees, 16 Kingfisher Close, Congleton, CW12 3FF
M6 GCE J. Berry, 245A Eaves Lane, Chorley, PR6 0AG
MU6 GCI J. Littlewood, Wayland Les Martin, L'islet, GY2 4XW
M6 GCJ G. Cooke, 25 Avill, Hockley, Tamworth, B77 5QE
M6 GCK G. Cornish, 78 Kerry Avenue, Ipswich, IP1 5LD
M6 GCM N. Baker, 56 Chalklands, Bourne End, SL8 5TJ
MW6 GCN G. Leedham, 15 Vale Park, Rhyl, LL18 2EN
M6 GCO S. Gilbert, 34 Ramsey Road, St Ives, PE27 5RD
M6 GCP M. Poulter, 32 Woburn Street, Hull, HU3 5LW
MW6 GCQ G. Jones, 6 Tre Ambrose, Holyhead, LL65 1LR
M6 GCS G. Smith, Glencairn, Gaston Lane, Malmesbury, SN16 0LY
M6 GCT G. Taylor, 24 Otter Close, Bletchley, Milton Keynes, MK3 7QP
M6 GCV G. Adey, 24 Burycroft, Welwyn Garden City, AL8 7AW
M6 GCW G. Wager, Dovecote, Turbary, Doncaster, DN9 1DY
M6 GCX M. Parnham, 9 The Close, Addington, West Malling, ME19 5BL
M6 GCY B. Thomson, 40 Northgate Road, Stockport, SK3 0LQ
M6 GDA J. Kemp, 9 Chequers Orchard, Stone Street, Canterbury, CT4 5PN
M6 GDB G. Buckley, 30 Saxon Road, Wheathampstead, St Albans, AL4 8NZ
M6 GDC D. Cowling, 11 Shakespeare Avenue, Scunthorpe, DN17 1SA
MI6 GDD Lord G. Drummond, 44 Camphill Park, Ballymena, BT42 2DJ
M6 GDI J. Williams, 41 Overton Lane, Hammerwich, Burntwood, WS7 0LQ
M6 GDJ G. Johnson, 31 Hillcrest Avenue, Grays, RM20 3DA
M6 GDL C. Foulkes, 5 Kennedy Close, Chaddesden, Derby, DE21 6LW
MI6 GDN G. Nelson, 65 Dernawilt Road, Annagolgan, Enniskillen, BT92 7FN
M6 GDO C. Stallard, 8 Holmeside Avenue, Trimdon, TS296HE
M6 GDP A. Parker, 18 Lincoln Close, Keynsham, Bristol, BS31 2LJ
M6 GDQ G. Whatmough, Stud Bungalow, Wakefield Lodge Estate, Towcester, NN12 7QX
M6 GDS G. Sole, 16 Beech Crescent, Hythe, Southampton, SO45 3QG
M6 GDT G. Thomas, 14 Lower Meadow, Cheshunt, Waltham Cross, EN8 0QU
M6 GDV G. Butterworth, 11 St. Davids Road, Robin Hood, Wakefield, WF3 3TG
MM6 GDY Lady D. Collins, 6, Blackberry Farm, Throntonloch, EH42 1QT
M6 GEA G. Abraham, 18 Harpenden Close, Bedford, MK41 9RG
M6 GEE B. Gee, 23 Kenmore Close, Gateshead, NE10 8WJ
M6 GEF G. Bridge, 16 Victoria Street, Ramsbottom, Bury, BL0 9ED
M6 GEG A. Gleave, 38 Barkus Way Stokenchurch, High Wycombe, HP14 3RE
M6 GEJ G. Johnson, 4 Wallace Close, Hullbridge, Hockley, SS5 6NE
M6 GEK S. Sissens, 20 Fallow Drive, Eaton Socon, St Neots, PE19 8QL
M6 GEP S. Ingledew, 34 Sunningbrook Road, Tiverton, EX16 6EB
M6 GEQ M. Emmett, 8 Coates Drove, Isleham, Ely, CB7 5SJ
M6 GEU G. Walker, 43 Wordsworth Crescent, Kidderminster, DY10 3EY
M6 GEV R. Heslop, 7 Fieldfare Close, Clanfield, Waterlooville, PO8 0NQ
M6 GEY A. Callaghan, 46 Highfields, Great Yeldham, Halstead, CO9 4QQ
M6 GFA C. Claypole, 1 Richmond Crescent, Leominster, HR6 8RX
M6 GFE G. Teale, Flat 3, 89-91 Barrack Road, Christchurch, BH23 2AJ
M6 GFF G. Gibbs, 2 Salesfrith Cottages, Bicknacre Road, East Hanningfield, Chelmsford, CM3 8AP
MM6 GFG D. Speirs, 45 Elmbank Crescent, Arbroath, DD11 4EZ
M6 GFH G. Molendijk, 47 Lodge Road, Scunthorpe, DN15 7EN
M6 GFM G. Cummins, 10 Baugh Gardens, Bristol, BS16 6PN
MI6 GFO G. Craig, 103 Moyle Parade, Larne, BT40 1ET
MW6 GFP G. Stephens, 3 Yr Hen Orsaf, Y Felinheli, LL56 4AB
M6 GFQ J. Cooper, 70 Rockhampton Close, Weymouth, DT3 6NG
MU6 GFR D. Robert, Nos Treis 7 Liberation Drive, Route Des Clos Landais, St. Saviour, Guernsey, GY7 9PH
M6 GFS G. Williams, 70 Southill Road, Poole, BH12 3AS
M6 GFV P. Dorman, 2 Moises Hall Road, Wombourne, Wolverhampton, WV5 0LF
M6 GFW G. Watson, The Dell, Nova Scotia Road, Great Yarmouth, NR29 3QD
M6 GFY G. Waring, 18 The Mead, Beaconsfield, HP9 1AW
M6 GFZ G. Linnett, 62 Melrose Avenue, Burtonwood, Warrington, WA5 4NW
M6 GGD G. Durham, 12 Beech Close, Sproatley, Hull, HU11 4XB
MM6 GGE G. Semple, 28 Newton Brae, Cambuslang, Glasgow, G72 7UW
MI6 GGF A. Dowling, 74 Ashmount Gardens, Lisburn, BT27 5DA
M6 GGG D. Rosenschein, 101 Christchurch Road, London, SW14 7AT
M6 GGI G. Ridley, 12 Garforth Avenue, Steeton, Keighley, BD20 6SP
M6 GGJ G. Jacks, 2 Corve View, Fishmore, Ludlow, SY8 2QD
M6 GGL S. Cooke, Flat 25, Attree Court, Brighton, BN2 0FZ
M6 GGM A. Mansfield, 20 High Street, Broughton, Kettering, NN14 1NG
MI6 GGN G. Scullion, 50 Tober Road, Pharis, Ballymoney, BT53 8NY
M6 GGO L. Jones, 44 Althorpe Drive, Loughborough, LE11 4QU
M6 GGQ G. Wilson, 2 Hill Row, Braunston, NN117HU
M6 GGT D. Turford, 51 Moorfield Avenue, Bolsover, Chesterfield, S44 6EJ
M6 GGV G. Guest, 90 Pearson Crescent Wombwell, Barnsley, S73 8SG
M6 GGW G. Warnock, 31 Greycote, Shortstown, Bedford, MK42 0XD
M6 GGX C. Ogidih, 89 West Road, Birmingham, B43 5PG
M6 GGZ T. Holman, 20 Green Drive, Wolverhampton, WV10 6DW
MI6 GHA E. Forde, 35 Torr Gardens, Larne, BT40 2JH
M6 GHB G. Brooks, 14 Chalton Crescent, Havant, PO9 4PT
M6 GHC L. Arenas Martinez, 22 Brickfield Road, Southampton, SO17 3AE
MW6 GHD A. Burleton, 9 Waghausel Close, Caldecott, Np26 4qr
MM6 GHF G. Farrer, 23 Upper Craigour, Edinburgh, EH17 7SE
MM6 GHH D. Livingstone, The Bungalow, Stewarton, Campbeltown, PA28 6PG
M6 GHI A. Haslam, 25 Lulworth Road Eccles, Manchester, M30 8WP
M6 GHJ J. O'Donnell, 24 Foxhill, Watford, WD24 6SY
M6 GHL G. Nicholls, 26 The Headlands, Darlington, DL3 8RP
M6 GHM G. Bell, 34 Manor Road, Eastham, Wirral, CH62 8BN
M6 GHN J. Oldham, 5 Amersham Rise, Nottingham, NG8 5QG
M6 GHP G. Pearson, 41 Myrica Grove, Hoole, Chester, CH2 3EW
M6 GHQ B. Gamble, 67 Queen Street, Burntwood, WS7 4QQ
M6 GHR J. Greenfield, 12 Firbeck Road, Nottingham, NG8 2FB
M6 GHS M. Roper, 26 Malpas Close, Bransholme, Hull, HU7 4HH
M6 GHT G. Hart, 11 Sadlers Ride, West Molesey, KT8 1SU
M6 GHV J. Harris, 108 Gresley Wood Road, Church Gresley, Swadlincote, DE11 9QN
M6 GHW G. Harrison-Webb, 2 Jubilee Close, Henlow, SG16 6FD
MM6 GHX R. James, The Garret, Alyth, Blairgowrie, PH11 8HQ
MW6 GHY C. Beamish, 15 Pen Yr Hwylfa, Harlech, LL46 2UW
MW6 GIA D. Owen, Tanrallt, Blaenpennal, Aberystwyth, SY23 4TP
M6 GIB R. Gibbs, 7 Thornhill, Eastfield, Scarborough, YO11 3LY
M6 GIC A. Oliveira, 13 A Lakefield Road, London, N22 6RR
MM6 GID W. Elder, 3 Mossgiel Place, Dundee, DD4 8AP
MI6 GIF G. Clarke, 12 Church Green, Dromore, BT25 1LL
M6 GIH G. Hutchinson, 128 Crescent Drive North, Woodingdean, Brighton, BN2 6SF
M6 GII C. Baines, 19 Kingfisher Road, Mountsorrel, Loughborough, LE12 7FG
M6 GIL G. Coleman, 5 Meeting Lane, Burton Latimer, Kettering, NN15 5LS
MI6 GIN B. Emerson1, 67 Castlemore Avenue, Belfast, BT6 9RH
M6 GIP P. Duffield, Flat 2, 4 Park Road West, Southport, PR9 0JS
M6 GIQ D. Lynch, M L & S Martin Lynch & Sons Ltd, Wessex House, Drake Avenue, Staines-upon-Thames, TW18 2AP
M6 GIS J. Goodman, 70 Bradford Road Eccles, Manchester, M30 9FT
MW6 GIU A. Sprott, 61 Brynymor, Three Crosses, Swansea, SA4 3PE
MW6 GIV A. Vowles, 232 Pentregethin Road, Gendros, Swansea, SA5 8AW
MM6 GIW C. Ferguson, 18 The Braes, Lochgelly, KY5 9QH
MW6 GIX A. Hodgson, Browns Holiday Park Towyn Road, Conwy, LL22 9HD
M6 GIY B. Smith, 26 Epperstone Court Patrick Road, West Bridgford, NG2 7QR
M6 GIZ M. Staley, 164 Rotherham Road Maltby, Rotherham, S66 8NA
M6 GJA A. Gallagher, 174 Queensway, West Wickham, BR4 9DZ
M6 GJB G. Baigent, 15 Soyland Town Road, Sowerby Bridge, HX6 4NB
M6 GJC G. Cregg, 17 Rake Way, Aylesbury, HP21 9AL
M6 GJD D. Toller, Field Cottage, Ash Lane, Etwall, Derby, DE65 6HT
M6 GJE G. Groves, 5 Beech Road, Ashurst, Southampton, SO40 7AY
M6 GJG S. Kelly, 4 Berrells Court, Olney, MK46 4AR
M6 GJH G. Henderson, 4 Castell Road, Loughton, IG10 2LT
M6 GJI T. Kelly, 50 Ivanhoe Road, Herne Bay, CT6 6EQ
M6 GJL J. Paine, 28 Laurel Way, Bottesford, Nottingham, NG13 0FP
M6 GJM G. Morris, 7 Rowley View, Bilston, WV14 8DE
M6 GJN G. Noble, 6 Sturrocks, Basildon, SS16 4PQ
M6 GJO G. Owens, 6, Khyber Close, Torpoint, PL11 2JB
M6 GJP G. Priest, Lilac Cottage, School Lane, Tamworth, B79 9JJ

M6 GJQ E. Wilkinson, 40 Charmouth Road, St Albans, AL1 4SN
M6 GJS G. Shaw, 6 Vickers Close, Woodley, Reading, RG5 4PA
M6 GJT G. Tyler, Crofton, Stoney Ley, Worcester, WR6 5NG
M6 GJV A. Wesselby, 16 Hudson Way, Grantham, NG31 7BX
M6 GJX A. Hunt, 76 Andrew Street, Bury, BL9 7HB
M6 GJY T. Garrett, 12 Poulders Gardens, Sandwich, CT13 0BE
MI6 GKB G. Black, 45 Meeting House Lane, Lisburn, BT27 5BY
M6 GKC G. Kapfunde, Gunnels Wood Road, Stevenage, SG1 2AS
MM6 GKE L. Scott, 5 Links Road, Saltcoats, KA21 6BE
M6 GKF G. Morris-Roe, 77 Bridlebank Way, Weymouth, DT3 5RP
M6 GKG G. Gordon, 40 Grange Crescent, Rubery, Birmingham, B45 9XB
M6 GKH R. Curant, 18 Ramley Road, Lymington, SO41 8GQ
M6 GKI J. Mcgowan, 2 Turnstone Drive, Liverpool, L26 7WR
M6 GKK A. Maclean, 10 Elizabeth Close, West Hallam, Ilkeston, DE7 6LW
M6 GKM J. Harrison, 20 Brafield Close, Belper, DE56 0EU
M6 GKN G. Davies, 6 Bayleys Close, Empingham, Oakham, LE15 8PJ
MW6 GKQ R. Buchan-Terrey, Godre'R Coed, Aberhosan, Machynlleth, SY20 8RA
M6 GKT Dr M. Gunn, Brynywawr, Talybont, SY24 5HJ
MW6 GKU P. Daniel, 71 Coed Isaf Road, Pontypridd, CF37 1EN
M6 GKV D. Crook, 72 Rushleigh Avenue, Cheshunt, Waltham Cross, EN8 8PS
M6 GKW G. Wade, 43 Green Park, Cambridge, CB4 1SX
M6 GKX E. York, Flat 23, Paul Stacey House, Coventry, CV1 5GU
MI6 GLC G. Crabbe, 39 Arran Avenue, Ballymena, BT42 4AP
M6 GLD R. Broughton, 3 The Row, Weeting, Brandon, IP27 0QG
MM6 GLI G. Irvine, 120A Shore Road, Innellan, Dunoon, PA23 7SS
M6 GLJ S. Solomon, 71 Ashby Road, Moira, Swadlincote, DE12 6DN
MW6 GLK S. Bateman, 26 Kenneth Treasure Court, Bethania Row, Cardiff, CF3 5UD
M6 GLM S. Weeks, Edithmead, Highbridge, TA9 4HE
M6 GLN G. Brittleton, 1 Littler Lane, Winsford, CW7 2NE
MM6 GLO G. Burns, 25 Dunskey Road, Kilmarnock, KA3 6FJ
M6 GLP G. Parker, 23 Woodrush Road, Purdis Farm, Ipswich, IP3 8RB
M6 GLS D. Hendricks, 60 Heath View, Leiston, IP16 4JP
M6 GLU M. Morgan-Lucas, 33 Stark Close, Diss, IP22 4BY
MW6 GLV C. Tanner, Pen Y Gogarth Llaneilian, Amlwch, LL68 9NH
M6 GLW G. Williams, 7 The Grove, Patchway, Bristol, BS34 6PE
M6 GLX Dr C. Cowen, Rosita, White Street Green, Sudbury, CO10 5JN
M6 GLZ G. Pardoe, 3 Bar Meadow, Shobdon, Leominster, HR6 9BZ
M6 GMA G. Marsden, 38 Sandhill Road, Rawmarsh, Rotherham, S62 5NT
M6 GMC T. Smith, 109 Ferriston, Banbury, OX16 1XA
M6 GMD M. Davison, 27 Ford Street, Consett, DH8 7AE
M6 GMF G. Fildes, 10 Windmill Gardens, St Helens, WA9 1EN
MW6 GMJ C. Jones, 28 Ivor Street, Maesteg, CF34 9AH
MI6 GMK G. Mckinley, 26 Newry Street, Warrenpoint, Newry, BT34 3JZ
M6 GMM M. Dickinson, 16 Shearwater Avenue, Newcastle upon Tyne, NE12 8PH
M6 GMO M. Boland, 24 Hallam Close, Moulton, Northampton, NN3 7LB
M6 GMP C. Crank, 12 Boundary Lane North, Cuddington, Northwich, CW8 2PL
M6 GMQ B. Healey, 14 Orchard Close, Ferring, Worthing, BN12 6QP
M6 GMR G. Mather, Shawdene House, Donnington, Newbury, RG14 3AJ
M6 GMS G. Moss, 125 Lavender Avenue, Mitcham, CR4 3RS
M6 GMT W. Pattison, 30 Main Street, Flixton, Scarborough, YO11 3UB
M6 GMU L. Lewis, 653 Main Road, Dovercourt, Harwich, CO12 4NF
M6 GMV M. Shepherd, 38 Pryors Lane, Bognor Regis, PO21 4LH
M6 GMY L. Allison, 25 Saddington Road, Fleckney, Leicester, LE8 8AX
MW6 GNA A. Jones, 23 Pinecroft Avenue, Aberdare, CF44 0HY
M6 GNC B. Chandler, 1 Rambridge Farm Cottages, Weyhill, Andover, SP11 0QF
M6 GND A. Holden, 4 Gilberts Drive, East Dean, Eastbourne, BN20 0DJ
MI6 GNF R. Simpson, 3 Portadown Road, Tandragee, Craigavon, BT62 2BB
M6 GNG C. Wilsher, 17 Milton Road, Blacon, Chester, CH1 5XE
M6 GNH G. Harris, 100 Bennett Lane, Batley, WF17 6DB
M6 GNJ G. Hance, 23 Catalina Avenue, Chafford Hundred, Grays, RM16 6RE
M6 GNM K. Markham, Hampers Cottage, Hampers Lane, Pulborough, RH20 3HZ
M6 GNN G. Norris, Trade Winds, Preston Road, Preston, PR4 0TT
M6 GNO A. Vanderahe, 28 Percival Close, Norwich, NR4 7EA
MI6 GNP M. Parke, 39 Crannog Park, Strathfoyle, Londonderry, BT47 6NF
M6 GNR P. Wollen, 41 Bernadette Close, Exeter, EX4 8DU
M6 GNS S. Hance, 23 Catalina Avenue, Chafford Hundred, Grays, RM16 6RE
M6 GNU P. Myszka, 1 The Winnows, Denton, M34 3QR
M6 GNV A. Stoll, 6 Dovestone Gardens, Littleover, Derby, DE23 4EJ
M6 GNW A. Land, 12 Beverley Close Holton Le Clay, Grimsby, DN36 5HG
M6 GNX J. Elliot, 82 Hollinside Road, Sunderland, SR4 8BG
M6 GNY A. Ali, 42 Blease Close, Staverton, Trowbridge, BA14 8WD
MI6 GOA J. Proctor, 16 Lisnavaragh Road, Scarva, Craigavon, BT63 6NX
M6 GOB I. Pashley, 45 Highfield View Road Newbold, Chesterfield, S41 7JZ
M6 GOC J. Adams, 10 Leckford Road, Oxford, OX2 6HY
M6 GOE P. Everton, 110 Sandstone Road, Sheffield, S9 1AG
M6 GOF M. Stitt, 17 Evans Street, Prescot, L34 6HU
M6 GOG M. Henman, 47 London Rd Kirton, Boston, pe201jj
M6 GOI E. Walsh, 12 Ock Meadow, Stanford In The Vale, Faringdon, SN7 8LN
M6 GOL M. Goldthorpe, 84 Park Lane Allerton Bywater, Castleford, WF10 2AP
MM6 GON J. Loughren, 16 Merlewood Road, Inverness, IV2 4NL
M6 GOO J. Goolden, 34 Grimsby Road, Louth, LN11 0DY
M6 GOQ J. Walsh, Flat 4, 14 Chantry Road, Bristol, BS8 2QD
MM6 GOR G. Campbell, 98 Netherton Road, East Kilbride, Glasgow, G75 9LB
M6 GOS S. Forshaw, 8 Stavesacre, Leigh, WN7 3LD
M6 GOU R. Harlow, 28 Dovecliff Crescent, Stretton, Burton-on-Trent, DE13 0JH
MM6 GOW N. Gowans, 1 Dunmuir Road, Castle Douglas, DG7 1LG
M6 GOX D. Chadwick, 19 Regent Crescent, Failsworth, Manchester, M35 0LR
MW6 GOY C. Rees, 5 Dare Villas, Aberdare, CF44 8AH
MW6 GOZ C. Gozzard, Craig Dulas, Rhydyfoel Road, Llanddulas, Abergele, LL22 8EG
M6 GPA M. Phillips, 36 Hyde Heath Court, Crawley, RH10 3UQ
M6 GPB G. Beacher, 22 Trowbridge Gardens, Luton, LU2 7JY
M6 GPC G. Coleman, 15 Redwood Drive, Ormskirk, L39 3NS
MM6 GPD L. Macdonald, 193 Den Walk, Buckhaven, Leven, KY8 1DJ
M6 GPE G. Davison, Lyncroft House, Swan Lane, Edenbridge, TN8 6AJ
M6 GPF G. Fleming, 17 Greenfield Close, Kippax, Leeds, LS25 7PX
M6 GPG G. Bates, 230 Brook Street, Erith, DA8 1DZ
M6 GPJ M. Stevens, 4 Wellwood Close, Horsham, RH13 6AL
M6 GPM G. Mccaffery, 7 Cliffe Court, Sunderland, SR6 9NT
M6 GPO P. Jones, 2 Herons Court, Doncaster, DN2 4GD
M6 GPP G. Powell, 7 Donstan Road, Highbridge, TA9 3LA
MI6 GPQ D. Boyd, 11 Abbey Gardens, Belfast, BT5 7HL
M6 GPS G. Stevens, 17 Manston Close, Ernesettle, Plymouth, PL5 2SN
M6 GPT R. Hampson, 12 Oakhays, South Molton, EX36 4DB
M6 GPU D. Mclean, 38 New Burlington Road, Bridlington, YO15 3HS
MI6 GPV G. Cassidy, 90 Ardmeen Green, Downpatrick, BT30 6JL
MI6 GPZ B. Cousins, 58 Bannview Heights, Banbridge, BT32 4NA
M6 GQA L. Warren, 85 Greenwood Avenue, Blackpool, FY1 6PR
M6 GQB S. Mccormack, 56 Careys Road, Pury End, Towcester, NN12 7NX
MM6 GQC B. Angus, Springfield Church Street, Dufftown, AB55 4AR
MW6 GQD B. Lee, 46 Knowling Mead, Tenby, SA70 8EB
M6 GQF D. Thomas, 18 Howard Close, Cambridge, CB5 8QU
M6 GQG E. Thresher, 18 Sandy Lane, Preesall, Poulton-le-Fylde, FY6 0EH
M6 GQH J. Fitzpatrick, 56 Littlehaven Lane, Horsham, RH12 4JB
MI6 GQI J. Steuerwald, 50 Myrtlefield Park, Belfast, BT9 6NF
M6 GQJ E. Lewis, 105 Wards Hill Road, Minster On Sea, Sheerness, ME12 2LH
M6 GQK E. Beckett, 4 Princes Avenue, Ramsgate, CT12 6DW
MM6 GQP P. Hunter, 3 Promenade, Leven, KY8 4HZ
M6 GQR J. Townsend, 47 Cosgrove Avenue, Sutton-in-Ashfield, NG17 3JY
M6 GQS S. Klee, 4 Abbots Road, Pershore, WR10 1LL
M6 GQT I. Alderman, 107 Manton Drive, Luton, LU2 7DL
M6 GQU A. Scott, 37 Pikestone Close, Hayes, UB4 9QT
M6 GQW H. Beavis, 8 Hookfield, Harlow, CM18 6QG
M6 GQX M. Martin, 19 Elmsleigh Road, Farnborough, GU14 0ET
M6 GQY L. Brearley, Ash Tree Lodge, Snaith Road, Goole, DN14 0AT
M6 GQZ P. Collier, Flat 9, Henry House, Wyvil Road, London, SW8 2TF
M6 GRC G. Clarke, 6 Coverdale Road, Scunthorpe, DN16 2RP
MD6 GRD C. Schofield, Rockside, Dreemskerry Road, Maughold, Isle Of Man, IM7 1BL
M6 GRH G. Hooper, 16 Trentham Drive, Orpington, BR5 2EP
M6 GRJ R. Davison, Lyncroft House, Swan Lane, Edenbridge, TN8 6AJ
M6 GRK G. Kenyon, 2 Langdale Terrace, Stalybridge, Sk151ex
M6 GRN A. Young, 48 Sussex Street, Cleethorpes, DN35 7NP
M6 GRP G. Priestley, 53 Millfield Gardens, Crowland, Peterborough, PE6 0HA
M6 GRQ D. Mcclurg, 40 Forfar Drive, Bletchley, Milton Keynes, MK3 7LS
M6 GRR G. Rodwell, 87 Park Avenue, Ruislip, HA4 7UL
M6 GRT J. Grant, 80 Fishbourne Road West, Chichester, PO19 3JL
M6 GRU J. Stewart, 43 Balcombe Gardens, Horley, RH6 9BY
M6 GRV C. Rawlin, 5 Japonica Hill, Immingham, DN40 1LT
M6 GRX S. Tomlinson, 8 Levett Road, Stanford-le-Hope, SS17 0BB
M6 GRZ G. Kober, 1 Sawston, King's Lynn, PE30 4XT
M6 GSC G. Casey, 10 Windermere Road, Dukinfield, SK16 4SJ
M6 GSD C. Riley, 2 Shottisham Hall Cottages, Alderton Road, Shottisham, Woodbridge, IP12 3EP
M6 GSF G. Ferns, Oxlea House, Meerbrook, Leek, ST13 8SL
MI6 GSG G. Gregg, 30 Claremont Avenue, Moira, Craigavon, BT67 0SS
M6 GSH P. Pritchard, Flat 148, Nine Acre Court, Salford, M5 3HU
M6 GSJ G. Sawyer, 432 Rowood Drive, Solihull, B92 9LQ
M6 GSK M. Silver, 52 Park Crescent, Elstree, Borehamwood, WD6 3PU
MW6 GSL A. Cassar, 48 High Street, Abergwynfi, Port Talbot, SA13 3YW
M6 GSO G. Wren, 14 Overdale, Triangle, Sowerby Bridge, HX6 3HZ
M6 GSP P. Lumb, 22 Lydney Road, Liverpool, L36 2LT
M6 GSQ J. Creese, Matrice, Church Road, Billericay, CM11 1RR
M6 GSR G. Rogers, 69 Leechcroft Avenue, Swanley, BR8 8AP
MW6 GSS G. Smith, Ty Clyd, Llanfihangel-Nant-Bran, Brecon, LD3 9NA
M6 GST G. Starling, 4 Three Corner Drive, Norwich, NR6 7HA
M6 GSX D. Tinkler, 19 Askew Dale, Guisborough, TS14 8JG
MM6 GSY N. Blampied, 6 Grosvenor Park, Riverview Country Park, Mundole, Forres, IV36 2UL
M6 GSZ D. Gray, 26 Highgate Road, Portsmouth, PO35as
M6 GTB G. Bolan, 82 Calve Croft Road, Manchester, M22 5FU
M6 GTC J. Tanner, 7 Larch Crescent Eastwood, Nottingham, NG16 3RB
M6 GTD D. Garratt, 215 Coalpool Lane, Walsall, WS3 1RF
M6 GTE G. Tomkins, 22 Arundel Drive, Orpington, BR6 9JG
M6 GTH G. Hofman, Brookside, High Street, Stockbridge, SO20 6EY
M6 GTK G. Tagg, Tinkers Cottage, Nevendon Road, Wickford, SS12 0QB
M6 GTO B. Thomas, 13 Killerton Close, Paignton, TQ3 3FT
M6 GTQ B. Partridge, 14 Whitewells Road, Bath, BA1 6NZ
M6 GTQ T. Haswell, 65 Eastlea Crescent, Seaham, SR7 8EE
M6 GTR M. Bailey, 17 Sparrowhawk Way, Hartford, Huntingdon, PE29 1XE
M6 GTT J. Owen, 8 Highridge Crescent, Bristol, BS13 8HN
M6 GTU R. Burton, 1 Broad Street, Long Eaton, Nottingham, NG10 1JH
MI6 GTY G. Mamijs, 33 Glenmore Walk, Lisburn, BT27 4RY
M6 GUA J. Hammond, 8 Rowntree Way, Saffron Walden, CB11 4DG
M6 GUB N. Emberson, Lion House, Audley End, Saffron Walden, CB11 4JB
M6 GUC R. Millen, 21 Sunnymead, Tyler Hill, Canterbury, CT2 9NW
MU6 GUE S. Tostevin, Hillside, Le Francais, Vale, Guernsey, GY3 5NL
M6 GUF A. Cook, 35 Park Road, Cheveley, Newmarket, CB8 9DF
M6 GUH D. Hughes, Flat 2, 13 Thorgam Court, Grimsby, DN31 2EU
M6 GUJ S. Noller, 3 Thor Road, Norwich, NR7 0JS
M6 GUM A. Bell, 36 Schneider Road, Barrow-in-Furness, LA14 5DW
M6 GUO Dr M. Wilson, The Avenue Care Home, 23 Avenue Road, Malvern, WR14 3AY
M6 GUS G. Frisholm, 99 Greenfell Mansions, Glaisher Street, London, SE8 3EX
M6 GUT G. Taljaard, Flat 140, 105 London Street, Reading, RG1 4QD
M6 GUU E. Shaw, Upper Floor Flat, 21 Norfolk Road, Littlehampton, BN17 5PW
M6 GUV D. Connolly, 2 Layton Close, Birchwood, Warrington, WA3 6PT
M6 GUW D. Hemp, 4 Wells Close, Hainford, Norwich, NR10 3NB
M6 GUX D. Wittering, 156 Langland Road, Netherfield, Milton Keynes, MK6 4HX
M6 GUY G. Westbrook, The Haven, St. Johns Road, Norwich, NR12 9BE
M6 GVA P. Kelly, 22 Manchester Place, Norwich, NR2 2SH
M6 GVC G. Clayton, The Forge, High Street, Moreton-in-Marsh, GL56 0LL
M6 GVD M. Mcsweeney, Barn House, Stone Quarry Road, Haywards Heath, RH17 7LP
M6 GVF G. Vailati Facchini, 8 Orchard Crescent, Edgware, HA8 9PW
M6 GVI J. Day, 120 Goring Road, Colchester, CO4 0DB
M6 GVJ D. Gough, 29 Belvedere Road, Biggin Hill, Westerham, TN16 3HX
M6 GVL B. Alston, 9 Central Avenue, Church Stretton, SY6 6EE

M6 GVM A. Spurling, 62 Swains Meadow, Church Stretton, SY6 6HT
M6 GVN G. Tilling, 1 Bellington Cottages, Worcester Road, Kidderminster, DY10 4NE
M6 GVP G. Haynes, 25 Ladbroke Road, Bishops Itchington, Southam, CV47 2RA
MW6 GVR G. Roberts, 14 Abercarn Fach, Cwmcarn, Newport, NP11 7EP
M6 GVS G. Greaves, 183 Wordsworth Avenue, Sheffield, S5 8NE
M6 GVT L. Edwards, 10 Marsh View, Beccles, NR34 9RT
M6 GVU E. Hammond, 14 Wannock Gardens, Polegate, BN26 5PA
M6 GVW G. Weston, 131 Ringwood Road, Eastbourne, BN22 8TQ
M6 GVY R. Kendall, 24 Scotland Road, Cambridge, CB4 1QG
M6 GWB G. Bunting, 31 Hardwick Avenue, Allestree, Derby, DE22 2LN
M6 GWE G. Wells, 1 Seagrove Way, Seaford, BN25 3QY
M6 GWF G. Waterfall, 18 Sandbed Lane, Belper, DE56 0SH
M6 GWG G. Weston, 94 Redhill Road, Northfield, Birmingham, B31 3LA
M6 GWI N. Shears, 1 The Old Rectory, Rectory Lane, Little Bookham, Leatherhead, KT23 4DY
MW6 GWK K. Jones, Gorswen Brynrefail, Caernarfon, LL55 3NT
MW6 GWL S. Gwillym, 5 Alder Close New Inn, Pontypool, NP4 0DF
MW6 GWM M. Martin, 1 Y Gorlan, Bryn Street, Newtown, SY16 2HN
M6 GWN G. Wynne, 19 Monks Orchard, Nantwich, CW5 5TX
MW6 GWO W. Jones, 39 Heolddu Grove, Bargoed, CF81 8UX
M6 GWQ M. Priest, 35 Albert Road, Chaddesden, Derby, DE21 6SJ
M6 GWS G. Salter, 9 Spring Gardens, Malvern Link, Malvern, WR14 1AP
M6 GWT G. Toomer, 7 Rosewood Drive, Barnby Dun, Doncaster, DN3 1BJ
M6 GWU T. Barrett, 43 The Ridgeway, Meols, Wirral, CH47 9RZ
MM6 GWW N. Robertson, Ladyburn, Port William, Newton Stewart, DG8 9QN
M6 GWZ K. Webb, 52 Princes Avenue, Walsall, WS1 2DH
M6 GXC T. Hales, Flat 6, St. Georges Court, Cambridge, CB1 7UP
M6 GXD J. Puddy, 9 Prospect Close, Easter Compton, Bristol, BS35 5SB
M6 GXF G. Fernando, 1 Rosemary Avenue, West Molesey, KT8 1QF
M6 GXH B. Gates, 3 Highfield, Taunton, TA1 5JE
MW6 GXI D. Burt, 2 Cae Masarn, Pentre Halkyn, Holywell, CH8 8JY
M6 GXJ D. Ashton, 13 Upstreet Flats Canterbury Road, Etchinghill, Folkestone, CT18 8BS
M6 GXK P. Blagden, 26 Duncroft Road, Hucclecote, Gloucester, GL3 3AS
M6 GXL A. Jones, 1A Invicta Road, Folkestone, CT19 6EY
M6 GXP H. Butcher, 12 Bath Road, Willesborough, Ashford, TN24 0BJ
M6 GXR S. Thresher, 4 Huntersway, Culmstock, Cullompton, EX15 3HJ
M6 GXS J. Allen, 29 Wood Cottage Lane, Folkestone, CT19 4QG
MW6 GXU S. Gordon, 8 Maesteg, Cymau, Wrexham, LL11 5EP
M6 GXV M. Butler, Longuenesse, Dodwell Lane, Southampton, SO31 1AD
M6 GXW A. Ashby, 3 The Caravan, Heather Bank, Maryport, CA15 6PB
M6 GXZ M. George, 26 Yew Tree Road, Ormskirk, L39 1NU
M6 GYA A. Bairstow, 12 Danesfield Avenue, Waltham, Grimsby, DN37 0QE
MW6 GYB L. Brown, 13 Station Road Loughor, Swansea, Sa46tr
M6 GYC G. Codrai, 27 Howard Avenue West Wittering, Chichester, PO20 8EX
M6 GYE D. Voak, 63 Green Lane, Crawley, RH10 8JX
M6 GYF S. Robottom-Scott, 73 St. Bernards Road, Solihull, B92 7DF
M6 GYG G. Christison, 9 Victoria Avenue, Market Harborough, LE16 7BQ
M6 GYH M. Thompson, 305 Highters Heath Lane, Birmingham, B14 4NX
M6 GYI A. Hofstedt, 16 Montana Road, London, SW17 8SN
M6 GYK S. Woodfield, 1 Kingsley Court, Church Road, Birmingham, B25 8XS
MM6 GYL M. Glasper, 1 Lindertis Cottages, Kirriemuir, DD8 5NT
M6 GYM J. Martin, 20 Hall Green Road, West Bromwich, B71 3LA
MM6 GYP A. Dobie, 201 High Street, Dalbeattie, DG54dw
M6 GYQ K. Holloway, Flat 3, Selwyn Court, Castle Street, Eccleshall, Stafford, ST21 6DG
MM6 GYR G. Brass, 114 Torbrex Rd Cumbernauld, Glasgow, G67 2JS
M6 GYS I. Cook, 93 Cathedral View, Houghton le Spring, DH4 4HN
M6 GYU D. Perry, Manor Garth, Wesley Road, Robin Hoods Bay, Whitby, YO22 4RW
M6 GYV D. Harris, Gatehouse 19, Skitfield Road, Dereham, NR20 5QN
M6 GYY T. Keep, Coombe Cottage, Coombe Lane, Cradley, Malvern, WR13 5JF
M6 GYZ D. Murray, 19 Andrew Allan Road, Rockwell Green, Wellington, TA21 9DY
M6 GZA G. Watson, 20 Windermere Drive, West Auckland, Bishop Auckland, DL149LF
M6 GZB R. Appleby, 20 Mcconnell Close, Aston Fields, Bromsgrove, B60 3SD
MW6 GZC C. Ezard, 59 Station Farm, Croesyceiliog, Cwmbran, NP44 2JW
M6 GZE R. Adams, 40 Lichfield Road, Gloucester, GL4 3AL
MI6 GZF V. Kinney, 49 Lanntara, Ballymena, BT42 3BE
M6 GZG J. Causer, 2 Kidd Croft, Tipton, DY4 0AF
M6 GZI B. Jones, 37 Sedgefield Close, Wirral, CH46 9RW
M6 GZJ N. Ngan, 6 Wynton Grove, Walton-on-Thames, KT12 1LW
M6 GZK N. Heywood, 38 Thurne Rise, Martham, Great Yarmouth, NR29 4PU
M6 GZL C. Snowden, 15 Rose Villas, Hebden Bridge, HX7 5ER
M6 GZN M. Armstrong, 5 Aireside, Cononley, Keighley, BD20 8LT
M6 GZO R. Bruce, 19 Spindle Beams, Rochford, SS4 1EH
M6 GZR G. Foster, 248 Harbour Lane, Milnrow, Rochdale, OL16 4EL
MM6 GZS G. Sinclair, 33 Keptie Road, Arbroath, DD11 3EF
M6 GZT T. Marshall, 63A Newport Road, Ventnor, PO38 1BD
M6 GZU C. Waters, 45 Elmdale Road, Bedminster, Bristol, BS3 3JF
M6 GZW C. Hughes, 41 Rotherham Road, Dinnington, Sheffield, S25 3RG
MW6 GZX L. Kurdi, 8 Gwel Afon, Penparcau, Aberystwyth, SY23 3PL
M6 GZY P. Garrett, April Cottage, Castle Avenue, Blandford Forum, DT11 0RY
MW6 GZZ G. Williams, Flat 137, Rosser, Aberystwyth, SY23 3LH
M6 HAC P. Selby, 24 Juniper Close, Guildford, GU1 1PA
MI6 HAD A. Mcmillen, 55 Northwood Road, Belfast, BT15 3QS
M6 HAE A. Watts, 175 Ber Street, Norwich, NR1 3HB
MI6 HAF A. Mckinley, 36 Grangewood Drive, Londonderry, BT47 5WN
M6 HAG S. Haigh, 17 Glebe Street, Swadlincote, DE11 9BW
MM6 HAH H. Halley, 1 Grant Crescent, Renton, Dumbarton, G82 4NH
M6 HAK R. Singer, 19 Rosalind Avenue, Bebington, Wirral, CH63 5JR
M6 HAM D. Trotter, 48 Swindon Road, Sunderland, SR3 4EE
MW6 HAR K. Harbour, Glaslwyn, Cwmoernant, Carmarthen, SA31 1EG
M6 HAS H. Searle, 2 Tukes Avenue, Gosport, PO13 0SE
M6 HAT R. Hatton, 18 Tangier Road, Guildford, GU1 2DF
M6 HAU A. Coghlan, 1 Bell View Cross Houses, Shrewsbury, SY5 6JR
MM6 HAV H. Venries, 24 Lady Place, Livingston, EH54 6TB
M6 HAY H. Harding, 1 Saddleton Grove, Saddleton Road, Whitstable, CT5 4LY
M6 HBB H. Buckley, 10 Lower Hey Lane, Mossley, Ashton-under-Lyne, OL5 9DE
M6 HBC J. Moore, 118 Heneage Road, Grimsby, DN32 9JQ
M6 HBD K. Mills, 72 Sycamore Road, Ecclesfield, Sheffield, S35 9YW
MM6 HBF J. Hawkins, 113 Meadowpark Avenue, Bathgate, EH48 2ST
M6 HBG H. Hutasuhut, Hawkridge, Warden Road, London, NW5 4SA
M6 HBH B. Wilson, 38 Cotleigh Drive, Sheffield, S12 4HU
MI6 HBI B. Huddleson, 4 Knightsbridge Court, Bangor, BT19 6SD
MW6 HBK A. Curry, 58 Greenfields, St. Martins, Oswestry, SY11 3AH
M6 HBN R. Tongs, 9 Woodland Drive, Winterslow, Salisbury, SP5 1SZ
M6 HBP H. Pascall, 60 Weyland Road, Witnesham, Ipswich, IP6 9ET
M6 HBQ I. Newton, 16 Cross Close, Newquay, TR7 3LB
MM6 HBR H. Ling, Leadburnlea, Leadburn, West Linton, EH46 7BE
M6 HBS J. Hobbs, 82 Perry'S Lane, Wroughton, Swindon, SN4 9AP
M6 HBT S. Hopkins, 3B Tolfa House Wellington Terrace, Truro, TR1 3JA
M6 HBU J. Mcdonald, 222 Bristol Avenue, Farington, Leyland, PR25 4QZ
M6 HBV J. Haynes, 16 Mountsfield, Frome, BA11 5AR
MM6 HBY J. Church, 4 Weir Place, Perth, Ph1 3Gp, Perth, PH1 3gp
M6 HCA C. Edwards, Apartment, Maiori, Downs Lane, West Looe, Looe, PL13 2HX
M6 HCB D. Mchugh, 38 Quarryhill Road, Wath. upon Dearne, S63 7TD
M6 HCD D. Humphries, 100 Sunnyside Avenue, Stoke-on-Trent, ST6 6EB
M6 HCE S. Ellis, 9 Deanwood Close, Whiston, Prescot, L35 3UX
M6 HCG C. Loud, 24 Harrington Avenue, Lowestoft, NR32 4JU
M6 HCI L. Call, 9 Hyperion Avenue, Polegate, BN26 5HT
M6 HCJ S. Shone, 22 Fenwick Road, Great Sutton, Ellesmere Port, CH66 4UF
MM6 HCK C. Northcott, 4/6 Castleview House, 2 Craigour Place, Edinburgh, EH17 7Rt
M6 HCO M. Moore, 46 Scholes Park Road, Scarborough, YO12 6QY
MI6 HCP S. Mawhinney, 14 Beech Park, Portadown, Craigavon, BT63 5ES
M6 HCR H. Rogers, 2 Broom Close, Waterlooville, PO7 8DP
MW6 HCS R. Hicks, 14 Carn Celyn Beddau, Pontypridd, cf38 2tf
M6 HCT J. Poulter, 71H, Highstreet, Canvey Island, SS8 7RD
M6 HCU P. Lockwood, 80 Falmouth Road Le5 4Wh, Leicester, Le5 4wh
M6 HCV D. Poulter, 1 Deacon Drive, Laindon, Basildon, SS15 5FY
M6 HCW C. Haynes, 4 Thorn Close, Rugby, CV21 1JN
MW6 HCY R. Porcher, 9 Blenheim Close, Oswestry, SY11 2UN
M6 HCZ C. Young, 21A Union Crescent, Margate, CT9 1NS
M6 HDA D. Harley, 39 Tweedale Crescent, Madeley, Telford, TF7 4EA
MW6 HDB H. Bancroft, Stop And Call, Goodwick, SA64 0EX
M6 HDD M. Hurst, 48 Radcot Close, Woodley, Reading, RG5 3BG
M6 HDE D. Edmondson, 21 Hawthorne Close, Heathfield, TN21 8HP
M6 HDG E. Majoch, 66 Boughton Green Road, Northampton, NN2 7SP
MI6 HDH G. Plunkett, 18 Carnhill Place, Carrickfergus, BT38 7RL
M6 HDI S. Ruddy, 27 Grove Park Walk, Harrogate, HG1 4BP
M6 HDK H. Karpuk, Clarence Drive, Harrogate, HG1 2QG
M6 HDM D. Money, 2 Lodge Farm Cottage, Black Horse Road, Norwich, NR10 5DJ
M6 HDO I. Johnson, 35 Church Parade, Canvey Island, SS8 9RQ
MW6 HDP M. Waldman, 9 David Street, Cwmbwrla, Swansea, SA5 8NX
M6 HDS D. Stapleton, 179 Woodcock Road, Norwich, NR3 3TQ
MW6 HDT D. Jenkins, 12 Llys Yr Onnen, Dafen, Llanelli, SA14 8PP
M6 HDU Dr S. Chipperfield, 3 Clayton Avenue, Upminster, RM14 2EZ
MW6 HDV J. Hodson, 6 Heol Pentwyn, Tonyrefail, Porth, CF39 8DF
M6 HDW D. Wilson, Flat 6, Saxon House, Draymans Way, Alton, GU34 1AY
M6 HDY D. Hardy, Flat 10, Bridport House, Hillwood Road, Birmingham, B31 1DN
MM6 HDZ H. De Zeeuw, Eastfield Farm, Ballater, AB35 5SH
MW6 HEA T. Heath, 16 Beacons Park, Brecon, LD3 9BR
M6 HEB S. Drakeley, 1 Council Houses, Churchtown, Wadebridge, PL27 7QA
MW6 HED M. Hedley, Windberry Top Begelly, Kilgetty, SA68 0XA
M6 HEE S. Marr, 49 Gallows Hill, Ripon, HG4 1RG
M6 HEF D. Robinson, Height End Farm, Kirk Hill Road, Haslingden, Rossendale, BB4 8TZ
M6 HEG P. Sellick, 1, The Gallery, Northwick Park, Blockley, GL56 9RJ
MW6 HEI R. Williams, 88 Parc Pendre, Kidwelly, SA17 4TE
M6 HEJ N. Lunden, Flat 15 Western House, 8 Woodfield Place, London, W9 2BJ
M6 HEK E. Wright, 351 Market Street, Droylsden, Manchester, M43 7EA
MM6 HEO L. Davis-Edmonds, 6 Barlockhart Park, Glenluce, Newton Stewart, DG8 0JQ
M6 HEP H. Purdy, 13 St. Cuthbert Street, Worksop, S80 2HN
MM6 HEQ R. Wilson, 3 St. Peters Park, Stromness, KW16 3EH
M6 HET I. Szabo, 95B Westgate, Grantham, NG31 6LE
M6 HEW W. Fearby, 165 Brandsfarm Way, Telford, TF3 2JJ
M6 HEX J. Ash, 47 Stein Road, Emsworth, PO10 8LB
MW6 HEY C. Hey, 84 Trefelin, Aberdare, CF44 8LF
MM6 HEZ W. Demczur, 25 Maitland Court, Helensburgh, G84 7EE
M6 HFA J. Wilson, Flat 5, Blake House, London, SE1 7DX
M6 HFF M. Collins, The Haven, Kettleby Lane, Wrawby, Brigg, DN20 8SW
M6 HFG H. Furniss, 3 Byron Avenue, Chapeltown, Sheffield, S35 1SQ
M6 HFI G. Mitchell, 49 Nash Road, Romford, RM6 5JP
M6 HFJ A. Cheung, 197 St. Lukes Avenue, Ramsgate, CT11 7HS
M6 HFK C. Boal, Chindits, Wagg Lane, Probus, Truro, TR2 4JX
M6 HFM H. Mcevoy, 18 Brookfield Gardens, Wirral, CH48 4EL
M6 HFN D. Kelly, 6 Sandham Walk, Bolton, BL3 6RA
M6 HFO G. Allen, 14 The Parsonage, Sixpenny Handley, Salisbury, SP5 5QJ
M6 HFP J. Darley, 159 Main Road, Hawkwell, Hockley, SS5 4EL
M6 HFQ A. White, 56 Seagrave Close, Oakwood, Derby, DE21 2HZ
M6 HFR R. Heffer, 36 Raven Avenue, Tibshelf, Alfreton, DE55 5NR
M6 HFT D. Merridale, The Granary, Falledge Lane, Upper Denby, HD8 8YH
M6 HFU J. Russell, Flat 1, Knapp Cottage, Roadwater, Watchet, TA23 0QY
M6 HFV M. Johnson, Charnwood, The Close, Ringwood, BH24 2PE
M6 HFX T. Crawshaw, 208 Ovenden Road, Halifax, HX3 5QG
MW6 HFY N. James, 8 Garth Lwyd, Caerphilly, CF83 3QB
MM6 HFZ P. Mcnally, 35 Crosshill Avenue, Glasgow, G42 8BZ
M6 HGA L. Brookhouse, 16 Clockmill Road, Walsall, WS3 4AH
M6 HGD L. Brookhouse, 16 Clockmill Road, Walsall, WS3 4AH
M6 HGE J. Webber, 14 Raleigh Street, Scarborough, YO12 7JZ
M6 HGF A. Highfield, 38 Brunswick Gardens, Garforth, Leeds, LS25 1HF
M6 HGG S. Entwisle, 30 Arden Mhor, Pinner, HA5 2HR
M6 HGH H. Hamilton, Flat B, 9 Cambridge Drive, London, SE12 8AG
M6 HGJ M. Hadley, 75 Glendower Avenue, Coventry, CV5 8BD
M6 HGM B. Holgate, 20 Milford Avenue, Bridlington, YO16 7AU
M6 HGN J. Allen, Flat 7, The Fairways, 35 The Esplanade, Knott End-On-Sea, Poulton-le-Fylde, FY6 0AD

M6 HGO P. Niewiadomski, Flat1 79A Dartmouth Road, London, SE23 3HT
M6 HGR S. Yardley, 22 Wedgwood Road, Clifton, Manchester, M27 8RT
MI6 HGS M. Doogan, 54 Birchdale Manor, Lurgan, Craigavon, BT66 7SY
M6 HGU V. Keith, Valentine Cottage, Frankton Road, Rugby, CV23 9QT
MI6 HGV E. Rainey, 22 Cherry Gardens, Ballymoney, BT53 7AS
M6 HGW G. Whitton, 40 Louville Avenue, Withernsea, HU19 2PB
M6 HGX E. Keith, Valentine Cottage, Frankton Road, Rugby, CV23 9QT
M6 HGY A. Flintoft, 71 Huntsman Lane, York, YO41 1ET
MW6 HGZ R. Taylor, 4 Castell Morgraig, Caerphilly, CF83 3JH
M6 HHA G. Coldham, 27 Welsby Road, Leyland, PR25 1JA
MM6 HHB D. Burgess, Quendale Farm, Quendale, Shetland, ZE2 9JD
M6 HHC K. Townsend, Flat 3, Drake House, Bexhill-on-Sea, TN39 3TS
M6 HHF J. Sim, 11 Haven Close Istead Rise, Gravesend, DA13 9JR
M6 HHH J. Hoare, Flat 3, 6 High Street, Watlington, OX49 5PR
MM6 HHJ D. Jappy, 21 Primrose Avenue, Grangemouth, FK3 8YG
M6 HHL J. Wang, 54 Mistle Thrush Drive, Cambridge, CB24 1BS
M6 HHM H. Mccarthy, 80 Vaughan Williams Way, Warley, Brentwood, CM14 5WT
M6 HHO D. Przybylski, 108 Windrows, Skelmersdale, WN8 8NW
MM6 HHP R. Jenkins, 54 Castle Bay Holiday Park, Portpatrick, Stranraer, DG9 9AA
M6 HHQ B. Harvey, Bennettshayes Barn, Awliscombe, Honiton, EX14 3PY
M6 HHR R. Joyce, 42 Long Close Station Road, Lower Stondon, Henlow, SG16 6JS
MD6 HHT R. Jefferies, The Old Police Station, Bay View Road, Port St Mary, Isle Of Man, IM9 5AW
M6 HHW C. Ogidih, 89 West Road, Birmingham, B43 5PG
MI6 HHX S. Gibson, 22 Station Road, Bangor, BT19 1HD
M6 HHY C. Taylor, Willow Court, Lincoln, LN4 1AS
MM6 HIA A. Crichton, 11 Baillie Court Sauchie, Alloa, FK10 3FG
M6 HIB D. Bedworth, 96 Balmoral Road, Stourbridge, DY8 5JB
MM6 HIG H. Glennie, 97 Smithfield Crescent, Blairgowrie, PH10 6UE
M6 HIJ M. Willis, 7 Belvawney Close, Chelmsford, CM1 4YR
M6 HIK A. Gretton, 67 Hawthorn Crescent, Arnold, Nottingham, NG5 8BE
M6 HIL H. Penfold, 15 Carmans Close, Loose, Maidstone, ME15 0DR
M6 HIM P. Green, 15 Dickenson Road, Chesterfield, S41 0RX
M6 HIP S. Admans, 77 Beaumont Road, Birmingham, B30 2EB
M6 HIT I. Petrie, 88 Vicarage Road, Henley-on-Thames, RG9 1JT
M6 HIU J. Leatherd, 115 Rothesay Road, Blackburn, BB1 2ER
MM6 HIZ D. Aitken, 64 Brunton Street Cathcart, Glasgow, G44 3NQ
M6 HJA K. Rouse, 42 Berkley Close Highwoods, Colchester, CO4 9RR
M6 HJB H. Barton, 86 Forge Lane, Kingswinford, DY6 0LG
M6 HJC H. Christie, 15 Heilsburg Road, Canvey Island, SS8 8HH
M6 HJD J. Hawley, 11 Upper Green Way, Tingley, Wakefield, WF3 1TA
M6 HJE M. Jessop, 24 Woodcroft Avenue, Hull, HU6 8lh
M6 HJF S. Rattley, 2 Burnt Cottages Beanacre, Melksham, SN12 7PT
M6 HJG A. Lawrence, 6 Beaver Court, Ashford, TN23 5QR
M6 HJH D. Money, Flat 1, Lewin Court 24B Plumstead High Street Plumstead, London, SE18 1SL
M6 HJI N. Crudgington, Appledore Blackness Lane, Keston, BR2 6HL
M6 HJK A. Scott, 19 Estuary Drive, Felixstowe, IP11 9TL
M6 HJM H. Mcneill, 44 Anglesey Road, Wirral, CH48 5EG
M6 HJQ G. Cairns Thomas, 121 London Road, Bagshot, GU19 5DH
M6 HJR D. Riman, 22 Princess Road, Hinckley, LE10 1EB
M6 HJT H. Hughes, 27 The Holt, Hailsham, BN27 3ND
M6 HJU C. Cairns Thomas, 121 London Road, Bagshot, GU19 5DH
M6 HJV J. Moore, 65 Hamsey Green Gardens, Warlingham, CR6 9RT
M6 HJX N. Fulcher, 33 Water Lane, London, SE14 5DN
M6 HJZ V. Lucock, 34 Wentworth Drive, Ipswich, IP8 3RX
M6 HKA J. Daniels, 27 Hammerwater Drive, Warsop, Mansfield, NG20 0DJ
M6 HKB D. Lock, 20 Jasmine Close Trimley St. Martin, Felixstowe, IP11 0UY
M6 HKC A. Chambers, 5 Blandford Road, Shepton Mallet, BA4 4FB
M6 HKI M. Davies, 2 Ellins Terrace, Normanton, WF6 1BL
M6 HKJ D. Killingley, 17 Colbert Drive, Leicester, LE3 2JB
MW6 HKL L. Chang, 14 Greenfield Gardens, Pentrebach, Merthyr Tydfil, CF48 4BQ
M6 HKN Dr D. Potts, 3A Westfield, Blean, Canterbury, CT2 9ER
M6 HKO S. Marden, 65 Hedley Way, Hailsham, BN27 3FZ
MI6 HKP S. Mccartney, 35 Blackstaff Road Clough, Downpatrick, BT30 8SR
M6 HKQ A. Lamont, 89 Newlands, Whitfield, Dover, CT16 3ND
M6 HKS D. Collier, 7 Compass Close, Ashford, TW15 1UT
M6 HKT B. Scott, 3 Chaplin Close, Basildon, SS15 4EJ
M6 HKU D. Holdbrook, 9 Johns Terrace, Harold Park, Romford, RM3 0AW
M6 HKV J. Robinson, Old Laira Road, Plymouth, PL3 6DH
M6 HKW H. Knowles, 16 Farmcote Road, Coventry, CV2 1SA
M6 HKX H. Khayer, 24 Clyde Road, Stoke-on-Trent, ST6 3DJ
M6 HKY A. Hickey, 144 Gisburn Road, Barnoldswick, BB18 5LQ
M6 HKZ J. Poole, 61 Lower Vickers Street, Miles Platting, Manchester, M40 7LX
M6 HLA A. Harvey, 20 Fellowes Place, Plymouth, PL1 5NB
MI6 HLC S. Savage, 469 Old Belfast Road, Bangor, BT19 1RQ
M6 HLE P. Bray, 24 Eldon Terrace, Bristol, BS3 4NZ
M6 HLF J. Taylor, 41 Waters View, Yarwell Mill, Yarwell, Peterborough, PE8 6EU
MW6 HLG H. Griffiths, 5 Heol-Y-Sarn, Llantrisant, CF72 8DA
M6 HLL Dr M. Beharrell, 110 Scotforth Road, Lancaster, LA1 4SQ
M6 HLP A. Wilson, 39 Rochford Garden Way, Rochford, SS4 1QH
MW6 HLQ L. Stevens, 32 St Daniels Drive, Pembroke, SA71 5QQ
M6 HLR L. Richardson, 35 Vidgeon Avenue, Hoo, Rochester, ME3 9DE
M6 HLS P. Hollis, 89 Longfield Lane, Cheshunt, Waltham Cross, EN7 6AN
M6 HLT A. Tomkins, 28 Newborough Close, Austrey, Atherstone, CV9 3EX
MW6 HLU A. Garner, 15 Midland Place, Llansamlet, Swansea, SA7 9QU
M6 HLW C. Robinson, 69 Sanger Avenue, Chessington, KT9 1BY
MI6 HLY R. Lawrence, 8 Wynford Park, Lisburn, BT27 5HJ
MM6 HLZ H. Ross, 16 Myreton Drive, Bannockburn, Stirling, FK7 8PX
MM6 HMB H. Brown, 7 Maple Avenue, Milton Of Campsie, Glasgow, G66 8BB
MM6 HMC H. Campbell, 10 Stewart Avenue, Linlithgow, EH49 6DQ
M6 HME W. Mayall, 17 Norrington Grove, Birmingham, B31 5NY
M6 HMG M. Hughes, 129 Jackson Road, Scunthorpe, DN15 8JT
M6 HMK H. Melhuish, 22 Mayflower Close, Glossop, SK13 8UD
M6 HML K. Lees, 24 Marks Road, Wokingham, RG41 1NN
MM6 HMM P. Bromley, Broadwood Treovis Upton Cross, Liskeard, G68 9JY
M6 HMQ J. Webb, 49 Perth Avenue, Leicester, LE3 6QQ
M6 HMR H. Pittard, 21 The Oaklands, Church Eaton, Stafford, ST20 0BA
M6 HMS M. Taylor, 35 Valley Drive, Wilnecote, Tamworth, B77 5FL
M6 HMU M. Gosi, 49 Elms Drive, Marston, Oxford, OX3 0NW
M6 HMV O. Wood, 26 Parkfield Crescent, Kimpton, Hitchin, SG4 8EQ
M6 HMW J. Walker, 5 Preston Avenue, Alfreton, DE55 7JX
M6 HMZ R. Dalziel, 24 Horringford Road, Liverpool, L19 3QX
M6 HNA M. Rosewell, 54 Alder Drive, Chelmsford, CM2 9EZ
M6 HNB H. Baird, 35 St. Peters Road, Wolvercote, Oxford, OX2 8AX
M6 HND A. Haylor, 33 Crimp Hill Road, Old Windsor, SL4 2QY
M6 HNF C. Seymour, 12 Silver Street Riccall, York, YO19 6PB
M6 HNG N. Kerry, 47 Harpley Dams Hillington, King's Lynn, PE31 6DP
M6 HNH S. Kozlowski, 28 Osney Crescent, Paignton, TQ4 5EY
M6 HNI M. Lawson, 131 Windermere Avenue, Ilkeston, DE7 4EZ
M6 HNJ E. Potter, 3 Thomson Court Chadwick Close, Crawley, RH11 9LH
MW6 HNN D. Gibson, 38 Newellhill, Tenby, Sa708en
M6 HNO P. Byrne, 18 St. Aidans Square, Bingley, BD16 2BN
MM6 HNQ A. Macintyre, 2 Memorial Square, Main Street, Castletown ?, Thurso, KW14 8TU
M6 HNS H. Stanley, 253 Brownley Road, Wythenshawe, M22 9UX
M6 HNT M. Hunt, 13 Pine Halt, Station Road, Cheltenham, GL54 4JX
M6 HNV P. Sargeant, 6 Meldon Way, Winlaton, Blaydon-on-Tyne, NE21 6HJ
MW6 HNW T. Fletcher, 15A Les Maisonette Penryhn Road, Colwyn Bay, LL29 8LG
M6 HNX A. Lorne, 8 Campbell Close, Grantham, Ng31 8aw
M6 HNZ M. Bruce, 28 Pheasants Way, Rickmansworth, WD3 7ES
M6 HOF S. Baines, 23 Gladstonbury Close, Belmont, Hereford, HR2 7YL
M6 HOG G. Snape, 3 Jasper Close, Barlaston, Stoke-on-Trent, ST12 9BL
MW6 HOH P. Gostelow, 6 Tree Field Caerau Farm, Llanidloes, SY18 6ll
M6 HOI M. Simkins, 11A Commercial Road, Southampton, SO15 1GF
M6 HOK K. Hough, 15 Moorside Road Endmoor, Kendal, LA8 0EN
M6 HOM S. Mirjalili Mohanna, Apartment 18B, White Croft Works, 69 Furnace Hill, Sheffield, S3 7AH
MM6 HOO G. Freeburn, 31 Courthill, Rosneath, Helensburgh, G84 0RN
M6 HOP R. Hope, 32 Winstanley Place, Rugeley, WS15 2QB
M6 HOQ D. Hampson, 29 Holywell Road Kilnhurst, Mexborough, S64 5UQ
M6 HOS B. Mulholland, 6 Burnside, Longhoughton, Alnwick, NE66 3JQ
M6 HOT D. Hubbard, 14 Parkfield Crescent, Kimpton, Hitchin, SG4 8EQ
M6 HOU R. Houlton, L11 (Ash 7) Woodlands Caravan Park, The Marshes Lane, Preston, PR4 6JS
M6 HOV N. Fox, Nik.Fox@Icloud.Com, Not Relevant, pe100fg
M6 HOY J. Wilson, 5 Queens Road Hoylake, Wirral, CH47 2AG
MW6 HPC H. Cooper, 3 Waunddu, Pontnewynydd, Pontypool, NP4 6QZ
M6 HPD R. Hodson, 99 Alcester Road, Hollywood, Birmingham, B47 5NR
M6 HPF R. Stringer, 9 Pershore Close, Walsall, WS3 2UQ
M6 HPG H. Bell, 7 Rosecomb Way, Haxby, York, YO32 3ET
M6 HPH H. Heathfield, Apartment 27, The Fitzgerald, 1 West Bar, Sheffield, S3 8PQ
MW6 HPK R. Hopkins, 132 Laurel Road Bassaleg, Newport, NP10 8PT
M6 HPL C. Gorse, Apartment 11, 30 Stockton Road, Hartlepool, TS25 1RY
MW6 HPN A. Hampson, 10 Coronation Terrace, Senghenydd, Caerphilly, CF83 4HU
M6 HPR I. Hooper, 25 Honey Lane, Buntingford, SG9 9BQ
M6 HPS H. Partridge, 19 Dickens Drive, Melton Mowbray, LE13 1HZ
M6 HPT S. Grubb, 28 Bury Road, Newmarket, CB8 7BT
M6 HPV C. Ward, 325 Cedar Road, Nuneaton, CV10 9DQ
M6 HPX W. Rainbow, 69 Abbotsweld, Harlow, CM18 6TG
M6 HQB J. Bristow, 55 Haldon Close, Bristol, BS3 5LR
MM6 HQC G. Fleming, 169 Oldtown Road, Inverness, Scotland, IV2 4QD
M6 HQD J. Hobbs, 1 Hotground Cottage Branfield, Hertford, SG14 2QG
MM6 HQE A. Mccormick, Flat 2 16 Marine Drive, Edinburgh, EH5 1FD
MM6 HQI W. Mcbain, 56 Scotstoun Park, South Queensferry, EH30 9PQ
MM6 HQK B. Mccabe, 173 Marmion Road Cumbernauld, Glasgow, G67 4AW
M6 HQL M. Dimmick, 16 Bushell Way, Kirby Cross, Frinton-on-Sea, CO13 0TW
M6 HQM R. Shulver, 63 Hill Farm Way, Southwick, Brighton, BN42 4YG
M6 HQN W. Hamlet, 18 Bridle Lane, Alfreton, DE55 1LG
M6 HQO R. Olive, Lorien, The Ridge, Thatcham, RG18 9HZ
M6 HQR P. Nolan, 14 Woodlands Road, Stafford, ST16 1QR
M6 HQW K. Kimura, Cornerways, Tennyson Road, Yarmouth, PO41 0PX
M6 HQZ H. Bamford, 14 Calf Close, Haxby, York, YO32 3NS
M6 HRC P. Humphreys, 30 The Chestnuts, Hinstock, Market Drayton, TF9 2SX
M6 HRD V. Greatwood, 11 The Green, Long Preston, Long Preston, Skipton, BD23 4PQ
M6 HRE S. Bligh-Wall, 81 Warmingham Road, Leighton, Crewe, CW1 4PS
MM6 HRF M. Reynolds, 22 Fergus Place, Dyce, Aberdeen, AB21 7DD
M6 HRG S. Probert, 10 The Green, Church Lawton, St73ed
M6 HRI H. Hughes, Cefn Glass, Clyro, Hereford, HR3 5JT
M6 HRJ R. Huelin, 15 Hill Chase, Walderslade, Chatham, ME5 9HE
MW6 HRK G. Hughes, 19 Gernant Braichmelyn Bethesda, Bangor, LL57 3RE
M6 HRL H. Russell, 4 Dearnsdale Close, Stafford, ST16 1SD
M6 HRM H. Mcbrien, Hamilton House, Hayes Lane, Wokingham, RG41 4TA
M6 HRN B. Hurren, 29 Chalk Lane, Ixworth, Bury St Edmunds, IP31 2JQ
M6 HRO J. Mills, 24 Charles Street, Ryhill, Wakefield, WF4 2BU
M6 HRQ S. Duncan, 5 Spring Vale Bilton, Hull, HU11 4DN
MW6 HRR H. Roberts, 12 Rheidol Terrace Ceredigion, Aberystwyth, SY23 1JU
M6 HRS H. Richardson, 103 Marys Mead, Hazlemere, High Wycombe, HP15 7DT
M6 HRT J. Krol, 40 Hampton Gardens, Southend-on-Sea, SS2 6RW
M6 HRV H. Harkishin, 2 Kingfisher Close, Bournemouth, BH6 5BB
M6 HRW H. Woolley, 84 Bowthorpe Road, Norwich, NR2 3TP
M6 HRX R. May, 6 Gordon Court Well Street, Loose, Maidstone, ME15 0QF
MM6 HRZ W. Mcewan, 240 Turriff Brae, Glenrothes, KY7 6UT
M6 HSA P. Sousa, 11 Broom Crescent, Ipswich, IP3 0EE
M6 HSB P. Henry, 22 Huddleston Close, Wirral, CH49 8JP
MM6 HSC H. Campbell, 8B Hawthorn Place, Uphall, Broxburn, EH52 5BX
M6 HSE H. Evans, 26 Peartree Court, Welwyn Garden City, AL7 3XN
M6 HSF F. Hayati, Apartment 19, 29 Longleat Avenue, Birmingham, B15 2DF
M6 HSH M. Williams, 21 Elmbrook Close, Basildon, SS14 2FH
M6 HSI J. Ostapiuk, 33 Kent Terrace, Haswell, DH62EL
MM6 HSK H. King, 19 Gleneagles Way, Deans, Livingston, EH54 8EW
MI6 HSL N. Davis, 19 Toberhewny Hall, Lurgan, BT66 8JZ
M6 HSP N. Hawkins, Deganwy Hardwick Road, King's Lynn, PE30 5BB
M6 HSQ R. Terry, 31 Barnwood Road, Birmingham, B32 2LY
M6 HSR H. Stewart-Roberts, Cinderfield House, Cornwells Bank, Lewes, BN8 4RH

MM6 HSS Prof. S. Skerratt, 8/11 Leslie Place, Edinburgh, EH4 1NH
M6 HST M. Ghost, 59 Kingswood Avenue, Park North, Swindon, SN3 2RB
M6 HSX H. Sati, 83 Edenfield Gardens, Worcester Park, KT4 7DX
M6 HSY G. Kokinis, 16 Wellesley Avenue, Beverley Road, Hull, HU6 7LW
M6 HTB S. Crane, 1 Nicol Mere Drive Ashton-In-Makerfield, Wigan, WN4 8DQ
M6 HTC A. Porter, 35 St. Andrews Crescent, Hindley, Wigan, WN2 3EQ
M6 HTF S. Griffiths, 8 Alldicks Road, Hemel Hempstead, HP3 9JJ
M6 HTG D. Quinney, 8 Crabwood Road, Southampton, SO16 9EZ
M6 HTI J. Trimmer, The Lodge, Horsemoor Lane Winchmore Hill, Amersham, HP7 0PL
M6 HTJ T. Hare, 2 Primula Drive, Norwich, NR4 7LZ
M6 HTK J. Ormston, 15 Thackeray End, Aylesbury, HP19 8JE
M6 HTM J. Clyne, Ravenswood, Green Lane, Wisbech, PE14 7BJ
M6 HTN A. Horton, 51 Walsingham Gardens, Epsom, KT19 0LS
MM6 HTS S. Spencer, 55 Blackwell Court, Inverness, IV2 7AR
MW6 HTT H. Thomas, 2 Ffordd Donaldson, Copper Quarter, Swansea, SA1 7FJ
M6 HTU T. Leatherbarrow, 17 Egerton, Skelmersdale, WN8 6AA
M6 HTW C. Brooks, 12 Beeches Court, Thornton-Cleveleys, FY5 4PZ
M6 HTX H. Banasiak, 16 St Christophers Close, Bath, BA2 6RG
M6 HTZ L. Shearson, 23 Thumpers, Hemel Hempstead, HP2 5SL
M6 HUC C. Lycett, 2 Royce Avenue, Hucknall, Hucknall, Nottingham, NG15 6FU
MM6 HUE J. Hughes, 2 Forest Place, Townhill, Dunfermline, KY12 0EP
M6 HUF S. Pennell, 272 Coal Clough Lane, Burnley, BB11 5BS
M6 HUH W. Lam, Clarence Drive, Harrogate, HG1 2QG
M6 HUI A. Windle, 10 Longshaw Street, Blackburn, BB2 4HS
M6 HUK A. Thomas, 6 Coniston Avenue, Grimsby, DN33 3EF
M6 HUL S. Lyon, 10 Sycamore Close, Preston, Hull, HU12 8TZ
M6 HUM D. Wood, School Farm, Brock Road, Preston, PR3 0XD
M6 HUN J. Collier, 133 Woodstock Road, Moston, Manchester, M40 0DG
M6 HUP M. Arif, 171 Henley Road, Bedford, MK40 4FZ
M6 HUQ R. Harwood, 5 Cloakham Drive, Axminster, EX13 5GT
M6 HUR J. Jefferies, Millfield Cottage, 1 Bolnhurst Road, Bedford, MK44 2LF
M6 HUS M. Hussain, 10 Mercia Crescent, Stoke-on-Trent, ST6 3JB
M6 HUV I. Barnes, 35 Copley Road, Stanmore, HA7 4PF
M6 HUW J. Mclean, 24 Durham Drive, Oswaldtwistle, BB5 3AT
M6 HUX M. Lisle, 16 Collegiate Crescent, Sheffield, S10 2BA
MW6 HUY D. Johnson, 12 Bro Dedwydd Dunvant, Swansea, SA2 7PR
M6 HUZ L. Hughes, 32 Calder Road, Blackpool, FY2 9TX
M6 HVA J. Wills, 1 Aberdeen Road, Harrow, HA3 7NF
M6 HVD L. Evans, 8 Meadowbrook, Blackpool, FY3 9UE
M6 HVE M. Simonsohn, 5 Pitt Close, Blandford St. Mary, Blandford Forum, DT11 9PS
M6 HVF S. Pearson, 3 Berkeley Road, Shirley, Solihull, B90 2HS
M6 HVH M. Rose, 149 Claremont Road, Blackpool, FY1 2QJ
M6 HVI A. Corbett, 122 Harrowby Road, Stoke-on-Trent, ST3 7AN
M6 HVK K. Orchard, 47 Trezaise Road Roche, St Austell, PL26 8HD
M6 HVL H. Ly, 11 Nodes Drive, Stevenage, SG2 8AL
M6 HVM P. Ashton, 32 Sycamore Road New Ollerton, Nr Newark, NG22 9PS
M6 HVO A. Jarvis, 10 West Park, Wadebridge, PL27 6AN
M6 HVP C. Harwood, 5 Cloakham Drive, Axminster, EX13 5GT
M6 HVR M. Murray, 31 Feeny Street Sutton Manor, St Helens, WA9 4BJ
M6 HVS B. Smithers, 4 Bidmead Court Kent Way, Surbiton, KT6 7SX
M6 HVU T. Raymond, 12 Mill Race Wolsingham, Bishop Auckland, DL13 3BW
M6 HVY E. Money, 4 Cromes Place, Badersfield, Norwich, NR10 5JT
M6 HWC H. Cheesman, 49 Front Street, Chirton, North Shields, NE29 7QN
M6 HWD J. Redgrave, 24 Burnham Close Trimley St. Mary, Felixstowe, IP11 0XG
M6 HWE A. Jepson, Edmonton Road, Mansfield, NG21 9ah
MM6 HWG C. Haws, 1/1 4 Fenella Street, Glasgow, G32 7JT
M6 HWH M. Milioto, 7 Bennett Green, Colchester, CO4 5ZR
M6 HWJ T. Armstrong, 12 Mayhouse Road, Burgess Hill, RH15 9RF
M6 HWL J. Dyson, Fy8 3Tl, Lytham St Annes, FY8 3TL
M6 HWM D. Hatchman, 14 Rudyard Close, Brighton, BN2 6UA
M6 HWN J. Laws, 47 Hampshire Place, Peterlee, SR8 2HE.
M6 HWO M. Wells, 14, Werrington Grove, Peterborough, pe46nt
M6 HWQ N. Williams, 245 Central Drive, Bilston, WV14 8JE
M6 HWT G. Scholey, Ln9 6Jh, Horncastle, LN9 6JH
MI6 HWV G. Osprey, 24 The Hollies Carrickfergus Bt388Ha, Carrickfergus, Bt388ha
M6 HWW A. Graham, 24 Primrose Bank, Wigton, CA7 9JW
M6 HWX G. Mcritchie, 21 St Marys Field, Colchester, CO3 3BP
M6 HXA P. Barden, 17 Chapel Fields, Charterhouse Road, Godalming, GU7 2BS
M6 HXB D. Jones, 91 Laburnum Road, High Wycombe, HP12 3LP
M6 HXC H. Chawdhry, Trinity College, Cambridge, CB2 1TQ
M6 HXD M. Shelley, 21 Ripley Close New Addington, Croydon, CR0 0RP
M6 HXE A. Moss, Winstons, Mayfield Lane Durgates, Wadhurst, TN5 6DG
M6 HXF J. Phillips, 10 Turners Avenue, Fleet, GU51 1DX
M6 HXG P. Stokes, 26 Ashford Road, Hastings, TN34 2HA
M6 HXI R. Bowen, 4 Crossley Gardens, Halifax, HX1 5PU
M6 HXK R. Calvert, 13 William Street, Tunbridge Wells, TN4 9RT
M6 HXN R. Hammond, 28 Birch Way, Hastings, TN34 2JZ
M6 HXO G. Dyson, 4 Davenport Avenue, Blackpool, FY2 9EP
M6 HXR D. Goodchild, Gravel Lane, Ringwood, Bh24 1xy
M6 HXT K. Ede, 7 Corner Garth Ferring, Worthing, BN12 5EL
M6 HXU A. Loader, Rowan Tree House, Crowfield, Brackley, NN13 5TW
M6 HXV H. Redington, Galeholm, Whitecroft, Gosforth, Seascale, CA20 1AY
M6 HXW B. Fitchett, 1 Hilly Fields Mews, Parsonage Estate, Rogate, Petersfield, GU31 5BF
M6 HXX A. Sayer, 23 Toronto Street Lincoln, Lincoln, LN2 5NN
MM6 HXY G. Towell, 265 Stirling Street, Denny, FK6 6QJ
M6 HYB G. Marfell, Springfields Bungalow, Drybrrok, GL17 9BW
M6 HYF R. Hubbard, Southbroom School House, Estcourt Street, Devizes, SN10 1LW
M6 HYH P. Thomas, 5 Mallard Close Chipping Sodbury, Bristol, BS37 6JA
M6 HYJ H. Jeram, 6 Lavender Lane, Rowledge, Farnham, GU10 4AX
M6 HYM A. Mears, Scotland Yard, Priors Leaze Lane, Hambrook, Chichester, PO18 8RQ
M6 HYQ G. Oliver, Bernard Road, Cromer, NR27 9AW
M6 HYR D. Spencer, 17 Rightup Lane, Wymondham, NR18 9NB
MW6 HYS C. Lowes, 3 Castle Close Creigiau, Cardiff, CF15 9NJ
M6 HYT P. Carr, 9 Hollydene Villas, Hythe, Southampton, SO45 4HU
M6 HYW H. Wong, Clarence Drive, Harrogate, HG1 2QG
M6 HYX S. Short, 16 Melrose Drive Fletton, Peterborough, pe2 9dn
M6 HYY S. Greenaway, 166 Marlborough Road, Swindon, SN3 1LU
M6 HYZ K. Evennett, 13 Elm Street, Dereham, NR20 3FN
MM6 HZA S. Gibb, 102 Whitehill Avenue, Musselburgh, EH21 6PE
M6 HZD H. Donovan, 2 All Saints, Weeting, IP270QH
M6 HZF J. Franklin, 31 Haining Gardens, Mytchett, GU16 6BJ
M6 HZJ L. Gynn-Burton, 1 Broad Street Long Eaton, Nottingham, NG10 1JH
M6 HZL H. Smith, 12 Lockgate East, Windmill Hill, Runcorn, WA7 6LB
M6 HZM D. Walker, 44 Albany Road Kilnhurst, Mexborough, S64 5UG
MI6 HZN D. Henderson, 16 Elliott Place, Enniskillen, BT74 7HQ
MM6 HZO A. Artamonovs, Clovenstone Gardens 23/5, Edinburgh, EH14 3ex
MW6 HZP T. Tilbrook, 17 Jubilee Close, Letterston, Haverfordwest, SA62 5SW
M6 HZQ J. Platt, 12 Tawny Grove Four Marks, Alton, GU34 5DU
M6 HZT S. Rickard, 66 Sinclair Drive, Basingstoke, RG21 6AD
M6 HZU J. Howarth, 19 Farnham Croft, Leeds, LS14 2HR
MM6 HZW P. Majumdar, 12 Richmond Avenue Clarkston, Glasgow, G76 7JL
M6 HZX B. Bailey, 215 Stretton Avenue, Coventry, CV3 3HQ
M6 HZZ S. Conway, 21 Milcote Avenue, Hove, BN3 7EJ
MM6 IAB J. Joyce, 7 Leitch Street, Greenock, PA15 2HJ
M6 IAC Z. Cole, 31 High Street, Kimpton, Hitchin, SG4 8RA
M6 IAF S. Shambhu, 34 Gascoigns Way, Patchway, Bristol, BS34 5BY
MW6 IAG J. Richards, 210A Pandy Road, Bedwas, Caerphilly, CF83 8EP
MM6 IAI I. Brown, 28 Garden Road, Cults, Aberdeen, AB15 9RE
M6 IAJ I. Jones, 21 Kennet Green, Worcester, WR5 1JQ
M6 IAL I. Lambert, 69 Anvil Crescent, Broadstone, BH18 9DZ
M6 IAN I. Shires, 19 Prince Charles Avenue, Sittingbourne, ME10 4NA
M6 IAO I. Phillips, 124 Brookwood Drive, Stoke-on-Trent, ST3 6LP
M6 IAQ A. Instone, 63 Larch Road, New Ollerton, Newark, NG22 9SX
MM6 IAR A. Rees, Mains Of Atherb, Maud, Peterhead, AB42 4RD
M6 IAS I. Scott, 21 Field Avenue, Shepshed, Loughborough, LE12 9SH
M6 IAT D. James, 53 Whittingham Road, Ilfracombe, EX34 9LL
M6 IAV I. Avery, 4 Southampton Drive, Liverpool, L19 2HE
MM6 IAW I. Ferguson, 5 Grahamsfield Court Kirkpatrick Fleming, Lockerbie, DG7 3BD
M6 IAX I. Lawton, 11 Goosewell Terrace, Plymstock, Plymouth, PL9 9HW
MM6 IAY A. Maclennan, 5 Baluachrach, Culbokie, Dingwall, IV7 8FP
M6 IAZ D. Goldthorpe, 6 West End Grove Haydock, St Helens, WA11 0AP
MM6 IBB I. Bainbridge, 2 Courthill Road Cottage, Arbroath, DD11 4UX
M6 IBC I. Barber, 8 Newlands Close, Lowestoft, NR33 7EY
MW6 IBD I. Daniel, 35 New Road, Upper Brynamman, Ammanford, SA18 1AF
M6 IBF I. Graham, 49 Eagle Close, Leighton Buzzard, LU7 4AT
M6 IBG I. Bruno-Gaston, 83 Althorne Gardens, London, E18 2DB
M6 IBH C. Tate, 87 Overdale Road, Middlesbrough, TS3 7NQ
M6 IBI A. Douglas, 3 Beech Avenue, Bilsborrow, Preston, PR3 0RH
M6 IBJ S. Moorby, 11 Vespasian Gardens Rooksdown, Basingstoke, RG24 9SH
M6 IBL Dr G. Eibl-Kaye, 1 Main Road, Littleton, Winchester, SO22 6PS
M6 IBO A. Lawrence, 5 Heighton Close, Bexhill-on-Sea, TN39 3UP
M6 IBP G. Stamp, 17 Harlow Manor Park, Harrogate, HG2 0EG
MM6 IBQ Dr J. Simpson, 18 Corrour Road, Glasgow, G43 2DX
MI6 IBR B. Rocks, 4 Millview, Randalstown, Antrim, BT41 3BA
M6 ICA C. Irons, 11 Elm Grove, Moira, Swadlincote, DE12 6HH
M6 ICB I. Bushnell, 4 Upper Tail, Watford, WD19 5DF
MI6 ICD I. Cairns, 18 Molyneaux Avenue, Larne, BT40 2TU
MW6 ICF K. Earnshaw, 5 Castle Mews George Street, Pontypool, NP4 6BU
M6 ICH M. Pullen, 2 Wyatts Lane, Little Cornard, Sudbury, CO10 0NT
M6 ICJ A. Lorentsen, 54 Kingsnorth Road, Gillingham, ME8 6QY
M6 ICK M. Ginty, 34 High Street, Branston, Lincoln, LN4 1NB
M6 ICL M. Dominguez, 52 St. Leonards Road, Amersham, HP6 6DR
MW6 ICM A. Moody, Perthiteg, Cwmhiraeth, Llandysul, SA44 5XJ
M6 ICO J. Stevenson, 18 Drakehouse Lane, Sheffield, S20 1FW
M6 ICP I. Pears, 10 Pixley Dell, Consett, DH8 7DB
M6 ICQ D. Banks, 41 East Road, Rotherham, S65 2UX
M6 ICR S. Barlow, Apartment 34, Jet Centro, Sheffield, S2 4AH
MW6 ICU A. Jones, 8 Arles Road, Cardiff, CF5 5AP
M6 ICX D. Lucock, 34 Wentworth Drive, Ipswich, IP8 3RX
M6 ICZ O. Barber, Homedale, St. Monicas Road Kingswood, Tadworth, KT20 6ET
M6 IDB M. Parker, 1 Ham Road, Wanborough, Swindon, SN4 0DF
M6 IDC I. Cosham, 54 Hawkins Crescent, Shoreham-by-Sea, BN43 6TP
M6 IDD P. Wilcox, 2 Merryhill Terrace, Belmont, Hereford, HR2 9RT
M6 IDE A. Currie, 31 Launceston Road, Bodmin, SO50 6AY
M6 IDF I. Firth, 124 Viking Road, Bridlington, YO16 6TB
M6 IDG I. Garrard, 33 Uplands Road, Hockley, SS5 4DL
MI6 IDJ K. Mclaverty, 123A Castle Road, Antrim, BT41 4NG
M6 IDK I. King, 7, Greenacres Avenue, Blythe Bridge, Stoke on Trent, ST11 9HU
M6 IDM E. Ogbua, 11 Coltness Crescent, London, SE2 0UY
M6 IDN I. Norfolk, Arwelfa, High Street, Uckfield, TN22 3LP
M6 IDO Y. Wang, 50 Sleaford Street, Cambridge, CB1 2PU
M6 IDP I. Nelson, 38 Warbro Road, Torquay, TQ1 3PW
M6 IDR I. Reeve, 36 Stone Pippin Orchard, Badsey, Evesham, WR11 7AA
M6 IDS A. Pulman, 3 Brunner Road, Billingham, TS23 1HW
M6 IDU M. Brannon, Leonard Cheshire Close, Salisbury, Sp47rn
M6 IDX J. Ledger, 39Eascroft Drive, Sheffield, S20 8JG
M6 IEA E. Aspden, 4 Lilac Close, Newcastle, ST5 7DH
M6 IEB I. Beales, 6 Edge Well Rise, Sheffield, S6 1FB
M6 IEC C. Williams, Flat 21, Holly House Holmes Street, Burnley, BB11 3BE
M6 IEI J. Lowe, 26 Rectory Drive Yatton, Bristol, BS49 4HF
M6 IEJ L. Thornton, 11 Seabourne Ave Blackpool Fy4 1Eh, Blackpool, FY4 1EH
M6 IEM J. Paine, 11 Ferndale Park Fifield Road, Bray, Maidenhead, SL6 2DZ
M6 IEO L. Metcalfe, 40 St. Anns Court, Hartlepool, TS24 7HY
M6 IEQ J. Skitt, 38 Bodiam Avenue Tuffley, Gloucester, GL4 0TJ
M6 IER A. Unwin, 152 Epsom Road, Guildford, GU1 2RP
M6 IEU R. Husher, 194 Leybourne Ave, Bournemouth, BH10 5NR
M6 IEW I. Williams, 23 Symons Close, Blackwater, Truro, TR4 8ER
M6 IEZ T. O'Gorman, Spencer Avenue, Birkenhead, CH42 2DN
M6 IFA J. Arnold, 17 Larch Road Roby, Liverpool, l369ty
M6 IFB M. Glen, 3 Mill Bungalows Catterick, Richmond, DL10 7LY
M6 IFC M. Harvey, May Tree Cottage, Kelvedon Road, Tiptree, Colchester, CO5 0LJ
MW6 IFE L. Pickering, 14 Bryn Wyndham Terrace, Cardiff, CF42 5NG
M6 IFF J. Sharrad, 52 Springwood Drive, Ashford, TN23 3LQ

M6 IFG A. Kreissl, 382 Oldfield Road, Altrincham, WA14 4QT
M6 IFH I. Harrison, 8 Jeffrey Avenue, Longridge, Preston, PR3 3TH
M6 IFI I. Iremonger, 2 Harbord Road, Cromer, NR27 0BP
M6 IFN T. Sargeant, 27 Digby Close Tilton On The Hill, Leicester, LE7 9LL
M6 IFO D. Harris, 35 Itchenor Road, Hayling Island, PO11 9SN
M6 IFQ M. Murray, Victor Court, Rainham, rm13 8el
M6 IFS G. Milner, 9 Lilydene Avenue Grimethorpe, Barnsley, S72 7AA
M6 IFT G. Saunders, 140 Highbridge Road, Burnham-on-Sea, TA8 1LW
M6 IFV P. Kearney, 22 Kingston Drive, Cheltenham, GL51 0UB
M6 IFW I. Warman, 8 Burley Road, Bishop's Stortford, CM23 3LR
M6 IFX S. Wade, 2 Esher Place, Beacon Lane, Cramlington, NE23 8JQ
M6 IFY M. Gratton, 4 Princes Road, Wisbech, PE13 2PG
M6 IFZ I. Huggett, 104 Hinchcliffe Orton Goldhay, Peterborough, PE2 5SS
M6 IGA E. O'Neill, 13 Goodwood Close, Market Harborough, LE168JF
MW6 IGC I. Curnock, 62 Heol Y Banc, Bancffosfelen, Llanelli, SA15 5DL
M6 IGH I. Garforth, 63 Upper Perry Hill, Bristol, BS3 1NJ
M6 IGJ I. Jackson, 22 Greenwood Avenue, Congleton, CW12 3HH
M6 IGK I. Wideman, Silcoates Lane, Wrenthorpe, Wakefield, WF2 0PD
M6 IGM G. Benford, 34 Victoria Gardens, Colchester, CO4 9YD
M6 IGP Dr J. Harmer, 1 Wynford Rise, Leeds, LS16 6HX
M6 IGQ S. Nelson, 25 Waterloo Place, North Shields, NE29 0NA
M6 IGR D. Drummond, 21 Beveland Road, Canvey Island, SS8 7QU
M6 IGS N. Hawkins, 17 Meddins Lane, Kinver, Stourbridge, DY7 6BZ
M6 IGT A. Davies, 6 Stonefield Close Shrivenham, Swindon, SN6 8DY
MI6 IGV R. Freeburn, 24Highfern Gardens, Belfast, bt133rd
M6 IGW A. Wilson, 38 Cotleigh Drive, Sheffield, S12 4HU
M6 IGZ A. Hogg, 17 Kedleston Close, Stretton, Burton-on-Trent, DE13 0FN
M6 IHA C. Haworth, Salisbury, Station Road Keyingham, Hull, HU12 9SZ
M6 IHC I. Clement, 24 Millais, Horsham, RH13 6BS
M6 IHD J. Booth, 27 Moorlands Scholes, Holmfirth, HD9 1SW
M6 IHE K. Harcombe, 29 Jurston Fields, Wellington, TA21 9FX
M6 IHH I. Hutchinson, Bridgend, Mill Lane, North Hykeham, Lincoln, LN6 9PA
M6 IHM A. Walker, 39 Delves Road Killamarsh, Sheffield, S21 1AW
M6 IHN N. Gooding, 41 The Crescent, Wolverhampton, WV6 8LA
M6 IHO B. Domigan, 156A Parkgate Road Holbrooks, Coventry, CV6 4GG
MM6 IHQ T. Rogers, Marypark Farm, Marypark, Ballindalloch, AB37 9BG
M6 IHR C. Witham, 162 Stockbridge Lane, Liverpool, L36 8EH
M6 IHS R. Hunter, Poplars, March Road Guyhirn, Wisbech, PE13 4DA
M6 IHT G. Lamming, 8 Meadow Lane Newport, Brough, HU15 2QN
M6 IHZ E. Pardoe, 2 Bar Meadow, Shobdon, HR6 9BZ
M6 IIB D. Faulkner, 21 Chestnut Way, Bromyard, HR7 4LG
M6 IIF D. Kerby-Collins, Wr158Qn, Worcester, wr158qn
M6 IIG R. Blair, Springfield, Pewsey Road, Rushall, Rushall, Pewsey, SN9 6EN
M6 IIJ I. Bailey, 33 Govett Road, St Helens, WA9 5NQ
M6 IIL L. Furr, 158 Eastern Avenue, Southend-on-Sea, SS2 4AZ
MM6 IIO R. Treble, 2 Baidland Meadow, Dalry, KA24 5HP
MM6 IIP P. Wilson, 2 Southerhouse Scousburgh, Dunrossness, Shetland, ZE2 9JE
M6 IIQ P. Baillie, 23 Brackens Drive Warley, Brentwood, cm145ue
MW6 IIU R. Selby, 7 West Hook Road Hook, Haverfordwest, SA62 4LS
M6 IIW J. Kelly, 25 Blunts Avenue Sipson, West Drayton, UB7 0DR
M6 IIX P. Hill, 83 Gladeside, Croydon, CR0 7RW
M6 IIY A. Wood, 85 Love Lane, Rayleigh, SS6 7DX
M6 IIZ A. Billett, 1 Tortoiseshell Drive, Attleborough, NR17 1GU
M6 IJB I. Boddy, 5 Boverton Avenue, Brockworth, Gloucester, GL3 4ER
MW6 IJD N. Dimonaco, 41 Brongwinau, Comins Coch, Aberystwyth, SY23 3BQ
M6 IJG D. Mowbray, 5 Heath Lane Leasingham, Sleaford, NG34 8JF
M6 IJH I. Holdford, 46 Hildreth Road, Prestwood, Great Missenden, HP16 0LY
MW6 IJJ I. Johns, Flat 16, Baker Street House James Street, Blaenavon, Pontypool, NP4 9EH
M6 IJK C. Birch, 5 Newport, Barton-upon-Humber, DN18 5QJ
M6 IJL B. Damazer, The Manse East Street, Crowland, PE6 0EN
M6 IJM I. Morgan, 30 Farm Road, Hutton, Weston-Super-Mare, BS24 9RH
M6 IJO S. Deere, 12 Leverton Road, Retford, Dn229he
MM6 IJP I. Purkis, Lochaber Croft, Tullynessle, Alford, AB33 8QQ
M6 IJQ R. Bedford, 29 Kent Road, Brookenby, Market Rasen, LN8 6EW
M6 IJR M. Elkins, 18 Scutts Close Lytchett Matravers, Poole, BH16 6HB
M6 IJT C. Partington, 55 Caldercroft, Elland, HX5 9AY
M6 IJV K. Thwaites, 65 Chessington Park Hill, Chessington, KT9 2BJ
MW6 IJW P. Williams, 52 Rhoslan, Tredegar, NP22 4PF
M6 IJX D. Wilderspin, 14 Tannery Court, North Street, Crewkerne, TA18 7AY
M6 IJZ S. Cross, 22 Park Avenue Washingborough, Lincoln, LN4 1DB
M6 IKA I. Abraham, 12 Graham Road, Halesowen, B62 8LJ
M6 IKD M. Draper, 160 Chanctonbury Road, Burgess Hill, RH15 9HA
M6 IKF M. Blyth, 9 High Grove, Ryton, NE40 3Jn
M6 IKH J. Jarvis, The Wheel House, Ellerton Hill Ellerton Upon Swale, Richmond, DL10 6AL
M6 IKI M. Anostalgia, 136 Avenue Road Extension, Leicester, LE2 3EH
M6 IKJ C. Smith, 2 Burley Gardens, Street, BA16 0SN
M6 IKL M. Mccallister, 13 Darvel Avenue Ashton-In-Makerfield, Wigan, WN4 0UA
M6 IKM M. Moffat, 7 Lingfell Avenue, Cockermouth, CA13 9BE
M6 IKN T. Caldwell, 29 The Vale, Coventry, CV3 1DW
MD6 IKR D. Cain, Flat, Ballavagher, Main Road, Isle of Man, IM4 4AR
MM6 IKS I. Stobo, 25 Greenfield Circle, Elgin, IV30 5NF
MM6 IKT Dr K. Brooks, 192/6 Causewayside, Edinburgh, EH9 1PN
M6 IKX D. Mielczarek, 7 Bryant Close, Camberley, GU16 8AD
M6 IKY M. Rawson, 34 Penlon Place, Abingdon, OX14 3QQ
M6 ILB R. Brocklehurst, 12 Harriers Close, Christchurch, BH23 4SL
M6 ILC C. Ng, 9A Romsey Road, Southampton, SO16 4BY
MI6 ILF I. Forsythe, 45 Kensington Park, Portadown, Craigavon, BT63 5PQ
M6 ILH I. Hobbs, 115 Adams Way, Croydon, CR0 6XR
M6 ILI N. Winfield, 1 Southview, School Lane, Stoke Row, Henley-on-Thames, RG9 5QX
M6 ILJ W. Fisher, 15A Trevenner Lane, Marazion, TR17 0BL
MW6 ILK T. Rathbone, 13 North Avenue, Rhyl, LL18 1HT
M6 ILM D. Sharpe, 31 Malyons, Basildon, SS13 1PJ
M6 ILN D. Bee, 25 Blatcher Close, Minster On Sea, Sheerness, ME12 3PG
M6 ILO M. Noblet, 1 Lingdale Road, Wirral, CH48 5DG
M6 ILP I. Patterson, 63 Orchard Road, South Ockendon, RM15 6HP
M6 ILR V. Seabright, 208 Park Way, Rubery, Birmingham, B45 9WA
M6 ILS J. Anthony, 21 Belgrave Street, Denton, Manchester, M34 3WP
M6 ILT P. Mccormick, Fieldview, Crown East Lane, Lower Broadheath, Worcester, WR2 6RH
M6 ILX M. Bennett, 33 Charles Street, Redditch, B97 5AA
M6 ILY D. Stuart, 6 Ross Crescent, Watford, WD25 0DB
MJ6 ILZ M. Thorpe, Dolphin Cottage, Union Road, Grouville, Jersey, JE3 9ER
M6 IMA I. Maley, 6 Meadow View Close, Newport, TF10 7NN
MM6 IMB M. Breimann, 5 Leaside, Mossbank, Shetland, ZE2 9TF
MM6 IMF I. Fairbairn, 2 The Steadings, Slackend, Buckie, AB56 5BS
M6 IMH I. Hickinbottom, Clover Cottage, Snead, Montgomery, SY15 6EB
M6 IMI A. Rayner, 5 Buttermere Close, Lincoln, LN6 0YD
MW6 IMM C. Milne, 18 Chapelfield, Deganwy, LL31 9BF
MM6 IMP C. Mcconnachie, Flat C, 5 Whitefaulds Crescent, Maybole, KA19 8AY
MI6 IMQ J. Crowe, 36 A Ballynahatty Road, Belfast, BT8 8LE
M6 IMR I. Rich, 39 Wren Close, Heathfield, TN21 8HG
M6 IMS M. Sims, 5 Sandy Leaze, Bradford-on-Avon, BA15 1LX
MW6 IMT I. Thornton, 31 Woodland Row, Cwmavon, Port Talbot, SA12 9LJ
MW6 IMU M. Mcdonald, Falcondale, Pleasant Valley, Stepaside, Wisemans Bridge, Narberth, SA67 8NT
MI6 IMV A. Brashaw, 3 The Straits, Lisbane, Comber, Belfast, BT23 6AQ
M6 IMW I. Walker, 24 Hawthorn Road, Norwich, NR5 0LP
MI6 IMY P. Calvin, 49 Gobhan Close Portadown, Craigavon, BT63 5QZ
M6 IMZ I. Ross, 48 Henry Drive, Leigh-on-Sea, SS9 3QF
M6 INA G. Allen, 13 Strathmore Avenue, Hull, HU6 7HJ
MI6 INB M. Gibson, 12 Drumcarn Gardens Portadown, Craigavon, BT62 4DH
MM6 INC A. Mcmath, 57 Hillhouse Avenue, Bathgate, EH48 4BB
M6 IND P. Ind, 30 Thompson Road, Stroud, GL5 1SY
M6 ING D. Singh, 28 Chadview Court, Chadwell Heath Lane, Romford, RM6 4BF
M6 INI M. Smith, Church Farm, Mucklestone, Market Drayton, TF9 4DN
M6 INM M. Davies, 91 Ameysford Road, Ferndown, BH22 9QD
MM6 INN A. Macinnes, 1 Brent Place, Glenrothes, KY7 6TA
M6 INP J. Walton, Walton Holt, Marsh Road Orby, Skegness, PE24 5HZ
MM6 INS C. Ralph, 37 Seaview Terrace, Edinburgh, EH15 2HE
M6 INT M. Pomfret, 5 Malvern Crescent, Ince, Wigan, WN3 4QA
M6 INV M. Basford, 4 Renoir Close, St Ives, PE27 3HF
M6 INW D. Inwood, 78 Lower Thrift Street, Northampton, NN1 5HP
M6 INX J. Porter, Flat 2, 14 Trafalgar Square, Scarborough, YO12 7PY
MW6 IOA I. Dumitrescu, The Rectory, Halkyn, Holywell, CH8 8BU
M6 IOD S. Mowbray, 5 Heath Lane Leasingham, Sleaford, NG34 8JF
M6 IOF J. Parton, 6 Windmill Road, Atherstone, CV9 1HP
M6 IOG F. Cooper, Needhams Farm House, Spittal Hill Road, Boston, PE22 0PA
M6 IOI L. Emanuel, 20 Wychwood Drive, Redditch, B97 5NW
M6 IOL M. Scott, 19 Saltburn Road, Sunderland, SR3 4DJ
M6 ION L. Rimington, 3 Amicombe, Wilnecote, Tamworth, B77 4JJ
M6 IOO J. Berrio, 29 Scalborough Close, Countesthorpe, Leicester, LE8 5XH
M6 IOQ G. Rowland, 52 Victoria Road, Mablethorpe, LN12 2AJ
M6 IOR S. Engelmann, 81 Tierney Road, London, SW2 4QH
M6 IOS G. Baker, 56 Chalklands, Bourne End, SL8 5TJ
M6 IOW F. Alfrey, 16 Walls Road, Bembridge, PO35 5RA
M6 IOY R. Melbourne, Barton Marina, Barton-under-Needwood, DE13 8DZ
MW6 IOZ M. Simons, 68 Harbour Village, Goodwick, SA64 0DZ
M6 IPA R. Howitt, 7 Badgers Close, Chelmsford, CM2 8QB
MI6 IPB D. Neill, 8 Castle Meadows Carrowdore, Newtownards, BT22 2TZ
M6 IPE R. Fulcher, 1 Edwards Close Hutton, Brentwood, CM13 1BU
M6 IPF A. Allen, 2 Robson Way, Blackpool, FY3 7PP
M6 IPG N. Head-Jenner, 5 Ponders Road, Colchester, CO63LX
M6 IPH J. Dunn, 39 Bramble Lane, Wye, TN25 5AB
M6 IPI A. Cook, 118 Avon Road, Chelmsford, CM1 2LA
M6 IPJ P. Allen, 45 Under Knoll, Peasedown St. John, Bath, BA2 8TY
M6 IPL I. Pilton, Caleril Barn, Pool Foot Farm Haverthwaite, Ulverston, LA12 8AA
MM6 IPP Lord I. Patterson, 8 Jane Street, Dunoon, PA23 7HX
M6 IPQ T. Baker, 92 Conway Avenue, Derby, DE723GR
M6 IPW A. Lee, 1 Lower Stoke Limpley Stoke, Bath, BA2 7FU
M6 IPY P. Sayer, 443 Sutton Road, Southend-on-Sea, SS2 5PJ
M6 IQA P. Patmore, 85 Frenchs Wells, Woking, GU21 3AU
M6 IQC R. Clark, 67 Seymour Street, Chorley, PR6 0RR
M6 IQE D. Hopwood, 64 Ridge Road, Stoke-on-Trent, ST6 5LP
MW6 IQF D. Jackson, 10 Hafod Wen Johnstown, Wrexham, LL14 2AT
M6 IQG J. Wooldridge, 6 Heskin Road, Lydiate, Liverpool, L31 0BS
M6 IQK C. Price, 270A Chorley New Road, Horwich, Bl6 5ny
M6 IQL N. Green, 44 Rushyford Drive, Chilton, Ferryhill, DL17 0EQ
M6 IQM A. Sharman, 8 Knowle Road, Biddulph, Stoke on Trent, ST8 6LH
M6 IQN J. Holford, 30 Meadow Avenue, Newcastle, ST5 9AE
M6 IQO P. Taylor, 32 Heliers Road, Liverpool, L13 4DH
M6 IQT K. Carey, 28 Honeysuckle Avenue Hailsham Bn274Fp, Hellingly, Hailsham, BN27 4FP
M6 IQU A. Pym, 58 Eastern Avenue, Chippenham, SN15 3LW
M6 IQW D. Parker, 51 Parsonage Road, Henfield, BN5 9HZ
M6 IQX D. Vaughan, 1 Boulnois Avenue, Poole, BH14 9NX
M6 IQY H. Fisher, Flat 15, Chaucer House, 25 Knightrider Street, Maidstone, ME15 6ND
M6 IQZ D. Cassidy, 186 Kingsman Drive, Boorley Green, Botley, SO32 2TG
M6 IRB R. Bewick, 357 Franklands Village, Haywards Heath, RH16 3RP
M6 IRC I. Crowson, 19 Burgoyne Road, Southsea, PO5 2JJ
MI6 IRE P. Mcaleer, 24 Wansbeck Street, Belfast, BT9 5FQ
M6 IRJ M. Impey, 20 Chalton Road, Luton, LU4 9ER
M6 IRK O. Kirton, Woodland, Moretonhampstead, Newton Abbot, TQ13 8SD
M6 IRL D. Lovejoy, 9 West View Close, Middlezoy, Bridgwater, TA7 0NP
M6 IRM M. Raine, 91 Lulworth Avenue, Jarrow, NE32 3SB
M6 IRP I. Pipe, 9 Sherlock Hoy Close, Broseley, Telford, Tf125jb
M6 IRU S. Airey, 22 Primrose Street, Lancaster, LA1 3BN
M6 IRW I. Wilkes, 7 Oriel Close, Dudley, DY1 2UW
M6 IRX G. Harman, 58 Laurence Avenue, Witham, CM8 1JB
M6 IRZ W. Drozdz, 29 Breedon Road, Birmingham, B30 2HT
M6 ISA I. Stone, 169 Booth Road, Wednesbury, WS10 0EW
M6 ISB S. Brown, 18 Goring Ave. Gorton., Manchester, M18 8WW
M6 ISD S. Dodd, 21 Sunningdale, Grantham, NG31 9PF

M6 ISE M. Petre, 52 Fremantle Road, Ilford, IG6 2AZ
M6 ISG T. Holland, 68 Church Street, Billericay, CM11 2TS
M6 ISH R. Kinder, 21 Oakdene, Chobham, Woking, GU24 8PS
M6 ISJ S. Jones, 5 Meadowlands, Kirton, Ipswich, IP10 0PP
M6 ISK N. Butler, 75 Rutland Street, Derby, DE23 8PR
M6 ISN A. Oakey, 53 Appledore Close, Margate, CT9 3RG
M6 ISR I. Robinson, 16, Stobarts Field New Ridley, Stocksfield, NE437RL
M6 ISU M. Dunstan, 5 Polgooth Close, Redruth, TR15 1QL
M6 ISW J. Atkinson, 1 St. Johns Crescent Bishop Monkton, Harrogate, HG3 3QZ
M6 ISX J. Rogerson, 8 Bonser Crescent Huthwaite, Sutton-in-Ashfield, NG17 2RE
MM6 ISY I. Phillips, 16 Seton Court, Port Seton, Prestonpans, EH32 0TU
M6 ISZ I. Townsend, 263D Stapleton Road, Bristol, BS5 0PQ
M6 ITD C. Knowles, 10 The Drive, Southwick, Brighton, BN42 4RR
M6 ITI C. Johnson, Suite 204, 33 Queen Street, Wolverhampton, WV13AP
M6 ITL J. White, 4 Kings Road, New Milton, BH25 5AY
M6 ITN R. Goodall, 76 Beaconfield Road, Plymouth, PL2 3LF
M6 ITP M. Baynes, 92 Belgrave Drive, Hull, HU4 6DW
M6 ITQ J. Swales, 90 Earlswood Road, Dorridge, Solihull, B93 8RN
M6 ITU P. James, Flat 2, Marigold House, 2 Ironbridge Road, Twigworth, Gloucester, GL2 9GS
M6 ITV I. Roberts, 15 Broadcroft, Hemel Hempstead, HP2 5YX
M6 ITX M. Reynolds, 24 Burton Close, Corringham, Stanford-le-Hope, SS17 7SB
M6 ITY A. Praveen, 28 Long Deacon Road, London, E4 6EG
M6 ITZ C. Mayer, 81 Bohelland Road, Penryn, TR10 8DY
MM6 IUE A. Young, 2 Dunvegan Place, Ellon, AB41 9TF
M6 IUF L. Barrett, 2 Barton Rise Feniton, Honiton, EX14 3HW
MD6 IUH K. Payne, 2 Dreeym Balley Cubbon, Ballacubbon, Isle of Man, IM9 4PR
M6 IUI D. Hendy, 12 Rumsam Gardens, Barnstaple, EX32 9EY
M6 IUK D. Grayson, 79 Errington Avenue, Sheffield, S2 2EA
M6 IUM R. Lincoln, 7 John Carrs Terrace, Bristol, BS8 1DW
MW6 IUN I. Jones, 21 Albert Street, Maesteg, CF34 0UF
MW6 IUP D. Flewin, Huntingdon Way, Swansea, SA2 9HN
MM6 IUR B. Findlay, 39 Barrhill Court Kirkintilloch, Glasgow, G66 3PL
M6 IVB C. Payne, 11 The Paddocks Great Chart, Ashford, TN23 3BE
M6 IVC I. Stephens, 14 Vardon Close Kingston Hill, Stafford, ST16 3YW
M6 IVE C. Cavalcante Pinheiro Filho, Flat 7, Hollybush House, Hollybush Gardens, London, E2 9QT
M6 IVG N. Simmons, 28 Odo Road, Dover, CT17 0DP
M6 IVI J. Knights, 18 Kenilworth Gardens, Blackpool, FY4 1JJ
MW6 IVK P. John, 136 Birchgrove Road Birchgrove, Swansea, SA7 9JT
M6 IVN I. Clarke, 140 Cashmere Drive, Andover, SP11 6SS
M6 IVO I. Yovchev, 11 Beverley Drive, Edgware, HA8 5NQ
MM6 IVP S. Morrison, 24 High Street, Fochabers, IV32 7DX
M6 IVQ R. Cutting, 19 Laburnum Close, Wisbech, PE13 3RJ
M6 IVR I. Goodman, 26 Vallansgate, Stevenage, SG2 8PY
M6 IVT C. Drake, 115, Woodlands Road, Gillingham, ME72DX
MW6 IVW D. Morgan, 28 Harbour Village, Goodwick, SA64 0DY
M6 IWA I. Wright, 1 Greaves Avenue, Old Dalby, Melton Mowbray, LE14 3QE
M6 IWB I. Bunting, 14 Mill Pightle, Aylsham, Norwich, NR11 6LX
M6 IWD A. Sirrell, 4 White Oaks North Pickenham, Swaffham, PE37 8LB
M6 IWE G. Cousins, 47 Nightingale Drive Taverham, Norwich, NR8 6LA
M6 IWF I. Francis, 34 Furlong Road, Bourne End, SL85AA
M6 IWI D. Gilbertson, 6 Lewens Lane, Wimborne, BH21 1LE
MM6 IWL C. Somerville, 112 Dickens Avenue, Glasgow, G813EP
M6 IWO C. Hargrave, 20 Gainsborough Road, Ashley Heath, Ringwood, BH24 2HY
M6 IWP I. Perkins, 49 Salters Lane, Tamworth, B79 8BH
M6 IWX A. James, 36 Cedar Close, Walton on the Naze, CO14 8NJ
MW6 IXD J. Harrington, 23 Heol Emrys, Fishguard, SA65 9EE
M6 IXF M. Cox, 17A Church End, Weston Colville, Cambridge, CB21 5PE
M6 IXG P. Gretton, 159 Birchfield Road Arnold, Nottingham, NG5 8BP
MM6 IXH R. Stewart, 31 Seatown, Lossiemouth, iv 31 6 jj
M6 IXI R. Forss, Ginger House, High Street, Blockley, Moreton-in-Marsh, GL56 9EX
M6 IXJ R. Reynolds, 3 Canna Park Drive, Highampton, Beaworthy, EX21 5AY
M6 IXM Dr D. Buckley, 7 Clary Meadow, Northwich, CW8 4XG
MM6 IXO K. Goggins, Farm Road, Glasgow, G81 6HH
M6 IXR D. Brotheridge, 14 Nuthatch Drive, Torquay, TQ2 7GF
M6 IXY A. Champkin, 17 Blundell Place, Bedford, MK42 9XB
MW6 IXZ C. Purviss, 45 Heol Camlas Gwersyllt, Wrexham, LL11 4HF
MM6 IYA D. Kean, 24 Morar Drive, Clydebank, G81 2YB
M6 IYC J. Mcintyre, 10 Fern Drive Dudley, Cramlington, NE23 7AF
M6 IYG G. Osborne, Dunelm Cottage, High Street Castle Camps, Cambridge, CB21 4SN
M6 IYJ M. Futcher, 46 Houldey Road, Birmingham, B313HJ
M6 IYL G. Bryant, 15 The Clock Inn Park Lydeway, Devizes, SN10 3PP
M6 IYM P. Hutchinson, 75 Windermere Avenue Orford, Warrington, WA2 0NB
M6 IYN A. Bailey, 52 Berkeley Road, Shirley, Solihull, B90 2HT
M6 IYS E. Crafter, 90 Connaught Avenue, Grays, RM162XT
M6 IYT M. Robino, 61 Mill Hill Little Hulton, Manchester, M38 9TN
M6 IYU S. Helm, 10 St. Annes Avenue, Middlewich, CW10 0AE
M6 IYV K. Smith, London Road, Grays, rm20 4aa
MW6 IYW P. Beech, 9 Wesley Court, Pembroke Dock, SA72 6NE
M6 IZA I. Brunt, 62 Greenwood Drive, Watford, WD25 0HX
M6 IZB A. Bowes, 178 Saxton Road, Abingdon, OX14 5HF
M6 IZE A. Hampson, 12 Oakhays, South Molton, EX36 4DB
M6 IZF C. Scouller, 72 Brookmans Avenue Brookmans Park, Hatfield, AL9 7QQ
MM6 IZH A. Sommerville, Smeaton Farm, Dalkeith, EH22 2NL
MD6 IZI I. Dorman, 1 Sprucewood Rise, Foxdale, Douglas, Isle Of Man, IM4 3JP
MM6 IZK I. Low, 124 Sheephousehill, Fauldhouse, Bathgate, EH47 9EL
M6 IZM J. Mccaffery, 56 Churchill Avenue Bulford, Salisbury, SP4 9HE
M6 IZN A. Mccaffery, 57 St.Leonards Road, Newton Abbot, TQ12 1JY
M6 IZO D. Drake, 16 Princess Street, Blackpool, FY1 5BZ
M6 IZP A. Jakstas, 36 Boxridge Avenue, Purley, CR8 3AQ
M6 IZQ W. Townsend, 8 Beech Avenue Kirkham, Preston, PR4 2UE
M6 IZT M. Scott, 51 Devona Avenue, Blackpool, FY4 4NU
M6 IZU B. Griggs, 61 Langdale Road, Blackpool, FY4 4RR
M6 IZW I. Whiteley, 2 The Meade, Manchester, M21 8FA
M6 IZY E. Coffey, 4 Castlehall, Tamworth, B77 2EG
M6 JAB J. Burrows, 78A Coronation Road, Earl Shilton, Leicester, LE9 7HJ
MI6 JAD S. Mullan, 135 Ballyavelin Road, Limavady, BT49 0QB
MM6 JAE J. Dalgety, 5 East Court, Edinburgh, EH16 4ED
M6 JAF D. Oliphant, 9 Portia Street, Ashington, NE63 9DL
M6 JAG R. Price, 92 Lincoln Road, Ingham, Lincoln, LN1 2XF
M6 JAJ J. Smith, 46 Mulberry Close, Goldthorpe, Rotherham, S63 9LB
M6 JAK J. King, 18 Ross Road, Wallington, SM6 8QB
M6 JAL M. Leggett, 7 Barley Way, Thetford, IP24 1LG
M6 JAM J. Hunter, 164 Grange Road, Newark, NG24 4PP
MM6 JAN J. Mcclure, Bridgefoot Croft, Udny, Ellon, AB41 6RT
M6 JAQ J. Snell, 32 Meadow Halt, Ogwell, Newton Abbot, TQ12 6FA
M6 JAR J. Revell, 63 Mountbatten Road, Bungay, NR35 1PP
M6 JAS J. Wells, 18 Roewood Road, Holbury, Southampton, SO45 2JH
M6 JAU J. Goldsmith, 20 Trinity Mews, Bury St Edmunds, IP33 3AT
M6 JAV M. Sharman, 33 Bungalow Estate, Lady Lane, Coventry, CV6 6BD
MM6 JAW J. White, 63 Allershaw Tower, Wishaw, ML2 0LP
M6 JAZ J. Cleeter, 49 Hunters Field, Stanford In The Vale, Faringdon, SN7 8LZ
M6 JBA J. Ashley, Rowborough Farm Cottages, Brading, Sandown, PO36 0BA
M6 JBB J. Berry, 31 New Hall Way, Flockton, Wakefield, WF4 4AX
M6 JBE J. Nicholson, 32 Home Orchard, Yate, Bristol, BS37 5XQ
M6 JBG J. Bethell, 19 Kiln Cottages, The Brickfields, Stowmarket, IP14 1RY
M6 JBI J. Innes, Trelawny, Marine Drive, Bude, EX23 0AH
M6 JBL B. Jedryka, 31 Backhold Lane, Halifax, HX3 9DR
M6 JBM J. Russell, 81 Chapman Street, Loughborough, Le11 1dd
MM6 JBN J. Breen, 26 Maxwelton Road, Glasgow, G33 1LR
M6 JBO J. Davies, 34 Station Road, Benton, Newcastle upon Tyne, NE129NQ
MW6 JBQ G. Smith, 49 Coed Cae, Caerphilly, CF83 1RU
M6 JBR J. Whitton, 11 Dursley Road, Bristol, BS11 9XB
M6 JBT J. Hawkins, 24 Green Lane, Stourbridge, DY9 7EW
M6 JBU J. Burnett, Hobart House, 16 Church Lane, Reepham, Lincoln, LN3 4DQ
M6 JBV M. Redmore, 8 Hambledon Rise, Northampton, NN4 8TT
M6 JBX J. Blower, 8 The Laundry, Seifton, Ludlow, SY8 2DH
MI6 JBZ J. Given, 15 Middle Road, Lisburn, BT27 6UU
M6 JCA J. Aubury, 27 Gravel Walk, Tewkesbury, GL20 5NH
M6 JCB C. Daniels, 19 Gerrard Street, Rochdale, OL11 2EB
M6 JCC J. Clare, Kimberley, Boston Road, Boston, PE20 3AP
M6 JCE J. Crewe, 22 Myrtle Tree Crescent Sandbay, Weston-Super-Mare, BS22 9UL
M6 JCF J. Faulkner, Mount Pleasant, Elkstones, Buxton, SK17 0LU
M6 JCG J. Anderson, The Firs, Crapstone, Yelverton, PL20 7PJ
MW6 JCH J. Hudson, 50 Dan Y Cwarre, Kidwelli, SA17 4JA
M6 JCJ J. Ward, 6 Fairfield, Telegraph Hill, Redruth, TR16 5AH
MM6 JCL J. Littlefair, 30 Maple Crescent, Cambuslang, Glasgow, G72 7NN
M6 JCN J. Barma, 28 Briarfield Road, Timperley, Altrincham, WA15 7DB
M6 JCP J. Preece, 42 Henderson Road, Widnes, WA8 7LR
M6 JCQ J. Stevens, 51 Cheddington Road, Pitstone, Leighton Buzzard, LU7 9AQ
M6 JCR J. Reynolds, 15 Chestnut Mead, Oxford Road, Redhill, RH1 1DR
M6 JCS J. Shettler, 504 Leeds Road, Huddersfield, HD2 1YW
M6 JCT J. Anderson, 139 Cromwell Road, Rushden, NN10 0EG
MM6 JCU J. Coubrough, 41 Bridge Court, Alexandria, G83 0BZ
MW6 JCV J. Baldwin, 94 King Street Abertridwr, Caerphilly, CF83 4BG
M6 JCW J. Wright, 19 Halstead Close, Woodley, Reading, RG5 4LD
M6 JCX J. Coxon, Flat 6, Neptune House, London, SE16 7AU
MM6 JCZ J. Mackenzie, 40 Deanswood Park, Deans, Livingston, EH54 8NX
M6 JDA J. Dale, Corydon, Church Street, Sevenoaks, TN14 7SW
M6 JDC J. Collier, Benallack Farm, Grampound Road, Truro, TR2 4BY
M6 JDD D. Riley, 14 Mond Road, Widnes, WA8 7NB
M6 JDF J. Firth, 7 Manor Avenue, Derby, DE23 6EB
M6 JDL J. Harding, 181 South Coast Road, Peacehaven, BN10 8NS
M6 JDM J. Briggs, 47 Greenland Avenue, Derby, DE22 4AQ
M6 JDQ J. Hole, 17 Cromford Street, Sheffield, S2 4BP
M6 JDU J. Clarke, 160 Hall Lane Estate, Willington, Crook, DL15 0PP
MW6 JDY J. Evans, 311 Delffordd, Rhos, Swansea, SA8 3ER
MM6 JEA J. Addison, Lambhill Bungalow, St. Katherines, Inverurie, AB51 8TS
M6 JED P. Carberry, 203 Fitzwilliam Road, Rotherham, S65 1NB
M6 JEF J. Marland, 34 Barnfield, Stoke-on-Trent, ST4 5JE
M6 JEG D. Stephens, 37 Pimpernel Close, Southampton, SO316TN
M6 JEH J. Harris, 45 Sleigh Road, Sturry, Canterbury, CT2 0HT
M6 JEJ J. Brennan, 42 Maryland Road, Thornton Heath, CR7 8DF
M6 JEK J. Durey, 7 Staplers Close, Great Totham, Maldon, CM9 8UN
MW6 JEL J. Christensen, 29 Caerphilly Close, Rhiwderin, Newport, NP10 8RF
M6 JEM J. Shaw, 2A Priory Avenue, Petts Wood, Orpington, BR5 1JF
M6 JEO J. Saunders, 123 Medway Road, Ferndown, BH22 8UR
M6 JEP D. Jepson, 104 Norris Street, Warrington, WA2 7RW
M6 JEQ J. Mcdermott, 2 Denton Terrace, Castleford, WF10 4LN
M6 JER J. Harley, 170 Windsor Road, Hull, HU5 4HH
M6 JET T. Edwards, 17 The Green, Woodbastwick, Norwich, NR13 6HH
M6 JEV J. Gourley, 19 Walter Nash Road East, Kidderminster, DY11 7EA
M6 JEY J. Johnson, 226 Preston New Road, Southport, PR9 8NY
M6 JEZ J. Mitchell, 11 Brookside Drive, Oadby, Leicester, LE2 4PB
M6 JFA J. Worsley, 3 Sheephouse Road, Hemel Hempstead, HP3 9LW
MM6 JFB J. Copland, Janefield, Colvend, Dalbeattie, DG5 4QN
M6 JFE D. Foley, 1 Hill Rise Close, Harrogate, HG2 0DQ
M6 JFF J. Carpenter, 20 Barton Grove, Kedington, Haverhill, CB9 7PT
M6 JFH J. Feltham, 12 Penrith Way, Eastbourne, BN23 8NS
MW6 JFI J. Power, 27 Seaview Crescent, Goodwick, SA64 0AZ
M6 JFJ J. Spratley, 10 The Willows, Jarrow, NE32 4QN
M6 JFK J. Knight, 30 Ash Meadow Lea, Preston, PR2 1RX
MW6 JFL J. Lewis, 9 Llwyn Bedw, Cefn Pennar, Mountain Ash, CF45 4DZ
M6 JFN S. Butler, 41 Angus Close, Chessington, KT9 2BN
MI6 JFO W. Forde, 35 Torr Gardens, Larne, BT40 2JH
M6 JFP J. Poole, 82 Merchants Way, Canterbury, CT2 8PN
M6 JFR A. Bass, 7 Woodlands, Horbury, Wakefield, WF4 5HH
M6 JFS R. Jones, 32 Remington Drive, Sheffield, S5 9AH
MM6 JFU J. Frew, 57 Ford Avenue, Dreghorn, Irvine, KA11 4BN
M6 JFV J. Hardingham, 11 All Saints Close, Weybourne, Holt, NR25 7HH
M6 JFW J. Leonard, 51 Molyneux Drive, Bodicote, Banbury, OX15 4AX
M6 JFY Dr C. Holmes, Old Vicarage Farmhouse, Course Lane, Wigan, WN8 7LA
MW6 JGC J. Sollis, 15 Llanwonno Road, Mountain Ash, CF45 3NB
M6 JGE J. Glover, 12 Willow Street, London, E4 7EG

M6 JGG J. Goodwin, 6 Worrall Street, Congleton, CW12 1DT
M6 JGH R. Hart, Jays, South Street, Kington Magna, Gillingham, SP8 5ET
M6 JGI J. Griffiths, 83 Golborne Road Ashton-In-Makerfield, Wigan, WN4 8XA
MI6 JGK G. Gardiner, 60 Limestone Meadows, Moira, Craigavon, BT67 0UT
M6 JGM J. Gould-Martin, 203 Maple Crescent, Leigh, WN7 5SW
M6 JGN J. Elsey, 6 Lewis Avenue, Walthamstow, London, E17 5BL
M6 JGP J. Paradi, 168 Castle Road, Northolt, UB5 4SG
M6 JGQ J. Chapman, Boundary Farm, Garlic Street, Diss, IP21 4RL
M6 JGR J. White, 62 Dalebrook Road, Burton-on-Trent, DE15 0AD
M6 JGS J. Seaton, 88 Stanley Road, Cambridge, CB5 8LB
M6 JGT J. Jeffryes, 15 Shirley Road, St Albans, AL1 5ES
M6 JGU J. Borrett, 84 Kingsway, Mapplewell, Barnsley, S75 6EX
M6 JGV J. Chown, 40 Kemps Green Road, Balsall Common, Coventry, CV7 7QF
M6 JGY J. Diez, 174 Humber Avenue, Coventry, CV1 2AR
M6 JGZ J. Gilbert, 101 Eastbrook Road, Lincoln, LN6 7EW
M6 JHB J. Burdett, 5 Winston Drive, Wainscott, Rochester, ME2 4LJ
M6 JHC M. Clark, 10 Priory Close, Sporle, King's Lynn, PE32 2DU
MI6 JHE J. Evans, 404 Foreglen Road, Dungiven, Londonderry, BT47 4PN
M6 JHF P. Attwater, 42 Danescourt Crescent, Sutton, SM1 3EA
M6 JHG J. Guess, Flat 257, Helen Gladstone House Nelson Square, London, SE1 0QB
MM6 JHH J. Hutchinson, 169 Glen Avenue, Largs, KA30 8QQ
M6 JHI A. Randall, 7 Blenheim Place Buckshaw Village Chorley Pr7 7Lt, Leyland, PR7 7LT
MW6 JHJ J. Cook, 40 Cemaes Crescent, Rumney, Cardiff, CF3 1TA
M6 JHK J. Hele Kergozou De La Boessiere, Lilac Cottage, The Street, Cheddar, BS27 3TH
M6 JHL J. Lumm, 25 Knowsley Way, Hildenborough, Tonbridge, TN11 9LG
M6 JHN J. Curwen, Corner Cottage, Tailors Green, Stowmarket, IP14 4LL
M6 JHO D. Carey, 7 Pelman House Pelman Way, Epsom, KT19 8HH
M6 JHP J. Palmer, 14 Linnet Close, Luton, LU4 0XJ
M6 JHQ J. Rimmer, 33 New Cut Lane, Southport, PR8 3DW
M6 JHR J. Ross, Foundry Cottage, Crowders Lane, Battle, TN33 9LP
M6 JHS J. Harris, 36 Northmoor Way, Wareham, BH20 4SJ
M6 JHT J. Hill, 45 Venus Street, Congresbury, Bristol, BS49 5HA
M6 JHU B. Jones, 64 Church Lane, Barwell, Leicester, LE9 8DG
M6 JHV J. Hancox, 10 Dunham Close, Newton-On-Trent, Lincoln, LN1 2LH
M6 JHZ J. Hazeltine, 21 Hassock Way, Wimblington, March, PE15 0PJ
M6 JIC R. Price, 17 Treleven Road, Bude, EX23 8SA
M6 JID J. Darwin, 14 Croftwood Grove, Whiston, Prescot, L35 3UT
MI6 JIE J. Evans, 404 Foreglen Road, Dungiven, Londonderry, BT47 4PN
M6 JIF J. Miazek, 2 Oak Grove, Armthorpe, Doncaster, DN3 2DJ
M6 JIG C. Ember, 35 Mattock Lane, Ealing, London, W5 5BH
M6 JIH Rev. J. Horsley, 33 Amalfi Tower, Sunderland, SR3 3AN
M6 JII J. Donovan, 18 Ellesmere Street, Eccles, Manchester, M30 0JN
M6 JIK B. Ferry, 1 New Front Street, Anfield Plain., Stanley, DH9 8JG
M6 JIN J. Ramachandran, 1 Keplerlaan, Estec, Tec-Sws, Noordwijk, Netherlands, 2201AZ
M6 JIQ J. Ritson, 38 Hawkshead Road, Burtonwood, Warrington, WA5 4PW
M6 JIR J. Barrett, 9 Hook Road, Goole, DN14 5JB
M6 JIS M. Furby, 14 Larch Avenue, Wickersley, Rotherham, S66 2PQ
M6 JIT J. Topping, 10 St. Pauls Road, Blackpool, FY1 2NY
M6 JIW J. Barker, Pearl Bungalow, Killerby Cliff, Cayton Bay, Scarborough, YO11 3NR
M6 JIX J. Packer, 20 Shipman Road, Market Weighton, Market Weighton, YO43 3RB
M6 JIY D. Baker, 22 Cleveland Road, Plymouth, PL4 9DF
M6 JIZ J. Watson, 82 Glendale Avenue, Washington, NE37 2JS
MM6 JJB J. Barclay, 24 Wellcroft Road, Hamilton, ML3 9SG
M6 JJD J. Smyth, 5 Lime Close, Lakenheath, Brandon, IP27 9AJ
M6 JJE S. Kelly, 7 Cedar Grove, Greetland, Halifax, HX4 8HT
M6 JJF J. Fradley, 9 Hagley Park Gardens, Rugeley, WS15 2GY
M6 JJG J. Greenhalgh, 7 Swynford Close, Kempsford, Fairford, GL7 4HN
M6 JJH J. Holbrook, 24 Birks Holt Drive, Maltby, Rotherham, S66 7JZ
M6 JJI J. Roberts, 27 Pike Purse Lane, Richmond, DL10 4PS
M6 JJK C. Webb, 16 Bexfield Road, Foulsham, Dereham, NR20 5SB
M6 JJL J. Jackson, 23 Hawkley Drive, Tadley, RG26 3YH
M6 JJM J. Marriott, 55 Ellis Avenue, Stevenage, SG1 3SL
M6 JJN J. Nicholls, 6 Marasca End, Holt Drive, Colchester, CO2 0DL
MW6 JJO J. Owens, 16 Blanche Street, Dowlais, Merthyr Tydfil, CF48 3PE
MM6 JJQ J. Mccrae, 50 Cuilmuir View, Croy, Glasgow, G65 9HQ
M6 JJS J. Stewart, 29 Comwall Close, Leamore, Walsall, WS3 2AR
M6 JJU K. Weston, 10 Hollow Hill Road Ditchingham, Bungay, NR35 2QZ
M6 JJV S. Virgo, 41 Lynch Road, Berkeley, GL13 9TE
MM6 JJW J. Wallace, 323 High Street, Dalbeattie, DG5 4DX
M6 JJX J. Adams, 18 Vanguard Court, Sleaford, NG34 7WL
MI6 JJZ J. Clark, Apartment 16, Pipers Field, 16B Comber Road, Dundonald, BT16 2AB
M6 JKA A. Kirby, 61 The Chase, Harlow, CM17 9JA
M6 JKB S. Lucas, 31 Lilian Close, Norwich, NR6 6RZ
M6 JKD J. Davies, 135 Silvercourt Gardens, Brownhills, Walsall, WS8 6EZ
M6 JKH J. Horry, 20 Churchgate, Sutterton, Boston, PE20 2NS
M6 JKM J. England, 15 Woodward Avenue Radford Gardens, Hereford, HR2 7FH
M6 JKQ R. Poulson, 9 Scattergate Green, Appleby-in-Westmorland, CA16 6SP
M6 JKR J. Krinks, 29 Swaledale Avenue, Congleton, CW12 2BY
M6 JKS J. Shaw, 25 High Street, Gorleston, Great Yarmouth, NR31 6RT
M6 JKT C. Baines, Donative Farm, Warton, Tamworth, B79 0JR
M6 JKW J. Gelder, 36 Westcombe Court, Wyke, Bradford, BD12 8PT
M6 JKX J. Kitto, 306 Lewisham Road, London, SE13 7PA
M6 JKY K. Young, 48 Sussex Street, Cleethorpes, DN35 7NP
M6 JLA J. Tayler, 22 Wheatley Road, Leicester, LE4 2HN
M6 JLB J. Blundell, 21 Walmsley Street, Fleetwood, FY7 6LJ
M6 JLD J. Drinkell, 11 Valley Walk, Kettering, NN16 0LY
M6 JLE J. Wiskow, 15 Ferndale Close, Sandbach, CW11 4HZ
M6 JLH J. Hughes, 27 Mitchell Street, Stoke-on-Trent, ST6 4EX
M6 JLL J. Caswell, 10 Beech Close, Scole, Diss, IP21 4EH
M6 JLM J. Mould, 30 Northfields, Dunstable, LU5 5AL
M6 JLQ M. Adams, 10 St. Johns Road, Wallasey, CH45 3LU
M6 JLR J. Ridley, 73 The Markhams, New Ollerton, Newark, NG22 9QY
M6 JLT J. Trunks, 37 Carlton Street, Haworth, Keighley, BD22 8JY
M6 JLU L. Adams, 10 St. Johns Road, Wallasey, CH45 3LU
M6 JLW J. Hart, 420 Butts Road, Southampton, SO19 1DD
MM6 JLX J. Leith, 13 Chesterhall Avenue, Macmerry, Tranent, EH33 1QJ
M6 JLY J. Lyall, 11 Bowmont, Ellington, Morpeth, NE61 5LT
M6 JLZ J. Beccles, 15 Lambert Avenue, Shepshed, Loughborough, LE12 9QH
MW6 JMB M. Bloore, Halfway House, Hyfrydle Road, Talysarn, Caernarfon, LL54 6HG
MI6 JMC J. Mccloskey, 19 Kinnyglass Road, Coleraine, BT51 3SN
MI6 JMD J. Mcdonald, 70 Allen Park Dunamanagh, Strabane, BT82 0PD
M6 JMF J. Fearn, 16 Sandringham Road, Retford, DN22 7QW
M6 JMG J. Mobbs, 6 School View, Banbury, OX16 4SD
MW6 JMH J. Hewitt, 1 Highfield, Gloucester Road, Chepstow, NP16 7DF
MM6 JMI J. Wilson, 1 West Long Cottages, Livingston, EH54 7AB
M6 JMJ M. Grimsley, 80 Holborn Avenue, Coventry, CV6 4FZ
M6 JMM J. Munton, 14 Serpells Meadow, Polyphant, PL157pr
M6 JMN J. Mason, 9 Little Warton Road, Warton, Tamworth, B79 0HR
M6 JMO C. Monahan, 48 Church Road, Earley, Reading, RG6 1HS
M6 JMP J. Pearse, 23 Buckingham Drive, Knutsford, WA16 8LH
M6 JMR J. Randall, 18 Therty First Avenue, Kingston Up on Hull, HU6 8DB
M6 JMX J. Matthews, 50A South Farm Road, Worthing, BN14 7AE
M6 JNA N. Watling, 8 Badger Close, Solihull, B90 4HR
MM6 JNB J. Beedie, 21 Marywell Village, Arbroath, DD11 5RH
M6 JNC J. Clay, Willow Brook, Bairstow Lane, Sowerby Bridge, HX6 2SY
M6 JNE J. Nobes, 22 Mansfield Road, Edwinstowe, Mansfield, NG21 9NJ
MM6 JNF J. Fyall, 56 Whins Road, Alloa, FK10 3RE
M6 JNH H. Blount, 70 Harrisons Lane, Ringmer, Lewes, BN8 5LJ
M6 JNI D. Storton, 50 Cromwell Road, Blackpool, FY1 2RG
M6 JNJ J. Dray, Rose Cottage, The Street East Brabourne, Ashford, TN25 5LR
M6 JNL S. Clarke, 59 Baden Powell Crescent, Pontefract, WF8 3QD
M6 JNM J. Nofrerias Mondejar, 31 Emmendingen Avenue, Newark, NG24 2FX
MM6 JNN J. Needham, 154 Ravenswood Rise, Dedridge, Livingston, EH54 6PQ
M6 JNO C. Pryke, 50 Raglan Gardens, Watford, WD19 4LL
MW6 JNP J. Powell, 31 Laburnum Close Gurnos Estate, Merthyr Tydfil, CF47 9SN
M6 JNQ J. Nicholas, 14 Poplar Close, Frome, BA11 2UH
M6 JNR Dr J. Reynolds, 38 Spring Lane, Hockley Heath, Solihull, B94 6QY
M6 JNS J. Shatford, 31 Pinner Park Avenue, Harrow, HA2 6LG
M6 JNV T. Wild, 31 Ivel Road, Shefford, SG17 5LB
M6 JNW J. Walker, 232 Bideford Green, Leighton Buzzard, LU7 2TS
M6 JNZ S. Jessup, 6 Fifth Ave, Catterick, DL9 4RJ
M6 JOA M. Hall, Ward 2 Berth 15, Royal Hospital, Chelsea Royal Hospital Road, London, SW3 4SR
M6 JOD J. O'Driscoll, 48 St. Ives Road, Coventry, CV2 5FZ
M6 JOE J. Bell, 8 Firsleigh Park, Roche, St Austell, PL26 8JN
M6 JOG S. Mayor, 12 Yealand Avenue, Heysham, Morecambe, LA3 2LT
MM6 JOH J. Hutton, 2 Watson Place, Dunfermline, KY12 0DR
M6 JOI W. Pickles, 31 Longfield Road, South Woodham Ferrers, CM3 5JL
MM6 JOK J. Stewart, 1 Barns Park, Dalgety Bay, Dunfermline, KY11 9XX
M6 JON J. Oakley, 59 Bewsey Street, Warrington, WA2 7JQ
M6 JOO J. Fairhall, 122 Thornhill Rise Portslade, Brighton, BN41 2YL
M6 JOP S. Gowers, High Elms Burnt House Lane, Dartford, DA2 7SP
M6 JOR J. Oliver, 16 Eastdale Road, Burgess Hill, RH15 0NH
MI6 JOS J. Millar, 3, Ahoghill, BT42 1JN
M6 JOU J. Best, 7 Lawns Court, Carr Gate, Wakefield, WF2 0UT
M6 JOV M. Dale, 50 St. Cuthberts Avenue, Colburn, DL9 4NT
M6 JOW J. Whitmore, 33 Wanton Lane Terrington St Clement, Kings Lynn, PE34 4NR
MM6 JOX J. Greer, 2/2 49 Strathcona Drive, Glasgow, G13 1JH
MI6 JOY J. Ruddell, 16 Beechfield Manor, Aghalee, Craigavon, BT67 0GB
M6 JPC P. Crossley, 4 Bartons Garth, Selby, YO8 9RR
M6 JPD D. Bache, 62 Whittingham Road, Halesowen, B63 3TP
M6 JPF J. Franks, 14 The Hamlet, Slades Hill, Templecombe, BA8 0HJ
M6 JPH J. Higginson, 187 Birmingham Road, Ansley Village, Warwickshire, CV10 9PQ
M6 JPI J. Pinder, 11 Andrews Close, Louth, LN11 0BP
M6 JPJ J. Leeson, 2 Hawthorn Road, Radstock, BA3 3NW
M6 JPK J. Knight, 16 Old Farm Lane, Norwich, NR3 2DR
M6 JPL J. Lynch, Beechway, Raddel Lane, Warrington, WA4 4EE
MM6 JPM J. Moffat, 6 Park Grove, Belhelvie, Aberdeen, AB23 8YG
M6 JPO J. Owen, 21 Marlborough Road, Luton, LU3 1EF
M6 JPR J. Raine, Braemar, Sandy Lane, Crawley, RH10 4HS
M6 JPS J. Potts, 103 Etherstone Street, Leigh, WN7 4HY
M6 JPT J. Thompson, 7 Hawthorne Terrace, Crosland Moor, Huddersfield, HD4 5RP
M6 JPU S. Stretton, 9 Kilton Road, Worksop, S80 2EG
M6 JPV C. Bone, 65 Walesmoor Avenue Kiveton Park, Sheffield, S26 5RF
M6 JPW P. Whitehead, 29 Cleveland, Bradville, Milton Keynes, MK13 7AZ
M6 JPX J. Powell, Temple Cottage, Monkey Island Lane, Maidenhead, SL6 2ED
MM6 JQA M. Batho, 25 Burghmuir Road, Perth, PH1 1LU
M6 JQC K. Swift, 1 Cedar Nook, Sheffield, S26 5HL
M6 JQE J. Drea, 25 Loxley Gardens, Burnley, BB12 6PW
MM6 JQF A. Aitken, Cherlor Mosscastle Road, Slamannan, FK13EL
M6 JQG D. Start, The Rise, Valley Lane Swaby, Alford, LN13 0BH
M6 JQK C. Gilbert, 3 Marston Close, Eastham, ch629ea
M6 JQL N. Eyers, Flat 9, Pennys Court, 450 New Road, Ferndown, BH22 8EX
M6 JQQ C. Murphy, 20 Bridge Close, Burgess Hill, RH15 8PD
M6 JQS J. Wort, 25 Renton Cottage, Ellery Grove, Lymington, SO41 9DX
M6 JQU C. Hodson, 64 Cradley, Widnes, WA8 7PL
M6 JQW J. Wang, Flat 31, 74 Arlington Avenue, London, N1 7AY
M6 JQZ G. Hardman, 12 Fernleigh Chorley New Road, Horwich, Bolton, BL6 6HD
M6 JRB J. Brown, 13 Waterlakes, Edenbridge, TN8 5BX
M6 JRC J. Clay, Riverside Mews, Nichols Yard, Sowerby Bridge, HX6 2EE
M6 JRE J. Redfern, Flat 11, Poplar Court, Poplar Street, Manchester, M34 5EJ
M6 JRH J. Harrison, 20 Somerton Avenue, Wilford, Nottingham, NG11 7FD
MW6 JRI J. Richards, 29 Kingsland Crescent, Barry, CF63 4JQ
M6 JRJ J. Rhodes, 13 Blake Hall Drive, Mirfield, WF14 9NL
M6 JRK J. De Vantier, 167 Ivy Road, Bolton, BL1 6EF
M6 JRL J. Peacock, 4 Luzley Cottages Luzley Road, Ashton-under-Lyne, OL6 9AL
M6 JRO J. Oldman, High Waters, Bentfield Green, Stansted, CM24 8HX
M6 JRS J. Sadler, 10 Spindle Warren, Havant, PO9 2PU
M6 JRT N. Wing, 39 Whittington Road, Hutton, Brentwood, CM13 1JX
M6 JRW J. Wallis, 20 Green Leys, West Bridgford, Nottingham, NG2 7RX

M6 JSH J. Horner, 36 Chadsfield Road, Rugeley, WS15 2QP
M6 JSI G. Harvey, 55 Trelawney Road Hainault, Ilford, IG6 2NJ
M6 JSJ M. Burrows, 1 Hedgemere, Taverham, Norwich, NR8 6GG
M6 JSM J. Mcclure, 25 Bluebell Avenue, Wigan, WN6 8NS
MM6 JSN J. Drummond, 1/1 2 Lethamhill Place, Glasgow, G33 2SD
M6 JSP J. Parris, 10, Wharfedale Grange, Ben Rhydding Drive, Ilkley, West Yorkshire, LS29 8AR
MI6 JSQ J. Quinn, 5 Woodhill Heights, Lurgan, Craigavon, BT66 7DJ
M6 JSR I. Smith, 2 George Street, Somercotes, Alfreton, DE55 4JT
M6 JSS J. Stones, Eads Astrium Ltd, Anchorage Road, Portsmouth, PO3 5PU
M6 JST J. Tidmarsh, 16 Birch Road, Wellington, TA21 8EP
M6 JSU J. Neal, 5 Shelley Close, Huntingdon, PE29 1NF
M6 JSV J. Slade, 199 Durham Road, Stevenage, SG1 4JP
M6 JSW J. Welsh, 16 Colton Crescent, Dover, CT16 2EP
M6 JSX Lord C. Burridge, 43 Rackenford Road, Tiverton, EX16 5AF
M6 JTA J. Allen, 15 Wessington Drive, Hereford, HR1 1AH
M6 JTC P. Askey, The Maltings, Brewery Yard, Kettering, NN14 3BT
M6 JTD D. Connor, 145 Welsby Road, Leyland, PR25 1JH
MM6 JTG J. Mcdowall, 50 Dailly Road, Maybole, KA19 7AU
M6 JTH J. Thomas, 14 Lower Meadow, Cheshunt, Waltham Cross, EN8 0QU
M6 JTI J. Smith, 8 Westbourne Terrace, Thirsk, YO7 1QD
M6 JTK D. Maitland, 125 Shottery Road, Stratford-upon-Avon, CV37 9QA
M6 JTL M. Simons, 123 Main Street, Little Harrowden, Wellingborough, NN9 5BA
MW6 JTM J. Morgan, The Lodge, Old Racecourse, Oswestry, SY10 7PQ
M6 JTN M. Chivers, 4 Hunters Lodge, Fareham, PO15 5NF
M6 JTO J. O'Reilly, 116 Coleridge Way, Crewe, CW1 5LF
M6 JTP J. Payne, 27 Newton Hall Gardens, Rochford, SS4 3EP
M6 JTQ J. Cook, 28 Wenny Estate, Chatteris, PE16 6UX
M6 JTU J. Bridge, 4 Knights Way, Camberley, GU15 1EQ
M6 JTW J. Waring, 12 Mary Street, Farnhill, Keighley, BD20 9AU
M6 JUB A. Casson, 7 Leebrook Court, Owlthorpe, Sheffield, S20 6QJ
M6 JUC B. Marsh, 21 Edward Road, Eynesbury, St Neots, PE19 2QF
MM6 JUE R. Cormack, 2 Coghill Street, Wick, KW1 4PN
M6 JUF J. Green, 26 Foxhill Road Burton Joyce, Nottingham, NG14 5DB
M6 JUG S. Jones, 25 Kents Lane, Crewe, CW1 4PX
M6 JUH M. Stanford-Taylor, Holly Flat, School Lane, St. Johns, Crowborough, TN6 1SE
M6 JUI Dr M. Ovenden, 59 Cemetery Road Woodlands, Doncaster, DN6 7RY
M6 JUJ W. Bruen, 25 Carlton Avenue Upholland, Skelmersdale, WN8 0AE
M6 JUK J. Rhodes, 73 Keats Way, Hitchin, SG4 0DP
M6 JUP A. Drewitt, 12 Kinross Crescent, Loughborough, le11 4uq
MW6 JUQ A. Morgan, 4 Courtney Street Manselton, Swansea, SA5 9NY
M6 JUT J. Thorne, 53 Elfleda Road, Cambridge, CB5 8NA
M6 JUU L. Hagan, 7 Betjeman Mews, Southend-on-Sea, SS2 5EJ
M6 JUW G. Rodriguez, 32 Mount Pleasant, Prestwich, Manchester, M25 2SD
M6 JUX A. Brown, 17 Quail Ridge, Ford, Shrewsbury, SY5 9LF
M6 JVB J. Cobb, 32 Dellmont Road, Houghton Regis, Dunstable, LU5 5HU
MI6 JVC J. Heyburn, 20 Victoria Street, Armagh, BT61 9DT
M6 JVD J. Boyd, 62 West Road, Shoeburyness, Southend-on-Sea, SS3 9DP
M6 JVF W. Terry, 11 Crescent Ave, Overhulton., Bolton, BL51EN
M6 JVK J. Kinsey, 1 Keystone Close, Goring-By-Sea, Worthing, BN12 6GA
MI6 JVM O. Hart, 18 Sandringham Court Portadown, Craigavon, BT63 5BF
M6 JVO J. Simmonds, Weir Cottage Lugwardine, Hereford, HR1 4AS
MM6 JVQ S. Cummings, 1 Inveresk Gardens, Dundee, DD4 0XZ
M6 JVS C. James, Ct93Nx, Margate, ct92nx
M6 JVW M. Fogerty, 25 Noel Street, Gainsborough, DN21 2RY
M6 JVX G. Cope, 171 Ockford Ridge, Godalming, GU7 2NN
M6 JVY P. Smythe, 15 Falcon Drive, Trowbridge, BA14 7GE
M6 JVZ E. Walker, St. Govans Ledbury Road, Ross-on-Wye, HR9 7BG
M6 JWC J. Cater, 5 Shady Grove, Hilton, Derby, DE65 5FX
M6 JWD J. Wilson, 35 Lawson Avenue, Jarrow, NE32 5UF
M6 JWE J. Loveday, 27 New Road, Chatteris, PE16 6BJ
M6 JWF A. Marsh, 140 Church Road, Redfield, Bristol, BS5 9HN
M6 JWG J. Whitworth, 31 Shirley Close, Chesterfield, S40 4RJ
M6 JWI D. Milne, 154 Cambridge Road, Ellesmere Port, CH65 5BW
MW6 JWL D. Jenkins, 16 Celtic Road, Whitchurch, Cardiff, CF14 1EG
M6 JWO J. Whalley, 28 Hatton Lane, Stretton, Warrington, WA4 4NG
M6 JWP J. Wade, 44 Newton Park Homes, Newton St. Faith, Norwich, NR10 3LP
M6 JWQ J. Oldfield, 2 Bailie Cross Cottages, Poole Road, Wimborne, BH21 4AE
M6 JWT J. Thompson, 19 Taverner Road, Boston, PE21 8NL
M6 JWW J. Whiteside, The Old Antique Shop, Bank Street Pulham Market, Diss, IP21 4TG
M6 JWX J. Wade, 105 Western Avenue, Woodley, Reading, RG5 3BL
M6 JWY J. Willby, 10 Sunbury Road, Birmingham, B31 4LJ
M6 JXD J. Darragh, 20 Templar Place, Hampton, TW12 2NE
M6 JXE G. Spiers, 2 Sponnes Road, Towcester, NN12 6ED
M6 JXF D. Wells, 96 Tennyson Avenue, Rugby, CV22 6JF
MM6 JXH J. Hinchcliffe, 27 Forest Glade, East Calder, Livingston, EH53 0FQ
M6 JXI J. Jago, 5 Horrell Court, Rushden, NN10 9EG
M6 JXJ C. Connell, 16 Woodman Drive, Bury, BL9 5HQ
M6 JXK J. Kelly, Flat 4, The Corner Place, 1 North Road, Harborne, Birmingham, B17 9PA
M6 JXN J. Culshaw, 13 Ravens Close, Knaphill, Woking, GU21 2LD
M6 JXO P. Mcnulty, 2 Belvedere Road Hanford, Stoke-on-Trent, ST4 8RL
M6 JXX J. Creaser, 8 Millwood Road, Hounslow, TW3 2HH
M6 JXY J. Yendole, 15 Borgie Place, Weston-Super-Mare, BS22 9HG
M6 JYB J. Brown, 37 Calshot Avenue, Chafford Hundred, Grays, RM16 6NS
M6 JYF A. Browning, 1 Stamford Ave, Blackpool, FY42BH
M6 JYI M. Piatkowski, 152 Anchorway Road, Coventry, CV3 6JG
M6 JYJ L. Richardson, 28 Durham Way Wyton, Huntingdon, PE28 2EQ
M6 JYL Y. Lai, Harrogate Ladies' College, Clarence Drive, Harrogate, HG1 2QG
M6 JYO S. Valsala, 12 Montague Street, Rainham Essex, RM13 8LW
M6 JYQ R. Cichocki, 12 Crossland Crescent, Wolverhampton, WV6 9JY
M6 JYR G. Hepworth, 12 Fourlands Gardens, Bradford, BD10 9SP
M6 JYS J. Pengilly, 8 Willard Close, Eastbourne, BN22 8SX
M6 JYT A. Lane, 26 Astral Gardens Sutton-On-Hull, Hull, HU7 4YS
M6 JYV K. Riddiough, 47 Brooke Street Hoyland, Barnsley, S74 9DP
M6 JYX N. Peppe, 5 Cherford Road, Bournemouth, BH11 8SU
M6 JYY R. Burton, 152 Hough Road, Walsall, WS2 9BQ
M6 JYZ J. Sanders, 2 Manor Court Manor Grove, Mangotsfield, Bristol, BS16 9LF
M6 JZC J. Clark, 26 Heron Way, Sandbach, CW11 3AU
M6 JZF D. Ward, 10 Royston Court, Potton, Sandy, SG19 2NJ
M6 JZH C. Garratt, 309, Brookles Mead, Harlow, CM19 4QD
M6 JZK J. Howlett, 29 Little London, Heytesbury, Warminster, BA12 0ES
M6 JZL J. Hope, 4 Beaumont Way, Prudhoe, NE42 6RA
M6 JZM K. Wood, 89 Oriel Road, Portsmouth, PO2 9EG
M6 JZN R. Rudelic, Flat 26 Neptune House, 1 Neptune Way, Southampton, SO14 3FN
M6 JZU P. Holmes, 82 Moore Avenue, Norwich, NR6 7LG
M6 JZV B. Shackleton, 7 Erringden Street, Todmorden, OL14 6AW
MM6 JZW S. Gillion, 13 Aignish, Isle of Lewis, HS20PB
M6 JZY J. Smale, 30 Gillian Close, Aldershot, GU12 4HU
MW6 JZZ J. Trahearn-O'Brien, 148 Gladstone Road, Barry, CF62 8ND
MW6 KAA K. Parry, 80 Cripps Avenue, Cefn Golau, Tredegar, NP22 3PB
MW6 KAB K. Boulter, 16 Danygraig, Pontlottyn, Bargoed, CF81 9RS
MW6 KAC M. Woodington, 44 Glas Y Gors, Aberdare, CF44 0BQ
M6 KAE J. Withall, Humphreys Cottage, Fords Green, Uckfield, TN22 3LJ
M6 KAG K. Goodyer, 5A Station Road Bow Brickhill, Milton Keynes, MK17 9JU
M6 KAH K. Holloway, 6 Britons Lane Close, Beeston Regis, Sheringham, NR26 8SH
M6 KAM K. Mcgill, 45 Fir Terraces, Esh Winning, Durham, DH7 9JQ
M6 KAN K. Sharpe, 18 Dudhill Road, Rowley Regis, B65 8HT
M6 KAO K. Gibbs, 30 George Road, Water Orton, Birmingham, B46 1PE
M6 KAQ K. Sidaway, 19 Larkwhistle Walk, Havant, PO9 4JA
M6 KAR K. Slotwinski, 51 Moorland Gate, Heathfield, Newton Abbot, TQ12 6TX
M6 KAS J. Gilhespy, 106 Durham Drive, Jarrow, NE32 4QY
M6 KAT C. Gibson, 16A Hillside Road, Wool, Wareham, BH20 6DY
MM6 KAU M. Krawczyk, 19 Wishart Archway, Dundee, DD1 2JA
M6 KAV K. Booth, 28 Farndale Gardens, Lingdale, Saltburn-by-the-Sea, TS12 3EW
M6 KAX J. Killman, 19 Moorland Avenue, Walkeringham, Doncaster, DN10 4LG
M6 KAZ K. Hamilton, 171 Jackmans Place, Letchworth Garden City, SG6 1RG
M6 KBA D. Reeve, 5 Antelope Avenue, Grays, RM16 6QT
MI6 KBB J. Elliott, 12 Drumbeg Drive, Lisburn, BT28 1NY
M6 KBC S. Williams, Flat 4, The Crown Mews, 24 Station Road, Midsomer Norton, Radstock, BA3 2FE
M6 KBE K. Bantock, 22 Deepdale Drive, Consett, DH8 7EH
M6 KBF B. Wiggins, The Wigwam 13 Hastings Road, Bromsgrove, B60 3NX
M6 KBG K. Glaysher, 66 Talbot Road, Farnham, GU9 8RR
M6 KBH K. Burt, 2 Mousehole Close Dalton, Rotherham, S65 4JF
M6 KBI M. Symmonds, 24 Woodville Grove, Stockport, SK5 7HU
M6 KBJ K. Jewell, 25 Park View, Liskeard, PL14 3EE
MW6 KBK B. Hopkins, 28 Ivor Street, Maesteg, CF34 9AH
M6 KBO R. Waterhouse, 19 Honeysuckle Close, St Leonards-on-Sea, TN37 7LX
M6 KBS K. Burness, 4 Fenwick Street, Boldon Colliery, NE35 9HU
M6 KBT R. Mundy, 9 Gable Court, Liverpool, L11 7DS
M6 KBV K. Clarke, 41 St. Hilda Street, Bridlington, YO15 3EE
MD6 KBW K. Whittle, 113 Ballaquark, Douglas, Isle of Man, IM2 2EU
M6 KBX J. Greenwood, 38 Baskerville Road, Sonning Common, Reading, RG4 9LS
M6 KBY A. Davies, 6 Gribble Road, Liverpool, L10 7NF
M6 KCA K. Millward, 133 Birchfield Way, Telford, TF3 5HN
MM6 KCB S. Gallagher, 29 Roslyn Drive Bargeddie, Baillieston, Glasgow, G69 7QZ
M6 KCC K. Cole, 16 Minster Road, Westgate-on-Sea, CT8 8BP
M6 KCD K. Crosby, 28 Exning Road, London, E16 4NA
M6 KCE K. Bradford, 3 Haven Close, Sutton-in-Ashfield, NG17 2DG
M6 KCF L. Bee, 25 Blatcher Close, Minster On Sea, Sheerness, ME12 3PG
M6 KCG K. Allen, 20 Rookery Road, Innsworth, Gloucester, GL3 1AT
M6 KCI D. Orme, 47 Poplar Avenue, Oldham, OL8 3TZ
MW6 KCJ K. Uprichard, 111 Fernhill, Mountain Ash, CF45 3EF
M6 KCK L. Wood, 2 The Bungalows, North Green, Woodbridge, IP13 9NP
M6 KCL J. Jopson, 12 Charing Close, Ringwood, BH24 1FA
MM6 KCM K. Mair, 92 Graham Street, Wishaw, ML2 8HR
M6 KCP A. Ostatek, 38 Coronation Way, Keighley, BD22 6HF
MW6 KCQ M. Cook, 40 Cemaes Crescent, Rumney, Cardiff, CF3 1TA
MM6 KCR K. Riddick, Davah, Port Road, Castle Douglas, DG7 3JW
M6 KCS A. Stooke, 128 Stamford Street, Grantham, NG31 7BP
M6 KCT K. Coton, 11 Pickers Way, Clacton-on-Sea, CO15 5RU
M6 KCU C. Pavey, 143 Queen Elizabeth Way, Colchester, CO2 8LT
MM6 KCV C. Vines, 49 Annandale Gardens, Glenrothes, KY6 1JD
M6 KCY R. Fisher, 14 Fraser Avenue, Caversham, Reading, RG4 6RT
M6 KCZ R. Robinson, 22 Riddings Court, Timperley, Altrincham, WA15 6BG
MW6 KDA D. Henderson, 1 Court Place, Tonypandy, CF40 2RE
M6 KDB K. Beech, 8 Gateford Drive, Worksop, S81 7HL
M6 KDC D. Stewart, 67 Little Moss Hey, Liverpool, L28 5RJ
M6 KDD K. Daniels, 64 Skelton Road, Norwich, NR7 9UH
M6 KDF J. Nash, 124 Tanhouse Avenue, Birmingham, B43 5AG
M6 KDG C. Harman, 58 Laurence Avenue, Witham, CM8 1JB
M6 KDJ K. Hall, 11 Hinton Villas, Hinton Charterhouse, Bath, BA2 7SS
M6 KDK K. Khan, 3 Marshall Close, Frimley, Camberley, GU16 9NY
M6 KDM K. Middlehurst, 7 Statham Drive, Lymm, WA13 9NW
M6 KDO K. Porter, 58 Panama Drive, Atherstone, CV9 3HJ
M6 KDP R. Hutton, 22A Victoria Road, Maldon, CM9 5HF
M6 KDQ A. Bowles, Golden Yews, Burnt Hill, Yattendon, Thatcham, RG18 0XD
M6 KDR C. Pearcey, 17 Peppercorn Close, Christchurch, BH23 3BL
M6 KDV M. Pulling, 21 Heathlands Avenue, West Parley, Ferndown, BH22 8RW
M6 KDW K. Dillow, 101 Martins Lane Hardingstone, Northampton, NN4 6DJ
M6 KDZ I. Whiteley, 29 Harvey Avenue, Wirral, CH49 1RT
M6 KEA A. Wheeler, 10 Handsacre Crescent, Rugeley, WS15 4DQ
M6 KEC D. Preston, 9 Daubney Street, Cleethorpes, DN35 7BB
MW6 KED S. Kedward, 9 Lawrence Avenue, Aberdare, CF44 9EW
M6 KEE M. Keeler, 15 Grove Park, Kingswinford, DY6 9AD
M6 KEF C. Thompson, 5 Oak Avenue, Charlton Kings, Cheltenham, GL52 6JG
M6 KEG P. Selwood, 27 Sharp Street, Warrington, WA2 7AP
M6 KEH K. Dalby, 52 Narborough Road South, Leicester, LE3 2FN
MW6 KEL K. Gemmell, 93 North Road, Ferndale, CF43 4RG
M6 KEM K. Morris, 1A Littlemoor, Queensbury, Bradford, BD13 1DB
M6 KEP K. Applegarth, 28 South View Gardens, Pontefract, WF8 2HW
MW6 KEQ K. Mogford, 49 Cefn Road, Rogerstone, Newport, NP10 9AQ

MM6 KER K. Hamilton, 25 Abbotsford Street, Falkirk, FK2 7NH
M6 KES M. Shore, 43 Daniel Fold, Rochdale, OL12 7JU
M6 KET J. Daws, 1157 Evesham Road, Astwood Bank, Redditch, B96 6DY
M6 KEU Y. Law, 59 Lake View, Edgware, HA8 7SA
M6 KEV K. Taylor, Heatherleigh B & B, 17 Queens Road, Shanklin, PO37 6AW
M6 KFB Dr F. Kuttikkate, 34 Shetland Crescent, Rochford, SS4 3FJ
MM6 KFE K. Farnington, 52 Glebe Park, Duns, TD11 3EE
M6 KFF S. Coupe, 47 Burn Street, Sutton-in-Ashfield, NG17 4LL
M6 KFG J. Stokes, 30 Anglers Reach, Grove Road, Surbiton, KT6 4EX
MD6 KFH K. Gascoyne, 22 Broogh Wyllin, Kirk Michael, Isle of Man, IM6 1HU
MM6 KFJ A. Keogh, 251 Main Street Plains, Airdrie, ML6 7JH
M6 KFK K. Kozlowski, Woking Homes / Flat 2, Oriental Road, Woking, GU22 7BE
M6 KFP K. Prentice, 24 Sulgrave Close, Liverpool, L16 6AD
MW6 KFQ J. Turner, 4 Imble Street, Uk, SA726QL
M6 KFS K. Fraser-Smith, Flat 3 Droffats House 69 Stafford Road, Southampton, SO15 5RS
M6 KFW K. Wells, 5 Cook Avenue, Newport, PO30 2LL
M6 KFX R. Stone, 63 Sands Lane, Lowestoft, NR32 3ER
M6 KGA K. Allgar, 13 Deacon Avenue, Kempston, Bedford, MK42 7DU
MW6 KGB W. Oliver, Pwllmeyric, Chepstow, NP16 6LE
M6 KGC K. Cossey, 34 Pinewood Road, Hordle, Lymington, SO41 0GP
M6 KGD M. Tabberer, 29 Chase Vale, Chasetown, Burntwood, WS7 3GD
M6 KGE K. Gibbins, 11A Wood End, Banbury, OX16 9ST
M6 KGF G. Flew, 58 Streamside, Mangotsfield, Bristol, BS16 9EA
M6 KGI D. Cooper, Shirley Villa Morley Road North, Sheringham, NR26 8JB
M6 KGK K. Goddard, 49 Chalet Hill, Bordon, GU35 0EF
M6 KGL M. Horwell, 3 Vanity Close, Oulton, Stone, ST15 8TZ
MM6 KGM K. Mitchell, 92 Whiteside Court, Bathgate, EH48 2TN
M6 KGN J. Badham, Comwillgur House Ross Road, Longhope, GL17 0LP
M6 KGO Dr B. Osborne, 15 Ferguson Avenue, Surbiton, KT5 8DS
M6 KGQ J. Loveland, 6 Wren Drive Bradwell, Great Yarmouth, NR31 8JW
M6 KGR K. Carter, 50 Elliman Avenue Bottom Flat, Slough, SL2 5BG
M6 KGS K. Stow, 39 Biverfield Road, Prudhoe, NE42 5ER
MI6 KGU D. Holloway, 3 Iniscarn Close, Lisburn, BT28 2BX
M6 KGV G. Carter, 54 Wood Street, Taunton, TA1 1UW
MM6 KHA K. Adrian, 1C Oliphant Court, Paisley, PA2 0DP
M6 KHB M. Heaton-Bentley, 65 Brookfield Road, Thornton-Cleveleys, FY5 4DR
M6 KHE K. Ballard, 34 Orange Street South Wigston, Leicester, LE184QB
M6 KHF R. Boden, 44 High Street, Blackpool, FY1 2BN
MM6 KHM K. Macrae, 2 Netherhill Avenue, Glasgow, G44 3XG
M6 KHP S. Smith, 2 Burnsall Avenue, Blackpool, FY3 7LQ
M6 KHQ C. Thompson, 12 Otho Way, North Hykeham, Lincoln, LN6 9ZD
M6 KHS K. Shaw, 3 The Grange, Woolley Grange, Barnsley, S75 5QP
M6 KHX K. Hewson, 48 Ruskin Road, Belvedere, DA17 5BB
M6 KIA A. Taylor, 130A Hazelwood Avenue, Eastbourne, BN22 0UX
M6 KIE K. Alejo-Blanco, 23 Southern Road, Thame, OX9 2EE
M6 KIF M. Chalkley, 36 Cowper Road, Bournemouth, BH9 2UJ
M6 KIG K. Greenfield, 49 Railway Street, Northfleet, Gravesend, DA11 9DU
M6 KII S. Hopper, Coombe Road, Brighton, BN24EA
M6 KIK W. Dover, Windcrest, Fox Lane, Basingstoke, RG23 7BB
M6 KIL B. Fryer, 16 Elston Place, Aldershot, GU12 4HY
M6 KIN M. Makin, 34 Carlton Gardens, Farnworth, Bolton, BL4 7TH
M6 KIO D. Bloomfield, 30 Wye Road, Clayton, Newcastle, ST5 4AZ
M6 KIP K. Davies, 29 Corser Street, Stourbridge, DY8 2DE
M6 KIR K. Halloway, 41 Trenoweth Estate, North Country, Redruth, TR16 4AQ
M6 KIT C. Ellis, 1 Rugby Road, Stockton-on-Tees, TS18 4AZ
M6 KIU S. Kato, 38-9-6, Kamikizaki6-Chome, Urawa-Ku, Saitama-City, Japan, 3300071
M6 KIV V. Ilic, Vlade Djordjevica 41, Leskovac, Serbia And Montenegro, 16000
MM6 KIW C. Ross, 4 Weir Place, Greenock, PA15 2JD
M6 KIX C. Hill, 3 Beechmount Rise, Stafford, ST17 4QR
M6 KJB K. Black, 1 Woodcock Close, Haxby, York, YO32 3NQ
M6 KJD K. Davison, 1, Barnsley, S71 2HS
M6 KJG K. Gallyot, 18, High Pines, St. Georges Close, Christchurch, BH23 4LN
M6 KJI M. Bidwell, 3 Walsingham Place, London, SW4 9RR
M6 KJK K. Kolesnik, 15 Steer Road, Swanage, BH19 2RU
M6 KJL N. Rasevic, Bulevar Despota Stefana 7, Novi Sad, Serbia And Montenegro, 21000
M6 KJM K. Mitchell, 5 Finch Crescent, Linslade, Leighton Buzzard, LU7 2PE
M6 KJP K. Parker, Flat 3, Gleadless Court, Sheffield, S2 3AE
M6 KJQ K. Freer, 1 Masefield Flats, Masefield Road, Rotherham, S63 6NQ
M6 KJR J. Cogman, 6 Stockholm Way, Dereham, NR19 1XF
MI6 KJW K. Watt, 2A Drumalane Road, Newry, BT35 8AP
MW6 KJX C. Hodson, 6 Heol Pentwyn Tonyrefail, Porth, CF39 8DF
M6 KJY K. Younger, 29 Eagle Walk, Bury St Edmunds, IP32 6RJ
MW6 KJZ M. Stevenson, 64 Leslie Terrace, Porth, CF39 9TE
M6 KKA R. Pallister, 15 High Row, Washington, NE37 2LZ
M6 KKD K. Edmands, 61 Somers Road, Keresley End, Coventry, CV7 8LE
M6 KKF F. Fitton, 6 Leaford Close, Denton, Manchester, M34 3QH
M6 KKG K. Komenan, 113 Wightman Road, Finsbury Park, London, N4 1RJ
M6 KKI R. Young, 10 Hareholme Lane, Rossendale, BB4 7JZ
M6 KKJ N. Dyer, 79 Saltmarsh Drive, Bristol, BS11 0NL
M6 KKL T. Cheng, Harrogate Ladies' College, Clarence Drive, Harrogate, HG1 2QG
M6 KKM D. Linnett, Flat19. Kestrel House. Osprey Close. Sinfin. Derby., Derby, De24 3dd
MI6 KKN A. Conboy, 16 Tullaghmurry Fold, St. Johns Close, Portstewart, BT55 7DT
M6 KKO T. Stack, 31A Chester Road South, Kidderminster, DY10 1XJ
MI6 KKU S. Mc Court, 17 Windermere Road, Lisburn, BT28 2WY
M6 KKW N. Webb, 27A Orchard Way, Bognor Regis, PO22 9HJ
M6 KKY M. Douglas, 26 Clumber Drive, Northampton, NN3 3NX
M6 KLA D. Bourne, 4 Market Street, Cheltenham, GL50 3NH
M6 KLC K. Chappuis, Priory Cottage, Priory Lane, Bridport, DT6 3RW
M6 KLD T. Lovell, 7 Victoria Wharf Victoria Street North, Grimsby, DN31 1PQ
M6 KLF K. Francis, 203 Colchester Road Lawford, Manningtree, CO11 2BU
M6 KLI J. Heng, 1 St. Giles Croft, Beverley, HU17 8LA
M6 KLK J. Li, Harrogate Ladies' College, Clarence Drive, Harrogate, HG1 2QG
MW6 KLL K. Linahan, 19 Heol Bryn Hebog, Merthyr Tydfil, CF48 1HH
M6 KLM K. Mason, 18 Goose Lane, Sutton, Norwich, NR12 9SE
M6 KLN C. Dzundza, 67 Sidegate Lane, Ipswich, IP4 4HY
M6 KLR S. St George, 37 Highfields, Bromsgrove, B61 7DA
MM6 KLT C. Sharp, 16 Wood Lane, Monifieth, Dundee, DD5 4HS
M6 KLV P. Ayre, 11 Paradise Road, Waltham Abbey, EN9 1RL
M6 KLW L. Woods, 193 Wimberley Street, Blackburn, BB1 8HU
M6 KLX A. Swain, 43 Stretton Road, Morton, Alfreton, DE55 6GW
MM6 KLZ N. Stewart, 35 Newbattle Gardens, Dalkeith, EH22 3DR
M6 KMA K. Machen, 18 Peveril Close, Whitefield, Manchester, M45 6NR
M6 KMC S. Day, 9 Lynton Street, Derby, DE22 3RW
M6 KMD K. Deans, 31 Northcroft, Sandy, SG19 1JJ
M6 KMF N. Powell, 21 Hitchen, Merriott, TA16 5QX
M6 KMG P. Matthew, 24 Jubilee Close, Pamber Heath, Tadley, RG26 3HP
M6 KMI N. Dobson, 2 Hills Road, Breaston, Derby, DE72 3DF
MW6 KMJ M. Beasley, Ffynnon Wen, Bontnewydd, Aberystwyth, SY23 4JJ
M6 KMK A. Platt, 43 The Butts, Frome, BA11 4AB
M6 KML K. Lavelle, 12 Shalewood Court, Atherton, Manchester, M46 0SN
M6 KMN K. Nilan, 15 Broomhall Road Pendlebury, Swinton, Manchester, M27 8XP
MM6 KMQ W. Donnelly, 10 Sentry Knowe, Selkirk, TD7 4BG
M6 KMR K. Riding, 8 Kingston View, Barton-upon-Humber, DN18 6DN
M6 KMS M. Harrison, 70 Hope Avenue, Goldthorpe, Rotherham, S63 9EA
MM6 KMU D. Cowe, 1 Pentland Road Bonnyrigg Midlothian, Edinburgh, EH192LG
M6 KMV A. Waller, 155 Bridgemary Road, Gosport, PO13 0ut
M6 KMW K. White, 12 Lakefields, West Coker, Yeovil, BA22 9BT
M6 KMX E. Rhodes, 10 New Road Cottages, Selborne Road, Selborne, Alton, GU34 3JA
M6 KMY K. Murray, 5 Princes Crescent, Basingstoke, RG22 6DP
M6 KMZ Dr K. Martinez, Forest View, Forest Road, Salisbury, SP5 2BP
M6 KNB M. Narayanankutty, Karikkattu House, Chelamattom, Ernakulam, India, 683550
M6 KNC J. Richardson, Berwick Cottage, Bailes Lane, Guildford, GU3 2AX
M6 KND D. Simmons, 23 Fairey Street, Cofton Hackett, Birmingham, B45 8GU
M6 KNG A. King, 23 Tower Crescent, Lincoln, LN2 5QF
MM6 KNJ J. Wilson, 16 Big Brigs Way, Newtongrange, EH22 4DG
M6 KNL G. Knowles, 29 Shepherds Cote Drive, Hepscott Park, Stannington, Morpeth, NE61 6FN
M6 KNM M. Hill, 109 Kitchener Street, St Helens, WA10 4LU
MM6 KNO T. Knox, 23 Hill Street, Alness, IV17 0QL
M6 KNP M. Lewis, 312 Twyford Avenue, Portsmouth, PO2 8NT
M6 KNS K. Fletcher, Beverley Hotel, 55 Old Brumby Street, Scunthorpe, DN16 2AJ
M6 KNU R. Weightman, 2 Bannister Grove, Winsford, CW7 1RJ
M6 KNV M. Brown, 99 Apprentice Drive, Colchester, CO45SE
M6 KNW K. Woolsey, 1 Park Farm Cottage, Westhorpe Road, Stowmarket, IP14 4SP
M6 KNY R. Dunnaker, 12 Dagger Lane, West Bromwich, B71 4BA
M6 KNZ P. Webb, 473 The Manor, Billing Garden Village, The Causeway, Great Billing, Northampton, NN3 9EX
M6 KOA L. Lemmon, 4 Honington Close, Wickford, SS11 8XB
MI6 KOB K. Boyle, 764 Springfield Road, Belfast, BT12 7JD
M6 KOD A. Adams, 27 George Street Stockton, Southam, CV47 8JS
MM6 KOE D. Robinson, 2 Brixwold Rise, Bonnyrigg, EH19 3FG
M6 KOH D. Elcock, 27 Harefield Road, Southampton, SO17 3TG
MW6 KOI C. Young, 13 Bryn Mawr Road, Holywell, CH8 7AP
MM6 KOJ C. Smith, 68 Craigmore Street, Dundee, DD3 0EA
M6 KOM D. Slater, 13 Longford Close, Rainham, Gillingham, ME8 8EW
M6 KON S. Rouse, 7 Cranbrook Road, Thurnby, Leicester, LE7 9UA
MM6 KOS J. Kosarzecki, 96 Colinton Mains Drive, Edinburgh, EH13 9BL
M6 KOV M. Burton, The Old Bakehouse, Lincoln Road, Goltho, Market Rasen, LN8 5NF
M6 KOZ R. Koziolek, 30 Lammas Beanhill, Milton Keynes, MK6 4LA
M6 KPA H. Ilie, Sos. Iancului Nr. 33 Bloc 105A Scara B Apt 63, Bucharest, Romania, 21717
M6 KPC K. Cooper, 12 Waverley Crescent, Brighton, BN1 7BG
M6 KPD K. Dance, 20 Harmers Hay Road, Hailsham, BN27 1SU
M6 KPE A. Donnabella, 124 Heron Way, Harwich, CO12 3FF
MW6 KPF K. Foster, Prince Of Wales House Short Bridge Street Sy18 6Ad, Llanidloes, SY18 6AD
M6 KPG K. Gallery, 89 Mavis Drive, Coppull, Chorley, PR7 5AE
M6 KPI J. Barnes, Landhill Farm, Halwill, Beaworthy, EX21 5TX
M6 KPJ M. Barrett, 57 Marlborough Avenue, Hornsea, HU18 1UA
M6 KPK K. Killingback, Flat 4 Shorland House, 6 Elm Grove Road, Dawlish, EX7 0BZ
M6 KPL A. Lloyd, 34 Diamond Way, Ellesmere, SY120FH
MW6 KPN A. Thomas, 10 Chapel Street, Gorseinon, Swansea, SA4 4DT
M6 KPP S. Hunter, 8 Clyde Street, Plymouth, PL2 1QQ
M6 KPQ S. Brooks, 68 Daws Heath Road, Benfleet, SS7 2TA
M6 KPR C. Eales, 7 Prima Cresent, Van Riebeeck Park, South Africa, 1619
M6 KPT K. Thrower, 19 Blyford Road, Lowestoft, NR32 4PZ
MW6 KPV D. Heywood, 6 Wentworth Close, Buckley, CH7 2QX
M6 KPX D. Cooper, Needhams Farm House, Spittal Hill Road, Boston, PE22 0PA
M6 KPZ R. Ballard, 11 Thurston, Skelmersdale, WN8 8QU
M6 KQA J. Miller, 21 Hedgerow Grove, Dunmow, CM6 4AS
M6 KQC H. Klettke, 13 Hastings Close, Wythall, Birmingham, B47 6AW
M6 KQD J. Bovey, 17 North Bank Road, Bingley, BD16 1UH
M6 KQF J. Casey, 81 Undercliffe Road, Bradford, BD2 3BP
MM6 KQH S. Dunn, 2 Glendaruel Avenue Bearsden, Glasgow, G61 2PR
M6 KQJ J. Clarke, 49 Brunel Close, Hartlepool, TS24 0UF
M6 KQK P. Constantine, 11 Lingwell Walk, Leeds West Yorkshire South Leeds, LS10 4TH
MW6 KQL M. Castle, 10 West Place, Gobowen, Oswestry, SY11 3NR
M6 KQM A. Green, 62 Briarwood Drive, Blackpool, FY2 0EB
M6 KQP L. Barlow, 51 Hare Hill Road, Littleborough, OL15 9HE
M6 KQR A. Kinsella, Apartment 28 241 Liverpool Road, Widnes, WA8 7HL
M6 KQS R. O'Sullivan, Woodhall Drive, Derby, De23 4RS
M6 KQW E. Almas, 10, Rindal, Norway, 6657
M6 KQZ P. Foster, 95 Yew Tree Drive, Bristol, BS15 4UF
M6 KRD D. Niggemann, 35 Holm Court, Twycross Road, Godalming, GU7 2QT
M6 KRE K. Knight, 11 Sweetbriar Lane Holcombe, Dawlish, EX7 0JZ
M6 KRF K. Furlong, 22 Swaisland Road, Dartford, DA1 3DE
M6 KRH K. Hall, 130 Aylestone Lane, Wigston, LE18 1BA
M6 KRI K. Irvin, 21 Fremantle Crescent, Middlesbrough, TS4 3HR

M6 KRK J. Birch, 3 Partridge Way, High Wycombe, HP13 5JX
M6 KRL P. Dudding, 14 St. Levan Close, Marazion, TR17 0BP
M6 KRM K. Mcleod, 7 Priory Place, Sporle, King's Lynn, PE32 2DT
MI6 KRP K. Pritchard, 18 Ashbrooke, Donaghadee, BT21 0EY
M6 KRQ M. Rotheram, 19 Oteley Avenue, Wirral, CH62 7DJ
M6 KRR G. Newton, 20 East Avenue, Syston, Leicester, LE7 2EH
M6 KRV K. Bott, 294 Walthall Street, Crewe, CW2 7LE
M6 KRW K. Wiles, 24 Cromwell Way, Witham, CM8 2ES
M6 KRX K. Rosema, Apartment 801, 25 Goswell Road, London, EC1M 7AJ
M6 KRZ K. Brazier, Elgin House, Seaside Lane, Easington Colliery, Peterlee, SR8 3JZ
M6 KSA C. Sibley, 57 Palatine Road, Thornton-Cleveleys, FY5 1EY
M6 KSC D. Clark, 34 Magdalene Road, Owlsmoor, Sandhurst, GU47 0UT
M6 KSD K. Darwin, 38 Springbank Road, Gildersome, Leeds, LS27 7DJ
M6 KSE A. Went, 25\\\, Wayletts\, Laindon, Basildon, SS16 6RN
M6 KSF D. Richards, 115 School Road Tilney St. Lawrence, King's Lynn, PE34 4RA
M6 KSG K. Stevens, 61A Main Road, Hoo, Rochester, ME3 9AA
M6 KSH K. Haywood, 126 Derby Street, Sheffield, S2 3NF
M6 KSI A. Thompson, 10 Belgrave Close, Hersham, Walton-on-Thames, KT12 5PH
MM6 KSJ J. Blick, Lower Adelaide House, 8 Mountstuart Road, Rothesay, Isle of Bute, PA20 9DY
MW6 KSL S. Davies, 5 Maldwyn Street, Cardiff, CF11 9JR
M6 KSP S. Pegg, 24 Fleetwood Close, Minster On Sea, Sheerness, ME12 3LN
M6 KSS S. Evennett, The Homestead, Pound Green Lane, Thetford, IP25 7LS
MM6 KSU G. Smith, 2 Croftfoot Place Gartcosh, Glasgow, G69 8EG
M6 KSV S. Karpukhina, 13 Rushmon Court, Barker Road, Chertsey, KT16 9HP
MW6 KSW M. Hancock, Halfway House, Hyfrydle Road, Talysarn, Caernarfon, LL54 6HG
M6 KSX B. Elms-Lester, Ferndale House Kerry'S Gate, Hereford, HR2 0AH
M6 KSZ P. Morris, Lindum, Dunholme Road Welton, Lincoln, LN2 3RS
M6 KTA K. Tilly, 14 Mcnally Place, Durham, DH1 1JE
M6 KTC I. Mcewen, 4 The Pantyles, Nightingale Lane, Sevenoaks, TN14 6BX
M6 KTH K. Thompson, 32 Westfield Harwell, Didcot, OX11 0LG
M6 KTI K. Raynor, 6 New Street, Castleford, WF10 2RN
M6 KTJ K. Jones, 7 Castlemans Cottages, Castlemans Lane, Hinton St. Mary, Sturminster Newton, DT10 1LY
M6 KTK N. Edwards, 42 Cambrai Avenue, Chichester, PO19 7UY
MW6 KTM P. Jones, 16 Severn View, Garndiffaith, Pontypool, NP4 7SN
M6 KTN K. Turner, 3 Beech Street, Sutton-in-Ashfield, NG17 3FL
M6 KTO K. Oxford, 5 Hazel Close, Rendlesham, Woodbridge, IP12 2UR
M6 KTQ Dr R. Pediani, Old School House Great Coxwell, Faringdon, SN7 7NB
MW6 KTS I. Williams, 5 Fron Goch, Llanberis, Caernarfon, LL55 4LE
M6 KTT K. Thistlethwaite, 140 Kingstown Road, Carlisle, CA3 0AY
M6 KTV B. Davies, 16 Pearmains Close, Orwell, Royston, SG8 5QY
M6 KTW K. Wise, 4 Cherrydown, Rayleigh, SS6 9ND
M6 KTX K. Taylor, 26 Elmbridge Road, Birmingham, B44 8AB
MM6 KTY K. Heron, 26 Lochancroft Lane, Wigtown, Newton Stewart, DG8 9JA
M6 KTZ K. Tasker, 16 Chopin Road, Basingstoke, RG22 4JN
M6 KUC C. Elliott, Vicarage Hill, Kingsteignton, TQ12 3BA
MW6 KUD L. Ansell, 114 Bowleaze, Greenmeadow, Cwmbran, NP44 4LG
M6 KUE I. Vernon, 7 Seaton Road Wick, Littlehampton, BN17 7LG
M6 KUG A. Jacobs, 41 Drake Ave, Sheerness, me123sa
M6 KUH S. Steinhoefel, 222 Stretford Road, Urmston, Manchester, M41 9NT
M6 KUI M. Diver, 2 West Lodge Lane Sutton, Ely, CB6 2NX
MI6 KUJ T. Calka, 71 Willowfield Street, Belfast, BT6 9AW
M6 KUK M. Decruz, 124 Antrim Road Woodley, Reading, RG5 3NY
M6 KUP A. Stain, Apartment 5, Bramble Road, Bridgwater, TA5 2FT
M6 KUR P. Fletcher, 14 Orchard Avenue, Aylesford, ME20 7LY
M6 KUT A. Roberts, 29 Manor Lane, Stourbridge, DY8 3ER
M6 KUU J. Tobin, 47C Crystal Palace Road East Dulwich, London, SE22 9EX
M6 KUY J. Moore, 9 Goldsmith Avenue, London, RM7 0EX
MW6 KVA P. Barnes, 95 Marcroft Road, Port Tennant, Swansea, SA1 8PN
M6 KVB K. Bushell, 4 Birch Grove, Harrogate, HG1 4HR
M6 KVD J. Miller, 5 Gilver Lane, Hanley Castle, Worcester, WR8 0AT
MM6 KVE M. Hodgson, 5 Jocks Loaning, Dumfries, DG2 0NQ
M6 KVF K. Emery-Ford, 48 Welham Grove, Retford, DN22 6TS
M6 KVG D. Miles, 133 Marston Lane, Nuneaton, CV11 4RE
M6 KVH S. Hepple, 20 Gower Road, Shaftesbury, SP7 8RU
MM6 KVI G. Hepburn, 33 Saxon Road, Glasgow, G13 2YQ
M6 KVJ D. Boyes, The Shute Fluxton, Ottery St Mary, EX11 1RL
M6 KVK G. Kirk, 15 Underwood Avenue, Ash, Aldershot, GU12 6PP
M6 KVL G. Dougherty, 22 Mayplace Avenue, Dartford, DA1 4PZ
M6 KVM Rev. D. Palmer, 14 Walcot Parade, Bath, BA1 5NF
M6 KVN K. Sewell, 12 Haylands Square, South Shields, NE34 0JB
MW6 KVO K. Owen, 27C Waen Fawr Estate, Holyhead, LL65 1LT
M6 KVP J. Wilkins, 3 Ganesfield Doulting, Shepton Mallet, BA4 4QA
M6 KVQ A. Taylor, 60 Wood Ride, Petts Wood, Orpington, BR5 1PY
MW6 KVS M. Kveksas, 25 Barnard Way, Church Village, Pontypridd, CF38 1DQ
M6 KVT R. Simpson, 48 Weatherhill Road, Horley, RH6 9LY
M6 KVU W. Cuddeford, 5 Rosevalley Threemilestone, Truro, TR3 6BH
M6 KVV G. Easton, Cowpen Road, Blyth, NE24 5TS
M6 KVW L. Sutherland, 198 Norbury Crescent, London, SW16 4JY
M6 KVX R. Timmons, 36 Worrall Road High Green, Sheffield, S35 3LP
MM6 KWA K. Mackenzie, Graceland, Crosshill, Duns, TD11 3UF
M6 KWC K. Chapman, 227 Raglan Street, Lowestoft, NR32 2LA
M6 KWG Dr W. Wightman, 36 Holyoake Avenue, Woking, GU21 4PW
M6 KWI K. Irwin, 11 The Crofts, Silloth, Wigton, CA7 4EU
M6 KWK D. Barnett, 12 Craig Walk Alsager, Stoke-on-Trent, ST7 2Rj
M6 KWL C. Henniker, 1 Brook House Drive, Fairfield, Buxton, SK17 7HW
M6 KWM D. Wright, 203 Winn Street, Lincoln, LN2 5EY
M6 KWP K. Pritchard, 124 Milburn Road, Ashington, NE63 0PQ
MM6 KWQ M. Brock, 18 Birchfield Place, Dumfries, DG1 4SD
M6 KWT J. Caulfield, 2 Thornley Road, Tow Law, Bishop Auckland, DL13 4ED
M6 KWU S. Penhaligan, 7 Trembath Crescent, Newquay, TR7 2DX
MM6 KWV I. Mcculloch, 4 Lady Street, Brydekirk, Annan, DG12 5LZ
M6 KWW B. Covill, Walnut Tree Cottage, Holt Street, Dover, CT15 4HX
M6 KWX K. Woodard, 4 Pingo Road, Watton, IP25 6ZB
M6 KWZ M. Cross, 43 Queens Park, Wadebridge, PL27 7PR
M6 KXA K. Adams, 57 Haddenham Road, Leicester, LE3 2BH
M6 KXE C. Howell, 47 Birch Park Coalway, Coleford, GL16 7RU
MM6 KXF D. White, 31 Dunmuir Road, Castle Douglas, DG7 1LQ
M6 KXH P. Hambly, Mendip Road, Stoke St Michael, BA35JU
M6 KXI D. Cook, 14 Filbridge Rise, Sturminster Newton, DT10 1AA
MI6 KXM K. Mitchell, 80 Markethill Road, Collone, Armagh, BT60 1LE
M6 KXN T. Canning, Whitegates, Mayfield Avenue New Haw, Addlestone, KT15 3AG
M6 KXO G. Barnard, 35 High Street, Hitchin, SG4 8AJ
M6 KXP C. Wooldridge, 26 Grieg Close, Basingstoke, RG22 4DU
M6 KXQ B. Wogden, 3 Gordonstoun Place, Blackburn, BB2 2PT
M6 KXR K. Golloghly, 7 Jenner Close, Bungay, NR35 1QR
M6 KXS R. Rigden, 36A Atherston, Bristol, BS30 8YB
M6 KXU B. Wood, 25 Chestnut Terrace, Lamerton, PL19 8RL
M6 KXV A. Jordan, 19 Arrow Lane, Newhaven, BN9 0FG
M6 KXW A. Jordan, 20 Normandy Close, Crawley, RH107XP
M6 KXX B. Cook, 11 Arbrook Lane, Esher, KT10 9EG
M6 KYB K. Greenshields, 3 Lovers Walk, Wells, BA5 2QL
M6 KYC M. Cook, 12 Davison Street, Lingdale, Saltburn-by-the-Sea, TS12 3DX
M6 KYE A. Letchford, 64 Medway Road, Sheerness, ME12 1DR
M6 KYI G. Virostek, 58 Hendon Street, Brighton, BN2 0EG
M6 KYK J. Slim, Troublesome Reach, Playford Road, Woodbridge, IP13 6ND
MW6 KYN K. O'Brien, Flat 8, 4 Market Street, Newport, NP20 1FU
M6 KYO A. Noriega, 363 Pracha Uthit Road, Don Mueang, Thailand, BANGKOK 10210
M6 KYP P. Higgins, 27 Middleton Avenue, Ross-on-Wye, HR9 5BD
M6 KYS E. Panayiotou, 13 Oakmere Close, Potters Bar, EN6 5JQ
MM6 KYV W. Hannah, 71 Herriot Avenue, Kilbirnie, KA25 7JB
M6 KYX A. Tandler, 6 Field View Cottages, Brimfield, Ludlow, SY8 4LB
M6 KYZ G. Amos, 9 Manor Road Sundridge, Sevenoaks, TN14 6DL
M6 KZB C. Wilkinson, 67 Middleton Park Grove, Leeds, LS10 4BG
M6 KZC J. Neal, 40 Channel View Road, Portland, DT5 2AY
M6 KZE P. Orwin, 108 Cordwell Avenue, Chesterfield, S41 8BN
M6 KZF E. Cook, 927 Manchester Road Linthwaite, Huddersfield, HD7 5NE
M6 KZH L. Gostelow, 7 Coronation Drive, Donhead St Mary, Shaftesbury, SP7 9NA
M6 KZI D. Ball, 28 Bryony Gardens, Gillingham, SP8 4TR
M6 KZK A. Whybrow, 64 Church Road Sevington, Ashford, TN24 0LF
M6 KZM K. Maddy, 56 Coachwell Close, Telford, TF3 2JB
M6 KZN J. Spence, Officers Mess, Royal Air Force, Brize Norton, Carterton, OX18 3LX
M6 KZQ M. Pott, 29 Lilliebrooke Crescent, Maidenhead, SL6 3XJ
M6 KZR D. Ellens, 150 Lumley Avenue, South Shields, NE34 7DJ
MI6 KZS K. Sullivan, 149 Largy Road Ahoghill, Ballymena, BT42 2RG
M6 KZU M. Macrae, 91 Chosen Way, Hucclecote, Gloucester, GL3 3BX
M6 KZW O. Dawkins, 47B Fieldside, Abingdon, OX14 1JT
M6 KZX G. Burtenshaw, The Blenheims, Keymer Road, Burgess Hill, RH15 0BA
M6 KZZ P. Keenan, 5 Downton Walk Tiptree, Colchester, CO5 0DH
M6 LAA L. Atkins, Mouse Hall, Low Row, Richmond, DL11 6PY
M6 LAC L. Ainger, 41 Gilbert Road, Camberley, GU16 7RD
MM6 LAD J. Cattigan, Lunan Home Farm Cottage, Lunan Bay, Arbroath, DD11 5ST
M6 LAE L. Thompson, 43 Manor Road, Horsham St. Faith, Norwich, NR10 3LF
M6 LAF S. Dooley, 1 Rosewood Close, Little Sutton, Ellesmere Port, CH66 4AJ
M6 LAG L. Burgess, 40 Sheridan Terrace, Hove, BN3 5AF
MM6 LAH L. Hail, 70 Nobleston Estate, Alexandria, G83 9DB
M6 LAI L. Lewczenko, 10 Saxon Court, Swaffham, PE37 7TP
MM6 LAK S. Ramsay, 6 Cross Road, Peebles, EH45 8DH
M6 LAL L. Cowley, 3 Park Villas, Keswick, CA12 5LQ
M6 LAM M. Quemby, 19 Oak Close, Coalville, LE67 4JU
MM6 LAO W. Neish, 29 Currievale Drive, Currie, EH14 5RN
M6 LAP S. Potts, 22 Valebridge Road, Burgess Hill, RH15 0QY
M6 LAQ L. Briggs, 5 Links Avenue, Cromer, NR27 0EQ
M6 LAS L. Scambell, 8 South Bank Road, East Cowes, PO32 6JE
M6 LAV L. Hand, 168 Barcroft Street, Cleethorpes, DN35 7DX
M6 LAY S. Lay, 7 Hunt Street, Swindon, SN1 3HW
M6 LAZ G. Sinclair, 23 Cummings Square, Wingate, TS28 5JF
M6 LBC L. Nicholas, 9 Lila Place, Swanley, br88jb
M6 LBE S. Blackburn, 20 Seascale Close, Blackburn, BB2 3TP
M6 LBI I. Hyde, 3 Hibbert Avenue, Denton, Manchester, M34 3NZ
M6 LBK I. Macdonald, Broomhill Mill Lane, Worthing, BN13 3DH
M6 LBL L. Bell, 56 Boyd Road, Wallsend, NE28 7SQ
M6 LBM L. Mason, 86 The Street, Rockland St. Mary, Norwich, NR14 7AH
M6 LBN P. Tolcher, 15 Langstone Close, Torquay, TQ1 3TX
M6 LBQ B. Li, Harrogate Ladies' College, Clarence Drive, Harrogate, HG1 2QG
M6 LBR S. Hayward, 20 Abbey Square, Walsall, WS3 2RJ
MM6 LBS W. Thomson, 20 Greenfield Road, Glasgow, G32 0LP
M6 LBT L. Thomas, 66 Sturdee Avenue, Great Yarmouth, NR30 4HL
M6 LBU L. Baldwin, 24 Rockrose Way, Portsmouth, PO6 4EZ
M6 LBV E. Price, Little Acre Eardisley, Hereford, HR3 6LX
M6 LBW E. White, 16 Illingworth Way, Foxton, Cambridge, CB22 6RY
M6 LBX A. Walker, 27 Fielding Avenue, Poynton, Stockport, SK12 1YX
M6 LBY A. Smith, 69 Keele Road, Newcastle under Lyme, ST52JT
M6 LCC P. Harris, 16 Laxton Gardens, Baldock, SG7 6DA
M6 LCF M. Macdonell, 54 Cinque Foil, Peacehaven, BN10 8DZ
M6 LCG E. Ayre, 1 Spring Gardens, Broadmayne, Dorchester, DT2 8PP
M6 LCH M. Balshaw, 16 East Avenue, Heald Green, Cheadle, SK8 3DL
M6 LCI L. O'Brien, 172 Cheriton High Street, Folkestone, CT19 4HN
M6 LCK L. Kerr, 24, Lindisfarne Street, Carlisle, CA1 2ND
MM6 LCL L. Clark, 9 Aileymill Gardens, Greenock, PA16 0QF
M6 LCP L. Bourn, 4 Fell Wilson Street, Warsop, Mansfield, NG20 0PT
M6 LCQ L. Man, 115 Northdown Park Road, Margate, CT9 3PX
MI6 LCR L. Robinson, 32 Corrycroar Road, Pomeroy, Dungannon, BT70 3DY
M6 LCT C. Lai, Clarence Drive, Harrogate, HG1 2QG
M6 LCU C. Keszei, 2 Blackmore Hill Farm Cottages, Calvert Road, Buckingham, MK18 2HA
M6 LCV A. Grant, 15A Elmdon Road Marston Green, Birmingham, B37 7BU
M6 LCW L. Weeks, 11 Brock Close, Deepcut, Camberley, GU16 6GA
MW6 LCX J. Hawkins, 104 Ty Fry, Aberdare, CF44 7PP
M6 LCY L. Heron, 301 Marton Road, Middlesbrough, TS4 2HG
M6 LDA M. Leggett, 55 Colomb Road, Gorleston, Great Yarmouth, NR31 8BU
M6 LDB L. Hoddinott, 30 Deans Mead, Bristol, BS11 0QX

M6 LDD E. Daniels, 2 Garstons Close, Titchfield Fareham, PO14 4EN
M6 LDE D. Simmons, 8 Lower Grange, Huddersfield, HD2 1RU
M6 LDF L. Ferguson, 67 Knowlton Road, Poole, BH17 9EE
M6 LDG R. Hazel, 2 Lynwood Grove Hull, Hull, HU5 2BE
M6 LDH O. Lai, Harrogate Ladies' College, Clarence Drive, Harrogate, HG1 2QG
MI6 LDI P. Dorrian, 12 Gortnamona Place, Belfast, BT11 8PP
M6 LDJ L. Jepson, 143 Walnut Avenue Weaverham, Northwich, CW8 3DX
M6 LDK L. Kelley, 51 Grasmere Street, Liverpool, L5 6RH
M6 LDL J. Hatton, 49 Buxton Street, Morecambe, LA4 5SR
MW6 LDM L. Martin, 2 Ael Y Glyn, Nant Road, Harlech, LL46 2UJ
M6 LDQ L. Dobinson, 20 Newholme Crescent, Evenwood, Bishop Auckland, DL14 9RY
M6 LDR L. Roworth, 27 Bury Road, Dagenham, RM10 7XR
MW6 LDS T. Jones, 13 Bond Street, Aberdare, CF44 7HA
M6 LDU L. Dumbleton, 9 Wareham Road, Rubery, Birmingham, B45 0JS
M6 LDW T. Evans, 70 Tremarle Home Park, North Roskear, Camborne, TR14 0AR
M6 LDY A. Collins, 15 North River Road, Great Yarmouth, NR30 1JY
MW6 LDZ T. Clapp, Crunns Farm. Coxhill, Narberth, SA67 8EH
M6 LEA J. Bridgehouse, 43 Age Croft, Oldham, OL8 2HG
M6 LEC L. Collinson, 26 Westway Avenue, Hull, HU6 9SA
M6 LEE L. Davies, 94 Trent Way Kearsley, , Bolton, BL4 8PS
M6 LEF L. Faulkner, Mount Pleasant, Elkstones, Buxton, SK17 0LU
M6 LEG K. Foulger, 89 Blaney Crescent, London, E6 6BB
M6 LEH L. Hargreaves, 32 Bank Road, Carrbrook, Stalybridge, SK15 3JX
M6 LEJ C. Calvert, 1 Moorsholme Avenue, Manchester, M40 9BW
M6 LEQ L. Pinkney, 18 Bridlington Road, Driffield, YO25 5HZ
MD6 LET D. Holohan, 22 Ballacannell Estate, Laxey, Isle of Man, IM4 7HH
M6 LEU L. Chadwick, 78 Blakemore, Telford, TF3 1PT
MM6 LEW M. Strachan, 62 Charleston Drive, Dundee, DD2 2EZ
M6 LEX K. Rowland, 9 Churchlands, North Bradley, Trowbridge, BA14 0TD
M6 LEY C. Harris, 4 Coronation Drive, Leigh, WN7 2UU
M6 LFB L. Bell, 42 Ocean Road, Walney, Barrow-in-Furness, LA14 3DX
M6 LFC D. Smith, 5 Verbena Close, Beechwood, Runcorn, WA7 3JA
M6 LFD L. Drew, 7 Bronte Court, Tamworth, B79 8DN
M6 LFE B. Lewin, 68 Brackley Square, Woodford Green, IG8 7LS
MW6 LFG J. Beavan, 21 Llannerch Road West, Rhos On Sea, Colwyn Bay, LL28 4AU
M6 LFJ M. Stevens, The Glen, Sunnyside Avenue, Sheerness, ME12 2RA
MI6 LFK L. Corr1, 6 Lime Park Balnamore, Ballymoney, Bt537qg
M6 LFL L. Theobold, 25 Aysgarth Road, Leicester, LE4 0ST
M6 LFM A. Campbell, 35 Goodwood Road, Gosport, PO12 4HN
MM6 LFN C. Bolton, 30 Brackenhill Drive, Hamilton, ML3 8AY
M6 LFO P. Noble, 30 Whitewater Rise, Dibden Purlieu, Southampton, SO45 4BY
M6 LFQ K. Blatch, 61 Linden Way, Haddenham, Ely, CB6 3UG
M6 LFR L. Davison, 58 Priestley Court, South Shields, NE34 9NQ
MM6 LFS A. Miles, 9 Buchanan Drive, Lenzie, Kirkintilloch, Glasgow, G66 5HS
M6 LFT Dr M. Cianni, 121 Springfield Park Avenue, Chelmsford, CM2 6EW
MI6 LFU C. Mccartney, 62 Lakeview, Crumlin, BT29 4YA
M6 LFW M. O'Donovan, Wyllsden House, Stroud, GL52PA
M6 LFX J. Lamb, 9A Matlock Road, Canvey Island, SS8 0EW
M6 LFY J. Wells, 27 Victoria Avenue, Camberley, GU15 3HT
M6 LFZ E. Fish, 16 Cartmel Place, Ashton-On-Ribble, Preston, PR2 1TY
MW6 LGA R. Samphire, Courtlands, Newport Road, Magor, Caldicot, NP26 3BZ
M6 LGB L. Brazier, 165 Avon Road, Worcester, WR4 9AH
M6 LGD L. Thomas, Fair View, Close Hill, Redruth, TR15 1EP
M6 LGF D. Riches, 48 Turner Road, Ipswich, IP3 0LX
M6 LGH J. Abraham, 18 Ferneley Crescent, Melton Mowbray, LE13 1RZ
M6 LGI L. Grover, 80 Hamilton Road, Bishopstoke, Eastleigh, SO50 6AN
M6 LGJ L. Sparkes, 40 Cambridge Road, Eastbourne, BN22 7BT
M6 LGL M. Rusu, 1 Turnbull Road, March, PE15 9RX
M6 LGM L. Mann, 14A Orchehill Avenue, Gerrards Cross, SL9 8PX
M6 LGQ M. Lee, 4 Cluny Court, Wavendon Gate, Milton Keynes, MK7 7TT
MM6 LGS C. Sloan, 7 Clova Street, Thornliebank, Glasgow, G468na
M6 LGT J. Leggett, 2 Getliff Mews North Quay, Great Yarmouth, NR30 1RP
M6 LGV L. Vickers, 24 Hearnes Meadow, Seer Green, Beaconsfield, HP9 2YJ
MI6 LGX C. Sheppard, 4 Fairview Drive, Whitehead, Carrickfergus, BT38 9NT
M6 LHA A. Crawford, 4 Trimpley Drive, Kidderminster, DY11 5LB
MW6 LHB S. Hillman, 8 Brigantine Grove, Duffryn, Newport, NP10 8ET
M6 LHD K. Newbould, 47 Old Barber, Harrogate, HG1 3DF
M6 LHE L. Heppenstall, 15 Gibraltar Road, Hemswell Cliff, Gainsborough, DN21 5XJ
M6 LHF M. Pickering, 30 Hotspur Avenue, Whitley Bay, NE25 8RP
M6 LHG L. Hagan, 7 Betjeman Mews, Southend-on-Sea, SS2 5EJ
M6 LHH B. Cummings, 8 Carshalton Way, Lower Earley, Reading, RG6 4EP
M6 LHI Dr M. Depardieu, 4 Belvedere Fff, Bath, BA1 5ED
M6 LHJ L. Smith, 17 Grove Street, Kirton Lindsey Lincs, DN21 4BY
M6 LHL J. Ablett, 33 Langley Road, Leeds, LS13 1AX
M6 LHM L. Mason, 2 Iris Close, Widnes, WA8 4GA
M6 LHN L. Hulse, 202A Shooters Hill Road, London, SE3 8RP
M6 LHO A. Rowan, 14 Craven Lea, Liverpool, L12 0NF
M6 LHP L. Pass, 14A Elm Avenue, Hucknall, Nottingham, NG15 6GE
M6 LHQ L. Daniels, 19 Redwell Grove, Kings Hill, West Malling, ME19 4BU
M6 LHR L. Rich, 39 Wren Close, Heathfield, TN21 8HG
M6 LHS H. Ren, 50 Highwoods Drive, Marlow, SL7 3PY
M6 LHV G. Mccarthy, 14 Cosedge Crescent, Croydon, CR0 4DN
M6 LHW H. Lau, 113 Ruxley Lane, Epsom, KT19 9EX
M6 LIB E. Bayliss, 19 Rugby Road, Dunchurch, Rugby, CV22 6PG
M6 LID J. Eades, 128 Russells Hall Road, Dudley, DY1 2NN
M6 LIE L. Gregory, 14 Anderson Road, Hemswell Cliff, Gainsborough, DN21 5XP
M6 LIH N. Hindl, Hill House Haulage 67 Yarmouth Road, Ellingham, Bungay, NR35 2PH
M6 LII A. James, 45 Knockholt Road Cliftonville, Margate, CT9 3HL
M6 LIJ J. Mason, 14A Brooke Avenue, Margate, CT9 5NG
M6 LIK T. Russell, 38 Speedwell Close, Witham, CM8 2XL
MM6 LIL L. Clark, 17 Lewis Rise, Broomlands, Irvine, KA11 1HH
M6 LIN L. Briggs, 20 Broad Lane, Pelsall, Walsall, WS4 1AP
M6 LIP S. Shakespeare, 36 Maitland Road, Russells Hall Estate, Dudley, DY1 2NU
M6 LIQ G. Pollard, 22 Girton Avenue, Ashton in Makerfield, WN4 9SA
M6 LIS M. Ramsbottom, 1 Barn Gill Close, Blackburn, BB2 3HU
MI6 LIT C. Brush, 56 Larchwood, Banbridge, BT32 3UT
M6 LIU D. Petrauskas, Ateities 7, Jurbarkas, Lithuania, LT74208
M6 LIV J. Hay, 13 Windsor Road, Workington, CA14 5BQ
MW6 LIW D. Evans, 23 Wellington Street, Aberdare, CF44 8EW
MW6 LIZ E. Martin, 62 Llwyn Ynn, Talybont, LL43 2AL
M6 LJA M. Duxbury, 32 Radford Street, Darwen, BB3 2PB
M6 LJB A. Billingham, 6 Kemble Close, Lincoln, LN6 0NR
M6 LJD L. Denham, 92 Windermere Avenue, Southampton, SO16 9GF
MM6 LJE L. Treble, 2 Baidland Meadow, Dalry, KA24 5HP
M6 LJF J. Bogdaniec, 3 Cavalry Chase, Okehampton, EX20 1GR
M6 LJG L. Goldsmith, Hunters Cottage, 61 Fengate Drove, Weeting, Brandon, IP27 0PW
M6 LJJ L. Jones, Brimham Lodge Farm, Harrogate, HG3 3HE
M6 LJK L. Kirkpatrick, The Cleave, Nine Oaks Estate, Yelverton, PL20 6ND
M6 LJM L. Marriott, 94 Lyndhurst Road, Worthing, BN11 2DW
MI6 LJO O. Conaghan, 94 Curlyhill Road, Strabane, BT82 8LS
M6 LJP L. Passam, Glanceiro, Llandre, Bow Street, SY24 5BS
M6 LJR C. Lonie Jr, 41 De La Hay Avenue, Plymouth, PL3 4HS
M6 LJS L. Smith, 177 Waterloo Road, Stoke-on-Trent, ST6 2ER
M6 LJT M. Rushton, 17 Highbury Gardens, Ramsgate, CT12 6QG
M6 LJU R. Gilliam, 7 Hamble Street B, London, SW6 2RT
M6 LJV S. Cilliers, 61 Heathwood Gardens, London, SE7 8ET
MI6 LJZ A. Porter, 8 Ballyregan Avenue Dundonald, Belfast, BT16 1JW
M6 LKA M. Wilkinson, 3 Balsams Close, Hertford, SG13 8BN
M6 LKD M. Dust, 1A Elm Road, Erith, DA8 2NN
M6 LKE L. Huddart, 1 Rydal Court, Penrith, CA11 8PN
M6 LKF L. Chapman, 4 Russell Road, Clacton-on-Sea, CO15 6BE
M6 LKI K. Lockstone, Oceana, The Parade Pevensey Bay, Pevensey, BN24 6LX
M6 LKK H. Freeman, 95 Raleigh Road, Wirral, CH46 2QY
M6 LKL G. Urban, 33 High Meadow, Hathern, Loughborough, LE12 5HW
M6 LKM L. Mcdonnell, 108 Long Lane, Garston, Liverpool, L19 6PQ
M6 LKS L. Spriggs, 19 Mackenzie Square, Shephall, Stevenage, SG2 9TT
M6 LKW L. Whitby, 45 Trent Road Shaw, Oldham, OL2 7YG
M6 LKY A. Lee, 27 Victoria Avenue, Camberley, GU15 3HT
M6 LLB L. Brigham, 42 Cayley Close, Clifton, York, YO30 5PT
M6 LLC L. Clark, 1 Brooklime Road, Liverpool, L11 2YH
M6 LLD A. Clark, 1 Brooklime Road, Liverpool, L11 2YH
M6 LLE R. Clark, 1 Brooklime Road, Liverpool, L11 2YH
MI6 LLG S. Horner, 10 Meadow Court, Bushmills, BT57 8SD
M6 LLH A. Hoyte, 43 Orchard Drive Mayland, Chelmsford, CM3 6EP
MI6 LLI R. Hetherington, 112 Screeby Road, Fivemiletown, BT75 0LG
M6 LLK Dr D. Roberts, Beggarwood House Ravensworth Park Estate, Gateshead, NE11 0HQ
M6 LLL L. Foulkes, 43 Mill Hayes Road, Stoke-on-Trent, ST6 4JB
M6 LLM L. Meredith, 131 Trimdon Avenue, Middlesbrough, TS5 8RY
M6 LLN A. Allen, 59 Bottels Road Warboys, Huntingdon, PE282RZ
M6 LLO S. Hamilton, 11 Jubilee Gardens, Telford, TF3 2BR
MI6 LLS B. Kelly, 153 Ardanlee, Ballynagard, Londonderry, BT48 8RT
M6 LLW H. Liu, Harrogate Ladies' College, Clarence Drive, Harrogate, HG1 2QG
MI6 LLZ K. Dorman, 25 Blackthorn Road, Newtownabbey, BT37 0GH
M6 LMB L. Bate, 87 Dunsheath, Telford, TF3 2BY
M6 LMC L. Chatt, Rosemary Cottage, Causeway End Road, Dunmow, CM6 3LU
M6 LMG J. Porter, 3 The Walks, Main Road, Woodbridge, IP12 3DZ
MM6 LMH L. Mitchell Hynd, Smithy House, Bruichladdich, Isle of Islay, PA49 7UN
M6 LMI L. Mikolka, Flat, 1 Scotney Court, Romney Marsh, TN29 9JP
M6 LMJ L. Mathlin, 29 Wagtail Drive, Stowmarket, IP14 5GH
M6 LMU J. Jarvis, 8 Wesleys Fold, Pinfold Street, Wednesbury, WS10 8UN
M6 LMW A. Thomson, Shire Jee Neevas, Cold Ash Hill, Thatcham, RG18 9PH
MM6 LNB S. Anderson, Middleton Mains, Gorebridge, EH23 4RL
M6 LNC D. Lancaster, Linkhill View, Frith Common, Eardiston, Tenbury Wells, WR15 8JX
M6 LND L. Dawson, 5 Harbour View, Roker, Sunderland, SR6 0NL
M6 LNE C. Bruce, Nunfield House, Bull Lane, Sittingbourne, ME9 7SL
M6 LNI G. Peters, Flat 6, Park Court, 46 North Park Road, Harrogate, HG1 5AD
M6 LNM W. Walters, 3 Heatherfields, Gillingham, SP8 4TT
MI6 LNP N. Scott, 5 Sharonmore Parade, Newtownabbey, BT36 6PR
M6 LNQ N. Sargent, 21 St. Michaels Road, Claverdon, Warwick, CV35 8NT
M6 LNR A. Evetts, 21 Instone Road, Halesowen, B63 4SA
M6 LNS D. Williams, Apartment 54, 7 Tiltman Place, London, N7 7EL
MW6 LNU R. Sullivan, 58 Ynyscynon Road, Tonypandy, CF40 2LN
M6 LNV R. Mayes, Crescent Avenue, London, RM17 6AZ
M6 LNX S. Wareham, 8 Simon Road, Longlevens, Gloucester, GL2 0ER
M6 LOB A. Brash, 44 Broadway East, Chester, CH2 2DP
M6 LOC L. Barker, 88 Cecilia Road, London, E8 2ET
MW6 LOD L. Broadhurst, 4 Lilburne Drive, Newport, NP19 0ET
M6 LOE N. Chaplin, 5 Maxwell Street, Bury, BL9 7QA
M6 LOF S. Mcilwaine, 70 Autumn Drive, Sutton, SM2 5BA
M6 LOG Dr I. Van Der Linde, 77 Port Vale, Hertford, SG14 3AF
M6 LOK B. Wolff, 7 Church Terrace, Reading, RG1 6AS
M6 LOL R. Cooper, 11A Ambleside, Gamston, Nottingham, NG2 6NA
MM6 LON L. Nicoll, 15 Redford Walk, Edinburgh, EH13 0AF
M6 LOQ P. Wright, 25, Paynes Meadow, Whitminster, Gloucester U. K., GL2 7PS
M6 LOS J. Hawthorn, 1 Tudor Close, Leigh-on-Sea, SS9 5AR
MI6 LOT L. Treanor, 1 Granemore Park, Keady, Armagh, BT60 2GP
M6 LOW J. Lowenthal, 133 Marshalswick Lane, St Albans, AL1 4UX
M6 LOZ L. Shaw, 21 Dyke Street, Stoke-on-Trent, ST1 2DF
M6 LPB Lord N. Petit-Brown, 6 Cope Avenue, Nantwich, CW5 5JE
M6 LPD D. Lester, 171 Glenavon Road, Birmingham, B14 5BT
M6 LPF A. Kuba, Flat 291-295, Jellicoe Court, Southampton, SO16 3UJ
M6 LPH F. Quinn, Harley Cottage Beech Grove Gardens, Carlisle, CA30LR
M6 LPI N. Stone, Flat, 97-99 Stoke Road, Gosport, PO12 1LR
M6 LPK L. Kiddell, 1 Sparham Hill, Sparham, Norwich, NR9 5QT
M6 LPN J. Leach, 9 Crown Point Drive, Ossett, WF5 8RQ
M6 LPO R. Cowperthwaite, 30 Glover Place, Bootle, L20 4QR
M6 LPP P. Petersen, 15 Kent Gardens, Birchington, CT7 9RS
M6 LPS D. Clement, 24 Millais, Horsham, RH13 6BS
MM6 LPT J. Mckinnon, 81 Willow Drive, Johnstone, PA5 0DA
M6 LPV D. Churchill, 39 East Street, Corfe Castle, Wareham, BH20 5EE
M6 LPW L. Walker, 7 Stroudley Close, Ashford, TN240TY

M6 LPX S. Kembrey, 101 Yew Tree Drive, Bristol, BS15 4UF
M6 LPY S. Taylor, 27 Anchorsholme Lane East, Thornton-Cleveleys, FY5 3QH
MW6 LPZ D. Stevens, 12 Firs Avenue Fairwater, Cardiff, CF53TH
M6 LQA J. Strickler, Friedlebenstraße 48, Frankfurt/Main, Germany, 60433
M6 LQC D. Tofield, 6 Church Grove, Barnstaple, EX32 9DJ
MM6 LQF J. Bennet, 24/11 Greenpark, Edinburgh, EH177TA
MW6 LQG S. Trump, 168 Brynglas Avenue, Newport, NP20 5LP
MW6 LQK M. Stanger, 66 Heather Court, Ty Canol, Cwmbran, NP44 6JR
M6 LQM B. Poulter, 1 Heathyfields Road, Farnham, GU9 0BN
M6 LQO J. Thompson, 32 Coult Avenue, North Hykeham, Lincoln, LN6 9RG
M6 LQP A. Clack, 3 Darwin Close, Swindon, SN3 3NF
M6 LQR M. Constantine, 82 The Oval, Brough, HU15 1DD
M6 LQT Dr P. Patureau, 4 Belvedere Fff, Bath, BA1 5ED
MM6 LQU S. Clayton, 19 Howe Park, Edinburgh, EH10 7HF
MW6 LQW M. Downward, 17 Beeston Terrace, Wrexham, ll139nn
M6 LQX A. Mcdonald, 11 Micklegate Murdishaw, Runcorn, WA76HT
M6 LQY I. Gilmore, 19 Green Sward Lane, Redditch, B98 0EN
M6 LRA L. Akred, 25 Kitchener Street, Walney, Barrow-in-Furness, LA14 3QW
M6 LRB D. Williamson, 4 King Edward Road, Northampton, NN1 5LU
M6 LRF R. Deller, Four Winds Farm, Buckworth Road, Huntingdon, PE28 4JX
M6 LRG M. Parker, Ridgeways, Mill Common, Westhall, Halesworth, IP19 8RQ
M6 LRH L. Hancock, 106 Hoyle Street, Warrington, WA5 0LW
MM6 LRK A. Lark, 20 Lawfield, Coldingham, Eyemouth, TD14 5PB
M6 LRL S. Vane, 17 Knights Walk, Abridge, Romford, RM4 1DR
MI6 LRN K. Bell, 3 Alexandra Crescent, Larne, BT40 1NE
MW6 LRO O. Thomas, Garth Celyn, St. Davids Road, Aberystwyth, SY23 1EU
M6 LRQ M. Bacon, 69 Doverdale Close, Redditch, B98 7SD
M6 LRR A. Nielsen, 29 Greygarth Close Bransholme, Hull, HU7 5AP
M6 LRU R. Walker, 125 Devereux Road, West Bromwich, B70 6RQ
M6 LRV H. Moore, 52 Limefield Street, Accrington, BB52AF
M6 LRW J. Welch, 49 Walshs Manor, Stantonbury, Milton Keynes, MK14 6BU
MM6 LRX G. Morrison, 11 Goodman Place, Maddiston, FK2 0NB
M6 LRZ S. Carr, 24 Park Road, Blyth, NE24 3DH
M6 LSA L. Allcock, 26 Castleton Grove, Inkersall, Chesterfield, S43 3HU
M6 LSB L. Catterall, 14 Dunham Drive, Whittle-Le-Woods, Chorley, PR6 7DN
M6 LSE C. Stoten, 12 Boyd Avenue, Dereham, NR19 1LU
M6 LSG L. Spong, 2 Strathmore Drive, Charvil, Reading, RG10 9QT
M6 LSH L. Shaw, 47 Beechfields, Eccleston, Chorley, PR7 5RF
M6 LSJ L. Sawkins, 20 Nye Close, Cheddar, BS27 3PB
M6 LSK C. Hare, 25 Southend Place, Sheffield, S2 5FQ
M6 LSL A. Wood, 4 St. Andrews, The Common, Cranleigh, GU6 8NX
M6 LSN P. Bamber, 15 Grantley Avenue Kingswood Oak, Shrewsbury, SY3 5LA
M6 LSO G. Coltman, The Oaks Rushden, Buntingford, SG9 0SN
M6 LSP L. Phillips, 5 Barnes Green, Wirral, CH63 9LU
M6 LST D. Lui, Clarence Drive, Harrogate, HG1 2QG
M6 LSU M. Pesendorfer, 13 Blake Road, London, N11 2AD
M6 LSV S. Ludziss, 82 Trinity Avenue, Mildenhall, Bury St Edmunds, IP28 7LS
MW6 LSW D. Johns, 35 Bronhaul, Talbot Green, Pontyclun, CF72 8HW
MI6 LSY L. Stock, 15 Mahon Drive, Portadown, Craigavon, BT62 3JB
M6 LTB L. Burke, Teviot, Malthouse Lane, Peasmarsh, TN31 6TA
MM6 LTC T. Craig, Cemetery Lodge, Lochmaben, Lockerbie, DG11 1RL
M6 LTD P. Asher, 124 Bath Street, Market Harborough, LE16 9JL
M6 LTL L. Layland, 3 Thirlmere Road, Golborne, Warrington, WA3 3HH
M6 LTM L. Simons, 123 Main Street, Little Harrowden, Wellingborough, NN9 5BA
MW6 LTN M. Ashford, 3 Candleston Place Bonymaen, City & County Of Swansea, Swansea, SA1 7JB
M6 LTO L. Trend, 140 Ardleigh, Basildon, SS16 5RW
M6 LTQ W. Coates, 36 The Crescent, Welwyn, AL6 9JQ
MW6 LTR C. Rowe, 21 Graig Terrace, Abercwmboi, Aberdare, CF44 6AH
M6 LTS L. Stant, 3 Uffa Fox Place, Cowes, PO31 7NX
M6 LTU M. Brazinskas, 25 Elswick Road, London, SE13 7SP
M6 LTV N. Griffin, 3 Gooseander House, Cirencester, GL7 5fh
M6 LUA T. Goddard, 217 Speedwell Road, Bristol, BS5 7SP
M6 LUC D. Boden, 249 Nottingham Road, Ilkeston, DE7 5AT
M6 LUD K. Willson, Ludpit Cottage, Ludpit Lane, Etchingham, TN19 7DB
M6 LUG P. Schoenmaker, 24 Greenheys Drive, London, E18 2HB
M6 LUH N. Porter, 114 Kingston Avenue, Worcester, WR3 8PP
M6 LUI G. Conboy, 7 Bell Clough Road, Droylsden, Manchester, M43 7NS
M6 LUJ S. Cottam, 14 Barnard Close, Rednal, Birmingham, B45 9SZ
M6 LUK L. Johnson, 7 Southover Way, Hunston, Chichester, PO20 1NY
M6 LUM L. Medza, 16 Huntroyde Avenue, Bolton, BL2 2ET
MI6 LUP M. Sullivan, 149 Largy Road Ahoghill, Ballymena, BT42 2RG
MW6 LUQ E. Williams, 56 Heol Llansantffraid Sarn, Bridgend, CF32 9NH
M6 LUR P. Hadley, 20 Merrybrook, Evesham, WR11 2QF
M6 LUT A. Lutley, Springfield, Rookery Hill, Ashtead, KT21 1HY
M6 LUW P. Steadman, 82 Hereford Road, Shrewsbury, SY3 7RA
M6 LUY G. Hinson, The Leys, Brierley Hill, DY5 3UJ
M6 LUZ G. Luscombe, 28 St. Giles Gate, Doncaster, DN5 8PQ
M6 LVA R. Scullion, 41 Myrica Grove, Hoole, Chester, CH2 3EW
M6 LVC I. Collins, 19 Peel Street, Kidderminster, DY11 6UG
MW6 LVD K. West, 58 Claerwen Gelligaer, Hengoed, CF82 8EX
M6 LVE C. Johnson, 25 Pelham Street, Worksop, S80 2TW
MW6 LVH E. Parker, Gadlys, London Road Valley, Holyhead, LL65 3DP
M6 LVJ I. Humphries, 13 Malvern Close, Banbury, OX16 9EL
M6 LVK K. Jones, 7 Fazan Court, Wadhurst, TN5 6BT
M6 LVM P. Molloy, 4 Tilt Meadow, Cobham, KT11 3AJ
M6 LVN Dr J. Clark, 14 Portobello Terrace Birtley, Chester le Street, DH3 2JS
M6 LVR A. Humphreys, 2 Harrop Place Ribbleton, Preston, PR2 6TD
M6 LVS E. Pasqual, Flat 203 2 South Ealing Road, London, W5 4BY
MM6 LVV R. Holtom, Old Post Office, Church Road, Laurencekirk, AB30 1YS
M6 LVW L. Walker, 17 Carr House Road, Halifax, HX3 7QY
M6 LVX T. Von Bergmann, 17 Seeley Crescent, Street, BA16 0RN
M6 LWA L. Alderson, 62 Rusland Park, Kendal, LA9 6AJ
MM6 LWB L. Bradley, Amon Sul, Kiltarlity, Beauly, IV4 7HT
M6 LWE A. Martin, 14 Wain Way North Wingfield, Chesterfield, S42 5PQ
MW6 LWF L. Fish, Iddon Cottage, Bronygarth, Oswestry, SY10 7NF
M6 LWG C. Rowland, 20 Bell Hill Park Lindale, Grange-over-Sands, LA11 6JZ
MI6 LWI A. Lewis, 8 Liester Park, Ballyclare, BT39 9RZ
M6 LWJ L. Webb, Fern Bank, Wood Lea, Rotherham, S66 8NN
M6 LWM C. Smithen, 10 High Street, Temple Ewell, Dover, CT16 3DU
M6 LWP A. Lawler, 58 Bahram Road, Costessey, Norwich, NR8 5EY
M6 LWQ G. Ford, The Lodge, Home Farm Lane, Rimpton, Yeovil, BA22 8AS
M6 LWR S. Ross, 51 Claypiece Road, Bristol, BS13 9DR
M6 LWS W. Sawyer, 20 Park Terrace Willington, Crook, DL15 0QL
M6 LWT M. Hennessey, 3 Northgate Cottage, Falmer Road, Rottingdean, Brighton, BN2 7DT
M6 LWV L. Brown, 99 Apprentice Drive, Colchester, CO4 5SE
M6 LWY G. Fenton, 40 High Street, Easington Lane, Houghton le Spring, DH5 0JN
M6 LWZ R. Marshall, 17 Haywards Place, Easterton, Devizes, SN10 4PP
M6 LXC Rev. L. Clark, 226 Philip Lane Tottenham, London, N15 4HH
M6 LXE A. Elena, 31 Larksfield Avenue, Bournemouth, BH9 3LW
M6 LXH S. Li, Clarence Drive, Harrogate, HG1 2QG
M6 LXI E. Pestano, 38 Third Avenue, Bexhill-on-Sea, TN40 2PA
M6 LXK A. Weston, 37-39 Clover Hill, Sunniside, Newcastle upon Tyne, NE16 5PT
M6 LXM A. Brighton, 67 Wilks Farm Drive, Sprowston, Norwich, NR7 8RG
M6 LXN R. Froggatt, The Hawthorns, 2 Laurels Drive, Barton St David, TA11 6AT
M6 LXP M. Ling, Flat 14, Rowan Court, London, SW20 0BA
M6 LXQ Dr G. Turner, 8 Scarborough Terrace, York, YO30 7AW
M6 LXR L. Rhodes, 91 Bedonwell Road, Bexleyheath, DA7 5PS
M6 LXS M. Konrad, 12 Dale View Silsden, Keighley, BD20 0JP
M6 LXU D. Walton, 45 Wells Hall Road Great Cornard, Sudbury, CO10 0NH
M6 LXW C. Stickley, 152 Fore Street, Pinner, HA5 2NE
M6 LXX A. Erlank, 28 Ashenden Road, Guildford, GU2 7XE
M6 LXY M. Oxley, 49 Dalton Crescent, Shildon, DL4 2LE
M6 LYA A. Polakovs, 76 Sandringham Crescent, Leeds, LS17 8DF
M6 LYB G. Turner, 43 Harding Avenue, Eastbourne, BN22 8PL
M6 LYC M. O'Connor, 5 Kesbrook Drive Ashwood Park, Overseal, DE126NS
M6 LYD R. Lyddall, 102 Chapel Road, Brightlingsea, Colchester, CO7 0HE
M6 LYF R. Downhill, 40 Collingwood Close, Eastbourne, BN23 6HW
M6 LYG D. Bache, 81 Westgate, Driffield, YO25 6TA
MM6 LYH R. Latimer, 6 Crawford Avenue Rosemarkie, Fortrose, IV10 8UX
M6 LYJ W. Bradley, 67 Bury Road, Leamington Spa, CV31 3JD
M6 LYN L. Groves, 3 Hudson Davies Close, Pilley, Lymington, SO41 5PA
M6 LYO O. Lyon, Splinters, Nelson Park Road, Dover, CT15 6HL
M6 LYP M. Jonusas, 70 Methuen Street, Southampton, SO14 6FR
M6 LYQ T. Mann, 6 Kenley Close, Wickford, SS11 8XL
M6 LYS A. Booth, 27 Shenstone Road, Rotherham, S65 2JR
M6 LYU J. Swanbrow, 7 Manor Crescent, Rookley, Ventnor, PO38 3NS
M6 LYY A. Allgood, 39 Eastwood, Chatteris, PE16 6RX
M6 LYZ S. Fearn, 79 Maudlin Drive, Teignmouth, TQ14 8SB
M6 LZA L. Jackson, 90 Horne Street, Bury, BL9 9HS
M6 LZC D. Cassidy, 172 Lyde Road, Yeovil, BA21 5PN
M6 LZD G. Kelley, 31 Cherry Park, Brandon, Durham, DH7 8TN
M6 LZF M. Clitheroe, Green End, North Street, Castle Acre, King's Lynn, PE32 2BA
M6 LZG T. Coldham, 4 Carr Meadow Bamber Bridge, Preston, PR5 8HS
M6 LZI J. Miller, 9 St. Nicholas Road Tillingham, Chelmsford, CM0 7SQ
MI6 LZL P. Robinson, 119 Avenue Road, Lurgan, Craigavon, BT66 7BD
M6 LZM M. Radulov, 14 Grove Road, Chatham, ME4 5HS
M6 LZN J. Marshall, 6 Foster Walk, Sherburn in Elmet, LS25 6EU
M6 LZP J. Lovelock, Sea Spray, Lighthouse Road The Lizard, Helston, TR12 7NU
M6 LZQ L. Hollingworth, 43 Wingfield Road, Hull, HU9 4PR
M6 LZS C. Dyson, 5 Welton Park Welton, Daventry, NN11 2JW
M6 LZT C. Plant, 6 Talsarn Grove, Stoke-on-Trent, ST4 8YL
MW6 LZU P. Matthews, The Chateau, Wynnstay Hall Estate, Ruabon, Wrexham, LL14 6LA
MW6 LZV K. Mccafferty, 19 Quarr Road, Pontardawe, SA8 4JD
MI6 LZW R. Lewis, 8 Liester Park, Ballyclare, BT39 9RZ
M6 LZX B. Siddle, 7 Farebrother Street, Grimsby, DN32 0NH
M6 LZY C. Hillcox, 2 New Hall Drive, Sutton Coldfield, B75 7UU
MW6 LZZ R. James, 168 Brynglas Avenue, Newport, NP20 5LP
M6 MAA M. Meadowcroft, 210 Dickinson Close, Blackburn, BB2 2LT
M6 MAD D. Binnall, 21 Appletree Road, Featherstone, Pontefract, WF7 5EA
M6 MAG C. Lavery, 2 Barley Mow Cottages, Malting Lane, Grundisburgh, Woodbridge, IP13 6TE
M6 MAH M. Hyett, 1 Darell Close, Quedgeley, Gloucester, GL2 4YR
M6 MAJ M. Kealey, 24 Ben Nevis Road, Birkenhead, CH42 6QY
M6 MAK P. Mcgrath, 24 Broadoak Drive, Lanchester, Durham, DH7 0QA
M6 MAL M. Wallace, 4 Edmund Street, Kettering, NN16 0HU
M6 MAM M. Mcdougall, 122 Lee Lane, Horwich, Bolton, BL6 7AF
M6 MAO I. Connors, 3 Wheatfield Way, Chelmsford, CM1 2QZ
M6 MAP M. Peters, 25 Windsor Court, Falmouth, TR11 3DZ
M6 MAS S. Shailes, 9 Ingham Street, Padiham, Burnley, BB12 8DR
M6 MAT M. Pye, 42, Colne, BB8 9QH
M6 MAW M. Ansell-Wood, Sanju, Old Lane, Nethertown, Bradford, BD11 1LU
M6 MAX M. Trivett, 36 Edward Street, Hartshorne, Swadlincote, DE11 7HG
M6 MAY M. Buist, 23 St. Chads Drive, Gravesend, DA12 4EL
M6 MBB M. Bennett-Blacklock, 46 Friern Road, London, SE22 0AX
MW6 MBC S. Cook, 114 Caerphilly Road, Bassaleg, Newport, NP10 8LJ
M6 MBD M. Bowell, 28 Jubilee Close, Byfield, Daventry, NN11 6UZ
M6 MBF M. Bailey, 34 Jephson Drive, Birmingham, B26 2HW
M6 MBG C. Collins, The Coppice, Old Coach Road, Sheffield, S6 6HX
M6 MBH M. Creedy, 25 Ryton Close, Redditch, B98 0EW
M6 MBI M. Collis, 35 Fishergate, Norwich, NR3 1SE
M6 MBK J. Seaman, 6 Gibbets, Hale Road, Thetford, IP25 7QX
M6 MBO M. Siddall, 10 Foston Drive, Chesterfield, S40 4SJ
M6 MBP N. Challis, 48 Brunsfield Close, Wirral, CH46 6HE
M6 MBQ M. Ball, 22 Wheatley Drive, Mirfield, WF14 8NW
M6 MBR M. Burr, 49 Knightsbridge Way, Hemel Hempstead, HP2 5ES
M6 MBS M. Smith, 78 New Croft, Weedon, Northampton, NN7 4RL
M6 MBU M. Burnett, 218 High Street, Clapham, Bedford, MK41 6BS
M6 MBX M. Bell, 68 Hereford Drive, Liverpool, L30 1PR
M6 MBY M. Cox, 120 Helmsdale, Bracknell, RG12 0TB
MW6 MBZ M. Argyle, 17 Heol Cae-Rhys, Cardiff, CF14 6AN
MM6 MCA J. Mcardle, 1 Queen Street, Hamilton, ML3 9JR
M6 MCB M. Cooper, 6 The Crescent, Cookley, Kidderminster, DY10 3RY
M6 MCC S. Allaker, 61 West Street, Winterton, Scunthorpe, DN15 9QG

M6 MCE G. Mcewen, 37 Malvern Way, Twyford, Reading, RG10 9PY
MI6 MCF J. Macfarlane, 1 Main Street, Uttony, Magheraveely, Enniskillen, BT92 6NB
M6 MCH M. Hill, 10 The Moorings, Littlehampton, BN17 6RG
M6 MCJ M. Coiley, 25 Spring Garden Street, Queensbury, Bradford, BD13 2AE
MI6 MCK S. Mckay, 27 Rathbeg Crescent, Limavady, BT49 0AT
M6 MCL M. Chaffey, 46 Bartlett Way, Poole, BH12 4FD
M6 MCM S. Mcmurtrie, 5 Hill Road, Carshalton, SM5 3RA
M6 MCO M. Denham, 2 Shorts Corner, Frithville, Boston, PE22 7EA
M6 MCP M. Pennington, 88 Gillsmans Hill, St Leonards-on-Sea, TN38 0SL
M6 MCR K. Wills, 24 Bitten Court, Northampton, NN3 8HH
M6 MCS M. Statham, Broad Oak Bungalow, Manston, Sturminster Newton, DT10 1EZ
MM6 MCT J. Leitch, 25 Lime Street, Grangemouth, FK3 8LZ
M6 MCU M. Barker, 18 Nickleby Road, Waterlooville, PO8 0RH
M6 MCW M. Wilson, 11A St. Julians Road, London, NW6 7LA
MM6 MCX B. Bannister, 28 Mansefield Crescent, Old Kilpatrick, Glasgow, G60 5JJ
M6 MCY M. Attree, 52 The Ridgeway, St Albans, AL4 9PS
M6 MCZ A. Davis, Old Malt Kiln House, Barden, Leyburn, DL8 5JS
M6 MDB T. Brown, 69 Lawn Closes Alt, Oldham, OL8 2HB
M6 MDC D. Smith, 193 Brooke Road, Oakham, LE15 6HQ
M6 MDG D. Green, 45 Highthorn Road, Kilnhurst, Mexborough, S64 5UP
M6 MDJ D. Jefferson, 74 Cloisters Avenue, Barrow-in-Furness, LA13 0BB
M6 MDL M. Luttrell, 116 Starkey Street, Heywood, OL10 4JH
M6 MDM S. Clarke, 27 Netherhouse Moor, Church Crookham, Fleet, GU51 5TZ
M6 MDN M. Norman, 28 Cumberland Close, Twickenham, TW1 1RS
M6 MDR M. Riley, 16 Dudley Avenue, Leicester, LE5 2EE
M6 MDT M. Taylor, 11 Holly Crescent, Sunnyside, Rotherham, S66 3PL
M6 MDU M. Duchar, 13 Thirlmere Avenue, Chester le Street, DH2 3ED
M6 MDX M. Abraham, 12 Graham Road, Halesowen, B62 8LJ
M6 MDZ M. Smith, 31 Atlantic Crescent, Sheffield, S8 7FW
M6 MEA M. Atfield, 42 Pauls Croft Cricklade, Swindon, SN6 6AJ
M6 MEB P. Shires, 30 Philip Garth, Wakefield, WF1 2LS
M6 MEC M. Carroll, 11 Old Hall Court, Old Hall Street, Malpas, SY14 8NE
M6 MED R. Muswell, 7 Stoneyfields Gardens, Edgware, HA8 9SP
M6 MEJ M. Bray, 13 Rosebay Close, Hartlepool, TS26 0ZL
M6 MEK A. Walters, 32 Lincoln Road, Tuxford, Newark, NG22 0HP
M6 MEL M. Lewis, 73 Addenbrooke Street, Wednesbury, WS108HJ
M6 MEN M. Lucas, 8 Greenwood Close, Bury, Ramsey, Huntingdon, PE26 2NZ
M6 MEO B. Clements, 129 Gunners Road, Southend-on-Sea, SS3 9SB
M6 MEP M. Lawton, 20 Wharfedale Walk, Stoke-on-Trent, ST3 2RS
M6 MEQ A. Riley, 35 Ross Avenue, Wirral, CH46 2SA
M6 MES J. Reeve, 5 Antelope Avenue, Grays, RM16 6QT
M6 MEU A. Sharif, 6 Buckle Rise, Seaford, BN25 2QN
M6 MEV M. Sanderson, 20 East View, Castleford, WF10 1PZ
MW6 MFB J. Murray, 17 Bro Dawel, Bodedern, Holyhead, LL65 3TB
M6 MFC C. Parkes, 3 Greenham Close, Middlesbrough, TS3 9NT
M6 MFD G. Mansfield, 2 School Street, Syston, Leicester, LE7 1HN
M6 MFF M. File, Flat 1, 4 Priory Courtyard, Ramsgate, CT11 9PW
M6 MFG M. Gainza, Stanhope, High Street, Saxmundham, IP17 3EP
M6 MFJ M. Coleman, 3 Tummon Road, Sheffield, S2 5FD
M6 MFK A. Powell, 37 Newnham Close, Mildenhall, Bury St Edmunds, IP28 7PD
M6 MFL M. Ross, 31 Duke Street, Oswaldtwistle, Accrington, BB5 3PN
M6 MFM M. Maheshwarappa, R43 Room 2 International House, University Of Surrey, Guildford, GU2 7JL
MW6 MFN A. Evans, Maesyronnen, Sarnau, Llanymynech, SY22 6QL
M6 MFO Lt. Col. M. Foster, 21 The Bourtons, Newton Road, Totnes, TQ9 6LS
MI6 MFR A. Morrow, 769 Farranseer Park, Macosquin, Coleraine, BT51 4NB
M6 MFS S. Finlayson, 41 Low Catton Road, Stamford Bridge, York, YO41 1DZ
M6 MFU K. Ledson, 202 Brodick Drive, Bolton, BL2 6UE
M6 MFZ M. Fitzgerald, Flat 35, Winterton House, London, E1 2QR
M6 MGA N. Valvona, 63 Vale Road, Ash Vale, Aldershot, GU12 5HR
MW6 MGC M. Carwardine, Buttington Lodge, Sedbury, Chepstow, NP16 7EX
M6 MGD L. Addison, 45 Fir Terraces Esh Winning, Durham, DH7 9JQ
M6 MGF K. Holman, 39 Trellech Court, Yeovil, BA21 3TE
M6 MGG R. Wenlock, 8 Dinchope Drive, Telford, TF3 2ES
M6 MGH M. Hopewell, 4 Cotes Crescent Bicton Heath, Shrewsbury, SY3 5AS
M6 MGJ B. Hardy, 10 Spring Farm Road, Burton-on-Trent, DE15 9BN
MW6 MGM M. Margetts, Central House, Llanfechain, SY22 6UJ
M6 MGN N. Cook, 210 Cemetery Road, Wath-Upon-Dearne, Rotherham, S63 6HZ
M6 MGO M. Wenlock, 3 Kennels Cottages, Hall Lane, Stone, ST15 0RD
MD6 MGP A. Breen, 1 Snugborough Close, Union Mills, Isle of Man, IM4 4NZ
M6 MGR M. Reeks, 33 Madresfield Village, Madresfield, Malvern, WR13 5AA
M6 MGT J. Dickenson, 2 Kirkleys Avenue North, Spondon, Derby, DE21 7FX
M6 MGU M. Golding, 11 Southwold Crescent, Broughton, Milton Keynes, MK10 7BW
M6 MGV M. Pike, 21 Watersmeet Close, Guildford, GU4 7NQ
M6 MGW M. Walker, 50 College Grove Road, Wakefield, WF1 3RL
M6 MGX M. Gillard, 66 West End Rd, Bradninch, ex5 4qp
MW6 MGY M. Gray, 15 The Circle, Cwmbran, NP44 7JP
MW6 MGZ M. Bannister, 45 Queens Drive, Llantwit Fardre, Pontypridd, CF38 2NT
M6 MHA M. Hersom, Room 303, 95-98 Talbot Street, Dublin, DO1 WR94
M6 MHD M. Jodrell, 2 Charlesworth Street, Crewe, CW1 4DE
M6 MHE T. Harvey, 12 Woodkirk Avenue, Tingley, Wakefield, WF3 1JL
MI6 MHI A. Menzies, 21 Woodview Park, Tandragee, Craigavon, BT62 2DD
M6 MHJ M. Jackson, 64 Main Road Moulton, Northwich, CW9 8PB
MI6 MHK C. Moonie, 21 Woodview Park, Tandragee, Craigavon, BT62 2DD
M6 MHL M. Lacey, 82 Bowerings Road, Bridgwater, TA6 6HF
M6 MHM M. Hashim, 1 Cheylesmore Drive, Frimley, Camberley, GU16 9BL
MM6 MHN J. Mulhern, 10 Fisher Court, Knockentiber, Kilmarnock, KA2 0DS
M6 MHO M. Hossell, 80 Murray Road, Sheffield, S11 7GG
M6 MHQ M. Rea, 15 Wensleydale Close, Royton, Oldham, OL2 5TQ
M6 MHU M. Humphries, 5 Coppice Mead, Stotfold, Hitchin, SG5 4JX
M6 MHV M. Clarke, 54 Stafford Grove, Shenley Church End, Milton Keynes, MK5 6AZ
M6 MHW M. Hall, 259 Lambourn Drive, Allestree, Derby, DE22 2ur
M6 MHY M. Williams, 37 Clarendon Road, Weston-Super-Mare, BS23 3EE
M6 MIA M. Andrews, 286 Huddersfield Road, Mirfield, WF14 9PY
MI6 MIB P. Cobain, 15 Drumfad Avenue, Millisle, Newtownards, BT22 2GS
M6 MIC M. Taylor, 24 Crowley Lane, Oldham, OL4 2PN
M6 MID I. Shiradski, 69 Masefield Avenue, Borehamwood, WD6 2HG
M6 MIE M. Johnson, 3 Bersted Mews Bersted Street, Bognor Regis, PO22 9RR
M6 MIF M. Vaughan, 2 Hendingham Close, Gloucester, GL4 0XS
M6 MIG M. Greenwood, Stone Fell Gate, Swallow Hill, Distington, Workington, CA14 4PR
MI6 MIH J. Mercer, 32 Templemore Avenue, Belfast, BT5 4FT
M6 MII M. Bell, 2 Fox Close, Dunton, Biggleswade, SG18 8RF
M6 MIK M. Harris, 27 Great Braitch Lane, Hatfield, AL10 9FD
M6 MIL J. Milbourne, 102 Bells Marsh Road, Gorleston, Great Yarmouth, NR31 6PR
M6 MIN D. Mitchell, Flat 2, Weavers Court, 51 Unwin Street, Sheffield, S36 6EH
M6 MIO F. Miocinovic, 14 Huxloe Rise, Northampton, NN3 8YA
M6 MIP M. Payne, 14 Linnell Road, Rugby, CV21 4AN
MW6 MIQ V. Mckendley, 9 Mary Street, Aberdare, CF44 7NF
M6 MIR H. Mir, 13-15 Wain Street, Stoke-on-Trent, ST6 4ES
MM6 MIS J. Marsh, 8 Hazelton Way, Broughty Ferry, Dundee, DD5 3BT
M6 MIT M. Tarling, 16 Cross Walk, Bristol, BS14 0RX
M6 MIU J. Marsh, 14 Eyam Road, Hazel Grove, Stockport, SK7 6HP
M6 MIV B. Davies, 60 Queensway, Blackburn, BB2 4QT
M6 MIY P. Billingham, 393 Landseer Road, Ipswich, IP3 9LT
M6 MIZ M. Kwiatkowski-Zelazny, 56 York Road, Hove, BN3 1DL
M6 MJA M. Austwick, 6 Worlaby Road, Grimsby, DN33 3JY
MM6 MJC M. Clifford, Bridgeton Castle, St. Cyrus, Montrose, DD10 0DN
M6 MJD P. Smith, 67 Gipsy Lane, Old Whittington, Chesterfield, S41 9JD
M6 MJF M. Fysh, 3 Jeffrey Close, Kings Lynn, PE30 2HX
MM6 MJG M. Gilbert, 3 Hoymansquoy, Stromness, KW16 3DR
M6 MJI M. Greensmith, 14 Fountain Road, Draycott-In-The-Clay, Ashbourne, DE6 5HP
M6 MJL M. Lawrance, 17 Wren Crescent, Scartho Top, Grimsby, DN33 3RA
M6 MJM M. Mccormack, Flat 13, 29 Stoneygate Road, Leicester, LE2 2AE
M6 MJN M. Neale, 41 Langford Road, Weston-Super-Mare, BS23 3PQ
M6 MJP S. Parker, 36 Eton Close, Lincoln, LN6 0YF
MM6 MJR M. Robertson, Tigh Jenny, Strath, Gairloch, IV21 2BX
M6 MJS S. Shields, 9 Berrington Drive, Newcastle upon Tyne, NE5 4BG
M6 MJV D. Morley, The Old Mill, Mill Lane, Loughborough, LE127UX
MM6 MJY M. Yarrow, Lomond Villa, Downies Village, Aberdeen, AB12 4QX
MW6 MJZ M. Shepley, 16 Heulwen Close, Hope, Wrexham, LL12 9PR
M6 MKB M. Buchanan, 36 Church Lane, Manby, Louth, LN11 8HL
M6 MKD D. Bell-Stephens, 6 Woodland Road, Warminster, BA12 8HJ
M6 MKE M. Gregory, 65 Nursery Crescent North Anston, Sheffield, S25 4BR
M6 MKH A. Freeth, 121 Highfield Road, Burntwood, WS7 9DA
M6 MKJ K. Juszczak, 68 College Road, Sandy, SG19 1RH
M6 MKK M. Kendall, 53 Ellerker Rise, Willerby, Hull, HU10 6EU
M6 MKM J. Mckie, 59 Leaholme Terrace, Blackhall Colliery, Hartlepool, TS27 4AB
M6 MKN N. Driscoll, 42 Adelaide Square, Shoreham-by-Sea, BN43 6LN
M6 MKO M. Kent, 7 Lockyers Drive, Ferndown, BH22 8AJ
M6 MKU D. Mcdonald, 29 Highfield Crescent, Rayleigh, SS6 8JP
M6 MKV M. Vardy, 60 Hucklow Avenue, North Wingfield, Chesterfield, S42 5PU
M6 MKW M. Wharton, Sea Cadets, Riverside Road, Great Yarmouth, NR31 6PX
M6 MKX S. Mcguckian, 71 Heathfield Drive, Tyldesley, Manchester, M29 8PJ
M6 MKY M. King, Flat 6, Derwent Court, Solihull, B92 7BU
M6 MLA M. Lovatt, 3 Withington Close, Atherton, Manchester, M46 0EZ
M6 MLE D. Pilkington, 197 Saltings Road, Snodland, ME6 5HP
M6 MLF J. Kiely, 35 Chestnut Avenue, Todmorden, OL14 5PH
M6 MLG P. Arnold, 25 Arliston Drive, Woodville, Swadlincote, DE11 8FS
M6 MLH M. Hoyland, 3 Telford Street, Barrow-in-Furness, LA14 2ER
M6 MLI R. Parker, 58 Bryncastell, Bow Street, SY24 5DF
M6 MLK M. Lake, 64 Womersley Road, Norwich, NR1 4QB
M6 MLL D. Neumunn, 92 Miner Street, Walsall, WS2 8QL
M6 MLM M. Milano, 35 Orion Road, Rochester, ME1 2UL
M6 MLN A. Mcdermid, 49 Jubilee Street, Irthlingborough, Wellingborough, NN9 5RL
M6 MLO M. Broyd, 16 George Downing House, Miles Mitchell Avenue, Plymouth, PL6 5XJ
M6 MLP M. Le-Petit, 3 Stanley Drive, Hatfield, AL10 8XX
M6 MLQ M. Byard, 1 Fieldside, Long Wittenham, Abingdon, OX14 4QB
M6 MLR L. Rolt, 40 Water Meadow Drive, Bradford, BD13 4EX
MM6 MLT P. Connon, 4 Highfield Court, Stonehaven, AB39 2PL
M6 MLU P. Rath, 60 Elstree Road, Bushey Heath, Hertfordshire, WD23 4GL
M6 MLV D. Malcolm, 66 Bracken Bank Grove, Keighley, BD22 7AU
M6 MLX M. Pacitti-Lamb, 41 Cowell Grove, Highfield, Rowlands Gill, NE39 2JQ
M6 MLY M. Livesey, 24 St. Marys Road, Bamber Bridge, Preston, PR5 6TD
M6 MMB M. Parkes, 2 Woodhouse Mount, Normanton, WF6 1BN
M6 MMC Dr M. Mcintyre, 18 Norlands Crescent, Chislehurst, BR7 5RN
MM6 MMG D. Anderson, Dail Darach, Monydrain Road, Lochgilphead, PA31 8LG
M6 MMH M. Houghton, 18 Leopold Way, Blackburn, BB2 3UE
M6 MMI R. Hewson, 10 Miriam Grove, Leigh, WN7 3EX
MM6 MML J. Lucas, 7 Rysland Avenue, Newton Mearns, Glasgow, G77 6EA
M6 MMM M. Hunter, 126 Turner Street, Stoke-on-Trent, ST1 2NE
M6 MMN M. Newbury, 2 Rowan Close, Clacton-on-Sea, CO15 2DB
M6 MMP M. Chu, Harrogate Ladies' College, Clarence Drive, Harrogate, HG1 2QG
M6 MMQ M. Majhail, 3 Poynders Hill, Hemel Hempstead, HP2 4PQ
M6 MMR S. Wellsted, 127 Goldthorn Hill, Wolverhampton, WV2 4PS
M6 MMS M. Strange, 101 Southbroom Road, Devizes, SN10 1LY
MI6 MMT M. Torley, 4 Yew Tree Park, Newry, BT34 2QP
MW6 MMU G. Edwards, 54 Old Street, Tonypandy, CF40 2AF
M6 MMX A. Iggulden, 78 Wrensfield Road, Stockton-on-Tees, TS19 0BD
M6 MMY M. Barber, 3 Baxter Road, Sunderland, SR5 4LH
M6 MNC A. Cockburn, 20 Hexham Avenue, Hebburn, NE31 2HN
M6 MND D. Rogers, 21 Belmont Park, Pensilva, Liskeard, PL14 5QT
MM6 MNE C. Macnee, 7 Church Street Chapelton, Strathaven, ML10 6SD
M6 MNG N. Giuliano, 13 Walton Drive, Littleover, Derby, DE23 1GN
M6 MNH M. Hunt, 37 Shortlands Avenue, Ongar, CM5 0BL
M6 MNI M. Walton, 38 Wingate Road, Grimsby, DN37 9DU
M6 MNK P. Roberts, 17 Cannon Hill, Prenton, CH43 4XR
MI6 MNL J. Johnston, 19 Killowen Grange, Lisburn, BT28 3HQ
M6 MNO A. Hood, 109 Trotters Field, Braintree, CM7 3NW
M6 MNP M. Killoran, 4 Victoria Road, Pudsey Leeds, LS28 7SR
MM6 MNQ R. Boan, 6 Philip Avenue, Newton Stewart, DG8 6HF
M6 MNS T. Hodson, 25 The Rise, Amersham, HP7 9AG
M6 MNT S. Hamer, Flat 5, 19 Frimley Road, Camberley, GU15 3EN
M6 MNU J. Williams, 1 Lower Meadow Drive, Congleton, CW12 4UX

M6 MNV F. Brunt, 74 Bardley Crescent, Tarbock Green, Prescot, L35 1RJ
MM6 MNW D. Ryan, 89 The Braes, Saltcoats, KA21 5EP
M6 MNX M. Norfolk, 185 Heath Road, Leighton Buzzard, LU7 3AD
M6 MNZ M. Bartlett, 34 Yarrow Drive, Birmingham, B38 9QR
M6 MOB P. Hawes, 6 Robert Street, Sunderland, SR4 6EY
M6 MOC T. Forss, Lower Conghurst Oast, Conghurst Lane, Cranbrook, TN18 4RW
M6 MOD M. O'Driscoll, 17 Petherton Gardens, Bristol, BS14 9BT
M6 MOF C. Moffitt, 5 Foxton Terrace, Horstead Avenue, Brigg, DN20 8QR
M6 MOG N. Saville, 4 Shannon Court, Downs Barn, Milton Keynes, MK14 7PP
M6 MOH M. Hinds, 10 Lustrells Close, Saltdean, Brighton, BN2 8AS
M6 MOI S. Lewis, 58 Ocean Close, Fareham, PO15 6QP
M6 MOJ A. Mckie, 5 Greenway Northenden, Manchester, M22 4LW
M6 MOK M. Lowin, 1A Burnside Avenue, Blackpool, FY4 4AF
M6 MOM J. Clare, Kimberley, Boston Road, Bicker, Boston, PE20 3AP
M6 MON P. Montgomery, 14 Saxon Crescent, Horsham, RH12 2HU
M6 MOP D. Pomeroy, 73 Pinewood Gardens, North Cove, Beccles, NR34 7PG
M6 MOQ K. Hamilton, 73 Leven Road, Stockton-on-Tees, TS20 1DB
M6 MOS M. Steele, 10 Green Lane, Houghton, Carlisle, CA3 0NT
M6 MOU D. Dormer, 69 Favell Drive, Furzton, Milton Keynes, MK4 1AX
M6 MOV C. Moverly, 17 Trefoil Road, Hailsham, bn27 4FR
M6 MOW J. Horry, 5 Donington Road, Bicker, Boston, PE20 3EF
M6 MOX M. Cox, 3 Cromwell Road, Hertford, SG13 7DP
MM6 MOY P. Mcdonald, 9 Partan Skelly Way, Cove Bay, Aberdeen, AB12 3PH
M6 MOZ M. Meadowcroft, 8 Lamlash Road, Blackburn, BB1 2AS
M6 MPB M. Bridger, 11, Beecham Close, Newcastle upon Tyne, NE15 6LG
M6 MPD R. Yarrow, 27 Staplers Road, Newport, PO30 2DB
M6 MPE M. Evans, 48 Paddock Lane, Aldridge, Walsall, WS9 0BP
M6 MPF J. Dunn, 3 Hobbs Way, Rustington, Littlehampton, BN16 2QU
MI6 MPH J. Martin, 22 Lisbane Road, Saintfield, Ballynahinch, BT24 7BS
M6 MPK J. Garwood, 4 Ryedale, Carlton Colville, Lowestoft, NR33 8TB
M6 MPL G. Clark, 65 Chyvelah Vale, Gloweth, Truro, TR1 3YJ
M6 MPM M. Mccall, 1 Jaunty Road, Sheffield, S12 3DT
M6 MPO M. Woolger, 25 Rookwood Park, Horsham, RH12 1UB
M6 MPP M. Michalowski, 39 Towan Avenue, Fishermead, Milton Keynes, MK6 2DS
MW6 MPQ P. Harris, 123 St. Georges Court, Tredegar, NP22 3DD
M6 MPR P. Rodgers, 25 Eagle Avenue, Barnsley, S75 1FE
M6 MPS S. O'Riordan, 46 Grange Road, London, HA20LW
M6 MPT M. Thompson, 35 Princes Avenue, Desborough, Kettering, NN14 2RQ
M6 MPV M. Varley, 50 Gorse Valley Road, Hasland, Chesterfield, S41 0JP
M6 MPW M. Whotton, 7 Heatherdale, Ibstock, LE67 6JU
M6 MPX M. Phillips, The Well House, Eastbury, Hungerford, RG17 7JL
M6 MPY R. Greenwood, The Oast, Hazel Street Farm, Spelmonden Road, Tonbridge, TN12 8EF
M6 MPZ M. Edmonds, 20 Tomline Road, Ipswich, IP3 8BZ
M6 MQB M. Bailey, 1 Oriel Drive, Glastonbury, BA6 9PA
M6 MQC M. Le Moine, 115 Rothesay Road, Blackburn, BB1 2ER
M6 MQD M. Oliver, 14 Ash Road, Ashurst, Southampton, SO40 7AT
M6 MQE M. Ayres, 32 Kinterbury Close, Hartlepool, TS25 1GQ
MI6 MQF D. Mulligan, 10 Seaview, Ardglass, Downpatrick, BT30 7SQ
M6 MQH A. Jovanovic, 33 Seward Road, London, W7 2JS
M6 MQJ M. Trathen, Gweal An Mayn Cottage, Newtown, St. Martin, Helston, TR12 6DP
M6 MQK D. Ferguson, 39 Fouracres, Maghull, Liverpool, l317bp
M6 MQL M. Boyle, 64 Spencerfield Crescent, Middlesbrough, TS3 9HD
M6 MQM M. Tozer, 4 The Grange Dousland, Yelverton, Pl206nn
M6 MQN J. Milner, 30 Rowena Drive, Thurcroft, Rotherham, S66 9HT
MM6 MQO S. Doonan, West Clanfin Farm Waterside, Kilmarnock, KA3 6JQ
M6 MQP P. Marlow, 59 Kinross Crescent, Beechdale, Nottingham, NG8 3FT
MM6 MQR Dr H. Erwood, Lunna House, Lunna Vidlin, Vidlin, ZE2 9QF
M6 MQT T. Mloduchowski, Flat 4, Gwynne House, London, E1 2AG
MW6 MQU R. Thomas, 49 Viscount Evan Drive, Newport, NP10 8HJ
MM6 MQV D. Robertson, Grindhus, Gonfirth, Voe, ZE2 9PY
M6 MQW J. Evans, Davies Lines 16A, Raf Cosford, Albrighton, Wolverhampton, WV7 3EX
MI6 MQX M. Elliott, 17 Milebush Road, Dromore, BT25 1RT
M6 MQY L. Andrews, 72 Grange Road, Alresford, SO24 9HF
M6 MRC M. Bridges, 176 Coombe Valley Road, Dover, CT17 0HE
M6 MRD P. Sephton, 11 Moss Avenue, Leigh, WN7 2HH
M6 MRG H. Martin, 27 Gordon Road, Fleetwood, FY7 6UE
M6 MRH M. Hayward, 8 Kiln Shaw, Langdon Hills, Basildon, SS16 6LE
MI6 MRI F. Rafferty, 96 Collinward Avenue, Newtownabbey, BT36 6DZ
MI6 MRJ J. Martin, 23 Winters Gardens, Omagh, BT79 0DZ
M6 MRK M. Mckenna, 54 Whickham Road, Hebburn, NE31 1QU
M6 MRM M. Barnfather, 2 Brazenhill Lane, Haughton, Stafford, ST18 9HS
M6 MRN P. Fisher, 26 Rydal Avenue, Barrow-in-Furness, LA14 4NW
M6 MRP P. Boxx, 19 Glenthorpe House, 54 West Stevenson Street, South Shields, NE33 4DL
M6 MRR M. Rutter-Dacosta, 144 Bellingdon Road, Chesham, HP5 2HF
M6 MRS C. Davies, 68 Wood Street, Castleford, WF10 1LN
M6 MRT A. Blamires, 2 Foldings Grove, Scholes, Cleckheaton, BD19 6DQ
M6 MRU A. Owen, 5 Kirkham Court, Goole, DN14 6JU
M6 MRV R. Martin, 21 Lonsdale Crescent, Dartford, DA2 6LQ
M6 MRW M. Willison, 8 Summervale Mews Wharf Lane, Ilminster, TA19 0BA
M6 MRX R. Brough, 36 Salstar Close, Aston, Birmingham, B6 4PP
M6 MRY B. Murray, 23 Tillotson Close, Crawley, RH10 7WQ
MM6 MSA M. Saddler, 12 Carnbee End, Edinburgh, EH16 6GJ
M6 MSC M. Colman, 4 Northmead Drive, North Walsham, NR28 0AU
MW6 MSE M. Roblin, 6 Gethin Street, Briton Ferry, Neath, SA11 2LU
M6 MSF M. Farnham, 94 Rochester Way, Crowborough, TN6 2DU
MI6 MSG G. Graham, Flat J 86 Sunningdale Gardens, Belfast, BT14 6SL
MM6 MSH M. Hatfull, Flat 2 1 Northumberland Place Lane, Edinburgh, EH3 6LD
M6 MSI M. Scott, 25 Smithburn Road, Gateshead, NE10 9DT
M6 MSJ C. Gibson, 17 Clyde Court, Grantham, NG31 7RB
M6 MSM M. Smith, 24 Fifth Avenue, Portsmouth, PO6 3PE
MW6 MSN P. Evans, Flat 1 Red Cow Annex, Lloyds Terrace, Adpar, Newcastle Emlyn, SA38 9EH
M6 MSO S. Marsh, 6 Sparsholt Road, Southampton, SO19 9NJ
M6 MSP S. Palmer, Sawrey Ground, Crosby, Maryport, CA15 6SH
MI6 MSR M. Ruddy, 34 Glenveagh Hilltown, Newry, BT34 5US
M6 MSS M. Smith, The Lawns Tylers Green, High Wycombe, HP108BH
M6 MST M. Streeter, Fairway, West Chiltington Road, Pulborough, RH20 2EE
M6 MSU S. Hodder, 19 Kingsclere, Huntington, York, YO32 9SF
MI6 MSV M. Steele, 134 Knock Road, Dervock, Ballymoney, BT53 8AB
M6 MSX M. Street, Flat 6, Derwent Court, Solihull, B92 7BU
M6 MSY M. Saysell, Gordano Valley Riding Centre, Moor Lane, Bristol, BS20 7RF
M6 MTA M. Ansari, 37 Lizmans Court, Silkdale Close, Oxford, OX4 2HF
M6 MTC D. Godfrey, 25 Bosworth Crescent, Romford, RM3 8JZ
M6 MTD P. Williams, 4 Red Gables, Shap, Penrith, CA10 3NL
MW6 MTE E. Thomas, 1 Cambrian Gardens Y Drenewydd, Newtown, SY16 2AW
MW6 MTG M. Blomfield, 99 Mountain Road, Upper Brynamman, Ammanford, SA18 1AN
M6 MTH M. Horton, 8 Liptraps Lane, Tunbridge Wells, TN2 3BS
M6 MTI M. Ilsley, 38 Coleridge Road, Ottery St Mary, EX11 1TD
M6 MTJ M. Jones, 18 Cleveleys Avenue, Heald Green, Cheadle, SK8 3RH
M6 MTK A. Morris, 4 Saunders Close, Uckfield, TN22 2BX
M6 MTL D. Baldwin, 19 Bramble Grove, Wigan, WN5 9PR
M6 MTM M. Tyler-Moore, 34 Guildford Road, Horsham, RH121LS
M6 MTN M. Banks, 37 Havelock Road, Southsea, PO5 1RU
MI6 MTO T. Mehaffey, 33 Lenaderg Road, Banbridge, BT32 4PT
M6 MTQ B. Turnbull, Brooklet Road, Heswall, CH601UL
M6 MTR P. March, 46 Christchurch Road, Tilbury, RM18 8XP
M6 MTS M. Smith, 5 Newland Avenue, Stafford, ST16 1NL
M6 MTT P. Crosweller, Flat 3, 18 Pelham Road, Seaford, BN25 1ES
M6 MTU D. Atkins, 32 Braybrook, Orton Goldhay, Peterborough, PE2 5SH
M6 MTV P. Bailey, 4 Roving Bridge Rise Clifton, Swinton, Manchester, M27 8AF
MW6 MTW M. Ellis, Heddwch Brithdir, Dolgellau, LL40 2SF
M6 MTZ R. Lenicker, 136 Nab Wood Drive, Shipley, BD18 4EW
MM6 MUA M. Mulligan, 18 Camperdown Road, Nairn, IV12 5AR
M6 MUB M. Masters, 49 St. Johns Avenue, Bridlington, YO16 4ND
M6 MUD C. Brock, 30 Cromer Road, Norwich, NR6 6LZ
M6 MUF W. Burridge, 20 Archer Close, Kingston upon Thames, KT2 5NE
MM6 MUH C. Fenton, 2 Rothiebrisbane Cottages Fyvie, Turriff, AB53 8LE
M6 MUJ I. Fores, 27 Southfield Lane, Whitwell, Worksop, S80 4NS
M6 MUK M. King, 7 Battismore Road, Morecambe, LA4 4QG
MM6 MUO A. Flett, Southview, Lein Road Kingston, Fochabers, IV32 7NW
M6 MUP C. Meredith, 6 North End, Shortstown, Bedford, MK42 0XB
MM6 MUR G. Murray, The Barn House, Springfield Farm, Carluke, ML8 4QZ
M6 MUS A. Sutton, 3 Grotes Buildings, London, SE3 0QG
M6 MUT M. Northwood, 34 Whitehead Drive, Wellesbourne, Warwick, CV35 9PW
M6 MUY F. Boyce, 277 Manor View, Par, PL24 2EP
M6 MUZ M. Colpman, 20 Rochford Road, Basingstoke, RG21 7TQ
M6 MVA M. Johnson, 143 Swan Lane, Wickford, SS11 7DG
M6 MVB M. Bradley, 13 Elizabeth Avenue, Bilston, WV14 8EA
M6 MVD J. Dilworth, 808 Liverpool Road, Southport, PR8 3QF
M6 MVE W. Cartledge, 20 Ariel Street, Newcastle, Ne639ez
M6 MVF M. Findon, 64 Fur Tree Grove, Birmingham, B735UN
M6 MVH R. Barker, 6 Trenoweth Cressent, Penzance, TR184RY
M6 MVK J. Hart, 57 Brentleigh Way, Stoke-on-Trent, ST1 3GX
MM6 MVM V. Mcgowan, 112 Oronsay Avenue, Port Glasgow, PA14 6EF
M6 MVN S. Forshaw, 22A Barley Hall Street, Heywood, OL10 4DH
MM6 MVQ I. Learmonth, 14 Deansloch Terrace, Aberdeen, AB16 5SN
M6 MVT M. Bell, 7 Shiregreen Lane, Sheffield, S5 6AA
M6 MVV M. Mcarthur, 25 Lingfield Road, Edenbridge, TN8 5DS
M6 MVW M. Ward, 13 Hyde Lane, Danbury, CM3 4QT
MM6 MVX D. Wilson, Rivendell Lodge, Glenkindie, Alford, AB33 8RN
M6 MVZ D. Money, Flat 1, Lewin Court 24B Plumstead High Street, London, SE18 1SL
M6 MWA S. Hyland-Davis, 34 Melody Close, Warden, Sheerness, ME12 4PU
M6 MWB M. Bryant, 284 Brantingham Road Chorlton Cum Hardy, Manchester, M21 0QU
M6 MWC M. Curwen, 40 Grange Street, Morecambe, LA4 6BW
M6 MWD M. White, 27 Winstone Close, Redditch, B98 8JS
MM6 MWE L. Stoneham, Kirkton Farm Balblair, Dingwall, IV7 8LG
MM6 MWF M. Flynn, 15 Riselaw Crescent, Edinburgh, EH10 6HN
MW6 MWG J. Davies, 1 South View, Pontycymer, CF32 8LE
MW6 MWN W. Noble, 2 Harriet Town, Troedyrhiw, Merthyr Tydfil, CF48 4HJ
M6 MWP M. Poole, 22 Padstow Gardens, Leeds, LS10 4NQ
M6 MWQ C. Cromie, 140 Whalley Road Wilpshire, Blackburn, BB1 9LJ
MW6 MWS A. Sneddon, 3 Marigold Close, Gurnos, Merthyr Tydfil, CF47 9DA
M6 MWT M. Tolmie, 28 Spencer Road, Rendlesham, Woodbridge, IP12 2TJ
M6 MWV K. Beal, 22 Harpers Road, Newhaven, BN9 9RR
M6 MWW M. Watts, 47 Westbury Crescent, Weston-Super-Mare, BS23 4RF
M6 MWZ M. Way, 9 Railton Avenue, Stoke on Trent, ST34BN
M6 MXA A. O'Reilly, 3B Summerleys, Edlesborough, Dunstable, LU6 2HR
M6 MXB M. Boddy, 26 Tulip Tree Road, Bridgwater, TA6 4XD
M6 MXC M. Craven, 78 Connaught Road, Brookwood, Woking, GU24 0HF
M6 MXD S. Ldr. M. Dalziel, 4 Meadow Close, St Albans, AL4 9TG
M6 MXF L. Watson, 616 Beverley Drive, Stoke-on-Trent, ST2 0RE
M6 MXH M. Head, 9 Kinsbourne Grreen, Dunscroft, Doncaster, DN7 4Bl
MM6 MXK M. Rawlings, Brownstone, 2 Springfield Terrace, Alness, IV17 0SP
M6 MXM M. Meehan, 14 Grosvenor Road, Walton, Liverpool, L4 5RB
M6 MXO V. Adedeji, 3 Royal Troon Mews, Wakefield, WF1 4JL
M6 MXP D. Burrows, 3 Novello St, Maltby Maltby, Rotherham, S66 7QB
M6 MXR M. Russell, Cowmans Cottage Spring Lane, Flintham, Newark, NG23 5LB
MM6 MXU A. Mcbain, 9/12 Tower Place, Edinburgh, EH6 7BZ
MW6 MXV W. Cooper, 1 Bedw Street, Maesteg, CF34 0TF
M6 MXX C. Morrow, 194 Cedar Road, Doncaster, DN4 9ET
M6 MXY C. Simpson, 2 Bradley Road, Haslington, Cheshire. Cw1 5Pw, Crewe, CW1 5PW
MI6 MXZ M. Masterson, 2 Pinley Drive, Banbridge, BT32 3TZ
M6 MYB M. Bannister, 110 Cotswold Crescent, Billingham, TS23 2QB
M6 MYC M. Beddall, 11 Sinodun Road, Wallingford, OX10 8AD
M6 MYD A. Middleton, 2 Moor View, Godshill, Ventnor, PO38 3HW
MW6 MYE A. Twose, 4 Bradford Street, Caerphilly, CF83 1GA
M6 MYF L. Thompson, 33 Dalton Crescent, Shildon Co Durham, DL4 2LE
M6 MYG M. Young, 72 Goddard Way, Saffron Walden, CB10 2EB
M6 MYH S. Elliott, 79 Somerton Road, Bolton, BL2 6LN

M6 MYI O. Makhura, H-1-6-S, The Maltings Haven Road, Colchester, CO2 8FU
MM6 MYK M. Cheetham, Plowvent, Muir Of Fowlis, Alford, AB33 8NX
M6 MYL J. Swann, 5 Lanark Close, Hazel Grove, Stockport, SK7 4RU
M6 MYM M. Beniston, Scowbuds, Camborne, TR14 8PB
M6 MYN R. Eaton, 31 Pinfold Lane Ruskington, Sleaford, NG34 9EU
M6 MYR M. Moss, 27 Dunn Side, Chelmsford, CM1 1DL
M6 MYS A. Smart, 7 Hinton Grove, Hyde, SK14 5ST
M6 MYT B. Hilton, 17 Bellwood, Westhoughton, Bolton, BL5 2RT
M6 MYU J. Othen, 149 Blumfield Crescent, Slough, SL1 6NN
M6 MYV R. Flux, 18 Ashdene Road, Ashurst, so40 7dp
MI6 MYW M. Mcwilliams, 84 Syerla Road, Dungannon, BT71 7ET
M6 MYY M. Ffrench, 24 Monkwood Road, Chesterfield, S41 8DG
M6 MZB M. Bauer, Flat 21, 5 Queensland Road, London, N7 7FE
M6 MZC G. Fox, 23 The Driveway, Canvey Island, SS8 0AB
M6 MZD B. Greenwood, 13 Mayflower Street, Blackburn, BB2 2RX
MM6 MZE C. Menzies, 0/2 47 Wellshot Road, Glasgow, G32 7XL
MM6 MZG K. Jones, 6 Traill Street Castletown, Thurso, KW14 8UG
M6 MZH M. Hubbard, 14 Parkfield Crescent, Kimpton, Hitchin, SG4 8EQ
M6 MZJ M. Juniper, 2 Cranbourne Drive, Hoddesdon, EN11 OQH
M6 MZL G. Johnson, 22 Beechwood Close, Blythe Bridge, Stoke-on-Trent, ST11 9RH
M6 MZN J. Forshaw, 22A Barley Hall Street, Heywood, OL10 4DH
M6 MZQ G. Peters, Curlew Court, Guys Head Road, Sutton Bridge, Spalding, PE12 9QQ
MI6 MZR M. Armstrong, 2 Thralcot Link, Carrickfergus, BT38 9RG
MW6 MZU T. Baldwin, Rose Cottage, High Street, Pontypool, NP4 6HE
M6 MZV A. Watson, 67 Alamein Drive, Romiley, Stockport, SK6 4JN
M6 MZX M. Pinter, Bercsenyi Str. 47, Totkomlos, Hungary, 5940
M6 MZY K. Phillips, 12 Copland Avenue, Minster On Sea, Sheerness, ME12 3PJ
M6 MZZ M. Driscoll, 14B Pretoria Road Hedge End, Southampton, SO30 0BS
MM6 NAA N. Mcdonald, 8 Newton Place, Perth, PH1 2QJ
MM6 NAB D. Bell, 82 Campbell Avenue, Stevenston, KA20 4BP
M6 NAC N. Carter, 25 Breachfield, Burghclere, Newbury, RG20 9HY
MM6 NAD N. Anderson, The Cedars, Church Street, Keith, AB55 4AR
MW6 NAG N. Berrall, 41 Nantgarw Road, Caerphilly, CF83 3FB
MM6 NAI A. Anderson, 18 Selkirk Street, Wishaw, ML2 8RA
M6 NAK N. Leech, 59 Lakeside Court, Brierley Hill, DY5 3RQ
M6 NAL N. Parry, 125 Lawsons Road, Thornton-Cleveleys, FY5 4PL
M6 NAM M. Cody, 139 Vicarage Road, Watford, WD18 0HA
M6 NAN Dr L. Alconcel, Top Lock Cottage, Stoke Pound Lane, Bromsgrove, B60 4LH
M6 NAO N. Griffiths, 67 Warstones Drive, Wolverhampton, WV4 4PF
M6 NAQ A. Cowley, 7 Harwood Road, Gosport, PO13 0TU
M6 NAS S. Nash, Ashmead, Hemel Hempstead, HP3 0BU
M6 NAT N. Jones, 5 Montgomery Crescent, Quarry Bank, Brierley Hill, DY5 2HB
M6 NAU N. Alders, 14 Forest Rise, Crowborough, TN6 2ES
M6 NAV J. Glicklich, 86 Ainsdale Road, Bolton, BL3 3ER
M6 NAW N. White, 5 Badgers Walk, Burgess Hill, RH15 0AE
MW6 NAX N. Jones, 7 Dyffryn, Burry Port, SA16 0TE
MW6 NAZ D. Tippett, 30 Berw Road, Tonypandy, CF40 2HD
M6 NBG J. Mccosh, The Mill House, Moorlands Road, Merriott, TA16 5NF
M6 NBH A. Barrett, 2 Friars Close, Clacton-on-Sea, CO15 4EU
MM6 NBI R. Mackay, 12 Robertson Square, Wick, KW1 5NF
M6 NBL Lady A. Mackendrick, Cecily Cottage, Brockhill, Wareham, BH20 7NH
M6 NBN R. Anderson, 26 Bowness Avenue, Warrington, WA2 9NQ
M6 NBO B. Vickers, 52 Edward Street, Grimsby, DN32 9HJ
M6 NBP N. Williams, 114 Essex Place Montague Street, Brighton, BN2 1LL
M6 NBQ Dr P. Fabrega, Flat 3, Post Office Court, Whitchurch, SY13 1QT
M6 NBS N. Barker, 17 Pippin Walk, Hardwick, Cambridge, CB23 7QD
MW6 NBU N. Bruetsch, The Firs, Penglais Road, Aberystwyth, SY23 2EU
M6 NBV D. Whitelock, 2 Shippards Road, Brighstone, Newport, PO30 4BG
M6 NBW N. Warden, 1 Forge House The Street, Earl Soham, Woodbridge, IP13 7RT
M6 NBX N. Cohen, 8 Henry Gepp Close, Adderbury, Banbury, OX17 3FE
M6 NBY N. Thompson, 10 Belgrave Close, Hersham, Walton-on-Thames, KT12 5PH
MW6 NCA N. Curry, 58 Greenfields, St. Martins, Oswestry, SY11 3AH
M6 NCB N. Bettridge, 37 Princess Avenue, Warsop, Mansfield, NG20 0PY
MI6 NCC N. Corbett, 10 Main Street, Rosslea, Enniskillen, BT92 7PP
M6 NCD D. Senior, 16 Cherry Tree Close, Billingshurst, RH14 9NG
M6 NCE C. Mclennan, 18 Loveridge Way, Eastleigh, SO50 9PW
M6 NCF N. Froude, 6 Park Road West, Chester, CH4 8BG
MI6 NCG N. Griffin, 327 Clonmeen, Drumgor, Craigavon, BT65 4AT
M6 NCI S. Ldr. B. Dowley, 120 Capel Street, Capel-Le-Ferne, Folkestone, CT18 7HB
M6 NCK N. Taylor, 212 Plantation Hill, Worksop, S81 0HD
M6 NCL C. Braddock, 22 Anncroft Road, Buxton, SK17 6UA
M6 NCM N. Mcveagh, 14 Chapel Close, Ravenstone, Coalville, LE67 2JT
M6 NCO K. Tonge, 25 Southcote Grove, Birmingham, B38 8ED
MM6 NCP N. Pollard, 30 Abbeyhill Crescent, Edinburgh, EH8 8DZ
M6 NCR G. Paton, 17 Blakeney Road, Stevenage, SG1 2LH
M6 NCU C. Johnson, 22 Carleton Close Great Yeldham, Halstead, CO9 4QJ
M6 NCV G. Killpack, 20 Fisher Close, Banbury, OX16 3ZW
M6 NCY J. Mooneapillay, 354 Upper Elmers End Road, Beckenham, BR3 3HG
M6 NDB N. Brown, 9 Devonshire Avenue, Wigston, LE18 4LP
M6 NDC N. Charlotte, 26 Nettleton Avenue, Mirfield, WF14 9AN
M6 NDE N. Evans, 46 Furzehill Road, Plymouth, PL4 7LA
M6 NDF D. Arnold, 15 Chiltern Drive, Verwood, BH31 6US
M6 NDG N. Graven, 33 Sheldrake Road, Broadheath, Altrincham, WA14 5LJ
M6 NDI A. Kaluarachchi, 103 Bentinck Road, Newcastle upon Tyne, NE4 6UX
M6 NDK C. Lucas, 15 Higher Moor, Ruan Minor, Helston, TR12 7JJ
M6 NDM N. Mason, 11 Chapel Court, Brierley Hill, DY5 2UT
M6 NDN K. Hutchens, 7 Lapwing Close, Thurston, Bury St Edmunds, IP31 3PW
M6 NDO A. Ashmore, 42 Holme Road, Chesterfield, S41 7JF
M6 NDP N. Plunkett, 11 Stoneleigh Gardens, Grappenhall, Warrington, WA4 3LE
M6 NDR N. Reeve, 4 Ash Grove, Swindon, SN2 1RX
M6 NDT R. Thatcher, 83 Westfield Drive, North Greetwell, Lincoln, LN2 4RE
M6 NDY M. Twitchen, 72 Finedon Road, Burton Latimer, Kettering, NN15 5QB
M6 NEA N. Asher, 17 Ashby Road, Cleethorpes, DN35 9PF
M6 NEC P. Bean, 14 St. Andrews Lane, Necton, Swaffham, PE37 8HY
MM6 NED E. Brophy, 44 Annieshill View, Plains, Airdrie, ML6 7NT
M6 NEE J. Todd, 20 Hexham Avenue, Hebburn, NE31 2HN
M6 NEF N. Chapman, 13 Clayton Grove, Bracknell, RG12 2PT
M6 NEG A. Brown, 19 Main Street Leconfield, Beverley, HU17 7NQ
M6 NEH Dr N. Hoare, 5 Kelsey Head, Port Solent, Portsmouth, PO6 4TA
M6 NEI N. Yorke, 30 Bramdene Avenue, Nuneaton, CV10 0DH
M6 NEJ N. Jones, 8 Regent Court Belvedere Close, Guildford, GU2 9GA
M6 NEL N. Price, 68 Powke Lane, Rowley Regis, B65 0AG
M6 NEM E. Chance, 33 Larkfield Avenue, Kirkby-In-Ashfield, Nottingham, NG17 9FE
M6 NEO C. Radford, 86 Flamstead Avenue Loscoe, Heanor, DE75 7RP
M6 NEQ J. Clarke, 24 Telford Court, East Howdon, NE280JH
M6 NES N. Preval, 63 Dudley Avenue, Leicester, LE5 2EF
M6 NET D. Bridges, 176 Coombe Valley Road, Dover, CT17 0HE
M6 NEV N. Chambers, 16A Hillside Road, Wool, Wareham, BH20 6DY
M6 NEW W. Chesworth, 28 Chapel Close, Gunnislake, PL18 9JB
M6 NEY M. Bullions, 25 Kirby Road, Dartford, DA2 6HE
M6 NEZ N. Jones, 46 Devon Street, Barrow-in-Furness, LA13 9PX
M6 NFB M. Gregory, 40 Withycombe Drive, Banbury, OX16 0SR
M6 NFC A. Cockburn, 52 Devon Road, Hebburn, NE31 2DW
M6 NFD P. Rickman, 2 Woodley Lane, Romsey, SO51 7JN
M6 NFE G. Singleton, 2 Rome Avenue, Burnley, bb115lq
MW6 NFG N. Terrell, 82 Baglan Street, Treherbert, Treorchy, CF42 5AR
MW6 NFH N. Holloman, The Cloisters, Llanvihangel Crucorney, Abergavenny, NP7 8DH
M6 NFI N. Mooney, 60 Rhyddings Street, Oswaldtwistle, Accrington, BB5 3EY
MW6 NFJ M. Woffindale, 108 College Road, Oswestry, SY11 2SB
M6 NFK N. Lee, 2 The Green Ormesby, Great Yarmouth, NR29 3JX
M6 NFL R. Wraith, 53 Links View Road, Croydon, CR0 8ND
MW6 NFN C. Davies, Foelallt, North Road, Aberystwyth, SY23 2EL
M6 NFP G. Robinson, 12 Donnington Street, Grimsby, DN32 9EN
M6 NFQ G. Brindle, 25 Chedworth Drive, Witney, OX28 5FS
M6 NFS J. Haigh, 1 Smithy Site, Farnborough, Wantage, OX12 8NS
M6 NFT N. Thomson, 92 Sutton Road Hull, Hull, HU67DT
M6 NFW P. Setter, 199 Southbourne Grove, Westcliff-on-Sea, SS0 0AN
M6 NFX N. Furneaux, Hill View, Highridge Road, Bristol, BS41 8JU
MW6 NFY R. Lannon, 14 Fairwood Close, Cardiff, CF5 3QP
M6 NFZ D. Creech, 29 Lake Road, Poole, BH15 4LE
M6 NGA S. Howell, 26 Greenbank Close, Grampound Road, Truro, Cornwall, Truro, TR24TD
M6 NGD N. Goode, 46 Robert Road, Tipton, DY4 9BJ
MW6 NGE G. Dixon, 9 The Glen, Bryncethin, Bridgend, CF32 9LX
M6 NGI P. Strachan-Buckley, 9 Short Street, Aldershot, GU11 1HA
M6 NGK G. Wheeler, 39 Woodbine Close, Newport, PO30 1AE
MI6 NGM A. Mckay, 17 Thorn Hill Road, Banbridge, BT32 3TL
M6 NGN G. Nelson, 4 Garnet Field, Yateley, GU46 6FN
M6 NGP A. Armstrong, 9 Cumberland Drive, Mansfield, NG19 6LS
M6 NGR N. Griffiths, 85 Foljambe Road, Chesterfield, S40 1NJ
M6 NGU R. Gwillym, Flat 3, Ady House 4A Alexandra Road, Farnborough, GU14 6DA
M6 NGW N. Lang, 1 Peartree Court, Old Orchards, Lymington, SO41 3TF
M6 NGX M. Royall, 7 Stacklands, Welwyn Garden City, AL8 6XW
M6 NGY R. Hawkes, Littleworth Lane, Littleworth, RH13 8JX
M6 NHA S. Ratcliff, 27 Furlong Road, Manchester, M22 1UD
MW6 NHC A. Taylor, 74 Fidlas Avenue, Cardiff, CF14 0NZ
M6 NHD D. Baker, 34 Farnham Road, Durham, DH1 5LA
M6 NHF A. Barford, 11 Tuffnells Way, Harpenden, AL5 3HJ
M6 NHG P. Spencer, 146, Winchester Road, Wolverhampton, WV10 6EZ
MI6 NHH E. Currid, 41 Mill Road, Portstewart, BT557PQ
M6 NHJ N. Jones, 18 Cleveleys Avenue, Heald Green, Cheadle, SK8 3RH
M6 NHK N. Bates, 40 First Street, Bradley Bungalows, Consett, DH8 6JT
MW6 NHL J. Keogh, 1 Dyffryn Ig Talley, Llandeilo, SA19 7YP
MM6 NHM N. Morris, 23 Sedgebank, Sedgebank, Livingston, EH54 6HE
M6 NHN N. Collins, 12 Fern Close, Eastbourne, BN23 8AQ
M6 NHO D. Leach, 35 Victoria Road, Alton, GU34 2DG
M6 NHP N. Pettitt, 2A The Oval, Bulford Road, Tidworth, SP9 7SB
M6 NHS N. Dodge, Crossways, Culford, Bury St Edmunds, IP28 6DT
M6 NHT M. Sheppard, 113 Chesterfield Road North, Mansfield, NG19 7JB
M6 NHU T. Butcher, Flat 1, Whitstone Orchard Whitstone Road, Paignton, TQ4 6EY
M6 NHV M. Mclaughlin, 6 Halsey Place, Watford, WD246BS
M6 NHW J. Foxall, 2 Millers Walk Pelsall, Walsall, WS3 4QS
M6 NHX N. Hurlock, 9 Little Meadow, Exmouth, EX8 4LU
M6 NHY M. Pittas, 89 Seddon Road, Morden, SM4 6ED
MM6 NIA N. Hague, 11 Auchriny Circle, Bucksburn, Aberdeen, AB21 9JJ
M6 NIB N. Bennett, 44 Glenmoor Road, Buxton, SK17 7DD
M6 NIC N. Bowker, 16 Farncombe Close, Wivelsfield Green, Haywards Heath, RH17 7RA
MI6 NID P. Moore, 32 Kinnegar Rocks, Donaghadee, BT21 0EZ
M6 NIE P. Martin, 26 Kingfisher Close, Chatteris, PE16 6TP
M6 NIK N. Armstrong, 1 Sea View Terrace, Churchtown, Helston, TR12 7BZ
M6 NIL J. Stacey, 47 Station Road, Foulsham, Dereham, NR20 5RD
MI6 NIM H. Mcallister, Ballycairn Drive, Belfast, BT8 8HG
M6 NIN J. Dewhirst, Flat 12, Lewis Court, Tamworth, B79 8BE
M6 NIQ N. Stokes, 618A Thorne Road, Netheravon, Salisbury, SP4 9QG
M6 NIS F. Nisar, 19A Cromwell Road, Basingstoke, RG21 5NR
MW6 NIT T. Dixon, Troedyrhiw, Abercych, Boncath, SA37 0EY
M6 NIU J. Coombs, 1 Beeston Avenue, Northampton, NN3 9UG
M6 NIV J. Snowden, 30 St. Christophers Walk, Wakefield, WF1 2UP
M6 NIX A. Wilkinson, Central House, Main Road, Hull, HU11 4DJ
M6 NJB N. Bennett, 35 West Shepton, Shepton Mallet, BA4 5UD
M6 NJD N. Dixon, 39 Urswick Green, Barrow-in-Furness, LA13 0BH
M6 NJE N. Spencer, 47 Tyne Road, Oakham, LE15 6SJ
M6 NJG N. Grey, 131 Links Avenue, Hellesdon, Norwich, NR6 5PQ
M6 NJH N. Hall, 24C Oakleigh Court, Church Hill Road, Barnet, EN4 8UX
M6 NJI N. Isherwood, 41 Livingstone Road, Blackburn, BB2 6NE
M6 NJK D. Hawes, 5 Valleyside Road, Hastings, TN35 5AD
MW6 NJM N. Morris, Fairview, Trefonen, Oswestry, SY10 9DP
M6 NJN K. Shaw, Flat 40, Bamford House, Bamford Road, Walsall, WS3 3SA
M6 NJO N. Owen, 32 Westfield Road, Dudley, DY2 8LE
M6 NJP N. Pipkin, 46 Charles Avenue, Albrighton, Wolverhampton, WV7 3LF
M6 NJS N. Sandy, 5 High Ercal Avenue, Brierley Hill, DY5 3QH
M6 NJT N. Jones, 63 Bernwell Road, London, E4 6HX
M6 NJV D. Villa, 33 North Street, Tywardreath, Par, PL24 2PW
M6 NJX Dr M. Nassau, 4A London Road, Liphook, GU30 7AN

M6 NKB N. Booth, 10 Games Walk, Wythenshawe, Manchester, M22 1SN
M6 NKC N. Carey, 28 Tremayne Road, St Austell, PL25 4NE
M6 NKE N. Rotherham, 4 Spenser Road, Cheltenham, GL51 7EA
M6 NKG J. Maddams, 18 Roche Way, Wellingborough, NN8 5YD
M6 NKH N. Hammond, 453 Smorrall Lane, Bedworth, CV12 0LD
M6 NKJ S. Teasdale, 63 Copley, Bishop Auckland, DL13 5LS
M6 NKK A. Hoad, 189 The Diplocks, Hailsham, BN27 3JZ
M6 NKM N. Morse, 33 Tower Close, Bassingbourn, Royston, SG8 5JX
M6 NKP N. Palin, 21 Ford Lane, Crewe, CW1 3EQ
M6 NKQ S. Gudgeon, Animals In Need, Pine Tree Farm London Road, Little Irchester, Wellingborough, NN8 2EH
M6 NKR R. Spearman, 36 Beechwood Road, Northampton, NN5 6JT
M6 NKY N. Brown, 12 Forest Close, Newport, PO30 5SF
MW6 NLA R. Williams, Bardsville, Porthdafarch Road, Holyhead, LL65 2LL
M6 NLB N. Burnet, 27 Mackenzie Way, Tiverton, EX16 4AW
M6 NLF D. Hydes, 31 Ridgehill Avenue, Sheffield, S12 2GL
MW6 NLG N. Gladding, 19 Laleston Close, Nottage, Porthcawl, CF36 3HW
M6 NLJ A. Stokes, 34 Watercress Close, Hartlepool, TS26 0QY
M6 NLK J. Cutter, 34 Greengate Lane, Knaresborough, HG5 9EL
M6 NLL G. Fores, 120 Whiteleas Avenue North Wingfield, Chesterfield, S42 5PW
M6 NLO P. Burt, 56 Winslade Road, Sidmouth, EX10 9EX
MM6 NLP M. Lawson, 23 Kirkfield View, Livingston Village, Livingston, EH54 7BP
M6 NLQ S. Elms-Lester, Ferndale House Kerrys Gate, Hereford, HR2 0AH
M6 NLR D. Miller, 11 Church Road, Hoveton, NR12 8UG
MM6 NLV J. Heywood, 2 Broaddykes Drive Kingswells, Aberdeen, AB15 8UE
M6 NLW S. Jones, 30 Crown Fields Close, Newton-le-Willows, WA12 0JW
M6 NLX P. Cooke, 26 Welby Way, Coxhoe, Durham, DH6 4BT
M6 NLZ J. Wainwright, 8 Mount Drive Purdis Farm, Ipswich, IP3 8UU
M6 NMD E. Martin, 61 Uffington Avenue, Lincoln, LN6 0AG
M6 NME E. Nudd, 14 Birkbeck Way, Norwich, NR7 0XZ
MW6 NMG M. Glenn, 31 South Drive, Rhyl, LL18 4SU
M6 NMH N. Hoddinott, 30 Deans Mead, Bristol, BS11 0QX
M6 NML K. Taylor, 9 Whitcliffe Grange, Richmond, DL10 4ES
M6 NMM S. Bell, 4 Dereham Terrace, Truro, TR1 3DE
M6 NMN N. Norsworthy, Pippins, Trusham, Newton Abbot, TQ13 0NW
MW6 NMP S. Quinn, 1334 Carmarthen Road Fforestfach, Swansea, SA5 4BR
M6 NMQ B. Clark, 46 Fraser Close, Laindon, Basildon, SS15 6SU
M6 NMR K. Wright, 6 Windsor Park, Dereham, NR19 2SU
M6 NMS J. Blackwell, 68 Scarborough Drive, Minster On Sea, Sheerness, ME12 2NQ
M6 NMT M. Afolabi, Clarence Drive, Harrogate, HG1 2QG
M6 NMU E. Riddle, 37B Stubbs Lane, Braintree, CM7 3NR
M6 NMW C. Marcus, 43 Townsend Square, Oxford, OX4 4BB
M6 NMZ R. Rutland, 53 Downs Avenue, Whitstable, CT5 1RR
M6 NNA L. Wilkinson, 20 Coniston Road, Chorley, PR7 2JA
M6 NNB A. Hopper, 7 Holmesdale Villas, Swallow Lane, Dorking, RH5 4EY
M6 NNC N. Hanson, 8 Oak Street, Skegby, Sutton-in-Ashfield, NG17 3FF
M6 NND N. Davies, 29 Burns Road, Congleton, CW12 3EE
M6 NNE D. Hanwell, 28 Chipperfield Road, Norwich, NR7 9RR
M6 NNF N. Ferenc, 2A Rosedene Ave, London, SW16 2LT
M6 NNJ P. Johnson, 25 Pelham Street, Worksop, S80 2TW
M6 NNK P. Banks, 110 Cherwell Drive, Walsall, WS8 7LL
M6 NNL D. Connolly, 5 Cotton Fold, Rochdale, OL16 5HJ
M6 NNM I. Friend, 12 Bladen Valley, Briantspuddle, Dorchester, DT2 7HP
M6 NNO P. Livermore, Cm31Pn, Chelmsford, CM3 1PN
M6 NNP J. Pinto, 21 Steed Crescent, Colchester, CO2 7SJ
MM6 NNS M. Juknevicius, 13/3 Wester Hailes Park, Edinburgh, EH14 3AE
M6 NNT D. Mallinson, 7 Abbots Way, Yeovil, BA21 3HX
M6 NNU D. Johnson, 25 Pelham Street, Worksop, S80 2TW
M6 NNV K. Jaroonsungnoen, 10 Stainsbury Street, London, E2 0NF
M6 NNW D. Salmon, Old School House, Ford Lane Alresford, Colchester, CO7 8AU
M6 NNX D. Austin, 1002 Marsden House Marsden Road, Bolton, BL1 2JX
M6 NNY D. Richardson, 10 Maple Drive, Penrith, CA11 8TU
M6 NOA J. Laszlo, 135 Willifield Way, London, NW11 6XY
M6 NOC C. Singfield, 35 Leamington Drive, Sutton-in-Ashfield, NG17 5BA
M6 NOD A. Smith, Flat 19, Crown Terrace, 10 High Street, Leamington Spa, CV31 3AN
MI6 NOE G. Mccann, 2 Mahon Close, Portadown, Craigavon, BT62 3JF
MM6 NOF K. Leiper, West Wing North Newseat Of Ardo Methlick, Aberdeen, AB417HQ
M6 NOH J. Fletcher, 74 Devonshire Road, Maltby, Rotherham, S66 7DQ
M6 NOI M. Richards, 13 Rosemary Gardens, Sudbury, CO10 1WL
M6 NOJ C. Denman, 16 Upper Oak Street, Windermere, LA23 2LB
M6 NOK N. Kibble, 5 Dunbar Drive, Thame, OX9 3YD
M6 NOL M. Roys, Flat 13, Gregory House, Lister Avenue, Rotherham, S62 7JA
M6 NOM N. O'Mahony, 23 Main Road, Broomfield, Chelmsford, CM1 7BU
MW6 NON A. Rosser, 25 Clos Tir-Y-Pwll, Newbridge, Newport, NP11 5GE
M6 NOQ P. Turner, 77 High Street St. Lawrence, Ramsgate, CT11 0QR
MM6 NOR N. Turnbull, 16 Barntongate Terrace, Edinburgh, EH4 8BA
MI6 NOS J. Baker, 324 Clonmeen, Drumgor, Craigavon, BT65 4AT
M6 NOT Dr S. Leask, 1 Collington Street, Beeston, Nottingham, NG9 1FJ
M6 NOY D. Noyek, 34 The Spinney, Sidcup, DA14 5NF
M6 NPD N. Dagger, 23 Vassall Road, Bristol, BS16 2LH
M6 NPE J. Perfect, 62 Warwick Close Holmwood, Dorking, RH5 4NL
M6 NPF N. Fairhurst, 7 Heatherlands, Sunbury-on-Thames, TW16 7QU
M6 NPL A. Lyman, 12 Chicheley Street, Newport Pagnell, MK16 9AR
M6 NPN R. Sandwell, Higher Hampson, Hampson Lane, Bow, Crediton, EX17 6LA
M6 NPO R. Chambers, ?Tylers?, Fitzhead, Taunton, Ta43jn
M6 NPV A. Price, 87 Derricke Road, Bristol, BS14 8NH
MW6 NPW N. Pitt, 33 Scotchwell View, Haverfordwest, SA61 2RD
M6 NPX N. Paxman, 128 Coggeshall Road, Braintree, CM7 9ES
M6 NPZ J. Freeman, 46 Wargrove Drive, Sandhurst, gu470du
M6 NQA J. Abrahart, Coombe Park, Bickington, Newton Abbot, TQ12 6NZ
M6 NQB D. Hickson, 19 Emerson Avenue Stainforth, Doncaster, DN7 5QL
M6 NQC R. Lane, 2 Dickiemoor Lane, Plymouth, PL5 3NU
MI6 NQD I. Taylor, 26 Shandon Park, Newry, BT34 1QD
M6 NQI P. Lounton, 107 Browning Hill Coxhoe, Durham, DH6 4SA
M6 NQJ A. Fraley, 28 Riverside Park Colehouse Lane, Clevedon, BS21 6TQ
M6 NQK T. Batchford, 22 Roberts Court.., Chester Road Erdington Chester Road Erdington, Erdington, B240BX
M6 NQL G. Norbury, 3 Sherard Croft, Birmingham, B36 0LS
M6 NQM P. Stevens, 36 Cooksey Lane, Birmingham, B44 9QN
M6 NQN S. Reason, 49 Highfield Road, Pudsey, LS28 7JW
M6 NQP K. Dibling, 23C Woodend Ave, Liverpool, L23 2TY
M6 NQQ P. Bridgwater, 18 Angelica, Amington, Tamworth, B77 3JZ
M6 NQR B. Somerville Roberts, 21 Regency Way, Ponteland, Newcastle upon Tyne, NE20 9AU
M6 NQT C. Gamble, 60 Hazelbury Road, Bristol, BS14 9ET
M6 NQU J. Brookhouse, 10 Davenport Drive Castle Vale, Birmingham, B35 7NT
M6 NQV T. Harris, The Lodge Chenies, Rickmansworth, WD3 6ER
M6 NQW J. Williams, 7 Spinney Lane, Nuneaton, CV10 9JA
M6 NRA R. Nagy, 40 Oakhampton Road, London, NW7 1NH
M6 NRB M. Beckett, 4 Sandcross Close, Orrell, Wigan, WN5 7AH
M6 NRC C. Willson, 17 Knightons Way, Brixworth, Northampton, NN6 9UE
MM6 NRE R. Hudson, 128 Neilston Road 2/2, Paisley, PA2 6EP
M6 NRF C. Halloway, 41 Trenoweth Estate, North Country, Redruth, TR16 4AQ
M6 NRG N. Genge, 21 Castle Mead, Washford, Watchet, TA23 0PZ
M6 NRH N. Holmes, 32 Spinney Close, Kidderminster, DY11 6DQ
MD6 NRI N. Rice, The Asters, 33 Upper Dukes Road, Douglas, Isle of Man, IM2 4AT
M6 NRJ N. Johnson, Belair, Western Road, Crediton, EX17 3NB
MM6 NRK G. Kerr, 43 Blairpark Avenue, Coatbridge, ML5 2ES
M6 NRM N. Miles, 28 Oakdale Road, Witney, OX28 1BJ
M6 NRN D. Reason, 3 Stanford Road, Thetford, IP24 1FH
M6 NRO N. Rostant, 14 Gas Street, Leamington Spa, CV31 3BY
MM6 NRQ A. Yates, 25 Keith-Hall Road, Inverurie, AB51 3UA
M6 NRR K. Bostock, 6 Jubilee Road, Daventry, NN11 9HB
M6 NRS J. Swain, 35 Heygate Close, Baildon, Shipley, BD17 6RT
MW6 NRT N. Tanner, 3 Maes Y Tyra, Resolven, Neath, SA11 4NN
M6 NRU D. Warden, 9 Chauntry Place, Coventry, CV1 1JR
M6 NRV I. Hall, 119 Linneaus, Hull, hu3 2qt
M6 NRW N. Waters, 9 Shirley Road, Droitwich, WR9 8NR
M6 NRX S. Cook, Deganwy, Hardwick Road, King's Lynn, PE30 5BB
M6 NSC C. Kemp, Forest Edge, Deer Park, Blandford Forum, DT11 0AY
M6 NSD C. Coppins, 17 Balmoral Road, Gillingham, ME7 4PY
M6 NSE H. Kennedy, 11 Green Road, High Wycombe, HP13 5BD
M6 NSG S. Green, 11 Lavender Walk, Beverley, HU17 8WE
M6 NSH B. Skelton, 1 Summer Hill Road, Bexhill-on-Sea, TN39 4LN
M6 NSI K. Maclean, Ch48Nz, Chester, ch48nz
M6 NSJ N. Inglis, 74 Runswick Avenue, Whitby, YO21 3UE
M6 NSK T. Pennell, 99 Westheath Avenue, Sunderland, SR2 9LQ
MM6 NSM C. Morris, 23 Sedgebank, Livingston, EH54 6HE
M6 NSO N. Osborne, 12 Spiller Road, Chickerell, Weymouth, DT3 4AX
M6 NSQ A. Skelton, 1 Summer Hill Road, Bexhill-on-Sea, TN39 4LN
M6 NSR E. Parrish, 89 Delamere Drive, Macclesfield, SK10 2PS
MD6 NSS N. Smith, 4 Cooil Farrane, Douglas, IM2 1NX
M6 NST K. Theobald, 21 Stirling Close, Rochester, ME1 3AJ
M6 NSW S. Wilson, 37 New Road, Minster On Sea, Sheerness, ME12 3PU
MW6 NSY B. Adamson, 16 Glan Garth, Wrexham, LL12 7DS
M6 NSZ K. Pesztranszki, Flat 5 72 Welldon Crescent, Harrow, HA1 1QR
M6 NTB N. Turrell, 30 Old Road, Longtown, Carlisle, CA6 5TH
M6 NTJ N. Jones, 1 Olaf Close, Andover, SP10 5NJ
M6 NTL B. Porter, 74 Whalley Road, Heywood, OL103JG
M6 NTM N. Mcniece, 17 Hempdyke Road, Scunthorpe, DN15 8LA
M6 NTN N. Hazlehurst, 4 Titchfield Close, Wolverhampton, WV10 8UN
MI6 NTP N. Prentice, 26 Claranagh Road, Claranagh, Enniskillen, BT94 3FJ
M6 NTR N. Reeve, 124 Greenhills Road, Eastwood, Nottingham, NG16 3FR
M6 NTT S. Black, 55 North Road, Hertford, SG14 1NE
M6 NTV Rev. N. Wood, 12 Spring Close, Verwood, BH31 6LB
MW6 NTW N. Davies, 99 Sandpiper Way, Duffryn, Newport, NP10 8WY
M6 NTY D. Balderson, 19 Valley Road, Wellingborough, NN8 2PH
M6 NUD N. Cull, 8 Eaton Road, Norwich, NR4 6PY
M6 NUF D. Metcalfe, 25 St. Abbs Walk, Hartlepool, TS24 7NW
MD6 NUG Capt. D. Cheadle, Bay House, College Green, Castletown, Isle of Man, IM9 1BE
M6 NUJ S. Burgess, 24 Read Avenue, Stafford, ST16 3NP
M6 NUL T. Moye, 33 Prince Charles Road, Colchester, CO2 8NS
MI6 NUM W. Millar, 55 Parklands, Antrim, BT41 4NH
MM6 NUO Dr C. Brown, 4 Damselfly View, Edinburgh, EH17 8XH
MM6 NUP J. Mack, 5 Glen Affric Court, Dumbarton, G82 2BN
M6 NUQ A. Louko, Gummeruksenkatu 3, JyvŽSkylŽ, Finland, 40100
M6 NUR D. Harrod, 44 Kenning Street, Clay Cross, Chesterfield, S45 9LE
M6 NUV D. Wright, 1 Prince Andrew Drive Dersingham, King's Lynn, PE31 6JW
MW6 NUW L. Newton, 45 Commercial Street, Risca, Newport, NP11 6AW
MI6 NUX J. Denvir, 14 Cleland Park Central, Bangor, BT20 3EP
M6 NUY W. Day, 137 Tuffley Lane, Gloucester, GL40NZ
M6 NVA M. Phillips, 2 Millstream Close, Goostrey, Crewe, CW4 8JG
M6 NVB N. Brown, 28 Broadway, Stockport, SK2 5SN
M6 NVG R. Adiga, 41 St. Pauls Road, Staines-upon-Thames, TW18 3HQ
M6 NVJ C. Wellman, 3 Church Street, Warwick, CV344AB
M6 NVL A. Wake, Wardens Bungalow, Wareham Forest Tourist Park, North Trigon, Wareham, BH20 7NZ
MI6 NVM N. Mccann, 3 Portadown Road, Tandragee, Craigavon, BT62 2BB
M6 NVO R. Marshall, 1 Hill Street Upper Gornal, Dudley, DY32DE
M6 NVR N. Ross, 38 Jamaica Road, Malvern, WR14 1TU
M6 NVT N. Tennant, 22 The Lizard, Wymondham, NR18 9BH
MM6 NVW V. Corcoran, Talland Brae, Beley Bridge, St Andrews, KY16 8LX
MM6 NVY M. Fitzgerald, 42 Rosetta Road, Peebles, EH45 8HJ
M6 NVZ D. O Brien, 5 Orson Leys, Rugby, CV22 5RG
M6 NWA N. Wong, Montefiore House, Wessex Lane, Southampton, SO18 2NU
M6 NWB J. Benbow, 44 Copthorne Park, Shrewsbury, SY3 8TJ
M6 NWC N. Clark, Denemead, Cromwell Road, Waltham Cross, EN7 6AS
M6 NWF A. Herbert, 42 Paulsgrove Orton Wistow, Peterborough, PE2 6YE
MM6 NWH N. Holgate, 12 Talisman Road, Glasgow, G13 3QN
M6 NWI A. Howard, 70 Victoria Road, Dukinfield, SK16 4UN
MM6 NWK I. Smith, 92 Malvina Place, Perth, PH1 5FJ
M6 NWL M. Webster, Summerfield Cottage, Walker Lane Wadsworth, Hebden Bridge, HX7 8SJ
MW6 NWM N. Miles, 16 Oakland Crescent, Cilfynydd, Pontypridd, CF37 4HD

M6 NWN J. Smith, 54 Sneinton Hermitage, Sneinton, Nottingham, NG2 4BS
M6 NWO D. Etherington, 20 Sandringham Drive, Leeds, LS17 8DA
M6 NWP C. Jeary, 2 Baker Close, North Walsham, NR28 9JE
M6 NWQ D. Colborne, 11 Martindale Court, Martindale Road, Weston-Super-Mare, BS22 8QQ
M6 NWS S. Wright, 139 Aberfield Drive, Leeds, ls10 3qa
M6 NWT J. Turner, 2 South Drive, Padiham, Burnley, BB12 8SH
M6 NWV J. German, 89 Sandham Grove Heswall, Wirral, CH60 1XW
M6 NWW N. Fields, 31 Oxford Drive, Wirral, CH63 1JG
MM6 NWX C. Mcintyre, 10 Castle Grove, Glasgow, G814AW
M6 NWY S. Newhouse, 28 Hillmorton Lane, Lilbourne, Rugby, CV23 0SS
M6 NWZ E. Griffin, 26 Pebworth Drive Hatton, Warwick, CV35 7UD
M6 NXA A. Milner Smith, 10 Clover Lane, Cricklade, Swindon, SN6 6FJ
M6 NXE P. Holroyd, 39 Ellerbeck, Tamworth, B77 4PP
M6 NXF W. Rowland, 5 Gwinear Downs, Leedstown, Hayle, TR27 6DJ
M6 NXH D. Gunter, 1 Clipstone Mews Bedford Road, Barton-Le-Clay, Bedford, MK45 4FN
M6 NXL P. Mcgee, 2 Fishers Close, Norwich, NR5 0QH
M6 NXO W. Byers, 37 Windermere Drive, Bletchley, Milton Keynes, MK2 3NR
M6 NXP L. Antins, 1 Green Lane Cottages, Penrith, CA11 0HN
M6 NXQ S. Donovan, 3 Speldhurst Court, Kents Hill, Milton Keynes, MK7 6JA
M6 NXR M. Rollason, 6 Shenton Road Barwell, Leicester, LE9 8AR
MW6 NXS A. Williams, Tyn Coed, Bodffordd, Llangefni, LL77 7LJ
M6 NXT L. Stevens, 55 Silverlands Road, St Leonards-on-Sea, TN37 7DF
MW6 NXV B. Williams, The Grange Llanddewi, Llandrindod Wells, LD1 6SF
M6 NXY M. Leach, 64 Grove Street, Wantage, OX12 7BG
M6 NXZ N. Dexter, 49 Kennedy Road, Horsham, RH13 5DB
M6 NYA N. Fill, 3 St. Albans Crescent, Flat B, London, N22 5NB
M6 NYB M. Matthews, 43 Nightingale Drive, Weymouth, DT3 5ST
MW6 NYE A. Minton, 22 Heol Serth, Caerphilly, CF83 2AN
M6 NYF D. Lamble, 4 Laburnum Road, Chorley, PR6 7BG
M6 NYG A. Oxtoby, 29 Masons Road, Hemel Hempstead, HP2 4QP
M6 NYJ A. Harvey, 17 Kerry Drive Smalley, Ilkeston, DE7 6ER
M6 NYM Prof. S. Braunstein, Crossways, Roundhay Park Lane, Leeds, LS17 8AR
M6 NYO J. Raybould, 33 Lincoln Road, Dorrington, Lincoln, LN4 3PT
M6 NYP J. Dyson, 77 Grantham Road, Southport, PR8 4LT
M6 NYR S. Hartley, 30 Coltman Avenue, Beverley, HU17 0EY
M6 NYT J. Woodcock, 31 Moorway Lane Littleover, Derby, DE23 2FR
M6 NYW A. Sim, 273A James Recketts Ave Hull, Hull, Hu8 8lq
M6 NYX S. Dyer, 24 Fossil Road, London, SE13 7DE
M6 NYY D. Paterson, 48 Sissons Road, Leeds, Ls104jt
M6 NYZ N. Mcgroarty, 8 Bakers Lane, Weldon, Corby, NN17 3LR
M6 NZB N. Woodruffe, 139 North Home Road, Cirencester, GL7 1DY
M6 NZE D. Mills, 8 Chestnut Avenue Bucknall, Woodhall Spa, LN10 5DU
M6 NZF K. Surtees, Flat 10 2 Beaufort Road, Bournemouth, BH6 5AL
M6 NZH B. Moore, 10 Shakespeare Drive, Leicester, LE3 2SP
M6 NZI S. Morriss, 10 Cherry Tree Avenue, London Colney, AL2 1RU
M6 NZJ B. Clark, 8 Langdale Close, Farnborough, GU14 0LQ
M6 NZK G. Jenkin, 24 Coronation Avenue, Camborne, TR14 7PE
M6 NZL P. Payne, 3 Queens Court, Woking, GU22 7NE
M6 NZM C. Mccallion-Gow, 70 Cheaton Close, Leominster, HR6 8EW
M6 NZN C. Naylor, 16 Boston Avenue, Norbreck Blackpool, fy29bz
MD6 NZO G. Wilby, Byways, Glenlough Circle, Glen Vine, Isle of Man, IM4 4AX
M6 NZQ G. Wood, 28 Marford Crescent, Sale, M33 4DH
M6 NZR R. Sadowski, 35 Fletcher Gardens, Bracknell, RG42 1FJ
M6 NZS A. Jackson, 13 Thorness Close Alvaston, Derby, DE24 0UY
M6 NZT T. Reseigh, 10 Higher Croft Parc, The Lizard, Helston, TR12 7RL
M6 NZV R. Todd, Lavender Cottage Hillcrest Road, Edenbridge, TN8 6JS
M6 OAB E. Colbert, 176 Carters Mead, Harlow, CM17 9EU
M6 OAD T. Waters, 1 Meadow Cottages, Church Lane, East Winch, King's Lynn, PE32 1NQ
M6 OAE M. Snow, 6 Evelyn Close, Doncaster, DN2 6PA
M6 OAF I. Parbery, 38 Moor End, Maidenhead, SL6 2YJ
MM6 OAG M. Scott, 71 Craigie Road, Perth, PH2 0BL
MM6 OAI S. Chacko, 25 Wemyss Court, Rosyth, Dunfermline, KY11 2LL
M6 OAJ A. Johnson, 19 Meadow Vale, Bristol, BS5 7RG
M6 OAL S. Chng, Cambridge, Cb3 9bb, UNITED KINGDOM
M6 OAN A. Neale, 4 Elizabeth Road, Rothwell, Kettering, NN14 6AJ
M6 OAO J. Wilkinson, 40 Station Road, Kenilworth, CV8 1JD
M6 OAP M. Deary, 7 Newbold Avenue, Sunderland, SR5 1LG
M6 OAQ D. Byng, 1 Ambell Close, Rowley Regis, B65 8PB
M6 OAS L. Rowlands, 6 St. Michaels Avenue, Clevedon, BS21 6LL
M6 OAT J. Bullock, 62 Hawthorn Close, Halstead, CO9 2SF
M6 OAU L. Boylan, 30 Pembrey Way, Liverpool, L25 9SN
M6 OAV J. Mennell, 16 Trafalgar Street West, Scarborough, YO12 7AU
M6 OAW A. Willis, 29 Bonington Drive, Three Elms Road, Hereford, HR4 0RU
M6 OAX A. Edmonds, 20 Tomline Road, Ipswich, IP3 8BZ
MI6 OAZ N. Armstrong, 1 Diamond Cottages, Ardmore Road, Crumlin, BT29 4QU
M6 OBB B. Beard, 8 Monks Close, Newcastle-under-Lyme, ST5 3QU
M6 OBC Dr C. Bridges, 23 Bramley Vale, Cranleigh, GU6 7FY
M6 OBH M. Lachs, 3 King Henrys Walk, Epping, CM16 6FH
M6 OBJ J. Bookham, 116 Clare Gardens, Petersfield, GU31 4EU
M6 OBK D. Cull, 6 Compass Way, Bromsgrove, B60 3GP
M6 OBO B. Bentham, 7 Maypole Crescent, Abram, Wigan, WN2 5YL
M6 OBP P. Percival, 2 St. Catherine Street, Ventnor, PO38 1HG
M6 OBQ D. Gaskell, 126 Higher Green Lane Astley, Tyldesley, Manchester, M29 7JB
M6 OBR B. O'Brien, 48 Wright Crescent, Bridlington, YO16 4RG
M6 OBS S. Martin, 77 Chatford Drive, Shrewsbury, SY3 9PH
MM6 OBU K. Moore, 26 Summerhill Place, Glasgow, G15 7JA
M6 OBX B. Gilbert, 34 Ramsey Road, StIves, pe275rd
M6 OBY P. Cairns, Thorverton, Exeter, EX5 5NB
M6 OBZ P. Dimes, 5 Meadowbrook, Oxted, RH8 9LT
M6 OCB M. Wilsher, Flat K, Wellington Court, Bedford, MK40 2HY
MW6 OCC B. Werrell, 26 Glynhafod Street, Cwmaman, Aberdare, CF44 6LD
M6 OCD Lord J. Gilbody, Winter Meadows, Puxton, Weston-Super-Mare, BS24 6TH
M6 OCH C. Howe, 21 Gotham Road, Birmingham, B26 1LB
M6 OCI J. Bruce, 17 Highcrest Grove Tyldesley, Manchester, M29 8GH
M6 OCJ O. Jones, 6 Pickworth, Stamford, PE9 4DJ
M6 OCK A. Mock, 60 Effra Road Wimbledon, London, SW19 8PP
M6 OCM I. Johnson, 9 Brook Road, Pontesbury, Shrewsbury, SY5 0QZ
MM6 OCP H. Adkins, 270 Cleeves Quadrant, Glasgow, G53 6NR
M6 OCR D. Meakin, 27 Spencer Road, Long Buckby, Northampton, NN6 7YP
M6 OCS O. Smith, 19 Tedder Road, Bournemouth, BH11 8BT
MW6 OCT A. Davies, 28 Y Fron, Cefneithin, Llanelli, SA14 7DN
M6 OCU E. Joynson, 15 Home Farm Lane, Bury St Edmunds, IP33 2QJ
M6 OCX D. Mccarthy-Stewart, 4 Hurst Hill, Chatham, ME5 9BX
M6 OCY L. Horne, 1268 Evesham Road Astwood Bank, Redditch, B96 6AX
M6 OCZ B. Salt, 1 Chantry Close, Harrow, HA3 9QZ
M6 ODA B. Naylor, 45 Meadowgrass Gardens, Worsley, Manchester, M28 1PS
M6 ODC G. Wellstead, 14A Hardy Road, West Moors, Ferndown, BH22 0EX
M6 ODD J. Saunders, 9 Capitol Close, Bolton, BL1 6LU
M6 ODE K. Hadley, 5 The Mead, Clutton, Bristol, BS39 5RF
M6 ODF L. Halloway, 82 Northwall Road, Deal, CT14 6PP
MI6 ODG D. Gallagher, 33 Mulnafye Road, Omagh, BT79 0PG
M6 ODH M. Vlachos, 2 Cumbrian Gardens, London, NW2 1EF
M6 ODK M. Nash, Unit 1 Bull Lane Ind Est Acton Sudbury Suffolk Co10 0Bd, Sudbury, CO100BD
M6 ODL M. Cavanagh, 97 Denecroft Crescent, Uxbridge, UB10 9HZ
M6 ODM J. Barrett, 5 Oakapple Drive, Dereham, NR19 2SR
MI6 ODN T. Pashler, 29 Clogher Road, Ballymena, BT43 6TD
M6 ODP Dr O. De Peyer, Flat 5, Molasses House, Clove Hitch Quay, London, SW11 3TN
M6 ODR M. Hornibrook, 16 Hamilton Road, Ryde, PO33 3QZ
M6 ODS A. Royds, 3A Fairfield Avenue, Rossendale, BB4 9TG
M6 ODY P. Lane, 5 Swan Court, Middle Watch, Swavesey, Cambridge, CB24 4AG
MM6 OEA D. Henretty, 52 Woodburn Avenue, Falkirk, FK2 9YG
M6 OEB D. James, 23 Hedingham Close Plympton, Plymouth, PL7 2FJ
MM6 OEC J. Maclean, 42B Coll, Isle of Lewis, HS2 0LR
M6 OEF S. Dickens, 73 Hengist Way Wallington, London, SM69BP
M6 OEG O. Blagrove, The Willows, Ford Heath, Shrewsbury, SY5 9GZ
M6 OEH K. Hillier, 8 Shuttle Close Rossington, Doncaster, DN11 0FR
MM6 OEI M. Gordon, 41 Jan Mayen Drive, Peterhead, AB42 3PX
MM6 OEJ R. Sutherland, 25 Burns Wynd, Maybole, KA19 8FG
M6 OEM M. Faulkner, 6 Stanley Avenue, Queenborough, ME11 5DT
M6 OEN T. Cummings, 45 Sutton Way, Shrewsbury, SY2 6EE
M6 OER K. Sadlik, 35 Tournament Road, Salisbury, SP2 9LQ
M6 OES I. Jones, 33 Cobham Avenue, Liverpool, L9 3BP
M6 OET B. Mead, 8 Wordsworth Road, Kettering, NN16 9LB
M6 OEU S. Fraser, Old Post Office, Mill Road Barton St. David, Somerton, TA11 6DF
M6 OEV M. Appleton, The Old School, Martin Street Baltonsborough, Glastonbury, BA6 8QS
M6 OEW L. Hudspith, 20 Low Wood Road, Denton, Manchester, M34 2PD
M6 OEZ C. Chandler, 19 West Cliff Park Drive, Dawlish, EX7 9EA
MM6 OFA J. Tervit, 94 Covington Road Thankerton, Biggar, ML12 6NE
M6 OFD G. Round, 128 Leicester Road, Shepshed, Loughborough, LE12 9DH
M6 OFE H. Bond, Flat 4, Athrington Court, First Avenue, Felpham, Bognor Regis, PO22 7LB
M6 OFF C. Gibson, 3 Conway Drive, Billinge, Wigan, WN5 7LH
M6 OFG D. Walker, 39 Spinkhill Avenue, Sheffield, S13 8FA
M6 OFH R. Williams, 16 Horseshoe Close, Cowes, PO31 8PZ
M6 OFJ D. Walker, 39 Spinkhill Avenue, Sheffield, S13 8FA
M6 OFL R. Seville, 29 Blakesley Lane, Portsmouth, PO3 5UG
M6 OFM J. Joyce, 2 Harold Road Cuxton, Rochester, ME2 1EE
MI6 OFN C. May, 20 Harryville Street, Drumgoon, Enniskillen, BT94 4QX
M6 OFQ C. Leece, 101 Wellstead Way, Hedge End, Southampton, SO30 2BH
MI6 OFZ A. Canning, 89 Drumachose Park, Limavady, BT49 0NY
M6 OGA C. Ingram, 17 Stackfield, Harlow, CM202LA
M6 OGC A. Chapman, 44 Malcolm Road, Hartlepool, TS25 3QR
M6 OGE T. Cullis, 98 Pendragon Road, Bromley, BR1 5LH
MI6 OGF J. Mcclarence, 10 Cloyfin Park, Coleraine, BT52 2BL
M6 OGH G. Holmes, 3 Honeysuckle Square, Wymondham, NR18 0FH
M6 OGJ J. Smith, 19 Primrose Terrace St. Michaels Street, Shrewsbury, SY1 2EY
M6 OGK V. Sim, 49 St. Julians Wells, Kirk Ella, Hull, HU10 7AF
M6 OGO G. O'Gorman, Potenza, Chapel Lane, Tewkesbury, GL20 8HS
M6 OGP A. Bhogal, 28 Leigham Drive, Isleworth, TW7 5LU
M6 OGQ P. Kelly, 30 Victoria Street, Maidstone, ME16 8HY
M6 OGS G. Saunders, Hastings Road, Battle, TN330TA
MW6 OGT T. Cording, 17 Maes Y Derwen, Llanrhaeadr Ym Mochnant, Oswestry, SY10 0LE
MM6 OGU H. Knox, Shalderha Holm, Orkney, KW17 2SA
M6 OGX G. Green, 90 Princes Way, Fleetwood, FY78DX
MM6 OHE P. Gush, 1 Gallanach Lochgair, Lochgilphead, PA31 8SD
M6 OHF K. Charlton, 14 Rubens Close, Aylesbury, HP19 8SW
M6 OHL M. O'Connor, 28 Cardigan Road, Southport, PR8 4SF
M6 OHN J. Jones, 3 Orchard Way, Oxted, RH8 9DJ
M6 OHO R. Owen, 41 Mount Road, Cosby, LE91SX
M6 OHP M. Watson, 59 Willowtree Avenue, Durham, DH1 1EA
M6 OHQ C. Smith, 8 The Spinney, Scarborough, YO12 5HQ
M6 OHS J. Hodgson Stokes, 40 Homewood Road, St Albans, AL1 4BQ
M6 OHT J. Holley, Lookers, Blenheim Road Littlestone, New Romney, TN28 8PR
M6 OHU T. Allen, 11 Church View Highworth, Swindon, SN6 7ER
M6 OHY A. Parkin, 22 Glorney Mead, Badshot Lea, Farnham, GU9 9NL
M6 OHZ N. Holden, Plum Cottage, Avon Dassett, Southam, CV47 2AP
M6 OIC G. Chinnappa, 34 Barker Road, Chertsey, KT16 9HX
MI6 OIM A. Mcguigan, 30 Cogry Hill, Ballyclare, BT39 0RY
MM6 OIR P. Riddiough, 1 Cedar Road, Ayr, KA7 3PE
MM6 OIT L. Elder, 18 Stewart Gardens, Airdrie, ML6 9AQ
M6 OIU J. Fox, 23 The Driveway, Canvey Island, SS8 0AB
MW6 OIV A. Prime, 106 Snowden Road, Cardiff, CF5 4PS
M6 OIW C. Thurling, 32 Birch Avenue, Chatteris, PE16 6JJ
M6 OIY S. Davis, 74 The Driveway, Canvey Island, SS8 0AD
M6 OIZ A. Papiewski, 42 Balmoral Avenue, Spalding, PE11 2RU
M6 OJB O. Beck, 11 Clarefield Drive, Maidenhead, SL6 5DW
M6 OJC A. Bruton, 29 Helyers Green, Wick, Littlehampton, BN17 7HB
M6 OJD K. Winton, 130 George V Avenue, Worthing, BN11 5RX
M6 OJF J. Brooks, Cromer Street, Middlesbrough, TS4 2DN

M6 OJI K. Schofield, 18 Berrow Walk, Bristol, BS3 5ES
MI6 OJK J. Kavanagh, 11 Pinewood Crescent, Claudy, BT47 4AB
M6 OJL L. Johnson, 33 Yetholm Place, Newcastle upon Tyne, NE5 4ED
M6 OJM J. Marriott, 66 Latimer Road, Cropston, Leicester, LE7 7GN
M6 OJO E. Mehmet, 8 Hailsham Road, Tooting, London, SW17 9EN
M6 OJP C. Mead, 17 Nelson Crescent, Ramsgate, CT11 9JF
M6 OJQ W. Tranter, 10 Wellpark Walk Newton Heath, Manchester, M40 1JE
M6 OJS A. Snow, 8 Summerhill Grove, Enfield, EN1 2HY
M6 OJT O. Trehearne, Claywood House, East Mascalls Lane Lindfield, Haywards Heath, RH16 2QJ
M6 OJU M. Ross, Whitewalls Off Golf Road, Mablethorpe, ln12 1lp
M6 OJZ J. Mees, 4 Heltwate Bretton, Peterborough, PE3 8RL
M6 OKC A. Armson, 2 Windmill Gardens, St Helens, WA9 1EN
M6 OKD J. Yates, 12 Iris Gardens, Gillingham, SP8 4QY
M6 OKH O. Hopkins, Apartment 17, White Croft Works, Sheffield, S3 7AH
M6 OKH K. Humble, 2 Woodford Close, Sunderland, SR5 5SA
M6 OKI O. Kelland, 16 Esher Place Avenue, Esher, KT10 8PY
MW6 OKJ J. Orchard, The Burrows, Spring Gardens, Whitland, SA34 0HL
M6 OKK G. Cooper, Holmfield, Chelmorton, Buxton, SK17 9SG
M6 OKL A. Gates, 46 Gloucester Place, Littlehampton, BN17 7AL
M6 OKM T. Banham, 16 Mallard Court, Oakham, LE15 6RQ
M6 OKO S. Neale, 13 Hilltop, St. Anns Chapel, Bigbury, Kingsbridge, TQ7 4HG
M6 OKP M. Kitching, 70 Harrington Street, Bourne, PE10 9HB
M6 OKQ M. Broum, 137 Culvers Avenue, Carshalton, SM5 2BA
M6 OKR M. Cole, 4 Park Manor, Britton Street, Gillingham, ME7 5EX
MI6 OKS A. Hanna, 21 Elms Park, Coleraine, BT52 2QF
M6 OKY C. Bietz, 7 Harry Watson Court, Norwich, NR3 3SX
M6 OKZ J. Moore, Flat 4, 219 Holland Road, Holland On Sea, Clacton-on-Sea, CO15 6NL
M6 OLB R. Carter, 39 Ambleside Court Marine Parade East, Clacton-on-Sea, CO15 6JL
M6 OLD G. Waters, 7 Roeselare Close, Torpoint, PL11 2LP
M6 OLE O. Ward, 33 Seventh Avenue, Oldham, OL8 3RY
M6 OLF M. Mills, 17 Hornby Street, Plymouth, PL2 1JD
M6 OLH P. Costall, 3 Gaynsford Place, Little Canfield, Dunmow, CM6 1WB
M6 OLI J. Barnett, 16 Beryl Avenue, Blackburn, BB1 9RR
MI6 OLJ J. Rankin, 36 Glenbank Place, Belfast, bt14 8an
MW6 OLL O. Booth, Oak Cottage, Knockin, Oswestry, SY10 8HQ
M6 OLP K. Baker, Flat 26, Redwing Court 58 High Street, Orpington, BR6 0LE
M6 OLQ R. Gaskell, Cottonwood California, Baldock, SG7 6NU
M6 OLR S. Jenkinson, 17A Britannia Road Burbage, Hinckley, LE10 2HE
M6 OLS J. Myers, 43 Boggart Hill Crescent, Leeds, LS14 1LF
M6 OLT A. Hicks, 15 West Road, Ruskington, Sleaford, NG34 9AL
M6 OLU J. Bugden, 57 Henley Meadows, Tenterden, TN30 6EN
M6 OLW O. Wood, 2 The Bungalows, North Green, Woodbridge, IP13 9NP
M6 OLY D. Goodman, 44 Roberts Road, Madeley, Telford, TF7 5JJ
MI6 OMA A. O'Brien, 15 Oldcastle Road, Newtownstewart, Omagh, BT78 4HX
M6 OME M. Edge, 15 Littlemoor Avenue, Kiveton Park, Sheffield, S26 5NZ
M6 OMF M. Flower, 1 Ridge Farm Cottages, Turnpike, Chilmark, Salisbury, SP3 5AD
M6 OMG N. Patel, 78 Wesley Close, South Harrow, Harrow, HA2 0QE
M6 OMH M. Hawkridge, 27 Northdale Road, Bradford, BD9 4HG
MM6 OMJ M. Reid, 16 Hayfield Road, Kirkcaldy, KY2 5DG
M6 OMK L. Flis, 15 Rifle Hill, Braintree, CM7 1DG
M6 OML M. Lewis, 1 Kingsmead Stretton, Burton on Trent, DE13 0FQ
M6 OMP O. Powell, 2 Little Bowden Manor Kettering Road, Market Harborough, LE16 8AW
M6 OMR B. Withers, 29 Yeoman Way, Trowbridge, BA14 0QL
M6 OMS M. Stradling, 25 Maple Court, Acacia Grove, New Malden, KT3 3BX
MM6 OMT C. Ingram, 22 The Square, Mintlaw, Peterhead, AB42 5EH
M6 OMX M. Amos, Willow Tree House, Deers Green, Clavering, Saffron Walden, CB11 4PX
M6 OMZ S. Smith, 8 Mill Field Close, South Kilworth, Leicester, LE17 6FE
M6 OND C. Howard, 75 Gordon Road, Herne Bay, CT6 5QX
M6 ONH A. Brown, 51 Towncroft, Chelmsford, CM1 4JX
M6 ONJ T. Roberts, 16 Malfort Road, London, SE5 8DQ
M6 ONL P. Destoop, 73, Oeselgemstraat, Wakken, Belgium, 8720
M6 ONM M. Kulakowski, Flat 4 Clement Mellish House East Stockwell St., Colchester, CO1 1GJ
M6 ONN T. Newberry, 2 Brodsworth Road, Peterbrough, pe28xf
M6 ONO R. Cobern, 21 Hopgarden Close, Lamberhurst, Tunbridge Wells, TN3 8DY
M6 ONP P. Maes, Schoolstraat 111, Sint-Niklaas, Belgium, 9100
M6 ONS J. Wood, 3 Onslow Mews, Cranleigh, GU6 8FD
M6 ONT M. Newman, 274 Long Drive, Ruislip, HA4 0HY
M6 ONU A. Kazmi, 39 Herga Road, Harrow, HA3 5AX
M6 ONW J. Gray, 7 Ruskin Avenue, Melksham, SN12 7NG
MW6 ONZ D. Cheeseman, 61 Ffordd Y Millenium, Barry, CF61 5BD
M6 OOA P. Ellis, 14 Mountfield Gardens, Newcastle upon Tyne, NE3 3DB
M6 OOB L. Atkinson, 55 Warkworth Crescent, Seaham, SR7 8JT
M6 OOC S. Kiely, 35 Chestnut Avenue, Todmorden, OL14 5PH
M6 OOD M. Holbrook-Bull, 66 Wayman Road, Corfe Mullen, Wimborne, BH21 3PN
M6 OOJ J. Edwards, 23 Rectory Lane, Great Ellingham, Attleborough, NR17 1LD
MM6 OOK A. Currie, 2/2 44 Robertson Street, Greenock, PA16 8QB
M6 OOL S. Latimer, 40 Petersham Road, Long Eaton, Nottingham, NG10 4DD
M6 OOM A. Shepherd, 33 Meadow Way, Turton, Bolton, BL7 0DE
M6 OOO P. France, Flat 39, Netley House, Birmingham, B32 2BT
MM6 OOP A. Brown, 11A Bowfield Road, West Kilbride, KA23 9LB
M6 OOQ M. Waite, 108A Chantry Gardens, Southwick, Trowbridge, BA14 9QS
M6 OOV G. Long, Maple House. 13 Welfen Lane, Claypole, Newark, NG23 5AL
M6 OOW W. Waterson, 43 Highland Road, Bath, BA2 1DY
M6 OOX M. Emans, 212 Hilldene Avenue, Romford, RM3 8DB
M6 OOZ J. Maclean, 6 Matford Mews Matford Alphington, Exeter, EX2 8XP
M6 OPA G. Mourelatos, 13 Stirling Road, London, N22 5BL
M6 OPB S. O'Loughlin, 1 Pine View, Headley Down, Bordon, GU35 8AX
M6 OPC O. Campbell, 3 Hillside Close, Helsby, Frodsham, WA6 9LB
MI6 OPD J. Reilly, 220 Ardmore Road, Londonderry, BT47 3TE
M6 OPE O. Evers, 7 Rosebury Street, Hull, HU3 6PQ
M6 OPF N. Prater, 100 Pitfold Road, London, SE12 9HY
M6 OPG A. Holmes, Bolehall Manor Club Ltd Amington Road, Tamworth, B77 3LH
M6 OPJ J. Viswambaran, 103 Inglehurst Gardens, Ilford, IG4 5HA
M6 OPL R. Topley, 85 Stuart Road, Aylsham, Norwich, NR11 6HW
M6 OPM O. Murphy, 7 Charlotte Street, Leamington Spa, CV31 3EB
M6 OPO O. Pochat, Hame, Bealswood Road, Gunnislake, PL18 9DA
M6 OPS B. Hillson, Flat 2, Heatherton Park House, Heatherton Park, Taunton, TA4 1EU
MI6 OPT A. Bailie, 130 John Street, Newtownards, BT23 4NA
M6 OQD L. Sproule, Dragonpits North Perrott, Crewkerne, TA18 7TH
M6 OQG G. Sampson, Green Head Cottage, More Hall Lane Bolsterstone, Sheffield, S36 3ST
M6 OQH C. Pendlebury, 73 Pool Street, Wigan, WN3 5BT
M6 OQL D. Cooper, 129 Church Street Eastwood, Nottingham, NG16 3HR
M6 OQN A. Tyrrell, Flat 1, 61 Vicarage Road, Eastbourne, BN20 8AH
M6 OQQ R. Webster, La Lade Caravan Park, Long Load, Langport, TA10 9JX
M6 OQR S. Dolby, 198 Cavendish Road, Worksop, S80 2SH
M6 OQT J. Rymell, 9 Carter Avenue Ruddington, Nottingham, NG11 6NP
M6 OQX M. Nicholson, 104 Harvesters Close Rainham, Gillingham, ME8 8PA
M6 OQZ R. Carpenter, 8 New Buildings Clee Hill, Ludlow, SY8 3PN
M6 ORB P. Thomas, 17 Vine Court, St. Pauls Road, Cheltenham, GL50 4LL
M6 ORC A. Cross, 12 Appleby Drive, Langdon Hills, Basildon, SS16 6NU
M6 ORE E. Moore, 33 Avon Drive, Congleton, CW12 3RQ
MW6 ORF S. Leigh, 27 Cae Clyd, Llandudno, LL30 1BL
MJ6 ORG R. Spencer, 1 Excelsior Villas, Le Mont Les Vaux, Jersey, JE3 8LS
MW6 ORH O. Hopkin, The Forge, Rock Road, St Athan, CF62 4PG
M6 ORI D. Dart, Ticklebelly Cottage Lower Charlton Trading Estate, Shepton Mallet, BA4 5QE
MW6 ORJ K. Collins, 27 Cae Clyd, Llandudno, LL30 1BL
MM6 ORK M. Herridge, The Hollies, Petticoat Lane, Orkney, KW17 2RP
M6 ORM R. Turner, 38 Cotman Road, Clacton-on-Sea, CO16 8YB
M6 ORO O. Gascoigne, 19 Shaftesbury Close, Nailsea, Bristol, BS48 2QH
M6 ORT Rev. G. Wellington, 94 Arlington Road, London, N14 5AT
MM6 ORU M. Drever, 7 Abbotswell Crescent, Aberdeen, AB12 5AQ
MW6 ORW O. Wilson, 45 Dol Helyg, Penrhyncoch, Aberystwyth, SY23 3GZ
M6 ORX D. Gibbeson, 33 Treefield Walk, Barnstaple, EX32 8PE
M6 OSC B. Wilson, 119 Fountains Close, Washington, NE38 7TQ
M6 OSF J. Clover, 174 Sturton Street, Cambridge, CB1 2QF
M6 OSI M. Kent, 56 Hill Street, Hilperton, Trowbridge, BA14 7RX
M6 OSJ S. Mcintosh, 12 Cuthwine Place, Lechlade, GL7 3EG
M6 OSL R. Kemish, Damons Cottage, Romsey, SO51 0GD
M6 OSM Dr E. Pavelin, 35 St. Marys Park, Ottery St Mary, EX11 1JA
M6 OSO J. Coombe, 21 Lesnewth, Par, PL24 2DE
M6 OSP D. Park, 29 Tresco Close, Blackburn, bb2 4rt
M6 OST R. Frost, 40 Greenwich Close, Downham Market, PE38 9TZ
M6 OSU O. Summerfield, 2 Walnut Close, Broughton Astley, Leicester, LE9 6PY
M6 OSW J. Halford, Flat 5, Lattice Court, 2 Leonora Walk, Campbell Park, Milton Keynes, MK9 4BA
M6 OSX M. Rawson, 71 Corn Mill Drive, Farnworth, Bolton, BL4 9EN
M6 OSY D. Coppenhall, 55 Vicarage Lane, Elworth, Sandbach, CW11 3BU
M6 OTB D. Tate, 14 Wordsworth Avenue, Hartlepool, TS25 5NG
M6 OTH M. Riches, Fransham Road Farm, Beeston, King's Lynn, PE32 2LZ
M6 OTI J. Bryan, 7 Riverside Apartments 43A Mowbray Street, Sheffield, S3 8EP
M6 OTJ A. Packer, 20 Shipman Road, Market Weighton, YO43 3RB
MW6 OTK J. Gilmour, Llyswen, Crick, Caldicot, NP26 5UW
M6 OTL A. Rawlins, 7 Kiln Hill, Slaithwaite, Huddersfield, HD7 5JS
M6 OTN T. Bluck, 2 Pendas Meadow, Pinvin, Pershore, Wr102dq
MI6 OTP T. Pearson, 20 Lammy Walk, Omagh, BT78 5JE
M6 OTR T. Rood, 5 Sunny View, Rise, Hull, HU11 5BW
M6 OTS M. Clapham, 12 Penkridge Grove Stechford, Birmingham, B33 9JX
MW6 OTT M. Locke, 18 Tan Y Capel Llanddaniel, Gaerwen, LL60 6EA
M6 OTU R. Hallam, 18 Trent Vale Walney, Barrow-in-Furness, la14 3nb
MI6 OTW P. Fallon, 18 Church View, Killough, Downpatrick, BT30 7RJ
M6 OTY N. Payne, 221 Orchard Avenue, Bridport, DT6 5RJ
M6 OUA A. Regmi, 29 Birchwood Avenue Littleover, Derby, DE23 1QA
M6 OUD A. Cattell, 2 St. James Close, Ruscombe, Reading, RG10 9LJ
M6 OUI B. Gimbert, 49 Ratcliffe Road, Loughborough, LE11 1LF
M6 OUL J. Nield, 24 Bagot Street South Shore Blackpool, Blackpool, FY1 6EZ
M6 OUN M. Johnson, 54 Birchwood Drive, Ulverston, LA12 9PN
M6 OUP A. Wheatley, 17 Cambridge Road, Southport, PR9 9NQ
M6 OUR J. Miall, 42 Garrard Road, Slough, SL2 2QW
M6 OUS M. Mckean, 45 Drakeloe Close Woburn, Milton Keynes, MK17 9QF
MM6 OUU C. Mcconnochie, 72 Duddingston Avenue, Kilwinning, KA13 6RS
MW6 OUW B. Davies, 8 Ferry Road, Kidwelly, SA17 5BJ
M6 OUY I. Boffey, 29 Elm Road Seaforth, Liverpool, L21 1BJ
M6 OVB R. Gowers, 43 Tungstone Way, Market Harborough, LE16 9GA
M6 OVE G. Whale, 1 Rokeby Gardens, Bradford, BD10 0DN
MW6 OVF M. Rowan, Flat 10, Albion Court, Wellington Street, Pembroke Dock, SA72 6JR
M6 OVH S. Langton, 44 Poppyfields Hesketh Bank, Preston, PR4 6TJ
M6 OVI Dr O. Rominger, 27 Paddock Close, Sixpenny Handley, Salisbury, SP5 5NZ
M6 OVN G. Ayres, Shannon, Gunville Road, Winterslow, Salisbury, SP51PP
M6 OVR M. Barry, 58 Bury Road, Radcliffe, Manchester, M26 2UU
M6 OVS R. Yilmaz, 2 Cumbrian Gardens, London, NW2 1EF
M6 OVT J. Priddin, 4, Priory Close Pen-Y-Ffordd Ch4 0Jb, Chester, Ch4 0jb
MW6 OVV D. Jonker, 8 Angle Village Angle, Pembroke, SA71 5AT
M6 OVW G. Sandell, 1 St. Margaret Road, Ludlow, SY8 1XN
M6 OVX M. Walton, 358 Old Heath Road, Colchester, CO2 8DD
M6 OVY H. Greenhill, 39 Pook Lane East Lavant, Chichester, PO18 0AH
M6 OWA J. Wright, 10 Whalley Road, Heskin, Chorley, PR7 5NY
M6 OWC S. Iles, Bigbury Bay Holiday Park, Challaborough, Kingsbridge, TQ7 4HS
MI6 OWD A. Kirkpatrick, 33D Ballyferris Walk, Bangor, Bt19 1ql
M6 OWG D. Burgfeld, 20 Wilson Row, Crowthorne, RG45 6WE
M6 OWI C. Pierce, 42 Staplehurst Road, Reigate, RH2 7PY
M6 OWJ D. Owen, 1 Nightingale Road, Malvern, WR14 2QA
MM6 OWL B. Ewart, 94 Kirkness Street, Airdrie, ML6 6ET
M6 OWM J. Thorpe, 57 Bratch Lane, Wombourne, Wolverhampton, WV5 8DL
M6 OWO A. Whitley, 37 Park Avenue Saltney, Chester, CH4 8TR
M6 OWP A. Stinton, 22 Durham Road. Ronkswood, Worcester, wr5 1nl
M6 OWS P. Blunden, 64 Sussex Drive, Banbury, OX16 1UN

M6 OWT R. Harwood, 24 Firle Crescent, Lewes, BN7 1QG
M6 OWU J. Woodcock, 1 Main Street Brandesburton, Driffield, YO258RL
MM6 OXA S. Waddington, Wester Lathallan, Leven, KY8 5QP
M6 OXB D. Whittaker, Flat 1, Clement Court, Manchester, M11 1EQ
M6 OXC I. Davison, 35 Newhouse Avenue, Esh Winning, Durham, DH7 9JH
M6 OXD S. Holt, 65 Brown Royd Avenue, Huddersfield, HD5 9QA
M6 OXF D. East, 43 Conduit Hill Rise, Thame, OX9 2EJ
M6 OXJ R. Low, 4A Bellevue Road, Romford, RM5 3AY
MI6 OXM C. O'Neill, 53A Coole Road, Dungannon, BT71 5DP
M6 OXR M. Thompson, 22 Churchfield Terrace, Barnsley, s728jt
M6 OXS C. Gibson, 20 Dordon Road Dordon, Tamworth, B78 1QN
M6 OXT R. Ingram, 17 Stackfield, Harlow, CM20 2LA
M6 OXU G. Eaton, 16 Holly Walk, Nuneaton, CV11 6UU
MW6 OXV S. Davies, 14 Elm Walk Mynydd Isa, Mold, CH7 6XZ
M6 OXW D. Ellis, 4 Vane Terrace, Darlington, DL3 7AT
MM6 OXX A. Berry, 8 Hill Street Striling Fk7 0Dh, Striling, FK7 0DH
M6 OXY A. Oxborrow, 24 Ickworth Crescent, Rushmere St. Andrew, Ipswich, IP4 5PQ
M6 OXZ A. Santese, 4 Vane Terrace, Darlington, DL3 7AT
M6 OYA K. Peakman, 32 Mob Lane, Walsall, WS4 1BB
MI6 OYB P. Magee, 11 Elm Corner, Dunmurry, Belfast, BT17 9PZ
M6 OYC S. Boyce, 15 Verbena Avenue Farnworth, Bolton, BL4 0EN
M6 OYD M. Boyd, 1 Harris Close, Brackley, NN13 6NS
M6 OYE A. Hatton, 270 Little Brays, Harlow, CM18 6HD
M6 OYF G. Bailey, 28 Station Road, Polesworth, Tamworth, B78 1BQ
M6 OYH C. Staples, 32 Browns Lane, Netherton, Bootle, L30 5RW
MM6 OYJ B. Smith, 88 Regent Court, Aberdeen, AB24 1ZS
MW6 OYL A. Owen, Bro Dawel, Brynrefail, Caernarfon, LL55 3NR
M6 OYM M. James, 50 Lambert Drive Acton, Sudbury, CO10 0BZ
M6 OYN D. Applewhite, 84 Eringden, Wilnecote, Tamworth, B77 4DB
MW6 OYO T. Puxley, 9 Wesley Court, Pembroke Dock, SA72 6NE
M6 OYP M. Curran, 32 Wedgewood Road, Lincoln, LN6 3LU
M6 OYU L. Rowe, 4 Allandale View, Lincoln, LN1 3RD
M6 OYX S. Rees, 24 Carlton Road, Helmsley, YO625HA
M6 OYY N. Weiner, 3 Bluebell Court Lower Mardyke Avenue, Rainham, RM13 8GF
M6 OYZ S. Bushnall, 7 Faraday Terrace Haswell, Durham, DH6 2DT
MW6 OZB D. Gardner, 1 Neath Court, Thornhill, Cwmbran, NP44 5UH
MW6 OZF C. Lyle, 41 Mount Pleasant Avenue, Llanrumney, Cardiff, CF3 5SY
M6 OZG K. Randle, 36 All Saints Road, Sittingbourne, ME10 3PB
M6 OZI S. Carpenter, 52 Mewstone Avenue, Wembury, Plymouth, PL9 0JZ
M6 OZL A. Vickers, 2 Overton Lane Hammerwich, Burntwood, WS7 0LH
M6 OZM M. Osborne, 6 Walnut Tree Way, Worthing, BN13 3QQ
M6 OZN R. Jacobs, 21 Queens Avenue, Ramsgate, CT12 6DQ
MW6 OZO S. Hayward, 22 Dewsland Street, Milford Haven, SA73 2AU
M6 OZQ N. Silvester, 34 The Burgage Eccleshall, Stafford, ST21 6DR
M6 OZR I. Plimmer, 120 Spitfire Road Castle Donington, Derby, DE74 2AU
M6 OZS T. Falkingham, 22 Almond Tree Avenue Carlton, Goole, DN149QQ
M6 OZT C. Moore, 2 Chapel Rise, Cross Hill, Filey, YO14 0JA
M6 OZW E. Duffy, 75 West Park, Selby, YO84JN
M6 OZX C. Russell, 9 Sandfields Road, St Neots, PE19 1PF
M6 OZY K. Johnson, 32 Redmire Close, Bransholme, Hull, HU7 5AQ
M6 OZZ M. Crockford, Centre Cottage Kelk, Kelk, YO258HL
M6 PAA P. Sadler, 52 Kent Avenue, Weston-Super-Mare, BS24 7FH
MW6 PAC P. Smith, 3 Islington Road, Bridgend, CF31 4QY
M6 PAD Dr P. Darlington, 8 Uplands Road, Urmston, Manchester, M41 6PU
M6 PAE P. Lobo-Kazinczi, Flat 1, 52 Park Road, Hull, HU5 2TA
M6 PAG W. Johnson, 15 Sanders Road, Hemel Hempstead, HP3 9UB
MW6 PAI P. Blackburne, 16 College View, Connah'S Quay, Deeside, CH5 4BY
M6 PAJ P. Moore, 112 Westbury Lane, Bristol, BS9 2PU
MW6 PAM G. Williams, 99 Maes Llwyn, Amlwch, LL68 9BG
M6 PAN P. Naylor, 38 Piggott Grove, Stoke-on-Trent, ST2 9BZ
M6 PAO P. Alley, 58 Osprey Close, Watford, WD25 9AR
M6 PAS P. Simcox, 67 Mervyn Road, Bilston, WV14 8DB
MI6 PAT P. Coogan, Flat B, 44 Ramoan Gardens, Belfast, BT11 8LL
MM6 PAU M. Scott, 28B Highfield Place, Birkhill, Dundee, DD2 5PZ
M6 PAV P. Villette, 62 Hallwood Road, Kettering, NN16 9RF
M6 PAW P. Wallis, 28 Munro Street, Stoke-on-Trent, ST4 5HA
M6 PAX S. Pashley, 45 Highfield View Road Newbold, Chesterfield, S41 7JZ
M6 PAY D. Potter, The Lines, Commonside, Boston, PE22 9PR
M6 PBE P. Carter, 16 Alexandra Park Paulton, Bristol, BS39 7QS
MW6 PBF P. Martin, 6 Herrick Place, Machen, Caerphilly, CF83 8TA
MI6 PBI L. Thompson, 28 Kirkdale, Newtownabbey, BT36 5BX
M6 PBK P. Browne, Ham Cottage, Hammingden Lane, Haywards Heath, RH17 6SR
M6 PBL P. Parkin, 29 Robinia Close, Lutterworth, LE17 4FS
M6 PBM P. Mills, 10 Laurel Estate, Cowes, PO31 7HW
MW6 PBO Dr W. Dickson, The Rowans, Pwllmeyric, Chepstow, NP16 6LA
M6 PBP S. Bradley, 90 Occaiso House, 90 Play House Square, Harlow, ME20 1AP
M6 PBR M. Massie, 3 East Mount, North Ferriby, HU14 3BX
M6 PBT P. Burgess, 61 Grosvenor Avenue, Torquay, TQ2 7JX
M6 PBV J. Brown, 26 Lynnes Close, Blidworth, Mansfield, NG21 0TU
MI6 PBW P. White, 46 Pine Cross, Dunmurry, Belfast, BT17 9QY
M6 PBY I. Maltby, 22 Fern Avenue, Doncaster, DN5 9QX
MI6 PBZ P. Bell, 3 Alexandra Crescent, Larne, BT40 1NE
M6 PCB C. Vincent, 81 Trethannas Gardens, Praze, Camborne, TR14 0LL
M6 PCC P. Cumbers, Azaiba, Fielden Road, Crowborough, TN6 1TP
M6 PCD P. Cheek, 55 Turpin Green Lane, Leyland, pr253ha
M6 PCE C. Fryer, 16 Elston Place, Aldershot, GU12 4HY
M6 PCF P. Faulkner, 32 Manvers Road, Beighton, Sheffield, S20 1AY
MI6 PCJ P. James, 14 Brookmount Crescent, Omagh, BT78 5HG
M6 PCL P. Lambert, 10 Cranmore, Brompton Road, Weston-Super-Mare, BS24 9BU
M6 PCM P. Mason, 1 Keepers Coombe, Bracknell, RG12 0TN
M6 PCN P. Newth, 3 Mulberry Court, Mulberry Street, Stratford-upon-Avon, CV37 6RT
M6 PCP P. Couch, 28 Polgover Way, St. Blazey, Par, PL24 2DL
M6 PCQ P. Morris, 14 Marina Road, Darlington, DL3 0AL
M6 PCR P. Raine, 29 Beech Gardens, Crawley Down, Crawley, RH10 4JB
MW6 PCT S. Gau, Disgwylfa, The Downs, Cardiff, CF5 6SB
M6 PCU T. Prince, 6 Hyperion Avenue, South Shields, NE34 9AE
M6 PCV B. Eastman, 49 Calshot Close, Calshot, SO451BP
M6 PCW P. Cullen, 77 Rising Brook, Stafford, ST17 9DH
M6 PCX P. Coombes, 2 Bissoe Cottages, Bissoe, Truro, TR4 8SU
M6 PCZ P. Colyer, 23 Florida Road, Torquay, TQ1 1JY
MM6 PDA S. Haynes, 86 Barrowfield Street, Coatbridge, ML5 4BJ
M6 PDB P. De Basto, 27 Cloudesley Square, London, N1 0HN
M6 PDC P. Chell, 174D South Road, Stourbridge, DY8 3RN
M6 PDD A. Coleman, 63 Newtown Green, Ashford, TN24 0PL
MW6 PDE P. Devlin, Brynteg, Fron Bache, Llangollen, LL20 7BP
MW6 PDG P. Gough, 92 Pendwyallt Road, Whitchurch, Cardiff, CF14 7EH
M6 PDI M. Lidster, 83 Stroma Gardens, Hailsham, BN27 3AZ
M6 PDJ P. Davies, 15 Carlyle Road, Wolverhampton, WV10 8SL
M6 PDK P. Coote, 251 Alfreton Road, Pye Bridge, Alfreton, DE55 4PB
M6 PDL M. James, 42 Doone Way, Ilfracombe, EX34 8HS
M6 PDM P. Moffitt, Wrawby Farm, Star Carr Lane, Brigg, DN20 8SG
M6 PDN D. Ebbs, 28 Betterton Court, Chapmangate, York, YO42 2ET
M6 PDO A. Dingwall, 48 Village Farm Caravan Site, Bilton Lane, Harrogate, HG1 4DL
MW6 PDP D. James, Coombe House, Coombe, Presteigne, LD8 2HL
M6 PDQ M. Summers, 22A Footners Lane, Burton, Christchurch, BH23 7NT
M6 PDR P. Roberts, 1 Winston Close, Eastleigh, SO50 4NS
M6 PDS P. Stean, 100 Faringford Road, London, E15 4DF
M6 PDT C. Devlin, 32 Kestrel Drive, Dalton-in-Furness, LA15 8QA
MW6 PDU L. Davies, 16 Pembrey Gardens, Pontllanfraith, Blackwood, NP12 2LR
M6 PDV F. Farrer, 16 High Street, Eagle, Lincoln, LN6 9DH
M6 PDW P. Willis, 21 Alton Close, Swindon, SN2 5HF
MM6 PDX W. Davison, 16 Millfore Court, Bourtreehill North, Irvine, KA11 1LT
MI6 PDY P. Davis, 35 Culross Drive, Dundonald, Belfast, BT16 2SQ
M6 PDZ P. Bromfield, 32 Mount Pleasant, Halesworth, IP19 8JF
MM6 PEA A. Pemberton, 111 Henderson Street, Bridge Of Allan, Stirling, FK9 4HH
M6 PEB P. Beckwith, 1 Whincroft Drive, Ferndown, BH22 9LH
M6 PEC B. Clayton, 1 Maude Crescent Sowerby Bridge, Halifax, HX6 1lb
M6 PED J. Pedley, 19 Ellis Drive, New Romney, TN28 8XH
M6 PEF P. Phipps, Meakers Cottage, Long Load, Langport, TA10 9JX
M6 PEG P. Boast, 104 Whittier Road, Nottingham, NG2 4AS
M6 PEI J. Brown, The Chapel, Heath Green, Leighton Buzzard, LU7 0AB
MI6 PEJ J. Beckett, 20 Old Forge, Banbridge, BT32 4AH
MW6 PEL P. Day, 15-16 Troedrhiw-Trwyn, Pontypridd, CF37 2SE
M6 PEM P. Metters, 137 Nevinson Avenue, South Shields, NE34 8NE
M6 PEN P. Kiley, 178 Kingfisher Drive, Woodley, Reading, RG5 3LQ
M6 PEO P. Owen, 17 Jameston, Bracknell, RG12 7WZ
M6 PEP S. Hill, 35 Longs Way, Wokingham, RG40 1QW
M6 PEQ P. Scarlett, 75 Ilfracombe Road, Southend-on-Sea, SS2 4PA
M6 PER P. Rimmington, 28 Skipton Road, Swallownest, Sheffield, S26 4NQ
M6 PEU G. Wale, 21A Pole Barn Lane, Frinton-on-Sea, CO13 9NH
M6 PEW P. Woodburn, 21 The Row, Silverdale, Carnforth, LA5 0UG
M6 PEX C. Rumball, 4 Hasted Close, Bury St Edmunds, IP33 2UA
MW6 PEY A. Fey, 28 Bryn Rhedyn, Caerphilly, CF83 3BT
M6 PEZ C. Perrin, 66 Central Avenue, Farnworth, Bolton, BL4 0AU
M6 PFA D. Rees, 16 Hopton Close, Hereford, HR1 4DQ
M6 PFB P. Payne, 5 Hurstwood Close, Bexhill-on-Sea, TN40 2TA
MI6 PFD P. Dallas, 12 Glendun Crescent, Coleraine, BT52 1UJ
M6 PFG P. Roberts, 1 Ballard Road, Wirral, CH48 9XU
M6 PFH P. Holmes, 18 Raleigh Avenue, Whiston, Prescot, L35 3PL
MI6 PFI P. Sweeney, 83 Rashee Park, Ballyclare, BT399AS
M6 PFK P. Lewis, 9 The Hill, Glapwell, Chesterfield, S44 5LX
M6 PFL P. Flatt, 8 Lingfield Crescent, Queensbury, Bradford, BD13 2SA
M6 PFM T. Sarosi, The Grange, Warwick Road, London, W5 3XH
M6 PFO P. Noble, 14 Park Street, Swallownest, Sheffield, S26 4UP
M6 PFP B. Robinson, 11 Wimbledon Drive, Stockport, SK3 9RZ
M6 PFQ J. Beeley, 47 Claremont Avenue, Hindley, Wigan, WN2 4JG
M6 PFR P. Reeves, 18 Ploughmans Lea, East Goscote, Leicester, LE7 3ZR
MM6 PFT D. Barclay, 45 Milton View, Gatehead, Kilmarnock, KA2 0AY
M6 PFU P. Ogle, 29 Patrick Street, Grimsby, DN32 0JQ
M6 PFV F. Popa, Flat 183, Edinburgh House, Edinburgh Gate, Harlow, CM20 2TJ
M6 PFW P. Woodmass, 29 Thirlmoor, Blackfell, Washington, NE37 1HT
M6 PFY P. Knox, 3 Woodside View, Benfleet, SS7 4PB
M6 PFZ P. Catterall, 117 Beech Hill Lane, Wigan, WN6 8PJ
M6 PGA P. Burrows, 31 Spencer Road, Lutterworth, LE17 4PG
M6 PGB P. Boultwood, 32 Makepiece Road, Bracknell, RG42 2HJ
MW6 PGC G. Shepherd, 48 Lasgarn View, Varteg, Pontypool, NP4 7RZ
MW6 PGD P. David, 5 The Bungalows, Oakfield Terrace, Bridgend, CF32 7SP
M6 PGF P. Bagely, 28 Tenacre Lane, Dudley, DY3 1XQ
M6 PGH P. Hill, 14 Drovers Way, Woodlands, Ivybridge, PL21 9XA
MI6 PGI P. Brown, 2 Legaterriff Road, Ballinderry Upper, Lisburn, BT28 2EY
M6 PGL P. Lewis, 154 Meadow Head, Sheffield, S8 7UF
M6 PGN D. Martin, 8 Garden City, Langport, TA10 9ST
M6 PGO J. Barnes, 6 Windsor Drive, Bredbury, Stockport, SK6 2EH
M6 PGP P. Pearce, 41 Tennyson Avenue, Boldon Colliery, NE35 9EP
M6 PGQ P. Challans, Flat 4, Sandringham Court, 2 Chandos Square, Broadstairs, CT10 1QN
MM6 PGT M. Paget, 40 Davaar Drive, Kilmarnock, KA3 2JG
M6 PGU P. Hall, 11 Middleton Court, Mansfield, NG18 3RN
M6 PGW P. Gonczarow, 25 Ribchester Avenue, Burnley, BB10 4PD
M6 PGX B. Williams, 8 Lindberg Way, Woodley, Reading, RG5 4XE
M6 PGY P. Chapman, 30 Roman Way Shrivenham, Swindon, SN6 8FA
M6 PGZ P. Barkley, 32 Tudor Road, Doncaster, DN2 6EN
M6 PHA P. Biggs, 4 Pendle Close, Southampton, SO16 4QT
M6 PHC P. Conduit, 16 Rectory Avenue, High Wycombe, HP13 6HW
M6 PHE P. Swinyard, 72 London Road, Wokingham, RG40 1YE
M6 PHF S. Shaw, Riverside, Calderbridge, Seascale, CA20 1DN
MM6 PHG P. Groundwater, Vollbekk, 14 Burnside, Kirkwall, KW15 1TF
MI6 PHH P. Kinney, 23 Glenariff Crescent, Ballymena, BT43 6ET
M6 PHJ D. Collier, 133 Woodstock Road, Moston, Manchester, M40 0DG
M6 PHK P. Bentley, The Vauce Farm, Langley-On-Tyne, Hexham, NE47 5NA
M6 PHL P. Hughes, 111 Wisbech Road, Littleport, Ely, CB6 1JJ
M6 PHM P. Meerman, 24 Horseshoe Crescent, Burghfield Common, Reading, RG7 3XW
MM6 PHO C. Hunter, 1 North Gate Lodge, Erines, Tarbert, PA29 6YL
M6 PHP J. Dando, 70 Lydgate Road, Southampton, SO19 6NG
MI6 PHQ P. Wilkin, 55 Clare Heights, Ballyclare, BT39 9SB

UK Callsigns

M6 PHS P. Swannick, 11 Derwent Road, Scunthorpe, DN16 2PA
M6 PHT P. Hart, 11 Sadlers Ride, West Molesey, KT8 1SU
M6 PHU J. Steel, 4 Station Terrace, Boroughbridge, York, YO51 9BU
MI6 PHV P. Lowe, 1 Ballymore Court, Tandragee, , BT62 2AG
M6 PHW P. Walsh, 55 Fore Street, St. Marychurch, Torquay, TQ1 4PU
M6 PHZ J. Smith, 51 Myrtle Avenue, Peterborough, PE1 4LR
M6 PIA M. Moriarty, 30 Longmead, Abingdon, OX14 1JQ
MM6 PIB M. Thiebaut, 1/1 Block A, 2 Barrack Street, Hamilton, ML3 0HZ
M6 PIC D. Berry, 27 Harcourt Terrace, Headington, Oxford, OX3 7QF
M6 PID P. Hodgson, 19 Raymonds Drive, Benfleet, SS7 3PL
MW6 PIH P. Humphreys, 14 Woodside, Oswestry, SY11 1EP
M6 PII S. Whitwell, 12 Lambton Road, Stockton-on-Tees, TS19 0ER
MM6 PIJ G. Wilson, Flat 7 29 Second Avenue, Clydebank, g813ab
M6 PIK D. Pike, 46 Haymans Close, Cullompton, EX15 1EH
M6 PIP P. Pritchard, 12 Easton Crescent, Billingshurst, RH14 9TU
MI6 PIR D. Hamilton, 41 Moss Lane, Ballynahinch, BT24 8FE
M6 PIT R. Peacock, 24 Vicarage Estate, Wingate, TS28 5BP
M6 PIU Dr A. Jackson, 22 Albert Avenue, Nottingham, NG8 5BE
M6 PIY B. Kay, 17 Upper Abbots Royd Caravan Park Barkisland, Halifax, HX4 0DE
M6 PJA P. Archer, 31 Stoney Bank Drive, Kiveton Park, Sheffield, S26 6SJ
M6 PJB P. Bacon, 124 Aughton Road, Swallownest, Sheffield, S26 4TH
M6 PJD P. Doyle, 27 Brockenhurst Avenue, Havant, PO9 4NS
M6 PJE P. Elmore, 8 Gray Street, Elsecar, Barnsley, S74 8JR
M6 PJH P. Higginson, 187 Birmingham Road, Ansley Village, Warwickshire, CV10 9PQ
M6 PJI P. Carne, 1 Curlew Close, Letchworth Garden City, SG6 4TG
MW6 PJJ P. Jones, 23 Pinecroft Avenue, Aberdare, CF44 0HY
M6 PJM P. Miller, 46 Great Brooms Road, Tunbridge Wells, TN4 9DH
MW6 PJN P. Noakes, 8 Ellis Avenue, Old Colwyn, Colwyn Bay, LL29 9LB
MW6 PJP P. Parsons, 31 Clos Tan-Y-Fron, Bridgend, CF31 2BY
MM6 PJR P. Russell, 21 St. Andrews Drive, Law, Carluke, ML8 5GB
M6 PJV K. Rutter, 72 Astbury Drive, Barnton, Northwich, CW8 4PX
M6 PJX S. Rodgers, 22 Crabtree Lane, Sutton-On-Sea, Mablethorpe, LN12 2RT
MM6 PJY K. Govan, 30 Glenwell Avenue, Stranraer, DG9 7BA
MM6 PKC I. Macnab, 4/4 27 St. Andrews Crescent, Glasgow, G41 5SD
M6 PKD C. Barker, 52 Hazelmoor, Hebburn, NE31 1DH
M6 PKH P. Holman, Brindle Lodge, Wilney Green Bressingham, Diss, IP22 2AJ
M6 PKL D. Cadet, 2 Paddockside, Middleton, Ludlow, SY8 3EB
M6 PKN P. Winterbottom, 25A Celandine Rise, Swinton, Mexborough, S64 8NZ
MM6 PKO G. Rothera, Greystones, Watten, Wick, KW1 5UG
M6 PKQ P. Robbins, 2 Bramble Drive Claremont Park, Berrow, Burnham-on-Sea, TA8 2NH
M6 PKR P. Karuppannan Rajan, 1 Jacquard Close, Coventry, CV3 5NG
M6 PKT P. Hall, 112 Osmaston Park Road, Derby, DE24 8EX
M6 PKV N. Beer, North View, 30 Wood End Bluntisham, Huntingdon, PE28 3LE
M6 PLB D. Smith, 13A, Barnes Road, Skelmersdale, WN8 8HN
MW6 PLC P. Carroll, 22 Llanbad, Brynna, Pontyclun, CF72 9QQ
M6 PLE P. Jenkins, 4 Boulton Avenue, West Kirby, Wirral, CH48 5HZ
M6 PLH P. Simmonds, 25 Dimlington Bungalows, Easington, Hull, HU12 0TH
M6 PLI P. Hartley, 99 Cyprus Street, Stretford, Manchester, M32 8BE
M6 PLK A. Jedryka, 71 West Royd Drive, Shipley, BD18 1HL
M6 PLO P. Thomas, 77 Hawthorn Avenue, Lowestoft, NR33 9BB
MW6 PLP P. Price, 80 Maesglas, Pontyates, Llanelli, SA15 5SH
M6 PLQ M. Kincaid, 40 Sorrel Close, Uttoxeter, ST14 8UP
M6 PLR P. Rasmussen, 7 Portal Drive North, Upper Heyford, Bicester, OX25 5TH
M6 PLS A. Hornby, 54 Amberley Road, Horsham, RH12 4LN
M6 PLV P. Le Vallois, 14 London Row, Arlesey, SG15 6RX
M6 PLZ L. Perry, 23 Victors Crescent, Hutton, Brentwood, CM13 2HZ
M6 PMA P. Ambrose, 61 Greenleas Road, Wallasey, CH45 8LR
M6 PMB P. Mcmullan-Bell, 57 Melford Avenue, Barking, IG11 9HS
M6 PMD P. Davies, 68 Sidmouth Avenue, Stafford, ST17 0HF
M6 PMF P. Fenney, 4 Frampton Crescent, Bristol, BS16 4JA
M6 PMG P. Gorman, 1 Hill Top, Stourbridge, DY9 9BZ
M6 PMJ P. Mullen, 14 Anderson Road, Hemswell Cliff, Gainsborough, DN21 5XP
M6 PMK T. Karpasitis, 4 East Crescent, Enfield, EN1 1BS
MI6 PML P. Mclarnon, 8C Greenview Way, Antrim, BT41 4EG
MI6 PMM P. Maguire, 139 Carrowshee Park, Drumhaw, Enniskillen, BT92 0FS
M6 PMP P. Parker, 3 Little Charlton, Basildon, SS13 2EJ
M6 PMR B. Rubery, 142 Chapel Street, Pensnett, Brierley Hill, DY5 4EQ
M6 PMT P. Troth, Beuna Vista, Hawford Wood, Droitwich, WR9 0EZ
M6 PMY P. Imm, 29 Naunton Crescent Leckhampton, Cheltenham, GL53 7BD
M6 PMZ P. Mansfield, 27 Popplechurch Drive, Swindon, SN3 5DE
MI6 PNB P. O'Kane, 31 Larchfield Gardens, Kilrea, BT51 5SB
M6 PNC P. Griffiths, 930 Cedar Point Cir, Rose Hill, USA, 67133
MM6 PND C. Lucking, 39 Overdale Street, Glasgow, G42 9PZ
M6 PNG D. Golik, Primrose Cottage, Newport Road Woodseaves, Stafford, ST20 0NP
MM6 PNH S. Nevins, 4 Ubbanford Norham, Berwick-upon-Tweed, TD15 2LA
M6 PNI M. Roe, Flat 20, 150 London Road, Guildford, Gu11uf
M6 PNJ P. Jackson, 1 Osbern Road, Preston, Paignton, TQ3 1HN
M6 PNK C. Taylor, 212 Plantation Hill, Worksop, S81 0HD
M6 PNL P. Neil, 1 High Graham Street, Sacriston, Durham, DH7 6LZ
M6 PNP A. Pepler, Flat 2, 5 Fore Street, Westbury, BA13 3AU
M6 PNR P. Nowak, 38 Brook Road, Craven Arms, SY7 9RF
M6 PNT A. Huby, 38 Ferryhill Road, Irlam, Manchester, M44 6DD
MM6 PNU I. Swanston, 6 Roberts Grove, Galashiels, TD1 2BJ
MW6 PNW P. White, 13 Bridge Street, Maesteg, CF34 9BJ
M6 PNX C. Peace, 49 Rossefield Way, Leeds, LS13 3RS
M6 PNY P. Sayles, 11 Malton Close, Monkston, Milton Keynes, MK10 9HR
MW6 PNZ C. Eckersley, 43 Bronantfer, Leyshon Road, Gwaun Cae Gurwen, Ammanford, SA18 1EN
M6 POA J. Lambert, 205 Reading Road, Wokingham, RG41 1LJ
M6 POB B. Campbell, 9 Granams Croft, Liverpool, l30 0ph
M6 POC Z. Schwingen, 1 Elizabeth Lockhart Way, Braintree, CM7 9RH
M6 POD D. Pochat, Hame, Bealswood Road, Gunnislake, PL18 9DA
M6 POE S. Barber, 31 Aysgarth Avenue, Crewe, CW1 4QE
MI6 POF A. Stewart, 35 West Wind Terrace, Hillsborough, BT26 6BS
M6 POG P. Gupta, 36 Slimmons Drive, St Albans, AL4 9AP
MI6 POH P. O'Hare, 15A Lisdoo Road, Clady, Strabane, BT82 9RQ
M6 POP P. Barrow, Beach View, Withernsea, HU19 2DS
M6 POQ H. Thomas, 5 Silver Drive, Frimley, Camberley, GU16 9QN
M6 POU C. Moye, 33 Prince Charles Road, Colchester, CO2 8NS
MM6 POV B. Barclay, 24 Hillcrest, Dalmellington, KA67ST
M6 POW L. Patton, 17 Hardwick Street, Weymouth, DT4 7HT
M6 POY M. Moore, 63 Homewater House Hulbert Road, Waterlooville, PO7 7JY
M6 POZ P. Osborne, 7 Bately Avenue, Gorleston, Great Yarmouth, NR31 6HJ
MW6 PPD J. Lawson, 138 Fosse Road, Newport, NP19 4TB
M6 PPE P. Wainwright, 30 Bank View Earlsheaton, Dewsbury, WF12 8HH
M6 PPH P. Holmquest, 6 Rhyme Hall Mews, Fawley, Southampton, SO45 1FX
M6 PPJ P. Pearson, 22 Norris Street, Darwen, BB3 3DR
M6 PPK P. Kay, 30 Broadway, Grange Park, St Helens, WA10 3RX
M6 PPL G. Smith, 36 Palma Park Homes, Shelly Street, Loughborough, LE11 5LB
M6 PPN T. Penberthy, 34 Aldrin Road, Exeter, EX4 5DN
M6 PPP R. Parker, 5 Hawksland Close, Gainsborough, DN21 5FF
M6 PPQ C. Rainbow, 13 Audley Road, Talke Pits, Stoke-on-Trent, ST7 1UG
M6 PPS J. Eyre, 13 Beech Close, Great Ayton, TS96NQ
M6 PPT G. Owen, 38 Trentham Drive, Bridlington, YO16 6ES
M6 PPV J. Smith, 48 London Road, Wymondham, NR18 9BP
M6 PPY J. Dales, 6 Woodfield Drive, Sawtry, Huntingdon, PE28 5TZ
M6 PPZ P. Lloyd, 8 Maydor Avenue, Saltney Ferry, Chester, CH4 0AH
M6 PQA P. Casado Arias, 24 Oldridge Road, London, SW12 8PJ
M6 PQD P. Coxon, 14 Barwick Street, Peterlee, SR8 3SA
M6 PQF P. Ferreira, Flat 48, Lambert Court, Bushey, WD23 2HF
M6 PQI K. Meakin, 8 Over Ross Street, Ross-on-Wye, HR9 7AS
M6 PQL L. Jansen, 45 Clive Road Failsworth, Manchester, M35 0NN
M6 PQO N. Morris, 19 Marshalswick Lane, St Albans, AL1 4UR
MW6 PQQ C. Kibblewhite, 2 Maes Ceidio, Llannerch-Y-Medd, LL71 7AE
M6 PQS R. Cowap, 53 Cross Lane, Newton-le-Willows, WA12 9QA
MW6 PQU E. Maher, 25 Ffordd Cadfan, Bridgend, CF31 2DP
M6 PQV B. Edwards, 34 Alderley Digmoor, Skelmersdale, WN89LZ
M6 PQW K. Matley, Fleetwood, Fleetwood, Fy77hz
M6 PRA A. Mcewen, 4 The Pantyles, Nightingale Lane, Sevenoaks, TN14 6BX
M6 PRB P. Bowes, 72 Keswick Road, Worksop, S81 7PS
M6 PRC L. Piercy, 35 Claremont Road, Grimsby, DN32 8NU
M6 PRE P. Evans, 67 Grenville Street, Stokport, SK3 9ER
M6 PRF J. Rowland-Stuart, 86 Wiltshire House, Lavender Street, Brighton, BN2 1LE
MM6 PRH P. Higgins, South Fallaws Farm, Strathmartine, Dundee, DD3 0PR
M6 PRM P. Munson, 8 Longley Lane, Spondon, Derby, DE21 7AT
M6 PRN R. Raine, 110 Stirling Avenue, Jarrow, NE32 4HS
M6 PRO A. Forrest, 1 Errington Bungalows, Sacriston, Durham, DH7 6NE
M6 PRP P. Punjabi, 62 Cleveland Road, London, W13 8AJ
M6 PRS P. Skidmore, 36 Princes Drive, Harrow, HA1 1XH
M6 PRV S. Bullock, 25 Greenfield Street, Haslingden, Rossendale, BB4 5TG
M6 PRW P. White, 63 Garden Drive, Brampton, Barnsley, S73 0TN
MM6 PRZ P. Misson, 95 Alexandra Street, Devonside, FK13 6JA
M6 PSC P. Croxford, 1 Meteor Close, Bicester, OX26 4YA
M6 PSG P. Griffith, Long Field Barn, Clarkes Lane, Beccles, NR34 8HR
M6 PSJ D. Williams, 2 Tyning Road, Peasedown St. John, Bath, BA2 8HT
M6 PSM P. Mansfield, 15 Earl Avenue, New Waltham, Grimsby, DN36 4NE
M6 PSN G. James, 28 Redcar Road, Romford, RM3 9PT
M6 PSO P. Szabo, 93 Norman Avenue, Southampton, SO16 6QB
M6 PSQ J. Anderson, 18 Blackwater Grove Alderholt, Fordingbridge, SP6 3AD
M6 PSR D. Rowe, 18 Burman Road, Wath-Upon-Dearne, Rotherham, S63 7ND
M6 PST P. Tugwell, 71 Dunkellin Way, South Ockendon, RM15 5ES
M6 PSV M. Beard, 4A St. Aidans Grove, Liverpool, L36 8JE
MM6 PSX A. Ross, 29 East Banks, Wick, KW1 5NL
M6 PSY M. O'Halloran, 7 Waver Close, Corby, NN18 8LL
M6 PSZ S. Macdonald, Woodside Cottage, Horton Way, Verwood, BH31 6JJ
MM6 PTE P. Davis, 4 Daisy Park, Baltasound, Unst, Shetland, ZE2 9EA
M6 PTF M. Haddleton, 42 Grove Road, Pontefract, WF8 2AB
M6 PTH P. Thomasson, 13 Ringway, Neston, CH64 3RS
M6 PTI S. Machon, 19 Houfton Road, Mansfield, NG18 2DG
M6 PTM P. Millington, 125 Telford Way, High Wycombe, HP13 5SZ
M6 PTO R. Morgan, 17 Forbes Close, Glenfield, Leicester, LE3 8LF
M6 PTP M. Taylor, 84 Elgar Crescent, Brierley Hill, DY5 4JJ
MM6 PTR P. Owen, 4 Doonhill Wood, Newton Stewart, DG8 6NU
M6 PTU Prof. P. Curtis, Cotswold, Salisbury Road, Abbotts Ann, Andover, SP11 7NX
M6 PTV C. Brough, 40 Denby Grange, Harlow, CM17 9PZ
M6 PTX P. Hutchins, 25 The Paddock, Maidenhead, SL6 6SD
M6 PTY P. Ryan, 3 St. Quentin Close, Derby, DE22 3JT
M6 PUA A. De Mora, 36 West Park, Minehead, TA24 8AN
M6 PUD J. Crudgington, 29 Wild Flower Way, Ditchingham, Bungay, NR35 2SF
MW6 PUF R. Bath, 10 Tan Rhiw, Penrhyn Bay, Llandudno, LL30 3RB
M6 PUG A. Ward, 29 Mainwaring Road, Wallasey, CH44 9DN
M6 PUH Dr M. Foster, 58 New Terrace, Staverton, Trowbridge, BA14 6NY
M6 PUL D. Pullen, 5 Weldon Close, Shotton Colliery, Durham, DH6 2YJ
M6 PUN J. Schofield, 21A Millgate, Thirsk, YO7 1AA
M6 PUP S. Nichols, 61B Norwich Common, Wymondham, NR18 0SW
M6 PUQ R. Vincent, 81 Trethannas Gardens, Praze, Camborne, TR14 0LL
M6 PUS A. Russell, 26 Diamond Ridge, Camberley, GU15 4LD
MW6 PUT K. Duggan, 97 Maesglas Grove, Newport, NP20 3DN
M6 PUU M. Coleman, 60 Church Rd, Fleet, GU514LY
MI6 PUX N. Jenkinson, 35 Scarvagh Heights, Scarva, Craigavon, BT63 6LY
M6 PUY E. Barrett, 54 The Parade, Greatstone, New Romney, TN28 8SU
M6 PUZ B. Titmus, 68 Hobart Road, Cambridge, CB1 3PT
MJ6 PVB A. Bertram, Roz-Den La Rue De La Guilleaumerie, Jersey, JE2 7HQ
M6 PVC C. Constantine, 257 Kings Road, Ashton-under-Lyne, OL6 9EG
M6 PVI J. Griffiths, Mill House, High Street, Waddingham, Gainsborough, DN21 4SW
M6 PVL P. Lansdown, 30 Bromley Road Kingsway Quedgeley, Gloucester, GL2 2JB
M6 PVM P. Massey, 11 Shakestone Close, Writtle, Chelmsford, CM1 3HS
M6 PVN P. Nicholls, 23 Bishops Gate, Birmingham, B31 4AJ
M6 PVR P. Richards, Hills & Hollows Barnwell Road Oundle, Peterborough, PE8 5PB
M6 PVV B. Woods, 193 Wimberley Street, Blackburn, BB1 8HU
M6 PVW M. Lawrence, Pheasant View, Church Lane Thwaite, Eye, IP23 7EJ
MM6 PVY C. English, 6 Torrance Court East Kilbride, Glasgow, G75 0RU
M6 PVZ D. Till, 157 Malkin Drive, Church Langley, Harlow, CM17 9HL
M6 PWB P. Barrett, 8 Durban Road, Portsmouth, PO1 5RR

M6 PWC P. Castle, 3 Wye Road, Brockworth, Gloucester, GL3 4PP
M6 PWD P. Deeprose, 2 Denehurst Gardens, Hastings, TN35 4PB
M6 PWE G. Corkett, 28 Underwood, Hawkinge, Folkestone, CT18 7NT
M6 PWG P. Green, 3 Beech Court, Long Stratton, Norwich, NR15 2WY
M6 PWH P. Whitworth, 34 Rye Crescent, Danesmoor, Chesterfield, S45 9HH
M6 PWK P. Martin, 5 Shropshire Drive, Wilpshire, Blackburn., BB1 9NF
M6 PWL M. Mynn, The Town House, Parsons Field, St. Mary'S, Hugh Town, TR21 0JJ
MW6 PWM J. Pattinson, 79 Waterloo Road, Penygroes, Llanelli, SA14 7PN
MW6 PWO P. Oseland, 6 Oaklands Close, Bridgend, CF31 4SJ
MI6 PWR P. Warriner, 25 St. Marys Road, Omagh, BT79 7JX
MW6 PWS P. Spong, 209 Neath Road, Briton Ferry, Neath, SA11 2BJ
M6 PWT J. Girard, 49 Beech Crescent, Hythe, Southampton, SO45 3QF
M6 PXD P. Donaghy, 67 Brockenhurst Way, Bicknacre, Chelmsford, CM3 4XN
M6 PXF P. Forbes, 55 The Henrys, Thatcham, RG18 4LS
M6 PXG P. Guest, 88 Windsor Drive, Wigginton, York, YO32 2YE
M6 PXI L. Willis, 12 Robartes Road, St. Dennis, St Austell, PL26 8DS
M6 PXK C. Smith, 83 Sea View Street, Cleethorpes, DN35 8HY
M6 PXL C. Hembrow, 5 Wetherby Road, Stoke-on-Trent, ST4 8AZ
M6 PXP J. Maudsley, Knight Stainforth Hall, Little Stainforth, Settle, BD24 0DP
M6 PXT L. O'Connor, Newton, Maldon Road, Witham, CM8 1HP
M6 PXW P. Wright, 4A Alma Street, Melbourne, Derby, DE73 8GA
M6 PXY A. Schofield, 24 Oldbrook, Bretton, Crowland, PE3 8SH
M6 PYD D. Robinson, 27 Abbeylea Drive, Westhoughton, Bolton, bl5 3zd
M6 PYE Dr D. Pye, 151 Smallbrook Lane, Leigh, WN7 5PZ
M6 PYF A. Whitmore, 12 Rutland Close, Congleton, CW12 1LT
M6 PYG G. Fielding, Chapel Court, Chapel Lane, Malvern, WR13 5HX
MW6 PYH G. Round, 3 Pen Yr Hwylfa, Harlech, LL46 2UW
MI6 PYL S. Begley, 46 Lany Road Moira, Craigavon, BT67 0NZ
M6 PYO P. Pritchard, 1 Winceby Road, Wolverhampton, WV6 7SY
MW6 PYP R. Hall, 8 Adam Street, Abertillery, NP13 1EX
M6 PYR P. Robinson, 15 Cornelius Drive, Wirral, CH61 9PY
MM6 PYX D. Ewart, 2 School Quadrant, Airdrie, ML6 6SP
M6 PYZ T. Brown, 69 Lawn Closes Alt, Oldham, Ol82hb
M6 PZA R. Hawes, 40 Nightingale Way, Thetford, IP24 2YN
M6 PZB P. Brown, 64 St. Johns Road, Swinton, Mexborough, S64 8QW
M6 PZF P. Fairbourn, 17 Perry'S Lane, Wroughton, Swindon, SN4 9AX
MM6 PZG Dr J. Wills, 2 Old Dalmore Gardens Auchendinny, Penicuik, EH26 0RR
M6 PZM A. Aslam, 101 The Oval, Guildford, GU2 7TP
M6 PZO S. Mitchell, 51 Chatsworth Avenue Tuffley, Gloucester, GL4 0SH
MW6 PZP S. Owen, 8 Old Tanymanod Terrace, Blaenau Ffestiniog, LL41 4BU
M6 PZW M. Heenan, 15 Woodacre Green, Bardsey, Leeds, LS17 9AB
M6 PZY C. Puzey, 17 Iverdale Close, Iver, SL0 9RJ
M6 PZZ L. Evans, 1A South View, Fryston, Castleford, WF10 2QF
M6 QFG F. Gardner, 67 Woodside Road, Tunbridge Wells, TN4 8PY
MI6 RAC R. Mcdaid, 29 Linen Green, Sion Mills, Strabane, BT82 9TL
MI6 RAD R. Todd, 58 Kilrea Road, Portglenone, Ballymena, BT44 8JB
M6 RAH R. Harriman, 9 Millers Close, Rushden, NN10 9RP
M6 RAL C. Jewell, 3 Marsh Gate, Clee St. Margaret, Craven Arms, SY7 9DU
MM6 RAM R. Law, 4A Burnside Court, Dundee, DD2 3AF
MU6 RAN R. Phibbs, Aqeb, St Martins, Guernsey, GY4 6AD
M6 RAO R. Rider, 25 Kimber Close, Lancing, BN15 8QD
MD6 RAQ R. Shooter, 14 Cooyrt Shellagh, Ballasalla, Isle of Man, IM9 2EU
M6 RAR R. Calleja, 3 Stoutsfield Close, Yarnton, Kidlington, OX5 1NX
MI6 RAS G. O'Neill, 46 Ashgrove Road, Newtownabbey, BT36 6LJ
MM6 RAV R. Nelson, 8 South Street Cambus Fk102Pa, Stirling, FK102PA
MW6 RAW A. Cotter, 3 George Street, Aberdare, CF44 6RY
M6 RAZ R. Booker, 6 Kipling Road, Dursley, GL11 4QB
M6 RBA R. Bristow, 6 Blackness Road, Crowborough, TN6 2LY
M6 RBB B. Plackett, 36 Dartmouth Crescent, Brinnington, Stockport, SK5 8BG
MM6 RBE J. Howison, 54 Whiteside, Bathgate, EH48 2RG
M6 RBF R. Fry, 20 St. Andrews Road, Ellesmere Port, CH65 5DG
M6 RBH R. Haynes, 28 Ridgeway View, Montgomery, SY15 6BF
MJ6 RBI R. Rumboll, Windsor House, La Grande Route De St. Laurent, Jersey, JE3 1NL
M6 RBK Dr B. Kalogerakis, Inglewood, Madingley Road, Cambridge, CB23 7PH
M6 RBO R. Axtell, 74 Elmshott Lane, Slough, SL1 5QZ
MI6 RBP R. Lyttle, 14 Rosehead Drive, County Antrim, Ireland, BT14 7BF
M6 RBQ R. Loseby, 22 Stanley Gardens, Doncaster, DN4 0JG
MM6 RBT R. Turpie, 11 Ashkirk Place, Dundee, DD4 0TN
M6 RBU R. Byard, Wynbourne, Wynolls Hill Lane, Coleford, GL16 8BP
M6 RBV R. Ford, 7 Briar Court, Wickersley, Rotherham, S66 1AF
M6 RBY R. Johnston-Stuart, 35 Robbins Close, Bradley Stoke, Bristol, BS32 8AS
MW6 RBZ R. Bishop, 96 Heol Homfray, Cardiff, CF5 5SB
M6 RCA R. Broadwith, Keepers Cottage, Rookwith, Ripon, HG4 4AY
M6 RCB R. Brignall, 53 Grange Crescent, St. Michaels, Tenterden, TN30 6DY
M6 RCD R. Coates, 123 Mansfield Crescent, Armthorpe, Doncaster, DN3 2AR
M6 RCE C. Mandville, Garth Hill College, Bull Lane, Bracknell, RG42 2AD
M6 RCF R. Goodier, 56 Chariot Street, Manchester, M11 1DP
M6 RCG P. Ballington, 7 Links Close, Sinfin, Derby, DE24 9PF
M6 RCI L. Rodgers, 27 Arden Houses Normanby-By-Spital, Market Rasen, LN8 2HE
MM6 RCK T. Mcgurk, 11 Palmer Road, Currie, EH14 5QH
M6 RCL R. Lynch, 2 Launceston Close, Oldham, OL8 2XE
M6 RCN S. Francis, 17 Garden Close, Rough Common, Canterbury, CT2 9BP
MM6 RCO R. Nicoll, 15 Redford Walk, Edinburgh, EH13 0AF
M6 RCP C. Edwards, 19 Bells Place, Coleford, GL16 8BX
MI6 RCR R. Reid, 13 Gelvin Grange, Londonderry, BT47 2LD
M6 RCS R. Codrai, 27 Howard Avenue West Wittering, Chichester, PO20 8EX
MI6 RCV H. Ashe, 74 Markville Portadown, Craigavon, BT63 5SZ
M6 RCW R. Weaver, 15 Sharps Field, Headcorn, Ashford, TN27 9UF
M6 RCY C. Radley, 7 Garforth Crescent, Droylsden, Manchester, M43 7SW
MM6 RCZ R. Taylor, Station Cottage, Main Street, Bathgate, EH48 3BU
M6 RDA A. Hill, 7 Knebworth Court, Congleton, CW12 3SW
MW6 RDC R. Cole, 14 Inner Loop Road, Beachley, Chepstow, NP16 7HF
M6 RDD B. Barwell, 48 Locke King Road, Weybridge, KT13 0TB
M6 RDF I. Sharman, 63 Duston Road, Northampton, NN5 5AR
M6 RDI E. Vecenans, 155 Upper Dale Road, Derby, DE23 8BP
M6 RDK R. Starr, 43 Newington Avenue, Southend-on-Sea, SS2 4RD
MW6 RDL R. Green, Hillside, Meinciau Road, Kidwelly, SA17 4RA
MM6 RDM R. Matthews, 5 Dollar Avenue, Falkirk, FK2 7LD
M6 RDN R. Vickers, 57 Cecil Road, Selly Park, Birmingham, B29 7QQ
M6 RDP A. Toynton, Flat 2, 7 Sea Lawn Terrace, Dawlish, EX7 0AD
M6 RDQ R. Ellison, 66 Coronation Road, Wingate, TS28 5JW
MI6 RDR Rev. R. Rowe, 31 Main Street, Brookeborough, BT94 4EZ
M6 RDS R. Sutton, Yew Tree Farm, Paddol Green, Shrewsbury, SY4 5QZ
MM6 RDT R. Tourish, 8 Linnpark Gardens, Johnstone, PA5 8LH
MM6 RDU R. Drummond, 11 Firwood Drive, Bo'ness, EH51 0NX
M6 RDV R. De Vries, Corner Cottage, Hillcrest Close, Sturminster Newton, DT10 2DL
M6 RDW I. Hartless, 32 Long Acre, Mablethorpe, LN12 1JF
M6 RDX A. Griffiths, 61 Hawarden Road, Penyffordd, Chester, CH4 0JD
M6 RDZ R. Sidwell, 27 Gressingham Drive, Lancaster, LA1 4RF
M6 REB R. Beardsley, 10 Moreton Close, Church Crookham, Fleet, GU52 8NS
M6 RED I. Sanderson, 9 Haigh Street, Cleethorpes, DN35 8QN
M6 REE R. Garvey, 2 Link Lane, Oldham, OL8 3AD
M6 REF C. Newsam, 15 Devonshire Avenue North, New Whittington, Chesterfield, S43 2DF
M6 REH R. Ashley, 15 Wimbourne Drive, Gillingham, ME8 9EN
M6 REI C. Reid, 28 Albion Road, London, N16 9PH
M6 REJ R. James, 31 Tehidy Gardens, Camborne, TR14 0ET
M6 REL R. Allen, 44 Overing Avenue, Great Waldingfield, Sudbury, CO10 0RJ
MI6 REM Dr R. Richey, 10 Crest Road, Enniskillen, BT74 6JJ
M6 REO D. Curry, 2 Norham Avenue South, South Shields, NE34 7LP
M6 REQ R. Saunders, 128 Foxcroft Drive, Wimborne, BH21 2LA
M6 RER R. Ridge, Roskellan House, Maenlay, Helston, TR12 7QR
M6 RES R. Strong, 2 Dean Avenue, Thornbury, Bristol, BS35 1JJ
MM6 REV T. Paterson, Free Church Manse, Church Street, Golspie, KW10 6TT
M6 REW R. Wells, 37 Water Meadows, Worksop, S80 3DF
M6 REX R. Moldoveanu, 17 Lynchford Road, Farnborough, GU14 6AR
M6 REY T. Surrey, 53 Central Avenue, North Shields, NE29 7JB
M6 RFA R. Allen, Milverton, Mill Road, Pulborough, RH20 2PZ
M6 RFB R. Woodrow, 190 Sedlescombe Road North, St Leonnards on Sea, TN377EN
M6 RFC R. Crockford, 17 Tadcroft Walk, Calcot, Reading, RG31 7JR
MI6 RFD A. Hudson, 31 Knollwood, Seapatrick, Banbridge, BT32 4PE
M6 RFE A. Braeman, 59 Freshbrook Road, Lancing, BN15 8DE
M6 RFF R. Hancock, 125 Fairham Road, Stretton, Burton-on-Trent, DE13 0BT
M6 RFG R. Morgan, 14 Ash Road, Ashurst, Southampton, SO40 7AT
M6 RFI G. Sullivan, 9 Hine Close, Gillingham, SP8 4GN
M6 RFJ D. Jacobs, 7 Coppice Close Ravenstone, Coalville, LE67 2NS
M6 RFK R. Brookes, 5 Radstock Road, Stretford, Manchester, M32 0AJ
M6 RFL R. Long, 45 Dean Street, Low Fell, Gateshead, NE9 5XL
M6 RFM R. Corney, Lavender Cottage, Worlds End, Hambledon, Waterlooville, PO7 4QU
M6 RFN S. Hinton, 33 Park Street, Kidderminster, DY11 6TP
M6 RFO A. Trohear, 3 Frampton Crescent, Brlstol, BS16 4JA
M6 RFP M. Bruce, 27 Blaenant, Emmer Green, Reading, RG4 8PH
M6 RFQ R. Newton, 38 Bedford Road, Denton, Northampton, NN7 1DR
M6 RFR R. Wheddon, 10 Penharget Close, Pensilva, Liskeard, PL14 5SA
MW6 RFS R. Shipman, 1 Lledfair Place, Heol Pentrerhedyn, Machynlleth, SY20 8DL
M6 RFV R. Manser, 39 Long Meadow, Markyate, St Albans, AL3 8JN
M6 RFW P. Wilson, 103 Thors Oak, Stanford-le-Hope, SS17 7BZ
M6 RFZ R. Finch, Garth Cottage North Cowton, Northallerton, DL7 0HL
MW6 RGA R. Anderson, 156 Cockett Road, Cockett, Swansea, SA2 0FQ
M6 RGC D. Clavey, 32 Apollo Close, Dunstable, LU5 4AQ
M6 RGD R. Yates, West Wing, Elsworth Manor, 20 Smith Street, Elsworth, CB23 4HY
M6 RGE W. Ridge, 91 Ridgeway, Rotherham, S65 3NL
M6 RGF R. Gott, 21 Broughton Road, Crewe, CW1 4NW
M6 RGI R. Griffiths, 2 Old Market Close Acle, Norwich, NR13 3EY
MW6 RGK T. Bourner, 23 St. Michaels Road, Pembroke, SA71 5JQ
MI6 RGM R. Murphy, 40 Stoneypath, Londonderry, BT47 2AF
M6 RGO R. Hunter, 9 Gelt Burn, Didcot, OX11 7TZ
M6 RGP R. Pearson, 12 Mere View, Great Livermere, Bury St Edmunds, IP31 1JU
M6 RGQ R. Hobbs, 2 Jenkyns Close Botley, Southampton, SO30 2UQ
M6 RGR R. Reeves, 20 Warren Close, Irchester, Wellingborough, NN29 7HF
M6 RGS R. Sneddon, 3 George Street, Dipton, Stanley, DH9 9HD
MM6 RGT R. Thomson, Inchully, Myreriggs Road, Blairgowrie, PH13 9HS
M6 RGU R. Emery, 115 Humberstone Road, Grimsby, DN32 8DR
MW6 RGW R. Woodland, Flat 14, Mays Court, Windsor Road, Neath, SA11 1NG
M6 RGX W. Morrison-Bates, 95 Beaulieu Close, Toothill, Swindon, SN5 8AJ
M6 RGY P. Jay, 113 Stanks Lane North, Leeds, LS14 5AS
M6 RGZ R. Hughes, 3 Wood Avens Close, Northampton, NN4 9TX
M6 RHC R. Cook11, 16 First Square, Stainforth, Doncaster, DN7 5RH
MW6 RHD R. Hughes-Burton, 6 Troed Y Garn, Llangybi, Pwllheli, LL53 6DQ
MM6 RHI A. Hunsley, 42 Waverley Place, Edinburgh, EH7 5SA
M6 RHK R. Kirby, 50 Harcourt Avenue, Harwich, CO12 4NT
MI6 RHL A. Calvin, 65 Tannaghmore Road, Markethill, Armagh, BT60 1TW
MD6 RHN D. Heaton, Rushen Vicarage Barracks Rd, Port St Mary, Isle Of Man, IM9 5LP
MM6 RHO R. Hendry, 3 Claddyburn Terrace, Cairnryan, Stranraer, DG9 8RD
MM6 RHQ R. Hepburn, 33 Saxon Road, Glasgow, G13 2YQ
MW6 RHR N. Oldham, Arhosfa, Carmel, Llannerch-Y-Medd, LL71 7DH
MM6 RHT Dr R. Harkness, Fernwood, 4 Cassalands, Dumfries, DG2 7NS
M6 RHU M. Ricketts, 45 Jesmond Road, Grays, RM16 2QS
M6 RHX A. Richardson, 18 Warkworth Road, Durham, DH1 5PB
MW6 RHY R. Cooper, 3 Waunddu, Pontnewynydd, Pontypool, NP4 6QZ
M6 RIA J. Aldersley, 6 Cavendish Close, Kingswinford, DY6 9PR
MM6 RIE E. Soames, 4 Johnstons Road, Rousay, Orkney, KW17 2PZ
M6 RIH R. Hayes, 32 Green Road, Newmarket, CB8 9BA
M6 RIJ S. Blades, 18 Hall Cliffe Road, Horbury, Wakefield, WF4 6BX
M6 RIL R. Studeny, 36 Maples Street, Nottingham, NG7 6AD
M6 RIN A. Barker, 3A Hillcrest Close, Castleford, WF10 3QS
M6 RIO J. Roberts, Worlds Wonder, Warehorne, Ashford, TN26 2LU
M6 RIQ R. Gallagher, 56 Chesterton Square, London, W8 6PJ
MI6 RIR E. Stevenson, 25 Woodlands Manor, Portadown, Craigavon, BT62 4JP
M6 RIS R. Sowden, Flat 8, Bramble Court Bramble Street, Derby, DE1 1HW
M6 RIT R. Williams, 25 Wolds Retreat, Brigg Road, Ln6 7Ru, Market Rasen, LN7 6RU
M6 RIU R. May, 18 The Glebe, Camborne, TR14 7EW

M6 RIV C. Jenkin, 64 Rivers Road, Yeovil, BA21 5RJ
MW6 RIY A. Holland, Four Winds, Nebo, Llanon, SY23 5LF
M6 RIZ R. Harris, 19 Nightingale Walk, Stevenage, SG2 0QE
M6 RJB R. Beardmore, 37 Fernhill Road, Solihull, B92 7RU
M6 RJE R. Edwards, 23 Queens Walk, Ruislip, HA4 0LX
M6 RJF G. Ropinski, 38 The Leys, Little Eaton, Derby, DE21 5AR
M6 RJH R. Harrison, 18-20 Hall Lane, Kirkburton, Huddersfield, HD8 0QW
M6 RJJ R. Pettigrew, 26 Middlewich Road, Northwich, CW9 7AN
M6 RJK R. Killner, Oak Cottage, Priors Byne Farm, Lock Lane, Horsham, RH13 8EF
M6 RJL J. Rose, 9 Denewood, New Barnet, Barnet, EN5 1LX
M6 RJM R. Meehan, 2A Pinfold Street, Macclesfield, SK11 6HA
MM6 RJN N. Robertson, 3 Wallach Brae, Dalbeattie, DG5 4GY
M6 RJP R. Parsons, Aintree, Hurn Bridge Road, Lincoln, LN4 4XT
M6 RJQ R. Munir, 29 Hamilton Drive, Guildford, GU2 9PL
M6 RJR S. Slater, 6 Hazel Dene, Hurstfarm Estate, Matlock, DE4 3TG
M6 RJS J. Statham, Oakwoods, School Lane Upper Basildon, Reading, RG8 8LT
M6 RJU R. Uzzell, 68 St.Michaels Road, Cirencester, GL7 1ND
M6 RJW R. Wakelam, 1 Cleeve Park Mews, Cleeve Park, Minehead, TA24 6JH
MW6 RJX M. Burns, 40 Matthysens Way, St. Mellons, Cardiff, CF3 0PS
MW6 RJY J. Reason, 158 Caerau Lane, Cardiff, CF5 5JS
M6 RJZ R. Duthie, 14 Kettles Close, Oakington, Cambridge, CB24 3XA
M6 RKA M. Halliburton, 7 Penrose Avenue West, Liverpool, L14 6UT
M6 RKB K. Bache, 49 Cheviot Way, Halesowen, B63 1HD
M6 RKC M. Callow, 4 The Firs, Canvey Island, SS8 9TW
M6 RKD M. Davey, Romea, Long Street, Attleborough, NR17 1LW
M6 RKE R. Elger, 3 Alexander Square, Eastleigh, SO50 4BW
MW6 RKF W. Lewis, 53 Leyshon Road, Gwaun Cae Gurwen, Ammanford, SA18 1EN
M6 RKI R. Hogan, 10 Forest Way, High Wycombe, HP13 7JF
M6 RKK R. Dick, 15 Havenwood, Arundel, BN18 0AH
M6 RKL L. Hall, 180 Moor Road, Chorley, PR7 2NT
M6 RKM R. Mason, 22 Coronation Close, Happisburgh, Norwich, NR12 0RL
M6 RKO R. Oczerklewicz, 4 Grisedale Place, Chorley, PR7 2JW
M6 RKP R. Page, 4 Vale Cottages, Kirmington Vale, DN38 6af
M6 RKQ N. Monaghan, 7 Hoyle Road, Wirral, CH47 3AG
M6 RKR G. Hirst, 94 Upper Brighton Road Sompting, Lancing, BN15 0LB
M6 RKS M. Swanston, 309 Main Road, Wharncliffe Side, Sheffield, S35 0DQ
MM6 RKT J. Browne, 33 Pilgrims Hill, Linlithgow, EH49 7LN
M6 RKU A. Rickwood, 10 Wardrop Road, Catterick Garrison, DL9 3BW
M6 RKW M. Ware, 20 Brentwood Avenue, Thornton-Cleveleys, FY5 3QR
M6 RKX M. Hemming, 11 Blackberry Way, Evesham, WR11 2AH
MW6 RLA M. Lewis, 7 Yew Tree Terrace, Cwmbran, NP44 3NU
M6 RLB R. Brown, 24 Malthouse Court, Wellington, Telford, TF1 1QJ
M6 RLC R. Lee, 28 Champion Way, Oxford, OX4 4NS
MM6 RLD D. Beeton, The Old School House, Sliddery, Isle of Arran, KA27 8PB
MW6 RLE R. Lewis, 5 Elizabeth Sparkes Close, Rogiet, Caldicot, NP26 3UT
M6 RLG J. Berrisford, 18 Trowels Ln, Derby, DE22 3LS
MM6 RLL J. Young, 101 Fleming Way, Hamilton, ML3 9QH
M6 RLN D. Carroll, 48 Foster Road, Trumpington, Cambridge, CB2 9JR
M6 RLP R. Palmer, New Haven, Stoneraise, Carlisle, CA5 7AX
M6 RLQ R. Skingley, Tregwin, The Commons, Helston, TR12 7HZ
M6 RLR R. Hanson, 664 Leeds Road, Huddersfield, HD2 1UB
M6 RLT R. Trim, 23 Coleman Road, Bournemouth, BH11 8EQ
MI6 RLU F. Burgess, 65 Sunnylands Avenue, Carrickfergus, BT38 8JT
M6 RLV R. Oliver, 3 Histon Road Cottenham, Cambridge, CB24 8UF
M6 RLW R. Wood, 7 Wishart Green, Old Farm Park, Milton Keynes, MK7 8QB
M6 RLX R. Lang, 31 Ewan Close, Barrow-in-Furness, LA13 9HU
M6 RLZ K. Lewis, 15 Gilbert Scott Way, Kidderminster, DY10 2EZ
M6 RMA R. Armstrong, 11 High Ditch Road, Fen Ditton, Cambridge, CB5 8TE
M6 RMD R. Dockray, 54 Kelsick Park, Seaton, Workington, CA14 1PY
MI6 RME R. Mcdonald, 6 Rose Park, Limavady, BT49 0BF
M6 RMG R. Grout, 5 Branton Close, Great Ouseburn, York, YO26 9SF
M6 RMI R. Le Feuvre, 14 Elm Drive, Brockworth, Gloucester, GL3 4DH
M6 RMK R. Keith-Hill, 32 Thornhill Way, Plymouth, PL3 5NP
M6 RML J. Hodson, 77 Waltham Road, Woodford Green, IG8 8DW
M6 RMN T. Ahern, 39 Essex Road, Romford, RM7 8BE
M6 RMO C. Larner, 98 Allandale, Hemel Hempstead, HP2 5AT
MW6 RMR R. Russell, 1 Horeb Cottages, Rhiw Road, Colwyn Bay, LL29 7TL
M6 RMS R. Somerville, 28 Yeathouse Road, Frizington, CA26 3QJ
MM6 RMT R. Titmarsh, Caberfeidh Clachan Na Luib, North Uist, HS6 5HD
M6 RMU R. Mitchell, 8 Prestwood Close, Benfleet, SS7 3LD
MM6 RMV J. Vine, Seaview, Half Of 4 Kilvaxter, Kilmuir, IV51 9YR
M6 RMW M. Willmott, 25 Clent Avenue, Kidderminster, DY11 7EH
M6 RNA P. Bannon, 73 London Road, Worcester, WR5 2DU
M6 RNC R. Blow, 131 Fairfax Avenue, Harrogate, HG2 7RU
M6 RND B. Starling, 4 Three Corner Drive, Norwich, NR6 7HA
M6 RNE S. Wilson, 2 Bydales Drive, Marske-By-The-Sea, Redcar, TS11 7HJ
M6 RNF D. Meekins, 30 Lorton Avenue, Workington, CA14 3JF
M6 RNI R. Hillier, 10 Buttermere Close, Folkestone, CT19 5JH
M6 RNM A. Atkinson, 2E Bagridge Road, Wolverhampton, WV3 8HW
M6 RNN N. Renny, 48 Tower Road Tividale, Oldbury, B69 1na
M6 RNO R. Noakes, 3 Firtree Road, Hastings, TN34 3TR
M6 RNQ R. Stevens, 7 Dunbreck Grove, Sunderland, SR4 7LL
M6 RNS J. Grint, 9 Mountbatten Drive, Leverington, PE13 5AF
M6 RNU M. Nutt, 110 Birkinstyle Lane, Shirland, Alfreton, DE55 6BT
M6 RNV Dr A. Green, 1 Buxton Road Dove Holes, Buxton, SK17 8DL
M6 RNZ R. Baldwin, 98 Rosemary Avenue, Braintree, CM7 2TA
M6 ROA R. Parrott, 37 Wren Place, Gillingham, SP8 4WE
M6 ROB R. Smith, 21 Canal Road Crossflatts, Bingley, BD16 2SR
M6 ROC C. Martin, Kiln Close, Main Road, Lincoln, LN4 4QH
M6 ROE P. Roe, 36 Rutland Crescent, Harworth, Doncaster, DN11 8HZ
M6 ROF A. Martin, 81 Langstone Road, Dudley, DY1 2NL
MM6 ROH R. Hutton, 2 Watson Place, Dunfermline, KY12 0DR
M6 ROI P. King, 18 Paddock Close, Wantage, OX12 7EQ
M6 ROJ R. Gudgeon, 6 The Choakles, Wootton, Northampton, NN4 6AP
M6 ROL R. Rollinson, Beauty Bank Farm, Six Ashes, Bridgnorth, WV15 6ER
M6 ROM C. Romocea, 14 Foxgrove Path, Watford, WD19 6YL
M6 ROP R. Pochat, Hame, Bealswood Road, Gunnislake, PL18 9DA
M6 ROU S. Courree, 4-6 Alhambra Road, Southsea, PO4 0RL
M6 ROV R. Van-Der-Wijst, 6 Willow Street, Romford, RM7 7LJ
M6 ROW C. Hartshorn, 63 Mill Road, Maldon, CM9 5HY
MW6 ROX S. Baker, Sunville, Village Road, Mold, CH7 6HT
M6 ROY R. Trent, 62 Dean Street, Radcliffe, Manchester, M26 3TZ
M6 RPC R. Cobb, 57 Adams Drive, Willesborough, Ashford, TN24 0FX
M6 RPD D. Rajanayagam, 87 Riffel Road, London, NW2 4PG
M6 RPE Dr A. Wallman, Oakwood, 30 Elmsway, Bramhall, Stockport, SK7 2AE
M6 RPH R. Hunt, 7 Tinsley Close, Luton, LU1 5QD
M6 RPL R. Lee, 13 Earle Street, Barrow-in-Furness, LA14 2PZ
M6 RPM D. Williams, 96 Bingley Court, Canterbury, CT1 2SX
MM6 RPN P. Toner, 1 Milton Mains Road, Clydebank, G81 3NF
M6 RPO W. Donnelly, 9 Old Laundry Mews, Laundry Lane, Ingleton, Carnforth, LA6 3GH
M6 RPP J. Anderson, 19A Buttermarket, Thame, OX9 3EP
M6 RPQ R. Back, 79 Aspin Close Wellington Home, Somerset, TA21 9EG
M6 RPR M. Roper, 13 St. Cuthbert Street, Worksop, S80 2HN
MW6 RPS C. Hurley, 14 Magazine Street, Maesteg, CF34 0TG
M6 RPU A. Smith, 20 Bishops Avenue, Worcester, WR3 8XA
M6 RPV R. Connolly, 29 Gayer Street, Coventry, CV6 7EU
M6 RPW C. Allan-Mcwilliams, 81 Skipper Way, Little Paxton, St Neots, PE19 6LT
M6 RPX R. Price, 1 Blakiston Close, Ashington, Pulborough, RH20 3GL
M6 RPY Dr L. Percival-Alwyn, 2 Sand Road, Great Gransden, Sandy, SG19 3AQ
MW6 RQC R. Chegwin, 17 Cyncoed Crescent, Cardiff, CF23 6SW
MM6 RQD R. Dean, 1 The Old Post Office Main Street, Golspie, KW10 6RA
M6 RQE R. Fakes, 24 Fairways, Weyhill, Andover, SP11 8DW
M6 RQH R. Hawkes, The Old Oak Bungalow, Crossway Green, Stourport-on-Severn, DY13 9SJ
MW6 RQM J. Morris, 1 Minafon, Newtown, SY16 1RH
M6 RQN J. Foster, 23 High Street, Cumnor, Oxford, OX2 9PE
M6 RRA R. Gibbons, 18 Langdale Road, Ribbleton, Preston, PR2 6AN
M6 RRC K. Robinson, 103 Recreation Street, Mansfield, NG18 2HP
M6 RRD R. Davies, 76 Berry Park, Saltash, PL12 6EN
M6 RRF T. Allman, 46 Belmont Road, Rugby, CV22 5NZ
M6 RRH S. Walters, 87 Fairbourne Close, Bransholme, Hull, HU7 5DH
M6 RRJ R. Jobber, 79 Falcon Way, Ashford, TN23 5UR
M6 RRK R. Rao, 3 Weller Mews, Enfield, EN2 8FG
M6 RRL K. Woodhams, 83 Langdale Place, Newton Aycliffe, DL5 7DY
M6 RRR R. Capon, 49 Littlemoor Road, Illingworth, Halifax, HX2 9EF
M6 RRV R. Oxley, 17 Hardhurst Road, Alvaston, Derby, DE24 0LF
MW6 RRW K. Hazlewood, 22 Lon Cwm, Llandrindod Wells, LD1 6BE
M6 RSB R. Blackman, 32 Kingfisher Road, Sprowston, NR7 8GX
M6 RSD R. Davies, 8 The Brambles, Bar Hill, Cambridge, CB23 8TA
M6 RSF A. Robson, 270 Calder Road, Lincoln, LN5 9TL
MI6 RSH M. Rush, 122 Cullaville Road Crossmaglen, Newry, BT35 9AQ
M6 RSI R. Best, 14 Charles Street, Bugle, St Austell, PL26 8PS
M6 RSJ S. Rebisz, 29 Rosecroft Drive, Nottingham, NG5 6EH
M6 RSL B. Leeson, 18 Georgeham Close, Wigston, LE18 2HZ
M6 RSO R. Oliver, 49 Albany Way, Skegness, PE25 2NB
MM6 RSP R. Phillips, 32 Parkview, Ayr, KA7 4QF
M6 RSR S. Richtering, 22 Darwin Drive, Driffield, YO25 5PF
M6 RSS R. Brooks, 52 Harebell Drive, Portslade, Brighton, BN41 2UZ
M6 RST M. O'Donoghue, 24 Whitelock Road, Abingdon, OX14 1NZ
MD6 RSV N. Hazell, 43 Allan Street, Douglas, Isle of Man, IM1 3DP
M6 RSY R. Sayre, 8 Lorne Road, Richmond, TW10 6DS
M6 RTA S. Himsworth, 17 Forber Road Saltersgill, Middlesbrough, TS4 3HJ
M6 RTC R. Churchill, 16 Mansfield Close, Worthing, BN11 2QR
M6 RTD R. Wilson, 18 Baldwin Road, Bewdley, DY12 2BP
M6 RTE S. Warde, 36 Cornwallis Road, London, E17 6NN
M6 RTG I. Sharpe, 4 Low Dowfold, Crook, DL15 9AE
M6 RTI R. Todd, 42 K D Tower, Cotterells, Hemel Hempstead, HP1 1AS
M6 RTM R. Townsend, 28 Steele Street Hoyland Common, Barnsley, S74 0ps
M6 RTN M. Fletcher, 8 Brigham Hill Mansion, Brigham Great Broughton, Brigham, Cockermouth, CA13 0TL
MM6 RTO M. Smith, 2 Richmond Court, Dundee, DD2 1BF
M6 RTP R. Paster, 8 Rachaels Lake View, Warfield, Bracknell, RG42 3XU
M6 RTQ Dr R. Tempo, 35 Warminster Road, Bath, BA2 6XG
M6 RTR A. Gill, 12 Andersons Court, Plymyard Avenue, Wirral, CH62 6ef
M6 RTU A. Loukes, 14 Batchwood View, St Albans, AL3 5TD
M6 RTX B. Purvis, 23 Deans Gardens, St Albans, AL4 9LS
M6 RTZ A. Lewis, 43 High Street, Colney Heath, St Albans, AL4 0NS
M6 RUB R. Beck, 13 Chaseside Avenue, Twyford, Reading, RG10 9BT
MI6 RUC W. Rafferty, 44 Lestannon Avenue, Whitehead, Carrickfergus, BT38 9NN
M6 RUH R. Plumley, 9 Haynes Road, Kettering, NN16 0NG
M6 RUI N. Ginns, 45 Gapstile Close, Desborough, Kettering, NN14 2TZ
M6 RUK M. Shockness, 38 Park Grange Court, Sheffield, S2 3SY
M6 RUM L. Rumbelow, 71 Anchorage Lane, Doncaster, DN5 8EB
M6 RUN M. Saunders, 68 Heywood Road, Prestwich, Manchester, M25 1FN
M6 RUP A. Tomlinson, 10 Beech Street, Hollingwood, Chesterfield, S43 2HN
M6 RUS R. Trickey, 19 Eastfields, Folkestone, CT19 5RU
M6 RUT R. Williams, 11 Berkeley Gardens, London, N21 2BE
M6 RUZ R. Brierley, 39 Hatfield Road, Alvaston, Derby, DE24 0BU
M6 RVA V. Davies, 49 Victoria Court, Birkdale, PR8 2DN
M6 RVD R. Davies, 4 Haven Villas, Ferry Road, Exeter, EX3 0JW
M6 RVE A. Mitchell, 121 Totteridge Lane, High Wycombe, HP13 7PH
M6 RVG R. Kobela, 32 Allen Road, Rushden, Nn10 0ft
M6 RVH R. Hughes, 19 Pendine Crescent, North Hykeham, Lincoln, LN6 8UW
M6 RVI R. Miranda, 33 Hill Rise, Luton, LU3 3EA
M6 RVJ J. Roa Vicens, 80 Queensway, London, W2 3RL
M6 RVM R. Moore, 22 Chase Close, Nuneaton, CV11 6AJ
M6 RVN I. Smith, 2 Frederick Street, Woodville, Swadlincote, DE11 8BX
M6 RVR R. Varasani, 34 Freemans Rd, Minster, Ramsgate, CT12 4EL
M6 RVY R. Harvey, 1 Bickley Moss, Whitchurch, SY13 4JF
M6 RWB W. Stobbs, 19 Stonecliffe Bank, Leeds, LS12 5BL
M6 RWE D. Woodhouse, 18 Kirk Close, Ripley, DE5 3RY
MM6 RWI R. Wilinski, 1 Keil Gardens Benderloch, Oban, PA37 1SY
M6 RWJ A. Nelson, 38 Warbro Road, Torquay, TQ1 3PW
M6 RWP R. Whalley, 65 Stanley Street, Nelson Lancashire, BB9 7ET
M6 RWS K. Saville, Flat 4, Birchitt Court, 152 Bradway Road, Sheffield, S17 4QX

M6 RWT R. Trueman, 83 Morton Street, Middleton, Manchester, M24 6AX
M6 RWV A. Durrant, 22 Supple Close, Norwich, NR1 4PP
M6 RWX C. Anderson, 191 Waveney Road, Hull, HU8 9NA
M6 RXA R. Arnold, 20 South View, Frampton Cotterell, Bristol, BS36 2HT
MI6 RXC T. Masterson, 2 Pinley Drive, Banbridge, BT32 3TZ
M6 RXE R. Ebdon, 6 Bankside, Headington, Oxford, OX3 8LT
M6 RXG M. Robson, 270 Calder Road, Lincoln, LN5 9TL
MI6 RXH R. Harvey, 464 Drumbeg South, Craigavon, BT65 5AQ
MM6 RXJ R. Johnstone, 14 Hoover Driver, Cambuslang, Glasgow, G72 7EF
M6 RXL H. Leach, 34 Forest Close, Crawley Down, Crawley, RH10 4LU
M6 RXR R. Rich, 54 Paddocks Way, Ferndown, BH22 9FW
M6 RXS R. Sanders-Hewett, 26 Verulam Road, Hitchin, SG5 1QE
M6 RXT R. Thomas, 43B Napier Road, Bromley, BR2 9JA
M6 RXU C. Attrill, 42 Purley Way, Clacton-on-Sea, CO16 8YX
M6 RXX A. Tomczynski, 66 Commercial Street, London, E1 6LT
MW6 RXZ J. Davies, 8 Coronation Road, Upper Brynamman, Ammanford, SA18 1BB
M6 RYA R. Tarr, 16 Stoneleigh Court, Frimley, Camberley, GU16 8XH
MI6 RYC R. Carmichael, 21 Stuart Park, Ballymoney, BT53 7BE
MW6 RYD R. Davies, 63 Maes Y Gwernen Road, Cwmrhydyceirw, Swansea, SA6 6LL
MM6 RYG R. Young, 22 Station Road Armadale, Bathgate, EH48 3LN
M6 RYK G. Mackay, 5 Furze Street, Carlisle, CA1 2DL
M6 RYL R. Harris, 183A Painswick Road, Gloucester, GL4 4AG
M6 RYN R. Clough, 8 Skeldyke Road, Kirton, Boston, PE20 1LR
M6 RYR R. Brown, 2 Westgate, Leominster, HR68SA
M6 RYZ J. Fryd, The Cottage, Old Blackalls Drive Cholsey, Wallingford, OX10 9HD
M6 RZA M. Box, 273 Broad Lane, Birmingham, B14 5AF
M6 RZD R. Dainton, 1 The Woodlands, Stroud, GL5 1QE
M6 RZE J. Stephenson, Flat 12, St. Clares Court, St. Clares Avenue, Havant, PO9 4JF
M6 RZI T. Hurst, 22 Coronation Street, Mansfield, NG18 2QL
MW6 RZL R. Higgins, 44 Maeshyfryd Road, Holyhead, LL65 2AL
M6 RZO D. Bardell, 32 Bridle Road, Watton, Thetford, IP25 6NA
M6 RZP R. Williams, 10 Bramley Close, Twickenham, TW2 7EU
M6 RZW R. Laye, 11 Hillhead Rise, Falmouth, TR11 5GZ
M6 RZX B. Forrest, 32 Idonia Road, Wolverhampton, WV6 7NQ
M6 RZY R. Howard, 8 Pennant Close Birchwood, Warrington, WA3 6RR
M6 RZZ R. Reynolds, 22A Beech Avenue, Shepton Mallet, BA4 5XW
MW6 SAA S. Llewellyn, 53 Cripps Avenue, Cefn Golau, Tredegar, NP22 3PF
M6 SAC M. Clark, 99 Cheylesmore Drive, Frimley, Camberley, GU16 9BW
M6 SAD S. Abraham, 12 Graham Road, Halesowen, B62 8LJ
M6 SAE S. Everett, 12 Broadlands, Netherfield, Milton Keynes, MK6 4HL
M6 SAF S. Fisher, 25 Tennyson Road, Diss, IP22 4PY
M6 SAH S. Painter, 47 Longmead Drive, Nottingham, NG5 6DP
M6 SAI T. Beighton, 97 Great North Road, Woodlands, Doncaster, DN6 7NH
M6 SAK S. Kendrick, 29 Waterside Silsden, Keighley, BD20 0LQ
M6 SAL S. Lester, 71 Ronelean Road, Surbiton, KT6 7LL
M6 SAM M. Sampson, 45 Carron Street, Stoke-on-Trent, ST4 3DT
M6 SAN S. Weightman, 131 Leeds Road, Birstall, Batley, WF17 0JZ
M6 SAQ S. Wills, 8 Amherst Road, Newcastle upon Tyne, NE3 2QQ
M6 SAS A. Speight, 112 Fern Avenue, Staveley, Chesterfield, S43 3RA
M6 SAU C. Saunders, 68 Heywood Road, Prestwich, Manchester, M25 1FN
M6 SAV S. Morrell, 20 Brocklebank Close, Bassingham, Lincoln, LN5 9LJ
MW6 SAW P. Campigli, 43 Waterloo Road, Penylan, Cardiff, CF23 9BJ
MW6 SAX J. Elsmore, 8 Clos Aberconway, Prestatyn, LL19 9HU
M6 SAY S. York, 1 The Cottage, Dogdyke Bank, Lincoln, LN4 4JQ
M6 SBB S. Jeffery, 79 Greenbank Road, Watford, WD17 4FJ
M6 SBD S. Radford, Littlehampton Marina, Ferry Road, Littlehampton, BN17 5DS
M6 SBE S. Bennett, 154 Humberstone Road, Grimsby, DN32 8HR
M6 SBF S. Raine, 91 Lulworth Avenue, Jarrow, NE32 3SB
M6 SBH T. Thomas, 269 Church Road, St. Annes, Lytham St Annes, FY8 3NP
M6 SBI S. Bassett, 3 Lower Merryfield, Anchor Road, Radstock, BA3 5PG
M6 SBJ S. Crombie, 8 Hazel Close, Haverhill, CB9 9LY
M6 SBK R. Colman-Whaley, 37 Suters Drive, Taverham, Norwich, NR8 6UU
M6 SBL S. Blaney, 213 Albert Road, Poole, BH12 2EZ
M6 SBM S. Brightmore, 36 Laburnum Road, Langold, Worksop, S81 9RR
M6 SBN S. Hanson, 59 Weetwood Avenue Ne71 6Af, Wooler, NE71 6AF
M6 SBO B. Mcglynn, 22 Bracken Bank Way, Keighley, BD22 7AB
M6 SBR S. Flowers, 56 Pilkington Road Braunstone, Leicester, LE3 1RA
M6 SBS S. Stiddard, 21 Studland Park, Westbury, BA13 3HQ
MW6 SBT A. Jones, 10 Dan Y Bryn, Caerau, Maesteg, CF34 0UW
M6 SBU S. De Chastelain, 31 Campion Court, Northampton, NN3 9BW
M6 SBV D. Hindle, 18 Haig Street, Selby, YO8 4BY
M6 SBW J. Stephenson, 4 Carlow Drive West Sleekburn, Choppington, NE62 5UT
M6 SBY S. Hollis, 24 Bosley View, Congleton, CW12 3TU
M6 SBZ D. Sibley, 85 Brick Crescent, Stewartby, MK43 9GG
M6 SCA S. Frampton, 4 Sussex Gardens, Fleet, GU51 2TL
M6 SCB S. Burnage, 124 Mayfields, Spennymoor, DL16 6TT
MI6 SCC S. Cole, 16 Otterbank Road, Strathfoyle, Londonderry, BT47 6YB
MW6 SCD S. Davies, 28Y Fron, Llanelli, SA14 7DN
M6 SCF S. Faulkner, Mount Pleasant, Elkstones, Buxton, SK17 0LU
MM6 SCG S. Greenland, 0/1 22 Linden Street, Anniesland, Glasgow, G13 1DQ
M6 SCH S. Hall, 50 Charnock Wood Road, Sheffield, S12 3HN
M6 SCI M. Pope, Drove Lodge 39 The Drove Barroway Drove, Downham Market, PE38 0AJ
M6 SCK J. Lynch, 19 Riverside, Clitheroe, BB7 2NP
M6 SCM S. Cumberland, 11 Cleveleys Road, Blackburn, BB2 3JS
M6 SCN R. Rothery, 24 Maes Y Dafarn, Carno, Caersws, SY17 5NG
M6 SCP S. Pettit, 24 Bickington Lodge Estate, Barnstaple, EX31 2LH
M6 SCQ A. Chamberlain, 35 Wexham Close, Luton, LU3 3TU
M6 SCU D. Willetts, Domus, Broyle Lane Ringmer, Lewes, BN8 5PG
M6 SCV S. Chandler, 7 Clinton Road, Redruth, TR15 2LL
M6 SCW S. Whittaker, 25 Cleveleys Road, Blackburn, BB2 3JS
M6 SCX S. Carpenter, Field View, Old Lyndhurst Road, Southampton, SO40 2NL
M6 SDA O. Suda, 52 Coltman Street, Hull, HU3 2SG
MW6 SDE D. Ezard, 59 Station Farm, Croesyceiliog, Cwmbran, NP44 2JW
M6 SDF A. Hulok, 3 Dickson Court, Sittingbourne, ME10 3LG
M6 SDG S. Mallows, 74 St Alban'S Close, Gillingham, me7 1tx
M6 SDH S. Hopkins, 61 Green Avenue, Astley, Tyldesley, Manchester, M29 7FF
M6 SDI S. Legg, Woodview, Clay Lane, Clacton-on-Sea, CO16 8HH
M6 SDL J. Layden, 51 Markham Road, Langold, Worksop, S81 9SH
M6 SDN A. Mcconkey, 7 Darwin Road, Stevenage, SG2 0DE
M6 SDP S. Porter, 18 The Crescent, Bircotes, Doncaster, DN11 8DT
M6 SDQ S. Dean, 39 Low Grange View, Leeds, LS10 3DT
M6 SDS S. Brown, 4 Bishop Fox Drive, Taunton, TA1 3HQ
M6 SDT D. Grace, 107 Bush Avenue, Little Stoke, Bristol, BS34 8NG
M6 SDU S. Seddon, 3 Kinsley Close, Ince, Wigan, WN3 4PQ
MW6 SDV S. Price, 27 Gilfach Road, Penygraig, Tonypandy, CF40 1EN
M6 SDW R. Wilson, 84 Sir Thomas Whites Road, Coventry, CV5 8DR
M6 SDX S. Page, Flat 25 The Riverfront, Eastern Esplanade, Canvey Island, SS8 7DN
M6 SDZ S. Dimopoulos, 160 West Wycombe Road, High Wycombe, HP12 3AE
M6 SEA S. Aspinall, 33 Covertside Road, Scarisbrick, Southport, PR8 5HB
M6 SEB S. Plowman, 7 Birkdale Close, Cudworth, Barnsley, S72 8EW
M6 SEC T. Summers, 5 Fairclough Place, Adlington, Chorley, PR7 4AN
M6 SEE A. Norton, 9 The Common, West Tytherley, Salisbury, SP5 1NS
MW6 SEF M. Williams, 2 St. Andrews Road, Wenvoe, Cardiff, CF5 6AF
M6 SEG S. Todd, 24 Pentland Avenue, Redcar, TS10 4HD
MM6 SEI D. Crane, Otterburn, Dervaig, Isle of Mull, PA75 6QL
M6 SEJ S. Jeronimo, 46 Pretoria Road, Chertsey, KT16 9AZ
M6 SEK A. Collins, 5 Grove Close, Basingstoke, RG21 3AS
M6 SEN A. Snelling, 6 Aylsham Road, Buxton, Norwich, NR10 5EX
M6 SEO A. Gardner, 17 Devonshire Court, Moorland Road, Weston-Super-Mare, BS23 4JA
MW6 SEP D. Jones, 46 East Avenue, Caerphilly, CF83 2SR
M6 SER S. Richards, 5 Bloomfield Drive, Bracknell, RG12 2JW
M6 SES S. Stoney, 2 Fishers Mead, Dulverton, TA22 9EN
M6 SEU M. Fairbairn, 36 Avebury Place Eastfield-Lea, Cramlington, NE23 2UR
M6 SEV S. Greaves, The Grange Farmhouse, Leicester Forest East, LE3 3GA
M6 SEY S. Borrell, Rose Cottage, Colchester Main Road, Colchester, CO7 8DD
MI6 SEZ Dr P. Wilson, 2 Tweskard Lodge, Belfast, BT4 2RH
MW6 SFB P. Latham, 20 Kenyon Avenue, Wrexham, LL11 2ST
M6 SFC J. Ellery, 7 Midanbury Crescent, Southampton, SO18 4FN
M6 SFC P. Craig, 4 Poolside, Burston, Stafford, ST18 0DR
MM6 SFF S. Ferguson, 17 Brown Street, Shotts, ML7 5HW
MI6 SFH S. Hand, 12 Church Street, Rosslea, Enniskillen, BT92 7DD
M6 SFJ A. Morton, 6 Chessar Ave, Chessar Ave Blakelaw, Newcastle, NE5 3RE
MI6 SFK S. Mcelmurray, 43 Main Street, Sixmilecross, Omagh, BT79 9NH
M6 SFL D. Webber, 11 Waghorn Road, Harrow, HA3 9ET
M6 SFM S. South, Ivy Cottage, Finkle Street Lane, Sheffield, S35 7DH
M6 SFQ B. Cross, 22 Park Avenue Washingborough, Lincolnshire, LN 4 1 DB
MM6 SFR A. Marshall, 6 Heron Way, Minnigaff, Newton Stewart, DG8 6PZ
M6 SFS S. Spooner, 51 Lewisham Court, Morley, Leeds, LS27 8QB
M6 SFT S. Foster, 16 Birchcroft Road, Ipswich, IP1 6PA
M6 SFV S. Fry, Piccadilly Farm, Aggs Hill, Cheltenham, GL54 4ET
M6 SFW J. Walker, 6 Wellington Terrace, Islip, Kettering, NN14 3LJ
M6 SGA S. Gregory, 9 Croftlands Road, Wythenshawe, M22 9YE
M6 SGC S. Latter, 1539 Great Cambridge Road, Enfield, EN1 4SY
M6 SGD S. Gibbs, 35 St. Michaels, Houghton le Spring, DH4 5NR
M6 SGE N. Pulev, Zapaden Park, 79-B, Flat 4, Sofia, Bulgaria, 1373
MM6 SGF S. Gray, 9 Caledonian Crescent, Prestonpans, EH32 9GF
M6 SGG S. Gillett, 38 Bury Road, Thetford, IP27 0Bt
M6 SGH S. Holt, 14 Fir Street, Cadishead, Manchester, M44 5AU
M6 SGJ S. Powell, Crosstrees, Main Road, Theberton, Leiston, IP16 4RX
M6 SGL S. Gillard, 1 Chevening Close, Stoke Gifford, Bristol, BS34 8NJ
M6 SGM S. Roberts, 17 East View Marsh, Huddersfield, HD1 4NU
M6 SGN B. Frith, 159 Milton Road, Grimsby, DN33 1DN
MM6 SGO S. Green, 4 Mid Avenue, Port Glasgow, PA14 6PL
M6 SGP S. Phipps, 53 Pioneer Avenue, Burton Latimer, Kettering, NN15 5LJ
MM6 SGQ S. Gilruth, 20 The Aspens, Carberry Crescent, Dundee, DD4 0XJ
M6 SGS G. Street, 105 Jeals Lane, Sandown, PO36 9NS
M6 SGT A. Williams, 11 Broadsmith Avenue, East Cowes, PO32 6QW
MI6 SGU A. Curran, 44 Ardboe Drive, Lurgan, Craigavon, BT66 8HP
M6 SGV M. Alcaino Pizani, Flat 45, Brian Redhead Court, 123 Jackson Crescent, Manchester, M15 5RR
M6 SGY S. Johnson, 2 North Square, Edlington, Doncaster, DN12 1ED
MM6 SGZ B. Fitzsimmons, 8 Firth View, Brodick, KA27 8FJ
M6 SHA M. Callaway, 19 The Oval, Dudley, DY1 2LN
MM6 SHB X. Bradshaw, 55 Headwell Avenue, Dunfermline, KY12 0JX
M6 SHC S. Corden, 59 Brindles Field, Tonbridge, TN9 2YR
M6 SHD G. Fisk, 22 Church Street, Wangford, Beccles, NR34 8RN
M6 SHF B. Thwaites, 109 Emerson Crescent, Sheffield, S5 7SW
M6 SHH R. Devos, 76 North Parade, Sleaford, NG348AW
M6 SHI S. Shillabeer, 6 Meadow Court, Nomans Heath, Malpas, SY14 8DU
MW6 SHJ S. Jones, 56 Ffordd Offa, Rhosllanerchrugog, Wrexham, LL14 2EY
M6 SHK A. Al-Shakarchi, 17 Fairfax Place, London, NW6 4EJ
M6 SHL S. Lalji, 76 Abbotts Drive, Wembley, HA0 3SG
MM6 SHM I. Mcewan, 40 Kirkland Road, Dunlop, Kilmarnock, KA3 4AQ
M6 SHP A. Sharp, 30 Burbidge Close, Calcot, Reading, RG31 7ZU
M6 SHQ J. Butcher, 14 Park Road, Lowestoft, NR32 1SW
M6 SHU S. Maton, 41 Bemerton Gardens, Kirby Cross, Frinton-on-Sea, CO13 0LQ
M6 SHV S. Roberts, 10 Morningside Way, Liverpool, L11 1BD
M6 SHW S. Wraith, 20 Alexandra Road Charlestown, Weymouth, DT4 9TL
M6 SHZ S. Dainton, 1 The Woodlands, Stroud, GL5 1QE
MW6 SID S. Jones, 25 The Crescent, Tredegar, NP22 3HN
M6 SIE B. Brown, 114 Woodhorn Road, Ashington, NE63 9EN
M6 SIF S. Frost, 4 Banister Way, Shipston-on-Stour, CV36 4JU
M6 SIH S. Tomkins, Flat 2, 18 Brandon Way, Birchington, CT7 9XE
M6 SII S. Gray, 12 Bell Street, Henley-on-Thames, RG9 2BG
M6 SIJ J. Simkins, 37 St. Andrews Meadow, Harlow, CM18 6BL
M6 SIK W. Henson, 24 Grimshaw Close, North Road, London, N6 4BH
MM6 SIM S. Beeson, Muir Cottage, The Muirs, Huntly, AB54 4GD
M6 SIP A. Davies, 32 Kinross Road, Wallasey, CH45 8LH
M6 SIR B. Greaves, 2 Peel Street, Padiham, Burnley, BB12 8RP
MM6 SIS G. Mccaughey, 24A, Williamson Place, Johnstone, PA5 9DW
M6 SIU S. Murdoch-Mckay, 19 Beachy Road, Crawley, RH11 9HN
MM6 SIV M. Sives, 4 Fir Grove, Livingston, EH54 5JP
M6 SIW S. Cartwright-Proctor, 448 Tuttle Hill, Nuneaton, CV10 0HR

M6 SIX B. North, 54 Parklands, Mablethorpe, LN12 1BY
M6 SIY S. Lilley, 34 Rye Crescent, Danesmoor, Chesterfield, S45 9HH
M6 SIZ W. Easdown, 11 Mulcaster Avenue, Kidlington, OX5 2HG
M6 SJA S. Bridges, 65 Abbots Gate, Bury St Edmunds, IP33 2GB
M6 SJB S. Buckley, 64 Wolseley Road, Rugeley, WS15 2ES
M6 SJC S. Charles, 29 Woolford Close, Winchester, SO22 4DN
MI6 SJD S. Dallas, 101 Coagh Road Stewartstown, Dungannon, BT71 5JL
M6 SJE J. Hill, 7 Knebworth Court, Congleton, CW12 3SW
M6 SJF S. Fox, 77 The Grove, London, N13 5JS
M6 SJH S. Hayden, 5 Blackbird Close, Thurston, Bury St Edmunds, IP31 3PF
M6 SJJ S. Jackson, 54 Sefton Avenue, Poulton-le-Fylde, FY6 8BL
MM6 SJK Dr D. Krauskopf, Cocklehaa, Urafirth, Shetland, ZE2 9RH
M6 SJN D. Lawson, 56 River Bank East, Stakeford, Choppington, NE62 5XA
M6 SJO S. Boniface, 19 Toronto Drive, Smallfield, Horley, RH6 9RB
M6 SJQ C. Gyngell, 54 Association Walk, Rochester, ME1 2XD
M6 SJR S. Ray, 18 Crescent Way, Cholsey, Wallingford, OX10 9NE
M6 SJU J. Searle, 18 The Avenue, Bloxham, Banbury, OX15 4QU
MI6 SJV D. Parkinson, 16 Beechwood Gardens, Moira, Moira, BT67 0LB
M6 SJW E. Watson, 4 Glenluce Drive, Preston, PR1 5TB
M6 SJX S. Bates, 6 Foxdell, Northwood, HA6 2BU
M6 SJZ S. Lampard, 63 Broadmayne Road, Poole, BH12 4EH
M6 SKF S. Froggett, 1 The Paddocks, Pilsley, Chesterfield, S45 8ET
M6 SKH S. Hemmings, 26 Austin Drive, Banbury, OX16 1DJ
M6 SKL S. Maqbool, 69 Waltham Close, West Bridgford, Nottingham, NG2 6LD
M6 SKN T. Watkins, 72A St. Clements Road, Keynsham, Bristol, BS31 1BA
M6 SKP S. Pryer, 16 Wayside Avenue, Worthing, BN13 3JU
M6 SKQ M. Berrisford, 5 Branwell Drive, Haworth, Keighley, BD22 8HG
MW6 SKR S. Rogers, 30 Coed Celynen Drive, Abercarn, Newport, NP11 5AU
M6 SKT T. Robinson, 39 Cemetery Road, Laceby, Grimsby, DN37 7ER
M6 SKU M. Baker, 14 Thornberry Drive, Dudley, DY1 2PL
M6 SKW S. Kneeshaw, 39 Cherrytree Walk East Ardsley, Wakefield, WF32HS
M6 SKX S. Key, 29 Ellesmere Crescent, Brackley, NN13 6BP
M6 SKY R. Coles, Bay Cottage, St. Catherines Road, Ventnor, PO38 2NE
M6 SKZ T. Barham, 88 Rundells, Harlow, CM18 7HD
M6 SLA S. Harvey, 40 Littlemoor Lane, Newton, Alfreton, DE55 5TY
M6 SLB S. Berry, 9 Magnolia Street, Winnington, Northwich, CW8 4EH
M6 SLC D. Morphew, 8 Blinco Lane, George Green, Slough, SL3 6RQ
M6 SLD S. Light, 61 Eversfield Road, Horsham, RH13 5JS
M6 SLE S. Lee, 360 Ringley Road, Stoneclough, Manchester, M26 1EP
M6 SLG S. Gash, 22 Wood View, Rugeley, WS15 1AT
M6 SLI S. Lisi, 60 Middlecroft Road, Staveley, Chesterfield, S43 3XH
M6 SLJ L. Jones, 137 Breck Road, Poulton-le-Fylde, FY6 7HJ
M6 SLK S. Lake, 64 Womersley Road, Norwich, NR1 4QB
M6 SLL L. Butler, 42 Station Road, Stanbridge, Leighton Buzzard, LU7 9JF
M6 SLR S. Perry, 44 Ashcombe Road, Dorking, RH4 1NA
M6 SLT S. Townsend, 28 Steele Street, Hoyland, Barnsley, S74 0PS
M6 SLU K. Cockburn, 20 Hexham Avenue, Hebburn, NE31 2HN
MM6 SLV K. Slaven, Drymuir, Drymuir, Peterhead, AB42 5RR
M6 SLZ S. Eloie, 26 Halsbrook Road, London, SE3 8QY
M6 SMA S. Hindmarsh, 7 George Street Murton, Murton, SR7 9BN
M6 SMB S. Bide, 9 Greenway, Watchet, TA23 0BP
M6 SMD S. Davies, 95 Lathkill Drive, Ashbourne, DE6 1TZ
M6 SME S. Elliott, 74 Preston Avenue, Alfreton, DE55 7JX
M6 SMF C. Smith, 21 Mill House Drive, Cheltenham, GL50 4RG
M6 SMG S. Hampshire, Rose Cottage, Heath Road, Norwich, NR12 0SU
M6 SMK S. Kendrick, 103A Latimer Street, Liverpool, L5 2RF
MM6 SML J. Macleod, 388 Garrynamonie, Lochboisdale, Isle of South Uist, HS8 5TX
MI6 SMM S. Maguire, 139 Carrowshee Park, Drumhaw, Enniskillen, BT92 0FS
MM6 SMN S. Nicoll, 15 Redford Walk, Edinburgh, EH13 0AF
M6 SMQ B. Mcgowan, 200 Thomas Drive, Liverpool, L14 3LE
M6 SMR G. Tudor, 31 Church Road, Little Sandhurst, Sandhurst, GU47 8HY
M6 SMX M. Feast, 30 Peveril Drive, Riddings, Alfreton, DE55 4AP
MM6 SMY S. Young, 103 Feorlin Way, Garelochhead, Helensburgh, G84 0EB
M6 SNC J. Sinclair, 19 Ridge Way, Edenbridge, TN8 6AU
M6 SNE S. Solanki, 2 Churchill Mews, Newcastle upon Tyne, NE6 1BH
M6 SNF S. Wall, Flat 1, 41 Alexandra Road, Cleethorpes, DN35 8LE
M6 SNG H. Pike, 14 Milton Avenue, Barnet, EN5 2EX
M6 SNH S. Hawkes, 2 St Mary'S Close, Aston, SG2 7EQ
M6 SNJ S. Briggs, 3 Chapel Road, Southrepps, Norwich, NR11 8UW
M6 SNN S. Ballard, 77 Lanchester Road, Birmingham, B38 9AG
M6 SNO X. Christofi, 19 Kingsland Avenue, Northampton, NN2 7PP
M6 SNP A. Alkhateb, 51 Mendip Crescent, Bedford, MK41 9ER
M6 SNQ S. Harrison, 38 Alma Road, Bournemouth, BH9 1AN
MM6 SNR S. Russell, 3 Rankin Road, Wishaw, ML2 8PG
M6 SNS S. Button, 6 Farfield, Retford, DN22 7TL
M6 SNU T. Lewis, 653 Main Road, Harwich, CO124NF
MD6 SNV A. Elliott, Round Table House, Ronague, Castletown, Isle of Man, IM9 4HJ
M6 SNW D. Snowden, 10 Woodcroft, Wakefield, WF2 7LS
MM6 SNY A. Dobie, 7 Urr Terrace, Castle Douglas, DG7 1BL
M6 SNZ S. Roshanmanesh, 123 The Vale, Edgbaston, Birmingham, B15 2RU
M6 SOC A. Wallace, 17 Dennis Road, Liskeard, PL14 3NS
M6 SOE P. Strickland, 100 Spitfire Road, Castle Donington, Derby, DE74 2AU
MW6 SOF D. Parry, 45 Taff Court Thornhill, Cwmbran, NP44 5UU
M6 SOG S. Jackson, 50 Leicester Road, Sharnford, Hinckley, LE10 3PR
M6 SOK J. Neary, 3 Wordsworth Close, Torquay, TQ2 6EA
MW6 SON G. Stewart, 62 Norton Road Penygroes, Llanelli, SA14 7RS
M6 SOO R. Landragin, 101 Linden Gardens, Enfield, EN1 4DY
MM6 SOR R. Ewing, Kildonan House, Caerlaverock Farm, Crieff, PH5 2BD
M6 SOT C. Scrivens, 28 Bank Hall Road, Stoke-on-Trent, ST6 7DL
M6 SOU A. Nur, 92 Bramble Avenue, Conniburrow, Milton Keynes, MK14 7AP
M6 SOV M. Mcdonald, 55 Bournemouth Avenue, Middlesbrough, TS3 0NN
M6 SOZ G. Knowles, 31 Gibb Lane, Catshill, Bromsgrove, B61 0JP
M6 SPA S. Etheridge, 50 Pond Road, Horsford, Norwich, NR10 3SW
M6 SPC J. Lyon, 16 Well Court, St. Anns Close, Andover, SP10 2FS
M6 SPG S. Coxon, 31 Marden Road, Staplehurst, Tonbridge, TN12 0NE
M6 SPH S. Harrison, 26 Lambert Road, Lancaster, LA1 2NA
M6 SPJ S. Johnston, 67 Eversfield Road, Horsham, RH13 5JS
M6 SPK S. Kay, 32 Glossop Street, Derby, DE24 8DU
M6 SPL S. Lycett, 58 Hazel Grove, Hucknall, Nottingham, NG15 6ED
M6 SPM S. Morris, 23 De Courtenai Close, Bournemouth, BH11 9PG
MW6 SPN S. Nelson, 13 Llwyn Briscoe, Holyhead, LL65 1HT
M6 SPP S. Fawcett, Hollins Farm, Marske, Richmond, DL11 7NH
M6 SPS S. Summers, 450 Baddow Road, Chelmsford, CM2 9RD
M6 SPT J. Gardner, Silverdale, Vicarage Lane, Ormskirk, L40 6HQ
M6 SPU G. Mason, 21 High Street, Carisbrooke, PO30 1NR
MI6 SPY J. Woods, 39 Shetland Street, Antrim, BT41 2TG
M6 SPZ S. Pendlebury, 6 Normanby Close, Bewsey, Warrington, WA5 0GJ
M6 SQB S. Stevens, 40 Heath Road, Exeter, EX2 5JX
M6 SQC S. Burton, 2 West Batter Law Farm Cottages, Hawthorn, Seaham, SR7 8RZ
M6 SQD P. Riding, 160 Capel Road, London, E7 0JT
M6 SQE S. Tudor, 201 Cruddas Park, Westmorland Road, Newcastle upon Tyne, NE4 7RG
M6 SQF Dr S. Pearce, 15 Hillfield Court Road, Gloucester, GL1 3QS
M6 SQI H. Blythe, 14 The Green, South Creake, Fakenham, NR21 9PD
M6 SQK C. Southey, The Limes, 21 Grove Road, Chelmsford, CM2 0EY
M6 SQL A. Burton, 12 Munden Grove, Watford, WD24 7EE
MI6 SQN A. Reid, 40 Hillfoot Street, Belfast, BT4 1PR
M6 SQO S. Macmurray, 21 Dymoke Green, St Albans, AL4 9LX
M6 SQU B. Turner, 35 Gosforth Lane, Watford, WD19 7AY
M6 SQW S. Waite, 61 King Edward Crescent, Leeds, LS18 4BE
M6 SRA S. Antill, Lodge Farm, Mansfield Road, Worksop, S80 3DL
M6 SRB S. Robinson, 2 Old Hall Crescent, Bentley, Doncaster, DN5 0DW
M6 SRC S. Cash, The Warren, Kent Hatch Road, Edenbridge, TN8 6SX
M6 SRD R. Strong, 82 Oakmount Road, Chandler'S Ford, Eastleigh, SO53 2LL
M6 SRE S. Bradley, 75 Weald Bridge Road North Weald, Epping, CM16 6ES
M6 SRF J. Cranston, 7 Cowen Gardens, Gateshead, NE9 7TY
M6 SRG S. Gibson, 66 Kinoulton Court, Grantham, NG31 7XP
M6 SRI J. Bibby, 12 James Street, Burton-on-Trent, DE14 3SB
MM6 SRL S. Leeman, 35 Foudland Court, Church Avenue, Insch, AB52 6JZ
M6 SRN R. Dunn, 15A Connaught Drive, Chapel St. Leonards, Skegness, PE24 5YS
M6 SRO C. King, 8A Barton Road, Bedford, MK42 0NA
M6 SRS S. Tait, 29 Hotspur Avenue, Bedlington, NE22 5TD
M6 SRV S. Vickers, 22 Thistle Green, Birmingham, B38 9TT
M6 SRZ P. Willetts, 223 Eaves Lane, Chorley, PR6 0AG
M6 SSA S. Aldersley, 245 St. Johns Road, Chesterfield, S41 8PE
M6 SSB M. Saywell, 8 Stanley Street North, Bristol, BS3 3LU
M6 SSC S. Azzaro, 5 Rye Hill Close, Bere Regis, Wareham, BH20 7LU
M6 SSD S. Shephard, 17 Grimsby Road, Laceby, Grimsby, DN37 7DF
MJ6 SSF S. Foot, 4 Aubin Place, Aubin Lane, Jersey, JE2 7PP
M6 SSM S. Millard, 44 South Bank Road, East Cowes, PO32 6JD
M6 SSN B. Woods, 28 Delph Drive, Burscough, Ormskirk, L40 5BE
MM6 SSO S. Jones-Martin, 14F High Street, Montrose, DD10 8JL
M6 SSP S. Pashley, 45 Highfield View Road Newbold, Chesterfield, S41 7JZ
M6 SSR S. Rymel, 3 Southview, Smalldale, Hope Valley, S33 9JQ
M6 SSS Dr S. Rowe, 9 Corfield Close, Finchampstead, Wokingham, RG40 4PA
M6 SST S. Slapper, 1 Standards Keep, Standards Road, Bridgwater, TA7 0EZ
M6 SSV K. Niendorf, 84 St. Marys Road, Stratford-upon-Avon, CV37 6XQ
M6 SSW B. Hood, 168 Shay Lane, Walton, Wakefield, WF2 6NP
M6 SSX J. Prout, 110 Dorothy Avenue North, Peacehaven, BN10 8DP
M6 SSY N. Hancock, 3 Market Street, Shipdham, Thetford, IP25 7LY
M6 SSZ A. Jarre, 8 Lingford Court Bishop Auckland County Durham Dl14 7Dz, Bishop Auckland, DL14 7DZ
M6 STA S. Ancill, 42 The Mill Pond, Holbury, Southampton, SO45 2QN
M6 STC S. Cousins, West Mill, Wareham Common, Wareham, BH20 6AA
M6 STF S. Briant, 1B College Road, Haywards Heath, RH16 1QN
M6 STH A. Huckle, 36 Weldbank Close, Beeston, Nottingham, NG9 5FU
M6 STI B. Richards, 58 Holm Lane, Oxton, Prenton, CH43 2HS
M6 STJ S. Sheppard, 34 Hardwicke Walk, Birmingham, B14 5XX
MU6 STK C. Stockwell, Fleurs Des Champs, La Colline Des Bas Courtils, St Saviours, Guernsey, GY7 9YQ
MM6 STM S. Mcmillan, 56 Mount Pleasant, Armadale, Bathgate, EH48 3HB
MI6 STN S. Nash, 5 Drumard Cottages, Dans Road, Ballymena, BT42 2PX
M6 STP S. Phythian, 24 Water Grove Road, Dukinfield, SK16 5QS
M6 STQ S. Totham, Cavaliers, Dane Court Manor, School Road, Tilmanstone, Deal, CT14 0JL
M6 STR M. Stroud, 1 Sefton Court, Welwyn Garden City, AL8 6WW
M6 STU S. Vzor, 40 Henlow Road, Birmingham, B14 5DS
M6 STV G. Wadsworth, 21 Appletree Road, Featherstone, Pontefract, WF7 5EA
M6 STY T. Fletcher, 9 Lawn Avenue, Kimpton, Hitchin, SG4 8QD
MM6 SUB S. Boyd, 269 Cloch Caravans Cloch Road, Gourock, PA19 1AZ
M6 SUC J. Murphy, Wessex House, Drake Avenue, Staines-upon-Thames, TW18 2AP
M6 SUD S. Griffiths, 37 Stourton Close, Knowle, Solihull, B93 9NP
M6 SUE S. Hadley, 60 Chapel Street, Pensnett, Brierley Hill, DY5 4EF
M6 SUF S. Fotheringham, 8 Ivanhoe Court, Ulrica Drive, Thurcroft, S66 9QP
M6 SUH S. Halewood, 12 Silver Street Riccall, York, YO19 6PB
M6 SUI S. Brooker, 4 Fleet End Close, Havant, PO9 5ED
M6 SUJ S. Jones, Marvin House, Ryhill Pits Lane, Cold Hiendley, Wakefield, WF4 2DU
M6 SUL R. Sullivan, 14 Colleton Drive, Twyford, Reading, RG10 0AU
M6 SUM P. Smith, 14 Highfield Crescent, Kettering, NN15 6JS
M6 SUN R. Beaumont, 61 Mitcham Road, Camberley, GU15 4AR
M6 SUP D. Markey, 7 Knightsway, Wakefield, WF2 7EG
MM6 SUR R. Fair, 6 Fairways, Stewarton, Kilmarnock, KA3 5DA
M6 SUU S. Coombes, 33 Clarence Park Road, Bournemouth, BH7 6LF
M6 SUV D. Arnold, The Chase, Rectory Road, Penzance, TR19 6BB
M6 SUW S. Urwin-Wright, 99 The Fairway, Midhurst, GU29 9JF
M6 SUX S. Harris, 17 Swindale, Wilnecote, Tamworth, B77 4LD
M6 SUY A. Fact, 50 Imison Street, Liverpool, L9 1EF
M6 SUZ S. Trigg, 2 Langley Common Road, Barkham, Wokingham, RG40 4TS
M6 SVF L. Garnett, 6 Tremaine Close, Norwich, NR6 5EL
M6 SVG J. Savage, 44 Hastings Road, Maidstone, ME15 7SP
M6 SVJ S. Carpenter, 18 Warwick Drive, Bury St Edmunds, IP32 6TF
MW6 SVM P. Edwards, Delfryn, Capel Dewi, Aberystwyth, SY23 3HU
M6 SVN S. Jackson, 18 Kentish Gardens, Tunbridge Wells, TN2 5XU
M6 SVT L. Tayler, 22 Wheatley Road, Leicester, LE4 2HN
M6 SVU S. Logan, Cedar House Reading Road North, Fleet, GU51 4AQ

MW6 SVV S. Berrow, 40 Priorsgate Oakdale, Blackwood, NP12 0EL
M6 SVY D. Southernwood, 24 Silver Gardens, Belton, Great Yarmouth, NR31 9PD
MW6 SVZ J. Pauline, 54 Laurel Road, Bassaleg, Newport, NP10 8NY
M6 SWA S. Worger, 6 Glendale Terrace, Mornington Road, Whitehill Bordon, GU35 9AJ
MI6 SWB S. Beatty, 132 Joanmount Gardens, Belfast, BT14 6NZ
MM6 SWC S. Caldwell, 45 Pappert, Alexandria, G83 9LE
M6 SWD C. Rich, Red Oak House, Summer Lane, Woodbridge, IP12 2QA
M6 SWE M. Nilsson, 81 Kingsway, Duxford, Cambridge, CB22 4QN
M6 SWG G. Wicks, 4 Bedford Street, Barnstaple, EX32 8JR
M6 SWH S. Hughes, 104 Thornley Road, Stoke-on-Trent, ST6 7BA
M6 SWI A. Waller, 64 Heaton Road, Billingham, TS23 3GP
M6 SWK K. Michael, 55 St Olaves Road, Norwich, NR3 4QB
M6 SWL P. Dekkers, 21 Nodens Way, Lydney, GL15 5NP
MW6 SWN S. Butler, 33 Heol Penlan, Neath, SA10 7LB
M6 SWO S. Woods, 92 Rubens Avenue, South Shields, NE34 8JT
M6 SWP S. Plume, 1A Handside Lane, Welwyn Garden City, AL8 6SE
M6 SWS I. Franklin, 23 Ingle Drive, Ashby-de-la-Zouch, LE65 2LW
MM6 SWT Z. Mckinnon, 8 Rowanlea Avenue, Paisley, PA2 0RP
M6 SXA N. Carter, 16 Swinburne Close Sutton Heights, Telford, TF7 4PZ
M6 SXB M. Comber, 9 Blackwell Road, East Grinstead, RH19 3HP
M6 SXC M. Revell, 13 Mount Pleasant, Framlingham, Woodbridge, IP13 9HQ
M6 SXD J. Brookes, 177 Charnwood Close, Rubery, Birmingham, B45 0JY
M6 SXG C. Champion, 155 Walton Road, Walton on the Naze, CO14 8NF
M6 SXI S. Robinson, 47 Platt Hill Avenue, Bolton, BL3 4JU
M6 SXM M. Cook, 30 St. Peters Crescent, Selsey, Chichester, PO20 0NA
M6 SXO A. Cooke, Cooke Towers, 9 Nyetimber Crescent, Pagham, PO213NN
M6 SXP S. Price, 43 Charter Road, Weston-Super-Mare, BS22 8LN
M6 SXT M. Dengate, 15 Barn Close, Pease Pottage, Crawley, RH11 9AN
M6 SXU K. Helm, 90 Horne Street, Bury, BL9 9HS
MM6 SXY L. Ritchie, Braiklaw, Blackhills, Peterhead, AB42 3LA
M6 SYB S. Baxby, 197 Cemetery Road, Wath-Upon-Dearne, Rotherham, S63 6LJ
M6 SYF H. Starling, 23 Lathom Road, Manchester, M20 4NX
M6 SYG S. Dibben, 2 Taynton Covert, Birmingham, B30 3QR
M6 SYH B. Hooper, 55 Gildas Avenue, Birmingham, B38 9HS
M6 SYI S. De Koster, 21 Normoor Road, Burghfield Common, Reading, RG7 3QG
M6 SYK P. Sykes, 16 Hill Fold, South Elmsall, Pontefract, WF9 2BZ
M6 SYW S. Ward, Wellbeck, Wheel Road, Alpington., Norwich, NR14 7NH
M6 SYX A. Davies, 4 Capella Path, Hailsham, BN27 2JY
MM6 SZA G. Inglis, The Shop, Roadside Skirza, Freswick, Wick, KW1 4XX
MM6 SZK T. Truesdale, 9 Cherry Lane, Banknock, Bonnybridge, FK4 1JY
MW6 SZL D. Phillips, 9 Baldwin Street, Newport, NP20 2LT
M6 SZP W. Davies, 61A Lightridge Road, Huddersfield, HD2 2HF
MW6 SZW S. Sewell, 53 Bryn Castell, Abergele, LL22 8QA
M6 TAA T. Akay, Caddebostan Mah. Plaj Yolu Sok. No25/15 Kad?Köy, ?Stanbul (Asya), Turkey, 34710
M6 TAD G. Tew, 66 St. Nicholas Estate, Baddesley Ensor, Atherstone, CV9 2EZ
M6 TAF A. Cornelius, 16 Crown House, North Street, Nailsea, Bristol, BS48 4SX
M6 TAG D. Cutter, 92 Hillcrest, Bar Hill, Cambridge, CB23 8TQ
M6 TAH T. Hutchinson, 134 Wingate Square, London, SW4 0AN
M6 TAJ D. Jarvice, 15 Meden Avenue, Warsop, Mansfield, NG20 0PS
MW6 TAK C. Smith, 7 Betws Avenue, Kinmel Bay, Rhyl, LL18 5BN
M6 TAL T. Lewis, 154 Meadow Head, Sheffield, S8 7UF
M6 TAO P. Washbrook, 1 Berry Hill Cottages, Berry Hill, Seaton, EX12 3BD
M6 TAU A. West, 33 Mundays Row, Waterlooville, PO8 0HF
MI6 TAW T. Wilmot, 4 Esdale Park, Bushmills, BT57 8RB
M6 TAY A. Ayres, Bryn Hyffryd, Phocle Green, Ross-on-Wye, HR9 7TW
MW6 TBC A. Doyle, 54 Bro Syr Ifor, Tregarth, Bangor, LL57 4AS
MW6 TBD A. Smith, In2Change 3 Palmyra Place, Newport, NP20 4EJ
M6 TBG A. Gravell, 21 Wickridge Close, Stroud, GL5 1ST
M6 TBH P. Bowman, 26 Albany Hill, Tunbridge Wells, TN2 3RX
M6 TBJ L. Perry, 56 Lambrook Road, Taunton, TA1 2AF
M6 TBK A. Green, Flat 19, 19-25 Marine Parade East, Clacton on Sea, CO15 1UX
MW6 TBL T. Lander, Glopa Cottage, Old Racecourse, Oswestry, SY10 7HP
M6 TBM T. Behan, 48 Montrose Avenue, Datchet, Slough, SL3 9NJ
M6 TBO M. Bradbury, 104 Lilly Hall Road Maltby, Rotherham, S66 8AT
M6 TBQ A. Trick, 2 Newell Close, Aylesbury, HP21 7FE
M6 TBR C. Herd, 10 Amethyst Close, Rainworth, Mansfield, NG21 0GH
M6 TBS D. Fletcher, 2 Hillside Close, Heddington, Calne, SN11 0PZ
M6 TBT A. Talbot, 30 Irwell Road, Walney, Barrow-in-Furness, LA14 3UZ
MW6 TBU M. Johns, 151 Somerset Street, Abertillery, NP13 1DR
M6 TBV R. Day, 152 Swievelands Road Biggin Hill, Biggin Hill Westerham, TN16 3QX
M6 TBX T. Burkinshaw, 78 Binsted Road, Sheffield, S5 8LL
M6 TCB P. Deluce, 114 Townsfield Road, Westhoughton, Bolton, BL5 2NT
M6 TCD T. Denny, 20 School Lane, Surbiton, KT6 7QH
M6 TCE R. Hannant, 24 Tower Hill Park Costessey, Norwich, NR8 5AT
M6 TCF I. Moore, 128 Dalestorth Street, Sutton-in-Ashfield, NG17 4EY
M6 TCG A. Gilberts, 22 Granby Road, Buxton, SK17 7TW
M6 TCH T. Hewlett, Tuckers Cottage, Alfold Road, Cranleigh, GU6 8NB
M6 TCI S. Friend, 1 Orford Road, Tunstall, Woodbridge, IP12 2JH
M6 TCK P. Pritchard, 11 Beacon Avenue, Dunstable, LU6 2AD
M6 TCL W. Hand, 168 Barcroft Street, Cleethorpes, DN35 7DX
M6 TCM R. Fowler, Windycroft, Whistlefield, Garelochhead, G840EY
M6 TCN D. Woodhams, 83 Langdale Place, Newton Aycliffe, DL5 7DY
MI6 TCO J. Allen, 3 Malwood Close, Belfast, BT9 6QX
MM6 TCS T. Moffat, 16 Drumellan Road, Ayr, KA7 4XQ
M6 TCX T. Atherton, 16 Steeple View, Ashton-On-Ribble, Preston, PR2 2PX
M6 TCY A. Todd, 27 Woodleigh Crescent, Ackworth, Pontefract, WF7 7JG
M6 TCZ T. Clarke, 168 Hykeham Road, Lincoln, LN6 8AP
M6 TDA M. Aubrey, 4 Broadleaf Close, Oakwood, Derby, DE21 2DH
MW6 TDB L. Brookes, Tyn Llidiart, Llanfairpwllgwyngyll, LL61 6EQ
MW6 TDC D. Cowling, Dorset Cottage, Alltyblacca, Llanybydder, SA40 9SU
MW6 TDE P. Phillips, Sedgemoor, Station Road, Kilgetty, SA68 0XS
M6 TDF T. Peck, 2 Primrose Lane, Miami Beach, Sutton on Sea, LN12 2JZ
M6 TDI A. Jones, 4 Park Close, Wymondham, NR18 9BA
M6 TDJ J. Stringer, 21 Ladywalk, Maple Cross, Rickmansworth, WD3 9YZ
M6 TDK I. Radford, 7 Eastmount Avenue, Hull, HU8 9EW
MW6 TDL T. Lawrence, 14 Railway Terrace, Caerau, Maesteg, CF34 0UE
M6 TDM S. Brown, 4 Dorado Gardens, Orpington, BR6 7TD
M6 TDO S. Hogg, 57 The Grange, Burton-on-Trent, DE14 2EX
M6 TDP P. Thorley, 57 Riverside Drive, Hambleton, Poulton-le-Fylde, FY6 9EH
M6 TDR T. Robertson, 6 Sandringham Court, Bircotes, Doncaster, DN11 8QU
MD6 TDU K. Dodds, 16 Second Avenue, Onchan, Isle of Man, IM3 4LE
M6 TDV T. Mcnamara, 19 Abbey Mews, Pontefract, WF8 1TD
M6 TDY E. Ashford, 56 Finch Close, Shepton Mallet, BA4 5GL
M6 TDZ A. Thomas, 5 Thornley Road, Wolverhampton, WV11 2HR
M6 TEE P. Hardacre1, 13 St. Johns Street, Bridlington, YO16 7NL
M6 TEF R. Ashman, 85A Fakenham Road, Great Ryburgh, Fakenham, NR21 7AQ
M6 TEG T. Guy, 16 Cogdeane Road, Poole, BH17 9AS
M6 TEJ T. Jones, Formula Cars, Wellington, TA21 9HW
M6 TEK A. Sood, Parima, Sewardstone Road, London, E4 7RA
M6 TEM J. Turner, 34 Vaughan Road, Stotfold, Hitchin, SG5 4EH
M6 TEO M. Prentice, Cherry Ash, Shaftesbury Road, Gillingham, SP8 4LL
M6 TEP T. Coles, 88C Dursley Road, Trowbridge Ba14 0Ns, Trowbridge, BA14 0NS
MM6 TEQ K. Ruchomski, 53A Barnton Avenue, Edinburgh, EH4 6JJ
M6 TER B. Lye, 7 Cider Way, South Molton, EX36 4FA
MW6 TES A. Bailey, 6 Trenos Gardens, Bryncae, Pontyclun, CF72 9SZ
M6 TEV T. Howson, 82 Bank End Avenue, Worsbrough, Barnsley, S70 4QN
MM6 TEW T. Warr, 407 Smerclate, Isle of South Uist, HS8 5TU
M6 TEY A. Brown, 3 Alston Road, New Hartley, Whitley Bay, NE25 0ST
M6 TEZ T. Archer, 1 Banks Road, Ashford, TN234NR
MW6 TFB T. Blanchard, 17 Heol Tyn-Y-Fron, Penparcau, Aberystwyth, SY23 3RP
M6 TFD D. Lyes, 2 Thelnetham Road, Blo Norton, Diss, IP22 2JQ
M6 TFE T. Fletcher, 3 Moorend Glade, Charlton Kings, Cheltenham, GL53 9AT
M6 TFF P. Ratcliffe, 61 Queens Avenue, Ilfracombe, EX34 9LS
MI6 TFG T. Mckee, 4 Earlford Heights, Newtownabbey, BT36 5WZ
M6 TFJ T. Johnsen, 5 Willow Lane, Billinghay, Lincoln, LN4 4FN
M6 TFK T. Kightly, Ferry Hill Farm, London Road, Chatteris, PE16 6SG
MW6 TFL T. Fletcher, 21 Sandpiper Road, Llanelli, SA15 4SG
MI6 TFN R. Taylor, 3 Austin Drive, Tandragee, Craigavon, BT62 2AR
M6 TFO A. Theobold, 25 Aysgarth Road, Leicester, LE4 0ST
M6 TFP T. Chambers, 104 Blue Hill Crescent, Leeds, LS12 4PB
M6 TFQ D. Thomson, 11 Uranus Road, Hemel Hempstead, HP2 5QF
M6 TFT H. Salter, 1 Rock Farm Cottages, Gibbs Hill, Maidstone, ME18 5HT
M6 TFV D. Forshaw, 14 Hope Carr Road, Leigh, WN7 3ET
M6 TFW F. Worrall, 297 Tamworth Road, Amington, Tamworth, B77 3DG
MM6 TFY T. Yates, 12 Fulmar Court, Newtonhill, Stonehaven, AB39 3QG
M6 TFZ E. Walton, 11 Parkfield Road, Northwich, CW9 7AR
MM6 TGB T. Bell, 22 Queens Road, Elderslie, Johnstone, PA5 9LJ
M6 TGC B. Grainger, 42 Madeira Avenue, Leigh-on-Sea, SS9 3EB
MM6 TGD I. Currie, 4 Greendyke Cottage, Falkirk, FK2 8PP
MW6 TGF M. Malone, 5 Brook Park Avenue, Prestatyn, LL19 7HH
M6 TGG T. Sutton, Yew Tree Farm, Paddol Green, Shrewsbury, SY4 5QZ
M6 TGH A. Hardy, 54 Trueway Drive Shepshed, Loughborough, LE12 9HG
M6 TGI T. Hunt, Mad Bess Cottage, Breakspear Road North, Uxbridge, UB9 6LZ
M6 TGJ T. Woolvin, 68 Jiggins Lane, Birmingham, B32 3LA
M6 TGK R. Willis, 12 Robartes Road St. Dennis, St Austell, PL26 8DS
M6 TGL C. Dolphin, 24 Saughall Road, Wirral, CH46 6DS
M6 TGM S. Thorpe, 4 Moredon Road, Swindon, SN25 3DQ
MM6 TGN A. Barclay, 21 Netherlea, Scone, Perth, PH2 6QA
M6 TGP R. Chambers, 3 Westby Way, Poulton-le-Fylde, FY6 8AD
M6 TGQ T. Webb, 24A Baldmoor Lake Road, Birmingham, B23 5QA
M6 TGR H. Mcphillips, 160 Pasley Street, Plymouth, PL2 1DT
M6 TGS A. Stubbs, 20 Mossfields, Crewe, CW1 4TD
M6 TGU M. Bostock, 86 Beauvale Drive, Ilkest on, DE7 8SJ
M6 TGV B. Mountford, 189 Lloyd Street, Stockport, SK4 1NH
M6 TGY T. Hopkins, Horrabridge Stores, Commercial Road, Horrabridge, Yelverton, PL20 7QB
M6 TGZ S. Richardson, 12 Good Road, Poole, BH12 3PJ
M6 THA K. Thacker, 2 New Cottages, Selham, Petworth, GU28 0PJ
M6 THB T. Barratt, 17 Main Road Collyweston, Stamford, PE9 3PF
M6 THC M. Hellon, 75 Brassey Street, Birkenhead, CH41 8BZ
M6 THI K. Barker, 10 Corn Mill Close, Rochdale, OL12 9UW
M6 THJ T. Hunt, 102 Gainsborough Close, Salisbury, SP2 9HJ
M6 THO Dr A. Thornett, Ty Bedwen, Birch Grove, Lichfield, WS13 6EP
MJ6 THP T. Pallot, Biltmore, La Grande Route De St. Laurent, Jersey, JE3 1NH
M6 THQ C. Taylor, 9 Cesson Close, Chipping Sodbury, Bristol, BS37 6NJ
M6 THS T. Shevchenko, 25 Kirby Road, Dartford, DA2 6HE
MM6 THU A. Gray, 28 Le Roux Drive, Oldmeldrum, Inverurie, AB51 0PJ
M6 THV T. Hunt, 14 Wenlock Road, South Shields, NE34 9BA
MW6 THW T. Woodley, 2 Parc Onen, Neath, SA10 6AA
M6 THY H. Tang, Flat 31, 74 Arlington Avenue, London, N1 7AY
M6 TIA C. Lyne, 4 Bridge Close, Catterick Garrison, DL9 4PG
M6 TIC R. Wainwright, 38 Stanway Close, Worcester, WR4 9XL
M6 TID A. Gilham, 38 Trojan Way, Waterlooville, PO7 8AL
M6 TIF C. Townsend, 64 Burnbridge Road Old Whittington, Chesterfield, S41 9LR
M6 TIH T. Mackenzie, 2 Newcastle Street, Carlisle, CA2 5UH
M6 TII T. Ince, 18 Holly Walk, Keynsham, Bristol, BS31 2TU
MI6 TIJ C. Thompson, 86 Huguenot Drive, Lisburn, BT27 4YD
M6 TIL P. Athersmith, 5 Aqueduct Lane Stirchley, Telford, TF3 1BW
M6 TIO T. Allsopp, 413 Boulton Lane, Derby, DE24 9DL
MW6 TIQ R. Baker, 20 Hawthorne Terrace, Aberdare, CF44 7HE
MM6 TIR S. Taylor, Pine Lodge, Easterton Of Auchleuchries, Hatton, Peterhead, AB420TQ
M6 TIU D. Firth, 5 Mowhay Gardens, Hatherleigh, EX20 3FE
M6 TIW T. Heath, 48 Brightholmlee Lane, Sheffield, S35 0DD
M6 TIY T. Cooper, 11 Warwick Road, Totton, Southampton, SO40 3QP
M6 TJA T. Atkin, 1 Wrangle Farm Green, Clevedon, BS21 5DR
M6 TJB T. Bexon, 51 Hookstone Drive, Harrogate, HG2 8PR
M6 TJD T. Davis, 11 Hinton Villas, Hinton Charterhouse, Bath, BA2 7SS
MW6 TJE G. Hodges, 79 Trefelin, Aberdare, CF44 8LF
MI6 TJF T. Ferguson, 3 Wheatfield Park, Ballybogy, Ballymoney, BT53 6NT
MW6 TJG D. Watts, 52 Ridgeway Avenue, Newport, NP20 5AH
MW6 TJH R. Orchard, The Burrows, Spring Gardens, Whitland, SA34 0HL
M6 TJI T. Dyer, 180 Seaton Lane, Hartlepool, TS25 1HF
M6 TJJ T. Hempsall, 16 Central Avenue, Warrington, WA2 8AJ

M6 TJL T. Lake, 64 Lytchett Drive, Broadstone, BH18 9LB
M6 TJN C. Austen, Fenbank House, Roman Bank, Holbeach Clough, Spalding, PE12 8DH
M6 TJO R. Austen, Fenbank House, Roman Bank, Holbeach Clough, Spalding, PE12 8DH
M6 TJQ D. Roberts, 61 Teign Bank Road, Hinckley, LE10 0ED
MW6 TJS T. Skerritt, 94 Turberville Street, Maesteg, CF34 0LU
M6 TJU T. Idiculla, 3 Stanhope Road, Slough, SL1 6JR
M6 TJV T. Humphreys, 93 Cornwall Crescent Yate, Bristol, BS37 7RU
M6 TJX T. Dalby, 52 Narborough Road South, Leicester, LE3 2FN
M6 TJY F. Grimley, 37 Reginald Road, Bexhill-on-Sea, TN39 3PH
M6 TJZ T. Johnson, 15 Tennyson Road, Creswell, Worksop, S80 4DW
M6 TKA K. Thompson, 184 Tickhill Road, Doncaster, DN4 8QS
M6 TKC T. Corcoran, 191 Queensway, Rochdale, OL11 2NA
M6 TKE N. Crabb, 1 Council Houses, Hall Lane, Crostwick, Norwich, NR12 7BB
M6 TKI J. Lewis, 93 Eastcliff, Portishead, Bristol, BS20 7AD
M6 TKK T. Kosteletos, 10 Church Lane, Southwick, Brighton, BN42 4GD
M6 TKN J. Atkinson, 8 Cragton Gardens, Blyth, NE24 5PR
M6 TKP A. Pickavance, 14 Loughrigg Avenue, St Helens, WA11 7AP
M6 TKR T. Berisford, 18 Cambridge Avenue, Winsford, CW7 2LL
M6 TKU S. Heys, 15 Heathfield Road, Fleetwood, FY7 7LY
MM6 TKV D. Hastings, Flat 38, Ericht Court, Upper Mill Street, Blairgowrie, PH10 6AE
M6 TKW T. Ward, 1 Darrismere Villas, Edinburgh Street, Hull, HU3 5AS
M6 TKX T. Keable, 5 Redhills Way, Hetton-Le-Hole, Houghton le Spring, DH5 0ES
M6 TLB R. Lawrence, 20 Coronation Drive, Forest Town, Mansfield, NG19 0AJ
M6 TLC T. Christopher, 27 Brinkhill Crescent, Nottingham, NG11 8GN
MW6 TLF T. Ford, Clwt Joli, Llanfrothen, Penrhyndeudraeth, LL48 6DU
MI6 TLG R. Greer, 11, Mullaghcarton Road, Lisburn, BT28 2TE
MM6 TLI A. Pates, 7/5 Sciennes, Edinburgh, EH9 1NH
M6 TLK A. Kyte, 42 Radford Street, Alvaston, Derby, DE24 8NS
M6 TLL A. Lear, 38 The Roundway, Claygate, Esher, KT10 0DW
M6 TLM I. Williams, 36 Telford Road, Tamworth, B79 8EY
MW6 TLN N. Weightman, 39 Brynamman Road, Lower Brynamman, Ammanford, SA18 1TR
M6 TLP T. Porter, 30 Woodbridge Close, Appleton, Warrington, WA4 5RD
MW6 TLR T. Rogers, 59 Park Place, Newport, NP11 6BN
M6 TLS J. Olver, 43 Heron Gardens, Portishead, Bristol, BS20 7DH
M6 TLX M. Egan, 4 Rutter Avenue, Warrington, WA5 0HP
M6 TLY A. Thompson, 25 Ardingly Road, Cuckfield, Haywards Heath, RH17 5HD
M6 TMA R. Smith, 3 Plane Tree Close Marple, Stockport, SK6 7RJ
M6 TMC S. Oram, 7 West Bank, Main Street, Old Weston, Huntingdon, PE28 5LJ
M6 TMD P. Kirby, 36 Durham Road, London, E12 5AX
M6 TMF D. Fletcher, 97 Wallace Crescent, Carshalton, SM5 3SU
M6 TMG M. Galloway, 2 Edendale Terrace, Horden, Peterlee, SR8 4RD
M6 TMH T. Hoyle, 60 Greenbank Crescent, Marple, Stockport, SK6 7PB
M6 TMJ M. Jones, 11 Lower Glen Park Pensilva, Liskeard, PL14 5PP
M6 TMK T. Walker, 11 Banburies Close, Bletchley, Milton Keynes, MK3 6JP
M6 TML T. Longmore, 3 Dairy Farm Cottages, Northlands Road, Gainsborough, DN21 5DN
M6 TMM T. Marsland, 34 Wakefield Road, Stalybridge, SK15 1AJ
M6 TMO T. Williams, Moor Farm, Moor Lane, Lincoln, LN3 4EG
M6 TMP T. James, 87A Beaconfield Road, Epping, CM16 5AT
MW6 TMQ D. Mottram, 35 Trem Arfon, Llanrwst, LL26 0BP
MI6 TMR M. Graham, 6 Sir Richard Wallace Drive, Lisburn, BT28 3BA
MM6 TMS S. Scott, 32 Dixon Terrace, Whitburn, EH47 0LH
M6 TMY M. Thackery, 1 Bockings Grove, Clacton-on-Sea, CO16 8DL
MI6 TMZ J. Allen, 192 Joanmount Gardens, Belfast, BT14 6PA
M6 TNA A. Golden, 3 Ampleforth Road, Middlesbrough, TS3 7PU
M6 TNB P. Bozikis, 336 Higham Hill Road, London, E17 5RG
M6 TNC J. Winstone, 31 Setterfield Way, Rugeley, WS15 1BJ
MI6 TND R. Dobson, 2 Mourne View, Crossgar, Downpatrick, BT30 9HW
M6 TNH L. Macdougall-Smith, 160 Thingwall Park, Bristol, BS16 2BU
MI6 TNI T. Nelson, 25 Monaghan Road Annashanco Rosslea, Belfast, bt927pt
M6 TNL C. Aston, 21 Hawks Drive, Winshill, Burton-on-Trent, DE15 0DL
MM6 TNO D. De Freitas, 14 York Street, Clydebank, G81 2PH
M6 TNR T. Robinson, 39 Cemetery Road, Laceby, Grimsby, DN37 7ER
M6 TNU J. Lambert, 69 Anvil Crescent, Broadstone, BH18 9DZ
M6 TNY A. Deeming, 16 Marshall Road, Exhall, Coventry, CV7 9BX
MI6 TNZ A. Mccullough, 45 Eglantine Park, Hillsborough, BT26 6HL
MI6 TOA T. Mcpolin, 40 Olympia Drive, Belfast, BT12 6NH
MI6 TOC T. Carvill, 18 Hospital Road, Newry, BT35 8PW
M6 TOD T. Sandham, 96 South Road, Morecambe, LA4 6JS
M6 TOE A. Smyth, 2 North Edge, Leigh, WN7 1HW
M6 TOF J. Morgan, Glas Y Dorlan, Pontrhydfendigaid, Ystrad Meurig, SY25 6EJ
M6 TOG P. Joyce, 2 Harold Road, Cuxton, Rochester, ME2 1EE
M6 TOK S. Halliday, 8 Newby Farm Road, Scarborough, YO12 6UN
M6 TOL C. Higgins, 27 Chetwynd Road, Birmingham, B8 2LB
M6 TON T. Ralph, 54 Clacton Road, Portsmouth, PO6 3QY
M6 TOT S. Dix, 8 Beaumont Road, Longlevens, Gloucester, GL2 0EJ
MW6 TOU W. Redbourn, 20 Radnor Drive Tonteg, Pontypridd, CF38 1LA
M6 TOV M. Murdoch, 19 Beachy Road, Crawley, RH11 9HN
M6 TOW A. Smith, 206 Essex Square, Salisbury, SP2 8HZ
M6 TOZ D. Torrance, 30 St. Norbert Drive, Ilkeston, DE7 4EH
M6 TPA A. Austin, 4 Cornwall Avenue, Oldbury, B68 0SW
MI6 TPC T. Crozier, 6 Garden Of Eden, Carrickfergus, BT38 7ls
M6 TPD T. Davies, 7 Crescent Road, Warley, Brentwood, CM14 5JR
M6 TPE A. Smith, 14 Queenstock Lane, Buxted, TN224AR
M6 TPG T. Gollins, 17 John Street, Stafford, ST16 3PJ
M6 TPI T. Kilfeather, Flat 1, 57 Chalk Hill, Watford, WD19 4DA
MW6 TPM T. Mckeown, 13 George Street, Treherbert, Treorchy, CF42 5AH
M6 TPO S. Bunworth, 21 Slattsfield Close, Selsey, Chichester, PO20 0EB
M6 TPR E. Butler, Tanglewood, Elms Lane, Wolverhampton, WV10 7JS
M6 TPS J. Bramley, 13 Bolton Street, Stockport, sk5 6be
M6 TPT T. Taylor, 35 Albany Way, Skegness, PE25 2nb
M6 TPV P. Watson, 6 Walnut Grove, Shepton Mallet, BA4 4HX
M6 TPX A. Patrick, 2 Beacon Grange, Malvern, WR14 3EU
M6 TPY T. Pollett, 10 Bridport Road, Poole, BH12 4BS
M6 TQF K. Taylor, 44 Main Street, Willoughby, Rugby, CV23 8BH
M6 TQW S. Moss, 27 Dunn Side, Chelmsford, CM1 1DL
M6 TRC N. Kent, Flat 1, Manor House, Redruth, TR15 1AX
M6 TRD M. Read, 133 Somersall Street, Mansfield, NG19 6EL
M6 TRE R. Evered, Ivy Cottage, Old Bristol Road, Wells, BA5 3AL
MI6 TRF D. Foy, 125 Drumbeg Tullygally, Craigavon, BT65 5AE
M6 TRH D. Constable, 7 Mill Field, Sutton, Ely, CB6 2QB
M6 TRI D. Freeman, 37 Bedale Road, Nottingham, NG5 3GL
MW6 TRK T. Keane, 19 Williton Road Llanrumney, Cardiff, CF3 5QE
MW6 TRL P. Terrell, 82 Baglan Street, Treherbert, Treorchy, CF42 5AR
MM6 TRO S. Troscheit, 20 James Street, St Andrews, KY16 8YA
M6 TRP T. Green, Bentley Country Park Ltd, Flag Hill, Great Bentley, Colchester, CO7 8RF
MW6 TRQ I. Troughton, Rhiwbina, Pentre Lane, Cwmbran, NP44 3AP
M6 TRS T. Steel, Barrowfield House, Much Hadham, SG10 6BD
M6 TRT J. Fry, Deal Cottage, Ipswich Road, Long Stratton, Norwich, NR15 2TF
M6 TRU G. Truman, 3 Mulberry Road, Rotherham, S254BH
M6 TRV T. Meakin, 33 The Markhams, New Ollerton, Newark, NG22 9QX
M6 TRW T. Wheatley, Three Chimneys, Ricket Lane Blidworth, Mansfield, NG21 0NA
MM6 TRX A. Mcneill, 13 Spinkhill, Laurieston, Falkirk, FK2 9JR
M6 TSA T. Stamp, 41 Glamorgan Close, Mitcham, CR4 1XG
MM6 TSC T. Couper, 10 Sclandersburn Road, Denny, FK6 5LP
MI6 TSH S. Hamilton, 38B Doagh Road Kells, Ballymena, BT42 3ND
M6 TSI S. Gilham, 32 Whitby Road, Lytham St Annes, FY8 3HA
M6 TSJ S. Mcbain, 13A St. Lukes Close, Cherry Willingham, Lincoln, LN3 4LY
M6 TSM T. Keay, 12 Quarryfields, Seahouses, NE68 7TB
MM6 TSN R. Thomson, Galloquhine Cottage, Auchenblae, AB30 1TT
M6 TSP C. Power, 23 Manor Road, Killamarsh, Sheffield, S21 1BU
M6 TSR T. Scopes, The Garden House, West Common Road, Keston, BR2 6AJ
M6 TST T. Tate, 339 West Dyke Road, Redcar, TS10 4PS
MD6 TSW D. Williamson, 45 Bluebell Close, Peel, Isle Of Man, IM5 1GH
M6 TSZ M. Wickens, St. James Vicarage, The Parade, Dudley, DY1 3JA
M6 TTB T. Bate, 87 Dunsheath, Telford, TF3 2BY
M6 TTE T. Battelle, 207 Alfreton Road, Blackwell, DE55 5JH
M6 TTF M. Lanham, 3 Park Cottages, New Common, Bishop's Stortford, CM22 7RT
M6 TTH T. Horsten, Kastelsvej 4, 2.Tv, Copenhagen E, Denmark, 2100
M6 TTK R. Rushton, 65 Wilton Bank, Saltburn-by-the-Sea, TS12 1PD
M6 TTM J. Aldwinckle-Day, 45 Felicia Way, Grays, RM16 4JF
M6 TTN M. Newham, 3 Laurel Drive, Brockworth, GL3 4GF
M6 TTO F. Armstrong, 38 Dovecote Drive, Haydock, St Helens, WA11 0SD
M6 TTP M. Mcgowan, 48 Alderley Road, Thelwall, Warrington, WA4 2JA
M6 TTR S. Chuck, 82 Redmayne Drive, Chelmsford, CM2 9AG
MW6 TTS M. Johns, 39 Tyla Coch, Llanharry, Pontyclun, CF72 9LT
M6 TTT A. Allen, 65 Hanbury Road, Dorridge, Solihull, B93 8DN
M6 TTV D. Hargrove, 24 Slade Road, Wolverhampton, WV10 6QS
M6 TTW M. Berlyn, 13 Hopping Jacks Lane, Danbury, Chelmsford, CM3 4PN
M6 TTY T. Tong, Lincoln House, Clarence Drive, Harrogate, HG1 2QD
M6 TTZ S. Marsh, 31A Broad Street, Stamford, PE9 1PJ
M6 TUD C. Barker, 18 Nickleby Road, Waterlooville, PO8 0RH
MM6 TUG A. Cairns, Tigh Bruadair, Achmore, Strome Ferry, IV53 8UX
M6 TUH R. Taylor, Penhawger Park, Liskeard, PL14 3LW
M6 TUK C. Yunnie, Lighthouse, Park Lane, Totnes, TQ9 7BD
MI6 TUM S. Tumilty, 18 Cluain-Air, Newry, BT34 1PW
M6 TUN A. Tunney, 79 Scott Street, Burnley, BB12 6NJ
M6 TUR D. Slee, Turnpike, Brearton, Harrogate, HG3 3BX
M6 TUT A. Mars, 14 Woodyard Close, London, NW5 4BX
MI6 TUV C. Bailie, 26 Moatview Park, Dundonald, Belfast, BT16 2BE
MI6 TUX E. Murray, Apartment 4, 20 Global Crescent, Belfast, BT6 8LN
M6 TVB S. Brett, 16 Walmsley Road, Broadstairs, CT10 2BH
M6 TVC L. Lovell, 28 Park Lane, Blunham, Bedford, MK44 3NJ
M6 TVH E. Grabham, 37 Redstart Road, Chard, TA20 1SD
M6 TVL T. Cannon, 242 Hall Road, Norwich, NR1 2PW
M6 TVM T. Melia, 6 Burnley Close, Blackburn, BB1 3HL
M6 TVN D. Royall, 7 Stacklands, Welwyn Garden City, AL86XW
M6 TVS N. Simmonds, 3 Noneley Hall Barns, Noneley, Wem, Shrewsbury, SY4 5SL
M6 TVW T. Hawksford, 24 Greenaway, Morchard Bishop, Crediton, EX17 6PA
M6 TVX D. Mills, 91 Harp Road, London, W7 1JQ
M6 TWA T. Alvey, 2 Sunny View, Back Road, Saxmundham, IP17 3NY
MM6 TWE S. Tweedie, 3 Edgar Road, Westruther, Gordon, TD3 6ND
MW6 TWH J. Hughes, Midfield Farm, Midfield Caravan Site, Aberystwyth, SY23 4DX
M6 TWI J. Truman, 9 Riddell Avenue, Langold, Worksop, S81 9SS
M6 TWL T. Willis, 143 Clarence House, Leeds, LS10 1LH
M6 TWQ J. Attwood, 13 John Winter Court, Euston Road, Great Yarmouth, NR30 1DU
MM6 TWS A. Gray, 73 Ash-Hill Drive, Aberdeen, AB16 5YR
M6 TWV T. Lamb, Marsh Green, Tn7 4Et, Colmans Hatch, TN7 4ET
M6 TWW T. Ward, 20 Ollerton Road, Edwinstowe, Mansfield, NG21 9QG
M6 TWY T. Wood, 26 Parkfield Crescent, Kimpton, Hitchin, SG4 8EQ
MM6 TWZ Dr T. Dodd, 2 White Wisp Gardens, Dollar, FK14 7BH
M6 TXG C. Whatmough, 11 Blackchapel Drive, Rochdale, OL16 4QU
M6 TXH J. Marriott-Levett, 37 Christleton Drive, Ellesmere Port, CH66 3NN
M6 TXI B. Cave, 2 Beaufort Close, Newcastle upon Tyne, NE5 3XL
M6 TXJ J. Hopkins, 53 Sprules Road, London, SE4 2NL
MM6 TXK Prof. T. Kerby, 1 St. Mark'S Lane, Edinburgh, EH15 2PX
M6 TXM A. Miller, Ashtree Farm, Rugby Road, Princethorpe, Rugby, CV23 9PN
MM6 TXN S. Moore, 35 Niddrie House Park, Edinburgh, EH16 4UH
M6 TXP P. Cassells, 49 Dodds Lane, Maghull, Liverpool, L31 0BD
M6 TXR R. Withers, 50 Coneygear Road, Hartford, Huntingdon, PE29 1QL
MI6 TXS R. Bates, 45 Upper Ballyboley Road, Ballyclare, BT39 9ST
M6 TXT M. Bookham, 116 Clare Gardens, Petersfield, GU31 4EU
M6 TXU A. Gillard, 58 Queens Road, Thame, OX9 3NQ
M6 TXX A. Gibbins, Flat 120, Greenhill, London, NW3 5TY
M6 TXZ R. Bolton, 18 Slimbridge Road, Tuffley, Gloucester, GL4 0NB
M6 TYB R. Thornett, Ty Bedwen, Birch Grove, Lichfield, WS13 6EP
M6 TYC T. Corcoran, 50 Grange Road, Bracebridge Heath, Lincoln, LN4 2PW
M6 TYE S. Connolly, 82 Cheswood Drive, Minworth, Sutton Coldfield, B76 1YE
MW6 TYG C. Morris, 17 Percy Road, Wrexham, LL13 7EA
M6 TYH T. Hill, 14 Hunters Mead, Motcombe, Shaftesbury, SP7 9QG
M6 TYK J. Goddard, 217 Speedwell Road, Bristol, BS5 7SP
M6 TYL T. Ward, Flat 26, Bassett Court, Bassett Avenue, Southampton, SO16 7DR

M6 TYN M. Tynan, 57 Alpine Drive, Wardle, Rochdale, OL12 9NY
MI6 TYR V. Brogan, 7 Richmond Park, Omagh, BT79 7SJ
M6 TYS S. Tyrrell, 14 Town Orchard, Southoe, St Neots, PE19 5YJ
M6 TYT D. Redhead, 113 Blythway, Welwyn Garden City, AL7 1DL
M6 TYU B. Grzelak, Flat 20, Cotterell Court, Southern Way, London, SE10 0DW
M6 TYZ G. Hudson, 74 Upperfield Drive, Felixstowe, IP11 9LS
M6 TZA A. Zaulincy Adams, 261 Nottingham Road, Derby, de21 6ap
M6 TZD N. Cripps, 66 Forest Road, London, E8 3BT
M6 TZE S. Johnston, 62 Charlecote Park, Telford, TF3 5HD
M6 TZI I. Tighe, 4 Beach Walk, Broadstairs, CT10 1FA
M6 TZO J. Morton, 44 Silam Road, Stevenage, SG1 1JJ
MI6 TZP C. Gault, 7 Gardenmore Place, Larne, BT40 1SE
M6 TZR T. Crouch, 2 Park Farm Close, Horsham, RH12 5EU
M6 TZU S. Moore, 32 Broadgreen Close, Leyland, PR25 2XA
MM6 TZX S. Mccorkell, 28 Leven Road, Hamilton, ML3 7WS
M6 TZY S. Crabb, 1 Council Houses, Hall Lane, Crostwick, Norwich, NR12 7BB
M6 TZZ P. Moore, 24 Plough Road, Dormansland, Lingfield, RH7 6PS
MI6 UAB A. Pritchard, 16 Ballymaconnell Road South, Bangor, BT19 6DQ
M6 UAD M. Kimber, 2 Church Road, Brandon, IP27 0EN
M6 UAE P. Potter, 21 Ulverston Crescent, Lytham St Annes, FY8 3RZ
M6 UAF J. Glynn, 106 Fairway Avenue, West Drayton, UB7 7AP
M6 UAJ A. Ibbotson, 323 Brincliffe Edge Road, Sheffield, S11 9DE
M6 UAL A. Laidlaw, 3 Litcham Close Litcham, King's Lynn, PE32 2QX
M6 UAO N. Smith, 2 Tibbett Close, Dunstable, LU6 3TT
M6 UAP Dr S. Ando, 84 New Hey Road, Cheadle, SK8 2AQ
M6 UAR D. Randles, 12 Wain Court. Rakeway, Saughall. Chester, CH16BF
M6 UAS A. Cudworth, 30 Compass Tower, Munnings Road, Norwich, NR7 9TW
M6 UAW A. Wragg, 14 Grizedale Avenue Sothall, Sheffield, S20 2DL
M6 UAX A. Macdonald, Woodside Cottage, Horton Way, Verwood, BH31 6JJ
M6 UBC B. Clifford, 7 Broke Court, Guildford, GU4 7HQ
MI6 UBE L. Calderwood, 43 Rathview Park, Mullybritt, Enniskillen, BT94 5EW
M6 UBH R. Tolman, 10 Woodcote Way, Abingdon, OX14 5NE
M6 UBI C. Cotton, 49 Cornwall Road, Portsmouth, PO1 5AR
M6 UBM B. Mcdowell, 10 Vineyard Lane, Kingswood, Wotton-under-Edge, GL12 8SB
M6 UBN B. Shephard, 13 Forest Street, Annesley Woodhouse, Kirkby In Ashfield, Nottingham, NG17 9HE
M6 UBR M. Moore, 3 St. Pauls Road Birkenshaw, Bradford, BD11 2JY
M6 UBS N. Evetts, 35 Wood End Road, Kempston, Bedford, MK43 9BB
MI6 UBT T. Mcwilliams, 221 Kings Road, Belfast, BT5 7EH
M6 UCE S. Roderick, 47 Cuckmans Drive, St Albans, AL2 3AY
M6 UCP D. Ali, 20 Millwall Close Gorton Lancshire, Manchester, M18 8LL
M6 UCR C. Rogers, 1 Tregurthen Close, Camborne, TR14 7EB
M6 UCS M. Coote, 4 Hunters Oak, Watton, Thetford, IP25 6HL
M6 UDA L. Symonds, 10 Cowper Court, Willunga, Australia, 5172
M6 UDB D. Hudson, 34 Upton Gardens, Upton upon Severn, WR8 0NU
M6 UDC D. Cole, 39 Hillside Road, Southminster, CM0 7AL
M6 UDM D. Meehan, 47 Clinton Road, Shirley, Solihull, B90 4RN
MI6 UDR R. Dunwoody, 16 Dernalea Road, Milford, Armagh, BT60 4DZ
M6 UDS G. Chapman, 253 St. Pauls Road, Preston, PR1 6NS
M6 UDX R. Berwick, 10 Hall Lane, Wacton, Norwich, NR15 2UH
M6 UDY D. Smout, Sunrays, Warbage Lane, Bromsgrove, B61 9BH
M6 UEA M. Robinson, 58 Haworth Road, Cross Roads, Keighley, BD22 9DL
M6 UEB J. Masters, 31 Lower Beeches Road, Birmingham, B31 5JB
M6 UED M. Tanji, Apartment 63, Westside One, Birmingham, B1 1LS
M6 UEE R. Goldsack, 5 Parc Dellen, Croft Farm Park, Luxulyan, Bodmin, PL30 5EW
M6 UEH Dr S. Smith, 557, Riverside Island Marina, Isleham, Ely, CB7 5SL.
MM6 UEN I. Macdonald, 20 Newbigging Terrace, Auchtertool, Kirkcaldy, KY2 5XL
M6 UES R. Hawes, 40 Nightingale Way, Thetford, IP24 2YN
M6 UET R. Taylor, 68 Charter Road, Chippenham, SN15 2RA
M6 UFC A. Hunt, 49 Freame Close, Stroud, GL68HG
M6 UFF T. George, 2 Dane Valley Road, Margate, CT9 3RX
M6 UFO C. Sims, 133 Canterbury Road, Hawkinge, Folkestone, CT18 7BS
M6 UFR S. Sinderbury, 40 Norfolk Gardens, Urmston, Manchester, M41 8RE
M6 UGA G. Amos, Willow Tree House, Deers Green, Clavering, Saffron Walden, CB11 4PX
MI6 UGE G. English, 232 Kernan Hill Manor, Portadown, Craigavon, BT63 5WU
M6 UGF S. Peare, 15 Clydesdale Gardens, Bognor Regis, PO22 9BE
M6 UGM G. Mountain, 34 Albert Road, Warlingham, CR6 9EP
M6 UGN S. Nesling, 64 Ruskin Avenue, Lincoln, LN2 4BT
M6 UGP G. Palin, 104 Nelson Street, Crewe, CW2 7LN
M6 UGX M. Street, 7 Salisbury Street, Sowerby Bridge, HX6 1EE
M6 UHF J. Smith, 24 Monk Road, Wallasey, CH44 1AJ
M6 UHM C. Imm, 29 Naunton Crescent, Cheltenham, GL53 7BD
M6 UHN B. Marks, 167 Linnet Drive, Chelmsford, CM2 8AH
M6 UHT P. Lunn, 179 Coventry Road, Nuneaton, CV10 7BA
M6 UHU J. Willetts, 102 Welch Road, Cheltenham, GL51 0EG
M6 UHV S. King, 29 Hall Road, Chatham, ME5 8PL
MI6 UIM F. Hand, 12 Church Street, Rosslea, Enniskillen, BT92 7DD
M6 UIR R. Muir, 4 Blandford Avenue, Worsley, Manchester, M28 2JE
M6 UIT A. Rackett, 151 Stoke Road, Gosport, PO12 1SE
M6 UJD J. Dixey, 108 Colchester Road West Bergholt, Colchester, CO6 3JS
M6 UJK J. Keddy, 63 High Street Kimpton, Hitchin, SG4 8PU
M6 UJM R. Banks, 3 Parkhayes, Woodbury Salterton, Exeter, EX5 1QS
M6 UJR J. Risby, 112 Stratton Heights, Cirencester, GL7 2RL
M6 UKA G. Curtis, 16 Chettell Way, Blandford St. Mary, Blandford Forum, DT11 9PW
M6 UKB J. Lightfoot, 23 Nichols Street Desborough, Desborough, nn142qu
M6 UKC P. Williams, 53 Lilac Avenue, Cannock, WS11 0AR
M6 UKF F. Hennigan, 50 Fairford Crescent, Downhead Park, Milton Keynes, MK15 9AE
M6 UKG J. English, 1 Niton Cottage Pound Lane, Meonstoke, Southampton, SO32 3NP
M6 UKH A. Shaw, 20 Hillcrest Close, Thrapston, Kettering, NN14 4TB
M6 UKI U. Trnjakov, 63 Welford Gardens, Abingdon, OX14 2BH
M6 UKJ J. Pithers, 77 Victoria Road, Laindon, Basildon, SS15 6RA
M6 UKM M. Reeves, 41 Hogarth Road, Whitwick, Coalville, LE67 5GF
M6 UKT G. Eason, Whitegates, Parsonage Road, Takeley, Bishop's Stortford, CM22 6QX
M6 UKX G. Radulescu, 41 Sherard Road, London, SE9 6EX
M6 ULA J. Van Der Elsen, 6 Kent Close, Churchdown, Gloucester, GL3 2HQ
MM6 ULL K. Mackenzie, Alderwood, Braes, Ullapool, IV26 2TB
MW6 ULX C. Griffiths, 21 Sandpiper Road, Sandy, Llanelli, SA15 4SG
M6 ULY B. Page, 12 Hitchens Close, Hemel Hempstead, HP1 2PP
M6 UMB M. Bramble, 133 Fane Way, Maidenhead, SL6 2TX
M6 UMD A. Slee, Foal Cottage Mare Hill Common, Pulborough, RH20 2DX
M6 UMI A. Smith, 83 Lottbridge Drive, Eastbourne, BN22 9PA
M6 UMM F. Colquhoun, 2 Heslop Road, London, SW12 8EG
M6 UMR U. Munir, Flat 6, Horton House, Field Road, London, W6 8HW
M6 UNA J. Taberner, 32 Bell Lane, Sutton Manor, St Helens, WA9 4BD
MI6 UNC C. Harper, 2 The Courtyard, Rathfriland, Newry, BT34 5PU
M6 UNI M. Knapton, The Orchard, Rucklers Lane, Kings Langley, WD4 9NA
M6 UNN D. Munn, 36 Moor Lea, Braunton, EX33 2PE
M6 UNS P. Unsworth, 83 Newbold Avenue, Sunderland, SR5 1LL
MW6 UNY L. Betts, 12A Maesgwyn, Pontnewydd, Cwmbran, NP44 1BQ
M6 UON J. Fernßndez Silva, 45 Veals Mead, Mitcham, CR4 3SB
M6 UPE R. Saddler, 41 Clapham Close, Swindon, SN2 2FL
MW6 UPH A. Williams, 8 Old Tanymanod Terrace, Blaenau Ffestiniog, LL41 4BU
M6 UPL M. Juuso, 32 Hart Road, St Albans, AL1 1NF
M6 UPS A. Morehen, 20 Castleton Grove, Inkersall, Chesterfield, S43 3HU
M6 UPT P. Turner, 59 Fairburn Avenue, Crewe, CW2 7SY
M6 UPU A. Stirk, 59 West Avenue, Lightcliffe, Halifax, HX3 8TJ
M6 UPW A. Wood, Berrybrook Northall Road, Eaton Bray, LU6 2DQ
M6 URA D. Kveksas, 36, Purley, CR8 3AQ
M6 URD J. Ratcliffe, 63 Dickens Lane, Poynton, Stockport, SK12 1NN
M6 URG S. Kitchen, 16 Crown Avenue, Cudworth, Barnsley, S72 8SE
M6 URH R. Roebuck, 469 East Bank Road, Sheffield, S2 2AE
MW6 URI M. Ouseley, Brynsiriol Manorowen, Fishguard, SA65 9PT
M6 URK M. Mcnamara, 50 Portsdown Road, Portsmouth, PO6 4QH
M6 URM A. Stansfield, 59 Prankerds Road, Milborne Port, Sherborne, DT9 5BX
MM6 URP S. Gray-Jones, Flat C, 7 Nelson Street, Aberdeen, AB24 5EP
MM6 URR M. Riddick, Davah, Port Road, Castle Douglas, DG7 3JW
M6 URS S. Banda, 45 High View Avenue, Grays, RM17 6RU
M6 URT B. Burnett, 44 Westoe Drive, South Shields, NE33 3EL
M6 URX M. Urban, 59 Sterling Gardens, London, SE14 6DU
M6 USA M. Dexter, 70 Rutland Street, Derby, DE23 8PR
MI6 USC W. Bradley, 14 Ardmore Grange, Ballygowan, Newtownards, BT23 5TZ
M6 USD J. Rufes, Flat 1 & 3-8 12 Smyrna Road, London, NW6 4LY
M6 USE D. Hardy, 186 Sneyd Hill, Stoke-on-Trent, ST6 1RA
M6 USG Dr B. Featherstone, 1 Hall Terrace, Bodinnick, Fowey, PL23 1LX
MW6 USK C. Moreton, 20 Millbrook Court, Little Mill, Pontypool, NP4 0HT
M6 USM J. Wright, 32 Carlton Road, Nottingham, NG10 3LF
M6 USO I. Usov, Flat 12, 48 The Avenue, Southampton, SO17 1XQ
M6 USP A. Brighton, 38 Greenfield Crescent, Nailsea, Bristol, BS48 1HR
MM6 USS E. Milligan, 46 Arkaig Drive Crossford, Dunfermline, KY12 8YW
M6 USV D. Soames, 40 Woodland Drive, North Anston, Sheffield, S25 4EP
M6 UTB E. Harrington, 53 Kingscroft Road, Banstead, SM7 3NA
M6 UTC A. Driscoll, 82 Station Road, Langford, Biggleswade, SG18 9PQ
M6 UTI J. Blay, 31 Woodcote House, Queen Street, Hitchin, SG4 9TL
M6 UTP C. Walker-Riley, 1 Farmcote Court, Hemlington, Middlesbrough, TS8 9LJ
M6 UTT C. Dyer, 74 Godstone Road, Lingfield, RH7 6BT
MI6 UTV T. Maclaine, 172 Moylagh Road, Seskanore, Omagh, BT78 2PN
M6 UTX P. Begg, 20 Wilsford Avenue, Uttoxeter, ST14 8XG
MM6 UTZ C. Mackie, 4/36 Gillsland Road, Edinburgh, EH10 5BW
M6 UUA D. Sealey, 6 Mizzen Road, Hull, HU6 7AG
M6 UUE N. Busley, Busley, South Drove, Spalding, PE11 3BD
M6 UUH I. Ashmore, 36 West End Avenue Royston, Barnsley, S71 4LQ
M6 UUU A. Castell, 35 Plantation Road, Chippenham, SN14 0EX
M6 UVB B. Axcell, 20 Sheerways, Faversham, ME13 8TP
M6 UVD D. Smith, 17 Stonepound Road, Hassocks, BN6 8PP
M6 UVF E. Goodwin, 55 Twickenham Road, Sunderland, SR3 4JN
MI6 UVS Q. Church, 51 Houston Park, Broughshane, Ballymena, BT42 4LB
M6 UWK J. Hadley, 75 Glendower Avenue, Coventry, CV5 8BD
M6 UWS M. Sharman, 3 Deben Crescent, Swindon, SN25 3QB
M6 UWU A. Siabi, 10 Ayloffs Walk, Hornchurch, RM11 2RJ
M6 UXB C. Hurley, 12 Grocott Road, Wednesbury, WS10 8RQ
M6 UYF E. Martin, 28 Mountford Close, Wellesbourne, Warwick, CV35 9QQ
M6 UYP A. Tomlinson, 45 Well Lane Willerby, Hull, HU10 6HB
M6 UZH U. Hussain, Cardwell Barn, Hostingley Lane, Dewsbury, WF12 0QH
M6 UZK A. Sweet, 3 Beechwood Grove, Blackpool, FY2 0DZ
M6 UZT L. Howe, 29 Abdy Avenue, Harwich, CO12 4QP
M6 UZZ D. Willis, 7 Shearman Avenue, Kimberworth Park, Rotherham, S61 3AG
MJ6 VAA V. Atherton, 4 Clos De La Mer, La Route De Noirmont, Jersey, JE3 8AL
MM6 VAB V. Julius, 100 Hogarth Drive, Cupar, KY15 5YU
M6 VAG K. Lane, 243 Edwin Road, Gillingham, ME8 0JL
M6 VAH V. Downes, 55 Ashfield Road Bromborough, Wirral, CH62 7EE
MI6 VAI L. Hambleton, 9 Springvale Road, Ballywalter, Newtownards, BT22 2PE
M6 VAJ A. Coomber, 3 Dolly Drove, Chard, TA20 1PF
M6 VAK D. Evans, 1 Copeland Street, Hyde, SK14 4TD
M6 VAM E. Driscoll, 42 Adelaide Square, Shoreham-by-Sea, BN43 6LN
M6 VAN E. Tsang, Harrogate Ladies' College, Clarence Drive, Harrogate, HG1 2QG
M6 VAP A. Mason, 13 Welton Gardens, Lincoln, LN2 2AY
M6 VAR Dr T. Varoudis, 179 Lymington Avenue, London, N22 6JL
M6 VAU S. Fairbourn, 17 Perry'S Lane, Wroughton, Swindon, SN4 9AX
M6 VAV A. Vasarhelyi, 1 Eldon Close, Langley Park, Durham, DH7 9FR
M6 VAW D. Vassie, 48 Greenfield Drive, Ridgewood, Uckfield, TN22 5SF
M6 VAY A. Vassie, Teddards, Filching, Polegate, BN26 5QA
M6 VAZ J. Bevan, 18 Martin Road, Diss, IP22 4HR
MI6 VBB D. Twaddle, 19 Lisanduff Park, Portballintrae, Bushmills, BT57 8RY
M6 VBD T. Hillier, 16 Priory Walk, Leicester Forest East, Leicester, LE3 3PP
MW6 VBE J. Jones, 7 St. Mary Street, Trelewis, Treharris, CF46 6AL
M6 VBF P. Staite, Chestnut Farm, Eastville, Boston, PE22 8LX
MI6 VBH V. Hazelton, 12 The Elms Bush, Dungannon, BT71 6UE
M6 VBJ J. Vanbesien, Flat 8, Suffolk House, Chester, CH1 3BZ
M6 VBN D. Melbourne, 160 Markfield Courtwood Lane, London, cr0 9hq
M6 VBP S. Vyner, 20 Ryde Lands, Cranleigh, GU6 7DD
M6 VBR J. Baughan, Chestnut Farm, Eastville, Boston, PE22 8LX
M6 VBS A. Cairns, 202 The Ridgeway, St Albans, AL4 9XJ

M6 VBX P. Otterwell, 50 Hythe Road, Staines-upon-Thames, TW18 3EE
M6 VBZ D. Scott, 13 Grey Close, Bredbury, Stockport, SK6 1HA
M6 VCA V. Chady, 12 Brent Place, Barnet, EN5 2DP
M6 VCB D. Bonney, 14 Jersey Close, Congleton, CW12 3TW
M6 VCC Dr O. Blacklock, 15 Kings Crescent, Lymington, SO41 9GT
M6 VCH C. Wood, 169 Bramerton Road, Nottingham, NG8 4NB
M6 VCK V. Corcoran, 50 Grange Road, Bracebridge Heath, Lincoln, LN4 2PW
M6 VCM V. Millson, Harrogate Ladies' College, Clarence Drive, Harrogate, HG1 2QG
M6 VCN D. Williams, 6 Homestone Gardens, Leicester, LE5 2LJ
M6 VCP C. Pinder, 70 Highfield Road, Beverley, HU17 9QR
M6 VCS V. Newell, 15 The Grove, Luton, LU1 5PE
M6 VCU K. Beswick, 29 Hops Lane, Halifax, HX3 5FB
M6 VCV Dr C. Jenkins, 52 Warden Abbey, Bedford, MK41 0SN
M6 VDC N. Fairbairn, 15 Hewitt Road, Dover, CT16 1TH
MW6 VDE D. Evans, 12 Caerfallwch, Rhosesmor, Mold, CH7 6PN
M6 VDF J. Roser, St Marys Terrace, Hastings, tn343lr
M6 VDH V. Harrington, 25 Victoria Street, Norwich, NR1 3QX
M6 VDJ D. Jones, 68 Dunelm Road Thornley, Durham, DH6 3HW
M6 VDL D. Lodwig, 15 Frithwood Park, Brownshill, Stroud, GL6 8AB
MM6 VDP D. Pounder, Birchbank, Lindean, Galashiels, TD1 3PA
M6 VDR C. Johnson, The Hollies, Belaugh Green Lane, Norwich, NR12 7AJ
M6 VDT C. Stephens, 8 New Molinnis, Bugle, St Austell, PL26 8QL
M6 VDX P. Martin, 108 Headlands Grove, Swindon, SN2 7HP
M6 VEC M. Broadhurst, 23 Hucklow Avenue, North Wingfield, Chesterfield, S42 5PX
M6 VED R. Dawson, 6 Harleys Field, Abbeymead, Gloucester, GL4 4RN
M6 VEG J. Waldron, 55 Sheringham Road, Poole, BH12 1NS
M6 VEL D. Rogers, 6 Guildford Street, Plymouth, PL4 8DS
M6 VEN A. Fisher, 15 Washdyke Lane, Osgodby, Market Rasen, LN8 3PB
M6 VEP E. Parker, 3 Dowgate Close, Tonbridge, TN9 2EH
M6 VET A. Williamson, 4 Garden End, Melbourn, Royston, SG8 6HD
M6 VFA A. Colcombe, 217 Church Drive, Quedgeley, Gloucester, GL2 4US
M6 VFC M. Chipperfield, 5 Lullingstone Close, Hempstead, Gillingham, ME7 3TS
M6 VFD H. Campbell, 50 Northwood Road, Whitstable, CT5 2ES
MM6 VFL B. Sonnet, 12 Winton Circus, Saltcoats, KA21 5DA
M6 VFN B. Bewick, 2 Grange Close, Hoveton, Norwich, NR12 8EA
MW6 VGA J. Duffain, 37 Mount Pleasant Street, Dowlais, Merthyr Tydfil, CF48 3AF
M6 VGB A. Green, 139 Clough Lane, Halifax, HX2 8SN
M6 VGC N. Lupton, 134 Ridgeway Road, Sheffield, S12 2SZ
M6 VGE S. Coates, 27 Primula Way, Chelmsford, CM1 6QT
M6 VGR I. Brodie, 38 Chatsworth Place, Stoke-on-Trent, ST3 7DP
MM6 VGS R. Adams, 1/R 61 Capelrig Street, Thornliebank, Glasgow, G46 8LP
M6 VGU K. Haythornwhite, 148 Nairne Street, Burnley, BB11 4NP
M6 VGV G. Venugopalan, 3 Southwater Close, London, E14 7TE
M6 VHA V. Gudgeon, 6 The Choakles, Wootton, Northampton, NN4 6AP
M6 VHB H. Blayer, 10 Riverside Close, Staines-upon-Thames, TW18 2LW
M6 VHM L. Mooney, Calde Cottage, Mannings Lane, Chester, CH2 2PB
M6 VHZ J. Telfer, 50 Agraria Road, Guildford, GU2 4LF
M6 VIA Dr L. Kirkcaldy, 17 Central Avenue, Exeter, EX4 8NG
M6 VID D. Mason, 94 Guessburn, Stocksfield, NE43 7QR
M6 VIE S. Croston, 12 Sefton Street, London, SW15 1LZ
M6 VIG C. Harvey, 1 New Pit Cottages, Bridge Place Road, Bath, BA2 0PE
M6 VIO S. Morgan, 8 Williams Drive, Braintree, CM7 5QJ
M6 VIT A. Vitiello, 8 Pegasus Road, Leighton Buzzard, LU7 3NJ
M6 VIV V. Lee, 56 Gordon Road, Fishersgate, Brighton, BN41 1PT
M6 VIX W. Reynolds, 4 Underwood Close, Stafford, ST16 1TB
M6 VJG V. Gallagher, 4 Kingsdown Way, Bromley, BR2 7PT
MI6 VJR R. Russell, 70 Allen Park, Dunamanagh, Strabane, BT82 0PD
M6 VJX B. Walker, 255 Packington Avenue, Birmingham, B34 7RU
M6 VKA S. Rush, 88 Mountview, Borden, Sittingbourne, ME9 8JZ
M6 VKB N. Morehen, 20 Castleton Grove, Inkersall, Chesterfield, S43 3HU
M6 VKE D. Lee, 26 Church Lane, Chalgrove, Oxford, OX44 7TA
M6 VKG M. Gibson, 6 Harrison Road, Mansfield, NG18 5RG
M6 VKH S. Hardy, 55 Westwood Road, East Peckham, Tonbridge, TN12 5DB
M6 VKK R. Cresswell, Meadow View, Hulver Road, Beccles, NR34 7UW
M6 VKM K. Minett, Rosedene, Honey Hill, Wimbotsham, King's Lynn, PE34 3QD
M6 VKN R. Reilly, 29 Burwell Close, Plymouth, PL6 8QD
M6 VKW V. Williams, Moor Farm, Moor Lane, Lincoln, LN3 4EG
M6 VKY V. Fleming, 17 Greenfield Close, Kippax, Leeds, LS25 7PX
M6 VKZ G. Delaforce, 29 Littlebridge Meadow Bridgerule, Holsworthy, EX22 7DU
M6 VLB K. Earwicker, Greenbank, Cripplestyle, Fordingbridge, SP6 3DU
M6 VLD V. Mereuta, 62 Sutherland Avenue, London, W9 2QU
M6 VLE V. Sayer, Observatory, Mount Bures Road Wakes Colne, Colchester, CO6 2AS
MM6 VLG W. Guy, 121 Stewarton Drive, Cambuslang, Glasgow, G72 8DH
M6 VLH D. Smith, 23 Old School House, Shotley Gate, Ipswich, IP9 1QP
MW6 VLO B. Henley, Rhewin Glas, Glynarthen, Llandysul, SA44 6PR
M6 VLQ D. Volkov, 2 Robert Close, Potters Bar, EN6 2DH
M6 VLR Rev. R. Blunden, 34 Stephenson Road Arborfield, Reading, RG2 9NP
MW6 VLS S. Morris, 21 Heritage Way Badgers Green, Llanymynech, SY226LL
M6 VMA C. Skupski, 57 Three Nooks, Bamber Bridge, Preston, PR5 8EN
MI6 VMC T. Mcelwee, Orchard Bank, 1A Dunover Road, Ballywalter, Newtownards, BT22 2LE
M6 VMF J. Jackson, 80A Clarence Road, Leighton Buzzard, LU7 3EL
M6 VMJ R. Atkinson, 75 Talbot Road, Penistone, Sheffield, S36 9ED
M6 VMK R. Whatling, 41 Lawrence House Camperdown Street, Bexhill on Sea, TN39 5EN
M6 VMO F. Deschacht, 28 Barnfield, Slough, SL1 5JW
M6 VMP I. Rotheram, 60 Whitewood Park, Liverpool, L9 7LG
M6 VMS M. Wickens, Haven Lea, Queens Drive, Windermere, LA23 2EL
M6 VMY V. Gunnoo, 76 Katherine Drive, Dunstable, LU5 4NU
M6 VNL M. Harris, 5 Lynmore Close, Northampton, NN4 9QU
MW6 VNP S. Beer, 67 Killan Road, Dunvant, Swansea, SA2 7TH
M6 VNT V. Sheppard, Flat 8, Riverside Court Cambridge Road, Harlow, CM20 2AD
M6 VNV D. Nicholls, 12 Northview, Tufnell Park Road, London, N7 0QB
MW6 VOC A. Parsons, 21 Rectory Drive, St. Athan, Barry, CF62 4PD
M6 VOD A. Motion, 45 Bradgate Lane, Asfordby, Melton Mowbray, LE14 3YB
MI6 VOF P. Mcfadden, 35 West Wind Terrace, Hillsborough, BT26 6BS
M6 VOG B. Rutkowski, 18 Squirrel Close, Coventry, CV2 1FP
M6 VOL S. Hill, 86 Gilbert Road, Chichester, PO19 3NL
M6 VOS S. Devos, Applecross Cottage, Main Road, Newark, NG23 7HR
MW6 VOW B. Moore, Lower Bulford Farm, Bulford Road, Haverfordwest, SA62 3ET
M6 VOX F. Riche, 1 Lovelstaithe, Norwich, NR1 1LW
M6 VOY M. Harrison, 2 Broad Farm Cottages, North Street, Hailsham, BN27 4DS
MI6 VOZ J. Kacprzyk, 80 The Brambles, Magherafelt, BT45 5RZ
MM6 VPF H. Phillips, Maplebank, Leithen Road, Innerleithen, EH446NJ
M6 VPH P. Haywood, 1 Scott Close Bidford-On-Avon, Alcester, B50 4HY
MW6 VPL S. Mccaffery, 5 Troedyrhiw, Caerlan, Abercrave, Swansea, SA9 1SX
MM6 VPM A. Cowan, 32 Esk Valley Terrace, Dalkeith, EH22 3FT
M6 VPS M. Partyka, Ridgebourne Cottage, Ridgebourne Road, Kington, HR5 3EG
M6 VPX S. Chapman, Wigrams Turn Marina, Shuckburgh Road, Southam, CV47 8NL
M6 VPZ T. Godfrey, 43B Broadbury Road, Bristol, BS4 1JT
M6 VQC C. Powell, 52 Primley Road, Sidmouth, EX10 9LF
M6 VQV T. Page, 19 Lamorna Drive, Callington, PL17 7QH
M6 VQX R. Olszewski, 24A Nightingale Road, South Croydon, CR2 8PT
M6 VRC A. Newbould, 9 Laburnum Road, Rudheath, Northwich, CW9 7JT
M6 VRD V. Bowen, 4 Crossley Gardens, Halifax, HX1 5PU
M6 VRE E. Vrentzos, 12 Floyer Close, Queens Road, Richmond, TW10 6HS
MM6 VRH C. Harvey, Hillside, Sarclet, Wick, KW1 5TU
M6 VRJ R. Jermy, 15 Oak Tree Close Martham, Great Yarmouth, NR29 4QN
M6 VRL R. Liszewski, 79 Malvern Drive, Warmley, Bristol, BS30 8XA
M6 VRO R. Sheen, 29 Aldermoor Avenue, Southampton, SO16 5GJ
M6 VRP I. Bell, 53 Shurland Avenue, Sittingbourne, ME10 4QT
M6 VRR K. Bird, Mount View, Hill Top Road Ashover, Chesterfield, S45 0BZ
M6 VRT A. Calvert, 5 Pond Cottages, Butts Pond, Sturminster Newton, DT10 1BE
MI6 VRX D. Todd, 2 Innishargie Gardens, Bangor, BT19 1SN
M6 VSA A. Chubb, 70 Goldcrest Road, Chipping Sodbury, Bristol, BS37 6XQ
M6 VSB B. Rhodes, 81 Witherston Way, Eltham, London, SE9 3JL
M6 VSC V. Schultz, 223 Long Down Avenue, Bristol, BS16 1GE
M6 VSE Dr S. Vellaichamy, 37 Briarswood, Chelmsford, CM1 6UH
MM6 VSI S. Irvine, 9 Pearce Grove, Edinburgh, EH12 8SP
M6 VSJ S. Jermy, 15 Oak Tree Close, Martham, Great Yarmouth, NR29 4QN
M6 VSL S. Lawrance, 69 Athelstan Gardens, Wickford, SS11 7EF
MM6 VSM S. Mulligan, 17 Crawfurd Gardens, Rutherglen, Glasgow, G73 4JP
M6 VSS M. Petrea-Goanta, 14 Oakdene Road, Watford, WD24 6RW
M6 VST J. Baker, 70 Vane Close, Norwich, NR7 0US
MW6 VTA J. Roberts, Hafan Llewelyn, Llanystumdwy, Criccieth, LL52 0SS
M6 VTB L. Rudge, 10 St. Johns Close, Aldingbourne, Chichester, PO20 3TH
M6 VTC S. Dillon, 33A Main Road, Cleeve, Bristol, BS49 4NS
M6 VTG D. Coles, 36 York Hill, Loughton, IG10 1HT
MW6 VTP M. Davies, 26 Ty Nant, Caerphilly, CF83 2RA
MM6 VTS S. Davidson, 6 Sutherland Walk, Mintlaw, Peterhead, AB42 5GT
M6 VTT D. Bedford, 55 Sycamore Avenue, Wickersley, Rotherham, S66 2NS
M6 VTZ H. Hatchard, 118 Marina, St Leonards on-Sea, TN380BN
M6 VUD A. Burr, 3 Larch Close Attleborough, Attleborough, NR 172HB
M6 VUK A. Buck, 10 Northfield Park, Mansfield Woodhouse, Mansfield, NG19 8PA
MM6 VUS C. Doherty, 22 Castlelaw Street, Glasgow, G32 0NF
MM6 VUV R. Fraser, 72 Ferguson Drive, Denny, FK6 5AG
M6 VVA A. Amos, Willow Tree House, Deers Green, Clavering, Saffron Walden, CB11 4PX
M6 VVB C. Harper, 21 Holly Close Farnham Common, Slough, SL2 3QT
M6 VVE R. Cook, 62 High Hazel Road, Moorends, Doncaster, DN8 4QN
M6 VVN V. Walker, 291 Park Road, Blackpool, FY1 6RR
M6 VVT M. Kenny, 5 Fallowfield, Hazlemere, High Wycombe, HP15 7RP
M6 VVX P. Buck, Staple Barton, Staple Fitzpaine, Taunton, TA3 5SW
M6 VWA W. Vinnicombe, 22 Victoria Park, Cambridge, CB4 3EL
M6 VWD J. Bartley, 29 Cheltenham Road, Blackburn, BB2 6HR
M6 VWE C. Brown, 123 Godinton Road, Ashford, TN23 1LN
MM6 VWH C. Robertson, Flat 3, Redheugh House, Redheugh Court, Kilbirnie, KA25 7JF
M6 VWN S. Mathew, 183 West End, Costessey, Norwich, NR8 5AW
M6 VWP E. Byrne, 859 Rochdale Road, Middleton, Manchester, M24 2RA
M6 VWR R. Hirons, 28 Crossfell Road, Hemel Hempstead, HP3 8RG
MM6 VWT W. Scott, 11 The Marches Armadale, Bathgate, EH48 2PG
M6 VWV J. Jackson, 65 Berkeley Close, Abbots Langley, WD5 0XD
M6 VWW A. Wilson, 28 Langham Road, Bristol, BS4 2LJ
M6 VWX G. Brownlow, 21 Church Lane Tittleshall, King's Lynn, PE32 2QD
MM6 VXB M. Al Saeed, 9 Appin Place, Edinburgh, EH14 1NJ
M6 VXI D. Holden, 24 Penny Gate Close, Hindley, Wigan, WN2 3DP
MM6 VXR A. Latto, 8 Aspen Avenue, Glenrothes, KY7 5TA
M6 VXT T. Searle, 18 Witton Lane, Little Plumstead, Norwich, NR13 5DL
M6 VYN M. Roberts, 11 Oakleigh Road, Pinner, HA5 4HB
M6 VZA Dr J. Lauxman Mccorkell, Suite 77 City House, 131 Friargate, Preston, PR1 2EF
M6 VZF E. Clarke, 1 Farm Drive, Croydon, CR0 8HX
M6 VZV M. O'Donovan, 10 Rockfield Close, Teignmouth, TQ14 8TS
MI6 VZW M. Mcpeake, 47 Dunclug Gardens, Ballymena, BT43 6NN
M6 WAA S. Myciunka, 67 Templeton Drive, Fearnhead, Warrington, WA2 0WR
MI6 WAB W. Mcdonald, 14 Edenmore Park, Limavady, BT49 0RG
M6 WAD P. Waddington, 98 Harrogate Street, Barrow-in-Furness, LA14 5LY
MI6 WAF W. Hawkes, 12 Meadow Court, Newtownards, BT23 8YE
MI6 WAG M. Barr, 2 Willowvale Close, Islandmagee, Larne, BT40 3SD
MM6 WAI W. Inglis, 13 Princes Street, California, Falkirk, FK1 2BX
M6 WAJ J. Watts, 31 Ingledene Close, Havant, PO9 1DG
MI6 WAM W. Mccormick, Dernawilt Road, Crocknaboghill, Rosslea, Enniskillen, BT92 7GG
MI6 WAN W. Nelson, 44 Lacky Road, Tattynageeragh, Rosslea, Enniskillen, BT92 7GA
M6 WAP W. Patterson, 6 Lea Drive, Nantwich, CW5 5JS
M6 WAQ J. Carn, 12 Woodcock Hill Estate Harefiled Road, Rickmansworth, WD3 1PQ
M6 WAS D. Simpson, Flat 4, Victoria Mansions, East Street, Harwich, CO12 3AS
M6 WAT P. Watson, 10 Whitelands Crescent, Baildon, Shipley, BD17 6NN
M6 WAV A. Snelson, 6 Rayleigh Close, Braintree, CM7 9TX
MM6 WAW M. Beaton, 76 Fergus Avenue, Howden, Livingston, EH54 6BG
M6 WAX P. Simmonds, 10 Castlewood Mobile Home Park, Hinckley Road, Leicester, LE9 4JZ
M6 WAY B. Waymark, 13 Beech Ave, Nottingham, NG7 7LJ
M6 WBA J. Edridge, 36 Conyngham Lane, Bridge, Canterbury, CT4 5JX

M6 WBB J. Whitehead, 12 Polkerris Road, Carharrack, Redruth, TR16 5RJ
M6 WBE D. Buckley, 99 Kings Drive, Bradwell, Great Yarmouth, NR31 8TF
M6 WBF I. Westwood, 1 Cook Avenue, Maltby, Rotherham, S66 8QZ
M6 WBH D. Thompson, 17 Sandpiper Close, Blyth, NE24 3QN
MM6 WBJ W. Jackson, 3 Annick Road, Dreghorn, Irvine, KA11 4EY
M6 WBK D. Beck, 13 Chipstead, Chalfont St. Peter, Gerrards Cross, SL9 9JZ
M6 WBM A. Buckley, 25 Queensway, Pilsley, Chesterfield, S45 8EJ
MI6 WBN M. Parke, 18 Brandywell Court, Londonderry, BT48 9HL
M6 WBO W. Beacroft, 3 Rose Farm Rise, Normanton, WF6 2PL
MM6 WBP W. Ferguson, 1D Macphail Drive, Kilmarnock, KA3 7EJ
M6 WBR J. Webber, 16 Marythorne Road, Bere Alston, Yelverton, PL20 7BZ
M6 WBS W. Bennison, 21 Ashdene Close, Chadderton, Oldham, OL1 2QG
MI6 WBT W. Turkington, 8A Drummullan Road, Moneymore, Magherafelt, BT45 7XS
MM6 WBU J. Mclelland, 24 Hyslop Street, Airdrie, ML6 0ES
M6 WBV W. Reeves, 33 Pond Bank, Blisworth, Northampton, NN7 3EL
M6 WBX D. Mills, 79 Eastbourne Avenue, Gosport, PO12 4NX
M6 WBZ K. Hunt, 11 De Marnham Close, West Bromwich, B70 6RJ
M6 WCA B. Neal, 6 Canterbury Street, Chaddesden, Derby, DE21 4LG
M6 WCB A. Williams, 74 Broadfield Road, Bristol, BS4 2UW
M6 WCC P. Homer, 20 Sunningdale Road, Middlesbrough, TS4 3HU
M6 WCE W. Steingold, 36 Hailsham Road, Romford, RM3 7SP
M6 WCF P. Wade, 85 Penymynydd Road, Penyffordd, Chester, CH4 0LF
M6 WCG C. Jones, 90 Station Rod, Wem, SY4 5BL
MW6 WCI W. Johns, 5 Heol Pantgwyn, Llanharry, Pontyclun, CF72 9HU
MM6 WCK D. Ross, 29 East Banks, Wick, KW1 5NL
M6 WCL O. Fallon, 26 Central Avenue, Corfe Mullen, Wimborne, BH21 3JD
M6 WCN D. Jones, 85 Black Butts Lane, Walney, Barrow-in-Furness, LA14 3JL
MW6 WCQ E. Edwards, 1 Brynhyfryd, Sarn, Bridgend, CF32 9UR
M6 WCQ J. Daly, 77 Broomfield Road, Swanscombe, DA10 0LU
M6 WCR C. Bavister, 10 Pheasant Grove, Wixams, Bedford, MK42 6AH
MW6 WCT M. Cooper, 3 Waunddu, Pontnewynydd, Pontypool, NP4 6QZ
MW6 WCU D. Crowe, 2 Severnside Cottages, Canal Road, Newtown, SY16 2JN
M6 WCV W. Marshall, 24 Union Road, Thorne, Doncaster, DN8 5EL
M6 WCX W. Dix, 21 Pine Vale Crescent, Bournemouth, BH10 6BG
M6 WCY A. Whyman, 2 Wilson Close, Thelwall, Warrington, WA4 2ET
M6 WCZ D. Howden, 46 Chestnut Avenue, Armthorpe, Doncaster, DN3 2EP
MI6 WDB W. Bradshaw, 206 Manor Street, Belfast, BT14 6ED
M6 WDC M. Moseley, 86 Sissons Terrace, Leeds, LS10 4LH
MI6 WDD D. Milligan, 30 Belgrano Ahoghill, Ballymena, BT42 2QQ
M6 WDF D. Leyland, 20 Newton Heath, Middlewich, CW10 9HL
M6 WDG D. Goodwin, 33 York Road Dunscroft, Doncaster, DN7 4LZ
M6 WDH W. Hunt, 128 Heath Road, Dudley, DY2 0AU
M6 WDI K. Lynch, Medindie, Woodside, Ryton, NE40 4SY
M6 WDL W. Leake, 61 Jubilee Road, Sutton-in-Ashfield, NG17 2DD
MI6 WDM W. Mccormick, 1 Mullagreenan Court, Rosslea, Enniskillen, BT92 7PR
M6 WDN Dr A. Brito Da Silva, 21 Grosvenor Gardens, Newcastle upon Tyne, NE2 1HQ
M6 WDO T. Smith, Chy Crowshensy, Clifton Road, Redruth, TR15 3UD
M6 WDQ W. Carey, 6 Gainsborough Road, Bexhill-on-Sea, TN40 2UL
M6 WDR A. Wallace, 46 Heathfield Road, Grantham, NG31 7NH
M6 WDW J. Summers, 5 The Meadow, Bosvigo Lane, Truro, TR1 3NG
MW6 WDX P. Dalton, 89 Hillcrest, Brynna, Pontyclun, CF72 9SL
M6 WDY C. Johnson, 45 Gordon Road, Chelmsford, CM2 9LN
MM6 WEB J. Webster, 6 Livingston Crescent, Winchburgh, Broxburn, EH52 6FX
MW6 WEE A. Morris, Fairview, Trefonen, Oswestry, SY10 9DP
M6 WEJ C. Buswell, 34 Hatchmore Road, Denmead, Waterlooville, PO7 6TF
M6 WEK K. Alabaster, 16 Butlers Road, Horsham, RH13 6AJ
M6 WEL D. Wells, 40 Barnham Broom Road, Wymondham, NR18 0DF
M6 WEN W. Jefferson, Mk451Lf, Flitwick, MK451LF
M6 WEO S. Kiel, 32 Weavers Avenue, Frizington, CA26 3AT
M6 WEP E. Raine, Stable Cottage, St. Martins, Richmond, DL10 4SJ
MI6 WEQ R. Skelton, 16 Demiville Avenue, Lisburn, BT27 5RE
MM6 WER J. Weir, 36 Main Rd, Ferguslie, Paisley, PA12QT
M6 WET D. Bilton, 9 Ashby Street, Allenton, Derby, DE24 8JR
M6 WEU C. Massimiani, Flat 2 37 York Road, Guildford, GU1 4DN
M6 WEV A. Weatherall, The Old Telephone Exchange, The Street, Canterbury, CT3 1ED
M6 WEW L. Wnekowski, 22 Simonds Grove, Spencers Wood, Reading, RG7 1BH
MI6 WEZ W. Todd, 2C Knockwood Crescent, Belfast, BT5 6GE
M6 WFA W. Durrant, 19 Rydal Rd, Gosport, PO12 4ES
MW6 WFB L. Bowman, Chanrick, Penderyn Road, Aberdare, CF44 9RU
M6 WFC P. Morgan, 25 Junction Street, Dudley, DY2 8XT
M6 WFE T. Mckenna, 67 Raynsford Road, Great Whelnetham, Bury St Edmunds, IP30 0TN
MW6 WFF W. Webb, 70 Beech Court, Bargoed, CF81 8NS
M6 WFG S. Woodruff, 172 Windsor Street, Wolverton, Milton Keynes, MK12 5DR
M6 WFH M. Sharp, Woodgate Farm, Livery Road, Salisbury, SP5 1RJ
M6 WFI I. Warnecke, 12 Caxton Road, Margate, CT9 5NP
MW6 WFJ T. Penman, 21 Maelog Road, Cardiff, CF14 1HP
MD6 WFK R. Kissack, 6 Falcon Cliff Court, Douglas, Isle of Man, IM2 4AQ
M6 WFL C. Coles, 2 Fern Square, Chickerell, Weymouth, DT3 4NZ
M6 WFR R. Pickwoad, 2 The Tannery, Barrowden, Oakham, LE15 8EA
M6 WFV J. Stevens, Parc-An-Lower Prospidnick, Helston, TR13 0RY
M6 WFW W. Fletcher-Wells, 46 London Road, Buxton, SK17 9NU
M6 WFX C. Giddens, 3 Neville Walk, Richmond, DL10 5AG
M6 WFY K. Payne, Eastern Esplanade, Broadstairs, CT10 1DR
M6 WFZ M. Toomey, Ravens Rock, 10A Castlebank, Northwich, CW8 1BL
M6 WGB G. Birch, 23 Hanson Street, Great Harwood, Blackburn, BB6 7LP
M6 WGC T. Buck, 6 Lynn Road, Terrington St. Clement, King's Lynn, PE34 4JX
MM6 WGD R. Ward, 35 Mill Street, Drummore, Stranraer, DG9 9PS
MI6 WGE G. Orderly, 30 Moatview Park, Dundonald, Belfast, BT16 2BE
M6 WGF G. Harris, 9 Penmeneth Trewennack, Helston, TR13 0PU
M6 WGG I. Wagstaff, 25 Bowbridge Gardens Bottesford, Nottingham, NG130AZ
M6 WGH I. Mccorquodale, 4 Wise Grove, Warwick, CV34 5JW
M6 WGJ W. Joyce, 2 Palmers Cottage, Main Street, Oakham, LE15 8DH
MI6 WGL W. Leonard, 57 Old Coach Road Mullanavehy, Enniskillen, BT92 2EW
MI6 WGM G. Mccusker, 10 Birchdale, Lurgan, Craigavon, BT66 7TR
M6 WGS C. Smart, 35B Church Lane, Melksham, SN12 7EF
MM6 WGT M. Gilchrist, 17 Colinslee Crescent, Paisley, PA2 6SD
M6 WGU G. Head, 45 Beechnut Road, Kendal, LA9 7FF
MI6 WGX J. Johnston, 16 Fort Road, Bangor, BT19 7BS
M6 WHA A. Whadcoat, 38 Edwin Panks Road, Hadleigh, Ipswich, IP7 5JL
M6 WHD R. Searby, 72 Polstain Road Threemilestone, Truro, TR3 6DH
M6 WHF D. Cracknell, 120 Woodhill, London, SE18 5JL
MD6 WHG W. Hogg, Medhamstead, Lhergydhoo, Isle of Man, IM5 2AE
MM6 WHI R. Leslie, 118 Western Avenue, Ellon, AB41 9EU
M6 WHN D. Donnelly, 33 Dunlin Drive, Ayton, Washington, NE38 0EB
M6 WHS D. Whitehouse, 6 Larch Close, Heathfield, TN21 8YW
M6 WHT K. Cullen, 29 Colman Road, London, E16 3JY
M6 WHU A. Hodgson, 515 Ashingdon Road, Rochford, SS4 3HE
MM6 WHW W. Woods, 1 Nursery Lane, Kilmacolm, PA13 4HP
MM6 WHX D. Strachan, 30 Belhaven Park, Muirhead, G69 9FB
M6 WHZ M. Underwood, 39 Westbury Crescent, Oxford, OX4 3SA
M6 WIB R. Wilson, 171 Cooden Drive, Bexhill-on-Sea, TN39 3AQ
M6 WIC Rev. B. Jackson, 46 Princess Road, Market Weighton, York, YO43 3BR
M6 WID G. Holland, 6 Moorfield Road, Widnes, WA8 3JE
M6 WIF M. Landragin, 101 Linden Gardens, Enfield, EN1 4DY
M6 WII W. Smith, 40 Westbury Rise, Harlow, CM17 9NS
M6 WIK W. Trofimiuk, 120A The Fairway, Northolt, UB5 4SW
M6 WIL W. Outram, 7 Bakewell Road, Baslow, Bakewell, DE45 1RE
M6 WIM A. Forrester, 14 Calder Close, Lytham St Annes, FY8 3NH
M6 WIR D. Daniels, 60 Hawthorn Way, Northway, Tewkesbury, GL20 8TQ
M6 WIS D. Wiskow, 15 Ferndale Close, Sandbach, CW11 4HZ
M6 WIT B. Witt, 20 Foxglove Close, Newton Aycliffe, DL5 4PF
M6 WIW C. Ring, Acorn Cottage Prospect Place, Helston, TR13 8RU
M6 WIX J. Unwin, 39 Whinfell Drive, Normanby, Middlesbrough, TS6 0BG
M6 WIZ S. Keen, 13 Ivy Road, Kettering, NN16 9TG
M6 WJA A. Woodhouse, 14 Wrens Gardens, Wath-Upon-Dearne, Rotherham, S63 7GD
M6 WJB W. Buss, 83 Gorse Avenue, Chatham, ME5 0UP
M6 WJF W. Foster, 55 Drake Avenue Minster On Sea, Sheerness, ME12 3SA
M6 WJJ K. Jones, 98 Common View, Stedham, Midhurst, GU29 0NU
MI6 WJK J. Kelso, 32 Old Park Manor, Ballymena, BT42 1RW
M6 WJL G. Hayers, 87 Bradleigh Avenue, Grays, RM17 5RH
M6 WJM J. Wheeler, 14 Doomgate Asby, Appleby-in-Westmorland, CA16 6RB
M6 WJN W. Naughton, 14 Carisbrooke, Frimley, Camberley, GU16 8XR
MM6 WJP W. Paterson, 1 Burnside Terrace, Stranraer, DG9 8HH
MM6 WJS S. Wilson, 2 Kinnear Court, Guardbridge, St Andrews, KY16 0UE
MI6 WJW W. Wilson, 9 Benbradagh Avenue, Limavady, BT49 0AP
M6 WKB K. Brett, 1 Tower Court, Peterborough, PE2 9AT
MM6 WKC M. Cormack, 2 Coghill Street, Wick, KW1 4PN
M6 WKD R. Hammond, Bye Road Cottage Pot Kiln Chase Co93Bh, Gestingthorpe, Co93bh
MI6 WKE R. Wilkinson, 11 Fairview Park, Dromore, BT25 1PN
M6 WKF A. Rowland-Stuart, 86 Wiltshire House, Lavender Street, Brighton, BN2 1LE
M6 WKG A. Cole, 104 Newport Road, Cowes, PO31 7PS
M6 WKI M. Hawker, Popes Corner Marina, Holt Fen, Little Thetford, Ely, CB6 3HR
M6 WKL P. Houghton, 6 Olivers Court, Calne, SN11 0FL
MI6 WKN W. Gamble, 11 Anderson Park, Limavady, BT49 0RH
M6 WKR G. Walker, 2 Cliffe Bank Cottages, Piercebridge, DL2 3SX
M6 WKS A. Wilkins, 124 Fullers Mead, Harlow, CM17 9AU
M6 WKT G. Williams, 18 Elmsleigh Road, Farnborough, GU14 0ET
M6 WKW K. Davies, 23 Egmanton Road, Meden Vale, Mansfield, NG20 9QN
M6 WKY C. Wilkinson, 40 Lumley Drive, Consett, DH8 7DT
M6 WKZ A. Norden, 10 School Lane, Watton At Stone, Hertford, SG14 3SF
M6 WLA S. Fearnhead, 1B Hampden Grove, Eccles, Manchester, M30 0QU
MW6 WLB K. Clarke, 13 Caerphilly Close, Dinas Powys, CF64 4PZ
M6 WLC W. Cogdon, 15 Strafford Avenue, Worsbrough, Barnsley, S70 6SU
M6 WLD W. Daley, 27 Rosebery Street, Manchester, M14 4UR
M6 WLE E. Williams, 24 Astbury Street, Congleton, CW12 4EQ
M6 WLF K. Smith, 3 Hawley Vale, Hawley Road, Dartford, DA2 7RL
M6 WLG D. Collins-Cubitt, 75 Anthony Drive, Norwich, NR3 4EW
M6 WLR A. Coats, 57 Mill Hill, Boulton Moor, Derby, DE24 5AF
M6 WLS W. Le Serve, 120 Cheam Road, Cheam, Sutton, SM1 2EB
MW6 WLY L. Lane-Wells, Gwaelod, Pool Quay, Welshpool, SY21 9LH
M6 WMA M. Wall, 227 Wayfield Road, Chatham, ME5 0HJ
M6 WMF N. Riggall, 41 Manor Drive, Waltham, Grimsby, DN37 0NS
M6 WMG M. Carpenter, The Retreat, High Lane Manaccan, Helston, TR12 6HT
M6 WMH W. Clarke, 1 Alvey Terrace, Nottingham, NG7 3DF
M6 WMJ M. Watts, 48 Tasburgh Street, Grimsby, DN32 9LB
MM6 WMK M. Walker, 1 Gilmerton Street, Glasgow, G32 7SQ
MM6 WMM W. Moodie, West Barcloy Farm Rockcliffe, Dalbeattie, DG5 4QL
MI6 WMN W. Mcmullen, 69 Lissize Avenue Rathfriland, Newry, BT34 5DE
M6 WMP M. Weaver, 16 Avocet Drive, Kidderminster, DY10 4JT
M6 WMS M. Smith, 12 Lincoln Drive Waddington, Lincoln, LN5 9NH
M6 WMV J. Hancock, 162 Preston Road, Yeovil, BA20 2EQ
M6 WMZ A. Law, 2 The Bank, Somersham, Huntingdon, PE28 3DJ
M6 WNB J. Hocking, 26 Musket Road, Heathfield, Newton Abbot, TQ12 6SB
M6 WNC C. Lindley, 187 Alexandra Road, Sheffield, S2 3EH
M6 WNE W. Gough, 42 Merevale Avenue, Hinckley, LE10 0PY
M6 WNF N. Waller, Olive Cottage, 6 Church Road Chelmondiston, Ipswich, IP9 1HS
M6 WNM W. Mccoo, Ivy House, West Drove South, Walpole Highway, Wisbech, PE14 7RA
M6 WNN W. Newey, 3 Forrester Close, Flanderwell, Rotherham, S66 2NL
M6 WNR N. Rapson, 15 School Close, Bampton, Tiverton, EX16 9NN
M6 WNS M. Tsun, Dyson's Farm, Long Row, Tibenham, NR16 1PD
M6 WNY N. Yeates, 149 Hallam Way West Hallam, Ilkeston, DE7 6LP
M6 WNZ A. Argent-Wenz, 6 Ashley Brake West Hill, Ottery St Mary, EX11 1TW
M6 WOB J. Blow, 23 Leaders Way, Lutterworth, LE17 4YS
MM6 WOC W. O'Rourke, 6 Busby Place, Kilwinning, KA13 7BA
MW6 WOD W. Davies, Foelallt, North Road, Aberystwyth, SY23 2EL
M6 WOE G. Needle, 195 Kingsley Avenue, Kettering, NN16 9ET
MI6 WOF A. Pulman, 69 Islandarragh Road, Cape Castle, Ballycastle, BT54 6HS
M6 WOH H. Pickett, 13 Clifden Close Mullion, Helston, TR12 7EQ
M6 WOI M. Jenkins, 5 Arthurs Hill, Shanklin, PO37 6EW
M6 WOK M. Smith, 7 Cherry Tree Close, Everton, Lymington, SO41 0ZG
M6 WOL M. Walters, 39 Portland Place, Coseley, Bilston, WV14 9TB

M6 WOM M. Wilkes, 1 Hawburn Close, Bristol, BS4 2PB
M6 WOO J. Wooldridge, 7 Heather Gardens, Belton, Great Yarmouth, NR31 9PP
M6 WOS W. Briggs, 88 Guild Avenue, Walsall, WS3 1LQ
M6 WOT D. Bowers-Edgley, 8 The Slopes, Lower Henley Road, Reading, RG4 5LE
MW6 WOV V. Lyall, 29 King Alfreds Road, Sedbury, Chepstow, NP16 7AQ
M6 WPA W. Armsden, 2 Rowan Close, Middlewich, CW10 0TA
MW6 WPB W. Thomas, 10 John Lewis Street Hakin, Milford Haven, SA73 3HS
M6 WPI T. Hobson-Smith, 15 Henconner Lane, Chapel Allerton, Leeds, LS7 3NX
M6 WPN S. Comper, 105 Westway, Copthorne, Crawley, RH10 3QS
MI6 WPP J. Quigley, 25 Rathfern Way, Newtownabbey, BT36 6BX
M6 WPS W. Phillips, 36 Beeches Road, Great Barr, Birmingham, B42 2HF
MI6 WPT W. Turkington, 230 Whitechurch Road, Ballywalter, Newtownards, BT22 2LB
MI6 WPW W. Wilson, 9 Benbradagh Avenue, Limavady, BT49 0AP
M6 WRB W. Bellamy, Flat 6, 26 Moremead Road Lewisham, London, SE6 3LP
MW6 WRC W. Lockyer, 31 Erw Las, Penywaun, Aberdare, CF44 9BG
M6 WRD W. Davis, 24 Strensall Road, Hull, HU5 5td
M6 WRE W. Rendell, 71 Hardwick Bank Road, Northway, Tewkesbury, GL20 8RP
M6 WRG M. Richardson, 64 Henry Road, West Bridgford, Nottingham, NG2 7ND
M6 WRH W. Hartley, 30 Coltman Avenue, Beverley, HU17 0EY
M6 WRJ W. Jones, 11 Wykeham Place, Fareham, PO16 0FA
M6 WRL S. Lacey, 2 Purbeck Cottages, Acton, Swanage, BH19 3LU
MI6 WRM W. Campbell, 9 Rochester Court, Coleraine, BT52 2JL
M6 WRN M. Wren-Hilton, 28 Arwenack Avenue, Falmouth, TR11 3JW
M6 WRO J. Wrobel, Flat 27, Tenney House, Curzon Drive, Grays, RM17 6SG
M6 WRR Dr C. Hall, 6 Browning Road Church Crookham, Fleet, GU52 0YJ
M6 WRS W. Sankey, 14 Carter Grove, Hereford, HR1 1NT
MI6 WRT R. Jameson, 64 Blackhill Road, Dromore, Omagh, BT78 3HL
M6 WRU W. Clough, 19 Cardigan Road, Bedworth, CV12 0LY
M6 WRV M. Wright, 4 Woodlands Road, Headington, Oxford, OX3 7RU
M6 WRW D. Hampton, 6 St. Georges Lane South, Worcester, WR1 1QZ
M6 WRX I. Scholey, 36 Longroyd Avenue, Leeds, LS11 5HA
M6 WRY L. Curtis, 39 Mount Stewart Street, Seaham, SR7 7NG
M6 WSB A. Mcallister-Bowditch, 24 Morse Close, Chippenham, SN15 3FY
M6 WSC D. Waterhouse, 10 Falconers Drive, Battle, TN33 0DT
MM6 WSG G. Greenwood, 92 Porterfield, Comrie, Dunfermline, KY12 9XG
M6 WSH Dr S. Halsey, 15 Rectory Drive, Yatton, Bristol, BS49 4HF
M6 WSK D. Clayton, 171 Warning Tongue Lane, Doncaster, DN4 6TU
M6 WSM W. Gee, 2 Milton Walk, Worksop, S81 0DH
M6 WSO D. Mathison, 74 Victoria Road, Tuebrook, Liverpool, L13 8AW
MI6 WSP T. Mullan, 35 Finn Park Rosslea, Enniskillen, Bt92 7lj
M6 WSR R. Emery, 15 Shaw Crescent, Hutton, Brentwood, CM13 1JD
M6 WSS S. Roberto-Southall, 121 Dunsheath, Telford, TF3 2DA
M6 WST R. West, 557 East Bank Road, Sheffield, S2 2AG
M6 WSU S. Blackburn, 36 Mardale Grove, Barrow-in-Furness, LA13 9QG
M6 WSW D. Rolph, 17 Moorlands Park Ashby Road, Sinope, Le67 3bd
M6 WTD T. Dallas, 74 Main Street, Asfordby, Melton Mowbray, LE14 3SA
M6 WTE A. White, 56 Raleigh Close, Churchdown, Gloucester, GL3 1NT
M6 WTF S. Searles-Bryant, 14 Canuden Road, Chelmsford, CM1 2SX
M6 WTG W. Jones, 8 Oakbrook Close, Ewyas Harold, Hereford, HR2 0NX
M6 WTL A. Page, 12 Hitchens Close, Hemel Hempstead, HP1 2PP
M6 WTM D. Parsloe, 13 Highview Park New Dover Road, Capel-Le-Ferne, Folkestone, CT18 7GA
M6 WTP W. Pascoe, 34 Jasper Road, London, E16 3TR
M6 WTR S. Whittaker, 115 Grange Close, Hoveton, Norwich, NR12 8EB
M6 WTT W. Barnard, 34 Forsyth Drive, Braintree, CM7 1AR
M6 WTW T. White, 9 Dashwood Court, Banbury, OX16 5DF
M6 WTX S. Jackson, 64 Main Road, Moulton, Northwich, CW9 8PB
MI6 WTZ T. Saunderson, 40 Hillfoot Street, Belfast, BT4 1PR
M6 WUB A. Best, 18 Chestnut Close Rendlesham, Woodbridge, IP12 2UW
M6 WUG C. Humphries, 44 Linksway, Folkestone, CT19 5LS
M6 WUH C. Hopkins, 24 Battle Road, Tewkesbury, GL20 5TZ
MW6 WUK G. Reason, 454 Cowbridge Road West, Cardiff, CF5 5BZ
MM6 WUL W. Fulton, 15 Staffa Avenue, Port Glasgow, PA14 6DT
M6 WUM J. Griffiths, 9 Cauldale Close, Middleton, Manchester, M24 5SU
M6 WUN H. Cheung, Clarence Drive, Harrogate, HG1 2QG
M6 WVB V. Brett, 1 Tower Court, Peterborough, PE2 9AT
MW6 WVC T. Rowlands, Caer Gog Farm, Bodffordd, LL77 7BX
M6 WVE M. Huggett, 12 West View Cottages Lewes Road, Lindfield, Haywards Heath, RH16 2LJ
M6 WVG D. Millward, 77A Meadowcroft, St Albans, AL1 1UG
M6 WVH P. Wood, 26 Wamil Way, Mildenhall, Bury St Edmunds, IP28 7JU
M6 WVK K. Andrews-Mead, 10 Alderlands Close Crowland, Peterborough, PE6 0BS
M6 WVV S. Legg, 4 Riverside Avenue, Fareham, PO16 8TF
M6 WWB W. Buczkowski, 54 Woodfield Heights, Wolverhampton, WV6 8PT
M6 WWC C. West, 112A Hawthorne Way Shelley, Huddersfield, HD8 8PX
M6 WWD R. Howard, 13 Top Common, East Runton, Cromer, NR27 9PW
M6 WWE S. George, 84 Kenilworth Crescent, Enfield, EN1 3RG
M6 WWF M. Johnson, 6 Swainson Road, Leicester, LE4 9DQ
M6 WWI B. Kuttikkate, 34 Shetland Crescent, Rochford, SS4 3FJ
M6 WWJ J. Harrison, 20 Kelcliffe Avenue, Guiseley, Leeds, LS20 9EW
M6 WWK M. Wilks, Flat 16, Overton Court, Cheltenham, GL50 3BW
M6 WWM M. Barnard, 47 Springfields, Dunmow, CM6 1BP
M6 WWR I. Coleman, 15 St. Andrews Close, Holme Hale, Thetford, IP25 7EH
MM6 WWS W. Stevenson, 28 Lightburn Road Cambuslang, Glasgow, G72 8UE
M6 WWT A. Walls, 7 Waveney Grove, York, YO30 6EQ
MW6 WWY K. Macleod, 11 Hillside Close, Goodwick, SA64 0AX
MW6 WWZ A. Harris, 4 Dol Elian, Old Colwyn, Colwyn Bay, LL29 8YZ
M6 WXD B. Wild, 1 Sunnymount, Midsomer Norton, Radstock, BA3 2AS
M6 WXF J. Alansbur-Warrior-Macbyrne, 4 Rayner Street, Stockport, SK1 4HX
M6 WXH D. Purnell, 30 Conygre Green Timsbury, Bath, BA2 0JU
M6 WXK P. Marsh, 30 Mount Pleasant, Aylesford, ME20 7BE
M6 WXN S. Harcourt, 71 Ingleby Road, Long Eaton, Nottingham, NG10 3DG
M6 WXP M. Howes, 8 Oat Hill Drive, Northampton, NN3 5AL
MM6 WXS A. Mccall, 1 Finlayson Drive, Airdrie, ML6 8LU
M6 WXX D. Watts, 3 Witney Road, Crawley, RH10 6GJ
M6 WXY G. Hogan, 58 Weeley Road, Little Clacton, Clacton-on-Sea, CO16 9EN
M6 WYD G. Holt, 6 Highfields, Holmfirth, HD9 2PZ
M6 WYG D. Griffiths, 5 New Grange Terrace, Pelton Fell, Chester le Street, DH2 2PB
MW6 WYH H. Williams, 174 Squirrel Walk, Pontarddulais, SA4 0UG
M6 WYM S. Koskinas, 50 Amderley Drive, Norwich, NR46HZ
M6 WYN D. Griffith, The Old Stables, Briantspuddle Dairy, Dorchester, DT2 7HT
M6 WYR P. Attwood, 17 Robins Corner, Evesham, WR11 4RJ
M6 WYT T. Webster, 1 Fen Close, Newton, Alfreton, DE55 5TD
M6 WYW M. Whatling, 34 Cannon Park Road, Coventry, CV4 7AY
M6 WYX N. Soane, 32 Pangdene Close, Burgess Hill, RH15 9UT
M6 WYY L. O'Neill, Flat 6, Freshwater Court, Lee-on-the-Solent, PO13 9BB
M6 WYZ A. Tanseli, 157 Warwick Road, Rayleigh, SS6 8SG
M6 WZB S. Burdis, 10 Johnston Avenue, Hebburn, NE31 2LJ
M6 WZF A. Downing, 67 Mayfield Drive Caversham, Reading, RG4 5JP
M6 WZK P. Goodhand, 2A Moor Road, Sutton, Norwich, NR12 9QN
M6 WZV T. Cole, 36 Fairlee Road, Newport, PO30 2EJ
MU6 WZY S. Kirkpatrick, Ste Helene Manor, St Andrew, Guernsey, GY6 8XN
M6 XAA J. Daniels, 33 Park View, Crewkerne, TA18 8HS
M6 XAB A. Clay, 4 Craiglands Park, Ilkley, LS29 8SX
MW6 XAC A. Oconnell, 31 North Avenue, Tredegar, NP22 3HF
M6 XAD J. Hyde11, 49 Chichester Close, Exmouth, EX8 2LR
MW6 XAE J. Smith, Llanstinan Fach, Letterston, Haverfordwest, SA62 5XD
M6 XAF T. Chapman, 12 Greenways Chilcompton, Radstock, BA3 4HT
MW6 XAG S. Broderick, 179 Malpas Road, Newport, NP20 5PP
M6 XAI A. Rumsby, 31 Howell Road, Drayton, Norwich, NR8 6BU
M6 XAK S. Ferguson, 80 London Road, Retford, DN22 7DX
M6 XAL A. Darlington, 92 Lancaster Road, St Albans, AL1 4ES
MI6 XAM M. Green, 7 Liester Park, Ballyclare, BT39 9RZ
M6 XAO A. Gray, 18 St. Botolphs Green, Leominster, HR6 8ER
M6 XAQ M. Dennis, 40 Windsor Road, Linton, Swadlincote, DE12 6PL
M6 XAS A. Smith, 91 Norcliffe Road, Blackpool, FY2 9EN
M6 XAT T. Whelan, 243 Duke Street, Barrow-in-Furness, LA14 1XU
M6 XAV B. Smith, 24 Oakwood Park, Leeds, LS8 2PJ
M6 XAW A. Wardle, 2 Deer Park Place, Sheffield, S6 5ND
M6 XAX K. Poulton, 21 East View, London, E4 9JA
M6 XAY A. Laxton, 7 St. Christophers Green, Broadstairs, CT10 2SS
MI6 XBA T. Agnew, 28 Knockdhu Park, Larne, BT40 2EJ
M6 XBB J. Clarke, 8 Waveney Heights, Brockdish, Diss, IP21 4LD
MM6 XBD S. Boyd, 1 St. Marks Lane, Edinburgh, EH15 2PX
M6 XBJ D. Mccarthy, 7 Shenley Close, Wirral, CH63 7QU
MI6 XBL D. Mooney, 12 Curragh Walk, Londonderry, BT48 8HX
M6 XBM B. Johnson, 6 Trevor Road, Swinton, Manchester, M27 0YH
M6 XBO S. Wilson, 7 Dalglish Drive, Blackburn, BB2 4FU
M6 XBP B. Pettit, 35 Lakeside Rise, Blundeston, Lowestoft, NR32 5BE
M6 XBQ M. Doherty, 151 College Croft, Eccles, Manchester, M30 0AN
M6 XBR J. Breward, 399 St. Margarets Road, Isleworth, TW7 7BZ
M6 XBS B. Scott, 8 Chelsea Close, Tilehurst, Reading, RG30 6EP
M6 XBV C. Somerville, Close House, Gretton, Cheltenham, GL54 5EP
M6 XBW B. Woollett, 23 Kinglake House, Southall, UB24FZ
M6 XBX M. Wheeler, 6 Severn Road, Melksham, SN12 8BQ
M6 XCA M. Jurisic, Flat 7, 27 Slade Way, Mitcham, CR4 2GA
M6 XCC S. Derner, 14 Elmhurst Drive, Huthwaite, Sutton-in-Ashfield, NG17 2NP
M6 XCF C. Fish, New House Farm, Peaton, Craven Arms, SY7 9DW
M6 XCJ C. Jones, 8 Henry Gepp Close Adderbury, Banbury, OX17 3FE
M6 XCK C. Kellers, 21 Aspen House West Terrace, Folkestone, CT20 1TH
M6 XCL L. Robertson, Clayhill Cottage Aldham, Ipswich, IP76NN
M6 XCO R. Kempton, Choices Folly, Marsh Road Gedney Drove End, Spalding, PE12 9PJ
M6 XCP C. Poole, 15 Devon Close, Macclesfield, SK10 3HB
M6 XCZ M. Hartley, 24 Burnham Avenue, Bognor Regis, PO21 2JU
MW6 XDA M. Herbert, 6 Yew Street Taffs Well, Cardiff, CF15 7PT
M6 XDB Dr D. Buttle, 29 Upthorpe Drive, Wantage, OX12 7DF
M6 XDG Dr G. Welch, Amazonas, Sandy Lane, Hightown, Liverpool, L38 3RP
M6 XDI I. Doolan-Tanner, 150 Milward Road, Hastings, TN34 3RT
M6 XDJ S. Scott, 13 Silver Close, Harrow, HA3 6JT
M6 XDO D. Witt, 20 Foxglove Close, Newton Aycliffe, DL5 4PF
M6 XDR C. Darby, 2 Lindsey Court, Alfred Street, Lincoln, LN5 7PZ
M6 XDT I. Hulme, 10 Coney Green, Bicton Heath, Shrewsbury, SY3 5AP
M6 XDW D. Tams, 2 Jacksons Lane, Hazel Grove, Stockport, SK7 6EL
M6 XDX S. James, 124 Alcock Avenue, Mansfield, NG18 2NF
M6 XDZ G. Parsons, 2 The Close, East Grinstead, rh191dq
M6 XEE S. Morton, 77 Hepworth Road, Stanton, Bury St Edmunds, IP31 2UA
M6 XEL L. Cook, Luckfield, Ellis Road Boxted, Colchester, CO4 5RN
MI6 XEM A. Mccusker, 41 The Granary, Waringstown, Craigavon, BT66 7TG
M6 XEP D. Heath, 3 Elm Road, Congleton, CW12 4PR
M6 XER S. Wood, 3 Lion Lane, Haslemere, GU27 1JF
M6 XEU L. Seddon, 53 Reedswood Crescent, Cramlington, NE23 6RW
M6 XEV N. Wright, 10 Olden Mead, Letchworth Garden City, SG6 2SP
MI6 XEX A. Rowan-Jenkins, 143 Glenkeen Avenue, Greenisland, Carrickfergus, BT38 8ST
M6 XEY J. Chadwick, 21 Horridge Street, Bury, BL8 1TN
M6 XFI F. Beckitt-Marshall, 13 Colegrave Street, Lincoln, LN5 8DR
MW6 XFM C. King, Nyth Dedwydd, Llywernog, Aberystwyth, SY23 3AB
MW6 XFU C. Jenkins, Flat 5, The Lawns, Usk, NP15 1BA
M6 XGB G. Bogg, The Shires, Main Street, Pickering, YO18 7PG
M6 XGC J. Beckitt-Marshall, 13 Colegrave Street, Lincoln, LN5 8DR
MW6 XGD G. Jukes, 52 Beach Road, Pyle, Cardiff, CF33 6AS
M6 XGF S. Wellborn, 36 Lidgett Park Court, Leeds, LS8 1ED
MM6 XGJ G. Jamieson, 6 Maryville Park, Aberdeen, AB15 6DU
MI6 XGN G. Houston, 51 Rockfield Heights, Connor, Ballymena, BT42 3LH
M6 XGS G. Stanley, 95 Old Vicarage, Westhoughton, Bolton, BL5 2EG
MM6 XGT G. Taylor, 21 Glenesk Avenue, Montrose, DD10 9AQ
M6 XGX J. Parry, 12 Kerrysdale Close Sutton, St Helens, WA9 3WA
MI6 XGZ J. Downing, 15 Knockburn Gardens, Lisburn, BT28 2QL
M6 XHB B. Hobbs, 2 Miller Court, Bedford, MK42 9PB
M6 XHF H. Beer, 12 Cross Street, Northam, Bideford, EX39 1BS
M6 XHI C. Smith, 17 Essex Street Wash Common, Newbury, RG14 6QJ
MI6 XHL S. Kerr, Bt71 6Sq, Dungannon, BT71 6SQ
MM6 XHU R. Giles, 304 Ellon Park Collydean, Glenrothes, KY7 6UY
M6 XIE S. Reveley, Neuadd Wen, Pontrhydygroes, Ystrad Meurig, SY25 6DQ
M6 XIP P. Armstrong, 10 Shirdley Avenue, Liverpool, L32 7QG

M6 XIT D. Bond, 53 Cotswold Grove, St Helens, WA9 2JD
M6 XJC J. Connett, 81 Holwick Close, Consett, DH8 7UJ
M6 XJD J. Cook, 10A East Street, Newton Abbot, Tq12 1ag
MW6 XJL J. Henley, Rhewin Glas, Glynarthen, Llandysul, SA44 6PR
M6 XJM J. Mcfarlane, 16 Lake View Houghton Regis, Dunstable, LU5 5GJ
M6 XJO J. Ochalek, 21 Fairfield Road, Buxton, SK17 7DN
M6 XJP C. Dennis, Hillsdene, Plex Lane, Ormskirk, L39 7JY
M6 XJR J. Redhead, 28 Sandfields, Frodsham, WA6 6PT
M6 XJS D. Washington, 13 Cheddar Waye, Hayes, UB4 0DZ
M6 XJW J. Wraith, 6 Ashfield Road, Chippenham, SN15 1QQ
MM6 XKC C. Robertson, 5 Broomlands Place, Irvine, KA12 0DU
MI6 XKE W. Duncan, 74 Bunderg Road, Douglas Bridge, Strabane, BT82 8QQ
M6 XKG K. Abel, 7 Foldgate View, Ludlow, SY8 1NB
M6 XKJ K. Cook, 4 Chelmscote Row, Upper Wardington, OX17 1SS
M6 XKM M. Koster, Stalworthy Manor Farm, Suton Lane, Suton, Wymondham, NR18 9JG
M6 XKN K. Northrop, 200 Nevells Road, Letchworth Garden City, SG6 4TZ
M6 XKT K. Todman, 12 Winscombe, Bracknell, RG12 8UD
M6 XKW A. Summerfield, Drovers Rest, Back Lane, Cross In Hand, Heathfield, TN21 0QA
M6 XKY K. Cokayne, 46A Roscoe Road, Irlam, Manchester, M44 6AR
M6 XLC L. Cockburn, 58 Priestley Court, South Shields, NE34 9NQ
M6 XLE J. Walker, 22 Lower Road, Malvern, WR14 4BX
M6 XLF L. Fairs, 140 Manchester Road, Chapel-En-Le-Frith, High Peak, SK23 9TP
M6 XLG L. Graham, Yews Avenue, Barnsley, S70 4BW
M6 XLN C. Beckitt-Marshall, 13 Colegrave Street, Lincoln, LN5 8DR
M6 XLP L. Pawson, 32 Cross Street, Upton, Pontefract, WF9 1EU
M6 XLR N. Sayle, 6 Glen View, Wigmore, Leominster, HR6 9UU
M6 XLT M. Roach, 31 Brumby Wood Lane, Scunthorpe, DN17 1AA
M6 XLX J. Gascoigne, 1 Mill Meadow, Aylesbury, HP19 8GW
M6 XMA D. Jasper, 38 Ingleway Avenue, Blackpool, FY3 8JJ
M6 XMB B. Wale, 23 Castleton Avenue, Bournemouth, BH10 7HW
M6 XMC M. Callis, 1 Webb Close, Letchworth Garden City, SG6 2TY
M6 XME D. Evans, 15 Cider Mill Court, Hereford, HR2 6RY
MI6 XMG A. Mcgarvey, 66A Scaddy Road, Downpatrick, BT30 9BS
M6 XMH M. Hoult, 43 Hutcliffe Wood Road, Sheffield, S8 0EY
M6 XMJ M. Jermy, 15 Oak Tree Close Martham, Great Yarmouth, NR29 4QN
M6 XMN M. O'Brien, Flat 9, Neilson Court, Manchester, M23 1LE
MW6 XMO J. Morris, 70 Forge Road, Port Talbot, SA13 1PF
MM6 XMQ L. Anderson, 6 St. Keiran Crescent, Stonehaven, AB39 2GQ
M6 XMR M. Robinson, 59 The Avenue Kennington, Oxford, OX1 5PP
M6 XMS P. Kentish, 17 Rose Drive, Walsall, WS8 7EB
M6 XMX D. Lyon, 1 Garnsgate Road, Long Sutton, Spalding, PE12 9BT
M6 XNC R. Jeffery, 126 Woodham Lane, New Haw, Addlestone, KT15 3NQ
M6 XNG J. Blackwell, 164 York Road Haxby, York, YO32 3EL
M6 XNL N. Lamerton, 51 High Street, Knaphill, Woking, GU21 2PX
M6 XNO J. Mirfield, 14 King Edward Avenue Horsforth, Leeds, LS18 4BD
M6 XNU S. Grace-Bolton, 26 Fairway Road, Blackpool, FY4 4AZ
M6 XNX M. Powell, 14 Moneta Rise, Leighton Buzzard, LU7 9SN
MM6 XOA Dr A. Onken, 19/7 Damside, Edinburgh, EH4 3BB
MI6 XOD H. Mcdowell, 51 Rockfield Heights, Connor, Ballymena, BT42 3LH
M6 XOJ J. Gunson, 66 Milward Crescent, Hastings, TN34 3RU
M6 XON T. Brownen, 43 Great Rea Road., Brixham., TQ5 9SW
M6 XOR M. Hauser, 23 Prince William Way Sawston, Cambridge, CB22 3SZ
MI6 XOS J. Mcgoldrick, 45 Stewarts Road, Dromara, Dromore, BT25 2AN
MI6 XOX S. Mccartan, 13 Lecale Park, Downpatrick, BT30 6ST
M6 XPB M. Partner, 22 Moordale Avenue, Bracknell, RG42 1RT
M6 XPC P. Cox, 75 Rosecroft Gardens, Swadlincote, DE11 9AF
M6 XPD P. Douglas, 76 Woodside Avenue, Benfleet, SS7 4NY
M6 XPH R. James, 159 Orton Avenue, Sutton Coldfield, B76 1JN
M6 XPK P. Davies, 2 Lynfords Drive, Runwell, Wickford, SS11 7PP
M6 XPN D. Rutter, 42 Great Coates Road, Grimsby, DN34 4ND
M6 XPT T. Parfitt, 5 Sheridan Road, Frimley, Camberley, GU16 7DU
M6 XPW P. Williams, 3 Swift Close, Cottingham, HU16 4DQ
M6 XQS P. Jarvis, 50 Northcott, Bracknell, RG12 7WR
M6 XQX S. Lowe, 14 Windmill Rise, York, YO26 4TX
MI6 XRC D. Mclaughlin, 134 Castleroe Road, Coleraine, BT51 3RW
M6 XRD R. Daniell, Wits End, Barley Mow Lane, Woking, GU21 2HY
M6 XRF R. Ford, 13 Green Street, Hereford, HR1 2QG
MM6 XRI R. Irvine, 9 Pearce Grove, Edinburgh, EH12 8SP
M6 XRL R. Lamerton, 51 High Street Knaphill, Woking, GU21 2PX
MW6 XRO A. Howard, 4 Fern Rise Neyland, Milford Haven, SA73 1RA
MM6 XRS C. Rankin, 2 Thomson Green, Livingston, EH54 8TA
M6 XRT Dr R. Tofts, Elmcroft, Redhill Road, Ross-on-Wye, HR9 5AU
M6 XRX R. Tooley, 18 Toronto Road, Petworth, GU28 0QX
M6 XSD C. Catlin, 27 Main Street, Frizington, CA26 3SA
M6 XSF S. Dockray, 54 Kelsick Park, Seaton, Workington, CA14 1PY
MW6 XSI N. Jones, 30 Cardiff Road, Pwllheli, LL53 5NU
M6 XSN S. Bolton, The Conifers, Methwold Road Methwold Hythe, Thetford, IP26 4QW
M6 XSO S. O'Brien, 15 Fair Ave, Bedlington, ne22 7br
M6 XSR S. Mansfield, The Old Piggery, Ham Lane, Compton Dundon, Somerton, TA11 6PQ
M6 XSS A. Corless, 4 Mayfield Road, Bentham, Lancaster, LA2 7LP
M6 XST S. Harper, 61 Elmhurst, Tadley, RG26 3LF
M6 XSZ B. Staszewski, 10 Priory Road, Stanford-le-Hope, SS17 7EW
M6 XTA T. Windle, 48 Sheridan Road, South Shields, NE34 9JJ
M6 XTB T. Beale, 28 Redhouse Way, Swindon, SN25 2AZ
M6 XTD S. Brookes, 71 Friends Road, Norwich, NR5 8HW
M6 XTF S. Ganti, 59 Fitzroy Road, Blackpool, FY2 0RJ
M6 XTK C. Walsh, 72 William Street, Totterdown, Bristol, BS3 4TX
M6 XTL D. Baker, 32 Richardson Crescent, Hethersett, Norwich, NR9 3HS
M6 XTM T. Mercer, Ralph Allen House, Railway Place, Bath, BA1 1SR
M6 XTR M. Moggeridge, 19 Sherwood Close, Fetcham, Leatherhead, KT22 9QT
M6 XTU S. Baker, 61 Victoria Road, Coleford, GL16 8DS
M6 XTY D. Stott, 63 Cowlishaw Lane, Shaw, Oldham, OL2 8SW
M6 XUI J. Ribbands, Dyson's Farm, Long Row, Tibenham, NR16 1PD
M6 XUM W. Webb, 84 Bruce Street, Swindon, SN2 2EN
M6 XUT D. Baker, 17 Woodroyd Gardens, Luddendenfoot, Halifax, HX2 6BG
M6 XUU P. Sanders, Alresford Road, Wivenhoe, CO7 9JX
M6 XUV S. Ellinor, 53 Hillside, Banstead, SM7 1HG
M6 XUX S. Andrews, 9 Southernhay Crescent, Bristol, BS8 4TT
M6 XVB J. Watson, Croft Holm New Market Street, Buxton, SK17 6LP
M6 XVC N. Booth, 30 Kingfisher Drive, Cheltenham, GL51 0WN
M6 XVJ C. Wallwork, 8 Daneswood Close, Whitworth, Rochdale, OL12 8UX
MJ6 XVL J. Bertram, Roz-Den, La Rue De La Guilleaumerie St. Saviour, Jersey, JE2 7HQ
M6 XVM E. Smith, Flat 29 Heron Court, Emscote Drive, Sutton Coldfield, B73 5NF
M6 XVS S. Cooksey, 22 Fitzgerald Place, Brierley Hill, DY5 2SZ
MW6 XVT C. Williams, 1 South View, Freeholdland Road, Pontypool, NP4 8LL
M6 XVX M. Lawrence, 16 Timson Close, Market Harborough, LE16 7UU
M6 XWB P. Forrest, 6 Scarisbrick Place, Liverpool, L11 7DJ
M6 XWD N. Bishop, 15 Katrine Road, Stourport-on-Severn, DY13 8QB
M6 XWG P. Matthewson, 30 Harvey Road, Rugeley, WS15 4HF
M6 XWR C. Wrobel, 33 Harsnett Road, Colchester, CO1 2HS
M6 XWS R. Molli Boulock, 3 Bray Lane, Telford, TF3 5HH
M6 XXB B. Bentham, 7 Maypole Crescent, Abram, Wigan, WN2 5YL
MW6 XXC D. Clark, 137 Llanedeyrn Road, Penylan, Cardiff, CF23 9DW
M6 XXD P. Jones, 222 York Road, Shrewsbury, SY13QE
M6 XXH C. Troughton, 20 Oakleigh Road, Uxbridge, UB10 9EL
M6 XXI B. Littlechild, 85 Long Road, Canvey Island, SS8 0JB
M6 XXS S. Harvey, Penleigh, Cherry Cross Totnes Down Hill, Totnes, TQ9 5EU
MM6 XXV J. Kerr, 2 Burngrange Cottages, West Calder, EH55 8EW
M6 XXZ T. O'Sullivan, Onibury Hedsor Road, Bourne End, SL8 5DH
M6 XYH D. Hatch, 18 Victory Park Road, Addlestone, KT15 2AX
M6 XYJ S. Jefferies, 28 Curlew Road Abbeydale, Gloucester, GL4 4TF
M6 XYL C. Harris, 50 The Warren, Horsham St Faith, Norwich, NR10 3JT
M6 XYY P. Liu, International Hall, University Of London, London, WC1N 1AS
M6 XZK E. Arthur, 7 Magpie Close, Coulsdon, CR5 1AT
M6 XZY A. Mclean, 141 Crawford Avenue, Tyldesley, Manchester, M29 8LS
M6 YAD G. Boam, 36 Merlin Way, Woodville, Swadlincote, DE11 7QU
MW6 YAE D. Bannister, 45 Queens Drive, Llantwit Fardre, Pontypridd, CF38 2NT
M6 YAF A. Garn, 5 Bassett Street, Walsall, WS2 9PZ
MW6 YAG A. Graham, 2 Heol Undeb, Beddau, Pontypridd, CF38 2LB
M6 YAH C. Bowden, 12 Butts Green, Stoke-on-Trent, ST2 8EH
M6 YAJ R. Harrison, 268 Central Drive, Blackpool, FY1 5JB
MI6 YAM J. Wilkinson, 11 Fairview Park, Dromore, BT25 1PN
M6 YAN Y. Watkins, 1 St. Saviour Close, Colchester, CO4 0PW
MM6 YAP C. Phillips, 8 The Square, Newtongrange, Dalkeith, EH22 4QD
M6 YAR R. Parker, 4 Hall Yard Metheringham, Lincoln, LN4 3BY
M6 YAS P. Savage, 53 Ashby Road, Braunston, Daventry, NN11 7HE
M6 YAT L. Jones, 8 Oakbrook Close, Ewyas Harold, Hereford, HR2 0NX
M6 YAV A. Vaile, 66 Grasmere Point, Old Kent Road, London, SE15 1DU
M6 YAW A. Nicholls, 2 Park End, Forsbrook, Stoke-on-Trent, ST11 9DR
M6 YAY C. Jack, 18 Swan Close, Martlesham Heath, Ipswich, IP5 3SD
M6 YBA M. Adams, 18 Vanguard Court, Sleaford, NG34 7WL
M6 YBB L. West, 153 Chartist House, Mount Street, Hyde, SK14 1RP
M6 YBC J. Bernard-Cooper, 12 River Court, Green Lane, Durham, DH1 3UA
M6 YBD R. Cook, 8 Kingsdown Way New Marske, Redcar, TS11 8JJ
MD6 YBE H. Jones, 77 Royal Park, Ramsey, Isle Of Man, IM8 3UH
MI6 YBH J. Mccourt, 70A Sessiagh Scott Road Rock, Dungannon, BT70 3JU
M6 YBK B. Watson, 173 Churchfield Lane, Darton, Barnsley, S75 5EA
M6 YBT M. Harrison, 2 Rosemount Court, Holly Bank Road, York, YO24 4EG
M6 YBV B. Young, 11 Gainsborough Avenue, Washington, NE38 7EF
MW6 YBZ P. Smith, 29 Heol Cwarrel Clark, Caerphilly, CF83 2NE
M6 YCA A. Young, 53 Arnold Grove, Newcastle, ST5 8LD
MM6 YCB J. Williamson, Clunie Cottage, Tullibardine Road, Auchterarder, PH3 1LX
M6 YCG C. Gouveia, 86A Spareacre Lane, Eynsham, Witney, OX29 4NP
M6 YCH C. Hohmann, 13 Woodbine Terrace Bensham, Gateshead, NE8 1RU
MM6 YCJ Prof. C. Jones, 11B Ettrick Road, Edinburgh, EH10 5BJ
M6 YCP C. Pearce, 51 Stanwick Drive, Cheltenham, GL51 9LG
M6 YCR W. Jones, 50 Bridge Place, Croydon, CR0 2BB
M6 YCT D. Ashton-Hilton, 14 Weetwood Road, Congresbury, Bristol, BS49 5BN
M6 YDA K. Abel, 7 Foldgate View, Ludlow, SY8 1NB
M6 YDB D. Bower, 4 Winsford Road, Sheffield, S6 1HT
M6 YDC D. Cook, 14 Alberta Avenue Selston, Nottingham, NG16 6GN
M6 YDD R. Howson, 37 Ghyllroyd Drive, Birkenshaw, Bradford, BD11 2ET
M6 YDF S. Gains, 14 Ainslie Street, Grimsby, DN32 0LU
M6 YDG D. Lyall, Royal Hospital Road, London, SW3 4SR
M6 YDN P. Norman, 11 Windmill Road, Irthlingborough, Wellingborough, NN9 5RJ
MW6 YDP P. Warburton, 71 Richards Terrace, Cardiff, CF24 1RW
M6 YDV D. Powell, 29 Kelston View, Bath, BA2 1NW
M6 YEB M. Clayton, 2 West Street, Wroxall, Ventnor, PO38 3BU
M6 YEF S. Shore, 8 Bromley Road, Colchester, CO4 3JE
M6 YEG J. Fenton, 4 Forest Hills, Newport, PO30 5NG
M6 YEH A. Bate, 16 East Avenue, Heald Green, Cheadle, SK8 3DL
M6 YEL J. Callaghan, 46 Highfields, Great Yeldham, Halstead, CO9 4QQ
M6 YEO A. Yeo, Bridge House, Tresillian, Truro, TR2 4AU
M6 YEQ G. Browning, 6 The Drive, Rickmansworth, WD3 4EB
M6 YES S. Crabb, 22 Mary Warner Road Ardleigh, Colchester, CO7 7RP
M6 YEU M. Read, 3613 Clary Ave, Fort Worth, USA, 76111
M6 YEY T. Windass, 58 Nicholas Gardens, High Wycombe, HP13 6JG
MM6 YEZ A. Ewing, Kildonan House, Caerlaverock Farm, Crieff, PH5 2BD
M6 YFN A. Boubaker, Citee Sonboula N5, Grombalia, Tunisia, 8030
M6 YFT M. Hubbard, Millstone, Highfield Road, Truro, TR4 8DZ
M6 YFU M. Coats, 3 Broadleaf Close, Exeter, EX1 3XA
M6 YFX I. Cox, 2 Welby Close Tadpole Garden Village, Swindon, SN25 2RJ
M6 YFZ N. Mason, 6 Maldern Avenue, Poulton-le-Fylde, FY6 7TL
M6 YGD D. Young, Thring Lodge, Granley Gardens, Cheltenham, GL51 6LQ
M6 YGL G. Hansell, 196 The Downs, Harlow, CM20 3RH
M6 YGN G. Nicholls, 65 Boleyn Way New Barnet, Barnet, EN5 5LH
M6 YGR G. Robertson, 12 Chester Close Ince, Wigan, WN3 4JP
MM6 YGT G. Christie, 10 Calcots Crescent, Elgin, IV30 6GL
M6 YHB H. Baker, 300 Lowerhouse Lane, Burnley, BB12 6LZ
M6 YHM H. Marchington, 30 Warwick Avenue Golcar, Huddersfield, HD7 4BX
MM6 YHO D. Chisholm, 111 Philips Wynd, Hamilton, ML3 8PH

M6 YHQ P. May, 1 Ballowall Terrace St. Just, Penzance, TR19 7BG
MW6 YHR R. Boulton, 41 St. Andrews Drive, Libanus Fields, Blackwood, NP12 2ET
M6 YHS D. Stevens, 67 New Road, East Hagbourne, Didcot, OX11 9JX
M6 YHV C. Strange, 18 Glenwood Court, Farnborough, GU14 7TB
M6 YHW R. Lloyd, 18 Newbury Grove Blurton, Stoke-on-Trent, ST3 3DD
MW6 YIG M. Graham, 60 Heol Seward Beddau, Pontypridd, CF38 2SR
M6 YIK F. Nisbet, 3 Fern Crescent, Seaham, SR7 7UJ
M6 YIN D. Horton, 27 Fishers Lane, Wirral, CH61 9NT
M6 YIT C. Peters, 15 Jasmine Crescent Trimdon Village, Trimdon Station, TS29 6QE
M6 YJB J. Wilson, 55 Chestnut Avenue Kirkby-In-Ashfield, Nottingham, NG17 8BA
M6 YJD D. Sanders, 1 Berkeley Grange, Carlisle, CA2 7PN
M6 YJH J. Harman, 1 Vernons Road, Newick, BN84NF
M6 YJK J. Williams, 56 Hillingdon Rise, Sevenoaks, TN13 3SB
M6 YKC P. Purnell, 89 St. Aubyns, Goldsithney, Penzance, TR20 9LS
MW6 YKI K. Griffiths, 11 Hatfield Meadow, Knighton, LD7 1RY
MW6 YKS B. Roberts, 6 Trem Y Moelwyn, Tanygrisiau, Blaenau Ffestiniog, LL41 3SS
MM6 YLA W. Lawson, 60 Inglis Avenue, Port Seton, EH32 0AQ
M6 YLC C. Rawlinson, Westfield Farm, Risden Lane, Cranbrook, TN18 5DU
M6 YLD D. Braithwaite, 300 Arnold Estate, Druid Street, London, SE1 2XN
M6 YLE A. Kelly, 62 Old Warren, Taverham, Norwich, NR8 6GA
MI6 YLG G. Mccormick, 24 Warren Park Drive, Lisburn, BT28 1HF
M6 YLH H. Earnshaw, 16 Estill Close, Cayton, Scarborough, YO11 3TA
M6 YLI E. Fisher, 1 Eaton Place, Kingswinford, DY6 8JU
M6 YLJ R. Whitehead, 29 Coulsons Road, Bristol, BS14 0NN
MM6 YLO G. Davis, 2 Virkie Cottages, Virkie, Shetland, ZE3 9JS
M6 YLP R. Campbell-Black, 10 Wren Close, Towcester, NN12 6RD
M6 YLR D. Walker, 290 Shannon Road, Hull, HU8 9RY
M6 YLS S. Woolley, Babbington House, Church Road, Crowborough, TN6 3LG
MI6 YLT S. Mccormick, 46 Lany Road, Moira, Craigavon, BT67 0NZ
M6 YLW T. Stromberg, 89 Sutton Road, Watford, WD17 2QG
M6 YLY F. Titeiu, Hillary Road, 18, Kidsgrove, st74dn
M6 YMA A. Andrew, 80 Hamble Drive, Abingdon, OX14 3TE
M6 YMB L. Latimer, 40 Petersham Road, Long Eaton, NG10 4DD
M6 YME M. Smith, 2 Tullig, Cahirciveen, County Kerry, Ireland, V23 V348
M6 YMF K. Mbachu, 46 Roseberry Gardens, Upminster, RM14 1NW
M6 YMM T. Daniel, 28 Polefield Circle Prestwich, Manchester, M25 2WP
M6 YMN M. Lee, 4 Addison Place, Bilston, WV14 7BD
M6 YMR L. Humphreys, Flat 5 Sabie Court, 21 Argyll Road, Bournemouth, BH5 1EB
M6 YMW M. Williamson, 5 John F Kennedy Walk, Tipton, DY4 0SF
M6 YMX M. Allen, 55 The Meadows, Skegness, PE25 2JA
M6 YMY K. Nelhams-Wright, The Gatehouse, Frogmore Lane, Aylesbury, HP18 9DZ
M6 YNA B. Wright, 5 Ridge Road, Kingswinford, DY6 9RB
MM6 YND A. Gillies, 43 Oak Wynd, Cambuslang, Glasgow, G72 7GS
MM6 YNG C. Young, 0/2 27 Cardwell Road, Gourock, PA19 1UW
M6 YNK S. Higton, 25 Shady Grove Hilton, Derby, DE65 5FX
M6 YNL O. Price, Queens Arms House 1 King Street Odiham, Hook, RG29 1NN
M6 YNN A. Green, 10 Coopers Hill, Eastbourne, BN20 9HX
M6 YNT A. Ashby, 5 New Street, Osbournby, Sleaford, NG34 0DL
M6 YNY K. Edwards, 352 Plessey Road, Blyth, NE24 3RD
M6 YOB C. Burns, 29 Foxglove Fold, Castleford, WF10 5UJ
M6 YOG S. Lee, 7 Lilac Avenue, Lower Quinton, Stratford upon Avon, CV37 8US
M6 YON A. Parkhouse, 3 St. Margarets Avenue, Ashford, TW15 1DR
M6 YOO M. Akena, 6 Concord Terrace, Coles Crescent, Harrow, HA2 0HJ
M6 YOS L. Shackleton, 1 Mayfield Road, Wooburn Green, High Wycombe, HP10 0HG
M6 YOT C. Watson, 51 Lodge Way, Weymouth, DT4 9UU
M6 YOU B. Roots, 212 Hunt Road, Tonbridge, TN10 4BJ
M6 YOW E. Murchie, 52 Kingswood Road, Manchester, M14 6SA
MM6 YOY J. Moir, 41 Brisbane Terrace, East Kilbride, Glasgow, G75 8DL
M6 YOZ S. Green, 28 Courtney Drive, Sunderland, SR3 1JR
M6 YPA A. Cattell, 40 Collyweston Road, Northampton, NN3 5ET
M6 YPD D. Hawkes, 62 Park Street, Wollaston, Wellingborough, NN29 7RR
M6 YPE J. Pye, 17 Langden Brook Mews, Morecambe, LA3 3SN
M6 YPG G. Evans, 19 Windsor Street, Thurnscoe, S63 0HB
MM6 YPN G. Lailvaux, 4 Oxenfoord Avenue, Pathhead, EH37 5QD
M6 YPS P. Smith, Meadows, Menthorpe Lane, North Duffield, Selby, YO8 5RL
M6 YPW Dr P. Woodfin, Laurel Cottage, Barrow Street, Much Wenlock, TF13 6EN
M6 YPX A. Gillard, 4 Horton Avenue, Thame, OX9 3NJ
MI6 YPY G. Kelly, 50 Tullyrain Road, Beagh, Enniskillen, BT94 2AS
M6 YPZ M. Meyer, 22 Orchard Grove, Newton Abbot, TQ12 1FZ
MM6 YQP A. Sharples, 49 Ramsay Road, Banchory, AB31 5TS
M6 YRB R. Braisby, Flat 3 Rose House Victoria Road, Woking, GU21 2AT
M6 YRC R. Curtis, 3 Coniston Close, Peterlee, SR8 5LW
M6 YRD S. Jones, 19 Tipnall Road, Castle Donington, Derby, DE74 2JY
M6 YRF Dr A. Saje, 72 Bedworth Road, Bulkington, Bedworth, CV12 9LL
M6 YRL R. Yerrell, 88 Woodcock Road, Norwich, NR3 3TD
MM6 YRO D. Kennedy, 7 Burns Terrace, Cowie, Stirling, FK7 7BS
M6 YRS J. Stallard, 6 Richmond Crescent, Leominster, HR6 8RX
M6 YRW R. Weber, 11 Harewood Road, Calstock, PL18 9QN
M6 YRY R. Lambert, 7 Templeway, Lydney, GL15 5HU
MW6 YSA B. Paterson, 35 Ynys Yr Afon, Neath, Sa114bp
M6 YSB P. Bunting, 29 Marion Avenue, Alverthorpe, Wakefield, WF20BJ
M6 YSD S. Daniels, 48 Gason Hill Road, Tidworth, SP9 7JX
M6 YSF S. Fletcher, Beverley Hotel, 55 Old Brumby Street, Scunthorpe, DN16 2AJ
M6 YSK A. Bragg, Buena Vista Low Moresby, Whitehaven, CA28 6RR
M6 YSM M. Young, 39 Hollins Lane, Sheffield, S6 5GQ
MM6 YST A. Thomson, 15 Waverley Road, Nairn, IV12 4RH
M6 YSU J. Wright, Paddock Views, Main Street, Leicester, LE7 9YB
MI6 YSW D. Lappin, 46 Grange Road, Kilmore, Armagh, BT61 8NX
MJ6 YSY G. Antcliff, Andylor Cottage, 6 Dicq Road, St Saviour, Jersey, JE2 7PD
M6 YTB R. Thomas, 9 Spa View Terrace, Sheffield, S12 4HG
M6 YTC T. Campbell, 3 Hillside Close, Helsby, Helsby, WA69LB
MM6 YTD T. Davidson, 141 Huron Avenue, Livingston, EH54 6LQ
M6 YTF T. Fox, 28 Parishes Mead, Stevenage, Stevenege, SG2 9QD
M6 YTI H. Cropp, 26 Coventry Street, Brighton, BN1 5PQ
M6 YTL J. Watkins, The Hundred, The Hundred Middleton On The Hill, Leominster, HR6 0HZ
M6 YTM N. Allison, 2 Gables Cottages Argos Hill, Rotherfield, Crowborough, TN6 3QH
M6 YTS J. Stacey, 12 Kimbridge Road, East Wittering, Chichester, PO20 8PE
M6 YTU A. Coward, 11 Oaklands, Ardingly, Haywards Heath, RH17 6UE
M6 YTX H. Taylor, 25 Northolme Avenue, Nottingham, NG6 9AP
M6 YUG G. Harmath, 101 Whitton Avenue East, Greenford, UB6 0QE
MM6 YUI I. Page, 37 Pine Crescent, Hamilton, ML3 8TZ
MM6 YUJ C. Duncan, 36 Commore Drive, Glasgow, G13 3TT
M6 YUK D. Barker, 12 The Weavers, Denstone, Uttoxeter, ST14 5DP
M6 YUL G. Szabo, 6 Stanway Road, Gloucester, GL4 4RE
M6 YUM T. Benson, 3 Tey Road, Coggeshall, Colchester, CO6 1SY
MM6 YUP B. Rankin, 5 Glen Affric Court, Dumbarton, G82 2BN
M6 YVN A. Lowe, 17 Burnside, Telford, TF3 1SS
MM6 YVO Y. Oldfield, 1 Gallanach, Lochgair, Lochgilphead, PA31 8SD
M6 YVR S. Preece, 14 Bettespol Meadows, Redbourn, St Albans, AL3 7EW
M6 YVT P. Van Staveren, 14 Fortune Green Road Flat 3, London, NW6 1UE
M6 YWA M. Smith, 51 Town Lane, Shepton Mallet, BA4 5LX
M6 YWG R. Holland, 431 Beverley Road, Newland, Hull, HU5 1LX
M6 YWO M. Bruyneel, 14 Riversmead, St Neots, PE19 1HA
M6 YWX D. Hartnell, 3 Fairmead, Sidmouth, EX10 9SU
M6 YXD D. Yeaman, Flat 3, 62 Madeley Road, Ealing, W5 2LU
M6 YXZ C. Cutting, 5A Clifton Mansions, Clifton Road, Folkestone, CT20 2EJ
MM6 YYB I. Nicolson, 1 Gullane Place, Dundee, DD2 3BF
M6 YYD A. Edge, 79 Waterside Drive, Stoke-on-Trent, ST3 3NU
M6 YYG D. Harwood, 32 John Reid Road, South Shields, NE34 9EB
M6 YYK S. Hoyle, 20 Brandsby Grove, Huntington, York, YO31 9HL
M6 YYL R. Ashton-Cox, 10 Princes Avenue, Benfleet, SS7 3AZ
M6 YYM Viscount M. Thomas, 13 Christchurch Gardens, Reading, RG2 7AH
M6 YYT G. Finney, 121 School Lane, Caverswall, Stoke-on-Trent, ST11 9EN
M6 YYU J. Underwood, 15 Fawsley Road, Northampton, NN4 8NR
M6 YYY S. Christofi, 19 Kingsland Avenue, Northampton, NN2 7PP
MW6 YZF G. Jones, 31 Liverpool Road, Buckley, CH7 3LH
M6 YZH B. Yu, 210 Ramsgate Road, Broadstairs, CT10 2EW
M6 YZM G. Brown, 134 Skipper Way, Lee-on-the-Solent, PO13 8HD
M6 YZO C. Gregory, 28 Rynal Street, Evesham, WR11 4QA
M6 YZR D. King, 170 Stepney Road, Scarborough, YO12 5NH
M6 YZX S. Forber, 32 Larch Avenue, Newton-le-Willows, WA12 8JF
M6 ZAB A. Fletcher, 74 Devonshire Road, Maltby, Rotherham, S66 7DQ
M6 ZAC M. Holmes, 6 Wells Court, Saxilby, Lincoln, LN1 2GY
M6 ZAD A. Ghafoor, 20 Maureen Avenue, Manchester, M8 5AR
M6 ZAF A. Barter, 4 Blackberry Way Kingsteignton, Newton Abbot, TQ12 3QX
M6 ZAH T. Hayden, The Ferns, Yarmouth Road, North Walsham, NR28 9LX
MW6 ZAN P. Smith, 3 Digby Street, Barry, CF63 4NP
M6 ZAQ Z. Gale, 7 Somerset Road Droylsden, Manchester, M43 7PF
M6 ZAS N. Zulkfli, Harrogate Ladies' College, Clarence Drive, Harrogate, HG1 2QG
M6 ZAU C. Sommers, 11 Westlands Grove, Fareham, PO16 9AA
M6 ZAV B. Appleby, 1E Wilfred Owen Close, London, SW19 8SW
M6 ZAW D. Packham, 5 The Avenue, Yate, Bristol, BS37 4PN
M6 ZAX M. Southgate, 107 Englands Lane, Loughton, IG10 2QL
M6 ZAY A. Carter, 19 Burn Lane, Newton Aycliffe, DL5 4HX
MM6 ZAZ G. Fyvie, Bridgefoot Of Gaval, Mintlaw, Peterhead, AB42 4HA
M6 ZBA J. Merritt, 41 Great Grove, Bushey, WD23 3BQ
M6 ZBB Dr M. Palmer, 116 Claverham Road, Yatton, Bristol, BS49 4LE
M6 ZBD I. Painter, 47 Longmead Drive, Nottingham, NG5 6DP
MW6 ZBE S. Berrow, 40 Priorsgate Oakdale, Blackwood, NP12 0EL
MM6 ZBG C. Halcrow, Hellia, Cunningsburgh, Shetland, ZE2 9HG
M6 ZBL Z. Raybould, 31 Meadow Road, Quinton, Birmingham, B32 1AY
M6 ZBM C. Abbott, 38 Foxcover, Linton Colliery, Morpeth, NE61 5SR
M6 ZBQ A. Kerr, 9 Martindale Way, Sawston, Cambridge, CB22 3BT
MW6 ZBR B. Morgan, 1 Y Dolydd, Aberdare, CF44 8EX
M6 ZBS K. Florence, 30 Lancaster Gardens, Ealing, London, W13 9JY
M6 ZBT B. Tomlinson, 7 Springwell Close, Crewe, CW2 6TX
M6 ZBW B. Waddingham, 11 Chandlers Court, Tidworth, SP9 7FN
M6 ZCA C. Archer, 31 Stoney Bank Drive Kiveton Park, Sheffield, S26 6SJ
M6 ZCB C. Burdett, 44 Emmett Carr Lane, Renishaw, Sheffield, S21 3UL
MM6 ZCD C. Docherty, 23 The Maltings, Haddington, EH41 4EF
M6 ZCL C. Lynch, 26 Warwick Street, Rochdale, OL129SW
M6 ZCM C. May, 18 Ravenscroft, Salisbury, SP2 8DL
M6 ZCP C. Palawinna, 3 Stirling Court Road, Burgess Hill, RH15 0PS
M6 ZCR C. Rhodes, 93 Southwark Close, Stevenage, SG1 4PH
MW6 ZCT C. Tsoi, Fairview, Llanbadarn Fawr, Aberystwyth, SY23 3QU
M6 ZDA D. Harris, 27 Ashley Road, Poole, BH14 9BS
M6 ZDB D. Brownsea, 47 Southill Road, Bournemouth, BH9 1SH
M6 ZDC Dr D. Cooke, Apartment 9, 27 Sheldon Square, London, W2 6DW
M6 ZDF D. Forbes, Flat 3, Arundel Court 1 Cherrywood Drive, London, SW15 6DS
MM6 ZDG D. Gillies, 43 Oak Wynd, Cambuslang, Glasgow, G72 7GS
M6 ZDL A. Abson, 117 Lysander Road, Rubery, Birmingham, B45 0EN
MM6 ZDW D. Woodford, Nordkette, Evnabrek, Levenwick, Shetland, ZE2 9GY
MM6 ZDY D. Young, Fada Fuireach, Erbusaig, Kyle, IV40 8BB
M6 ZDZ M. Clark, 52 Redan Road, Aldershot, GU12 4ST
M6 ZEA C. Scott, 40 The Close, Skipton, BD23 2BZ
M6 ZEB A. Hulme, 6 Asmall Close, Ormskirk, L39 3PX
MD6 ZEE C. Glaister, Balleigh Villa, Jurby Road, Isle of Man, IM8 3NZ
M6 ZEF J. Ferry, 7, Pangbourne Road Thurnscoe, Rotherham, S63 0LQ
M6 ZEJ M. Jeffrey, 9 Stoney Lands, Plymouth, Pl125df
M6 ZEK E. Short, 2 Richmond Street, Kings Sutton, Banbury, OX17 3RS
M6 ZEL S. King, 105 Bluebell Street, Plymouth, PL6 8EQ
M6 ZEN J. Charters-Reid, Woodlands Farm, Flaxton, York, YO60 7RJ
M6 ZEP S. Redford, 78 Boyds Walk, Dukinfield, SK16 4AU
M6 ZES G. Dimitrov, Flat 6, Blenholme Court, Hampton, TW12 2BL
M6 ZET A. Balmer, 11 China Street, Darlington, DL3 0EJ
M6 ZEW W. Woodhead, 1 Hayle Road, Oldham, OL1 4NP
M6 ZFE J. Mcinnes-Boylan, 17 Hindsford Bridge Mews, Atherton, Manchester, M46 9QZ
MM6 ZFG S. Street, Tangaroa, Fairfield Gardens, Kilcreggan, Kilcreggan, G84 0HS
M6 ZGR C. Panaitescu, 131 Stafford Road, Croydon, CR0 4NN
MM6 ZGS G. Sneddon, 26 Linkwood Road, Airdrie, ML6 6GP
M6 ZGY T. Talbot-Humphries, 22 Vicar Street, Wednesbury, WS10 9HF
M6 ZGZ N. Glazzard, Snitton Gate Cottage, Snitton, Ludlow, SY8 3JX
M6 ZIA Z. Sati, 27 Hanover Road, London, SW19 1EB

M6 ZIB S. Barker-Mawjee, 23 Raleigh Road, London, N8 0JB
M6 ZIP A. Bradley, Flat 3, 56 Chase Green Avenue, Enfield, EN2 8EN
M6 ZIY Dr J. Medland, 34 Allderidge Avenue, Hull, HU5 4EQ
MW6 ZIZ A. Morgan, 2 Parc Cambria, Old Colwyn, Colwyn Bay, LL29 9AJ
M6 ZJB J. Brownsea, 47 Southill Road, Bournemouth, BH9 1SH
M6 ZJH J. Hughes, Herons Way, Munslow, Craven Arms, SY7 9ET
M6 ZJP J. Povall, Elsich Barn Farm, Seifton, Craven Arms, SY7 9LF
M6 ZJT T. Heyes, The Coach House, Mossley Hall, Congleton, CW12 3LZ
M6 ZJW Z. Wilcoxen, 5 Douglas Lane, Wraysbury, Staines-upon-Thames, TW19 5NF
M6 ZKA D. Owings, 11 Thingwall Road East, Thingwall, Wirral, CH61 3UY
MD6 ZKK A. Kissack, 6 Falcon Cliff Court, Douglas, Isle of Man, IM2 4AQ
M6 ZLA Z. Lucas, 31 Lilian Close, Norwich, NR6 6RZ
M6 ZLC S. Rice, 30 Oveton Way, Bookham, Leatherhead, KT23 4ND
M6 ZLD P. Frost, 26 Hollies Court, Britannia Road, Banbury, OX16 5DR
M6 ZLL T. Wiggins, 158 Prince Charles Avenue, Derby, DE22 4LQ
M6 ZLM M. Swindells, 35 Rivington Street, St Helens, WA10 4BL
M6 ZLN Z. Newell, 15 The Grove, Luton, LU1 5PE
M6 ZLP Z. Pepper, 2 Greenways, Walton On The Hill, Tadworth, KT20 7QE
M6 ZLR A. Zeller, Flat 1, 57 Chalk Hill, Watford, WD19 4DA
M6 ZMC M. Clayton, 21 Green Leys, Maidenhead, SL6 7EZ
M6 ZMI T. Kelly, 2 Weaver House, Chester Road, Runcorn, WA7 3EG
M6 ZMR M. Richards, 33 Daleside, Dewsbury, WF12 0PJ
M6 ZNF N. Farrington, The Old Hall, Main Street, Elmley Castle, Pershore, WR10 3HS
M6 ZNZ Dr G. Richardson, Berwick Cottage, Bailes Lane, Guildford, GU3 2AX
M6 ZOC M. Cozens, 17 Ash Grove, Norwich, NR3 4BE
MW6 ZOD C. Williams, 1 Lawrence Terrace, Llanelli, SA15 1SW
M6 ZOG S. Littlechild, 539 Tyburn Road, Birmingham, B24 9RX
MW6 ZOL R. Bowen, 25 Maendu Terrace, Brecon, LD3 9HH
M6 ZOM S. Corbishley, 86 Boundary Lane, Congleton, CW12 3JA
M6 ZOO N. Weale, Flat 85, Redbridge Tower, Southampton, SO16 9AW
M6 ZPB P. Bannon, Rose House, 73 London Road, Worcester, WR5 2DU
M6 ZPE M. Hills, 19 Wilkes Road, Broadstairs, CT10 2HL
M6 ZPH C. Ward, 1 Granary Close, Codford, Warminster, BA12 0PR
M6 ZPL R. Manning, 21 Whitethorn Way, Oxford, OX4 6ER
M6 ZPS Z. Svanda, 5 Red River Country Park, Hullbridge, Hockley, SS56EP
M6 ZPT A. Pattison-Turner, 47 Jefferson Way, Bannerbrook Grange, Coventry, CV4 9AN
M6 ZPY R. Pyner, 1 Avon Court, 63 Shakespeare Road, Bedford, MK40 2DS
M6 ZPZ P. Zhang, Harrogate Ladies' College, Clarence Drive, Harrogate, HG1 2QG
M6 ZRG G. Reynolds, 43 Orchard Drive, Watford, WD17 3DX
M6 ZRJ S. Rogers, 28 Damian Way, Hassocks, BN6 8BJ
M6 ZRL G. Lockett, 15 Deepdale Street, Hetton-Le-Hole, Houghton le Spring, DH5 0DQ
M6 ZRO M. Brough, 40 Denby Grange, Harlow, CM17 9PZ
M6 ZRT T. Cooper, 9 Websters Close, Shepshed, Loughborough, LE12 9AT
MM6 ZRX K. Scollay, Nirvana, Orphir, Orkney, KW17 2RB
M6 ZRZ D. Gill, 60 Woodcock Court, Three Mile Cross, RG7 1BZ
M6 ZSA S. Evans, Le3 0Ea, Leicester, LE3 0EA
M6 ZSB S. Bache, 62 Whittingham Road, Halesowen, B63 3TP
M6 ZSD S. Adams, 31 Northway, Dudley, DY3 3PH
M6 ZSE P. Holmes, 53 Bishops Hull Road, Bishops Hull, Taunton, TA1 5EP
M6 ZSH M. Gupta, Flat 26, Bassett Court, Southampton, SO16 7DR
M6 ZSK S. Kneller, 12A Richard Street, Crewe, CW1 3AF
M6 ZST S. Harris, 61 Monks Park Avenue, Bristol, BS7 0UA
M6 ZSV S. Trevor, 8 Aldin Grange Terrace, Bearpark, Durham, DH7 7AN
M6 ZTA H. Glendinning, 9 Spinners Avenue, Wakefield, WF1 3QD
M6 ZTB R. Janezko, 10A Lens Road, Allestree, Derby, DE22 2NB
M6 ZTC S. Broom, 128 Springhill Road, Wolverhampton, WV11 3AQ
M6 ZTD A. Maiden, 79 Green End Road, Manchester, M19 1LE
MI6 ZTM M. Meagher, 42 Mourne View Park, C.Down, Newry, BT35 6BZ
M6 ZTO K. Balls, 7 Rowan Close, Holbeach, Spalding, PE12 7BT
M6 ZTP E. Pritchard, Lilliput, Doctors Commons Road, Berkhamsted, HP4 3DR
M6 ZTS G. Harris, Sunnyside Lodge, Mongeham Road, Deal, CT14 8JW
M6 ZUB R. Landragin, 101 Linden Gardens Enfield En1 4Dy United Kingd101, England, EN1 4DY
M6 ZUF P. Norman, Four Acres Bungalow Farm, Winwick Gated Road West Haddon, Northampton, NN6 7BH
M6 ZUT M. Beyoglu, Flat 22, Globe House, 30 Southall Street, Manchester, M3 1LP
MM6 ZUY J. Woods, 33A Dalrymple Street, Girvan, KA26 9EU
M6 ZVD M. Ratcliffe, 30 Newton Cross Road, Newton In Furness, Barrow-in-Furness, LA13 0NB
M6 ZVL V. Lynch, 16 Okehampton Crescent, Sale, M33 5HR
M6 ZWB M. Wilson, 10 Willow Lane, Langar, Nottingham, NG13 9HL
M6 ZWP J. Scarcliffe, 17 Fillingfir Drive, Leeds, LS16 5EG
M6 ZWT A. Gow, Flat 203, Viotti Heights Sandy Hill Road, London, SE18 6PA
M6 ZXB P. Holloway, Flat 26, Swan House, Romford, RM7 8AZ
M6 ZXC T. Hunter, 5 Regal Court, Dewsbury, WF12 7DE
M6 ZXH P. Baker, 69 Cavendish Place, Walsall, WS27HH
M6 ZXI J. Clark, 54 Coleshill Place, Bradwell Common, Milton Keynes, MK13 8DP
M6 ZXL C. Cox, 47 Thurlstone Road, Penistone, Sheffield, S36 9EF
MW6 ZXO R. Zlotnicki, Gwynfryn, Goginan, Aberystwyth, SY23 3PD
M6 ZXQ A. Cockle, 32 Old Coach Road, Playing Place, Truro, TR3 6ES
M6 ZXR D. Thomas, 10 Priory Drove, Great Cressingham, Thetford, IP25 6NJ
MI6 ZXT C. Mckinley, 36 Grangewood Drive, Londonderry, BT47 5WN
M6 ZXZ J. Richards, 41 Mount Pleasure, Camborne, TR14 7RR
M6 ZYE G. Winnett, 24 Underleys, Beer, Seaton, EX12 3LT
M6 ZYG C. Richardson, Heathercroft, Kirkby Mills, Kirkbymoorside, York, YO62 6NN
M6 ZYH Z. Hill, 101 Abbeydale Road, Sheffield, S7 1FE
M6 ZYK E. Matwiejczyk, 3 Tippett Avenue, Swindon, SN25 2GQ
MW6 ZYQ J. Moorhouse, 33 Goylands Close, Llandrindod Wells, LD1 5RB
M6 ZYX W. Coburn, 42 Hinton Wood Avenue, Christchurch, BH23 5AH
MM6 ZYZ R. Mcglynn, 35 Barrie Quadrant, Clydebank, G81 3EH
M6 ZZA J. Isles, 128 Ditton Lane, Fen Ditton, Cambridge, CB5 8SS
M6 ZZB S. Bird, Milestone Cottage, Station Road Kimberley, Wymondham, NR18 9HQ
M6 ZZC S. Alexander, 13 Padgate, Thorpe End, Norwich, NR13 5DG
M6 ZZD J. Nixon, East View, Cloudside, Congleton, CW12 3QG
M6 ZZE C. Hogan, 26 Mowbray Avenue, St Helens, WA11 9JD
MM6 ZZG G. Runcie, Smithy Cottage, Brotherton, Montrose, DD10 0HW
M6 ZZK J. Gooch, 38 Wards Crescent, Bodicote, Banbury, OX15 4DY
M6 ZZQ S. Collins, 5 Fernleigh Gardens, Stafford, ST16 1HA
M6 ZZS A. Barrett, 24 Burhill Way, St Leonards-on-Sea, TN38 0XP
M6 ZZV A. Siddall, 12 Russell Gardens, Sipson, West Drayton, UB7 0LS
M6 ZZY R. Soar, 31 Sidlesham Close, Hayling Island, PO11 9ST

M7

MI7 AAC A. Cairns, 14 Otterbank Road, Strathfoyle, Londonderry, BT47 6YB
MM7 AAE G. Mann, Halcyon, Strachan, Banchory, AB31 6NL
M7 AAH B. Lavery, 113 Ena Crescent, Leigh, WN75ET
M7 AAK E. Blackford, 55 Wellington Street Lemington, Newcastle upon Tyne, NE15 8RH
M7 AAL D. Hollbrook, 64 Stonycroft Albany, Washington, NE37 1UN
M7 AAP G. Walton, 141 Lunsford Lane Larkfield, Aylesford, ME20 6HP
M7 AAQ S. Glover, 20 Lilly Street, Bolton, BL1 3AU
M7 AAT Dr D. Lester, 87 Lime Road Normanby, Middlesbrough, TS6 0BZ
M7 AAU R. Lucas, 4 Midwinter Gardens, Swindon, SN3 4NZ
M7 AAV I. Blofield, 13 Wimborne Avenue, Ipswich, IP3 8QW
M7 AAW P. Mabbott, Flat 16, Tintagel Court, 1-3 Arthur Street, Hove, BN3 5EY
M7 AAX O. Pollard, Flat 1, Kent Mansions Brighton Road, Worthing, BN11 3EH
M7 AAY S. Britt, 2 West Terrace West End, Herstmonceux, Hailsham, BN27 4NT
M7 AAZ P. Hutchinson, Corby Steps, The Street, Woodbridge, IP12 2QG
M7 ABA M. Voros, Flat 4, 5 Spencer Road, Eastbourne, BN21 4PB
M7 ABB P. Chadwick, 1 Latrigg Crescent Middleton, Manchester, M24 4LU
M7 ABC S. Dearne, 26 Dilly Lane Barton On Sea, New Milton, BH25 7DQ
M7 ABD A. O'Reilly, 45 Commercial Street, Southampton, SO18 6LY
M7 ABE A. Hyndman, 7 Field View, Castleford, WF10 5TU
M7 ABG W. Watson, 54 Pannal Ash Grove, Harrogate, Hg2 0hz
M7 ABM C. White, Flat 2 55 Church Road, London, SE19 2TE
M7 ABN R. Fletcher, Riversdale Drive, Goole, dn14 5lh
M7 ABP R. Miller, 175 Warrington Road, Widnes, WA8 0BA
M7 ABQ D. Pacey, 35 Regent Court, Lord Street, Southport, PR9 0QQ
M7 ABR A. Brook, Inkerman House 113 Clovelly Road, Bideford, EX39 3BY
M7 ABT A. Ryan, Flat 8/91 Westward Road, Stroud, GL5 4LF
MI7 ABU D. Clarke, 117 Gregg Street, Lisburn, BT27 5AW
MI7 ABV M. Robinson, 31 Glenavon Crescent, Lurgan, Craigavon, BT66 8JR
M7 ABW K. Nelson, 185 Headlands, Fenstanton, Huntingdon, PE28 9LP
M7 ABX R. Culley, 6 The Hollies, Straight Road, Foxhall, Ipswich, IP10 0FN
MM7 ABY I. Mcgowan, Carbeth Road, Glasgow, G62 7PT
M7 ACA J. Smith, 44 Old Station Way, Goldalming, GU7 3HA
MI7 ACD J. Miskimmin, 47 Shore Road, Kircubbin, Newtownards, BT22 2RP
M7 ACE V. Foster, 8 Skirlington Grove, Hull, HU9 3RH
M7 ACG M. Mann, Woodlands, Coston, Norwich, NR9 4DT
M7 ACH M. Norfield, 122 Huntingdon Road, Upwood, Ramsey, Huntingdon, PE26 2QQ
MW7 ACK B. Davies, 10 Long Acre Court, Bishopston, Swansea, SA3 3AY
MI7 ACL J. Mcpeake, 47 Dunclug Gardens, Ballymena, BT436NN
M7 ACM B. Chamberlin, 349A Hungerdown Lane, Chippenham, SN14 0JW
M7 ACN M. Wandby, 44 Windrush Road, Hollywood, Birmingham, B47 5QA
MM7 ACS C. Chroston, 6 Queeness Road, Vidlin, Shetland, ZE2 9UB
M7 ACV T. Abel, 20 Falcon Fields, Maldon, CM9 6YA
M7 ACW R. Gower, 30 Westerham Road, Sittingbourne, ME10 1XF
M7 ACX V. Oliveira, 60 Hermine House Moselle Street, London, N17 8DE
M7 ACY P. Young, 11 St. Andrews Avenue, Washington, NE37 1AH
M7 ADC A. Capon, 1 Windermere Way, North Common, BS305XN
M7 ADF N. Barrett, 7 Tannersfield Shalford, Guildford, GU4 8JW
MM7 ADI A. Mcmillan, 13 Roxburgh, Greenock, Pa15 4pu
MW7 ADL C. Walker, Perthi, Llaneilian, Amlwch, LL68 9LY
M7 ADN A. Thomas, Penberthy Road, Portreath, Tr164lu
M7 ADO S. Steppens, 21 Keynes Close, London, N2 9NE
MW7 ADT P. Hoath, 8 Liverpool Terrace, Llithfaen, Pwllheli, LL53 6NN
M7 ADU I. Jackson, 40 Highgate, Cleethorpes, DN35 8NT
M7 ADW R. Naden, 10 Suffolk Close, Holland On Sea, Clacton-on-Sea, CO15 5SQ
M7 ADX M. Lindup, 87 Oakwood Avenue, West Mersea, CO5 8BD
M7 AEA R. Reed, 27 Howard Close, Braintree, CM7 3DT
M7 AEB A. Winton, 10 Old Parsonage Court Guithavon Street, Withem, CM8 1XP
M7 AED J. Searl, Studio 1 34-36 Crown Street, Reading Uk, RG1 2SE
M7 AEF K. Duncan, 72 Elmswood Road, Tranmere, CH42 7HR
MM7 AEI R. Kemp, 14 Marshall Road, Luncarty, Perth, PH1 3UT
M7 AEJ L. Marsh, 11 Woodlands Road, Haywards Heath, RH16 3JU
M7 AEL G. Mayell, Flat 11, Eagle House, Goldsmiths, Grays, RM17 6PX
MW7 AEN C. Trigg, 23 Crystal Glen, Cardiff, CF14 5QH
M7 AER K. Weston, 34 Alexandra Terrace, Lincoln, LN1 1JE
M7 AES R. Carter-Sutherland, 74 Legsby Avenue, Grimsby, DN320NE
MM7 AEU B. Weir, 22 Belvedere Road, Bathgate, EH48 4AX
MI7 AEW G. Colgan, 42 Loughview Village, Carrickfergus, BT38 7PD
M7 AEX P. King, 29 Newbury Road, Manchester, Sk83pa
M7 AEY A. Clarke, Little Acre, Pound Lane, Hardwicke, Gloucester, GL2 4RJ
M7 AEZ C. Bown, 19 Victory Villas, Hatherop Road, Fairford, GL7 4JU
M7 AFA L. Carmon, 21 Wyncroft Road, Widnes, WA88QE
M7 AFD T. Quinlan, 29 Seklton Road, Diss, Ip22 4pw
M7 AFE G. Ede, 53 Mayfield Close, Bognor Regis, PO21 3PS
M7 AFH S. Ali, 13 Stonehill Drive, Rochdale, OL12 7JN
M7 AFI N. Cheetham, 39 Burns Avenue, Church Crookham, Fleet, GU52 6BN
M7 AFK B. Fox, Heath Gardens, Stone, ST15 0AW
M7 AFL G. Clarke, High Gables, Sandstone Close, Dudley, DY3 2EQ
M7 AFO O. Phillips, 54 Marshall Road, Cambridge, CB1 7TY
MW7 AFP S. Cawsey, 8 Hickman Road, Penarth, CF64 2AJ
M7 AFR P. Gianrossi, 2 Limetree Close, Cambridge, CB1 8PF
M7 AFS S. Williamson, 3 Mill Green, Warboys, Huntingdon, PE28 2SA
M7 AFU R. Cardus, 3 Washington Avenue Chaddesden, Derby, DE216JS
M7 AFW G. Appleton, Hawthorn Lodge, Ludborough Road, North Thoresby, Grimsby, DN36 5RF
M7 AFZ C. Pike, 47 Whitstone Rise, Shepton Mallet, BA4 5QA
M7 AGB A. Boocock, 25 Smallwood Road, Dewsbury, WF12 7RU
M7 AGD G. Bonny, 6 Highfield Court, Highfield Avenue, Harwich, CO12 4JG

MM7 AGE J. Groundwater, Newhall, St Margarets Hope, KW17 2RW
M7 AGG A. Scotchmer, 207 Cornelian Street, Blackburn, BB1 9QN
M7 AGH A. Hill, 24 Keelham Drive, Leeds, LS19 6SG
M7 AGJ S. Eardley, 20 Nash Peake Street Tunstall, Stoke on Trent, ST6 5BT
MM7 AGL E. Barton, Old Bakery Cottage, Main Road St. Cyrus, Montrose, DD10 0BA
M7 AGN M. Quinlan-Sandy, 173 Melfort Road, Thornton Heath, CR7 7RU
M7 AGO J. Maxwell, The Bungalow, Latton Bush Centre, Southern Way, Harlow, CM18 7BL
M7 AGQ T. Tofts, 5 Woodcroft, Harlow, CM18 6XX
M7 AGS B. Gojkovic, 42B Hogarth Road, London, SW5OPU
M7 AGU S. Mcdermott, 19 Walpole Street, Wolverhampton, WV6 0AT
M7 AGV D. Hart, 7 Penrose Road, Ferndown, BH22 9JF
M7 AGX V. Fox, 48 Grenham Avenue, Manchester, M15 4HD
MM7 AGY K. Walsh, 12B Humbie Holdings, Kirknewton, Eh27 8ds
M7 AHA R. Bishop, 12A Goseley Avenue, Hartshorne, Swadlincote, DE11 7EZ
M7 AHE L. Rawlings, Gladstone Road, Kingswood, Kingswood, BS151SW
M7 AHG D. Sherlock, London Road, Swanley, BR8 7HA
M7 AHH J. Kent, 28 Connaught Drive, Chapel St. Leonards, Skegness, PE24 5YS
MW7 AHI R. King, 29 Heol Y Waun, Seven Sisters, Neath, SA10 9BL
MM7 AHK J. Martin, 3F Kirkgate, Irvine, KA12 0DF
MW7 AHM K. Burrows, 25 Mill Road, Pontllanfraith, Blackwood, NP12 2GE
M7 AHN F. Tombling, 48 Venables Road, Guisborough, TS14 6LQ
M7 AHT S. Horne, Beaucroft, Keswick Road, Benfleet, SS7 3HU
M7 AHU D. Newton, 22 Cleves Road, Richmond, TW10 7LD
M7 AIE R. Ward, 45 Lumbards, Welwyn Garden City, AL7 1PJ
MW7 AIF A. Hatch, Pembrey Gardens, Blackwood, NP12 2LR
MW7 AIH J. Harris, Heol Y Groes, Cwmbran, Np44 7lt
M7 AII A. Ifrim, 46 Northdown Road, Solihull, B91 3NB
MM7 AIK Dr J. Emery-Barker, 3 Main Street, Newmills, Dunfermline, KY12 8SR
M7 AIM I. Sproson, Flat 15, Olive Standring House, Todmorden Road, Littleborough, OL15 9AH
MM7 AIP R. Stewart, 14 Marshall Court, Queen Street, Dunoon, PA23 8BA
M7 AIQ G. Hunt, 21 Needham Street, Codnor, Ripley, de5 9rr
M7 AIR L. Sanduly, 1 Archford Croft, Emerson Valley, Milton Keynes, MK4 2EZ
M7 AIW W. Lancaster, 39 Chester Grove, Seghill, Cramlington, NE23 7TR
M7 AIZ A. Bailey, 69 New Street, Baddesley Ensor, CV9 2DN
M7 AJA A. Ashurst, 66 Arcot Avenue, Nelson Village, Cramlington, NE23 1EY
M7 AJB A. Burgess, 5 Wilkie Road, Birchington, CT7 9HE
MI7 AJD D. Hurst, 85 Market Street, Ballycastle, BT54 6DS
M7 AJF M. Soden, 24 Barn Rise, Seaford, BN25 3BY
M7 AJG A. Golding, 44 Blendon Drive, Andover, SP10 3NG
M7 AJH A. Holmes, Lodge 7, Arrowbank Caravan Park, Leominster, HR6 9BG
MM7 AJJ M. Mersinis, Flat 30, 88 Albion Street, Glasgow, G1 1NY
M7 AJK M. Collins, Purcell Walk, Bristol, Bs4 1xt
M7 AJM A. Mauger, 51 Pethertons, Halberton, Tiverton, EX16 7AZ
M7 AJO C. Doggrell, 16 Coronation Road, Frome, BA11 2BJ
M7 AJP A. Pickrell, 22 Fosseway, Lichfield, WS14 0AD
MM7 AJQ J. Brummell, 1 Moorpark Cottages, Kirkinner, Newton Stewart, DG8 9BY
M7 AJR G. Huff, 135 Westminster Road Toothill, Swindon, SN5 8JE
M7 AJT B. Starkey, 109 Parc An Tansys, Pengegon, Camborne, TR14 7PH
M7 AJU Maj. A. Taylor-Roberts, 38 Old Coach Road, Bulford, Salisbury, SP4 9DA
M7 AJX S. Evans, 7 The Lampreys, Gloucester, GL4 6QD
M7 AJZ O. Payne, Holland Hall, Clydesdale Road, Exeter, EX4 4SA
M7 AKA J. Harding, 28 Berecroft, Harlow, cm18 7sa
M7 AKC T. Farr, 125 Rochford Way ., Walton on the Naze, CO14 8SP
M7 AKG M. Chapman, 21 Connaught Street, Northampton, NN1 3BP
M7 AKM S. Hey, 3 Queens Square, Kirkby Lonsdale, Carnforth, LA6 2AZ
M7 AKN W. Clarke, 30 Swalecliffe Court Drive, Whitstable, CT5 2LZ
M7 AKP J. Hockridge, 7 Queen Street, Great Oakley, Harwich, CO12 5AS
M7 AKQ S. Stothard, 84 Guntons Close, Soham, CB7 5DN
M7 AKS D. Jackson, 22 Trent Road, Walsall, WS3 4DQ
M7 AKU R. Mortlock, 11 Thompson Avenue, Richmond, Tw9 4jp
M7 AKW D. Kryukov, Verno, Hollow Lane, Virginia Water, GU25 4LP
M7 AKX F. Cairns, 20 St. Davids Close, Maidenhead, SL6 3BB
M7 AKZ A. Frago, Psc 37 Box 3828, Bury St Edmunds, IP28 8NG
M7 ALC A. Bell, 15 Kininvie Close, Redcar, ts104jq
M7 ALF A. Hawkes, 9 Ferneley Avenue, Hinckley, LE100FE
MM7 ALI A. Oakes, 55 Northfield Meadows Longridge, Bathgate, EH47 8SA
MM7 ALL G. Allathan, 1A Irvine Road, Kilmarnock, KA1 2JN
M7 ALN A. James, 36 Westcote Close, Solihull, B92 8PL
MM7 ALO A. Chambers, 35 Echline Grove, South Queensferry, EH30 9RU
M7 ALR A. Rudgley, The New House, Plymouth Road, Buckfastleigh, TQ11 0DB
M7 ALS A. Stanley, 75.Birmingham Road, Kidderminster, DY10 2SR
MW7 ALT D. Clarke, 8 East Walk, Barry, CF62 8DA
MM7 ALW S. Doyle, 19 Cairns Gardens, Balerno, EH14 7HJ
M7 AMA A. Martinavarro, Twyford Court, University Campus, Guildford, GU2 7JP
M7 AMB A. Bennett, 112 Vicarage Crescent, Redditch, B97 4RP
M7 AMC K. Conlon, 136 Chart Downs, Dorking, RH5 4DG
MJ7 AME P. Greenaway, Plat Douet Road, St Saviour S, Jersey, Je27pn
M7 AMJ M. Napiorkowski, 74 Wigley Road, Feltham, TW13 5HE
M7 AMK Dr T. Hoban, 28 Cranley Road, Guildford, GU1 2JS
M7 AMM C. Quinn, 3 Hunters Ride, Stafford, ST17 9HU
M7 AMP R. Barker, 33 Milton Brow, Weston-Super-Mare, BS22 8DB
M7 AMQ A. Wilkinson, 67 Middleton Park Grove, Leeds, ls10 4bg
M7 AMR A. Rowland, 5, Gwinear Downs, Leedstown, Hayle, TR27 6DJ
M7 AMS A. Sonnex, 49 Salisbury Road, Tonbridge, TN10 4PD
M7 AMT R. Pickard, 23 Blackberry Way, Midsomer Norton, Radstock, BA3 2RN
M7 AMW P. Hind, 19 Egmont Drive, Ringwood, BH24 2BN
M7 AMX J. Wright, 149 Wroslyn Road, Freeland, Witney, OX29 8HR
M7 ANA A. Gimenez, Flat 27, Lamb Court, 69 Narrow Street, London, E14 8EJ
M7 ANB D. Smith, 42 Channel View Road, Portland, DT5 2AY
M7 AND A. Goncalves, Flat 20, Brecon Lodge, 2 Wintergreen Boulevard, West Drayton, UB7 9GJ
M7 ANE A. Simpson, 18 Clavell Road, Liverpool, L19 4TR
M7 ANG G. Gordon, 1 Waverland Terrace, Gillingham, SP8 4NT
M7 ANH J. Birkinshaw, 57 Walnut Tree Avenue, Hereford, HR2 7JU
M7 ANI D. Rockey, 113 Bradwell Lane, Newcastle, ST5 8qd
M7 ANJ N. Moorby, 5 Nightingale Close Lower Tean, Stoke on Trent, ST10 4LX
M7 ANK C. Tate, 18 Spenser Walk, South Shields, NE34 9NF
M7 ANM P. Ashworth, 3 Hardman Close, Rossendale, BB4 7DL
M7 ANN L. King, 2 Ebenezer Cottages, Thorney Road, Eye, Peterborough, PE6 7UB
M7 ANO P. Kilkenny, 11 Ash Close, Rochdale, OL12 9an
M7 ANP S. Langton, 44 Poppyfields, Hesketh Bank, Preston, PR4 6TJ
M7 ANR A. Rawson, 64 Dukes Mead, Fleet, GU51 4HE
MW7 ANT A. Lockwood, Church Road, Newport, np197el
M7 ANU S. Archer, 1 Waudby Garth South Cave, Brough, HU15 2GT
M7 ANV A. Young, 22 Martin Close, Cambridge, PE292WA
M7 ANX A. Matsell, 1 Anglia Close, Quarrington, Sleaford, NG34 8WX
M7 ANY N. Glass, 79 Marsh Lane Farndon, Newark, NG24 4TA
M7 AOC S. James, 14 Dene Walk, West Parley, Ferndown, BH22 8PQ
M7 AOD L. Weston, 7 Chestnut Grove, Arnold, Nottingham, NG5 8BD
M7 AOI M. Firth, 12 Pendle Street West, Sabden, Clitheroe, BB7 9EG
M7 AOL S. Wright, 18 Garratts Road, Bushey, wd24 4la
M7 AOM R. Phillips, 148 Northborough Road, Slough, SL2 1TA
M7 AOO A. Watling, 1 Dovendale Cottage Horncastle Road, Tathwell, LN11 9SA
M7 AOR P. Rollason, 68 Heathfield Lane West, Wednesbury, WS10 8QP
MI7 AOU S. Sheridan, 90 The Meadows Randalstown, Randalstown, BT412JB
M7 AOV N. Austerfield, Princes Avenue, Withernsea, HU192JA
MM7 AOY A. Liddell, 22 Tassie Court, Leargan, Leven, KY8 5FL
M7 AOZ D. Hughes, 14 Holts Lane, Clayton, Bradford, BD14 6BL
MM7 APB J. Quinn, 2 Manse Road, Shotts, ML7 5EL
M7 APC R. Nicholson, 27 Bishopton Way, Hexham, NE46 2LR
M7 APF P. Richardson, 91 Park Road, Blackpool, FY1 4JE
M7 APH A. Harcourt, Flat 6, 93 Westridge Road, Southampton, SO17 2HJ
M7 APM G. Booton, 69 The Street, Deal, CT14 0AJ
M7 APN T. Gore, 22 Stoppard Road, Burnham-on-Sea, TA8 1QB
M7 APO L. Martin, 51 Meadow Close, Stretton On Dunsmore, Rugby, CV23 9NL
M7 APV C. West, 11 Peterswood, Harlow, CM18 7RJ
MM7 APW A. Ward, 20 Schoolbrae, Lhanbryde, Iv308qg
M7 APY D. Appleton, Field Farm, Mumby Road, Hogsthorpe, Skegness, PE24 5PD
M7 APZ B. Hender, 11 Cambridge Road, Walton-on-Thames, KT12 2DP
M7 AQC R. Pearce, 5 Jubilee House, Chapel Street, Bicester, OX26 6FE
M7 AQF R. Tew, 61 Magna Road, Bournemouth, BH11 9ND
M7 AQG K. Wallis-Gare, 40 The Forge The Green, Upton, Norwich, NR13 6AY
M7 AQI M. Pownall, 43A Hughes Ave, Warrington, wa29ew
M7 AQL M. Turnier, Unit F, Winston Business Park, Churchill Way, Chapeltown, Sheffield, S35 2PS
M7 AQM S. Cleaver, 77 Main Street, Cockermouth, CA13 9JS
M7 AQO G. Shaw, 1 Platts, Lydlinch, DT10 2HX
M7 AQP C. Watts, 17 Hollycroft Road, Emneth, Wisbech, PE14 8AY
M7 AQQ I. Dodds, 54 Philip Road, Newark, NG24 4PD
M7 AQR S. Salisbury, 6 Ryan Close, Leyland, PR25 2XW
M7 AQS M. Davys, 9 Littlecombe Close, Kersfield Road, London, SW15 3HR
M7 AQY M. Parker, 113 Burringham Road, Ashby, Scunthorpe, DN17 2DF
M7 AQZ T. O'Donnell, 911A New Hey Road, Halifax, HD3 3FH
M7 ARA M. Hicks, 108 Northorpe, Thurlby, Bourne, PE10 0HZ
M7 ARC M. Johnson, 24 Stjames'S Park, Wakefield, WF1 4EU
MW7 ARE J. Browne, 4 Bro Walker, Ammanford, SA18 3EF
M7 ARH C. Pilling, 28 Waverley Rd, Exmouth, EX8 3HJ
M7 ARL E. Hartopp, 38 Lanchester Road, Middlesbrough, TS6 7HG
M7 ARN M. Head, 99 Bell Hill Road, St George, Bristol, BS5 7LY
M7 ARO Lord K. Powell, 57 Valletts Lane, Bolton, BL1 6DW
M7 ARP A. Pomfret, 40 Alwyne Grove, York, YO30 5RT
M7 ARQ M. Feakins, 29 Whitesfield Road Nailsea, Bristol, BS48 2DY
M7 ART A. Fourie, 5 Old Farm Road, Bexhill-on-Sea, TN39 4DN
M7 ARU A. Bird, 18 Welbeck Drive, Spalding, PE11 1PD
M7 ARW D. White, 15 Ropehaven Road, St Austell, PL25 4DU
M7 ARZ D. Trickett, 22 Spring Croft, Rotherham, S61 3RF
M7 ASA B. Chislett, Seavale Road, Clevedon, BS217QB
M7 ASB K. Strykowski, Flat 2, Leyla House, 2 Dunn Street, London, E8 2DB
M7 ASE M. Wilson, 11 Rushdene Avenue, Barnet, EN4 8EN
M7 ASF A. Axtell, Flat 11, Owen House, 6 College Road, Ripon, HG4 2AP
M7 ASH S. Ashfield, 9 Penny Hill Close, Spalding, PE11 2DE
M7 ASJ A. Reader, 4 Hollicombe Close, Tilehurst, Reading, RG30 4PA
M7 ASK J. Postlethwaite, 9 Grange Street, Morecambe, LA4 6BW
MW7 ASM P. Hodgkiss, 35 Princess Avenue, Buckley, CH72LN
MM7 ASN D. Macfarlane, 8 Kingsmill Drive, Kennoway, Leven, KY8 5LX
M7 ASO K. Hedges, 28 Hill Park, Congresbury, Bristol, BS49 5BT
MM7 ASP A. Aspinall, 4 The Maples, Dundee, DD40XQ
M7 ASU R. Crosby, 85A Mendip Road, Portishead, Bristol, BS20 6DF
M7 ASW A. Ward, Westways, Rudston Road, Burton Agnes, Driffield, YO25 4NE
MU7 ASX R. Best, Route De St Andre, St Andrew, Guernsey, GY6 8TU
M7 ASY R. Samson, 20 Chichester Park, Westbury, BA133AN
M7 ASZ K. Burt, 38 Church Lane, Moldgreen, Huddersfield, HD5 9EB
M7 ATA A. Atack, 601 Long Cross, Bristol, bs110tx
M7 ATB Z. Berces, 22 Breydon Walk, Crawley, RH10 6RE
M7 ATC A. Hayler, 19 Newgate Road, St Leonards-on-Sea, TN37 6SA
M7 ATD I. Leonard, 10 Bouchers Mead, Chelmsford, CM1 6PJ
M7 ATE C. Stubbs, High Street, Stoke on Trent, st71py
MM7 ATF J. Shaver, 1 Glendale Crescent, Ayr, KA7 3SQ
M7 ATG M. Cowell, 3 Selborne Mews, Blackburn, BB2 2SQ
M7 ATH A. Higginbottom, 2 Claremont Gardens, Waterlooville, PO7 5LL
M7 ATI A. Toth, 14 Runcorn Road, Sunderland, SR5 5ET
M7 ATJ A. Jemmett, 46 Barkway Road, Royston, SG8 9EB
M7 ATL G. Des Jardins, 4 Bettoney Vere, Bray, SL6 2BA
M7 ATN A. Nelson, 125 Brighton Road, Crawley, RH10 6TE
M7 ATO Dr J. Hunt, 10 Couzens Close, Chippenham, SN15 1US
MW7 ATR T. Brown, 12 George Lansbury Drive, Newport, NP19 9DS
M7 ATT N. Watson, 37 Riber Avenue, Barnsley, S71 3PU
M7 ATV J. Robinson, 12 Rising Side, Barrow-in-Furness, LA13 9ES
M7 ATW A. Wiseman, 16 Hartford Close, Essex, SS69DQ
M7 ATX D. Pyke, 38 Mayflower Close, South Killingholme, Immingham, DN40 3HF
M7 AUB D. Bingham, 132 Gipsy Road, Welling, DA16 1JG
M7 AUC S. Cooney, 1 The Moorings, 254A Astley Street, Dukinfield, SK16 4QW
M7 AUD A. Valvona, 26 Craigbank Court, Fareham, PO14 1AQ
MW7 AUE M. Hallett, 9 Chapel Close, Monmouth, NP25 3NN

M7 AUF I. Edwards, 9 Sanctuary Road, Holsworthy, EX22 6DQ
MM7 AUH K. Stankiewicz, 8 Victoria Place, Cullen, AB56 4TU
M7 AUI S. Mauer, Flat 18, Channing Court, Osborne Road, London, W3 8SY
M7 AUL B. Hatwell, 15 Northey Road, Bodmin, PL31 1JE
M7 AUM J. Shaughnessy, 37 Belton Road, London, NW2 5QE
M7 AUO D. Tonks, 54 Pageant Drive, Aqueduct, Telford, TF4 3RF
M7 AUP J. Matthews, 2 Farm Close, Bungay, NR35 1JG
M7 AUR M. Gibbon, 14 Hillside Way, West Lutton, Malton, YO17 8TE
MW7 AUU M. Waller, 9 Tre Rhoser, Newbourgh, ll61 6tg
MW7 AUX A. Jenkins, 13 Church Street, Bargoed, CF81 8RN
M7 AUY D. Higton, 1 Allendale Road, Caister-on-Sea, NR30 5ES
M7 AUZ G. Muir, The King William Iv, Station Road, Wanstrow.., Shepton Mallet, BA4 4SZ
M7 AVA I. Titchener, 18 King Edgar Close, Ely, CB6 1DP
M7 AVB M. Bareham, 136 Coppins Road, Clacton on Sea, co153la
M7 AVC K. Spowage, 3 Arcadia Avenue, Mansfield, NG20 8JS
M7 AVE S. Wright, 115 Marlborough Road, Derby, de24 8ds
M7 AVF A. Hooper, 13 Westend Avenue, Coppull, Pr7 5db
M7 AVG A. Thompson-Dale, 6 Emsley Avenue, Cudworth, Barnsley, S72 8HU
M7 AVH S. Ackerley, Pierremont Crescent, Darlington, DL3 9PB
MM7 AVI B. Horan, 82 Alloway Road, Lochside, Dumfries, DG2 9LT
M7 AVL A. Hayter, 50 Hamilton Road, Alford, LN13 9AX
M7 AVN T. Ballam, 28 Lannett Road, Gloucester, GL1 5DE
M7 AVO D. Eccles, 16 Pasture Close, Kelsall, Tarporley, CW6 0PN
M7 AVP G. Prenas, 58 Losinga Road, King's Lynn, PE30 2DH
M7 AVQ J. West, 61 Teign Bank Road, Hinckley, LE10 0ED
M7 AVR D. Singleton, 3 Swansbury Drive, Bournemouth, BH80LB
M7 AVS P. Smith, 431 New Street, Biddulph Moor, Stoke-on-Trent, ST8 7NG
M7 AVT A. Velveris, 32 Harrison Street, Carlisle, CA2 4EP
M7 AVW M. Simpkins, 67 Kirkdale, London, SE26 4BL
M7 AWD A. Walkden, 5 Ivy Close, Shaw, Oldham, Ol27tq
M7 AWE S. Vijaysampath, 19 Wallbrook Avenue, Macclesfield, Sk10 3gl
M7 AWF S. Allen, Bar-Point, Carpalla, Foxhole, St Austell, PL26 7TY
M7 AWG A. Goldsmith, 13 Skinners Lane, Galleywood, Chelmsford, CM2 8RH
M7 AWH B. Knight, 5 Lapwing, Tamworth, B77 5NW
M7 AWJ I. Parkin, 93 Lakeside, Isleham Marina, Fen Bank, Isleham, Ely, CB7 5ZD
M7 AWK A. Lowe, 91 Maidenway Road, Paignton, TQ3 2AQ
MM7 AWL M. Love, 17 Lindsay Road, East Kilbride, Glasgow, G74 4HZ
M7 AWM S. Baker, 76 Malvern Road, Bournemouth, BH9 3AJ
M7 AWN J. Coles, 4 Carloggas Close, St Mawgan. Newquay, Truro, TR8 4HJ
MW7 AWO Dr R. Stoyle, 37 Geiriol Road, Townhill, Swansea, SA1 6QP
M7 AWQ A. Williams, 13 Birch Close Colden Common, Colden Common, SO21 1XE
M7 AWW B. Andrews, 18 Parkside, Wilnecote, Tamworth, B77 2JU
M7 AXA B. Thomas, 33 Pelham Road, Southsea, PO5 3DT
MM7 AXB A. Locke, Cotswold 40 Craigton Av, Inverness, iv3 8az
M7 AXC K. Davies, 15 Sandpiper Drive, Weston-Super-Mare, BS22 8UH
M7 AXD J. Warburton, 82 Hampton Drive, Newport, TF10 7RF
M7 AXE J. King, 140 Valley Drive, Harrogate, HG2 0JS
M7 AXF D. Entwisle, 3 Ainley Street, Elland, HX5 0AJ
M7 AXJ S. Brett, 51 West End, Wirksworth, DE44EG
M7 AXM D. White, 11 Seymour Place, Canterbury, CT1 3SF
M7 AXU R. Mills, 6 Gould Close, Bristol, BS13 0BJ
M7 AXV S. Kelly, 12 Carmarthen Way, Rushden, NN10 0TN
MM7 AXX M. Ronaldson, 59B Glasgow Street, Ardrossan, KA22 8EP
M7 AXY J. Cooper, 42 St. Augustines Crescent, Chesterfield, S40 2SD
M7 AYF M. Young, 155 Arctic Road, Cowes, PO31 7XS
MM7 AYG P. Payne, 116 Main Street, West Kilbride, KA23 9AR
M7 AYJ V. Sajanani, University Of Warwick, Post Room, Gibbet Hill Road, 13 F Whitefields, Coventry, CV4 7ES
M7 AYP T. Hobbs, 10 Chervil Close, Chandler'S Ford, Eastleigh, SO53 4JL
M7 AYR S. Hoare, 47 Bright Street, Darlington, DL1 4EY
M7 AYS A. Sullivan, 155 Kings Road, Chelmsford, CM12BA
M7 AYU P. Hasney, 22 Newburn Court, South Shields, NE33 4HR
M7 AYW R. Chandler, 48 Middletons Road, Yaxley, Peterborough, PE7 3NU
M7 AYZ S. Benniman, 5 Round Hill Lane, Shrewsbury, SY1 2NE
M7 AZA K. Brooks, 7 Watermill Close, Polegate, BN26 5JU
M7 AZG D. Smith, 27A Priory Road, Newbury, RG14 7QS
M7 AZH P. Maxfield, 1 Lynchet Lane, Worksop, S817AN
M7 AZJ A. Clarke, 48 East Street, Okehampton, EX20 1AU
M7 AZK M. Wills, 38 New Park House, New Park Road, Shrewsbury, SY1 2RT
M7 AZN Dr A. Rennie, 80 Bartons Drive, Yateley, GU46 6DP
M7 AZP D. I?Onn, 1 Bligh Road, Westhoughton, Bolton, BL5 3TR
M7 AZQ A. Williams, 11 The Fairway, Farnham, GU9 9BB
M7 AZR R. White, 7 Cornwell Close, Redditch, B98 7TG
M7 AZS M. Hughes, 16 Marlborough Road, Urmston, Manchester, M41 5QG
M7 AZV W. Mcintosh, 69A Park Hill, London, SW4 9NS
M7 AZW J. Hartley, 9 Westfield, Bradninch, Exeter, EX5 4QU
M7 BAA T. Stockings, 51 The Mount, Driffield, YO25 5JQ
M7 BAC N. Bacala, Flat 3, 232A Seven Sisters Road, London, N4 3NX
M7 BAE S. Stenning, 18 Knights Mead, Chudleigh Knighton, tq13 0re
M7 BAF P. Symon, 8 Glebe Close, St Columb, TR9 6TA
M7 BAG R. Banks, 1 Holly Meadows, Ashford, TN23 3QR
M7 BAJ A. Carbonell, 2 Wolvesmere Woolmer Green, Knebworth, SG3 6JW
M7 BAK Dr A. Baker, Victoria Cottage, Swan Lane, Aughton, Ormskirk, L39 6SU
M7 BAN A. Wilson, 4 Oxford Street, Doe Lea, Chesterfield, S44 5PH
MW7 BAO B. Williams, Bryn Celyn, Hendre Road, Conwy, LL32 8RJ
M7 BAQ D. Wills, 37 Hawden Road, Bournemouth, BH118RP
M7 BAR D. Saunders, 68 Heywood Road, Prestwich, Manchester, M25 1FN
M7 BAS B. Rice, 178 Hungerhill Road, Nottingham, NG3 3LL
M7 BAV K. Bavin, Browns Lane, Tamworth, B798TA
M7 BAW B. Withington, 20 Bond Way, Hednesford, Cannock, WS12 4SN
M7 BAY B. Young, 1 Bugle Place, Newton Abbot, TQ12 1GZ
MM7 BBG B. Gordon, 125 Caird Street, Hamilton, ML3 0AL
M7 BBI R. Rudling, 18 Bournemouth Avenue, Gosport, PO12 4NP
M7 BBK A. Hedgecock, 45 Gordon Road, High Wycombe, HP13 6EQ
M7 BBM A. Uren, 12 Frewin Gardens, Plymouth, PL6 6PN
M7 BBN D. Click, 3 Brean Gardens, Bristol, BS3 5ET
M7 BBO E. Hanson, 44 Dovehouse Road, Haverhill, CB9 0BZ
M7 BBS S. Dean, 154 Broad Lane, Walsall, WS3 2TQ
M7 BBT J. Mcnaught, Flat 2, 29 Prowse Place, London, NW1 9PN
M7 BBV C. Butler, Wv14 7Np, Wolverhampton, WV14 7NP
M7 BBW G. Davison, 29 Lansbury Way, Sunderland, SR5 3DD
M7 BBY D. Lovick, 328 Cranbrook Avenue, Hull, HU6 9PH
M7 BBZ N. Granville, Honeygarston Road, Bristol, bs139ly
M7 BCA T. Pearson, 4 Oakwood Drive, Barrow-in-Furness, LA13 0UB
M7 BCE D. Davison, 213 Haworth Road, Bradford, BD9 6NX
M7 BCH S. Bailey, 39 Aintree Drive, Balby, Doncaster, DN4 8TU
M7 BCI A. Leonard, 109 Webb Crescent, Dawley, Telford, TF4 3DX
MM7 BCL P. Siviter, 17 Cairndhuna Terrace, Wick, KW1 5BJ
MW7 BCM T. Jones, Gowan Bank Aberarth, Aberaeron, sa460lp
M7 BCN B. Nicholson, 5 Arundel Road, Billingham, TS23 2DJ
M7 BCV B. Chadwick, 44 Glendale Drive, Mellor, Blackburn, BB2 7HD
M7 BCW S. Ferguson, Flat 6, 38 Springfield Road, Blackpool, FY1 2AU
MW7 BCY E. Jones, 34 Llys Charles, Towyn, ll229nt
M7 BCZ D. Williams, Little Benifold, Headley Hill Road, Headley, Bordon, GU35 8DU
M7 BDB D. Bandiera, 1 Broadhill Road, Kegworth, Derby, DE74 2DQ
MM7 BDC P. Chant, Hazelbank Dunrossness, Shetland, ZE2 9JB
M7 BDE L. Kingaby, 5 School Lane, Aldford, Chester, CH3 6HZ
M7 BDF W. Sinnott, 33 Mayfield Street, Atherton, Manchester, M46 0AQ
M7 BDG D. Dheerasinghe, 29 Sharman Close, Stoke-on-Trent, ST4 7LS
M7 BDO J. Middlemiss, Kenmore, Marlpits Lane, Ninfield, Battle, TN33 9LD
M7 BDP D. Taft, 16 Briar Place, Eastbourne, BN23 8DB
MW7 BDT R. Bowen, 16 Lon Hywel, Whitland, SA34 0BE
M7 BDV H. Woolston, 32 Brodrick Road, Eastbourne, BN22 9NR
MM7 BDW A. Mccubbin, 1 Clark Place, Saltcoats, KA21 6JU
M7 BDX S. Cooper, 19 West End Road, Silsoe, MK45 4DU
M7 BDZ B. Hickey, 15 Balmoral Drive, Leicester, LE3 3AD
M7 BEE R. Yaxley, 10 South Rise, North Walsham, NR280EE
MW7 BEF M. Hopkins, 95 Heol-Y-Parc, North Cornelly, Bridgend, CF33 4LY
M7 BEG K. Greenwood, 1 Eastfield Close, Clipstone Village, Mansfield, NG21 9AZ
MM7 BEI R. Hoey, 12 Chamberlain Street, St Andrews, KY16 8JF
MW7 BEK A. Kendrick, Cilgwyn Isaf, Carmarthen, SA33 6NE
M7 BEL M. Tielemans, 2 Parr Close, Grange Park, Swindon, SN5 6JY
M7 BEM S. Christie, 124 Bickershaw Lane, Abram, Wigan, WN2 5PP
M7 BEO M. Carter, 32 Victoria Avenue, Brighouse, HD6 1QT
M7 BEP D. Clarke, 22 Hornsby Avenue, Harley-Goodacre, Worcester, WR4 0PN
M7 BET M. Wu, 1 Cookson Close, Newcastle upon Tyne, NE4 5RY
M7 BEV P. Cheshire, 43 Fremantle Avenue, St Helens, WA95SN
M7 BEX A. Sluijters, 9 George Street, Ashford, TN23 7AF
M7 BFA M. Ashford, 11 Pond Close, Felixstowe, IP112JW
M7 BFB J. Hunter, 121 Longacre, Bamberbridge, PR5 8HQ
M7 BFG A. Rogers, 60 Wellington Road, Bury, BL9 9BQ
M7 BFJ P. Bragg, 10 Wheelers Rise, Brackley, Nn135nd
M7 BFK J. Marchant, 129 Highbury Grove, Clapham, Bedford, MK41 6DU
MM7 BFL L. Robertson, 3 Greenhead Street, Dailly, KA26 9SN
M7 BFN A. Threlfall, 148 Windlehurst Road High Lane, Stockport, SK6 8AG
MI7 BFP T. Smyth, 139 Tullyreagh Road, Gorteen, Tempo, Enniskillen, BT94 3PH
M7 BFQ J. Jenkins, 28, Fitzgerald Avenue, Herne Bay, CT6 8NA
M7 BFV S. Thornton, Low Fold Barn, Orton, Penrith, CA10 3RX
M7 BFX N. Daley, 1 Lewis Road, Chipping Norton, OX7 5JT
M7 BFY A. Evans, 44 Wordsworth Ave Headless Cross, Redditch, B97 5bh
M7 BFZ I. Cooper, Fettlers Wharf Marina Ltd, Station Road, Rufford, Ormskirk, L40 1TB
M7 BGB M. Wanless, 3 Bromlow Hall Barns, Bromlow, Minsterley, Shrewsbury, SY5 0DX
MW7 BGD B. Davies, 3 Yr Hafan, St. Davids, Haverfordwest, SA62 6RA
MI7 BGE J. Mccluskey, 10 Arosa Parade, Belfast, BT15 3JF
M7 BGF B. Grospe, 8, Lincombe Road, Manchester, M22 1GA
M7 BGG A. De Kock, Cartwright Road, London, RM9 6JL
M7 BGH G. Cockle, 2 Beccles Road, Lowestoft, NR33 8QX
M7 BGO P. Chinn, 14 Gwelmeneth, Albion Road, Helston, TR13 8JH
M7 BGR P. Gardner, 7 Great Blakelands, Marston Moretaine, Bedford, MK43 0WY
M7 BGU N. Rendell, 95 Parkstone Road, Poole, BH15 2NZ
M7 BGV T. Vines, 14, Finningham Road, Old Newton, Stowmarket, IP14 4EG
MM7 BGW J. Galbraith, 16 Cameron Way, Prestonpans, EH32 9FH
MW7 BGX P. Harding, 6 Ridgeway View, Newport, NP205AW
M7 BHC R. Mcmanus, 125 Whitaker Road, Derby, DE23 6AQ
M7 BHK C. Fox, 209 Sulgrave Road, Washington, NE37 3DE
M7 BHL S. Gillen, 3 Shetland Road, Dronfield, S18 1WB
M7 BHM R. Billington, 77 Clacton Road St Osyth, Clacton on Sea, co168pd
M7 BHN P. Dawson, 2 Harman Walk, Clacton on Sea, CO16 8UN
M7 BHO J. Dobrucki, 43 Priory Lane, Macclesfield, SK10 3HJ
M7 BHQ P. Green, 26 All Saints Close, Doddinghurst, Brentwood, CM15 0NH
MM7 BHR D. Mcnelis, 35 Parkinch, Erskine, PA8 7HZ
M7 BHT S. Dickson, 11 Benfield, Grasmere, Ambleside, LA22 9RD
M7 BHU D. Cruz, 16 Addison Close, Winchester, SO22 4ED
MW7 BHZ P. Stagg, 8 Canal Terrace, Ystalyfera, SA9 2LP
MM7 BIC A. Knox, 5 Blackdales Avenue, Largs, KA30 8HU
M7 BID G. Russ, 244 Goresbrook Road, Dagenham, RM9 6XU
M7 BIE B. Whitfield, 39 Moralee Close, Gateshead, NE404QE
MW7 BIH D. Grewar, 19 Holland Close, Rogerstone, Newport, NP10 0AU
M7 BIL W. Hughes, The Old Cider House, Well Farm Lane, Alford, BA7 7PW
M7 BIO B. Leeson-Earle, 14 Sedgewaite House, Willowbourne, Fleet, GU51 5BJ
M7 BIP J. Turner, 12 Southfield Lane, Addingham, Ilkley, LS29 0SS
M7 BIQ R. Tuthill, 27 Woodbury House Woodbury Rise, Salisbury, SP2 8FG
M7 BIS C. Lewis, 55 Glynn Road West, Peacehaven, BN10 7SL
MI7 BIU I. Kondrashenkov, 44 Forge Manor, Magheralin, Craigavon, BT67 0XP
MM7 BIW J. Campbell, 113 Brent Field Circle, Ellon, AB41 9DB
M7 BIX S. Eggleson, 17 Fourth Ave, Raf Scampton, Lincoln, LN1 2UP
MM7 BJA B. Anderson, 22 Gladstone Gardens, Fettercairn, Laurencekirk, AB30 1FR
MM7 BJC B. Corkindale, 8 Rockland Park, Largs, KA30 8HB
MM7 BJG P. Ling, 5 Whinstone Place, Ratho, Newbridge, EH28 8AD
M7 BJH B. Havers, 81 Cheviot Road, Slough, SL3 8UE
M7 BJI A. Edwards, 15 Hindsford Bridge Mews, Atherton, Atherton, M46 9QZ
M7 BJK B. Kemble, 3 Red House Close, Chudleigh Knighton, Chudleigh, Newton Abbot, TQ13 0RH

MD7 BJL B. Lerigo, Flat 5, Ballure Court, Queens Drive West, Ramsey, Isle of Man, IM8 2JD
M7 BJN K. Smith, 31 Greenfinch Road, Easington Lane, Houghton le Spring, DH5 0GG
MW7 BJQ A. Jones, 12 Uwch Y Nant, Mynydd Isa, Mold, CH7 6YP
MM7 BJR J. Burton, 6 Kilfinnan Lodges, Spean Bridge, PH34 4EB
M7 BJS B. Slater, Flaxman Rise, Greater Manchester, OL1 4QB
M7 BJT P. Edwards, 4 Stodart Road, London, SE20 8ET
MM7 BJU B. Marshall, 13 Alder Grove, Westquarter, FK2 9SU
M7 BJV D. Shepherd, 10 Luscombe Crescent, Paignton, TQ3 3TW
MW7 BJW C. Weaver, 2 Ty Cerrig, Llanddoged, Llanrwst, LL26 0TY
M7 BJY A. Dawkins, 12 Salisbury Road, Chatham, ME4 5NW
MM7 BKA A. Stanley, 22 Braid Mount, Edinburgh, EH10 6JJ
M7 BKB A. Page, 49 Chiltern Close, Shoreham-by-Sea, BN43 6LE
M7 BKE A. Siani, Flat 3, Windsor Lodge, High Street, Brighton, BN2 1RP
M7 BKG T. Ward, 65 Birtles Road, Macclesfield, SK10 3JG
M7 BKH J. Gray, 5 Oldfield Grove, London, SE16 2NA
M7 BKI S. Manley, The Cottage, Crow Ash Road, Berry Hill, Coleford, GL16 7RB
M7 BKJ M. Mcmullen, 11 Benlaw Grove, Felton, Morpeth, NE65 9NG
M7 BKL J. Baxter, 6 Clayton Mead, Godstone, RH9 8NX
M7 BKM Dr S. Shales, 47 Bradley Avenue, Winterbourne, Bristol, BS36 1HX
M7 BKN J. Forrester, 4 Hingley Street, Cradley Heath, B64 5LA
M7 BKO P. Gregson, 41 Newhouse Road, Blackpool, FY4 4JJ
M7 BKP D. Scrivens, Apartado De.Correos 306, Albox, Spain, 4800
M7 BKS M. Gorrill, 30 Higher Dunscar Egerton, Bolton, BL7 9TF
M7 BKT P. Smith, 3 Crown Cottages School Road, King's Lynn, PE32 1UX
M7 BKU R. Lowe, 31 Chapel Road, Habrough, Immingham, DN40 3AE
M7 BKX M. Gannon, 13 Holyoake Avenue, Woking, GU21 4PW
M7 BKZ A. Prince, 1A Greenacres, Crewe, CW1 4JU
M7 BLC J. North, 7 Kingsbridge Road, Newbury, RG14 6DY
M7 BLD T. Bartholomew, 6 Kilmaine Road, Harwich, CO12 4UZ
M7 BLG V. Robinson, Rectory Cottage Cuxham, Watlington, OX49 5NQ
M7 BLI R. Thompson, 1 Moat Farm Cottage Ashford Road Great Chart, Ashford, TN23 3DH
M7 BLJ J. Utley, 43 Wheatcrofts, Barnsley, S70 6BZ
M7 BLL B. Louder, Flat 1, 30 Mill Street, Bideford, EX39 2JJ
M7 BLM A. Harris, 14 All Saints Road, Bedworth, CV12 0BL
M7 BLO B. Wright, 60 The Ryde, Leigh-on-Sea, SS9 4TN
MM7 BLP R. Mcdonald, 135 Feorlin Way, Garelochhead, Helensburgh, G84 0EB
M7 BLR J. Fitzsimmons, 42 Handcross Road, Luton, LU2 8JF
MW7 BLU D. Bluer, 12 Llys Dewi Sant, Bangor, LL57 2UJ
M7 BLX D. Cooke, 2 Wiffens Loke, Hethersett, Norwich, NR9 3RH
M7 BLY A. Woolhouse, 70 Ditmas Avenue, Kempston, Bedford, MK42 7DW
M7 BMC N. Rijckmans, 1 Nurseries Avenue, Brundall, Norwich, NR13 5NS
M7 BME R. Hale, 19 Glen Mobile Home Park, Colden Common, Winchester, SO21 1TE
M7 BMF B. Hudson, 19 The Ropery, Whitby, Yo224ey
M7 BMG R. Huntrod, 64 Escallond Drive, Seaham, SR7 7JZ
M7 BMH B. Hobbs, 14 Durleigh Hill, Bridgwater, TA5 2AG
M7 BMJ B. Josyfon, 25 Norrice Lea, London, N2 0RD
MM7 BMK K. Biegun, 27 School Walk, Aberdeen, AB24 1XX
M7 BMO L. Taylor, 84 Victoria Way, Stafford, ST17 0NX
M7 BMR B. Royles, 18 Crown Avenue, Cudworth, Barnsley, S72 8SE
MW7 BMU N. Parker, 18 Woodland Terrace, Abercarn, Newport, NP11 4SQ
MW7 BMW J. Angeles, 3 Tolpath, Coed Eva, Cwmbran, NP44 6UD
M7 BMX V. Cepraga, 9 Kynon Close, Gosport, PO12 4LW
MM7 BMY R. Fairman, 52 Broomhill Crescent, Alexandria, G83 9PW
MI7 BMZ K. Mcshane, 40 Oar Building Annadale Crescent, Belfast, BT7 3NB
M7 BNB D. Gerrard, 268A Whelley, Wigan, WN2 1DA
M7 BND R. Clarke, 110 Fairmile Road, Halesowen, B63 3QD
M7 BNE C. Lees, Flat 8, Atlantic House, 57 Sandylands Promenade, Heysham, Morecambe, LA3 1DW
M7 BNG M. Byng, 177 Rosefield Road, Smethwick, B67 6DY
M7 BNH D. Daly, 9 Oxford Street Finedon Nn95Ez, Northants, NN95EZ
M7 BNI R. Dovey, Green Tiles, 138 Frinton Road, Holland-On-Sea, Clacton-on-Sea, CO15 5PN
M7 BNK P. Amos, Flat 2, 3 North Terrace, Birstall, Batley, WF17 9EU
M7 BNL S. Jones, 19 Runshaw Lane, Euxton, Chorley, PR7 6AU
M7 BNN D. Gelkin, Flat 5, Rock House, Station Hill, Chudleigh, Newton Abbot, TQ13 0EE
M7 BNP N. Mason, 4 Berry Hill Close, Mansfield, NG18 4RS
M7 BNR K. Roast, 11 Bracken Road, Keighley, bd227df
M7 BNS D. Croft, Wn49Xa, Wigan, WN4 9XA
MI7 BNV D. Thompson, 45 Ardnabrocky Drumahoe, Londonderry, BT47 3BF
M7 BNW G. Mustoe, 171 Buckingham Road, Aylesbury, HP19 9QF
M7 BNZ A. Sanders, 24 Leeming Lane South, Mansfield Woodhouse, Mansfield, NG19 9AB
M7 BOA A. Woodhead, The Briggs, Ingbirchworth Road, Thurlstone, Sheffield, S36 9QN
M7 BOC J. Woodford, 58 Benedict Close, Romsey, SO51 8PN
M7 BOD G. Stafford, 25 Chilton Gardens, Houghton le Spring, DH4 6LD
M7 BOE A. Beveridge, Kernick Cottage, Sparry Bottom, Carharrack, Redruth, TR16 5SH
M7 BOF K. Clegg, 11 Birchwood Drive, Fulwood, Preston, PR2 9UJ
M7 BOJ C. Eggleton, 20 Arkwright Court, Leominster, HR6 8NF
M7 BOK G. Boka, B Flat, 41 Kimberley Gardens, London, N4 1LB
M7 BON D. Harris, 12A Clevelands Park, Northam, Bideford, EX39 3QH
M7 BOO H. Boocock, 25 The Arches, Claremount Road, Halifax, HX3 6LD
M7 BOP T. Gambrell, 23 Birkside Way, Blackpool, FY4 4TZ
M7 BOQ C. Carr, 52 Blundell Lane, Penwortham, Preston, PR1 0AX
M7 BOS M. Fairchild, Brooklyn Caravan Park 21 Almond Brow Gravel Lane Banks, Southport, PR98BU
M7 BOU A. Clothier, 57 Middle Road, Southampton, SO19 8FT
M7 BOW D. Taylor, 62 Floyds Lane, Walsall, WS4 1LE
M7 BOZ S. Ghent, 86 Tamworth Road, Long Eaton, Nottingham, NG10 3NA
M7 BPA M. Pr³stel, 2 Market Hill Apartments, Market Hill, St Ives, PE27 5AL
M7 BPC B. Coley, 17 Livingstone Road Handsworth, Birmingham, B20 3LS
M7 BPD G. Glover, 834 Moss Bank Way, Johnson Fold, Bolton, BL1 5TB
M7 BPE W. Smith, 69 Victoria Road Netley Abbey, Southampton, SO315DQ
M7 BPH B. Heley, 7 Ringwood Grove, Weston-Super-Mare, BS23 2UA
M7 BPJ A. Odedra, 8 Littlecote Grove, Peterborough, PE4 6BJ
M7 BPN R. Wardley, 218 Chester Road, Watford, WD18 0LJ
M7 BPO T. Lacey, 5 Pilgrim Way, Hastings, TN34 2LF
MW7 BPQ Maj. S. Jha, 19 Armoury Drive, Cardiff, CF14 4NP
M7 BPS P. Ross, 12 Haig Avenue, Great Sankey, Warrington, WA5 2TG
M7 BPW K. Bonfield, 625A Burton Road, Midway, Swadlincote, DE11 0DH
M7 BPX C. Esposito, 1 Chestnut Avenue Wicken Green Village, Fakenham, NR21 7QL
MW7 BPZ S. Hughes, 7 Lower Row, Dowlais, Merthyr Tydfil, CF48 3ND
M7 BQB Rev. P. Callway, 7 Heath Road, Langley, Maidstone, ME17 3LH
M7 BQC P. Van-Embden, 14 Maurice Cullen Close Shotton Colliery, Durham, DH6 2FA
MM7 BQF D. Michael, 24, Taylor Court, Aberlour, AB389LA
MM7 BQG Sir R. Cherry, 1/2 282 Royston Rd, Glasgow, G212JB
M7 BQH A. Gambrell, 7 St. Michaels Close, Shipton-Under-Wychwood, Chipping Norton, OX7 6BE
M7 BQI T. Searchfield, 2 Wares Field, Ridgewood, Uckfield, TN22 5SG
M7 BQJ S. Porter, 8 Brickwork Avenue, Liphook, GU30 7wp
MW7 BQK T. Edge, Llys Y Gwynt, Rhoscefnhir, Pentraeth, LL75 8YU
M7 BQL A. Arcia, Cosmopolitan Tower#400 Apt 2D, Panamµ, Panama
MW7 BQM G. Edwards, 9 Dalton Close, Merthyr Tydfil, CF47 0TE
MM7 BQP C. Duncan, 22 Dalgleish Avenue, Cumnock, KA18 1QU
M7 BQU E. Bray, 56 The Links, Gosport, PO13 0DX
M7 BQX A. Toal, 29 Highland Drive, Oakley, Basingstoke, RG23 7LF
M7 BQY M. Smith, 52 Hungerford Gardens, Bristol, BS4 5HB
M7 BQZ G. Manzetti, 98 Mold Crescent Banbury Ox160Ex, Banbury, OX16 0EX
M7 BRA S. Bracegirdle, Flat 28, West Fryerne, Parkside Road, Reading, RG30 2BY
M7 BRC D. Dobbie, 24 Harrow Road, Leighton Buzzard, LU7 4UQ
MM7 BRG A. Brown, 4 Damselfly View, Edinburgh, Eh17 8xh
M7 BRH B. Harwood, 9 Wyndham Rd, Newstead, Stoke on Trent, ST3 3LX
M7 BRI B. Freeman, 11 Rachel Grove, Stoke-on-Trent, ST4 3QX
M7 BRJ D. Withnall, 35 Dalton Bank, Warrington, WA1 3AH
M7 BRM J. Holden, 47 Copse Hill, London, SW20 0NJ
M7 BRO S. Finch, 2 De Grey Road, Colchester, CO4 5YE
MW7 BRQ J. Mann, 11 Hilltop Close, Port Talbot, SA12 8YH
M7 BRR M. Ding, 130 Windermere Avenue, Warrington, WA2 0NE
M7 BRX D. Reed, 12 George Trollope Road, Watton, Thetford, IP25 6AS
M7 BRY B. Bell, 101 Victoria Road, Gateshead, NE8 2SY
M7 BRZ D. Bishop, 19 Cambrian Way, Winsford, CW7 1QT
M7 BSF D. Westlake, 63 Carswell, Kingskeswell Newton Abbot, TQ125BL
M7 BSH A. Gladosz, 45 Bushey Ley, Welwyn Garden City, AL7 3HB
M7 BSI A. Stockton, 31 Jones Close, Brackley, NN13 6JD
MW7 BSK M. Watts, 2 Cysgod-Y-Cwm, St. Dogmaels, Cardigan, SA43 3DS
M7 BSL E. Crobu, 16 York Street, Flat 6, London, W1U 6PS
M7 BSM B. Smith, Bees Corner, Ideford, Chudleigh, Newton Abbot, TQ13 0AZ
M7 BSN B. Shaw, 9 Pheasant Close, Mulbarton, Norwich, NR14 8BL
M7 BSS A. Jackson, 11 New Drake Green, Westhoughton, Bolton, BL5 2RF
M7 BSU S. Ash, 10 Anchor Court 132 Bury Old Road, Manchester, M85DR
M7 BSV W. Baron, 10 Hembury Close, Middleton, Manchester, M24 2SX
M7 BSX B. Shepard, 3 The Leys, Oxhill, CV35 0QX
M7 BSZ A. Watson, 14 Linlithgow Close, Papworth Everard, Cambridge, CB23 3RX
M7 BTD D. Sampson, 116 South Mossley Hill Road, Liverpool, L19 9BJ
MW7 BTF S. Nutting, 9 Lugg View, Presteigne, LD8 2DG
M7 BTG W. Trott, 8 Dunlop Crescent, South Shields, NE34 6QG
MI7 BTH R. Hannon, 10 Malone Meadows, Belfast, BT9 5BG
M7 BTL S. Parrott, 39 Chichester Rd, Cleethorpes, DN35 0HY
M7 BTO L. Tod, 6 Middleham Close, Sandy, SG19 1TU
M7 BTQ L. Jones, Warren Farm Road, Birmingham, B44 0QB
M7 BTS M. Evans, Carlton Manor Caravan Park, Carlron Coleville, Chapel Road, NR33 8BL
M7 BTU O. Schonrock, 1 Powell Road, Poole, BH14 8SG
M7 BTV G. Hodkinson, 14 Mitton Close, Culcheth, Warrington, WA3 4EU
M7 BTW B. Wyatt, 2 Birchwood Gardens, Southampton, SO30 2AR
MM7 BTX I. Morrison, Coach House, Olivers Brae, Stornoway, HS12SX
M7 BTY R. Chandler, 56 Marsh Lane, Addlestone, KT15 1UN
M7 BTZ W. Pearce, 6A Havelock Street, Desborough, Kettering, NN14 2LU
M7 BUB G. Veale, 8 Duchy Cottages, Stoke Climsland, Callington, PL17 8PA
M7 BUE I. Shepherd, 61 Chatsworth Place Meir, Stoke on Trent, st37dp
M7 BUG D. Coyle, 7 The Willows Bilsborrow, Preston, PR3 0SG
M7 BUH G. Butler, 4 Mercia Close Worksop, Worksop, S81 0SS
M7 BUI D. Nixon, 20 Oakdale Road, Retford, DN22 7GX
M7 BUK A. Brown, 17 Harley St, Coventry, CV2 4EZ
M7 BUL N. Bull, Eldoret, Castle Street Bampton, Tiverton, EX16 9NS
M7 BUN D. Shingleton, 6 Newsham Walk, Manchester, M12 5QB
M7 BUR D. Burton, 16 Cage Lane, Great Staughton, St Neots, PE19 5DB
M7 BUS G. Busacca, Flat 6, Leigh Court, Tavistock Place, Bedford, MK40 2XU
M7 BUV A. Harding, 32 North Falls Road, Canvey Island, Ss87qg
M7 BUY K. Hodgson, 411 Moss Bay Road, Workington, CA14 5AB
M7 BVA I. Gourlay, 1 Littlemoor Road, Mirfield, wf149al
M7 BVC A. Marvel, Dean House Farm, Nordan, Leominster, HR6 0AW
M7 BVF C. Morris, 4 Wroxeter, Shrewsbury, SY5 6PH
M7 BVH D. Clews, 25 Lea Green Lane, Wythall, Birmingham, B47 6HE
M7 BVI J. Lambert, 7 Fawn Close, Huntington, Cannock, WS12 4UP
M7 BVJ J. Fairbrother, 6, The Dene, Bethersdon, Ashford, TN26 3AR
M7 BVK P. Macfarlane, 26 Whittingham Drive Ramsbottom, Bury, BL0 9LZ
MM7 BVN R. Bevan, 46 Oak Drive, Portlethen, Aberdeen, AB12 4XF
M7 BVO D. Blencowe, 25 Harrington Road, Kelmarsh, Northampton, NN6 9LX
M7 BVQ D. Lamrhari, 16 The Close, Potters Bar, EN6 2HY
M7 BVS D. Szmytkowski, Van Crombrugghe, Brussels, Belgium, 1150
M7 BVT R. Morris, 45 St. Kildas Road, Bath, BA2 3QL
M7 BVV P. Pullinger, 10 Cowley Drive Lancing, Lancing West Sussex, BN158DJ
M7 BVZ B. Mockford, 322 High Street, Eastleigh, SO50 5ND
M7 BWE N. Hughes, Hawes Bank Pow Lane, Shap, CA10 3NB
M7 BWG B. Gentry, 5 Whitlock Drive, Great Yeldham, Halstead, CO9 4EE
M7 BWH J. Lane, 9 Bere Court Old Mill Lane, Crewkerne, TA18 7BN
M7 BWJ W. Jacques, 3 The Butts, Crudwell, SN16 9HF
MM7 BWK M. Borthwick, Shepherds Cottage, Sunnyside, Hawick, TD9 9SS

M7 BWL B. Langford, North Street House, North Street, Langport, TA10 9RL
M7 BWM K. Brennan, 14 West View, Silsden, Keighley, BD20 9JY
MM7 BWN G. Scott, 29 Kennedy Drive Dunure, Dunure, ka7 4lr
M7 BWO R. Yarnall, 279 Longfield, Falmouth, TR11 4SS
M7 BWP A. Gornall, 29 Clifton Street, Darwen, BB3 0BR
MM7 BWT N. Mansfield, Blencathra House, Township Road, Auckengill, KW14xp
M7 BWW T. Smith, Grange De Lings, Lincoln, LN2 2LX
MM7 BWZ A. Marshall, 46 Langton Road, Westquarter, Falkirk, FK2 9SZ
M7 BXA R. Angus, 48 Castle Avenue, Epsom, KT17 2PH
M7 BXC P. Wilbraham, 122 Prenton Hall Road, Prenton, CH43 3BJ
M7 BXE P. Martin, 46 Melody Road, Biggin Hill, Westerham, TN16 3PH
M7 BXF G. Payne, 508 Crownhill Road, Plymouth, PL5 2QU
MM7 BXL B. Longmuir, 12 Cloverfield Gardens, Bucksburn, Aberdeen, AB21 9AY
M7 BXM K. Scott, Flat 16, Quayside Court, The Quay, Harwich, CO12 3HH
MW7 BXN P. Lang, 3 Heol-Yr-Enfys, Bridgend, CF31 4UQ
M7 BXO P. Wade, 89 Top Fair Furlong, Redhouse Park, Milton Keynes, MK14 5FQ
M7 BXT J. Stanley, 57 Walterbush Road, Chipping Norton, OX7 5DP
M7 BYC R. Bagnall, 9 Anson Road, Stoke-on-Trent, ST3 7AT
M7 BYD K. Potter, 7 Leesfield Road, Meadowfield, Durham, DH7 8NJ
M7 BYE G. Frost, The Old Chestnut, 51 Knightcott Gardens, Banwell, BS29 6HD
M7 BYF S. Austin, 35 Flordon, Skelmersdale, WN8 6PA
MM7 BYH W. Gordon, 5 Balgarvie Road, Cupar, KY15 4AH
MM7 BYK M. Weeks, 109 East Princes Street, Helensburgh, G84 7DN
M7 BYL M. Kimberley, 45 Harbord Close, North Walsham, NR28 0TA
M7 BYM G. Young, 65 Boyton Road, Ipswich, IP3 9PD
MM7 BYQ G. Nicholson, 22Kilbowie Rd South Carbrain, Cumbernauld, G67 2PX
M7 BYS M. Bosman, Lindridge, Church Lane, Baschurch, Shrewsbury, SY4 2ED
MM7 BYT Lord M. Bryant, 46 Waverley Drive Glenrothes, Glenrothes, KY6 2LU
M7 BYU M. Annan, 6 Whiteleaze Westbury On Trym, Bristol, BS10 5ED
MW7 BYV S. Richards, 62 Glyn Bedw, Caerphilly, CF833PG
MW7 BYW P. Whitby, 28 Bryn Dyrys, Bagillt, CH6 6BX
M7 BZA J. Fagan, 20 O'Sullivan Crescent, Blakbrook, St Helens, WA11 9RB
M7 BZB G. Woodham, 5 Colne Rise Rowhedge, Colchester, Rowhedge, CO5 7EH
M7 BZC T. Mather, 40 Sandown Road, Haslingden, Rossendale, BB4 6PL
M7 BZK J. Hammond, 11 Cornish Houses Buckshaft Rd, Cinderford, GL143DX
M7 BZL N. Ormrod, 13 Boscombe Ave, Peel Green Peel Green, Manchester, M307DU
M7 BZO D. Neill, Burrs Shoes, 7-9 Leys Avenue, Letchworth Garden City, SG6 3EA
M7 BZP S. Tudge, 11 Boar Croft, Coventry, CV4 9SJ
MI7 BZR A. Bandeja, 16, Craigavon Crescent, Dungannon, BT71 7BD
M7 BZU A. Moore, 178 Park Road, Bedworth, CV12 8LA
M7 BZV K. Houghton, 26 Heathfield Road, Grantham, NG31 7NH
M7 BZZ A. Busby, 24 Hall Close, Bourn, Cambridge, CB23 2SW
M7 CAA D. Morland, Greystone, Cliburn, Penrith, CA10 3AL
M7 CAD G. Jones, 7 Wheatlands Close, Cannock, WS12 3XL
MW7 CAJ N. Hewelt, 17 Saint Marys Road, Llandudno, LL30 2UB
M7 CAK Dr L. Walker, 12 Priory Gardens, Corbridge, NE45 5HZ
M7 CAL J. Groombridge, 57 Lower Road, Swanley, BR8 7RY
MM7 CAN J. Cannon, 31 Ellismuir Street, Coatbridge, ML5 5BH
M7 CAQ D. Martin, 179 Dartford Road, Dartford, DA1 3EW
MI7 CAT C. Todd, 75 Suffolk Road, Belfast, BT11 9PU
M7 CAW C. Gray, The Pigsty, Cleverton, Chippenham, SN15 5BT
M7 CAZ M. Ball, 16 King Street, Burntwood, WS7 4QJ
MW7 CBF D. Holdcroft, 7 Llys Mai, Buckley, Ch72gz
M7 CBG A. Rizvi, 41 Clarence Close, Manchester, BL9 6HE
M7 CBH C. Heath, 130 Holyfields, West Allotment, Newcastle upon Tyne, NE27 0EY
M7 CBL A. Evans, 45 Bonington Road, Mansfield, NG19 6QH
M7 CBP D. Mcdonald, 20 Montague Road, Bournemouth, BH5 2EP
MW7 CBT P. Rogers, 109 North Road, Pontywaun, Cross Keys, Newport, NP11 7FS
M7 CBX C. Barber, 32 Ashcroft Road, Ipswich, IP1 6AB
M7 CBY N. Porter, 37 Windermere Park, Lowestoft, NR32 4UD
MM7 CCB C. Boal, 20 Fairinsfell, Broxburn, EH52 6AL
MM7 CCC B. Cook, 2 Marlfield Farm Cottages, Kelso, TD5 8ED
M7 CCF S. Greenan, 1 Bramley Drive, Frome, bA112EY
M7 CCG A. Jones, 15 Wheatfields, Bradeley, Stoke-on-Trent, ST6 7QD
M7 CCH C. Chowdhury-Hanscombe, 35 Clavering Avenue, London, SW13 8DX
M7 CCK G. Allen, 22 Acrefield, Padiham, Burnley, BB12 8HN
M7 CCL C. Crawford, 31 Marston Drive, Newbury, RG14 2SQ
M7 CCM N. Pilling, 12 Brooke Close, Baxenden, BB5 2QX
M7 CCO B. Wyatt, 20 Britannia Gardens, Westcliff-on-Sea, SS0 8BN
M7 CCQ A. Woosnam, 15 Saddlers Square, Northampton, NN3 5AY
M7 CCR I. Read, 8 The Moors Cressage, Shrewsbury, SY5 6DA
MW7 CCS C. Smith, Flat 15, Llys Morfa, Sunny Crescent, Cymmer, Port Talbot, SA13 3PQ
M7 CCU F. Richardson, 68 Peregrine Road, Brockworth, Gloucester, GL3 4ZE
M7 CCW C. Wright, 11 Trinity Fields, Lower Beeding, Horsham, RH13 6GH
M7 CCZ B. Crudgington, 1 Sharmford Meadows, Barham, Ipswich, IP6 0QY
M7 CDA C. Ashley, 22 Pasture Close, Lower Earley, Reading, RG6 4UY
M7 CDB P. Wood, 11 Honey Hill Lees Oldham 0L4 5Dp Lancashire, Oldham, ol4 5dp
M7 CDF C. Forrester, 21 The Crescent, Northfleet, Gravesend, DA11 7EB
M7 CDH Dr C. Haley, Cold Norton Farm Ockham Lane Hatchford, Cobham, KT11 1LW
M7 CDK I. Casbon, 117A Pasture Street, Grimsby, DN32 9EE
M7 CDO W. Harvey, 17 Saunton Road, Braunton, EX331HB
M7 CDQ J. Mastros, 41 Radnor Street, Swindon, SN1 3PR
M7 CDS C. Sole, 2 Shepley Street, Manchester, M359dy
M7 CDT C. Smith, 17 Midwinter Avenue, Milton, Abingdon, OX14 4XA
M7 CDU A. Zahid Begum, 3 Mercury House, 4-8 Cheam Road, Epsom, KT17 1SN
M7 CDW C. Woolf, 127 Lavender Avenue Mitcham, Mitcham, CR4 3RS
M7 CDX C. Dibben, 30 Harty Road, Haydock, WA11 0YY
M7 CEB C. Brennan, 19 The Furrow, Littleport, Ely, CB6 1GL
M7 CEF A. Lendrum, 35 Hardfield Road, Alkrington, Middleton, Manchester, M24 1JA
MM7 CEH C. Harris, 27 Avon Road, Bathgate, EH48 4AH
MI7 CEK J. Harvey, 6 Montrose Street, Belfast, BT5 4HY
M7 CEL C. Dench, Thornby, Down St. Mary, Crediton, EX17 6DU
MW7 CEO D. Hughes, 6 Cambrian House, Old School Lane, Pontypridd, CF37 2DD
M7 CEP S. Dunklin, 2 High Street, Boxworth, Cb234ly
M7 CER C. Storer, 90 Midland Road, Ellistown, Coalville, LE67 1EH
MM7 CES D. Woods, 3 Walnut Crescent, Johnstone, PA5 9QJ
M7 CEV A. Peet, 49 Staveley Road, Poolsbrook, Chesterfield, S43 3JX
MW7 CEW C. Worka, 9 Bertha Street, Pontypridd, Cf37 1ts
M7 CEY I. Rehacevs, Flat 13, Rowan Court, 19 The Avenue, Beckenham, BR3 5LH
M7 CEZ C. Neagu, Flat 4, 17 Wastbourne Gardens, Folkestone, CT20 2HZ
MM7 CFC S. Smith, 10 Munro Street, Stenhousemuir, Larbert, FK5 4QF
MM7 CFE G. Pow, 29 Seafar Drive, Kelty, KY4 0JX
M7 CFF D. Adelodun, 128 Olney Road, London, SE17 3HR
M7 CFG P. Luck, Durban House, Merrymeet, Liskeard, PL14 3LP
M7 CFL C. Locock, 12 Park Road, Moseley, Birmingham, B13 8AB
M7 CFR C. Ralston, 11 Back Gillmoss Lane, Liverpool, L11 0AY
MM7 CFU H. Mackenzie, 31 Ulladale Crescent, Strathpeffer, IV14 9AQ
M7 CFW D. Cadden, 24 Lindenbrook Vale, Stafford, ST17 4QN
MI7 CFX M. Douglas, 7 Beechill Crescent Tandragee, Craigavon, BT62 2BN
M7 CGD P. Miller, 21 Golborne Road, Lowton, Warrington, WA3 2DP
M7 CGH C. Heys, 32 Wolseley Road, Sale, M33 7AU
M7 CGI M. Gilliver, 24 Fritchley Lane, Fritchley, Belper, DE56 2FN
M7 CGK G. Steel, 4 Doublegates Close, Ripon, HG4 2TU
M7 CGM A. Davies, 4 Erpingham Road, Poole, BH12 1EX
M7 CGO M. Flint, 22A High Street, Hinxton, Saffron Walden, CB10 1QY
M7 CGQ K. Slater, 4 Ridgewell Close, Lincoln, LN63QQ
MM7 CGS C. Stephens, Blairmains Farm, Harthill, Shotts, ML7 5TJ
M7 CHA M. Charlesworth, 18 Bushell Close, Leighton Buzzard, LU7 4TQ
M7 CHB M. Bell, 11 Greaves Road, Sheffield, S5 9DB
M7 CHC D. Woodward, 21 Northolt Avenue, Bishop's Stortford, CM23 5DR
M7 CHD C. Copley, 18 Pavement Lane, Halifax, HX2 9JJ
M7 CHE R. Drage, 4 Bruce'S Close Conington, Peterborough, PE73QW
M7 CHF S. Brown, 313 Milton Road, Cambridge, CB4 1XQ
M7 CHG A. Salerno, 1A Woodland Way, Marlow, SL7 3LD
M7 CHH C. Haycock, 45 Barden Road, Wakefield, WF1 4HP
M7 CHI S. Speak, 32 Noyes Avenue, Laxfield, Woodbridge, IP13 8EB
M7 CHJ O. Burr, 6 Curson Road, Tasburgh, Norwich, NR15 1NH
M7 CHN I. Bott, 22 Middlefield Road, Cossington, Leicester, LE7 4UT
M7 CHP R. Chappell, 24 Woodend Drive, Shipley, BD18 2BW
M7 CHR C. Pascoe, Treleven, Primrose Hill, Goldsithney, Penzance, TR20 9JR
M7 CHS C. Schroth, Flat 3, 15B Cavendish Road, Bournemouth, BH1 1QX
MW7 CIE M. Thomas, 88 Dafodil Court Ty Canol, Cwmbran, np44 6jf
MW7 CIH M. Knight, 1 Bessemer Close, Newport, NP20 6PE
M7 CII T. Bickerstaff, 43 Riverway, Durrington, Salisbury, SP48ES
M7 CIJ S. Mcdermott, 48 Horseshoe Crescent, Bordon, GU35 0DP
M7 CIL S. Perera, 33 Lockington Croft, Halesowen, B62 9BP
MU7 CIM R. Price, Les Vaurioufs, St Martins, Guernsey, GY4 6TE
M7 CIN K. Darker, 19, The Horseshoe, Hemel Hempstead, HP3 8QT
M7 CIO C. Green, 36 Wren Gardens Alderholt, Fordingbridge, sp63pj
M7 CIS R. Scott, 34 Moorfield Road, Birmingham, B34 6QY
M7 CIT A. Jacobs, 11 Office Road, Cinderford, GL14 2HZ
M7 CIU J. Weber, 9 Orpwood Way, Abingdon, OX14 5PX
M7 CIX A. Williamson, 15 Montgomery Road, Gosport, PO13 0UZ
M7 CIZ P. Upton, 27 Pavilion Way, Little Chalfont, Amersham, HP6 6PZ
MI7 CJA J. Coles, 20 Upper Cairncastle Road, Larne, BT40 2DT
M7 CJB C. Braiden, Flat 5, Lansdowne House, 12 Twickenham Close, Swindon, SN3 3FQ
M7 CJC M. Criscione, 46 South Molton Street, London, W1K 5RX
M7 CJE M. Hill, 75 Orpen Road, Sholing, Southampton, SO19 0EH
M7 CJG C. Gosling, 34 Coldhams Crescent, Huntingdon, PE29 1UG
M7 CJH S. Seddon, 3 Riverside Wharf, Dartford, DA1 5TN
M7 CJL C. Davey, 1 Hillend Cottages, Eastnor, HR8 1RF
M7 CJM J. Mccarthy, 60 Milborough Crescent, London, SE12 0RN
MM7 CJS C. Sutton, 29 Hazel Dene, Methil, Leven, KY8 2JL
M7 CJU C. Longster, 136 140 Old Shoreham Road, Hove, Bn37bd
M7 CJV D. Cooke, 42 Branston Road, Uppingham, Oakham, LE15 9RS
M7 CJX P. Jones, 8 Normanton Close, Edwinstowe, Mansfield, NG21 9PF
M7 CJY C. Yates, 10 Grosvenor Avenue, Torquay, TQ2 7LA
M7 CKA M. Angus, 4 Irlam Road, Sale, M33 2BH
M7 CKC J. Winsor, 8 Sages Lea, Woodbury Salterton, Exeter, EX5 1RA
M7 CKF B. Morrissey, Fiboard House, 5 Oakleigh Gardens, London, N20 9AB
M7 CKG L. Staples, 4 Halidon Terrace, St Columb, TR96GQ
M7 CKN G. Harris, 18 Waggoners Way Bugbrooke, Northampton, NN7 3QT
M7 CKR A. Shilling, 18 Simmons Way, Okehampton, EX2O 1PY
M7 CKT Rev. S. Perry, 8 Collingwood Avenue, March, pe15 9ef
M7 CKV M. Sanger, 45 Pier Plain, Gorleston, Great Yarmouth, NR31 6PS
M7 CKW C. Wise, 32 Commercial Street, Willington, DL15 0AD
M7 CKX C. Whittaker, Little Polmarth Farm Polmarth Carnminellis Nr Redruth Cornwall, Turo, tr166nt
M7 CKY M. Logan, 14 Long Ley, Langley Upper Green, Saffron Walden, CB11 4RX
M7 CLC G. Walsh, 42 St. Huberts Road, Great Harwood, Blackburn, BB6 7AR
MW7 CLE P. Roberts, 18 Maes Mawr, Llanrwst, Ll260hw
MM7 CLH C. Hornby, 2 Galabank Gardens, Annan, DG12 5FA
M7 CLI C. Davies, 51 Holland Street, Crewe, CW1 3TT
M7 CLJ C. Streader, 56 Lowtown Street, Worksop, S80 2JR
M7 CLK D. Rigby, 4 Avalon Close Tottington, Bury, BL8 3LW
M7 CLM C. Morse, Redhouse, The Green, Dauntsey, Chippenham, SN15 4JH
MM7 CLN G. Mcara, 24 Balfour Street, Alloa, FK10 1RU
M7 CLP W. Pryce, 6 Banbury Close, Shrewsbury, SY2 6TE
M7 CLQ R. Davies, 32 Saddleback Road, Swindon, SN5 5RL
M7 CLU D. Pollard, 6 Amble Road, Callington, PL17 7QE
M7 CLW C. Wishart, Flat 22, Lake View, Alcove Road, Bristol, BS16 3AG
M7 CLX C. Greenhalgh, 2 Acredales, Abbeytown, Wigton, CA7 4RT
MM7 CMB C. Mcconnell, 4 Doune Road, Bragar, HS2 9DF
MM7 CMC C. Mcintyre, 70 Oldwood Place, Livingston, EH54 6US
M7 CMD E. Hauser, 23 Prince William Way Sawston, Cambridge, CB22 3SZ
M7 CME A. Banwell, Hazel Cottage, Chapel Lane, Hinton, Chippenham, SN14 8HD
M7 CMI A. Castle, 12 Churchfield House, Guessens Road, Welwyn Garden City, AL8 6RJ
MW7 CMJ G. Hillman, 5 Avondale Street, Abercynon, Abercynon, CF454YU
M7 CML C. Cresswell, 23 Wellfield Close, Newcastle upon Tyne, NE15 9JL
MM7 CMM M. Duncan, 72 Scalloway Road, Gartcosh, Glasgow, G69 8LH
M7 CMS M. Cocciu, 92 Dowdeswell Close, London, SW15 5RL
M7 CMW C. Wardrop, 5, Highstreet, Bidborough, Tunbridge Wells, TN3 0UJ

MM7 CMY A. Mccall Smith, 16 Napier Road, Edinburgh, EH10 5AY
M7 CNA M. Higginbottom, 6 Stuart Close, Strensall, York, YO32 5ZP
M7 CNB J. Gulliver, 32 The Row, Welford, Newbury, RG20 8HS
M7 CND W. Johns, 19 Kingston Close, Seaford, BN25 4NF
M7 CNE A. Clarkson, 9 St. Andrews Square, Stoke-on-Trent, ST4 7GA
M7 CNI S. Arnull, 1 Limetree Cottage, The Street, Erpingham, Norwich, NR11 7QD
M7 CNJ T. Grear, 8 Petworth Close, Frimley, Camberley, GU16 8XS
M7 CNK K. Robinson, 1 Berry Close, Earls Barton, Northampton, NN6 0HU
M7 CNL M. Mcneil, 249 Surbiton Road, Fairfield, Stockton-on-Tees, TS19 7SF
M7 CNM F. Sheane-Smith, 9 Tagwell Grange, Droitwich Spa, WR9 7FD
M7 CNN R. Sissons, 57 Sedgehill Road, London, SE6 3QR
MM7 CNQ H. Brown, 6 Rupert Street 1/2, Glasgow, G4 9AR
MM7 CNX L. Day, 25 Meiklerig Court, Glasgow, G53 5UY
M7 CNY A. Taylor, 23 Park Lane, Southwick, Brighton, BN42 4DL
M7 CNZ J. Greaves, Petherton Preston New Road, Nr. Blackburn, BB2 7PU
M7 COB P. Cobb, 166 Milton Street, Southport, PR9 7AP
M7 COD M. Ray, 60 Albemarle Road, Gorleston, Great Yarmouth, NR31 7AS
M7 COF P. Perkins, 47 Ranulf Road, Flitch Green, Dunmow, CM6 3GR
M7 COH S. O'Hanlon, 113 Taunton Drive, Liverpool, L108JN
M7 CON D. Wilkinson, 8 Brecon Close, Newcastle upon Tyne, ne54td
M7 COP C. Speak, 32 Noyes Avenue, Laxfield, Woodbridge, IP13 8EB
M7 COQ J. Trajdos, 43 Mount Road, Manchester, M18 7BX
M7 COR D. Jones, Ch658Hn, Ellesmere Port, CH65 8HN
M7 COS C. Rowantree, 25 Hillside, Briston, Melton Constable, NR24 2DW
M7 COU M. Cartwright, 7 Old Mill Close, Grove, Wantage, OX12 7LD
M7 COV M. Higginson, 190 St. Georges Road, Coventry, CV1 2DF
M7 COX G. Cox, 2 Poulteney Road, Stansted, CM24 8ED
M7 COY J. Voak, Flat 3, Dove Court, Packers Lane, Ramsgate, CT11 8QA
M7 CPA C. Wilkins, Alma Place, North Road, Barnstaple, EX31 1PA
M7 CPC C. Clark, 90 Castlebridge Gardens, Wolverhampton, WV11 3NQ
M7 CPG D. Cooper, 21 Staley Drive, Glapwell, Chesterfield, S44 5QG
M7 CPH P. Hazelton, 20 Greenways, Chelmsford, CM1 4EF
MM7 CPI J. Devlin, 200 Second Avenue, Clydebank, Glasgow, G81 3LE
M7 CPJ G. Pawson, 9 Ivan Clarks Corner Abington, Cambridge, CB216XR
M7 CPO R. Tullock, 22A Tunstall Avenue, Billingham, TS23 3SP
M7 CPQ R. Pink, 75 Greenfinches, Hempstead, Gillingham, ME7 3PW
M7 CPT J. Cressey, 32 Shakespeare Road, Bredbury, Stockport, SK6 2HS
M7 CPW A. Wells, 8 Ward Street, Earls Barton, Northampton, NN6 0JW
MM7 CPZ F. Dove, 5 Cairnorchies, Mintlaw, Peterhead, AB42 4LH
M7 CQA S. Jones, 36 Woodbury, Lambourn, Hungerford, RG17 7LT
M7 CQD D. Ransome, Ashleigh, Coombe Lane, Shepton Mallet, BA4 5UY
M7 CQE D. Harmer, 98 King Georges Avenue, Coventry, CV6 6FF
M7 CQF D. Owens, Lucas Green Nurseries, Lucas Green, West End, Woking, GU24 9LY
M7 CQJ D. Nicoll, Dingle Lane Farm, Dingle Lane, Hilderstone, Stone, ST15 8SG
M7 CQK R. Baber, 38 Sinope Street, Gloucester, GL1 4AR
M7 CQL Z. Gong, Room 1111, Scape Wembley, Fulton Rd, Wembley Park, Wembley, London, HA90TF
M7 CQM J. Law, 58 Westfields, Zeals, BA126PW
M7 CQS S. Peters, 7 Greenfield Avenue, Chatburn, BB7 4AJ
M7 CQU V. Rotaru, 15 Leicester Drive, Glossop, SK13 8SH
M7 CQX N. Horton, Parkside Farm, Shripney Lane, Bognor Regis, PO22 9NU
M7 CQY M. Blencowe, 11 Gendalls Way, Launceston, PL15 8SE
MM7 CQZ P. Addison, 29 Clachnaharry Road, Inverness, IV3 8RA
M7 CRA B. Tilley, 46 Longway Avenue, Charlton Kings, Cheltenham, GL53 9JJ
MM7 CRB C. Balmain, 331 Dunecht Court, Glenrothes, KY7 6UQ
M7 CRC C. Clark, 48 Church Drove, Wisbbech, PE148RH
M7 CRD C. Davis, 47 Mendip Road, Weston-Super-Mare, BS23 3HB
M7 CRF Dr C. Fletcher, 7 Highfield Crescent, Baildon, Shipley, BD17 5NR
M7 CRG C. Fryer, 35 Palmer Avenue Abbeymead, Gloucester, GL4 5BH
M7 CRK C. Hughes, 120 Treningle View, Bodmin, PL311PD
M7 CRM C. Marsh, 75 Priorswood Road, Taunton, TA2 7PT
MW7 CRP D. Crisp, 5 Gyrnosfa, Lower Cwmtwrch, Swansea, SA9 1DR
M7 CRQ L. Adams, 38 St. Albans Road, Kingston upon Thames, KT2 5HQ
M7 CRR M. Ketteringham, 4 Westgate Street, Downham Market, PE38 0PA
M7 CRS C. Oyitch, 69 Melrose Road, Gainsborough, DN21 2SA
M7 CRU L. Oconnor, 7 Wentworth Way, St Leonards-on-Sea, TN38 0XG
M7 CRV Z. Mohammad, 35 Fairlands Ave, London, CR7 6HD
M7 CRW C. Smith, 6 Sawley Avenue Lowton, Warrington, wa32ew
MM7 CRY M. Napier-Holford, 26 Craig Street Rosyth., Dunfermline, KY112ng
M7 CRZ C. Sharples, 35 Ennerdale Drive, Barrow-in-Furness, LA14 4PN
M7 CSA C. Afonso, 4 Adams Forge, Littleport, Ely, CB6 1FA
MM7 CSC G. Porter, Brae, Port William, Newton Stewart, DG8 9RT
M7 CSF D. Hensman, 31 Farmer Ward Road, Kenilworth, cv8 2dj
M7 CSI P. Johnson, 135 Addison Road, Bilton, Rugby, CV22 7HB
M7 CSK C. Kent, 20 Cherrytree Grove, Spalding, PE11 2NA
M7 CSL C. Lowe, Keynsham Water Front, Stidham Lane, Keynsham, Bristol, BS31 1GB
MI7 CSM Y. Kyle, 2 Wattstown Crescent, Coleraine, BT52 1SP
M7 CSN L. Beynon, Flat 3, 2 Gloucester Mews, Weymouth, DT4 7DA
MW7 CSO J. Healey, 1 Park Crescent, Brynmawr, Ebbw Vale, NP23 4HR
M7 CSP C. Pyne, Blenheim Road, Bridgwater, TA6 4HE
M7 CSR C. Reeves, 36 Oakfield Avenue, Wrenbury, Nantwich, CW5 8ER
M7 CSU R. Morley, 7 Eliot Road, St Austell, PL25 4NL
M7 CSW P. Williams, 14 Springfield Road, Flat 3, Ilfracombe, EX34 9JW
M7 CSX R. Whiting, Swanwick, Church Green, Churchfields, Wellington, TA21 8SF
MW7 CTA A. Smedley, 27 West Place, Gobowen, Oswestry, SY11 3NR
M7 CTB C. Thorne, 67 Devon Road, Cadishead, Manchester, M44 5HB
M7 CTF Dr R. Kisiel, 98 Iolanthe Drive, Exeter, EX4 9EA
MM7 CTG T. Gallagher, 18 Argyle Drive, Hamilton, ML3 9EB
M7 CTI S. Stokes, 2C The Street, Holywell Row, Bury St Edmunds, IP28 8LS
MI7 CTL G. Goodwin, 29 Glenfield Road, Lurgan, Craigavon, BT66 8ER
M7 CTM C. Hoult, 11 Queen Street, Alnwick, NE66 1RD
M7 CTN C. Cantral, 38 Langtree Avenue, Old Whittington, Chesterfield, S41 9HP
M7 CTP C. Parkyn, 3 Walnut Road, Honiton, EX14 2UG
M7 CTS T. Jones, 7 Sedum Close, Huntington, Chester, CH3 6BL
M7 CTT M. Steeples, 44 Trunch Road, Mundesley, Norwich, NR11 8JX
M7 CTV C. Thompson, 75A Park Lane, Bonehill, Tamworth, B78 3HZ
M7 CTY P. Murton, 7 Stony Close, Long Meadow, Worcester, WR4 0JY
M7 CUA J. Woollett, 3 Wishingtree Close, St Leonards-on-Sea, TN38 9JG
M7 CUD C. Dewhirst, 40 Mountain Road Thornhill, Dewsbury, wf120bw
M7 CUG D. Parsons, 12 Heraldry Way, Exeter, EX2 7RA
MM7 CUH C. Mitchell, 25 Shoulderigg Road, Coalburn, ML110EL
MW7 CUI T. Tansley, 7 Edward Street, Oswestry, SY112BL
M7 CUJ P. Kane, The Snuggle Up, Sandhutton, YO7 4RW
M7 CUN A. Newman, 25 Pensford Court, Craydon Road, Bristol, BS14 8EQ
MW7 CUO C. Evans, 16 Oak Road, Llanharry, Pontyclun, CF72 9HT
M7 CUP C. Hatton, 54 St. Josephs Avenue, Birmingham, B31 2XQ
M7 CUQ P. Budgen, 9 Hornfair Road, London, se77be
M7 CUR R. Jones, St. Fillans, The Warren, East Horsley, Leatherhead, KT24 5RH
M7 CUS A. Otton, 21 Marshall Avenue, Bognor Regis, PO21 2TJ
M7 CUU C. Lloyd, 53 Gleneagles Road, Birmingham, B26 2HT
M7 CUW P. Beckett, 19 Greave Clough Close, Bacup, OL13 9HS
M7 CUZ A. Sequeira, 15 Bernstein Court, Garibaldi Street, London, SE18 1DP
M7 CVC C. Chisholm, Hill Top Kellah, Haltwhistle, NE49 0JL
MI7 CVD D. Harrison, Carnmoney Road, Belfast, BT366HT
M7 CVH A. Pickering, 4 Simpson Close, Chaple St Leonards, PE245JU
M7 CVJ A. Thornton, 4 Riley Lane, Kirkburton, Huddersfield, HD8 0RX
M7 CVM D. Wesley, 14 Boylan Road, Coalville, LE67 3JG
M7 CVQ I. Giles, 4 Dainton Mews Fisher Street, Paignton, TQ4 5UA
MW7 CVT C. Thorley, Helston, The Mountain, Holyhead, LL65 1YR
M7 CVW R. Nesbit, 7 Teal Court Juniper Close, St Leonards-on -Sea, TN38 9RW
M7 CVX J. Miller, 62 Sheffield Road, Penistone, Sheffield, S36 6HE
M7 CVZ R. Perry, 24 Conrad Road, Stanford le Hope, SS170AT
M7 CWA C. Walson, 30 West Crescent, Duckmanton, Chesterfield, S44 5HE
M7 CWB C. Bell, 56 Derby Drive Moorside, Durham, DH8 8DX
M7 CWD C. Dabrowski, Overton, Salisbury Terrace, Mytchett, Camberley, GU16 6DB
M7 CWK A. Wade, 6 Elliott Grove, Brixham, TQ5 8RT
M7 CWL K. Horigan, 21 Churchtown, Gwinear, Hayle, TR27 5JL
M7 CWM C. Mercer, 31 Downside Avenue, Plymouth, PL6 5SD
M7 CWN C. Woodall, 73 Ridyard Street, Wigan, WN5 9QD
M7 CWS I. Handley, 110 Longslow Road, Market Drayton, TF9 3BW
M7 CWT M. Crighton, Flat 5 Sheridan Court, Rochester, ME1 3BL
M7 CWX C. Wright, 151 Atkinson Road, Sunderland, SR6 9AY
M7 CXB C. Beachell, 19 Oldfield Avenue, Stannington, Sheffield, S6 6DQ
M7 CXD J. Tunstall, 46 The Hawthorns, West Kyo, Stanley, DH9 8TX
M7 CXE W. Softley, 15 Hare Law Gardens, Stanley, DH9 8DE
M7 CXF S. Higginson, 14 Hans Price Close, Weston-Super-Mare, BS23 1NG
M7 CXJ D. Garside, 14 Roche Avenue, York, YO31 9BB
M7 CXL P. Needham, 29 Sandhill Close, Bedford, MK45 2JD
M7 CXP A. Jackson, 21 Hibson Avenue Norden, Rochdale, OL12 7RU
M7 CXQ D. Cumming, 47 The Thicket, Fareham, PO16 8QA
MM7 CXR M. Irons, 70 Fairlie Street, Camelon, Falkirk, FK1 4NL
M7 CXT G. Boxer, 31 Barnfield Road Flat 2, Orpington, BR5 3LP
M7 CXU R. Gilbert, 51 Maudslay Road, London, SE9 1LH
M7 CXV R. Jaya Chandran, 14 Egerton Road, Reading, RG2 8HQ
M7 CXW E. Mcgee, 44 May Tree Road, Radstock, BA33TU
M7 CXY S. Stray, 9 Highfields Mews, Great Gonerby, Grantham, NG31 8XA
M7 CXZ J. Turner, 2 Dukewood Road, Clayton West, Huddersfield, HD8 9HF
M7 CYA C. Yates, 7 Plessey Terrace, Newcastle upon Tyne, NE7 7DJ
M7 CYC M. Harrison, Oldcastle Avenue, Newcastle under Lyme, ST5 8HF
M7 CYE A. Done, 3 Home Farm Cottages, Old Warden Park, Old Warden, Biggleswade, SG18 9DU
M7 CYG P. Leftwich, 43 Norman Drive, Old Sarum, Salisbury, SP4 6FP
M7 CYH I. Sweet, 70 Moor Park Drive, Ilkley, LS290PT
M7 CYI R. Hawley, Anni Healey Close, Woodbridge, IP12 1GZ
MI7 CYL L. Miskimmin, 47 Shore Road, Kircubbin, Newtownards, BT22 2RP
MW7 CYN C. Vugts, Coed Coch, Llangammarch Wells, LD4 4BS
M7 CYQ T. Considine, Signal House Jacklyns Lane, Alresford, SO24 9JJ
M7 CYR C. Dagga, 12 Columbine Road, Rochester, ME2 2XZ
M7 CYS S. Cole, 8 Wayfarer Close, Warsash, Southampton, SO31 9AU
M7 CYU A. Todd, 16 Watling Close, Lincoln, LN42BD
M7 CYV D. Bradbury, 26 Sandbanks Way, Hailsham, BN27 3LL
MI7 CYX T. Wojciechowski, Flat 7, Easton House, 69-71 Cliftonville Road, Belfast, BT14 6JP
M7 CYY L. Hawkins, 11 Coronation Place, St. Budeaux, Plymouth, PL5 1UP
M7 CZA R. Myrcha, 100 Third Street, Horden, Peterlee, SR8 4EH
M7 CZB R. Hyland, 35 Brodrick Road, Eastbourne, BN22 9NR
MM7 CZD M. Rostant, 5 Baird Crescent, Alexandria, G83 0TX
M7 CZF Y. Chadun, 77 John Archer Way, London, SW18 2TS
M7 CZM Dr D. Shephard, 2 Charleville Mews, Isleworth, TW7 7BW
M7 CZQ C. Ballard, 3 Pope Court The Galleries Warley, Brentwood, CM14 5FR
M7 CZR S. Cartwright, Bs30 5Qu, Bristol, BS 30 5QU
M7 CZU A. Durucz, 7 Quayside, St. Marks Square, Lincoln, LN5 7EX
M7 CZW E. Clothier, 15 Trafalgar Road, Newport, PO30 1QD
M7 CZZ T. Brown, Bridge House, Redenhall Road, Harleston, IP20 9QN
M7 DAB D. Beard, 1 Bond Close, Leonard Stanley, Stonehouse, GL10 3GQ
M7 DAC A. Capitan, 41 Cunningham Drive, Runcorn, WA7 4DL
M7 DAD M. Bland, 10 Pennycross Square, Sunderland, SR40HR
M7 DAE D. Elley, 34 Nash Way, Coleford, GL16 8RQ
M7 DAF R. Jones, 34 Lytchett Way, Poole, BH16 5LS
MM7 DAH D. Hancox, 52 High Main Street, Dalmellington, Ayr, KA6 7QN
M7 DAJ D. Jarman, 62 Combe Drive, Dunstable, LU6 2AE
M7 DAK N. Goult, 5 Colbourne Close, Bransgore, Christchurch, BH23 8BW
MM7 DAM D. Mcintosh, Crunes Way, Greenock, PA15 2WH
M7 DAN D. Savage, Rufford, Barnes Lane Milford On Sea, Lymington, SO41 0RR
M7 DAP D. Pope, 176 Innsworth Lane, Gloucester, GL3 1DX
M7 DAR D. Rockall, 24 Alvington Road, Newport, PO30 5AR
M7 DAY D. Allen, 19 Brooklands Close, Uttoxeter, ST148UH
M7 DAZ D. Walton, 106 Cumberworth Lane, Lower Cumberworth, Huddersfield, HD8 8PG
M7 DBB C. Poole, 128 Ansley Common, Nuneaton, CV10 0QA
M7 DBD N. Dyer, Five Rochester Close, Chichester, PO195DS
MW7 DBF M. Bugby, Derwen Deg, Fron Goch, Bala, LL23 7NT
M7 DBG A. Lydiate, 89 Berrylands Road, Wirral, CH46 7TY
M7 DBH B. Taylor, 43 Melbourne Way, Bush Hill Park, Enfield, EN1 1XG

M7 DBJ A. Millington, 8 Bede Road, Nuneaton, CV10 8HP
M7 DBL O. Cable, 116A London Road, Hadleigh, SS7 2PG
M7 DBM D. Edwards, 20 Keelson Pointe, Carl Street, Walsall, WS2 7DB
M7 DBN N. Dinsdale, 3 Pond Cottages, Faulkland, Radstock, BA3 5XB
M7 DBP D. Pickles, 27 Keelham Lane, Keighley, BD20 6DE
MI7 DBR D. Bennett, 15 Elgin Street, Belfast, BT7 3AG
M7 DBX S. Ranger, 137 Warren Avenue, Southampton, SO16 6AF
M7 DBZ M. Webb, 10 Sausthorpe Street, Lincoln, LN5 7XW
M7 DCA D. Acton, 31 Harrogate Terrace, Murton, SR7 9PQ
MM7 DCD D. Clark, 9D West Court Clydebank, Glasgow, G81 4PG
M7 DCK D. Sutton, 18 Knights Hill Severn Stoke, Worcester, wr8 9jd
M7 DCL P. Willingale, 10 Therfield Walk Houghton Regis, Dunstable, LU55QB
M7 DCM D. Clarke, 34 Kimberlee Avenue, Cookley, Kidderminster, DY10 3TN
M7 DCN D. Newman, Common Mead Lane, Gillingham, SP8 4RE
M7 DCP D. Potter, 33 Wellington Grove, Doncaster, DN5 9RN
M7 DCR D. Ross, Flat A, 174 Estcourt Road, London, SW6 7HD
M7 DCT D. Thomas, 8 Cedar Avenue, Weston-Super-Mare, BS22 8HL
M7 DCV M. Dutton, 10 Braemore Close Shaw Ol2 7Na, Oldham, OL2 7NA
M7 DCX C. Dunn, 29 Concord Street, Leeds, LS2 7QB
M7 DDA D. Daniel, 38 Netherthorpe Lane, Killamarsh, Sheffield, S21 1DA
M7 DDC R. Williams, 4 Bluebell Close, Barlborough, Barlborough, S43 4WT
M7 DDE S. Kirby, 4 Tittesworth Avenue, Leek, st13 6ps
M7 DDF J. Sivyour, 16, Highlands Way., Salisbury, SP52SZ
M7 DDK P. Cowling, 11 Palmerston Road, Shanklin, PO376AU
M7 DDM P. Hennings, 12 Pendleway, Pendlebury, Swinton, Manchester, M27 8QR
M7 DDN D. Reed, Ivy Cottage, Foundry Yard, Ridsdale, Hexham, NE48 2TG
MW7 DDP R. Mitchell, The Hayloft, Plas Devon Court, Rossett Road, Commonwood, Holt, Wrexham, LL13 9SY
M7 DDQ S. Brunsden, 43 Porthcawl Drive, Washington, NE37 2LT
M7 DDR D. Rushton, 4 Avocet Drive Irlam, Salford, M44 6PJ
M7 DDT A. Dryburgh, 201 Watling Street, Grendon, Atherstone, CV9 2PJ
MM7 DDV J. Mcintosh, 1 Marchburn Court Northfield, Aberdeen, AB16 7PQ
M7 DDW D. Wilde, 12 Windhill Road Walker, Newcastle upon Tyne, NE6 3TQ
MM7 DDX D. Cunningham, 3 Dallerie, Crieff, PH7 4JH
M7 DDY C. Westlotorn, 84 Alverstone Road, Whippingham, East Cowes, PO32 6NX
M7 DEA D. Sanders, 37 Maple Avenue, Little Sutton, Ellesmere Port, CH66 3QU
M7 DEE D. Edwards, 2 Dingwall Road, London, SW18 3AZ
M7 DEI D. Taylor, 18 Moorside Place, Dewsbury, wf136qj
M7 DEJ P. Goodwin, 60 Redburn Drive, Shipley, BD18 3AZ
M7 DEK D. Howard, 31 White Mullein Drive, Redlodge, ip288xp
M7 DEL M. Sprague, 11 The Orchards, Witcham, Ely, CB6 2LR
M7 DEP D. Damodar, 11, Tudor Court, Church Lane, Rickmansworth, WD3 8PX
M7 DER D. Jacobson, 17 Robinia Walk Whitchurch, Bristol, bs14 0sh
M7 DET D. Tipping, Laurel Close, Mountsorrel, LE12 7JN
M7 DEW D. Wade, 2 Springwood Square, Back Spring Street, Huddersfield, HD1 4AJ
M7 DEX D. Flood, 506 Preston Old Road, Blackburn, BB2 5LY
M7 DEY S. Kitchener, 14A Lower Road, Malvern, WR14 4BX
MM7 DFC D. Clark, 106 Braes Avenue, Clydebank, G81 1DP
M7 DFD D. Ford, 9 The Crofts, Hatch Warren, Basingstoke, RG22 4RE
MW7 DFE K. Brown, 1 Heol Arthur Fear, Abertillery, NP13 3JQ
M7 DFF P. Huber, 15 Chelford Road, Bromley, BR1 5QT
M7 DFI P. Robins, 20 Saffron Close, Chineham, Basingstoke, RG24 8XQ
M7 DFK C. Latham, Hunter Road, Wigan, Wn5 0qd
M7 DFM D. Mccarron, 22 Peel Close, Woodley, Reading, RG5 4SR
M7 DFO J. Ball, 11Lydieth Lea, Liverpool, L278YL
M7 DFP M. Cavill, 4 Corner Farm, Luke Lane, Brailsford, Ashbourne, DE6 3BQ
MW7 DFX D. Fowler, 5 Cotswold Way, Risca, Newport, NP11 6QT
M7 DFY D. Ford, 17 Adelaide Crescent, Burton-on-Trent, DE15 0PA
M7 DGB U. Decruz, Victoria Road, Lowestoft, NR33 9LR
MM7 DGC D. Chappelle, 10 Whitefield Terrace, Lennoxtown, Glasgow, G66 7JT
M7 DGE L. Dauphin, 22 Lincoln Place, Wigan, WN5 0RA
M7 DGF D. Bailey, 1 The Magpies, Maulden, MK45 2EG
M7 DGG K. Mann, 29, Wisteria Avenue, Branston Branston, Lincoln, LN4 1QE
M7 DGH D. Henderson, 75 Laburnum Road, Waterlooville, PO7 7EW
MM7 DGK M. Greig, 22 Rowanhill Close, Port Seton, Prestonpans, EH32 0SY
MI7 DGL D. Gordon, 48 Dunsy Way, Comber, Newtownards, BT23 5DF
M7 DGN D. Rose, 3 Iona Way, Urmston, Manchester, M41 7EY
M7 DGP D. Potter, 1 Upper Park Road, Colchester, CO7 0JP
MM7 DGS T. Sneddon, Fourwinds, Northfield, Denny, FK66RB
M7 DGU S. Finn, 12 Longfield Place, Maidstone, ME15 9AJ
M7 DGW D. Warriner, 1 St. Johns Avenue, North Hykeham, Lincoln, LN6 8QR
M7 DGZ D. Godfrey, 7 Austin Close, Dudley, DY1 2ST
M7 DHA D. Cook, 20 The Chine, London, N10 3PY
MM7 DHB M. Kinnon, 45D Blackthorn Place, Blairgowrie, PH10 6FH
MM7 DHC D. Canavan, 9 Ormelie Terrace, Edinburgh, EH15 2EX
M7 DHD R. Dodge, 4 Haunts Cottages Middle Chinnock, Crewkerne, TA18 7PW
MM7 DHE N. Milne, 83 Dalfarson Avenue, Dalmellington, Ayr, KA6 7TX
M7 DHF R. Wildman, 2 St. Leonards Way, Barnsley, S71 5BS
MW7 DHG R. Hood, Coed Bach, Morgan Street, Abercrave, Swansea, SA9 1TS
M7 DHL D. Watson, 30 Campbell Street, Tow Law, Bishop Auckland, DL13 4DX
M7 DHM R. Freeman, 14 Reeves Road, Devizes, SN10 3UB
M7 DHP C. Taylor, Innismore, Old Road, Liskeard, PL14 6DL
M7 DHT R. Groves, 34 Clover Way, Paddock Wood, TN12 6BQ
M7 DHZ S. Harriss, 30 Chatsworth Place, Harrogate, HG1 5HR
M7 DIF C. Walters, 10 Grampian Crescent, Chesterfield, S40 4QB
MJ7 DIJ D. Mashev, Flat 23, 1875 Wesley Street, St. Helier, Jersey, JE2 4DA
M7 DIL P. Bartlett, 58 Ashdown Road, Chandler'S Ford, Eastleigh, SO53 5QJ
M7 DIM G. Jackson, 14 Southdown Crescent, Cheadle Hulme, Cheadle, SK8 6EQ
M7 DIO K. Grzywaczewski, Flat 2, Abbot House, Smythe Street, London, E14 0HD
M7 DIP J. Grainger, 42 Jackson Road, South Cerney, Cirencester, GL7 6JB
M7 DIQ R. Duffy, 27A Mortimer Close, Totton, Southampton, SO40 2QE
M7 DIS Prof. R. Houlston, 51 Adelaide Road, Surbiton, KT6 4SR
M7 DIU B. Sancto, 133 Hawthorne Avenue, Gillingham, ME8 6YE
MM7 DIV D. Frazer, 32 Loudoun Avenue, Kilmarnock, KA1 3RZ
M7 DIW W. Lockley-Gardiner, 193 Westbourne, Telford, TF7 5QP
MW7 DIZ C. Cooper, 26 Dunlin Court, Barry, CF63 4JY
M7 DJA A. Deakin, 69 Hammonds Ridge, Burgess Hill, RH15 9QW
M7 DJB D. Bowles, 3 Iris Road, Southampton, SO16 3GU
M7 DJC D. Cleak, 8 St Davids Road North, Lytham St Annes, FY8 2BL
M7 DJD D. De-Steunder, 67 Mow Lane, Gillow Heath, Stoke-on-Trent, ST8 6QB
M7 DJG D. Gorst, 8 Devonshire Road, Padgate, Warrington, WA1 3JS
M7 DJK D. Kearley, 101 Ashdale, Bishop's Stortford, CM23 4EB
M7 DJL C. Simon, 55 Shirecroft Road, Weymouth, DT4 0NH
M7 DJM D. Moseley, 50 Chestnut Road, Walsall, WS3 1BE
M7 DJN D. Norry, 58 Wealdstone, Telford, TF7 5PT
M7 DJP D. Peagram, 1 Larke Rise, Southend-on-Sea, SS2 6GQ
MW7 DJQ D. Cooper, 1 Bedw Street, Caerau, Maesteg, Cardiff, cf340tf
M7 DJS D. Spear, 72 Hunters Road, Leyland, PR25 5TT
MI7 DJT E. Morgan, 12A Garden Street, Magherafelt, BT45 5DD
M7 DJW D. Wyld, 33 Pinewood Drive, Accrington, BB5 6UG
M7 DJY M. Cowsill, 23 Laithes Croft, Earlsheaton, Dewsbury, WF12 8BN
M7 DKA D. Allen, 28 Fourth Avenue, Wickford, SS11 8RQ
MI7 DKB J. Shaw, 14 Milewater Terrace, Onewtownabbey, BT36 5UY
M7 DKC G. Bell, 1 Bailey Close, Haverhill, CB9 0LH
M7 DKE D. Kane, 6 Bentry Road, Dagenham, RM8 3PA
MM7 DKH D. Hancock, Ellensgeo, Ulbster, Lybster, KW2 6AA
MU7 DKI J. Donaldson, 62 Le Banquage Rue De Beaumont, Alderney, Guernsey, GY9 3YP
M7 DKJ J. Bailey, The Green, Lower Brailes, Banbury, OX15 5HZ
M7 DKK D. Kirk, 57 Sandringham Avenue, London, SW20 8JY
M7 DKL T. Lee, 6 Kerry Close, Upminster, RM14 1JD
M7 DKM J. Lyth, 170 Robin Hood Road, Coventry, cv33au
M7 DKU P. Piper, 53 Garfield Avenue, Bournemouth, BH1 4QT
MM7 DKX M. Jakob, 11 Dunnet Place, Thurso, KW14 8JE
MM7 DKY S. Drummond, 45C Dunbeth Avenue, Coatbridge, ML5 3JD
M7 DLB D. Barton, 30 Mayfield Close, Bognor Regis, PO213PL
MI7 DLD D. Donley, 64 Glenbank Place, Belfast, BT14 8AN
M7 DLJ S. Harvey, 110 Vicarage Road, Wednesfield, Wolverhampton, WV11 1SF
M7 DLK D. Swann, Flat 4, 70 Hamstead Road, Handsworth, B191DG
M7 DLO J. Miles, 14 Westville Road, Bexhill on Sea, TN393QB
M7 DLQ R. Clark, 7 The Square, Chipping, Buntingford, SG9 0PJ
M7 DLR B. Bleier, Whipley Manor Cottage, Palmers Cross, Bramley, Guildford, GU5 0LL
M7 DLS D. Salt, Christmas Cottage, Ram Hill, Coalpit Heath, Bristol, BS36 2TX
M7 DLV A. Yakimov, 4 Erin Close, Bromley, BR1 4NX
M7 DLX L. Austin, 2 List Meadows, Littlebourne, Canterbury, CT3 1XW
M7 DLZ S. Cockerton, Hall Lodge, Wood Dalling Road, Wood Dalling, Norwich, NR11 6SG
M7 DMA D. Andrijauskas, 7 Hawthorn Grove, Enfield, EN2 0DU
M7 DMB D. Buschbaum, 42 Cordery Road, Leicester, LE56DE
M7 DMF A. Fielding, 4 Poot Hall, Rochdale, OL120AS
M7 DMH R. Hennefer, 2 Stirling Avenue, Ince, Wigan, WN2 2JG
M7 DMI M. Tyson, 93 Wroxall Drive Grantham, Grantham, Ng317eg
M7 DMJ D. Jones, 53 Bradley Road, Stourbridge, DY8 1UX
M7 DMK J. Davidson, 78 Old Heath, Shrewsbury, SY1 4SE
M7 DMP S. Buckley, 70 Rockbourne Road Manor Farm, Bristol, BS105AS
M7 DMS D. Steel, 36 Silver Street, South Petherton, TA13 5AL
M7 DMV M. Chequer, 14 Pasture Lane, Lazenby, Middlesbrough, TS6 8EG
MM7 DMW D. White, 9 Tweedsyde Park, Kelso, TD5 7RF
M7 DMY D. Westwood, 28 Weybridge Mead, Yateley, GU46 7UY
MM7 DNE K. Thomson, Inverluther, Laurencekirk, AB30 1QT
MI7 DNG N. Foster, 48 Beatrice Villas Bellaghy, Magherafelt, BT45 8JA
M7 DNH D. Horne, Alias, Lowthorpe, Southrey, Lincoln, LN3 5TD
M7 DNM D. Simpson, 12 Guildford Avenue, Midway, Swadlincote, DE11 7LN
M7 DNN C. Norris, Heather View, Forest Front, Hythe, Southampton, SO45 3RJ
M7 DNO N. Wright, 61 High Street, Arlesey, SG15 6SW
MW7 DNQ T. Allen, 53 Pendre Close, Brecon, LD3 9EH
M7 DNW J. Armour, Room 8, Beechwood Grove Care Home 42-44 East Dean Road, Eastbourne, BN0 8EH
M7 DOD R. Dodd, 82 Dickens Road, Coppull, Nr Chorley, Preston, PR7 5BH
M7 DOG J. Warman, 6 Down View Road, Denbury, Newton Abbot, TQ12 6ER
M7 DOK D. Kelly, 23A Baldwyns Road, Bexley, DA5 2AB
M7 DON D. Buck, 9 Ashwood Close, Mansfield Woodhouse, NG19 9HD
M7 DOS S. Redmond, 12 Hemlock Way, Manchester, M9 7GR
M7 DOX A. Pinder, 13 Hatfield Road, Bradford, BD2 4QX
M7 DOZ S. Edwards, 87 Garden Village, Micklefield, L5254AD
M7 DPB D. Bristoll, 138 Chatham Road, Northfield, Birmingham, B31 2PL
M7 DPC D. Cardoso, 11 Coppice Way, Aylesbury, HP20 1XG
M7 DPD D. Street, 18 Meadow Gardens, Bath, BA1 3RY
M7 DPE P. Ashton, 5 New Barns Avenue, Mitcham, CR4 1LG
MW7 DPG T. Pigott, 16A East Road, Tylorstown, Tylorstown, CF43 3HF
M7 DPH D. Hozman, 12 Bober Court, Holt Drive, Colchester, CO2 0DR
MM7 DPL D. Lambert, 76 Martin Crescent, Ballingry, Lochgelly, KY5 8QA
MW7 DPO R. Davis, 19 Ael-Y-Bryn, Caerphilly, CF83 2QX
MM7 DPP S. Prime, Torcroft Lodges, Balnain, Drumnadrochit, IV63 6TJ
M7 DPT M. Perescu, 8 Marauder Road, Norwich, NR66HD
M7 DPU M. Birkett, 5 King Oswald Drive, Blaydon-on-Tyne, NE21 4FD
M7 DPW D. Watmough, 2 Alston Road, Wigan, WN21AU
M7 DPX D. Palacionis, 7 Swallow Road, Crawley, RH11 7RF
M7 DPY D. Skinner, Albany House, 5 Butler Road, Shrewsbury, SY3 7AJ
M7 DPZ K. Harte, 44 Mowrick Road, North Shields, Ne29 8jb
M7 DQC S. Tait, 29 Hillside View Sherburn Village, Durham, dh61dz
M7 DQD S. Nelson-Smith, Welbeck, Wickham Road, Fareham, PO17 5BU
M7 DQF M. Johnson, 2 Himley Close, Willenhall, WV12 4LX
M7 DQG M. Mccarthy, 1 Scargells Yard, High Street, March, PE15 9la
M7 DQI C. Male, 4 Orton Lane, Twycross, Atherstone, CV9 3HA
M7 DQJ M. Hine, 5 Tamar Close, Spalding, PE11 3GZ
M7 DQO D. Coward, R F Welding Ltd, Jekils Bank, Holbeach St. Johns, Holbeach, Spalding, PE12 8RQ
M7 DQQ C. Horne, 54 Bridgnorth Road, Wollaston, Stourbridge, DY8 3QG
M7 DQU D. Sharp, 197 Preston Down Road, Preston, Paignton, TQ3 1DL
MW7 DQV D. Rhodes, 110 Swansea Road, Trebanos, Pontardawe, Swansea, SA8 4BN
MW7 DQW N. Sparkes, Oxford Lodge, 15 Leg Street, Oswestry, SY11 2NL
M7 DQY S. Bone, 20 Keld Avenue, Uckfield, TN22 5BN
M7 DRB Dr B. Deane, 2 Seymour Road, Preston, PR2 2EU
M7 DRC D. Chalk, Garretts Close, Southampton, so19 9rw

M7 DRE B. Owen, 39 Bernard Avenue Appleton, Warrington, WA4 3BA
MM7 DRG D. Gray, 23 South End, Stromness, KW16 3DJ
M7 DRH Dr J. Horlock, 7 Laurel Drive, Prestbury, Cheltenham, GL52 3DE
M7 DRJ D. Jones, 20 Nightingale Road, Guisborough, TS14 8HA
M7 DRL D. Lamb, Unit 7, The Craft Workshops, Hutton-Le-Hole, York, YO62 6UA
M7 DRR D. Robson, 8 The Vale, Stockton-on-Tees, TS19 0XL
M7 DRS Dr M. Sinclair, 40 Grotto Road, South Shields, NE34 7AH
MW7 DRT R. Thomas, 46 Home Farm Way, Penllergaer, Swansea, SA4 9HF
MM7 DRV S. Paton, 31 Woodlands Place, Inverbervie, DD100SL
M7 DRW A. Harcourt, 12 Woodside Close, Finchampstead, Wokingham, RG40 4EY
M7 DRY D. Young, 1 Bugle Place, Newton Abbot, TQ12 1GZ
MM7 DSA D. Sewell, 51Main Street, Swinton, Duns, Berwickshire Main Street, Swinton, Duns, TD11 3JJ
M7 DSH D. Shaw, 81 Chesterfield Road Tibshelf, Alfreton, DE55 5NJ
M7 DSL D. Livingstone, 4 Hatfield Road, Southport, PR8 2PE
MW7 DSP D. Parfitt, 16, Gladstone Street, Blaina, NP13 3HJ
M7 DSQ D. Chalton, 36 Coronation Road, Wingate, TS28 5JN
M7 DSR D. Rudgley, The New House, Plymouth Road, Buckfastleigh, TQ11 0DB
MM7 DSV G. Crawford, 4 Windram Road Chirnside, Duns, TD11 3UT
M7 DSZ C. Terry, 65 Intrepoid Close, Hartlepool, TS25 1GF
M7 DTA T. Cocks, 87 Wyatt Road, Sutton Coldfield, B75 7NH
M7 DTD D. Diiorio, Aspley Close, Chesterfield, S404HG
M7 DTF D. Fincham, 2 Glebe Close, St Columb Major, TR9 6TA
M7 DTG D. Gryta, 5 Sefton House, Tenbury Fold, Bradford, BD4 0BD
MM7 DTL E. Blakeway, 25 Allanton Grove, Wishaw, ML2 7LL
M7 DTR B. Million, 53 Chestnut Avenue Rossington, Doncaster, DN11 0DF
M7 DTW D. Twaites, 23 Church Street, Hadfield, SK13 2AD
M7 DTZ P. Coulson, 52 Milbank Terrace, Station Town, Ts285ef
M7 DUB D. Wood, 3 Ripley Close, Wakefield, WF3 2FG
M7 DUC D. Garland, 18 Sandgate Road, Tipton, DY4 0SX
M7 DUD I. Atanasov, 12 Seymour Drive, Eaglescliffe, Stockton-on-Tees, TS16 0LQ
MM7 DUF A. Duff, 170 Saint Leonard Street, Lanark, ml11 7du
M7 DUH J. Edwards, 23 Glossop Brook View, Glossop, SK13 8BF
MW7 DUI M. Bennewith, 1 Leisure Centre House, Chester Road West, Queensferry, Deeside, CH5 1SA
M7 DUJ D. Mcclelland, 4 Rectory Drive, Coppull, Chorley, PR7 4QE
M7 DUM R. Walker, 36, Lindrick Rd, New Marske, TS11 8HT
M7 DUP S. Miller, 39 Busticle Lane Sompting, Lancing, BN15 0DJ
M7 DUR P. Stagg, 70 Turin Court Roman Way Estate, Andover, SP10 5LD
M7 DUT P. Ajayi, 149 Albert Road, London, N22 7AQ
M7 DUU A. Munson, The Priory, Flowton, Ipswich, IP8 4LH
M7 DVB J. Sandon, 461 Archer Road, Pin Green, Stevenage, SG1 5QP
M7 DVD D. Van Dijk, 76 High Street, Tetsworth, Thame, OX9 7AE
MM7 DVE D. Thompson, 19 Tulloch Terrace, Perth, PH1 2PF
M7 DVJ P. Calvert, 9 Stanley Road, Keighley, BD22 7DE
M7 DVK D. Kirk, 26 Ingham Grove, Cramlington, NE23 3LH
MM7 DVM B. Webb, 14A Greendykes Road, Dundee, DD4 7NA
M7 DVO M. Welsh, Ch41 8Jf, Merserside, Ch41 8jf
M7 DVP D. Ward, 6 Hillhead Bungalows Hillhead, Colyton, EX24 6NN
M7 DVQ A. Vickers, 16 Birdham Road, Brighton, BN2 4RF
M7 DVS C. Lord, 25 Thurso Close, Romford, RM3 0YR
M7 DVT J. Harrison, 20 Kelcliffe Avenue, Guiseley, Leeds, LS20 9EW
M7 DVX R. Evmez, Luscot, North Allington, Bridport, DT6 5EQ
M7 DVY R. Werner, 7 Summer Road, Kidderminster, DY11 7JR
M7 DWD A. Storr, 4 Shipley Road, Southwater, Horsham, RH13 9BD
M7 DWE D. Waller, 4 Cymens Ora, Keynor Lane, Sidlesham, Chichester, PO20 7NL
M7 DWF D. Fear, 38 Axbridge Road, Bristol, BS4 2RX
MM7 DWK D. Keay, Parkhill, Cromwell Park, Almondbank, Perth, PH1 3LW
M7 DWP D. Page, Tamar Way, Slough, SL3 8SY
MM7 DWQ W. Henderson, 106 Gilberstoun, Edinburgh, EH15 2QZ
M7 DWT A. Dykes, 16 Paterson Drive, Stafford, ST16 1WH
MM7 DWV L. Davidson, 106 Fintry Drive, Dundee, DD4 9HH
M7 DWW D. Woodall, 9 Calmington Lane, Sandymoor, Runcorn, WA7 1QE
M7 DX Newton Le Willows ARC, c/o K. Horsfield, 59 Queens Drive, Newton-le-Willows, WA12 0LY
MM7 DXC A. Mcintyre, 14 Wavell Street, Grangemouth, FK3 8TG
M7 DXF M. Broad, 7 Long Ley, Harlow, CM20 3NH
M7 DXH N. Mason, 8 Barrowby Gate, Grantham, NG31 7LT
M7 DXJ K. Coleman, Nepturn Cottage, School Lane, North Newington, Banbury, OX15 6AQ
MI7 DXL D. Lyons, 16 William Street, Donaghcloney, BT66 7LS
M7 DXP C. Swift, 10 Sankey Road, Haydock, St Helens, WA11 0DD
M7 DXQ R. Foulds, 40 Broome Avenue, Swinton, Mexborough, S64 8QQ
M7 DXS C. Bassett, 2 Culverden Square, Tunbridge Wells, TN4 9NS
M7 DXT A. Wickins, Flat 8, Victoria Court, Milton Green, Langdon Hills, Basildon, SS16 6GB
MI7 DXU X. Qin, 9 Glenbrook Avenue, Belfast, BT5 5JP
M7 DXX P. Kilby, 1 Home Farm Barns, South End, Milton Bryan, Milton Keynes, MK17 9HS
M7 DXZ J. Cheeran, 37 Farnol Road, Dartford, DA1 5NG
MM7 DYA C. Buchanan, 4 Smiddy Place Letham, Forfar, DD8 2SD
M7 DYB N. Abdallahi, 66 West Street, Stratford-upon-Avon, CV37 6DR
M7 DYG Dr M. Ali, 4 Woolcombers Way, Bradford, BD4 8JF
MW7 DYH D. Haden, 1 Barham Road, Trecwn, Haverfordwest, SA62 5XX
M7 DYI A. Giles, Apartment 437, Holden Mill, Blackburn Road Blackburn Road, Bolton, BL1 7QJ
M7 DYM S. Medhurst, 50 Robson Drive, Hoo, Rochester, ME3 9EA
M7 DYO G. Shaw, 12 Aged Miners Homes, Boldon Colliery, NE35 9JE
M7 DYP R. Forrest, 5 Harmony Hill, Milnthorpe, LA7 7QA
M7 DYR A. Loveday, 50 Tenderah Road, Helston, TR13 8NT
M7 DYS D. Sandland, 54 Bishopdale Drive, Rainhill, Prescot, L35 4QH
M7 DYU A. Styles, 6 Morpeth Avenue, Darlington, DL1 2QG
MD7 DYW S. St.John, 8 Slieau Ree Apartments, Main Road, Union Mills, Isle of Man, IM4 4ND
M7 DYZ B. Humble, Linden Close, Dewsbury, wf128pl
M7 DZC J. Baker, 19 Littlebrook Lane, Bedford, MK43 9RA
M7 DZE D. Simmons, 23 Parker Road, Ashmore Park, Wednesfield, Wolverhampton, WV11 2HL
M7 DZF K. Phillips, 89 Cochrane Road, Dudley, DY2 0RU
M7 DZG D. Martin, Fawcett House 34 Forton Road, Chard, TA20 2HL
M7 DZI G. Hales, Flat 19, The Cross, 101-107 Commercial Road, Poole, BH14 0DL
M7 DZK D. Pyott, 64 Carlton Avenue, Worksop, S81 7JZ
M7 DZL D. Silsbury, 35A Churchill Close, Wimborne, BH21 4BH
M7 DZM D. Fox, 183 Grange Road, Letchworth Garden City, SG6 4LP
M7 DZO L. Pearson, 45 Clementine Drive Mapperley, Nottingham, NG3 5UX
M7 DZP R. Newbon, Rose Cottage, Sowters Lane, Burton-On-The-Wolds, Loughborough, LE12 5AL
M7 DZQ I. Coleby, Engine House Cottage, Creech St. Michael, Taunton, TA3 5RA
MM7 DZS K. Smillie, 1 Blaven Court, Baillieston, Glasgow, G69 7HY
M7 DZT S. Bradley, Eden Cottage, Salters Lane, Shotton Colliery, DH6 2PZ
M7 DZU S. Clements, Church Farm, South Scarle, NG237JH
M7 DZV R. Darby, 39 Leonard Road Wollaston, Stourbridge, DY8 3LU
M7 DZW H. Harclerode, The Bungalow, Silverlace Green, Parham, Woodbridge, IP13 9AD
MM7 EAB R. Boyd, 9 Abbey Road, Scone, PH2 6LWL
M7 EAC Dr R. Brown, 46 Mandarin Way, Derby, DE24 8YE
M7 EAE E. Barrett, 4 Corinth Walk, Worsley, Manchester, M28 3EB
M7 EAH C. Taylor, Pond Farm, Sandy Lane, East Tuddenham, Dereham, NR20 3JF
M7 EAK D. Smith, 3 Littleton Crescent, Penkridge, Penkridge, ST19 5BQ
M7 EAM E. Millett, Church Road, Skeyton, NR10 5AX
M7 EAN I. Crossley, 81 Quinton Drive, Bradwell, Milton Keynes, MK13 9HP
M7 EAO M. Lake, 18 Keppel Road, Dagenham, RM9 5LT
M7 EAP G. Watkins, 10 Maple Drive, Newport, PO30 5QP
MM7 EAR J. Eardley, 8 Greenbraes Crescent, Gourdon, Montrose, DD10 0NG
M7 EAS J. Chidley, 6 Rhodes Avenue, London, N22 7UT
M7 EBC A. Turner, 602 Manchester Road, Sheffield, S10 5PT
M7 EBI P. Watkins, 10 Lea Lane, Cookley, Kidderminster, DY10 3TA
M7 EBJ Z. Mcmillan, 2 Burgess Close, Whitfield, Dover, CT16 3NP
M7 EBL C. Edwards, 57 Shawbrooke Road, Eltham, SE96AL
M7 EBO T. Evans, 68 Wildbrook Road, Little Hulton, Manchester, M38 0FU
M7 EBU A. Crombie, 22 Withdean Court, London Road, Preston, Brighton, BN1 6RN
M7 EBW L. Culshaw, Barnsole Farm, Fleming Road Staple, Canterbury, CT3 1LG
M7 EBY P. Norwell, Russet, Litcham Road, Great Dunham, King's Lynn, PE32 2LJ
M7 EBZ C. Parrott, 15 Park Lane Snitterfield, Stratford upon Avon, CV370LT
M7 ECF S. Phillipo, 16 Leach Close, Chelmsford, CM2 7DS
M7 ECG F. Ehlers, North Lodge, Stanford Hall Estate, Stanford On Soar, Loughborough, LE12 5QW
M7 ECM A. Warr, 27 Blundell Place, Bedford, MK42 9XB
M7 ECN S. Titley, 20 Tower Road, Portishead, BS20 8RE
M7 ECP Lord A. Aspinall, 27 Swanswell Street, Coventry, CV1 5FZ
M7 ECT E. Tattersall, 17 Malt Kiln Way, Sandbach, CW11 1JL
MM7 ECV A. Brand, 6 Uraghag .Garenin, Stornoway, Hs29aj
M7 ECW R. Holmes, 21 Castle Hill Close, Eckington, Sheffield, S21 4BJ
M7 ECY Lord D. Giles, 22 Boston Close, Chaddesden, Derby, DE21 6WB
M7 ECZ R. Jobson, 25A Admirals Way, Thetford, IP24 2LB
M7 EDE D. Edwards, 21 Burton Old Road, Streethay, Lichfield, WS13 8LJ
M7 EDK D. Kennett, Twinstead Cottage, Sudbury, CO10 7NA
M7 EDL E. Long, 18 Prunus Road, Crewe, CW1 4HB
M7 EDM R. Edminson, 22 Southwold Place, Cramlington, NE23 8HE
MI7 EDO E. Mclaughlin, 1 Stewarts Terrace, Derry, BT487LH
M7 EDS E. Stewart, Flat 1, 33 Robert Street, Harrogate, HG1 1HP
M7 EDW E. Wright, 2 Wulfrath Way, Ware, SG12 0DN
M7 EDY S. Clarkson, 14 Hambleton View, Tollerton, York, YO61 1QW
M7 EDZ K. Edwards, 24 Abbey Road, Halesowen, B63 2HE
M7 EEC C. Johnson, 40 Springfield Elstead, Godalming, GU8 6EG
M7 EEE R. Vickery, 17 Plain-An-Gwarry, Redruth, TR15 1JB
M7 EEG M. Gee, 97 Lincolns Mead, Lingfield, RH76TA
M7 EEH J. Butters, Overlands, Green Lane, Hyde Lea, Stafford, ST18 9AY
M7 EEL L. Smith, 10 Lee Road, Dewsbury, WF13 3AX
M7 EEM V. Gibson, 9 Dorchester Drive, Hartlepool, TS24 9QY
MD7 EEO S. Williamson, 45 Bluebell Close, Peel, IM5 1GH
MM7 EEQ R. Hamlet, 4 Broaddykes Avenue, Kingswells, Aberdeen, AB15 8UH
M7 EEU W. Wren, 9 Church Road Terrace, High Harrington, Workington, CA14 5PL
M7 EEW N. Robson, Roseland House, Pidney, Hazelbury Bryan, Sturminster Newton, DT10 2EB
M7 EFA A. Corbett, 3 Tretawn Gardens, Selsey, Chichester, PO20 0DW
M7 EFB J. Armstrong, Flat 2, 8 Thornfield Gardens, Tunbridge Wells, TN2 4RZ
M7 EFC S. Walmsley, 45 Johnston Avenue, Bootle, L20 6HF
M7 EFH A. Phillips, 44 Hunton Hill, Birmingham, B23 7NE
MM7 EFJ L. Yip, 70 Kenley Road, Renfrew, PA4 8YW
M7 EFM D. Compton, Pin Lan, Far Vallens, Hadley, Telford, TF1 5SE
M7 EFO A. Johnson, 25 Eton Avenue, East Barnet, Barnet, EN4 8TU
MM7 EFP E. Peat, 1 Blackwood Drive New Cumnock, Cumnock, KA18 4BW
M7 EFR M. Lane, Kerity, Northchurch Common, Berkhamsted, HP4 1LR
M7 EFS P. Ransom, 15 Moat Drive, Harrow, HA1 4RY
M7 EFT C. Towle, 23 More Avenue, Aylesbury, HP21 8JY
M7 EFU P. Jones, Sussex Cottage, The Limes, Felbridge, East Grinstead, RH19 2QY
M7 EFV N. Bonnar, 5 Lavant Close, Waterlooville, PO8 8BQ
M7 EFX G. Abson, 62 New Hall Street, Burnley, BB10 1PT
M7 EFZ D. Thomas, 43 Buckland Road, Taunton, TA2 8EW
M7 EGA N. Allsworth, 6 Woodcroft, Oxford, Ox15nh
M7 EGB J. Smith, 14 Alexander Avenue, Droitwich, WR9 8NH
M7 EGE R. Jarvis, Postal Point 428, Manchester, M8 2FS
M7 EGJ A. Prime, 68 Langley Drive, Crewe, CW2 8LN
M7 EGK D. Frost, 169 West Auckland Road, Darlington, DL3 0SP
MM7 EGM E. Mackay, 10 Blackpark Avenue, Invergordon, IV18 0HY
M7 EGP O. Lazar, 84 Bruce Street, Swindon, SN2 2EN
MI7 EGS E. Paulikas, 33 Clarefield, Dungannon, BT71 6TQ
M7 EGT E. Travers, 24 Mittell Court, Lydd, Romney Marsh, TN29 9BJ
M7 EGW P. Farman, 84, Hawthorn Road., Exeter, EX2 6ED
M7 EGY J. Beer, 92 Ashby Road, Hull, HU4 7JT
M7 EHB M. Smith, 30 Pitfold Road, London, SE12 9HX
M7 EHC C. Cannon, 8 Piltdown Rise, Uckfield, TN22 1UH
M7 EHD I. Scriviner, 32 Mountbatten Way, Plymouth, PL9 9EJ
M7 EHG E. Hodgkiss, 190 Ulverely Green Road, Solihull West Midlands, B92 8AD
M7 EHL C. Talbot, 24A Calder Road, Stourport-on-Severn, DY13 8QD

M7 EHQ S. Al-Hajri, 806 Witham Wharf Brayford Street, Lincoln, LN5 7DL
MW7 EHV R. Scammells, 11 Gower Green, Croesyceiliog, Cwmbran, NP44 2QL
M7 EHW E. White, 16 Sisley Way, Hinckley, LE10 0GJ
MM7 EID D. Fitzgibbon, 1 Walton Road, Kirkpatrick Durham, Castle Douglas, DG7 3HG
MW7 EIJ R. Ramsay, 29 Cardigan Close, Tonteg Tonteg, Pontypridd, CF38 1LB
M7 EIK M. Rowlands, 69 Greenvale Road, London, SE91PB
M7 EIP D. Stanyer, 37 Heron Crescent, Melton Mowbray, LE13 1FT
M7 EIR I. Steele, 4 Godwin Close, Colchester, CO3 4BU
MM7 EIT D. Sipetan, 10 North Square, Coatbridge, ML5 2HB
M7 EIU M. Williams, 70 Barton Drive, Barton On Sea, New Milton, BH25 7JL
M7 EIW J. Clinton, Flat 2, Watson Court, 12A The Crescent, Netley Abbey, Southampton, SO31 5HP
M7 EJB P. Barnett, 6 Flintsham Grove, Stoke-on-Trent, ST1 5QS
MW7 EJC Dr E. Harries, Ty Traeth, Caerwedros, Llandysul, SA44 6BS
MW7 EJE E. Evans, 10 Bryn Drive, Coedpoeth, Wrexham, LL11 3LJ
M7 EJH L. Batten, Y Felin Barn, Llawr-Y-Glyn, Caersws, SY17 5RH
M7 EJI J. Earley, White Dymes, Winford Road, Newchurch, Sandown, PO36 0NE
MW7 EJJ E. Jenkins, 41 Park Street, Bridgend, CF31 4AX
M7 EJK R. Watkins, 31 Talbot Road, Maidstone, ME160HB
M7 EJM E. Mcmillan, 12 Longendale Road, Standish, Wigan, WN6 0UE
M7 EJN H. Williams, 4 Linden Court Station Road, Forest Hall, Newcastle-upon-Tyne, NE12 9BQ
M7 EJP M. Zuber, 36 Cashs Lane, Coventry, CV1 4DS
M7 EJS E. Stammers, 40 Tillingbourne Road, Shalford, Guildford, GU4 8EY
MI7 EJV W. Moorhead, 3 Ballyminstragh Road, Killinchy, Newtownards, BT23 6PE
M7 EJW E. Weatherill, Northend Cottage, North End Road, Yatton, Bristol, BS49 4AS
M7 EJY J. Tarr, 33 Easebourne Road, Dagenham, RM8 2DW
M7 EKA G. Leach, 43 Aldbanks, Dunstable, LU6 1Ah
M7 EKD S. Brown, 56 Ashwood Avenue, Abram, Wigan, WN2 5YE
M7 EKK G. Neagle, 51 Kingston Lane, Southwick, Brighton, BN42 4SJ
M7 EKL R. Slater, 23 Lawn Road, Eastleigh, SO50 4GT
MM7 EKP J. M Voinot, 30 Argyll Street, Lochgilphead, PA31 8NE
MM7 EKQ G. Gallagher, 5 Mcnay Crescent, Saltcoats, KA21 6AX
M7 EKR P. Clarke, 19 Woodlands Close, Denby Dale, Huddersfield, HD8 8RH
M7 EKS E. Schonrock, 1 Powell Road, Poole, BH14 8SG
M7 EKT N. Thorpe, Dove Cottage, Kithurst Park, Storrington, Pulborough, RH20 4JH
M7 EKU M. Butterworth, 228 High Street, Leeds, LS23 6AD
M7 EKV T. Cooper, 14 Hawthorn Drive, Brackley, NN13 6PA
MM7 EKZ S. Strachan, 8 Pitforthie Place, Brechin, DD9 7AX
M7 ELE E. Chauvelaine, 292 Mount Pleasant, Redditch, B97 4JL
M7 ELG K. Pollard, 1 Ruskin Close, Selsey, PO20 0TE
MW7 ELL P. Ellis, Flat 4, 32 Lawson Road, Colwyn Bay, LL29 8HE
M7 ELN A. Miller, East Lyn, Lazonby, Penrith, CA10 1BX
M7 ELO J. Myers, 3 Cavendish Street, Leigh, WN7 1SG
M7 ELS E. Sanders, 36 Queens Road, Waterlooville, PO7 7SB
M7 ELU M. Morton, 40 Mulgrave Road, Woolwich, London, SE18 5TY
M7 ELZ E. Hase, 23 Southdown Road, Thatcham, RG19 3BF
M7 EMC J. Reed, 3 Kings Croft, Ealand, Scunthorpe, DN17 4GA
MM7 EMD P. Turner, 7 Gartcows Drive, Falkirk, FK1 5QQ
M7 EME J. Smethurst, 57 Willoughby Close, Warrington, WA5 9QP
M7 EMH C. Holmes, Plot 5 Van 8 Hygrove Residential Park, Hygrove Lane, Minsterworth, Gloucester, GL2 8LE
M7 EMI J. Isaacs, 20 Vine Street, Worcester, WR3 7DY
M7 EMJ E. Callow, Orchardside, Lampley Road, Kingston Bridge, Clevedon, BS21 6TY
MW7 EMK H. Parker, Dolifor, Llanwrthwl, Llandrindod Wells, LD1 6NU
M7 EMO M. Hall, 6 Sturgess Street, Newton-le-Willows, WA12 9HS
M7 EMU D. Lambert-Barnett, Fordlands Farm, Exeter, EX6 7ST
M7 EMV M. Blain, 63 Walmington Fold, London, N12 7LD
M7 EMW D. Pearson, 74 Arena Park, Exeter, EX4 8RD
M7 EMY D. Emerton, 15 Swallow Close, Stoke-on-Trent, ST3 7FN
M7 ENB M. Glen, 134 Spen Lane, Leeds, LS6 3NA
M7 ENC J. Hoang, 2 Chichester Avenue, Ruislip, HA4 7EH
MM7 ENF D. Line, 55 Burn Brae Westhill, Inverness, IV2 5RH
M7 ENH I. Hodgkiss, 190 Ulverley Green Road, Solihull, B92 8AD
M7 ENI V. Curtis, 52 Nairn Road, Lancaster, LA1 5UY
M7 ENJ M. Faben, 15 Monks Close, Doncaster, DN7 4QL
MW7 ENN Maj. I. Wright, Wernelli, Llannon, Llanelli, SA14 6AP
MM7 ENQ S. Wright, 67 Park Terrace, Broxburn, EH52 6AP
M7 ENU M. Dadswell, Avaeya Villa, Winchester Road, Waltham Chase, Southampton, SO32 2LX
M7 ENV M. Savage, 5 Mere Hall Barns, Mere Lane, Enville, Stourbridge, DY7 5JL
M7 ENW Dr M. Heywood, 17 Lavenham Close, Macclesfield, SK10 2TS
M7 EOA M. Allen, 8 Oakdene Road, Watford, WD246RW
M7 EOB S. Tyler-Murphy, 63 Sedlescombe Gardens, St. Leonards-On-Sea, St Leonards on Sea, TN38 0YT
M7 EOD A. Parker, 51 Smithy Lane, Tingley, Wakefield, WF3 1QH
M7 EOE T. Fowler, 16 Walsingham Drive, Taverham, Norwich, NR8 6FZ
M7 EOH E. Crisp-Mcneil, Jacaranda, Gandish Road, East Bergholt, CO7 6TP
M7 EOK Z. Downer, 16 Windmill Lane Freshwater Isle Of Wight, Freshwater, PO40 9DX
M7 EOL M. Williams, 19 Cotebrook Drive, Upton, Chester, CH2 1RA
M7 EOQ J. Kirk, 108 Lodge Road, Heacham, King's Lynn, PE31 7SZ
M7 EOS S. Allsopp, 5 Grebe Road, Banbury, OX16 9YZ
M7 EOT A. Highfield, 54 Parkfield Street, Rowhedge, Colchester, CO5 7EL
M7 EOU E. Pereira, 15 Burgess Green Close, St. Annes Park, Bristol, BS4 4DG
M7 EOV A. Beauchamp, Nightingale Lane, Burgess Hill, rh15 9jj
M7 EOZ Z. Howling, Sysonby Knoll Hotel Asfordby Road Melton Mowbray, Melton Mowbray, LE13 0HP
MM7 EPD K. Macandie, 33 Jordanhill Crescent, Glasgow, G13 1UN
M7 EPF C. Miller, Whitechapel Mission, 212 Whitechapel Road, London, E1 1BJ
M7 EPG E. Gresty, 7 Chester Road, Winsford, CW7 2NG
MW7 EPH C. Howells, 22 Hafod Arthen Estate, Brynithel, Abertillery, NP13 2HX
M7 EPI T. Flude, 32 Borough Street, Brighton, BN1 3BG
M7 EPK N. Shave, 52 Da Volls Court, Gorleston, Great Yarmouth, NR31 6NH
M7 EPM R. Abeykoon, 20, Holmewood Crescent, Nottingham, NG5 5JH
M7 EPN M. Levett, 16 Firs Avenue, Waterlooville, PO8 8RS
M7 EPS B. Irving, 25 Doniford Road, Williton, Taunton, TA4 4SG
M7 EPT C. Duncombe, 17 South Lynn Crescent, Bracknell, RG12 7JU
M7 EPV S. Mawn, 1 Blue Water Drive, Elborough, Weston-Super-Mare, BS24 8PF
MM7 EPW A. Philip, 10 Gordon Brown Place, Mallaig, PH41 4RL
M7 EPY C. Ward, 10 Fairfield Link, Sherburn In Elmet, Leeds, LS25 6LT
M7 EQC E. Coates, 3 Glenview Close, Crawley, RH10 8AS
MM7 EQF O. Nicol, 34 Torvean Crescent, Kirkcaldy, KY2 6FT
M7 EQG A. Ratowski, 18B Hurstbrook Drive, Stretford, Manchester, M32 9JQ
M7 EQI J. Mc Elhinney, 16 Chaucer Road, London, E17 4BE
M7 EQM S. Cartlidge, 45 Forest Road, Market Drayton, TF9 3HX
M7 EQN C. Evans, Lilly Cottage, Wood Enderby, Boston, PE22 7PQ
M7 EQP K. Knight, 72 Beaconsfield Villas, Brighton, BN1 6HE
M7 EQQ M. Steed, 38 Rivelands Road Swindon Village, Cheltenham, GL51 9RF
M7 EQR F. Liput, 8 Willow Way, Horsford, Norwich, NR10 3GE
M7 EQS K. Whitman, The Little Oyster Care Home Seaside Ave, Minster on Sea, me12 2nj
MW7 EQT M. Rumball, Crex, Tregynon, Newtown, SY16 3PY
MW7 EQU D. Adams, 1 Parkhall Caravan Park Penycwm, Haverfordwest, SA626LS
MM7 EQX S. Wallace, 28 Laurel Bank Terrace, Castle Douglas, DG7 1BP
M7 EQZ A. Cartwright, 154 Camp Hill Road, Nuneaton, CV10 0JJ
M7 ERB E. Russell-Brown, 2 Churn Hill, North Cerney, Cirencester, GL7 7DN
M7 ERE C. Dodds, 22 Cambridge Street, Wolverton, Milton Keynes, MK12 5AJ
M7 ERF I. Thomson, 23B Vicarage Gardens, Netheravon, Salisbury, SP4 9RW
M7 ERG E. Earle, 20 Queens Crescent, Gorleston, Great Yarmouth, NR31 7NN
M7 ERI D. Conlan, 33 Urmson Road, Wallasey, CH45 7LE
M7 ERJ C. Whittenbury, Little Orchard, The Common, Abberley, Worcester, WR6 6AY
M7 ERK E. Keeley, 10 Farringdon Close, Peterborough, PE1 4RQ
MW7 ERP T. Clode, 31 Heol Rhos, Caerphilly, CF83 2BE
M7 ERQ M. Ward, 81 Houldey Road, Birmingham, B31 3HH
M7 ERS S. Stephens, 65 Summerlin Drive, Milton Keynes, Buckinghamshire, MK17 8GP
M7 ERT A. Eastwood, 3 Bowes Nook, Bradford, BD6 2BJ
M7 ERU A. Usher, 3 Meadow View, Patrington, HU12 0QG
M7 ERV M. Smith, 34 Winnycroft Lane Matson, Gloucester, GL4 6EJ
M7 ESB D. Wilson, 7 Wisteria Grove, Birmingham, B44 9AX
M7 ESC S. Christie, 21 Lovat Lane, London, EC3R 8EB
M7 ESE S. Ambrosio, 8 The Pines, Long Lane, Chester, CH2 2QF
MM7 ESG A. Xenos, 8 Hayburn Crescent Main Door, Glasgow, G11 5AX
M7 ESH E. Sheremetyeva, 6 Pixey Close, Yarnton, Kidlington, OX5 1FY
M7 ESI A. Clark, 62 New Road, London, SE2 0QG
M7 ESK S. Shields, 28 Bank Field Westhoughton, Bolton, BL5 2QG
M7 ESR Dr N. Rajendran, 1 Sharp?S Yard, London Road, Six Mile Bottom, Newmarket, Cb8 0uh
M7 ESS M. Taylor, 168 Long Furrow, East Goscote, Leicester, LE7 3SU
MM7 EST E. Thorpe, 27 Castle Street, Dunbar, EH42 1EX
M7 ESW J. Roberts, 34 Gloucester Road, Consett, DH8 7LL
M7 ESX K. Polston, 10 Marsh Farm Road, South Woodham Ferrers, Chelmsford, CM3 5WP
M7 ESZ E. Zaralli, 53A Winchester Street, London, SW1V 4NY
M7 ETB E. Ball, 49 Wunburra Circle, Pacific Pines, Gold Coast, Australia, 4211
M7 ETH E. Wiggins, 32 Prestbury Road, Duston, Northampton, NN5 6XP
MM7 ETI J. Stokroos, 20 Ochlochy Park, Dunblane, FK15 0DU
M7 ETN R. Sheffield, 7 Elysium Terrace, Northampton, NN2 6EN
M7 ETZ I. Maddock, 6 Flintsham Grove, Hanley, ST1 5QS
M7 EUS M. Allington, Flat 16, Isambard Court, Damers Road, Dorchester, DT1 1RQ
M7 EVM E. Merritt, 35 Jocelyn Drive, Wells, BA5 2ER
M7 EWB E. Beckett, 55 Schoolfield, Preston, Pr58bh
M7 EWT A. Newton, 4 Acre Close, Market Rasen, LN8 3DL
M7 EXT C. Newman, 3 Orchard Way, Northstowe, Cambridge, CB24 1AG
MM7 EYO E. Young, 31 Hilton Road, Milngavie, Glasgow, G62 7DN
M7 EYP S. Wilkinson, 16 Chaucer Street, Hull, HU8 8NA
M7 FAB F. Brazier, 277 Feltham Hill Road, Ashford, TW15 1LT
M7 FAF A. Davidson, 2 Neil Street, Widnes, WA86RH
M7 FAL M. Roberts, Bowerbank Hall, Scotland Road, Penrith, CA11 9HL
M7 FAM J. Jordaan, 41 Bentham Place, Standish, Wigan, WN6 0NB
M7 FAN R. Fanyinka, 9 Orchard Rise, Chard, TA20 1FX
M7 FAV Dr S. Favell, Lodge Farm, Moor Lane, Reepham, Lincoln, LN3 4EE
M7 FAX N. Tarrant, 87 Bideford Green, Linslade, Leighton Buzzard, LU7 2TJ
M7 FBC J. Gittins, 9 Beechfield, Grasscroft, Oldham, OL4 4EN
M7 FBM C. Evans, 7 Westfield View, Wakefield, WF1 3RU
M7 FBR C. Wilson, Collingwood Avenue, Tolworth, KT59PT
M7 FBW D. Bevan, Flat, The Roost, 30 Treverbyn Road, Stenalees, St Austell, PL26 8TJ
M7 FCA D. Grant, 22 Alba Gardens, London, NW11 9NR
M7 FCB C. Barrett, 40 Hillside Road, Dover, CT17 0JQ
M7 FCH C. Hoyle, 2 Westfield Street, Chadderton, Oldham, OL9 6RQ
M7 FDC P. Ainsworth, 4 St. Johns Road, Wirral, CH62 0BN
M7 FDD F. Done, 3 Home Farm Cottages, Old Warden Park, Old Warden, Biggleswade, SG18 9DU
MW7 FDI S. Turner, 9 Baldwin Street, Newport, NP20 2LT
M7 FDS F. Da Silva, Flat 72, Vesta House, 4 Liberty Bridge Road, London, E20 1AN
M7 FDW F. Woolley, 10 Hazelmoor Fold, Elland, HX5 0DR
MW7 FEC J. Williams, 12 Pantycelyn, Fishguard, SA65 9EH
MM7 FEM J. Small, Gemmell Crescent, Ayr, Ka8 0jr
M7 FET D. Purslow, 169 Hoo Road, Kidderminster, DY10 1LP
M7 FEU G. Godfrey, 11 Defrene Road, London, SE26 4AB
MM7 FEW J. Frew, 125 Minnoch Crescent, Maybole, KA19 8dr
M7 FGB F. Byrne, 144 Meridian Road, Boston, PE21 0NF
M7 FGC F. Chapman, 77 Cedar House, Spelthorne Grove, Sunbury-on-Thames, TW16 7DD
M7 FGH D. Barwick, 2 Sutton Close, Bury St Edmunds, Ip327ep
MM7 FGL A. Hall, 94 Muirside Avenue, Kirkintilloch, Glasgow, G66 3PH
M7 FGS P. Pendlebury, 79 Pinfold Close, Westhoughton, Bolton, BL5 2RN
M7 FGT F. Tozzi, 120 Wellhouse Lane, Barnet, EN5 3DP
M7 FGY W. Evenden, 15 Knights Garden, Hailsham, BN27 3JR
M7 FHG J. Akhtar, 986 London Road, Alvaston, Derby, DE24 8PY
MI7 FHM F. Mcalister, 24 Ballyhone Road Gleno, Larne, BT403LW
M7 FIN P. Webber, 17 Lilleshall Drive, Elstow, Bedford, MK42 9FG

M7 FJS F. Simpson, 36 Lyndhurst Ave, Blackpool, FY4 3AX
M7 FKR A. Ferrer, Flat 8, 372 Prince Avenue, Westcliff-on-Sea, SS0 0NB
M7 FLD S. Muller, 426A St. John Street, London, EC1V 4NJ
M7 FLG F. Gardner, 2 Austen Road, Guildford, GU1 3NP
M7 FLJ J. Cross, 18 Brickly Road, Luton, LU4 9EU
MW7 FLK R. Sawle, Westering, Stonewall Hill, Presteigne, LD8 2HB
M7 FLR C. Day, 14 Windsor Drive, Ramsey Forty Foot, Ramsey, Huntingdon, PE26 2XX
M7 FLY W. Mcguire, 16 Piel View Grove, Barrow-in-Furness, LA13 0EF
MM7 FLZ C. Coutts, 1 Springfield Avenue, Duns, TD11 3BF
M7 FMA F. Allen, 109 Barston Road, Oldbury, B680PU
M7 FMB F. Mclean-Brown, 101 Mayfield Avenue, Southend-on-Sea, SS2 6NR
MW7 FMD F. Mcdermott, Maesnewydd Talsarn, Lampeter, SA48 8RE
M7 FMG Dr F. Morales Gundin, Corner Close Broadmead Road Send, Woking, GU23 7AD
M7 FMM C. Wood, 17 Chestnut Close, Greetland, Halifax, HX4 8HX
M7 FMR R. Clark, Stone Ends, Hesket Newmarket, Wigton, CA7 8JS
M7 FMT F. Tang, 78A Church Lane, Mill End, Rickmansworth, WD3 8HE
MM7 FNG E. Boardman, 86 James Park, Burntisland, KY3 9EP
MI7 FNY M. Feeney, Leamington Place, Lisburn, bt27 4ul
M7 FOK I. Wilkinson, The Beeches, Whitchurch Road, Aston, Nantwich, CW5 8DJ
MW7 FON G. Pritchard, 3 Llwyn, Rhosgadfan, Caernarfon, LL54 7HN
M7 FQZ D. Garner, 1 Rowland Avenue, Field Street, Hull, HU9 1HR
M7 FRD R. Haughton, 5 Fairfield Way Wesham, Preston, PR4 3EP
M7 FRE P. Freeman, Grove Orchard, Knapp Lane Coaley, Dursley, GL11 5AR
M7 FRG H. Glover, 26 Mersey Walk, Birkenhead, CH42 3UN
MW7 FRN G. White, 5 Ramson Close, Penpedairheol, Hengoed, CF82 8LB
M7 FRR F. Ruddick, 12 Davids Avenue, Great Sankey, Warrington, WA5 1LN
M7 FRS B. Horne, 31 Grampian Way, Oulton, Lowestoft, NR32 3EP
M7 FRT P. Smith, 11 Brine Road, Nantwich, CW5 7BA
M7 FRW L. Finlayson, 44 School Lane, Ashurst Wood, East Grinstead, RH19 3QP
MM7 FRZ F. Laidlaw, 12 Limeview Avenue, Paisley, PA2 8NB
MM7 FSR R. Hamilton Brodie, 18 Torcraik Crescent, North Middleton, EH23 4SU
M7 FTD S. Lord, 18 Caunsall Road, Kidderminster, Dy11 5 yb
M7 FTF A. Redshaw, 155 Wainbody Avenue South, Coventry, CV3 6BY
MM7 FTH G. Hill, 21 Affric Drive, Falkirk, FK27UF
M7 FTL Dr S. Frankau, 79 Boyne Road, London, SE13 5AN
M7 FTQ K. Gonulkirmaz, 14 Ashbourne Avenue, London, NW11 0DR
M7 FTS P. Johnson, 12A Beechroyd, Pudsey, LS28 8BH
M7 FTX J. Hole, 40 Campkin Road, Wells, BA5 2DG
M7 FTZ P. Sharpe, 23 Sparrows Mead, Redhill, RH12EJ
M7 FUN D. Hills, 4 Birchglade, Totton, SO40 2GP
MM7 FUR Q. Misell, Flat 59 St Peter'S Studios, Aberdeen, AB24 3HQ
M7 FVS P. Squires, 3 Kenan Drive, Attleborough, NR17 2RJ
M7 FWA M. Hurson, Hq Company Clive Barracks, Ternhill, TF9 3QE
M7 FWB D. Lewis, 23 Malvern Road, Southampton, SO16 6PZ
M7 FWH F. West, Harrington Lodge, Orton, Kettering, NN14 1LN
MM7 FWK F. Kinloch, 53 Belmont Street, Newtyle, Blairgowrie, PH12 8UB
M7 FWR W. Rogerson, 37 Victoria Park, Stockport, SK1 4EP
M7 FWW A. Crocker, 32 Godmanston Close, Poole, BH17 8BU
MJ7 FXC F. Chesnay, 3 Santa Rosa, La Ruette De Patier, St. Saviour, Jersey, JE2 7LQ
M7 FXD S. Pallister, 32 Greensnook Lane, Bacup, OL13 9DQ
M7 FYF K. Fyfe, 40 Hampden Road, Ashford, TN23 6JL
M7 FYL P. Smith, 29 Fowlmere Road Foxton, Foxton, Cambridge, CB226RT
M7 FYM B. Friedman, St. Peters House, Broad Hinton, Swindon, SN4 9PA
MI7 FYR G. Cleary, 2 Loanda Crescent, Newry, Ireland, BT35 8EZ
M7 FZA B. Mellor, Flat 1, 138 Street Lane, Gildersome, Morley, Leeds, LS27 7JB
MM7 GAC G. Cattanach, 14 Stewart Park Place, Aberdeen, AB24 4GA
MW7 GAH G. Howard, Sport Y Gwynt, Middle Road, Llandegfan (Nr Menai Bridge), LL595YD
MW7 GAN S. Ganley, 1 Bron Wern Llanddulas, Abergele, LL22 8JD
M7 GAP P. Grange, 14 Westley Avenue, Whitley Bay, NE26 4PA
MW7 GAR G. Williams, 10 Strawberry Place Morriston, Swansea, SA6 7AG
M7 GAT J. Reynolds, 7A New Inn Lane, Trentham, Stoke-on-Trent, ST4 8HA
M7 GAW T. Stephenson, 10 Lower Heyshott, Petersfield, GU31 4PZ
M7 GAX G. Rowntree, 86 Clog Mill Gardens, Selby, YO8 3EH
MU7 GBG G. De Putron, Le Murier Cottage, Le Murier, St. Sampson, Guernsey, GY2 4HQ
M7 GBH G. Hodgson, 6 Gilbert Road, Sunderland, SR4 9QU
MW7 GBJ G. Britten-Jones, 101 Mill View Estate, Maesteg, CF34 0DE
MM7 GBL G. Lennox, 49 Sorbie Road, Ardrossan, KA22 8AP
M7 GBO A. Woods, 4 Sedgwick Close, Atherton, M46 9EG
M7 GBP M. Finch, 186 Hookfield, Harlow, CM18 6QW
M7 GBQ H. Rowland, 5, Gwinear Downs, Leedstown, Hayle, TR27 6DJ
M7 GBR E. Wilson, 51 Pincroft Wood, Longfield, DA3 7HB
M7 GBS D. Hockley, Green Farm, Mendlesham Green, Stowmarket, IP14 5RE
MW7 GBW M. Williams, 42 Parkside, Overton, Wrexham, LL13 0HA
MI7 GBX S. Golemboski-Byrne, 79 Lackan Road, Ballyroney, Banbridge, BT32 5HR
M7 GBZ P. Pearce, Criggion Mw Radio Station, Back Lane, Criggion, Shrewsbury, SY5 9BE
M7 GCA G. Allison, 20 Joseph Johnson Road, Sandbach, CW11 3TE
MI7 GCC G. Castorio, 1A Marlborough Park North, Belfast, BT9 6HJ
M7 GCH G. Chlapoutakis, 10A Thorpes Avenue, Denby Dale, HD8 8SP
M7 GCL T. Cox, 74 Dean Way, Storrington, Pulborough, RH20 4QS
M7 GCR G. Roberts, 10 Haslam Close, Ickenham, UB10 8TJ
MM7 GCS G. Stuart, 12 Todshaugh Gardens, Kirkliston, EH29 9GE
M7 GCT P. Hughes, 17 Worcester Park, Bath, BA1 6QU
MI7 GCV L. Moore, 83 Cloyfin Road, Coleraine, BT52 2NZ
M7 GCW D. Booth, 19 Oak Avenue, Elloughton, Brough, HU15 1LA
M7 GCX G. Churcher, 4 The Crescent, Tonbridge, TN9 1JH
M7 GCY G. Chedgy, 67 Studland Park, Westbury, BA13 3HN
M7 GDF D. Coombs-Farnsworth, 27 Stork Lane, Rothwell, NN14 6GE
M7 GDG G. Gibson, 4 Swan Close, Swannington, Norwich, NR9 5NL
MM7 GDI G. Irvine, 5 Carlingnose Court, North Queensferry, Inverkeithing, KY11 1EP
M7 GDW G. Wills, 15 Markham Avenue, Bournemouth, BH10 7HL
M7 GDY A. Good, 19 Milfields, Barton-upon-Humber, Dn185na
M7 GEB J. Nel, 12 Jodrell Close, Waterlooville, PO8 9NH
M7 GEF G. Barden, 9 Nene Close, Wellingborough, NN8 5WB
M7 GEG G. Kendall, 39 Foundry Gate, Wombwell, Barnsley, S73 0LF
MW7 GEH G. Herd, Cartref, Penisarwaun, Caernarfon, LL55 3BS
MW7 GEL N. Shepherd, Prospect, Newchapel, Llanidloes, SY18 6JY
M7 GEN T. Gentry, 175 Huddersfield Road, Halifax, HX3 0AS
M7 GEO G. Salt, Christmas Cottage, Ram Hill, Bristol, BS36 2TX
M7 GEP P. Griffiths, 12 Plant Farm Crescent, Waterlooville, PO7 3DB
M7 GER G. Tilley, 24 Viador, Chester le Street, DH3 3TP
M7 GET R. French, South Cottage, Barns Green Farm, Chapel Road, Barns Green, Horsham, RH13 0PR
M7 GEZ G. Hever, 15 Juniper Close, Reigate, RH2 7NQ
M7 GFA A. Grime, 13 Yarrow Gate, Chorley, PR7 3AZ
M7 GFB G. Bingley, 18 Seaton Close, Nuneaton, CV11 6YX
M7 GFC B. Capewell, 18 Westminster Road, Kidderminster, DY11 6HG
M7 GFR N. Ramsey, 16 Sunnycroft Close Bishops Cleeve, Cheltenham, GL52 8AU
M7 GFY D. Ward, 20 Norbury Road, Wolverhampton, wv109rl
MM7 GGD G. Sneddon, 34 Myreside Avenue, Kennoway, KY8 5EN
M7 GGF G. Farmer, 5 Atlas Road Earls Colne, Colchester, CO62LU
M7 GGH Dr G. Howling, Sysonby Knoll Hotel, Asfordby Road, Melton Mowbray, LE13 0HP
M7 GGK G. Kerekes, 10 Brookfield Road, Bury, BL9 5JZ
MM7 GGL G. Ledgerwood, 28 Cannerton Park, Milton Of Campsie, Glasgow, G66 8HR
M7 GGM G. Mariner, 6 Richmond Road, Manton, Worksop, S80 2TP
M7 GHB G. Brinklow, 45 Gillsway, Northampton, NN2 8HT
M7 GHE M. Ghent, 76 Blue Hill Crescent, Wortley, Leeds, LS12 4PB
MM7 GHF J. Hazle, Grange Hill Farm, Kinghorn, Burntisland, KY3 9YF
M7 GHI G. Hewitt, 5 Station Road, Dinnington, Sheffield, S25 3RW
M7 GHJ K. Selvaratnam, 15 Mark Close, Southall, UB1 3QJ
M7 GHK G. Kerr, 6 Cressages Close, Felsted, Dunmow, CM6 3NW
M7 GHO S. Wheelhouse, 30 Highfield Drive, Gildersome, Morley, Leeds, LS27 7DW
M7 GHS A. Jurd, 54 Caesar Close, Andover, SP10 5JR
M7 GHW G. Heaton, Crooke Village Marina, Crooke Road, Standish Lower Ground, Wigan, WN6 8LR
M7 GHZ S. Rankin, 2 Thompson Road, Broadbridge Heath, Horsham, RH12 3TT
MW7 GIA G. Jacobsen, 98 Laburnum Drive, Newport, NP19 9AN
M7 GIB P. Gibbons, 2A Gipsy Castle Estate, Hay-On-Wye, Hereford, HR3 5EG
M7 GIF C. Arthur, 25 Pump Hollow Lane, Mansfield, NG18 3DU
M7 GIL S. Gilmartin, 6 Newport Crescent, Leeds, LS6 3BY
M7 GIN W. Pilling, 12 Brooke Close, Accrington, BB5 2QX
M7 GIP P. Malinowski, 12 Lodge Road, Southampton, SO14 6RN
MM7 GIR W. Girdwood, 95 Parkhead Crescent, West Calder, EH55 8BE
M7 GIS Dr D. Rodriguez, 23 Thorpe Way, Cambridge, CB5 8UJ
M7 GJA G. Finnigan, 33 Birchwood Drive, Ulverston, LA12 9PN
M7 GJF G. Flatman, Devenport Cresent, Manchester, OL26JX
M7 GJG G. Graham, Todcroft, Elsdon, Newcastle upon Tyne, NE19 1AA
M7 GJJ G. Johnson, 4 Delta Park Drive, Hesketh Bank, Preston, PR4 6SE
M7 GJL G. Larner, 58 Heather Close, Carterton, Ox181th
M7 GJP A. Bonwitt, 60 Wellhouse Road, Beech, Alton, GU34 4AG
M7 GKA H. Griffiths, The Old Church Hall, Christmas Common, Watlington, OX49 5HL
M7 GKB A. Hewitt, 64 Ridgeway, Clowne, Chesterfield, S43 4BD
MM7 GKD Rev. J. Maclennan, 87 Glenfruin Road, Blantyre, Glasgow, G72 9RJ
M7 GKL A. Gkelios, 25 Newlands Avenue Norton, Stocton on Tees, TS202PQ
M7 GKR D. Radulescu, 8 White Street, Hull, HU3 5PS
MW7 GKT G. Thomas, 37 Broadstairs Road, Cardiff, CF11 8DE
M7 GKX R. Newell, 44 New Street, Chagford, Newton Abbot, TQ13 8BB
M7 GLB G. Briggs, 37 Knutton Crescent, Sheffield, S5 9NW
M7 GLE G. Porritt, 23 Greenlands Road Pickering, Pickering, YO18 8BQ
MW7 GLF N. Jones, 11 Wood Green, Mold, CH7 1UG
M7 GLK D. Hancock, 7 Runswick Close Silksworth, Sunderland, SR3 2YG
M7 GLN G. Strong, 11 Tamar Walk, Scunthorpe, DN17 1UE
M7 GLO J. Elton, 63 Bazeley Road, Gloucester, GL4 6JE
M7 GLS G. Sangwell, 34 Gatcombe Close, Calcot, Reading, RG31 4XQ
M7 GLW G. Williams, 329 Longford Lane Longford, Gloucester, GL2 9ES
M7 GLY G. Freeman, 116 Madeira Way, Eastbourne, BN23 5UJ
M7 GMA C. Clunn, 38 Byfleet Avenue, Basingstoke, RG247HR
M7 GMC G. Mcloughlin, 12 Harwood Road, Norwich, NR1 2NG
M7 GMD G. Doggett, 68 Pilgrims Way, Harleston, IP20 9QE
M7 GMF Dr G. Fahy, 40 Mayfair Close, Great Sankey, Warrington, WA5 3PL
M7 GMG A. Beckett, 23 Lingcrest Close, Manchester, M19 2WJ
M7 GMH G. Heath, 2 Lower Drake Fold, Westhoughton, Bolton, BL5 2RE
M7 GMJ I. Matthews, Astral, South Dock Marina, Rope Street, London, SE16 7SZ
M7 GMK G. Madgwick, 5 Plantation Drive, Walkford, Christchurch, BH23 5SE
M7 GMN D. Jowle, 6 Newton Ave, Sheffield, s36 1el
M7 GMP G. Preston, 3 Thirlwall Avenue, Conisbrough, Doncaster, DN12 3JZ
M7 GMR G. Mylonas, 68 Lancaster Gate, Cambourne, CB236AT
MM7 GMS G. Smith, 37 Kilmorie Drive, South Lanarkshire, G732ER
MM7 GMW G. Mcfarlane, 18 Stanistone Road, Carluke, ML8 4DY
M7 GNA G. Gray, 9 Bevan Way, Market Drayton, TF9 3US
M7 GNB G. Baguley, 58 Chestnut Avenue, Great Notley, Braintree, CM77 7YJ
M7 GNC G. Chalklin, 18 Trinity Close, West Mersea, Colchester, CO5 8RW
M7 GNH G. Harris, 6 Penling Close, Cookham, Maidenhead, SL6 9NF
MI7 GNI G. Baine, 13 Cherrymount, Newtownabbey, BT36 5NH
MW7 GNN S. Horan, 26 Ffordd Ysgubor Goch, Caernarfon, LL55 2RU
M7 GNR B. Ashcroft, Ash Crescent, Seaham, SR7 7UE
MM7 GNX A. Cybulski, 46 Waulkmill Avenue, Barrhead, Glasgow, G78 1DD
M7 GOB J. Holden, 262 St. Margarets Road, Twickenham, TW1 1PR
M7 GOG D. Rolf, 28 Titchener Way, Hook, RG27 9GB
MW7 GOM G. O'Meara, 30 West View, Chirk, LL14 5HN
M7 GOO A. Goose, 12 Brown Street, Rainham, Gillingham, ME8 7JN
M7 GOT A. Croot, 13 Bickford Road, Wolverhampton, WV10 0NH
M7 GOU P. Gould, 9 Crowden Way Thamesmead, London, SE28 8HE
M7 GOW K. Laing, 16 Cherrywood Drive Gonerby Hill Foot, Grantham, NG31 8QL
M7 GOZ S. Gosbee, Water Hall Farm, Wixoe, Stoke By Clare, Sudbury, CO10 8UA
M7 GPH G. Hawes, 59 Hartlands, Bedlington, NE22 6JG
M7 GPI G. Pipkin, 22 Orchard Way, Sutton, SM1 3QQ
M7 GPL G. Lomas, 28 Moat Bank, Bretby, Burton-on-Trent, DE15 0QJ
M7 GPM G. Manning, 8 Devonshire Row, Princetown, Yelverton, PL20 6QD
M7 GPP G. Phillips, Labourn Fell Farm, Chopwell, NE17 7AY
MI7 GPT P. Hegarty, 26 Ashveagh Benburb, Dungannon, BT717TS

M7 GPW C. Godfrey, 3 Moor Park, Neston, Corsham, SN13 9YJ
M7 GPX R. Haines, 47 Upshire Road, Waltham Abbey, EN9 3NZ
M7 GPY D. Twells, Preston Lodge 3 Willoughby Court, Norwell, Newark, NG23 6JJ
M7 GRD G. Downing, Fort Road, Gosport, Po12 2dt
MM7 GRG G. Manning, Snipefield, Culsalmond, Insch, AB526TU
M7 GRH I. Justice, 4 Saxon Street, Droylsden, Manchester, M43 7FR
M7 GRK J. Honeyands, 154 Burnham Road, Hullbridge, Hockley, SS5 6HJ
M7 GRM Dr G. Lioliou, Flat 1 22 Grand Parade, Brighton, BN2 9QB
M7 GRP G. Purser, 1 Lichfield Gardens, Aldwick, Bognor Regis, PO21 3RB
M7 GRR A. Moorhouse, 66 Wicor Mill Lane, Fareham, PO16 9EG
M7 GRS G. Smith, 148 Firhill Road, London, SE6 3SQ
M7 GSA K. Hemmings, 90 Grovewood Avenue, Leigh-on-Sea, SS9 5EG
M7 GSD A. Smith, 30 Knutsford Road, Alderley Edge, SK9 7SD
MI7 GSI R. Nelson, 61 Ballycrune Road, Hillsborough, BT26 6NH
M7 GSP G. Patterson, 3 Sunte Close, Haywards Heath, RH16 1QT
M7 GSS A. Netting, 7 Manor Close, Ivybridge, PL21 9BQ
M7 GST R. Budinger, Kings Arms, Bexhill Road, Ninfield, Battle, TN33 9JB
M7 GTA G. Ashcroft, 8 Launceston Close, Winsford, CW7 1LY
M7 GTC G. Bradley, Greentree Cottage, Town End, Broadclyst, Exeter, EX5 3HW
M7 GTD R. Dunnigan, 21 Coldstone Drive Garswood, Ashton - in - Makerfield, WN40RW
M7 GTE P. Bleasdale, 12 Malvern Ave, Padiham, bb127dt
M7 GTI S. Lifely, The Farm, Moreton Jefferies, Hereford, HR1 3QY
M7 GTJ G. Lawton, 11 Highlands Drive, Stockport, SK2 5HX
M7 GTL Dr I. Chessell, Pondfield House, Pondfield, Great Dunmow, CM6 1FX
M7 GTM G. Moulds, Flat 88, Barkis House, Brownlow Close, Portsmouth, PO1 4ES
MM7 GTS G. Gault, Ka73Sb, Ayr, ka73sb
MW7 GTW G. Whitley, 73 Penrhos Avenue, Llandudno Junction, LL31 9EL
M7 GTX R. Twyning, 64 Hornbeam Way, Kirkby-in-Ashfield, NG17 8RL
M7 GTZ C. Aldous, 342 Feltham Hill Road, Ashford, TW151IW
M7 GVM A. Miles, The Cider House Weston Street East Chinnock, Yeovil, BA22 9EJ
M7 GVP J. Hunt, 18 Oakville Close, Worcester, WR2 4XL
MW7 GVR P. Perrett, 1 Llys Y Wennol Northop Hall, Mold, CH7 6GE
M7 GWC G. Castledine, 11 Cam Road, Cheltenham, GL52 5QS
M7 GWD A. Dainty, 6 St Nicholas Way Wygate Park Spalding, Spalding, PE11 3GF
M7 GWF R. Lewis, 46 Graydon Avenue, Chichester, PO19 8RG
M7 GWH J. Housley, Lesede Cottage, The Town, Carsington, Matlock, DE4 4PX
M7 GWJ S. Joyner, 34 Hycliffe Gardens, Chigwell, IG7 5HJ
M7 GWM G. Goddard, 13 Rosslyn Avenue, Coventry, CV6 1GL
MW7 GWT C. Marrs, 49 Abraham Court, Lutton Close, Oswestry, SY11 2TH
M7 GWW G. Worthington, 190 Higher Road, Liverpool, L26 1UW
M7 GYT C. Clark, 2B Hayling Rise High Salvington, Worthing, BN13 3AL
M7 GYZ M. Rothery, 24 Maes Y Dafarn, Caersws, SY17 5NG
M7 GZA J. Gresty, 81 Old Hall Road, Sale, M33 2HU
MI7 GZO J. Montgomery, 27 Fir Park Broughshane, Ballymena, BT42 4DH
MW7 GZR N. Davies, 24 Tal Y Fan, Glan Conwy, Colwyn Bay, LL28 5NG
M7 GZV P. Fell, 1 Victoria Terrace, Church Walk, Melksham, SN12 6NA
MI7 GZX P. Brennan, 23 Ardchrois, Donaghmore, Dungannon, BT70 3LB
MW7 HAD J. Hadley, Rhianfryn, Caergelach, Llandegfan, Menai Bridge, LL59 5UF
M7 HAF H. Fletcher, Flat 4, Chase Court, High Street, Theale, Reading, RG7 5AR
M7 HAL M. Iddon, 40 Winchester Avenue, Waterloo, Liverpool, L22 2AT
M7 HAM N. Parker, 18 Sandown Road Bishops Cleeve, Cheltenham, GL52 8BZ
MW7 HAW A. Hawkins, 5 Keats Road, Caldicot, NP26 4LH
MW7 HAY M. Hay, Helyg, Abernant, Carmarthen, SA33 5RR
M7 HBA K. Anand, 19 Tempest Road, Upper Cambourne, Cambridge, CB23 6HW
M7 HBB T. Delsi, 42 Westergate Avenue, Brooklands, Milton Keynes, MK10 7LQ
M7 HBG H. Gruen, Allfrey House, Herstmonceux, Hailsham, BN27 4RS
MM7 HCM H. Mcintosh, 43 Woodlands Avenue, Cults, Aberdeen, AB15 9DE
M7 HDC D. Horridge, Flat 8, Blenheim Court, Marlborough Road, London, N19 4HR
M7 HDJ J. Denton, 7 Cedda Place, Sandbach, CW11 3SW
M7 HDM Z. Harmath, 2 Colyton Close, Wembley, HA0 2HG
M7 HDR I. Hammond, 88 Great Innings North, Watton at Stone, SG14 3TD
M7 HDS D. Samson, 20 Chichester Park, Westbury, BA13 3AN
M7 HDX P. Matthews, 6, Willow Road, Sunderland, dh45qf
M7 HED M. Skipper, 21 Shire Close, Kettering, NN15 6NN
M7 HEF J. Kelly, 14 Quarry Close, Bloxham, Banbury, OX15 4HA
M7 HEL S. Helyer, 21 Staddons View, Bovey Tracey, Newton Abbot, TQ13 9HN
MW7 HEN D. Edkins, Penhower Newydd Caerhun, Bangor, LL57 4DT
M7 HES Dr F. Sedgemore, 12 The Brambles, Royston, SG8 9NQ
MW7 HEW I. Hewitt, Gwern Borter Cottage Rowen, Conwy, LL32 8YL
M7 HEX K. King, Flat 18, St. Peters House, Jacobs Wells Road, Bristol, BS8 1DY
M7 HFB Y. Sun, Apartment 131, 29F Hope Street 29F Hope Street, Liverpool, L1 9BR
MM7 HFC S. Woodburn, 6 Roman Place, Bellshill, ML4 2AU
M7 HFD R. Beare, 246 Elgar Avenue, Malvern, WR142YB
M7 HFY S. Spencer, 9 Racecourse Marina, Boroughbridge Road, Ripon, HG4 1UG
M7 HGF H. Fernandes, 118 Bushfield Drive, Redhill, RH1 5BW
M7 HGK P. Morris, 5 Pebble Court 129-131 Henver Road, Newquay, TR7 3DT
M7 HGM J. Moran, 33 Chedington Close, Poole, BH17 9EQ
M7 HGN H. Nixon, 24 Lingfield Road, Worcester Park, KT4 8TG
M7 HGO P. Holpin, 32 Friars Way, Bushey, WD23 2BT
M7 HGV S. Jordan, 74 Soane Gardens, South Shields, Ne34 8nn
M7 HHH H. Hughes, 24 Broadley Green, Windlesham, GU20 6AL
M7 HHK H. Khoo, Flat 2, 59 Lilac Road, Southampton, SO16 3DA
M7 HHS D. Sutch, 2 Elizabeth Cottages, Lutterworth Road, Shawell, Lutterworth, LE17 6AE
M7 HHY M. Marsh, 64 Hewitt Avenue, St Helens, WA10 4EW
MM7 HIC Dr L. Campbell, 21 Fordyce Way, Auchterarder, PH3 1BE
M7 HIL G. Lawry, 5 Trevithick Road, Pool, Redruth, TR15 3NW
MM7 HIM A. Harker, 29 Overmills Crescent, Ayr, KA7 3LN
MM7 HIR M. Joglekar, 21 Lovat Avenue, Bearsden, Glasgow, G61 3LQ
M7 HJB H. Bingham, 14 Slaters Close, Rushden, NN10 0EE
MW7 HJE H. Evans, 20 South Hook Road, Gelliswick, Milford Haven, SA73 3RU
M7 HJH H. Hughes, 346 Griffiths Drive, Wolverhampton, WV11 2LB
M7 HJM H. Mclean, 3 Yarnton Close, Royton, Oldham, OL2 6PF
M7 HJO J. Holroyd, 31A Cross Street, Tenbury Wells, WR15 8EF
M7 HJP H. Piper, 14 Ashwood Close, Worthing, BN11 2AF
M7 HJR H. Robinson, 7 Higney Road, Hampton Vale, Peterborough, PE7 8LZ
M7 HJS A. Stocks, 82 Carr Street, Huddersfield, HD3 4BQ
MI7 HJW J. Watson, 74 Bann Meadows, Bann Road, Ballymoney, BT53 7RN
M7 HKA J. Durkin, 18 Woodlands Road, Harrow, HA1 2RS
M7 HKP D. Cotton, 53 Brookdale Road, Hartshorne, Swadlincote, DE11 7HH
M7 HKY J. Hockey, 5 Old Farm Road, Birmingham, B33 9HH
M7 HLA P. Coleman-Smith, Rowley Bank, Quarry Bank, Utkinton, Tarporley, CW6 0LA
MM7 HLD A. Mac An Iasgair, 3 Poppy Court, Abbey Road, Scone, Perth, PH2 6GD
M7 HLG G. Williams, 40 Manion Avenue, Liverpool, L31 4ED
M7 HLK S. Fisher, 37 Lancaster Crescent, St. Eval, Wadebridge, Pl277TP
M7 HLN H. Fox, 24 Northwood Street, Stapleford, NG9 8GH
M7 HLR M. Hasler, 15 Priory View, Cornworthy, Totnes, TQ9 7HN
M7 HMB D. Williams, 6 Swettenham Close Alsager, Stoke-on-Trent, ST72XG
M7 HMC Dr H. Colman, 16 Gleneagle Road, Plymouth, PL3 5HJ
MI7 HMD D. Hargreaves, 19 Laurel Park, Ahoghill, Ballymena, BT42 1LN
M7 HME S. Swinton, 9 Carnoustie, Bolton, BL3 4TF
M7 HMO R. Jones, 43 Walton Avenue, Penwortham, Preston, PR1 0XR
M7 HMR W. Andes, 141 Sunningdale Avenue, Hanworth, Feltham, TW13 5JS
M7 HMS A. Draycott, 56 Jacks Lane, Marchington, Uttoxeter, ST14 8LW
MW7 HMT H. Todd, 29 Maes Maelor, Penparcau, Aberystwyth, sy23 1sz
MI7 HNC C. Oliver, 46E Reilly Park, Banbridge, BT32 3DJ
M7 HNK J. Henkins, 69 Verdayne Gardens, Warlingham, CR6 9RP
M7 HNU B. Gannamani, 13 St. Anthony Road, Basingstoke, RG24 9XP
M7 HOG S. Hogg, 38 The Gables, Widdrington, Morpeth, NE61 5RA
M7 HOM S. Mcgaughey, 60 Newington Street, Hull, HU35LX
M7 HOP Dr J. Hopcroft, 18 Heron Road, Honiton, EX14 2GL
M7 HOU G. Hoult, 47 Woodborough Road, Mansfield, NG19 6NN
M7 HPS J. Sykes, 79 Owler Park Road, Ilkley, LS29 0BG
MM7 HPX N. Taylor, 9 Tigh Na Mara, Main Street, Gaza, Portmahomack, Tain, IV20 1YS
MM7 HQS H. Matthews, 88 Nevis Crescent, Alloa, FK10 2BN
M7 HRT R. Hart, 26 Bell Road, Norwich, NR3 4RA
M7 HSG G. Slater, 38 Bradstock Road, Epsom, KT17 2LH
M7 HSI D. Markland, 1 Andromeda Way, Brackley, NN13 6GU
M7 HTC H. Castree, 31 Fairview Thickwood, Colerne, Chippenham, SN14 8BS
M7 HTL S. Pelc, 44 Frithwood Crescent, Kents Hill, Milton Keynes, MK7 6BQ
M7 HTR R. Hopkins - Esteris, 1 Edendale Terrace, Horden, Peterlee, SR8 4RD
M7 HTS J. Hogarth, 23 Willow Road East, Darlington, DL3 6PY
M7 HTU E. Lewis, 49 Ridgebourne Road, Shrewsbury, SY3 9AB
M7 HUB D. Hubbard, 66 Parkfield Road, Rainham, ME8 7SZ
M7 HUD D. Hudson, 30 Sarmatian Fold, Ribchester, Preston, PR3 3YG
MM7 HUE Dr Z. Yang, 48 Foxglove Road, Newton Mearns, Glasgow, G77 6FP
M7 HUG M. Townsend, 35 Church Way, Hungerford, RG17 0JP
M7 HUL A. Shakesby, 14 Dawnay Road, Bilton, Hull, HU11 4HB
M7 HUX D. Robertson, Lingley, Cold Pool Lane, Badgeworth, Cheltenham, GL51 4UP
M7 HVA H. Valentine, 117 Great North Road, New Barnet, Barnet, EN5 1AW
M7 HVH C. Sidaway, 6 Rookery Chase, Deepcar, Sheffield, S36 2NF
M7 HVL K. Howell, 4 Wisteria Gardens Harlow Green, Crag Lane, Harrogate, HG3 1FP
M7 HWB D. Maloney, 21 Poplar Court, Northampton, NN3 6SE
M7 HWK A. Hawkins, Marellis, North Lane, Swaby, Alford, LN13 0BD
M7 HWL D. Laker, 85 The Gore, Basildon, SS14 2DB
M7 HWO Dr H. Orridge, Stonecross Cottage, Wadworth Hall Lane, Wadworth, Doncaster, DN11 9BH
MM7 HXI M. Moir, 21 Primrosehill Gardens, Aberdeen, AB24 4EQ
M7 HXO R. James, 22 Moss Road, Watford, WD25 0EN
M7 HYB M. Greenway, 28 Purbrook, Tamworth, B77 2NB
M7 HYE A. Dandy, Longridge, Oakridge Lynch, Stroud, GL6 7NZ
M7 HYL F. Stephenson, 67 Geneva Avenue, Lincoln, LN2 4EB
M7 HYM C. Mears, Waterways, Scotland Yard, Priors Leaze Lane Hambrook, Chichester, PO18 8RQ
M7 HYP D. Cook, 5 Pendennis Court, Red Lane, Bugle, St Austell, PL268QP
MM7 IAB I. Black, 30 Victoria Street, Dumbarton, G82 1HP
MM7 IAM I. Mort, 5 Rockfield Road, Tobermory, Isle of Mull, PA75 6PN
M7 IAP I. Pennington, 2 St. Lukes Place, Preston, PR1 5DE
M7 IAZ I. Lowbridge, 4802 East Ray Road #23-513, Phoenix, USA, 85044
M7 ICD J. Caine, 16 Wheatfield Drive, Shifnal, TF11 8HL
M7 ICG I. Gilbert, 84 Carlyle Road, London, W5 4BJ
M7 ICH C. Healey, 37 Ennerdale Drive, Manchester, M33 5NF
M7 ICK M. Percival, 10 Binder Close, Higham Ferrers, NN108PH
M7 ICL I. Larman, 30 Kiln Drive, Tydd St Mary, Wisbech, PE13 5RA
M7 ICP A. Moran, 451 Haworth Road, Allerton, Bradford, BD15 9LL
M7 ICQ D. Mason, 25 Culm Valley Way, Uffculme, Cullompton, EX15 3XZ
MM7 ICR I. Crosbie, West Gates Avenue, Lochwinnoch, PA12 4HG
M7 ICS I. Sherratt, 52 Corran Close, Northampton, NN5 7AL
M7 ICY S. Brace, 2 Greenfields Cottages, Brockhampton Estate, Bringsty, WR6 5TB
MW7 IDI I. Williams, 72A Stryt Issa, Pen-Y-Cae, Wrexham, LL14 2PN
M7 IDM I. Mills, Garfield Station Road, Ganton, YO12 4PB
M7 IDP I. Eaton, 12 St. Christopher Road, Uxbridge, UB8 3SG
M7 IDS I. Stanley, 11 Reading Road, Burghfield Common, RG7 3PY
M7 IDY I. Duxbury, 8 Eden Gardens, Longridge, Preston, PR33WF
M7 IED M. Belch, The Nook, Lower North Street, Cheddar, BS27 3HA
MM7 IEF A. Mclennan, 10 Trefoil Place, Ayr, KA7 3XG
MM7 IFC I. Collins, 22 Ninian Quadrant, Glenrothes, KY7 4HP
M7 IFR I. Rudge, 24 Granville Road, Cradley Heath, B64 7QH
MW7 IFZ I. Jones, 27 Cae Bach Aur Estate, Bodffordd, Llangefni, LL77 7JS
M7 IGG G. Gray, 20 Philip Road, Staines-upon-Thames, TW18 1PW
M7 IGN Prof. I. Neal, 8 Rushey Gill, Brandon, Durham, DH7 8BL
M7 IGT S. Mucklow, Blithbury House, Blithbury Road, Rugeley, WS15 3HR
M7 IGW I. Westland, 123 Eton Hill Road Radcliffe, Manchester, M26 2XQ
MM7 IIE H. Chen, Iq Fountainbridge Room 118E 114 Dundee Street, Edinburgh, EH11 1AD
MM7 IIH J. Jarvie, Berryhill Farm, Tak-Ma-Doon Road, Kilsyth, Glasgow, G65 0RY
M7 IIO T. Groves, 34 Maple Way, Coulsdon, Cr53rn
M7 IIZ C. Bohoris, Waltana, Fallowmead, Stag Close, Bishopstoke, Eastleigh, SO50 8AN
M7 IJB I. Bentley, 34 Knapton Close, Chelmsford, CM1 6UL
M7 IJC I. Chilley, 24 Blackheath Road, Lowestoft, NR33 7JG
MM7 IJK I. Killoh, 49 May Place, Perth, PH1 3BJ
M7 IJM I. Mcadam, 43 Hightown Road, Ringwood, BH24 1NQ
M7 IJP S. Holmes, Plot 5 Van 8 Highgrove Residential Park, Highgrove Lane,

Minsterworth, Gloucester, GL2 8LE
M7 IKA J. Hawkes, 142 Mercian Way, Slough, SL1 5LE
M7 IKB I. Bourne, The Granary, Pincott Farm, Nuthill, Brockworth, GL3 4RL
M7 IKC I. Colebourn, 21 Well Close, Addingham, Ilkley, LS29 0SH
MW7 IKH I. Hooper, 43 Meadow Lane, Porthcawl, CF365EY
M7 IKO B. Giraud, 1 Carlton Terrace, Dewsbury, WF12 9LD
M7 ILA A. Nicholson, 2 Broom Close, Leyland, PR25 5RQ
MM7 ILB Capt. B. Anderson, 1 New Houses, East Fortune, North Berwick, EH39 5JZ
MW7 ILL R. Bolger, 49 St. Benedict Crescent Heath, Cardiff, CF14 4DP
MM7 ILY S. Urquhart, Sornbank, Bridgend, Isle of Islay, PA44 7PQ
MM7 IMC I. Campbell, Fossoway Lodge, Kinross, KY13 0PD
MM7 IME S. Boal, 20 Fairinsfell, Broxburn, EH526AL
M7 IMH I. Mahmood, Loughborough University, B919111 Faraday Hall, Ashby Road, Loughborough, LE11 3TY
MW7 IMJ I. Jones, 4 Cwm Tecaf, Rhosybol, Amlwch, LL68 9PU
MW7 IML P. Lowe, 9 Felin Uchaf, Dolgellau, LL40 1NS
MW7 IMO I. Wright, The Barn, Penfeidr Farm, Treffgarne, Haverfordwest, SA62 5PL
M7 INA M. Mitchell-Hardy, Northcote Road, Norwich, NR3 4QF
M7 INC G. Jones, 158 Withington Lane, Aspull, Wigan, WN2 1JE
M7 IND A. Tree, 7 Blackthorne Road Biggin Hill, Westerham, TN16 3SH
M7 INE J. Miller, Manor Farm House, 16 Cause End Road, Bedford, MK43 9DB
M7 ING R. Smith, Guyatt Court, Burwell, CB25 0DP
M7 INN A. Seal, 33 Daleham Mews, London, NW3 5DB
M7 INS S. Coppins, 33 Honywood Road, Lenham, Maidstone, ME17 2HH
M7 INT G. Farmer, 288 Margate Road, Ramsgate, Ct12 6aj
MM7 INV I. Hamilton, 27 Cameron Avenue, Balloch, Inverness, IV2 7JT
M7 INX A. Henderson, Higginstown, Ballyshannon, Ireland, F94 ND35
MM7 IOD J. Jarvis, 39 Ladyacre Wynd, Irvine, Ka112fy
MW7 IOL I. Roberts, Llwydiarth, 38 Tal Y Cae, Tregarth, Bangor, LL57 4AE
M7 ION T. Heyes, 11 Roughtor View Planet Park, Delabole, PL33 9BX
M7 IOR I. Rutherford, 12 Madox Brown End, College Town, Sandhurst, GU47 0GJ
M7 IOS P. Lisewski, 40D Gledstanes Road, London, W14 9HU
MM7 IOW S. Mason, 31 Clifton Road, Lossiemouth, IV31 6DJ
M7 IPL I. Leyland, 61 Stacey Avenue, Milton Keynes, MK12 5DN
M7 IRB I. Bell, 11 Cherry Blossom Close, Northampton, NN3 9DN
M7 IRC N. Dean, 24 Haddon Road, Leamington Spa, CV32 7QX
MM7 IRG Dr M. Sutcliffe, 11 Low Borland Way, Eaglesham, Glasgow, G76 0BP
M7 IRJ J. Abel-Moir, 6 Egypt Hill, Leckhampstead, Newbury, RG20 8QF
MM7 IRM Dr I. Maoileoin, Wester Blackfold, Abriachan, Inverness, IV3 8LB
MM7 IRQ G. Stirling, 45 Mains Ave, Invergordon, IV18 0JT
M7 IRW R. Irwin, 2 Sutton Close, Redditch, B98 0JR
M7 ISA M. Davies, 40 Lake Crescent, Daventry, NN11 9EB
M7 ISB B. Isac, 66 Dacre Gardens, Borehamwood, WD6 2JW
MI7 ISF I. Forde, 101 Princess Way, Portadown, BT635EJ
M7 ISG I. Siomadis, 15 Maidstone Road, Norwich, NR1 1EA
M7 ISM P. Lyba, 6 Ambassadors Way, North Shields, NE29 8ST
MM7 ITA C. Di Bona, 294 Broadholm Street, Glasgow, G22 6DN
M7 ITD D. Ward, Flat 15, Barnes Court, Whitley Mead, Stoke Gifford, Bristol, BS34 8XT
M7 ITI D. Foxcroft, 8 Cowlard Close, Launceston, PL15 7EQ
M7 ITR I. Robinson, 220 Collis Street, Stourbridge, DY8 4EQ
M7 IVA I. Lucas, Orde House, Whitchurch, Ross-on-Wye, HR9 6DQ
M7 IVN I. Peikov, 41 Wellfield Road, London, SW16 2BT
M7 IVY A. Gilbert, 31 Elizabeth Road, Leamington Spa, CV31 3LJ
M7 IWF D. Smith, 3 Skipton Close, Corby, NN18 0NS
M7 IWN I. Antoniw, 128 Rosemary Rd Beighton, Sheffield, S20 1DA
M7 IWT I. Thompson, 26 Countrymans Way, Shepshed, Loughborough, LE12 9RB
M7 IWW I. Watts, The Walled Cottage 9 Holland Road, Frinton-on-Sea, CO13 9DH
M7 IXH S. Simon, 4 Mina Close, Peterborough, PE2 8TG
MW7 JAB J. Byast, 10 Chapel Street, Amlwch Port, Amlwch, LL68 9HT
M7 JAC J. Parkes, 4 Round Saw Croft, Rubery, Rednal, Birmingham, B45 9TT
M7 JAD J. Davis, Wakefield, The Street, Tirley, Gloucester, GL19 4ES
M7 JAG J. Stone, 55 Austin Waye, Uxbridge, UB8 2RQ
M7 JAJ J. Jobling, 152 Derwent Way, York, YO31 0RQ
M7 JAP P. Whiteley, 16 Fustian Avenue, Heywood, OL10 3FN
M7 JAQ J. Quinlan, 45 Mill Road, Beccles, NR349UT
M7 JAR J. Robinson, 6 Oswin Place, Walsall, WS3 1PU
M7 JAS J. Singh, Lynx House, School Lane, Hadlow Down, TN22 4JE
M7 JAT J. Tesseyman, 31 Tennyson Street, Guiseley, Leeds, LS20 9LN
M7 JAV J. Singh, 35 Eaton Rise, London, W5 2HE
M7 JAW J. Withington, 20 Bond Way, Hednesford, Cannock, WS12 4SN
M7 JAY J. Buttery, 11 Warton Lane, Austrey, Atherstone, CV9 3EJ
M7 JAZ P. Pennington, 10 Rosina Street, Manchester, M11 1HX
M7 JBB B. Butler, The Mount, Cobmoor Road, Kidsgrove, Stoke-on-Trent, ST7 4DF
M7 JBE J. Etheridge, 34 Westonzoyland Road, Bridgwater, TA6 5BN
M7 JBL J. Booth, 2 Fairfax Mews, London, E161TY
M7 JBM J. Minterne, 32A Schole Road, Willingham, Cambridge, CB24 5JD
M7 JBO J. Boslem, 14 Morrins Close, Great Wakering, SS3 0DY
M7 JBP P. Kinch, 22 Westcott Way, Honiton, EX14 1JE
M7 JBS J. Stone, 47 Alma Road, Weymouth, DT4 0AJ
M7 JBT B. Thompson, 301 Pleck Road, Walsall, WS2 9HA
M7 JBW J. King, 10 Holly Park View, Gateshead, NE10 9NH
M7 JBY J. Youel, 18 Park Avenue, Penistone, Sheffield, S36 6DN
M7 JCA J. Carter, 4 Pewter Court, Wilnecote, Tamworth, B77 5FX
MI7 JCB J. Bradshaw, 51 Albany Drive, Carrickfergus, BT38 8BF
M7 JCC J. Clark, 7 Moorland Avenue, Blackburn, BB2 5EQ
M7 JCD J. Diss, 18 Rutland Avenue, Thornton-Cleveleys, FY5 2DU
M7 JCE J. Clark, 4 Exeter Street, North Tawton, EX20 2HB
M7 JCG J. Groves, 350 Middle Deal Road, Deal, CT14 9SN
M7 JCH J. Howarth, 67A Crawford Ave, Manchester, M298ET
M7 JCK J. Macnally, 3 Sydenham Terrace, Covington Road, Westbourne, Emsworth, PO10 8SZ
M7 JCL J. Campion, 5 Sylvester Court, Beverley, HU17 8DZ
M7 JCM J. Mcleod, 11 Ryebank Close, Birmingham, B30 1SN
M7 JCP J. Pablo, 22 Nightingale Lane, London, N8 7QU
M7 JCS J. Sydney, 17 Streamside, Tonbridge, TN10 3PU
M7 JCX J. Clare, 5 Highbury Close, Westhoughton, Bolton, BL5 2QU
M7 JDB J. De Beer, 5 Kyte Close, Warminster, BA12 8GE
M7 JDC J. Cumming, Farley Court, 100 Homer Close, Gosport, PO13 9TL
M7 JDH J. Hunter, 53 Grove Road, Tiptree, Colchester, CO5 0JJ
M7 JDI J. Elliott, 22 Bath Crescent, Huntingdon, PE28 2EH
MM7 JDL J. Lockhart, 41 Westfield Ave, Cupar, KY15 5AA
M7 JDM D. Moyse, 2 Station Road, Princetown, Yelverton, PL20 6QX
M7 JDP J. Powell, The Gables, Whites Lane, Kessingland, Lowestoft, NR33 7TF
M7 JDW J. Wilkinson, 8 Hunters Point, Chinnor, OX39 4TG
M7 JEE S. Mukherjee, 28 Culverlands Close, Stanmore, HA7 3AG
M7 JEF J. Sparrow, 12 St. Huberts Close, Gerrards Cross, SL9 7EN
M7 JEH J. Hartley, 20A Park Homer Drive, Wimborne, BH21 2SR
MW7 JEL J. Slade, Flat 1, 29 Broad Street, Welshpool, SY21 7RW
M7 JEN J. Swift, 16 Lime Tree Avenue, Kiveton Park, Sheffield, S26 5NY
M7 JEP D. Palmer, Ts37Hp, Middlesbrough, Ts37hp
M7 JEX C. Lawton, 15 Grenville View, Cotford St. Luke, Taunton, TA4 1JH
MM7 JFA J. Faulds, 123 East Main Street, Darvel, KA17 0JG
MM7 JFB J. Burns, 191 Grahams Road, Falkirk, FK2 7BU
M7 JFF F. Gutai, Flat 1-082, Arthur Sanctuary House, Sandfield Road, Headington, Oxford, OX3 7RH
M7 JFH J. Fisk, 2 Garden Cottage, Hartsholme Park, Lincoln, LN6 0EY
M7 JFJ J. Forte, 109 Lower Galdeford, Ludlow, Shropshire, Sy8 1Ru, Ludlow, SY81RU
M7 JFK J. King, 99 Devizes Road, Salisbury, SP27LQ
M7 JFL D. Searley, 18 Plantagenet Crescent, Bournemouth, BH11 9PJ
M7 JFR B. Roberts, Lynwood, Wrigglebrook, Kingsthorne, Hereford, HR2 8AR
M7 JFX J. Fox, 2 Rhodes Avenue, Rotherham, S61 3LG
M7 JGB J. Davies, 166 Victoria Avenue, Princes Avenue, Hull, HU5 3DY
M7 JGD J. Green, 42 Longfield Ave, Newbarn, DA3 7LA
M7 JGP J. Pearce, 53 Kettlewell Close, Woking, GU21 4HY
M7 JGS J. Southan, 39 Hardy Way, Enfield, EN2 8NW
M7 JGW J. Weatherill, Northend Cottage, North End Road, Yatton, Bristol, BS49 4AS
M7 JGX J. Greenwood, 45, Billington Gardens Hedge End, Southampton, so30 2ax
MM7 JGY G. Yang, 48 Foxglove Road, Newton Mearns, Glasgow, G77 6FP
M7 JGZ C. Millar, 195 Worsley Road, Frimley, Camberley, GU16 9BH
M7 JHA J. Ratcliffe, 7 De Havilland Drive, Hazlemere, HP15 7FP
M7 JHD J. Haswell, 15 Oaklands, Swalwell, Newcastle upon Tyne, NE16 3EJ
MW7 JHN J. Hardman, Rosedale, Abergele Road, Old Colwyn, Colwyn Bay, LL29 8AS
M7 JHR J. Ratty, 12 Challoners Close, East Molesey, KT8 0DW
M7 JHT J. Turrell, 24 Barnfield, East Allington Totnes Devon, Totnes, TQ9 7QR
M7 JHW J. Wheaton, Flat 4, 25 Greenside, Waterbeach, Cambridge, CB25 9HW
MW7 JHX J. Humphreys, 134 Mill Street, Tonyrefail, Porth, CF39 8AF
M7 JHY J. Hurley, 64 Carleton Road, Chorley, PR6 8UB
M7 JIA R. Jia, 48 Boston Av, Reading, RG1 6JU
M7 JIB G. Ogram, 42 Juniper Way, Bradley Stoke, Bristol, BS32 0BR
M7 JIM J. Austin, 25 Cotmore Gardens, Thame, OX9 3LZ
M7 JIZ J. Cook, 20 Waverley Avenue, Kidlington, OX5 2NA
M7 JJC J. Claridge, 31 Conifer Crest, Newbury, RG14 6RS
M7 JJG J. Glass, 8 Marlborough Close, Ramsbottom, Bury, BL0 9YU
M7 JJI J. Iseton, 23 Heddon Grove, Ingleby Barwick, Stockton-on-Tees, TS17 0FT
M7 JJK J. Page, 22 Coppersmith, Combs, Stowmarket, IP14 2FD
M7 JJO D. Simpson, 1 Bridge Cottage, Blackpool, Dartmouth, TQ6 0RG
M7 JJT J. Turner, 24 Edith Wood Close, Alvaston, Derby, DE24 0HJ
MW7 JJY J. Young, Rhos Owen, Llangristiolus, Bodorgan, LL62 5RD
M7 JKA Dr J. Kinrade, 20 Harrier Court, Fenton Street, Lancaster, LA1 1AE
M7 JKC D. Langford, 1 Peveril Street Allenton, Derby, De248dg
M7 JKD J. Dalton, 166 Ennerdale Road, Cleator Moor, CA25 5LF
M7 JKF J. Korzybski, 24 Brabazon Close, Shortstown, Bedford, MK42 0FF
M7 JKM J. Moore, 19 Exe View, Exminster, Exeter, EX6 8AL
M7 JKN M. Jackson, Paragon Hotel 123 North Marine Road, Scarborough, YO12 7HU
M7 JLC J. Cole, 68 Northfield Road, Gloucester, GL4 6TX
M7 JLD J. Dickins, 78 Gobions, Basildon, SS16 5AY
M7 JLG S. Raneses Grospe, 8 Lincombe Road, Manchester, M22 1GA
MW7 JLM J. Simons, 24 Church Terrace, Porth, CF39 0ET
M7 JLW J. Walkey, 1 Elm Avenue, Poulton-le-Fylde, FY6 7SP
M7 JLX J. Laud, 2F Manor Road, Atherstone, CV9 1QJ
M7 JMA J. Mclachlan, 3 Wentworth Road, Oxford, OX2 7TG
M7 JMC J. Mcdonald, 18 Ridgeway, Gateshead, NE10 8DD
M7 JMD J. Davidson, 4 Willows Avenue, Alfreton, DE55 7ER
M7 JME J. Earle, 24 Brook Vale, Charlton Kings, Cheltenham, GL52 6JD
M7 JMF J. Marvel, Dean House Farm, Nordan, Leominster, HR6 0AW
M7 JMH M. Hill, Buzon 169, Avd De La Condomina 53, Local 3, Alicante, Spain, 3540
M7 JMJ J. Jimenez Sevilla, 14Bentley Close Bentley Close, Loughborough, LE11 1SY
M7 JMO J. Partovi, 3 Willowburn, Morpeth, NE61 2FY
MM7 JMP C. Jump, 15 The Lade, Bonhill, Alexandria, G83 9JR
MM7 JMR J. Rennie, 39 Birdland Avenue, Bonesss, EH519LX
MI7 JMS J. Mcshane, 14 Annahugh Road Loughgall, Armagh, BT61 8PQ
MW7 JMW J. Whale, 60 Thornhill Road, Cardiff, CF14 6PF
M7 JMX J. Mullings, 112 Vicarage Lane, Great Baddow, Chelmsford, CM2 8JD
MW7 JNC J. Craig, 32 The Uplands, Port Talbot, SA13 2EW
M7 JNH S. Johnston, 33 Angus Crescent, North Shields, NE29 6UF
MI7 JNJ J. Agg, 70C Cloghoge Road, Tandragee, Craigavon, BT62 2HB
MM7 JNK J. Park, 6 Robsland Avenue, Ayr, KA7 2RW
M7 JNQ A. North, 37 Cromwell Road, Southend, SS25NG
M7 JNR A. Primo, 68 Menzies Avenue, Basildon, SS15 6SY
M7 JNU J. Nuttall, 73 Severn Drive, Walton-Le-Dale, Preston, PR5 4TD
M7 JNY J. Curtis, Juniper, Broadway Lane, Fladbury, Pershore, WR10 2QF
M7 JOC J. O'Connell, 23 Halstead Road, Gosfield, CO9 1PG
M7 JOD J. Pollard, 9 Chillum Drive, Manchester, ol9 0ly
MI7 JOE J. Hannon, 75 Suffolk Road, Belfast, BT11 9PU
M7 JOL J. Wainwright, 83 Bronte Paths, Stevenage, SG20PJ
MW7 JON J. Loisz, 46 Fairview Avenue, Risca, Newport, NP11 6HW
M7 JOP A. Jopson, 43 Lower Landedmans, Westhoughton, Bolton, BL5 2QL
M7 JOS J. Stringer, 2 Westend Marston Magna, Somerset, BA22 8BW
M7 JOX K. Brailsford, 4 Maple Drive Gedling, Nottingham, NG44AF
M7 JOY J. Organ, 58 Sycamore Close, Taunton, TA1 2QJ
M7 JOZ A. Jozwik, 68 Pipit Rise, Bedford, MK41 7JT
M7 JPC J. Christiaens, 6 Richard Hicks Drive, Scarning, Dereham, NR19 2TN
M7 JPF J. Fulcher, 2 Bagot Grove, Stoke-on-Trent, ST1 6JD

M7 JPM J. Meredith, 26 Hallam Drive, Shrewsbury, SY1 4YE
M7 JPP J. Parish, 60 New Drove, Wisbech, PE13 2RZ
M7 JPR J. Prince, Wellfield Avenue, Luton, LU33AT
M7 JPS N. Mantle, St36Qj, Stoke-on-Trent, St36qj
M7 JPX J. Walters-Pennell, Clopton Grange, Clopton, Woodbridge, IP13 6QR
M7 JRA Dr J. Addis, 28 Shaw Street Culcheth, Warrington, WA3 5EX
M7 JRB J. Baines, 34 Crompton Road, Stone, ST15 8NL
M7 JRG J. Godbold, 22 Lark Hill Rise, Winchester, SO22 4LX
M7 JRH J. Howling, Sysonby Knoll Hotel, Asfordby Road, Melton Mowbray, LE13 0HP
M7 JRK J. Kermode, 17A Vancouver Road, London, SE23 2AG
M7 JRM J. Bailey, 97 Cherry Tree Ave, Cowplain, PO88AX
MM7 JRS J. Scott, 81 Kinellar Drive, Glasgow, G14 0EU
MM7 JRX J. Hutton, 66/2 Craigmount Brae, Edinburgh, EH12 8XF
M7 JSA J. Stephens, 16 Northfield Road, Harborne, Birmingham, B17 0SU
M7 JSC J. Scully, 22, Lansdowne Terrace West North Shields, Newcastle upon Tyne, NE29 0RW
M7 JSD J. Dale, 123 Rosedale Gardens Belton, Great Yarmouth, NR31 9PL
M7 JSF J. Fletcher, 3 Moorend Glade Charlton Kings, Cheltenham, GL53 9AT
MW7 JSG E. Phillips, 60 Gittin Street, Oswestry, SY111DS
M7 JSH J. Shears, 7 Tinderley Grove, Huddersfield, HD5 8PE
M7 JSI B. Wolfe, 60 Lindley Street, Rotherham, S65 1RT
M7 JSK J. King, 1 Kennion Road, Wells, BA5 2NP
M7 JSL A. Cramp, 30 Park Drive, Hastings, TN34 2PR
M7 JSP J. Pope, 5 Cotswold Rd Little Sandhurst, Sandhurst, GU47 8NA
M7 JSS W. Fisher, 24 Lisle Road, Weston-Super-Mare, BS22 7UA
M7 JST J. Talbot, Hales Road, Wednesbury, ws109bs
M7 JSW J. Spooner, 82 Reedswood Lane, Walsall, WS2 8QP
M7 JSY J. Yapp, 6 Avondale, Dawley Bank, Telford, TF4 2LW
MW7 JTH J. Turobin-Harrington, Michaelmas Barn, Sawmills, Kerry, Newtown, SY16 4LL
M7 JTN T. Evans, 69 Lincoln Street, Liverpool, L19 8LF
MM7 JTR J. Robertson, Pinewood Place, Edinburgh, EH225JA
M7 JTZ J. Skinner, 35 Holbein Close, Bedworth, CV12 8TA
MW7 JUE J. Burford, 13 Maes Ceidio, Llannerch-Y-Medd, LL71 7AE
M7 JUK S. Jukes, 147 Vicarage Road, West Bromwich, B71 1AE
M7 JUL J. Brewer, 1 Bentley Road, Forncett St. Peter, Norwich, NR16 1LH
MM7 JVA J. Howes, 22 Criffel Drive, Lincluden, Dumfries, DG2 0PE
M7 JVC L. Shaw, 23 Church Street, Buckingham, MK18 1BY
M7 JVH J. Henry, 32 Kendricks Fold, Rainhill, L35 9LX
MM7 JVR J. Rowlands, 3 Earls Gate, Slackbuie, Inverness, IV2 6FF
M7 JVS V. Jagannathan, 19 Wallbrook Avenue, Macclesfield, SK10 3GL
M7 JWE J. Emery, Flat 3, 12 Buxton Road, Ashbourne, DE6 1EX
M7 JWF J. Willis-Fisher, Linden House, Portman Road Pimperne, Blandford Forum, DT11 8UJ
M7 JWH J. Hunter, 53 Grove Road, Tiptree, Colchester, CO5 0JJ
MM7 JWK J. Keogh, The Mound, Dornoch, IV25 3JE
M7 JWL J. Limer, 86 Elldawn Avenue Milton, Stoke-on-Trent, st68xe
M7 JWM Dr M. Moss, 2 Crosby Avenue, Worsley, Manchester, M28 3FQ
M7 JWP J. Plant, Longacres, Station Road, Old Leake, Boston, PE22 9RF
M7 JWR J. Wright-Roberts, 19 Field Lane, Dursley, GL116JF
M7 JWW J. Walker-Wilson, Rest Harrow Southside, Scorton, Richmond, DL10 6DN
M7 JWZ J. Marek, 91 Dial House Road, Sheffield, S6 4WU
M7 JXR J. Richmond, 18 Ullswater Drive, Bradford, BD6 2TE
M7 JXW J. Warrington, 22 Knights Close, Buntingford, SG9 9SE
M7 JXX J. Thompson, 11 Foxbury Drive, Orpington, BR66EJ
M7 JYP J. Capstick, 186 Forest Lane, Harrogate, HG2 7EE
MM7 JYS J. Steven, 34 Queens Drive, Troon, KA10 6SE
M7 JZN J. Nichols, 4 Elmlee Close, Chislehurst, BR7 5DU
M7 KAD N. Jaggs, 15 St. Anthonys Road, Kettering, NN15 5HT
M7 KAE K. Barnett, 22 Highclere Road, Southampton, SO16 7AW
M7 KAG K. Gillkerson, Flat 14, Target Place, 489 Butts Road, Southampton, SO19 1AD
M7 KAK K. Kariraman, 26 Tythe Barn Lane, Shirley, Solihull, B90 1RW
MW7 KAP K. Gamwasam, 01, Churton Drive, Wrexham, LL13 8RU
M7 KAS K. Fenton, 40 High Street, Easington Lane, Houghton le Spring, DH5 0JN
MI7 KAT K. Nesbitt, 2 Church Road, Gracehill, Ballymena, BT42 2NL
M7 KAU A. Mahal, Kingsley Hall Stretton Road, Great Glen, Leicester, LE8 9GP
M7 KAV P. Walmsley, 24 Camton Road, Middleleaze, Swindon, SN5 5TP
M7 KAW K. Welch, Jacobs Leys, Falcon Lane, Ledbury, HR8 2JS
M7 KBO T. Menzies, 4 Meadow Road, Muxton, Telford, TF2 8JH
M7 KBT K. Tang, Mail Boxes Etc Suite 88 1 Hanley Street, Nottingham, NG1 5BL
M7 KCB K. Beckett, 29 Park Road, Camberley, GU15 2SP
M7 KCD C. Daly, 2 Withen Cottages, Withen Lane, Aylesbeare, Exeter, EX5 2JQ
M7 KCE B. Toombs, 21 Hawthorn Crescent, Shepton Mallet, BA4 5XR
M7 KCP K. Pugh, 4 Salt Boxes, Pinvin, Pershore, WR10 2LB
M7 KCR K. Charkseliani, 2, Raymond Crescent, Guildford, GU2 7SX
M7 KCY C. Jones, 4 Guestwick, Tonbridge, TN10 4HU
MW7 KDA K. Smith, 48 Ffordd Gryffydd, Llay, Wrexham, LL12 0RT
M7 KDB K. Brough, 161 Gloucester Road, Kidsgrove, Stoke-on-Trent, ST7 1EH
M7 KDH K. Dhun, 2 Sparvell Road Knaphill, Woking, GU21 2RR
M7 KDR D. Brook, 24 Botany Avenue, Bradford, BD2 1EU
M7 KDU S. Callen, 3 Pendleton Gardens, Blackfield, Southampton, SO45 1DQ
M7 KDV P. Field, 63 Hartford Road, Davenham, Northwich, CW9 8JE
M7 KDZ K. Davenport, 6 Wolfe Close, Walton, Chesterfield, S40 2DF
M7 KEC J. Crawley, 65 Wrentham Avenue, Herne Bay, CT6 7UX
MW7 KEE S. Keeble, Mynachlog, Tyn Y Gongl, LL74 8SG
MM7 KEF K. Sterling, 1 Glamis Road, Kinghorn, Burntisland, KY3 9UR
MM7 KEH J. Hunter, 22 Watson Place, Dunfermline, ky12 0dr
M7 KEI K. Evans, 10 Wharfedale Close, Blackburn, BB2 5EY
M7 KEJ S. Jordan, 8 Averham Close, Swadlincote, DE11 9SG
MM7 KEL S. Kelly, 15 Shinwell Avenue, Linnvale, Clydebank, G81 2RA
M7 KEN K. Spilsbury, 81 Leacroft Avenue, Wolverhampton, WV10 9DB
M7 KEP P. Kenderes, 9 Waverton Road, London, SW18 3BY
M7 KEV K. Allison, 4 Ashwood Grove, Sunderland, Sr53bu
M7 KEZ K. Buckwell, 31 West Green, Stokesley, TS95BE
M7 KFC P. Roppa, 7 Kilowna Close, Charvil, Reading, RG10 9QU
M7 KFG C. Freebody, 94 Fingringhoe Road, Colchester, CO2 8EE
M7 KGA K. Gracey, Gwendoline Close, Merseyside, Ch611dl
MM7 KGC R. Condie, 48 Carleton Avenue, Glenrothes, KY7 5AJ
MM7 KGT K. Thomson, 26 Canal Court, Linlithgow, EH49 6LZ
M7 KGW K. Wheaton, 3 Walnut Tree Crescent, Fenstanton, PE289LE
M7 KHA H. Khan, 167 Desborough Avenue, High Wycombe, HP11 2ST
M7 KHB K. Bridge, 43 Highbank, Blackburn, BB1 9SX
M7 KHX K. Heydon, 34 Charles Road, Amble, Morpeth, NE65 0SQ
MW7 KID B. Ryan, 8 Beechfield Close, Garnlydan, Ebbw Vale, NP23 5EN
MM7 KIE R. Mackie, 72 Rosemount Crescent, Carstairs, ML118QW
M7 KIK B. Brightwell, 3 Longedryve, Off Wavell Avenue, Colchester, CO2 7HH
M7 KIM S. Kimber, 8 Westfield View, Wakefield, WF1 3RU
M7 KIT A. Snell-Pym, 50 Newton Avenue, Gloucester, GL4 4NU
MW7 KJC K. Cox, 61 Folly View, Penygarn, Pontypool, NP4 8BU
M7 KJD J. Drew, Slough Hall, Slough Lane, Little Cornard, Sudbury, CO10 0NY
M7 KJG K. Green, 9 Oakfield Road, Bishops Cleeve, Cheltenham, GL52 8LA
M7 KJL K. Lever, 121 Greenwater Court Mainway, Lancaster, la12ay
M7 KJM C. Mcmahon, 42 St. Mildreds Avenue, Broadstairs, CT10 2BX
MW7 KJN K. Nicholls, 35 Partridge Road, Cardiff, CF24 3QW
M7 KJP K. Price, 17 Oakmoore, Sandymoor, Runcorn, WA7 1NR
M7 KJS K. Smith, 30 Azalea Drive, Swanley, BR8 8HZ
MM7 KJV K. Vielhaber, Nether Littlefold, Crieff, PH7 3NY
M7 KJW K. Webster, 17 Cardinal Close, Tonbridge, TN9 2EN
M7 KKB K. Bennett, 24 The Quadrant, Uppingham, Oakham, LE15 9QP
M7 KKG K. Gibson, 44 Churchville, Micklefield, Leeds, LS25 4AP
M7 KKM N. Gregory, 8 Firman Close, New Malden, KT3 4BN
M7 KKO K. Oliver, 50 Tewkes Road, Canvey Island, SS8 8HG
MW7 KKW K. Kirkpatrick, 151 Bridgend Road, Aberkenfig, CF32 9AE
M7 KKY Y. Jo, 103-503 Lg Apt., Jure2-Dong, Busan, Korea, Republic Of, 46997
M7 KLA K. Adcock, 37 Flansham Park, Bognor Regis, PO22 6QH
M7 KLB K. Bragg, Flat 5 126 Harrington Road, Workington, CA14 2UW
MW7 KLC K. Evans, 21 Orchard Street, Pontardawe, Swansea, SA8 4ER
M7 KLF J. Rosewarn, 47 Bond Street, Trowbridge, BA14 0AS
M7 KLJ P. Jones, 2 Whitley Place, Stoneley Park, Crewe, CW1 4GH
M7 KLM N. Winchcombe, 1 Talbot Villa, Talbot Street, Glossop, SK13 7DG
M7 KLR S. Mccarthy, 1 Fishmoor Drive, Blackburn, bb2 3tj
M7 KLT K. Louth, 36 Copelea Cheswardine, Market Drayton, TF9 2RX
M7 KLW S. Walker, 2 Williamson Avenue, Stoke-on-Trent, ST6 8AB
MW7 KMB K. Byast, 10 Chapel Street, Amlwch Port, Amlwch, LL68 9HT
M7 KMF K. Fell, 211 Lower Way, Thatcham, RG19 3TN
M7 KMO O. Glover, 9 Lea Road, Grays, RM16 4DD
MI7 KMQ K. Mchugh, 31 Moyle Avenue, Ballycastle, BT546NX
M7 KMS K. Singleton, 2 Rome Avenue, Burnley, bb115lq
M7 KMV J. Varkevisser, Stocks Road, Wittersham, tn307ey
M7 KMW J. Nihill, 14 Hereford Avenue, Clayton, Newcastle, ST5 3ED
MD7 KNE W. Kneale, 2 The Crescent, Douglas, Isle of Man, IM2 5ET
M7 KNG A. King, 6 Martin Court Ashby De Launde Lincoln Lincolnshire Ln4 3Tw, Lincoln, LN4 3TW
M7 KNI R. Knight, 10 Hargreaves Close, Basingstoke, RG24 9SS
M7 KNO R. Massie, 13 Noel Coward Gardens, Aldington, Ashford, TN25 7EU
M7 KNR D. Tickner, 26 Wye Road, Borough Green, Sevenoaks, TN15 8DX
M7 KOF S. Kofler, 29 Fownes Street, London, SW11 2TJ
M7 KOL K. Dowling, 8 Chathill Close, Morpeth, NE61 2TH
M7 KOP S. Allen, 26 Gloucester Avenue, Rayleigh, Essex, SS68XR
M7 KOR D. Korzeniewicz, 6 Barley Down Drive, Winchester, SO22 4LS
M7 KOW B. Staniforth, The Mill House, Chantry Mill, Chantry Lane, Storrington, Pulborough, RH20 4AB
M7 KOY T. Headley, 73 Tudor Green, Jaywick, Clacton-on-Sea, CO15 2PB
M7 KPB K. Birkett, 17 Torridon Crescent., Bradford, bd62ty
M7 KPM J. Ward, 66 Abbey Road, Astley, Tyldesley, Manchester, M29 7WG
M7 KPS K. Snipe, 5 Draycott Road, Chiseldon, Swindon, SN4 0LT
M7 KPW K. Watson, 43 Rectory Place, Weyhill, Andover, SP11 0PZ
M7 KRA K. Hiltz, 35 Highview Road, London, W13 0HA
M7 KRD K. Dodwell, 14 Montacute Road, Houndstone, Yeovil, BA22 8ZD
M7 KRE K. Mills, 3 The Seasons, Summerway, Exeter, EX4 8DQ
M7 KRH J. Burton, 2A Empire Road, Salisbury, SP2 9DF
M7 KRK C. Guri Comallonga, 10 Collingwood Road, Sutton, SM1 2RZ
M7 KRN K. Ayre, St. Andrews Farm East, Woodbridge Lane, Bedchester, Shaftesbury, SP7 0BF
M7 KRW K. Windle, 44 The Park, Kirkburton, Huddersfield, HD8 0NP
M7 KSD K. Driver, 91 Springfield Gardens, Ilkeston, DE7 8JA
MW7 KSE T. Bancroft, 2 Bryn Heli, Old Colwyn, ll299er
M7 KSF A. Brooking, 1 West Villas Cotford St Luke, Taunton, TA4 1DF
MD7 KSL C. Larkham, Monte Rosa, 7 Ballaughton Close, Douglas, IM2 1JE
M7 KST K. Turner, 40 Walsingham Close, Eastbourne, BN22 0UD
M7 KTL J. Boutros, 15 Navigation House, Whiting Way, London, SE16 7EG
MM7 KTO K. Todd, 11 Hillfield Road, Inverkeithing, KY11 1BP
MW7 KTP P. Joyce, 73 Glenwood Close, Coychurch, Bridgend, CF35 5EU
M7 KTR K. Railton, 102 Shieldfield House, Barker Street, Newcastle upon Tyne, NE2 1BQ
M7 KTT P. Trent, 3.Berkshire Close, Leigh-On-Sea, SS9 4RT
MM7 KTW R. Rae, 1 Jura Drive, Tweedbank, Galashiels, TD1 3ST
MI7 KTY R. Maguire, 45 Dorchester Park Portadown, Craigavon, BT62 3EB
M7 KVM S. Ahmed, 445 Ashingdon Road, Rochford, SS4 3EN
M7 KVS K. Slaney, 69A Victoria Street, Littleport, Ely, CB6 1NA
M7 KVZ R. Elles, 8 Dilton Close, Trowbridge, BA140FS
MM7 KWA K. Waitz, C/O Waitz-Rainey, 16 Inverkeithing Road, Aberdour, Burntisland, KY3 0RS
M7 KWB K. Buckley, 44 Greenfields, Gnosall, Stafford, ST20 0HR
M7 KWG W. Green, 48 Springfield Rd Lofthouse, Wakefield, WF33FN
M7 KWI K. Wilczynski, 45 Caroline Street, Preston, PR1 5UY
M7 KWS F. Stepien, 11A Wirral Gardens, Wirral, CH63 3BD
M7 KWT K. Tyler, Ground Floor Flat 20 Brighton Road Newhaven, Newhaven, BN9 9NB
M7 KXD K. Davis, 38 South Lynn Crescent, Easthampstead, Bracknell, RG127JY
MM7 KXI A. Stewart, 5 Mosswater Wynd, Cumbernauld, Glasgow, G68 9JU
M7 KXL K. Lee, 16 Longfield Close, Barnsley, S738DX
MW7 KYL K. Jones, 24 Crown Street, Swansea, SA6 8BD
M7 KZE J. Taylor, 97 George Street, Cleethorpes, DN35 8PL
M7 KZT L. Kinzett, 52 Thistle Drive, Peterborough, PE2 8HX

M7 LAA A. Lester, Flat 7 Wyre View 28 Queens Terrace, Fleetwood, FY7 6BT
M7 LAE F. Khatir, 12 Ashburn Garth, Hightown, Ringwood, BH24 3DS
M7 LAG L. Gutierrez, Flat 2, St. Albans, Elmwood Avenue, Feltham, TW13 7AA
M7 LAH L. Horsfall, 48 Orchard Drive, Grimsby, DN32 7AW
MI7 LAJ J. O'Shaughnessy, 17 Hood Court, Antrim, BT41-4HW
M7 LAK L. Kroon, Flat 102, The Courtyard, Circus Street, Brighton, BN2 9AL
M7 LAN C. Langman, 24 St. Marys Road, Nuneaton, CV11 5AU
MM7 LAP P. Lapworth, Meadow View, Borgue, Kirkcudbright, DG6 4SH
M7 LAT L. Thompson, 3 Leasway, Grays, RM16 2HD
M7 LAW L. Wright, 16 Sheffield Close, Potton, SG19 2NY
M7 LBI I. Brown, 11 Snowdon Close, Warrington, WA5 3HD
M7 LBM L. Munckton, 20 Malvern Close, Bishops Waltham, Southampton, SO32 1AY
M7 LBX M. Dickinson, 16A Horsebrook Park, Calne, SN11 8EY
M7 LBY C. Lazenby, 33 Towning Close Deeping St. James, Peterborough, PE68HS
M7 LCA M. Piatek, 3 Chestnut Close, Wilmslow, SK9 2NT
M7 LCB L. Baker, 1B The Parade, Moss Road, Askern, Doncaster, DN6 0LF
M7 LCF Prof. M. Atkinson, 21 Dennington Crescent, Fryerns, Basildon, SS14 2FF
M7 LCJ C. Ludlow, M33 3An, Manchester, M33 3AN
M7 LCL J. Phypers, Cranborne Farm Bungalow, Blandford Road, Coombe Bissett, Salisbury, SP5 4LF
M7 LCP L. Pickering, 7 Ashmore Terrace, Sunderland, SR2 7DE
M7 LCR L. Richardson, 33 Dingley Road, Wednesbury, WS10 9PU
M7 LCS T. Bayley, 67 Shaftesbury Avenue, Feltham, TW14 9LN
M7 LCW L. Wilkin, 14 Queens Gardens, Panfield, Braintree, CM7 5AH
M7 LCY A. Gee, 18 St. James Avenue West, Stanford-le-Hope, SS17 7BB
M7 LDJ D. Loveridge, 2 Abbots Mead, Cholsey, OX10 9RJ
M7 LDS L. Sawyer, 85, Beechwood Road, Sheffield, S6 4LQ
M7 LDU G. Burridge, 88 Eaton Road, Kempston, Bedford, MK42 7RX
M7 LDV P. Copeland, 15 Station Gardens, Woodford Halse, Daventry, NN11 3PX
M7 LDW D. Whewell, 41 Fordway Avenue, Blackpool, Fy3 8jl
M7 LDX G. Taylor, 64 St. Elphins Close, Warrington, WA1 2EH
M7 LED M. Clayton, 16 Westwood Close, Scarborough, YO11 2JB
M7 LEL L. Lovell, 28 Park Lane, Blunham, MK44 3NJ
M7 LEM E. Lemaire, 238 Inskip, Skelmersdale, WN8 6JX
M7 LEN L. Wharram, 8 Bourne Way, Sutton, Sm12en
MM7 LEO L. Taylor, 22 Briggies Wynd, Kintore, Inverurie, AB51 0TX
M7 LEP L. Potter, 201, Shadyside, Doncaster, DN4 0HE
M7 LEV L. Levene, 4 Worthing Street, Wyke, Bradford, BD12 9PS
M7 LEW S. Freestone, 47 Salisbury Street, Gainsborough, DN21 2RS
M7 LEX J. Lewthwaite, 12 Fullers Close, Melksham, SN12 7BX
MM7 LEY R. Farley, 16 Greenbank Road, Darvel, KA17 0NN
M7 LEZ M. Jakes, Crimond, Coles Lane, Kingskerswell, TQ12 5BQ
M7 LFF A. Buck, 1 Sandbeck Cottage, Stragglethorpe, Lincoln, LN5 0QZ
M7 LFG M. Sykes, 5 Greenfield Cottages, Barnsley, S71 3LF
M7 LFI C. Sotiriou, Studio 210, 23 Station Street, Nottingham, NG2 3ER
M7 LGC Dr C. Roe-Bullion, 96A Nunnery Street, Castle Hedingham, Halstead, CO9 3DP
M7 LGD A. Bent, Lime Garth, Sherburn Road, Durham, DH1 2JR
M7 LGG L. Goodfellow, 19 Quarry Bank, Newcastle under Lyme, ST5 5AG
M7 LGV O. Vieira Leite, 60 Cardigan Street, Derby, DE21 6DX
M7 LGW N. Westhead, 33 Theobalds Way, Frimley, Camberley, GU16 9RF
M7 LHC A. Walrond, Leigh Hill Cottage, Lowton, Taunton, TA3 7SU
M7 LHR D. Langley, 2 Fulwood Close, Hayes, UB3 2NF
M7 LIE E. Wall, 7 Fearnham Close, Leigh, WN7 3LB
M7 LIN L. Chadwick, 19Regent Crescent, Manchester, M350LR
M7 LIP Dr P. Lidstone, Maddox, Bradworthy, Holsworthy, EX22 7QY
M7 LIU H. Liu, Saunderites, Queens Drive, Godalming, GU7 2GA
M7 LJA D. Mccorrie, 145 Mildenhall Road, Fordham, Ely, CB7 5NW
MM7 LJD L. Dickson, 8 Hailes Park, Edinburgh, EH13 0NG
M7 LJH L. Healy, 31 Roach Road, Sheffield, S11 8UA
M7 LJM L. Moyle, 14 St. Lukes Close, Kettering, NN15 5HD
MI7 LJR L. Roberts, Clonduff Drive, Belfast, BT6 9NT
M7 LKS A. Melchior, 12 Midge Hall Drive, Rochdale, OL11 4AX
M7 LKY M. Hackney, 15 Station Drive, Winchester, SO21 3FS
M7 LKZ A. Frydrych, Flat 1, Ashford Court, Bodenham Road, Hereford, HR1 2TS
M7 LLB F. Cunliffe, 33 Moor View Road, Sheffield, S8 0HH
M7 LLE A. Miller, 69 Rawmarsh Hill, Parkgate, Rotherham, S62 6DN
MM7 LLK D. Tanner, 14 Kincaldrum Place, Dundee, DD3 7HG
MM7 LLM J. Rae, 4 Hillside Crescent, Langholm, DG13 0EE
M7 LLR T. Gregory, 50 Croft Street Lincoln England Ln25Az, Lincoln, Ln25az
M7 LLS M. Duckett, 17 High Way, Lingwood, Norwich, NR13 4BU
M7 LLY C. Salisbury, 9 Oakville Road, Heysham, Morecambe, LA3 2TB
M7 LMA L. Afonso, Flat 1, 59 Croham Road, South Croydon, CR2 7HF
M7 LMB R. Barwell, 13 Pembroke Avenue, Orton Waterville, Peterborough, PE2 5EY
M7 LMC L. Coupe, 51 Leamington Drive, Sutton-in-Ashfield, NG17 5BA
M7 LMF W. Turner, Low Farm, Bishop Norton, Market Rasen, LN8 2AJ
M7 LMG S. Crowther, 4 Brigsteer Close, Clayton Le Moors, Accrington, BB5 5GE
M7 LMK L. Mery-Kennedy, 69 Alexandea Rd, Ipswich, IP42RN
MM7 LMM L. Mullaney, 8 School Road, Wellbank, Broughty Ferry, Dundee, DD5 3PL
M7 LMN M. Horwich, H Horwich Motor Engineers, Unit 2, Peel Green Trading Estate, Green Street, Eccles, Manchester, M30 7HF
M7 LMP I. Jabegu, 51 South Crescent, Blandford Camp, Blandford Forum, DT11 8AJ
M7 LMQ R. Lewis, 20 Hillary Road, Rugby, CV22 6EU
M7 LMS L. Scully, 17 Fenwick Close, Woking, GU21 3BY
M7 LNA N. Arnold, 44 Small Drove Weston, Spalding, PE12 6HS
M7 LND N. Legall, 122 Cheddon Road, Taunton, TA2 7DN
M7 LNG A. Laing, 37 Crossways, South Croydon, CR2 8JQ
M7 LOC Prof. J. Stochl, 23 Hardy Close, Huntingdon, pe29 1rr
MW7 LOI L. Illingworth, 54 Swansea Road, Garden Village, Gorseinon, Swansea, SA4 4HE
MM7 LOL A. Bennett, 60 Rosemount Buildings, Edinburgh, EH3 8DD
M7 LOT R. Coates, 1 Audrey Road, Sheffield, S13 8DQ
MW7 LOV M. Chacon-Dawson, 57 Fair View, Blackwood, NP12 3NR
M7 LOW M. Rook, 116 Ollerton Road, Athersley North, Barnsley, S71 3DL
MM7 LOX M. Macfarlane, 9 Dreghorn Park, Colinton, Edinburgh, EH13 9PH
M7 LOZ L. Woodward, Oaklands, Ashford Lane, Hockley Heath, Solihull, B94 6RH
M7 LPL P. Stanton, 12 St. Lawrence Close, Liverpool, L8 4XP
M7 LPP Dr A. Holt, Ellon House, Church Road, Sutton, Norwich, NR12 9SG
M7 LRB L. Bennett, North Cottage Frilsham, Thatcham, RG18 9UZ
M7 LRJ L. Jarrett, 7 Valley View Road, Plymouth, PL3 6QJ
M7 LRM R. Millington, 34 Grosvenor Road, Rotherham, S65 1QP
M7 LRP L. Paston, 138A Forest View Road, Manor Park, London, E12 5HX
M7 LRQ E. Rowe, Allandale View Ermine West, Lincoln, ln13rd
M7 LRY L. Young, 5A Fernlea Road, Burnham-on-Crouch, CM0 8EJ
M7 LSA S. Deakin, 20 Riccat Lane, Stevenage, SG1 3XY
M7 LSB P. Curtis, 112 Goring Road, Colchester, CO4 0DA
M7 LSE S. Evans, 2A Manor Road, Merstham, RH1 3LT
M7 LSL Rev. L. Williams, 36 Royd Court, Mirfield, WF14 9DJ
M7 LST L. Turnbull, 56 Wyngates, Leighton Buzzard, LU7 2LE
MW7 LTE L. Evans, 32 Breakwater House, Cardiff, CF11 0JQ
M7 LTH L. Hopkins, 4 Deepfield Way, Coulsdon, CR5 2SY
M7 LTM L. Marrable, 6 Piccadilly Close, Northampton, NN4 8RU
M7 LTR R. Latter, Longreach, Chapel End, Broxted, Dunmow, CM6 2BW
MW7 LTS C. Morris, Hideaway Bettws Cedewain, Newtown, Powys, SY16 3DS
MD7 LTT R. Talbot, 43 Harcroft Meadow, New Castletown Road, Douglas, Isle Of Man, IM2 1JT
M7 LTY G. Alty, 20 Long Meadows, Everton, Doncaster, DN10 5BL
MM7 LUB L. Balmforth, 1 Laburnum Avenue, Beith, KA15 1BQ
M7 LUF S. Luffingham, 25 Taylor Close, Tonbridge, TN9 2FE
M7 LUK L. Patalon, 54A Castle Road, Bournemouth, BH9 1PJ
M7 LUT A. Moss, 2 Saxted Close, Luton, LU2 9SQ
M7 LUV C. Harrod, 99 Bergholt Road, Colchester, CO4 5AF
MM7 LWC L. Copland, 67 Naughton Road, Wormit, Newport-on-Tay, DD6 8NG
M7 LWI L. Wilde, 33 Watermouth Road, Ilfracombe, EX34 9QY
M7 LWP L. Pearson, 136 Nottingham Road, Stapleford, Nottingham, NG9 8AR
M7 LWR L. Russell, 11 Hartbushes Station Town, Wingate, TS28 5GA
MM7 LWZ L. Macinally, 3 Aglath, Stenness, Stromness, KW16 3HA
M7 LXB R. Thompson, Croft Michael Farm, Croft Mitchell, Troon, Camborne, TR14 9JJ
M7 LXD L. Davis, 9 High Street, Chapel-En-Le-Frith, High Peak, SK23 0HD
MM7 LXI L. Hunter, 9 Glenview, Dalmally, PA33 1BE
M7 LXT D. Ingram, 19A Parkville Road, London, SW6 7DA
M7 LYD L. Gill, Sixhills Street, Grimsby, DN32 9HT
M7 LYF Capt. J. Pearce, Clematis Cottage, 52 Cheselbourne, Dorchester, DT2 7NP
M7 LYM S. Owens, 18 Harvester Way, Lymington, SO41 8YD
M7 LYN L. Walsh, 4 Musbury Crescent, Rossendale, BB4 6AY
MM7 LYS R. Howell, 338 Albert Drive, Glasgow, G41 5HH
MM7 LYZ L. Johnston, 61 Stafford Road, Greenock, PA16 0TG
M7 LZT L. Telfer, 7 Beech Street, Failsworth, Manchester, M35 0BE
M7 MAA G. Sangiorgi, Flat 3, Windsor Lodge, 84 High Street, Brighton, BN2 1RP
MM7 MAB S. Kent, 7 Craigie Place, Galston, KA4 8AX
M7 MAD F. Nixon, 16 Westwood Close, Scarborough, YO11 2JB
M7 MAE M. Parsons, 16 Latimer Road, Teddington, TW11 8QA
MW7 MAI C. O'Connor, Vaenor Park House, Vaenor Park, Bryndu Road, Llanidloes, SY186DN
M7 MAM M. Matis, Unit 15, Old Cement Works, Newhaven, BN9 0HS
MD7 MAN Dr J. Daniels, 24 King Orry Road Glen Vine, Isle of Man, IM4 4ES
MW7 MAP M. Pluke, 3 Maes Y Ficerdy, Rhosllanerchrugog, Wrexham, LL14 2EJ
M7 MAS M. Szpytko, 21 California Avenue, Scratby, NR29 3PE
MW7 MAT M. Burnell, Ty Talwyn Farm, Cefn Cribwr, Bridgend, CF32 0BP
M7 MAW A. Whyatt, 30 Vale Road Midway, Swadlincote, DE11 0HA
MM7 MBB M. Burke, 103 Seaforth Road, Falkirk, FK2 7TQ
M7 MBC D. Cooper, 38 Barnsdale Road, Leicester, LE4 1AR
M7 MBE R. Baldwin, 10 Queen Elizabeth Avenue, South Ferriby, Barton-upon-Humber, DN18 6HJ
M7 MBK M. Booker, 19 The Bentleys, Southend-on-Sea, SS2 6UJ
M7 MBO M. Booth, 59 Kenmore Road, Sale, M33 4LG
M7 MBT T. Youp, 123 Tanfields, Skelmersdale, WN8 8NS
MM7 MCA C. Anderson, 57 Fa'Side Avenue South, Wallyford, Musselburgh, EH21 8AN
M7 MCB M. Crosby, 40 Galtres Drive, Easingwold, York, YO61 3DJ
MI7 MCC W. Mcclean, 6 Alveston Drive Carryduff, Belfast, BT8 8RL
MM7 MCD A. Mcdonald, Kettlehills, Cupar, KY15 7TW
M7 MCF M. Lambert, 311 Manchester Road, Manchester, M34 5GR
M7 MCG M. Godfrey, 2 Park Street, Barnoldswick, BB18 5BT
M7 MCI M. Inseal, 63 Woodside Avenue, Benfleet, SS7 4NX
M7 MCJ M. Johnson, 27 Tyndall Walk, Birmingham, B32 3UN
MI7 MCK A. Mckenzie, 32 Abbot Gardens, Newtownards, BT23 8UL
MM7 MCM A. Mcmillan, 275 Saughs Drive Robroystone, Glasgow, G33 1BN
MM7 MCN J. Mclaren, 33 Foulford Street, Cowdenbeath, KY4 9NB
M7 MCO M. Onassis, 12 Gogle Close, Mattishall, Dereham, NR20 3SY
M7 MCQ T. Mcquiggan, 28 Stocks Park Drive, Horwich, Bolton, BL6 6DD
M7 MCR T. Carroll, 37 Mayswood Gardens, Dagenham, RM10 8UU
MI7 MCS A. Davidson, 8 Rashee Court, Ballyclare, BT39 9SE
MM7 MCT F. Genolini, 29 Primrosehill Drive, Aberdeen, AB24 4ER
M7 MCU J. Whitworth, 5 Hilltop Close, Shrewton, SP3 4EB
M7 MCY M. Charity, 36A Beeston Street, Northwich, CW8 1ER
M7 MDA M. Andrijauskas, 7 Hawthorn Grove, Enfield, EN2 0DU
M7 MDB M. Burden, 17 Water Meadow, Cullompton, EX15 1QS
M7 MDD J. Dee, 4 Marlborough Way, Kennington, Ashford, TN24 9HH
M7 MDH M. Haffenden, 147 The Diplocks, Hailsham, BN27 3JZ
M7 MDR M. Read, 4 Farleigh Road, Shrivenham, Swindon, SN6 8BD
M7 MDY A. Waters, 21 Kings Road, Lee-on-the-Solent, PO13 9NU
M7 MDZ P. Wilkinson, 11, St Marys Row Aldeby, Beccles, NR34 0AL
MD7 MEB M. Behrman, 60 Scarlett Road, Castletown, Isle Of Man, IM9 1PN
M7 MEC M. Crook, 25 Race Hill, Launceston, PL15 9BD
M7 MED C. Babb, 35 James Crescent, Werrington, Stoke-on-Trent, ST9 0DZ
MW7 MEJ E. Jones, 8 Ty Newydd Court, High Street, Ruabon, Wrexham, LL14 6BF
M7 MEK G. James, 19 Fairway, Chatteris, PE16 6SX
M7 MEO W. Beckett, 20 St Fabians Close Newmarket, Suffolk, CB8 0EJ
M7 MEP M. Price, 31 Covert Mead Handcross, Haywards Heath, RH17 6DL
M7 MET A. Morel, Flat 67, Chadbrook Crest, Richmond Hill Road, Birmingham, B15 3RN
M7 MEX A. Delchini, 5 Hillside, Harefield, Uxbridge, UB9 6AU
M7 MEZ Dr R. Mackie, 31C Champion Hill, London, SE5 8BS
M7 MFD M. Dougan, Flat 7 Brooke Court 300 Kilburn Lane, London, W10 4BW
M7 MFH Dr M. Al-Naday, 39 Quayside Drive, Colchester, CO2 8GE

MM7 MFR M. Ritchie, 97 Woodburn Crescent, Bonnybridge, fk4 2dl
M7 MFV M. Frolov, 135 Sapphire Road, Bishops Cleeve, Cheltenham, GL52 7YT
MI7 MFW D. Hall, 7 Kingsbury Gardens, Coleraine, BT52 2JE
M7 MGB M. Gardiner, 21 Bell Chase, Aldershot, GU11 3GY
M7 MGD M. Drury, 15 Averingcliffe Road, Bradford, BD10 9HQ
M7 MGG M. Garth, 77 Centenary Way, Chelmsford, CM1 6AU
M7 MGH M. Harvey, 4 Scilly Close, Ellesmere Port, CH65 9JU
M7 MGL C. Mcglinchey, 9 Spring Rise, Glossop, SK13 6US
M7 MGM Y. Weng, 20 Kendal Grove, Leeds, LS3 1NS
M7 MGU M. Guidolin, Flat 3, 3 Grimston Gardens, Folkestone, CT20 2PT
M7 MGW M. Waterman, 3 School Farm Cottages, Graveney Road, Faversham, ME13 8UR
M7 MGX S. Swinton, 4 Fallow Road, Newton Aycliffe, DL5 4SU
M7 MHA M. Harper, 67 Ludsden Grove, Thame, OX9 3BY
M7 MHB M. Balyuzi, 48 Cleveland Gardens, London, NW2 1DY
MW7 MHC D. Roberts, 25 Peibio Close, Holyhead, LL65 2EG
MI7 MHG M. Hughes, 96 White Rise, Dunmurry, Belfast, BT17 0XD
MM7 MHJ M. Hancox, 33 Caledonian Road Brechin, Brechin, DD9 6BG
M7 MHM Capt. R. Massingham, 6 Larch Crescent, Holt, NR25 6TU
M7 MHS M. Smith, 32 Charity William Way, Stanton, Bury St Edmunds, IP31 2FB
M7 MIQ A. Gardner, 16A Spring Lane, Shepshed, LE12 9JE
M7 MIR G. Buaras, 16 Potterswood Close, Bristol, BS15 8LW
M7 MIT T. Haynes, Jeetex Marine, Riverside Estate, Brundall, Norwich, NR13 5PL
M7 MJB J. Brittain, 66 High Street Swavesey, Cambridge, CB24 4QU
M7 MJC M. Crees, Flat 3 Fox House, Fox Lane North, Chertsey, KT16 9GY
M7 MJG M. Gardiner, 6 Garston Grove, Hartlepool, TS25 1HL
M7 MJH M. Hall, 88 Kingscote Yate, Bristol, BS378YE
M7 MJK M. Keay, 8 Coronation Road, Stoke-on-Trent, ST4 6BH
MW7 MJL M. Ladd, 50 Brynmelyn Avenue, Llanelli, SA15 3RT
M7 MJO M. Jolly, 7 Castle Road Colne Lancashire, Pendle, BB8 7AR
M7 MJR M. Rigby, 28A Lower Boston Rd Flat A, Hanwell, W7 2NR
MW7 MJT M. Thomson, 27 Sovereign Gardens, Miskin, Pontyclun, CF72 8SZ
M7 MKB M. Bradshaw, 70 Brook Gardens, Emsworth, PO10 7LB
MI7 MKD Rev. M. Donald, Station Road, Garvagh, BT51 5LA
M7 MKJ M. Jarkiewicz, 22 Lower Anchor Street, Chelmsford, CM2 0AS
M7 MKO M. Conlin, 11 James Street, Sunderland, SR5 2DJ
MM7 MKV W. Macleod, 26 Upper Barvas, Isle of Lewis, HS2 0QX
M7 MKY D. Curtis, 62 Hartley Road, Birmingham, B44 0RD
M7 MKZ B. Monks, 7 Springwell Road, Bootle, L20 6LU
M7 MLB L. Bennett, 19 Campion Crescent, Cranbrook, TN17 3QJ
M7 MLC M. Coombes, 33 Woodside Road North Baddesley, Southampton, SO52 9NB
M7 MLD M. Dudley, 88 King Street, Dawley, Telford, TF4 2AH
M7 MLE E. Groves, 350 Middle Deal Road, Deal, CT14 9SN
M7 MLK M. Kubiak, 132 Evistone Gardens, Newcastle upon Tyne, NE63RU
M7 MLM A. Stevens, 5 Ifield Mill Close, Stone Cross, BN245PF
M7 MLP M. Pullen, 8 Athena Avenue, Waterlooville, PO7 8AE
M7 MLY Dr E. Billinge, 23 Lamb Drive, Sheffield, S5 8LT
M7 MLZ D. Mullett, 17 Green Crescent, Bucklesham, Ipswich, IP10 0EA
MJ7 MMA M. Aubert, Flat, Deansway La Grande Route De St. Martin, St. Saviour, Jersey, Jersey, JE2 7GR
MM7 MMB A. Hood, 26 Annan Avenue, East Killbride, G75 8XT
M7 MMC M. Mccarthy, Flat 45, Beech, Beresford Road, Brighton, BN2 5DD
MW7 MME E. Holmes, 111 Llanfabon Drive, Trethomas, Caerphilly, CF83 8GX
M7 MML M. Larke, 16 Linton Road, Tamerton Foliot, Plymouth, PL5 4PG
M7 MMM P. Mocker, Flat 24 The Gardens Clapton Common, London, E5 9AZ
MM7 MMO S. O'Neill, Flat 5-8 460 Sauchiehall Street, Glasgow, G2 3JW
M7 MMP Q. Cope, 102 Moreton Road, Buckingham, MK18 1PW
M7 MMT M. Truman, 3 Mulberry Road, North Anston, Rotherham, S254BH
MM7 MMW M. Wylde, 73 Glendale Crescent, Ayr, KA7 3RZ
M7 MND M. Soles, 70 Mitchell Road, Bedworth, CV12 9HP
MM7 MNY M. Cowie, Larachmohr, Hawkhill, Keiss, Wick, KW1 4XF
MW7 MOC M. Angell, Fronheulog, Cloth Hall Lane, Cefn Coed, Merthyr Tydfil, CF48 2NT
M7 MOE M. Moen, 221 Buckhurst Way, Buckhurst Hill, IG9 6JB
M7 MOH H. Singh, 19 Elmore Road, Peterborough, PE3 9PS
M7 MOI S. Schonrock, 1 Powell Road, Poole, BH14 8SG
M7 MOK M. Swaby, 12 Market Square, Westerham, TN16 1AW
M7 MOL H. Molyneux, 16 Millbrook Close, Shaw, Oldham, OL2 8QA
MW7 MON J. Byast, Ty Arbennig Bull Bay Road, Amlwch, LL68 9EA
M7 MOR M. Robson, Induno, Hallgate, Moulton, Spalding, PE12 6QG
M7 MOT T. Wilcox, 17 Westminster Avenue, Royton, Oldham, OL2 5XY
M7 MOU P. Thomas, 2 Darwin Close, Stamford, PE9 1JL
M7 MOW M. Mower, 83 Sipson Road, West Drayton, UB7 9DH
M7 MOY M. Moyse, 37 Kingfisher Road, Buckingham, MK18 7EX
MW7 MOZ B. Morris, 20, Poets Field Road, Barry, CF62 9TY
M7 MPD M. Daryanani, 20 Blossom Way, Uxbridge, UB10 9LN
M7 MPL M. Loveridge, 63 Guildford Road, Portsmouth, PO1 5HU
MW7 MPM M. Moyse, 10 Clifton Rise, Abergele, LL22 7DN
M7 MPQ Dr A. Gair-Harris, Osterley, White Lackington, Piddletrenthide, Dorchester, DT2 7QU
M7 MPS M. Spikings, 84 Northampton Road, Broughton, Kettering, NN14 1NS
MI7 MQX M. Colhoun, 102 Dowland Road, Limavady, BT49 0HR
M7 MRA R. Ayres, 11 Sunningdale Crescent, Bournemouth, BH10 5LL
M7 MRB A. Bugler, 2 Roman Barn, Worth Matravers, Swanage, BH19 3LZ
M7 MRG G. Walker, Flat 74, Riversley Court, 205 Wensley Road, Reading, RG1 6ED
M7 MRJ M. Jones, 4 Farmland Way, Hailsham, BN27 1SP
MW7 MRK M. Morgan, 30 Hafan Werdd, Mornington Meadows, Caerphilly, CF83 3BU
M7 MRL M. Lee, 2 Rupert Road Chaddesden, Derby, de214nd
M7 MRO M. Roffe, 27 Athena Close, Southend on Sea, SS2 4GL
M7 MRP L. Pares, 22 Hawthorn Road, Bourne, PE10 9SN
M7 MRQ G. Newell, 12 Firecrest Close, Wymondham, NR18 9FA
M7 MRR M. Rogers, 26 Boldrewood, Swindon, SN3 6JP
M7 MRT M. Twyman, Hazeldene, St. Marys Road, Great Bentley, Colchester, CO7 8NN
M7 MRX A. Stokes, 175 Exhall Close, Church Hill South, Redditch, B98 9JA
M7 MSA M. Simpson, Elm Crescent, Peterborough, PE67LE
M7 MSB M. Bridges, 11 Fontana Close, Worth, Crawley, RH10 7SE
M7 MSD M. Dawson, 26 Goodwin Close, Wellingborough, NN8 4BS
M7 MSF P. Reynolds, 1 Aspen Grove, Newcastle upon Tyne, NE16 6QP
M7 MSJ S. Smith, 14 Benbow Close, Malvern, WR14 4JJ
MM7 MSL M. Craig, 20B Duke Street, Coldstream, TD12 4BW
M7 MSM M. Boisriveau-Mitchell, 16 Pochard Close, Quedgeley, Gloucester, GL2 4LL
M7 MSR M. Ramsay, 28A Strathmore Drive, Charvil, Reading, RG10 9QT
M7 MSV A. Varney, 11 Redburn Drive, Birmingham, B14 5XA
M7 MSW M. Wood, 44 Harveyfields, Waltham Abbey, EN9 1HN
MI7 MSX M. Silva, Apartment 205, 2 William Street South, Belfast, BT1 4FJ
MM7 MTB A. Campbell, 10 Crathes Close, Glenrothes, KY7 4SS
M7 MTE P. Polanyk, 44 Lowther Hill Honor Oak, London, SE23 1PY
M7 MTH M. Thompson, 49 Beechburn Park, Crook, DL15 8NA
MD7 MTM D. Wilson, 23 Snugborough Avenue, Union Mills, Braddan, Isle of Man, IM4 4LT
M7 MTS M. Shepard, 3 The Leys, Oxhill, Warwick, CV35 0QX
M7 MTV M. Venters, 18 Foxcovert Road, Werrington, Peterborough, PE4 6RF
MI7 MTY M. Thompson, 25 Ivy Mead Mews, Altnagelvin, Londonderry, BT47 3FH
M7 MUF P. Dilly, 4 Marsh Farm Cottages, Marsh Farm Lane, Alresford, Colchester, CO7 8BQ
M7 MUJ B. Isaac, 19 East Lodge, Fareham, PO15 5LZ
M7 MUS D. Mcdean, 2 Rigbys Row, Nantwich, CW5 5RX
M7 MUT L. Lewin, 83 Fulbourn Road, Cambridge, CB1 9AJ
M7 MVB M. Boldero, 1 Grasleigh Avenue, Allerton, Bradford, BD15 9AR
M7 MVD D. Matthews, 13 East Road, Egremont, CA22 2ED
M7 MVF Rev. P. Campbell, 116 Kingsway Park, Urmston, Manchester, M41 7FH
MM7 MVR D. Hood, 70 Kirkfield Gardens, Renfrew, PA4 8JE
M7 MVS A. Lammiman, 177 Obelisk Rise, Northampton, NN2 8TX
M7 MWG M. Gittins, 4 Old Mill Close, East Knoyle, Salisbury, SP3 6EX
MM7 MWL M. Lowson, Old Leslie Farmhouse, Leslie, Insch, AB52 6NS
M7 MWM M. Warwick, 63 The Crescent, Irlam, Manchester, M44 6FG
M7 MWT G. Griffiths, 23 Kilnwick Court, Mill Lane, Northallerton, DL7 8XS
M7 MWW M. Weekes, Lorraine, Winchester Road, Hawkhurst, Cranbrook, TN18 4DQ
M7 MWY M. Wilson, 35 Allen Close, York, YO10 3TS
MM7 MXB M. Butterworth, 97A Dunearn Drive, Kirkcaldy, KY2 6AL
M7 MXM C. Ford, 23 Seaton Orchard, Sparkwell, PL75HX
M7 MXY J. Crewe-Read, The Branches, Mill Lane, Hartley Wespall, Hook, RG27 0BQ
M7 MYA M. Joseph, 53 Wilman Road, Tunbridge Wells, TN4 9AL
M7 MYB M. Blicharz, 58 Park Barn Drive, Guildford, GU2 8ES
M7 MYC D. Mycroft, 11 Paisley Walk, Church Gresley, Swadlincote, DE11 9FF
M7 MYK M. Nicholas, 1 Buckland Barton Cottages, Newton Abbot, TQ12 4SA
M7 MYM M. Malik, School Of Mechanical, Aerospace & Civil Engineering University Of Manchester Oxford Road, Manchester, M13 9PL
M7 MYU K. Jenner, 24 The Willows, Nailsea, Bristol, BS48 1JQ
M7 MZM R. Hopkins, 26 High Firs Road, Romsey, SO51 5PZ
M7 MZN M. Nabi, 59 Sunninghill Avenue, Hove, BN3 8JB
M7 MZO D. Wilson, 3 Primrose Court, Moreton-in-Marsh, GL56 0JG
M7 MZT J. Maertens, 23 Cottingham Street, Goole, DN14 5RR
M7 NAB M. Bradbury, 7 Lulworth Avenue, Ipswich, IP38RW
M7 NAC A. Campbell, 21 Sherbrook Gardens, London, N21 2NX
M7 NAG N. Nicholls, 16 King Edwards Way, Edith Weston, Oakham, LE15 8EZ
M7 NAI I. Goldsmith, Foxwood Lane, York, Yo24 3lt
M7 NAJ N. Musgrave, 14 Clarendon Green, Orpington, BR5 2PA
M7 NAL A. Davies, 5 Worths Way, Stratford-upon-Avon, CV37 0RR
M7 NAM Dr N. Mirza, 21 Knoll Road, London, SW18 2DF
M7 NAN S. Morozov, Haroldene, Towpath, Shepperton, TW17 9LL
M7 NAS N. Stratford, 108 Nottage Crescent, Braintree, CM7 2TX
MW7 NAT N. Morris, Llwyn-Y-Gorras, Castlemorris, SA62 5ES
M7 NAV R. Matthewman, 44 Huntwick Crescent Featherstone, Pontefract, WF75JQ
M7 NAW C. Smith, 3 Park Rd Sunderland, Sunderland, SR2 8HR
M7 NAZ R. Messen, 45 Church Lane North Bradley, Trowbridge, BA14 0TE
M7 NBA C. Sims, 52 Station Road, Royston, SG8 0NP
MW7 NBC N. Conway, 7, Heatherdale Close, Stansty Park. Gwersyllt, LL11 4SZ
M7 NBX N. Brown, 22 Cherrytree Road, Bristol, BS16 4EX
M7 NBZ S. Johnson, 43 Rutherford Street Howdon, Wallsend, NE28 0AY
M7 NCA N. Atkins, 41 Church Lane, Bardsey, Leeds, LS17 9DR
M7 NCB N. Brooks, 105B Upper Woodcote Road, Caversham, Reading, RG4 7JZ
M7 NCC A. Whiteley, 16 Fustian Avenue, Heywood, OL10 3FN
MM7 NCD N. Coulter, 17 Gartinny, Coalsnaughton, Tillicoultry, FK13 6LF
M7 NCM N. Moore, 2 The Paddocks, Off The Croft, Longhoughton, NE66 3DD
M7 NCN N. Nurse, 3 Portnall Place, Cranfield, mk430jw
M7 NCO N. Cooke, 10 Westbury Road, Shrewsbury, SY1 3HF
M7 NCV C. Nutu, 37 Brading Road, London, SW2 2AP
M7 NCW N. Worthington, 190 Higher Road, Liverpool, L26 1UW
MI7 NCZ N. Conway, 30 Greenvale Manor, Antrim, BT41 1SB
M7 NDA N. Ayre, 58 Burford Avenue, Swindon, SN3 1BN
M7 NDB J. Jones, 227 George Lane, Bredbury, Stockport, SK6 1DJ
M7 NDI A. Roe, 3A Fairfield Avenue, Rossendale, BB4 9TG
M7 NDK D. Needham, 6 Apley Drive, Wellington, Telford, TF1 3DU
M7 NDL D. Callery, 15 Cliff Road, Holland-On-Sea, Clacton-on-Sea, CO15 5QQ
M7 NDM A. Mackey, Kings Barrow, North Bovey, Newton Abbot, TQ13 8RS
M7 NDO N. Downes, 17 Knightswood Close, Rosliston, Swadlincote, DE12 8JJ
M7 NDS N. Spencer, 11 Hitchcock Close, Shepperton, TW17 0QT
M7 NDT A. Tunley, 5 Camborne Close, Lower Earley, Reading, RG6 4EN
M7 NEA N. Morris, 31 Endeavour Place, Stourport-on-Severn, DY13 9RL
M7 NED N. Hambly, 172 Wareham Road Lytchett Matravers, Poole, BH16 6DT
M7 NEJ J. Adeniji, 28 Wigeon Path London, London, SE28 0DS
M7 NEL D. Nelson, Chapman Way, Hayward?S Heath, Rh164ul
M7 NEN N. Nowell, 12 Townhead Court, Settle, BD24 9HY
M7 NES N. Smith, 17 Lady Margaret Gardens, Woodbridge, IP12 4EZ
M7 NET D. Picker, 12 Crown Close, Barnsley, S70 4DB
M7 NEW A. Fleck, 108 Hillsview Avenue, Newcastle upon Tyne, NE3 3LA
M7 NEZ L. Ponton, 3 Capon Hill, Brampton, CA8 1QJ
M7 NFL S. Bradley, Leacroft Road, Penkridge, ST19 5BX
M7 NFX N. Friend, 4 Bartholomew Street, Dover, CT16 2LH
M7 NFZ N. Fenn, 7 Pottery Road, Tilehurst, Reading, RG30 6BA
M7 NGC N. Connolly, 32 St. Oswald Road, Bridlington, YO16 7SD
M7 NGL N. Bennetts, 1 The Moor, Falmouth, TR11 3QA
MM7 NGZ N. Grainger, 4 Hillhead, Inverfarigaig, Inverness, IV2 6XS
M7 NHF N. Frost, 5 George Street Elworth, Sandbach, CW11 3BL
M7 NHG N. Groom, 4 Burley Hill, Allestree, Derby, DE22 2ET

M7 NHJ T. Sidebottom, 256 Rolleston Road, Burton-on-Trent, DE13 0AY
M7 NHS Dr R. De Silver, 37 Hopeville Avenue, Broadstairs, CT10 2TR
M7 NIB N. Beresford, 2 Meadow View, Great Addington, Kettering, NN14 4BN
MI7 NIH H. Drinkwater, 26 Hillview Place, Holywood, BT18 9DL
M7 NIK Dr N. Crisp, Apartment 88, Advent House, 2 Isaac Way, Manchester, M4 7EP
M7 NJD N. Dowe, 47 Reachfields, Hythe, CT21 6LS
M7 NJJ N. Foreman, 10 Oak Road, Stilton, Peterborough, PE7 3RB
M7 NJM N. Manning, 9 Hawthorne Grove, Stockport, SK6 2PJ
M7 NJR N. Roche, 4 Kennedy Close, Rayleigh, SS6 8UW
M7 NKA N. Aldrich, 11 Town Mead, West Green, Crawley, RH11 7EG
M7 NKE Dr J. Holmes, Sexeys Farmhouse, Spring Street, Wool, Wareham, BH20 6DB
M7 NKN N. Newens, 5 Lillian Place, Gosport Road, Lower Farringdon, Alton, GU34 3DH
MM7 NLH G. Frame, 27 Oswald Court, Ayr, KA8 8NL
MJ7 NLK N. King, 15 Le Clos De Noirmont, La Route De Noirmont, St. Brelade, Jersey, JE3 8AP
M7 NLP N. Pereira, 55B North Street, Carshalton, SM5 2HG
M7 NLS N. Andre, Churchill College, Storey'S Way, Cambridge, CB3 0DS
MM7 NLW N. Wilson, 92 Inveroran Drive, Bearsden, Glasgow, G61 2AT
MI7 NMA M. Dowie, 16 Castle Meadow Park, Cloughey, Newtownards, BT22 1GB
M7 NMC N. Cardoso, 24 Northcroft Road, Gosport, PO12 3DR
M7 NMD L. Lynch, 10 Blackthorn Grove, Woburn Sands, Milton Keynes, MK17 8PZ
M7 NMM Dr H. Chen, Flat 37, Castel Mill, Rodger Dudman Way, Oxford, OX1 1AD
MM7 NMO C. Nimmo, 6 Raith Avenue, Prestwick, KA9 1DL
MW7 NMT N. Thomas, 82 Fairhill, Fairwater, Cwmbran, NP44 4RB
MM7 NNM N. Milne, 27 Brechin Road, Arbroath, DD11 1ST
MW7 NNR U. Miranovich, Flat 1, 35 Queens Road, Aberystwyth, SY23 2HN
M7 NNY W. Heaney, 9 Copeland Avenue Egremont, Cumbria, CA222QT
M7 NOC P. Parkin, 1 Swinley Chase, Wilmslow, SK9 2LZ
M7 NOL M. Nolan, 7 Freston Gardens, Barnet, EN4 9LX
M7 NOO J. Coakley, 44 Haydock Close, Birmingham, B36 8UN
M7 NOV J. Boase, 28 Broad Lane, Rochdale, OL16 4PG
M7 NOW N. Wadsley, 5 Paget Close, Needham Market, Ipswich, IP6 8XF
MM7 NOZ S. Wallace, 8 Gullane Avenue, Dundee, DD2 3BU
M7 NPC N. Cox, 182 North Tenth Street, Milton Keynes, MK9 3AY
M7 NPH N. Haigh, High Farm Cottage, Blindcrake, Cockermouth, CA13 0QP
MM7 NPT I. Stewart, Douneside, Allanfearn, Inverness, IV2 7HX
M7 NRC N. Cooper, 3 Dalwood Gardens, Benfleet, SS7 2NN
M7 NRH A. Ball, 7 Primrose Avenue The Drive, Horley, Rh67jw
M7 NRJ N. Raj, Apartment 702 Metis, 1 Scotland Street, Sheffield, S3 7AQ
M7 NRM N. May, 37 Brightling Avenue, Hastings, TN35 5EG
MM7 NRN T. Laurenson, Flat G, 4 South Lodge Court, Ayr, KA7 2TA
M7 NRO N. Olsson, 60 Bebington Road, Birkenhead, CH42 6PX
M7 NRR R. Levitt, 4 Ashley Row, Aylesbury, HP20 1HJ
M7 NRT M. Norton, 56 Melrose Avenue Vicars Cross, Chester, CH3 5JB
MI7 NRW R. Bestek, 111 Cloughwater Road, Ballymena, BT43 6SZ
M7 NSC N. Cooper, Romer Cottage, Long Reach, Ockham, Woking, GU23 6PF
M7 NSP M. Pinder, 3 Hatherop, Cirencester, GL73NA
M7 NSS V. De Souza, 52 Newhall Street, Swindon, SN1 5QS
MI7 NSU Dr A. Fogarty, 18 Oakleigh Park, Portadown, Craigavon, BT62 3QF
MM7 NSY N. Young, 61 Park Terrace, Broxburn, EH52 6AP
M7 NTD N. Tate, 418 Broad Lane, Leeds, LS13 3DF
M7 NTH N. Howells, 1 Eunice Way, Newdale, Telford, TF3 5FH
M7 NTT S. Bishop, 5 Mulberry Court, 266 Goring Road, Goring-By-Sea, Worthing, BN12 4PF
M7 NUK L. Colley, 31 Wharton Drive, North Walsham, Nr280ug
M7 NUT T. Gifford, 53 Haydn Avenue, Purley, CR8 4AJ
M7 NVM D. Mills, 31 Gorseway, Hatfield, AL10 9GS
M7 NVY J. Brodie, 10 Hillhead Rise, Falmouth, TR11 5GZ
MI7 NWA A. Carlisle, 21 Agnes Close, Belfast, Bt13 1Dj
MW7 NWJ Dr N. Jones, 54 Glanrhyd, Coed Eva, Cwmbran, NP44 6TY
M7 NWL R. Heaton, 19 Douglas Drive, Ormskirk, L39 1LJ
M7 NWT N. Trice, 4 Atkins Crescent, Maldon, CM9 6JB
M7 NWV N. Wright, 12 Hartington Road, Oakhurst, Swindon, SN25 2EF
M7 NWZ N. Willis, 3 Harpur-Crewe Cottages, Alstonefield, DE6 2GD
M7 NXC E. Walley, 4 Langley Mill Close, Sutton Coldfield, B75 7BY
MW7 NYE A. Price, 1 Graig Park Villas, Newport, NP20 6GU
M7 NYH N. Hodson, 163 Lincoln Road, Branston, Lincoln, LN4 1NS
MI7 NYK N. Montgomery, 50 Somerset Park, Coleraine, BT51 3LH
M7 NYY R. Coleman, 41 Lawrence House, Camperdown Street, Bexhill-on-Sea, TN39 5EN
M7 NZL M. Francis, 11 Barton Avenue, Paignton, TQ3 3JQ
MW7 NZP Z. Pugh, 28 Hill View Road, Llandudno, LL30 1SL
M7 OAD D. Drynski, Flat 22, Nicholas Court, Corney Reach Way, Chiswick, London, W4 2TS
M7 OAK D. Dalton, 8 Moorland Crescent, Pudsey, LS28 8EW
M7 OAN J. Robinson, 96 White Lodge Park, Shawbury, Shrewsbury, SY4 4NU
M7 OBI D. Goatcher, 5 Coopers Close, Burgess Hill, RH158AN
M7 OBJ D. Evans, 134 Somerton Road, Bolton, BL2 6LW
M7 OBX A. Wilkinson, 88 Main Street, Balderton, Newark, NG24 3NU
M7 OBY G. Inman, 3 Paisley Road, Bournemouth, BH6 5ET
M7 OCB C. Burls, 86 All Saints Road, Kings Heath, Birmingham, B14 7LN
M7 OCJ O. Cox, 110 Maidstone Road, Borough Green, Sevenoaks, TN15 8HG
M7 OCM M. Thirkettle, 320 Raglan Street, Lowestoft, NR32 2LB
M7 OCN E. Robinson, 7 Carlton Terrace, Lipson, Plymouth, PL4 8PR
M7 OCO K. O'Connell, 63 Hazelton Road, Colchester, CO4 3DS
MI7 OCP S. Spallen, 65 Tirgarvil Road, Upperlands, Maghera, BT46 5UW
M7 OCT S. Collins, 15 Twiss Grove, Hythe, Ct21 5pa
MM7 OCX A. Mather, 12 An Creagan Place, Port Charlotte, Pa48 7uf
M7 ODD S. Aynsley, 24 Grange Avenue, Bedlington, NE22 7EW
M7 ODG A. Cleveland, The Flat, 2 Church Street, Wells-Next-the-Sea, NR23 1JA
M7 ODH D. Harris, 381 Littleworth Road, Cannock, WS12 1HY
M7 ODO C. Dawson, 9 Telford Road, Sunderland, SR3 4HZ
M7 ODY Dr M. Venables, 50 Church View Close, Norwich, NR7 8QA
M7 OEL I. Robinson, 12 Osprey Road, Flitwick, MK45 1RU
M7 OEN M. Nicholson, 10 Beechfield Road, Cheadle Hulme, Cheadle, SK87DS
M7 OFG S. Wilson, 64 Deer Park Close, Sheffield, S6 5NA
MM7 OFH A. Gajdos, 262 Clement Rise, Livingston, EH54 6LR
M7 OFM J. Joyce, 2 Harold Road Cuxton, Rochester, ME2 1EE
MW7 OFS A. Williams, 49 Maesheli, Penparcau, Aberystwyth, SY23 1TB
M7 OGB D. Fixter, 68 Southwood, Coulby Newham, Middlesbrough, TS8 0UF
M7 OGC S. Grey, 40 Almery Drive, Carlisle, CA2 4EX
M7 OHM S. Hagan, 5 High Street, Brampton, Huntingdonshire, PE28 4TG
M7 OJA O. Acton, 93 Braunespath Estate, New Brancepeth, Durham, DH7 7JQ
M7 OJO C. Strikes, American Church In London, 79A Tottenham Court Road, London, W1T 4TD
M7 OJS J. Siddall, 16 Alandale Avenue, Langwith Junction, Mansfield, NG20 9RU
M7 OJV O. Verity, 2 Mill Hill Cottage Masham, Ripon, HG4 4BP
M7 OJW O. Williams, 6 Millfield Drive, Bristol, BS30 5NR
M7 OKG K. Gregson, 14 Tarragon Drive, Stoke-on-Trent, ST3 7YE
MM7 OKO J. Jasiewski, 5 Katrine Road, Shotts, ML7 4JA
M7 OKR K. Rowsell, Crofts Edge, Loud Hill, Stanley, DH9 8PL
M7 OKV P. Skermer, 23 Lea Grove Bardney, Lincoln, LN35XN
M7 OLA S. Coombe, 66 Tungstone Way, Market Harborough, LE16 9GG
M7 OLM R. Holtom, 79 Bowling Green Rd, Stourbridge, DY8 3RZ
M7 OLO K. Elms-Lester, Ferndale House, Kerrys Gate, Hereford, HR2 0AH
M7 OLY O. Botwright, 16 Upper Whistler Walk, World'S End Estate, London, SW10 0ER
M7 OMA T. Anable, The Old Court House, South Street, Winterton, Scunthorpe, DN15 9RP
M7 OMB P. Cummings, 12 Odile Mews, Gilstead, Bingley, BD16 3QL
M7 OMD C. Penny, 9 Winifred Road, Dartford, DA1 3BL
M7 OMG S. Orme, 78 Castle Road, Northolt, UB5 4SE
M7 OMM R. Reid, Postbridge, Yelverton, PL206TJ
M7 OMO P. Hickey, 11 Border Brook Lane, Boothstown, M281xj
M7 OMP O. Phillips, 1 Gate Foot Lane, Shepley, Huddersfield, HD8 8AZ
M7 OMW O. Williams, Apartment 20, 104 Pensby Road, Heswall, CH60 7RF
M7 OMY T. Sanduly, 1 Archford Croft, Emerson Valley, Milton Keynes, MK4 2EZ
M7 ONO M. Winkle, 75 Primrose Ave Haslington, Crewe, cw15ny
M7 ONR K. Mair, 12 Cheltenham Crescent, Moreton, Wirral, CH46 1PU
MM7 OOF B. Blyth, 58 Spencerfield Road, Inverkeithing, KY11 1PG
M7 OOI A. Wood, Chicago House, The Bowjey Hill, Newlyn, Penzance, TR18 5LW
M7 OOW J. Early, 6 Goverseth Hill, Foxhole, St Austell, PL26 7UZ
M7 OPB C. Mann, Swanton Road, Dereham, NR20 4PS
M7 OPH S. Richards, 60 Brooklyn Avenue, Loughton, IG10 1BN
M7 OPM P. Mcmillan, 44 Wheatfield Grove, Benton, NE12 8DP
M7 OPY P. Hope, 22 Ayrshire Close, Salisbury, SP2 9PF
M7 OQN T. Schonrock, 1 Powell Road, Poole, BH14 8SG
M7 ORC M. Minett, 82 Lower Market Street, Penryn, TR10 8BH
MM7 ORE J. Moore, 43 Alexander Grove, Kilmardinny Grange, Bearsden, G61 3EF
M7 ORG A. Angell, Nightingales, Greys Green, Rotherfield Greys, Henley-on-Thames, RG9 4QQ
MM7 ORR K. Orr, 9 Wallace Avenue, Stevenston, ka20 4bn
M7 OSD S. Mccallister, 44 Chester Street, Accrington, BB5 0SD
M7 OSH O. Hirst, 7 The Crescent Charles Street, Elland, HX5 0HR
M7 OSM O. Morris, Linden House Church End Weston Colville Cambridge Cb21 5Pe, Weston Coville, CB215PE
M7 OSS C. Donachie, 22 Glasgow Street, Hull, HU3 3PR
M7 OST A. Fuller, 44 Carbeile Road, Torpoint, PL11 2HR
MW7 OTK O. Knowles, 4 Church Street, Pen-Y-Cae, Wrexham, LL14 2RL
M7 OTO P. Chamberlain, 37 Curzon Street, Colne, BB8 0HE
M7 OTZ J. Bithell, 118 Hassam Parade, Newcastle, ST5 9DN
M7 OUT K. Mason, 214 Carfield, Skelmersdale, WN8 9DW
M7 OVM J. Harley, 33 Miranda Rd Archway, London, N193RA
M7 OWA A. Oliver, 49 Wharf Close, St. Georges, Telford, TF2 9PX
M7 OWD V. Hodge, 37 Monkswood Crescent, Tadley, RG26 3UE
M7 OWJ O. Jackson, 38 Bishop Close, Poole, BH12 5HT
M7 OWL W. Burgess, 241 King Avenue, Bootle, L20 0BY
MM7 OWN O. Mcdermid, 38 Steading Drive, Alexandria, G83 9EB
MM7 OWO L. Tarvit, 1/10 Chapel Lane, Leith, Edinburgh, EH6 6ST
M7 OXO J. Bliszko, 34 Hambledines, Redhouse Park, Milton Keynes, MK14 5FS
M7 OXS O. Spearing, 2 Goldcroft Road, Weymouth, DT4 0DZ
M7 OXY D. Oxendale, Flat 1, Quay House, Middle Wharf, Mevagissey, St Austell, PL26 6UP
M7 OYR S. Roberts, 77 Lambwath Road, Hull, HU8 0HB
M7 OZA A. Ozarski, 86 Howards Way, Newton Abbot, TQ124HX
M7 OZC O. Sloane-Hase, 23 Southdown Road, Benham Hill, Thatcham, RG19 3BF
M7 OZE O. Martin, 22 Frankton Avenue, Hayward's Heath, RH16 3QX
M7 PAA P. Rossi, Tantivy, Lower Street, Eastry, Sandwich, CT13 0HX
M7 PAC P. Cann, 47 Drury Lane, Houghton Regis, LU55ED
M7 PAG P. Goodier, Macaulay Road, Hartlepool, TS25 4NE
MW7 PAJ P. Jewell, 27 Overdale Avenue, Mynydd Isa, Mold, CH7 6US
MW7 PAN S. Carter, The Stables, Penllyn, Cowbridge, CF71 7RQ
M7 PAR P. Pankhania, 20 Dean Road, Hounslow, TW3 2EZ
M7 PAU J. Hamilton, 60 Woodlands Road, Surbiton, KT6 6PY
M7 PAV P. Venables, 7 Lawton Road, Rushden, NN10 0DX
M7 PAY S. Payne, 82 Lower Barnes Street, Clayton Le Moors, Accrington, BB5 5SW
M7 PBD P. Bowers-Davis, 28 Muchelney Road, Morden, SM4 6HU
M7 PBN D. Newman, Finsbury Ave, Blackpool, fy16qn
M7 PBO A. Bogg, 13, Emerald Grove, Hull, hu35ae
M7 PBS P. Baskerville, 4 Fosse Close, Swindon, SN2 2BP
M7 PBT P. Townsend, 80 Uplands Road, West Moors, Ferndown, BH22 0BT
M7 PBW P. Winn, Glen Lyon, Hoggs Drove, Marham, King's Lynn, PE33 9JW
M7 PBX P. Barnett, 18 Cloverdale, Northwich, CW8 4UE
M7 PCB P. Cridland, 49 North Drive, Thornton Cleveleys, FY5 3AJ
M7 PCC P. Culmer, 56 Main Road Washingborough, Lincoln, LN4 1AU
M7 PCD P. Delaney, 32 Netherfield View, Rotherham, S65 3RB
M7 PCE P. Clark, 8 Kyreside, Tenbury Wells, WR15 8BX
M7 PCF P. Finch, 6 Holmer Manor Close, Hereford, HR4 9QZ
MW7 PCG P. Gorton, Bradnant, Tylwch, Llanidloes, SY18 6JZ
M7 PCL J. Perry, 198 Totteridge Road, High Wycombe, HP13 7LF
M7 PCS P. Sanders, 6 Primrose Hill, Warwick, CV34 5HW
M7 PCW S. Cheetham Ward, 21 Scarborough Close, Walsall, WS2 9TN
M7 PCX P. Eley, 172 New Bristol Road, Weston Super Mare, BS226BG
M7 PDK P. Thomas, 6 Middleton Road, Acomb, York, YO24 3AS
M7 PDQ L. Pooler, 56 Jacks Lane, Marchington, Uttoxeter, ST14 8LW

M7 PDR P. Ryder, 5 Weyhouse Close, Stroud, GL5 2JJ
M7 PDS P. Steels, 18 Moss Close, Arnold, Nottingham, NG5 8SD
MM7 PDW P. Whitbread, Mains Of Whitehill, Banff, AB453ER
M7 PDZ R. Attan, 45 Village Road, London, N3 1TJ
M7 PEB P. Brennan, 2 Ethel Road, Birmingham, B17 0EL
M7 PEJ J. Puzhiveliparambil James, 279 New Road, Dagenham, RM10 9ND
M7 PEL D. Pelling, 8 Fowler Close Earley, Reading, RG6 7SS
M7 PEN J. Penn, 8 Doulton Close, Harlow, CM17 9RG
M7 PES P. Dale, 14 Locksley Drive, Rotherham, S66 9NU
M7 PEV D. Maxwell, 28 High Street, Wilden, MK44 2PB
M7 PEW A. Pewsey, Pineholm, High Close, Bovey Tracey, Newton Abbot, TQ13 9EX
M7 PFB P. Billington, 10 East Crescent, Accrington, BB5 5BS
M7 PFC P. Colman, 16 Gleneagle Road, Plymouth, PL3 5HJ
M7 PFD Prof. P. Davis, 83 South Street, Crowland, Pe6 0ah
M7 PFF P. Francis, 67 Longs View, Charfield, Wotton-under-Edge, GL12 8HZ
M7 PFK P. Kusiak, 17 Kingsway Black, Camberley, GU17 0JW
MD7 PFM M. Squires, Eaglehurst, The Crescent, Ramsey, Isle Of Man, IM8 2JN
M7 PFT A. Richardson, Valley Farm, Long Lane, Dunston Heath, Stafford, ST18 9FB
M7 PGA P. Greenough, 26 Beech Grove, Liverpool, L21 1BP
M7 PGE P. Glaze, 28 Meddins Lane, Kinver, DY7 6BY
M7 PGF P. Featherstone, 122 Addiscombe Court Road, Croydon, CR0 6TS
MM7 PGG P. Greig, 22 Rowanhill Close, Port Seton, Prestonpans, EH32 0SY
M7 PGH P. Hodgson, 32 Main Road, Seaton, Workington, CA14 1HS
MM7 PGL M. Saunders, 8D Springhill Road, Port Glasgow, PA14 5QP
M7 PGM P. Millington, 81 Kilpin Green, North Crawley, Newport Pagnell, MK16 9LZ
M7 PGP S. Hunt, 12 Exmoor Drive, Leamington Spa, CV32 7BB
MM7 PGR P. Riggs, 106 Braes Avenue, Clydebank, G81 1DP
M7 PGY P. Gray, 3 Herrington Court, Newton Aycliffe, DL5 4RA
M7 PHA P. Alves, 2 Sherwood Court High Road Leavesden, Watford, WD25 7PA
M7 PHB P. Baines, 11 Orchid Avenue, Farnworth, Bolton, BL4 0ES
M7 PHE K. Munson, Birchwood, Green Lane, North Kelsey, Market Rasen, LN7 6FH
MW7 PHI P. Marks, 61 Trefelin Crescent, Port Talbot, SA13 1DX
MW7 PHY P. Williams, 14 Penyfan Road, Llanelli, SA15 1JP
M7 PHZ P. Hunt, 21 Lawrence Drive, Cobham, Gravesend, DA12 3BU
M7 PIX T. Bacala, Flat 3, 22 A, Seven Sisters Rd, London, N4 3NX
M7 PJB P. Black, 7 Birch View, Manchester, Ol129pz
M7 PJD P. Dee, 17 William Way, Lawford Manningtree, CO11 2GE
M7 PJF P. Foster, 34 Straight Bit, Flackwell Heath, High Wycombe, HP10 9LT
MW7 PJH P. Houghton, 48, Woodlands Caravan Park, Tyn Y Morfa, Holywell, CH8 9JN
M7 PJI Dr D. Mcconnell, 2 Hancock Court, Norwich, NR5 9NN
M7 PJK P. Keightley, 2 Spring Meadow, Upton Bishop, Ross-on-Wye, HR9 7SS
M7 PJM P. Mackwell, Wistaria, Hawthorn Lane, Pickering, YO18 7EA
M7 PJO P. Offord, 1 Adare Close, Dunmow, CM6 2GR
M7 PJS P. Lawrence, 13 Woodcock Close, Haverhill, CB9 0JP
M7 PJT P. Johnson, 16 Peregrine Close, Hythe, CT21 6QZ
M7 PJY P. Abraham, 211B Hornby Road, Blackpool, Lancashire, FY1 4JA
M7 PKA E. Fergus, 98A West Vale, Neston, CH64 0TL
M7 PKK P. Kitching, 83 Birchlands Avenue Wilsden, Bradford, BD15 0HB
MW7 PKT Dr K. Trimmis, 142 Arran Street, Cardiff, CF24 3HU
M7 PKW D. Woodward, 178A Worcester Road, Malvern, WR14 2HL
M7 PLA P. Larcombe, 55 Forest Drive, Weston-Super-Mare, BS23 2UG
M7 PLB P. Brown, 28 Farley Way, Stockport, SK5 6JD
M7 PLD C. Abbott, 1 Milton Road, Ellesmere Port, CH65 5AT
MM7 PLG D. Johnson, 8 Sandmartin Grove, Lenzie, Kirkintilloch, Glasgow, G66 3WF
M7 PLH P. Hilton, Shankly Cottage 161 Highgate, Jennings Yard, Kendal, LA9 4EN
MW7 PLL J. Jillings, Cane Garden, Dolau, Llandrindod Wells, LD1 5TE
M7 PLN P. Neal, 65 Van Diemans, Stanford in the Vale, SN7 8HW
M7 PLO P. Osborne, 28 Elizabeth Way, Rushden, nn108jr
M7 PLP P. Dingsdale, 51 Adelaide Avenue, Thatto Heath, St Helens, WA9 5RU
M7 PLS P. Shakespeare, 12 Brickley Lane, Devizes, SN10 3BQ
M7 PLT G. Lunt, 9 Hereford Way, Aylesbury, HP19 9GY
M7 PLW P. Walker, 33 Bradshaw Street, Manchester, M24 2AG
MW7 PLY D. Evans, Bleddfa, Maes Meyrick, Heolgerrig, Merthyr Tydfil, CF48 1RZ
MM7 PMB P. Barton, 2/2 5 Park Quadrant, Glasgow, G3 6BS
M7 PMF P. Fotherby, Furndale Foxhill Court, Leeds, LS16 5PL
M7 PMH P. Harston, 70 Dairymans Walk, Burpham, Guildford, GU4 7FF
M7 PMK P. Mckay, 18 Dalewood Crescent, Elton, Chester, CH2 4PR
M7 PML P. Long, 206A Henver Road, Newquay, TR7 3EH
M7 PMY Dr P. Ylioja, 12 Nelson Crescent, Cambridge, CB24 3GN
M7 PNG P. Garbett, 827C Warwick Road, Birmingham, B11 2EL
M7 PNO M. Troscianczyk, 5 Haldane Court, Hull, HU4 6ST
M7 PNR P. Whiterod, 9 Brewhouse Lane, Soham, Ely, CB7 5JD
M7 PNS N. Wheeler, 6 Fishermens Court, Attleborough, NR172QW
M7 PNY P. Yarrow, 54 Wansbeck Road, Jarrow, NE32 5SS
MJ7 POA L. Coates, The Manse Clos De Carrel, St Brelade, Jersey, JE3 8LJ
M7 POC P. O'Callaghan, 220 Chapel House Road, Nelson, BB90QR
M7 POE Lord J. Williams, 7 Southrop Road Kingsway, Quedgeley, Gloucester, GL2 2HN
M7 POG Dr J. Pogmore, Sunnybanks, West End, Barlborough, Chesterfield, S43 4HE
M7 POI R. Lanham, 16 Pinewood Road, Ferndown, BH22 9RW
MM7 POL P. Mccandlish, 2 Dumbarton Road, Glasgow, G11 6PB
M7 POW J. Pownall, 75 Park Barn Drive, Guildford, GU2 8ER
M7 PPF R. Mayo, 8 Station Terrace, Kempley Road, Dymock, GL18 2BD
M7 PPG A. Wilson, 196 Spring Lane Lambley, Nottingham, NG4 4PE
M7 PPL P. Latham, 135 Ashgate Road, Chesterfield, S40 4AN
M7 PPM P. Marks, 4 Belmont Drive, Lymington, SO41 3AE
M7 PPR P. Rush, 8 Ashworth Street, Radcliffe, Manchester, M26 2XU
MM7 PPW P. Wilkie, 1 Kinnaber Road, Hillside, Montrose, DD10 9HE
M7 PRA R. Park, 53 Sheepwash Avenue, Choppington, NE62 5NN
M7 PRE A. Hill, 39 Powis Road, Ashton-On-Ribble, Preston, PR2 1AD
M7 PRF B. Fulford, 43 St. Peters Court, Whitby, YO22 4JQ
M7 PRG P. Gill, 102 Leighton Road, Sheffield, S14 1SS
M7 PRH Dr P. Holford, 52 Coldstream Close, Warrington, WA20ll
M7 PRI P. Acharya, 61 Rose Court Nursery Road Pinner, London, HA5 2AR
M7 PRN D. Crisan, Flat A, 9 Fortis Green Avenue, London, N2 9LY
M7 PRW P. West, Stour House, Church Street, West Stour, Gillingham, SP8 5RL
M7 PRZ A. Prozapas, 149 Osborne Road, Wisbech, PE13 3JP
M7 PSB P. Browning, 38 Moorsfield, Houghton le Spring, DH4 5PF
M7 PSC Capt. P. Cunningham, Flat 1, Torvesco House, Knowles Hill Road, Newton Abbot, TQ12 2PW
MM7 PSG P. Anderson, 76 Devonway, Clackmannan, FK10 4LF
M7 PSI J. Stokes, Orchard House, Old Smithy Close, Pattingham, Wolverhampton, WV6 7AZ
M7 PSJ S. Jackson, 8 Buttree Court, South Kirkby, Pontefract, WF9 3NB
M7 PSM S. Miles, 21A Bearwood Road, Barkham, Wokingham, RG41 4TB
M7 PSS P. Sanderson, 26 Ponsford Road, Minehead, TA24 5DY
M7 PSX A. Faulkner, 13 Cherry Orchard, Woodchurch, Ashford, TN26 3QX
MI7 PSY S. Redmond, 79A Bridge Street, Portadown, Co Armagh, Belfast, Portadown, BT63 5AA
M7 PSZ S. Marek, 91 Dial House Road, Sheffield, S6 4WU
M7 PTP C. Laycock, 35A High Street, Henlow, SG16 6AA
MW7 PTW P. Williams, Gorn Newydd, The Gorn, Llanidloes, SY18 6LA
M7 PTY P. Taylor, 89 Sunningdale Gardens, Bognor Regis, PO22 9LE
M7 PUF D. Williams, 11 Friars Place, Littleport, Ely, CB6 1LG
M7 PUL P. Parkinson, 5 Gail Close, Failsworth, Manchester, M35 0TG
M7 PUP R. Strutt, Collingwood, 39 Henley Court, Ipswich, IP1 3SD
M7 PUZ N. Titmus, 68 Hobart Road, Cambridge, CB1 3PT
M7 PVP R. Armshaw, 11 Collingwood Crescent, Matlock, DE4 3TB
M7 PVS P. Sloper, 51 Obelisk Road, Southampton, SO19 9BL
M7 PVW M. Vaites, 19 Campion Drive, Sheffield, S21 1TG
M7 PVX A. Mount, 48 Cow Lane, Wakefield, WF4 1BA
M7 PWB P. Bone, 11 Fox Hill Drive, Stalybridge, SK15 2RP
MW7 PWC P. Colwill, 85 Davis Avenue, Bryncethin, Bridgend, CF32 9JL
M7 PWD J. Savage, 7 Elmsleigh Drive, Swadlincote, De110et
MW7 PWH P. Hughes, Awel Menai Penmaen Park, Llanfairfechan, LL33 0RN
M7 PWK S. Padwick, 37 The Fieldings, Southwater, RH13 9LZ
M7 PWL R. Sparrow, 97 Lower Street, Horning, Norwich, NR12 8PF
M7 PWW P. Wilmore, 1 Highfield Terrace, Lower Bentham, Lancaster, LA2 7EP
M7 PXM P. Melichar, 10 Thistlemead, Chislehurst, BR7 5RF
M7 PYA P. Yates, 7 Plessey Terrace, Newcastle upon Tyne, NE7 7DJ
MW7 PYM P. Man, 17A Cradock Street, Swansea, SA1 3HE
MW7 PYS C. Sandever, 22 Cae Melin, Little Mill, Pontypool, NP4 0HX
M7 PZG P. Gahagan, 28 Brow Hey, Bamber Bridge, Preston, PR5 8DS
M7 RAB R. Bolger, Little Annaside, Bootle, LA19 5XL
M7 RAD R. Day, Flat 22, Jefferies Lodge, 48-60 Footscray Road, London, SE9 2SU
M7 RAE R. Evans, 11 Swane Road Stockwood, Bristol, BS14 8NQ
M7 RAJ R. Kumar, 1 Stilton Close, Aylesbury, HP19 8JH
M7 RAS R. Statham, 47 Essington Way, Stoke-on-Trent, ST6 5EE
MW7 RAT R. Couper, 58 York Place, Newport, NP20 4GD
M7 RAV A. Ravenscroft, 35 Adelaide Street, Fleetwood, FY7 6AD
M7 RAW R. Wallace, Killerton Road, Bude, ex238en
M7 RAX R. Hawkins, 1 Northbank Close, The Reddings, Cheltenham, GL51 6UA
M7 RAY R. Lovell, Formby, Formby, Livepool, L37 4BP
M7 RAZ A. Smith, 55 Van Diemans Road, Thame, OX9 2DH
M7 RBC R. Coomer, 30 Torquay Road, Kingskerswell, Newton Abbot, TQ12 5EZ
M7 RBE R. Beardmore, 28 Broadway, Ilkeston, DE7 8TD
M7 RBF R. Blackmann, Flat 1, Merrywood House, Merrywood Lane, Thakeham, Pulborough, RH20 3HD
MM7 RBT R. Telford, 4 Cumming Avenue, Carluke, ML8 4RL
M7 RBY A. Fox, 8 Giles Road, Swindon, SN25 1QD
M7 RCE J. Beniston, 100 The Ridgeway, Croydon, CR0 4AF
M7 RCG M. Hiscock, 12 Pickwick Close, Basildon, SS155SW
M7 RCH R. Lounton, 107 Browning Hill, Durham, DH6 4SA
M7 RCI A. Cottell, 3 Honeylight View, Swindon, SN25 4XS
M7 RCM J. Lingard, 28 Pendine Crescent, North Hykeham, Lincoln, LN6 8UR
M7 RCR R. Carter, 5 Englefield Close, Crewe, CW1 3YN
M7 RCS J. Mclean, 3 Burgess Field, Woodlands, Wimborne, BH21 8LQ
MM7 RCT T. Bray, 2/1 West Grange Gardens, Edinburgh, EH9 2RA
MM7 RCU R. Abringe, 5 Suisnish Place, Broadford, Isle of Skye, IV49 9BZ
MM7 RCW R. Walker, 8 Laverock Braes Road, Grandhome, Aberdeen, AB22 9AE
MM7 RCX R. Caves, West Lamberkin Farmhouse, Tibbermore, Perth, PH1 1QA
M7 RCZ J. Haskell, 60 Blenheim Drive, Witney, OX28 5LJ
MI7 RDA G. Mulholland, 62 Castleward Road, Downpatrick, BT30 7JT
M7 RDB R. Boag, Sudeley Rd, Nuneaton, CV10 7AJ
M7 RDC R. Cunningham, 130 Station Road, Shimpling, Diss, IP21 4UA
M7 RDH R. Holder, 34 Looseleigh Lane, Plymouth, PL6 5HQ
M7 RDN A. Robinson, 10A Denver Court Stapleford, Nottingham, NG9 8LN
M7 RDS C. Dannemann, Bracelands, The Headlands, Stroud, GL5 5PS
M7 RDZ R. Diamond, 51 Huddersfield Road, Diggle, Oldham, OL3 5NT
M7 RED A. Rodionov, 29 West Parkside, London, SE10 0JT
M7 REL R. Lee, 80 Bailey Drive Bootle, Bootle, l20 6hb
M7 REO D. Cooper, Seinna Bryne Lane, Padbury, MK18 2AL
M7 REP S. Roberts, 21 Hednesford Road, Walsall, WS8 7LS
M7 RER R. Dixon, Coach House, Chapel Lane, Ellel, Lancaster, LA2 0PN
MW7 RES P. Leitch, 57 Stephen Road, Prestatyn, LL19 7EH
M7 RET P. Barrett, 58 Friars Avenue, Northampton, NN4 8PX
M7 REV P. Howard, 4 Wayside, Telford, TF7 5NF
M7 REX A. Palmer, 7 Oak Street, Lechlade, GL7 3AX
M7 REY J. Stanesby, 21 Godwin Close, Wokingham, RG41 2AH
M7 RFA M. Culley, 99 Mill Lane, Stockport, SK6 1QL
M7 RFD G. Davies, 11, Liverpool, L12 3HS
M7 RFM R. Fox, 8 Croome Close, Drakes Broughton, Pershore, WR10 2BH
MM7 RFP M. Stephenson, 20 Balmoral Road, Galashiels, TD1 1JL
MM7 RFW D. Wells, 17. Henry Street, Kirriemuir, DD85DL
M7 RGB R. Bamber, 28 Market Street, Cheltenham, GL50 3NH
M7 RGD R. Dukes, 40 Eastern Road, Brightlingsea, Colchester, CO7 0HU
M7 RGF R. Gallagher, 8B Eden Close, Beverley, hu177he
M7 RGG R. Hearn, 39 John Gaskell Court, Hensingham, Whitehaven, CA28 8PH
M7 RGJ I. Harding, 7 Hawthorne Close, River, Dover, CT17 0NG
M7 RGK G. Clarke, 1 Huntley Crescent, Sheringham, NR26 8QQ
MM7 RGM R. Mcbride, 7 Garnock Road, Stevenston, KA20 3BA
MI7 RGN A. Fleming, 7 Laurel Hill Gardens, Coleraine, BT51 3GW
M7 RGP G. Peck, 90 Chadwell Avenue, Cheshunt, Waltham Cross, EN8 0ER
M7 RGR R. Robinson, 4 Limetree Court, Taverham, Norwich, NR8 6QY
M7 RGS R. Selby, 1 Wakeling Close, Southwell, NG25 0JF
M7 RHA R. Ashitaka, 148 Coppice Lane, Basildon, SS15 4JS

M7 RHB R. Butt, Waterloo Cottage, Barrack Hill, Little Birch, Hereford, HR2 8AX
MW7 RHC H. Mckirdy, 7 Pritchard Court, Albany Road, Cardiff, CF24 3RW
MW7 RHF P. Passmore, 298 Dinas, Newtown, SY16 1NW
M7 RHH Dr R. Hopper, 255 Wellbrook Way, Girton, Cambridge, CB3 0GL
M7 RHI R. Harris, 3 Bramble Close, Guildford, GU3 3BQ
M7 RHM R. Ham, 232 Moor Road, Chorley, PR7 2NT
MW7 RHS R. Jones, 38 High Street, Pwllheli, LL53 5RT
M7 RHW R. Haines-White, 111 Glendower Crescent, Orpington, BR6 0UP
M7 RIA A. Edwards, 124 Ardrossan Gardens Worcester Park, Worcester Park, KT47AY
MM7 RIK R. Maccormack, 90 Wardlaw Crescent, East Kilbride, Glasgow, G75 0PY
M7 RIL J. Riley, 67 Moss Bank Road, St Helens, WA11 7DE
MM7 RIN I. Gubaciova, 38/25 South Clerk Street, Edinburgh, EH8 9PS
M7 RIS R. Stone, 374 Bourne Road, Spalding, PE11 3LL
M7 RIZ M. Heywood-Macdonald, 12 Burton Road, Kendal, LA9 7JA
M7 RJB Sir R. Lindeman, 26 Seattle Avenue, Blackpool, FY2 0PW
M7 RJD R. Denby, 168 Nottingham Road, Nuthall, Nottingham, NG16 1AB
M7 RJH R. Herbert, 24 Green Acres Road, Birmingham, B38 8NH
M7 RJI R. Ibbotson, 33 St. Peters Avenue, Caversham, Reading, RG4 7DH
M7 RJJ R. Jarrott, 11 Stewkins, Stourbridge, DY8 4YW
M7 RJL R. Land, 31 Winifred Road Cale Green, Stockport, SK2 6HF
M7 RJR R. Riley, 117 Woodgrove Road, Burnley, BB11 3EJ
M7 RJT I. Valasakis, 16 Seaton Square, London, NW7 1GB
M7 RKA R. Arnold, 68 Springfield Drive, Calne, SN11 0UG
M7 RKC M. Coull, 20A Queen Street, Warley, Brentwood, CM14 5JZ
M7 RKD R. King, 169 Morley Road, Oakwood, Derby, DE21 4QY
M7 RKE D. Roake, 9 Falcondale Walk, Westbury On Trym, Bristol, BS93JG
MM7 RKJ K. Roy, 21 Irvine Place Stirling, Stirling, FK8 1BZ
M7 RKL R. Kelleher, 65Rugby Road, Leigh, WN73HB
M7 RKN M. Corristine, 16B Daventry Road, Bristol, BS4 1DG
M7 RKS M. Stockdale, 3 Manor Close, Sproatley, Hull, HU11 4PY
M7 RKW M. Wallis, 3 Lawson Court, Boldon Colliery, NE35 9NH
M7 RKY M. Jordan-Reed, Pump Corner, High Street Green, Sible Hedingham, Halstead, CO9 3LG
M7 RKZ R. Keene, 33 Jasmine Gardens, Bradwell, Great Yarmouth, NR31 8HU
M7 RLA R. Adams, N1 Creeksea Place Caravan Park, Ferry Road, Burnham on Crouch, cm08pj
MI7 RLC R. Calvert, 29 Coachmans Way, Hillsborough, BT26 6HQ
M7 RLD R. Lapinskas, 38 Forest Road, Enfield, EN3 6ST
M7 RLL R. Lester, Finkle Street, Doncaster, DN5 0RP
M7 RLS R. Sharpe, 23 Sparrowsmead, Redhill, RH1 2EJ
MW7 RLT A. Noakes, Westra Holt, Westra, Dinas Powys, CF64 4HA
M7 RLW R. Walker, Pilgrims, Gore Road, Burnham-on-Sea, TA8 2HL
MW7 RMC R. Cooke, 6 The Forestry, Trecastle, Brecon, LD3 8YA
M7 RMF R. Fellows, 41 Dalesford Road, Aylesbury, HP21 9XE
M7 RML R. Little, 296 Welford Road Leicester, Leicester, LEG6EG
M7 RMP R. Pearson, 14 Highfields, Lakenheath, Brandon, IP27 9DZ
M7 RMU R. Underwood, 32 Lowlands Close, Rectory Farm, Northampton, NN3 5EP
M7 RMW R. Ambler, 21 Whitley Spring Road, Ossett, WF5 0QA
M7 RMZ S. Lilley, 1 Ormonde Road, Chester, CH2 2AH
M7 RNE R. Evans, 364 Aldermans Green Road, Coventry, CV2 1NN
M7 RNI P. Whitehouse, 52 Hagley Road, Rugeley, WS15 2AW
M7 RNL A. Scoggins, 7 North Road, Southwold, IP18 6BG
M7 RNO R. Fisher, 6 Armitstead Road, Wheelock, Sandbach, CW11 3LP
M7 RNS A. Atkinson, 21 Dennington Crescent, Basildon, SS14 2FF
M7 RNU I. Eliade, 44 Brookside Road, Stratford-upon-Avon, CV37 9PH
M7 RNX R. Nixon, 32 Albany Road, Hornchurch, RM12 4AF
M7 ROC C. Rockey, 10 Doulton Drive Porthill, Newcastle under Lyme, ST5 8SE
M7 ROI A. Ingham, Rosemount, Church Whitfield Road, Whitfield, Dover, CT16 3HZ
M7 ROO M. Ratcliffe, 76 Churchill Road, Stone, ST15 0DY
M7 ROT F. Sacca, 78 St. Johns Avenue, London, NW10 4EG
M7 ROV J. Mckeown, 5 Sidney Avenue, Hesketh Bank, Preston, PR4 6SU
M7 ROX E. Hauser, 23 Prince William Way, Sawston, Cambridge, CB22 3SZ
MW7 ROY D. Underwood, 1 Bro Hafan, Cross Inn, Llandysul, SA44 6NQ
M7 RPA R. Parker, 21 Broadbridge Lane, Smallfield, Horley, RH6 9RE
M7 RPI A. Wynne, 7 Ruskin Close, Oxford, OX2 9FU
M7 RPK R. Pike, 44 Falkirk Close Bransholme, Hull, HU7 5BX
M7 RPL R. Lyons, 103A Oxney Road, Peterborough, PE1 5NG
MI7 RPP M. Burns, 1 Ashley Close, Armagh, BT60 1EX
M7 RPR P. Shingler, 4 Lawnswood Avenue, Burntwood, WS74YD
M7 RPY R. Young, 23 Brewton Drive, Deeping St. James, Peterborough, PE6 8GR
M7 RQQ D. Cockram, 4 Greenleaze, Marston Meysey, Swindon, SN6 6LJ
M7 RRG S. Brown, 48 Austerby, Bourne, PE10 9JG
MW7 RRW T. Williams, Sea View House, Calon Fawr, Lon Masarn, Tycoch, Sketty, Swansea, SA2 9EX
M7 RSC J. Lawrence, 42 Alma Street, Weston-Super-Mare, BS23 1RD
M7 RSD J. Stannard, Rose Cottage, Hereford, HR69SY
M7 RSH R. Davies, Lamb Cottage 3 Manor Barns, Snowshill, Broadway, WR12 7JR
M7 RSJ S. Riley, 51 Kenilworth Avenue, Reading, RG30 3DL
M7 RSM P. Collins, 7 Fitzmaurice Square, Calne, SN11 8NL
M7 RST R. Upton, 139 Withens Lane, Wallasey, CH45 7NF
M7 RTB J. Blundell, 17 Griffin Close, Chester, CH1 5TX
MI7 RTD R. Donaghy, 35 Garvagh Road, Dungiven, Dungiven, BT47 4LU
M7 RTE R. Logue, 21 Hammers Lane, London, NW7 4BY
M7 RTK R. Kelly, 48 Strathearn Drive, Bristol, bs10 6tj
M7 RTO A. Rutson-Edwards, 89 Bemerton Gardens, Kirby Cross, Frinton-on-Sea, CO13 0LQ
M7 RTS P. Craske, 1 Hythe Lane, Burwell, Cambridge, CB25 0EH
M7 RTT A. Tait, The Reddings, Dirty Lane, Beausale, Warwick, CV35 7AQ
M7 RTU C. Rees, 14 Elizabeth Court, Droitwich Spa, WR9 8RU
MW7 RTY M. Hayden, 17 Haulfryn, Kenfig Hill, Bridgend, CF33 6EJ
M7 RUE P. Brindley, 70 Atherfield Road, Reigate, RH2 7PS
M7 RUI R. Dos Santos Amorim, 20 Urswick Road, Dagenham, RM9 6EA
M7 RUK A. Johnson, 7 Island Green, Stafford, ST17 0QB
M7 RUS R. Downham, 7 Cherry Garden Road, Maldon, CM9 6ES
M7 RUT I. Rutlidge, 42 Crispin Close, Locks Heath, Southampton, SO31 6TD
MM7 RVC C. Cooper, Beech Tree House, Bogriffie, Fintray, Aberdeen, AB21 0YQ
M7 RVD W. Williams, White Gates, Skinners Lane, Churchill, Winscombe, BS25 5PW
M7 RVF D. Grear, 8 Petworth Close, Frimley, Camberley, GU16 8XS
M7 RVH C. Goodall, 253 Conway Road, Fordbridge, Birmingham, B37 5LG
MM7 RVI F. Liddell, 36 Ladeside, Newmilns, KA16 9BE
MI7 RVJ R. Johnston, 4 Fennel Drive, Antrim, BT41 4FN
MM7 RVN N. Fox, 9 Springfield Terrace, Alness, IV17 0SP
M7 RVT C. Patching, 28 Chenies Drive, Basildon, SS15 4AE
MJ7 RVV W. Stewart, 9 Rue Verte Villas La Grande Route De Saint Laurent, St Lawrence, Jersey, JE3 1NJ
M7 RWA R. Ambrose, 217A Woodgrange Drive, Southend-on-Sea, SS1 2SG
M7 RWE R. Evans, 14 Doncaster Drive, Upton, Wirral, CH494NX
M7 RWF P. Baker, 3 Carrion View, Gateford, Worksop, S81 8UZ
M7 RWH R. Hart, 15 West Street, Rottingdean, BN2 7HP
M7 RWJ R. Jones, 31 Mary Mead, Warfield, Bracknell, RG42 3SZ
MI7 RWT R. Thompson, 18 Parklands, Ballymena, BT43 6FD
M7 RWW Capt. R. Wilkins, 27 Northside Drive, Sutton Coldfield, B74 3QQ
MI7 RXD R. Kearney, 4 Flat P Clarendon Street, Derry, BT487ES
MM7 RXH R. Haddow, Pointhouse, King'S Cross, Isle of Arran, KA27 8RG
M7 RXK R. Kilcommons, Haven, Meadow View Drive, Ravenfield, Rotherham, S65 4RJ
M7 RXM J. Jordan, 8 Leven Road, Tamworth, B77 2TX
M7 RXW R. Winkley, 109 Furze Park Road, Bratton Fleming, EX31 4TA
M7 RYD D. Ryder, 2 The Birches Canterbury Road, Herne Common, CT6 7LE
MW7 RYK Dr A. Rykala, 20 Mount Pleasant, Blaina, Abertillery, NP13 3DD
MI7 RYP G. English, 15 Murrays Hollows, Ballyroney, Banbridge, BT32 5ES
M7 RYV C. Ryves, 21 Augustus Way, Chandler'S Ford Chandler'S Ford, Eastleigh, Eastleigh, SO53 2BD
M7 RZA A. Keen, 54 Leigh Road, Hale, Altrincham, WA15 9BD
M7 RZE D. Rezaie, 25 Coach Road Green, Gateshead, NE10 0EH
M7 RZO A. Brzosko, 32 Lower Park Street, Cambridge, CB5 8AR
MM7 RZT J. Baugh-Clark, 10 Deveron Street, Huntly, AB54 8BY
M7 SAB S. Bodsworth, 51 Berry Way, Andover, SP10 3RZ
M7 SAC S. Carro, 149 Ruston Rd, London, SE18 5QY
MM7 SAH S. Hancox, 45 Park Road, Brechin, DD9 7AE
M7 SAK P. Phizackerley, 19 Fir Court Avenue, Churchstoke, Montgomery, SY15 6BA
M7 SAP S. Elton, 43 Glenville Road, Bournemouth, BH10 5DD
M7 SAV S. Sedgwick, Flat 2/A, St. Georges Court 44 Thorne Road, Doncaster, DN1 2JA
M7 SAZ S. Ziya, 32 Latchford Road, Wirral, CH60 3RW
M7 SBA A. Shayes, 4 Vicarage Road Deopham, Wymondham, NR18 9DR
M7 SBB S. Swallow, 16 Quarry Lane, Chesterfield, S40 3AS
MW7 SBD S. Davies, Llanbadarn Fynydd, Llandrindod Wells, Ld16ye
M7 SBM S. Brackstone, 75 Mozart Close, Basingstoke, RG22 4HZ
M7 SBN Dr S. Georgescu, 42A Denbigh Road, London, W13 8NH
M7 SBO S. Bickham, 114 Parkway, Bridgwater, TA6 4HT
M7 SBY C. Sobey, 9 Waterhouse Lane, Chelmsford, CM1 2TE
M7 SBZ S. Ball, 23 High Heath Close, Birmingham, B30 1HU
M7 SCB S. Buckley, 341 Manchester Road, Northwich, CW9 7NL
M7 SCC S. Cartwright, 12 Stirling Court, Stafford, ST16 3YG
M7 SCF S. Fadhley, 36 Torrington Gardens, London, N112AB
M7 SCG S. Groves, 152 Rundells, Harlow, CM18 7HF
M7 SCH S. Head, 42 Bitterne Drive, Woking, GU21 3JU
M7 SCN S. Nuttall, 17 Redgate, Northwich, CW8 4TQ
MM7 SCO D. Raeburn, 168 Stonylee Road, North Carbrain, Cumbernauld, G67 2LU
M7 SCP S. Palmer, 173 Sorrel Bank, Linton Glade, Croydon, CR0 9LZ
M7 SCR S. Ride, 61 Pepper Street, Sutton-in-Ashfield, NG17 5GD
M7 SCT G. Nightingale, 21 Turners Close, Ongar, CM5 9HH
M7 SCW S. Wormleighton, 4 Selwyn Avenue, Wick, Littlehampton, BN17 7NF
M7 SCY B. Tocknell, 16 Calder Avenue, Walsall, WS1 2BQ
M7 SDB S. Hartshorne, 143 Winchester Drive Chelmsley Wood, Birmingham, B37 5QL
M7 SDD S. Dawe, Whiterocks House, St Anns Chapel, PL18 9HN
M7 SDE S. Evans, 20 Edgbaston Way, Edlington, Doncaster, DN12 1SQ
M7 SDF S. Forshaw, 32 Fox Way, Eastfield, YO11 3PH
M7 SDI S. Unsworth, Heath Terrace, Towcester, NN128UP
MW7 SDJ S. Crabbe, 58 Severn Crescent, Chepstow, NP16 5EA
M7 SDX J. Moore, 45 Northgate, Wiveliscombe, Taunton, TA4 2LF
M7 SEA S. Hodgson, 35 Hamilton Drive, Guildford, GU2 9PL
M7 SEG S. Green, Flat 12, Canute House, Strand Street, Poole, BH15 1EJ
M7 SEH S. Hodds, 4 Alconbury Way, Middlesbrough, TS3 9QW
M7 SEJ S. Adams, 1 Byford Way, Winslow, Buckingham, MK18 3RJ
M7 SEK R. Sek, 6 Tanners Road, Cheltenham, GL51 7LH
M7 SES S. Smith, 133 Radcliffe New Road, Whitefield, Manchester, M45 7RP
M7 SEW I. Hunt, 17 Robartes Road, St Austell, PL26 8DS
M7 SEY B. Ramsey, 2 Hamilton Walk, Great Yarmouth, NR29 4TB
M7 SFB S. Brown, 1 Boundary Cottage, Bures Road, Little Cornard, Sudbury, CO10 0NN
M7 SFD S. Doust, 109 Athol Mount, Halifax, HX3 5RH
M7 SGA M. Riches, 7 Leighton Square, Ipswich, IP3 0LL
M7 SGB S. Gee, 17 Newchurch Street, Rochdale, OL11 2TA
M7 SGF S. Fowler, 74 Osborne Road, Wisbech, PE13 3JW
M7 SGM M. Stringer, Castle Street, Keinton Mandeville, TA11 6DX
M7 SGT R. Hunter, 134 Casttoecoombe Drive London Wimbledon Sw196Rt, London, SW19 6RT
M7 SHF Rev. T. Cundy, 82 Queensbury Close, Bedford, MK40 4RE
M7 SHN S. Gannaway, 8 Stubbington End, Evesham, WR11 2SF
M7 SHO S. Blackwell, 16A Drummond Roadbournemouth, Bournemouth, BH1 4DS
M7 SHP R. Shepherd, 19 Elford Avenue, Newcastle upon Tyne, NE13 9AP
M7 SHQ D. Stone, 6 Beechwood Gardens, Bournemouth, BH5 1NF
MM7 SHT J. Short, 7 Macinnes Drive Newarthill, Motherwell, ML1 5TY
M7 SHW S. Webster, 22, St Michael'S Gardens, South Petherton, TA13 5BD
M7 SHZ S. Baines, 65 Milner St, York, YO24 4NJ
M7 SIB S. Blight, 18 Sophia Way, Newcastle, ST5 8TB
M7 SID Dr P. Bunyan, 228 Aldwick Road, Bognor Regis, PO21 3QH
MW7 SIF S. French, 2 Cambrian Road, Tywyn, LL36 0AG
M7 SIJ S. Smart, 26 Wycote Road, Gosport, PO13 0TG
M7 SIN D. Murphy, 29 Beverley Crescent, Northampton, NN3 2PY
M7 SIS S. Issaias, 35 Cowper Road, London, N14 5RR
M7 SIW S. Wallace, 6 Nuttall Gardens, Cranleigh, GU67FQ

M7 SJB S. Bagnall, 30 Bideford Avenue, Stafford, ST17 0HB
MM7 SJF S. Ferguson, 1 Rhindmuir Grove, Baillieston, Glasgow, G69 6NE
M7 SJG S. Gardner, 137 Allerburn Lea, Alnwick, NE66 2QP
M7 SJH S. Hearne, 8 Morgan?S Road, Calne, SN11 0FH
M7 SJL S. Lovell, 20 Courtenay Walk, Weston-Super-Mare, BS22 7TQ
M7 SJM S. Martin, 6 Cherry Tree Drive, Sedgefield, Stockton-on-Tees, TS21 3DN
M7 SJT J. Thirlwell, 6A Brook Street, Warminster, BA12 8DN
M7 SJW S. Wilson, 21 Earlham Grove, Weston-Super-Mare, BS23 3JH
M7 SKA L. Steingold, 20 Spenser Avenue, Exeter, EX2 6BW
M7 SKB Capt. S. Bahulayan, 23, Pennine Way, Farnborough, GU14 9HT
MM7 SKE M. Chomentowski, 24 Arbroath Lane, Aberdeen, AB12 5BY
M7 SKI H. Allen, 8 Winchcombe Gardens, South Cerney, Cirencester, GL7 5WJ
MM7 SKL M. Johnston, 10 Montgomerie Drive, Skelmorlie, PA17 5AG
M7 SKN C. Hodgskin, 2 Roberta Walk, Bridgwater, TA5 2GW
M7 SKO S. Ochot, 6 Ufford Close, Harrow, HA3 6PP
M7 SKP Capt. M. Williams, 2 Addington Court, Madeira Road, Weston-Super-Mare, BS23 2EY
M7 SKS K. Stevens, 32 Christchurch Road Hucknall, Nottingham, NG156SA
M7 SKT S. Turner, 326 Middle Park Way, Havant, PO9 5DS
M7 SKV L. Paplauskas, 19 Watery Lane, Coventry, CV6 2GE
M7 SKW S. Wickham, 45 Nicholson Drive, Beccles, NR34 9UX
M7 SLA S. Smith, 103 Charles Road Quarry Bank Brierley Hill, Dudley, DY5 1AE
M7 SLB S. Appleby, 21 Osborne Park, Scarborough, YO12 5QF
M7 SLD Capt. M. Bazzocchi, 29 Turners Avenue, Fleet, GU51 1DU
M7 SLE S. Griffin, 54 Marine Drive, Hartlepool, TS24 0dY
M7 SLG S. Graham, 23 North Hill, Fareham, PO16 7HN
M7 SLH S. Hodgkiss, 14 Dales Close, Biddulph Moor, Stoke-on-Trent, ST8 7LZ
M7 SLI S. Lane, 12 Ashenden Road, Guildford, GU2 7UU
M7 SLK B. Slack, 3 The Ridge, Broad Road, Braintree, CM7 9RX
M7 SLO R. Mallinson, 20 Moorside Court, Moorends, Doncaster, DN8 4SL
M7 SLP M. Jones, Flat, 2 The Anchor Centre, Bridge Street, Kingsbridge, TQ7 1SB
M7 SLR S. Bennett, 7 Holme Park, High Bentham, LA2 7ND
M7 SLW S. Whitfield, 20 Cherrytree Drive, Langley Park, DH7 9FX
MM7 SLY M. Sisley, 3 Harley Terrace, Portlethen, Portlethen Village Portlethen, Aberdeen, AB12 4NS
M7 SMC P. Harding, 16 Wightman Road, London, N4 1SQ
M7 SME R. Hauk, 6 Hall Road, Stowmarket, IP14 1TN
MM7 SMG S. Mcgill, 39 Dudley Drive, Townhead, Coatbridge, ML5 2PJ
M7 SMH S. Hammond, Ellsworth, Thrigby Road, Filby, Great Yarmouth, NR29 3HJ
M7 SMN S. Trenchard, 1 The Willows Pitch Place Worplesdon Road, Guildford, GU3 3LL
M7 SMP S. Pressland, 14 Conyerd Road Borough Green, Sevenoaks, TN15 8RJ
M7 SMX S. Mansfield, 545 Portswood Road, Southampton, SO17 3SA
M7 SNB S. Brown, 5 Overbrook Nursery, Green End, Landbeach, Cambridge, CB25 9FD
MM7 SNP R. Cosh, Lyndhurst, Kingston Road, Neilston, Glasgow, G78 3DY
M7 SNR S. Nolan, 24 West Avenue, Chelmsford, CM1 2DE
M7 SNS S. Sampson, 5 Healdwood Close, Castleford, WF10 3AR
M7 SOC S. O'Connor, 4 Charlton Grove, Bradford, bd20 0qg
M7 SOE C. Haffenden, 147 The Diplocks, Hailsham, BN27 3JZ
M7 SON S. O'Neill, 33 Norlands Park, Widnes, WA8 5BH
M7 SOP P. Weaver, 2 Almer Road, Poole, BH15 4JR
M7 SOT D. Baddeley, The Old Cottage, Sandy Lane, Baldwins Gate, Newcastle, ST5 5DP
M7 SOX R. Keeys, 5 Little Fields, High Street, Hadlow, Tonbridge, Tn11 0ed
M7 SPA M. Caddick, 97 Hollemeadow Avenue, Walsall, WS3 1JB
M7 SPC S. Curtis, Flat 10, Maurice House, Southmill Road, Bishop's Stortford, CM23 3DH
M7 SPE A. Campbell, 26 Castle Grove, Horbury, Wakefield, WF4 5DX
M7 SPH S. Harris, Lackington Drove, Dorchester, DT2 7QU
M7 SPJ S. Johnson, 8 Foster Close, Seaford, BN25 2JL
M7 SPM S. Manon, 30 Glycena Road, London, SW11 5TR
M7 SPO N. Booth, Spook Hollow, Spook Hill, North Holmwood, Dorking, RH5 4JP
M7 SPP S. Parkes, 19 Keats Road Harden, Walsall, WS3 1DS
M7 SPS S. Polap, 42 Woodland Drive, Leicester, LE3 3EB
M7 SPW S. Worsley, 2 Cassidy Close Pennington Wharf, Leigh, WN7 4GB
M7 SPX S. Page, 99 Masons Hill, Bromley, BR29HT
M7 SPY N. Ashton, 33 Paxford Road, Wembley, HA0 3RQ
M7 SRB B. Briggs, Blackthorn Road, Northallerton, Dl7 8WB
M7 SRE S. Holroyd, 31A Cross Street, Tenbury Wells, WR15 8EF
M7 SRH S. Harvey, 3 North Holmes Close, Horsham, RH12 4HB
M7 SRP S. Robinson, 7 Higney Road, Hampton Vale, Peterborough, PE7 8LZ
MM7 SRR S. Rastocky, Dudley Dr, Glasgow, g129sb
M7 SRT T. Keloglou, 52 Nutcroft Road, London, SE15 1AF
M7 SRV A. Insarov, Broadway Chambers, 20 Hammersmith Broadway, London, W6 7AF
M7 SRW S. Wood, 18 Eller Drive, West Winch, King's Lynn, PE33 0NN
M7 SRZ S. Roberts, 16 Corelli Road, Basingstoke, RG22 4NB
M7 SSB J. Virgo, 41 Lynch Road, Berkeley, GL13 9TE
M7 SSC S. Compagno, 26 Barnet Road Erdington Birmingham B23 6Jl, Birmingham, B23 6jl
M7 SSE S. Evans, 78 High Brooms Road, Tunbridge Wells, TN4 9BN
M7 SSF S. Farnsworth, 26 Burrows Close, Headington, Oxford, OX3 8AN
MI7 SSJ S. Spallen, 65 Tirgarvil Road, Upperlands, Maghera, BT46 5UW
M7 SSO S. Thompson, 19 Tewkesbury Place Nether Street, Beeston, Nottingham, NG9 2BA
M7 SSS S. Tierney, 40 Gypsy Lane, Watford, Wd48pr
M7 SSW S. Webb, Wa103Ht, St Helens, WA10 3HT
M7 SSY A. Parrott, 24 The Bungalows, Savory Road, Wallsend, NE28 7HX
MI7 SSZ S. Stewart, 7 Orangefield Crescent, Armagh, BT60 1DS
MM7 STF S. Ferguson, Stewards House, Ardeer Golf Club, Stevenston, KA20 4LB
M7 STH S. Hoyle, 7 Horderns Road, Chapel-En-Le-Frith, High Peak, SK23 9ST
MM7 STK T. Storkey, Flat E, 29 Herbert Street, Glasgow, G20 6NB
MM7 STL S. Laing, 57 Rosebery Avenue, South Queensferry, EH30 9JQ
MM7 STM S. Mitchell, 97 Barbieston Road, Auchinleck, KA18 2ED
MM7 STS S. Sim, 3 Deveron View, Glass, Huntly, AB54 4XP
MM7 STT S. Thom, 47 Muirhouse Green, Edinburgh, EH4 4RB
M7 STU S. Dingle, 18 Parsons Mead, Abingdon, OX14 1LS
M7 STV S. Elliott, Lower Ground Flat, 34 Maxwell Road, Fulham, London, SW6 2HR
M7 SUE S. Braisby, 4 Langmans Way, Woking, GU21 3QY
M7 SUM S. Dossa, 24 Warwick Drive, Cheshunt, Waltham Cross, EN8 0BW
M7 SUN L. Bell, 21 Great North Road, New Barnet, Barnet, EN5 1EJ
M7 SUP N. Paul, Enfield, Gunton Road, Wymondham, NR18 0QP
MW7 SUS B. Williams, 25 Park Avenue, Flint, CH6 5DW
M7 SUZ P. Hiles, 209 Oldbrook Boulevard, Oldbrook, Milton Keynes, MK6 2QB
M7 SVA A. Mystikis, 23 Barrier Point Road, London, E16 2SB
M7 SVB C. Bolton, 1A Salterns Terrace, Bideford, EX39 4AG
M7 SVE S. Coles, 5 Pinfold Road, Giltbrook, Nottingham, NG16 2FT
MM7 SVI A. Robertson, 1 Greenbank Street, Galashiels, TD1 3BL
M7 SVK D. Durkot, 15 Thornhill Road, Chorley, PR60JB
M7 SVP V. Sterea, 8 Hamilton Street, Stalybridge, SK15 1LL
MM7 SVR W. Mcglone, 90 Nevis Park, Inverness, IV3 8PP
M7 SVT S. Purcell, 39 Broadfield Road, Stoke-on-Trent, ST65PW
M7 SWA S. Allison, 14 Thameside Court, Northmoor, Witney, OX29 5BL
MM7 SWC S. Copland, 91 Nimmo Avenue, Perth, PH1 2PU
M7 SWD J. Marcelino, 18 Blayney Row, Newcastle upon Tyne, NE15 8QD
M7 SWE C. Sjostedt, 14 New Road, Marlow, SL7 3NG
M7 SWK M. Sedgwick, 44 Cundall Road, Hartlepool, TS26 8LG
MM7 SWM S. Mackie, Flat 115, 70 Kennishead Avenue, Thornliebank, Glasgow, G46 8RS
MM7 SWO L. Mackay, 16 Swordale, Isle of Lewis, HS2 0BP
M7 SWR S. Prandoczky, 8 South View, Kimblesworth, Chester le Street, DH2 3QN
M7 SWS S. Walters-Smith, 83 Chesterfield Road, Tibshelf, Alfreton, DE55 5NJ
M7 SWT S. Tredwell, 9 Ferry Road, Iwade, Sittingbourne, ME9 8RG
M7 SWW W. Taylor, 10 Miller Close, Redbourn, St Albans, AL3 7BG
M7 SWX S. Wilson, 7 Boreham Road, Great Leighs, Chelmsford, CM3 1NH
M7 SWY J. Yeo, 10, Bradgate Close, Leicester, Le79np
M7 SXH A. Dawson, 30 Dukes Road, Lindfield, Haywards Heath, RH16 2JQ
M7 SYB D. Bunting, 19 Hornscroft Park, Kingswood, Hull, HU7 3GS
M7 SYD J. Sydenham, 23 Lawrence Way, Bicester, OX26 2FP
M7 SYE S. Stuart, 18 Park Avenue, Castle Donington, Derby, DE74 2JT
MW7 SYK D. Burns, Old Tavern, Llangeinor, Bridgend, CF32 8PE
MI7 SYL C. Waldie, 47 Oakfield Road, Oakfield, Letterbreen, Enniskillen, BT92 2GJ
M7 SYW S. Yem, 8 Beechwood Avenue, Wallasey, CH45 8NX
M7 SYZ S. Morfitt, 10 Greenway, Ryde, PO33 3SD
M7 SZD S. Dossa, 24 Warwick Drive, Cheshunt, Waltham Cross, EN8 0BW
M7 SZO Z. Szot, 45 Ealing Park Gardens, London, W5 4EX
M7 SZU M. Tarrant, Holly Cottage Nanstallon, Bodmin, PL30 5JZ
M7 TAA T. Aavola, 25 Honiton Road, Southend-on-Sea, SS1 2RT
M7 TAD D. Prior, Holly House, Cross Street Drinkstone, Bury St Edmunds, IP30 9TP
MW7 TAF J. Watkins, 6 High Street Abersychan, Pontypool, NP4 7AB
M7 TAG B. Murtagh, 4 Sandcroft Court, 76 Garlands Road, Redhill, RH1 6GZ
MW7 TAH T. Hughes, 1 Craig Y Don, Amlwch, LL68 9DN
M7 TAS T. Jones, 2 Woodcote, Hanham, Bristol, BS15 8QS
M7 TAT M. Humphries, Meadow View 40 Main Road, Cleeve, Bristol, BS49 4NR
M7 TAW A. Walker, 5 Bepton Down, Petersfield, GU31 4PR
M7 TAY K. Taylor, 56 Gibraltar Lane, Haughton Green Denton, Manchester, M34 7GG
M7 TAZ A. Allen, 5 Davison Court, Longhorsley, Ne65 8ld
M7 TBA A. Greff, 28 Merrylands Road, Great Bookham, Leatherhead, KT23 3HW
M7 TBC R. Boland, 10 Kenilworth Drive, Kidderminster, DY10 1YD
M7 TBD D. Harvey, 5 Tithe Barn Drive Bray, Maidenhead, SL6 2DF
M7 TBF F. Grande, Restrup Engvej 26, Aalborg, Denmark, 9000
M7 TBI T. Baker, 23 Ferrars Road, Tinsley, Sheffield, S9 1RX
M7 TBM T. Mannion, 31 Lockwell Road, Dagenham, RM107RE
M7 TBO T. Wright, 161 Golf Road, Mablethorpe, LN12 1EZ
M7 TBR Dr D. Elliott, Maynestone Road, Chinley, SK23 6AQ
M7 TBZ B. Miller, 1A Crowthorne Close, Cambridge, CB1 9LZ
M7 TCA I. Daraban, 2 Bassenthwaite, Huntingdon, PE29 6UL
M7 TCC C. Cleggett, 20 Albany Street, Maidstone, ME14 5AJ
M7 TCF J. Ashcroft, 104 Deneside, Seghill, Cramlington, NE23 7EU
M7 TCK R. Wicks, 30 Ashridge Close, Rushden, NN10 9HS
M7 TCM P. Twyman, 2 Bronte Cottages, Blackmore End, Braintree, CM7 4DG
M7 TCS J. Mutter, 27 Snowdonia Way, Huntingdon, PE29 6XP
M7 TCT M. Burrows, 17 Jessie Road Bedhampton Havant, Portsmouth, po93th
M7 TCV M. Vincent, Clouds Hill, White Pit, Shillingstone, Blandford Forum, DT11 0SZ
M7 TCW I. Sykes, 31 The Park, Warrington, WA5 2sg
M7 TDA A. Danks, Flat 6, Cartwright House, Wolverhampton Road, Bloxwich, Walsall, WS3 2HD
M7 TDK T. Duke, 49 Stanley Webb Close, Sawston, Cambridge, CB22 3FE
M7 TDL T. Large, 4 Blackfirs Lane Somerford, Congleton, cw12 4qq
M7 TDS J. Rushton, 7 Torpoint Walk, New Most On, Manchester, M400FZ
M7 TDV N. Dorling, 51 Haygarth Close, Cirencester, GL7 1WY
MI7 TDY J. Field, 9 Broughton Park, Belfast, BT6 0BD
M7 TEG T. Goddard, 60 Lake Road, Bristol, BS10 5JF
M7 TEJ J. Jewitt, 148 Headley Way Headington, Oxford, OX3 7SZ
M7 TEO T. Orzechowski, 15 Rothesay Terrace, Northampton, NN2 7ER
M7 TER T. Hedger, 1 Berry Terrace, Acton Square, Sudbury, CO10 1HT
M7 TEU Dr S. Hill, 36 The Woodlands, Market Harborough, LE16 7BW
MW7 TEW M. Tew, 2 Allen Street Caegarw, Mountain Ash, CF45 4BD
M7 TEY A. Jordan, Flat 2, The Roveries, Hardwick Close, Epsom, KT19 9NR
M7 TFA A. Brown, 56 Ashwood Avenue, Wigan, WN2 5YE
M7 TFD D. Johnson, 5 Keats Walk, Hutton, Brentwood, CM13 2RY
M7 TFG T. Gallagher, 67 Denfield Crescent, Halifax, HX3 5NQ
M7 TFN M. Rodgers, 53 Shaw Avenue, Normanton, WF6 2TS
M7 TFT C. Walker, 14 Oaks Meade, Carterton, OX18 1JX
M7 TFW W. Johnson, 5 Keats Walk, Hutton, Brentwood, CM13 2RY
M7 TFX T. Forshaw, 2A, Victoria Road, Fallowfield, Manchester, M14 6AP
MM7 TGA T. Goodenough, 83 Craufurdland Road, Kilmarnock, KA3 2HU
M7 TGD D. Edwards, 8 Carey Avenue, Higher Bebington, CH63 8LU
M7 TGE I. Cartwright, 6 Overwood Place, Stoke-on-Trent, ST6 6XD
M7 TGF T. Griffiths, 6 Witney Lane, Edge, Malpas, SY148JJ
M7 TGH A. Walker, 196 High Street Worsbrough Dale, Barnsley, S704SQ
M7 TGJ T. Jones, 27 Chamberlain Grove, Fareham, PO14 1HH
M7 TGR T. Greenbank, 13 Three Stiles Road, Farnham, GU9 7DE
M7 THF T. Fulbrook, 2 Cob Place, Westbury, BA13 3GS
M7 THO A. Swallow, 16 Quarry Lane, Chesterfield, S40 3AS

M7 THR J. Summers, 263 Stroud Road, Gloucester, GL1 5JZ
MM7 THU A. Ridley, 12 Meadow Court, Thurso, KW14 8DD
M7 THX L. Frohock, Flat 3, Parkham Court, Shortlands Road, Bromley, BR2 0JF
M7 TIA N. Truchla, 15 Portchester Close, Peterborough, PE2 8UP
M7 TIC J. Sanders, 11 Butler Close, Plymouth, PL6 6PL
M7 TIE B. Shearer, 22 Linkscroft Avenue, Ashford, TW15 2BE
M7 TIM T. Eilec, 5 Balata Way, Basingstoke, RG24 9YP
M7 TIN D. Smith, 82 Armstead Road, Beighton, Sheffield, S20 1ET
M7 TJB T. Bowron, 1, Broughton Avenue. Lu5 6Bq, Toddington, Nr. Dunstable, LU5. 6BQ
M7 TJC T. Tredwell-Coleman, 22 The Quadrant, Hull, HU6 8NX
M7 TJD T. Dix, Willow Cottage 31 London Road, Woolmer Green, SG3 6JE
M7 TJE J. Edwards, The Granary Old Hall Farm School Road Tunstall, Woodbridge, IP122JQ
M7 TJF T. Franklin, Flat 57 The Maltings Clifton Road, Gravesend, DA11 0AH
M7 TJM T. Mcgoun, 64 Buttfield Lane, Howden, Goole, DN14 7DS
M7 TJN T. Newton, 4 Manor Close, Bradford Abbas, Sherborne, DT9 6RN
M7 TJS T. Sherwin, 17A London Road, Aston Clinton, Aylesbury, HP22 5HG
M7 TJX T. Jinkerson, 104 Foxcote, Finchampstead, Wokingham, RG40 3PE
M7 TKB T. Bishop, 12 Athelstan Road, Folkestone, CT19 6EU
M7 TKD R. Mackay, 7 Darwin Close, Lee-on-the-Solent, PO13 8LS
M7 TKH T. Howell, London Southend Airport Southend Airport, Southend-on-Sea, SS2 6YF
M7 TKI T. King, 83 Westbury Rise, Harlow, CM17 9NT
M7 TKL T. Klimczak, 3 Mariner Walk, Chorley, PR6 9FF
M7 TKR P. Dulac, 22 Colliery Green Drive, Little Neston, Neston, CH64 0UA
M7 TKV L. Clinton, 16A Melrose Avenue, London, NW2 4JS
M7 TKW B. Whaley, 14 Kings Way, Harrow, HA1 1XU
M7 TLA P. Barrows, 5A Magdalen Road, Willoughby, Rugby, CV23 8BJ
M7 TLC J. Phillips, 1 Wath Cottages, Cundall, York, YO61 2RL
M7 TLG J. Turner, 3 Lulworth Close, Winsford, CW7 1LZ
M7 TLJ R. Mcwilliam, 3 Fountains Close, Riccall, York, YO19 6QN
M7 TLM T. Miller, 24 Waveney Residential Park Pound Road, Beccles, NR34 9BJ
MW7 TLS M. Comerford, 5 Oak Avenue, Penley, LL13 0NW
MW7 TLX M. Blandford, 33 Greenmeadow Drive, Parc Seymour, Penhow, Caldicot, NP26 3AW
MW7 TMA T. Allen, 32 Tir Morfa Road, Port Talbot, SA12 7PF
M7 TMC C. Marris, 16 Cedar Close, Louth, LN11 0EH
M7 TMD M. Mcdowell, 22 Paddock Close, Clapham, Bedford, MK41 6BD
M7 TMH T. Hudson, Mimosa Cottage, Ashley Gardens, Mayfield, TN20 6DU
M7 TMM A. Mira Quinones, 14, Langwood House, 63-81 High Street, Rickmansworth, WD3 1EQ
MM7 TMO T. Oakes, 55 Northfield Meadows Longridge, Bathgate, EH47 8SA
M7 TMP M. Firestone, 1 Reeds Close, Rossendale, BB4 8ND
M7 TMR T. Reaney, Odessa Marine, Little London, Newport, PO30 5BS
M7 TMW F. Pereira Da Silva, 19 Hatton Park Road, Wellingborough, NN8 5BA
MW7 TNM T. Feierabend, 140 Pearl Street, Cardiff, CF24 1PN
M7 TNY T. Holden, 132 Sutherland Street, Barrow in Furness, LA14 2BJ
M7 TOG G. Tant, 61 Western Way, Sandy, SG19 1DU
M7 TON A. Naylor, 2 Byron Close Huyton Merseyside, Liverpool, l360uh
M7 TOT S. Turner, 4 Bridge Close, Shoeburyness, Southend-on-Sea, SS3 9PE
M7 TOW A. Martin, 17 Suffolk Road, Bexhill-on-Sea, TN39 5BH
M7 TOY T. Littlebury, 71 Hampden Drive, Kidlington, OX5 2LT
M7 TPD T. Davis, 37 Curzon Avenue, Northwich, CW8 4YU
M7 TPO P. Davis, 7 Rothbury Green, Cannock, WS12 2TR
M7 TPT P. Taylor, 20 Burlingham Avenue, Worcestershire, WR113EE
M7 TPU A. Usher, White Cottage, Bratoft, Skegness, PE24 5DD
M7 TPW P. Watt, 22 Schofield Road, Eccles, Manchester, M30 7LG
M7 TRA T. Alexander, 6 Moselle Close, Farnborough, GU14 9YB
M7 TRB A. Naylor, 1 Harvington Close, Kidderminster, DY11 5LP
M7 TRG R. Gill, Peartree Cottage, Chapel Road, Necton, Swaffham, PE37 8JA
M7 TRH T. Hawkins, Deganwy, Hardwick Road, King's Lynn, PE30 5BB
M7 TRM S. Amos, 1 Philbye Mews, Cippenham, Slough, SL1 5US
M7 TRV R. Woodward, 8 Fulwell Grove, Birmingham, B44 0EF
M7 TRW K. Dunleavy, 22 Gloucester Court, Newcastle upon Tyne, NE3 2XJ
MW7 TRY S. Stupple, 7 Glan Preseli, Llanddewi Velfrey, Narberth, SA67 7PG
M7 TSA P. Markiewicz, 7 Parkfields, Abram, Wigan, WN2 5XR
M7 TSB A. Stanford-Beale, 2A Albert Road, Caversham, Reading, RG4 7PE
M7 TSG A. Rice, 23 Chilver Drive, Tong, Bradford, BD4 0TS
M7 TSH T. Harrand, 3 Chapel Lane, Sykehouse, Goole, DN14 9BN
M7 TSO G. Smith, 37 The Crescent, Bracebridge Heath, Lincoln, LN4 2NP
M7 TSZ F. Holmes, 2 Station Lane, Hartlepool, TS25 1AX
MW7 TTA A. Gunn, 63 Ffordd Tudur, Holyhead, LL65 2DU
M7 TTB M. Britton, Butterwick Low, Hales Street Tivetshall St. Margaret, Norwich, NR15 2EE
M7 TTS A. Thorpe, 8 Syke Avenue, Tingley, Wakefield, WF3 1LU
M7 TUC M. Tucker, 42 Martock Avenue, Southend on Sea, SS00HH
M7 TUF A. Hawkins, 111 Overbrook, Swindon, SN3 6AT
M7 TUX D. Parker, 51 Ruskin Road, Congleton, CW12 4EA
MI7 TVA R. Moore, 34 Rathvarna Walk, Lisburn, BT28 2UD
MW7 TVH A. Holland, 1 Llys Colwyn, Fairmount, Old Colwyn, Colwyn Bay, LL29 9NF
M7 TVS L. Jenkins, 40 Apollo Drive, Bordon, GU35 0DZ
M7 TVT M. Burge, 24 Oakdale Road, North Anston, Sheffield, S25 4EY
M7 TWA S. Waterson, 7 West Farm Road, Newcastle upon Tyne, NE6 4JA
M7 TWC T. Stevens, 6 West Street, Ilchester, Yeovil, BA22 8NN
MW7 TWG J. Pritchard, 193 Main Road Bryncoch, Neath, SA10 7TT
M7 TWK T. Kerswill, 11A Vernon Drive, Prestwich, Manchester, M25 9RA
MM7 TWN J. Keymer, 2 Crunklaw Farm Cottage, Duns, TD11 3RA
M7 TWO J. Watts, 3 Witney Road, Crawley, RH10 6GJ
M7 TWW D. Hughes, 10 Kingscroft Court, Northampton, NN3 9BH
M7 TXA E. Rippingale-Combes, 3 Leyfield, Albourne, Hassocks, BN6 9DA
M7 TXB A. Blews, 57 Highfield Grove, Stafford, ST17 9RA
M7 TXC A. Collins, 32 Upper Weybourne Lane, Farnham, GU9 9DF
M7 TXM T. Lockyer, 5A Eastgate, Deeping St. James, Peterborough, PE6 8HH
M7 TXR B. Roberts, 20 Mitchell Road, Enderby, LE19 4NX
MM7 TXY A. Dunlop, 12 Kendal Avenue 2/1, Glasgow, G120DL
MW7 TYB J. Procter, Awelfa, Ffordd Glan Mor, Talybont, LL43 2AR
MW7 TYD R. Grimes, Tyddyn Iolyn, Pentrefelin, Criccieth, LL52 0RB
M7 TYG C. Sandbrook, 24 Greensome Lane, Stafford, ST16 1HE
M7 TYM T. Boniface, 51 Harebeating Crescent, Hailsham, BN27 1JL
MW7 TYR N. Griffiths, 30 Offas Green, Norton, Presteigne, LD82NX
M7 TYS P. Tysoe, 51 Grendon Road Polesworth, Tamworth, b78 1nx
M7 TZF C. Cooper, 2 Brook View, Railway Street, Slingsby, YO62 4AN
M7 TZG T. Grabiec, 16 Jubilee Crescent, Clowne, Chesterfield, S43 4ND
M7 TZW A. Saxton, Clifton, Nottingham, NG11 9HB
M7 UAM Capt. G. Story, 7 Wheelock Close, Northwich, CW9 8TQ
M7 UAP P. Richards, 22 Waterpump Court, Northampton, NN3 8US
M7 UCA P. Wheeldon, 6 The Crescent, Brimington, Chesterfield, S43 1AZ
M7 UCD S. Cole, 13 Boscobel Road North, St Leonards-on-Sea, TN38 0NY
M7 UCL T. Sun, Orchard Heights, 31 Frogmore Street, Bristol, BS1 5BY
M7 UDV G. Lee, Princess Beatrice House, 192 Finborough Road, London, SW10 9BA
M7 UFC J. Ross, 42 Brosscroft, Hadfield, Glossop, SK13 1HF
M7 UFF K. Simkin, 19 Summercourt, Hailsham, BN273AW
M7 UFL Dr S. Sadashivajois, Flat 131, Aviator House, 227 Station Road, Addlestone, KT15 2US
M7 UGL Dr P. Hall, 120 Miswell Lane, Tring, HP23 4EU
M7 UHU C. Matthews, Manor Farm Knedlington, Goole, DN14 7EU
M7 UKD G. Lowe, Stone Cottage, Esplanade Lane, Watchet, TA23 0AH
M7 UKN K. Mcdonald, 11 Micklegate Murdishaw, Runcorn, WA7 6HT
M7 UKP P. Csorba, 59 Brightgreen Street, Stoke on Trent., ST35DG
M7 UKU M. Fowler, 191 Rayne Road, Braintree, CM7 2QE
M7 UKW A. Wearne, 58 Mainstone Avenue, Princerock, Plymouth, PL4 9NB
M7 UKX N. Radulescu, 41 Sherard Road, London, SE9 6EX
M7 UKZ C. Conghos, Flat 30, The Dutch Quarter Ap. West Stockwell Street, Colchester, CO11FQ
M7 ULV S. Price, 41 Beech Drive., Ulverston, LA129EX
MI7 UNI C. Mccrea, 20 Kilmore Park, Kilmore, Armagh, BT61 8NT
M7 UOK R. Dutton, 3 Kilkenny Road, Guisborough, TS14 7LE
M7 UPS A. Upstone, 19 Calfe Fen Close, Soham, Ely, CB7 5GD
M7 UPT B. Fellowes, 35 Bolburn, Gateshead, NE10 8XB
MW7 URL C. Cowling, 14 Shelley Drive, Bridgend, CF31 4QA
M7 USA A. Carter, 36 Marriotts Close, Ramsey Mereside, Ramsey, Huntingdon, PE26 2TX
M7 USB J. Hunter, 30 Glebelands, Pulborough, RH20 2JJ
M7 USE J. Everatt, 50 Barnsley Road, Thorpe Hesley, Rotherham, S61 2RR
M7 USL A. Lee, 76 Sherborne Road, Stockport, SK3 0SN
M7 USV C. Carter, Bungalow 7, Higher Ingsdon Quarry, Liverton, Newton Abbot, TQ12 6JA
M7 UTD P. Brice-Bullows, 3 Ashcroft, Chard, TA20 2JH
M7 UTG D. Balch, 41 Crosscombe Drive, Bristol, BS13 0DE
M7 UTM P. Richmond, 34 Garfield Road, Scarborough, YO12 7LJ
M7 UTT C. Little, 28 Cecil Avenue, Warmsworth, Doncaster, DN4 9QW
MM7 UVI H. Yang, Flat3/2 7 Cooperage Place, Glasgow, G3 8QP
M7 UWL R. Scott, 157 Walton Back Lane, Walton, Chesterfield, S42 7LT
M7 UXO J. Coleman, 67 Church Meadows, Deal, CT14 9QZ
M7 VAC T. Easton, 62 Gunton Drive, Lowestoft, NR32 4QB
M7 VAD A. Dsouza, 13A Sellons Avenue, London, NW10 4HJ
M7 VAL G. Hall, 124 Borfard Road, Liverpool, l250pr
M7 VAT P. Cartwright, 43 Tarragon Drive Bispham, Blackpool, fy20wl
M7 VBO M. Apps, 15 Green Court, Bridge, Canterbury, CT4 5LU
M7 VBR A. Riddle, 19 Cottey Crescent, Exeter, EX4 9DT
M7 VCC S. Cairney, 42 Gorse Avenue, Stretford, Manchester, M32 0UE
M7 VCF D. Peacock, 41 Oxford Meadow, Sible Hedingham, Halstead, CO9 3QW
M7 VCH I. Vernall, 1 Owen Place, Bridge Street, Kington, HR5 3DH
M7 VCK K. Vickers, 11 Kendal Drive, Rainhill, Prescot, L35 9JQ
M7 VCL V. Forsyth, 4 Charmouth Walk, Manchester, M22 5BF
MW7 VCM R. Poyser, Glandwr, Snowdon Street, Porthmadog, LL49 9DF
M7 VCT D. Cooper, 58 Serpentine Road, Widley, Waterlooville, PO7 5EF
M7 VCZ P. Simcox, 29 Oakamoor Road, Cheadle, Stoke-on-Trent, ST10 1BS
MD7 VDL R. Langstaff, 17A Elm Drive, Onchan, Onchan, Isle Of Man, IM3 4EH
M7 VDO C. Seal, 65 Sandpits, Leominster, HR6 8HT
M7 VEG S. Goode, 135 Parkside Drive, Watford, WD17 3BA
M7 VEN G. Venn, 14 School Road, Wychbold, Droitwich, WR9 7PU
MM7 VES E. Smith, 15 School Wynd, Quarrier's Village, PA11 3NL
M7 VET M. Williams, 6 Wellhouse Avenue, West Mersea, Colchester, CO5 8GF
MI7 VFR W. Mccandless, 35 Whinsmoor Park, Broughshane, Ballymena, BT42 4JG
M7 VGG G. Gribbin, 96 Old Church Lane, Stanmore, HA7 2RR
MM7 VGR V. Grauso, 31 Curlew Brae, Livingston, EH54 6UG
MW7 VHT J. Banfield, 188 Penderry Road Penlan, Swansea, SA5 7ER
MW7 VIB A. Jones, 5 Harlech Road, Llandudno, LL30 1RQ
M7 VIK A. Cretu, 14 Ashton Gate, Flitwick, MK45 1AG
M7 VIN V. Brindle, 185 Brunshaw Road, Burnley, BB10 4DL
M7 VIP G. Matthews, 255 Coach Road Estate, Washington, NE37 2EU
M7 VIV V. Roy, 12 Salisbury Road, Penenden Heath, Maidstone, ME14 2TX
M7 VJH V. Hayes, 68 Billingsley Road, Birmingham, B26 2EA
M7 VJM N. Marchant, 18, Allington Crescent, Allington Road, Newick, bn84nt
M7 VJT V. Thayil, 18 Bargrove Avenue, Hemel Hempstead, HP1 1QP
MM7 VKR V. Lang, Darnaway, Castleton Place, Braemar, Ballater, AB35 5ZQ
M7 VLA V. Oleinicenco, 91A Brent Street, London, NW4 2DY
MM7 VLC A. Cruickshank, 58 Cavalry Park Drive, Edinburgh, EH15 3QG
M7 VLD G. Baciu, 7A Holmstall Avenue, Edgware, HA8 5JQ
M7 VLQ Dr A. Volkov, 2 Robert Close, Potters Bar, EN6 2DH
M7 VMB S. Howard, 57 Mill Lane, Huthwaite, Sutton-in-Ashfield, NG17 2SJ
MM7 VMC V. Mccutcheon, 30 Belhaven Park, Muirhead, Glasgow, G69 9FB
M7 VMR V. Roberts, Cobden Avenue, Southampton, SO18 1FW
M7 VMS S. Spurr, 12 Millom Way, Grimsby, DN32 7EJ
M7 VNX A. Wright, 32 Temple Grove, Leeds, LS15 0HT
M7 VOD J. Naylor, 16 Eaton Street, Prescot, L34 6HD
M7 VOT A. Hopper, 38 Mitchell Avenue, Thornaby, TS17 9AF
M7 VOX D. Fincham, 2 Glebe Close, St Columb, TR9 6TA
M7 VPM V. Mann, Ss0 9Rj, Westcliff on Sea, SS0 9RJ
M7 VRA K. Cope, 9 Amber Heights, Ripley, DE5 3SP
M7 VRM V. Mccullagh, 4 Sholebroke Mount, Leeds, LS7 3HG
M7 VSD S. Pester, 85 Lytham Close, Great Sankey, Warrington, WA5 2GH
M7 VSZ F. Zanchi, Flat 2 4 Helios Road, London, SM6 7BZ
M7 VTT T. Turner, 39 High Street, Brampton, PE284TG

M7 VUK F. De Meira Lins, 90 Nora Street, South Shields, NE34 0RB
M7 VUX D. Contractor, 43 The Spinney, High Wycombe, HP11 1QE
M7 VVB D. Elliott, 21A The Broadway, Swindon, SN25 3BN
M7 VVS S. Swingler, 41 Mount Pleasant, East Lulworth, Wareham, BH20 5QW
M7 VVV H. Horan, 5 Hunters Ride, Stafford, ST17 9HU
M7 VVZ R. Hall, 7 St Saviours Road, Totland Bay, PO390EZ
MM7 VWG D. Morrison, Osbourne Cottage, Benderloch, Oban, PA37 1QP
M7 VWS S. Broomberg, 31 Crundwell Road, Tunbridge Wells, TN4 0LL
MM7 VWT G. Henderson, 2 Monks Moss, Ladybank, Cupar, KY15 7NN
M7 VXE S. Holden, 21 Warren Crescent, East Preston, Littlehampton, BN16 1BH
M7 VXU M. Paddock, 22 The Lant, Shepshed, Loughborough, LE12 9PD
M7 VXW J. Halsall, 53 Grasmere Ave Ince, Wigan, WN2 2NN
M7 VYN M. Taylor, 20 King Richard Drive, Bournemouth, BH11 9PE
MM7 WAB P. Scott, 62 Kerse Terrace, Rankinston, Ayr, KA6 7HG
M7 WAC S. Preston, 105 Lodge Road Thackley, Bradford, BD10 0RF
MW7 WAD W. Dabrowski, 2 Underhill Crescent, Knighton, LD7 1DG
M7 WAF E. Rogers, Manor Farm Close, Westont Turville, HP225SD
M7 WAI D. Waite, The Haven, Kemsley Road, Tatsfield, Westerham, TN16 2BH
M7 WAM I. Hamilton, Grenfell House, Kirkby Stephen, Ca17 4hl
M7 WAW C. Coath, 17 Ashville Road, Bristol, BS3 2AP
M7 WAY W. Hind, 6 Whinfield Avenue, Shotton Colliery, Co Durham, DH6 2HE
M7 WBB W. Bevan, 25 Lymbrook Close South Leigh, Witney, OX29 6XL
M7 WBF A. Smith, 3 Marion Avenue, Middlesbrough, TS5 5JG
M7 WBN N. Nelson, 40 Delph Mount, Leeds, LS6 2HS
M7 WBT C. Elliott, 51 Lower Street Quainton, Quainton, HP22 4BL
M7 WBX E. Smart, 81 Hunt Close, London, w11 4jx
M7 WBY P. Mulder, 8 Chapel Close, Little Gaddesden, Berkhamsted, HP4 1QG
M7 WCC C. Watson, 30 Moorfield Parade, Irlam, Manchester, M44 6FY
M7 WCF W. Coombs-Farnsworth, 6 Burditt Close, Rothwell, NN14 6LD
MM7 WCT W. Forbes, Tillathrowie, Gartly, Huntly, AB54 4SB
M7 WCW A. Abeka, 35 Goldbeaters Grove, Edgware, HA8 0QE
M7 WDC J. Bell, 9 Somerville Green, Leeds, LS14 6AY
MW7 WDP D. Wheatley, 10 Castle Street, Pennar, Pembroke Dock, SA72 6RH
MI7 WDT D. Tagg, 23 Cloverhill Park Coleraine, Coleraine, BT51 3RF
M7 WDW W. Wakeman, Blunts Hall Cottage, Blunts Hall Road, Witham, CM8 1LX
M7 WDX L. Baker, 3 Oak Court, Peterborough, PE7 3FS
MW7 WDY C. Wood, 50 Heather Court, Cwmbran, NP446JR
M7 WEA M. Speller, 38 Brockley Crescent, Romford, RM5 3JX
M7 WEB M. Webb, 60 Waddens Brook Lane, Wednesfield, Wolverhampton, WV11 3SF
M7 WEL S. Weller, 25A Orchard Way, East Grinstead, RH19 1AY
M7 WEP G. Brearley, 8 Albert Road Eston, Middlesbrough, TS6 9QW
M7 WER W. Rees, 69 Pewley Way, Guildford, gu13pz
M7 WET J. Smith, 39 Kemp Road, North Walsham, NR28 0FP
M7 WEZ I. Bird, 2 Church Street, Wiveliscombe, Taunton, TA4 2LR
M7 WFD J. Kershaw, 17 The Sycamores, Wakefield, Horbury, WF4 5QG
M7 WFG W. Gola, 128 Bournville Road, Weston-Super-Mare, BS23 3RS
M7 WFR G. Kirby, East Road, West Mersea, CO5 8EB
MM7 WGB W. Barclay, Bressa Waterside, Strathdon, AB36 8XA
MM7 WGC A. Woodwark, 35/1 Darnell Road, Edinburgh, EH5 3PH
M7 WGF D. Bullen, 5 Westview, Long Sutton, Langport, TA10 9LT
M7 WGM A. Wigham, 15 Owens Way, Croxley Green, Rickmansworth, WD3 3PS
M7 WGR R. Woodford, 7 Steer Road, Swanage, BH19 2RU
MI7 WGS W. Giles, 11 Old Antrim Mews, Ballymena, BT42 2SP
M7 WGX M. Chenery, 27 Dunstan Road, Glastonbury, BA6 8EE
M7 WHA Rev. W. Hackman, Kynance, Barden Road, Speldhurst, Tunbridge Wells, TN3 0QB
M7 WHC Q. Wang, 14, Marquis House, 45 Beadon Road, London, W6 0BT
M7 WHI P. Whitehouse, 2 Nans Rosen, Threemilestone, Truro, TR3 6FW
MM7 WIC C. Aitken, Windybraes, Upper Gills, Canisbay, Caithness, Canisbay, KW1 4YB
M7 WID J. Butler, 112 Netherfield, Widnes, WA8 8BZ
M7 WIG S. Biggs, 81 St Abbs Drive, Bradford, BD6 1EJ
M7 WIM W. Mckechnie, 49 Bedeswell Close, Hebburn, NE31 2GB
M7 WIS W. Stone, 63 Sands Lane, Lowestoft, NR32 3ER
M7 WIT V. Vaznais, 35 Lynwood Drive, London, KT4 7AA
M7 WIV R. Laidler, Swallow Cottage, Wiverton, Plympton, Plymouth, PL7 5AA
M7 WIX S. Wix, 43 Radburn Close, Harlow, CM18 7EE
M7 WJA W. Auger, Augernik Fruit Farm, Hopton Wafers, Kidderminster, DY14 0HH
M7 WJB J. Blackall, 2 Ryson Avenue, Blackpool, FY4 4DN
M7 WJC C. Andrews, 10 Hartford Road, Hartley Wintney, Hook, RG27 8QW
MI7 WJD W. Burrowes, 208 Ballynure Road, Ballyclare, BT39 9AJ
M7 WJG W. Garvey, 254 Bury Road, Tottington, Bury, BL8 3DT
M7 WJH W. Hoar, 46 Pendean Avenue, Liskeard, PL14 6DA
MI7 WJL W. Little, 87 Meadowvale Park, Limavady, BT49 0RD
MM7 WJM M. Wigham, 5 Milton Park, Auchtertool, Kirkcaldy, KY2 5QX
M7 WJR Dr W. Rickford, Rushcroft, Green End, Weston, Hitchin, SG4 7AL
M7 WKD A. Brown, 11 Newlands Close, Chandler'S Ford, Eastleigh, SO53 4PD
MM7 WKK G. Hamilton, 45 Argyle Square, Wick, kw1 5aj
M7 WKS S. Wilkins, Colroger Close, Mullion, TR12 7DZ
M7 WLF R. Flores De Guirior, 17 Keyhaven Close, Derby, DE21 4SQ
M7 WLH W. Hughes, 346 Griffiths Drive, Wednesfield, Wolverhampton, WV11 2LB
M7 WLL C. Norman, 24 Jasmine Grove, Paignton, TQ3 3TH
M7 WLT S. Connor, 2 Swedish Houses, Shalbourne, Marlborough, SN8 3PX
M7 WLZ D. Walter, 13 Hill Top Road, Little Harrowden, Wellingborough, NN9 5BP
MM7 WMG W. Mcguigan, 1 West Drip Farm Cottages, Stirling, FK9 4UJ
M7 WMK M. Bare, 4 Elenors Grove, Ryde, PO33 4HE
M7 WML Rev. M. Legg, 30 Canonbury Road, Enfield, EN1 3LW
M7 WMM T. Cross, Piccotts End Farm 117 Piccotts End Road, Hemel Hempstead, HP13AU
MJ7 WMQ R. Raynes, Mon Plaisir Rue De Samares, St Clement, Jersey, JE2 6LZ
MW7 WMS P. Williams, 127 Penchwintan Road, Bangor, LL57 2YG
M7 WMY A. Henslok, 31 Pritchard Street, Wednesbury, WS10 9EW
MI7 WNY P. Ffitch, 1 Lisburn Road, Ballynahinch, BT24 8BL
MW7 WOC O. Crump, 26 Stanly Road, Wrexham, LL14 1HH
MM7 WOK H. Cameron, 6 North Steadings Torlundy, Fort William, PH33 6SW
M7 WOL P. Wallis, 24 Goldsdown Road, Enfield, EN37QZ
M7 WOZ W. Walton, 1 West Hall, Yeadon, Leeds, LS19 7AJ
MM7 WPC J. Pearson, 4 Falsidehill Farm Cottages, Kelso, TD5 7TT
M7 WPE W. Edmondson, 9 Mayfield Avenue, Clitheroe, BB7 1LB
M7 WPH W. Hearn, 16 The Cobbins, Burnham on Crouch, Essex, CM0 8QL
M7 WPL P. Conyers, 112 Weston Park, Crouch End, London, N8 9PN
MI7 WPX R. Arbuckle, 1 Greenfield Park, Strathfoyle, Derry, BT47 6XE
MW7 WRG G. Williams, 8 Fron Deg, Pantymwyn, CH7 5EU
M7 WRI S. Wright, 4 Cassino Road, Watchet, TA23 0TX
M7 WRM W. Maudsley, Knight Stainforth Hall Little Stainforth, Settle, BD24 0DP
MW7 WRZ D. Rees, 19 Waunfawr Road, Cross Keys, Newport, NP11 7PG
MM7 WSI B. Inglis, 51 Blackcraig Brae, Blantyre, Glasgow, G72 0TZ
M7 WSJ D. Cooper, 58 Serpentine Road, Widley, Waterlooville, PO7 5EF
M7 WSM S. Reed, 101 Milton Road, Weston Super Mare, Bs23 2ux
M7 WSP G. Lebond-Carroll, 9 Hawthorn Way Carlton In Lindrick, Worksop, S81 9HN
MM7 WSS W. Shaw, Shaws Farm, Selkirk, TD7 4PR
M7 WST C. West, 12 St. Georges Road, Wallington, SM6 0AS
M7 WSX G. Kinder, 6 Ten Acres, Shaftesbury, SP7 8PP
M7 WTG W. Grimes, 17 Handley Close, Stockport, SK3 8NQ
M7 WTJ T. Wright, 26 Stamford Rd, Carrington, M31 4BA
MI7 WTS B. Rollins, 2 Tarragon Park, Antrim, BT41 4PF
MM7 WTX S. Thomson, Woodbine Cottage, Mosstodloch, Fochabers, IV32 7HZ
MI7 WTZ J. Jordan, 2 Mill House, Crumlin, BT29 4XN
M7 WUT Dr D. Hall, 53 Elfleda Road, Cambridge, CB5 8NA
M7 WVC W. Clancy, 58 Beatrice Place, Hitchin, SG5 4RZ
M7 WVM M. Lomas, 33 Carnation Road, Oldham, OL4 5QD
MW7 WWF M. Ransom, 16 Glyn Garfield Close, Neath, SA11 2JR
M7 WWL A. Walker, 50 Parkstone Crescent, Hellaby, Rotherham, S66 8HD
MM7 WWM J. Hale, Glenburn Rd, Drynoch, Lochgilphead, PA308EU
M7 WWR R. Wall, 3 Stag Drive, Huntington, Cannock, WS12 4UJ
M7 WWT M. Howard, Flat 10, Oakdale, 6 Westgate Road, Beckenham, BR3 5DY
M7 WWX S. Hammond, 5 Nursery Gardens, Sturminster Marshall, Wimborne, BH21 4AX
M7 WXD R. Springall, 7 The Spinney Grange Park, Northampton, NN4 5BT
M7 WXS S. Thomas, 2820 Winding Creek Rd, Prosper, USA, 75078
M7 WYD R. Jenkinson, 8 Newstead Avenue Off Newstead Street, Hull, HU5 3NE
MW7 WYN A. Williams, 5 Brynymor, Three Crosses, Swansea, SA4 3PE
MM7 WYZ W. Zaczyk, 19 Parkgrove Street, Edinburgh, EH4 7NT
M7 WZM R. Turner, 2 Martins Mews, Haverhill, CB9 7FU
M7 WZO D. Mulder, 217 Cloes Lane, Clacton-on-Sea, CO16 8AG
MW7 WZQ G. Roberts, Ynys 3 Rhodfa Gwilym, Four Mile Bridge, LL65 2TX
M7 WZY W. Cooke, 3 Torrington Close, Potton, Sandy, SG19 2SD
MM7 XAC A. Campbell, Margreig, Laghall Court, Kingholm Quay, Dumfries, DG1 4SX
MI7 XAG A. Gregson, 16 Dunkirk Road, Waringstown, Craigavon, BT66 7SW
M7 XAM A. Moore, 11 Glen View, Wigmore, Leominster, HR6 9UU
M7 XAN A. Nasseri, 4 Kenbury Close, Ickenham, Uxbridge, UB10 8HU
M7 XAP A. Parham, 2 Finch Gardens North Bersted, Bognor Regis, po229eq
MW7 XAW R. Brindle, 10 Brynderw, Stanley Road, Aberystwyth, SY23 1LB
M7 XBS B. Saunders, 60 Mountford Close, Wellesbourne, Wellesbourne, CV35 9QQ
M7 XBX Dr R. Bland, 5 Seafield Road, Lytham St Annes, FY8 5PY
M7 XCB D. Bayley, 13 St. Patrick'S Drive Poolfields, Newcastle-under-Lyme, ST5 2NS
M7 XCS C. Shepherdson, 149 Scarborough Road, Norton, Malton, YO17 8AD
M7 XDE D. Edwards, 69 Park Street, Ripley, DE5 3ES
M7 XDS D. Soulsby, 93 Stonecross Road, Hatfield, AL10 0HW
M7 XHC R. Williams, 98B, The Manor, Billing Garden Village, The Causeway, Great Billing, Northampton, NN3 9EX
M7 XHF C. Collins, 27 Arnold Close, Castle Gresley, Swadlincote, DE11 9HF
M7 XIA P. Mayne, 56 Palmerston Street, Plymouth, PL1 5LJ
M7 XIX P. Cammish, 18 Hertford Close, Eastfield, Scarborough, YO11 3HJ
MW7 XJO J. Mcneil, Mosaic House, Beachley Road, Tutshill, Chepstow, NP16 7EG
M7 XJP A. Griffiths, 13 Hill Place, Knowlwood Road, Todmorden, OL14 6PN
MW7 XJQ D. Smith, Tyddyn Bach, Bethel, Caernarfon, LL55 1YD
M7 XJW J. Graham, 17 Valley Road, Wotton-under-Edge, GL12 7NP
M7 XLA S. Mayes, 14 Prior Street, Hereford, HR4 9LB
M7 XLF T. Evans, 36A Swanmore Road, Ryde, PO33 2TQ
M7 XLH T. Kidds, Laurels Garth, York, YO606SE
M7 XLR P. Ffitch, 22 Beeching Close, Halwill Junction, EX21 5XY
M7 XLX A. Freeman, 122 Tandra, Beanhill, Milton Keynes, MK6 4LL
M7 XMB P. Foster, 100 Howe Road, Norton, Malton, YO17 9BL
MW7 XMD J. Rossiter, 20 Bennions Road, Wrexham, LL13 7AW
M7 XMG M. Griffiths, 23 Gawdy Close, Harleston, IP20 9ET
M7 XNG M. Galbraith, Oaklands, Didcot Road Harwell, Didcot, OX11 0DP
M7 XNY K. Blowing, 194 Cheltenham Road East, Gloucester, GL3 1AL
M7 XON H. Tranter, Flat 6 Oak Apple Court 25 Acorn Road Catshill, Bromsgrove, B61 0TR
M7 XPG P. Goldsmith, 20 Lichfield Drive, Prestwich, Manchester, M25 0HX
M7 XPT L. Cowles, 5 Chalvedon Square, Pitsea, Basildon, SS13 3QX
M7 XPX J. Porter, 45A Wealden Way, Bexhill-on-Sea, TN39 4NZ
M7 XQS D. Dalik, 26 Hanson Street, London, W1W 6UH
MW7 XRE G. Phillips, 109 Mount Pleasant Road, Ebbw Vale, NP23 6JL
M7 XRN A. Craven, 14 Deverel Road, Charlton Down, Dorchester, DT2 9UD
M7 XRO J. Roberts, 11 Vulcan Way, Castle Donington, Derby, DE74 2UJ
MM7 XRR J. Caldwell, 35, Clunie Drive, Larbert, FK5 4UA
M7 XRS R. Spencer, 41B Angerstein Close, Weeting, Brandon, IP27 0RL
M7 XSB S. Bishop, 168 Mansfield Road, Clipstone Village, Mansfield, NG21 9AE
M7 XSE S. Edwards, 24 Philips Lane Great Sutton, Ellesmere Port, CH66 4TP
M7 XST S. Rouse, 20 Chepstow Road, Corby, NN18 8QR
M7 XTB M. Machyna, 12 Victoria Street, Grantham, NG31 7BW
M7 XTW P. Bryant, 4 Neptune Drive, Hemel Hempstead, HP2 5QQ
M7 XTX T. Jones, 3 Langford Road, Liverpool, L19 3RA
M7 XUN A. Southern, 89 Purlewent Drive Weston, Bath, BA1 4BD
M7 XUP S. Davison, Shinwell House, Central Road, Dearham, Maryport, CA15 7HD
M7 XXS E. Savill, 61 Lower Road, Orpington, BR5 4AH
MM7 XYF J. Major, 14 Younger Gardens, St Andrews, KY16 8AB
M7 XYX R. Franklin, 7 Harvest Way, Thornbury, Bristol, BS35 1AL
M7 XYZ I. Gould, 2 Harps Avenue Minster, Sheerness, ME12 3PF
MI7 XZD M. Alsallal, Flat 3 Tollgate House Bradbury Place, Belfast, BT7 1PH
M7 XZW Z. Williamson, 9 Vine Avenue, Cleckheaton, BD19 3DW
M7 YAC A. Cross, 8 Briar Close, Lowestoft, NR324SU
M7 YAG P. Rogers, 9 William Burt Close, Weston Turville, Aylesbury, HP22 5QX
M7 YAM G. Smith, 47 Percy Road, Carlisle, CA2 6ER

M7 YAN J. Wojcik, 3 Whitehall, Lidlington, Bedford, MK43 0RS
M7 YAU A. Yau, Pembury Road, Tunbridge Wells, TN2 4ND
M7 YAZ R. Dabrowski, Overton, Salisbury Terrace, Mytchett, Camberley, GU16 6DB
M7 YBB R. Walker, 25 Hedges Close, Hatfield, AL10 0HZ
M7 YBC C. Walker, 8 Foxhollows, Hatfield, AL10 0HX
M7 YBZ I. Bradley, 41 Beckford Avenue, Bracknell, RG12 7ND
M7 YCA E. Cooper, 20 Manor Close, Baston, Peterborough, PE6 9PH
MI7 YDB D. Boucher, 52 Mourneview Park, Kilkeel, BT34 4NB
M7 YDX A. Crudele, 5 Greatfield Road, Winchester, SO22 6HN
MM7 YEA D. Yeaman, 7 Brimmond Crescent, Westhill, AB32 6RD
M7 YET S. Reed, 69 Kingfisher Drive, Lydney, GL155FX
M7 YEW P. Cannings, 30 Graham Gardens, Luton, LU3 1NQ
M7 YGI R. Skuse, 17 Church Way, Gloucester, GL4 4NW
M7 YGK G. Kane, 32 Gidleys Meadow, Dartington, Totnes, TQ9 6JZ
M7 YGM G. Meadows, 36 Belbroughton Road Blakedown, Kidderminster, DY10 3JG
M7 YGR G. Riccardi, 25 Montefiore Road Flat 2, Hove, BN3 1RD
M7 YHZ J. Hiltz, 35 Highview Road, London, W13 0HA
M7 YIB R. Kershaw, 40 The Butts, Frome, BA11 4AA
M7 YIF J. Allen, 23 Amwell Court, Waltham Abbey, en9 3ea
M7 YIN M. Cavallaro, 51 Mccreery Road, Sherborne, DT9 4DT
M7 YKD N. Stevens, 8 Moorfields, West Moor Lane, Raskelf, York, YO61 3UZ
M7 YKE A. Wood, 19 The Shepherdies, Ripon, hg43hu
M7 YKK R. Pas, 12 Monument Way, Bodmin, PL31 1NZ
M7 YLB A. Doyle, 2 The Greenways Paddock Wood, Tonbridge, TN12 6LS
M7 YMA A. Thomas, 20 Coronation Road, Ludgershall, Andover, SP11 9NN
M7 YMC D. Ruscoe, Wayside, Chester High Road, Neston, CH64 7TT
M7 YNC R. Dalumpines, 98 Green Hills, Harlow, CM20 3SZ
MM7 YNK N. Kirtley, 10 Millcroft Road, Auldearn, Nairn, IV12 5TW
M7 YOB M. Miller, 1 Tonbridge Road, West Molesey, kt82el
M7 YOE G. Yeomans, 98 Well Street, Biddulph, Stoke-on-Trent, ST8 6HY
M7 YOL A. Fresta, Flat 2, Dobson House, John Williams Close, London, SE14 5XF
M7 YOR R. Bertie, 14 St. Giles Street, Padiham, Burnley, BB12 8HL
M7 YOY M. Cooper, 2 May Tree Close, Waterthorpe, Sheffield, S20 7JB
MI7 YOZ M. Kelly, 19 Gowanvale Drive, Banbridge, BT32 3GD
MM7 YRN R. Thomson, Meadow Steading Tornaveen, Torphins, Banchory, AB31 4PJ
M7 YSE R. Moyse, 22 Sugden Road Wandsworth, London, SW11 5EF
M7 YSU D. Farrell, 15 Woodcote Way, Bexhill-on-Sea, TN39 4GP
M7 YTO S. Scorer, Flat15, Hanover House, Tewin Well Avenue, Harrogate, HG28AP
MM7 YTR M. Sandor, 77 Montgomery Avenue, Coatbridge, ML5 1QT
M7 YVR R. Minarro Escalona, 134 Leopold Road, Kensington, Liverpool, L7 8SS
M7 YXZ A. Manczuk, Unit 15 Old Cement Works The Hollow, South Heighton, Newhaven, BN9 0HS
M7 YYF T. Bradley, 23 Grafton Road, Gloucester, GL2 0QP
M7 YYQ H. Lee, 71 Easemore Road, Redditch, B98 8EY
M7 YYT P. Boulding, 16 Higher Croft Road, Lower Darwen, Darwen, BB3 0QR
M7 YYZ N. Williams, 1 African Row Barripper, Camborne, TR14 0QT
MW7 YZF A. Parry, 2 Moorfield Avenue, Clarbeston Road, SA63 4UU
M7 YZP N. Doyle, 32 Donald Road, Bristol, BS13 7BU
M7 YZY M. Davies, 12 Hopewell Road, Baldock, SG7 5AA
M7 YZZ D. Green, 84 Clarondale, Hull, HU7 4AR

United Kingdom

'Details withheld'

The callsigns in this list are those of active licences for which the owner has requested details not be given.

2#0

2#0 AAD
2#0 AAE
2#0 AAG
2#0 AAH
2#0 AAL
2#0 AAM
2#0 AAP
2#0 AAR
2#0 AAS
2#0 ABW
2#0 ACO
2#0 ACS
2#0 ADD
2#0 ADE
2#0 ADK
2#0 ADM
2#0 ADO
2#0 AER
2#0 AFA
2#0 AFO
2#0 AGT
2#0 AHO
2#0 AHX
2#0 AIB
2#0 AIH
2#0 AIV
2#0 AJS
2#0 AJZ
2#0 AKA
2#0 AKG
2#0 AKT
2#0 ALC
2#0 ALE
2#0 ALI
2#0 ALK
2#0 AOO
2#0 APR
2#0 AQH
2#0 AQQ
2#0 AQR
2#0 AQX
2#0 ARE
2#0 ARI
2#0 ARO
2#0 ARW
2#0 ASJ
2#0 ATF
2#0 AUH
2#0 AWA
2#0 AWF
2#0 AWL
2#0 AWM
2#0 AWO
2#0 AWP
2#0 AWX
2#0 AXX
2#0 AYA
2#0 AYD
2#0 AYR
2#0 AZC
2#0 AZN
2#0 AZQ
2#0 AZS
2#0 AZT
2#0 AZV
2#0 BAE
2#0 BAS
2#0 BAW
2#0 BBB
2#0 BBC
2#0 BBD
2#0 BBP
2#0 BBV
2#0 BCA
2#0 BCI
2#0 BCN
2#0 BCP
2#0 BCU
2#0 BCV
2#0 BCY
2#0 BDA
2#0 BDC
2#0 BDF
2#0 BDG
2#0 BDL
2#0 BDW
2#0 BDX
2#0 BDY
2#0 BDZ
2#0 BEJ
2#0 BEK
2#0 BEN
2#0 BES
2#0 BEU
2#0 BEX
2#0 BEY
2#0 BEZ
2#0 BFB
2#0 BFH
2#0 BFI
2#0 BFK
2#0 BFL
2#0 BFP
2#0 BFR
2#0 BFU
2#0 BFX
2#0 BFY
2#0 BGA
2#0 BGB
2#0 BGF
2#0 BGG
2#0 BGR
2#0 BHE
2#0 BHI
2#0 BHK
2#0 BHL
2#0 BHM
2#0 BHO
2#0 BHR
2#0 BHV
2#0 BHX
2#0 BIA
2#0 BIF
2#0 BIJ
2#0 BIV
2#0 BIX
2#0 BIZ
2#0 BJC
2#0 BJG
2#0 BJJ
2#0 BJN
2#0 BJO
2#0 BKC
2#0 BKF
2#0 BKR
2#0 BKS
2#0 BKW
2#0 BLB
2#0 BLE
2#0 BLH
2#0 BLO
2#0 BLP
2#0 BLU
2#0 BLV
2#0 BMA
2#0 BMC
2#0 BML
2#0 BMV
2#0 BMX
2#0 BMZ
2#0 BNL
2#0 BNM
2#0 BNN
2#0 BNP
2#0 BNQ
2#0 BNR
2#0 BNS
2#0 BNU
2#0 BOA
2#0 BOE
2#0 BOO
2#0 BOQ
2#0 BOU
2#0 BOW
2#0 BPA
2#0 BPB
2#0 BPC
2#0 BPD
2#0 BPQ
2#0 BPZ
2#0 BQS
2#0 BRA
2#0 BRG
2#0 BRO
2#0 BRV
2#0 BSD
2#0 BSL
2#0 BSV
2#0 BSZ
2#0 BTH
2#0 BTK
2#0 BTL
2#0 BTY
2#0 BUB
2#0 BUG
2#0 BUL
2#0 BUR
2#0 BUS
2#0 BUW
2#0 BVA
2#0 BVC
2#0 BVM
2#0 BVP
2#0 BVX
2#0 BVY
2#0 BWB
2#0 BWC
2#0 BWE
2#0 BWF
2#0 BWG
2#0 BWM
2#0 BWR
2#0 BWS
2#0 BWW
2#0 BWZ
2#0 BXH
2#0 BXL
2#0 BXP
2#0 BXR
2#0 BYE
2#0 BYP
2#0 BYS
2#0 BYV
2#0 BYX
2#0 BYZ
2#0 BZD
2#0 BZK
2#0 BZR
2#0 BZY
2#0 CAB
2#0 CAC
2#0 CAH
2#0 CAI
2#0 CAM
2#0 CAM
2#0 CAN
2#0 CAY
2#0 CAZ
2#0 CBD
2#0 CBK
2#0 CBM
2#0 CBP
2#0 CBR
2#0 CBS
2#0 CBW
2#0 CBY
2#0 CCM
2#0 CCN
2#0 CCP
2#0 CCS
2#0 CCV
2#0 CCX
2#0 CDB
2#0 CDC
2#0 CDD
2#0 CDG
2#0 CDH
2#0 CDN
2#0 CDP
2#0 CDX
2#0 CEC
2#0 CEL
2#0 CET
2#0 CEW
2#0 CEZ
2#0 CFF
2#0 CFR
2#0 CFS
2#0 CFU
2#0 CGA
2#0 CGC
2#0 CGF
2#0 CGY
2#0 CHB
2#0 CHD
2#0 CHF
2#0 CHG
2#0 CHI
2#0 CHJ
2#0 CHM
2#0 CHO
2#0 CHP
2#0 CHS
2#0 CIC
2#0 CIL
2#0 CIP
2#0 CIW
2#0 CIY
2#0 CIZ
2#0 CJC
2#0 CJH
2#0 CJN
2#0 CJR
2#0 CJT
2#0 CJU
2#0 CJY
2#0 CKA
2#0 CKD
2#0 CKF
2#0 CKG
2#0 CKK
2#0 CKU
2#0 CKZ
2#0 CLA
2#0 CLC
2#0 CLG
2#0 CLK
2#0 CLM
2#0 CLO
2#0 CLQ
2#0 CLR
2#0 CMB
2#0 CMG
2#0 CMI
2#0 CML
2#0 CMM
2#0 CMR
2#0 CMS
2#0 CMU
2#0 CMV
2#0 CMW
2#0 CMX
2#0 CNF
2#0 CNJ
2#0 CNK
2#0 COO
2#0 COP
2#0 COR
2#0 COS
2#0 COU
2#0 CPA
2#0 CPI
2#0 CPJ
2#0 CPM
2#0 CPU
2#0 CPW
2#0 CQA
2#0 CQE
2#0 CQP
2#0 CQV
2#0 CQW
2#0 CQY
2#0 CRE
2#0 CRF
2#0 CRG
2#0 CRJ
2#0 CRK
2#0 CRL
2#0 CRP
2#0 CRT
2#0 CRZ
2#0 CSA
2#0 CSC
2#0 CSI
2#0 CSR
2#0 CSS
2#0 CTG
2#0 CTH
2#0 CTP
2#0 CTS
2#0 CTX
2#0 CUF
2#0 CUN
2#0 CUP
2#0 CUQ
2#0 CUR
2#0 CUZ
2#0 CVA
2#0 CVB
2#0 CVI
2#0 CVT
2#0 CWD
2#0 CWG
2#0 CWN
2#0 CWP
2#0 CWT
2#0 CWU
2#0 CWX
2#0 CWY
2#0 CXC
2#0 CXR
2#0 CXS
2#0 CXY
2#0 CXZ
2#0 CYA
2#0 CYB
2#0 CYD
2#0 CYF
2#0 CYH
2#0 CYJ
2#0 CYK
2#0 CYN
2#0 CYQ
2#0 CYV
2#0 CZE
2#0 CZH
2#0 CZO
2#0 CZQ
2#0 CZV
2#0 CZX
2#0 CZY
2#0 DAD
2#0 DAF
2#0 DAG
2#0 DAK
2#0 DAM
2#0 DAS
2#0 DAV
2#0 DAZ
2#0 DBC
2#0 DBD
2#0 DBE
2#0 DBJ
2#0 DBR
2#0 DBT
2#0 DBU
2#0 DBV
2#0 DBX
2#0 DCB
2#0 DCC
2#0 DCE
2#0 DCG
2#0 DCO
2#0 DCQ
2#0 DCU
2#0 DDE
2#0 DDT
2#0 DDV
2#0 DDY
2#0 DEB
2#0 DED
2#0 DEL
2#0 DET
2#0 DFK
2#0 DFR
2#0 DFU
2#0 DFW
2#0 DFY
2#0 DGE
2#0 DGF
2#0 DGN
2#0 DGO
2#0 DGP
2#0 DGX
2#0 DGY
2#0 DGZ
2#0 DHH
2#0 DHL
2#0 DHN
2#0 DHU
2#0 DIA
2#0 DIR
2#0 DIW
2#0 DIY
2#0 DIZ
2#0 DJB
2#0 DJD
2#0 DJE
2#0 DJG
2#0 DJN
2#0 DJO
2#0 DJP
2#0 DKC
2#0 DKH
2#0 DKJ
2#0 DKK
2#0 DKL
2#0 DKN
2#0 DLG
2#0 DLI
2#0 DLL
2#0 DLM
2#0 DLQ
2#0 DLS
2#0 DLX
2#0 DLY
2#0 DMF
2#0 DMH
2#0 DML
2#0 DMO
2#0 DMT
2#0 DNA
2#0 DNK
2#0 DNN
2#0 DNQ
2#0 DNT
2#0 DNY
2#0 DNZ
2#0 DOA
2#0 DOB
2#0 DOK
2#0 DOO
2#0 DOS
2#0 DOT
2#0 DOU
2#0 DPE
2#0 DPG
2#0 DPJ
2#0 DPK
2#0 DPP
2#0 DPQ
2#0 DPU
2#0 DPV
2#0 DQE
2#0 DQF
2#0 DQI
2#0 DQK
2#0 DQM
2#0 DQR
2#0 DQV
2#0 DQW
2#0 DR
2#0 DRD
2#0 DRH
2#0 DRL
2#0 DRM
2#0 DRP
2#0 DRR
2#0 DRS
2#0 DRU
2#0 DRX
2#0 DSA
2#0 DSD
2#0 DSF
2#0 DSM
2#0 DSR
2#0 DSX
2#0 DTA
2#0 DTI
2#0 DTK
2#0 DTU
2#0 DUC
2#0 DUG
2#0 DUK
2#0 DUT
2#0 DUX
2#0 DUY
2#0 DVA
2#0 DVB
2#0 DVE
2#0 DVJ
2#0 DVL
2#0 DVQ
2#0 DVR
2#0 DVS
2#0 DVT
2#0 DVZ
2#0 DWH
2#0 DWI
2#0 DWN
2#0 DWO
2#0 DWQ
2#0 DWX
2#0 DWY
2#0 DXB
2#0 DXD
2#0 DXE
2#0 DXG
2#0 DXH
2#0 DXI
2#0 DXM
2#0 DXN
2#0 DXP
2#0 DXQ
2#0 DXR
2#0 DXS
2#0 DXT
2#0 DXU
2#0 DXV
2#0 DXX
2#0 DXY
2#0 DYA
2#0 DYC
2#0 DYE
2#0 DYF
2#0 DYH
2#0 DYO
2#0 DYR
2#0 DYS
2#0 DYT
2#0 DYW
2#0 DZB
2#0 DZD
2#0 DZG
2#0 DZH
2#0 DZI
2#0 DZL
2#0 DZP
2#0 DZU
2#0 DZW
2#0 EAB
2#0 EAE
2#0 EAG
2#0 EAH
2#0 EAJ
2#0 EAK
2#0 EAP
2#0 EAQ
2#0 EBC
2#0 EBE
2#0 EBM
2#0 EBT
2#0 EBY
2#0 ECA
2#0 ECB
2#0 ECE
2#0 ECF
2#0 ECH
2#0 ECL
2#0 ECM
2#0 ECN
2#0 ECR
2#0 ECS
2#0 ECT
2#0 ECX
2#0 EDB
2#0 EDJ
2#0 EDK
2#0 EDO
2#0 EDT
2#0 EDU
2#0 EDW
2#0 EDZ
2#0 EEA
2#0 EEC
2#0 EEE
2#0 EEG
2#0 EEH
2#0 EEL
2#0 EEN
2#0 EEP
2#0 EEY
2#0 EFF
2#0 EFL
2#0 EFM
2#0 EFQ
2#0 EFT
2#0 EGD
2#0 EGF
2#0 EGG
2#0 EGH
2#0 EGJ
2#0 EGU
2#0 EGV
2#0 EGW
2#0 EGX
2#0 EGY
2#0 EGZ
2#0 EHC
2#0 EHE
2#0 EHF
2#0 EHG
2#0 EHI
2#0 EHM
2#0 EHU
2#0 EHY
2#0 EIA
2#0 EIC
2#0 EIF
2#0 EIH
2#0 EII
2#0 EIJ
2#0 EIK
2#0 EIL
2#0 EIM
2#0 EIN
2#0 EIO
2#0 EIP
2#0 EIQ
2#0 EIR
2#0 EIS
2#0 EIT
2#0 EIV
2#0 EIY
2#0 EJB
2#0 EJE
2#0 EJF
2#0 EJG
2#0 EJH
2#0 EJI
2#0 EJN
2#0 EJS
2#0 EJX
2#0 EJY
2#0 EKF
2#0 EKG
2#0 EKL
2#0 EKQ
2#0 EKV
2#0 EKY
2#0 EKZ
2#0 ELF
2#0 ELG
2#0 ELH
2#0 ELJ
2#0 ELL
2#0 ELM
2#0 ELN
2#0 ELS
2#0 ELU
2#0 ELW
2#0 ELX
2#0 ELZ
2#0 EMA
2#0 EMC
2#0 EMH
2#0 EMJ
2#0 EMO
2#0 EMR
2#0 EMT
2#0 EMU
2#0 EMW
2#0 ENA
2#0 ENH
2#0 ENK
2#0 ENL
2#0 ENR
2#0 ENS
2#0 ENT
2#0 ENU
2#0 ENX
2#0 ENY
2#0 EOA
2#0 EOB
2#0 EOE
2#0 EOG
2#0 EOJ
2#0 EOO
2#0 EOQ
2#0 EOR
2#0 EOW
2#0 EOY
2#0 EPD
2#0 EPI
2#0 EPL
2#0 EPO
2#0 EPQ
2#0 EPS
2#0 EPV
2#0 EPX
2#0 EPY
2#0 EQA
2#0 EQF
2#0 EQM
2#0 EQV
2#0 EQX
2#0 EQY
2#0 ERC
2#0 ERH
2#0 ERI
2#0 ERW
2#0 ERY
2#0 ESD
2#0 ESE
2#0 ESF
2#0 ESG
2#0 ESH
2#0 ESK
2#0 ESM
2#0 ESN
2#0 ESQ
2#0 ESR
2#0 EST
2#0 ESV
2#0 ESZ
2#0 ETK
2#0 ETM
2#0 ETO
2#0 ETR
2#0 ETS
2#0 EUA
2#0 EUC
2#0 EUG
2#0 EUL
2#0 EUT
2#0 EUU
2#0 EUX
2#0 EVD
2#0 EVG
2#0 EVH
2#0 EVJ
2#0 EVL
2#0 EVN
2#0 EVO
2#0 EVT
2#0 EVY
2#0 EVZ
2#0 EWC
2#0 EWD
2#0 EWG
2#0 EWI
2#0 EWJ
2#0 EWK
2#0 EWN
2#0 EWO
2#0 EWR
2#0 EWU
2#0 EWV
2#0 EWW
2#0 EWZ
2#0 EXF
2#0 EXT
2#0 EYA
2#0 EYD
2#0 EYL
2#0 EYM
2#0 EYN
2#0 EYO
2#0 EYR
2#0 EYT
2#0 EYV
2#0 EYW
2#0 EYZ
2#0 EZB
2#0 EZD
2#0 EZE
2#0 EZH
2#0 EZJ
2#0 EZK
2#0 EZM
2#0 EZN
2#0 EZP
2#0 EZQ
2#0 EZR
2#0 EZT
2#0 EZU
2#0 EZW
2#0 FAD
2#0 FAF
2#0 FAI
2#0 FAL
2#0 FAO
2#0 FAW
2#0 FAX
2#0 FAZ
2#0 FBB
2#0 FBG
2#0 FBK
2#0 FBP
2#0 FBT
2#0 FBV
2#0 FBW
2#0 FBX
2#0 FBZ
2#0 FCA
2#0 FCC
2#0 FCD
2#0 FCI
2#0 FCK
2#0 FCL
2#0 FCN
2#0 FCP
2#0 FCR
2#0 FCU
2#0 FCV
2#0 FCX
2#0 FCY
2#0 FDA
2#0 FDF
2#0 FDK
2#0 FDL
2#0 FDM
2#0 FDN
2#0 FDO
2#0 FDP
2#0 FDR
2#0 FDV
2#0 FDX
2#0 FDY
2#0 FEA
2#0 FED
2#0 FEF
2#0 FEG
2#0 FEH
2#0 FEI
2#0 FEK
2#0 FEL
2#0 FEN
2#0 FES
2#0 FET
2#0 FEW
2#0 FFA
2#0 FFD
2#0 FFE
2#0 FFG
2#0 FFQ
2#0 FFT
2#0 FFU
2#0 FFX
2#0 FFZ
2#0 FGB
2#0 FGC
2#0 FGD
2#0 FGE
2#0 FGG
2#0 FGJ
2#0 FGK
2#0 FGL
2#0 FGP
2#0 FGU
2#0 FGX
2#0 FGZ
2#0 FHF
2#0 FHI
2#0 FHL
2#0 FHO
2#0 FHP
2#0 FHS
2#0 FHT
2#0 FHU
2#0 FHW
2#0 FHY
2#0 FHZ
2#0 FID
2#0 FIE
2#0 FIG
2#0 FIH
2#0 FII
2#0 FIN
2#0 FIQ
2#0 FIS
2#0 FIZ
2#0 FJB
2#0 FJF
2#0 FJH
2#0 FJL
2#0 FJO
2#0 FJR
2#0 FKA
2#0 FKI
2#0 FKL
2#0 FKO
2#0 FKP
2#0 FKQ
2#0 FKR
2#0 FKX
2#0 FKZ
2#0 FLB
2#0 FLC
2#0 FLD
2#0 FLE
2#0 FLP
2#0 FLS
2#0 FLT
2#0 FLX
2#0 FLY
2#0 FLZ
2#0 FMC
2#0 FMD
2#0 FMJ
2#0 FMM
2#0 FMU
2#0 FMW
2#0 FMZ
2#0 FNC
2#0 FND
2#0 FNE
2#0 FNF
2#0 FNK
2#0 FNL
2#0 FNP
2#0 FNU
2#0 FNW
2#0 FNX
2#0 FNZ
2#0 FOA
2#0 FOB
2#0 FOJ
2#0 FOM
2#0 FON
2#0 FOQ
2#0 FOS
2#0 FOU
2#0 FOV
2#0 FOW
2#0 FOY
2#0 FPE
2#0 FPG
2#0 FPH
2#0 FPL
2#0 FPN
2#0 FPP
2#0 FPR
2#0 FPS
2#0 FPU
2#0 FPV
2#0 FPZ
2#0 FQB
2#0 FQD
2#0 FQF
2#0 FQH
2#0 FQK
2#0 FQL
2#0 FQN
2#0 FQO
2#0 FQQ
2#0 FQV
2#0 FQX
2#0 FQY
2#0 FQZ
2#0 FRE
2#0 FRH
2#0 FRI
2#0 FRM
2#0 FRN
2#0 FRT
2#0 FRW
2#0 FSJ
2#0 FSL
2#0 FSO
2#0 FST
2#0 FSU
2#0 FSW
2#0 FSY
2#0 FSZ
2#0 FTE
2#0 FTG
2#0 FTJ
2#0 FTK
2#0 FTN
2#0 FTR
2#0 FTY
2#0 FUB
2#0 FUI
2#0 FUM
2#0 FUU
2#0 FUV
2#0 FUY
2#0 FVA
2#0 FVG
2#0 FVH
2#0 FVI
2#0 FVN
2#0 FVP
2#0 FVQ
2#0 FVU
2#0 FVW
2#0 FVY
2#0 FWA
2#0 FWE
2#0 FWF
2#0 FWG
2#0 FWH
2#0 FWI
2#0 FWJ
2#0 FWL
2#0 FWM
2#0 FWP
2#0 FWQ
2#0 FWU
2#0 FWV
2#0 FWX
2#0 FXG
2#0 FXI
2#0 FXK
2#0 FXN
2#0 FXQ
2#0 FXR
2#0 FXS
2#0 FXT
2#0 FXZ
2#0 FYB
2#0 FYH
2#0 FYI
2#0 FYJ
2#0 FYK
2#0 FYM
2#0 FYV
2#0 FYY
2#0 FZH
2#0 FZI
2#0 FZL
2#0 FZP
2#0 FZQ
2#0 FZR
2#0 FZV
2#0 FZX
2#0 GAB
2#0 GAD
2#0 GAE
2#0 GAG
2#0 GAI
2#0 GAT
2#0 GAZ
2#0 GBC
2#0 GBL
2#0 GBQ
2#0 GBR
2#0 GBX
2#0 GBY
2#0 GBZ
2#0 GCA
2#0 GCK
2#0 GCR
2#0 GCU
2#0 GCV
2#0 GCZ
2#0 GDC
2#0 GDD
2#0 GDI
2#0 GDJ
2#0 GDK
2#0 GDP
2#0 GDQ
2#0 GDR
2#0 GDS
2#0 GDU
2#0 GDX
2#0 GEA
2#0 GED
2#0 GEH
2#0 GEI
2#0 GEP
2#0 GEQ
2#0 GEU
2#0 GEW
2#0 GEY
2#0 GFA
2#0 GFD
2#0 GFG
2#0 GFH
2#0 GFI
2#0 GFL
2#0 GFM
2#0 GFN
2#0 GFS
2#0 GGB
2#0 GGC
2#0 GGD
2#0 GGE
2#0 GGF
2#0 GGH
2#0 GGK
2#0 GGL
2#0 GGR
2#0 GGS
2#0 GGX
2#0 GHI
2#0 GHL
2#0 GHM
2#0 GHN
2#0 GHO
2#0 GHQ
2#0 GHU
2#0 GHW
2#0 GIB
2#0 GIC
2#0 GID
2#0 GIG
2#0 GII
2#0 GIJ
2#0 GIK
2#0 GIL
2#0 GIM
2#0 GIN
2#0 GIO
2#0 GIP
2#0 GIQ
2#0 GIR
2#0 GIU
2#0 GJA
2#0 GJC
2#0 GJD
2#0 GJF
2#0 GJH
2#0 GJI
2#0 GJL
2#0 GJM
2#0 GJU
2#0 GJV
2#0 GJW
2#0 GJX
2#0 GJY
2#0 GJZ
2#0 GKC
2#0 GKE
2#0 GKH
2#0 GKI
2#0 GKK
2#0 GKO
2#0 GKV
2#0 GKW
2#0 GKY
2#0 GLB
2#0 GLE
2#0 GLF
2#0 GLG
2#0 GLH
2#0 GLK
2#0 GLN
2#0 GLO
2#0 GLP
2#0 GLX
2#0 GLZ
2#0 GME
2#0 GMH
2#0 GMI
2#0 GMJ
2#0 GMK
2#0 GMP
2#0 GMQ
2#0 GMT
2#0 GMV
2#0 GMX
2#0 GMY
2#0 GNA
2#0 GNB
2#0 GND
2#0 GNE
2#0 GNF
2#0 GNK
2#0 GNM
2#0 GNR
2#0 GNT
2#0 GNX
2#0 GNY
2#0 GNZ
2#0 GOA
2#0 GOB
2#0 GOC
2#0 GOF
2#0 GOH
2#0 GOI
2#0 GOJ
2#0 GOT
2#0 GOU
2#0 GOV
2#0 GOX
2#0 GPM
2#0 GPN
2#0 GPO
2#0 GPP
2#0 GPW
2#0 GQB
2#0 GQG
2#0 GQJ
2#0 GQQ
2#0 GQV
2#0 GQX
2#0 GQY
2#0 GRB
2#0 GRC
2#0 GRD
2#0 GRG
2#0 GRJ
2#0 GRO
2#0 GRT
2#0 GRU
2#0 GSD
2#0 GSM
2#0 GSQ
2#0 GSU
2#0 GSV
2#0 GSX
2#0 GSZ
2#0 GTD
2#0 GTF
2#0 GTG
2#0 GTH
2#0 GTI
2#0 GTK
2#0 GTP
2#0 GTQ
2#0 GTS
2#0 GTV
2#0 GTW
2#0 GTY
2#0 GUB
2#0 GUD
2#0 GUE
2#0 GUK
2#0 GUO
2#0 GUP
2#0 GUQ
2#0 GUW
2#0 GUX
2#0 GVA
2#0 GVB
2#0 GVD
2#0 GVE
2#0 GVF
2#0 GVH
2#0 GVJ
2#0 GVM
2#0 GVQ
2#0 GVS
2#0 GVT
2#0 GVU
2#0 GVV
2#0 GVX
2#0 GVY
2#0 GWG
2#0 GWL
2#0 GWN
2#0 GWQ
2#0 GWW
2#0 GWY
2#0 GXA
2#0 GXJ
2#0 GXL
2#0 GXS
2#0 GXV
2#0 GYA
2#0 GYE
2#0 GYG
2#0 GYI
2#0 GYO
2#0 GYR
2#0 GYU
2#0 GZB
2#0 GZC
2#0 GZE
2#0 GZG
2#0 GZI
2#0 GZK
2#0 GZL
2#0 GZR
2#0 GZY
2#0 GZZ
2#0 HAD
2#0 HAI
2#0 HAO
2#0 HAQ
2#0 HAR
2#0 HAU
2#0 HAX
2#0 HAY
2#0 HBA
2#0 HBC
2#0 HBF
2#0 HBG
2#0 HBH
2#0 HBI
2#0 HBR
2#0 HBU
2#0 HBZ
2#0 HCA
2#0 HCB
2#0 HCD
2#0 HCE
2#0 HCH
2#0 HCI
2#0 HCJ
2#0 HCN
2#0 HCP
2#0 HCR
2#0 HCT
2#0 HCU
2#0 HDJ
2#0 HDK
2#0 HDO
2#0 HDR
2#0 HDT
2#0 HDV
2#0 HED
2#0 HEE
2#0 HEH
2#0 HEJ
2#0 HEK
2#0 HEL
2#0 HEN
2#0 HER
2#0 HEV
2#0 HEW
2#0 HFB
2#0 HFC
2#0 HFF
2#0 HFH
2#0 HFL
2#0 HFM
2#0 HFP
2#0 HFQ
2#0 HFR
2#0 HFU
2#0 HFV
2#0 HFW
2#0 HFX
2#0 HFY
2#0 HFZ
2#0 HGB
2#0 HGC
2#0 HGH
2#0 HGJ
2#0 HGL
2#0 HGS
2#0 HGV
2#0 HGY
2#0 HGZ
2#0 HHD
2#0 HHH
2#0 HHI
2#0 HHL
2#0 HHM
2#0 HHO
2#0 HHQ
2#0 HHR
2#0 HHS
2#0 HHV
2#0 HHY
2#0 HHZ
2#0 HIA
2#0 HID
2#0 HIF
2#0 HIH
2#0 HII
2#0 HIM
2#0 HIN
2#0 HIU
2#0 HIX
2#0 HJA
2#0 HJD
2#0 HJF
2#0 HJH
2#0 HJL
2#0 HJM
2#0 HJO
2#0 HJQ
2#0 HJT
2#0 HJU
2#0 HJW
2#0 HJX
2#0 HKA
2#0 HKB
2#0 HKD
2#0 HKF
2#0 HKJ
2#0 HKM
2#0 HKO
2#0 HKQ
2#0 HKS
2#0 HKT
2#0 HKV
2#0 HKX
2#0 HLA
2#0 HLB
2#0 HLD
2#0 HLE
2#0 HLI
2#0 HLJ
2#0 HLL
2#0 HLN
2#0 HLQ
2#0 HLU
2#0 HLV
2#0 HLX
2#0 HLY
2#0 HLZ
2#0 HMB
2#0 HME
2#0 HMF
2#0 HMI
2#0 HMK
2#0 HMN
2#0 HMO
2#0 HMT
2#0 HMW
2#0 HMX
2#0 HMY
2#0 HMZ
2#0 HNA
2#0 HNB
2#0 HND
2#0 HNE
2#0 HNI
2#0 HNL
2#0 HNN
2#0 HNO
2#0 HNS
2#0 HNT
2#0 HNU
2#0 HNV
2#0 HNW
2#0 HNY
2#0 HOC
2#0 HOD
2#0 HOE
2#0 HOF
2#0 HOI
2#0 HOJ
2#0 HOP
2#0 HOZ
2#0 HPC
2#0 HPG
2#0 HPH
2#0 HPN
2#0 HPP
2#0 HPQ
2#0 HPS
2#0 HPT
2#0 HPU
2#0 HPV
2#0 HPW
2#0 HPX
2#0 HPY
2#0 HPZ
2#0 HQC
2#0 HQE
2#0 HQG
2#0 HQH
2#0 HQI
2#0 HQK
2#0 HQL
2#0 HQM
2#0 HQN
2#0 HQP
2#0 HQR
2#0 HQS
2#0 HQT
2#0 HQW
2#0 HQX
2#0 HQZ
2#0 HRA
2#0 HRF
2#0 HRK
2#0 HRO
2#0 HRQ
2#0 HRW
2#0 HRX
2#0 HRZ
2#0 HSA
2#0 HSD
2#0 HSF
2#0 HSH
2#0 HSI
2#0 HSJ
2#0 HSK
2#0 HSM
2#0 HSN
2#0 HSO
2#0 HSQ
2#0 HSR
2#0 HSS
2#0 HSX
2#0 HSZ
2#0 HTA
2#0 HTD
2#0 HTE
2#0 HTH
2#0 HTL
2#0 HTP
2#0 HTT
2#0 HTU
2#0 HTY
2#0 HTZ
2#0 HUA
2#0 HUC
2#0 HUE
2#0 HUF
2#0 HUI
2#0 HUP
2#0 HUQ
2#0 HUS
2#0 HUV
2#0 HUW
2#0 HUY
2#0 HVA
2#0 HVC
2#0 HVI
2#0 HVJ
2#0 HVK
2#0 HVN
2#0 HVO
2#0 HVP
2#0 HVR
2#0 HVX
2#0 HWA
2#0 HWB
2#0 HWF
2#0 HWH
2#0 HWL
2#0 HWR
2#0 HWT
2#0 HWU
2#0 HWV
2#0 HWX
2#0 HWZ
2#0 HXA
2#0 HXB
2#0 HXD
2#0 HXE
2#0 HXI
2#0 HXK
2#0 HXM
2#0 HXO
2#0 HXR
2#0 HXS
2#0 HXT
2#0 HXV
2#0 HXX
2#0 HXZ
2#0 HYA
2#0 HYB
2#0 HYL
2#0 HYN
2#0 HYQ
2#0 HYR
2#0 HYT
2#0 HYV
2#0 HYX
2#0 HZD
2#0 HZF
2#0 HZG
2#0 HZJ
2#0 HZM
2#0 HZN
2#0 HZT
2#0 HZV
2#0 HZX
2#0 HZY
2#0 HZZ
2#0 IAA
2#0 IAB
2#0 IAD
2#0 IAE
2#0 IAI
2#0 IAP
2#0 IAR
2#0 IAU
2#0 IAW
2#0 IAY
2#0 IBC
2#0 IBE
2#0 IBF
2#0 IBG
2#0 IBH
2#0 IBJ
2#0 IBK
2#0 IBL
2#0 IBO
2#0 IBQ
2#0 IBS
2#0 IBV
2#0 IBX
2#0 IBZ
2#0 ICA
2#0 ICG
2#0 ICH
2#0 ICJ
2#0 ICL
2#0 ICM
2#0 ICO
2#0 ICQ
2#0 ICS
2#0 ICV
2#0 ICZ
2#0 IDD
2#0 IDH
2#0 IDI
2#0 IDP
2#0 IDQ
2#0 IDS
2#0 IDU
2#0 IDV
2#0 IDW
2#0 IDZ
2#0 IEF
2#0 IEG
2#0 IEH
2#0 IEJ
2#0 IEN
2#0 IES
2#0 IEU
2#0 IEV
2#0 IEY
2#0 IEZ
2#0 IFD
2#0 IFE
2#0 IFH
2#0 IFI
2#0 IFJ
2#0 IFL
2#0 IFM
2#0 IFQ
2#0 IFR
2#0 IFX
2#0 IFY
2#0 IGB
2#0 IGD
2#0 IGF
2#0 IGG
2#0 IGI
2#0 IGJ
2#0 IGK
2#0 IGS
2#0 IGU
2#0 IGV
2#0 IGZ
2#0 IHA
2#0 IHC
2#0 IHD
2#0 IHF
2#0 IHG
2#0 IHJ
2#0 IHK
2#0 IHP
2#0 IHQ
2#0 IHX
2#0 IHY
2#0 IHZ
2#0 IIA
2#0 IIB
2#0 IIC
2#0 IIE
2#0 IIG
2#0 IIK
2#0 IIL
2#0 IIN
2#0 IIO
2#0 IIS
2#0 IIW
2#0 IIX
2#0 IJA
2#0 IJC
2#0 IJE
2#0 IJF
2#0 IJI
2#0 IJJ
2#0 IJK

Withheld

2#0 IJM
2#0 IJN
2#0 IJO
2#0 IJS
2#0 IJU
2#0 IJV
2#0 IJY
2#0 IKA
2#0 IKD
2#0 IKF
2#0 IKI
2#0 IKJ
2#0 IKK
2#0 IKL
2#0 IKP
2#0 IKR
2#0 IKY
2#0 IKZ
2#0 ILA
2#0 ILB
2#0 ILG
2#0 ILH
2#0 ILI
2#0 ILK
2#0 ILP
2#0 ILR
2#0 ILT
2#0 ILU
2#0 ILV
2#0 ILY
2#0 IMA
2#0 IME
2#0 IMF
2#0 IMI
2#0 IMK
2#0 IML
2#0 IMQ
2#0 IMR
2#0 IMU
2#0 IMY
2#0 INB
2#0 INF
2#0 ING
2#0 INH
2#0 INI
2#0 INL
2#0 INM
2#0 INN
2#0 INR
2#0 INV
2#0 IOA
2#0 IOC
2#0 IOC
2#0 IOP
2#0 IOT
2#0 IPD
2#0 IPE
2#0 IPH
2#0 IPM
2#0 IPP
2#0 IPR
2#0 IPS
2#0 IPZ
2#0 IQD
2#0 IQI
2#0 IQJ
2#0 IQS
2#0 IRM
2#0 IRP
2#0 IRQ
2#0 IRR
2#0 IRS
2#0 IRV
2#0 IRY
2#0 ISC
2#0 ISD
2#0 ISG
2#0 ISH
2#0 ISJ
2#0 ISL
2#0 ISR
2#0 IST
2#0 ISV
2#0 ISW
2#0 ITA
2#0 ITE
2#0 ITK
2#0 ITL
2#0 ITT
2#0 IUB
2#0 IUZ
2#0 IVA
2#0 IVG
2#0 IVM
2#0 IVT
2#0 IWA
2#0 IWC
2#0 IWH
2#0 IWK
2#0 IWO
2#0 IWW
2#0 IWX
2#0 IXD
2#0 IXN
2#0 IYX
2#0 IZC
2#0 IZL
2#0 JAD
2#0 JAG
2#0 JAH
2#0 JAS
2#0 JAU
2#0 JAV
2#0 JBB
2#0 JBG
2#0 JBH
2#0 JBN
2#0 JBO
2#0 JBP
2#0 JBR
2#0 JBU
2#0 JBV

2#0 JCJ
2#0 JDG
2#0 JDJ
2#0 JDN
2#0 JDT
2#0 JDZ
2#0 JEB
2#0 JEF
2#0 JEL
2#0 JER
2#0 JEU
2#0 JEX
2#0 JFA
2#0 JFB
2#0 JFC
2#0 JFG
2#0 JFH
2#0 JFM
2#0 JFQ
2#0 JFS
2#0 JFT
2#0 JFX
2#0 JFZ
2#0 JGA
2#0 JGC
2#0 JGR
2#0 JHB
2#0 JHL
2#0 JHM
2#0 JHR
2#0 JHW
2#0 JHX
2#0 JHY
2#0 JII
2#0 JIJ
2#0 JIK
2#0 JIN
2#0 JJC
2#0 JJD
2#0 JJE
2#0 JJL
2#0 JJM
2#0 JJO
2#0 JJT
2#0 JJX
2#0 JKA
2#0 JKC
2#0 JKE
2#0 JKJ
2#0 JKK
2#0 JKM
2#0 JKO
2#0 JKV
2#0 JKZ
2#0 JLA
2#0 JLC
2#0 JLG
2#0 JLH
2#0 JLL
2#0 JLN
2#0 JLO
2#0 JLP
2#0 JLY
2#0 JMA
2#0 JMG
2#0 JMM
2#0 JMN
2#0 JMP
2#0 JMQ
2#0 JMS
2#0 JMT
2#0 JMU
2#0 JMZ
2#0 JNC
2#0 JND
2#0 JNI
2#0 JNK
2#0 JNL
2#0 JNQ
2#0 JNS
2#0 JNT
2#0 JNX
2#0 JNZ
2#0 JOB
2#0 JOJ
2#0 JOO
2#0 JOT
2#0 JOZ
2#0 JPG
2#0 JPI
2#0 JPJ
2#0 JPK
2#0 JPN
2#0 JPV
2#0 JQQ
2#0 JQY
2#0 JQZ
2#0 JRC
2#0 JRF
2#0 JRG
2#0 JRH
2#0 JRM
2#0 JRO
2#0 JRX
2#0 JSA
2#0 JSC
2#0 JSL
2#0 JSP
2#0 JSW
2#0 JTC
2#0 JTD
2#0 JTF
2#0 JTI
2#0 JTL
2#0 JTP
2#0 JTR
2#0 JTU
2#0 JTZ
2#0 JUB
2#0 JUE
2#0 JUG
2#0 JUK

2#0 JUS
2#0 JUZ
2#0 JVB
2#0 JVK
2#0 JVL
2#0 JVW
2#0 JVZ
2#0 JWA
2#0 JWF
2#0 JWI
2#0 JWK
2#0 JWO
2#0 JWR
2#0 JWT
2#0 JWX
2#0 JWZ
2#0 JXE
2#0 JXI
2#0 JXJ
2#0 JXL
2#0 JXP
2#0 JXR
2#0 JXV
2#0 JXW
2#0 JYK
2#0 JYL
2#0 JYT
2#0 JYX
2#0 JYY
2#0 JZA
2#0 JZB
2#0 JZF
2#0 JZL
2#0 JZN
2#0 JZR
2#0 JZX
2#0 JZZ
2#0 KAA
2#0 KAD
2#0 KAE
2#0 KAH
2#0 KAI
2#0 KAJ
2#0 KAM
2#0 KAO
2#0 KAQ
2#0 KAR
2#0 KAV
2#0 KAW
2#0 KAY
2#0 KAZ
2#0 KBH
2#0 KBI
2#0 KBT
2#0 KBV
2#0 KBW
2#0 KBY
2#0 KCA
2#0 KCC
2#0 KCE
2#0 KCI
2#0 KCK
2#0 KCM
2#0 KCS
2#0 KDA
2#0 KDD
2#0 KDK
2#0 KDL
2#0 KDM
2#0 KDN
2#0 KDO
2#0 KDP
2#0 KDU
2#0 KDW
2#0 KDX
2#0 KDY
2#0 KEB
2#0 KEE
2#0 KEF
2#0 KEH
2#0 KEL
2#0 KEM
2#0 KER
2#0 KET
2#0 KEX
2#0 KFC
2#0 KFF
2#0 KFU
2#0 KGG
2#0 KGI
2#0 KGL
2#0 KGO
2#0 KGS
2#0 KGX
2#0 KHB
2#0 KHD
2#0 KHF
2#0 KHH
2#0 KHR
2#0 KHS
2#0 KHT
2#0 KHZ
2#0 KIB
2#0 KIC
2#0 KID
2#0 KIE
2#0 KIF
2#0 KIK
2#0 KIP
2#0 KIR
2#0 KIW
2#0 KIX
2#0 KJB
2#0 KJC
2#0 KJM
2#0 KJN
2#0 KJP
2#0 KJR
2#0 KJS
2#0 KJU
2#0 KJV
2#0 KJW

2#0 KKA
2#0 KKB
2#0 KKD
2#0 KKP
2#0 KKS
2#0 KKU
2#0 KKV
2#0 KKW
2#0 KKX
2#0 KLG
2#0 KLO
2#0 KLU
2#0 KLX
2#0 KLZ
2#0 KMC
2#0 KME
2#0 KMH
2#0 KMJ
2#0 KMK
2#0 KMM
2#0 KMO
2#0 KMR
2#0 KMV
2#0 KMW
2#0 KMX
2#0 KNG
2#0 KNJ
2#0 KNO
2#0 KNR
2#0 KNS
2#0 KNY
2#0 KNZ
2#0 KOG
2#0 KOJ
2#0 KOL
2#0 KON
2#0 KOO
2#0 KOT
2#0 KOW
2#0 KOY
2#0 KPG
2#0 KPJ
2#0 KPK
2#0 KPY
2#0 KRA
2#0 KRJ
2#0 KRL
2#0 KRM
2#0 KRO
2#0 KRV
2#0 KRW
2#0 KRZ
2#0 KSB
2#0 KSE
2#0 KSF
2#0 KSS
2#0 KTA
2#0 KTB
2#0 KTC
2#0 KTE
2#0 KTF
2#0 KTO
2#0 KTT
2#0 KTY
2#0 KTZ
2#0 KUB
2#0 KUL
2#0 KUP
2#0 KUW
2#0 KUX
2#0 KVD
2#0 KVG
2#0 KVL
2#0 KVP
2#0 KVT
2#0 KVW
2#0 KVX
2#0 KWD
2#0 KWH
2#0 KWI
2#0 KWK
2#0 KWN
2#0 KWR
2#0 KWT
2#0 KWX
2#0 KXO
2#0 KXQ
2#0 KXT
2#0 KXY
2#0 KXZ
2#0 KYA
2#0 KYC
2#0 KYE
2#0 KYF
2#0 KYG
2#0 KYM
2#0 KYN
2#0 KYO
2#0 KYR
2#0 KYS
2#0 KZA
2#0 KZP
2#0 KZZ
2#0 LAA
2#0 LAF
2#0 LAG
2#0 LAJ
2#0 LAM
2#0 LAN
2#0 LAP
2#0 LAT
2#0 LAU
2#0 LAZ
2#0 LBC
2#0 LBD
2#0 LBF
2#0 LBO
2#0 LBT
2#0 LBX
2#0 LCB
2#0 LCF
2#0 LCG

2#0 LCL
2#0 LCO
2#0 LCS
2#0 LCT
2#0 LCX
2#0 LDA
2#0 LDB
2#0 LDK
2#0 LDO
2#0 LDS
2#0 LDT
2#0 LDW
2#0 LEC
2#0 LEH
2#0 LEJ
2#0 LEK
2#0 LER
2#0 LES
2#0 LFA
2#0 LFB
2#0 LFD
2#0 LFJ
2#0 LFO
2#0 LFP
2#0 LFV
2#0 LGA
2#0 LGC
2#0 LGD
2#0 LGK
2#0 LGM
2#0 LGN
2#0 LGP
2#0 LGX
2#0 LHB
2#0 LHG
2#0 LHM
2#0 LHP
2#0 LHT
2#0 LHW
2#0 LIA
2#0 LIB
2#0 LID
2#0 LIE
2#0 LIG
2#0 LIK
2#0 LIL
2#0 LIN
2#0 LIP
2#0 LIS
2#0 LIU
2#0 LJA
2#0 LJM
2#0 LJU
2#0 LJW
2#0 LJZ
2#0 LKA
2#0 LKI
2#0 LKL
2#0 LKN
2#0 LKP
2#0 LKQ
2#0 LKY
2#0 LLB
2#0 LLH
2#0 LLP
2#0 LLR
2#0 LLS
2#0 LMA
2#0 LMB
2#0 LMC
2#0 LMF
2#0 LMJ
2#0 LMN
2#0 LMX
2#0 LMZ
2#0 LNA
2#0 LND
2#0 LNE
2#0 LNL
2#0 LNS
2#0 LNT
2#0 LNY
2#0 LOB
2#0 LOD
2#0 LOJ
2#0 LOK
2#0 LOM
2#0 LOT
2#0 LOV
2#0 LPA
2#0 LPB
2#0 LPF
2#0 LPI
2#0 LPK
2#0 LPL
2#0 LPT
2#0 LPU
2#0 LQX
2#0 LRF
2#0 LRH
2#0 LRM
2#0 LRS
2#0 LRT
2#0 LRW
2#0 LRX
2#0 LRY
2#0 LSD
2#0 LSJ
2#0 LSM
2#0 LSN
2#0 LSP
2#0 LSQ
2#0 LSY
2#0 LTA
2#0 LTE
2#0 LTN
2#0 LTY
2#0 LTZ
2#0 LUA
2#0 LUB
2#0 LUC
2#0 LUH

2#0 LUI
2#0 LUP
2#0 LUV
2#0 LUX
2#0 LUZ
2#0 LVA
2#0 LVI
2#0 LVO
2#0 LVX
2#0 LVZ
2#0 LWC
2#0 LWF
2#0 LWG
2#0 LWH
2#0 LWI
2#0 LWP
2#0 LWS
2#0 LWX
2#0 LXE
2#0 LXH
2#0 LXL
2#0 LXM
2#0 LXN
2#0 LYL
2#0 LYO
2#0 LYT
2#0 LZF
2#0 LZH
2#0 LZO
2#0 LZR
2#0 LZS
2#0 LZZ
2#0 MAG
2#0 MAI
2#0 MAK
2#0 MAM
2#0 MAR
2#0 MAT
2#0 MAU
2#0 MAW
2#0 MBC
2#0 MBF
2#0 MBH
2#0 MBL
2#0 MBP
2#0 MBU
2#0 MBY
2#0 MCE
2#0 MCO
2#0 MCU
2#0 MCX
2#0 MCZ
2#0 MDA
2#0 MDD
2#0 MDI
2#0 MDJ
2#0 MDL
2#0 MDM
2#0 MDQ
2#0 MDV
2#0 MDW
2#0 MDX
2#0 MDY
2#0 MEB
2#0 MEC
2#0 MEE
2#0 MEF
2#0 MEJ
2#0 MEM
2#0 MEN
2#0 MEP
2#0 MER
2#0 MEW
2#0 MFE
2#0 MFH
2#0 MFL
2#0 MFM
2#0 MFO
2#0 MFP
2#0 MFQ
2#0 MFV
2#0 MFW
2#0 MGB
2#0 MGD
2#0 MGF
2#0 MGG
2#0 MGH
2#0 MGK
2#0 MGN
2#0 MGO
2#0 MGS
2#0 MGZ
2#0 MHA
2#0 MHB
2#0 MHD
2#0 MHF
2#0 MHH
2#0 MHK
2#0 MHO
2#0 MHQ
2#0 MHR
2#0 MHS
2#0 MHW
2#0 MHX
2#0 MHY
2#0 MHZ
2#0 MIA
2#0 MIE
2#0 MII
2#0 MIN
2#0 MIP
2#0 MIR
2#0 MIT
2#0 MJB
2#0 MJF
2#0 MJI
2#0 MJN
2#0 MJR
2#0 MJU
2#0 MJV
2#0 MK
2#0 MKD

2#0 MKL
2#0 MKM
2#0 MKN
2#0 MKO
2#0 MKP
2#0 MKQ
2#0 MKR
2#0 MKS
2#0 MKU
2#0 MKY
2#0 MLM
2#0 MLN
2#0 MLO
2#0 MLP
2#0 MLW
2#0 MLZ
2#0 MMA
2#0 MMK
2#0 MMN
2#0 MMR
2#0 MMS
2#0 MMV
2#0 MMW
2#0 MMY
2#0 MMZ
2#0 MNE
2#0 MNI
2#0 MNK
2#0 MNM
2#0 MNN
2#0 MNO
2#0 MNR
2#0 MNT
2#0 MNV
2#0 MOA
2#0 MOC
2#0 MOD
2#0 MOE
2#0 MOG
2#0 MOH
2#0 MOI
2#0 MOP
2#0 MOS
2#0 MOV
2#0 MOX
2#0 MPD
2#0 MPH
2#0 MPI
2#0 MPK
2#0 MPL
2#0 MPM
2#0 MPQ
2#0 MPU
2#0 MPW
2#0 MPZ
2#0 MQK
2#0 MQM
2#0 MQN
2#0 MRB
2#0 MRE
2#0 MRH
2#0 MRI
2#0 MRK
2#0 MRL
2#0 MRN
2#0 MRR
2#0 MRS
2#0 MRU
2#0 MRV
2#0 MRW
2#0 MRX
2#0 MSD
2#0 MSF
2#0 MSG
2#0 MSJ
2#0 MSK
2#0 MSP
2#0 MST
2#0 MSV
2#0 MSW
2#0 MSX
2#0 MSY
2#0 MTF
2#0 MTG
2#0 MTI
2#0 MTJ
2#0 MTK
2#0 MTP
2#0 MTU
2#0 MTV
2#0 MTW
2#0 MUC
2#0 MUF
2#0 MUG
2#0 MUK
2#0 MUL
2#0 MUO
2#0 MUP
2#0 MUX
2#0 MVA
2#0 MVC
2#0 MVF
2#0 MVG
2#0 MVK
2#0 MVM
2#0 MVR
2#0 MVY
2#0 MVZ
2#0 MWM
2#0 MWR
2#0 MWU
2#0 MXD
2#0 MXE
2#0 MXH
2#0 MXI
2#0 MXJ
2#0 MXS
2#0 MXV
2#0 MXX
2#0 MXY
2#0 MXZ
2#0 MYC

2#0 MYF
2#0 MYJ
2#0 MYM
2#0 MYN
2#0 MYO
2#0 MYP
2#0 MYR
2#0 MZA
2#0 MZF
2#0 MZG
2#0 MZH
2#0 MZI
2#0 MZJ
2#0 MZO
2#0 MZT
2#0 MZW
2#0 NAC
2#0 NAH
2#0 NAK
2#0 NAL
2#0 NAO
2#0 NAV
2#0 NAW
2#0 NBA
2#0 NBL
2#0 NBP
2#0 NBS
2#0 NBT
2#0 NBW
2#0 NBY
2#0 NCH
2#0 NCL
2#0 NCP
2#0 NCW
2#0 NCX
2#0 NDB
2#0 NDC
2#0 NDD
2#0 NDM
2#0 NDS
2#0 NDX
2#0 NEA
2#0 NEB
2#0 NEE
2#0 NEG
2#0 NEK
2#0 NEP
2#0 NES
2#0 NET
2#0 NEW
2#0 NEX
2#0 NFA
2#0 NFG
2#0 NFJ
2#0 NFL
2#0 NFO
2#0 NFR
2#0 NFT
2#0 NFU
2#0 NFY
2#0 NGA
2#0 NGE
2#0 NGI
2#0 NGP
2#0 NGQ
2#0 NGS
2#0 NGU
2#0 NGV
2#0 NGX
2#0 NGY
2#0 NHD
2#0 NHT
2#0 NHX
2#0 NIC
2#0 NID
2#0 NIH
2#0 NII
2#0 NIK
2#0 NIM
2#0 NIN
2#0 NIQ
2#0 NIS
2#0 NIU
2#0 NIV
2#0 NIY
2#0 NJA
2#0 NJD
2#0 NJH
2#0 NJP
2#0 NJR
2#0 NKB
2#0 NKD
2#0 NKJ
2#0 NKL
2#0 NKV
2#0 NKX
2#0 NKY
2#0 NLC
2#0 NLD
2#0 NLF
2#0 NLG
2#0 NLH
2#0 NLJ
2#0 NME
2#0 NMG
2#0 NMH
2#0 NMI
2#0 NMJ
2#0 NML
2#0 NMM
2#0 NMO
2#0 NMP
2#0 NMR
2#0 NMT
2#0 NNA
2#0 NNB
2#0 NNS
2#0 NOG
2#0 NOO
2#0 NPB
2#0 NPC
2#0 NPD

2#0 NPH
2#0 NPI
2#0 NPJ
2#0 NPK
2#0 NPQ
2#0 NPV
2#0 NQP
2#0 NRC
2#0 NRD
2#0 NRM
2#0 NRO
2#0 NRP
2#0 NRY
2#0 NRZ
2#0 NSD
2#0 NSD
2#0 NSE
2#0 NSM
2#0 NSO
2#0 NST
2#0 NSU
2#0 NSV
2#0 NSX
2#0 NSZ
2#0 NTA
2#0 NTB
2#0 NTE
2#0 NTO
2#0 NTX
2#0 NUE
2#0 NUF
2#0 NUK
2#0 NUM
2#0 NUN
2#0 NUR
2#0 NVD
2#0 NVG
2#0 NVH
2#0 NVM
2#0 NVR
2#0 NVS
2#0 NVY
2#0 NWG
2#0 NWI
2#0 NWM
2#0 NXJ
2#0 NXK
2#0 NXN
2#0 NXR
2#0 NXS
2#0 NXT
2#0 NYH
2#0 NYR
2#0 NYY
2#0 NZL
2#0 NZS
2#0 NZW
2#0 NZY
2#0 NZZ
2#0 OAD
2#0 OAE
2#0 OAF
2#0 OAL
2#0 OAN
2#0 OAT
2#0 OAV
2#0 OAW
2#0 OBA
2#0 OBJ
2#0 OBR
2#0 OBW
2#0 OBX
2#0 OBY
2#0 OCA
2#0 OCD
2#0 OCE
2#0 OCK
2#0 OCO
2#0 OCR
2#0 OCX
2#0 OCZ
2#0 ODC
2#0 ODD
2#0 ODE
2#0 ODH
2#0 ODY
2#0 OED
2#0 OEI
2#0 OEO
2#0 OER
2#0 OEW
2#0 OFC
2#0 OFG
2#0 OFK
2#0 OFN
2#0 OFP
2#0 OFR
2#0 OFS
2#0 OFV
2#0 OFX
2#0 OFZ
2#0 OGC
2#0 OGI
2#0 OGL
2#0 OGM
2#0 OGP
2#0 OGS
2#0 OGW
2#0 OHG
2#0 OHH
2#0 OHI
2#0 OHM
2#0 OHR
2#0 OIC
2#0 OID
2#0 OII
2#0 OJA
2#0 OJP
2#0 OJR
2#0 OJW
2#0 OJX
2#0 OKE

2#0 OKF
2#0 OKI
2#0 OKM
2#0 OKW
2#0 OLA
2#0 OLB
2#0 OLD
2#0 OLI
2#0 OLL
2#0 OLN
2#0 OLO
2#0 OLS
2#0 OLW
2#0 OLX
2#0 OMB
2#0 OMC
2#0 OMK
2#0 OML
2#0 OMM
2#0 OMO
2#0 OMY
2#0 OMZ
2#0 OND
2#0 ONF
2#0 ONK
2#0 ONN
2#0 ONO
2#0 ONR
2#0 ONX
2#0 ONZ
2#0 OOA
2#0 OOB
2#0 OOD
2#0 OOE
2#0 OOF
2#0 OOK
2#0 OOL
2#0 OOP
2#0 OOS
2#0 OOZ
2#0 OPA
2#0 OPG
2#0 OPP
2#0 OPQ
2#0 OPR
2#0 OPV
2#0 OPW
2#0 OPZ
2#0 OQU
2#0 ORA
2#0 ORB
2#0 ORG
2#0 ORN
2#0 ORO
2#0 ORU
2#0 ORZ
2#0 OSB
2#0 OSL
2#0 OSM
2#0 OSO
2#0 OSR
2#0 OSW
2#0 OSZ
2#0 OTD
2#0 OTH
2#0 OTJ
2#0 OTK
2#0 OTL
2#0 OTO
2#0 OTQ
2#0 OTS
2#0 OTV
2#0 OUC
2#0 OUE
2#0 OUN
2#0 OUT
2#0 OUZ
2#0 OVA
2#0 OVE
2#0 OVG
2#0 OVK
2#0 OVO
2#0 OWI
2#0 OWN
2#0 OWW
2#0 OWX
2#0 OWZ
2#0 OXE
2#0 OXH
2#0 OXI
2#0 OXJ
2#0 OXM
2#0 OXN
2#0 OXS
2#0 OXW
2#0 OYF
2#0 OYI
2#0 OZC
2#0 OZP
2#0 OZT
2#0 OZV
2#0 OZZ
2#0 PAI
2#0 PAM
2#0 PAQ
2#0 PAR
2#0 PAW
2#0 PAZ
2#0 PBA
2#0 PBD
2#0 PBG
2#0 PBI
2#0 PBQ
2#0 PCA
2#0 PCB
2#0 PCH
2#0 PCJ
2#0 PCK
2#0 PCM
2#0 PCS
2#0 PCY
2#0 PDA

2#0 PDC
2#0 PDD
2#0 PDE
2#0 PDF
2#0 PDI
2#0 PDK
2#0 PDN
2#0 PDS
2#0 PDW
2#0 PDY
2#0 PEA
2#0 PEB
2#0 PEK
2#0 PEO
2#0 PER
2#0 PES
2#0 PFC
2#0 PFI
2#0 PFM
2#0 PFX
2#0 PGA
2#0 PGD
2#0 PGE
2#0 PGF
2#0 PGN
2#0 PGW
2#0 PGX
2#0 PGY
2#0 PGZ
2#0 PHD
2#0 PHH
2#0 PHI
2#0 PHJ
2#0 PHN
2#0 PHT
2#0 PHW
2#0 PHY
2#0 PIC
2#0 PIE
2#0 PIF
2#0 PII
2#0 PIR
2#0 PIV
2#0 PIZ
2#0 PJA
2#0 PJB
2#0 PJG
2#0 PJP
2#0 PJW
2#0 PJX
2#0 PJZ
2#0 PKC
2#0 PKD
2#0 PKE
2#0 PKG
2#0 PKO
2#0 PKP
2#0 PKT
2#0 PKW
2#0 PKX
2#0 PKY
2#0 PKZ
2#0 PLD
2#0 PLF
2#0 PLG
2#0 PLL
2#0 PLM
2#0 PLO
2#0 PLW
2#0 PLZ
2#0 PMF
2#0 PMH
2#0 PMJ
2#0 PMK
2#0 PMQ
2#0 PMT
2#0 PMY
2#0 PNE
2#0 PNF
2#0 PNO
2#0 PNS
2#0 PNT
2#0 PNX
2#0 PNY
2#0 POA
2#0 POC
2#0 POF
2#0 POG
2#0 POI
2#0 POJ
2#0 POM
2#0 POP
2#0 POS
2#0 POV
2#0 POW
2#0 POY
2#0 PPC
2#0 PPE
2#0 PPG
2#0 PPN
2#0 PPP
2#0 PPQ
2#0 PPS
2#0 PPT
2#0 PPU
2#0 PPX
2#0 PQP
2#0 PQV
2#0 PRB
2#0 PRF
2#0 PRH
2#0 PRI
2#0 PRJ
2#0 PRK
2#0 PRN
2#0 PRP
2#0 PRT
2#0 PRU
2#0 PRW
2#0 PRX
2#0 PRY
2#0 PS

2#0 PSB
2#0 PSI
2#0 PSJ
2#0 PSL
2#0 PST
2#0 PSU
2#0 PSV
2#0 PSX
2#0 PSY
2#0 PTD
2#0 PTH
2#0 PTK
2#0 PTL
2#0 PTP
2#0 PTQ
2#0 PTT
2#0 PTU
2#0 PTV
2#0 PTW
2#0 PTX
2#0 PUC
2#0 PUD
2#0 PUK
2#0 PUP
2#0 PUR
2#0 PUX
2#0 PVC
2#0 PVM
2#0 PVO
2#0 PVP
2#0 PVR
2#0 PVT
2#0 PVX
2#0 PWA
2#0 PWE
2#0 PWH
2#0 PWN
2#0 PWS
2#0 PWW
2#0 PWX
2#0 PWZ
2#0 PXB
2#0 PXC
2#0 PXG
2#0 PXK
2#0 PXM
2#0 PXQ
2#0 PXT
2#0 PXX
2#0 PYB
2#0 PYE
2#0 PYG
2#0 PYO
2#0 PYT
2#0 PYU
2#0 PYW
2#0 PYX
2#0 PZA
2#0 PZL
2#0 PZO
2#0 PZR
2#0 PZS
2#0 PZW
2#0 RAC
2#0 RAE
2#0 RAN
2#0 RAO
2#0 RAR
2#0 RAT
2#0 RAV
2#0 RAX
2#0 RAY
2#0 RAZ
2#0 RBB
2#0 RBD
2#0 RBE
2#0 RBF
2#0 RBJ
2#0 RBL
2#0 RBM
2#0 RBS
2#0 RCH
2#0 RCK
2#0 RCQ
2#0 RCS
2#0 RCY
2#0 RDB
2#0 RDM
2#0 RDY
2#0 RED
2#0 REE
2#0 REF
2#0 REI
2#0 REK
2#0 RET
2#0 REV
2#0 RFA
2#0 RFB
2#0 RFJ
2#0 RFQ
2#0 RFR
2#0 RFV
2#0 RFX
2#0 RFZ
2#0 RGE
2#0 RGF
2#0 RGG
2#0 RGP
2#0 RGQ
2#0 RGV
2#0 RGX
2#0 RHD
2#0 RHG
2#0 RHK
2#0 RHR
2#0 RHU
2#0 RHV
2#0 RHX
2#0 RHY
2#0 RHZ
2#0 RIB
2#0 RIC

2#0 RIF
2#0 RIG
2#0 RIJ
2#0 RIP
2#0 RIV
2#0 RIX
2#0 RJC
2#0 RJF
2#0 RJI
2#0 RJK
2#0 RJN
2#0 RJS
2#0 RJT
2#0 RKA
2#0 RKI
2#0 RKJ
2#0 RKL
2#0 RKN
2#0 RKT
2#0 RKW
2#0 RKY
2#0 RLD
2#0 RLF
2#0 RLI
2#0 RLL
2#0 RLM
2#0 RLT
2#0 RLZ
2#0 RMA
2#0 RME
2#0 RMG
2#0 RMX
2#0 RNC
2#0 RNG
2#0 RNL
2#0 RNN
2#0 RNY
2#0 ROA
2#0 ROB
2#0 ROF
2#0 ROH
2#0 ROL
2#0 ROT
2#0 ROW
2#0 ROZ
2#0 RPG
2#0 RPI
2#0 RPJ
2#0 RPK
2#0 RPS
2#0 RPT
2#0 RPX
2#0 RPZ
2#0 RQB
2#0 RQF
2#0 RQO
2#0 RQR
2#0 RQT
2#0 RRB
2#0 RRG
2#0 RRJ
2#0 RRM
2#0 RRN
2#0 RRR
2#0 RRS
2#0 RRV
2#0 RRW
2#0 RRZ
2#0 RSF
2#0 RSJ
2#0 RSK
2#0 RSL
2#0 RSO
2#0 RSP
2#0 RSQ
2#0 RSR
2#0 RSS
2#0 RST
2#0 RSU
2#0 RSW
2#0 RSX
2#0 RSY
2#0 RTB
2#0 RTR
2#0 RTS
2#0 RTT
2#0 RTV
2#0 RTX
2#0 RUB
2#0 RUE
2#0 RUF
2#0 RUH
2#0 RUK
2#0 RUM
2#0 RUN
2#0 RVC
2#0 RVL
2#0 RVN
2#0 RVO
2#0 RVP
2#0 RVR
2#0 RVS
2#0 RVX
2#0 RVY
2#0 RWD
2#0 RXB
2#0 RXD
2#0 RXF
2#0 RXI
2#0 RXJ
2#0 RXM
2#0 RXN
2#0 RXP
2#0 RXS
2#0 RYA
2#0 RYC
2#0 RYE
2#0 RYK
2#0 RYM
2#0 RYO
2#0 RYT
2#0 RYU

2#0 RYW
2#0 RYZ
2#0 RZA
2#0 RZB
2#0 RZG
2#0 RZK
2#0 RZR
2#0 RZZ
2#0 SAB
2#0 SAC
2#0 SAE
2#0 SAJ
2#0 SAL
2#0 SAM
2#0 SAO
2#0 SAQ
2#0 SAV
2#0 SBA
2#0 SBE
2#0 SBF
2#0 SBG
2#0 SBI
2#0 SBQ
2#0 SBT
2#0 SBU
2#0 SBV
2#0 SBY
2#0 SCI
2#0 SCT
2#0 SCY
2#0 SCZ
2#0 SDB
2#0 SDF
2#0 SDH
2#0 SDI
2#0 SDL
2#0 SDN
2#0 SDX
2#0 SEI
2#0 SEM
2#0 SEN
2#0 SEQ
2#0 SEU
2#0 SEV
2#0 SFI
2#0 SFK
2#0 SFM
2#0 SFO
2#0 SFR
2#0 SFU
2#0 SGA
2#0 SGC
2#0 SGE
2#0 SGF
2#0 SGJ
2#0 SGN
2#0 SGP
2#0 SGR
2#0 SGT
2#0 SGU
2#0 SGX
2#0 SGZ
2#0 SHD
2#0 SHE
2#0 SHL
2#0 SHQ
2#0 SHS
2#0 SHT
2#0 SIE
2#0 SIG
2#0 SII
2#0 SIN
2#0 SIP
2#0 SIR
2#0 SJC
2#0 SJD
2#0 SJL
2#0 SJW
2#0 SJY
2#0 SKB
2#0 SKC
2#0 SKD
2#0 SKF
2#0 SKJ
2#0 SKO
2#0 SKS
2#0 SKT
2#0 SKW
2#0 SKX
2#0 SLB
2#0 SLC
2#0 SLE
2#0 SLF
2#0 SLX
2#0 SMC
2#0 SME
2#0 SMG
2#0 SMH
2#0 SMP
2#0 SMQ
2#0 SMR
2#0 SMU
2#0 SMW
2#0 SNH
2#0 SNI
2#0 SNN
2#0 SNO
2#0 SNQ
2#0 SNR
2#0 SNT
2#0 SNX
2#0 SNY
2#0 SOA
2#0 SOC
2#0 SOG
2#0 SOH
2#0 SOL
2#0 SOM
2#0 SON
2#0 SOO
2#0 SOU
2#0 SOY

2#0 SPC
2#0 SPI
2#0 SPK
2#0 SPO
2#0 SPQ
2#0 SQA
2#0 SQJ
2#0 SQM
2#0 SQP
2#0 SQR
2#0 SQW
2#0 SQZ
2#0 SRE
2#0 SRK
2#0 SRL
2#0 SRM
2#0 SRN
2#0 SRS
2#0 SRU
2#0 SRW
2#0 SSH
2#0 SSR
2#0 SSS
2#0 SSV
2#0 SSW
2#0 STC
2#0 STD
2#0 STE
2#0 STF
2#0 STG
2#0 STJ
2#0 STM
2#0 STO
2#0 STR
2#0 STU
2#0 STW
2#0 STY
2#0 SUG
2#0 SUJ
2#0 SUL
2#0 SUM
2#0 SUN
2#0 SUP
2#0 SUQ
2#0 SUR
2#0 SUV
2#0 SVA
2#0 SVD
2#0 SVE
2#0 SVJ
2#0 SVL
2#0 SVM
2#0 SVO
2#0 SVR
2#0 SVS
2#0 SVX
2#0 SVY
2#0 SWC
2#0 SWD
2#0 SWI
2#0 SWL
2#0 SWW
2#0 SWX
2#0 SWY
2#0 SXA
2#0 SXB
2#0 SXD
2#0 SXG
2#0 SXH
2#0 SXL
2#0 SXN
2#0 SXO
2#0 SXR
2#0 SXX
2#0 SYR
2#0 SYZ
2#0 SZD
2#0 SZE
2#0 SZN
2#0 SZO
2#0 SZP
2#0 SZS
2#0 SZW
2#0 SZX
2#0 SZY
2#0 TAP
2#0 TBB
2#0 TBC
2#0 TBG
2#0 TBK
2#0 TBM
2#0 TBN
2#0 TBO
2#0 TBP
2#0 TBT
2#0 TBU
2#0 TBY
2#0 TCE
2#0 TCH
2#0 TCS
2#0 TDA
2#0 TDB
2#0 TDG
2#0 TDK
2#0 TDM
2#0 TDQ
2#0 TDS
2#0 TDW
2#0 TDX
2#0 TDY
2#0 TDZ
2#0 TEA
2#0 TEB
2#0 TEC
2#0 TEL
2#0 TEM
2#0 TEO
2#0 TEQ
2#0 TER
2#0 TES
2#0 TET
2#0 TEV

2#0 TEX
2#0 TFC
2#0 TFG
2#0 TFL
2#0 TFP
2#0 TFR
2#0 TFS
2#0 TFU
2#0 TFZ
2#0 TGE
2#0 TGH
2#0 TGP
2#0 TGR
2#0 TGS
2#0 TGW
2#0 TGX
2#0 TGY
2#0 TGZ
2#0 THB
2#0 THC
2#0 THD
2#0 THH
2#0 THI
2#0 THK
2#0 THL
2#0 THM
2#0 THO
2#0 THV
2#0 THW
2#0 THX
2#0 THY
2#0 TIB
2#0 TIC
2#0 TIE
2#0 TIG
2#0 TII
2#0 TIK
2#0 TIM
2#0 TIP
2#0 TIX
2#0 TJA
2#0 TJB
2#0 TJH
2#0 TJL
2#0 TKA
2#0 TKB
2#0 TKC
2#0 TKH
2#0 TKI
2#0 TKK
2#0 TKL
2#0 TKM
2#0 TKN
2#0 TKO
2#0 TKP
2#0 TKR
2#0 TKT
2#0 TKW
2#0 TKZ
2#0 TLK
2#0 TLL
2#0 TLS
2#0 TLV
2#0 TLZ
2#0 TMG
2#0 TMI
2#0 TMJ
2#0 TMP
2#0 TMR
2#0 TMT
2#0 TMU
2#0 TMZ
2#0 TNA
2#0 TNF
2#0 TNG
2#0 TNJ
2#0 TNK
2#0 TNL
2#0 TNR
2#0 TNS
2#0 TNT
2#0 TNX
2#0 TNY
2#0 TOC
2#0 TOD
2#0 TOE
2#0 TOH
2#0 TOZ
2#0 TPB
2#0 TPJ
2#0 TPK
2#0 TPS
2#0 TPW
2#0 TPX
2#0 TPZ
2#0 TQR
2#0 TQW
2#0 TQX
2#0 TRB
2#0 TRC
2#0 TRF
2#0 TRG
2#0 TRN
2#0 TRQ
2#0 TRT
2#0 TRV
2#0 TSB
2#0 TSD
2#0 TSF
2#0 TSG
2#0 TSH
2#0 TSL
2#0 TSN
2#0 TSS
2#0 TST
2#0 TSW
2#0 TSZ
2#0 TTI
2#0 TTK
2#0 TTL
2#0 TTO
2#0 TTQ

Withheld

2#0 TTR
2#0 TTV
2#0 TTX
2#0 TTZ
2#0 TUP
2#0 TUZ
2#0 TVA
2#0 TVC
2#0 TVE
2#0 TVI
2#0 TVK
2#0 TVO
2#0 TVP
2#0 TVV
2#0 TWA
2#0 TWB
2#0 TWC
2#0 TWE
2#0 TWF
2#0 TWG
2#0 TWH
2#0 TWM
2#0 TWN
2#0 TWR
2#0 TXF
2#0 TXN
2#0 TXO
2#0 TXS
2#0 TXT
2#0 TXX
2#0 TXZ
2#0 TYB
2#0 TYD
2#0 TYF
2#0 TYK
2#0 TYM
2#0 TYN
2#0 TYO
2#0 TYP
2#0 TYU
2#0 TYV
2#0 TZA
2#0 TZC
2#0 TZH
2#0 TZJ
2#0 TZN
2#0 TZT
2#0 TZX
2#0 UAF
2#0 UAG
2#0 UAI
2#0 UAJ
2#0 UBA
2#0 UBD
2#0 UBS
2#0 UBZ
2#0 UCK
2#0 UCW
2#0 UDG
2#0 UDL
2#0 UDP
2#0 UDV
2#0 UDX
2#0 UEM
2#0 UEX
2#0 UFJ
2#0 UFS
2#0 UFT
2#0 UFX
2#0 UGD
2#0 UGS
2#0 UGX
2#0 UHD
2#0 UHH
2#0 UHU
2#0 UHV
2#0 UHZ
2#0 UIO
2#0 UIR
2#0 UKC
2#0 UKE
2#0 UKF
2#0 UKH
2#0 UKI
2#0 UKL
2#0 UKO
2#0 UKP
2#0 UKQ
2#0 UKR
2#0 UKS
2#0 UKV
2#0 UKZ
2#0 ULA
2#0 ULF
2#0 ULL
2#0 UMB
2#0 UMC
2#0 UMF
2#0 UMI
2#0 UMM
2#0 UMS
2#0 UMT
2#0 UND
2#0 UNE
2#0 UNK
2#0 UNM
2#0 UNO
2#0 UNX
2#0 UOG
2#0 UOH
2#0 UOO
2#0 UOS
2#0 UPG
2#0 UPN
2#0 UPP
2#0 URL
2#0 URM
2#0 URN
2#0 URS
2#0 USE
2#0 USP
2#0 USQ
2#0 USR
2#0 UST
2#0 UTM
2#0 UTO
2#0 UTS
2#0 UTU
2#0 UUC
2#0 UUF
2#0 UUK
2#0 UUN
2#0 UUO
2#0 UUP
2#0 UUU
2#0 UVA
2#0 UVV
2#0 UWB
2#0 UWU
2#0 UWX
2#0 UXA
2#0 UXL
2#0 UXP
2#0 UYI
2#0 UYP
2#0 UYU
2#0 UZA
2#0 UZE
2#0 VAB
2#0 VAD
2#0 VAE
2#0 VAK
2#0 VAL
2#0 VAP
2#0 VAR
2#0 VAX
2#0 VAZ
2#0 VBC
2#0 VBD
2#0 VBE
2#0 VBG
2#0 VBO
2#0 VBQ
2#0 VBV
2#0 VCF
2#0 VCH
2#0 VCJ
2#0 VCN
2#0 VCO
2#0 VCT
2#0 VCX
2#0 VDD
2#0 VDF
2#0 VDK
2#0 VDR
2#0 VDS
2#0 VDX
2#0 VEC
2#0 VED
2#0 VEE
2#0 VEI
2#0 VEL
2#0 VEM
2#0 VEN
2#0 VEP
2#0 VER
2#0 VEV
2#0 VEY
2#0 VFM
2#0 VFS
2#0 VFZ
2#0 VGD
2#0 VGG
2#0 VGH
2#0 VGI
2#0 VGM
2#0 VGN
2#0 VGO
2#0 VGP
2#0 VGW
2#0 VGX
2#0 VHB
2#0 VIC
2#0 VIE
2#0 VII
2#0 VIK
2#0 VIM
2#0 VIP
2#0 VIR
2#0 VIX
2#0 VIY
2#0 VJC
2#0 VJM
2#0 VJT
2#0 VKC
2#0 VKD
2#0 VKF
2#0 VKH
2#0 VKJ
2#0 VKP
2#0 VKT
2#0 VKV
2#0 VKX
2#0 VLC
2#0 VLK
2#0 VLP
2#0 VLS
2#0 VLU
2#0 VLV
2#0 VMB
2#0 VME
2#0 VMF
2#0 VMI
2#0 VMK
2#0 VML
2#0 VMM
2#0 VMO
2#0 VMR
2#0 VMS
2#0 VMX
2#0 VMZ
2#0 VNA
2#0 VNB
2#0 VNC
2#0 VNG
2#0 VNU
2#0 VOE
2#0 VOL
2#0 VOM
2#0 VON
2#0 VOO
2#0 VOS
2#0 VOV
2#0 VOW
2#0 VOX
2#0 VPC
2#0 VPR
2#0 VPV
2#0 VPW
2#0 VPX
2#0 VPY
2#0 VPZ
2#0 VQD
2#0 VRG
2#0 VRH
2#0 VRJ
2#0 VRM
2#0 VRN
2#0 VRS
2#0 VSA
2#0 VSR
2#0 VSS
2#0 VST
2#0 VSV
2#0 VSX
2#0 VTD
2#0 VTE
2#0 VTG
2#0 VTI
2#0 VTL
2#0 VTM
2#0 VTN
2#0 VTO
2#0 VTX
2#0 VTY
2#0 VUE
2#0 VUX
2#0 VVD
2#0 VVG
2#0 VVL
2#0 VVN
2#0 VVQ
2#0 VVR
2#0 VVV
2#0 VWA
2#0 VWB
2#0 VWC
2#0 VWI
2#0 VWP
2#0 VWR
2#0 VWS
2#0 VWV
2#0 VXC
2#0 VXN
2#0 VXP
2#0 VXV
2#0 VXY
2#0 VYD
2#0 VYG
2#0 VYL
2#0 VYM
2#0 VYV
2#0 VYZ
2#0 VZC
2#0 VZJ
2#0 VZK
2#0 VZM
2#0 VZT
2#0 VZV
2#0 VZW
2#0 VZZ
2#0 WAB
2#0 WAC
2#0 WAM
2#0 WAN
2#0 WAR
2#0 WAW
2#0 WAZ
2#0 WBA
2#0 WBC
2#0 WBK
2#0 WBM
2#0 WBN
2#0 WBP
2#0 WBY
2#0 WCA
2#0 WCD
2#0 WCF
2#0 WCH
2#0 WCK
2#0 WCS
2#0 WCT
2#0 WCW
2#0 WDC
2#0 WDF
2#0 WDJ
2#0 WDK
2#0 WDT
2#0 WDV
2#0 WDX
2#0 WDZ
2#0 WEA
2#0 WEH
2#0 WEM
2#0 WEN
2#0 WEP
2#0 WER
2#0 WEW
2#0 WEY
2#0 WFA
2#0 WFL
2#0 WFM
2#0 WFT
2#0 WFX
2#0 WFZ
2#0 WGA
2#0 WGH
2#0 WGK
2#0 WGR
2#0 WGW
2#0 WGX
2#0 WGY
2#0 WGZ
2#0 WHC
2#0 WHE
2#0 WHG
2#0 WHI
2#0 WHJ
2#0 WHL
2#0 WHM
2#0 WHN
2#0 WHQ
2#0 WHR
2#0 WHY
2#0 WIA
2#0 WIB
2#0 WII
2#0 WIM
2#0 WIN
2#0 WIT
2#0 WIW
2#0 WIY
2#0 WJG
2#0 WKA
2#0 WKC
2#0 WKD
2#0 WKF
2#0 WKH
2#0 WKI
2#0 WKO
2#0 WKQ
2#0 WKS
2#0 WLB
2#0 WLC
2#0 WLF
2#0 WLG
2#0 WLJ
2#0 WLL
2#0 WLR
2#0 WLW
2#0 WLZ
2#0 WMA
2#0 WMF
2#0 WMI
2#0 WMR
2#0 WMS
2#0 WMV
2#0 WMX
2#0 WND
2#0 WNF
2#0 WNJ
2#0 WOE
2#0 WOJ
2#0 WOK
2#0 WOM
2#0 WON
2#0 WOO
2#0 WOT
2#0 WPA
2#0 WPG
2#0 WPL
2#0 WPM
2#0 WPR
2#0 WPW
2#0 WQE
2#0 WQZ
2#0 WRA
2#0 WRB
2#0 WRC
2#0 WRD
2#0 WRG
2#0 WRJ
2#0 WRN
2#0 WRP
2#0 WRZ
2#0 WSA
2#0 WSB
2#0 WSD
2#0 WSE
2#0 WSF
2#0 WSL
2#0 WSO
2#0 WTB
2#0 WTC
2#0 WTF
2#0 WTK
2#0 WTL
2#0 WTO
2#0 WTV
2#0 WTW
2#0 WTX
2#0 WUE
2#0 WUH
2#0 WUS
2#0 WUT
2#0 WUV
2#0 WVA
2#0 WVB
2#0 WVX
2#0 WVY
2#0 WVZ
2#0 WWU
2#0 WWW
2#0 WXC
2#0 WXV
2#0 WXX
2#0 WYB
2#0 WYD
2#0 WYI
2#0 WYK
2#0 WYL
2#0 WYP
2#0 WYR
2#0 WYY
2#0 WZC
2#0 WZD
2#0 WZG
2#0 WZL
2#0 WZO
2#0 WZP
2#0 WZV
2#0 XAB
2#0 XAC
2#0 XAD
2#0 XAF
2#0 XAJ
2#0 XAK
2#0 XAQ
2#0 XAT
2#0 XBF
2#0 XBH
2#0 XBI
2#0 XBJ
2#0 XBS
2#0 XCE
2#0 XCG
2#0 XCK
2#0 XCQ
2#0 XCW
2#0 XCX
2#0 XDB
2#0 XDE
2#0 XDJ
2#0 XDL
2#0 XDO
2#0 XDP
2#0 XDV
2#0 XDW
2#0 XEC
2#0 XED
2#0 XEQ
2#0 XEX
2#0 XFC
2#0 XFK
2#0 XFL
2#0 XFS
2#0 XFU
2#0 XGK
2#0 XGL
2#0 XGM
2#0 XGN
2#0 XGP
2#0 XGR
2#0 XHA
2#0 XHD
2#0 XHF
2#0 XHK
2#0 XHP
2#0 XHX
2#0 XID
2#0 XIE
2#0 XIH
2#0 XIM
2#0 XIO
2#0 XIX
2#0 XIZ
2#0 XJA
2#0 XJB
2#0 XJC
2#0 XJS
2#0 XJX
2#0 XKB
2#0 XKF
2#0 XKG
2#0 XKI
2#0 XKN
2#0 XKV
2#0 XKZ
2#0 XLA
2#0 XLC
2#0 XLD
2#0 XLF
2#0 XLL
2#0 XLV
2#0 XMA
2#0 XMB
2#0 XMD
2#0 XME
2#0 XMJ
2#0 XML
2#0 XMM
2#0 XMR
2#0 XMX
2#0 XMZ
2#0 XNA
2#0 XNB
2#0 XNG
2#0 XNI
2#0 XNP
2#0 XNR
2#0 XNV
2#0 XNX
2#0 XOB
2#0 XOF
2#0 XOM
2#0 XOO
2#0 XOS
2#0 XOW
2#0 XOX
2#0 XPB
2#0 XPE
2#0 XPL
2#0 XPN
2#0 XPO
2#0 XPR
2#0 XPW
2#0 XPX
2#0 XQB
2#0 XQS
2#0 XQT
2#0 XQU
2#0 XQZ
2#0 XRA
2#0 XRB
2#0 XRC
2#0 XRE
2#0 XRF
2#0 XRI
2#0 XRJ
2#0 XRK
2#0 XRP
2#0 XRT
2#0 XRW
2#0 XRY
2#0 XSE
2#0 XSH
2#0 XSI
2#0 XSM
2#0 XSQ
2#0 XSR
2#0 XSS
2#0 XSX
2#0 XTE
2#0 XTJ
2#0 XTK
2#0 XTN
2#0 XTO
2#0 XTW
2#0 XTX
2#0 XTY
2#0 XUB
2#0 XUE
2#0 XUP
2#0 XUT
2#0 XUX
2#0 XVS
2#0 XWB
2#0 XWF
2#0 XWH
2#0 XWM
2#0 XWP
2#0 XWT
2#0 XWV
2#0 XWZ
2#0 XXA
2#0 XXD
2#0 XXG
2#0 XXH
2#0 XXJ
2#0 XXL
2#0 XXN
2#0 XXR
2#0 XXS
2#0 XXV
2#0 XXZ
2#0 XYE
2#0 XYL
2#0 XYN
2#0 XYR
2#0 XZC
2#0 XZF
2#0 XZL
2#0 XZQ
2#0 XZT
2#0 XZY
2#0 YAA
2#0 YAK
2#0 YAN
2#0 YAT
2#0 YAU
2#0 YAZ
2#0 YBC
2#0 YBG
2#0 YBJ
2#0 YBK
2#0 YBM
2#0 YBO
2#0 YBQ
2#0 YBT
2#0 YBV
2#0 YBW
2#0 YCB
2#0 YCC
2#0 YCK
2#0 YCL
2#0 YCM
2#0 YCP
2#0 YCS
2#0 YCT
2#0 YCU
2#0 YCX
2#0 YCZ
2#0 YDE
2#0 YDL
2#0 YDS
2#0 YDW
2#0 YDZ
2#0 YED
2#0 YEE
2#0 YFA
2#0 YFL
2#0 YFM
2#0 YFO
2#0 YFT
2#0 YFY
2#0 YGA
2#0 YGG
2#0 YGN
2#0 YGW
2#0 YHB
2#0 YHQ
2#0 YIC
2#0 YIP
2#0 YJH
2#0 YJM
2#0 YKF
2#0 YKG
2#0 YKM
2#0 YKS
2#0 YLK
2#0 YLM
2#0 YLO
2#0 YLW
2#0 YLY
2#0 YMC
2#0 YMD
2#0 YMI
2#0 YML
2#0 YMN
2#0 YMO
2#0 YNC
2#0 YNE
2#0 YNG
2#0 YNM
2#0 YNV
2#0 YNW
2#0 YOB
2#0 YOC
2#0 YOR
2#0 YOS
2#0 YOT
2#0 YOU
2#0 YOW
2#0 YOX
2#0 YPA
2#0 YPC
2#0 YPD
2#0 YPL
2#0 YPM
2#0 YPO
2#0 YPT
2#0 YPZ
2#0 YQF
2#0 YQN
2#0 YRA
2#0 YRD
2#0 YRG
2#0 YRK
2#0 YRN
2#0 YRO
2#0 YSD
2#0 YSH
2#0 YSW
2#0 YTK
2#0 YTO
2#0 YTS
2#0 YTV
2#0 YTX
2#0 YUB
2#0 YUC
2#0 YUE
2#0 YUK
2#0 YUM
2#0 YUS
2#0 YUV
2#0 YVE
2#0 YVN
2#0 YVO
2#0 YVS
2#0 YVV
2#0 YVX
2#0 YVZ
2#0 YWF
2#0 YWH
2#0 YWK
2#0 YXP
2#0 YXS
2#0 YYO
2#0 YYX
2#0 YZE
2#0 YZF
2#0 YZG
2#0 YZV
2#0 ZAD
2#0 ZAK
2#0 ZAN
2#0 ZAR
2#0 ZAT
2#0 ZAV
2#0 ZAW
2#0 ZAX
2#0 ZAY
2#0 ZAZ
2#0 ZBA
2#0 ZBG
2#0 ZBU
2#0 ZCL
2#0 ZCR
2#0 ZDF
2#0 ZDK
2#0 ZDY
2#0 ZEC
2#0 ZEM
2#0 ZEO
2#0 ZEY
2#0 ZFM
2#0 ZFS
2#0 ZFU
2#0 ZFX
2#0 ZGB
2#0 ZGN
2#0 ZIG
2#0 ZIL
2#0 ZIM
2#0 ZIX
2#0 ZKX
2#0 ZLG
2#0 ZLK
2#0 ZLR
2#0 ZLT
2#0 ZMX
2#0 ZNH
2#0 ZOG
2#0 ZON
2#0 ZPM
2#0 ZPS
2#0 ZQR
2#0 ZRA
2#0 ZRN
2#0 ZRO
2#0 ZSP
2#0 ZTP
2#0 ZTX
2#0 ZUA
2#0 ZUT
2#0 ZWH
2#0 ZWZ
2#0 ZXA
2#0 ZXP
2#0 ZXW
2#0 ZXX
2#0 ZYF
2#0 ZYQ
2#0 ZZI
2#0 ZZM
2#0 ZZR
2#0 ZZW
2#0 ZZX

2#1

2#1 AAH
2#1 ACX
2#1 ADB
2#1 ADW
2#1 AED
2#1 AFH
2#1 AFO
2#1 AFT
2#1 AJK
2#1 AJL
2#1 AJM
2#1 AKC
2#1 AKO
2#1 AKU
2#1 ALF
2#1 ALQ
2#1 ANP
2#1 AOQ
2#1 AOR
2#1 APT
2#1 AQG
2#1 AQO
2#1 ART
2#1 ARV
2#1 ATB
2#1 AVX
2#1 AWM
2#1 AWW
2#1 AWX
2#1 AXG
2#1 AXN
2#1 AXX
2#1 AYY
2#1 AZD
2#1 AZE
2#1 AZL
2#1 BBN
2#1 BCW
2#1 BDF
2#1 BDL
2#1 BDM
2#1 BDU
2#1 BDX
2#1 BDZ
2#1 BEI
2#1 BFD
2#1 BFO
2#1 BGC
2#1 BHS
2#1 BJF
2#1 BKA
2#1 BKU
2#1 BKW
2#1 BMD
2#1 BMM
2#1 BNA
2#1 BNB
2#1 BNF
2#1 BNP
2#1 BOF
2#1 BOS
2#1 BPC
2#1 BPJ
2#1 BRE
2#1 BRM
2#1 BRT
2#1 BRY
2#1 BTD
2#1 BTH
2#1 BTS
2#1 BTX
2#1 BUC
2#1 BUF
2#1 BVM
2#1 BVS
2#1 BVX
2#1 BWZ
2#1 BZF
2#1 BZU
2#1 CAD
2#1 CBC
2#1 CBL
2#1 CCB
2#1 CCM
2#1 CDJ
2#1 CDL
2#1 CEL
2#1 CES
2#1 CFA
2#1 CFU
2#1 CGT
2#1 CIC
2#1 CIH
2#1 CII
2#1 CJK
2#1 CKE
2#1 CLU
2#1 CMH
2#1 CMP
2#1 CNT
2#1 CQG
2#1 CSN
2#1 CTB
2#1 CTF
2#1 CTR
2#1 CTV
2#1 CTZ
2#1 CUK
2#1 CUT
2#1 CUZ
2#1 CVG
2#1 CXB
2#1 CYH
2#1 CYR
2#1 CZI
2#1 CZY
2#1 DAA
2#1 DAG
2#1 DAH
2#1 DCF
2#1 DCH
2#1 DCJ
2#1 DCL
2#1 DCU
2#1 DDD
2#1 DDH
2#1 DDN
2#1 DDT
2#1 DDU
2#1 DEU
2#1 DEV
2#1 DEW
2#1 DFD
2#1 DFI
2#1 DFR
2#1 DHH
2#1 DHL
2#1 DHN
2#1 DIK
2#1 DIQ
2#1 DJM
2#1 DJN
2#1 DJR
2#1 DJZ
2#1 DKI
2#1 DKT
2#1 DMM
2#1 DMW
2#1 DNH
2#1 DNZ
2#1 DOX
2#1 DPJ
2#1 DPU
2#1 DQR
2#1 DQU
2#1 DSD
2#1 DSG
2#1 DSM
2#1 DSV
2#1 DTI
2#1 DTR
2#1 DTT
2#1 DUE
2#1 DUN
2#1 DUU
2#1 DUZ
2#1 DVE
2#1 DVS
2#1 DWY
2#1 DWZ
2#1 DXY
2#1 DYR
2#1 DZF
2#1 DZR
2#1 DZY
2#1 EAR
2#1 EAS
2#1 EBK
2#1 EBO
2#1 EBS
2#1 ECQ
2#1 ECY
2#1 EDF
2#1 EER
2#1 EEU
2#1 EFH
2#1 EFZ
2#1 EGK
2#1 EGQ
2#1 EIA
2#1 EIL
2#1 EJI
2#1 EJN
2#1 EKU
2#1 EKW
2#1 ELH
2#1 ELM
2#1 ELR
2#1 ELY
2#1 EMM
2#1 EMS
2#1 EMT
2#1 ENF
2#1 EOW
2#1 EQM
2#1 EQY
2#1 ESU
2#1 ESX
2#1 ESY
2#1 ETP
2#1 EUB
2#1 EUG
2#1 EUT
2#1 EUW
2#1 EVB
2#1 EWF
2#1 EXB
2#1 EXC
2#1 EXG
2#1 EYI
2#1 EYO
2#1 EZF
2#1 FAV
2#1 FCN
2#1 FCV
2#1 FDZ
2#1 FEK
2#1 FEX
2#1 FFC
2#1 FGK
2#1 FGY
2#1 FHJ
2#1 FIY
2#1 FJU
2#1 FKA
2#1 FKE
2#1 FMS
2#1 FNP
2#1 FOJ
2#1 FOK
2#1 FOP
2#1 FPD
2#1 FPH
2#1 FPR
2#1 FQP
2#1 FQU
2#1 FTP
2#1 FTU
2#1 FTW
2#1 FUE
2#1 FUF
2#1 FUN
2#1 FUO
2#1 FUZ
2#1 FVC
2#1 FVG
2#1 FWL
2#1 FWR
2#1 FXO
2#1 FYA
2#1 FYE
2#1 FYI
2#1 FZF
2#1 FZK
2#1 FZP
2#1 GAD
2#1 GAJ
2#1 GCT
2#1 GCX
2#1 GDC
2#1 GEF
2#1 GEH
2#1 GEN
2#1 GEO
2#1 GEV
2#1 GEW
2#1 GFD
2#1 GFH
2#1 GFL
2#1 GFY
2#1 GGW
2#1 GGX
2#1 GHW
2#1 GIA
2#1 GIB
2#1 GIP
2#1 GIR
2#1 GJN
2#1 GJW
2#1 GKA
2#1 GLA
2#1 GLO
2#1 GMD
2#1 GMH
2#1 GMS
2#1 GNW
2#1 GOG
2#1 GON
2#1 GOO
2#1 GOQ
2#1 GOS
2#1 GPF
2#1 GPR
2#1 GPT
2#1 GQA
2#1 GQK
2#1 GQL
2#1 GQM
2#1 GQX
2#1 GQZ
2#1 GRD
2#1 GRE
2#1 GRY
2#1 GSH
2#1 GSJ
2#1 GSK
2#1 GSN
2#1 GSV
2#1 GSY
2#1 GTU
2#1 GTZ
2#1 GUA
2#1 GUB
2#1 GUT
2#1 GUU
2#1 GWB
2#1 GWI
2#1 GXD
2#1 GXN
2#1 GXP
2#1 GXT
2#1 GYJ
2#1 GYP
2#1 GZB
2#1 GZE
2#1 GZU
2#1 GZW
2#1 HAA
2#1 HAB
2#1 HAF
2#1 HAK
2#1 HAT
2#1 HAV
2#1 HBI
2#1 HBV
2#1 HCJ
2#1 HCL
2#1 HDL
2#1 HDU
2#1 HDW
2#1 HEB
2#1 HEC
2#1 HEH
2#1 HEL
2#1 HEN
2#1 HIB
2#1 HIE
2#1 HIT
2#1 HJK
2#1 HJP
2#1 HJQ
2#1 HJT
2#1 HJY
2#1 HKO
2#1 HKQ
2#1 HKR
2#1 HKT
2#1 HLB
2#1 HLR
2#1 HMH
2#1 HMW
2#1 HNK
2#1 HOJ
2#1 HOL
2#1 HOM
2#1 HOW
2#1 HPE
2#1 HPN
2#1 HPX
2#1 HQB
2#1 HQM
2#1 HQN
2#1 HQS
2#1 HRA
2#1 HRC
2#1 HRT
2#1 HRZ
2#1 HSF
2#1 HSK
2#1 HSQ
2#1 HSU
2#1 HSW
2#1 HTG
2#1 HTW
2#1 HTX
2#1 HVG
2#1 HWD
2#1 HWE
2#1 HXG
2#1 HYA
2#1 HYG
2#1 HYK
2#1 HYP
2#1 HYT
2#1 HYU
2#1 HYV
2#1 HZA
2#1 HZE
2#1 HZH
2#1 HZW
2#1 IAK
2#1 IAV
2#1 IBG
2#1 IBQ
2#1 IBW
2#1 IBY
2#1 ICA
2#1 ICB
2#1 ICH
2#1 ICK
2#1 ICY
2#1 ICZ
2#1 IDA
2#1 IDD
2#1 IDJ
2#1 IDK
2#1 IDL
2#1 IDN
2#1 IDP
2#1 IDQ
2#1 IDT
2#1 IDU
2#1 IEC
2#1 IED
2#1 IEH
2#1 IEU
2#1 IEW
2#1 IEY
2#1 IFC
2#1 IFG
2#1 IFH
2#1 IFK
2#1 IFO
2#1 IFP
2#1 IFS
2#1 IGD
2#1 IGE
2#1 IGH
2#1 IGN
2#1 IGW
2#1 IHC
2#1 IHH
2#1 IHI
2#1 IHL
2#1 IHP
2#1 IHZ
2#1 IIH
2#1 IIK
2#1 IIN
2#1 IIQ
2#1 IIS
2#1 IIY
2#1 IIZ
2#1 IJB
2#1 IJZ
2#1 IKC
2#1 IKD
2#1 IPA
2#1 IPS
2#1 IXG
2#1 JAJ
2#1 JAN
2#1 JAY
2#1 JBA
2#1 JBC
2#1 JDT
2#1 JFJ
2#1 JJE
2#1 JMK
2#1 JTL
2#1 KAC
2#1 KGB
2#1 KLM
2#1 KLO
2#1 LBH
2#1 LCY
2#1 LMW
2#1 LNJ
2#1 LOK
2#1 LOM
2#1 LOU
2#1 LTW
2#1 MCT
2#1 MDD
2#1 MDM
2#1 MEX
2#1 MHH
2#1 MJE
2#1 MJG
2#1 MMH
2#1 MNP
2#1 MOG
2#1 MPC
2#1 MPH
2#1 MTA
2#1 NAI
2#1 NBC
2#1 NBY
2#1 NOJ
2#1 NOP
2#1 NOW
2#1 NRJ
2#1 OAT
2#1 OCT
2#1 ODB
2#1 OKR
2#1 ONÿ
2#1 OOO
2#1 PAK
2#1 PAN
2#1 PAT
2#1 PEF
2#1 PJE
2#1 PJT
2#1 PKY
2#1 PRK
2#1 RAN
2#1 RBS
2#1 RCL
2#1 RCM
2#1 RDX
2#1 REK
2#1 RIC
2#1 RNY
2#1 ROY
2#1 RSX
2#1 RUS
2#1 RWH
2#1 SAM
2#1 SIM
2#1 SKJ
2#1 SLE
2#1 SPS
2#1 SRM
2#1 SSB
2#1 SWW
2#1 TAF
2#1 TEC
2#1 TER
2#1 TFJ
2#1 THX
2#1 TJH
2#1 TMP
2#1 USC
2#1 VAL
2#1 VHS
2#1 VIN
2#1 VIX
2#1 WCB
2#1 WND
2#1 WOW
2#1 WSR
2#1 WTG
2#1 WWC
2#1 YAP
2#1 YOT
2#1 ZFA

G0

G0 AAR
G0 AAV
G0 AAW
G0 ABY
G0 ABZ
G0 ACE
G0 ADR
G0 AEI
G0 AFB
G0 AFD
G0 AGH
G0 AGM
G0 AGQ
G0 AHN
G0 AJC
G0 AJQ
G0 AKA
G0 AKI
G0 AKQ
G0 ALD
G0 ALV
G0 AMB
G0 ANC
G0 ANS
G0 ANV
G0 AOU
G0 APE
G0 APM
G0 APQ
G0 APW
G0 AQL
G0 AQP
G0 ARC
G0 ARW
G0 ASC
G0 ASE
G0 ATR
G0 ATV
G0 AUO
G0 AUQ
G0 AVO
G0 AVR
G0 AVW
G0 AWD
G0 AWL
G0 AWV
G0 AXN
G0 AYE
G0 AYN
G0 AYR
G0 AYU
G0 AYZ
G0 AZS
G0 AZZ
G0 BBH
G0 BBS
G0 BBU
G0 BBZ
G0 BCK
G0 BDX
G0 BDY
G0 BEA
G0 BEH
G0 BEO
G0 BEQ
G0 BEU
G0 BFO
G0 BGC
G0 BGD
G0 BGN
G0 BGQ
G0 BGW
G0 BHD
G0 BHX
G0 BHY
G0 BIS
G0 BJF
G0 BJH
G0 BJO
G0 BJV
G0 BJW
G0 BJX
G0 BKO
G0 BLD
G0 BLN
G0 BMA
G0 BMB
G0 BMX
G0 BNC
G0 BND
G0 BNI
G0 BOB
G0 BPB
G0 BPI
G0 BQN
G0 BQT
G0 BRC
G0 BRP
G0 BRV
G0 BRY
G0 BSG
G0 BSS
G0 BSU
G0 BTO
G0 BTP
G0 BUS
G0 BUY
G0 BVH
G0 BVX
G0 BVZ
G0 BWS
G0 BWT
G0 BXT
G0 BYI
G0 BZI
G0 BZK
G0 BZY
G0 BZZ
G0 CAA
G0 CAC
G0 CAF
G0 CAR
G0 CAU
G0 CAW
G0 CBE
G0 CBH
G0 CCI
G0 CCK
G0 CCP
G0 CCR
G0 CCT
G0 CCW
G0 CDD
G0 CDF
G0 CDX
G0 CED
G0 CET
G0 CFL
G0 CFR
G0 CFS
G0 CFV
G0 CFY
G0 CGL
G0 CHT
G0 CIB
G0 CID
G0 CIF
G0 CIQ
G0 CIU
G0 CKG
G0 CKP
G0 CKT
G0 CKW
G0 CLB
G0 CLK
G0 CLN
G0 CLP
G0 CME
G0 CMF
G0 CMQ
G0 CNE
G0 CNH
G0 CNN
G0 CNR
G0 CNS
G0 COD
G0 CPK
G0 CPW
G0 CRH
G0 CRI
G0 CRQ
G0 CTO
G0 CTX
G0 CUD
G0 CUK
G0 CUQ
G0 CUT
G0 CVL
G0 CVT
G0 CVZ
G0 CWC
G0 CWM
G0 CWT
G0 CWV
G0 CXF
G0 CXM
G0 CXN
G0 CXP
G0 CXQ
G0 CXT
G0 CYA
G0 CYP
G0 CYT
G0 CZA
G0 CZB
G0 CZE
G0 CZP
G0 CZT
G0 DBH
G0 DCA
G0 DCB
G0 DCD
G0 DCK
G0 DCN
G0 DDD
G0 DDX
G0 DFK
G0 DFP
G0 DGV
G0 DGX
G0 DHN
G0 DIN
G0 DJE
G0 DJF
G0 DJH
G0 DJR
G0 DKL
G0 DLD
G0 DLO
G0 DLV
G0 DLY
G0 DMI
G0 DML
G0 DMO
G0 DNC
G0 DNE
G0 DNJ
G0 DNM
G0 DOI
G0 DON
G0 DOT
G0 DPB
G0 DPH
G0 DPZ
G0 DQL
G0 DQX
G0 DRC
G0 DTU
G0 DTX
G0 DUU
G0 DVF
G0 DVR
G0 DVV
G0 DVZ
G0 DWT
G0 DXA
G0 DXC
G0 DXD
G0 DXQ
G0 DXS
G0 DYB
G0 DYI
G0 DYZ
G0 EAC
G0 EAD
G0 EAK
G0 EAR
G0 EAZ
G0 EBB
G0 EBR
G0 EBU
G0 ECE
G0 ECO
G0 EED
G0 EES
G0 EFX
G0 EGX
G0 EGY
G0 EHF
G0 EHG
G0 EHH
G0 EHU
G0 EHY
G0 EIK
G0 EIO
G0 EIU
G0 EIX
G0 EJG
G0 EJW
G0 EKJ
G0 EKL
G0 EKP
G0 EKQ
G0 EKR
G0 ELE
G0 ELS
G0 ELT
G0 EME
G0 EMG
G0 ENC
G0 ENE
G0 ENG
G0 ENH
G0 ENR
G0 ENS
G0 EOA
G0 EOB
G0 EOC
G0 EOQ
G0 EOU
G0 EPG
G0 EPH
G0 EPX
G0 EPZ
G0 EQA
G0 EQG
G0 EQP
G0 EQZ
G0 ERI
G0 ERM
G0 ESB
G0 ESC
G0 ESE
G0 ESV
G0 ETB
G0 ETC
G0 ETN
G0 ETO
G0 ETY
G0 EUB
G0 EUF
G0 EUH
G0 EUI
G0 EUT
G0 EWM
G0 EWN
G0 EWS
G0 EXF
G0 EXH
G0 EXJ
G0 EXX
G0 EXZ
G0 EYJ
G0 EZA
G0 EZN
G0 EZP
G0 FAF
G0 FAI
G0 FAQ
G0 FAY
G0 FBI
G0 FBR
G0 FBY
G0 FCD
G0 FCP
G0 FDM
G0 FDN
G0 FEA
G0 FEF
G0 FEX
G0 FFH
G0 FFR
G0 FHM
G0 FHR
G0 FHV
G0 FII
G0 FIK
G0 FIL
G0 FIM
G0 FIR
G0 FIZ
G0 FJI
G0 FJT
G0 FJV
G0 FJY
G0 FKH
G0 FKL
G0 FKM
G0 FLB
G0 FLC
G0 FLN
G0 FLO
G0 FMM
G0 FMY
G0 FMZ
G0 FNK
G0 FNL
G0 FNY
G0 FOA
G0 FOD
G0 FOJ
G0 FOQ
G0 FOX
G0 FOZ
G0 FPD
G0 FPV
G0 FQB
G0 FQN
G0 FQT
G0 FRE
G0 FRJ
G0 FRK
G0 FRQ
G0 FSO
G0 FTD
G0 FTM
G0 FUQ
G0 FVG
G0 FVV
G0 FVX
G0 FWK
G0 FWL
G0 FWO
G0 FWU
G0 FWV
G0 FWX
G0 FXU
G0 FXW
G0 FXZ
G0 FYG
G0 FYN
G0 FYQ
G0 FYS
G0 FYV
G0 FZP
G0 GAC
G0 GAD
G0 GBD
G0 GBO
G0 GBX
G0 GCB
G0 GCD
G0 GCH
G0 GCL
G0 GCP
G0 GCV
G0 GCW
G0 GCY
G0 GDA
G0 GDO
G0 GDP
G0 GDU
G0 GDW
G0 GDX
G0 GDY
G0 GDZ
G0 GEM
G0 GEY
G0 GFX
G0 GGF
G0 GGU
G0 GGX
G0 GHI
G0 GHQ
G0 GIC
G0 GIG
G0 GIP
G0 GIQ
G0 GIV
G0 GIW
G0 GIY
G0 GJK
G0 GJP
G0 GJU
G0 GJZ
G0 GKW
G0 GLM
G0 GLT
G0 GMF
G0 GMG
G0 GMR
G0 GMT
G0 GMU
G0 GNL
G0 GNM
G0 GNR
G0 GNX
G0 GOA
G0 GPZ
G0 GQF
G0 GQM
G0 GQU
G0 GQZ
G0 GRA
G0 GRK
G0 GRN
G0 GRP
G0 GRT
G0 GRT
G0 GSD
G0 GTK
G0 GTM
G0 GTO
G0 GTR
G0 GTS
G0 GTU
G0 GUE
G0 GUH
G0 GUK
G0 GUP
G0 GVC
G0 GVI
G0 GVM
G0 GVQ
G0 GVY
G0 GWK
G0 GXL
G0 GYK
G0 GYZ
G0 GZS
G0 HAD
G0 HAF
G0 HAH
G0 HCO
G0 HCS
G0 HCT
G0 HDA
G0 HDN
G0 HEG
G0 HEH
G0 HFH
G0 HFN
G0 HFW
G0 HGK
G0 HGX
G0 HHG
G0 HHS
G0 HIA
G0 HIO
G0 HIQ
G0 HIS
G0 HIX

G0 HJQ
G0 HKA
G0 HKP
G0 HKS
G0 HKU
G0 HLD
G0 HLM
G0 HLN
G0 HLO
G0 HLR
G0 HLT
G0 HLX
G0 HMW
G0 HMY
G0 HMZ
G0 HNA
G0 HNB
G0 HNF
G0 HOK
G0 HOL
G0 HON
G0 HOO
G0 HOR
G0 HPO
G0 HPZ
G0 HQB
G0 HQD
G0 HQE
G0 HQJ
G0 HQW
G0 HRA
G0 HSO
G0 HSS
G0 HSZ
G0 HUO
G0 HUP
G0 HUS
G0 HUU
G0 HUZ
G0 HVL
G0 HVM
G0 HVQ
G0 HVU
G0 HVV
G0 HVW
G0 HWJ
G0 HWN
G0 HWW
G0 HXO
G0 HXT
G0 HXW
G0 HXX
G0 HXY
G0 HXZ
G0 HYF
G0 HYI
G0 HYY
G0 HZP
G0 HZR
G0 HZU
G0 HZV
G0 HZW
G0 HZX
G0 IAM
G0 IAN
G0 IAT
G0 IBB
G0 IBD
G0 IBP
G0 ICF
G0 ICI
G0 ICX
G0 ICZ
G0 IDA
G0 IEX
G0 IFB
G0 IFG
G0 IFI
G0 IFM
G0 IFP
G0 IFZ
G0 IHB
G0 IIR
G0 IIS
G0 IIZ
G0 IJJ
G0 IJM
G0 IJO
G0 IKA
G0 IKG
G0 IKH
G0 IKS
G0 IKV
G0 ILF
G0 ILG
G0 ILS
G0 ILV
G0 IMN
G0 IMT
G0 INP
G0 IOX
G0 IPS
G0 IPU
G0 IPY
G0 IQX
G0 IRG
G0 IRN
G0 IRO
G0 IRV
G0 IRW
G0 ISD
G0 ISV
G0 ISW
G0 ITA
G0 ITB
G0 IUE
G0 IUQ
G0 IVA
G0 IVM
G0 IWA
G0 IWM
G0 IWT
G0 IXB
G0 IXE
G0 IXI
G0 IXP
G0 IYF
G0 IYZ
G0 IZA
G0 IZM
G0 IZU
G0 IZX
G0 JAY
G0 JBD
G0 JBK
G0 JBW
G0 JCM
G0 JDF
G0 JDI
G0 JDJ
G0 JDK
G0 JDP
G0 JDU
G0 JED
G0 JEI
G0 JFN
G0 JGP
G0 JGZ
G0 JHA
G0 JHC
G0 JHF
G0 JHV
G0 JHY
G0 JIP
G0 JIZ
G0 JJL
G0 JKB
G0 JKN
G0 JKQ
G0 JKV
G0 JKX
G0 JKZ
G0 JLO
G0 JLW
G0 JLY
G0 JMC
G0 JMP
G0 JMT
G0 JMV
G0 JMX
G0 JNH
G0 JNL
G0 JNN
G0 JNP
G0 JNS
G0 JOH
G0 JOO
G0 JOW
G0 JOY
G0 JPA
G0 JPN
G0 JPO
G0 JPS
G0 JQG
G0 JQJ
G0 JQL
G0 JQM
G0 JQN
G0 JQO
g0 jrk
G0 JRP
G0 JSB
G0 JSH
G0 JSS
G0 JSV
G0 JTB
G0 JUU
G0 JUW
G0 JUX
G0 JVA
G0 JVY
G0 JWB
G0 JWQ
G0 JXF
G0 JXM
G0 JYB
G0 JYM
G0 JYO
G0 JYP
G0 JYT
G0 JZG
G0 JZI
G0 JZP
G0 KAC
G0 KAD
G0 KAE
G0 KAF
G0 KAG
G0 KAI
G0 KAJ
G0 KAL
G0 KAO
G0 KAR
G0 KAW
G0 KBE
G0 KCQ
G0 KDJ
G0 KEH
G0 KEN
G0 KET
G0 KFA
G0 KFO
G0 KFX
G0 KGH
G0 KGK
G0 KGO
G0 KGV
G0 KHB
G0 KHC
G0 KHD
G0 KHG
G0 KHO
G0 KHV
G0 KHW
G0 KIJ
G0 KIS
G0 KIW
G0 KIZ
G0 KJV
G0 KJY
G0 KKB
G0 KKI
G0 KKJ
G0 KLH
G0 KLI
G0 KLM
G0 KLR
G0 KLX
G0 KME
G0 KMX
G0 KNK
G0 KOA
G0 KOR
G0 KPK
G0 KPM
G0 KPR
G0 KQL
G0 KQZ
G0 KRA
G0 KSB
G0 KSE
G0 KST
G0 KTM
G0 KUM
G0 KUN
G0 KUO
G0 KVH
G0 KVL
G0 KVT
G0 KVY
G0 KVZ
G0 KWT
G0 KWV
G0 KXA
G0 KXE
G0 KXM
G0 KYC
G0 KZC
G0 KZZ
G0 LAH
G0 LAI
G0 LAT
G0 LBK
G0 LBW
G0 LCL
G0 LCQ
G0 LCW
G0 LDD
G0 LDM
G0 LDW
G0 LEG
G0 LET
G0 LFK
G0 LFU
G0 LFW
G0 LGD
G0 LGN
G0 LGS
G0 LGT
G0 LGU
G0 LHF
G0 LHK
G0 LHT
G0 LIF
G0 LIJ
G0 LIL
G0 LIX
G0 LJS
G0 LJT
G0 LJX
G0 LKM
G0 LKQ
G0 LKZ
G0 LLW
G0 LMF
G0 LMI
G0 LMM
G0 LMO
G0 LMT
G0 LNR
G0 LNY
G0 LOG
G0 LOQ
G0 LOU
G0 LOV
G0 LPO
G0 LPS
G0 LQP
G0 LQY
G0 LRD
G0 LRH
G0 LRX
G0 LRY
G0 LSB
G0 LSH
G0 LTG
G0 LTN
G0 LTU
G0 LTW
G0 LUJ
G0 LUS
G0 LUZ
G0 LVB
G0 LVN
G0 LVQ
G0 LVV
G0 LVZ
G0 LWH
G0 LWK
G0 LWS
G0 LWX
G0 LWY
G0 LWZ
G0 LXA
G0 LXJ
G0 LXS
G0 LXZ
G0 LYL
G0 LYV
G0 LZB
G0 LZO
G0 MAB
G0 MAO
G0 MAV
G0 MBE
G0 MBT
G0 MCU
G0 MDA
G0 MDG
G0 MDI
G0 MDL
G0 MEG
G0 MEJ
G0 MEK
G0 MEL
G0 MEM
G0 MEP
G0 MES
G0 MEU
G0 MGA
G0 MGE
G0 MGF
G0 MGK
G0 MHR
G0 MIN
G0 MIQ
G0 MIY
G0 MJI
G0 MJM
G0 MJN
G0 MJQ
G0 MJS
G0 MJW
G0 MKB
G0 MKI
G0 MKJ
G0 MKM
G0 MKT
G0 MLD
G0 MLG
G0 MLS
G0 MMN
G0 MMR
G0 MNE
G0 MNM
G0 MOD
G0 MOP
G0 MPD
G0 MPG
G0 MPX
G0 MQA
G0 MQF
G0 MQG
G0 MQP
G0 MQS
G0 MRV
G0 MSL
G0 MSP
G0 MTR
G0 MTS
G0 MTX
G0 MTZ
G0 MUO
G0 MUT
G0 MUU
G0 MUV
G0 MUW
G0 MUY
G0 MWR
G0 MXF
G0 MXN
G0 MYW
G0 MYX
G0 MYY
G0 MZB
G0 MZS
G0 MZT
G0 MZV
G0 MZX
G0 NAB
G0 NAN
G0 NAV
G0 NBF
G0 NBM
G0 NBO
G0 NCF
G0 NCM
G0 NCN
G0 NCR
G0 NDN
G0 NDT
G0 NEH
G0 NEW
G0 NEY
G0 NFQ
G0 NGL
G0 NGT
G0 NGU
G0 NGV
G0 NGX
G0 NHF
G0 NHN
G0 NHY
G0 NIH
G0 NII
G0 NIM
G0 NIS
G0 NIV
G0 NJF
G0 NJM
G0 NJY
G0 NKE
G0 NKR
G0 NKV
G0 NLE
G0 NLR
G0 NLS
G0 NME
G0 NMK
G0 NML
G0 NMQ
G0 NMT
G0 NND
G0 NNY
G0 NOK
G0 NOS
G0 NOT
G0 NOY
G0 NPH
G0 NQZ
G0 NRG
G0 NRL
G0 NTE
G0 NTK
G0 NTM
G0 NTO
G0 NTP
G0 NTQ
G0 NTX
G0 NVF
G0 NVN
G0 NVP
G0 NVW
G0 NWB
G0 NWX
G0 NYG
G0 NZO
G0 NZQ
G0 NZW
G0 OAH
G0 OAI
G0 OAK
G0 OAN
G0 OAQ
G0 OAY
G0 OBS
G0 OCH
G0 OCI
G0 ODF
G0 ODL
G0 ODV
G0 OEF
G0 OEP
G0 OFC
G0 OFJ
G0 OGO
G0 OGT
G0 OGU
G0 OHB
G0 OHE
G0 OHI
G0 OIL
G0 OJA
G0 OJD
G0 OJH
G0 OJI
G0 OJZ
G0 OKB
G0 OKR
G0 OKS
G0 OKW
G0 OLB
G0 OLH
G0 OLI
G0 OMG
G0 OMK
G0 OMP
G0 OMS
G0 OMV
G0 ONC
G0 OOG
G0 OOH
G0 OOM
G0 OOT
G0 OOW
G0 OOY
G0 OPD
G0 OPD
G0 OPH
G0 OPR
G0 OPS
G0 OPW
G0 OQB
G0 OQC
G0 OQD
G0 OQJ
G0 OQM
G0 OQU
G0 OQV
G0 OQW
G0 ORA
G0 ORH
G0 ORI
G0 OSY
G0 OTA
G0 OTO
G0 OUF
G0 OUI
G0 OUQ
G0 OVI
G0 OVU
G0 OXI
G0 OXJ
G0 OXO
G0 OXQ
G0 OYB
G0 OZK
G0 OZU
G0 OZW
G0 OZX
G0 OZY
G0 OZZ
G0 PAT
G0 PCL
G0 PDL
G0 PDR
G0 PDS
G0 PDU
G0 PFC
G0 PGE
G0 PHS
G0 PHU
G0 PIF
G0 PIG
G0 PIH
G0 PIM
G0 PIP
G0 PJK
G0 PJX
G0 PKH
G0 PKS
G0 PKU
G0 PLM
G0 PLN
G0 PLP
G0 PLV
G0 PMA
G0 PMH
G0 PMJ
G0 PNY
G0 PNZ
G0 POH
G0 POI
G0 PPB
G0 PPF
G0 PPN
G0 PPO
G0 PPP
G0 PQE
G0 PQK
G0 PQQ
G0 PQT
G0 PRT
G0 PSM
G0 PSN
G0 PSU
G0 PTJ
G0 PTS
G0 PUA
G0 PUB
G0 PUF
G0 PUH
G0 PUU
G0 PVK
G0 PWR
G0 PWZ
G0 PYB
G0 PYN
G0 PYT
G0 PYZ
G0 PZA
G0 PZE
G0 PZI
G0 RAB
G0 RAH
G0 RAP
G0 RAY
G0 RAZ
G0 RBE
G0 RCC
G0 RCK
G0 RCM
G0 RCZ
G0 RDI
G0 RDQ
G0 RDW
G0 REJ
G0 REM
G0 RES
G0 REW
G0 RFD
G0 RFO
G0 RGB
G0 RGF
G0 RGR
G0 RGV
G0 RGY
G0 RHH
G0 RHT
G0 RIN
G0 RIO
G0 RIT
G0 RIV
G0 RIV
G0 RJP
G0 RKF
G0 RKL
G0 RLR
G0 RMA
G0 RME
G0 RMQ
G0 RMZ
G0 RNE
G0 RNG
G0 RNR
G0 RNT
G0 RNU
G0 ROG
G0 ROI
G0 ROJ
G0 ROK
G0 ROV
G0 RPH
G0 RPP
G0 RPR
G0 RQK
G0 RRA
G0 RSH
G0 RSN
G0 RSP
G0 RSQ
G0 RTD
G0 RTG
G0 RTK
G0 RTX
G0 RUE
G0 RUJ
G0 RUP
G0 RVA
G0 RVF
G0 RVG
G0 RVJ
G0 RVL
G0 RVN
G0 RVT
G0 RVU
G0 RVY
G0 RWB
G0 RWD
G0 RWF
G0 RWK
G0 RWR
G0 RXG
G0 RXH
G0 RXL
G0 RXN
G0 RXT
G0 RXW
G0 RXY
G0 RYC
G0 RYF
G0 RYH
G0 RYN
G0 RYY
G0 RZC
G0 RZE
G0 RZF
G0 RZJ
G0 RZR
G0 RZV
G0 SAA
G0 SAG
G0 SAM
G0 SAN
G0 SAP
G0 SAQ
G0 SAS
G0 SAW
G0 SBD
G0 SBE
G0 SBN
G0 SBS
G0 SBW
G0 SCF
G0 SDI
G0 SDN
G0 SDO
G0 SDP
G0 SDV
G0 SDY
G0 SEA
G0 SEM
G0 SEV
G0 SFF
G0 SFH
G0 SFL
G0 SFR
G0 SGB
G0 SGN
G0 SGO
G0 SHE
G0 SHQ
G0 SHW
G0 SIC
G0 SID
G0 SIF
G0 SIP
G0 SIT
G0 SIZ
G0 SJI
G0 SJK
G0 SJY
G0 SKB
G0 SKE
G0 SKO
G0 SKP
G0 SKT
G0 SKU
G0 SLO
G0 SLV
G0 SMA
G0 SMB
G0 SMD
G0 SMG
G0 SMI
G0 SML
G0 SMX
G0 SNC
G0 SNE
G0 SNJ
G0 SOB
G0 SOB
G0 SOJ
G0 SOL
G0 SOM
G0 SOO
G0 SOP
G0 SOR
G0 SOT
G0 SOZ
G0 SP
G0 SPE
G0 SPG
G0 SPR
G0 SQZ
G0 SRM
G0 SSB
G0 SSI
G0 SSO
G0 STE
G0 STG
G0 STN
G0 STQ
G0 STU
G0 STV
G0 SUG
G0 SUJ
G0 SUM
G0 SVG
G0 SVT
G0 SWG
G0 SWP
G0 SWQ
G0 SWR
G0 SXB
G0 SXC
G0 SXF
G0 SXH
G0 SYA
G0 SYJ
G0 SYM
G0 SYO
G0 SYU
G0 SYW
G0 SZP
G0 SZR
G0 SZS
G0 TAP
G0 TAU
G0 TBA
G0 TBK
G0 TBN
G0 TBR
G0 TBT
G0 TBX
G0 TCD
G0 TCX
G0 TCZ
G0 TDB
G0 TDD
G0 TDF
G0 TDI
G0 TDY
G0 TEC
G0 TEQ
G0 TER
G0 TET
G0 TEY
G0 TEZ
G0 TFH
G0 TFS
G0 TFW
G0 TGC
G0 TGD
G0 TGI
G0 TGJ
G0 TGK
G0 TGT
G0 TGW
G0 TGY
G0 TGZ
G0 THG
G0 THM
G0 THZ
G0 TIK
G0 TIM
G0 TIQ
G0 TIY
G0 TJB
G0 TJZ
G0 TKD
G0 TKO
G0 TKP
G0 TKV
G0 TLB
G0 TLC
G0 TLD
G0 TLE
G0 TLG
G0 TLK
G0 TLM
G0 TLY
G0 TMB
G0 TMC
G0 TMD
G0 TMP
G0 TMX
G0 TMY
G0 TNI
G0 TNJ
G0 TNR
G0 TNT
G0 TNV
G0 TOG
G0 TOR
G0 TOU
G0 TOV
G0 TOW
G0 TPC
G0 TPS
G0 TPT
G0 TPU
G0 TPV
G0 TPX
G0 TPZ
G0 TQM
G0 TQN
G0 TQO
G0 TRF
G0 TRT
G0 TRX
G0 TRZ
G0 TSM
G0 TSP
G0 TSW
G0 TTA
G0 TTD
G0 TTJ
G0 TTU
G0 TTV
G0 TTX
G0 TUA
G0 TUB
G0 TUD
G0 TUH
G0 TUK
G0 TUY
G0 TVE
G0 TVF
G0 TVH
G0 TVI
G0 TVJ
G0 TVK
G0 TVP
G0 TVY
G0 TWN
G0 TWW
G0 TXE
G0 TXI
G0 TYH
G0 TYX
G0 TZA
G0 TZE
G0 TZK
G0 TZN
G0 TZS
G0 UAE
G0 UAL
G0 UAM
G0 UAN
G0 UAT
G0 UAX
G0 UAZ
G0 UBC
G0 UBD
G0 UBE
G0 UBQ
G0 UBV
G0 UCR
G0 UDA
G0 UDC
G0 UDE
G0 UDF
G0 UDK
G0 UDM
G0 UDX
G0 UEI
G0 UEJ
G0 UEN
G0 UER
G0 UES
G0 UEX
G0 UFH
G0 UFK
G0 UGC
G0 UGE
G0 UGF
G0 UGO
G0 UGT
G0 UHA
G0 UHE
G0 UHY
G0 UIB
G0 UII
G0 UIK
G0 UIM
G0 UIN
G0 UIS
G0 UJA
G0 UJK
G0 UKB
G0 UKI
G0 UKN
G0 UKR
G0 ULE
G0 ULK
G0 ULR
G0 ULV
G0 UMF
G0 UNJ
G0 UNO
G0 UNU
G0 UOA
G0 UOG
G0 UOH
G0 UON
G0 UOO
G0 UOR
G0 UPA
G0 UPC
G0 UPQ
G0 UPX
G0 UQN
G0 UQR
G0 UQX
G0 URE
G0 URG
G0 URJ
G0 URM
G0 URR
G0 URY
G0 USD
G0 USH
G0 USL
G0 USP
G0 USR
G0 USU
G0 USX
G0 UTL
G0 UTS
G0 UTY
G0 UUK
G0 UVB
G0 UVM
G0 UWG
G0 UWM
G0 UWY
G0 UXA
G0 UXC
G0 UXE
G0 UXM
G0 UYB
G0 UYL
G0 UYN
G0 UZC
G0 UZG
G0 UZJ
G0 UZL
G0 VAK
G0 VAW
G0 VBA
G0 VBC
G0 VBI
G0 VBJ
G0 VBS
G0 VCA
G0 VCE
G0 VCG
G0 VCH
G0 VCL
G0 VCM
G0 VCP
G0 VCQ
G0 VCS
G0 VDG
G0 VDI
G0 VEA
G0 VFG
G0 VFI
G0 VFK
G0 VFR
G0 VGF
G0 VGQ
G0 VGS
G0 VGX
G0 VGZ
G0 VHB
G0 VHM
G0 VHS
G0 VHZ
G0 VIH
G0 VIN
G0 VIP
G0 VIR
G0 VIU
G0 VIW
G0 VIZ
G0 VJD
G0 VJG
G0 VJV
G0 VJW
G0 VKP
G0 VKQ
G0 VKT
G0 VLG
G0 VLH
G0 VLN
G0 VLO
G0 VMG
G0 VML
G0 VMX
G0 VNG
G0 VNP
G0 VNR
G0 VNS
G0 VNZ
G0 VOM
G0 VOW
G0 VOY
G0 VOZ
G0 VPF
G0 VPL
G0 VPM
G0 VPR
G0 VRD
G0 VRI
G0 VRN
G0 VRR
G0 VSV
G0 VTH
G0 VTR
G0 VTU
G0 VTX
G0 VTZ
G0 VUF
G0 VUO
G0 VUV
G0 VUW
G0 VUZ
G0 VVE
G0 VVM
G0 VVN
G0 VVS
G0 VWC
G0 VWG
G0 VWL
G0 VWR
G0 VWS
G0 VWY
G0 VXN
G0 VXO
G0 VYH
G0 VYO
G0 VYS
G0 VZG
G0 VZJ
G0 VZM
G0 VZW
G0 VZY
G0 WAF
G0 WAG
G0 WAI
G0 WAJ
G0 WAK
G0 WAP
G0 WAU
G0 WAV
G0 WBE
G0 WBJ
G0 WBX
G0 WBY
G0 WCA
G0 WCF
G0 WCG
G0 WCP
G0 WCW
G0 WCY
G0 WDA
G0 WDD
G0 WDM
G0 WEC
G0 WEE
G0 WEH
G0 WEJ
G0 WEN
G0 WFJ
G0 WFS
G0 WFW
G0 WFY
G0 WGC
G0 WGF
G0 WGK
G0 WGQ
G0 WHA
G0 WHE
G0 WHF
G0 WHG
G0 WHI
G0 WHJ
G0 WHM
G0 WHP
G0 WHS
G0 WHX
G0 WIU
G0 WIV
G0 WJE
G0 WJQ
G0 WKB
G0 WKC
G0 WKD
G0 WKG
G0 WKK
G0 WKR
G0 WKV
G0 WLB
G0 WLL
G0 WLO
G0 WLS
G0 WLW
G0 WMF
G0 WMM
G0 WNA
G0 WNC
G0 WNH
G0 WNO
G0 WNQ
G0 WNV
G0 WOD
G0 WOE
G0 WOK
G0 WOZ
G0 WPA
G0 WPB
G0 WPG
G0 WPJ
G0 WPN
G0 WPP
G0 WPR
G0 WQB
G0 WQI
G0 WQJ
G0 WQJ
G0 WQK
G0 WQL
G0 WQS
G0 WQT
G0 WQU
G0 WQV
G0 WRB
G0 WRF
G0 WRJ
G0 WRO
G0 WRP
G0 WSE
G0 WSG
G0 WSM
G0 WSO
G0 WSQ
G0 WST
G0 WSV
G0 WSW
G0 WTH
G0 WTT
G0 WTZ
G0 WUD
G0 WUE
G0 WUQ
G0 WUT
G0 WVK
G0 WVL
G0 WVU
G0 WVX
G0 WWG
G0 WWV
G0 WWW
G0 WXB
G0 WXI
G0 WXN
G0 WXS
G0 WXT
G0 WXU
G0 WYL
G0 WZI
G0 WZR
G0 XAX
G0 XBI
G0 XBJ
G0 XBU
G0 XEL
G0 XEM
G0 XGK
G0 XJB
G0 XRO
G0 XTV
G0 XXL
G0 XYS
G0 YAF
G0 YAK
G0 YAP
G0 YEW
G0 YLM
G0 YNM
G0 YOY
G0 ZAA
G0 ZEN
G0 ZGB
G0 ZIG

G1

G1 AAA
G1 AAE
G1 AAS
G1 ABR
G1 ABY
G1 ACH
G1 ADD
G1 AED
G1 AEG
G1 AES
G1 AEV
G1 AGI
G1 AGT
G1 AHA
G1 AIH
G1 AIV
G1 AJH
G1 AJI
G1 AJR
G1 AKN
G1 AKR
G1 AKT
G1 AKX
G1 ALT
G1 AMI
G1 AMV
G1 AMW
G1 ANT
G1 AOU
G1 AOY
G1 APV
G1 AQQ
G1 ARG
G1 ARI
G1 ARN
G1 ARV
G1 ASE
G1 AST
G1 ATI
G1 AUJ
G1 AUN
G1 AUQ
G1 AVI
G1 AVQ
G1 AYL
G1 BAD
G1 BAF
G1 BCL
G1 BCP
G1 BCR
G1 BCZ
G1 BDK
G1 BED
G1 BEI
G1 BEM
G1 BFQ
G1 BFU
G1 BFZ
G1 BGY
G1 BHM
G1 BIQ
G1 BIR
G1 BJC
G1 BJH
G1 BJJ
G1 BKH
G1 BLD
G1 BMF
G1 BND
G1 BNH
G1 BOE
G1 BOP
G1 BOZ
G1 BPH
G1 BPR
G1 BQC
G1 BRM
G1 BRV
G1 BSN
G1 BSP
G1 BSQ
G1 BSR
G1 BTU
G1 BUO
G1 BVI
G1 BWE
G1 BWT
G1 BXB
G1 BXG
G1 BXL
G1 BYY
G1 BZB
G1 BZG
G1 CAB
G1 CAE
G1 CAJ
G1 CAT
G1 CBD
G1 CCC
G1 CDV
G1 CEB
G1 CEH
G1 CFM
G1 CHB
G1 CIK
G1 CKL
G1 CLG
G1 CLH
G1 CLL
G1 CMB
G1 CMU
G1 COR
G1 COT
G1 CPP
G1 CPQ
G1 CQB
G1 CQL
G1 CQN
G1 CQO
G1 CQP
G1 CQV
G1 CSC
G1 CSL
G1 CSM
G1 CSU
G1 CUX
G1 CWC
G1 CWN
G1 CWO
G1 CXA
G1 CXD
G1 CXG
G1 CYB
G1 DBC
G1 DBD
G1 DCG
G1 DDH
G1 DDY
G1 DEA
G1 DET
G1 DFL
G1 DHD
G1 DHE
G1 DIQ
G1 DIW
G1 DJA
G1 DJG
G1 DJT
G1 DKA
G1 DLC
G1 DLL
G1 DMQ
G1 DMX
G1 DOK
G1 DOS
G1 DOW
G1 DQR
G1 DQX
G1 DRF
G1 DRP
G1 DRX
G1 DSC
G1 DST
G1 DTV
G1 DUE
G1 DUH
G1 DUM
G1 DVA
G1 DVW
G1 DVX
G1 DWG
G1 DWH
G1 DWN
G1 DX
G1 DXS
G1 DYD
G1 DYE
G1 DYX
G1 EAG
G1 EAY
G1 ECA
G1 ECL
G1 ECO
G1 ECZ
G1 EEL
G1 EEP
G1 EEQ
G1 EEU
G1 EFI
G1 EFY
G1 EGC
G1 EHN
G1 EHT
G1 EIQ
G1 EJE
G1 EJX
G1 EJY
G1 EKH
G1 EKW
G1 ELG
G1 EMU
G1 ENI
G1 ENX
G1 EQT
G1 ERJ
G1 ERW
G1 ESV
G1 ETH
G1 EUZ
G1 EWA
G1 EWP
G1 EWQ
G1 EWV
G1 EYO
G1 EZG
G1 EZV
G1 FBH
G1 FBK
G1 FCG
G1 FDB
G1 FDK
G1 FDY
G1 FEV
G1 FGB
G1 FGT
G1 FHB
G1 FJQ
G1 FLQ
G1 FMD
G1 FML
G1 FNR
G1 FOX
G1 FPS
G1 FPU
G1 FQE
G1 FQR
G1 FQV
G1 FSH
G1 FTF
G1 FTO
G1 FTQ
G1 FUA
G1 FUD
G1 FXE
G1 FXJ
G1 FYG
G1 FYX
G1 GAB
G1 GBL
G1 GBO
G1 GBQ
G1 GCT
G1 GCU
G1 GCX
G1 GDS
G1 GFJ
G1 GGC
G1 GGH
G1 GHF
G1 GHM
G1 GIB
G1 GJC
G1 GJF
G1 GJM
G1 GKU
G1 GKY
G1 GLA
G1 GOE
G1 GRD
G1 GRG
G1 GRP
G1 GRW
G1 GTI
G1 GTO
G1 GTV
G1 GUE
G1 GUT
G1 GVE
G1 GVI
G1 GWH
G1 GWL
G1 GWN
G1 GWV
G1 GXM
G1 GZY
G1 HAG
G1 HAK
G1 HAN
G1 HEH
G1 HFC
G1 HFD
G1 HHL
G1 HIZ
G1 HJH
G1 HJW
G1 HKE
G1 HLF
G1 HLW
G1 HMA
G1 HMO
G1 HMP
G1 HNI
G1 HOK
G1 HOS
G1 HPD
G1 HRE
G1 HRI
G1 HRW
G1 HSN
G1 HSS
G1 HUI
G1 HUL
G1 HUP
G1 HUW
G1 HWH
G1 HWN
G1 HXM
G1 HYD
G1 HZA
G1 HZW
G1 IAH
G1 IAR
G1 IAT
G1 IBH
G1 ICO
G1 IDN
G1 IFL
G1 IGK
G1 IGZ
G1 IID
G1 IIE
G1 IIN
G1 IKO
G1 ILB
G1 ILN
G1 ILW
G1 IMH
G1 IMJ
G1 IML
G1 IMN
G1 IMQ
G1 INC
G1 INF
G1 INH
G1 IOJ
G1 IOS
G1 IOV
G1 IPM
G1 IQZ
G1 IRB
G1 IRP
G1 ISG
G1 ISO
G1 IUN
G1 IVE
G1 IWX
G1 IXY
G1 IYM
G1 IYU
G1 IZF
G1 IZQ
G1 JAJ
G1 JAK
G1 JAN
G1 JAR
G1 JAS
G1 JBD
G1 JDY
G1 JET
G1 JEV
G1 JFV
G1 JFW
G1 JFX
G1 JGQ
G1 JHI
G1 JHL
G1 JJJ
G1 JLE
G1 JME
G1 JMO
G1 JPL
G1 JPV
G1 JQI
G1 JQJ
G1 JQW
G1 JRG
G1 JRH
G1 JRT
G1 JSA
G1 JSF
G1 JSH
G1 JSZ
G1 JTD
G1 JTG
G1 JTO
G1 JTY
G1 JUX
G1 JWZ
G1 JXI
G1 JZK
G1 JZQ
G1 KAJ
G1 KAN
G1 KAW
G1 KBU
G1 KBV
G1 KCE
G1 KDB
G1 KDC
G1 KDQ
G1 KDU
G1 KEA
G1 KEJ
G1 KEK
G1 KEX
G1 KFC
G1 KFD
G1 KHL
G1 KIK
G1 KMA
G1 KMO
G1 KNP
G1 KOO
G1 KPS
G1 KQJ
G1 KRG
G1 KRI
G1 KTJ
G1 KTN
G1 KUL
G1 KVD
G1 KVM
G1 KVN
G1 KWJ
G1 KXF
G1 KXW
G1 KXY
G1 KYO
G1 KZL
G1 LAJ
G1 LCX
G1 LCZ
G1 LDI
G1 LDM
G1 LFU
G1 LGA
G1 LGX
G1 LHQ
G1 LIE
G1 LIF
G1 LII
G1 LIQ
G1 LIS
G1 LIW
G1 LIZ
G1 LJA
G1 LJB
G1 LJD
G1 LJR
G1 LJU
G1 LJY
G1 LLB
G1 LNT
G1 LNY
G1 LOA
G1 LOP
G1 LOR
G1 LOT
G1 LPW
G1 LQN
G1 LRO
G1 LRO
G1 LSJ
G1 LUS
G1 LUV
G1 LVN
G1 LWA
G1 LXI
G1 LXP
G1 LXR
G1 LXW
G1 MAK
G1 MBT
G1 MCD
G1 MCP
G1 MCX
G1 MDK
G1 MEZ
G1 MFF
G1 MFG
G1 MFN
G1 MHU
G1 MIB
G1 MIC
G1 MII
G1 MIK
G1 MIL
G1 MIR
G1 MIS
G1 MJD
G1 MKH
G1 MKJ
G1 MLK
G1 MLO
G1 MLR
G1 MME
G1 MMS
G1 MNP
G1 MPH
G1 MSE
G1 MSU
G1 MTC
G1 MTT
G1 MTY
G1 MUK
G1 MVL
G1 MWI
G1 MWP
G1 MXJ
G1 MXJ
G1 MXS
G1 NAF
G1 NAG
G1 NAR
G1 NBQ
G1 NCI
G1 NCS
G1 NDA
G1 NDG
G1 NEJ
G1 NES
G1 NFF
G1 NFT
G1 NGH
G1 NHB
G1 NHH
G1 NHW
G1 NIB
G1 NID
G1 NIG
G1 NIS
G1 NKC
G1 NKU
G1 NLD
G1 NLI
G1 NLV
G1 NLX
G1 NMH
G1 NNP
G1 NOC
G1 NOD
G1 NOL
G1 NON
G1 NOX
G1 NPT
G1 NPV
G1 NQG
G1 NQP
G1 NQW
G1 NRL
G1 NRS
G1 NRT
G1 NRU
G1 NRV
G1 NSA
G1 NSU
G1 NTE
G1 NTH
G1 NTY
G1 NUC
G1 NUJ
G1 NUN
G1 NUR
G1 NUU
G1 NVQ
G1 NYN
G1 NZB
G1 NZM
G1 NZO
G1 NZT
G1 OAC
G1 OAH
G1 OAT
G1 OAV
G1 OAY
G1 OBD
G1 OBF
G1 OBY
G1 OCA
G1 OCJ
G1 OCN
G1 OEN
G1 OEX
G1 OFA
G1 OFZ
G1 OGU
G1 OHN
G1 OHS
G1 OHT
G1 OHZ
G1 OIF
G1 OIN
G1 OIP
G1 OJJ
G1 OJP
G1 OKD
G1 OKT
G1 OLJ
G1 OMK
G1 OMO
G1 OMQ
G1 ONL

G1 ONN
G1 ONY
G1 OOE
G1 OOV
G1 OPE
G1 OPH
G1 OPK
G1 ORY
G1 OTR
G1 OUR
G1 OUT
G1 OVA
G1 OVR
G1 OWP
G1 OXD
G1 OXL
G1 OXX
G1 OYQ
G1 OYS
G1 OYW
G1 OYY
G1 OZF
G1 OZT
G1 OZU
G1 OZZ
G1 PAC
G1 PBO
G1 PCM
G1 PCS
G1 PCW
G1 PCX
G1 PDZ
G1 PEP
G1 PFP
G1 PFQ
G1 PGF
G1 PHL
G1 PHQ
G1 PHZ
G1 PIB
G1 PIC
G1 PIE
G1 PIR
G1 PJW
G1 PKY
G1 PKZ
G1 PLB
G1 PLM
G1 PLT
G1 PLY
G1 PMN
G1 PMR
G1 PMS
G1 PNP
G1 POS
G1 PPA
G1 PPC
G1 PPH
G1 PPP
G1 PRA
G1 PRB
G1 PRX
G1 PSP
G1 PSP
G1 PTH
G1 PTO
G1 PVO
G1 PVY
G1 PWD
G1 PWW
G1 PXH
G1 PXJ
G1 PYA
G1 PYT
G1 RAH
G1 RBK
G1 RBM
G1 RCD
G1 RCL
G1 RCP
G1 RCQ
G1 RCT
G1 REQ
G1 RFH
G1 RFO
G1 RHV
G1 RIJ
G1 RKF
G1 RKO
G1 RLP
G1 RMH
G1 RNA
G1 ROI
G1 ROL
G1 RRD
G1 RRM
G1 RSB
G1 RSH
G1 RSL
G1 RSW
G1 RSZ
G1 RTG
G1 RTJ
G1 RTN
G1 RUN
G1 RUV
G1 RUX
G1 RUY
G1 RVD
G1 RVE
G1 RVU
G1 RVV
G1 RVX
G1 RWA
G1 RWB
G1 RWK
G1 RWS
G1 RWU
G1 RXC
G1 RXR
G1 RYO
G1 RZT
G1 RZV
G1 SAA
G1 SAN
G1 SAO
G1 SBB
G1 SBG
G1 SBO
G1 SBV
G1 SCP
G1 SCX
G1 SDC
G1 SDD
G1 SDT
G1 SE
G1 SEN
G1 SFI
G1 SFN
G1 SFO
G1 SFS
G1 SGC
G1 SGX
G1 SHC
G1 SHM
G1 SHV
G1 SHX
G1 SJF
G1 SKL
G1 SLN
G1 SLW
G1 SMA
G1 SMG
G1 SMX
G1 SNH
G1 SNJ
G1 SNL
G1 SNW
G1 SOA
G1 SOU
G1 SPC
G1 SPO
G1 SPQ
G1 SQB
G1 SQH
G1 SQY
G1 SRX
G1 SRZ
G1 SSO
G1 SSR
G1 SST
G1 STE
G1 SUI
G1 SUT
G1 SUV
G1 SWN
G1 SWY
G1 SXO
G1 SXV
G1 SXW
G1 SXZ
G1 SYG
G1 SYS
G1 TAQ
G1 TBA
G1 TBC
G1 TBL
G1 TCM
G1 TDB
G1 TDG
G1 TDM
G1 TED
G1 TEJ
G1 TFI
G1 TFV
G1 TGC
G1 TGJ
G1 TGM
G1 TKO
G1 TLB
G1 TLF
G1 TLG
G1 TLQ
G1 TLY
G1 TMJ
G1 TMN
G1 TNG
G1 TNL
G1 TOA
G1 TOP
G1 TOU
G1 TOX
G1 TPA
G1 TQC
G1 TQP
G1 TQW
G1 TRB
G1 TRD
G1 TSE
G1 TSJ
G1 TSQ
G1 TSY
G1 TSZ
G1 TTM
G1 TTR
G1 TTW
G1 TTY
G1 TUK
G1 TUO
G1 TVG
G1 TVJ
G1 TVL
G1 TXC
G1 TXD
G1 TXY
G1 TYH
G1 TYV
G1 TYY
G1 TZC
G1 TZG
G1 TZJ
G1 UAI
G1 UAJ
G1 UAN
G1 UAS
G1 UAU
G1 UBA
G1 UBB
G1 UBW
G1 UCA
G1 UCP
G1 UCW
G1 UDA
G1 UDH
G1 UDK
G1 UDL
G1 UDO
G1 UEB
G1 UEG
G1 UFN
G1 UFQ
G1 UGM
G1 UGY
G1 UHD
G1 UHE
G1 UHI
G1 UHQ
G1 UHU
G1 UHY
G1 UHZ
G1 UIF
G1 UJB
G1 UJO
G1 UJU
G1 UJV
G1 UJW
G1 UKE
G1 UKI
G1 UKQ
G1 UKR
G1 UKU
G1 UKX
G1 UKY
G1 ULN
G1 ULS
G1 UME
G1 UMH
G1 UML
G1 UMR
G1 UMV
G1 UMX
G1 UND
G1 UNF
G1 UNL
G1 UNP
G1 UNW
G1 UOJ
G1 UOM
G1 UOZ
G1 UPC
G1 UPH
G1 UPK
G1 UPR
G1 UPU
G1 UPY
G1 UQO
G1 URC
G1 USC
G1 USS
G1 UTA
G1 UTI
G1 UVQ
G1 UVY
G1 UWF
G1 UWG
G1 UWL
G1 UWR
G1 UWW
G1 UWY
G1 UXA
G1 UXB
G1 UXC
G1 UXG
G1 UXH
G1 UXK
G1 UXL
G1 UXP
G1 UXQ
G1 UYA
G1 UYE
G1 UYM
G1 UYP
G1 UZL
G1 VAC
G1 VAE
G1 VAF
G1 VBI
G1 VBM
G1 VBR
G1 VBW
G1 VBX
G1 VCA
G1 VCD
G1 VCN
G1 VCP
G1 VCT
G1 VDA
G1 VDB
G1 VDY
G1 VEB
G1 VEC
G1 VEQ
G1 VFD
G1 VFE
G1 VFG
G1 VFJ
G1 VFS
G1 VGE
G1 VGF
G1 VGT
G1 VGZ
G1 VHD
G1 VHG
G1 VHQ
G1 VIE
G1 VIK
G1 VIM
G1 VJW
G1 VKP
G1 VKZ
G1 VLB
G1 VLC
G1 VLM
G1 VLR
G1 VLV
G1 VLW
G1 VMB
G1 VMG
G1 VMH
G1 VMJ
G1 VML
G1 VMS
G1 VMT
G1 VMU
G1 VMW
G1 VMY
G1 VNA
G1 VOA
G1 VOK
G1 VPB
G1 VPK
G1 VPL
G1 VPP
G1 VPR
G1 VPV
G1 VPW
G1 VPX
G1 VQN
G1 VQO
G1 VQU
G1 VQW
G1 VRG
G1 VRH
G1 VSE
G1 VSL
G1 VSV
G1 VSW
G1 VTA
G1 VTD
G1 VTI
G1 VUE
G1 VUH
G1 VUL
G1 VUS
G1 VUT
G1 VUU
G1 VUX
G1 VVP
G1 VWV
G1 VWW
G1 VXJ
G1 VXK
G1 VXL
G1 VXQ
G1 VYK
G1 VYZ
G1 VZC
G1 VZM
G1 VZP
G1 WAT
G1 WCB
G1 WCL
G1 WCO
G1 WCR
G1 WDF
G1 WDM
G1 WDW
G1 WEA
G1 WEB
G1 WEU
G1 WEX
G1 WFK
G1 WFV
G1 WGH
G1 WGJ
G1 WGP
G1 WGV
G1 WHC
G1 WHH
G1 WHN
G1 WHP
G1 WIA
G1 WIH
G1 WIO
G1 WJZ
G1 WKP
G1 WKR
G1 WKV
G1 WLM
G1 WMD
G1 WMJ
G1 WNS
G1 WOB
G1 WOQ
G1 WPD
G1 WRK
G1 WRM
G1 WSA
G1 WTF
G1 WUJ
G1 WUP
G1 WVF
G1 WVN
G1 WVO
G1 WWV
G1 WWX
G1 WXA
G1 WXB
G1 WXI
G1 WXR
G1 WXZ
G1 WZA
G1 XAD
G1 XAE
G1 XBH
G1 XBU
G1 XCC
G1 XCS
G1 XCW
G1 XCZ
G1 XDA
G1 XDC
G1 XDH
G1 XDM
G1 XDT
G1 XEL
G1 XEM
G1 XFQ
G1 XGF
G1 XHW
G1 XIC
G1 XIG
G1 XIJ
G1 XIP
G1 XIX
G1 XJP
G1 XJS
G1 XJV
G1 XKM
G1 XKR
G1 XKT
G1 XLC
G1 XLP
G1 XLX
G1 XMF
G1 XMK
G1 XMV
G1 XNS
G1 XOL
G1 XOX
G1 XPC
G1 XPR
G1 XQT
G1 XQX
G1 XSF
G1 XSZ
G1 XTB
G1 XTP
G1 XWJ
G1 XWU
G1 XWX
G1 XWY
G1 XYL
G1 XYU
G1 XYX
G1 XZL
G1 YAI
G1 YAI
G1 YAR
G1 YBJ
G1 YBN
G1 YBQ
G1 YCB
G1 YCC
G1 YDH
G1 YDN
G1 YDT
G1 YEF
G1 YEX
G1 YEY
G1 YFF
G1 YFR
G1 YFU
G1 YFV
G1 YFW
G1 YFX
G1 YFY
G1 YGD
G1 YGJ
G1 YHR
G1 YIB
G1 YIF
G1 YIJ
G1 YIK
G1 YIO
G1 YJE
G1 YKV
G1 YLU
G1 YMI
G1 YMU
G1 YMW
G1 YMY
G1 YNE
G1 YNG
G1 YNI
G1 YNL
G1 YNW
G1 YNY
G1 YNZ
G1 YOK
G1 YOM
G1 YOR
G1 YPA
G1 YPB
G1 YPG
G1 YPQ
G1 YQH
G1 YQJ
G1 YQK
G1 YRV
G1 YRW
G1 YSR
G1 YST
G1 YSZ
G1 YUM
G1 YUO
G1 YUV
G1 YVF
G1 YVH
G1 YVR
G1 YVT
G1 YWC
G1 YWR
G1 YXF
G1 YXM
G1 YYC
G1 YYE
G1 YYN
G1 YYV
G1 YZB
G1 YZR
G1 YZS
G1 YZY
G1 ZAF
G1 ZAL
G1 ZAT
G1 ZCV
G1 ZCW
G1 ZDB
G1 ZDC
G1 ZDI
G1 ZDK
G1 ZDM
G1 ZEB
G1 ZEF
G1 ZEO
G1 ZEZ
G1 ZFO
G1 ZGC
G1 ZGL
G1 ZGL
G1 ZGU
G1 ZGX
G1 ZGZ
G1 ZHM
G1 ZHS
G1 ZHX
G1 ZIE
G1 ZIF
G1 ZIT
G1 ZIY
G1 ZJC
G1 ZJM
G1 ZJV
G1 ZKC
G1 ZKJ
G1 ZLG
G1 ZLH
G1 ZLM
G1 ZLU
G1 ZMA
G1 ZMM
G1 ZMS
G1 ZMV
G1 ZNE
G1 ZNI
G1 ZNL
G1 ZOI
G1 ZOM
G1 ZOV
G1 ZQA
G1 ZQC
G1 ZQI
G1 ZQZ
G1 ZRF
G1 ZRN
G1 ZRZ
G1 ZSJ
G1 ZUG
G1 ZUK
G1 ZVB
G1 ZVL
G1 ZVN
G1 ZVQ
G1 ZWA
G1 ZWT
G1 ZXA
G1 ZXH
G1 ZXQ
G1 ZXT
G1 ZYB
G1 ZYC
G1 ZYH
G1 ZYX
G1 ZZX
G1 ZZY

G2

G2 AA
G2 AP
G2 AVV
G2 AXQ
G2 BC
G2 BDV
G2 BFC
G2 BGI
G2 CWO
G2 DA
G2 DPQ
G2 DS
G2 DVP
G2 DX
G2 FQS
G2 FZ
G2 GS
G2 GT
G2 HC
G2 HPG
G2 IV
G2 JP
G2 LO
G2 LV
G2 ML
G2 MT
G2 NM
G2 NV
G2 OT
G2 PLX
G2 PY
G2 RW
G2 SM
G2 TT
G2 TV
G2 WW
G2 XV
G2 XYL
G2 ZE

G3

G3 AGW
G3 AHD
G3 AHW
G3 AJY
G3 AQT
G3 ARC
G3 ARK
G3 ARO
G3 ARX
G3 ASP
G3 ATA
G3 ATI
G3 BA
G3 BK
G3 BRK
G3 BRW
G3 BSA
G3 BSQ
G3 BXF
G3 BYV
G3 BZW
G3 CAR
G3 CDR
G3 CEN
G3 CKO
G3 CNX
G3 CPS
G3 CQQ
G3 CSO
G3 CXX
G3 DAM
G3 DDI
G3 DIH
G3 DJR
G3 DKJ
G3 DKN
G3 DLS
G3 DNE
G3 DRK
G3 DX
G3 DZE
G3 EFB
G3 EFU
G3 EGY
G3 EHG
G3 EIG
G3 EJS
G3 EKT
G3 EML
G3 EMZ
G3 ENG
G3 EPN
G3 ESP
G3 EWE
G3 EYB
G3 FAB
G3 FKB
G3 FRR
G3 FRW
G3 FSO
G3 FVO
G3 GFA
G3 GNF
G3 GNX
G3 GRM
G3 GUL
G3 GVB
G3 GWI
G3 GZJ
G3 HCM
G3 HCN
G3 HCU
G3 HDS
G3 HEL
G3 HER
G3 HGI
G3 HHT
G3 HIJ
G3 HPP
G3 HRL
G3 HRZ
G3 HSR
G3 HUE
G3 HZX
G3 IGQ
G3 IGR
G3 II
G3 IIA
G3 IJN
G3 IMR
G3 IMR
G3 INF
G3 ITK
G3 IVB
G3 IZV
G3 JBQ
G3 JEV
G3 JFH
G3 JFL
G3 JHB
G3 JJM
G3 JMC
G3 JMT
G3 JPL
G3 JRT
G3 JVB
G3 JW
G3 JXB
G3 KCJ
G3 KER
G3 KFN
G3 KLB
G3 KMZ
G3 KNE
G3 KNM
G3 KOU
G3 KPP
G3 KRC
G3 KSG
G3 KSU
G3 KSY
G3 KUM
G3 KUS
G3 KWA
G3 KXU
G3 KYG
G3 LBL
G3 LCT
G3 LFZ
G3 LHE
G3 LHI
G3 LKZ
G3 LLL
G3 LMD
G3 LMK
G3 LMO
G3 LPP
G3 LSS
G3 LX
G3 LXI
G3 LYE
G3 LYK
G3 LZG
G3 MA
G3 MAL
G3 MAN
G3 MDB
G3 MDC
G3 MDX
G3 MEJ
G3 MGM
G3 MGP
G3 MGV
G3 MHW
G3 MII
G3 MJO
G3 MMH
G3 MMP
G3 MMR
G3 MTM
G3 MUI
G3 MXG
G3 MZK
G3 NAC
G3 NAS
G3 NCM
G3 NCR
G3 NFO
G3 NGD
G3 NKB
G3 NKR
G3 NLN
G3 NLZ
G3 NNF
G3 NNH
G3 NPI
G3 NQV
G3 NRB
G3 NRK
G3 NRT
G3 NSB
G3 NSM
G3 NTK
G3 NUE
G3 NUF
G3 NWD
G3 NYY
G3 OBX
G3 ODR
G3 OFA
G3 OFL
G3 OGF
G3 OGY
G3 ORH
G3 OSB
G3 OSW
G3 OU
G3 OUF
G3 OUK
G3 OUR
G3 OVM
G3 OVQ
G3 OWF
G3 OWS
G3 OXQ
G3 OYE
G3 OYP
G3 OYS
G3 OZQ
G3 OZS
G3 PBA
G3 PBS
G3 PBU
G3 PCN
G3 PCQ
G3 PCR
G3 PD
G3 PDG
G3 PDK
G3 PGD
G3 PGV
G3 PHA
G3 PHR
G3 PHU
G3 PJD
G3 PJI
G3 PKM
G3 PKW
G3 PLH
G3 PLS
G3 PLU
G3 PMM
G3 PNY
G3 PPG
G3 PQU
G3 PRN
G3 PTD
G3 PTU
G3 PUK
G3 PUV
G3 PVS
G3 PVW
G3 PWH
G3 PXQ
G3 PXT
G3 PYA
G3 PYU
G3 RAS
G3 RAT
G3 RAX
G3 RAY
G3 RBA
G3 RBG
G3 RBX
G3 RCU
G3 REA
G3 REM
G3 REN
G3 RFF
G3 RFL
G3 RFS
G3 RFT
G3 RHN
G3 RIE
G3 RJQ
G3 RLB
G3 RLU
G3 RME
G3 RNQ
G3 ROI
G3 RRX
G3 RSS
G3 RSY
G3 RTG
G3 RTU
G3 RVU
G3 RWA
G3 RWP
G3 RXF
G3 RXL
G3 RXQ
G3 RZA
G3 SAZ
G3 SCU
G3 SDQ
G3 SEV
G3 SEW
G3 SIP
G3 SJZ
G3 SKC
G3 SKP
G3 SLR
G3 SMP
G3 SMX
G3 SPB
G3 SPE
G3 SPX
G3 SPZ
G3 SQV
G3 SR
G3 SRA
G3 SRO
G3 SSC
G3 SSG
G3 SSO
G3 STR
G3 STS
G3 SUF
G3 SUY
G3 SVL
G3 SWF
G3 SWM
G3 SYF
G3 SZO
G3 SZT
G3 TAH
G3 TAM
G3 TAN
G3 TBA
G3 TCA
G3 TCE
G3 TCR
G3 TDJ
G3 TFT
G3 TGP
G3 TGW
G3 TIB
G3 TIC
G3 TIQ
G3 TKV
G3 TKW
G3 TLE
G3 TMG
G3 TMH
G3 TMZ
G3 TOE
G3 TQH
G3 TRF
G3 TSH
G3 TSK
G3 TTK
G3 TVA
G3 TVQ
G3 TWE
G3 TWZ
G3 TXA
G3 TYB
G3 TYY
G3 TZH
G3 UBU
G3 UBW
G3 UCM
G3 UCV
G3 UDU
G3 UFK
G3 UFP
G3 UGG
G3 UGK
G3 UGY
G3 UIE
G3 UKU
G3 ULF
G3 UMK
G3 UMW
G3 UNC
G3 UNH
G3 UNR
G3 UPK
G3 USB
G3 UUW
G3 UWI
G3 UXB
G3 VAA
G3 VBS
G3 VCU
G3 VDX
G3 VEQ
G3 VFP
G3 VFU
G3 VFV
G3 VGF
G3 VGG
G3 VGU
G3 VHB
G3 VHG
G3 VHJ
G3 VHY
G3 VII
G3 VIV
G3 VLP
G3 VLT
G3 VMB
G3 VMH
G3 VMS
G3 VNC
G3 VOD
G3 VPD
G3 VRL
G3 VSS
G3 VUF
G3 VUP
G3 VUQ
G3 VW
G3 VWF
G3 VWL
G3 VWP
G3 VXI
G3 VXP
G3 VXX
G3 VZZ
G3 WA
G3 WCS
G3 WFB
G3 WFK
G3 WGC
G3 WGM
G3 WGQ
G3 WHL
G3 WHR
G3 WIC
G3 WIE
G3 WIR
G3 WKD
G3 WKU
G3 WM
G3 WMM
G3 WMR
G3 WNU
G3 WOF
G3 WOX
G3 WPO
G3 WQO
G3 WQP
G3 WQT
G3 WRS
G3 WRU
G3 WRV
G3 WSN
G3 WTA
G3 WVY
G3 WVZ
G3 WW
G3 WWW
G3 WXL
G3 WXX
G3 WZD
G3 WZN
G3 XAD
G3 XAJ
G3 XCR
G3 XDA
G3 XDE
G3 XDH
G3 XDQ
G3 XDV
G3 XEB
G3 XEK
G3 XFA
G3 XFE
G3 XFV
G3 XFZ
G3 XGM
G3 XGQ
G3 XGS
G3 XHF
G3 XJF
G3 XJK
G3 XJO
G3 XKQ
G3 XKR
G3 XKT
G3 XLF
G3 XNH
G3 XNO
G3 XNT
G3 XPE
G3 XPO
G3 XQX
G3 XRE
G3 XSA
G3 XSH
G3 XSJ
G3 XTX
G3 XUA
G3 XVF
G3 XWJ
G3 XWM
G3 XXT
G3 XYS
G3 XYX
G3 XZS
G3 XZT
G3 YAM
G3 YAP
G3 YAS
G3 YBD
G3 YBT
G3 YC
G3 YCA
G3 YCR
G3 YDL
G3 YEH
G3 YFJ
G3 YFN
G3 YGS
G3 YIP
G3 YIS
G3 YIV
G3 YJO
G3 YJR
G3 YL
G3 YLE
G3 YLG
G3 YLO
G3 YMP
G3 YMX
G3 YNH
G3 YNT
G3 YOK
G3 YPP
G3 YPQ
G3 YPX
G3 YQD
G3 YRE
G3 YRJ
G3 YRO
G3 YRW
G3 YRZ
G3 YSC
G3 YSH
G3 YTH
G3 YTK
G3 YUE
G3 YUV
G3 YVF
G3 YWQ
G3 YXX
G3 YXZ
G3 YYD
G3 YYH
G3 YYO
G3 YYT
G3 YZG
G3 YZJ
G3 YZY
G3 ZAF
G3 ZCC
G3 ZCU
G3 ZDD
G3 ZDR
G3 ZDV
G3 ZEH
G3 ZEV
G3 ZFJ
G3 ZHH
G3 ZHI
G3 ZHW
G3 ZHX
G3 ZKF
G3 ZLB
G3 ZLW
G3 ZMC
G3 ZMD
G3 ZMS
G3 ZMU
G3 ZMY
G3 ZOD
G3 ZOE
G3 ZPF
G3 ZPQ
G3 ZQX
G3 ZRK
G3 ZRV
G3 ZSE
G3 ZSK
G3 ZSP
G3 ZSV
G3 ZTO
G3 ZUD
G3 ZWI
G3 ZXH
G3 ZXL
G3 ZXP
G3 ZXX
G3 ZYN
G3 ZYS
G3 ZYW
G3 ZZO

G4

G4 AAO
G4 AAW
G4 ABD
G4 ABF
G4 ABP
G4 ABT
G4 ADN
G4 ADX
G4 AFG
G4 AHS
G4 AJC
G4 AJK
G4 AJM
G4 AKU
G4 AKZ
G4 ALG
G4 ALH
G4 ALI
G4 ANA
G4 ANH
G4 ANS
G4 AOG
G4 AON
G4 AOV
G4 APT
G4 APV
G4 AQU
G4 AQV
G4 AQY
G4 ARG
G4 ARP
G4 ASE
G4 AST
G4 ATI
G4 ATX
G4 AVB
G4 AVD
G4 AVU
G4 AWC
G4 AWP
G4 AXH
G4 AXN
G4 AXP
G4 AYF
G4 AYJ
G4 AYM
G4 AYT
G4 AYV
G4 AZF
G4 AZR
G4 AZZ
G4 BAZ
G4 BBS
G4 BBZ
G4 BCK
G4 BDE
G4 BDO
G4 BDP
G4 BDY
G4 BEE
G4 BEG
G4 BEN
G4 BEW
G4 BFJ
G4 BFP
G4 BFY
G4 BFZ
G4 BGQ
G4 BGX
G4 BIJ
G4 BIP
G4 BJM
G4 BKC
G4 BKM
G4 BLF
G4 BLH
G4 BLI
G4 BLJ
G4 BLM
G4 BLN
G4 BLQ
G4 BLX
G4 BMF
G4 BMH
G4 BNU
G4 BOY
G4 BPQ
G4 BSU
G4 BSX
G4 BTH
G4 BTS
G4 BUD
G4 BVH
G4 BWD
G4 BWQ
G4 BXO
G4 BXP
G4 BYY
G4 BZD
G4 BZO
G4 CAO
G4 CAW
G4 CBR
G4 CCX
G4 CDE
G4 CDM
G4 CDT
G4 CEH
G4 CFF
G4 CFJ
G4 CGS
G4 CIG
G4 CIL
G4 CIN
G4 CIV
G4 CJC
G4 CJF
G4 CJS
G4 CJX
G4 CJZ
G4 CKN
G4 CKR
G4 CLK
G4 CLQ
G4 CLT
G4 CLU
G4 CND
G4 CNE
G4 CNF
G4 CNG
G4 COK
G4 COO
G4 CPK
G4 CPR
G4 CQJ
G4 CQU
G4 CRF
G4 CRJ
G4 CRV
G4 CSN
G4 CSS
G4 CTG
G4 CTL
G4 CTP
G4 CTR
G4 CTS
G4 CUJ
G4 CUS
G4 CVY
G4 CXH
G4 CXM
G4 CYE
G4 CYQ
G4 CZE
G4 CZJ
G4 CZS
G4 DAA
G4 DAE
G4 DAR
G4 DBI
G4 DBL
G4 DBW
G4 DCQ
G4 DCR
G4 DCV
G4 DDR
G4 DDS
G4 DED
G4 DEE
G4 DGK
G4 DGU
G4 DHA
G4 DHJ
G4 DHO
G4 DHP
G4 DIF
G4 DIW
G4 DJW
G4 DKB
G4 DKF
G4 DKK
G4 DKO
G4 DLB
G4 DLX
G4 DMA
G4 DMD
G4 DME
G4 DMN
G4 DMU
G4 DNI
G4 DOI
G4 DPE
G4 DPN
G4 DPX
G4 DPY
G4 DQY
G4 DRK
G4 DRM
G4 DRY
G4 DSB
G4 DST
G4 DTD
G4 DTF
G4 DUC
G4 DUH
G4 DUK
G4 DUZ
G4 DVF
G4 DVP
G4 DVQ
G4 DVS
G4 DVW
G4 DWQ
G4 DWV
G4 DX
G4 DXA
G4 DXG
G4 DXX
G4 DYA
G4 DYK
G4 DYT
G4 DYW
G4 DZE
G4 DZF
G4 DZL
G4 DZR
G4 EAA
G4 EAR
G4 ECM
G4 ECY
G4 EDE
G4 EDS
G4 EEI
G4 EEM
G4 EER
G4 EFT
G4 EFW
G4 EGH
G4 EGT
G4 EHA
G4 EHH
G4 EHZ
G4 EIB
G4 EIS
G4 EIT
G4 EJB
G4 EJN
G4 EKK
G4 ELD
G4 ELN
G4 EMM
G4 EMO
G4 EMR
G4 EOZ
G4 EPJ
G4 EQB
G4 EQH
G4 EQQ
G4 ETH
G4 ETQ
G4 ETV
G4 EUE
G4 EUM
G4 EUQ
G4 EWC
G4 EWN
G4 EWS
G4 EWX
G4 EXA
G4 EXQ
G4 EXW
G4 EYL
G4 EYR
G4 EZR
G4 EZT
G4 EZV
G4 EZW
G4 EZZ
G4 FBA
G4 FBJ
G4 FBW
G4 FBX
G4 FCR
G4 FDB
G4 FDL
G4 FDO
G4 FFJ
G4 FHA
G4 FHI
G4 FID
G4 FIK
G4 FIS
G4 FIX
G4 FJQ
G4 FJR
G4 FKK
G4 FLH
G4 FLN
G4 FLQ
G4 FLU
G4 FMG
G4 FOC
G4 FOK
G4 FP
G4 FPH
G4 FPK
G4 FPS
G4 FRE
G4 FRS
G4 FRU
G4 FSC
G4 FSJ
G4 FTC
G4 FTF
G4 FTJ
G4 FUB
G4 FUF
G4 FUH
G4 FUN
G4 FUT
G4 FVG
G4 FVN
G4 FWC
G4 FWG
G4 FWX
G4 FXB
G4 FXO
G4 FXQ
G4 FYA
G4 FYF
G4 FYX
G4 FZE
G4 GA
G4 GAD
G4 GAS
G4 GAU
G4 GAV
G4 GBG
G4 GCB
G4 GCD
G4 GCG
G4 GCM
G4 GCP
G4 GCW
G4 GCZ
G4 GEF
G4 GER
G4 GES
G4 GFK
G4 GFX
G4 GGS
G4 GHD
G4 GHJ
G4 GHP
G4 GHU
G4 GIF
G4 GII
G4 GJC
G4 GJH
G4 GJL
G4 GJN
G4 GJW
G4 GKA
G4 GLT
G4 GMA
G4 GME
G4 GMJ
G4 GMO
G4 GMP
G4 GMV
G4 GNC
G4 GNK
G4 GOC
G4 GOH
G4 GOQ
G4 GOY
G4 GPK
G4 GPU
G4 GPZ
G4 GQB
G4 GRI
G4 GRW
G4 GSQ
G4 GSV
G4 GTA
G4 GTF
G4 GTG
G4 GTI
G4 GTP
G4 GTR
G4 GTT
G4 GTW
G4 GTY
G4 GUF
G4 GUI
G4 GUJ
G4 GVB
G4 GVC
G4 GVD
G4 GVN
G4 GVX
G4 GWD
G4 GWO
G4 GYB
G4 GYO
G4 GZ
G4 GZJ
G4 HAD
G4 HAQ
G4 HBJ
G4 HCF
G4 HCL
G4 HCW
G4 HDK
G4 HEL
G4 HEQ
G4 HFB
G4 HFF
G4 HFT
G4 HGI
G4 HGQ
G4 HGU
G4 HHI
G4 HHT
G4 HHV
G4 HIP
G4 HJG
G4 HJM
G4 HKF
G4 HKK
G4 HKM
G4 HKT
G4 HLP
G4 HLR
G4 HLU
G4 HLU
G4 HMG
G4 HMI
G4 HMJ
G4 HNB
G4 HNH
G4 HNY
G4 HOV
G4 HPF
G4 HPG
G4 HPI
G4 HPL
G4 HPW
G4 HRK
G4 HRM
G4 HRO
G4 HRV
G4 HSO
G4 HSR
G4 HSU
G4 HSY
G4 HTP
G4 HUK
G4 HVE
G4 HVK
G4 HVX
G4 HXD
G4 HXT
G4 HYC
G4 HYX
G4 HZA
G4 HZB
G4 IAL
G4 IAZ
G4 IBK
G4 IBZ
G4 ICD
G4 ICT
G4 ICV
G4 IDK
G4 IDO
G4 IDY
G4 IDZ
G4 IEJ
G4 IFB
G4 IFC
G4 IFU
G4 IGO
G4 IGW
G4 IGZ
G4 IHH
G4 IHL
G4 IIG
G4 IIQ
G4 IIU
G4 IJC
G4 IJE
G4 IJL
G4 IKH
G4 IKK
G4 IKP
G4 IKR
G4 IKZ
G4 IMD
G4 IMN
G4 IMO
G4 IMZ
G4 INK
G4 INT
G4 IOX
G4 IPP
G4 IPQ
G4 IPW
G4 IQE
G4 IQI
G4 IQP
G4 IRN
G4 IRQ
G4 IRW
G4 IRX
G4 ISI
G4 ISP
G4 ITF
G4 IUB
G4 IUG
G4 IUR
G4 IUW
G4 IVH
G4 IVK
G4 IVS
G4 IVV
G4 IWB
G4 IXL
G4 IXN
G4 IYJ
G4 IYL
G4 IYU
G4 IZC
G4 JBC
G4 JBI
G4 JBX
G4 JCI
G4 JCN
G4 JCO
G4 JCP
G4 JDL
G4 JEC
G4 JEG
G4 JEN
G4 JEQ
G4 JET
G4 JFB
G4 JFK
G4 JFO
G4 JFT
G4 JGL
G4 JGP
G4 JGR
G4 JHD
G4 JHM
G4 JIB
G4 JIN
G4 JIS
G4 JJI
G4 JKO
G4 JLB
G4 JLQ
G4 JLT
G4 JLU
G4 JMH
G4 JMJ
G4 JMN
G4 JMP
G4 JMV
G4 JNC
G4 JNN
G4 JNW
G4 JNY
G4 JOJ
G4 JOK
G4 JON
G4 JOP
G4 JPI
G4 JQE
G4 JQM
G4 JQP
G4 JQT
G4 JRC
G4 JRH
G4 JRL
G4 JRN
G4 JRO
G4 JRT
G4 JRV
G4 JSL
G4 JTJ
G4 JTY
G4 JUT
G4 JUY
G4 JVF
G4 JVG
G4 JWB
G4 JWG
G4 JWT
G4 JXL
G4 JXS
G4 JXX
G4 JXY
G4 JZJ
G4 KAG
G4 KAJ
G4 KAN
G4 KAO
G4 KAW
G4 KAY
G4 KBC
G4 KBD
G4 KBF
G4 KCS
G4 KCW
G4 KDG
G4 KDX
G4 KEU
G4 KEV
G4 KFE
G4 KFK
G4 KFO
G4 KHA
G4 KHV
G4 KIJ
G4 KIO
G4 KIV

Withheld

G4 KJQ
G4 KJW
G4 KLG
G4 KLH
G4 KLL
G4 KLW
G4 KMC
G4 KMV
G4 KND
G4 KNK
G4 KNM
G4 KNP
G4 KOB
G4 KOH
G4 KOP
G4 KPB
G4 KPJ
G4 KPT
G4 KPW
G4 KQU
G4 KRB
G4 KRM
G4 KRO
G4 KRV
G4 KSD
G4 KSE
G4 KSW
G4 KTE
G4 KTS
G4 KUI
G4 KVA
G4 KWG
G4 KXN
G4 KYA
G4 KYC
G4 KYD
G4 KYF
G4 KYN
G4 KYV
G4 KZS
G4 KZY
G4 LAC
G4 LAH
G4 LAX
G4 LBB
G4 LBI
G4 LCD
G4 LCZ
G4 LDC
G4 LDG
G4 LDY
G4 LDZ
G4 LEB
G4 LEF
G4 LEL
G4 LFA
G4 LFM
G4 LFU
G4 LGD
G4 LGE
G4 LGF
G4 LGK
G4 LGZ
G4 LHN
G4 LHU
G4 LHX
G4 LIB
G4 LIP
G4 LJJ
G4 LJZ
G4 LKI
G4 LKV
G4 LLX
G4 LMJ
G4 LMS
G4 LMZ
G4 LNA
G4 LND
G4 LNF
G4 LOU
G4 LPB
G4 LQY
G4 LRR
G4 LRS
G4 LRX
G4 LTF
G4 LTX
G4 LUG
G4 LUL
G4 LUP
G4 LUZ
G4 LVM
G4 LWR
G4 LXE
G4 LXI
G4 LXK
G4 LXL
G4 LXQ
G4 LXS
G4 LXT
G4 LXZ
G4 LZG
G4 LZL
G4 LZM
G4 LZW
G4 MAD
G4 MAM
G4 MAP
G4 MAW
G4 MBN
G4 MBS
G4 MBU
G4 MCL
G4 MCN
G4 MDI
G4 MDP
G4 MDX
G4 MEC
G4 MEZ
G4 MFU
G4 MGL
G4 MHN
G4 MHU
G4 MIN
G4 MIU
G4 MIY
G4 MJH
G4 MJO
G4 MJS
G4 MJZ
G4 MKS
G4 MLA
G4 MLJ
G4 MLP
G4 MNO
G4 MOF
G4 MOF
G4 MOM
G4 MOX
G4 MPD
G4 MPU
G4 MPV
G4 MPX
G4 MQ
G4 MQB
G4 MQE
G4 MR
G4 MRE
G4 MRG
G4 MRH
G4 MRM
G4 MRV
G4 MSF
G4 MSH
G4 MSO
G4 MTK
G4 MUF
G4 MUR
G4 MVC
G4 MVH
G4 MVK
G4 MVL
G4 MVN
G4 MVR
G4 MVU
G4 MVV
G4 MWC
G4 MWK
G4 MWM
G4 MWR
G4 MXA
G4 MXQ
G4 MXR
G4 MYC
G4 MYI
G4 MZJ
G4 NAB
G4 NAD
G4 NAR
G4 NAZ
G4 NBZ
G4 NCE
G4 NCN
G4 NCT
G4 NDH
G4 NDR
G4 NDW
G4 NEC
G4 NEV
G4 NEX
G4 NFB
G4 NFC
G4 NFD
G4 NFJ
G4 NFU
G4 NGE
G4 NGH
G4 NHJ
G4 NHS
G4 NIA
G4 NIS
G4 NIW
G4 NJE
G4 NJH
G4 NKL
G4 NKM
G4 NKT
G4 NKV
G4 NLP
G4 NLX
G4 NLY
G4 NMI
G4 NMX
G4 NNA
G4 NNC
G4 NNU
G4 NNW
G4 NOV
G4 NOW
G4 NPJ
G4 NPX
G4 NQF
G4 NQK
G4 NQX
G4 NRL
G4 NSG
G4 NSI
G4 NSU
G4 NTB
G4 NTF
G4 NTM
G4 NTN
G4 NUL
G4 NVB
G4 NVF
G4 NVO
G4 NVR
G4 NVZ
G4 NWP
G4 NXH
G4 NXK
G4 NXM
G4 NYS
G4 NZF
G4 NZV
G4 OAC
G4 OAO
G4 OAO
G4 OAQ
G4 OAT
G4 OBI
G4 OBL
G4 OCO
G4 OCP
G4 ODC
G4 OEL
G4 OFY
G4 OGD
G4 OGE
G4 OGF
G4 OGI
G4 OGP
G4 OGT
G4 OHK
G4 OHN
G4 OIP
G4 OIX
G4 OIZ
G4 OJC
G4 OJO
G4 OKK
G4 OKT
G4 OLN
G4 OLT
G4 OLV
G4 OMA
G4 OME
G4 OMF
G4 OMH
G4 OMX
G4 ONE
G4 ONK
G4 ONL
G4 ONW
G4 OOP
G4 OOR
G4 OOW
G4 ORF
G4 ORM
G4 ORR
G4 OSW
G4 OTK
G4 OTQ
G4 OUK
G4 OUL
G4 OVK
G4 OVP
G4 OVU
G4 OWF
G4 OWI
G4 OWO
G4 OWW
G4 OXC
G4 OXF
G4 OXW
G4 OXY
G4 OYD
G4 OYF
G4 OYS
G4 OZH
G4 OZZ
G4 PAB
G4 PAG
G4 PAN
G4 PAO
G4 PAP
G4 PBL
G4 PBV
G4 PC
G4 PCA
G4 PCI
G4 PCS
G4 PDF
G4 PDL
G4 PDS
G4 PDV
G4 PEB
G4 PEM
G4 PF
G4 PFC
G4 PFD
G4 PFN
G4 PGI
G4 PGP
G4 PHJ
G4 PHO
G4 PHU
G4 PHW
G4 PIO
G4 PIV
G4 PIX
G4 PJA
G4 PJC
G4 PJF
G4 PJG
G4 PLC
G4 PLN
G4 PLP
G4 PLT
G4 PM
G4 PMN
G4 PMQ
G4 PMX
G4 PNF
G4 PNJ
G4 PNW
G4 POA
G4 POQ
G4 PPM
G4 PQA
G4 PQL
G4 PQT
G4 PRV
G4 PSA
G4 PSB
G4 PSF
G4 PSN
G4 PTB
G4 PTI
G4 PTJ
G4 PTS
G4 PTX
G4 PUF
G4 PUJ
G4 PUR
G4 PVA
G4 PVB
G4 PWO
G4 PXI
G4 PXL
G4 PXO
G4 PXW
G4 PXZ
G4 PYM
G4 PYR
G4 PZE
G4 PZK
G4 RAD
G4 RAF
G4 RBI
G4 RBO
G4 RBV
G4 RCA
G4 RDK
G4 REA
G4 REL
G4 REQ
G4 REV
G4 RFM
G4 RFQ
G4 RFT
G4 RFX
G4 RFZ
G4 RGK
G4 RGV
G4 RHD
G4 RHN
G4 RHO
G4 RIC
G4 RII
G4 RIT
G4 RIW
G4 RKJ
G4 RKY
G4 RLQ
G4 RMI
G4 RMK
G4 RMM
G4 RMR
G4 RNL
G4 RNS
G4 ROE
G4 ROO
G4 RPF
G4 RPX
G4 RQE
G4 RQY
G4 RRC
G4 RRZ
G4 RS
G4 RSA
G4 RSB
G4 RSE
G4 RSH
G4 RSQ
G4 RSR
G4 RSY
G4 RTB
G4 RUV
G4 RVB
G4 RVM
G4 RVR
G4 RWA
G4 RWC
G4 RWO
G4 RWU
G4 RXN
G4 RXT
G4 RYD
G4 RYF
G4 RYN
G4 RYU
G4 RYX
G4 RZG
G4 RZK
G4 RZS
G4 SAE
G4 SAP
G4 SAQ
G4 SBC
G4 SBT
G4 SBX
G4 SCP
G4 SDG
G4 SDM
G4 SDN
G4 SDP
G4 SDR
G4 SDW
G4 SFI
G4 SFM
G4 SFU
G4 SFX
G4 SGB
G4 SGM
G4 SGO
G4 SGZ
G4 SHD
G4 SHP
G4 SHS
G4 SHZ
G4 SIO
G4 SIU
G4 SIV
G4 SJS
G4 SJX
G4 SJY
G4 SKH
G4 SKK
G4 SKN
G4 SKS
G4 SLU
G4 SLX
G4 SMN
G4 SMY
G4 SNK
G4 SNS
G4 SNT
G4 SON
G4 SOX
G4 SPG
G4 SPK
G4 SPO
G4 SPX
G4 SQU
G4 SRA
G4 SRB
G4 SRG
G4 SRH
G4 SRK
G4 SRS
G4 SSG
G4 SSI
G4 SSK
G4 SST
G4 SSX
G4 STA
G4 STM
G4 STR
G4 SUH
G4 SUI
G4 SVF
G4 SVN
G4 SWI
G4 SWL
G4 SWW
G4 SWX
G4 SXR
G4 SXS
G4 SYJ
G4 SZD
G4 SZM
G4 TAA
G4 TAI
G4 TAS
G4 TAX
G4 TBA
G4 TBD
G4 TBL
G4 TBW
G4 TBZ
G4 TCG
G4 TCJ
G4 TCN
G4 TCU
G4 TCX
G4 TCZ
G4 TDS
G4 TEI
G4 TET
G4 TEV
G4 TFE
G4 TFN
G4 TFR
G4 THD
G4 THE
G4 THJ
G4 THL
G4 THR
G4 TIN
G4 TIO
G4 TJB
G4 TJC
G4 TJE
G4 TJJ
G4 TJT
G4 TKA
G4 TKG
G4 TKI
G4 TKJ
G4 TKN
G4 TKR
G4 TKV
G4 TLP
G4 TLV
G4 TMT
G4 TMW
G4 TNI
G4 TNN
G4 TNT
G4 TNW
G4 TNX
G4 TOC
G4 TOJ
G4 TOK
G4 TPA
G4 TQE
G4 TQF
G4 TQG
G4 TQJ
G4 TQM
G4 TQM
G4 TRK
G4 TSH
G4 TSM
G4 TSR
G4 TTI
G4 TTU
G4 TTW
G4 TUE
G4 TUG
G4 TUS
G4 TUT
G4 TUY
G4 TUZ
G4 TVH
G4 TVK
G4 TVP
G4 TVR
G4 TWE
G4 TWI
G4 TWJ
G4 TXB
G4 TXP
G4 TXR
G4 TXS
G4 TXX
G4 TYB
G4 TYC
G4 TZJ
G4 TZP
G4 TZT
G4 UAO
G4 UCD
G4 UCO
G4 UDC
G4 UDJ
G4 UDR
G4 UDT
G4 UEM
G4 UEQ
G4 UHC
G4 UHD
G4 UHN
G4 UHP
G4 UIN
G4 UIX
G4 UJD
G4 UJG
G4 UKF
G4 UKP
G4 ULA
G4 ULF
G4 ULS
G4 UMX
G4 UNG
G4 UNP
G4 UNR
G4 UNT
G4 UNU
G4 UOQ
G4 UPE
G4 UPF
G4 UPZ
G4 UQB
G4 UQC
G4 URH
G4 URZ
G4 USB
G4 USG
G4 USI
G4 UTT
G4 UTW
G4 UVE
G4 UVP
G4 UWA
G4 UWD
G4 UWQ
G4 UXR
G4 UXU
G4 UYL
G4 UZX
G4 VAD
G4 VAJ
G4 VBT
G4 VBW
G4 VCD
G4 VCH
G4 VCK
G4 VCT
G4 VCY
G4 VDE
G4 VDU
G4 VDV
G4 VEK
G4 VFA
G4 VFM
G4 VFS
G4 VFT
G4 VGE
G4 VGH
G4 VHF
G4 VHU
G4 VIN
G4 VJG
G4 VJM
G4 VJU
G4 VJW
G4 VJX
G4 VJY
G4 VKB
G4 VLC
G4 VLM
G4 VLO
G4 VLR
G4 VLY
G4 VOM
G4 VOQ
G4 VOX
G4 VPO
G4 VPT
G4 VQG
G4 VQO
G4 VQQ
G4 VRQ
G4 VRR
G4 VSF
G4 VSZ
G4 VTF
G4 VTH
G4 VTK
G4 VTX
G4 VUB
G4 VUJ
G4 VUO
G4 VUT
G4 VUX
G4 VVG
G4 VVU
G4 VXC
G4 VXO
G4 VXQ
G4 VXY
G4 VYM
G4 VYS
G4 VYW
G4 VYX
G4 VZG
G4 VZN
G4 VZO
G4 VZV
G4 VZZ
G4 WAD
G4 WAU
G4 WBB
G4 WBC
G4 WBM
G4 WBR
G4 WCS
G4 WDL
G4 WDU
G4 WEA
G4 WEO
G4 WFQ
G4 WHM
G4 WHO
G4 WHR
G4 WHX
G4 WID
G4 WIF
G4 WIX
G4 WJI
G4 WJT
G4 WKA
G4 WKH
G4 WKJ
G4 WKX
G4 WKY
G4 WLC
G4 WMI
G4 WMT
G4 WNC
G4 WNE
G4 WNM
G4 WNO
G4 WNT
G4 WOC
G4 WOF
G4 WOV
G4 WOY
G4 WPD
G4 WPN
G4 WPS
G4 WQG
G4 WQI
G4 WQR
G4 WQV
G4 WQX
G4 WRI
G4 WRT
G4 WRW
G4 WRY
G4 WRZ
G4 WSM
G4 WSQ
G4 WSZ
G4 WTF
G4 WTV
G4 WVE
G4 WVG
G4 WVU
G4 WVX
G4 WWD
G4 WWK
G4 WWS
G4 WXM
G4 WXN
G4 WXP
G4 WXZ
G4 WYJ
G4 WZE
G4 WZK
G4 XAK
G4 XBT
G4 XCA
G4 XCP
G4 XCS
G4 XDA
G4 XDD
G4 XDF
G4 XDZ
G4 XEC
G4 XEH
G4 XEM
G4 XEN
G4 XFB
G4 XFD
G4 XGB
G4 XGU
G4 XHJ
G4 XHS
G4 XIB
G4 XID
G4 XIF
G4 XIH
G4 XIV
G4 XIY
G4 XJA
G4 XJB
G4 XJC
G4 XJP
G4 XJW
G4 XKB
G4 XKH
G4 XKX
G4 XKY
G4 XLS
G4 XMG
G4 XMW
G4 XNG
G4 XNH
G4 XNJ
G4 XOB
G4 XOO
G4 XOT
G4 XOU
G4 XOX
G4 XPC
G4 XPE
G4 XPL
G4 XPM
G4 XQL
G4 XQS
G4 XRI
G4 XRS
G4 XRZ
G4 XSD
G4 XSH
G4 XTN
G4 XUR
G4 XUT
G4 XWG
G4 XWJ
G4 XWV
G4 XYE
G4 XYT
G4 XZL
G4 XZU
G4 XZV
G4 YAD
G4 YAR
G4 YBC
G4 YBO
G4 YBU
G4 YCB
G4 YCC
G4 YCI
G4 YCK
G4 YCT
G4 YDK
G4 YDU
G4 YEA
G4 YED
G4 YEH
G4 YEN
G4 YEY
G4 YEZ
G4 YFE
G4 YFN
G4 YFP
G4 YFQ
G4 YHE
G4 YHF
G4 YHV
G4 YHW
G4 YIB
G4 YIJ
G4 YIL
G4 YJO
G4 YKC
G4 YKJ
G4 YKS
G4 YKU
G4 YLA
G4 YLB
G4 YLJ
G4 YLV
G4 YMK
G4 YMV
G4 YNA
G4 YNP
G4 YOB
G4 YOE
G4 YOH
G4 YOM
G4 YOQ
G4 YOU
G4 YOW
G4 YPX
G4 YPY
G4 YQY
G4 YRB
G4 YRG
G4 YRK
G4 YRS
G4 YTW
G4 YUM
G4 YUP
G4 YUQ
G4 YWL
G4 YWW
G4 YWY
G4 YXG
G4 YXL
G4 YXT
G4 YY
G4 YYB
G4 YZI
G4 YZX
G4 ZAJ
G4 ZAQ
G4 ZAT
G4 ZAU
G4 ZAZ
G4 ZBA
G4 ZBR
G4 ZBV
G4 ZBY
G4 ZCZ
G4 ZDK
G4 ZED
G4 ZEK
G4 ZEO
G4 ZEP
G4 ZFL
G4 ZFZ
G4 ZGC
G4 ZGH
G4 ZIA
G4 ZIC
G4 ZIO
G4 ZJF
G4 ZJN
G4 ZJW
G4 ZKK
G4 ZKU
G4 ZLH
G4 ZLZ
G4 ZMF
G4 ZMG
G4 ZMI
G4 ZOM
G4 ZOW
G4 ZQP
G4 ZQZ
G4 ZQZ
G4 ZRE
G4 ZRL
G4 ZRP
G4 ZSF
G4 ZSN
G4 ZU
G4 ZUF
G4 ZUP
G4 ZUQ
G4 ZUR
G4 ZVC
G4 ZVE
G4 ZVY
G4 ZWH
G4 ZWL
G4 ZWP
G4 ZXE
G4 ZXX
G4 ZYG
G4 ZYJ
G4 ZYT
G4 ZYU
G4 ZZB
G4 ZZR

G5

G5 AAA
G5 AIB
G5 AJK
G5 AKB
G5 ALF
G5 ALG
G5 BAY
G5 BMH
G5 BSD
G5 CAZ
G5 CHC
G5 CL
G5 CLR
G5 CTH
G5 DEM
G5 DKT
G5 DOC
G5 DSG
G5 DWS
G5 DX
G5 DXC
G5 EA
G5 ECO
G5 EEK
G5 EPC
G5 FLY
G5 FS
G5 GEI
G5 HRS
G5 HRY
G5 HVH
G5 HWB
G5 IOM
G5 IPX
G5 JDA
G5 JET
G5 JNX
G5 KEN
G5 KNY
G5 KSO
G5 KV
G5 LB
G5 LBS
G5 LDR
G5 LEA
G5 LO
G5 LOW
G5 MAT
G5 MCL
G5 MDR
G5 MHS
G5 MUP
G5 NG
G5 NI
G5 ODB
G5 OPW
G5 OTA
G5 PH
G5 PIR
G5 PJH
G5 PJR
G5 RMH
G5 ROC
G5 RPG
G5 RRY
G5 RSD
G5 SKY
G5 SRO
G5 TA
G5 TM
G5 UK
G5 UM
G5 VHF
G5 VK
G5 VU
G5 WCB
G5 WCG
G5 WCW
G5 WZY
G5 XV
G5 XW
G5 YTT
G5 ZL
G5 ZX

G6

G6 AAD
G6 AAL
G6 AAQ
G6 AAW
G6 ABF
G6 ABQ
G6 ABR
G6 ABU
G6 ACA
G6 ACF
G6 ACT
G6 ACY
G6 ADK
G6 ADP
G6 ADW
G6 AEG
G6 AEN
G6 AFY
G6 AGC
G6 AHY
G6 AIC
G6 AJY
G6 AKE
G6 AKJ
G6 ALA
G6 ALI
G6 ALM
G6 AMP
G6 AMU
G6 AMY
G6 ANF
G6 ANG
G6 ANM
G6 ANN
G6 AOU
G6 APR
G6 APT
G6 APU
G6 AQF
G6 AQK
G6 AQN
G6 AQP
G6 AQQ
G6 ARB
G6 ARN
G6 ARU
G6 ASB
G6 ASI
G6 ASP
G6 ASV
G6 ASW
G6 ATJ
G6 ATO
G6 AWC
G6 AWI
G6 AWK
G6 AWS
G6 AWT
G6 AWV
G6 AXF
G6 AXL
G6 AXM
G6 AXW
G6 AYK
G6 AZA
G6 AZF
G6 AZJ
G6 AZS
G6 AZU
G6 AZV
G6 BAI
G6 BAJ
G6 BAK
G6 BAR
G6 BBC
G6 BBE
G6 BBM
G6 BCC
G6 BCT
G6 BDA
G6 BDE
G6 BDV
G6 BDX
G6 BDZ
G6 BEX
G6 BEZ
G6 BFD
G6 BFT
G6 BFW
G6 BGD
G6 BGT
G6 BGW
G6 BHK
G6 BHM
G6 BHW
G6 BIR
G6 BJM
G6 BKH
G6 BLL
G6 BNB
G6 BOD
G6 BOJ
G6 BOL
G6 BPC
G6 BPM
G6 BQI
G6 BQJ
G6 BRE
G6 BRH
G6 BRL
G6 BRO
G6 BSE
G6 BSK
G6 BSU
G6 BTL
G6 BTM
G6 BTN
G6 BUB
G6 BVN
G6 BWR
G6 BWX
G6 BXA
G6 BZA
G6 BZB
G6 BZO
G6 BZX
G6 CAF
G6 CAO
G6 CAT
G6 CAU
G6 CBR
G6 CBX
G6 CCA
G6 CCM
G6 CCM
G6 CGE
G6 CHH
G6 CHO
G6 CIW
G6 CJF
G6 CJK
G6 CJM
G6 CJN
G6 CKT
G6 CLE
G6 CLI
G6 CLO
G6 CMG
G6 CMK
G6 COQ
G6 COW
G6 CPC
G6 CPI
G6 CPZ
G6 CQA
G6 CQD
G6 CQM
G6 CQO
G6 CRB
G6 CRL
G6 CRP
G6 CRV
G6 CSY
G6 CVC
G6 CVI
G6 CVL
G6 CW
G6 CWD
G6 CWL
G6 CWR
G6 CWS
G6 CXB
G6 CYR
G6 CYZ
G6 CZC
G6 CZF
G6 CZQ
G6 DAZ
G6 DBE
G6 DBT
G6 DCB
G6 DCB
G6 DCM
G6 DDE
G6 DDQ
G6 DDS
G6 DEH
G6 DEN
G6 DEQ
G6 DES
G6 DFD
G6 DFF
G6 DGT
G6 DHJ
G6 DHS
G6 DHY
G6 DIK
G6 DIV
G6 DIY
G6 DJA
G6 DJI
G6 DJT
G6 DKB
G6 DKC
G6 DKQ
G6 DLA
G6 DMA
G6 DMR
G6 DNM
G6 DNN
G6 DOM
G6 DOP
G6 DPC
G6 DPP
G6 DPR
G6 DPW
G6 DQM
G6 DQN
G6 DRR
G6 DSM
G6 DTB
G6 DTM
G6 DTX
G6 DUG
G6 DUS
G6 DUZ
G6 DUZ
G6 DVU
G6 DVZ
G6 DWI
G6 DWX
G6 DX
G6 DXH
G6 DXM
G6 DYA
G6 DYC
G6 DYD
G6 DYE
G6 DYJ
G6 DYX
G6 DZC
G6 DZD
G6 DZH
G6 DZM
G6 DZW
G6 EAL
G6 EAS
G6 EAV
G6 EBN
G6 ECF
G6 ECH
G6 EDL
G6 EEA
G6 EEY
G6 EFH
G6 EFN
G6 EFX
G6 EFY
G6 EGG
G6 EGM
G6 EGV
G6 EHB
G6 EHO
G6 EHU
G6 EIA
G6 EIG
G6 EIJ
G6 EIL
G6 EIW
G6 EJA
G6 EJG
G6 EJP
G6 EJV
G6 EJY
G6 EKG
G6 EKJ
G6 ELL
G6 ELO
G6 ELQ
G6 ELR
G6 ELY
G6 EMH
G6 EMT
G6 ENC
G6 ENG
G6 ENH
G6 EOC
G6 EOF
G6 EOL
G6 EOY
G6 EPQ
G6 EPW
G6 EQJ
G6 ERC
G6 ERV
G6 ETE
G6 ETT
G6 EVD
G6 EVT
G6 EWV
G6 EXF
G6 EXO
G6 EXY
G6 EYB
G6 EYD
G6 EYX
G6 EYY
G6 EZS
G6 FAB
G6 FBI
G6 FCW
G6 FCX
G6 FDN
G6 FEO
G6 FEP
G6 FGK
G6 FGO
G6 FGP
G6 FHW
G6 FID
G6 FIQ
G6 FIR
G6 FIU
G6 FIV
G6 FIY
G6 FJT
G6 FKQ
G6 FLX
G6 FMZ
G6 FNE
G6 FNF
G6 FNI
G6 FNL
G6 FOM
G6 FOP
G6 FQJ
G6 FQW
G6 FRT
G6 FRU
G6 FSB
G6 FSH
G6 FSS
G6 FTN
G6 FTP
G6 FUB
G6 FUD
G6 FUW
G6 FVK
G6 FWA
G6 FWC
G6 FZU
G6 GAX
G6 GBH
G6 GCX
G6 GDG
G6 GDT
G6 GEA
G6 GEC
G6 GEI
G6 GEJ
G6 GEY
G6 GGE
G6 GGP
G6 GGU
G6 GIR
G6 GIY
G6 GJQ
G6 GJT
G6 GLP
G6 GMC
G6 GN
G6 GNN
G6 GOU
G6 GPS
G6 GQP
G6 GRE
G6 GRG
G6 GRX
G6 GTD
G6 GTN
G6 GTR
G6 GUN
G6 GUP
G6 GUR
G6 GUX
G6 GVG
G6 GVK
G6 GWR
G6 GXH
G6 GYJ
G6 GZH
G6 HAC
G6 HAE
G6 HAG
G6 HAL
G6 HBK
G6 HBN
G6 HBX
G6 HCL
G6 HCU
G6 HDP
G6 HEE
G6 HEW
G6 HFI
G6 HHB
G6 HHP
G6 HIL
G6 HIZ
G6 HJD
G6 HJO
G6 HJP
G6 HJR
G6 HJX
G6 HKQ
G6 HKR
G6 HLZ
G6 HMM
G6 HNF
G6 HNJ
G6 HNM
G6 HNU
G6 HOL
G6 HOM
G6 HOQ
G6 HOU
G6 HPC
G6 HPO
G6 HRN
G6 HSF
G6 HSM
G6 HSV
G6 HTL
G6 HUA
G6 HUH
G6 HUL
G6 HUU
G6 HVI
G6 HVP
G6 HXK
G6 HXV
G6 HYN
G6 HYU
G6 HZW
G6 IA
G6 IBC
G6 IBH
G6 IBI
G6 IBZ
G6 ICF
G6 ICJ
G6 ICP
G6 IDI
G6 IDP
G6 IDZ
G6 IEA
G6 IEJ
G6 IEZ
G6 IFD
G6 IFF
G6 IFX
G6 IFZ
G6 IGA
G6 IGB
G6 IHN
G6 IIG
G6 IJJ
G6 IKT
G6 IKW
G6 ILE
G6 IMF
G6 IMR
G6 IMT
G6 IMU
G6 IPU
G6 IQA
G6 IQL
G6 IRB
G6 IRS
G6 ISC
G6 ISD
G6 ISO
G6 ISQ
G6 ISW
G6 ITD
G6 ITF
G6 ITH
G6 ITK
G6 ITO
G6 IVT
G6 IWY
G6 IXS
G6 IZC
G6 IZI
G6 IZN
G6 IZS
G6 JAD
G6 JDZ
G6 JEA
G6 JEG
G6 JEJ
G6 JEL
G6 JEQ
G6 JEV
G6 JFS
G6 JFT
G6 JFY
G6 JGD
G6 JGS
G6 JHA
G6 JHD
G6 JHE
G6 JHR
G6 JID
G6 JIE
G6 JIV
G6 JKG
G6 JKL
G6 JKQ
G6 JKT
G6 JLH
G6 JLM
G6 JME
G6 JMF
G6 JMS
G6 JON
G6 JOX
G6 JPA
G6 JPJ
G6 JQV
G6 JQW
G6 JR
G6 JRC
G6 JRV
G6 JSB
G6 JSK
G6 JSO
G6 JSV
G6 JTB
G6 JTN
G6 JUU
G6 JUZ
G6 JVD
G6 JVS
G6 JVV
G6 JVY
G6 JWE
G6 JWR
G6 JXE
G6 JXN
G6 JYC
G6 JYG
G6 JYQ
G6 JZZ
G6 KAE
G6 KAR
G6 KAX
G6 KBG
G6 KBJ
G6 KBR
G6 KDF
G6 KDQ
G6 KEC
G6 KFN
G6 KFW
G6 KFY
G6 KGI
G6 KGW
G6 KGZ
G6 KHK
G6 KIL
G6 KJR
G6 KJU
G6 KLS
G6 KLW
G6 KMA
G6 KME
G6 KMM
G6 KMT
G6 KMV
G6 KMY
G6 KNA
G6 KNF
G6 KNH
G6 KNX
G6 KOA
G6 KOI
G6 KOS
G6 KOY
G6 KOZ
G6 KPM
G6 KQI
G6 KQP
G6 KRD
G6 KRF
G6 KRL
G6 KSM
G6 KSN
G6 KUA
G6 KUC
G6 KUG
G6 KUS
G6 KUT
G6 KWJ
G6 KWM
G6 KWN
G6 KWO
G6 KXO
G6 KYD
G6 KYX
G6 KZD
G6 KZJ
G6 KZT
G6 KZX
G6 KZZ
G6 LAM
G6 LBA
G6 LBC
G6 LBD
G6 LCC
G6 LDI
G6 LDK
G6 LDL
G6 LDN
G6 LDV
G6 LDZ
G6 LFN
G6 LGF
G6 LGZ
G6 LHD
G6 LHJ
G6 LHO
G6 LHW
G6 LIQ
G6 LIS
G6 LJJ
G6 LLB
G6 LLO
G6 LLX
G6 LMZ
G6 LOH
G6 LPF
G6 LPY
G6 LPZ
G6 LQU
G6 LQW
G6 LRE
G6 LRQ
G6 LTG
G6 LTI
G6 LTO
G6 LTS
G6 LTT
G6 LTZ
G6 LUG
G6 LUH
G6 LUW
G6 LUZ
G6 LVW
G6 LWY
G6 LXK
G6 LXQ
G6 LXX
G6 LYY
G6 LZU
G6 MAF
G6 MCI
G6 MCJ
G6 MDD
G6 MDT
G6 MDU
G6 MEK
G6 MEN
G6 MFD
G6 MGB
G6 MGS
G6 MHH
G6 MIF
G6 MII
G6 MIT
G6 MIW
G6 MIY
G6 MJG
G6 MJN
G6 MJQ
G6 MLN
G6 MMC
G6 MND
G6 MNM
G6 MNO
G6 MNQ
G6 MOG
G6 MOY
G6 MPG
G6 MQW
G6 MRF
G6 MRI
G6 MRJ
G6 MSQ
G6 MSS
G6 MTZ
G6 MVC
G6 MVO
G6 MWJ
G6 MWP
G6 MWR
G6 MXJ
G6 MXR
G6 MXT
G6 MYW
G6 MZK
G6 MZS
G6 MZX
G6 NAT
G6 NCF
G6 NCN
G6 NCQ
G6 NCR
G6 NCU
G6 NDI
G6 NDR
G6 NDR
G6 NEB
G6 NEM
G6 NER
G6 NET
G6 NFF
G6 NGO
G6 NHR
G6 NIL
G6 NIS
G6 NJF
G6 NJI
G6 NJM
G6 NJP
G6 NKD
G6 NKG
G6 NKM
G6 NLL
G6 NLW
G6 NNN
G6 NNV
G6 NOS
G6 NOT
G6 NOZ
G6 NPL
G6 NPQ
G6 NPR
G6 NQF
G6 NQO
G6 NQP
G6 NRC
G6 NSB
G6 NSY
G6 NTI
G6 NTK
G6 NUA
G6 NUD
G6 NUM
G6 NUN
G6 NUT
G6 NVA
G6 NVT
G6 NWF
G6 NWO
G6 NWR
G6 NWV
G6 NXN
G6 NXR
G6 NYC
G6 NZB
G6 NZP
G6 OAJ
G6 OAX
G6 OAY
G6 OBW
G6 OCD
G6 OCE
G6 OCL
G6 OCQ
G6 OCT
G6 OCW
G6 OCX
G6 OEY
G6 OGY
G6 OHB
G6 OI
G6 OIL
G6 OJB
G6 OJU
G6 OLS
G6 OMT
G6 OMY
G6 ONC
G6 ONN
G6 OOO
G6 OPB
G6 OQK
G6 ORA
G6 ORC
G6 ORN
G6 OTI
G6 OTJ
G6 OTM
G6 OTU
G6 OVL
G6 OVN
G6 OVQ
G6 OVR
G6 OVU
G6 OVY
G6 OVZ
G6 OXV
G6 OXW
G6 OXY
G6 OYE
G6 OZV
G6 OZW
G6 PAF
G6 PAI
G6 PBN
G6 PBS
G6 PBT
G6 PCA
G6 PCD
G6 PCU
G6 PDH
G6 PDV
G6 PDW
G6 PER
G6 PFE
G6 PHA
G6 PHB
G6 PHD
G6 PHJ
G6 PHQ
G6 PJA
G6 PJT
G6 PJV
G6 PJZ
G6 PLN
G6 PLP
G6 PLV
G6 PMN
G6 PMY
G6 PNB
G6 PNC
G6 PNH
G6 PNX
G6 POL
G6 POQ
G6 POY
G6 PPJ

G6 PPX
G6 PQW
G6 PRB
G6 PRT
G6 PRV
G6 PSI
G6 PTJ
G6 PTL
G6 PTY
G6 PUB
G6 PUC
G6 PUX
G6 PUX
G6 PVF
G6 PVQ
G6 PWO
G6 PWT
G6 PWY
G6 PXI
G6 PXK
G6 PXT
G6 PZP
G6 PZW
G6 RBH
G6 RBK
G6 RCA
G6 RCP
G6 RDG
G6 RFG
G6 RFQ
G6 RHZ
G6 RIB
G6 RIK
G6 RIO
G6 RJT
G6 RKE
G6 RKP
G6 RMS
G6 RNE
G6 RNM
G6 ROP
G6 RPM
G6 RPX
G6 RQO
G6 RRL
G6 RSC
G6 RSE
G6 RST
G6 RSY
G6 RTV
G6 RUK
G6 RUQ
G6 RVB
G6 RVW
G6 RWI
G6 RWT
G6 RWV
G6 RYC
G6 RZA
G6 SAF
G6 SAG
G6 SAS
G6 SBF
G6 SBR
G6 SCD
G6 SCR
G6 SDJ
G6 SDP
G6 SEA
G6 SEA
G6 SEM
G6 SET
G6 SFB
G6 SFN
G6 SFP
G6 SHB
G6 SIC
G6 SIY
G6 SJF
G6 SJH
G6 SJJ
G6 SJO
G6 SJY
G6 SKC
G6 SKL
G6 SKN
G6 SKO
G6 SLK
G6 SLM
G6 SLU
G6 SMZ
G6 SNJ
G6 SNL
G6 SNO
G6 SNR
G6 SPC
G6 SPE
G6 SQC
G6 SQD
G6 SSF
G6 SSX
G6 SUD
G6 SVB
G6 SVN
G6 SXW
G6 SYC
G6 SYT
G6 SZF
G6 SZK
G6 SZQ
G6 TAT
G6 TAU
G6 TBA
G6 TBT
G6 TCO
G6 TDC
G6 TDL
G6 TDP
G6 TDR
G6 TEC
G6 TFB
G6 TFC
G6 TFU
G6 TFY
G6 TGD
G6 TGO
G6 TGP
G6 TGV
G6 TGZ
G6 THB
G6 THR
G6 THX
G6 TIH
G6 TIM
G6 TIX
G6 TJS
G6 TKI
G6 TKK
G6 TKS
G6 TKX
G6 TLH
G6 TLR
G6 TME
G6 TMK
G6 TMM
G6 TMO
G6 TMW
G6 TNO
G6 TNZ
G6 TOB
G6 TOC
G6 TOZ
G6 TPK
G6 TPV
G6 TQO
G6 TQT
G6 TQU
G6 TRD
G6 TRE
G6 TRS
G6 TRT
G6 TRU
G6 TSH
G6 TTL
G6 TTW
G6 TUH
G6 TUW
G6 TUZ
G6 TXW
G6 TXZ
G6 TYJ
G6 TYQ
G6 TYS
G6 TZB
G6 UAD
G6 UAU
G6 UAX
G6 UAY
G6 UBB
G6 UBD
G6 UBL
G6 UBM
G6 UBP
G6 UCJ
G6 UDE
G6 UDN
G6 UEB
G6 UEJ
G6 UEO
G6 UFJ
G6 UFT
G6 UGD
G6 UGL
G6 UIA
G6 UID
G6 UIP
G6 UIU
G6 UJB
G6 UJH
G6 UKB
G6 UKK
G6 UKV
G6 ULP
G6 ULX
G6 UNM
G6 UOD
G6 UQB
G6 UQC
G6 UQQ
G6 URA
G6 URB
G6 URE
G6 URY
G6 USH
G6 UT
G6 UTC
G6 UTD
G6 UTP
G6 UTQ
G6 UUM
G6 UVE
G6 UVH
G6 UWH
G6 UXD
G6 UXN
G6 UYG
G6 UZF
G6 UZH
G6 UZT
G6 VAH
G6 VAM
G6 VAM
G6 VCY
G6 VDH
G6 VDP
G6 VGW
G6 VHC
G6 VHF
G6 VIR
G6 VIW
G6 VIX
G6 VJE
G6 VLB
G6 VLK
G6 VME
G6 VND
G6 VNH
G6 VQU
G6 VQX
G6 VRD
G6 VRH
G6 VRT
G6 VTR
G6 VTU
G6 VUI
G6 VUL
G6 VVA
G6 VVC
G6 VVM
G6 VVT
G6 VVX
G6 VWC
G6 VXB
G6 VXE
G6 VXM
G6 VXX
G6 VYS
G6 VYT
G6 WBJ
G6 WBV
G6 WDE
G6 WDV
G6 WEX
G6 WEZ
G6 WFQ
G6 WGD
G6 WGK
G6 WGP
G6 WHI
G6 WHR
G6 WII
G6 WIM
G6 WJP
G6 WKW
G6 WLA
G6 WLL
G6 WLN
G6 WLZ
G6 WNA
G6 WNO
G6 WNW
G6 WOE
G6 WOL
G6 WOM
G6 WOV
G6 WOZ
G6 WQB
G6 WQU
G6 WRA
G6 WRB
G6 WRI
G6 WRQ
G6 WRW
G6 WSI
G6 WTA
G6 WTX
G6 WVK
G6 WVX
G6 WWO
G6 WXC
G6 WYN
G6 WZH
G6 WZO
G6 XAC
G6 XAI
G6 XBC
G6 XBF
G6 XBW
G6 XBX
G6 XCF
G6 XDP
G6 XDQ
G6 XEG
G6 XFD
G6 XFI
G6 XFP
G6 XFW
G6 XGL
G6 XHM
G6 XIP
G6 XIQ
G6 XJU
G6 XLU
G6 XLV
G6 XMI
G6 XMO
G6 XMY
G6 XNC
G6 XOI
G6 XPM
G6 XQW
G6 XRA
G6 XRQ
G6 XRX
G6 XSP
G6 XTL
G6 XTN
G6 XTO
G6 XTR
G6 XTW
G6 XUB
G6 XVN
G6 XVO
G6 XVT
G6 XVW
G6 XWE
G6 XWK
G6 XWN
G6 XWO
G6 XWT
G6 XWU
G6 XXW
G6 XYE
G6 XYT
G6 XYZ
G6 XZL
G6 XZV
G6 YAF
G6 YAL
G6 YBM
G6 YBY
G6 YCK
G6 YDD
G6 YGD
G6 YGG
G6 YHT
G6 YHU
G6 YHV
G6 YJN
G6 YKT
G6 YLJ
G6 YMP
G6 YMV
G6 YNZ
G6 YON
G6 YOS
G6 YOV
G6 YPU
G6 YQY
G6 YRK
G6 YRM
G6 YTC
G6 YTZ
G6 YUH
G6 YUI
G6 YVE
G6 YVX
G6 YWB
G6 YWC
G6 YWF
G6 YWX
G6 YYV
G6 YYX
G6 ZAB
G6 ZBB
G6 ZBI
G6 ZC
G6 ZDQ
G6 ZFJ
G6 ZFY
G6 ZGD
G6 ZGO
G6 ZIM
G6 ZIS
G6 ZKH
G6 ZKT
G6 ZLA
G6 ZLI
G6 ZMC
G6 ZMF
G6 ZMK
G6 ZMY
G6 ZMZ
G6 ZOD
G6 ZOI
G6 ZOO
G6 ZOV
G6 ZPD
G6 ZQE
G6 ZQQ
G6 ZQX
G6 ZQY
G6 ZSD
G6 ZSN
G6 ZSR
G6 ZTA
G6 ZTV
G6 ZVE
G6 ZWQ
G6 ZXU
G6 ZYD
G6 ZYF
G6 ZYG
G6 ZYP
G6 ZZP
G6 ZZX
G6 ZZZ

G7

G7 AAC
G7 AAF
G7 AAQ
G7 AAW
G7 ACE
G7 ACU
G7 ACW
G7 AEB
G7 AED
G7 AEJ
G7 AEO
G7 AEP
G7 AER
G7 AEU
G7 AFK
G7 AFM
G7 AGZ
G7 AHE
G7 AHH
G7 AHJ
G7 AHQ
G7 AHS
G7 AHT
G7 AHY
G7 AIE
G7 AII
G7 AIJ
G7 AIS
G7 AIT
G7 AIV
G7 AJC
G7 AJH
G7 AJM
G7 AKF
G7 AKL
G7 ALG
G7 ALN
G7 ALW
G7 ALZ
G7 AME
G7 AMF
G7 AMG
G7 AMH
G7 AMJ
G7 AMK
G7 AMO
G7 AMU
G7 ANF
G7 ANJ
G7 ANZ
G7 AOO
G7 AOP
G7 AOR
G7 AOZ
G7 APC
G7 AQR
G7 AQU
G7 AQX
G7 AQY
G7 ARC
G7 ASJ
G7 ASX
G7 ATQ
G7 ATV
G7 AUG
G7 AVO
G7 AWN
G7 AXB
G7 AXC
G7 AXF
G7 AXG
G7 AXJ
G7 AXU
G7 AXV
G7 BAP
G7 BBF
G7 BBL
G7 BBP
G7 BBQ
G7 BCB
G7 BCE
G7 BCL
G7 BCS
G7 BDP
G7 BDQ
G7 BEA
G7 BED
G7 BEH
G7 BEX
G7 BFD
G7 BFK
G7 BFO
G7 BFP
G7 BGA
G7 BHB
G7 BHH
G7 BHN
G7 BHQ
G7 BIU
G7 BJA
G7 BJH
G7 BJJ
G7 BJQ
G7 BJU
G7 BJY
G7 BKI
G7 BKK
G7 BKU
G7 BKV
G7 BKX
G7 BLR
G7 BMA
G7 BMS
G7 BNV
G7 BPE
G7 BPO
G7 BQC
G7 BRC
G7 BRH
G7 BRV
G7 BSV
G7 BSZ
G7 BTA
G7 BUG
G7 BUQ
G7 BUT
G7 BUZ
G7 BVK
G7 BVY
G7 BWP
G7 BZN
G7 BZW
G7 BZX
G7 BZZ
G7 CAK
G7 CAN
G7 CAQ
G7 CAR
G7 CAW
G7 CAY
G7 CBG
G7 CBL
G7 CBM
G7 CCI
G7 CCZ
G7 CDA
G7 CDK
G7 CDZ
G7 CFJ
G7 CFO
G7 CFR
G7 CFU
G7 CFV
G7 CFY
G7 CGG
G7 CGK
G7 CGV
G7 CHO
G7 CHT
G7 CHX
G7 CHY
G7 CIB
G7 CJY
G7 CKD
G7 CKF
G7 CKR
G7 CKU
G7 CKX
G7 CLP
G7 CLR
G7 CLZ
G7 CMJ
G7 CMK
G7 CMS
G7 CNF
G7 CNH
G7 CNQ
G7 COI
G7 COU
G7 COV
G7 COW
G7 CPB
G7 CPC
G7 CRO
G7 CRX
G7 CTA
G7 CTD
G7 CTX
G7 CUG
G7 CUH
G7 CUI
G7 CVH
G7 CVJ
G7 CWA
G7 CWH
G7 CWQ
G7 CWY
G7 CXP
G7 CXP
G7 CXR
G7 CYK
G7 CYO
G7 CYP
G7 CYX
G7 CZH
G7 CZO
G7 CZT
G7 CZZ
G7 DAA
G7 DAK
G7 DBA
G7 DBE
G7 DBJ
G7 DBK
G7 DBL
G7 DBM
G7 DBS
G7 DCD
G7 DCG
G7 DCO
G7 DDD
G7 DDY
G7 DEL
G7 DEM
G7 DEQ
G7 DER
G7 DET
G7 DFM
G7 DFO
G7 DGH
G7 DGR
G7 DGU
G7 DGW
G7 DGZ
G7 DHL
G7 DIH
G7 DIQ
G7 DJP
G7 DJW
G7 DKE
G7 DKJ
G7 DKL
G7 DKM
G7 DKQ
G7 DKW
G7 DKX
G7 DLC
G7 DLH
G7 DLM
G7 DLS
G7 DLV
G7 DLW
G7 DMD
G7 DMJ
G7 DMO
G7 DOH
G7 DOI
G7 DOM
G7 DOO
G7 DOT
G7 DPJ
G7 DPP
G7 DPY
G7 DQC
G7 DQD
G7 DQF
G7 DQG
G7 DQM
G7 DQR
G7 DQS
G7 DRI
G7 DRK
G7 DRS
G7 DSG
G7 DSI
G7 DSJ
G7 DSK
G7 DSX
G7 DUO
G7 DUP
G7 DUR
G7 DUW
G7 DVG
G7 DVY
G7 DWD
G7 DWK
G7 DXP
G7 DXS
G7 DXY
G7 DYA
G7 DYG
G7 DYH
G7 DYI
G7 DYO
G7 DYQ
G7 DZI
G7 DZS
G7 DZX
G7 EAG
G7 EAK
G7 EAP
G7 EBK
G7 EBP
G7 EBY
G7 ECK
G7 ECL
G7 ECM
G7 ECV
G7 EDC
G7 EDM
G7 EDQ
G7 EDT
G7 EDU
G7 EEP
G7 EET
G7 EEY
G7 EFC
G7 EFE
G7 EFN
G7 EGI
G7 EGK
G7 EGN
G7 EGZ
G7 EIM
G7 EIO
G7 EIT
G7 EIX
G7 EJG
G7 EKI
G7 EKY
G7 ELF
G7 ELK
G7 ELM
G7 ELR
G7 EMF
G7 EMK
G7 EML
G7 EOB
G7 EOL
G7 EOO
G7 EOQ
G7 EOR
G7 EOT
G7 EOX
G7 EOZ
G7 EPC
G7 EPG
G7 EQM
G7 EQU
G7 EQZ
G7 ERB
G7 ERM
G7 ERO
G7 ERU
G7 ERX
G7 ESB
G7 ESK
G7 ESP
G7 ESR
G7 ETA
G7 ETI
G7 ETN
G7 ETR
G7 ETT
G7 ETU
G7 ETW
G7 ETX
G7 ETZ
G7 EUA
G7 EUH
G7 EUM
G7 EUW
G7 EVA
G7 EVE
G7 EVW
G7 EWG
G7 EXN
G7 EYX
G7 EZG
G7 EZM
G7 EZP
G7 EZR
G7 EZT
G7 EZW
G7 EZY
G7 FAL
G7 FAN
G7 FAX
G7 FBD
G7 FBX
G7 FCT
G7 FCV
G7 FDN
G7 FDP
G7 FDR
G7 FDT
G7 FDU
G7 FDZ
G7 FEK
G7 FEW
G7 FEX
G7 FEZ
G7 FFA
G7 FFE
G7 FFL
G7 FGF
G7 FHC
G7 FHI
G7 FHJ
G7 FHK
G7 FHN
G7 FHW
G7 FHY
G7 FIF
G7 FIG
G7 FIT
G7 FJA
G7 FJI
G7 FKB
G7 FKE
G7 FKH
G7 FLO
G7 FLR
G7 FLT
G7 FLW
G7 FMF
G7 FMM
G7 FMN
G7 FMX
G7 FNA
G7 FNE
G7 FNS
G7 FOF
G7 FOJ
G7 FOK
G7 FPE
G7 FPL
G7 FQF
G7 FQG
G7 FQH
G7 FQW
G7 FRA
G7 FRB
G7 FRI
G7 FSI
G7 FTB
G7 FTL
G7 FTN
G7 FTP
G7 FTQ
G7 FTR
G7 FTY
G7 FUJ
G7 FVB
G7 FVD
G7 FVF
G7 FVG
G7 FVT
G7 FVV
G7 FWL
G7 FWO
G7 FWR
G7 FWT
G7 FWV
G7 FXM
G7 FXV
G7 FYO
G7 FYR
G7 FZV
G7 FZX
G7 GAD
G7 GAF
G7 GAM
G7 GAN
G7 GAS
G7 GAV
G7 GAX
G7 GBA
G7 GBC
G7 GCK
G7 GCO
G7 GCQ
G7 GCR
G7 GDH
G7 GDR
G7 GEQ
G7 GFS
G7 GFU
G7 GFW
G7 GGS
G7 GGU
G7 GGV
G7 GGW
G7 GGX
G7 GGZ
G7 GHA
G7 GHK
G7 GHS
G7 GHV
G7 GJF
G7 GJH
G7 GKO
G7 GKV
G7 GLE
G7 GLK
G7 GLN
G7 GLO
G7 GLT
G7 GMF
G7 GMG
G7 GMJ
G7 GMN
G7 GNB
G7 GNF
G7 GNG
G7 GNK
G7 GNL
G7 GNM
G7 GOC
G7 GOP
G7 GPR
G7 GQF
G7 GQP
G7 GRB
G7 GRK
G7 GSH
G7 GSI
G7 GSL
G7 GSN
G7 GSP
G7 GSQ
G7 GSW
G7 GTB
G7 GTI
G7 GTO
G7 GTV
G7 GUD
G7 GUE
G7 GUF
G7 GUJ
G7 GUN
G7 GVK
G7 GVV
G7 GVZ
G7 GWB
G7 GWM
G7 GWR
G7 GWV
G7 GXF
G7 GXG
G7 GXK
G7 GXY
G7 GYB
G7 GYP
G7 GZE
G7 GZG
G7 GZH
G7 HAA
G7 HAT
G7 HAV
G7 HAW
G7 HCD
G7 HCE
G7 HDA
G7 HDB
G7 HDE
G7 HDI
G7 HDJ
G7 HDO
G7 HDX
G7 HEL
G7 HEM
G7 HER
G7 HEU
G7 HEX
G7 HGP
G7 HIF
G7 HIP
G7 HJY
G7 HKD
G7 HKF
G7 HKK
G7 HKM
G7 HKP
G7 HKY
G7 HLF
G7 HLH
G7 HLJ
G7 HLL
G7 HLO
G7 HMC
G7 HMJ
G7 HNH
G7 HNH
G7 HNK
G7 HNN
G7 HNO
G7 HNQ
G7 HNX
G7 HOB
G7 HOD
G7 HOI
G7 HOW
G7 HPB
G7 HPE
G7 HPG
G7 HPH
G7 HPK
G7 HPM
G7 HPP
G7 HPS
G7 HPW
G7 HQB
G7 HQH
G7 HRC
G7 HRQ
G7 HRU
G7 HRW
G7 HSE
G7 HSG
G7 HST
G7 HTP
G7 HTS
G7 HUD
G7 HUH
G7 HUR
G7 HUW
G7 HVT
G7 HVW
G7 HWC
G7 HWD
G7 HWF
G7 HWJ
G7 HWK
G7 HWL
G7 HWT
G7 HWU
G7 HWW
G7 HXC
G7 HXJ
G7 HXM
G7 HXQ
G7 HXS
G7 HXV
G7 HXW
G7 HXX
G7 HYF
G7 HYK
G7 HZI
G7 IBI
G7 IBO
G7 IBY
G7 ICH
G7 ICQ
G7 ICS
G7 ICV
G7 ICW
G7 ICX
G7 ICY
G7 IDB
G7 IDD
G7 IDF
G7 IDJ
G7 IDL
G7 IEE
G7 IEI
G7 IFD
G7 IFF
G7 IGB
G7 IGC
G7 IGH
G7 IGZ
G7 IHA
G7 IHS
G7 III
G7 IIM
G7 IIU
G7 IJK
G7 IJM
G7 IJR
G7 IJS
G7 IJT
G7 IKR
G7 IKT
G7 IKU
G7 IKW
G7 IKY
G7 ILK
G7 ILV
G7 ILZ
G7 ILZ
G7 IMF
G7 IND
G7 INM
G7 INN
G7 INO
G7 INP
G7 INS
G7 INV
G7 INX
G7 IOV
G7 IOW
G7 IOY
G7 IPB
G7 IPK
G7 IPK
G7 IPT
G7 IQJ
G7 IQK
G7 IQT
G7 IQW
G7 IRV
G7 ISO
G7 ISY
G7 ISZ
G7 ITD
G7 ITL
G7 IUJ
G7 IUK
G7 IUO
G7 IUU
G7 IUV
G7 IVC
G7 IVH
G7 IVJ
G7 IVP
G7 IVR
G7 IVY
G7 IWG
G7 IWS
G7 IXA
G7 IXJ
G7 IXL
G7 IXM
G7 IXO
G7 IXU
G7 IYB
G7 IYD
G7 IYR
G7 IYS
G7 IYT
G7 IYW
G7 IYZ
G7 IZD
G7 IZG
G7 IZH
G7 IZI
G7 IZQ
G7 IZR
G7 IZX
G7 IZZ
G7 JAB
G7 JAC
G7 JAD
G7 JAH
G7 JAK
G7 JBB
G7 JBE
G7 JBQ
G7 JBT
G7 JCC
G7 JCJ
G7 JCP
G7 JDD
G7 JDE
G7 JDV
G7 JDW
G7 JEW
G7 JEX
G7 JFF
G7 JFH
G7 JFS
G7 JGD
G7 JGL
G7 JHF
G7 JHJ
G7 JHL
G7 JHO
G7 JHP
G7 JHQ
G7 JIC
G7 JID
G7 JIQ
G7 JIW
G7 JJF
G7 JJI
G7 JJZ
G7 JKB
G7 JKO
G7 JKV
G7 JKX
G7 JKZ
G7 JLI
G7 JLM
G7 JLR
G7 JLZ
G7 JMG
G7 JMI
G7 JND
G7 JNT
G7 JNY
G7 JOB
G7 JOG
G7 JOK
G7 JON
G7 JOP
G7 JOQ
G7 JOV
G7 JOX
G7 JOY
G7 JPD
G7 JPE
G7 JPH
G7 JPU
G7 JPW
G7 JPY
G7 JQE
G7 JQK
G7 JQL
G7 JRE
G7 JRV
G7 JRW
G7 JRZ
G7 JSF
G7 JSP
G7 JTQ
G7 JTS
G7 JTT
G7 JUE
G7 JUG
G7 JUS
G7 JUY
G7 JVL
G7 JVW
G7 JWB
G7 JWR
G7 JYD
G7 JYE
G7 JYF
G7 JYM
G7 JYR
G7 JYS
G7 JYU
G7 JZD
G7 JZE
G7 JZL
G7 JZO
G7 JZP
G7 JZR
G7 JZU
G7 JZW
G7 KAD
G7 KAH
G7 KAT
G7 KAW
G7 KBF
G7 KBL
G7 KBP
G7 KBV
G7 KBY
G7 KCD
G7 KCF
G7 KCK
G7 KCS
G7 KCW
G7 KCZ
G7 KDA
G7 KDC
G7 KDP
G7 KDZ
G7 KEH
G7 KEJ
G7 KES
G7 KET
G7 KEZ
G7 KFA
G7 KFC
G7 KFD
G7 KFG
G7 KFH
G7 KFI
G7 KFL
G7 KFT
G7 KFW
G7 KFX
G7 KFY
G7 KGB
G7 KGC
G7 KHN
G7 KIG
G7 KIM
G7 KJF
G7 KJG
G7 KKH
G7 KKJ
G7 KKL
G7 KKN
G7 KKP
G7 KKR
G7 KKS
G7 KKV
G7 KLA
G7 KLD
G7 KLI
G7 KMJ
G7 KMX
G7 KMZ
G7 KNE
G7 KNF
G7 KNI
G7 KNO
G7 KNP
G7 KNV
G7 KNZ
G7 KOA
G7 KOQ
G7 KOU
G7 KPB
G7 KPN
G7 KPO
G7 KPP
G7 KPQ
G7 KPR
G7 KPS
G7 KPT
G7 KPU
G7 KPX
G7 KPY
G7 KPZ
G7 KQA
G7 KQC
G7 KQF
G7 KQG
G7 KQI
G7 KQK
G7 KQM
G7 KQO
G7 KQY
G7 KSI
G7 KSW
G7 KTE
G7 KTK
G7 KTM
G7 KTN
G7 KTO
G7 KUF
G7 KUJ
G7 KUZ
G7 KVE
G7 KVK
G7 KVO
G7 KWC
G7 KWI
G7 KWX
G7 KWZ
G7 KXB
G7 KXD
G7 KXE
G7 KXM
G7 KXW
G7 KXX
G7 KXY
G7 KYB
G7 KYC
G7 KYM
G7 KYM
G7 KYN
G7 KZA
G7 KZF
G7 KZX
G7 KZZ
G7 LAE
G7 LBG
G7 LBS
G7 LBY
G7 LCB
G7 LCC
G7 LCG
G7 LCL
G7 LCM
G7 LCU
G7 LCY
G7 LDE
G7 LDF
G7 LEC
G7 LEE
G7 LEU
G7 LFR
G7 LFW
G7 LGD
G7 LGK
G7 LHA
G7 LHB
G7 LHE
G7 LHF
G7 LHG
G7 LHM
G7 LIM
G7 LIN
G7 LIZ
G7 LJN
G7 LJV
G7 LJW
G7 LJY
G7 LKB
G7 LKN
G7 LLQ
G7 LLS
G7 LMC
G7 LMF
G7 LMH
G7 LMJ
G7 LMK
G7 LMN
G7 LMP
G7 LMY
G7 LNC
G7 LOF
G7 LOH
G7 LOI
G7 LOM
G7 LOP
G7 LOR
G7 LOS
G7 LOU
G7 LQE
G7 LQH
G7 LQR
G7 LQY
G7 LRD
G7 LRE
G7 LRF
G7 LRJ
G7 LRK
G7 LRO
G7 LRQ
G7 LRR
G7 LRS
G7 LSH
G7 LSL
G7 LSR
G7 LSS
G7 LST
G7 LSV
G7 LSY
G7 LTQ
G7 LUC
G7 LUD
G7 LUS
G7 LUU
G7 LUV
G7 LVF
G7 LVO
G7 LVV
G7 LVW
G7 LVX
G7 LVY
G7 LVZ
G7 LWC
G7 LWI
G7 LWK
G7 LWM
G7 LWN
G7 LWT
G7 LWV
G7 LXC
G7 LXL
G7 LXP
G7 LXT
G7 LXX
G7 LYP
G7 LYQ
G7 LYR
G7 LZI
G7 LZK
G7 LZW
G7 MAJ
G7 MAX
G7 MAZ
G7 MCI
G7 MCP
G7 MDH
G7 MDR
G7 MDU
G7 MEI
G7 MFK
G7 MFT
G7 MGK
G7 MGL
G7 MGN
G7 MHK
G7 MHM
G7 MHN
G7 MIJ
G7 MIL
G7 MIW
G7 MIX
G7 MJA
G7 MJG
G7 MJK
G7 MJL
G7 MJN
G7 MJO
G7 MJW
G7 MJZ
G7 MKK
G7 MKO
G7 MKR
G7 MKS
G7 MLF
G7 MLZ
G7 MMA
G7 MMF
G7 MMM
G7 MMS
G7 MMT
G7 MNH
G7 MNM
G7 MNN
G7 MNY
G7 MOI
G7 MOT
G7 MPW
G7 MQB
G7 MQD
G7 MQL
G7 MQT
G7 MQV
G7 MRC
G7 MRD
G7 MRG
G7 MRI
G7 MRK
G7 MRN
G7 MRP
G7 MRT
G7 MRV
G7 MRY
G7 MSD
G7 MTK
G7 MTX
G7 MUI
G7 MUL
G7 MUQ
G7 MVC
G7 MVP
G7 MVR
G7 MWF
G7 MWN
G7 MXB
G7 MXJ
G7 MYB
G7 MYL
G7 MYR
G7 MYZ
G7 MZH
G7 MZN
G7 MZO
G7 MZR
G7 MZT
G7 NAD
G7 NAG
G7 NAM
G7 NAQ
G7 NAS
G7 NAU
G7 NBA
G7 NBK
G7 NBN
G7 NBW
G7 NCK
G7 NCN
G7 NCO
G7 NCT
G7 NDF
G7 NDG
G7 NDH
G7 NDJ
G7 NDL
G7 NDY
G7 NEI
G7 NEL
G7 NEO
G7 NFJ
G7 NFP
G7 NFS
G7 NFU
G7 NGG
G7 NGS
G7 NGV
G7 NGZ
G7 NHA
G7 NHG
G7 NIE
G7 NIG
G7 NIP
G7 NIS
G7 NJC
G7 NJF
G7 NJK
G7 NJO
G7 NKK
G7 NKL
G7 NKP
G7 NKT
G7 NLB
G7 NLQ
G7 NLV
G7 NLX
G7 NMF
G7 NMG
G7 NMO
G7 NNE
G7 NNF
G7 NNG
G7 NNY
G7 NOA
G7 NOF
G7 NOJ
G7 NOO
G7 NOT
G7 NOV
G7 NOX
G7 NOY
G7 NOZ
G7 NPK
G7 NPP
G7 NPW
G7 NQH
G7 NQI
G7 NQL
G7 NQN
G7 NQO
G7 NQS
G7 NRE
G7 NRH
G7 NRI
G7 NSA
G7 NSC
G7 NSR
G7 NSU
G7 NSY
G7 NTB
G7 NTC
G7 NTY
G7 NUK
G7 NVA
G7 NVE
G7 NVK
G7 NVO
G7 NVQ
G7 NWA
G7 NWB
G7 NWI
G7 NWK
G7 NWR
G7 NYC
G7 NYK
G7 NZK
G7 NZT
G7 OAE
G7 OAK
G7 OAU
G7 OBH
G7 OBI
G7 OBK
G7 OBO
G7 OBU
G7 OBY
G7 OBZ
G7 OCD
G7 OCG
G7 OCM
G7 OCO
G7 OCS
G7 OCT
G7 ODA
G7 ODD
G7 ODE
G7 ODJ
G7 ODO
G7 ODQ
G7 ODX
G7 ODY
G7 OEL
G7 OEM
G7 OEP
G7 OEX
G7 OFC
G7 OFR
G7 OFX
G7 OGA
G7 OGG
G7 OGX
G7 OHA
G7 OHK
G7 OHT
G7 OII
G7 OIJ
G7 OIW
G7 OJH
G7 OJN
G7 OJR
G7 OJW
G7 OKD
G7 OKE
G7 OKU
G7 OKZ
G7 OLA
G7 OLD
G7 OLI
G7 OLP
G7 OLS
G7 OML
G7 OML
G7 OMT
G7 OMZ
G7 OND
G7 ONG
G7 ONK
G7 ONM
G7 ONS
G7 ONX
G7 ONY
G7 OOA
G7 OOD
G7 OOJ
G7 OOR
G7 OOW
G7 OOX
G7 OOZ
G7 OPC
G7 OPO
G7 OPZ
G7 OQH
G7 OQJ
G7 OQK
G7 OQN
G7 OQS
G7 OQV
G7 OQZ
G7 ORH
G7 ORI
G7 OSG
G7 OSL
G7 OTY
G7 OUJ
G7 OUL
G7 OUN
G7 OVA
G7 OVC
G7 OVV
G7 OWE
G7 OWH
G7 OXM
G7 OXX
G7 OYC
G7 OYO
G7 OZF
G7 OZO
G7 OZY
G7 PAB
G7 PAJ
G7 PAL
G7 PAS
G7 PAV
G7 PAZ
G7 PBX
G7 PCA
G7 PCJ
G7 PCL
G7 PCM
G7 PCN
G7 PCQ
G7 PDL
G7 PDP
G7 PDY
G7 PDZ
G7 PEA
G7 PEK
G7 PEM
G7 PEQ
G7 PET
G7 PEY
G7 PFB
G7 PFE
G7 PFN
G7 PFO
G7 PFU
G7 PFX
G7 PGB
G7 PGS
G7 PGZ
G7 PHN
G7 PHU
G7 PHZ
G7 PIA
G7 PID
G7 PIM
G7 PIQ
G7 PJC
G7 PJH
G7 PJH
G7 PJT
G7 PJV
G7 PKF
G7 PKO
G7 PKS
G7 PKU
G7 PKV
G7 PKW
G7 PKX
G7 PLF
G7 PLG
G7 PLN
G7 PLT
G7 PLX
G7 PLY
G7 PLZ
G7 PME
G7 PMJ
G7 PMO
G7 PMS
G7 PNQ
G7 PNR
G7 PNU
G7 POG
G7 POJ
G7 POL
G7 POU
G7 PPB
G7 PPE
G7 PPF
G7 PPM
G7 PPT
G7 PQF
G7 PQN
G7 PQR
G7 PQU
G7 PQV
G7 PRA
G7 PRH
G7 PRJ
G7 PRN
G7 PRR
G7 PRY
G7 PSA
G7 PSQ
G7 PSX
G7 PSY
G7 PTL
G7 PTQ
G7 PTU
G7 PUD
G7 PUQ
G7 PUV
G7 PVA
G7 PVJ
G7 PVR
G7 PVS
G7 PVV
G7 PWE
G7 PWH
G7 PXA
G7 PXB
G7 PXD
G7 PXE
G7 PXF
G7 PXK
G7 PXP
G7 PXT
G7 PYH
G7 PYI
G7 PZI
G7 PZP
G7 PZS
G7 PZV
G7 PZX
G7 RAC
G7 RAD
G7 RAP
G7 RAR
G7 RBG
G7 RBV
G7 RBW
G7 RC
G7 RCD
G7 RCI
G7 RCQ
G7 RDC
G7 RDI
G7 RDM
G7 RDO
G7 RDV
G7 RDX
G7 REB
G7 RFG
G7 RFI
G7 RFR
G7 RGB
G7 RGH
G7 RGQ

Withheld

G7 RGW
G7 RHU
G7 RID
G7 RIK
G7 RIQ
G7 RIY
G7 RJA
G7 RJC
G7 RJL
G7 RJT
G7 RJU
G7 RJV
G7 RJZ
G7 RKI
G7 RKK
G7 RKM
G7 RKY
G7 RLI
G7 RMB
G7 RMH
G7 RMI
G7 RMN
G7 RMV
G7 RNE
G7 ROF
G7 ROS
G7 RPF
G7 RPG
G7 RPH
G7 RPO
G7 RPR
G7 RPS
G7 RPX
G7 RQA
G7 RQB
G7 RQE
G7 RQG
G7 RQJ
G7 RQM
G7 RQN
G7 RQS
G7 RQT
G7 RQU
G7 RQX
G7 RRF
G7 RRI
G7 RRP
G7 RRQ
G7 RRT
G7 RSD
G7 RSI
G7 RSO
G7 RST
G7 RSV
G7 RSZ
G7 RTK
G7 RUA
G7 RUG
G7 RUL
G7 RUM
G7 RUT
G7 RUV
G7 RVJ
G7 RVL
G7 RVS
G7 RWA
G7 RWH
G7 RWI
G7 RWJ
G7 RWO
G7 RWQ
G7 RWU
G7 RWX
G7 RXG
G7 RXH
G7 RXP
G7 RXR
G7 RXT
G7 RXV
G7 RXY
G7 RYE
G7 RYR
G7 RYV
G7 RZA
G7 RZD
G7 RZF
G7 RZM
G7 RZT
G7 RZU
G7 SAB
G7 SBH
G7 SBQ
G7 SBU
G7 SBV
G7 SBW
G7 SCC
G7 SCF
G7 SCY
G7 SDJ
G7 SDV
G7 SEH
G7 SEI
G7 SET
G7 SFC
G7 SFQ
G7 SFR
G7 SFV
G7 SFW
G7 SGB
G7 SGG
G7 SGS
G7 SHG
G7 SHW
G7 SHX
G7 SIC
G7 SII
G7 SIV
G7 SJP
G7 SJV
G7 SKE
G7 SKI
G7 SKJ
G7 SLD
G7 SLK
G7 SLM
G7 SLO
G7 SLS
G7 SLT
G7 SMA
G7 SMK
G7 SMO
G7 SMW
G7 SMY
G7 SOG
G7 SOI
G7 SOJ
G7 SOQ
G7 SPG
G7 SPI
G7 SPK
G7 SPR
G7 SPS
G7 SPT
G7 SPU
G7 SPV
G7 SPW
G7 SQC
G7 SQK
G7 SQL
G7 SQR
G7 SQS
G7 SQU
G7 SQX
G7 SRE
G7 SRG
G7 SSD
G7 SSE
G7 SSZ
G7 STA
G7 STB
G7 SUD
G7 SUH
G7 SUI
G7 SUP
G7 SUR
G7 SUZ
G7 SVB
G7 SVG
G7 SVK
G7 SVW
G7 SWC
G7 SWP
G7 SXA
G7 SXE
G7 SXH
G7 SXQ
G7 SXR
G7 SXX
G7 SYB
G7 SYK
G7 SYO
G7 SYP
G7 SYW
G7 SYZ
G7 SZD
G7 SZJ
G7 SZN
G7 SZR
G7 SZT
G7 TAC
G7 TAG
G7 TAH
G7 TAI
G7 TAJ
G7 TAO
G7 TAR
G7 TAS
G7 TBE
G7 TBH
G7 TBT
G7 TCI
G7 TCK
G7 TCL
G7 TCM
G7 TDO
G7 TDP
G7 TDW
G7 TDX
G7 TDY
G7 TDZ
G7 TED
G7 TEF
G7 TEM
G7 TER
G7 TFD
G7 TFH
G7 TFJ
G7 TFP
G7 TGE
G7 TGL
G7 TGO
G7 TGR
G7 TGV
G7 TGY
G7 THC
G7 THD
G7 THL
G7 THT
G7 THX
G7 TIA
G7 TIC
G7 TII
G7 TIS
G7 TIZ
G7 TJE
G7 TJK
G7 TJS
G7 TJX
G7 TJY
G7 TKE
G7 TKJ
G7 TKK
G7 TKL
G7 TLG
G7 TLN
G7 TLP
G7 TMG
G7 TMJ
G7 TMW
G7 TNA
G7 TNC
G7 TND
G7 TNF
G7 TNL
G7 TOC
G7 TOG
G7 TOP
G7 TOW
G7 TOX
G7 TPI
G7 TPL
G7 TPP
G7 TPR
G7 TPV
G7 TPX
G7 TQG
G7 TQJ
G7 TQS
G7 TQY
G7 TQZ
G7 TRE
G7 TRJ
G7 TRN
G7 TRW
G7 TRZ
G7 TSA
G7 TSL
G7 TSX
G7 TTR
G7 TTT
G7 TUA
G7 TUJ
G7 TUY
G7 TVB
G7 TVD
G7 TVH
G7 TVI
G7 TVP
G7 TVU
G7 TVX
G7 TWI
G7 TWP
G7 TWR
G7 TWX
G7 TWZ
G7 TXH
G7 TXJ
G7 TXM
G7 TXU
G7 TYA
G7 TYV
G7 TYY
G7 TZE
G7 TZH
G7 TZY
G7 UAA
G7 UAU
G7 UBF
G7 UBJ
G7 UBL
G7 UBT
G7 UBU
G7 UCC
G7 UCF
G7 UCX
G7 UDH
G7 UDP
G7 UDQ
G7 UDS
G7 UDT
G7 UEB
G7 UED
G7 UEF
G7 UEH
G7 UEM
G7 UEO
G7 UEQ
G7 UEY
G7 UFA
G7 UFG
G7 UFJ
G7 UFO
G7 UFP
G7 UFS
G7 UGE
G7 UGF
G7 UGG
G7 UGK
G7 UGN
G7 UGP
G7 UGS
G7 UGT
G7 UGV
G7 UGX
G7 UHF
G7 UHN
G7 UHV
G7 UIE
G7 UIL
G7 UIQ
G7 UJD
G7 UJG
G7 UJI
G7 UJM
G7 UJN
G7 UKG
G7 UKT
G7 ULE
G7 ULT
G7 UMB
G7 UME
G7 UMP
G7 UMR
G7 UMU
G7 UMX
G7 UND
G7 UNE
G7 UNF
G7 UNH
G7 UNN
G7 UNQ
G7 UOV
G7 UOZ
G7 UPC
G7 UPE
G7 UQF
G7 UQN
G7 UQP
G7 URA
G7 URB
G7 URD
G7 URN
G7 URU
G7 USP
G7 UTK
G7 UTL
G7 UTN
G7 UUV
G7 UUV
G7 UVC
G7 UVD
G7 UVG
G7 UVH
G7 UVT
G7 UWA
G7 UWF
G7 UWR
G7 UXG
G7 UXI
G7 UXV
G7 UXW
G7 UXX
G7 UYD
G7 UYH
G7 UZM
G7 UZP
G7 UZX
G7 VAJ
G7 VAK
G7 VAU
G7 VAV
G7 VAX
G7 VBB
G7 VBC
G7 VBP
G7 VBQ
G7 VBR
G7 VBS
G7 VBX
G7 VCC
G7 VCD
G7 VCQ
G7 VCX
G7 VDM
G7 VDR
G7 VEO
G7 VEQ
G7 VEZ
G7 VFD
G7 VFK
G7 VFW
G7 VFZ
G7 VGG
G7 VGP
G7 VGU
G7 VGW
G7 VGZ
G7 VHP
G7 VHV
G7 VIA
G7 VJC
G7 VJL
G7 VJN
G7 VJR
G7 VJX
G7 VKF
G7 VKK
G7 VKP
G7 VKQ
G7 VKU
G7 VKZ
G7 VLG
G7 VLN
G7 VLY
G7 VMC
G7 VMG
G7 VMI
G7 VMX
G7 VNB
G7 VNR
G7 VNU
G7 VNY
G7 VOF
G7 VOZ
G7 VPI
G7 VPK
G7 VPO
G7 VPZ
G7 VQD
G7 VQQ
G7 VQU
G7 VQV
G7 VRC
G7 VRD
G7 VRI
G7 VRT
G7 VSH
G7 VSQ
G7 VTF
G7 VTO
G7 VTU
G7 VTV
G7 VUB
G7 VUF
G7 VUR
G7 VUV
G7 VUZ
G7 VVE
G7 VVG
G7 VVU
G7 VWH
G7 VXV
G7 VXX
G7 VYC
G7 VYH
G7 VYJ
G7 VYL
G7 VZE
G7 VZJ
G7 VZN
G7 VZT
G7 VZW
G7 VZX
G7 WAJ
G7 WAM
G7 WAP
G7 WAT
G7 WBB
G7 WBI
G7 WBX
G7 WCA
G7 WCC
G7 WCI
G7 WCM
G7 WCO
G7 WDK
G7 WDR
G7 WDT
G7 WEF
G7 WEL
G7 WEX
G7 WFA
G7 WFE
G7 WFY
G7 WGC
G7 WGF
G7 WGJ
G7 WGK
G7 WHC
G7 WHF
G7 WHJ
G7 WHK
G7 WHN
G7 WIA
G7 WIB
G7 WIE
G7 WII
G7 WIJ
G7 WIK
G7 WIP
G7 WIR
G7 WIT
G7 WIU
G7 WIW
G7 WIZ
G7 WJI
G7 WKD
G7 WKE
G7 WKN
G7 WKX
G7 WKY
G7 WLI
G7 WLS
G7 WLW
G7 WLZ
G7 WMA
G7 WMC
G7 WOS
G7 WOT
G7 WRO
G7 WST
G7 XAL
G7 XJS
G7 ZZZ

G8

G8 AAG
G8 ACE
G8 ACM
G8 ACZ
G8 ADH
G8 ADM
G8 AFC
G8 AFZ
G8 AGR
G8 AHK
G8 AIT
G8 AJF
G8 AJN
G8 ALB
G8 AMP
G8 AMV
G8 AMZ
G8 ANJ
G8 AOA
G8 AQX
G8 ARG
G8 ARP
G8 ARV
G8 ASZ
G8 ATH
G8 ATO
G8 ATU
G8 ATV
G8 AUC
G8 AUT
G8 AWO
G8 AXA
G8 AXK
G8 AXW
G8 AXX
G8 AXZ
G8 AZC
G8 AZI
G8 BBC
G8 BBI
G8 BBS
G8 BDA
G8 BDU
G8 BEG
G8 BFT
G8 BGG
G8 BHB
G8 BIJ
G8 BJG
G8 BJY
G8 BKS
G8 BMD
G8 BNL
G8 BOU
G8 BPR
G8 BPV
G8 BQR
G8 BTS
G8 BTT
G8 BTW
G8 BTW
G8 BUJ
G8 BUN
G8 BUR
G8 BVI
G8 BXU
G8 BZY
G8 CAP
G8 CAR
G8 CBL
G8 CCG
G8 CCR
G8 CDF
G8 CDJ
G8 CHB
G8 CHP
G8 CHQ
G8 CHW
G8 CIL
G8 CJS
G8 CJZ
G8 CKF
G8 CKT
G8 CLE
G8 CME
G8 CMJ
G8 CML
G8 CNE
G8 COE
G8 COJ
G8 COV
G8 CPB
G8 CPZ
G8 CQA
G8 CQK
G8 CQP
G8 CTZ
G8 CUV
G8 CVL
G8 CVY
G8 CWM
G8 CZV
G8 CZZ
G8 DAZ
G8 DBN
G8 DCZ
G8 DDR
G8 DEN
G8 DGZ
G8 DID
G8 DJK
G8 DJM
G8 DJN
G8 DKC
G8 DLA
G8 DLW
G8 DMP
G8 DMS
G8 DNW
G8 DOG
G8 DQQ
G8 DQX
G8 DRL
G8 DVG
G8 DVO
G8 DVR
G8 DWA
G8 DWN
G8 DXT
G8 DYF
G8 DYK
G8 DZP
G8 EBH
G8 EBJ
G8 EDE
G8 EDH
G8 EDJ
G8 EEZ
G8 EFP
G8 EFV
G8 EGT
G8 EHK
G8 EIS
G8 EJG
G8 EJN
G8 EKO
G8 ELA
G8 ELR
G8 ELS
G8 ELZ
G8 EMO
G8 EPA
G8 EPM
G8 EPO
G8 EQA
G8 ERJ
G8 ERT
G8 ESB
G8 ESZ
G8 ETC
G8 ETQ
G8 EUD
G8 EUK
G8 EUQ
G8 EVY
G8 EWG
G8 EWJ
G8 EWV
G8 EYG
G8 EYH
G8 EZO
G8 FAW
G8 FBC
G8 FBO
G8 FBZ
G8 FCW
G8 FDO
G8 FDS
G8 FEC
G8 FED
G8 FEG
G8 FEL
G8 FGD
G8 FGF
G8 FGK
G8 FHF
G8 FHM
G8 FHN
G8 FIW
G8 FMQ
G8 FPN
G8 FQE
G8 FQW
G8 FRF
G8 FRL
G8 FSO
G8 FTH
G8 FTI
G8 FTO
G8 FUP
G8 FUQ
G8 FUR
G8 FVY
G8 FWF
G8 FXB
G8 FXD
G8 FYE
G8 FYH
G8 FZA
G8 FZC
G8 FZH
G8 GBW
G8 GCU
G8 GD
G8 GDF
G8 GDS
G8 GEC
G8 GFG
G8 GFH
G8 GFI
G8 GGG
G8 GGY
G8 GHD
G8 GHF
G8 GHX
G8 GIJ
G8 GIN
G8 GIY
G8 GKA
G8 GKQ
G8 GKS
G8 GKU
G8 GLM
G8 GLO
G8 GLW
G8 GMC
G8 GMD
G8 GME
G8 GMT
G8 GNC
G8 GND
G8 GNL
G8 GNW
G8 GOI
G8 GOW
G8 GQT
G8 GRA
G8 GSQ
G8 GSR
G8 GTC
G8 GUV
G8 GVK
G8 GWD
G8 GWL
G8 GWT
G8 GXP
G8 GYW
G8 GYZ
G8 HAJ
G8 HAK
G8 HAX
G8 HBA
G8 HBM
G8 HBS
G8 HCB
G8 HCO
G8 HCU
G8 HCY
G8 HED
G8 HEM
G8 HFI
G8 HFP
G8 HGT
G8 HHV
G8 HHX
G8 HIK
G8 HIM
G8 HIP
G8 HJT
G8 HKF
G8 HLU
G8 HLW
G8 HNB
G8 HNQ
G8 HOD
G8 HOH
G8 HOM
G8 HON
G8 HOP
G8 HOV
G8 HQJ
G8 HQV
G8 HRH
G8 HRK
G8 HRR
G8 HRZ
G8 HSO
G8 HTH
G8 HTX
G8 HUA
G8 HUE
G8 HUI
G8 HUL
G8 HVR
G8 HVY
G8 HWG
G8 HWZ
G8 HXT
G8 HYU
G8 HZW
G8 IAY
G8 IB
G8 ICJ
G8 IDX
G8 IEJ
G8 IFR
G8 IFW
G8 IGP
G8 IGS
G8 IGV
G8 IGY
G8 IIC
G8 IIV
G8 IJA
G8 IJF
G8 IJN
G8 IJP
G8 IJZ
G8 IKL
G8 IKP
G8 ILF
G8 ILM
G8 ILS
G8 IMD
G8 IML
G8 IMN
G8 IMP
G8 IMR
G8 INE
G8 INI
G8 INN
G8 INP
G8 INU
G8 INV
G8 IQG
G8 IQK
G8 IQL
G8 IQP
G8 ISX
G8 IUV
G8 IVC
G8 IVI
G8 IWL
G8 IWN
G8 IXA
G8 IXE
G8 IYP
G8 IZU
G8 IZZ
G8 JAO
G8 JAS
G8 JAU
G8 JAV
G8 JAZ
G8 JBE
G8 JC
G8 JCT
G8 JCW
G8 JEF
G8 JFJ
G8 JFW
G8 JGC
G8 JGI
G8 JGO
G8 JGQ
G8 JIX
G8 JKP
G8 JKP
G8 JKQ
G8 JKR
G8 JLE
G8 JLZ
G8 JNF
G8 JNJ
G8 JNS
G8 JOD
G8 JOJ
G8 JOR
G8 JPZ
G8 JQK
G8 JTJ
G8 JTY
G8 JUF
G8 JUX
G8 JUZ
G8 JVD
G8 JXA
G8 JXJ
G8 JYJ
G8 JZA
G8 JZE
G8 JZY
G8 KAI
G8 KBC
G8 KBI
G8 KBO
G8 KBV
G8 KBZ
G8 KCG
G8 KCM
G8 KCQ
G8 KCS
G8 KCV
G8 KDG
G8 KDI
G8 KFA
G8 KFB
G8 KFW
G8 KGA
G8 KGC
G8 KGH
G8 KHJ
G8 KHR
G8 KHS
G8 KHW
G8 KHZ
G8 KIO
G8 KIT
G8 KJC
G8 KJV
G8 KKS
G8 KLD
G8 KLQ
G8 KLR
G8 KMH
G8 KNV
G8 KOE
G8 KPM
G8 KPS
G8 KQH
G8 KQW
G8 KRA
G8 KSP
G8 KTO
G8 KUE
G8 KUH
G8 KUJ
G8 KUW
G8 KVM
G8 KVP
G8 KWC
G8 KWI
G8 KWQ
G8 KWR
G8 KWX
G8 KXM
G8 KYT
G8 KYU
G8 KZU
G8 LAW
G8 LBD
G8 LBV
G8 LCO
G8 LDP
G8 LEC
G8 LGI
G8 LGX
G8 LIB
G8 LIL
G8 LIR
G8 LIT
G8 LIY
G8 LJC
G8 LKD
G8 LKM
G8 LLB
G8 LLK
G8 LMD
G8 LMN
G8 LMS
G8 LMX
G8 LNR
G8 LNT
G8 LOE
G8 LOL
G8 LOY
G8 LSV
G8 LVI
G8 LVK
G8 LVZ
G8 LWY
G8 LXI
G8 LXO
G8 LYS
G8 LYT
G8 LZI
G8 LZR
G8 LZU
G8 MAC
G8 MAG
G8 MAQ
G8 MAU
G8 MBI
G8 MCD
G8 MCG
G8 MCV
G8 MCZ
G8 MDA
G8 MEJ
G8 MES
G8 MHK
G8 MHR
G8 MJW
G8 MKK
G8 MKM
G8 MLQ
G8 MMD
G8 MMH
G8 MML
G8 MNK
G8 MNZ
G8 MNZ
G8 MOA
G8 MOU
G8 MPP
G8 MQH
G8 MQN
G8 MRB
G8 MRJ
G8 MSA
G8 MSG
G8 MSR
G8 MSW
G8 MTM
G8 MTQ
G8 MUE
G8 MUQ
G8 MVF
G8 MVP
G8 MVV
G8 MWF
G8 MWK
G8 MWQ
G8 MXE
G8 MXZ
G8 MZG
G8 NAX
G8 NBD
G8 NBR
G8 NCF
G8 NCW
G8 NCX
G8 NDA
G8 NDL
G8 NDP
G8 NEH
G8 NEZ
G8 NHN
G8 NHR
G8 NIC
G8 NJD
G8 NJJ
G8 NJW
G8 NKD
G8 NKV
G8 NKX
G8 NKY
G8 NLY
G8 NMC
G8 NMI
G8 NML
G8 NMU
G8 NMW
G8 NNT
G8 NOC
G8 NPF
G8 NPV
G8 NQA
G8 NRY
G8 NSS
G8 NSU
G8 NSV
G8 NT
G8 NTI
G8 NTU
G8 NUC
G8 NUG
G8 NUP
G8 NVF
G8 NVJ
G8 NVY
G8 NWR
G8 NXI
G8 NXL
G8 NXV
G8 NYS
G8 NYW
G8 NZE
G8 NZH
G8 NZM
G8 NZU
G8 OBV
G8 OCN
G8 ODF
G8 ODP
G8 OEB
G8 OEF
G8 OEL
G8 OEM
G8 OER
G8 OES
G8 OFJ
G8 OFK
G8 OFS
G8 OFT
G8 OGI
G8 OGJ
G8 OGO
G8 OHL
G8 OHR
G8 OIS
G8 OIV
G8 OLD
G8 OLT
G8 OMG
G8 OML
G8 OMU
G8 ONK
G8 ONX
G8 OPV
G8 OQF
G8 OQN
G8 OQV
G8 ORE
G8 ORZ
G8 ORZ
G8 OSA
G8 OSD
G8 OSE
G8 OSN
G8 OTE
G8 OTJ
G8 OUA
G8 OUQ
G8 OUU
G8 OUX
G8 OVM
G8 OVT
G8 OWB
G8 OWL
G8 OWO
G8 OWX
G8 OZG
G8 OZJ
G8 OZV
G8 OZW
G8 OZZ
G8 PAF
G8 PAF
G8 PBJ
G8 PCB
G8 PCF
G8 PEF
G8 PEV
G8 PFE
G8 PFG
G8 PFS
G8 PGD
G8 PGL
G8 PHN
G8 PHP
G8 PIB
G8 PID
G8 PIZ
G8 PJH
G8 PKL
G8 PKY
G8 PLE
G8 PLL
G8 PLR
G8 PLW
G8 PLZ
G8 PME
G8 PMQ
G8 PMU
G8 PNC
G8 PND
G8 PNO
G8 PNX
G8 POC
G8 POW
G8 PPE
G8 PPS
G8 PRF
G8 PRI
G8 PRR
G8 PRU
G8 PSE
G8 PSP
G8 PTD
G8 PTJ
G8 PTP
G8 PTR
G8 PUO
G8 PUX
G8 PVB
G8 PVD
G8 PVI
G8 PVJ
G8 PVT
G8 PWA
G8 PWC
G8 PWY
G8 PXB
G8 PXL
G8 PY
G8 PYP
G8 PYT
G8 PYX
G8 PZJ
G8 PZL
G8 PZP
G8 PZR
G8 PZT
G8 QR
G8 RAE
G8 RAK
G8 RAL
G8 RCP
G8 RDE
G8 RDH
G8 RDO
G8 RDX
G8 REM
G8 REU
G8 REY
G8 RFN
G8 RGQ
G8 RHT
G8 RIA
G8 RIE
G8 RJA
G8 RJD
G8 RJU
G8 RJX
G8 RKC
G8 RLJ
G8 RLZ
G8 RMM
G8 ROD
G8 ROM
G8 ROZ
G8 RPH
G8 RPK
G8 RPV
G8 RQP
G8 RQY
G8 RTC
G8 RTW
G8 RTZ
G8 RUM
G8 RVC
G8 RVG
G8 RVX
G8 RWE
G8 RWT
G8 RWT
G8 RWX
G8 RXK
G8 RXL
G8 RXM
G8 RXP
G8 RXU
G8 RYC
G8 RZA
G8 RZJ
G8 SAE
G8 SAJ
G8 SAM
G8 SBE
G8 SBF
G8 SCH
G8 SCK
G8 SCU
G8 SDC
G8 SDT
G8 SEJ
G8 SEZ
G8 SFR
G8 SFU
G8 SHU
G8 SIJ
G8 SIQ
G8 SJP
G8 SKU
G8 SLN
G8 SLX
G8 SNG
G8 SNH
G8 SOZ
G8 SQF
G8 SQM
G8 SRD
G8 STU
G8 SUP
G8 SUT
G8 SVC
G8 SVE
G8 SVF
G8 SVL
G8 SVO
G8 SVU
G8 SWG
G8 SWM
G8 SWZ
G8 SXC
G8 SXN
G8 SYR
G8 SYU
G8 SYZ
G8 SZY
G8 TA
G8 TAM
G8 TBI
G8 TBK
G8 TCM
G8 TDB
G8 TDL
G8 TDW
G8 TEE
G8 TET
G8 TFH
G8 TFI
G8 TFK
G8 TFO
G8 TFV
G8 TGR
G8 THS
G8 TIC
G8 TIJ
G8 TIM
G8 TIS
G8 TJQ
G8 TKR
G8 TKV
G8 TLQ
G8 TLX
G8 TLZ
G8 TMY
G8 TNC
G8 TOK
G8 TOU
G8 TPA
G8 TPT
G8 TPX
G8 TPY
G8 TQY
G8 TRB
G8 TSE
G8 TSJ
G8 TT
G8 TTK
G8 TV
G8 TVH
G8 TVT
G8 TWH
G8 TWJ
G8 TWL
G8 TXM
G8 TXO
G8 TYI
G8 TYV
G8 TZT
G8 UAW
G8 UCB
G8 UCM
G8 UDH
G8 UEC
G8 UEM
G8 UEP
G8 UEQ
G8 UFD
G8 UFK
G8 UFN
G8 UFR
G8 UGB
G8 UGC
G8 UGC
G8 UGD
G8 UGM
G8 UGT
G8 UGZ
G8 UHN
G8 UHV
G8 UIB
G8 UJB
G8 UJP
G8 UKN
G8 UKT
G8 ULG
G8 ULO
G8 ULV
G8 UMX
G8 UNZ
G8 UOD
G8 UON
G8 UOQ
G8 UQI
G8 UQQ
G8 URC
G8 URE
G8 URP
G8 USX
G8 UUA
G8 UUG
G8 UUH
G8 UUL
G8 UVC
G8 UVD
G8 UVE
G8 UVK
G8 UVR
G8 UVS
G8 UWF
G8 UWJ
G8 UWO
G8 UWU
G8 UWW
G8 UXE
G8 UXU
G8 UYA
G8 UYC
G8 UYN
G8 UZJ
G8 UZL
G8 UZO
G8 UZV
G8 UZZ
G8 VAU
G8 VAZ
G8 VBO
G8 VC
G8 VCB
G8 VCW
G8 VDO
G8 VEB
G8 VFH
G8 VFN
G8 VFS
G8 VGC
G8 VHC
G8 VIQ
G8 VIW
G8 VJ
G8 VJH
G8 VJS
G8 VKB
G8 VKD
G8 VKT
G8 VKU
G8 VKV
G8 VLQ
G8 VNA
G8 VNE
G8 VNU
G8 VNW
G8 VOC
G8 VOP
G8 VOV
G8 VPQ
G8 VPS
G8 VQM
G8 VRI
G8 VSK
G8 VSU
G8 VTF
G8 VTN
G8 VUR
G8 VUW
G8 VVK
G8 VVQ
G8 VWN
G8 VXG
G8 VXV
G8 VXX
G8 VYD
G8 VZA
G8 VZE
G8 WAA
G8 WBB
G8 WBH
G8 WBI
G8 WCK
G8 WED
G8 WEG
G8 WEN
G8 WHS
G8 WHZ
G8 WID
G8 WIG
G8 WIO
G8 WIO
G8 WJZ
G8 WKU
G8 WMQ
G8 WMZ
G8 WND
G8 WNX
G8 WOF
G8 WOW
G8 WPD
G8 WQB
G8 WQD
G8 WQM
G8 WQQ
G8 WSG
G8 WT
G8 WTR
G8 WTV
G8 WVW
G8 WWB
G8 WWS
G8 WWZ
G8 WY
G8 WYZ
G8 WZB
G8 XAC
G8 XAH
G8 XBA
G8 XBC
G8 XBI
G8 XBP
G8 XBW
G8 XBZ
G8 XCH
G8 XCK
G8 XDU
G8 XEV
G8 XFA
G8 XFP
G8 XFX
G8 XGA
G8 XGE
G8 XGI
G8 XGM
G8 XGP
G8 XGR
G8 XHO
G8 XIJ
G8 XIT
G8 XIX
G8 XJR
G8 XJU
G8 XKB
G8 XKR
G8 XKV
G8 XLM
G8 XLV
G8 XLY
G8 XMP
G8 XNI
G8 XNQ
G8 XNV
G8 XOH
G8 XOT
G8 XPA
G8 XPF
G8 XPL
G8 XPV
G8 XQF
G8 XQI
G8 XQQ
G8 XTK
G8 XTX
G8 XUJ
G8 XUX
G8 XVA
G8 XVC
G8 XVD
G8 XVV
G8 XWG
G8 XWI
G8 XWJ
G8 XWK
G8 XXE
G8 XXF
G8 XXK
G8 XYY
G8 XZD
G8 XZO
G8 YAJ
G8 YBL
G8 YCR
G8 YCV
G8 YDU
G8 YFG
G8 YHH
G8 YHM
G8 YHS
G8 YHW
G8 YIH
G8 YIJ
G8 YJM
G8 YKF
G8 YKW
G8 YLB
G8 YLU
G8 YLX
G8 YMV
G8 YNJ
G8 YNM
G8 YOJ
G8 YOV
G8 YPC
G8 YPE
G8 YPJ
G8 YQD
G8 YQK
G8 YQQ
G8 YQT
G8 YRI
G8 YRJ
G8 YRY
G8 YSK
G8 YSO
G8 YTC
G8 YUQ
G8 YVR
G8 YXE
G8 YXL
G8 YXY
G8 YYH
G8 YZD
G8 YZH
G8 YZZ
G8 ZAZ
G8 ZBT
G8 ZBW
G8 ZBX
G8 ZCQ
G8 ZCV
G8 ZDF
G8 ZDW
G8 ZEC
G8 ZEG
G8 ZEZ
G8 ZGA
G8 ZGI
G8 ZGX
G8 ZHC
G8 ZHD
G8 ZHJ
G8 ZHL
G8 ZHP
G8 ZID
G8 ZJD
G8 ZJE
G8 ZJJ
G8 ZJP
G8 ZKK
G8 ZKV
G8 ZKY
G8 ZKZ
G8 ZLO
G8 ZMP
G8 ZNL
G8 ZNU
G8 ZNW
G8 ZOH
G8 ZPL
G8 ZPU
G8 ZQO
G8 ZQR
G8 ZRL
G8 ZRO
G8 ZSG
G8 ZSH
G8 ZSN
G8 ZTQ
G8 ZUD
G8 ZVE
G8 ZVJ
G8 ZWE
G8 ZWQ
G8 ZWS
G8 ZXG
G8 ZXW
G8 ZYO
G8 ZZE
G8 ZZF
G8 ZZZ

M0

M0 AAI
M0 AAL
M0 AAO
M0 AAT
M0 AAX
M0 AAY
M0 ABE
M0 ABL
M0 ABM
M0 ABR
M0 ABS
M0 ACD
M0 ACE
M0 ACF
M0 ACG
M0 ACH
M0 ACL
M0 ACQ
M0 ACS
M0 ADC
M0 ADI
M0 ADJ
M0 ADM
M0 ADP
M0 ADS
M0 ADX
M0 AED
M0 AEG
M0 AEH
M0 AEI
M0 AEY
M0 AFA
M0 AFE
M0 AFH
M0 AFL
M0 AFO
M0 AGC
M0 AGG
M0 AGK
M0 AGN
M0 AGX
M0 AHN
M0 AHP
M0 AHX
M0 AIA
M0 AII
M0 AIO
M0 AIR
M0 AIU
M0 AJL
M0 AJM
M0 AJN
M0 AJO
M0 AJW
M0 AJY
M0 AJZ
M0 AKB
M0 AKH
M0 AKO
M0 AKT
M0 AKW
M0 ALA
M0 ALC
M0 ALJ
M0 ALL
M0 ALW
M0 AMO
M0 ANG
M0 ANT
M0 AOP
M0 AOV
M0 APG
M0 APO
M0 APS
M0 APT
M0 APX
M0 AQC
M0 AQM
M0 AQS
M0 AQY
M0 ARF
M0 ARM
M0 ARP
M0 ARW
M0 ASA
M0 ASH
M0 ASP
M0 ASW
M0 ASZ
M0 ATE
M0 ATM
M0 AUE
M0 AUI
M0 AUQ
M0 AUV
M0 AVC
M0 AVD
M0 AVE
M0 AVG
M0 AWA
M0 AWF
M0 AWR
M0 AWS
M0 AWV
M0 AWW
M0 AXB
M0 AXF
M0 AXH
M0 AXK
M0 AXM
M0 AXP
M0 AXQ
M0 AXT
M0 AXY
M0 AYD
M0 AYJ
M0 AYP
M0 AYR
M0 AYW
M0 AZA
M0 AZH
M0 AZI
M0 AZM
M0 AZO
M0 AZQ
M0 BAN
M0 BAS
M0 BBA
M0 BBB
M0 BBJ
M0 BBP
M0 BBX
M0 BBY
M0 BBZ
M0 BCD
M0 BCY
M0 BDG
M0 BDP
M0 BEF
M0 BEG
M0 BEN
M0 BEW
M0 BFE
M0 BFF
M0 BFN
M0 BFO
M0 BFU
M0 BFY
M0 BGA
M0 BGC
M0 BGD
M0 BGL
M0 BGM
M0 BGP
M0 BGQ
M0 BGV
M0 BHB
M0 BHG
M0 BHL
M0 BHU
M0 BHY
M0 BIM
M0 BIR
M0 BIU
M0 BIW
M0 BJA
M0 BJH
M0 BJW
M0 BJY
M0 BKB
M0 BKQ
M0 BKR
M0 BKW
M0 BKZ
M0 BLB
M0 BLE
M0 BLK
M0 BLP
M0 BLQ
M0 BMH
M0 BMK
M0 BMO
M0 BNA
M0 BNE
M0 BNH
M0 BNI
M0 BNJ
M0 BNK
M0 BNY
M0 BOG
M0 BOJ

Withheld

M0 BOP
M0 BOT
M0 BOV
M0 BOZ
M0 BPH
M0 BPI
M0 BPJ
M0 BPK
M0 BPL
M0 BPO
M0 BQA
M0 BQP
M0 BQU
M0 BQV
M0 BQW
M0 BRC
M0 BRK
M0 BRV
M0 BSA
M0 BSR
M0 BTQ
M0 BUD
M0 BUJ
M0 BUL
M0 BUO
M0 BUQ
M0 BUS
M0 BUU
M0 BUX
M0 BVL
M0 BWR
M0 BXH
M0 BXR
M0 BXW
M0 BYH
M0 BYQ
M0 BYW
M0 BZG
M0 BZT
M0 BZU
M0 BZW
M0 CAA
M0 CAH
M0 CAL
M0 CAP
M0 CBC
M0 CBE
M0 CBH
M0 CBJ
M0 CBU
M0 CBV
M0 CBW
M0 CCE
M0 CCO
M0 CDD
M0 CDI
M0 CDP
M0 CDX
M0 CEE
M0 CEF
M0 CEL
M0 CEM
M0 CEN
M0 CEP
M0 CET
M0 CEV
M0 CFI
M0 CFM
M0 CFN
M0 CFO
M0 CFU
M0 CFV
M0 CFW
M0 CGC
M0 CGD
M0 CGI
M0 CGL
M0 CGM
M0 CGQ
M0 CHK
M0 CHW
M0 CHZ
M0 CIJ
M0 CIL
M0 CIN
M0 CIQ
M0 CIU
M0 CIV
M0 CJX
M0 CKQ
M0 CKR
M0 CLA
M0 CLC
M0 CLF
M0 CLX
M0 CMB
M0 CMD
M0 CMM
M0 CMR
M0 CMV
M0 CMX
M0 CMY
M0 CNQ
M0 COG
M0 COH
M0 COK
M0 COX
M0 COY
M0 CPA
M0 CPJ
M0 CPP
M0 CPR
M0 CPX
M0 CPY
M0 CQA
M0 CQE
M0 CQG
M0 CQK
M0 CQL
M0 CQU
M0 CQY
M0 CRX
M0 CSH
M0 CSI
M0 CSL
M0 CSM
M0 CSN
M0 CSS
M0 CSX
M0 CTW
M0 CUB
M0 CUV
M0 CUX
M0 CVL
M0 CVM
M0 CVX
M0 CWL
M0 CWM
M0 CWO
M0 CXC
M0 CXF
M0 CXG
M0 CXM
M0 CXN
M0 CXR
M0 CXT
M0 CXX
M0 CYF
M0 CYP
M0 CYQ
M0 CYS
M0 CYW
M0 CYZ
M0 CZI
M0 CZJ
M0 CZV
M0 CZW
M0 DAF
M0 DAJ
M0 DAO
M0 DAQ
M0 DAU
M0 DAV
M0 DBL
M0 DBN
M0 DBP
M0 DBQ
M0 DCJ
M0 DCK
M0 DCX
M0 DDD
M0 DDJ
M0 DDL
M0 DDM
M0 DDN
M0 DDP
M0 DDR
M0 DDS
M0 DED
M0 DEG
M0 DEM
M0 DFC
M0 DFG
M0 DFI
M0 DFK
M0 DFM
M0 DFS
M0 DFT
M0 DFU
M0 DFV
M0 DGD
M0 DGF
M0 DGG
M0 DGL
M0 DGP
M0 DGW
M0 DGY
M0 DGZ
M0 DHA
M0 DHD
M0 DHG
M0 DHH
M0 DHJ
M0 DHL
M0 DHR
M0 DHS
M0 DHZ
M0 DIA
M0 DIF
M0 DIP
M0 DIR
M0 DIU
M0 DIY
M0 DIZ
M0 DJE
M0 DJJ
M0 DJK
M0 DJM
M0 DJN
M0 DJP
M0 DJS
M0 DJU
M0 DJV
M0 DJY
M0 DJZ
M0 DKA
M0 DKE
M0 DKH
M0 DKI
M0 DKK
M0 DKO
M0 DKQ
M0 DKW
M0 DKZ
M0 DLD
M0 DLJ
M0 DLK
M0 DLQ
M0 DLT
M0 DLV
M0 DLW
M0 DMC
M0 DMG
M0 DMH
M0 DML
M0 DMM
M0 DMP
M0 DMQ
M0 DMW
M0 DNA
M0 DNC
M0 DNE
M0 DNG
M0 DNI
M0 DNL
M0 DNQ
M0 DNS
M0 DNT
M0 DOE
M0 DOF
M0 DOG
M0 DOI
M0 DOJ
M0 DOQ
M0 DOU
M0 DOV
M0 DPA
M0 DPB
M0 DPC
M0 DPD
M0 DPE
M0 DPM
M0 DPN
M0 DPO
M0 DPP
M0 DPR
M0 DPU
M0 DPX
M0 DPZ
M0 DQE
M0 DQJ
M0 DQM
M0 DQR
M0 DQT
M0 DQV
M0 DQW
M0 DQX
M0 DQY
M0 DRC
M0 DRH
M0 DRP
M0 DRV
M0 DRW
M0 DRX
M0 DSB
M0 DSD
M0 DSH
M0 DSJ
M0 DSK
M0 DSQ
M0 DST
M0 DSU
M0 DSV
M0 DTC
M0 DTG
M0 DTM
M0 DTP
M0 DTQ
M0 DTT
M0 DTU
M0 DTV
M0 DTX
M0 DUC
M0 DUE
M0 DUF
M0 DUG
M0 DUH
M0 DUI
M0 DUJ
M0 DUK
M0 DUL
M0 DUN
M0 DUS
M0 DUW
M0 DUZ
M0 DVA
M0 DVC
M0 DVE
M0 DVO
M0 DVP
M0 DVS
M0 DVV
M0 DVY
M0 DWA
M0 DWJ
M0 DWN
M0 DWV
M0 DWY
M0 DXB
M0 DXE
M0 DXK
M0 DXL
M0 DXO
M0 DXU
M0 DXW
M0 DXY
M0 DYD
M0 DYE
M0 DYF
M0 DYI
M0 DYK
M0 DYL
M0 DYM
M0 DYN
M0 DZI
M0 DZJ
M0 DZP
M0 DZQ
M0 DZR
M0 DZS
M0 DZU
M0 DZY
M0 DZZ
M0 EAA
M0 EAG
M0 EAH
M0 EAJ
M0 EAP
M0 EAV
M0 EAW
M0 EBA
M0 EBB
M0 EBC
M0 EBE
M0 EBF
M0 EBH
M0 EBK
M0 EBL
M0 EBM
M0 EBS
M0 EBW
M0 EBZ
M0 ECA
M0 ECB
M0 ECD
M0 ECE
M0 ECG
M0 ECH
M0 ECI
M0 ECO
M0 ECS
M0 ECU
M0 EDB
M0 EDC
M0 EDD
M0 EDG
M0 EDH
M0 EDI
M0 EDJ
M0 EDK
M0 EDT
M0 EDV
M0 EDW
M0 EDY
M0 EDZ
M0 EEA
M0 EEC
M0 EED
M0 EEE
M0 EEF
M0 EEI
M0 EEJ
M0 EEY
M0 EGH
M0 EGM
M0 EIB
M0 EIG
M0 EIP
M0 EIQ
M0 EJN
M0 ELG
M0 ELI
M0 ELT
M0 EM
M0 EMS
M0 ENA
M0 ENG
M0 EOG
M0 EPV
M0 EPW
M0 ERA
M0 ESE
M0 ESG
M0 ETM
M0 ETO
M0 EUG
M0 EVT
M0 EWS
M0 EXV
M0 EYB
M0 FAB
M0 FAQ
M0 FBI
M0 FBT
M0 FDA
M0 FDB
M0 FDE
M0 FDF
M0 FDO
M0 FDS
M0 FDY
M0 FET
M0 FEW
M0 FFF
M0 FGM
M0 FHB
M0 FHN
M0 FIN
M0 FMA
M0 FMC
M0 FME
M0 FMF
M0 FNJ
M0 FNR
M0 FOS
M0 FOZ
M0 FRE
M0 FRZ
M0 FSD
M0 FWD
M0 FWZ
M0 FXP
M0 FYF
M0 FZQ
M0 FZY
M0 GAA
M0 GAD
M0 GAF
M0 GAJ
M0 GAK
M0 GAM
M0 GAP
M0 GAU
M0 GAW
M0 GAZ
M0 GBP
M0 GBR
M0 GBS
M0 GBV
M0 GBX
M0 GBY
M0 GCE
M0 GCG
M0 GCJ
M0 GCK
M0 GCL
M0 GCM
M0 GCO
M0 GCP
M0 GCW
M0 GCZ
M0 GDA
M0 GDB
M0 GDE
M0 GDF
M0 GDK
M0 GDN
M0 GDO
M0 GDQ
M0 GDR
M0 GDW
M0 GDY
M0 GDZ
M0 GEE
M0 GEG
M0 GEH
M0 GEJ
M0 GER
M0 GET
M0 GEV
M0 GFB
M0 GFQ
M0 GFS
M0 GFT
M0 GFU
M0 GFV
M0 GFY
M0 GFZ
M0 GGA
M0 GGE
M0 GGI
M0 GGR
M0 GGS
M0 GGY
M0 GHB
M0 GHD
M0 GHF
M0 GHG
M0 GHH
M0 GHP
M0 GHQ
M0 GHS
M0 GHU
M0 GIC
M0 GIH
M0 GIK
M0 GIO
M0 GIR
M0 GIV
M0 GIX
M0 GJE
M0 GJF
M0 GJO
M0 GJQ
M0 GJR
M0 GJW
M0 GJY
M0 GJZ
M0 GKC
M0 GKH
M0 GKI
M0 GKM
M0 GKX
M0 GKY
M0 GLC
M0 GLD
M0 GLH
M0 GLJ
M0 GLM
M0 GLN
M0 GLO
M0 GLR
M0 GLW
M0 GLZ
M0 GMB
M0 GML
M0 GMM
M0 GMV
M0 GMX
M0 GMY
M0 GND
M0 GNE
M0 GNT
M0 GNV
M0 GNZ
M0 GOE
M0 GOJ
M0 GOR
M0 GPA
M0 GPI
M0 GPM
M0 GPR
M0 GPS
M0 GPT
M0 GQA
M0 GQC
M0 GQH
M0 GQK
M0 GQL
M0 GQN
M0 GQQ
M0 GQY
M0 GRK
M0 GRL
M0 GRQ
M0 GRV
M0 GSA
M0 GSE
M0 GSF
M0 GSG
M0 GSJ
M0 GSM
M0 GSU
M0 GTA
M0 GTC
M0 GTD
M0 GTF
M0 GTK
M0 GTS
M0 GTW
M0 GUB
M0 GUI
M0 GUP
M0 GUQ
M0 GUS
M0 GUT
M0 GUV
M0 GUY
M0 GVA
M0 GVD
M0 GVG
M0 GVH
M0 GVO
M0 GVR
M0 GVS
M0 GVU
M0 GVV
M0 GWG
M0 GWI
M0 GWJ
M0 GWP
M0 GWS
M0 GWW
M0 GWX
M0 GXA
M0 GXD
M0 GXF
M0 GXJ
M0 GXL
M0 GXP
M0 GXR
M0 GXS
M0 GXT
M0 GXX
M0 GXZ
M0 GYD
M0 GYE
M0 GYQ
M0 GYT
M0 GZG
M0 GZO
M0 GZP
M0 GZQ
M0 GZR
M0 GZS
M0 GZT
M0 GZU
M0 GZV
M0 GZY
M0 HAA
M0 HAE
M0 HAI
M0 HAJ
M0 HAK
M0 HAQ
M0 HAV
M0 HAX
M0 HAY
M0 HBA
M0 HBB
M0 HBD
M0 HBF
M0 HBI
M0 HBK
M0 HBP
M0 HBQ
M0 HBR
M0 HBS
M0 HBZ
M0 HCB
M0 HCD
M0 HCF
M0 HCH
M0 HCJ
M0 HCK
M0 HCL
M0 HCS
M0 HCU
M0 HDB
M0 HDD
M0 HDF
M0 HDH
M0 HDL
M0 HDO
M0 HDX
M0 HDY
M0 HEA
M0 HEC
M0 HEE
M0 HEH
M0 HEI
M0 HEK
M0 HEL
M0 HEO
M0 HEQ
M0 HER
M0 HES
M0 HEU
M0 HEV
M0 HEZ
M0 HFD
M0 HFJ
M0 HFK
M0 HFL
M0 HFM
M0 HFN
M0 HFP
M0 HFT
M0 HGB
M0 HGC
M0 HGE
M0 HGH
M0 HGI
M0 HGL
M0 HGP
M0 HGQ
M0 HGR
M0 HGT
M0 HGU
M0 HGW
M0 HGX
M0 HGZ
M0 HHJ
M0 HHK
M0 HHL
M0 HHN
M0 HHO
M0 HHQ
M0 HHS
M0 HHY
M0 HHZ
M0 HIB
M0 HIF
M0 HII
M0 HIK
M0 HIR
M0 HIS
M0 HIU
M0 HJA
M0 HJH
M0 HJM
M0 HJS
M0 HJT
M0 HJV
M0 HJX
M0 HJZ
M0 HKD
M0 HKN
M0 HKO
M0 HKQ
M0 HKY
M0 HKZ
M0 HLA
M0 HLG
M0 HLK
M0 HLL
M0 HLO
M0 HLT
M0 HMA
M0 HMC
M0 HMD
M0 HMG
M0 HMH
M0 HMK
M0 HML
M0 HMM
M0 HMN
M0 HMP
M0 HMQ
M0 HMW
M0 HNB
M0 HNP
M0 HNR
M0 HNS
M0 HNU
M0 HNV
M0 HNZ
M0 HOC
M0 HOE
M0 HOG
M0 HOH
M0 HOS
M0 HOW
M0 HOX
M0 HPA
M0 HPD
M0 HPI
M0 HPO
M0 HPY
M0 HQH
M0 HQK
M0 HQN
M0 HQS
M0 HQT
M0 HQV
M0 HQW
M0 HQX
M0 HRE
M0 HRF
M0 HRJ
M0 HRK
M0 HRR
M0 HRS
M0 HRU
M0 HRX
M0 HSA
M0 HSD
M0 HSE
M0 HSF
M0 HSK
M0 HSL
M0 HSM
M0 HSN
M0 HSP
M0 HST
M0 HSY
M0 HTD
M0 HTH
M0 HTM
M0 HTN
M0 HTP
M0 HTT
M0 HTZ
M0 HUB
M0 HUC
M0 HUD
M0 HUE
M0 HUK
M0 HUO
M0 HUP
M0 HUT
M0 HUX
M0 HVH
M0 HVJ
M0 HVT
M0 HVY
M0 HWA
M0 HWB
M0 HWE
M0 HWK
M0 HWX
M0 HWY
M0 HXD
M0 HXJ
M0 HXP
M0 HXQ
M0 HXR
M0 HXT
M0 HXU
M0 HXW
M0 HXY
M0 HYB
M0 HYF
M0 HYO
M0 HYR
M0 HYS
M0 HYT
M0 HYU
M0 HYV
M0 HYW
M0 HYY
M0 HYZ
M0 HZG
M0 HZH
M0 HZQ
M0 HZS
M0 HZZ
M0 IAB
M0 IAI
M0 IAN
M0 IAO
M0 IAP
M0 IAQ
M0 IAR
M0 IAU
M0 IAV
M0 IAW
M0 IAY
M0 IBA
M0 IBB
M0 IBF
M0 IBG
M0 IBP
M0 IBS
M0 IBU
M0 ICC
M0 ICD
M0 ICF
M0 ICH
M0 ICM
M0 ICN
M0 ICR
M0 ICV
M0 ICW
M0 ICX
M0 ICY
M0 IDA
M0 IDB
M0 IDD
M0 IDF
M0 IDH
M0 IDP
M0 IDQ
M0 IDS
M0 IDU
M0 IDV
M0 IDX
M0 IDZ
M0 IEC
M0 IEE
M0 IEF
M0 IEG
M0 IEI
M0 IEN
M0 IEU
M0 IEV
M0 IEX
M0 IFE
M0 IFR
M0 IGA
M0 IGD
M0 IGE
M0 IGI
M0 IGK
M0 IGN
M0 IGQ
M0 IGS
M0 IGU
M0 IGV
M0 IGZ
M0 IHA
M0 IHD
M0 IHF
M0 IHG
M0 IHH
M0 IHI
M0 IHK
M0 IHO
M0 IHP
M0 IHQ
M0 IHV
M0 IHZ
M0 IIB
M0 IIC
M0 IID
M0 IIH
M0 IIJ
M0 IIK
M0 IIL
M0 IIN
M0 IIO
M0 IIP
M0 IIS
M0 IIT
M0 IIV
M0 IIW
M0 IIX
M0 IIY
M0 IJD
M0 IJH
M0 IJK
M0 IJM
M0 IJN
M0 IJS
M0 IJV
M0 IJW
M0 IJX
M0 IJZ
M0 IKF
M0 IKL
M0 IKO
M0 IKP
M0 IKQ
M0 IKR
M0 IKU
M0 IKX
M0 IKY
M0 IKZ
M0 ILB
M0 ILD
M0 ILE
M0 ILL
M0 ILQ
M0 ILR
M0 ILS
M0 ILV
M0 ILW
M0 IMA
M0 IMB
M0 IMC
M0 IMG
M0 IMH
M0 IMK
M0 IMO
M0 IMX
M0 INA
M0 ING
M0 INJ
M0 INK
M0 INN
M0 INO
M0 INQ
M0 INR
M0 INV
M0 IOD
M0 IOF
M0 IOG
M0 IOH
M0 IOJ
M0 ION
M0 IOP
M0 IOR
M0 IOX
M0 IOY
M0 IPA
M0 IPC
M0 IPF
M0 IPG
M0 IPK
M0 IPM
M0 IPN
M0 IPT
M0 IPU
M0 IPW
M0 IPX
M0 IPY
M0 IQA
M0 IQB
M0 IQD
M0 IQE
M0 IQI
M0 IQJ
M0 IQP
M0 IQS
M0 IQU
M0 IQV
M0 IQW
M0 IRM
M0 IRN
M0 IRQ
M0 IRR
M0 IRY
M0 ISA
M0 ISB
M0 ISH
M0 ISM
M0 ISO
M0 ISR
M0 ISS
M0 ISV
M0 ISX
M0 ITD
M0 ITF
M0 ITH
M0 ITJ
M0 ITK
M0 ITL
M0 ITM
M0 ITN
M0 ITP
M0 ITQ
M0 ITR
M0 ITU
M0 IUF
M0 IUH
M0 IUL
M0 IUO
M0 IUP
M0 IUS
M0 IUW
M0 IUX
M0 IUY
M0 IVA
M0 IVB
M0 IVG
M0 IVI
M0 IVP
M0 IVQ
M0 IVR
M0 IVT
M0 IVY
M0 IWD
M0 IWE
M0 IWF
M0 IWH
M0 IWK
M0 IWL
M0 IWR
M0 IWY
M0 IXA
M0 IXE
M0 IXI
M0 IXK
M0 IXO
M0 IXS
M0 IYA
M0 IYB
M0 IYI
M0 IYL
M0 IYR
M0 IYW
M0 IYY
M0 IZB
M0 IZD
M0 IZE
M0 IZH
M0 IZL
M0 IZO
M0 IZT
M0 IZY
M0 JAA
M0 JAB
M0 JAH
M0 JAL
M0 JAS
M0 JAU
M0 JBE
M0 JBJ
M0 JBL
M0 JBM
M0 JBN
M0 JBO
M0 JBP
M0 JBQ
M0 JBR
M0 JBX
M0 JCA
M0 JCF
M0 JCI
M0 JCJ
M0 JCN
M0 JCO
M0 JCU
M0 JCW
M0 JCX
M0 JDC
M0 JDH
M0 JDI
M0 JDK
M0 JDM
M0 JDN
M0 JDO
M0 JDR
M0 JED
M0 JEE
M0 JEN
M0 JEV
M0 JEX
M0 JEY
M0 JFA
M0 JFC
M0 JFF
M0 JFG
M0 JFN
M0 JFP
M0 JFQ
M0 JFS
M0 JFT
M0 JFV
M0 JFZ
M0 JGA
M0 JGC
M0 JGD
M0 JGI
M0 JGJ
M0 JGL
M0 JGN
M0 JGO
M0 JGQ
M0 JGT
M0 JGW
M0 JGX
M0 JGY
M0 JGZ
M0 JHA
M0 JHE
M0 JHI
M0 JHJ
M0 JHK
M0 JHQ
M0 JHR
M0 JHS
M0 JHU
M0 JHX
M0 JHZ
M0 JIF
M0 JIG
M0 JIH
M0 JII
M0 JIK
M0 JIN
M0 JIR
M0 JIT
M0 JIV
M0 JIW
M0 JIY
M0 JJF
M0 JJG
M0 JJL
M0 JJO
M0 JJS
M0 JJT
M0 JJY
M0 JKA
M0 JKC
M0 JKD
M0 JKK
M0 JKO
M0 JKR
M0 JKT
M0 JKZ
M0 JLA
M0 JLD
M0 JLG
M0 JLJ
M0 JLK
M0 JLL
M0 JLQ
M0 JLX
M0 JLZ
M0 JMA
M0 JMG
M0 JMM
M0 JMO
M0 JMR
M0 JMX
M0 JMZ
M0 JNA
M0 JNB
M0 JNC
M0 JND
M0 JNE
M0 JNG
M0 JNH
M0 JNI
M0 JNK
M0 JNN
M0 JNO
M0 JNR
M0 JNT
M0 JNW
M0 JNZ
M0 JOA
M0 JOE
M0 JOF
M0 JOG
M0 JOQ
M0 JOS
M0 JOT
M0 JOU
M0 JOV
M0 JOZ
M0 JPF
M0 JPH
M0 JPI
M0 JPJ
M0 JPQ
M0 JPU
M0 JPY
M0 JPZ
M0 JQA
M0 JQC
M0 JQE
M0 JQF
M0 JQH
M0 JQI
M0 JQM
M0 JQN
M0 JQQ
M0 JQR
M0 JQT
M0 JQU
M0 JQX
M0 JQZ
M0 JRB
M0 JRC
M0 JRD
M0 JRG
M0 JRK
M0 JRM
M0 JRS
M0 JRT
M0 JRV
M0 JRY
M0 JSB
M0 JSC
M0 JSF
M0 JSI
M0 JSM
M0 JSO
M0 JSQ
M0 JST
M0 JTA
M0 JTF
M0 JTI
M0 JTK
M0 JTO
M0 JTP
M0 JTR
M0 JTS
M0 JTV
M0 JTZ
M0 JUD
M0 JUE
M0 JUF
M0 JUG
M0 JUH
M0 JUM
M0 JUO
M0 JUR
M0 JUS
M0 JUY
M0 JVA
M0 JVB
M0 JVD
M0 JVF
M0 JVI
M0 JVK
M0 JVN
M0 JVP
M0 JVS
M0 JWB
M0 JWF
M0 JWG
M0 JWK
M0 JWN
M0 JWO
M0 JWS
M0 JXA
M0 JXB
M0 JXC
M0 JXH
M0 JXJ
M0 JXK
M0 JXL
M0 JXN
M0 JXO
M0 JXP
M0 JXQ
M0 JXT
M0 JXV
M0 JXW
M0 JXZ
M0 JYA
M0 JYD
M0 JYE
M0 JYH
M0 JYI
M0 JYL
M0 JYT
M0 JYU
M0 JYX
M0 JZA
M0 JZC
M0 JZD
M0 JZH
M0 JZJ
M0 JZL
M0 JZQ
M0 JZX
M0 JZY
M0 KAA
M0 KAF
M0 KAH
M0 KAL
M0 KAQ
M0 KAS
M0 KAT
M0 KAV
M0 KAY
M0 KBE
M0 KBF
M0 KBI
M0 KBM
M0 KBO
M0 KBQ
M0 KBS
M0 KBV
M0 KCB
M0 KCG
M0 KCM
M0 KCN
M0 KCQ
M0 KCR
M0 KCT
M0 KCU
M0 KCX
M0 KDB
M0 KDD
M0 KDF
M0 KDI
M0 KDJ
M0 KDK
M0 KDN
M0 KDP
M0 KDQ
M0 KDS
M0 KEA
M0 KEC
M0 KEH
M0 KEI
M0 KEK
M0 KEM
M0 KER
M0 KEX
M0 KEZ
M0 KFE
M0 KFG
M0 KFM
M0 KFN
M0 KFQ
M0 KFS
M0 KFT
M0 KFX
M0 KFZ
M0 KGB
M0 KGC
M0 KGD
M0 KGH
M0 KGJ
M0 KGL
M0 KGN
M0 KGO
M0 KGR
M0 KGX
M0 KGZ
M0 KHF
M0 KHH
M0 KHI
M0 KHJ
M0 KHL
M0 KHP
M0 KHR
M0 KHT
M0 KHU
M0 KHV
M0 KIA
M0 KIE
M0 KIH
M0 KII
M0 KIK
M0 KIS
M0 KIT
M0 KIU
M0 KIV
M0 KIX
M0 KJA
M0 KJB
M0 KJI
M0 KJP
M0 KJR
M0 KJV
M0 KJW
M0 KKI
M0 KKR
M0 KKS
M0 KKT
M0 KKU
M0 KKV
M0 KKX
M0 KLC
M0 KLF
M0 KLO
M0 KLP
M0 KLY
M0 KMA
M0 KMD
M0 KMF
M0 KMM
M0 KMP
M0 KMV
M0 KMY
M0 KMZ
M0 KNA
M0 KNC
M0 KNG
M0 KNO
M0 KNR
M0 KNT
M0 KNZ
M0 KOA
M0 KOP
M0 KOS
M0 KOX
M0 KPG
M0 KPH
M0 KPL
M0 KPN
M0 KPR
M0 KPS
M0 KPZ
M0 KQA
M0 KRB
M0 KRF
M0 KRG
M0 KRI
M0 KRJ
M0 KRT
M0 KSD
M0 KSH
M0 KSJ
M0 KSM
M0 KSN
M0 KSP
M0 KSW
M0 KTA
M0 KTB
M0 KTC
M0 KTH
M0 KTK
M0 KTM
M0 KTX
M0 KTY
M0 KTZ
M0 KUB
M0 KUC
M0 KUN
M0 KVC
M0 KVG
M0 KVI
M0 KVL
M0 KVP
M0 KVW
M0 KVX
M0 KW
M0 KWD
M0 KXO
M0 KXS
M0 KXT
M0 KXX
M0 KYB
M0 KYE
M0 KYG
M0 KYJ
M0 KYM
M0 KYO
M0 KYT
M0 KYW
M0 KZL
M0 KZR
M0 KZX
M0 LAC
M0 LAH
M0 LAJ
M0 LAK
M0 LAM
M0 LAN
M0 LAP
M0 LAX
M0 LBC
M0 LBT
M0 LBV
M0 LBW
M0 LCB
M0 LCD
M0 LCE
M0 LCF
M0 LCG
M0 LDA
M0 LDB
M0 LDF
M0 LDI
M0 LDM
M0 LDN
M0 LDU
M0 LDW
M0 LEG
M0 LEJ
M0 LEU
M0 LFA
M0 LFD
M0 LFG
M0 LFO
M0 LFR
M0 LFV
M0 LFX
M0 LGH
M0 LGJ
M0 LGO
M0 LGV
M0 LHD
M0 LHH
M0 LHM
M0 LHT
M0 LIH
M0 LIL
M0 LIN
M0 LIP
M0 LIT
M0 LJB
M0 LJE
M0 LJM
M0 LJP
M0 LKA
M0 LKB
M0 LKC
M0 LKQ
M0 LKW
M0 LLB
M0 LLC
M0 LLF
M0 LLL
M0 LLP
M0 LLT
M0 LLZ
M0 LMD
M0 LMF
M0 LMK
M0 LMX
M0 LMZ
M0 LNA
M0 LNG
M0 LNT
M0 LNZ
M0 LOK
M0 LOM
M0 LON
M0 LOR
M0 LOT
M0 LOV
M0 LOY
M0 LPD
M0 LPE
M0 LPH
M0 LRE
M0 LRF
M0 LRI
M0 LRQ
M0 LRT
M0 LRV
M0 LRX
M0 LSE
M0 LSK
M0 LSP
M0 LTC
M0 LTF
M0 LTH
M0 LTM
M0 LTV
M0 LTY
M0 LUG
M0 LUI
M0 LUX
M0 LVV
M0 LWB
M0 LWC
M0 LWG
M0 LWI
M0 LWO
M0 LWQ
M0 LXB
M0 LXM
M0 LXQ
M0 LXR
M0 LXT
M0 LXX
M0 LYC
M0 LYM
M0 LYT
M0 LYX
M0 LZA
M0 LZP
M0 LZR
M0 LZY
M0 MAA
M0 MAD
M0 MAE
M0 MAK
M0 MAM
M0 MAS
M0 MAV
M0 MBF
M0 MBJ
M0 MBK
M0 MBN
M0 MBP
M0 MBQ
M0 MBU
M0 MBX
M0 MBY
M0 MCD
M0 MCF
M0 MCK
M0 MCN
M0 MCR
M0 MCU
M0 MCX
M0 MCZ
M0 MDA
M0 MDB
M0 MDD
M0 MDI
M0 MDK
M0 MDL
M0 MDM
M0 MDR
M0 MDS
M0 MDU
M0 MDW
M0 MDX
M0 MDY
M0 MEC
M0 MEE
M0 MEM
M0 MES
M0 MEU
M0 MEZ
M0 MFD
M0 MFF
M0 MFM
M0 MFO
M0 MFR
M0 MFZ
M0 MGC
M0 MGD
M0 MGG
M0 MGH
M0 MGN
M0 MGO
M0 MGR
M0 MGT
M0 MGV
M0 MGY
M0 MGZ
M0 MHA
M0 MHB
M0 MHS
M0 MIA
M0 MII
M0 MIK
M0 MIN
M0 MIO
M0 MIP
M0 MIS
M0 MIW
M0 MIX
M0 MJA
M0 MJC
M0 MJE
M0 MJJ
M0 MJL
M0 MJO
M0 MJP
M0 MJR
M0 MJU
M0 MKA
M0 MKB
M0 MKD
M0 MKF
M0 MKI
M0 MKK
M0 MKL
M0 MKM
M0 MKS
M0 MKT
M0 MKU
M0 MKW
M0 MKX
M0 MLC
M0 MLN
M0 MLP
M0 MLQ
M0 MLX
M0 MMA
M0 MMB
M0 MME
M0 MMH
M0 MMK
M0 MML
M0 MMM
M0 MMQ
M0 MMW
M0 MMY
M0 MMZ
M0 MNE
M0 MNF
M0 MNJ
M0 MNK
M0 MNM
M0 MNT
M0 MOA
M0 MOE
M0 MOF
M0 MOG
M0 MOM
M0 MOZ
M0 MPC
M0 MPG
M0 MPH
M0 MPJ
M0 MPL
M0 MPR
M0 MPX
M0 MPZ
M0 MQB
M0 MQM
M0 MQS
M0 MQT
M0 MQV
M0 MRA
M0 MRD
M0 MRN
M0 MRR
M0 MRU
M0 MRZ
M0 MSC
M0 MSD
M0 MSK
M0 MSL
M0 MSN
M0 MST
M0 MSU
M0 MSW
M0 MTB
M0 MTE
M0 MTH
M0 MTL
M0 MTM
M0 MTP
M0 MTT
M0 MTU
M0 MTV
M0 MTY
M0 MTZ
M0 MUD
M0 MUF
M0 MUG
M0 MUJ
M0 MUS
M0 MUT
M0 MUX
M0 MVA
M0 MVC
M0 MVK
M0 MVR
M0 MVV
M0 MWC
M0 MWD
M0 MWF
M0 MWG
M0 MWH
M0 MWI
M0 MWM
M0 MWV
M0 MWZ
M0 MXD
M0 MXH
M0 MXJ
M0 MXL
M0 MXO
M0 MXR
M0 MXV
M0 MXW
M0 MXY
M0 MYO
M0 MYR
M0 MYT
M0 MZB
M0 MZF
M0 MZG
M0 MZH
M0 MZL
M0 MZM
M0 MZR
M0 MZW
M0 MZZ
M0 NAC
M0 NAD
M0 NAH
M0 NAN
M0 NAT
M0 NAV
M0 NBB
M0 NBG
M0 NBR
M0 NBT
M0 NBW
M0 NBX
M0 NBY
M0 NCB
M0 NCF
M0 NCH
M0 NCP
M0 NCS
M0 NCT
M0 NCW
M0 NCY
M0 NDB
M0 NDN
M0 NDS
M0 NDW
M0 NEA
M0 NEB
M0 NED
M0 NEE
M0 NEI
M0 NEL
M0 NEN
M0 NES
M0 NEY
M0 NFA
M0 NFC
M0 NFE
M0 NFJ
M0 NFL
M0 NFP
M0 NGA
M0 NGD
M0 NGE
M0 NGG
M0 NGO
M0 NGT
M0 NHC
M0 NHG
M0 NHQ
M0 NHR
M0 NHS
M0 NIA
M0 NIK
M0 NIM
M0 NIN
M0 NIQ
M0 NIR
M0 NIT
M0 NIU
M0 NIX
M0 NIY
M0 NJA
M0 NJB
M0 NJC
M0 NJH
M0 NJM

Withheld

M0 NJN
M0 NKA
M0 NKC
M0 NKD
M0 NKG
M0 NKI
M0 NKK
M0 NKL
M0 NKM
M0 NKP
M0 NLB
M0 NLG
M0 NLM
M0 NLO
M0 NME
M0 NML
M0 NMP
M0 NMQ
M0 NMS
M0 NNE
M0 NNK
M0 NNN
M0 NNO
M0 NNQ
M0 NNR
M0 NOD
M0 NOP
M0 NOT
M0 NOX
M0 NPE
M0 NPK
M0 NPN
M0 NPP
M0 NPT
M0 NPW
M0 NQN
M0 NQP
M0 NRA
M0 NRD
M0 NRE
M0 NRF
M0 NRO
M0 NRT
M0 NRX
M0 NRZ
M0 NSA
M0 NST
M0 NSU
M0 NSZ
M0 NTD
M0 NTE
M0 NTL
M0 NTM
M0 NTO
M0 NTP
M0 NTR
M0 NTS
M0 NTV
M0 NTW
M0 NTX
M0 NUE
M0 NUM
M0 NUR
M0 NVA
M0 NVH
M0 NVS
M0 NWB
M0 NWF
M0 NWH
M0 NWR
M0 NWX
M0 NXN
M0 NXS
M0 NXT
M0 NYE
M0 NYL
M0 NYW
M0 NZG
M0 NZH
M0 NZO
M0 NZX
M0 NZY
M0 NZZ
M0 OAF
M0 OAH
M0 OAK
M0 OAP
M0 OAS
M0 OAV
M0 OBI
M0 OBJ
M0 OBO
M0 OBS
M0 OCB
M0 OCD
M0 OCH
M0 OCN
M0 OCS
M0 OCW
M0 ODC
M0 ODH
M0 ODN
M0 ODP
M0 ODV
M0 ODY
M0 OEE
M0 OEI
M0 OEY
M0 OFA
M0 OFC
M0 OFI
M0 OFP
M0 OFR
M0 OFV
M0 OFX
M0 OFY
M0 OGG
M0 OGY
M0 OHC
M0 OHM
M0 OII
M0 OIL
M0 OJD
M0 OJE
M0 OJM
M0 OJR
M0 OJY
M0 OKE
M0 OKO
M0 OKW
M0 OLA
M0 OLF
M0 OLI
M0 OLO
M0 OLP
M0 OLT
M0 OLY
M0 OML
M0 OMM
M0 OMN
M0 OMO
M0 OMR
M0 OMY
M0 ONA
M0 ONE
M0 ONM
M0 ONN
M0 ONP
M0 OOA
M0 OOB
M0 OOE
M0 OOJ
M0 OOM
M0 OON
M0 OOP
M0 OOX
M0 OPA
M0 OPE
M0 OPO
M0 OPR
M0 OPT
M0 OPV
M0 OPW
M0 OPX
M0 OQU
M0 ORA
M0 ORB
M0 ORD
M0 ORG
M0 ORH
M0 ORO
M0 OSA
M0 OSC
M0 OSG
M0 OSK
M0 OSS
M0 OTM
M0 OTP
M0 OTQ
M0 OTV
M0 OTW
M0 OTX
M0 OTY
M0 OUF
M0 OUK
M0 OUZ
M0 OVA
M0 OVE
M0 OVG
M0 OVL
M0 OVR
M0 OVY
M0 OWE
M0 OWN
M0 OWW
M0 OWZ
M0 OXF
M0 OXH
M0 OXI
M0 OXN
M0 OXS
M0 OYG
M0 OZZ
M0 PAB
M0 PAD
M0 PAE
M0 PAH
M0 PAN
M0 PAP
M0 PAS
M0 PAT
M0 PAU
M0 PBA
M0 PBD
M0 PBK
M0 PBL
M0 PBM
M0 PBQ
M0 PBS
M0 PCF
M0 PCI
M0 PCM
M0 PCP
M0 PDJ
M0 PDM
M0 PDT
M0 PEK
M0 PEL
M0 PEY
M0 PEZ
M0 PFD
M0 PFM
M0 PFX
M0 PGA
M0 PGE
M0 PGF
M0 PGG
M0 PGR
M0 PGT
M0 PGZ
M0 PHE
M0 PHG
M0 PHI
M0 PHS
M0 PHT
M0 PIH
M0 PIO
M0 PIU
M0 PIW
M0 PJB
M0 PJE
M0 PJH
M0 PJO
M0 PJW
M0 PKB
M0 PKD
M0 PKG
M0 PKM
M0 PKO
M0 PKP
M0 PKR
M0 PKT
M0 PKZ
M0 PLA
M0 PLD
M0 PLF
M0 PLK
M0 PLL
M0 PLM
M0 PMD
M0 PME
M0 PMK
M0 PMQ
M0 PMS
M0 PMT
M0 PMZ
M0 PNF
M0 PNG
M0 PNP
M0 PNX
M0 POJ
M0 POL
M0 POW
M0 PPD
M0 PPE
M0 PPF
M0 PPK
M0 PPL
M0 PPQ
M0 PPT
M0 PRJ
M0 PRL
M0 PRU
M0 PRY
M0 PSH
M0 PSL
M0 PSP
M0 PSQ
M0 PSU
M0 PSX
M0 PTC
M0 PTD
M0 PTH
M0 PTK
M0 PTM
M0 PTP
M0 PTU
M0 PTV
M0 PUB
M0 PUG
M0 PUN
M0 PUP
M0 PUX
M0 PUZ
M0 PVA
M0 PVB
M0 PVG
M0 PVM
M0 PVX
M0 PWA
M0 PWG
M0 PWH
M0 PWJ
M0 PWK
M0 PWN
M0 PWW
M0 PWX
M0 PWZ
M0 PXB
M0 PXG
M0 PXH
M0 PXS
M0 PYU
M0 PZB
M0 PZG
M0 PZL
M0 PZS
M0 PZT
M0 PZW
M0 QVE
M0 RAA
M0 RAH
M0 RAJ
M0 RAO
M0 RAQ
M0 RAV
M0 RAY
M0 RBS
M0 RBW
M0 RBZ
M0 RCB
M0 RCO
M0 RCQ
M0 RCW
M0 RDE
M0 RDF
M0 RDH
M0 RDJ
M0 RDM
M0 RDU
M0 RDW
M0 REE
M0 REK
M0 REL
M0 REN
M0 REO
M0 RER
M0 RES
M0 RET
M0 REU
M0 REW
M0 REY
M0 RFC
M0 RFD
M0 RFF
M0 RFI
M0 RFL
M0 RFN
M0 RFR
M0 RFT
M0 RFT
M0 RFX
M0 RGA
M0 RGG
M0 RGJ
M0 RGM
M0 RGT
M0 RGU
M0 RGZ
M0 RHF
M0 RHJ
M0 RHK
M0 RHV
M0 RHZ
M0 RIE
M0 RIF
M0 RIJ
M0 RIP
M0 RIQ
M0 RIT
M0 RIX
M0 RJA
M0 RJC
M0 RJD
M0 RJF
M0 RJG
M0 RJL
M0 RJP
M0 RJV
M0 RKJ
M0 RKL
M0 RKS
M0 RLA
M0 RLB
M0 RLE
M0 RLG
M0 RLH
M0 RLK
M0 RLR
M0 RLX
M0 RMA
M0 RMC
M0 RMR
M0 RMX
M0 RNA
M0 RNE
M0 RNJ
M0 RNN
M0 RNS
M0 RNT
M0 RNY
M0 ROA
M0 ROB
M0 ROD
M0 ROE
M0 ROF
M0 ROI
M0 ROK
M0 ROQ
M0 ROS
M0 ROT
M0 ROU
M0 ROZ
M0 RPA
M0 RPC
M0 RPH
M0 RPP
M0 RPQ
M0 RPS
M0 RPW
M0 RPX
M0 RPY
M0 RRA
M0 RRB
M0 RRD
M0 RRI
M0 RRK
M0 RRO
M0 RRQ
M0 RRS
M0 RRU
M0 RRV
M0 RRW
M0 RSK
M0 RSL
M0 RSS
M0 RTB
M0 RTG
M0 RTI
M0 RTR
M0 RTX
M0 RTZ
M0 RUD
M0 RUF
M0 RUN
M0 RUS
M0 RVA
M0 RVF
M0 RVI
M0 RVM
M0 RVP
M0 RVR
M0 RVW
M0 RVX
M0 RVZ
M0 RWX
M0 RWY
M0 RXJ
M0 RXL
M0 RXO
M0 RXP
M0 RXQ
M0 RXT
M0 RXW
M0 RXY
M0 RYC
M0 RYL
M0 RYN
M0 RYP
M0 RYU
M0 RYW
M0 RYZ
M0 RZA
M0 RZF
M0 RZK
M0 RZM
M0 RZW
M0 SAE
M0 SAF
M0 SAG
M0 SAN
M0 SAS
M0 SAT
M0 SAW
M0 SBE
M0 SBG
M0 SBI
M0 SBM
M0 SBS
M0 SBU
M0 SBV
M0 SBW
M0 SCF
M0 SCH
M0 SCI
M0 SCJ
M0 SCK
M0 SCL
M0 SCM
M0 SCQ
M0 SCZ
M0 SDI
M0 SDJ
M0 SDL
M0 SDN
M0 SDV
M0 SDX
M0 SDZ
M0 SEE
M0 SEG
M0 SEH
M0 SEI
M0 SEN
M0 SEQ
M0 SES
M0 SEU
M0 SFB
M0 SFU
M0 SFZ
M0 SGB
M0 SGC
M0 SGE
M0 SGI
M0 SGL
M0 SGM
M0 SGN
M0 SGO
M0 SGP
M0 SGR
M0 SHL
M0 SHW
M0 SHX
M0 SIB
M0 SIE
M0 SIG
M0 SIM
M0 SIS
M0 SIZ
M0 SJA
M0 SJB
M0 SJC
M0 SJF
M0 SJH
M0 SJM
M0 SJP
M0 SJT
M0 SJY
M0 SKB
M0 SKE
M0 SKJ
M0 SKL
M0 SKS
M0 SKU
M0 SLA
M0 SLJ
M0 SLK
M0 SLM
M0 SLS
M0 SLT
M0 SLW
M0 SLX
M0 SMI
M0 SMK
M0 SMM
M0 SMO
M0 SMR
M0 SMU
M0 SMX
M0 SNC
M0 SNE
M0 SNF
M0 SNI
M0 SNO
M0 SNR
M0 SNW
M0 SNX
M0 SOB
M0 SOM
M0 SON
M0 SOY
M0 SPF
M0 SPG
M0 SPI
M0 SPO
M0 SPR
M0 SPU
M0 SPW
M0 SPY
M0 SQJ
M0 SQL
M0 SQO
M0 SQZ
M0 SRC
M0 SRF
M0 SRH
M0 SRI
M0 SRQ
M0 SRS
M0 SRT
M0 SRW
M0 SRY
M0 SSI
M0 STB
M0 STD
M0 STE
M0 STG
M0 STH
M0 STP
M0 STQ
M0 STR
M0 STW
M0 STZ
M0 SUC
M0 SUE
M0 SUL
M0 SUT
M0 SUV
M0 SVG
M0 SVK
M0 SVL
M0 SVM
M0 SVU
M0 SWA
M0 SWG
M0 SWJ
M0 SWP
M0 SWS
M0 SWW
M0 SWX
M0 SWY
M0 SXD
M0 SXF
M0 SXP
M0 SXR
M0 SXY
M0 SXZ
M0 SYC
M0 SYP
M0 SYR
M0 SZ
M0 SZA
M0 SZE
M0 SZM
M0 SZP
M0 SZT
M0 TA
M0 TAH
M0 TAI
M0 TAU
M0 TBF
M0 TBL
M0 TBM
M0 TBO
M0 TBP
M0 TBT
M0 TBU
M0 TBX
M0 TCG
M0 TCH
M0 TCW
M0 TCX
M0 TDA
M0 TDI
M0 TDN
M0 TDR
M0 TEC
M0 TED
M0 TEE
M0 TEH
M0 TEM
M0 TEO
M0 TEP
M0 TEU
M0 TEV
M0 TEY
M0 TFD
M0 TFG
M0 TFR
M0 TFT
M0 TFW
M0 TGA
M0 TGH
M0 TGK
M0 TGO
M0 TGR
M0 TGZ
M0 THC
M0 THD
M0 THI
M0 THL
M0 THR
M0 THW
M0 THX
M0 THZ
M0 TIB
M0 TIC
M0 TID
M0 TIG
M0 TII
M0 TIS
M0 TJH
M0 TJK
M0 TJP
M0 TKK
M0 TKL
M0 TKO
M0 TLA
M0 TLB
M0 TLE
M0 TLI
M0 TLS
M0 TLT
M0 TLW
M0 TLZ
M0 TME
M0 TMK
M0 TML
M0 TMR
M0 TMU
M0 TMW
M0 TNA
M0 TND
M0 TNK
M0 TNM
M0 TNN
M0 TNR
M0 TNW
M0 TNY
M0 TOA
M0 TOC
M0 TOE
M0 TOT
M0 TOX
M0 TOY
M0 TPM
M0 TPN
M0 TPO
M0 TPR
M0 TPS
M0 TPV
M0 TPX
M0 TPZ
M0 TQM
M0 TQR
M0 TRA
M0 TRD
M0 TRI
M0 TRJ
M0 TRQ
M0 TRR
M0 TSE
M0 TSF
M0 TSG
M0 TSH
M0 TSK
M0 TSL
M0 TSP
M0 TSR
M0 TSS
M0 TST
M0 TSU
M0 TSV
M0 TSX
M0 TTA
M0 TTD
M0 TTJ
M0 TTM
M0 TTP
M0 TTQ
M0 TTS
M0 TTT
M0 TTV
M0 TTW
M0 TTZ
M0 TUC
M0 TUG
M0 TUL
M0 TVE
M0 TVI
M0 TVJ
M0 TVM
M0 TVO
M0 TVQ
M0 TVZ
M0 TWA
M0 TWB
M0 TWD
M0 TWE
M0 TWI
M0 TWP
M0 TWT
M0 TWZ
M0 TXE
M0 TXG
M0 TXI
M0 TXR
M0 TXT
M0 TXU
M0 TYA
M0 TYB
M0 TYL
M0 TYM
M0 TYY
M0 TZX
M0 UAB
M0 UAF
M0 UAL
M0 UAN
M0 UAW
M0 UAX
M0 UBA
M0 UBC
M0 UBS
M0 UBT
M0 UCB
M0 UCQ
M0 UCU
M0 UCW
M0 UCX
M0 UDP
M0 UDR
M0 UEA
M0 UED
M0 UEM
M0 UFM
M0 UFO
M0 UFP
M0 UFX
M0 UGA
M0 UGB
M0 UGS
M0 UHF
M0 UHR
M0 UIA
M0 UIB
M0 UIO
M0 UIT
M0 UKB
M0 UKD
M0 UKE
M0 UKG
M0 UKK
M0 UKL
M0 UKP
M0 UKR
M0 ULF
M0 ULT
M0 UMZ
M0 UNC
M0 UND
M0 UNG
M0 UNO
M0 UNX
M0 UOC
M0 UOS
M0 UPM
M0 URC
M0 URH
M0 URM
M0 URP
M0 USA
M0 USE
M0 USH
M0 USP
M0 USR
M0 USS
M0 USX
M0 UTC
M0 UTU
M0 UUF
M0 UVA
M0 UVK
M0 UVM
M0 UVR
M0 UVU
M0 UXD
M0 UXE
M0 UXH
M0 UXP
M0 UZE
M0 VAE
M0 VAF
M0 VAK
M0 VAX
M0 VAZ
M0 VBA
M0 VBD
M0 VBH
M0 VBM
M0 VBP
M0 VBX
M0 VCD
M0 VCH
M0 VCK
M0 VCM
M0 VCO
M0 VCT
M0 VCW
M0 VCZ
M0 VDC
M0 VDD
M0 VDL
M0 VDT
M0 VDU
M0 VDZ
M0 VEA
M0 VEB
M0 VED
M0 VEE
M0 VEI
M0 VEM
M0 VEN
M0 VEV
M0 VFA
M0 VFB
M0 VFT
M0 VFU
M0 VFX
M0 VGB
M0 VGP
M0 VGW
M0 VHF
M0 VHP
M0 VHR
M0 VHS
M0 VIA
M0 VIC
M0 VID
M0 VIE
M0 VIM
M0 VIP
M0 VIV
M0 VIX
M0 VJL
M0 VJM
M0 VJS
M0 VKA
M0 VKD
M0 VKH
M0 VKM
M0 VKO
M0 VKU
M0 VKV
M0 VLE
M0 VLF
M0 VMA
M0 VME
M0 VMF
M0 VMK
M0 VML
M0 VMO
M0 VMP
M0 VMT
M0 VMZ
M0 VNA
M0 VNL
M0 VNP
M0 VOA
M0 VOC
M0 VOD
M0 VOO
M0 VOP
M0 VOR
M0 VOW
M0 VOX
M0 VPH
M0 VPN
M0 VPS
M0 VPT
M0 VPZ
M0 VRA
M0 VRC
M0 VRE
M0 VRF
M0 VRH
M0 VRL
M0 VRR
M0 VRX
M0 VRY
M0 VSA
M0 VSI
M0 VSN
M0 VSS
M0 VST
M0 VSV
M0 VTB
M0 VTF
M0 VTM
M0 VTO
M0 VTT
M0 VTX
M0 VTZ
M0 VUG
M0 VUP
M0 VVF
M0 VVJ
M0 VVK
M0 VVS
M0 VVZ
M0 VWA
M0 VWB
M0 VWG
M0 VWH
M0 VWV
M0 VWX
M0 VXA
M0 VXK
M0 VXN
M0 VXO
M0 VXR
M0 VXT
M0 VYD
M0 VZO
M0 VZZ
M0 WAL
M0 WAN
M0 WAW
M0 WBA
M0 WBH
M0 WBN
M0 WBT
M0 WBX
M0 WCC
M0 WCF
M0 WCN
M0 WCQ
M0 WCY
M0 WDC
M0 WDF
M0 WDM
M0 WDN
M0 WDR
M0 WDT
M0 WDV
M0 WDW
M0 WDX
M0 WEA
M0 WEF
M0 WEH
M0 WEM
M0 WEZ
M0 WFD
M0 WGD
M0 WGG
M0 WGH
M0 WGN
M0 WGT
M0 WGX
M0 WGZ
M0 WHD
M0 WHH
M0 WHI
M0 WHU
M0 WHW
M0 WID
M0 WIX
M0 WJD
M0 WJE
M0 WJI
M0 WJJ
M0 WJN
M0 WJO
M0 WKC
M0 WKD
M0 WKS
M0 WKW
M0 WLC
M0 WLM
M0 WLO
M0 WLR
M0 WLW
M0 WLZ
M0 WMC
M0 WMD
M0 WME
M0 WMF
M0 WMM
M0 WNH
M0 WNN
M0 WNP
M0 WNU
M0 WOE
M0 WOK
M0 WOL
M0 WOO
M0 WOZ
M0 WPD
M0 WPF
M0 WPG
M0 WPH
M0 WQD
M0 WQX
M0 WRB
M0 WRF
M0 WRL
M0 WRM
M0 WRN
M0 WRT
M0 WSO
M0 WSS
M0 WTA
M0 WTB
M0 WTD
M0 WTF
M0 WTH
M0 WTM
M0 WTP
M0 WTR
M0 WTS
M0 WTT
M0 WTU
M0 WTV
M0 WUE
M0 WUF
M0 WUK
M0 WUT
M0 WVR
M0 WVV
M0 WVX
M0 WWA
M0 WWG
M0 WWJ
M0 WWS
M0 WWW
M0 WXB
M0 WXG
M0 WXT
M0 WYA
M0 WZC
M0 WZD
M0 WZF
M0 WZW
M0 WZY
M0 XAA
M0 XAF
M0 XAG
M0 XAH
M0 XAI
M0 XAL
M0 XAN
M0 XAP
M0 XAX
M0 XAY
M0 XAZ
M0 XBC
M0 XBE
M0 XBF
M0 XBO
M0 XBP
M0 XCB
M0 XCD
M0 XCE
M0 XCG
M0 XCK
M0 XCM
M0 XCV
M0 XCW
M0 XDB
M0 XDD
M0 XDG
M0 XDI
M0 XDM
M0 XDP
M0 XDR
M0 XDW
M0 XDY
M0 XEC
M0 XEN
M0 XEQ
M0 XET
M0 XFD
M0 XFJ
M0 XFM
M0 XFN
M0 XFR
M0 XFT
M0 XGA
M0 XGC
M0 XGD
M0 XGM
M0 XGP
M0 XHE
M0 XHM
M0 XHO
M0 XHT
M0 XIE
M0 XII
M0 XIM
M0 XIN
M0 XIO
M0 XIP
M0 XIT
M0 XIX
M0 XJA
M0 XJB
M0 XJC
M0 XJR
M0 XKC
M0 XKF
M0 XKG
M0 XKI
M0 XKN
M0 XKR
M0 XKV
M0 XLD
M0 XLE
M0 XLF
M0 XLG
M0 XLO
M0 XLS
M0 XLZ
M0 XMA
M0 XME
M0 XMM
M0 XMR
M0 XNA
M0 XNI
M0 XNP
M0 XOF
M0 XOJ
M0 XPC
M0 XPO
M0 XPU
M0 XPX
M0 XQZ
M0 XRB
M0 XRD
M0 XRE
M0 XRG
M0 XRJ
M0 XRK
M0 XRO
M0 XRS
M0 XSI
M0 XSP
M0 XSR
M0 XSW
M0 XSX
M0 XTA
M0 XTC
M0 XTE
M0 XTF
M0 XTF
M0 XTH
M0 XTO
M0 XTR
M0 XTY
M0 XUE
M0 XUT
M0 XVM
M0 XVR
M0 XVT
M0 XVW
M0 XWB
M0 XWE
M0 XWF
M0 XWN
M0 XWR
M0 XXA
M0 XXF
M0 XXQ
M0 XXR
M0 XXS
M0 XXU
M0 XXY
M0 XXZ
M0 XYB
M0 XYD
M0 XYF
M0 XYG
M0 XZA
M0 XZF
M0 XZI
M0 XZL
M0 XZZ
M0 YAA
M0 YAF
M0 YAP
M0 YAT
M0 YAX
M0 YAZ
M0 YBG
M0 YBM
M0 YBO
M0 YBX
M0 YCC
M0 YCL
M0 YCM
M0 YCP
M0 YCT
M0 YDA
M0 YDC
M0 YDD
M0 YDG
M0 YDR
M0 YDT
M0 YED
M0 YEH
M0 YEL
M0 YEW
M0 YFL
M0 YFR
M0 YGF
M0 YGH
M0 YGR
M0 YHB
M0 YHC
M0 YHF
M0 YHS
M0 YIK
M0 YIP
M0 YJB
M0 YJG
M0 YJR
M0 YKK
M0 YKW
M0 YMC
M0 YME
M0 YMK
M0 YNE
M0 YNG
M0 YNW
M0 YOH
M0 YOK
M0 YOR
M0 YOU
M0 YOY
M0 YPA
M0 YPT
M0 YPZ
M0 YQV
M0 YRK
M0 YRS
M0 YRU
M0 YRX
M0 YSB
M0 YSC
M0 YSL
M0 YSO
M0 YSU
M0 YTB
M0 YTE
M0 YTF
M0 YTH
M0 YTK
M0 YTL
M0 YTM
M0 YTT
M0 YTX
M0 YTY
M0 YUI
M0 YUL
M0 YUP
M0 YVE
M0 YWG
M0 YXB
M0 YXD
M0 YYC
M0 YYO
M0 YYZ
M0 YZA
M0 YZC
M0 YZL
M0 YZX
M0 YZZ
M0 ZAD
M0 ZAG
M0 ZAH
M0 ZAN
M0 ZAO
M0 ZAT
M0 ZAX
M0 ZAZ
M0 ZBB
M0 ZBU
M0 ZCC
M0 ZCH
M0 ZCR
M0 ZDF
M0 ZDR
M0 ZDZ
M0 ZEP
M0 ZER
M0 ZES
M0 ZFA
M0 ZFJ
M0 ZFR
M0 ZFX
M0 ZGD
M0 ZGN
M0 ZGY
M0 ZHA
M0 ZHN
M0 ZHX
M0 ZIE
M0 ZIX
M0 ZLB
M0 ZLR
M0 ZLW
M0 ZMF
M0 ZNH
M0 ZNK
M0 ZOC
M0 ZOD
M0 ZOL
M0 ZOS
M0 ZPK
M0 ZQI
M0 ZQR
M0 ZRA
M0 ZRN
M0 ZRQ
M0 ZRT
M0 ZSD
M0 ZSP
M0 ZSU
M0 ZTP
M0 ZTT
M0 ZTX
M0 ZUN
M0 ZUU
M0 ZVX
M0 ZWE
M0 ZXA
M0 ZXP
M0 ZXT
M0 ZXZ
M0 ZZB
M0 ZZK
M0 ZZM
M0 ZZR
M0 ZZW
M0 ZZX
M0 ZZY

M1

M1 AAA
M1 AAD
M1 AAG
M1 AAK
M1 AAM
M1 AAO
M1 AAT
M1 ABD
M1 ABE
M1 ABK
M1 ABP
M1 ACI
M1 ACM
M1 ACP
M1 ACS
M1 ACU
M1 ACZ
M1 ADA
M1 ADC
M1 ADG
M1 ADL
M1 ADW
M1 AEC
M1 AEM
M1 AES
M1 AFD
M1 AFM
M1 AFR
M1 AGB
M1 AGF
M1 AGJ
M1 AGN
M1 AHB
M1 AHE
M1 AHO
M1 AHS
M1 AHZ
M1 AIF
M1 AIS
M1 AIU
M1 AJB
M1 AJJ
M1 AJK
M1 AJR
M1 AJV
M1 AKE
M1 AKJ
M1 AKS
M1 ALI
M1 ALN
M1 ALY
M1 AMF
M1 AMH
M1 AMK
M1 AMN
M1 AMT
M1 AMY
M1 ANH
M1 ANV
M1 ANW
M1 ANX
M1 AOA
M1 AOC
M1 AOH
M1 APN
M1 APU
M1 APX
M1 AQF
M1 AQH
M1 AQK
M1 AQM
M1 AQO
M1 AQR
M1 AQV
M1 AQW
M1 ARA
M1 ARI
M1 ARN
M1 ARW
M1 ARY
M1 ARZ
M1 ASF
M1 ASO
M1 ASW
M1 ASY
M1 ATF
M1 ATH
M1 ATM
M1 ATW
M1 ATZ
M1 AUD
M1 AUS
M1 AUU
M1 AVE
M1 AVG
M1 AVP
M1 AWG
M1 AWH
M1 AWZ
M1 AXA
M1 AXI
M1 AXL
M1 AXN
M1 AXO
M1 AXV
M1 AXY
M1 AYE
M1 AYI
M1 AYM
M1 AYX
M1 AYZ
M1 AZS
M1 AZT
M1 AZZ
M1 BAT
M1 BAU
M1 BAW
M1 BAZ
M1 BBM
M1 BBN
M1 BBQ
M1 BBT
M1 BBV
M1 BCF
M1 BCJ
M1 BCV
M1 BCX
M1 BDB
M1 BDI
M1 BDK
M1 BDM
M1 BEA
M1 BEG
M1 BEK
M1 BEY
M1 BFH
M1 BFP
M1 BFZ
M1 BGB
M1 BGC
M1 BGQ
M1 BGU
M1 BGV
M1 BHD
M1 BHR
M1 BHS
M1 BIA
M1 BIJ
M1 BIN
M1 BIR
M1 BIZ
M1 BJH
M1 BJI
M1 BJL
M1 BJP
M1 BJR
M1 BKC
M1 BKD
M1 BKM
M1 BKU
M1 BLM
M1 BLV
M1 BMJ
M1 BMN
M1 BNE
M1 BNJ
M1 BNL
M1 BNM
M1 BNP
M1 BNW
M1 BNZ
M1 BOC
M1 BOI
M1 BPJ
M1 BPT
M1 BRH
M1 BRQ
M1 BRR
M1 BSA
M1 BSC
M1 BSZ
M1 BTF
M1 BTJ
M1 BTY
M1 BUD
M1 BUV
M1 BUZ
M1 BVA
M1 BWB
M1 BWO
M1 BWP
M1 BWT
M1 BWW
M1 BWY
M1 BXE
M1 BXG
M1 BXI
M1 BXT
M1 BXW
M1 BXY
M1 BYA
M1 BYB
M1 BYN
M1 BYO
M1 BYY
M1 BZM
M1 BZN
M1 BZV
M1 CAF
M1 CAK
M1 CAM
M1 CAN
M1 CBG
M1 CBP
M1 CBQ
M1 CBW
M1 CCC
M1 CCO
M1 CCP
M1 CDE
M1 CDG
M1 CDH
M1 CDM
M1 CDU
M1 CDZ
M1 CEB
M1 CEC
M1 CEE
M1 CEH
M1 CEK
M1 CEL
M1 CEN
M1 CEU
M1 CEV
M1 CEX
M1 CFJ
M1 CFK
M1 CFL
M1 CFN
M1 CFV
M1 CGE
M1 CGG
M1 CGH
M1 CGX
M1 CHG
M1 CHN
M1 CHO
M1 CHV
M1 CHW
M1 CIA
M1 CID
M1 CIW
M1 CIX
M1 CIY
M1 CJK
M1 CJP
M1 CJU
M1 CJY
M1 CKC
M1 CKG
M1 CKM
M1 CKP
M1 CLF
M1 CLK
M1 CLN
M1 CLQ
M1 CLV
M1 CMC
M1 CMP
M1 CMS
M1 CMV
M1 CNG
M1 CNO
M1 CNU
M1 CNW
M1 COI
M1 COK
M1 CPE
M1 CPH
M1 CPM
M1 CPO
M1 CPU
M1 CPV
M1 CPW
M1 CPX
M1 CQO
M1 CRB
M1 CRG
M1 CRN
M1 CRS
M1 CRV
M1 CRX
M1 CSB
M1 CSK
M1 CSO
M1 CSQ
M1 CSS
M1 CTA
M1 CTN
M1 CTP
M1 CUE
M1 CUI
M1 CUJ
M1 CUW
M1 CUZ
M1 CVA
M1 CVE
M1 CVJ
M1 CVR
M1 CVZ
M1 CWM
M1 CWR
M1 CXB
M1 CXE
M1 CXF
M1 CXI
M1 CXJ
M1 CXX
M1 CYE
M1 CYI
M1 CYS
M1 CYV
M1 CYW
M1 CYY
M1 CZB
M1 CZC
M1 CZR
M1 CZT
M1 CZW
M1 DAJ
M1 DAT
M1 DAZ
M1 DBB
M1 DBE
M1 DBG
M1 DBS
M1 DBT
M1 DBZ
M1 DCG
M1 DCJ
M1 DCN
M1 DCO
M1 DCW
M1 DDD
M1 DDE
M1 DDK
M1 DDL
M1 DDN
M1 DDV
M1 DEC
M1 DED
M1 DEI
M1 DEL
M1 DEN
M1 DEO
M1 DFG
M1 DFX
M1 DGA
M1 DGC
M1 DGD
M1 DGF
M1 DGH
M1 DGN
M1 DGO
M1 DGZ
M1 DHB
M1 DHD
M1 DHH
M1 DHL
M1 DHP
M1 DIC
M1 DID
M1 DIG
M1 DII
M1 DIP
M1 DJE
M1 DJF
M1 DJK
M1 DJR
M1 DJU
M1 DKC
M1 DKU
M1 DKV
M1 DLB
M1 DLC
M1 DLF
M1 DLJ
M1 DLL
M1 DLW
M1 DLY
M1 DMC
M1 DMJ
M1 DMQ
M1 DMS
M1 DMX
M1 DNF
M1 DNL
M1 DNO
M1 DNS
M1 DNV
M1 DOC
M1 DOJ
M1 DOP
M1 DOW
M1 DPB
M1 DPD
M1 DPM
M1 DPV
M1 DPZ
M1 DQA
M1 DQF
M1 DQJ
M1 DQY
M1 DRC
M1 DRN
M1 DRQ
M1 DRT
M1 DRX
M1 DSB
M1 DSL
M1 DSR
M1 DST
M1 DTC
M1 DTI
M1 DTJ
M1 DTX
M1 DUL
M1 DUM
M1 DUN
M1 DUQ
M1 DUX
M1 DUY
M1 DUZ
M1 DVM
M1 DWM
M1 DWO
M1 DWX
M1 DXF
M1 DXK
M1 DXN
M1 DXP
M1 DXZ
M1 DYP
M1 DYR
M1 DYX
M1 DZA
M1 DZI
M1 DZI
M1 DZK
M1 DZO
M1 DZU
M1 EAC
M1 EAG
M1 EAR
M1 EAU
M1 EAX
M1 EBM
M1 EBO
M1 ECE
M1 ECF
M1 ECK
M1 ECN
M1 ECP
M1 ECR
M1 ECY
M1 EDB
M1 EDG
M1 EDN
M1 EDT
M1 EDU
M1 EDX
M1 EEM
M1 EEU
M1 EEV
M1 EFI
M1 EFL
M1 EFQ
M1 EFY
M1 EGF
M1 EGH
M1 EGI
M1 EGJ
M1 EGO
M1 EHL
M1 EHM
M1 EHU

M1 EHW
M1 EHY
M1 EIB
M1 EIK
M1 EIV
M1 EJC
M1 EJH
M1 EJK
M1 EJU
M1 EKG
M1 EKI
M1 EKT
M1 EKV
M1 EKX
M1 EKZ
M1 ELC
M1 ELF
M1 ELJ
M1 ELK
M1 ELT
M1 ELY
M1 EME
M1 EMF
M1 EMH
M1 EMI
M1 EMJ
M1 EMT
M1 EMZ
M1 ENG
M1 ENH
M1 ENI
M1 ENO
M1 ENR
M1 ENU
M1 ENV
M1 ENW
M1 ENY
M1 EOA
M1 EOE
M1 EOG
M1 EOM
M1 EON
M1 EOX
M1 EPH
M1 EPL
M1 EPM
M1 EPO
M1 EPQ
M1 EPW
M1 EPY
M1 EPZ
M1 EQN
M1 EQP
M1 EQS
M1 ERG
M1 ERM
M1 ESA
M1 ESC
M1 ESG
M1 ESL
M1 ESR
M1 EST
M1 ESU
M1 ESW
M1 ETI
M1 EUA
M1 EUI
M1 EUS
M1 EUW
M1 EUZ
M1 EVA
M1 EVG
M1 EVK
M1 EVV
M1 EVY
M1 EWH
M1 EWK
M1 EWZ
M1 EXA
M1 EXE
M1 EXL
M1 EXP
M1 EXR
M1 EXX
M1 EXZ
M1 EYB
M1 EYE
M1 EYK
M1 EYW
M1 EYX
M1 EYY
M1 EZA
M1 EZF
M1 EZS
M1 EZV
M1 EZW
M1 EZY
M1 FAM
M1 FAU
M1 FAV
M1 FBC
M1 FBE
M1 FBG
M1 FBH
M1 FBJ
M1 FBR
M1 FBT
M1 FBU
M1 FBV
M1 FBX
M1 FBY
M1 FBZ
M1 FCD
M1 FCP
M1 FDB
M1 FDI
M1 FDJ
M1 FDQ
M1 FDS
M1 FDT
M1 FDU
M1 FDZ
M1 FEB
M1 FEF
M1 FEV
M1 FFK
M1 FFQ
M1 FFZ
M1 FGA
M1 FGE
M1 FGF
M1 FGG
M1 FGI
M1 FGK
M1 FGL
M1 FGN
M1 FGQ
M1 FGR
M1 FGS
M1 FGU
M1 FGY
M1 FGZ
M1 FHD
M1 FHG
M1 FHI
M1 FHK
M1 FHM
M1 FHU
M1 FID
M1 FIF
M1 FIH
M1 FIJ
M1 FIM
M1 FIN
M1 FIO
M1 FIQ
M1 FIU
M1 FIW
M1 FJE
M1 FJI
M1 FJN
M1 FJO
M1 FJU
M1 FJV
M1 FMJ
M1 FMW
M1 FRH
M1 FXY
M1 FZL
M1 GAC
M1 GAZ
M1 GGY
M1 GLA
M1 GLO
M1 GMJ
M1 GMW
M1 GNB
M1 GPD
M1 GPO
M1 GRY
M1 GSE
M1 GTR
M1 HAC
M1 HAT
M1 HAX
M1 HCO
M1 HDD
M1 HIS
M1 HOG
M1 HRQ
M1 HSB
M1 HTG
M1 HUD
M1 HUL
M1 HVA
M1 HVD
M1 HWH
M1 HZJ
M1 IJJ
M1 ISR
M1 IVA
M1 IWB
M1 JCM
M1 JDK
M1 JED
M1 JEN
M1 JJK
M1 JKL
M1 JMH
M1 JMR
M1 JMW
M1 JNC
M1 JPF
M1 JPM
M1 JRC
M1 JRG
M1 JRV
M1 JSC
M1 JUL
M1 JUN
M1 JWT
M1 KAB
M1 KAH
M1 KCS
M1 KDA
M1 KE
M1 KEN
M1 KET
M1 KFM
M1 KGB
M1 KGC
M1 KLS
M1 KTM
M1 KTO
M1 LCR
M1 LEE
M1 LIV
M1 LKY
M1 LNB
M1 MAR
M1 MAS
M1 MAX
M1 MBZ
M1 MDH
M1 MHS
M1 MHZ
M1 MIS
M1 MJT
M1 MOW
M1 MOZ
M1 MPR
M1 MSD
M1 MSG
M1 MWD
M1 MZX
M1 NCD
M1 NCP
M1 NCR
M1 NDF
M1 NED
M1 NIC
M1 NIX
M1 NJJ
M1 NRT
M1 NSP
M1 NTP
M1 NWT
M1 OAK
M1 OEY
M1 OMW
M1 OOO
M1 PDJ
M1 PEP
M1 PGC
M1 PHS
M1 PJC
M1 PLA
M1 PMO
M1 POM
M1 POP
M1 PPP
M1 PRO
M1 PRV
M1 PSH
M1 PSY
M1 PUB
M1 PUP
M1 PVS
M1 RCA
M1 RDB
M1 REW
M1 RGN
M1 RGZ
M1 RIP
M1 RKH
M1 RMB
M1 ROK
M1 RON
M1 RPM
M1 RPW
M1 RPY
M1 RWN
M1 SAG
M1 SBH
M1 SCH
M1 SDH
M1 SJE
M1 SJM
M1 SJP
M1 SMH
M1 SMJ
M1 SPH
M1 SSY
M1 STE
M1 STX
M1 TAY
M1 TCH
M1 TDM
M1 TEX
M1 TGQ
M1 TIG
M1 TLD
M1 TLK
M1 TNT
M1 TNY
M1 TRV
M1 TTT
M1 TTY
M1 TUT
M1 TWO
M1 TXI
M1 TZR
M1 USA
M1 VAW
M1 VCG
M1 VFO
M1 VFR
M1 VKJ
M1 VOX
M1 VSA
M1 VTR
M1 VWB
M1 WCJ
M1 WEB
M1 WML
M1 WPB
M1 WRG
M1 XCQ
M1 XDX
M1 XED
M1 XER
M1 XJC
M1 XMP
M1 XOR
M1 XUP
M1 XWD
M1 XXX
M1 YAP
M1 YJG
M1 ZRP

M3

M3 ACK
M3 ACV
M3 ADF
M3 ADV
M3 AEX
M3 AFV
M3 AFW
M3 AFZ
M3 AGC
M3 AGJ
M3 AGO
M3 AGQ
M3 AGS
M3 AHK
M3 AHM
M3 AHT
M3 AIX
M3 AJB
M3 AJG
M3 AKD
M3 AKU
M3 ALP
M3 ALR
M3 ALT
M3 AMG
M3 AMN
M3 AMY
M3 ANX
M3 ANY
M3 AOB
M3 AOI
M3 APF
M3 APK
M3 APL
M3 APM
M3 APP
M3 APR
M3 ARC
M3 ARV
M3 ARX
M3 ASQ
M3 ASU
M3 ATA
M3 ATH
M3 ATK
M3 ATM
M3 ATX
M3 AUD
M3 AUE
M3 AUH
M3 AUR
M3 AUV
M3 AUW
M3 AVB
M3 AVC
M3 AVG
M3 AVN
M3 AVP
M3 AVT
M3 AWA
M3 AWF
M3 AWJ
M3 AWL
M3 AWV
M3 AXJ
M3 AYG
M3 AYK
M3 AYR
M3 AZC
M3 AZD
M3 AZW
M3 BAF
M3 BAP
M3 BAV
M3 BAX
M3 BBE
M3 BBJ
M3 BBM
M3 BBT
M3 BBV
M3 BBW
M3 BCN
M3 BCO
M3 BCT
M3 BDE
M3 BDF
M3 BDI
M3 BDL
M3 BDM
M3 BDO
M3 BDS
M3 BEH
M3 BEI
M3 BEO
M3 BEY
M3 BFA
M3 BFS
M3 BFW
M3 BGO
M3 BGP
M3 BHN
M3 BHV
M3 BHX
M3 BIJ
M3 BIL
M3 BIM
M3 BIN
M3 BJC
M3 BJD
M3 BJI
M3 BJU
M3 BJX
M3 BKF
M3 BKL
M3 BKR
M3 BKZ
M3 BLL
M3 BLS
M3 BLV
M3 BLW
M3 BMD
M3 BMY
M3 BNH
M3 BNI
M3 BNO
M3 BNP
M3 BNT
M3 BOM
M3 BOQ
M3 BOX
M3 BOY
M3 BPB
M3 BPE
M3 BPK
M3 BQA
M3 BQR
M3 BQU
M3 BQV
M3 BQW
M3 BRP
M3 BSA
M3 BSL
M3 BSQ
M3 BSX
M3 BTC
M3 BTD
M3 BTE
M3 BTV
M3 BTY
M3 BUB
M3 BUG
M3 BUK
M3 BUO
M3 BVC
M3 BVN
M3 BVZ
M3 BWM
M3 BWO
M3 BXA
M3 BXI
M3 BXL
M3 BYG
M3 BYH
M3 BYY
M3 BYZ
M3 BZG
M3 BZO
M3 CBD
M3 CBI
M3 CCR
M3 CDC
M3 CDD
M3 CDF
M3 CDH
M3 CDS
M3 CDZ
M3 CEC
M3 CED
M3 CEE
M3 CEF
M3 CEH
M3 CEI
M3 CEK
M3 CEY
M3 CFC
M3 CFE
M3 CFH
M3 CFK
M3 CFL
M3 CFN
M3 CFQ
M3 CFS
M3 CFW
M3 CFY
M3 CGB
M3 CGV
M3 CGW
M3 CGY
M3 CHA
M3 CHC
M3 CHD
M3 CHH
M3 CHO
M3 CIB
M3 CIN
M3 CJQ
M3 CJV
M3 CKB
M3 CKG
M3 CKS
M3 CKV
M3 CLF
M3 CLN
M3 CLR
M3 CLV
M3 CLZ
M3 CMD
M3 CMT
M3 CNU
M3 CNX
M3 COH
M3 COO
M3 COR
M3 COS
M3 COX
M3 COY
M3 CPA
M3 CPB
M3 CPJ
M3 CPP
M3 CPT
M3 CPU
M3 CPV
M3 CQE
M3 CQG
M3 CQI
M3 CQJ
M3 CQK
M3 CQL
M3 CQM
M3 CQZ
M3 CRB
M3 CRE
M3 CRG
M3 CRJ
M3 CRU
M3 CSP
M3 CSW
M3 CTF
M3 CTG
M3 CTH
M3 CTJ
M3 CTL
M3 CTW
M3 CUD
M3 CUG
M3 CUI
M3 CUO
M3 CUP
M3 CUQ
M3 CUS
M3 CVF
M3 CVG
M3 CVI
M3 CVT
M3 CWB
M3 CWD
M3 CWM
M3 CWR
M3 CWS
M3 CWT
M3 CWU
M3 CWW
M3 CXA
M3 CXL
M3 CXS
M3 CXW
M3 CYP
M3 CYR
M3 CYY
M3 CZC
M3 CZD
M3 CZH
M3 CZO
M3 CZZ
M3 DAC
M3 DAJ
M3 DAT
M3 DAY
M3 DBD
M3 DBI
M3 DBM
M3 DBO
M3 DBQ
M3 DCA
M3 DCB
M3 DCG
M3 DCO
M3 DCQ
M3 DCT
M3 DDL
M3 DDN
M3 DDO
M3 DDP
M3 DDV
M3 DEG
M3 DEK
M3 DEO
M3 DEQ
M3 DEU
M3 DFE
M3 DFI
M3 DGC
M3 DGE
M3 DHC
M3 DHL
M3 DHP
M3 DID
M3 DIF
M3 DII
M3 DJS
M3 DJT
M3 DKD
M3 DKF
M3 DKQ
M3 DKR
M3 DKX
M3 DKY
M3 DLA
M3 DLL
M3 DLS
M3 DLV
M3 DLY
M3 DMB
M3 DMC
M3 DMH
M3 DMV
M3 DMX
M3 DNK
M3 DNL
M3 DNZ
M3 DOE
M3 DOI
M3 DOQ
M3 DOV
M3 DPD
M3 DPE
M3 DPL
M3 DPO
M3 DPR
M3 DPU
M3 DPW
M3 DQC
M3 DQF
M3 DQO
M3 DQT
M3 DRL
M3 DRP
M3 DRQ
M3 DRT
M3 DSN
M3 DSO
M3 DTF
M3 DTS
M3 DTW
M3 DUA
M3 DUF
M3 DUG
M3 DUV
M3 DUW
M3 DUX
M3 DUY
M3 DVE
M3 DVK
M3 DVS
M3 DVX
M3 DWJ
M3 DWN
M3 DWS
M3 DWT
M3 DWU
M3 DWY
M3 DXB
M3 DXC
M3 DXH
M3 DXJ
M3 DXP
M3 DYD
M3 DYG
M3 DYH
M3 DYI
M3 DYJ
M3 DYP
M3 DYS
M3 DYX
M3 DZJ
M3 DZY
M3 EAC
M3 EAG
M3 EAJ
M3 EAM
M3 EAO
M3 EAV
M3 EBN
M3 ECA
M3 ECH
M3 ECL
M3 ECM
M3 ECT
M3 ECX
M3 EDB
M3 EDE
M3 EDF
M3 EDM
M3 EEA
M3 EEB
M3 EEY
M3 EFE
M3 EFF
M3 EFM
M3 EFY
M3 EFZ
M3 EGN
M3 EGO
M3 EGP
M3 EHE
M3 EHN
M3 EHO
M3 EHQ
M3 EHW
M3 EHX
M3 EIE
M3 EII
M3 EIR
M3 EIS
M3 EJA
M3 EJG
M3 EJT
M3 EJZ
M3 EKB
M3 EKQ
M3 EKX
M3 ELE
M3 ELH
M3 ELJ
M3 ELU
M3 EMF
M3 EMH
M3 EMP
M3 EMT
M3 EMY
M3 ENB
M3 END
M3 ENQ
M3 ENT
M3 ENW
M3 ENX
M3 EOA
M3 EOG
M3 EOK
M3 EPA
M3 EPB
M3 EPE
M3 EPN
M3 EPP
M3 EPS
M3 EPY
M3 EPZ
M3 EQC
M3 EQI
M3 EQJ
M3 EQN
M3 EQR
M3 EQV
M3 EQX
M3 EQZ
M3 ERG
M3 ERK
M3 ERL
M3 ERW
M3 ESC
M3 ESR
M3 ETD
M3 ETP
M3 ETY
M3 EUC
M3 EUO
M3 EUT
M3 EVL
M3 EVZ
M3 EWC
M3 EWJ
M3 EWL
M3 EXE
M3 EXO
M3 EXQ
M3 EXU
M3 EYE
M3 EYG
M3 EYU
M3 EYV
M3 EZA
M3 EZD
M3 EZT
M3 EZW
M3 FAD
M3 FAQ
M3 FAV
M3 FBE
M3 FBH
M3 FBT
M3 FBZ
M3 FCC
M3 FCD
M3 FCE
M3 FCH
M3 FCI
M3 FCJ
M3 FDD
M3 FDE
M3 FDI
M3 FDK
M3 FDP
M3 FDT
M3 FDW
M3 FDY
M3 FEE
M3 FEH
M3 FEP
M3 FER
M3 FEV
M3 FEZ
M3 FFG
M3 FFW
M3 FGD
M3 FGE
M3 FGJ
M3 FHN
M3 FID
M3 FIF
M3 FIN
M3 FIR
M3 FIV
M3 FJU
M3 FKA
M3 FKF
M3 FKH
M3 FKJ
M3 FKT
M3 FKX
M3 FLF
M3 FLH
M3 FLQ
M3 FLW
M3 FLY
M3 FMA
M3 FMG
M3 FMJ
M3 FML
M3 FMO
M3 FMT
M3 FMX
M3 FMZ
M3 FNF
M3 FNK
M3 FNP
M3 FNU
M3 FNW
M3 FOB
M3 FOI
M3 FON
M3 FOT
M3 FOU
M3 FPP
M3 FPX
M3 FPY
M3 FQB
M3 FQE
M3 FQF
M3 FQH
M3 FQK
M3 FQL
M3 FQP
M3 FQV
M3 FQW
M3 FRA
M3 FRF
M3 FRI
M3 FRW
M3 FSA
M3 FSJ
M3 FSM
M3 FSO
M3 FST
M3 FTL
M3 FTO
M3 FTS
M3 FUE
M3 FUL
M3 FUU
M3 FVF
M3 FVI
M3 FVK
M3 FVQ
M3 FVT
M3 FVU
M3 FVV
M3 FWB
M3 FWG
M3 FWQ
M3 FXH
M3 FXN
M3 FXR
M3 FXS
M3 FXV
M3 FYD
M3 FYH
M3 FYO
M3 FYP
M3 FYT
M3 FYU
M3 FYX
M3 FZG
M3 FZH
M3 FZK
M3 FZN
M3 FZP
M3 FZQ
M3 FZU
M3 FZW
M3 FZY
M3 GAI
M3 GAJ
M3 GAR
M3 GAS
M3 GAU
M3 GBG
M3 GBL
M3 GBQ
M3 GBU
M3 GBV
M3 GCA
M3 GCE
M3 GCG
M3 GCL
M3 GCO
M3 GCQ
M3 GCU
M3 GCV
M3 GDA
M3 GDB
M3 GDJ
M3 GDP
M3 GDQ
M3 GDS
M3 GDW
M3 GDZ
M3 GED
M3 GEG
M3 GEO
M3 GEV
M3 GFC
M3 GFK
M3 GFN
M3 GFU
M3 GFV
M3 GFX
M3 GFY
M3 GGC
M3 GGG
M3 GGI
M3 GGK
M3 GGL
M3 GGP
M3 GGT
M3 GGU
M3 GGY
M3 GGZ
M3 GHP
M3 GHQ
M3 GHV
M3 GIA
M3 GIB
M3 GIN
M3 GIO
M3 GIP
M3 GIU
M3 GJA
M3 GJF
M3 GJJ
M3 GJL
M3 GJP
M3 GJS
M3 GJZ
M3 GKC
M3 GKF
M3 GKP
M3 GKQ
M3 GKS
M3 GKV
M3 GLB
M3 GLE
M3 GLF
M3 GLG
M3 GLU
M3 GLW
M3 GLX
M3 GME
M3 GMH
M3 GMN
M3 GMO
M3 GMQ
M3 GMZ
M3 GNC
M3 GNE
M3 GNH
M3 GNL
M3 GNO
M3 GNQ
M3 GNS
M3 GNV
M3 GNW
M3 GNX
M3 GOC
M3 GOF
M3 GOG
M3 GON
M3 GOO
M3 GOP
M3 GPF
M3 GPO
M3 GPT
M3 GPU
M3 GPZ
M3 GQF
M3 GQG
M3 GQH
M3 GQQ
M3 GQU
M3 GRD
M3 GRE
M3 GRQ
M3 GRU
M3 GRW
M3 GRX
M3 GRZ
M3 GSA
M3 GSN
M3 GSS
M3 GSV
M3 GTI
M3 GTS
M3 GUA
M3 GUE
M3 GUK
M3 GUL
M3 GUP
M3 GUT
M3 GUW
M3 GVI
M3 GVP
M3 GVR
M3 GVX
M3 GVY
M3 GVZ
M3 GWE
M3 GWL
M3 GWM
M3 GWO
M3 GWX
M3 GWY
M3 GXC
M3 GXF
M3 GXS
M3 GXT
M3 GXV
M3 GXW
M3 GYG
M3 GYS
M3 GYX
M3 GZA
M3 GZB
M3 GZH
M3 HAA
M3 HAH
M3 HAO
M3 HAR
M3 HAY
M3 HBA
M3 HBI
M3 HBO
M3 HBZ
M3 HCZ
M3 HDB
M3 HDO
M3 HDR
M3 HDY
M3 HEF
M3 HEN
M3 HEP
M3 HES
M3 HEW
M3 HEZ
M3 HFB
M3 HFK
M3 HFM
M3 HFV
M3 HFY
M3 HGB
M3 HGD
M3 HGF
M3 HGS
M3 HGY
M3 HHJ
M3 HHT
M3 HHW
M3 HHY
M3 HIJ
M3 HIY
M3 HJH
M3 HJL
M3 HJN
M3 HJR
M3 HKG
M3 HKI
M3 HKK
M3 HKO
M3 HKS
M3 HKX
M3 HLO
M3 HMX
M3 HNC
M3 HNH
M3 HNI
M3 HNY
M3 HNZ
M3 HOB
M3 HOF
M3 HOI
M3 HOJ
M3 HOO
M3 HOS
M3 HOZ
M3 HPG
M3 HPJ
M3 HPL
M3 HPP
M3 HPR
M3 HPS
M3 HPX
M3 HQG
M3 HQI
M3 HQO
M3 HRC
M3 HRI
M3 HRP
M3 HRQ
M3 HRR
M3 HSA
M3 HSB
M3 HSN
M3 HSP
M3 HSY
M3 HTB
M3 HTC
M3 HTD
M3 HTT
M3 HTU
M3 HTV
M3 HTZ
M3 HUA
M3 HUC
M3 HUF
M3 HUH
M3 HUI
M3 HUJ
M3 HVB
M3 HVF
M3 HVG
M3 HVI
M3 HVR
M3 HVT
M3 HWB
M3 HWE
M3 HWF
M3 HWG
M3 HWH
M3 HWK
M3 HWT
M3 HWZ
M3 HXA
M3 HXE
M3 HXI
M3 HXP
M3 HXT
M3 HXX
M3 HYA
M3 HYC
M3 HYK
M3 HYM
M3 HYW
M3 HYZ
M3 HZF
M3 HZG
M3 HZV
M3 HZY
M3 IAB
M3 IAJ
M3 IAN
M3 IAS
M3 IAT
M3 IBB
M3 IBD
M3 IBF
M3 IBH
M3 IBL
M3 IBN
M3 IBO
M3 IBW
M3 IBX
M3 ICJ
M3 ICX
M3 ICY
M3 ICZ
M3 IDJ
M3 IDL
M3 IDP
M3 IDS
M3 IDU
M3 IEO
M3 IES
M3 IEY
M3 IFC
M3 IFH
M3 IFX
M3 IGI
M3 IGK
M3 IGL
M3 IGP
M3 IGW
M3 IHE
M3 IHG
M3 IHH
M3 IHT
M3 IIB
M3 IIE
M3 IIF
M3 IIQ
M3 IIS
M3 IIY
M3 IJB
M3 IJH
M3 IKA
M3 IKF
M3 IKL
M3 IKX
M3 IKY
M3 ILE
M3 ILF
M3 ILI
M3 ILL
M3 ILO
M3 ILP
M3 ILX
M3 IMN
M3 IMT
M3 IMU
M3 IMW
M3 INB
M3 INI
M3 INK
M3 INV
M3 INX
M3 INZ
M3 IOB
M3 IOE
M3 IOP
M3 IOY
M3 IPB
M3 IPI
M3 IPN
M3 IPU
M3 IQH
M3 IQK
M3 IQL
M3 IQR
M3 IRD
M3 IRI
M3 IRK
M3 IRL
M3 IRN
M3 ISF
M3 ISV
M3 ITE
M3 ITF
M3 ITJ
M3 ITP
M3 ITQ
M3 ITR
M3 IUB
M3 IUG
M3 IUL
M3 IUQ
M3 IUR
M3 IUY
M3 IVE
M3 IVH
M3 IVJ
M3 IVL
M3 IVP
M3 IWH
M3 IWS
M3 IWY
M3 IXI
M3 IXS
M3 IXV
M3 IXW
M3 IYC
M3 IYD
M3 IYF
M3 IYJ
M3 IYL
M3 IYN
M3 IYS
M3 IYT
M3 IYU
M3 IYZ
M3 IZC
M3 IZR
M3 IZS
M3 IZX
M3 JAH
M3 JAJ
M3 JAN
M3 JAT
M3 JBA
M3 JBC
M3 JBD
M3 JBL
M3 JBN
M3 JBO
M3 JBU
M3 JBX
M3 JCJ
M3 JCM
M3 JDC
M3 JDD
M3 JDE
M3 JDK
M3 JDV
M3 JDY
M3 JDZ
M3 JEG
M3 JEL
M3 JEO
M3 JEU
M3 JEV
M3 JEX
M3 JEZ
M3 JFG
M3 JFI
M3 JFJ
M3 JFM
M3 JFO
M3 JFU
M3 JGD
M3 JHD
M3 JHF
M3 JHI
M3 JHK
M3 JHO
M3 JHQ
M3 JHU
M3 JIB
M3 JIE
M3 JIO
M3 JIY
M3 JJA
M3 JJF
M3 JJK
M3 JJO
M3 JJR
M3 JJX
M3 JKC
M3 JKF
M3 JKO
M3 JKQ
M3 JKV
M3 JLC
M3 JLG
M3 JLN
M3 JLO
M3 JLP
M3 JLU
M3 JLY
M3 JLZ
M3 JMB
M3 JME
M3 JMF
M3 JMH
M3 JMN
M3 JNF
M3 JNO
M3 JNS
M3 JOB
M3 JOG
M3 JOK
M3 JOX
M3 JOZ
M3 JPE
M3 JPR
M3 JPT
M3 JPW
M3 JQA
M3 JQE
M3 JQL
M3 JQO
M3 JQP
M3 JQQ
M3 JQZ
M3 JRB
M3 JRP
M3 JSC
M3 JSP
M3 JSR
M3 JSU
M3 JSW
M3 JSX
M3 JSY
M3 JTD
M3 JTG
M3 JTI
M3 JTL
M3 JTP
M3 JTQ
M3 JTR
M3 JTW
M3 JTY
M3 JUA
M3 JUG
M3 JUS
M3 JUT
M3 JUU
M3 JVE
M3 JVL
M3 JVN
M3 JVO
M3 JVU
M3 JVZ
M3 JWH
M3 JWI
M3 JWL
M3 JWO
M3 JWU
M3 JWX
M3 JWY
M3 JXB
M3 JXT
M3 JYJ
M3 JYQ
M3 JYU
M3 JZB
M3 JZH
M3 JZQ
M3 JZS
M3 JZU
M3 JZZ
M3 KAA
M3 KAO
M3 KAT
M3 KAZ
M3 KBA
M3 KBI
M3 KBK
M3 KBO
M3 KBU
M3 KBV
M3 KBW
M3 KCK
M3 KCN
M3 KCW
M3 KDB
M3 KDD
M3 KDT
M3 KEP
M3 KEX
M3 KFA
M3 KFD
M3 KFF
M3 KFX
M3 KFY
M3 KFZ
M3 KGA
M3 KGC
M3 KGH
M3 KGM
M3 KGR
M3 KGX
M3 KHD
M3 KHN
M3 KHO
M3 KHS
M3 KHV
M3 KIC
M3 KIH
M3 KII
M3 KIJ
M3 KIK
M3 KIV
M3 KIW
M3 KIX
M3 KJI
M3 KJN
M3 KJO
M3 KJX
M3 KJZ
M3 KKJ
M3 KKR
M3 KKV
M3 KLA
M3 KLD
M3 KLF
M3 KLG
M3 KLI
M3 KLQ
M3 KLR
M3 KLW
M3 KLX
M3 KMG
M3 KMI
M3 KMJ
M3 KMM
M3 KMV
M3 KNG
M3 KNI
M3 KNQ
M3 KNX
M3 KOD
M3 KOE
M3 KOH
M3 KOL
M3 KOO
M3 KOQ
M3 KOV
M3 KPA
M3 KPH
M3 KPN
M3 KPR
M3 KPS
M3 KPV
M3 KPX
M3 KPY
M3 KQA
M3 KQE
M3 KQH
M3 KQJ
M3 KQQ
M3 KQZ
M3 KRG
M3 KRJ
M3 KRU
M3 KRV
M3 KRW
M3 KSB
M3 KSF
M3 KSL
M3 KSR
M3 KST
M3 KSU
M3 KSY
M3 KSZ
M3 KTF
M3 KTN
M3 KTQ
M3 KTU
M3 KUA
M3 KUF
M3 KUL
M3 KUR
M3 KUT
M3 KUW
M3 KVF
M3 KVS
M3 KWA
M3 KWK
M3 KWN
M3 KWO
M3 KWQ
M3 KWW
M3 KXN
M3 KXO
M3 KXQ
M3 KXW
M3 KXX
M3 KYF
M3 KYN
M3 KYS
M3 KYT
M3 KYU
M3 KYY
M3 KZG
M3 KZL
M3 KZM
M3 KZN
M3 KZO
M3 KZU
M3 LAA
M3 LAC
M3 LAE
M3 LAH
M3 LAI
M3 LAL
M3 LAM
M3 LAR
M3 LBF
M3 LBH
M3 LBI
M3 LBU
M3 LCA
M3 LCB
M3 LCK
M3 LCM
M3 LCQ
M3 LCV
M3 LCX
M3 LDG
M3 LDN
M3 LDP
M3 LDV
M3 LDZ
M3 LEA
M3 LEM
M3 LEQ
M3 LEU
M3 LEY
M3 LFA
M3 LFD
M3 LFF
M3 LFJ
M3 LFK
M3 LFM
M3 LFS
M3 LFT
M3 LFW
M3 LFY
M3 LGA
M3 LGG
M3 LGQ
M3 LGT
M3 LGV
M3 LHD
M3 LHH
M3 LHJ
M3 LHL
M3 LHS
M3 LHT
M3 LHY
M3 LIA
M3 LIF
M3 LII
M3 LIJ
M3 LIQ
M3 LIR
M3 LJD
M3 LJN
M3 LJO
M3 LJU
M3 LJV
M3 LKA
M3 LKG
M3 LKI
M3 LKK
M3 LKL
M3 LKX
M3 LKZ
M3 LLE
M3 LLI
M3 LLJ
M3 LLO
M3 LLR
M3 LLY
M3 LMO
M3 LMP
M3 LMT
M3 LND
M3 LNK
M3 LNS
M3 LNV
M3 LOD
M3 LOH
M3 LOJ
M3 LOM
M3 LON
M3 LOV
M3 LOW
M3 LPD
M3 LPO
M3 LPV
M3 LPX
M3 LPY
M3 LQG
M3 LQH
M3 LQM
M3 LQQ
M3 LQR
M3 LQS
M3 LQT
M3 LRA
M3 LRB
M3 LRD
M3 LRL
M3 LSD
M3 LSJ
M3 LSM
M3 LSN
M3 LSP
M3 LSQ
M3 LSY
M3 LSZ
M3 LTC
M3 LTD
M3 LTE
M3 LTF
M3 LTJ
M3 LTK
M3 LTM
M3 LTN
M3 LTO
M3 LTQ
M3 LTZ
M3 LUA
M3 LUB
M3 LUE
M3 LUF
M3 LUH
M3 LUI
M3 LUJ
M3 LUM
M3 LUQ
M3 LUR
M3 LUT
M3 LVE
M3 LVG
M3 LVH
M3 LVL
M3 LVO
M3 LVQ
M3 LVS
M3 LVU
M3 LVV
M3 LVW
M3 LWA
M3 LWB
M3 LWD
M3 LWE
M3 LWH
M3 LWI
M3 LWL
M3 LWM
M3 LWN
M3 LWS
M3 LWY
M3 LXC
M3 LXG
M3 LXM
M3 LXO
M3 LXQ
M3 LXX
M3 LXY
M3 LYB
M3 LYD
M3 LYF
M3 LYJ
M3 LYK
M3 LYO
M3 LYT
M3 LYW
M3 LYY
M3 LZB
M3 LZE
M3 LZG
M3 LZI
M3 LZJ
M3 LZM
M3 LZN
M3 LZP
M3 LZQ
M3 LZW
M3 LZX
M3 LZZ
M3 MAF
M3 MAH
M3 MAI
M3 MAL
M3 MAQ
M3 MAW
M3 MAX
M3 MBN
M3 MBP
M3 MBW
M3 MCH
M3 MCJ
M3 MCO
M3 MCP
M3 MCZ
M3 MDL
M3 MDQ
M3 MDT
M3 MDZ
M3 MEQ
M3 MEX
M3 MFI
M3 MFM
M3 MFQ
M3 MFW
M3 MGG
M3 MGM
M3 MGX
M3 MHB
M3 MHJ
M3 MHM
M3 MHO
M3 MHT
M3 MIB
M3 MIL
M3 MIY
M3 MJB
M3 MJC
M3 MJP
M3 MJQ
M3 MJT
M3 MJU
M3 MJW
M3 MKL
M3 MKR
M3 MKU
M3 MKX
M3 MLC
M3 MLD
M3 MLE
M3 MLN
M3 MLO
M3 MLS
M3 MLU
M3 MLY
M3 MMH
M3 MMR
M3 MMT
M3 MMX
M3 MNA
M3 MNE
M3 MNF
M3 MNH
M3 MNM
M3 MNN
M3 MNO
M3 MNP
M3 MNZ
M3 MOA
M3 MOI
M3 MOM
M3 MOS
M3 MOU
M3 MPD
M3 MPF
M3 MPG
M3 MPN
M3 MPQ
M3 MPU
M3 MPV
M3 MPX
M3 MPY
M3 MQD
M3 MQE
M3 MQF
M3 MQG
M3 MQK
M3 MQN
M3 MQQ
M3 MQS
M3 MQU
M3 MQV

Withheld

M3 MQW
M3 MQZ
M3 MRP
M3 MRR
M3 MRT
M3 MSE
M3 MSI
M3 MSM
M3 MSS
M3 MTA
M3 MTG
M3 MTH
M3 MTJ
M3 MTN
M3 MTV
M3 MTY
M3 MUC
M3 MUH
M3 MUJ
M3 MUM
M3 MUR
M3 MUV
M3 MUY
M3 MUZ
M3 MVA
M3 MVC
M3 MVD
M3 MVF
M3 MVG
M3 MVH
M3 MVQ
M3 MVS
M3 MVZ
M3 MWF
M3 MWK
M3 MWL
M3 MWP
M3 MWU
M3 MWY
M3 MXB
M3 MXD
M3 MXI
M3 MXS
M3 MXT
M3 MXY
M3 MYA
M3 MYB
M3 MYJ
M3 MYL
M3 MYN
M3 MYO
M3 MYP
M3 MYR
M3 MYU
M3 MYV
M3 MYX
M3 MYY
M3 MZH
M3 MZI
M3 MZL
M3 MZQ
M3 MZS
M3 MZU
M3 MZY
M3 NAA
M3 NAM
M3 NBA
M3 NBC
M3 NBF
M3 NBM
M3 NBO
M3 NBR
M3 NBT
M3 NBW
M3 NCF
M3 NCR
M3 NCU
M3 NCV
M3 NCW
M3 NCX
M3 NCY
M3 NCZ
M3 NDD
M3 NDH
M3 NDP
M3 NDV
M3 NDY
M3 NEJ
M3 NEK
M3 NEU
M3 NEZ
M3 NFC
M3 NFD
M3 NFI
M3 NGI
M3 NGL
M3 NGQ
M3 NGT
M3 NGW
M3 NHF
M3 NHJ
M3 NHL
M3 NHO
M3 NHQ
M3 NHT
M3 NHX
M3 NIB
M3 NIH
M3 NIL
M3 NIM
M3 NIO
M3 NIU
M3 NIY
M3 NJG
M3 NJL
M3 NJT
M3 NJZ
M3 NKA
M3 NKD
M3 NKH
M3 NKI
M3 NKJ

M3 NKM
M3 NKS
M3 NKT
M3 NKV
M3 NKY
M3 NLB
M3 NLC
M3 NLL
M3 NLO
M3 NLS
M3 NLU
M3 NLV
M3 NMA
M3 NMC
M3 NME
M3 NMF
M3 NMH
M3 NMN
M3 NMO
M3 NMT
M3 NMW
M3 NMY
M3 NMZ
M3 NNB
M3 NNC
M3 NND
M3 NNF
M3 NNK
M3 NNL
M3 NNR
M3 NNX
M3 NOC
M3 NOG
M3 NOK
M3 NOT
M3 NOU
M3 NPJ
M3 NPL
M3 NPP
M3 NPQ
M3 NPV
M3 NPY
M3 NQC
M3 NQD
M3 NQF
M3 NQG
M3 NQJ
M3 NQP
M3 NQQ
M3 NQV
M3 NQW
M3 NQX
M3 NQZ
M3 NRD
M3 NRO
M3 NRS
M3 NRU
M3 NRY
M3 NRZ
M3 NSE
M3 NSW
M3 NSY
M3 NSZ
M3 NTF
M3 NTM
M3 NTO
M3 NTS
M3 NTU
M3 NUA
M3 NUD
M3 NUF
M3 NUJ
M3 NUK
M3 NUR
M3 NUS
M3 NUZ
M3 NVB
M3 NVI
M3 NVM
M3 NVT
M3 NVU
M3 NVZ
M3 NWB
M3 NWG
M3 NWI
M3 NWM
M3 NWN
M3 NXB
M3 NXG
M3 NXI
M3 NXL
M3 NXM
M3 NXN
M3 NXP
M3 NXU
M3 NXV
M3 NXW
M3 NXX
M3 NYE
M3 NYJ
M3 NYL
M3 NYO
M3 NYR
M3 NYT
M3 NYU
M3 NYV
M3 NZB
M3 NZC
M3 NZD
M3 NZF
M3 NZL
M3 NZP
M3 NZQ
M3 NZS
M3 OAD
M3 OAF
M3 OAN
M3 OAO
M3 OAU
M3 OAV
M3 OBE
M3 OBH

M3 OBL
M3 OBP
M3 OBY
M3 OCB
M3 OCD
M3 OCF
M3 OCH
M3 OCI
M3 OCM
M3 OCT
M3 OCU
M3 ODD
M3 ODF
M3 ODG
M3 ODI
M3 ODJ
M3 ODN
M3 ODQ
M3 ODS
M3 ODU
M3 ODW
M3 ODZ
M3 OEA
M3 OEK
M3 OEP
M3 OET
M3 OEU
M3 OEW
M3 OEX
M3 OEZ
M3 OFG
M3 OFK
M3 OFL
M3 OFM
M3 OFP
M3 OFQ
M3 OFT
M3 OFY
M3 OFZ
M3 OGA
M3 OGE
M3 OGF
M3 OGH
M3 OGI
M3 OGJ
M3 OGO
M3 OGP
M3 OGQ
M3 OGW
M3 OGX
M3 OGZ
M3 OHA
M3 OHH
M3 OHM
M3 OHT
M3 OHU
M3 OHV
M3 OHW
M3 OIA
M3 OID
M3 OIE
M3 OIF
M3 OIG
M3 OIJ
M3 OIO
M3 OIP
M3 OIQ
M3 OIR
M3 OIT
M3 OIU
M3 OIV
M3 OIW
M3 OJA
M3 OJB
M3 OJH
M3 OJI
M3 OJM
M3 OJQ
M3 OJT
M3 OKD
M3 OKK
M3 OKL
M3 OKM
M3 OKN
M3 OKQ
M3 OKV
M3 OKW
M3 OLB
M3 OLG
M3 OLH
M3 OLJ
M3 OLL
M3 OLP
M3 OLR
M3 OLS
M3 OMB
M3 OMJ
M3 OMK
M3 OML
M3 OMM
M3 OMN
M3 OMO
M3 OMP
M3 OMQ
M3 OMR
M3 OMV
M3 OMW
M3 ONA
M3 ONJ
M3 ONL
M3 ONQ
M3 ONR
M3 ONS
M3 ONT
M3 ONU
M3 ONW
M3 ONZ
M3 OOB
M3 OOJ
M3 OOK
M3 OON
M3 OOO

M3 OOS
M3 OOV
M3 OOW
M3 OPE
M3 OPF
M3 OPH
M3 OPJ
M3 OPQ
M3 OPR
M3 OPY
M3 OQA
M3 OQE
M3 OQF
M3 OQM
M3 OQO
M3 OQP
M3 OQU
M3 OQW
M3 OQX
M3 OQY
M3 ORA
M3 ORF
M3 ORH
M3 ORI
M3 ORJ
M3 ORW
M3 ORX
M3 OSD
M3 OSH
M3 OSL
M3 OSM
M3 OSR
M3 OSX
M3 OTC
M3 OTD
M3 OTH
M3 OTK
M3 OTL
M3 OTN
M3 OTO
M3 OTX
M3 OTY
M3 OUA
M3 OUB
M3 OUD
M3 OUE
M3 OUJ
M3 OUK
M3 OUM
M3 OUO
M3 OUP
M3 OUR
M3 OUW
M3 OUX
M3 OUZ
M3 OVB
M3 OVH
M3 OVI
M3 OVL
M3 OVN
M3 OVP
M3 OVQ
M3 OVS
M3 OVU
M3 OVW
M3 OVY
M3 OWA
M3 OWC
M3 OWD
M3 OWJ
M3 OWK
M3 OWL
M3 OWM
M3 OWR
M3 OWT
M3 OWV
M3 OWW
M3 OWX
M3 OWY
M3 OXA
M3 OXF
M3 OXG
M3 OXI
M3 OXJ
M3 OXL
M3 OXM
M3 OXS
M3 OXT
M3 OXU
M3 OXW
M3 OXX
M3 OYA
M3 OYF
M3 OYG
M3 OYH
M3 OYI
M3 OYK
M3 OYM
M3 OYO
M3 OYT
M3 OYV
M3 OYX
M3 OYY
M3 OZA
M3 OZL
M3 OZQ
M3 OZR
M3 OZS
M3 OZV
M3 OZX
M3 OZZ
M3 PAA
M3 PAB
M3 PAJ
M3 PAK
M3 PAO
M3 PAQ
M3 PBD
M3 PBF
M3 PBH
M3 PBJ
M3 PBM

M3 PBZ
M3 PCA
M3 PCB
M3 PCM
M3 PCU
M3 PCZ
M3 PDA
M3 PDI
M3 PDS
M3 PEG
M3 PEI
M3 PEJ
M3 PEK
M3 PEO
M3 PEP
M3 PEW
M3 PEX
M3 PFD
M3 PFG
M3 PFI
M3 PFO
M3 PFP
M3 PFQ
M3 PFS
M3 PFT
M3 PFV
M3 PFW
M3 PFZ
M3 PGJ
M3 PGM
M3 PGQ
M3 PGR
M3 PGV
M3 PGW
M3 PGX
M3 PGZ
M3 PHB
M3 PHI
M3 PHJ
M3 PHK
M3 PHN
M3 PHV
M3 PIF
M3 PII
M3 PIJ
M3 PIR
M3 PJA
M3 PJB
M3 PJD
M3 PJY
M3 PKA
M3 PKD
M3 PKF
M3 PKG
M3 PKJ
M3 PKK
M3 PKN
M3 PKU
M3 PKV
M3 PKX
M3 PKY
M3 PLC
M3 PLF
M3 PLG
M3 PLK
M3 PLM
M3 PLQ
M3 PLR
M3 PLY
M3 PMA
M3 PMB
M3 PMC
M3 PMD
M3 PMG
M3 PMJ
M3 PMP
M3 PMQ
M3 PMV
M3 PND
M3 PNJ
M3 PNM
M3 PNP
M3 PNS
M3 PNU
M3 PNX
M3 POA
M3 POG
M3 POS
M3 POU
M3 POY
M3 POZ
M3 PPF
M3 PPJ
M3 PPP
M3 PPS
M3 PPT
M3 PPV
M3 PQA
M3 PQC
M3 PQD
M3 PQH
M3 PQR
M3 PQW
M3 PQX
M3 PRC
M3 PRF
M3 PRI
M3 PRO
M3 PRQ
M3 PRT
M3 PRW
M3 PRX
M3 PSK
M3 PSP
M3 PSQ
M3 PST
M3 PSX
M3 PTJ
M3 PTK
M3 PTS
M3 PTW
M3 PTY

M3 PUG
M3 PUK
M3 PUM
M3 PUS
M3 PUX
M3 PUY
M3 PVA
M3 PVD
M3 PVH
M3 PVJ
M3 PVK
M3 PVL
M3 PVM
M3 PVN
M3 PVO
M3 PVS
M3 PVW
M3 PWA
M3 PWB
M3 PWF
M3 PWG
M3 PWI
M3 PWJ
M3 PWQ
M3 PWR
M3 PWU
M3 PWV
M3 PWX
M3 PXC
M3 PXD
M3 PXH
M3 PXI
M3 PXJ
M3 PXS
M3 PYA
M3 PYF
M3 PYH
M3 PYN
M3 PYP
M3 PYU
M3 PYZ
M3 PZA
M3 PZD
M3 PZG
M3 PZH
M3 PZI
M3 PZM
M3 PZP
M3 PZT
M3 PZU
M3 RAO
M3 RAP
M3 RAX
M3 RAY
M3 RBV
M3 RBZ
M3 RCA
M3 RCF
M3 RCK
M3 RCN
M3 RCP
M3 RCU
M3 RDE
M3 RDG
M3 RDI
M3 RDN
M3 RDT
M3 RDZ
M3 REE
M3 REF
M3 RER
M3 RES
M3 RFP
M3 RFT
M3 RFU
M3 RFY
M3 RFZ
M3 RGI
M3 RGM
M3 RGQ
M3 RGT
M3 RGX
M3 RGY
M3 RHE
M3 RHM
M3 RHU
M3 RHV
M3 RHX
M3 RHY
M3 RIB
M3 RIH
M3 RIN
M3 RIR
M3 RIY
M3 RJA
M3 RJD
M3 RJQ
M3 RJR
M3 RJT
M3 RJV
M3 RJW
M3 RKA
M3 RKM
M3 RKP
M3 RKU
M3 RKX
M3 RLI
M3 RLL
M3 RLQ
M3 RLR
M3 RLU
M3 RLY
M3 RLZ
M3 RMC
M3 RME
M3 RMK
M3 RMM
M3 RMO
M3 RMR
M3 RNB
M3 RND
M3 RNE

M3 RNF
M3 RNI
M3 RNJ
M3 RNL
M3 RNP
M3 RNR
M3 RNV
M3 ROE
M3 ROH
M3 ROJ
M3 ROT
M3 RPC
M3 RPG
M3 RPL
M3 RPM
M3 RPN
M3 RPU
M3 RPY
M3 RQF
M3 RQL
M3 RQS
M3 RQT
M3 RQU
M3 RQZ
M3 RRD
M3 RRK
M3 RRL
M3 RRO
M3 RRS
M3 RRX
M3 RSK
M3 RSL
M3 RSP
M3 RSQ
M3 RSW
M3 RTF
M3 RTG
M3 RTX
M3 RUA
M3 RUD
M3 RUJ
M3 RUN
M3 RUU
M3 RUX
M3 RVA
M3 RVF
M3 RVG
M3 RVL
M3 RVU
M3 RVW
M3 RVY
M3 RVZ
M3 RWE
M3 RWG
M3 RWL
M3 RWT
M3 RWU
M3 RWY
M3 RXC
M3 RXI
M3 RXR
M3 RXS
M3 RXV
M3 RXX
M3 RYH
M3 RYK
M3 RYL
M3 RYM
M3 RYQ
M3 RYV
M3 RYX
M3 RZH
M3 RZK
M3 RZR
M3 RZS
M3 SBG
M3 SBH
M3 SBI
M3 SBK
M3 SBL
M3 SBU
M3 SBV
M3 SBZ
M3 SCE
M3 SCK
M3 SCS
M3 SCT
M3 SCY
M3 SCZ
M3 SDM
M3 SDO
M3 SDT
M3 SEH
M3 SEP
M3 SEQ
M3 SER
M3 SEW
M3 SFB
M3 SFE
M3 SFF
M3 SFH
M3 SFI
M3 SFO
M3 SFQ
M3 SFR
M3 SFU
M3 SFV
M3 SGC
M3 SGN
M3 SGO
M3 SGY
M3 SHC
M3 SHD
M3 SHE
M3 SHH
M3 SHM
M3 SIA
M3 SIF
M3 SIH
M3 SIJ
M3 SIU
M3 SIV

M3 SJC
M3 SJG
M3 SJZ
M3 SKE
M3 SKG
M3 SKO
M3 SKS
M3 SLE
M3 SLR
M3 SLU
M3 SLV
M3 SMO
M3 SMS
M3 SMU
M3 SNC
M3 SND
M3 SNM
M3 SNP
M3 SNT
M3 SNU
M3 SNV
M3 SOH
M3 SOI
M3 SOJ
M3 SOK
M3 SOM
M3 SOO
M3 SOR
M3 SOU
M3 SOX
M3 SPB
M3 SPD
M3 SPF
M3 SPM
M3 SPO
M3 SPV
M3 SPW
M3 SPX
M3 SQB
M3 SQC
M3 SQD
M3 SQF
M3 SQR
M3 SRE
M3 SRN
M3 SRX
M3 SRZ
M3 SSE
M3 SSN
M3 SSP
M3 SST
M3 SSY
M3 STB
M3 STE
M3 STF
M3 STG
M3 STM
M3 STO
M3 STX
M3 STZ
M3 SUA
M3 SUH
M3 SUN
M3 SUQ
M3 SUZ
M3 SVA
M3 SVE
M3 SVG
M3 SVI
M3 SVQ
M3 SVR
M3 SVS
M3 SVV
M3 SVX
M3 SWJ
M3 SWL
M3 SWN
M3 SWR
M3 SWX
M3 SXA
M3 SXD
M3 SXG
M3 SXN
M3 SXO
M3 SYA
M3 SYE
M3 SYG
M3 SYK
M3 SZH
M3 SZJ
M3 SZL
M3 SZN
M3 SZP
M3 SZR
M3 SZU
M3 SZV
M3 SZW
M3 SZX
M3 SZZ
M3 TAI
M3 TAJ
M3 TAO
M3 TAQ
M3 TBB
M3 TBD
M3 TBI
M3 TBN
M3 TBR
M3 TBY
M3 TCA
M3 TCH
M3 TCL
M3 TCM
M3 TCO
M3 TCV
M3 TCZ
M3 TDA
M3 TDF
M3 TDI
M3 TDN
M3 TDQ
M3 TDV

M3 TDW
M3 TDZ
M3 TEA
M3 TEC
M3 TEJ
M3 TEO
M3 TER
M3 TEU
M3 TFQ
M3 TFU
M3 TGF
M3 TGG
M3 TGH
M3 TGN
M3 TGS
M3 TGX
M3 TGY
M3 THG
M3 THL
M3 THT
M3 THU
M3 THZ
M3 TIH
M3 TIO
M3 TIP
M3 TIR
M3 TIW
M3 TJA
M3 TJD
M3 TJN
M3 TJP
M3 TJV
M3 TJY
M3 TKA
M3 TKB
M3 TKC
M3 TKD
M3 TKF
M3 TKK
M3 TKL
M3 TKY
M3 TKZ
M3 TLA
M3 TLE
M3 TLK
M3 TMD
M3 TMH
M3 TMI
M3 TMS
M3 TMV
M3 TNI
M3 TNR
M3 TNU
M3 TOH
M3 TOQ
M3 TOU
M3 TOX
M3 TPC
M3 TPL
M3 TPP
M3 TPX
M3 TQC
M3 TQE
M3 TQL
M3 TQQ
M3 TQR
M3 TQW
M3 TQZ
M3 TRD
M3 TRN
M3 TRQ
M3 TRS
M3 TRU
M3 TSC
M3 TSK
M3 TSL
M3 TSM
M3 TSS
M3 TSU
M3 TTG
M3 TTJ
M3 TTQ
M3 TTV
M3 TTW
M3 TUE
M3 TUK
M3 TUT
M3 TUV
M3 TVE
M3 TVS
M3 TVX
M3 TWJ
M3 TWO
M3 TWU
M3 TWX
M3 TXB
M3 TXD
M3 TXE
M3 TXN
M3 TXO
M3 TXX
M3 TXZ
M3 TYD
M3 TYF
M3 TYN
M3 TYY
M3 TZA
M3 TZH
M3 TZJ
M3 TZL
M3 TZM
M3 TZT
M3 TZV
M3 TZX
M3 TZZ
M3 UAB
M3 UAD
M3 UAF
M3 UAU
M3 UBA
M3 UBI
M3 UBJ

M3 UBM
M3 UBO
M3 UBP
M3 UBV
M3 UBW
M3 UBZ
M3 UCB
M3 UCG
M3 UCK
M3 UCM
M3 UCQ
M3 UCT
M3 UCW
M3 UDE
M3 UDG
M3 UDH
M3 UDM
M3 UDP
M3 UEM
M3 UEO
M3 UES
M3 UFB
M3 UFM
M3 UFN
M3 UFO
M3 UFP
M3 UFR
M3 UFV
M3 UGN
M3 UGP
M3 UHA
M3 UHD
M3 UHE
M3 UHM
M3 UHR
M3 UHU
M3 UHZ
M3 UIH
M3 UIZ
M3 UJG
M3 UJI
M3 UJK
M3 UJM
M3 UJU
M3 UJW
M3 UKA
M3 UKE
M3 UKI
M3 UKL
M3 UKZ
M3 ULA
M3 ULF
M3 ULG
M3 ULH
M3 ULK
M3 ULR
M3 ULV
M3 UME
M3 UMI
M3 UMK
M3 UMO
M3 UMP
M3 UMS
M3 UMT
M3 UNA
M3 UNE
M3 UNM
M3 UNS
M3 UNV
M3 UNX
M3 UOA
M3 UOG
M3 UOH
M3 UOI
M3 UOQ
M3 UOT
M3 UOU
M3 UPC
M3 UPD
M3 UPE
M3 UPH
M3 UPM
M3 UPR
M3 UPU
M3 UPW
M3 UQS
M3 UQU
M3 UQV
M3 UQX
M3 URC
M3 URI
M3 URJ
M3 URM
M3 URP
M3 URY
M3 USE
M3 USG
M3 USI
M3 USL
M3 USO
M3 USQ
M3 USY
M3 UTA
M3 UTC
M3 UTD
M3 UTF
M3 UTG
M3 UTL
M3 UTW
M3 UUB
M3 UUC
M3 UUI
M3 UUM
M3 UUQ
M3 UUU
M3 UVE
M3 UVI
M3 UVL
M3 UVR
M3 UVZ
M3 UWC
M3 UWG

M3 UWK
M3 UWQ
M3 UXA
M3 UXD
M3 UXJ
M3 UXP
M3 UXT
M3 UXV
M3 UXW
M3 UYB
M3 UYN
M3 UYR
M3 UZC
M3 UZD
M3 UZG
M3 UZR
M3 UZS
M3 UZY
M3 VAI
M3 VAL
M3 VAN
M3 VAO
M3 VAP
M3 VAZ
M3 VBD
M3 VBK
M3 VBO
M3 VBS
M3 VBX
M3 VCH
M3 VCL
M3 VCU
M3 VDJ
M3 VDK
M3 VDR
M3 VDS
M3 VEB
M3 VED
M3 VEE
M3 VEI
M3 VEO
M3 VEU
M3 VEV
M3 VFA
M3 VFD
M3 VFG
M3 VFH
M3 VFT
M3 VGC
M3 VGI
M3 VGO
M3 VGQ
M3 VGR
M3 VGU
M3 VGY
M3 VID
M3 VIE
M3 VIF
M3 VIK
M3 VIT
M3 VIX
M3 VIY
M3 VJQ
M3 VJR
M3 VJY
M3 VJZ
M3 VKC
M3 VKE
M3 VKH
M3 VKI
M3 VKK
M3 VKW
M3 VKZ
M3 VLC
M3 VLK
M3 VLR
M3 VLW
M3 VLY
M3 VLZ
M3 VMB
M3 VMG
M3 VMK
M3 VMO
M3 VMX
M3 VNA
M3 VNB
M3 VND
M3 VNE
M3 VNY
M3 VOD
M3 VOE
M3 VOF
M3 VOK
M3 VOM
M3 VOO
M3 VOV
M3 VPC
M3 VPF
M3 VPI
M3 VPM
M3 VPU
M3 VPZ
M3 VQB
M3 VQO
M3 VQT
M3 VQU
M3 VQW
M3 VRE
M3 VRG
M3 VRH
M3 VRJ
M3 VRS
M3 VRW
M3 VSA
M3 VSD
M3 VSM
M3 VSP
M3 VSR
M3 VSV
M3 VSY
M3 VTD
M3 VTI

M3 VTM
M3 VTY
M3 VTZ
M3 VUB
M3 VUE
M3 VUF
M3 VUG
M3 VUM
M3 VVG
M3 VVM
M3 VVP
M3 VWA
M3 VWB
M3 VWI
M3 VWL
M3 VWM
M3 VWN
M3 VWS
M3 VWT
M3 VWU
M3 VWX
M3 VWZ
M3 VXD
M3 VXQ
M3 VXT
M3 VXU
M3 VXW
M3 VXZ
M3 VYG
M3 VYI
M3 VYL
M3 VYW
M3 VYZ
M3 VZA
M3 VZB
M3 VZE
M3 VZG
M3 VZK
M3 VZM
M3 VZR
M3 VZT
M3 VZW
M3 VZY
M3 WAA
M3 WAG
M3 WAH
M3 WAI
M3 WAO
M3 WAQ
M3 WAU
M3 WAZ
M3 WBS
M3 WBW
M3 WBX
M3 WCC
M3 WCD
M3 WCF
M3 WCH
M3 WCJ
M3 WCK
M3 WCL
M3 WCT
M3 WCU
M3 WDD
M3 WDG
M3 WDM
M3 WDQ
M3 WDS
M3 WDW
M3 WDX
M3 WEC
M3 WED
M3 WEK
M3 WEL
M3 WEN
M3 WEO
M3 WER
M3 WEX
M3 WFD
M3 WFI
M3 WFV
M3 WFW
M3 WFX
M3 WFZ
M3 WGF
M3 WGG
M3 WGH
M3 WGN
M3 WGQ
M3 WGT
M3 WGX
M3 WHD
M3 WHE
M3 WHF
M3 WHM
M3 WHN
M3 WHO
M3 WIB
M3 WIE
M3 WIF
M3 WIS
M3 WIY
M3 WJB
M3 WJC
M3 WJH
M3 WJJ
M3 WJQ
M3 WJS
M3 WJX
M3 WKA
M3 WKD
M3 WKG
M3 WKQ
M3 WKS
M3 WKT
M3 WKU
M3 WLC
M3 WLE
M3 WLF
M3 WLH
M3 WLJ

M3 WLM
M3 WLN
M3 WLP
M3 WLQ
M3 WLR
M3 WLT
M3 WMA
M3 WMF
M3 WMN
M3 WMT
M3 WMZ
M3 WND
M3 WNL
M3 WNO
M3 WNU
M3 WNW
M3 WNY
M3 WOE
M3 WOH
M3 WOJ
M3 WOM
M3 WON
M3 WOO
M3 WOU
M3 WPD
M3 WPR
M3 WPZ
M3 WQD
M3 WQH
M3 WQJ
M3 WQK
M3 WQP
M3 WQQ
M3 WQU
M3 WQZ
M3 WSA
M3 WSD
M3 WSG
M3 WST
M3 WSU
M3 WSX
M3 WSY
M3 WTJ
M3 WTQ
M3 WTX
M3 WTZ
M3 WUC
M3 WUF
M3 WUT
M3 WUY
M3 WVA
M3 WVE
M3 WVM
M3 WVS
M3 WVU
M3 WWB
M3 WWE
M3 WWF
M3 WWI
M3 WWL
M3 WWT
M3 WWX
M3 WXK
M3 WXM
M3 WXO
M3 WXQ
M3 WXV
M3 WXY
M3 WYD
M3 WYE
M3 WYH
M3 WYL
M3 WYP
M3 WYU
M3 WYX
M3 WZC
M3 WZD
M3 WZI
M3 WZK
M3 WZO
M3 WZQ
M3 WZU
M3 WZX
M3 XAA
M3 XAD
M3 XAL
M3 XAP
M3 XAQ
M3 XBA
M3 XBD
M3 XBI
M3 XBJ
M3 XBP
M3 XBQ
M3 XBU
M3 XBV
M3 XBW
M3 XBX
M3 XCC
M3 XCD
M3 XCG
M3 XCK
M3 XCL
M3 XCM
M3 XCO
M3 XCQ
M3 XCZ
M3 XDA
M3 XDC
M3 XDN
M3 XDO
M3 XDR
M3 XDT
M3 XDY
M3 XEC
M3 XED
M3 XEE
M3 XEH
M3 XEM
M3 XEV
M3 XEW
M3 XEZ

M3 XFE
M3 XFF
M3 XFG
M3 XFJ
M3 XFL
M3 XFQ
M3 XFU
M3 XFV
M3 XFW
M3 XFY
M3 XGE
M3 XGF
M3 XGJ
M3 XGK
M3 XGM
M3 XGO
M3 XHA
M3 XHD
M3 XHI
M3 XHJ
M3 XHP
M3 XHR
M3 XHS
M3 XHV
M3 XIB
M3 XII
M3 XIR
M3 XIX
M3 XJD
M3 XJN
M3 XJR
M3 XJU
M3 XJV
M3 XJY
M3 XKB
M3 XKC
M3 XKK
M3 XKQ
M3 XKR
M3 XKS
M3 XKT
M3 XKV
M3 XKW
M3 XKZ
M3 XLA
M3 XLD
M3 XLF
M3 XLI
M3 XLN
M3 XLY
M3 XMF
M3 XML
M3 XMM
M3 XMR
M3 XMV
M3 XMX
M3 XND
M3 XNG
M3 XNI
M3 XNJ
M3 XNS
M3 XNY
M3 XNZ
M3 XOB
M3 XOF
M3 XOM
M3 XOO
M3 XOP
M3 XPA
M3 XPB
M3 XPC
M3 XPE
M3 XPO
M3 XPQ
M3 XPV
M3 XPZ
M3 XQA
M3 XQD
M3 XQR
M3 XQU
M3 XRA
M3 XRB
M3 XRF
M3 XRL
M3 XRR
M3 XRU
M3 XSB
M3 XSH
M3 XSL
M3 XSO
M3 XSQ
M3 XSS
M3 XSW
M3 XTB
M3 XTH
M3 XTO
M3 XTQ
M3 XTW
M3 XTX
M3 XUB
M3 XUK
M3 XUL
M3 XUM
M3 XUQ
M3 XUZ
M3 XVA
M3 XVE
M3 XVN
M3 XVV
M3 XWA
M3 XWD
M3 XWH
M3 XWJ
M3 XWL
M3 XWT
M3 XXC
M3 XXD
M3 XXF
M3 XXH
M3 XXN
M3 XYD
M3 XYF

M3 XYG
M3 XYL
M3 XYQ
M3 XYR
M3 XYY
M3 XZI
M3 XZL
M3 XZM
M3 XZO
M3 XZQ
M3 XZU
M3 XZZ
M3 YAF
M3 YAH
M3 YAP
M3 YAR
M3 YAU
M3 YAY
M3 YAZ
M3 YBE
M3 YBH
M3 YBO
M3 YBP
M3 YBS
M3 YBV
M3 YBY
M3 YBZ
M3 YCA
M3 YCC
M3 YCF
M3 YCH
M3 YCP
M3 YCY
M3 YDC
M3 YDE
M3 YDG
M3 YDO
M3 YDR
M3 YEB
M3 YED
M3 YEF
M3 YEH
M3 YEI
M3 YEO
M3 YEV
M3 YEX
M3 YEY
M3 YFC
M3 YFK
M3 YFO
M3 YFP
M3 YFU
M3 YGE
M3 YGH
M3 YGN
M3 YGX
M3 YHB
M3 YHE
M3 YHI
M3 YHS
M3 YHX
M3 YIA
M3 YIB
M3 YIN
M3 YIP
M3 YIW
M3 YJC
M3 YJO
M3 YJQ
M3 YJX
M3 YKB
M3 YKD
M3 YKG
M3 YKJ
M3 YKQ
M3 YKU
M3 YKX
M3 YKY
M3 YLC
M3 YLF
M3 YLI
M3 YLY
M3 YMO
M3 YMP
M3 YMR
M3 YMT
M3 YMZ
M3 YNF
M3 YNR
M3 YNU
M3 YNV
M3 YNW
M3 YOA
M3 YOD
M3 YOJ
M3 YON
M3 YPC
M3 YPL
M3 YPM
M3 YPO
M3 YPT
M3 YPV
M3 YPZ
M3 YQB
M3 YQE
M3 YQF
M3 YQI
M3 YQK
M3 YQN
M3 YQO
M3 YQS
M3 YQV
M3 YQY
M3 YRD
M3 YRG
M3 YRI
M3 YRK
M3 YRL
M3 YRN
M3 YRQ
M3 YRU
M3 YRY

M3 YSB
M3 YSG
M3 YSH
M3 YSI
M3 YSK
M3 YSO
M3 YSP
M3 YSR
M3 YST
M3 YSX
M3 YTD
M3 YTJ
M3 YTK
M3 YTO
M3 YTT
M3 YTU
M3 YTY
M3 YUG
M3 YUJ
M3 YUL
M3 YUO
M3 YUZ
M3 YVH
M3 YVI
M3 YVL
M3 YVM
M3 YVO
M3 YVP
M3 YVQ
M3 YVX
M3 YWA
M3 YWB
M3 YWN
M3 YWQ
M3 YWR
M3 YWS
M3 YWV
M3 YWX
M3 YWY
M3 YWZ
M3 YXG
M3 YXI
M3 YXO
M3 YXY
M3 YYA
M3 YYC
M3 YYL
M3 YYP
M3 YYY
M3 YZD
M3 YZF
M3 YZG
M3 YZL
M3 YZT
M3 YZY
M3 ZAB
M3 ZAG
M3 ZAH
M3 ZAJ
M3 ZAX
M3 ZBA
M3 ZBE
M3 ZBK
M3 ZBN
M3 ZBP
M3 ZBU
M3 ZBY
M3 ZCK
M3 ZCL
M3 ZCP
M3 ZCV
M3 ZCZ
M3 ZDO
M3 ZDP
M3 ZDY
M3 ZDZ
M3 ZEA
M3 ZEF
M3 ZEG
M3 ZEK
M3 ZEP
M3 ZEU
M3 ZEZ
M3 ZFA
M3 ZFE
M3 ZFM
M3 ZFP
M3 ZFQ
M3 ZFT
M3 ZFU
M3 ZFX
M3 ZGB
M3 ZGQ
M3 ZGV
M3 ZGY
M3 ZGZ
M3 ZHA
M3 ZHK
M3 ZHN
M3 ZHO
M3 ZHR
M3 ZHT
M3 ZIB
M3 ZIC
M3 ZIG
M3 ZIJ
M3 ZIP
M3 ZIQ
M3 ZIW
M3 ZJC
M3 ZJI
M3 ZJR
M3 ZJT
M3 ZJU
M3 ZJW
M3 ZJZ
M3 ZKB
M3 ZKC
M3 ZKG
M3 ZKH
M3 ZKK
M3 ZKM

Withheld

M3 ZKN
M3 ZKU
M3 ZKV
M3 ZLB
M3 ZLC
M3 ZLG
M3 ZLH
M3 ZLM
M3 ZLN
M3 ZLQ
M3 ZLT
M3 ZMA
M3 ZMF
M3 ZMG
M3 ZMH
M3 ZMK
M3 ZND
M3 ZNE
M3 ZNH
M3 ZNI
M3 ZOF
M3 ZOS
M3 ZOT
M3 ZOW
M3 ZOX
M3 ZPA
M3 ZPF
M3 ZPH
M3 ZPI
M3 ZPK
M3 ZPN
M3 ZPP
M3 ZPQ
M3 ZPS
M3 ZPU
M3 ZPV
M3 ZPX
M3 ZQD
M3 ZQE
M3 ZQH
M3 ZQI
M3 ZQQ
M3 ZQR
M3 ZQT
M3 ZQY
M3 ZQZ
M3 ZRD
M3 ZRI
M3 ZRJ
M3 ZRN
M3 ZRT
M3 ZRU
M3 ZSN
M3 ZSP
M3 ZSR
M3 ZSS
M3 ZSX
M3 ZSZ
M3 ZTA
M3 ZTE
M3 ZTG
M3 ZTO
M3 ZTQ
M3 ZTY
M3 ZUE
M3 ZUH
M3 ZUQ
M3 ZUX
M3 ZVM
M3 ZVO
M3 ZVP
M3 ZVQ
M3 ZVZ
M3 ZWV
M3 ZWZ
M3 ZXB
M3 ZXJ
M3 ZXM
M3 ZXP
M3 ZXS
M3 ZXV
M3 ZXW
M3 ZYA
M3 ZYB
M3 ZYC
M3 ZYD
M3 ZYH
M3 ZYJ
M3 ZYL
M3 ZYN
M3 ZZB
M3 ZZC
M3 ZZZ

M5

M5 AAD
M5 AAE
M5 AAO
M5 AAS
M5 ABP
M5 ABV
M5 ACM
M5 ACU
M5 ACZ
M5 ADU
M5 ADX
M5 ADZ
M5 AED
M5 AEJ
M5 AEK
M5 AFD
M5 AFK
M5 AFO
M5 AFU
M5 AGC
M5 AGH
M5 AGT
M5 AHQ
M5 AIF
M5 AIL
M5 AIM
M5 AIR
M5 AJF
M5 AJT
M5 AJV
M5 AKA
M5 ALI
M5 AML
M5 ASF
M5 AX
M5 BRY
M5 CDS
M5 CJW
M5 CSM
M5 DBH
M5 DEN
M5 DWI
M5 EEK
M5 EPA
M5 ESE
M5 ET
M5 EVT
M5 FET
M5 FLY
M5 FUN
M5 GB
M5 GTI
M5 GUM
M5 IA
M5 ICB
M5 IEP
M5 IIT
M5 IJH
M5 IPX
M5 IXG
M5 JAB
M5 JAN
M5 JB
M5 JDW
M5 JDZ
M5 JFS
M5 JSW
M5 JTS
M5 KAW
M5 KEV
M5 KR
M5 KRG
M5 LAM
M5 LLT
M5 LOE
M5 MAN
M5 MAT
M5 MD
M5 MDC
M5 MKW
M5 MOW
M5 MPC
M5 MSX
M5 NCW
M5 OFV
M5 PD
M5 PMJ
M5 RAO
M5 RIO
M5 RJL
M5 RS
M5 SPY
M5 ST
M5 STC
M5 TAW
M5 TGW
M5 TKA
M5 UF
M5 UFO
M5 UK
M5 URR
M5 UTC
M5 WMF
M5 WOB
M5 WRG
M5 XT
M5 YRG
M5 ZZR

M6

M6 AAB
M6 AAI
M6 AAN
M6 AAS
M6 AAT
M6 ABA
M6 ABC
M6 ABH
M6 ABI
M6 ABJ
M6 ABL
M6 ABU
M6 ABW
M6 ABY
M6 ACG
M6 ACK
M6 ACL
M6 ACN
M6 ACO
M6 ACT
M6 ACU
M6 ADA
M6 ADH
M6 ADK
M6 ADN
M6 ADO
M6 ADP
M6 ADR
M6 ADW
M6 AEA
M6 AEH
M6 AEO
M6 AEP
M6 AEV
M6 AFD
M6 AFE
M6 AFJ
M6 AFO
M6 AFP
M6 AFR
M6 AFT
M6 AFV
M6 AFZ
M6 AGB
M6 AGC
M6 AGE
M6 AGM
M6 AGN
M6 AGP
M6 AGU
M6 AGW
M6 AGX
M6 AHC
M6 AHI
M6 AHM
M6 AIG
M6 AIX
M6 AIY
M6 AJE
M6 AJG
M6 AJK
M6 AJR
M6 AJS
M6 AJT
M6 AJU
M6 AJV
M6 AJW
M6 AJY
M6 AJZ
M6 AKB
M6 AKL
M6 AKM
M6 AKR
M6 AKS
M6 AKU
M6 AKX
M6 ALA
M6 ALF
M6 ALN
M6 ALR
M6 ALS
M6 ALV
M6 AMB
M6 AMH
M6 AMP
M6 AMQ
M6 AMS
M6 AMU
M6 AMY
M6 AMZ
M6 ANA
M6 ANC
M6 ANF
M6 ANG
M6 ANH
M6 ANJ
M6 ANK
M6 ANL
M6 ANQ
M6 ANU
M6 ANW
M6 ANZ
M6 AOA
M6 AOC
M6 AOD
M6 AOI
M6 AOM
M6 AOO
M6 AOS
M6 APE
M6 APF
M6 APH
M6 APJ
M6 APK
M6 APL
M6 APN
M6 APP
M6 APQ
M6 APT
M6 APV
M6 APX
M6 APY
M6 APZ
M6 AQB
M6 AQC
M6 AQF
M6 AQQ
M6 AQS
M6 AQX
M6 AQZ
M6 ARA
M6 ARE
M6 ARG
M6 ARZ
M6 ASA
M6 ASB
M6 ASG
M6 ASR
M6 ATE
M6 ATN
M6 ATO
M6 ATT
M6 ATZ
M6 AUK
M6 AUN
M6 AUR
M6 AUS
M6 AUT
M6 AUV
M6 AUW
M6 AUY
M6 AVB
M6 AVC
M6 AVI
M6 AVN
M6 AVP
M6 AWF
M6 AWH
M6 AWK
M6 AWM
M6 AWT
M6 AWX
M6 AXE
M6 AXK
M6 AXN
M6 AXO
M6 AXQ
M6 AXV
M6 AXY
M6 AXZ
M6 AYB
M6 AYF
M6 AYO
M6 AYT
M6 AYX
M6 AZB
M6 AZI
M6 AZJ
M6 AZL
M6 AZW
M6 AZZ
M6 BAE
M6 BAF
M6 BAL
M6 BAO
M6 BAP
M6 BAT
M6 BBI
M6 BBN
M6 BBV
M6 BBW
M6 BBY
M6 BBZ
M6 BCA
M6 BCD
M6 BCE
M6 BCI
M6 BCL
M6 BCM
M6 BCO
M6 BCP
M6 BCR
M6 BCS
M6 BCT
M6 BCX
M6 BDB
M6 BDC
M6 BDE
M6 BDF
M6 BDH
M6 BDK
M6 BDP
M6 BDT
M6 BDW
M6 BDY
M6 BDZ
M6 BEA
M6 BEO
M6 BEP
M6 BER
M6 BET
M6 BEV
M6 BEW
M6 BEY
M6 BFA
M6 BFD
M6 BFE
M6 BFT
M6 BFW
M6 BFY
M6 BGN
M6 BGQ
M6 BGR
M6 BGU
M6 BGX
M6 BGY
M6 BHG
M6 BHL
M6 BHR
M6 BHV
M6 BHW
M6 BHX
M6 BIC
M6 BIK
M6 BIL
M6 BIQ
M6 BIW
M6 BIZ
M6 BJD
M6 BJH
M6 BJR
M6 BJS
M6 BJU
M6 BJV
M6 BKB
M6 BKG
M6 BKW
M6 BKX
M6 BKY
M6 BKZ
M6 BLD
M6 BLJ
M6 BLL
M6 BLR
M6 BLU
M6 BMA
M6 BMD
M6 BMG
M6 BMH
M6 BML
M6 BMS
M6 BMU
M6 BMX
M6 BND
M6 BNE
M6 BNI
M6 BNR
M6 BNZ
M6 BOH
M6 BON
M6 BOO
M6 BOU
M6 BPF
M6 BPN
M6 BPP
M6 BPT
M6 BPZ
M6 BQI
M6 BQT
M6 BQU
M6 BQX
M6 BQY
M6 BRA
M6 BRC
M6 BRD
M6 BRE
M6 BRF
M6 BRG
M6 BRM
M6 BRW
M6 BRX
M6 BSD
M6 BSE
M6 BSH
M6 BSJ
M6 BSM
M6 BSN
M6 BTD
M6 BTI
M6 BTJ
M6 BTL
M6 BTO
M6 BTU
M6 BTV
M6 BTW
M6 BTX
M6 BTZ
M6 BUK
M6 BUN
M6 BUO
M6 BUP
M6 BUQ
M6 BUT
M6 BUV
M6 BUZ
M6 BVA
M6 BVB
M6 BVF
M6 BVH
M6 BVL
M6 BVO
M6 BVT
M6 BVY
M6 BWA
M6 BWD
M6 BWI
M6 BWM
M6 BWS
M6 BWT
M6 BWX
M6 BXC
M6 BXE
M6 BXG
M6 BYB
M6 BYC
M6 BYD
M6 BYG
M6 BYI
M6 BYW
M6 BZJ
M6 BZL
M6 BZO
M6 BZR
M6 BZS
M6 BZZ
M6 CAB
M6 CAC
M6 CAL
M6 CAM
M6 CAO
M6 CAX
M6 CBB
M6 CBK
M6 CBR
M6 CBS
M6 CBT
M6 CBV
M6 CBX
M6 CBZ
M6 CCC
M6 CCD
M6 CCK
M6 CCL
M6 CCP
M6 CCT
M6 CCZ
M6 CDA
M6 CDF
M6 CDI
M6 CDM
M6 CDP
M6 CDS
M6 CDX
M6 CDY
M6 CEC
M6 CEK
M6 CER
M6 CEV
M6 CFB
M6 CFC
M6 CFF
M6 CFG
M6 CFJ
M6 CFK
M6 CFM
M6 CFP
M6 CGA
M6 CGG
M6 CGI
M6 CGN
M6 CGP
M6 CGR
M6 CGU
M6 CHA
M6 CHE
M6 CHK
M6 CHO
M6 CHR
M6 CIB
M6 CIC
M6 CID
M6 CIF
M6 CIH
M6 CII
M6 CIL
M6 CIN
M6 CIO
M6 CIT
M6 CIV
M6 CJB
M6 CJO
M6 CKA
M6 CKB
M6 CKG
M6 CKH
M6 CKK
M6 CKN
M6 CKS
M6 CKW
M6 CKX
M6 CLD
M6 CLE
M6 CLH
M6 CLN
M6 CLR
M6 CLS
M6 CLU
M6 CLY
M6 CMA
M6 CMD
M6 CME
M6 CMH
M6 CMJ
M6 CMN
M6 CMP
M6 CMR
M6 CMS
M6 CMV
M6 CMX
M6 CNB
M6 CNE
M6 CNF
M6 CNH
M6 CNJ
M6 CNM
M6 CNU
M6 CNW
M6 COA
M6 COG
M6 COJ
M6 COO
M6 COQ
M6 COR
M6 COS
M6 COT
M6 COZ
M6 CPD
M6 CPF
M6 CPG
M6 CPI
M6 CPL
M6 CPY
M6 CQA
M6 CQC
M6 CQG
M6 CQI
M6 CQK
M6 CQM
M6 CQQ
M6 CQT
M6 CQW
M6 CQZ
M6 CRB
M6 CRD
M6 CRK
M6 CRL
M6 CRM
M6 CRN
M6 CRS
M6 CRV
M6 CRY
M6 CSB
M6 CSH
M6 CSQ
M6 CST
M6 CSV
M6 CSY
M6 CTF
M6 CTM
M6 CTP
M6 CTR
M6 CTU
M6 CUB
M6 CUG
M6 CUH
M6 CUI
M6 CUK
M6 CUO
M6 CUT
M6 CUY
M6 CVA
M6 CVD
M6 CVG
M6 CVI
M6 CVL
M6 CVR
M6 CVS
M6 CVT
M6 CVX
M6 CVZ
M6 CWE
M6 CWM
M6 CWS
M6 CWT
M6 CWY
M6 CXE
M6 CXG
M6 CXK
M6 CXL
M6 CXR
M6 CXS
M6 CXT
M6 CXU
M6 CXX
M6 CXY
M6 CYH
M6 CYI
M6 CYJ
M6 CYK
M6 CYZ
M6 CZE
M6 CZG
M6 CZK
M6 CZM
M6 CZP
M6 CZQ
M6 CZT
M6 CZX
M6 CZY
M6 CZZ
M6 DAA
M6 DAH
M6 DAJ
M6 DAV
M6 DBC
M6 DBM
M6 DBO
M6 DBR
M6 DBY
M6 DBZ
M6 DCC
M6 DCF
M6 DCG
M6 DCV
M6 DCX
M6 DDA
M6 DDE
M6 DDJ
M6 DDK
M6 DEB
M6 DEN
M6 DEQ
M6 DET
M6 DEU
M6 DEW
M6 DEX
M6 DFB
M6 DFF
M6 DFN
M6 DFO
M6 DFR
M6 DFV
M6 DGB
M6 DGE
M6 DGF
M6 DGL
M6 DGT
M6 DGV
M6 DGX
M6 DHA
M6 DHC
M6 DHD
M6 DHE
M6 DHH
M6 DHJ
M6 DHL
M6 DHM
M6 DHN
M6 DHO
M6 DHP
M6 DHQ
M6 DHR
M6 DHY
M6 DIA
M6 DID
M6 DIN
M6 DIP
M6 DIR
M6 DIS
M6 DIW
M6 DIX
M6 DIY
M6 DJB
M6 DJD
M6 DJE
M6 DJF
M6 DJH
M6 DJM
M6 DJO
M6 DJP
M6 DJR
M6 DJU
M6 DJV
M6 DJY
M6 DKB
M6 DKD
M6 DKE
M6 DKG
M6 DKH
M6 DKJ
M6 DKP
M6 DKQ
M6 DKV
M6 DKZ
M6 DLE
M6 DLF
M6 DLG
M6 DLK
M6 DLN
M6 DLS
M6 DLZ
M6 DMI
M6 DMJ
M6 DML
M6 DMM
M6 DMR
M6 DMT
M6 DMU
M6 DMV
M6 DMX
M6 DMZ
M6 DNA
M6 DNE
M6 DNH
M6 DNK
M6 DNT
M6 DNU
M6 DNW
M6 DNY
M6 DOE
M6 DOH
M6 DOI
M6 DOK
M6 DOL
M6 DOO
M6 DOP
M6 DOQ
M6 DOS
M6 DOT
M6 DOU
M6 DOV
M6 DOY
M6 DPB
M6 DPD
M6 DPE
M6 DPK
M6 DPM
M6 DPS
M6 DPT
M6 DQA
M6 DQE
M6 DQF
M6 DQI
M6 DQM
M6 DQX
M6 DRA
M6 DRB
M6 DRC
M6 DRD
M6 DRE
M6 DRL
M6 DRN
M6 DRQ
M6 DRS
M6 DRT
M6 DRX
M6 DRY
M6 DSA
M6 DSB
M6 DSF
M6 DSG
M6 DSI
M6 DSK
M6 DSL
M6 DSM
M6 DSX
M6 DTB
M6 DTI
M6 DTK
M6 DTW
M6 DTZ
M6 DUA
M6 DUC
M6 DUC
M6 DUQ
M6 DUS
M6 DUU
M6 DVB
M6 DVD
M6 DVG
M6 DVI
M6 DVJ
M6 DVK
M6 DVL
M6 DVX
M6 DWB
M6 DWD
M6 DWH
M6 DWK
M6 DWL
M6 DWN
M6 DWQ
M6 DWR
M6 DWU
M6 DWX
M6 DXB
M6 DXD
M6 DXE
M6 DXM
M6 DXP
M6 DXR
M6 DXT
M6 DXU
M6 DXX
M6 DXY
M6 DYE
M6 DYJ
M6 DYK
M6 DYQ
M6 DYS
M6 DYT
M6 DYZ
M6 DZD
M6 DZE
M6 DZG
M6 DZI
M6 DZJ
M6 DZK
M6 DZL
M6 DZM
M6 DZO
M6 DZU
M6 DZW
M6 EAM
M6 EAQ
M6 EAR
M6 EAU
M6 EAV
M6 EAY
M6 EBF
M6 EBH
M6 EBK
M6 EBM
M6 EBN
M6 EBT
M6 EBV
M6 ECA
M6 ECF
M6 ECG
M6 ECI
M6 ECK
M6 ECL
M6 ECN
M6 ECQ
M6 ECS
M6 ECU
M6 ECY
M6 EDE
M6 EDF
M6 EDT
M6 EDU
M6 EDX
M6 EEE
M6 EEI
M6 EES
M6 EEV
M6 EEY
M6 EFA
M6 EFN
M6 EFO
M6 EFQ
M6 EFT
M6 EFZ
M6 EGI
M6 EGQ
M6 EGR
M6 EGY
M6 EHG
M6 EHJ
M6 EHN
M6 EHS
M6 EHT
M6 EIP
M6 EIQ
M6 EIT
M6 EIV
M6 EJD
M6 EJE
M6 EJH
M6 EJN
M6 EJQ
M6 EJZ
M6 EKG
M6 EKH
M6 EKJ
M6 EKR
M6 EKU
M6 EKZ
M6 ELA
M6 ELG
M6 ELK
M6 ELM
M6 ELO
M6 ELP
M6 ELT
M6 EMB
M6 EMD
M6 EMN
M6 EMR
M6 EMU
M6 ENG
M6 ENO
M6 ENT
M6 EOD
M6 EOF
M6 EOG
M6 EOH
M6 EOI
M6 EOJ
M6 EOL
M6 EOO
M6 EPB
M6 EPC
M6 EPG
M6 EPI
M6 EPK
M6 EPP
M6 EPR
M6 EPU
M6 EPY
M6 EQP
M6 EQU
M6 EQX
M6 ERB
M6 ERE
M6 ERG
M6 ERJ
M6 ERK
M6 ERM
M6 ERV
M6 ERX
M6 ERY
M6 ESB
M6 ESC
M6 ESD
M6 ESE
M6 ESF
M6 ESG
M6 ESI
M6 ESK
M6 ESM
M6 ESN
M6 ESQ
M6 ESU
M6 ESX
M6 ETB
M6 ETF
M6 ETI
M6 ETK
M6 ETM
M6 ETO
M6 ETQ
M6 ETT
M6 ETV
M6 ETX
M6 EUD
M6 EUF
M6 EUG
M6 EUJ
M6 EUM
M6 EUN
M6 EUQ
M6 EUR
M6 EUS
M6 EUT
M6 EVF
M6 EVG
M6 EVJ
M6 EVN
M6 EVP
M6 EVQ
M6 EVT
M6 EVV
M6 EWB
M6 EWE
M6 EWF
M6 EWG
M6 EWM
M6 EWS
M6 EXM
M6 EYC
M6 EYE
M6 EYN
M6 EZB
M6 EZD
M6 EZH
M6 EZI
M6 EZJ
M6 EZL
M6 EZQ
M6 EZU
M6 EZW
M6 FAA
M6 FAD
M6 FAI
M6 FAL
M6 FAO
M6 FAP
M6 FAQ
M6 FAR
M6 FAT
M6 FAV
M6 FAZ
M6 FBC
M6 FBG
M6 FBI
M6 FBL
M6 FBX
M6 FCK
M6 FCL
M6 FCO
M6 FCP
M6 FCU
M6 FCY
M6 FDA
M6 FDB
M6 FDC
M6 FDE
M6 FDJ
M6 FDM
M6 FDO
M6 FDP
M6 FDQ
M6 FEG
M6 FES
M6 FET
M6 FEV
M6 FEW
M6 FEY
M6 FFA
M6 FFC
M6 FFG
M6 FFH
M6 FFL
M6 FFX
M6 FGB
M6 FGD
M6 FGF
M6 FGL
M6 FGO
M6 FGP
M6 FGT
M6 FGU
M6 FGV
M6 FGX
M6 FHC
M6 FHL
M6 FHT
M6 FHU
M6 FHW
M6 FHZ
M6 FIE
M6 FIF
M6 FIH
M6 FIM
M6 FIP
M6 FIQ
M6 FIS
M6 FJB
M6 FJG
M6 FJK
M6 FJP
M6 FJR
M6 FJY
M6 FKK
M6 FKL
M6 FKM
M6 FKT
M6 FKX
M6 FLD
M6 FLI
M6 FLM
M6 FLP
M6 FLT
M6 FLV
M6 FMA
M6 FMB
M6 FMD
M6 FMH
M6 FMK
M6 FMQ
M6 FMR
M6 FMX
M6 FNC
M6 FND
M6 FNH
M6 FNI
M6 FNJ
M6 FNN
M6 FNS
M6 FNW
M6 FOB
M6 FOH
M6 FOI
M6 FOM
M6 FON
M6 FOP
M6 FOQ
M6 FOS
M6 FOU
M6 FOV
M6 FOW
M6 FPA
M6 FPB
M6 FPR
M6 FPS
M6 FPT
M6 FPZ
M6 FQB
M6 FQI
M6 FQQ
M6 FQS
M6 FQT
M6 FQV
M6 FRA
M6 FRB
M6 FRH
M6 FRL
M6 FRM
M6 FRU
M6 FRX
M6 FSF
M6 FSH
M6 FSK
M6 FSL
M6 FSS
M6 FSX
M6 FSY
M6 FTB
M6 FTD
M6 FTE
M6 FTH
M6 FTJ
M6 FTL
M6 FTM
M6 FTR
M6 FTS
M6 FUB
M6 FUG
M6 FUI
M6 FUP
M6 FUS
M6 FUV
M6 FUZ
M6 FVF
M6 FVG
M6 FVI
M6 FVQ
M6 FVU
M6 FVX
M6 FWA
M6 FWF
M6 FWH
M6 FWI
M6 FWJ
M6 FWK
M6 FWN
M6 FWR
M6 FWU
M6 FXC
M6 FXF
M6 FXH
M6 FXK
M6 FXN
M6 FXR
M6 FXS
M6 FXT
M6 FXV
M6 FYC
M6 FYI
M6 FYO
M6 FYS
M6 FYU
M6 FYX
M6 FYZ
M6 FZA
M6 FZH
M6 FZI
M6 FZO
M6 FZP
M6 FZU
M6 GAJ
M6 GAY
M6 GAZ
M6 GBT
M6 GCA
M6 GCF
M6 GCG
M6 GCH
M6 GCR
M6 GCU
M6 GCZ
M6 GDF
M6 GDG
M6 GDK
M6 GDM
M6 GDR
M6 GDU
M6 GDW
M6 GDX
M6 GEB
M6 GEC
M6 GED
M6 GEH
M6 GEI
M6 GEM
M6 GEN
M6 GEO
M6 GER
M6 GES
M6 GEZ
M6 GFB
M6 GFC
M6 GFD
M6 GFG
M6 GFI
M6 GFJ
M6 GFL
M6 GFT
M6 GFU
M6 GFX
M6 GGA
M6 GGB
M6 GGC
M6 GGK
M6 GGP
M6 GGR
M6 GGU
M6 GGY
M6 GHE
M6 GHG
M6 GHK
M6 GHO
M6 GHU
M6 GHZ
M6 GIE
M6 GIG
M6 GIJ
M6 GIK
M6 GIO
M6 GIR
M6 GJF
M6 GJK
M6 GJR
M6 GJU
M6 GJW
M6 GKA
M6 GKD
M6 GKJ
M6 GKL
M6 GKO
M6 GKP
M6 GKR
M6 GKY
M6 GLA
M6 GLB
M6 GLE
M6 GLF
M6 GLG
M6 GLL
M6 GLQ
M6 GLR
M6 GLT
M6 GLY
M6 GMB
M6 GME
M6 GMG
M6 GMH
M6 GML
M6 GMN
M6 GMW
M6 GMX
M6 GMZ
M6 GNB
M6 GNE
M6 GNI
M6 GNK
M6 GNL
M6 GNQ
M6 GNT
M6 GOH
M6 GOJ
M6 GOK
M6 GOP
M6 GOT
M6 GOV
M6 GPH
M6 GPN
M6 GPW
M6 GPY
M6 GQE
M6 GQL
M6 GQM
M6 GQN
M6 GQO
M6 GQQ
M6 GRB
M6 GRE
M6 GRF
M6 GRG
M6 GRI
M6 GRL
M6 GRM
M6 GRO
M6 GRS
M6 GRW
M6 GSA
M6 GSB
M6 GSE
M6 GSI
M6 GSM
M6 GSN
M6 GSU
M6 GSV
M6 GTA
M6 GTG
M6 GTJ
M6 GTL
M6 GTN
M6 GTP
M6 GTV
M6 GTX
M6 GTZ
M6 GUD
M6 GUG
M6 GUI
M6 GUK
M6 GUL
M6 GUN
M6 GUP
M6 GUQ
M6 GUR
M6 GUZ
M6 GVB
M6 GVE
M6 GVG
M6 GVH
M6 GVK
M6 GVO
M6 GVQ
M6 GVV
M6 GVX
M6 GVZ
M6 GWC
M6 GWP
M6 GWR
M6 GWV
M6 GWY
M6 GXB
M6 GXE
M6 GXG
M6 GXM
M6 GXN
M6 GXQ
M6 GXT
M6 GXX
M6 GXY
M6 GYD
M6 GYJ
M6 GYN
M6 GYO
M6 GYT
M6 GYW
M6 GYX
M6 GZC
M6 GZH
M6 GZM
M6 GZP
M6 GZQ
M6 GZV
M6 HAA
M6 HAB
M6 HAI
M6 HAJ
M6 HAL
M6 HAN
M6 HAO
M6 HAP
M6 HAQ
M6 HAW
M6 HAX
M6 HAZ
M6 HBA
M6 HBL
M6 HBO
M6 HBW
M6 HBX
M6 HBZ
M6 HCC
M6 HCF
M6 HCH
M6 HCN
M6 HCX
M6 HDF
M6 HDJ
M6 HDL
M6 HDQ
M6 HDR
M6 HEC
M6 HEL
M6 HEM
M6 HEN
M6 HES
M6 HEU
M6 HEV
M6 HFD
M6 HFH
M6 HFL
M6 HFS
M6 HFW
M6 HGB
M6 HGC
M6 HGK
M6 HGL
M6 HGP
M6 HGT
M6 HHD
M6 HHI
M6 HHK
M6 HHS
M6 HHU
M6 HHV
M6 HHZ
M6 HIC
M6 HIE
M6 HIF
M6 HIH
M6 HIN
M6 HIO
M6 HIQ
M6 HIR
M6 HIS
M6 HIW
M6 HIX
M6 HIY
M6 HJJ
M6 HJL
M6 HJN
M6 HJO
M6 HJP
M6 HJW
M6 HJY
M6 HKD
M6 HKE
M6 HKF
M6 HKG
M6 HKH
M6 HKK
M6 HKM
M6 HKR
M6 HLB
M6 HLD
M6 HLH
M6 HLI
M6 HLJ
M6 HLK
M6 HLM
M6 HLN
M6 HLO
M6 HLV
M6 HLX
M6 HMA
M6 HMF
M6 HMH
M6 HMI
M6 HMJ
M6 HMN
M6 HMO
M6 HMT
M6 HMX
M6 HMY
M6 HNE
M6 HNK
M6 HNL
M6 HNM
M6 HNP
M6 HNR
M6 HNU
M6 HNY
M6 HOB
M6 HOC
M6 HOD
M6 HOE
M6 HOJ
M6 HOL
M6 HOW
M6 HOX
M6 HPA
M6 HPB
M6 HPE
M6 HPI
M6 HPJ
M6 HPM
M6 HPO
M6 HPP
M6 HPQ
M6 HPU
M6 HPW
M6 HPY
M6 HQA
M6 HQF
M6 HQG
M6 HQH
M6 HQJ
M6 HQP
M6 HQQ
M6 HQS
M6 HQT
M6 HQU
M6 HQV
M6 HQY
M6 HRA
M6 HRB
M6 HRH
M6 HRP
M6 HRU
M6 HRY
M6 HSD
M6 HSG
M6 HSJ
M6 HSN
M6 HSO
M6 HSU
M6 HSW
M6 HSZ
M6 HTA
M6 HTD
M6 HTE
M6 HTL
M6 HTO
M6 HTP
M6 HTR
M6 HTV
M6 HTY
M6 HUA
M6 HUB
M6 HUD
M6 HUG
M6 HUJ
M6 HUO
M6 HUT
M6 HVB
M6 HVC
M6 HVG
M6 HVJ
M6 HVN
M6 HVQ
M6 HVT
M6 HVV
M6 HVW
M6 HVX
M6 HVZ
M6 HWA
M6 HWB
M6 HWF
M6 HWK
M6 HWP
M6 HWR
M6 HWS
M6 HWU
M6 HWY
M6 HWZ
M6 HXH
M6 HXJ
M6 HXL
M6 HXM
M6 HXP
M6 HXQ
M6 HXS
M6 HXZ
M6 HYA
M6 HYC
M6 HYD
M6 HYE
M6 HYG
M6 HYI
M6 HYK
M6 HYL
M6 HYN
M6 HYO
M6 HYP
M6 HYU
M6 HYV
M6 HZB
M6 HZC
M6 HZG
M6 HZH
M6 HZK
M6 HZS
M6 HZV
M6 HZY
M6 IAD
M6 IAE
M6 IAM
M6 IAP
M6 IAU
M6 IBA
M6 IBE
M6 IBK
M6 IBM
M6 IBN
M6 IBS
M6 IBT
M6 IBU
M6 IBV
M6 IBW
M6 IBX
M6 IBY
M6 IBZ
M6 ICC
M6 ICG
M6 ICI
M6 ICN
M6 ICS
M6 ICT
M6 ICV
M6 ICW
M6 ICY
M6 IDA
M6 IDH
M6 IDL
M6 IDT
M6 IDW
M6 IDY
M6 IED
M6 IEE
M6 IEG
M6 IEH
M6 IEK
M6 IEL
M6 IEN
M6 IEP
M6 IES
M6 IET
M6 IEX
M6 IEY
M6 IFD
M6 IFJ
M6 IFK
M6 IFL
M6 IFM
M6 IFP
M6 IFR
M6 IFU
M6 IGB
M6 IGD
M6 IGE
M6 IGF
M6 IGG
M6 IGI
M6 IGL
M6 IGN
M6 IGO
M6 IGU
M6 IGX
M6 IGY
M6 IHB
M6 IHF
M6 IHG
M6 IHI
M6 IHJ
M6 IHK
M6 IHU
M6 IHV
M6 IHW
M6 IHX
M6 IHY
M6 IIA
M6 IIC
M6 IID
M6 IIE
M6 IIH
M6 III
M6 IIK
M6 IIM
M6 IIN
M6 IIS
M6 IIT
M6 IIV
M6 IJA
M6 IJC
M6 IJE
M6 IJF
M6 IJI
M6 IJS
M6 IJU
M6 IJY
M6 IKB
M6 IKC
M6 IKE
M6 IKG
M6 IKK
M6 IKO
M6 IKP
M6 IKQ
M6 IKU
M6 IKV
M6 IKW
M6 IKZ
M6 ILA
M6 ILD
M6 ILE
M6 ILG
M6 ILL
M6 ILU
M6 ILV
M6 ILW
M6 IMD
M6 IME
M6 IMG
M6 IMJ
M6 IMK
M6 IML
M6 IMN
M6 IMO
M6 IMX
M6 INE
M6 INF
M6 INH
M6 INJ
M6 INK
M6 INL
M6 INO
M6 INQ
M6 INR
M6 INY
M6 INZ
M6 IOB
M6 IOC
M6 IOE
M6 IOH
M6 IOJ
M6 IOK
M6 IOM
M6 IOP
M6 IOT
M6 IOU
M6 IOX
M6 IPC
M6 IPD
M6 IPK
M6 IPN
M6 IPO
M6 IPR
M6 IPS
M6 IPT
M6 IPU
M6 IPV
M6 IPX
M6 IQB
M6 IQD
M6 IQI
M6 IQJ
M6 IQP
M6 IQQ
M6 IQR
M6 IQS
M6 IQV
M6 IRD
M6 IRF
M6 IRG
M6 IRH
M6 IRN
M6 IRO
M6 IRQ
M6 IRR
M6 IRS
M6 IRT
M6 IRV
M6 IRY
M6 ISC
M6 ISL
M6 ISO
M6 ISP
M6 ISQ
M6 ISS
M6 IST
M6 ISV
M6 ITA
M6 ITB
M6 ITC
M6 ITE
M6 ITF
M6 ITG
M6 ITH
M6 ITJ
M6 ITK
M6 ITO
M6 ITS
M6 ITT
M6 IUA
M6 IUB
M6 IUC
M6 IUG
M6 IUJ
M6 IUL
M6 IUO
M6 IUQ
M6 IUS
M6 IUT
M6 IUU
M6 IUV
M6 IUW
M6 IUX
M6 IUY

Withheld

M6 IUZ
M6 IVA
M6 IVD
M6 IVF
M6 IVH
M6 IVM
M6 IVS
M6 IVU
M6 IVV
M6 IVX
M6 IVY
M6 IVZ
M6 IWC
M6 IWG
M6 IWH
M6 IWJ
M6 IWK
M6 IWN
M6 IWR
M6 IWS
M6 IWT
M6 IWU
M6 IWV
M6 IWW
M6 IWY
M6 IWZ
M6 IXA
M6 IXB
M6 IXC
M6 IXE
M6 IXK
M6 IXL
M6 IXN
M6 IXQ
M6 IXS
M6 IXT
M6 IXU
M6 IXV
M6 IXW
M6 IXX
M6 IYB
M6 IYD
M6 IYE
M6 IYF
M6 IYI
M6 IYK
M6 IYO
M6 IYP
M6 IYQ
M6 IYR
M6 IYX
M6 IZC
M6 IZD
M6 IZG
M6 IZJ
M6 IZL
M6 IZR
M6 IZS
M6 IZV
M6 IZX
M6 IZZ
M6 JAA
M6 JAC
M6 JAH
M6 JAI
M6 JAO
M6 JAT
M6 JAX
M6 JAY
M6 JBD
M6 JBF
M6 JBJ
M6 JBK
M6 JBP
M6 JBS
M6 JBW
M6 JBY
M6 JCI
M6 JCK
M6 JCO
M6 JCY
M6 JDB
M6 JDE
M6 JDG
M6 JDH
M6 JDI
M6 JDJ
M6 JDO
M6 JDP
M6 JDR
M6 JDS
M6 JDT
M6 JDW
M6 JDZ
M6 JEE
M6 JEI
M6 JEN
M6 JER
M6 JES
M6 JEU
M6 JEX
M6 JFC
M6 JFD
M6 JFM
M6 JFQ
M6 JFT
M6 JFX
M6 JFZ
M6 JGA
M6 JGB
M6 JGD
M6 JGF
M6 JGJ
M6 JGL
M6 JGO
M6 JGW
M6 JGX
M6 JHA
M6 JHM
M6 JHW
M6 JHX
M6 JHY
M6 JIA
M6 JIB
M6 JIJ
M6 JIL
M6 JIM
M6 JIO
M6 JIP
M6 JIU
M6 JIV
M6 JJA
M6 JJC
M6 JJP
M6 JJR
M6 JJT
M6 JJY
M6 JKC
M6 JKE
M6 JKF
M6 JKG
M6 JKI
M6 JKJ
M6 JKK
M6 JKL
M6 JKN
M6 JKO
M6 JKP
M6 JKU
M6 JKZ
M6 JLC
M6 JLF
M6 JLG
M6 JLI
M6 JLK
M6 JLN
M6 JLO
M6 JLP
M6 JLS
M6 JLV
M6 JMA
M6 JME
M6 JMK
M6 JMQ
M6 JMS
M6 JMT
M6 JMU
M6 JMV
M6 JMY
M6 JMZ
M6 JND
M6 JNG
M6 JNK
M6 JNT
M6 JNX
M6 JNY
M6 JOB
M6 JOC
M6 JOF
M6 JOJ
M6 JOL
M6 JOM
M6 JOQ
M6 JOT
M6 JOZ
M6 JPA
M6 JPE
M6 JPG
M6 JPN
M6 JPQ
M6 JPY
M6 JQB
M6 JQD
M6 JQH
M6 JQI
M6 JQJ
M6 JQM
M6 JQN
M6 JQP
M6 JQR
M6 JQT
M6 JQX
M6 JQY
M6 JRA
M6 JRD
M6 JRG
M6 JRM
M6 JRP
M6 JRQ
M6 JRR
M6 JRU
M6 JRV
M6 JRX
M6 JRY
M6 JRZ
M6 JSA
M6 JSB
M6 JSC
M6 JSD
M6 JSE
M6 JSF
M6 JSG
M6 JSK
M6 JSL
M6 JSO
M6 JSY
M6 JSZ
M6 JTB
M6 JTF
M6 JTJ
M6 JTR
M6 JTS
M6 JTT
M6 JTX
M6 JTY
M6 JTZ
M6 JUA
M6 JUD
M6 JUL
M6 JUM
M6 JUN
M6 JUO
M6 JUR
M6 JUS
M6 JUY
M6 JVA
M6 JVE
M6 JVH
M6 JVI
M6 JVJ
M6 JVL
M6 JVN
M6 JVP
M6 JVR
M6 JVT
M6 JWA
M6 JWB
M6 JWJ
M6 JWK
M6 JWM
M6 JWN
M6 JWR
M6 JWS
M6 JWV
M6 JWZ
M6 JXA
M6 JXB
M6 JXC
M6 JXG
M6 JXL
M6 JXM
M6 JXP
M6 JXQ
M6 JXR
M6 JXS
M6 JXT
M6 JXU
M6 JXV
M6 JXW
M6 JXZ
M6 JYA
M6 JYC
M6 JYG
M6 JYH
M6 JYK
M6 JYM
M6 JYN
M6 JYP
M6 JYU
M6 JYW
M6 JZA
M6 JZD
M6 JZE
M6 JZG
M6 JZJ
M6 JZO
M6 JZP
M6 JZQ
M6 JZR
M6 JZS
M6 JZX
M6 KAD
M6 KAI
M6 KAJ
M6 KAK
M6 KAL
M6 KAP
M6 KAW
M6 KAY
M6 KBL
M6 KBM
M6 KBP
M6 KBQ
M6 KBR
M6 KBU
M6 KBZ
M6 KCH
M6 KCN
M6 KCW
M6 KCX
M6 KDE
M6 KDH
M6 KDI
M6 KDL
M6 KDN
M6 KDS
M6 KDT
M6 KDU
M6 KDY
M6 KEB
M6 KEI
M6 KEJ
M6 KEK
M6 KEN
M6 KEO
M6 KEW
M6 KEX
M6 KEY
M6 KEZ
M6 KFA
M6 KFD
M6 KFI
M6 KFL
M6 KFN
M6 KFR
M6 KFU
M6 KFV
M6 KFY
M6 KFZ
M6 KGG
M6 KGH
M6 KGJ
M6 KGP
M6 KGT
M6 KGW
M6 KGY
M6 KGZ
M6 KHC
M6 KHD
M6 KHH
M6 KHI
M6 KHJ
M6 KHK
M6 KHL
M6 KHN
M6 KHO
M6 KHR
M6 KHT
M6 KHU
M6 KHV
M6 KHW
M6 KHY
M6 KHZ
M6 KIB
M6 KIC
M6 KID
M6 KIH
M6 KIJ
M6 KIM
M6 KIQ
M6 KIS
M6 KIY
M6 KIZ
M6 KJC
M6 KJE
M6 KJH
M6 KJN
M6 KJO
M6 KJS
M6 KJT
M6 KJU
M6 KJV
M6 KKC
M6 KKE
M6 KKH
M6 KKP
M6 KKQ
M6 KKR
M6 KKS
M6 KKT
M6 KKV
M6 KKX
M6 KKZ
M6 KLB
M6 KLG
M6 KLJ
M6 KLO
M6 KLQ
M6 KLS
M6 KLY
M6 KMB
M6 KME
M6 KMH
M6 KMM
M6 KMO
M6 KMT
M6 KNA
M6 KNE
M6 KNH
M6 KNK
M6 KNN
M6 KNQ
M6 KNR
M6 KNX
M6 KOF
M6 KOG
M6 KOL
M6 KOO
M6 KOP
M6 KOQ
M6 KOR
M6 KOT
M6 KOU
M6 KOW
M6 KOX
M6 KOY
M6 KPB
M6 KPH
M6 KPS
M6 KPW
M6 KPY
M6 KQB
M6 KQE
M6 KQG
M6 KQI
M6 KQN
M6 KQO
M6 KQQ
M6 KQT
M6 KQU
M6 KQV
M6 KQX
M6 KQY
M6 KRA
M6 KRB
M6 KRC
M6 KRG
M6 KRJ
M6 KRN
M6 KRO
M6 KRS
M6 KRT
M6 KRU
M6 KRY
M6 KSB
M6 KSK
M6 KSN
M6 KSO
M6 KSQ
M6 KSR
M6 KSY
M6 KTB
M6 KTD
M6 KTE
M6 KTF
M6 KTG
M6 KTL
M6 KTP
M6 KTR
M6 KTU
M6 KUA
M6 KUB
M6 KUF
M6 KUL
M6 KUM
M6 KUN
M6 KUO
M6 KUS
M6 KUW
M6 KUX
M6 KUZ
M6 KVC
M6 KVR
M6 KVY
M6 KVZ
M6 KWB
M6 KWD
M6 KWE
M6 KWF
M6 KWJ
M6 KWO
M6 KWR
M6 KWS
M6 KWY
M6 KXB
M6 KXC
M6 KXD
M6 KXG
M6 KXJ
M6 KXK
M6 KXL
M6 KXT
M6 KXY
M6 KXZ
M6 KYA
M6 KYD
M6 KYF
M6 KYG
M6 KYH
M6 KYJ
M6 KYM
M6 KYQ
M6 KYR
M6 KYT
M6 KYU
M6 KYW
M6 KYY
M6 KZA
M6 KZD
M6 KZG
M6 KZJ
M6 KZO
M6 KZP
M6 KZT
M6 KZV
M6 KZY
M6 LAJ
M6 LAN
M6 LAT
M6 LAU
M6 LAW
M6 LAX
M6 LBA
M6 LBB
M6 LBF
M6 LBG
M6 LBH
M6 LBJ
M6 LBO
M6 LBZ
M6 LCA
M6 LCB
M6 LCE
M6 LCJ
M6 LCM
M6 LCO
M6 LCS
M6 LDC
M6 LDN
M6 LDO
M6 LDP
M6 LDT
M6 LDX
M6 LED
M6 LEI
M6 LEL
M6 LEM
M6 LEN
M6 LEO
M6 LER
M6 LES
M6 LEZ
M6 LFA
M6 LFF
M6 LFP
M6 LFV
M6 LGE
M6 LGG
M6 LGK
M6 LGN
M6 LGO
M6 LGP
M6 LGU
M6 LGW
M6 LGY
M6 LGZ
M6 LHC
M6 LHK
M6 LHT
M6 LHU
M6 LHX
M6 LHY
M6 LHZ
M6 LIA
M6 LIF
M6 LIG
M6 LIR
M6 LIX
M6 LIY
M6 LJC
M6 LJH
M6 LJI
M6 LJL
M6 LJN
M6 LJQ
M6 LJW
M6 LJX
M6 LJY
M6 LKB
M6 LKC
M6 LKG
M6 LKH
M6 LKJ
M6 LKN
M6 LKO
M6 LKP
M6 LKQ
M6 LKR
M6 LKT
M6 LKU
M6 LKX
M6 LKZ
M6 LLA
M6 LLF
M6 LLJ
M6 LLP
M6 LLQ
M6 LLR
M6 LLT
M6 LLV
M6 LLX
M6 LMA
M6 LMD
M6 LME
M6 LMF
M6 LML
M6 LMM
M6 LMN
M6 LMO
M6 LMP
M6 LMR
M6 LMS
M6 LMT
M6 LMV
M6 LMX
M6 LMY
M6 LNA
M6 LNF
M6 LNH
M6 LNJ
M6 LNK
M6 LNL
M6 LNN
M6 LNO
M6 LNR
M6 LNT
M6 LNW
M6 LNY
M6 LNZ
M6 LOA
M6 LOH
M6 LOI
M6 LOJ
M6 LOM
M6 LOP
M6 LOR
M6 LOU
M6 LOV
M6 LOX
M6 LOY
M6 LPA
M6 LPE
M6 LPJ
M6 LPL
M6 LPM
M6 LPQ
M6 LPU
M6 LQB
M6 LQD
M6 LQE
M6 LQH
M6 LQI
M6 LQJ
M6 LQL
M6 LQQ
M6 LQS
M6 LQV
M6 LQZ
M6 LRC
M6 LRD
M6 LRE
M6 LRJ
M6 LRM
M6 LRP
M6 LRS
M6 LRT
M6 LRY
M6 LSC
M6 LSD
M6 LSF
M6 LSI
M6 LSM
M6 LSQ
M6 LSR
M6 LSS
M6 LSX
M6 LSZ
M6 LTA
M6 LTE
M6 LTF
M6 LTG
M6 LTH
M6 LTI
M6 LTP
M6 LTT
M6 LTW
M6 LTX
M6 LTY
M6 LTZ
M6 LUB
M6 LUE
M6 LUF
M6 LUL
M6 LUN
M6 LUO
M6 LUS
M6 LUU
M6 LUV
M6 LUX
M6 LVB
M6 LVF
M6 LVG
M6 LVI
M6 LVL
M6 LVO
M6 LVP
M6 LVQ
M6 LVT
M6 LVU
M6 LVY
M6 LVZ
M6 LWC
M6 LWD
M6 LWH
M6 LWK
M6 LWL
M6 LWN
M6 LWO
M6 LWU
M6 LWW
M6 LWX
M6 LXA
M6 LXB
M6 LXF
M6 LXG
M6 LXJ
M6 LXL
M6 LXO
M6 LXT
M6 LXV
M6 LXZ
M6 LYE
M6 LYI
M6 LYK
M6 LYL
M6 LYM
M6 LYR
M6 LYT
M6 LYV
M6 LYW
M6 LYX
M6 LZB
M6 LZE
M6 LZH
M6 LZJ
M6 LZK
M6 LZO
M6 LZR
M6 MAC
M6 MAE
M6 MAI
M6 MAN
M6 MAQ
M6 MAR
M6 MAU
M6 MAV
M6 MAZ
M6 MBA
M6 MBJ
M6 MBL
M6 MBM
M6 MBT
M6 MBW
M6 MCD
M6 MCG
M6 MCI
M6 MCN
M6 MCV
M6 MDA
M6 MDD
M6 MDE
M6 MDF
M6 MDI
M6 MDK
M6 MDP
M6 MDQ
M6 MDS
M6 MDV
M6 MDW
M6 MDY
M6 MEE
M6 MEF
M6 MEH
M6 MEI
M6 MEM
M6 MER
M6 MET
M6 MEW
M6 MEX
M6 MEY
M6 MEZ
M6 MFH
M6 MFP
M6 MFQ
M6 MFT
M6 MFV
M6 MFW
M6 MFX
M6 MFY
M6 MGB
M6 MGE
M6 MGI
M6 MGK
M6 MGL
M6 MGQ
M6 MGS
M6 MHB
M6 MHC
M6 MHG
M6 MHP
M6 MHR
M6 MHS
M6 MHX
M6 MHZ
M6 MIJ
M6 MIW
M6 MIX
M6 MJB
M6 MJE
M6 MJH
M6 MJJ
M6 MJK
M6 MJQ
M6 MJU
M6 MJW
M6 MJX
M6 MKA
M6 MKC
M6 MKF
M6 MKG
M6 MKI
M6 MKL
M6 MKP
M6 MKQ
M6 MKR
M6 MKS
M6 MKT
M6 MKZ
M6 MLB
M6 MLC
M6 MLD
M6 MLJ
M6 MLS
M6 MLW
M6 MLZ
M6 MMA
M6 MMD
M6 MME
M6 MMF
M6 MMJ
M6 MMK
M6 MMO
M6 MMV
M6 MMW
M6 MMZ
M6 MNA
M6 MNB
M6 MNF
M6 MNJ
M6 MNM
M6 MNN
M6 MNY
M6 MOA
M6 MOE
M6 MOL
M6 MOR
M6 MOT
M6 MPA
M6 MPC
M6 MPG
M6 MPJ
M6 MPN
M6 MPU
M6 MQA
M6 MQG
M6 MQI
M6 MQQ
M6 MQZ
M6 MRA
M6 MRB
M6 MRE
M6 MRF
M6 MRL
M6 MRO
M6 MRQ
M6 MRZ
M6 MSB
M6 MSD
M6 MSK
M6 MSL
M6 MSQ
M6 MSW
M6 MSZ
M6 MTB
M6 MTF
M6 MTP
M6 MTX
M6 MTY
M6 MUC
M6 MUE
M6 MUG
M6 MUI
M6 MUM
M6 MUN
M6 MUQ
M6 MUV
M6 MUW
M6 MUX
M6 MVC
M6 MVG
M6 MVJ
M6 MVO
M6 MVP
M6 MVR
M6 MVS
M6 MVU
M6 MWH
M6 MWK
M6 MWL
M6 MWM
M6 MWR
M6 MWU
M6 MWX
M6 MWY
M6 MXG
M6 MXI
M6 MXJ
M6 MXL
M6 MXN
M6 MXQ
M6 MXS
M6 MXT
M6 MXW
M6 MYA
M6 MYJ
M6 MYP
M6 MYQ
M6 MYX
M6 MYZ
M6 MZA
M6 MZF
M6 MZI
M6 MZK
M6 MZM
M6 MZO
M6 MZP
M6 MZS
M6 MZT
M6 MZW
M6 NAE
M6 NAF
M6 NAH
M6 NAR
M6 NBA
M6 NBB
M6 NBC
M6 NBD
M6 NBE
M6 NBF
M6 NBJ
M6 NBK
M6 NBT
M6 NBZ
M6 NCJ
M6 NCN
M6 NCQ
M6 NCT
M6 NCW
M6 NCX
M6 NDA
M6 NDD
M6 NDH
M6 NDJ
M6 NDL
M6 NDQ
M6 NDS
M6 NDU
M6 NDV
M6 NDW
M6 NDX
M6 NEB
M6 NEK
M6 NEN
M6 NEP
M6 NER
M6 NEX
M6 NFA
M6 NFF
M6 NFO
M6 NFR
M6 NFU
M6 NFV
M6 NGB
M6 NGC
M6 NGG
M6 NGH
M6 NGJ
M6 NGL
M6 NGO
M6 NGQ
M6 NGS
M6 NGT
M6 NGV
M6 NGZ
M6 NHB
M6 NHE
M6 NHI
M6 NHQ
M6 NHZ
M6 NIF
M6 NIH
M6 NII
M6 NIO
M6 NIR
M6 NIW
M6 NIY
M6 NIZ
M6 NJA
M6 NJC
M6 NJF
M6 NJJ
M6 NJL
M6 NJQ
M6 NJR
M6 NJU
M6 NJW
M6 NJY
M6 NKD
M6 NKF
M6 NKI
M6 NKL
M6 NKN
M6 NKO
M6 NKS
M6 NKT
M6 NKU
M6 NKV
M6 NKW
M6 NKX
M6 NKZ
M6 NLC
M6 NLD
M6 NLE
M6 NLH
M6 NLI
M6 NLM
M6 NLN
M6 NLT
M6 NLU
M6 NLY
M6 NMA
M6 NMB
M6 NMC
M6 NMF
M6 NMI
M6 NMJ
M6 NMK
M6 NMO
M6 NMV
M6 NMX
M6 NMY
M6 NNG
M6 NNH
M6 NNI
M6 NNN
M6 NNQ
M6 NNR
M6 NNZ
M6 NOG
M6 NOO
M6 NOP
M6 NOU
M6 NOV
M6 NOW
M6 NOZ
M6 NPA
M6 NPB
M6 NPC
M6 NPG
M6 NPH
M6 NPI
M6 NPJ
M6 NPK
M6 NPM
M6 NPP
M6 NPQ
M6 NPR
M6 NPT
M6 NPU
M6 NPY
M6 NQE
M6 NQF
M6 NQG
M6 NQH
M6 NQO
M6 NQS
M6 NQX
M6 NQY
M6 NQZ
M6 NRD
M6 NRL
M6 NRY
M6 NRZ
M6 NSA
M6 NSB
M6 NSF
M6 NSL
M6 NSN
M6 NSP
M6 NSU
M6 NSV
M6 NSX
M6 NTA
M6 NTC
M6 NTD
M6 NTE
M6 NTF
M6 NTG
M6 NTH
M6 NTI
M6 NTO
M6 NTQ
M6 NTS
M6 NTU
M6 NTX
M6 NTZ
M6 NUA
M6 NUB
M6 NUE
M6 NUH
M6 NUI
M6 NUK
M6 NUN
M6 NUS
M6 NUT
M6 NUU
M6 NUZ
M6 NVC
M6 NVD
M6 NVE
M6 NVF
M6 NVH
M6 NVI
M6 NVK
M6 NVN
M6 NVP
M6 NVQ
M6 NVU
M6 NVV
M6 NVX
M6 NWD
M6 NWE
M6 NWG
M6 NWJ
M6 NWR
M6 NWU
M6 NXB
M6 NXC
M6 NXD
M6 NXG
M6 NXI
M6 NXJ
M6 NXK
M6 NXM
M6 NXN
M6 NXU
M6 NXW
M6 NXX
M6 NYC
M6 NYH
M6 NYI
M6 NYK
M6 NYN
M6 NYQ
M6 NYS
M6 NYU
M6 NYV
M6 NZA
M6 NZC
M6 NZD
M6 NZG
M6 NZP
M6 NZU
M6 NZW
M6 NZX
M6 NZZ
M6 OAA
M6 OAC
M6 OAH
M6 OAK
M6 OAM
M6 OAY
M6 OBA
M6 OBD
M6 OBE
M6 OBF
M6 OBG
M6 OBI
M6 OBL
M6 OBM
M6 OBN
M6 OBT
M6 OBV
M6 OBW
M6 OCA
M6 OCE
M6 OCF
M6 OCG
M6 OCL
M6 OCN
M6 OCO
M6 OCQ
M6 ODI
M6 ODJ
M6 ODQ
M6 ODT
M6 ODU
M6 ODV
M6 ODW
M6 ODX
M6 ODZ
M6 OED
M6 OEE
M6 OEL
M6 OEO
M6 OEQ
M6 OEX
M6 OEY
M6 OFB
M6 OFC
M6 OFI
M6 OFO
M6 OFP
M6 OFR
M6 OFS
M6 OFT
M6 OFU
M6 OFV
M6 OFW
M6 OFX
M6 OFY
M6 OGB
M6 OGD
M6 OGG
M6 OGI
M6 OGL
M6 OGM
M6 OGN
M6 OGR
M6 OGV
M6 OGW
M6 OGY
M6 OGZ
M6 OHA
M6 OHB
M6 OHC
M6 OHD
M6 OHG
M6 OHH
M6 OHI
M6 OHJ
M6 OHK
M6 OHM
M6 OHR
M6 OHV
M6 OHW
M6 OHX
M6 OIA
M6 OIB
M6 OID
M6 OIE
M6 OIF
M6 OIG
M6 OIH
M6 OII
M6 OIJ
M6 OIK
M6 OIL
M6 OIN
M6 OIO
M6 OIP
M6 OIQ
M6 OIS
M6 OIX
M6 OJA
M6 OJG
M6 OJH
M6 OJJ
M6 OJR
M6 OJV
M6 OJW
M6 OJX
M6 OJY
M6 OKA
M6 OKE
M6 OKF
M6 OKG
M6 OKN
M6 OKT
M6 OKU
M6 OKV
M6 OKW
M6 OKX
M6 OLA
M6 OLC
M6 OLK
M6 OLM
M6 OLN
M6 OLO
M6 OLV
M6 OLZ
M6 OMB
M6 OMC
M6 OMD
M6 OMI
M6 OMM
M6 OMN
M6 OMO
M6 OMQ
M6 OMU
M6 OMV
M6 OMW
M6 OMY
M6 ONA
M6 ONB
M6 ONC
M6 ONE
M6 ONF
M6 ONG
M6 ONI
M6 ONK
M6 ONQ
M6 ONR
M6 ONV
M6 ONX
M6 ONY
M6 OOE
M6 OOF
M6 OOH
M6 OOI
M6 OON
M6 OOR
M6 OOS
M6 OOT
M6 OOU
M6 OPH
M6 OPI
M6 OPK
M6 OPN
M6 OPP
M6 OPQ
M6 OPR
M6 OPU
M6 OPV
M6 OPW
M6 OPX
M6 OPY
M6 OPZ
M6 OQA
M6 OQC
M6 OQE
M6 OQF
M6 OQI
M6 OQJ
M6 OQK
M6 OQM
M6 OQO
M6 OQP
M6 OQS
M6 OQU
M6 OQV
M6 OQW
M6 OQY
M6 ORA
M6 ORD
M6 ORL
M6 ORP
M6 ORQ
M6 ORR
M6 ORS
M6 ORV
M6 ORY
M6 ORZ
M6 OSA
M6 OSB
M6 OSD
M6 OSE
M6 OSG
M6 OSH
M6 OSK
M6 OSQ
M6 OSR
M6 OSS
M6 OSV
M6 OSZ
M6 OTA
M6 OTC
M6 OTD
M6 OTE
M6 OTF
M6 OTO
M6 OTQ
M6 OTV
M6 OTZ
M6 OUB
M6 OUC
M6 OUE
M6 OUG
M6 OUH
M6 OUJ
M6 OUK
M6 OUM
M6 OUO
M6 OUQ
M6 OUT
M6 OUV
M6 OUX
M6 OUZ
M6 OVA
M6 OVC
M6 OVD
M6 OVG
M6 OVJ
M6 OVK
M6 OVL
M6 OVM
M6 OVO
M6 OVP
M6 OVQ
M6 OVU
M6 OVZ
M6 OWB
M6 OWE
M6 OWF
M6 OWH
M6 OWK
M6 OWN
M6 OWQ
M6 OWR
M6 OWV
M6 OWW
M6 OWX
M6 OWY
M6 OWZ
M6 OXG
M6 OXH
M6 OXI
M6 OXK
M6 OXL
M6 OXN
M6 OXO
M6 OXP
M6 OXQ
M6 OYG
M6 OYI
M6 OYK
M6 OYQ
M6 OYR
M6 OYS
M6 OYT
M6 OYV
M6 OYW
M6 OZA
M6 OZC
M6 OZD
M6 OZE
M6 OZH
M6 OZJ
M6 OZK
M6 OZP
M6 OZU
M6 OZV
M6 PAB
M6 PAH
M6 PAK
M6 PAL
M6 PAQ
M6 PAR
M6 PBA
M6 PBB
M6 PBC
M6 PBG
M6 PBH
M6 PBQ
M6 PBS
M6 PBX
M6 PCA
M6 PCG
M6 PCH
M6 PCI
M6 PCK
M6 PCO
M6 PCS
M6 PCY
M6 PDF
M6 PDH
M6 PEE
M6 PEH
M6 PEK
M6 PEV
M6 PFC
M6 PFE
M6 PFF
M6 PFS
M6 PFX
M6 PGE
M6 PGG
M6 PGJ
M6 PGK
M6 PGM
M6 PGR
M6 PGS
M6 PGV
M6 PHB
M6 PHD
M6 PHI
M6 PHN
M6 PHR
M6 PHY
M6 PIE
M6 PIF
M6 PIO
M6 PIQ
M6 PIV
M6 PIW
M6 PIX
M6 PIZ
M6 PJC
M6 PJF
M6 PJG
M6 PJL
M6 PJO
M6 PJQ
M6 PJU
M6 PJW
M6 PJZ
M6 PKA
M6 PKB
M6 PKE
M6 PKF
M6 PKG
M6 PKI
M6 PKJ
M6 PKM
M6 PKP
M6 PKS
M6 PKW
M6 PKX
M6 PKY
M6 PKZ
M6 PLA
M6 PLD
M6 PLF
M6 PLG
M6 PLJ
M6 PLL
M6 PLM
M6 PLU
M6 PLX
M6 PMC
M6 PME
M6 PMI
M6 PMN
M6 PMO
M6 PMQ
M6 PMS
M6 PMU
M6 PMV
M6 PMX
M6 PNA
M6 PNE
M6 PNF
M6 PNM
M6 PNO
M6 PNQ
M6 PNS
M6 PNV
M6 POI
M6 POJ
M6 POM
M6 POS
M6 PPA
M6 PPB
M6 PPC
M6 PPF
M6 PPG
M6 PPI
M6 PPM
M6 PPO
M6 PPR
M6 PPU
M6 PPW
M6 PPX
M6 PQB
M6 PQC
M6 PQE
M6 PQG
M6 PQH
M6 PQJ
M6 PQN
M6 PQR
M6 PQT
M6 PQX
M6 PQY
M6 PQZ
M6 PRD
M6 PRG
M6 PRI
M6 PRJ
M6 PRK
M6 PRL
M6 PRQ
M6 PRR
M6 PRU
M6 PRX
M6 PRY
M6 PSA
M6 PSB
M6 PSD
M6 PSE
M6 PSF
M6 PSH
M6 PSI
M6 PSK
M6 PSL
M6 PSP
M6 PSS
M6 PSU
M6 PSW
M6 PTA
M6 PTB
M6 PTC
M6 PTD
M6 PTG
M6 PTJ
M6 PTK
M6 PTL
M6 PTN
M6 PTQ
M6 PTS
M6 PTT
M6 PTW
M6 PTZ
M6 PUB
M6 PUC
M6 PUE
M6 PUK
M6 PUR
M6 PVA
M6 PVG
M6 PVP
M6 PVX
M6 PWA
M6 PWI
M6 PWJ
M6 PWW
M6 PWX
M6 PWY
M6 PWZ
M6 PXB
M6 PXQ
M6 PXS
M6 PYA
M6 PYB
M6 PYK
M6 PYM
M6 PYS
M6 PYT
M6 PYU
M6 PZD
M6 PZL
M6 PZT
M6 PZX
M6 QCB
M6 RAA
M6 RAB
M6 RAE
M6 RAF
M6 RAG
M6 RAI
M6 RAJ
M6 RAK
M6 RAP
M6 RAT
M6 RAU
M6 RAY
M6 RBC
M6 RBD
M6 RBG
M6 RBJ
M6 RBL
M6 RBM
M6 RBR
M6 RBS
M6 RBX
M6 RCH
M6 RCM
M6 RCQ
M6 RCT
M6 RCX
M6 RDB
M6 RDE
M6 RDG
M6 RDH
M6 RDJ
M6 RDO
M6 RDY
M6 REA
M6 REC
M6 REK
M6 REN
M6 REP
M6 RET
M6 REU
M6 REZ
M6 RFT
M6 RFU
M6 RFX
M6 RFY
M6 RGB
M6 RGG
M6 RGJ
M6 RGL
M6 RGV
M6 RHE
M6 RHG
M6 RHH
M6 RHJ
M6 RHM
M6 RHP
M6 RHS
M6 RHV
M6 RHW
M6 RHZ
M6 RIC
M6 RIF
M6 RIG
M6 RII
M6 RIP
M6 RIW
M6 RIX
M6 RJ
M6 RJA
M6 RJC
M6 RJD
M6 RJG
M6 RJI
M6 RJT
M6 RJV
M6 RKG
M6 RKH
M6 RKJ
M6 RKN
M6 RKV
M6 RKY
M6 RKZ
M6 RLF
M6 RLH
M6 RLI
M6 RLJ
M6 RLK
M6 RLM
M6 RLO
M6 RLS
M6 RLY
M6 RMB
M6 RMC
M6 RMH
M6 RMM
M6 RMP
M6 RMQ
M6 RMY
M6 RMZ
M6 RNB
M6 RNG
M6 RNH
M6 RNJ
M6 RNL
M6 RNR
M6 RNW
M6 RNY
M6 ROK
M6 ROO
M6 ROQ
M6 ROR
M6 ROS
M6 ROT
M6 ROZ
M6 RPA
M6 RPG
M6 RPI
M6 RPK
M6 RQB
M6 RQI
M6 RQL
M6 RRB
M6 RRE
M6 RRI
M6 RRM
M6 RRO
M6 RRP
M6 RRS
M6 RRT
M6 RRU
M6 RRX
M6 RRY
M6 RRZ
M6 RSC
M6 RSG
M6 RSK
M6 RSM
M6 RSN
M6 RSQ
M6 RSW
M6 RSX
M6 RSZ
M6 RTB
M6 RTH
M6 RTJ
M6 RTK
M6 RTL
M6 RTS
M6 RTV
M6 RTW
M6 RTY
M6 RUA
M6 RUD
M6 RUE
M6 RUF
M6 RUU
M6 RVB
M6 RVC
M6 RVF
M6 RVK
M6 RVL
M6 RVP
M6 RVT
M6 RVV
M6 RWC
M6 RWD
M6 RWH
M6 RWK
M6 RWN
M6 RWQ
M6 RWR
M6 RWW
M6 RXD
M6 RXF
M6 RXI
M6 RXM
M6 RXN
M6 RXO
M6 RXP
M6 RYB
M6 RYE
M6 RYO
M6 RYS
M6 RYT
M6 RYW
M6 RZB
M6 RZJ
M6 RZM
M6 RZR
M6 RZS
M6 RZT
M6 SAB
M6 SAJ
M6 SAO
M6 SAP
M6 SAR
M6 SAT
M6 SBA
M6 SBC
M6 SBQ
M6 SBX
M6 SCE
M6 SCJ
M6 SCL
M6 SCO
M6 SCR
M6 SCS
M6 SCT
M6 SCY
M6 SDB
M6 SDC
M6 SDD
M6 SDM
M6 SDO
M6 SDY
M6 SED
M6 SEH
M6 SEL
M6 SEM
M6 SEQ
M6 SET
M6 SEW
M6 SEX
M6 SFA
M6 SFD
M6 SFE
M6 SFG
M6 SFI
M6 SFN
M6 SFO
M6 SFU
M6 SFX
M6 SFY
M6 SFZ
M6 SGI
M6 SGK
M6 SGR
M6 SGW
M6 SGX
M6 SHE
M6 SHN
M6 SHO
M6 SHR
M6 SHS
M6 SHT
M6 SHX
M6 SHY

M6 SIB
M6 SIG
M6 SIN
M6 SIO
M6 SIQ
M6 SJI
M6 SJL
M6 SJM
M6 SJP
M6 SJS
M6 SJT
M6 SJY
M6 SKA
M6 SKB
M6 SKC
M6 SKD
M6 SKE
M6 SKG
M6 SKI
M6 SKJ
M6 SKK
M6 SKO
M6 SKS
M6 SLF
M6 SLH
M6 SLO
M6 SLQ
M6 SLS
M6 SLW
M6 SLX
M6 SLY
M6 SMC
M6 SMJ
M6 SMO
M6 SMP
M6 SMS
M6 SMT
M6 SMU
M6 SMV
M6 SMW
M6 SMZ
M6 SNA
M6 SNB
M6 SND
M6 SNI
M6 SNK
M6 SNL
M6 SNM
M6 SNT
M6 SNX
M6 SOA
M6 SOB
M6 SOH
M6 SOJ
M6 SOL
M6 SOM
M6 SOP
M6 SOQ
M6 SOY
M6 SPB
M6 SPD
M6 SPE
M6 SPF
M6 SPI
M6 SPO
M6 SPQ
M6 SPR
M6 SPV
M6 SPW
M6 SPX
M6 SQA
M6 SQJ
M6 SQM
M6 SQR
M6 SQS
M6 SQT
M6 SQY
M6 SQZ
M6 SRH
M6 SRJ
M6 SRK
M6 SRP
M6 SRT
M6 SRU
M6 SRW
M6 SRX
M6 SRY
M6 SSE
M6 SSG
M6 SSH
M6 SSJ
M6 SSK
M6 SSL
M6 SSQ
M6 STB
M6 STE
M6 STG
M6 STL
M6 STO
M6 STS
M6 STT
M6 STW
M6 STX
M6 SUA
M6 SUG
M6 SUK
M6 SUS
M6 SVA
M6 SVB
M6 SVC
M6 SVD
M6 SVE
M6 SVK
M6 SVL
M6 SVO
M6 SVR
M6 SVS
M6 SVW
M6 SVX
M6 SWM
M6 SWR
M6 SWW
M6 SWX
M6 SWY
M6 SWZ
M6 SXE
M6 SXH
M6 SXJ
M6 SXL
M6 SXR
M6 SXS
M6 SXW
M6 SXX
M6 SYA
M6 SYD
M6 SYL
M6 SYM
M6 SYN
M6 SYP
M6 SYR
M6 SYS
M6 SYU
M6 SYV
M6 SYY
M6 SYZ
M6 SZE
M6 SZF
M6 SZR
M6 SZS
M6 SZT
M6 SZY
M6 TAB
M6 TAC
M6 TAE
M6 TAI
M6 TAM
M6 TAN
M6 TAP
M6 TAQ
M6 TAR
M6 TAS
M6 TAT
M6 TAV
M6 TAZ
M6 TBB
M6 TBE
M6 TBI
M6 TBP
M6 TBW
M6 TBZ
M6 TCC
M6 TCJ
M6 TCP
M6 TCQ
M6 TCR
M6 TCT
M6 TCU
M6 TCW
M6 TDG
M6 TDH
M6 TDN
M6 TDQ
M6 TDS
M6 TDT
M6 TDW
M6 TDX
M6 TEA
M6 TEB
M6 TEC
M6 TED
M6 TEH
M6 TEI
M6 TET
M6 TEX
M6 TFA
M6 TFC
M6 TFI
M6 TFM
M6 TFR
M6 TFS
M6 TFU
M6 TFX
M6 TGA
M6 TGE
M6 TGW
M6 TGX
M6 THD
M6 THE
M6 THF
M6 THG
M6 THH
M6 THK
M6 THL
M6 THM
M6 THN
M6 THR
M6 THT
M6 THX
M6 THZ
M6 TIB
M6 TIG
M6 TIK
M6 TIM
M6 TIN
M6 TIP
M6 TIS
M6 TIV
M6 TIX
M6 TIZ
M6 TJC
M6 TJK
M6 TJM
M6 TJR
M6 TJT
M6 TJW
M6 TKB
M6 TKD
M6 TKF
M6 TKH
M6 TKJ
M6 TKL
M6 TKM
M6 TKO
M6 TKT
M6 TLA
M6 TLD
M6 TLE
M6 TLH
M6 TLJ
M6 TLT
M6 TLU
M6 TLV
M6 TLW
M6 TLZ
M6 TMB
M6 TME
M6 TMI
M6 TMN
M6 TMT
M6 TMV
M6 TMW
M6 TMX
M6 TNE
M6 TNF
M6 TNG
M6 TNK
M6 TNP
M6 TNS
M6 TNW
M6 TNX
M6 TOB
M6 TOH
M6 TOJ
M6 TOP
M6 TOR
M6 TOX
M6 TOY
M6 TPF
M6 TPJ
M6 TPK
M6 TPL
M6 TPN
M6 TPP
M6 TPW
M6 TPZ
M6 TQB
M6 TQM
M6 TQP
M6 TQY
M6 TRA
M6 TRB
M6 TRG
M6 TRJ
M6 TRM
M6 TRN
M6 TRY
M6 TSB
M6 TSD
M6 TSE
M6 TSF
M6 TSG
M6 TSL
M6 TSO
M6 TSQ
M6 TSS
M6 TSU
M6 TTA
M6 TTC
M6 TTD
M6 TTG
M6 TTI
M6 TTJ
M6 TTQ
M6 TUB
M6 TUC
M6 TUE
M6 TUI
M6 TUP
M6 TUS
M6 TUZ
M6 TVA
M6 TVE
M6 TVG
M6 TVI
M6 TVO
M6 TVR
M6 TVT
M6 TVV
M6 TVZ
M6 TWB
M6 TWC
M6 TWD
M6 TWF
M6 TWG
M6 TWO
M6 TWP
M6 TWR
M6 TWT
M6 TWX
M6 TXA
M6 TXB
M6 TXC
M6 TXL
M6 TXO
M6 TXW
M6 TYD
M6 TYF
M6 TYM
M6 TYO
M6 TYP
M6 TYW
M6 TYX
M6 TYY
M6 TZF
M6 TZG
M6 TZH
M6 TZL
M6 TZN
M6 TZT
M6 TZW
M6 UAA
M6 UAC
M6 UAH
M6 UAM
M6 UAQ
M6 UAT
M6 UAV
M6 UBA
M6 UBJ
M6 UBU
M6 UBZ
M6 UCC
M6 UCH
M6 UCK
M6 UCO
M6 UCW
M6 UCY
M6 UDD
M6 UDE
M6 UDF
M6 UDG
M6 UDH
M6 UDI
M6 UDK
M6 UDP
M6 UDQ
M6 UDW
M6 UDZ
M6 UEC
M6 UEX
M6 UFI
M6 UGB
M6 UGD
M6 UGG
M6 UGT
M6 UGV
M6 UHD
M6 UHH
M6 UHK
M6 UHZ
M6 UIK
M6 UIN
M6 UIX
M6 UJB
M6 UJC
M6 UJI
M6 UKD
M6 UKE
M6 UKK
M6 UKL
M6 UKO
M6 UKP
M6 UKR
M6 UKS
M6 UKV
M6 UKW
M6 UKY
M6 UKZ
M6 ULF
M6 ULN
M6 ULP
M6 ULT
M6 ULU
M6 UMA
M6 UME
M6 UMF
M6 UMH
M6 UML
M6 UMP
M6 UNE
M6 UNG
M6 UNK
M6 UNO
M6 UNX
M6 UOJ
M6 UOK
M6 UOO
M6 UOS
M6 UPD
M6 UPP
M6 URB
M6 URC
M6 URF
M6 URL
M6 URZ
M6 USB
M6 USH
M6 USL
M6 USR
M6 UST
M6 UTA
M6 UTD
M6 UTE
M6 UTH
M6 UTM
M6 UTR
M6 UTS
M6 UTU
M6 UUB
M6 UUC
M6 UUF
M6 UUG
M6 UUL
M6 UUN
M6 UVE
M6 UVL
M6 UVR
M6 UWE
M6 UWF
M6 UWM
M6 UXH
M6 UXL
M6 UXO
M6 UXP
M6 UXY
M6 UYT
M6 UZI
M6 VAC
M6 VAD
M6 VAF
M6 VAO
M6 VAS
M6 VBA
M6 VBC
M6 VBG
M6 VBM
M6 VBO
M6 VBY
M6 VCG
M6 VCJ
M6 VCO
M6 VCR
M6 VCT
M6 VCW
M6 VCX
M6 VCZ
M6 VDM
M6 VDN
M6 VDQ
M6 VDS
M6 VEE
M6 VEI
M6 VEK
M6 VEM
M6 VER
M6 VES
M6 VEU
M6 VEX
M6 VEY
M6 VEZ
M6 VFM
M6 VFO
M6 VFR
M6 VFT
M6 VFX
M6 VGF
M6 VGG
M6 VGH
M6 VGL
M6 VGM
M6 VGT
M6 VGW
M6 VGX
M6 VHC
M6 VHD
M6 VHF
M6 VHG
M6 VHP
M6 VHR
M6 VHS
M6 VII
M6 VIK
M6 VIM
M6 VIN
M6 VIP
M6 VIR
M6 VIS
M6 VJB
M6 VJK
M6 VJM
M6 VJP
M6 VJS
M6 VJT
M6 VKC
M6 VKD
M6 VKL
M6 VKO
M6 VKP
M6 VKR
M6 VKV
M6 VLC
M6 VLF
M6 VLJ
M6 VLK
M6 VLL
M6 VLM
M6 VLP
M6 VMB
M6 VME
M6 VML
M6 VMM
M6 VMR
M6 VMW
M6 VMZ
M6 VNA
M6 VNB
M6 VNC
M6 VNE
M6 VNO
M6 VNR
M6 VNS
M6 VNY
M6 VOB
M6 VOE
M6 VOJ
M6 VOK
M6 VON
M6 VOO
M6 VOP
M6 VOV
M6 VPC
M6 VPN
M6 VPR
M6 VPT
M6 VPW
M6 VRA
M6 VRB
M6 VRS
M6 VRV
M6 VSG
M6 VSP
M6 VSY
M6 VTD
M6 VTF
M6 VTL
M6 VTO
M6 VTR
M6 VTV
M6 VTW
M6 VTX
M6 VTY
M6 VUB
M6 VUP
M6 VVC
M6 VVD
M6 VVK
M6 VVM
M6 VVO
M6 VVP
M6 VVV
M6 VVW
M6 VWG
M6 VWM
M6 VWY
M6 VXC
M6 VXD
M6 VXH
M6 VXL
M6 VXN
M6 VXO
M6 VXX
M6 VYD
M6 VYK
M6 VYL
M6 VYZ
M6 VZG
M6 VZM
M6 VZN
M6 VZO
M6 VZT
M6 VZY
M6 WAC
M6 WAE
M6 WAH
M6 WAK
M6 WAL
M6 WAO
M6 WAR
M6 WAU
M6 WAZ
M6 WBC
M6 WBD
M6 WBG
M6 WBL
M6 WCD
M6 WCH
M6 WCJ
M6 WCM
M6 WCO
M6 WCS
M6 WDA
M6 WDE
M6 WDJ
M6 WDK
M6 WDP
M6 WDT
M6 WDU
M6 WDV
M6 WDZ
M6 WEA
M6 WEC
M6 WEF
M6 WEG
M6 WEH
M6 WEI
M6 WEY
M6 WFD
M6 WFM
M6 WFN
M6 WFS
M6 WGA
M6 WGP
M6 WGW
M6 WGY
M6 WHB
M6 WHC
M6 WHE
M6 WHH
M6 WHK
M6 WHL
M6 WHM
M6 WHO
M6 WHQ
M6 WHR
M6 WHV
M6 WHY
M6 WIA
M6 WIQ
M6 WJD
M6 WJG
M6 WJH
M6 WJO
M6 WJR
M6 WJT
M6 WJV
M6 WJX
M6 WKA
M6 WKM
M6 WLJ
M6 WLK
M6 WLL
M6 WLM
M6 WLT
M6 WLZ
M6 WMC
M6 WMD
M6 WMI
M6 WMO
M6 WMQ
M6 WMR
M6 WMT
M6 WMU
M6 WMX
M6 WMY
M6 WNA
M6 WND
M6 WNT
M6 WNX
M6 WON
M6 WOZ
M6 WPF
M6 WPJ
M6 WPK
M6 WPL
M6 WPM
M6 WPX
M6 WPZ
M6 WQI
M6 WQP
M6 WQW
M6 WRA
M6 WRF
M6 WRI
M6 WRK
M6 WRP
M6 WRZ
M6 WSA
M6 WSD
M6 WSE
M6 WSF
M6 WSL
M6 WSN
M6 WSX
M6 WSZ
M6 WTA
M6 WTB
M6 WTH
M6 WTI
M6 WTK
M6 WTN
M6 WTO
M6 WTS
M6 WTV
M6 WTY
M6 WUF
M6 WUS
M6 WUT
M6 WUV
M6 WUZ
M6 WVD
M6 WVI
M6 WVL
M6 WVM
M6 WVU
M6 WVW
M6 WVZ
M6 WWA
M6 WWP
M6 WWW
M6 WWX
M6 WXA
M6 WXC
M6 WXG
M6 WXM
M6 WXW
M6 WXZ
M6 WYE
M6 WYK
M6 WYL
M6 WYP
M6 WYS
M6 WZA
M6 WZG
M6 WZH
M6 WZL
M6 WZP
M6 WZS
M6 WZT
M6 WZZ
M6 XAH
M6 XAJ
M6 XAN
M6 XAZ
M6 XBE
M6 XBF
M6 XBG
M6 XBH
M6 XBK
M6 XBN
M6 XBT
M6 XCB
M6 XCD
M6 XCE
M6 XCG
M6 XCM
M6 XCS
M6 XCT
M6 XCV
M6 XCX
M6 XDD
M6 XDE
M6 XDF
M6 XDH
M6 XDK
M6 XDL
M6 XDM
M6 XDN
M6 XDP
M6 XDS
M6 XDV
M6 XDY
M6 XEB
M6 XED
M6 XEG
M6 XEH
M6 XEN
M6 XEO
M6 XEQ
M6 XES
M6 XET
M6 XEW
M6 XFC
M6 XFD
M6 XFF
M6 XFK
M6 XFT
M6 XFV
M6 XFZ
M6 XGE
M6 XGK
M6 XGM
M6 XGP
M6 XGR
M6 XHA
M6 XHK
M6 XHT
M6 XHV
M6 XHX
M6 XII
M6 XIJ
M6 XIM
M6 XIO
M6 XIS
M6 XIV
M6 XJA
M6 XJB
M6 XJF
M6 XJH
M6 XJK
M6 XJT
M6 XJX
M6 XKA
M6 XKB
M6 XKD
M6 XKP
M6 XKR
M6 XKX
M6 XLA
M6 XLB
M6 XLD
M6 XLL
M6 XLM
M6 XLV
M6 XLW
M6 XLY
M6 XLZ
M6 XMD
M6 XMK
M6 XML
M6 XMM
M6 XMP
M6 XMT
M6 XNI
M6 XOC
M6 XOF
M6 XOL
M6 XOM
M6 XOO
M6 XOP
M6 XOV
M6 XOZ
M6 XPA
M6 XPE
M6 XPG
M6 XPM
M6 XPR
M6 XPS
M6 XPY
M6 XQD
M6 XQE
M6 XRA
M6 XRB
M6 XRE
M6 XRG
M6 XRH
M6 XRJ
M6 XRK
M6 XRM
M6 XRW
M6 XRY
M6 XSB
M6 XSG
M6 XSJ
M6 XSM
M6 XSP
M6 XSQ
M6 XSW
M6 XSX
M6 XSY
M6 XTC
M6 XTE
M6 XTG
M6 XTH
M6 XTN
M6 XTP
M6 XTS
M6 XTV
M6 XTW
M6 XTX
M6 XUD
M6 XUE
M6 XUK
M6 XUL
M6 XUP
M6 XVE
M6 XVN
M6 XVZ
M6 XWF
M6 XWH
M6 XWM
M6 XWP
M6 XWT
M6 XWX
M6 XWZ
M6 XXA
M6 XXE
M6 XXF
M6 XXG
M6 XXK
M6 XXO
M6 XXP
M6 XXR
M6 XXX
M6 XXY
M6 XYA
M6 XYE
M6 XYM
M6 XYO
M6 XYS
M6 XYT
M6 XYX
M6 XYZ
M6 XZA
M6 XZQ
M6 XZV
M6 XZX
M6 XZZ
M6 YAA
M6 YAB
M6 YAC
M6 YAI
M6 YAK
M6 YAL
M6 YAQ
M6 YAU
M6 YAX
M6 YAZ
M6 YBF
M6 YBG
M6 YBM
M6 YBP
M6 YBU
M6 YBW
M6 YBX
M6 YCC
M6 YCE
M6 YCL
M6 YCN
M6 YCQ
M6 YCS
M6 YCZ
M6 YDJ
M6 YDM
M6 YDS
M6 YDT
M6 YDX
M6 YDZ
M6 YEA
M6 YEC
M6 YED
M6 YEE
M6 YEM
M6 YEN
M6 YEP
M6 YER
M6 YEW
M6 YFC
M6 YFJ
M6 YFL
M6 YFS
M6 YFY
M6 YGC
M6 YGG
M6 YGH
M6 YGI
M6 YGK
M6 YGM
M6 YGW
M6 YGZ
M6 YHA
M6 YHF
M6 YHT
M6 YHZ
M6 YIF
M6 YIP
M6 YJC
M6 YJE
M6 YJM
M6 YJP
M6 YJS
M6 YKB
M6 YKK
M6 YKL
M6 YKV
M6 YKZ
M6 YLB
M6 YLF
M6 YLL
M6 YLM
M6 YLX
M6 YLZ
M6 YMC
M6 YMD
M6 YMG
M6 YMJ
M6 YMK
M6 YML
M6 YMO
M6 YMZ
M6 YNE
M6 YNF
M6 YOA
M6 YOK
M6 YOL
M6 YOM
M6 YOP
M6 YOR
M6 YOX
M6 YPC
M6 YPF
M6 YPH
M6 YPK
M6 YPL
M6 YPM
M6 YPT
M6 YQB
M6 YQY
M6 YRA
M6 YRG
M6 YRK
M6 YRM
M6 YRP
M6 YRT
M6 YRU
M6 YSL
M6 YSO
M6 YSZ
M6 YTE
M6 YTH
M6 YTK
M6 YTN
M6 YTY
M6 YUC
M6 YUR
M6 YVM
M6 YVX
M6 YWF
M6 YWK
M6 YWY
M6 YWZ
M6 YXA
M6 YXS
M6 YXX
M6 YXY
M6 YYA
M6 YYC
M6 YYP
M6 YYZ
M6 YZA
M6 YZG
M6 YZP
M6 YZV
M6 YZW
M6 YZZ
M6 ZAE
M6 ZAG
M6 ZAI
M6 ZAJ
M6 ZAM
M6 ZAO
M6 ZAP
M6 ZAT
M6 ZBC
M6 ZBF
M6 ZBI
M6 ZBK
M6 ZBO
M6 ZBP
M6 ZCC
M6 ZCJ
M6 ZDD
M6 ZDK
M6 ZDM
M6 ZDR
M6 ZDX
M6 ZEC
M6 ZED
M6 ZEY
M6 ZEZ
M6 ZFS
M6 ZFX
M6 ZGP
M6 ZHC
M6 ZIG
M6 ZIM
M6 ZIN
M6 ZIX
M6 ZJC
M6 ZJO
M6 ZJS
M6 ZKC
M6 ZKG
M6 ZKL
M6 ZKM
M6 ZKO
M6 ZKX
M6 ZKZ
M6 ZLG
M6 ZMD
M6 ZNQ
M6 ZOA
M6 ZOE
M6 ZON
M6 ZOT
M6 ZOZ
M6 ZPC
M6 ZPG
M6 ZPM
M6 ZQQ
M6 ZRA
M6 ZRB
M6 ZSC
M6 ZSR
M6 ZSU
M6 ZSY
M6 ZTE
M6 ZTT
M6 ZTZ
M6 ZUJ
M6 ZUK
M6 ZUM
M6 ZUU
M6 ZVV
M6 ZWD
M6 ZXA
M6 ZXJ
M6 ZXY
M6 ZZI
M6 ZZO
M6 ZZR
M6 ZZT
M6 ZZW
M6 ZZZ

M7

M7 AAD
M7 AAF
M7 AAG
M7 AAI
M7 AAJ
M7 AAM
M7 AAN
M7 AAO
M7 AAR
M7 AAS
M7 ABF
M7 ABH
M7 ABI
M7 ABJ
M7 ABK
M7 ABL
M7 ABO
M7 ABS
M7 ACB
M7 ACC
M7 ACF
M7 ACI
M7 ACJ
M7 ACO
M7 ACP
M7 ACQ
M7 ACR
M7 ACT
M7 ACU
M7 ACZ
M7 ADB
M7 ADD
M7 ADE
M7 ADG
M7 ADH
M7 ADJ
M7 ADK
M7 ADM
M7 ADP
M7 ADQ
M7 ADR
M7 ADV
M7 ADY
M7 AEC
M7 AEE
M7 AEG
M7 AEH
M7 AEK
M7 AEM
M7 AEO
M7 AEP
M7 AEQ
M7 AET
M7 AEV
M7 AFB
M7 AFC
M7 AFF
M7 AFG
M7 AFJ
M7 AFM
M7 AFN
M7 AFQ
M7 AFT
M7 AFV
M7 AFX
M7 AFY
M7 AGA
M7 AGC
M7 AGF
M7 AGI
M7 AGM
M7 AGP
M7 AGR
M7 AGT
M7 AGW
M7 AGZ
M7 AHB
M7 AHC
M7 AHF
M7 AHJ
M7 AHL
M7 AHO
M7 AHP
M7 AHQ
M7 AHS
M7 AHV
M7 AHW
M7 AHX
M7 AHY
M7 AHZ
M7 AIA
M7 AIB
M7 AIC
M7 AIJ
M7 AIL
M7 AIN
M7 AIO
M7 AIS
M7 AIT
M7 AIU
M7 AIV
M7 AIX
M7 AIY
M7 AJC
M7 AJE
M7 AJI
M7 AJL
M7 AJS
M7 AJV
M7 AJW
M7 AJY
M7 AKB
M7 AKD
M7 AKE
M7 AKF
M7 AKH
M7 AKI
M7 AKJ
M7 AKK
M7 AKL
M7 AKO
M7 AKR
M7 AKT
M7 AKV
M7 AKW
M7 ALA
M7 ALB
M7 ALD
M7 ALE
M7 ALG
M7 ALH
M7 ALJ
M7 ALK
M7 ALP
M7 ALQ
M7 ALU
M7 ALV
M7 ALX
M7 ALY
M7 ALZ
M7 AMD
M7 AMF
M7 AMG
M7 AMI
M7 AML
M7 AMN
M7 AMO
M7 AMU
M7 AMV
M7 AMY
M7 AMZ
M7 ANC
M7 ANF
M7 ANL
M7 ANQ
M7 ANS
M7 ANW
M7 ANZ
M7 AOA
M7 AOB
M7 AOE
M7 AOF
M7 AOG
M7 AOJ
M7 AOK
M7 AON
M7 AOP
M7 AOQ
M7 AOS
M7 AOT
M7 AOW
M7 AOX
M7 APA
M7 APD
M7 APE
M7 APG
M7 API
M7 APK
M7 APL
M7 APP
M7 APQ
M7 APR
M7 APS
M7 APT
M7 APU
M7 APX
M7 AQA
M7 AQB
M7 AQD
M7 AQE
M7 AQH
M7 AQJ
M7 AQK
M7 AQN
M7 AQT
M7 AQU
M7 AQV
M7 AQW
M7 AQX
M7 ARB
M7 ARD
M7 ARF
M7 ARG
M7 ARI
M7 ARJ
M7 ARK
M7 ARM
M7 ARR
M7 ARS
M7 ARV
M7 ARX
M7 ARY
M7 ASC
M7 ASD
M7 ASG
M7 ASI
M7 ASL
M7 ASQ
M7 ASR
M7 ASV
M7 ATK
M7 ATM
M7 ATP
M7 ATQ
M7 ATS
M7 ATU
M7 ATY
M7 ATZ
M7 AUA
M7 AUG
M7 AUJ
M7 AUK
M7 AUN
M7 AUQ
M7 AUS
M7 AUV
M7 AUW
M7 AVD
M7 AVJ
M7 AVK
M7 AVM
M7 AVU
M7 AVV
M7 AVX
M7 AVY
M7 AVZ
M7 AWA
M7 AWB
M7 AWC
M7 AWI
M7 AWP
M7 AWR
M7 AWS
M7 AWT
M7 AWU
M7 AWV
M7 AWX
M7 AWY
M7 AWZ
M7 AXG
M7 AXH
M7 AXI
M7 AXK
M7 AXL
M7 AXN
M7 AXO
M7 AXP
M7 AXQ
M7 AXR
M7 AXS
M7 AXT
M7 AXW
M7 AXZ
M7 AYA
M7 AYC
M7 AYD
M7 AYE
M7 AYH
M7 AYI
M7 AYK
M7 AYL
M7 AYM
M7 AYN
M7 AYO
M7 AYQ
M7 AYT
M7 AYX
M7 AYY
M7 AZB
M7 AZC
M7 AZD
M7 AZE
M7 AZF
M7 AZI
M7 AZL
M7 AZM
M7 AZT
M7 AZU
M7 AZX
M7 AZY
M7 AZZ
M7 BAB
M7 BAD
M7 BAH
M7 BAI
M7 BAL
M7 BAM
M7 BAP
M7 BAT
M7 BAU
M7 BAX
M7 BAZ
M7 BBA
M7 BBB
M7 BBC
M7 BBE
M7 BBF
M7 BBH
M7 BBL
M7 BBP
M7 BBQ
M7 BBR
M7 BBU
M7 BBX
M7 BCB
M7 BCC
M7 BCD
M7 BCF
M7 BCG
M7 BCJ
M7 BCK
M7 BCO
M7 BCQ
M7 BCR
M7 BCS
M7 BCT
M7 BCU
M7 BCX
M7 BDA
M7 BDD
M7 BDH
M7 BDI
M7 BDJ
M7 BDK
M7 BDK
M7 BDL
M7 BDM
M7 BDN
M7 BDR
M7 BDS
M7 BDU
M7 BDY
M7 BEA
M7 BEB
M7 BEC
M7 BED
M7 BEH
M7 BEJ
M7 BEN
M7 BEQ
M7 BER
M7 BES
M7 BEU
M7 BEW
M7 BEY
M7 BEZ
M7 BFC
M7 BFD
M7 BFE
M7 BFF
M7 BFH
M7 BFI
M7 BFM
M7 BFO
M7 BFR
M7 BFS
M7 BFT
M7 BFU
M7 BFW
M7 BGA
M7 BGC
M7 BGI
M7 BGJ
M7 BGK
M7 BGL
M7 BGM
M7 BGN
M7 BGP
M7 BGQ
M7 BGS
M7 BGT
M7 BGY
M7 BGZ
M7 BHA
M7 BHB
M7 BHD
M7 BHE
M7 BHF
M7 BHG
M7 BHH
M7 BHI
M7 BHJ
M7 BHP
M7 BHS
M7 BHV
M7 BHW
M7 BHX
M7 BHY
M7 BIA
M7 BIB
M7 BIF
M7 BII
M7 BIJ
M7 BIK
M7 BIM
M7 BIN
M7 BIR
M7 BIV
M7 BIY
M7 BIZ
M7 BJB
M7 BJD
M7 BJE
M7 BJF
M7 BJM
M7 BJO
M7 BJP
M7 BJX
M7 BJZ
M7 BKC
M7 BKD
M7 BKF
M7 BKK
M7 BKQ
M7 BKR
M7 BKV
M7 BKW
M7 BKY
M7 BLA
M7 BLB
M7 BLE
M7 BLF
M7 BLH
M7 BLK
M7 BLN
M7 BLQ
M7 BLS
M7 BLT
M7 BLV
M7 BLW
M7 BLZ
M7 BMA
M7 BMB
M7 BMI
M7 BML
M7 BMM
M7 BMN
M7 BMP
M7 BMQ
M7 BMS
M7 BMT
M7 BMV
M7 BNA
M7 BNC
M7 BNF
M7 BNM
M7 BNO
M7 BNQ
M7 BNT
M7 BNU
M7 BNX
M7 BNY
M7 BOB
M7 BOH
M7 BOI
M7 BOR
M7 BOT
M7 BOV
M7 BOX
M7 BOY
M7 BPF
M7 BPG
M7 BPK
M7 BPL
M7 BPM
M7 BPP
M7 BPR
M7 BPT
M7 BPU
M7 BPV
M7 BPY
M7 BQA
M7 BQD
M7 BQE
M7 BQN
M7 BQO
M7 BQR
M7 BQS
M7 BQT
M7 BQV
M7 BQW
M7 BRB
M7 BRD
M7 BRE
M7 BRF
M7 BRK
M7 BRL
M7 BRN
M7 BRP
M7 BRT
M7 BRU
M7 BRV
M7 BRW
M7 BSA
M7 BSB
M7 BSC
M7 BSD
M7 BSE
M7 BSG
M7 BSJ
M7 BSO
M7 BSP
M7 BSQ
M7 BSR
M7 BST
M7 BSW
M7 BSY
M7 BTA
M7 BTB
M7 BTC
M7 BTE
M7 BTI
M7 BTJ
M7 BTK
M7 BTM
M7 BTN
M7 BTP
M7 BTR
M7 BTT
M7 BUA
M7 BUC
M7 BUD
M7 BUF
M7 BUJ
M7 BUO
M7 BUP
M7 BUQ
M7 BUT
M7 BUU
M7 BUW
M7 BUX
M7 BUZ
M7 BVB
M7 BVD
M7 BVE
M7 BVG
M7 BVL
M7 BVM
M7 BVP
M7 BVR
M7 BVU
M7 BVW
M7 BVX
M7 BVY
M7 BWA
M7 BWC
M7 BWD
M7 BWF
M7 BWI
M7 BWQ
M7 BWR
M7 BWS
M7 BWU
M7 BWV
M7 BWX
M7 BWY
M7 BXB
M7 BXD
M7 BXG
M7 BXH
M7 BXI
M7 BXJ
M7 BXK
M7 BXP
M7 BXQ
M7 BXR
M7 BXS
M7 BXU
M7 BXV
M7 BXW
M7 BXX
M7 BXY
M7 BXZ
M7 BYA
M7 BYB
M7 BYG
M7 BYI
M7 BYJ
M7 BYN
M7 BYO
M7 BYP
M7 BYR
M7 BYX
M7 BYY
M7 BYZ
M7 BZD
M7 BZE
M7 BZF
M7 BZG
M7 BZH
M7 BZI
M7 BZJ
M7 BZM
M7 BZN
M7 BZQ
M7 BZS
M7 BZT
M7 BZW
M7 BZX
M7 BZY
M7 CAB
M7 CAC
M7 CAE
M7 CAF
M7 CAG
M7 CAH
M7 CAI
M7 CAM
M7 CAO
M7 CAP
M7 CAR
M7 CAS
M7 CAU
M7 CAV
M7 CAX
M7 CAY
M7 CBA
M7 CBB
M7 CBC
M7 CBD
M7 CBE
M7 CBI
M7 CBJ
M7 CBK
M7 CBM
M7 CBN
M7 CBO
M7 CBQ
M7 CBR
M7 CBS
M7 CBU
M7 CBV
M7 CBW
M7 CBZ
M7 CCA
M7 CCD
M7 CCE
M7 CCI
M7 CCN
M7 CCP
M7 CCT
M7 CCV
M7 CCX
M7 CCY
M7 CDC
M7 CDD
M7 CDE
M7 CDG
M7 CDI
M7 CDJ
M7 CDL
M7 CDM
M7 CDN
M7 CDR
M7 CDV
M7 CDY
M7 CDZ
M7 CEA
M7 CEC
M7 CED
M7 CEE
M7 CEI
M7 CEJ
M7 CEN
M7 CEQ
M7 CET
M7 CEU
M7 CEX
M7 CFA
M7 CFB
M7 CFD
M7 CFH
M7 CFI
M7 CFJ
M7 CFK
M7 CFN
M7 CFO
M7 CFP
M7 CFQ
M7 CFS
M7 CFT
M7 CFV
M7 CFY
M7 CFZ
M7 CGA
M7 CGB
M7 CGC
M7 CGE
M7 CGG
M7 CGJ
M7 CGL
M7 CGN
M7 CGP
M7 CGR
M7 CGT
M7 CGU
M7 CGV
M7 CGW
M7 CGX
M7 CGY
M7 CGZ
M7 CHK
M7 CHL
M7 CHM
M7 CHO
M7 CHQ
M7 CHT
M7 CHU
M7 CHV
M7 CHW
M7 CHX
M7 CHY
M7 CHZ
M7 CIA
M7 CIB
M7 CIC
M7 CID
M7 CIF
M7 CIG
M7 CIK
M7 CIP
M7 CIQ
M7 CIR
M7 CIV
M7 CIW
M7 CIY
M7 CJD
M7 CJF
M7 CJI
M7 CJJ
M7 CJK
M7 CJN
M7 CJO
M7 CJP
M7 CJQ
M7 CJR
M7 CJT
M7 CJW
M7 CJZ
M7 CKB
M7 CKD
M7 CKE
M7 CKH
M7 CKI
M7 CKJ

Withheld

M7 CKK
M7 CKL
M7 CKM
M7 CKO
M7 CKP
M7 CKQ
M7 CKS
M7 CKU
M7 CKZ
M7 CLA
M7 CLB
M7 CLD
M7 CLF
M7 CLG
M7 CLL
M7 CLO
M7 CLR
M7 CLS
M7 CLT
M7 CLV
M7 CLY
M7 CLZ
M7 CMA
M7 CMF
M7 CMG
M7 CMH
M7 CMK
M7 CMN
M7 CMO
M7 CMP
M7 CMQ
M7 CMR
M7 CMT
M7 CMU
M7 CMV
M7 CMX
M7 CMZ
M7 CNC
M7 CNF
M7 CNG
M7 CNH
M7 CNO
M7 CNP
M7 CNR
M7 CNS
M7 CNU
M7 CNV
M7 CNW
M7 COA
M7 COE
M7 COG
M7 COI
M7 COJ
M7 COM
M7 COO
M7 COT
M7 COZ
M7 CPB
M7 CPD
M7 CPE
M7 CPF
M7 CPK
M7 CPL
M7 CPM
M7 CPN
M7 CPP
M7 CPR
M7 CPS
M7 CPU
M7 CPV
M7 CPX
M7 CPY
M7 CQB
M7 CQC
M7 CQG
M7 CQH
M7 CQI
M7 CQN
M7 CQO
M7 CQP
M7 CQQ
M7 CQR
M7 CQT
M7 CQV
M7 CQW
M7 CRE
M7 CRH
M7 CRI
M7 CRJ
M7 CRL
M7 CRN
M7 CRO
M7 CRT
M7 CRX
M7 CSB
M7 CSD
M7 CSE
M7 CSG
M7 CSH
M7 CSJ
M7 CSQ
M7 CSS
M7 CST
M7 CSV
M7 CSY
M7 CSZ
M7 CTC
M7 CTD
M7 CTE
M7 CTH
M7 CTJ
M7 CTK
M7 CTO
M7 CTQ
M7 CTR
M7 CTU
M7 CTW
M7 CTZ
M7 CUB
M7 CUC
M7 CUE
M7 CUF

M7 CUK
M7 CUL
M7 CUT
M7 CUV
M7 CUX
M7 CUY
M7 CVA
M7 CVB
M7 CVE
M7 CVF
M7 CVG
M7 CVI
M7 CVK
M7 CVL
M7 CVN
M7 CVO
M7 CVP
M7 CVR
M7 CVS
M7 CVU
M7 CVV
M7 CVY
M7 CWC
M7 CWE
M7 CWF
M7 CWG
M7 CWH
M7 CWI
M7 CWJ
M7 CWO
M7 CWP
M7 CWQ
M7 CWR
M7 CWU
M7 CWV
M7 CWW
M7 CWY
M7 CWZ
M7 CXA
M7 CXC
M7 CXG
M7 CXH
M7 CXI
M7 CXK
M7 CXM
M7 CXN
M7 CXO
M7 CXS
M7 CXX
M7 CYB
M7 CYD
M7 CYF
M7 CYJ
M7 CYK
M7 CYM
M7 CYO
M7 CYP
M7 CYT
M7 CYW
M7 CYZ
M7 CZC
M7 CZE
M7 CZG
M7 CZH
M7 CZI
M7 CZJ
M7 CZK
M7 CZL
M7 CZN
M7 CZO
M7 CZP
M7 CZS
M7 CZT
M7 CZV
M7 CZX
M7 CZY
M7 DAA
M7 DAG
M7 DAI
M7 DAL
M7 DAO
M7 DAQ
M7 DAS
M7 DAU
M7 DAV
M7 DAW
M7 DAX
M7 DBA
M7 DBC
M7 DBE
M7 DBI
M7 DBK
M7 DBO
M7 DBQ
M7 DBS
M7 DBT
M7 DBU
M7 DBV
M7 DBW
M7 DBY
M7 DCB
M7 DCC
M7 DCE
M7 DCF
M7 DCG
M7 DCH
M7 DCI
M7 DCJ
M7 DCO
M7 DCQ
M7 DCS
M7 DCU
M7 DCW
M7 DCY
M7 DCZ
M7 DDB
M7 DDD
M7 DDG
M7 DDH
M7 DDI
M7 DDJ
M7 DDL

M7 DDO
M7 DDS
M7 DDU
M7 DDZ
M7 DEB
M7 DEC
M7 DED
M7 DEF
M7 DEG
M7 DEH
M7 DEM
M7 DEN
M7 DEO
M7 DEQ
M7 DES
M7 DEU
M7 DEV
M7 DEZ
M7 DFA
M7 DFB
M7 DFG
M7 DFH
M7 DFJ
M7 DFL
M7 DFN
M7 DFQ
M7 DFR
M7 DFS
M7 DFT
M7 DFU
M7 DFV
M7 DFW
M7 DFZ
M7 DGA
M7 DGD
M7 DGI
M7 DGJ
M7 DGM
M7 DGO
M7 DGQ
M7 DGR
M7 DGT
M7 DGV
M7 DGX
M7 DGY
M7 DHH
M7 DHI
M7 DHJ
M7 DHK
M7 DHN
M7 DHO
M7 DHQ
M7 DHR
M7 DHS
M7 DHU
M7 DHV
M7 DHW
M7 DHX
M7 DHY
M7 DIA
M7 DIB
M7 DID
M7 DIG
M7 DIH
M7 DII
M7 DIN
M7 DIR
M7 DIT
M7 DIX
M7 DIY
M7 DJF
M7 DJH
M7 DJI
M7 DJJ
M7 DJO
M7 DJR
M7 DJU
M7 DJV
M7 DJX
M7 DJZ
M7 DKD
M7 DKF
M7 DKG
M7 DKN
M7 DKO
M7 DKP
M7 DKQ
M7 DKR
M7 DKS
M7 DKT
M7 DKV
M7 DKW
M7 DKZ
M7 DLA
M7 DLC
M7 DLE
M7 DLF
M7 DLG
M7 DLH
M7 DLI
M7 DLL
M7 DLM
M7 DLN
M7 DLP
M7 DLT
M7 DLU
M7 DLW
M7 DLY
M7 DMC
M7 DME
M7 DMG
M7 DML
M7 DMM
M7 DMN
M7 DMO
M7 DMQ
M7 DMR
M7 DMT
M7 DMU
M7 DMX
M7 DMZ
M7 DNA

M7 DNB
M7 DNC
M7 DND
M7 DNF
M7 DNI
M7 DNJ
M7 DNK
M7 DNL
M7 DNP
M7 DNR
M7 DNS
M7 DNT
M7 DNU
M7 DNV
M7 DNX
M7 DNY
M7 DNZ
M7 DOA
M7 DOB
M7 DOC
M7 DOE
M7 DOF
M7 DOH
M7 DOI
M7 DOJ
M7 DOL
M7 DOM
M7 DOO
M7 DOP
M7 DOQ
M7 DOT
M7 DOU
M7 DOV
M7 DOW
M7 DOY
M7 DPA
M7 DPF
M7 DPI
M7 DPJ
M7 DPK
M7 DPM
M7 DPN
M7 DPQ
M7 DPR
M7 DPS
M7 DPV
M7 DQA
M7 DQB
M7 DQE
M7 DQH
M7 DQK
M7 DQL
M7 DQM
M7 DQN
M7 DQP
M7 DQS
M7 DQT
M7 DQX
M7 DQZ
M7 DRA
M7 DRD
M7 DRF
M7 DRI
M7 DRK
M7 DRM
M7 DRN
M7 DRO
M7 DRP
M7 DRQ
M7 DRU
M7 DRX
M7 DRZ
M7 DSB
M7 DSD
M7 DSE
M7 DSF
M7 DSG
M7 DSI
M7 DSJ
M7 DSK
M7 DSM
M7 DSN
M7 DSO
M7 DSS
M7 DST
M7 DSU
M7 DSW
M7 DSX
M7 DSY
M7 DTB
M7 DTC
M7 DTE
M7 DTJ
M7 DTK
M7 DTM
M7 DTN
M7 DTO
M7 DTP
M7 DTQ
M7 DTS
M7 DTT
M7 DTU
M7 DTV
M7 DTX
M7 DTY
M7 DUA
M7 DUE
M7 DUG
M7 DUK
M7 DUL
M7 DUN
M7 DUO
M7 DUQ
M7 DUS
M7 DUV
M7 DUW
M7 DUX
M7 DUY
M7 DUZ
M7 DVA
M7 DVC
M7 DVF

M7 DVG
M7 DVH
M7 DVI
M7 DVL
M7 DVN
M7 DVR
M7 DVU
M7 DVV
M7 DVW
M7 DVZ
M7 DWA
M7 DWB
M7 DWC
M7 DWG
M7 DWH
M7 DWI
M7 DWJ
M7 DWL
M7 DWM
M7 DWN
M7 DWO
M7 DWR
M7 DWS
M7 DWU
M7 DWX
M7 DWY
M7 DWZ
M7 DXA
M7 DXB
M7 DXD
M7 DXE
M7 DXG
M7 DXI
M7 DXK
M7 DXM
M7 DXN
M7 DXO
M7 DXR
M7 DXV
M7 DXW
M7 DXY
M7 DYC
M7 DYD
M7 DYE
M7 DYF
M7 DYJ
M7 DYK
M7 DYL
M7 DYN
M7 DYQ
M7 DYT
M7 DYV
M7 DYX
M7 DZA
M7 DZB
M7 DZD
M7 DZH
M7 DZJ
M7 DZN
M7 DZR
M7 DZX
M7 DZY
M7 DZZ
M7 EAA
M7 EAD
M7 EAF
M7 EAG
M7 EAI
M7 EAJ
M7 EAL
M7 EAQ
M7 EAT
M7 EAU
M7 EAV
M7 EAW
M7 EAX
M7 EAY
M7 EAZ
M7 EBB
M7 EBD
M7 EBE
M7 EBF
M7 EBG
M7 EBH
M7 EBK
M7 EBM
M7 EBN
M7 EBQ
M7 EBR
M7 EBS
M7 EBT
M7 EBV
M7 EBX
M7 ECA
M7 ECB
M7 ECC
M7 ECD
M7 ECE
M7 ECH
M7 ECI
M7 ECJ
M7 ECK
M7 ECL
M7 ECO
M7 ECQ
M7 ECR
M7 ECS
M7 ECU
M7 ECX
M7 EDA
M7 EDB
M7 EDC
M7 EDD
M7 EDF
M7 EDG
M7 EDH
M7 EDI
M7 EDJ
M7 EDN
M7 EDP
M7 EDQ
M7 EDT

M7 EDU
M7 EDV
M7 EDX
M7 EEA
M7 EEB
M7 EED
M7 EEF
M7 EEI
M7 EEJ
M7 EEK
M7 EEN
M7 EEP
M7 EER
M7 EES
M7 EET
M7 EEV
M7 EEX
M7 EEY
M7 EEZ
M7 EFD
M7 EFE
M7 EFF
M7 EFG
M7 EFI
M7 EFK
M7 EFL
M7 EFN
M7 EFQ
M7 EFW
M7 EFY
M7 EGC
M7 EGD
M7 EGF
M7 EGG
M7 EGH
M7 EGI
M7 EGL
M7 EGN
M7 EGO
M7 EGQ
M7 EGR
M7 EGU
M7 EGV
M7 EGX
M7 EGZ
M7 EHA
M7 EHE
M7 EHF
M7 EHH
M7 EHI
M7 EHJ
M7 EHK
M7 EHM
M7 EHN
M7 EHO
M7 EHP
M7 EHR
M7 EHS
M7 EHT
M7 EHU
M7 EHX
M7 EHY
M7 EHZ
M7 EIA
M7 EIB
M7 EIC
M7 EIE
M7 EIF
M7 EIG
M7 EIH
M7 EII
M7 EIL
M7 EIM
M7 EIN
M7 EIQ
M7 EIS
M7 EIV
M7 EIX
M7 EIY
M7 EIZ
M7 EJA
M7 EJD
M7 EJF
M7 EJG
M7 EJL
M7 EJO
M7 EJQ
M7 EJR
M7 EJT
M7 EJU
M7 EJX
M7 EJZ
M7 EKB
M7 EKC
M7 EKE
M7 EKF
M7 EKG
M7 EKH
M7 EKI
M7 EKJ
M7 EKM
M7 EKN
M7 EKO
M7 EKW
M7 EKX
M7 EKY
M7 ELA
M7 ELB
M7 ELC
M7 ELD
M7 ELF
M7 ELH
M7 ELI
M7 ELJ
M7 ELK
M7 ELM
M7 ELP
M7 ELQ
M7 ELR
M7 ELT
M7 ELV
M7 ELW

M7 ELX
M7 ELY
M7 EMA
M7 EMB
M7 EMF
M7 EMG
M7 EML
M7 EMM
M7 EMN
M7 EMP
M7 EMQ
M7 EMS
M7 EMT
M7 EMX
M7 EMZ
M7 ENA
M7 END
M7 ENE
M7 ENG
M7 ENK
M7 ENL
M7 ENM
M7 ENO
M7 ENP
M7 ENR
M7 ENS
M7 ENT
M7 ENX
M7 ENY
M7 ENZ
M7 EOC
M7 EOF
M7 EOG
M7 EOI
M7 EOJ
M7 EOM
M7 EON
M7 EOO
M7 EOP
M7 EOR
M7 EOW
M7 EOX
M7 EOY
M7 EPA
M7 EPB
M7 EPC
M7 EPE
M7 EPJ
M7 EPL
M7 EPO
M7 EPP
M7 EPQ
M7 EPR
M7 EPU
M7 EPX
M7 EQA
M7 EQB
M7 EQD
M7 EQE
M7 EQH
M7 EQJ
M7 EQK
M7 EQL
M7 EQO
M7 EQV
M7 EQW
M7 EQY
M7 ERA
M7 ERC
M7 ERD
M7 ERH
M7 ERL
M7 ERM
M7 ERN
M7 ERO
M7 ERR
M7 ERW
M7 ERX
M7 ERY
M7 ERZ
M7 ESA
M7 ESD
M7 ESL
M7 ESN
M7 ESO
M7 ESP
M7 ESU
M7 ETC
M7 ETD
M7 ETE
M7 ETR
M7 ETT
M7 EUC
M7 EUL
M7 EUN
M7 EUR
M7 EVC
M7 EVD
M7 EVE
M7 EVN
M7 EVO
M7 EVS
M7 EVX
M7 EVY
M7 EWC
M7 EWE
M7 EWK
M7 EWO
M7 EWS
M7 EWZ
M7 EXA
M7 EXC
M7 EXD
M7 EXE
M7 EXM
M7 EXO
M7 EXP
M7 EXU
M7 EXY
M7 EYE
M7 EZA
M7 EZE

M7 EZF
M7 EZH
M7 EZT
M7 EZY
M7 FAC
M7 FAE
M7 FAK
M7 FAQ
M7 FAR
M7 FAS
M7 FAT
M7 FAY
M7 FAZ
M7 FBB
M7 FBI
M7 FBT
M7 FCC
M7 FCK
M7 FCR
M7 FDK
M7 FDP
M7 FDX
M7 FED
M7 FEE
M7 FEK
M7 FEL
M7 FEN
M7 FER
M7 FES
M7 FEZ
M7 FFB
M7 FFC
M7 FFF
M7 FFM
M7 FFS
M7 FFY
M7 FGR
M7 FHA
M7 FHP
M7 FHT
M7 FIG
M7 FIH
M7 FII
M7 FIL
M7 FIR
M7 FIS
M7 FIT
M7 FIX
M7 FJA
M7 FJB
M7 FJH
M7 FKB
M7 FKS
M7 FKU
M7 FKW
M7 FLB
M7 FLC
M7 FLF
M7 FLH
M7 FLI
M7 FLM
M7 FLO
M7 FLS
M7 FLT
M7 FLW
M7 FLX
M7 FMC
M7 FMJ
M7 FML
M7 FMY
M7 FNM
M7 FNT
M7 FOA
M7 FOE
M7 FOG
M7 FOL
M7 FOR
M7 FOS
M7 FOX
M7 FOY
M7 FOZ
M7 FPI
M7 FPR
M7 FPV
M7 FRA
M7 FRB
M7 FRC
M7 FRL
M7 FRM
M7 FRO
M7 FRU
M7 FRX
M7 FRY
M7 FSA
M7 FSB
M7 FSH
M7 FSM
M7 FSS
M7 FST
M7 FSW
M7 FTA
M7 FTC
M7 FTE
M7 FTP
M7 FTW
M7 FTY
M7 FUD
M7 FUM
M7 FUQ
M7 FVE
M7 FVJ
M7 FWS
M7 FXJ
M7 FXO
M7 FXT
M7 FYA
M7 FYI
M7 FZP
M7 FZR
M7 FZS
M7 FZZ
M7 GAB

M7 GAD
M7 GAF
M7 GAG
M7 GAJ
M7 GAK
M7 GAL
M7 GAM
M7 GAO
M7 GAS
M7 GAV
M7 GAZ
M7 GBA
M7 GBB
M7 GBC
M7 GBD
M7 GBE
M7 GBI
M7 GBK
M7 GBM
M7 GBT
M7 GBY
M7 GCB
M7 GCF
M7 GCG
M7 GCK
M7 GCN
M7 GCP
M7 GDA
M7 GDB
M7 GDC
M7 GDD
M7 GDE
M7 GDH
M7 GDJ
M7 GDK
M7 GDL
M7 GDM
M7 GDO
M7 GDP
M7 GDV
M7 GDX
M7 GDZ
M7 GEA
M7 GEE
M7 GEI
M7 GEM
M7 GEV
M7 GEW
M7 GFH
M7 GFK
M7 GFM
M7 GFS
M7 GFT
M7 GGB
M7 GGC
M7 GGG
M7 GGP
M7 GGR
M7 GGX
M7 GGY
M7 GGZ
M7 GHG
M7 GHM
M7 GHQ
M7 GHR
M7 GHT
M7 GHV
M7 GIC
M7 GIG
M7 GIK
M7 GIO
M7 GIX
M7 GIZ
M7 GJB
M7 GJC
M7 GJH
M7 GJM
M7 GJR
M7 GJW
M7 GJZ
M7 GKC
M7 GKH
M7 GKK
M7 GKM
M7 GKP
M7 GKS
M7 GKU
M7 GLA
M7 GLC
M7 GLD
M7 GLG
M7 GLH
M7 GLJ
M7 GLM
M7 GLP
M7 GLR
M7 GLT
M7 GLX
M7 GLZ
M7 GMB
M7 GME
M7 GMM
M7 GMO
M7 GMT
M7 GMX
M7 GMZ
M7 GND
M7 GNG
M7 GNO
M7 GNS
M7 GNT
M7 GNU
M7 GNZ
M7 GOL
M7 GON
M7 GOR
M7 GOS
M7 GOV
M7 GPC
M7 GPE
M7 GPF

M7 GPK
M7 GPO
M7 GPR
M7 GPS
M7 GPU
M7 GPZ
M7 GRA
M7 GRB
M7 GRC
M7 GRI
M7 GRJ
M7 GRL
M7 GRN
M7 GRO
M7 GRT
M7 GRU
M7 GRW
M7 GRX
M7 GRY
M7 GRZ
M7 GSB
M7 GSC
M7 GSE
M7 GSF
M7 GSK
M7 GSL
M7 GSM
M7 GSO
M7 GSR
M7 GSW
M7 GSX
M7 GSY
M7 GTB
M7 GTF
M7 GTG
M7 GTH
M7 GTO
M7 GTP
M7 GTR
M7 GTT
M7 GTV
M7 GTY
M7 GUB
M7 GUN
M7 GUP
M7 GUR
M7 GUS
M7 GUT
M7 GUV
M7 GUY
M7 GVB
M7 GVE
M7 GVS
M7 GVZ
M7 GWA
M7 GWB
M7 GWN
M7 GWR
M7 GWS
M7 GWX
M7 GWY
M7 GXB
M7 GXL
M7 GXN
M7 GXX
M7 GYB
M7 GYH
M7 GYN
M7 GZM
M7 GZZ
M7 HAB
M7 HAC
M7 HAH
M7 HAI
M7 HAK
M7 HAN
M7 HAQ
M7 HAR
M7 HAS
M7 HAT
M7 HAX
M7 HAZ
M7 HBD
M7 HBF
M7 HBK
M7 HBL
M7 HBR
M7 HBX
M7 HBY
M7 HCA
M7 HCB
M7 HCC
M7 HCF
M7 HCG
M7 HCI
M7 HCK
M7 HCL
M7 HCN
M7 HCY
M7 HDB
M7 HDD
M7 HDG
M7 HDI
M7 HDL
M7 HDP
M7 HDZ
M7 HEC
M7 HEG
M7 HEJ
M7 HEM
M7 HEY
M7 HEZ
M7 HFE
M7 HGI
M7 HGS
M7 HHA
M7 HHC
M7 HIG
M7 HIJ
M7 HIN
M7 HIP
M7 HIT

M7 HIX
M7 HJF
M7 HJL
M7 HJT
M7 HKE
M7 HKG
M7 HKN
M7 HKR
M7 HKS
M7 HKW
M7 HLI
M7 HLL
M7 HLM
M7 HLO
M7 HLP
M7 HLS
M7 HLV
M7 HLW
M7 HLY
M7 HMF
M7 HMG
M7 HMJ
M7 HMM
M7 HNA
M7 HNM
M7 HNR
M7 HNT
M7 HOB
M7 HOE
M7 HOL
M7 HOO
M7 HOT
M7 HOW
M7 HPG
M7 HPL
M7 HPR
M7 HQA
M7 HRA
M7 HRC
M7 HRD
M7 HRE
M7 HRH
M7 HRO
M7 HRV
M7 HRX
M7 HRY
M7 HRZ
M7 HSA
M7 HSE
M7 HSK
M7 HSL
M7 HSQ
M7 HST
M7 HSW
M7 HSZ
M7 HTB
M7 HTD
M7 HTG
M7 HTM
M7 HTO
M7 HTP
M7 HTW
M7 HTY
M7 HUF
M7 HUK
M7 HUM
M7 HUN
M7 HUT
M7 HUW
M7 HUZ
M7 HVC
M7 HWA
M7 HWD
M7 HWF
M7 HWM
M7 HWY
M7 HXL
M7 HXR
M7 HXS
M7 HXX
M7 HZA
M7 IAH
M7 IAI
M7 IAN
M7 IAO
M7 IAS
M7 IBB
M7 IBC
M7 IBJ
M7 IBL
M7 IBM
M7 IBW
M7 ICA
M7 ICB
M7 ICC
M7 ICE
M7 ICJ
M7 ICM
M7 ICO
M7 ICT
M7 ICW
M7 ICX
M7 ICZ
M7 IDA
M7 IDB
M7 IDC
M7 IDE
M7 IDH
M7 IDK
M7 IDX
M7 IEG
M7 IEL
M7 IFA
M7 IFB
M7 IFS
M7 IGK
M7 IGR
M7 IGX
M7 IHB
M7 IHH
M7 IHR
M7 IHS

M7 III
M7 IJD
M7 IJF
M7 IJH
M7 IJL
M7 IJO
M7 IJV
M7 IKD
M7 IKE
M7 ILF
M7 ILI
M7 ILK
M7 ILN
M7 ILP
M7 ILS
M7 ILW
M7 IMA
M7 IMD
M7 IMF
M7 IMG
M7 IMM
M7 IMP
M7 IMR
M7 IMS
M7 IMT
M7 INF
M7 INK
M7 IOB
M7 IOM
M7 IOO
M7 IOP
M7 IOT
M7 IOU
M7 IPA
M7 IPD
M7 IPH
M7 IPO
M7 IPS
M7 IPT
M7 IPX
M7 IQQ
M7 IRE
M7 IRF
M7 IRH
M7 IRL
M7 IRN
M7 IRV
M7 ISI
M7 ISK
M7 ISO
M7 ISP
M7 ISS
M7 IST
M7 ISW
M7 ITB
M7 ITC
M7 ITG
M7 ITH
M7 ITM
M7 ITO
M7 ITS
M7 ITV
M7 ITX
M7 ITY
M7 IVL
M7 IVP
M7 IWH
M7 IXN
M7 IZD
M7 IZI
M7 IZY
M7 JAA
M7 JAF
M7 JAH
M7 JAI
M7 JAK
M7 JAL
M7 JAM
M7 JAN
M7 JAO
M7 JAU
M7 JAX
M7 JBA
M7 JBC
M7 JBD
M7 JBG
M7 JBR
M7 JBX
M7 JBZ
M7 JCF
M7 JCI
M7 JCJ
M7 JCO
M7 JCT
M7 JCU
M7 JCW
M7 JCY
M7 JDA
M7 JDD
M7 JDE
M7 JDG
M7 JDJ
M7 JDK
M7 JDN
M7 JDO
M7 JDR
M7 JDS
M7 JDT
M7 JDV
M7 JDZ
M7 JEA
M7 JEB
M7 JEC
M7 JED
M7 JEG
M7 JEK
M7 JEM
M7 JER
M7 JES
M7 JET
M7 JEY
M7 JEZ

M7 JFC
M7 JFM
M7 JFS
M7 JFT
M7 JGA
M7 JGG
M7 JGJ
M7 JGK
M7 JGL
M7 JGR
M7 JGT
M7 JHB
M7 JHC
M7 JHG
M7 JHH
M7 JHP
M7 JHS
M7 JIN
M7 JIP
M7 JIT
M7 JJA
M7 JJB
M7 JJD
M7 JJF
M7 JJH
M7 JJM
M7 JJP
M7 JJR
M7 JJS
M7 JJW
M7 JKB
M7 JKI
M7 JKK
M7 JKL
M7 JKP
M7 JKR
M7 JKS
M7 JKT
M7 JKV
M7 JKX
M7 JLA
M7 JLB
M7 JLE
M7 JLH
M7 JLO
M7 JLP
M7 JLR
M7 JLT
M7 JLY
M7 JLZ
M7 JMB
M7 JMG
M7 JMI
M7 JMK
M7 JML
M7 JMM
M7 JMN
M7 JMQ
M7 JMT
M7 JMU
M7 JMV
M7 JMY
M7 JMZ
M7 JNA
M7 JND
M7 JNL
M7 JNM
M7 JNP
M7 JNS
M7 JNT
M7 JNX
M7 JNZ
M7 JOA
M7 JOB
M7 JOG
M7 JOH
M7 JOJ
M7 JOK
M7 JOR
M7 JOT
M7 JOW
M7 JPA
M7 JPB
M7 JPD
M7 JPG
M7 JPH
M7 JPK
M7 JPL
M7 JPN
M7 JPO
M7 JPT
M7 JPW
M7 JPZ
M7 JQI
M7 JQN
M7 JRC
M7 JRD
M7 JRE
M7 JRF
M7 JRJ
M7 JRL
M7 JRO
M7 JRP
M7 JRR
M7 JRT
M7 JRV
M7 JRW
M7 JSB
M7 JSM
M7 JSN
M7 JSO
M7 JSR
M7 JTA
M7 JTB
M7 JTC
M7 JTD
M7 JTK
M7 JTL
M7 JTM
M7 JTP
M7 JTQ
M7 JTS

M7 JTT
M7 JTW
M7 JTX
M7 JUD
M7 JUG
M7 JVB
M7 JVE
M7 JVF
M7 JVL
M7 JVM
M7 JWA
M7 JWB
M7 JWN
M7 JWS
M7 JWT
M7 JWX
M7 JXC
M7 JXJ
M7 JXK
M7 JXN
M7 JXP
M7 JXS
M7 JXT
M7 JYD
M7 JYM
M7 JYN
M7 KAB
M7 KAC
M7 KAH
M7 KAI
M7 KAJ
M7 KAL
M7 KAM
M7 KAN
M7 KAO
M7 KAQ
M7 KAR
M7 KAX
M7 KAY
M7 KAZ
M7 KBE
M7 KBL
M7 KBM
M7 KBR
M7 KBX
M7 KCC
M7 KCI
M7 KCL
M7 KCN
M7 KCS
M7 KCT
M7 KCW
M7 KCX
M7 KDC
M7 KDF
M7 KDL
M7 KDM
M7 KDS
M7 KDW
M7 KDX
M7 KEB
M7 KED
M7 KEG
M7 KEK
M7 KER
M7 KES
M7 KET
M7 KEY
M7 KFH
M7 KFM
M7 KGB
M7 KGI
M7 KGR
M7 KGS
M7 KHI
M7 KHK
M7 KHT
M7 KHZ
M7 KIA
M7 KIB
M7 KIG
M7 KIL
M7 KIN
M7 KIP
M7 KIS
M7 KIZ
M7 KJA
M7 KJB
M7 KJH
M7 KJK
M7 KJO
M7 KJR
M7 KKA
M7 KKH
M7 KKJ
M7 KKP
M7 KLE
M7 KLH
M7 KLI
M7 KLL
M7 KLN
M7 KLO
M7 KLS
M7 KLU
M7 KLV
M7 KLX
M7 KLY
M7 KLZ
M7 KMA
M7 KMC
M7 KMD
M7 KMH
M7 KMK
M7 KML
M7 KMM
M7 KMR
M7 KMT
M7 KMX
M7 KMZ
M7 KNB
M7 KNJ
M7 KNS

M7 KNT
M7 KNZ
M7 KOI
M7 KOJ
M7 KOM
M7 KON
M7 KOO
M7 KOQ
M7 KOS
M7 KOT
M7 KPD
M7 KPF
M7 KPH
M7 KPK
M7 KPO
M7 KPP
M7 KPR
M7 KPT
M7 KPX
M7 KPZ
M7 KRB
M7 KRC
M7 KRI
M7 KRM
M7 KRO
M7 KRS
M7 KRT
M7 KSA
M7 KSH
M7 KSI
M7 KSK
M7 KSP
M7 KSR
M7 KSS
M7 KTC
M7 KTE
M7 KTF
M7 KTG
M7 KTH
M7 KTI
M7 KTM
M7 KTN
M7 KTU
M7 KTX
M7 KTZ
M7 KUC
M7 KUH
M7 KUL
M7 KVA
M7 KVH
M7 KVJ
M7 KVK
M7 KVN
M7 KVP
M7 KVX
M7 KWK
M7 KWM
M7 KWP
M7 KWZ
M7 KXA
M7 KXP
M7 KYN
M7 KYO
M7 KZB
M7 KZI
M7 KZN
M7 KZZ
M7 LAB
M7 LAC
M7 LAD
M7 LAF
M7 LAL
M7 LAR
M7 LAS
M7 LAV
M7 LAX
M7 LAY
M7 LAZ
M7 LBC
M7 LBK
M7 LBL
M7 LBR
M7 LBS
M7 LCC
M7 LCE
M7 LCG
M7 LCH
M7 LCM
M7 LCO
M7 LCT
M7 LCX
M7 LDA
M7 LDB
M7 LDC
M7 LDF
M7 LDG
M7 LDN
M7 LDP
M7 LDQ
M7 LDR
M7 LDT
M7 LDY
M7 LEA
M7 LEB
M7 LEC
M7 LEE
M7 LEG
M7 LEI
M7 LEK
M7 LER
M7 LES
M7 LET
M7 LFC
M7 LFO
M7 LFS
M7 LFX
M7 LGJ
M7 LGN
M7 LGO
M7 LGR
M7 LGS
M7 LGX

M7 LHG
M7 LHJ
M7 LHL
M7 LHM
M7 LHN
M7 LHO
M7 LHS
M7 LHT
M7 LIA
M7 LIG
M7 LIJ
M7 LIL
M7 LIS
M7 LIT
M7 LIZ
M7 LJB
M7 LJC
M7 LJF
M7 LJG
M7 LJN
M7 LJO
M7 LJP
M7 LJS
M7 LJW
M7 LKA
M7 LKB
M7 LKC
M7 LKD
M7 LKG
M7 LKL
M7 LKM
M7 LKR
M7 LLA
M7 LLC
M7 LLF
M7 LLG
M7 LLH
M7 LLI
M7 LLL
M7 LLT
M7 LMD
M7 LMH
M7 LML
M7 LMO
M7 LMR
M7 LMT
M7 LMW
M7 LMY
M7 LMZ
M7 LNC
M7 LNE
M7 LNR
M7 LNS
M7 LNX
M7 LNZ
M7 LOB
M7 LOD
M7 LOG
M7 LOK
M7 LOM
M7 LON
M7 LOP
M7 LOR
M7 LOS
M7 LOU
M7 LOY
M7 LPD
M7 LPE
M7 LPG
M7 LPM
M7 LPO
M7 LPS
M7 LPW
M7 LQE
M7 LRA
M7 LRD
M7 LRG
M7 LRH
M7 LRI
M7 LRN
M7 LRO
M7 LRS
M7 LRW
M7 LSC
M7 LSD
M7 LSH
M7 LSI
M7 LSK
M7 LSM
M7 LSP
M7 LSR
M7 LSW
M7 LSY
M7 LTB
M7 LTD
M7 LTG
M7 LTN
M7 LTU
M7 LTW
M7 LTX
M7 LTZ
M7 LUC
M7 LUD
M7 LUM
M7 LUX
M7 LVE
M7 LVN
M7 LVR
M7 LVS
M7 LWA
M7 LWK
M7 LWS
M7 LWT
M7 LXF
M7 LXL
M7 LXP
M7 LXX
M7 LYK
M7 LYO
M7 LZA
M7 LZB
M7 LZE

M7 LZW
M7 LZZ
M7 MAC
M7 MAF
M7 MAH
M7 MAJ
M7 MAK
M7 MAL
M7 MAO
M7 MAQ
M7 MAR
M7 MAU
M7 MAV
M7 MAX
M7 MAY
M7 MAZ
M7 MBA
M7 MBD
M7 MBF
M7 MBG
M7 MBH
M7 MBL
M7 MBM
M7 MBN
M7 MBR
M7 MBS
M7 MBW
M7 MBX
M7 MBZ
M7 MCE
M7 MCH
M7 MCL
M7 MCP
M7 MCV
M7 MCW
M7 MCX
M7 MCZ
M7 MDF
M7 MDG
M7 MDK
M7 MDL
M7 MDM
M7 MDO
M7 MDP
M7 MDS
M7 MDT
M7 MDV
M7 MDW
M7 MDX
M7 MEA
M7 MEE
M7 MEG
M7 MEH
M7 MEI
M7 MEL
M7 MEM
M7 MER
M7 MES
M7 MEW
M7 MFA
M7 MFC
M7 MFJ
M7 MFP
M7 MFS
M7 MFT
M7 MGA
M7 MGC
M7 MGE
M7 MGI
M7 MGK
M7 MGN
M7 MGO
M7 MGP
M7 MGR
M7 MGT
M7 MGV
M7 MGY
M7 MHL
M7 MHN
M7 MHT
M7 MHX
M7 MHY
M7 MHZ
M7 MIA
M7 MIB
M7 MIC
M7 MID
M7 MIF
M7 MIG
M7 MIH
M7 MIJ
M7 MIK
M7 MIL
M7 MIM
M7 MIP
M7 MIS
M7 MIW
M7 MIX
M7 MIZ
M7 MJA
M7 MJD
M7 MJE
M7 MJF
M7 MJJ
M7 MJM
M7 MJP
M7 MJS
M7 MJW
M7 MJY
M7 MJZ
M7 MKA
M7 MKE
M7 MKL
M7 MKN
M7 MKP
M7 MKR
M7 MKS
M7 MKT
M7 MKU
M7 MLA
M7 MLG
M7 MLH
M7 MLJ
M7 MLN
M7 MLO
M7 MLR
M7 MLS
M7 MLT
M7 MLW
M7 MLX
M7 MMD
M7 MMG
M7 MMH
M7 MMR
M7 MMS
M7 MMX
M7 MMY
M7 MMZ
M7 MNA
M7 MNE
M7 MNG
M7 MNK
M7 MNL
M7 MNM
M7 MNO
M7 MNP
M7 MNS
M7 MNT
M7 MNX
M7 MOA
M7 MOB
M7 MOD
M7 MOF
M7 MOG
M7 MOJ
M7 MOM
M7 MOP
M7 MOS
M7 MOV
M7 MOX
M7 MPA
M7 MPB
M7 MPC
M7 MPE
M7 MPF
M7 MPG
M7 MPH
M7 MPK
M7 MPP
M7 MPR
M7 MPT
M7 MPW
M7 MPX
M7 MQS
M7 MQT
M7 MRC
M7 MRD
M7 MRE
M7 MRF
M7 MRH
M7 MRM
M7 MRS
M7 MRU
M7 MRW
M7 MRZ
M7 MSC
M7 MSN
M7 MSP
M7 MSS
M7 MST
M7 MSY
M7 MTA
M7 MTC
M7 MTD
M7 MTF
M7 MTJ
M7 MTK
M7 MTL
M7 MTN
M7 MTO
M7 MTP
M7 MTR
M7 MTT
M7 MTX
M7 MTZ
M7 MUD
M7 MUG
M7 MUK
M7 MUM
M7 MUN
M7 MUP
M7 MUR
M7 MUX
M7 MUZ
M7 MVC
M7 MVK
M7 MVN
M7 MVP
M7 MWA
M7 MWB
M7 MWC
M7 MWH
M7 MWI
M7 MWJ
M7 MWK
M7 MWP
M7 MWR
M7 MWS
M7 MWV
M7 MWZ
M7 MXA
M7 MXC
M7 MXD
M7 MXE
M7 MXG
M7 MXJ
M7 MXP
M7 MXR
M7 MXT
M7 MXX
M7 MXZ
M7 MYD
M7 MYN
M7 MYR
M7 MYY
M7 MZB
M7 MZC
M7 MZD
M7 MZU
M7 MZX
M7 MZZ
M7 NAD
M7 NAF
M7 NAK
M7 NAP
M7 NAR
M7 NAU
M7 NAY
M7 NBB
M7 NBD
M7 NBI
M7 NBL
M7 NBQ
M7 NBS
M7 NBY
M7 NCE
M7 NCF
M7 NCG
M7 NCH
M7 NCJ
M7 NCK
M7 NCL
M7 NCT
M7 NCX
M7 NDC
M7 NDP
M7 NDR
M7 NDW
M7 NDY
M7 NDZ
M7 NEB
M7 NEC
M7 NEE
M7 NEG
M7 NEO
M7 NER
M7 NEV
M7 NFD
M7 NFG
M7 NFI
M7 NFJ
M7 NFM
M7 NFO
M7 NFT
M7 NGA
M7 NGB
M7 NGD
M7 NGF
M7 NGO
M7 NGR
M7 NGY
M7 NHC
M7 NHK
M7 NHL
M7 NHM
M7 NIA
M7 NIC
M7 NIF
M7 NII
M7 NIN
M7 NIO
M7 NIU
M7 NIX
M7 NJA
M7 NJB
M7 NJC
M7 NJH
M7 NJL
M7 NJN
M7 NJP
M7 NJS
M7 NJW
M7 NKD
M7 NKI
M7 NKJ
M7 NKO
M7 NKT
M7 NKY
M7 NLD
M7 NLM
M7 NLX
M7 NME
M7 NMR
M7 NNN
M7 NNQ
M7 NOG
M7 NOK
M7 NOM
M7 NOP
M7 NOR
M7 NOS
M7 NOT
M7 NOX
M7 NPB
M7 NPF
M7 NPN
M7 NPS
M7 NPY
M7 NQL
M7 NRD
M7 NRF
M7 NRG
M7 NSA
M7 NSD
M7 NSG
M7 NSM
M7 NSN
M7 NSR
M7 NSW
M7 NTA
M7 NTB
M7 NTS
M7 NTX
M7 NTY
M7 NUC
M7 NUG
M7 NUL
M7 NUS
M7 NUX
M7 NVA
M7 NVK
M7 NWC
M7 NWH
M7 NWO
M7 NWR
M7 NWW
M7 NXF
M7 NXT
M7 NXX
M7 NYA
M7 NYX
M7 NZZ
M7 OAH
M7 OAP
M7 OAR
M7 OAT
M7 OAX
M7 OBE
M7 OBG
M7 OBK
M7 OBN
M7 OBS
M7 OCA
M7 OCC
M7 OCD
M7 OCE
M7 OCS
M7 OCW
M7 ODB
M7 ODE
M7 ODN
M7 ODP
M7 ODR
M7 OED
M7 OEM
M7 OEW
M7 OEY
M7 OFB
M7 OGA
M7 OGD
M7 OGG
M7 OGT
M7 OGV
M7 OGY
M7 OGY
M7 OHC
M7 OHS
M7 OIL
M7 OIO
M7 OIR
M7 OJB
M7 OJC
M7 OJG
M7 OJH
M7 OJT
M7 OKA
M7 OKE
M7 OKI
M7 OKL
M7 OKW
M7 OKY
M7 OLD
M7 OLE
M7 OLG
M7 OLH
M7 OLI
M7 OLK
M7 OLL
M7 OLT
M7 OLW
M7 OLX
M7 OMC
M7 OML
M7 OMR
M7 OMT
M7 OND
M7 ONE
M7 ONG
M7 ONI
M7 ONK
M7 ONS
M7 ONX
M7 ONY
M7 OOH
M7 OOK
M7 OOL
M7 OOM
M7 OON
M7 OOO
M7 OOP
M7 OOS
M7 OOT
M7 OOZ
M7 OPC
M7 OPG
M7 OPL
M7 OPN
M7 OPO
M7 OPS
M7 ORA
M7 ORB
M7 ORD
M7 ORF
M7 ORK
M7 ORM
M7 ORS
M7 ORT
M7 ORX
M7 OSC
M7 OSI
M7 OSJ
M7 OSL
M7 OSO
M7 OSU
M7 OTA
M7 OTC
M7 OTG
M7 OTI
M7 OTJ
M7 OTP
M7 OTR
M7 OTT
M7 OUV
M7 OUZ
M7 OWC
M7 OWE
M7 OWW
M7 OXF
M7 OXG
M7 OXN
M7 OXT
M7 OXW
M7 OYD
M7 OYZ
M7 OZD
M7 OZI
M7 OZN
M7 OZO
M7 OZX
M7 OZY
M7 OZZ
M7 PAB
M7 PAD
M7 PAE
M7 PAF
M7 PAH
M7 PAK
M7 PAL
M7 PAM
M7 PAP
M7 PAS
M7 PAT
M7 PAW
M7 PAX
M7 PAZ
M7 PBA
M7 PBC
M7 PBK
M7 PBM
M7 PBP
M7 PBR
M7 PBU
M7 PBZ
M7 PCH
M7 PCK
M7 PCM
M7 PCO
M7 PCR
M7 PCT
M7 PCZ
M7 PDB
M7 PDC
M7 PDD
M7 PDF
M7 PDH
M7 PDM
M7 PDP
M7 PDY
M7 PEA
M7 PEC
M7 PED
M7 PEE
M7 PEF
M7 PEG
M7 PEK
M7 PEM
M7 PEP
M7 PER
M7 PET
M7 PEZ
M7 PFO
M7 PGB
M7 PGC
M7 PGD
M7 PGS
M7 PGV
M7 PGW
M7 PGZ
M7 PHD
M7 PHL
M7 PHM
M7 PHO
M7 PHP
M7 PHS
M7 PHW
M7 PHX
M7 PIA
M7 PIC
M7 PIE
M7 PIL
M7 PIP
M7 PIR
M7 PIT
M7 PJA
M7 PJC
M7 PJE
M7 PJG
M7 PJL
M7 PJP
M7 PJR
M7 PJW
M7 PJX
M7 PKB
M7 PKD
M7 PKE
M7 PKJ
M7 PKL
M7 PKM
M7 PKR
M7 PKS
M7 PKY
M7 PLC
M7 PLE
M7 PLF
M7 PLJ
M7 PLM
M7 PLR
M7 PLU
M7 PLX
M7 PLZ
M7 PMA
M7 PMC
M7 PMD
M7 PMG
M7 PMI
M7 PMJ
M7 PMM
M7 PMN
M7 PMO
M7 PMP
M7 PMQ
M7 PMR
M7 PMS
M7 PMV
M7 PMX
M7 PMZ
M7 PNK
M7 PNW
M7 PNX
M7 POB
M7 POD
M7 POH
M7 POM
M7 POP
M7 PPC
M7 PPK
M7 PPP
M7 PPS
M7 PPY
M7 PQR
M7 PRB
M7 PRC
M7 PRD
M7 PRJ
M7 PRK
M7 PRL
M7 PRM
M7 PRO
M7 PRP
M7 PRS
M7 PRT
M7 PRU
M7 PRX
M7 PRY
M7 PSA
M7 PSD
M7 PSH
M7 PSK
M7 PSL
M7 PSO
M7 PSP
M7 PSR
M7 PST
M7 PSW
M7 PTA
M7 PTB
M7 PTC
M7 PTE
M7 PTG
M7 PTK
M7 PTL
M7 PTM
M7 PTN
M7 PTO
M7 PTR
M7 PTS
M7 PTT
M7 PTX
M7 PTZ
M7 PUB
M7 PUD
M7 PUG
M7 PUH
M7 PUN
M7 PUR
M7 PVB
M7 PVD
M7 PVL
M7 PVM
M7 PVT
M7 PWE
M7 PWG
M7 PWJ
M7 PWM
M7 PWN
M7 PWO
M7 PWP
M7 PWR
M7 PWS
M7 PWZ
M7 PXE
M7 PXK
M7 PXP
M7 PXR
M7 PXT
M7 PXX
M7 PYB
M7 PYF
M7 PYX
M7 PZK
M7 PZT
M7 PZW
M7 RAA
M7 RAC
M7 RAG
M7 RAH
M7 RAK
M7 RAL
M7 RAM
M7 RAN
M7 RAP
M7 RAQ
M7 RAR
M7 RBA
M7 RBB
M7 RBD
M7 RBH
M7 RBJ
M7 RBN
M7 RBS
M7 RCA
M7 RCB
M7 RCD
M7 RCJ
M7 RCK
M7 RCL
M7 RCN
M7 RCO
M7 RCY
M7 RDE
M7 RDF
M7 RDG
M7 RDJ
M7 RDL
M7 RDM
M7 RDO
M7 RDP
M7 RDR
M7 RDT
M7 RDV
M7 RDW
M7 RDX
M7 RDY
M7 REA
M7 REB
M7 REC
M7 REE
M7 REG
M7 REH
M7 REJ
M7 REM
M7 REN
M7 RFC
M7 RFJ
M7 RFQ
M7 RFS
M7 RFT
M7 RFX
M7 RGA
M7 RGC
M7 RGL
M7 RGW
M7 RGX
M7 RHE
M7 RHK
M7 RHL
M7 RHO
M7 RHT
M7 RHU
M7 RIB
M7 RIC
M7 RIF
M7 RIG
M7 RIO
M7 RIP
M7 RIR
M7 RIX
M7 RJA
M7 RJE
M7 RJF
M7 RJK
M7 RJM
M7 RJO
M7 RJP
M7 RJS
M7 RJV
M7 RJW
M7 RJX
M7 RJY
M7 RKF
M7 RKG
M7 RKK
M7 RKM
M7 RKO
M7 RKR
M7 RKT
M7 RLB
M7 RLH
M7 RLK
M7 RLM
M7 RLX
M7 RLY
M7 RMA
M7 RMB
M7 RMD
M7 RME
M7 RMG
M7 RMJ
M7 RMK
M7 RMM
M7 RMR
M7 RMS
M7 RMT
M7 RMX
M7 RMY
M7 RNA
M7 RND
M7 RNF
M7 RNG
M7 RNJ
M7 RNM
M7 RNN
M7 RNR
M7 RNT
M7 RNW
M7 ROA
M7 ROB
M7 ROE
M7 ROF
M7 ROJ
M7 ROK
M7 ROL
M7 ROM
M7 ROP
M7 ROR
M7 ROS
M7 ROU
M7 ROW
M7 ROZ
M7 RPB
M7 RPC
M7 RPE
M7 RPG
M7 RPH
M7 RPJ
M7 RPM
M7 RPN
M7 RPS
M7 RPX
M7 RPZ
M7 RRA
M7 RRB
M7 RRD
M7 RRJ
M7 RRK
M7 RRL
M7 RRM
M7 RRR
M7 RRS
M7 RRY
M7 RSA
M7 RSB
M7 RSF
M7 RSG
M7 RSI
M7 RSL
M7 RSN
M7 RSR
M7 RSS
M7 RSU
M7 RSW
M7 RSX
M7 RSZ
M7 RTC
M7 RTF
M7 RTG
M7 RTH
M7 RTI
M7 RTJ
M7 RTL
M7 RTM
M7 RTN
M7 RTP
M7 RTR
M7 RTX
M7 RTZ
M7 RUB
M7 RUD
M7 RUF
M7 RUG
M7 RUM
M7 RUX
M7 RVB
M7 RVP
M7 RVR
M7 RVS
M7 RVX
M7 RWB
M7 RWC
M7 RWD
M7 RWI
M7 RWK
M7 RWM
M7 RWP
M7 RWR
M7 RWS
M7 RWX
M7 RWZ
M7 RXA
M7 RXB
M7 RXT
M7 RXX
M7 RXY
M7 RYA
M7 RYB
M7 RYE
M7 RYG
M7 RYN
M7 RZL
M7 RZZ
M7 SAA
M7 SAD
M7 SAE
M7 SAF
M7 SAG
M7 SAI
M7 SAL
M7 SAM
M7 SAN
M7 SAS
M7 SAT
M7 SAW
M7 SAX
M7 SAY
M7 SBC
M7 SBF
M7 SBG
M7 SBH
M7 SBJ
M7 SBL
M7 SBR
M7 SBS
M7 SBT
M7 SBX
M7 SCA
M7 SCE
M7 SCI
M7 SCJ
M7 SCL
M7 SCM
M7 SCU
M7 SCZ
M7 SDA
M7 SDC
M7 SDH
M7 SDK
M7 SDL
M7 SDM
M7 SDN
M7 SDO
M7 SDP
M7 SDR
M7 SDS
M7 SDT
M7 SDU
M7 SDV
M7 SDY
M7 SEB
M7 SEC
M7 SED
M7 SEE
M7 SEI
M7 SEL
M7 SEN
M7 SEO
M7 SEP
M7 SER
M7 SET
M7 SFA
M7 SFC
M7 SFK
M7 SFM
M7 SFN
M7 SFR
M7 SFS
M7 SFV
M7 SFX
M7 SGC
M7 SGE
M7 SGG
M7 SGH
M7 SGK
M7 SGL
M7 SGP
M7 SGR
M7 SGS
M7 SGW
M7 SGX
M7 SGZ
M7 SHA
M7 SHD
M7 SHE
M7 SHG
M7 SHH
M7 SHI
M7 SHL
M7 SHS
M7 SHX
M7 SIA
M7 SIE
M7 SIL
M7 SIM
M7 SIO
M7 SIP
M7 SIR
M7 SIU
M7 SIX
M7 SJA
M7 SJC
M7 SJD
M7 SJE
M7 SJI
M7 SJK
M7 SJO
M7 SJP
M7 SJQ
M7 SJR
M7 SJS
M7 SKC
M7 SKD
M7 SKK
M7 SKM
M7 SKR
M7 SKX
M7 SKY
M7 SLF
M7 SLJ
M7 SLL
M7 SLM
M7 SLS
M7 SLX
M7 SMA
M7 SMB
M7 SMD
M7 SMF
M7 SMI
M7 SMJ
M7 SMK
M7 SMM
M7 SMO
M7 SMR
M7 SMS
M7 SMT
M7 SMU
M7 SMW
M7 SMY
M7 SNA
M7 SNC
M7 SNE
M7 SNF
M7 SNG
M7 SNK
M7 SNM
M7 SNO
M7 SNT
M7 SNW
M7 SOA
M7 SOK
M7 SOL
M7 SOM
M7 SOO
M7 SOR
M7 SOU
M7 SOV
M7 SOY
M7 SPB
M7 SPD
M7 SPF
M7 SPG
M7 SPI
M7 SPK
M7 SPN
M7 SPR
M7 SPT
M7 SPV
M7 SQD
M7 SQF
M7 SQL
M7 SQN
M7 SQT
M7 SRA
M7 SRC
M7 SRD
M7 SRI
M7 SRJ
M7 SRK
M7 SRM
M7 SRO
M7 SRS
M7 SRX
M7 SRY
M7 SSA
M7 SSD
M7 SSH
M7 SSL
M7 SSM
M7 SSP
M7 SSR
M7 SST
M7 STA
M7 STB
M7 STC
M7 STD
M7 STE
M7 STG
M7 STI
M7 STJ
M7 STN
M7 STO
M7 STP
M7 STR
M7 STW
M7 STX
M7 STY
M7 STZ
M7 SUB
M7 SUC
M7 SUD
M7 SUI
M7 SUK
M7 SUV
M7 SUX
M7 SVF
M7 SVG
M7 SVH
M7 SVN
M7 SVS
M7 SVV
M7 SWF
M7 SWG
M7 SWH
M7 SWI
M7 SWJ
M7 SWL
M7 SWN
M7 SWP
M7 SXA
M7 SXB
M7 SXC
M7 SXI
M7 SXP
M7 SXY
M7 SYG
M7 SYM
M7 SYN
M7 SYR
M7 SZA
M7 SZM
M7 SZR
M7 SZY
M7 SZZ
M7 TAB
M7 TAC
M7 TAJ
M7 TAK
M7 TAL
M7 TAM
M7 TAN
M7 TAO
M7 TAP
M7 TAR
M7 TAU
M7 TAV
M7 TAX
M7 TBE
M7 TBG
M7 TBJ
M7 TBL
M7 TBN
M7 TBS
M7 TBT
M7 TBU
M7 TBW
M7 TBY
M7 TCB
M7 TCD
M7 TCG
M7 TCH
M7 TCJ
M7 TCO
M7 TCP
M7 TCX
M7 TCY
M7 TCZ
M7 TDB
M7 TDC
M7 TDD
M7 TDG
M7 TDI
M7 TDM
M7 TDN
M7 TDP
M7 TDW
M7 TDX
M7 TDZ
M7 TEA
M7 TEC
M7 TED
M7 TEE
M7 TEH
M7 TEK
M7 TEL
M7 TEM
M7 TEN
M7 TES
M7 TET
M7 TEV
M7 TEX
M7 TEZ
M7 TFF
M7 TFH
M7 TFJ
M7 TFO
M7 TGB
M7 TGI
M7 TGL
M7 TGS
M7 TGZ
M7 THC
M7 THD
M7 THE
M7 THH
M7 THK
M7 THM
M7 THS
M7 THT
M7 THW
M7 THZ
M7 TIF
M7 TIG
M7 TIK
M7 TIL
M7 TIO
M7 TIP
M7 TIR
M7 TIS
M7 TJA
M7 TJG
M7 TJH
M7 TJJ
M7 TJK
M7 TJL
M7 TJP
M7 TJR
M7 TJW
M7 TKA
M7 TKE
M7 TKF
M7 TKK
M7 TKM
M7 TKS
M7 TLB
M7 TLD
M7 TLF
M7 TLH
M7 TLK
M7 TLP
M7 TLR
M7 TLT
M7 TLW
M7 TLY
M7 TLZ
M7 TMB
M7 TME
M7 TMG
M7 TMI
M7 TMJ
M7 TMK
M7 TML
M7 TMN
M7 TMQ
M7 TMS
M7 TMT
M7 TMU
M7 TMY
M7 TNC
M7 TND
M7 TNE
M7 TNG
M7 TNI
M7 TNK
M7 TNN
M7 TNO
M7 TNQ
M7 TNS
M7 TNT
M7 TNX
M7 TNZ
M7 TOB
M7 TOC
M7 TOD
M7 TOE
M7 TOF
M7 TOL
M7 TOM
M7 TOP
M7 TOR
M7 TOZ
M7 TPA
M7 TPB
M7 TPC
M7 TPE
M7 TPG
M7 TPH
M7 TPP
M7 TPR
M7 TPS
M7 TPX
M7 TQJ
M7 TRE
M7 TRF
M7 TRI
M7 TRJ
M7 TRK
M7 TRL
M7 TRN
M7 TRO
M7 TRP
M7 TRS
M7 TRT
M7 TRU
M7 TRX
M7 TSC
M7 TSD
M7 TSE
M7 TSF
M7 TSK
M7 TSL
M7 TSM
M7 TSR
M7 TSW
M7 TSX
M7 TTC
M7 TTE
M7 TTF
M7 TTG
M7 TTI
M7 TTJ
M7 TTL
M7 TTM
M7 TTR
M7 TTW
M7 TTX
M7 TTY
M7 TTZ
M7 TUD
M7 TUG
M7 TUI
M7 TUK
M7 TUR
M7 TVB
M7 TVC
M7 TVG
M7 TVK
M7 TVM
M7 TVN
M7 TVR
M7 TVZ
M7 TWB
M7 TWE
M7 TWH
M7 TWI
M7 TWM
M7 TWP
M7 TWR
M7 TWS
M7 TWT
M7 TWX
M7 TWY
M7 TWZ
M7 TXK
M7 TXL
M7 TXP
M7 TXV
M7 TYK
M7 TYN
M7 TYO
M7 TYT
M7 TZB
M7 TZM
M7 TZN
M7 TZR
M7 TZT
M7 TZX
M7 TZZ
M7 UAE
M7 UAN
M7 UAS
M7 UAT
M7 UAV
M7 UAX
M7 UCF
M7 UCH
M7 UDP
M7 UDR
M7 UDT
M7 UDX
M7 UEI
M7 UEL
M7 UFO
M7 UHC
M7 UHF
M7 UID
M7 UKA
M7 UKC
M7 UKE
M7 ULF
M7 ULI
M7 ULS
M7 UMA
M7 UMM
M7 UMP
M7 UNE
M7 UNK
M7 UNO
M7 UNX
M7 UPH
M7 UPP
M7 URC
M7 URG
M7 URM
M7 URR
M7 URW
M7 USK
M7 USS
M7 UTB
M7 UTC
M7 UTE
M7 UTH
M7 UTO
M7 UTW
M7 UVW
M7 UWU
M7 UXA
M7 UXB
M7 UXT
M7 UZT
M7 VAA
M7 VAG
M7 VAM
M7 VAN
M7 VAP
M7 VAR
M7 VAS
M7 VAX
M7 VAZ
M7 VBC
M7 VCB
M7 VCG
M7 VCO
M7 VCR
M7 VCW
M7 VDK
M7 VEC
M7 VED
M7 VEE
M7 VER
M7 VEV
M7 VFA
M7 VFD
M7 VFS
M7 VGA
M7 VGD
M7 VGO
M7 VHF
M7 VHS
M7 VIC
M7 VID
M7 VIE
M7 VIM
M7 VIR
M7 VIS
M7 VIX
M7 VJE
M7 VJS
M7 VJW
M7 VKA
M7 VKG
M7 VKK
M7 VKV
M7 VLF
M7 VLL
M7 VLM
M7 VLT
M7 VLX
M7 VMD
M7 VMG
M7 VMX
M7 VNA
M7 VNC
M7 VND
M7 VNE
M7 VNG
M7 VNS
M7 VNU
M7 VOB
M7 VOE
M7 VOH
M7 VOL
M7 VOM
M7 VON
M7 VOR
M7 VOS
M7 VOZ
M7 VPB
M7 VPG
M7 VPH
M7 VPI
M7 VPN
M7 VPR
M7 VPT
M7 VQQ
M7 VRF
M7 VRJ
M7 VRS
M7 VRT
M7 VSI
M7 VSM
M7 VST
M7 VSW
M7 VTR
M7 VTX
M7 VUP
M7 VVR
M7 VVT
M7 VWB
M7 VWC
M7 VWV
M7 VXR
M7 VXT
M7 VXX
M7 VXZ
M7 VYZ
M7 WAG
M7 WAH
M7 WAK
M7 WAL
M7 WAN
M7 WAP
M7 WAR
M7 WAS
M7 WAT
M7 WAV
M7 WAX
M7 WAZ
M7 WBC
M7 WBD
M7 WBR
M7 WBS
M7 WBV
M7 WCB
M7 WCD
M7 WCJ
M7 WCL
M7 WCM
M7 WCS
M7 WDG
M7 WDH
M7 WDM
M7 WDS
M7 WEC
M7 WEE
M7 WEN
M7 WEO
M7 WES
M7 WEY
M7 WFC
M7 WFK
M7 WFS
M7 WFT
M7 WGA
M7 WGG
M7 WGK
M7 WGL
M7 WHB
M7 WHE
M7 WHL
M7 WHM
M7 WHN
M7 WHO
M7 WHP
M7 WHU
M7 WHW
M7 WHX
M7 WHY
M7 WIB
M7 WII
M7 WIL
M7 WIN
M7 WIO
M7 WIZ
M7 WJF
M7 WJI
M7 WJQ
M7 WJS
M7 WJT
M7 WJW
M7 WJX
M7 WKC
M7 WKM
M7 WKN
M7 WKO
M7 WKP
M7 WLA
M7 WLC
M7 WLE
M7 WLG
M7 WLK
M7 WLM
M7 WLS
M7 WMA
M7 WMC
M7 WMD
M7 WMR
M7 WMT
M7 WMZ
M7 WND
M7 WOA
M7 WOB
M7 WOD
M7 WOF
M7 WOM
M7 WON
M7 WOO
M7 WOT
M7 WOW
M7 WPA
M7 WPD
M7 WPJ
M7 WPS
M7 WPT
M7 WPW
M7 WQE
M7 WQW
M7 WRC
M7 WRH
M7 WRJ
M7 WRL
M7 WRS
M7 WRT
M7 WRX
M7 WRY
M7 WSE
M7 WSL
M7 WSN
M7 WSO
M7 WSR
M7 WSZ
M7 WTC
M7 WTF
M7 WTH
M7 WTM
M7 WTN
M7 WTP
M7 WTR
M7 WTT
M7 WUB
M7 WUD
M7 WUF
M7 WVF
M7 WWB
M7 WWC
M7 WWD
M7 WWI
M7 WWW
M7 WWZ
M7 WXM
M7 WXW
M7 WXY
M7 WXZ
M7 WYA
M7 WYB
M7 WYC
M7 WYE
M7 WYO
M7 WZA
M7 WZD
M7 WZE
M7 WZF
M7 WZL
M7 WZT
M7 XAB
M7 XAD
M7 XAJ
M7 XAK
M7 XAL
M7 XAR
M7 XAT
M7 XAV
M7 XAX
M7 XAZ
M7 XBD
M7 XBP
M7 XBR
M7 XBT
M7 XBW
M7 XCD
M7 XCG
M7 XCH
M7 XCQ
M7 XCR
M7 XCT
M7 XCV
M7 XCX
M7 XCZ
M7 XDB
M7 XDC
M7 XDD
M7 XDG
M7 XDH
M7 XDL
M7 XDM
M7 XDR
M7 XDV
M7 XDX
M7 XDZ
M7 XEA
M7 XED
M7 XEH
M7 XEL
M7 XEM
M7 XEN
M7 XER
M7 XFH
M7 XFM
M7 XFT
M7 XFX
M7 XGA
M7 XGB
M7 XGP
M7 XGR
M7 XGW
M7 XHL
M7 XIM
M7 XIN
M7 XIO
M7 XIV
M7 XJB
M7 XJD
M7 XJI
M7 XJM
M7 XJR
M7 XJT
M7 XJX
M7 XKL
M7 XKP
M7 XKX
M7 XLT
M7 XLV
M7 XLW
M7 XLY
M7 XMR
M7 XMS
M7 XMX
M7 XNC
M7 XNW
M7 XOC
M7 XOF
M7 XOR
M7 XOX
M7 XOZ
M7 XPC
M7 XPP
M7 XPS
M7 XPZ
M7 XQU
M7 XRC
M7 XRF
M7 XRM
M7 XRP
M7 XRT
M7 XRV
M7 XRY
M7 XSD
M7 XSM
M7 XSS
M7 XSW
M7 XTC
M7 XTF
M7 XTL
M7 XTN
M7 XTR
M7 XTS
M7 XTT
M7 XTZ
M7 XUK
M7 XVF
M7 XVK
M7 XVT
M7 XVZ
M7 XWK
M7 XWL
M7 XWT
M7 XWZ
M7 XXD
M7 XXL
M7 XXT
M7 XYG
M7 XYL
M7 XYM
M7 XZT
M7 XZX
M7 YAK
M7 YAP
M7 YAR
M7 YAS
M7 YAT
M7 YAY
M7 YBA
M7 YBE
M7 YBJ
M7 YBN
M7 YBS
M7 YCB
M7 YDC
M7 YDT
M7 YDU
M7 YED
M7 YEP
M7 YES
M7 YEY
M7 YFR
M7 YFY
M7 YGG
M7 YGN
M7 YGX
M7 YHI
M7 YIO
M7 YIY
M7 YKL
M7 YKS
M7 YKX
M7 YLE
M7 YMT
M7 YOG
M7 YOT
M7 YOU
M7 YPE
M7 YRC
M7 YRK
M7 YRU
M7 YSC
M7 YSL
M7 YST
M7 YSX
M7 YTH
M7 YTI
M7 YTT
M7 YTX
M7 YUK
M7 YUM
M7 YUU
M7 YVZ
M7 YWB
M7 YWH
M7 YXY
M7 YYX
M7 YYY
M7 YZC
M7 YZV
M7 YZX

Special Contest Calls

G0A	GW4SKA
G0B	M0BUL
G0C	G0CER
G0H	G0HEU
G0V	GM0OQV
G0W	G0VDZ
G0Z	G1RVD
G1A	M0NKR
G1B	G1PPA
G1C	GM1BSG
G1D	M0AQM
G1E	G1TPA
G1G	G4KIV
G1J	MM0BQI
G1K	M0RTQ
G1M	G0PZE
G1N	G0URR
G1P	M0IEP
G1V	G4CZP
G1W	M0HAO
G1X	GM0HBF
G2A	GJ3IT
G2B	G8VVY
G2C	M0VCB
G2D	M0HRF
G2E	M0ORD
G2G	MM0DFV
G2K	M0ICR
G2L	G3SVJ
G2M	G0HDB
G2N	G4ARN
G2O	G0UVX
G2P	G0JCC
G2R	GW4BVE
G2T	MM0CPS
G2U	G0UGO
G2V	GM3WOJ
G2W	G4DBW
G2X	G0DCK
G2Y	MM0DXH
G2Z	G8JYV
G3A	MM0JOM
G3B	G2DPQ
G3C	GM0WED
G3D	G4PDS
G3F	G4AFF
G3G	G3XLG
G3J	G4RMV
G3K	G3SVK
G3L	G3LHJ
G3M	G5LK
G3N	G3OTK
G3P	G3WPH
G3Q	G3RXQ
G3R	G3CKR
G3S	GM3SEK
G3T	G3VGZ
G3U	G3UJE
G3V	G3VER
G3W	GM4RIV
G3X	GM3POI
G3Y	G3YBY
G3Z	G3ZME
G4A	G4TSH
G4C	G1FCW
G4D	GW4OKT
G4E	G4ENZ
G4G	M0RRG
G4H	GI4JTF
G4J	GW0ETF
G4L	G4LDL
G4M	GM4UBJ
G4N	G4ZVB
G4O	GM0IIO
G4P	G3YPP
G4Q	G3PRI
G4R	G4RFR
G4S	G3TXF
G4T	MI0SMK
G4U	G4SGX
G4V	MM0CCC
G4W	GW4EVX
G4X	GM4WZG
G4Y	G0CCT
G4Z	GM4ZUK
G5A	GM8VL
G5B	G4SIV
G5C	G4OGB
G5D	M0TTG
G5F	GD4RFZ
G5G	MM0TGH
G5H	G8TRF
G5I	GI4DOH
G5K	G0BNR
G5L	GW4ZAR
G5M	MS0TJT
G5N	G0NWM
G5O	G6UQ
G5P	GW0EGH
G5Q	G3SVL
G5R	GW3YDX
G5T	G3XSD
G5U	G3RXP
G5V	G3KAF
G5W	G3BJ
G5X	GM4YXI
G5Z	G3HRH
G6A	G3VDB
G6C	M0ITR
G6K	G0EAK
G6M	G4BYG
G6N	G0GDU
G6T	G4MKP
G6X	M0KLO
G7A	GM0ADX
G7C	M5ARC
G7G	M0XAR
G7H	G7SYW
G7L	G7LRQ
G7M	M0SDC
G7N	GW4MVA
G7R	GM0NAI
G7T	M0VSE
G7V	GM2MP
G7W	GC0VPR
G7X	G0MCV
G8A	G3XSV
G8B	G3UBX
G8C	M0WLY
G8D	G3SJJ
G8K	GW4BRS
G8L	G8LZE
G8N	G8LED
G8P	G4LIP
G8S	G4IDF
G8T	M0BAA
G8W	M0VCT
G8X	G4FJK
G9A	GM4FDM
G9C	MM0GHM
G9D	G6NHU
G9F	G4BVY
G9J	GW0GEI
G9P	M0NCG
G9T	GW4WXM
G9V	M0VSQ
G9W	M0DXR
G9X	M1LCR
G9Y	M0YHC
G9Z	GW1YQM
M0A	G8APB
M0B	G4KZD
M0C	MM0CEZ
M0H	MI0KOA
M0I	MI0RRE
M0K	G8FMC
M0M	MN0NID
M0N	M0NVS
M0P	M0RYB
M0T	GM3WUX
M0X	M0RTI
M1A	MI0ULK
M1B	G1YBB
M1C	G1XOW
M1E	MM0GOR
M1F	M0NVK
M1G	G0UWS
M1L	G0LGS
M1M	MI0LLG
M1N	M1DST
M1R	MM0ZBH
M1T	M0KYB
M1U	M0UTD
M1V	M1VPN
M1W	M0ICK
M1X	G0CKP
M2A	G3SDC
M2C	MD0CCE
M2D	G4NVR
M2E	G0RPM
M2F	G0OOG
M2G	G4RCG
M2I	G0FRE
M2J	G4NBS
M2K	GU3HFN
M2L	M0BJL
M2N	MM0GPZ
M2P	G1ZJP
M2R	GM0GNK
M2S	G0MGM
M2T	GM0LIR
M2U	M0DHP
M2W	G0MIN
M2Y	MW0YVK
M3A	M0UKR
M3C	G4GFI
M3D	G3XTT
M3E	G4CWH
M3F	G3WZD
M3M	G3PLE
M3N	GM4SID
M3P	G3PIA
M3R	G3RTU
M3S	G0MFR
M3T	MM0LCG
M3W	M0HDG
M3X	M0IHT
M4A	G6UW
M4C	G0FCT
M4D	G8DYT
M4I	GI4SJQ
M4J	G0DVJ
M4K	GD0EMG
M4M	M0PNN
M4N	G4IZZ
M4R	GD4XUM
M4T	M0BEW
M4U	G0RGH
M4W	G8SRC
M4X	G3SZU
M4Y	M5ESE
M5A	G0AAA
M5B	M0ZIP
M5C	G3CO
M5D	G4WQI
M5E	GU4YOX
M5G	M0ROA
M5I	GI0RQK
M5K	MI0SLE
M5M	G4BRK
M5N	G0GJV
M5O	G3LET
M5P	M5BIR
M5R	MW0EDX
M5S	G4IRN
M5T	M0RIU
M5W	M0HMJ
M5X	G3RLE
M5Y	G3UES
M5Z	M0CFW
M6A	G4KZY
M6C	M0HFC
M6M	GW4BVJ
M6N	M0NPK
M6O	G3WGN
M6T	G4MRS
M6W	G3WW
M7A	G4HVC
M7C	M5RIC
M7K	M0SDV
M7N	G3RWF
M7O	M0VKY
M7Q	G4PIQ
M7R	G0TPH
M7T	G3YYD
M7V	M0VAA
M7W	G3TBK
M7X	G0TSM
M7Z	G4BWP
M8A	M0HDF
M8C	G3RCV
M8K	M0HMO
M8M	G0JJG
M8P	MW0OMB
M8R	GW4SHF
M8T	MM0CWJ
M8Z	GM7VSB
M9A	G3ZVW
M9B	M0LKW
M9I	GM0OPS
M9K	M0SIY
M9N	G7WHI
M9T	M0MCG
M9W	GW0KRL
M9X	M0PGX

Permanent Special Event Callsigns

GB0MWM	47 Oakfield Road, Kidderminster,United Kingdom, DY11 6PL
GB0SMA	STOW MARIES AERODROME, Hackmans Lane, Purleigh, Chelmsford,United Kingdom, CM3 6RN
GB100RSM	Blandford Camp, Blandford, Dorset,United Kingdom, DT11 8RH
GB1900HA	14 Captain's Wharf, South Shields, Tyne and Wear,United Kingdom, NE33 1JQ
GB1900HW	27 Bishopton Way, Hexham, Northumberland,United Kingdom, NE46 2LR
GB1NHS	52 Manor Rise, Lichfield,United Kingdom, WS14 9RF
GB1PBL	Adjacent Portland Bill Lighthouse, Portland Bill Road,United Kingdom, DT5 2JT
GB20WM	Kiln Corner, Kirkwall, Orkney,United Kingdom, KW15 1LB
GB2CWP	East Kirkby Airfield,, Spilsby , Lincolnshire,United Kingdom, PE23 4DE
GB2EVR	Warcop Station, Appleby, Cunbria,UK, DL16 6PR
GB2GM	The Marconi Centre, Poldhu, Cornwall,United Kingdom, TR12 7JB
GB2HAM	Harrington Aviation Museum, Sunnyvale Farm, Lamport Road, Harrington,United Kingdom, NN6 9PF
GB2MOP	Museum of Power, Castell Pridd, Tan-y-Groes, Ceredigion,,United Kingdom, SA43 2JS
GB2SJ	Souter Lighthouse, Coast Road,, Sunderland, Tyne and Wear,United Kingdom, SR6 7NH
GB2SPY	Rosemount, Church Whitfield, Dover,United Kingdom, CT16 3HZ
GB4GCT	Greenham Common Control Tower, Burys Bank Road, Thatcham,United Kingdom, RG19 8BZ
GB4HRH	Osyerber Farm, Low Bentham, Lancaster,United Kingdom, LA2 7ET
GB4LD	The Old Marconi Hut, Lloyds Lane, Lizard, Cornwall,United Kingdom, TR12 7AP
GB4SMH	Signals Museum Building 104 RAF Henlow, Bedfordshire,United Kingdom, SG16 6DN
GB4UAS	Maze Long Kesh, 94-b Halftown Road, Lisburn,United Kingdom, BT27 5RF
GB6GEO	1 Clavering Court, Lincombe Drive, Torquay,United Kingdom, TQ1 2HH
GB6SL	12, Tramayne Avenue, Brough,United Kingdom, HU15 1BL

Postcode Index

AL
(St Albans)

B
(Birmingham)

BA
(Bath)

BB
(Blackburn)

BD
(Bradford)

BH
(Bournemouth

BL
(Bolton)

BN
(Brighton)

BR
(Bromley)

BS
(Bristol)

BT
(Belfast)

CA
(Carlisle)

CB
(Cambridge)

CF
(Cardiff)

CH
(Chester)

CM
(Chelmsford)

CO
(Colchester)

CR
(Croydon)

CT
(Canterbury)

CV
(Coventry)

CW
(Crewe)

D
(Dartford)

DD
(Dundee)

DE
(Derby)

DG
(Dumfries)

DH
(Durham)

DL
(Darlington)

DN
(Doncaster)

DT
(Dorset)

DY
(Dudley)

E
(East London)

EC
(East Central London)

EH
(Edinburgh)

EN
(Enfield)

EX
(Exeter)

FK
(Falkirk)

FY
(Fylde)

G
(Glasgow)

GL
(Gloucester)

GU
(Guildford)

GY
(Guernsey)

HA
(Harrow)

HD
(Huddersfield)

HG (Harrogate)

HR
(Hereford)

HS
(Scottish Islands)

HU
(Hull)

HX
(Halifax)

IG
(Ilford)

IM
(Isle of Man)

IP
(Ipswich)

IV
(Inverness)

JE
(Jersey)

KA
(Kilmarnock)

KT
(Kingston Upon Thames)

KW
(Kirkwall)

KY
(Kirkcaldy)

L
(Liverpool)

LA
(Lancaster)

LD
(Llandrindod)

LE
(Leicester)

LL
(Llandudno)

LN
(Lincoln)

LS
(Leeds)

LU
(Luton)

M
(Manchester)

ME
(Medway)

MK (Milton Keynes)

ML
(Motherwell)

N
(North London)

NE
(Newcastle Upon Tyne)

NG
(Nottingham)

NN
(Northampton)

NP
(Newport)

NR
(Norwich)

NW
(North West London)

OL
(Oldham)

OX
(Oxford)

PA
(Paisley)

PE
(Peterborough)

PH
(Perth)

PL
(Plymouth)

PO
(Portsmouth)

PR
(Preston)

RG
(Reading)

RH
(Redhill)

RM
(Romford)

S
(Sheffield)

SA
(Swansea)

SE
(South East London)

SG
(Stevenage)

SK
(Stockport)

SL
(Slough)

SM
(Sutton)

SN
(Swindon)

SO
(Southampton)

SP
(Salisbury)

SR
(Sunderland)

SS
(Southend on Sea)

ST
(Stoke on Trent)

SW
(South West London)

SY
(Shrewsbury)

TA
(Taunton)

TD
(Tweed)

TF
(Telford)

TN
(Tonbridge)

TQ
(Torquay)

TR
(Truro)

TS
(Teeside)

TW
(Twickenham)

UB
(Uxbridge)

W
(West London)

WA
(Warrington)

WC
(West Central London)

WD
(Watford)

WF
(Wakefield)

WN
(Wigan)

WR
(Worcester)

WS
(Walsall)

WV
(Wolver-hampton)

YO
(York)

ZE
(Zetland)

AB10 6ED 2M0EMM
AB10 6ED MM0MUN
AB10 6HP GM4BAP
AB10 6JD GM4HTU
AB10 6JU GM4VRE
AB10 6QH GM0NRT
AB10 6RA GM1MCN
AB10 6SB GM6GFQ
AB10 7BS MM3OMI
AB10 7JE GM3KJE
AB11 6XL MM0GAL
AB11 7WD 2M0RND
AB11 7WD MM0ROV
AB11 7WD MM3MVY
AB11 7WD MM3YBG
AB11 7WF GM7VZV
AB11 7WF MM0AOF
AB11 8FX 2M0AKS
AB11 8FX MM0GDG
AB11 8FX MM3JJC
AB12 3DZ 2M1IIW
AB12 3JJ MM1ABA
AB12 3NG GM1THS
AB12 3PH MM6MOY
AB12 3RL GM3NUU
AB12 3RN GM8FFX
AB12 3SH GM7BYB
AB12 4NS MM7SLY
AB12 4NY GM4RAZ
AB12 4NY GM4RGS
AB12 4QA MM0AOY
AB12 4QX 2M0MJY
AB12 4QX MM0MJY
AB12 4QX MM6MJY
AB12 4TF GM3ZEU
AB12 4XF MM7BVN
AB12 4XL GM8GDN
AB12 4XT 2M0DWC
AB12 4XT MM3YXN
AB12 4XT MM3ZGK
AB12 5AB GM0DBW
AB12 5AQ 2M0FPI
AB12 5AQ MM6ORU
AB12 5BY MM7SKE
AB12 5DQ GM4UWN
AB12 5LJ MM0DBR
AB12 5QT 2M1HRS
AB12 5XT GM0VFY
AB13 0EF GM7WGM
AB13 0ER GM0OPX
AB13 0JB GM3TLA
AB14 0AB GM3CIO
AB14 0AB GM4EMX
AB14 0LN GM0FRT
AB14 0LN GM4ZUK
AB14 0NX GM4TEF
AB14 0TU GM0CQV
AB15 4BP GM1MYF
AB15 4UE GM1KBZ
AB15 4UH GM8ADK
AB15 6AN GM3OUU
AB15 6BH 2M1HTR
AB15 6DU 2M0GEJ
AB15 6DU MM0TGG
AB15 6DU MM6XGJ
AB15 7QA GM4EKC
AB15 7SX GM3VEY
AB15 7XP GM7RNJ
AB15 8LQ GM3WIJ
AB15 8LQ GM4GTV
AB15 8LT GM3HGA
AB15 8LX GM4FBP
AB15 8SF GM3BSQ
AB15 8SF GM4GVK
AB15 8UE 2M0JUH
AB15 8UE MM6NLV
AB15 8UH MM7EEQ
AB15 9AR GM4NVI
AB15 9BU 2M0RVM
AB15 9DE MM7HCM
AB15 9JX GM4IXH
AB15 9PL GM8MFZ
AB15 9QF GM3VAP
AB15 9RE MM6IAI
AB15 9RH GM4EGX
AB16 5QG MM3RBJ
AB16 5RP 2M0RYL
AB16 5RP 2M0SYL
AB16 5RP MM0BCR
AB16 5RR MM6AIK
AB16 5SB MM3KQI
AB16 5SN 2M0NSA
AB16 5SN 2M0STB
AB16 5SN MM0ROR
AB16 5SN MM6MVQ
AB16 5UY MM0HZO
AB16 5YR MM6TWS
AB16 6FN GM6JBF
AB16 6FU GM0MYQ
AB16 6NX GM7KBK
AB16 6PD GM8MHU
AB16 6XY MM6CIA
AB16 7DQ 2M0FYG
AB16 7PQ MM7DDV
AB16 7SL 2M0PKA
AB16 7YR MM0KSS
AB21 0JZ GM4NNK
AB21 0LH MM0ALY
AB21 0NG GM8BZP
AB21 0NS MM6AGL
AB21 0QF 2M0MRO
AB21 0RJ GM8GHV
AB21 0RS GM1BNP
AB21 0TW GM0IPV
AB21 0TW GM4SID
AB21 0TY MM1FDF
AB21 0TZ GM6GMZ
AB21 0YQ MM7RVC
AB21 7DD MM6HRF
AB21 7FN GM3WSR
AB21 9AY MM7BXL
AB21 9HJ MM0CZH
AB21 9HS GM0PTY
AB21 9JJ 2M0NIA
AB21 9JJ MM0CUG
AB21 9JJ MM6NIA
AB21 9PQ MM0DOT
AB21 9QS 2M0BPV
AB21 9QS MM3WMQ
AB21 9QX GM7NHS
AB21 9RH GM0SZA
AB214NR MM3SWW
AB22 8LJ GM7MWL
AB22 8RW GM7MMI
AB22 8TP GM1AUZ
AB22 8WF 2M0HIB
AB22 8WF MM3ZRZ
AB22 8WG GM0MCJ
AB22 8WY GM0JOV
AB22 8XB GM6MJY
AB22 8YJ 2M0OQR
AB22 8YJ MM0OQR
AB22 8YJ MM3OQR
AB22 8ZB MM5ISS
AB22 8ZH 2M0DHI
AB22 8ZH MM6DHI
AB22 9AE MM7RCW
AB23 8BD GM8BNH
AB23 8EH GM1FSU
AB23 8PL GM4PMT
AB23 8PW 2M0RMH
AB23 8PW MM3NXY
AB23 8QD GM4YWV
AB23 8QN GM1LKD
AB23 8UT 2M1ENI
AB23 8XY GM7SPB
AB23 8YG MM6JPM
AB24 1WT GM4BFX
AB24 1XX 2M0OKO
AB24 1XX MM7BMK
AB24 1ZS MM6OYJ
AB24 2RP GM0AUL
AB24 2XS MM6BNN
AB24 2XS MM6BNO
AB24 3HQ MM7FUR
AB24 3JL GM8BSU
AB24 4EQ MM7HXI
AB24 4ER MM7MCT
AB24 4GA MM7GAC
AB24 4HX GM4CAU
AB24 4NH GM4TVB
AB24 4NJ GM6SDV
AB24 4NT GM7RXL
AB24 5BE MM0TJR
AB24 5EP 2M0URP
AB24 5EP GM6URP
AB24 5EP MM6URP
AB25 1DQ GM4HQF
AB25 1LR MM6EUA
AB25 1LW MM5SHA
AB25 2JX 2M1CUS
AB25 2YF MM0DBF
AB25 3XX MM6CJY
AB3 1NX GM7IHR
AB30 1FR MM7BJA
AB30 1QT MM7DNE
AB30 1TT MM6TSN
AB30 1YS MM6LVV
AB31 4AF GM4JLZ
AB31 4EN GM6EUC
AB31 4HG GM0TCU
AB31 4JU MM0IWS
AB31 4LS GM8GJI
AB31 4NB GM6UHC
AB31 4PJ 2M0OYS
AB31 4PJ MM7YRN
AB31 4QA MM3XAF
AB31 4QL 2M0BZL
AB31 4QL MM0XPT
AB31 4QL MM6CRW
AB31 4RY GM0SXQ
AB31 5AU GM4MFB
AB31 5HA MM0ACR
AB31 5HA MM0ACT
AB31 5TS MM6YQP
AB31 5UY 2M0DJT
AB31 5XA GM0PKQ
AB31 5XW MM0DRT
AB31 5YG 2M0CXI
AB31 5YG MM0HVW
AB31 5ZF GM7BDD
AB31 6BL GM0AZV
AB31 6DT GM0RAO
AB31 6NL MM0IYD
AB31 6NL MM7AAE
AB32 6RD 2M0FTI
AB32 6RD MM7YEA
AB32 6TH 2M0YMA
AB32 6WS GM8BSQ
AB32 6WW GM1XEA
AB32 6XH MM0JOM
AB32 6XY GM0GAT
AB32 6YE MM3HLG
AB32 7EQ GM4NHI
AB33 8BH GM8ZFW
AB33 8ER 2M0PTE
AB33 8ER MM3UDI
AB33 8HQ GM0MHD
AB33 8HQ MM1MHD
AB33 8NX MM6MYK
AB33 8QA MM6ELU
AB33 8QQ MM6IJP
AB33 8QW MM5CFA
AB33 8RN 2M0MVX
AB33 8RN MM0MVX
AB33 8RN MM6MVX
AB33 8SH MM1DZW
AB33 8SH MM3RGH
AB33 8UB GM7IEU
AB34 4TA GM8AT
AB34 4UH MM3VNU
AB34 5JF GM4JXP
AB34 5JZ GM0KDP
AB34 5PQ 2M0EFI
AB34 5PQ MM0EFI
AB35 5QH GM4RLV
AB35 5SF GM8FVN
AB35 5SH MM6HDZ
AB35 5UT GM0DZW
AB35 5ZQ MM6CHN
AB35 5ZQ MM7VKR
AB36 8UJ 2M0COT
AB36 8UJ GM3ALZ
AB36 8UJ MM0ODL
AB36 8UJ MM6ALZ
AB36 8XA MM7WGB
AB37 9BG MM6IHQ
AB37 9ET MM3WFU
AB37 9HW MM3WQO
AB38 7BD MM3YDK
AB38 7QZ GM0OSJ
AB38 9NU GM1TGY
AB38 9SQ MM3XMT
AB389LA MM7BQF
AB39 2AD GM1JPJ
AB39 2BZ 2M0DRY
AB39 2BZ MM6BRV
AB39 2EG GM4PVQ
AB39 2GF GM4HVS
AB39 2GQ 2M0XMQ
AB39 2GQ MM0XMQ
AB39 2GQ MM6XMQ
AB39 2JA MM0CFE
AB39 2LU GM4VQY
AB39 2PL 2M0YZT
AB39 2PL MM6MLT
AB39 2UA GM1JHM
AB39 3PF GM4KOI
AB39 3PG GM6OSZ
AB39 3QG MM6TFY
AB39 3SY MM0ALM
AB39 3UL 2M0MGM
AB39 3UL MM3RLG
AB39 3XW MM0GQF
AB4 5SE GM1VSR
AB41 6BJ GM4MBG
AB41 6RT MM6JAN
AB41 7DF GM4LTL
AB41 7DH GM7LDU
AB41 7DU GM7KRQ
AB41 7HS GM4IBI
AB41 7JY GM1WKH
AB41 7NZ MM3OZW
AB41 7PH MM6DXJ
AB41 8BA GM0PKF
AB41 8BH GM4FVS
AB41 8QS GM3UAG
AB41 8QW GM4GNR
AB41 8TF MM5AJN
AB41 8UJ GM7LAC
AB41 8UJ MM6BOS
AB41 8UW 2M0DDS
AB41 8UW MM0TDS
AB41 8UW MM3DDS
AB41 8YH GM2MP
AB41 8YH GM4YXI
AB41 9DB 2M0IAQ
AB41 9DB MM7BIW
AB41 9EU MM6WHI
AB41 9HF GM0JEF
AB41 9JB GM6AOR
AB41 9LW GM4OBD
AB41 9NF GM0MIS
AB41 9TF 2M0IUE
AB41 9TF MM6IUE
AB417HQ MM6NOF
AB42 0LX GM5JDG
AB42 0NG GM7LJE
AB42 0NY GM1TBW
AB42 0PP GM4PXB
AB42 0PZ MM3TAV
AB42 0RE MM1BMK
AB42 0RE MM6BOQ
AB42 0TB GM8AGM
AB42 0TQ MM6FYR
AB42 1HB MM0CJF
AB42 1HL GM1JNS
AB42 1NX GM4UFD
AB42 1RD GM6KAM
AB42 2UF GM7FYB
AB42 3AT MM3PDM
AB42 3AY GM1GCB
AB42 3DD 2M0CBE
AB42 3DD MM6CBE
AB42 3DN 2M0BVN
AB42 3DN MM0IHE
AB42 3DN MM3XIZ
AB42 3LA MM6SXY
AB42 3PX MM6OEI
AB42 4HA MM6ZAZ
AB42 4HX GM7NNS
AB42 4JU GM1XIN
AB42 4JU GM6FDQ
AB42 4LH MM0MPW
AB42 4LH MM7CPZ
AB42 4NL GM4EHP
AB42 4QG MM1EYZ
AB42 4RD MM6IAR
AB42 5AY GM7OJJ
AB42 5AY GM7OTT
AB42 5BL MM0XTW
AB42 5DE GM3ZMA
AB42 5DG MM3WKF
AB42 5EH MM6OMT
AB42 5EJ GM6WTT
AB42 5ES GM0RSI
AB42 5GT MM6VTS
AB42 5HR GM1KUI
AB42 5LR GM0HAN
AB42 5RR MM6SLV
AB42 5WE MM0AOQ
AB42 7TB GM8ZKU
AB420TQ MM6TIR
AB423HZ GM0DYU
AB43 6NE GM1KZG
AB43 6NN MM0CAE
AB43 6NQ GM7OGS
AB43 6SY MM3FYN
AB43 6TJ MM0DVZ
AB43 7ED MM0JGP
AB43 7JS GM4PMH
AB43 7JT MM3LNT
AB43 7JT MM3LOF
AB43 7NW MM1DAK
AB43 8RU GM7FQZ
AB43 8WA 2M1VXB
AB43 9DH MM1CAC
AB43 9NL GM1INS
AB44 1RP GM3UBJ
AB44 1UL 2M0KIA
AB45 1DB GM1HRY
AB45 1DB GM4VVY
AB45 1DZ GM4TOE
AB45 1HS MM0HKU
AB45 2BQ GM0FHD
AB45 2BQ GM0JLJ
AB45 2JR GM1CCI
AB45 2JT MM0BUH
AB45 2PJ GM4PSJ
AB45 3BR GM6JOA
AB45 3UD GM1ROX
AB451FB GM4LFE
AB453ER MM7PDW
AB5 2BJ GM1VBD
AB51 0AB GM4ZEX
AB51 0ES GM4DZM
AB51 0JT GM0APN
AB51 0PJ MM6THU
AB51 0SN GM0PKX
AB51 0TF GM7NQP
AB51 0TX MM7LEO
AB51 0XA GM0TGE
AB51 0XA MM0DTL
AB51 0XX MM6ACM
AB51 3UA MM6NRQ
AB51 3WJ GM0TFQ
AB51 4FL GM0VGI
AB51 4RQ GM4DIN
AB51 4RY GM7NUQ
AB51 4TB GM1RDG
AB51 5BY MM6FVD
Ab51 5HE GM7UPD
Ab51 5LY GM4YAU
AB51 5QT GM3XOQ
AB51 5QZ GM0FIQ
AB51 5RH GM1FSZ
AB51 7HH MM3MXN
AB51 7QP 2M0GMB
AB51 8TQ GM1MKC
AB51 8TS MM6JEA
AB52 6HX MM5FWD
AB52 6JY GM1KHU
AB52 6JY MM6FYW
AB52 6JZ MM6SRL
AB52 6LH GM8UPC
AB52 6NS MM7MWL
AB52 6PD MM0TXO
AB52 6UN MM3ERP
AB52 6ZD GM0UYS
AB526TU MM7GRG
AB53 4GN GM3OXX
AB53 4GS MM0UDI
AB53 4LX MM0ECV
AB53 4UD GM0EMQ
AB53 5QP GM4OCA
AB53 5QP GM6KMK
AB53 5RN GM0EQS
AB53 5TQ GM7CPJ
AB53 5WH 2M0TRA
AB53 5WH GM7SZA
AB53 5WH MM3OGU
AB53 5YH 2M0FTH
AB53 5YH MM3SZM
AB53 6SL GM4TRS
AB53 6TE GM0NGJ
AB53 6UP MM6ACI
AB53 8ED GM1LXA
AB53 8HD GM0BNQ
AB53 8LE MM6MUH
AB54 4GD MM3VSX
AB54 4GD MM6SIM
AB54 4JB GM8NNM
AB54 4NN MM3VSU
AB54 4NN MM6EQM
AB54 4PF 2M0IOK
AB54 4SB MM7WCT
AB54 4XP MM7STS
AB54 5NY GM0WPU
AB54 5NY GM8JWQ
AB54 6AT GM6KDB
AB54 6HA GM1XLH
AB54 7LQ MM3TPF
AB54 7LR GM7ESM
AB54 7LR MM3IOM
AB54 7SY GM8REG
AB54 7XR GM4NXT
AB54 8BY MM7RZT
AB55 4AR 2M0TPE
AB55 4AR MM6GQC
AB55 4AR MM6NAD
AB55 4AZ MM0SMD
AB55 4AZ MM3JIN
AB55 4EF GM7TWM
AB55 5AG GM0EIT
AB55 5AP GM0LVK
AB55 5AP GM1MRS
AB55 5BX GM1LOZ
AB55 5EB MM1DVC
AB55 5ED MM1DTN
AB55 5ED MM3TQI
AB55 5EG GM7DXT
AB55 5EH GM4FGL
AB55 5EQ MM0DVB
AB55 5FX MM3AWC
AB55 6LP GM1HNZ
AB55 6LQ MM0IEL
AB55 6QU GM4WJA
AB55 6UR GM8PSV
AB56 1DD GM3UKG
AB56 4LD 2M0HJS
AB56 4LD MM3UDQ
AB56 4NB MM3TLQ
AB56 4NR GM4UWX
AB56 4QA GM4OEZ
AB56 4SD GM8JCF
AB56 4TU MM7AUH
AB56 5AL GM1KWG
AB56 5BS MM6IMF
AB56 5BW GM0ARY
AB56 5EP GM3KHH
AB56 5HB GM1OXB
AB56 5HP 2M0FSP
AB56 5HP MM6FSP
AB56 5YD GM4YGS
AL1 1DW 2E1BDB
AL1 1NF M6UPL
AL1 1UG 2E0WVG
AL1 1UG G8AMC
AL1 1UG M0TPT
AL1 1UG M6WVG
AL1 2PG G6BDW
AL1 2QS G6CMD
AL1 3AG G8VMP
AL1 4BQ M6OHS
AL1 4ES M6XAL
AL1 4JZ M6CYP
AL1 4PZ G3XXF
AL1 4RD G3YCY
AL1 4SN M6GJQ
AL1 4TT G0MVY
AL1 4UR M6PQO
AL1 4UX M6LOW
AL1 4XZ G0LQU
AL1 5AE M6ADI
AL1 5BX 2E0MRM
AL1 5ES M6JGT
AL1 5EX G4BIX
AL1 5LF G7ARF
AL1 5NJ G4DDV
AL1 5NS M0GTJ
AL1 5QB M0CZX
AL10 0HW M7XDS
AL10 0HX M7YBC
AL10 0HZ M7YBB
AL10 0UB G1ROH
AL10 8DQ G4RMD
AL10 8HE G6INV
AL10 8QF G1IHE
AL10 8XX 2E0FWS
AL10 8XX M6MLP
AL10 9FD M6MIK
AL10 9GS 2E0MVN
AL10 9GS M7NVM
AL10 9GW M0TMS
AL10 9LE G3VIX
AL10 9NT M1JES
AL10 9PB G4SNI
AL10 9PE 2E0BXF
AL10 9RW G3NQN
AL10 9SB G6UBH
AL10 9WN G4KDS
AL2 1HL M1FCX
AL2 1RU M6NZI
AL2 2AH G1ARL
AL2 2AR G8GYK
AL2 2QL G8UCP
AL2 2RD G6URU
AL2 3AY 2E0UCE
AL2 3AY M6UCE
AL2 3DD G7GJN
AL2 3ND G0BEE
AL2 3PU G1WVD
AL2 3QA G8NGR
AL2 3SE G7SDG
AL2 3SJ G0HTL
AL2 3SJ G0KVA
AL2 3SR G1IHS
AL2 3ST G6CKW
AL2 3TD G4LMN
AL2 3TD G8KGG
AL3 4HH G3GEX
AL3 4JZ G4WSL
AL3 4NJ G4HVG
AL3 4TL G6PWL
AL3 5RE G3PZF
AL3 5TD 2E0RTY
AL3 5TD M0JZG
AL3 5TD M6RTU
AL3 5TU 2E1HHE
AL3 5TX 2E0YXX
AL3 5UB 2E0FYZ
AL3 6HP G8SGF
AL3 6LR 2E0BIY
AL3 6LR M3KIZ
AL3 7BG M7SWW
AL3 7EW 2E0CIM
AL3 7EW M6YVR
AL3 7HD 2E1EWN
AL3 7JB G0EHO
AL3 7NL G4YSJ
AL3 7PF G6USO
AL3 8EF G7MSG
AL3 8HW M3MNY
AL3 8JB G1NTJ
AL3 8JN G3LFV
AL3 8JN M6RFV
AL4 0DH G0LZW
AL4 0DH G8BNR
AL4 0DR 2E1IDI
AL4 0EN G8UPX
AL4 0EX 2E0WAP
AL4 0EX M6APW
AL4 0EY M6GBF
AL4 0GE G6PFP
AL4 0JG G4HHJ
AL4 0NS M6RTZ
AL4 0QA G6DYM
AL4 0QR G0MDR
AL4 0QS 2E0KHA
AL4 0QS M3KHA
AL4 0XA G4XJS
AL4 8JD G4WNP
AL4 8NZ M6GDB
AL4 8PE G4BOU
AL4 8PR G0HGO
AL4 8RY G6AHE
AL4 8TP G0XBC
AL4 8TP M0ABY
AL4 9AF G4ZRA
AL4 9AP M6POG
AL4 9DW G4IXY
AL4 9EH G6PWS
AL4 9HD M6APR
AL4 9JU G4BEO
AL4 9JX G4AUE
AL4 9LS M6RTX
AL4 9LX M6SQO
AL4 9NZ G4CZA
AL4 9PS G7IFJ
AL4 9PS M6MCY
AL4 9PW G7PKH
AL4 9QH M6BVS
AL4 9QN M6ACD
AL4 9SJ G3XYH
AL4 9TG G4DJX
AL4 9TG M0SCY
AL4 9TG M6MXD
AL4 9XB G4ZES
AL4 9XJ M6VBS
AL4 9XQ 2E0IOZ
AL4 9XQ M0IZA
AL5 1EF 2E1BZB
AL5 1EZ G7HDR
AL5 1HD G4DOC
AL5 1JQ G3UHN
AL5 1LL G1XEH
AL5 1RF 2E0IHB
AL5 1RF M0RBK
AL5 1RF M6AKY
AL5 1RP G8YNB
AL5 1SR G8CLY
AL5 1TD G8KEK
AL5 2AL G0DFF
AL5 2HJ M0AYS
AL5 2QJ G4GEZ
AL5 3AS G1BRP
AL5 3HJ M6NHF
AL5 3LJ 2E0JSG
AL5 3NX G4OGZ
AL5 4BT G1ZYJ
AL5 4DB G0CPN
AL5 4DB G8ASP
AL5 4HN G4KCN
AL5 4JU M6FJI
AL5 4LP G3YWA
AL5 4TE M5WIZ
AL5 5BT G4OAV
AL5 5DW G3MSW
AL5 5DW G8XOB
AL5 5HR G4RIK
AL5 5LY 2E0SUU
AL5 5LY M0SUU
AL5 5QS G1ZGF
AL5 5QU M0MLT
AL5 5SJ G7EFL
AL5 5ST M1CBV
AL5 5SU G0RVH
AL5 5SU G3SDG
AL6 0DB G1MMZ
AL6 0DH G0URC
AL6 0DZ G0NYY
AL6 0EN G3AQF
AL6 0PS G7UOU
AL6 0RU G8TRU
AL6 0RW G3KOX
AL6 9BB G6KJA
AL6 9JQ M6LTQ
AL6 9NW G3XDM
AL6 9PX G0WTF
AL6 9SZ G0IPJ
AL6 9TF G4SZI
AL7 1DL M6TYT
AL7 1DU G0IIG
AL7 1DW 2E0XYT
AL7 1NL G7BDS
AL7 1PD 2E0JMJ
AL7 1PJ M7AIE
AL7 1QP G0KLU
AL7 1SD G0WAT
AL7 2BW G6BUP
AL7 2HF G6POE
AL7 2JS M1ACE
AL7 2QJ G5KW
AL7 2QJ G8FXM
AL7 2QJ M3FXM
AL7 3FA 2E0YUN
AL7 3FA M0LSY
AL7 3HB M7BSH
AL7 3LR G0NJS
AL7 3RJ G8TVW
AL7 3RL G0MKN
AL7 3TD G1PGJ
AL7 3TE G3TNE
AL7 3TW G0HVX
AL7 3XN M6HSE
AL7 4AG 2E1PDM
AL7 4AQ G0JJM
AL7 4JT G3HLN
AL7 4LR M0HSQ
AL7 4TG M1FEK
AL8 6HT G4JVA
AL8 6RF G4LWV
AL8 6RH G1LLW
AL8 6RJ M7CMI
AL8 6SE M6SWP
AL8 6TA G0CYC
AL8 6WW 2E0HOG
AL8 6WW M6STR
AL8 6XW M6NGX
AL8 7AW M6GCV
AL8 7BA M1PWT
AL8 7BG G4MDC
AL8 7DQ G0UBO
AL8 7EE M0GJK
AL8 7EN G0VDT
AL8 7HX M0JLW
AL8 7HZ G4POB
AL8 7LF G6CXY
AL8 7LH G4DRI
AL8 7NN G8TVZ
AL8 7SG G4TIM
AL8 7SN G6FVF
AL8 7ST G8CYX
AL86XW M6TVN
AL9 7AY M0BMU
AL9 7QQ 2E0HFI
AL9 7QQ M6IZF
AL95HQ G7CMB
B1 1LS M6UED
B1 1UJ G3OTY
B1 1UJ G4ACF
B10 0LE 2E0XTV
B10 0LE M3XTV
B10 9TA G7LOZ
B10 9TB G6HLR
B11 2EL M7PNG
B11 3QJ G8EMX
B11 4NJ G0IVD
B12 9ER G0NHP
B12 9JQ G1YIL
B13 0BA G6DSD
B13 0BH G7UGC
B13 0BH G8ENA
B13 0BH G8GKR
B13 0DL G4KZV
B13 0DL G4OHM
B13 0EJ G4PUD
B13 0JL M3JPP
B13 0NR G6DRN
B13 0PR G4GIG
B13 0RQ G4BJS
B13 8AB M7CFL
B13 8JZ G0EBW
B13 8NZ M0AEJ
B13 9AD M6AKA
B13 9EN G4DFO
B13 9QR 2E0BTU
B13 9QR M3YGR
B13 9TY G3YHF
B13 9YP M0JMT
B14 4DS G4ETZ
B14 4LS G1POR
B14 4NX M6GYH
B14 4TE G8MYK
B14 4TG G1LZH
B14 4TJ 2E0MSE
B14 4TJ M0MSE
B14 4TW M3LTH
B14 5AF M6RZA
B14 5BT 2E0LPD
B14 5BT M0WYH
B14 5BT M6LPD
B14 5DS 2E0NYC
B14 5DS M0NYP
B14 5DS M6STU
B14 5EF G3PQP
B14 5HD G4MTG
B14 5HD G8TXL
B14 5LX M1CPC
B14 5XA M7MSV
B14 5XX 2E0SDD
B14 5XX M6STJ
B14 6AD G8YQC
B14 6DE G2BBC
B14 6DE G3YXM
B14 6DE M3YXM
B14 6HH G4ZVS
B14 6TN G1FDD
B14 7AS G4JGH
B14 7DB M0DGQ
B14 7DB M3ZCR
B14 7LN 2E0IKN
B14 7LN M7OCB
B14 7QA G4WPT
B15 2AA G7ARP
B15 2DF M6HSF
B15 2RU M6SNZ
B15 2TT G7AFQ
B15 2XA G3UOA
B15 2XA G8CQH
B15 3JB G3NMW
B15 3RL G4NMC
B15 3RN M7MET
B16 0EF G1JDE
B16 0EF G6AUO
B16 8HU M0LTL
B16 8UQ G6ZYZ
B16 9EY G1XFL
B17 0AQ G6DMQ
B17 0AT G3SJH
B17 0EL 2E0GXP
B17 0EL M7PEB
B17 0NW G0KOI
B17 0QT G4NZY
B17 0SU M7JSA
B17 8BN G4HWN
B17 8PY G4KXV
B17 8QH 2E1BHB
B17 8QH G4XDM
B17 9EB M0BZY
B17 9JU G4LQF
B17 9JX M6EFM
B17 9LE G4YDO
B17 9PA 2E0EXV
B17 9PA M6JXK
B17 9QX G1KUQ
B17 9RE M5PIP
B17 9TB G6PGO
B17 9TZ 2E1DLA
B18 4NE G1UBT
B18 6JQ 2E0HRS
B18 6JQ M3NBH
B19 1EH G6KEX
B191DG M7DLK
B20 1EQ 2E1TBW
B20 1EQ M3LTV
B20 2AQ G6EEB
B20 2AS M3OCQ
B20 2HN G6KHN
B20 2NX G7WKG
B20 3LS 2E0GJP
B20 3LS M0JNF
B20 3LS M7BPC
B21 9NU G1AJK
B21 9QD M3EWV
B23 5DF 2E0FOX
B23 5LL M3UGB
B23 5QA M6TGQ
B23 5RG G7RYA
B23 5XE G8LIX
B23 5XR G8NZO
B23 6BT G0RAR
B23 6DX G8NDT
B23 6jl M7SSC
B23 6NN G6IHW
B23 6PJ M3PKQ
B23 7JJ 2E0LPJ
B23 7JJ M3XIP
B23 7JR G0VBQ
B23 7LG G8ROU
B23 7LT G6JAC
B23 7NE M7EFH
B23 7PB G0LNE
B23 7XL G7EKJ
B24 0SN M0KRW
B24 0TQ 2E0GDM
B24 0TQ G4JDC
B24 8JP 2E0GOG
B24 8LX G4ZCR
B24 9BY 2E0XEB
B24 9BY M0DID
B24 9EQ M0YJL
B24 9LX G6GBT
B24 9NE M6CKT
B24 9NF G8YJT
B24 9NF M0YJT
B24 9RJ M3UTV
B24 9RX M6ZOG
B240BX M6NQK
B25 8EJ 2E0DJV
B25 8EJ M3ZQN
B25 8JF G8AFI
B25 8LH 2E0IZJ
B25 8LQ G8XXZ
B25 8NE M3TBF
B25 8NZ G7PSC
B25 8TJ M6AUO
B25 8XS M6GYK
B26 1DY G3NXC
B26 1LB M6OCH
B26 1PR G3KYE
B26 2AF G3JTT
B26 2AW G1VSD
B26 2EA 2E0FMV
B26 2EA M0KBG
B26 2EA M7VJH
B26 2HT G6KHM
B26 2HT M7CUU
B26 2HW 2E0MCH
B26 2HW M6MBF
B26 2NY M6DFG
B26 2PP M0GNP
B26 2PP M0SAL
B26 3BX M0WSN
B26 3EL G7HNR
B26 3HY G8DHI
B26 3LA 2E0EUN
B26 3LA M0TAX
B26 3LS G8KGS
B26 3PW G4NPG
B26 3ST G0LBQ
B27 6BB G1OOW
B27 7LA M6CYY
B27 7PB 2E0DHZ
B27 7PB M3UNH
B27 7RU G6HFZ
B28 0DQ M6CUP
B28 0HH G1BLK
B28 0HR M6EQS
B28 0LA G8ADD
B28 0QX G6KVR
B28 0TE G7IJI
B28 0TW G3RGD
B28 0TZ M6EXQ
B28 8PF 2E0TBR
B28 8PF M6AUL
B28 9EQ G3KLD
B28 9NT G1SWU
B29 4AG G8GDZ
B29 4AH G3ZKQ
B29 4LT 2E1GYZ
B29 4LT M1EBK
B29 4NH G4KLA
B29 4RD G7EJN
B29 5DG 2E0MBR
B29 5DG M0PTG
B29 5DG M6FLE
B29 5DR M3ONM
B29 5LE G1LWH
B29 5NX G0HDF
B29 5PU G0ETJ
B29 6NH M0JZT
B29 6PX M0BCN
B29 7QJ G4RCZ
B29 7QQ M6RDN
B29 7SG G4NCY
B30 1DR G4NPA
B30 1DR G4NPB
B30 1DU G8LKW
B30 1HU M7SBZ
B30 1LY G8ORR
B30 1NG G0BWP
B30 1PU 2E0GAH
B30 1QN M3SIW
B30 1RF M1EAN
B30 1SN M7JCM
B30 1TH G8DYH
B30 2AX G6WPD
B30 2BA G4YUI
B30 2DY G0UYT
B30 2EB M6HIP
B30 2HT M6IRZ
B30 2LS M3VCB
B30 2SH G4OMP
B30 3BZ G1EAX
B30 3LG M3YIY
B30 3NG 2E0GLM
B30 3QJ G7PIR
B30 3QN G0MQC
B30 3QR M6SYG
B30 3RP M6BJN
B31 1AL G6VGT
B31 1DJ M6DJW
B31 1DN 2E0HBD
B31 1DN M6HDY
B31 1DX G4RPV
B31 1NE G7DNX
B31 1UQ 2E0DOQ
B31 1XH 2E0AZL
B31 1XH M3HTG
B31 2AX G8XIZ
B31 2BJ G8ZXY
B31 2EB G1FVC
B31 2EE M0RTE
B31 2EJ G4YTO
B31 2FG G0FPN
B31 2FS G8VBE
B31 2FT G8AHE
B31 2HD G6IRJ
B31 2HW M0PGD
B31 2JG G3XVW
B31 2LY G1JRF
B31 2PL M7DPB
B31 2PP G8VNL
B31 2QB G8KAE
B31 2QY M0DTR
B31 2QZ G2AZM
B31 2RP M0VAT
B31 2XQ M7CUP
B31 3GG G7BZM
B31 3HH M7ERQ
B31 3HU 2E0PGT
B31 3LA M6GWG
B31 3NL M6EBP
B31 3SY M3XHM
B31 3XE M6FBO
B31 4AJ 2E0PVN
B31 4AJ M0PVN
B31 4AJ M6PVN
B31 4AU G7FUW
B31 4BS M0BAZ
B31 4LJ 2E0JWY
B31 4LJ M0WBY
B31 4LJ M6JWY
B31 4LN M0ALH
B31 4PA M3XOQ
B31 4QN G1JON
B31 4QN G3OHM
B31 4QN G6BV
B31 4QN G8SH
B31 4SS G6TDG
B31 4SU 2E0INC
B31 4SU M0LWT
B31 4SU M6ERA
B31 4TA M3UTM
B31 4TQ G8MIT
B31 5AU 2E1BHF
B31 5DP M0BSP
B31 5HH 2E0UGO
B31 5HH M3UGO
B31 5JB M6UEB
B31 5NT M3HNE
B31 5NY 2E0REN
B31 5NY M6HME
B313HJ M6IYJ
B32 1AY M6ZBL
B32 1DR G4SWA
B32 1EG G4XUA
B32 1EL G4CGU
B32 1HE G4XEJ
B32 1JA 2E0KEG
B32 1LB G1MAR
B32 1LB G3MAR
B32 1LB G8BHE
B32 1LB G8OHM
B32 1PG 2E1AIT
B32 1PG G0PWQ
B32 1QT G0LGO
B32 2AA G7TPB
B32 2BA G4JCZ
B32 2BS 2E0PPH
B32 2BT M6OOO
B32 2LY M6HSQ
B32 2NW G0PPX
B32 2SB G1NHX
B32 2TT G4ZTM
B32 2UX G4CKK
B32 3LA M6TGJ
B32 3NL M6EYH
B32 3TA G1PKG
B32 3UN 2E0ROX
B32 3UN M7MCJ
B32 4LW 2E0DLF
B32 4LW M6EPF
B33 0LH G4KII
B33 0NR G4SAS
B33 0RL G7DQQ
B33 0YL G6BNO
B33 0YP G4FGF
B33 8AB M0DZO
B33 8JP 2E0MJX
B33 8SL 2E0UOK
B33 8SL M3UOK
B33 8ST G6LKZ
B33 9BU G7FPZ
B33 9DL G8KCB
B33 9HH M7HKY
B33 9JX M6OTS
B33 9NZ G6SFW
B34 6JL G4LTT
B34 6PZ G6NOW
B34 6QY 2E0HMP
B34 6QY M0JVL
B34 6QY M7CIS
B34 7AY 2E1SJG
B34 7BU M0BEM
B34 7LP G4ZPI
B34 7QX G0FEO
B34 7RS G4LAJ
B34 7RU 2E0VJX
B34 7RU M0VJX
B34 7RU M6VJX
B35 7NT M6NQU
B35 7PD G8TBW
B35 7PF M3VVH
B36 0AD G6WYQ
B36 0BX G6VMR
B36 0EH M1NPH
B36 0HH M3VVA
B36 0JT 2E0DEZ
B36 0LG G3OKH
B36 0LS 2E0GTN
B36 0LS M0XGN
B36 0LS M6NQL
B36 0PB G4OOX
B36 0UH G7UEJ
B36 8HA G0OTF
B36 8QG G3NQA
B36 8UN M7NOO
B36 9DY M3VVB
B36 9DY M3ZAU
B36 9HX G4SGA
B36 9JD G4IMB
B36 9JD G8PXU
B36 9JD M3SZY
B36 9LL G6VXL
B36 9LL G6YLL
B36 9LL G7OZE
B36 9LL M3OZE
B36 9SN G3DID
B36 9ST G4LRN
B36 9TS M3JYA
B36 9TW G1GFW
B36 9TY G4XPV
B37 5HS M3CIP
B37 5HX G0FCQ
B37 5LG M7RVH
B37 5QL M7SDB
B37 6DL G4SPY
B37 6JQ G0WFX
B37 6SB G6APQ
B37 7BU M0CDQ
B37 7BU M6LCV
B37 7EE G4HPT
B37 7HS G4TDF
B37 7PA M3XJP
B37 7QT M6FMI
B37 7RD G1PRF
B38 0AB G1BUJ

B38 0AL M0BTO
B38 0DN G3YKO
B38 0EP G0MTN
B38 0EP G0WRC
B38 0EP G7WAC
B38 8AJ G1UOD
B38 8DA G8NLK
B38 8DT G4AEG
B38 8ED 2E0NCO
B38 8ED M6NCO
B38 8LB G0MKU
B38 8LS G1XPD
B38 8NH M7RJH
B38 8PH G4KRT
B38 8PW G3TGL
B38 8PW G8XUN
B38 8TH G0AHC
B38 9AG M6SNN
B38 9HS M6SYH
B38 9LA G0BOT
B38 9LW M6DCW
B38 9PA G0WYT
B38 9QR M6MNZ
B38 9QT G1MZM
B38 9RH M3UZV
B38 9RY G0OKI
B38 9TT 2E0SRV
B38 9TT M0SSV
B38 9TT M6SRV
B38 9UW M3VXB
B42 1EU G7TVL
B42 1HF G0FBQ
B42 1LP G1HUM
B42 1LY G4WSI
B42 1PL G0DOG
B42 1PZ G6UGA
B42 1RY G1EDT
B42 2BH M0MLS
B42 2BX G1MOV
B42 2DT G4NTV
B42 2EA M3EYK
B42 2EH M3PHX
B42 2HF 2E0WPS
B42 2HF M0WPS
B42 2HF M6WPS
B42 2HJ G0GPF
B42 2HT G8ASW
B42 2JW G8SFQ
B42 2LX M0GYA
B42 2NU G1LTE
B42 2PQ G6NHW
B42 2QB G6UGZ
B42 2QJ G4HOM
B42 2RL G6YQW
B42 2SQ G4NBW
B43 5AG M6KDF
B43 5EH M0ANK
B43 5HH 2E1ILM
B43 5JR G7IKG
B43 5LN M3OHX
B43 5ND 2E0XAW
B43 5ND M6ARW
B43 5PG 2E0EHX
B43 5PG M6GGX
B43 5PG M6HHW
B43 6BB G3FIA
B43 6HX G0TEM
B43 6NA M3EMO
B43 6QE G6YSB
B43 7HG G6LPB
B43 7HX 2E0NVP
B43 7HX M3NVP
B43 7JW M3NVR
B43 7PG G4PFK
B43 7PQ 2E0CVU
B43 7PQ M0TVU
B44 0AL G6WOI
B44 0AY 2E0VMV
B44 0AY M0VMV
B44 0AY M3VMV
B44 0EF M7TRV
B44 0LB G1BJK
B44 0LF G0FOC
B44 0NF M3WMU
B44 0QB M7BTQ
B44 0QS 2E0OAM
B44 0RD M7MKY
B44 0SN M0WET
B44 8AB M6KTX
B44 8EN M1CSC
B44 8ER G7CWO
B44 8JB G4DIP
B44 8LQ G6WIG
B44 8RG 2E0ELD
B44 8RG M3ELD
B44 8RL G0WMU
B44 8RS G7OPD
B44 8SW 2E0GOM
B44 8SW G8CHA
B44 9AX M7ESB
B44 9BY G4RZD
B44 9DB G1XKB
B44 9NY G8TZW
B44 9QH G7ORT
B44 9QN M6NQM
B44 9RP G6NZW
B44 9RR G3URV
B44 9SS G4BTK
b440ng M3INC
B45 0DA 2E1IAT
B45 0DD 2E0GCB
B45 0DD M3GKH
B45 0DD M3HIO
B45 0EN 2E0SVZ
B45 0EN M6ZDL
B45 0JB G8VXY
B45 0JS M6LDU
B45 0JY G0HPG
B45 0JY G0HPH
B45 0JY M6BNP
B45 0JY M6SXD
B45 0NB G6ZSG
B45 8EH G0NJT
B45 8GU 2E0AAI
B45 8GU M6KND
B45 8HP G4TSB
B45 8NG M0VGG
B45 8NZ G7UMY
B45 8QU 2E0VUK
B45 8QU M3VUK
B45 8SJ M3JYE
B45 8TQ 2E1ANG
B45 8TQ 2E1ANH
B45 9DA G1XKY
B45 9EU 2E0AQE
B45 9EU M0JTB
B45 9HW G6YXW
B45 9HY G7RLV
B45 9LR G4YTJ
B45 9LW M0BYJ
B45 9RU M3KCQ
B45 9SZ 2E0LUJ
B45 9SZ G0HVN
B45 9SZ G3IUB
B45 9SZ G8IUB
B45 9SZ M6LUJ
B45 9TT 2E0FXJ
B45 9TT M6FXJ
B45 9TT M7JAC
B45 9WA M6ILR
B45 9XB G8IFT
B45 9XB M6GKG
B46 1EP G4OPE
B46 1HL G6NTY
B46 1NP 2E0TJE
B46 1PD G3UPA
B46 1PE 2E0KHG
B46 1PE M6KAO
B46 1RU G8MWE
B46 1SA 2E0MEY
B46 1SA M6CKR
B46 1SN G0CKE
B46 1SN G3ZUM
B46 1TW G6VNC
B46 1UF G1KAT
B46 2NH 2E0YOM
B46 2NH M0YOM
B46 2NH M3YOM
B46 3EH G0BUV
B46 3LZ G4YQD
B46 3NE G4FTY
B47 5EN G3OIC
B47 5HJ G8TWZ
B47 5HY G1WAC
B47 5HY G4VPD
B47 5HY G4WAC
B47 5NR 2E0HPD
B47 5NR M0IFT
B47 5NR M6HPD
B47 5PX G6JYO
B47 5QA 2E0YZW
B47 5QA M7ACN
B47 5QE G0NES
B47 6AW 2E0HKL
B47 6AW M6KQC
B47 6HE M7BVH
B47 6HP G4OJL
B47 6LX G0ICJ
B48 7NE G1PHN
B48 7PL G3UYX
B48 7TB G0BNZ
B48 7TB G4SVL
B49 5DD G1YFD
B49 5DD G4GYI
B49 5EG 2E0BOT
B49 5EG M0OJG
B49 5HA G0UGY
B49 5HY G6MOZ
B49 5LJ G0UMV
B49 5LJ G3UMV
B49 5LJ G4UMV
B49 5LJ G6LRT
B49 6AP M1EZD
B49 6DR G1IKL
B49 6HQ G3USA
B49 6LF G3EVT
B49 6LF G4ACZ
B49 6LX G1WLU
B49 6LX G1ZUU
B49 6PG G0RGW
B49 6PG G8EXF
B49 6QY G6BAY
B5 7NE G6DWS
B50 4AN G4OHJ
B50 4AN M3MEB
B50 4AP G0VBZ
B50 4AP G3VRF
B50 4HY 2E0VPH
B50 4HY M6VPH
B50 4NP 2E1MGB
B50 4NP M0XLX
B50 4NX M0BHM
B50 4PH G7BRF
B6 4PP M6MRX
B6 5HW 2E0DHY
B60 1AD G1BHO
B60 1AL G1BIU
B60 1AW G1GDR
B60 1BN M0FRA
B60 1BP G1FPY
B60 1DG G8DEC
B60 1DS G4LVK
B60 1DS G6VGG
B60 1DW G8VFP
B60 1DY G4OJS
B60 1DY G4WZA
B60 1DZ G6NQB
B60 1HE G0KHK
B60 1HN G4BBU
B60 1HN M3RXP
B60 1HW G4AHK
B60 1pj M0WOD
B60 2DB 2E1GNE
B60 2EB G0BLT
B60 2EB G0HAW
B60 3AY G0EVY
B60 3EP G4UIW
B60 3GP 2E0OBK
B60 3GP M0OBK
B60 3GP M6OBK
B60 3HB G0BIR
B60 3HB G0EHQ
B60 3NX 2E0KBF
B60 3NX M6KBF
B60 3SD M6GZB
B60 4EB 2E1GXI
B60 4EB G1TCK
B60 4EB M3CBC
B60 4LH M6NAN
B60 4LS G6OPY
B60 4NF G4FNQ
B61 0DX G6OIF
B61 0EL G4AHO
B61 0ER G6EET
B61 0ER M1CLZ
B61 0JP M6SOZ
B61 0JT G4LRL
B61 0JT M1ESD
B61 0LQ G0AQF
B61 0LU G4NZK
B61 0LU G6LPS
B61 0PA G8XUW
B61 0TR 2E0XON
B61 0TR M7XON
B61 0TT M3KGO
B61 7BE G1IEC
B61 7DA M6KLR
B61 7EB G4CQS
B61 7EB G8GEB
B61 7JG G0BGA
B61 7JT M1JSS
B61 7PR M0LLS
B61 8HY M3SKU
B61 8NB 2E0MJT
B61 8NQ G0SUQ
B61 8PE G4OTC
B61 8PN M3UHW
B61 8UA G4PKW
B61 9BH 2E0MIS
B61 9BH M0LIJ
B61 9BH M6UDY
B61 9JN M3XNE
B61 9JT G4CAF
B61 9JW M3AXT
B61 9LH M3OJS
B62 0HU G1WTH
B62 0HW 2E0UAE
B62 0HW M3ULS
B62 0JL M0JFW
B62 0LN 2E0ZXJ
B62 0LN G4OHB
B62 0LN M0ZXJ
B62 0LN M3XSJ
B62 0NE G0EVM
B62 0PG M1CGB
B62 8EU G7DIZ
B62 8EU G7IIS
B62 8EZ M0VCR
B62 8JA G7AXW
B62 8JS G2FXZ
B62 8JS G6GQJ
B62 8LJ M3SAR
B62 8LJ M6IKA
B62 8LJ M6MDX
B62 8LJ M6SAD
B62 8LR 2E0KPP
B62 8LU M6EFJ
B62 8QE M0CWX
B62 8SH G6IJQ
B62 8TB G4YAH
B62 8TH G1IAL
B62 9AW G3VBV
B62 9BP M7CIL
B62 9DX 2E0DSS
B62 9DX M6DSS
B62 9HJ M3CPY
B62 9NQ G4VPZ
B62 9NR G4DPZ
B62 9PP G1EBB
B62 9QJ 2E0MKB
B62 9QJ M0FCW
B62 9QJ M0KMW
B62 9QW G4JSV
B62 9RF G4PVP
B62 9SX G6EVW
B62 9TF 2E0SLJ
B629LP M0BRE
B63 1BB G1FET
B63 1BZ G4OQX
B63 1DU G8TXA
B63 1EQ 2E1WIN
B63 1EQ G1PPZ
B63 1HD M6RKB
B63 1JQ G4LWF
B63 1JQ G4MEB
B63 1JY G4AYK
B63 1JY G4NQW
B63 2AY G7VJT
B63 2DW G6XYO
B63 2HE 2E0XSZ
B63 2HE M7EDZ
B63 2JA G7FIJ
B63 2JJ G0GUG
B63 2PP G4EHR
B63 2PR M3XMJ
B63 2PY G1VQV
B63 2SY G8TIA
B63 2SZ G8ZPD
B63 2SZ M0ZPD
B63 2TB G6LDA
B63 2UY G1KII
B63 2XH M0YIM
B63 3DN G8XSA
B63 3JX G4UMY
B63 3QD 2E0RWC
B63 3QD M7BND
B63 3QR G4FAV
B63 3QR M3EFL
B63 3RZ M3FFA
B63 3RZ M3FFE
B63 3ST M6EUX
B63 3TJ G1BEG
B63 3TP 2E0CUY
B63 3TP 2E0JTM
B63 3TP G7JMZ
B63 3TP M6EDA
B63 3TP M6JPD
B63 3TP M6ZSB
B63 3TZ M6ARK
B63 4BX M6EMS
B63 4HD M0CTK
B63 4HG G6ALW
B63 4HQ G1VOB
B63 4HQ G3DTU
B63 4PB G0NLA
B63 4QG G1BRF
B63 4RN M1BZG
B63 4SA M6LNR
B63 4TJ M1EVF
B632BD G7VIE
B64 5BG G8MUV
B64 5EX G7TUV
B64 5LA M7BKN
B64 6DU G0PPJ
B64 6EA M0EHS
B64 6QX G0BHR
B64 6RB G4JFF
B64 7EZ G3VPX
B64 7EZ G7BRA
B64 7HH 2E0FGT
B64 7HJ G6NNO
B64 7JS G4HPB
B64 7LE G4ISQ
B64 7QH M7IFR
B65 0AQ M6NEL
B65 0HF G6LOR
B65 0NP G0WYA
B65 0QE M1FFG
B65 0QE M3FXU
B65 0RR 2E0JAM
B65 0RR 2E0LEE
B65 0RR 2E1HXM
B65 0RR 2E1HXN
B65 8DT G4YIV
B65 8DW M3VBI
B65 8DW M3VRP
B65 8HT 2E0KAS
B65 8HT M0KAU
B65 8HT M6KAN
B65 8NX M0GSX
B65 8PB M6OAQ
B65 9BQ G1OOZ
B65 9DZ G0PAI
B65 9DZ M3XNK
B65 9HZ G0OYF
B65 9JN G0KJG
B65 9JX 2E0HAZ
B65 9JX M3GHH
B65 9LG G0MJT
B65 9NJ M0IMM
B65 9NT G3TRG
B65 9NT G8TRG
B65 9QN M3FGU
B65 9SD G4FJJ
B67 5AY G1JVM
B67 5AY M0JVM
B67 5DH G1UCO
B67 5PD G7UNB
B67 6DA M6DKK
B67 6DY M7BNG
B67 6HU M0BBW
B67 6LA G1UUZ
B67 6PR G1FFO
B67 6QS M3JJH
B67 6QX G0TVR
B67 6QX M3DVU
B67 7BX G4KVC
B68 0BJ M0MEL
B68 0NA G8DEM
B68 0NU G8PTF
B68 0PU 2E0PYA
B68 0PU M0PYA
B68 0SW 2E0XYX
B68 0SW M6TPA
B68 8AQ G4SFG
B68 8BE M1AED
B68 8BP M3LXV
B68 8HY M3EFQ
B68 8LT G4OJJ
B68 8LT M3TOF
B68 8NG G0UFJ
B68 8PP M1WEH
B68 8PR G4OYT
B68 8PT G2BXP
B68 9DP G1XFO
B68 9DP G6IQY
B68 9DU M3XVZ
B68 9ES G8MKE
B68 9LQ G1IOU
B68 9LU G4VRX
B68 9PW G8RAO
B68 9TB G7RTQ
B68 9UL G0WXA
B680PU M7FMA
B69 1AJ G7IYY
B69 1BA 2E0TAM
B69 1BA G0NJZ
B69 1BA G7IYX
B69 1JU 2E0TAT
B69 1na M6RNN
B69 1NP G4JVH
B69 1NP G8FCO
B69 1NT G3NZS
B69 1PW G4YFT
B69 1QA G1WHY
B69 1SE G6EOR
B69 1SW 2E0SOX
B69 1TP G8SPU
B69 1UD G8TFB
B69 4QA G7LOE
B69 6RN G3VJX
B70 0SL G4DVM
B70 6RJ 2E0WBZ
B70 6RJ M0WBZ
B70 6RJ M6WBZ
B70 6RQ 2E1LJW
B70 6RQ M6LRU
B70 7LS G1TBK
B70 8JY 2E0KVE
B70 8JY M3GUM
B70 8LD G0BZP
B70 8PL M3LMD
B70 8QR M3LLC
B70 9ES 2E0ZGL
B70 9ES M0VGL
B70 9ES M3NUE
B71 1AE M7JUK
B71 1DQ M6BSI
B71 1DX M3DIU
B71 1NJ G0WCK
B71 1RU G4PMW
B71 1RU G6FPN
B71 2AA 2E0SJS
B71 2AA M3MFS
B71 2DY G7RCP
B71 2DY M3APE
B71 2QJ 2E0BMU
B71 2QW 2E0RWN
B71 2QW G4LWN
B71 2QW G8UID
B71 2QW M0RWN
B71 2QW M3RWN
B71 3BT G6VFO
B71 3DA 2E0DEM
B71 3DA M6EED
B71 3EE M1DMT
B71 3HX G6MPT
B71 3LA 2E0LIT
B71 3LA 2E0XAV
B71 3LA M0PPY
B71 3LA M6GYM
B71 3LX G0RCX
B71 3NE 2E1IHE
B71 4BA 2E0LET
B71 4BA M6KNY
B71 4HN 2E0BCD
B71 4LQ M3OFU
B71 4LR M1GAP
B71 4LR M3TYZ
B72 1AG G1AAL
B72 1HB G7ETS
B72 1HP G3PLP
B72 1JP G0UQJ
B72 1JU G1NFN
B72 1LG G5GWH
B72 1YE G4TLR
B72 1YF G0VXK
B72 1YZ G8NFD
B73 5AR G7FBY
B73 5EA G4TYR
B73 5EH M3FSU
B73 5EL G8EZB
B73 5JY G1KYK
B73 5LD M3TGE
B73 5LF G0VZL
B73 5LT G0GEP
B73 5LT G1LTG
B73 5NF M6XVM
B73 5SP G7EKW
B73 6NZ G4OFN
B73 6NZ G7ORV
B73 6PG G8AMD
B73 6QA G4LBT
B73 6QR G4JJN
B73 6UQ G1NPA
B735UN M6MVF
B74 2AE M3LIY
B74 2BS G6HOC
B74 2BU M3IZN
B74 2DA 2E1EQE
B74 2DG M0CUD
B74 2EA G7JJW
B74 2JE G8KFF
B74 2LA G4DDD
B74 2PS G3MCB
B74 2QA G3JZF
B74 2TB G4ABW
B74 2UH G6AYI
B74 3JU G6VPL
B74 3LR G6RBO
B74 3NP G1MGZ
B74 3NP G1OBA
B74 3PG G4IWF
B74 3PQ G7OBC
B74 3QE G0PJG
B74 3QE M0AZK
B74 3QP G1EYT
B74 3QQ M7RWW
B74 3TU G1XDV
B74 3UD 2E0XCN
B74 3UD M3XCN
B74 4AN G0MBI
B74 4BL M3YOH
B74 4DQ G1AZE
B74 4DY G6REY
B74 4HL G8EFU
B74 4HL M0DBM
B74 4JS G3WWL
B74 4LQ 2E0CLI
B74 4SG G8XMH
B74 4SJ G7NJW
B74 4UG 2E0MAJ
B74 4UG M0MAJ
B74 4UG M3OFJ
B74 4XG G8OHS
B74 4XR G1JXX
B74 4YD G1GFA
B74 4YD G7SER
B75 5AQ M0AZE
b75 5ej G7PZQ
B75 5EX G7LHK
B75 5LD G1NSG
B75 5LH G0KLK
B75 5PQ G6VIY
B75 5TJ G7VBJ
B75 6AU G0EVH
B75 6AX M3FQX
B75 6DB G6HNS
B75 6DH G7UCG
B75 6DW G3AVE
B75 6EN 2E0GTL
B75 6EN G1AZZ
B75 6EN G3MYC
B75 6EN M3PNF
B75 6SN G8TBB
B75 6TF G5RGS
B75 7AA G6AGO
B75 7AA G6DGR
B75 7BL G6BBR
B75 7BY M7NXC
B75 7HN M7DTA
B75 7TH G6HSR
B75 7UU M6LZY
B756AA M0BUC
B76 1EA G7IXH
B76 1FN M3WAP
B76 1HU G8VUU
B76 1HY 2E0ROJ
B76 1HY M3HOD
B76 1JN 2E0FDE
B76 1JN M6XPH
B76 1JQ G1BUQ
B76 1JR G1XKN
B76 1LQ G7PHF
B76 1LZ G8IOS
B76 1PJ G6KPX
B76 1QZ G8ERN
B76 1XR G6UED
B76 1YD G0HID
B76 1YE 2E0CCB
B76 1YE M6TYE
B76 1YR 2E1DSU
B76 2PT G3VNY
B76 2SY G1FQX
B77 1AB 2E0HZS
B77 1BT G0WKI
B77 1BY G6EOO
B77 1DF M6BTN
B77 1HB G4GYA
B77 1JD G4MFN
B77 1NY 2E0CCC
B77 1NY M3FEG
B77 1PE G1MGN
B77 1QR G4ORW
B77 1QT G6RZY
B77 2EG M6IZY
B77 2HQ G3YUX
B77 2JA G0UYP
B77 2JU M7AWW
B77 2LD G0GUD
B77 2NA G0PFQ
B77 2NB M7HYB
B77 2QD G6SRU
B77 2RS M3JWN
B77 2RY G8SCG
B77 2TX M7RXM
B77 3BT G4LZV
B77 3BT G4NWM
B77 3BT G4NWN
B77 3BT G8UXD
B77 3DG M6TFW
B77 3JH G4NWO
B77 3JW M5ACJ
B77 3JZ 2E0KPB
B77 3JZ G8RFL
B77 3JZ M6NQQ
B77 3LA G6LPX
B77 3LH M1FEQ
B77 3LH M6OPG
B77 3NB M3FLB
B77 3PH G8MCT
B77 3PP M1FER
B77 3PR M6AQA
B77 3QG M1CVH
B77 4AA G4INA
B77 4AB 2E0DPS
B77 4AB M6DSZ
B77 4AQ G0SKK
B77 4BZ G0LHR
B77 4DB M6OYN
B77 4DF M6APU
B77 4DL G6NHG
B77 4EJ G0KFF
B77 4EJ G4ZPJ
B77 4EN 2E1DAT
B77 4EN G3DAT
B77 4EU G0LTR
B77 4EZ M3FLZ
B77 4HT G0IKQ
B77 4JJ 2E0LCR
B77 4JJ M6ION
B77 4JL G0TRB
B77 4JW M3ASN
B77 4LD M6SUX
B77 4LU M3RLT
B77 4LX 2E0BME
B77 4LX M3UWV
B77 4NA G1UZW
B77 4NA M3UZW
B77 4NA M3YII
B77 4PP M6NXE
B77 4QY 2E0EHB
B77 4QY M0NPL
B77 4QY M6EHB
B77 5EY G0FXL
B77 5FL M6HMS
B77 5FX M7JCA
B77 5GG M6DSO
B77 5JD 2E1FDY
B77 5JE 2E0ZAP
B77 5JE M3RJB
B77 5NW M7AWH
B77 5PL G1GAW
B77 5PQ G8VPH
B77 5QE M6GCJ
B77 5QF G6MOD
B77 5TD G6FBH
B78 1BQ M6OYF
B78 1DE G7ODM
B78 1JY 2E0RPU
B78 1JY M3ZQG
B78 1LS 2E0LLE
B78 1LS M0XVK
B78 1LS M6CKO
B78 1LW M6AMG
B78 1NJ G8NAP
B78 1NU G6DYR
b78 1nx M7TYS
B78 1QN M6OXS
B78 1QZ G7UKF
B78 1SY G2BZR
B78 2DR G4BKA
B78 2EP G4NRY
B78 2ER 2E0JBM
B78 2ER M1DNE
B78 2ER M3SWV
B78 2ER M6CSM
B78 2ET M6FAJ
B78 2JR M3OHJ
B78 2JU G0FEO
B78 2LA G0GEF
B78 2LG 2E0GEV
B78 2LG M3YFH
B78 2LH G4ORY
B78 2LW G6SJD
B78 2LX G6VWI
B78 3DE G3LGW
B78 3HS M6DEJ
B78 3HW G8OSX
B78 3HZ M7ECV
B78 3NW M6CRJ
B78 3QD G0NXY
B78 3RA G4NRX
B78 3SE G7HJG
B78 3SW G7UHG
B78 3SZ M1APQ
B78 3TJ G4ICI
B78 3YA G4SBS
B79 0DJ G6VBD
B79 0HR M6JMN
B79 0JR M6JKT
B79 0LD M1EDF
B79 7BE G0BFC
B79 7BQ G0FGK
B79 7EW G0FZM
B79 7RA M6FAH
B79 7SQ M6EEP
B79 8BE M6NIN
B79 8BH M6IWP
B79 8BZ 2E1SHE
B79 8DE G8KRV
B79 8DN M6LFD
B79 8EJ G8NCK
B79 8EY 2E0TLM
B79 8EY 2E1JMW
B79 8EY M6TLM
B79 8HZ 2E0JRW
B79 8JB G8JOB
B79 8JB M3SQU
B79 8LB G8TRS
B79 8NB M3SQV
B79 8NF 2E0FAJ
B79 8PE M0TSD
B79 8QB G7WFK
B79 8UA M3SUY
B79 9BP G3ENO
B79 9JA 2E0CLP
B79 9JA M0RGH
B79 9JA M6BHB
B79 9JG G8RVZ
B79 9JJ M6GJP
B798TA M7BAV
B8 2AU G3RSC
B8 2AU M1CQN
B8 2AU M3ASZ
B8 2AU M3BJV
B8 2AU M3LMA
B8 2AU M3SVN
B8 2EA G7GGT
B8 2LB M6TOL
B8 2PD G0NFR
B8 3LL M3AOM
B8 3LL M3REZ
B80 7HD G1YFA
B80 7HN G4YJC
B80 7JJ G7APL
B80 7LX G4LMF
B80 7PG G4STE
B80 7QL M3VDL
B80 7RD G6FDO
B80 7RR G3CON
B80 7SH G0WDU
B807BZ M3RKE
B9 5NG G7LTG
B9 5RY G7LXV
B90 1DS G0MLY
B90 1DS G7DDN
B90 1DS M6FAB
B90 1LF G4AQJ
B90 1RW 2E0KNK
B90 1RW G7JHX
B90 1RW M0KNK
B90 1RW M7KAK
B90 2BB G3XBY
B90 2BQ G7GFP
B90 2DR G7WBJ
B90 2HB G4ZVZ
B90 2HS 2E0HVF
B90 2HS M6HVF
B90 2HT 2E0EPA
B90 2HT M0IQM
B90 2HT M6IYN
B90 2HY G4RTI
B90 2LN G7IMQ
B90 2PR G7IMR
B90 2PU M3UKX
B90 2QF M0GWM
B90 2QW G6AQW
B90 3DF G0IHU
B90 3DQ M3IBS
B90 3HX G4KOR
B90 3JE M1DJG
B90 3JE M3DJG
B90 3JF M3FAE
B90 3JR G8ZUL
B90 3JZ G6FIO
B90 3LG G4TBJ
B90 3LJ 2E0DBL
B90 3LJ M0PYT
B90 3LJ M6GAL
B90 3PL G8XQA
B90 3RE G6IHB
B90 3SA G3MRZ
B90 4BU G6BWT
B90 4BU M3DGF
B90 4BX G1ZQE
B90 4DX G4LCH
B90 4HR M6JNA
B90 4JT G4LPE
B90 4PH G4BQW
B90 4PH G6HSD
B90 4PN G6HNR
B90 4RN 2E0UDM
B90 4RN M6UDM
B90 4RU G6UUR
B90 4UU G4HOR
B90 4YB G0NFZ
B902TA M0DOH
B91 1BS G6VUJ
B91 1DD G4MPG
B91 1DD G7KZV
B91 1DQ G7MAB
B91 1DX G4AMI
B91 1DY M3ZAR
B91 1LN 2E1GYC
B91 1LN 2E1GYD
B91 1LN G4WMH
B91 1PR G4LOE
B91 1QB G1FMW
B91 1TJ G3LUA
B91 1TQ G8GBM
B91 1TZ M0AEC
B91 2EG G3DEJ
B91 2EH G4EET
B91 2PL G6JEB
B91 3GA G1RBX
B91 3HH G0NOB
B91 3JY G1BHB
B91 3LL G6KMQ
B91 3LL G8UXW
B91 3NB 2E0XZM
B91 3NB M7AII
B91 3ND G4NOG
B91 3PW G1ZLC
B91 3UD G3WZI
B91 3XR G4NYG
B92 0BS G8SAN
B92 7BU 2E0XMS
B92 7BU M0LRO
B92 7BU M6MKY
B92 7BU M6MSX
B92 7DF 2E0ZSR
B92 7DF M0SRZ
B92 7DF M6GYF
B92 7ES G8GWR
B92 7EY G6BJG
B92 7HB G6DFH
B92 7HE G1STK
B92 7HE G6MCY
B92 7HH G4FJB
B92 7JA G8YMU
B92 7JB G0HLW
B92 7JH G4VMO
B92 7NU G4KQV
B92 7PN M1LCL
B92 7PN M3CTO
B92 7RU M6RJB
B92 8AD 2E0YZY
B92 8AD G4ENH
B92 8AD M7EHG
B92 8AD M7ENH
B92 8AL G6EDF
B92 8DP G7LED
B92 8DX G4CEX
B92 8EE G3GEI
B92 8EE G4BBT
B92 8HQ 2E0AXB
B92 8NB G4MVB
B92 8NN G4PCE
B92 8PL M0PPX
B92 8PU M7ALN
B92 8QA M1CXA
B92 8QS G8ULQ
B92 9BJ G1ASU
B92 9DB G1FUG
B92 9DQ G0PHR
B92 9EY M0WAO
B92 9HH G7OJO
B92 9LQ 2E0SSL
B92 9LQ M6GSJ
B92 9ND G4KWO
B92 9NW G7JVF
B92 9PQ G6VXN
B92 9PT G7GQX
B92 9QH G4EIG
B93 0BZ G4AOJ
B93 0HX G0IZQ
B93 0PT G3OIF
B93 8AN G8KBG
B93 8AW M0LYQ
B93 8DN G4RWG
B93 8DN M6TTT
B93 8NN G7FFS
B93 8QP G3UFQ
B93 8RA G8IK
B93 8RN 2E0ITQ
B93 8RN G1DWU
B93 8RN M0JSL
B93 8RN M6ITQ
B93 8SY M0XXT
B93 9AW G6VKS
B93 9EQ G3KEK
B93 9JL G4MAU
B93 9LA G3TZM
B93 9LG M0RXV
B93 9LQ G1JYR
B93 9LQ G3VPE
B93 9NP 2E0THE
B93 9NP M0SUD
B93 9NP M6SUD
B93 9PA G1VIW
B93 9PP G4CVM
B94 5DP G3GBS
B94 5EB G8NOF
B94 5LP G7OKF
B94 5LP M3JKZ
B94 5RZ G6HSD
B94 5SD G4NSB
B94 6LN M0HRT
B94 6PA 2E0DTV
B94 6PA M0REM
B94 6QH M3KKO
B94 6QY 2E0RNJ
B94 6QY M0RJH
B94 6QY M6JNR
B94 6RH 2E0LQZ
B94 6RH M7LOZ
B95 5BA G0EPL
B95 5HU G8JMY
B95 5LR G0CRB
B95 5NN G4PIP
B95 6AB 2E1FRC
B95 6BH G3UOC
B95 6BH G4CGR
B95 6JA G3XTI
B96 6AX 2E0LHE
B96 6AX M6OCY
B96 6BB G7VTR
B96 6DE 2E0EOU
B96 6DE M0EOU
B96 6DY 2E0XET
B96 6DY M6KET
B96 6EB M0CUS
B96 6ED G1DCY
B96 6ED G3RZI
B96 6LT G6CUQ
B96 6NG 2E0EGP
B96 6NG M6EGP
B97 4JL 2E0MYD
B97 4JL G0MYD
B97 4JL G0TOX
B97 4JL M7ELE
B97 4LX G3WF
B97 4NE G1JJA
B97 4NP G3KWK
B97 4PL G1RFI
B97 4PN G8MGK
B97 4RL G8DSM
B97 4RP 2E0GUF
B97 4RP M7AMB
B97 4SP M0JEM
B97 5AA 2E0ILX
B97 5AA G2FXJ
B97 5AA M6ILX
B97 5AS G6EAY
B97 5AY G3TNI
B97 5bh M7BFY
B97 5DF G4ZWR
B97 5EP G4SGV
B97 5FP G4NTG
B97 5LX G0TPG
B97 5NG G0ORE
B97 5NG G8XUB
B97 5NW 2E0LRV
B97 5NW G0RMG
B97 5NW M0RGP
B97 5NW M6IOI
B97 5PZ G1XVY
B97 5QT G1TQU
B97 5RX G0EYO
B97 5RX G6SL
B97 5TB G4NRP
B97 5UA G3OXL
B97 5UN G4SRV
B97 5WB G4NYZ
B97 5WG G8RCE
B97 5XS G0NRF
B97 6JH M3PSF
B97 6LP M3JFS
B97 6NJ G2HHH
B97 6NS G0TNH
B97 6PG M1ERO
B97 6PH G1AJV
B97 6SE G3XHY
B97 6TS G0CLM
B97 6UF G7RBQ
B975YR G4TJS
B98 0AG G7GFM
B98 0AS G4MUV
B98 0BH G4SWR
B98 0BJ 2E1FPM
B98 0BJ M3DVN
B98 0BJ M3FPM
B98 0BJ M3HLN
B98 0BJ M5DRW
B98 0BX G1AVW
B98 0EN 2E0LQY
B98 0EN M0LQY
B98 0EN M6LQY
B98 0EW M6MBH
B98 0EY 2E0PWM
B98 0EY M3PWM
B98 0HH M3TXQ
B98 0HJ G3CKE
B98 0HJ G6WPE
B98 0JQ G0PLA
B98 0JQ G7HSL
B98 0JR M7IRW
B98 0LA M3VNM
B98 0NA G1DOA
B98 0NF G1RAG
B98 0NF M5AFV
B98 0PJ G0CAX
B98 0PP 2E0ULC
B98 0QT G7AZH
B98 0SB G4DFC
B98 0TQ G6DVP
B98 7NH G1DHM
B98 7NR G0NVV
B98 7PD G6FVL
B98 7PL M6AGA
B98 7PW G3TBW
B98 7QA G3WJN
B98 7QF 2E0ETH
B98 7QF M0NAK
B98 7QF M6BNQ
B98 7RF M3EFX
B98 7SD M6LRQ
B98 7SZ G6OUX
B98 7TG M7AZR
B98 7UT G4UDK
B98 7XE G0UQE
B98 7XT G1MZT
B98 7YD G6HCW
B98 7YL G4LDB
B98 7YR G3SAH
B98 7YZ M0AMF
B98 8EY M7YYQ
B98 8HT 2E0HOL
B98 8HT 2E0MNC
B98 8JS M6MWD
B98 8PL G0KWQ
B98 8QL G0LFY
B98 8QL G0LFZ
B98 8RL G7KMW
B98 8RW G6VUN
B98 8SQ G4KNX
B98 9JA M7MRX
B98 9JH G6ERJ
B98 9JX G4TRI
B98 9JX G8EOJ
B98 9LE M6DWY
BA1 1SR M6XTM
BA1 2BL G4YNM
BA1 2UU G0EJR
BA1 2XH M3XUR
BA1 3HH M0RCR
BA1 3LN G8DWW
BA1 3RB G6MZW
BA1 3RY M7DPD
BA1 4BD 2E0XUN
BA1 4BD M7XUN
BA1 4DZ G3UMM
BA1 4NR G1BCB
BA1 5ED 2E0TGA
BA1 5ED M6LHI
BA1 5ED M6LQT
BA1 5JU M3TCG

BA1 5NF 2E0EOL
BA1 5NF M6KVM
BA1 5SW G6UTK
BA1 5SY G1ZUC
BA1 5TB M0VNR
BA1 5TB M6EUL
BA1 5TL G7SDQ
BA1 5TW G6UZG
BA1 6EF M3GZP
BA1 6JR G1MDC
BA1 6NA G8DRK
BA1 6NP G7MZY
BA1 6NZ M6GTQ
BA1 6QN G7VSJ
BA1 6QU 2E1XGX
BA1 6QU M7GCT
BA1 7BA G0WZY
BA1 7EE G3PWX
BA1 7NX 2E1DAR
BA1 7SB G1OPW
BA1 7TJ G3RVX
BA1 7TT G4JQB
BA1 8AD G0WAW
BA1 8AD G7STQ
BA10 0BS G3KZR
BA10 0HR G0WRL
BA10 0JD G7VNC
BA10 0RJ G4WTX
BA11 1AQ G3XBW
BA11 1RR M3IKD
BA11 2BD G1JPK
BA11 2BJ M7AJO
BA11 2DZ 2E0BHQ
BA11 2DZ M3SXZ
BA11 2QD M0IBW
BA11 2TN G4OWH
BA11 2UH M6JNQ
BA11 3BL G0JDM
BA11 3DP G8VGI
BA11 3DY M3BPG
BA11 3LR M0EDA
BA11 3LR M3LPF
BA11 4AA 2E0YIB
BA11 4AA M7YIB
BA11 4AB M6KMK
BA11 4JA G4XAG
BA11 4JA G4XAH
BA11 4JB M0NUH
BA11 4NR G7VQX
BA11 5AR 2E0HBV
BA11 5AR M0JCP
BA11 5AR M6HBV
BA11 6PR G1JAL
bA11 2EY M7CCF
BA12 0AE G7AZV
BA12 0ES 2E0JZK
BA12 0ES M0JZK
BA12 0ES M6JZK
BA12 0JW G0BGI
BA12 0PR 2E0FDH
BA12 0PR M6ZPH
BA12 0PW G7FHA
BA12 0RN G0AYD
BA12 6BP G0WSJ
BA12 6HH G3MAZ
BA12 6JX G4SSP
BA12 6LR G4KDK
BA12 6QZ G7DDV
BA12 7AG G7VHC
BA12 7AP G3NFJ
BA12 7BB M6FRP
BA12 7HE M0RSG
BA12 7PA G1ACY
BA12 8BU M3HNL
BA12 8DN 2E0HGD
BA12 8DN M0JUA
BA12 8DN M7SJT
BA12 8EB G0GGG
BA12 8EZ G4ILF
BA12 8GE M7JDB
BA12 8HJ M6MKD
BA12 8JR M0BYL
BA12 8JR M0HZT
BA12 8LY G7COA
BA12 8TB G6MTV
BA12 9HS G4YMG
BA12 9DU M0SJG
BA12 9EF M1SRP
BA12 9HX G4NKP
BA12 9PN G3MHV
BA12 9PN G4WHV
BA12 9PN M0ZYF
BA12 9PT 2E0PYC
BA12 9PT M0XAW
BA12 9PT M6BNG
BA126PW 2E0IEQ
BA126PW M7CQM
BA13 2GR 2E0TMX
BA13 3AE 2E0DNC
BA13 3AE M0SFT
BA13 3AE M6DNC
BA13 3AN M7HDS
BA13 3AQ M1BPW
BA13 3AU M6PNP
BA13 3ES G4IQZ
BA13 3GS M0WZT
BA13 3GS M3WZT
BA13 3GS M7THF
BA13 3HL 2E0JWJ
BA13 3HL M3JWJ
BA13 3HN M7GCY
BA13 3HP G0VFS
BA13 3HQ G4YSE
BA13 3HQ M6SBS
BA13 3HW G3ZQC
BA13 3JW G7LND
BA13 3LQ M0ADW
BA13 3QL M5EAY
BA13 3RW G6ZDE
BA13 3SH G1WFO
BA13 3UE G6ESJ
BA13 3UH G0GHW
BA13 3UH G8CPF
BA13 3UW G7PVY
BA13 3XG M0XBI
BA13 4AT G4YXS
BA13 4BH G3WZH
BA13 4EA G0PVN
BA13 4LG G4JQN
BA13 4NX G4JQX
BA13 4NY G4CLC
BA13 4PT M1FHB
BA13 4QQ G0MZJ
BA13 4TH G0JYL
BA133AN M7ASY
BA133XQ G8FAS
BA14 0AS M7KLF
BA14 0HD G0HAS
BA14 0HG G0RKB
BA14 0HG G0TFX
BA14 0HS M1WAZ
BA14 0LH G4KHK
BA14 0LH G6CKL
BA14 0LJ G7FXY
BA14 0NA M0LAZ
BA14 0NS 2E0BKU
BA14 0NS G0BKU
BA14 0NS M6TEP
BA14 0QL 2E0OMR
BA14 0QL M6OMR
BA14 0QP M3DHV
BA14 0QP M3DHW
BA14 0QS G7OIB
BA14 0RE G7NGI
BA14 0RX G7KNU
BA14 0TD G1UGV
BA14 0TD M6LEX
BA14 0TE 2E0XLI
BA14 0TE M0WIZ
BA14 0TE M7NAZ
BA14 0TX G0KCZ
BA14 0UJ G6PAE
BA14 0UX G6HQX
BA14 6EH M1CKJ
BA14 6EH M3ACA
BA14 6EZ G0HEL
BA14 6JG G6ZRL
BA14 6JG G7RAF
BA14 6JQ G0LJG
BA14 6JZ G3VMZ
BA14 6NA G4UJJ
BA14 6NY 2E0PUH
BA14 6NY M0PUH
BA14 6NY M6PUH
BA14 6QP G7JQW
BA14 6SA G4SPE
BA14 7GE 2E0JVY
BA14 7GE M0JVY
BA14 7GE M6JVY
BA14 7LE G5BBL
BA14 7PE G8GUA
BA14 7PG 2E0BUA
BA14 7PG G1HFY
BA14 7PG M3PVU
BA14 7PH G4GFJ
BA14 7PR G0VYU
BA14 7RS G3AZW
BA14 7RX M6OSI
BA14 7TX G6MQH
BA14 7UN G0KHQ
BA14 8QP G1TST
BA14 8RY G0GKH
BA14 8RY G2BQY
BA14 8RY M1BQY
BA14 8WD 2E0DUQ
BA14 8WD M6GNY
BA14 9BQ G0WPL
BA14 9DA G6POW
BA14 9GG M6ALE
BA14 9HH G0BQG
BA14 9HL G0BNG
BA14 9HS M6DNO
BA14 9JZ M3WAY
BA14 9LQ G3PYF
BA14 9PH G7PQW
BA14 9PW G0HFX
BA14 9PW G8BYI
BA14 9QS 2E0FPF
BA14 9QS M6OQQ
BA14 9RB G0TOE
BA14 9RB G6YDP
BA14 9TB G0EUR
BA14 9TP M3FMI
BA140FS M7KVZ
BA15 1AX G0IAK
BA15 1EN M0IJT
BA15 1HZ G7KBD
BA15 1JF G4YPE
BA15 1LL G8MGQ
BA15 1LX 2E0MSI
BA15 1LX M0IMS
BA15 1LX M6IMS
BA15 1RJ G0OFT
BA15 1SE G4VVZ
BA15 1SE G4ZAP
BA15 1SJ M3OKE
BA15 1TB M3VAT
BA15 1TJ G3BBX
BA15 1UD G7SPM
BA15 2BH G1LCN
BA15 2HL M0AXW
BA15 2SB G3YIQ
BA152RZ G0JYF
BA16 0BY G4PLY
BA16 0HX G7SDD
BA16 0HX M0BRH
BA16 0HY G4JBW
BA16 0HY G8SCO
BA16 0RE G4XRM
BA16 0RL G4LWQ
BA16 0RN M6LVX
BA16 0SA G1ZTM
BA16 0SN 2E0WRT
BA16 0SN M0VUW
BA16 0SN M6IKJ
BA16 0TE G1FGK
BA16 0TE G4XLY
BA16 9PE G7NCE
BA16 9PF G0BNF
BA16 9QN G6YZF
BA16 9QX 2E0RWW
BA16 9QX M6FCT
BA16 9RJ G1OOB
BA2 0AP G3WEY
BA2 0DH G0LTE
BA2 0DH G8LJY
BA2 0DZ G4BHP
BA2 0EA G4OMG
BA2 0HB M1LRX
BA2 0HB M3LRX
BA2 0JU M6WXH
BA2 0PE M6VIG
BA2 1AE G4YTN
BA2 1DY M6FZF
BA2 1DY M6OOW
BA2 1HS G6VIF
BA2 1NW G4YCE
BA2 1NW G7ANB
BA2 1NW M6YDV
BA2 1PY 2E0CXK
BA2 2DL 2E0ICP
BA2 2DL M0ICP
BA2 2DL M3XKP
BA2 2DR G4CVD
BA2 2LZ G0KTN
BA2 2NG M1ZZY
BA2 2NG M3BFY
BA2 2PG G8CJT
BA2 2PS 2E0PBL
BA2 2QB G6ZKU
BA2 2TB 2E0TFO
BA2 2TB M0TFO
BA2 2TB M3TFO
BA2 2UD G1EEZ
BA2 3BS G0FUW
BA2 3BS M3MOB
BA2 3JL G8XZB
BA2 3JW 2E0PPD
BA2 3NN M0JFJ
BA2 3PT G5RET
BA2 3QL 2E0HGQ
BA2 3QL M0VIS
BA2 3QL M7BVT
BA2 4DH M6FWZ
BA2 4HS G3NAW
BA2 4LP G0CQC
BA2 4RJ G4BBD
BA2 5AL G7NFN
BA2 5JE G8FRI
BA2 5NF G4NDT
BA2 5NF G4NDU
BA2 5PL M3ZJE
BA2 5PU 2E0MIZ
BA2 5PU M3VWR
BA2 6AL G4GON
BA2 6DE 2E1CZO
BA2 6DE G3VTO
BA2 6DE M5SUE
BA2 6DF G3VWC
BA2 6PJ G7AYL
BA2 6RG 2E0HTB
BA2 6RG M0HTB
BA2 6RG M6HTX
BA2 6UE G6MBF
BA2 6XG 2E0RTK
BA2 6XG M0RTK
BA2 6XG M6RTQ
BA2 7AF G4NBG
BA2 7AX 2E0DUF
BA2 7AX M0IDJ
BA2 7AX M6FYJ
BA2 7AY M0HFO
BA2 7BA G4FEA
BA2 7FU M0OLS
BA2 7FU M6IPW
BA2 7GR G3TTJ
BA2 7SS M6KDJ
BA2 7SS M6TJD
BA2 7TJ G3TXF
BA2 8AF G0LIB
BA2 8EF G1WFU
BA2 8EQ M0WYB
BA2 8HT M0UAS
BA2 8HT M6PSJ
BA2 8SA G4PVX
BA2 8TG M6FIZ
BA2 8TY M6IPJ
BA2 9AF M6DBW
BA2 9DZ M1TAP
BA20 1DQ M1AZF
BA20 2AZ G7SFY
BA20 2BD G4JBH
BA20 2BD G8TCQ
BA20 2BD M3VBH
BA20 2DB G8AFN
BA20 2ED 2E0JFW
BA20 2ED M3PIW
BA20 2EH G1FZL
BA20 2EQ M6WMV
BA20 2PD G4GNV
BA21 3AH G0UMS
BA21 3BE M0BAO
BA21 3BT G7LNJ
BA21 3BT M3LNJ
BA21 3HX 2E0EVU
BA21 3HX M6NNT
BA21 3JB G7AIB
BA21 3LF M6FMT
BA21 3QP M0ZFF
BA21 3TB G0TIJ
BA21 3TE 2E0KNH
BA21 3TE M0KNH
BA21 3TE M6MGF
BA21 3TW G7MSK
BA21 3UA G7KBE
BA21 4AW G7GGJ
BA21 4BA G8WBT
BA21 4DD G4EVI
BA21 4EX M3RKZ
BA21 4HF G7EAQ
BA21 4JF G3CMH
BA21 4JF M0WOB
BA21 4NN G3KCV
BA21 4NX 2E0MJM
BA21 4NX M3OWO
BA21 4PG G4KKG
BA21 4RA M3VQF
BA21 4RJ M0BHO
BA21 5FP M3SJW
BA21 5FQ M0SCA
BA21 5HA G6FHR
BA21 5JB G0AIL
BA21 5JE G3MYM
BA21 5NY G7AIC
BA21 5PN 2E0XIG
BA21 5PN M0VLX
BA21 5PN M6LZC
BA21 5RJ M6RIV
BA21 5SH G3ZLQ
BA21 5SU G6IUQ
BA21 5XA G3OBL
BA21 5XQ G6LLP
BA213FJ G8VCN
BA22 7AP G8UED
BA22 7NQ M1FFO
BA22 7QZ G8EGG
BA22 8AS 2E0LWQ
BA22 8AS M6LWQ
BA22 8BW G0LNI
BA22 8BW M7JOS
BA22 8HF G4SMD
BA22 8JY 2E0FFW
BA22 8NN M7TWC
BA22 8NS G1XNK
BA22 8NY G0WRK
BA22 8RB G7WBE
BA22 8RB M3ARS
BA22 8UR G3YPL
BA22 8ZD M7KRD
BA22 9BT M3OQJ
BA22 9BT M6KMW
BA22 9EJ M7GVM
BA22 9EN G3AST
BA22 9HF 2E0BHH
BA22 9HF M0KRP
BA22 9LY G4BMO
BA22 9QW G1GAN
BA22 9RR G2AMG
BA228RB M0DZH
BA25BN M0HLV
BA25DN G8HKK
BA3 2AS 2E0WXD
BA3 2AS M0WLD
BA3 2AS M6WXD
BA3 2AX G8HKP
BA3 2FE M6KBC
BA3 2FG G7IRF
BA3 2JH G4KVI
BA3 2JH M3FED
BA3 2NB G7IYA
BA3 2PR G1IHI
BA3 2RA M3YOB
BA3 2RN M7AMT
BA3 2RZ G6VJP
BA3 2SD 2E0DFG
BA3 2SD M0HWP
BA3 2SD M6EQK
BA3 2SL 2E1DOZ
BA3 3JZ M3XSA
BA3 3NW 2E0EFU
BA3 3NW M6JPJ
BA3 4AN G1ORL
BA3 4BB G6EYI
BA3 4BR G3RHU
BA3 4BR G4PSP
BA3 4BZ G0TJP
BA3 4EX G0HKB
BA3 4GT G4VVP
BA3 4HA M6BEB
BA3 4HG G7SSA
BA3 4HT 2E0CSE
BA3 4HT M6XAF
BA3 4JL G4MKE
BA3 4QH G1ORN
BA3 4QH G8WKK
BA3 4QH G8WKL
BA3 4RW M0SKV
BA3 4SS G6KPD
BA3 5AL 2E1FPV
BA3 5ED G7LKI
BA3 5FP M6EPO
BA3 5PG M0SBY
BA3 5PG M6SBI
BA3 5PJ G7VCY
BA3 5PP G7KEP
BA3 5PT G0HIC
BA3 5XB M0PDN
BA3 5XB M7DBN
BA3 5XU M3VQG
BA33TU M7CXW
BA34SF M3WNC
BA35JU M6KXH
BA4 4AZ M1ANL
BA4 4FB 2E0HKC
BA4 4FB M6HKC
BA4 4FD G4JJP
BA4 4FD M0WSX
BA4 4HX M6TPV
BA4 4LS M6AXR
BA4 4PS 2E0TUW
BA4 4PS M0TUW
BA4 4QA M6KVP
BA4 4QN G8EDQ
BA4 4SN G0WKM
BA4 4SZ M7AUZ
BA4 4TL G4YJH
BA4 5DH M1ERH
BA4 5GL M6DGA
BA4 5GL M6TDY
BA4 5JW M3BVK
BA4 5JW M3CGS
BA4 5JW M3FES
BA4 5LE M0PRF
BA4 5LG G7FPW
BA4 5LX M6YWA
BA4 5PX G4YJX
BA4 5QA M7AFZ
BA4 5QE 2E0ORI
BA4 5QE M0ORI
BA4 5QE M6ORI
BA4 5UD 2E0FGQ
BA4 5UD M6NJB
BA4 5UY M7CQD
BA4 5XR G8KKA
BA4 5XR M7KCE
BA4 5XW M6RZZ
BA4 5YG G6RRY
BA4 6BB G7ORK
BA4 6LZ G8DMN
BA4 6NG G4STH
BA4 6NG G4YLO
BA4 6NG G6SIG
BA4 6PN M3ULZ
BA5 1AH M5ENM
BA5 1DG G3OJL
BA5 1JA M1FJL
BA5 1PD G3XTT
BA5 1SA G2BJK
BA5 1UD G1WQU
BA5 2DG M7FTX
BA5 2EN G4KQQ
BA5 2ER M7EVM
BA5 2FE M3MYM
BA5 2FF G6SIM
BA5 2FN G3IJU
BA5 2GA M6BJK
BA5 2HZ G4ZDR
BA5 2JU G3ZJF
BA5 2NP M7JSK
BA5 2QL G4FSU
BA5 2QL M6KYB
BA5 2XG G3IUZ
BA5 3AL 2E0FFB
BA5 3AL G2RE
BA5 3AL M6EVE
BA5 3AL M6TRE
BA5 3ED G4SFS
BA5 3FG G0AQA
BA5 3HY G4XWE
BA5 3QN 2E0JNM
BA52DG G8BFV
BA6 8AW G6UVO
BA6 8EE M7WGX
BA6 8EJ G0MYH
BA6 8QS M6OEV
BA6 8RG M0BYZ
BA6 8UB G4FOB
BA6 9AN G0GJX
BA6 9JW 2E0FNT
BA6 9JW M3MMZ
BA6 9PA M6MQB
BA6 9PH G5FM
BA6 9PH G6HIQ
BA6 9PH G7SWB
BA6 9PH M0IOA
BA6 9PH M0MSA
BA6 9PQ M6DBP
BA6 9PU G7LVN
BA6 9SH M6AWB
BA6 9SP G8JDN
BA6 9TT 2E0WWF
BA7 7EH M1FFP
BA7 7FE G1YWY
BA7 7HE G0ANP
BA7 7JY G6ZJK
BA7 7LA G7PBT
BA7 7LA G8YEO
BA7 7LB M3UED
BA7 7LT G7VCJ
BA7 7PW M7BIL
BA8 0BP G0ENW
BA8 0DN G6NDA
BA8 0ED M0GOT
BA8 0HJ 2E0RNO
BA8 0HJ M0VIT
BA8 0HJ M6JPF
BA8 0HR M1EYT
BA8 0JG G4KHU
BA8 0JP G4WJW
BA8 0LR G0GRI
BA8 0LR G6ZXN
BA8 0TH G3BCE
BA9 8AL G6UZR
BA9 8LY 2E0CYC
BA9 8LY M0HQP
BA9 8LY M6BOE
BA9 9BS G8GJA
BA9 9BZ G4YXX
BA9 9EJ M0FZR
BA9 9LS G4BIN
BA9 9LT G7SSG
BA9 9PB G3TZT
BA9 9RB G6IRX
BA9 9SB G6RCT
BB1 2AS 2E0MOZ
BB1 2AS M6MOZ
BB1 2DR G3YWH
BB1 2ER 2E0MQC
BB1 2ER M6HIU
BB1 2ER M6MQC
BB1 2HB G0PXH
BB1 2HY G4XHZ
BB1 2JQ G0DTI
BB1 2NN G4FSD
BB1 3HL M6TVM
BB1 3LP G1DJQ
BB1 4BH G7WJC
BB1 4EE M3TIQ
BB1 4ES G6SUV
BB1 4JX G7WBZ
BB1 4ND G7IJL
BB1 4NP G6SZS
BB1 5HQ G4CDR
BB1 6LF 2E0DIL
BB1 6LF M0YKB
BB1 6LF M6DIL
BB1 7EX G4BJJ
BB1 7EX G8EMB
BB1 8HU 2E0KLW
BB1 8HU M6KLW
BB1 8HU M6PVV
BB1 8LP G1BBC
BB1 8NS 2E0BPP
BB1 8QZ M0SKA
BB1 9DP G3SQO
BB1 9LJ 2E0EUJ
BB1 9LJ M0IQY
BB1 9LJ M6MWQ
BB1 9NF 2E0PME
BB1 9NF M0NWI
BB1 9NF M6PWK
BB1 9NH M3JRI
BB1 9NS 2E0JCM
BB1 9NS M6AEK
BB1 9PE G8LTC
BB1 9PW G0GSN
BB1 9QN M7AGG
BB1 9QT G7MOB
BB1 9QY G0HTD
BB1 9RR M6OLI
BB1 9SA G0VOF
BB1 9SX M0KHB
BB1 9SX M7KHB
BB10 1BA G0KMK
BB10 1EU G4EOX
BB10 1HU 2E1FUH
BB10 1HZ M3LIU
BB10 1PT M7EFX
BB10 2HB G8YYX
BB10 2JT G0ECG
BB10 2LG M3OAT
BB10 2NP M3CFJ
BB10 2QP G8FLS
BB10 3AG G4YMQ
BB10 3DS G3KJY
BB10 3JF G6OHK
BB10 3JF M3JAB
BB10 3JG G8ZGF
BB10 3NG G4OKZ
BB10 3NU G6SYI
BB10 3PS M0CGB
BB10 3QH G3XAB
BB10 4AJ G6PPY
BB10 4BH 2E0FHR
BB10 4DL 2E0VNY
BB10 4DL M0VNY
BB10 4DL M7VIN
BB10 4HX 2E0LFI
BB10 4HX M0WBM
BB10 4JA G3MBU
BB10 4LA G8BEK
BB10 4LA M0BEK
BB10 4LB G0MLL
BB10 4NR M6CXW
BB10 4PD 2E0GON
BB10 4PD M6PGW
BB10 4QH G0OCK
BB10 4QH G0OCL
BB10 4TE G4JSK
BB11 1UG G0FNJ
BB11 3BE M6IEC
BB11 3EJ M7RJR
BB11 3NX M3HLV
BB11 3PR M0AIS
BB11 4BJ M3RFR
BB11 4BP G8XCE
BB11 4DN G0UAA
BB11 4LH 2E0YFR
BB11 4LH M3YFR
BB11 4NP M6VGU
BB11 4QG M3PFR
BB11 5BS M6HUF
BB11 5EA 2E0CSG
BB11 5HP G7BRS
BB11 5HX M3KVL
BB11 5QL M0CHJ
BB11 5RB G4OPN
bb115lq 2E0EVQ
bb115lq M0OGZ
bb115lq M6NFE
bb115lq M7KMS
BB12 0DF M3HLA
BB12 0DX G4GOU
BB12 0EF G0DZC
BB12 0EF G1ZBP
BB12 0JG G0MUH
BB12 0JJ G6YRC
BB12 6AA G8IUQ
BB12 6DT 2E0CKT
BB12 6ET 2E1CTU
BB12 6JT G6PRA
BB12 6LZ 2E0EXM
BB12 6LZ M0ISU
BB12 6LZ M6YHB
BB12 6NG G6RIM
BB12 6NJ 2E0TUN
BB12 6NJ G6IFV
BB12 6NJ M6TUN
BB12 6NQ G3YGC
BB12 6NZ G7DMS
BB12 6PW M6JQE
BB12 6QE 2E0DWK
BB12 7AU 2E0BLZ
BB12 7AU M3SZS
BB12 7DB G4RQA
BB12 7HT G3ROS
BB12 7HY G0DLT
BB12 7QG G3XAC
BB12 7QH G4NYL
BB12 8DR 2E0SBD
BB12 8DR 2E1CXF
BB12 8DR M6MAS
BB12 8HL M7YOR
BB12 8HN M7CCK
BB12 8JB M0MMX
BB12 8NP G1JCW
BB12 8RP M6SIR
BB12 8SH M0NWT
BB12 8SH M6NWT
BB12 9LW M0IEY
BB12 9QA G0RTU
bb127dt 2E0PRV
bb127dt M7GTE
BB18 5BS G1BDY
BB18 5BT M7MCG
BB18 5JB G4GQZ
BB18 5LB M1DHA
BB18 5LP 2E0BYM
BB18 5LQ 2E0CKS
BB18 5LQ M6HKY
BB18 5NH G7WAW
BB18 5NW G4LWG
BB18 5PD G7SNQ
BB18 5PD G7WEK
BB18 5PR M6EYY
BB18 6DD G6FEQ
BB18 6LX 2E0ZTM
BB18 6NA M3VPQ
BB18 6NA M3VQD
BB18 6WA G1VAL
Bb2 2hl M1FJF
BB2 2LT M6MAA
BB2 2NQ G1WYB
BB2 2PT M6KXQ
BB2 2RX 2E0MZD
BB2 2RX M0MZD
BB2 2RX M6MZD
BB2 2SQ M7ATG
BB2 2TX G0TPE
BB2 3EH G4HYT
BB2 3HU M6FFD
BB2 3HU M6LIS
BB2 3JS 2E0SCW
BB2 3JS M0WSW
BB2 3JS M6SCM
BB2 3JS M6SCW
BB2 3JZ 2E0JQI
BB2 3LQ M3WOX
BB2 3NZ G1BWI
BB2 3ST M0CAS
bb2 3tj M7KLR
BB2 3TP M6LBE
BB2 3UE M6MMH
BB2 4AU 2E0MPO
BB2 4AU M3NJO
BB2 4EU 2E0MFC
BB2 4EU M3JQM
BB2 4FT 2E0BGO
BB2 4FT M3OFB
BB2 4FU M6XBO
BB2 4HS M6HUI
BB2 4lx M0BXM
BB2 4NG G6ZAL
BB2 4QJ G4UCC
BB2 4QT 2E0MIV
BB2 4QT M6MIV
BB2 4RQ M0MCW
bb2 4rt 2E0FSD
bb2 4rt M6OSP
BB2 4TD G7PZT
BB2 4TY 2E0TRH
BB2 4TY M6ARH
BB2 5AH M0BHP
BB2 5DT G7VIY
BB2 5DU G6OWI
BB2 5EJ G6ZKZ
BB2 5EQ M3ZWF
BB2 5EQ M7JCC
BB2 5ER G7DEC
BB2 5EY M7KEI
BB2 5LE G3UUM
BB2 5LQ M0YCX
BB2 5LY 2E0FNH
BB2 5LY M0JFH
BB2 5LY M7DEX
BB2 5NN G3OZC
BB2 6ET G7VNK
BB2 6HR M6VWD
BB2 6LW G4TMY
BB2 6NE 2E0NJI
BB2 6NE G0ELR
BB2 6NE M0NJI
BB2 6NE M6NJI
BB2 7DS G3JXC
BB2 7HA G6ILH
BB2 7HD 2E0XUK
BB2 7HD M7BCV
BB2 7PA G4HUQ
BB2 7PN G1JHP
BB2 7PU M7CNZ
BB2 7QS G8CFD
bb25ej 2E0BFN
bb25ej M3JIT
BB3 0AJ G6UXU
BB3 0AQ 2E1HXP
BB3 0AQ G3PJV
BB3 0AY G1VLS
BB3 0BR M7BWP
BB3 0EH G8WZW
BB3 0HW G8YLM
BB3 0HZ G4JBY
BB3 0JB G6WGA
BB3 0JW G4IAT
BB3 0LU G6ETP
BB3 0QR 2E0YYM
BB3 0QR M7YYT
BB3 0QT G1ZEX
BB3 0QT G1ZFD
BB3 0QY G7DNM
BB3 0RG G1FDN
BB3 1EF M1EXS
BB3 1LQ G1JBE
BB3 1NP M3OYR
BB3 1NS G3KWO
BB3 2BS G4JGF
BB3 2DT G1BKZ
BB3 2HP G6CIE
BB3 2JH G6FKR
BB3 2LW G6RXP
BB3 2PB M6LJA
BB3 2SA G7VZS
BB3 2SF G1XNX
BB3 2SQ G0NPJ
BB3 2SQ G1ECC
BB3 2SQ G4PSE
BB3 2SS G4IBS
BB3 3AG G0BPQ
BB3 3AG M0DOC
BB3 3DR M0PJP
BB3 3DR M6PPJ
BB3 3EB 2E0FRC
BB3 3EB M0TAQ
BB3 3EB M6FRC
BB3 3GZ G0KXD
BB3 3JH G3YTI
BB3 3QA G8TJG
BB4 4AJ G4LPO
BB4 4AN G0OCW
BB4 4DZ G7JXD
BB4 4EA M3HZE
BB4 4EE G4NGV
BB4 4FN G7BXL
BB4 4HN M6FBR
BB4 4PB G0KIM
BB4 5BQ G0LRR
BB4 5BQ G4TWG
BB4 5EF M3ILV
BB4 5NA G1BPS
BB4 5NG G0PGW
BB4 5TE G8FGB
BB4 5TG M6PRV
BB4 6AY 2E0GQE
BB4 6AY G7IAW
BB4 6AY M7LYN
BB4 6BE G4ZLJ
BB4 6DS G8JCN
BB4 6EE G3XWB
BB4 6PL M7BZC
BB4 6QN G7WGO
BB4 6RX G4UUE
BB4 6TH G6GLT
BB4 7DL M7ANM
BB4 7HN G0RFY
BB4 7HN G6MBV
BB4 7JA G0MEX
BB4 7JA M0AKS
BB4 7JZ M6KKI
BB4 7TH G7CZL
BB4 7TT G6VZZ
BB4 8AG M3HJG
BB4 8JG M3FHK
BB4 8LY 2E0ETA
BB4 8LY 2E0WIE
BB4 8LY M0GLI
BB4 8LY M3RYO
BB4 8ND M7TMP
BB4 8PY G4VVK
BB4 8QH G0VYX
BB4 8QL G0ZAP
BB4 8TZ 2E0HEF
BB4 8TZ M0LMN
BB4 8TZ M6HEF
BB4 8UW G0FCA
BB4 9BT 2E0PLA
BB4 9HG G6TKB
BB4 9JE G8GNO
BB4 9PX G0FQF
BB4 9SD G4MKT
BB4 9TG 2E0HFE
BB4 9TG G5GRG
BB4 9TG M0IXW
BB4 9TG M6ODS
BB4 9TG M7NDI
BB4 9TQ G4LNE
BB4 9TR G6OIN
BB5 0FN G0JSA
BB5 0HQ G0SVP
BB5 0JJ M0BZS
BB5 0NA 2E0BBL
BB5 0NA M3HJV
BB5 0ND M3GMY
BB5 0NT G7MYJ
BB5 0SD M7OSD
BB5 0SQ G3ILE
BB5 1SL G4ZMB
BB5 2AF M0MMD
BB5 2AF M6DQT
BB5 2AS G6MKQ
BB5 2JD 2E0PCV
BB5 2JD M3LYV
BB5 2NF G1OPV
BB5 2NY G1ZED
BB5 2NY M3UOL
BB5 2PA G1BQQ
BB5 2PF G4DPU
BB5 2PR G6EYS
BB5 2QP G6VJM
BB5 2QX 2E0VIB
BB5 2QX 2E0XIA
BB5 2QX M7CCM
BB5 2QX M7GIN
BB5 2TQ G7PQL
BB5 3AT G0BMH
BB5 3AT M0HFI
BB5 3AT M0HIP
BB5 3AT M6HUW
BB5 3AU 2E0LGV
BB5 3BL M0CZU
BB5 3EY 2E0NFI
BB5 3EY G4ELR
BB5 3EY M0NFI
BB5 3EY M6NFI
BB5 3LH G4GQP
BB5 3PN M6MFL
BB5 3TA G0VGN
BB5 4AL G7DHD
BB5 4AR 2E0LLX
BB5 4AR G0UNK
BB5 4AR M3LLX
BB5 4AR M3UNK
BB5 4BZ M0JMY
BB5 4DX G4GHK
BB5 4HL G0SVH
BB5 4HL M3LPE
BB5 4JQ G4GQV
BB5 4JR G7ILY
BB5 4LY M1CEM
BB5 4PH G7WJK
BB5 5AU G2CJK
BB5 5BS M7PFB
BB5 5DH G0GBU
BB5 5GE 2E0UKU
BB5 5GE M7LMG
BB5 5GG G1CTQ
BB5 5GH G3RFN
BB5 5LA G4XEI
BB5 5NQ G4FQW
BB5 5QD G3KEG
BB5 5QH G6ZFO
BB5 5QJ G6IXP
BB5 5RB G0LQO
BB5 5RB M1CVX
BB5 5RU G8JIE
BB5 5RX G3PUO
BB5 5SW M7PAY
BB5 5XF M0AFC
BB5 5XG G6LVM
BB5 5XP G4FRF
BB5 6BD G4BYL
BB5 6BJ G1BQI
BB5 6BS G4GLW
BB5 6JG G1VVM
BB5 6JQ G0OVP
BB5 6NB M3BLO
BB5 6PL G0TZY
BB5 6QU M6CND
BB5 6TD G8RBV
BB5 6UG M7DJW
BB52AF 2E0HTR
BB52AF M0WON
BB52AF M6LRV
BB6 7AR M7CLC
BB6 7AS G0PXI
BB6 7HU G8UQY
BB6 7JE G6PXX
BB6 7JS G0TPP
BB6 7JS G8MEC
BB6 7JU G8VJO
BB6 7LP 2E0WBI
BB6 7LP M6WGB
BB6 7NF M0CRH
BB6 7NL G6LXU
BB6 7NL M0CKO
BB6 7PH G3SXC
BB6 7QF G6ENQ
BB6 7ST M6EIA
BB6 7UQ G8CZG
BB6 8BQ G6HSG
BB6 8DD 2E0VXR
BB6 8DD M0YFT
BB6 8ET G4VEY
BB6 8HB G3DMO
Bb67dj G7DSV
BB7 1DG M0BXC
BB7 1LA M1DGL
BB7 1LB M7WPE
BB7 1ND G0AZH
BB7 1PD G7RDP
BB7 2LS G8XSU
BB7 2NP M6SCK
BB7 2NX M0BXF
BB7 2QD G4JJD
BB7 2QH M1BOZ
BB7 2QN G0RKP
BB7 2QW G4LKZ
BB7 3DA G0AZG
BB7 3JD G4UMB
BB7 3JW 2E0EAW
BB7 3JW M6AZH
BB7 3LB G4WJG
BB7 4AJ M7CQS
BB7 4ES G4LDD
BB7 4lj M3HJD
BB7 4RR M0DFH
BB7 4RS G2FLW
BB7 9BJ G3YFG
BB7 9EG M7AOI
BB7 9JN G4XSG
BB7 9JW M1CLX
BB7 9TJ 2E0EXY
BB7 9XR G8IEZ
BB8 0HE M7OTO
BB8 0ND G6PDM
BB8 0TX G6BXS
BB8 7AA M3TBP
BB8 7AR G4UCU
BB8 7AR M7MJO
BB8 7HW G1WAP
BB8 7JY G3ZCX
BB8 7NH G4HXL
BB8 8BF G0JMR
BB8 8BW G6HMN
BB8 8DP G0OKV
BB8 8DT G3KLN
BB8 8JX G4DUI
BB8 8NR G4JMO
BB8 8QS G0IQM
BB8 8SA G0RFQ
BB8 8TB G0HBN
BB8 9AA M1EAZ
BB8 9AZ M0FIL
BB8 9DF G6DFV
BB8 9ED 2E0WLN
BB8 9ED M6EUP
BB8 9PE G0UGM
BB8 9QH M6MAT
BB8 9QJ G4RTS
BB8 9QR G0VQJ
BB8 9RS G7VZM
BB88JD 2E0MMP
BB9 0AP G0TFK
BB9 0EG G4ZFD
BB9 0EZ G0BQC
BB9 0HZ G4ZLU
BB9 0LE G0AQZ
BB9 0QD G0BPR
BB9 0RP G4MJX
BB9 0SB G4UEA
BB9 0ST G4MLB
BB9 0TW M1BSV
BB9 0YH G4KFF
BB9 5BN M0ETY
BB9 5DR 2E0DLZ
BB9 5DR 2E0ZLD
BB9 5DR G0ECB
BB9 5DR M0WOW
BB9 5DR M0ZOE
BB9 5DR M3LHF
BB9 5DR M3LHI
BB9 5DR M3OFC
BB9 5DR M3OFD
BB9 5DR M3VSB
BB9 5DT G4HCC
BB9 5DT M0YRC
BB9 5QB 2E0MUN
BB9 5QB M6APM
BB9 5RS G4MYU
BB9 5RX M6AZG
BB9 6BL G1LQE
BB9 6BQ G0LLL
BB9 6BT G3YGD
BB9 6EX G6JAL
BB9 6HE 2E0PIO
BB9 6HE M0JPN
BB9 6HE M3PIO
BB9 6LD G4WZM
BB9 6LZ G4SEG
BB9 7BD G4TSV
BB9 7BD G4UUA
BB9 7ET 2E0RWB
BB9 7ET M6RWP
BB9 7RA G0ABV
BB9 8AB G2IF
BB9 8DF G7PZU
BB9 8EE G6RUY
BB9 8SD G0JAF
BB9 8SQ M3NSJ
BB9 9HR G0DFO
BB9 9JL G4UNB
BB9 9LL M6ENS
BB9 9LN G3TRK
BB9 9RH 2E0VEK
BB9 9RH M3VYD
BB9 9RR G4KTR
BB90QR M7POC
BD10 0DN M6OVE
BD10 0LP G4ONZ
BD10 0NX M0AMM
BD10 0RF M7WAC

BD10 0RJ G4RSF
BD10 0UR G7ELE
BD10 8PU G1CHQ
BD10 8RZ M3UAV
BD10 9HQ M7MGD
BD10 9JW G0LJM
BD10 9SP 2E0FYR
BD10 9SP M6JYR
BD11 1AL G7SLP
BD11 1ET 2E1GVD
BD11 1HE G4TCT
BD11 1HR G4YJM
BD11 1JL G3YXH
BD11 1LU M6MAW
BD11 2EE G8CHN
BD11 2EF M3LHW
BD11 2ET M6YDD
BD11 2JE M0BCE
BD11 2JY M6UBR
BD11 2LG G7OQO
BD11 2NN G0FOI
BD11 2NN G3SVC
BD11 2PG G4UNJ
BD12 0JQ G6BIU
BD12 0UX G0WJC
BD12 8DN G0BVQ
BD12 8PT M6JKW
BD12 9DA 2E0YZQ
BD12 9DA M0XAK
BD12 9DA M6ALK
BD12 9HL 2E0OIR
BD12 9HL M0IIQ
BD12 9LS G7HSS
BD12 9NR G0LGB
BD12 9PS M7LEV
BD13 1DB M6KEM
BD13 1LD 2E0WBQ
BD13 1LD M6EHU
BD13 1LP G4XGN
BD13 1NE G7NEC
BD13 1PL G4NTA
BD13 2AE M6MCJ
BD13 2BE M6UBX
BD13 2FF M0AAM
BD13 2HQ 2E0FKU
BD13 2HQ M3AIE
BD13 2JN G0KVM
BD13 2LJ G4UKW
BD13 2QA M3JNR
BD13 2SA M6PFL
BD13 3BE G0SNV
BD13 3BE G7DFC
BD13 3BG M0NAP
BD13 3DQ G8FJR
BD13 3PQ G4JHS
BD13 4DX G0JBC
BD13 4EX M6MEH
BD13 4EY G0IBS
BD13 4LZ G4WFC
BD13 5BA G7HLV
bd133at M6EHQ
BD14 6BL 2E0GQC
BD14 6BL M6NAL
BD14 6BL M7AOZ
BD14 6DF M1EKK
BD14 6JZ G3HMV
BD14 6PJ G1IEP
BD14 6PJ G1UHB
BD14 6RY G6CMN
BD15 0BE G3UCK
BD15 0HB G7HEN
BD15 0HB M7HHB
BD15 0HH G4EZX
BD15 7AU G3NYR
BD15 7AU G4TIV
BD15 7PR G1OLQ
BD15 7QZ G1FKM
BD15 9AR 2E0MQB
BD15 9AR M7MVB
BD15 9BD G0IFT
BD15 9BD G4ZFT
BD15 9BT G6OSJ
BD15 9LD 2E0KEI
BD15 9LL M7ICP
BD15 9LP G7EVR
BD16 1AD 2E0PJH
BD16 1AD M0IRK
BD16 1BD M0EAD
BD16 1DA G8ESK
BD16 1HW G4MOY
BD16 1LN G3UOI
BD16 1LZ G0CEF
BD16 1PU M3HHC
BD16 1QB M0CRD
BD16 1QD G7CKG
BD16 1RB G4SMK
BD16 1UF G0MPP
BD16 1UH 2E0FOP
BD16 1UH M0JQB
BD16 1UH M6KQD
BD16 2BN M6HNO
BD16 2EY G0MDO
BD16 2SR 2E0ERT
BD16 2SR M0XOM
BD16 2SR M6ROB
BD16 3AY G3ZXZ
BD16 3BX G3RXS
BD16 3DF G6KJT
BD16 3DH G4YCP
BD16 3LG G8UPK
BD16 3PL G3NN
BD16 3QL M7OMB
BD16 3QN 2E0PMM
BD16 4ED G8NTZ
BD16 4EE G8PZF
BD16 4EJ M0HVX
BD16 4HL G4BLL
BD16 4LB G7UTR
BD16 4QD G4YTI
BD16 4RN G0HUK
BD17 5DW G8FTX
BD17 5NR 2E0FXM
BD17 5NR G0RJC
BD17 5NR G7HKU
BD17 5NR M0JIX
BD17 5NR M7CRF
BD17 5RL M6DER
BD17 5RS G0YKK
BD17 5TJ M1RAL
BD17 6DR G0FVO
BD17 6NN 2E0PAK
BD17 6NN M0PKW
BD17 6NN M6WAT
BD17 6RT M6NRS
BD17 7PQ 2E0HTS
BD17 7PQ M0YKS
Bd17 7pq M3TLL
BD18 1EH 2E0CDR
BD18 1EH M3RZE
BD18 1HL 2E0PLK
BD18 1HL M6PLK
BD18 2BW 2E0FKC
BD18 2BW M7CHP
BD18 2EY G1XJO
BD18 2JB 2E1GES
BD18 2JB G7KHE
BD18 2LT G0OEJ
BD18 2NT G0GME
BD18 2NT M3VFU
BD18 3AZ M7DEJ
BD18 3JB G1LZF
BD18 3LW M0IZV
BD18 4AR G8ZMG
BD18 4AW G0VJB
BD18 4DY G8ORM
BD18 4EJ G0PCM
BD18 4EW M6MTZ
BD18 4JZ M0TAN
BD19 3BY G0DPX
BD19 3DG G1SKE
BD19 3DW M7XZW
BD19 3EJ 2E0ZDW
BD19 3EJ M0WIT
BD19 3EJ M6DFW
BD19 3PX M3ZPT
BD19 4JT G4OTL
BD19 4LG G1GQB
BD19 4NX M3TAN
BD19 4RU G0MZZ
BD19 5RX G3WYP
BD19 6DQ 2E0MBQ
BD19 6DQ M6MRT
BD19 6DU M0WRE
BD19 6JH G4OSO
BD19 6JH G4OSP
BD19 6LJ G3HVP
BD2 1BQ G8AQH
BD2 1EU M7KDR
BD2 1JP G4XYR
BD2 3BP M6KQF
BD2 3HD G0PVB
BD2 3RB G1OCS
BD2 4DP 2E0VPI
BD2 4HX G8MVD
BD2 4HY G4TFD
BD2 4QX M7DOX
BD2 4SA G6PAR
BD2 4SG G1FVS
BD20 0JP M6LXS
BD20 0LD M3HOU
BD20 0LQ 2E0SJK
BD20 0LQ M0KWS
BD20 0LQ M0PLN
BD20 0LQ M6DXA
BD20 0LQ M6SAK
BD20 0ND G6IPB
BD20 0QG 2E0HZH
BD20 0QG M3VGH
bd20 0qg M7SOC
BD20 0QQ G0JFC
BD20 5AN G3WZZ
BD20 5DB G0HOT
BD20 5HN M3YBB
BD20 5JT G7BUR
BD20 5NW G4AFS
BD20 5TD M1EKD
BD20 5TD M5LAR
BD20 5TL G0SUI
BD20 6DE M7DBP
BD20 6HP G4YXB
BD20 6SP 2E0DVI
BD20 6SP M0JTM
BD20 6SP M6GGI
BD20 6SZ G1OTA
BD20 6SZ G7KDH
BD20 7BH M0PWF
BD20 7DN G4CPA
BD20 7DN M0BCQ
BD20 7EU M0AEZ
BD20 9QR G0STK
BD20 8LT M6GZN
BD20 9AU 2E0JCB
BD20 9AU M0AGN
BD20 9AU M6JTW
BD20 9JY M7BWM
BD20 9LL G0LVT
BD205HN G0KLN
BD21 1HY G3VDK
BD21 2QJ G0OPT
BD21 2QJ G1ONJ
BD21 2RX M0DIT
BD21 3HY M1PAC
BD21 4NP G8VPX
BD21 4TA 2E1HFV
BD21 4TD M0CEX
BD21 4TD M1DSV
BD21 4TF G0AJF
BD21 4TL G0BWY
BD21 4YG G0MJB
BD21 4YG G7BQU
BD21 5BD M0CAN
BD22 0AP G7SFJ
BD22 0HA G3TFF
BD22 0JQ M0MSS
BD22 0NU M0HSS
BD22 0QY G4AEE
BD22 6DD G0MEA
BD22 6ET 2E0RIZ
BD22 6EU M6AXH
BD22 6FF G4IUP
BD22 6HF M6KCP
BD22 7AB 2E0CQN
BD22 7AB M6SBO
BD22 7AU M6MLV
BD22 7BP G1AVA
BD22 7DE M7DVJ
BD22 7DH 2E1FON
BD22 7DH M1BVX
BD22 7DN G0MMC
BD22 7DW M1BZD
BD22 7EX G4TFT
BD22 7NQ G0VXS
BD22 7PD G0BQVE
BD22 7QS G3UBD
BD22 7RH G0BZH
BD22 7SH G4XNF
BD22 8BJ G0RLY
BD22 8BJ G0TSJ
BD22 8BY G6BAD
BD22 8HG 2E0MFI
BD22 8HG M6SKQ
BD22 8JY M6JLT
BD22 8PL G1NQU
BD22 8PR 2E0DOG
BD22 8PR M0UEZ
BD22 9DL M6UEA
BD22 9LE G1SRA
BD22 9LE G7KRG
BD22 9SS G3OPW
bd227df M7BNR
BD23 1BB G0XDL
BD23 1BP 2E0VRX
BD23 1BP G0RAX
BD23 1BP M6BEQ
BD23 1DQ M1DER
BD23 1RX M0IYF
BD23 1TL 2E0RCI
BD23 2BE 2E0BWN
BD23 2BE M3EHH
BD23 2BZ M6ZEA
BD23 2ET 2E0PHL
BD23 2ET M3NYF
BD23 2ET M3SKN
BD23 2LE M3CWC
BD23 2PY 2E0JBC
BD23 2PY M0YBC
BD23 2PY M3CSZ
BD23 2QE M0NTT
BD23 2RR G0UCD
BD23 2SP M3SBY
BD23 3DW G8LKQ
BD23 3DW M0DRF
BD23 3DW M0YCG
BD23 3NT G7FCD
BD23 3RY M0XLT
BD23 3SE G7CYN
BD23 3TH 2E0XLG
BD23 3TP G1LNQ
BD23 4HJ 2E0PIW
BD23 4HJ M0FTL
BD23 4HJ M3HLD
BD23 4JE G6LYE
BD23 4JW G6RIY
BD23 4LB G6LFC
BD23 4PQ 2E0HRD
BD23 4PQ M0VHG
BD23 4PQ M6HRD
BD23 5EN G6LHA
BD23 5EZ M1DEY
BD23 5JA G8UHL
BD23 6SD G4YQA
BD24 0AG G4GLC
BD24 0AH 2E0PFL
BD24 0AH M6DLQ
BD24 0DP 2E0PXP
BD24 0DP M0PXP
BD24 0DP M6PXP
BD24 0DP M7WRM
BD24 0HG G4HOK
BD24 0LH G4ZVD
BD24 9DA 2E0RSM
BD24 9HL G7EKH
BD24 9HY M7NEN
BD24 9JP M5CMO
BD24 9QR M0ANP
BD3 7BU M3IAO
BD3 7BY G7FDW
BD3 9JT G0GNU
BD3 9NU G0JTP
BD4 0BD M7DTG
BD4 0JJ 2E1HZM
BD4 0QH G4HAG
BD4 0SJ G1ANA
BD4 0TS M7TSG
BD4 6BY G0BZV
BD4 6JY M3TZS
BD4 6PJ M0CXL
BD4 7JD M1BGY
BD4 7JJ 2E1GDD
BD4 7JJ M3DXI
BD4 7JT 2E1GVC
BD4 8EN M0LMO
BD4 8JF M7DYG
BD4 8PB 2E0RAS
BD4 8PB M3RSX
BD4 9JH G1XGM
BD4 9JJ 2E0NOK
BD4 9LX G4KZW
BD5 8NX G4JRW
BD5 9AN M3XWN
BD5 9HB G0ITU
BD6 1EJ 2E0WGG
BD6 1EJ M0WIG
BD6 1EJ M7WIG
BD6 1ET 2E0JPA
BD6 1ET G6FDG
BD6 1ET M0JPA
BD6 1HL M6BRS
BD6 1PQ G0FUY
BD6 1QU G0IQH
BD6 1RP G4RFO
BD6 1RP M0AMB
BD6 1UL M6FKP
BD6 1UU G4YOR
BD6 2BJ M7ERT
BD6 2BT G6RFL
BD6 2TE M7JXR
BD6 3DJ G4UNH
BD6 3JQ G0VVK
BD6 3SW G7JZM
BD6 3SW M3XKY
BD6 3XE G0BTA
BD6 3YN M3AHL
BD62AB 2E0DMN
BD62AB M0ZUB
BD62AB M6ATB
bd62ty M7KPB
BD7 2LX G7RTA
BD7 4AS M3RQO
BD7 4BG 2E1HEM
BD7 4DB G6CGI
BD8 0AA G0OWI
BD8 0EN M0GEN
BD8 0EN M3CAP
BD8 0JZ 2E0GYW
BD8 7SZ G4VFU
BD8 9EX G7MMJ
BD9 4AX G8ZMQ
BD9 4ES M3FTU
BD9 4HG M6OMH
BD9 5PA M1DJL
BD9 6DQ G0TVM
BD9 6EX G1KWK
BD9 6EZ G7PHW
BD9 6EZ G7PHY
BD9 6NX M7BCE
BD9 6PU M3XJE
BF19DB G7GGF
BFPO59 G1VJE
BH1 1QX 2E0GQK
BH1 1QX M0VKP
BH1 1QX M7CHS
BH1 2PG G3PUZ
BH1 3DX G4LXY
BH1 3SS G4UGV
BH1 3TE G3XFD
BH1 4DS M7SHO
BH1 4PH G0OFE
BH1 4QT M7DKU
BH10 4BB 2E0UXV
BH10 4EE G1INB
BH10 4EY G7IWW
BH10 4HP G3WZP
BH10 4HP M0LWZ
BH10 5AW G4ZLT
BH10 5DD 2E0GXM
BH10 5DD M7SAP
BH10 5EP G1OQM
BH10 5JT G8CYT
BH10 5JT M0GIU
BH10 5LF G8UAD
BH10 5LG G4OEB
BH10 5LL M7MRA
BH10 5NR M6IEU
BH10 6BG 2E0WCX
BH10 6BG M6WCX
BH10 7AA G1XES
BH10 7EU G3ZWD
BH10 7HL M7GDW
BH10 7HW M6XMB
BH11 8AE M0KHA
BH11 8AW G8SLE
BH11 8BN M3MYZ
BH11 8BT M6OCS
BH11 8EE M0VCA
BH11 8EQ 2E0RAI
BH11 8EQ M6RLT
BH11 8NN G1WAW
BH11 8PS G4DKM
BH11 8RB 2E0KSC
BH11 8RB M6CJW
BH11 8RH G3XBZ
BH11 8RH G4PRS
BH11 8RH M3BDC
BH11 8SL M3ZWN
BH11 8SU 2E0EQD
BH11 8SU M6JYX
BH11 9EQ G7TKG
BH11 9EU M0AUY
BH11 9HJ M3ZKE
BH11 9HN G7VJI
BH11 9JD G7VJJ
BH11 9LJ G4XOH
BH11 9ND 2E0WMT
BH11 9ND M7AQF
BH11 9PE M7VYN
BH11 9PG 2E0SPM
BH11 9PG M0SXM
BH11 9PG M6SPM
BH11 9PJ M7JFL
BH11 9SJ G7JNS
bh118rp 2E0BAQ
BH118RP M7BAQ
BH119BP G1ERS
BH12 1DU M3XCH
BH12 1EX 2E0JNR
BH12 1EX M7CGM
BH12 1NS G1MXD
BH12 1NS M6VEG
BH12 1QE G0UCX
BH12 2EZ M6SBL
BH12 2HE G0ROZ
BH12 2HT 2E0TMN
BH12 2HT M3HCE
BH12 2HT M3KMN
BH12 2JF G4FDS
BH12 2JQ M0MMC
BH12 2LQ M3SUK
BH12 3AQ M6ANV
BH12 3AS M6GFS
BH12 3AW M0NMO
BH12 3BT M6GBH
BH12 3DA G4EOE
BH12 3HB G0CRY
BH12 3HE G6AHC
BH12 3HF M3KCA
BH12 3JW G7CAA
BH12 3LP G0GZN
BH12 3LP G0ISO
BH12 3PJ M6TGZ
BH12 4BD G1UIO
BH12 4BS 2E0TPY
BH12 4BS M0TPY
BH12 4BS M6TPY
BH12 4DH G0AEP
BH12 4DY 2E1BVY
BH12 4EH M6SJZ
BH12 4FD M6MCL
BH12 4JH G0LOU
BH12 4JH M6EKB
BH12 4JQ G4JQW
BH12 4LD M0GUH
BH12 4LT G0ISC
BH12 4PJ M0BTJ
BH12 5AY M3TFW
BH12 5ET M3OEF
BH12 5HT M7OWJ
Bh121ja 2E0WEG
BH13 6AU M0GWU
BH13 6EF G0MJO
BH13 6EN 2E0MIL
BH13 6EN M0IHM
BH13 7BE G4PVY
BH14 0DB G6EJH
BH14 0DL M7DZI
BH14 0LL G6CGQ
BH14 0PA M1PJH
BH14 0PF G8YYA
BH14 0PJ G3RRP
BH14 0PP G3MXF
BH14 0QS G3WJJ
BH14 0RZ G0RPA
BH14 8AD G6ABA
BH14 8DL M0PTR
BH14 8EG G3RGJ
BH14 8EY G8GAT
BH14 8NW G8IWE
BH14 8QY G3LUH
BH14 8SD M0BVI
BH14 8SG M7BTU
BH14 8SG M7EKS
BH14 8SG M7MOI
BH14 8SG M7OQN
BH14 8UE G0KKL
BH14 9AU G1NNR
BH14 9BS 2E0MVT
BH14 9BS M6ZDA
BH14 9EL G1TEX
BH14 9HU M6BLC
BH14 9JH 2E0DUD
BH14 9JH M3YVT
BH14 9LW G4UWS
BH14 9NX M0ILX
BH14 9NX M6IQX
BH14 9QP G0KCC
BH15 1EH G3YEK
BH15 1EJ 2E1STU
BH15 1EJ G5STU
BH15 1EJ M7SEG
BH15 1JY G8FDZ
BH15 1TT G6OFM
BH15 1TU G4UTG
BH15 1UA M0LEZ
BH15 1UP M0GPV
BH15 1UR 2E0DMY
BH15 1UR M6EYL
BH15 1XL G7FKS
BH15 2BZ G1YHE
BH15 2DB G0CBP
BH15 2DW G0PRS
BH15 2DW G4WCK
BH15 2DW G6OAI
BH15 2EF M6CDL
BH15 2ES G4CQW
BH15 2ES M1EZG
BH15 2EX G0JII
BH15 2EX G0KZD
BH15 2HQ G6NIO
BH15 2JQ G3WCU
BH15 2NZ M7BGU
BH15 2QG 2E0GVP
BH15 3AG M0IDL
BH15 3AQ G3NIL
BH15 3EE G6TEL
BH15 3ET G3HUO
BH15 3ET G3UEE
BH15 3NE G3FTK
BH15 3NJ G1ZXD
BH15 3NR G3JTK
BH15 3PH G0WTG
BH15 3PU M0PXR
BH15 3QU G0IBJ
BH15 3QZ G3RZV
BH15 4DQ G3PLR
BH15 4DQ G8EPK
BH15 4DX M3YYO
BH15 4HP G1UEQ
BH15 4HP G4CVX
BH15 4JD G1FPZ
BH15 4JR M7SOP
BH15 4JS G0JJI
BH15 4JX M6BUW
BH15 4JZ G7TZB
BH15 4JZ M3TZB
BH15 4LE 2E0DJW
BH15 4LE M6NFZ
BH15 4LT G0KQH
BH16 5BZ G0UPG
BH16 5ED G8MCW
BH16 5EJ M0ATB
BH16 5EQ G1TQN
BH16 5HE G4NFT
BH16 5LA G2HKQ
BH16 5LS M7DAF
BH16 5LY M6EKE
BH16 5PB G0JBZ
BH16 5PP G3VOB
BH16 5QT G6AKG
BH16 5RA G0IWZ
BH16 5RT G6DUN
BH16 5RX G7MUT
BH16 6AP G0ICG
BH16 6DP G0FCV
BH16 6DP M3WHQ
BH16 6DT M3EZB
BH16 6DT M7NED
BH16 6EQ G0FRR
BH16 6EQ G3PFM
BH16 6EQ G6SFR
BH16 6EQ M1DKY
BH16 6ET G3PKL
BH16 6HB M6IJR
BH16 6HJ G1JUI
BH16 6JH G1YHI
BH16 6JS G0FAJ
BH16 6NB G4MPQ
BH16 6NB G7BIK
BH17 7AH G3ZPR
BH17 7AX 2E0CHN
BH17 7DS M0RWB
BH17 7DS M1DYS
BH17 7DW 2E1HBJ
BH17 7DW G7TEZ
BH17 7DW M3GCM
BH17 7EH M0AIJ
BH17 7EU G3MEC
BH17 7HA 2E1DBS
BH17 7SB G6MXL
BH17 7UP G0FVH
BH17 7UP M3RLX
BH17 7XT 2E1FDK
BH17 7YJ 2E0LGR
BH17 8BP G7SVI
BH17 8BU 2E0EXO
BH17 8BU G1ZSE
BH17 8BU M6EXO
BH17 8BU M7FWW
BH17 8DB G1HBF
BH17 8PW G4UKR
BH17 8SQ G7MZX
BH17 8SR 2E0AOL
BH17 8SR M0AKY
BH17 8SU G0RVI
BH17 9AH M1EEQ
BH17 9AS 2E0TCF
BH17 9AS M0TGY
BH17 9AS M6TEG
BH17 9BF G4IBM
BH17 9BF G8RSE
BH17 9BG G7OAX
BH17 9DU M5PDL
BH17 9EE M6LDF
BH17 9EQ M7HGM
BH17 9WE G7GNU
BH18 8AE M0HZU
BH18 8DY 2E0UEL
BH18 8DY M0HTY
BH18 8DY M6DWV
BH18 8HZ M3WBM
BH18 8JS 2E0AMW
BH18 8JS M0CJC
BH18 8LN M3FSB
BH18 8LN M3FSD
BH18 9AE G4BYO
BH18 9AJ G1YHJ
BH18 9AX G4SZO
BH18 9BD G3PSV
BH18 9DF G8HNA
BH18 9DZ G8TNU
BH18 9DZ M5ADL
BH18 9DZ M6IAL
BH18 9DZ M6TNU
BH18 9ED G8JMB
BH18 9EH G7EYA
BH18 9HQ G4BKE
BH18 9HY G6PIM
BH18 9JG G3SGX
BH18 9JG M0BHH
BH18 9LB M6TJL
BH18 9NR G0WKH
BH18 9NR G0BRQ
BH18 9QZ 2E1DFZ
BH19 1HY G1HEJ
BH19 1HY G4ZPB
BH19 1JT 2E0HVU
BH19 1LG G7FXO
BH19 1NF G6CAC
BH19 1PQ M3SGZ
BH19 1QF G8CLJ
BH19 1QU G7OVM
BH19 2DE G3MBM
BH19 2DL M0HRZ
BH19 2ET G3NR
BH19 2LB G4ZYY
BH19 2PX G4CEK
BH19 2PZ G7OYX
BH19 2RB G0WZK
BH19 2RG M1DCH
BH19 2RU 2E0KJK
BH19 2RU 2E0WRO
BH19 2RU M0KJK
BH19 2RU M6KJK
BH19 2RU M7WGR
BH19 2SL G0TOT
BH19 3LU 2E0WDR
BH19 3LU 2E0WVL
BH19 3LU M0WVL
BH19 3LU M6WRL
BH19 3LX G6BOF
BH19 3LZ M7MRB
BH2 5JT G8CJH
BH2 5NR G4FNI
BH2 5RA M3XZR
BH2 5RA M3XZS
BH2 6LT M0IJY
BH20 4AQ G0WFE
BH20 4DR G7VXQ
BH20 4EL G4GHA
BH20 4HY G0BAO
BH20 4JA G3TIG
BH20 4QX G0TGP
BH20 4SD M3KQP
BH20 4SG G1HXR
BH20 4SJ M6JHS
BH20 5BB M0IXC
BH20 5EE M6LPV
BH20 5ES G0MWV
BH20 5QW M7VVS
BH20 5RN G1LZZ
BH20 5RQ G6HCH
BH20 5RY G1EQJ
BH20 6AA M6STC
BH20 6DB M7NKE
BH20 6DY M6KAT
BH20 6DY M6NEV
BH20 6EQ G7MND
BH20 6HB G7EWY
BH20 6JZ M3MUF
BH20 6NN M0HXA
BH20 7BD 2E0XAM
BH20 7BD M0XAM
BH20 7BD M3XAM
BH20 7BX G1AAH
BH20 7DH G1PJB
BH20 7JF G7BRU
BH20 7LU M6BEE
BH20 7LU M6SSC
BH20 7NH M6NBL
BH20 7NZ 2E0NVL
BH20 7NZ M6NVL
BH20 7PA G6XEL
BH20 7PE 2E0JPD
BH20 7PE 2E0VGC
BH20 7PE M3ZNP
BH20 7PE M6DKT
BH20 7QA 2E0CTE
BH20 7QA M6CCE
BH21 1JJ G3VQR
BH21 1LE 2E0IWI
BH21 1LE M0IWI
BH21 1LE M6IWI
BH21 1PJ M0JLE
BH21 1PJ M0TAK
BH21 1PL G4RLM
BH21 1QJ G4FUP
BH21 1QT G6NZN
BH21 1RG G6EOK
BH21 1RX M1EBH
BH21 1SL G8CEZ
BH21 1SR G6JKK
BH21 1SX M0CGT
BH21 1TB G6SHS
BH21 1TE 2E0JWS
BH21 1TP G0GOX
BH21 1TU G4WEY
BH21 1TZ M3ONK
BH21 1UQ G8MXW
BH21 2AA M1ETT
BH21 2AU G0EEF
BH21 2BL M0MJS
BH21 2DH G0UPS
BH21 2EA G0MTV
BH21 2HR G4NHE
BH21 2JB G7MQU
BH21 2JU G0DVE
BH21 2JZ G4RFV
BH21 2LA M1BAI
BH21 2LA M6REQ
BH21 2LD G3BHM
BH21 2LE G0MDK
BH21 2NG G8YLR
BH21 2NZ G0WAL
BH21 2PQ G3ZXW
BH21 2QB 2E0AKK
BH21 2QB M0GJV
BH21 2QB M3GTH
BH21 2SF G0SKN
BH21 2SR M7JEH
BH21 2ST G8CHI
BH21 2UW G0UGS
BH21 3AF M6CVF
BH21 3BG G3RJH
BH21 3BJ G4VZT
BH21 3DS G1VIP
BH21 3DS G8UCY
BH21 3EZ G0API
BH21 3EZ G3ZTY
BH21 3EZ G6NLC
BH21 3EZ G7MHO
BH21 3EZ G7PRO
BH21 3HL G0PRH
BH21 3HW M0EYT
BH21 3JD 2E0WCL
BH21 3JD M0WTL
BH21 3JD M6WCL
BH21 3LG G0FKY
BH21 3LW M0FBB
BH21 3LZ G1YHV
BH21 3NH G7AUF
BH21 3NN 2E0BNA
BH21 3PN 2E0ZXR
BH21 3PN M6OOD
BH21 3PU G8NVC
BH21 3PZ M1EZL
BH21 3QG M3YZP
BH21 3SB G0ODP
BH21 3SL G8KNS
BH21 3SP G0KFM
BH21 3SU G3NUN
BH21 3TB M1PMR
BH21 3TF G7AGB
BH21 3TL M0MRP
BH21 3TW M0GUJ
BH21 3UA G8CRV
BH21 3UW 2E0CMT
BH21 4AE M6JWQ
BH21 4AX M7WWX
BH21 4BH G4WUK
BH21 4BH M7DZL
BH21 4DS M6FRV
BH21 4HD G8ZEK
BH21 5AF G6CHX
BH21 5BH M3RLB
BH21 5ET G4GXZ
BH21 5RD G7MYI
BH21 5RD G7KJI
BH21 6RR M3XQM
BH21 6SG G4MHF
BH21 7AN G3OAF
BH21 7AZ G7WBU
BH21 7EZ G4MSA
BH21 7JZ M0MAI
BH21 8LQ M7RCS
BH211XQ G4YGM
BH22 0BQ G4PIJ
BH22 0BT M7PBT
BH22 0BW G1GRB
BH22 0BZ G1JBM
BH22 0DW G8TEO
BH22 0EE G0KLQ
BH22 0EX M6ODC
BH22 0EY 2E1FSZ
BH22 0HN G4FOL
BH22 0LG G4VLP
Bh22 8AA G1BBI
BH22 8AJ M6MKO
BH22 8AL G0UJI
BH22 8AT G1LDJ
BH22 8BA G4VHI
BH22 8BY G1BRS
BH22 8BY G2BRS
BH22 8BY G3CPN
BH22 8DZ G8UJS
BH22 8EX M6JQL
BH22 8PA 2E1FSY
BH22 8PQ M7AOC
BH22 8QS M0WPR
BH22 8QS G8DJL
BH22 8QT M6FCD
BH22 8QW G8TSZ
BH22 8RG 2E1JME
BH22 8RS G0AEL
BH22 8RW M6KDV
BH22 8RZ 2E0ADN
BH22 8SB G3ZCL
BH22 8SF M0EWG
BH22 8SF M1EWF
BH22 8SF M5BMW
BH22 8UR 2E0DQX
BH22 8UR M6JEO
BH22 8XA M5ACX
BH22 8XN M0DOS
BH22 9EQ G8YHF
BH22 9ES G4LJN
BH22 9FW 2E0RXR
BH22 9FW M6RXR
BH22 9HN G1RSE
BH22 9HP G4ULQ
BH22 9HW G4CFW
BH22 9JE G3WPB
BH22 9JF 2E0XOE
BH22 9JF M0WEO
BH22 9JH M7AGV
BH22 9JH G0NLM
BH22 9JP G8DQE
BH22 9JQ G0CDY
BH22 9LA G4PAI
BH22 9LH M6PEB
BH22 9LJ G3WEG
BH22 9LY M3PIL
BH22 9NB G8XKD
BH22 9PD G0ENA
BH22 9QB M3ORN
BH22 9QD M6INM
BH22 9QE G3EBP
BH22 9QE M0DOY
BH22 9QQ G1DNK
BH22 9RW M7POI
BH22 9SD G4ACJ
BH22 9SG G4XEE
BH22 9UX M0VWP
BH220BU M0CPU
BH23 1FA G0JIL
BH23 1JU 2E0IYY
BH23 1QL 2E1CDS
BH23 1QL G0RCN
BH23 1QT G4AWA
BH23 1QX 2E0XAZ
BH23 2AF G8ZPW
BH23 2AG G4HEC
BH23 2AJ 2E0GFE
BH23 2AJ M6GFE
BH23 2AU 2E1CEQ
BH23 2DF G4XWT
BH23 2EH M3VRM
BH23 2HJ M0CDY
BH23 2LL 2E1FTI
BH23 2NG M3LAQ
BH23 2NP M1EGM
BH23 2NW G6MUQ
BH23 2NW M3TJO
BH23 2RP G3YPT
BH23 2SP G3UXR
BH23 2TW G6MYT
BH23 3BJ G4SEU
BH23 3BJ G6LUK
BH23 3BL 2E0KFR
BH23 3BL M6KDR
BH23 3EW G6SPB
BH23 3JA G0AHK
BH23 3JA G0MUD
BH23 3JA G7MUD
BH23 3NB 2E0CSF
BH23 3NB M3CRL
BH23 3NB M3CSF
BH23 3NS 2E1DHX
BH23 3QN G4MUU
BH23 3SN G4SNR
BH23 4BP G6FAH
BH23 4DD G0NIQ
BH23 4ED G0TVS
BH23 4HB G1SLO
BH23 4HB G6MEH
BH23 4LN 2E0KJG
BH23 4LN M6KJG
BH23 4QQ 2E0CXM
BH23 4SL M6ILB
BH23 5AH 2E0CPS
BH23 5AH M0YAH
BH23 5AH M6ZYX
BH23 5BA G8GZN
BH23 5DB G7UWZ
BH23 5DF G3NDS
BH23 5DX G1HQG
BH23 5HD 2E0DBB
BH23 5HD M6DAG
BH23 5PN G3YNC
BH23 5QD M3EBA
BH23 5RF M0MRW
BH23 5SE M7GMK
BH23 5SG M6BCU
BH23 6BE G0IXT
BH23 7BE G1LLA
BH23 7LD G0IKN
BH23 7LE M0DEP
BH23 7NJ G4TKP
BH23 7NT M6PDQ
BH23 7NW G4FND
BH23 8BS G8YBZ
BH23 8BT G4OXK
BH23 8BW 2E0LYZ
BH23 8BW M7DAK
BH23 8DU G3PIY
BH23 8DU G3SGL
BH23 8HG G1IOP
BH23 8HG G8JEM
BH23 8NA G0HJL
BH238BU G1BLO
BH24 1FA 2E0KCL
BH24 1FA M6KCL
Bh24 1ll 2E0EIE
Bh24 1ll M3FSS
BH24 1LS G1RRR
BH24 1LS G1SVJ
BH24 1NE G6SUC
BH24 1NQ M7IJM
BH24 1NX M3ZER
BH24 1PH G4ZLI
BH24 1PQ G4ZPW
BH24 1PU G8LHZ
BH24 1PX G8BDF
BH24 1QU G3OMY
BH24 1QX G6PBO
BH24 1UY M1EHF
BH24 1UY M3ANH
BH24 1XD G6PLU
BH24 1XL G1LOU
Bh24 1xy 2E0EID
Bh24 1xy M6HXR
BH24 2BH 2E1HGE
BH24 2BL G1MTA
BH24 2BN M7AMW
BH24 2BT M6DTY
BH24 2DU 2E0YKX
BH24 2HH G6EDU
BH24 2HH G6IMH
BH24 2HR G1MMT
BH24 2HS 2E0DWS
BH24 2HS M0WCS
BH24 2HS M6BUI
BH24 2HY G5HY
BH24 2HY M6IWO
BH24 2JA G0SQH
BH24 2JP G0FUR
BH24 2JP G1VBA
BH24 2PE M6HFV
BH24 2PF G8SXJ
BH24 2PH G3YNJ
BH24 2QJ G3NWL
BH24 2QS M6DOB
BH24 2QW M5AGG
BH24 2QZ 2E0OEV
BH24 2QZ M3OEV
BH24 3BE M0HVO
BH24 3BP 2E0RFG
BH24 3BP M3RFG
BH24 3DS M7LAE
BH24 3DT G3ZCI
BH24 3DX 2E0IKC
BH24 3ER G1OCH
BH24 3HT G0MYL
BH24 3NQ G1WSN
BH24 3NQ M3WSN
BH24 4EL G0DBI
BH25 5AE M3WRN
BH25 5AY M6ITL
BH25 5BQ M0JRE
BH25 5ED M0AYC
BH25 5EJ M0EVI
BH25 5GX G4ZKT
BH25 5JF M0DBX
BH25 5JG G7EPE
BH25 5JP 2E1GQN
BH25 5JP 2E1HAW
BH25 5JP 2E1HFA
BH25 5JP G1ZEC
BH25 5JP G1ZVC
BH25 5JP G8CON
BH25 5JP M3NBB
BH25 5NA G0RIX
BH25 5NB G0TMZ
BH25 5NL M6CFY
BH25 5NQ 2E1IHW
BH25 5NQ G4MVP
BH25 5NS M0IVZ
BH25 5PE G1HHO
BH25 5PW G0HOQ
BH25 5SD 2E1DUW
BH25 5SD G4DUW
BH25 5XE M3RLH
BH25 5XP G1IAV
BH25 6AB G3DCO
BH25 6ES G4ODM
BH25 6ES G6JDP
BH25 6EY G4HFQ
BH25 6NW 2E0VDD
BH25 6NW 2E0VDE
BH25 6NW M0NFR
BH25 6NW M0RBF
BH25 6NW M3VDE
BH25 6NW M3YGF
BH25 6NZ G6LVC
BH25 6NZ G8WQW
BH25 6QE G1UWV
BH25 6SL G3UEZ
BH25 6TA G4XYU
BH25 7AJ 2E0TFE
BH25 7AJ M0ORC
BH25 7AJ M3TFE
BH25 7AW G7ILD
BH25 7DF M0ORE
BH25 7DQ 2E0GMF
BH25 7DQ M3IZQ
BH25 7DQ M7ABC
BH25 7HR 2E1IGY
BH25 7HR G6RFG
BH25 7JL M7EIU
BH25 7JT M1HHL
BH25 7LU G4BJB
BH25 7LU G7MER
BH25 7NX M0ABI
BH3 7AF G7HEY
BH3 7AF M3TIE
BH3 7AG M0MEY
BH3 7DG G3YUZ
BH3 7DY M1GPE
BH31 6BL G8YZL
BH31 6BL M0SPC
BH31 6BS G4GTH
BH31 6DX M6DYY
BH31 6EY G0SWF
BH31 6HX G8NAG
BH31 6JA M6FFM
BH31 6JJ 2E0EHV
BH31 6JJ 2E0PSZ
BH31 6JJ G4WHY
BH31 6JJ M0PSZ
BH31 6JJ M0VLT
BH31 6JJ M6PSZ
BH31 6JJ M6UAX
BH31 6JP 2E1HVM
BH31 6LB 2E0NTV
BH31 6LB M6NTV
BH31 6LH G0WXZ
BH31 6LQ G4TMF
BH31 6QB G0VPJ
BH31 6US M6NDF
BH31 6XA G6SRV
BH31 7LE G3XYE

BH31 7PG G0DEJ
BH31 7PW G4HJH
BH4 8AB G8PKG
BH4 8AD M0IPJ
BH4 8AL M0DZT
BH4 8AL M3DUL
BH4 8AT G1HDO
BH4 8BX G7FPR
BH5 1EB 2E0YMR
BH5 1EB M6YMR
BH5 1LT M1BFG
BH5 1LY G8PVK
BH5 1NF M7SHQ
BH5 2DJ G1PFY
BH5 2EP M7CBP
BH5 2HT M6FCN
BH6 3DU G3WAL
BH6 3HJ G8ARA
BH6 3JR M0TRF
BH6 3LU G7SUM
BH6 3ND M0MPS
BH6 3NN G4YNO
BH6 3PZ M3ZNZ
BH6 4AE G6EZM
BH6 4DT G4YRY
BH6 4DX G4RGP
BH6 4DX M3GPP
BH6 4HQ G4WLP
BH6 4HU G3IQX
BH6 4NB G8CRZ
BH6 5AL M1BSB
BH6 5AL M6NZF
BH6 5BB M6HRV
BH6 5ET M7OBY
BH6 5JL G8GLY
BH6 5LD G6INW
BH6 5NW G4JYH
BH6 5PY G8ASX
BH7 6LF 2E0SUZ
BH7 6LF G4ERV
BH7 6LF M0SUZ
BH7 6LF M6SUU
BH7 7AA G4ELC
BH7 7AS G4BNO
BH7 7BD G0OPI
BH7 7BD G4JDW
BH8 0BT 2E0RIS
BH8 0BT M0JAX
BH8 0BT M6BCY
BH8 0DS M1EGL
BH8 0JP 2E0DFA
BH8 0JQ 2E1ICM
BH8 0NL G4STB
BH8 8NP G8FXL
BH8 8PS 2E0BKE
BH8 8PS M0KUP
BH8 8PS M3UHX
BH8 8SF M5BUF
BH8 8SR G1BNG
BH8 8SR G4WFZ
BH8 8ST M3MNR
BH8 8UA M1ERU
BH8 9AE G8OWZ
BH8 9HW G0EJO
BH8 9HY G0EGR
BH8 9PJ G7HLG
BH8 9SG G8MQT
BH8 9UD G3IWV
BH80LB M7AVR
BH9 1AN M6SNQ
BH9 1DB G4KLB
BH9 1DB G8SWO
BH9 1LH G7ADH
BH9 1LR M1ELM
BH9 1PJ M7LUK
BH9 1SH 2E0ZDB
BH9 1SH M0ZDB
BH9 1SH M6ZDB
BH9 1SH M6ZJB
BH9 1SJ M6BPV
BH9 1TX G8RAJ
BH9 2BZ G4HBD
BH9 2JE G3JAU
BH9 2JQ G6CML
BH9 2ND G3AWP
BH9 2QJ G0UAP
BH9 2UD G6ZAX
BH9 2UJ M0KIF
BH9 2UJ M6KIF
BH9 3AJ M7AWM
BH9 3BB M3HDV
BH9 3BX M6FCI
BH9 3EH M3VUN
BH9 3LP G7VIK
BH9 3LP M3JER
BH9 3LW G1HRH
BH9 3LW M6LXE
BH9 3NS G0URB
BH9 3PZ M1PKB
BH9 3PZ M3BVA
BH9 3SB G3BHA
BL0 0DP G4PAS
BL0 0LD G4YVO
BL0 0LD G6ZVO
BL0 0QA 2E1EPD
BL0 0QA 2E1EPE
BL0 0RY G4PVS
BL0 9BR M3IMR
BL0 9DF 2E0CCD
BL0 9ED 2E0GAO
BL0 9ED M0YYA
BL0 9ED M6GEF
BL0 9JX G7CER
BL0 9LZ M7BVK
BL0 9ND G3XPZ
BL0 9ND G4SVA
BL0 9QG G6SHD
BL0 9RE G6JJF
BL0 9SB 2E0IFP
BL0 9UF G0UKM
BL0 9UN G7BYS
BL0 9UT G4APJ
BL0 9XE G6CHC
BL0 9YN G0JOG
BL0 9YP 2E0YLR
BL0 9YU M7JJG
BL1 2JP G0WWH
BL1 2JX 2E0NNX
BL1 2JX M6NNX
BL1 2XP G1ITV
BL1 3AU M7AAQ
BL1 3DJ 2E0FNB
BL1 3EZ G8JIT
BL1 3LD G0IZJ
BL1 3PE G6VFB
BL1 3RH M3YWG
BL1 4HW G7ROM
BL1 4LZ G6TWX
BL1 4NJ G6YEA
BL1 4PA G0VUX
BL1 4RQ G0BWC
BL1 4RQ G6GVI
BL1 4RQ G6MRY
BL1 4SA G4NTC
BL1 4UA G0FRL
BL1 5FJ M0HZX
BL1 5RG 2E1HLA
BL1 5RG M0BGU
BL1 5TB M7BPD
BL1 5UN 2E1IAB
BL1 6AA G0PVP
BL1 6AH M3ZHP
BL1 6AJ M6BRP
BL1 6BL G1KQZ
BL1 6DW M7ARO
BL1 6EF M6JRK
BL1 6HZ M0CSP
BL1 6LU M6ODD
BL1 6NT M3TEY
BL1 6RR G7KON
BL1 7QJ M7DYI
BL1 7RJ G6ZBV
BL1 7RJ M3ZBV
BL1 8PA G4RWS
BL1 8RW G1PSL
BL1 8SD G4YNK
BL1 8SF M1AWC
BL2 1AD G7DAZ
BL2 1PA G1OUG
BL2 2DE 2E1EKQ
BL2 2ET M6LUM
BL2 2SS M3DPI
BL2 2TA G0JFE
BL2 2TA G0TCY
BL2 2TA G1IOO
BL2 2UR G1JMV
BL2 3AY G1YQI
BL2 3AY G4XPU
BL2 3BW M3UNT
BL2 3JJ M0ZVF
BL2 3NG G4CMH
BL2 3NS G0HYG
BL2 3QN G6MVR
BL2 3QT 2E0RIT
BL2 4AQ G1RRE
BL2 4DU G1ZWQ
BL2 4ET G0EWV
BL2 4LN G1CCD
BL2 4LZ 2E0MLK
BL2 4LZ M0MLK
BL2 4LZ M6FYL
BL2 4NU 2E0JAC
BL2 4NU M0JBC
BL2 5ED 2E1GNZ
BL2 5LY G0GRX
BL2 5LY G4MUQ
BL2 5NL G7HVO
BL2 5NS G4TZG
BL2 5QU G6ERK
BL2 5RJ G0KEX
BL2 6BJ G1VFW
BL2 6EU G1EVR
BL2 6JJ G3TVT
BL2 6LN 2E0MYX
BL2 6LN M6MYH
BL2 6LQ M3SVT
BL2 6LT G4AQB
BL2 6LW M7OBJ
BL2 6NX G8YOY
BL2 6RQ G4CIC
BL2 6TU M6BNB
BL2 6UD M3OPN
BL2 6UE M6MFU
BL2 6US G3ZIK
BL26HB M3ZPJ
BL3 1BA 2E0BVJ
BL3 1BA M3VPY
BL3 1BA M6BWC
BL3 1JU G4YKB
BL3 1NX G1BIF
BL3 1PN G6VPH
BL3 1PW G1EDM
BL3 1QE M3KKZ
BL3 1RG 2E0DAH
BL3 1RG 2E0RIN
BL3 1RG M0LSW
BL3 1RG M0TEG
BL3 1RG M3PYO
BL3 1RG M3YRH
BL3 1RN G8ROS
BL3 1TE G8RGN
BL3 1TR M3YTA
BL3 1UB G7GLS
BL3 1UD G6CTH
BL3 3AU G7MGC
BL3 3ER 2E0IRN
BL3 3ER M6NAV
BL3 3JD G1YYH
BL3 3JH G6KUJ
BL3 3LG M1FJJ
BL3 3NU M0IZG
BL3 3RB 2E0NSG
BL3 4BR M3GPC
BL3 4BW G0RIP
BL3 4JU M6SXI
BL3 4LF G4TNA
BL3 4PH 2E0JSR
BL3 4PP G4DFP
BL3 4QR G7RZW
BL3 4QZ G1AEQ
BL3 4QZ G1ONE
BL3 4QZ M3DEE
BL3 4TF M7HME
BL3 5HX G4PMV
BL3 5NU G0KAB
BL3 5PJ 2E0CAV
BL3 5PJ M0TJV
BL3 5PQ G7SKV
BL3 6QP M6EGJ
BL3 6RA M6HFN
BL3 6SJ M1FAF
BL3 6YH 2E0SIA
BL3 6YH M6ARL
BL33DT M0IWN
BL4 0AU M6PEZ
BL4 0AY M1TOM
BL4 0DS G6KQN
BL4 0EN M6OYC
BL4 0ES M7PHB
BL4 0HQ G1RWX
BL4 0RG 2E1AMW
BL4 0RN M0BOC
BL4 7DF M3KVR
BL4 7DQ G0BMS
BL4 7DQ M3LUP
BL4 7HS M0JDE
BL4 7QD G3BRS
BL4 7QD M0OWS
BL4 7TG M0CPT
BL4 7TH M6KIN
BL4 8DW G1FGE
BL4 8NR G1YCK
BL4 8NT G6QA
BL4 8PS M6LEE
BL4 9EN M6OSX
BL4 9HT G3XUM
BL4 9LX G0KHJ
BL4 9NE G6SNZ
BL4 9PT G4HYG
BL4 9QW G0TBW
BL40QH 2E1HDF
BL5 1BA G1JZX
BL5 1BL 2E0SXP
BL5 1BL M3SXP
BL5 1DL M0YOT
BL5 2AF M3BBC
BL5 2AF M3LMH
BL5 2DT G6SSH
BL5 2EG 2E0XGS
BL5 2EG M0XGS
BL5 2EG M6XGS
BL5 2EU G7BKL
BL5 2GR G6OBE
BL5 2HR G2ANC
BL5 2LE G0AIM
BL5 2LY G6GVR
BL5 2NT M6TCB
BL5 2PJ G7CGN
BL5 2PJ M0SPH
BL5 2QG M0KSE
BL5 2QG M7ESK
BL5 2QL G7GKC
BL5 2QL M7JOP
BL5 2QU M7JCX
BL5 2RE 2E0HGM
BL5 2RE M7GMH
BL5 2RF M7BSS
BL5 2RN M7FGS
BL5 2RT M6MYT
BL5 2SD G6UVS
BL5 2SL M6ABV
BL5 3AX 2E1CKH
BL5 3DN G6GYC
BL5 3DP G0LBE
BL5 3HQ G4RIE
BL5 3TR M7AZP
BL5 3UZ G8IZR
BL5 3YB G4EGG
bl5 3zd 2E0JBQ
bl5 3zd M6PYD
BL51EN 2E0FMI
BL51EN M6JVF
Bl53js G0IUA
BL6 4AZ G4PDD
BL6 4BB M0WHB
BL6 4BB M3BXC
BL6 4FA G4IAD
BL6 4HX G0MXH
BL6 4JF G1AQP
BL6 4LG G8WLL
BL6 4PJ G3JNM
BL6 4RQ G1VHW
BL6 5BG G0MRL
BL6 5BG M3LBQ
BL6 5GZ 2E0HLG
Bl6 5ny M6IQK
BL6 5QX 2E0EBJ
BL6 5RH G7CIQ
BL6 5TE G1VON
BL6 5TR 2E0DDN
BL6 5TR M6DNQ
BL6 5UG G4CFP
BL6 5UG G4WBO
BL6 6DD M7MCQ
BL6 6DJ 2E0HHT
BL6 6DJ M6CRR
BL6 6HD 2E0GXC
BL6 6HD M0MHW
BL6 6HD M6JQZ
BL6 6HN 2E1BKT
BL6 6JE G0UXF
BL6 6NR G6YQI
BL6 6PZ G7GFK
BL6 6QG G3KMS
BL6 7AF M6MAM
BL6 7BE G0WBT
BL6 7ED G0KGI
BL6 7HZ G7ETK
BL6 7JU G4FSN
BL6 7LF G6CLX
BL65SZ G1EFP
BL7 0DE M0XAS
BL7 0DE M6OOM
BL7 0HS G0RGO
BL7 0HS G0TWD
BL7 0HS G7PSU
BL7 0PA G3HCZ
BL7 0PW G4XTG
BL7 9BN G1HSG
BL7 9DX G8UQV
BL7 9LL G6MAJ
BL7 9RR M0HDQ
BL7 9SP 2E0LXF
BL7 9SP M0PSY
BL7 9SP M3NPO
BL7 9TF 2E0HBP
BL7 9TF M7BKS
BL8 1BJ G0HUQ
BL8 1DW G6DEG
BL8 1JA M0IWV
BL8 1JA M0IWW
BL8 1JB G0TGR
BL8 1RT G7SPL
BL8 1TN M6XEY
BL8 1UE G1RAO
BL8 1XB G0ROC
BL8 1XX G6BIT
BL8 2BQ G0BZU
BL8 2HD G4GSY
BL8 2HF M1BNH
BL8 2QG M3IXO
BL8 2RE G7NLP
BL8 2TS G6DGV
BL8 2UL 2E0DQJ
BL8 2UL M6EVH
BL8 3AU M3TIF
BL8 3BL G0VYP
BL8 3DB G6AMX
BL8 3DB G6PZE
BL8 3DB G8AEN
BL8 3DB G8EZU
BL8 3DT 2E0FPW
BL8 3DT M0WJG
BL8 3DT M7WJG
BL8 3DY 2E0DNB
BL8 3DY M3DNB
BL8 3DY M3DNC
BL8 3JA G0WVY
BL8 3JE G3DQQ
BL8 3LW M7CLK
BL8 4BG G3BQT
BL8 4EP M0BSB
BL9 5BT G0KAY
BL9 5DL G1PKO
BL9 5DL G4KLT
BL9 5DQ G0GPH
BL9 5HQ M6JXJ
BL9 5JG G3UGC
BL9 5JZ M7BJG
BL9 6HE M7CBG
BL9 6JH 2E0PDG
BL9 6JH M3PDG
BL9 6NE M3HGH
BL9 6PP G4YYD
BL9 6RN G1YUS
BL9 6RN G4UYJ
BL9 7BU G4KQZ
BL9 7EL G8AJZ
BL9 7HB M6GJX
BL9 7QA 2E0LOE
BL9 7QA G1AKE
BL9 7QA M6LOE
BL9 7SG G7JUL
BL9 8DN G3TNQ
BL9 8HN G6MHO
BL9 8PD G1ADB
BL9 8PD G7DAL
BL9 9BQ M7BFG
BL9 9DQ G0LVX
BL9 9ET M6DSV
BL9 9EY M1EKU
BL9 9EY M3EKU
BL9 9HS M3SXU
BL9 9HS M6LZA
BL9 9HS M6SXU
BL9 9HZ G7TYO
BL9 9PD M3OAZ
BL9 9QB 2E0UTD
BL9 9QB M3OEH
BL9 9QE M1AGA
BL9 9SD M0LZH
BL9 9TA G4XHT
BL9 9UD G0RAN
BN0 8EH M7DNW
BN1 3BG M7EPI
BN1 3LS G6LFJ
BN1 3PS G3XCT
BN1 3RU G6YRJ
BN1 4AB G0CIT
BN1 5AG G8KTG
BN1 5EL G3TDL
BN1 5EP G4CFB
BN1 5FA G0CJZ
BN1 5FN G4FNL
BN1 5GB G3UWZ
BN1 5HH G3CUY
BN1 5NH G6CXO
BN1 5PQ M6YTI
BN1 6EB G3WBK
BN1 6FB G4HWF
BN1 6HE M7EQP
BN1 6RN M7EBU
BN1 6RZ 2E0WMY
BN1 6SB G4UNX
BN1 6WG G7KIL
BN1 7BG M6KPC
BN1 7EG G3GZT
BN1 7EJ G6LMB
BN1 7GH G6JNW
BN1 7HP G3EWT
BN1 7HX G7UDE
BN1 8AH G0SFV
BN1 8HA G7PYT
BN1 8HF G8GEZ
BN1 8LF M6FMJ
BN1 8LP M3HXG
BN1 8NE G3WXG
BN1 8NL G2FSR
BN1 8NL G8VYP
BN1 8NL M1CTK
BN1 8NP G8VIC
BN1 8QW G8TTP
BN1 8RE G1GGB
BN1 8RH G1OYG
BN1 8SH G4WCP
BN1 8TQ G4ZWD
BN1 8UE M6FXI
BN1 8WE G3XBN
BN1 8XQ G0CKA
BN1 8YG G1GDJ
BN1 9AD M3TJT
BN1 9EB M6EWC
BN10 7EF G0MIB
BN10 7JA G1KZI
BN10 7LA M3RTR
BN10 7LS M0HEP
BN10 7NS 2E0JAW
BN10 7PS G0RNS
BN10 7RL M3UZN
BN10 7RS G4ASX
BN10 7SL M7BIS
BN10 8AB M3CEZ
BN10 8BA G7FJC
BN10 8DP M6SSX
BN10 8DS G4XLM
BN10 8DZ 2E0XCO
BN10 8DZ M0XCO
BN10 8DZ M6LCF
BN10 8GF G7UBB
BN10 8HX G8KHH
BN10 8HX M5KHH
BN10 8JB G0KVF
BN10 8JT G0TEL
BN10 8NS G7RUQ
BN10 8NS M6JDL
BN10 8PJ G8ZTG
BN10 8RP G4PNP
BN10 8SA M0AZT
BN10 8TE M6CWK
BN10 8TH M3ZZA
BN10 8TJ G0AWG
BN11 1LB G6GPF
BN11 1RD M6CJU
BN11 1UG 2E0WGB
BN11 1UG M0WGB
BN11 1UG M3ORE
BN11 2AF M7HJP
BN11 2DW 2E0LJK
BN11 2DW G4FFE
BN11 2DW M0LJK
BN11 2DW M6LJM
BN11 2JG G7LLD
BN11 2JG M3ZNM
BN11 2LT G4DYI
BN11 2PH M0GMT
BN11 2PH M1RAD
Bn11 2pn M0JKW
BN11 2PN M6ARQ
BN11 2QE M0GEK
BN11 2QR M6RTC
BN11 2SS M3ZAM
BN11 3EH M7AAX
BN11 3LE M6FWS
BN11 3NE M6DTX
BN11 3RB G1IJY
BN11 3RU G8RKG
BN11 4JB G0HKN
BN11 5EF G3SPN
BN11 5HB G6JEY
BN11 5LP G0UZD
BN11 5QS G8RJB
BN11 5QY G8MAD
BN11 5RX 2E0OJD
BN11 5RX M0MCL
BN11 5RX M6OJD
BN12 4BH 2E1CPP
BN12 4BH M0CQQ
BN12 4HG G3YZT
BN12 4LD G3SZM
BN12 4LH G4TWK
BN12 4PF 2E0KYT
BN12 4PF M7NTT
BN12 4PN G8ZNK
BN12 4PU M3RIP
BN12 4QA G4TRP
BN12 4TD M1KSB
BN12 4TQ 2E0VBY
BN12 4TQ M0VBY
BN12 4TQ M3VBY
BN12 4TU 2E0KZM
BN12 4TU M0KZM
BN12 4UB 2E0SRB
BN12 4UB M0HBJ
BN12 5BQ G8RHU
BN12 5BU G6IEQ
BN12 5DJ G1DKX
BN12 5EL M6HXT
BN12 5JF G7RQD
BN12 5JF M0REG
BN12 5JG M6CCI
BN12 5JW M6CCJ
BN12 5PL G4GUO
BN12 5QA G4KHM
BN12 5RD G0PBV
BN12 6AB G6JVT
BN12 6DY 2E0CHK
BN12 6EY G4UWM
BN12 6GA M6JVK
BN12 6HR G4FZF
BN12 6JE M6FQN
BN12 6LA G4GOT
BN12 6LH 2E0GPX
BN12 6LH M6DCU
BN12 6QA G3HPB
BN12 6QP 2E0DYU
BN12 6QP M6GMQ
BN12 6QS G8MMF
BN13 1AE G4HNU
BN13 1BH M3NHU
BN13 1BH M3NHV
BN13 1DA G1OLE
BN13 1DA G3YSW
BN13 1DA M3YSW
BN13 1DG G0MVP
BN13 1DX 2E0EKB
BN13 1DX M0EKB
BN13 1DZ G4CZX
BN13 1ET M0HVD
BN13 1JS G6HXW
BN13 1LQ G1ZZC
BN13 1LX G3SKI
BN13 1PQ G6BWP
BN13 1QX M6BMO
BN13 1RE M3DZQ
BN13 2AU G1YVI
BN13 2BH G7GOA
BN13 2DH G0WSS
BN13 2DH G1KIZ
BN13 2DH G4LKW
BN13 2EN 2E0TGQ
BN13 2EN M0MEW
BN13 2LP M3GXI
BN13 2NT G3UEQ
BN13 2PW G4BAQ
BN13 2QX G4OHC
BN13 2QY G3YHM
BN13 2SU G7EWA
BN13 2TT 2E0LDY
BN13 3AG G0JXX
BN13 3AG G4GFN
BN13 3AG G8GRQ
BN13 3AL M7GYT
BN13 3AT G3XIA
BN13 3BH G3SXE
BN13 3BU M0STC
BN13 3BZ G0SBB
BN13 3DG G4ALZ
BN13 3DH 2E0IAL
BN13 3DH M0IAD
BN13 3DH M6LBK
BN13 3dn 2E0BHP
BN13 3dn M0GJS
BN13 3DS G7EQR
BN13 3HB G6FCS
BN13 3HG G4RRU
BN13 3HH G6FJL
BN13 3HZ M6EEA
BN13 3JD G8YGK
BN13 3JG G4XHE
BN13 3JU 2E0XBT
BN13 3JU M6SKP
BN13 3LG G4FPM
BN13 3PG G7NUE
BN13 3QG M3UJS
BN13 3QQ M6OZM
BN133BH G4DXO
BN14 0AH G3OUA
BN14 0EY 2E1AMB
BN14 0EZ G7FCU
BN14 0HR G7GSD
BN14 7 EG G8WOK
BN14 7AE M6JMX
BN14 7BL M0AIB
BN14 7DB M6FMN
BN14 7EE G0ELN
BN14 7EG 2E0DSK
BN14 7EG M6DCQ
BN14 7HE G4NSJ
BN14 7HH G8MIH
BN14 7HW G8ZWC
BN14 7NE G7JRM
BN14 7PE G1KKH
BN14 7PY G6MBL
BN14 7QU 2E1EOO
BN14 7QU M3YEA
BN14 7QW 2E1HEG
BN14 7QW G8DHE
BN14 7QY M0PCR
BN14 7SB G0SWH
BN14 8AH G7PIK
BN14 8AZ G6MMJ
BN14 8DG G3UQD
BN14 8EL G6KHG
BN14 8EN G6PLL
BN14 8ET G6GAW
BN14 9AU G2ALM
BN14 9BB M3VEM
BN14 9DG M3CWH
BN14 9EA M3ZUJ
BN14 9JE G8RDK
BN14 9JT M0GHO
BN14 9NJ G3VXJ
BN14 9RJ G3UDH
BN15 0AE G4GPW
BN15 0DJ G0WAM
BN15 0DJ M7DUP
BN15 0DX G4JEI
BN15 0HU G1EOM
BN15 0LB 2E0RKR
BN15 0LB M0RKR
BN15 0LB M6RKR
BN15 0LU 2E0EOI
BN15 0LU M3YFY
BN15 0ND G7FGZ
BN15 0NG G4ZZS
BN15 0NN G8XEU
BN15 0NZ G4OEH
BN15 8AS G3JSU
BN15 8DE G1UGB
BN15 8DE M1DNZ
BN15 8DE M6RFE
BN15 8EN G3IUE
BN15 8EQ M6BJI
BN15 8HX G6ENT
BN15 8JN G1OCL
BN15 8LF 2E0FVV
BN15 8LF M0WTO
BN15 8LF M6FVV
BN15 8LW G4JEY
BN15 8LZ G1BGC
BN15 8LZ G6BWN
BN15 8NY G0WMG
BN15 8PA G4RTX
BN15 8QD 2E0LNG
BN15 8QD M6RAO
BN15 8RL G0TLU
BN15 8RL G4GQR
BN15 8RL M0AAN
BN15 9BS G7NVI
BN15 9DD G1JPI
BN15 9PZ G3UFS
BN15 9QD G1DLB
BN15 9QL M0THT
BN15 9QL M3CCF
BN15 9RQ G4FIG
BN158DJ M7BVV
BN159SY G8XRX
BN16 1AJ G4HNX
BN16 1AJ M1BXO
BN16 1BH M7VXE
BN16 1DG M6CFZ
BN16 1DT G7KAV
BN16 1DU G8JVE
BN16 1HE G6MIC
BN16 1HE G7IWZ
BN16 1HF G0LOF
BN16 1JS M0KEL
BN16 1QS G3WMU
BN16 1QS M0MNG
BN16 1QS M3DNX
BN16 1SQ M0JIQ
BN16 2AB G8FNH
BN16 2EF G4TMG
BN16 2HH G3ZBG
BN16 2NY G0BHK
BN16 2QU M0RUX
BN16 2QU M6MPF
BN16 2RZ G4IWU
BN16 2UB M3RBQ
BN16 3ET G0GMC
BN16 3HW M0HJF
BN16 3JP M0RJK
BN16 3JQ M0GVB
BN16 3NP G8XDV
BN16 3PA M1BSP
BN16 3QG G8NXG
BN16 3QU G6WHH
BN16 3RU G4KOU
BN16 3RZ G4EJG
BN16 3TS G1LTI
BN16 4AF G8MFO
BN16 4FH M1DPY
BN16 4HE G4CHE
BN16 4HF G7AIF
BN16 4HF G8VCH
BN17 5BX M0IMP
BN17 5BX M3PGI
BN17 5DB M3RPQ
BN17 5DP G1WXC
BN17 5DS M6SBD
BN17 5DW G6TXY
BN17 5EY G0AMP
BN17 5HA G4XWD
BN17 5HE G7HNF
BN17 5HN M0FRH
BN17 5NU 2E0VKZ
BN17 5NU M0TVV
BN17 5PW 2E0GKL
BN17 5PW M6GUU
BN17 5QQ 2E0BDQ
BN17 5QQ M0WJT
BN17 5QQ M3KJE
BN17 6AS G7KEA
BN17 6AU 2E0KCB
BN17 6AU M3TKO
BN17 6BG G1AFK
BN17 6JN G0NGP
BN17 6LT G0MOU
BN17 6ND G4ZKE
BN17 6NN G1VYM
BN17 6PA G4ZFV
BN17 6QB G0XAG
BN17 6QD M0RDV
BN17 6QX G4RBC
BN17 6RG M6MCH
BN17 6RY G8DOW
BN17 6UP G3ZTZ
BN17 6UT M3CZJ
BN17 7AL 2E0OKL
BN17 7AL M0OKL
BN17 7AL M6OKL
BN17 7AT G0SPZ
BN17 7BW M3EPQ
BN17 7DF G1MOW
BN17 7HB 2E0TCB
BN17 7HB M6OJC
BN17 7HN 2E0CQZ
BN17 7HN M0JSR
BN17 7HN M6BXK
BN17 7HY M6DXI
BN17 7JS G3GGN
BN17 7LG 2E0KUE
BN17 7LG M6KUE
BN17 7ND 2E0RCN
BN17 7NE M1BSO
BN17 7NE M3UJX
BN17 7NF M7SCW
BN17 7PJ G6FPQ
BN17 7PJ M0FPQ
BN17FB G7UCO
BN18 0AH 2E0WWD
BN18 0AH M6RKK
BN18 0DH G4YCS
BN18 0DJ G0JZU
BN18 0EG 2E0RKO
BN18 0FD 2E0MKJ
BN18 0JA G8DIU
BN18 0JE G4TSQ
BN18 0JE G4WYL
BN18 0JH G0LPF
BN18 0PA G8KFJ
BN18 0PW G7JXB
BN18 0RP G0RBQ
BN18 0SD G6FKY
BN18 0SX 2E0DEG
BN18 0SX M0MRI
BN18 0SX M6ELD
BN18 0TG M0CKM
BN2 0EG M6KYI
BN2 0FZ M6GGL
BN2 0HP M1ATU
BN2 0JH G4SLW
BN2 0JL G3XDK
BN2 0JL G4AQG
BN2 1HE 2E0CKX
BN2 1JZ G7WBO
BN2 1LE 2E0CWW
BN2 1LE M6PRF
BN2 1LE M6WKF
BN2 1LL M6NBP
BN2 1RP M7BKE
BN2 1RP M7MAA
BN2 2EX G4AGQ
BN2 2EX G8AAC
BN2 3BG G7MRH
BN2 3DA G6NCE
BN2 3EQ G8ROF
BN2 3RH G1EXG
BN2 4LT G1KKF
BN2 4LU G4YLW
BN2 4RF M7DVQ
BN2 4TD G4YLG
BN2 5DA G0VAJ
BN2 5DD M7MMC
BN2 5GH G1BCG
BN2 5JA M3TGZ
BN2 5JS G8JFT
BN2 5PD M0IVW
BN2 5PU G0JKP
BN2 5TD M0DKD
BN2 5UP G3UON
BN2 6BE G6JNV
BN2 6DJ G1SBZ
BN2 6LH G8VTZ
BN2 6PF G4NQC
BN2 6RG G0ODI
BN2 6RH M3KXZ
BN2 6SF M0OTA
BN2 6SF M6GIH
BN2 6UA M6HWM
BN2 6UE 2E0ABD
BN2 6UE G7PTZ
BN2 7DP G3ITF
BN2 7DT 2E0SPG
BN2 7DT M0MBV
BN2 7DT M6LWT
BN2 7GG G8ZRM
BN2 7HA G4GDP
BN2 7HH G1RSC
BN2 7HH M0ATQ
BN2 7HP M7RWH
BN2 8AF G2CHI
BN2 8AS M6MOH
BN2 8DH G8LGS
BN2 8DJ G0ENJ
BN2 8EJ G4CBZ
BN2 8EZ G7NKU
BN2 8EZ M3NKU
BN2 8FQ G0ECW
BN2 8FR M0EBI
BN2 8PF M3DVB
BN2 8PH G6CIT
BN2 9AL M7LAK
BN2 9QB M7GRM
BN2 9YE G4MFR
BN20 0AS G4CUG
BN20 0DJ M6GND
BN20 0EU G8GEA
BN20 7EU G6TIQ
BN20 7HD G4DRV
BN20 7HD M0AEK
BN20 7HS G8CRC
BN20 7QE G4NYA
BN20 7TT M0NZL
BN20 7TZ 2E1OLI
BN20 7XR G0UGD
BN20 8AH 2E0TIZ
BN20 8AH M0TIZ
BN20 8AH M6OQN
BN20 8AU M0JAO
BN20 8AU M3XTE
BN20 8DQ G3YXW
BN20 8HY G4MHK
BN20 8LP G4YJW
BN20 8LP G7ULN
BN20 8NT M6BVP
BN20 8SD G1KSN
BN20 8SY 2E0IGO
BN20 8UG G1AKV
BN20 8XD G0USA
BN20 8XD G7ALC
BN20 9DJ 2E1HKY
BN20 9DJ M3DVG
BN20 9DL M3NIC
BN20 9DP 2E1LGV
BN20 9DY M1EHJ
BN20 9HX M6YNN
BN20 9JA G1RMN
BN20 9NJ G0EDS
BN20 9NJ G1KAR
BN20 9NJ G3WQK
BN20 9NJ G3ZQB
BN20 9NS G1UUP
BN20 9PP G7VAE
BN20 9QD G3ZHZ
BN20 9QJ G1FNF
BN20 9QJ M0WNT
BN20 9RD G4UPY
BN20 9RJ G7GSC
BN20 9RQ 2E0WFC
BN20 9RQ M0WTC
BN20 9RQ M3WFC
BN20 9SQ 2E0JXU
BN20 9SQ M6EYO
BN21 1HB G3MHF
BN21 1LW G7NIZ
BN21 1PS M0RMT
BN21 1SH M3SPP
BN21 1UD G7GMB
BN21 1UJ G6GVM
BN21 1UJ M3OYU
BN21 2AH 2E0VHC
BN21 2AH M3WUK
BN21 2BX M3TBW
BN21 2DN 2E0BVZ
BN21 2DN M0GUR
BN21 2DN M3YGV
BN21 2HR G4WLV
BN21 2LQ G1SSS
BN21 2LQ G8LQZ
BN21 2LQ M3BBA
BN21 2SD G0GZB
BN21 2TL M0LDH
BN21 2UR G1VGK
BN21 3DD G3UCW
BN21 3JG M0IJC
BN21 3QD G0RSG
BN21 3QD M0THM
BN21 4HR G6KJM
BN21 4PB M7ABA
BN22 0AB G8NVB
BN22 0AJ G0JPY
BN22 0JP G4ZWE
BN22 0PZ M0DJQ
BN22 0TT G4YGJ
BN22 0UD M7KST
BN22 0UT G7DWI
BN22 0UX 2E0DIG
BN22 0UX M6KIA
BN22 0UZ G1MNX
BN22 0XH G4NBQ
BN22 0XH G7UWG
BN22 7BT M6LGJ
BN22 7JF M6CHG
BN22 7JG G8DXU
BN22 7JL M3EVM
BN22 8EH G0EBF
BN22 8EH G7PGH
BN22 8LW 2E1PJJ
BN22 8LW G6TFJ
BN22 8LW M3MJJ
BN22 8PL M6LYB
BN22 8RS G7MJI
BN22 8SX M6JYS
BN22 8TQ 2E0DMV
BN22 8TQ M6GVW
BN22 8UH G8HGM
BN22 8UQ G4SHM
BN22 9EJ G0MUC
BN22 9EZ G6RGA
BN22 9EZ M3IJE
BN22 9HP G0DAX
BN22 9HQ 2E0JRP
BN22 9JL 2E1IKB
BN22 9NR M7BDV
BN22 9NR M7CZB
BN22 9PA M6UMI
BN22 9QQ G4RUL
BN22 9RH G7JVO
BN22 9RN M1DRB
BN22 9RN M1HLG
BN22 9RR M3ENE
BN22 9RR M3ENF
BN22 9SF 2E0SPU
BN22 9SF G0UAI
BN22 9SF M3SPU
bn227sz M3YHR
BN23 5AJ M3HIP
BN23 5BN G1SGR
BN23 5BN M3NJM
BN23 5PG G4XXM
BN23 5PL G3IAZ
BN23 5RA G0GZP
BN23 5TH G0FIP
BN23 5TH G0LTD
BN23 5UB 2E0WGO
BN23 5UB M0WGO
BN23 5UB M3WGO
BN23 5UH G8CCO
BN23 5UJ M7GLY
BN23 6AB G0OZG
BN23 6AF G4EKS
BN23 6AL G6GVU
BN23 6HQ M3UDS
BN23 6HW M3OWF
BN23 6HW M6LYF
BN23 6JP G7TSP
BN23 6LN G0EYE
BN23 6LN G6UNU
BN23 6TQ G7PMY
BN23 7AQ M3PWZ
BN23 7BE G6GVL
BN23 7BH G1LEO
BN23 7BT M0DUP
BN23 7DS M0CHO
BN23 7NY G1FSF
BN23 7PF G4PRJ
BN23 7PF G8XCY
BN23 7QP G1FHY
BN23 7QT G4TWP
BN23 7QT G4ZQS
Bn23 7RW M6DPL
BN23 7TP G8ANT
BN23 8AL M1APT
BN23 8AN M3LHU
BN23 8AQ M6NHN
BN23 8BR M3GZQ
BN23 8BX M0PHL
BN23 8DB M7BDP
BN23 8DX G0IGU
BN23 8FB G0GFP
BN23 8JD M3GUQ
BN23 8JQ 2E0SCR
BN23 8JQ M6CJP
BN23 8NS M6JFH
BN24 5AW M3TZE
BN24 5DT M0RFW
BN24 5DY M1ABM
BN24 5HW M0CUY
BN24 5HW M0RJO
BN24 5NL M3RVM
BN24 5NN M6COH
BN24 5QX G4PGA
BN24 6BE 2E0DWW
BN24 6DE G1NEZ
BN24 6DE G6GVS
BN24 6LX M0KIL
BN24 6LX M6LKI
BN24 6RT M6DQH
BN24 6RW G0RPJ
BN24 6SL G0HVO
BN245LB 2E0BRL
BN245LB M3UYO
BN245PF M7MLM
BN24EA M6KII
BN25 1ES M6MTT
BN25 1QF G8DWX
BN25 1SP G6FYA
BN25 1SW G3DQT
BN25 2EB G1OIO
BN25 2EW G3JKF
BN25 2EW G6GAF
BN25 2HA G0GJH
BN25 2HZ G8WHD
BN25 2JL M7SPJ
Bn25 2jq 2E0WCO
BN25 2JZ G4BMK
BN25 2NE G4MGN
BN25 2NZ G6OHR
BN25 2NZ G6VDD
BN25 2QN 2E0MEU
BN25 2QN G6MAR
BN25 2QN M6MEU
BN25 2RT M0RKF
BN25 2RU G0KDD
BN25 2RU M0ABT
BN25 2TX G1VAY
BN25 2UL G3AGF
BN25 2UN 2E1IIJ
BN25 2UN M3GLC
BN25 2XG 2E0FJK
BN25 2XG M0USC
BN25 2XW G7AFZ
BN25 3AD G3KXF
BN25 3AX G3ZUN
BN25 3BY M7AJF
BN25 3EZ G8XXJ
BN25 3HH 2E0BXE
BN25 3HH 2E0RBN

BN25 3HW G4BQA
BN25 3JB 2E0PQR
BN25 3JF 2E1MPN
BN25 3JR G6FLH
BN25 3LB M1EFP
BN25 3QY M6GWE
BN25 3ST M3TPY
BN25 3ST M3WCO
BN25 3TN G0IOF
BN25 3UE 2E0DMX
BN25 4DU G8FGN
BN25 4LZ G1BAB
BN25 4NF M7CND
BN25 4NU G8KUV
BN25 4PD 2E0MEZ
BN25 4PD M6DZP
BN25 4QL G7TOI
BN25 4QL M3PJG
BN25 4QR G1KVR
BN25 4QT 2E1JAC
BN25 4QT 2E1MJM
BN25 4QT M3JRM
BN25 4QT M3MJM
Bn252rt M6BEL
BN26 5HT M6HCI
BN26 5JU M7AZA
BN26 5NS G3CMU
BN26 5PA M6GVU
BN26 5QA M6VAY
BN26 6DA G4BJP
BN26 6DQ M3VEJ
BN26 6LU G6NIX
BN26 6NE G0UOI
BN27 1DS 2E0MED
BN27 1DS M0EBQ
BN27 1JL M7TYM
BN27 1NA M1BSY
BN27 1PJ G4ZKQ
BN27 1SP 2E0KFX
BN27 1SP M7MRJ
BN27 1SR G4RWM
BN27 1SU M6KPD
BN27 1TU G6YFF
BN27 1TW G6JTK
BN27 1UN G0UZY
BN27 1UX M6CQX
BN27 2BT G8JLD
BN27 2BU G8NFZ
BN27 2JY 2E0CVF
BN27 2JY M0NAZ
BN27 2JY M6SYX
BN27 2LX G1GLG
BN27 2PE M3YNE
BN27 3AH G1RFW
BN27 3AQ M6EHY
BN27 3AT M3TEG
BN27 3AZ M6FJM
BN27 3AZ M6PDI
BN27 3BZ G3JYG
BN27 3DJ G4BLS
BN27 3EG 2E0ERF
BN27 3EG M6BUJ
BN27 3FU G7JWX
BN27 3FZ M6HKO
BN27 3GP 2E0SEB
BN27 3GP M0STT
BN27 3HU G0RFF
BN27 3JR M7FGY
BN27 3JZ M6NKK
BN27 3JZ M7MDH
BN27 3JZ M7SOE
BN27 3LJ G8MAF
BN27 3LL M7CYV
BN27 3LS G1ATL
BN27 3ND 2E0BXG
BN27 3ND M0HJL
BN27 3ND M6HJT
BN27 3NP G4KAR
BN27 3NP G8TNH
BN27 3NX M1BMR
BN27 3QG 2E1GPE
BN27 3RA M3YDA
BN27 3TL G4LYU
BN27 3TP M0AWY
BN27 4BG G3RXO
BN27 4DS M6VOY
BN27 4FP M6IQT
bn27 4FR M6MOV
BN27 4JL G0CRD
BN27 4LN G8URG
BN27 4LN M0CTC
BN27 4NH 2E1BUR
BN27 4NT M7AAY
BN27 4PP M3UWF
BN27 4RA 2E1GTT
BN27 4RS 2E0HBX
BN27 4RS M0TXH
BN27 4RS M7HBG
BN27 4TS M0IRF
BN27 4TX G4HEJ
BN27 4UT 2E0ACA
BN27 4UT G3SGR
BN273AW 2E0PUF
BN273AW M0ABU
BN273AW M7UFF
BN3 1DL M6MIZ
BN3 1PP G4HZR
BN3 1RD M7YGR
BN3 1RL M1BIL
BN3 1RN G3LAZ
BN3 2BP G0CIM
BN3 2FF M0DTB
BN3 2LH G6ONI
BN3 2RT M0EEK
BN3 3RE G3LRU
BN3 3WJ G4GUX

BN3 3WW M6CZC
BN3 5AF M6LAG
BN3 5BE G3VBE
BN3 5DF G3MOL
BN3 5EY M7AAW
BN3 5HP G7WHX
BN3 5HP M5ABJ
BN3 5ND G3ECM
BN3 5NY 2E0BHS
BN3 5NY M3XKF
BN3 5SQ G6JAY
BN3 6FZ M6ETL
BN3 6HP G1VUP
BN3 6HP G3WOR
BN3 6NT G3HZT
BN3 6PJ G4DGW
BN3 6PL G3VLC
BN3 6UJ M6CJK
BN3 6UQ G1NKN
BN3 6WQ G0IGA
BN3 7AQ G0UCF
BN3 7EJ M6HZZ
BN3 7FW G7MGG
BN3 7GX G1GSK
BN3 7NS G4KIL
BN3 7QB M0LHB
BN3 7QX G3LCF
BN3 8BA G4ORP
BN3 8EE G1JYH
BN3 8EN G8BZL
BN3 8EQ G1KGL
BN3 8FE G0FUI
BN3 8JB M7MZN
BN3 8LQ G6FFH
BN3 8PP G8IQX
Bn37bd M7CJU
BN41 1GG G7UTH
BN41 1PT M6VIV
BN41 1SA M6CWV
BN41 1SJ G1FZS
BN41 1SJ M3LPN
BN41 1SW 2E0MRZ
BN41 2DN G4XKF
BN41 2DN M6FTX
BN41 2HN G1GID
BN41 2LS G7OBD
BN41 2PJ G0SJG
BN41 2PN M5AJZ
BN41 2RD G0NDF
BN41 2UZ M6RSS
BN41 2WU M3OHN
BN41 2YB G1ELE
BN41 2YE G7NLZ
BN41 2YF G0XAN
BN41 2YF M3TII
BN41 2YF M3TIJ
BN41 2YG G0SLW
BN41 2YL M6JOO
BN42 4AT M6BBQ
BN42 4DL M7CNY
BN42 4GD M6TKK
BN42 4LA G4YRT
BN42 4NE G0DWZ
BN42 4NH M3HXF
BN42 4QS M3XEX
BN42 4RR M6ITD
BN42 4SJ M7EKK
BN42 4YG 2E0MME
BN42 4YG M6HQM
BN43 5AA M1IRB
BN43 5AN G4ZHK
BN43 5AY M3GQB
BN43 5ES G8AIP
BN43 5GD G4XUZ
BN43 5HL G4OCZ
BN43 5JH M3ITK
BN43 5LG G7VML
BN43 5LN G0FJJ
BN43 5LW G1RDU
BN43 5NE G6YIK
BN43 5NN M3RQR
BN43 5NQ G4EUK
BN43 5TH G0FIG
BN43 5WY G4JBA
BN43 6BH G4GNX
BN43 6GH G7GMD
BN43 6GP G0FZB
BN43 6GR G0AQH
BN43 6HF G2DPY
BN43 6HY G3LLD
BN43 6LE M7BKB
BN43 6LN M3MZV
BN43 6LN M6MKN
BN43 6LN M6VAM
BN43 6NN G0SIU
BN43 6PF G6WDR
BN43 6PJ 2E0CCA
BN43 6RA G0SON
BN43 6TP 2E0IDC
BN43 6TP G3XOI
BN43 6TP G6ENS
BN43 6TP G7GFX
BN43 6TP M6IDC
BN43 6YB G0IFQ
BN44 3AG M1PKW
BN44 3FJ M5NEV
BN44 3FP G3REP
BN44 3GD G4GZH
BN44 3GF G0VLC
BN44 3HF M0FVV
BN44 3LH M0MBH
BN44 3LN G6AIK
BN44 3LR G0BAF
BN44 3RL G0VQD
BN44 3RQ G1HWY
BN44 3SN M6FLR

BN44 3TB G4RDH
BN44 3TF G4AND
BN44 3WE G3EUE
BN44 3WH G0CPR
BN44 3WH G0NKN
BN44 3WH G1WOR
BN44 3WH G4UDU
BN44 3WH M3ULO
BN44 3WH M3UPO
BN44 3WJ G0UQY
BN44 3WJ G7MGT
BN44 3WJ M3UQY
BN5 9DG G8RDG
BN5 9FR M6EWT
BN5 9HZ 2E0IQW
BN5 9HZ M0LDP
BN5 9HZ M6IQW
BN5 9SB G8XNB
BN5 9UX G3WMY
BN5 9YU G4XFV
BN6 8BB G4VTD
BN6 8BJ G8YKV
BN6 8BJ M6ZRJ
BN6 8DD G3XTH
BN6 8HR G6MJW
BN6 8HR G6YPY
BN6 8HZ G1JNX
BN6 8JL G4HJS
BN6 8NB G0DKY
BN6 8NS G7VNG
BN6 8NU G8XYR
BN6 8PD G0GZE
BN6 8PD G4LYX
BN6 8PD M0NJD
BN6 8PE G4GDT
BN6 8PP M6UVD
BN6 8SB G4CYZ
BN6 9BZ G3SGF
BN6 9DA M7TXA
BN6 9RT 2E0LBW
BN6 9RT M5REW
BN6 9RZ G3NYX
BN6 9TR M3NIF
BN7 1BT M0GKP
BN7 1EN G0VPX
BN7 1HT G7IIH
BN7 1HY M1ALO
BN7 1LT 2E1CCI
BN7 1LT G4XBG
BN7 1NP G3IIO
BN7 1QG G4OWT
BN7 1QG G6CWH
BN7 1QG M6OWT
BN7 1SP G8OYM
BN7 1SP M6AIT
BN7 1XU G8UND
BN7 2BE G6RXK
BN7 2DL G1WSE
BN7 2DL M1CNS
BN7 2EJ G4TBM
BN7 2EN G7MOW
BN7 2HY G4PZU
BN7 2SH G6GOS
BN7 2TX G1MXC
BN7 2TX G7PUW
BN7 2UB G4YGE
BN7 3JL G8BYC
BN7 3PB M0YDM
BN8 4DS G1RKR
BN8 4GD 2E1BWT
BN8 4HP G7JKW
BN8 4NA G6DGK
BN8 4NA M0SXE
BN8 4PG 2E1JBJ
BN8 4PG M3BER
BN8 4PG M3BJB
BN8 4PG M3IMB
BN8 4PS G2FQZ
BN8 4QX G0XAM
BN8 4RH M6RSH
BN8 5HX G3YXO
BN8 5JD G7SMT
BN8 5LJ M6JNH
BN8 5PA M3IQG
BN8 5PG M6SCU
BN8 6DR 2E1GJG
BN8 6DR 2E1HQA
BN8 6DR M3BOB
BN8 6DR M3HAT
BN8 6EQ M6CNZ
BN8 6HW G4SVB
BN8 6LS G4MJC
BN8 6PS G3EKJ
BN84NF M6YJH
bn84nt M7VJM
BN9 0FG 2E0KXV
BN9 0FG M6KXV
BN9 0HS M7MAM
BN9 0HS M7YXZ
BN9 0JU G8ZVX
BN9 0ND G4KZX
BN9 0PS G8NQK
BN9 0QU G0CCX
BN9 0RD G4YRA
BN9 9AH G1AAR
BN9 9DJ G8KAS
BN9 9DY M6ACB
BN9 9ER 2E0CAE
BN9 9ER M3XGU
BN9 9EZ M1UKC
BN9 9LJ G6FYR
BN9 9NB M7KWT
BN9 9NH G0TJH
BN9 9NR M3VQI
BN9 9QB G7WHU
BN9 9RR M6MWV

BN9 9SP M1DNC
BR1 2AT G4XGP
BR1 2NF G0JBP
BR1 2PR M3EFW
BR1 2UD 2E0JXB
BR1 3BL G4NCS
BR1 3PU G4FXR
BR1 4DB G4OIM
BR1 4LP G8OTG
BR1 4NX M7DLV
BR1 4QY G1GYF
BR1 5AZ G3YQN
BR1 5BT G4AHT
BR1 5EA 2E0RTM
BR1 5EG G7CRK
BR1 5EW G0UBA
BR1 5LH M6OGE
BR1 5LR M3RHB
BR1 5LR M6BHZ
BR1 5NA G0WYG
BR1 5NA G1WYG
BR1 5QT G4EPV
BR1 5QT M7DFF
BR2 0DN M0HOV
BR2 0EE G3IUJ
BR2 0JF M7THX
BR2 0LS G1ROK
BR2 0RW G3IZA
BR2 0TN 2E0PAF
BR2 6AJ M6TSR
BR2 6BF G0REE
BR2 6BF G7HCC
BR2 6DG G6YRY
BR2 6HL 2E0EHQ
BR2 6HL M0ITG
BR2 6HL M6HJI
BR2 7AL M3UMB
BR2 7DY G3XDL
BR2 7DY G8ZYH
BR2 7EH M3FLV
BR2 7HE G3TSC
BR2 7HE G4XAT
BR2 7HE M3LSU
BR2 7HE M3TVC
BR2 7HU G4TYT
BR2 7HX G1OJO
BR2 7JA G4WJQ
BR2 7JS G4IPF
BR2 7LX G8SXU
BR2 7PT M6VJG
BR2 7QJ G8BVQ
BR2 7QN G3OQD
BR2 7QT G3NGK
BR2 7QT G8TSD
BR2 8EY G4MSK
BR2 8FG G3WOE
BR2 8JB G3WIM
BR2 8LT G1BYS
BR2 8PF G1ZDR
BR2 8PP G3ZJX
BR2 8QQ G8ITB
BR2 8QQ M0XBY
BR2 9EN G0ELU
BR2 9JA M6RXT
BR2 9NW G1ZQN
BR2 9NW M0HTJ
BR2 9NW M0TZO
BR2 9UL M0HUN
BR29HT M7SPX
BR3 1LE G7RUS
BR3 1NB M3UNP
BR3 1NN M6AWG
BR3 1NY G4HZX
BR3 3AY G7TOU
BR3 3BJ G7HNG
BR3 3HG 2E0NCY
BR3 3HG M6NCY
BR3 3JT G1HIG
BR3 3JW G7EWS
BR3 3QU G7RUX
BR3 3RL G7KWO
BR3 3SB G7CXO
BR3 4AE G7HYM
BR3 4JJ G4VJT
BR3 4JT M6DOM
BR3 4PZ M6FZC
BR3 4RU G8NKM
BR3 4SS G3SZR
BR3 4XS G4RND
BR3 5AA G3EFS
BR3 5DY 2E0FPX
BR3 5DY M7WWT
BR3 5HJ G3OSR
BR3 5LH M7CEY
BR3 6LJ G4RTV
BR3 6LX M0ZVB
BR3 6SN 2E0JFL
BR3 6SN M0JKM
BR3 6SN M6BXL
BR4 0HA G8VZS
BR4 0HB M1ENE
BR4 0QH G1PGS
BR4 0SG G8OMQ
BR4 0TA G4IKL
BR4 9AH G3ZQF
BR4 9DZ M6GJA
BR4 9JA M1LES
BR4 9JA M1PAM
BR5 1AB G6MYZ
BR5 1AL 2E0BBG
BR5 1AL M3IXU
BR5 1BZ G8IVB
BR5 1EZ G3NAT
BR5 1EZ G4WGZ
BR5 1EZ M0UYR
BR5 1JF G8YTZ

BR5 1JF M6ELY
BR5 1JF M6JEM
BR5 1NF G7URM
BR5 1PY 2E0KVV
BR5 1PY M0KVB
BR5 1PY M6KVQ
BR5 2DL G8KPY
BR5 2EH G1FAA
BR5 2EP M6GRH
BR5 2JF G1BPD
BR5 2JR G7RUY
BR5 2JX G7KWM
BR5 2LZ G6LGR
BR5 2NT G7RFM
BR5 2NY G1UPT
BR5 2PA M7NAJ
BR5 2SB G7CEW
BR5 2SH M0LRH
BR5 2TF G0RRL
BR5 3AN G7BIP
BR5 3AT G0OMH
BR5 3AT M6AIZ
BR5 3LF G4DOE
BR5 3LG G6DEV
BR5 3LP M7CXT
BR5 3LQ M3TKT
BR5 3LY G7FRH
BR5 4AH M7XXS
BR5 4EQ G6XYD
BR5 4JP G1EYS
BR5 4JZ G7SAT
BR5 4LA G1VOJ
BR5 4LU G3BVA
BR5 4NS G0LXF
BR5 4PF G0JPM
BR5 4PF M1AHJ
BR5 4PN G4KMF
BR6 0AQ G7PJD
BR6 0BH G1VII
BR6 0BT G7HOL
BR6 0EJ G0TZR
BR6 0EQ G4NMT
BR6 0ER G8LSC
BR6 0LE M6OLP
BR6 0QE G1PJI
BR6 0QJ G0VPC
BR6 0TD G8LSI
BR6 0UP M7RHW
BR6 6BE G7DRW
BR6 6HW M1DPO
BR6 6JF M3XFB
BR6 6JU G3SIU
BR6 7QG M0GZI
BR6 7QJ M3XEF
BR6 7QN M0HHE
BR6 7RT G7RGJ
BR6 7RT M0HAL
BR6 7SD G3VFD
BR6 7TD 2E0CGG
BR6 7TD M0REZ
BR6 7TD M6TDM
BR6 8BL G8KZJ
BR6 8DJ 2E0BWI
BR6 8DJ M0YRG
BR6 8HJ G3ZOH
BR6 8HP G3MCA
BR6 8JB G6GKL
BR6 8JP G4TQO
BR6 9AA 2E0AYQ
BR6 9BS G0BZX
BR6 9BZ G8HKN
BR6 9EW M0HRY
BR6 9JG M6GTE
BR6 9LE M3HVE
BR6 9LL 2E0PAT
BR6 9LL M6EWD
BR6 9LY G1JTX
BR6 9NF G8IKA
BR6 9PN G1EOH
BR6 9PN G6XVQ
BR6 9RS G7AQK
BR6 9RS M3TYX
BR66EJ 2E0JTT
BR66EJ M7JXX
BR7 5DU M0JZN
BR7 5DU M7JZN
BR7 5RF M7PXM
BR7 5RN M6MMC
BR7 6AG G7CXU
BR7 6BQ G7UQQ
BR7 6JD G7ULL
BR7 6JR 2E0SPW
BR7 6JR M0GXN
BR7 6JR M6AEI
BR7 6LA G0GZV
BR7 6QT G6DNL
BR7 6SJ M1FAA
BR7 7HA M7AHG
BR8 7JA G1PUO
BR8 7JY G4IKQ
BR8 7LS M1LYE
BR8 7QT M3GML
BR8 7QZ M3KGG
BR8 7QZ M3KSG
BR8 7RJ G7LIT
BR8 7RY M7CAL
BR8 7SB G3IKQ
BR8 7SE G6SYA
BR8 7TW G7AKM
BR8 7UR G7PEB
BR8 8AP M6GSR
BR8 8DD 2E0SSI
BR8 8DQ 2E1DQQ
BR8 8DQ M3DQQ
BR8 8HZ 2E0FVB
BR8 8HZ M7KJS

BR8 8JL M0RAW
BR8 8LP G8VJG
BR8 8NZ G4AGC
BR8 8TN M1AEP
br88jb M6LBC
BS 30 5QU M7CZR
BS1 4AU M0HKH
BS1 5BY M7UCL
BS1 5JR G4FAZ
BS1 6EH M3NWQ
BS1 6EH M3NYG
BS10 5ED M7BYU
BS10 5JF M7TEG
BS10 5SR M3HXQ
BS10 6BP G0PID
BS10 6JN G0JKU
bs10 6tj M7RTK
BS10 7PH G0HNO
BS10 7TE G4SQQ
BS10 7XE G1EIR
BS105AS M7DMP
BS11 0AG M6EVR
BS11 0EF G4HHL
BS11 0NL M6KKJ
BS11 0QX 2E0LHD
BS11 0QX M6LDB
BS11 0QX M6NMH
BS11 9AD M0KGA
BS11 9JD M0HXI
BS11 9RT G0OKL
BS11 9RY M0GMW
BS11 9SR G1PBX
BS11 9SR G8OTH
BS11 9TX G3YHV
BS11 9TX G4AHG
BS11 9TX M3YHV
BS11 9UF G0XAK
BS11 9UP G3LWF
BS11 9XB 2E0JCC
BS11 9XB G1RWT
BS11 9XB M6JBR
bs110tx M7ATA
BS12 3LQ G4KXO
BS13 0BB 2E0SZH
BS13 0BB M3AMB
BS13 0BJ M7AXU
BS13 0DE M7UTG
BS13 0HS M0WLF
BS13 0JG G8VUM
BS13 0NL M0PCH
BS13 0QA G7VOK
BS13 7BU M7YZP
BS13 7BW G0RCU
BS13 7DB G3XED
BS13 7DE G3WCL
BS13 7LY G6FMN
BS13 7SA G7NQJ
BS13 8AQ G6SDW
BS13 8DB M0AKF
BS13 8EF 2E0WSW
BS13 8EF M3VIU
BS13 8EF M3WZS
BS13 8HN 2E0VTT
BS13 8HN M0JFO
BS13 8HN M6GTT
BS13 8HQ G1AVB
BS13 8HQ M3RWD
BS13 8HU G8GRS
BS13 8PY M1KEV
BS13 8SA M3NRJ
BS13 9DR M6LWR
BS13 9HS M1BOB
BS13 9RB G0CEM
BS13 9RL M3HBP
bs139ly M7BBZ
BS14 0DS G4OJU
BS14 0EF 2E0LRL
BS14 0EG G4YZR
BS14 0EH M0DIL
BS14 0EH M3SWS
BS14 0EJ 2E0WIL
BS14 0EJ M3WPW
BS14 0HH G4JFX
BS14 0HS G3IOI
BS14 0NN 2E0JIL
BS14 0NN G3YLJ
BS14 0NN M0DXV
BS14 0NN M3YLJ
BS14 0NN M6BEX
BS14 0NN M6YLJ
BS14 0PB G4EPH
BS14 0PP G0DRX
BS14 0PZ G4WUB
BS14 0QG G0CCA
BS14 0RX M6MIT
bs14 0sh M7DER
BS14 8DY G4RAB
BS14 8EQ M7CUN
BS14 8LQ G0JLI
BS14 8NH 2E0FNV
BS14 8NH M6NPV
bs14 8nq 2E0UGE
bs14 8nq M0UGE
BS14 8NQ M7RAE
BS14 8PZ G0KDS
BS14 8SS G7PVG
BS14 8SZ G1SLU
BS14 8TT M0CRO
BS14 9BD 2E0HCC
BS14 9BD M6MOD
BS14 9ED G4XML
BS14 9ET M6NQT
BS14 9LW G4NFS
BS14 9ND M3CFU
BS14 9SF M3VRN
BS14 9TB M6FZM

BS15 1DE M0ORN
BS15 1EP G1HPZ
BS15 1HF G4CQI
BS15 1JR M0PHN
BS15 1TA G0VBK
BS15 1UW M0EAE
BS15 1XB G0TDV
BS15 3BY G3KZC
BS15 3JR G7MWJ
BS15 3JX G8DBP
BS15 3JZ G7IRP
BS15 3PW 2E0BYQ
BS15 3PW M0MMO
BS15 3PW M6BGT
BS15 3RA M1MVX
BS15 3TB G0JYX
BS15 3TJ G4EQP
BS15 4HN G6NQM
BS15 4HT G7BME
BS15 4JA G7KCC
BS15 4LT G7IPH
BS15 4PA G6XEB
BS15 4QB M3VAE
BS15 4QJ G4UBI
BS15 4RT G1FUJ
BS15 4UF 2E0EYU
BS15 4UF M0UTK
BS15 4UF M6KQZ
BS15 4UF M6LPX
BS15 8AA G8DQD
BS15 8DQ G4OPO
BS15 8ES G4BVK
BS15 8LW M7MIR
BS15 8NT G4WOD
BS15 8NX G4FJH
BS15 8NZ G1DFM
BS15 8NZ G1ODD
BS15 8QR M1BEP
BS15 8QS M7TAS
BS15 9NN G4DCX
BS15 9PU M1MAD
BS15 9QJ G1XYR
BS15 9QP G1IHL
BS15 9SH M0VVM
BS15 9UE G8ZEX
BS15 9ZE G0CFM
BS151SW M7AHE
BS16 1DQ G3EWF
BS16 1FB G3TEX
BS16 1FD M0DEY
BS16 1GE M6VSC
BS16 1RD G8TYD
BS16 1WL G3UHK
BS16 2AH M5AJO
BS16 2BU M6TNH
BS16 2LH M6NPD
BS16 2SW G4CXW
BS16 2UB G3RUJ
BS16 2UD G4TVD
BS16 3AG M7CLW
BS16 3DR G1IXE
BS16 3DR G1IXF
BS16 3NG G1FWF
BS16 3QY G7NZZ
BS16 3TL M0KBB
BS16 4EX M7NBX
BS16 4JA 2E0BZU
BS16 4JA M6PMF
BS16 4JA M6RFO
BS16 4JD G7UUC
BS16 4PQ M0HDJ
BS16 4QE G4OPQ
BS16 4SQ M1BGK
BS16 5AA G7HVL
BS16 5AU M6CJN
BS16 5BL G0PDV
BS16 5DE G7BLT
BS16 5EH 2E0FEP
BS16 5LE G3IZM
BS16 5LG G6YB
BS16 5QS G4DEU
BS16 5RU G4FHN
BS16 5RU G8LMC
BS16 5TN G4UGU
BS16 5UP G1LBM
BS16 6EG 2E0BLW
BS16 6EG M0LHS
BS16 6EG M3VHI
BS16 6HN G6CWF
BS16 6JG G4OEQ
BS16 6PN M6GFM
BS16 7BP G2BAR
BS16 7BP G4VKO
BS16 7BW M3MBZ
BS16 7DN G0FMJ
BS16 7DQ M0NIC
BS16 7EB 2E0TBS
BS16 7EB M6CXB
BS16 7HB G4JQK
BS16 7JL G6AWZ
BS16 9AZ G0KGL
BS16 9BQ G8HSR
BS16 9BZ G0XAF
BS16 9DR G0NQJ
BS16 9DW M0MAT
BS16 9EA M6KGF
BS16 9EY M0DSB
BS16 9EY M6DFS
BS16 9HN G4RXF
BS16 9LB 2E0BFF
BS16 9LF M6JYZ
BS16 9NH G7CWN
BS16 9QF G7DSQ
BS162UF G8JUT
BS17 1RH G4YCD
BS2 9TB G1ODE

BS2 9TL G8OTA
BS2 9UB G3TVV
BS20 0DE G7IEB
BS20 0DR G6PHM
BS20 0EQ G0SNP
BS20 0EY M0BJS
BS20 0JR G4VYK
BS20 0JT 2E0DQQ
BS20 0JT M3GNM
BS20 0JX G4XDP
BS20 0JX G4YSZ
BS20 0LF G6TSE
BS20 0QB G4WRQ
BS20 6BB M6BGW
BS20 6DF M7ASU
BS20 6JR G7AQF
BS20 6JX G4DRZ
BS20 6LA G4KMB
BS20 6LD G8EZR
BS20 6LU G4YNV
BS20 6PF G4UGT
BS20 6QY G4USQ
BS20 6RG G7UTS
BS20 6SR G0RWI
BS20 6YT G6EQZ
BS20 6YT M0DRD
BS20 7AD M6TKI
BS20 7AW M0JXR
BS20 7DH M6TLS
BS20 7DY G4EJH
BS20 7FG G0XZT
BS20 7FW G1UPP
BS20 7FW G4CDU
BS20 7HH M0RBH
BS20 7RF M6MSY
BS20 7TQ G6ETL
BS20 8AJ G4NXI
BS20 8DD G0FKJ
BS20 8EG G1SWX
BS20 8EG M3CPH
BS20 8JB G6TNP
BS20 8JQ G4KPM
BS20 8JQ G6ZPV
BS20 8LG G4NVV
BS20 8LG G6ZOE
BS20 8RE M7ECN
BS20 8RQ G4VEH
BS20 8RW G1SEW
BS21 5AQ G0ELO
BS21 5DR M6TJA
BS21 5HB G4ONS
BS21 5HN G7ENR
BS21 6AY G6IQF
BS21 6EH M3YMX
BS21 6HR M1EPX
BS21 6JE 2E0CFY
BS21 6JJ G1CZW
BS21 6JU G1DAX
BS21 6LE G4IWQ
BS21 6LL 2E0LXR
BS21 6LL M6OAS
BS21 6LQ G0IMA
BS21 6LQ G1YDJ
BS21 6QS G4OCF
BS21 6RD G0AZE
BS21 6RG G4NAQ
BS21 6SN 2E0LZT
BS21 6SN M3LZT
BS21 6TQ 2E0KKG
BS21 6TQ M6NQJ
BS21 6TY M7EMJ
BS21 6UQ 2E0IVO
BS21 6UQ M0IVO
BS21 6UQ M3IVO
BS21 7AJ G4ULP
BS21 7DY G0AKS
BS21 7DY G1EFS
BS21 7LU G3WBA
BS21 7QZ G8SPC
BS21 7RL G4BBL
BS21 7RL G8CUW
BS21 7TN G6BGY
BS21 7UP G6AEC
BS21 7US G7RKT
BS21 7YJ M3GYA
BS21 7YJ M5AFH
BS21 7YN G8OUG
BS217QB M7ASA
BS22 6DJ G1FWZ
BS22 6DQ G6AMV
BS22 6EN G8AVK
BS22 6NX M0RPZ
BS22 6NY 2E0HOQ
BS22 6NY M3HOQ
BS22 6NY M3UEQ
BS22 6PW 2E0JQF
BS22 6PW M3JQF
BS22 6RA 2E0DAR
BS22 6RA M0RGO
BS22 6RA M3PWO
BS22 6RL G0ATD
BS22 6XP G8PRP
BS22 7DR M3VUI
BS22 7FA G1OOJ
BS22 7FW G6OPD
BS22 7HP G4ANZ
BS22 7LU M1EYD
BS22 7PA M1NEW
BS22 7PF G8VZI
BS22 7TJ G1PEI
BS22 7TQ 2E0GXQ
BS22 7TQ M0GXQ
BS22 7TQ M7SJL
BS22 7TS 2E0WSM
BS22 7UA M7JSS

BS22 7YB G0IHE
BS22 8AD M0ONS
BS22 8DB M7AMP
BS22 8DD 2E0UWI
BS22 8DD G7UWI
BS22 8DD M3UWI
BS22 8DD M3UWJ
BS22 8DD M6ABZ
BS22 8HL 2E0PHE
BS22 8HL M0OOR
BS22 8HL M7DCT
BS22 8JX G7HYS
BS22 8LN M6SXP
BS22 8PS 2E0BAP
BS22 8PT G4AYD
BS22 8PT G4YQG
BS22 8QH M1BQD
BS22 8QN G0SVA
BS22 8QQ M6NWQ
BS22 8SE M1JMB
BS22 8UH M7AXC
BS22 8UY G8GYV
BS22 8XR 2E0AUV
BS22 8XR 2E1HTM
BS22 8XS M3KUS
BS22 9AL G4WAZ
BS22 9AQ G7OPJ
BS22 9AY G0WMW
BS22 9BD G1XRO
BS22 9BD G4JWV
BS22 9DZ M3NKB
BS22 9HG M6JXY
BS22 9HT G4ZUX
BS22 9HT M1BQC
BS22 9HT M1BQE
BS22 9HU G3PLT
BS22 9LG M0FXB
BS22 9LW G4DUQ
BS22 9QS G0ADW
BS22 9SP G3GMC
BS22 9UL 2E0JCO
BS22 9UL M0JCE
BS22 9UL M6JCE
BS22 9UU M0CAZ
BS22 9YD G7HVN
BS226BG M7PCX
BS229HA G0LOJ
BS23 1NG M7CXF
BS23 1RD 2E0RSC
BS23 1RD M7RSC
BS23 1RQ G3PLJ
BS23 2BH G3LJD
BS23 2EY M7SKP
BS23 2HE G3RNX
BS23 2JR G3TJE
BS23 2SY G6LQI
BS23 2UA G4UPR
BS23 2UA M7BPH
BS23 2UG 2E0VPA
BS23 2UG M7PLA
BS23 2US M3HPZ
Bs23 2ux 2E0WXU
Bs23 2ux M7WSM
BS23 3BX G0TCH
BS23 3DE G1CMZ
BS23 3DF G4CXQ
BS23 3DF G8WSM
BS23 3EE M6MHY
BS23 3HB M7CRD
BS23 3JH M7SJW
BS23 3LY M3NFJ
BS23 3PQ M6MJN
BS23 3RR G4RNZ
BS23 3RS M7WFG
BS23 3RT M3NVC
BS23 3UY G3JLK
BS23 3XQ G4CDI
BS23 3XX M3LJX
BS23 4BG G0VJM
BS23 4DH 2E1KIP
BS23 4HR M6EDH
BS23 4HU G6YEK
BS23 4JA M6SEO
BS23 4JR G0PZB
BS23 4JY G8JNO
BS23 4JZ G0VAZ
BS23 4LQ M6CTZ
BS23 4QQ G1VSX
BS23 4QS G4NMV
BS23 4RF M6MWW
BS23 4TU G7LXA
BS23 4TU M0LXA
BS23 4YH G0IYE
BS24 0AB G0TAT
BS24 0AN M3TQX
BS24 6TH M6OCD
BS24 7AS G7BRX
BS24 7DZ M7FIX
BS24 7EF G7USJ
BS24 7FH M6PAA
BS24 7GT G0WWE
BS24 7HS G8JMK
BS24 7SB M1EPN
BS24 7TJ 2E0ZDJ
BS24 8AD G6FEX
BS24 8AG M3PLV
BS24 8BH G4HJF
BS24 8DS G3LAA
BS24 8PA G8WAL
BS24 8PF M7EPV
BS24 8RZ G6XNQ
BS24 9BU M6PCL
BS24 9DY G7UUK
BS24 9EB M3UPP
BS24 9EH G0BMT
BS24 9JF G6EQI

BS24 9JG G0BOR
BS24 9JW G4PWP
BS24 9LH G3WXH
BS24 9LW G3RUD
BS24 9NJ M1BDO
BS24 9QX G0DKM
BS24 9RH M0KYX
BS24 9RH M6IJM
BS25 1AL G1HHQ
BS25 1AR 2E1EMH
BS25 1AT G7SMH
BS25 1HB G3GTA
BS25 1HD G7PRI
BS25 1HL G0CHJ
BS25 1HQ G8TTX
BS25 1JX 2E0GHA
BS25 1JX M0HAG
BS25 1JX M3ZWJ
BS25 1NB G1CGJ
BS25 1NH G3YOL
BS25 1SA G3RXG
BS25 1TG G4XYH
BS25 1TR G4RCY
BS25 1UE G7DRO
BS25 5PD G7NJX
BS25 5PE M0GAC
BS25 5PW M7RVD
BS26 2AX G4MCE
BS26 2EH G3XLX
BS26 2QW G4YKG
BS27 3BW M3HPF
BS27 3DH M1DIR
BS27 3HA M7IED
BS27 3HS G8OGP
BS27 3HZ G0VSS
BS27 3JH 2E0VHA
BS27 3JH M3VHA
BS27 3LE 2E0AFT
BS27 3NY G6OSR
BS27 3PB M6LSJ
BS27 3TH M6CHD
BS27 3TH M6JHK
BS27 3UB G0VVR
BS27 3XN G4KAE
BS28 4BZ G3RBJ
BS28 4HL G4FXM
BS28 4SW G0HVB
BS29 6AZ G3KVR
BS29 6DG M6BJL
BS29 6DG M6BJP
BS29 6HD 2E0GFJ
BS29 6HD M7BYE
BS29 6JD 2E1ADP
BS3 1NJ M6IGH
BS3 1XB G0VJN
BS3 2AP M7WAW
BS3 2BP G0FXI
BS3 3DY G7PXR
BS3 3EA G4XED
BS3 3EA G8BDZ
BS3 3JF M6GZU
BS3 3LU M6SSB
BS3 3PW G1HYQ
BS3 4LG M0SYJ
BS3 4NZ 2E0EKM
BS3 4NZ M6HLE
BS3 4TX M6XTK
BS3 5BT G8SUV
BS3 5EG G4EGR
BS3 5ES G4OJI
BS3 5ES M6OJI
BS3 5ET M7BBN
BS3 5LN G3IUO
BS3 5LR M6HQB
BS3 5ND 2E0RUU
BS30 5JH M3OPG
BS30 5NR 2E0IHR
BS30 5NR M7OJW
BS30 5PN G8PQA
BS30 5PP G4VRP
BS30 5PP G6AUR
BS30 5PW G7VVO
BS30 5WJ G1AIG
BS30 6DY G4SVS
BS30 6EZ G8VZB
BS30 6JA G0VDJ
BS30 6RH G0JJS
BS30 7BS G1ABA
BS30 8EJ G4MQF
BS30 8UT G4MCQ
BS30 8UT G8KGE
BS30 8XA M6VRL
BS30 8YB 2E0RBG
BS30 8YB M6KXS
BS30 9DU G4YOC
BS30 9PX G8ZFL
BS30 9QQ G7IYK
BS30 9UE G2BTZ
BS30 9UE G4ULV
BS30 9XB G4SNU
BS30 9XB G7JVK
BS30 9YQ G4EXZ
BS305XN 2E0TPL
BS305XN M0TPL
BS305XN M7ADC
BS31 1AS 2E0DUH
BS31 1AS M6FKG
BS31 1BA M6SKN
BS31 1DB G1PCA
BS31 1GB M7CSL
BS31 1JX 2E1GOZ
BS31 1QE G0JZH
BS31 1QF M0HBT
BS31 1QF M6BDQ
BS31 1QG G0DPW
BS31 1QW G7CBI

BS31 1XD 2E1CCG
BS31 1XD G0DSB
BS31 1XG G3XAW
BS31 2EQ G7PHE
BS31 2LJ M6GDP
BS31 2NL M3PKE
BS31 2NL M3PKL
BS31 2TR M6DYA
BS31 2TU M6TII
BS31 3BG G3TKF
BS31 3BG G4HTV
BS31 3BZ 2E0CTK
BS31 3BZ M6CDT
BS31 3DU G7TFU
BS31 3DX G8VPG
BS31 3LA G6AYY
BS32 0AP G6TJZ
BS32 0BH G4GOA
BS32 0BR M7JIB
BS32 0DA G7IYM
BS32 0DW G1YHN
BS32 0DZ G4GSA
BS32 4HH 2E1CYP
BS32 4LQ G4OST
BS32 8AF M0ZLI
BS32 8AS G4FUA
BS32 8AS M0GBH
BS32 8AS M6RBY
BS32 8AU M0JOB
BS32 8BB G4BOL
BS32 8BP 2E0IFG
BS32 8BP G0RVM
BS32 8DP M0SUG
BS32 9AR G0RAT
BS34 5AU G0HTS
BS34 5BG 2E0CNV
BS34 5BY M6IAF
BS34 5HH G0UMP
BS34 5HX G4IYE
BS34 5LF M6DJJ
BS34 5NP G0NFH
BS34 5NP G1DOX
BS34 5PW G3ZZU
BS34 5PY G6ZRS
BS34 5RN G0ECM
BS34 5SA G6RUP
BS34 6EB G4DVV
BS34 6EF G1YXA
BS34 6PE M6GLW
BS34 7LJ G7MNO
BS34 7LU M3WLO
BS34 7RD G4YQQ
BS34 8GD G6TVJ
BS34 8GD G7RHT
BS34 8GD G8EKZ
BS34 8NG G3VM
BS34 8NG M6SDT
BS34 8NJ 2E0CCL
BS34 8NJ M0GYP
BS34 8NJ M6SGL
BS34 8PJ G0SXU
BS34 8QH G8IMB
BS34 8RZ G4LAW
BS34 8TG M0PLO
BS34 8UD G4AEL
BS34 8XA 2E0PGS
BS34 8XA G8YMM
BS34 8XA M0NBC
BS34 8XA M3PGS
BS34 8XN G4CJV
BS34 8XT M7ITD
BS35 1AL M7XYX
BS35 1AY G1USV
BS35 1DP G4YZD
BS35 1HB G7FCO
BS35 1HX M0HFH
BS35 1JF G3XNN
BS35 1JH G4AGH
BS35 1JJ 2E0RES
BS35 1JJ M6RES
BS35 1SR G8AZT
BS35 1SX M3EQQ
BS35 1TB G0JPU
BS35 1UE 2E0DFV
BS35 1UL G4EDQ
BS35 2DN G8MXD
BS35 2EJ G7NVZ
BS35 2FH G6FLE
BS35 2JE G4ZVQ
BS35 2LL M6CPA
BS35 2LX G1PXW
BS35 2QX G0RYM
BS35 2YD G6OLJ
BS35 2YE G3XIY
BS35 3JG G0WMB
BS35 3JL G1FND
BS35 3LQ G4UGO
BS35 3LZ G4GMW
BS35 3NJ G0WOI
BS35 3NJ G4RAE
BS35 3PR G3HTJ
BS35 3RR G0HKC
BS35 3RR G6RAZ
BS35 3SB G6YNL
BS35 3TA G0NBP
BS35 3TL G6FFL
BS35 4AQ G0EBZ
BS35 4DX G0MGC
BS35 4DZ G4YHG
BS35 4HA M6ATI
BS35 4HJ G0CYD
BS35 4LZ M0LDG
BS35 4LZ M5JON
BS35 4PF G0WRN
BS35 5RE G3ZUT
BS35 5SB M6GXD

BS36 1BY G4PHK
BS36 1EP G3FYX
BS36 1HQ 2E0FKE
BS36 1HQ M0IYQ
BS36 1HQ M6EZX
BS36 1HX M7BKM
BS36 1NA G3GBD
BS36 2AT M3PSO
BS36 2EN G0LXC
BS36 2FD G4DEM
BS36 2HL G0NQG
BS36 2HT M6RXA
BS36 2NB G4FKA
BS36 2NQ G4RKG
BS36 2SB G4NBP
BS36 2TX M7DLS
BS36 2TX M7GEO
BS37 4DJ M6EIY
BS37 4EG G0IUD
BS37 4EY G3YAD
BS37 4GB G4FBK
BS37 4JX G7IBF
BS37 4LL 2E0JEZ
BS37 4LL M0JEZ
BS37 4LL M3YSQ
BS37 4LR G4IGU
BS37 4LS M3ZGM
BS37 4PF G6FFB
BS37 4PN M6ZAW
BS37 5DS 2E0EZZ
BS37 5DY M0TPW
BS37 5EX G1WVM
BS37 5PJ G6VEJ
BS37 5TE G0JMD
BS37 5TE G1YEH
BS37 5UR M0SVR
BS37 5XQ M6JBE
BS37 6HE G0LXL
BS37 6JA M6HYH
BS37 6JB 2E0PSM
BS37 6JB G4ABC
BS37 6JB M0ZMB
BS37 6JB M6AVL
BS37 6JF G4EIJ
BS37 6JU G4FPI
BS37 6LA G1ZFF
BS37 6LD M6FUA
BS37 6NJ M6THQ
BS37 6XA G0JYN
BS37 6XB G7BYN
BS37 6XF M0EEP
BS37 6XF M1EEP
BS37 6XJ G1JOR
BS37 6XQ M6VSA
BS37 7AH M6DTT
BS37 7LL G7TZO
BS37 7NA G7GNS
BS37 7NA M3BUU
BS37 7RU 2E0TJV
BS37 7RU G7SGR
BS37 7RU M0TJX
BS37 7RU M6TJV
BS37 8SA G7CJD
BS37 8UA G0JTR
BS37 8YW G6HKZ
BS37 9XN M0XXX
BS37 9XN M0YYY
BS375TF G0JZF
BS377SJ M0GTT
BS378YE M7MJH
BS39 4BH G4SND
BS39 4NT G1ZFG
BS39 5ED M3NEE
BS39 5PB G1FZV
BS39 5PL M6CBO
BS39 5RF M6ODE
BS39 5RJ G0BLB
BS39 5SA G4OTJ
BS39 5UP G0KMV
BS39 6TW 2E0VHV
BS39 6TW M3VHV
BS39 6UD M5AKY
BS39 7LU G0RWT
BS39 7LW M1ALR
BS39 7NL M3LDS
BS39 7PN G1OWD
BS39 7PX G1DOJ
BS39 7PX G4LAF
BS39 7QB G7AEQ
BS39 7QS 2E0GHT
BS39 7QS G4BAD
BS39 7QS G4FZV
BS39 7QS M6PBE
BS39 7RP M0ALZ
BS39 7SE 2E0DMA
BS39 7XA G8OEU
BS39 7YX G1GFD
BS4 1BN G4YTH
BS4 1DG M7RKN
BS4 1JT M6VPZ
BS4 1PL G7LPP
BS4 1QE M0RCE
BS4 1RN G4RZY
BS4 1SL 2E0WXT
BS4 1UQ G7BYK
Bs4 1xt M7AJK
BS4 2DL G4KUQ
BS4 2DL G8SHR
BS4 2DX G6HMV
BS4 2LA G7WFH
BS4 2LJ 2E0MXR
BS4 2LJ M6VWW
BS4 2PB M6WOM
BS4 2QP G4KVT
BS4 2RX M7DWF
BS4 2UP G0EXU

BS4 2UW 2E0WCB
BS4 2UW G3ATI
BS4 2UW M6WCB
BS4 2XN M3KPZ
BS4 3BY M0IPH
BS4 3EY G0CCU
BS4 3JD M5AXA
BS4 3JF G6LPG
BS4 3QP G4KKU
BS4 4AF G1XXE
BS4 4BN 2E0JPH
BS4 4DG G7FTS
BS4 4DG M7EOU
BS4 4HN G4XCB
BS4 4JL G3XOD
BS4 4JT G6VPW
BS4 4LP G1XZV
BS4 4NX G7AES
BS4 4QX M1ACQ
BS4 4RA G4GTD
BS4 5DJ G0LTB
BS4 5DZ M6FWL
BS4 5ES G0NVJ
BS4 5HB M7BQY
BS4 5HQ M6EAN
BS4 5JA G0SYI
BS40 5EB G3XSV
BS40 5EG 2E0DDB
BS40 5HD M6DCP
BS40 5LD 2E1DTN
BS40 5PX G6KID
BS40 5QG M0FCP
BS40 6AD G4MKX
BS40 6AP G1KTY
BS40 6BJ G4AXX
BS40 6BJ M0CAM
BS40 6BZ G0PCQ
BS40 6HF G3XMC
BS40 6JE G3SDH
BS40 7TL G3ZMH
BS40 8BG M0TTI
BS40 8SS G8ZOO
BS41 8JA M0KEE
BS41 8JU M6NFX
BS41 9AQ G3KOS
BS41 9AZ 2E0EKT
BS41 9FE G7HMQ
BS41 9NA 2E0AKJ
BS41 9NA M3OQL
BS41 9NF G3YQV
BS48 1BB G8XXG
BS48 1HR M6USP
BS48 1JD G0CCB
BS48 1JL G7PVL
BS48 1JQ G4WAW
BS48 1JQ G7KNA
BS48 1JQ M7MYU
BS48 1LT 2E0GWF
BS48 1PS G8DEX
BS48 2AG G6ATS
BS48 2BH G0IFF
BS48 2BL G7RUR
BS48 2BN M0HCT
BS48 2DS G6WLX
BS48 2DY 2E0GKF
BS48 2DY M7ARQ
BS48 2DZ G1ODB
BS48 2EE 2E1IHO
BS48 2EE M1FFN
BS48 2QH M6ORO
BS48 2XD G0LHD
BS48 3BG G3ATX
BS48 3NR G7UTT
BS48 3PB G3PRH
BS48 3RX G4DEQ
BS48 4RA G0GHM
BS48 4RT G0CEN
BS48 4SX 2E0CFN
BS48 4SX M6TAF
BS48 4YD 2E0DBW
BS48 4YD G8NQO
BS48 4YD M3NQO
BS48 4YD M3ZDW
BS48 4YH G4LSX
BS48 4YH G8HVT
bs49 4 rg G4HBT
BS49 4AS G0LCX
BS49 4AS M7EJW
BS49 4AS M7JGW
BS49 4DA G4DPH
BS49 4DF G0JLF
BS49 4EB G4FZV
BS49 4EB G6GVH
BS49 4ER G6DMC
BS49 4HF M6IEI
BS49 4HF M6WSH
BS49 4JU M6AEZ
BS49 4LE 2E0ZBB
BS49 4LE M6ZBB
BS49 4LN G6ANJ
BS49 4LS G4BWR
BS49 4NR M7TAT
BS49 4NS M6VTC
BS49 4RB M0SWH
BS49 4RG G6BZW
BS49 5BN 2E0YCD
BS49 5BN M3FPZ
BS49 5BN M6YCT
BS49 5BT 2E0FYO
BS49 5BT M0JVE
BS49 5BT M7ASO
BS49 5ES G4DYM
BS49 5ES G4FDK
BS49 5EX M0DWC
BS49 5EX M3DWD
BS49 5EY G4KWL

BS49 5EY G4RSC
BS49 5HA 2E0JHT
BS49 5HA M0XOX
BS49 5HA M6JHT
BS49 5HB G0ALI
BS49 5HB G4TCI
BS49 5HL G4REH
BS49 5HQ G3SWH
BS5 0DL G4BWO
BS5 0PQ M6ISZ
BS5 0SE G1JOO
BS5 6RJ G4TPV
BS5 6SY G8BIR
BS5 6TN M0SVA
BS5 7BQ G4GGE
BS5 7EJ G0SCK
BS5 7LY M7ARN
BS5 7NE G0GJN
BS5 7QT 2E0GHE
BS5 7QZ M3WMC
BS5 7RG 2E0OAJ
BS5 7RG M0OAJ
BS5 7RG M6OAJ
BS5 7RL 2E0KAC
BS5 7RL M3KAC
BS5 7SP 2E0TUK
BS5 7SP M6LUA
BS5 7SP M6TYK
BS5 8DX G3ZKI
BS5 8DX G4JPS
BS5 8DX G8XAA
BS5 8HF G7GLQ
BS5 8JU 2E0LJT
BS5 8JU M0LJT
BS5 8JU M6CQJ
BS5 8LN G4ZBL
BS5 8RH G4EIA
BS5 8ST G8CKK
BS5 8ST M0BUV
BS5 8ST M6EZG
BS5 8SZ G0JLE
BS5 8TA G6YFG
BS5 8TW G0KWF
BS5 9HN M3UIC
BS5 9HN M6JWF
BS5 9HW M0IIU
BS5 9JT M3MCU
BS5 9QN G4TAH
BS6 5AH 2E0PFB
BS6 5AH M0HWQ
BS6 5BQ M1BBR
BS6 5HX G8NNU
BS6 6BE G3ORV
BS6 6DU G1YOY
BS6 6PD G4TRN
BS6 7LG G4LOX
BS6 7SU G1UGO
BS6 7XS G8UXB
BS66NS G4FVX
BS7 0HQ G8OQG
BS7 0HS G0NZT
BS7 0RG G0NZU
BS7 0RH G6ZBG
BS7 0RH G8ZRN
BS7 0RP G8VYT
BS7 0RT G3XPJ
BS7 0SA G7UHS
BS7 0SG G7ISR
BS7 0SR G7DRU
BS7 0TT G4CSE
BS7 0UA 2E0ZST
BS7 0UA M6ZST
BS7 0UH G3FPY
BS7 0UQ G3IUV
BS7 0US G4CGF
BS7 8EX G3MGS
BS7 8JJ G8MHD
BS7 8LT G7AGI
BS7 8LU G0KKC
BS7 8QG G4YQH
BS7 9DW M0HCV
BS7 9JW G6RNZ
BS7 9ST G4ROX
BS7 9XS 2E0RBK
BS7 9YW G7PKJ
BS8 1DW M6IUM
BS8 1DY M7HEX
BS8 2HF G6GGN
BS8 2QD M6GOQ
BS8 3EG G7RLQ
BS8 3PE G6DUC
BS8 4DL G8OOQ
BS8 4TT M6XUX
BS9 1DR G4NYK
BS9 1NG G8HJD
BS9 1NP G3YER
BS9 1PL M1PUW
BS9 1QP G4FRO
BS9 1SN G0SDW
BS9 2BA 2E0KMI
BS9 2BA M0KMI
BS9 2BW G3TCO
BS9 2JF G0OBT
BS9 2LU G4WBV
BS9 2PU M6PAJ
BS9 2QP G0KJM
BS9 2QQ G8XIM
BS9 2QR G4ZBQ
BS9 2QT G0OER
BS9 2QU G8PSO
BS9 3DQ G3XOB
BS9 3RN M0AWH
BS9 3SX G8GRD
BS9 3TU 2E0CES
BS9 3TU M0HVR
BS9 3UU G4HCB

BS9 3UW G3OWX
BS9 4BU G3SJI
BS9 4DW G4NBN
BS9 4EL G2BQP
BS9 4QW G8FNR
BS9 4RH G0CJG
BS9 4RS G4KSR
BS9 4TB G3TZA
BS9 4TF G4OEP
BS93JG 2E0FJX
BS93JG M0RKE
BS93JG M7RKE
BS99 5LG G0TEI
BS99 5LG G1IBO
BT1 4FJ MI7MSX
BT10 0AN GI0JRI
BT10 0AS GI6JOP
BT10 0JQ GI0JRD
BT10 0QE MI0GPB
BT11 8BP GI0DPV
BT11 8LL MI6PAT
BT11 8LP GI7OMY
BT11 8PP MI6LDI
BT11 9ED 2I0TJM
BT11 9ED MI0TJM
BT11 9ED MI3FSX
BT11 9GF GI3UIH
BT11 9LU MI6CVO
BT11 9PU 2I0JHH
BT11 9PU MI7CAT
BT11 9PU MI7JOE
BT119ER MI1DAW
BT119ER MI5DAW
BT12 4NH 2I0GCC
BT12 4NH MI6GAQ
BT12 4QB GI7PJU
BT12 5NR MI1RDR
BT12 5NR MI3BRX
BT12 6GF GI0VWU
BT12 6JS MI0SAM
BT12 6JT MI6AHR
BT12 6NH MI6TOA
BT12 7JD 2I0KBS
BT12 7JD MI6KOB
BT12 7LG GI0XBH
Bt13 1Dj MI7NWA
BT13 2DR GI4XFN
BT13 2SB GI4GOL
BT13 3DZ GI0CDM
BT13 3LG GI4CFQ
BT13 3LR GI4PXM
BT13 3LR MI0MFI
BT13 3LR MI6EAI
BT13 3PQ MI6DCH
BT13 3PS MI3RLA
BT13 3XN GI0ZAK
bt133rd MI6IGV
BT14 6BT MI0HPE
BT14 6ED MI6WDB
BT14 6ES MI0ILJ
BT14 6ES MI6DND
BT14 6JP MI7CYX
BT14 6NZ MI6SWB
BT14 6PA 2I0CVR
BT14 6PA MI6TMZ
BT14 6RZ GI4OZI
BT14 6SL MI6MSG
BT14 6TE GI4IKF
BT14 7GD GI7TPO
BT14 7NF MI3MPL
bt14 8an MI6OLJ
BT14 8AN MI7DLD
BT14 8FP MI0BDZ
BT14 8HD GI4EIZ
BT14 8JX GI4CUV
BT14 8JY GI7MBP
BT14 8QB MI3GEI
BT14 8RE 2I0ZFZ
BT14 8RE MI6FNZ
BT146BE MI6FRQ
BT15 2BP MI0ISY
BT15 3FY GI0RWO
BT15 3JF MI7BGE
BT15 3LA MI5HNA
BT15 3NP MI0PCJ
BT15 3QS MI6HAD
BT15 4EP GI1KDS
BT15 4GR GI4MNN
BT15 4JU GI7IFW
BT15 5AJ 2I0EBB
BT15 5AJ MI6EDP
BT15 5GL GI3RQU
BT16 1JD 2I0DHR
BT16 1JD MI3DHR
BT16 1JW MI6LJZ
BT16 1SL MI5AMO
BT16 1UU GI4SAM
BT16 1XD GI4NBO
BT16 1XU GI8SJS
BT16 2AB MI6JJZ
BT16 2BB GI4OCL
BT16 2BE 2I0TUV
BT16 2BE GI4TUV
BT16 2BE MI6TUV
BT16 2BE MI6WGE
BT16 2HB MI3MDV
BT16 2NT GI7KHR
BT16 2PQ MI3CST
BT16 2SQ MI6PDY
BT17 0AF GI1JQP
BT17 0JG GI4CSO
BT17 0NZ MI3OIB
BT17 0NZ MI6BFJ
BT17 0XD MI7MHG
BT17 9PY GI4KCO
BT17 9PZ MI6OYB

BT17 9QY 2I0DHC
BT17 9QY MI6PBW
BT18 0DY GI3GTR
BT18 0HG GI0TDP
BT18 0HH GI4JTF
BT18 0NB GI5SBZ
BT18 0PJ GI4ZTU
BT18 0PL GI3USK
BT18 9DL MI7NIH
BT18 9EL GI3WUO
BT18 9EU GI0USQ
BT18 9JQ MI0RJW
BT18 9JX GI8YJV
BT18 9NX 2I0VOQ
BT19 1AA GI4MUE
BT19 1BJ MI3SEV
BT19 1DQ GI6ATD
BT19 1EQ GI4POC
BT19 1FG MI3CCA
BT19 1GH MI3LXZ
BT19 1HD 2I0HHX
BT19 1HD MI0WHX
BT19 1HD MI6HHX
BT19 1HG MI3LXN
BT19 1HQ MI0CBX
BT19 1HQ MI3OHF
BT19 1JA MI3XOI
BT19 1LB GI6UFU
BT19 1NP MI3XIU
BT19 1NS MI0NYC
BT19 1NT GI4OCK
Bt19 1ql MI6OWD
BT19 1RQ 2I0DUR
BT19 1RQ MI0OBR
BT19 1RQ MI6DUR
BT19 1RQ MI6HLC
BT19 1SN MI6VRX
BT19 1SR MI1EVD
BT19 1YG GI6IHM
BT19 1YN MI3JSH
BT19 1YU GI3XRQ
BT19 1YU GI4DOH
BT19 1YU MI3FSW
BT19 1ZQ MI3FBX
BT19 6AE GI3TZB
BT19 6AF GI4TPY
BT19 6AF GI7DIT
BT19 6AR GI3OTU
BT19 6AY GI1VPA
BT19 6AY GI4NSS
BT19 6BA GI3MBB
BT19 6DG GI6BDN
BT19 6DQ 2I0UAD
BT19 6DQ MI6UAB
BT19 6DT GI7VIW
BT19 6DU GI3YMY
BT19 6EB MI6CBG
BT19 6FN GI3MMG
BT19 6HW GI4TCR
BT19 6HY GI6KJC
BT19 6JF GI6JGB
BT19 6LX MI3SRG
BT19 6LZ MI3PPI
BT19 6NJ 2I0POD
BT19 6NR GI7FOD
BT19 6NX GI1SZC
BT19 6SD GI3VAF
BT19 6SD MI6HBI
BT19 6XG GI6BNI
BT19 6ZH MI3FSR
BT19 6ZW GI6PLO
BT19 7BS MI6WGX
BT19 7EY GI4JNS
BT19 7FE 2I0DTE
BT19 7FE MI0OBC
BT19 7FE MI6DTE
BT19 7GB GI7GXZ
BT19 7GY MI0ILH
BT19 7HL GI0UOL
BT19 7HR GI4MRZ
BT19 7QB GI3WFP
BT19 7RB GI4WYE
BT20 3BF MI0RSO
BT20 3EP GI4LZS
BT20 3EP MI6NUX
BT20 3ER GI4JJF
BT20 3HA GI0HSB
BT20 3JD GI0POB
BT20 3PP GI3TZX
BT20 3PU GI8JPF
BT20 3TP GI7JEB
BT20 4DF GI0WPV
BT20 4HS GI6IVJ
BT20 4JD GI0FZT
BT20 4NN MI3BLN
BT20 4NP GI4OPH
BT20 4NS MI6EAC
BT20 4PE GI0HHV
BT20 4PP 2I0LNZ
BT20 4PT GI4SPU
BT20 4PT GI6SBW
BT20 4PX GI7ISX
BT20 4PX MI3ISX
BT20 4QB GI6FZI
BT20 4RS GI4TMB
BT20 4TG GI8YJD
BT20 4TX GI4MRN
BT20 4US GI0BEY
BT20 4UX GI7TVV
BT20 5LT MI3CBJ
BT20 5NT MI3FEO
BT20 5NU GI8ZHW
BT20 5PN GI3UBA
BT20 5RF GI4FLG
BT20 5RQ GI4TPI
BT21 0AX MI3CBL

BT21 0BN GI1SYM
BT21 0BQ GI7FNP
BT21 0DR 2I1HNZ
BT21 0ES GI8VTK
BT21 0EY MI6KRP
BT21 0EZ 2I0ETW
BT21 0EZ GI0SSA
BT21 0EZ MI0HWG
BT21 0EZ MI1EIH
BT21 0EZ MI3BPS
BT21 0EZ MI3CQO
BT21 0EZ MI6NID
BT21 0GA GI4BXB
BT21 0HL GI1WGK
BT21 0HZ MI6CUZ
BT21 0LN GI3UPG
BT21 0NR MI3VFJ
BT21 0NS GI7PIZ
BT21 0PU MI3FVW
BT21 0PY GI3TJM
BT21 0QR GI0BCP
BT21 0SH GI4PQV
BT21 0SH GI6IOU
BT210FB GI4OVN
BT22 1AF MI0AEX
BT22 1AF MI0JSJ
BT22 1AJ MI0BSU
BT22 1AU MI3MJI
BT22 1BW GI0PNP
BT22 1DZ MI0SRM
BT22 1EL GI0AQD
BT22 1GB MI7NMA
BT22 1HP GI0VTS
BT22 1HP GI4XSF
BT22 1JX GI0UAG
BT22 1LL GI4TTL
BT22 1ND GI4NAE
BT22 1NE GI6IES
BT22 1NQ 2I0GTO
BT22 1NQ GI3UZJ
BT22 1NQ MI3GTO
BT22 1QT MI0GLG
BT22 1QT MI3VFZ
BT22 1RB GI4YPR
BT22 2BG GI7NMK
BT22 2BN GI3OBO
BT22 2GS MI6MIB
BT22 2HW MI3CSS
BT22 2HY GI4PBS
BT22 2HY GI4PBT
BT22 2JQ MI0RPT
BT22 2LA GI4HCX
BT22 2LB MI6WPT
BT22 2LE 2I0TMH
BT22 2LE MI0TMP
BT22 2LE MI6VMC
BT22 2NG MI0UST
BT22 2NQ MI0WST
BT22 2NQ MI3PYX
BT22 2NW MI3VEQ
BT22 2PE MI6VAI
BT22 2QF GI4PGN
BT22 2RP 2I0FIP
BT22 2RP MI7ACD
BT22 2RP MI7CYL
BT22 2TH GI4NKK
BT22 2TR GI4EQN
BT22 2TZ 2I0FPB
BT22 2TZ GI7VCR
BT22 2TZ MI0IGL
BT22 2TZ MI6IPB
BT23 3BN GI4AFH
BT23 4AN MI0JPC
BT23 4BN 2I0BXJ
BT23 4BN MI3LZF
BT23 4LY GI7ALH
BT23 4NA MI6OPT
BT23 4ND MI3WJO
BT23 4NT MI3ZMJ
BT23 4PD MI0RSN
BT23 4SF GI7KTU
BT23 4SQ GI4MCW
BT23 4TB GI1WLJ
BT23 4TN MI6BAI
BT23 4TP GI4JTS
BT23 4TQ GI4PGH
BT23 4UW GI0PFL
BT23 5DF MI7DGL
BT23 5EW GI6EGE
BT23 5HA GI7FGQ
BT23 5HE GI0SMU
BT23 5HR GI4GST
BT23 5JJ GI4TSK
BT23 5LD MI1FAR
BT23 5LN GI6CAG
BT23 5LP GI4MBM
BT23 5LT GI6VLY
BT23 5QP GI3NYJ
BT23 5RJ GI4OSG
BT23 5RN GI0EEO
BT23 5TQ GI4OWB
BT23 5TZ 2I0WWB
BT23 5TZ GI5TKA
BT23 5TZ MI0WWB
BT23 5TZ MI6USC
BT23 5XA MI3SXI
BT23 5YX GI4IYO
BT23 6AQ MI6IMV
BT23 6BB GI4MQA
BT23 6DE MI3VFF
BT23 6EN GI3XZM
BT23 6ES GI3VHM
BT23 6PB MI1RGL
BT23 6PE MI7EJV
BT23 6RZ MI3ZMP
BT23 7AR GI0LTT

BT23 7BN 2I0LJQ
BT23 7BN MI3LJQ
BT23 7BW MI0ABD
BT23 7BZ GI0OHT
BT23 7ED MI6CMQ
BT23 7GQ GI6JMD
BT23 7QP GI4OYI
BT23 7RE GI0WYK
BT23 8NN MI0MGJ
BT23 8QS GI3AMY
BT23 8QT GI0HHZ
BT23 8RF MI3PMR
BT23 8RS MI0BVG
BT23 8RT GI0DUP
BT23 8SN MI5ALO
BT23 8TE GI3WHA
BT23 8UA MI3LVZ
BT23 8UL 2I0FZO
BT23 8UL MI7MCK
BT23 8XD MI5CFM
BT23 8YE 2I0WMH
BT23 8YE MI6WAF
BT24 7BE GI4BJK
BT24 7BS MI6MPH
BT24 7DP MI3EZF
BT24 7FQ GI0VIF
BT24 7HU MI3NSF
BT24 7PR 2I0LXS
BT24 7PR MI6CVW
BT24 8BL 2I0WKY
BT24 8BL MI7WNY
BT24 8FE MI6PIR
BT24 8GA GI4JOR
BT24 8HU GI8TME
BT24 8HW GI4MUN
BT24 8LB GI4TUJ
BT24 8LF GI4XLB
BT24 8LF GI6ATZ
BT24 8PT GI0WJI
BT24 8QQ GI4AXV
bt24 8qy GI6DCX
BT24 8QZ GI7MDP
BT24 8QZ MI3NJU
BT24 8YS GI4SZP
BT25 1BD 2I0YMF
BT25 1BD MI0YMF
BT25 1BD MI3YMF
BT25 1DD MI0BAT
BT25 1LL 2I0GIF
BT25 1LL MI6GIF
BT25 1NP MI0SRR
BT25 1NP MI3AJK
BT25 1PN 2I0WKE
BT25 1PN MI6WKE
BT25 1PN MI6YAM
BT25 1RT MI6MQX
BT25 2AF MI1CUS
BT25 2AN 2I0EIG
BT25 2AN 2I0ESA
BT25 2AN MI6EQA
BT25 2AN MI6XOS
BT25 2EQ GI0UXD
BT25 2HN GI6SFO
BT25 2HY GI7VXC
BT25 2NT GI0TSS
BT25 2PN MI0BTK
BT26 6BH GI4KSO
BT26 6BL MI0DGX
BT26 6BS 2I0VOF
BT26 6BS MI6POF
BT26 6BS MI6VOF
BT26 6BX GI4OZJ
BT26 6DJ 2I0LPG
BT26 6EG MI3VPO
BT26 6ES GI3VPV
BT26 6HL MI6TNZ
BT26 6HQ MI1BRS
BT26 6HQ MI7RLC
BT26 6HU GI7NFB
BT26 6LJ GI4TCS
BT26 6NH MI7GSI
BT26 6NS GI6UUT
BT26 6PW GI4RXM
BT27 4BH GI7TFK
BT27 4BS GI4NFH
BT27 4DA GI4ERM
BT27 4DA GI8KEP
BT27 4EF GI6ETQ
BT27 4EW GI6CMA
BT27 4JA GI6WFX
BT27 4PL 2I0SHM
BT27 4QA GI4NLQ
BT27 4QX GI4NKY
BT27 4RY MI6GTY
bt27 4ul MI7FNY
BT27 4YD MI6TIJ
BT27 5AW MI7ABU
BT27 5BF GI4AHP
BT27 5BT GI6FOR
BT27 5BY 2I0GKB
BT27 5BY MI6GKB
BT27 5BZ MI0CZF
BT27 5DA 2I0EIU
BT27 5DA MI6GGF
BT27 5DB GI4XIR
BT27 5HJ MI6HLY
BT27 5LF MI0OPM
BT27 5LF MI1OPM
BT27 5LF MI3TEM
BT27 5LQ GI7PBQ
BT27 5LS GI1GKI
BT27 5LW MI1FRM
BT27 5PD GI0LWO
BT27 5PD GI8TSI
BT27 5RE 2I0EUV
BT27 5RE MI6WEQ

BT27 5RF 2I1AXH
BT27 6UU 2I0ITY
BT27 6UU MI0TUB
BT27 6UU MI6JBZ
BT27 6YW MI1ERL
Bt276ts MI3GSW
BT28 1EX GI0RDJ
BT28 1HE GI0GDF
BT28 1HF 2I0GYL
BT28 1HF GI7PWQ
BT28 1HF MI6YLG
BT28 1LD GI0RYK
BT28 1NY 2I0KBB
BT28 1NY MI6KBB
BT28 1QD MI0JLC
BT28 1QD MI3OHG
BT28 1SQ GI0UVD
BT28 1TY GI0OKM
BT28 1YJ GI0PGC
BT28 2BX MI6KGU
BT28 2DN GI0TJV
BT28 2DR GI3LQY
BT28 2DW MI0BME
BT28 2EY MI6PGI
BT28 2GZ 2I0JCH
BT28 2GZ GI7SBF
BT28 2GZ MI6EAS
BT28 2HU MI3SYF
BT28 2HX MI3MOT
BT28 2LH GI8DGB
BT28 2PL MI0MSB
BT28 2QF MI0MOD
BT28 2QL MI6XGZ
BT28 2TE MI0TGL
BT28 2TE MI0TLG
BT28 2TE MI6TLG
BT28 2UD MI7TVA
BT28 2UN MI3NSR
BT28 2WY MI6KKU
BT28 2XU GI4PES
BT28 3BA MI6TMR
BT28 3BT GI4DNW
BT28 3DG GI7ROB
BT28 3DS GI4XTC
BT28 3HD MI3MRG
BT28 3HG GI4LKG
BT28 3HN 2I0PRL
BT28 3HN MI6FIU
BT28 3HQ MI6MNL
BT28 3JH GI4MEQ
BT28 3LL GI8YWE
BT28 3LP GI6GNA
BT28 3PD GI3ECQ
BT28 3QB GI4RKC
BT28 3QX MI0ASV
BT28 3QY MI0ABN
BT28 3QY MI3EOD
BT28 3RE GI4SNA
BT28 3RR GI0DVU
BT28 3TQ GI1YSG
BT28 3TR GI7GSB
BT28 3UT GI8UUN
BT28 3YB MI3TBL
BT29 4FX MI0IXH
BT29 4JL 2I0EQC
BT29 4JL 2I0EQR
BT29 4JL 2I0EQS
BT29 4JL GI4WTT
BT29 4JL MI3EQS
BT29 4JL MI3WTT
BT29 4JQ GI0AIJ
BT29 4JW MI5AJH
BT29 4QU 2I0OAZ
BT29 4QU MI6OAZ
BT29 4RH GI7IRJ
BT29 4SF GI7ULG
BT29 4SG GI7UBV
BT29 4WT GI7PJF
BT29 4XN MI7WTZ
BT29 4YA MI1DJW
BT29 4YA MI6LFU
BT30 6JL MI6GPV
BT30 6NS GI1XTK
BT30 6PZ GI4NJQ
BT30 6ST MI6XOX
BT30 7AZ MI3LXE
BT30 7DA GI8UIU
BT30 7JT MI7RDA
BT30 7LY GI4RMA
BT30 7NU MI5AFL
BT30 7RJ 2I0OTW
BT30 7RJ MI6OTW
BT30 7SA GI4HYM
BT30 7SQ 2I0NEJ
BT30 7SQ GI6MQF
BT30 8SR MI6HKP
BT30 9BS 2I0TXM
BT30 9BS MI0HNQ
BT30 9BS MI0TXM
BT30 9BS MI6XMG
BT30 9BU GI4MMJ
BT30 9BW MI3AIN
BT30 9EH GI4MAC
BT30 9EW GI4MHD
BT30 9GB GI3TAC
BT30 9HW MI6THD
BT30 9PD GI3HNM
BT30 9PX GI1RXM
BT31 9SJ GI0LAM
BT32 3AW MI0DNM
BT32 3BQ MI0MAP
BT32 3DJ MI7HNC
BT32 3GD MI7OYZ
BT32 3RD GI0USS
BT32 3TL 2I0NGM
BT32 3TL MI6NGM

BT32 3TZ MI6MXZ
BT32 3TZ MI6RXC
BT32 3UT MI6LIT
BT32 3YA GI0WZW
BT32 4AH MI6PEJ
BT32 4BN GI7INR
BT32 4HF GI0HXH
BT32 4JL 2I0BSH
BT32 4JL MI0TFK
BT32 4LF GI3WEM
BT32 4NA MI6GPZ
BT32 4NU MI6EFD
BT32 4PE MI6RFD
BT32 4PT MI6MTO
BT32 4PY GI4GUH
BT32 4PZ 2I0SXM
BT32 4PZ MI3SXM
BT32 4RA GI8RQI
BT32 4RE GI0UQK
BT32 5ES 2I0GQR
BT32 5ES MI0SNY
BT32 5ES MI7RYP
BT32 5HR MI7GBX
BT32 5JF MI0CGV
BT32 5NN GI4GPC
BT32 5RD GI1JXE
Bt323gd 2I0PBM
Bt323gd MI3GJI
BT33 0AR GI6DWZ
BT33 0DT MI6AGV
BT33 0FA GI1VKJ
BT33 0HW MI1FCB
BT33 0NQ GI3FJX
BT34 1HL MI3PMW
BT34 1JG 2I0BIR
BT34 1JG MI3RIF
BT34 1JW MI0PYN
BT34 1NZ GI0LRZ
BT34 1PW MI6TUM
BT34 1QD MI6NQD
BT34 2BQ MI0AQX
BT34 2BQ MI0BES
BT34 2JB MI5AHG
BT34 2NA GI6JJR
BT34 2ND GI0THR
BT34 2NY GI0WAH
BT34 2PG GI0UYY
BT34 2PG GI1YEA
BT34 2PJ GI8ZFZ
BT34 2QP 2I0MMT
BT34 2QP MI6MMT
BT34 3AN MI6AOR
BT34 3BJ GI6RBD
BT34 3BL GI1WFP
BT34 3DX GI8YJF
BT34 3EP MI0ZKX
BT34 3HL GI4MBQ
BT34 3JZ MI6GMK
BT34 3NL MI3EOH
BT34 3PG GI0VVC
BT34 3SA GI1RAA
BT34 3SR MI0DDW
BT34 4AQ GI4BBE
BT34 4DA GI7IVX
BT34 4FT 2I0TXB
BT34 4JJ GI6EBX
BT34 4LP GI4XFY
BT34 4NB MI7YDB
BT34 4NU GI3HJH
BT34 4XW MI3SXQ
BT34 5DE 2I0WMN
BT34 5DE MI6WMN
BT34 5EL 2I1ALE
BT34 5EL GI0UTE
BT34 5LS 2I0SEC
BT34 5LS MI0NLY
BT34 5LS MI6EEC
BT34 5PU MI6UNC
BT34 5TJ GI4WAH
BT34 5US MI6MSR
BT35 0JG GI4DQT
BT35 0PJ GI0VGL
BT35 6BZ 2I0ZXM
BT35 6BZ MI6ZTM
BT35 6DD 2I0DBK
BT35 6EH GI6FXY
BT35 6LF MI3RIV
BT35 6NA GI4OVE
BT35 6NS 2I0FEX
BT35 6NS 2I0JAP
BT35 6NS 2I0RRE
BT35 6NS MI0KAG
BT35 6NS MI0RRE
BT35 6NS MI3FEX
BT35 6NS MI3RRE
BT35 6NS MI3WBU
BT35 6SD 2I0BYL
BT35 6SD 2I0FUT
BT35 6SD MI0GQG
BT35 6SD MI0GQI
BT35 6SD MI3GMI
BT35 6SD MI3LXJ
BT35 6TW 2I0MBI
BT35 6TW MI3SYI
BT35 7FD MI0YCK
BT35 7GA MI6CQS
BT35 7HD GI4SZW
BT35 7HD MI0JWY
BT35 8AP MI6KJW
BT35 8PP MI6DVN
BT35 8PW GI1ANG
BT35 8PW MI6UCM
BT35 8XA MI3CQX
BT35 9AQ 2I0EIB
BT35 9AQ MI6RSH
BT35 9RR MI3CQR

BT35 9RR MI3NCC
BT35 9TX GI0CTI
BT36 4QT GI4OXO
BT36 4TP GI4BUJ
BT36 4WL GI4BTG
BT36 4WQ GI4SQL
BT36 5BX MI6PBI
BT36 5FG MI1AUI
BT36 5GD GI7AQO
BT36 5GZ GI4RXX
BT36 5JU GI7JYK
BT36 5JU GI8UCS
BT36 5JU MI5JYK
BT36 5JZ GI1WYZ
BT36 5LA 2I0BID
BT36 5LA MI0GTI
BT36 5LA MI3OEQ
BT36 5NH MI7GNI
BT36 5ST GI4KEQ
BT36 5ST MI1ASN
BT36 5UF GI5CEO
BT36 5UY MI7DKB
BT36 5WQ MI3LPH
BT36 5WR GI4JIW
BT36 5WZ 2I0TCJ
BT36 5WZ MI0TBV
BT36 5WZ MI6TFG
BT36 6BA GI4OTG
BT36 6BD 2I0EPC
BT36 6BX MI6WPP
BT36 6DZ MI6MRI
BT36 6LE MI3DNN
BT36 6LJ 2I0CLS
BT36 6LJ MI6RAS
BT36 6LS GI4KSH
BT36 6PA MI0NAJ
BT36 6PR MI6LNP
BT36 6QQ GI0BEB
BT36 6SP GI8RPT
BT36 6TZ GI8LUR
BT36 6UA MI1BOE
BT36 6UN GI4JWW
BT36 7DG MI3CXM
BT36 7DR MI1EZZ
BT36 7DR MI3STW
BT36 7HA GI0USW
BT36 7SU GI4GID
BT36 7TG 2I0TUI
BT36 7TG MI0HRO
BT36 7YP GI0JFF
BT366HT MI7CVD
BT37 0AZ GI4MCH
BT37 0BE GI4RNP
BT37 0BE MI0JQS
BT37 0EL GI6KVS
BT37 0GH 2I0KFD
BT37 0GH MI0KKD
BT37 0GH MI6LLZ
BT37 0JN GI0BDU
BT37 0LH MI0XLK
BT37 0N MI0OIM
BT37 0RR GI0IQA
BT37 0TD GI0HHE
BT37 0UL GI4KBW
BT37 0XH GI4SFE
BT37 0XY GI4VWC
BT37 0ZY 2I0LOR
BT37 0ZY MI6BAJ
BT37 9JZ MI3FPN
BT37 9PD GI4RYL
BT37 9sh 2I0SUB
BT37 9SH GI6ANC
BT37 9SQ MI0TBE
BT379QZ 2I0RWG
BT379QZ MI0RGX
BT379QZ MI6BJG
BT38 7BL 2I0CYW
BT38 7BL MI3ZCY
BT38 7DJ MI3CGT
BT38 7DJ MI3CGU
BT38 7EH GI0USC
BT38 7EP MI3KIL
BT38 7HG MI0PQR
BT38 7HN MI3JTB
BT38 7HQ MI3WMK
BT38 7JT GI8LCJ
BT38 7LD MI0KMJ
BT38 7LL MI0PCW
BT38 7ls 2I0TPC
BT38 7ls MI6TPC
BT38 7LZ 2I1EXU
BT38 7NG GI3YRL
BT38 7NG GI4SBA
BT38 7PD 2I0FKM
BT38 7PD MI7AEW
BT38 7QD GI4DAV
BT38 7RB GI6ROI
BT38 7RL MI6HDH
BT38 7RT MI3FCK
BT38 7RU MI0ZSC
BT38 7UE MI0PJL
BT38 7XU 2I0TLT
BT38 7XU MI0TIP
BT38 8BF GI7VGR
BT38 8BF MI0SYN
BT38 8BF MI7JCB
BT38 8BQ 2I0LBS
BT38 8BQ MI0AFT
BT38 8BQ MI0LBS
BT38 8BY GI0IBC
BT38 8BY GI4GVS
BT38 8DP MI3VQH
BT38 8EW 2I0HRV
BT38 8EW MI0HRV
BT38 8EW MI6AOX
BT38 8FB GI4RVF
BT38 8FG GI4OYG
BT38 8GP MI3FBW
BT38 8GQ 2I0SMD
BT38 8HZ GI4GCN
BT38 8JT MI6RLU
BT38 8NE 2I0IYH
BT38 8NE MI3IYH
BT38 8NN 2I0CEI
BT38 8NN MI0HHV
BT38 8NN MI3YVB
BT38 8ps GI0XYZ
BT38 8SN GI0IJB
BT38 8ST MI6XEX
BT38 8SY GI6EJW
BT38 8TX GI0RUC
BT38 9AG 2I0UCS
BT38 9AG MI0GJN
BT38 9AG MI3UCS
BT38 9AP GI0DFD
BT38 9BB GI4FUE
BT38 9DL GI8SKN
BT38 9DL MI6EXU
BT38 9DN GI8KYI
BT38 9EA GI6TFF
BT38 9EG GI8WBZ
BT38 9EG MI0AWL
BT38 9GZ GI0PCU
BT38 9HE GI8KFG
BT38 9JD GI7NOW
BT38 9LF GI4IZF
BT38 9ND GI6GRV
BT38 9NN MI6RUC
BT38 9NT MI6LGX
BT38 9NZ GI4XHO
BT38 9RG MI6MZR
BT38 9RL GI1TFC
BT38 9SU MI0JPL
BT38 9XA GI7JEM
Bt388ha MI6HWV
BT39 0BW GI0SRL
BT39 0BW GI3PDN
BT39 0FA GI4BWM
BT39 0HZ GI3ZVZ
BT39 0JP GI4KQA
BT39 0PH GI4SZY
BT39 0PZ 2I0BAD
BT39 0QB GI4PRH
BT39 0QB GI6PAZ
BT39 0RY 2I0INA
BT39 0RY MI0IRX
BT39 0RY MI6OIM
BT39 0SB GI7JKA
BT39 0SQ 2I0BAC
BT39 0SQ GI7GJX
BT39 0SQ MI3IIH
BT39 0SQ MI3IIL
BT39 0TN GI3XDD
BT39 0TQ GI0LGV
BT39 9AJ 2I0SGK
BT39 9AJ MI7WJD
BT39 9DD MI3FXE
BT39 9FL 2I0GOK
BT39 9GN GI7FCM
BT39 9HE GI1CKU
BT39 9HT GI7GUT
BT39 9HZ GI4TAJ
BT39 9HZ GI4XGO
BT39 9HZ MI6ALL
BT39 9JS GI7DBZ
BT39 9PS GI7GVI
BT39 9QW GI4JYJ
BT39 9QW GI6UUC
BT39 9RZ 2I0LXW
BT39 9RZ MI3LXW
BT39 9RZ MI6LWI
BT39 9RZ MI6LZW
BT39 9RZ MI6XAM
BT39 9SB MI6PHQ
BT39 9SE 2I0KUN
BT39 9SE MI0UFL
BT39 9SE MI7MCS
BT39 9ST MI6TXS
BT39 9TS GI8VKA
BT39 9WE GI4NXJ
BT399AS MI6PFI
BT4 1LJ MI3IAI
BT4 1NA GI7PUG
BT4 1ND GI0RBC
BT4 1PR MI6SQN
BT4 1PR MI6WTZ
BT4 1QT GI7RAM
BT4 1QU GI4IOO
BT4 1RJ GI7HVC
BT4 2BH GI2BX
BT4 2BH GI4NKB
BT4 2BH GI6DEY
BT4 2BL GI4JLF
BT4 2BY GI3TNK
BT4 2DX GI4LGP
BT4 2EH GI4GOS
BT4 2HH GI7RAH
BT4 2HS GI4FZD
BT4 2HT GI6FEN
BT4 2JZ GI4ILZ
BT4 2PA GI4WRJ
BT4 2RB 2I0OHE
BT4 2RB MI3OHE
BT4 2RH MI6SEZ
BT4 3DE GI0VAB
BT4 3DJ GI6DRK
BT40 1BD GI1TRZ
BT40 1BD MI0CNI
BT40 1EB GI8WHP
BT40 1ET 2I0GFO
BT40 1ET MI6GFO
BT40 1EW MI3BKA
BT40 1NE 2I0LRN
BT40 1NE 2I0PBZ
BT40 1NE MI6LRN
BT40 1NE MI6PBZ
BT40 1QL GI6EWO
BT40 1SE 2I0RPM
BT40 1SE MI3TUZ
BT40 1SE MI6TZP
BT40 1UB GI6VCG
BT40 1UL GI7DZE
BT40 2DF GI4MXV
BT40 2DT 2I0HBL
BT40 2DT MI7CJA
BT40 2EG GI4MVQ
BT40 2EJ MI6XBA
BT40 2ER MI3CIV
BT40 2ER MI3CIW
BT40 2JE MI3CIZ
BT40 2JH 2I0JFO
BT40 2JH MI6GHA
BT40 2JH MI6JFO
BT40 2JR MI0BOU
BT40 2PA GI8WFA
BT40 2PH GI4MAJ
BT40 2QA GI6DNI
BT40 2TU 2I0EPG
BT40 2TU MI6ICD
BT40 2WF 2I0TAA
BT40 2WF MI3VHW
BT40 3DU MI3SRL
BT40 3FF GI7USA
BT40 3HL GI7MDK
BT40 3JG GI4UPC
BT40 3NF 2I0XDR
BT40 3NT MI3UMC
BT40 3SD GI1CET
BT40 3SD MI6WAG
BT40 3SJ MI0IZP
BT40 3SQ GI4RVT
BT40 3TT GI4MTZ
BT40 3TX GI0HWO
BT40 3TX GI7WCS
BT40 3TX GI8IZB
BT40 3TX MI3JCB
BT40 3UG GI0TWX
BT40 3UG GI0XAC
BT403LW MI7FHM
BT41 1AY MI3MRF
BT41 1HH GI3RNO
BT41 1HP MI0GFE
BT41 1LN 2I0RZT
BT41 1LN MI0GIJ
BT41 1LN MI3RZT
BT41 1SB 2I0EHW
BT41 1SB MI6FHG
BT41 1SB MI7NCZ
BT41 2DP MI3TUS
BT41 2DR GI0MQN
BT41 2EU GI6OQL
BT41 2EY GI1EOS
BT41 2HJ GI3YDM
BT41 2JD MI6CAD
BT41 2JH GI0NNK
BT41 2PN MI0DFO
BT41 2QT GI4PID
BT41 2RF MI6DUP
BT41 2TG MI6SPY
BT41 2TR MI1DPL
BT41 2TS GI7IPO
BT41 3BA MI6IBR
BT41 3BJ GI4VJZ
BT41 3DX GI8TWB
BT41 3NH 2I0LPO
BT41 3NH MI0LPO
BT41 3NX GI0NQC
BT41 3NX GI0WOW
BT41 3RT GI4DCC
BT41 4EG MI6PML
BT41 4EQ GI0PML
BT41 4FG MI0HHU
BT41 4FG MI6BZC
BT41 4FN MI7RVJ
BT41 4HD GI4KUM
BT41 4HD MI3NOH
BT41 4JQ GI4IHY
BT41 4NG 2I0IDJ
BT41 4NG MI6IDJ
BT41 4NH MI6NUM
BT41 4NP GI0VLE
BT41 4NS MI6FNO
BT41 4PF MI7WTS
BT41 4QQ GI4DWZ
BT41 4SA GI7EBM
BT41 4SB GI4FUM
BT41 4SB GI4SIW
BT41 4SB GI8MIV
BT412JB MI7AOU
BT41-4HW MI7LAJ
BT42 1AX GI4SFZ
BT42 1AX MI3FHM
BT42 1DE GI3FFF
BT42 1DE GI4HCN
BT42 1GW GI1XIB
BT42 1JA GI0JEV
BT42 1JL MI3AVJ
BT42 1JN 2I0JOS
BT42 1JN MI6JOS
BT42 1JW MI0CUN
BT42 1LN MI7HMD
BT42 1NZ MI3PZV
BT42 1PU GI6KBX
BT42 1RW 2I0EOS
BT42 1RW MI6WJK
BT42 1RX MI3PQM
BT42 1SB GI8MOV
BT42 2AU GI0AYG
BT42 2BJ GI4OYL
BT42 2DG GI4TOR
BT42 2DJ MI0LGD
BT42 2DJ MI6GDD
BT42 2DQ GI4OGQ
BT42 2LR MI6AJO
BT42 2NL GI1XPV
BT42 2NL MI7KAT
BT42 2PX 2I0STN
BT42 2PX MI0WWF
BT42 2PX MI6STN
BT42 2QH GI0THO
BT42 2QQ 2I0WDD
BT42 2QQ MI6WDD
BT42 2RG MI6KZS
BT42 2RG MI6LUP
BT42 2RJ MI1DEZ
BT42 2RP GI4KUZ
BT42 2SP MI7WGS
BT42 3BE MI6GZF
BT42 3DF MI3DSM
BT42 3GH 2I0RBV
BT42 3JE 2I0RGT
BT42 3LD GI0OHG
BT42 3LH MI6XGN
BT42 3LH MI6XOD
BT42 3ND MI6TSH
BT42 4AP 2I0GLC
BT42 4AP MI6GLC
BT42 4DH MI3LNC
BT42 4DH MI7GZO
BT42 4JG MI7VFR
BT42 4LB MI6UVS
BT42 4NJ MI3LRR
BT43 5HE GI4ESI
BT43 5NP GI0RBO
BT43 5PR GI3XDX
BT43 5PY GI0ITJ
BT43 6DT 2I1JMC
BT43 6DT MI0JML
BT43 6DT MI3JMC
BT43 6ET MI6PHH
BT43 6FD MI7RWT
BT43 6JG MI3OHP
BT43 6NF GI6OJC
BT43 6NN MI0GVC
BT43 6NN MI6VZW
BT43 6PB GI4NNM
BT43 6QE 2I0IPB
BT43 6QE MI3NUW
BT43 6SX MI0IOU
BT43 6SZ 2I0GQA
BT43 6SZ MI0REI
BT43 6SZ MI7NRW
BT43 6TA GI0LMR
BT43 6TD MI6ODN
BT43 6TL GI1KHF
BT43 7ap GI1GME
BT436NN MI7ACL
BT44 0NS GI4PCQ
BT44 0QZ GI1LGM
BT44 8AD GI4XFX
BT44 8AR GI6CXD
BT44 8EF GI1BSJ
BT44 8EF GI6VCL
BT44 8EF GI7UPQ
BT44 8HH GI4RXS
BT44 8JB 2I0RVT
BT44 8JB MI6RAD
BT44 8QH MI3DZD
BT44 9BN GI0WYB
BT44 9DL GI4SZU
BT44 9DT 2I0RHQ
BT44 9HZ MI0PJS
BT44 9JJ MI3XUC
BT44 9NA MI3SIL
BT44 9PE MI6FHJ
BT44 9QA MI0BBF
BT44 9RH MI3UHL
BT448NZ 2I0WAI
BT448NZ MI3UKW
BT45 5DD MI7DJT
BT45 5JF GI0EUG
BT45 5JZ GI4SOY
BT45 5LY GI7CTW
BT45 5QA GI0SRP
BT45 5QG GI4NGP
BT45 5RP MI0UTY
BT45 5RZ MI6VOZ
BT45 6DN MI0GRN
BT45 6DN MI3XEY
BT45 6DS MI3UTY
BT45 6EX GI7MWA
BT45 6HW GI1BZT
BT45 6JU MI0GOZ
BT45 6ly MI3LQN
BT45 6ND MI3UFD
BT45 6NH GI4WNH
BT45 6PF MI0JPD
BT45 6PU GI0PJH
BT45 6PY GI4EQA
BT45 7DT 2I0CKB
BT45 7DT 2I0GQZ
BT45 7LY GI4BGB
BT45 7NU GI3TIJ
BT45 7QF GI4NFW
BT45 7TW GI7JKM
BT45 7XS 2I0WBF
BT45 7XS MI6WBT
BT45 8AL GI3RXV
BT45 8BZ GI4BDR
BT45 8JA MI7DNG
BT45 8JE MI0LBA
BT45 8JE MI0TLF
BT45 8JE MI3EAQ
BT45 8NQ GI6OCC
BT45 8NQ MI3OYP
BT45 8PJ MI3OLM
BT457ht 2I0PND
BT46 5JR GI3ZTL
BT46 5NX MI3EYB
BT46 5SH 2I0ZCM
BT46 5SH MI3BCM
BT46 5TU MI3MIE
BT46 5UW MI7OCP
BT46 5UW MI7SSJ
BT47 2AF 2I0RGM
BT47 2AF MI0RYM
BT47 2AF MI6RGM
BT47 2BY GI7FJY
BT47 2ES GI8AFS
BT47 2HA MI3TXI
BT47 2HW GI4YWT
BT47 2HW MI0KQU
BT47 2HW MI3YWT
BT47 2LD MI0SDR
BT47 2LD MI3TPR
BT47 2LD MI6RCR
BT47 2NL GI3FTT
BT47 2PL 2I0HKW
BT47 2QR MI3RNN
BT47 2RD GI6AUI
BT47 2RD MI0AKU
BT47 2RD MI5TCC
BT47 2RY GI0AYB
BT47 2RY MI3NYB
BT47 2RY MI3XEB
BT47 2RY MI3XYB
BT47 2SL MI3LGL
BT47 3BF MI7BNV
BT47 3DD 2I0MRY
BT47 3FH MI7MTY
BT47 3JW GI0NCA
BT47 3NY MI6FAU
BT47 3PR GI6BVQ
BT47 3PW GI0LDI
BT47 3PW MI0NWG
BT47 3PW MI3MFD
BT47 3QS MI5ROY
BT47 3SF MI3STY
BT47 3TE MI3CCN
BT47 3TE MI3GER
BT47 3TE MI3PPD
BT47 3TE MI6OPD
BT47 3TR MI3FGK
BT47 3YE 2I0TDL
BT47 3YE MI0HXB
BT47 4AB 2I0OJK
BT47 4AB MI6OJK
BT47 4AD MI3OZT
BT47 4AQ GI4AHD
BT47 4AQ GI8RPP
BT47 4AS MI3TFF
BT47 4ff MI3KKP
BT47 4GA 2I0VTZ
BT47 4HQ MI3EGJ
BT47 4JN MI3RXF
BT47 4LU MI7RTD
BT47 4PJ GI0EFW
BT47 4PN 2I0JIE
BT47 4PN GI0AZA
BT47 4PN GI0AZB
BT47 4PN GI4MFT
BT47 4PN GI5ALP
BT47 4PN MI0OPC
BT47 4PN MI6JHE
BT47 4PN MI6JIE
BT47 4PP 2I0DMC
BT47 4PP MI3RUV
BT47 4ST MI6CEJ
BT47 4TJ GI4HDJ
BT47 4TT 2I0EGN
BT47 5HE GI0RJO
BT47 5JP GI0AWK
BT47 5PS MI3TJM
BT47 5QY GI0IOT
BT47 5SZ GI8ZDB
BT47 5WN MI6HAF
BT47 5WN MI6ZXT
BT47 5XS GI4OWA
BT47 5YD GI4FHB
BT47 6DU MI0UNA
BT47 6DU MI0WJM
BT47 6DU MI1JOE
BT47 6DU MI1UNA
BT47 6DU MI3CEM
BT47 6DU MI3SVM
BT47 6DU MI3XRT
BT47 6FE GI0TSA
BT47 6FE MI3TWM
BT47 6HY MI6CGQ
BT47 6NF MI6GNP
BT47 6SE GI0WYO
BT47 6UG GI0IUP
BT47 6UW MI3IFO
BT47 6XA MI3DBB
BT47 6XE MI7WPX
BT47 6XL MI3CXD
BT47 6YB MI6SCC
BT47 6YB MI7AAC
BT47 6YU MI3WQC
BT48 0AD GI4MJD
BT48 0AD GI8YBU
BT48 0AD MI0DVH
BT48 0AU GI3TJJ
BT48 0BY GI0NWN
BT48 0JX MI3DKN
BT48 0QA GI4ZLD
BT48 0RL MI3CGA
BT48 0RS GI0HDO
BT48 7ER GI0PXS
BT48 7JN GI4DGI
BT48 7PN GI4ZCH
BT48 7UA MI3GJG
BT48 8AQ GI0GGY
BT48 8BA 2I0HWW
BT48 8BA MI0ITS
BT48 8BA MI3PCF
BT48 8EF GI0LRB
BT48 8GQ GI3TME
BT48 8HN MI3IYP
BT48 8HX 2I0NAT
BT48 8HX MI3WFT
BT48 8HX MI3WVL
BT48 8HX MI6XBL
BT48 8JD GI0SFT
BT48 8JD GI7KVR
BT48 8JW GI0TQD
BT48 8JW GI1BEU
BT48 8JW GI4EXI
BT48 8NT GI7UPU
BT48 8PF GI4EPK
BT48 8RT MI6LLS
BT48 8TH GI0DSG
BT48 8UJ MI6BLW
BT48 9DU GI0OZQ
BT48 9HL MI6WBN
BT48 9LA GI1PVE
BT487ES MI7RXD
BT487LH MI7EDO
BT49 0AE GI4EBS
BT49 0AP MI6WJW
BT49 0AP MI6WPW
BT49 0AS MI3VTJ
BT49 0AT MI6MCK
BT49 0BB MI3YXX
BT49 0BD GI0RBS
BT49 0BF 2I0NWO
BT49 0BF MI0NWO
BT49 0BF MI3NWO
BT49 0BF MI6EJK
BT49 0BF MI6RME
BT49 0BH MI0GKL
BT49 0BH MI0JBT
BT49 0DH GI1ELP
BT49 0DQ MI3JTM
BT49 0EQ GI4OKU
BT49 0HR MI7MQX
BT49 0HS MI3XWW
BT49 0HS MI3YYT
BT49 0NF 2I0KEW
BT49 0NF MI3IHZ
BT49 0NT GI4PMP
BT49 0NY GI4MAZ
BT49 0NY MI6OFZ
BT49 0QB MI6JAD
BT49 0QF GI0BZM
BT49 0QF GI3KVD
BT49 0QF MI3UIW
BT49 0RD MI7WJL
BT49 0RG 2I0WBD
BT49 0RG MI6WAB
BT49 0RH MI6WKN
BT49 0RW MI3MBM
BT49 0SF GI3VAW
BT49 0SH MI0DWD
BT49 0SH MI3BIE
BT49 0UF 2I0BHT
BT49 0UF MI0GGB
BT49 0UF MI3IHY
BT49 9AD GI4YDP
BT49 9BQ GI7JRG
BT49 9BS GI6VWS
BT49 9BS MI3INS
BT49 9BS MI3JDQ
BT49 9EN MI3XGR
BT49 9EW MI3RYJ
BT49 9HQ GI4GNT
BT49 9HT 2I0BTT
BT49 9LR GI4SIZ
BT49 9LY GI4ODT
BT49 OAT MI3UIV
BT5 4FE 2I0WSH
BT5 4FT 2I0HRM
BT5 4FT MI6MIH
BT5 4HY MI7CEK
BT5 4NT GI7RCH
BT5 5HS GI6BDI
BT5 5JP MI7DXU
BT5 5LB 2I0HBO
BT5 5LB MI0HZD
BT5 5LB MI6EMI
BT5 5LT 2I0CGZ
BT5 5LT MI6AOZ
BT5 5NT GI8YTH
BT5 6AD GI0UTV
BT5 6AL GI7IMU
BT5 6BT GI4FNU
BT5 6DD GI7UQW
BT5 6ED MI1DQB
BT5 6FN GI7URC
BT5 6GE MI6WEZ
BT5 6NG GI0BRO
BT5 6NG GI0UEG
BT5 6PU GI0WCE
BT5 7DH GI4CBG
BT5 7EH MI6UBT
BT5 7EQ GI3MMF
BT5 7EQ GI6YM
BT5 7EY GI0OUM
BT5 7EZ GI4SSF
BT5 7GX GI6HKE
BT5 7HL 2I0GPQ
BT5 7HL GI8YYM
BT5 7HL MI0RUR
BT5 7HL MI6GPQ
BT5 7HW GI4ZOS
BT5 7JB GI7KMC
BT5 7JP GI7MDJ
BT5 7LY MI0ALS
BT5 7NR GI0BDZ
BT5 7PS GI4MYT
BT5 7RL MI3SXR
BT51 3GW MI7RGN
BT51 3LH MI7NYK
BT51 3QZ GI4OHH
BT51 3RA MI0RTY
BT51 3RD GI8LTB
BT51 3RE GI4GPA
BT51 3RF MI7WDT
BT51 3RW MI6XRC
BT51 3RZ GI4VIZ
BT51 3SN MI6JMC
BT51 3TN 2I0SEK
BT51 3TN MI0PLC
BT51 3TN MI3SEK
BT51 4AR MI1BNO
BT51 4BD GI7LCQ
BT51 4DN MI0PMX
BT51 4LZ GI7SOB
BT51 4NB 2I0GCN
BT51 4NB MI0ULK
BT51 4NB MI6MFR
BT51 4NE GI1LBI
BT51 4NG MI3VXI
BT51 4RA GI4ZAH
BT51 4SD GI3PLL
BT51 4SZ MI3CON
BT51 4TS GI4HVI
BT51 4US GI8OLH
BT51 5BJ MI1CSA
BT51 5JP MI0BYR
BT51 5LA 2I0LOP
BT51 5LA MI0JQY
BT51 5LA MI7MKD
BT51 5RZ MI0HYQ
BT51 5RZ MI0SLE
BT51 5SB MI0DMT
BT51 5SB MI0DNB
BT51 5SB MI0MSM
BT51 5SB MI6PNB
BT51 5TA GI6JXG
BT51 5YR GI6GBK
BT51 5YR MI0AAZ
BT52 1EN 2I0TAN
BT52 1EN MI0TBD
BT52 1EN MI3ZYU
BT52 1EW GI4JFP
BT52 1JR GI7TEB
BT52 1JW GI6IBL
BT52 1NG GI1RXL
BT52 1QH GI1JHQ
BT52 1SP 2I0BKI
BT52 1SP MI0WGW
BT52 1SP MI3UHI
BT52 1SP MI7CSM
BT52 1TL GI4WWF
BT52 1TU MI3GFA
BT52 1TW GI6EIR
BT52 1TY MI3BRJ
BT52 1UJ 2I0PAC
BT52 1UJ MI3UIA
BT52 1UJ MI6PFD
BT52 1WN MI3BYJ
BT52 2BL 2I0UDR
BT52 2BL MI3VCI
BT52 2BL MI6OGF
BT52 2DY MI6BEF
BT52 2ES MI0RJN
BT52 2EU MI0HMY
BT52 2EX MI3NWU
BT52 2HD MI0HPX
BT52 2HW MI0JZZ
BT52 2JE MI7MFW
BT52 2JL 2I0WAS
BT52 2JL MI0WJC
BT52 2JL MI6WRM
BT52 2ND 2I0ETB
BT52 2ND MI6ETE
BT52 2NY GI7TTO
BT52 2NZ MI7GCV
BT52 2QA MI3CKF
BT52 2QE 2I0IMO
BT52 2QE MI3IMO
BT52 2QF MI6CWC
BT52 2QF MI6OKS
BT521WN MI0CIB
BT53 6BZ MI0JAY
BT53 6EQ MI0ENR
BT53 6ES 2I0BPO
BT53 6ES MI0JAR
BT53 6ES MI3FPE
BT53 6EX MI0JPO
BT53 6JG MI3MMC
BT53 6NT 2I0TJF
BT53 6NT MI6TJF
BT53 6PT MI6GAU
BT53 6QB GI6GIE
BT53 6QF GI3OYG
BT53 7AH GI6GAQ
BT53 7AQ MI3FOJ
BT53 7AS 2I0EAR
BT53 7AS MI6HGV
BT53 7BB MI3DCM
BT53 7BE MI0MCC
BT53 7BE MI6RYC
BT53 7BX MI0VAC
BT53 7DE MI0GHI
BT53 7QL GI8DHW
BT53 7QS MI0OBE
BT53 7RN MI7HJW
BT53 8AB MI6MSV
BT53 8DL MI3LMR
BT53 8HD 2I0FMN
BT53 8HD GI4OSF
BT53 8JT GI4JRA
BT53 8NL MI0JTE
BT53 8NL MI3CQB
BT53 8NL MI3TRR
BT53 8NY MI6GGN
BT53 8PS MI1WWG
BT53 8QQ MI0MCB
BT53 8YH 2I1IKL
Bt537qg MI6LFK
BT54 6AN MI3YJE
BT54 6BW MI3BUT
BT54 6BW MI3BYQ
BT54 6DQ 2I1SWD
BT54 6DQ MI0CRQ
BT54 6DQ MI1DOG
BT54 6DQ MI3DMM
BT54 6DQ MI3SWD
BT54 6DS GI8WIU
BT54 6DS MI7AJD
BT54 6DZ 2I0RMK
BT54 6DZ 2I0TJK
BT54 6DZ MI0JBK
BT54 6DZ MI0RMK
BT54 6DZ MI3TJK
BT54 6HS MI6WOF
BT54 6JH MI3WES
BT54 6JL GI0JQQ
BT54 6LE MI3WDO
BT54 6LP MI3POB
BT54 6PF MI0MRV
BT54 6PF MI3DDK
BT54 6PF MI3TXT
BT54 6QZ MI0CRR
BT54 6QZ MI0MRG
BT54 6RT MI1BLZ
BT546NX MI7KMQ
BT55 7AU 2I0PHA
BT55 7DL 2I0ELR
BT55 7DL MI3CDA
BT55 7DT MI6KKN
BT55 7EG GI7JLD
BT55 7HA MI0KAM
bt55 7hs 2I0IOI
bt55 7hs MI0OUT
bt55 7hs MI3CAB
BT557PQ MI6NHH
BT56 8AS GI4OYM
BT56 8GP GI4CZO
BT56 8HN GI7TMQ
BT56 8JN GI4OQY
BT56 8NJ GI7SLN
BT56 8QA GI8PGJ
BT56 8SP GI0OTC
BT56 8ST MI3DOD
BT57 8AE MI0GBU
BT57 8AE MI0RUC
BT57 8AE MI3NRB
BT57 8RB MI6TAW
BT57 8RY MI6VBB
BT57 8SD 2I0LLG
BT57 8SD MI0LLG
BT57 8SD MI6LLG
BT57 8UX GI0ISQ
BT57 8YX GI8AIR
BT6 0AD MI3PUH
BT6 0AE GI7CMC
BT6 0AE MI1BRA
BT6 0BD MI7TDY
BT6 0DG GI4IBV
BT6 0DZ GI7KEC
BT6 0ER GI0DHW
BT6 0FN MI0AAW
BT6 0NH GI4FGH
BT6 0NR GI3YDH
BT6 8BH GI4RCK
BT6 8LN MI6TUX
BT6 8NL GI3LEG
BT6 9AW 2I0KUJ
BT6 9AW MI0KUJ
BT6 9AW MI6KUJ
BT6 9DS 2I0IZI
BT6 9DS MI6AHO
BT6 9GJ GI3UEX
BT6 9JF GI4LZR
BT6 9NT MI7LJR
BT6 9NU MI3CID
BT6 9PJ GI6NTP
BT6 9RB MI5WJB
BT6 9RH 2I0OTC
BT6 9RH MI6GIN
BT6 9RP GI0URN
BT6 9RP MI0NOR
BT6 9RX GI4GOV
BT6 9SE GI3XEQ
BT6 9SF GI4MDD
BT6 9SF GI7JAM
BT6 9SR MI3KDR
BT60 1DS GI4XFE
BT60 1DS MI7SSZ
BT60 1EX MI7RPP
BT60 1HP GI4RYP
BT60 1JU MI1EQI
BT60 1LE 2I0KXM
BT60 1LE MI6KXM
BT60 1LP GI4SRQ
BT60 1LQ 2I0NYL
BT60 1LQ GI3NYL
BT60 1LQ MI3WLW
BT60 1LZ MI0DBK
BT60 1NT GI8OCR
BT60 1QR GI6TBC
BT60 1TW GI7WLA
BT60 1TW MI6RHL
BT60 2BH 2I0FPT
BT60 2BN GI0DWN
BT60 2GP MI3ZJN
BT60 2GP MI6LOT
BT60 2JF GI0MSI
BT60 2JQ MI3RST
BT60 2NA 2I0GWA
BT60 2NA GI0KOW
BT60 2UP MI3WBL
BT60 3AA GI0MSH
BT60 3JS 2I0CBV
BT60 3JS GI0NWG
BT60 3JS MI3YTH
BT60 3LJ MI3LWU
BT60 3TS GI0VHG
BT60 4AS GI6NNP
BT60 4BL GI0GPG
BT60 4BU GI1VMF
BT60 4BU MI3LFE
BT60 4DZ 2I0RHN
BT60 4DZ MI6UDR
BT60 4NZ 2I0KPA
BT60 4NZ MI0KPA
BT60 4NZ MI0WHG
BT61 7DF GI8RLG
BT61 7JB GI8RNG
BT61 7JD GI0NYI
BT61 7PE MI6DOD
BT61 7QU GI0LTF
BT61 7SA GI0MSG
BT61 7SA GI7IEZ
BT61 8BU MI0CLP
BT61 8JD GI6NFK
BT61 8ND MI0GRG
BT61 8NP GI0MTE
BT61 8NT MI7UNI
BT61 8NX GI0OND
BT61 8NX MI0VFW
BT61 8NX MI6YSW
BT61 8PQ MI7JMS
BT61 8QN GI3WTG
BT61 8RW GI4SYM
BT61 9BB GI0MSK
BT61 9DT MI6JVC
BT61 9HA GI7DWF
BT61 9HA MI3BSN
BT61 9JX GI4FFL
BT61 9LD GI0ADD
BT61 9LD GI8RLE
BT61 9LT GI7GRY
BT61 9RW 2I0NTH
BT61 9RW MI6BZI
BT62 1EH GI0KUH
BT62 1EW MI3IRV
BT62 1EY GI0RYU
BT62 1JN GI7FCP
BT62 1JX GI0AIQ
BT62 1RN GI0UJG
BT62 2AG MI6PHV
BT62 2AR MI6TFN
BT62 2BB MI6GNF
BT62 2BB MI6NVM
BT62 2BH GI1RSR
BT62 2BL MI0SAP
BT62 2BL MI3SEO
BT62 2BN MI7CFX
BT62 2DD 2I0PRM
BT62 2DD MI0IZN
BT62 2DD MI0PRM
BT62 2DD MI3WWJ
BT62 2DD MI6MHI
BT62 2DD MI6MHK
BT62 2EJ MI0IRZ
BT62 2HB MI7JNJ
BT62 2NE MI3VWH
BT62 2NF MI6ECC
BT62 2NJ GI3ONF
BT62 2NJ GI4IVI
BT62 3BN 2I0EJT
BT62 3BN MI6AJA
BT62 3EB MI7KTY
BT62 3HH MI3IFI
BT62 3HY MI6CAV
BT62 3JB 2I0UCY
BT62 3JB MI6LSY
BT62 3JF MI6NOE
BT62 3QF MI7NSU
BT62 3QH GI4SJQ
BT62 3QU GI8TAX
BT62 3QX GI0DQJ
BT62 3RN GI4BQI
BT62 3TD 2I0SAI
BT62 3TD 2I0WAH
BT62 3TD MI0NID
BT62 3TD MI0SAI
BT62 3TD MI3XUS
BT62 3TD MI3YBI
BT62 3TF 2I0EBS
BT62 3TF MI6EBS
BT62 4DH MI6INB
BT62 4HL MI0TRC
BT62 4HL MI3DFR
BT62 4HP GI4MDO
BT62 4HX 2I0NIO
BT62 4HX MI6FWD
BT62 4JP 2I0RIR
BT62 4JP MI6RIR
BT622BL 2I0SGM
BT63 5AA MI7PSY
BT63 5AR 2I0JPP
BT63 5AR GI8XSB
BT63 5AR MI0UYD
BT63 5BF MI0INL
BT63 5BF MI6JVM
BT63 5DQ GI1ACN
BT63 5ES MI6HCP
BT63 5FY GI1PXX
BT63 5HR 2I0TME
BT63 5LT GI6FTM
BT63 5PF GI0NOX
BT63 5PQ MI6ILF
BT63 5QZ MI6IMY
BT63 5RJ MI3KFI
BT63 5RS GI7TGJ
BT63 5RX MI3ISC
BT63 5SW GI4YCZ
BT63 5SZ 2I0MFJ
BT63 5SZ MI6RCV
BT63 5TB MI6ENR
BT63 5TL GI6EEH
BT63 5UU 2I0TJR
BT63 5UU MI3TJR
BT63 5WU MI6UGE
BT63 5XT MI3IRY
BT63 5YD GI0PVG
BT63 5YD GI6FHD
BT63 5YD GI6NAQ
BT63 5YD GI7GHC
BT63 5YH GI4MXW
BT63 5YJ GI8JOA
BT63 6AT MI1AYL
BT63 6BH GI3WWY
BT63 6FA GI6MTL
BT63 6LY 2I0SFA
BT63 6LY MI6PUX
BT63 6NX MI6GOA
BT635EJ MI7ISF
BT64 3AN MI3PDN
BT65 4AB GI6GAG
BT65 4AL GI0STS
BT65 4AT 2I0ZXD
BT65 4AT MI0NWA
BT65 4AT MI6NCG
BT65 4AT MI6NOS
BT65 5AE MI6TRF
BT65 5AQ MI6RXH
BT65 5DH GI0SZH
BT65 5JF MI3LZA
BT66 6LD GI0MHB
BT66 6PS MI0TBN
BT66 6QW GI6FQT
BT66 7AN 2I0TRM
BT66 7AN MI0LRC
BT66 7AN MI3RYD
BT66 7BD 2I0HEZ
BT66 7BD MI6LZL
BT66 7DJ 2I0JSQ
BT66 7DJ MI6JSQ
BT66 7EB GI3POS
BT66 7HD GI8GZM
BT66 7HH GI4BDL
BT66 7LE MI1FIS
BT66 7LP 2I0RGD
BT66 7LP MI6BGD
BT66 7LS MI7DXL
BT66 7NE GI4GEL
BT66 7NE GI8LFY
BT66 7PD MI3RBM
BT66 7PP MI0BPB
BT66 7PQ GI4YRP
BT66 7QY GI0OHU
BT66 7RU GI3WEL
BT66 7SW MI7XAG
BT66 7SY 2I0OUI
BT66 7SY MI6HGS
BT66 7TG 2I0GLY
BT66 7TG MI6XEM
BT66 7TR 2I0WGM
BT66 7TR MI0WGM
BT66 7TR MI6WGM
BT66 7TY MI0MEV
BT66 7UA GI4ISR
BT66 7UT 2I0ROC
BT66 7UT MI6DED
BT66 8DF 2I0SNG
BT66 8DF MI0SNG
BT66 8EE 2I0FVX
BT66 8ER MI7CTL
BT66 8HP MI6SGU
BT66 8JP MI0EDF
BT66 8JR MI7ABV
BT66 8JZ 2I0HSL
BT66 8JZ MI6HSL
BT66 8LE GI3SCM
BT66 8LW GI7OOM
BT66 8QT MI3WQT
BT66 8RH GI0OUZ
BT67 0AU 2I0FNN
BT67 0AU MI6FIK
BT67 0ED MI0BMM
BT67 0FA MI0SMV
BT67 0FQ MI0GCV
BT67 0FQ MI3OFX
BT67 0GB GI4MWA
BT67 0GB MI6AJN
BT67 0GB MI6JOY
BT67 0GJ 2I0YDF
BT67 0GJ MI0YAM
BT67 0GJ MI3YDF
BT67 0HP GI4AIO
BT67 0HW MI6GBC
BT67 0JB GI4MNF
BT67 0JZ 2I0FBY
BT67 0LB 2I0SJV
BT67 0LB GI0OEH
BT67 0LB MI6SJV
BT67 0LN MI3UGI
BT67 0NL GI1FSJ
BT67 0NZ 2I0YLT
BT67 0NZ MI0YLT
BT67 0NZ MI6PYL
BT67 0NZ MI6YLT
BT67 0QT GI0VGV
BT67 0RN MI5HIL

BT67 0SS 2I0GSG
BT67 0SS MI6GSG
BT67 0SZ GI6IRL
BT67 0TU 2I0MOU
BT67 0TU MI3GAM
BT67 0TU MI5BUG
BT67 0UT 2I0DKQ
BT67 0UT MI6JGK
BT67 0XP 2I0IGX
BT67 0XP MI0LTX
BT67 0XP MI7BIU
BT67 9BS MI3PJM
BT67 9HN MI5LYN
BT67 9JN GI4ELQ
BT7 1BL MI3BCR
BT7 1PH 2I0FQW
BT7 1PH MI7XZD
BT7 2GF GI8HUD
BT7 2GJ MI0BOK
BT7 3AG MI7DBR
BT7 3HA GI6KDN
BT7 3NB MI7BMZ
BT70 1SZ GI3OZW
BT70 1TH GI4XKI
BT70 1UH GI0RDM
BT70 2EL GI4VCZ
BT70 3BU GI7FCW
BT70 3DY GI0ZER
BT70 3DY GI7TDA
BT70 3DY MI6LCR
BT70 3JU 2I0YBH
BT70 3JU MI0YBH
BT70 3JU MI6YBH
BT70 3JY GI4XAA
BT70 3LB 2I0GZX
BT70 3LB MI0KBL
BT70 3LB MI7GZX
BT70 3LU GI4LDN
BT70 3PU GI0STM
BT71 4AA MI6FQY
BT71 4AG GI7NEB
BT71 4AJ GI7RTB
BT71 5DP MI6OXM
BT71 5GA MI0CAC
BT71 5JL 2I0SMY
BT71 5JL MI0SMY
BT71 5JL MI6SJD
BT71 5LY GI0JHS
BT71 6DD GI0URI
BT71 6DR GI3NFM
BT71 6HX MI6EQD
BT71 6PJ GI4CSP
BT71 6PW 2I0EXP
BT71 6PW MI6EXP
BT71 6QG GI3NPP
BT71 6QN GI3OQR
BT71 6SQ MI6XHL
BT71 6TN GI0IVJ
BT71 6TQ 2I1WBC
BT71 6TQ MI7EGS
BT71 6UE 2I0VBH
BT71 6UE MI6VBH
BT71 7AY MI3IOH
BT71 7BD 2I0IIQ
BT71 7BD MI7BZR
BT71 7BH GI0KPF
BT71 7BN MI6EQC
BT71 7ER GI0KVQ
BT71 7ER MI0AVI
BT71 7ET MI6MYW
BT71 7HT GI0VIB
BT71 7JN GI8ITD
BT71 7NH 2I0BIQ
BT71 7NY 2I0WBU
BT71 7PY MI0CXE
BT71 7QF GI4TED
BT71 7QF GI4XJD
BT71 7SQ GI4SLQ
BT717TS MI7GPT
BT74 4DY 2I0NIE
BT74 4DY MI3NIE
BT74 4RB MI3TMN
BT74 5ND GI7NET
BT74 5NQ GI4CZW
BT74 6AZ MI6FEB
BT74 6BQ MI3GHW
BT74 6EP GI0BQX
BT74 6ET MI0BTM
BT74 6EU GI0BFD
BT74 6HD MI1FHE
BT74 6JJ MI6REM
BT74 6JP 2I0EAS
BT74 6JP MI3YUT
BT74 7HQ MI6HZN
BT74 7JN GI4UHA
BT75 0LG 2I0EPN
BT75 0LG MI0LLM
BT75 0LG MI6LLI
BT76 0HJ GI0EWE
BT78 1ES MI6AUU
BT78 1QS 2I0CKN
BT78 1QS MI3MWA
BT78 1QZ GI4OHW
BT78 2DD MI6BJZ
BT78 2DD MI6BKD
BT78 2DL GI0LXN
BT78 2EH GI0KDH
BT78 2PN GI7SZV
BT78 2PN MI6UTV
BT78 2PP GI7THY
BT78 3AW 2I0CFW
BT78 3AW MI3DWQ
BT78 3HL MI6WRT
BT78 3LR GI0OXK
BT78 4HX GI1CAI
BT78 4HX MI6OMA
BT78 4JT GI4KJC
BT78 4JZ GI0HVJ
BT78 4LG GI0TMS
BT78 5AL MI1FCQ
BT78 5BB MI3KRL
BT78 5BQ MI6PCJ
BT78 5JE MI6OTP
BT79 0DW 2I0JXO
BT79 0DW MI3JXO
BT79 0DZ 2I0OMA
BT79 0DZ MI0UFT
BT79 0DZ MI6MRJ
BT79 0EB GI4RSI
BT79 0HF GI3NVW
BT79 0HX MI1CTQ
BT79 0JX GI0STC
BT79 0PG MI6ODG
BT79 0PH GI0GQG
BT79 0PL GI4UDI
BT79 0RZ GI6EGJ
BT79 0WA MI6BVN
BT79 0XS GI3XCZ
BT79 7GL 2I0LDC
BT79 7GL MI6BII
BT79 7GL MI6ECV
BT79 7GL MI6EJT
BT79 7JJ GI4LIF
BT79 7JX MI6PWR
BT79 7LA GI7UCS
BT79 7LB MI6EDK
BT79 7PQ GI0EJT
BT79 7PQ GI1RIB
BT79 7RA 2I0DJM
BT79 7RA GI4SXV
BT79 7RA MI3DJM
BT79 7SJ GI7AUY
BT79 7SJ GI7FHZ
BT79 7SJ MI6TYR
BT79 7TJ GI0EJU
BT79 7WJ GI3YBZ
BT79 7WJ GI8BNC
BT79 7XG 2I0MSO
BT79 7XG MI0IFG
BT79 7XG MI0MSO
BT79 9AY MI3DLO
BT79 9NH MI6SFK
BT79 9PH MI1CCU
BT79 9PH MI6CCU
BT79 9QA GI4IWP
BT8 6HU GI0BFA
BT8 6JX MI5EEM
BT8 6LF 2I0JXA
BT8 6LF MI3JXA
BT8 6RG GI3YMT
BT8 6SB GI4KIX
BT8 6WL GI4JIC
BT8 7DB GI4WME
BT8 7RJ GI1VAZ
BT8 8BQ GI4TAP
BT8 8HG MI6NIM
BT8 8HQ GI4TAV
BT8 8JZ GI4TGR
BT8 8LE MI6IMQ
BT8 8LX GI8PDK
BT8 8NL MI0BDX
BT8 8RL 2I0WJM
BT8 8RL MI7MCC
BT8 8TG GI4SIP
BT80 0JL MI6CAK
BT80 8AS MI0RYU
BT80 8BW GI8XSY
Bt80 8ez MI0LNL
BT80 8JH MI0AHH
BT80 8JH GI7BET
BT80 8PL GI3SOO
BT80 8PL GI8YGG
BT80 8PW MI0AHI
BT80 8PW MI0BML
BT80 8PZ 2I0HGI
BT80 8QE GI6JRY
BT80 8QJ GI0JHR
BT80 8XW GI8JRE
BT80 9LB GI6WHZ
BT80 9RN MI0AIH
BT808RS GI6NDM
BT81 7DF MI6DAY
BT81 7DR MI3SLT
BT81 7NA MI3NVX
BT81 7NW MI3JQD
BT81 7QF 2I0NDJ
BT81 7QF MI0NDK
BT81 7QY MI0JAT
BT81 7RF MI0HRG
BT81 7RF MI1CCT
BT81 7TJ MI3CCT
BT81 7TJ MI3FCA
BT82 0AE GI0VJE
BT82 0AE MI3BXQ
BT82 0AE MI3JVJ
BT82 0AE MI3JVV
BT82 0AE MI5ALJ
BT82 0AJ MI3EZK
BT82 0BD GI4DXK
BT82 0DN MI3RXU
BT82 0DP GI3CFH
BT82 0DP GI4OUN
BT82 0DP MI3OUN
BT82 0DW MI0RIB
BT82 0DW MI3ASH
BT82 0DW MI3BEG
BT82 0DZ MI3CGZ
BT82 0DZ MI3FPB
BT82 0DZ MI3IGO
BT82 0GZ GI0EWP
BT82 0HN MI3NEN
BT82 0HN MI3TLV
BT82 0LZ 2I0BSA
BT82 0LZ 2I0FLO
BT82 0LZ MI3ATT
BT82 0PD MI6EAF
BT82 0PD MI6JMD
BT82 0PD MI6VJR
BT82 0PP GI4XGQ
BT82 0QQ MI3RIL
BT82 0QQ MI5RJS
BT82 8EL GI7THH
BT82 8LD GI4VQK
BT82 8LF MI3NMG
BT82 8LS MI0HOZ
BT82 8LS MI6LJO
BT82 8NW GI4VKS
BT82 8QQ MI6XKE
BT82 9AJ 2I0VFO
BT82 9AJ MI0GTM
BT82 9AJ MI3OZK
BT82 9DT 2I0IMB
BT82 9DT MI3KMB
BT82 9LA MI1AWM
BT82 9PN MI1DRP
BT82 9PN MI3AGR
BT82 9QL GI4DYE
BT82 9QS MI3PDL
BT82 9RQ MI6POH
BT82 9TL MI6RAC
BT820DB GI4OUP
BT9 5AA GI3ZCK
BT9 5AQ GI1KGZ
BT9 5BG MI7BTH
BT9 5EL MI0EFM
BT9 5FQ MI6IRE
BT9 5GZ MI3WDI
BT9 5PG GI1MJJ
BT9 5QL GI4OMK
BT9 5QL GI4VIP
BT9 5QP GI4WVN
BT9 5QP GI4XFS
BT9 6HA GI3VYY
BT9 6HJ 2I0IGA
BT9 6HJ MI7GCC
BT9 6JX MI3GWQ
BT9 6LJ GI8SKR
BT9 6NF MI6GQI
BT9 6PJ GI7JUH
BT9 6QR GI3SG
BT9 6QX 2I0NGK
BT9 6QX MI6TCO
BT9 6TJ GI3PSQ
BT9 6UH GI3VCI
BT9 7AX GI4ISH
BT9 7JD GI4DOM
BT92 0EQ MI6AFI
BT92 0EQ MI6AFM
BT92 0FS MI1RAY
BT92 0FS MI6PMM
BT92 0FS MI6SMM
BT92 0FW MI3LDO
BT92 2DL GI0LEC
BT92 2DL GI6ZIR
BT92 2EW 2I0WGL
BT92 2EW MI0WGL
BT92 2EW MI6WGL
BT92 2GJ MI7SYL
BT92 3BU GI0CAH
BT92 3BU MI6CAH
BT92 6NB 2I0PPW
BT92 6NB MI0PPW
BT92 6NB MI6MCF
BT92 7DD 2I0WMC
BT92 7DD MI6BMC
BT92 7DD MI6SFH
BT92 7DD MI6UIM
BT92 7FN MI6GDN
BT92 7GA 2I0XAN
BT92 7GA MI6WAN
BT92 7GG MI6WAM
BT92 7HP MI6COM
Bt92 7lj MI6WSP
BT92 7PP MI6NCC
BT92 7PR MI6WDM
BT92 7QY GI7JGT
BT92 8AE MI6CAY
bt927pt 2I0RVH
bt927pt MI0RVH
bt927pt MI6TNI
BT93 0FQ GI1OMD
BT93 1UH GI4SZQ
BT93 3EZ MI5IMB
BT93 3FT GI0PTQ
BT93 5BR MI0BHX
BT93 6FR 2I1GIH
BT93 7DN GI6PYP
BT93 8DF GI3WBR
BT94 1HQ MI3JXG
BT94 1RN MI5MTC
BT94 2AS MI6YPY
BT94 2JA GI1RBI
BT94 3DB 2I0GHY
BT94 3DB MI0IIG
BT94 3DB MI3GHY
BT94 3EH GI8WJN
BT94 3EQ 2I0EKN
BT94 3FJ 2I0NTP
BT94 3FJ MI0NPR
BT94 3FJ MI6NTP
BT94 3PH 2I0HGF
BT94 3PH MI7BFP
BT94 4EZ 2I0RDR
BT94 4EZ MI6RDR
BT94 4QX MI6OFN
BT94 5BU 2I0GFR
BT94 5BX GI6RMO
BT94 5EW 2I0UBE
BT94 5EW GI4VHO
BT94 5EW MI0UBE
BT94 5EW MI6UBE
BT94 5FB GI6JPO
BT94 5FB MI0RCF
BT94 5FL MI0DWE
BT94 5HJ GI6UFO
BT941je GI7UIP
CA1 1LX G4RQW
CA1 2DL M6RYK
CA1 2ND M6LCK
CA1 2QE M6ELH
CA1 2QJ G3XWA
CA1 2SZ G3WTO
CA1 2TA M0SHD
CA1 2WF M3OLQ
CA1 3AS M3TUC
CA1 3BP M0YDP
CA1 3HF G4KVQ
CA1 3HT G1COY
CA1 3JQ M3ZYF
CA1 3QB G1FBE
CA10 1AJ G7UNZ
CA10 1AT M3OKP
CA10 1BA G1AEI
CA10 1BX 2E0AND
CA10 1BX M7ELN
CA10 1DT G4WKG
CA10 1PD G0IYQ
CA10 1QF G4JHV
CA10 1QR G4EXD
CA10 1QX G4GXO
CA10 1TA G4XET
CA10 2AW G4UWG
CA10 2LL G0VRZ
CA10 3AL G0TNF
CA10 3AL M7CAA
CA10 3HD G3WGV
CA10 3JE G7VVX
CA10 3LT M3ITH
CA10 3NB M7BWE
CA10 3NL M6MTD
CA10 3PF G0VMP
CA10 3PF G7DEG
CA10 3RX M7BFV
CA10 3RZ G8OZT
CA10 3TW G0NAS
CA10 3UE G3LRA
ca11 0ay 2E0FQI
CA11 0DT G1XTD
CA11 0HN M6NXP
CA11 0LS M1XCG
CA11 0XB G0AUG
CA11 0XB G0AUH
CA11 7JY G0MDV
CA11 7JY G1FNP
CA11 7QT M1EUM
CA11 7UW G0TDM
CA11 7UW G7GQL
CA11 7UW G7GQM
CA11 8AX G0OFR
CA11 8BJ G3YEP
CA11 8EH G4GLQ
CA11 8LX G4YTD
CA11 8PN M6LKE
CA11 8SD G1AQV
CA11 8TU M6NNY
CA11 8TW G4GQZ
CA11 8UF G4FUI
CA11 8UF G8LBT
CA11 9HL M7FAL
CA11 9HN G1VKG
CA11 9NA M5TNT
CA11 9PD M6FWW
CA11 9PF M3WBR
CA11 9SJ M6CTO
CA11 9SS G6XWD
CA11 9TR G4YLI
CA12 4AP G3ZED
CA12 4AZ M0AOT
CA12 4HS G0UQC
CA12 4NN G4ERL
CA12 5LQ M6LAL
CA12 5SA G0END
CA13 0AR 2E1APW
CA13 0LA M0GCU
CA13 0QP M7NPH
CA13 0QR G1GWE
CA13 0TJ M0AYB
CA13 0TL M6RTN
CA13 0UL M6BUU
CA13 9BE 2E0IKM
CA13 9BE M6IKM
CA13 9EQ G7OHM
CA13 9JD 2E0LDF
CA13 9JS M7AQM
CA13 9UQ G0JSO
CA13 9UQ M0WLA
CA14 1AA G3WCM
CA14 1DT 2E0OMT
CA14 1DT M3OMT
CA14 1DW M0XAT
CA14 1EE G1OAE
CA14 1HE G3RRI
CA14 1HS 2E0KKT
CA14 1HS M7PGH
CA14 1LL G0HCE
CA14 1LP G3WJH
CA14 1PY G0EMM
CA14 1PY M6RMD
CA14 1PY M6XSF
CA14 1XJ 2E0MAX
CA14 1XJ G4ZFX
CA14 1XJ M0WCR
CA14 2PT 2E0IFF
CA14 2PT G8RZ
CA14 2PT M3IFF
CA14 2QY 2E0DNW
CA14 2QY M6CVP
CA14 2TP G6CSL
CA14 2UG G0OMB
CA14 2UW M7KLB
CA14 2XZ M3HWP
CA14 3EH 2E0BEG
CA14 3EH M0XMD
CA14 3EH M3LCI
CA14 3EY G7NMB
CA14 3HR G6TAK
CA14 3JF M6RNF
CA14 3PE M1RIC
CA14 4HY G1LEH
CA14 4JA M3LDL
CA14 4NU G1GTA
CA14 4NX G4CJP
CA14 4PR M6MIG
CA14 4PU G6PTF
CA14 5AB M7BUY
CA14 5AF G1LZL
CA14 5BQ M6LIV
CA14 5HP G0DPE
CA14 5JA G1AQI
CA14 5LB 2E0YHI
CA14 5LZ M3SDN
CA14 5PL M7EEU
CA14 5QR 2E0NWT
CA14 5QR G0MTD
CA14 5UJ G1FYE
CA15 6BQ G0NNN
CA15 6BQ G4HLZ
CA15 6HR M3XVK
CA15 6PB M6GXW
CA15 6SH M6MSP
CA15 7BU G0TII
CA15 7BU M0BCV
CA15 7DB M3UVA
CA15 7DX 2E0GSR
CA15 7DX M6EZA
CA15 7EG G0MWE
CA15 7ER M1EBC
CA15 7ER M1EBD
CA15 7HD M7XUP
CA15 7JU M3SDK
CA15 7JW M1CIE
CA15 7QJ 2E1BMB
CA15 7QN G0ORM
CA15 7RG M0GDH
CA15 7SS G0MTQ
CA15 7SS G7DIS
CA15 8RX 2E0ENP
CA15 8RX G2KS
CA15 8RX M0WRC
CA15 8RX M6ENP
CA15 8ST M0VLA
CA157LW 2E0IKX
CA16 6BD G0KDB
CA16 6BD G0VGJ
CA16 6DA G7ITT
CA16 6JS G1ZBW
CA16 6JZ 2E0AOK
CA16 6JZ 2E1EIO
CA16 6JZ G4XTA
CA16 6PX G0ANT
CA16 6PX G8TXJ
CA16 6PX M5TXJ
CA16 6QE G0OPM
CA16 6RB M6WJM
CA16 6RD G0NYQ
CA16 6RS G4PZN
CA16 6SP M0UKR
CA16 6SP M3ZOP
CA16 6SP M6JKQ
CA16 6TA G7THI
CA16 6TB G7NRG
CA16 6XU G1AOZ
CA17 4AU G4TUA
CA17 4BZ G8OCA
CA17 4EX G7BNB
Ca17 4hl M7WAM
CA17 4HR G4SPR
CA174sn 2E0TIN
CA174sn M0XXL
CA174sn M3ZFW
CA18 1RG 2E0GTM
CA18 1RG M6CWA
CA18 1SL 2E0DAL
CA18 1SL M3ZCX
CA18 1SW G1KGU
CA19 1UD M1PAF
CA19 1UL 2E0NEN
CA19 1XH G3PNU
CA2 4EP M7AVT
CA2 4EX 2E0BKV
CA2 4EX M3OZI
CA2 4EX M7OGC
CA2 4LD M1LSD
CA2 4PZ G0NUD
CA2 5DZ M0AOH
CA2 5UH 2E0TMS
CA2 5UH M6TIH
CA2 6AG M0ADG
CA2 6DH M5FAB
CA2 6ER 2E0UKY
CA2 6ER M0XOD
CA2 6ER M7YAM
CA2 6JJ G8CAU
CA2 6LZ G4FQM
CA2 6LZ G8KIZ
CA2 7DA M6COB
CA2 7LD M3VZQ
CA2 7LU G4ARS
CA2 7LU G6LSO
CA2 7LU M0SOL
CA2 7PN M6YJD
CA2 7RD G0JXO
CA2 7TB M3IRH
CA2 7XD G8GUH
CA2 7XY M3BSF
CA20 1AY 2E0EHA
CA20 1AY G0SVK
CA20 1AY M6HXV
CA20 1DN M6PHF
CA20 1EF M3IUX
CA20 1LB M3MLI
CA20 1PZ G1HZJ
CA21 2YY G1MPG
CA22 2AT 2E0CBB
CA22 2DN M0KHZ
CA22 2DT M0HCI
CA22 2ED M7MVD
CA22 2EH G4VFL
CA22 2NA G1GWF
CA22 2NA G3BJD
CA22 2NN 2E0OZM
CA22 2PW 2E0XUH
CA22 2PW M0XUH
CA22 2PW M3XUH
CA22 2QQ G1HDX
CA22 2RL M0EJB
CA22 2SG M5GVY
CA22 2TP 2E0DNX
CA222QT M7NNY
CA23 3DY G1LMW
CA23 3DY M0LMS
CA24 3JL G3YSD
CA25 5EU G4RVH
CA25 5LF M7JKD
CA25 5LR M0ALR
CA26 3AT 2E0WEO
CA26 3AT M0LVL
CA26 3AT M6WEO
CA26 3QJ M6RMS
CA26 3QS G0RZI
CA26 3RU M0ELC
CA26 3SA 2E0XSD
CA26 3SA M0XSD
CA26 3SA M6EPW
CA26 3SA M6XSD
CA26 3XN M1DZR
CA27 0AN G4DSI
CA27 0AR G7KSE
CA27 0AR M0LCW
CA27 0AS G0ESI
CA27 0EG G3WIN
CA27 0EG G4PEW
CA27BS M0PTT
CA28 6AQ M3USF
CA28 6EF M0SHP
CA28 6JZ G6WLP
CA28 6LS 2E0XGO
CA28 6PE M6FYM
CA28 6RR 2E0YSK
CA28 6RR M6YSK
CA28 6SG M0FWM
CA28 6SW G7RKV
CA28 6SX G8IMM
CA28 6XL 2E0MIX
CA28 6XL M6DEG
CA28 7AW G7SGO
CA28 7XG G4VIA
CA28 8EP G0LJB
CA28 8HS G0HYP
CA28 8JP G7MRL
CA28 8JP M0BEE
CA28 8JP M0CRM
CA28 8PH M7RGG
CA28 8PX M3YRR
CA28 8PZ G0DIG
CA28 8XJ G4OVF
CA28 8YG M0SKY
CA28 9AH G4FBC
CA28 9PB M3AQP
CA28 9PD G4MTR
CA28 9PD G8ORO
CA28 9TG G0CZU
CA28 9UG 2E0BEE
CA3 0AW G0KEE
CA3 0AY M6KTT
CA3 0DB M1ETM
CA3 0LF G0TFC
CA3 0NT M6MOS
CA3 0NU G1FVA
CA3 9RE M1ERP
CA3 9RE M3MIR
CA3 9SN G0DHI
CA30LR 2E0FDW
CA30LR M6LPH
CA4 0JF G7SZZ
CA4 0RP M3BWT
CA4 8BD G8UTQ
CA4 8ER M3OLF
CA4 8FG G7DXB
CA4 8LD G4BUI
CA4 8PL G4ISS
CA4 9PX G7JKD
CA4 9QR G4IIB
CA4 9QY M3FZV
CA4 9QY M3ZTR
CA4 9RR 2E0TXG
CA4 9RR M3TXA
CA5 6TB M3WGV
CA5 6EN G1OFY
CA5 6EN G6PPD
CA5 6NS G0NVA
CA5 6NS G7EKM
CA5 6NS M3GMG
CA5 6PB M3OLN
CA5 6QW G4SCE
CA5 7AX 2E0MXP
CA5 7AX M6DLP
CA5 7AX M6RLP
CA5 7DP G8KAP
CA5 7NS M3DLJ
CA5 7NW G7IVG
CA5 7NW M3VYT
CA5 7QH G8GJW
CA6 4EA G0WFH
CA6 4LY G6WAM
CA6 4NN G8ITX
CA6 4PZ G0OYI
CA6 5SS G4LIL
CA6 5TH M6NTB
CA6 5TS M1JWR
CA6 5UJ M1JON
CA6 5UJ M3CFM
CA6 5UJ M3JDG
CA6 5UJ M3MDK
CA6 5XX M3FBJ
CA6 5XX M3HFX
CA6 6LH G4HZP
CA6 6LJ 2E0NGT
CA6 6PA M3WLD
CA7 0BZ G3KEL
CA7 0DF 2E0COB
CA7 0DF M3OSC
CA7 0DF M3OSF
CA7 1JF G4NEH
CA7 1JF G8DVW
CA7 1JG M1ACO
CA7 2JF G1GDB
CA7 3AH M0IHX
CA7 3DJ M0DXT
CA7 3QX G0ZMH
CA7 3QX M0AMX
CA7 3RJ G7ATW
CA7 4EU M6KWI
CA7 4JJ G0TNL
CA7 4QH G7KWQ
CA7 4RT M7CLX
CA7 5AN G0TRW
CA7 5AS M0PMR
CA7 5AS M3AQN
CA7 8AU G0HNQ
CA7 8JS M7FMR
CA7 8PF 2E0BKP
CA7 8PF 2E1BKP
CA7 8PF G0ORO
CA7 9DR G8DNH
CA7 9HQ M3JMY
CA7 9JL 2E0PGR
CA7 9JL M0PEG
CA7 9JW M6HWW
CA7 9QW M3TEE
CA7 9RE M3UIK
CA8 1EN G3VUS
CA8 1EX G0LEJ
CA8 1LE G6KXD
CA8 1QJ M7NEZ
CA8 1RZ M3TFM
CA8 1SR G0JVU
CA8 1SR G7ANH
CA8 1TT 2E0ACE
CA8 1TY M3XUO
CA8 2BB G4IIY
CA8 2NJ G3OHL
CA8 7AL G4RHC
CA8 7LA M1DIN
CA8 9AA G8IBK
CA8 9DH G4LQM
CA8 9HF M6BAX
CA8 9JA G7NZR
Ca81bn M6BJW
CA9 3DY G4ZXF
CA9 3LH G1RLT
CA9 3LH M3HVQ
CA9 3LZ G0NXS
CB1 1DP 2E0BPE
CB1 1DP M0GJU
CB1 1DP M3XQZ
CB1 1LE M0DVK
CB1 2DW 2E0BOV
CB1 2EW 2E0XZW
CB1 2EW M0XZW
CB1 2FJ M1DUO
CB1 2LG G6TIU
CB1 2LP G3ROP
CB1 2PU 2E0IDO
CB1 2PU M6IDO
CB1 2QF M6OSF
CB1 3AJ M3KWR
CB1 3EF G7PSF
CB1 3LR M0GXM
CB1 3LW 2E0EFW
CB1 3PT 2E0CDV
CB1 3PT 2E0OIK
CB1 3PT 2E0PUZ
CB1 3PT M0TFB
CB1 3PT M6PUZ
CB1 3PT M7PUZ
CB1 3PW G0IJZ
CB1 3QL M0GLT
CB1 3RS M6RSS
CB1 3RY G4EZN
CB1 3RY G6UW
CB1 3UB G0SFA
CB1 3UB G1OKI
CB1 3UF G1KUN
CB1 6HS G0DEU
CB1 7TY 2E0FLQ
CB1 7TY M0JEI
CB1 7TY M7AFO
CB1 7UP M6GXC
CB1 7UR G0GVZ
CB1 7UR G3INR
CB1 8PF M7AFR
CB1 8QR G6ZEY
CB1 8RG G8OUS
CB1 8RX G0XAS
CB1 8SG G4EAG
CB1 8SH G4GLI
CB1 9AJ M7MUT
CB1 9AL M3UBS
CB1 9AZ G6EUI
CB1 9GH 2E0BLI
CB1 9GH M0SKI
CB1 9GH M3XSK
CB1 9JA G0PNQ
CB1 9LN G3UPJ
CB1 9LZ M7TBZ
CB1 9YL G7VCE
CB10 1LZ G6KDW
CB10 1NR M3IAQ
CB10 1QD M0ECC
CB10 1QY M7CGO
CB10 2AH G1WSF
CB10 2AJ G4YHN
CB10 2AP G4UHR
CB10 2EB 2E0MYG
CB10 2EB G6YBH
CB10 2EB M0MYG
CB10 2EB M6MYG
CB10 2HG G8SZR
CB10 2RW G4FBY
CB10 2TZ G0TCI
CB11 3BJ G8XLB
CB11 3DN G0GWM
CB11 3ED G0ICB
CB11 3ER G0JML
CB11 3HD G0IMG
CB11 3HU 2E0PKM
CB11 3PU G4FON
CB11 3RU G3WMS
CB11 3SL G7PNP
CB11 3SP G4HSN
CB11 4AU M1EMO
CB11 4BS M1CKQ
CB11 4DE G0CLJ
CB11 4DE G1RBO
CB11 4DG 2E0GUA
CB11 4DG M1JDH
CB11 4DG M6GUA
CB11 4DJ 2E1HLH
CB11 4DJ G0HEM
CB11 4DJ G1LTL
CB11 4DW 2E0NOZ
CB11 4DW M0NOZ
CB11 4DW M3TPN
CB11 4JB G8ULM
CB11 4JB M6GUB
CB11 4JU G3YZO
CB11 4LX G0OJR
CB11 4PX 2E0GGA
CB11 4PX 2E0MXA
CB11 4PX 2E0VVA
CB11 4PX M0ONL
CB11 4PX M0VVA
CB11 4PX M6OMX
CB11 4PX M6UGA
CB11 4PX M6VVA
CB11 4RX M7CKY
CB11 4SJ G3TXC
CB11 4SJ G6FEM
CB11 4TH G8EBX
CB11 4UU G3NOX
CB11 4XG G3XHZ
CB2 0QZ G7PLE
CB2 1RL 2E0ION
CB2 1RL M0POE
CB2 1ST M0XUB
CB2 1ST M6BXU
CB2 1TA M3HZC
CB2 1TJ M0FFX
CB2 1TP M0ERY
CB2 1TQ 2E0TRY
CB2 1TQ M0WCK
CB2 1TQ M6HXC
CB2 3BU 2E0FJG
CB2 3BU M0JDY
CB2 3BU M3VUS
CB2 3HU M0RSW
CB2 3JU M0TMN
CB2 3JU M3TNM
CB2 6SP M5AKW
CB2 9AQ M3VFC
CB2 9BG M0IPQ
CB2 9HR G3NHB
CB2 9HS G1YPU
CB2 9JR 2E0BUV
CB2 9JR M6RLN
CB2 9JW 2E0CFL
CB2 9NJ 2E0BCK
CB2 9NJ M0EAQ
CB21 4AD G4PDI
CB21 4EG G4STV
CB21 4EG G6BOX
CB21 4EG M0BOX
CB21 4EU M0RSP
CB21 4HS G7KYI
CB21 4HS M1BEC
CB21 4JQ 2E1HLF
CB21 4JQ G7SRK
CB21 4JT G1VCU
CB21 4LY G1CQA
CB21 4RY G0CHC
CB21 4SN M6IYG
CB21 5BG M6FQO
CB21 5BT G4YJK
CB21 5DT G8RYL
CB21 5EY G4VTU
CB21 5JY G4HJW
CB21 5LD G3UDI
CB21 5LR G3JQS
CB21 5LR G4FZM
CB21 5LU M0RBU
CB21 5LY G4FTW
CB21 5PE G3ZUK
CB21 5PE M0JJJ
CB21 5PE M6IXF
CB215PE M7OSM
CB216XR M7CPJ
CB22 3BN G1GFC
CB22 3BT 2E0WWA
CB22 3BT M6ZBQ
CB22 3DH G4WEZ
CB22 3DU G7KRT
CB22 3DU G8TMV
CB22 3DW M1AQP
CB22 3DW M1CKO
CB22 3EA G8LVU
CB22 3FE M7TDK
CB22 3HJ G6PUO
CB22 3LP M1DBK
CB22 3LP M3NQS
CB22 3SP G6ASH
CB22 3SZ 2E0XOR
CB22 3SZ M0XOR
CB22 3SZ M6XOR
CB22 3SZ M7CMD
CB22 3SZ M7ROX
CB22 3UR M6BSL
CB22 4LX G6BBN
CB22 4NE G4LCE
CB22 4NG G0PGQ
CB22 4QN 2E0BKQ
CB22 4QN M0ZDG
CB22 4QN M3UYC
CB22 4QN M6SWE
CB22 4RR M0ZZE
CB22 4RT G6WJX
CB22 5AE G8MEI
CB22 5AW G0ETP
CB22 5BP M1KTA
CB22 5BP M3KTA
CB22 5BX G3WLD
CB22 5BX M0LAT
CB22 5EG G3UUT
CB22 5EG G5PI
CB22 5JJ G6WKQ
CB22 5JW G3TVM
CB22 5JZ M0GCX
CB22 5LG G1IMM
CB22 5LW M6FCG
CB22 6RT G6UFL
CB22 6RY M6LBW
CB22 6SP M0DCV
CB22 7QF M1HJE
CB22 7RP M3IQN
CB22 7RZ G8SHC
CB223HZ G3WFF
CB225JF G7UUT
CB226RT M7FYL
CB23 1LS G6KWA
CB23 2NG G6FSU
CB23 2NL G4VHK
CB23 2RE G3TAG
CB23 2RE G4VYG
CB23 2RE M3CMX
CB23 2RY G7SPZ
CB23 2SJ G7ASZ
CB23 2SW M7BZZ
CB23 2TH G4IPM
CB23 2TU G0BOM
CB23 3RX M7BSZ
CB23 3XP 2E0GGW
CB23 3XP M0HHB
CB23 4HY M6RGD
CB23 4JQ G4UGQ
CB23 4JQ G4VWS
CB23 5ED G0ULA
CB23 5EF M1AQR
CB23 5FN M6CWJ
CB23 5HY G0RQN
CB23 6FN M3WCX
CB23 6FP G0REU
CB23 6GA 2E0DWV
CB23 6GA M0IMZ
CB23 6GA M6FVZ
CB23 6GU 2E0LCM
CB23 6GU M0LCM
CB23 6HW 2E0VUL
CB23 6HW M7HBA
CB23 7AE G8MEX
CB23 7AZ G8GEV
CB23 7BU G4DGL
CB23 7DY G3PJT
CB23 7NX G3GGK
CB23 7PG M0YCS
CB23 7PH 2E0VBK
CB23 7PH M6RBK
CB23 7PS G3RUH
CB23 7PW G3XZP
CB23 7QD 2E0NHB
CB23 7QD M0HZR
CB23 7QD M6NBS
CB23 7QL G4JZQ
CB23 7RY G8EYP
CB23 7UZ G7GCW
CB23 7XA 2E0BKD
CB23 7XA M3FHO
CB23 7XJ G4NBS
CB23 7XJ G4NXO
CB23 7XL G3YMW
CB23 7XR G6HFS
CB23 7XR G6NIW
CB23 8BT G4AKD
CB23 8BW 2E0JJA
CB23 8BW 2E0JOA
CB23 8BW M6CVB
CB23 8BX G7BIX
CB23 8EP 2E1AEJ
CB23 8EP G0NJJ
CB23 8EQ G3LOD
CB23 8HH G6KIB
CB23 8SA M0JMQ
CB23 8SF G0KRB
CB23 8SJ G8RPI
CB23 8SU G6VTN
CB23 8SZ G6TUS
CB23 8TA M6RSD
CB23 8TD G6HSW
CB23 8TD M1MPW
CB23 8TQ M6TAG
CB23 8TR G1ZCC
Cb234ly M7CEP
CB236AT 2E0GTJ
CB236AT M0SVP
CB236AT M7GMR
CB24 1AG M7EXT
CB24 1BS M6HHL
CB24 3AY G8NPR
CB24 3EH G7LBL
CB24 3GN M7PMY
CB24 3XA 2E0RJZ
CB24 3XA M6RJZ
CB24 4AG 2E0OTY
CB24 4AG M6ODY
CB24 4QU M7MJB
CB24 4RT G7CPQ
CB24 4RT G8JDQ
CB24 4SF G7NBL
CB24 4SR G3XIP
CB24 4TR 2E0BPX
CB24 4TR M0GUM
CB24 4TR M3VGZ
CB24 5HY G1JZN
CB24 5HY G1XAA
CB24 5JD M7JBM
CB24 5JU G4NEY
CB24 5JX G4XXX
CB24 5LS G6XRY
CB24 5NB G0DAH
CB24 5NJ G8GJU
CB24 5PZ G4BHJ
CB24 5TZ G1MKP
CB24 5UR G1JKF
CB24 5UU G8FTE
CB24 5UX G8IWQ
CB24 6AD G7DJA
CB24 6BS G0OIR
CB24 6DA G1NWM
CB24 6DG G8CRB
CB24 6DT G4LNY
CB24 6EB G0WZD
CB24 6ET 2E0HWK
CB24 6ET M0HKW
CB24 6ET M3UFS
CB24 8AW G0MBD
CB24 8QG G1KIJ
CB24 8QR G0GKP
CB24 8SA G7WFV
CB24 8SD G0EUZ
CB24 8TB M0BLF
CB24 8TG G1DEQ
CB24 8TQ G3XXX
CB24 8TR M0VFC
CB24 8TS M1EMG
CB24 8TX G3XPW
CB24 8UF 2E0RLV
CB24 8UF M0RLV
CB24 8UF M6RLV
CB24 8UJ M6ESO
CB24 8UL G2CR
CB24 8UL M0VMC
CB24 9JL G8XPQ
CB24 9JU G0THI
CB24 9PE G7KDG
CB24 9PP G8HSS
CB24 9PP M0AQF
CB25 0BA G4PZW
CB25 0BB M3MKZ
CB25 0DP M7ING
CB25 0DT G8NEO
CB25 0DT M0CNP
CB25 0DW M3HWW
CB25 0EH M7RTS
CB25 0EL G8DQF
CB25 0ER M0DRK
CB25 0ES G7SCO
CB25 0ES M1BIB
CB25 0HD G3TFX
CB25 0HL G3XBM
CB25 0HT G8BXA
CB25 0JU G4FVA
CB25 0JX G0SBK
CB25 0LD G6ALB
CB25 0LD M0BXT
CB25 9BQ G3PTQ
CB25 9DW 2E0HGW
CB25 9FD M7SNB
CB25 9HB G7BNM
CB25 9HW G4VME
CB25 9HW M7JHW
CB25 9JW 2E0SDY
CB25 9JW M0SND
CB25 9NB G8XLE
CB25 9NE G8ZRD
CB25 9PX G4BAO
CB3 0BN 2E0TJG
CB3 0BN M0JSH
CB3 0DG M0HFQ
CB3 0DG M6BHK
CB3 0DS M0XOS
CB3 0DS M7NLS

CB3 0GL M0RHC
CB3 0GL M7RHH
CB3 0GP G0VDE
CB3 0HX G4XIL
CB3 0LT M1MAJ
CB3 0LT M3ZCB
CB3 0NP G1DAZ
CB3 0NS 2E1FLN
CB3 0PE M0HOT
CB3 0PP G0ECL
CB3 6GA G3HZP
CB3 9DY G8KDO
CB3 9LQ G3MRX
CB4 1DG G3VGX
CB4 1ET G6AIG
CB4 1FD G0RDB
CB4 1LW G4DNG
CB4 1NP G6FKS
CB4 1NX M0ZCJ
CB4 1QG M6GVY
CB4 1SQ G3ZAY
CB4 1SX M6GKW
CB4 1TT G3RQS
CB4 1TZ G1DVD
CB4 1TZ G8CTX
CB4 1XG G3URX
CB4 1XQ M7CHF
CB4 1YZ G8NNA
CB4 2AJ G3OWB
CB4 2AS G7AZJ
CB4 2AU G1TPV
CB4 2BJ G8UDV
CB4 2DB G3RSE
CB4 2DB G8UDG
CB4 2HH M0HOJ
CB4 2HT G1VJG
CB4 2LD G7KHL
CB4 2NP G6KSV
CB4 2NS M6EBG
CB4 2QR M5CHH
CB4 2RA G7DWU
CB4 2SH G7NZY
CB4 2UF G4NUA
CB4 2UH G8XYS
CB4 2UN G1AQX
CB4 3BW G4OSX
CB4 3EL M6VWA
CB4 3ND G3HRE
CB4 3PA G4AJW
CB4 3PA G8KNN
CB5 0JF G1YAS
CB5 8AR M0RZO
CB5 8AR M7RZO
CB5 8LB 2E0JGS
CB5 8LB M0JGS
CB5 8LB M0WBJ
CB5 8LB M6JGS
CB5 8NA M6JUT
CB5 8NA M7WUT
CB5 8QU 2E0DVY
CB5 8QU M6FVM
CB5 8QU M6GQF
Cb5 8se G7SWW
CB5 8SS 2E0IHI
CB5 8SS M0RTW
CB5 8SS M6FHI
CB5 8SS M6ZZA
CB5 8TE M6RMA
CB5 8UB M6AWR
cb5 8uj 2E0IFN
cb5 8uj M0KGE
CB5 8UJ M7GIS
CB5 8US M6CFX
CB5 8XJ 2E0FPQ
CB5 8XJ M6FPQ
CB58DL M1CWA
CB6 1AB G0KFQ
CB6 1AY G0HCP
CB6 1BS G7VGH
CB6 1DA M3UPB
CB6 1DP 2E0XCF
CB6 1DP G8GLB
CB6 1DP M0XCF
CB6 1DP M7AVA
CB6 1DZ G6YTO
CB6 1FA M7CSA
CB6 1FH G7HMI
CB6 1GL 2E0LTP
CB6 1GL M0OEB
CB6 1GL M7CEB
CB6 1HU G3KKC
CB6 1JD M1BFY
CB6 1JJ 2E0DPH
CB6 1JJ M6PHL
CB6 1JS G6YQJ
CB6 1JW G0MGI
CB6 1LG M7PUF
CB6 1NA M7KVS
CB6 1RJ 2E0TTN
CB6 1RJ M6DXL
CB6 1RQ G7BNL
CB6 1SB G4HWJ
CB6 1TL 2E0RDF
CB6 1TL M0MTW
CB6 1TL M6DFK
CB6 1TP G4BUH
CB6 2AQ G6OIB
CB6 2EQ G1HMT
CB6 2HH G1DUI
CB6 2HS G1PEK
CB6 2LR 2E0FHA
CB6 2LR M7DEL
CB6 2NW G4FIT
CB6 2NX M6KUI
CB6 2PD G0PPL
CB6 2QB M0TVT
CB6 2QB M6TRH
CB6 2QF M0DBJ
CB6 2RL 2E0DNF
CB6 2RL G0PZJ
CB6 2RL M0ZEB
CB6 2RL M6DNF
CB6 2RX G3IUY
CB6 2ST G4WMZ
CB6 2TQ M1ZEM
CB6 2UH M0BWH
CB6 3AL G0ICE
CB6 3AL G1XCY
CB6 3DA G4GXY
CB6 3DL G3MXA
CB6 3DN G1CQR
CB6 3HR M0RKA
CB6 3HR M6WKI
CB6 3JT G8OHP
CB6 3PP G4GND
CB6 3PW G0PPI
CB6 3PW G4KNS
CB6 3PW G7IEY
CB6 3SA G7VNP
CB6 3ST G4BEL
CB6 3UE 2E1NPH
CB6 3UE G0PYE
CB6 3UE G4NPH
CB6 3UE G8NPH
CB6 3UE M3NPH
CB6 3UG M6LFQ
CB6 3XE G1VLD
Cb61dj G0OKK
CB7 4BU G1UAF
CB7 4HU G6IOB
CB7 4PQ G6YPD
CB7 4SS M0CGR
CB7 4TH G4KPX
CB7 4UN G0LQV
CB7 4YJ G1AAQ
CB7 5AL M0TIX
CB7 5AL M3DCS
CB7 5AT G3UCL
CB7 5AT M0SNB
CB7 5AT M1GEO
CB7 5BL G8LWQ
CB7 5DH G6BUU
CB7 5DN M7AKQ
CB7 5DS G7BRP
CB7 5EL G8PAG
CB7 5EX 2E0VAS
CB7 5EX M0VAS
CB7 5EX M3VAS
CB7 5fn 2E0GXB
CB7 5GD M7UPS
CB7 5HJ M3YEJ
CB7 5HL 2E0FBE
CB7 5HL M6FTF
CB7 5JD 2E0RPN
CB7 5JD M6DZV
CB7 5JD M7PNR
CB7 5NR G4UJV
CB7 5NW 2E0LWA
CB7 5NW M7LJA
CB7 5QX G8WAV
CB7 5RU G0PIK
CB7 5SH G0TLQ
CB7 5SJ M6GEQ
CB7 5SL. 2E0UEH
CB7 5SL. M0UEH
CB7 5SL. M6UEH
CB7 5TN G4ERO
CB7 5TW G3FCM
CB7 5UU G0EIM
CB7 5XE G7FMV
CB7 5XH M0GUO
CB7 5XW M6BSF
CB7 5ZD M7AWJ
CB8 0BY 2E0BLR
CB8 0DJ G6GQI
CB8 0EJ M7MEO
CB8 0QD 2E0NEV
CB8 0SE G3LUN
CB8 0SE G8IDL
CB8 0SQ G1ACB
Cb8 0uh M7ESR
CB8 7AZ G8WUR
CB8 7BQ G0SPK
CB8 7BT M6HPT
CB8 7DU G6JJK
CB8 7RT M3VSF
CB8 7SD G6BBK
CB8 7SD G8BBK
CB8 7YP G0ELK
CB8 7YP G1NHG
CB8 8AD G7JAX
CB8 8DT G4DSN
CB8 8HU G7VGN
CB8 8RL M6AWY
CB8 8SW M3YDY
CB8 8YH 2E0TUG
CB8 8YH M6EZN
CB8 9BA M6RIH
CB8 9DF M6GUF
CB8 9DG G3NUL
CB8 9DG G3SZY
CB8 9DQ G0ADZ
CB8 9NB G0FRU
CB8 9NR G4IQR
CB8 9PD M1BJE
CB8 9RX G4HVV
CB8 9RX G8HVV
CB8 9RY G6JDO
CB8 9SA G0VQH
CB9 0AP G7MVU
CB9 0AU M1DQU
CB9 0BZ M7BBO
CB9 0ED M3ZCA
CB9 0EQ M3LHD
CB9 0JA G0OJB
CB9 0JP M7PJS
CB9 0JR G0EQG
CB9 0LH M7DKC
CB9 0PH M3DMG
CB9 7DL 2E0ELA
CB9 7ED G4PQY
CB9 7EE G7LVM
CB9 7EW G7MQC
CB9 7FU G1WZM
CB9 7FU M7WZM
CB9 7GH M0JDD
CB9 7GH M1RTT
CB9 7NA G6MQN
CB9 7NN G4KVU
CB9 7PT M6JFF
CB9 7RF M3ZKI
CB9 7SG M6EZT
CB9 7ST G8FFF
CB9 7WG G8UBX
CB9 7XW G1DNO
CB9 8EH M6BEJ
CB9 8LX M5PYE
CB9 9DD M3XCA
CB9 9LE G4LYF
CB9 9LY M6SBJ
CB9 9ND 2E0PCI
CF1 8BZ GW3KXX
CF11 0AP MW1CQP
CF11 0JQ MW7LTE
CF11 7BG GW8WEY
CF11 7HF MW3ZYQ
CF11 8AJ 2W0HRL
CF11 8AJ MW3LLV
CF11 8DE MW7GKT
CF11 8HQ GW4ISJ
CF11 9JR 2W0TKS
CF11 9JR MW0KST
CF11 9JR MW6KSL
CF11 9NQ MW0CCK
CF14 0NZ MW6NHC
CF14 0RF GW8VRS
CF14 0RW MW3KNR
CF14 1EG 2W0DSJ
CF14 1EG GW4LFO
CF14 1EG MW0MPD
CF14 1EG MW6JWL
CF14 1EH MW6CJS
CF14 1EH MW6ESW
CF14 1HP MW6WFJ
CF14 1NQ GW4LFW
CF14 1UE GW4VGB
CF14 1UH GW6RCK
CF14 1UN GW4TJQ
CF14 2BR MW7AWH
CF14 2EH GW4NHB
CF14 2FU GW6LSL
CF14 2LR GW7ASL
CF14 2RY 2W0PCN
CF14 2TE GW4NHH
CF14 3DU GW0KRQ
CF14 3DU GW1WZI
CF14 3PL MW3FPF
CF14 3QA GW8NCN
CF14 4AQ 2W1CYC
CF14 4DJ GW7EMV
CF14 4DP MW7ILL
CF14 4HN MW3XIJ
CF14 4HS GW4GZX
CF14 4JQ GW8LKX
CF14 4NP MW7BPQ
CF14 4NS GW0GDI
CF14 4SB GW1AWH
CF14 4SP GW8ARC
CF14 5HJ GW4JOG
CF14 5JW MW3HZB
CF14 5LR
MW3WWO
CF14 5PW MW3DEM
CF14 5QA GW6UAS
CF14 5QH MW7AEN
CF14 6AE GW4LXO
CF14 6AN MW6MBZ
CF14 6BJ GW1BCI
CF14 6DF GW6DFX
CF14 6JZ GW0BBC
CF14 6JZ GW8RWX
CF14 6NS MW6FKC
CF14 6PE GW0WMT
CF14 6QF MW7JMW
CF14 6QH MW3PJU
CF14 6RL GW6ZHM
CF14 6SS GW6ITB
CF14 6SW GW3WLN
CF14 7EH 2W0CDZ
CF14 7EH GW3UZS
CF14 7EH MW6PDG
CF14 7TP GW0TPL
CF14 9AF GW4LWA
CF14 9AH GW7EYP
CF14 9LA GW3TKH
CF15 7PT MW6XDA
CF15 7SE MW3ZKX
CF15 8AL GW8EHO
CF15 8DD GW6MNC
CF15 8FE GW6VTZ
CF15 8RA GW8FNO
CF15 8RE GW0HCB
CF15 9JP GW1CUQ
CF15 9NJ MW6HYS
CF15 9NN MW0ZAP
CF15 9NZ GW7HOM
CF15 9PB GW3TOB
CF15 9QQ 2W0JYI
CF15 9QQ MW0WSH
CF15 9QQ MW6EDQ
CF15 9QR GW3VOL
CF15 9SJ GW4LFV
CF15 9SJ GW6IOA
CF15 9TJ GW8YYF
CF2 6JN GW3UMD
CF23 5BY GW8EHQ
CF23 5EZ GW0WLI
CF23 5HE MW6EPX
CF23 5NN GW4ZAW
CF23 5SB MW3PGN
CF23 6EJ GW6MHV
CF23 6EQ GW3XXB
CF23 6HN GW4JJV
CF23 6HN GW4JJW
CF23 6LF GW6AGS
CF23 6PA GW3YGH
CF23 6QU GW0MLN
CF23 6SW 2W0RQC
CF23 6SW GW0RQC
CF23 6SW MW6RQC
CF23 6SZ 2W0EUO
CF23 6SZ MW6EUO
CF23 7DA GW3MKT
CF23 7HN GW0VMD
CF23 7HQ GW4KOE
CF23 9AZ GW7JDX
CF23 9BJ MW6SAW
CF23 9DW 2W0FCF
CF23 9DW MW6XXC
CF23 9HX MW0AQT
CF23 9NR 2W0YIF
CF23 9NR GW6CUR
CF23 9NR GW8NP
cf238xj GW0HYU
CF24 0HS GW0DGJ
CF24 1PN MW7TNM
CF24 1RW GW7HOC
CF24 1RW MW3YKL
CF24 1RW MW5HOC
CF24 1RW MW6YDP
CF24 2EL GW1SMT
CF24 2NW GW7EMO
CF24 3HU MW7PKT
CF24 3QW 2W0KXN
CF24 3QW MW0KXN
CF24 3QW MW7KJN
CF24 3RW MW7RHC
CF24 4BZ GW8AHB
CF24 4LS MW3KHH
CF24 4RU GW1XZI
CF24 4RU GW4SGR
CF24 4RU GW7BTC
CF3 0AQ GW6YML
CF3 0BY GW7KDU
CF3 0DF MW6FPL
CF3 0DQ GW0FVC
CF3 0DU MW6ATC
CF3 0NB 2W0OCF
CF3 0PS MW6RJX
CF3 1RE MW3IHD
CF3 1TA MW6JHJ
CF3 1TA MW6KCQ
CF3 2TX MW6ETY
CF3 2WL MW3ASG
CF3 3AG MW3UFH
CF3 3HQ GW4IUN
CF3 3LX 2W0BJE
CF3 3LX MW3RXK
CF3 3LZ GW0POA
CF3 5QE MW6TRK
CF3 5SY MW6OZF
CF3 5UD MW6GLK
CF3 6XW GW4BNC
CF3 7LL GW4OES
CF31 1HD 2W1GKJ
CF31 1PF GW0SVX
CF31 1PQ GW0PYU
CF31 1QN MW6DHK
CF31 1RL 2W0ZAA
CF31 1RL MW3USX
CF31 2AT MW3KHY
CF31 2BY MW6YPF
CF31 2DP 2W0PQU
CF31 2DP 2W0YAB
CF31 2DP MW3YDS
CF31 2DP MW6PQU
CF31 2EA GW0RFM
CF31 2ED GW0FJP
CF31 2HL GW8PJE
CF31 2JA MW3THI
CF31 2JG MW3VEL
CF31 2JR MW0AXA
CF31 2LA GW1ERA
CF31 2LJ GW6HMJ
CF31 2ND GW4MTE
CF31 2ND GW8XJC
CF31 2NE GW0NKG
CF31 2NG GW3ZTH
CF31 2PF GW1ZNC
CF31 2PR MW0LCH
CF31 2QF MW6AHT
CF31 2SS GW8SKB
CF31 3BA GW7RRS
CF31 3DA MW0GRZ
CF31 3LN GW4SQD
CF31 3LN GW8ZIL
CF31 4AX 2W0JKN
CF31 4AX GW7RQV
CF31 4AX MW7EJJ
CF31 4BB GW0SZU
CF31 4EY GW3SRF
CF31 4HD GW0SIS
CF31 4HD GW4NBY
CF31 4LU MW3OEJ
CF31 4NR GW1KQV
CF31 4PY MW0AEL
CF31 4QA 2W0HQA
CF31 4QA MW7URL
CF31 4QX GW1FKY
CF31 4QY 2W0GNG
CF31 4QY MW6PAC
CF31 4RN MW0FDG
CF31 4SJ 2W0PWO
CF31 4SJ MW0GYV
CF31 4SJ MW6PWO
CF31 4SR GW6MAB
CF31 4UQ MW7BXN
CF31 5BA GW6VET
CF31 5BT GW7DVJ
CF32 0BJ GW0MNP
CF32 0BP 2W0SYT
CF32 0BP MW7MAT
CF32 0EW 2W0SNW
CF32 0EW MW3SNW
CF32 0HL GW0TTF
CF32 0LH GW1XJJ
CF32 0PJ GW4ZEA
CF32 0SL GW4SKP
CF32 7BB 2W0PEH
CF32 7BB MW0PEH
CF32 7BB MW3PEH
CF32 7SP MW6PGD
CF32 7SY GW1CXT
CF32 8AH GW7VBE
CF32 8BL MW1CFE
CF32 8HY GW8BWX
CF32 8LE MW6MWG
CF32 8PE MW7SYK
CF32 8SA MW0DCQ
CF32 8SD GW0GQC
CF32 8SE MW3BBQ
CF32 8UU MW3SLL
CF32 8YB GW1OUP
CF32 8YB GW1YHA
CF32 8YB MW3DIL
CF32 9AE MW7KKW
CF32 9AF GW4VKG
CF32 9AQ GW7TNS
CF32 9BJ MW3LEW
CF32 9DH GW3SSK
CF32 9DH MW3MRL
CF32 9JL MW7PWC
CF32 9LU GW0ACH
CF32 9LX MW6NGE
CF32 9NH MW6LUQ
CF32 9RJ MW0WNL
CF32 9SB GW0KAX
CF32 9SB GW4WVO
CF32 9SX MW6CLP
CF32 9UE MW3BOC
CF32 9UE MW3UBY
CF32 9UG GW4THK
CF32 9UR 2W0WCQ
CF32 9UR MW6WCQ
CF32 9YH GW8CNS
CF32 9YN GW4SKA
CF33 4EU MW3OEC
CF33 4LT MW0COE
CF33 4LT MW0COF
CF33 4PL MW7BEF
CF33 4RQ GW4IFE
CF33 6AJ GW7PEX
CF33 6AP MW0AGE
CF33 6AS MW6XGD
CF33 6EJ MW7RTY
CF33 6EU GW0RTP
CF33 6HT GW6FNB
CF33 6HY MW0PDR
CF33 6HY MW0PRC
CF34 0AB 2W0EJP
CF34 0AB GW0CYK
CF34 0BB GW3WRE
CF34 0BP MW3LSG
CF34 0DE MW0WGR
CF34 0DE MW1WYN
CF34 0DE MW7GBJ
CF34 0DP GW6OMV
CF34 0HT 2W1AID
CF34 0HY MW3MVT
CF34 0LU MW6TJS
CF34 0NG GW7DUI
CF34 0NT GW0TOM
CF34 0NT GW0VMS
CF34 0RU MW0COD
CF34 0SW 2W0JAI
CF34 0TF 2W0ETZ
CF34 0TF MW6MXV
CF34 0TG 2W0ENV
CF34 0TG MW6RPS
CF34 0UE MW6TDL
CF34 0UF 2W0DRK
CF34 0UF 2W0IUN
CF34 0UF MW0HXX
CF34 0UF MW0IUN
CF34 0UF MW6DRM
CF34 0UF MW6IUN
CF34 0UR GW0IRP
CF34 0UW 2W0LFY
CF34 0UW 2W0LLY
CF34 0UW MW6SBT
CF34 0YG MW0GWT
CF34 0YP GW4HCA
CF34 9AH MW6GMJ
CF34 9AH MW6KBK
CF34 9BJ MW6PNW
CF34 9HP GW3XJC
CF34 9JL 2W0FAR
CF34 9JL MW6BAU
CF34 9JL MW6FAN
CF34 9ST MW0AWO
CF34 9TB 2W0DMG
CF34 9TB MW3GPG
CF340TB MW3NCS
cf340tf MW7DJQ
CF35 5BH GW4ZXG
CF35 5EG MW1MFY
CF35 5EU MW7KTP
CF35 5HX MW1BUN
CF35 5NA GW6ZGY
CF35 5NA MW0WIL
CF35 5NF MW0JAN
CF35 5QA GW6CZE
CF35 5RH GW7NJT
CF35 6ED GW8HWL
CF35 6FA GW8TGS
CF35 6HE MW0SWR
CF35 6JR GW0KFL
CF35 6PD GW3FSP
CF35 6SD GW3RVG
CF35 6TF MW0JLN
CF35 6TY GW1YBF
CF35 6TY MW0PSG
CF35 6YA 2W1FUD
CF35 6YD MW3NXD
CF35 6YD MW3WQE
CF35 6YF MW5MWR
CF35 6YG 2W0DJS
CF35 6YG MW3NUX
CF36 3HR GW0BAH
CF36 3HW MW6NLG
CF36 3PR GW7SAQ
CF36 3QH GW0VST
CF36 3SW GW6JFV
CF36 3UN GW7NUL
CF36 3YS GW0BZA
CF36 5AJ GW7DTB
CF36 5AN GW3RIY
CF36 5AU GW4BKG
CF36 5HN GW1MNC
CF36 5LP GW4ZVO
CF36 5SG GW3ARS
CF363EP GW4BCF
CF365EY MW7IKH
CF37 1EN MW6GKU
CF37 1HT GW0ONU
CF37 1HY GW0SQY
CF37 1NH GW8HF
CF37 1NW GW0RQP
CF37 1SS GW1FKL
Cf37 1ts 2W0FMR
Cf37 1ts MW7CEW
CF37 2DA GW4ZVV
CF37 2DD MW7CEO
CF37 2HF GW3YBN
CF37 2LG 2W0BAO
CF37 2SE 2W0PCD
CF37 2SE MW0HKA
CF37 2SE MW6PEL
CF37 3BW GW8LRO
CF37 3DF GW0SLC
CF37 3EU GW0JHH
CF37 4AN 2W0KKR
CF37 4DP MW0ATG
CF37 4HD MW6NWM
CF37 4NR MW3HBF
CF37 5DA MW0MWL
CF37 5ER MW6ARX
CF37 5HH 2W0LMM
CF38 1AU GW0MOQ
CF38 1BW GW1KQY
CF38 1BY GW4CNL
CF38 1DQ 2W0VBT
CF38 1DQ MW0VBT
CF38 1DQ MW6KVS
CF38 1DW MW3ZSO
CF38 1LA MW6TOU
CF38 1LB MW7EIJ
CF38 1LF 2W0BMR
CF38 1LF MW3PBV
CF38 1LY 2W0CBJ
CF38 1RP GW0WLN
CF38 1RY GW8RLI
CF38 1UE GW3ZXI
CF38 1UF GW4NPC
CF38 2AL MW3CHZ
CF38 2EN 2W0KPH
CF38 2EN MW6CDV
CF38 2LB 2W0OAG
CF38 2LB MW0YAG
CF38 2LB MW6YAG
CF38 2LN 2W0CWL
CF38 2LN MW0HUY
CF38 2LN MW6CDZ
CF38 2LN MW6CWL
CF38 2NH GW4PAF
CF38 2NT 2W0RLB
CF38 2NT MW6MGZ
CF38 2NT MW6YAE
CF38 2SR 2W0YMG
CF38 2SR MW6YIG
cf38 2tf 2W0MEX
cf38 2tf MW6HCS
CF39 0EG MW3ZSM
CF39 0ET MW0DVM
CF39 0ET MW7JLM
CF39 8AF MW7JHX
CF39 8AL GW4RKZ
CF39 8AL GW6MIH
CF39 8DF MW6HDV
CF39 8DF MW6KJX
CF39 8DU GW7CSK
CF39 8PL GW0UHJ
CF39 8PL GW7PIN
CF39 8PP GW0WLQ
CF39 8TW 2W0GER
CF39 8UG 2W0BKA
CF39 8UG MW0HAB
CF39 9EL MW6BUH
CF39 9HT MW0GWL
CF39 9TE 2W0YMS
CF39 9TE MW6KJZ
CF4 5SR GW8NNF
CF4 6HD GW0NIY
CF4 6HY GW1YKT
CF4 7PD GW4WQC
CF40 1BW MW6BZP
CF40 1BY GW7RKC
CF40 1DQ GW1BDF
CF40 1DQ GW1BDG
CF40 1DW MW0SBJ
CF40 1EN MW6SDV
CF40 1EQ MW0SKD
CF40 1EU 2W0BED
CF40 1HF MW0SWB
CF40 1HP 2W0DLE
CF40 1HR GW2FOF
CF40 1HR GW4BUZ
CF40 1JD MW0MWA
CF40 1JT GW0KQU
CF40 1LX GW6ZYI
CF40 1LX MW0WRQ
CF40 1NG MW6FIN
CF40 1PJ MW0PRP
CF40 1QY GW4YCO
CF40 1TA 2W0SCL
CF40 1TA MW3ZDQ
CF40 2AF MW6MMU
CF40 2AH MW6CGH
CF40 2DZ MW3FVH
CF40 2HD MW6NAZ
CF40 2HJ MW6AEX
CF40 2HN GW4SYO
CF40 2JD GW0LYF
CF40 2LN MW6LNU
CF40 2NX 2W1FVH
CF40 2RE MW6KDA
CF40 2UN MW3MKN
CF41 7HF GW3XHG
CF41 7JT MW1DTT
CF41 7NJ GW0DLA
CF41 7PW MW6CUA
CF41 7PW MW6FBA
CF41 7TW GW0FJQ
CF41 7UD GW1NZF
CF41 7UE MW1COB
CF41 7UG GW4NOS
CF42 5AH GW1JNH
CF42 5AH MW6TPM
CF42 5AR 2W0KSA
CF42 5AR MW6DJT
CF42 5AR MW6NFG
CF42 5AR MW6TRL
CF42 5DR GW4XAZ
CF42 5DT 2W0YAE
CF42 5DT MW0YAE
CF42 5DT MW3YAE
CF42 5HL MW3TWQ
CF42 5NG 2W0EQQ
CF42 5NG MW6IFE
CF42 6AW GW7CYT
CF42 6DF GW0NDZ
CF42 6DT 2W0VDW
CF42 6DT 2W0VEH
CF42 6DT MW3VEH
CF42 6ED MW0TJD
CF42 6RL GW1VAW
CF42 6RL MW6AOY
CF42 6SN 2W0PHP
CF42 6SN MW3XZP
CF42 6TF GW8UAP
CF43 3BD MW0DCM
CF43 3HF MW7DPG
CF43 3LF GW7VSO
CF43 3LJ MW1BTM
CF43 3NW 2W0CJG
CF43 3NW MW3YHW
CF43 3RN MW0DRU
CF43 4EU GW3SUH
CF43 4HE 2W0VKA
CF43 4HE MW3IPK
CF43 4HE MW3VKA
CF43 4PB MW0HVL
CF43 4RG MW6KEL
CF44 0BQ 2W0TMB
CF44 0BQ MW6KAC
CF44 0EJ 2W0XMG
CF44 0EJ MW0XMG
CF44 0EJ MW3USS
CF44 0EJ MW3XMG
CF44 0ER GW1HNG
CF44 0HR GW0UKC
CF44 0HY 2W0PJJ
CF44 0HY MW0HPH
CF44 0HY MW0PJJ
CF44 0HY MW6GNA
CF44 0HY MW6PJJ
CF44 0PB GW3SFC
CF44 0PF GW4UAJ
CF44 0PW MW0KAX
CF44 0PY GW4SUN
CF44 0TP MW6AUX
CF44 0TU GW6MKR
CF44 6AH MW6LTR
CF44 6DD GW0OUH
CF44 6EN 2W0DTR
CF44 6LD GW3ASW
CF44 6LD GW4RUX
CF44 6LD MW0TTR
CF44 6LD MW6OCC
CF44 6NG 2W0XTP
CF44 6NG MW0TMI
CF44 6NL 2W0KOP
CF44 6RY MW6ALC
CF44 6RY MW6RAW
CF44 6SH 2W1HFZ
CF44 6UW 2W0LJD
CF44 6UW MW6EFV
CF44 6YG MW1AND
CF44 7HA MW6LDS
CF44 7HE MW6TIQ
CF44 7NF MW6MIQ
CF44 7PP MW6LCX
CF44 8AH MW6GOY
CF44 8BW GW6YCT
CF44 8EW MW6BQA
CF44 8EW MW6LIW
CF44 8EX MW6ZBR
CF44 8HA GW4OPW
CF44 8LA GW4ZUA
CF44 8LA MW6DRV
CF44 8LF 2W0PPL
CF44 8LF MW6AZP
CF44 8LF MW6HEY
CF44 8LF MW6TJE
CF44 8TE GW4SXA
CF44 8TH MW0BWM
CF44 8TL MW3EQE
CF44 8UB GW0KZE
CF44 8YB GW4KCY
CF44 9BG MW6WRC
CF44 9EW 2W0KED
CF44 9EW MW0JZM
CF44 9EW MW6KED
CF44 9JU MW3FJW
CF44 9LY GW0UHX
CF44 9LY GW1SVV
CF44 9ND GW4VEI
CF44 9NG MW0DAR
CF44 9RU 2W0WFB
CF44 9RU MW0WFB
CF44 9RU MW6WFB
CF44 9SA MW6BHD
CF44 9TX 2W0DSO
CF44 9TX MW6CQL
CF44 9YJ 2W0XBC
CF45 3AB GW0WXO
CF45 3BS GW0UHO
CF45 3BS MW3VGJ
CF45 3BY MW6BVG
CF45 3EF MW6KCJ
CF45 3NB MW6JGC
CF45 3ND GW4LWD
CF45 3NU 2W0SKG
CF45 3YA GW0VQZ
CF45 3YW GW4ZUJ
CF45 3YW MW6ENY
CF45 4BD MW7TEW
CF45 4BG 2W0RAD
CF45 4BG MW0PIC
CF45 4BG MW3ZLX
CF45 4BL 2W0IWM
CF45 4BX MW3IQE
CF45 4DZ MW6JFL
CF45 4SY GW7HVA
CF454YU MW7CMJ
CF46 5AF MW0UKX
CF46 5HG GW0JTJ
CF46 5HQ GW1IQS
CF46 5LA GW7SXN
CF46 5LB GW0OPP
CF46 5LN GW7HDS
CF46 5RL 2W0BGS
CF46 5RL MW3ONG
CF46 6AB GW0PSV
CF46 6AL MW6VBE
CF46 6AW GW0TOI
CF46 6HJ MW3NKG
CF46 6HQ 2W0AZP
CF46 6HQ MW3XVB
CF46 6JF GW4ZCM
CF46 6LF MW3MOJ
CF46 6NS GW0OUV
CF46 6NT GW6OVD
CF46 6RY MW6EGH
CF46 6SD MW6DAI
CF46 6UD MW0DYS
CF47 0TA MW0KDL
CF47 0TB GW0UZX
CF47 0TE MW7BQM
CF47 0TG MW6CIY
CF47 0UX GW4PWZ
CF47 0YU 2W1SRB
CF47 8HJ GW3RYR
CF47 8LU GW1MTH
CF47 8NJ GW0VMZ
CF47 8RH MW3YCL
CF47 9DA 2W0MJA
CF47 9DA MW0DOZ
CF47 9DA MW0EQL
CF47 9DA MW0MUM
CF47 9DA MW6MWS
CF47 9SN MW6JNP
CF47 9TP GW3PYD
CF48 1EE GW1WRV
CF48 1EN GW0UXJ
CF48 1EN MW0IBI
CF48 1HH MW6KLL
CF48 1HW GW6AYR
CF48 1JH 2W0DTM
CF48 1JH MW0KIJ
CF48 1JS GW4GJA
CF48 1JU 2W1CLC
CF48 1LR 2W0BPJ
CF48 1LR MW3WFF
CF48 1LU MW3BOP
CF48 1PH GW4CGZ
CF48 1RZ 2W0HHN
CF48 1RZ MW7PLY
CF48 1TP GW6VRN
CF48 1YY GW0KIG
CF48 1YY MW3IXZ
CF48 2BB GW3LNR
CF48 2DG MW3TPJ
CF48 2NT MW7MOC
CF48 2SA MW3AFX
CF48 3AF 2W0GFX
CF48 3AF MW6VGA
CF48 3HE GW0BYZ
CF48 3ND MW7BPZ
CF48 3NY 2W0BOK
CF48 3NY GW3YUC
CF48 3PE MW3DTO
CF48 3PE MW6JJO
CF48 3PU GW0JTE
CF48 4BQ MW6HKL
CF48 4EE GW3RDB
CF48 4EE GW4ZRW
CF48 4HD GW4BIS
CF48 4HJ MW6MWN
CF48 4NG GW6RUE
CF48 4RN GW0WUL
CF5 1DH 2W0FCM
CF5 1GH GW1VRR
CF5 1GH GW1YXR
CF5 1HL MW3MHG
CF5 1JP MW0WZX
CF5 1QS MW1ABZ
CF5 1RF GW8HWS
CF5 2AU GW8REV
CF5 2DH 2W0BTP
CF5 2JF 2W0USN
CF5 2JN GW2DHM
CF5 2JR MW0CCL
CF5 2QP GW8RAS
CF5 2QS GW8THM
CF5 2QT GW3WCV
CF5 3EW 2W0UZO
CF5 3EW MW0UZO
CF5 3EW MW3UZO
CF5 3QP MW6NFY
CF5 3SS GW6NHB
CF5 4AP GW4KQ
CF5 4FG GW8VFF
CF5 4FH MW0DYZ
CF5 4HJ MW3VNZ
CF5 4PG MW3HAS
CF5 4PH GW6RDV
CF5 4PR 2W0LEN
CF5 4PS 2W0FFN
CF5 4PS MW6OIV
CF5 4QW MW3KCL
CF5 4QY MW6AYA
CF5 5AH MW6CTE
CF5 5AP MW6ICU
CF5 5AQ GW4FOM
CF5 5BZ 2W0GJR
CF5 5BZ MW0IHW
CF5 5BZ MW6WUK
CF5 5DA 2W0SWO
CF5 5DA MW3SWO
CF5 5DD GW7ESF
CF5 5JS 2W0PEG
CF5 5JS MW6RJY
CF5 5NW MW3LGS
CF5 5PB 2W0YFC
CF5 5PL MW6BSR
CF5 5SB MW3HSZ
CF5 5SB MW6AWV
CF5 5SB MW6RBZ
CF5 6AF MW6SEF
CF5 6BQ 2W0SHX
CF5 6BQ GW1CHS
CF5 6BQ GW5NIC
CF5 6BQ MW6EEJ
CF5 6DA GW0WDV
CF5 6EP GW1OZW
CF5 6LQ GW3RIH
CF5 6SB 2W0PCT
CF5 6SB MW0PCT
CF5 6SB MW6PCT
CF5 6SS GW4CSY
CF5 6TN GW4JCK
CF53TH MW6LPZ
cf54ln 2W0ZBC
CF61 1GX GW8KEV
CF61 1GX GW8MFQ
CF61 1TA GW4RER
CF61 1TA GW6GSR
CF61 1YA 2W0DUN
CF61 1YD GW4LFF
CF61 2GS GW4MOG
CF61 2SG 2W0IDT
CF61 2SG GW7HFZ
CF61 2SG MW0IDT
CF61 2UF GW3NCT
CF61 2UG GW0CNJ
CF61 3XR MW3FYA
CF61 5BD MW6ONZ
CF62 3DZ GW4ZCL
CF62 3EA GW6ZMN
CF62 3FT GW8YTO
CF62 3HQ GW0LTC
CF62 3LB MW6AQJ
CF62 3LQ GW6OGD
CF62 4JD MW6ADS
CF62 4PD 2W0VOC
CF62 4PD MW6VOC
CF62 4PG 2W0ORH
CF62 4PG MW6ORH
CF62 5AB 2W0ODS
CF62 5AB MW6FNV
CF62 5BD GW1ABX
CF62 5US GW0TSL
CF62 6RA GW1JVB
CF62 6RA GW1XFB
CF62 6RA GW8LJJ
CF62 7HX GW1VRW
CF62 7TG GW1ZLL
CF62 8BJ GW1XUD
CF62 8BL GW6CNS
CF62 8BR GW8KZA
CF62 8BU MW3DJV
CF62 8BU MW3DJZ
CF62 8BY GW0JCB
CF62 8DA MW7ALT
CF62 8HF MW3JDA
CF62 8ND GW4BRS
CF62 8ND GW6BRC
CF62 8ND MW0VRQ
CF62 8ND MW6JZZ
CF62 9DE MW6BVR
CF62 9DF 2W1SDR
CF62 9DR GW7PFK
CF62 9DW 2W0NCA
CF62 9HD GW0SQT
CF62 9HJ GW3WSU
CF62 9HJ GW4GSH
CF62 9SW GW3CBA
CF62 9TE MW6FYY
CF62 9TG MW0RUH
CF62 9TG MW3RUH
CF62 9TY MW7MOZ
CF63 1DW GW7AUQ
CF63 1FP MW0CWF
CF63 1FP MW1COY
CF63 1QJ GW7HQL
CF63 1RQ GW0NXW
CF63 2FE MW3AWA
CF63 3QD GW4ANK
CF63 4EF GW6ATT
CF63 4EF GW7VST
CF63 4JQ MW6JRI
CF63 4JY MW7DIZ
CF63 4NP MW6ZAN
CF63 4PQ GW0GRQ
CF64 1DD GW3VKL
CF64 1DD MW0DHF
CF64 1TN GW3MOV
CF64 1TR GW1LKG
CF64 1WW GW3LQE
CF64 2AJ 2W0VOG
CF64 2AJ MW7AFP
CF64 2JS MW3LZC
CF64 2PG GW0GHF
CF64 2QZ GW3WBU
CF64 2RX GW0PZZ
CF64 2RX MW3WXN
CF64 2SA GW0OTY
CF64 2TZ GW4UGI
CF64 3BE GW3UWL
CF64 3DD GW3SPA
CF64 3DH MW1EAA
CF64 3DH MW3RRW
CF64 3EN GW3FZV
CF64 3HF GW3VNZ
CF64 3HH GW1MBV
CF64 3HY MW0HAT
CF64 3JD GW4ILU
CF64 3PP MW3NHC
CF64 3QX 2W1YEG
CF64 3QX MW3YEG
CF64 3RB GW3PYX
CF64 4HA 2W0INY
CF64 4HA MW7RLT
CF64 4LR GW3SFQ
CF64 4PS GW1BAI
CF64 4PZ MW6WLB
CF64 5BW GW1DQV
CF64 5GA GW4EUA
CF64 5GA GW8JWL
CF64 5JW 2W0ICD
CF64 5QD MW0GBW
CF64 5QE MW6FTC
CF64 5SX MW3UDO
CF64 5TQ GW4BEQ
CF64 5UG GW0CKK
CF71 7EX GW4MOZ
CF71 7LE GW7KQN
CF71 7LX GW0ANA
CF71 7PB 2W1ETN
CF71 7RA GW7NJQ
CF71 7RQ MW7PAN
CF72 8BG 2W0YYP
CF72 8DA GW0DIV
CF72 8DA MW6HLG
CF72 8DY 2W0CGM
CF72 8EG 2W0ITH
CF72 8EG 2W0BMO
CF72 8EG MW0OMB
CF72 8EG MW3RXH
CF72 8HQ GW4SML
CF72 8HR GW4VMT
CF72 8HW MW6LSW
CF72 8SZ MW7MJT
CF72 9BP 2W0HPM
CF72 9DB GW4HDZ
CF72 9ER GW1DPV
CF72 9FZ GW6PFK
CF72 9HT MW7CUO
CF72 9HU MW6WCI
CF72 9JD 2W1ACM
CF72 9LT 2W0RTJ
CF72 9LT MW6TTS
CF72 9NR GW7VSF
CF72 9QQ MW6PLC
CF72 9SB MW0PVW
CF72 9SB MW6DYL
CF72 9SJ GW0VWD
CF72 9SL MW6WDX
CF72 9SZ MW6TES

CF72 9TF GW7VOO
CF72 9TX GW4DVB
CF72 9TX MW0MOX
CF8 4AU GW4ZBN
CF8 4DL GW1SRB
CF8 8NA 2W0IVZ
CF81 8JS GW0LBI
CF81 8JX 2W1EKR
CF81 8NB GW0WGW
CF81 8NS MW6WFF
CF81 8NT 2W0WMB
CF81 8NW MW1AFW
CF81 8PP 2W0EGK
CF81 8PP MW3EGZ
CF81 8QJ MW3PKC
CF81 8RN MW7AUX
CF81 8TG MW0TAF
CF81 8US MW3FFL
CF81 8UU 2W0HAK
CF81 8UX MW6GWO
CF81 9AB GW8GOC
CF81 9DZ MW3IWC
CF81 9LQ 2W1JOL
CF81 9PT MW3XNW
CF81 9RS MW6KAB
CF82 7DB 2W0UAA
CF82 7DB MW0MAU
CF82 7HF MW3ZTH
CF82 7JB MW0ATK
CF82 7JP MW0ICE
CF82 7JP MW3CCE
CF82 7NP GW4TPG
CF82 7QQ 2W1EUR
CF82 7RH GW0PCJ
CF82 7RZ GW6JSJ
CF82 8BR MW1FEU
CF82 8EG MW3NQH
CF82 8ET GW0TKX
CF82 8EW MW0IQZ
CF82 8EW MW6EOP
CF82 8EX MW6LVD
CF82 8FN MW0XMI
CF82 8LA GW4XMU
CF82 8LB MW7FRN
CF828EW 2W0KMU
CF828EW MW3KMU
CF83 1AN MW6CRU
CF83 1GA MW6MYE
CF83 1JW MW6EXV
CF83 1LA MW3ZTB
CF83 1NQ MW6FNT
CF83 1RF 2W0RDD
CF83 1RF MW0NNX
CF83 1RF MW3DQB
CF83 1RU MW6JBQ
CF83 1SP GW4HZM
CF83 1SQ GW6IPR
CF83 1SZ GW0IRC
CF83 2AN MW6NYE
CF83 2BE MW7ERP
CF83 2JL 2W0BUQ
CF83 2JL MW3ZWO
CF83 2JU MW0MEX
CF83 2LA 2W0PCE
CF83 2LA MW0WTK
CF83 2LL MW3USP
CF83 2NE 2W0CYY
CF83 2NE 2W0YBZ
CF83 2NE MW0YBZ
CF83 2NE MW6CVC
CF83 2NE MW6YBZ
CF83 2NJ 2W0RFT
CF83 2NJ MW3IQY
CF83 2NR GW4YKM
CF83 2NZ GW0MXG
CF83 2QX 2W0HZC
CF83 2QX MW7DPO
CF83 2RA 2W0VTP
CF83 2RA MW3HAQ
CF83 2RA MW6VTP
CF83 2RX GW8UAM
CF83 2SR MW6SEP
CF83 2TX GW0OSB
CF83 2TX GW1MAX
CF83 2UN MW3TJG
CF83 2UW GW4HDF
CF83 3BT 2W0FEY
CF83 3BT MW6PEY
CF83 3BU 2W1MRK
CF83 3BU GW6WTK
CF83 3BU MW7MRK
CF83 3FB 2W0MSL
CF83 3FB MW6NAG
CF83 3FT GW0KWO
CF83 3HF GW4VEB
CF83 3JH MW6HGZ
CF83 3QB MW6HFY
CF83 3RT GW6DOC
CF83 4BG 2W0JCV
CF83 4BG MW6JCV
CF83 4BN GW4SLI
CF83 4DD MW3WLB
CF83 4DH MW6ECR
CF83 4EQ GW0CKL
CF83 4GG GW0PUP
CF83 4HJ 2W0BGQ
CF83 4HJ MW3IUS
CF83 4HU MW6HPN
CF83 8EP MW6IAG
CF83 8FY GW0MTI
CF83 8GX MW7MME
CF83 8TA MW6PBF
CF83 8TT 2W0NQE
CF83 8TT MW0GEI
CF83 8TT MW3NQE

CF833PG MW7BYV
CH1 2NW G3XIR
CH1 3BZ M6VBJ
CH1 3HG G8WLD
CH1 3ND 2E0BSR
CH1 4AN G3SES
CH1 4BE 2E0IFW
CH1 4BE M0JXE
CH1 4BE M3JXE
CH1 5AF G6LRU
CH1 5AZ G0ULM
CH1 5DP 2E1CRA
CH1 5DZ M6BNK
CH1 5JQ G1URW
CH1 5NW G7VPQ
CH1 5PP G7JGS
CH1 5QX M1EQD
CH1 5RU M3LIN
CH1 5SW M6DOA
CH1 5SY M1DAP
CH1 5TX M7RTB
CH1 5XE M6GNG
CH1 6HQ G0JCG
CH1 6BF M0IVM
CH1 6JS G4SMM
CH1 6LU G4WXO
CH1 6QJ G0CGD
CH16BF 2E0DDR
CH16BF M0AUT
CH16BF M6UAR
CH2 1BB 2E0UPT
CH2 1BB M3UPT
CH2 1BS G6SJV
CH2 1DG G8GWX
CH2 1HF G6VGV
CH2 1HT G8BMH
CH2 1JG M0NUG
CH2 1LX G0NLQ
CH2 1LX G7HGB
CH2 1NF G3ZVH
CH2 1NF G7GFC
CH2 1NW M0AUA
CH2 1PF G0HEJ
CH2 1RA M7EOL
CH2 1RD G8RRS
CH2 2AH 2E0IIV
CH2 2AH M7RMZ
CH2 2AQ G3KJS
CH2 2BX G0OIV
CH2 2DH M3UZA
CH2 2DP M6LOB
CH2 2HL M3IHA
CH2 2LA G0ROY
CH2 2PB M6VHM
CH2 2PL G0NMD
CH2 2QF M7ESE
CH2 3EW 2E0GPE
CH2 3EW M0PEA
CH2 3EW M6GHP
CH2 3EW M6LVA
CH2 3HT M3OME
CH2 3JT M0CVP
CH2 3NL G0MTA
CH2 3RE G6SKP
CH2 4DQ M1RWB
CH2 4PR M7PMK
CH2 4QJ G7IDE
CH2 4RP M3HWV
CH3 5DY G1YNJ
CH3 5HA G1LML
CH3 5HD M3OAS
CH3 5HE 2E0WVD
CH3 5HE M3WVD
CH3 5HE M3WVF
CH3 5HE M3WVT
CH3 5HF M0PWS
CH3 5JB G7HEK
CH3 5JB M7NRT
CH3 5JF 2E0OBZ
CH3 5JF M3OBZ
CH3 5LA G8KKN
CH3 5LE G1XPW
CH3 5LU M3JAL
CH3 5LY G1CZU
CH3 5LY G7BQY
CH3 5PT G8ZRE
CH3 5RR G4JMF
CH3 5RR G8GIZ
CH3 5RR M3XSD
CH3 5UH G8XDY
CH3 5XN G1DIA
CH3 6BL 2E0HQY
CH3 6BL G7REV
CH3 6BL M7CTS
CH3 6HZ M7BDE
CH3 6NE G0VAH
CH3 6QX M3CZX
CH3 7EJ G8KWP
CH3 7EL G6FMS
CH3 7NT G0DUB
CH3 7NT G0RCW
CH3 7NT G4NRC
CH3 7NT M0LSA
CH3 7QD G6FDK
CH3 7QD M3DVA
CH3 7QQ 2E0IBA
CH3 7QQ M3OCS
CH3 8EQ G6VUX
CH3 8LP G7NEH
CH3 8LR G0DYL
CH3 8LR G6MYL
CH3 9BE G7EEN
CH4 0AH 2E0LUD
CH4 0AH M6PPZ
CH4 0AL M0CSO
CH4 0AQ G6FKB

CH4 0FT M6AQT
CH4 0FY M6FIB
Ch4 0jb M6OVT
CH4 0JD M6RDX
CH4 0LF M6WCF
CH4 0NH G0EDC
CH4 0NZ 2E0BTA
CH4 0NZ 2E0BTD
CH4 0NZ M3LXB
CH4 0NZ M3LXK
CH4 0PT G8VNN
CH4 0PU M3LBC
CH4 0QE M6AKF
CH4 0QT G6OKC
CH4 0RJ M3KXV
CH4 0RL G7TZX
CH4 0RN G1LDY
CH4 0SG G7HXF
CH4 0SQ G8ZBC
CH4 0SZ G0ISE
CH4 7LE G6XKJ
CH4 7LU G8VAD
CH4 7LZ G1LED
CH4 7NF M1BZI
CH4 7PQ G1NCD
CH4 8BB G7KLN
CH4 8BG M6NCF
CH4 8BY M3IUV
CH4 8DE M1FJP
CH4 8DH 2E0JWE
CH4 8DH M3YYW
CH4 8FB 2E0CKL
CH4 8FB M0UFA
CH4 8FB M6BHF
CH4 8JG G8SLP
CH4 8LB G4UXD
CH4 8PA G8DOF
CH4 8PJ 2E0SEY
CH4 8SJ 2E0TWI
CH4 8SS M3JUY
CH4 8TR M6OWO
CH4 9DG G4YNS
CH4 9NG G0CTH
CH4 9NG G1SVN
CH4 9NN G0UVT
CH4 9NN G0UWA
CH4 9NU G6YCW
CH41 0AX M1AAS
CH41 3LF 2E0LGZ
CH41 4FF G4NCI
CH41 5LD G7EED
CH41 8BZ M6THC
CH41 8HJ 2E0MTH
CH41 8HJ M5TT
CH41 8HJ M6FTT
Ch41 8jf M7DVO
CH416JA 2E0DEQ
CH42 0LB G8UWG
CH42 2BU G6RYM
CH42 2DN M6IEZ
CH42 3UN M7FRG
CH42 3XQ G6CHD
CH42 3YE 2E0OQH
CH42 3YE M3OQH
CH42 3YH 2E1SOX
CH42 3YH M3SHJ
CH42 4PD G3OTW
CH42 4QX M0BZZ
CH42 5LH M5HFJ
CH42 5LR M3ELV
CH42 6PX M7NRO
CH42 6QY 2E0YJY
CH42 6QY G4EWJ
CH42 6QY M6MAJ
CH42 6RG 2E0OOY
CH42 7HR M7AEF
CH42 7JA G4WVC
CH42 8JZ M0HGG
CH42 8QA G6JZE
CH42 9LJ G0KXL
CH42 9NZ G6RVH
CH42 9NZ M3RGJ
CH42 9PH G7CQZ
CH42 9QA 2E1PGB
CH42 9QA M3PGB
CH43 0TT G6HGI
CH43 0XB G1MHF
CH43 2 HE 2E1GMA
CH43 2HE G0PZO
CH43 2HS 2E0WED
CH43 2HS G6XHF
CH43 2HS M6CUL
CH43 2HS M6STI
CH43 2JL M3LXH
CH43 2LZ G0LZI
CH43 2NQ M1ARH
CH43 3AS 2E0GLR
CH43 3BB M3XCB
CH43 3BJ M7BXC
CH43 3DW 2E0ENZ
CH43 4UL G6VDX
CH43 4XR 2E0MFN
CH43 4XR M6MNK
CH43 5RP M0GNA
CH43 6TA G4ASM
CH43 6UT G0UCP
CH43 6UY G0MHF
CH43 7NX G4MUP
CH43 7SE G0WMY
CH43 7SQ G6PNB
CH43 7YQ G6EWK
CH43 8SU G8FPU
CH43 9YT M3LGX
CH43 9YT M3LGY
CH44 0AQ G1RCI
CH44 0BQ G4NXG

CH44 0BQ G6MSY
CH44 0BQ M0AJB
CH44 0BQ M3PSB
CH44 0BQ M3YJB
CH44 0EB 2E1GXL
CH44 0EB G0VAX
CH44 1AJ M6UHF
CH44 1AT M0AXJ
CH44 2EY G1BYI
CH44 3AF G1PMJ
CH44 3DJ M6AOQ
CH44 3DZ G8NNS
CH44 3DZ G8TFW
CH44 3ED G0JQK
CH44 4DL G0DPO
CH44 5RQ G4ZCA
CH44 5RQ G7KOS
CH44 5UH 2E0AAN
Ch44 5uh G8AA
CH44 5XB 2E0DHQ
CH44 5XB M6EKW
CH44 6PZ G7VQM
CH44 6QA M1MPB
CH44 7BJ 2E1TAB
CH44 7BJ M3ATB
CH44 8AB G7OKR
CH44 8AZ G4LWC
CH44 9BR G4PGO
CH44 9DN 2E0WAE
CH44 9DN M6PUG
CH44 9DX G4KVP
CH44 9DZ G1NXB
CH45 0LG M6AAC
CH45 1HA G0NBD
CH45 1HD G6ABO
CH45 1JG G0WMQ
CH45 1JT G1JEZ
CH45 2LZ G4KRF
CH45 2NG M0NBJ
CH45 2NN M6EXS
CH45 2PE G8STF
CH45 3HA G6RUU
CH45 3HJ G0GQW
CH45 3LU M6JLQ
CH45 3LU M6JLU
CH45 4PN M6BTM
CH45 5AY G4ENK
CH45 5EJ 2E0PSO
CH45 5HB G0AWW
CH45 5HN G1FEP
CH45 6TD G3SEJ
CH45 6TD G8STD
CH45 6TE G3XCD
CH45 6TF M0ATZ
CH45 7LE M7ERI
CH45 7LQ 2E0PJD
CH45 7LQ M3IRU
CH45 7NF M7RST
CH45 7NZ G1PDA
CH45 7NZ G3VEB
CH45 7QF G8TRY
CH45 7QF G8WDC
CH45 7QG G4VQS
CH45 7QQ G4KOY
CH45 8JN M3DBX
CH45 8LH M6SIP
CH45 8LR G0PIT
CH45 8LR M6PMA
CH45 8NX 2E0SYA
CH45 8NX G8JUV
CH45 8NX G8UUC
CH45 8NX M7SYW
CH45 9JA G1EYJ
CH45 9LJ G6IIN
CH45 9LW G6IHD
CH46 0QT G3PEZ
CH46 0RS M6DXZ
CH46 0SN G6UFI
CH46 0TN G7NOS
CH46 0TP 2E0RAA
CH46 0TU G3LCI
CH46 1PU M7ONR
CH46 1QH G0PTE
CH46 1QZ G7NHR
CH46 1RQ M3PHU
CH46 1RW 2E0NLP
CH46 1RW M6ATS
CH46 1RX M3HQP
CH46 1RX M3PAP
CH46 2QR G6HBF
CH46 2QY M6LKK
CH46 2QZ G0WUO
CH46 2SA 2E0WBL
CH46 2SA M6MEQ
CH46 2SB G0PVY
CH46 3RR M6BXW
CH46 3RX 2E0AZK
CH46 3SG 2E1HOU
CH46 5NE M6GAW
CH46 6AX G3YHB
CH46 6BJ G6USU
CH46 6DH G4WUA
CH46 6DH M0MTC
CH46 6DS M6TGL
CH46 6EL 2E0DLJ
CH46 6EN 2E0BDO
CH46 6EN M3KZW
CH46 6ET M1ELR
CH46 6FL G4VWL
CH46 6HB G4UJP
CH46 6HB G6WGG
CH46 6HE 2E0MTX
CH46 6HE M0WBG
CH46 6HE M6MBP
CH46 6HQ G3XJZ
CH46 7SJ G0CLV
CH46 7TY M7DBG

CH46 7UP G7FND
CH46 7UT G0DNQ
CH46 8TT M0SPD
CH46 8TX G0EQE
CH46 9PF G0LHN
CH46 9PS 2E1RSB
CH46 9RW M6GZI
CH46 9RY M3VYK
CH47 0LB G3OKA
CH47 0LQ G4ELA
CH47 2AE 2E0LGH
CH47 2AE M3LGH
CH47 2AG 2E0YOZ
CH47 2AG M6HOY
CH47 3AG M6RKQ
CH47 3AJ M3FLL
CH47 3BZ G4FPB
CH47 3DD G0KQY
CH47 3DD G8THZ
CH47 5AW G3VDS
CH47 7DA G1ZZA
CH47 9RW 2E0CDM
CH47 9RW M3VZC
CH47 9RZ M6GWU
CH48 3HE M0BFV
CH48 4EL 2E0ECJ
CH48 4EL M6HFM
CH48 5DG 2E0ILO
CH48 5DG M0ILO
CH48 5DG M6ILO
CH48 5EG M6HJM
CH48 5HQ G8ODK
CH48 5HZ M6PLE
CH48 6DA G0VAX
CH48 6DJ G1UHO
CH48 6ET M3IYO
CH48 7EU M0SJK
CH48 7EX G3OVE
CH48 7HR G8LEM
CH48 9UF G7NIR
CH48 9UL G0SSG
CH48 9UX G0GSF
CH48 9XP 2E0ETD
CH48 9XP M6ETD
CH48 9XS M5AOI
CH48 9XU 2E0PCU
CH48 9XU M6PFG
CH486EJ M1AOR
ch48nz M6NSI
CH49 0TD G4BKF
CH49 1RT 2E0TFM
CH49 1RT M6KDZ
CH49 1SS G6PRE
CH49 2RQ G6KBQ
CH49 2RZ G8OBT
CH49 3AG G3YSM
CH49 3AW G3MAJ
CH49 3NG G0MQJ
CH49 3QH G0LPU
CH49 4GD G8WWD
CH49 4GS G8ZJH
CH49 4GY G3VUY
CH49 4QP G0TYN
CH49 4QP M3LDH
CH49 6JD G7MIZ
CH49 6LA G3ZIC
CH49 6PN 2E1JEF
CH49 6PW G8CVF
CH49 7LH M3TDM
CH49 7NJ 2E0BJW
CH49 7NT M6DYG
CH49 8EE G7GRU
CH49 8JP 2E0SQK
CH49 8JP M6HSB
CH49 9AW G6WCW
CH494NX 2E0ARC
CH494NX M7RWE
CH5 1EH GW7UVO
CH5 1EZ GW7GKX
CH5 1HH GW4KDI
CH5 1NU GW1ATZ
CH5 1NU GW3ATZ
CH5 1SA MW7DUI
CH5 1XE MW0SZR
CH5 1XU MW3JIJ
CH5 2BR MW3RPX
CH5 2DP GW7SBJ
CH5 2FX GW4MVA
CH5 3AL GW0PBJ
CH5 3BF GW8UEK
CH5 3BF MW3UEK
CH5 3EX GW0TCL
CH5 3HS GW4EYO
CH5 3JG GW4SII
CH5 3RW GW0DRS
CH5 3RW GW0HRG
CH5 4AL GW1RVC
CH5 4AU MW3XZV
CH5 4BY MW6PAI
CH5 4GG GW6POO
CH5 4HN 2W0DPI
CH5 4HN MW6AFK
CH5 4HW GW4ZAG
CH5 4JP GW0DSP
CH5 4JP GW1LFX
CH5 4LG GW3KNZ
CH5 4LG MW3JLK
CH5 4LG MW3MRK
CH5 4PA GW1LEL
CH5 4QA MW3FLI
CH5 4RE GW4RQS
CH5 4RE GW4VBM
CH5 4SH GW6NVJ
CH5 4SN GW7AAU
CH5 4SN GW7AAV
CH5 4TF MW3SUF

CH5 4TN GW0UEO
CH5 4WP GW0SXE
CH52HX GW1GBH
CH6 5AN GW0UGQ
CH6 5DW MW7SUS
CH6 5HU GW0UIP
CH6 5HU MW0GPP
CH6 5JP GW6LMI
CH6 5JU MW1CRE
CH6 5LN MW1LSG
CH6 5LP 2W0CCG
CH6 5LT MW1RES
CH6 5QA GW4ZAR
CH6 5QB GW4VFE
CH6 5RQ MW0GRJ
CH6 5RX GW4RYJ
CH6 5SB GW0GZR
CH6 5SB GW7FZW
CH6 5YE GW7CMM
CH6 6BX MW7BYW
CH6 6JE GW0POG
CH6 6JS 2W1RSS
CH6 6JS MW0RKD
CH6 6LS GW7CCR
CH60 0BP G3WIO
CH60 1UH G6LKG
CH60 1XW M6NWV
CH60 1YD G6AGP
CH60 2SG G0MQR
CH60 2TT G7SFM
CH60 2TT M0CKV
CH60 3RW 2E0SSZ
CH60 3RW M7SAZ
CH60 5RY G4EIC
CH60 5RZ M3YKZ
CH60 6RD M3REX
CH60 6TE G7FVR
CH60 7RA G3UVR
CH60 7RA G4MGR
CH60 7RF M7OMW
CH60 9JT G3RLA
CH601UL M6MTQ
CH61 0HD G4QBF
CH61 3UY 2E0EFC
CH61 3UY M6ZKA
CH61 3XH G7AAI
CH61 3XS G0WAB
CH61 4UA G1RIX
CH61 5UG G6KIV
CH61 6UP G1KJX
CH61 6UZ G4NOY
CH61 6XT G8DTS
CH61 6YH G8HLJ
CH61 7UL G7IIF
CH61 8SU G4ILA
CH61 8SX G8REQ
CH61 9NT M6YIN
CH61 9PR M0CNN
CH61 9PY 2E0PYR
CH61 9PY M6PYR
CH61 9QA 2E0XGA
CH61 9QA M3XGA
Ch611dl 2E0IKS
Ch611dl 2E0KGA
Ch611dl M7KGA
CH62 0BD G3ODC
CH62 0BN M7FDC
CH62 1BJ M0BQH
CH62 1DR M3IRJ
CH62 1DY M3YWJ
CH62 3LH M0HVS
CH62 5DB M3TFI
CH62 5DB M3TXU
CH62 6AN G6HKY
CH62 6AW G7IGU
CH62 6BE G6PYL
CH62 6BE M3FMV
CH62 6BE M3ODC
CH62 6BR G8WHR
CH62 6BR G8WSR
CH62 6BT G0GBN
CH62 6DT G8IWJ
CH62 6DX G8LAU
CH62 6EA M0WAD
CH62 6ef M6RTR
CH62 7BA 2E0HLP
CH62 7BA M3HLP
CH62 7DJ M6KRQ
CH62 7EE 2E0DUP
CH62 7EE M0VWD
CH62 7EE M6VAH
CH62 7EX G3NWR
CH62 7EX G4YWD
CH62 7JY G3NNV
CH62 8BB G6AVY
CH62 8BB M0CMI
CH62 8BN M6GHM
CH62 8BS G3YGL
CH62 8BS G0VEO
CH62 8DG G1WAS
CH62 8DL G4UCE
CH62 8EB G1NPN
CH62 8EQ G3TRR
CH62 8EQ G4XKL
CH62 9AA G3AVL
CH62 9AZ 2E0ICE
CH62 9DF G0GFG
ch629ea M6JQK
CH63 0JA G6TPE
CH63 0JJ G8TSG
CH63 0JW G7NHD
CH63 0LA G8EYA
CH63 0QF M0OFB
CH63 1JG M6NWW
CH63 1JS G7TIY
CH63 2JU G6NOI

CH63 2LW M6CXH
CH63 3BD 2E0HZE
CH63 3BD M0KSX
CH63 3BD M7KWS
CH63 5JR G4NSA
CH63 5JR M6HAK
CH63 5PA G0OXA
CH63 6HN M3PRU
CH63 7LR 2E0OWL
CH63 7QU M6XBJ
CH63 8LP G5YSS
CH63 8LU M7TGD
CH63 9AH G6BWO
CH63 9FW G8DRB
CH63 9LP G0PXO
CH63 9LU M6LSP
CH63 9NG M0HWO
CH63 9YJ M1BZZ
CH64 0TL M7PKA
CH64 0UA M7TKR
CH64 0UY G8VSI
CH64 0UZ G4EEQ
CH64 0XP G1NUH
CH64 1TN G4JZR
CH64 2XG M0PER
CH64 2XQ G8OKZ
CH64 3RS M6PTH
CH64 3TJ 2E0IGQ
CH64 4AN G6HQ
CH64 4AT G4HGL
CH64 4BJ G3NTI
CH64 5TJ 2E0CPB
CH64 5TJ M3YPB
CH64 6QB G0KQS
CH64 6RB 2E0MLD
CH64 6RB G6PTL
CH64 6SF G4ELK
CH64 7TR G6ADO
CH64 7TR G8MMM
CH64 7TR M3JLI
CH64 7TT M7YMC
CH64 9SW G7DEY
CH65 2BD 2E0GWI
CH65 2BE G1AOF
CH65 2BE M0COC
CH65 4BB G0OJC
CH65 5AD M1SUM
CH65 5AT M7PLD
CH65 5BW M6JWI
CH65 5DG M6FEK
CH65 5DG M6RBF
CH65 5DX G0HFE
CH65 6PB G7JKH
CH65 6RW 2E0RFF
CH65 6RW G0RGG
CH65 6RW M6EBC
CH65 6SB 2E0BJX
CH65 6SB G3YCO
CH65 6SB M0GPG
CH65 6SB M3UQJ
CH65 6SE G4JTK
CH65 6TD G6ALN
CH65 7AZ M3LCL
CH65 7BW G0WVW
CH65 7DX M6BLV
CH65 7DZ 2E1LES
CH65 8EP 2E0IHO
CH65 8EP M3IHO
CH65 8HN M7COR
CH65 9DZ G0PJY
CH65 9EN G0CTR
CH65 9EY G8RVY
CH65 9JU M7MGH
CH66 1JF G1PRL
CH66 1JG M3UHB
CH66 1JJ G3VZM
CH66 1JW G0JZJ
CH66 1JX M3AIS
CH66 1NY M3UQZ
CH66 1QT G5MUN
CH66 1RL M0AMS
CH66 1RW G0AKU
CH66 1TY G4STZ
CH66 2BG M0BAU
CH66 2BJ G4DBG
CH66 2BL G4YBJ
CH66 2GU G8XPB
CH66 2GX G6IIM
CH66 2LH G1GWS
CH66 2SX G4ZKG
CH66 2TD 2E0FWZ
CH66 2US G0IEQ
CH66 2WL M3LYU
CH66 2YJ G3CSA
CH66 2YJ G4WSE
CH66 3LL G0COJ
CH66 3NN M6TXH
CH66 3PF G7GQW
CH66 3PH G4UXL
CH66 3QU M7DEA
CH66 3SQ G3TXH
CH66 4AJ M6LAF
CH66 4NA G7NHV
CH66 4PD G0RCY
CH66 4QY M1ZXZ
CH66 4SG G5CH
CH66 4SG G8IPT
CH66 4TP M7XSE
CH66 4UF M6HCJ
CH66 5NB M3YQH
CH66 5PB M1AOX
CH66DP G6WGY
CH7 1HH 2W0GDA
CH7 1HH MW0WML
CH7 1HH MW6AQU

CH7 1LD GW0TCV
CH7 1SU GW7MVG
CH7 1TH MW3RAA
CH7 1UG MW7GLF
CH7 2AG GW3UOO
CH7 2AG GW4HER
CH7 2AX GW7MHB
CH7 2BS 2W1HUX
CH7 2BS GW3RIB
CH7 2EQ GW0EGQ
CH7 2HQ GW7AIY
CH7 2JA GW0GXQ
CH7 2JL GW0FHL
CH7 2JS MW6BPH
CH7 2LJ 2W0NAD
CH7 2LJ MW0JRX
CH7 2PA GW3HDF
CH7 2PB 2W0IBM
CH7 2QP MW0HLW
CH7 2QX MW6KPV
CH7 3BL GW3UVA
CH7 3BR GW7BOY
CH7 3DU MW0IEH
CH7 3EL GW1OTI
CH7 3ER 2W0VMC
CH7 3ER MW3NQK
CH7 3ET GW4GTE
CH7 3HN GW4HBZ
CH7 3JU MW6BNH
CH7 3LH 2W0RSV
CH7 3LH MW0RSV
CH7 3LH MW6YZF
CH7 3NB MW0RHT
CH7 3NG GW0PJA
CH7 3QA GW0OHJ
CH7 3QF MW6FEF
CH7 4NE GW0MDQ
CH7 4NE GW3YHR
CH7 4NS MW3JNX
CH7 4PY GW4GDM
CH7 4QD GW1PKW
CH7 4SB GW7APM
CH7 4SN GW6UTF
CH7 4SS GW3TMP
CH7 4TU GW4SDO
CH7 5AE GW1FWE
CH7 5DZ GW8EQI
CH7 5EL GW7VFJ
CH7 5EU MW7WRG
CH7 5HY MW0ISC
CH7 5NU GW0LNO
CH7 5RF GW8NZN
CH7 5UY MW0JFX
CH7 6DU GW6RNV
CH7 6ED GW7NFM
CH7 6EE 2W0CCK
CH7 6EE MW0GYF
CH7 6EE MW0TTK
CH7 6EE MW3MTB
CH7 6EF GW4EVX
CH7 6GE MW7GVR
CH7 6HT MW6ROX
CH7 6JF GW3FPH
CH7 6PN MW6VDE
CH7 6RT 2W0VAC
CH7 6RT MW0ZAQ
CH7 6RT MW3YCR
CH7 6RW GW8ICT
CH7 6SL GW3IVK
CH7 6SW GW4BTW
CH7 6TD GW4NEI
CH7 6TF GW0PZU
CH7 6TR MW0CVW
CH7 6TU MW3KML
CH7 6US GW6FKP
CH7 6US MW7PAJ
CH7 6UZ GW3LWU
CH7 6WD MW0RPE
CH7 6XZ MW6OXV
CH7 6YP GW3KJN
CH7 6YP MW7BJQ
CH7 6YU GW4GSG
Ch72gz MW7CBF
CH72LN MW7ASM
CH75BE MW0HCC
CH8 7AP MW6KOI
CH8 7AU MW3UDA
CH8 7DF GW0PFZ
CH8 7DR GW0HKQ
CH8 7EQ 2W0EUY
CH8 7PJ GW1CGD
CH8 7QR GW4MOK
CH8 7SJ GW4TIZ
CH8 7UG GW6WQJ
CH8 7UJ GW6VZB
CH8 7UR GW7KIS
CH8 7XG GW4VUH
CH8 8AX GW0UDJ
CH8 8BU MW6IOA
CH8 8DL GW4JUN
CH8 8DU MW3PZC
CH8 8ES GW0FEU
CH8 8HA MW0CIH
CH8 8HE GW1PJL
CH8 8JY 2W0GXI
CH8 8JY MW0XRT
CH8 8JY MW6GXI
CH8 8NG MW3CBX
CH8 9BZ MW0LCK
CH8 9HJ MW0ATT
CH8 9HY GW6RCX
CH8 9JB MW1ARM
CH8 9JB MW3ARM
CH8 9JN 2W0FXF
CH8 9JN MW7PJH
CH8 9JQ GW7KBI

CH8 9NZ MW3NGN
CH8 9NZ MW3NGP
CH8 9PQ MW3LBX
CH88ES MW0CRA
CM0 7AH 2E1AEU
CM0 7AL 2E0XDM
CM0 7AL M6UDC
CM0 7BA M3TCR
CM0 7HJ 2E0CUS
CM0 7NF G4JDH
CM0 7QE G6VQV
CM0 7QT G7JDR
CM0 7QT M0RBI
CM0 7RD 2E1EDV
CM0 7RD 2E1EDW
CM0 7RY M6EOV
CM0 7SQ 2E0EYG
CM0 7SQ M6LZI
CM0 8BT G1IGN
CM0 8DP G7RGR
CM0 8EH M0CZC
CM0 8EJ M7LRY
CM0 8EY G4POP
CM0 8EY G6CNQ
CM0 8HS G8CCL
CM0 8HS M3CCY
CM0 8PZ G6EUW
CM0 8QL M7WPH
CM0 8RB G1LUX
cm08pj M7RLA
CM1 1DL 2E0NKR
CM1 1DL M6MYR
CM1 1DL M6TQW
CM1 1QB G7KOI
CM1 2AR G8UVY
CM1 2AT 2E0TXL
CM1 2AT M0TXL
CM1 2AT M6FIL
CM1 2BW 2E1GTQ
CM1 2DE M7SNR
CM1 2DZ M1EZE
CM1 2JA G3XRC
CM1 2JA G4YTG
CM1 2LA M6IPI
CM1 2NH 2E1HRB
CM1 2NJ G7RFT
CM1 2NQ G0NVM
CM1 2PR G4BNE
CM1 2PT G3SLT
CM1 2QZ 2E0CBX
CM1 2QZ M6MAO
CM1 2RR G0NAX
CM1 2SE G7GBZ
CM1 2SS G3NWG
CM1 2SX M6WTF
CM1 2SY 2E0FSX
CM1 2SY M6FNR
CM1 2TE M7SBY
CM1 3EA G4IMS
CM1 3GW M3ZJD
CM1 3HS M6PVM
CM1 3JB M0HBY
CM1 3JZ G4RAP
CM1 3NU G8GWB
CM1 4DA M3DIM
CM1 4DG 2E0JCS
CM1 4DG G1CSR
CM1 4DG G3CSR
CM1 4DG G8GFF
CM1 4DG M3AAY
CM1 4DG M3JCS
CM1 4DN G6ZVV
CM1 4DU G4TRF
CM1 4EF G3VCG
CM1 4EF M7CPH
CM1 4EJ G8AEU
CM1 4HJ G1NZD
CM1 4JQ 2E0KFM
CM1 4JQ M0KVR
CM1 4JQ M3GTV
CM1 4JX 2E0ONH
CM1 4JX M0ONH
CM1 4JX M6ONH
CM1 4JY G4FKH
CM1 4JY G4TOO
CM1 4NW M0ROO
CM1 4PG M0CSV
CM1 4SJ G6ZWC
CM1 4UN G4HSK
CM1 4UQ G6DJS
CM1 4XU G8SGM
CM1 4XZ M0BQC
CM1 4YG M3YNM
CM1 4YR M6HIJ
CM1 6AU M7MGG
CM1 6BF 2E0KIL
cm1 6bf M3HNK
CM1 6BU G0URK
CM1 6FD G0LLB
CM1 6HY G4JJH
CM1 6JF 2E0RMI
CM1 6JF M3RMI
CM1 6JX G4DJC
CM1 6PJ M7ATD
CM1 6QR G3VSI
CM1 6QT M6VGE
CM1 6TX G7RCL
CM1 6UH 2E0VSE
CM1 6UH M6VSE
CM1 6UJ M3NOJ
CM1 6UJ M6ERR
CM1 6UL M7IJB
CM1 6UW G8AAE
CM1 6UX G1EFL
CM1 6UX M0VAM

CM1 6UX M3VAM
CM1 6XY 2E0HDW
CM1 6YP G8PFL
CM1 7AN G8CDB
CM1 7BU M6NOM
CM1 7PG G8ETN
CM1 7RD M3GUO
CM1 7RX 2E0BWU
CM1 7RX M3OJN
CM1 7RY G6GOV
CM11 1AU G4EVZ
CM11 1ET G0ORT
CM11 1ET G4HTG
CM11 1HB G0NGD
CM11 1LH G0PXA
CM11 1QN G8CCN
CM11 1RR M6GSQ
CM11 2LL G1DHB
CM11 2NX G0TST
CM11 2NZ M0BZC
CM11 2PD G8URI
CM11 2QN G3UTC
CM11 2QN G4SIS
CM11 2RQ G8HGN
CM11 2RY G6KVY
CM11 2TS M6ISG
CM11 2XA G3MVV
CM11 2XE G4XTS
CM11 2YX M6EZP
CM12 0HH G3TTB
CM12 0JF G7PMI
CM12 0PF G8GZV
CM12 0PY M3MQB
CM12 0SU G1UKA
CM12 0SU M3CAD
CM12 0UD M6FVR
CM12 0YE M0DKS
CM12 9DS M0HSX
CM12 9JN G1DSG
CM12 9LX G8DQN
CM12 9NT M1EMU
CM12 9SN G5PMJ
CM129LH M3VUC
CM12BA M7AYS
CM13 1BT G3JWI
CM13 1BU 2E0EKC
CM13 1BU M0IPE
CM13 1BU M6IPE
CM13 1JD M6WSR
CM13 1JX 2E0JRT
CM13 1JX M6JRT
CM13 1JZ M0TIU
CM13 1LW G6YUY
CM13 1RH G0UBL
CM13 1SJ G0PNO
CM13 2AG G7RCW
CM13 2HZ G8NYD
CM13 2HZ M6PLZ
CM13 2LA G7HRL
CM13 2NF G8CUB
CM13 2RY M7TFD
CM13 2RY M7TFW
CM13 2UF G4WYI
CM13 3AL G0FMU
CM13 3DD M0OGS
CM13 3DZ G0LKY
CM13 3RJ M6DII
CM13 3TR 2E0RKB
CM14 4HE M0TKU
CM14 4UU G7HFL
CM14 5AZ G8APZ
CM14 5AZ G8PUB
CM14 5DB M0BAW
CM14 5DG G4XFZ
CM14 5FR M7CZQ
CM14 5JR 2E0DTD
CM14 5JR M6TPD
CM14 5JZ M7RKC
CM14 5NS M0GNC
CM14 5UE G6PZF
CM14 5WT M6HHM
cm145ue M6IIQ
CM14DU G1IBP
CM15 0BQ G4SYR
CM15 0LX G6TNJ
CM15 0NB G8DXV
CM15 0NH M7BHQ
CM15 0PP 2E0XDC
CM15 0PP M0XDC
CM15 0PP M3NUU
CM15 0PY G4ZON
CM15 0QX G8PVM
CM15 8BW G3ZRH
CM15 8JL G4NGS
CM15 8PF G4YVW
CM15 8SA G8DWP
CM15 9ND M0MPY
CM15 9ND M0ZNP
CM15 9QA M3NRW
CM15 9QS G4BLD
CM15 9SG 2E0KWG
CM16 5AT M6TMP
CM16 5DL G4LGY
CM16 6BP M3WEA
CM16 6EF M5AGV
CM16 6ES M6SRE
CM16 6EW G0PYV
CM16 6FH M6OBH
CM16 6HA G6OIX
CM16 6JA G7CWX
CM16 6JQ M3EYY
CM16 6JQ M3JXX
CM16 6LF G4WWY
CM16 6PJ M3WYT
CM16 6RR M6ARR
CM16 7BB G1WXS

CM16 7ET G8DZH
CM16 7HT M3GKJ
CM16 7JX G4ACL
CM17 0EX M3MER
CM17 0HD G3UUY
CM17 0HN G8CUA
CM17 0HQ G8MHO
CM17 0JY G0BXL
CM17 0LH G3UEG
CM17 0LQ G0CPU
CM17 0QR G8DNP
CM17 0SB G7ULS
CM17 9AU G6DMF
CM17 9AU M6DMF
CM17 9AU M6WKS
CM17 9EU M0OAB
CM17 9EU M6OAB
CM17 9HL M6PVZ
CM17 9JA M6JKA
CM17 9NL 2E1HJW
CM17 9NS M6WII
CM17 9NT M7TKI
CM17 9PA M0DXR
CM17 9PL G0AWY
CM17 9PL G6MZV
CM17 9PR G4XUQ
CM17 9PZ M6PTV
CM17 9PZ M6ZRO
CM17 9QG G0KNJ
CM17 9RG M7PEN
CM18 6BL 2E0SIJ
CM18 6BL G6BUT
CM18 6BL G7OBS
CM18 6BL M3OBS
CM18 6BL M6SIJ
CM18 6EB G7IZW
CM18 6ES G0HDZ
CM18 6EY G3RYK
CM18 6HD M6OYE
CM18 6HN M6BUY
CM18 6QE M0JPG
CM18 6QE M3ZBX
CM18 6QN M6GQW
CM18 6QN G8VQE
CM18 6QT M6FKI
CM18 6QW M7GBP
CM18 6QY 2E0AUU
CM18 6ST G8POE
CM18 6TG M6HPX
CM18 6XL 2E0TCI
CM18 6XL M6EJA
CM18 6XX M7AGQ
CM18 7BL M7AGO
CM18 7BY G3WRO
CM18 7BY M3ZSL
CM18 7EE M7WIX
CM18 7HD M6SKZ
CM18 7HF M7SCG
CM18 7PG G4JQU
CM18 7QD G0HRR
CM18 7QD G1FRL
CM18 7QZ 2E0VAO
CM18 7RE 2E0YND
CM18 7RE M3YND
CM18 7RF G1OSI
CM18 7RJ 2E0VLT
CM18 7RJ M0PHO
CM18 7RJ M3VLT
CM18 7RJ M7APV
cm18 7sa M7AKA
CM18 7SY G1FIM
CM19 4DB M3GOV
CM19 4DQ M6BUE
CM19 4HE M6FXL
CM19 4NJ G1SFU
CM19 4NW G3GWE
CM19 4QD M6JZH
CM19 4RD G7RFZ
CM19 4RR G0AGO
CM19 5LJ G0MGU
CM19 5NU G6GXO
CM19 5NZ G7AZA
CM19 5RN G0REF
CM19 5SA G6TPO
CM2 0AS M7MKJ
CM2 0ES G1JIJ
CM2 0EY M6SQK
CM2 0FJ G8NWL
CM2 0RX 2E0DNU
CM2 0RX M6DNV
CM2 0SA G8MYJ
CM2 0TY G0MWT
CM2 0TY G1FCW
CM2 0TY G4PVM
CM2 0TY G8CW
CM2 0TY M5BAE
CM2 0UA G6BZQ
CM2 6AP 2E1GQB
CM2 6AQ G4MDB
CM2 6AQ M3HPY
CM2 6AQ M3ZAN
CM2 6AZ G8DET
CM2 6BA G7RVY
CM2 6DH G6CNK
CM2 6DJ G1AJQ
CM2 6DN M0MVO
CM2 6DX M1SNM
CM2 6EB G1INA
CM2 6EW 2E0SES
CM2 6EW M0OER
CM2 6EW M6LFT
CM2 6HA G4JDS
CM2 6QE G6WFM
CM2 6QL G4KGL
CM2 6QN G4LAE
CM2 6QW G8UDD
CM2 6TN 2E0NNZ
CM2 6UH G4LQE
CM2 6XN M3LOT
CM2 7AU G6JYB
CM2 7AX M3IWO
CM2 7DD G4OLU
CM2 7DS M7ECF
CM2 7JL G1NLZ
CM2 7JN G3PMW
CM2 7LP G0KOM
CM2 7LP G2AD
CM2 7LT G8BOS
CM2 7PP 2E0IWT
CM2 7PP M3IWT
CM2 7SF G3JOX
CM2 7TD G3FMO
CM2 8AH 2E0WHB
CM2 8AH M6UHN
CM2 8AL 2E0LZI
CM2 8AL M0LZI
CM2 8AL M6DNR
CM2 8HN G1GKN
CM2 8HY G7KLV
CM2 8JD M7JMX
CM2 8NF G0RLI
CM2 8NT G3YDY
CM2 8NT G8BGV
CM2 8NY G0BDS
CM2 8PD G8GNZ
CM2 8PT G0KOU
CM2 8QB M6IPA
CM2 8QP G3WGE
CM2 8RH M7AWG
CM2 8RQ M6ERF
CM2 8RR M3OQI
CM2 8UR 2E0YYA
CM2 8UR M0MYM
CM2 8XJ G3VUO
CM2 8YY G6FJE
CM2 8YY G6ZNJ
CM2 9AG M6TTR
CM2 9BD 2E1RMD
CM2 9BD M3OGM
CM2 9BZ G8VU
CM2 9DD G3PEM
CM2 9DX G1MVI
CM2 9DZ G3UCD
CM2 9DZ G4BMW
CM2 9EZ M6HNA
CM2 9GJ 2E1GHE
CM2 9HA G0POK
CM2 9LF G1NNN
CM2 9LN 2E0WDY
CM2 9LN M6WDY
CM2 9LW G0SQP
CM2 9NY 2E1IDH
CM2 9NY M3IDH
CM2 9PW G0SXK
CM2 9RD M6SPS
CM2 9RE G3PBT
CM2 9SN G0NEM
CM2 9SN G8VYK
CM2 9XQ G6ZGI
CM20 1SW M0SLC
CM20 2AD 2E0VNT
CM20 2AD M6VNT
CM20 2JG M0TJB
CM20 2LA 2E0OXT
CM20 2LA M6OXT
CM20 2PA M3WMV
CM20 2PZ G4GGX
CM20 2TJ M6PFV
CM20 3BW M3APA
CM20 3DP G8CHO
CM20 3DW 2E0JOP
CM20 3DW M3ZLJ
CM20 3DY G4MQG
CM20 3EF G6UXX
CM20 3EY G7UFF
CM20 3EY M0NRY
CM20 3EY M3LEN
CM20 3HB G4YBN
CM20 3HL M6FQX
Cm20 3jp M0XID
CM20 3LE G8XGO
CM20 3LY 2E0MBD
CM20 3LY M3NBD
CM20 3NH M7DXF
CM20 3RH M6YGL
CM20 3RU G4ADG
CM20 3SZ M7YNC
CM202LA 2E0OGA
CM202LA M6OGA
CM21 0AU G4PGB
CM21 0BT G8WUG
CM21 0DZ G1XRT
CM21 0ER G4SNL
CM21 0HX G4ERS
CM21 9ED G7NBQ
CM21 9HG G8FUL
CM21 9NF G0UEU
CM21 9NF G4DMI
CM21 9NN G3NWX
CM21 9NT G3WOO
CM22 6BG M0KIB
CM22 6LG G6IWT
CM22 6NT M3HKV
CM22 6QX 2E1UKT
CM22 6QX M6UKT
CM22 6SW G4PFJ
CM22 7AJ G7VAB
CM22 7AS G8BPH
CM22 7FF M6DJQ
CM22 7LT G4PJD
CM22 7NB G6ART
CM22 7PA M1SGA
CM22 7RT M6TTF
CM22 7SJ G4WWP
CM22 7SQ G7VRX
CM22 7TP G7VJE
Cm226ut 2E0WWK
Cm226ut M6BLB
CM23 1BD M0DXZ
CM23 1HX G6ECS
CM23 2BL G3WYD
CM23 2BQ G0CGH
CM23 2HU M0GCA
CM23 2HX G4CWH
CM23 3DH M7SPC
CM23 3EU G7CQX
CM23 3EX G3KCG
CM23 3EX M0AQO
CM23 3JN G7TET
CM23 3LR M6IFW
CM23 3NG M0AZR
CM23 3NP G4SUA
CM23 3NQ G4WPO
CM23 3PU 2E1WGB
CM23 3PU M5REG
CM23 3PX G1NRK
CM23 3PX G1OSH
CM23 3QH M3XBC
CM23 3QN M6BUA
CM23 3RJ G3PRI
CM23 3YW 2E0WAG
CM23 4AD G0PQF
CM23 4AD G1ARU
CM23 4AD G5ZG
CM23 4BL G4SNJ
CM23 4DU G0PLS
CM23 4EB M7DJK
CM23 4EY G0GZU
CM23 4HU G0XAD
CM23 4JL G3SUS
CM23 4JP G4OBV
CM23 4LL G6LIB
CM23 4LL G8THH
CM23 4PD G6YMU
CM23 5AP G3ZXF
CM23 5DE G8ZPH
CM23 5DR M7CHC
CM23 5DS G4HXH
CM23 5HN M0ITY
CM23 5HY G4GIS
CM23 5JJ G7WJV
CM23 5LS G4LSE
CM23 5NU G8SEY
CM23 5QH G0ODE
CM24 8ED M7COX
CM24 8HX 2E0EAL
CM24 8HX M0GZM
CM24 8HX M6JRO
CM24 8LR G3PWR
CM3 1AS G6ANO
CM3 1DA G1UZC
CM3 1DA M1BWN
CM3 1ND 2E0GZD
CM3 1NH M7SWX
CM3 1NP G8LVW
CM3 1NR G6YJH
CM3 1PN M6NNO
CM3 1QS G3WSV
CM3 1RB G6OTE
CM3 1RT G6EEE
CM3 2AY G4VVQ
CM3 2DF G8RSX
CM3 2JA M3AAS
CM3 2JE 2E0NOD
CM3 2JE M3NOD
CM3 2LB G8LYV
CM3 2LJ G4OAU
CM3 2LP 2E0DEO
CM3 2LU M0VRS
CM3 2QS G4KTX
CM3 3AD G6CMS
CM3 3BY G4DBM
CM3 3EF M3MFX
CM3 3JJ 2E0BOB
CM3 4DP G3JCM
CM3 4HT G7KSQ
CM3 4HT M6BDL
CM3 4LS G3PJC
CM3 4LS G4EOG
CM3 4PG M0DTA
CM3 4PH G7UVP
CM3 4PN M6TTW
CM3 4PS G0TRM
CM3 4QT M6MVW
CM3 4RB M6BAK
CM3 4RP G4TNB
CM3 4SD G8CYK
CM3 4XL M3DRH
CM3 4XN 2E0PXD
CM3 4XN M0PXD
CM3 4XN M6PXD
CM3 5FP M0DVD
CM3 5FS G7HBV
CM3 5FS M3EQY
CM3 5GH G4GYJ
CM3 5GY G3STP
CM3 5JF G4DUJ
CM3 5JJ M0AMZ
CM3 5JL 2E0JOI
CM3 5JL M0XJL
CM3 5JL M6JOI
CM3 5NJ G0AUR
CM3 5NL 2E0RFH
CM3 5NL M3RFH
CM3 5NP 2E0MYZ
CM3 5NY G7VEE
CM3 5PF G7NBR
CM3 5PQ G3XMB
CM3 5SB G0BCW
CM3 5TS G0VGC
CM3 5TS G7MOY
CM3 5TY G1YJF
CM3 5WH G8NAM
CM3 5WP 2E0FEQ
CM3 5WP M0XYM
CM3 5WP M7ESX
CM3 5YG G6KWY
CM3 5ZT G7BHE
CM3 5ZU 2E0JPU
CM3 5ZU M0GGX
CM3 5ZU M3JPU
CM3 6AQ G4TWL
CM3 6BY G0IJN
CM3 6BY G0UTT
CM3 6DD G0LWL
CM3 6DF G4THV
CM3 6DP G7EHS
CM3 6DQ G6TXV
CM3 6DT G8WRB
CM3 6DU G3ZES
CM3 6EJ 2E0GRA
CM3 6EP 2E0IWZ
CM3 6EP 2E0THT
CM3 6EP M0IWZ
CM3 6EP M0UHH
CM3 6EP M3IWZ
CM3 6EP M6LLH
CM3 6EU G1SNU
CM3 6EU G7LIK
CM3 6JE G3HMQ
CM3 6LF G6AHD
CM3 6NF G4BUP
CM3 6NZ 2E0DJQ
CM3 6PZ G0MLO
CM3 6RX 2E0DVH
CM3 6TP G7TAV
CM3 7AH M1GUS
CM3 7BD 2E0FJJ
CM3 7BD M3FHI
CM3 8AP M6GFF
CM3 8EL G3XYI
CM3 8EW G6BCL
CM3 8HJ G8CRX
CM3 8UT M3USH
CM3 8XA G4OGL
CM3 8XA G8EGE
CM4 0AL G3FJO
CM4 0HA G8YQO
CM4 0RS G8GIG
CM4 9DT G8LAB
CM4 9QA G7WKP
CM5 0BL M6MNH
CM5 0ES M6BUC
CM5 0QG G0KKH
CM5 9BG G4KVR
CM5 9HH M7SCT
CM5 9HW M3YIF
CM5 9QN G6NAX
CM5 9QW G4TOX
CM6 1AJ G7JXL
CM6 1BP M6WWM
CM6 1BY G7DPR
CM6 1BY M6DWO
CM6 1FX M7GTL
CM6 1SD M0TRE
CM6 1WB 2E0RHA
CM6 1WB M0RXA
CM6 1WB M6OLH
CM6 1WT G8OPI
CM6 1WU M6DVS
CM6 1XQ M6BCW
CM6 2AP M5BIL
CM6 2BS M1SUE
CM6 2BW M7LTR
CM6 2DL G0PTG
CM6 2GR G1SDN
CM6 2GR M0PXO
CM6 2GR M7PJO
CM6 2JR 2E0MAA
CM6 2JR M0MAF
CM6 2JR M3RXQ
CM6 2PD M6DVY
CM6 2PU G1ILG
CM6 2PU M3ILG
CM6 2QZ 2E0XGY
CM6 2QZ G8GSY
CM6 2SQ M3ZIF
CM6 3AQ G6WYH
CM6 3AX G7TKI
CM6 3BP M0DUT
CM6 3GR M7COF
CM6 3LU M6LMC
CM6 3NW M7GHK
CM6 3PP G8HUT
CM6 4AS M6KQA
CM7 1AB M3TPU
CM7 1AR M6WTT
CM7 1DG M6OMK
CM7 1DL M3RTI
CM7 1DR G1JSK
CM7 1EB G4TVX
CM7 1XF G6HSC
CM7 2LH 2E0PAB
CM7 2LR M3XOL
CM7 2QE M7UKU
CM7 2RZ G6XCU
CM7 2TA M6RNZ
CM7 2TX M7NAS
CM7 2WB G6OIY
CM7 3DT M7AEA
CM7 3JR M3TXH
CM7 3JX 2E0XRS
CM7 3JX M3YFW
CM7 3LG G4ICP
CM7 3NR 2E0JSE
CM7 3NR M0JBY
CM7 3NR M6NMU
CM7 3NW M6MNO
CM7 3PE G3TGB
CM7 4BY G8PLO
CM7 4DG M7TCM
CM7 4JZ G4ONH
CM7 4LE G4YQL
CM7 5AH M7LCW
CM7 5EG G3WVR
CM7 5FS 2E1IAI
CM7 5HN G4YWN
CM7 5HU G6DIO
CM7 5NB G0HWQ
CM7 5PY G0EMK
CM7 5PY G3XG
CM7 5PY M0CLO
CM7 5QJ M6VIO
CM7 5SD 2E0TKY
CM7 5SH G0OSI
CM7 5UE G7IRK
CM7 5UQ G1XWN
CM7 9DZ 2E0XLS
CM7 9DZ M3OIL
CM7 9ES 2E0NPX
CM7 9ES M6NPX
CM7 9LL G0DEC
CM7 9LL G1YYY
CM7 9LL G8XVO
CM7 9LR G3ITL
CM7 9NF G0DCR
CM7 9NJ G1WRH
CM7 9RH M6POC
CM7 9RX M7SLK
CM7 9TX 2E0WAV
CM7 9TX M0WAV
CM7 9TX M6WAV
CM7 9UG 2E1HYX
CM7 9UJ G7PCG
CM7 9UR G1TZZ
CM75LW G8DJO
CM75LW M0IAE
CM77 6BY G4WXT
CM77 6DE G7SLJ
CM77 7JW G6EJF
CM77 7PX G8ZIW
CM77 7UA G6IOW
CM77 7YJ M7GNB
CM77 8BP G1GNQ
CM77 8BP M0VZA
CM77 8BP M0XG
CM77 8HN G7HHN
CM8 1HG 2E1IGJ
CM8 1HP M6PXT
CM8 1HR G6CSK
CM8 1JB 2E0IRX
CM8 1JB 2E0KDG
CM8 1JB M6IRX
CM8 1JB M6KDG
CM8 1JZ G3MMA
CM8 1LX M7WDW
CM8 1QJ G0FWA
CM8 1TJ M3RYT
CM8 1XP 2E0FJC
CM8 1XP M7AEB
CM8 1XY G8NGM
CM8 2AQ M0IYZ
CM8 2ES M6KRW
CM8 2LH G6GXG
CM8 2LJ G8WSC
CM8 2LL G0DJK
CM8 2NP 2E0NRH
CM8 2NP M0NRH
CM8 2NP M3ILM
CM8 2NP M3TSA
CM8 2PA M6ENM
CM8 2PE G1SDN
CM8 2PT 2E0XIS
CM8 2SZ G1NNB
CM8 2SZ G7BIQ
CM8 2SZ G7LEY
CM8 2TY 2E0CSL
CM8 2TY M0GGL
CM8 2TY M3JYP
CM8 2UQ G7BND
CM8 2UQ M3BAW
CM8 2XG G3KXI
CM8 2XL M6LIK
CM8 2XQ G8CMO
CM8 3JR G1UZD
CM8 3LN G3LVL
CM8 3LT G7NAI
CM8 3LZ G3VYD
CM8 3NR G6AUD
CM8 3NS G3XAX
CM8 3NZ M0PWB
CM8 3PH G3XVV
CM8 3QY 2E1IIV
CM8 3QY M0UTH
CM8 3RZ 2E0RNS
CM8 3RZ M3RNS
CM8 3SN M0TCT
CM8 3SP 2E1HGG
CM8 3SP G4KQE
CM8 3SP G8WOZ
CM8 3SP G8WQZ
CM8 3SY G8UUO
CM9 4BL G6RKF
CM9 4BW M0RAX
CM9 4LF M1OCN
CM9 4PB G1WFA
CM9 4RQ G0AAN
CM9 4SE 2E0WAA
CM9 4SE M0WAG
CM9 4SE M3WPI
CM9 4US G1YKZ
CM9 4YN G0MBY
CM9 4YN G8DHV
CM9 4YX G8JAH
CM9 5DP G3VNP
CM9 5EE G1BQG
CM9 5HF 2E0KCP
CM9 5HF M0KCP
CM9 5HF M6KDP
CM9 5HY M6ROW
CM9 5JJ G8GLP
CM9 5JQ M6BZW
CM9 6BE 2E0OVF
CM9 6BE M0SHQ
CM9 6BE M6AEE
CM9 6DJ G4HGH
CM9 6ES M7RUS
CM9 6EW G1NMN
CM9 6EW G4MOV
CM9 6JB M7NWT
CM9 6JF G7RFC
CM9 6JF G7SRC
CM9 6JF G8PYD
CM9 6JF M1BDR
CM9 6JH G6LJR
CM9 6LR G6PCP
CM9 6QP G3UYC
CM9 6UH G4MNT
CM9 6UQ G8WTM
CM9 6YA M7ACV
CM9 6YX G7PIX
CM9 7TY G1PMU
CM9 8AY M1GFE
CM9 8AZ G8UVG
CM9 8BW G0UHD
CM9 8BY M1DOZ
CM9 8EH M3YWO
CM9 8HA M6EAP
CM9 8LL G0VDV
CM9 8PB G4EMB
CM9 8PB G6OXQ
CM9 8PJ G0VDP
CM9 8PR G3ZOX
CM9 8QA G0ASN
CM9 8RJ G6CYE
CM9 8SB G3GLL
CM9 8SR G4IIH
CM9 8UA G1FXM
CM9 8UD G1FWY
CM9 8UF G6IFH
CM9 8UN 2E1DET
CM9 8UN M3DVP
CM9 8UN M6JEK
CM9 8XB G0IBN
CO1 1GJ 2E0ONM
CO1 1GJ M6ONM
CO1 1PR G0VKY
CO1 1UR 2E0BYF
CO1 1UR M3BYF
CO1 2GQ 2E1FRI
CO1 2GQ M1CUY
CO1 2HS 2E0ETY
CO1 2HS M0IQR
CO1 2HS M6XWR
CO1 2HU G7LYH
CO1 2JJ M6AWP
CO1 2LF 2E0FTQ
CO1 2LF M6FTQ
CO1 2SN M0LCX
CO1 2UT M1FJD
CO1 2WA M0WNF
CO1 2WA M3WNF
CO10 0BZ M6OYM
CO10 0DJ G0PAO
CO10 0DX M3LEL
CO10 0EP G0VAS
CO10 0FH G4LSK
CO10 0HH G1DEU
CO10 0JN G4GGC
CO10 0LJ M6DNX
CO10 0NH M6LXU
CO10 0NN M7SFB
CO10 0NT M6ICH
CO10 0NY M7KJD
CO10 0QP G8AAR
CO10 0RJ 2E0REL
CO10 0RJ M6REL
CO10 0RN G6RKG
CO10 0YE 2E0BYC
CO10 0YF G7SRA
CO10 0YF G8LTY
CO10 0YU G4JAA
CO10 1HT 2E0ORX
CO10 1HT M7TER
CO10 1JB G6DKE
CO10 1NT M0JMV
CO10 1PJ G7HMF
CO10 1QX G7MLK
CO10 1WL M6NOI
CO10 1YU G3PZU
CO10 2BU G7NZV
CO10 2PP G7UTC
CO10 2TP G1IUW
CO10 2TU M3TFP
CO10 2XZ M0JGK
CO10 5JA G1YJL
CO10 5JN 2E0TBZ
CO10 5JN M0TBZ
CO10 5JN M6GLX
CO10 6JB G1PMF
CO10 7AR 2E0CNE
CO10 7AR M0HGS
CO10 7AR M6AWL
CO10 7EZ G6VSM
CO10 7JX G4DHU
CO10 7LS G1EUF
CO10 7NA M7EDK
CO10 7PH G4OBB
CO10 7PN M0COJ
CO10 7PS M0AUR
CO10 7PS M0CEQ
CO10 7QZ M1COQ
CO10 7RL G3YAI
CO10 7RS G4SWQ
CO10 7RW G3WRD
CO10 7SB G0HUG
CO10 7SJ G0PGS
CO10 7SJ G7KSH
CO10 8AS G7MTQ
CO10 8AS M3MTQ
CO10 8DA G4IMM
CO10 8EQ G4HGH
CO10 8HA G1BWX
CO10 8JN G4HTZ
CO10 8JP G6FCL
CO10 8JS G8GRL
CO10 8LQ 2E1MJH
CO10 8LQ 2E1ORT
CO10 8LQ M0MJH
CO10 8LQ M3MJH
CO10 8NN G8CDG
CO10 8PD G6UQZ
CO10 8UA M7GOZ
CO10 9LF G6CHJ
CO10 9QD G1TWY
CO10 9RU G0SNU
CO10 9SD M3NIT
CO100BD M6ODK
CO11 1DL M0CGE
CO11 1NS G1XMH
CO11 1RN G1NNA
CO11 2AL G0NXH
CO11 2AL G4JVM
CO11 2BU 2E0KLF
CO11 2BU M0JVC
CO11 2BU M6KLF
CO11 2DT G7EPM
CO11 2GE M7PJD
CO11 2HE M0BGE
CO11 2HE M6EWJ
CO11 2HX G3YAJ
CO11 2HZ 2E1RFS
CO11 2LH M0AFX
CO11 2QP M1DVO
CO11 2RB G7VFC
CO11 2SU M3JIG
co112sp G6UWK
CO11FQ 2E0TLW
CO11FQ M7UKZ
CO12 3AS M6WAS
CO12 3BD G4WHK
CO12 3DS M3DZT
CO12 3FF M6KPE
CO12 3HH M7BXM
CO12 3HS G0OEY
CO12 3JA G1PJJ
CO12 3JZ M0NWW
CO12 3NR 2E0TUD
CO12 3PL G8XAX
CO12 3PP M0DSY
CO12 3PP M6DSY
CO12 4BW G3UJB
CO12 4DW M0RIC
CO12 4DW M0TYG
CO12 4EQ G1VCZ
CO12 4HR G7HMN
CO12 4JG M7AGD
CO12 4JU M6BLO
CO12 4LA M0BVQ
CO12 4NF 2E0NRN
CO12 4NF G4NDR
CO12 4NF M6GMU
CO12 4NT M6RHK
CO12 4QP M6UZT
CO12 4QU 2E0GYY
CO12 4QU G7OWB
CO12 4TS G0STW
CO12 4UZ 2E0HEI
CO12 4UZ M7BLD
CO12 4XN G3VTT
CO12 4XN M0GVP
CO12 5AS M7AKP
CO12 5BQ G0RDX
CO12 5BS G2ABR
CO12 5DH G4JIG
CO12 5EE G4YJQ
CO12 5EJ G3YYZ
CO12 5HE G4GBY
CO12 5JF G4EYE
CO12 5NE G8PAI
CO12 5NN 2E0NOM
CO12 5NN G4TZM
CO124NF M6SNU
CO13 0AB 2E0UPM
CO13 0AN G6CDU
CO13 0AQ G0KCD
CO13 0BH G0NLG
CO13 0ET G1WJR
CO13 0EW M0LWM
CO13 0JL G0MJA
CO13 0JS M0TLR
CO13 0JS M3UGT
CO13 0LQ 2E0XLH
CO13 0LQ G6NHU
CO13 0LQ M6DHU
CO13 0LQ M6SHU
CO13 0LQ M7RTO
CO13 0QH G3OMB
CO13 0SF G4OAX
CO13 0SW 2E0MGP
CO13 0SW M0MGP
CO13 0SW M3NGC
CO13 0TG G4ZDY
CO13 0TH M3LIX
CO13 0TQ M6FCV
CO13 0TW M6HQL
CO13 9DH M7IWW
CO13 9DW G4PBD
CO13 9HR G3OJS
CO13 9JE G6WHT
CO13 9JZ 2E0NTR
CO13 9JZ M6BDO
CO13 9LY G0HPM
CO13 9NH M6PEU
CO14 8BL G0WWR
CO14 8BU G0GYY
CO14 8DF G8HSI
CO14 8EW G0OFF
CO14 8LE G6GJV
CO14 8LT M6ALY
CO14 8NF 2E0FYA
CO14 8NF M0LTG
CO14 8NF M3MGP
CO14 8NF M6SXG
CO14 8NJ M6IWX
CO14 8PE 2E0DPX
CO14 8PE G3VLD
CO14 8RE G1ISS
CO14 8RG G3CO
CO14 8RG G6XOU
CO14 8RG M1COL
CO14 8RG M3LFO
CO14 8RL G3TUU
CO14 8RL G7BWF
CO14 8RR 2E1GXH
CO14 8SP M3ZLY
CO14 8SP M7AKC
CO14 8SR G4MVX
CO14 8SS G0KKT
CO14 8SS G0OIF
CO14 8UD G1HZD
CO14 8UE 2E1FPU
CO15 1EJ G3XRN
CO15 1EJ G4AQZ
CO15 1HB 2E1HGU
CO15 1HB 2E1HLP
CO15 1HB G7PMU
CO15 1LA G6DLZ
CO15 1NQ 2E1HHB
CO15 1TX 2E1HAQ
CO15 1TX M0GWE
CO15 1TX M3KEV
CO15 1UW G6EAR
CO15 1UX M6TBK
CO15 2DA G1JOA
CO15 2DB M6MMN
CO15 2DD G6AGN
CO15 2JN G6WIL
CO15 2JW M6EVD
CO15 2NN G0KFV
CO15 2NU 2E1IGU
CO15 2PA G6JIR
CO15 2PB M7KOY
CO15 2PE G4PQU
CO15 2QB 2E0DKM
CO15 2QB M6DKM
CO15 2QT G8DMT
CO15 2RW M0HTF
CO15 3DW 2E0IKH
CO15 3HS G6CYA
CO15 3JR M0CHL
CO15 3LA G1YJJ
CO15 3LA M0RBC
CO15 3LA M3ZSC
CO15 3NF G7AJG
CO15 3QX G0UHU
CO15 3SB M0DHG
CO15 3SE G6XJC
CO15 3SR G0DGH
CO15 4EU 2E0VBJ
CO15 4EU M6NBH
CO15 4JE G3ZEZ
CO15 4LH G8ZZL
CO15 4ND G0LQT
CO15 4NN G7UHX
CO15 4PA G4AEB
CO15 4PJ G8TOI
CO15 4RJ G4ZXS
CO15 4RJ G8ATL
CO15 4RU G0VAL
CO15 4RZ G0HWC
CO15 5DG 2E0FBQ
CO15 5DG M6EML
CO15 5DJ G3LZR
CO15 5JF 2E0COI
CO15 5JF M6BEZ
CO15 5LA G4ZJK
CO15 5NA G0SMJ
CO15 5PN M7BNI
CO15 5QQ M7NDL
CO15 5RH 2E1HIO
CO15 5RU M6KCT
CO15 5SQ 2E0TNO
CO15 5SQ G0NAD
CO15 5SQ M7ADW
CO15 5SR G7KME
CO15 5XH M3TLO
CO15 5XY M0DNO
CO15 6BE M6LKF
CO15 6DP G6CTY
CO15 6EN M0IWQ
CO15 6JL M6OLB
CO15 6LN G4TTQ
CO15 6NL 2E0XTA
CO15 6NL M6OKZ
CO15 6NX G7MAV
co153la M7AVB
CO16 0AX G0UBJ
CO16 0BU M0PJK
CO16 0EG 2E0CWZ
CO16 0EG M0KEB
CO16 0EG M6FGG
CO16 0EP G4CYF
CO16 0HT G1JCP
CO16 7HF G0MBA
CO16 7HF G0PKT
CO16 7HQ G4OJH
CO16 7HQ G8JPJ
CO16 8AG M7WZO
CO16 8BN G4RUJ
CO16 8DB G4WHZ
CO16 8DG G8LID
CO16 8DL M6TMY
CO16 8EP G1FWR
CO16 8FB G7RIA
CO16 8FE G7HJK
CO16 8FQ G0DZZ
CO16 8FQ M0JXY
CO16 8HB G0HCY
CO16 8HH M6SDI
CO16 8PH G3CRC
CO16 8PH M1EAK
CO16 8QE G1NQB
CO16 8RQ G0CKD
CO16 8RQ G0RIU
CO16 8TH M0CNL
CO16 8UA G6XHI
CO16 8UN M7BHN
CO16 8US G4FJT
CO16 8YB 2E0NNG
CO16 8YB M6ORM
CO16 8YD G6TNQ
CO16 8YE G0GEQ
CO16 8YU G4RJQ
CO16 9AN G3VMP
CO16 9EN M6WXY
CO16 9ES G3PQM
CO16 9NE G8CVP
CO16 9PS G0FIW
CO16 9PS G1OGY
CO16 9PU G1HBW
CO16 9QP G0EBI
CO16 9RS G7BVZ
CO167DX 2E1PAW
CO168BX M0CHS
co168pd M7BHM
CO169EN G1IPU
CO2 0DL 2E0JJN
CO2 0DL M0JJN
CO2 0DL M6JJN
CO2 0DR M7DPH
CO2 0JE G4YJN
CO2 0NQ G4YOF
CO2 7EN G6DNA
CO2 7HH M7KIK
CO2 7LD G7WLC
CO2 7LQ G4NNN
CO2 7PE M3WWR
CO2 7SJ 2E0NNP
CO2 7SJ M0IUI
CO2 7SJ M6NNP
CO2 7UG G7EQX
CO2 7UX G6FBB
CO2 8AR G6XWY
CO2 8BH G1XAJ
CO2 8BP G7PST
CO2 8DD 2E0OVX
CO2 8DD M6OVX
CO2 8EE 2E0KFG
CO2 8EE M7KFG
CO2 8EG M3EUM
CO2 8FU M6MYI
CO2 8GE M7MFH
CO2 8GY M3RPA
CO2 8LT M6KCU
CO2 8NS G7TAT
CO2 8NS M6NUL
CO2 8NS M6POU
CO2 8NZ G8UBU
CO2 8PA G7UFT
CO2 8QF G7RWF
CO2 8QY 2E0LSI
CO2 8QY M0LSI
CO2 8QY M6EPZ
CO2 8SJ G4DIS
CO2 8UB G7ELH
CO2 8UD G6FLW
CO2 8UD M0VLL
CO2 8UJ M6EBW
CO2 8UP 2E0JYE
CO2 8WW G6RVP
CO2 9DT M0KTR
CO2 9DZ M0KAP
CO2 9HN M3WMO
CO2 9JR M1CZY
CO2 9JR M3CZY
CO2 9NE M6ASK
CO3 0HP G4CTA
CO3 0HP G6WYL
CO3 0HY G8VJR
CO3 0NP G7CGC
CO3 0RZ M1BPY
CO3 0YA M0BCK
CO3 0YJ G0NHB
CO3 3BP M6HWX
CO3 3HE M0LFS
CO3 3HW G4UTJ
CO3 3QE M6ALU
CO3 3RS G0KOY
CO3 4BU M7EIR
CO3 4JJ G0WZV
CO3 4JP G4JKC
CO3 4NQ G6DFZ
CO3 4NU M6CRX
CO3 4PS 2E0JBK
CO3 4PS M3JBK
CO3 4QN G4SOB
CO3 9AQ G6RDD
CO3 9DP G0HHC
CO3 9TR M1XXT
CO3 9XJ 2E0YTZ
CO4 0AJ G0OWK
CO4 0DA M7LSB
CO4 0DB M6GVI
CO4 0EA G6IIF
CO4 0HX G3MQR
CO4 0LD 2E0BMG
CO4 0LD M3LFQ
CO4 0LP G4OYH
CO4 0PJ G4AUG
CO4 0PW G8RFC
CO4 0PW M0YNK
CO4 0PW M6YAN
CO4 0QJ G6AQI
CO4 3AS G8SGB
co4 3ds 2E0KFV
co4 3ds M0KFV
CO4 3DS M7OCO
CO4 3EY G4HKC
CO4 3FD G4HKB
CO4 3FE G0VYC
CO4 3FN G4RKB
CO4 3HB G4JHP
CO4 3JA M0ASG
CO4 3JE M6YEF
CO4 3JH G7NID
CO4 3JP G0PFM
CO4 3JR M1AMI
CO4 3JR M3DFC
CO4 3LX G7EHY
CO4 3NF G0VAE
CO4 3NF M0DEQ
CO4 3NF M3ZZN
CO4 3NL G1DTE
CO4 3SQ M0UOE
CO4 3SQ M3PQI
CO4 3SQ M3TFA
CO4 3SZ M3SCH
CO4 3UP M3JKB
CO4 3UT 2E0CEA
CO4 3UT M6CEA
CO4 3XN M0CVH
CO4 3YA G6CLA
CO4 3YH G0SLB
CO4 3YP G3ZOL
CO4 5AF M7LUV
CO4 5BD M3KMS
CO4 5DU 2E0TFH
CO4 5JT G8MKN
CO4 5JX 2E1DXB
CO4 5JX M3AJD
CO4 5LQ 2E1EAG
CO4 5PE G0AHE
CO4 5PY G8WBY
CO4 5RD G6KTR
CO4 5RN M6XEL
CO4 5SE M6LWV
CO4 5YA M3YAL
CO4 5YE M7BRO
CO4 5ZR M6HWH
CO4 9EA G4ZOR
CO4 9RE M3JBE
CO4 9RJ M1BQZ
CO4 9RR M6HJA
CO4 9SL 2E1EWK
CO4 9ST 2E0GMM
CO4 9ST M3OEG
CO4 9WQ G7BIV
CO4 9YD 2E0IGM
CO4 9YD M6IGM
CO4 9YU G1GWO
CO45SE 2E0EMZ
CO45SE M0ITB
CO45SE M6KNV
CO5 0AA G6XIF
CO5 0AE G6KVI
CO5 0AY G4TFI
CO5 0BN G7JMW
CO5 0DH M6KZZ
CO5 0DT G4KTB
CO5 0EF G4TFP
CO5 0EF G7NMI
CO5 0EN M0JJH
CO5 0HJ M1ASV
CO5 0HN M0HIG
CO5 0JG M6CHP
CO5 0JG M6FSC
CO5 0JJ 2E0VVB
CO5 0JJ M7JDH
CO5 0JJ M7JWH
CO5 0JU G4ZZL
CO5 0LJ 2E0MNH
CO5 0LJ M0JBV
CO5 0LJ M6IFC
CO5 0LQ M0ECL
CO5 0LR G0IZK
CO5 0QD G3GML
CO5 0QT G6DBY
CO5 0RH G6LTD
CO5 0RP G6YWU
CO5 0SL M0XDA
CO5 5AD 2E0NOW
CO5 5AD M0HKG
CO5 5AD M3HHB
CO5 7AS G4PZX
CO5 7EH M7BZB
CO5 7EL M7EOT
CO5 7ET G6KNM
CO5 7HF G3KHK

CO5 7LB G4YK
CO5 7LJ G4PBR
CO5 8BD M7ADX
CO5 8DR G8SOI
CO5 8EB 2E0WFR
CO5 8EB M1RJS
CO5 8EB M7WFR
CO5 8GF 2E0DPW
CO5 8GF M7VET
CO5 8GG M5RHG
CO5 8GJ G8IFN
CO5 8JU M3BSH
CO5 8LN G0JDE
CO5 8PX M1EVN
CO5 8QP G0APP
CO5 8RW 2E0FIO
CO5 8RW M7GNC
CO5 8RY G6OOT
CO5 8SL M1CAY
CO5 8SS M0PUC
CO5 9AG G8ADZ
CO5 9AH G8AFQ
CO5 9TD G3NXK
CO5 9TH G4PYG
CO5 9TU M3VXK
CO6 1BP G0NMB
CO6 1DB G3YEC
CO6 1NY G1GMM
CO6 1RQ G1NPX
CO6 1SG G0WJD
CO6 1SY M6YUM
CO6 1UL G7VGC
CO6 1XG G1IUD
CO6 1XG G6LJF
CO6 1XG G8IUD
CO6 1YN 2E0JPR
CO6 1YN M0JAI
CO6 1YN M6ETU
CO6 1YP M1REH
CO6 2AS M6VLE
CO6 2DZ G8LHF
CO6 2NH 2E0VAN
CO6 2NH M0YEP
CO6 2NH M3VXC
CO6 2PF G7LNU
CO6 3DB G0VHF
CO6 3DB G4ZTR
CO6 3DB M1CRO
CO6 3DX G8VAF
CO6 3EP G8YMZ
CO6 3HY 2E0FBA
CO6 3JS M6UJD
CO6 3LX 2E0CGV
CO6 3LX M0HBV
CO6 3LX M6ALT
CO6 3NJ 2E0WDM
CO6 3NJ M3VZV
CO6 3RY 2E0ORP
CO6 4BJ G6XMM
CO6 4EH 2E0ALB
CO6 4EH M3XMA
CO6 4LT G4DKX
CO6 4PQ G7DTT
CO6 4QH G3LST
CO6 4UN G1AWK
CO6 5AG M0EJG
CO6 5AR G0IBZ
CO62LU M7GGF
CO63LX 2E0IPG
CO63LX M6IPG
CO7 0AQ 2E0KKC
CO7 0DU G6AEB
CO7 0HE 2E0DPZ
CO7 0HE M6LYD
CO7 0HU M7RGD
CO7 0JP M7DGP
CO7 0LA G4IZX
CO7 0LB G4CIA
CO7 0NA G1XUU
CO7 0NZ G6PDR
CO7 0PE G3ELS
CO7 0PR G3MGW
CO7 0QR G4JAC
CO7 0RH G7KRO
CO7 0RP G7TBU
CO7 0RS G0SOX
CO7 0RU G4EUW
CO7 0SJ G1BFF
CO7 0SJ M0UEI
CO7 0ST G8DRE
CO7 6ES G4PZL
CO7 6QY G1VZT
CO7 6RF G0VEI
CO7 6SJ G8CJD
CO7 6TP G1CHN
CO7 6TP M7EOH
CO7 6TX G6VJK
CO7 6UX M6AZX
CO7 7AS M0PDE
CO7 7AS M0PDF
CO7 7EG G7USX
CO7 7GA G3YVW
CO7 7PQ 2E0HCO
CO7 7PQ M0OKD
CO7 7RP 2E0HUM
CO7 7RP M6YES
CO7 7RY G8GML
CO7 7SD 2E0BZM
CO7 7SD M3ZMQ
CO7 7SY G6IGU
CO7 8AU M6NNW
CO7 8BQ M7MUF
CO7 8DD 2E0DDU
CO7 8DD M0XLB
CO7 8DD M6SEY
CO7 8DH G0GYI
CO7 8HS 2E0WMG
CO7 8HS G8EWC
CO7 8JA G4AZD
CO7 8JB G4URA
CO7 8JH G1TWH
CO7 8JH M3DEB
CO7 8LH M6BAR
CO7 8LJ G4ZKS
CO7 8NN G0PJZ
CO7 8NN M7MRT
CO7 8RF M6TRP
CO7 9EH G0BEP
CO7 9JX 2E0WIV
CO7 9JX M6XUU
CO7 9LD G0GGM
CO7 9LG 2E0AIK
CO7 9LG G4SDI
CO7 9LS 2E0PXY
CO7 9LS 2E0PXZ
CO7 9LS M0PXZ
CO7 9LS M3PXY
CO7 9LS M3PXZ
CO7 9NH G4FTP
CO7 9QH M1BQF
CO7 9QQ G4NXR
CO7 9QZ G7TWA
CO7 9RP M3XGB
CO7 9SD M1DYW
CO8 5BN G6WPJ
CO9 1AS G1ZLD
CO9 1BJ M1CJF
CO9 1ED G6LMC
CO9 1EH G7UUA
CO9 1EH G7UUD
CO9 1JU G4YAX
CO9 1JU G6NEK
CO9 1LB G0GWN
CO9 1NY G4FJC
CO9 1PD G4BCV
CO9 1PD G7ORE
CO9 1PD M0NAS
CO9 1PD M3JDS
CO9 1PG 2E0JTO
CO9 1PG M0JOC
CO9 1PG M7JOC
CO9 1PY M3PSI
CO9 1QG G4LBJ
CO9 1TD G1CCW
CO9 1UB G6OXJ
CO9 1UG G4TEB
CO9 1YB M3IRX
CO9 2HF G0GII
CO9 2HJ M3JEP
CO9 2NW G7UVY
CO9 2SF M6OAT
CO9 2TA G6LHG
CO9 2TB G0CQI
CO9 2TB G1GRZ
CO9 2UA 2E0MKW
CO9 3AZ G0LZY
CO9 3BZ G4ZUL
CO9 3DP M7LGC
CO9 3HX G0GZF
CO9 3LG 2E0SLG
CO9 3LG M7RKY
CO9 3NX G1OFX
CO9 3QN G4WVH
CO9 3QW 2E0FIW
CO9 3QW M0OBM
CO9 3QW M7VCF
CO9 3RN M5AJB
CO9 4AB 2E0AAF
CO9 4AB 2E0FLR
CO9 4AB M0JNQ
CO9 4AB M0VJR
CO9 4AB M6DTR
CO9 4EE 2E0GFZ
CO9 4EE M7BWG
CO9 4JG G3PGN
CO9 4LN G6WHY
CO9 4LX G8JVV
CO9 4QJ M6NCU
CO9 4QQ M6GEY
CO9 4QQ M6YEL
Co93bh 2E0COM
Co93bh G1OIS
CR0 0AA G0DDT
CR0 0BL G7VTH
CR0 0DN M3XNR
CR0 0JE G0VQM
CR0 0NU G0HSH
CR0 0PF M0RXZ
CR0 0RP M6HXD
CR0 1JS G7SFA
CR0 1XL G1ALR
CR0 2BB 2E0WBO
CR0 2BB M6YCR
CR0 2HX G1OIS
CR0 2LP G4NHA
CR0 2LW 2E0DIS
CR0 2LW M6EEW
CR0 2PF 2E0BPU
CR0 2PF M3FUB
CR0 3AD G4MZK
CR0 3JF G0GFY
CR0 3NF G1IEY
CR0 3QP G8GYM
CR0 3SW G6FGY
CR0 4AF M7RCE
CR0 4DN M6LHV
CR0 4EG G4DAF
CR0 4HG G7LTP
CR0 4JD M6BGG
CR0 4NN 2E0IPL
CR0 4NN M0UPL
CR0 4NN M6ZGR
CR0 4PU G4FFX
CR0 5BA G3WBN
CR0 5BP G8TBL
CR0 5HX G4STD
CR0 5PS G6DAY
CR0 5QA M0CCF
CR0 5QA M0VOG
CR0 5QA M1ACF
CR0 5QA M1CCF
CR0 5QA M3ACF
CR0 5ST G4KQO
CR0 5TB G3SCB
CR0 6AP M3WXH
CR0 6AQ M6CGF
CR0 6DE G4DPO
CR0 6TS M7PGF
CR0 6XR M6ILH
CR0 7EB G4AVV
CR0 7HY G4RWW
CR0 7PR G3ZIO
CR0 7QP M0CVZ
CR0 7RW 2E0EQU
CR0 7RW M6IIX
CR0 7SH M3WRO
CR0 8HJ M3OSA
CR0 8HX M6VZF
CR0 8LG G7ODG
CR0 8ND M6NFL
CR0 8PN G3RMN
CR0 8SB G3ALG
CR0 8YQ M3YRV
CR0 9DG G7MSF
CR0 9HE G4UIO
cr0 9hq 2E0IDM
cr0 9hq M6VBN
CR0 9LN G4FUU
CR0 9LZ M7SCP
CR06SW G7NQW
CR2 0BL M3YDJ
CR2 0DZ G0HWY
CR2 0EF G3WMT
CR2 0LB G4SXY
CR2 0RB M0CGF
CR2 6EE G0VVX
CR2 6EE G3WDS
CR2 6EJ G6CID
CR2 6NE G1WFG
CR2 7EF G0DJT
CR2 7ER G8MNY
CR2 7GE G3TCZ
CR2 7HF M7LMA
CR2 7HH G3MCX
CR2 7JJ G4DAC
CR2 7RE G4GFC
CR2 8AR 2E0DCZ
CR2 8AR M6CNX
CR2 8AR M6DCZ
CR2 8HR 2E1DRV
CR2 8JQ M7LNG
CR2 8PH G4LZE
CR2 8PT M6VQX
CR2 8RA G3ZYZ
CR2 8SL M3YGO
CR2 8SL M3ZGO
CR2 9DU M1ZAR
CR2 9ES G3KKZ
CR2 9HY M0PCC
CR2 9HY M6API
CR2 9JR G0TCE
CR2 9JY G8DNL
CR3 0AB G8IYS
CR3 0AJ G7VFX
CR3 0EP G4BWG
CR3 5DL G1RCE
CR3 5EL G0SCR
CR3 5EL G4APL
CR3 5EL G7BSF
CR3 5EL M3BYL
CR3 5JN G6LTK
CR3 5LJ G4LXR
CR3 5QH G8LMI
CR3 5RB G7BWE
CR3 5RT G0SYR
CR3 5RT G8DTQ
CR3 5SG 2E0FSC
CR3 5SH M3SOF
CR3 5TG G4ECS
CR3 5XN G7APO
CR3 5ZU G7ONI
CR3 6AD G8PRK
CR3 6BA G3ODX
CR3 6DQ G6CDW
CR3 6HN G4DTC
CR3 6NJ M3YTG
CR3 6QX G0TNM
CR3 7EH G8HDP
CR36SA G7NGB
CR4 1JF M3WVG
CR4 1LF G0WCR
CR4 1LG M7DPE
CR4 1NY G7JMQ
CR4 1XG 2E0TSA
CR4 1XG M6TSA
CR4 1XJ M3MZP
CR4 1XN G1IZN
CR4 2EN M0UCI
CR4 2GA M6XCA
CR4 2LF G0LUL
CR4 2LT M3URZ
CR4 3DW G7OZJ
CR4 3DZ G0VTM
CR4 3JS G4RBH
CR4 3LL 2E1GOE
CR4 3LW G1KGO
CR4 3RQ G7OLH
CR4 3RS 2E0YGS
CR4 3RS M0IYN
CR4 3RS M6GMS
CR4 3RS M7CDW
CR4 3SB M6UON
CR4 4HD G3LCH
CR4 4LZ G1FOF
CR5 1AT M6XZK
CR5 1BB G3MZA
CR5 1DF G8EIN
CR5 1HR G0KZT
CR5 1HR G4FUR
CR5 1HR M1FUR
CR5 1JS G1KGA
CR5 1NF M1DOR
CR5 1NL G8EDN
CR5 1QP G3CQU
CR5 1QS G0GZM
CR5 1RF G1TLW
CR5 2BL G4CJR
CR5 2EG G7OKY
CR5 2EJ G4FVL
CR5 2JF G8GAR
CR5 2LD G8RWG
CR5 2LF G3ZPB
CR5 2SY M7LTH
CR5 3BP G0KUE
CR5 3DD G0FUH
CR5 3DE G2LW
CR5 3DE G3OOU
CR5 3PH G6UVB
CR5 3SJ M6EBE
CR5 3SX G3GWC
Cr53rn M7IIO
CR6 9EP 2E0UGM
CR6 9EP M0UGM
CR6 9EP M0USM
CR6 9EP M6UGM
CR6 9HZ G3WZK
CR6 9LB G4LLM
CR6 9LP G8JAC
CR6 9RP M7HNK
CR6 9RT 2E0HJV
CR6 9RT M6HJV
CR6 9TD 2E0FJA
CR6 9TD M0IDM
CR6 9TD M6FJA
CR7 6BX G1POM
CR7 6EB M0HNC
CR7 6HD M7CRV
CR7 7AF G3UFY
CR7 7HQ G4KRD
CR7 7JE G3JAL
CR7 7NP G1HER
CR7 7NQ G7HLW
CR7 7RU M7AGN
CR7 8DF G0IOO
CR7 8DF M6JEJ
CR7 8NY G4REK
CR7 8RP G1OTN
CR8 1BB G0DLP
CR8 1JA G7JAQ
CR8 1JB G4BUK
CR8 1JL G0UCT
CR8 1JQ G7PWV
CR8 2DY 2E0BCF
CR8 2DY M0GNM
CR8 2DY M3IHV
CR8 2HQ G8FOT
CR8 2LR G6GFJ
CR8 3AQ 2E0LTU
CR8 3AQ M0PZD
CR8 3AQ M6IZP
CR8 3AQ M6URA
CR8 3PE G0UQO
CR8 4AJ M7NUT
CR8 4HH G3TWJ
CR8 4NB G8KWV
CR8 5DG M0CUP
CR8 5DG M3IDY
CR8 5JJ G3ZXV
CT1 1GY 2E1TAG
CT1 1NR M3KRE
CT1 1PZ 2E1EHM
CT1 1QG G7RBB
CT1 1SJ M3AJA
CT1 1TS G6EGU
CT1 1TS G6YZU
CT1 1WX M0HWJ
CT1 1YG 2E0WPJ
CT1 1YG M0WPJ
CT1 1YH M0XED
CT1 2AA G1AUI
CT1 2SX 2E0DTW
CT1 2SX M6RPM
CT1 3DH G4KGY
CT1 3JL M0ELS
CT1 3LD G3MMJ
CT1 3SF M7AXM
CT1 3UF M0PAM
CT1 3UP 2E0DBP
CT1 3UP M3AYQ
CT10 1DR 2E0RDP
CT10 1DR 2E0WFY
CT10 1DR G8FME
CT10 1DR M6WFY
CT10 1FA M6TZI
CT10 1HN M3TDP
CT10 1HR M0ATS
CT10 1PG 2E1EKM
CT10 1QN 2E0PGC
CT10 1QN M5PGC
CT10 1QN M6PGQ
CT10 1RP G0LGW
CT10 1SR G6SFC
CT10 1TL G4GAT
CT10 2BH M6TVB
CT10 2BX M7KJM
CT10 2DT M0ASC
CT10 2EW M6YZH
CT10 2HA M0AQA
CT10 2HG G8DZN
CT10 2HL M6ZPE
CT10 2HY M3UVJ
CT10 2HY M6DWM
CT10 2JG G4AQE
CT10 2JL G8XJE
CT10 2ND G7FNU
CT10 2NG G0LEU
CT10 2PE 2E1HPT
CT10 2PE M0CAG
CT10 2PL M0TFC
CT10 2SD G0RJJ
CT10 2SS M6XAY
CT10 2TR M7NHS
CT10 2TX G0KCA
CT10 2XN G7OHO
CT10 2XN M0DLI
CT10 2XU 2E0NJJ
CT10 2XU M0GIG
CT10 2XU M3NJJ
CT10 3AA G1VID
CT10 3AH G4PTE
CT10 3AZ G0VUT
CT10 3DE M3JKJ
CT10 3DE M3JKP
CT10 3DL M3MQP
CT10 3DR G1KMJ
CT10 3ES G4GAP
CT10 3EY G0GUW
CT10 3HN G8WQT
CT10 3LS M3LQC
CT10 3NE G0RXU
CT10 3QP G4VBI
CT10 3SD G4VPI
CT102XN 2E1CPF
CT11 0BH M3UCF
CT11 0DF M6DJA
CT11 0HT M3WOD
CT11 0JJ M3XJZ
CT11 0LL G3LHS
CT11 0LQ M1CCL
CT11 0ND G7MGZ
CT11 0PB G7FMI
CT11 0PF 2E0CVJ
CT11 0PF M0ZRF
CT11 0PL G7SYE
CT11 0PX G0UAK
CT11 0QR 2E0OOW
CT11 0QR M6NOQ
CT11 0RN G1YZT
CT11 0RR G7DNQ
CT11 0RY G0AHA
CT11 7AS M0IJA
CT11 7DP M6FFU
CT11 7EF G6ENA
CT11 7HS M6HFJ
CT11 7HT G4ADS
CT11 7LP 2E0XDY
CT11 7NW G7ORS
CT11 8JP M3TWY
CT11 8QA 2E1GDO
CT11 8QA M7COY
CT11 9BP G6MWQ
CT11 9DE G3TVD
CT11 9JF M6OJP
CT11 9LP G1ZFB
CT11 9LP M6AVQ
CT11 9LU G0JIF
CT11 9PB G4HAK
CT11 9PW M6MFF
CT11 9QZ G4BXI
CT11 9RT G8GP
CT110ND G0MBL
CT12 4AG 2E1HNF
CT12 4BE G6UXG
CT12 4EL M3UGH
CT12 4EL M6RVR
CT12 4EP 2E0MRJ
CT12 4EW G1UFJ
CT12 4HR G6OKB
CT12 5AW G0LFM
CT12 5ED G3YCV
CT12 5ED M1DEJ
CT12 5LD G0WWP
CT12 5LX G7IFU
Ct12 6aj M7INT
CT12 6DD 2E0IAJ
CT12 6DD M3INL
CT12 6DE G0TFB
CT12 6DQ M6OZN
CT12 6DW M6GQK
CT12 6DX 2E1HPS
CT12 6DX G7ITS
CT12 6EZ G3ZBF
CT12 6JQ G7SFD
CT12 6NX M0DIJ
CT12 6NX M3EOX
CT12 6NZ G1ODK
CT12 6QG M6LJT
CT12 6RY M6EAJ
CT12 6SW G8UHJ
CT12 6UG M3TNH
CT13 0AQ 2E0PBY
CT13 0AQ 2E1ABW
CT13 0AQ 2E1ATH
CT13 0AS 2E1GTF
CT13 0BE M6GJY
CT13 0BG G7BPZ
CT13 0BG G7BQA
CT13 0DP G7VQJ
CT13 0DW M3SOQ
CT13 0EU G4RXG
CT13 0HX M7PAA
CT13 0JZ G1EHK
CT13 0PE G0JBA
CT13 9AS G4KPF
CT13 9JB 2E0ALH
CT13 9JB M3IYY
CT13 9JB M3KRM
CT13 9JB M3VKJ
CT13 9JE G4JYU
CT13 9JE G8MLB
CT13 9JF G0BVA
CT13 9JF G1PJR
CT13 9NY G0MAZ
CT13 9NY G7BVH
CT13 9QA 2E0BZV
CT13 9QA M0GUG
CT13 9QA M3ZNT
CT13NN 2E0WJF
CT14 0AJ 2E0GYJ
CT14 0AJ M0JSK
CT14 0AJ M7APM
CT14 0BT G0LGK
CT14 0JH G4WOS
CT14 0JL 2E0EWF
CT14 0JL M6STQ
CT14 0LA G4SUS
CT14 6PP 2E0LJH
CT14 6PP M0PKH
CT14 6PP M6CEH
CT14 6PP M6ODF
CT14 7BL M3BWZ
CT14 7DA 2E0VWW
CT14 7DA M3VWW
CT14 7EZ G0RDN
CT14 7PX M3OZC
CT14 7QB 2E0AAK
CT14 7QB G0OXX
CT14 7SY M1AJU
CT14 8BT G8YXR
CT14 8DD G4YWA
CT14 8EB G0DQI
CT14 8JW G1IUA
CT14 8JW G7MSC
CT14 8JW M6ZTS
CT14 9AA M3WGZ
CT14 9AT 2E1ACK
CT14 9DQ G3MZI
CT14 9EE G1LLU
CT14 9EF 2E0JEN
CT14 9EF G8SOU
CT14 9EF M0EEH
CT14 9EW G3FBU
CT14 9HW M6DWI
CT14 9JF G3VIR
CT14 9JF M3VIR
CT14 9JR M1DNA
CT14 9LL G4FJF
CT14 9LS G7HIX
CT14 9NB G0DUK
CT14 9NJ G0SRR
CT14 9NL M0ARX
CT14 9NP G4RXH
CT14 9QN M0SOE
CT14 9QZ M7UXO
CT14 9RG M1DEG
CT14 9RG M3AJU
CT14 9SN 2E0VRI
CT14 9SN M0VRI
CT14 9SN M7JCG
CT14 9SN M7MLE
CT14 9TW G7MSS
CT15 4AW 2E1DFB
CT15 4AW G7SGH
CT15 4BT G6IOM
CT15 4HE M0HHI
CT15 4HH M0WPL
CT15 4HR G6BNW
CT15 4HX M6KWW
CT15 4HZ M3OVC
CT15 5BY G7FCL
CT15 5JD G8SMZ
CT15 5JW G6HIG
CT15 6AH G8YMD
CT15 6AH M1BKI
CT15 6BS M3VFM
CT15 6DD G7SXJ
Ct15 6dl 2E1BTK
CT15 6ED G6PSQ
CT15 6EJ G4OPR
CT15 6HL 2E1BTG
CT15 6HL M6LYO
CT15 6JL G6WPK
CT15 7ES G4OJG
CT15 7HF G8MBV
CT15 7HJ G8UMB
CT15 7HR G3ZHU
CT15 7JN G4MIX
CT15 7JS G4AWW
CT15 7LJ G0AXD
CT15 7LP G4VRB
CT16 1EX 2E0CUV
CT16 1LH G0SET
CT16 1PX G4DDB
CT16 1TH 2E0VDC
CT16 1TH M0KID
CT16 1TH M6VDC
CT16 2AR G1PQX
CT16 2AX G4BBH
CT16 2EP M6JSW
CT16 2JW 2E1ABN
CT16 2LH 2E0NFX
CT16 2LH M7NFX
CT16 2NQ 2E1NFD
CT16 2QP G6APW
CT16 2QW G7MBU
CT16 2SG 2E0LYD
CT16 2SG M0IML
CT16 2SG M6BDV
CT16 3BA M6FEN
CT16 3DU M6LWM
CT16 3EE M1EJO
CT16 3EJ G4HHX
CT16 3HA G4RLX
CT16 3HZ G0ROO
CT16 3HZ G3ROO
CT16 3HZ M7ROI
CT16 3LQ G4UHT
CT16 3LT G7JOW
CT16 3ND 2E0EHR
CT16 3ND M6HKQ
Ct16 3ng M3WZP
CT16 3NP G4XDW
CT16 3NP M7EBJ
CT17 0BD 2E0BEH
CT17 0BD M0GGO
CT17 0BD M3JOM
CT17 0BS G8IYN
CT17 0DP M6IVG
CT17 0DY M6BAQ
CT17 0ER G4WIY
CT17 0FP G6JHT
CT17 0HE M3VPD
CT17 0HE M6MRC
CT17 0HE M6NET
CT17 0JQ M7FCB
CT17 0NG 2E0FHK
CT17 0NG M7RGJ
CT17 0NP 2E1ABY
CT17 0NT G8YNG
CT17 0NX G0ADK
CT17 0PR G7BRM
CT17 0PS G0KOK
CT17 0PS G3YMD
CT17 0PS G6USA
CT17 0RX M1CJB
CT17 9ES M3IWX
CT17 9JZ G6TRM
CT17 9LL 2E0CPL
CT17 9LL M0UGR
CT17 9LL M3UGR
CT17 9LQ G4FXE
CT17 9PN G4SMX
CT17 9QF M6ELZ
CT17 9QQ 2E0BHJ
CT17 9QQ 2E0OIN
CT17 9QQ G8WLY
CT17 9QQ M3KYH
CT17 9QQ M3OIN
CT17 9RB 2E0FQR
CT17 9RB M0IER
CT17 9RB M6FQR
CT17 9RQ G0CCG
CT18 7AP G4IMP
CT18 7AP G4RJZ
CT18 7AP M3PNA
CT18 7BS 2E0UGF
CT18 7BS M0WHO
CT18 7BS M3UGF
CT18 7BS M6UFO
CT18 7DS G6RRS
CT18 7DS M0KBC
CT18 7DZ G4TGP
CT18 7FA M6AKH
CT18 7FU G0BPS
CT18 7GA M6WTM
CT18 7GZ G4MDZ
CT18 7HB 2E0NCI
CT18 7HB M6NCI
CT18 7JS G0IXV
CT18 7LB G3OJZ
CT18 7LH 2E1DYT
CT18 7LH M1CVF
CT18 7LQ G0VHL
CT18 7LS 2E0DTS
CT18 7LW 2E0WUN
CT18 7LW M3RVQ
CT18 7LZ M0DRN
CT18 7NT M6PWE
CT18 7QL G4DUE
CT18 7TG M3FAK
CT18 8BS M6GXJ
CT18 8BY G8KSD
CT18 8BY M0CFH
CT18 8DF 2E0JKG
CT18 8DF M0JKG
CT18 8DF M3XJG
CT18 8DS G0VAI
CT18 8LX M3PMN
CT19 4AF G1RQI
CT19 4BX M3RWI
CT19 4EQ G0NOH
CT19 4HN M6LCI
CT19 4JA 2E0DJF
CT19 4JA M3PQB
CT19 4JA M3ZEJ
CT19 4JT 2E0MTR
CT19 4JW 2E0FLV
CT19 4JW M6AZQ
CT19 4QG M6GXS
CT19 4QH G0VWB
CT19 4RL 2E0DLR
CT19 4RL M6EQT
CT19 5AT G8YXQ
CT19 5BZ M1CMN
CT19 5DD M3HOM
CT19 5EL M3JBF
CT19 5HG G4IVL
CT19 5HG G7IWV
CT19 5JF G6GXE
CT19 5JH M6RNI
CT19 5LS 2E0NEC
CT19 5LS M6WUG
CT19 5NA G8SUQ
CT19 5NB G4EQJ
CT19 5NE 2E0VYW
CT19 5NE M0VYW
CT19 5NR M1DFB
CT19 5NR M3DFB
CT19 5PW M1AZO
CT19 5QY M0TUX
CT19 5RA M3MSN
CT19 5RU M6RUS
CT19 5SH G8YNH
CT19 5SH M3BNU
CT19 5TA M3IZP
CT19 5UB G8EKD
CT19 6BA G3XVY
CT19 6BP M3OZP
CT19 6EU M7TKB
CT19 6EY M6GXL
CT19 6NE 2E0ALL
CT19 6NX G4IDX
CT19 6PR G4MHS
CT2 0HR G8WMK
CT2 0HR M3CSR
ct2 0ht 2E0EKR
ct2 0ht M0GRB
CT2 0HT M3EKR
CT2 0HT M6JEH
CT2 0HX G1LHL
CT2 0JZ G7NIN
CT2 0LA M1DFM
CT2 0LA M3DFM
CT2 0LA M3TTA
CT2 0LL M6DYV
CT2 0QB G6ZEQ
CT2 0QH G3FMU
CT2 7AH M3ZIZ
CT2 7BX 2E0MNY
CT2 7BX M0RDP
CT2 7BX M6FQU
CT2 7HH 2E0DFQ
CT2 7JZ G3HTT
CT2 7TA M3DKO
CT2 7TF M0DCO
CT2 8BA M0KKB
CT2 8DR G6WDS
CT2 8PN M6JFP
CT2 8PQ G8UIV
CT2 9BL G4PCN
CT2 9BP M0SGF
CT2 9BP M6RCN
CT2 9BU M3WTU
CT2 9DB 2E0IMJ
CT2 9DB M0YMJ
CT2 9DB M3IMJ
CT2 9DL G8BRD
CT2 9ER 2E0HNP
CT2 9ER M0HKP
CT2 9ER M6HKN
CT2 9JX 2E0XXK
CT2 9JX G2DLX
CT2 9JX M0XXK
CT2 9JX M3XXK
CT2 9NW M6GUC
CT20 1DA G6ZNW
CT20 1DF G6WGM
CT20 1DL M3FSC
CT20 1ET 2E0SDJ
CT20 1HY G7HIY
CT20 1JU 2E1BYH
CT20 1TH M6XCK
CT20 2EJ 2E0LCK
CT20 2EJ M6YXZ
CT20 2HZ M7CEZ
CT20 2PQ G8IWX
CT20 2PT 2E0MGU
CT20 2PT M7MGU
CT20 2RP M6ECZ
CT20 2TY G0SLJ
CT20 2TY G2FA
CT20 3BE M0DRO
CT20 3EJ M6ASH
CT20 3JY M6BOK
CT20 3LA M3BJL
CT20 3LH 2E0RDE
CT20 3QJ 2E1GTI
CT20 3SA 2E0AYY
CT20 3SA G4EGQ
CT20 3SA M0RNP
CT20 3TA G3XHW
CT20 3TL M0GCB
CT21 4EA G7SZO
CT21 4JP G3LWD
CT21 4JQ G3ZHT
CT21 4SE G4SSZ
Ct21 5pa M7OCT
CT21 5RY M3PLI
CT21 6AG G3VYW
CT21 6DG 2E0JXT
CT21 6LS M7NJD
CT21 6QZ M7PJT
CT3 1DJ M6BAC
CT3 1ED 2E0TWA
CT3 1ED M6WEV
CT3 1JX G3JIX
CT3 1LG M7EBW
CT3 1LL G7MZA
CT3 1LN G3TAJ
CT3 1LY G4FLR
CT3 1RQ M6FHQ
CT3 1SY G3KFG
CT3 1TZ G3KTZ
CT3 1UA G3RWF
CT3 1UH G4KJS
CT3 1XW 2E0XFD
CT3 1XW M7DLX
CT3 2LP G1MXM
CT3 2NH G8DBU
CT3 3AQ G6EPL
CT3 3HZ G6WSF
CT3 4DB G4NEE
CT3 4HL 2E1JEH
CT3 4JN G7VQW
CT3 4LN G4PLS
CT34DS G0TXA
CT4 5EX G4ZHN
CT4 5JX M6WBA
CT4 5LU M7VBO
CT4 5PN M6GDA
CT4 6DN G3MLO
CT4 6HX G4AZG
CT4 6JB G6GES
CT4 6JQ G7LNB
CT4 6NP G4GWI
CT4 6RN M0IBX
CT4 6RT G4BQS
CT4 7AH G4ZIF
CT4 7ND G0VRW
CT4 7ND G7MPV
CT4 7NN G3XAQ
CT4 7QL G0ONB
CT5 1EL G0IAA
CT5 1JQ M3PPU
CT5 1JQ M3RZY
CT5 1JZ 2E0GTB
CT5 1NS G4SIL
CT5 1QF G4RIS
CT5 1QF M3BQN
CT5 1RR 2E0RUT
CT5 1RR M6NMZ
CT5 2DH G7EOE
CT5 2DS G8BNK
CT5 2DY G8WIR
CT5 2ES 2E0VTF
CT5 2ES M0IUG
CT5 2ES M6VFD
CT5 2LA 2E0TDO
CT5 2LA M0ZTD
CT5 2LA M6CWX
CT5 2LB G8NXQ
CT5 2LE G4LTS
CT5 2LE M6BIE
CT5 2LZ M7AKN
CT5 2NW G4PYA
CT5 2PH G0FAE
CT5 3AR M0KFJ
CT5 3EJ G3KXB
CT5 3JZ G0IFS
CT5 3PE G4ELP
CT5 3PE G4ICM
CT5 3PJ G0AUW
CT5 3QD G4SEJ
CT5 3RF G8YMN
CT5 4BZ G0BEX
CT5 4DN G4GRJ
CT5 4DT G4WZZ
CT5 4DT G4XAB
CT5 4EL G0PPY
CT5 4EQ M1DHI
CT5 4HX G4CZU
CT5 4LA G1PVA
CT5 4LA M3MRS
CT5 4LQ 2E0BEI
CT5 4LY G7UWW
CT5 4LY M6HAY
CT5 4NL G8BUI
CT5 4ST G0LJD
CT5 4ST M6NHS
CT5 4TE M3PNB
CT5 4TH G0NBB
CT5 4TH G1PRH
CT6 5QX 2E0TUT
CT6 5QX G8NEL
CT6 5QX M6OND
CT6 6BH G1KQE
CT6 6DX G7EIA
CT6 6EQ 2E0TDJ
CT6 6EQ M0TKW
CT6 6EQ M6GJI
CT6 6GZ G8ELP
CT6 6HG G0ANK
CT6 6HU G4ZIH
CT6 6JA G4WZQ
CT6 6JS M6AFC
CT6 6NT G0ILO
CT6 6PP G8PPQ
CT6 6QZ G1FNN
CT6 6RE 2E0AIT
CT6 6RF G0HVC
CT6 6RF M1DHY
CT6 6SB G0LXB
CT6 6SB G0OXE
CT6 6SB G8ESW
CT6 6SE G2FMW
CT6 6SR G4SUK
CT6 6SS M3BBY
CT6 6UQ M0UCJ
CT6 7DW G4EVD
CT6 7EQ G8FEZ
CT6 7EW G1TQH
CT6 7HB G0ALJ
CT6 7HD G1AGW
CT6 7HG G6CVD
CT6 7LE G1EDK
CT6 7LE M7RYD
CT6 7NA M3BOV
CT6 7QA M3CUH
CT6 7RS G0IHI
CT6 7RS G1VNM
CT6 7RS G4MKI
CT6 7TA M6BNY
CT6 7TB M3PSE
CT6 7UW G0ETI
CT6 7UX M7KEC
CT6 7XB M3SEY
CT6 7XF 2E0MDC
CT6 7XF G1DSZ
CT6 7XF M3YMG
CT6 7XG G6VRI
CT6 8AD G8KOL
CT6 8AE G4XOE
CT6 8AN G7MIF
CT6 8HG G4URD
CT6 8HX G4VMZ
CT6 8HX G4ZZK
CT6 8JA M0HWM
CT6 8JS G6PKS
CT6 8JS M3KTV
CT6 8LN G6RMA
CT6 8LP 2E0ATZ
CT6 8LS G0LAA
CT6 8LT G0HMK
CT6 8LZ G7LFQ
CT6 8NA M7BFQ
CT6 8QW G8KDU
CT6 8QW M6CHZ
CT6 8RX G1DKY
CT6 8SD G4RKV
CT6 8TU G0NFG
CT6 8UQ G0EAG
CT7 0BT M0PJG
CT7 0DX G8NAV
CT7 0EL G0ANW
CT7 9AJ M6AYS
CT7 9AS G1UEA
CT7 9AZ G0INT
CT7 9DF G0CIX
CT7 9ED G1JEH
CT7 9HE 2E1AJB
CT7 9HE M7AJB
CT7 9NA 2E1EHY
CT7 9QD G3OND
CT7 9QH G0CTQ
CT7 9QN M1TCI
CT7 9RS 2E0EPP
CT7 9RS M6LPP
CT7 9SR G3VID
CT7 9UQ G0MUZ
CT7 9XE M6SIH
CT79TY M3TGJ
CT8 8AN G4SBD
CT8 8AN G4SEK
CT8 8AN G8NZA
CT8 8AP 2E1DDZ
CT8 8AP G1NLQ
CT8 8AP G7MEZ
CT8 8BP M6KCC
CT8 8BT G4KPV
CT8 8BX 2E1CQP
CT8 8BX G0TBU
CT8 8HR 2E0CNU
CT8 8HR M6BOA
CT8 8LW G3PNT
CT8 8RJ G0CEY
CT88AH G6KKW
CT9 1FE G0WVA
CT9 1HU G0GUE
CT9 1NS 2E0EMN
CT9 1NS M6HCZ
CT9 1TR G4YGB
CT9 2EJ M0AJC
CT9 2EN G3YEQ
CT9 2EN M0MVS
CT9 2NH M6EBL
CT9 2NS G0SJP
CT9 2PS G0GNQ
CT9 2SL M3CAE
CT9 2SW G0HRS
CT9 2SW G0PDZ
CT9 2SW G7HTI
CT9 2SW M5IC
CT9 2TD M6DGJ
CT9 3BH 2E0JFK
CT9 3DU G3YWF
CT9 3EF 2E1COV
CT9 3ES G8WMW
CT9 3HD G3SVJ
CT9 3HD G8ATD
CT9 3HJ 2E0XPY
CT9 3HL G0LGE
CT9 3HL M6LII
CT9 3JB G0DFI
CT9 3JB G1DUJ
CT9 3PT M0RST
CT9 3PT M1CBT
CT9 3PX G6KNU
CT9 3PX M6LCQ
CT9 3RG M6ISN
CT9 3RX G0CHN
CT9 3RX M6UFF
CT9 3SL 2E1PDQ
CT9 3XJ G0NEP
CT9 3XT 2E0EMB
CT9 3XT M6AKJ
CT9 3YA M1BBS
CT9 4DH G3YUH
CT9 4HA G3SHX
CT9 4LL G6ILU
CT9 4NE G7NOR
CT9 5DT G7EYE
CT9 5EH G3ZZZ
CT9 5EH G8GHH
CT9 5HT G3TRX
CT9 5JA G4BSW
CT9 5JZ G0TBO
CT9 5LN G1AOC
CT9 5NE 2E1HQH
CT9 5NG 2E1HUC
CT9 5NG M6LIJ

CT9 5NP 2E0DUE
CT9 5NP M6WFI
CT9 5NT 2E1DNC
CT9 5PA G7MKF
CT9 5PF 2E0ZDE
CT9 5PF M0ZDE
CT9 5PF M6DES
CT9 5PN M6FZB
CT9 5QA G8DHJ
ct92nx M6JVS
CV1 1FZ G1XLL
CV1 1JR M6NRU
CV1 2AA G6AJC
CV1 2AL M6FKW
CV1 2AR M6JGY
CV1 2DF M7COV
CV1 2PQ M0WFI
CV1 4DS M7EJP
CV1 4EB M3YVZ
CV1 4HL M1DCV
CV1 5EA G6XMT
CV1 5FZ M7ECP
CV1 5GU M6GKX
CV10 0BA G1JWY
CV10 0BA G1WRN
CV10 0BX G4UQU
CV10 0DF G0MDN
CV10 0DH 2E0NEI
CV10 0DH M0NKE
CV10 0DH M6NEI
CV10 0DW G0VZO
CV10 0DW G3VDU
CV10 0DW G7SKL
CV10 0DZ G4NCV
CV10 0HP G8VHI
CV10 0HR G1VVL
CV10 0HR M6SIW
CV10 0JJ M7EQZ
CV10 0LB G0FBG
CV10 0LY M3TID
CV10 0NH M3FRU
CV10 0NL G7ANQ
CV10 0QA M7DBB
CV10 0SL G0LLP
CV10 0SN G4KQL
CV10 0TD G8SYE
CV10 7AJ M7RDB
CV10 7AW G1FTH
CV10 7BA M6UHT
CV10 7DE G3ZSQ
CV10 7EE G4ZUE
CV10 7ES G4AEH
CV10 7ES G7BBC
CV10 7ET G3ZOY
CV10 7HG G7USI
CV10 7HP G4YOZ
CV10 7JX M6AIU
CV10 7PW G3YTW
CV10 7SW G0NZI
CV10 7SW M3BCW
CV10 8BB 2E0FQG
CV10 8BD G6DDA
CV10 8DL M3VCO
CV10 8EG G4BEH
CV10 8HL M3CBW
CV10 8HL M3CCB
CV10 8HP M7DBJ
CV10 8JQ G8ITJ
CV10 8NH G4NHF
CV10 8NU G4BBQ
CV10 8NW M6FSJ
CV10 8PA G0REO
CV10 8PT G7FSC
CV10 8PT M6CFI
CV10 8QA M6DZY
CV10 9AR 2E0NAG
CV10 9AR M3NWY
CV10 9DQ M6HPV
CV10 9DY G0SZG
CV10 9DY M3ETQ
CV10 9DY M3EYR
CV10 9EL G0BYF
CV10 9GS 2E0BNO
CV10 9HU 2E0CJF
CV10 9HU M6AHU
CV10 9HY G0PYL
CV10 9JA M6NQW
CV10 9JH G7JXF
CV10 9JZ G1VOR
CV10 9PB G6OQV
CV10 9PQ M6JPH
CV10 9PQ M6PJH
CV10 9QJ G8OMB
CV10 9QJ M0NDA
CV10 9SD G1FLW
CV10 9SG G7CEN
cv107nh 2E0DRN
CV11 4LH G4SHY
CV11 4RE M6KVG
CV11 4XG G7LSD
CV11 4XQ G4APS
CV11 5AT G1VDP
CV11 5AT G3VLG
CV11 5AU M7LAN
CV11 5HP G8AWI
CV11 5LU G1EXV
CV11 5PD G6MTG
CV11 5RR M0GNB
CV11 5UB G1JVL
CV11 6AJ M6RVM
CV11 6BQ G4LMK
CV11 6DW M6EMF
CV11 6DZ G4NQZ
CV11 6EE G0SBC
CV11 6EZ G8GXN
CV11 6FF G6SFY
CV11 6GA G4ASQ
CV11 6HE G6UHF
CV11 6HN G3SLK
CV11 6JA G4HXC
CV11 6JR G3YTT
CV11 6UU 2E0GXG
CV11 6UU G0NCQ
CV11 6UU M6OXU
CV11 6YX M7GFB
CV12 0AX G0DCZ
CV12 0AX G0OBE
CV12 0BL M7BLM
CV12 0DE G0JUN
CV12 0EL G8GMU
CV12 0JH M0HSJ
CV12 0LD M6NKH
CV12 0LY G0LDP
CV12 0LY M6WRU
CV12 0NS M3HBM
CV12 8AL M3HSH
CV12 8BE G0IYW
CV12 8BQ M3ZNV
CV12 8DA G1LBH
CV12 8DG G1OJD
CV12 8DU M1BNI
CV12 8JG M3JOS
CV12 8LA M7BZU
CV12 8QU M6ENZ
CV12 8TA M7JTZ
CV12 8TB 2E0VWM
CV12 9AG G1VNB
CV12 9BX G0BVS
CV12 9DZ 2E0BGL
CV12 9EL M1JAK
CV12 9HP M7MND
CV12 9JB G0BIN
CV12 9LL 2E0YRF
CV12 9LL M0SLO
CV12 9LL M6YRF
CV12 9ND G3YQZ
CV12 9PR M3SYH
CV12 9PS G1ORG
CV12 9SG G6AD
CV12 9SG M0LFF
CV13 0AU M3MFU
CV13 0HQ G6CLC
CV13 0NT G1LQP
CV13 0QH G3RHZ
CV13 0QH G4IQF
CV13 6BB G1SWI
CV13 6EP M0CJS
CV13 6HN G0OEM
CV2 1AF G0BJA
cv2 1dl G1HOP
CV2 1EF G1NAT
CV2 1FP M6VOG
CV2 1NN 2E0OGB
CV2 1NN M6DTL
CV2 1NN M7RNE
CV2 1PP M3XXU
CV2 1SA G1AVF
CV2 1SA M6HKW
CV2 2JD G6FXE
CV2 2JG G4HRY
CV2 2JS G4WKB
CV2 2JS G7DBO
CV2 2NE G6FYL
CV2 2QF G7PEN
CV2 3BH G1BNX
CV2 3DX G7JSQ
CV2 3NG G8JSC
CV2 4EU G0MTP
CV2 4EZ M7BUK
CV2 4GL G0UGX
CV2 4HT M5BIR
CV2 4HT M6BIR
CV2 4JW G8MIE
CV2 4QY M0HWI
CV2 5BH G2ASF
CV2 5BH G7ASF
CV2 5BH G8SEQ
CV2 5BH M3MKV
CV2 5BT M3TSE
CV2 5EH G4GEE
CV2 5FZ 2E0FNG
CV2 5FZ M0ZAI
CV2 5FZ M6FNG
CV2 5FZ M6JOD
CV2 5GJ G6WVM
CV2 5GL G8TGD
CV2 5jw G8ZJO
CV2 5LG M0COV
CV2 5NQ G1OPA
CV2 5NU G4ZXN
CV21 1DE M6EEN
Cv21 1fb M3JUO
CV21 1HW G3OQO
CV21 1HW M0RTM
CV21 1JB G6GND
CV21 1JN 2E0HCW
CV21 1JN M0WCH
CV21 1JN M6HCW
CV21 1NJ G1VAJ
CV21 1PG G0ZMC
CV21 1PG G7TMC
CV21 1PZ 2E0PPK
CV21 1PZ M0GWK
CV21 1PZ M3PPK
CV21 2QU G1FLI
CV21 2SZ 2E0SND
CV21 2SZ G0REQ
CV21 2SZ M3REQ
CV21 2TE G1ICA
CV21 3AB M0BBH
CV21 3BD G7BNI
CV21 3LH G8VTY
CV21 3NQ 2E1LOZ
CV21 3QH G3IKL
CV21 3SZ G0COY
CV21 3TS G0FIN
CV21 4AN M6MIP
CV21 4BP G8EYY
CV21 4EF M0DEF
CV21 4HG G7KRE
CV21 4HJ M0HVV
CV21 4JY G7NFO
CV21 4LT G7SRL
CV21 4LT M0ASD
CV22 5BG G0UFW
CV22 5ET G7APD
CV22 5ET G8LYB
CV22 5HN G0DLB
CV22 5HN G0RLV
CV22 5JN M0CNE
CV22 5NY M3SWK
CV22 5NZ 2E0MTA
CV22 5NZ M6RRF
CV22 5QJ G8AQN
CV22 5RG 2E0EXA
CV22 5RG G4GZS
CV22 5RG M6NVZ
CV22 5RW G8FFZ
CV22 5RW M3AEA
CV22 6BG M3VIG
CV22 6BG M3XRP
CV22 6DD M1CIZ
CV22 6EH G1BIM
CV22 6EU 2E0LMQ
CV22 6EU M0LMQ
CV22 6EU M7LMQ
CV22 6HB G3OBV
CV22 6JF 2E0JXF
CV22 6JF M0WCZ
CV22 6JF M6JXF
CV22 6LG G4ACY
CV22 6NS M0ANH
CV22 6PG M6LIB
CV22 6PX G8HF
CV22 7AP G6CYT
CV22 7BZ G8DLX
CV22 7EG G7RGO
CV22 7EW G4SSW
CV22 7GN M6EIE
CV22 7HB M7CSI
CV22 7PT G7GAB
CV22 7PT G7LIH
CV22 7RE G4AUB
CV22 7RG G8BLK
CV22 7RG M1AXE
CV22 7TN G1EBW
CV22 7TT G7JBW
CV23 0DB G1BIN
CV23 0DE G1UUV
CV23 0EE G3TQF
CV23 0JA G4JSX
CV23 0NN G4CFG
CV23 0NZ G0GZL
CV23 0PE G0GSA
CV23 0PH G0TZH
CV23 0PH G4AZH
CV23 0SP G7DDF
CV23 0SS 2E0NWY
CV23 0SS M0NWY
CV23 0SS M6NWY
CV23 0TS M0CNH
CV23 0UH G4IXT
CV23 1BS 2E0HAB
CV23 1BS M0HOU
CV23 8BH 2E0HOV
CV23 8BH M0KNV
CV23 8BH M6TQF
CV23 8BJ 2E0VEZ
CV23 8BJ M0WDK
CV23 8BJ M7TLA
CV23 8DN G3ZJK
CV23 8HF M0GVW
CV23 8TR G3TYP
CV23 8UF G6IQM
CV23 8XQ G4VHW
CV23 9NL M7APO
CV23 9PN 2E0KWA
CV23 9PN M6TXM
CV23 9QA G8CDC
CV23 9QT M6HGU
CV23 9QT M6HGX
CV230LB 2E0CJW
CV230LB M6CJT
CV238UB G8AWY
CV3 1AN M0BVV
CV3 1AU G0DKR
CV3 1DW M6IKN
CV3 1EQ G0RDU
CV3 1GG M6DZF
CV3 1GW G8MJX
CV3 1LY M6DIU
CV3 1NZ G1HGD
CV3 1QU 2E0VRZ
CV3 2AJ G4GDY
CV3 2AX G8YKY
CV3 2HA G3HLI
CV3 2HF G8OHG
CV3 2LA M3JMI
CV3 2LZ 2E0PDP
CV3 2LZ M3PDP
CV3 2NQ G4ZDG
CV3 2NQ M6FKA
CV3 2NW G1COV
CV3 2NW G8HPV
CV3 2NW M0HPV
CV3 2PE G4ZCP
CV3 2QG G0LXI
CV3 3EQ G7MNT
CV3 3HH G1AMS
CV3 3HH M3WCQ
CV3 3HQ M6HZX
CV3 4BQ M1ECM
CV3 4FS G7LXY
CV3 5AG G0JKJ
CV3 5AU G4ROA
CV3 5DA M0IRT
CV3 5DS G6CTC
CV3 5DS G8ISJ
CV3 5HH M0BBQ
CV3 5JT G3LMQ
CV3 5LF G3TZG
CV3 5NG M6PKR
CV3 5PJ 2E0BLX
CV3 5PJ M0GKO
CV3 5PJ M3RVE
CV3 6BT G1EOK
CV3 6BT M3BMQ
CV3 6BT M3XGV
CV3 6BY M7FTF
CV3 6EB G7RWY
CV3 6EW M0BVW
CV3 6JB M0AZY
CV3 6JG M6JYI
CV3 6JH G6RZS
CV3 6JT G7VCZ
CV3 6NY M5ZAP
CV3 6RD G4BBM
CV31 1DQ G4DXN
CV31 1GE M0JJR
CV31 1LA M5ALS
CV31 1RJ G3XKE
CV31 1WA 2E0CWC
CV31 1WA M0SER
CV31 1WA M3ZCW
CV31 2JX 2E0HFK
CV31 2JX M6EFW
CV31 2JZ G0GIL
CV31 2PB G0RTI
CV31 2RB G4GEP
CV31 2TQ G3SPL
CV31 3AN M6NOD
CV31 3BD M1ERJ
CV31 3BY 2E0TRI
CV31 3BY M6NRO
CV31 3EB M6OPM
CV31 3JD M6LYJ
CV31 3JE G0KWC
CV31 3LJ 2E0JTG
CV31 3LJ M7IVY
CV32 5AJ G4TYA
CV32 5JG G4PKT
CV32 5LU G6ISG
CV32 5NU G0KPH
CV32 6BA G8UIO
CV32 6BA G8WRG
CV32 6ES M6AVR
CV32 6HE G0BNE
CV32 6HY 2E1FQY
CV32 6LR G7DIW
CV32 6PB G3ZCG
CV32 6PW G3YJA
CV32 6QG G0LXG
CV32 7BB M7PGP
CV32 7DN G0FEH
CV32 7DY M1AUK
CV32 7EY G4CYG
CV32 7HT G6GSI
Cv32 7JD G8AIM
CV32 7JD M3JHT
CV32 7JT G3HTB
CV32 7QX M7IRC
CV32 7RX G0HCR
CV32 7SN G0DWJ
CV32 7TF G6NMA
CV33 9HQ G6VQW
CV33 9HQ G1EAN
CV33 9LN G3YJN
CV33 9NP G8ESK
CV33 9PX M1CZL
CV33 9QL G1PVR
CV33 9RN G1APQ
CV33 9RZ G0ASX
CV33 9TE G4USP
CV33 9TE M3EBK
CV33 9TH G1FEO
cv33au M7DKM
CV34 4SS G6BVR
CV34 5BX G0FGC
CV34 5FE G8GEE
CV34 5HW 2E0NEM
CV34 5HW M0NEM
CV34 5HW M7PCS
CV34 5HX G6VGZ
CV34 5JW 2E0JGM
CV34 5JW M6WGH
CV34 5NJ M3NYM
CV34 5NL G7RMW
CV34 5NL G8XDL
CV34 5PB G1FIP
CV34 5PU G0OYR
CV34 5RQ G4TIF
CV34 5SN G6YPK
CV34 5TG G1HRL
CV34 5TS G4JFS
CV34 5UJ G6YLB
CV34 5UL M0GWB
CV34 5XD G0GLU
CV34 5XD G1WXF
CV34 5YQ M0LCA
CV34 6FQ G1ABM
CV34 6JD G4RCR
CV34 6JQ G3UDN
CV34 6JQ G6WAR
CV34 6JQ M0OAE
CV34 6PF G3VRW
CV34 6QA G3TTC
CV34 6QJ G0GRM
CV34 6QX M3RYZ
CV34 6QY G3PXU
CV34 6XB G1XWM
CV344AB M6NVJ
CV345FE G3GYQ
CV35 0HH M0AYA
CV35 0QX M7BSX
CV35 0QX M7MTS
CV35 0RE G1YYP
CV35 0SB G6NGF
CV35 0SS G8GDC
CV35 0SS M0BXA
CV35 0UE M0BBT
CV35 7AQ G5ART
CV35 7AQ M7RTT
CV35 7HF G1WUM
CV35 7HF M3WUM
CV35 7TL G8NTQ
CV35 7TQ G1OUY
CV35 7UA G0OMN
CV35 7UD M6NWZ
CV35 8ED G1UTJ
CV35 8ER G8CLI
CV35 8JX M0CLM
CV35 8NT 2E0RTU
CV35 8NT M6LNQ
CV35 8QP M0XTD
CV35 8TY G6XCC
CV35 9HQ G6HOB
CV35 9LY G6VGH
CV35 9NZ G3XMQ
CV35 9PW M6MUT
CV35 9QQ 2E0UYF
CV35 9QQ M6UYF
CV35 9QQ M7XBS
CV359AR 2E0IMG
CV36 4DJ M1DCX
CV36 4FD G0CHO
CV36 4FD G0SOA
CV36 4FD G1SWR
CV36 4HS G1RLD
CV36 4JU 2E0SIF
CV36 4JU M6SIF
CV36 4NG G8PYU
CV36 4NR G1AWJ
CV36 4PG G0CXJ
CV36 5DA G8HGG
CV36 5LN G0UPY
CV37 0AP G1HPB
CV37 0DN G8JXP
CV37 0DZ G3YED
CV37 0JD M6BAH
CV37 0PP G4RYM
CV37 0PP G6MMD
CV37 0RR M7NAL
CV37 0TT G1UNN
CV37 6DD G0TVC
CV37 6DR M7DYB
CV37 6RT 2E0WTH
CV37 6RT M6PCN
CV37 6ST G6SGE
CV37 6TD G0REP
CV37 6TF G0JJY
CV37 6XG G0EDT
CV37 6XQ M6SSV
CV37 7DD G0CCJ
CV37 7DW M0LGW
CV37 7EX 2E0MDX
CV37 7EX M3YOE
CV37 7JF G4IQW
CV37 7LW G8RIM
CV37 7PL G1EIX
CV37 7TL G3XZO
CV37 8EU M3XWC
CV37 8HJ M0HGA
CV37 8RA 2E1DMH
CV37 8US G3GXG
CV37 8US M6YOG
CV37 8XT 2E1DFS
CV37 8XT M0VKB
CV37 8XY G3XGV
CV37 9ED 2E0NDZ
CV37 9ED M3NDZ
CV37 9FZ G4YGY
CV37 9JN G0JUQ
CV37 9PH 2E0YOI
CV37 9PH G6JNY
CV37 9PH M7RNU
CV37 9QA M6JTK
CV37 9QL G0BKA
CV37 9QL G0BKB
CV37 9QW G8OCS
CV37 9SR 2E0VAV
CV37 9ST 2E1FCC
CV37 9ST 2E1FCD
CV37 9ST G8RDT
CV37 9XN 2E0BZA
CV37 9XN M3MXV
CV37 9XR G0WKT
CV370LT M7EBZ
CV377PL 2E0EKW
CV377PL M6AGT
CV4 7AY M6WYW
CV4 7DU G1KCR
CV4 7ED G6GLZ
CV4 7ES M7AYJ
CV4 8HF G0KNX
CV4 9AA G0OTT
CV4 9AN M6ZPT
CV4 9AU G6UCT
CV4 9AZ G8BZT
CV4 9BW G6PSO
CV4 9EX M0URX
CV4 9EZ G4VCX
CV4 9JH G1YFI
CV4 9LA G6WLM
CV4 9RN G1IUL
CV4 9SJ M7BZP
CV4 9TD G6JPR
CV47 0ES M3BOU
CV47 0HW G8OKN
CV47 1AS G1VBB
CV47 1EX G1CNY
CV47 1GH M3LQJ
CV47 1NF G3TFA
CV47 1PH G0AFZ
CV47 2AP 2E0OHN
CV47 2AP M0NPH
CV47 2AP M6OHZ
CV47 2AT G0CDO
CV47 2AT G0SDX
CV47 2BN G3ZST
CV47 2FJ G4EWW
CV47 2QS G0GNF
CV47 2RA M6GVP
CV47 2XY G1GAS
CV47 2XY G4YIE
CV47 8JS 2E0KOD
CV47 8JS M6KOD
CV47 8LG M0ZOM
CV47 8LG M6FZY
CV47 8LX G3SNO
CV47 8NE G1KCS
CV47 8NL M6VPX
CV47 8NN G3OJI
CV47 8NN G6EVC
CV471TA M0TOZ
CV47AL M3WRK
CV5 6AR 2E0CBG
CV5 6AY G3VWD
CV5 6BB G4RUN
CV5 6DJ G3PUR
CV5 6DR G3WCQ
CV5 6FF G0EMX
CV5 6GH 2E1IAX
Cv5 6lf G1MSA
CV5 6QH M6CVJ
CV5 7AD G8BHY
CV5 7AH G6MQP
CV5 7BY G8WYI
CV5 7DA M6BHT
CV5 7EH G3MAU
CV5 7FR G1KFQ
CV5 7GP G1PPB
CV5 7HP G7DTK
CV5 7JU G3NAP
CV5 7LJ G3UKD
CV5 7ND 2E0UAC
CV5 7NE G8AHR
CV5 7NR G4HOC
CV5 7QE G4MWP
CV5 7QE G4UVW
CV5 8BD 2E0UWK
CV5 8BD M6HGJ
CV5 8BD M6UWK
CV5 8DA M0XKW
CV5 8DE M1EQV
CV5 8DR 2E0RRC
CV5 8DR M0DTJ
CV5 8DR M6SDW
CV5 8FL G4DXH
CV5 9AS G0ENY
CV5 9EA M0NOI
CV5 9EN G8AZL
CV5 9HL G3JYS
CV5 9JT G4BCG
CV5 9JT G8GET
CV5 9JY G3NAY
CV5 9LN G4STI
CV5 9LQ G8RHM
CV5 9NA G4JDO
CV5 9NS G4KUE
CV6 1BE G7LAW
CV6 1BY G0EWZ
CV6 1GL 2E0LXC
CV6 1GL M7GWM
CV6 1LF G8KTX
CV6 1LH G1UUO
CV6 1NW G6KHA
CV6 1PW G7TFZ
CV6 2GE M7SKV
CV6 2GQ G1UWQ
CV6 2HB G8LGT
CV6 2NN G4KZU
CV6 2PG M3ORU
CV6 3GD 2E0COV
CV6 3GD M0HIN
CV6 3GD M0TAV
CV6 3GF G1ZNK
CV6 3GJ M3ECW
CV6 3GL G6IHO
CV6 3GS G7IZV
CV6 4EY G4XML
CV6 4FH G3CWH
CV6 4FZ M6JMJ
CV6 4GG M6IHO
CV6 4HL M3VYN
CV6 6BD M6JAV
CV6 6FA G1NML
CV6 6FF 2E0UPK
CV6 6FF G0UPK
CV6 6FF M7CQE
CV6 6GY G4OQV
CV6 6JZ G6ZTF
CV6 6NB M3JST
CV6 7EU 2E0TSP
CV6 7EU M6RPV
CV6 7JJ M6EXH
CV6 7LB M3LVP
CV7 7AB G4LMV
CV7 7FX G4IEV
CV7 7FX G6SCM
CV7 7GG G6OES
CV7 7LU G0EPV
CV7 7LW G8XUU
CV7 7NB 2E0TUX
CV7 7NB M0YMM
CV7 7NB M3POV
CV7 7NQ 2E1ACG
CV7 7NQ G4KUR
CV7 7NQ G7UBX
CV7 7QF G3UKM
CV7 7QF M6JGV
CV7 7QL M0CAR
CV7 7QN G1UUS
CV7 7QW G4JYE
CV7 7QW M3HYF
CV7 8AQ M3UYF
CV7 8AU G4IAG
CV7 8BF G4RWH
CV7 8BY M6DAD
CV7 8ER G3SZE
CV7 8HL M0BVX
CV7 8HL M0BVZ
CV7 8HY G6GYN
CV7 8JJ G3SPY
CV7 8JJ G3UVW
CV7 8JJ G3ZFR
CV7 8JX M3CVO
CV7 8LB M0NIE
CV7 8LB M0SQC
CV7 8LE M6KKD
CV7 8PR G1UUK
CV7 8PR G6KXN
CV7 9AQ M0BVU
CV7 9AU G6VJR
CV7 9AU G7AMQ
CV7 9BJ G4LML
CV7 9BX M6TNY
CV7 9FH M0DVG
CV7 9JE G1BXQ
CV7 9LR G0AIZ
CV7 9LT G1OPJ
CV7 9NJ G4WXK
CV7 9NQ M0OCV
CV8 1AH G7RYO
CV8 1BE G1UUT
CV8 1DG G6XVZ
CV8 1DJ M0AZJ
CV8 1EZ G0KPG
CV8 1FL G4GJY
CV8 1GH M0LDY
CV8 1GL M3UKV
CV8 1HX G6GQF
CV8 1JD M6OAO
CV8 1JW G3VTL
CV8 1RB G8CYA
CV8 2AF M0BNC
CV8 2BB M0JDB
CV8 2BE G6XRI
CV8 2BX G4YMT
cv8 2dj M7CSF
CV8 2DU G0DKV
CV8 2EE G0JMW
CV8 2JZ M6ENC
CV8 2PP G8YHO
CV8 2XH G7JYY
CV8 3ET G6WTD
CV8 3FR G0HOV
CV8 3HH M0JCT
CV8 3JF G3JZL
CV9 1ET M0WHR
CV9 1ET M3WHR
CV9 1HN G1SIM
CV9 1HP 2E0EKA
CV9 1HP G8LQF
CV9 1HP M6IOF
CV9 1JY M0VEX
CV9 1NZ G4IWA
CV9 1PS G0LKA
CV9 1QJ M7JLX
CV9 2AA G7TYJ
CV9 2DA G4XDE
CV9 2DN M7AIZ
CV9 2ET G4YNX
CV9 2EZ M1CNL
CV9 2EZ M6TAD
CV9 2PJ M7DDT
CV9 2PL G8SEI
CV9 2RP G6SGY
CV9 3BP M3GTQ
CV9 3EH 2E0NDW
CV9 3EJ G0VZK
CV9 3EJ M7JAY
CV9 3EX G0HUW
CV9 3EX M6HLT
CV9 3HA M7DQI
CV9 3HJ M6KDO
CV9 3LT G4KMT
CV9 3PJ G8NDB
CV9 3RE G2FXQ
CV91NX G6MUW
CV91QA G0KKV
CW1 3AF 2E0ZSK
CW1 3AF M6ZSK
CW1 3AX G8DTT
CW1 3AX M6FUJ
CW1 3EG M0MTF
CW1 3EQ 2E0NKP
CW1 3EQ M6NKP
CW1 3JN 2E1EHP
CW1 3JN M3EHP
CW1 3LE 2E0BDP
CW1 3NW 2E0FQC
CW1 3NW M6FQC
CW1 3PJ 2E0AZZ
CW1 3PJ 2E1UJE
CW1 3PJ M5WGD
CW1 3QD G7PEH
CW1 3SG 2E0KNC
CW1 3TT M7CLI
CW1 3XJ G6WZY
CW1 3XN M0MBG
CW1 3XN M6BCV
CW1 3YN M7RCR
CW1 4AL 2E0RMS
CW1 4AL 2E1RMS
CW1 4AL M0TDC
CW1 4AL M3AHS
CW1 4AL M3RMS
CW1 4AU M3KQW
CW1 4BS G7BMT
CW1 4DA 2E1RAO
CW1 4DE M3DMJ
CW1 4DE M6MHD
CW1 4DF 2E0NPP
CW1 4DF M3ZJF
CW1 4DF M3ZMV
CW1 4DU M3USB
CW1 4DY G0TWH
CW1 4EL 2E1FIV
CW1 4ES G1DBI
CW1 4GH 2E0HBK
CW1 4GH M7KLJ
CW1 4HB M7EDL
CW1 4HL M3ISG
CW1 4HZ 2E0HST
CW1 4HZ M0KDA
CW1 4HZ M0WFX
CW1 4JU M7BKZ
CW1 4JZ M3VFL
CW1 4NW M6RGF
CW1 4PS M6HRE
CW1 4PX M6JUG
CW1 4QE 2E0FTV
CW1 4QE M6FTV
CW1 4QE M6POE
CW1 4RA G4DBX
CW1 4RP M0UTD
CW1 4RS 2E0CKM
CW1 4RS M6AVU
CW1 4TD M6TGS
CW1 4TT G6INM
CW1 4TX M6CSI
CW1 4TY G0HIZ
CW1 5LF 2E0FFM
CW1 5LF G7UUG
CW1 5LF M6JTO
CW1 5PB 2E0DCP
CW1 5PB M6DEO
CW1 5PW M6MXY
CW1 5PX 2E0CPE
CW1 5PX 2E0ECO
CW1 5PX M0HPW
CW1 5PX M0MRH
CW1 5PX M6EXC
CW1 5QE G6OBT
CW1 5RA G0CGQ
CW1 5SY G3ZFC
CW1 5YF G0AVE
CW1 5YQ M0CJK
CW1 6ES M3HQS
CW1 6HD G4DQG
CW1 6HN 2E0RAL
CW10 0AE 2E0SHK
CW10 0AE M0IUV
CW10 0AE M6IYU
CW10 0DG M3LVY
CW10 0DL M0BMR
CW10 0EA M6DMA
CW10 0HR G3MCC
CW10 0JS G4CFK
CW10 0PF 2E0CDQ
CW10 0PF M0WJH
CW10 0PF M6AFY
CW10 0PJ G4ZKR
CW10 0TA M6WPA
CW10 9AU 2E1GDA
CW10 9BY G0NTG
CW10 9HG G0GKN
CW10 9HL 2E0DWF
CW10 9HL M6WDF
CW10 9NW M3KKQ
CW10 9QY M1BUR
CW11 1BL M3NNZ
CW11 1BN M0CJI
CW11 1BS G6MPN
CW11 1BZ G6ZGB
CW11 1EB G0BRZ
CW11 1EP G6RXD
CW11 1FW G4RXB
CW11 1FY G4GVE
CW11 1GF G7JHE
CW11 1JL 2E0NRL
CW11 1JL M7ECT
CW11 1RS M3TAE
CW11 1RW 2E0ZAC
CW11 1RW M0BZI
CW11 1SG G7OOI
CW11 1SQ G4MXE
CW11 1WN M3LIV
CW11 1XP 2E0FUH
CW11 1XP M0EEG
CW11 1XP M3FUH
CW11 1XP M3NBX
CW11 1XU M3SIS
CW11 3AR M6DBQ
CW11 3AS G0JNE
CW11 3AU 2E0JZC
CW11 3AU M6JZC
CW11 3BL 2E0CXN
CW11 3BL M6DVA
CW11 3BL M7NHF
CW11 3BU 2E0OSY
CW11 3BU G4OHP
CW11 3BU G6IIU
CW11 3BU M6OSY
CW11 3FZ G1XRJ
CW11 3JF G7HOA
CW11 3JF M0OBW
CW11 3JF M1CNY
CW11 3LP M7RNO
CW11 3QL G4UCT
CW11 3RD G6BMG
CW11 3SW M7HDJ
CW11 3TA G7CTE
CW11 3TE M7GCA
CW11 4BP M0JEH
CW11 4HL G4VIF
CW11 4HS G7SHI
CW11 4HZ 2E0WIS
CW11 4HZ M6JLE
CW11 4HZ M6WIS
CW11 4PQ G6VPJ
CW11 4RE G4PMY
CW11 4RZ G1GRN
CW11 4SP G3HUD
CW11 4SP G3JOE
CW11 4TN G6UKN
CW12 1AG 2E0GRS
CW12 1AG M7FIH
CW12 1DT M6JGG
CW12 1FS G3TVX
CW12 1HE G3WNW
CW12 1LT M6PYF
CW12 1NU G0REB
CW12 1NU G4LTZ
CW12 1NY G7PZF
CW12 1QW M6CLX
CW12 1SD G7VTQ
CW12 1SE G4JYK
CW12 1SE G4NVN
CW12 1SH G6DZX
CW12 2BH 2E0FGI
CW12 2BH M6EGB
CW12 2BS 2E0SOT
CW12 2BS M3SOT
CW12 2BY M6JKR
CW12 2DX G6TDJ
CW12 2EP M3KXE
CW12 2EQ M0OWO
CW12 2HN G3WRK
CW12 2HQ G7HIO
CW12 2JJ G7RMJ
CW12 2JU M3JCH
CW12 3DB M0AVQ
CW12 3DE G4GMZ
CW12 3DZ M3KWZ
CW12 3EE M6NND
CW12 3EP 2E1CPQ
CW12 3FF M6GCD
CW12 3HH M6IGJ
CW12 3HQ M6DEI
CW12 3HS G4UJO
CW12 3JA M6ZOM
CW12 3JJ M6BSO
CW12 3JP G7NCD
CW12 3JY G8JPU
CW12 3JY M3ZBL
CW12 3LD M0TCL
CW12 3LL G6VMV
CW12 3LZ M6ZJT
CW12 3PL G1LCE
CW12 3PL G4FSH
CW12 3QG G0DOC
CW12 3QG M6ZZD
CW12 3RB G3ZAU
CW12 3RH G4TAG
CW12 3RQ 2E0ICY
CW12 3RQ M0RZY
CW12 3RQ M6ORE
CW12 3RX M6AMX
CW12 3SW M6RDA
CW12 3SW M6SJE
CW12 3TD 2E1CAT
CW12 3TD G8WWF
CW12 3TU M6SBY
CW12 3TW M1EER
CW12 3TW M6VCB
CW12 3TY G8RXY
CW12 4BS 2E0CJD
CW12 4BS M3SQT
CW12 4DY G4WTZ
CW12 4EA 2E0TXE
CW12 4EA M7TUX
CW12 4EQ M6WLE
CW12 4FR G4HMX
CW12 4FR M6EAL
CW12 4HL G8UVZ
CW12 4JL M0VVG
CW12 4JL M1LOL
CW12 4JL M1REK
CW12 4JN M0GVE
CW12 4JR M3ZQB
CW12 4NE G0UXO
CW12 4NN G6ORT
CW12 4PH G0CQY
CW12 4PR M6XEP
CW12 4PS M3AVZ
CW12 4PY G7CPN
CW12 4QH G4IWV
cw12 4qq M7TDL
CW12 4QR M6EHC
CW12 4QU G6XYV
CW12 4QX G6MAT
CW12 4RF G1VIY
CW12 4RR G6LAO
CW12 4RS G6WXI
CW12 4SN M3HXM
CW12 4TQ G4BOH
CW12 4UX M6MNU
Cw13tl 2E0NNH
Cw13tl M0NCC
Cw13tl M0NNH
cw15ny M7ONO
CW15PN M3MVK
CW2 5AW G0NWF
CW2 5AW G4BZI
CW2 5AZ 2E1GDB
CW2 5AZ G8SIG
CW2 5AZ M3MAR
CW2 5AZ M3ZTT
CW2 5BJ G0TMK
CW2 5BY 2E1EGV
CW2 5ez 2E0PWJ
CW2 5EZ M6EFP
CW2 5GP G4VXX
CW2 5GS G0FAU
CW2 5LX G0GRO
CW2 6HN 2E0ZHG
CW2 6HN M3ZHG
CW2 6JB G8DHQ
CW2 6RD G0DRM
CW2 6SQ M0XPA
CW2 6TG G1LSN
CW2 6TJ G6IGW
CW2 6TX M0VFR
CW2 6TX M0YVX
CW2 6TX M6ZBT
CW2 6XP G3NKC
CW2 7DP M3BMI
CW2 7HD M3KKF
CW2 7JY M3KPF
CW2 7LE M6KRV
CW2 7LN 2E0UGP
CW2 7LN M6UGP
CW2 7LQ G0DRN
CW2 7NY 2E0DYB
CW2 7NY M0SWC
CW2 7NY M6DYB
CW2 7SY M6UPT
CW2 8AP G0MMH
CW2 8AP G4RRM
CW2 8AP G6TW
CW2 8BD M1LEO
CW2 8BD M3AGB
CW2 8BN G4YFO
CW2 8BT G7IFI
CW2 8DS G1GRM
CW2 8EX G7GZZ
CW2 8HG G4MAG
CW2 8HJ G4LVR
CW2 8HP G0SIE
CW2 8HS G3ZBM
CW2 8LN 2E0IHL
CW2 8LN M7EGJ
CW2 8NA G8ZSK
CW2 8NH 2E0JDD
CW2 8NH M3LZK
CW2 8NS M3EUU
CW2 8PB G7PWU
CW2 8PL M3NVF
CW2 8QQ M3SDH
CW2 8RE M0RCT
CW2 8RL 2E0MCG
CW2 8RS G7OMN
CW2 8SH G6LVT
CW2 8SZ G8BTD
CW27JS 2E0BCJ
CW3 0AH 2E0KZL
CW3 0AH M3LYP
CW3 0BA G3YMU
CW3 0BE 2E0COX
CW3 0BE M3MIO
CW3 0JB G0NBH
CW3 0JB G1XHR
CW3 9BS M3KEF
CW3 9DN G8FNG
CW3 9EG G4YPQ
CW3 9EQ G3JJR
CW3 9EU G1VTN
CW3 9SZ M3YHQ
CW3 9SZ M3ZHY
CW4 7AS G3GMM
CW4 7BT G0CTP
CW4 7DR M3IFE
CW4 7HA 2E0MHT
CW4 7HA M0MHT
CW4 7LA G1DBL
CW4 8AS M3HYV
CW4 8DT G0CSY
CW4 8HX M0SUN
CW4 8JB G4NVA
CW4 8JG M0ESR
CW4 8JG M0ZED
CW4 8JG M6DNM
CW4 8JG M6NVA
CW4 8JU G4ZVA
CW4 8LJ M0YIJ
CW4 8LL 2E0OCG
CW4 8PG M3JWZ
CW4 8PP 2E0SKA
CW4 8PP M3BMV
CW5 5JE M6LPB
CW5 5JS M6WAP
CW5 5NW G0EEI
CW5 5QJ M0ELO
CW5 5RX M7MUS
CW5 5TX M6GWN
CW5 6AL G3KMV
CW5 6AL G8UHT
CW5 6DX G4XNR
CW5 6ED G0RBA

CW5 6HJ G4PCR
CW5 6HL G0FUZ
CW5 6JJ G6SNI
CW5 6NL G0RBW
CW5 6QF G0VJR
CW5 6SQ M0JEG
CW5 7BA M7FRT
CW5 7BB G3HEH
CW5 7BJ G8AVO
CW5 7EJ G3PFE
CW5 7ER G0OEW
CW5 7FL G4SHC
CW5 7GU 2E0JCD
CW5 7GU M0JCD
CW5 7HY G4DBD
CW5 7JA G0CZD
CW5 7NN G1BDQ
CW5 7NX G4MHX
CW5 7PN G4PUM
CW5 7QD G4PNC
CW5 8AQ M3FJN
CW5 8AQ M3FJR
CW5 8AQ M3HAK
CW5 8BG G6PGG
CW5 8DJ M7FOK
CW5 8ER M7CSR
CW5 8JD G0ZPV
CW6 0BA 2E0RJM
CW6 0BA M0RJM
CW6 0BA M3RZV
CW6 0DA G0TQJ
CW6 0GJ G8NXY
CW6 0HJ G3TRL
CW6 0LA M7HLA
CW6 0NL G4VPC
CW6 0PN M7AVO
CW6 0PU G3NHP
CW6 0QR M3SHK
CW6 0SD G7VOI
CW6 0SX G3VUN
CW6 9AJ M0SOA
CW6 9BP M1EIU
CW6 9JT G4KZO
CW6 9QX G4XMX
CW6 9RT M6AYI
CW6 9TF G0MMT
CW7 1DZ M3RVX
CW7 1EU G1ILH
CW7 1FE M3JIC
CW7 1HA M0RDR
CW7 1LU G0GZI
CW7 1LU G6DSA
CW7 1LU G7DSA
CW7 1LU M3NLJ
CW7 1LY 2E0FZE
CW7 1LY M7GTA
CW7 1LZ M7TLG
CW7 1NE G7RYN
CW7 1NE M1ATI
CW7 1NJ M6CAG
CW7 1QL M1ARS
CW7 1QT M7BRZ
CW7 1QZ M3HGP
CW7 1RE G4TEZ
CW7 1RJ 2E0KNU
CW7 1RJ M6KNU
CW7 1SW G4JVX
CW7 2BJ G0SPH
CW7 2ED 2E1HEV
CW7 2ED G7GRO
CW7 2LJ M1DJP
CW7 2LL M6TKR
CW7 2NE M6GLN
CW7 2NG M7EFG
CW7 2QD G0EOL
CW7 2SE G0FOU
CW7 2SE M0TGS
CW7 2UE M3XFD
CW7 2UW G6WEL
CW7 2YR 2E0PXW
CW7 2YR M3PXW
CW7 3DE G7KFZ
CW7 3EN G0ADU
CW7 3EN G6WZZ
CW7 3JB 2E0RJD
CW7 3JB M0JFD
CW7 3JB M3LKD
CW7 3JU M1BXM
CW7 3LE G1DBR
CW7 3NG G8UYB
CW7 4DP M3YUA
CW8 1BL M6WFZ
CW8 1BW G0VOK
CW8 1ER M7MCY
CW8 1HX 2E0HGE
CW8 1HX M3HGE
CW8 1JR M3YET
CW8 1LA 2E0SDK
CW8 1LA M3KDK
CW8 1PL G6HXU
CW8 1PY G3ZTT
CW8 1PY G4XUV
CW8 1RD G1GYJ
CW8 2BQ G0JIT
CW8 2JB G3YWU
CW8 2LW G4MIH
CW8 2PB G1YJR
CW8 2PL M6GMP
CW8 2PP G6AFG
CW8 2XJ G4JYP
CW8 2XN G6DQO
CW8 3DX 2E0LDJ
CW8 3DX M0NEX
CW8 3DX M1ALA
CW8 3DX M6LDJ
CW8 3EZ M0XEY
CW8 3HD G4YVI
CW8 3HH G1GOP
CW8 3JD G7VDQ
CW8 3LP M3KJB
CW8 3PN G4KCZ
CW8 3PN G6NBP
CW8 3PT G4IAB
CW8 3RH G6LCS
CW8 4AA G4XDG
CW8 4BA G0DQQ
CW8 4DF G0HXD
CW8 4EH M6SLB
CW8 4GS M0WKG
CW8 4GS M6CSS
CW8 4HR G6GAK
CW8 4JU G0WOA
CW8 4JX G4ROK
CW8 4PU G0BKH
CW8 4PU G1MCG
CW8 4PX M6PJV
CW8 4TQ M0SCN
CW8 4TQ M7SCN
CW8 4UE M7PBX
CW8 4XA G8NCS
CW8 4XG 2E0IXM
CW8 4XG M6IXM
CW8 4YU M7TPD
CW81SH M6ENX
CW9 5JL 2E0DHJ
CW9 5JL M6EXB
CW9 5LJ G0DND
CW9 5LJ M3CRV
CW9 5NT 2E0RNI
CW9 5NT M3CUU
CW9 5PS G1LDC
CW9 5QR G4OJF
CW9 6DA G8BJA
CW9 6DR G4XQB
CW9 6DS G1MVE
CW9 6EB G1URR
CW9 6ED M3GYH
CW9 6EZ G6LDM
CW9 6EZ M3OUQ
CW9 6EZ M3OVG
CW9 6HJ G0THJ
CW9 6LN 2E0XOJ
CW9 6LN M3XOJ
CW9 6PP G8PHZ
CW9 6PX G8VNX
CW9 7AN M6RJJ
CW9 7AR 2E0DHV
CW9 7AR M6EQO
CW9 7AR M6TFZ
CW9 7AS M3ISH
CW9 7BF M3GHG
CW9 7EP M3LZR
CW9 7JA G0OGJ
CW9 7JB 2E1GYG
CW9 7JB G1GOQ
CW9 7JQ M3XRQ
CW9 7JT 2E0VRC
CW9 7JT M6VRC
CW9 7LD M3CGI
CW9 7NL M7SCB
CW9 7QD M3XQV
CW9 7QD M3XQY
CW9 7QH 2E0HGX
CW9 7QH G4PKX
CW9 7QH M3HGX
CW9 7QJ M0CRP
CW9 7RG G0WPO
CW9 7TY M6EWU
CW9 8AR G6LJX
CW9 8BN M3NDJ
CW9 8GF G7KTH
CW9 8GG G6ZTT
CW9 8GG G7LQD
CW9 8GP G4MUA
CW9 8JE 2E0OJN
CW9 8JE G8XIN
CW9 8PB 2E0STX
CW9 8PB M0WTX
CW9 8PB M0XSJ
CW9 8PB M6EAX
CW9 8PB M6MHJ
CW9 8PB M6WTX
CW9 8QA G0LBO
CW9 8QQ G4CAX
CW9 8RQ M0FCT
CW9 8TQ M7UAM
CW95PX M3KKN
DA1 1ND G4ZHX
DA1 1QT M3YJU
DA1 1TR G6UML
DA1 1YY M3ALB
DA1 2QN M1CXK
DA1 2RZ G6MVN
DA1 2TX G8PHJ
DA1 3AJ M1CYN
DA1 3AN G3ZPS
DA1 3AN M3RZX
DA1 3AP M0JVT
DA1 3BA G7KAO
DA1 3BL M7OMD
DA1 3BP G0RJL
DA1 3DE M6CAF
DA1 3DE M6GAF
DA1 3DE M6KRF
DA1 3EE M6BPC
DA1 3EW M7CAQ
DA1 3JU G4XKZ
DA1 3JX M0TGV
DA1 3NX G1RCV
DA1 3NX G3RCV
DA1 4PZ M6KVL
DA1 4RY G8ZHR
DA1 4SN G7WLL
DA1 4SU G4APB
DA1 5HT G6CMB
DA1 5JW G6TSC
DA1 5JW G8XJB
DA1 5NF M3YZN
DA1 5NG 2E0GLT
DA1 5NG M0HPF
DA1 5NG M7DXZ
DA1 5TN M7UCH
DA10 0LP G0KDV
DA10 0LT G0NZR
DA10 0LU M6WCQ
DA11 0AH M7TJF
DA11 0PH G4WTE
DA11 7AQ G7MIE
DA11 7BN G8JAD
DA11 7BN M1ENQ
DA11 7BN M3JUL
DA11 7EB 2E0IJH
DA11 7EB G8TRF
DA11 7EB M0JKX
DA11 7EB M0JUQ
DA11 7EB M7CDF
DA11 7EZ G0DWS
DA11 7LB G8IEA
DA11 7LG G0RNP
DA11 7NY M3UUZ
DA11 7QG G3NPS
DA11 7QG G8LDV
DA11 8NH M1DUV
DA11 8NQ G4ALD
DA11 9DU M1CWG
DA11 9DU M6KIG
DA11 9DX G1YXY
DA11 9FF M0HZF
DA12 2LP G7EOC
DA12 3BU M7PHZ
DA12 4AR M3TJI
DA12 4BJ G3DCV
DA12 4EL 2E0OOM
DA12 4EL M6FDY
DA12 4HD G7CJS
DA12 4HJ G3RE
DA12 4HJ G4IYK
DA12 4NA G4BBJ
DA12 4NJ G8XIR
DA12 5BD G1PHZ
DA12 5BD G1PNL
DA12 5DN G1OFL
DA13 0BW G3KBH
DA13 0EA G0OAT
DA13 0LS G3BAC
DA13 0QA G0RJN
DA13 0SQ G4YGU
DA13 0TQ M6DDF
DA13 0TX 2E0RCL
DA13 0TX M0RBX
DA13 0TX M6DBI
DA13 0UD G0AFH
DA13 0UD G0FBB
DA13 0XB G1FCN
DA13 9DS G3PBF
DA13 9EJ G0SSN
DA13 9EJ G6CBY
DA13 9JR 2E0EDA
DA13 9JR M6HHF
DA14 4BE G8MCA
DA14 4ET G4YMF
DA14 4LJ G7EDA
DA14 4PS G6INU
DA14 4QU G0BAX
DA14 5LT G8PJF
DA14 5NF 2E0YEK
DA14 5NF G4FAA
DA14 5NF M0NOY
DA14 5NF M6NOY
DA14 5NG G6GTC
DA14 6JQ G3BNE
DA14 6JT G8XIN
DA14 6SG M1BXU
DA15 7NL G4ILH
DA15 7PA 2E0JQB
DA15 7PA M0KUK
DA15 8AT 2E0SDW
DA15 8AT M0SJW
DA15 8AT M3NHI
DA15 8ER G1TXO
DA15 8JN G8JTG
DA15 8QL G1HBR
DA15 8QY G4NHP
DA15 8RX M0DDV
DA15 8SZ G8KAM
DA15 8TA G1PRZ
DA15 9DZ G8OPA
da158rd 2E0ERX
DA16 1DE M1TAD
DA16 1EJ G0KPZ
DA16 1JG M7AUB
DA16 2AW G8SWC
DA16 2BN G6NRH
DA16 2BP G0KTS
DA16 2HJ G2DZH
DA16 2HX G8GKC
DA16 2LH G1OAZ
DA16 2QD G0GKI
DA16 2RU G6ODW
DA16 3AW G0LZF
DA17 5BB 2E0KHX
DA17 5BB M0KBX
DA17 5BB M6KHX
DA17 5BG G8MIF
DA17 5BT G7UHW
DA17 5EW G3CUR
DA17 6HB G4WJH
DA17 6JE M0PBR
DA17 6LP G0VWF
DA2 6BN G7FIA
DA2 6DN G1JNQ
DA2 6HD G0IPC
DA2 6HE M6NEY
DA2 6HE M6THS
DA2 6HZ G1ZAW
DA2 6JS G4OFU
DA2 6JX G6ZAY
DA2 6LQ M6MRV
DA2 6NB G4CVC
DA2 7ES G4IQQ
DA2 7LP G0HRD
DA2 7NW G0DVL
DA2 7PB G0KUU
DA2 7QQ G4LGU
DA2 7RL M6CMO
DA2 7RL M6WLF
DA2 7SP G3XVC
DA2 7SP M6JOP
DA2 7WB G1EWE
DA2 8BH G1UXZ
DA2 8BH G7JYL
DA2 8BZ M0DPS
DA26HZ G7PXM
DA3 7HB M7GBR
DA3 7HD G3FGP
DA3 7LA 2E1JGD
DA3 7LA M7JGD
DA3 7NS G6XND
DA3 8DD G1FJJ
DA3 8EU G4NRV
DA3 8LG G8RW
DA3 8LG M0XPS
DA3 8LN 2E0RBI
DA3 8LN M0RJT
DA3 8LN M3RBI
DA4 0BE 2E0IAG
DA4 0BE G7FUV
DA4 0BE M0UAT
DA4 0BE M3YBJ
DA4 0BE M3ZBQ
DA4 0HA G0WMC
DA4 0HA M1ABU
DA4 9AR M1CGP
DA4 9DQ 2E0WAF
DA4 9DQ G4BWV
DA4 9DQ M3WPK
DA4 9EX G6YLD
DA4 9HS M3YOU
DA5 1LX G6IRP
DA5 2AB M7DOK
DA5 2ER G3XEW
DA5 2ES G4CW
DA5 2HN M3VSO
DA5 3AH G0FDZ
DA5 3BT G6YZB
DA6 7LA M1DYD
Da6 7ny G6AVP
DA6 7PA G4MB
DA6 8HU G4RPP
DA6 8JS G1MAV
DA7 4 G4DDP
DA7 4 G8BXC
DA7 4AJ M6ASM
DA7 4JL G4XKV
DA7 4JZ G4KOW
DA7 4LY M3JPG
DA7 4PG G6CUE
DA7 4PQ 2E0WUF
DA7 4QE M6BZG
DA7 4RL G0FAS
DA7 4ST G8CXI
DA7 4TZ M0HFR
DA7 4UE M1CZA
DA7 4UX G7MJJ
DA7 5BT G7UUB
DA7 5BT M3FTI
DA7 5DG G6ENO
DA7 5DZ G3KHR
DA7 5HG M3USW
DA7 5JU G0PXT
DA7 5JU M3KVJ
DA7 5NP G8JZT
DA7 5NP G8TNK
DA7 5PS 2E0LRR
DA7 5PS M6LXR
DA7 5SL G7GPI
DA7 6AF G4DUM
DA7 6NL G0UAO
DA7 6QU G4GZN
DA7 6SG M6BLT
DA8 1DZ 2E0GPG
DA8 1DZ M0YUG
DA8 1DZ M6GPG
DA8 1JQ M3JLV
DA8 1LU G4LSU
DA8 1NN G4MHJ
DA8 1NN M1AJT
DA8 1PZ M0JHP
DA8 2AQ M1CXN
DA8 2DR 2E0XUZ
DA8 2JD G4EGU
DA8 2JD M3VNX
DA8 2JG 2E0WTZ
DA8 2JG M1CBK
DA8 2NN M6LKD
DA8 3BA M0LHA
DA8 3BL G3NRZ
DA8 3LZ G6NNU
DA8 3NG 2E0JET
DA8 3QT M0SHI
DA8 3SN G8YLC
DA9 9PG G8MHI
DD1 2AP MM0GDI
DD1 2JA 2M0KAU
DD1 2JA MM6KAU
DD1 3AW GM7TTU
DD1 4LY GM0TOF
DD10 0BA MM7AGL
DD10 0DN MM6MJC
DD10 0HW MM6ZZG
DD10 0HX GM4PKJ
DD10 0HX GM8ZEQ
DD10 0NG MM7EAR
DD10 0PH 2M0BXN
DD10 0RY MM3NWF
DD10 0SB GM4TXN
DD10 0SL GM0ENQ
DD10 0SR GM1MLS
DD10 0SW GM3YAO
DD10 0TG GM1TDU
DD10 0TT GM0WRV
DD10 8AZ MM1COS
DD10 8JL MM6SSO
DD10 8PW MM3DFG
DD10 8TW GM3COQ
DD10 9AQ MM6XGT
DD10 9BX MM3PFA
DD10 9DD GM5CGA
DD10 9EJ GM0ARH
DD10 9HE MM7PPW
DD10 9RR MM0BIX
DD100SL MM7DRV
DD11 1ST MM7NNM
DD11 2LZ GM4SXJ
DD11 2NX GM0MIW
DD11 2NX GM4YWU
DD11 3BZ MM0CXA
DD11 3EF 2M0GXZ
DD11 3EF MM0KGS
DD11 3EF MM6GZS
DD11 4BH MM3NJV
DD11 4EZ 2M0DOI
DD11 4EZ MM6GFG
DD11 4HT 2M0ULD
DD11 4RA GM4YWS
DD11 4RH GM0WNS
DD11 4SR GM1XHZ
DD11 4SX GM0SHD
DD11 4SX GM3GBZ
DD11 4UX MM6IBB
DD11 5FG MM0HRI
DD11 5RH MM6CWB
DD11 5RH MM6JNB
DD11 5SS GM1MON
DD11 5ST 2M0MAV
DD11 5ST MM0LBX
DD11 5ST MM6LAD
DD11 5SY GM3OBG
DD2 1BF 2M0RTO
DD2 1BF MM6RTO
DD2 1JT MM0DNH
DD2 1PJ GM8JVZ
DD2 1QN MM6DBJ
DD2 2BU GM8UMN
DD2 2BU MM3XGS
DD2 2EZ 2M0LEW
DD2 2EZ MM0LWS
DD2 2EZ MM6LEW
DD2 2FB 2M0MFK
DD2 2NH MM6AKQ
DD2 2PW GM3OQI
DD2 2PW MM0AKX
DD2 3AF MM6RAM
DD2 3BF 2M0NYK
DD2 3BF MM6YYB
DD2 3BU MM7NOZ
DD2 3JR GM3WPA
DD2 4EG GM3WJE
DD2 4HP MM6BXH
DD2 4TT 2M0XDS
DD2 4UA GM4JCM
DD2 4XB GM4VXM
DD2 5AH 2M0SEF
DD2 5DL GM0KKE
DD2 5PX GM4LUD
DD2 5PZ 2M0INE
DD2 5PZ MM6PAU
DD2 5QJ GM1CMF
DD2 5QN GM0TAY
DD2 5QN GM4FLP
DD2 5RY GM8LON
DD3 0BN GM4ZFS
DD3 0EA 2M1KOJ
DD3 0EA MM6KOJ
DD3 0JR MM3NPG
DD3 0LT MM6AAR
DD3 0PH GM4API
DD3 0PR MM6PRH
DD3 0QN 2M0CVK
DD3 6AQ MM0DTW
DD3 6DD GM3ZXB
DD3 7HG MM7LLK
DD3 8AF 2M0XZX
DD3 8EJ GM3ZTP
DD3 8PY GM5BDW
DD3 9LH 2M0MGY
DD3 9PL GM0ROU
DD3 9RG GM0FSW
DD4 0AD GM1KCH
DD4 0NR GM4AAF
DD4 0NR MM0DXD
DD4 0NR MM3JFW
DD4 0PP GM7ONJ
DD4 0PW GM0GYQ
DD4 0QU GM0NLU
DD4 0TN 2M0RTA
DD4 0TN MM0RTT
DD4 0TN MM6RBT
DD4 0UJ GM7OWU
DD4 0XJ MM6SGQ
DD4 0XZ MM6JVQ
DD4 7NR GM8LUK
DD4 7EL GM0PIV
DD4 7NA MM7DVM
DD4 8AP MM6GID
DD4 8NN MM3PXG
DD4 8RP MM6DDX
DD4 9EX GM1VWA
DD4 9HH MM7DWV
DD4 9HQ GM0DNH
DD4 9LP GM0ISA
DD4 9ND MM0TMG
DD40XQ MM7ASP
DD5 1QT GM3NHQ
DD5 1QW GM1JTK
DD5 1QW GM7GOE
DD5 2EU GM4CAB
DD5 2RE GM1MUY
DD5 2RJ GM6BML
DD5 3AT GM0URU
DD5 3BN GM0RKU
DD5 3BT 2M0DOL
DD5 3BT MM6MIS
DD5 3DQ MM3SYU
DD5 3EF GM0CQL
DD5 3JG GM4JPZ
DD5 3LE GM3YVX
DD5 3PD GM8OEG
DD5 3PL 2M0LMY
DD5 3PL MM7LMM
DD5 3QN GM4UGF
DD5 3RE MM3NLH
DD5 3WN MM3HKE
DD5 4AA GM7DPI
DD5 4HG GM7AUX
DD5 4HS MM6KLT
DD5 4HT GM4MUZ
DD5 4LD MM1CFC
DD5 4SW GM0VOL
DD5 4TS MM0DRA
Dd51dy GM3LIW
DD6 8AP GM0TGG
DD6 8DT GM4FSB
DD6 8HL GM3MOR
DD6 8HL GM4RXW
DD6 8JH MM0BTD
DD6 8LF GM0CNW
DD6 8ND GM8RPE
DD6 8NG MM7LWC
DD6 9DT 2M0CLN
DD6 9EX MM3ZUP
DD6 9JD 2M1IBX
DD6 9LG GM4RDI
DD6 9NE MM3FZI
DD6 9NX GM4OLH
DD7 6DQ MM1ELE
DD7 6EX GM4FEI
DD7 6HT MM1EDY
DD7 6HW GM6RAK
DD7 6JY 2M0NGO
DD7 6JY MM3XLO
DD7 7BR MM0GGD
DD7 7BR MM1DSD
DD7 7PJ GM3XJE
DD7 7QF MM0EQE
DD7 7QF MM1EQE
DD7 7QQ MM3XFP
DD7 7SQ GM4JEJ
DD7 7TB GM0EFT
DD8 1AD GM3KEZ
DD8 1DQ GM1JWJ
DD8 1JR GM0DGK
DD8 1UE MM6BHS
DD8 1UF GM3ZBR
DD8 1UF GM4GOW
DD8 1UN GM0BTK
DD8 2AT 2M0YEQ
DD8 2AT MM0YEQ
DD8 2AT MM3YEQ
DD8 2PQ 2M0EDV
DD8 2PQ GM8RSC
DD8 2PQ MM3IZO
DD8 2SD MM7DYA
DD8 2SP GM0TWB
DD8 2UZ GM4RWE
DD8 3AG GM4XKP
DD8 3ES GM4WZY
DD8 3EY GM6AAJ
DD8 3EY MM0BSX
DD8 3TU GM1TCN
DD8 5ab MM3KVV
DD8 5ET MM3KVX
DD8 5LD GM4YHS
DD8 5NT MM6GYL
DD85DL MM7RFW
DD9 2PJ MM3HAF
DD9 6BG MM7MHJ
DD9 6JD GM4LFL
DD9 6RQ GM0JZV
DD9 6SD GM7DAP
DD9 6UH GM3KC
DD9 6UH MM0ERK
DD9 7AE 2M0SYH
DD9 7AE MM7SAH
DD9 7AX MM7EKZ
DD9 7QE MM0BDA
DD9 7SZ GM6RGD
DD9 7UX GM4YRE
DE1 1HW M6RIS
DE1 3RW M1CWY
DE11 0BH G0VVF
DE11 0DH M7BPW
DE11 0DS G1PKR
DE11 0EB G3OMT
DE11 0HA M7MAW
DE11 0JR G4WPE
DE11 0LY M0APK
DE11 0LZ G6DCS
DE11 0NB G1PKV
DE11 0RX G4EWK
DE11 0RX M1FBF
DE11 0SL M6CGM
DE11 0SP 2E1FLD
DE11 0SR M3THN
DE11 0TG G7EJO
DE11 0UU 2E0KPO
DE11 0UU 2E0NBR
DE11 0UU G8UZQ
DE11 0UU M0KPO
DE11 0UU M0SCW
DE11 0UU M3KPO
DE11 7EZ 2E0FNO
DE11 7EZ M0JEO
DE11 7EZ M0URJ
DE11 7EZ M7AHA
DE11 7HG M6MAX
DE11 7HH M7HKP
DE11 7LN M7DNM
DE11 7QU G6NZL
DE11 7QU M6YAD
DE11 8AA 2E0EWL
DE11 8AA G4RJO
DE11 8AD G8VBC
DE11 8BG G0WHC
DE11 8BX M6RVN
DE11 8FS M6MLG
DE11 8HD 2E0XPK
DE11 8HD M3XPK
DE11 8LX G1ZMW
DE11 9AF M6XPC
DE11 9AQ M6EZF
DE11 9AT G7AYB
DE11 9BQ G7LAS
DE11 9BW 2E0FGM
DE11 9BW M0OSI
DE11 9BW M6HAG
DE11 9DZ G1OWK
DE11 9EJ 2E0TOX
DE11 9FB G7JHW
DE11 9FB M3JQY
DE11 9FF 2E0MYA
DE11 9FF G4UWA
DE11 9FF M7MYC
DE11 9FH G1SNQ
DE11 9FH G4ZDE
DE11 9HD G8FVC
DE11 9HF M7XHF
DE11 9PH M1BFV
DE11 9PJ 2E0BGV
DE11 9PJ M3FKV
DE11 9QN M6GHV
DE11 9QU M6DYP
DE11 9SG 2E0KEJ
DE11 9SG M0TRZ
DE11 9SG M7KEJ
De110et M7PWD
DE12 6BL G4AYS
DE12 6DN M6GLJ
DE12 6EX 2E0DAQ
DE12 6EX M3VWD
DE12 6HH M6ICA
DE12 6HQ G1YBM
DE12 6JJ M3KBY
DE12 6JJ M3KRB
DE12 6JL M3XWP
DE12 6LU G4CRT
DE12 6PL M6XAQ
DE12 6PZ M3UDF
DE12 7AS G1OAX
DE12 7BD G1MVQ
DE12 7EG G6JTV
DE12 7JG M3XUT
DE12 7JH M0SDP
DE12 7JZ G7MGX
DE12 7JZ M0PCA
DE12 7JZ M1LYN
DE12 7NA G7KVZ
DE12 7PW G6MYO
DE12 7QG G7EIK
DE12 7QU G0HYR
DE12 8BJ G4YBP
DE12 8DL M0APD
DE12 8ES G1VQH
DE12 8JJ 2E0IDY
DE12 8JJ M0NXD
DE12 8JJ M7NDO
DE12 8JW G7EHU
DE12 8LB G0OES
DE12 8NJ G4HIW
DE126NS 2E0EZO
DE126NS M6LYC
DE128HG G6DIC
DE13 0AL G6NSQ
DE13 0AY M7NHJ
DE13 0BT M3BJE
DE13 0BT M6RFF
DE13 0DQ G4YFZ
DE13 0EE G4GKZ
DE13 0FD M6BSW
DE13 0FN M6IGZ
DE13 0FQ 2E0EZG
DE13 0FQ M0XMX
DE13 0FQ M6OML
DE13 0JD G4RAC
DE13 0JD G8VBW
DE13 0JH 2E0PKS
DE13 0JH G5RJH
DE13 0JH M6GOU
DE13 0JP G1ATA
DE13 0LD G1OYZ
DE13 0LD G1OZD
DE13 0LR M1CBO
DE13 0ND M1CDT
DE13 0NN M3GPM
DE13 0NT M3RCW
DE13 0NU G4XKL
DE13 0NY G4IPV
DE13 0RD M3FKN
DE13 0RD M3IEW
DE13 0RF G4HBY
DE13 0SQ M3HRV
DE13 0UU G4VPF
DE13 0WJ G1PRW
DE13 7AJ G3CRH
DE13 7EB G6ZVR
DE13 7EZ G1DQU
DE13 7EZ G4UJW
DE13 8AB G7VSM
DE13 8DS G7KIF
DE13 8DZ M6IOY
DE13 8JR G4RWD
DE13 8NT G4CHI
DE13 8PT G4ICZ
DE13 9AA G6EIH
DE13 9AG G8VBA
DE13 9AR G0JLU
DE13 9AR G4LIR
DE13 9BJ G0FNH
DE13 9BJ G1OYU
DE13 9DE G4SXE
DE13 9DS G4KBP
DE13 9DY 2E1IIL
DE13 9EG G0SHO
DE13 9JR G0AOD
DE13 9JR M3RHK
DE13 9LE G3BZB
DE13 9NH G8OZP
DE13 9QD G0JEE
DE13 9RT G3TOY
DE14 1BS G6TJK
DE14 1BW M3WJY
DE14 1QS M6FDY
DE14 2EX 2E0SMX
DE14 2EX M6TDO
DE14 2FL G8OWA
DE14 2JS G4TBQ
DE14 2NB G8WSY
DE14 2NP G1NBO
DE14 2SN M3KQR
DE14 2SW M3VLH
DE14 3BY M0HFA
DE14 3DA G6FLY
DE14 3DY M3AUF
DE14 3DZ G8HZI
DE14 3FL G0HFL
DE14 3FQ 2E0GSB
DE14 3FQ M3VDF
DE14 3GJ M3OTS
DE14 3JE 2E0CGU
DE14 3JE M3HVA
DE14 3Jf G6AZL
DE14 3LR G0FSF
DE14 3SB 2E0JBY
DE14 3SB M6SRI
DE14 3TX 2E0KLR
DE15 0AB G4DIH
DE15 0AD M6JGR
DE15 0AR G6OVA
DE15 0BA G7RUC
DE15 0BW 2E0CGJ
DE15 0BW M0KAJ
DE15 0BY G0VUC
DE15 0DL M6TNL
DE15 0EJ G0IDD
DE15 0PA M7DFY
DE15 0PT G0CQS
DE15 0QJ G4PGJ
DE15 0QJ M7GPL
DE15 0QW G0TZM
DE15 0RY G8EBM
DE15 0SL G4ELG
DE15 0TT G6JPQ
DE15 9BJ M0AGS
DE15 9BN 2E0ZAI
DE15 9BN M0ISZ
DE15 9BN M6MGJ
DE15 9EB G0HJR
DE15 9EN M1EWB
DE15 9EY G4PGX
DE15 9FR G6BRY
DE15 9QN G3NFC
DE15 9QN G8EKG
DE15 9QW M3XXG
DE15 9SE G8UUR
DE21 2DH M6TDA
DE21 2HY 2E0ZCG
DE21 2HY M1BUQ
DE21 2HY M3ZCG
DE21 2HZ M6HFQ
DE21 2LH M1CFZ
DE21 2UG 2E0RCB
DE21 2UG M3LFP
DE21 4DX M0NMD
DE21 4GA 2E0TGV
DE21 4GB M3DGN
DE21 4HQ G3HCS
DE21 4HQ G8HCS
DE21 4JL G1VQB
DE21 4JY M6FFI
DE21 4LG 2E0TCU
DE21 4LG M6WCA
DE21 4LY 2E0JNH
DE21 4LY M6DDY
DE21 4QY 2E0KGY
DE21 4QY M7RKD
DE21 4RF G6VFA
DE21 4RL 2E0OES
DE21 4RL M0TRY
DE21 4RL M3XUG
DE21 4SQ M7WLF
DE21 5AP G4AOA
DE21 5AR 2E1ICU
DE21 5AR G4IBW
DE21 5AR M6RJF
DE21 5AX G4FQZ
DE21 5BN G4RAR
DE21 5DY M0IMF
de21 6ap M6TZA
DE21 6DX M7LGV
DE21 6DZ M3JVW
DE21 6FW M0CNY
DE21 6FX G4TBK
DE21 6FZ M1CWV
DE21 6LW M6GDL
DE21 6NX G1NZN
DE21 6NX G7AFW
DE21 6PH G1DHY
DE21 6PL G1UJX
DE21 6SH G4EYN
DE21 6SJ 2E0EBQ
DE21 6SJ M0IEQ
DE21 6SJ M6GWQ
DE21 6TR G1XFE
DE21 6TZ 2E0BFA
DE21 6TZ M3MVV
DE21 6WB M7ECY
DE21 6WX G8BFA
DE21 7AT 2E0PET
DE21 7AT M0KEF
DE21 7AT M6PRM
DE21 7DS M5GAC
DE21 7DZ G4UMT
DE21 7EN G3OCA
DE21 7EN G3ZBI
DE21 7ES G3JFT
DE21 7FX M6MGT
DE21 7GL G1UYZ
DE21 7GT M0BJT
DE21 7JW G6WYD
DE21 7JZ G3OXN
DE21 7LR 2E1GOC
DE21 7LR G0IDL
DE21 7LX G6JPS
DE21 7QA G6FVJ
DE21 7RG 2E0SYD
DE21 7RG M3IIP
DE21 7TP M3NTZ
de214nd 2E0MXL
de214nd M7MRL
DE216JS M7AFU
DE22 1EQ G4OMT
DE22 1EY G8RBK
DE22 1GE M3AWQ
DE22 2BJ G3SZJ
DE22 2BJ G4PGV
DE22 2EH G6NLE
DE22 2ET M7NHG
DE22 2GN 2E1DIA
DE22 2GR M0ADL
DE22 2GW G6DNH
DE22 2HA 2E1FBS
DE22 2HF G1BEJ
DE22 2HF G4FPY
DE22 2HG G0RII
DE22 2HH M3WMS
DE22 2JR G5GIH
DE22 2LN 2E0GWB
DE22 2LN G3YOO
DE22 2LN M0HLP
DE22 2LN M6BJM
DE22 2LN M6GWB
DE22 2NA G6WXJ
DE22 2NB M0MPO
DE22 2NB M6ZTB
DE22 2RE G0GDL
DE22 2SN M3ULN
DE22 2TA M6FHX
DE22 2ur 2E0WCE
DE22 2ur M6MHW
DE22 2XP G4LOF
DE22 3AS G0VVA
DE22 3EF G6EBL
DE22 3GL G7ODZ
DE22 3JP G4CPQ
DE22 3JT M6PTY
DE22 3LS 2E0RLG
DE22 3LS M6RLG
DE22 3RW M6KMC
DE22 3UE G4CZH
DE22 4AQ M6JDM
DE22 4BW G3SMV
DE22 4FL G7RWW
DE22 4HN G3PWS
DE22 4JU 2E1ADJ
DE22 4JX 2E1EDD
DE22 4LQ M6ZLL
DE23 1BZ G4CMZ
DE23 1DN 2E0KDH
DE23 1GN 2E0MNG
DE23 1GN M6MNG
DE23 1PR M3RCI
DE23 1QA M6OUA
DE23 1RH G6XOR
DE23 1WN G0IOZ
DE23 1XF 2E1GDF
DE23 2FR 2E0NYT
DE23 2FR M6NYT
DE23 2UB G3HAG
DE23 3RL G4COR
DE23 3XU M1MPK
DE23 4BJ M3JBZ
DE23 4EJ M6GNV
De23 4RS M6KQS
DE23 6AQ 2E0NPF
DE23 6AQ M7BHC
DE23 6BL G7SJS
DE23 6BW 2E1DBT
DE23 6DA G7GXE
DE23 6DL G3URU
DE23 6EB M6JDF
DE23 6EN G4BGD
DE23 6FB 2E0OJB
DE23 6FB M3NZK
DE23 6FT 2E0KRS
DE23 6GA M0HLS
DE23 6GA M6BFO
DE23 6HB G0WQY
DE23 6HB G8MXT
DE23 6JD G8TNE
DE23 6JD M3JBB
DE23 6JD M3TNE
DE23 6JY M0DVQ
DE23 6NF M0OTS
DE23 6NF M5EHG
DE23 6NQ G6CWW
DE23 6PF G1VAB
DE23 6PS 2E1CZJ
DE23 6TE M0GPK
DE23 6TG G3RLO
DE23 6TG G6XGF
DE23 6XL G3FUJ
DE23 8AX 2E0DLP
DE23 8AX M0OAC
DE23 8BP 2E0GCG
DE23 8BP M6GBE
DE23 8BP M6RDI
DE23 8LH M6EPJ
DE23 8PR 2E0SIK
DE23 8PR M6ISK
DE23 8PR M6USA
DE24 0AQ G8YCK
DE24 0BP G7LPV
DE24 0BU 2E0RUZ
DE24 0BU M0RUZ
DE24 0BU M6RUZ
DE24 0DJ G0PRY
DE24 0ER G0PHC
DE24 0FJ G6SOX
DE24 0FT G0DKX
DE24 0GH G1HEU
DE24 0GQ G4FAE
DE24 0HJ M7JJT
DE24 0HZ G6KUI
DE24 0JU G4YVV
DE24 0LF 2E0VRR
DE24 0LF M6RRV
DE24 0LH G0BIE
DE24 0LP G7SZG
DE24 0LQ G3ERD
DE24 0LQ G4AKE
DE24 0PX G8BFC
DE24 0SA G3KWY
DE24 0TA G4LNQ
DE24 0UQ 2E1HVZ
DE24 0UY M1EDL
DE24 0UY M6NZS
DE24 3AA G3WSM
DE24 3AG M1WTL
DE24 3AG M3WTL
DE24 3BH 2E0WHO
DE24 3BH G0WAC
DE24 3BS G1LKK
De24 3dd M6KKM
DE24 3DT M0LKN
DE24 3DZ G0IWF
DE24 3EA G4MVD
DE24 3HE G0CPO
DE24 3JA G4PTZ
DE24 5AF 2E0BNH
DE24 5AF 2E0WBX
DE24 5AF M0XFX
DE24 5AF M3SKV
DE24 5AF M6WLR
DE24 8AZ G7BJB
DE24 8BD G4PDE
DE24 8DA M3PRS
de24 8ds M7AVE
DE24 8DU M6SPK
DE24 8EX M6PKT
DE24 8GT G0RUR
DE24 8JR M6WET
DE24 8NF G0TJT
DE24 8NP G6UGT
DE24 8NS M6TLK
DE24 8PY M7FHG
DE24 8RN G4SNN
DE24 8WD G7VNO
DE24 8YE M7EAC
DE24 9BT G1IWT
DE24 9DL M6TIO
DE24 9FQ 2E0VEW
DE24 9GY 2E0FOR
DE24 9GY M6FOR
DE24 9LZ M1AJQ
DE24 9NG G6CXI
DE24 9PD G4YYI
DE24 9PF 2E0RTE
DE24 9PF M6RCG
DE24 9PT G6XEN
de248at 2E0JWW
de248at M3NVG
De248dg M7JKC
DE24SD 2E0FMO
DE3 0DF G4HJL
DE3 0DU M3JIK
DE3 0ED M0WSB
DE3 0EE G4YIZ
DE3 0PA G7BHY

DE3 0PH M3LFG
DE3 0QQ G4ZHD
DE3 0QT G4MBH
DE3 0RD G0MSS
DE3 0RD G7SVQ
DE3 0RG M1FAJ
DE3 0RR G3TOV
DE3 0RS M3JKG
DE3 0TB M3JXI
DE3 4LQ G7BJC
DE3 9AH G8GBU
DE3 9AQ G6MWS
DE3 9BD M1MTV
DE3 9FJ G4KRW
DE3 9FL G7WGP
DE3 9FP G6WZE
DE3 9FT G8YNK
DE3 9FY G1UZS
DE3 9GH G7GJI
DE3 9GH G8QZ
DE3 9GH M3DPS
DE3 9GQ G0DMK
DE3 9GT G0CRN
DE3 9HB G4KOJ
DE3 9HZ G7GEX
DE3 9JT G1KKD
DE3 9JW G2DAN
DE3 9LN G4BKO
DE3 9LU M3JZE
DE4 2AH M3FUV
DE4 2GG G7RBY
DE4 2GG G7VWN
DE4 2GL G0KKD
DE4 2HF G3RBP
DE4 2HF M5AA
DE4 2HP G0OTJ
DE4 2JJ G1CJC
DE4 2JW G4XTK
DE4 2PW G6HWR
DE4 3BA 2E0VKS
DE4 3BA M0VKS
DE4 3BA M3VKS
DE4 3BT G4ZEY
DE4 3EU G7NBJ
DE4 3EW 2E1AWZ
DE4 3GP G7NLJ
DE4 3JW G3YGB
DE4 3JW G8GGR
DE4 3QU G0FSB
DE4 3SD G3RLX
DE4 3SN G1TCH
DE4 3TB M7PVP
DE4 3TE G0BJD
DE4 3TG M6RJR
DE4 4AD G4UIQ
DE4 4AD G6NMU
DE4 4EX M0CVS
DE4 4FR G1SGP
DE4 4PG G0MMO
DE4 4PX 2E0HSY
DE4 4PX M7GWH
DE4 5AR M6BOR
DE4 5BH G8DLP
DE4 5DG G4HTW
DE4 5DJ G0FVU
DE4 5DJ G6JNZ
DE4 5DJ M3NVA
DE4 5DJ M3WTC
DE4 5EG G3VLF
DE4 5FF G8RNU
DE4 5FS G0ACZ
DE4 5GT G7EJK
DE4 5HN G4UUQ
DE4 5LA G0OWC
DE4 5LT M3OGD
DE42DT 2E0GHJ
DE42DT M0JKS
DE44EG 2E0VMG
DE44EG M7AXJ
DE45 1BH G4JIK
DE45 1DD G6EAZ
DE45 1DX G7ABR
DE45 1DX M3DXD
DE45 1FG M0PKV
DE45 1JB G4GWX
DE45 1NJ G3VOT
DE45 1QQ G3SPV
DE45 1RE M6WIL
DE45 1TP G6CSC
DE5 3ES M7XDE
DE5 3FG G0WWO
DE5 3FG G1OLT
DE5 3FL G7GZU
DE5 3HD G1YPT
DE5 3JL 2E1HFN
DE5 3LJ G0NYM
DE5 3PJ 2E0GRM
DE5 3PY G4GGL
DE5 3RE G3NJX
DE5 3RR G3ZYC
DE5 3RR G8ZYC
DE5 3RY M6RWE
DE5 3SP 2E0VRA
DE5 3SP G5VRA
DE5 3SP M7VRA
DE5 3TR 2E1EOK
DE5 3TR G8ERV
DE5 8JG G0NNU
DE5 8JG G7OWZ
DE5 8PQ G4DXY
DE5 8PQ G6TFL
DE5 8RF 2E1AZA
DE5 9QN G8HNT
DE5 9RB G8ZIY
DE5 9RN M0XTX
de5 9rr M7AIQ
DE5 9SP G3YQL
DE55 1LG M6HQN
DE55 1LN G6KPW
DE55 1RL 2E0IPA
DE55 2AJ G7DGF
DE55 2AJ M0GNU
DE55 2AS G1UAZ
DE55 2BJ G0DGQ
DE55 2BR M5RST
DE55 2DN G4CHM
DE55 2EJ G1BFK
DE55 2EP M0DAG
DE55 2ER M0HXO
DE55 2HE 2E0CND
DE55 2HE M3CND
DE55 2HE M6BYS
DE55 2HS G8VSN
DE55 2JD G0VBX
DE55 2JD G7SJD
DE55 2JD M3IQM
DE55 3AP G6TRQ
DE55 3AW M3DAF
DE55 3NW M3FYM
DE55 4AA M6EPQ
DE55 4AF G4KBI
DE55 4AG G1NWH
DE55 4AG G7ANO
DE55 4AP 2E0MFS
DE55 4AP M0IEM
DE55 4AP M6SMX
DE55 4BL G1OGC
DE55 4BP G3JWQ
DE55 4EX M3TFK
DE55 4JT M6JSR
DE55 4JY 2E1BIT
DE55 4LA 2E1DBQ
DE55 4LA M0SBA
DE55 4LH M1GHT
DE55 4LT G4OHV
DE55 4NF G6TLA
DE55 4PB M6PDK
DE55 5AJ M0TRP
DE55 5HD G4AMF
DE55 5HL M3HUN
DE55 5HT G7BHW
DE55 5HU M3EAT
DE55 5JH M6TTE
DE55 5JJ M3EUW
DE55 5JL G6LSD
DE55 5JL M5SRE
DE55 5JN M1DDI
DE55 5LL G4VNG
DE55 5LQ G4PCL
DE55 5LT M3VLO
DE55 5LU G0DXT
DE55 5NA G4THI
DE55 5NJ 2E0FUO
DE55 5NJ 2E1SWS
DE55 5NJ G1SWS
DE55 5NJ M0XSH
DE55 5NJ M7DSH
DE55 5NJ M7SWS
DE55 5NR M6HFR
DE55 5QH G4EKD
DE55 5TD 2E0WYT
DE55 5TD M0WYT
DE55 5TD M6WYT
DE55 5TT G6ZDB
DE55 5TY 2E0SBK
DE55 5TY G6AZE
DE55 5TY M0BAY
DE55 5TY M0BMY
DE55 5TY M6SLA
DE55 6BT 2E0RNU
DE55 6BT M0RNU
DE55 6BT M6RNU
DE55 6DX M0SMG
DE55 6EH G3LGK
DE55 6GW M6KLX
DE55 6JG G1AII
DE55 6JW G4DTP
DE55 6LD M3ZHW
DE55 6LH G7GLL
DE55 7DG G4NID
DE55 7DG G4UTN
DE55 7ER G1XJM
DE55 7ER M0YJD
DE55 7ER M7JMD
DE55 7FL M3TXV
DE55 7HT G7SMZ
DE55 7HX G0GHD
DE55 7JN G4LSV
DE55 7JP G3GRL
DE55 7JX 2E0SBX
DE55 7JX M6HMW
DE55 7JX M6SME
DE55 7JY G7SMD
DE55 7JY G7TYP
DE55 7JY M3SUJ
DE55 7LP G4EWZ
DE55 7NH 2E0CEM
DE55 7NH M0ILM
DE55 7PL G3TIX
DE555QL G8XZX
DE56 0EA G1RYQ
DE56 0EU M6GKM
DE56 0HD M0WMX
DE56 0HL M3FXQ
DE56 0HN G0ORC
DE56 0HN M0KBZ
DE56 0NS G4JTX
DE56 0PF 2E0CTM
DE56 0PF G7EMZ
DE56 0PF M3UJY
DE56 0PG G4ANV
DE56 0PQ G0RLJ
DE56 0PY G8IHA
DE56 0QH G7HHZ
DE56 0QR G4RVL
DE56 0SH M6GWF
DE56 0TG G4NXL
DE56 0TZ G1NMI
DE56 0UB G6SKK
DE56 1AP G1BZE
DE56 1BX G8VYO
DE56 1EE G0MGX
DE56 1EJ G3ZYD
DE56 1FQ M0TVX
DE56 1GH 2E0YPG
DE56 1JG G4VSI
DE56 1LX 2E1HES
DE56 1NE G0ROD
DE56 1PB G8GJC
DE56 1PD G0FRY
DE56 1RE M6FED
DE56 1UF G7NAE
DE56 1UT M0XXI
DE56 1UT M1EAI
DE56 2AL G3PDD
DE56 2BA G4UBR
DE56 2FN M7CGI
DE56 2GR G0UOV
DE56 2GR G8GTD
DE56 2GS G0JNK
DE56 2HP G6JFE
DE56 2LH 2E0FGY
DE56 2LH M6FGY
DE56 2TR M0NOV
DE56 2UW G0AEU
DE56 4BH G3SNR
DE56 4DF G4RVU
DE56 4DP G4DJP
DE56 4DR G3ROD
DE56 4FJ G7UCB
DE56 4FX G6TQC
DE6 1AT G4HIJ
DE6 1BB G8PKM
DE6 1BJ G1NDL
DE6 1BR G3TVU
DE6 1DF G1MUT
DE6 1EX 2E0HWE
DE6 1EX G0DLS
DE6 1EX M7JWE
DE6 1LZ G0BPZ
DE6 1QB G6YWV
DE6 1TZ M6SMD
DE6 1UD G7NZO
DE6 2AS G0LAU
DE6 2EH G3KAG
DE6 2FB G0DJQ
DE6 2GD M7NWZ
DE6 2GL G6MQD
DE6 2GY G8NOP
DE6 2LP G0CQP
DE6 2LP G7KRM
DE6 3AG M3IJF
DE6 3AJ G3OQT
DE6 3BQ M7DFP
DE6 3EN G4DKV
DE6 3FB G0CXD
DE6 3FE G6RBM
DE6 4JS G3UBS
DE6 4LP M0BAJ
DE6 4NG G7DDR
DE6 4NJ G7EUT
DE6 5BE M3MQI
DE6 5HP 2E0MGA
DE6 5HP M6MJI
DE6 5JZ G6YWN
DE6 5NY G3LKV
DE6 6NB G4MIT
DE65 5DL G1PUU
DE65 5ER G4PYI
DE65 5FD 2E1BUM
DE65 5FX 2E0JWC
DE65 5FX G0WOC
DE65 5FX G7RIS
DE65 5FX M0KCA
DE65 5FX M6JWC
DE65 5FX M6YNK
DE65 5FY G1JJT
DE65 5HP G7BJE
DE65 5JG M3LFV
DE65 5LP G4RJA
DE65 5QQ G1TKY
DE65 6DW G7ILL
DE65 6EB G0HLI
DE65 6ES G6CZD
DE65 6FX M3XHN
DE65 6GP G0MRY
DE65 6GR G0DMS
DE65 6GT G8VAN
DE65 6GT G8WSQ
DE65 6HD G4PWY
DE65 6HL G0SUT
DE65 6HT 2E0DWE
DE65 6HT M6GJD
DE65 6HY G3ZIF
DE7 4DA G1ZHZ
DE7 4DF M0AIT
DE7 4DL G6CDT
DE7 4EH 2E0GBU
DE7 4EH M6TOZ
DE7 4EZ M6HNI
DE7 4HD M3IZH
DE7 4HD M3IZM
DE7 4HQ M0CQF
DE7 4JY G3XRD
DE7 4LE 2E1GOK
DE7 4NE M0BCJ
DE7 4PW G0ITL
DE7 5AP 2E0CTC
DE7 5AT G0LUI
DE7 5AT M6LUC
DE7 5DW G6IQP
DE7 5EF G3IFX
DE7 5EX G4XCK
DE7 5HR G4AIB
DE7 5HT G8NBI
DE7 5NZ M0BCC
DE7 5RB M0BQT
DE7 6AU G4UFX
DE7 6AW G3RKZ
DE7 6AX 2E0VGK
DE7 6AX M3VGK
DE7 6DD G1GNP
DE7 6DE G8DKV
DE7 6EG G4NOB
DE7 6ER M6NYJ
DE7 6EZ G3NNN
DE7 6FR G4LPF
DE7 6GB 2E1GMV
DE7 6GB G4DAM
DE7 6GB G4EDD
DE7 6GU G8KSW
DE7 6HW G8RJF
DE7 6LP M6WNY
DE7 6LW G1GXW
DE7 6LW M6GKK
DE7 6PE M3ZYV
DE7 6XB G7UUP
DE7 8AL 2E1BFP
DE7 8AW G8WRY
DE7 8EY G1DMH
DE7 8JA M7KSD
DE7 8NL G0JVB
DE7 8PW 2E0FSH
DE7 8PX G6KBC
DE7 8SJ 2E0DTF
DE7 8SJ M0WCA
DE7 8SJ M6TGU
DE7 8TD 2E0HGA
DE7 8TD M0RLF
DE7 8TD M7RBE
DE7 8WB G0KBN
DE7 9HE G1YEV
DE7 9HJ M0IVL
DE7 9HJ M3AQK
DE7 9HJ M6EPD
DE7 9JW G3ZDK
DE7 9LF M1DYC
DE72 2AU G0JKC
DE72 2AU G0NHR
DE72 2BJ G0PBM
DE72 2BJ G4NYB
DE72 2DF G4RBZ
DE72 2DQ 2E0GEN
DE72 2DQ M0TDY
DE72 2DQ M6CSX
DE72 2GE M6CDD
DE72 2GL 2E0AVW
DE72 2GL M0LJC
DE72 2GS G3ZOW
DE72 3AQ 2E0RRF
DE72 3DF M6KMI
DE72 3DZ G4EHX
DE72 3EE G4EAX
DE72 3GN G8KEA
DE72 3HB G7TLL
DE72 3JJ M0SDS
DE72 3LN G8YZC
DE72 3NP G0DPQ
DE72 3PN G4OQL
DE72 3PN G6CHI
DE72 3RP G1DGY
DE72 3TE G1SGZ
DE72 3TF G1PWH
DE723GR 2E0EIZ
DE723GR M6IPQ
DE73 1AG M3IHU
DE73 1BW G3NYZ
DE73 5as 2E1BHC
DE73 5BX M1BGT
DE73 5SJ G1SPT
DE73 5SU G7CYF
DE73 6PB 2E0PFT
DE73 6PB M0PFT
DE73 6PF G7DZR
DE73 6RD G4LPZ
DE73 6RP G4XYD
DE73 6SB M0CAT
DE73 6XH G1PER
DE73 7AA M0JVR
DE73 7GY G0JCK
DE73 7HD M3IIN
DE73 7JJ G3XER
DE73 8AG G4FQI
DE73 8AG G8DBO
DE73 8BD G8MZA
DE73 8BX G8NFM
DE73 8EB 2E0GRL
DE73 8FG M3ZZL
DE73 8GA 2E0WRI
DE73 8GA M6PXW
DE73 8JH M3RBT
DE74 2AU 2E0CPX
DE74 2AU M6OZR
DE74 2AU M6SOE
DE74 2DN G0SOU
DE74 2DN G4IRH
DE74 2DQ M7BDB
DE74 2DS G0WUG
DE74 2DT G0RDY
DE74 2EJ G4TCA
DE74 2GQ G3JRS
DE74 2JG 2E0MPJ
DE74 2JG M3WQX
DE74 2JT M7SYE
DE74 2JY M6YRD
DE74 2LF G0EKK
DE74 2NG 2E1CPI
DE74 2NG G8BAG
DE74 2PG 2E0RJU
DE74 2PJ G1ZEK
DE74 2PQ G3KTP
DE74 2QQ G8PTW
DE74 2RX G4UVG
DE74 2SR G7PNG
DE74 2UJ M7XRO
DE74 2XA G0VNQ
DE74 2XB G1LCR
DE74 2XB G1SPA
DE75 7AN G7LGY
DE75 7BN 2E1FAT
DE75 7BN G6XTD
DE75 7FW G7MLT
DE75 7GQ G4DMF
DE75 7HB G1IQG
DE75 7HF G7JJC
DE75 7HG G3LZC
DE75 7JX M1EUR
DE75 7LY G8IOW
DE75 7NE M6BIA
DE75 7NJ G4GBC
DE75 7NJ M3MBI
DE75 7PQ G0FEZ
DE75 7PQ G4TYY
DE75 7PQ G4ZNI
DE75 7PZ G1GKK
DE75 7QB G8UCC
DE75 7RL M3UFX
DE75 7RP M6NEO
DE75 7TJ G1SEA
DE75 7UN 2E0CID
DE75 7UN M6BCZ
DG1 1BT GM1OXQ
DG1 1PP GM4JKB
DG1 1QN GM1JVU
DG1 1RB MM5AKZ
DG1 1RT GM7TUD
DG1 1RZ GM2AJW
DG1 1UZ 2M0RDG
DG1 1UZ MM3RSR
DG1 2LU GM0PFH
DG1 3HE GM4JR
DG1 3HE GM6EPU
DG1 3HE GM7WJP
DG1 3LH MM6BXF
DG1 3RJ GM3OXK
DG1 3SL MM6BNT
DG1 4BU GM4TNJ
DG1 4DN GM0WIZ
DG1 4DW GM4RJF
DG1 4EW GM0BQQ
DG1 4HN MM6EIX
DG1 4HU GM8VBX
DG1 4LU MM3OYW
DG1 4SD MM6KWQ
DG1 4SX MM7XAC
DG10 9DY GM1KBJ
DG10 9JU GM3OIT
DG11 1JL 2M0NEO
DG11 1JL MM0NEO
DG11 1JL MM3VRX
DG11 1NF MM3MID
DG11 1RL 2M0NJS
DG11 1RL MM0NJS
DG11 1RL MM6LTC
DG11 1SA GM8UPI
DG11 2BA MM0WHA
DG11 3AE GM3UHT
DG11 3AT GM0WMH
DG11 3BA GM1VLA
DG11 3DU GM0CBC
DG11 3JH GM0BVG
DG12 5FA MM7CLH
DG12 5LN 2M0ETJ
DG12 5LZ MM6KWV
DG12 5QL GM3YLU
DG12 6BU GM6LYJ
DG12 6dy GM0PMW
DG12 6HX GM4WHA
DG12 6HX GM6SMW
DG12 6QX GM0FQV
DG12 6QX GM6TVR
DG12 6UH MM0JZI
DG13 0AT GM7IHH
DG13 0AX GM4BDJ
DG13 0BG GM8ETJ
DG13 0EE 2M0GQS
DG13 0EE MM0JYR
DG13 0EE MM7LLM
DG13 0HJ GM6NJL
DG13 0JW GM1XGI
DG13 0JW GM1PGP
DG13 0PS MM6AEB
DG14 0RZ GM6LJE
DG14 0UY GM1CUC
DG16 5JS GM3MRV
DG2 0AJ GM4RPO
DG2 0NQ MM6KVE
DG2 0PE 2M0HDL
DG2 0PE MM7JVA
DG2 0QX GM0RJG
DG2 0QX GM1EOA
DG2 7AS 2M0EBW
DG2 7AS MM3RBF
DG2 7DH GM1SVQ
DG2 7DL GM8RQW
DG2 7LW GM6PFJ
DG2 7NS 2M0RHT
DG2 7NS MM6RHT
DG2 8DA MM6BLE
DG2 8LN GM8XKW
DG2 8NR GM0OHD
DG2 8QR 2M1SCO
DG2 9BP GM4XUJ
DG2 9DZ MM3ELT
DG2 9LT MM7AVI
DG2 9NN MM3MFR
DG2 9NR GM0AVB
DG2 9NR GM1PKB
DG3 4HD GM1YRD
DG3 5HR GM1SQA
DG3 5HR GM7ENM
DG4 6JZ MM0CIK
DG4 6LD GM4NTL
DG4 6LD GM8ZXQ
DG4 6SN GM7AWK
DG5 4DX 2M0HOS
DG5 4DX MM6JJW
DG5 4EB GM6VVG
DG5 4EE GM4BRB
DG5 4EN GM3LJR
DG5 4FA MM1BHO
DG5 4FA MM3PBQ
DG5 4GY MM6RJN
DG5 4QL MM6WMM
DG5 4QN MM6JFB
DG54dw MM6GYP
DG6 4HL GM4DSO
DG6 4LH MM3ZJL
DG6 4PE GM4KOO
DG6 4QD MM3SLD
DG6 4RN GM6RXQ
DG6 4SH MM7LAP
DG6 4TT MM3UJQ
DG6 4XT MM0KBT
DG7 1AH 2M0WUL
DG7 1BL MM6SNY
DG7 1BP MM7EQX
DG7 1EG MM6ACV
DG7 1GB GM6ZLY
DG7 1HS MM6ACY
DG7 1HX GM1BZR
DG7 1HX GM4VIS
DG7 1LG MM6GOW
DG7 1LQ MM6KXF
DG7 1LW GM4TPR
DG7 1LW GM6WNX
DG7 1LW MM6FNL
DG7 1TN MM6DPV
DG7 2EA MM0GOF
DG7 2JE MM6DMC
DG7 2JJ GM4DON
DG7 2NG MM3WDT
DG7 3BD 2M0ETL
DG7 3BD MM6IAW
DG7 3BF GM3JKS
DG7 3HG MM7EID
DG7 3JW MM6KCR
DG7 3JW MM6UYZ
DG7 3LD MM6DYH
DG7 3UR MM6DPJ
DG7 3XJ MM0YOS
DG72PW GM8CJG
DG8 0JU 2M0HEO
DG8 0JQ MM6HEO
DG8 0PN GM6FXZ
DG8 6AB GM6AQL
DG8 6BH GM3MZX
DG8 6HF MM6MNQ
DG8 6NU MM6PTR
DG8 6PZ MM6SFR
DG8 7AE GM3ZYE
DG8 7AZ MM6DGZ
DG8 7HU 2M0TKE
DG8 7HU MM3TKE
DG8 7JB GM3VNA
DG8 8BA GM4DUX
DG8 8DQ 2M0NSW
DG8 8DQ GM3SEK
DG8 8DQ GM5RP
DG8 8DQ MM0WNW
DG8 8DU GM4FZH
DG8 8LD GM0TFF
DG8 8LD GM3AUE
DG8 8NZ GM0ECU
DG8 8NZ GM0IDJ
DG8 8PP GM6FSG
DG8 9AH GM4GDF
DG8 9AQ GM4UQK
DG8 9BY MM7AJQ
DG8 9EE GM8AVM
DG8 9EE MM5WIG
DG8 9HX 2M0GKD
DG8 9JA MM6KTY
DG8 9JG GM4XAW
DG8 9LJ GM7NFF
DG8 9LL MM6FKB
DG8 9QN 2M0NIX
DG8 9QN MM0WWH
DG8 9QN MM6GWW
DG8 9RT MM7CSC
DG9 0AU GM6WAZ
DG9 0DX GM0AGV
DG9 0DX GM0HPL
DG9 0LJ GM0UWV
DG9 0LN GM0AJT
DG9 0LW GM0VYB
DG9 0RY GM7OJA
Dg9 7ba 2M0FBU
DG9 7BA MM6PJY
DG9 7BX GM1FMX
DG9 7LL MM3NWZ
DG9 7TA GM0AGN
DG9 8BQ GM4LPT
DG9 8DB GM3TGG
DG9 8ET MM3MZO
DG9 8EY GM6KXP
DG9 8HH 2M0WJP
DG9 8HH MM6WJP
DG9 8LX 2M0BOY
DG9 8RD MM6DGY
DG9 8RD MM6RHO
DG9 8TF GM8IWR
DG9 9AA MM6HHP
DG9 9AW 2M0WML
DG9 9AW MM6FJV
DG9 9BZ GM0HPK
DG9 9BZ GM4RIV
DG9 9BZ MM3MCG
DG9 9EX GM1ZIV
DG9 9HF MM6DPO
DG9 9NE GM4DLG
DG9 9PS MM6WGD
DG9 9PT GM1HZL
DG9 9QE MM6EDB
DG97HR GM0CXY
DH1 1EA M6OHP
DH1 1EB G7VIL
DH1 1HA G0PMX
DH1 1HW 2E0BYO
DH1 1HW M3YWU
DH1 1JE M6KTA
DH1 1LT G1GXW
DH1 2JR 2E0FYT
DH1 2JR 2E0OGZ
DH1 2JR M7LGD
DH1 2JU G7BYE
DH1 3RH M0BXP
DH1 3TY M0RCL
DH1 3UA M6YBC
DH1 4DS 2E0WTQ
DH1 4DS M0HNN
DH1 4DS M6CPQ
DH1 4EN G6RMJ
DH1 5DQ G4EBN
DH1 5DR G1SLP
DH1 5EH G7DGD
DH1 5GL G8PSS
DH1 5HP G7VJH
DH1 5HZ M0CKC
DH1 5JN G0DGB
DH1 5JN G0DRJ
DH1 5LA M6NHD
DH1 5LN M6EUB
DH1 5NL G3JQL
DH1 5PB M6RHX
DH1 5PN M0HAM
DH1 5PT 2E0UTC
DH1 5PT M0SGZ
DH1 5PT M6BBA
DH1 5PU G4PFE
DH1 5RJ G4LOM
DH1 5ZG G0VLF
DH1 5ZG G4EUZ
DH11JA M0DIN
DH11JA M3ADB
DH2 1JJ G3ZIJ
DH2 1NR G1NCL
DH2 1QH G4OCQ
DH2 1QY M1ELW
dh2 1rs M6BRL
DH2 1TA 2E0DEK
DH2 1TA M6HCM
DH2 1UF G4PSS
DH2 2BY 2E0DGQ
DH2 2BY M6EOC
DH2 2LD G7TFX
DH2 2LQ 2E0KMT
DH2 2LQ M0KMX
DH2 2LX G4ODE
DH2 2PB M6WYG
DH2 2SD G7OES
DH2 3DD G8YWK
DH2 3EA G6LCL
DH2 3ED 2E0LOW
DH2 3ED M6MDU
DH2 3EU G8APW
DH2 3HY G7DPZ
DH2 3JS G4MJA
DH2 3LX M3NJD
DH2 3QN M7SWR
DH3 1AR G4RXQ
DH3 1HP G1JGE
DH3 2EY G0UEC
DH3 2JG G0CGW
DH3 2JL M0CBP
DH3 2JS M6LVN
DH3 2NW M0CZO
DH3 2PT M0CSF
DH3 3ED M0BPM
DH3 3PR G4OSK
DH3 3PR M3GQP
DH3 3TP M7GER
DH3 4HU G1REO
DH3 4LU G6GLR
DH3 4LU G8FBQ
DH4 4BW M0FCY
DH4 4BW M0HRA
DH4 4EF M1ADK
DH4 4FB G8KLT
DH4 4HN M6CLO
DH4 4HN M6GYS
DH4 4HU 2E0BMW
DH4 4HU M3OVO
DH4 4PE M3EKP
DH4 4XR 2E0GRI
DH4 4XR M0IAS
DH4 4XR M3XNC
DH4 5BB G2DWC
DH4 5BB M0EXM
DH4 5ND M3JEE
DH4 5NR M6SGD
DH4 5PF M7PSB
DH4 6EA G0AAU
DH4 6GG G0FRZ
DH4 6JG G0FDA
DH4 6JG G4VOK
DH4 6JG G6KVA
DH4 6LA M3IVI
DH4 6LD M7BOD
DH4 6PA G3NMD
DH4 6PA M0RZE
DH4 7FB 2E0ODB
DH4 7FB M0ZXW
DH4 7HP 2E0JDH
DH4 7HP M0GDV
DH4 7HP M0HJD
DH4 7LP G0FBL
DH4 7RD G0OLO
DH4 7RZ M1FBI
DH4 7TD G0EHV
DH45JZ 2E0EYI
dh45qf M7HDX
DH5 0BA M0AHT
DH5 0BH M0ETP
DH5 0BH M0ETQ
DH5 0DB 2E0AAC
DH5 0DQ M6ZRL
DH5 0EF G4KOT
DH5 0ES M6TKX
DH5 0GG M7BJN
DH5 0JN 2E0HIS
DH5 0JN M0ODZ
DH5 0JN M6LWY
DH5 0JN M7KAS
DH5 0NR M0EGA
DH5 0NW M3BYG
DH5 0SE G6BIM
DH5 8AG M0GNG
DH5 8DP 2E0HNJ
Dh5 8eq M3VQC
DH5 8EX G4WJV
DH5 8GB M0RNR
DH5 8HN M3JMQ
DH5 8HR G0CKI
DH5 9LB 2E0IAS
DH5 9LJ M3KUV
DH5 9LX G7GJV
DH5 9QX 2E0DLB
DH5 9QX M6DLB
DH5 9RD M6FMS
DH5 9RR G4HBR
DH5 9RW G8VLS
DH6 1LB 2E0MBB
DH6 1LS G6WML
DH6 1PA G4SYD
DH6 1QB G6CQH
DH6 1QN G4WOT
DH6 1RH G4RHL
DH6 1RJ G0WPM
DH6 2DT M6OYZ
DH6 2FA M7BQC
DH6 2HE 2E0WWY
DH6 2HE M7WAY
DH6 2NF G0WIC
DH6 2PZ M7DZT
DH6 2QD M0IOC
DH6 2RG M6BLK
DH6 2TR G0PXQ
DH6 2YJ 2E0PUL
DH6 2YJ M6PUL
DH6 3DB M3ZAL
DH6 3HW M6VDJ
DH6 3LH G4RNI
DH6 4BE 2E0NMK
DH6 4BE M6EHF
DH6 4BT 2E0NLX
DH6 4BT M6NLX
DH6 4DG G7DIG
DH6 4DN G7MVX
DH6 4NN G4WUI
DH6 4NP G0DUG
DH6 4QW M6CAT
DH6 4SA 2E0PIU
DH6 4SA M0OPL
DH6 4SA M6NQI
DH6 4SA M7RCH
DH6 5ES G6LLD
DH6 5QA G7NFG
DH6 5QB G0SLP
dh61dz M7DQC
DH61LB M3GBB
DH62EL 2E0DZQ
DH62EL M6HSI
DH7 0BA G7UXQ
DH7 0BQ G6CQC
DH7 0HL G0GGB
DH7 0QA 2E0MPG
DH7 0QA M6MAK
DH7 6LZ M6PNL
DH7 6NE 2E0WET
DH7 6NE M6PRO
DH7 6SG G8FWY
DH7 6SQ G1WSZ
DH7 6TF G0CUO
DH7 6TW G3SNT
DH7 7AN G7AYF
DH7 7DP M3ZGE
DH7 7FD G1RVT
DH7 7JQ M7OIA
DH7 7NN G7MBH
DH7 7RL M0BAR
DH7 8BL 2E0INX
DH7 8BL M0KEO
DH7 8BL M7IGN
DH7 8NJ M7BYD
DH7 8PQ G7UKA
DH7 8RY M0DOW
DH7 8RY M3DPX
DH7 8SQ G6YAR
DH7 8TG G1WRF
DH7 8TN 2E0LZD
DH7 8TN M6LZD
DH7 9AG M6FSZ
DH7 9AU G0MNH
DH7 9FR 2E0JVA
DH7 9FR M6VAV
DH7 9FX M7SLW
DH7 9JH 2E0FEJ
DH7 9JH M6OXC
DH7 9JQ 2E0LLC
DH7 9JQ M6KAM
DH7 9JQ M6MGD
DH7 9LY G7TLR
DH7 9NF M6ESH
DH7 9TB M0NIW
DH7 9UW G6XCO
DH7 9YA G6PTT
DH8 0RF G0XVC
DH8 5JF G1EJQ
DH8 5LS G0JRT
DH8 5LS G4PFQ
DH8 5PA G0OHA
DH8 6AF M6FOX
DH8 6EZ M0HTV
DH8 6HZ 2E1EPA
DH8 6JG M3UJN
DH8 6JT M6NHK
DH8 6JZ 2E0GBB
DH8 6JZ M6AQO
DH8 6TL G0GKK
DH8 7AE G0SBU
DH8 7AE M6GMD
DH8 7DB M6ICP
DH8 7DT M6WKY
DH8 7DY G3UTS
DH8 7EH M0MFL
DH8 7EH M6KBE
DH8 7HS 2E0IGL
DH8 7JB 2E0CIN
DH8 7JB M6GAM
DH8 7JE M0WDJ
DH8 7LL M7ESW
DH8 7QT G1NVN
DH8 7SB G8IAJ
DH8 7SJ G3EGF
DH8 7UJ M6XJC
DH8 8DX M7CWB
DH8 8TU G6WDM
DH8 9AP G3KMG
DH8 9JU G1BET
DH8 9JU M3BET
DH8 9RF G8KZN
Dh88tq 2E0FTT
DH9 0QD G1NTI
DH9 0QP G1HYG
DH9 0QR G7LQK
DH9 6BE G1HEX
DH9 6ET 2E0DYD
DH9 6ET M6EKI
DH9 6HX 2E0GBN
DH9 6HX M6GBN
DH9 6JD M3TSO
DH9 6JU M0BZX
DH9 6XE G7OVB
DH9 7AJ G6FLK
DH9 7HP G4APP
DH9 7JW M3EAS
DH9 7PH M5ABT
DH9 7TR G1AIF
DH9 7TZ M0AYI
DH9 8DE M7CXE
DH9 8DG G1KRX
DH9 8DG G6UEV
DH9 8EQ G7GJU
DH9 8JG M6JIK
DH9 8JW M1FIV
DH9 8PL M7OKR
DH9 8QY G1OGH
DH9 8TX M7CXD
DH9 8UH G4MHA
DH9 9HB G0KYL
DH9 9HD M6RYS
DH9 9LZ G0LEI
DH9 9PA G0IVX
DH9 9PT G0VII
DH9 9PT M3KQB
DH9 9UH M0CLD
DL1 1EQ 2E0AES
DL1 1HG G8BWH
DL1 1JG M6AAJ
DL1 2DX G0FGG
DL1 2HY G0LUY
DL1 2QF G7CIU
DL1 2QG M7DYU
DL1 2TA G6VPV
DL1 2TU G4NYJ
DL1 3HH G6SPQ
DL1 3ND M0WLY
DL1 3ND M6DLY
DL1 4EY M7AYR
DL1 4LG G8OWS
DL1 4XG G4KEL
DL1 5BJ 2E0NYM
DL1 5BJ M3CRD
DL1 5DF G0TJC
DL1 5DF G1LEN
DL1 5DX G4AUN
DL1 5EA G6XCD
DL1 5LH G0JHD
DL1 5LX 2E0LRP
DL1 5LX M3HUG
DL10 4ES 2E0UXM
DL10 4ES M6NML
DL10 4ET M6ABR
DL10 4PQ G3XHB
DL10 4PQ G8YPN
DL10 4PS M6JJI
DL10 4QY M0OJC
DL10 4SJ 2E0YEA
DL10 4SJ M1SDE
DL10 4SJ M6WEP
DL10 5AG M6WFX
DL10 5BJ 2E0BNC
DL10 5BJ M3WUS
DL10 5DB G3EKL
DL10 5DB G3RLV
DL10 5DJ G0MQV
DL10 5PE M0BYU
DL10 5QD M0DBB
DL10 6AL M6IKH
DL10 6DN 2E0JTS
DL10 6DN G2MN
DL10 6DN M7JWW
DL10 6EE M0JWR
DL10 6EE M5JWR
DL10 6SB 2E1GUN
DL10 6SB G4FZN
DL10 6SB G8HQW
DL10 6SB M0RRG
DL10 6SB M1ECH
DL10 7AE G8ZNB
DL10 7AG G4VGY
DL10 7BG M3NPZ
DL10 7BQ G0RYS
DL10 7DL G4DBY
DL10 7DX M3VON
DL10 7JE 2E0WGC
DL10 7JE M0WGC
DL10 7JE M6CGW
DL10 7JP 2E0LXA
DL10 7JP G7LLY
DL10 7JP M6AXL
DL10 7LQ G6XJJ
DL10 7LY 2E0MXG
DL10 7LY M6IFB
DL10 7PB G6CUK
DL10 7QD G6RNR
DL10 7RT 2E0BEA
DL10 7SP M1EAB
DL10 7SS M3UBU
DL11 6HH G0JTA
DL11 6PY M6LAA
DL11 6QX G8TOQ
DL11 6TE G1FCX
DL11 7AJ M3CDI
DL11 7JH G6IET
DL11 7NH M6SPP
DL11 7RD G3XWV
DL12 0AH 2E0BWP
DL12 0AH M6DLM
DL12 0JD M6EVM
DL12 0QU G0ADO
DL12 0RP G4IPB
DL12 0UU G4AFE
DL12 8EB G0NRK
DL12 8LA G4OUB
DL12 8LF 2E0BFM
DL12 8LQ 2E0BPF
DL12 9PQ G8LIE
DL13 1JD G6FZC
DL13 2AY G4VZS
DL13 2NB G7GLP
DL13 2QR G4DWM
DL13 2XY G6OTL
DL13 3BW 2E0FTD
DL13 3BW M6HVU
DL13 4DS 2E0AIM
DL13 4DU G4TMI
DL13 4DX M7DHL
DL13 4ED M6KWT
DL13 4EE M6CZW
dl13 4js G4OJY
DL13 5LS M6NKJ
DL13 5NF G0LLC
DL13 5RH G4KUX
DL14 0DH G0OCB
DL14 0DH G1XNI
DL14 0DH M6AMD
DL14 0RN M3WIT
DL14 0RP 2E1EYS
DL14 0RX G0EQD
DL14 0TG G1ASR
DL14 0TP 2E0MRQ
DL14 0TP M3ZJG
DL14 6AR M3RLO
DL14 6ET M3UOM
DL14 6NT M5CJH
DL14 6TU M0BLN
DL14 6UB M3BAH
DL14 6UB M3EMA
DL14 7DS G0KIC
DL14 7DZ M6SSZ
DL14 7GH M3FSV
DL14 7LZ M0ESZ
DL14 9EJ G4LRG
DL14 9LG G4TTF
DL14 9LG M0ACV
DL14 9LQ G1YNO
DL14 9PY G0JQP
DL14 9PY M3UOB
DL14 9QH M0DQD
DL14 9RY 2E0LDQ
DL14 9RY M6LDQ
DL149LF 2E0WGE
DL149LF M6GZA
DL15 0AD 2E0WKM
DL15 0AD M0WCW
DL15 0AD M7CKW
DL15 0BJ 2E0BJA
DL15 0BJ M0BPC

DL15 0BJ M3NAR
DL15 0DA G4STP
DL15 0DS M3XWM
DL15 0HN G6LQR
DL15 0LQ 2E0CEG
DL15 0LQ G1YUB
DL15 0LQ M3ALX
DL15 0NS M0MEG
DL15 0PP 2E0CWQ
DL15 0PP M6JDU
DL15 0QL 2E0LIW
DL15 0QL M6LWS
DL15 8AG G7VNM
DL15 8NA M7MTH
DL15 8NN G4FZS
DL15 8QU G4PSI
DL15 9AE 2E0RTG
DL15 9AE M0RRF
DL15 9AE M6RTG
DL15 9DB G7OCK
DL15 9DY G7PTT
DL15 9PT M0LYI
DL15 9QX G0HCD
DL15 9SE G4VIO
DL15 9SF M1GSM
DL15 9UT G7BPN
DL16 6DW G7ESY
DL16 6ER M3AHR
DL16 6LY G0TSR
DL16 6NB G0RNY
DL16 6QU M6DTJ
DL16 6RN 2E0XVF
DL16 6RN G0KNN
DL16 6RN M0XVF
DL16 6RN M3XVF
DL16 6TJ M0PJM
DL16 6TT M6SCB
DL16 6TZ G7MKQ
DL16 7DZ G8NSX
DL16 7HF G7VYZ
DL16 7HN G1LPS
DL16 7HU G7KJR
DL16 7HZ G0JEC
DL16 7QS G0BVO
DL16 7TA M3AHO
DL16 7TU 2E0HDF
DL16 7UG G0PWK
DL166HP G7ESX
DL17 0EQ 2E0NGL
DL17 0EQ M0NGN
DL17 0EQ M6IQL
DL17 0RW G0BNY
DL17 0RW G6CAY
DL17 8DA G4MQV
DL17 8DA G6CII
DL17 8DA G8UYM
DL17 8DL G4JIX
DL17 8NG G4XHP
DL17 8QE G1LGQ
DL2 1AU G0AIH
DL2 1HF G3SGY
DL2 1JG 2E0UCH
DL2 1JG 2E0ZOT
DL2 1JG M0VLI
DL2 1JG M0XEE
DL2 1JG M3VFS
DL2 1JG M3VLI
DL2 1JG M3ZVA
DL2 2AH 2E0VVF
DL2 2AL 2E1DPK
DL2 2GL G0SBP
DL2 2LN M3KRX
DL2 2PX G4NLL
DL2 2SL G8MTV
DL2 2SQ M1SAZ
DL2 3JJ G4MIJ
DL2 3NS 2E1HHL
DL2 3NS M1DYJ
DL2 3RY G3GJJ
DL2 3SX M0OAT
DL2 3SX M6WKR
DL3 0AH G3UHJ
DL3 0AL 2E0PCQ
DL3 0AL M6PCQ
DL3 0EJ M6ZET
DL3 0HX G8HZS
DL3 0JQ G1LEC
DL3 0JY G2HMK
DL3 0NW G4YFS
DL3 0SP M7EGK
DL3 6ES G7TCD
DL3 6HN G4JIR
DL3 6HT M1CUB
DL3 6PY M7HTS
DL3 7AT M0OXW
DL3 7AT M6OXW
DL3 7AT M6OXZ
DL3 7HP G1DMN
DL3 7SJ G1BIA
DL3 8BH G4OXU
DL3 8BH G8YBO
DL3 8HU G6KMG
DL3 8HY G4FVP
DL3 8HY M0NFD
DL3 8LD G0ARZ
DL3 8NH M3ONH
DL3 8RP M6GHL
DL3 9AT 2E0GBA
DL3 9AT M3XNM
DL3 9DT G6BCG
DL3 9PB G3CDM
DL3 9PB M7AVH
DL3 9PF G0MSZ
DL3 9SA G0VQG
DL3 9SR M3MCA
DL3 9XS G8TJR
DL4 1AP G7BPR
DL4 1BH G3LUC
DL4 1EB G0RYW
DL4 1EB M0SDE
DL4 1QA G1DFN
DL4 2AS M0MYE
DL4 2AZ M3NFU
DL4 2DN 2E0SUX
DL4 2DN M0SGA
DL4 2LE 2E0XLY
DL4 2LE M0XLY
DL4 2LE M6LXY
DL4 2LE M6MYF
DL5 4HX M6ZAY
DL5 4PF M6WIT
DL5 4PF M6XDO
DL5 4RA M7PGY
DL5 4SU 2E0VNE
DL5 4SU M7MGX
DL5 5EQ G0IDS
DL5 6RF G3OAL
DL5 6TA G1ZEU
DL5 7BD M0PST
DL5 7BQ M0WRI
DL5 7DR. G1AZC
DL5 7DX G8LCP
DL5 7DX M0BUG
DL5 7DY 2E0RRL
DL5 7DY 2E0WPD
DL5 7DY M0RRL
DL5 7DY M0UDL
DL5 7DY M6RRL
DL5 7DY M6TCN
DL5 7LN G3FBT
DL5 7NP M3VRY
DL5 7NT M3UVY
DL5 7PS G4GMB
DL6 1DZ G0GBG
DL6 1ED G0JQA
DL6 1ED G3KJX
DL6 1EE G8FLV
DL6 1HP G1XRE
DL6 1QQ G0LEL
DL6 1QT G8IHT
DL6 1RQ G0GCK
DL6 1SF G1RQK
DL6 1SJ G7NKJ
DL6 2AA G7HHK
DL6 2BD G1JER
DL6 2BE G3MAE
DL6 2LB 2E0FBC
DL6 2LB M6FOY
DL6 2RD G0WWF
DL6 2RD G2PB
DL6 2SN G1HLV
DL6 3QA M0BBK
DL7 0HL 2E0DNH
DL7 0HL M6RFZ
DL7 8AD M3XYJ
DL7 8BN 2E1CQQ
DL7 8BN G7COC
DL7 8JF M0MFT
Dl7 8WB M7SRB
DL7 8XS M7MWT
DL7 9RA G8CEE
DL8 1SX G7HSN
DL8 2DR G3GEJ
DL8 2HT G1WWH
DL8 2JE G4AFU
DL8 2QF G4NUY
DL8 3HN G8AMJ
DL8 3RH G3ZMO
DL8 4LU M3LOA
DL8 4NA G1BLV
DL8 4QN 2E1AZW
DL8 4QN G0OLR
DL8 4TS M1BTI
DL8 5BH G4HVF
DL8 5EL M1FGH
DL8 5HU G0BAQ
DL8 5JQ G3XPQ
DL8 5JS 2E0CQX
DL8 5JS M6MCZ
DL8 5QN G8LNQ
DL8 5QN M0KXD
DL9 3BW M6RKU
DL9 3NJ 2E1COG
DL9 3NJ G1YQY
DL9 3NP M6FGM
DL9 3RA G7WBW
DL9 4HX M0DIG
DL9 4NT M6JOV
DL9 4PG 2E0TCN
DL9 4PG M0TCN
DL9 4PG M6TIA
DL9 4RJ M6JNZ
DN1 2JA 2E0MLC
DN1 2JA M7SAV
DN1 2NP G0UQU
DN1 2PZ G4OEM
DN1 2QU G1PBY
DN10 4BE G4OZN
DN10 4BU G0OKZ
DN10 4DP G8EGL
DN10 4EF M1FGM
DN10 4HT M1WDK
DN10 4LG 2E0CTO
DN10 4LG M6KAX
DN10 4NG G0EJD
DN10 5BL M7LTY
DN10 5BS G6GWP
DN10 6HE G8JFC
DN10 6NW G3RCW
DN10 6NW M0LLW
DN10 6QF M1CTB
DN10 6RT M3ZFS
DN10 6SW 2E0XAH
DN11 0AA G0PXD
DN11 0BT M0BZQ
DN11 0DF M7DTR
DN11 0FR G4ZLP
DN11 0FR M6OEH
DN11 0JG M0YDB
DN11 0JP 2E0DMJ
DN11 0JP 2E0KPI
DN11 0LU G7JDH
DN11 0NQ G0PXX
DN11 0NQ G6USL
DN11 0PU M3EGM
DN11 0QF M0CTN
DN11 0QF M3CTN
DN11 0QX M0CTQ
DN11 0RW M3EYW
DN11 0RW M3EYX
DN11 0RZ M3BZQ
DN11 0SB 2E1IOS
DN11 0SB 2E1RAD
DN11 0SH M3WXB
DN11 0TS M3ZLS
DN11 0UP G4NQM
DN11 0UP G6DIE
DN11 0UR G8HIQ
DN11 8DT M6SDP
DN11 8FP G1BAL
DN11 8FP M0DRM
DN11 8FP M0MHY
DN11 8HP G4ZVP
DN11 8HT G0FVD
DN11 8HZ M6ROE
DN11 8LL G4BZG
DN11 8LU M3IOK
DN11 8QJ 2E0ZCB
DN11 8QJ M0YCB
DN11 8QP M6FYD
DN11 8QU M6TDR
DN11 8SN M6ENE
DN11 9BH 2E0HWO
DN11 9BH M0YZT
DN11 9BH M7HWO
DN11 9BW G0TTL
DN11 9ET G6HMS
DN11 9EW 2E0DFT
DN11 9PW M0BOI
DN11 9TE G7BGV
DN11 9UE M6BGP
DN11 9UH G7EDF
DN11 9UL M3XPN
DN12 1BD 2E0DKS
DN12 1BD 2E0IKB
DN12 1ED 2E0SEJ
DN12 1ED M0SBK
DN12 1ED M6SGY
DN12 1NQ M3SMD
DN12 1NW 2E0XOO
DN12 1NW M3KAK
DN12 1SQ M7SDE
DN12 2JG 2E0XPM
DN12 2JG M0XPM
DN12 2JW G3NXZ
DN12 3DF G3WBG
DN12 3DG M6EMJ
DN12 3JZ M7GMP
DN12 3LB 2E0RDX
DN12 4EN M0TOR
DN12 4LR M3LTG
DN14 0AT 2E0DXZ
DN14 0AT M6GQY
DN14 0BU M0ROW
DN14 0JX G4FPO
DN14 0LN G4III
DN14 0NW G6LUE
DN14 0PD G8DML
DN14 0PG M0JKP
DN14 0QY G4AFZ
DN14 0RD G8FCT
DN14 5HQ M3HVU
DN14 5JB M6JIR
dn14 5lh M7ABN
DN14 5NY G4GNP
DN14 5RR M7MZT
DN14 5XF G8FWC
DN14 6DH G0PQX
DN14 6DY M3BRW
DN14 6EB M6BPX
DN14 6EL M3NVE
DN14 6HJ 2E0EWX
DN14 6HJ M0JDU
DN14 6JU M6MRU
DN14 6JX G3LYZ
DN14 6LG M5RHS
DN14 6NA G0GLZ
DN14 6QW G8VHL
DN14 6SH 2E0CAD
DN14 6XN G4GRP
DN14 7DS 2E0TUM
DN14 7DS M7TJM
DN14 7EN G3ZGT
DN14 7EU M7UHU
DN14 7FD G4NLG
DN14 7HD G0KHZ
DN14 7HE G4DBN
DN14 7JL G1RFC
DN14 7LZ G6COE
DN14 7NT M6BIB
DN14 7TP M3KQS
DN14 7YA 2E0UIP
DN14 8DE 2E0XAG
DN14 8DE M3WQF
DN14 8DW M6AFH
DN14 8ER M1AKQ
DN14 8ET G0FRX
DN14 8ET M0UVZ
DN14 8EX G4BDX
DN14 8HL G4LKD
DN14 8HL M0NTG
DN14 8RP G6PXJ
DN14 8RW G7UJT
DN14 8RW M3UJT
DN14 9BN M7TSH
DN14 9JG G7OYD
DN14 9JH M3SRI
DN14 9NG M3GUG
DN14 9NG M3PKH
DN14 9QR G0UYF
DN149QQ M6OZS
DN15 0AD G7MTJ
DN15 0BE M3OZB
DN15 6SN M6FJW
DN15 7AT G0RMD
DN15 7BT G0OQX
DN15 7EG G4KWW
DN15 7EH G1HKM
DN15 7EN 2E0XNL
DN15 7EN M6GFH
DN15 7EQ M0ADH
DN15 7HT 2E0PFY
DN15 7HT M6FXU
DN15 7PJ G0DLL
DN15 7PJ M6BYQ
DN15 8AB G3KFB
DN15 8AD 2E1HHJ
DN15 8AU G0EFY
DN15 8BP G0RKQ
DN15 8JT M6HMG
DN15 8LA M6NTM
DN15 8NS 2E1FVS
DN15 8NS G0GHK
DN15 8NS G7HAH
DN15 8NS M0HOM
DN15 8PE M0PDC
DN15 8UG G6JVO
DN15 9AG G0DVG
DN15 9DH G4IPE
DN15 9EW G4CDC
DN15 9EX M3IPM
DN15 9HH G0JRR
DN15 9NQ G1RUE
DN15 9NQ M0GBB
DN15 9NR G4YZH
DN15 9NW 2E0AWT
DN15 9NW M0GIQ
DN15 9QG M6MCC
DN15 9QP M0JYZ
DN15 9QY G8TLL
DN15 9RP M7OMA
DN15 9RT G6CMX
DN15 9RU 2E0FTW
DN15 9RU M1DOS
DN158LL M3VTP
DN16 1EB G4JJY
DN16 1EY G0GWY
DN16 1HT 2E1MIN
DN16 1HT M1FFM
DN16 1NA G0OKF
DN16 1QH 2E0FQT
DN16 1RW 2E0MBO
DN16 2AJ 2E0LYF
DN16 2AJ M0LYF
DN16 2AJ M6KNS
DN16 2AJ M6YSF
dn16 2be G4JRY
DN16 2ES G4HFZ
DN16 2LQ G1YNQ
DN16 2LR 2E0CQD
DN16 2PA M6PHS
DN16 2RP M6GRC
DN16 3DE G4MGD
DN16 3EN G8GIH
DN16 3LQ G0TNS
DN16 3PH G0PQY
DN16 3SA G8LIP
DN16 3SW G4NJA
DN16 3TT 2E0GYK
DN16 3TT M0XRV
DN17 1AA M6XLT
DN17 1AF G4CBY
DN17 1AS G0RUH
DN17 1DU G1ZRT
DN17 1SA 2E0DGC
DN17 1SA G0UTM
DN17 1SA M0HDV
DN17 1SA M6GDC
DN17 1SW 2E1CZB
DN17 1TS 2E1GXY
DN17 1TS M3BAT
DN17 1UE M7GLN
DN17 1YB 2E0CZA
DN17 1YB M6DQR
DN17 1YL 2E1HHA
DN17 2BD G8YVC
DN17 2DF M7AQY
DN17 2EQ G3KNU
DN17 2LW 2E0RNP
DN17 2NQ G7LAL
DN17 2QY G0JRB
DN17 2TP G0ECS
DN17 3HZ G7URL
DN17 3JR G8JET
DN17 3PB M0ACB
DN17 3RA G6PAJ
DN17 3SB G0JEU
DN17 3SB G4ZGE
DN17 3SZ G0PAB
DN17 3TT G4EQD
DN17 3TU G4STW
DN17 4AT G6DBC
DN17 4DG G4FZR
DN17 4ET G8MOF
DN17 4GA M7EMC
DN17 4HU G8ZGY
DN17 4LH G4LSL
DN17 4NB M3ZPO
DN17 4PP G4VOV
DN17 4PQ M1LMO
DN17 4RZ G0NVD
DN18 5BS 2E0GDN
DN18 5BS M0IIE
DN18 5DP G1ZQG
DN18 5DZ G3RHQ
DN18 5HH G0HMD
DN18 5LE M0AEP
DN18 5LN 2E0RWX
DN18 5NH G4IWR
DN18 5QA G0PHP
DN18 5QJ M6IJK
DN18 5TD G8RQH
DN18 5TW M3USJ
DN18 6AE G4TGE
DN18 6DN M6KMR
DN18 6HJ M7MBE
DN18 6JS M0IUZ
Dn185na M7GDY
DN19 7AU G0KAT
DN19 7AX G8FWE
DN19 7EE G4JPK
DN19 7EE M1TAT
DN19 7EY G1CUM
DN19 7HG G4CFD
DN19 7QB G7PXX
DN19 7QG M3BEE
DN19 7QG M3PWS
DN19 7TL 2E0MPR
DN19 7TL M3JZI
DN2 4BB M1BUX
DN2 4DA M3KJY
DN2 4GD M6GPO
DN2 4JS M3LKY
DN2 4LF M3WPV
DN2 4QA G0IDZ
DN2 4RF M6DJL
DN2 4RJ 2E0PBP
DN2 4RJ M6DPJ
DN2 5HN G4PJL
DN2 5RF G3XGU
DN2 5TR 2E0IER
DN2 6AN G4LKX
DN2 6DT G4ZGM
DN2 6EN M6PGZ
DN2 6HF G0VID
DN2 6HF G7GJO
DN2 6HF M0BDL
DN2 6JQ M0JIC
DN2 6LG M1CHM
DN2 6LN 2E0ECG
DN2 6PA M6OAE
DN20 0AU M3UPN
DN20 0BB G3MHT
DN20 0NN G8YAU
DN20 0PP G0AOJ
DN20 0SE 2E0SXF
DN20 8PW M0ICJ
DN20 8QR M6MOF
DN20 8SG M6PDM
DN20 8SH G0CMK
DN20 8SW M6HFF
DN20 9BE G3MPW
DN20 9DG G4ROC
DN20 9FN M3YUN
DN20 9HU G8YYW
DN20 9JG G6GNE
DN200SJ 2E1HHK
DN200SJ G4TYL
DN208RR 2E0NXM
DN21 1BQ M3VYF
DN21 1BT G3SCJ
DN21 1DD G1HKF
DN21 1DH G7AVU
DN21 1NA G7OLF
DN21 1PA G7GMR
DN21 1QZ M1EBN
DN21 1QZ M3DXR
DN21 1SS 2E0RKS
DN21 1SS M0ZRS
DN21 1SS M6AYP
DN21 1TT G8VRN
DN21 1WB G7MDM
DN21 1XQ M3LAJ
DN21 2HQ G6FOW
DN21 2JA G3YPS
DN21 2JZ G3UHS
DN21 2RS 2E1LEW
DN21 2RS M7LEW
DN21 2RY M6JVW
DN21 2SA M7CRS
DN21 2TH M3TNL
DN21 2TU G6YVS
DN21 3DL 2E0FAM
DN21 3DL G1CIM
DN21 3DL M0CGS
DN21 3DL M6FLC
DN21 3HY M3CVD
DN21 3JY G4XHC
DN21 3QZ G4YET
DN21 3RR G3ZOT
DN21 3RU G7WGY
DN21 3RU M3WGY
DN21 3SL G7JQT
DN21 3SR G3TMD
DN21 3TZ G0PZW
DN21 4AF G4GZA
DN21 4BA 2E0BCX
DN21 4BA M0JAE
DN21 4BB G4SXZ
DN21 4BY M6LHJ
DN21 4DF M3FWU
DN21 4ER G4MBK
DN21 4FE M6CGX
DN21 4NE 2E1HXY
DN21 4NE 2E1HXZ
DN21 4NE G4CCH
DN21 4NF G0MPK
DN21 4SW G6PHC
DN21 4SW M6PVI
DN21 5BZ G3RAU
DN21 5DN 2E0LQR
DN21 5DN M0LQR
DN21 5DN M6TML
DN21 5FF M6PPP
DN21 5JU G7UGY
DN21 5LB G8SFF
DN21 5LY G7IYQ
DN21 5NL M0AYF
DN21 5PT G6WVO
DN21 5TQ G3SET
DN21 5UT G0SQS
DN21 5UT G7JVC
DN21 5XJ M6LHE
DN21 5XP 2E0CJK
DN21 5XP M0PMJ
DN21 5XP M6LIE
DN21 5XP M6PMJ
DN22 0AS G7SRI
DN22 0BY G6NRL
DN22 0FL G7KYL
DN22 0HU M3BAL
DN22 0JX G0TGB
DN22 0LG G4SOI
DN22 0LN G1EWH
DN22 0QQ G0RAF
DN22 0QQ G3VCA
DN22 6LR 2E1FRW
DN22 6NH G0MYA
DN22 6NW G4XTU
DN22 6QH G0CEB
DN22 6QU 2E0LDV
DN22 6QU M3MZG
DN22 6RT G7IBD
DN22 6SF G4YWX
DN22 6SQ M0PAL
DN22 6TS 2E0KVF
DN22 6TS M0KVF
DN22 6TS M6KVF
DN22 6UB G0HLU
DN22 6UF G3DXZ
DN22 7AD 2E1HYI
DN22 7BT G6GKK
DN22 7DX M6XAK
DN22 7GX M7BUI
DN22 7LJ G3NTD
DN22 7NX G4ZDN
DN22 7QW M6JMF
DN22 7TL 2E0SNS
DN22 7TL M6SNS
DN22 7UW G4WBH
DN22 8AJ G0IAS
DN22 8NL G4VEO
DN22 8PQ G0NPG
DN22 8RS G4AWU
DN22 9NQ G0VUL
Dn229he M6IJO
DN3 1AN G1MCT
DN3 1AW M3XYN
DN3 1BJ M6GWT
DN3 1BS G4NZX
DN3 1DP G0UZW
DN3 1DP G7RXE
DN3 1DP M0CJZ
DN3 1DS G3UWT
DN3 1JU G4KKJ
DN3 1LY G0EZY
DN3 1LY G8CDV
DN3 1NY M6DMO
DN3 2AR M6RCD
DN3 2AZ 2E0PKB
DN3 2AZ G7MEX
DN3 2AZ M0BOH
DN3 2AZ M0EVE
DN3 2AZ M0MEO
DN3 2AZ M1DAH
DN3 2AZ M3DYF
DN3 2AZ M3HJB
DN3 2AZ M3PLB
DN3 2AZ M3WQI
DN3 2DE G1IND
DN3 2DE M3IND
DN3 2DJ M6AMC
DN3 2DJ M6JIF
DN3 2EP M6WCZ
DN3 2ES M3MSC
DN3 2ET G6XVH
DN3 2HE M3YJG
DN3 2HN G4PNT
DN3 2JF G7GEI
DN3 2PE G4MDE
DN3 2QF M0CRT
DN3 3AJ G7JTF
DN3 3HE G7GZK
DN3 3HE G8XKI
DN3 3HG M3PTV
DN3 3HS G0GOH
DN3 3PB G0WKA
DN31 1PQ M0ZLF
DN31 1PQ M6KLD
DN31 2DN G6NDH
DN31 2EU 2E0GUH
DN31 2EU M6GUH
DN31 2NB G1WSC
DN31 2PW M3YAV
DN32 0HN 2E0OZW
DN32 0HN M3HVV
DN32 0JQ M6PFU
DN32 0LU M6YDF
DN32 0NG G8BMZ
DN32 0NH M6EJP
DN32 0NH M6LZX
DN32 0NQ M3WJA
DN32 7AW M7LAH
DN32 7EJ M7VMS
DN32 7JE G1HCI
DN32 7JE G1HCJ
DN32 7NJ M3DKW
DN32 7PL G0CJV
DN32 7SB G7HJR
DN32 8DR M6RGU
DN32 8DZ G7VKY
DN32 8HR M6CBP
DN32 8HR M6SBE
DN32 8NU M6PRC
DN32 8NX M0DDY
DN32 8QS G7LIW
DN32 9EE M7CDK
DN32 9EN M6NFP
DN32 9HJ M6NBO
DN32 9HT M7LYD
DN32 9JQ M6HBC
DN32 9LB M6WMJ
DN32 9NL 2E0GYX
DN32 9NL M0XRF
DN32 9NP M6EPS
DN32 9NU M3XPU
DN32 9NW G1CWJ
DN32 9PQ M1BYQ
DN32 9QY G1KBL
DN32 9QG M0DMY
DN320NE M7AES
DN328BJ M3TYW
DN33 1BG G4HZF
DN33 1DE G3DAE
DN33 1DN M6SGN
DN33 1EZ G8XXI
DN33 1JB M0AJT
DN33 1LU 2E0KQV
DN33 1NL M3UIP
DN33 1SB M0SDG
DN33 2BB G0KUD
DN33 2BY 2E0UVO
DN33 2BY M3UVO
DN33 2EA G4NPS
DN33 2HF G3DSZ
DN33 2JS G1HZN
DN33 2LG G6EXZ
DN33 2NW G1FVE
DN33 2PL G0CGZ
DN33 2PL G0IOR
DN33 3BL G0CSV
DN33 3BS G1BKJ
DN33 3BW G4VIM
DN33 3EE G3PLN
DN33 3EF M6HUK
DN33 3JP G0JNT
DN33 3JT G3SWU
DN33 3JY M6MJA
DN33 3NG G4MSY
DN33 3RA M6MJL
DN34 4AD 2E1FHZ
DN34 4DP 2E0OZQ
DN34 4ND M6XPN
DN34 4NJ M6AVY
DN34 4PN G0IGC
DN34 4PN G0IGH
DN34 4PW M0FIS
DN34 4QH G0HXL
DN34 4QN G0RIR
DN34 4SG M5ACD
DN34 4TY G8RHN
DN34 5DB G0RUS
DN34 5JP 2E0GCI
DN34 5JP M0KIN
DN34 5JP M3LAP
DN34 5LX M1NCC
DN34 5PE G6RIQ
DN34 5RE G7EOG
DN34 5RE M3LEO
DN34 5TG 2E1HIL
DN34 5UP 2E1GCB
DN34 5UQ G4JMY
DN34 5UQ G4VUP
DN35 0HY M7BTL
DN35 0JQ G8ULH
DN35 0NN M1BYI
DN35 0SE M0CDU
DN35 0SF G4YSO
DN35 7BB M6KEC
DN35 7DX G0LFA
DN35 7DX M6LAV
DN35 7DX M6TCL
DN35 7JJ 2E0YQT
DN35 7JJ M3YQT
DN35 7NG G4YVJ
DN35 7NP 2E0GRN
DN35 7NP 2E0JKY
DN35 7NP M6FFS
DN35 7NP M6GRN
DN35 7NP M6JKY
DN35 7SD G4ZVX
DN35 7UR M0KWK
DN35 8HY M6PXK
DN35 8HZ G0MNI
DN35 8HZ G6MHY
DN35 8JA G8DKS
DN35 8LE M6SNF
DN35 8NT M7ADU
DN35 8PD G0GSY
DN35 8PL G6VJA
DN35 8PL M7KZE
DN35 8PR G7FFW
DN35 8QN G0PBR
DN35 8QN M6RED
DN35 9DE G4SRF
DN35 9HF G4ZRV
DN35 9JR M0SMT
DN35 9PF M6NEA
DN35 9PP G7CHB
DN35 9PP G7PSV
DN35 9PP M3RHP
DN36 4BA M6CVK
DN36 4DE G1INK
DN36 4DF G8ZLF
DN36 4DS G0BBT
DN36 4EA G7DEU
DN36 4JA G0KFD
DN36 4LJ G1JZL
DN36 4NE 2E0PBJ
DN36 4NE M6PSM
DN36 4NS 2E0WNW
DN36 4NS M6DQK
DN36 4QY G8ECI
DN36 4RB M0ANU
DN36 4ST G6CCB
DN36 4TT G7IMZ
DN36 4UF G3ZSF
DN36 4WH M3UII
DN36 5AD G4HOF
DN36 5AQ G3VIP
DN36 5BG 2E1CAW
DN36 5BG G0NQW
DN36 5BH 2E0CTZ
DN36 5BH M6FBB
DN36 5BQ G4PYD
DN36 5DS G4GAB
DN36 5HG 2E0GNW
DN36 5HG M6GNW
DN36 5JE G4OII
DN36 5LP M0DER
DN36 5NF M1CYP
DN36 5NJ G4SEP
DN36 5PL G4GOJ
DN36 5QS G4LOY
DN36 5RF M7AFW
DN36 5SQ M6FJU
DN36 5TP G1IZB
DN37 0EE M1CDQ
DN37 0JR G3TVC
DN37 0NB G7NCV
DN37 0NS M6WMF
DN37 0QE 2E0VBM
DN37 0QE M6GYA
DN37 0UA G4WOH
DN37 0UZ G5ATT
DN37 0YJ G0TWE
DN37 7BA G0IIQ
DN37 7DB G0ATW
DN37 7DF M6DCS
DN37 7DF M6SSD
DN37 7ER M6SKT
DN37 7ER M6TNR
DN37 8LB G0OZO
DN37 9DU M6MNI
DN37 9EE M0COI
DN37 9HA G7PKP
DN37 9QJ G4EBK
DN37 9QL G3RGC
DN37 9QZ 2E0SFT
DN37 9RH G1WHT
DN37 9RN G8RIW
DN37 9RX G4DXB
DN37 9RZ G1BRB
DN37 9SQ M1RMO
DN38 6af M6RKP
DN38 6AP G8HDJ
DN38 6AU G7TRG
DN38 6BD G4FHF
DN38 6HF G7IJC
DN38 6HJ M6FOK
DN39 6SG G0EFZ
DN39 6UQ G0FEQ
DN4 0EP 2E0SOF
DN4 0EP M3JZT
DN4 0HE M7LEP
DN4 0JG M6RBQ
DN4 0UD 2E0NDH
DN4 5EE G8XTU
DN4 5EG G6RQJ
DN4 5EQ 2E1BGN
DN4 5EY M1EUF
DN4 6AL M3OYZ
DN4 6AU G0RFV
DN4 6DQ G3NYB
DN4 6DX G3AAF
DN4 6DX G7PUL
DN4 6EZ 2E0IMC
DN4 6EZ G4FYE
DN4 6HE G1HZR
DN4 6HX G6ZKS
DN4 6LF G0LOL
DN4 6PD 2E1DBZ
DN4 6PD M3DBZ
DN4 6QB G0GTI
DN4 6RE M3CMP
DN4 6SN G8GHB
DN4 6SQ G4OEK
DN4 6TB M3WSI
DN4 6TU G0DAM
DN4 6TU G3TCQ
DN4 6TU M6WSK
DN4 6UR M3KFH
DN4 6UT G7KEK
DN4 7AX G4WTR
DN4 7EG G7CDI
DN4 7HU G3VHZ
DN4 7JH G7KMO
DN4 7JQ G7TMO
DN4 7JY G0NMJ
DN4 7JY G4DMH
DN4 7RB M0MCH
DN4 8QS M6TKA
DN4 8SX G8RFZ
DN4 8TN G6MWD
DN4 8TN M5ADA
DN4 8TU 2E0TTP
DN4 8TU M7BCH
DN4 9AJ G3OFP
DN4 9AR G3TTU
DN4 9DA M0KAI
DN4 9DB M0AGJ
DN4 9DQ G0SSC
DN4 9DT G6DJX
DN4 9EL M3KEW
DN4 9ET M6MXX
DN4 9LA G0LJV
DN4 9QR G0LHM
DN4 9QW 2E0GZJ
DN4 9QW M7UTT
DN40 1AR G7BTP
DN40 1DW G4YAM
DN40 1EE 2E1HWU
DN40 1HH G0NBC
DN40 1LT 2E0CMZ
DN40 1LT M0PEM
DN40 1LT M6GRV
DN40 1NR G2HLB
DN40 1PT G0KTX
DN40 1PW G0DOB
DN40 1RB G7PLS
DN40 2AS G7GOV
DN40 2AZ G8XFY
DN40 2DQ G0HDS
DN40 2EQ G7BRZ
DN40 2JH G3VCX
DN40 2SG G1RSK
DN40 3AE M7BKU
DN40 3AX M0PLY
DN40 3DA G4YTQ
DN40 3HF M7ATX
DN40 3PS M3KYV
DN40 3PT G3YKK
DN402EU G0MAF
DN41 7JB G1STW
DN41 7RD G4YHP
DN41 7RE G1LWL
DN41 7SR G7RLK
DN41 8AN G8FKF
DN41 8ED G1OPD
DN41 8EL M0MAL
DN41 8ER G6PEH
DN41 8HE G4KAL
DN41 8HG G6SXN
DN41 8JB G1EXU
DN5 0DW M6SRB
DN5 0ED G0RLH
DN5 0EG M0BDU
DN5 0LH G7VYY
DN5 0NT G0IBG
DN5 0PQ M3GFO
DN5 0RP M7RLL
DN5 7BL G7ANA
DN5 7EN G4OVS
DN5 7LH G3VLL
DN5 7LW G0UQQ
DN5 7QF G8LHT
DN5 7RE G7ONV
DN5 7RY G1SCO
DN5 7SD 2E0HNF
DN5 7TE G0GNP
DN5 8EB M6RUM
DN5 8EH G8JJR
DN5 8ER G6NBF
DN5 8NQ G0DQB
DN5 8PQ M6LUZ
DN5 8QA G0PXE
DN5 8QN G4ANP
DN5 8QP G3IGU
DN5 8QW G0HLJ
DN5 8RR G6GNO
DN5 9DY M3KKA
DN5 9HD G7ACO
DN5 9QX M6PBY
DN5 9RN M7DCP
DN5 9ST G4PCD
DN50RA M6BMW
DN6 0AL G4YTR
DN6 0AR G6GYV
DN6 0EZ G0TSH
DN6 0EZ G7AYO
DN6 0LF M7LCB
DN6 0LU 2E0BSJ
DN6 0LU M3LCS
DN6 7AX G7CTG
DN6 7BT M0RWA
DN6 7EA G4JII
DN6 7EE G0GBQ
DN6 7JL M3LQB
DN6 7JQ M3GRC
DN6 7JQ M3YEZ
DN6 7JZ G6HWI
DN6 7JZ G8KYS
DN6 7LX M3WXP
DN6 7NH M6SAI
DN6 7QQ 2E0GRC
DN6 7QQ 2E0AXA
DN6 7QQ M6DGG
DN6 7RY M6JUI
DN6 7SL G7BGZ
DN6 7TT M0FCI
DN6 8EB G3UWR
DN6 8EH G1JTZ
DN6 8JE G1ILF
DN6 8JL M0CCD
DN6 8JQ G4MRU
DN6 8PD G0AHV
DN6 8RJ G8WCX
DN6 9BY G3WZW
DN6 9HJ G7UBO
DN6 9HX G7JWY
DN6 9HX G7VMO
DN6 9JX G0DQM
DN67NW M3JVK
DN69JU 2E0UBT
DN69JU M3UBT
DN7 4AH M0DLY
DN7 4AH M3GCN
DN7 4AH M3SJK
DN7 4AS M1EJX
DN7 4BI M6MXH
DN7 4BS G3VZE
DN7 4BU G1HOU
DN7 4EJ M1ALU
DN7 4HB G0TJR
DN7 4HB M3NVS
DN7 4HS 2E1SKA
DN7 4LZ M6WDG
DN7 4QL M7ENJ
DN7 5BS G8TVU
DN7 5PE G4VVE
DN7 5PT G6OSK
DN7 5PU M0ASJ
DN7 5QL M6NQB
DN7 5RH 2E0RHC
DN7 5RH M6RHC
DN7 5TQ G4ALF
DN7 6AB G3OZD
DN7 6AH G0LCT
DN7 6DA G4ZR5
DN7 6EH M6BTP
DN7 6ER G8PRH
DN7 6JX M3IKJ
DN7 6QQ G7BPF
DN7 6RY M0YRF
DN8 4NG M3RRV
DN8 4NU G0BKD
DN8 4QN M6VVE
DN8 4QT G0LNT
DN8 4SF M1EMR
DN8 4SL 2E0WUK
DN8 4SL M7SLO
DN8 4ST 2E0VVE
DN8 5BS G4HZN
DN8 5EA G7MGQ
DN8 5EL M6WCV
DN8 5JG G3KPU
DN8 5QS M0VYC
DN8 5QU M0GIW
DN8 5TL G4SZX
DN8 5YW G7MZW
DN84ST 2E1GLT
DN9 1DP G4WJE
DN9 1DY M6GCW
DN9 1HA G7ABT
DN9 1JL G4HOY
DN9 1JL M0XRC
DN9 1LB 2E0SWF
DN9 1LB G1IAB
DN9 1LB M3SWF
DN9 1LJ G7IMD
DN9 1LL G4NSM
DN9 1LR G4GZC
DN9 1LT 2E0CMY
DN9 1LT M6FAS
DN9 1MB G4NEG
DN9 1NA G3OS
DN9 1NW 2E0EDI
DN9 1PD 2E0EZX
DN9 1PD M3EZX
DN9 1PG G0OPA
DN9 1RG G4ZWQ
DN9 1RG G6ONE
DN9 1RT G8EVI
DN9 1RT M0DMZ
DN9 1RT M0NLR
DN9 1SD G8NDK
DN9 1SY M0RTQ
DN9 2DG G4YUK
DN9 2FB G4OGB
DN9 2HL 2E0LYK
DN9 2HP M3LQX
DN9 2HX G8BPS
DN9 2JQ G0UQV
DN9 2JX 2E1PMT
DN9 2JX G0TTR
DN9 2JX M1IFT
DN9 2LH G6BWA
DN9 2LR G8BAQ
DN9 2NX G4XRO
DN9 2PE 2E0CRS
DN9 3AE 2E0GWP
DN9 3AE M3RCC
DN9 3AJ G4RHY
DN9 3AJ M3RYN
DN9 3BA M0CAX
DN9 3ED 2E0MRG
DN9 3EG M3XPW
DN9 3JR G4RHZ
DN9 3PE G0RON
DN9 3PQ G0TUU
DO1 WR94 2E0CBT
DO1 WR94 M6MHA
DT1 1LH M0AFR
DT1 1LL G1DZZ
DT1 1NZ G1GMV
DT1 1RQ M7EUS
DT1 2BA G2DGB
DT1 2BS M0EJW
DT1 2BY G6KXW
DT1 2DY M6DDB

DT1 2EF G4CFY
DT1 2EL M0MZX
DT1 2JJ 2E0VWK
DT1 2JJ M0VWK
DT1 2JJ M3VWK
DT1 2LZ G3YUD
DT1 2NR G4EAS
DT1 2PE G7JIM
DT1 2PE G8SDS
DT1 2PF G4ZPO
DT1 2PS G3XIG
DT1 2SB G1RUL
DT1 3RJ G6PBG
DT1 3RR M1CSG
DT1 3SU M1ALT
DT1 3WP G4EDK
DT10 1AA M6KXI
DT10 1AR M0GID
DT10 1BE 2E0VXT
DT10 1BE M0XAJ
DT10 1BE M6VRT
DT10 1BN 2E0OEK
DT10 1BN M0OEK
DT10 1DD G4JBL
DT10 1EW M6FKJ
DT10 1EZ 2E0MSA
DT10 1EZ M6MCS
DT10 1HQ G4CEL
DT10 1HQ G4LRV
DT10 1LQ G8MNO
DT10 1LY M6KTJ
DT10 1PS G1SDJ
DT10 2BE G3RTD
DT10 2DL 2E0SDP
DT10 2DL M6RDV
DT10 2DT M3PAI
DT10 2EB M7EEW
DT10 2HF G7GPJ
DT10 2HX M7AQO
DT10 2LZ G0UXZ
DT11 0AA G3ZDQ
DT11 0AY 2E0RHS
DT11 0AY G4RFR
DT11 0AY G7RDT
DT11 0AY M0LXS
DT11 0AY M0ODE
DT11 0AY M0RHS
DT11 0AY M0VYB
DT11 0AY M6NSC
DT11 0DQ G6XAG
DT11 0EF G8RGU
DT11 0HF G7RMG
DT11 0HF M0RMG
DT11 0HF M3RMG
DT11 0HU G0NXQ
DT11 0JG G3VOO
DT11 0PP 2E0DKR
DT11 0PP M6DKR
DT11 0RY M6GZY
DT11 0SN M3GNY
DT11 0SS G6DAI
DT11 0SZ M7TCV
DT11 7DE G3KWN
DT11 7DL G1XCK
DT11 7GF 2E0NUQ
DT11 7GF M3NUQ
DT11 7HH G4GKX
DT11 7LW G3TPH
DT11 7LX 2E1EJC
DT11 7LZ G7PCE
dt11 7pa G7JTB
DT11 7RT M1LXM
DT11 7RU G3XGY
DT11 7UQ M0GKD
DT11 7XG G3PCW
DT11 7XQ M6ECX
DT11 7XU M0WAM
DT11 8AJ 2E0MHJ
DT11 8AJ M7LMP
DT11 8BP M6FKH
DT11 8JD 2E1ZPR
DT11 8JQ G1RRW
DT11 8JY G4TEN
DT11 8RH G3RGE
DT11 8TA G7NUN
DT11 8UJ M7JWF
DT11 9BU G4TRM
DT11 9BY G8LSS
DT11 9DW G1YHG
DT11 9HG G6JSN
DT11 9NB G3RVS
DT11 9NN G8BXQ
DT11 9NP G6AOF
DT11 9PH G4ZLX
DT11 9PS 2E0HVE
DT11 9PS M6HVE
DT11 9PW M6UKA
DT11 9QP G6PJD
DT12fe M6EKF
DT2 0AQ G8DJW
DT2 0AQ M3DOR
DT2 0BP G0OJP
DT2 0ER G0LOH
DT2 7HA G8GYL
DT2 7HP G8BAZ
DT2 7HP M6NNM
DT2 7HT M6WYN
DT2 7HX M1NAD
DT2 7NP 2E0LOA
DT2 7NP G8SEA
DT2 7NP M7LYF
DT2 7QQ G8XBY
DT2 7QS M3OPW
DT2 7QU 2E1MPQ
DT2 7QU 2E1SPH
DT2 7QU M7MPQ
DT2 7QU M7SPH
DT2 8AE G3KKJ
DT2 8BG G4DRS
DT2 8BG G4JQL
DT2 8BG M3WEZ
DT2 8BQ G1HYX
DT2 8EB G4PQX
DT2 8EG G8JUG
DT2 8EW M1EZX
DT2 8FJ 2E0PWP
DT2 8FJ M0PAA
DT2 8FJ M3UDN
DT2 8HS M0TRW
DT2 8JW G1WPG
DT2 8PE G0HVA
DT2 8PP 2E0SDZ
DT2 8PP G4TFF
DT2 8PP M0ZGT
DT2 8PP M6LCG
DT2 8QJ G0FIT
DT2 8TS G1RPT
DT2 8TX G7TNO
DT2 9DD G4AXU
DT2 9DS M6BPO
DT2 9ES G3IVC
DT2 9HZ 2E1HSA
DT2 9HZ G0PGT
DT2 9HZ M3TEN
DT2 9JN G0KYH
DT2 9JT G8DTE
DT2 9JU G6DOR
DT2 9LT G1LUC
DT2 9QB G0TZO
DT2 9QB M0HUV
DT2 9QX G3ZGN
DT2 9RE M3FRS
DT2 9RG G4GUV
DT2 9RY G4JQS
DT2 9TW M0JOL
DT2 9UD 2E0FZT
DT2 9UD M7XRN
DT3 4AX M6NSO
DT3 4BA 2E1LEC
DT3 4BA 2E1NRQ
DT3 4BA M3PSC
DT3 4BB G3SBT
DT3 4GU G0ROT
DT3 4HG G3VPF
DT3 4JP G7LPN
DT3 4JW M3YAD
DT3 4LD G0ZEP
DT3 4LD G2EC
DT3 4LD G4PTU
DT3 4LD M6DUK
DT3 4NZ M0XFL
DT3 4NZ M6WFL
DT3 5AG G8XAJ
DT3 5BG G3ZGP
DT3 5BP G3RAF
DT3 5BP G6RAF
DT3 5BP M0MJG
DT3 5HE G3OWE
DT3 5HF 2E0SEA
DT3 5HF M0ESU
DT3 5JS M0MCE
DT3 5LF G4XSC
DT3 5NG G3YWW
DT3 5PB G4RSL
DT3 5PF G3SDO
DT3 5QA G3RZG
DT3 5RP M6GKF
DT3 5SA M0ACC
DT3 5ST M6NYB
DT3 6AA M0XDL
DT3 6AA M6DLO
DT3 6AS G4TRW
DT3 6AS G6ZMU
DT3 6BG G0LQI
DT3 6DE G0NEV
DT3 6HS G4XRR
DT3 6JW G1KDO
DT3 6LE G0FBS
DT3 6LF G0BMQ
DT3 6LF G3JRL
DT3 6LZ G6GYG
DT3 6NE G3EGV
DT3 6NG M6GFQ
DT3 6NL G2CO
DT3 6PD G3XCY
DT3 6PT G3LAG
DT3 6PT G6LBJ
DT3 6PT M0BQO
DT3 6QL G1HTO
DT3 6RB M3UYL
DT3 6RD G0SEC
DT3 6RE 2E0GNJ
DT3 6RH G6XSK
DT3 6RP G3PGK
DT3 6SG M3NMJ
DT4 0AJ M7JBS
DT4 0AS 2E1DNX
DT4 0AS G0RWL
DT4 0BE G0VLJ
DT4 0DS G1YHB
DT4 0DZ M7OXS
DT4 0ET G4ZXP
DT4 0EZ 2E1OZY
DT4 0EZ M0WIK
DT4 0EZ M3OZY
DT4 0FE G0EVW
DT4 0FE G3SDS
DT4 0FE G8WQ
DT4 0JX G0ROX
DT4 0LN G4OWY
DT4 0LN G6AUW
DT4 0NH 2E0IJP
DT4 0NH M7DJL
DT4 0NJ G0ECX
DT4 0PW M0BND
DT4 0QF M3LUU
DT4 0QL 2E0PBS
DT4 0QY G1CAN
DT4 0SA 2E1TNE
DT4 0UE G6BSP
DT4 7DA M7CSN
DT4 7HQ M3WJU
DT4 7HT M6POW
DT4 7JH G3YIC
DT4 7LX M3ZUZ
DT4 7PS G4SDJ
DT4 8RF G3UHU
DT4 8SG G0LVF
DT4 8SQ 2E0TPH
DT4 8SQ M3TPH
DT4 8SQ M3YTQ
DT4 8TX M3KIO
DT4 9AL G4FYT
DT4 9AL M3MHZ
DT4 9AU M0SGV
DT4 9BH G0HOS
DT4 9EZ 2E1HFS
DT4 9HJ G7OLW
DT4 9JU 2E1HSJ
DT4 9LE G8ZVS
Dt4 9ly G7WAQ
DT4 9NB M5RIC
DT4 9NU G4BEI
DT4 9PA 2E0HIO
DT4 9RN G3VYX
DT4 9RR G7DOW
DT4 9RX M3XPI
DT4 9SA G4MFS
DT4 9SG G6LNF
DT4 9TL M6SHW
DT4 9UE G1BBT
DT4 9UU M6YOT
DT5 1AS 2E1RBH
DT5 1AS G0RYL
DT5 1BD G4WWH
DT5 1FR G2FHF
DT5 1JP 2E0RJB
DT5 1JP M3KKS
DT5 1NH G0PNG
DT5 2AA 2E1DSX
DT5 2AA G4DOL
DT5 2AA G4ZIY
DT5 2AB G0BZW
DT5 2AR G4YAA
DT5 2AY 2E0ERB
DT5 2AY G0ACQ
DT5 2AY M6KZC
DT5 2AY M7ANB
DT5 2DE G0CAE
DT5 2EE G4CFZ
DT5 2HJ 2E1DQZ
DT5 2JG G7NIU
DT5 2JX M3VWP
DT5 2JZ 2E1NRL
DT5 2JZ G7EIS
DT5 2NQ G4RAK
DT6 3EB G7KLZ
DT6 3PW G0JUV
DT6 3RW M6KLC
DT6 3UB G4PPZ
DT6 4AN G3PXH
DT6 4DW G8JCC
DT6 4EH M0CYG
DT6 4ES G0EZJ
DT6 4ET G4MZL
DT6 4HX G4JFG
DT6 4NQ G4XMZ
DT6 4NU M6CIS
DT6 4NY G6LEU
DT6 4NY M5ABC
DT6 4QL M0AXC
DT6 4QN G8IUN
DT6 4RG G4JWA
DT6 4RY M3BPF
DT6 5EQ M7DVX
DT6 5HX G3MTP
DT6 5LS G6RMV
DT6 5QY G4UHU
DT6 5RF G4MZY
DT6 5RF G7BYG
DT6 5RJ M6OTY
DT6 6AH M0IJL
DT6 6AL M3FRE
DT6 6DF G0HET
DT6 6DF G1EFX
DT6 6DU G3ZUE
DT6 6dy G4HNG
DT6 6JE G6VSE
DT6 6LR M0DRB
DT6 6PE G8ZTN
DT6 6SA G1EFG
DT7 3QY G8FIF
DT7 3SL G1GTM
DT7 3SX G0AYY
DT7 3UR G1ENA
DT8 3BQ G1SCN
DT8 3ES G6BTP
DT8 3HD M6BRZ
DT8 3HU G8JBQ
DT8 3JT G8LOJ
DT8 3PJ G1XHA
DT8 3RA G4PAC
DT8 3RF G4BQN
DT9 3EQ G0WDG
DT9 3RT G7AUU
DT9 3RT M3AHJ
DT9 4AT G3OPJ
DT9 4DT M7YIN
DT9 4NH M0MRB
DT9 4PZ M1EZT
DT9 4SZ M1BUP
DT9 5BD M0ARO
DT9 5BX M6URM
DT9 5DD G8DXO
DT9 5ED G7UPP
DT9 5FD G4INX
DT9 5PD G8CIG
DT9 6AF M3DIB
DT9 6BU G4GHI
DT9 6DN G7BYV
DT9 6EJ M1FFA
DT9 6HL G3UCT
DT9 6HX M6DGM
DT9 6NQ G8AFA
DT9 6RG G2UH
DT9 6RN 2E0TJN
DT9 6RN M0TJN
DT9 6RN M7TJN
DY1 1SL G8ZZT
DY1 2AZ G0NXD
DY1 2EF M0MGS
DY1 2ET 2E0NZD
DY1 2EU M3LWG
DY1 2GG G0OIY
DY1 2LN M6SHA
DY1 2NL M6ROF
DY1 2NN M6LID
DY1 2NU M6LIP
DY1 2NZ 2E0LSB
DY1 2NZ M3OLU
DY1 2PD 2E0SWS
DY1 2PD M3NUI
DY1 2PL M0EAF
DY1 2PL M6SKU
DY1 2QZ G1BTV
DY1 2RT G7EDZ
DY1 2RX M6EVI
DY1 2SL G6VJC
DY1 2SN G4XME
DY1 2SN G6PPA
DY1 2ST M7DGZ
DY1 2UN G6ZSU
DY1 2UW M6IRW
DY1 3DF G6IRG
DY1 3ED G7OLC
DY1 3JA M6TSZ
DY1 3JU M3UWU
DY1 3LE G0OWU
DY1 3RE M3XOV
DY1 4DZ G4ZLK
DY1 4LU G4OUH
DY1 4LU G4OUI
DY1 4NE 2E0COJ
DY1 4NE M0KCC
DY1 4NE M6BXO
DY1 4NP G7UHY
DY10 1LP M7FET
DY10 1LQ M3NGK
DY10 1LR G0NFO
DY10 1LR G1ZNV
DY10 1SE 2E0DUB
DY10 1SE M6DUB
DY10 1SS M0JMP
DY10 1XJ 2E0KKO
DY10 1XJ M0KKO
DY10 1XJ M1DTO
DY10 1XJ M6KKO
DY10 1YD 2E0XBB
DY10 1YD M0XBB
DY10 1YD M7TBC
DY10 1YH G7ABZ
DY10 2BZ G0MBG
DY10 2EZ 2E0RLX
DY10 2EZ M6RLZ
DY10 2HB G0MJX
DY10 2QD M0UCK
DY10 2RA 2E0ZVR
DY10 2RA M0ZVR
DY10 2RH G7EZE
DY10 2RY M3XNV
DY10 2SR M7ALS
DY10 2ST G0ISG
DY10 2SX M1DCY
DY10 2TH G1XJZ
DY10 2TP G4XCX
DY10 2UN G6BAM
DY10 2UT M0ORR
DY10 2XT G3SZG
DY10 2YB G0MJY
DY10 2YB G0WVT
DY10 3AG 2E1DLR
DY10 3AP G8WOX
DY10 3AQ G4TXV
DY10 3BH G4OBC
DY10 3DG G4ILQ
DY10 3DT 2E0DGG
DY10 3DT M0MCO
DY10 3DT M3TIC
DY10 3EG M0TTX
DY10 3EY M6GEU
DY10 3JG G0RQO
DY10 3JG M7YGM
DY10 3JR G0PPU
DY10 3LH G7ILG
DY10 3ND 2E1IFW
DY10 3ND M0NYX
DY10 3ND M1EJG
DY10 3ND M3ZOG
DY10 3QR G8OXG
DY10 3QS G0EYW
DY10 3QS G4AFY
DY10 3QZ G6DQZ
DY10 3RY M6MCB
DY10 3TA M7EBI
DY10 3TL G8TPM
DY10 3TN M7DCM
DY10 3TU G1HOL
DY10 3UA M6DTD
DY10 3UB G0UQB
DY10 3XL G4OIK
DY10 3XP M3WVJ
DY10 3XS M6BXP
DY10 4HG G4ZLN
DY10 4JR M1FDY
DY10 4JT 2E0WMP
DY10 4JT M6WMP
DY10 4NE M6GVN
DY10 4RS G4ROJ
Dy11 5 yb M7FTD
DY11 5BB M0JXF
DY11 5DL G8AKX
DY11 5DW G0FJD
DY11 5HJ G6BDY
DY11 5JN G0GPS
DY11 5JP G1AJU
DY11 5LB 2E0VEX
DY11 5LB M0MKV
DY11 5LB M6LHA
DY11 5LP M7TRB
DY11 5LU G8TLH
DY11 5LZ G0VBP
DY11 5NE G6ORM
DY11 5RU G0UUZ
DY11 5ST G8UEF
DY11 5TZ G4ZIB
DY11 5UA G0HRL
DY11 5UJ M0BRU
DY11 6AA G3TAW
DY11 6AU M3HXH
DY11 6BX G4CTU
DY11 6DQ G4PRD
DY11 6DQ M6NRH
DY11 6EE M6CEY
DY11 6HG 2E0HZQ
DY11 6HG M7GFC
DY11 6JU G0EOF
DY11 6PL G4BXD
DY11 6RL G3ZQQ
DY11 6RL G8DQP
DY11 6TE G8GYI
DY11 6TP M6RFN
DY11 6UG M6LVC
DY11 7BY G7DCJ
DY11 7DR G8SKA
DY11 7EA M6JEV
DY11 7EH M6RMW
DY11 7EW G0GFE
DY11 7HD G4ALT
DY11 7JR M7DVY
DY11 7LA G4NPN
DY11 7XS G1UNU
DY11 7XU G1JRX
DY11 7YL G7RDQ
DY12 1AT G6EAM
DY12 1BY M6BGZ
DY12 1DB G0UDI
DY12 1JH M0PGS
DY12 1RB G1YFG
DY12 1TR G0ESH
DY12 2BP M6RTD
DY12 2ED G6DFB
DY12 2HT G4OIL
DY12 2JQ G1OZB
DY12 2JX G0LOZ
DY12 2PU G4SPZ
DY12 2QG G4OCH
DY12 2QZ G7RBZ
DY12 3AA G4SNO
DY12 3AE M0ABF
DY13 0AR G4DKP
DY13 0EB G0TUW
DY13 0EL G7ESI
DY13 0EQ G7JWL
DY13 0EW G4OPV
DY13 0HE G7PLP
DY13 0HJ G7KZJ
DY13 0JR G0TLP
DY13 0JT G0IBT
DY13 0LL G1LBK
DY13 0LT M6CAA
DY13 0NU G7SCZ
DY13 0NU G7WDC
DY13 0NY G0PMF
DY13 0RH G8SPD
DY13 0RU M3MGO
DY13 0RX G0PJM
DY13 8EL 2E0MOL
DY13 8EL G3GJL
DY13 8EL M3YRM
DY13 8JB M3YCT
DY13 8JG G7WFQ
DY13 8JG M1DRZ
DY13 8LP G6CBB
DY13 8LR G3VHL
DY13 8NT G6EHG
DY13 8QB M6XWD
DY13 8QD M7EHL
DY13 8QP G0PWC
DY13 8RZ G0LJK
DY13 8SR G8DUW
DY13 8SU G6CPO
DY13 8TA 2E0TEH
DY13 8TA M6BHA
DY13 8TF G0MKL
DY13 8TF G6YIS
DY13 8TH 2E1CVE
DY13 8UQ G7KPF
DY13 8XW G0MAL
DY13 9EU G8BKL
DY13 9LR G7SAI
DY13 9NA 2E1DHJ
DY13 9NZ 2E0VXX
DY13 9NZ M0VXX
DY13 9NZ M3VXX
DY13 9PB G7IEO
DY13 9RL M7NEA
DY13 9RY G7RXX
DY13 9RZ G6PZH
DY13 9SJ M6RQH
DY14 0HH M7WJA
DY14 0QB G8IYE
DY14 0UA G4EQK
DY14 8AG G0EZT
DY14 8EB G8DFI
DY14 8LS G1WWA
DY14 8QQ G4EMD
DY14 9DB G4ZKD
DY14 9HP G0EXB
DY14 9HX G0IMK
DY14 9JY G3GGL
DY14 9LH 2E0BTR
DY14 9LH M3SFN
DY14 9LJ G3WBL
DY14 9NR G3WPQ
DY14 9NT G7JSG
DY14 9SF G6UMS
DY14 9TJ G8ETR
DY14 9XT G1DFR
DY2 0AU M6WDH
DY2 0BN M0MWT
DY2 0DG G7WLV
DY2 0EE G0AMW
DY2 0EE G0BFZ
DY2 0HZ G4WBP
DY2 0JY M3AYP
DY2 0NW G7ROP
DY2 0NW M0AKZ
DY2 0RU M7DZF
DY2 7QT M3PKR
DY2 7RA G4GSB
DY2 7RA G7CRS
DY2 7TS G6NTQ
DY2 8BB 2E0SLR
DY2 8BT M0PPC
DY2 8DH 2E0MTB
DY2 8DH M3FWO
DY2 8ER G0GUC
DY2 8EY 2E0PPY
DY2 8HP 2E0NTC
DY2 8HP M0NTC
DY2 8LE M6NJO
DY2 8XT M6WFC
DY2 9EU G7IZM
DY2 9EZ M3RWR
DY2 9HE G8TEK
DY2 9HR M3YZA
DY2 9JN M3NCL
DY2 9JT M0GQE
DY2 9LA M0HRC
DY20NW G1HRU
dy29ln M0LPF
DY3 1AL G6DKM
DY3 1LB G0BZT
DY3 1LF 2E0VPO
DY3 1NG G0DPJ
DY3 1PD G1GST
DY3 1RL G4ICU
DY3 1TG G3XEV
DY3 1UU G7RXZ
DY3 1XQ M6PGF
DY3 1XW G0CSW
DY3 1YA M6FBM
DY3 2AU M6DTS
DY3 2AX 2E0GOS
DY3 2AX M3GOZ
DY3 2BB G4DFE
DY3 2BB G8FUI
DY3 2DF G7HVF
DY3 2DQ G2HDF
DY3 2DQ M0BLT
DY3 2DQ M1AEA
DY3 2DQ M5HDF
DY3 2EQ M7AFL
DY3 2HJ M3IDD
DY3 2JF G7GEU
DY3 2LR G8WSB
DY3 2LR M1BVT
DY3 2PS G4HCZ
DY3 2QQ G7OIE
DY3 2QQ M0SJV
DY3 2RX G4IFM
DY3 2SH M6ARI
DY3 2UU M3FGO
DY3 3AS G7GEL
DY3 3BH M1AZG
DY3 3EF G8BOP
DY3 3LB G0OCR
DY3 3LH G6DAQ
DY3 3LN G7JIN
DY3 3NA 2E0KPD
DY3 3NP M0JYN
DY3 3PH M6ZSD
DY3 3QJ M0IMT
DY3 3TH M0KSA
DY3 3YQ G3IIV
DY3 4NF 2E0BJY
DY3 4NF M0VKY
DY3 4NF M3NFB
DY3 4NQ G0TMF
DY3 4NT G1BLJ
DY32DE M6NVO
DY32JU G0CUZ
DY4 0AF M6GZG
DY4 0AJ G6VAA
DY4 0SF 2E0YMW
DY4 0SF M5SW
DY4 0SF M6YMW
DY4 0SU M3HAJ
DY4 0SX M7DUC
DY4 4AU G3ZQS
DY4 4AU M0BPT
DY4 7PU 2E0SMZ
DY4 7PU M3OAL
DY4 7TP 2E0XEA
DY4 7TP M6CDB
DY4 8AJ G6WJJ
DY4 8HZ G7JCD
DY4 8UQ M6DYR
DY4 8XY G7CBW
DY4 9BJ M6NGD
DY4 9QS M3SMZ
DY5 1AE M7SLA
DY5 1DB G1YMH
DY5 1EQ G4PQI
DY5 2AY G6BEB
DY5 2BS M0DZD
DY5 2BU G1DIG
DY5 2DA G4NUS
DY5 2DH M3IUZ
DY5 2EL G7EVI
DY5 2HB 2E0NJO
DY5 2HB M6NAT
DY5 2HT G4YVA
DY5 2LQ G1YRQ
DY5 2LY G8JTL
DY5 2NG M1CWD
DY5 2PH G7FAZ
DY5 2QF M6CFE
DY5 2QG G4FAH
DY5 2SZ M6XVS
DY5 2UT M6NDM
DY5 2XY G6HZK
DY5 3BU M3LTP
DY5 3DP G4TGM
DY5 3DP G6IWD
DY5 3JE G6LUM
DY5 3JE G6XKE
DY5 3NT M3IFK
DY5 3NY G4TDB
DY5 3QH M6NJS
DY5 3RQ M6NAK
DY5 3UJ 2E0ETQ
DY5 3UJ M0IUQ
DY5 3UJ M6LUY
DY5 3YY G4NLW
DY5 4AW 2E0MLA
DY5 4AW M3HHN
DY5 4EF G0JKY
DY5 4EF G6JKY
DY5 4EF M6SUE
DY5 4EQ M0RSD
DY5 4EQ M0WYZ
DY5 4EQ M3FNC
DY5 4EQ M6PMR
DY5 4EX G0OWJ
DY5 4HA M3LKO
DY5 4JF 2E0TSC
DY5 4JJ M6PTP
DY5 4LB M3LTW
DY5 4LG M3CGM
DY5 4ND 2E0VCD
DY5 4ND M6DJX
DY5 4QL G8WSF
DY5 4QN G8TMQ
DY6 0BG 2E0KAB
DY6 0BG M0KAB
DY6 0DA G8PXW
DY6 0HL M1DRM
DY6 0HL M1EGG
DY6 0HX G8NWS
DY6 0LG M6EMM
DY6 0LG M6HJB
DY6 0LL G4CVU
DY6 0LN G0SRY
DY6 7AA G1MTU
DY6 7HG 2E1FPW
DY6 7QE 2E0ARA
DY6 7QE G0IPX
DY6 7QE G6ZUV
DY6 7QE M0CDL
DY6 7RP G4NRA
DY6 7RQ G4LVA
DY6 8BT G1WFJ
DY6 8DJ 2E0PPZ
DY6 8EE G6PYI
DY6 8JU M6YLI
DY6 8LL G4IEB
DY6 8LW 2E0BLT
DY6 8LW M3VDN
DY6 8PD G8YZF
DY6 8RQ 2E0RLY
DY6 8RZ G8KPG
DY6 8SP G4YBT
DY6 8SP G6GTB
DY6 8XP M3YVJ
DY6 9AD M6KEE
DY6 9DX G3ITH
DY6 9DY M3ZWH
DY6 9ET G4KEB
DY6 9PR M6RIA
DY6 9RB M6YNA
DY6 9RE G3APL
DY6 9RJ G6PSZ
DY6 9RJ M6AKG
DY6 9RP G4VNE
DY6 9TQ G1WJK
DY68JT 2E0BVT
DY68JT M0GUC
DY68JT M3UPF
DY7 5HU G4CYB
DY7 5JL 2E0HXP
DY7 5JL M0XIC
DY7 5JL M7ENV
DY7 6AD G4WDS
DY7 6BN G0TBI
DY7 6BN G8RAF
DY7 6BN M0VMW
DY7 6BY M7PGE
DY7 6BZ M6IGS
DY7 6DR G4ACS
DY7 6DU G4FTN
DY7 6DX G3TDC
DY7 6ED G4IES
DY7 6ED G8MXR
DY7 6EE 2E0LYY
DY7 6EE G0TDR
DY7 6HF G0AOW
DY7 6HF G4WRA
DY7 6HW G3KZG
DY7 6LG M6EVK
DY7 6LS G6SRS
DY7 6LS G8UAE
DY7 6RR G0DBJ
DY7 6RW G0AOX
DY7 6SP G7FXZ
DY8 1AJ G4XNW
DY8 1AX M3UFF
DY8 1HD G4MD
DY8 1NW M3ZWX
DY8 1QX M6FBY
DY8 1UX M7DMJ
DY8 2BE M1DBC
DY8 2DE G8XNL
DY8 2DE M6KIP
DY8 2HL G3ORI
DY8 2LH G4OLL
DY8 3BD G1RDJ
DY8 3DZ G7RCS
DY8 3EF G6BDH
DY8 3EH 2E0ITC
DY8 3EP G0EZX
DY8 3ER M6KUT
DY8 3JE G3NDN
DY8 3JH G7DTV
DY8 3JR G7JWJ
DY8 3LU 2E0IJT
DY8 3LU M7DZV
DY8 3NA M6AHW
DY8 3PH G4HWH
DY8 3PJ G4FYQ
DY8 3PJ G4NQB
DY8 3QG M7DQQ
DY8 3RN M6PDC
DY8 3RP G0MWS
DY8 3RZ M7OLM
DY8 3SY M3TFG
DY8 3UF G8OKB
DY8 3UT M1CMX
DY8 3UX M3UUE
DY8 3UZ G7FMQ
DY8 3XU M3ZJM
DY8 4BW M0KMT
DY8 4EQ M7ITR
DY8 4JQ M3PML
DY8 4QF G6WMR
DY8 4QF G8RLN
dy8 4qg G7HEZ
DY8 4QS G0FLW
DY8 4RN G4OLS
DY8 4SF G8PIP
DY8 4XS G0ASK
DY8 4XS G0ASL
DY8 4XS G4TFB
DY8 4XS G4TFC
DY8 4XU G0NLT
DY8 4XX G1IHJ
DY8 4YW M6ATW
DY8 4YW M7RJJ
DY8 5AU G0WBA
DY8 5HR G0ONH
DY8 5JB M6HIB
DY8 5JJ G4WIG
DY8 5LR G1VHN
DY8 5PH G0KVK
DY8 5PU G0OYY
DY8 5RF M6DIH
DY8 5YJ G0OVV
Dy84pa G1DCU
Dy84pa M3KTT
DY9 0JH G7RLX
DY9 0LX G6YAK
DY9 0RE G4VPE
DY9 0RT G6DDO
DY9 0SD G7VFL
DY9 0UP G0TZV
DY9 0YB 2E0ZDH
DY9 0YB G4IIC
DY9 0YB M0ZDH
DY9 0YE G3PWJ
DY9 0YH G1NZZ
DY9 7AZ G0WWA
DY9 7DT G7JDK
DY9 7EW M6JBT
DY9 7LA M1BQM
DY9 7PS G3XOV
DY9 8DE G1OHV
DY9 8SG G8JBP
DY9 8SG M3LKU
DY9 8YG M3KJM
DY9 9AE M3FFV
DY9 9BZ M6PMG
DY9 9DE G7EZH
DY9 9DL G7OJZ
DY9 9DL M0CHD
DY9 9DL M3OJZ
DY9 9EH G4KZB
DY9 9EL G0EWH
DY9 9EL G0KZM
DY9 9EL G4XOM
DY9 9HH G7JYZ
DY9 9HT G6RSI
DY9 9HX G6NME
E1 0EE G6USX
E1 0EE M3FLC
E1 1BJ M7EPF
E1 2AG 2E0MTM
E1 2AG M6MQT
E1 2QR 2E0SAW
E1 2QR M0WSA
E1 2QR M6MFZ
E1 4TD G1POJ
E1 6LT 2E0FBJ
E1 6LT M6RXX
E1 8PW M0HZY
E10 5QD 2E0OON
E10 6DX 2E0WJB
E10 6DX M6EHV
E10 6EE G4VXP
E10 6JD M3SJV
E10 6JL 2E0COL
E10 6QT G8VZD
E10 7EB G0TVV
E10 7JS M0GZK
E10 7LQ M0GHV
E11 1BA G7JRJ
E11 1JX G0VEH
E11 1JX G4ONP
E11 1LP M3YSA
E11 2JD G4GTZ
E11 2LD M3HTF
E11 2PP G4GGT
E11 2QQ M0CMQ
E11 4AA G7LAF
E12 5AX 2E0TMD
E12 5AX M6TMD
E12 5DZ G1SEF
E12 5EF G6RVZ
E12 5HD 2E1FZC
E12 5HD M3FZC
E12 5HX 2E0RLP
E12 5HX M7LRP
E12 5NB 2E0EWM
E12 5NB M3ZEW
E12 5PQ M3FNM
E12 6AA M0EDP
E12 6RE M3GVN
E14 0FN M6EBI
E14 0HD M7DIO
E14 0SL M3YSM
E14 3AJ G8XCJ
E14 3JB 2E0CPR
E14 3JB M6BZH
E14 3UG M1MOD
E14 6HP G6DFY
E14 6HU M1HMP
E14 6LS G1GLN
E14 7TE 2E0VGV
E14 7TE M0VGV
E14 7TE M6VGV
E14 8EJ M7ANA
E14 8EY G0KBO
E14 8LH G4LXH
E14 8SR G1RYS
E15 3AL M0PSS
E15 4DF M6PDS
E16 1GJ G0TDG
E16 1LQ 2E0NLM
E16 1LQ M3NLM
E16 1QW M6AUM
E16 1UT G8PUY
E16 2SB M7SVA
E16 3JY 2E0IJZ
E16 3JY M6WHT
E16 3NJ M3TGL
E16 3NJ M3UVS
E16 3QL 2E0CXO
E16 3RY M0GYU
E16 3TA G0MXE
E16 3TR M6WTP
E16 4NA M6KCD
E161TY G7GRM
E161TY M7JBL
E17 3RJ G3PKQ
E17 4BD G1HLS
E17 4BE 2E0HYF
E17 4BE M7EQI
E17 4HH G8PWK
E17 4PR M0TNL
E17 4QY G4IWD
E17 5AZ M1ADZ
E17 5BL M6JGN
E17 5EY G8KNF
E17 5RG 2E0PNB
E17 5RG M0PNB
E17 5RG M6TNB
E17 6NN M6RTE
E17 7ER M1EUE
E17 7ER M3EUE
E17 7HD M6EKY
E17 7JB M0VBE
E17 8AF M1CQM
E17 9EH G8BVY
E18 1AP G0FMB
E18 1BN G8ZHS
E18 1DG 2E0RCW
E18 1DG M6ECW
E18 1TP 2E0IGH
E18 2AB G3XXC
E18 2AW G6HIU
E18 2DA G4PSR
E18 2DB M6IBG
E18 2HB 2E0LGS
E18 2HB M0PDS
E18 2HB M6LUG
E18 2NJ G6SPN
E18 2PL G0KRX
E18 2PZ G6XBD
E18 2QA G4LKT
E1W 2BG M0IBD
E1W 2PJ G3UKI
E1W 2QW G1ETD
E1W 3EE M0GFD
E1W 3EE M3XXA
E2 0HE G6EUU
E2 0NF M6NNV
E2 8LP 2E1HBT
E2 9QT 2E0LUK
E2 9QT M6IVE
E20 1AN M7FDS
E3 2BB G0NXM
E3 3EG M0CVA
E4 6DR G6EXN
E4 6EG M6ITY
E4 6HX 2E0NJT
E4 6HX M6NJT
E4 6NA G4CGA
E4 6RP M6DZQ
E4 6RT G4RMQ
E4 7AG G7GCI
E4 7BP G3JPG
E4 7DX G8IUC
E4 7EG 2E0JGE
E4 7EG M0JGR
E4 7EG M6JGE
E4 7HN G2XG
E4 7HX G2HR
E4 7HX G8CSA
E4 7JG M1BLJ
E4 7LG G8JYX
E4 7ND G6XRF
E4 7NR G7ITX
E4 7PZ G0DPT
E4 7PZ G1MBW
E4 7PZ M0MBW
E4 7PZ M3OAX
E4 7RA 2E0TKX
E4 7RA M0TKX
E4 7RA M6TEK
E4 8BG M3WPC
E4 8DZ G4NEL
E4 8DZ G8MJH
E4 8HD G4RSX
E4 9DU G0LWN
E4 9EA G4HWK
E4 9HE G6SKR
E4 9JA 2E0LDN
E4 9JA M0VXY
E4 9JA M6XAX
E4 9NW G8XAO
E4 9RE G5YC
E4 9RE M0BPQ
E4 9RR G0PRK
E4 9SJ G7IQM
E5 0LY M6BKF
E5 0RG G4SHO
E5 9AZ M0IVN
E5 9AZ M7MMM
E6 2BS G1SJU
E6 3BW M0GTO
E6 3RY G0JYJ
E6 5XX 2E0TBW
E6 5XX M0LGN
E6 5XX M6CXP
E6 6BB M6LEG
E6 6HG 2E0CKR
E6 6HG M0HFZ
E6 6HG M6BKU
E7 0DN M0GHE
E7 0HN G7KWS
E7 0JS G0LLE
E7 0JT M6SQD
E7 9PR G7FRR
E8 2AD M0DYW
E8 2DB M7ASB
E8 2ET M6LOC
E8 2HE G7SSU
E8 3BT M6TZD
E8 3LE G0IFD
E8 4LN G4UAV
E9 7HU G3WUB
EC1M 7AJ 2E0KRX
EC1M 7AJ M0KRX
EC1M 7AJ M6KRX
EC1R 5XB G3WIP
EC1V 1LR G0MZF
EC1V 3AJ M3XMS
EC1V 4NJ M7FLD
EC1V 7NS M6BTH
EC1Y 8NL G0JKM
EC2Y 8BD 2E0EEM
EC3R 8EB M7ESC
EC4V 3EJ G0PCE
EC4Y 7EX G8FBK
EH1 2HR GM7ORX
eh1 2pb 2M0XRX
eh1 2pb MM0TEI
eh1 2pb MM6BBX
EH10 4EQ GM1THR
EH10 5AY MM7CMY
EH10 5BJ 2M0YCJ
EH10 5BJ MM0YCJ
EH10 5BJ MM6YCJ
EH10 5BW 2M0UTZ
EH10 5BW MM6UTZ
EH10 5DS GM4DTH
EH10 5DS GM8HHC
EH10 5JY 2M0CFA
EH10 5JY MM0GXY
EH10 5JY MM3RKF
EH10 5LW MM0AKM

EH10 5NU MM3FKO
EH10 5SJ GM7GNO
EH10 5SL GM0VMV
EH10 5SN GM7RMF
EH10 6ER GM8OTI
EH10 6HN GM8BJF
EH10 6HN MM6MWF
EH10 6JJ 2M0GWJ
EH10 6JJ MM7BKA
EH10 6NN GM4ZOA
EH10 6PS MM0YMG
EH10 6XE MM0KCS
EH10 6XE MM3MDB
EH10 7AQ GM3PSP
EH10 7BU GM3HAM
EH10 7BU GM4BYF
EH10 7BU GM6KAY
EH10 7BX MM0CPZ
EH10 7DG GM7REG
EH10 7HF GM7WFT
EH10 7HF MM6LQU
EH11 1AD 2M0IQR
EH11 1AD MM7IIE
EH11 1JZ GM8LYQ
EH11 1LR MM6EVO
EH11 1RP GM0AXY
EH11 1RP GM4YMM
EH11 1TT MM0ZBD
EH11 3GZ GM4AJV
EH11 3PF GM4DIZ
EH11 3SY MM3UDV
EH111HJ 2M0NQT
EH12 5AY GM6AXZ
EH12 5HG 2M0CLF
EH12 5HG 2M0JCF
EH12 5LG GM7JGR
EH12 5NG GM3HNE
EH12 5RF GM4HYR
EH12 6HN MM3DVD
EH12 6NB GM0BPT
EH12 6NB GM0CMO
EH12 6XB GM4DIJ
EH12 7EB GM7UJJ
EH12 7EW GM1CNH
EH12 7EW GM4HAM
EH12 7EW MM0BMG
EH12 7HP MM6CFA
EH12 7JW 2M0CBC
EH12 7JW MM3ZQW
EH12 7PD GM1PEL
EH12 7PU GM8CSE
EH12 8DH GM0PXV
EH12 8DL GM0SSQ
EH12 8DL GM3VJY
EH12 8DL GM4CUX
EH12 8EA GM0CDW
EH12 8EE GM8DKB
EH12 8LD 2M0DGJ
EH12 8LD MM0HVF
EH12 8LD MM6ERO
EH12 8NQ GM7VLZ
EH12 8QA GM4ZRR
EH12 8QU MM0DMU
EH12 8QW GM3RVL
EH12 8SP 2M0RWH
EH12 8SP MM0RHU
EH12 8SP MM0XRI
EH12 8SP MM6VSI
EH12 8SP MM6XRI
EH12 8SY GM1CHT
EH12 8TB GM4DHN
EH12 8UW MM3LWZ
EH12 8XF MM7JRX
EH12 8XH 2M0EGE
EH12 8XH MM6FXW
EH12 8XW GM3XUW
EH12 9JF MM3GOI
EH13 0AF 2M0RMN
EH13 0AF 2M0SMN
EH13 0AF MM6LON
EH13 0AF MM6RCO
EH13 0AF MM6SMN
EH13 0JH MM3UKG
EH13 0NA 2M0WRX
EH13 0NA MM0KJG
EH13 0NA MM3KJG
EH13 0NG MM7LJD
EH13 0RA GM8CVN
EH13 9AG GM8IIO
EH13 9BL MM6KOS
EH13 9LF GM0FMW
EH13 9PH 2M0LPX
EH13 9PH MM0MGM
EH13 9PH MM7LOX
EH130EW GM4ZQH
EH14 1EF GM3SRV
EH14 1HE MM1ICE
EH14 1HG GM4RZW
EH14 1HW GM4EZJ
EH14 1JW GM6NIA
EH14 1NJ 2M0VXB
EH14 1NJ MM6VXB
EH14 1NS 2M0TFF
EH14 1UH MM0AWU
EH14 2LB GM0AXX
EH14 3AE MM6NNS
EH14 3ER GM4YLN
EH14 3ex MM6HZO
EH14 5DW GM3OWU
EH14 5DX GM6MUZ
EH14 5EZ GM4XZN
EH14 5JN GM8LKL
EH14 5JN GM8NZL
EH14 5PP GM4REN
EH14 5QH MM6RCK
EH14 5RN MM6LAO
EH14 5RR MM0WAK
EH14 5SA GM7RYT
EH14 5SS MM6BXQ
EH14 5SX GM3TFY
EH14 5SY 2M0PSA
EH14 5SY MM0PSA
EH14 7ED GM1GDO
EH14 7EJ GM4EAU
EH14 7HJ MM7ALW
EH14 7HS GM4RAH
EH14 7JE GM1JKJ
EH14 7JE MM0MGB
EH15 1JP MM6BMQ
EH15 1NB GM8BHR
EH15 1NH GM7CZC
EH15 1RA GM3DIE
EH15 2BD GM1HJX
EH15 2EX MM7DHC
EH15 2HE 2M0INS
EH15 2HE MM0INS
EH15 2HE MM6INS
EH15 2PX 2M0TXK
EH15 2PX 2M0XBD
EH15 2PX MM0SBO
EH15 2PX MM0TKE
EH15 2PX MM6TXK
EH15 2PX MM6XBD
EH15 2QZ MM7DWQ
EH15 3QG GM4BHU
EH15 3QG MM7VLC
EH16 4ED MM6JAE
EH16 4LX MM3VVI
EH16 4TG GM7IFX
EH16 4UH 2M0WMU
EH16 4UH MM0TXN
EH16 4UH MM6TXN
EH16 5DD 2M1GFG
EH16 5RY MM0RHL
EH16 5UW GM8LFI
EH16 6GJ MM6MSA
EH16 6NX GM4WZP
EH16 6YE GM4HWO
eh165ee 2M0BUX
eh165ee MM0GSQ
eh165ee MM6ART
EH17 7Rt 2M0FFY
EH17 7Rt MM6HCK
EH17 7SE 2M0SRF
EH17 7SE MM0VTV
EH17 7SE MM3SRF
EH17 7SE MM6GHF
EH17 8AW GM4DTJ
EH17 8DX MM3MTM
EH17 8SR GM6OUL
EH17 8UF 2M0DTP
EH17 8UF MM6FYF
EH17 8XH 2M0NUO
EH17 8XH MM0NUO
EH17 8XH MM6NUO
Eh17 8xh MM7BRG
EH177TA 2M0LQF
EH177TA MM0LQF
EH177TA MM6LQF
EH18 1AB GM4DCL
EH18 1LN GM6SEV
EH19 2EH GM0OTU
EH19 2EU GM8HSY
EH19 2LD GM4SUR
EH19 2PQ MM0FQS
EH19 3FG MM6KOE
EH19 3JG 2M0SRY
EH19 3JG MM6DKI
EH19 3JY GM0QK
EH19 3JY MM3TQK
EH19 3LQ GM4GGF
EH19 3NF 2M0DXC
EH192LG MM6KMU
EH2 3NS MM3LGU
EH2 4AN GM1CZE
EH20 9EE GM3BST
EH20 9JT 2M0SWA
EH20 9JT MM3SWA
EH20 9LU GM7LTX
EH20 9PA GM6JGH
EH20 9RR GM0PQV
EH20 9SN GM6WMA
EH21 6AR GM7VLC
EH21 6EX GM0VYL
EH21 6JQ 2M0EZA
EH21 6PE MM6HZA
EH21 6RR MM0JXI
EH21 6SB GM6KFO
EH21 6TU 2M1DZW
EH21 6TU GM7PPN
EH21 7RD GM0EFQ
EH21 8AA 2M0BEB
EH21 8AA MM0FZV
EH21 8AA MM3ZVY
EH21 8AN MM7MCA
EH21 8JT GM1FAF
EH21 8JT GM4GVJ
EH22 1EE GM4CMI
EH22 2LW MM0CHV
EH22 2NL MM6IZH
EH22 2NS GM1CCN
EH22 2NZ GM0LOD
EH22 3BN GM4HQU
EH22 3DB GM0SUF
EH22 3DR 2M0KPZ
EH22 3DR MM6KLZ
EH22 3FT 2M0VPM
EH22 3FT MM0VPM
EH22 3FT MM6VPM
EH22 3LT GM0TFE
EH22 4AR MM3JPS
EH22 4BW MM1CMU
EH22 4BW MM3ALG
EH22 4BW MM3BHD
EH22 4DG 2M0EPK
EH22 4DG MM6KNJ
EH22 4QD 2M0YSR
EH22 4QD MM0YAB
EH22 4QD MM6YAP
EH22 4SJ MM0GFP
EH22 5AX MM3KVY
EH22 5DT GM7DRY
EH22 5DT MM3ZDG
EH22 5ER 2M1BYW
EH22 5HX GM0PKP
EH225JA MM7JTR
EH23 4BT GM0KDC
EH23 4EQ MM0BAC
EH23 4EU MM3DHN
EH23 4GX MM1BJO
EH23 4JF 2M1HIN
EH23 4LD GM4WCE
EH23 4NT GM0IMZ
EH23 4RL MM6LNB
EH23 4RS MM0MOK
EH23 4SU MM7FSR
EH25 9NP GM4LHW
EH25 9QJ GM4VZI
EH25 9SB GM6BRU
EH25 9TD MM0AJQ
EH26 0DR MM6DUV
EH26 0EL GM0MJR
EH26 0RR 2M0ITI
EH26 0RR MM0JTX
EH26 0RR MM6PZG
EH26 8BW MM6DWC
EH26 8DF 2M0CEX
EH26 8DF GM1GEQ
EH26 8DF MM0VPR
EH26 8DF MM6ANB
EH26 8EF GM0LUF
EH26 8HB MM6GBS
EH26 9AN GM1OVW
EH26 9EE GM6GFL
EH26 9HS GM1PSZ
EH27 8AN GM7KZL
EH27 8AU 2M0DIF
EH27 8AU MM6DHZ
EH27 8BH GM7HHB
EH27 8BH MM3RIX
EH27 8DD MM1RAH
EH27 8DQ GM1ROL
EH27 8DQ GM1ROM
Eh27 8ds MM7AGY
EH28 8AD MM7BJG
EH28 8PD GM4XHQ
EH28 8PD MM0DLH
EH29 9AN GM1YFO
eh29 9du MM3TUR
EH29 9DX MM0SFM
EH29 9DZ GM6BEY
EH29 9GE MM7GCS
EH3 5LP MM0WXD
EH3 5NA GM8EKF
EH3 5PD GM7PNX
EH3 6HB GM6KOR
EH3 6LD 2M0CNS
EH3 6LD MM6MSH
EH3 6RP 2M0BDN
EH3 7TZ GM6CZM
EH3 7TZ GM6NIC
EH3 8DD MM0OOF
EH3 8DD MM7LOL
EH3 9PL 2M0FSV
EH30 9JQ MM7STL
EH30 9PL GM7RAK
EH30 9PQ 2M0EEV
EH30 9PQ GM4XWL
EH30 9PQ MM0IGO
EH30 9PQ MM6HQI
EH30 9RU 2M0FZA
EH30 9RU MM0JKY
EH30 9RU MM7ALO
EH30 9UP GM8ZMF
EH30 9XG GM0SUY
EH32 0AD MM0VBC
EH32 0AN GM4LRU
EH32 0AQ 2M0WLA
EH32 0AQ MM0MLD
EH32 0AQ MM6YLA
EH32 0BB 2M1IBH
EH32 0BP GM0TCC
EH32 0BY MM3FJX
EH32 0EE GM4UYZ
EH32 0EE MM0CPS
EH32 0HZ 2M1HVR
EH32 0PG 2M1DZS
EH32 0PG 2M1EUV
EH32 0PG 2M1GLD
EH32 0SY 2M0BUY
EH32 0SY 2M0ILZ
EH32 0SY 2M0PGO
EH32 0SY MM3YMN
EH32 0SY MM7DGK
EH32 0SY MM7PGG
EH32 0TA GM4TAL
EH32 0TU 2M0ISY
EH32 0TU MM6ISY
EH32 9AT MM0CGZ
EH32 9DD 2M1IGO
EH32 9FB GM4YWI
EH32 9FH MM7BGW
EH32 9GF MM6SGF
EH32 9HA 2M0FZD
EH32 9HA MM0SHB
EH32 9HA MM6FZD
EH32 9HF MM3MFN
EH32 9PX GM7PSH
EH32 9RD 2M1IBE
EH32 9RD MM0IBE
EH33 1HU MM3XDW
EH33 1QE 2M0EAC
EH33 1QE MM3EWI
EH33 1QE MM3XVD
EH33 1QJ MM6JLX
EH33 2AL GM0NTL
EH33 2AL MM0WCG
EH33 2AL MM3ITA
EH33 2AS MM0CCC
EH33 2EB MM0DXC
EH33 2EE 2M1ETM
EH33 2HR MM6AUJ
EH33 2PL MM1ATY
EH34 5BB GM7MAG
EH34 5EH GM3GKJ
EH34 5EH MM1BVW
EH34 5HA 2M0KWP
EH37 5QB 2M0KLL
EH37 5QB MM0LBF
EH37 5QB MM3ISA
EH37 5QB MM3LWJ
EH37 5QD 2M0GRE
EH37 5QD 2M0GZA
EH37 5QD MM0GZA
EH37 5QD MM3YPN
EH37 5QD MM6YPN
EH37 5RA 2M0PMR
EH37 5RL MM3EYM
EH37 5RL MM3EYN
EH37 5RS GM8MNG
EH37 5SQ GM3OIB
EH37 5UJ MM3DZW
EH37 5UP GM8MJV
EH38 5YD 2M0NOP
EH39 4BP GM0EHL
EH39 4NR GM0WPW
EH39 4PS GM4KGZ
EH39 5JE MM0JMI
EH39 5JZ MM7ILB
EH4 1AX MM3KRR
EH4 1LW MM0VTA
EH4 1NH 2M0GUL
EH4 1NH 2M0PCW
EH4 1NH MM0DXH
EH4 1NH MM3JHS
EH4 1NH MM6HSS
EH4 1PJ GM8SOK
EH4 2AU GM4SVM
EH4 2JU GM4DMQ
EH4 2ND MM0MRO
EH4 2TT GM1CQC
EH4 2UE GM1PHD
EH4 3BB 2M0XOA
EH4 3BB MM0XAO
EH4 3BB MM6XOA
EH4 3DQ GM4GIO
EH4 3JU GM8MST
EH4 3TP GM0AXM
EH4 4RB MM7STT
EH4 5AW GM0RMV
EH4 5BF GM4HCE
EH4 5BQ GM7IRI
EH4 5JA GM3SBC
EH4 5NE GM7FIE
EH4 5PY GM6PYD
EH4 6EB GM6UNQ
EH4 6JJ MM6TEQ
EH4 7AH GM3KBP
EH4 7BJ GM7KTY
EH4 7BJ GM7PZH
EH4 7DF GM0WFB
EH4 7HN GM1UWE
EH4 7NT MM7WYZ
EH4 7RQ MM3BUZ
EH4 8BA MM6NOR
EH4 8DL MM3ICD
EH40 3BH GM4HJQ
EH40 3EB GM8PEB
EH41 3BE 2M0WDG
EH41 3BE MM0INH
EH41 3BE MM6FXZ
EH41 3DU GM3LGU
EH41 3LS MM1DSX
EH41 4EF 2M0VOZ
EH41 4EF MM0WCD
EH41 4EF MM6ZCD
EH41 4LW 2M0YUP
EH41 4QA 2M0TJO
EH41 4QA MM0HZI
EH41 4QA MM6FXM
EH42 1AY GM4FQG
EH42 1BA GM0JPG
EH42 1EX MM7EST
EH42 1GJ 2M0FLJ
EH42 1GJ MM3SQJ
EH42 1NW GM6KEV
EH42 1QT MM6GDY
EH42 1RT GM1SYC
EH42 1UF MM3KZD
EH42 1UG GM3UIN
EH42 1YR GM3VEI
EH44 6JT MM0ABJ
EH44 6LZ 2M0NMD
EH446NJ 2M0HJP
EH446NJ MM0VPF
EH446NJ MM6VPF
EH45 8DH 2M0LAO
EH45 8DH MM6LAK
EH45 8HB GM6JFP
EH45 8HJ MM6NVY
EH45 8JE GM1RKI
EH45 8NU GM0UTD
EH45 8PP GM4WTK
EH45 8PW GM4LDX
EH45 8QZ MM6DAT
EH45 9DB GM1MSS
EH45 9HX GM8NAL
EH45 9LX GM1OPO
EH45 9LX MM3NTX
EH46 7BA GM4OGM
EH46 7BE 2M0SPL
EH46 7BE MM0SPL
EH46 7BE MM3PZJ
EH46 7BE MM6HBR
EH47 0AJ MM3MPK
EH47 0AP MM3YTI
EH47 0BH 2M0WEV
EH47 0BH MM3WEV
EH47 0DN MM0HSR
EH47 0DN MM0PTE
EH47 0LH 2M0HIE
EH47 0LH MM6TMS
EH47 0NL MM0SMB
EH47 0NS 2M0HZL
EH47 0NS MM0HZL
EH47 7QB 2M0SXJ
EH47 7QB GM1SXJ
EH47 7QB MM3SXJ
EH47 7QJ MM6CMM
EH47 8AT MM6DSC
EH47 8EW MM0WKJ
EH47 8SA MM7ALI
EH47 8SA MM7TMO
EH47 9AZ MM6DGC
EH47 9BW 2M0DGB
EH47 9BW MM0GOG
EH47 9BW MM3NVD
EH47 9EL MM6IZK
EH470QA 2M0FCQ
EH48 1AS GM8ZKN
EH48 1DD GM0KBR
EH48 1DF GM4LHJ
EH48 1DU MM0LGS
EH48 2AF MM6GBX
EH48 2BD MM6DDN
EH48 2BU MM3UPY
EH48 2GZ GM1WMU
EH48 2JL MM0IQN
EH48 2JS MM3HTY
EH48 2LT MM0IEJ
EH48 2PB MM5AES
EH48 2PG 2M0CTB
EH48 2PG MM6VWT
EH48 2RG MM6RBE
EH48 2RR MM0DIS
EH48 2ST MM0HVA
EH48 2ST MM6HBF
EH48 2TD GM1PSU
EH48 2TN MM6KGM
EH48 3BU MM6RCZ
EH48 3HB MM6STM
EH48 3HG MM0HVU
EH48 3JB GM0DXB
EH48 3JU GM0EWF
EH48 3LA MM6EOU
EH48 3LN 2M0RYG
EH48 3LN MM6RYG
EH48 3PP MM5AHO
EH48 4AH MM7CEH
EH48 4AX MM7AEU
EH48 4BB MM6INC
EH48 4DP 2M0XFM
EH48 4DP MM3XFM
EH48 4HG GM0EDQ
EH48 4LD GM4LZO
EH48 4NU GM4BQD
EH48 4NW MM1DQV
EH482HH GM8IOL
EH49 6BP 2M0MIF
EH49 6BP MM6DCM
EH49 6BS GM3PDX
EH49 6DD GM1FAI
EH49 6DQ MM6HMC
EH49 6HA GM4LUS
EH49 6HA GM4XUS
EH49 6JS 2M0JXC
EH49 6JT 2M1ADM
EH49 6JT MM0GTU
EH49 6JT MM3ADM
EH49 6LW GM7VPT
EH49 6LZ MM7KGT
EH49 6SD 2M0GAN
EH49 6SF MM3TZP
EH49 6SH GM0MWJ
EH49 7AP GM4CAQ
EH49 7BP GM4YPL
EH49 7BP MM0ZCT
EH49 7BS GM0FTH
EH49 7JU GM3VTH
EH49 7LD GM0RUW
EH49 7LN 2M0CQI
EH49 7LN MM6RKT
EH49 7ND 2M0APX
EH49 7ND MM0ASB
EH49 7PL MM6AHX
EH49 7SR MM6EOZ
EH5 1EX MM0XXW
EH5 1FD 2M0DZX
EH5 1FD MM6HQE
EH5 1FT GM8YIK
EH5 2ET GM0BKS
EH5 2GU MM3KLO
EH5 2HP GM1FMV
EH5 2JJ GM0CII
EH5 2JN 2M0XTH
EH5 3AR GM0TEA
EH5 3NH GM4AOR
EH5 3PH MM7WGC
EH51 0AE 2M0PID
EH51 0AE 2M0VTB
EH51 0AE GM4FH
EH51 0AE MM0PID
EH51 0AE MM3VAH
EH51 0AE MM3VTB
EH51 0BQ 2M0IRC
EH51 0BQ MM3RQC
EH51 0DD GM0CQQ
EH51 0NX 2M0DRO
EH51 0NX GM5RAB
EH51 0NX MM6RDU
EH51 9EG GM7LFX
EH51 9ER GM0IKY
EH51 9LT GM0WKZ
EH51 9PE MM0FWG
EH51 9QD GM4HRL
EH51 9QN GM0HSC
EH519LX MM7JMR
EH52 5BX MM6HSC
EH52 5HQ 2M1GBG
EH52 5NS 2M0MMG
EH52 5NS MM0MMG
EH52 5NS MM3IQD
EH52 5PN GM4VDG
EH52 6AL MM7CCB
EH52 6AP MM7ENQ
EH52 6AP MM7NSY
EH52 6FB 2M0ZEB
EH52 6FX 2M0JXK
EH52 6FX MM6WEB
EH52 6HA MM0FFC
EH52 6LY GM0UHC
EH52 6PL GM0DHD
EH52 6QB GM3OIV
EH52 6TH MM0IRJ
EH52 6UW MM3MRX
EH526AL 2M0OOR
EH53 0BT MM7IME
EH53 0EG GM7GIF
EH53 0FQ MM6JXH
EH53 0NG MM0PWM
EH53 0NT 2M0CRR
EH53 0NT MM3LYH
EH53 0RN 2M1AOL
EH53 0SJ GM0ERT
EH53 0TE GM7RYK
EH54 5AD GM0NAZ
EH54 5JP GM6OGN
EH54 5JP MM6SIV
EH54 5LE MM6FFQ
EH54 5LP GM0REZ
EH54 5NA GM0ALS
EH54 6BG MM6WAW
EH54 6BZ 2M0FTA
EH54 6BZ MM6FTA
EH54 6DB MM0DMK
EH54 6EE 2M0AWY
EH54 6HE 2M0DIB
EH54 6HE 2M0MHN
EH54 6HE MM0LIV
EH54 6HE MM0NHM
EH54 6HE MM6NHM
EH54 6HE MM6NSM
EH54 6JD 2M0GGY
EH54 6JD MM3FOE
EH54 6JJ GM0CBA
EH54 6LQ 2M0YTD
EH54 6LQ MM6YTD
EH54 6LR MM7OFH
EH54 6LT GM7GTS
EH54 6PG 2M0HMV
EH54 6PG GM4UQD
EH54 6PG MM0RYE
EH54 6PG MM6FVK
EH54 6PQ MM6JNN
EH54 6RJ GM4HML
EH54 6RP GM3UCN
EH54 6TB MM6HAV
EH54 6UG MM7VGR
EH54 6US 2M0KRG
EH54 6US MM7CMC
EH54 6UX MM1EYI
EH54 7AB MM6JMI
EH54 7BP 2M0TZB
EH54 7BP MM6NLP
EH54 7BZ GM7AHA
EH54 8AP MM0KFR
EH54 8EW MM0HOL
EH54 8EW MM6HSK
EH54 8HW GM8MYO
EH54 8JB GM1PZT
EH54 8JG 2M0LAS
EH54 8JG MM3GTF
EH54 8JN 2M0OSC
EH54 8LQ MM0GEO
EH54 8NX MM6JCZ
EH54 8QP MM6EQY
EH54 8RW MM3TWG
EH54 8TA MM6XRS
EH54 9AA MM0DEC
EH54 9AA MM0NEW
EH54 9AT MM0SZC
EH54 9BZ MM3DYT
EH54 9DT GM0LOT
EH54 9DW 2M0CNZ
EH54 9DW MM0JRR
EH54 9DW MM6BTR
EH54 9JE GM3ZZA
EH54 9JG 2M1AVZ
EH54 9LD MM3IDR
EH55 8BE MM7GIR
EH55 8EW MM6XXV
EH55 8LT GM7IKB
EH55 8NL 2M0CPV
EH55 8NL MM0JHL
EH6 4DU 2M0BUT
EH6 4DU MM0WAP
EH6 4DU MM3RQP
EH6 4RU GM7GAE
EH6 4RZ. MM0PHD
EH6 4TE GM6HVY
EH6 4TR GM4HQZ
EH6 6ST MM7OWO
EH6 7BZ MM6MXU
EH6 7JH GM4IUS
EH6 7QQ MM6AUG
EH6 8QD MM6BYJ
EH6 8QP GM7AUW
EH6 8QP MM3AUX
EH65RY MM3WBV
EH7 4AY GM8GHQ
EH7 4HN 2M0TOK
EH7 4HJ GM3XQP
EH7 5JA GM1YME
EH7 5NG 2M0AYZ
EH7 5PS GM3HSO
EH7 5SA 2M0LBH
EH7 5SA MM6RHI
EH7 5TT GM7IHZ
EH7 5UF GM0HLK
EH7 6AU GM0LVL
EH7 6TH MM1FHL
EH7 7DW MM0BQI
EH8 7JL GM4JEM
EH8 7LZ MM0CJT
EH8 7PU GM0WTP
EH8 7SP 2M0ECK
EH8 7SP MM3ZQX
EH8 8DZ MM6NCP
EH8 8HA MM0HLU
EH8 9PS MM7RIN
EH8 9UW 2M0GZU
EH9 1AY MM0HQI
EH9 1NH MM6TLI
EH9 1PN MM6IKT
EH9 1QG GM7OCU
EH9 1UF GM6RQU
EH9 2EH GM4XNQ
EH9 2EL GM3FRU
EH9 2JL GM4HAO
EH9 2JY 2M1GYX
EH9 2NZ MM3NQT
EH9 2RA MM7RCT
EH9 3AR MM0GYG
EN1 1BS M6PMK
EN1 1NE 2E0DGD
EN1 1NE M0HLX
EN1 1NE M6DGD
EN1 1XG M7DBH
EN1 2BZ G4OBE
EN1 2HY M6OJS
EN1 2JU G3YJZ
EN1 3AE G0NEU
EN1 3DE G1GTS
EN1 3DW G4RTC
EN1 3LW M7WML
EN1 3NQ G3RWL
EN1 3NT G8PSF
EN1 3NU G8TWR
EN1 3RG M6WWE
EN1 3UP 2E0DHE
EN1 3UP 2E0JKS
EN1 3UP M0HVC
EN1 3UT M5AJK
EN1 3UU G0NQV
EN1 4AD M1EJQ
EN1 4BD G3ATC
EN1 4BD M5AGW
EN1 4DY 2E0RHL
EN1 4DY 2E0RIA
EN1 4DY 2E0WOB
EN1 4DY M6SOO
EN1 4DY M6WIF
EN1 4DY M6ZUB
EN1 4HR G4BUB
EN1 4PS G4IPR
EN1 4SY M6SGC
EN1 4TX G7MNK
EN10 6EE G6XAR
EN10 6JA G4UOI
EN10 6LD G4GAK
EN10 6NY G4XVY
EN10 6QE 2E1IBS
EN107NR M6CKJ
EN11 0BB G1SDK
EN11 0JS G8HTA
EN11 0LP M3MRJ
EN11 8RW G0JXR
EN11 9DG G1MRE
EN11 9DG G4ARY
EN11 9DL 2E0EHD
EN11 9DL M6EHD
EN11 9DX 2E0CHW
EN11 9DX M3VCW
EN11 9EP G3VSJ
EN11 9JR G1PZP
EN11 9JS G4YUZ
EN11 9NR G1CAY
EN11 9QH M0XDS
EN11 9QN G7HYG
EN11 9QS 2E0RNT
EN11 OQH M6MZJ
EN2 0DU M7DMA
EN2 0DU M7MDA
EN2 0DZ M0NOE
EN2 6NQ G6ODA
EN2 7BY G4XIN
EN2 7DB M1MUS
EN2 7EL G8AQO
EN2 7EL M0DDC
EN2 7EL M0HCY
EN2 7EN G8MOL
EN2 7PD M3FDM
EN2 8EN M6ZIP
EN2 8FG 2E0KRR
EN2 8FG M6RRK
EN2 8HS G4KZD
EN2 8HS G4ODR
EN2 8NW M7JGS
EN2 8QL G7MNE
EN3 4EF G0HAK
EN3 4EF G1PWO
EN3 4QE G3RKJ
EN3 5SE M0DJI
EN3 6EB M3VYE
EN3 6HE 2E1FQE
EN3 6NT G7VVF
EN3 6ST M7RLD
EN3 7HA G3JNJ
EN3 7JF G4GJO
EN3 7QA 2E1HVI
EN3 7RW M3XFA
EN37QZ M7WOL
EN4 0BB G6DLM
EN4 0EX G3UJV
EN4 8DQ G0CEQ
EN4 8EA 2E1DLS
EN4 8EN M7ASE
EN4 8JS G6PKX
EN4 8NJ G3LSX
EN4 8PU G8DVJ
EN4 8TU M7EFO
EN4 8UX M6NJH
EN4 9AS G7MGM
EN4 9DS G0LCS
EN4 9DS G5SIX
EN4 9DS G8VR
EN4 9LX M7NOL
EN4 9PA G8OQC
EN4 9QT G3GMY
EN4 9RL G1MPU
EN4 9TX G0NSI
EN5 1AW M7HVA
EN5 1EJ M7SUN
EN5 1LX M6RJL
EN5 1RJ G7VFY
EN5 2BP G8FVE
EN5 2BW G1ALA
EN5 2DB M6DUT
EN5 2DP M6VCA
EN5 2EX M6SNG
EN5 2JL G0SNO
EN5 2LE G4BWY
EN5 2LE G8AIE
EN5 2LS G4GPR
EN5 2NQ G3IQY
EN5 2QU 2E1JOD
EN5 3BB G0TOY
EN5 3BP 2E1HAS
EN5 3BP M3BCH
EN5 3DP M7FGT
EN5 3JR G6NVI
EN5 4JG M6FUT
EN5 5LH M6YGN
EN5 5QH G3YFW
EN5 5RH G8TAU
EN5 5TN G4WRD
EN6 1LG G3PRK
EN6 1QW G3RTE
EN6 1QW G6AY
EN6 1XW G1END
EN6 2BD G3WEA
EN6 2BQ G4EZU
EN6 2DH 2E0KSP
EN6 2DH M6VLQ
EN6 2DH M7VLQ
EN6 2HY M7BVQ
EN6 3AJ M0ATL
EN6 3DL G3DEN
EN6 3DZ G7BHR
EN6 4DZ G0ICK
EN6 4DZ G4JIJ
EN6 4JE G4ZSC
EN6 5HE G4HTE
EN6 5JQ M6KYS
EN6 5LX G8ZCK
EN6 5NB G7RHI
EN7 5EQ 2E0GSN
EN7 5EW G3PID
EN7 5JT G3SSZ
EN7 5NL G3MVM
EN7 6AN 2E1HRJ
EN7 6AN G7OYP
EN7 6AN M6HLS
EN7 6AQ 2E0ROM
EN7 6AQ M0ROM
EN7 6AQ M3XYA
EN7 6AQ M6DAP
EN7 6AS M6NWC
EN7 6DF G8KKH
EN7 6HB 2E1JGM
EN7 6LG M1GDB
EN7 6TF G8XTR
EN8 0BW 2E0SSA
EN8 0BW M0SAO
EN8 0BW M6AHD
EN8 0BW M7SUM
EN8 0BW M7SZD
EN8 0ER M7RGP
EN8 0EW M5BXB
EN8 0QU M6GDT
EN8 0QU M6JTH
EN8 7JJ G4GRN
EN8 8DA 2E1EGI
EN8 8NH G4SGN
EN8 8PS M6GKV
EN8 8UX G8VLP
EN8 9AP G6YNT
EN8 9BZ G0PXY
EN8 9JJ G6XBS
EN8 9QY G0NLV
EN8 9RQ G4WIP
EN8 9RQ G6AFX
EN9 1HN M7MSW
EN9 1PS M3UBK
EN9 1RL M6KLV
EN9 1SZ M3TEP
EN9 2AL G4XVM
EN9 2LB G3HYG
EN9 3DS 2E0DDF
EN9 3DS M6DFD
en9 3ea M7YIF
EN9 3HP 2E0AWC
EN9 3LD G4MOI
EN9 3LD G7OKV
EN9 3LQ M0MHO
EN9 3NS M3HSM
EN9 3NZ M7GPX
EX1 2BG G8NEI
EX1 2SP 2E0RGH
EX1 2SR G7EQO
EX1 3AH M0GJX
EX1 3AH M0GMO
EX1 3EQ 2E0DHX
EX1 3GA G3WVM
EX1 3JQ 2E0CRD
EX1 3JQ M3DGD
EX1 3NP M3DDY
EX1 3NT 2E0GHR
EX1 3NT M3GHR
EX1 3RA G6HUO
EX1 3WU 2E0TSY
EX1 3XA M6YFU
EX1 3XP G4KXR
EX10 0AY M6FSM
EX10 0ER G6IDG
EX10 0LR M3YFN
EX10 9BW 2E0ZZT
EX10 9BW M3HUS
EX10 9EW G8GTV
EX10 9EX 2E0BZS
EX10 9EX M0PBC
EX10 9EX M6NLO
EX10 9JA G3DCE
EX10 9JJ G4UBB
EX10 9LF M6VQC
EX10 9LS G6RUM
EX10 9NY G6BHY
EX10 9SJ G4IFQ
EX10 9SU M6YWX
EX10 9TJ G0AXC
EX10 9TJ G6XUV
EX10 9TJ G7AGO
EX10 9TJ M0KSO
EX11 1AR G4BNP
EX11 1BX G1OEF
EX11 1BY G8LAY
EX11 1DJ 2E0ITF
EX11 1DT G0WGH
EX11 1EN G7DYB
EX11 1EP G7NBZ
EX11 1FQ 2E1VAR
EX11 1FQ G1ARM
EX11 1JA G8NGE
EX11 1JA M6OSM
EX11 1PF G0WYF
EX11 1PF G6LAW
EX11 1PR G4VTQ
EX11 1RG G4TIG
EX11 1RL 2E0KVJ
EX11 1RL M6KVJ
EX11 1SY G8NVT
EX11 1TD G4UUH
EX11 1TD M6MTI
EX11 1TW M6WNZ
EX11 1UY G4BOP
EX12 2DB G6JJA
EX12 2DJ G4XXK
EX12 2EQ 2E0EBL
EX12 2EQ M6EVX
EX12 2HH G2BSW
EX12 2HN G8VXU
EX12 2HX M0TGC
EX12 2NJ G7VCB
EX12 2NT G8WCQ
EX12 2PD G1NFQ
EX12 2SB G0CTF
EX12 2SS M3ECS
EX12 2TP M3GPR
EX12 2TU G3LVB
EX12 2UH M0HSH
EX12 2XW G6WZP
EX12 3BD M6TAO
EX12 3BL 2E0TWO
EX12 3BL G4APG
EX12 3BL M3HKM
EX12 3LT M6ZYE
EX13 5BL G1KNU
EX13 5BS G4VHG
EX13 5GT M6HUQ
EX13 5GT M6HVP
EX13 5JT G8MCC
EX13 5RW G0AOS
EX13 5SQ G6WWY
EX13 5SQ M3WWY
EX13 5SX G3ZVW
EX13 5SZ G7CMP
EX13 5TD G4KAM
EX13 5TX M3ZII
EX13 7AF G8CYG
EX13 7LU G3GOS
EX13 7PB G0HRH
EX13 7PS G8SHF
EX13 7RW G4WNU
EX13 7SQ M3YKH
EX13 7ST G3CFR
EX13 8AH M1AVB
EX13 8AP G3CYX
EX13 8AQ G4FVU
EX13 8TT G0GHH
EX13 8TT G8CA
EX14 1AX G4PQS
EX14 1AX G7RMX
EX14 1JB G3GRQ
EX14 1JE M7JBP
EX14 1QZ G0BCO
EX14 1QZ M3VTA
EX14 1QZ M3WDY
EX14 2DF M3ZDS
EX14 2GL M7HOP
EX14 2GP G0SWV
EX14 2GP G3USE
EX14 2TT G0KJJ
EX14 2UG M7CTP
EX14 2YP G0WWL
EX14 3EX G0JSC
EX14 3EX G0JST
EX14 3EX G7JST
EX14 3HW M6IUF
EX14 3NL G4ENJ
EX14 3PY M6HHQ
EX14 3UP G4TRU
EX14 4QS G3OLB
EX14 4UL M1VPN
EX14 4UZ 2E1GFS
EX14 9AL G0TEB
EX14 9TA G0BVC
EX14 9TQ 2E0OSE
EX149RY 2E1HMB
EX15 1BQ G4TLL
EX15 1DW G4HMA
EX15 1EH 2E0PAO
EX15 1EH M6PIK
EX15 1FJ 2E1FWX
EX15 1PA G4FDG
EX15 1QD M0AKJ
EX15 1QS M7MDB
EX15 1SS G0SXN
EX15 1UD G0KOF
EX15 2LW M0CVR
EX15 2RN G8HPS
EX15 2RN G8IYD
EX15 2RN M0BVM
EX15 2SH G0IFA
EX15 2TT G6VDA
EX15 3AU G6TJJ
EX15 3BA G3WNV
EX15 3DN G0HPS
EX15 3DS G0OED
EX15 3ES G6EED
EX15 3HJ M6GXR
EX15 3LH G4SCV
EX15 3QG M3YNI
EX15 3RD M6FJE
EX15 3SE G6FVD
EX15 3XA G0TAZ
EX15 3XZ M7ICQ
EX16 4AW 2E0IFC
EX16 4AW M6NLB
EX16 4BE G6XWM
EX16 4BN G6SWD
EX16 4BW G7JGZ
EX16 4ER G4XXD
EX16 4ET G4ZAL
EX16 4LF M0PJY
EX16 4LN M3IEG
EX16 4PY G8MFF
EX16 5AF M6JSX
EX16 5FH G1HEA
EX16 5JE M3TME
EX16 5PA G8URB
EX16 5PE 2E0TIV
EX16 6BU M6CPZ
EX16 6EB 2E0CZC
EX16 6EB M6GEP
EX16 6EE G0IFC
EX16 6EE G4TSW
EX16 6FJ M0HMJ
EX16 6HE 2E0MEL
EX16 6JU G0MAS
EX16 6PL G4DKC
EX16 6RB G3VGY
EX16 6RG 2E0AYI
EX16 6RG M3CLO
EX16 6RJ G4YCV
EX16 6XJ G4PYU
EX16 7AZ M7AJM
EX16 7BS G0GLQ
EX16 7HT G3ZQL
EX16 7HT G8PXO
EX16 7NA G3PCY
EX16 8AZ M0GMS
EX16 8LQ G3ZRJ
EX16 8PP 2E0KCF
EX16 9DW G6YWZ
EX16 9JQ G6ASK
EX16 9JQ M0KSB
EX16 9JQ M0LPR
EX16 9JU G0LOW
EX16 9NE G8MBE
EX16 9NN 2E0RNW
EX16 9NN M6WNR
EX16 9NS 2E0PLJ
EX16 9NS M7BUL
EX16 9QH G3HXK
EX16 9QH G4FJK
Ex166ar G6MFX
EX17 2DH G4YAQ
EX17 2DH G6TWD
EX17 2DN G1JXA
EX17 2EJ M3LLK

EX17 2EJ M3VVY
EX17 2EP G3HTA
EX17 2HA G4WJJ
EX17 3HE M3MEF
EX17 3NB 2E0NRJ
EX17 3NB M0NRJ
EX17 3NB M6NRJ
EX17 4AE 2E0CTQ
EX17 4QX M0GEB
EX17 4RU M3TBU
EX17 5AL G4RRA
EX17 5HR M0PXI
EX17 5NA G0JUM
EX17 6AY G6DLJ
EX17 6DH M1AEI
EX17 6DU M7CEL
EX17 6LA M6NPN
EX17 6NA G4MQQ
EX17 6NA G8RYX
EX17 6PA M6TVW
EX18 7BQ 2E0SXS
EX18 7BQ M6DVT
EX18 7QX G0ABI
EX18 7SN M1DAS
EX19 8DP G8JRW
EX19 8HD G8RIR
EX19 8JW M6BPE
EX19 8JX G3LBM
EX19 8SL G3ZQR
EX2 4DG G8LKA
EX2 4JS G3AJK
EX2 4LR M6EUV
EX2 4SJ G6ZTP
EX2 5DU M0IHT
EX2 5EP 2E0SJM
EX2 5ES 2E0RQK
EX2 5ES M0RQK
EX2 5JT M0REB
EX2 5JX 2E0SQB
EX2 5JX M6SQB
EX2 5NH G6BJL
EX2 5QJ G7WLM
EX2 5QN G8RCZ
EX2 5RG M3LQI
EX2 6AN G0BNW
EX2 6BW M7SKA
EX2 6EA G7UIU
EX2 6ED M7EGW
EX2 6EG M0CPF
EX2 6JJ G4BQH
EX2 6LE M3RPS
EX2 6LS M3YDV
EX2 7AQ M0TCF
EX2 7BD M3IEA
EX2 7BQ M0NJE
EX2 7RA M7CUG
EX2 8LA M6AJL
EX2 8XN 2E0ECW
EX2 8XN G0JQS
EX2 8XP M6OOZ
EX2 9AH 2E1GZZ
EX2 9AH M3MPC
EX2 9JF G4CPN
EX20 1AU M7AZJ
EX20 1BD G1DKB
EX20 1EA G3PSZ
EX20 1EA G7VMQ
EX20 1GR 2E0SPS
EX20 1GR M6LJF
EX20 1HZ G7CIH
EX20 1JF G0MOH
EX20 1NU 2E0VFT
EX20 1PL M6BPU
EX20 1PR G0FJA
EX20 1PR G0FYL
EX20 1QQ G4NJK
EX20 1QX G8JZZ
EX20 2HB 2E0WLQ
EX20 2HB M0PIR
EX20 2HB M7JCE
EX20 2HX G3YGA
EX20 2JF G3VDL
EX20 2NG G1BOB
EX20 3AJ G4NFP
EX20 3EZ G0FUV
EX20 3EZ G6LWT
EX20 3FE 2E0GLL
EX20 3FE M0SLL
EX20 3FE M6TIU
EX20 3LE G3ZLS
EX20 3LU G4BTN
EX20 3NR G8WKA
EX20 3NX 2E0CER
EX20 3PL G8TAY
EX20 3PP G4VQP
EX20 3QW G0GHT
EX20 4DQ G4XIX
EX20 4LL G0IVO
EX21 5AY M6IXJ
EX21 5BP G4FIV
EX21 5BQ G0OAW
EX21 5EF G0ASG
EX21 5eu G0ESY
EX21 5HA G0LSI
EX21 5LT G8UIL
EX21 5LT M3WYJ
EX21 5PX M3AFF
EX21 5TX M6KPI
EX21 5UD G0IKC
EX21 5UF G3BYG
EX21 5XY 2E0XLO
EX21 5XY M7XLR
EX22 6AY 2E0SAT
EX22 6AY M6AIR
EX22 6DA 2E1GQU
EX22 6DA G7VTE
EX22 6DA M1BNR
EX22 6DQ M7AUF
EX22 6HB G0CLC
EX22 6HX G0WTI
EX22 6QN G8MWW
EX22 6RS G1GZI
EX22 6RS M3GZI
EX22 6TB 2E0MYB
EX22 6TB M0MYB
EX22 6UU G4WWR
EX22 6UU G6WWR
EX22 6UU M0BKV
EX22 6XX G3VDH
EX22 7AA G6LAU
EX22 7BQ M0OZJ
EX22 7DU M6VKZ
EX22 7ED G6REW
EX22 7EN M3VJW
EX22 7EW G4CQM
EX22 7NB M1AOB
EX22 7NB M6DEM
EX22 7NY G0RQL
EX22 7NY G1DIF
EX22 7NY M0OMC
EX22 7QY M7LIP
EX22 7RN G0KDY
EX22 7RT G6ILD
EX22 7YG G3ZGI
EX23 0AH M6JBI
EX23 0AU M0LSS
EX23 0BE G6LSB
EX23 0BN G0OKN
EX23 0DT G4FWT
EX23 0ES G0TDE
EX23 0HH M0HEN
EX23 0HY G4NCJ
EX23 8AP G1IQN
EX23 8DD G3XNE
EX23 8EU 2E0FWC
EX23 8EU G4YTC
EX23 8FF G8YWL
EX23 8HZ G3KMQ
EX23 8JB G8YRW
EX23 8LN M6EJV
EX23 8LR G6MAA
EX23 8LR G8DHA
EX23 8NA G0DBD
EX23 8PD G8ULJ
EX23 8PG G1RLB
EX23 8QF M0MJW
EX23 8RY G4IDU
EX23 8RY G4MMT
EX23 8SA M6JIC
EX23 8SB G7TRM
EX23 8SP M0BUI
EX23 9AF M0SMJ
EX23 9AG G1IJQ
EX23 9BB G4NLH
EX23 9BJ G4ASF
EX23 9BP 2E0RCM
EX23 9BP M0RBM
EX23 9ES M6CHW
EX23 9JA G8MEH
EX23 9JR G4OJQ
EX23 9LE 2E1EOQ
EX23 9LE M3EOQ
EX23 9RE G4MYD
ex238en M7RAW
EX24 6JU G6LUJ
EX24 6NN M7DVP
EX24 6PN G0GIN
EX24 6QZ G7EHR
EX2O 1PY M7CKR
EX3 0DY G1DKE
EX3 0DY M6AZM
EX3 0JW M6RVD
EX3 0LG G3IMW
EX3 0NA G4ETO
EX3 0PE G8ATC
EX3 0RA G1WYM
EX31 1PA M7CPA
EX31 1QA G3XYG
EX31 1QF G3AKJ
EX31 1QX G4RVJ
EX31 1RZ G0OQZ
EX31 2BH G6RNT
EX31 2HL M6DHW
EX31 2HY M0DJH
EX31 2JG M0IGF
EX31 2LH 2E0TVD
EX31 2LH M6SCP
EX31 2NJ M1FWD
EX31 3AF M0DWX
EX31 3AP G8NBO
EX31 3BS G3PGA
EX31 3BS G6GZZ
EX31 3DA G4PEK
EX31 4FL 2E0TGC
EX31 4FL M3TGC
EX31 4HY G6TAP
EX31 4JD M6EQL
EX31 4JQ G4RZI
EX31 4JR G0BUJ
EX31 4LP G0RVS
EX31 4PR G4YCW
EX31 4TA M7RXW
Ex313ap M3YBN
EX32 0DF G4SGW
EX32 0EG 2E0GZM
EX32 0eg M3MSU
EX32 0JN G4SJP
EX32 0JN G8KQB
EX32 0ND M3UKF
EX32 0NJ G0WUA
EX32 0TX G4MZH
EX32 7AS G3ZFV
EX32 7AY G7FJZ
EX32 7AY M3VJM
EX32 7HN G6USZ
EX32 7HW G0EYF
EX32 8DN G4TNE
EX32 8EJ G0DNV
EX32 8ET M6FEQ
EX32 8JR M6SWG
EX32 8LA M3HJF
EX32 8NW G0NPV
EX32 8PE 2E0ITX
EX32 8PE M6ORX
EX32 8QH G4KXQ
EX32 9BG G8VCQ
EX32 9BW G4XWQ
EX32 9DG M1CVK
EX32 9DJ 2E0LVC
EX32 9DJ M6LQC
EX32 9DP M3XZT
EX32 9EY 2E0EON
EX32 9EY M6IUI
EX32 9JX M3BOR
EX33 1AT G1LLI
EX33 1BB G4XZF
EX33 1DH G0DUH
EX33 1HE G3OIP
EX33 1PP 2E0FEO
EX33 1PP M0TLO
EX33 1PP M6FEO
EX33 2EH G0SHP
EX33 2EL G3MTD
EX33 2EZ M3IXF
EX33 2HL G8EZZ
EX33 2HN G1MJI
EX33 2LT G0LKI
EX33 2PE 2E0PEX
EX33 2PE M0PEX
EX33 2PE M6UNN
EX33 2PF 2E0IXX
EX33 2PF M3IXX
EX331HB M7CDO
EX34 0HQ M0AME
EX34 0NL 2E0JSX
EX34 7BX 2E1NJC
EX34 7BX G4NGB
EX34 7BX M3NJC
EX34 8DT G0HGM
EX34 8DW G8RPD
EX34 8DZ 2E1GBN
EX34 8EW G5HBM
EX34 8HS 2E0PBV
EX34 8HS M6PDL
EX34 8LG M3MXK
EX34 8LR 2E0IHW
EX34 8NH G4RWK
EX34 9HP G3JUW
EX34 9JW M7CSW
EX34 9LG G1GBC
EX34 9LL M6IAT
EX34 9LN G0DIZ
EX34 9LN M3MRU
EX34 9LS 2E0JQK
EX34 9LS M3NCH
EX34 9LS M6TFF
EX34 9QY M7LWI
EX34 9TA G1JWO
EX35 6EW M6BNA
EX36 3FL G6TWR
EX36 3HJ G8VEQ
EX36 3HL G0EOP
EX36 4AL G0HIW
EX36 4BH G4JBR
EX36 4DA M6FEJ
EX36 4DB 2E0GPT
EX36 4DB M6GPT
EX36 4DB M6IZE
EX36 4EW G4RVG
EX36 4FA M6TER
EX36 4HJ G0FGI
EX36 4HJ G8BXO
EX36 4HJ G8SBH
EX36 4HJ G8SSS
EX36 4JU G1KBF
EX36 4JW M0DYV
EX36 4PX G4YUO
EX37 9HY 2E0CNB
EX37 9HY M0JAP
EX37 9JG G1EEO
EX37 9QE G8RIS
EX37 9ST 2E0DYM
EX37 9TT G7TVQ
EX38 7BE G4INI
EX38 7NZ M1AEO
EX38 8AL G0OKA
EX38 8AS G4UFK
EX38 8EE G6KEH
EX38 8EX G0CVI
EX38 8HZ G8UZY
EX38 8NG M1GRA
EX38 8NR G8TBU
EX388BY G8VDD
EX39 1BD G4NUJ
EX39 1BS G0XCF
EX39 1BS M6XHF
EX39 1DF G6YSL
EX39 1GU G0VXB
EX39 1HG G0EDU
EX39 1LS M3KZS
EX39 1LT G0AYM
EX39 1LT G0BYL
EX39 1LT G6XYL
EX39 1NW G4FRV
EX39 1NX G0UKF
EX39 1NX G4XHK
EX39 1PA G1ZUS
EX39 1PE G0PGK
EX39 1PW G7MWI
EX39 1QY 2E0TUC
EX39 1QY M3TUH
EX39 1RD G4SRP
EX39 1SG G0UNB
EX39 1UT 2E0RGK
EX39 1UT M3RGK
EX39 1UW G6WZN
EX39 1XE G4CHD
EX39 1XN G4TCK
EX39 2BP M3VUY
EX39 2EA 2E0AJX
EX39 2EA M0OCJ
EX39 2JJ 2E0FTZ
EX39 2JJ M7BLL
EX39 2LL G1ZTJ
EX39 2LU 2E0HVZ
EX39 2LU M0WWD
EX39 2RR G3MNV
EX39 3BN G8SFD
EX39 3BY 2E0FQE
EX39 3BY M0MFS
EX39 3BY M7ABR
EX39 3DJ 2E0WWG
EX39 3DJ G3YGJ
EX39 3DJ M6DBV
EX39 3EB 2E1GJY
EX39 3EQ M3KWS
EX39 3ET G4SOF
EX39 3LZ G2FKO
EX39 3LZ G3JKL
EX39 3NE G4DXT
EX39 3QH M7BON
EX39 3QZ G1ZVZ
EX39 3RE G6LPV
EX39 3RW G1SVP
EX39 3RW G3YIN
EX39 3SF M0PIP
EX39 4AG 2E0IFA
EX39 4AG M7SVB
EX39 4AP G3YBP
EX39 4DX G0HAU
EX39 4HA 2E0AZJ
EX39 4HA M0VMH
EX39 4HA M3FXX
EX39 4HG G0GFK
EX39 4QR G0VWQ
EX39 4RS G3APU
EX39 5HN G0JZT
EX39 5HY G3RSF
EX39 5PL G3VYS
EX39 5PL G8FXG
EX39 5RH G8VNO
EX39 5SF M6ENI
EX39 5SH G8CPN
EX39 5ST M3WLL
EX39 5XY 2E0EUZ
EX39 5XY M6CCH
EX39 6BQ M3VYB
EX39 6JA G7VJA
EX4 1FD 2E0HOO
EX4 1FD M3UNF
EX4 1NH 2E0CNL
EX4 1NH M3VNL
EX4 1NJ G0FGE
EX4 1NX G4EDG
EX4 1PQ M3GSR
EX4 1SW M0KFW
EX4 1TA G3YKB
EX4 2AP G7BAE
EX4 2DF G3YRX
EX4 2EF M0UWD
EX4 2EG M1BAS
EX4 2EU G1YPM
EX4 2JP G4PCB
EX4 2JP G8NZB
EX4 2LR G4FYJ
EX4 2PN G0UTA
EX4 2QD 2E0JSK
EX4 4PW 2E1GMT
EX4 4RR G3SBP
EX4 4SA M7AJZ
EX4 4SJ G0TQS
EX4 4SJ G4BPV
EX4 5AJ G4PXE
EX4 5DN G4SJU
EX4 5DN M6PPN
EX4 6EY G3SXH
EX4 6HG G6IKC
EX4 6ND M0BBV
EX4 6NG G1VNH
EX4 7DY 2E0WRS
EX4 7DY M0ZZT
EX4 7DY M3WRS
EX4 7EA G3YBK
EX4 8DQ 2E0HYY
EX4 8DQ M7KRE
EX4 8DU M6GNR
EX4 8HB G3ZVI
EX4 8NG 2E0VIA
EX4 8NG M0VSD
EX4 8NG M6VIA
EX4 8QD G0BTQ
EX4 8RD M7EMW
EX4 9DT 2E0HVY
EX4 9DT M7VBR
EX4 9DY G4VDY
EX4 9EA M7CTF
EX4 9ES 2E0HVW
EX4 9ES M6QWB
EX4 9HP M0CQW
EX4 9HP M1AZY
EX4 9JT G3TLH
EX5 1EE G3XYO
EX5 1HP M0DKV
EX5 1JD G3VRV
EX5 1LL G4TDV
EX5 1PP G4INF
EX5 1QS 2E0UJM
EX5 1QS G3UJM
EX5 1QS M6UJM
EX5 1RA M3HXS
EX5 1RA M7CKC
EX5 2BY G3RUV
EX5 2EX G7NBU
EX5 2JQ M7KCD
EX5 2TJ G7VNQ
EX5 3DX G4AP
EX5 3DX G4ARE
EX5 3HW 2E0SPE
EX5 3HW G6ECG
EX5 3HW M7GTC
EX5 4BB G3ORN
EX5 4HT M0WPN
EX5 4JZ 2E0NXB
EX5 4JZ M0NXB
ex5 4qp M6MGX
EX5 4QU 2E0MRC
EX5 4QU M0CHE
EX5 4QU M3CHE
EX5 4QU M7AZW
EX5 5DA G7NUC
EX5 5NB M6OBY
EX5 7AZ M0DPK
EX5 7GB G4VMF
EX6 6AP 2E0KMP
EX6 6AP M0ZCM
EX6 6AP M6AZA
EX6 6ET G1PHU
EX6 7AD G4WTU
EX6 7SR G4BZE
EX6 7ST M7EMU
EX6 7TZ G3KQG
EX6 7UB M0IOS
EX6 7XS G4WVT
EX6 8AL M7JKM
EX6 8AT 2E0HHJ
EX6 8AT M0KOB
EX6 8DW M3FTE
EX6 8NS G0XRC
EX6 8NS M0THJ
EX6 8SD G7UEK
EX7 0AB G0KYS
EX7 0AB M0MCG
EX7 0AD M6RDP
EX7 0AN M0GSI
EX7 0AN M3GSI
EX7 0AN M3XOR
EX7 0BP G0JIS
EX7 0BZ M6KPK
EX7 0EA G4ONV
EX7 0EJ M3DOM
EX7 0HB G4KPP
EX7 0JT M3SGE
EX7 0JZ 2E0KRE
EX7 0JZ G1ZVO
EX7 0JZ M0KRE
EX7 0JZ M6KRE
EX7 0NA G4VXH
EX7 0NA G7URW
EX7 0QL G3AVN
EX7 0RP G1MVT
EX7 9BS G3WEJ
EX7 9BU G7CNP
EX7 9BU M3HUW
EX7 9DY G0WGJ
EX7 9DZ G3WWT
EX7 9EA M6OEZ
EX7 9FA G0BBO
EX7 9FD G6MTB
EX7 9HP M6AYJ
EX7 9JB G0CAS
EX7 9NQ G4WLA
EX7 9QH G8NVS
EX7 9RS G4DUT
EX8 1DT G4SET
EX8 1QN G4EBO
EX8 1QS G8DQK
EX8 2AF G0RKC
EX8 2JW M3LPU
EX8 2LR M6XAD
EX8 2PU M3LHQ
EX8 2PU M3UIL
EX8 2QR G4VQL
EX8 2RP G8VSO
EX8 2SG G1YOF
EX8 3ba M3GUJ
EX8 3DT G7BWO
EX8 3EJ G1GZG
EX8 3EJ G1MQC
EX8 3HJ M7ARH
EX8 3HY G4YRM
EX8 3JX G4UUW
EX8 3LA G4DPJ
EX8 3NB G1ILY
EX8 3NE G3UZM
EX8 4DN M1CML
EX8 4EU G8EKW
EX8 4EY G3ZDG
EX8 4LF G8GON
EX8 4LF G8UWE
EX8 4LU M6NHX
EX8 4NA M6AFF
EX8 5AP G0GKH
EX8 5BU G1BMZ
EX8 5EE G0ETZ
EX8 5JD M6AFA
EX8 5PN G0GHO
EX8 5PQ G0AVJ
EX8 5RF G0NKC
EX8 5SP G0XJS
EX9 6JN G6UEI
EX9 6LY G0KNY
EX9 7AS G4WJZ
EX9 7BP G4JBF
EX9 7JE G7AKJ
EX9 7JL G4ZQF
FK1 1RL GM0TXJ
FK1 2AG MM1CKW
FK1 2AP GM6PKP
FK1 2BX MM0RSI
FK1 2BX MM6WAI
FK1 2EA 2M0RZE
FK1 2EA MM0GSS
FK1 2EA MM3JRK
FK1 2RP GM6JUA
FK1 2RP MM0DBC
FK1 3BA MM6BOD
FK1 4BJ 2M0MMF
FK1 4BJ MM0PQN
FK1 4BJ MM0RDT
FK1 4ED GM1BVA
FK1 4JF GM0KWL
FK1 4LY MM6EJO
FK1 4NL MM7CXR
FK1 4PB GM6RGY
FK1 4PB MM1CLR
FK1 4PG MM3ZTS
FK1 5BQ 2M1ENK
FK1 5DS GM0KUP
FK1 5HU MM0RAM
FK1 5JX GM1LUZ
FK1 5JX MM5AGM
FK1 5NZ MM0KCD
FK1 5QQ MM7EMD
FK1 5UE GM4MIG
FK10 1DD GM0WWX
FK10 1NQ GM4DGT
FK10 1PT 2M0CSZ
FK10 1PT MM6CNO
FK10 1QZ MM0OBT
FK10 1RU 2M0HMU
FK10 1RU MM7CLN
FK10 2BN GM0UUB
FK10 2BN MM7HQS
FK10 2DT GM8FMR
FK10 2EG GM0UGH
FK10 2ER 2M0BCL
FK10 2ER MM0GKU
FK10 2JG GM0LWD
FK10 2JU GM4WQH
FK10 2RX GM4ARJ
FK10 2TT 2M0WSK
FK10 2TT MM0WSK
FK10 3AN GM0IIO
FK10 3AW GM0FRC
FK10 3AW GM8GAX
FK10 3BG GM8WGU
FK10 3DZ GM4WGC
FK10 3FG MM6HIA
FK10 3JN GM8JJN
FK10 3PD GM0PEI
FK10 3RE MM6JNF
FK10 4LF MM7PSG
FK102PA 2M0RMD
FK102PA MM0RMD
FK102PA MM6RAV
FK11 7DG GM4FXL
FK12 5AL GM4HUX
FK12 5BL 2M0WDP
FK12 5BL MM6DWP
FK12 5DS GM0NAQ
FK12 5HU GM1PST
FK12 5JU GM1BVT
FK12 5NN GM4XHV
FK13 6HB 2M0FXL
FK13 6HB 2M0KTL
FK13 6HB MM0KTL
FK13 6HF GM0TTY
FK13 6HT GM8CIF
FK13 6JA MM6PRZ
FK13 6LF MM7NCD
FK13 6NH MM0GNX
Fk136pu 2M0FKF
Fk136pu MM6DVM
FK13EL MM6JQF
FK14 7AE MM0MSH
FK14 7BA MM0GJC
FK14 7BH MM6TWZ
FK14 7HP GM4WBU
FK14 7JZ GM1FTG
FK14 7NT MM6FKD
FK14 7PT MM0PTX
FK15 0DU 2M0SWM
FK15 0DU MM7ETI
FK15 0EB GM3YTS
FK15 0EB GM5CX
FK15 0EB GM8VL
FK15 0HQ GM1SRR
FK15 0JS GM0DUX
FK15 9AR 2M1IGQ
FK15 9AT 2M0PXH
FK15 9AT MM0AXR
FK15 9AT MM3ZNN
FK15 9ES GM6UCN
FK15 9HB GM0GRL
FK15 9JL GM4VZY
FK15 9NA GM0GMD
FK15 9PX GM0MXP
FK15 9RA GM1RIG
FK17 8LA GM8SAP
FK2 0BJ GM7GLJ
FK2 0EJ MM3KUU
FK2 0FN 2M0DKX
FK2 0FN MM0PRB
FK2 0GU GM7KMM
FK2 0HB MM0VWR
FK2 0LP MM0DSM
FK2 0LP MM3DEC
FK2 0LY MM3OIX
FK2 0NB MM6LRX
FK2 0QW MM3ZQP
FK2 0RE MM0MPA
FK2 0SU GM4MIM
FK2 0TF GM0KLO
FK2 0TJ GM6KDD
FK2 0UP GM7DZK
FK2 0UX GM3JJQ
FK2 0XT 2M0BKL
FK2 0XT MM3ZXL
FK2 7AY 2M1CJL
FK2 7AY GM0DEM
FK2 7BJ MM3YQP
FK2 7BQ MM6DBT
FK2 7BU 2M0HJG
FK2 7BU MM7JFB
FK2 7LD MM6RDM
FK2 7LF MM3ZSF
FK2 7NH MM6KER
FK2 7NY MM7MBB
FK2 7TQ MM3KAQ
FK2 8AN GM0FTG
FK2 8DX GM0ZAM
FK2 8EE GM7UJO
FK2 8GD GM8VYZ
FK2 8GE GM1IEL
FK2 8JP GM0HZI
FK2 8NP 2M0TGD
FK2 8PP MM0OKG
FK2 8PP MM6TGD
FK2 9JR 2M0TXR
FK2 9JR MM6TRX
FK2 9QQ GM4XQJ
FK2 9SU MM7BJU
FK2 9SZ MM7BWZ
FK2 9UF GM0PGD
FK2 9UT GM0NJL
FK2 9YS MM6OEA
FK20 8RQ GM0EWW
FK27UF MM7FTH
FK3 0DE GM3IWX
FK3 0JJ 2M0OAB
FK3 0JJ MM0PSM
FK3 8LZ 2M0JLS
FK3 8LZ MM6MCT
FK3 8NT MM0CNV
FK3 8NZ GM7GTX
FK3 8TG 2M1ANY
FK3 8TG MM7DXC
FK3 8YG MM6HHJ
FK3 9AT GM3YKA
FK3 9JD GM6JWH
FK3 9JD MM0GXU
FK3 9JN 2M0WWX
FK39EQ MM3TEQ
FK4 1AR MM0OZY
FK4 1GD GM6MHC
FK4 1HE GM0AEY
FK4 1JY MM6SZK
FK4 1NB GM0IYA
FK4 1TP MM1BFE
fk4 2dl MM7MFR
FK5 3JU GM3UCH
FK5 3LH GM4EJX
FK5 4DX MM0INT
FK5 4DX MM0NDX
FK5 4LT GM0KMJ
FK5 4QF MM0SAJ
FK5 4QF MM7CFC
FK5 4UA 2M0XRV
FK5 4UA MM7XRR
FK5 4UF GM0OKJ
FK6 5AG 2M0VUV
FK6 5AG MM0VUV
FK6 5AG MM6VUV
FK6 5EG GM4XRP
FK6 5HX GM0AZC
FK6 5JF GM6PNJ
FK6 5LP 2M0LNR
FK6 5LP MM6TSC
FK6 6PN GM0DKK
FK6 6QJ 2M0VVS
FK6 6QJ MM0IAL
FK6 6QJ MM3VVS
FK6 6QJ MM6HXY
FK6 6RB GM4NTX
FK66RB MM7DGS
FK7 0DH 2M0OXX
FK7 0DH MM0OXX
FK7 0DH MM6OXX
FK7 0DT GM0VRP
FK7 0HJ GM0EGI
FK7 0HJ MM0AIK
FK7 0LH GM4YMD
FK7 0LJ GM4VGR
FK7 0NP GM0WUP
FK7 0NP GM7VDL
FK7 0NP MM3WUP
FK7 0NQ 2M0ANE
FK7 0NQ 2M0BYT
FK7 0NQ GM0WUR
FK7 0NQ MM3RGZ
FK7 0QD 2M0OCC
FK7 0QP GM4UYE
FK7 0QP GM6NX
FK7 7BS MM6YRO
FK7 7QZ 2M0GEK
FK7 7QZ MM6CJZ
FK7 7UL GM0NWI
FK7 7XA GM0FNE
FK7 8NQ GM8MOI
FK7 8PX GM1BSG
FK7 8PX MM6HLZ
FK7 8UF MM3RCR
FK7 9BY GM7EHN
FK7 9LP GM1SBD
FK8 1BZ MM7RKJ
FK8 1EP GM0RMT
FK8 1SA MM0JMK
FK8 1TA 2M0FXV
FK8 1TA MM6AJI
FK8 2JE MM3RHA
FK8 2JS GM0KQB
FK8 3JY GM0HZO
FK8 3PD MM0PDD
FK8 3SZ MM0WCT
FK9 4HH MM6PEA
FK9 4PS GM4PJR
FK9 4SA GM0SVS
FK9 4SA GM1XJE
FK9 4UJ MM7WMG
FK9 5BE MM0BSM
FK9 5EU GM4PYJ
FK9 5LR GM4RCN
FK9 5LR GM4RTN
FK9 5LS GM0FSV
FK9 5PX MM1CCR
FY1 2AQ G1UGG
FY1 2AU M7BCW
FY1 2BN M3SBT
FY1 2BN M6KHF
FY1 2EG M3LDT
FY1 2JS 2E0MHE
FY1 2JS G3CTQ
FY1 2JS M3SZQ
FY1 2NY M6JIT
FY1 2QJ 2E0EEI
FY1 2QJ M0IHJ
FY1 2QJ M6HVH
FY1 2QL G7HEJ
FY1 2RG 2E0EKS
FY1 2RG M6JNI
FY1 2RW G1YMR
FY1 2SF G1KJQ
FY1 3NN G1TIH
FY1 3RB G1MET
FY1 4JA M7PJY
FY1 4JE 2E0FXA
FY1 4JE M7APF
FY1 4JQ 2E0DAX
FY1 4JQ M6BIF
FY1 4NB G6ZFT
FY1 4QP G0UHS
FY1 5BZ M6IZO
FY1 5HP G4GOR
FY1 5JB M6YAJ
FY1 5NA G1OLM
FY1 6DW M3CLT
FY1 6DZ G7FIK
FY1 6EZ M6OUL
FY1 6JP G4EYX
FY1 6LP G0NBW
FY1 6LY M1EOZ
FY1 6NP M6BJC
FY1 6NW G4ZPN
FY1 6PR M6GQA
FY1 6RR M6VVN
fy12pw G1NGR
fy16qn M7PBN
FY2 0AW G7MXN
FY2 0DZ 2E0UZK
FY2 0DZ M6UZK
FY2 0EB 2E0DYY
FY2 0EB M0LJH
FY2 0EB M6KQM
FY2 0EN G1BFG
FY2 0LW M0TUT
FY2 0NQ G0ETV
FY2 0PD M3BAA
FY2 0PW M7RJB
FY2 0RJ M6XTF
FY2 0SD G1ZRE
FY2 0SR G6WWV
FY2 0TR G8ATG
FY2 0TU G4WYF
FY2 0UG M1ALF
FY2 0XH 2E1DZL
FY2 0XH M3RKN
FY2 9AQ G4TUM
FY2 9EN M6XAS
FY2 9EP M6HXO
FY2 9EQ G6XNU
FY2 9LQ 2E0OYG
FY2 9NU M3XUW
FY2 9QT G3LLE
FY2 9QW G1NQN
FY2 9TB G4RRJ
FY2 9TX M6HUZ
fy20wl M7VAT
fy29bz 2E0MOW
fy29bz M6NZN
FY3 0AL G3UIT
FY3 0BU G3WPT
FY3 7BQ M6DFU
FY3 7BQ M6FTU
FY3 7LQ 2E0SGS
FY3 7LQ M6KHP
FY3 7PP M6IPF
FY3 7PW G7EPY
FY3 7QS M3KDY
FY3 7RR M3MEW
FY3 7RS G1CFE
FY3 7UB M3VKF
FY3 7UE M6BQP
FY3 7UJ G1CWQ
FY3 7UJ M3UUO
FY3 8DP G4OMS
FY3 8HS G6FCI
FY3 8JJ M6XMA
Fy3 8jl M7LDW
FY3 8LX 2E0MLL
FY3 8LX M6CLL
FY3 8PF 2E0EZY
FY3 9TN G4VAL
FY3 9UE M6HVD
FY4 1EH M0DKR
FY4 1EH M6IEJ
FY4 1JJ M6IVI
FY4 1JP M3MGZ
FY4 1LB G4MPT
FY4 1PW G0LLX
FY4 1QG G1EEA
FY4 1QR G4AKC
FY4 1QS G4EZM
FY4 1SX G6IRU
FY4 2BT G4XPI
FY4 2BW G4JAJ
FY4 2DH G7PTM
FY4 2NF M1CDX
FY4 3AX 2E0TZE
FY4 3AX M7FJS
FY4 3JQ G6IGV
FY4 3LH G6BXO
FY4 3LY M3RPR
FY4 3NH G0FYD
FY4 3PQ M6DSR
FY4 3QA G6BGH
FY4 3QA M3BPL
FY4 3QQ G6VEZ
FY4 4AB G0EUP
FY4 4AF M6MOK
FY4 4AR 2E1DAP
FY4 4AX G1NQO
FY4 4AZ M6XNU
FY4 4DN 2E0YJB
FY4 4DN M7WJB
FY4 4JJ G1HSO
FY4 4JJ M7BKO
FY4 4LQ M3DCI
FY4 4NA G6GJD
FY4 4NU M6IZT
FY4 4PY M0JEB
FY4 4QS G4YVQ
FY4 4QS M0BOL
FY4 4RL M1IMC
FY4 4RQ G4XQF
FY4 4RR M6IZU
FY4 4TZ M7BOP
FY4 4UD G4RXK
FY4 4UZ G6XNI
FY4 4YF M1BUG
FY4 5BT G4SZB
FY4 5BX G6AOS
FY4 5DS G0BGR
FY4 5DW G6IFQ
FY4 5EZ G0SVQ
FY4 5HG G1ZME
FY4 5HT G4DJY
FY4 5RE G6VRF
FY4 5RF G0HZL
FY42BH M6JYF
FY5 1BG G6PGV
FY5 1BU G4BCT
FY5 1DL G3RFH
FY5 1EY M6KSA
FY5 1HQ M3DEH
FY5 1JD G3HKA
FY5 1JW G4KQC
FY5 1LP G0PIL
FY5 1LP G7URJ
FY5 1PR G7SUV
FY5 1SP 2E0RWT
FY5 1SU G3ZGZ
FY5 2AR 2E0UHJ
FY5 2AR M3UHJ
FY5 2DU M7JCD
FY5 2HX M3XGY
FY5 2JS G1PPK
FY5 2LG G1NFE
FY5 2LL G3WBB
FY5 2NW G6VCR
FY5 2PH G0NXR
FY5 2QX M3SBP
FY5 2RT G4ANE
FY5 2UG G4ATH
FY5 2UG G4FRK
FY5 2UG G6GMW
FY5 2ZA G0MWM
FY5 3AJ M7PCB
FY5 3DH 2E0EDX
FY5 3DQ M3OYJ
FY5 3EY G6SRJ
FY5 3HZ G8ZSM
FY5 3QA M3YVW
FY5 3QB G0GER
FY5 3QB G0LSA
FY5 3QB G7FCR
FY5 3QH M6LPY
FY5 3QR G1DEP
FY5 3QR M1AMZ
FY5 3QR M6RKW
FY5 3RQ G4RDY
FY5 3SE G1EPL
FY5 3UH G6HMX
FY5 4DL G0VVG
FY5 4DR 2E0MKH
FY5 4DR M6KHB
FY5 4NB G3SNH
FY5 4NG M3TWV
FY5 4PL 2E0CQJ
FY5 4PL 2E0TOG
FY5 4PL M0NAG
FY5 4PL M6FAC
FY5 4PL M6NAL
FY5 4PZ M6HTW
FY5 5AE G8MKQ
FY5 5AY G4UHI
FY5 5BH M6CDH
FY5 5DR M3NXO
FY5 5EA M0EAL
FY5 5HH 2E1ADR
FY5 5HP G3ZNG
FY5 5NS M3NDF
FY6 0AD M6HGN
FY6 0AJ 2E0CVG
FY6 0BS 2E0MYT
FY6 0BS M6CHC
FY6 0BT G7IED
FY6 0BT G7PDR
FY6 0DD G0JZT
FY6 0DF G0AJW
FY6 0DF G4RND
FY6 0EH 2E0EEW
FY6 0EH M6GQG
FY6 0EW G7NYD
FY6 0HA G0HIU
FY6 0HG 2E0DQB
FY6 0HG G8FFW
FY6 0HG M6FKQ
FY6 0LD G6KHD
FY6 0LZ M0HOI
FY6 0PB G7UWE
FY6 0PW G1OCK
FY6 0QE G4DRX
FY6 0QE M3YRX
FY6 0QW M6FUE
FY6 0RS G6NJE
FY6 7BL G6NLQ
FY6 7DB M3TVJ
FY6 7DJ G4KHJ
FY6 7DY M3UUN
FY6 7EG G3UIS
FY6 7EW 2E1EIU
FY6 7EW M3UGA
FY6 7FY 2E0SEP
FY6 7FY M6FUQ
FY6 7HJ M6SLJ
FY6 7NE G0LQN
FY6 7PB G6KOE
FY6 7PE M0HZP
FY6 7PN G3BDT
FY6 7PW G3XCE
FY6 7QX G4TMA
FY6 7QY G0RNA
FY6 7RD 2E0OMG
FY6 7RD M3XZN
FY6 7SP M7JLW
FY6 7SR G6MQG
FY6 7TL M6YFZ
FY6 7UB G8KBH
FY6 7UB G8YOK
FY6 7UQ M3NCE
FY6 7UX G4LOR
FY6 7UX G4PNI
FY6 7UX M3PNI
FY6 8AD G3MCE
FY6 8AD M6TGP
FY6 8BL G0BRX
FY6 8BL M6SJJ
FY6 8BZ G7FED
FY6 8EB G3VDO
FY6 8ED G4BVW
FY6 8ED G4VYL
FY6 8EH M3SJM
FY6 9AP G4XKR
FY6 9AW G0RHF
FY6 9EA G6UDA
FY6 9EE G6AEK
FY6 9EH 2E0TDH
FY6 9EN M6TDP
FY7 6AD M7RAV
FY7 6BG M0BAL
FY7 6BT M7LAA
FY7 6DA 2E0FFS
FY7 6DA G4NIU
FY7 6FG M0GNJ
FY7 6FG M0NAU
FY7 6JR G8IMJ
FY7 6LJ M0JME
FY7 6LJ M6JLB
FY7 6QA M3AYT
FY7 6QJ M3ZUA
FY7 6UE M6MRG
FY7 7ar G8YPY
FY7 7DU G3SSN
FY7 7EA G0CUX
FY7 7HA 2E1FUJ
FY7 7HA G0LRK
FY7 7HA G0NCY
FY7 7HH G4TTG
FY7 7HJ M0HMU
FY7 7HJ M0JFE
FY7 7HW M3FPS
FY7 7HY M0BLS
FY7 7JF 2E0XHG
FY7 7JF M3XHG
FY7 7JH M6EFR
FY7 7LF M0DLB
FY7 7LF M3MRN
FY7 7LJ M3IPY
FY7 7LJ M3TCY
FY7 7LY M3UJE
FY7 7LY M6TKU
FY7 7NH G0IAE
FY7 7PY M3WWH
FY7 8BH 2E0KNF
FY7 8BQ G4GYF
FY7 8EG G4FWM
FY7 8EG G8RDP
FY7 8EG M3RAE
FY7 8NW M6BHE

Fy77hz M6PQW
FY78DX 2E0GRZ
FY78DX M0RGV
FY78DX M6OGX
FY8 1DW G4JFV
FY8 1ET G8LCZ
FY8 1HW G2HFP
FY8 1HZ G7CWI
FY8 1LZ G1CDQ
FY8 1PQ G1TJR
FY8 1PU G4OPT
FY8 1TN M3RZG
FY8 2BL M7DJC
FY8 2DA 2E0BKZ
FY8 2DA M0GMN
FY8 2DA M3UGX
FY8 2HE 2E1DZV
FY8 2HE M3JMK
FY8 2HW G0HLB
FY8 2JF G7HJQ
FY8 2LT G4GFE
FY8 2QN G3WBI
FY8 2QX G1TBN
FY8 2SG G7DIE
FY8 2SG M3SGS
FY8 3BG G0AIN
FY8 3DB G6KTO
FY8 3HA 2E0TSI
FY8 3HA M6TSI
FY8 3HY G7FDS
FY8 3NH M6WIM
FY8 3NP M3ZOI
FY8 3NP M6SBH
FY8 3PS G8SND
FY8 3QF G0JBS
FY8 3RZ M6FRZ
FY8 3RZ M6UAE
FY8 3SL G3IEJ
FY8 3TL M6HWL
FY8 4AR G8LMF
FY8 4BJ G4NXW
FY8 4PF M1FHJ
FY8 4QR G4WLE
FY8 4UE G7CBY
FY8 4UE G7CBZ
FY8 5PX M6AVD
FY8 5PY M7XBX
FY83AG G8SNR
G1 1HD M6BPY
G1 1LH MM0IBO
G1 1NY MM3MMO
G1 1NY MM7AJJ
G11 5AP GM0EDR
G11 5AX MM7ESG
G11 5EA GM8TXC
G11 5PE 2M0HZO
G11 6PB MM7POL
G11 6QP 2M0YYU
G11 7LG GM4ACM
G11 7PP GM7KVU
G11 7QX 2M0TLE
G12 0AS 2M0CTN
G12 0ER 2M0SCG
G12 0LE GM8DKG
G12 0PB GM0ILQ
G12 8TL GM3RUP
G12 9DZ GM7DLY
G12 9JU MM0MHP
G120DL MM7TXY
g129sb MM7SRR
G13 1DQ MM6SCG
G13 1JH MM6JOX
G13 1UN MM7EPD
G13 2LA GM0EFH
G13 2RJ GM0EAH
G13 2XX 2M0GFT
G13 2XX MM0NNA
G13 2YQ MM6KVI
G13 2YQ MM6RHQ
G13 3AQ GM0GYT
G13 3EP 2M0IPO
G13 3EP MM0JOX
G13 3QN MM6NWH
G13 3TT MM6YUJ
G13 3YE MM6CIJ
G13 4HL GM0FHJ
G13 4QE MM6CTL
G13 4QN GM4NWK
G14 0EU 2M0VJS
G14 0EU MM7JRS
G15 6AU 2M0RCD
G15 6AU MM0HCO
G15 6AU MM6AOH
G15 7JA MM6OBU
G15 7QE GM7NZI
G2 3AU MM6AXT
G2 3JW 2M0SGO
G2 3JW MM7MMO
G20 0HJ 2M0UTH
G20 0HJ MM3UTH
G20 0HT 2M0NTY
G20 6AG MM0CXB
G20 6AG MM0PYS
G20 6NB 2M0STK
G20 6NB MM0WSG
G20 6NB MM1STK
G20 6NB MM7STK
G20 6XZ GM0FWY
G20 7YQ 2M0TTF
G20 7YQ MM0GUE
G20 7YQ MM6AYR
G20 8LF GM0USI
G20 8LF GM4FVQ
G20 9JQ GM0KVD
G21 2AP MM3RDP
G21 2DE 2M0WFN
G21 2DE MM6AHJ
G21 2QF MM3UHK
G21 3HY GM0KUJ
G21 3JS MM6AKZ
G21 4XY GM4ENN
G212JB MM7BQG
G22 6DN MM7ITA
G22 7RA MM3SNB
G22 7RG 2M0GCF
G22 7RG MM0GCF
G22 7RG MM0KHG
G23 5DJ GM0JHE
G3 6BS MM7PMB
G3 6QN GM3VBT
G3 8QP MM7UVI
G31 3JL GM0KTO
G31 4RT MM0TBH
G31 5RP GM3PUY
G32 0LP MM6LBS
G32 0NF 2M0VUS
G32 0NF MM6VUS
G32 6TA GM3SPT
G32 7JT 2M0HWG
G32 7JT MM0YCH
G32 7JT MM6HWG
G32 7SQ MM6WMK
G32 7XL MM6MZE
G32 9BW GM3JKC
G33 1BN MM7MCM
G33 1BU MM0SIL
G33 1LR MM6JBN
G33 1RS GM7GXI
G33 2DD GM4PCT
G33 2DP MM3VZI
G33 2LD GM7BRL
G33 2SD MM6JSN
G33 5HU GM1POA
G33 5JJ GM4EWL
G33 5JJ GM8DRA
G33 6NJ MM0XEA
G34 9AR 2M0FYF
G34 9AR MM0JRF
G4 0PH 2M1EZA
G4 0TQ MM3LFI
G4 9AR MM7CNQ
G4 9JX GM6AQR
G40 3LE GM8LYO
G41 2AF GM3DIN
G41 3BS GM8PIV
G41 4PW GM0PAC
G41 4QN GM0OAA
G41 5HH MM7LYS
G41 5RL MM0TWX
G41 5SD MM6PKC
G42 0DW MM6FBU
G42 8BZ MM6HFZ
G42 8DR GM3WUX
G42 8DW 2M0FRA
G42 8DW MM6CFH
G42 8RT 2M0IEC
G42 8RT MM3IEC
G42 9PZ MM6PND
G43 1BW 2M0HYZ
G43 1BW MM3UTU
G43 2BW MM3SQM
G43 2DX MM6IBQ
G43 2DY MM0GYD
G43 2SY MM0GUW
G43 2YL MM3FYF
G44 3NQ MM6HIZ
G44 3XG MM6KHM
G44 4PA GM0WRR
G44 4TJ MM1BJT
G44 5JU 2M0MTO
G44 5JU MM0MTO
G44 5PF 2M0RRO
G44 5PF GM7WED
G44 5PF MM3LKV
G45 9QD 2M0PBC
G46 6DB GM3NGW
G46 6LA GM6FIK
G46 6QB GM4IYZ
G46 7AE GM3COB
G46 7LU GM0WRH
G46 8AB GM0LIM
G46 8LP MM6VGS
G46 8RS MM7SWM
G46 8UR GM7UTD
G468na MM6LGS
G51 1HG GM3ZDH
G52 2TP MM6FEX
G52 3HA GM4BGS
G52 3HA MM6AHY
G52 3JY GM1JNC
G52 3PU MM0CDK
G52 4HJ MM3WHS
G53 5UY MM7CNX
G53 6BI MM1DEE
G53 6NQ 2M0AZW
G53 6NR MM6OCP
G53 6QW GM4FFF
G53 6QW MM0DGR
G53 6QW MM0EPC
G53 7UP 2M0SAX
G53 7UP MM0SAX
G60 5HR MM0TCQ
G60 5JJ MM6MCX
G60 5LE GM3YCB
G60 5LE GM6AQB
G60 5LJ MM0CZM
G61 1EJ GM0GCO
G61 1RE GM3RQQ
G61 2AT MM7NLW
G61 2LT GM7SCJ
G61 2PR MM6KQH
G61 3EF MM7ORE
G61 3EZ GM8ZGC
G61 3HD 2M1HFE
G61 3HD 2M1MIC
G61 3HD 2M1SJB
G61 3HD GM3VTB
G61 3HD GM4VTB
G61 3HD MM1VTB
G61 3HD MM3CVB
G61 3HQ GM0ATL
G61 3JX GM3GTQ
G61 3LG GM0UPE
G61 3LQ MM7HIR
G61 3LX GM0NUQ
G61 4EH GM4AWB
G62 7DN MM7EYO
G62 7DT GM1XEB
G62 7HA 2M0IBW
G62 7HA MM0SUS
G62 7JP 2M0VXL
G62 7JP MM3RXM
G62 7JP MM3VXL
G62 7PT MM7ABY
G62 7RA GM4KAV
G62 7RL GM4JRF
G62 7RR GM3MAS
G62 8BE GM8ZKF
G62 8HD MM0GDL
G62 8HX MM0HLQ
G63 0EX GM3ODP
G63 0NP GM0GIB
G63 0PF MM1FHO
G63 0XE MM3AYS
G64 1QW MM1PTT
G64 2EU GM8ZCS
G64 2HP GM0GMO
G64 2NS GM3HOM
G64 3AD GM8ZAK
G64 3AQ MM6BSK
G65 0EE GM0BRJ
G65 0EX 2M0RCZ
G65 0EX MM0POD
G65 0EX MM3RCZ
G65 0RY 2M0IIH
G65 0RY GM8IIH
G65 0RY MM0OJJ
G65 0RY MM7IIH
G65 9EA MM3SAK
G65 9EJ MM1FZR
G65 9HF MM0GUX
G65 9HJ MM0GFA
G65 9HJ MM1DHU
G65 9HQ MM6JJQ
G65 9UL GM0BUE
G66 1AX GM3GRG
G66 2BD GM4LYV
G66 2BH 2M1EKI
G66 2BH MM3XOK
G66 2QE 2M0OMS
G66 2QE MM0OMS
G66 3AS GM3WYL
G66 3BW GM0VEK
G66 3HJ MM3HYG
G66 3JL GM1BTL
G66 3JW MM3RHT
G66 3NX GM7JDS
G66 3PH MM7FGL
G66 3PL MM6IUR
G66 3RY GM8WWY
G66 3WF 2M0VOB
G66 3WF MM7PLG
G66 4BF GM7SPA
G66 4DF MM1AHL
G66 4DF MM3DIZ
G66 4EL GM6TIB
G66 4EN GM4RUP
G66 4RE GM3SER
G66 4SR MM1BXF
G66 5HS 2M0LFS
G66 5HS MM0TMZ
G66 5HS MM6LFS
G66 5NG GM0BEL
G66 5NG GM4ZRX
G66 7EP GM7FLG
G66 7JT MM7DGC
G66 8AY GM0HNV
G66 8AY GM0PEX
G66 8AY GM3JMM
G66 8BB GM0HJU
G66 8BB MM6HMB
G66 8ER GM7GBD
G66 8ET GM0HIM
G66 8ET GM0RED
G66 8ET GM7HIR
G66 8HG GM0ATA
G66 8HR 2M0LWD
G66 8HR GM5LWD
G66 8HR MM7GGL
G661nu MM1DEE
G67 1JE MM3VBF
G67 1LR MM6ATR
G67 2AY MM1BJZ
G67 2BL GM1RRJ
G67 2DA GM7FIS
G67 2DW MM0MIJ
G67 2JS 2M0IVS
G67 2JS MM6GYR
G67 2LU MM7SCO
G67 2NP 2M0UGL
G67 2NP MM0SEY
G67 2NP MM3UGL
G67 2NT MM1RIK
G67 2PX MM7BYQ
G67 2QW 2M0VKO
G67 2QW GM6FLL
G67 2QW MM3VKO
G67 3AX 2M0LSG
G67 3LU GM4XLU
G67 4AD GM7RQK
G67 4AJ 2M0BYI
G67 4AJ MM0MYL
G67 4AJ MM3YUW
G67 4AJ MM3YUX
G67 4AW MM6HQK
G67 4ES MM0GXQ
G68 0AF GM8NFT
G68 0EP MM3BSC
G68 0JB GM0PLH
G68 0JB MM0LMC
G68 0JR MM0HUF
G68 9BJ GM4EIW
G68 9DZ GM1VFR
G68 9DZ GM7JUX
G68 9EG GM7MZZ
G68 9ER MM1DEA
G68 9JD MM6DBN
G68 9JU MM7KXI
G68 9JY 2M0PLB
G68 9JY MM0PLB
G68 9JY MM6HMM
G68 9NG GM1TFF
G68 9NW GM4VWV
G69 0JW GM1MRY
G69 0LZ MM0ODI
G69 0PH MM3CBO
G69 6JB MM6FTG
G69 6LQ GM4RPE
G69 6ND MM3DOP
G69 6NE MM7SJF
G69 6qp MM0DNX
G69 7BH GM8YGI
G69 7HW GM3SAN
G69 7HY MM7DZS
G69 7QZ MM6KCB
G69 8AG MM3DQV
G69 8AG MM3DQX
G69 8EG MM6KSU
G69 8LE MM0OWL
G69 8LH MM7CMM
G69 9FB 2M0WHX
G69 9FB GM5DAV
G69 9FB MM6WHX
G69 9FB MM7VMC
G69 9ND GM0LBR
G71 6ED GM7FGH
G71 6LG 2M0MBE
G71 6LG MM3VPK
G71 7BQ GM0IMW
G71 7ET GM7VYR
G71 7ET MM3VYR
G71 8AR GM1VBE
G71 8BP 2M0SSO
G71 8BP MM0SSG
G71 8BP MM3YNP
G72 0RQ GM0XFK
G72 0TR MM0GPZ
G72 0TZ MM7WSI
G72 7EF MM6RXJ
G72 7GS MM6YND
G72 7GS MM6ZDG
G72 7NN MM6JCL
G72 7SQ MM3UMY
G72 7SQ MM3UMZ
G72 7TP GM7SFE
G72 7UW MM6GGE
G72 8DH MM6VLG
G72 8NL GM4NDV
G72 8RD GM4TTC
G72 8UE 2M0WWC
G72 8UE MM6WWS
G72 9DT GM7WBP
G72 9NX GM0AZU
G72 9RJ MM7GKD
G72 9UP MM3XNP
G73 3BG GM4REF
G73 3EN GM0LBN
G73 3EU MM3WJZ
G73 3QP GM7AYW
G73 3QP GM8SNB
G73 4AE GM0JVV
G73 4DX MM3CKP
G73 4EA 2M0DKV
G73 4EA MM6EGC
G73 4JP 2M0FCT
G73 4JP MM6VSM
G73 5RG MM0AYE
G732ER MM7GMS
G74 1BQ GM4DLU
G74 1DR MM0BMA
G74 2HU GM0TPI
G74 3DL MM0ZXI
G74 3DN MM6CHY
G74 3HZ GM1BNA
G74 4HZ 2M0FZW
G74 4HZ MM0JJU
G74 4HZ MM7AWL
G74 4TZ GM7LOK
G75 0BX MM3YVU
G75 0PY 2M0HRN
G75 0PY MM7RIK
G75 0RU MM6PVY
G75 8DL 2M0YOY
G75 8DL MM6YOY
G75 8LH 2M0EEQ
G75 8LH MM6EWR
G75 8RZ MM1EGS
G75 8SA MM1DME
G75 8TN GM1SQZ
G75 8XT 2M0HUJ
G75 8XT GM7GDE
G75 8XT MM0YEK
G75 8XT MM7MMB
G75 8YG GM0UET
G75 8YG MM3CLA
G75 8YG MM3UET
G75 9FS GM8NET
G75 9LB MM6GOR
G75 9NW MM3UQT
G76 0BP MM0IRG
G76 0BP MM0KFP
G76 0BP MM7IRG
G76 0DG GM1PUR
G76 0EU GM0VIY
G76 0EU MM3PEY
G76 0HA GM0PHG
G76 7AD GM6ANZ
G76 7DU GM4VBE
G76 7HG GM0SEP
G76 7HG GM0UIG
G76 7HG GM4SRL
G76 7JL MM6HZW
G76 7PL GM4XGY
G76 7XT GM0IVQ
G76 7XT GM6FOT
G76 9BN GM0DVO
G77 5LS GM4KNU
G77 5PP GM1NEW
G77 5QJ GM7SKB
G77 5TQ GM4KLO
G77 6EA MM6MML
G77 6FP 2M0GQH
G77 6FP GM6PZY
G77 6FP MM7HUE
G77 6FP MM7JGY
G77 6HP GM3NEQ
G77 6LQ GM1NET
G77 6LQ GM6OQN
G77 6PZ MM6DVR
G77 6UG GM4AGU
G77 6UG GM5AUG
G77 6UG GM5KCC
G77 6UJ 2M1EDM
G77 6UJ GM0UKZ
G77 6UU MM0HFU
G77 6XX GM3KXQ
G77 6YG GM6LIN
G78 1DD MM7GNX
G78 2DH GM0GMI
G78 2DH GM1TFZ
G78 2LR MM6DAQ
G78 3DY MM7SNP
G78 3JA GM6JIL
G78 3JX GM3AEI
G78 3PZ GM4LCP
G78 3QP GM4OSV
G78 3QP MM3USV
G78 4AA GM0DJG
G78 4AA GM3USL
G81 1DP 2M0LZY
G81 1DP GM2UWX
G81 1DP MM7DFC
G81 1DP MM7PGR
G81 1EH MM3OYL
G81 1ER GM4BLO
G81 1RF MM0HJC
G81 1RF MM0IXJ
G81 1RF MM0YZE
G81 2LL GM0GYM
G81 2PH 2M0NIT
G81 2PH MM6TNO
G81 2RA MM7KEL
G81 2ST GM4PLI
G81 2YB MM6IYA
G81 3EH MM6ZYZ
G81 3LE 2M0HNQ
G81 3LE GM5MAJ
G81 3LE MM7CPI
G81 3LH GM4KHE
G81 3NF MM6RPN
G81 3RR GM1BNS
G81 4HH MM6AQM
G81 4HJ GM7OMU
G81 4PG MM7DCD
G81 4QD MM6ECO
G81 5BS MM3YTB
G81 5EA 2M0UFO
G81 5EG GM4ZMK
G81 5HJ 2M1EDT
G81 5PD GM6KTP
G81 6AW GM3KCY
G81 6HH MM6IXO
G81 6HX MM3TAM
G81 6LW GM3RGU
G81 6NU GM3ITN
G81 6PX GM4AGG
G81 6PX GM4TOQ
g813ab 2M0PIJ
g813ab MM6PIJ
G813EP MM6IWL
G814AW MM6NWX
G82 1HP MM7IAB
G82 2BN MM6NUP
G82 2BN MM6YUP
G82 2JA MM3ODV
G82 2PF MM6DHF
G82 2PN GM7NHU
G82 2QB 2M0IXT
G82 2QB MM0IXT
G82 2RZ MM0CXZ
G82 3ER GM0KZX
G82 3JU 2M0MSB
G82 3JU MM3MGK
G82 3NU 2M1HSG
G82 3NU MM3HSG
G82 3PB MM1JAC
G82 3QW GM8FFH
G82 4BG MM5AON
G82 4JL MM0LGT
G82 4NH MM6HAH
G82 4QA MM3FMB
G82 5HD GM1UIR
G82 5HD MM3DMZ
G82 5HP GM1MLW
G83 0BZ 2M0JBZ
G83 0BZ MM6JCU
G83 0DW GM7MYF
G83 0LL GM6JWF
G83 0PU GM4UPN
G83 0RJ MM3OFE
G83 0RZ GM4LGM
G83 0TX MM7CZD
G83 0UF MM1DMU
G83 0US MM1LJB
G83 7DB 2M0OVV
G83 7DB MM3OVV
G83 8BL 2M0LNF
G83 8BL MM3LNF
G83 8EX MM6EOE
G83 8RP GM0RTY
G83 8RU MM3OCY
G83 8SD MM0HDA
G83 9BJ MM3XDP
G83 9BU 2M0BIL
G83 9BU MM0ELF
G83 9BU MM3FCG
G83 9DB MM3XKH
G83 9DB MM6LAH
G83 9EB MM7OWN
G83 9JR 2M0JUM
G83 9JR MM7JMP
G83 9LE MM6SWC
G83 9LG 2M0TXY
G83 9LG MM6FCA
G83 9PW MM7BMY
G83 9QL GM7OBM
G83 9QT MM3DDQ
G84 0DW MM6BGV
G84 0EB 2M0VFV
G84 0EB MM6SMY
G84 0EB MM7BLP
G84 0HS 2M0ZFG
G84 0HS MM0ZFG
G84 0HS MM6ZFG
G84 0JN MM0EAI
G84 0QD GM1KWX
G84 0QR GM1FTZ
G84 0QX 2M0CTI
G84 0QX MM6CNV
G84 0QZ GM4MFO
G84 0RL 2M1HKA
G84 0RN MM6HOO
G84 7DN MM7BYK
G84 7EE MM6HEZ
G84 7PL GM8VAM
G84 8JP GM7OAF
G84 8NN 2M0BRH
G84 8NN MM0GPL
G84 8NN MM3FPI
G84 9DA 2M0BJU
G84 9DA MM0GON
G84 9DA MM3OZU
G84 9DN GM4RJX
G84 9DX MM3ECO
G84 9DX MM3ERD
G84 9DX MM3YOL
G84 9JD GM1VYF
G84 9JZ MM0JWH
G84 9LL GM0SIM
G84 9QP GM4ZWJ
G840EY M6TCM
GL1 1DH G7UWP
GL1 2AR M6BBS
GL1 2PB G1ZSZ
GL1 2QZ M3TYI
GL1 2RX 2E0LGW
GL1 3DE 2E0HBE
GL1 3QE G2CIW
GL1 3QH M3CGC
GL1 3QS 2E0SFQ
GL1 3QS M0SPZ
GL1 4AD M6SQF
GL1 4AR M7CQK
GL1 5BY G8VUO
GL1 5HU M7AVN
GL1 5ER G3RKH
GL1 5HL 2E0BNF
GL1 5HL G1ZXC
GL1 5HL G6GLO
GL1 5JU M6CZD
GL1 5JZ 2E0THR
GL1 5JZ M7THR
GL1 5QD G7BPX
GL10 2DG G1USZ
GL10 2DH G7MLW
GL10 2PZ G7MWC
GL10 2QH G4CMY
GL10 3GA 2E0CMC
GL10 3GA M6BBO
GL10 3GQ 2E0UDB
GL10 3GQ M0UDB
GL10 3GQ M7DAB
GL10 3HS G4XWZ
GL10 3HX G4EXF
GL10 3JN G3TBF
GL10 3LD G4EDY
GL10 3LD G4YYR
GL10 3LD G8ILN
GL10 3LJ G0RUY
GL10 3NA G0PDE
GL10 3NL G4UBC
GL10 3PJ M6AKT
GL10 3QW G4SHB
GL10 3QW M0JWQ
GL10 3RT G4CRG
GL10 3SN G4FRR
GL10 3TU G4CIO
GL11 4AP G6HKL
GL11 4AS 2E0KLD
GL11 4EP M0LEP
GL11 4EW G4JXC
GL11 4QB 2E0RBZ
GL11 4QB M6RAZ
GL11 5AR 2E0HHA
GL11 5AR M7FRE
GL11 5DA G4KYI
GL11 5EL G4VZR
GL11 5JQ G8ZTM
GL11 5SG G3ZKN
GL11 5SW G1GDT
GL11 6HB G4FQH
GL11 6HY G4ETS
GL11 6JE G0DQS
GL11 6LF G0BRW
GL116JF M7JWR
GL12 7BJ G4VCQ
GL12 7HA G4ARW
GL12 7LQ G0SYF
GL12 7NP M7XJW
GL12 7PD G4YIC
GL12 7RH G7FEQ
GL12 8AS 2E0EUW
GL12 8AS G1HXT
GL12 8AS M0JEA
GL12 8AS M6EUW
GL12 8DA G0MIG
GL12 8HZ M7PFF
GL12 8NB G1USW
GL12 8SB 2E0UBM
GL12 8SB M6UBM
GL12 8SF M0IPZ
GL12 8SG G7JWE
GL12 8TJ G0CBK
GL12 8TN G1VNL
GL13 9BU G0NVX
GL13 9BU G4HQX
GL13 9DF G8ZHN
GL13 9EB M6EAT
GL13 9NP M6DZZ
GL13 9PL G6JWO
GL13 9PY G0UGR
GL13 9TE M6JJV
GL13 9TE M7SSB
GL13 9TG G7OPB
GL13 9TQ G7FPU
GL13 9TQ M6DCL
GL13 9UA M3YXF
GL13 9UT G0IHC
GL14 1BW G3XBX
GL14 1NB G6GUC
GL14 1QX G0XAE
GL14 2BH G4KRJ
GL14 2BH G6FGA
GL14 2DE G8PGH
GL14 2DW G0DZA
GL14 2DW G0WBS
GL14 2EB G0PBB
GL14 2EB G0SNB
GL14 2EB G7KXN
GL14 2EF G0ODN
GL14 2EF G6CHT
GL14 2HJ M5AFX
GL14 2HZ M7CIT
GL14 2QU G4THC
GL14 3AX G0HUX
GL14 3BL G1AXW
GL14 3DZ G0DAB
GL143DX M7BZK
GL15 4AJ M3DCL
GL15 4BS G8TMA
GL15 4HR G7VQI
GL15 4JX G4NIF
GL15 4NY G4ULG
GL15 4QD G4UHJ
GL15 5AZ 2E1IFM
GL15 5AZ 2E1IFN
GL15 5AZ 2E1IJL
GL15 5AZ 2E1KID
GL15 5AZ G7VHJ
GL15 5GB G4NNJ
GL15 5HU M6YRY
GL15 5JD G0ENF
GL15 5LP G0FDD
GL15 5NP 2E0FOD
GL15 5NP M0SSJ
GL15 5NP M6SWL
GL15 5QS M3YJM
GL15 5QS M3YJP
GL15 5QS M6BPG
GL15 5RE 2E0SYS
GL15 5SL G4EPW
GL15 5TA 2E0NSS
GL15 5TA M3NSS
GL15 6BZ M1ZZA
GL15 6DN G4IKX
GL15 6HG 2E0DWP
GL15 6HG M3BRV
GL15 6JQ G0SDD
GL15 6LQ G3CZL
GL15 6NA 2E0NBC
GL15 6NT G7EEG
GL15 6PE G8BXD
GL15 6PF M0GLQ
GL15 6TN G6UXY
GL15 6TN M3FTJ
GL15 6UN G4PWA
GL155FX M7YET
GL16 7AG G3NOC
GL16 7AQ G1EHX
GL16 7BL G7GOK
GL16 7EB G0PKV
GL16 7LG 2E0CJP
GL16 7LG M3FHQ
GL16 7LR G8AOJ
GL16 7PU G0KWG
GL16 7QB G0OOF
GL16 7QB M0ACW
GL16 7QB M0ATX
GL16 7QD M1ART
GL16 7RB 2E0MNL
GL16 7RB G1KUG
GL16 7RB M7BKI
GL16 7RG 2E1IFL
GL16 7RU 2E0EQE
GL16 7RU M0TXB
GL16 7RU M6KXE
GL16 8AY M1AYN
GL16 8AY M1BKE
GL16 8AZ G3KTI
GL16 8BG G1EDP
GL16 8BJ M6BLG
GL16 8BN M6CMZ
GL16 8BP M6RBU
GL16 8BS 2E0SDR
GL16 8BX M6RCP
GL16 8BY G7AEF
GL16 8BY M1EKM
GL16 8DE M6CNA
GL16 8DN G1BWP
GL16 8DS M6XTU
GL16 8PQ G8WGD
GL16 8PT G3TLD
GL16 8RQ M7DAE
GL168EZ 2E0EFK
GL168EZ M0IQQ
GL168EZ M6APS
GL17 0DF M0NOS
GL17 0DQ M3FHP
GL17 0JE G0NQI
GL17 0LP 2E0KGN
GL17 0LP M6KGN
GL17 0PH G0FGZ
GL17 0SB G7MWW
GL17 9AU M0HIY
GL17 9AU M0RHR
GL17 9BW M0OED
GL17 9BW M6HYB
GL17 9QS M1AVU
GL17 9SB G0RMX
GL17 9SB M6FUR
GL17 9UD G0ORD
GL17 9XR G7TUS
GL17 9YN M3MYG
GL17 9YN M3MYI
GL18 1HN G6PMJ
GL18 1PS M6DDD
GL18 1PZ G3WVQ
GL18 1QB M3DGJ
GL18 1TL G1NRX
GL18 2BD M7PPF
GL18 2BW G4WUH
GL18 2EJ G8SQH
GL19 3BN G8CQZ
GL19 3BP G1MHM
GL19 3QY G8TTJ
GL19 4BT G0ECJ
GL19 4BT G6VKA
GL19 4DE G6BOQ
GL19 4DS G1KQD
GL19 4ES M7JAD
GL19 4NX G8DYG
GL19 4NY G4CIB
GL19 4NY G4RHK
GL2 0DP G4NVY
GL2 0DZ G4FLS
GL2 0EJ 2E0TOT
GL2 0EJ M6TOT
GL2 0ER M6LNX
GL2 0HA G1JMF
GL2 0JG M1DKA
GL2 0JH G7WAY
GL2 0LX G4MGW
GL2 0NA G0UJT
GL2 0NB M3NVO
GL2 0NQ G3XMM
GL2 0PS M3CCJ
GL2 0PX G4CHS
GL2 0PX G4HJV
GL2 0QP M7YYF
GL2 0RX G0EEA
GL2 2BF G4WXF
GL2 2BH G4LEN
GL2 2HN 2E0POE
GL2 2HN M7POE
GL2 2JB M6PVL
GL2 4GS G4MOH
GL2 4JH G4IVD
GL2 4LL 2E0MSN
GL2 4LL M7MSM
GL2 4LZ G1GAT
GL2 4NP G4OIN
GL2 4PB G0DBM
GL2 4QE G0JQX
GL2 4QQ G8RMI
GL2 4RJ 2E0UPS
GL2 4RJ M7AEY
GL2 4RT 2E0CMD
GL2 4RT M0HFY
GL2 4RT M6CBY
GL2 4SY G8MMG
GL2 4SY M0OMD
GL2 4UF M3NCQ
GL2 4US 2E0VFA
GL2 4US M6VFA
GL2 4WF G8OTD
GL2 4WJ G0KVO
GL2 4YR M6MAH
GL2 4YY G1DNT
GL2 5BJ M6DDT
GL2 5NZ G7NGN
GL2 7DF G8IEW
GL2 7DJ G7JWQ
GL2 7DN 2E0JCR
GL2 7ED G1WZO
GL2 7HR G4LZQ
GL2 7LH G3STZ
GL2 7LW G1XAL
GL2 7PS 2E0EQI
GL2 7PS M6LOQ
GL2 7pt G3ILO
GL2 8EB G7DMZ
GL2 8EH G1ISY
GL2 8JP G4PJJ
GL2 8LE M7EMH
GL2 8LE M7IJP
GL2 8LJ G4ZYR
GL2 8NH G3PJQ
GL2 9ES M7GLW
GL2 9GS 2E0ITU
GL2 9GS M6ITU
GL2 9HB G1IDV
GL2 9NW G4BGW
GL2 9PS G7GVJ
GL2 9RB G0HTO
GL2 9RD M0ASO
GL20 5DG G6VAR
GL20 5FB G7AEE
GL20 5FB G7VTL
GL20 5NH 2E0JCA
GL20 5NH M6JCA
GL20 5NN G8YMR
GL20 5PD G1KNX
GL20 5RF G0UHG
GL20 5RL G0MMA
GL20 5RX M5ADE
GL20 5TW G7AEC
GL20 5TZ G0VFZ
GL20 5TZ G8AZN
GL20 5TZ M6WUH
GL20 6BB G4CRN
GL20 6DL G0CLR
GL20 6DL G0NFV
GL20 6DW G0NXA
GL20 6JW G6AHX
GL20 7AT G3FHG
GL20 7AU G3XGW
GL20 7EH G1XYF
GL20 7EP 2E0UHF
GL20 7EP G6LJU
GL20 7EP M0RAR
GL20 7QL G0HDB
GL20 7RS G0NUL
GL20 7RW G8WWC
GL20 7WL G0COZ
GL20 8AQ M3SZO
GL20 8AS M1AMA
GL20 8AT G4EAZ
GL20 8BB M0WMR
GL20 8BT G1NFB
GL20 8HS G3OLW
GL20 8HS M6OGO
GL20 8NN 2E0CME
GL20 8NN M6BBU
GL20 8NT G7OHW
GL20 8PJ M3XIV
GL20 8PR G8JXS
GL20 8QP G8VSH
GL20 8QY G0FCM
GL20 8RB G1NVS
GL20 8RE G7CVC
GL20 8RP M6WRE
GL20 8SN G0PTR
GL20 8TQ M6WIR
GL29ED G4DCK
GL3 1AA G0JVH
GL3 1AL M7XNY
GL3 1AT M6KCG
GL3 1BL M3WHG
GL3 1DB 2E0GKN
GL3 1DX M7DAP
GL3 1LL M3TMY
GL3 1NT M6WTE
GL3 2AU G4ENZ
GL3 2BA G4BCA
GL3 2BT G0VIG
GL3 2DW G6XQO
GL3 2HQ 2E0UPA
GL3 2HQ M0UPA
GL3 2HQ M6ULA
GL3 2HT G1CMH
GL3 2HW M3SFL
GL3 2JS G0MTW
GL3 2LD G3SUA
GL3 2PN G4FRI
GL3 2PU G6UGW
GL3 2RY G6OTP
GL3 2TR G4MYW
GL3 3AS 2E0GXK
GL3 3AS M6GXK
GL3 3BX 2E0KZU
GL3 3BX M6KZU
GL3 3DH G0VWH
GL3 3JE G4HBV
GL3 3JF G3UUL
GL3 3JH G3ZLM
GL3 3QS 2E0TFN
GL3 3QS M6DLV
GL3 3SX G8IWB
GL3 3TZ G6USD
GL3 3TZ G8DHF
GL3 3UL 2E0YIN
GL3 4BD M6ECT
GL3 4DH M6RMI
GL3 4ER 2E0JIA
GL3 4ER M6IJB
GL3 4ES G1NVO
GL3 4GF M6TTN
GL3 4NP M3IXD
GL3 4PF 2E0DDQ
GL3 4PF G0FNF
GL3 4PF M0HWT
GL3 4PF M6ECE
GL3 4PP 2E0PWC
GL3 4PP M6PWC
GL3 4RL M7IKB
GL3 4ZE M7CCU
GL4 0AB 2E0UED
GL4 0AL G3XUC
GL4 0BX G0RGJ
GL4 0DA G3HXN
GL4 0HP G8WRI
GL4 0JT M3UEY
GL4 0NB M6TXZ
GL4 0NZ 2E0OKK
GL4 0NZ M0IRD
GL4 0PG G0MGG
GL4 0QN G0HBB
GL4 0QY G1DIM
GL4 0RA G1IFF
GL4 0SH M6PZO
GL4 0SR G6UER
GL4 0TD G4CLR
GL4 0TJ M6IEQ
GL4 0TS G1AET
GL4 0TT G4TDG
GL4 0XS M6MIF
GL4 0XW M0PCB
GL4 0XW M0VSQ
GL4 0YQ 2E0IXC
GL4 0YQ M3IXC
GL4 3AG G0UWU
GL4 3AG G0UWW
GL4 3AH G7RHM
GL4 3AJ G7OUG
GL4 3AL 2E0IEI
GL4 3AL M6GZE
GL4 3AX G1BWH
GL4 3JL G1NJI
GL4 3TQ G4PTW
GL4 3TX G1UGX
GL4 4AG G0ULH
GL4 4AG G2HX
GL4 4AG M6RYL
GL4 4NU M0KTN
GL4 4NU M7KIT
GL4 4NW M7YGI
GL4 4PU M6FOD
GL4 4RB G0GAJ
GL4 4RE M6YUL
GL4 4RN M6VED
GL4 4TF M6XYJ
GL4 4UA G7CSM
GL4 4WA G7BVS
GL4 4WH G0FBX
GL4 4WH G0FHK
GL4 4WH G1FHK
GL4 4WP G0JBV
GL4 4WS M6BHI
GL4 4WZ 2E0NZA
GL4 4WZ M0XGL
GL4 4XD G4KXK
GL4 4XH G6BXT
GL4 4XJ G3TDT
GL4 5BH M7CRG
GL4 5DG G3VTS
GL4 5FD G4BNW
GL4 5FJ G7AEA
GL4 5FJ M1AGK
GL4 5FQ 2E0PMA
GL4 5GD G7GQC
GL4 5GD G7JUP
GL4 5GD M0XAC
GL4 5XT 2E0ADY
GL4 5XT G5ADY
GL4 6DW G0MIE
GL4 6EJ M7ERV
GL4 6JE M7GLO
GL4 6PB G0IMQ
GL4 6QD M7AJX
GL4 6QU M0JJA
GL4 6TX M7JLC
GL4 6UY 2E1FXN
GL4 6WE 2E1CAF
GL4 8AL G0NRZ
GL4 8DZ M0GCN
GL4 8HB G1SCV
GL4 8HB M6AND
GL4 8LD G3YJE
GL40NZ M6NUY
GL5 1ES M3BGE
GL5 1HD M3UGV
GL5 1HS G6SQT
GL5 1PL G8VLY
GL5 1QE 2E0RKD
GL5 1QE M6RZD
GL5 1QE M6SHZ
GL5 1RD G0MZK
GL5 1ST 2E0TPG
GL5 1ST G0NUN
GL5 1ST M0HZ
GL5 1ST M0TPG
GL5 1ST M6TBG
GL5 1SY 2E0IND
GL5 1SY M0IND
GL5 1SY M6IND
GL5 1US G0FCO
GL5 2DG M0BJP
GL5 2DG M3CIJ
GL5 2JJ 2E0RYD

GL5 2JJ	M7PDR
GL5 2RF	G3REB
GL5 2RG	G6UTB
GL5 2UG	G8AER
GL5 3PJ	G0OIS
GL5 3QT	M1FCV
GL5 3QZ	G4GXB
GL5 3TS	G4GWZ
GL5 3TS	G4RNK
GL5 4DF	G4RJG
GL5 4JW	G1JCT
GL5 4LF	M7ABT
GL5 4PX	G0TNG
GL5 4PX	G1OEM
GL5 4SA	G0IUW
GL5 4UN	2E0BKJ
GL5 4UN	2E0UUW
GL5 4UN	M3UGJ
GL5 4UN	M3UUW
GL5 4UW	G0FCJ
GL5 5JS	G3ICA
GL5 5JY	G3YIE
GL5 5LN	G4ENA
GL5 5PS	M7RDS
GL50 2AW	G3DNS
GL50 2LU	G8JJK
GL50 2NG	G4GVZ
GL50 2RD	M0WEB
GL50 3BW	M6WWK
GL50 3FJ	2E0JDO
GL50 3NH	2E0KXL
GL50 3NH	M6KLA
GL50 3NH	M7RGB
GL50 4BY	G8GKH
GL50 4HS	G4PBY
GL50 4LL	M6ORB
GL50 4NU	G3SMD
GL50 4NX	G8DTA
GL50 4NY	M3JIL
GL50 4PX	G8XRS
GL50 4QB	G4FZG
GL50 4RG	2E0SMF
GL50 4RG	G0NDU
GL50 4RG	M0SMF
GL50 4RG	M6SMF
GL50 4RJ	M0DCB
GL50 4SA	G0UTR
GL50 4SA	G0WJA
GL50 4SB	G4VTA
GL50 4SE	G6GJN
Gl504NW	G3RMD
GL51 0AE	G4IOA
GL51 0EG	2E0GEL
GL51 0EG	M6UHU
GL51 0JN	2E0OCH
GL51 0JN	M0UCH
GL51 0LZ	G8CYU
GL51 0NY	G0SMM
GL51 0NZ	G6AFE
GL51 0PP	G6MBH
GL51 0QH	2E0ETU
GL51 0QH	M6FYK
GL51 0QT	M6ECJ
GL51 0TU	G1CQK
GL51 0TW	G6UYN
GL51 0UB	2E0IFV
GL51 0UB	M6IFV
GL51 0UP	M0SYG
GL51 0UP	M1SYG
GL51 0WN	M3JQV
GL51 0WN	M6XVC
GL51 0XE	G8ZEE
GL51 3BB	G6TRY
GL51 3BN	G4DVX
GL51 3JF	G4BZU
GL51 3LA	G0EYP
GL51 3LL	G6VQN
GL51 3LL	M6EPL
GL51 3LX	G4FUJ
GL51 3LX	G6FTR
GL51 3PD	G4HQC
GL51 3QL	G0ALA
GL51 3RA	G3NKS
GL51 3RR	G4SGI
GL51 3RZ	M1ELN
GL51 3WA	G6DPS
GL51 4GQ	M3ZZX
GL51 4SB	G0TPO
GL51 4SN	G3KII
GL51 4SZ	G7URT
GL51 4TG	G4ERR
GL51 4UP	M7HUX
GL51 4UW	G6RII
GL51 5RX	G6XSC
GL51 6AL	G0EVX
GL51 6AL	G4XDL
GL51 6BA	G8TWS
GL51 6BU	G6FPK
GL51 6GR	G3LHU
GL51 6JE	G0SDG
GL51 6JE	G4BSC
GL51 6JQ	2E1GKY
GL51 6JQ	M3GKY
GL51 6JS	G0WIG
GL51 6LQ	2E0DYG
GL51 6LQ	M0YAY
GL51 6LQ	M6YQD
GL51 6NP	G1EDX
GL51 6NX	G4WUJ
GL51 6QE	G1WVK
GL51 6QP	M0GPC
GL51 6QZ	G8DOB
GL51 6RL	G3LVP
GL51 6RL	G8APY
GL51 6RT	G8JAY
GL51 6RT	M3YAJ
GL51 6RW	G4ILI
GL51 6SN	G1AEB
GL51 6UA	G5ROB
GL51 6UA	M7RAX
GL51 7BQ	2E0GBG
GL51 7BQ	M6DMG
GL51 7DQ	G8PZD
GL51 7EA	2E0NKE
GL51 7EA	M6NKE
GL51 7LH	M7SEK
GL51 8AF	G3YEU
GL51 8DZ	2E0CLY
GL51 8DZ	M0JLY
GL51 8DZ	M6BBP
GL51 8HR	G3JJT
GL51 8NX	G6BHS
GL51 9DQ	G1NMW
GL51 9LG	M6YCP
GL51 9RD	M1AUO
GL51 9RF	M7EQQ
GL51 9RW	M6CJM
GL51 9TG	G0LRI
GL52 2JR	M0IPB
GL52 2NR	G4NSZ
GL52 2NR	G8NSZ
GL52 2QR	G4YSF
GL52 3BB	G3XPR
GL52 3DA	M1BFR
GL52 3DE	M7DRH
GL52 3DF	G8ZFT
GL52 3DG	G3XKD
GL52 3DS	M0JVV
GL52 3DT	G4CWM
GL52 3DT	G6UJI
GL52 3DU	G3UYL
GL52 3DU	G4LCM
GL52 3DU	G4PKO
GL52 3DX	G7NCW
GL52 3DY	G0LFP
GL52 3DY	G7DEI
GL52 3EH	G3SNN
GL52 3EH	G5BK
GL52 3ES	G4PDQ
GL52 3HF	G8LHP
GL52 3LB	G6EEU
GL52 3PS	G8ENW
GL52 3QB	M0DQB
GL52 5AN	G0COI
GL52 5EG	G3LME
GL52 5GH	M0XQX
GL52 5JQ	G4KFT
GL52 5QS	M0YZG
GL52 5QS	M7GWC
GL52 6JD	2E0RLE
GL52 6JG	M7JME
GL52 6JJ	M6KEF
GL52 6NZ	G1KYV
GL52 6RF	G8UJF
GL52 6SX	G3VKV
GL52 6TX	G6DXD
GL52 6YA	G7GQA
GL52 7UZ	M6FRG
GL52 7YT	M7MFV
GL52 7YU	M3WXD
GL52 8AG	G3XKH
GL52 8AU	M7GFR
GL52 8BG	G6LTB
GL52 8BN	G3PYI
GL52 8BY	M1CBY
GL52 8BY	M1CBZ
GL52 8BZ	2E0GWV
GL52 8BZ	M7HAM
GL52 8LA	M7KJG
GL52 8NA	G3AKI
GL52 8NU	G7PTV
GL52 8NX	G0IYO
GL52 8NZ	G4GKK
GL52 8PA	G4SYW
GL52 8PF	M0RON
GL52 8SA	G4ERP
GL52 8SS	G0VSY
GL52 8SS	G0VSZ
GL52 8TG	G0MBZ
GL52 8TG	G4MPJ
GL52 8UL	G4ZRD
GL52 8US	G1LMU
GL52 8XP	G4CKX
GL52 8XQ	G8TQP
GL52 9HN	2E0DKZ
GL52 9HN	M0HLI
GL52 9HU	G0SFE
GL52 9PY	G3KGT
GL52 9QP	G3NOI
GL52 9QS	G3ZTI
GL52 9QU	G4IZZ
GL52 9RJ	M0JNV
GL52 9TZ	G8MGD
GL52 9UD	G4BMP
GL525GJ	M6BEM
GL525LG	2E0VHF
Gl526da	G7VDS
GL52PA	2E0IRE
GL52PA	M6LFW
GL53 0AZ	G4RFU
GL53 0BH	G0VCD
GL53 0HE	G4NOE
GL53 0LU	G4WGK
GL53 0NQ	G6VVL
GL53 0NU	G8BTV
GL53 0PQ	G7AKI
GL53 0PU	G6PYM
GL53 7BD	M6PMY
GL53 7BD	M6UHM
GL53 7BQ	G3ZQJ
GL53 7BQ	M0DMF
GL53 7DB	G3XTP
GL53 7DX	G8KMR
GL53 7JJ	G3CGD
GL53 7RT	G3PCL
GL53 7RU	M0DKP
GL53 7RY	M0IBQ
GL53 8AR	M6BCC
GL53 8NQ	G4XXA
GL53 8NQ	G4XXB
GL53 8NS	G0LSQ
GL53 9AT	2E0KVS
GL53 9AT	M6TFE
GL53 9AT	M7JSF
GL53 9DQ	G0VNJ
GL53 9HU	G4VQE
GL53 9JJ	M7CRA
GL53 9LB	G4INB
GL53 9QG	G8NSO
GL530AD	G8CQX
GL54 1DJ	M3RXD
GL54 1JD	G3RWI
GL54 1LG	G8FMZ
GL54 2EW	G0BXS
GL54 2NZ	G1NAP
GL54 2NZ	G1WQX
GL54 2QX	G3NGZ
GL54 2QX	G3TSO
GL54 3JJ	M1BPS
GL54 3QJ	G1VXS
GL54 4DD	G1XJT
GL54 4ET	M6SFV
GL54 4HP	G4VLH
GL54 4HP	M0URF
GL54 4HZ	G7IHN
GL54 4JX	M6HNT
GL54 4LL	M0AUW
GL54 5EP	M6XBV
GL54 5JX	G4MUW
GL54 5LQ	G0TPD
GL54 5QR	G3DPM
GL54 5QX	G7AWP
GL54 5XH	M0BAM
GL55 6AG	G3DWI
GL55 6BY	G1TFY
GL55 6JL	M0PHV
GL55 6TD	G0LSK
GL55 6XP	G0TPA
GL56 0FX	G8OYA
GL56 0JG	M7MZO
GL56 0LL	2E0GVC
GL56 0LL	M6GVC
GL56 9EX	2E0IXI
GL56 9EX	M6IXI
GL56 9JZ	G8DCJ
GL56 9RJ	2E0HEG
GL56 9RJ	M6HEG
GL56 9TE	G4PQB
GL6 0AP	G0FDE
GL6 0DR	G3WMQ
GL6 0LD	G0DZM
GL6 0NJ	G3ZTX
GL6 0PJ	M0HMR
GL6 0PY	G3OYX
GL6 0RT	G6CYO
GL6 0TB	G0PPM
GL6 6AD	G7KUU
GL6 6AG	G0ISH
GL6 6EY	G4TCO
GL6 6NJ	G4HXQ
GL6 6PH	M0BAP
GL6 6RQ	G4VLV
GL6 6SA	G0VPT
GL6 6TJ	G3OAD
GL6 7DA	G6PKG
GL6 7EF	G0PKJ
GL6 7NY	G4SJN
GL6 7NZ	M7HYE
GL6 7QA	G3SPO
GL6 8AA	G3XYZ
GL6 8AB	2E0VDL
GL6 8AB	M6VDL
GL6 8DG	G1VDE
GL6 8FB	G3VBQ
GL6 8LH	M5OTA
GL6 8LZ	G3NQF
GL6 8ND	G3TEV
GL6 8NX	M0HNH
GL6 8NY	G4OHA
GL6 9BA	G4GUG
GL6 9BA	G5TAM
GL6 9BZ	G4BSM
GL6 9BZ	G7EXX
GL6 9EP	G6KKV
GL6 9HB	G4IJV
GL6 9HF	G3UDD
GL67LE	G4MQL
GL68HG	M6UFC
GL7 1AU	G0UIX
GL7 1BJ	M1TCP
GL7 1BJ	M6FXG
GL7 1BL	M1BHZ
GL7 1BT	G0AZD
GL7 1BT	G7AEH
GL7 1DY	2E0FWK
GL7 1DY	M0NYY
GL7 1DY	M6NZB
GL7 1GJ	G4GPV
GL7 1HF	2E0ARJ
GL7 1HF	G5HI
GL7 1JT	M0AYX
GL7 1ND	M6RJU
GL7 1PR	2E0LLW
GL7 1PS	2E0KYD
GL7 1RU	M3IAA
GL7 1TG	G0RYR
GL7 1TG	M3GVJ
GL7 1TG	M3WHA
GL7 1UG	G8BAS
GL7 1WY	2E0HFI
GL7 1WY	M7TDV
GL7 2EJ	M0ANO
GL7 2JW	G8FBS
GL7 2LS	G4EVE
GL7 2NG	G1FRD
GL7 2PZ	G0WIY
GL7 2RH	G0NKK
GL7 2RL	2E0UJR
GL7 2RL	M6UJR
GL7 3AR	G3UJZ
GL7 3AX	M7REX
GL7 3EG	M6OSJ
GL7 3JG	G3MNJ
GL7 3NN	M6CAW
GL7 3SD	G4JTO
GL7 3SD	G7AYE
GL7 4EQ	G3ZVC
GL7 4FZ	2E0KSR
GL7 4HN	G6RPW
GL7 4HN	M6JJG
GL7 4JU	M0ONY
GL7 4JU	M7AEZ
GL7 5fh	M6LTV
GL7 5HQ	G1ELF
GL7 5PR	2E0DIU
GL7 5PR	M6EOY
GL7 5QS	G1MYQ
GL7 5ST	G2OU
GL7 5ST	M1BEO
GL7 5WJ	M7SKI
GL7 6BE	G6EMB
GL7 6BS	G8KKD
GL7 6JB	M7DIP
GL7 7BA	G7ORG
GL7 7BB	G3UK
GL7 7DL	G0IZP
GL7 7DN	M7ERB
GL7 7JU	G3TA
GL73NA	M7NSP
GL77JY	G4ABY
GL8 8BU	2E0XVZ
GL8 8BU	M0GVQ
GL8 8BU	M3LLN
GL8 8HA	G0RXQ
GL8 8JE	G1TJW
GL8 8JU	2E0AYX
GL8 8JU	M0OSM
GL8 8SN	G4IHT
GL8 8YU	G3MAV
GL9 1HH	G1KKS
GL9 1HP	G0XAY
GL9 1HP	G4GCT
GL9 1HT	2E1IJK
GL9 1HT	M0TFY
GR - 72053	G3URA
GU1 1BT	2E0BZH
GU1 1EP	G0SWC
GU1 1EP	G0SWT
GU1 1FR	G7MZS
GU1 1HX	G1MDS
GU1 1HX	G8UKH
GU1 1NA	2E1CML
GU1 1PA	M6HAC
GU1 1QL	G0KTV
GU1 1RQ	G4CDX
GU1 2DF	2E0HAT
GU1 2DF	M6HAT
GU1 2FB	M0RCN
GU1 2FJ	G0LPG
GU1 2JQ	G0DRT
GU1 2JS	2E0GHC
GU1 2JS	M7AMK
GU1 2QF	G1RNV
GU1 2RP	M6IER
GU1 2TY	G7UCR
GU1 3JR	G4TJK
GU1 3NP	G6AFK
GU1 3NP	M7FLG
GU1 3PZ	G3PQC
GU1 3PZ	M0SBT
GU1 4DD	M1TOD
GU1 4DN	M6WEU
GU1 4NL	G4WSW
GU1 4NP	G8TZN
GU1 4NQ	G7VNL
GU10 1BY	G0VYQ
GU10 1LE	G4GNO
GU10 2JG	G1RUG
GU10 2JG	M0CJO
GU10 2QU	2E1JIM
GU10 2QU	G4IFX
GU10 2QU	M3CPX
GU10 4AX	M6HYJ
GU10 4BJ	M0PSI
GU10 4DW	G0LVR
GU10 4EX	2E0BOD
GU10 4JW	G8RYJ
GU10 4LU	G4ZKJ
GU10 4PQ	2E0JLW
GU10 4PQ	M6FZG
GU10 4RJ	M6AGQ
GU10 4RL	G4AEW
GU10 5EL	G4KWX
GU10 5HZ	G0NFA
GU10 5LP	G4ROM
GU10 5LS	G4LES
GU10 5PA	G8IXL
GU10 5QE	M0HUT
GU10 5RD	G1CEI
GU11 1HA	M0NGI
GU11 1HA	M6NGI
GU11 1YY	G6RQZ
GU11 3BW	M0GPW
GU11 3DB	M0AHJ
GU11 3EL	G4GVV
GU11 3EP	M0PSE
GU11 3ET	G4GMI
GU11 3GY	2E0HEU
GU11 3GY	M7MGB
GU11 3JX	G4GGZ
GU11 3SL	G7EXO
Gu11uf	M6PNI
GU12 4AG	G1XIE
GU12 4AT	G8YKM
GU12 4EU	2E1FHQ
GU12 4HU	M6JZY
GU12 4HY	M6KIL
GU12 4HY	M6PCE
GU12 4HZ	G4YFU
GU12 4LD	G8OLY
GU12 4RD	G3KND
GU12 4SF	G4SPD
GU12 4ST	M6ZDZ
GU12 5AY	G7ITM
GU12 5ED	G7TEP
GU12 5HP	G8PDP
GU12 5HR	M6MGA
GU12 5HS	G4WEL
GU12 5HS	G4WEM
GU12 5JG	G8ZAJ
GU12 5JP	G1LKJ
GU12 5JP	M3LKJ
GU12 5NQ	G6XNP
GU12 5PR	2E1HAC
GU12 5PR	G8CXF
GU12 5QW	G8NEF
GU12 5SF	G4CDH
GU12 6HP	2E0DFI
GU12 6HP	M6EHA
GU12 6LX	G3VKI
GU12 6LX	G4YCL
GU12 6LY	G7PWA
GU12 6NZ	G6TSX
GU12 6PL	M0DGT
GU12 6PP	M6KVK
GU12 6QB	G1IYE
GU12 6QN	G7QXT
GU12 6SG	G7CUF
GU12 6ST	M1EGZ
GU124HY	G1OQG
GU124HY	M0COM
GU125JT	G8BCO
GU125JT	M0FRS
gu13pz	2E0RWR
gu13pz	M7WER
GU14 0DD	G0EMR
GU14 0DW	G6MRW
GU14 0ET	2E0WKT
GU14 0ET	M0WKT
GU14 0ET	M6GQX
GU14 0ET	M6WKT
GU14 0HL	M6AAO
GU14 0LQ	2E0FCE
GU14 0LQ	M0IYM
GU14 0LQ	M6NZJ
GU14 0PB	G1AAP
GU14 0RF	G8IBC
GU14 0RW	M6AOP
GU14 6AR	2E0TRX
GU14 6AR	M6REX
GU14 6AX	M6BQK
GU14 6BW	M0CMP
GU14 6DA	M6NGU
GU14 6DS	G6RIZ
GU14 6JS	2E0DFL
GU14 6JS	M6EHI
GU14 6JT	M6BDM
GU14 6LX	G1NTX
GU14 6NU	G4JFN
GU14 6NU	G4VAH
GU14 6QA	G4KCC
GU14 6RF	G3WOS
GU14 6TH	M0HAU
GU14 7AE	G3SRR
GU14 7AR	G3PQC
GU14 7DA	G0DWM
GU14 7EU	G3UHW
GU14 7EY	G1OER
GU14 7HH	G0HOD
GU14 7PP	G7VTS
GU14 7TB	M6YHV
GU14 8AG	2E1CYZ
GU14 8AG	M3MKB
GU14 8BJ	M0RCD
GU14 8DG	2E0GFF
GU14 8DG	M0YMA
GU14 8DG	M6ADB
GU14 8DJ	2E0EOV
GU14 8DJ	M6FKN
GU14 8NN	G4PDR
GU14 8PH	G1FVH
GU14 8QJ	G3PQF
GU14 8SR	G7HGT
GU14 9EA	G1KXQ
GU14 9HT	M7SKB
GU14 9JQ	G0MPI
GU14 9PJ	G0MCQ
GU14 9PW	M3IHS
GU14 9QH	M3NFW
GU14 9RL	G4BNK
GU14 9RP	2E0HHB
GU14 9RY	2E0LHS
GU14 9RY	M6FKO
GU14 9UH	G0HUT
GU14 9UR	M3XSU
GU14 9XU	G4NSN
Gu14 9xu	M3HCL
GU14 9XY	M6ASQ
GU14 9YB	M7TRA
GU15 1BF	G3YTX
GU15 1DE	G8GQS
GU15 1DG	G3HEJ
GU15 1DL	G6FYC
GU15 1DL	M6CHX
GU15 1EF	G8BVB
GU15 1EQ	M6JTU
GU15 1LF	G4UQE
GU15 1NP	G6IXM
GU15 1PS	G0RGX
GU15 1PS	G8DXZ
GU15 1RD	2E0CIX
GU15 1RU	G8EAD
GU15 2AJ	G0IZY
GU15 2BH	G3RKK
GU15 2BU	M0HUI
GU15 2DE	G0DSQ
GU15 2JQ	G6JTO
GU15 2LD	G4REE
GU15 2LT	G4ELJ
GU15 2LY	M6BTB
GU15 2NR	G6LWA
GU15 2NT	G0WZX
GU15 2RQ	2E1AFA
GU15 2SP	G6ENU
GU15 2SP	M6ENU
GU15 2SP	M7KCB
GU15 3EB	M6EYA
GU15 3EN	M6MNT
GU15 3HT	M0KWY
GU15 3HT	M0LKY
GU15 3HT	M0RCK
GU15 3HT	M6DBK
GU15 3HT	M6ELI
GU15 3HT	M6LFY
GU15 3HT	M6LKY
GU15 3NP	G4GLP
GU15 3NR	G3LPN
GU15 3XB	G4YHK
GU15 4AR	G8OUY
GU15 4AR	M6SUN
GU15 4JR	M6BTQ
GU15 4LD	M6PUS
GU15 4ZF	M0CMF
GU16 6BJ	M6HZF
GU16 6BZ	G1OSO
GU16 6DB	M7CWD
GU16 6DB	M7YAZ
GU16 6DG	2E0SEE
GU16 6DG	M6AZE
GU16 6DT	G6WZD
GU16 6ER	G6BLU
GU16 6EY	M1CIS
GU16 6GA	M6LCW
GU16 6HG	M6BMJ
GU16 6JJ	G4SYB
GU16 6PH	G6MJB
GU16 6SD	M6CTW
GU16 7DU	2E0SKZ
GU16 7DU	2E0XAI
GU16 7DU	2E0XPT
GU16 7DU	M3SKZ
GU16 7DU	M3XAI
GU16 7DU	M6XPT
GU16 7EB	M6BQR
GU16 7RD	M6LAC
GU16 8AD	M0ILA
GU16 8AD	M6IKX
GU16 8LP	G7KNK
GU16 8LP	M3KNK
GU16 8LR	G3PPU
GU16 8NA	G8XNC
GU16 8PS	G7LTW
GU16 8RT	G8IRM
GU16 8RW	G4MEA
GU16 8SA	G3KFU
GU16 8XH	M6RYA
GU16 8XQ	G4DJB
GU16 8XR	M6WJN
GU16 8XS	M7CNJ
GU16 8XS	M7RVF
GU16 8XX	G3TJI
GU16 8YN	G7WDD
GU16 8YT	M6ABB
GU16 9BH	2E0JXG
GU16 9BH	M7JGZ
GU16 9BL	M6MHM
GU16 9BW	M6SAC
GU16 9NY	M6KDK
GU16 9QN	M6POQ
GU16 9QU	G3YAG
GU16 9RB	G7AQL
GU16 9RF	M7LGW
GU17 0DH	G3WCY
GU17 0DJ	G0ADT
GU17 0DU	G6KRY
GU17 0EN	G1FOE
GU17 0HB	G8JZO
GU17 0JW	M7PFK
GU17 0LG	2E0CUL
GU17 0LG	M0KUL
GU17 0NJ	G3TMU
GU17 9BP	G8HIO
GU17 9DL	G4CGW
GU17 9DP	M6DMQ
GU17 9DX	G3ZYX
GU17 9DZ	G3REL
GU17 9DZ	G3SFU
GU17 9ET	G4EMV
GU17 9HA	G8HUF
GU17 9HH	G6AAB
GU18 5RZ	G3TJS
GU18 5TA	G0KDL
GU18 5TE	G8JMP
GU18 5TJ	G4MBZ
GU18 5TP	G3UKE
GU18 5TR	G8HXD
GU18 5TS	G8NYJ
GU18 5UJ	G4SUO
GU18 5UW	M1AKZ
GU18 5XH	G4SJH
GU19 5DH	M6HJQ
GU19 5DH	M6HJU
GU19 5JR	M3XTA
GU19 5JX	G4ZRB
GU19 5NU	G8MZD
GU2 4AZ	G2DBH
GU2 4EL	G4PEA
GU2 4JF	G8UHK
GU2 4LA	G6FKN
GU2 4LF	2E0VHZ
GU2 4LF	M6VHZ
GU2 5XH	G7DQE
GU2 7JG	M0WAC
GU2 7JH	M3RZU
GU2 7JL	M3UJO
GU2 7JL	M3VKT
GU2 7JL	M6MFM
GU2 7JN	M3ZRV
GU2 7JP	M7AMA
GU2 7QN	G0SUL
GU2 7RT	G7JVQ
GU2 7SA	G1WTX
GU2 7SR	G4OWL
GU2 7SR	G8DPQ
GU2 7SX	M7KCR
GU2 7TP	M6PZM
GU2 7TS	M3MBF
GU2 7UU	M7SLI
GU2 7XE	M6LXX
GU2 7XH	M0LTS
GU2 7YW	2E0VNN
GU2 7YW	M3VNN
GU2 8AL	M0DGT
GU2 8AP	G8LOZ
GU2 8AR	G8FUH
GU2 8AU	M6ASD
GU2 8BL	G1MGF
GU2 8DE	G3KMI
GU2 8DE	M0DNY
GU2 8ER	2E0JIP
GU2 8ER	M0JIP
GU2 8ER	M7POW
GU2 8ES	M7MYB
GU2 8JD	M0SFI
GU2 8JU	G4PMZ
GU2 8LX	2E0CZS
GU2 8LX	M0PVI
GU2 8LX	M3PVI
GU2 8UT	2E1AOG
GU2 9GA	2E0SUY
GU2 9GA	M0SNJ
GU2 9GA	M6NEJ
GU2 9NP	G8TZU
GU2 9PG	G1ZDG
GU2 9PJ	G0HNL
GU2 9PJ	G8NOB
GU2 9PL	M6RJQ
GU2 9PL	M7SEA
GU2 9PW	G7UPN
GU2 9TY	M3UIU
GU2 9WA	2E0JRN
GU2 9WA	G1YJY
GU2 9WA	M0JRL
GU20 6AL	2E0HUZ
GU20 6AL	M7HHH
GU20 6BU	G8WSP
GU21 2AT	M6YRB
GU21 2DD	G1OFW
GU21 2DD	M0RDK
GU21 2fz	G0NGI
GU21 2HY	M6XRD
GU21 2JN	G8FZV
GU21 2LD	M6JXN
GU21 2LH	G4HWI
GU21 2NG	G0JXP
GU21 2PX	M6XNL
GU21 2PX	M6XRL
GU21 2QP	2E0SSJ
GU21 2RR	M7KDH
GU21 2TA	G1AOE
GU21 3AU	M6IQA
GU21 3BP	G4PRW
GU21 3BY	2E0FQA
GU21 3BY	M7LMS
GU21 3BZ	M0RDB
GU21 3DB	M0IRI
GU21 3DD	G8JMU
GU21 3JU	M7SCH
GU21 3NZ	G8BTL
GU21 3PJ	G7WIY
GU21 3QS	M0VPC
GU21 3QY	2E0XSG
GU21 3QY	M0XSG
GU21 3QY	M7SUE
GU21 4BG	G3USX
GU21 4HY	M7JGP
GU21 4JG	G8SSY
GU21 4PW	2E0DRW
GU21 4PW	M6KWG
GU21 4PW	M7BKX
GU21 5EG	G0NIX
GU21 6DQ	M6FBT
GU21 7PN	G7CSJ
GU21 7QT	G4RHJ
GU21 8AW	M0PES
GU21 8SS	M0KFU
GU21 8SY	M6CEI
GU22 0AA	G4GGI
GU22 0JB	2E0WFS
GU22 0NT	G3ZHB
GU22 0NT	G6BQC
GU22 7BE	2E0KFK
GU22 7BE	M6KFK
GU22 7NE	M6NZL
GU22 7NW	2E0CLH
GU22 7NW	M6BGE
GU22 7SZ	2E0BQN
GU22 7UE	G3GAQ
GU22 7UR	G8LSA
GU22 8PE	G0AAA
GU22 8PE	G3WVG
GU22 8PE	M0ZIP
GU22 8QZ	G0AFP
GU22 9AJ	M6EMO
GU22 9AU	G0NRJ
GU22 9AU	M3YSC
GU22 9BQ	G4YPC
GU22 9BQ	G5RS
GU22 9ND	G7USO
GU22 9PH	G7PVU
GU22 0NU	G4SLL
GU220NQ	G7HHI
GU23 6DQ	G0JOP
GU23 6EN	G0DKN
GU23 6PF	2E0GJT
GU23 6PF	M0RNZ
GU23 6PF	M7NSC
GU23 7AD	M0BUF
GU23 7AD	M7FMG
GU23 7AP	G4EML
GU23 7ET	G8AFU
GU23 7HR	G3IUW
GU24 0HF	2E0MXC
GU24 0HF	M0MXC
GU24 0HF	M6MXC
GU24 8PR	G7CSI
GU24 8PS	M6ISH
GU24 8RL	M6AXS
GU24 8AR	G3KMA
GU24 9DG	G0RVK
GU24 9LY	M7CQF
GU24 9QE	G0DER
GU25 4EZ	G0TTG
GU25 4LD	G8MLK
GU25 4LP	M7AKW
GU25 4NG	M6ACC
GU26 6LR	G0NIN
GU26 6NY	G3ZFT
GU26 6PA	2E1JON
GU26 6PA	G4NOT
GU26 6PW	G7VBL
GU26 6PZ	G7UEI
GU26 6PZ	G7WBM
GU26 6RP	G4CMG
GU26 6SX	G3WLH
GU27 1AZ	M1CZX
GU27 1JD	G1OTZ
GU27 1JF	M6DFE
GU27 1JF	M6XER
GU27 1LA	G8EOV
GU27 2BX	G7NDN
GU27 2FD	2E0CVQ
GU27 2FD	M6CTN
GU27 2NY	G1MAL
GU27 3DZ	2E0URJ
GU27 3JL	G0XTL
GU27 3JS	M0EBX
GU27 3ND	G4DQZ
GU27 3RG	G0NON
GU27 3SN	2E0SJI
GU27PH	2E0IFO
GU28 0EQ	G6NMQ
GU28 0EQ	G8WSX
GU28 0EQ	M3UQE
GU28 0NY	G0HJZ
GU28 0PJ	M6THA
GU28 0QX	M6XRX
GU28 9DH	2E0BQZ
GU28 9DH	M0PUT
GU28 9DH	M3UNB
GU28 9DH	M6BBE
GU28 9DP	M0GPD
GU29 0BP	G4BGM
GU29 0DJ	G4YEI
GU29 0NQ	G3VXF
GU29 0NU	M6WJJ
GU29 9BU	G1TLH
GU29 9JF	M6SUW
GU29 9JG	G0DDY
GU29 9TE	M1STI
GU3 1AZ	G0SJH
GU3 2AX	2E0ZNZ
GU3 2AX	M0ZNZ
GU3 2AX	M6KNC
GU3 2AX	M6ZNZ
GU3 2EN	G4HZV
GU3 2JW	G3VCY
GU3 3AU	G4PNB
GU3 3BE	G6GPR
GU3 3BQ	M7RHI
GU3 3BX	G3TQC
GU3 3LL	M7SMN
GU3 3NE	G8IMS
GU3 3PQ	M1EBL
GU3 3PR	G4MPW
GU30 7AN	2E0MTT
GU30 7AN	M0NJX
GU30 7AN	M6NJX
GU30 7BS	G4VLF
GU30 7BS	G6JQX
GU30 7BY	M0RGI
GU30 7GB	G8VG
GU30 7GW	G4TTJ
GU30 7HG	G3TSR
GU30 7HR	G4VKC
GU30 7SH	G8ISI
GU30 7wp	M7BQJ
GU30 7XA	G8UYW
GU30 7XD	G4AER
GU31 4EU	2E0DFF
GU31 4EU	2E0JLB
GU31 4EU	M6OBJ
GU31 4EU	M6TXT
GU31 4NX	G1MGU
GU31 4PN	G1LQH
GU31 4PR	M7TAW
GU31 4PZ	M7GAW
GU31 4QA	G6SLZ
GU31 5AX	G3TZL
GU31 5BF	2E0WEI
GU31 5BF	M6HXW
GU31 5HJ	G0BUZ
GU31 5NZ	2E1JJM
GU31 5NZ	M3JJM
GU31 5SB	G1THA
GU32 1DD	G4AEP
GU32 2AN	G4VQT
GU32 2AX	G6XJB
GU32 3AX	G8LWC
GU32 3AZ	G4VRC
GU32 3BT	G4SBU
GU32 3BW	G0MBQ
GU32 3LS	G0BTU
GU32 3LS	G8XDD
GU32 3PJ	G4ACW
GU33 6DA	M3TQP
GU33 6HG	G8XJO
GU33 7BF	G4CFS
GU33 7BP	2E0MPA
GU33 7BP	M3ZQF
GU33 7DB	M1CQU
GU33 7DL	G4ELM
GU33 7JE	M0PDH
GU33 7JE	M3EMX
GU34 1AY	M6HDW
GU34 1JA	G4ASL
GU34 1NU	G4FOY
GU34 1PW	G0JMI
GU34 1QY	G0JVL
GU34 1QY	G1LAO
GU34 1RA	G3RSI
GU34 1RR	G8MOS
GU34 2BJ	G1EML
GU34 2DG	M6NHO
GU34 2ED	G8XWR
GU34 2EE	G0DBS
GU34 2EW	2E0BTW
GU34 2EW	M0GOK
GU34 2EW	M3XRK
GU34 2fj	2E0BUF
GU34 2fj	M0VVQ
GU34 2HP	G8YFH
GU34 2PB	M3KOA
GU34 2PF	G1JUP
GU34 2QT	G0IBR
GU34 2QT	G7LWH
GU34 2RD	G4JOU
GU34 2TL	G8IRL
GU34 2TP	G8UML
GU34 3DH	M7NKN
GU34 3EJ	G6BXV
GU34 3JA	M6KMX
GU34 3NP	G0DWR
GU34 4AG	G4AJB
GU34 4AG	M7GJP
GU34 4AN	G8EAN
GU34 4AQ	G0EFP
GU34 4LJ	G4DNK
GU34 4LJ	G8DDH
GU34 5AD	G4JCX
GU34 5AX	G8LES
GU34 5BX	G0RMR
GU34 5BY	G4IPI
GU34 5BY	G8WSZ
GU34 5DU	2E0VVX
GU34 5DU	M0VVX
GU34 5DU	M6HZQ
GU34 5EF	G3PRQ
GU34 5JF	M0BDF
GU34 5LG	G6UEQ
GU34 5NP	G3YVI
GU34 5PB	G0BHA
GU34 5PB	G4OBN
GU34 5PB	G6BHH
GU34 5PB	G6BHI
GU34 5PF	G8CKN
GU34 5PR	G3SJX
GU35 0DG	G0PLZ
GU35 0DP	M7CIJ
GU35 0DZ	M7TVS
GU35 0EF	M6KGK
GU35 0FJ	G8OXS
GU35 0HB	G3ZRM
GU35 0PP	G0OYN
GU35 0QB	G8AZB
GU35 0RA	M6EYF
GU35 0SG	2E0BII
GU35 0TH	G8NLF
GU35 0XA	G8BGI
GU35 8AJ	G0PVJ
GU35 8AX	M6OPB
GU35 8DU	M7BCZ
GU35 8EP	G1CBB
GU35 8HQ	G6SNA
GU35 8HU	G4ZEL
GU35 8JX	2E0FCB
GU35 8NA	G4LJB
GU35 8NR	G6VFI
GU35 9AJ	2E0SNJ
GU35 9AJ	M6SWA
GU35 9DZ	G8CYL
GU35 9ED	2E0CSN
GU35 9ED	M6BYZ
GU35 9EX	2E0DVD
GU35 9EX	M0EUY
GU35 9EX	M3EUY
GU35 9PJ	G8SXB
GU35 9QN	2E0CCW
GU35 9TA	G1TQY
GU359BA	M0MGA
GU4 7FF	M6FSD
GU4 7FF	M7PMH
GU4 7HQ	M6UBC
GU4 7NP	M1RJJ
GU4 7NQ	M6MGV
GU4 7PD	2E1ICW
GU4 7PD	G0KUF
GU4 7QB	G6MQZ
GU4 7TJ	M1ACK
GU4 7UG	M0AGP
GU4 7XR	G3PIZ
GU4 7XR	M0ECW
GU4 8AY	M0DHN
GU4 8EY	2E0WKW
GU4 8EY	M0REQ
GU4 8EY	M7EJS
GU4 8HZ	G0JRE
GU4 8JG	G8NTH
GU4 8JW	M7ADF
GU4 8LL	G6ZAC
GU4 8LQ	G1HTN
GU4 8PH	G3WBQ
GU46 6BX	G3WUL
GU46 6DP	M7AZN
GU46 6DW	G4URP
GU46 6FA	G8PJQ
GU46 6FN	2E0NGN
GU46 6FN	M6NGN
GU46 6GZ	2E0PTG
GU46 6GZ	M0URL
GU46 6JT	G4VDF
GU46 6NE	G8IFH
GU46 6NT	G6FTY
GU46 6PB	G1WNL
GU46 6PE	G7CFW
GU46 6YP	G7AOA
GU46 6YW	G0YOU
GU46 7AD	G4RYV
GU46 7QP	M0RSB
GU46 7RD	G8INZ
GU46 7SY	G4ZHZ
GU46 7TT	G4TTZ
GU46 7UY	2E0HZP
GU46 7UY	M0KFK
GU46 7UY	M7DMY
GU47 0GJ	M7IOR
GU47 0QT	M3BVQ
GU47 0ST	G7VHZ
GU47 0UP	G8EII
GU47 0UT	2E0DCL
GU47 0UT	M6CLK
GU47 0UT	M6KSC
GU47 0YQ	G8LMY
GU47 0YQ	M5LMY
GU47 0YY	G6ACJ
GU47 8HT	G0FGA
GU47 8HT	G3ZJQ
GU47 8HY	M6SMR
GU47 8JH	G1FNS
GU47 8JL	G8FXU
GU47 8LD	G8NMH
GU47 8NA	M7JSP
GU47 8QS	G4XYW
GU47 8QS	G6GS
GU470DU	2E0WXY
gu470du	M6NPZ
GU470HH	2E0KAL
GU470HH	M0KAR
GU5 0BL	G0UNF
GU5 0HZ	G4NMD
GU5 0HZ	G6APJ
GU5 0HZ	G8KVU
GU5 0HZ	M3CIS
GU5 0JP	G0DRR
GU5 0LL	M7DLR
GU5 0SA	G4VQF
GU5 0SE	G6ZPR
GU5 0SX	G8WMF
GU5 0TJ	G8XAK
GU5 9RS	G0FCU
GU50EG	G1UNB
GU51 1AU	G8VIV
GU51 1DU	M7SLD
GU51 1DX	M6HXF
GU51 2TE	G4KTZ
GU51 2TL	M6SCA
GU51 3AH	M3ZCM
GU51 3BS	G7BQT
GU51 3DY	G3MSL
GU51 3EB	G1FRS
GU51 3EB	G4OQZ
GU51 3EL	G7MCS
GU51 3LY	G8YVM
GU51 3LY	M0NAM
GU51 4AG	2E0CKO
GU51 4AQ	2E0DOZ
GU51 4AQ	M0SLG
GU51 4AQ	M6DOZ
GU51 4AQ	M6SVU
GU51 4HA	G4EFY
GU51 4HB	G8JNI
GU51 4HE	2E0FVZ
GU51 4HE	M0FAR
GU51 4HE	M7ANR
GU51 4HN	G4FTK
GU51 4SW	G0EFL
GU51 5BJ	M7BIO
GU51 5DP	G1CNV
GU51 5NJ	2E0BBS
GU51 5NP	G3NVM

GU51 5NR G7KIT
GU51 5SU M0DHO
GU51 5TZ M6MDM
GU514LY M6PUU
GU52 0TE M3WRJ
GU52 0UR G3KOB
GU52 0YJ 2E0EVC
GU52 0YJ M6WRR
GU52 6AS G3HUK
GU52 6AZ M0RHE
GU52 6BN G4AFI
GU52 6BN M7AFI
GU52 6JD 2E0BBY
GU52 6JD M3KFR
GU52 6LQ G3WKW
GU52 6PW G3RCD
GU52 7LE G3LGT
GU52 7LN G1JPP
GU52 7UG G3BRQ
GU52 7XE G4NYY
GU52 8NS 2E0ETE
GU52 8NS 2E0REB
GU52 8NS M6FFF
GU52 8NS M6REB
GU52 8XG G0NWL
GU526AY G8SUJ
GU6 7AA M3BSM
GU6 7DD M6VBP
GU6 7DS G7PRZ
GU6 7EN M0AWE
GU6 7ET G4CJT
GU6 7FY 2E0OBC
GU6 7FY M0IEB
GU6 7FY M6OBC
GU6 7HZ G5OD
GU6 7HZ G6XN
GU6 7HZ M0GJH
GU6 7JB G3OHC
GU6 7JB G4GRG
GU6 7RW G8TIO
GU6 8BZ G4YIF
GU6 8FD 2E0XJW
GU6 8FD M6ONS
GU6 8NB M6TCH
GU6 8NX 2E0LSL
GU6 8NX M0LSL
GU6 8NX M6LSL
GU6 8PQ G1DSM
GU6 8UD 2E1BEB
GU6 8UD M0RUT
GU67FQ M7SIW
GU7 1AH M0DLX
GU7 1QQ G8IBL
GU7 1RL G6WOT
GU7 1RL M0GKK
GU7 1SB 2E0DPO
GU7 1SB M0HZM
GU7 1SB M6FLS
GU7 1SY G4CWP
GU7 1YA 2E1IAC
GU7 1YB 2E0XEN
GU7 1YB M0MOW
GU7 1YB M0NGY
GU7 1YB M3XEN
GU7 2AA M3KVU
GU7 2BS M6HXA
GU7 2GA M7LIU
GU7 2LD G3TCU
GU7 2LJ 2E0ERK
GU7 2LJ 2E0YAP
GU7 2NE G1WTW
GU7 2NN M6JVX
GU7 2NT G4VRN
GU7 2QT 2E0DTN
GU7 2QT M0KRD
GU7 2QT M6KRD
GU7 2RW G1SWZ
GU7 3EU G8ZMC
GU7 3HA M7ACA
GU7 3HG G3KZB
GU7 3NN G8YEF
GU7 3NT G3MBK
GU7 3NZ G1FJF
GU7 3QJ G6NAV
GU7 3SH M0IAM
GU8 4AB M3TGK
GU8 4BD G4JEF
GU8 4ND G0SUA
GU8 4NT G8EBT
GU8 4RG 2E1AFS
GU8 5AB G7CND
GU8 5AB M3FJB
GU8 5DN G4GIX
GU8 5NR G4XBF
GU8 5TU G4DUF
GU8 6DE G0GNE
GU8 6DE G6RLM
GU8 6DU G3YZN
GU8 6DZ 2E1AFR
GU8 6EG M7EEC
GU9 0BN M6LQM
GU9 0BS G6FRS
GU9 0BS G7NMT
GU9 0BS G8ATK
GU9 0HY G8FQN
GU9 0NU M5AMN
GU9 0RZ G0FRS
GU9 0RZ G0YYY
GU9 7BP G4WUV
GU9 7BP G4WUW
GU9 7BT G7HOE
GU9 7BX G3VYI
GU9 7BX G4ALE
GU9 7DE M7TGR
GU9 7DH M0BSD
GU9 7RQ G4UEL
GU9 8BZ G3TAX
GU9 8JQ G8OGR
GU9 8LF G0CTZ
GU9 8RR 2E0KBG
GU9 8RR M6KBG
GU9 8SA M0PRA
GU9 8SA M6DPR
GU9 8TA G0CXU
GU9 8TW G0EYA
GU9 9AY G7JUR
GU9 9BB M7AZQ
GU9 9DF M7TXC
GU9 9EA G8UMA
GU9 9EY M3OSP
GU9 9FT M3CJH
GU9 9HJ G6BEN
GU9 9HJ G7FAQ
GU9 9NL 2E0EXQ
GU9 9NL M0IVK
GU9 9NL M6OHY
GY1 1TS GU7MXZ
GY1 1WJ GU0RAG
GY1 1WU GU4EON
GY1 2DZ MU0MXF
GY1 2ES MU0GSY
GY2 4AH 2U1EKH
GY2 4EX MU6GBG
GY2 4HJ GU4HUY
GY2 4HQ MU7GBG
GY2 4JS MU6CPV
GY2 4RN 2U0WGE
GY2 4RN MU6FXB
GY2 4XW 2U0EJL
GY2 4XW MU6GCI
GY3 5JG GU4YBW
GY3 5SE MU0IIF
GY3 5TZ 2U0PTB
GY4 6JW GU4WMG
GY4 6NB GU6RWD
GY4 6QJ GU4NYT
GY4 6QJ GU4SYQ
GY4 6SF GU3YVV
GY5 7FQ GU0SUP
GY5 7FQ GU3HFN
GY5 7XD GU4XIT
GY6 8EZ 2U1EJF
GY6 8LN GU1IIW
GY7 9LD GU0PSP
GY7 9NN GU7CQN
GY7 9NN MU0CHN
GY7 9PH 2U0EFR
GY7 9PH MU0EFR
GY7 9PH MU6GFR
GY7 9QF 2U0LRB
GY9 3JD GU3LPV
GY9 3TZ GU7APA
GY9 3UY GU0UVH
H52 0EU G0PKW
HA0 2HG M7HDM
HA0 2NX 2E0LHR
HA0 3QG G6MLV
HA0 3RQ M7SPY
HA0 3SG M6SHL
HA1 1QR M6NSZ
HA1 1XH 2E0SRP
HA1 1XH M0SRP
HA1 1XH M6PRS
HA1 1XU M7TKW
HA1 2QT M0JZO
HA1 2RS M7HKA
HA1 3HT G0UHK
HA1 3PR M0EVT
HA1 4AJ G3VFX
HA1 4AL G4IQD
HA1 4AL G4MSE
HA1 4BH 2E1HVL
HA1 4BW M0EGL
HA1 4DJ M6CIX
HA1 4EF M6EGA
HA1 4RY M7EFS
ha11qr M3NGO
HA2 0HJ M6YOO
HA2 0PU G7OZA
HA2 0QE 2E0GEE
HA2 0QE M6OMG
HA2 0QJ G7FSJ
HA2 0SX G4TPM
HA2 6AP G4JMG
HA2 6AQ G1RYF
HA2 6HE G0CAG
HA2 6HE G5FRG
HA2 6HF G3LGQ
HA2 6LG 2E0LBG
HA2 6LG M0JNS
HA2 6LG M6JNS
HA2 7EJ G4PKV
HA2 7JJ M6FFP
HA2 7NU G4IUM
HA2 7RB G4IRP
HA2 7RP M3KHW
HA2 9EA G4KEW
HA2 9EF G8DKW
HA2 9JL G4WVW
HA2 9LD G4MMA
HA2 9LJ G7IEF
HA2 9QF G4UKD
HA20LW 2E0NSY
HA20LW M6MPS
HA3 0PB G3FKI
HA3 0XQ G4KAB
HA3 5AX 2E0EYF
HA3 5AX M6ONU
HA3 6AJ G0IPK
HA3 6BX G3VHK
HA3 6HY M0HWZ
HA3 6JT 2E0DJX
HA3 6JT M0ZDJ
HA3 6JT M6XDJ
HA3 6PP M0GFK
HA3 6PP M7SKO
HA3 6QL 2E0HBT
HA3 6QL M3HBT
HA3 6TN G0BSP
HA3 7AX M0GCI
HA3 7HU M6ABT
HA3 7NF M6HVA
HA3 8HA M6FQF
HA3 8JB G3BQE
HA3 8NF G3ASR
HA3 8NF G4KEP
HA3 8NQ G4ZPR
HA3 9EJ G8FAT
HA3 9ET M6SFL
HA3 9LB G7HIH
HA3 9NB G1TMW
HA3 9QZ 2E0RDA
HA3 9QZ 2E0UTL
HA3 9QZ M0AAG
HA3 9QZ M0UMS
HA3 9QZ M6OCZ
HA4 0AU G3KRT
HA4 0DS G0ACK
HA4 0DW G0DVC
HA4 0HL G8FTY
HA4 0HY 2E0FGR
HA4 0HY M6ONT
HA4 0LX 2E0EJR
HA4 0LX M0RJE
HA4 0LX M6RJE
HA4 6AJ G0RNF
HA4 6BA M3TLG
HA4 6ED G0DZH
HA4 6ED G1SWF
HA4 6ND M0MAR
HA4 6QZ M3PYB
HA4 6SX G0NRN
HA4 7EH M7ENC
HA4 7JB G3EFX
HA4 7JB G4AUF
HA4 7LZ G0JIM
HA4 7LZ G0VTC
HA4 7LZ G4XIT
HA4 7LZ G6JIM
HA4 7QD G8UOZ
HA4 7RD G6TDX
HA4 7TQ M0XRA
HA4 7UL G4LHT
HA4 7UL M6GRR
HA4 7UN G4UZE
HA4 7XR G0TAN
HA4 8AJ G4FFA
HA4 8PH G8UMY
HA4 8RY G3OEC
HA4 8SB G0JCF
HA4 8SB G1DES
HA4 8SD G7KUG
HA4 8UA G6SYB
HA4 9AG G3JVM
HA4 9BY M6FGZ
HA4 9DL G6ETZ
HA4 9HD G6PLR
HA4 9JT G1LQV
HA4 9LF G6SPI
HA4 9SA M0HJR
HA5 1JH G0IRM
HA5 1NB G4PXY
HA5 1NJ G1RGT
HA5 1NL 2E0CGB
HA5 1NL M3VUQ
HA5 1PH G7HSO
HA5 1TL G0KSL
HA5 1TN G0CHQ
HA5 2AR M7PRI
HA5 2AU M0DLC
HA5 2DA G8CAB
HA5 2EH G3LFD
HA5 2HR 2E0ZYX
HA5 2HR G0HGG
HA5 2HR M6HGG
HA5 2NB G8NTR
HA5 2NE M6LXW
HA5 3UP G6LFD
HA5 3UP M3YGK
HA5 3YF 2E0XLJ
HA5 3YF G4AZL
HA5 3YF G6XUD
HA5 3YF G8DAI
HA5 3YF M3XLJ
HA5 4HB 2E0VYN
HA5 4HB M6VYN
HA5 4LN G3UAS
HA5 4NJ G0WHV
HA5 4RA G0NTA
HA5 4TX G1VIO
HA5 5SP M0SEW
HA54RA G0NTB
HA6 1DW G4MEX
HA6 2BU 2E0SEO
HA6 2BU M0AAV
HA6 2BU M6SJX
HA6 2FY G4AGM
HA6 2HN G7NBF
HA6 2UJ G8MAA
HA6 2YP G7KWT
HA7 1JH M3THQ
HA7 2RR 2E0IIF
HA7 2RR M7VGG
HA7 3AG M7JEE
HA7 3HT 2E0UAY
HA7 3HT M3XSG
HA7 3PB M3WCA
HA7 3PL G1DQQ
HA7 4EP G1MYM
HA7 4FP G0MOX
HA7 4LD G3NDC
HA7 4PF 2E0PPM
HA7 4PF M6HUV
HA7 4TQ M6DAU
HA7 4UQ M0ALQ
HA8 0QE M7WCW
HA8 0SA G4DRO
HA8 0TW M3BCS
HA8 0TW M3OAK
HA8 5JQ M7VLD
HA8 5NQ 2E0PBW
HA8 5NQ M6IVO
HA8 5RJ G0SQL
HA8 6DL G8RNM
HA8 6HL G7MHQ
HA8 6JE 2E0ODZ
HA8 7SA M6KEU
HA8 7UH 2E0CQT
HA8 7UH M6FEA
HA8 8AE M6EZM
HA8 8JU G4XEW
HA8 8PS G4GLM
HA8 8RH G4RFI
HA8 8SH G0STR
HA8 8TR G0KCL
HA8 8TR G4ZOD
HA8 8UA G4SCG
HA8 8XJ M0HOF
HA8 9HE G1GXC
HA8 9JD M0CSG
HA8 9PW M6GVF
HA8 9SP M6MED
HA9 7DQ M0TAT
HA9 8NX G4DNP
HA9 8TP G3ZZF
HA9 9HA M1FFR
HA9 9TQ G3SRN
HA90TF 2E0HYS
HA90TF M7CQL
HD1 3SL G0BFJ
HD1 3SL G6LD
HD1 4AJ M7DEW
HD1 4NU M6SGM
HD1 4PW 2E0KUK
HD1 4UR G1MOZ
HD1 5DY G7BNS
HD2 1AS G1MSD
HD2 1BN G1JJR
HD2 1DA G1CYY
HD2 1DH G3WUI
HD2 1ED G8YNE
HD2 1EJ M0GDJ
HD2 1JT G1GPT
HD2 1LB G0ODY
HD2 1LL 2E0ALJ
HD2 1LL M0OYZ
HD2 1LL M3SVZ
HD2 1QH G4YDI
HD2 1QH G6KKA
HD2 1QP G1AOR
HD2 1QY G8YOC
HD2 1RN G3XXR
HD2 1RU 2E0EDL
HD2 1RU M0RIU
HD2 1RU M6LDE
HD2 1TH M6BJO
HD2 1UB 2E0RLR
HD2 1UB M6RLR
HD2 1UY G3WLW
HD2 1XT 2E0LOG
HD2 1XT M0LAG
HD2 1YW 2E0RNR
HD2 1YW M6LOS
HD2 2EB G0PRF
HD2 2EH G4XTE
HD2 2EW G7EOD
HD2 2HF M6SZP
HD2 2JP G7RAI
HD2 2NE G0BWQ
HD2 2NF G4BWC
HD2 2NF M0AGR
HD2 2NH G4BYW
HD2 2PE M0TKA
HD2 2PQ G4MEK
HD2 2YD 2E0WBH
HD3 3DB G3BKJ
HD3 3EJ G0IRY
HD3 3FH M7AQZ
HD3 3QH G0MKP
HD3 3QX G3TFO
HD3 3RJ G8MAR
HD3 3YW M3ZVU
HD3 3ZF G4KDM
HD3 4BN G8DTX
HD3 4BQ M7HJS
HD3 4GH G4LRD
HD3 4HQ 2E0BOR
HD3 4HQ M3SVJ
HD3 4LD G1AGA
HD3 4LD G1AGB
HD3 4RS G8RWN
HD3 4SW 2E0BWJ
HD3 4SW M3YLK
HD3 4SW M3YLL
HD4 5DR G6VBA
HD4 5LL 2E0VVK
HD4 5NS G4GNA
HD4 5RP M6JPT
HD4 6HN M3VHU
HD4 6PJ M3GDX
HD4 6PL M1DQX
HD4 6QX G6TGE
HD4 6RA G1GTH
HD4 6SS M0AFS
HD4 6SZ M0PIE
HD4 6TE G4TML
HD4 7AS M3LQE
HD4 7DA G0BWB
HD4 7DA G0LPV
HD4 7DA G4CDD
HD4 7EP G8TOT
HD4 7HE G7UID
HD4 7HS G1BGM
HD4 7JS G0NMH
HD4 7JS G8XSF
HD4 7JX G0SVU
HD4 7RA G3UEU
HD4 7RJ G0TPM
HD4 7SR G4JLO
HD5 0AT G4RLA
HD5 0JB G7MHQ
HD5 0JD G0ISX
HD5 0LJ G7UZO
HD5 0NJ M3CMI
HD5 8EX G4KMK
HD5 8LD M0DXA
HD5 8LS 2E0RSB
HD5 8LS G4BQC
HD5 8LS M0RBG
HD5 8LU G6PJL
HD5 8PE M7JSH
HD5 8QB 2E0VET
HD5 8QB M6ARC
HD5 8QG G6GSV
HD5 8RP M0AOB
HD5 8UJ G4ITV
HD5 8UP G4RAJ
HD5 8XT G7OOH
HD5 8XU G4OTE
HD5 9EB M7ASZ
HD5 9HG G1FYS
HD5 9HX G0BWO
HD5 9JW G0LUU
HD5 9QA 2E0STH
HD5 9QA M6OXD
HD5 9RG G7FBT
HD5 9UW G6ZVU
HD5 9XT M3OAQ
HD5 9XT M3RTP
HD6 1HH G0DIU
HD6 1QT 2E0GSP
HD6 1QT M7BEO
HD6 2AT G1GTQ
HD6 2AY G1GTR
HD6 2BJ G8AUL
HD6 2EP G4EEJ
HD6 2NA G7LBD
HD6 2NW M0ARK
HD6 2RS G8MLD
HD6 2RU G4SDX
HD6 2RU M3JQJ
HD6 3AH M3JLX
HD6 3JS G4WAP
HD6 3LD G7KXS
HD6 3NP G3WAH
HD6 3QX G8RFF
HD6 3SR G4GMT
HD6 3XF M3RFF
HD6 4AP G3JWN
HD6 4EQ G0TJQ
HD6 4JQ G4IPH
HD6 4JZ G4REG
HD7 2SQ G7KPH
HD7 4BT G4FSQ
HD7 4BX G7AXL
HD7 4BX M6YHM
HD7 4JR G1MQQ
HD7 4JR G4LYY
HD7 4JU G1HEP
HD7 4JY G0HRF
HD7 4LA 2E0PAD
HD7 4LA M3WUG
HD7 4NS G4IDR
HD7 4NU 2E1CRI
HD7 4QP G4UFL
HD7 4QQ G4UFJ
HD7 4RA G3XWN
HD7 4RE G8VMF
HD7 5BW G0PLD
HD7 5JS M6OTL
HD7 5LS G0JTM
HD7 5NE M6KZF
HD7 5NF G6HKA
HD7 5QU G7JMU
HD7 5QX 2E1BYI
HD7 5QX M0HVP
HD7 5TY M3ZNC
HD7 6BJ M3VQN
HD7 6JX M6BBT
HD8 0DE G3TSA
HD8 0GD 2E0SCJ
HD8 0GD M3SCJ
HD8 0JA 2E0RUD
HD8 0JA M3PBE
HD8 0JB G4AHJ
HD8 0JG G4XJG
HD8 0JG G7NDC
HD8 0NP M7KRW
HD8 0QW 2E0RJH
HD8 0QW M0RJX
HD8 0QW M6RJH
HD8 0RX M7CVJ
HD8 0SW M1AYG
HD8 8AG G3LUK
HD8 8AG M0ALT
HD8 8AP G4OPY
HD8 8AZ M7OMP
HD8 8DS G7UKK
HD8 8HP G1DEN
HD8 8HP G3SDY
HD8 8HP G8KMK
HD8 8HP M3OEN
HD8 8JZ M3OUS
HD8 8LX G0PHI
HD8 8PG 2E0FEV
HD8 8PG M7DAZ
HD8 8PX M6WWC
HD8 8QF G6UJR
HD8 8QW G7EXD
HD8 8RH M7EKR
HD8 8SG M1ELQ
HD8 8SP M7GCH
HD8 8UA M3WYV
HD8 8UB 2E0MPB
HD8 8UB M6CNI
HD8 8XP G6VIQ
HD8 8YH 2E0HFT
HD8 8YH M0ODM
HD8 8YH M6FFT
HD8 8YH M6HFT
HD8 9AB M0RGC
HD8 9BS G7VBU
HD8 9BT G4HKY
HD8 9BW G4BYD
HD8 9DB M0IQC
HD8 9DS G4TKO
HD8 9EH G4JCL
HD8 9ET G8PRN
HD8 9HF M7CXZ
HD8 9PA M1DLR
HD8 9QP G4EMQ
HD9 1EH G0HNW
HD9 1EN G1LFD
HD9 1ES G3RJT
HD9 1EU M0LUS
HD9 1EU M6AIN
HD9 1HL G7VDH
HD9 1HQ G8SNJ
HD9 1LH M3CGP
HD9 1SW 2E0EGS
HD9 1SW M6IHD
HD9 1XP M1EGD
HD9 1XU G8MQK
HD9 2LF G7UBQ
HD9 2PQ G3YPE
HD9 2PS G1JOW
HD9 2PX G8LIK
HD9 2PZ M6WYD
HD9 3ES G8PUT
HD9 3ES M3PUT
HD9 3ET G6WSX
HD9 3XZ G4UFZ
HD9 4AG G8GFY
HD9 4AJ G3NAK
HD9 4AJ G7GKQ
HD9 4DR G3LDJ
HD9 4DY 2E0DAW
HD9 4DY M3VIJ
HD9 4ED M3XOW
HD9 5LT 2E0NVK
HD9 5LT M3NVK
HD9 5PG G8TYB
HD9 5PL M0GBK
HD9 6DD G4LLZ
HD9 6DS G0JTL
HD9 6EG 2E1FWA
HD9 6HX G0BAK
HD9 6ND G0DBP
HD9 7ER G0SJB
HD9 7FB 2E1IAE
HD9 7SG G1VLU
HD9 7TJ G8ZML
HG1 1HP M7EDS
HG1 1TY G0MKK
HG1 2AS G6HAT
HG1 2BW G0NHG
HG1 2DP 2E0UAB
HG1 2DP M6CBC
HG1 2HA G4CWB
HG1 2JF M6FJJ
HG1 2JN 2E0WAY
HG1 2JN M3WDZ
HG1 2QD M3YXR
HG1 2QD M6TTY
HG1 2QG M3EFV
HG1 2QG M3JYW
HG1 2QG M3KHT
HG1 2QG M3MIU
HG1 2QG M3OGC
HG1 2QG M3OHQ
HG1 2QG M3OHY
HG1 2QG M3OIY
HG1 2QG M3OSY
HG1 2QG M3SVF
HG1 2QG M3TNK
HG1 2QG M3TWK
HG1 2QG M3UCL
HG1 2QG M3UFE
HG1 2QG M3UYK
HG1 2QG M3VHC
HG1 2QG M3VHO
HG1 2QG M3WSW
HG1 2QG M3WTO
HG1 2QG M3XOY
HG1 2QG M3XUF
HG1 2QG M3XZF
HG1 2QG M3YCK
HG1 2QG M3YCU
HG1 2QG M3YEE
HG1 2QG M3YHL
HG1 2QG M3YKF
HG1 2QG M3YKO
HG1 2QG M3YLM
HG1 2QG M3YTL
HG1 2QG M3YUH
HG1 2QG M3ZCJ
HG1 2QG M3ZJJ
HG1 2QG M3ZKO
HG1 2QG M3ZNO
HG1 2QG M6AYE
HG1 2QG M6BPW
HG1 2QG M6BTF
HG1 2QG M6CSW
HG1 2QG M6CYL
HG1 2QG M6FIA
HG1 2QG M6HDK
HG1 2QG M6HUH
HG1 2QG M6HYW
HG1 2QG M6JYL
HG1 2QG M6KKL
HG1 2QG M6KLK
HG1 2QG M6LBQ
HG1 2QG M6LCT
HG1 2QG M6LDH
HG1 2QG M6LLW
HG1 2QG M6LST
HG1 2QG M6LXH
HG1 2QG M6MMP
HG1 2QG M6NMT
HG1 2QG M6VAN
HG1 2QG M6VCM
HG1 2QG M6WUN
HG1 2QG M6ZAS
HG1 2QG M6ZPZ
HG1 2QL M3XOZ
HG1 2QL M3YHM
HG1 2QL M3YSZ
HG1 3BU 2E0BUI
HG1 3BW G1ZEA
HG1 3DF M6LHD
HG1 3EA G1VQK
HG1 3LT G8MJF
HG1 3NA G4KAT
HG1 4BP 2E0EFG
HG1 4BP M6HDI
HG1 4DL 2E0PDO
HG1 4DL G6GMU
HG1 4DL M0TUV
HG1 4DL M6PDO
HG1 4EG G0UIW
HG1 4EH G6VKP
Hg1 4es 2E0BLK
Hg1 4es M3NUL
HG1 4HR 2E0BWK
HG1 4HR M6KVB
HG1 4JN G7BBJ
HG1 4QR G0KIY
HG1 4RH 2E0WKZ
HG1 4RH M3WKZ
HG1 4SG G1JLB
HG1 5AD 2E0GGP
HG1 5AD M6LNI
HG1 5HR 2E0SXI
HG1 5HR G0SDR
HG1 5HR M7DHZ
HG1 5HS G4LGX
HG1 5JS G3ZTB
HG1 5JU G1WUC
HG2 0BJ G4ZIU
HG2 0DA M0VGA
HG2 0DQ 2E0CTU
HG2 0DQ M6JFE
HG2 0EG M6IBP
HG2 0ES G7FTS
Hg2 0hz M7ABG
HG2 0JS M7AXE
HG2 0LL G3OGZ
HG2 7AJ 2E0LMH
HG2 7AJ G8XZC
HG2 7AJ M0LMH
HG2 7AJ M3YUV
HG2 7DP G3XPM
HG2 7EE 2E0JYP
HG2 7EE G4RCD
HG2 7EE M7JYP
HG2 7ER G4TJI
HG2 7ES 2E0RHH
HG2 7ES M0HYJ
HG2 7ES M0RDY
HG2 7HQ G1UIB
HG2 7LJ G1ARH
HG2 7LJ G1ROD
HG2 7NJ M0AVF
HG2 7RU M6RNC
HG2 8HN M3YIT
HG2 8HW G4MNE
HG2 8LB G4EKJ
HG2 8LE G4KCR
HG2 8LX G4HOX
HG2 8PR G4LQG
HG2 8PR M6TJB
HG2 8QF M1DFW
HG2 9BS G1BPE
HG2 9HP G3YHC
HG2 9LD M0MCI
HG2 9NL G1CCX
HG2-7JU 2E0FDU
HG28AP M7YTO
HG29PG G8DMU
HG3 1BH 2E0PAE
HG3 1BH G8NMT
HG3 1FP M7HVL
HG3 1HY M3URS
HG3 1JR G4IUF
HG3 1LY G6WKO
HG3 1NF G3NXL
HG3 1NY M0DYO
HG3 1PB M1CTG
HG3 1QX G0JNG
HG3 1RJ G0DOZ
HG3 2AA G4MEM
HG3 2DG 2E0AGI
HG3 2DG 2E1BDV
HG3 2DG G1ZBO
HG3 2DG M5ACF
HG3 2DJ 2E1CJC
HG3 2DS G3FMW
HG3 2QE G6EQT
HG3 2RF G0UFL
HG3 2SU M3UTP
HG3 2SU M3YHP
HG3 2UT G1JWL
HG3 2WE G1XKJ
HG3 2WE G6WTM
HG3 3BX 2E0IVY
HG3 3BX G3RHR
HG3 3BX M6TUR
HG3 3DS G4PAH
HG3 3HE G5VO
HG3 3HE G6MC
HG3 3HE M0WKR
HG3 3HE M3UZL
HG3 3HE M3YPG
HG3 3HE M6LJJ
HG3 3JR G4EEV
HG3 3JR G8IKS
HG3 3LA G4VNA
HG3 3PB G4AFT
HG3 3QZ M6ISW
HG3 4EW G3RYP
HG3 4HA 2E0JDM
HG3 5LH G3ZCY
HG3 5LY G3NNO
HG3 5PG G1TBT
HG3 5QX M6FXO
HG4 1JU 2E0NAE
HG4 1NG M0TJW
HG4 1NW G7OFM
HG4 1RG 2E0GEB
HG4 1RG M6HEE
HG4 1RZ G4DSC
HG4 1UA G3AB
HG4 1UG M7HFY
HG4 1UJ G0ORJ
HG4 1UP M3ZAZ
HG4 1US G1WRE
HG4 2AP M7ASF
HG4 2LJ G4VMY
HG4 2LJ G6MGH
HG4 2LN G1VKC
HG4 2NB M0HQJ
HG4 2PG G3UVY
HG4 2TU M7CGK
HG4 3LL G1PWM
HG4 4AU G0UFI
HG4 4AU G4VUN
HG4 4AY M6RCA
HG4 4BP M7OJV
HG4 4LH G7LYN
HG4 4PW 2E0EEJ
HG4 4PW M0RBY
HG4 5BU G1JYK
HG4 5LF G3LEO
hg43hu M7YKE
HG5 0BW G0EYL
HG5 0DX M0DAL
HG5 0LW M0AQH
HG5 0QA G8BNE
HG5 0SW 2E1EJX
HG5 0TH M0RMN
HG5 8EL M0CVG
HG5 8HD G0KGT
HG5 8HQ G7IPX
HG5 8HX G7HGD
HG5 8JN G6HEB
HG5 8JT G3DUW
HG5 8LH G1BKU
HG5 9BJ G4YBD
HG5 9BZ G3XQJ
HG5 9DE G3XZV
HG5 9EL G3UNA
HG5 9EL G6CP
HG5 9EL M6NLK
HG5 9HN 2E0DPL
HG5 9HN M0HYH
HG5 9LD G0GYU
HP1 1AS M6RTI
HP1 1LB G3GBN
HP1 1NS M6FTN
HP1 1QP M7VJT
HP1 1XQ G8FXC
HP1 2BP M0ZJV
HP1 2BQ G0TIW
HP1 2HY 2E1EOI
HP1 2LN M1ABC
HP1 2PP M0IHY
HP1 2PP M6ULY
HP1 2PP M6WTL
HP1 2QR G0EOK
HP1 3BN G4WGA
HP1 3BN G7NUT
HP1 3BU M5AEO
HP1 3EW G7SGM
HP1 3HY M1EQO
HP1 3JU G7IPA
HP10 0BP G4KCX
HP10 0BP G8NMT
HP10 0DU G1FEF
HP10 0HG M6YOS
HP10 0HJ G8ZRG
HP10 0PZ 2E1TWB
HP10 8BL G4XVP
HP10 8DL G1SHT
HP10 8HG M3MVO
HP10 8JJ G8FBW
HP10 8LH G4DDM
HP10 9AX G3PWY
HP10 9DF G4VLL
HP10 9DZ G8XQT
HP10 9EQ G4PIE
HP10 9LH G3WQG
HP10 9LT M7PJF
HP10 9PL G3INZ
HP10 9RH G8JQV
HP108BH 2E0MSS
HP108BH M0VWS
HP108BH M6MSS
HP11 1GL G3SRQ
HP11 1PA G4TJM
HP11 1QA M3OJW
HP11 1QE M7VUX
HP11 1QT G8DJF
HP11 1TX M6FNE
HP11 1UB G1UCT
HP11 2JL G0OCY
HP11 2JP M5SSB
HP11 2PL G1EUI
HP11 2ST M7KHA
HP11 2UA G4CJY
HP11NE 2E0LFW
HP12 3AE M6SDZ
HP12 3AS M0HCN
HP12 3DS G7GUO
HP12 3JQ G0IZC
HP12 3LP M6HXB
HP12 3NG M0MIQ
HP12 3NG M6BQZ
HP12 3NN G1LQX
HP12 3PA G8VJW
HP12 4AA G8WCH
HP12 4AU G7OST
HP12 4BB G6BFP
HP12 4DA M0YOJ
HP12 4DD G0ANO
HP12 4DD G4ZOC
HP12 4DD G8ANO
HP12 4PG G8DLZ
HP12 4QD M1EBY
HP12 4QR G0IKE
HP12 4SP G3ZGQ
HP12 4TY G3PVJ
HP13 5BD 2E0RPY
HP13 5BD M6NSE
HP13 5JN G3FNO
HP13 5JN G4TZQ
HP13 5JX 2E0KRK
HP13 5JX 2E0YGB
HP13 5JX G4EBY
HP13 5JX M0YGB
HP13 5JX M3YGB
HP13 5JX M6KRK
HP13 5NG G8HDL
HP13 5NW G7RTI
HP13 5QA G8AAU
HP13 5RH G0CTD
HP13 5RH G1EYG
HP13 5SP M1BDL
HP13 5SS G4LMM
HP13 5SZ 2E0KNA
HP13 5SZ 2E0PTM
HP13 5SZ M6PTM
HP13 5TA G4MUI
HP13 5TZ G4TBG
HP13 5UD G6HBJ
HP13 5UN G8JAW
HP13 5XN G1CPC
HP13 5XT G4COS
HP13 5XY G0NCL
HP13 6AU 2E0BKT
HP13 6AU M0GGW
HP13 6AU M3TUU
HP13 6EQ M7BBK
HP13 6HW M6PHC
HP13 6JG M6ELF
HP13 6JG M6YEY
HP13 6SN G3OUV
HP13 6SW M0HMB
HP13 6TJ G1GAR
HP13 6XW G6GTZ
HP13 7EF G3GAA
HP13 7JF M6RKI
HP13 7LF M7PCL
HP13 7PH G0OQK
HP13 7PH M6RVE
HP13 7UE G1MIY
HP13 7XN G4YAN
HP13AU M7WMM
HP14 3BN G4KBB
HP14 3JN G4PUO
HP14 3JW G3LYP
HP14 3RE M6GEG
HP14 3RP G0FUN
HP14 3RP G4WJS
HP14 3SJ G1OLY
HP14 3TF G0GGL
HP14 4JG G4YBH
HP14 4JN G4YKQ
HP14 4QB G6XQB
HP14 4RA G8GLV
HP14 4SG G0HND
HP14 4UH G3SRJ
HP14 4UY G0IOU
HP15 6BW G4XJN
HP15 6EG G6AHN
HP15 6EY G7MLX
HP15 6JR G3VCT
HP15 6LJ G3WNS
HP15 6LJ G8AKU
HP15 6SL G4PFA
HP15 6TN G0RBB
HP15 6UT G4GED
HP15 6XA G0TRE
HP15 6XD G7DZD
HP15 6XF G1RDX
HP15 7AT G3XZK
HP15 7BX M3YXC
HP15 7DT M6HRS
HP15 7ED G6UDF
HP15 7FP 2E0SVB
HP15 7FP M7JHA
HP15 7PH M6BEH
HP15 7RE G3TYG
HP15 7RP M6VVT
HP15 7TF G4PPK
HP15 7TF G6GUD
HP15 7TP G6EYJ
HP156NT M0TPE
HP157BY G7DPE
HP16 0DF G4SNQ
HP16 0HB G6YOZ
HP16 0HW M3HEI
HP16 0LY 2E0DUU
HP16 0LY M6IJH
HP16 0NA G7TQA
HP16 0NJ G8XTJ
HP16 0NL M0IEA
HP16 0NL M6FZK
HP16 0QE 2E0MPF
HP16 0QE G4CLM
HP16 0SL M1CVT
HP16 9DS G1MDG
HP16 9DS G3MDG
HP16 9DS G3ZNU
HP16 9EJ G4DHQ
HP16 9JH G6MJA
HP16 9RE G0AUK
HP17 8AN G3OCP
HP17 8DH G7JLF
HP17 8DH M0SOC
HP17 8DJ M3AWD
HP17 8DL M6ERI
HP17 8EU G1MZG
HP17 8EY G3WUN
HP17 8HN G0IKP
HP17 8JG G1TRL
HP17 8QN G7AAR
HP17 8RG G3YSR
HP17 8XQ G8WDX
HP17 8XU M0SFD
HP18 0BL G6JRM
HP18 0BN M5AJP
HP18 0DN G6IFE
HP18 0EZ G1ATY
HP18 0HJ G6OIH
HP18 0LH G0MMI
HP18 0LY G0KMC
HP18 0RX G4CYR
HP18 9AU M3BLF
HP18 9DP 2E1HNS
HP18 9DZ M6YMY
HP18 9HH G0RAS
HP18 9HP G7TPD
HP18 9JZ G1ULQ
HP18 9LD G6DAP
HP18 9QZ G0UUI
Hp189pj 2E1FZY
Hp189pj M0BSQ
HP19 7BG G6YLV
HP19 7FY G1IJM
HP19 7HR M0TOL
HP19 7HR M1EIO
HP19 7HW G8UEI
HP19 7QJ G0MHZ
HP19 8GW 2E0XLX
HP19 8GW M6XLX
HP19 8GZ M0WDU
HP19 8JE M6HTK
HP19 8JH 2E0RJY
HP19 8JH M7RAJ
HP19 8SW 2E0HAF
HP19 8SW 2E0OHF
HP19 8SW M6OHF
HP19 8WA G8AQP
HP19 9GY M7PLT
HP19 9LN G0TGM
HP19 9QA 2E0PNP
HP19 9QA M6BEU
HP19 9QF M7BNW
HP19 9QT 2E0TBQ
HP19 9QT G7HIU
HP19 9QY G7GQH
HP2 4BA G0IAL
HP2 4BN G8RWU
HP2 4HE G1URZ
HP2 4LZ M3AZF
HP2 4LZ M3RDH
HP2 4PB G8UCK
HP2 4PQ 2E0MMJ
HP2 4PQ M0MMJ
HP2 4PQ M6MMQ
HP2 4PX G1EIZ
HP2 4QP M6NYG
HP2 4RP G7HCQ
HP2 4UG G1PAK
HP2 5AT 2E0MOB
HP2 5AT G0BLU
HP2 5AT M0RMO
HP2 5AT M6RMO
HP2 5EL G8VVB
HP2 5ES M6MBR
HP2 5JG G8CBE
HP2 5JL G1DDF
HP2 5QF G8SYD
HP2 5QF M6TFQ
HP2 5QQ M7XTW
HP2 5SL M6HTZ
HP2 5WR G6FGV
HP2 5WR M0CZB
HP2 5XT M0GDS
HP2 5YX 2E0VTV
HP2 5YX M6ITV

HP2 6DS G4BJN
HP2 6DX G8ZGQ
HP2 6JB M0SNZ
HP2 6LA G1PWF
HP2 7AR M3XHT
HP2 7JP G7WDO
HP20 1HJ M7NRR
HP20 1HW G7IMY
HP20 1JW G8ALQ
HP20 1pz M3OOE
HP20 1QE M3VRL
HP20 1RB 2E0PNR
HP20 1XG 2E0MHC
HP20 1XG M7DPC
HP20 2AB M3PIA
HP20 2AJ 2E0MJH
HP20 2AJ M3CTT
HP20 2AS G4BKS
HP20 2EX M5ASR
HP20 2NU M3MMP
HP20 2QT G1NMP
HP21 7BD G4FXI
HP21 7EP G1FNA
HP21 7FE M6TBQ
HP21 7NS G8VVZ
HP21 8BN G8MBM
HP21 8EH G1BHW
HP21 8JG G1MZW
HP21 8JY M7EFT
HP21 8LY G4CGE
HP21 8NW 2E1PPM
HP21 8TW M0SVB
HP21 9AL M6GJC
HP21 9DS G8AYM
HP21 9DT G8MEM
HP21 9EF M3HER
HP21 9EF M3HIM
HP21 9LY 2E0DSV
HP21 9LY M6FIW
HP21 9NX G7ITO
HP21 9SL M6BFU
HP21 9UB M0ILK
HP21 9XE M7RMF
HP22 4BL M7WBT
HP22 4BX M3FKW
HP22 4EF G8BQH
HP22 4LH G3XOP
HP22 4LT G4HFU
HP22 4PA G0WZH
HP22 5AB M3UCC
HP22 5EG G1MRI
HP22 5HG M7TJS
HP22 5HH G4VRS
HP22 5HH G6GDI
HP22 5LA G1ZDU
HP22 5ND G8TJI
HP22 5QX M7YAG
HP22 5RX 2E0CNM
HP22 5RX M6CBH
HP22 5YQ G0BZN
HP22 6AR G1DDR
HP22 6AW G4PFR
HP22 6PH G1RAF
HP22 6PH G8ZGK
HP22 6QH G8ZME
HP225SD M7WAF
HP225TX M0TIF
HP23 4DS G6DZJ
HP23 4DS G7UYW
HP23 4EA G3NBZ
HP23 4EU M7UGL
HP23 4HZ G0SEY
HP23 4LX M0ZKK
HP23 4LX M6DSQ
HP23 4QG G4SSE
HP23 4TP G8MKX
HP23 5DG G1NUO
HP23 5HG M3MGQ
HP23 5HL G7PWI
HP23 5PB G3WXM
HP23 5PJ G3RWV
HP23 5PL M0HXV
HP23 6BX G1IGA
HP23 6ED G8GYY
HP23 6EN G3MEH
HP23 6HH G4PMG
HP23 6HH G8MKW
HP23 6HH M3PTG
HP23 6JB G1ODZ
HP23 6LE G0BKP
HP23 6NE G6HGM
HP27 0BQ G0SNF
HP27 0EB G0SPB
HP27 0EB G1DAE
HP27 0JA G7AYS
HP27 0LG G0NLJ
HP27 0LG G4HMC
HP27 0NB G3SNP
HP27 0PG G0MBB
HP27 0PW G0SKA
HP27 0QY G7DPF
HP27 9AY G1THD
HP27 9FZ 2E0PJY
HP27 9HE G0WQQ
HP27 9HU G3OZK
HP27 9JY G7CUD
HP27 9RP G3NBS
HP3 0BS G3VSQ
HP3 0BT M3UST
HP3 0BU M6NAS
HP3 0DS G6HIA
HP3 0EA G0VFW
HP3 0EG G8MFH
HP3 0HG G1OEB
HP3 0QN G4VCO
HP3 8EQ G8OYQ
HP3 8JL 2E0HNC
HP3 8NN G1JHD
HP3 8NW G8HZQ
HP3 8PR G6EAX
HP3 8QS G8LPC
HP3 8QT M7CIN
HP3 8RG M6VWR
HP3 8SH 2E0NDR
HP3 8SH G1BSZ
HP3 8SH G6EXX
HP3 8SH M0NGL
HP3 8SH M3NDR
HP3 9GS G6SVL
HP3 9HB G7JHM
HP3 9HS M3YYE
HP3 9JA M0SCS
HP3 9JA M1SCS
HP3 9JJ M6HTF
HP3 9JY G4MSW
HP3 9LW M6JFA
HP3 9NG 2E0BVU
HP3 9NG M0GPE
HP3 9NG M3VUX
HP3 9PD G0GVN
HP3 9PW G2KG
HP3 9UB M6PAG
HP3 9UE G1KOG
HP3 9UQ G0EVS
HP3 9WU G6HGE
HP4 1FG G1ABW
HP4 1HF 2E0CIJ
HP4 1HF M0HDT
HP4 1HF M6AZV
HP4 1LR M7EFR
HP4 1QG G7WBY
HP4 1QG M7WBY
HP4 2PD G3RPA
HP4 3BS G7VJD
HP4 3BS M0AYG
HP4 3DR M6ZTP
HP4 3JZ 2E0FGW
HP4 3JZ M6FGW
HP4 3PW G1WMN
HP5 1QP G4UXY
HP5 1RL M1BPU
HP5 1RW G0LZG
HP5 1SL M0IWG
HP5 1SQ G8YDE
HP5 1SS G3WPD
HP5 1SS M1BAA
HP5 1TA G3XZG
HP5 2BU G6HPK
HP5 2BU G7PEU
HP5 2EB G0ABP
HP5 2EB G6LBG
HP5 2HF M6MRR
HP5 2HL G8BLB
HP5 2PG M6EWZ
HP5 2QR M6BLF
HP5 2RY G7GLZ
HP5 2UQ G4XRV
HP5 2XL G3VZF
HP5 2XW G1CUZ
HP5 3BT G1GMQ
HP5 3DJ G3VRY
HP5 3JH G6CDV
HP5 3JQ 2E1LGJ
HP5 3RJ G4CYY
HP6 5PY G8ETV
HP6 5RN G6FUT
HP6 5RW G6XSL
HP6 6DR 2E0EJC
HP6 6DR M6ICL
HP6 6HG M0FJS
HP6 6HL G7MMC
HP6 6HP G7EAT
HP6 6JF G1JNG
HP6 6JF M0BTX
HP6 6LD G0NLX
HP6 6PZ M7CIZ
HP6 6QL G0XDI
HP6 6QU M0GUL
HP6 6RJ G4GYN
HP6 6RP M3ZAI
HP6 6RR G4TNU
HP6 6SX M6DJG
HP6 6SX M6DWZ
HP7 0EF G0RTF
HP7 0PL M6HTI
HP7 0PN G4WQB
HP7 0PY G2SH
HP7 0RQ G1HPU
HP7 9AG M6MNS
HP7 9AU G3WLO
HP7 9BN M1DTG
HP7 9DZ G0VRV
HP7 9EP G1KTS
HP7 9EP G7STD
HP8 4AR G3DTX
HP8 4BG M1DYO
HP8 4HQ G3LPY
HP8 4HU M0KHN
HP8 4LG G3PBI
HP9 1AE M0GPY
HP9 1AW G0AFT
HP9 1AW M3MVR
HP9 1AW M6GFY
HP9 1BD G8FIG
HP9 1YB G0BKR
HP9 2AE G7ABQ
HP9 2BA G6DOV
HP9 2LA G3BII
HP9 2QU G3OHX
HP9 2XH G4LHE
HP9 2XP G3BEX
HP9 2YJ M6LGV
HR1 1AH M6JTA
HR1 1BY G8OHH
HR1 1HE 2E0DQZ
HR1 1HE M6DUW
HR1 1HQ G7KVT
HR1 1HT G4BJD
HR1 1NH G1HWP
HR1 1NT M6WRS
HR1 1QL G6KWZ
HR1 1QX M1EIR
HR1 1RB G3WRA
HR1 1SS M1CAR
HR1 1TJ 2E1GFW
HR1 1XQ M5RJC
HR1 1XY G8IVO
HR1 2HB G4OJP
HR1 2QG M6XRF
HR1 2RQ G3LZM
HR1 2TS M7LKZ
HR1 3BJ G1AYP
HR1 3BJ M3KAL
HR1 3DF G0SVY
HR1 3DF M3SVY
HR1 3DR G3RDC
HR1 3DR G3YQC
HR1 3DU G7JUN
HR1 3ED G3NDM
HR1 3EH G7LSG
HR1 3EP G0SBA
HR1 3ET G8RAC
HR1 3LP G3IXZ
HR1 3LR G1JWD
HR1 3QY M7GTI
HR1 3RF G6PXQ
HR1 4AF G0PMS
HR1 4AS M6JVO
HR1 4BT M0MVL
HR1 4DL G6TID
HR1 4DQ M6PFA
HR1 4EN M3UKO
HR1 4EU M1BDH
HR1 4EU M3BDH
HR1 4JP G3ESY
HR1 4LR G7LVG
HR1 4PR G8XMS
HR1 4RQ G1YBG
HR1 4UL G0LJP
HR1-1DG G0JWJ
HR1-1DG G3YDD
HR2 0AH 2E0KSX
HR2 0AH M0SWV
HR2 0AH M6KSX
HR2 0AH M6NLQ
HR2 0AH M7OLO
HR2 0EQ 2E0DJH
HR2 0EQ M0RNI
HR2 0EQ M6DAW
HR2 0HP G4ASR
HR2 0JA G0NMC
HR2 0NX 2E0YAV
HR2 0NX 2E0YBL
HR2 0NX M0YAV
HR2 0NX M6WTG
HR2 0NX M6YAT
HR2 0QG G0JCC
HR2 0QG G7KTP
HR2 0QR M1FDK
HR2 0SW M0DRL
HR2 6NG G8UVU
HR2 6RU G1YFC
HR2 6RY M6XME
HR2 7DF G1YBB
HR2 7EG G1LSX
HR2 7FH M6JKM
HR2 7JU 2E0HDI
HR2 7JU M7ANH
HR2 7LJ M0JDP
HR2 7LS G0DYM
HR2 7LU G4MBJ
HR2 7NT 2E1MIB
HR2 7UG G1DYQ
HR2 7UG G7BTI
HR2 7YL M6HOF
HR2 7ZD M1EKB
HR2 8AR M7JFR
HR2 8AX 2E0IKG
HR2 8AX M7RHB
HR2 8HJ G0SYT
HR2 8JT G4OGW
HR2 8LU G4VSO
HR2 8LU M0MUU
HR2 8NA G0VKC
HR2 8PU 2E0AFP
HR2 9AN G7RLO
HR2 9BE G4XTF
HR2 9EW M3VHB
HR2 9HD G4ZSY
HR2 9LT M6DZR
HR2 9QN G0DHB
HR2 9RH G1RLF
HR2 9RT M6IDD
HR26LJ G0VHT
HR3 5DW G0RHK
HR3 5EG 2E0OCY
HR3 5EG M7GIB
HR3 5JT M0ECX
HR3 5JT M6HRI
HR3 5ND M0TUI
HR3 5NF G7EAA
HR3 5RX G4CZK
HR3 6BY G0IAH
HR3 6BY G1UMS
HR3 6HT G7IFR
HR3 6LF G3IUC
HR3 6LX M6LBV
HR4 0JY G1DRW
HR4 0LP G4LNZ
HR4 0NE M6BDI
HR4 0PN G4MET
HR4 0RU M6OAW
HR4 0SR 2E0PTA
HR4 0SR M3PTA
HR4 7AZ G0RQF
HR4 7EH G6PFX
HR4 7PN G3YYC
HR4 7QP M0BHR
HR4 7RW G3NFP
HR4 7SW G4GOG
HR4 8AJ G4FAD
HR4 8AN 2E0OKG
HR4 8AN M6DVF
HR4 8FJ M1AGE
HR4 8NN G4ZXQ
HR4 8NT 2E1HOF
HR4 8NT G6TVC
HR4 8QG G4RGB
HR4 8RQ G7HII
HR4 8SW G0UVE
HR4 8TH G4RCB
HR4 9HP G4WZT
HR4 9LA G0AAG
HR4 9LB M7XLA
HR4 9NE G4HZT
HR4 9NF G4DTB
HR4 9NJ G6DXC
HR4 9PF G7BLL
HR4 9QZ M7PCF
HR4 9TJ G3WAG
HR4 9TJ G6XPY
HR4 9TJ G8FFA
HR4 9XF G0OGX
HR5 3BA 2E0EVF
HR5 3BA M0WJA
HR5 3BA M3EVF
HR5 3DH M7VCH
HR5 3DJ G4SOG
HR5 3DJ G5PAT
HR5 3EG M6VPS
HR5 3HZ G3ZUS
HR5 3JS M6FTI
HR6 0AW 2E0YHF
HR6 0AW M0WYP
HR6 0AW M7BVC
HR6 0AW M7JMF
HR6 0AZ G4KOK
HR6 0BG G4EBL
HR6 0EF G7UUN
HR6 0HZ M6YTL
HR6 0JL M3HCP
HR6 0JZ G7AZC
HR6 0NG G3VST
HR6 0PF G8BPN
HR6 0SS G1JKN
HR6 8BJ G8GVO
HR6 8ER M6XAO
HR6 8EW M6NZM
HR6 8HE G4YPF
HR6 8HT G6OZU
HR6 8HT M0VDO
HR6 8HT M3AZK
HR6 8HT M7VDO
HR6 8NF M7BOJ
HR6 8QQ M0PSK
HR6 8RE M0XGA
HR6 8RX 2E0JRS
HR6 8RX G6UQI
HR6 8RX M6GFA
HR6 8RX M6YRS
HR6 8SA M6AUF
HR6 8SD G0BRH
HR6 8SS G0JRV
HR6 8TQ G6GQG
HR6 9AY G8TMH
HR6 9BG 2E0FIK
HR6 9BG M7AJH
HR6 9BN G0ARF
HR6 9BZ M6GLZ
HR6 9BZ M6IHZ
HR6 9DN G4ASP
HR6 9EE G6FYE
HR6 9HB G4IIA
HR6 9HB G8NOD
HR6 9je G0ODG
HR6 9JQ G4VOY
HR6 9QN G0OEA
HR6 9QR G4BOF
HR6 9UF G3XJP
HR6 9UU M6XLR
HR6 9UU M7XAM
HR68SA 2E0RYR
HR68SA M6RYR
HR69SY M7RSD
HR7 4BD G4CPL
HR7 4DZ G4ZWY
HR7 4ES M1BPK
HR7 4EY 2E0TOP
HR7 4LB G1MUQ
HR7 4LG M6IIB
HR7 4LY G3OGK
HR7 4UL G4FHQ
HR8 1BE G4FTX
HR8 1HG G3WPN
HR8 1ND G0TWT
HR8 1NX G3VZR
HR8 1PP G0WHN
HR8 1RF M7CJL
HR8 1SG 2E0JAO
HR8 1SU G1JHY
HR8 1SU G1JHZ
HR8 2EG G0GFR
HR8 2FE G4PHX
HR8 2FR G8AVB
HR8 2JS 2E0KTQ
HR8 2JS M7KAW
HR8 2QB G8ZVZ
HR8 2UU G8EZE
HR8 2XD G1GCJ
HR9 5AU 2E0CUW
HR9 5AU M0HNK
HR9 5AU M6XRT
HR9 5BD M6KYP
HR9 5BL G1ZND
HR9 5BL G8RZS
HR9 5JL G4EFH
HR9 5JY G3DYO
HR9 5PN G0FGS
HR9 5RL G1OKP
HR9 5ST G4JNZ
HR9 5TA G6XLR
HR9 5TE M0GZF
HR9 6DH G0TRI
HR9 6DQ M7IVA
HR9 6GJ G4AWF
HR9 7AS M6PQI
HR9 7BE G0IMV
HR9 7BE G6TSL
HR9 7BG M6JVZ
HR9 7DN G7RVI
HR9 7DR M1AHN
HR9 7GP G0TME
HR9 7HR M6EQB
HR9 7HT G0RIJ
HR9 7JH G0GQI
HR9 7JP M0HZE
HR9 7JR G3EFL
HR9 7NF G7RIU
HR9 7PW G8PFR
HR9 7QW 2E0LTX
HR9 7QW M0RPI
HR9 7QW M6EQN
HR9 7SS M7PJK
HR9 7TL G8CMU
HR9 7TW 2E0WYE
HR9 7TW G7AEG
HR9 7TW M0TAY
HR9 7TW M1EJJ
HR9 7TW M6TAY
HR9 7UF G0EHT
HS1 2DS GM0DRU
HS1 2DS GM6BGJ
HS1 2JN GM4PTQ
HS1 2NP GM0IPW
HS1 2PG GM7JED
HS1 2QN MM3IIT
HS1 2TL 2M0APB
HS1 2UB MM0JKH
HS1 2UP MM6CTQ
HS1 2UR MM6BIO
HS12SX 2M0HRP
HS12SX MM7BTX
HS2 0AA MM0KGK
HS2 0BG GM7FAS
HS2 0BP MM7SWO
HS2 0DD GM6TYX
HS2 0DZ GM0LZE
HS2 0EZ GM4BMS
HS2 0HD 2M1EOV
HS2 0HD GM0HMM
HS2 0HE GM0WZO
HS2 0HE MM0BPV
HS2 0HU GM0NHT
HS2 0LP MM6BIP
HS2 0LR MM6OEC
HS2 0PD MM0TYR
HS2 0PL MM0KNN
HS2 0QB GM1MQA
HS2 0QB GM4KGK
HS2 0QW GM0FSY
HS2 0QX MM0IOL
HS2 0QX MM1RMS
HS2 0QX MM3SYB
HS2 0QX MM7MKV
HS2 0RT MM0TSB
HS2 0SW 2M0AKI
HS2 0SW GM7PBB
HS2 0SW MM3OHD
HS2 0XF GM7JFN
HS2 9AU GM7KUN
HS2 9DF MM7CMB
HS2 9DH 2M0FNI
HS2 9HR MM0UIG
HS2 9JA GM6VHA
hs2 9jt MM6ELW
HS2 9PJ 2M0DNM
HS2 9PJ 2M0WNB
HS2 9PN 2M0OSK
HS2 9PN MM3OSK
HS2 9QE GM0TVT
HS20PB MM6JZW
Hs29aj MM7ECV
HS3 3JA MM3WVP
HS5 6BP GM1MLY
HS6 5EU 2M1LPT
HS6 5EU GM0HBF
HS6 5EU MM0IRC
HS6 5HD 2M0RBQ
HS6 5HD GM8SAU
HS6 5HD MM0SIA
HS6 5HD MM6RMT
HS6 5HR GM7VEC
HS8 5SS MM6AVH
HS8 5SY GM7DMN
HS8 5TE MM3BCA
HS8 5TS 2M0TNM
HS8 5TU MM0CWJ
HS8 5TU MM6TEW
HS8 5TX MM6SML
HS9 5XU MM1FHZ
HS9 5XX GM0OGN
HS9 5XX MM3BRR
HS9 5YD MM0IOB
HS9 5ZE GM6SEL
HU1 2AJ M1EJL
HU1 2EP M0GPN
HU1 2EP M3LOY
HU10 6AJ G8DCD
HU10 6AR G1RVH
HU10 6AW G3TEU
HU10 6BJ G0KLG
HU10 6DD G4NJB
HU10 6EU M6MKK
HU10 6HB M6UYP
HU10 6HS G4LNR
HU10 6LW 2E1AGO
HU10 6SA G0ULN
HU10 6SF G0UGA
HU10 6SG M0DGU
HU10 6ST G0VQO
HU10 7AF 2E0KVK
HU10 7AF M0KVK
HU10 7AF M6EYR
HU10 7AF M6OGK
HU10 7DT G4CGG
HU10 7DT G4EKT
HU10 7HD 2E0OKP
HU10 7JE G4AQA
HU10 7JJ G8JIU
HU10 7JN 2E0FZM
HU10 7JN M3FZM
HU10 7JU G8NIE
HU10 7TL G4LBC
HU10 7UX G7PER
HU106ST G4BYG
HU106ST M0ARC
HU11 4AX G0GBW
HU11 4BN G4FEJ
HU11 4BN G6ZEN
HU11 4DJ M6NIX
HU11 4DN 2E0EPB
HU11 4DN M6HRQ
HU11 4ER G7CVA
HU11 4HB 2E0FYW
HU11 4HB G0AJX
HU11 4HB M0JJI
HU11 4HB M7HUL
HU11 4NR G5GX
HU11 4PR M3KUH
HU11 4PY 2E0GZP
HU11 4PY M7RKS
HU11 4QH G6JFU
HU11 4RA M0BTN
HU11 4RN G4LZJ
HU11 4RN G7JZY
HU11 4RY G1UCZ
HU11 4SD G0NAA
HU11 4UW G6ZDV
HU11 4XB M6GGD
HU11 4XF M0IOK
HU11 5BH G4YTV
HU11 5BH G6EKT
HU11 5BW 2E0OTR
HU11 5BW M6OTR
HU11 5QP M0VEY
HU11 5RN 2E0HYC
HU11 5RN 2E0RIO
HU11 5RN G3BOK
HU11 5RN M0HYC
HU11 5RN M3UMR
Hu115fb M1EHB
HU12 0DP 2E0THS
HU12 0DP G0THS
HU12 0DX M1CAV
HU12 0HH G4PMJ
HU12 0JF M3XSV
HU12 0JU G0UMK
HU12 0NE G3ULD
HU12 0PJ G4KFA
HU12 0PT G1RON
HU12 0QE G4PYW
HU12 0QG M7ERU
HU12 0RY G3UIF
HU12 0SF G7ONR
HU12 0TE G6AWP
HU12 0TH M6PJH
HU12 0TJ M1DCE
HU12 0US 2E0EYC
HU12 0US M0PAY
HU12 0UX M1COV
HU12 8BG M0MFP
HU12 8ET G6AXC
HU12 8FP M6FPF
HU12 8GF G6ZJI
HU12 8GT M0MGF
HU12 8HA M0OZO
HU12 8HQ G1BGO
HU12 8LB G7WDM
HU12 8LH G0VAY
HU12 8LX G0NYK
HU12 8NH G7SCT
HU12 8NH M0SCT
HU12 8NT G4CJJ
HU12 8QH G0LUP
HU12 8QH G4ZJC
HU12 8TZ 2E0XKX
HU12 8TZ M0SLY
HU12 8TZ M6HUL
HU12 8UB G7MFO
HU12 8XF G1EFO
HU12 9AF G7RPW
HU12 9BP M1VRC
HU12 9GB 2E1TAP
HU12 9HP G4LHF
HU12 9JZ 2E0CJA
HU12 9LD M3BJJ
HU12 9LD M3JMJ
HU12 9NG G3YBU
HU12 9QG G1EQU
HU12 9RG G1EQF
HU12 9SZ M6IHA
HU13 0AS G7USB
HU13 0HB 2E0DVW
HU13 0JH G6GKT
HU13 0JP 2E0VDQ
HU13 0JP M0VDQ
HU13 0JW 2E0CIK
HU13 0ND M1CBC
HU13 0NS G6MPK
HU13 0PY G3NPT
HU13 0RL G4JIO
HU13 0RT G0KWE
HU13 0RT M3WJK
HU13 9BP G0CNV
HU13 9HU G0OYX
HU14 3AE M3PLP
HU14 3AN G1EIB
HU14 3BG G8FEK
HU14 3BX M6PBR
HU14 3DW G3YNO
HU14 3DW G3YQA
HU14 3HZ G0GVT
HU14 3JS G3RSB
HU14 3NL M0REC
HU14 3PR G7UJY
HU14 3RH G4XWA
HU15 1BA 2E0XYM
HU15 1BA M3XYM
HU15 1BL G0SWL
HU15 1BL G4VHM
HU15 1DA G7CLY
HU15 1DD 2E0EXK
HU15 1DD M6LQR
HU15 1EN G6KIA
HU15 1EN G6UQA
HU15 1FF G3VMV
HU15 1HY G4FGO
HU15 1JT M3HZW
HU15 1LA 2E0FSS
HU15 1LA G6PIB
HU15 1LA M1FCW
HU15 1LA M7GCW
HU15 1NL G8FDR
HU15 1NL M3UAG
HU15 1PE G8LZG
HU15 1QL G0GDS
HU15 1RE M1ASR
HU15 1RJ 2E0DQP
HU15 1RJ G4DQP
HU15 1RJ M6DQP
HU15 2AL G4MJT
HU15 2AL G4SGU
HU15 2AL G4UVF
HU15 2GT M7ANU
HU15 2HR G6POZ
HU15 2JG G3VLR
HU15 2LT G0WJK
HU15 2NU 2E0DKI
HU15 2NU M0GWH
HU15 2QN M6IHT
HU15 2TU 2E0GDL
HU15 2TY G4YEB
HU15 2UH G0LHV
HU15 2XU G0JNQ
HU15 2ZU G3VDE
HU16 4AL M3PXL
HU16 4DQ M6XPW
HU16 4HN G0PAS
HU16 4NF G0KBP
HU16 4SA G7WGN
HU16 4SD G4EBT
HU16 5EH M0INM
HU16 5JD M0JBA
HU16 5JG G3NOP
HU16 5TF G0SXW
HU16 5TR G8EAH
HU16 5UG G4SEF
HU17 0BG G7TYT
HU17 0DF 2E1FIQ
HU17 0EY 2E0DYP
HU17 0EY M6NYR
HU17 0EY M6WRH
HU17 0HP 2E0JMR
HU17 0HP M0BNR
HU17 0ND G7FLS
HU17 0PA G8SGP
HU17 0PA M3NZG
HU17 0RU G4HBL
HU17 0SA 2E1BDC
HU17 0TH G4EIM
HU17 5NR 2E0SJE
HU17 5NR G7EFV
HU17 5NU G3RMX
HU17 7AB G8NJI
HU17 7BP G4HYD
HU17 7BW M6DIO
HU17 7HD G6UGS
HU17 7HP G4GPY
HU17 7NQ M6NEG
HU17 7NX G0TAS
HU17 7PN G8OEK
HU17 7RJ G7FDD
HU17 8DZ M7JCL
HU17 8FZ G8MEA
HU17 8GE G3MJM
HU17 8HY G4UOZ
HU17 8JL G4ESY
HU17 8LA 2E0EPJ
HU17 8LA 2E0KLI
HU17 8LA M6KLI
HU17 8LL M1EZK
HU17 8PF G0DMP
HU17 8PJ G4CMK
HU17 8PU G3NVX
HU17 8QA G4DEA
HU17 8QP G7UUO
HU17 8SD G3ZRS
HU17 8SD M0GWA
HU17 8WE 2E0USG
HU17 8WE M6NSG
HU17 8XB G6VCF
HU17 8YB M0HVQ
HU17 8YB M6BOP
HU17 9GL G4CQN
HU17 9LH G0KVR
HU17 9PG G6EUF
HU17 9QR 2E0VCP
HU17 9QR M6VCP
HU17 9RH G1AJD
HU17 9RH G1AJE
HU17 9RH G8UAI
hu177he M7RGF
HU18 1AF 2E0FLF
HU18 1AF M0FLF
HU18 1AF M6FLF
HU18 1AL G6GHE
HU18 1BF G0PFD
HU18 1BP G3OGE
HU18 1BX G0PSJ
HU18 1BX G7JZS
HU18 1EF G4FCT
HU18 1EF G8PZX
HU18 1ES G0VET
HU18 1ES G6DKF
HU18 1EU 2E0LKE
HU18 1EU M0LKE
HU18 1EW G4ADE
HU18 1JG G0IVP
HU18 1JU G7KHV
HU18 1LX M0CZR
HU18 1PB G8LJQ
HU18 1QD G1CFA
HU18 1QU 2E0EQK
HU18 1QU M6FFW
HU18 1RE G0JWY
HU18 1TP G8KFK
HU18 1TT G0GXX
HU18 1UA 2E0EQG
HU18 1UA M0KPJ
HU18 1UA M6KPJ
HU18 1XU M6DGP
HU19 2AY 2E0BQR
HU19 2AY M0SGJ
HU19 2AY M3FGG
HU19 2BB 2E0ENW
HU19 2BB M3EYH
HU19 2DB G6ZBT
HU19 2DB M0EGC
HU19 2DB M0ZBT
HU19 2DS M6POP
HU19 2DT G2DPA
HU19 2DT G6YAS
HU19 2DT M3HRT
HU19 2DT M5EXY
HU19 2DW G1KST
HU19 2EW G1OSG
HU19 2EW M0KBH
HU19 2JA M6EUE
HU19 2LN M3ZHV
HU19 2LP 2E0MWW
HU19 2LZ G4SJI
HU19 2LZ G6YMI
HU19 2PB M6HGW
HU19 2QD G3RJM
HU19 2QU M0VES
HU19 2RG 2E0JCL
HU19 2RG M0UTX
HU19 2RG M0VQJ
HU192JA M7AOV
HU2 8JR G8ZFX
HU2 9BL G1WFS
HU20 3UU G1WQL
HU20 3UX G8HUG
HU20 3XH G6KTK
HU3 1NS M0AYO
HU3 1SA G0RUF
HU3 1SA G0UBY
HU3 1SA G6CTE
HU3 1YD M6BYP
HU3 2LT M0AAR
hu3 2qt M6NRV
HU3 2SG M6SDA
HU3 2TL 2E0LKH
HU3 2TT M6CLF
HU3 2TZ G0IRJ
HU3 3PR 2E0FOH
HU3 3PR M7OSS
HU3 5AS 2E0CNC
HU3 5AS M0GOC
HU3 5AS M6TKW
HU3 5LW M6GCP
HU3 5PP G1FPC
HU3 5PS M7GKR
HU3 6PQ M6OPE
HU3 6QY G0SWO
HU3 6QY G3AMW
HU3 6QY G3JDY
HU3 6QY G8GBY
HU3 6RD M3OPX
HU3 6SL M3IDK
HU3 6SL M3PHR
HU35AE 2E0JOV
hu35ae M7PBO
HU35LX M7HOM
HU4 6AP M1ADN
HU4 6DW 2E0NMA
HU4 6DW M6ITP
HU4 6EF 2E0LDR
HU4 6EF M0LDR
HU4 6EF M3LRZ
HU4 6EW G1LAN
HU4 6LJ M0BXB
HU4 6LU M0BEO
HU4 6NG G6DLT
HU4 6PL G0IER
HU4 6PL M0HND
HU4 6RD M3UKH
HU4 6RW G1KNZ
HU4 6SN G1MZP
HU4 6SQ G1PGX
HU4 6ST M7PNO
HU4 6TL G3TKA
HU4 6UG G1DQD
HU4 6XJ G6HVE
HU4 7AH G0POQ
HU4 7AR M0GOY
HU4 7HB G4ZPQ
HU4 7JT 2E0HKZ
HU4 7JT M7EGY
HU4 7PL G0TRH
HU4 7QE M6ACF
HU4 7QG G4WCD
HU4 7QH G6VVZ
HU4 7rh M3UNI
HU4 7SW G3GJA
HU4 7SW G8EQZ
HU4 7SW M0BMB
HU5 1LX M6YWG
HU5 1ND G0OVT
HU5 2AB M3GIK
HU5 2BE M6LDG
HU5 2TA M6PAE
HU5 2TP M0KPE
HU5 3DW G3KDU
HU5 3DY M7JGB
HU5 3ER M0JDT
HU5 3HH M6CKY
HU5 3JY G0UNC
HU5 3NE M7WYD
HU5 3PF M3PBR
HU5 3SQ G1PCQ
HU5 3TW M6APA
HU5 3TW M6EKA
HU5 4AD M3KIU
HU5 4AG M1EIE
HU5 4ED G1PLU
HU5 4ED G4HJD
HU5 4EQ M6ZIY
HU5 4HH M6JER
HU5 4HN G1MCI
HU5 4HN G1SQW
HU5 4JZ 2E0BWL
HU5 4JZ M0RWL
HU5 4JZ M3USZ
HU5 4LP G3RDP
HU5 4PF G1EYY
HU5 5DJ G6VWF
HU5 5ES 2E0BVE
HU5 5HS G4CBL
HU5 5HW G0TGX
HU5 5HZ G0GXZ
HU5 5JW G7JGY
HU5 5LE G0VNH
HU5 5LN 2E0ZRB
HU5 5LN M3ZRB
HU5 5NH 2E0SCA
HU5 5NX G6ABG
HU5 5QG M6ACX
HU5 5td M6WRD
hu5 5xx 2E0SHW
HU5 5YH G7BAC
HU5 5YY G4COT
HU55JD 2E0EHN
HU6 7AG M6UUA
HU6 7AS G0CHP
HU6 7BA G4ERZ
HU6 7BW G1IUF
HU6 7DD G1LSZ
HU6 7DD M0TTL
HU6 7DZ G3ZZP
HU6 7HJ M6INA
HU6 7JT G0GXF
HU6 7LW M6HSY
HU6 7QB 2E0VCY
HU6 7QB M3VCY
HU6 7XE G4MPL
HU6 8DB M6JMR
HU6 8DG M6BZM
HU6 8DG M6BZN
HU6 8lh M6HJE
HU6 8NX M7TJC
HU6 8SA G3PQY
HU6 8SB G4AUQ
HU6 8SB G6LNV
HU6 8SB M0HFC
HU6 8SB M1DWT
HU6 9EB G0LPT
HU6 9ER G0RSB
HU6 9ER M3SBB
HU6 9EU G8VHG
HU6 9EZ G1TDO
HU6 9EZ M3TYS
HU6 9HA M3GZU
HU6 9HG M0SRN
HU6 9JH M3VPH
HU6 9LS 2E0JRL
HU6 9PH M7BBY
HU6 9RA G8ZFD
HU6 9RW 2E0TWZ
HU6 9SA G1TDN
HU6 9SA M6LEC
HU67DT M6NFT
HU7 0DU G0UKS
HU7 0DU G0VTD
HU7 3AF G0OYQ
HU7 3AL G1XGE
HU7 3FA G4KXU
HU7 3GS M7SYB
HU7 4AL M3MPT
HU7 4AR M7YZZ
HU7 4BX 2E0NYE
HU7 4HH 2E0DXW
HU7 4HH 2E0NSQ
HU7 4HH M0MRX
HU7 4HH M1BLO
HU7 4HH M3NSQ
HU7 4HH M6GHS
HU7 4HN G0WEA
HU7 4PQ G0BAP
HU7 4PY G0EPA
HU7 4RF G1ARD
HU7 4RP M3VGP
HU7 4RU G7POW
HU7 4ST 2E0DIX
HU7 4ST 2E0ELO
HU7 4ST 2E0HAL
HU7 4ST M3XHY
HU7 4ST M3ZZD
HU7 4ST M3ZZW
HU7 4UY G8WVB
HU7 4YE G1FQI
HU7 4YS 2E0EPF
HU7 4YS M6JYT
HU7 4YZ G6KRJ
HU7 4ZA M0BOM
HU7 5AP M6LRR
HU7 5AQ 2E0OAH
HU7 5AQ M6OZY
HU7 5AS G7UCZ
HU7 5BU G0PTL
HU7 5BX M0KUH
HU7 5BX M0RKK
HU7 5BX M7RPK
HU7 5DH 2E0GRH
HU7 5DH M6RRH
HU7 5FB M3RIE
HU7 5XD 2E0VTC
HU7 5XD M0PLR
HU7 5XD M3XMQ
HU7 5YY G0VVP
HU7 6DE G0MKZ
HU7 6EE G0WDX
HU7 6EL G6WVR
Hu74xz 2E0JYM
HU8 0AN M1DRK
HU8 0HB 2E0OYR
HU8 0HB M0OYR
HU8 0HB M7OYR
HU8 0RU 2E0FIY
HU8 0ST G0WFF
HU8 7RY 2E0BRQ
HU8 7RY M3OOQ
HU8 7RY M3SHW
HU8 7TJ G6WQH
Hu8 8lq M6NYW
HU8 8NA M7EYP
HU8 8QR M3KLB
HU8 8TQ M3GWW
HU8 8TQ M3HME
HU8 9BN G1ADE
HU8 9DG G0UTX
HU8 9EW M6TDK
HU8 9LY M3MWM
HU8 9NA 2E0LTC
HU8 9NA M6RWX
HU8 9RY 2E0FLA
HU8 9RY M6YLR
HU9 1EJ G1CFB
HU9 1HR 2E0IDL
HU9 1HR M7FQZ
HU9 1JB M3LSL
HU9 1ST 2E0GMW
HU9 2AS 2E0HKR
HU9 2PF 2E1XXX
HU9 2QR G7IWE
HU9 2TE G0TUJ
HU9 3JQ G0PSL
HU9 3JU G7UGW
HU9 3LB G0NAU
HU9 3PF M6BIG
HU9 3PN M0BYI
HU9 3PR M0GLV
HU9 3RG M1DHJ
HU9 3RH M7ACE
HU9 4BE M3MPM
HU9 4DR 2E0KCW
HU9 4DR M0PKE
HU9 4DR M6BWP
HU9 4HH M3LUD
HU9 4PG 2E0UVZ
HU9 4PG M0HBO
HU9 4PR 2E0LZQ
HU9 4PR M0XLH
HU9 4PR M6LZQ
HU9 4QR 2E0RWP
HU9 4UE G1ZSV
HU9 5JZ 2E0HDX
HU9 5LH G0PCD
HX1 2BJ G8WFP
HX1 2EP M1RIG
HX1 3EA G6FPF
HX1 3LE M0CRG
HX1 3LE M0GQB
HX1 4QG M3BXY
HX1 4TA G0CVC
HX1 5PU 2E0VRD
HX1 5PU M6HXI
HX1 5PU M6VRD
HX2 0HT M3BXX
HX2 0NP G3TAY

Postcode

HX2 0PA G7ELX
HX2 0PJ G0OZP
HX2 0PL G1RZZ
HX2 0QP M3DWA
HX2 0RB G8EXN
HX2 0RH G7OMF
HX2 0RL G4EHD
HX2 6BG 2E0WSG
HX2 6BG G2UG
HX2 6BG M6XUT
HX2 6EX G8RKX
HX2 6HQ M6DJN
HX2 6UX G6CNL
HX2 7HG G0INK
HX2 7HP G8SDE
HX2 7PN G6YGV
HX2 7RB G7TDN
HX2 7SG M3HTE
HX2 7TX G4NSH
HX2 8AA G3UI
HX2 8DL G1GWX
HX2 8DL G1VVB
HX2 8RE M1CMR
HX2 8RL M6BKN
HX2 8SG G8SJA
HX2 8SN M6VGB
HX2 9AZ G4XYS
HX2 9EF M6RRR
HX2 9JD G6GGV
HX2 9JH G0JAQ
HX2 9JJ M7CHD
HX2 9JQ G2SU
HX2 9JQ G3TQA
HX2 9LY G8XZZ
HX2 9PQ G0CBI
HX2 9PZ G6AVN
HX2 9SJ G0IJI
HX2 9SQ M6DKS
HX3 0AG G6COG
HX3 0AH M0HVI
HX3 0AL G4OYZ
HX3 0AL M3OOX
HX3 0AL M3OOY
HX3 0AL M3OPC
HX3 0AL M3OPD
HX3 0AS 2E0TGO
HX3 0AS M0TGE
HX3 0AS M7GEN
HX3 0BD G7SWZ
HX3 0DL M6EMH
HX3 0HA M3IDA
HX3 0JQ 2E1SKR
HX3 0SR G0SPX
HX3 0SU G7IHX
HX3 0TH M0RWD
HX3 5FB M6VCU
HX3 5LU 2E1BOO
HX3 5LU 2E1BVJ
HX3 5LU G3UGF
HX3 5NL M0DEJ
HX3 5NQ M7TFG
HX3 5PA 2E1AHK
HX3 5PE G1IYA
HX3 5QF G8REO
HX3 5QG M6HFX
HX3 5QU G1WHU
HX3 5RH M7SFD
HX3 5SZ G1HCU
HX3 6LD 2E0HTW
HX3 6LD M7BOO
HX3 6PL M0PDW
HX3 7BT G1NDQ
HX3 7EB M6AHS
HX3 7EP G6MDC
HX3 7LB G8UTW
HX3 7NA G0PMU
HX3 7NE G8ZXZ
HX3 7NH G0HRJ
HX3 7NH M3KJK
HX3 7PW G4KEX
HX3 7QY M0HFX
HX3 7QY M6LVW
HX3 7RN 2E0BAB
HX3 7RN M0TKD
HX3 8HB G7FZJ
HX3 8NJ M3PNH
HX3 8QF G4BZS
HX3 8TJ 2E0UPU
HX3 8TJ G1ILO
HX3 8TJ M0UPU
HX3 8TJ M3JJU
HX3 8TJ M6UPU
HX3 8XP M6FYN
HX3 9BH M6FAX
HX3 9DR M6JBL
HX3 9DT G0GVB
HX3 9EE G6XTT
HX3 9PS G7DWY
HX3 9PS M0INB
HX3 9QZ G8NQN
HX3 9SE M3YEM
HX4 0BW 2E1HUE
HX4 0DE M6PIY
HX4 0EW G8HTB
HX4 0HD G3PSB
HX4 8HT M6JJE
HX4 8HX M7FMM
HX4 8JF G6APB
HX4 8NU G3YGZ
HX4 8PA M3MQY
HX4 8QF G1MWT
HX4 9AS G4LIX
HX4 9DN M3JQT
HX4 9HZ G4MSP
HX5 0AJ M7AXF
HX5 0BB G4FMM

HX5 0BB M3LAG
HX5 0DR 2E0MWT
HX5 0DR M7FDW
HX5 0EE M0XRL
HX5 0HR M7OSH
HX5 0JH 2E0BJQ
HX5 0LA 2E0HQJ
HX5 0LS G6XNJ
HX5 0PF 2E0PHU
HX5 0QA G3YCJ
HX5 9AY M6IJT
HX5 9JF 2E0ZXV
HX6 1AD G6DQU
HX6 1AH M3SNR
HX6 1BX G8UTH
HX6 1EE M0UGX
HX6 1EE M6UGX
HX6 1lb 2E0PEC
HX6 1lb M6PEC
HX6 1NL G4AES
HX6 1NS G0DUN
HX6 2EE M6JRC
HX6 2EF 2E1FBY
HX6 2QP M3GSM
HX6 2RP G1MBM
HX6 2RU G0FOB
HX6 2SY M6JNC
HX6 3HZ M6GSO
HX6 4AG G0WFG
HX6 4AG G7NSN
HX6 4LQ G8EWN
HX6 4NB M6GJB
HX6 4NU M3JQS
HX6 4PA M1EYG
HX6 4RZ G6AUC
HX7 5ER M6GZL
HX7 5HS M6DTP
HX7 7AL 2E0KPT
HX7 7ED G0TOB
HX7 7HD G0TVB
HX7 7HD G6TRG
HX7 7JA G4EFX
HX7 7JP 2E0JCK
HX7 7JP M0JCK
HX7 7NX G6XPZ
HX7 7PB G7CLX
HX7 7PG G3MLS
HX7 7PH G8BBZ
HX7 8SJ 2E0FUJ
HX7 8SJ M6NWL
IG1 2ER G6ZQJ
IG1 2UF G0IQK
IG1 4RY M0NSP
IG1 4RY M6FDW
IG10 1BN M7OPH
IG10 1HT 2E0DQL
IG10 1HT M6VTG
IG10 1SB G0LWM
IG10 1SB G7EEE
IG10 1TS G0SGX
IG10 2AD G4WTQ
IG10 2AJ G4IMU
IG10 2LT M6GJH
IG10 2QL 2E0ZAU
IG10 2QL M6ZAX
IG10 2RE G0TOC
IG10 2RE G7KJV
IG10 2SA G3NQT
IG10 3BY M1EZC
IG10 3EP G0PFY
IG10 3JT G1DJI
IG10 3PL G1IZA
IG10 3PR G1IVP
IG10 3QY G7VKB
IG11 0DS G6MPJ
IG11 0LD M0BDQ
IG11 0NL G0POY
IG11 7EB M3ZBS
IG11 7QE M3WXW
IG11 7JW G1UAL
IG11 9DD G3OHS
IG11 9EE G0BIW
IG11 9HS M6PMB
IG11 9NY M1EEY
IG11 9PS M3DVH
IG11 9PS M3EKY
IG11 9XB G0OVE
IG11 9XB G6JMJ
IG2 6AS G3CAZ
IG2 6EQ G4PQW
IG2 6QA G7EMH
IG2 6QY G6RAH
IG2 6YG G1OYM
IG2 7DL G6EEF
IG2 7NQ G4JRD
IG2 7SF G7JUZ
IG2 7SF M3JUZ
IG3 8NN G0VKH
IG3 8XE G7VBF
IG3 9TB G0MEW
IG3 9XJ M0JLI
IG4 5AE G3ALK
IG4 5HA M6KFH
IG4 5HN G3WPR
IG4 5JX M0CTF
IG5 0DL M1EMX
IG5 0HN 2E0HTI
IG5 0HP G1ICK
IG5 0NP G8JFL
IG5 0RZ G6XSB
IG5 0TG G8VBK
IG5 0XN G8YAE
IG6 1BJ G0ATP
IG6 1PJ G4YUF
IG6 2AZ M6ISE
IG6 2BN M0OIA

IG6 2NJ 2E0XBA
IG6 2NJ M0XBA
IG6 2NJ M6JSI
IG6 2PH G4GQL
IG6 2QE 2E0DDH
IG6 2QE M0MBD
IG6 2QE M6DLH
IG6 2QU G8HST
IG6 2QU G8PRJ
IG6 3AH G8BUF
IG6 3SR M1EDO
IG6 3TF G1HEQ
IG7 4DG 2E0CFT
IG7 4EA M1CPD
IG7 4HQ 2E0EER
IG7 4HQ M6AGD
IG7 5AU G2CD
IG7 5ED G4GMN
IG7 5HJ M7GWJ
IG7 5HZ G4NZB
IG7 5NX M0XHS
IG7 5QZ G7VGJ
IG7 6AD G8LGU
IG8 7LS 2E0DOW
IG8 7LS M0CEO
IG8 7LS M6LFE
IG8 7QU G4AJG
IG8 8DW 2E0RMM
IG8 8DW M6RML
IG8 8JN 2E1ICT
IG8 9AA G6SWT
IG8 9AA G8FRH
IG8 9HY G4BNB
IG8 9QZ M1ETW
IG9 5QE G4BMU
IG9 5QF 2E0EFO
IG9 5RU G8MCR
IG9 5TZ G4YOA
IG9 6AQ G3VGR
IG9 6BY G0LWI
IG9 6JB M7MOE
IM1 3DP MD6RSV
IM2 1JE 2D0VMN
IM2 1JE 2D0WNY
IM2 1JE MD0LRK
IM2 1JE MD3VMN
IM2 1JE MD7KSL
IM2 1JT MD3JTT
IM2 1NX 2D0EYK
IM2 1NX MD0NSS
IM2 1NX MD6NSS
IM2 1QX MD3LEP
IM2 2AD GD1JNB
IM2 2AR MD0IOM
IM2 2EU MD6KBW
IM2 2LA MD3MAN
IM2 4AH MD0RKI
IM2 4AQ 2D0WFK
IM2 4AQ MD0WRK
IM2 4AQ MD6WFK
IM2 4AQ MD6ZKK
IM2 4AR 2D0IMN
IM2 4AR MD3FMN
IM2 4AT MD6NRI
IM2 4PE GD4FMB
IM2 4PE GD4ZAB
IM2 5ET MD7KNE
IM2 6AG GD0KQE
IM2 6EY 2D0OMN
IM3 2JF GD6JHP
IM3 3BU GD4RVQ
IM3 3HY GD0LQL
IM3 4AT GD0OUD
IM3 4AT GD7HTG
IM3 4LE MD6TDU
IM4 1BB GD0TFO
IM4 2DN GD0SFI
IM4 3HE GD4EIP
IM4 4AR MD6IKR
IM4 4AX 2D0PEY
IM4 4AX MD6NZO
IM4 4ES 2D0KTH
IM4 4ES GD4FWQ
IM4 4ES GD7MAN
IM4 4FE GD0IFU
IM4 4LT 2D0NXK
IM4 4LT MD0NKX
IM4 4LT MD7MTM
IM4 4LU MD3LEG
IM4 4ND MD7DYW
IM4 4NZ MD6MGP
IM4 5EA GD4XOD
IM4 7HH MD6LET
IM5 1GH MD7EEO
IM5 1HP GD4MCR
IM5 1JJ GD4NTR
IM5 2AE 2D0WFH
IM5 2AE GD4UHB
IM5 2AE MD6WHG
IM5 3BA GD6TWF
IM6 1AF GD1GHK
IM6 1AF MD6AGF
IM6 1HU MD6KFH
IM7 2EY GD4RGR
IM7 2HB GD1XMA
IM7 3HP GD3RFK
IM7 3HP MD3BZA
IM8 1LJ GD7BOJ
IM8 1LJ MD0BJM
IM8 2EG MD3OIS
IM8 2JD MD7BJL
IM8 2PT GD8BUE
IM8 3DA 2D0JBE
IM8 3NZ 2D0VES
IM8 3NZ MD6ZEE
IM8 3PU MD0FIX

IM8 3UH 2D1GCC
IM9 1BE MD6NUG
IM9 1HP MD3KSN
IM9 1HW 2D0VJK
IM9 1HW MD0RLS
IM9 1HY GD7DUZ
IM9 1HY MD3XDD
IM9 1TQ GD4GSR
IM9 2EU MD1DXW
IM9 2EU MD3DXW
IM9 2EU MD6RAQ
IM9 3AY 2D0JKW
IM9 4EF GD6OXG
IM9 4EP GD6XHG
IM9 4HJ 2D0TSE
IM9 4HJ MD0OOH
IM9 4HJ MD6SNV
IM9 4NB GD4VBA
IM9 4NR 2D0TRL
IM9 4PR MD6IUH
IM9 5ED GD0MWL
IM9 5ED MD3AAI
IM9 5LH 2D0NAU
IM9 5LH MD0YAU
IM9 5LX GD4RAG
IM9 5PR GD1USI
IM9 6DB GD4IZL
IM9 6DB GD8PPU
IM9 6EL GD0VIK
IM9 6EL MD3EEW
IM9 6HA MD6DYM
IM9 6HR GD8EUH
IM9 6LS GD3ZZN
IM9 6ND GD4ZZN
IM9 6QU GD7KAM
IP1 1RS M0IOV
IP1 1RS M0IOZ
IP1 2PH G7SMN
IP1 3QF G3XCO
IP1 3QZ G0UBM
IP1 3SA G7LKY
IP1 3SD M7PUP
IP1 4DP M3HGL
IP1 4JY G1YAB
IP1 4JY G4YUG
IP1 4LD G4LRB
IP1 4LU M3ZBR
IP1 4NZ G8MUF
IP1 5DR G0AKC
IP1 5EA G7OCH
IP1 5HD G1WQY
IP1 5HS G8FTW
IP1 5JX G3YWM
IP1 5LD 2E0GCY
IP1 5LD M6GCK
IP1 5LR G7PLV
IP1 5NJ 2E0AXZ
IP1 6AB 2E0UCB
IP1 6AB G7VLJ
IP1 6AB M7CBX
IP1 6BD G1HGF
IP1 6DU G0SAR
IP1 6DU G8XOR
IP1 6EW G7UQV
IP1 6HP M3UAQ
IP1 6JB G0HJK
IP1 6LB G0JNY
IP1 6PA M6SFT
IP1 6PQ G8VNP
IP1 6PQ M0AKK
IP1 6PS M3YGZ
IP1 6RE M0JAJ
IP1 6RG G4YGV
IP1 6RL G8ZQA
IP1 6RL M3ZQA
IP1 6SS G7FNN
IP10 0DE G4LXX
IP10 0EA M7MLZ
IP10 0EU G4SYG
IP10 0EW G6FRY
IP10 0FN M7ABX
IP10 0JX G4XQD
IP10 0PF G4LYD
IP10 0PF G4RHR
IP10 0PL G4DYH
IP10 0PP M0JSA
IP10 0PP M6ALX
IP10 0PP M6ISJ
IP10 0PZ G8BHC
IP10 0QF G0KBM
IP10 0QU 2E0VAA
IP10 0QU G4DDK
IP10 0QU G8EMY
ip100np G8MHA
IP11 0RG G8BHK
IP11 0UU G3OJ
IP11 0UU G3XIX
IP11 0UY 2E0EDR
IP11 0UY M0LOC
IP11 0UY M6HKB
IP11 0XG 2E0SYB
IP11 0XG M6HWD
IP11 0XL G8MXV
IP11 0YQ M0DNJ
IP11 2BG M0OKB
IP11 2EA M3YKN
IP11 2NS G8EAX
IP11 2NT M0LVW
IP11 2NY G1JRL
IP11 2PN G8BBV
IP11 2PQ G0CFT
IP11 2UH G4ZFR
IP11 2UH G7OPS
IP11 2UL G7TUG
IP11 7EG G3WDE

IP11 7JR M1EJS
IP11 7LG G4FAW
IP11 7NR G0UPD
IP11 7NX 2E0VJK
IP11 7PR G4BTX
IP11 7QW M0HLY
IP11 7RL G7ILA
IP11 7RP G3XFF
IP11 7RR G0PPS
IP11 7RR G0PRU
IP11 7RR G4DNX
IP11 7SP G0DUS
IP11 9AL G7MNS
IP11 9BS 2E0IPX
IP11 9BU G6MKO
IP11 9DE M3BDQ
IP11 9DS M3JFA
IP11 9EE M6BFF
IP11 9HS G4IVC
IP11 9JJ M0HQL
IP11 9JT G0GSL
IP11 9LG G3TRD
IP11 9LR G4FBV
IP11 9LS M6TYZ
IP11 9NN G3ZKZ
IP11 9NP G8NYC
IP11 9PJ G8RRN
IP11 9PJ M6FLX
IP11 9PL G3MVE
IP11 9PS G8JDS
IP11 9SL 2E1CZS
IP11 9SS G6MCG
IP11 9TJ M0MWR
IP11 9TL 2E0HJK
IP11 9TL G8XUL
IP11 9TL M6HJK
IP11 9TS M3MFG
IP11 9TX G0XEG
IP112JW 2E0XQA
IP112JW M7BFA
IP12 1AH G8VZZ
IP12 1BE M0JPS
IP12 1GZ M7CYI
IP12 1HB G8AXO
IP12 1JL 2E1GXU
IP12 1JQ G7JXR
IP12 1JS G0OZR
IP12 1LB 2E1FSR
IP12 1LD G1AGK
IP12 1LQ G0DDZ
IP12 1PE M0BPW
IP12 1RL M6BHY
IP12 1RN G7DWN
IP12 1RW M1ADV
IP12 2AJ G3RHP
IP12 2ED G6FS
IP12 2ED M0BCT
IP12 2GA M6DON
IP12 2GL M3VDO
IP12 2HZ G4AVS
IP12 2JE G4HHA
IP12 2JH 2E0TDN
IP12 2JH M6TCI
IP12 2PL G3ZYP
IP12 2PL G8BZJ
IP12 2QA M6SWD
IP12 2QG G0JFM
IP12 2QG G1YOU
IP12 2QG M7AAZ
IP12 2TE M6AZN
IP12 2TJ M6MWT
IP12 2UD G1SAR
IP12 2UD G3ADZ
IP12 2UD G7CIY
IP12 2UR M6KTO
IP12 2UT G3RTB
IP12 2UW M6WUB
IP12 3DZ M6LMG
IP12 3EP 2E0DBY
IP12 3EP M6GSD
IP12 3JT 2E0ROP
IP12 3JT M0NOC
IP12 3JY G0ILZ
IP12 3JY M6AOL
IP12 3LL G0THW
IP12 3QL M1AGP
IP12 3QU G1XWS
IP12 3TP G0VWE
IP12 4DU 2E0YZC
IP12 4DU M0GQR
IP12 4DU M3YZC
IP12 4EZ M7NES
IP12 4HE 2E0TYL
IP12 4HE M3TYL
IP12 4HR G3PFH
IP12 4HR G7SCR
IP12 4HW G3TLY
IP12 4JP G4FSG
IP12 4JR G4AKW
IP12 4JW G8BRL
IP12 4JZ G4WAG
IP12 4JZ G6MAD
IP12 4NN G7RMQ
IP12 4NY M1CGM
IP12 4NZ G8ONH
IP12 4PT G1DIK
IP12 4PT G3WTS
IP12 4RR M3STQ
IP12 4SL G1MOS
IP12 4ST G4DFD
IP12 4SU G4SYA
IP122JQ M7TJE
IP13 0AT M6BID
IP13 0BP 2E0EVE
IP13 0BP 2E0MIG
IP13 0BP M0SDY

IP13 0BP M3EBZ
IP13 0BP M3KIT
IP13 0BP M3LDC
IP13 0BP M3MIG
IP13 0BP M3VDT
IP13 0ES G4TRE
IP13 0HE G0RZG
IP13 0LN 2E0BEW
IP13 0LN M3IKR
IP13 0LR G3RPB
IP13 0LR G6KQ
IP13 0ND 2E1EJD
IP13 0QS G7ABO
IP13 0RQ 2E0FIR
IP13 0RQ M0MJF
IP13 0RQ M6FIR
IP13 0SL G6AYX
IP13 6DP M0BVT
IP13 6EB G3RVC
IP13 6ES G8VCU
IP13 6JF G8XYQ
IP13 6LA 2E0SUF
IP13 6LA M0SUF
IP13 6ND M6ADY
IP13 6ND M6KYK
IP13 6PL G0VQS
IP13 6QR 2E0XFX
IP13 6QR M7JPX
IP13 6SE 2E0LFR
IP13 6SU G0AOY
IP13 6TE M6MAG
IP13 6TH 2E0CQQ
IP13 6TH M0IAH
IP13 6UP 2E0CQR
IP13 6UP M6BTT
IP13 7AW M6EGZ
IP13 7JP M6CCF
IP13 7NU M6EBX
IP13 7PP G4RSD
IP13 7RT M6NBW
IP13 8BB 2E0STQ
IP13 8BB M0LIE
IP13 8BB M6BZU
IP13 8DT G7JCF
IP13 8EB M7CHI
IP13 8EB M7COP
IP13 8LZ G3PYW
IP13 9AD 2E0DZA
IP13 9AD M7DZW
IP13 9FD G0IVV
IP13 9HA G8JGF
IP13 9HP G7MIP
IP13 9HQ M6SXC
IP13 9LY M3DFP
IP13 9NP M6KCK
IP13 9NP M6OLW
IP13 9PQ G3LQR
IP13 9QH G4KXF
IP13 9SL 2E1GXV
IP13 9TE G6HTT
IP14 1BT M1CGO
IP14 1GH G0HEV
IP14 1LP G3TAQ
IP14 1RJ G1HNH
IP14 1RY 2E0JBI
IP14 1RY M6JBG
IP14 1TD G3TZE
IP14 1TN 2E0OME
IP14 1TN M7SME
IP14 1TS 2E1BRC
IP14 1TS G6SUR
IP14 1UQ G0NFE
IP14 2AB G3SUK
IP14 2EY G0BEC
IP14 2FD M7JJK
IP14 2JE M0IZR
IP14 2JT G6PGM
IP14 2LT G6RAQ
IP14 2NA M0JKI
IP14 2NZ G0BCN
IP14 2RE G1UBH
IP14 3AB G7ONB
IP14 3AG G2TO
IP14 3DJ G4AFX
IP14 3GA G3MXH
IP14 3HE M6EQV
IP14 3NF G6JZV
IP14 3NW M1ALH
IP14 3PA G4BIY
IP14 3RQ M3XQX
IP14 4DL G3XVL
IP14 4EG M7BGV
IP14 4EJ 2E0BBM
IP14 4EZ M5AEC
IP14 4LL M6JHN
IP14 4LP G0NMS
IP14 4NR G8SED
IP14 4NT M6DLA
IP14 4RF 2E0DIQ
IP14 4RF M6DIQ
IP14 4SP M6KNW
IP14 4TP G6XGT
IP14 5AN G4BJO
IP14 5AX M6EVC
IP14 5BB G1YRJ
IP14 5DS G7BUL
IP14 5ET G3XLG
IP14 5ET G4THN
IP14 5GH 2E0DEH
IP14 5GH M6LMJ
IP14 5HB G3ZEQ
IP14 5JL G3ZQU
IP14 5LP G0ERL
IP14 5LS M3IFB
IP14 5LX 2E1GRT
IP14 5PE G0OZS

IP14 5RE M7GBS
IP14 5SH M3XGL
IP14 5SN G6SYW
IP14 5UG 2E0XDD
IP14 5UG G0JJG
IP14 5UG M3VUH
IP14 5UG M3XDD
IP14 6AJ G4GBA
IP14 6BU 2E1HWJ
IP14 6DJ G4VPA
IP14 6HD G8ZZS
IP14 6LB G3ONL
IP14 6LX G3VNT
IP14 6LX G4EVN
IP14 6RN G8FMI
IP15 5EE M6CZI
IP15 5QE 2E0ALD
IP16 4AR 2E1HBF
IP16 4AR G1YRF
IP16 4AR G3DBJ
IP16 4BB 2E0IAF
IP16 4BB M0VCP
IP16 4BB M3VCP
IP16 4BY G3MYA
IP16 4DJ G0MLF
IP16 4DT G0UEA
IP16 4JP M6GLS
IP16 4QZ G0CFI
IP16 4RX 2E0JDF
IP16 4RX 2E0XOK
IP16 4RX M0ABG
IP16 4RX M0XOK
IP16 4RX M6EKK
IP16 4RX M6SGJ
IP16 4TA G6LKA
ip16 4ta M5ADQ
IP16 4UQ G4ZWB
ip16et M3PBK
IP17 1BA M1DQE
IP17 1BU M6CUD
IP17 1DP M3DTD
IP17 1EA 2E0KOS
IP17 1EA M0IPS
IP17 1EY G0CJX
IP17 1FB G0OIO
IP17 1HR G2FKZ
IP17 1HR M0RSE
IP17 1HR M1ACB
IP17 1PP G1BFV
IP17 1UX G0GDJ
IP17 1UX M3TIL
IP17 1WB G4ZJH
IP17 1XW M0LAY
IP17 2AA G0EGW
IP17 2AS G0KDR
IP17 2JP G1MJV
IP17 2NW G4IIK
IP17 2PU M0BJR
IP17 2RA G4CXT
IP17 3AH M6EGL
IP17 3ED G4XVE
IP17 3EP M6MFG
IP17 3LW M6CSR
IP17 3NY M6TWA
IP18 6BG M7RNL
IP18 6LR G6BPY
IP18 6RE G7BLK
IP18 6UJ G0BYK
IP18 6UL M1DNG
IP186UZ 2E0DFO
IP186UZ M0HWV
IP19 0BB 2E0FIB
IP19 0BB M3FIB
IP19 0BB M3OJU
IP19 0EA G4PUQ
IP19 0PY G0CFB
IP19 8DB G8TBV
IP19 8EE 2E1CPJ
IP19 8EG G0JJE
IP19 8JF M6PDZ
IP19 8JT M6DRP
IP19 8RP G3LXJ
IP19 8RQ 2E0LRG
IP19 8RQ M0SAZ
IP19 8RQ M6LRG
IP19 8JT G8KWN
IP19 9BH M0ARY
IP19 9DY G4IHI
IP19 9DZ G8EUE
IP2 0EH 2E0EKD
IP2 0NQ M0TMB
IP2 0QG G8LBS
IP2 8AZ G4BVI
IP2 8GA M0ULR
IP2 8GA M3ULQ
IP2 8HA 2E0BCM
IP2 8JZ G4AVL
IP2 9AS G3WJS
IP2 9JE M0HDK
IP2 9JF G4FSK
IP2 9PA G7PQX
IP2 9SX G3SXV
IP2 9TA G4BQR
IP2 9UE M6CQD
IP20 0NY G6IUF
IP20 9ET M7XMG
IP20 9HY G3JPZ
IP20 9JP G4RAV
IP20 9JU G3ICG
IP20 9NQ G4PFG
IP20 9PU G6ZYM
IP20 9QE M7GMD
IP20 9QN M7CZZ
IP21 4DW G7UNU
IP21 4EE G4YFV
IP21 4EH 2E0EPM

IP21 4EH M6JLL
IP21 4LD M6XBB
IP21 4NG G4DML
IP21 4PH M3YLZ
IP21 4QY 2E0CQG
IP21 4QY M0HIW
IP21 4RL M6JGQ
IP21 4TG 2E0HPJ
IP21 4TG M0HIJ
IP21 4TG M0HPJ
IP21 4TG M6JWW
IP21 4TQ G3IPG
IP21 4UA M7RDC
IP21 4XP G8OCV
IP21 4YJ 2E0XGW
IP21 4YJ M0XGW
IP21 4YJ M3XGW
IP21 5JH M1DPX
IP21 5LS G0DFC
IP21 5LS M0WRR
IP22 1AR G0OJJ
IP22 1RN M3XIY
IP22 1RU G4IYS
IP22 1RW G8IBR
IP22 1RZ M6FHO
IP22 2AJ 2E0PKH
IP22 2AJ M6PKH
IP22 2HS G8JPA
IP22 2JQ M6TFD
IP22 2PS G6WJW
IP22 2PS M3JNY
IP22 2PX G8STW
IP22 2PY M6CZL
IP22 2QR G3MFO
IP22 4BY M6GLU
IP22 4DF M0UGG
IP22 4DJ G0RPY
IP22 4EG G6RTM
IP22 4GJ G4MMI
IP22 4GL M0SDW
IP22 4HL G8BOB
IP22 4HR M3JLB
IP22 4HR M6VAZ
IP22 4NA G0WCJ
IP22 4NF G3XLL
IP22 4NP G6ILT
Ip22 4pw M7AFD
IP22 4PY M3XBF
IP22 4PY M6SAF
IP22 4QW 2E1LME
IP22 5RG G7OCQ
IP22 5RU G8RML
IP22 5SP 2E0TUF
IP22 5SP M3TUF
IP23 7DE G4UWW
IP23 7EE G3XAP
IP23 7EJ M6PVW
IP23 7JX M3PFM
IP23 8DY G1TIJ
IP23 8DY M3GFW
IP23 8EF M0ITX
IP23 8HP M1BLX
IP24 1BZ G0GSZ
IP24 1DG M6FBW
IP24 1FH M6NRN
IP24 1JJ 2E0PFG
IP24 1JJ M3VFE
IP24 1LG M6JAL
IP24 1LQ 2E0FCH
IP24 1LQ G0IFL
IP24 1NG G4LXD
IP24 1NP G3SXP
IP24 1PB G4RKK
IP24 1PF G0DCJ
IP24 1RR M3LGJ
IP24 1TQ G0CLT
IP24 2JH G7ACN
IP24 2LB M7ECZ
ip24 2lf M0CNM
IP24 2LY M3SPL
IP24 2QA 2E0GPS
IP24 2QA M3SKY
IP24 2QS G6CWP
IP24 2QX G3ZZQ
IP24 2UU G4OQK
IP24 2YN M6PZA
IP24 2YN M6UES
IP24 3HQ G4VEL
IP24 3NF G1LRK
IP24 3NP M3OCC
IP24 3PT M3UQL
IP25 6AS M7BRX
IP25 6BZ G3GIB
IP25 6BZ G7JRP
IP25 6DB G4GRT
IP25 6DN G4OZY
IP25 6DW G1UKH
IP25 6EA G4AED
IP25 6EU G8XOU
IP25 6EU M0XOU
IP25 6EY G3DOV
IP25 6FF G0NMP
IP25 6FF G7DXN
IP25 6HB M6DYC
IP25 6HE M6FHF
IP25 6HL M6DBH
IP25 6HL M6UCS
IP25 6LG G1NAN
IP25 6LL G8XOC
IP25 6NA M6RZO
IP25 6NJ M6ZXR
IP25 6NL G6YTV
IP25 6SD 2E0EKU
IP25 6SD M0VCX
IP25 6SD M3ZVW
IP25 6SR G0ALQ

IP25 6SS M6EKO
IP25 6XA 2E0FLN
IP25 6XA M0NUX
IP25 6XA M6FLN
IP25 6XB G8YRG
IP25 6ZB M6KWX
IP25 7EH G0KZI
IP25 7EH M6WWR
IP25 7EW G4TJY
IP25 7EZ G0SCT
IP25 7FD G1APL
IP25 7HW G4VBX
IP25 7LH M0HUA
IP25 7LS G0LGF
IP25 7LS M6KSS
IP25 7LX M3LRW
IP25 7LX M5IMI
IP25 7LY M6SSY
IP25 7NJ G4TUK
IP25 7PJ G0FMI
IP25 7QN M0BHW
IP25 7QX M6MBK
IP25 7RA G6BLA
IP25 7RA M6GBB
IP26 4BD G0CZR
IP26 4JG G4THP
IP26 4LD G7LVA
IP26 4QJ G0CLH
IP26 4QW 2E0XSB
IP26 4QW M6XSN
ip26 4rf G6YPJ
IP26 5EG 2E0CNN
IP26 5EG M6BKM
IP26 5EQ 2E0BXS
IP26 5EQ M0GQP
IP26 5EQ M3YZI
IP26 5JA 2E0BCO
IP26 5JA M0FZX
IP27 0BS G0WON
IP27 0Bt M6SGG
IP27 0DN G4KZK
IP27 0DX G8VMZ
IP27 0EN M6UAD
IP27 0LJ G8ENY
IP27 0NR 2E0EAA
IP27 0NR M0SAA
IP27 0NR M3ZPW
IP27 0PW 2E0GOL
IP27 0PW 2E0LJG
IP27 0PW M0LJD
IP27 0PW M0NKR
IP27 0PW M3UYG
IP27 0PW M6LJG
IP27 0QG M6GLD
IP27 0RL M7XRS
IP27 9AJ M6JJD
IP27 9AU G8XQD
IP27 9DZ M0JRP
IP27 9DZ M7RMP
IP27 9ES G6LOJ
IP27 9EU M6EKL
IP27 9EU M6EMX
IP27 9EZ G6EUO
IP27 9EZ G6XYX
IP27 9EZ M3XYX
IP27 9HF G6DFR
IP27 9HR G1EYZ
IP27 9HS G6MGN
IP27 9LA M0JQL
IP27 9SA G0BBN
IP27 9SA G0SDE
IP27 9SA G6TSP
IP270PU G7CNX
IP270QH M6HZD
IP28 6DT G6NNS
IP28 6DT M6NHS
IP28 6ER G3HGE
IP28 6ES G3PFJ
IP28 6TQ G6PCC
IP28 6UA G0FIU
IP28 6UG G4PXC
IP28 6UG G4XRK
IP28 6XF G3YFP
IP28 7BT G8DIY
IP28 7DP G4XTW
IP28 7FP G7PHD
IP28 7HX G8BCA
IP28 7JU M6WVH
IP28 7LN G0DZY
IP28 7LS M6LSV
IP28 7PD M3LWV
IP28 7PD M3ZWK
IP28 7PD M6MFK
IP28 7PR G6VAZ
IP28 8JF G0BRM
IP28 8JF G1SXY
IP28 8LQ G4BWP
IP28 8LQ G4MBC
IP28 8LS M7CTI
IP28 8NG M7AKZ
IP28 8QB G7OYF
IP28 8QD 2E0BJP
IP28 8SF M3PTB
ip288xp 2E0HBJ
IP288XP M0JUV
ip288xp M7DEK
IP29 4DL G7RKU
IP29 4PH G4KNO
IP29 4PL G3VTR
IP29 4SD M6AXM
IP29 5AD G1XAM
IP29 5AD G8KMM
IP29 5AG G4CGV
IP29 5AP M0CWN
IP29 5DD G0WVE
IP29 5DX G4ERF

IP29 5HE G7WKC
IP29 5HR G4VBS
IP29 5QL G8CRM
IP29 5RD M3WOL
IP29 5RP G3JPM
IP29 5SE G0JRM
IP3 0BW M6BGB
IP3 0EE M6HSA
IP3 0EJ M3FCS
IP3 0EW G4LVG
IP3 0LF G7DNT
IP3 0LF M0XON
IP3 0LJ M3ESK
IP3 0LL M7SGA
IP3 0LX M6LGF
IP3 0LZ M3KGJ
IP3 0LZ M3SMY
IP3 0LZ M3ULW
IP3 0PU G3PAI
IP3 0QB G8EHS
IP3 0QH G1HSL
IP3 0RG M3TGP
IP3 8AT G6MMT
IP3 8BW M1BOP
IP3 8BW M5BOP
IP3 8BZ 2E0DWU
IP3 8BZ 2E0SER
IP3 8BZ M6MPZ
IP3 8BZ M6OAX
IP3 8EU M0JFR
IP3 8JQ 2E1HKB
IP3 8JQ M0HKB
IP3 8JW G0BPU
IP3 8NX M6CLV
IP3 8PD M3MXM
IP3 8QW M7AAV
IP3 8RB M6GLP
IP3 8RZ G8OYX
IP3 8SP G3TJU
IP3 8UT M6DME
IP3 8UU M1AOF
IP3 8UU M6NLZ
IP3 9DE G4POU
IP3 9GY G8BEH
IP3 9LT 2E0MIY
IP3 9LT M6MIY
IP3 9NJ G3XKU
IP3 9PD M0CSQ
IP3 9PD M7BYM
IP3 9QF M6AIB
IP3 9RE G4LBU
IP30 0AU M3FDB
IP30 0DG G4ICH
IP30 0JP M6FII
IP30 0QB 2E1BLP
IP30 0QD M6FYB
IP30 0RW M3MSP
IP30 0SZ G7MPF
IP30 0TN M6WFE
IP30 0TS G7MLO
IP30 9AF G1FTD
IP30 9BS G8XOM
IP30 9DQ M3TWS
IP30 9HJ G6AWO
IP30 9PX G3MWO
IP30 9RL 2E0FBD
IP30 9RL M0STI
IP30 9RL M6FBD
IP30 9TP M7TAD
Ip30la G4LVD
IP31 1JJ G4UDD
IP31 1JU M6RGP
IP31 1NG G4IXQ
IP31 1TE G1FVU
IP31 2AY G0AHL
IP31 2BN G8YYC
IP31 2EE G0MEV
IP31 2EE G0MEZ
IP31 2EF G4SBW
IP31 2EP M3FRX
IP31 2EW G4LIG
IP31 2FB M7MHS
IP31 2JQ M6HRN
IP31 2LB G8JSL
IP31 2NJ G6TDW
IP31 2PL G1CFK
IP31 2PP G4GVW
IP31 2QU G4BSA
IP31 2RP G0DUA
IP31 2TF M6FJZ
IP31 2UA M6XEE
IP31 2UQ M3SII
IP31 3EL G7UXD
IP31 3EN G3LPT
IP31 3HX G3GIH
IP31 3PD G4MID
IP31 3PF M3EGV
IP31 3PF M3EGY
IP31 3PF M6DSJ
IP31 3PF M6SJH
IP31 3PW M6NDN
IP31 3PY M6AWI
IP31 3RZ M0DFL
IP31 3SP G4UZF
IP32 6DF G8VQJ
IP32 6ED 2E1FZH
IP32 6ED G7SDC
IP32 6PF M6AZD
IP32 6PU G6LPD
IP32 6RJ M6KJY
IP32 6RR 2E0PUG
IP32 6RR M6CHL
IP32 6RU M6BLN
IP32 6TF M6SVJ
IP32 7AZ M6ERS
IP32 7HR M0PDA

IP32 7HR M6DDP
Ip327ep 2E0OBM
Ip327ep M7FGH
IP33 1JH G3OWQ
IP33 1RF G3TMX
IP33 1SW 2E0CHQ
IP33 1SW M6AUB
IP33 1YP M6AWO
IP33 2GB M6SJA
IP33 2JA G6IGK
IP33 2LS G0DVT
IP33 2LT G3TQX
IP33 2LT M6FVE
IP33 2NJ M6FAK
IP33 2QB M6AUQ
IP33 2QJ M6OCU
IP33 2QL 2E0KEA
IP33 2QL M3THY
IP33 2SA G8XGT
IP33 2SN G7KFP
IP33 2SY G4VSB
IP33 2UA M6PEX
IP33 2UB M1AKN
IP33 3AT M6JAU
IP33 3DU G1UGH
IP33 3EG G6EPD
IP33 3QF G0ARU
IP33 3QJ G1VGI
IP33 3SD M6AUP
IP33 3UB 2E0CWI
IP33 3UB M6CHU
IP33 3XQ G4XSM
IP38RW M7NAB
IP4 1PU M6AWQ
IP4 2BX M0KHX
IP4 2PJ G4SUU
IP4 2TL G8GCO
IP4 2TS G1SWK
IP4 2UB 2E0CZJ
IP4 2UB M0JBZ
IP4 2UB M0NBA
IP4 2UB M6CZJ
IP4 3AB M0RLC
IP4 3AH G4KGO
IP4 3AS M3ZSA
IP4 3BT G3YUJ
IP4 3JH M3UOC
IP4 3LJ G3ZIN
IP4 3NG 2E1CWN
IP4 3PP G6HNN
IP4 4AD G1BEK
IP4 4AD M6EGN
IP4 4BS G8XKT
IP4 4BU M1NIZ
IP4 4HY M6KLN
IP4 4JJ G4ILN
IP4 4JN G8TPC
IP4 4JX G6PDE
IP4 4LP G4UCX
IP4 4QN G4TVT
IP4 4QP G8WVO
IP4 4SF G6FIL
IP4 5AX G4WMF
IP4 5BP M0CCZ
IP4 5DW G6BEH
IP4 5PP G4YVK
IP4 5PQ 2E0XAE
IP4 5PQ M6OXY
IP4 5QZ G6RHK
IP4 5RH G4ROH
IP4 5RZ G0CWW
IP4 5SA G4BRL
IP4 5SD G4FWA
IP4 5SX G4ETC
IP4 5TF G7UWB
IP4 5UQ G7UWC
IP4 5UQ G8CJL
IP4 5UU G3PWB
IP42RN M7LMK
IP5 1AQ G1YLE
IP5 1AS M6AAK
IP5 1AY G3WIU
IP5 1EB G4AEY
IP5 1ED G3UKW
IP5 1ED G3XDY
IP5 1ED G4FZZ
IP5 1EL G8VVR
IP5 1EN M1BMD
IP5 1JZ M1DLE
IP5 1LU G4VBQ
IP5 1NQ M3YCD
IP5 2DE 2E0GVZ
IP5 2DE M6DQD
IP5 2DQ M6EGM
IP5 2DQ M6ENJ
IP5 2DQ M6FCX
IP5 2DS M3KUM
IP5 2EP 2E0UOM
IP5 2EP M3YAW
IP5 2EQ 2E0SOR
IP5 2EQ M0SOR
IP5 2FB G0EBQ
IP5 2FB M3ZVF
IP5 2FX M6BFV
IP5 2GB 2E0VKQ
IP5 2GB G6GSG
IP5 2GB M1ADT
IP5 2GB M3ADT
IP5 2GB M3AKQ
IP5 2GP G8LQB
IP5 2GW G4LSQ
IP5 2GW M6CRG
IP5 2XF M3IBZ
IP5 2YN G4FNR
IP5 2YN M3XBL
IP5 2YN M3XPL
IP5 2YR M1DGY
IP5 2YT G0BQO
IP5 2YU M3MXG
IP5 2YU M3PNO
IP5 3QR G3NYK
IP5 3SD M6YAY
IP5 3SN G8VQH
IP5 3ST G6KWH
IP5 3SU 2E1BIM
IP5 3SU G0NIK
IP5 3SU G3ZID
IP5 3SU G4DWF
IP5 3SY M1ATJ
IP5 3SY M6EDN
IP5 3TZ M1CVB
IP5 3TZ M3BTZ
IP5 3TZ M3BUA
IP5 3UA G4EQE
IP5 3UF 2E1FWD
IP5 3UQ G4OPB
IP6 0DZ M6EFC
IP6 0HQ G7VGE
IP6 0HZ G3TXE
IP6 0JQ G4XDK
IP6 0ND M5AEH
IP6 0PY G8ZYM
IP6 0QY M7CCZ
IP6 0RH M1CVG
IP6 0RJ M0DFQ
IP6 8DP G7VXS
IP6 8ER G4ZYZ
IP6 8FJ G0PEY
IP6 8JH G6PHU
IP6 8JS 2E0FUD
IP6 8JS G3EKE
IP6 8JS M0NIL
IP6 8JS M6DQC
IP6 8TF 2E0JSS
IP6 8TF 2E0ZBD
IP6 8XF M7NOW
IP6 9AJ M0DNZ
IP6 9AR G0TCP
IP6 9AR M0HDR
IP6 9ET M6HBP
IP6 9EU M6CDQ
IP6 9HG G1YRE
IP6 9HR G4BAV
IP6 9HR G4IRC
IP6 9LP G4MRS
IP6 9LP G4PIQ
IP6 9NG G3IRQ
IP6 9NX G7LNI
IP6 9TD G0IIA
IP7 5BS G0MGN
IP7 5JL 2E0WBS
IP7 5JL M6WHA
IP7 5LJ G8HCZ
IP7 5RQ G1HIU
IP7 5RQ G3CQL
IP7 5SQ G7JVE
IP7 6AW G1XVD
IP7 6FE G3OPB
IP7 6FE G8AAI
IP7 6JH G0PEC
IP7 6NN G0ORG
IP7 6NN G0WTR
IP7 6PN G0NQK
IP7 7AL G0FJB
IP7 7BZ G4MGO
IP7 7DE G6HWT
IP7 7EJ 2E0CNG
IP7 7EJ M6AWN
IP7 7JH G7NGF
IP7 7LR G6MGQ
IP7 7LU 2E0XFH
IP7 7LU M3XFH
IP76NN M6XCL
IP8 3AP G0TJY
IP8 3DX G8ENM
IP8 3EY G0DVJ
IP8 3EY G0RGH
IP8 3EY M0GIM
IP8 3LF G4IJJ
IP8 3NH G4KYU
IP8 3RR G4TTB
IP8 3RX 2E0EDN
IP8 3RX 2E0ICX
IP8 3RX M0VLK
IP8 3RX M6HJZ
IP8 3RX M6ICX
IP8 4AU G3MYY
IP8 4HD G4FDF
IP8 4LH M7DUU
IP8 4LR 2E0YSP
IP8 4PE G8FFU
IP8 4PP G1XZX
IP8 4PQ G3XOK
IP8 4PU G3JFW
IP8 4SP G3WJI
IP8 4SR G4DMT
IP9 1AR G4MRD
IP9 1BA M3TWB
IP9 1BP M0GFJ
IP9 1DX G8PPD
IP9 1HS G0PJO
IP9 1HS M6WNF
IP9 1HX G7VPN
IP9 1LL 2E0XKC
IP9 1NP 2E1HKC
IP9 1NW M3GLM
IP9 1QH G0DTC
IP9 1QH G4BXZ
IP9 1QH M0HKC
IP9 1QP M6VLH
IP9 2DR M3TQD
IP9 2HT G4XOJ
IP9 2HW G8XRW
IP9 2JB G3WRT
IP9 2NF M3FDQ
IP9 2SW G3HMB
IP9 2TT G0LIN
IP9 2XD M1SRH
iv 31 6 jj MM6IXH
IV1 1NS GM7RJG
IV1 1XG GM7MWX
IV1 3XG GM3MUA
IV1 3XU GM4LXM
IV1 3XU MM6BDA
IV1 3YG MM0FQQ
IV10 8RA 2M0ALS
IV10 8RA GM0UDL
IV10 8RA MM3UDL
IV10 8SQ GM6IYJ
IV10 8UX MM6LYH
IV10 8XA GM4KLN
IV11 8XF GM4OHY
IV12 4RH MM6YST
IV12 5AR MM6MUA
IV12 5AZ MM1EHO
IV12 5BX MM1HWB
IV12 5EW GM7BCC
IV12 5EW MM3BCC
IV12 5LF 2M0TAS
IV12 5LF MM3UBD
IV12 5PJ GM0RML
IV12 5QG GM0WFS
IV12 5QP GM4WWU
IV12 5RA GM1VAD
IV12 5SD 2M0RUP
IV12 5SE GM3PIL
IV12 5SE MM3YOC
IV12 5TW 2M0YNK
IV12 5TW MM7YNK
IV13 7YE MM1MOY
IV13 7YQ MM6EPV
IV13 7YR GM3LVA
IV14 9AQ MM7CFU
IV15 9NJ MM0ACN
IV15 9PG GM8LEA
IV15 9RB GM4YGN
IV15 9RL 2M0AYU
IV16 9XH MM6CCW
IV16 9XH MM6GBP
IV17 0QL MM6KNO
IV17 0SP MM6MXK
IV17 0SP MM7RVN
IV17 0SZ MM5ALX
IV17 0TR GM0JOL
IV17 0TR GM0SFQ
IV17 0TR GM3PTR
IV17 0TU GM4TTD
IV17 0XL MM3ZRF
IV17 0XN GM0CAD
IV17 0YQ GM6IDF
IV18 0BG GM3SAE
IV18 0EY MM0AAJ
IV18 0GF 2M0WTE
IV18 0HY MM7EGM
IV18 0JT MM7IRQ
IV18 0LU GM4FWF
IV18 0PE GM3WOJ
IV18 0PR GM4FDT
IV18 0PR GM4MFL
IV19 1AZ MM3ONI
IV19 1ED MM0TWK
IV19 1LA GM4SUF
IV19 1QH GM3YQG
IV2 3DT GM0TNK
IV2 3ET GM4XWS
IV2 3EW GM7BOW
IV2 3HT GM0OMC
IV2 3HT GM1MYR
IV2 3HW MM6CEM
IV2 3LX GM0IQD
IV2 4EN GM0GEE
IV2 4EX MM0BAG
IV2 4LD GM0SXP
IV2 4LD GM8DFC
IV2 4NL MM6GON
IV2 4QD 2M0GHF
IV2 4QD MM6HQC
IV2 4XJ GM7RVR
IV2 5AQ MM0BGW
IV2 5DU GM4GKH
IV2 5EP MM0DHY
IV2 5EQ GM7IBM
IV2 5ES GM4UOD
IV2 5PN GM4VAC
IV2 5RG MM1ATR
IV2 5RH MM7ENF
IV2 6AX GM4VIK
IV2 6EQ GM6JOS
IV2 6FF MM7JVR
IV2 6XJ MM6AAA
IV2 6XS MM7NGZ
IV2 7AR 2M0NJM
IV2 7AR MM6HTS
IV2 7HB MM1AEL
IV2 7HN GM4ZIT
IV2 7HX MM7NPT
IV2 7JL GM4OIJ
IV2 7JT MM7INV
IV2 7ND MM3MEH
IV2 7RW GM1BQP
IV2 7SR MM0BQL
IV2 7ST MM3IAG
IV20 1UP GM7PFQ
IV20 1XP MM6DJC
IV20 1YS MM7HPX
IV21 2BX MM6MJR
IV21 2DH GM4CHX
IV24 3DW GM6VRU
IV25 3JE MM7JWK
IV26 2TB 2M0UAL
IV26 2TB MM0ULL
IV26 2TB MM6ULL
IV26 2TX GM6AJA
IV27 4AD GM7WHQ
IV27 4DG MM1YAM
IV27 4ED GM0IYP
IV27 4ED GM4VVX
IV27 4ED MM3MMB
IV27 4EG 2M1ECF
IV27 4JB GM6ZCX
IV27 4JB GM6ZCY
IV27 4JD GM4BYT
IV27 4JQ GM8IEM
IV27 4NL MM0ACN
IV27 4RP 2M0AMD
IV27 4RP MM0URN
IV27 4TG GM6JHH
IV27 4XQ MM3SCO
IV28 3XE MM1WKD
IV3 5AH MM3HJC
IV3 5BD 2M0HUD
IV3 5ES GM4UZR
IV3 5HE MM0IPD
iv3 8az MM7AXB
IV3 8LB GM7HQW
IV3 8LB MM7IRM
IV3 8LH GM0AKJ
IV3 8PF GM7GVD
IV3 8PP MM7SVR
IV3 8RA MM7CQZ
IV3 8SD GM4JNB
IV3 8SJ 2M0JAT
IV3 8SR GM8RMR
Iv3 8te 2M1HLE
IV30 1SF GM6WLJ
IV30 1TB GM0UYZ
IV30 2YR GM0LOK
IV30 2YR GM7CPY
IV30 4BU MM3GLH
IV30 4DJ GM0MZD
IV30 4HJ MM1FAS
IV30 4HL GM7FPN
IV30 4HN GM7LSI
IV30 4HX 2M0CFB
IV30 4HX GM0MGO
IV30 4HX MM0GYX
IV30 4HX MM6AON
IV30 4JH GM0DQV
IV30 4JH GM8FFK
IV30 4LY GM4ILS
IV30 4NB GM8YKT
IV30 4NS GM0LPD
IV30 5NF MM6IKS
IV30 5PQ MM3FJA
IV30 5PT MM3FJB
IV30 5PY 2M1DHG
IV30 5QN MM1GBS
IV30 5RN MM6DZC
IV30 5SU GM0OTS
IV30 5XY 2M0REH
IV30 5XY MM0TQH
IV30 5XY MM3TQH
IV30 5YD MM6CPE
IV30 5YE GM7LWA
IV30 5YN 2M0CXG
IV30 5YN MM6CXD
IV30 6BP 2M0BZB
IV30 6EJ GM4EWM
IV30 6EJ MM0JBM
IV30 6ER GM1ASY
IV30 6GL MM6YGT
IV30 6XF MM0JMB
IV30 8FE GM0GNY
IV30 8JU GM0ONN
IV30 8JZ GM0RVA
IV30 8NY 2M1HCP
IV30 8NY 2M1HCQ
IV30 8NY MM3BWV
IV30 8PE GM0AEG
IV30 8PE MM0HAR
IV30 8PP 2M0SQL
IV30 8PP MM3PHP
IV30 8SY GM0EMC
IV30 8SY GM6TUE
IV30 8TB 2M0WMJ
IV30 8TB MM0KKF
IV30 8UA GM6IKN
IV30 8UR GM3VBY
Iv308qg MM7APW
IV31 6BA GM4PGM
IV31 6BA MM3ZSK
IV31 6DJ 2M0FUS
IV31 6DJ MM7IOW
IV31 6NA GM4HMN
IV31 6QH GM4MKU
IV31 6TP GM6PLG
IV32 7DX MM6IVP
IV32 7EH MM0DAT
IV32 7HF 2M0JVR
IV32 7HZ MM7WTX
IV32 7JH MM3TVQ
IV32 7JY 2M0BPM
IV32 7JY MM0LER
IV32 7JY MM3WLA
IV32 7LU 2M0GYM
IV32 7LU MM3UBR
IV32 7NL MM0CQT
IV32 7NW 2M0DTJ
IV32 7NW MM6CYQ
IV32 7NW MM6MUO
IV36 1BQ 2M0MMM
IV36 1JB GM8DPV
IV36 2SG GM0KNT
IV36 2UF GM0GHN
IV36 2UL MM6GSY
IV36 2WE GM1FRG
IV36 3UA GM6NOO
IV4 7AB MM0LOZ
IV4 7AB MM3PSL
IV4 7AE GM7DXE
IV4 7AH 2M0XCT
IV4 7EY GM4TJD
IV4 7EY GM4UMA
IV4 7GL 2M0DAC
IV4 7HT 2M0LWB
IV4 7HT MM0VSU
IV4 7HT MM6LWB
IV4 7HZ GM7ORJ
IV4 7HZ MM3SRK
IV4 7JQ GM4GZD
IV40 8BB MM6ZDY
IV418PT GM6DBJ
IV42 8PY GM8RBR
IV44 8RD GM0HBK
IV44 8RF GM0MAC
IV44 8RQ GM3YAC
IV44 8RQ GM4YAC
IV45 8RU MM3VXP
IV47 8SH MM6DTV
IV47 8SJ GM0FHF
IV47 8SN GM0PRQ
IV49 9AQ GM1RLV
IV49 9BZ MM7RCU
IV51 9DN GM3SWK
IV51 9JX MM0LUP
IV51 9PE GM0EWX
IV51 9PW GM0DXE
IV51 9YN MM3ZDI
IV51 9YR 2M0RMV
IV51 9YR MM6RMV
IV52 8TN MM0GWO
IV53 8UX MM6TUG
IV55 8GL GM4TRH
IV55 8WF GM4AEK
IV55 8WL GM4RXD
IV55 8WP GM3MTW
IV55 8WS GM0CVD
IV55 8WS GM4WZD
IV56 8FJ GM4PUS
IV56 8FL GM6UCX
IV6 7PX MM3LQK
IV6 7QG GM6WEI
IV6 7QG MM0GKB
IV6 7SB GM0OGZ
IV6 7XN GM4RRP
IV63 6TJ MM7DPP
IV63 6TN GM0BZS
IV63 6UW GM4EKI
IV63 7YJ GM6JJN
IV7 8AW 2M1DZX
IV7 8AW GM4SFW
IV7 8BH MM3SES
IV7 8FP MM0YIA
IV7 8FP MM6IAY
IV7 8JS GM0NAI
IV7 8JS MM0APF
IV7 8LB GM3WED
IV7 8LG MM6MWE
IV7 8LH GM4FIZ
IV7 8LL GM0JFK
IV8 8PA GM7JYW
IV8 8PG GM0TQB
IV8 8PJ GM0JFL
IV8 8PR 2M1EBJ
IV9 8QL GM4MAI
JE2 3RX MJ6ADQ
JE2 3SY GJ7AOG
JE2 3ZA 2J0RZD
JE2 3ZA MJ0RZD
JE2 3ZA MJ3RZD
JE2 4DA MJ7DIJ
JE2 4RY MJ1COO
JE2 7AH 2J0SZI
JE2 7AH MJ3SZI
JE2 7HQ GJ8PVL
JE2 7HQ MJ6PVB
JE2 7HQ MJ6XVL
JE2 7LQ MJ7FXC
JE2 7PP MJ6SSF
JE2 7TA GJ7FGS
JE2 7TZ GJ8RRP
JE3 1GL GJ6WDK
JE3 1NH MJ6THP
JE3 1NL MJ6RBI
JE3 5AA GJ7DNJ
JE3 6AS GJ4YLP
JE3 6ER 2J0ODX
JE3 6ER MJ0PMA
JE3 6ER MJ6AUD
JE3 8AL MJ6VAA
JE3 8AP MJ7NLK
JE3 8AQ MJ0IEW
JE3 8GB GJ4JVP
JE3 8GQ GJ6EFW
JE3 8LS MJ6ORG
JE3 9AP 2J0CZD
JE3 9AP MJ6CQF
JE3 9AY 2J1CWG
JE3 9AY 2J1CWH
JE3 9AY GJ7UIT
JE3 9AY MJ0JIS
JE3 9EF GJ8CEY
JE3 9EP GJ3YLN
JE3 9ER 2J0ELQ
JE3 9ER MJ0THP
JE3 9ER MJ6ILZ
KA1 1UH GM7KIY
KA1 2HP GM0AAX
KA1 2HP MM3TYA
KA1 2JN MM7ALL
KA1 2LA GM6KPL
KA1 2NU 2M0YAF
KA1 3AR MM3LWT
KA1 3SE 2M0BSE
KA1 3LQ GM0ADX
KA1 3LQ GM3OZB
KA1 3QL MM3OJV
KA1 3RY GM0FKP
KA1 3RZ MM7DIV
KA1 3SJ GM0NTY
KA1 4EB MM3FET
KA1 4LQ GM4UYP
KA1 4PB GM0FQQ
KA1 4PB MM3VSC
KA1 4QT GM0LYH
KA1 5AB GM7AAJ
KA1 5NA 2M0XND
KA1 5NA MM0XNC
KA1 5ND GM4CAM
KA1 5SF GM4JMU
KA10 6AW 2M0JST
KA10 6DA 2M0CMA
KA10 6DA MM0YET
KA10 6DA MM6AMA
KA10 6NJ 2M0BMN
KA10 6NJ MM0CKF
KA10 6NJ MM3NNO
KA10 6QW MM6AAM
KA10 6SD GM4XRY
KA10 6SE MM7JYS
KA10 6SH MM0ODX
KA10 6TB GM3NIG
KA10 6TT GM7VXR
KA10 6TZ GM4SLY
KA10 6UG GM6BAO
KA10 6UP GM4IGS
KA10 6XE GM8GIQ
KA10 7AH MM0KJM
KA10 7DY GM6JEP
KA10 7EE GM4CXF
KA10 7GZ 2M0SOE
KA10 7HA GM4BIT
KA11 1BD GM0SEI
KA11 1BD GM7KSA
KA11 1BD MM0JJV
KA11 1BW GM7UAC
KA11 1BY MM3WNP
KA11 1DB MM0ZIF
KA11 1HH MM6LIL
KA11 1JD GM0DWH
KA11 1LR GM6JOD
KA11 1LT MM6PDX
KA11 1NJ GM3JOB
KA11 1PN GM0JQE
KA11 2BY MM3PEV
KA11 2DR MM0UKW
KA11 2EB GM1VJD
KA11 2EU GM0DNG
KA11 3AL MM1DQW
KA11 4BN MM6JFU
KA11 4EB MM3OIZ
KA11 4ER GM0WUX
KA11 4EY 2M0WBJ
KA11 4EY MM0HYM
KA11 4EY MM6WBJ
KA11 4FB 2M0NLA
Ka112fy MM7IOD
KA12 0DF MM7AHK
KA12 0DU 2M0XTS
KA12 0DU MM0XKA
KA12 0DU MM6XKC
KA12 0QE GM7VFR
KA12 0QG GM7CZU
KA12 0QS MM1CXO
KA12 0SE GM3NYG
KA12 0SE GM4PGV
KA12 0TR GM7KHA
KA12 0XA MM3UOE
KA12 0XH 2M0BMF
KA12 8DR GM0FCI
KA12 8PR MM0XPP
KA12 8RZ MM0CTT
KA12 8RZ MM0CTU
KA12 8RZ MM3TNG
KA12 9DN GM1ASA
KA12 9DT MM3TWA
KA12 9ER 2M0WUI
KA12 9ER GM5PEB
KA12 9ER MM3WUI
KA12 9JJ 2M0HXQ
KA12 9JJ MM1LDR
KA12 9JJ MM3LDR
KA12 9LU MM6DAF
KA12 9PA MM3UOS
KA13 6BE MM0DHQ
KA13 6BE MM0KLR
KA13 6EB GM1MSN
KA13 6JJ MM0HLN
KA13 6JJ MM3DKA
KA13 6LU MM3YUU
KA13 6PL GM3YQK
KA13 6QN MM0BRG
KA13 6RS 2M0OUU
KA13 6RS MM0OUU
KA13 6RS MM6OUU
KA13 6WH MM3YUY
KA13 7AR GM0TAE
KA13 7BA MM6WOC
KA13 7DT GM0UGG
KA13 7HE GM1MSO
KA13 7JN MM0CKK
KA13 7LQ MM1FHR
KA13 7ND MM1LBA
KA13 7PT GM4KQS
KA13 7QY GM7EGQ
KA14 3AJ GM8MMW
KA14 3BQ GM1AXI
KA15 1AR GM6OHF
KA15 1BQ MM7LUB
KA15 1JE GM7DTC
KA15 2AJ GM4VHZ
KA15 2BA GM7DHA
KA15 2DZ MM3OGS
KA15 2HG GM4LIS
KA15 2LN GM6UHE
KA15 2LN GM8NVG
KA16 9BE MM7RVI
KA16 9HZ MM0OVK
KA17 0DQ GM4ZNS
KA17 0JB MM3OJE
KA17 0JG MM7JFA
KA17 0LP 2M0UEA
KA17 0LP GM0VBE
KA17 0LP GM1FNX
KA17 0LP MM3UEA
KA17 0NN MM7LEY
KA18 1HA MM0JOK
KA18 1HA MM3EJB
KA18 1HN GM7ANE
KA18 1HN MM3GQT
KA18 1NP MM3PHC
KA18 1PU GM0BKX
KA18 1PU GM0NET
KA18 1PU GM1MDO
KA18 1QU MM7BQP
KA18 2ED GM3STM
KA18 2ED MM7STM
KA18 2LL GM1TCP
KA18 2LL GM1YKE
KA18 2NZ MM1ANP
KA18 2RE GM7OKX
KA18 3AG GM0GRD
KA18 3AT GM4XFU
KA18 3GA GM0DBK
KA18 3HS GM4XMD
KA18 4BW MM7EFP
KA19 7AE 2M0BET
KA19 7AE MM3BOJ
KA19 7AU MM6JTG
KA19 7HZ 2M0GYN
KA19 7HZ MM6EBD
KA19 7PW MM0KMK
KA19 7RE MM0TFU
KA19 8AY MM6IMP
KA19 8dr MM7FEW
KA19 8EN MM6BNS
KA19 8FG 2M0MSU
KA19 8FG MM6OEJ
KA19 8LR GM3YDN
KA197AE MM0LGR
KA2 0AX GM4ATA
KA2 0AX GM8DYT
KA2 0AY MM6PFT
KA2 0DS MM6MHN
KA2 0ES 2M0DTZ
KA2 0ES MM0GKN
KA2 0LD MM3DHG
KA2 9DX MM3EJV
KA2 9EW GM7IUF
KA2 9EX GM4VKI
KA2 9EZ GM8JUY
KA2 9HZ MM3DHE
KA20 3BA MM7RGM
KA20 3DG GM4WNQ
KA20 3LG MM3RTH
KA20 3PU MM3VFK
KA20 4BG 2M0YIO
KA20 4BG MM3YIO
ka20 4bn MM7ORR
KA20 4BP MM6NAB
KA20 4EF MM3PXO
KA20 4HN MM0CWB
KA20 4HN MM6CCS
KA20 4JE MM3UVF
KA20 4LB MM7STF
KA21 5DA MM6VFL
KA21 5EP MM6MNW
ka21 5lj MM0BQJ
KA21 5NH GM0XAV
KA21 5PS MM3AQM
KA21 6AB 2M0JCG
KA21 6AB MM6AVE
KA21 6AQ GM0OTB
KA21 6AQ MM6CKC
KA21 6AX MM7EKQ
KA21 6HE MM6GKE
KA21 6HY GM4UEH
KA21 6JU MM7BDW
KA21 6LN MM6DHS
KA22 7AJ MM3YBD
KA22 7BZ MM6BXT
KA22 7DY 2M0OVD
KA22 7DY MM0OVD
KA22 7ER MM3YFT
KA22 7NQ GM7VSB
KA22 7NQ MM6ARN
KA22 7PT GM7FPX
KA22 8AP MM7GBL
KA22 8EP MM7AXX
KA22 8LW 2M0RGU
KA22 8LW MM3NMI
KA22 8PH GM7SXI
KA228HA MM3UOR
KA23 9AR MM7AYG
KA23 9BX GM0EFC
KA23 9BX GM0EFD
KA23 9BY GM4YMC
KA23 9DA 2M0HCF
KA23 9LB GM0AYT
KA23 9LB MM6OOP
KA23 9LE GM3LLP
KA23 9LX GM3JIG
KA23 9NF GM0DEX
KA24 4AF GM0NHL
KA24 4DJ MM1EVJ
KA24 4HP MM0SVE
KA24 5HP MM6IIO
KA24 5HP MM6LJE
KA25 6AB MM3XXI
KA25 6DB GM7VCV
KA25 6EY GM0ODB
KA25 7JB MM6CCO
KA25 7JB MM6KYV
KA25 7JF MM6VWH
KA25 7LB GM0HQF
KA26 0AA 2M0NCM
KA26 0AA MM3GZG
KA26 0AA MM3NCM
KA26 0AA MM6CNC
KA26 0AE MM3GIR
KA26 0BP GM0FSZ
KA26 0BX GM0NBG
KA26 0BY MM6CTH
KA26 0EA GM0UDY
KA26 0EB MM1DWU
KA26 0EF MM0CBL
KA26 0NQ GM4ZTO
KA26 0PA GM4WEW
KA26 0QT 2M0YRB
KA26 0QT GM1ROB
KA26 9AH GM0JMO
KA26 9AH GM7SAK
KA26 9AH MM0SAK
KA26 9AH MM3UDB
KA26 9DJ GM0INC
KA26 9DZ GM6OFB
KA26 9EL GM0KCY
KA26 9EU MM6ZUY
KA26 9JH GM0KWW
KA26 9LP GM4TPQ
KA26 9SN MM7BFL
KA27 8FJ MM6SGZ
KA27 8PB MM6RLD
KA27 8PG GM8LQL
KA27 8PR GM4BVZ
KA27 8QH GM3HEN
KA27 8QH GM4JFH
KA27 8QH GM4XJF
KA27 8RG MM7RXH
KA27 8RL GM3UA
KA27 8SA GM1ISJ
KA27 8SF GM4PWR
KA28 0DL MM3WWQ
KA28 0DP GM8BJJ
KA29 0DG GM3DOD
KA29 0DP MM0JBS
KA3 1TU GM4ZNC
KA3 1TU MM0TJT
KA3 2AS 2M0OOT
KA3 2AS MM3OOT
KA3 2DA GM0MZH
KA3 2DW GM4UYK
KA3 2EQ GM0JFH
KA3 2EU 2M0ONW
KA3 2GN MM3LZU
KA3 2GP MM0GNS
KA3 2HU 2M0GZQ
KA3 2HU MM7TGA
KA3 2JG GM0ONX
KA3 2JG GM6JIC
KA3 2JG MM3ONX
KA3 2JG MM6PGT
KA3 2RS 2M0VNW
KA3 2RS MM0ABB
KA3 2RS MM3VNW
KA3 3BT 2M0TSR
KA3 3DQ GM3AXX
KA3 3HG GM4VAY
KA3 3HT GM4OSS
KA3 4AQ MM6SHM
KA3 4BP GM0GOV
KA3 4BP MM3LZD
KA3 5AT MM3WGW
KA3 5DA 2M0FRJ
KA3 5DA MM6SUR
KA3 5JT GM0PDQ
KA3 5JT GM4HCO
KA3 6DB GM0OXS
KA3 6FJ MM6GLO
KA3 6HJ GM0LYO
KA3 6JQ 2M0FDT
KA3 6JQ GM5TIM
KA3 6JQ MM6MQO
KA3 6JT GM0DVH
KA3 7DT GM4SQM
KA3 7DT GM5JNJ
KA3 7DT GM6JNJ
KA3 7EJ 2M0AAQ
KA3 7EJ MM6WBP
KA3 7RB MM3JGT
KA30 8EF GM8EUG
KA30 8ER GM0BFW
KA30 8HB GM5BDX
KA30 8HB MM7BJC
KA30 8HU GM1RJS
KA30 8HU MM7BIC
KA30 8PG GM4ZZW
KA30 8QQ MM6JHH
KA30 8RH GM0NUI
KA30 8RQ MM3XIW
KA30 8SJ MM0RFA
KA30 8SN GM3MWX
KA30 8SN GM6XW
KA30 9BH GM1LXM
KA30 9BZ GM1BKR
KA30 9DN MM0TCP
KA30 9EQ GM0DWY
KA30 9ER MM0MNS
KA30 9ET GM0GMN
KA30 9ET GM4IPA
KA30 9ET MM3YPH
KA30 9EX GM0RYD
KA4 8AX MM7MAB
KA4 8EA 2M0GCS
KA4 8EA MM0GHM
KA4 8EA MM0RAI
KA4 8EA MM3GDC
KA4 8EJ GM0KAZ
KA4 8HT GM8MNR
KA4 8NA MM3WZL
KA5 5AE GM1XBK
KA5 5QU GM7RPT
KA5 5RP MM3LDK
KA5 6DT GM7OPN
KA5 6EH GM0GRW
KA5 6HU MM0HSV
KA5 6HY MM3GSL
KA6 5BE MM0XCP
KA6 6BH GM8KXF
KA6 6EL GM1VFQ
KA6 6GJ MM6EBZ
KA6 6HB MM3ENP
KA6 6LB GM4PPT
KA6 6ND GM4SFA
KA6 6NL MM3WNH
KA6 6QG GM0POD
KA6 7AU 2M1FQI
KA6 7AU MM3FQI
KA6 7AU MM3WVQ
KA6 7DG MM0CNF
KA6 7EF GM1VXE
KA6 7HG MM7WAB
KA6 7LQ MM6CHM
KA6 7ND GM1OST
KA6 7QN MM7DAH
KA6 7RN GM0SDS
KA6 7SJ MM3JGR
KA6 7TX GM1VDZ
KA6 7TX MM7DHE
KA67LR GM7USC
KA67ST MM6POV
KA7 2LW GM4FGS
KA7 2NF 2M0LXX
KA7 2RW MM7JNK
KA7 2TA MM7NRN
KA7 3JB GM0TBH
KA7 3JJ MM0MLO
KA7 3LN MM7HIM
KA7 3NF GM7OIN
KA7 3PE GM4SQO
KA7 3PE MM6OIR
KA7 3QH MM6CLM
KA7 3QJ MM0BPF
KA7 3RL MM6CPK
KA7 3RZ MM7MMW
KA7 3SQ MM7ATF
KA7 3XG MM7IEF
KA7 4DA GM3PMB
KA7 4DA GM5VG
KA7 4EQ GM0EPO
KA7 4ET MM0RAG
ka7 4lr MM7BWN
KA7 4PB MM1JAS
KA7 4PW GM4WFV
KA7 4QF MM6RSP
KA7 4QN GM6BHR
KA7 4UB GM6MD
KA7 4XA MM3HQL
KA7 4XQ 2M0MOF
KA7 4XQ MM6TCS
ka73sb MM7GTS
KA8 0EF MM0CHX
Ka8 0jr MM7FEM
KA8 8NL MM7NLH
KA8 8NX MM6BXR
KA8 9BW MM6BCF
KA9 1DL MM7NMO
KA9 1JW MM3WVN
KA9 1LT MM0GAI
KA9 2DB GM4TPX
KA9 2ED GM0DEQ
KA9 2EQ MM3EDW
KA9 2HE GM3MQO
KA9 2HE GM4OOU
KA9 2HY GM4DOZ
KA9 2NN MM1DPH
KA9 2PW GM4LVW
KT1 2RU G3RLT
KT1 3PS G8VXB
KT1 3QB M0BEJ
KT1 3RW G8SIK
KT10 0AZ G0EYT
KT10 0BJ M0XDV
KT10 0DT G4IUA
KT10 0DW G6FWO
KT10 0DW M6TLL
KT10 0HS G3RIM
KT10 0JZ G1PQT
KT10 0QE G4GPB
KT10 0RZ G8YXZ
KT10 8DL M0PLP
KT10 8PY M6OKI
KT10 9EG 2E0KXX
KT10 9EG M6KXX
KT10 9HF G0IDB
KT11 1AU G6LHQ
KT11 1LW M7CDH
KT11 2AJ M0TWM
KT11 2AQ G1LKH
KT11 2BB G0OAS
KT11 2BH M1CNH
KT11 2RU G8ORX
KT11 2SX G4YKH
KT11 3AJ 2E0ETF
KT11 3AJ M5PM
KT11 3AJ M6LVM
KT12 1BB G4ARO
KT12 1LE G3LFX
KT12 1LW M6GZJ
KT12 1NF G8GTR
KT12 2AT G6AYS
KT12 2DP 2E0SYO
KT12 2DP M7APZ
KT12 2HS G0HJD
KT12 2LB G6CLK
KT12 2SU M3FJE
KT12 2SZ G2HS
KT12 2SZ M1GWZ
KT12 3AW M1IAN
KT12 3BA G0UHQ
KT12 3DA G4AWZ
KT12 3EQ G8DXP
KT12 3QH M1AEJ
KT12 3SQ G6INX
KT12 4HQ G1JKV
KT12 5BT M6ALJ
KT12 5HF G1KEV
KT12 5LW G8ZXL
KT12 5PH M6KSI
KT12 5PH M6NBY
KT12 5QG 2E1RJS
KT13 0BL G7SQH
KT13 0JW G0CKV
KT13 0NR G4CXL
KT13 0NR G8HCL
KT13 0TB M6RDD
KT13 0XE G1JHS
KT13 8AS G4BSV
KT13 8AS G4CPM
KT13 8PE G3NMH
KT13 8PU G6CUV
KT13 8RA G3YMN
KT13 8UP G4LQD
KT13 9AR 2E0SKK
KT13 9AR M0RND
KT13 9AR M6AGG
KT13 9AT M3XSZ
KT13 9LS M0ZCO
KT13 9NU G8IPN
KT13 9RU G0JRH
KT14 6AE M0HMF
KT14 6AP G0RWM
KT14 6DT G6JRZ
KT14 6DY M6AQW
KT14 7BZ M3EJH
KT14 7EF G6XAN
KT14 7HX G4EVA
KT14 7NG G0PBL
KT15 1DH G4WMP
KT15 1SR G7HIN
KT15 1UN M7BTY
KT15 2AX M6XYH
KT15 2DE G8HFW
KT15 2DN G0KEY
KT15 2US M7UFL
KT15 3AA G4ZMU
KT15 3AA M6CUS
KT15 3AG 2E0EQT
KT15 3AG M0TKC
KT15 3AG M6KXN
KT15 3AQ M3WQA
KT15 3DJ 2E0ABT
KT15 3DJ G4CCZ
KT15 3DL G1SMB
KT15 3DZ M0DJW
KT15 3DZ M1VPL
KT15 3ET G8JSN
KT15 3EY G1IDE
KT15 3EY G3UES
KT15 3EY M0SAR
KT15 3HP G6MJT
KT15 3HS G0PVF
KT15 3LX G3XSZ
KT15 3NQ 2E0XNC
KT15 3NQ M6XNC
KT15 3NR M0XMC
KT15 3SE G8CUG
KT15 3TU G6YHE
KT15 3UA G0RMU
KT151NX M6FDX
kt153lh 2E0BSM
kt153lh M0GPH
kt153lh M3YFI
KT16 0BP G0EMT
KT16 0ER G1UMY
KT16 0HD G6KJE
KT16 0LG G4LQI
KT16 0PH G7DUK
KT16 0RE G0GPT
KT16 8AP M0DXP
KT16 8BU G8BWA
KT16 8EP G0PCZ
KT16 8LE M0BSF
KT16 8QB G1AEA
KT16 8RA G7MYO
KT16 8RW G3ZZL
KT16 9AZ M6SEJ
KT16 9ED G6YRI
KT16 9GY 2E0FLH
KT16 9GY M7MJC
KT16 9HP M6KSV
KT16 9HX M6OIC
KT16 9JJ G8NYB
KT16 9PH G3OLH
KT16 9QR M0JFI
KT17 1EJ 2E0WOL
KT17 1EJ M3XEI
KT17 1QP G8HPF
KT17 1SN 2E0HIY
KT17 1SN M7CDU

KT17 2EF G4VHJ
KT17 2HD G0DGF
KT17 2LH M7HSG
KT17 2NG G7PBC
KT17 2NW G0ECK
KT17 2PH M7BXA
KT17 3LN G1ZVE
KT17 3NL G3LQP
KT17 4EN G8EPS
KT18 5JL G3TVL
KT18 5PB 2E0NOU
KT18 5PB M0LKJ
KT18 5PY G4IQV
KT18 5QT G4GFI
KT18 7BU 2E0DQO
KT18 7BU M6EZS
KT18 7JE 2E0DZF
KT18 7JE G6KVG
KT18 7JE M3GQL
KT18 7SL 2E1FYC
KT19 0AU G4ZSW
KT19 0HD M0AAF
KT19 0HF G8MTI
kt19 0jp G0WZM
KT19 0LB G4ROI
KT19 0LS G6QN
KT19 0LS M0ZEY
KT19 0LS M6HTN
KT19 0PQ G8EBQ
KT19 0RJ G6YIJ
KT19 0SZ G0PFA
KT19 7EW G6REH
KT19 7LD G1DPW
KT19 8BW M3KQC
KT19 8HG M3IDB
KT19 8HH M6JHO
KT19 8JX G3YGG
KT19 8RP G1LQC
KT19 8RP G4SYT
KT19 9DP G4WFL
KT19 9DP G6LFW
KT19 9EE G0KAS
KT19 9EX M6LHW
KT19 9HD 2E0BJL
KT19 9HD M3ZLL
KT19 9HH G1IKG
KT19 9LB M0GDU
KT19 9LD M0TBS
KT19 9LU M0DSL
KT19 9NR M7TEY
KT19 9QR G0OXZ
KT19 9UL G4RSU
KT2 5GG G4PPN
KT2 5HH G6OBA
KT2 5HQ M7CRQ
KT2 5NE M6MUF
KT2 5QB G8VOY
KT2 5TL M3XXE
KT2 5TU G6RFU
KT2 5TU M0ETA
KT2 6AH M3WEJ
KT2 6ES 2E0HMA
KT2 6PN G6JTD
KT2 6RA G8YJQ
KT2 7AL G6ZLS
KT2 7AN M0EIJ
KT2 7ES M0JZV
KT2 7JT G0TRD
KT2 7QD G4GQA
KT20 5EW G6HVQ
KT20 5JF G4WNV
KT20 5JQ G7PHK
KT20 5PS M0JHM
KT20 5RZ G3SUX
KT20 5SQ G0XTM
KT20 5UA G4XXI
KT20 5UT G6ALG
KT20 6BS G3NPC
KT20 6ET G0SWN
KT20 6ET G6CJR
KT20 6ET G8WGP
KT20 6ET M3URA
KT20 6ET M6ICZ
KT20 6XE G4HVO
KT20 6XE G6GGY
KT20 7AD G7KWF
KT20 7BA G4BYZ
KT20 7LN 2E0WZA
KT20 7LS G8GGS
KT20 7LZ G4NLB
KT20 7QE M6ZLP
KT21 1BE G4JUM
KT21 1EB G3KWJ
KT21 1HY 2E0DKP
KT21 1HY M0LUT
KT21 1HY M6LUT
KT21 1NA G6TZE
KT21 1NE G1HWR
KT21 1NE G4KGE
KT21 1NN M0JSN
KT21 1PD G7LSB
KT21 1RH G3OKU
KT21 1SG G7RSK
KT21 2HB G8ZZR
KT21 2HU G4CLY
KT21 2HY G4XQE
KT21 2JG G8NXB
kt21 2pg G4NHO
KT21 2TL G8FXN
KT21 2UF G6BQQ
KT212EG G6DTW
KT212EG M0KCI
KT22 0HR G8UYL
KT22 7EE M0GXV
KT22 7SL 2E0ZEH
KT22 7SL M0ZEH

KT22 7SL M3ZEH
KT22 8JT G3ZQT
KT22 8LR M0CDJ
KT22 8LZ M1KOS
KT22 8NU G8MHN
KT22 8RD G4EYB
KT22 8UJ G4XYC
KT22 9AZ G3NFV
KT22 9AZ G8GWM
KT22 9EF M0CDB
KT22 9JR G3UZX
KT22 9NB G6BTC
KT22 9NT G3XKG
KT22 9QT M6XTR
KT22 9TL M3FZE
KT22 9XA 2E0GBK
KT22 9XA M6BMB
KT23 3DA 2E0CDK
KT23 3DU G1FCU
KT23 3EY G4FDU
KT23 3HW M7TBA
KT23 3JF M3CSI
KT23 3PY G3LHN
KT23 4BJ G1RJN
KT23 4BQ G4HXN
KT23 4BS G8IWT
KT23 4BS G8KW
KT23 4DB G3UZW
KT23 4DY M6GWI
KT23 4HP 2E0ESU
KT23 4JX G3SIA
KT23 4ND 2E0LTR
KT23 4ND M6ZLC
KT23 4RR G4CVN
KT24 5DY G6GEK
KT24 5RH 2E0HQF
KT24 5RH M0JWU
KT24 5RH M7CUR
KT24 5SD G0RAL
KT24 6ED G7NJI
KT24 6QJ G3NIW
KT24 6QN G4PED
KT3 3BX M6OMS
KT3 3DY G4JUZ
KT3 3HP G3TON
KT3 3HP G4FJP
KT3 3HR 2E0WEE
KT3 3HR M6FRO
KT3 3HZ G8BPY
KT3 3HZ M3BPY
KT3 3LB M3WZR
KT3 4BN M7KKM
KT3 4HR G8GGI
KT3 4LD 2E0WNI
KT3 4LD M0WNI
KT3 4LD M3WNI
KT3 4LE G3XXH
KT3 5BZ G7UVF
KT3 6BS 2E0GRF
KT3 6BS M0HWS
KT3 6BS M3ZJV
KT36LX M0OOZ
KT4 7AA 2E0JRE
KT4 7AA M0UPS
KT4 7AA M7WIT
KT4 7DX G4ADM
KT4 7DX G7IQZ
KT4 7DX M6HSX
KT4 7JQ G7LCS
KT4 7PH M3XWY
KT4 7SJ 2E0FMY
KT4 7SJ G0KRT
KT4 8EZ G4VLN
KT4 8NB G4BQY
KT4 8SN G7RBC
KT4 8TG M7HGN
KT4 8UD M0BMZ
KT4 8XG G6VAX
KT4 8YA G1LGB
KT4 8YA M0CLG
KT47AY M7RIA
KT5 8DS M6KGO
KT5 8JJ G6DKS
KT5 8JZ G8KEJ
KT5 8TS G0IRK
KT5 9BS G6SMJ
KT5 9DX G0JOS
KT5 9EA G7EKC
KT5 9EA M3EKC
KT5 9JH G4PPU
KT5 9LJ 2E0OYN
KT5 9LJ G1KAG
KT5 9LJ M3OYN
KT5 9LW G7MLJ
KT5 9PH M3VXH
KT5 9RJ G0BQV
KT59PT 2E0HIW
KT59PT M7FBR
KT6 4DH G2SR
KT6 4DH G7JYQ
KT6 4DH M1SRC
KT6 4EX M6KFG
KT6 4SR 2E0HYO
KT6 4SR M0KBK
KT6 4SR M7DIS
KT6 4SW M0SHA
KT6 5AF G0PCY
KT6 5JD G3JVC
KT6 5RE 2E0MJO
KT6 6LJ G0SAC
KT6 6LJ G4WGE
KT6 6PY M7PAU
KT6 7JT G6LFQ
KT6 7LL G8UBJ
KT6 7LL M6SAL
KT6 7NA G1OEP

KT6 7NR G0KEB
KT6 7PP G8KFS
KT6 7QH M6TCD
KT6 7SX 2E0CEB
KT6 7SX M6HVS
KT6 7SZ G7STL
KT6 7UN M0DWG
KT6 7UN M3HZK
KT7 0BL M0DHP
KT7 0YH G7WEP
KT7 0YN G4CPV
KT8 0DW 2E0LXQ
KT8 0DW G5LUX
KT8 0DW M7JHR
KT8 1QF 2E0GXF
KT8 1QF M6GXF
KT8 1SU 2E0GPH
KT8 1SU M0PGH
KT8 1SU M6GHT
KT8 1SU M6PHT
KT8 1TE G7HMU
KT8 2ET G1DZB
KT8 2ET G4YKK
KT8 2EX G8CLW
KT8 2HY G0KDI
KT8 2PY G1YVV
KT8 2QJ 2E0JIZ
KT8 2QL M1CEA
KT8 9AN M0CGW
KT8 9DG 2E0LEO
KT8 9DG 2E0TJS
KT8 9DG M3NCP
kt82el M7YOB
KT82EX M5AAT
KT9 1BW G0DOE
KT9 1BY M6HLW
KT9 1EF M0AWN
KT9 1JY G0NCH
KT9 1NL G4HJY
KT9 1PL G6XKF
KT9 1PN G0WIT
KT9 2BJ M6IJV
KT9 2BN M6JFN
KT9 2BU G1SIP
KT9 2DE G1RMC
KT9 2DE G4LBM
KT9 2DE G7GLR
KT9 2DE G8RWH
KT9 2DT G6DFM
KT9 2EY G1JRR
KT9 2EY G1NYP
KT9 2HG G6KIE
KT9 2JE G6YOG
KT9 2LA G0SDF
KT9 2NN G6HMG
KT9 2QD G3SXW
KT9 2SA G4OFO
KT92QS 2E0RBH
KT92QS M3VYS
KW1 4DN GM7BNF
KW1 4DN GM7CLM
KW1 4DN GM7REF
KW1 4NT GM4JUE
KW1 4PF MM6EER
KW1 4PF MM6EEX
KW1 4PN MM6DRJ
KW1 4PN MM6JUE
KW1 4PN MM6WKC
KW1 4QT GM1GXH
KW1 4RG GM4MPR
KW1 4RX GM8EXU
KW1 4UD GM4JYB
KW1 4UD GM6KON
KW1 4XF 2M0HNM
KW1 4XF MM7MNY
KW1 4XP GM4OFI
KW1 4XP GM7UWL
KW1 4XX MM6CXJ
KW1 4XX MM6SZA
KW1 4YB 2M0SKR
KW1 4YB MM0WIC
KW1 4YB MM7WIC
kw1 5aj MM7WKK
KW1 5AS GM4XLN
KW1 5BJ MM7BCL
KW1 5EY MM0HDW
KW1 5HQ GM4RAI
KW1 5HW GM8LFB
KW1 5NE MM6EFH
KW1 5NF 2M0WIC
KW1 5NF MM0WKY
KW1 5NF MM6NBI
KW1 5NL 2M0WTN
KW1 5NL MM6PSX
KW1 5NL MM6WCK
KW1 5SS GM7JGH
KW1 5TF GM4DZX
KW1 5TU MM0DUR
KW1 5TU MM5AJW
KW1 5TU MM6VRH
KW1 5UG MM6PKO
KW1 5YJ GM6WFP
KW10 6RA MM6RQD
KW10 6TL 2M0SOP
KW10 6TT 2M0WTT
KW10 6TT MM6REV
KW12 6UZ MM3YLP
KW12 6XJ 2M0CEE
KW12 6XJ GM4NHX
KW12 6XJ MM0RJR
KW12 6XJ MM6AHK
KW12 6YE GM0MST
KW14 7ES GM6TMH
KW14 7ND GM6TJD
KW14 7NW GM6YQA
KW14 7QA MM3FMY

KW14 7RG GM3POT
KW14 7TB GM6LDG
KW14 7UE MM0BDJ
KW14 7XB GM8YRE
KW14 7XB GM8YRX
KW14 7XH GM4RKH
KW14 7YJ GM8NYV
KW14 8DD MM7THU
KW14 8JE MM7DKX
KW14 8JP MM6EWX
KW14 8SY MM3PPA
KW14 8TG GM6JRX
KW14 8TU 2M0FRR
KW14 8TU MM6HNQ
KW14 8UG GM6JNQ
KW14 8UG MM6FNQ
KW14 8UG MM6MZG
KW14 8UT GM4EFR
KW14 8XN GM0TKB
KW14 8YT GM1BAN
KW14 8YT MM0PFH
KW14 8YT MM3YHA
KW14 8YT MM3YMU
KW14 8YT MM6FGR
KW14xp 2M0HZR
KW14xp MM7BWT
KW15 1EE GM1RQD
KW15 1FF GM0GSG
KW15 1FF GM0HIG
KW15 1FP GM0OYU
KW15 1FP GM7FLZ
KW15 1FU MM0RDD
KW15 1LN GM4LNN
KW15 1NA GM4ZZH
KW15 1NA GM6WOF
KW15 1QR MM1DXU
KW15 1RL GM0MHS
KW15 1SL GM8NFG
KW15 1SX GM0OWM
KW15 1SX GM0WED
KW15 1SX GM3IBU
KW15 1SX
MM0MWW
KW15 1SX MM0ORK
KW15 1SX MM3SUS
KW15 1SX MM3TLH
KW15 1TF MM6PHG
KW15 1UU GM0EEY
KW15 1UU MM0SLB
KW15 1XE 2M0DES
KW15 1XE MM3XUI
KW15 1XJ GM1VZG
KW15 1XR GM1MWK
KW16 3DJ GM3PLO
KW16 3DJ MM7DRG
KW16 3DR MM6MJG
KW16 3EH MM6HEQ
KW16 3EP GM4IOB
KW16 3EX GM0PMO
KW16 3HA MM7LWZ
KW16 3LB GM6GJW
KW16 3NP GM0BXR
KW16 3NP MM3UBB
KW16 3NZ MM3JKX
KW16 3PQ GM8OFQ
KW16 3PQ MM3SDP
KW163NP GM1BZD
KW17 2AN GM1XPE
KW17 2AN GM3HDT
Kw17 2AN MM0BYB
KW17 2BA GM0CVP
KW17 2BA GM4WUR
KW17 2EB GM7RDY
KW17 2EH MM0EAX
KW17 2EZ GM4WMM
KW17 2HR GM0HTT
KW17 2HW MM3VIV
KW17 2JS 2M0LLU
KW17 2JS MM3LLU
KW17 2JU GM0HTH
KW17 2LQ GM7RDH
KW17 2NN GM0IDV
KW17 2NS MM3WJD
KW17 2NS MM5DWW
KW17 2NY GM7VKN
KW17 2PJ GM6BNS
KW17 2PZ MM6RIE
KW17 2QL GM3POI
KW17 2QL MM0DGI
KW17 2QL MM3POI
KW17 2RB MM6ZRX
KW17 2RD GM1MXE
KW17 2RF GM0HQG
KW17 2RL GM8WJK
KW17 2RP 2M0ORK
KW17 2RP MM0MDH
KW17 2RP MM6ORK
KW17 2RW MM7AGE
KW17 2SA 2M0URT
KW17 2SA MM6OGU
KW17 2ST MM3GQY
KW17 2SU GM0KTH
KW17 2TR GM7GJM
KW172TT 2M0ESL
KW172TT MM0OMG
KW172TT MM6ESL
KW2 6AA MM7DKH
KW3 6AS GM6YGW
KW3 6AT MM3SCQ
KW3 6BD MM0PFN
KW3 6BX GM4XRT
KW6 6EA GM3YKP
KW8 6HH MM6BQG
KW8 6HP GM0HLV
KW9 6NN GM4SJB
KY1 2AT GM0TKE

KY1 2HS 2M0DQY
KY1 2HS MM0SNK
KY1 2HS MM6DQY
KY1 2JG GM7DAJ
KY1 2XP GM7CNW
KY1 2XP MM0CWI
KY1 4LB GM0RSE
KY1 4LB GM4PSL
KY1 4LB
GBM0SSB
KY1 4LB MM0TGB
KY10 3AU GM8YUM
KY10 3HU GM4EOU
KY10 3UD GM0KDO
KY10 3UJ GM4FQE
KY11 1BP MM7KTO
KY11 1EP MM7GDI
KY11 1ET GM3KJZ
KY11 1LD GM4UKG
KY11 1PG MM7OOF
KY11 2EZ GM7PXJ
KY11 2HJ GM8YAQ
KY11 2LL 2M0DSY
KY11 2LL MM6OAI
KY11 2LX GM0PUN
KY11 2QW MM3GVE
KY11 2SL MM0CZK
KY11 2SS GM0DYD
KY11 2SS GM0GBH
KY11 2TJ 2M0CLU
KY11 2TJ MM0TIE
KY11 2TJ MM6BKP
KY11 3HS GM8SZS
KY11 3JG GM4YMI
KY11 3JY 2M0SZC
KY11 3JY MM0SKX
KY11 3JY MM3ZCS
KY11 3LG GM3TAL
KY11 3LH GM3UDK
KY11 3LJ GM4JKT
KY11 4AX MM0AOL
KY11 4BW MM3RUZ
KY11 4NA GM0OVD
KY11 4PE 2M0LRO
KY11 4PE GM3HYX
KY11 7HZ GM4FAU
KY11 8BF GM1ZVJ
KY11 8BH GM0TKC
KY11 8BT MM3LWO
KY11 8DH 2M0CSX
KY11 8DH MM0KZJ
KY11 8DH MM3CSX
KY11 8DH MM3UUT
KY11 8DP GM8IVR
KY11 8GE MM0MLB
KY11 8JL GM0SXO
KY11 8RA GM4HVM
KY11 9AA GM8PAH
KY11 9GT GM0VUY
KY11 9LF GM3OZJ
KY11 9LH GM4GZW
KY11 9LR GM8IXZ
KY11 9NG MM0KOZ
KY11 9SJ GM7IHJ
KY11 9TD MM1CHQ
KY11 9UA GM3SHR
KY11 9UQ GM4BRN
KY11 9UQ MM1CIR
KY11 9UW GM4GRC
KY11 9UW MM0DYX
KY11 9XR GM0LLJ
KY11 9XX 2M0JOK
KY11 9XX MM6JOK
KY11 9YE GM3VYJ
KY11 9YE GM4WZG
KY11 9YU GM3RXZ
KY111AZ 2M0XCS
KY111AZ MM0XCS
KY112ng MM7CRY
KY12 0AH GM8VKN
KY12 0BY MM0TIR
KY12 0DR 2M0RDH
KY12 0DR MM0RBR
KY12 0DR MM6JOH
KY12 0DR MM6ROH
ky12 0dr MM7KEH
KY12 0EP MM6HUE
KY12 0HN GM0IXO
KY12 0JX MM6SHB
KY12 0NL MM0TBY
KY12 0NL MM3XSF
KY12 0RL MM0KTE
KY12 0XW MM3EXW
KY12 0YX GM0RLZ
KY12 7EF GM4FSF
KY12 7RG GM1YGW
KY12 7TB GM3NVQ
KY12 8PT GM3NCO
KY12 8QS GM8SNE
KY12 8SR MM7AIK
KY12 8SS GM0VOU
KY12 8XL 2M0FLG
KY12 8XL MM6FLG
KY12 8YD GM0SRQ
KY12 8YL GM4ZNX
KY12 8YW MM6USS
KY12 9BE GM3ZSH
KY12 9BX GM4MPC
KY12 9JQ GM0RHP
KY12 9LP GM4NUU
KY12 9LU GM3YND
KY12 9UQ GM7LFT
KY12 9US MM0ONX
KY12 9XG MM6WSG
KY12 9YA GM0URD
KY12 9YD GM6CMQ

KY12 9YG 2M0BRD
KY12 9YG MM0XBD
KY12TS GM6VIU
KY13 0PD MM0MMN
KY13 0PD MM7IMC
KY13 8DN GM4HJO
KY13 8QY MM0GNH
KY13 8QZ GM0GTY
KY13 8QZ GM0GUU
KY13 8RT GM4VUG
KY13 8RU GM1ZOX
KY13 8TA GM0KVE
KY13 9JE 2M0YTA
KY13 9JE MM0YTA
KY13 9JE MM6FXQ
KY13 9TB GM3OXA
KY13 9XG GM0VCN
KY13 9XG GM7FTK
KY13 9XU GM3VAL
KY14 7BE MM6DOC
KY14 7DX MM0IQK
KY14 7ES MM3OJR
KY14 7HB 2M0EWY
KY14 7HB MM6EWY
KY146DN 2M0IQU
KY146DN MM3IQU
KY15 4AE MM6AUI
KY15 4AH MM7BYH
KY15 4HS GM0SEF
KY15 4NA GM7STI
KY15 4NB MM3TOV
KY15 4PT GM3WKB
KY15 4SS GM0OFL
KY15 4UG GM3LAW
KY15 4UG GM4ENF
KY15 5AA MM7JDL
KY15 5BS GM6ZAK
KY15 5BW GM6SZJ
KY15 5DA GM4ZCV
KY15 5JF MM5AMV
KY15 5JS MM0DWF
KY15 5LH GM0JRQ
KY15 5PJ MM0MUL
KY15 5QB GM0UZV
KY15 5RF GM4IEF
KY15 5SE GM7CPL
KY15 5SH 2M0ZNQ
KY15 5SH MM3ZNQ
KY15 5SQ MM3ZJY
KY15 5UF GM4RGU
KY15 5WN GM4BUA
KY15 5YU MM6VAB
KY15 7BW GM3SJY
KY15 7NN MM7VWT
KY15 7PT 2M0SGQ
KY15 7PT MM0SGQ
KY15 7PT MM3SGQ
KY15 7QD MM6BMP
KY15 7QX GM0AIR
KY15 7TF GM6GPH
KY15 7TW 2M0GOY
KY15 7TW MM7MCD
KY15 7TY GM0OFM
KY15 7UH GM3VPN
KY15 7UR 2M0BEC
KY15 7UR MM3IOF
KY155DY 2M0ITO
KY16 0AQ MM0IBL
KY16 0FG MM0BGO
KY16 0UE 2M0WJS
KY16 0UE MM6WJS
KY16 0XE GM1MZZ
KY16 8AB MM7XYF
KY16 8JF MM7BEI
KY16 8JP GM6BGL
KY16 8LX MM6NVW
KY16 8NJ MM0HRL
KY16 8NS GM3YOI
KY16 8NT GM4FVO
KY16 8QR GM4FFP
KY16 8RA GM3VWY
KY16 8RB GM0WFA
KY16 8YA 2M0BYK
KY16 8YA MM0TRS
KY16 8YA MM6TRO
KY16 9DA GM4FPZ
KY16 9ND MM0LJA
KY16 9NE GM4ENP
KY16 9NQ GM4HKV
KY16 9NQ MM0DQP
KY16 9UA GM0GKR
KY16 9YB GM1FEM
KY2 5AD GM4XND
ky2 5db MM0IMU
KY2 5DG MM6OMJ
KY2 5PH GM1BLX
KY2 5PX MM3CWO
KY2 5PZ MM0JSG
KY2 5QX MM7WJM
KY2 5RF MM0GSW
KY2 5SD 2M0NAN
KY2 5SD MM0YSK
KY2 5SD MM6CZH
KY2 5SE MM3DFZ
KY2 5SQ GM3YBQ
KY2 5SQ GM7UQM
KY2 5SQ MM0MOC
KY2 5TU 2M0IWD
KY2 5TU GM4DOF
KY2 5TU GM8CZU
KY2 5TX MM6FJS
KY2 5XL 2M0IAH
KY2 5XL MM6UEN
KY2 5YT MM1AEZ
KY2 5YX MM0SAH
KY2 6AL 2M0FVJ

KY2 6AL MM0JJQ
KY2 6AL MM7MXB
KY2 6DR GM0NXO
KY2 6ES GM0LKT
KY2 6FT MM7EQF
KY2 6HR GM0SNT
KY2 6JE 2M0DSU
KY2 6JE MM3YKR
KY2 6UD MM0IKG
KY2 6XP GM0LOO
KY2 6YB GM8ZTV
KY2 6YB MM3MMI
KY3 0AZ MM0AGV
KY3 0BT MM6DQG
KY3 0DQ GM0RDA
KY3 0JW GM0IJR
KY3 0RS 2M0KRI
KY3 0RS MM0WTZ
KY3 0RS MM7KWA
KY3 0RZ GM4TRZ
KY3 0XE 2M0ZBH
KY3 0XE MM0ZBH
KY3 0XH GM0CFC
KY3 0XJ GM4GUL
KY3 9EP MM7FNG
KY3 9EU GM0BUI
KY3 9JQ GM0VFD
KY3 9UR MM7KEF
KY3 9XS GM0CNP
KY3 9YF MM7GHF
KY4 0BG GM1SZM
KY4 0EQ 2M0GTR
KY4 0EQ GM8KSJ
KY4 0FH GM4UPB
KY4 0JF MM0DAA
KY4 0JX MM7CFE
KY4 0NL GM8ZOW
KY4 8AW GM8XUK
KY4 8AZ GM8NVE
KY4 8ER GM7OQE
KY4 8EU GM1ATW
KY4 8EU GM6OOA
KY4 9EP GM7KVB
KY4 9NB 2M0JMI
KY4 9NB MM7MCN
KY5 0AX GM0KLP
KY5 0DG GM8JGB
KY5 0DG MM0WLL
KY5 0EN MM0OVV
KY5 0EN MM3NGJ
KY5 0ND GM3ZGH
KY5 0ND GM4NNH
KY5 0NW MM0RBN
KY5 0NW MM3YEC
KY5 8AL MM0CBI
KY5 8NS MM0PMW
KY5 8NS MM1DPC
KY5 8QA MM7DPL
KY5 9EU GM0CFW
KY5 9JS MM3VNT
KY5 9LT GM1COF
KY5 9QH MM6GIW
KY6 1DU GM0EZR
KY6 1HW GM4SBP
KY6 1JD MM6KCV
KY6 1JP GM0MRJ
KY6 1JR MM3IQA
KY6 1LZ 2M0KPE
KY6 1NA GM0CFK
KY6 2BQ GM4DEX
KY6 2JU 2M0SXT
KY6 2JU MM3SXT
KY6 2LU MM7BYT
KY6 2PA GM0IJD
KY6 2QA GM1JZM
KY6 3AX GM7BOZ
KY6 3LH GM7CQQ
KY6 3PJ GM1CEJ
KY6 3QQ MM0AWJ
KY6 3QR GM8ZQY
KY7 4ES GM3NMN
KY7 4HP MM7IFC
KY7 4HS GM4TNP
KY7 4HS GM7FRC
KY7 4HS MM3OYB
KY7 4PH GM4ZJI
KY7 4RF MM3KMX
KY7 4RU GM6AOJ
KY7 4RU MM3BHG
KY7 4RW GM0GTL
KY7 4SJ GM3VVF
KY7 4SS GM6HGW
KY7 4SS MM7MTB
KY7 4UE GM7FWA
KY7 5AJ MM7KGC
KY7 5AX GM1ADI
KY7 5AX GM4XAV
KY7 5DF MM3EHM
KY7 5TA 2M0DSL
KY7 5TA MM0KDY
KY7 5TA MM6VXR
KY7 5TD GM7GKT
KY7 5TD MM3LYS
KY7 5TE GM3XMY
KY7 6BN GM7CTV
KY7 6HX MM3ATI
KY7 6LA GM3OBC
KY7 6TA MM6INN
KY7 6UP MM3XGP
KY7 6UQ MM7CRB
KY7 6US 2M0SOV
KY7 6UT MM6HRZ
KY7 6UY MM6XHU
KY7 6XE GM0IOA
KY7 6XX GM0MNV
KY7 6YL 2M0AGS

KY7 6YL GM3KMF
KY74EA MM3VYY
KY8 1BG MM6FSV
KY8 1DJ MM6GPD
KY8 1DW MM6CFS
KY8 1EZ MM3KYO
KY8 2EQ GM4PWQ
KY8 2HA GM6NYT
KY8 2JL MM7CJS
KY8 3AR MM0HSB
KY8 3DH GM4HBQ
KY8 3QG MM6EHL
KY8 4DF GM7VQB
KY8 4HZ MM6GQP
KY8 4LH GM0GUJ
KY8 4QE MM6BBK
KY8 4QN GM3RUI
KY8 5BD MM6CHV
KY8 5BP MM3WWP
KY8 5DL MM3TWW
KY8 5DZ GM7KPE
KY8 5EN MM7GGD
KY8 5FA GM4FEO
KY8 5FL MM7AOY
KY8 5HA 2M1JKG
KY8 5JH MM3ZKQ
KY8 5LT GM1EAH
KY8 5LX MM7ASN
KY8 5QP GM4FYH
KY8 5QP MM1DJJ
KY8 5QP MM6OXA
KY8 5SW GM3AHR
KY8 5TL GM0IET
KY8 5TR GM4EJI
KY8 5XA GM0IWX
KY8 6HH GM6BGQ
L N12 G4NMP
L1 9BR M7HFB
L10 4XL G4VKV
L10 4YW M0RDC
L10 7NF 2E0LAR
L10 7NF M6KBY
L10 8LU M0EBD
L108JN M7COH
L11 0AY M7CFR
L11 1AN G7SZF
L11 1AY G6NRK
L11 1AZ 2E0PUE
L11 1AZ M3PUE
L11 1BD M6BDR
L11 1BD M6SHV
L11 1DF G0WRE
L11 1DF M3XPS
L11 2YH 2E0JPC
L11 2YH M3JNQ
L11 2YH M6LLC
L11 2YH M6LLD
L11 2YH M6LLE
L11 3BQ G8WAW
L11 4UW G6OPV
L11 7DJ M6XWB
L11 7DS M6KBT
L11 7DW M3ZYG
L11 8LR G0JSZ
L11 8LR G1TIQ
L11 9AG G4XAR
L11 9DP G4WWB
L12 0AW M6DKC
L12 0HX G8XTD
L12 0NF 2E0HLO
L12 0NF M0SGG
L12 0NF M6LHO
L12 0QN M6BQQ
L12 3HQ G6NFR
L12 3HS 2E0FXW
L12 3HS M7RFD
L12 4YR G4VYR
L12 6RE M0YYV
L12 7HE G6WIO
L12 8RN G0MXW
L12 8RW G1AVC
L12 9EQ 2E0BZE
L12 9EQ M3YVD
L12 9GY G0SMR
L12 9NE G8YGM
L13 0AY M3GYY
L13 2BA M0HQQ
L13 3DY G6BFM
L13 4DH 2E0DKO
L13 4DH M6IQO
L13 7DJ G3OSI
L13 8AW M6WSO
L13 9AA M3SPY
L139DF M3UFY
L14 0LG G3RBD
L14 0NU G7DUY
L14 3LE M6SMQ
L14 3LF M3HSC
L14 3LT G7NHQ
L14 6UT M6RKA
L14 9PA G0SXA
L14 9PA M5CAB
L15 0HN M3GAP
L15 3JG G4KRN
L15 4JX M0BFM
L15 4ND G7DOD
L15 4PW G0STF
L15 5AX G0OSC
L15 6TG G6VNW
L15 8HX G6KXJ
L16 3NE G6IMJ
L16 5ET G3ZIM
L16 5EU G8HTF
L16 6AD M6KFP
L16 7PA G0VMT
L16 7PQ G0KBS

L16 7QR G7GEE
L16 8NZ G3RWW
L17 0DP G6EQL
L17 1AB G4IHS
L17 5DB G8FXA
L17 6AD G0ADP
L17 7AL G1JHG
L17 7DT G8JJF
L17 8TT 2E0WJT
L17 8TT M0TWJ
L17 8UD G0EVB
L17 8UH G1VAA
L18 1EY G3PDC
L18 1JZ G0MIT
L18 1JZ G7ACK
L18 3EX G8IPK
L18 4PZ M5RMF
L18 4QE G1YEP
L18 4QP M3XSN
L18 7HQ M0NVJ
L18 7HU G1YUX
L18 7HY G5JIM
L18 7HY G6WRC
L18 7JG G7VCN
L18 8AH M3XVU
L18 9SJ G0JUA
L18 9TB G0RBJ
L18 9TN G0MSO
L19 0LS G1RUZ
L19 0NA G4DKQ
L19 1QJ M3XNT
L19 1RB 2E0CFD
L19 2HE M6IAV
L19 2QZ G1BES
L19 2QZ G1KOP
L19 3QX M6HMZ
L19 3RA 2E0XRQ
L19 3RA M7XTX
L19 4TB G4KVL
L19 4TR M3PBB
L19 4TR M7ANE
L19 4UG 2E1PGA
L19 4UR M3OBD
L19 6PQ 2E0LKM
L19 6PQ M6LKM
L19 6PW M3LPJ
L19 6PZ M0CEW
L19 8LF M7JTN
L19 9AN G1ZBU
L19 9BJ 2E0HFJ
L19 9BJ M7BTD
L19 9BX M0BAI
L19 9DG G2BGG
L19 9EA G0GQX
L20 0AL M3DYY
L20 0BY M3AKE
L20 0BY M7OWL
L20 4QR M6LPO
L20 4UU G0WMJ
L20 5AN 2E0IOG
L20 5AN 2E0MTC
L20 5AN M3MTC
L20 5BG G7OLG
L20 5EA G1LOV
L20 6DJ G7GHT
L20 6EJ G0OCC
L20 6ES M3GTT
l20 6hb M7REL
L20 6HF M7EFC
L20 6LU M7MKZ
L20 6NN M3JID
L203HN M3EFC
L21 1BJ M3LNN
L21 1BJ M6OUY
L21 1BP M7PGA
L21 2PE G3YXS
L21 3UB M5SE
L21 4PJ G4BM
L21 7NN G0DBE
L21 7NN M3JIH
L21 7NN M3JII
L21 7NN M6DX
L21 7NX 2E0FQJ
L21 7QH G1DFP
L21 7QH G4VIL
L21 8HG G7OEA
L21 8NQ 2E0HGG
L22 1RJ M0CEC
L22 2AT M7HAL
L22 3YX G1PJK
L22 3YX G1SWE
L22 4RF 2E0ZOM
L22 4RF M3ZZS
L22 8QB G4XRX
L22 onx 2E0TTE
L23 0RF G1MTJ
L23 0RG G6FDI
L23 0RQ G8JYV
L23 0SG M0CMW
L23 0TT M6EMV
L23 1US G3XMG
L23 2RD M1BTD
L23 2TY M6NQP
L23 2XF G4OJK
L23 3BN M6BFB
L23 3BZ G3IZF
L23 7UL G1KLK
L23 8SS M3IHR
L23 9UF G8WMG
L23 9XE M0TER
L23 9XE M1NIS
L23 9XE M3BMW
L23 9XJ G3VXK
L23 9XS G6GOG
L23 9XY G0MJG
L23 9YU G3LNL

L23 9YU G8BVU
L24 3TH G4CMP
L24 5RR M0NOW
L24 5RR M1SPW
L25 0NY G7RYW
L25 0QD G0AJL
L25 1PR M1CEW
L25 2RB G0MFH
L25 2RP M3NVL
L25 3PY G0KWD
L25 3QE G8NNX
L25 3SE G1CPU
L25 4SQ G3XSN
L25 4TJ G6BOK
L25 5PN M3XCS
L25 6DG M0BGT
L25 6DH M0PMO
L25 6DW G3KRX
L25 7UH G1CZH
L25 7UJ M3AVF
L25 8QG 2E1KLT
L25 8QG M0KLT
L25 9QY M3PPZ
L25 9SN 2E0OAU
L25 9SN M0OAU
L25 9SN M6OAU
l250pr M7VAL
L26 1UN M0JYM
L26 1UW M7GWW
L26 1UW M7NCW
L26 6LW G0RFI
L26 7WR M6GKI
L26 7YW G7NHF
L26 9TL M1DHM
L26 9uf G0ARV
L26 9XZ M0BJD
L278YL M7DFO
L28 3QB M0LPL
L28 3QB M3HAM
L28 5RJ M6KDC
L3 4BP M3SBQ
L3 6JF M3KXG
L3 8PZ G4DLY
l30 0ph 2E0ILF
l30 0ph M6POB
L30 1PR M6MBX
L30 1PX M6BFP
L30 1PZ 2E0ECU
L30 1PZ M6FYE
L30 1SG 2E0XTC
L30 1SG M0TNX
L30 1SG M0WYR
L30 5QD G7IKM
L30 5RW 2E0FCG
L30 5RW G2OA
L30 5RW M0OYH
L30 5RW M6OYH
L30 7RN M0GSO
L31 0BD 2E0TXP
L31 0BD M0TXP
L31 0BD M6TXP
L31 0BS 2E0EJO
L31 0BS M6IQG
L31 0BW G8ZTF
L31 0DA G4OVD
L31 0DB 2E0FKD
L31 0DB M3LPR
L31 1DY G4WGN
L31 2JS G1ZEW
L31 2NN G0PWL
L31 3DN 2E1IIE
L31 3DN M3AZR
L31 4ED M7HLG
L31 5JJ G0DOR
L31 5NL G1DGW
L31 5PD G7FMW
L31 5PD M0AJJ
L31 6AS G1ACD
L31 6BS G4HDU
L31 6BY G3WDD
L31 7AN G8IIS
L31 8AE M3RJF
L31 8AX 2E0TRJ
L31 8BZ 2E0PRO
L31 8BZ M3AIR
L31 8EG G8GTI
L31 9AL G4AGY
L31 9DE M1BWZ
L31 9DG M6CUX
l316dy 2E0DLK
L316DY M0HVE
L317BP 2E0KEY
L317BP M0YOB
l317bp M6MQK
L32 0TT M0TMC
L32 0TT M3FOD
L32 0TT M3OPM
L32 0TT M3XNB
L32 1TP M1BWH
L32 7QG 2E0PVW
L32 7QG M6XIP
L32 7QH G0XBL
L32 7QR G6OWS
L33 1UW G0JIB
L34 1LW G3YCX
L34 1NH M3IJD
L34 1PX 2E0SJR
L34 1PX M0SJR
L34 1PX M3ITI
L34 2RS G4WCO
L34 2RY G6CKM
L34 6HD M7VOD
L34 6HU M6GOF
L34 9EH M6ARY
L35 0QH G1BEB
L35 0QR G1YCN

L35 0QR G6FIT
L35 1QG G8MKO
L35 1RJ M0FCB
L35 1RJ M6MNV
L35 1SF G4XNV
L35 2XX G1NVY
L35 2YG G0LPQ
L35 3JL G6MMG
L35 3PL 2E0PFH
L35 3PL M6PFH
L35 3PP G0IEW
L35 3RW G0ILC
L35 3SF G0RPD
L35 3SG G1CKV
L35 3UT M6JID
L35 3UX M6HCE
L35 4PA G3RJE
L35 4QH 2E0FZZ
L35 4QH G4TTO
L35 4QH M7DYS
L35 6PJ G3WOH
L35 7NE M3LXA
L35 7NG G4VCP
L35 8NE G0SLK
L35 8PY 2E0YAR
L35 8PY M3ZVI
L35 9JQ 2E0VCK
L35 9JQ M7VCK
L35 9LX 2E0MJZ
L35 9LX M7JVH
L36 0TU M3KAY
L36 0XA M3RSN
L36 1TA M3OHR
L36 1TH G4EGM
L36 1UD G0NYR
L36 1XX 2E1CYD
L36 2LT M6GSP
L36 2LX 2E0IXS
L36 5SF G1ATG
L36 5TL M3CYS
L36 5TL M3TNY
L36 5TN G3TVN
L36 6ER M0ANN
L36 8BN 2E1CYS
L36 8BN G0LSV
L36 8DA M0HHR
L36 8EH G7AWW
L36 8EH M6IHR
L36 8JE M6PSV
l360uh M7TON
l369ty 2E0EMQ
l369ty M6IFA
L37 0AH G4PKP
L37 2DG G3MIP
L37 2JD G6OEM
L37 3JL G6CIP
L37 3LB M0CPL
L37 3LX G3POG
L37 3LY 2E0GMU
L37 3NH 2E0BZG
L37 3NH M0HUZ
L37 3NH M3ZIH
L37 4BP 2E0MXN
L37 4BP M0MXN
L37 4BP M7RAY
L37 6AD G8DUF
L37 6AY 2E0WFG
L37 6AY M3WFG
L37 6BF G7CVZ
L37 6BF M0DGA
L37 6BG M6CJA
L37 6BQ G4MZZ
L37 6DF G4SMT
L37 6FA G0CTC
L37 7ER M1DLM
L37 7EX M0COO
L38 0BG G0OWP
L38 3RP 2E0XDG
L38 3RP M0XZG
L38 3RP M6XDG
L39 0EB G6HIO
L39 0HW G6BZL
L39 1LJ M7NWL
L39 1LR G0OXV
L39 1NN G3SZV
L39 1NU M6GXZ
L39 2BA 2E0ZMO
L39 2BA M3ZMO
L39 3LD G4ATU
L39 3NS 2E0GPC
L39 3NS M0YIG
L39 3NS M6GPC
L39 3PX M6ZEB
L39 4RE G8TUN
L39 4TF G7LFC
L39 4TF M0NCI
L39 4TF M0NLI
L39 4TF M0ORM
L39 4TF M3LFC
L39 4TF M6BUB
L39 4TF M6COV
L39 4UD G8FUB
L39 4XP G4TAK
L39 5AP G6ODU
L39 5AT G0BLW
L39 5AY G3FIC
L39 5BG G3NNT
L39 5QG 2E0ORM
L39 5QJ G4UVB
L39 5QJ G6EXC
L39 6RF G8SAX
L39 6SU 2E0GKU
L39 6SU M7BAK
L39 6SY M6EEG
L39 7JL G7SWV
L39 7JU G7HSB
L39 7JY 2E0XJP
L39 7JY M6XJP
L39 7LD 2E0OLG
L39 7LD M3BHP
L39 7LD M5DR
L4 1UL 2E0VWL
L4 2UN G0OIU
L4 3SX M3KCC
L4 5RB 2E0MXM
L4 5RB M0HHA
L4 5RB M6MXM
L4 6TJ 2E0EPU
L4 7TL 2E0HYU
L4 8TY M0WHC
L4 8UL M1SWB
L4 9UG G7UZA
L40 0RA G6ITU
L40 1TB M7BFZ
L40 1TT G4CLE
L40 2QJ M0RDA
L40 2QS G0UCS
L40 3TF G0SGI
L40 3TG 2E0EUR
L40 3TG M6EKP
L40 5AA G7EVY
L40 5BE 2E0SSN
L40 5BE M0SSN
L40 5BE M6SSN
L40 5TL G1JZZ
L40 5TN G0DDU
L40 5TU M0HNG
L40 5TU M6CUV
L40 5UA G6YGP
L40 6HQ 2E0WHD
L40 6HQ M6SPT
L40 6JG G1XKL
L40 6JQ G6NJJ
L40 6JR G4LTI
L40 7RB G0AFQ
L40 7RP G0CBB
L40 9QG M0DWZ
L40 9RS G4IQJ
L42 2BR G0DYG
L5 0TD G0ELZ
L5 0TD M0LBL
L5 2RF 2E0GNN
L5 2RF 2E0SMK
L5 2RF M3GNN
L5 2RF M6SMK
L5 3PD M0JCZ
L5 6RH M6LDK
L5 6SE 2E0KHI
L5254AD M7DOZ
L6 0BZ G1KON
L6 2NB G7JCQ
L6 2NG M3FGX
L6 6BN G0EDF
L60 0BD G4XCM
L60 6BB G0KTY
L7 0JE M0RAT
L7 3EN M6BZB
L7 8SS M7YVR
L8 0UF G8DEY
L8 4TH 2E0SNA
L8 4TH M0SET
L8 4TH M3YRO
L8 4XP M7LPL
L8 4XR G1MCW
L8 6QL G8ADA
L8 8NE M3WUN
L9 0NA G4XCY
L9 0NG G7VGK
L9 1AG G1EKM
L9 1AN G8CCD
L9 1EF M6SUY
L9 1JZ 2E0LPR
L9 1JZ M3MTL
L9 3BL M3ZUC
L9 3BP 2E0IRJ
L9 3BP M0OES
L9 3BP M6OES
L9 7LG 2E0CLD
L9 7LG M0JPW
L9 7LG M0MSY
L9 7LG M0ODD
L9 7LG M0OGI
L9 7LG M6VMP
L9 8HF G4YPD
L9 9AY G0DHJ
L9 9BZ G0OIX
L9 9EJ M3DLP
L9 9HQ G7JZK
LA1 1AE M7JKA
LA1 2AS G0FZA
LA1 2HA M3LCZ
LA1 2HE M6CPH
LA1 2HR G8XQS
LA1 2NA M6SPH
LA1 2NW M1JHG
LA1 2QP M3KHM
LA1 2QP M3KIQ
LA1 2QP M3KNV
LA1 2QW M3APO
LA1 2QX M3UFT
LA1 2TZ G3VWJ
LA1 2UQ G1OHH
LA1 3AW G6ZHL
LA1 3BN M6IRU
LA1 3DS G7BBN
LA1 3DU G8EXJ
LA1 3DY G8UHO
LA1 3FA G0VGP
LA1 3HJ M0BUR
LA1 3ND G1HFA
LA1 3SY 2E0FYE
LA1 3SY M3ZFY
LA1 4BD G1YYD
LA1 4ER G0AWM
LA1 4HX M1GDE
LA1 4HX M3GDE
LA1 4LQ G8TZJ
LA1 4LZ G4UGR
LA1 4NT M1FHQ
LA1 4PZ G3NNA
LA1 4QZ G4EKZ
LA1 4QZ M0RGS
LA1 4RF M6RDZ
LA1 4SQ M6HLL
LA1 4TS M3BIO
LA1 4UU 2E1GPV
LA1 5BD G4DLP
LA1 5BY 2E1YES
LA1 5EB G3XEN
LA1 5EB G8KED
LA1 5FS G6BJB
LA1 5JA M3VZH
LA1 5JD G0LLG
LA1 5JP M0AFF
LA1 5LB G6PHF
LA1 5RS 2E0MBA
LA1 5UY M7ENI
LA10 5AB M1ONE
LA10 5AB M3ONE
LA10 5AY 2E0CXU
LA10 5AY M6CYG
LA10 5AZ G1GVJ
LA10 5AZ G4MIS
LA10 5DE G1EKU
LA10 5JE 2E0CXJ
LA10 5JE M6CQV
LA10 5PH G4MWG
LA10 5PJ G6NQL
LA10 5QB G3LUO
LA10 5TQ G6LZZ
LA10 5TQ G6SSQ
LA11 6AZ G0KDG
LA11 6DL G4ATB
LA11 6DP G3LZZ
LA11 6EB G8JAG
LA11 6EB G8KPL
LA11 6EX M0GKJ
LA11 6JZ M6LWG
LA11 6PH M0DCZ
LA11 7LJ G4CEJ
LA11 7PF G0WHD
LA12 0AE M0KLJ
LA12 0DB G6ZGH
LA12 0JB 2E0FZK
LA12 0JB G1DXM
LA12 0JF G1BMB
LA12 0LL M0RBE
LA12 0PZ M0TES
LA12 0QH G8ALE
LA12 0RE 2E0AYW
LA12 0TB M3MEU
LA12 7ES G4UCL
LA12 7EU M0KYL
LA12 7JA M6DUI
LA12 8AA 2E0LCE
LA12 8AA M6IPL
LA12 8AW G0AOM
LA12 8NQ G3XTN
LA12 9EA G6NVF
LA12 9JP M1RDX
LA12 9LS M0POS
LA12 9NU G7GUB
LA12 9PD 2E0HHC
LA12 9PD G4EOB
LA12 9PD G6LKB
LA12 9PN 2E0URK
LA12 9PN M0MJX
LA12 9PN M6OUN
LA12 9PN M7GJA
LA12 9QZ G4PXR
LA12 9RL M3WTA
LA129EX M7ULV
la12ay M5TUE
LA13 0AU G0SKJ
LA13 0BB M6MDJ
LA13 0BH 2E0TCC
LA13 0BH M6AMO
LA13 0BH M6NJD
LA13 0BX G4AFR
LA13 0EE G3ZFZ
LA13 0EF M7FLY
LA13 0EQ M5RAG
LA13 0HE 2E0SLK
LA13 0HE M6BLZ
LA13 0HU G4ZEG
LA13 0NB 2E0HZK
LA13 0NB M6ZVD
LA13 0UB M7BCA
LA13 9EA G4ZOI
LA13 9ES M7ATV
LA13 9HJ M6AGY
LA13 9HU M6RLX
LA13 9PX M6NEZ
LA13 9QG G7UKR
LA13 9QG M6WSU
LA13 9RG G1GBF
LA13 9SF M3TXP
LA13 9TD M0MOL
LA14 1BQ G0MOM
LA14 1BQ M3FUR
LA14 1SZ 2E0FAS
LA14 1SZ M3CYW
LA14 1XU M6XAT
LA14 2AD M0FGB
LA14 2AD M1WDX
LA14 2BG 2E0HTC
LA14 2BG M6AUE
LA14 2BJ 2E0TXH
LA14 2BJ M7TNY
LA14 2BU G0TUE
LA14 2ER M6MLH
LA14 2HH M6AHF
LA14 2JG M6FWO
LA14 2PZ M6RPL
LA14 2RX G6UMN
LA14 2UT M3PAU
LA14 3AN G4RQJ
LA14 3BP M1MIJ
LA14 3DE G1DNI
LA14 3DE G4NFR
LA14 3DX 2E0LBI
LA14 3DX M6LFB
LA14 3EJ G0GSJ
LA14 3JL M6WCN
LA14 3LP G1AGM
LA14 3LP G1CGP
la14 3nb M6OTU
LA14 3NU M3HQW
LA14 3NU M3UFU
LA14 3NU M3UQF
LA14 3PR M6DHV
LA14 3PS M1MCW
LA14 3PS M3YBL
LA14 3QW 2E0NGC
LA14 3QW M0IGG
LA14 3QW M6LRA
LA14 3SF M6FWT
LA14 3TS M1AVV
LA14 3UD 2E0MYK
LA14 3UD M0MYK
LA14 3UD M6CET
LA14 3UZ M6TBT
LA14 4AH G0GJM
LA14 4LE M0REJ
LA14 4LE M0SJJ
LA14 4LE M3GEN
LA14 4LR M1SKY
LA14 4NH G4USW
LA14 4NW M6MRN
LA14 4PA G0UXH
LA14 4PH G4BZR
LA14 4PN M7CRZ
LA14 4PZ G7OXB
LA14 5DW 2E0TEI
LA14 5DW 2E0TEJ
LA14 5DW G2DT
LA14 5DW G4ARF
LA14 5DW M0TEB
LA14 5DW M6DTF
LA14 5DW M6GUM
LA14 5HW G6YXX
LA14 5LH M3MWV
LA14 5LY M6WAD
LA14 5QQ G0LSU
LA14 5RL M3YNS
LA14 5RQ M6ANX
LA14 5SE M6EEO
LA14 5TP M3UDJ
LA14 5TX G7UAK
LA14 5XP G3IZD
LA14 5XP M3ZZV
LA15 8BW G7RNX
LA15 8EU M1CQF
LA15 8LZ G4VEK
LA15 8NE M3UBE
LA15 8NR 2E0MND
LA15 8NR M3MND
LA15 8QA G4SPW
LA15 8QA M0SSD
LA15 8QA M6PDT
LA15 8SE G7MCE
LA16 7EW M3HVL
LA16 7EY G0HIK
LA16 7EY M3MGU
LA16 7JG 2E0CPG
LA16 7JG G3DCT
LA16 7JG M0KPW
LA16 7JG M6AYY
LA17 7XJ G4ATK
LA18 4LS G7PKQ
LA18 5DB G1NBY
LA18 5HB G8JHG
LA18 5HE G1ZIM
LA19 5UE G6HEF
LA19 5UE G6HFB
LA19 5XL M1AOD
LA19 5XL M7RAB
LA19 5XN M3TZI
LA2 0EG G4UJI
LA2 0PN M0SWI
LA2 0PN M7RER
LA2 6AG G4TOT
LA2 6ER G0VPU
LA2 6HJ G0NTT
LA2 6LB M0BZR
LA2 6NH G4DPT
LA2 6NL G7JPT
LA2 6QB G0UWX
LA2 6QJ G4ZEZ
LA2 7AB M0DGK
LA2 7AG G0OIE
LA2 7BA M0DWK
LA2 7BF G7IHD
LA2 7EP G4LIY
LA2 7EP M7PWW
LA2 7ET G1JYB
LA2 7ET M3YOG
LA2 7HX G8CSQ
LA2 7JN G1KLZ
LA2 7JZ 2E0CBL
LA2 7JZ M3ZMX
LA2 7LP M6XSS
LA2 7ND 2E0XRL
LA2 7ND M0YYT
LA2 7ND M1BUU
LA2 7ND M7SLR
LA2 8ER M0XPL
LA2 8NP G1LAT
LA2 9AN G4KWJ
LA2 9AN G4KWK
LA2 9BD M6AHE
LA2 9DL 2E0ZDX
LA2 9DL M6DCB
LA2 9LF G4NEQ
LA2 9NH 2E0MAF
LA2 9NH M0MOS
LA2 9NH M3WCZ
LA2 9NY G1CFJ
LA2 9PB G3UVQ
LA2 9PJ G4ZJE
LA2 9PJ M0CLR
LA2 9PJ M1TCQ
LA21 8EQ G8YGO
LA21 8EQ M0DCY
LA22 9BD G7IFB
LA22 9EX M3CQP
LA22 9RD 2E0GVI
LA22 9RD M7BHT
LA23 1EU M3LQO
LA23 1PX 2E0NFS
LA23 2EL 2E0KMN
LA23 2EL M0NOM
LA23 2EL M6VMS
LA23 2EU G3VGD
LA23 2LA G7IJW
LA23 2LB M6NOJ
LA23 3PW G8IRN
LA26QW G6WBS
LA3 1DW M7BNE
LA3 1ET G0LRE
LA3 1JH G0TOO
LA3 1jn M1BSF
LA3 1LR G7UBP
LA3 1TS G6BTM
LA3 1UR G1RLK
LA3 2BB M0SED
LA3 2JF M6GBL
LA3 2LR G3KOQ
LA3 2LT 2E0TFT
LA3 2LT M6JOG
LA3 2LX G3RCZ
LA3 2PW M3LDF
LA3 2PW M3LQP
LA3 2QJ 2E0DXO
LA3 2TB 2E0HCS
LA3 2TB G4ZJL
LA3 2TB M7LLY
LA3 2UR 2E0TBF
LA3 2UR M6DBE
LA3 3AA G4UME
LA3 3AJ G8BME
LA3 3EF G1HIB
LA3 3HU G6XVY
LA3 3JA G0LWU
LA3 3LF 2E0LJL
LA3 3LF 2E0SFZ
La3 3Qa M0LCY
LA3 3RA M3OBM
LA3 3RA M3UAY
LA3 3RZ G1DXN
LA3 3SH G0JHG
LA3 3SN M6YPE
LA32FH 2E1HER
LA4 4EF M3IAE
LA4 4EW 2E0AZA
LA4 4EW 2E0AZB
LA4 4EW G1CCL
LA4 4HS G4LDS
LA4 4LN M3HAZ
LA4 4LQ G6OUT
LA4 4LQ G7NLR
LA4 4NN 2E0BND
LA4 4NY G0IYU
LA4 4PZ G3CSY
LA4 4QD G0EBP
LA4 4QG 2E0GBO
LA4 4QG M6MUK
LA4 4RE G1ONH
LA4 4TJ G3VSE
LA4 4TU G7CAF
LA4 5LE 2E1DGL
LA4 5LJ G6ZJN
LA4 5LT G1UKS
LA4 5NF M3REM
LA4 5QD G3KSP
LA4 5RA G1PWY
LA4 5RU 2E0SBC
LA4 5RU M0SBC
LA4 5RU M3FRQ
LA4 5SR 2E0JOH
LA4 5SR M0YJO
LA4 5SR M6ALO
LA4 5SR M6LDL
LA4 5SU G6FKE
LA4 5UJ 2E1GNU
LA4 5UJ G0RDH
LA4 5UJ M0SCG
LA4 5XP G4XLC
LA4 6BN 2E0BQG
LA4 6BN M0GRR
LA4 6BT G7OOE
LA4 6BW M6MWC
LA4 6BW M7ASK
LA4 6DD G0FYW
LA4 6DQ M3ZTP
LA4 6DX 2E0XYZ
LA4 6DX M3SNZ
LA4 6DX M5XYZ
LA4 6EJ G0CHV
LA4 6EN 2E0WXZ
LA4 6HD G0GEU
LA4 6HS G0IFN
LA4 6JS 2E0VVJ
LA4 6JS M6TOD
LA4 6LA G1VTE
LA4 6LT G4IFT
LA4 6LU G4ZJO
LA4 6PD G6HMA
LA4 6PS 2E0CKC
LA4 6PS M6APB
LA4 6QG G0AUF
LA4 6QL G8FJA
LA4 6QR G0ASQ
LA4 6TB G0FYH
LA4 6TD G1GZT
LA46NU G4BWX
LA5 0BH G6OOK
LA5 0BX 2E0DVV
LA5 0BX M6EYV
LA5 0UG 2E0EET
LA5 0UG M0HPG
LA5 0UG M0PWD
LA5 0UG M6PEW
LA5 8AQ G1PEE
LA5 8AW G4RHB
LA5 8BL G4YBS
LA5 8BL G6ZET
LA5 8DJ M0FSK
LA5 8EN G0CWP
LA5 8HA G4TRY
LA5 8HE 2E0DTG
LA5 8HJ G0VLV
LA5 8HQ G0SCU
LA5 8HR G6IKE
LA5 8LY G3SZU
LA5 9BE G7MEU
LA5 9EQ 2E0WIO
LA5 9EQ M0WIO
LA5 9HS G4ABE
LA5 9LD M3LCE
LA5 9LG M6AAG
LA5 9NU M0XOB
LA5 9QR G4UGB
LA5 9QZ G4VAP
LA5 9SA M0RGB
LA6 1AX G8EUF
LA6 1DE 2E1GYR
LA6 1DJ G7IER
LA6 1HY G0KSS
LA6 1JE M6EOR
LA6 1QN G6POI
LA6 2AF G3HGD
LA6 2AZ M7AKM
LA6 2JL 2E1TIM
LA6 2JL M1JCB
LA6 2LJ G0JSK
LA6 2LJ G8JAI
LA6 2LJ M0IJU
LA6 3DZ G4VPS
LA6 3GH 2E0RPO
LA6 3GH M6RPO
LA7 7BB G0BBE
LA7 7BB G0KRG
LA7 7BB G0KRS
LA7 7BB G0RLO
LA7 7BB G7KRC
LA7 7BL G7OEW
LA7 7DZ M0DZW
LA7 7EJ G8TTD
LA7 7EJ M0CYR
LA7 7EW G6BKY
LA7 7JB G3JGP
LA7 7JP G1YMV
LA7 7ND M0PRN
LA7 7NL G6UPL
LA7 7NL G6UPM
LA7 7PT G4UXH
LA7 7QA M7DYP
LA7 7QF G6NPC
LA8 0DF 2E0TUE
LA8 0DF M5TUE
LA8 0EN 2E0HOK
LA8 0EN M0KOH
LA8 0EN M6HOK
LA8 0NG M0PJX
LA8 8AT G8VTX
LA8 8DJ G4ZBW
LA8 8HP G3PLX
LA8 8QE G8MGG
LA8 9BT M1IKE
LA8 9HL G4GLH
LA8 9NJ M3OPS
LA9 4EN 2E0SGL
LA9 4EN M7PLH
LA9 4LN G3GEF
LA9 4QR G3VVT
LA9 4QR M3RIA
LA9 4UU G3HMR
LA9 5DU M3RBP
LA9 5HN M3ZOO
LA9 5QE G7OWV
LA9 5QE M0RIS
LA9 6AJ M6LWA
LA9 6AU G1IGC
LA9 6AU G6UYK
LA9 6AU G6YCM
LA9 6AU M0RMH
LA9 6BN G3VJI
LA9 6EA G3WRI
LA9 6HT G4ZBO
LA9 6LB M1KMC
LA9 7ED G7NJD
LA9 7EY 2E0BXM
LA9 7EY M3VHH
LA9 7FE G4WJR
LA9 7FF M6WGU
LA9 7HA G6IPN
LA9 7JA M7RIZ
LA9 7JB M3HSE
LA9 7JG G0KHA
LA9 7JG G0RTM
LA9 7NA G0HIF
LA9 7NT M3ZWG
LA9 7PE G3XJI
LA9 7PL M3TYV
LA9 7PN G0RSL
LA9 7QB 2E0BNJ
LA9 7SN M3XVJ
LA9 8LH G4ETD
LD1 5BH 2W0OZY
LD1 5BH MW0OZI
LD1 5BH MW6CQO
LD1 5LW GW4XXJ
LD1 5NL GW0RBH
LD1 5RB 2W0XOT
LD1 5RB MW0XOT
LD1 5RB MW6ZYQ
LD1 5RD GW4LUX
LD1 5SE GW1PXM
LD1 5TE 2W0JRJ
LD1 5TE MW0MWJ
LD1 5TE MW7PLL
LD1 5UR GW7UNV
LD1 6BE MW6RRW
LD1 6BU MW0MVM
LD1 6DT GW6OJK
LD1 6EW MW0EBT
LD1 6EW MW3GUH
LD1 6NU GW6BDM
LD1 6NU MW7EMK
LD1 6PD GW4YCJ
LD1 6PH GW0BOE
LD1 6SF 2W0NXV
LD1 6SF MW6NXV
LD1 6SL GW8LQH
Ld16ye MW7SBD
LD2 3EN GW7JYJ
LD2 3UL MW6BUF
LD2 3UU GW4MUJ
LD2 3UY MW0LLY
LD3 0AT GW6DQH
LD3 0AT GW8YPR
LD3 0DU GW0BKJ
LD3 0HN GW1SMJ
LD3 0HY MW3RRU
LD3 0PW GW7RFA
LD3 0UR GW4NKR
LD3 7JF GW0IVT
LD3 7NY GW4BVT
LD3 7RT 2W0BOC
LD3 7RT MW0GMH
LD3 7RT MW3BOO
LD3 7UW GW0ABT
LD3 8DJ 2W0MCB
LD3 8DJ MW3BTN
LD3 8PD GW0XAP
LD3 8PD MW6ELX
LD3 8UP GW6JJX
LD3 8YA MW0OWM
LD3 8YA MW7RMC
LD3 9BR 2W0MFD
LD3 9BR MW6HEA
LD3 9EH MW7DNQ
LD3 9ES GW1BXX
LD3 9HH 2W0HSD
LD3 9HH 2W0YLL
LD3 9HH 2W0ZJA
LD3 9HH MW0HTO
LD3 9HH MW0UAA
LD3 9HH MW3VUJ
LD3 9HH MW3YVK
LD3 9HH MW6ZOL
LD3 9HT GW0GIH
LD3 9LH 2W0RHI
LD3 9LH MW3YWC
LD3 9NA MW6GSS
LD3 9NG MW6FVA
LD3 9PW GW4KCV
LD3 9RL GW6NSK
LD3 9RN GW8BCL
LD3 9SZ MW0HXC
LD4 4BS 2W0HSU
LD4 4BS GW3CPM
LD4 4BS MW7CYN
LD4 4DR GW7KIV
LD5 6BN GW0KQX
LD6 5HA GW7BNC
LD6 5HA MW6EQF
LD6 5LD GW1FXL
LD7 1AG GW7HDC
LD7 1DG 2W0LKX
LD7 1DG MW0LKX
LD7 1DG MW7WAD
LD7 1EF MW3TMR
LD7 1HY MW3RGD
LD7 1PT GW0KIR
LD7 1RY MW6YKI
LD7 1SW GW0JEQ
LD7 1YL GW8ARR
LD7 1YT GW4GAF
LD8 2AN GW3RMJ
LD8 2DG MW7BTF
LD8 2EG GW1ENP
LD8 2EP GW8CAK
LD8 2HB MW7FLK
LD8 2HL 2W0TLG
LD8 2HL MW0OTG
LD8 2HL MW6PDP
LD8 2LB GW0CTG
LD8 2LH GW4VVF
LD8 2PA GW7IZA
LD82NX MW7TYR
LD82RE 2W0GZN
LD82RE MW6DRW
LE1 5XR G6ANA
LE1 7RH G4ION
LE10 0DY G1CJI
LE10 0ED 2E0TJQ
LE10 0ED M6TJQ
LE10 0ED M7AVQ
LE10 0GB G8SUN
LE10 0GJ M7EHW
LE10 0HP M0BWF
LE10 0LA G0GSH
LE10 0LG G1CIV
LE10 0LY M6CJF
LE10 0NH G4CNZ
LE10 0PY M6WNE
LE10 0SE M3DZK
LE10 0TW G3XWD
LE10 0WU M3AQJ
LE10 1DY 2E0HTM
LE10 1DY M6AYN
LE10 1EB 2E0SCC
LE10 1EB M6HJR
LE10 1HE G1WJO
LE10 1LS G6UX
LE10 1LT G3MFK
LE10 1SJ G4NNI
LE10 1SL G4YFI
LE10 1SS M3IRQ
LE10 1TH M3IRP
LE10 1TJ 2E1GOP
LE10 1TJ G8JSE
LE10 2AS G4CLB
LE10 2AU G7FSD
LE10 2EG G8KFD
LE10 2HE 2E0MTE
LE10 2HE M6OLR
LE10 2HL M6ATM
LE10 2LY G4ALB
LE10 2LZ G6HVD
LE10 2ND G3XPU
LE10 2NX G8CTJ
LE10 2PS 2E0BMJ
LE10 2SW G4PJP
LE10 2TN 2E0DTL
LE10 2TN M0LTD
LE10 2UD G1KFB
Le10 2uh M1EGX
LE10 3LF G4ZXA
LE10 3PE G4TUH
LE10 3PR M6SOG
LE100FE 2E0FFV
LE100FE M7ALF
LE100PH G0GGA
LE11 1BA M3XMO
LE11 1BU M0WAI
LE11 1DD 2E0CGT
Le11 1dd M6JBM
LE11 1JW M0RWM
LE11 1LF 2E0SLV
LE11 1LF M6OUI
LE11 1LP 2E0FCZ
LE11 1LP M0WFC
LE11 1LP M6FCZ
LE11 1NX M0JPP
LE11 1PU 2E1IBT
LE11 1RD G7IQO
LE11 1SL G4YJT
LE11 1SN G4SGY
LE11 1SU M0JVZ
LE11 1SY M7JMJ
LE11 2AA G3ABG
LE11 2AA G4IAQ
LE11 2AA G4IAR
LE11 2AA G4LAB
LE11 2JG G0VAU
LE11 2JL G6OUJ
LE11 2PA G0LCU
LE11 2PB G0DMH
LE11 2PG 2E1HGJ
LE11 2PH G7HBO
LE11 2PH M0CQV
LE11 2PH M3HZH
LE11 2QZ G7WIQ
LE11 2RU G6AK
LE11 2RU G8IXX
LE11 3JP G1WLW
LE11 3JP G3MKE
LE11 3JS G0WTA
LE11 3JT G3OMK
LE11 3LT G7RAL
LE11 3LT G8SNF
LE11 3LT M3SNF
LE11 3NL G6GBU
LE11 3PS G7LTU
LE11 3PT G3ZUB
LE11 3RA G7NII
LE11 3RL G4SBM
LE11 3RX G7DST
LE11 3SS G7WCP
LE11 3ST G8UIW
LE11 3TB G7BDR
LE11 3TY M7IMH
LE11 4LF G4AOP
LE11 4LF G8CPK
LE11 4LJ M3EKZ
LE11 4LL M3ROQ
LE11 4ND M3AUC
LE11 4PP G4JQV
LE11 4PU G4CCI
LE11 4PX G7ACM
LE11 4QB M3OOH
LE11 4QD G0MMJ
LE11 4QU 2E0DUW
LE11 4QU M6GGO
LE11 4QW G4OYP
LE11 4SA G8APF
LE11 4SN G4OKY
le11 4uq M6JUP
LE11 5AN 2E0MWJ
LE11 5EZ G4NTJ
LE11 5HB G4UAT
LE11 5JW G1VVT
LE11 5LB 2E0JOG
LE11 5LB M6PPL
LE11 5UJ G7OXA
LE11 5UU G8HMA
LE11 5UW G8LVL
LE11 5YB G7SCL
LE11 5YZ G4VCN
LE11 5YZ G6VAW
LE12 5AL M7DZP
LE12 5DF G4MPH
LE12 5EE G1KBC
LE12 5HA 2E0OCN
LE12 5HA M6FHV
LE12 5HQ G4DCI
LE12 5HW 2E0GUR
LE12 5HW M0URB
LE12 5HW M6LKL
LE12 5NQ 2E0FXO
LE12 5QW M7ECG
LE12 5RQ M6BTK
LE12 6JF G8HYP
LE12 6LD G4ZMA
LE12 6NN G8HVF
LE12 6PL M0IUB
LE12 6PP 2E0LEA
LE12 6PP G3ZJV
LE12 6PW G8JHE
LE12 6ST G4BZP
LE12 6ST G8DTM
LE12 6TX M1GUR
LE12 6UB G3XYC
LE12 6UZ G0INJ
LE12 7BN G6HVD
LE12 7BP G0LMA
LE12 7FG M6GII
LE12 7HB G1IFX
LE12 7HY M3JNU
LE12 7JN M7DET
LE12 7ND G1DHQ
LE12 7NX M0BWU
LE12 7PX G4MIV
LE12 7PZ 2E0DGH
LE12 7RQ G6MOI
LE12 7SB G1HGA
LE12 7SH G0AUV
LE12 7SX G0JHJ
LE12 7TG M3OMU
LE12 7UH 2E0MFT
LE12 7UH M3MFT
LE12 8BB G4ERT
LE12 8BF G6TZO
LE12 8EJ G8IQT
LE12 8HT M3URH
LE12 8JZ G7UGA
LE12 8LA G4ZSP
LE12 8LG M6GAO
LE12 8LR 2E0MCW
LE12 8RP M3HAI
LE12 8SG G3THW
LE12 9AD G1HOD
LE12 9AT 2E0ZRT
LE12 9AT M0ZRR
LE12 9AT M6ZRT
LE12 9BZ M3ZTL
LE12 9DA M0TMM
LE12 9DG M6DJS
LE12 9DH 2E0EVR
LE12 9DH M0XGX
LE12 9DH M6OFD
LE12 9DY G7HDW
LE12 9HG 2E0HDY
LE12 9HG M6TGH
LE12 9HP M0ZAK
LE12 9HY G4JVQ
LE12 9JE M0ITZ
LE12 9JE M7MIQ
LE12 9LS G0UGI
LE12 9LS G4HTH
LE12 9LW M0TAB
LE12 9LW M1CJT
LE12 9NL G7BMM
LE12 9PD M7VXU
LE12 9PU G3MKU
LE12 9QH M6JLZ
LE12 9RB 2E0GQP
LE12 9RB M7IWT
LE12 9RW G4JCH
LE12 9SG M3OAB
LE12 9SH M6IAS
LE12 9SQ G1ETZ
LE12 9SS G8BUB
LE127UX G0MCV
LE127UX M6MJV
LE13 0BY G4LWB
LE13 0DS G0WYM
LE13 0DU M6CJI
LE13 0EU G1DPN
LE13 0EW 2E0CKE
LE13 0EW M3KSP
LE13 0GE G1JMC
LE13 0HP 2E0GZH
LE13 0HP G5HOW
LE13 0HP M7EOZ
LE13 0HP M7GGH
LE13 0HP M7JRH
LE13 0LU G0SCO
LE13 0LU G4CAZ
LE13 0NB G3MLQ
LE13 0NN G8AKE
LE13 0RA G7PCT
LE13 0RA M0MKE
LE13 0SR G4LAU
LE13 0TF M6CXZ
LE13 1ES M3UOX
LE13 1FT M7EIP
LE13 1HG G4NNZ
LE13 1HY G4YSP
LE13 1HZ 2E0HMG
LE13 1HZ M0WES
LE13 1HZ M6HPS
LE13 1JZ G4FOX
LE13 1JZ G4PTK
LE13 1JZ G7FOX
LE13 1LJ M0LBM
LE13 1RT G1MZH
LE13 1RZ M6LGH
LE13 1SE G8RBY
LE13 1SH G0LUB
LE13 1TT M3PTQ
LE13 1UH G6FDD
LE13 1UR G8NYH
LE14 2AN G6PZN
LE14 2AP G8YEJ
Le14 3af G2UT
LE14 3DT G4AMN
LE14 3EW G7UDJ
LE14 3EX G6PQI
LE14 3QA G4CWC
LE14 3QE M6IWA
LE14 3QG G3UOD
LE14 3QH G0PAG
LE14 3QH M0VGK
LE14 3RY G4HEE
LE14 3SA M6WTD
LE14 3YB M6VOD
LE14 4BU 2E0MDT
LE14 4BU M0VAR
LE14 4BU M6CJQ
LE14 4DD G4EVK
LE144BX M3MFZ
LE15 6BE G4BZV
LE15 6BE M6ELB
LE15 6EB G3MCK
LE15 6HQ M6MDC
LE15 6LT G4FDP
LE15 6LZ M1BAN
LE15 6ND 2E0HYK
LE15 6PH G6LEI
LE15 6QA G0CWU
LE15 6RQ M6OKM
LE15 6SJ G3WMA
LE15 6SJ M6NJE
LE15 7DB G3PKD
LE15 7DZ G3ZDW
LE15 7DZ G8FC
LE15 7HP M6ELS
LE15 7JL G4PKZ
LE15 7JL G8HLM
LE15 7NJ G4BOQ
LE15 7NX G7FGR
LE15 7PS G1VIG
LE15 7PX M0HQC
LE15 7RL G7JBD
LE15 8AD G3XQZ
LE15 8DH 2E0OCW
LE15 8DH M6WGJ
LE15 8EA M6WFR
LE15 8EZ M7NAG
LE15 8HY M3SNN
LE15 8JS G3OKB
LE15 8PH M0ILZ
LE15 8PJ 2E0XAL
LE15 8PJ M0OGD
LE15 8PJ M6GKN
LE15 8QB M6EOT
LE15 8RH G4OSJ
LE15 8SD G3USR
LE15 8SU G8TTE
LE15 9PG G4GWH
LE15 9QP M7KKB
LE15 9RL M6EII
LE15 9RS M7CJV
LE15 9TS G0KAQ
LE16 7BQ G4JOV
LE16 7BQ M6GYG
LE16 7BW 2E0TEU
LE16 7BW M0SVH
LE16 7BW M7TEU
LE16 7DD G8GMB
LE16 7DD M3GMB
LE16 7DE G0SFJ
LE16 7EH G0BHS
LE16 7FP G3VUH
LE16 7JF G4XRA
LE16 7JJ G4WVR
LE16 7JJ G8PAN
LE16 7LG M3HZN
LE16 7LW M3KXI
LE16 7NY G4XEX
LE16 7PQ G3YKS
LE16 7TA M3HZA
LE16 7UU 2E0XVX
LE16 7UU M6XVX
LE16 7XE 2E0GAP
LE16 7XE G1IVG
LE16 7XE M3YVF
LE16 7XE M3YVG
LE16 7XE M6AJP
LE16 8AW M6OMP
LE16 8BH G0VLR
LE16 8BH G4ISN
LE16 8BH G7VLR
LE16 8BH G8LVM
LE16 8BQ G4DIA
LE16 8EX G6OFZ

LE16 8LD G1TTH
LE16 8LD G4LXA
LE16 8QJ G1FJH
LE16 8QJ G3ZSU
LE16 8QJ M3UXE
LE16 8QJ M3UXM
LE16 8SJ G1NQH
LE16 8SS M6BCN
LE16 8XP G5UI
LE16 8XR M3MRA
LE16 8XS G7KRS
LE16 9DP M6BWL
LE16 9DX G7PZB
LE16 9GA 2E0OVB
LE16 9GA M0OVB
LE16 9GA M6OVB
LE16 9GG M7OLA
LE16 9JL M6LTD
LE16 9JS 2E0HFG
LE16 9JS M6BWJ
LE16 9LW G3SFV
LE16 9NA G4JYT
LE16 9NW G1IVF
LE16 9NW G1WVR
LE16 9RZ G1IDR
LE16 9RZ G1ZHD
LE168JF 2E0HRB
LE168JF M6IGA
Le169jw G1PHV
LE17 4DD G6OGZ
LE17 4FS 2E0PBK
LE17 4FS M6PBL
LE17 4JJ G0LAP
LE17 4NW G1ICI
LE17 4NW G4XPD
LE17 4PG M6PGA
LE17 4PS G3ORY
LE17 4PS G3SDC
LE17 4SP 2E0EEU
LE17 4SP M0IIA
LE17 4SP M6EEU
LE17 4SQ G0TTW
LE17 4TR G0OLS
LE17 4TR G7FFZ
LE17 4TU M0RVD
LE17 4TU M3RDV
LE17 4US G1IRQ
LE17 4UT 2E0EEZ
LE17 4UT M0IKV
LE17 4XB G3RIR
LE17 4XB G8TKQ
LE17 4XF G0JEW
LE17 4XF G4APD
LE17 4XJ G7DMG
LE17 4XJ M5WAH
LE17 4YS M6WOB
LE17 5AG G6EDC
LE17 5AS G4VSJ
LE17 5DE G4VOZ
LE17 5DE G8LM
LE17 5DL G6TZT
LE17 5DL G8ZUF
LE17 5HX G4NGF
LE17 5HY G4CLA
LE17 5QA G8PGI
LE17 5RP G1DPT
LE17 6AE M7HHS
LE17 6AZ G4ABX
LE17 6EQ 2E1BKX
LE17 6EQ G7LCK
LE17 6EQ M3TDB
LE17 6FE 2E0EXX
LE17 6FE M0XAE
LE17 6FE M6BSP
LE17 6FE M6OMZ
LE17 6HD M6DMD
LE17 6JW G4VTM
LE17 6NT G6YQU
LE18 1BA M6KRH
LE18 1BH M3XFC
LE18 1BR G6IFN
LE18 1FQ G1UFT
LE18 1FX G6HPR
LE18 1GD G8SSX
LE18 1HU G4OAI
LE18 1HY G0HLL
LE18 1JZ G4KKS
LE18 2EF 2E0FOK
LE18 2EF M6FOJ
LE18 2EP G6LTR
LE18 2FP G7CEC
LE18 2FU G4XDX
LE18 2HQ M3IHN
LE18 2HZ M6RSL
LE18 2JB G1VNS
LE18 2JH M0GTE
LE18 2QX G0TCJ
LE18 2RF G6HVJ
LE18 3RL G8EVD
LE18 3TY G4EMW
LE18 3WD G4OKD
LE18 3XW G1YQP
LE18 4LP 2E0NLB
LE18 4LP M6NDB
LE18 4LY M0KVN
LE18 4LY M1KVN
LE18 4NA 2E0VOD
LE18 4NA M3UFW
LE18 4TH 2E0APY
LE18 4UH G1IAQ
LE18 4WH G6YII
LE18 4WL M3ZPE
LE184QB 2E0KFL
LE184QB M6KHE
LE19 2FX G8RFY
LE19 2FY M6CBM
LE19 2HT G0AIG
LE19 2RA G7PPL
LE19 3DY 2E0PMU
LE19 3EZ G3TWY
LE19 3PS G0WBC
LE19 4NQ G7VUU
LE19 4NQ M0AZW
LE19 4NX 2E0TXQ
LE19 4NX M7TXR
LE19 4QD G8NTD
LE19 4QX G1VIN
LE19 4QZ G0TIZ
LE192AW G0PBY
LE193EW M0IOO
LE2 1PN M3MBV
LE2 1WQ G8TYF
LE2 1YD M6BHJ
LE2 2AE M6MJM
LE2 3EA G1JAG
LE2 3EH 2E0OTP
LE2 3EH M6IKI
LE2 3RJ G6EHJ
LE2 3WR M6FUD
LE2 4NT G1IGP
LE2 4NY 2E1FEC
LE2 4PB 2E0RMJ
LE2 4PB M6JEZ
LE2 4QA 2E0BQE
LE2 4QA M3XCU
LE2 4QD 2E0BQF
LE2 4QD M6MOM
LE2 4QD M3UFG
LE2 4RJ G0HNI
LE2 4RQ 2E0BQM
LE2 4RQ M0NCK
LE2 4RQ M3VWY
LE2 4RS M3WQU
LE2 5FH G7SEU
LE2 5HF G6PFN
LE2 5PF G3XKX
LE2 5TR G1HWO
LE2 5UE G3HAN
LE2 5YF 2E0EEF
LE2 5YF M0NDF
LE2 5YF M6FZL
LE2 6AD G0TTE
LE2 6FF M6EPN
LE2 6FN G3MCP
LE2 6HQ G3LQW
LE2 6JE M6AER
LE2 6NT M3CBY
LE2 6TS 2E1CPC
LE2 8DH G8FCQ
LE2 8DJ G6WMU
LE2 8DL G1YFT
LE2 8HW G4NBH
LE2 8NJ G1PPQ
LE2 8QA G6ITW
LE2 8SF M0HOP
LE2 8SF M1HOP
LE2 8UH G4MQS
LE2 8UH M1CSL
LE2 8UP G5YH
LE2 9DD G1ZYN
LE2 9GA G6WZM
LE2 9GY G0WUS
LE2 9NS G6HSI
LE2 9NS G8RFE
LE2 9QJ G0JQZ
LE2 9TH G0PBP
LE2 9TJ G4TQR
LE2 9TT G6VLT
LE3 0EA M6ZSA
LE3 0PB G4OHF
LE3 0SA G3ZCT
LE3 0TN G0EVI
LE3 1AU M0ZKA
LE3 1AX G6RRJ
LE3 1FG G6EJU
LE3 1GR G0FPU
LE3 1JF M3VJE
LE3 1PA 2E0DSQ
LE3 1PA G6XRS
LE3 1PA M0VSE
LE3 1PA M6ESV
LE3 1RA M3CGH
LE3 1RA M6SBR
LE3 2BH M6KXA
LE3 2BW G3WQL
LE3 2EJ G0FZC
LE3 2FN M6KEH
LE3 2FN M6TJX
LE3 2JB 2E0GXT
LE3 2JB M6HKJ
LE3 2JU G0FRV
LE3 2SP G0ORY
LE3 2SP G3LRS
LE3 2SP M6NZH
LE3 2UU G1AEZ
LE3 2XQ M1BSE
LE3 3AD G4NUK
LE3 3AD M7BDZ
LE3 3AE 2E0MZE
LE3 3AP 2E0TPP
LE3 3DQ G4DJK
LE3 3EB M7SPS
LE3 3EY G1CKY
LE3 3FF G4SDZ
LE3 3GA 2E0XAY
LE3 3GA M6ACG
LE3 3GN G7GLA
LE3 3HB G3HYH
LE3 3LA G7AYI
LE3 3PP G1EBV
LE3 3PP M6VBD
LE3 3PS M3NNM
LE3 3SW G0MTF
LE3 5RA G6WSN
LE3 6BD G6JOL
LE3 6FG G0UHI
LE3 6FG G7OCY
LE3 6LU M3SAO
LE3 6NA G4KGX
LE3 6NF G4RLC
LE3 6nr M1EZR
LE3 6PL 2E0DLD
LE3 6PL G1GEV
LE3 6PL M3PIY
LE3 6PL M3WBQ
LE3 6QQ M6HMQ
LE3 8AF G1YEZ
LE3 8AG G6BBI
LE3 8BN M0FRD
LE3 8GF G8ZQG
LE3 8GH M0GPO
LE3 8LF M6PTO
LE3 8LU 2E0KFH
LE3 8LU M6CDK
LE3 8QX M1GAS
LE3 9HG M6DPA
LE3 9ND M3PYG
LE3 9ND M3UNL
LE3 9NH M6FPD
LE3 9NH M6FPE
LE3 9PS M6CMW
LE3 9QB M3BXW
LE4 0BJ G3OCH
LE4 0GS G3KQV
LE4 0LL G4AGN
LE4 0PP G7DHJ
LE4 0QY 2E0BQC
LE4 0QY M3XEA
LE4 0SF M3PIK
LE4 0ST M6LFL
LE4 0ST M6TFO
LE4 0SU G0AZM
LE4 0UG M6FPG
LE4 0UR G3UST
LE4 1AR M7MBC
LE4 1BL G6ICV
LE4 1BX G8ATE
LE4 2BJ G4ITP
LE4 2HN 2E0SVT
LE4 2HN M6JLA
LE4 2HN M6SVT
LE4 2LH G4ILT
LE4 3FF M0IVU
LE4 3GW M0GAH
LE4 3HB G4ZCJ
LE4 3HQ G4BJF
LE4 4DA G4MEF
LE4 4EH 2E1HMQ
LE4 4FU G6TIW
LE4 5LH M3BYX
LE4 7AE M3FPT
LE4 7SU M3EVP
LE4 8AY G0VTL
LE4 8BD G7DOA
LE4 8BP G4VWI
LE4 8GL M1BUJ
LE4 8JP 2E1SDI
LE4 8NB M6EIH
LE4 8NR M3HVS
LE4 9DQ M6WWF
LE4 9JP M0LRG
LE48FP G7IUI
LE5 0LF G5MY
LE5 0PH M3WZV
LE5 1EA G4EOF
LE5 1ED G6WAY
LE5 1FA 2E0LDE
LE5 1FA M0LDE
LE5 1PA G3FJL
LE5 2BF M0ESB
LE5 2DE G7VDU
LE5 2DE M3VDU
LE5 2EE M6MDR
LE5 2EF G1ECK
LE5 2EF G4ZTD
LE5 2EF G6MAM
LE5 2EF M6NES
LE5 2GP G3ZZW
LE5 2HQ M6EIF
LE5 2LJ M6VCN
LE5 2RL G0ZIP
LE5 2RL G7VIP
LE5 4EN G1KOH
LE5 4LU 2E0KOI
LE5 4QL 2E0BIC
LE5 4QL M3BIC
Le5 4wh M6HCU
LE5 5UD M0COQ
LE5 6AH G1GKA
LE5 6HL M3UOZ
LE5 6JB G4ZDQ
LE5 6PT G4MGG
LE5 6SA G1WZQ
LE5 6SY G4IGL
LE5 6XT G4YTF
LE56DE M7DMB
LE6 0BA G3UAA
LE6 0BN G1ZQV
LE6 0BT G4TFO
LE6 0EX 2E0GSA
LE6 0LH G6PGP
LE6 0LH M5MUF
LE6 0YL G8POS
LE6 0YP G0HZG
LE6 1HT G1XDS
LE6 1HT G1YBK
LE65 1ES G4WYN
LE65 1EU G4LAI
LE65 1EW G6EQB
LE65 1HT 2E0RKK
LE65 1HT M0GNO
LE65 1HT M3RKK
LE65 1TR G4JNH
LE65 1UL M0BPS
LE65 1WA G3UBB
LE65 2FQ G7SEK
LE65 2HL G0NXT
LE65 2JR G7NBE
LE65 2JR M3BBB
LE65 2JZ G6CLP
LE65 2LW 2E0SQN
LE65 2LW M6SWS
LE65 2QQ G1CWW
LE65 2QY 2E1LGA
LE65 2QY M0JIO
LE651LY G4SGD
LE67 1BR M6CEF
LE67 1EH M7CER
LE67 1GB G1JOJ
LE67 2GB G1ULP
LE67 2GL G1ODQ
LE67 2GP G8YSA
LE67 2HE G4PLK
LE67 2HU G1YQU
LE67 2JT M6NCM
LE67 2ND M1DHO
LE67 2NS 2E0DFJ
LE67 2NS M6RFJ
LE67 2QL G0HME
LE67 2SS M0HMZ
LE67 2XD 2E1HGF
LE67 3AF G4AEO
Le67 3bd 2E0SLO
LE67 3BD M0MCY
Le67 3bd M6WSW
LE67 3BE G8YGT
LE67 3JG M7CVM
LE67 3PB M3KOF
LE67 3PL G8VFM
LE67 3PW G7WDG
LE67 4AN M0MHQ
LE67 4BF G8ZZY
LE67 4DD G1NNF
LE67 4DT G0IXS
LE67 4HZ G7KMK
LE67 4JU M6LAM
LE67 4TG G4JDP
LE67 4TG G8OZQ
LE67 5AY G8JMG
LE67 5AZ G1IWE
LE67 5BP G4DCE
LE67 5BP G6UZJ
LE67 5FD G0PXK
LE67 5GF M6UKM
LE67 5GN 2E0FSK
LE67 5GN M6GAH
LE67 5PA M1AYC
LE67 5PA M1BAC
LE67 5PF M0DJB
LE67 5PT G1KSC
LE67 5PT G3RAL
LE67 5PT G7SEG
LE67 6HP G8YOG
LE67 6JT G3ZJG
LE67 6JU M6MPW
LE67 6JW G0FZE
LE67 6LF G4JKG
LE67 6LP G7NVS
LE67 6LP M0LCN
LE67 6LP M0PFC
LE67 6LW M0JSD
LE67 6QD G0BAI
LE67 6QG M6CZR
LE67 8LY M3DFL
LE67 8NU G6IHU
LE67 8RJ G4IPX
LE67 9PY M1FJB
LE67 9RF G6OWB
LE67 9RJ G4RKD
LE67 9SN M1EXO
LE67 9TX G4ARI
LE67 9UJ G3SBF
LE67 9WA G4CQQ
LE67 9WH G4BWF
LE7 1GQ G1VBQ
LE7 1HJ G8MZY
LE7 1HN M6MFD
LE7 1HX G0PVE
LE7 1LY G0IPB
LE7 2EH M6KRR
LE7 2EN G6FPO
LE7 2JH 2E0ODF
LE7 2JN G6NGV
LE7 2JW G8TLC
Le7 3dn G7OOT
LE7 3RJ G8DBK
LE7 3SU M7ESS
LE7 3TU G3DAQ
LE7 3ZJ G0ORL
LE7 3ZR M6PFR
LE7 4UT M7CHN
LE7 4YN G6ABP
LE7 7BH G7UYJ
LE7 7BQ M5CAH
LE7 7DU G4LRO
LE7 7EU G6TMQ
LE7 7FA G7NQU
LE7 7GN M6OJM
LE7 7PU G8LDB
LE7 7QG M0XMZ
LE7 7RL G4NWS
LE7 7RL G4WRC
LE7 9AN M3MRZ
LE7 9DA G4ZJR
LE7 9GW G6ISM
LE7 9HD 2E1CJZ
LE7 9HD G7OEY
LE7 9HD M0DBG
LE7 9JS G3URQ
LE7 9LL M6IFN
LE7 9PP G3VOV
LE7 9PQ G4DR
LE7 9UA 2E0XZZ
LE7 9UA M6KON
LE7 9UR G4XXZ
LE7 9YB M6YSU
LE79NP 2E0TJY
Le79np M7SWY
LE8 0AP G8DGH
LE8 0JJ G0UFP
LE8 0JJ M0BNS
LE8 0NP M3JZX
LE8 0RA G4EEL
LE8 3NE G3RUO
LE8 4AB M6DXS
LE8 4BE G4WGU
LE8 4DL G6KJH
LE8 4FQ M0CDF
LE8 4FU G3KYF
LE8 4HF G8OBP
LE8 5RH G7UOS
LE8 5SB M6ELN
LE8 5SU G1IPP
LE8 5TB G3MXV
LE8 5TG G4YJU
LE8 5TL G4FIE
LE8 5TP G1HEN
LE8 5WJ G8BTU
LE8 5WJ G8CJA
LE8 5WS M0HKM
LE8 5XH 2E0RCJ
LE8 5XH M6IOO
LE8 6HY 2E0KTV
LE8 6HY M0KTV
LE8 6HY M6BCH
LE8 6JQ G6ZZE
LE8 6NF G8ZQB
LE8 6NT G7FMJ
LE8 6YB G4CPY
LE8 8AP G4EQL
LE8 8AX M6GMY
LE8 8BP G1PPU
LE8 8TE 2E0LTD
LE8 8TR M6BUR
LE8 8TX G4IHR
LE8 8TX G8RBI
LE8 9FH G8HMJ
LE8 9FP G8FWA
LE8 9GF 2E1IDG
LE8 9GF 2E1JMG
LE8 9GP M7KAU
LE9 1SE G8VEN
LE9 1SX G1IUT
LE9 1UW G1OOM
LE9 2BL G4ZIZ
LE9 2BN G4ECO
LE9 2BW G1YYL
LE9 2DD G0DMB
LE9 2DE G4FSS
LE9 2DR M0MRJ
LE9 2EN G3ONV
LE9 2EU G3OVH
LE9 3EB G3WTD
LE9 3GE G8XRG
LE9 3GX G8PEA
LE9 4BW G4EPN
LE9 4BW G4RCC
LE9 4DY M1ANN
LE9 4DZ G0CND
LE9 4DZ G0JSE
LE9 4JZ M6WAX
LE9 4LG G8PGO
LE9 4LL G7WHI
LE9 6EE 2E1BAD
LE9 6HJ G4OZD
LE9 6PT G0LDO
LE9 6PY M6OSU
LE9 6QE G0CRT
LE9 6RL 2E0IMH
LE9 6YU G8OPX
LE9 7AF G0BSN
LE9 7AW G8EHM
LE9 7AY M6CRO
LE9 7DA G8BSM
LE9 7DU G1LTK
LE9 7EY M6BGL
LE9 7FF G1MPD
LE9 7FP 2E0SBM
LE9 7FY G2AXO
LE9 7GT G4VMM
LE9 7HJ 2E0SHB
LE9 7HJ M3NXE
LE9 7HJ M6JAB
LE9 7HQ G6MKL
LE9 7HW G3TFV
LE9 7PD 2E0RBU
LE9 7PH M6BYL
LE9 7TF G3PVG
LE9 8AP M1EYA
LE9 8AR M6NXR
LE9 8DG M6JHU
LE9 8DN G8CGW
LE9 8DR G6XXN
LE9 8DR M3TLJ
LE9 8EH G4CAJ
LE9 8EW G1WIW
LE9 8FS G0UIF
LE9 8FS G7TZU
LE9 8FS G8LUV
LE9 8GP G6FNQ
LE9 8JH 2E0OFT
LE9 8JH 2E1FTA
LE9 8JH G0SKR
LE9 8JX G7RXO
LE9 9JG 2E1GXQ
LE9 9JJ G6AYE
LE9 9JR G0TPH
LE9 9LG G0EDO
LE9 9LG G8LMW
LE9 9LG G8RYE
LE9 9LQ G1WSW
LE9 9LQ G4UUG
LE9 9QF G3LMR
LE91SX M6OHO
LEG6EG M7RML
LL11 1EF MW3PPQ
LL11 2BG GW6WVD
LL11 2EU GW4AVC
LL11 2LU GW6NSG
LL11 2ST 2W0SFB
LL11 2ST MW6SFB
LL11 2YF GW7UMW
LL11 3ED MW6EMK
LL11 3LJ MW7EJE
LL11 3NN GW8UPJ
LL11 3PB GW7KNN
LL11 3PG 2W0PWR
LL11 3PG MW3VSG
LL11 3PZ MW0ARL
LL11 3RY MW6FVT
LL11 3TW GW1IAW
LL11 3YT GW1LHV
LL11 4AF GW8ASD
LL11 4BY MW6DCN
LL11 4EB MW3WCE
LL11 4EE MW6EGD
LL11 4HF MW6IXZ
LL11 4PF MW1VCD
LL11 4SZ MW7NBC
LL11 4UE 2W0YDK
LL11 4UE MW0ECF
LL11 5AH GW8KSE
LL11 5AH GW8KSF
LL11 5AS MW6DFC
LL11 5AW GW7CEA
LL11 5DH GW6IWC
LL11 5EP 2W0MLG
LL11 5EP MW0XAD
LL11 5EP MW6GXU
LL11 5EU GW8VEE
LL11 5EU MW3SAI
LL11 5HH GW6JMC
LL11 5LE GW6PVK
LL11 5NG 2W0OSG
LL11 5SH MW6EYU
LL11 5UF GW4TJN
LL11 5UH GW4GSS
LL11 5UN GW6MRO
LL11 5YF GW4OVH
LL11 5YP GW6NLP
LL11 6BP MW3IGZ
LL11 6DN 2W0KGP
LL11 6DN 2W0KGQ
LL11 6DN GW2HFR
LL11 6DN MW0KGP
LL11 6DN MW3KGP
LL11 6DN MW3KGQ
LL11 6HR MW3HQV
LL11 6HR MW3HRE
LL11 6NS GW0VMR
LL11 6PD GW6FUY
LL11 6RG MW3NDO
LL11 6RH GW1ZHI
LL12 0AU GW0MMY
LL12 0BP GW4IGF
LL12 0DA GW3TOW
LL12 0HN GW8HPL
LL12 0NW GW0KPU
LL12 0NW MW0RCH
LL12 0NW MW0YLS
LL12 0PW 2W1VMR
LL12 0RT GW6MPX
LL12 0RT MW7KDA
LL12 0UW GW3YTL
LL12 7DS MW6NSY
LL12 7EP MW0GXC
LL12 7PD GW0ABE
LL12 7PP GW6SBD
LL12 7SG MW1DBA
LL12 7TW 2W0TNB
LL12 7TW MW6BYX
LL12 7UG GW6GMF
LL12 7UH 2W1DAO
LL12 8AF GW0EMB
LL12 8BE GW6FED
LL12 8BN MW1DAU
LL12 8DD MW3WPN
LL12 8HW 2W0RBX
LL12 8LB 2W0WWR
LL12 8LD GW8IJT
LL12 8RN 2W0JCP
LL12 8RN MW3JWV
LL12 8RN MW3XLR
LL12 8SJ GW6JTX
LL12 8ST GW0CES
LL12 8YL MW0BNB
LL12 9DH GW7MQE
LL12 9PE GW0NOO
LL12 9PE GW0NOP
LL12 9PE GW7COB
LL12 9PJ GW6IGY
LL12 9PR MW6MJZ
LL13 0AY GW4VAG
LL13 0BL GW6ILY
LL13 0BQ GW1IPJ
LL13 0HA MW7GBW
LL13 0LJ GW4XDR
LL13 0LN GW6AJK
LL13 0NW MW7TLS
LL13 0PQ GW3HEU
LL13 0UJ GW6FES
LL13 7AW MW3NIA
LL13 7AW MW7XMD
LL13 7DX GW4GDB
LL13 7EA 2W0TYG
LL13 7EA MW0TBB
LL13 7EA MW6TYG
LL13 7QE GW0GWE
LL13 7QJ GW8MGF
LL13 7QW 2W0ISZ
LL13 8RU 2W0KAP
LL13 8RU MW7KAP
LL13 9BA MW0CQR
LL13 9HR 2W0AZM
LL13 9LY GW6ZHY
LL13 9NQ GW3XQO
LL13 9QH GW4TNF
LL13 9SY MW7DDP
LL13 9TA GW4ZYM
ll139nn MW6LQW
LL14 1BB MW0ARV
LL14 1HH MW7WOC
LL14 1NF 2W0MDG
LL14 1NF MW0MDT
LL14 1ST GW4WVB
LL14 1UA GW6GTS
LL14 1UA GW8GGW
LL14 2AT MW6IQF
LL14 2DT MW6BYV
LL14 2EA GW6ZCR
LL14 2EA MW3JAP
LL14 2EJ MW7MAP
LL14 2EY 2W0JYN
LL14 2EY MW3WZZ
LL14 2EY MW6SHJ
LL14 2HT GW7PCX
LL14 2HT MW3PCX
LL14 2ND MW0ASL
LL14 2PN MW7IDI
LL14 2RL GW0WZZ
LL14 2RL GW8WYW
LL14 2RL MW0MRS
LL14 2RL MW7OTK
LL14 2SR MW3OXV
LL14 3EE GW0WER
LL14 3GB MW0ZDX
LL14 3JH GW7TUQ
LL14 3JH MW0IKH
LL14 4DA 2W1FJG
LL14 4JD MW1BJB
LL14 5BG MW6BYT
LL14 5DW GW7MHF
LL14 5HN GW7GOM
LL14 5PF GW1CTO
LL14 6AH GW0EHA
LL14 6AN 2W0WXM
LL14 6AT MW3ZCI
LL14 6BF 2W0RUA
LL14 6BF MW7MEJ
LL14 6DD GW1MKV
LL14 6DP 2W1PGL
LL14 6EG GW0MMB
LL14 6HS GW1GXQ
LL14 6LA 2W0LZU
LL14 6LA MW0PTY
LL14 6LA MW6LZU
LL14 6TD MW6EAO
LL15 1AQ GW0GLI
LL15 1HB 2W0PJM
LL15 1JA GW4ITQ
LL15 1JA MW0GWY
LL15 1QN MW0IBT
LL15 1RR MW0ZUS
LL15 1YP GW3ODB
LL15 2DW 2W1LCO
LL15 2DW GW6XRE
LL15 2EY GW6JWL
LL15 2HR GW7NUU
LL16 3BE MW3AFR
LL16 3BE MW6FMW
LL16 3EU GW3RUE
LL16 3HE GW4CQZ
LL16 3PE MW0CKT
LL16 3PW MW3IIJ
LL16 3TD MW3CMG
LL16 3YA GW8FSN
LL16 4BD GW4IEZ
LL16 4BS 2W0BLG
LL16 4BS MW3PNR
LL16 4BU GW0WWQ
LL16 4DT GW3GZX
LL16 4HH GW8RHP
LL16 4NN GW8SIE
LL16 4PQ 2W0LJC
LL16 4PQ MW6CLJ
LL16 4RL GW4RWR
LL16 5PA MW3GQS
LL16 5YL MW3XHL
LL17 0AD GW6RNA
LL17 0BH GW7NIW
LL17 0BP GW6ONZ
LL17 0RG MW3KIS
LL17 0SX GW1TFB
LL17 0SX MW0YDX
LL17 0SX MW3GMB
LL17 0UH GW6VEI
LL17 0UH MW0WIW
LL17 0UP MW3DBF
LL170DB MW0MFB
LL18 1DA GW4ZPM
LL18 1HT MW6ILK
LL18 1PN MW6BME
LL18 1TF 2W1EPL
LL18 1TF 2W1EPO
LL18 2BX MW1EOK
LL18 2EN MW6GCN
LL18 2LW GW4UJF
LL18 2NS MW6BVQ
LL18 2NU GW4XEF
LL18 2RS GW4XWN
LL18 2RY GW0DFY
LL18 2TP GW6IYP
LL18 2UL GW8WUM
LL18 2UL MW1EOO
LL18 2UL MW3TOB
LL18 3BG GW3OIN
LL18 3EE GW3UTG
LL18 3ER GW4NLD
LL18 3PE GW4ARC
LL18 3PE GW4HDR
LL18 3US GW0AYP
LL18 4AD GW1NGN
LL18 4AF GW0MOF
LL18 4EH 2W0FAP
LL18 4EH MW0HAP
LL18 4HH GW3FJI
LL18 4JF GW8OYT
LL18 4JH GW8XLL
LL18 4NF MW3CYM
LL18 4NF MW5CYM
LL18 4PN MW0CCN
LL18 4SH MW6EUC
LL18 4SU GW8TBY
LL18 4SU MW6NMG
LL18 4TN GW7VQA
LL18 4TY MW3HFO
LL18 5AG GW7EXH
LL18 5AG GW7NFY
LL18 5BN MW6TAK
LL18 5BP MW6ATG
LL18 5EW GW6VEN
LL18 5EY GW0HBZ
LL18 5HW GW4VLU
LL18 5HY 2W0FKW
LL18 5JE GW4DMR
LL18 5JE MW0CCS
LL18 5LS GW6XXY
LL18 5LS MW3XXY
LL18 5NH GW8FVI
LL18 5NS MW3XMY
LL18 5PP MW3CDL
LL18 5RB MW6FJQ
LL18 5RF GW6SLO
LL18 5RW GW0JSX
LL18 5TB 2W1HTK
LL18 5TB MW3EGB
LL18 5TL GW0DSJ
LL18 5TL GW0NWR
LL18 5TT 2W1BYK
LL18 5UP GW4JOT
LL18 5UP MW1JAN
LL18 5UP MW3GNB
LL18 5YF MW3GVF
LL18 6HS GW4TYH
LL18 6HS MW3SZD
LL18 6HT GW4UIR
LL19 7DF 2W1CEE
LL19 7DS GW0ULP
LL19 7EH MW7RES
LL19 7HH MW6TGF
LL19 7HL MW3HCW
LL19 7LP GW7RZN
LL19 7LP GW7TEO
LL19 7PF GW7AMS
LL19 7PF MW6FRW
LL19 7TS GW4DTQ
LL19 8DG GW7PEO
LL19 8HQ MW3FTP
LL19 8JP GW6RUO
LL19 8LU GW4XXP
LL19 8RD 2W0CIV
LL19 8RD MW0MIE
LL19 8RD MW3YMY
LL19 8RN GW6YDT
LL19 8RN MW3YDT
LL19 8RU 2W0ALZ
LL19 8SG GW3JGE
LL19 8TS GW6PMC
LL19 8YQ MW6EUK
LL19 9DU GW6ITJ
LL19 9HL GW3JGA
LL19 9HU 2W0JPE
LL19 9HU MW6SAX
LL19 9NN GW0CNK
LL19 9NX MW6FXA
LL19 9PB GW0KZW
LL19 9SH GW4VHP
LL19 9SU MW1LLL
LL199DT GW0ADC
LL199DT GW3CF
LL20 7AD 2W1FJZ
LL20 7AR 2W1EEP
LL20 7AS 2W0CJJ
LL20 7AT 2W1FJN
LL20 7AT GW7GHE
LL20 7BP 2W0PFD
LL20 7BP MW0PDV
LL20 7BP MW6PDE
LL20 7DD MW3WAL
LL20 7UH GW0TBM
LL20 8AS MW0ICO
LL20 8AS MW3XQJ
LL20 8DD GW6EWX
LL20 8RD 2W0XKL
LL20 8RD MW0XKL
LL20 8RD MW3XKL
LL20 X GW1UYW
LL21 0DL GW6WAG
LL21 0EH MW6BQM
LL21 0HA GW4URG
LL21 0PE GW6NNB
LL21 0SB GW7KDI
LL21 0SB MW5DAD
LL21 9AA GW6UMU
LL21 9EG GW1ZKN
LL21 9NA MW0PPM
LL21 9NP MW0DFY
LL21 9NP MW3SLI
LL21 9TB GW6KAV
LL22 7AR MW3FLU
LL22 7BU GW1PYY
LL22 7BU MW3PYY
LL22 7DF MW3MRC
LL22 7DG GW4IDV
LL22 7DN 2W0NKS
LL22 7DN MW7MPM
LL22 7DU MW0IBH
LL22 7DX MW0KGU
LL22 7HE GW7NTP
LL22 7JF GW7EVG
LL22 7JF MW0GWV
LL22 7ND 2W0HUU
LL22 7UD GW0BCR
LL22 7UQ GW0NSZ
LL22 7UU MW1TAF
LL22 8EG 2W0CEO
LL22 8EG MW0CDG
LL22 8EG MW6DIG
LL22 8EG MW6GOZ
LL22 8JA GW0SPY
LL22 8JD GW3WEQ
LL22 8JD MW7GAN
LL22 8QA GW1CDH
LL22 8QA MW6SZW
LL22 8SS GW4NJL
LL22 8UL MW3YHC
LL22 9HD 2W0GIX
LL22 9HD MW0OTH
LL22 9HD MW6GIX
LL22 9LP MW1COJ
LL22 9ND GW3YIH
LL22 9NE GW4VHS
LL22 9PH MW3IXJ
LL22 9YH GW4DLC
ll229nt MW7BCY
LL23 7NL GW4WPH
LL23 7NT MW7DBF
LL23 7RF GW4NNO
LL23 7SF GW4UFQ
LL24 0DS GW8HJC
LL26 0BP MW6TMQ
LL26 0EF 2W0TIR
LL26 0EF GW0GTW
LL26 0EP MW6ACZ
LL26 0EP MW6ADF
LL26 0RG GW0ETU
LL26 0RG GW4TSQ
LL26 0TY 2W0AMN
LL26 0TY MW6CVN
LL26 0TY MW6DEB
Ll260hw 2W0IEP
Ll260hw MW7CLE
LL27 0JE GW0MHC
LL27 0JJ MW0BTI
LL28 4AJ GW0UTC
LL28 4AU MW6LFQ
LL28 4NR GW8SXI
LL28 4QA GW4AMX
LL28 4RS GW6MLI
LL28 4SL GW4SYI
LL28 4TF GW6GKP
LL28 4UU GW1OIK
LL28 4UU MW3OIK
LL28 4YG GW7VHD
LL28 4YH MW3FVC
LL28 5HS 2W0CVE
LL28 5NG MW7GZR
LL28 5NJ GW3DZJ
LL28 5NN GW8YDR
LL284TL 2W0ENQ
LL284TL MW3ENQ
LL29 6AF GW4GJI
LL29 6DG GW0GGW
LL29 6DH GW7KGD
LL29 6DH MW6BAS
LL29 6DL MW0AFD
LL29 6DS MW3YYQ
LL29 7AX GW8BQK
LL29 7BU MW0HSI
LL29 7EY GW8VPP
LL29 7HB GW6STK
LL29 7NB GW0EVE
LL29 7NB GW0EVG
LL29 7ND MW6ECB
LL29 7SD MW3JUM
LL29 7TL 2W0DFM
LL29 7TL MW6RMR
LL29 7TT 2W0ITM
LL29 7TT 2W0RWF
LL29 7TT MW0TMH
LL29 8AS MW7HAS
LL29 8ER GW0LUA
LL29 8EU 2W1HNH
LL29 8EU MW3RHI
LL29 8EU MW3UGW
LL29 8EX GW0PRM
LL29 8HE MW7ELT
LL29 8LE GW1SAM
LL29 8LG MW6HNW
LL29 8PW GW4NNL
LL29 8SA GW1CJJ
LL29 8TA GW0PNE
LL29 8UT GW0VOG
LL29 8UY GW0OGL
LL29 8YZ MW6WWZ
LL29 8ZA GW8VUG
LL29 9AJ MW6ZIZ
LL29 9DS GW8JJP
LL29 9EL GW0DYH
LL29 9HP GW4RCM
LL29 9HT MW1DGR
LL29 9LA 2W1MWS
LL29 9LA 2W1SWB
LL29 9LB MW3GPJ
LL29 9LB MW6PJN
LL29 9LJ GW3FLU
LL29 9LJ GW4AMZ
LL29 9LQ GW0VMW
LL29 9NF MW7TVH
LL29 9NL GW0SKC
LL29 9NL GW0TWI
LL29 9PY GW1KHH
LL29 9RH MW3DEL
ll299er MW7KSE
LL30 1BG MW0JYW
LL30 1BL GW4WBT
LL30 1BL MW6ORF
LL30 1BL MW6ORJ
LL30 1ES GW1BBH
LL30 1ES GW6HVA
LL30 1HQ GW0GST
LL30 1HZ GW1RVP
LL30 1JJ GW4HBS
LL30 1LP GW6WFW
LL30 1LT MW1AWT
LL30 1NG 2W0AMB
LL30 1PF GW0EZB
LL30 1PF GW3UTL
LL30 1PY GW8WFS
LL30 1RQ MW7VIB
LL30 1SL GW0TPF
LL30 1SL MW7NZP
LL30 1TU GW4PFL
LL30 1UH GW0AEZ
LL30 1UH GW7UUH
LL30 1UW GW7UIZ
LL30 2AA GW8VVX
LL30 2BU GW3FMR
LL30 2PQ GW2DNJ
LL30 2QA GW7TDQ
LL30 2RA GW3GJQ
LL30 2TY MW3EVN
LL30 2UB 2W0NTS
LL30 2UB MW7CAJ
LL30 2UF GW0IQC
LL30 2YE GW6NYR
LL30 2YP GW8FOY
LL30 2YQ GW7AHR
LL30 3BY GW0NEC
LL30 3EY MW0HCW
LL30 3LL MW0BET
LL30 3LN GW4PVU
LL30 3NL 2W0DFN
LL30 3NL MW0HTG
LL30 3NL MW6EIK
LL30 3RB MW6PUF
LL31 9AN MW3MMJ
LL31 9BF 2W0IMM
LL31 9BF MW6IMM
LL31 9BJ GW7RKQ
LL31 9BZ GW0SFP
LL31 9EL GW3LCQ
LL31 9EL MW7GTW
LL31 9ER GW1BDH
LL31 9HF GW6IVY
LL31 9HQ GW6IYA
LL31 9HS 2W0DDZ
LL31 9HS MW6DAR
LL31 9LP GW4KFY
LL31 9PF MW1AHU
LL31 9QE 2W0CRB
LL31 9QE 2W0TYE
LL31 9QE MW0HYK
LL31 9QF MW0CPN
LL31 9RG GW1PSW
LL31 9RR MW3NFF
LL31 9UG GW0TLS
LL31 9UG GW3XKB
LL31 9UG MW0NAB
LL31 9UT GW1KKJ
LL32 8EU GW6ZDH
LL32 8NB MW3NAQ
LL32 8NP GW3YQP
LL32 8NR GW8JKC
LL32 8NS GW6SIX
LL32 8RJ 2W0GYQ
LL32 8RJ MW0JTG
LL32 8RJ MW7BAO
LL32 8SB GW6WKU
LL32 8WB MW6EYI
LL32 8YL MW7HEW
LL32 8YY GW4HHD
LL33 0LL GW0WUM
LL33 0RN MW7PWH
LL33 0rp MW3XME
LL33 0SN GW7JLG
LL33 0SR GW0FPY
LL33 0SS GW0LBA
LL330TN MW3RAU
LL34 6ER GW1ACV
LL34 6TE MW0RXD
LL34 6UA GW1SGG
LL346LY GW8XAS
LL35 0PT GW0TWO
LL35 0RF MW0HYA
LL36 0AG GW4JPP
LL36 0AG MW7SIF
LL36 0AY GW3XSR
LL36 0BL MW0DWU

LL36 0TA GW6INF
LL36 9DE GW4XXF
LL36 9EG MW3MWO
LL36 9EP GW4KYK
LL36 9HG GW6JDF
LL36 9LY GW0COH
LL36 9ND GW3LTX
LL36 9SB GW4MSI
LL36 9SL MW0ISF
LL36 9SL MW6BXV
LL36 9UY GW4XZJ
LL37 2QB GW6UWW
LL37 2UZ GW6NXL
LL38 2BX MW3WMI
LL40 1GA MW5CKN
LL40 1NS 2W0IMX
LL40 1NS MW7IML
LL40 2AH GW6AZG
LL40 2AS
MW0WMW
LL40 2EE MW3AXW
LL40 2EF MW1CKK
LL40 2RP GW0UCA
LL40 2SF 2W0MCQ
LL40 2SF MW6MTW
LL40 2YG GW1URD
LL41 3AF GW5DRE
LL41 3DL GW0WGM
LL41 3PU GW1EKC
LL41 3SS 2W0ORT
LL41 3SS MW0DOR
LL41 3SS MW6YKS
LL41 4BH 2W0DUL
LL41 4BH MW0WYN
LL41 4BH MW6DUL
LL41 4BU 2W0OSH
LL41 4BU 2W0UPH
LL41 4BU MW0UPH
LL41 4BU MW3BSJ
LL41 4BU MW6PZP
LL41 4BU MW6UPH
LL41 4DD GW8ZLT
LL41 4PN GW7TTX
LL42 1DX GW0HHD
LL42 1EN GW6DDF
LL42 1PL MW1KDP
LL42 1PT MW3HGR
LL43 2AG 2W0VTK
LL43 2AG GW8WNB
LL43 2AG MW0VTK
LL43 2AG MW3VTK
LL43 2AL MW6LIZ
LL43 2AN 2W0LLL
LL43 2AR MW7TYB
LL43 2BB GW0AYQ
LL43 2BB GW4LZP
LL44 2BG MW1WEJ
LL44 2BU MW6CEG
LL44 2RQ GW3FDZ
LL45 2HH GW6JBN
LL45 2HT GW0AGL
LL45 2LA MW6DSU
LL45 2LU GW0UKF
LL45 2ND GW4ZTG
LL46 2SS GW1KVI
LL46 2UJ 2W0LGG
LL46 2UJ MW6LDM
LL46 2UW MW6GHY
LL46 2UW MW6PYH
LL46 2YW MW6CHQ
LL47 6YA MW3UKK
LL48 6AL GW3JSG
LL48 6AY GW4VWO
LL48 6BH GW8AAF
LL48 6DU MW6TLF
LL48 6EF MW1BQO
LL48 6EG GW6IMS
LL48 6LS GW4PHB
LL48 6LS MW3WMP
ll48 6ps GW7RLZ
LL48 6SR GW1EWY
LL48 6ST MW3SZF
LL49 9AT MW3RVN
LL49 9AT MW5CAD
LL49 9BU GW4HLO
LL49 9DF 2W0SAK
LL49 9DF MW0GSR
LL49 9DF MW3OYC
LL49 9DF MW6BOZ
LL49 9DF MW7VCM
LL49 9NB GW6AZX
LL49 9UH MW3DEI
LL49 9YA MW6BQL
LL49 9YD GW4OWQ
LL51 9SX MW3WSC
LL51 9UJ GW7KIO
LL52 0EF MW3EOY
LL52 0EG MW3ZHU
LL52 0RB MW7TYD
LL52 0SR GW4IDC
LL52 0SR GW8PKB
LL52 0SS MW6VTA
LL53 5AP GW0HGC
LL53 5LF GW0FOL
LL53 5NU 2W0VJL
LL53 5NU MW3VJL
LL53 5NU MW6XSI
LL53 5RT MW7RHS
LL53 6DQ MW0GQO
LL53 6DQ MW0RHD
LL53 6DQ MW3CYQ
LL53 6DQ MW3CYU
LL53 6DQ MW6RHD
LL53 6DW GW6KLC
LL53 6EA GW7UOH
LL53 6EG GW3NNB
LL53 6NN 2W0XMW
LL53 6NN MW0XTK
LL53 6NN MW7ADT
LL53 6RQ GW1ROE
LL53 6UB GW0RMB
LL53 6UY GW0IWD
LL53 7ED 2W0UUA
LL53 7PA 2W0BVK
LL53 7PB GW8WNK
LL53 7PF MW0AEV
LL53 7PG GW4UKU
LL53 7PG GW4WKQ
LL53 7RL MW3VVJ
LL53 8AE GW0PZT
LL53 8AE GW3KJW
LL53 8EH GW8BZN
LL53 8ND GW4UIL
LL53 8NW GW1IEB
LL53 8UG GW7KJO
LL536NT GW6NGR
Ll538ng GW7NJM
LL54 5EY GW1CFM
LL54 5HG GW4XRW
LL54 5HG GW4XZP
LL54 5SF MW3UWZ
LL54 6EW GW0NDA
LL54 6HD GW1OXJ
LL54 6HG 2W0HRH
LL54 6HG MW0PNC
LL54 6HG MW6JMB
LL54 6HG MW6KSW
LL54 6PT GW0WZL
LL54 6RT GW4CAT
LL54 6SN MW3ZKY
LL54 6SS MW3ZVD
LL54 7BH MW3PZO
LL54 7ED GW0GLX
LL54 7HN MW7FON
LL54 7HU MW0LPG
LL54 7LB GW8HQM
LL54 7PU MW3QJK
LL54 7TW GW7BZY
LL54 7YE GW0ENT
LL55 1BE 2W0SVW
LL55 1EY 2W0CYM
LL55 1EY GW6MUP
LL55 1HF GW4MEI
LL55 1LL GW0AQR
LL55 1UP GW6TGR
LL55 1YB GW1PIH
LL55 1YB GW6REQ
LL55 1YD 2W0XJQ
LL55 1YD MW7XJQ
LL55 1YL GW4KAZ
LL55 1YT 2W0LLA
LL55 1YT MW0KLW
LL55 1YT MW6GBA
LL55 2AG GW7CEQ
LL55 2EF 2W0OVT
LL55 2EF MW3OVT
LL55 2LW 2W0GVK
LL55 2RL GW6BUW
LL55 2RU MW7GNN
LL55 2UR GW4RLP
LL55 2UW MW5DJO
LL55 3BN GW4ZCY
LL55 3BN GW4ZPL
LL55 3BN GW6AKS
LL55 3BN GW6UKO
LL55 3BS MW7GEH
LL55 3HD GW1JIE
LL55 3LE 2W0AWW
LL55 3LG MW0INW
LL55 3LU GW1PCD
LL55 3ND GW0HHW
LL55 3NR 2W0OYL
LL55 3NR MW6OYL
LL55 3NT 2W0GWK
LL55 3NT MW6GWK
LL55 3PW GW4PUX
LL55 3PW GW7UQJ
LL55 3PW GW8XUM
LL55 4AZ MW3EWR
LL55 4HB MW0VCC
LL55 4LE 2W0GAQ
LL55 4LE MW1EYU
LL55 4LE MW6KTS
LL55 4RR MW0PPO
LL55 4TW MW0ANV
LL55 4YY MW3DTC
LL56 4AB MW6GFP
LL56 4HQ MW0LEF
LL56 4PB GW4WKZ
LL56 4QY GW7TSO
LL57 1HF MW3YNK
LL57 1LU 2W0JCN
LL57 1LU MW0IWU
LL57 1LU MW0JWP
LL57 1LU MW6CTB
LL57 1NA 2W0HMS
LL57 1NG GW0DXO
LL57 1NH 2W1AZU
LL57 2NS GW4KTQ
LL57 2UD GW0HGN
LL57 2UJ MW7BLU
LL57 2YG MW0WLP
LL57 2YG MW7WMS
LL57 3AY MW0EDS
LL57 3RE MW6HRK
LL57 3SP GW6MYY
LL57 3SP MW0GOX
LL57 3SP MW3AWI
LL57 3TD GW8GOO
LL57 3YJ MW3VQJ
LL57 4AE MW7IOL
LL57 4AF 2W0GVN
LL57 4AS 2W0YLE
LL57 4AS MW6TBC
LL57 4AX GW7EWD
LL57 4DJ GW0FQZ
LL57 4DP GW3VFZ
LL57 4DR GW0ETF
LL57 4DT MW7HEN
LL57 4LD GW4FQU
LL57 4LX 2W0CZU
LL57 4LX MW6BKS
LL57 4LX MW6CTG
LL57 4NS GW4HMR
LL57 4PA GW6ORE
LL57 4TG 2W0CHV
LL57 4TG GW5AMS
LL57 4TG MW0NWA
LL57 4TG MW3SET
LL57 4UA MW0BEA
LL57 4UR MW0BYT
LL58 8AF MW3GKI
LL58 8DB 2W0GVO
LL58 8HH GW4EMS
LL58 8HW GW0UWD
LL58 8LH MW1DOU
LL58 8PG GW4MBL
LL58 8PH GW3GUX
LL58 8PT GW8PZS
LL58 8RG GW8TOX
LL58 8RW GW6TOX
LL58 8SU GW4JUI
LL58 8UE GW4UCV
LL58 8YG GW0BDW
LL59 5DA GW8ZEI
LL59 5LR GW3JBJ
LL59 5LR GW4RQQ
LL59 5NB GW4CFC
LL59 5NG GW0GFO
LL59 5PP 2W0CEF
LL59 5PP MW0HCA
LL59 5PW GW1SUK
LL59 5PW MW5ADD
LL59 5RD GW3XRM
LL59 5UA GW4HAT
LL59 5UF MW7HAD
LL59 5YA MW3MEY
LL59 5YH GW6KIW
LL59 5YH MW0DNK
LL595YD MW7GAH
LL60 6AL GW3TLP
LL60 6DG MW3YQL
LL60 6EA MW6OTT
LL60 6HL MW1BTA
LL60 6HL MW3GWH
LL60 6HL MW6FNK
LL60 6JN 2W0DER
LL60 6JN MW0MON
LL60 6JN MW3UUY
LL60 6JW GW7BZR
LL60 6NW GW2OG
LL61 5AX GW4JXN
LL61 5JB GW4GTC
LL61 5JB GW8UTK
LL61 5JF 2W0DIV
LL61 5JF 2W0SLP
LL61 5JF MW5ADW
LL61 5JR GW0IQZ
LL61 5JY 2W0DAA
LL61 5JY GW3VVC
LL61 5JY GW4JKR
LL61 5JY GW4TTA
LL61 5SZ GW6IUK
LL61 5TZ GW6DOK
LL61 6EQ MW6TDB
LL61 6SY GW0PZS
LL61 6TG MW1DKM
ll61 6tg MW7AUU
LL61 6TZ GW3PRL
LL61 6TZ GW4NBM
LL62 5AS 2W0MTD
LL62 5AS GW7FNQ
LL62 5AS MW3UPX
LL62 5AW GW3EIZ
LL62 5BG 2W1FPK
LL62 5BG MW1DAM
LL62 5LF GW4HKX
LL62 5RD 2W0JJY
LL62 5RD MW7JJY
LL63 5PQ GW0GHG
LL63 5TW GW0EAW
LL64 5XB GW3YTC
LL64 5XB MW0JHC
LL65 1AL GW0AVD
LL65 1BG GW6VIC
LL65 1ES GW4XAU
LL65 1EU GW0LIS
LL65 1HT MW6SPN
LL65 1LR MW6GCQ
LL65 1LT MW6KVO
LL65 1NW GW4IEU
LL65 1SN GW6JCK
LL65 1YR 2W0PYL
LL65 1YR GW4CVT
LL65 1YR MW0KBN
LL65 1YR MW0TMJ
LL65 1YR MW1CFA
LL65 1YR MW7CVT
LL65 2AL 2W0RZL
LL65 2AL MW6RZL
LL65 2AZ GW1HHM
LL65 2DU MW7TTA
LL65 2EF GW0KPV
LL65 2EG MW7MHC
LL65 2EX GW0NRW
LL65 2HD MW0BLU
LL65 2LL 2W0RBW
LL65 2LL MW6NLA
LL65 2LU MW0AQZ
LL65 2NF GW0IJY
LL65 2NN MW0CLB
LL65 2NS MW3YQU
LL65 2PN GW4MHR
LL65 2PN GW4XYI
LL65 2TX MW7WZQ
LL65 2UG MW0REH
LL65 2WA MW3GKB
LL65 2YH MW0KGY
LL65 2YW GW4VDP
LL65 3BT MW3ZVB
LL65 3DP MW6LVH
LL65 3ES GW0ENU
LL65 3ES MW0GNF
LL65 3HS GW7JKK
LL65 3LU MW0OGI
LL65 3LU MW0MAH
LL65 3PL GW0ONY
LL65 3RD GW0VYG
LL65 3RD GW4VEQ
LL65 3RR GW6TMD
LL65 3SY GW0MHK
LL65 3TB MW6MFB
LL65 4AG GW6EGX
LL65 4AG GW6NHL
LL65 4AG GW8JEI
LL65 4AH GW4ZMQ
LL65 4BG GW4CWU
LL65 4EA MW1EPI
LL65 4PY MW3NDU
LL65 4SL GW7JRT
LL65 4SW 2W0MON
LL65 4SW MW6AAW
LL65 4UY GW4WZS
LL65 4UY GW4YVX
LL65 4YL GW4YQL
LL65 4YL GW8FOL
LL653yd 2W0GDB
LL67 0NP MW3UPK
LL68 0RE GW0FEM
LL68 0RE GW8FDI
LL68 9AT MW6DYF
LL68 9BG 2W0GIW
LL68 9BG MW1EWJ
LL68 9BG MW6PAM
LL68 9BT GW3SRM
LL68 9DL GW6OAW
LL68 9DN 2W0IJD
LL68 9DN MW7TAH
LL68 9DS MW0SSB
LL68 9EA 2W0JYB
LL68 9EA MW7MON
LL68 9HT 2W0FNA
LL68 9HT MW7JAB
LL68 9HT MW7KMB
LL68 9LY 2W0LKJ
LL68 9LY GW1KGW
LL68 9LY MW7ADL
LL68 9NE GW6HYI
LL68 9NH 2W0GLV
LL68 9NH MW0LLK
LL68 9NH MW6GLV
LL68 9PD GW1XAS
LL68 9PU MW7IMJ
LL68 9RX GW0ESK
LL69 9BZ GW4ZWN
LL69 9YB
GW1WWW
LL70 9DQ 2W0GGG
LL70 9DQ MW0RKB
LL70 9DQ MW3LTU
LL71 7AE MW6PQQ
LL71 7AE MW7JUE
LL71 7BU MW0SEC
LL71 7DH MW6RHR
LL71 8BG MW0BER
LL71 8ED GW0TPR
LL71 8EJ MW6FJX
LL72 8HB GW1OKY
LL72 8HN GW4EXE
LL72 8LN GW7JGF
LL73 8PE GW8YUJ
LL73 8PP GW0KRL
LL73 8PP GW7APP
LL74 8RD 2W0VSW
LL74 8RD MW3XQE
LL74 8RP GW4EOW
LL74 8SG 2W0YMP
LL74 8SG MW7KEE
LL74 8SR GW0BTB
LL75 8LN GW7SBO
LL75 8NQ MW0WEY
LL75 8UR 2W0CBZ
LL75 8UR 2W0LTG
LL75 8UR MW0GMZ
LL75 8UR MW3ZVH
LL75 8UT 2W0SJT
LL75 8UT MW0GCT
LL75 8YE MW0JDW
LL75 8YG GW1ISK
LL75 8YU MW7BQK
LL77 7BX 2W0RTZ
LL77 7BX MW6WVC
LL77 7DZ GW6YMS
LL77 7EJ MW1DNY
LL77 7JS MW6FJT
LL77 7JS MW7IFZ
LL77 7LJ GW3TWN
LL77 7LJ MW6ZAA
LL77 7NA GW6BMP
LL77 7QD 2W0ZZF
LL77 7QD MW6ADU
LL77 7QE GW3PRW
LL77 7SJ MW0BTU
LL77 7SU GW0ESU
LL77 7UR GW6HZJ
LL77 7WD MW3TOI
LN4 1DB 2E0DKY
LN4 1DB M6SFQ
LN1 1JE M7AER
LN1 2AS G4STO
LN1 2AS G5FZ
LN1 2AS G6COL
LN1 2BU M0KEY
LN1 2DY G4ZKW
LN1 2EP G7OJU
LN1 2EP M6DDQ
LN1 2FT 2E1GYB
LN1 2FT M0GYB
LN1 2FT M3GYB
LN1 2GY 2E0ZLO
LN1 2GY M0ZLE
LN1 2GY M6ZAC
LN1 2HA G8PK
LN1 2HZ G7KIE
LN1 2JD M3EJR
LN1 2LH M6JHV
LN1 2NH G0BUC
LN1 2NJ 2E0HXJ
LN1 2PE M6DUJ
LN1 2QA M3WVO
LN1 2QB G1HTM
LN1 2QL G3UPI
LN1 2QN G6BPH
LN1 2RG M3WBN
LN1 2SG G4GDC
LN1 2TL 2E0LEY
LN1 2TL 2E0VWG
LN1 2TL M6EGX
LN1 2UP M7BIX
LN1 2XF M6JAG
LN1 3JS G4SSV
LN1 3LB G0VWW
LN1 3LU G4OEF
LN1 3NN G0RQQ
LN1 3QL M3XCX
LN1 3QP G7KYJ
LN1 3QP M0XYZ
LN1 3RD 2E0WWP
LN1 3RD M6OYU
LN1 3SZ 2E0SIY
LN1 3SZ M0SIY
LN1 3SZ M3SIY
LN10 5DU M6NZE
LN10 5DY G6NKL
LN10 6RL M6ARB
LN10 6RT 2E0JHO
LN10 6RT M0JHO
LN10 6RT M6DKF
LN10 6RW G1VMX
LN10 6SD G3JKB
LN10 6TH G3QBX
LN10 6UE G3MDI
LN10 6UR 2E0TPD
LN10 6UR M0VLC
LN10 6UR M3WUV
LN10 6UU G4IJA
LN11 0AP M3ZWI
LN11 0AZ 2E1GIK
LN11 0AZ M5COL
LN11 0BD G8SQY
LN11 0BP M6JPI
LN11 0DF 2E0PJS
LN11 0DN G7YNA
LN11 0DW M3CDV
LN11 0DY G0LWG
LN11 0DY M6UDZ
LN11 0EH M7TMC
LN11 0JD M3REL
LN11 0PB G0KKR
LN11 0PN G4MXI
LN11 0XF G1VUG
LN11 7AD G3PRR
LN11 7AG 2E1GZF
LN11 7BZ G8OOS
LN11 7JR G3ZPM
LN11 7QG M0GHA
LN11 7QH G7VIG
LN11 7QH M0VIG
LN11 7QH M6AVS
LN11 7QU M3LRP
LN11 7QW G7MWH
LN11 7SA G3NRQ
LN11 8AS 2E0EDS
LN11 8AS M6EDS
LN11 8BS G6XFB
LN11 8HL 2E0MDT
LN11 8HL M6MKB
LN11 8HY G3KRZ
LN11 8JD G6XXL
LN11 8LJ G3NHR
LN11 8RP 2E1GJL
LN11 8RY G4HUD
LN11 8SG G1PRS
LN11 8SP M1BTR
LN11 8TA G0DPY
LN11 8UJ G0VRU
LN11 8UN G4XFG
LN11 8YN M6DIF
LN11 9AL 2E1FEF
LN11 9DG G8AJM
LN11 9HT G0GNW
LN11 9RQ G7AJP
LN11 9SA M7AOO
LN11 9TF G4YZC
LN11 9UE G0WIW
LN11 9UP G1GMX
LN12 1BQ G0SWS
LN12 1BQ G6TAN
LN12 1BQ M0ERG
LN12 1BY 2E0BWQ
LN12 1BY G4ECE
LN12 1BY M6SIX
LN12 1DA G0FMG
LN12 1DA M3INQ
LN12 1EZ M7TBO
LN12 1JF 2E0PGI
LN12 1JF M0PGI
LN12 1JF M6BHH
LN12 1JF M6RDW
LN12 1LL G3SLS
ln12 1lp 2E0FHX
ln12 1lp M6OJU
LN12 1NX G3WNQ
LN12 1QB G0KED
LN12 1QE G1XWD
LN12 1QE 2E0YYD
LN12 1QR M0XDX
LN12 2 BA G0LAN
LN12 2 BA G4ETM
LN12 2AJ 2E0GBH
LN12 2AJ G3VDV
LN12 2AJ G7BUK
LN12 2AJ M0DZA
LN12 2AJ M3FEY
LN12 2AJ M6IOQ
LN12 2AS G1IEX
LN12 2AX M3YUB
LN12 2AY G0DXK
LN12 2BF M6DRK
LN12 2BP M3FQT
LN12 2BS 2E0LRK
LN12 2BS M3NQL
LN12 2DF G7ANK
LN12 2EJ 2E0FHV
LN12 2EL M3RDY
LN12 2HP G6NZG
LN12 2JU G8EWF
LN12 2JZ 2E0JSJ
LN12 2JZ M6TDF
LN12 2LN G8PPA
LN12 2NP G1BVV
LN12 2NU G1AAC
LN12 2NU G4BFV
LN12 2PT M3TMG
LN12 2QS G3LSA
LN12 2QY G4NJW
LN12 2QZ G4HHM
LN12 2RE 2E1FYZ
LN12 2RE G0CBM
LN12 2RT G0NQA
LN12 2RT M6PJX
LN12 2TX G1HEY
LN122GY G3MNS
LN13 0BD M7HWK
LN13 0BH 2E0JQG
LN13 0BH M0JQD
LN13 0BH M6JQG
LN13 0EH G4CPI
LN13 0JP G0DVY
LN13 0JP G3SYB
LN13 0JP G4WEC
LN13 0JW G3ZUI
LN13 0LR G1NVV
LN13 0NP 2E0KIM
LN13 0PN G0KBZ
LN13 9AX M7AVL
LN13 9HJ G4LOP
LN13 9NG G6RHB
LN13 9PH 2E1FFJ
LN13 9PH G7RAG
LN13 9PT G8HSV
LN13 9RF M3NUI
LN13 9RL 2E0CDI
ln13rd M7LRQ
LN2 1JD G8OPP
LN2 1PY 2E0TTM
LN2 1PY M0TMA
LN2 2AE G0WQA
LN2 2AY M6VAP
LN2 2JL G0CEG
LN2 2JL G4RUE
LN2 2JR G0PKN
LN2 2JU M1FBW
LN2 2JW G6BWM
LN2 2LX M7BWW
LN2 2NE G3NIC
LN2 2NQ G0HEE
LN2 2PH G6BD
LN2 2PH G7FAR
LN2 2QU G3RRN
LN2 2WB M3HGW
LN2 3JS G1LQB
LN2 3JX G0NXF
LN2 3RS M6KSZ
LN2 3SR G4YQK
LN2 3TR 2E0DAT
LN2 3TR M3NKL
LN2 3TT G7PMQ
LN2 3TU G0JOD
LN2 3US 2E1RWC
LN2 3US M3RWC
LN2 4BT 2E0XXO
LN2 4BT M6UGN
LN2 4EB M7KTG
LN2 4EW G3VEV
LN2 4LT 2E0LRJ
LN2 4LT M0TEF
LN2 4LT M1DYI
LN2 4LT M3LRJ
LN2 4LX G6EWP
LN2 4LX G6SWZ
LN2 4PT G4KSA
LN2 4QH M5RFD
LN2 4RE 2E0NDT
LN2 4RE M6NDT
LN2 4TX M0EJL
LN2 4WS G7JRD
LN2 5EY 2E0KWM
LN2 5EY M0KWM
LN2 5EY M6KWM
LN2 5LB 2E0OYQ
LN2 5LB M0IYE
LN2 5LP G7RQO
LN2 5NN M6HXX
LN2 5QF 2E0KVR
LN2 5QF M6KNG
LN2 5RU G3NGJ
Ln25az M7LLR
LN3 4AN G6DKK
LN3 4AX G0VKF
LN3 4BE G3PTZ
LN3 4BG M0CZT
LN3 4BH M1AYR
LN3 4BJ G3AWK
LN3 4BZ G8FPA
LN3 4DH G6RFR
LN3 4DQ 2E0OAK
LN3 4DQ M3CTB
LN3 4DQ M6JBU
LN3 4EB G7NSK
LN3 4EE 2E0SFV
LN3 4EE M0VEL
LN3 4EE M7FAV
LN3 4EG 2E0TMO
LN3 4EG 2E0VKW
LN3 4EG G0OSO
LN3 4EG M6TMO
LN3 4EG M6VKW
LN3 4EJ G0TQV
LN3 4HT G3NSL
LN3 4JS G4XMQ
LN3 4LS 2E0CTL
LN3 4LS M6BGA
LN3 4LY 2E0SSM
LN3 4LY M0PTA
LN3 4LY M5SJM
LN3 4LY M6TSJ
LN3 4NA G4SLG
LN3 5AB M0BKK
LN3 5TD 2E0GFY
LN3 5TD M7DNH
LN3 5XS G7PHR
LN35XN M7OKV
LN35XT G7OBX
LN4 1AE G7JQT
LN4 1AS M6HHY
LN4 1AU 2E0HGU
LN4 1AU M7PCC
LN4 1DB M6IJZ
LN4 1DX M5YEX
LN4 1DZ 2E1BME
LN4 1DZ G4URX
LN4 1EG G4DTL
LN4 1EQ G4UAL
LN4 1EY G4JQJ
LN4 1LH 2E0FZY
LN4 1LH G1SVI
LN4 1LH M3GUU
LN4 1LJ G4DBS
LN4 1NB M6ICK
LN4 1NN G1OQF
LN4 1NS M7NYH
LN4 1NW G4YYE
LN4 1PG G4RYB
LN4 1PU G7PLJ
LN4 1QB G7PKG
LN4 1QE M7DGG
LN4 1QT G1ATU
LN4 1TT G4JNL
LN4 1TU G0GTV
LN4 2EX M0ONQ
LN4 2LE G3WOK
LN4 2NP 2E1TSO
LN4 2NP M7TSO
LN4 2PW 2E0TYC
LN4 2PW M6TYC
LN4 2PW M6VCK
LN4 2QG G7JBZ
LN4 2RD M6FYV
LN4 2RP G3WTT
LN4 2TD M6CDC
LN4 2TY M6CXV
LN4 3AP 2E1GJP
LN4 3BQ M0RDS
LN4 3BY M6YAR
LN4 3DS G7VSN
LN4 3DT G3NXT
LN4 3NP M0EAU
LN4 3PT 2E0FBO
LN4 3PT M0IVJ
LN4 3PT M6NYO
LN4 3RN G8IJC
LN4 3SL G3LAS
LN4 3TW M7KMG
LN4 3YG M0HIL
LN4 4AU G6AWY
LN4 4EA G3MON
LN4 4FN M6TFJ
LN4 4JJ M3GUF
LN4 4JQ 2E0CZK
LN4 4JQ M6SAY
LN4 4PB G0MPQ
LN4 4QA G1EFK
LN4 4QH 2E0WOW
LN4 4QH M6ROC
LN4 4QU G4DDI
LN4 4RG G6ZTL
LN4 4RG M0HAZ
LN4 4RQ G7SKW
LN4 4SP G4MWJ
LN4 4XA 2E0NAI
LN4 4XA M6CIK
LN4 4XT M6RJP
LN4 4YJ 2E1HGM
LN4 4YJ M1ECB
LN42BD M7CYU
LN5 0ER G4PWM
LN5 0ER G4UNF
LN5 0FD G6PMW
LN5 0QZ M7LFF
LN5 0RN G7OVS
LN5 0RP 2E0MFF
LN5 0RP M6FFY
LN5 0SJ G3VHN
LN5 7DL M7EHQ
LN5 7EX M7CZU
LN5 7PZ 2E0XRD
LN5 7PZ M6XDR
LN5 7QF M0MNV
LN5 7SJ G8TSV
LN5 7UT G0MRB
LN5 7XW 2E0HYI
LN5 7XW M7DBZ
LN5 8DA G0UOZ
LN5 8DA M3VOZ
LN5 8DR M6XFI
LN5 8DR M6XGC
LN5 8DR M6XLN
LN5 8EL G0LEN
LN5 8EL G7LEN
LN5 8ND 2E0MRF
LN5 8QS M6CAJ
LN5 8QX G7RUP
LN5 8RL M0DIW
LN5 8RX G0FVI
LN5 8SH M0CES
Ln5 8sj 2E0BVG
LN5 8SN 2E0COF
LN5 8TF G7CNM
LN5 8TG 2E0LUY
LN5 8TG G8YMW
LN5 8TG M0LUY
LN5 9DA 2E0GSJ
LN5 9DR M3ULU
LN5 9DT G4LQH
LN5 9FW G0WFV
LN5 9JD G0WAY
LN5 9LJ M6SAV
LN5 9NE M3GEK
LN5 9NH 2E0MSZ
LN5 9NH M0WMS
LN5 9NH M6WMS
LN5 9PY G7PRB
LN5 9QR G0MQD
LN5 9SW G4ZOX
LN5 9SX 2E0TRP
LN5 9TL 2E0RXG
LN5 9TL M0RXG
LN5 9TL M6RSF
LN5 9TL M6RXG
LN5 9UF M6AGR
LN5 9UT G0MEY
LN6 0AG 2E0XMO
LN6 0AG M6NMD
LN6 0EY M7JFH
LN6 0JB M6EBA
LN6 0LZ G1XZG
LN6 0NR 2E0EOD
LN6 0NR M6LJB
LN6 0QN G4JZV
LN6 0RH G6FYX
LN6 0SS G4OMD
LN6 0SY G8HMZ
LN6 0XQ G0BEN
LN6 0YD M6IMI
LN6 0YF 2E0SBL
LN6 0YF M6MJP
LN6 0YZ M6DTO
LN6 3LU M6OYP
LN6 3NR G7JQZ
LN6 3NU G0EUN
LN6 3RQ G6OYU
LN6 5TF G0EJV
LN6 5TW G4HNQ
LN6 5UF 2E0CJM
LN6 5UF M0KAN
LN6 5UF M6AVZ
LN6 5UU M0CQH
LN6 5UW G6KSR
LN6 7DR G7ERS
LN6 7DR M0ERS
LN6 7EW M6JGZ
LN6 7LQ G1ZBY
LN6 7NL G0VRE
LN6 7NL G8VEM
LN6 7PA G6HUP
LN6 7PJ M6CKE
LN6 7PN G3PVU
LN6 7PY M6DKO
LN6 7RD G7UAV
LN6 7RQ M6CMG
LN6 7TG M3FKL
LN6 7TG M3FKM
LN6 7UP M6CYW
LN6 8AP M6FCZ
LN6 8BT M3GFZ
LN6 8BX G0LZS
LN6 8LY G4OSB
LN6 8QR 2E0WGD
LN6 8QR M7DGW
LN6 8QS 2E0DDC
LN6 8QS M3XOE
LN6 8SD G6GLW
LN6 8SN G0IMP
LN6 8TZ M3OVA
LN6 8UB M1DJS
LN6 8UR M7RCM
LN6 8UW 2E0GAL
LN6 8UW 2E0KAG
LN6 8UW M3YZV
LN6 8UW M6RVH
LN6 9AZ M6BXX
LN6 9BG G8BGL
LN6 9BS G4HIV
LN6 9DH 2E0OVR
LN6 9DH G7JFI
LN6 9DH M6PDV
LN6 9EX G3IYF
LN6 9JE G8SGI
LN6 9PA 2E0IHH
LN6 9PA M6IHH
LN6 9PJ G7VTN
LN6 9RG M3TPD
LN6 9RG M6LQO
LN6 9SB M0KED
LN6 9SZ G0UND
LN6 9TB M3XBO
LN6 9TG 2E0DZV
LN6 9ZD M6KHQ
LN63QQ M7CGQ
LN69XA G4JES
LN7 6EL G0CVN
LN7 6FH M7PHE
LN7 6GN G4JA
LN7 6NQ 2E0FML
LN7 6PA G4FQP
LN7 6RB G3RXP
LN7 6RN G7KPM
LN7 6RU G6IYS
LN7 6RU M6RIT
LN7 6SW 2E0TYY
LN8 2AJ M7LMF
LN8 2HE 2E0FHD
LN8 2HE M6RCI
LN8 2HL G4CLL
LN8 3AA G7JXX
LN8 3AA M3LMY
LN8 3AG G4UHZ
LN8 3AP M6EOX
LN8 3BE M6DVZ
LN8 3DL M7EWT
LN8 3DS 2E0RPD
LN8 3DS M0RPD
LN8 3DS M3RPD
LN8 3EE G0GMA
LN8 3EE G6MVD
LN8 3EW G4OER
LN8 3EW M3OER
LN8 3JA G0FLG
LN8 3LD G1YFQ
LN8 3LD G3KDY
LN8 3LE G4PWF
LN8 3NQ M0JJE
LN8 3PB M6VEN
LN8 3QP G4XFC
LN8 3UB M3GEZ
LN8 3UR M3JVX
LN8 3US M0OOO
LN8 3US M3ZSW
LN8 3UT M0DBY
LN8 3XL G1PRM
LN8 3XL G8KLC
LN8 3XZ M1BRY
LN8 3YL G8THR
LN8 3YL M3PCC
LN8 5NF M6KOV
LN8 5RB M1DHV
LN8 5RB M3WRA
LN8 5RF G6TVP
LN8 5RF M5ZZZ
LN8 5RY G8KJI
LN8 5SL G7UEV
LN8 6AD G4KIZ
LN8 6AJ G3YFU
LN8 6AN M0BIC
LN8 6AZ G8UDI
LN8 6DE G8JCS
LN8 6DH G4LAY
LN8 6DX 2E0EDG
LN8 6DX M6EDG
LN8 6EW 2E0IJQ
LN8 6EW 2E0LJB
LN8 6EW 2E0ODL
LN8 6EW M3ODK
LN8 6EW M3ODL
LN8 6EW M6IJQ
LN8 6EX G4RBP
LN9 5AW G3VUE
LN9 5HF G0NDY
LN9 5JE M3VUV
LN9 5JF M1MSF
LN9 5LP 2E0ZRQ
LN9 5LP M3ZRQ
LN9 5NH G0SWW
LN9 5NH G7MES
LN9 5NW G6SWO
LN9 5PT G1JZG
LN9 6AF G6ZQS
LN9 6AW G3ZPU
LN9 6AW M0TCM
LN9 6BE G4FOT
LN9 6BH 2E0PKR
LN9 6EY M0AJD
LN9 6HG G1EGR
LN9 6JH M6HWT
LN9 6LB G3OUT
LN9 6LD G7NLA
LN9 6PQ G4WMO
LN9 6PQ G4ZWA
LN9 6QH G8RYO
LN9 6RE G4AIE
LN9 6RR 2E0XOD
LN9 6RR M3XOD
LS10 1LH 2E0CJQ
LS10 1LH M6TWL
LS10 3DT 2E0USD
LS10 3DT M6SDQ
LS10 3EJ M3IBJ
ls10 3qa M6NWS
LS10 3TB 2E0CRM
LS10 3TB M3BZC
LS10 3TJ G0CYB
LS10 4BG 2E0FNJ
LS10 4BG M6KZB
ls10 4bg M7AMQ
LS10 4LH M6WDC
LS10 4NQ M6MWP
LS10 4QN 2E0SDT
LS10 4QN M0SDT
LS10 4SN G1AUH
LS10 4TH M6KQK
Ls104jt M6NYY
LS11 0HZ M0VBV
LS11 0HZ M3VBT
LS11 5HA M6WRX
LS11 5RP G4RCH
LS11 5RX G4MPO
LS11 5SG G4IDT
LS11 5SG G8UAF
LS11 6NN G7HUJ
LS11 7LS 2E0BKY
LS11 7LS M3TQT
LS11 8AH M6ARF
LS11 8BG M3BTG
LS11 8HR G7POS
LS11 8TP G1KSW
LS11 9RE G7OCX
LS12 2PG G7ECA
LS12 2RE G0WXG
LS12 2RL M3XFI
LS12 3LB G7RME
LS12 3QJ G8SWK
LS12 3SN G8PKJ
LS12 4HQ G4ICF
LS12 4HQ G8KZY
LS12 4JX G0RFA
LS12 4LA G6YMH
LS12 4PB M6TFP
LS12 4PB M7GHE
LS12 4RD 2E0ZRL
LS12 4RR G4IDD
LS12 4UW G6SFH
LS12 5BL M6RWB
LS12 5EA G4LXW
LS12 5LP G4YQW
LS12 5QS G4RYE
LS12 5SU G7HRP
LS12 5SZ 2E0GNU
LS12 5SZ M3XGI
LS12 6AY G1OUX
LS13 1AX M6LHL
LS13 1BL G4SQG
LS13 1EA G7LBH
LS13 1ED M6EHO
LS13 2BJ 2E0JAQ
LS13 2BJ M3JTO
LS13 2BX G1DWT
LS13 2LE G4HLI
LS13 2LH G4OFA
LS13 3DF M0RSF
LS13 3DF M7NTD
LS13 3DH G7BXA
LS13 3EH 2E0XPP
LS13 3EH M3XPP
LS13 3JX G0URF
LS13 3RQ G6XGK
LS13 3RS M6PNX
LS13 3SG 2E1ODG
LS13 4PW G1IKW
LS13 4QY G4OOJ
LS13 4RN M0AVZ
LS13 4SU 2E1RON
LS13 4TT 2E0ERM
LS14 1AP G0JUT
LS14 1HH G1DZC
LS14 1JP G0TGH
LS14 1LF M6OLS
LS14 1LJ 2E0KZH
LS14 1LJ M0KZH
LS14 2EP G0VTJ
LS14 2EQ G8TBX
LS14 2HR 2E0HZU
LS14 2HR M0JKE
LS14 2HR M0SHO
LS14 2HR M6HZU
LS14 3AU G1ZUB
LS14 3BU G3YRU
LS14 3HU M3XSI
LS14 5AS M6RGY
LS14 5BQ M3NPW
LS14 5NR 2E0WFD
LS14 5PD G7VEB
LS14 6AA G7POQ
LS14 6AE M0CVC
LS14 6AY M7WDC
LS14 6JL G4JJS
LS14 6JZ G7PMX
LS14 6RH G4VRW
LS14 6SY M0MLH
LS15 0EZ G7ELS
LS15 0EZ G7KJT
LS15 0HT 2E0VNX
LS15 0HT G5VNX
LS15 0HT M7VNX
LS15 0LW G4XZI
LS15 0NQ G7NXV
LS15 0NQ M5FOX
LS15 4EJ 2E0JOE
LS15 4EJ M3JOW

LS15 4EZ G3XVP
LS15 4HJ G4ZRF
LS15 4JD G3GXQ
LS15 4NY G4OOB
LS15 7DN M1CAE
LS15 7HA G6FFQ
LS15 7HD 2E1GLR
LS15 7HD M1DIE
LS15 7QE G1IUZ
LS15 7QW G0HOB
LS15 7SQ G7DCT
LS15 8AY G0BTD
LS15 8ED M0RAP
LS15 8JJ G8ZXT
LS15 8LW G4HSZ
LS15 8QZ M1ATB
LS15 8SD G7MTI
LS15 8SW G6INO
LS15 8UP M3CJI
LS15 9JD G3RZY
LS150HQ G7OZQ
LS16 5DN 2E0JAF
LS16 5DN M0JAF
LS16 5DY M3ZTN
LS16 5EG 2E0TMF
LS16 5EG M6AMT
LS16 5EG M6ZWP
LS16 5PL M7PMF
LS16 6BU G8MDG
LS16 6BU M0DSI
LS16 6BX G0FWP
LS16 6DU 2E1ANQ
LS16 6EF G1SGM
LS16 6HX 2E0IGP
LS16 6HX M0RVB
LS16 6HX M6IGP
LS16 6JA M0DBD
LS16 6NA M0APY
LS16 7AB G0PAN
LS16 7AS M3WTD
LS16 7ES M0RCP
LS16 7ES M3OOL
LS16 7HF G4PHP
LS16 7PG G4TSF
LS16 7PN G0WRT
LS16 7PP G0SCL
LS16 7QX G4HAJ
LS16 7RB G8TEL
LS16 7SL G4BNL
LS16 8BL G3NRM
LS16 8DE G1SBK
LS16 8EJ G4CSZ
LS16 8JQ G1CXQ
LS16 9DQ G3MFJ
LS16 9DR 2E1CNO
LS16 9LE G4HGT
LS16 9LF G4BNM
LS17 5AX 2E1EMN
LS17 5AX M1APF
LS17 5BH G8UHW
LS17 5ET G4UZN
LS17 5HW M1AZM
LS17 5JP M1AHR
LS17 5NS G4THX
LS17 5PQ G3WNR
LS17 6DB M0SDU
LS17 6FD G4RYS
LS17 6LL M0BGS
LS17 6RY G3AAS
LS17 7DU G0SBZ
LS17 7EE G4BDC
LS17 7EP 2E1COM
LS17 7NH G3KVJ
LS17 7PD G4LYM
LS17 7QB G4MSN
LS17 7RB G7RBA
LS17 8AR M6NYM
LS17 8BG 2E0LSE
LS17 8BG M0SME
LS17 8BG M0YKR
LS17 8BS G4XZG
LS17 8BY G4NCK
LS17 8DA M6NWO
LS17 8DF 2E0LBA
LS17 8DF M6LYA
LS17 8LX G0PFT
LS17 8LX G8ASG
LS17 8TB G3HJP
LS17 9AB 2E0SFS
LS17 9AB M0XXG
LS17 9AB M6PZW
LS17 9AD M0JRO
LS17 9DR M7NCA
LS17 9EG G1MPT
LS17 9ER G3TAF
LS17 9NL G1HHT
LS18 4BD M6XNO
LS18 4BE 2E0NCC
LS18 4BE M3NCT
LS18 4BE M6SQW
LS18 4HL 2E0CUA
LS18 4HL M0ABO
LS18 4HL M3IDW
LS18 4HS G7RKJ
LS18 4PJ G0KVC
LS18 4RL G4OVT
LS18 4RL G4YDW
LS18 5AW G7VHU
LS18 5DT M0ALD
LS18 5HB G4LAD
LS18 5HB G8WYR
LS18 5JL G1ELX
LS18 5JP G4XQV
LS18 5JS G4GYL
LS18 5LD G4GOP
LS18 5PP G4RVO
LS18 5QE G4OOH
LS18 5QU M1DIB
LS18 5RN G0LHU
LS18 5RQ G4YBA
LS18 5SJ G6OTS
LS18 5UN G4FKY
LS19 6AD G1BQV
LS19 6AR G3ZNK
LS19 6BS G4BZL
LS19 6EH G7BHG
LS19 6JX G3KWT
LS19 6NE 2E0KDV
LS19 6NE M0KDV
LS19 6NE M3KDV
LS19 6PU G4AYH
LS19 6RJ G3RMQ
LS19 6SG 2E0VCA
LS19 6SG M7AGH
LS19 7AJ 2E0GNO
LS19 7AJ M7WOZ
LS19 7AS G4FBB
LS19 7AU G4REC
LS19 7ED G1PFZ
LS19 7NL G0VXV
LS19 7SQ G8RTK
LS19 7TE G7KHT
LS19 7XE G0HQU
LS2 7QB M7DCX
LS20 8AY G4GGR
LS20 8EJ M0JSE
LS20 8HA G1NEV
LS20 8LD G4JVZ
LS20 8NX G3KKP
LS20 9AU M1BZF
LS20 9DY M0CSR
LS20 9EF G7PFY
LS20 9EW M6WWJ
LS20 9EW M7DVT
LS20 9HN M0JAM
LS20 9LN M7JAT
LS20 9LW M0TEX
LS21 1DH G0CLD
LS21 1JZ G6XOG
LS21 2AA G8HPY
LS21 2AL G7OXH
LS21 2BY G0NIG
LS21 2DB M1FHX
LS21 2DP G0KDQ
LS21 2DT G0AEX
LS21 2DT G3XEP
LS21 2DT G8LVQ
LS21 2RS G1RRG
LS21 3DS 2E0LAL
LS21 3DS M3FZJ
LS21 3HY 2E0SNZ
LS21 3HY M6FFK
LS21 3LE G0SNW
LS21 3LE M0HWL
LS21 3LN G1TVW
LS21 3LW G7RDJ
LS21 3NW G7LNT
LS21 3NZ G4GDL
LS21 3NZ G4JZT
LS21 3PN G4MWQ
LS212BQ 2E1COD
LS22 4BB M3XVW
LS22 4EH G7OAV
LS22 4ET 2E0OST
LS22 4ET M0KOV
LS22 5BL G1EGK
LS22 5BW G4YZN
LS22 6NP G1UDB
LS22 6RN G1SBN
LS22 7PU G3YTN
LS22 7QU M0DCD
LS22 7QY G4IEC
LS22 7RA G4WWL
LS22 7UD 2E0PUB
LS22 7UE G1RPE
LS23 6AD M7EKU
LS23 6AY G4JVC
LS23 6BW M0BRL
LS23 6EJ M3HTO
LS23 6HU G6RDO
LS23 6PU G8XMU
LS23 6PX G1VAG
LS23 6PX M0VAG
LS23 6RN G4XYY
LS23 6RP G3VMW
LS24 8AG G0PEV
LS24 8BS G0AUE
LS24 8JD G1GCF
LS24 9BR G4FFM
LS24 9DL G0HXU
LS24 9HP M3PWE
LS24 9HR M3BLG
LS24 9HS G0RCL
LS24 9PQ G4TDC
LS24 9QW M0EBR
LS24 9RL G4GIY
LS25 1BN G1XPI
LS25 1BZ G4HKR
LS25 1DD G7CCL
LS25 1EF M6FVP
LS25 1EN G4DTT
LS25 1HE G4CRP
LS25 1HF 2E0LZB
LS25 1HF M6HGF
LS25 1HN G8MWX
LS25 1NS M0UPK
LS25 1PW M1BYT
LS25 2EG G1VGP
LS25 2EN G4DVZ
LS25 2EP G7ADW
LS25 2HA M1DGE
LS25 2JR G1FDL
LS25 2LF G0MYM
LS25 2LY G4ZOT
LS25 2NW G4JKH
LS25 2QB M0NOA
LS25 4AP 2E0LHF
LS25 4AP M7KKG
LS25 4AU G4ZSS
LS25 5AY G0FXY
LS25 5BT G4ZSA
LS25 5PJ G8INL
LS25 6AF G4NCB
LS25 6BY M3TFB
LS25 6BY M3VXN
LS25 6EU 2E0LZN
LS25 6EU M0LZN
LS25 6EU M6LZN
LS25 6LT M7EPY
LS25 6LW M3PMO
LS25 6QA M3SXV
LS25 7DY G6JSR
LS25 7ET G1UVI
LS25 7ET G4TXK
LS25 7PB G3KEP
LS25 7PB G4DMP
LS25 7PX 2E0GPF
LS25 7PX 2E0VKY
LS25 7PX M0HPP
LS25 7PX M6GPF
LS25 7PX M6VKY
LS252BN M3STJ
ls254ap 2E0KJD
LS257BX M3VSQ
LS257EX 2E1AYS
LS26 0DR G4MAK
LS26 0HL G4OVR
LS26 0PP G8TMD
LS26 0QZ 2E0MEH
LS26 0QZ M0TCB
LS26 0QZ M6AXC
LS26 8DL M3JEH
LS26 8ET M3YDI
LS26 8FJ G4NMO
LS26 8QF M3YTV
LS26 8QG G0RDR
LS26 8QJ G3ZRL
LS26 8RY G7NZU
LS26 8SQ G4JMT
LS26 8UN M0NDO
LS26 9AJ M3ZAA
Ls260sw 2E0JRZ
Ls260sw M0NOK
LS27 0DP G0BVT
LS27 0DS 2E1HYD
LS27 0JA G7NAP
LS27 0SG G1MJT
LS27 7BT G7AJN
LS27 7BT M3OCL
LS27 7DJ G0JVI
LS27 7DJ G7PAF
LS27 7DJ M6KSD
LS27 7DN M0LGA
LS27 7DW M7GHO
LS27 7JB M7FZA
LS27 7RT M0FJM
LS27 7RT M3MSL
LS27 8NS G6ZGF
LS27 8QB M6SFS
LS27 8TD G0PTI
LS27 8UH 2E1GDK
LS27 8UL M0NJW
LS27 8UL M0SSW
LS27 8US M3ZRH
LS27 9DS M1EWV
LS27 9NP 2E0JBA
LS27 9NP M3FOQ
LS27 9NP M3FOS
LS27 9PL G4MBE
LS28 5AA G6FMF
LS28 5BF M0BNO
LS28 5DY G4XZK
LS28 5DZ G0DPS
LS28 5DZ G4PTM
LS28 5HW G7OQB
LS28 5LW G4BXQ
LS28 5PZ G7AQA
LS28 5QH M0STS
LS28 5QH M3BVL
LS28 5RE G4NZN
LS28 6AA G1III
LS28 6AA M0XXV
LS28 6AA M3WYR
LS28 6BG G7RRO
LS28 6DJ G4SKO
LS28 7EP G1NSD
LS28 7JW 2E0FOT
LS28 7JW M6NQN
LS28 7SR 2E0FCO
LS28 7SR M6MNP
LS28 7SS M0LRA
LS28 7SS M0SGS
LS28 8AX 2E0WJC
LS28 8AX M3KFO
LS28 8BH 2E0FTS
LS28 8BH M7FTS
LS28 8DG G4TZV
LS28 8EF G3VXE
LS28 8EQ 2E0HGR
LS28 8EW M7OAK
LS28 8NH 2E1ACZ
LS28 8NH G0SSZ
LS28 8QD G4MOT
LS28 9AD 2E1DNB
LS28 9NQ G8WCT
LS29 0BG 2E0SOJ
LS29 0BG M7HPS
LS29 0LE G8KSX
LS29 0QQ G8HZJ
LS29 0QZ G1RPP
LS29 0SH M7IKC
LS29 0SP 2E0GNL
LS29 0SS M7BIP
LS29 6HE G0SNX
LS29 6HH G8SRN
LS29 7BX 2E0DHK
LS29 7BX M0HYN
LS29 7BX M6DUZ
LS29 7EN G3ZNR
LS29 7HS G0LXR
LS29 7LL M3HSV
LS29 7NP G6MHR
LS29 7NP G8JTD
LS29 7NR G4BSS
LS29 7PA G7VUH
LS29 7QB G0KEV
LS29 7QB G8AWN
LS29 7QB M3AWN
LS29 7QH G0CMU
LS29 7RG G6NWK
LS29 8AH G6JWM
LS29 8AH M0TIN
LS29 8AR 2E0CRI
LS29 8AR M6JSP
LS29 8LU 2E0KBJ
LS29 8LU M3SBJ
LS29 8PH G8AXR
LS29 8QP G7LYS
LS29 8SX M0DXN
LS29 8SX M6XAB
LS29 9BP G8WBK
LS29 9DB G4TGJ
LS29 9JN G4DCY
LS29 9SN G0AJB
LS290PT M7CYH
LS3 1NS 2E0NXP
LS3 1NS M0MMP
LS3 1NS M7MGM
LS4 2LN M6AYG
LS4 2PP 2E0LDL
LS4 2PP M0LOF
LS4 2SU M6CBA
LS4 2TQ M1APL
LS4 2TT G0ETL
LS5 3HX G4TFV
LS6 2AP G1JTC
LS6 2HS M7WBN
LS6 2RD G0FXK
LS6 3BY M7GIL
LS6 3DB G4YES
LS6 3HB G4MXM
LS6 3HB G4OOI
LS6 3NA M7ENB
LS6 3NX G3FLV
LS6 4EX G0TUM
LS6 4HD G4IJI
LS6 4HD M0UOY
LS6 4JT G0JUI
LS6 4PZ 2E0CSD
LS7 1SL G7VCK
LS7 2DY G1EZF
LS7 2NN G4UPD
LS7 3HG M7VRM
LS7 3LY G4TCB
LS7 3NX 2E0WPI
LS7 3NX M0WPI
LS7 3NX M6WPI
LS8 1ED M6XGF
LS8 1EW G4ZOB
LS8 1NS G3PKC
LS8 1RZ G1DIR
LS8 1RZ G4GCX
LS8 2BS G4KAX
LS8 2EE G4FTI
LS8 2LA G3VFH
LS8 2PJ M0TLL
LS8 2PJ M6XAV
LS8 2RH G0KQP
LS8 2SX G8HBQ
LS8 2TA G6MTF
LS8 4AG G0DIY
LS8 4DH G7OOS
LS8 5BY G1UVE
LS8 5DB 2E1DLX
LS9 0AE G4HJS
LS9 0LL M3UEX
LS9 6BY M1AUR
LS9 6RG M1AFV
LS9 7SY G4FVV
LS9 8FB M0DJD
LU1 4AN G8ADC
LU1 4DJ G8OBB
LU1 4DU G6CQB
LU1 4DU G6NDJ
LU1 4EN G0GJA
LU1 4EN G0JIK
LU1 4EN G3VER
LU1 4EN G4VER
LU1 4ER G6WVS
LU1 5LB G1OAM
LU1 5PE G1AYH
LU1 5PE M6VCS
LU1 5PE M6ZLN
LU1 5PF G8XSD
LU1 5QD M6RPH
LU2 0DN G6OUO
LU2 0FD G7PJG
LU2 0JF G0BWK
LU2 0PJ M0GQU
LU2 0PL M0HHM
LU2 0RE G1OUA
LU2 0RE G6OUA
LU2 0RE G6UCQ
LU2 0RP G6IAT
LU2 7BG G7FRW
LU2 7BY G4UVD
LU2 7DL 2E0GQT
LU2 7DL M0XIA
LU2 7DL M6GQT
LU2 7EU G3IWH
LU2 7EU M0KHW
LU2 7JL G4IET
LU2 7JY 2E0GPB
LU2 7JY M0PPG
LU2 7JY M6GPB
LU2 7NL G1FGC
LU2 7PJ M3HSW
LU2 7PP G7BNW
LU2 7SA G8NXS
LU2 7TE G4ENB
LU2 7TE G8DDC
LU2 7TX G1PZD
LU2 7UH G3NVL
LU2 7UU G3TAZ
LU2 8AF G4LBH
LU2 8DP G4FAX
LU2 8EJ G4NMY
LU2 8EP G1BDI
LU2 8HQ G8LXY
LU2 8JF M7BLR
LU2 8JH G0KUC
LU2 8JH G3LQS
LU2 8PP G0LQZ
LU2 8QL 2E0LUT
LU2 8RU G8IZW
LU2 8TS M0BHA
LU2 8TZ G8FFM
LU2 9AX G4BMM
LU2 9BL G0TNO
LU2 9ET G7RKW
LU2 9HL G0KYJ
LU2 9HL G4OIS
LU2 9JZ G7VWA
LU2 9RA G0ALB
LU2 9RA G7VYB
LU2 9SQ M7LUT
LU3 1EF 2E0JPO
LU3 1EF G8IXK
LU3 1EF M6JPO
LU3 1HB G8JHM
LU3 1NQ 2E0ONI
LU3 1NQ M7YEW
LU3 1PY G8UOJ
LU3 2AP G4HPY
LU3 2DW G3WBC
LU3 2HJ G6PPV
LU3 2JJ G1SCA
LU3 2LZ G8CBU
LU3 2NY G6DWM
LU3 2PY G1SVL
LU3 2TH 2E0YFZ
LU3 2TH M3YFZ
LU3 2UA G0DCS
LU3 2UR G8CPQ
LU3 3BH G8UYF
LU3 3EA 2E0RVI
LU3 3EA M6RVI
LU3 3JZ G7HBN
LU3 3LA G6MUJ
LU3 3PY G4CPE
LU3 3SU M0CAJ
LU3 3TU M6SCQ
LU3 3TW G1NWZ
LU33AT 2E0JJZ
LU33AT G5JPR
LU33AT M7JPR
LU4 0BU G0WKJ
LU4 0HX M0IQF
LU4 0QW G7JAS
LU4 0UN 2E0OTI
LU4 0UN M3OTI
LU4 0XJ M6JHP
LU4 8ER M3HQB
LU4 8NU G6RPD
LU4 8PY G8EZV
LU4 9AL 2E0DHA
LU4 9AL M3JZL
LU4 9AN 2E0EBA
LU4 9AN M0TMT
LU4 9AN M6EST
LU4 9AS G1XZW
LU4 9EN G7BPI
LU4 9EN G7DBT
LU4 9EN M3XQT
LU4 9ER M6IRJ
LU4 9EU M7FLJ
LU4 9FD M0TSZ
LU4 9HG G3WLM
LU4 9JE G6XFR
LU4 9LT G1JIG
LU5 4AL G1FXB
LU5 4AQ 2E0CQC
LU5 4AQ M0OBY
LU5 4AQ M6RGC
LU5 4BG G1GPL
LU5 4BN M0JUX
LU5 4EA G8NZD
LU5 4LF G3ZNT
LU5 4NU 2E0TVG
LU5 4NU M6VMY
LU5 4PR G1ANS
LU5 4PW G4UOO
LU5 4SH G0LJU
LU5 4SH G7UAL
LU5 5AL M6JLM
LU5 5EP M0SKM
LU5 5GJ M6XJM
LU5 5HU M0JBF
LU5 5HU M6JVB
LU5 5HZ M5IGE
LU5 5JG G1AYI
LU5 5NX M3FMP
LU5 5NX M3FMQ
LU5 5PD M0WHP
LU5 5PN M0LTN
LU5 5TJ M6ACP
LU5 6AL G8HJK
lu5 6aw G1VHC
LU5 6AW G1ZWB
LU5 6AZ G4DUL
LU5 6EL G0WFT
LU5 6LE G1PII
LU5 6LG G8AZR
LU5 6NF G3NRW
LU5 6NT G1ZNX
LU5. 6BQ M7TJB
LU55ED M7PAC
LU55HB M0ZMO
LU55QB M7DCL
LU6 1Ah M7EKA
LU6 1AJ M0MNO
LU6 1BN M0KDH
LU6 1BN M3ODH
LU6 1HW 2E0EXB
LU6 1HW M3FRJ
LU6 1NQ M6ALM
LU6 1NX M0BZN
LU6 1PN M0DZG
LU6 1QJ G0ODU
LU6 1TS M3WGM
LU6 1UG M3WFO
LU6 2AD 2E0TCK
LU6 2AD M6TCK
LU6 2AE 2E0FYC
LU6 2AE M0JJX
LU6 2AE M7DAJ
LU6 2AG G0WTL
LU6 2AH G0HWK
LU6 2AW G3ZFP
LU6 2BJ G8DBH
LU6 2DD G7TVT
LU6 2DF G3SUL
LU6 2DQ 2E0IFB
LU6 2DQ M6UPW
LU6 2FH G0SJR
LU6 2HL 2E0ASW
LU6 2HL 2E1GYH
LU6 2HL G4CVS
LU6 2HL M0CLW
LU6 2HL M0UKC
LU6 2HR G4XLG
LU6 2HR M6MXA
LU6 2JQ G3PTG
LU6 2JU M3XJH
LU6 3AS 2E0SBZ
LU6 3AS M0SBZ
LU6 3DD G6IAN
LU6 3JT 2E0EAZ
LU6 3JT G1HIJ
LU6 3JT M6ETH
LU6 3JZ 2E0NBZ
LU6 3JZ M3NBZ
LU6 3LJ G3NWH
LU6 3NS G4ILR
LU6 3PT G4EKB
LU6 3QQ M1CYT
LU6 3RG G8KGV
LU6 3RS G4WYO
LU6 3RS G8PJD
LU6 3TT M6UAO
LU7 0AB M6PEI
LU7 0EP G4FVZ
LU7 0EP M5EI
LU7 0ER G4IVZ
LU7 0EX G3XTQ
LU7 0HZ G8GIK
LU7 0HZ G8ZVI
LU7 0LB 2E0HJJ
LU7 0LJ G8GZX
LU7 0NT G1HMW
LU7 0QE G6JFN
LU7 0RY 2E0BFG
LU7 0RY M0PJD
LU7 0RY M3LLZ
LU7 0SQ G6FZW
LU7 0SY G6NAG
LU7 1AH M3MIV
LU7 1DJ G8KKU
LU7 1FQ 2E0OBI
LU7 1FQ M3UHN
LU7 1HT M6BFK
LU7 1SE G1LYX
LU7 2LE M7LST
LU7 2PE M3MHP
LU7 2PE M6KJM
LU7 2QQ G0KRR
LU7 2QR 2E0RZM
LU7 2QR M3RZM
LU7 2QW G0FRD
LU7 2RE M6ACQ
LU7 2RH G4CQA
LU7 2SJ 2E0OOV
LU7 2TJ M7FAX
LU7 2TS G6POV
LU7 2TS M6JNW
LU7 2TY G3YZW
LU7 2XD G4DDC
LU7 2XD G8XTW
LU7 2YG G3YYG
LU7 3AB M6CMF
LU7 3AD M6MNX
LU7 3DE M0NTY
LU7 3DN G1MMI
LU7 3DS G0AYC
LU7 3EG M0HSG
LU7 3EL 2E0VPG
LU7 3EL M6VMF
LU7 3GE M3KEJ
LU7 3LH G4ATQ
LU7 3LR M3GHA
LU7 3LU G3MBO
LU7 3NJ 2E0VIT
LU7 3NJ M0SPV
LU7 3NJ M6VIT
LU7 3UJ G6IXH
LU7 4AT 2E0GFK
LU7 4AT M6IBF
LU7 4BS M0POA
LU7 4HY G0DYW
LU7 4HY G1EOJ
LU7 4JG M6ADJ
LU7 4RF M6CWO
LU7 4SU M6CLZ
LU7 4TG M6BQN
LU7 4TQ M7CHA
LU7 4UQ 2E0HQB
LU7 4UQ M7BRC
LU7 4YU 2E0FEB
LU7 4YU M3SQP
LU7 9AQ 2E0JCQ
LU7 9AQ M0JCQ
LU7 9AQ M6JCQ
LU7 9DZ 2E0PGM
LU7 9DZ M0PAO
LU7 9DZ M6AZS
LU7 9GF M0VWW
LU7 9JF M6BAM
LU7 9JF M6SLL
LU7 9SN M6XNX
M1 2FA G7HFW
M11 1DP M6RCF
M11 1EQ M6OXB
M11 1HQ M3VNH
M11 1HQ M3YMH
M11 1HX M7JAZ
M11 4FB M3XGC
M12 5PU G0WGL
M12 5QB 2E0GQN
M12 5QB M0WOM
M12 5QB M7BUN
M12 5RP M3YPI
M13 9PL M7MYM
M14 4UR 2E0WLD
M14 4UR M6WLD
M14 5JS M6FML
M14 6AH G3YTZ
M14 6AP M7TFX
M14 6BZ M3YSF
M14 6PE G0BML
M14 6RW M6EGU
M14 6RX G4FFW
M14 6SA M6YOW
M14 6SB 2E0WQK
M14 6SB M0WQK
M14 7FN M0TCC
M14 7FZ G0EIZ
M14 7FZ G4JS
M14 7PX G7OQU
M14 7WT G1ZST
M15 4FZ M6EQR
M15 4HD 2E0XOH
M15 4HD M7AGX
M15 5AF M3NMR
M15 5RR M0VAP
M15 5RR M6SGV
M16 0HD M0HXS
M16 8BB M3MBQ
M16 8LX M3NXZ
M16 8PW G3SMN
M16 8QQ M3OAG
M16 9GQ M6CLQ
M18 7BX M7COQ
M18 8BP G6YFL
M18 8LL M6UCP
M18 8WW 2E0ISB
M18 8WW G7NAL
M18 8WW M0MWB
M18 8WW M6ISB
M19 1LE 2E0TZD
M19 1LE M0TZD
M19 1LE M6ZTD
M19 1QY M0FVD
M19 2FA G4GDG
M19 2LQ G3JRK
M19 2NL 2E0MQT
M19 2NL G0PXZ
M19 2NP G8OYF
M19 2WJ M7GMG
M19 3EH G4BVQ
M19 3FL M6EGW
M20 1DH M0IGY
M20 1EN 2E1HZI
M20 2QT G6BYR
M20 2SQ 2E0CFM
M20 3DD M3IYE
M20 4GH G7AZW
M20 4NX 2E0HFS
M20 4NX M6SYF
M20 5GJ M1DFO
M20 5LQ 2E0BDV
M20 5LQ M3MVM
M20 5ND G8NNZ
M20 5PL G0OVQ
M20 5QS M3PGO
M20 5WN G0UXI
M20 6BB G7JME
M20 6BB G7MIT
M20 6JA G3YHQ
M20 6SU 2E0BTB
M20 6TG 2E1HLW
M21 0QU 2E0MWB
M21 0QU M0UFC
M21 0QU M6MWB
M21 0UX G7EPL
M21 7LE M3OJK
M21 7QG M3BLX
M21 7TG G4RYT
M21 8FA 2E0IZW
M21 8FA M6IZW
M21 8XT G6BHA
M21 9DP G4OCR
M21 9JU G0CRF
M21 9JU G6CRF
M22 1AF G1KEW
M22 1AH 2E1IHS
M22 1AH G0PXG
M22 1AH M3AHZ
M22 1GA 2E0LHY
M22 1GA M7BGF
M22 1GA M7JLG
M22 1NN G1UTM
M22 1SN 2E0EWQ
M22 1SN M6NKB
M22 1UD M3NHA
M22 1UD M6NHA
M22 1UR G0JYS
M22 4DP G1OJB
M22 4DZ G0PWA
M22 4GG G1PNC
M22 4LW M6MOJ
M22 4QS G0JPT
M22 4WB M3XAO
M22 5AQ G8OZD
M22 5BF M7VCL
M22 5DH M3DAV
M22 5DJ G7SYJ
M22 5FU M6GTB
M22 5JY G4RWN
M22 5QG G7TEA
M22 5QG M0TEA
M22 9PL 2E1GNK
M22 9UX M6BBL
M22 9UX M6HNS
M22 9YE 2E0SGY
M22 9YE M6SGA
M23 1HF G0DNY
M23 1JX G0SAY
M23 1JY G6BMY
M23 1LD 2E0PLH
M23 1LE M6XMN
M23 1LP G0NID
M23 2AT G7VSE
M23 2UL G0LNA
M23 2YX G0LNB
M23 9AW G8LPX
M23 9BT G8LUL
M23 9NZ G7PXS
M24 1HE G6CVW
M24 1JA M7CEF
M24 1QD G1XGN
M24 1TJ G7VGX
M24 2AG M7PLW
M24 2AH M0LGP
M24 2EQ G0TRK
M24 2NN G6GCY
M24 2RA M6VWP
M24 2RF G1OZG
M24 2SX M7BSV
M24 2TQ G4VVT
M24 2WB G6EJI
M24 4AE G4WLI
M24 4AE G8IKW
M24 4JJ G4WVP
M24 4LU M7ABB
M24 4RU G4ZQL
M24 5LS M3TRC
M24 5RL G0FCZ
M24 5SU 2E0YMM
M24 5SU M5JPG
M24 5SU M6WUM
M24 6AX M6RWT
M24 6BN M1EKH
M24 6FF 2E0EBN
M24 6FF M0PTB
M25 0FZ G7PYN
M25 0HE G1HHU
M25 0HR G8JWC
M25 0HX G3ZVT
M25 0HX M7XPG
M25 1FN M6RUN
M25 1FN M6SAU
M25 1FN M7BAR
M25 1JY M6AAX
M25 1NB G4VUK
M25 2RD 2E0DIJ
M25 2RD M6DIJ
M25 2SD 2E0EVI
M25 2SD M0ODB
M25 2SD M6JUW
M25 2WP 2E0TFD
M25 2WP M6YMM
M25 3BT G6BVF
M25 3DY M6AUC
M25 3HZ G0CSU
M25 3JJ G7VKJ
M25 3JL G1ULB
M25 9GW G1PNB
M25 9GW M3WWD
M25 9LB M0MBS
M25 9NT G0UAY
M25 9RA 2E0HWY
M25 9RA M7TWK
M25 9RE G8OFI
M25 9RU 2E0OAI
M25 9RU M6DWS
M25 9TG G1JZY
M26 1EP M6SLE
M26 1HN G1JUD
M26 1HN G6RXF
M26 1HY G0BMM
M26 1JA G1PKP
M26 1JA G6CLW
M26 1LE M3ZFJ
M26 1YF G3ZCH
M26 1YF G6OTQ
M26 2GF G6YGB
M26 2QS G0WEF
M26 2RB M3LJB
M26 2RS G0HEA
M26 2UU M6OVR
M26 2UX G0ADL
M26 2XQ M7IGW
M26 2XU M7PPR
M26 3GL G4UAU
M26 3RU M0ANQ
M26 3TZ M6ROY
M26 3UA 2E0GPY
M26 3UJ G1ESW
M26 4FA M0BOQ
M26 4FS G8ZIC
M26 4NS M0XCT
M26 4PT G1ZOQ
M26 4QF G6DGQ
M27 0JT M6CIW
M27 0YH 2E0XBN
M27 0YH G0MKA
M27 0YH G6YIO
M27 0YH M0XBN
M27 0YH M3OTQ
M27 0YH M6XBM
M27 4FA G8RAX
M27 5GZ G4IRB
M27 5GZ G8SMR
M27 5NA G0KLF
M27 5QY M3OVE
M27 6AQ G3STJ
M27 6BU G5SRC
M27 6BU G7WAE
M27 6BU M0KHQ
M27 6BU M0NTI
M27 6PT M1KCB
M27 6WH G4FXA
M27 6WY G1SYV
M27 8AF M6MTV
M27 8FS G0CVH
M27 8QR M7DDM
M27 8QS G4VDJ
M27 8RE G0MZY
M27 8RT M6HGR
M27 8XP G0LVG
M27 8XP M6KMN
M27 9QP G7DWQ
M27 9RD G4EFD
M28 0SL G8KOQ
M28 0SX G3PMJ
M28 0TP G4NTY
M28 1BE 2E0ENE
M28 1BE M6FVL
M28 1DE G7ELA
M28 1DF 2E1GPO
M28 1FY M0SEB
M28 1HW G4SXG
M28 1LP G7DPW
M28 1ND G8DTF
M28 1PS M6ODA
M28 2JE M6UIR
M28 2RJ G0TEE
M28 2RW G1VJQ
M28 2TX 2E0YJL
M28 2TX M3YJL
M28 3DH 2E0ENB
M28 3DH M6ASZ
M28 3EB M7EAE
M28 3FQ M7JWM
M28 3JW M1VIP
M28 3WD G2LK
M28 7EN M0CAV
M28 7EX 2E1WJB
M28 7EX G4UPU
M28 7JD G7FFK
M28 7JW G4TJU
M28 7QF G4JLG
M28 7TS G8JOC
M28 7XS G7FPS
M28 7XX G6YGC
M281xj M7OMO
M29 7AT M6BSS
M29 7AX 2E0REM
M29 7AX M6AVT
M29 7BH G4JHA
M29 7BJ M6BST
M29 7DB 2E0IZR
M29 7DB G0IZR
M29 7DB M0SLR
M29 7DB M3RXT
M29 7EJ G8NSE
M29 7FF M6SDH
M29 7HH G4GKT
M29 7HS G4DZK
M29 7HS G8FQZ
M29 7JB 2E0FMH
M29 7JB M6OBQ
M29 7PE 2E1GSM
M29 7PH G0TFP
M29 7PZ M6ERH
M29 7RE G7CQW
M29 7RT G8SYM
M29 7WF M3GAV
M29 7WG M7KPM
M29 8GH M6OCI
M29 8LQ M0AVA
M29 8LS M6XZY
M29 8PG M3RFI
M29 8PH 2E1HZV
M29 8PH M3EGC
M29 8PJ M6MKX
M29 8WQ M6AEQ
M29 9DN M3LFH
M298ET 2E0HJC
M298ET M7JCH
M3 1LP M6ZUT
M30 0AN M6XBQ
M30 0FS M3LUZ
M30 0HW G6YAQ
M30 0JN M6JII
M30 0LT G1ASN
M30 0PZ M6CPC
M30 0QU M6WLA
M30 7HF M7LMN
M30 7LG M7TPW
m30 8ag G4VPW
M30 8AJ M1ABV
M30 8BP G0BVM
M30 8DD G0FPO
M30 8JR 2E0FDD
M30 8JR M6FDD
M30 8UW G8FYK
M30 8WB M1NUS
M30 8WP 2E0FWW
M30 8WP M3GHI
M30 8WP M6GHI
M30 9DL 2E0PCX
M30 9DL M0PCO
M30 9DL M6BGM
M30 9EE G8YPH
M30 9FT 2E0EFY
M30 9FT M0XJG
M30 9FT M6GIS
M30 9LY M0HTS
M307DU M7BZL
M31 4BA M7WTJ
M31 4DB 2E0HVD
M31 4DL G8BRU
M31 4DY G3VYA
M31 4GW G7GAK
M31 4NP G0TRU
M31 4PT G0LZL
M32 0AJ M6RFK
M32 0AN G7KRI
M32 0RG M3OVX
M32 0RY M6APC
M32 0UE M7VCC
M32 8BE M6PLI
M32 8DQ G1JHX
M32 8NA 2E0JED
M32 9AF 2E0MUA
M32 9AF M3MUA
M32 9BS G7FKJ
M32 9BX G4LVI
M32 9JB G6YOP
M32 9JQ M7EQG
M32 9QA M0NTH
M32 9QF G0BJK
M32 9RT 2E0RDU
M32 9SJ G1UFM
M33 2AR M0WRA
M33 2BH M7CKA
M33 2BJ G1NVL
M33 2DQ G7FQY
M33 2DY 2E1ICI
M33 2EA G4URV
M33 2HU M0YIT
M33 2HU M7GZA
M33 2JT M0DHE
M33 2LD G8LQO
M33 2PJ M6CGS
M33 2XF G0OVY
M33 2XP M0UQB
M33 2XP M0WKL
M33 2XP M0XQB
M33 3AN M7LCJ
M33 3GT G2HW
M33 3GT G3SVW
M33 3LD 2E1IBP
M33 3LH G0RFG
M33 3LQ G0TKL
M33 3LQ M3PSR
M33 3LW 2E0CPF
M33 3LW M6BQF
M33 3QP G1BOO
M33 3TN G0KKF
M33 3WS G4AOK
M33 4AE G4JZZ
M33 4DH M0IYP
M33 4DH M6NZQ
M33 4ER 2E1IGP
M33 4FS G4EDM
M33 4FW M0KWV
M33 4HL M0IXX
M33 4HP G4KGU
M33 4HR G7TUH
M33 4LG M7MBO
M33 4NS G8ZZF
M33 4QL M0HIH
M33 4RF G4AUR
M33 4RG G0NWJ
M33 4WG G4MYB
M33 5BE 2E0MTQ
M33 5BE M0MTQ
M33 5DL G8ILW
M33 5FB G3KCB
M33 5HF G0SSV
M33 5HR 2E0ZVL
M33 5HR M0LCR
M33 5HR M0SII
M33 5HR M6ZVL
M33 5NF M7ICH
M33 5NQ M3CVM
M33 5NY G0MRK

M33 5PR G6ISA
M33 5QJ G0BHP
M33 5RP M3XAU
M33 5UF G3FVA
M33 6EZ G3ZDM
M33 6GS M3VRU
M33 6NF G3WFT
M33 6NW G3SUI
M33 6QW G4FPA
M33 6WU G0FRB
M33 7AU M7CGH
M33 7EN G8RSI
M34 2DQ G0ROW
M34 2ER 2E0PBE
M34 2PD M6OEW
M34 2WP G0WYP
M34 3GL G7GXR
M34 3HU 2E1YRK
M34 3NS G7TYB
M34 3NX G3XVG
M34 3NZ M6LBI
M34 3QH 2E0KKF
M34 3QH M6KKF
M34 3QR M6GNU
M34 3TE G0BOH
M34 3TH G4WGR
M34 3WP 2E0TFK
M34 3WP M6ILS
M34 5BD G8HTN
M34 5EJ M6JRE
M34 5FB M3IFG
M34 5GB M3OWN
M34 5GR M7MCF
M34 5LJ G1JDT
M34 5LZ M0KCJ
M34 5QX G0NHO
M34 5QX G7DNR
M34 5UH G7GTU
M34 6AF G1BBK
M34 6EA G6NVU
M34 6EA G7HRZ
M34 6FT G7DOY
M34 6JY G0YRT
M34 6LR G1EJK
M34 7GG 2E0GKT
M34 7GG M7TAY
M34 7LB M3PBU
M34 7TF M6EKD
M34 7TX G1UTS
M34 7TX M0GMA
M34 7TX M3ZPM
M342DJ G0HUH
M35 0BE M7LZT
M35 0LH G0KTR
M35 0LR M6GOX
M35 0NN M6PQL
M35 0PX G7BAB
M35 0TG M7PUL
M35 9EJ G3PYH
M350LR M7LIN
M359dy 2E0OAS
M359dy M7CDS
M38 0BU M3WTN
M38 0FQ G1SPM
M38 0FU 2E0HVB
M38 0FU M7EBO
M38 0LW M0APZ
M38 9GL G6MIU
M38 9GQ 2E0SCD
M38 9GQ M6FPC
M38 9HB M3IRF
M38 9TN 2E0EMK
M38 9TN M6IYT
M38 9UB M0IRV
M38 9UH M0KEJ
M38 9XU G6PKV
M4 5LB M0DAH
M4 6GB G7FPJ
M4 7BB 2E0NTW
M4 7BB M3NTW
M4 7EP M7NIK
M40 0DG G6WGZ
M40 0DG M6EAE
M40 0DG M6HUN
M40 0DG M6PHJ
M40 0GB G0IUK
M40 1JE M6OJQ
M40 1LF 2E0GHS
M40 1LF M0SJS
M40 1LF M3GHS
M40 1QE M0FRG
M40 2SY M3GAF
M40 3GG 2E0AVS
M40 3GG M3AJN
M40 3NL G7TZV
M40 3SE G6IRF
M40 3TD M3VNP
M40 3WP G7ORW
M40 5HL G7VCG
M40 5HL M3IAF
M40 5QD G0VRQ
M40 5QD G0VXW
M40 5QW 2E0YES
M40 5QW M0TLY
M40 5QW M3GBC
M40 7LX 2E0JKP
M40 7LX M0JPE
M40 7LX M6HKZ
M40 9BW 2E0WCC
M40 9BW M6LEJ
M40 9LF G7USQ
M400FZ M7TDS
M41 0GF G7HKQ
M41 0PY G6PSA
M41 0RJ G3JIS
M41 0SB M3IRM
M41 0TY G3XGE
M41 0TY M0RML
M41 0XY G0ONF
M41 5AT 2E0DJY
M41 5BZ G7LKL
M41 5HA G7TIR
M41 5QG 2E0GNQ
M41 5QG M7AZS
M41 6FJ G1SCL
M41 6HP G1NYJ
M41 6LE G3XXG
M41 6NQ G7OTE
M41 6PU M0XPD
M41 6PU M6PAD
M41 6WA G8YUK
M41 7BT G4SDL
M41 7DY 2E0FPA
M41 7DY M0FPA
M41 7DY M3FPA
M41 7EY M7DGN
M41 7FH 2E0ILJ
M41 7FH M7MVF
M41 7WH G4WLJ
M41 7WX G4MFI
M41 8PB M0UUU
M41 8QT 2E0DCS
M41 8QT M0WXX
M41 8RE M6UFR
M41 8RN G8HXE
M41 8RN M0TWC
M41 8RN M3KMH
M41 8UE G4RYI
M41 8UE M3OPV
M41 8UE M3RYI
M41 8UT G1VZB
M41 9JZ G7HMW
M41 9NG G7AGC
M41 9NJ G1LWE
M41 9NT 2E0KUH
M41 9NT G0SYP
M41 9NT M6KUH
M41 9PT G4NCZ
M41 9QE G0IMB
M43 6DS G1WPL
M43 6DX G1YCM
M43 6EF G8CZE
M43 6FS G4HBI
M43 6HL G6KRS
M43 6QG G6GXY
M43 7EA M6HEK
M43 7FR 2E0RNH
M43 7FR M0RNH
M43 7FR M7GRH
M43 7JX G8OXX
M43 7LJ M6BXN
M43 7NS M6LUI
M43 7PF M6ZAQ
M43 7PL 2E0IMS
M43 7SW M6RCY
M43 7UQ G4IFI
M43 7YG M6EIL
M44 5AS 2E0YDJ
M44 5AS M0YDJ
M44 5AU 2E0SGH
M44 5AU G0GSK
M44 5AU M0MBE
M44 5AU M6SGH
M44 5HB 2E0UCT
M44 5HB G5CTX
M44 5HB M7CTB
M44 5JQ G6LBQ
M44 6AQ G8TXW
M44 6AR M6XKY
M44 6DD M6PNT
M44 6EH M3RYG
M44 6EH M3YOZ
M44 6EN G0JYK
M44 6FG M7MWM
M44 6FY M7WCC
M44 6HD G6SPG
M44 6HH 2E0JIR
M44 6HH M3JIR
M44 6LR G0IGT
M44 6PJ M7DDR
M44 6QA M0IPR
M44 6QA M0SRA
M44 6TD M3UJP
M44 6TE G6MMA
M44 6WH G1UKZ
M44 6WJ M0JWE
M44 6WY G0NPI
M45 6DP G8PAL
M45 6EP G8AOG
M45 6NR 2E0KMA
M45 6NR M6KMA
M45 6TJ G8DUT
M45 7LR G1COX
M45 7NT G3GXI
M45 7NT G8KRG
M45 7PR M6BPL
M45 7RP 2E0FYS
M45 7RP M7SES
M45 7SS G8NKJ
M45 7TS G4UGW
M45 8FR G4FZP
M45 8GG M1CRZ
M45 8LH 2E0GRX
M45 8NT G1NSB
M45 8PP M0OBU
M45 8WY G0AHR
M45 8XR M1APV
M46 0AQ M7BDF
M46 0EZ 2E0LTT
M46 0EZ M0LTT
M46 0EZ M6MLA
M46 0LQ G6SKM
M46 0PG G6LCX
M46 0QE G0FTO
M46 0SN M6KML
M46 9AB G6YBC
M46 9EG M7GBO
M46 9LN G6TFV
M46 9LQ 2E0ENC
M46 9LQ G4HZJ
M46 9LQ G7LDR
M46 9NQ 2E0HBY
M46 9PQ G0OMF
M46 9PY M0GNY
M46 9QB G0KBA
M46 9QB G4GBK
M46 9QZ 2E0JMC
M46 9QZ M0NWC
M46 9QZ M0OBZ
M46 9QZ M6ZFE
M46 9QZ M7BJI
M46 9RQ 2E1IHB
M46 9RQ M3WAC
M46 9RR 2E1HLL
M46 9RR M1HLL
M46 9TZ G6HFW
M5 3HU M6GSH
M5 5HX G7KQL
M5 5LF 2E0DMD
M5 5LF M0WDD
M50 3RB 2E1MPB
M6 5YB G0HQN
M6 6DW G0GIF
M6 7LA G0PLX
M6 7PF M0JPR
M6 7PQ G6FEI
M6 7WP M3UDC
M6 8QQ G0IYM
M7 1PR 2E0YWN
M7 2GB 2E0PDX
M7 3NX 2E1BLG
M7 3NZ G1JHB
M7 3TF M0LKS
M7 4JA M0ASU
M7 4TH M3OSU
M7 4TH M3OSW
M8 2FS M7EGE
M8 4GG M3VVQ
M8 4GT M6BLS
M8 4WQ G4YNI
M8 5AR M6ZAD
M8 5AW G5JCB
M8 5AW G5MS
M85DR M7BSU
M9 0RQ G6OGT
M9 4EX G4XBX
M9 4PZ 2E1IEM
M9 5XY 2E0TRK
M9 5XY M3ZAV
M9 6HR G3KIQ
M9 6QX M3YPR
M9 7AR G3NSW
M9 7EQ G6MDG
M9 7GR M7DOS
M9 8AT G1XIH
M9 8AT G2ALN
M9 8JD G0SNZ
ME1 1FB M3FPU
ME1 1HY G3NRU
me1 1jg G0NCE
ME1 1NB G7FZN
ME1 1QH G8WKX
ME1 2BL G7VFU
ME1 2DL G0WFM
ME1 2HZ G0WYV
ME1 2NN G4SAW
ME1 2QF M3VJX
ME1 2RE M3JHW
ME1 2SS M3RMZ
ME1 2ST G4NRT
ME1 2UA G2FSH
ME1 2UH G3ZJP
ME1 2UL M6MLM
ME1 2XD M0PLG
ME1 2XD M6SJQ
ME1 3AJ M6NST
ME1 3BL M7CWT
ME1 3EF G7KCK
ME1 3JU G0DKO
ME1 3NJ G1ERZ
ME1 3QE M0SAC
ME1 3QR G0BIX
ME1 3RJ G6AIC
ME10 1BY G3ISD
ME10 1EH G8LDJ
ME10 1ER M3ZKD
ME10 1ET M0HZW
ME10 1EY G1HHG
ME10 1JD M3DSI
ME10 1JF G0TNC
ME10 1LB G3EHW
ME10 1QT G0UAS
ME10 1QY G7GSR
ME10 1RD G3VPA
ME10 1RF G4TWC
ME10 1TJ M1DUB
ME10 1TJ M3PJS
ME10 1TJ M3PZX
ME10 1TS G6DUI
ME10 1XF M7ACW
ME10 1XJ 2E1AUN
ME10 1XJ M3PHF
ME10 1YL G1BOX
ME10 1YL G8ONY
ME10 2DP M6BAZ
ME10 2EA G6AAC
ME10 2EY G4IBH
ME10 2HR G1JUO
ME10 2JF G0AXQ
ME10 2LR G1YTO
ME10 2LZ M0AZP
ME10 2PX M0PIA
ME10 3BL 2E1JRS
ME10 3LG M6SDF
ME10 3NR G1VJN
ME10 3PB 2E0OZG
ME10 3PB M6OZG
ME10 3QY M6EWH
ME10 3TG 2E1CLG
ME10 4HS G0VTV
ME10 4HS G4LBS
ME10 4LF M3RXG
ME10 4NA G1SPX
ME10 4NA M6IAN
ME10 4QD G0DRQ
ME10 4QD G1DLJ
ME10 4QE G4VEC
ME10 4QT 2E0VRP
ME10 4QT M0VRP
ME10 4QT M6VRP
ME10 4RX G7BWV
ME10 4SL G0RNQ
ME10 4TB 2E0EWH
ME10 5AN G7MEE
ME10 5AS G6PGT
me102te G7AEY
ME11 5AG G8DPW
ME11 5AW G1JCL
ME11 5BU G6ILN
ME11 5DT 2E0RNF
ME11 5DT M6OEM
ME12 1BW M1WAW
ME12 1DR M6KYE
ME12 1UT G4KSU
ME12 2AT M3TMQ
ME12 2BS G1AFW
ME12 2DJ G0OLX
ME12 2DJ G0PEH
ME12 2DJ G8CUX
ME12 2HG G0JYH
ME12 2HX M0CSU
ME12 2LH 2E0WGF
ME12 2LH M0WGF
ME12 2LH M6GQJ
ME12 2NH G0DRT
me12 2nj M7EQS
ME12 2NL G7NJB
ME12 2NQ 2E0XCC
ME12 2NQ M6NMS
ME12 2RA M6LFJ
ME12 3BB G7LZY
ME12 3BQ G3ZYQ
ME12 3DH G8XGB
ME12 3DH M0XGB
ME12 3EZ M3RAK
ME12 3HX G3GDH
ME12 3JX G6NUI
ME12 3JX M1DJN
ME12 3LN G8JNZ
ME12 3LN M6KSP
ME12 3NF 2E0BPY
ME12 3NF M3UIJ
ME12 3NJ G1SGS
ME12 3NR G2HKU
ME12 3PF 2E0FHE
ME12 3PF M7XYZ
ME12 3PG 2E0ILN
ME12 3PG M0ILN
ME12 3PG M6ILN
ME12 3PG M6KCF
ME12 3PJ M6MZY
ME12 3PT M6BRK
ME12 3PU M6NSW
ME12 3QR 2E0HSV
ME12 3QR G7ECQ
ME12 3QR M3ILB
ME12 3RZ M6DTN
ME12 3SA 2E0ELT
ME12 3SA M0KLK
ME12 3SA M6WJF
ME12 3SQ G7IAU
ME12 3UA M3ZZF
ME12 4EJ M3VMU
ME12 4JU G4WPB
ME12 4PU M3YGD
ME12 4PU M6MWA
ME12 4QY M3XIL
ME121YA M0CST
me123sa M6KUG
ME13 0EF M3MUX
ME13 7DL M1ARU
ME13 7ES G0PZX
ME13 7JG 2E0BRJ
ME13 7JZ G4VXB
ME13 7PD M3VWF
ME13 7TA M3MFF
ME13 8DE M6BOI
ME13 8DG M3NPS
ME13 8DZ G1HDK
ME13 8HP G1EUQ
ME13 8JA G8MVH
ME13 8QL M3ZFB
ME13 8RU G7AFS
ME13 8TP M6UVB
ME13 8UR M7MGW
ME13 8XN G4UVX
ME13 9LF 2E1BAE
ME13 9NR M3XEQ
ME13 9SE G4AWG
ME14 2BX M1BWJ
ME14 2EA G0REN
ME14 2ER G3YJS
ME14 2JN G3PCX
ME14 2QR G4UAQ
ME14 2QW M0CZN
ME14 2TX M7VIV
ME14 3JN G3ZAL
ME14 4BH G0KYM
ME14 4HG G7MOX
ME14 4JB G0BUW
ME14 4NJ G0SPQ
ME14 5AJ M7TCC
ME14 5BH G6VVS
ME14 5HJ G8HLE
ME14 5PG G3WXU
ME15 0AJ G7DJT
ME15 0DR M6HIL
ME15 0QF M6HRX
ME15 0QH G4XMS
ME15 6AA G1VWP
ME15 6FF G6XAK
ME15 6ND M6IQY
ME15 6PX G7AZT
ME15 6QB 2E0BAI
ME15 6TH G4CCQ
ME15 6UL M6FLB
ME15 6UU G8LEG
ME15 6YN M3XSP
ME15 7AU G7VOM
ME15 7AU G7VON
ME15 7EL G6ZDY
ME15 7PL M0HLD
ME15 7RR 2E0GTA
ME15 7RT G3WWI
ME15 7SP 2E0SVG
ME15 7SP M6SVG
ME15 8DY G7OPG
ME15 8EG G4IWN
ME15 8EP M0HBN
ME15 8JP G0VAR
ME15 8JZ G7GAZ
ME15 8RA G0DBY
ME15 8TN G8LGA
ME15 9AB G6LUD
ME15 9AE G6ALJ
ME15 9AJ M7DGU
ME15 9JB G8VPD
ME15 9PD G0NCW
ME15 9QR G7HON
ME15 9RA G3RJF
ME15 9RP G0LCH
ME15 9RS G4WBA
ME16 0DL G3ORP
ME16 0JL M0PMC
ME16 0JS G6RVS
ME16 0JS M0KMR
ME16 0JS M0KRM
ME16 0JS M3RVS
ME16 0NF G3VNI
ME16 0PQ 2E0JIT
ME16 0QP G0UED
ME16 0QW 2E0KKQ
ME16 0RX M1AEX
ME16 8BS M1BGS
ME16 8DR G4AXD
ME16 8EN G8PWT
ME16 8HY M6OGQ
ME16 8NP G1XJK
ME16 8PA G0FKK
ME16 8QP M1AJM
ME16 8QU 2E0JHG
ME16 8QU M0JHG
ME16 9HF 2E0HIZ
ME16 9HF G4HIZ
ME16 9HF M3SIZ
ME16 9HH G6HTH
ME16 9JG M0DHM
ME160HB M7EJK
ME17 1AD G3YIF
ME17 1AY M1DDW
ME17 1BS G3YUU
ME17 1SR 2E0LAH
ME17 1SR M6BEC
ME17 1SX G1BLB
ME17 1SZ 2E0VPL
ME17 1TJ G3UOJ
ME17 1TZ G6KTG
ME17 2AN G6USR
ME17 2AX G0IJK
ME17 2EJ G3LRX
ME17 2ET G4XSI
ME17 2EU G8WHB
ME17 2HH 2E0KWE
ME17 2HH M7INS
ME17 2NH 2E0PSR
ME17 2NH M0NHL
ME17 2NH M6GAE
ME17 2NS G1HQJ
ME17 3LH M7BQB
ME17 3NP G1YQN
ME17 4DN G1ZMG
ME17 4EE G4LJT
ME17 4JY G6HTS
ME17 4JZ G3KOM
ME17 4LG G4OPP
ME17 4QS M3XZK
ME18 5HT M6TFT
ME18 5PR G3DEY
ME18 5PX G4LNM
ME18 5QF G6NAL
ME18 5QH G8NVH
ME18 6DX M0GZH
ME185BL G7BNZ
ME19 4BS G6FMU
ME19 4BU M6LHQ
ME19 4DL G8LDU
ME19 4TT G0OLT
ME19 5BL M6GCX
ME19 5BW G6BGA
ME19 5EH G3PAG
ME19 5EN G3XJS
ME19 5EN M0VLP
ME19 5JX G0VUN
ME19 5QY G1AJY
ME19 6ES G7LMT
ME19 6QS G1JJQ
ME19 6SJ M1CMW
ME2 1DJ G7DSU
ME2 1EE 2E0OFM
ME2 1EE M0OFM
ME2 1EE M6OFM
ME2 1EE M6TOG
ME2 1EE M7OFM
ME2 2QB 2E1IEK
ME2 2QY M0AZS
ME2 2TZ G7UTB
ME2 2XZ M7CYR
ME2 2YE G4ZFC
ME2 2YT M1DNJ
ME2 2YW G4TVW
ME2 3BA G0LAD
ME2 3BZ M3OKC
ME2 3EP G0GIR
ME2 3JH G7LAN
ME2 3LW G1EWM
ME2 3NF G6CKK
ME2 3PN G0VMQ
ME2 3QG G0MIF
ME2 3RE M3BFX
ME2 3SP G3FNZ
ME2 4BB G6HIV
ME2 4BB M3LYZ
ME2 4EB G0VKL
ME2 4JD M0BPN
ME2 4LJ 2E0JAX
ME2 4LJ M6JHB
ME2 4NE G0GQT
ME2 4NL G4RFN
ME2 4NU G4CWV
ME2 4PG G1DOT
ME2 4PN G3NZR
ME2 4UR M0OLW
ME2 4XF G8CAM
ME20 1AP M6PBP
ME20 6AE M1BJS
ME20 6BB M3VMQ
ME20 6EF G0PWX
ME20 6ET G3PYO
ME20 6HP M7AAP
ME20 6LA G4RIU
ME20 6QZ G7LJQ
ME20 6RE M3MDY
ME20 7BE 2E0WXK
ME20 7BE M6WXK
ME20 7LY 2E0KUR
ME20 7LY M6KUR
ME20 7SF G1XUE
ME206EF 2E0WGP
ME206EF M3WZG
ME3 0BJ M6AAE
ME3 0BS M0ZMX
ME3 7BA G7AJT
ME3 7EH G0WAN
ME3 7JR 2E0BBA
ME3 7RN M3UIF
ME3 7SH G3OHP
ME3 7TN G1HRV
ME3 8BE G6KSK
ME3 8DW G8MIN
ME3 8HT G1GGI
ME3 8LR G3BSN
ME3 8NE G4FVB
ME3 8TR G4FJW
ME3 9AA 2E0KSG
ME3 9AA M6KSG
ME3 9DE 2E0HLR
ME3 9DE G0PEK
ME3 9DE M0KRO
ME3 9DE M6HLR
ME3 9DF 2E0BPG
ME3 9DF M3XFT
ME3 9EA M3XBS
ME3 9EA M7DYM
ME3 9EU G4TAM
ME3 9GF 2E0CYP
ME3 9GF M0HSU
ME3 9GF M6DVW
ME3 9JP G0AUN
ME3 9PR G0PCA
ME3 9PR G0PCB
ME3 9ST G1ALK
ME3 9TA G6IYD
ME3 9TG 2E0REY
ME3 9TG M3UVC
ME39TG 2E0HEM
ME4 4XJ M0DKT
ME4 4XJ M3DKT
ME4 5BZ 2E0MTR
ME4 5HS 2E0YRM
ME4 5HS M0YRM
ME4 5HS M6LZM
ME4 5LE G6TXP
ME4 5NN M3FOR
ME4 5NW M7BJY
ME4 5SJ G6TOT
ME4 5SJ M3PQG
ME4 5TW G7CVY
ME4 6HL G1YXH
ME4 6LY G4EHQ
ME4 6LY G7COY
ME4 6QD 2E0CZR
ME4 6QD M6BYF
ME4 6RE G4YLT
ME4 6UU G8CJM
ME5 0BH G4HJE
ME5 0DD M1BIK
ME5 0DQ G1EDH
ME5 0DY G7MFW
ME5 0HJ 2E0KVB
ME5 0HJ M6ETA
ME5 0HJ M6WMA
ME5 0HU M5LRO
ME5 0JS G2FJA
ME5 0JS G8VJU
ME5 0SB G8TGB
ME5 0UP M6WJB
ME5 7AA M3MDS
ME5 7HH 2E0KLY
ME5 7HH M3SJY
ME5 7NG 2E0AXN
ME5 7NG M3DSU
ME5 7PT G0ABN
ME5 7QD G4YTU
ME5 8ER M3ZIM
ME5 8JD G6FHK
ME5 8NY G1LHD
ME5 8NY G7IBN
ME5 8PL M6UHV
ME5 8SR G6TXB
ME5 8TN G0VWV
ME5 8TN G7KNM
ME5 8UL G3PSR
ME5 9AA G7MPZ
ME5 9AG M3MZN
ME5 9AR 2E0JSH
ME5 9AR M0FSH
ME5 9BN 2E0EPH
ME5 9BN M0JCG
ME5 9BX M6OCX
ME5 9DJ M0IXN
ME5 9EX M0KIQ
ME5 9HA G8IXC
ME5 9HB M0DTK
ME5 9HE 2E0HRJ
ME5 9HE M0GYI
ME5 9HE M0HRP
ME5 9HE M6HRJ
ME5 9JG G6BLK
ME5 9LA G4DQN
ME5 9LF G6SQS
ME5 9LU M6DAK
ME5 9QP 2E1GTB
ME5 9RD G0AKR
ME5 9SP G6YHP
ME5 9TD G3JHU
ME6 5HP 2E0MLE
ME6 5HP M0MZE
ME6 5HP M6MLE
ME6 5HX G0TUZ
ME6 5PA G6ZMG
ME6 5PY G8CCJ
ME6 5RL G0JAR
ME7 1DZ G0GJW
ME7 1EW 2E0RNE
ME7 1JB 2E0WAK
ME7 1JB M0WKO
ME7 1JB M3WYQ
ME7 1ND G6OLV
me7 1tx G8JMS
me7 1tx M6SDG
ME7 1UE M3SML
ME7 1UG 2E1IJX
ME7 1UG M0PPS
ME7 2DN 2E0DCH
ME7 2HL M3MEI
ME7 2LN G3TCI
ME7 2RE 2E0BAY
ME7 2RE M3IQQ
ME7 2RE M6BAY
ME7 2SW G0JLP
ME7 2SX G0VQB
ME7 2UN M0OOT
ME7 2UW G3JRD
ME7 2YH G5ECQ
ME7 3AQ M0CPG
ME7 3AY G6ITM
ME7 3EH M3NBK
ME7 3EX M1BDS
ME7 3JE G7AJK
ME7 3LS G4ZTF
ME7 3NA 2E1EVM
ME7 3PT G5MW
ME7 3PT G6IVP
ME7 3PT G8MWA
ME7 3PW M7CPQ
ME7 3QQ G0TAR
ME7 3QQ M0ECK
ME7 3TJ G7NJG
ME7 3TS 2E1EVJ
ME7 3TS G3VFC
ME7 3TS M6VFC
ME7 4BA M3EJL
ME7 4EP G3ZYV
ME7 4HB 2E0JEF
ME7 4HB G6SXC
ME7 4HB M0FMY
ME7 4JB G7FTG
ME7 4NN G1XIO
ME7 4PY M6NSD
ME7 4RN M1ETC
ME7 4RR M0DYG
ME7 5EX M6OKR
ME7 5HY G8STY
ME7 5QG 2E1HRY
ME7 5QG 2E1RIO
ME7 5QG M3ANE
ME7 5QG M3HRY
ME7 5QH M1ALM
ME72DX M6IVT
ME8 0AQ G1IBJ
ME8 0DZ M3HWX
ME8 0JD G0VGR
ME8 0JG 2E0ICI
ME8 0JL M6VAG
ME8 0JN G6OCO
ME8 0JX G7MNG
ME8 0LZ G8VVM
ME8 0NR G6YLW
ME8 0PE G7MIM
ME8 0PE M1CQT
ME8 0RD 2E1DMU
ME8 0TH 2E1IJS
ME8 0TL G4HZI
ME8 6JE G0TCW
ME8 6LX M0PFW
ME8 6ND G1XNC
ME8 6QY M6ICJ
ME8 6XP G0RBV
ME8 6YE M7DIU
ME8 7JN 2E0PKF
ME8 7JN M7GOO
ME8 7QB G8MNL
ME8 7SZ M7HUB
ME8 7TA G1CBK
ME8 8EW 2E0HIP
ME8 8EW M0HLB
ME8 8EW M6KOM
ME8 8HE G3PQD
ME8 8JU G4ZRY
ME8 8PA M6OQX
ME8 8TA G0DWF
ME8 8TL 2E0WSS
ME8 8TL M3WSS
ME8 9EN 2E0RPA
ME8 9EN M0HNE
ME8 9EN M6REH
ME8 9ES G0APB
ME8 9HX G1ZBH
ME8 9JQ M6GYS
ME8 9PP M0AAK
ME9 0LR G1PGI
ME9 0PL G4MKG
ME9 0TY G4JJX
ME9 7AG G4TQS
ME9 7AS M3DWV
ME9 7EX G3UIB
ME9 7JU G4IOG
ME9 7QA G4LUO
ME9 7RH G3PSS
ME9 7SL M0LNE
ME9 7SL M6LNE
ME9 7SY G3UKL
ME9 8DX M6COL
ME9 8JZ 2E0IKV
ME9 8JZ M6VKA
ME9 8RG M7SWT
ME9 8SB G7HTN
ME9 9HU G4MPA
ME9 9LH G6SIQ
ME9 9PL M0CUT
ME9 9SP G6HOS
ME9 9SW M3MXO
MK1 1BL G8ABB
MK10 7BW M6MGU
MK10 7LQ M7HBB
MK10 9AF G8SJK
MK10 9AY 2E0ROY
MK10 9HG G8OSH
MK10 9HR 2E0CVV
MK10 9HR 2E0EBR
MK10 9HR M6EBR
MK10 9HR M6PNY
MK10 9HZ M0MPB
MK10 9JL G6KGK
MK11 1AT M3DXL
MK11 1ET G8XJL
MK11 1EX G4ZNY
MK11 1HX G6AZR
MK11 1RB G4NNX
MK11 2AA 2E1CKQ
MK11 2AF G0GMB
MK11 2BQ 2E0RKV
MK11 2BQ M0VTE
MK11 2BQ M3RKV
MK12 5AJ 2E0ORE
MK12 5AJ M7ERE
MK12 5AY G6WXM
MK12 5BE G8AQB
MK12 5DN 2E0IHU
MK12 5DN G6WZL
MK12 5DN M7IPL
MK12 5DR M6WFG
MK12 5HB M0BZK
MK12 6AL G0TGU
MK12 6AN G3NEH
MK12 6LR G3UQL
MK13 0PN G0RKV
MK13 0PN M3ROW
MK13 0QP G6INI
MK13 7AJ G8KFN
MK13 7AR 2E0GXX
MK13 7AR M0JTL
MK13 7AR M3GXX
MK13 7AY G4LIJ
MK13 7AY G8CCV
MK13 7AZ 2E0KOA
MK13 7AZ M6JPW
MK13 7DA G1NRY
MK13 7DY G8NKN
MK13 7EP M0GPJ
MK13 7HS G3SHD
MK13 7PZ G7OOU
MK13 8AT G8EPQ
MK13 8DP M6ZXI
MK13 9BT 2E0EBP
MK13 9HP M7EAN
MK13 9HS G0BYH
MK14 5AX G3YYN
MK14 5AX M0ALO
MK14 5BZ M1SHE
MK14 5BZ M3LEK
MK14 5EP G0TAI
MK14 5FQ M7BXO
MK14 5FS M7OXO
MK14 5JQ G4WBG
MK14 5JY G7AFL
MK14 5PA 2E0CKJ
MK14 5PA M3PWW
MK14 5QA G3JVR
MK14 5QA G8KYP
MK14 6AY G6ALU
MK14 6BU 2E0XJL
MK14 6BU M3RZO
MK14 6BU M6LRW
MK14 6EJ G7UTY
MK14 6EJ M0CMS
MK14 6ER G6HKS
MK14 6ER G7NRB
MK14 6GX M0AWI
MK14 7AP M6SOU
MK14 7BB G0FMN
MK14 7DY G4TOH
MK14 7LL G1DMR
MK14 7LU G1LRU
MK14 7PJ 2E0TFI
MK14 7PP 2E1HSP
MK14 7PP G1RNZ
MK14 7PP M3CMK
MK14 7PP M6MOG
MK14 7QN G7VOA
MK14 7QS G1SYU
MK14 7QS G7AUE
MK14 7RR G7KGR
MK15 9AE 2E0FMK
MK15 9AE M0UKF
MK15 9AE M6UKF
MK15 9HP 2E1HCT
MK15 9HP M3HKS
MK16 0EE G1BAQ
MK16 0LH G0FWD
MK16 0LL M0ATY
MK16 0LT G4MDR
MK16 0NG M3YJA
MK16 8BB M0HXN
MK16 8JB 2E0RTC
MK16 8JB M6AIE
MK16 8JG G3WKL
MK16 8PG M0GYK
MK16 8PL G4FIA
MK16 8PP G4VMX
MK16 8RF G1CYQ
MK16 8SR G6GCM
MK16 8TZ M6BEI
MK16 9AR 2E0NPL
MK16 9AR M6NPL
MK16 9BG G4GJR
MK16 9DN G4MTF
MK16 9LZ G3ZCJ
MK16 9LZ M7PGM
MK16 9NQ G6ZQU
MK17 0BX G3ZLX
MK17 0DQ G1FXT
MK17 0EP G0OUR
MK17 0EP M0ANS
MK17 0HP M0BZE
MK17 0LL M0RBD
MK17 0LW G4UCJ
MK17 0QR G0AGR
MK17 0QR G4NUG
MK17 0SB G7MFZ
MK17 0SB G8FAK
MK17 8AW G4XWM
MK17 8EA M0RPJ
MK17 8GP M7ERS
MK17 8HX G0TUL
MK17 8HX G8RCK
MK17 8PZ M7NMD
MK17 8QG G0GQH
MK17 8TN G3VZV
MK17 9AG M3ZCN
MK17 9AJ G0NGQ
MK17 9AJ G4UQR
MK17 9ED G0LAX
MK17 9ED G6VNI
MK17 9HS 2E0HCZ
MK17 9HS M0JQP
MK17 9HS M7DXX
MK17 9JU M6KAG
MK17 9QF M6OUS
MK18 1BY M7JVC
MK18 1DZ M1EUX
MK18 1PW M7MMP
MK18 2AL M0YAW
MK18 2AL M7REO
MK18 2ar M3ZKT
MK18 2FD G4HIF
MK18 2FG M3RHR
MK18 2FS G0GLG
MK18 2GS M0MBO
MK18 2HA 2E0EMP
MK18 2HA M6LCU
MK18 2JH G6JXC
MK18 2LZ G7PMW
MK18 2NY M3WRB
MK18 2PF G1UEO
MK18 2QD G0GGN
MK18 2QR G2BSJ
MK18 2QS G1MYO
MK18 3AB G8GLZ
MK18 3BN 2E0KVA
MK18 3DY G4FYO
MK18 3HX G8FMC
MK18 3JU G8MBS
MK18 3LZ M0THN
MK18 3NA G7JHV
MK18 3NT 2E0TAC
MK18 3NT M3TPI
MK18 3NT M5REV
MK18 3RJ 2E0WNL
MK18 3RJ M0SEJ
MK18 3RJ M7SEJ
MK18 4BX G8RDB
MK18 4EL G3SHZ
MK18 4HL G4UUF
MK18 4LH G0ICC
MK18 4LX M3KHI
MK18 5BJ G4YTL
MK18 7EQ G0BLQ
MK18 7EW G3THS
MK18 7EX M7MOY
MK18 7HR G6TSJ
MK19 6BT M3JHR
MK19 6BY G4MDU
MK19 6BY M3LJF
MK19 6JD 2E0ECI
MK19 6JD M6CIE
MK19 7AG M0JSZ
MK19 7BE G3LFR
MK19 7BT G8CXT
MK19 7JQ G4XJE
MK19 7LD G4KST
MK19 7LF 2E0FSG
MK19 7LF G4FZA
MK19 7LF M6AMJ
MK19 7LQ G4MGV
MK19 7LU G3XLN
MK19 7NF G8UEZ
MK19 7NH G1WWP
MK19 7NY 2E0CFQ
MK19 7NY M0HCP
MK2 2EZ G0EYZ
MK2 2FB G7ESL
MK2 2HT G0UUF
MK2 2LA G0GQP
MK2 2QD G0LRW
MK2 2RN G0OQE
MK2 2XP 2E1FCO
MK2 2XP 2E1GRG
MK2 3AD G3CPT
MK2 3EB G8GIL
MK2 3EB M3IDQ
MK2 3FH G1GXX
MK2 3LS G4JGG
MK2 3NB G3YJD
MK2 3NG 2E1OUJ
MK2 3NR 2E0FEE
MK2 3NR M6NXO
MK3 5AF 2E0EAF
MK3 5AF M3UHY
MK3 5AF M3UXL
MK3 5AF M3ZYT
MK3 5DQ 2E0APN
MK3 5DQ G3PPO
MK3 5DR M6AYU
MK3 5EN G0TUO
MK3 5HA G3NAI
MK3 6BY G6ZJD
MK3 6HX M3ZFI
MK3 6JP 2E0MKT
MK3 6JP M0TSW
MK3 6JP M6TMK
MK3 6PE M3ELN
MK3 6PJ M3GGA
MK3 6PL 2E1GJO
MK3 7BG M3UMV
MK3 7BP 2E1BEV
MK3 7EU G3KQQ
MK3 7LD M0HFB
MK3 7LS M6GRQ
MK3 7PS M1ELI
MK3 7QP M6GCT
MK3 7RQ G4GCJ
MK3 7TA G0EFN
MK4 1AX M6MOU
MK4 1BJ 2E0BGP
MK4 1BQ G7NDO
MK4 1DP G1TTC
MK4 1EF G0UCK
MK4 1HY M0AEQ
MK4 1JH M1ESM
MK4 2AU G0OGM
MK4 2BT G1PCN
MK4 2EZ 2E0USS
MK4 2EZ M0BOY
MK4 2EZ M7AIR
MK4 2EZ M7OMY
MK4 2GF G7LXB
MK4 3EN 2E0TWK
MK4 3EN G8LOF
MK4 3EN M0WSE
MK4 4AX M3EQL
MK4 4AX M3SSI
MK4 4DH G4UFS
MK40 1TF M6CYD
MK40 1ZL M0LRS
MK40 2BE G6GCO
MK40 2DS 2E0ZPY
MK40 2DS M6ZPY
MK40 2HY M6OCB
MK40 2TR G5BH
MK40 2XU M7BUS
MK40 3DF G3USD
MK40 3PP M6CSF
MK40 3RJ M0IMV
MK40 3RJ M5DHB
MK40 3SA 2E0ACR
MK40 3UG G0BMG
MK40 3UG M3TIY
MK40 4FZ M6HUP
MK40 4GW G7UOD

Postcode

MK40 4HJ G6GIH
MK40 4PZ G0ADB
MK40 4RE M7SHF
MK40 4SB M3IKN
MK403by M3OQD
MK41 0HL G8CZP
MK41 0LB M0GGT
MK41 0LB M3NKZ
MK41 0SN 2E0VCZ
MK41 0SN M6VCV
MK41 0TF G1MKR
MK41 0TF M0MKR
MK41 0TX M3XAK
MK41 0UP G0BKN
MK41 0UP G0MBR
MK41 0UP G6MBR
MK41 6BD M7TMD
MK41 6BS M6MBU
MK41 6DA G6UPA
MK41 6DU 2E0GXD
MK41 6DU G0RJM
MK41 6DU M7BFK
MK41 6DY M6FPM
MK41 7BS G4GDX
MK41 7BT G4KWH
MK41 7BT G4KYX
MK41 7DH G3UED
MK41 7HY G8CGM
MK41 7JP G4KJU
MK41 7JT M7JOZ
MK41 7LS G7RVC
MK41 7ST G3UBV
MK41 7ST G4EZQ
MK41 7TU G1BYT
MK41 7UH G6CVE
MK41 8AP G4UCY
MK41 8AS G4AQS
MK41 8AS G4GIR
MK41 8AW 2E1EMI
MK41 8BT G3RPL
MK41 8DR G4XDB
MK41 8DR G8HJH
MK41 8EL G0HER
MK41 8JS G6PAA
MK41 8JS G7RDA
MK41 8NX G6YNW
MK41 8QD G0VJJ
MK41 8QY G8HGL
MK41 9AL G8SDN
MK41 9AL G8XEF
MK41 9AS G7LEB
MK41 9DD G4YYC
MK41 9ER M6SNP
MK41 9LD M1PGH
MK41 9LD M3PGH
MK41 9LG G8RTN
MK41 9NE G3WBP
MK41 9QR G1JZT
MK41 9QW M3UDU
MK41 9RB M6FLK
MK41 9RG M6GEA
MK42 0AF G3XDU
MK42 0AF G8HXW
MK42 0FF M7JKF
MK42 0NA 2E0OAA
MK42 0NA M0TXS
MK42 0NA M6SRO
MK42 0PX G4IMH
MK42 0RS M3GHE
MK42 0SE G0GBI
MK42 0SJ G0OXY
MK42 0UJ 2E0CEP
MK42 0UJ M0HBX
MK42 0UJ M3SDV
MK42 0XB M6MUP
MK42 0XD M6GGW
MK42 6AH M6WCR
MK42 7BE G4OOQ
MK42 7DP M3DPQ
MK42 7DU M1MPA
MK42 7DU M3ZKA
MK42 7DU M6KGA
MK42 7DW M7BLY
MK42 7EH G7MIS
MK42 7EN G4ZPH
MK42 7EX G8YZT
MK42 7FR G1NRE
MK42 7JN 2E0GLD
MK42 7JN M3ORB
MK42 7JX G0DZB
MK42 7PE G8MGP
MK42 7RX M7LDU
MK42 7SX M6FAF
MK42 8BU G7SPE
MK42 8DT G0BDK
MK42 8HS G3WTP
MK42 8HS M0ZAE
MK42 8HS M1BED
MK42 8HS M3ZIE
MK42 8NL G7SQM
MK42 8RU G8WGQ
MK42 8RX G4OGG
MK42 8SY M0CMK
MK42 9EH G1CJL
MK42 9FG M7FIN
MK42 9PB 2E0XHB
MK42 9PB M6XHB
MK42 9RG G0EYG
mk42 9sh M3LFZ
MK42 9TH G6HCQ
MK42 9TZ 2E0RML
MK42 9UX G4NGD
MK42 9XB 2E1LEX
MK42 9XB M6IXY
MK42 9XB M7ECM
MK42 9YX G0VME

MK43 0DF G4DCW
MK43 0EL 2E0CQB
MK43 0EL M6CQB
MK43 0HA G7BBD
MK43 0HP G4PMB
MK43 0HP G6EZR
MK43 0HS G7KLR
MK43 0HX M6BFI
MK43 0JA G4JXZ
MK43 0LA G8OQK
MK43 0LW G4PMS
MK43 0PN G7UYB
MK43 0PN M0BZO
MK43 0QR G3YQO
MK43 0QZ G0FGJ
MK43 0RS M7YAN
MK43 0RU M6CYB
MK43 0RW G0RAV
MK43 0SD G8XER
MK43 0WY M7BGR
MK43 7BB G4ABQ
MK43 7BG M0NGS
MK43 7BG M0PSR
MK43 7BG M5RPT
MK43 7DR G4CEC
MK43 7ED G3INY
MK43 7HJ M6EMQ
MK43 7HN G8SZG
MK43 7HN M3HML
MK43 7JL G3LDG
MK43 7JT M3MSJ
MK43 7JX G3SVQ
MK43 7LP G8ADY
MK43 7LT G6EDT
MK43 7NJ M0KGI
MK43 7QF G4PNK
MK43 7SG 2E0ACQ
MK43 7SG G0GPV
MK43 7TD G4XPJ
MK43 8JA G4MBA
MK43 8LF G4UMW
MK43 8LG G7TTP
MK43 8PR G4OZP
MK43 8QJ G3XYV
MK43 8QJ M3ZLE
MK43 9BB 2E0LUN
MK43 9BB G0XIT
MK43 9BB M6UBS
MK43 9BT 2E1SPY
MK43 9BT M3CLP
MK43 9BT M3OLI
MK43 9BT M5HYD
MK43 9DB M7INE
MK43 9DW G8BAK
MK43 9EZ G0DIM
MK43 9GG 2E0SIB
MK43 9GG M6SBZ
MK43 9JL G6SNV
MK43 9JW G3YUQ
MK43 9JW G6IEE
MK43 9LB G8XNN
MK43 9RA M7DZC
mk430jw M7NCN
MK44 1DY G1BUV
MK44 1EN M1EBS
MK44 1EX G6HMF
MK44 1HB G6IPT
MK44 1JP G8IIG
MK44 1PL G0MOR
MK44 1SE G8ACQ
MK44 2AU G4FFC
MK44 2BA G0CZY
MK44 2DT G4TGV
MK44 2ER G4WDZ
MK44 2ES M0SFR
MK44 2EW G4MSQ
MK44 2HP G4OCX
MK44 2JX G1PSH
MK44 2LA G4HUE
MK44 2LF 2E0HUR
MK44 2LF M6HUR
MK44 2PB M7PEV
MK44 2RP G4HTX
MK44 2RP G6JJT
MK44 3 LR 2E0XFF
MK44 3 LR M0ZSS
MK44 3 LR M6ESS
MK44 3BH G4TXG
MK44 3BQ G4RLR
MK44 3NJ M6TVC
MK44 3NJ M7LEL
MK44 3NT M0DLM
MK44 3QW G4ACP
MK44 3UJ G3VHF
MK45 1AG M7VIK
MK45 1AQ G8LLD
MK45 1BU G0COG
MK45 1GW M0WAS
MK45 1JY G6JZW
MK45 1LF G1JCC
MK45 1LF M0JCC
MK45 1LN M1AFP
MK45 1LN M3ADL
MK45 1PJ G8CTB
MK45 1QA G4OWN
MK45 1QD G6PHZ
MK45 1RU 2E0OEL
MK45 1RU M7OEL
MK45 1TB G0WZA
MK45 1TQ G7WAS
MK45 1XP M0OYQ
MK45 2AD G7PCF
MK45 2AE G0BVW
MK45 2AQ 2E1FDJ
MK45 2BT G4AHM
MK45 2DJ G0WAS

MK45 2EG 2E0VUM
MK45 2EG M7DGF
MK45 2EY G3VES
MK45 2JD M7CXL
MK45 2JY G8PBM
MK45 2QL G8VHX
MK45 2RS G7NBI
MK45 2RS M0CKA
MK45 2SP G0RNI
MK45 2SP G4FKI
MK45 2SP G8GPO
MK45 2TP G0JJK
MK45 3BU G1WDQ
MK45 3EF G6AJX
MK45 3GD G3ZHJ
MK45 3LQ 2E0PSW
MK45 3LQ M0PSW
MK45 3LQ M3PPG
MK45 3PJ G8HRW
MK45 3QB G4USN
MK45 3WE G4MGX
MK45 3WE G4XQQ
MK45 3WH G7STG
MK45 4AS G8LQM
MK45 4BG G3FJE
MK45 4BG G4LOO
MK45 4DA G4VLA
MK45 4DU M7BDX
MK45 4FE G7HZQ
MK45 4FN M6NXH
MK45 4LN 2E1FDC
MK45 4LT M3IHQ
MK45 4NE G4DAQ
MK45 4NE M1EAJ
MK45 4NY M6BWF
MK45 4PF G4EIY
MK45 4TB G6BJO
MK45 5JN G0TFI
MK45 5LB M3YQG
MK45 5LP 2E1GXE
MK45 5LP M3IEM
MK451LF M6WEN
MK46 4AR M6GJG
MK46 4DN G0FQU
MK46 4ER G4TXT
MK46 4EY G7IMB
MK46 5BJ G4NEO
MK46 5ES G0HMF
MK46 5ES M6BJX
MK46 5FB G0SXY
MK46 5HT G4LHP
MK5 6AS M3JSQ
MK5 6AS M3LNQ
MK5 6AZ 2E0MEV
MK5 6AZ M6MHV
MK5 6HT G0UVG
MK5 6JE 2E0BZO
MK5 6JE M3XPY
MK5 6JL G4ETK
MK5 7AF G3HIU
MK5 7AF G3ZPA
MK5 7AF G8MKC
MK5 7AJ G4CLG
MK5 7AX G4LNC
MK5 7HA G6DOI
MK5 7HB G8XXM
MK5 8BL G4NIV
MK5 8EB G7HRH
MK6 2DS M6MPP
MK6 2HG G1WVP
MK6 2HT 2E1CJN
MK6 2HT G0FTI
MK6 2QB M7SUZ
MK6 3AY G3LMX
MK6 3ES G6TTX
MK6 3JJ G4ETX
MK6 3LQ G7SYU
MK6 3LQ M5MK
MK6 3LU G0RDG
MK6 4HL M6SAE
MK6 4HX M6GUX
MK6 4LA M6KOZ
MK6 4LL M7XLX
MK6 5AE 2E1DAK
MK6 5DA M3HAW
MK6 5LB M6FIO
MK7 6BQ M7HTL
MK7 6JA M6NXQ
MK7 7EG G6VTE
MK7 7FH G3XVA
MK7 7HH G0HRT
MK7 7TB G8XKH
MK7 7TT M6LGQ
MK7 8DH M3XKI
MK7 8LR M0MUZ
MK7 8LR M3YZU
MK7 8PL G1OWZ
MK7 8QB 2E0RLW
MK7 8QB M0UKA
MK7 8QB M6RLW
MK8 0AT G7THZ
MK8 0BB G1OWJ
MK8 0DB G4TZR
MK8 0EJ 2E0MGL
MK8 0EJ M3GLT
MK8 0EJ M3LTA
MK8 0EJ M3MGL
MK8 8AY G1WLX
MK8 8AY M0GOA
MK8 8BX 2E0DHF
MK8 8BX M6EHM
MK8 9EN 2E0BQB
MK8 9EN M3VPB
MK9 1LR G8VQN
MK9 3AY 2E0NYG
MK9 3AY M0NYG

MK9 3AY M7NPC
MK9 4BA M6OSW
ML1 1LQ MM3TRZ
ML1 2LB GM0ARD
ML1 2RL GM6OXL
ML1 3JW GM4UBJ
ML1 3NA GM0WNR
ML1 3QU GM0HWB
ML1 3TH 2M0YCG
ML1 3TH MM3YCG
ML1 4JL 2M0MMB
ML1 4JL MM3BQK
ML1 4RD 2M0FDZ
ML1 4RD MM3BPR
ML1 4ZL 2M0CRQ
ML1 4ZL GM1XOI
ML1 4ZL MM0RKN
ML1 4ZL MM0RWJ
ML1 4ZL MM6CRQ
ML1 5EF MM0SEK
ML1 5LB MM3XJA
ML1 5LQ GM0VXA
ML1 5NU GM4LQR
ML1 5RZ GM0EEH
ML1 5TY MM7SHT
ML10 6DL GM0SYV
ML10 6FW MM3SHT
ML10 6GN 2M0MCD
ML10 6GN MM3NZX
ML10 6SD MM6MNE
ML11 0FF GM0DQC
ML11 0FX GM0GBV
ML11 0HY GM6YRH
ML11 0NZ 2M0DSG
ML11 0NZ MM0TDB
ML11 0QA MM0MOT
ml11 7du MM7DUF
ML11 7PS GM4IIR
ML11 7SE GM3YXY
ML11 8AP GM4HBG
ML11 8ES GM0VYY
ML11 8HA 2M0AIE
ML11 8HB MM0WHM
ML11 8LH MM3GYU
ML11 8NE MM3XIA
ML11 8SU MM1BJG
ML11 9AE GM3NNZ
ML11 9QA GM6BIG
ML11 9SR GM4VGU
ML11 9XS GM0GYN
ML110EL MM7CUH
ML118QW MM7KIE
ML12 6FW GM4XHH
ML12 6GJ MM0FCM
ML12 6HQ GM0SCA
ML12 6LB GM7GIO
ML12 6NB GM0PJD
ML12 6NE MM6OFA
ML12 6PQ MM0THE
ML12 6PU GM4KKV
ML12 6RJ GM0DXI
ML12 6RR MM6EQW
ML12 6TP MM6FPY
ML2 0AH 2M0DQN
ML2 0AQ MM6AHB
ML2 0LP MM6JAW
ML2 7HG GM0WRU
ML2 7JX GM0LIR
ML2 7LL 2M0VGY
ML2 7LL GM5TDX
ML2 7LL MM7DTL
ML2 8BZ MM6BIY
ML2 8HR 2M0KVM
ML2 8HR MM6KCM
ML2 8PG 2M0SRX
ML2 8PG MM6SNR
ML2 8QR MM3VIS
ML2 8RA 2M0NAX
ML2 8RA MM6NAI
ML2 8SE GM0VXZ
ML2 9BL GM3TCW
ML2 9QL GM8KIQ
ML3 0AL MM7BBG
ML3 0HZ MM6PIB
ML3 6BF MM0KZA
ML3 6PD MM6CWN
ML3 6PN MM0RKT
ML3 6PP GM4UBF
ML3 7DD GM8NBV
ML3 7FB GM0STB
ML3 7FB GM4UQG
ML3 7FR MM3JLS
ML3 7HB GM0KDF
ML3 7HJ 2M0ELP
ML3 7HJ GM0ELP
ML3 7HJ MM0ELP
ML3 7HN 2M0RTD
ML3 7HN MM0RTD
ML3 7JY MM0GHN
ML3 7LL GM8LBC
ML3 7LZ 2M0JHN
ML3 7LZ MM0JZB
ML3 7PE GM4ILE
ML3 7PN 2M0JSB
ML3 7PN MM3JSB
ML3 7PW GM0MDX
ML3 7WS MM6TZX
ML3 8AY MM6LFN
ML3 8JS 2M0VLF
ML3 8NZ GM4WFE
ML3 8PH MM6YHO
ML3 8QH GM0CSN
ML3 8TZ MM6YUI
ML3 9AF MM0HQD
ML3 9EB MM7CTG
ML3 9JR MM6MCA

ML3 9QH MM6RLL
ML3 9SG MM6JJB
ML3 9UX GM4BVU
ML4 1JB MM0GGG
ML4 1LE MM6EEQ
ML4 2AU MM7HFC
ML4 2EQ 2M0RVF
ML4 2RT GM6PVA
ML5 1QT MM7YTR
ML5 2ES MM6NRK
ML5 2HB MM7EIT
ML5 2PJ MM7SMG
ML5 2SX GM4CHF
ML5 3JD MM7DKY
ML5 4BJ MM6PDA
ML5 4FN 2M0CWV
ML5 4FN MM6DKN
ML5 4JQ GM4SZG
ML5 4LL MM0CEZ
ML5 4NE MM3VHM
ML5 5BH MM7CAN
ML5 5LR MM0ZYT
ML5 5LR MM3XWB
ML6 0ES 2M0RMP
ML6 0ES MM6WBU
ML6 0QQ GM4AUP
ML6 0QQ GM5DVS
ML6 6ET 2M0WLX
ML6 6ET MM6OWL
ML6 6GP MM6ZGS
ML6 6PA GM0VXQ
ML6 6SP MM6PYX
ML6 7DH GM1MMK
ML6 7DH GM8HBY
ML6 7JF GM7CPR
ML6 7JH 2M0EFD
ML6 7JH MM6KFJ
ML6 7NT MM6NED
ML6 7QB MM3VYA
ML6 7SZ GM0ART
ML6 8BL GM4WTS
ML6 8FX GM0EEG
ML6 8LU 2M0WXS
ML6 8LU MM0WXS
ML6 8LU MM6WXS
ML6 8QG GM0VWZ
ML6 8SF GM4PRO
ML6 8XL GM7SWX
ML6 9AQ MM6OIT
ML6 9DF GM3ZNC
ML6 9ND MM0GHT
ML6 9ND MM1XJS
ML6 9ND MM3KNY
ML7 4HP MM3NFX
ML7 4JA MM7OKO
ML7 5AX GM0PHW
ML7 5EL MM7APB
ML7 5HW MM6SFF
ML7 5TJ 2M0EGI
ML7 5TJ MM0WEI
ML7 5TJ MM3WEI
ML7 5TJ MM7CGS
ML8 4AF MM3VYU
ML8 4DY MM7GMW
ML8 4NR GM4ARU
ML8 4QT GM4XXO
ML8 4QY MM5AHM
ML8 4QZ 2M0MUR
ML8 4QZ MM0MUR
ML8 4QZ MM0RRM
ML8 4QZ MM6MUR
ML8 4RL MM7RBT
ML8 5GB 2M0ZEE
ML8 5GB MM6PJR
ML8 5HB GM3UCI
ML8 5HB GM4GM
ML8 5HB MM3UCI
ML8 5HB MM3VRI
ML8 5HR GM4COX
ML8 5HR GM4UXX
ML8 5JG 2M0ISA
ML8 5JG GM1OQT
ML8 5JG MM0TPD
ML8 5LT 2M0LAW
ML8 5LT MM3SUV
ML8 5PH GM4LAO
ML8 5RX MM3OQV
ML8 5TS GM8FHK
ML8 5UW 2M0ZBF
ML8 5UW MM6EZO
ML9 1EH 2M0XXP
ML9 1EH MM0XXP
ML9 1EH MM3XXP
ML9 1JX 2M0MMO
ML9 1JX GM3PXK
ML9 1JX MM0MOB
ML9 1JX MM3MAO
ML9 1QT GM3MXN
ML9 1RA GM3HVK
ML9 1UR GM1WYV
ML9 2DA GM7AOM
ML9 2LG GM3NKG
ML9 2TD GM4ISM
ML9 3AJ GM1DLS
ML9 3BB 2M0TGM
ML9 3BB MM6FPX
ML9 3BT 2M0VPU
ML9 3EN GM6HFH
ML9 3NR 2M0TOR
MS2 9BD MM3JNJ
N1 0HN M6PDB
N1 0HQ M0ZAY
N1 1QN G1GBX
N1 1TN G0TJD
N1 2BN M6CYT
N1 2PJ G4FIH

N1 3NW M3ZUT
N1 4NU G0OOS
N1 4NU M0ESP
N1 7AR G0BQI
N1 7AY 2E0JQW
N1 7AY M0JQW
N1 7AY M0THY
N1 7AY M6JQW
N1 7AY M6THY
N1 7NW 2E0GKG
N1 8LX 2E1GPG
N1 8UA G4GJV
N1 9LS 2E0EZV
N10 1EG M0TPH
N10 1LX G3ASX
N10 2DE G3KOD
N10 2HS G4KPH
N10 2JU M0PCS
N10 2PD G8HHZ
N10 3AA G4DFJ
N10 3PB G7FJU
N10 3PY M7DHA
N11 1AX 2E1HAM
N11 1AX M3GTA
N11 1RD M3FWA
N11 1TA M6DSH
N11 2AD 2E0DTY
N11 2AD M0OTO
N11 2AD M6LSU
N11 2DE G8YNC
N11 2JL G6FDX
N11 2JS M0IRO
N11 2JY G0NSO
N11 2LT M0ULC
N11 2RL G1IHA
N11 3HJ M0PKL
N11 3NN M6DLD
N11 3NQ G3GGI
N112AB M7SCF
N12 0LS G8XRP
N12 0NX G6ABJ
N12 7AD G8LSH
N12 7JG G4CCT
N12 7LD M7EMV
N12 8RB 2E0KSO
N12 9AU G1YSA
N12 9BG G4RIH
N13 4HD 2E1HWV
N13 4RG 2E1HXD
N13 5BS G6NTW
N13 5BT G3YTY
N13 5JS M6SJF
N14 4DH G0KUX
N14 4DH G1YJI
N14 4PG G1BPU
N14 4RP G4IEH
N14 4XN 2E1FMW
N14 5AT 2E1GW
N14 5AT M6ORT
N14 5RR M7SIS
N14 6QU G4BAN
N14 6RR G1DEY
N14 7NJ G6BLC
N15 4HH 2E0GSL
N15 4HH M0LAE
N15 4HH M6LXC
N15 4JN G8OJR
N15 4LA M6COI
N15 4SQ M6ATA
N16 6DJ 2E1PPK
N16 6DJ M3GKG
N16 6EP G0PIU
N16 9PH 2E0SDM
N16 9PH M6REI
N17 0JD G3YBS
N17 0TE G0CEU
N17 6BP G8ASC
N17 6EH G0FWF
N17 7HT G3OVX
N17 8DE M7ACX
N18 1QD M0TUK
N19 3DJ M3FWJ
N19 3HF G7LGI
N19 3SJ G7VDI
N19 4HR M7HDC
N19 5BQ G8WPA
N19 5NJ M0HPZ
N19 5NJ M6DYI
N19 5TR G7RJO
N193RA M7OVM
N2 0HP 2E0AQI
N2 0HP G0UKA
N2 0HP M0DAN
N2 0HP M1DAN
N2 0QB G7TWC
N2 0QB M0HWC
N2 0RD 2E0JSY
N2 0RD M0JSY
N2 0RD M7BMJ
N2 8AY M1KAZ
N2 8JT G7LZB
N2 8JW 2E0MCA
N2 9LY M7PRN
N2 9NE M7ADO
N2 9NS G7ULM
N20 0AL G4XYK
N20 0DS G0ESF
N20 0HT G4LCB
N20 0HT M6FKU
N20 0QN G0EIQ
N20 8QH G1FLX
N20 9AB M7CKF
N21 1BA G6XIW
N21 1EH G3TIE
N21 1NP G3SFG
N21 1NP G4DFB

N21 1NP G6QM
N21 1QP G8DWF
N21 2BE G0RPM
N21 2BE M6RUT
N21 2NX 2E0FXU
N21 2NX M7NAC
N21 3AU M5AGZ
N21 3LP M0VKJ
N22 5BL M6OPA
N22 5NB M6NYA
N22 6JL M6VAR
N22 6LH 2E1FVJ
N22 6NT 2E1FVK
N22 6QD M0SKF
N22 6RG M6FCQ
N22 6RR 2E0VIS
N22 6RR M6GIC
N22 7AQ M7DUT
N22 7EX G4KKT
N22 7UT M7EAS
N22 7XG 2E0OJS
N22 8NN M3HMT
N22 8NN M3RGE
N22 8YQ M0KSR
N3 1NA 2E0YAQ
N3 1QL M1MRS
N3 1QL M3SMR
N3 1TJ M7PDZ
N3 2DD G0WCZ
N3 3HP G7UNW
N3 3NJ G0ODM
N3 3TX 2E0NAZ
N4 1HN G0PLC
N4 1LB M7BOK
N4 1RJ M6KKG
N4 1SQ M7SMC
N4 2LN M3XLK
N4 2ZT M0INF
N4 3DW G0JYU
N4 3NX 2E0YLZ
N4 3NX M7BAC
N4 3NX M7PIX
N5 0AD G4HOJ
N5 2DE G8ZQJ
N6 4BA G4GRS
N6 4BA G8MIC
N6 4BH M6SIK
N6 4EU G3XKY
N6 4QD G7PZM
N6 5AU G0RIC
N6 5PJ G0OOD
N6 5TS G4OJW
N6 6JR G4HDD
N7 0EL G1NWG
N7 0QB 2E0VNV
N7 0QB M6VNV
N7 0SH M0RXB
N7 0SN G0AKM
N7 7AY M3MUQ
N7 7EL M6LNS
N7 7FE M6MZB
N7 8HY G4IZU
N8 0JB M6ZIB
N8 0NA G4DVG
N8 7QU M7JCP
N8 7SJ G1WNZ
N8 8AA 2E0RUI
N8 8AA M0RLM
N8 8HD 2E0GES
N8 8NX G4KMM
N8 8QH 2E0CAO
N8 9BP G6RZR
N8 9PN M7WPL
N9 0HJ 2E1FNB
N9 0HJ 2E1HRM
N9 0HJ M3HRM
N9 7QG G1ALD
N9 8HD G4YGH
N9 8LJ G4ZIS
N9 8QL G1RFS
N9 9EQ G7COQ
N9 9HP G4UTR
N9 9HU G8SZZ
NE10 0DR 2E0RLJ
NE10 0EH M7RZE
NE10 8DD M7JMC
NE10 8QS M6ETG
NE10 8SW G7CIT
NE10 8UE G7JVG
NE10 8WJ M6GEE
NE10 8XB M7UPT
NE10 9AB M3YTZ
NE10 9BQ 2E1DRX
NE10 9DH G7EKG
NE10 9DT M6MSI
NE10 9NB G1OMX
NE10 9NH M7JBW
NE10 9ST M6DJK
NE10 9TZ G4VTN
NE11 0BS G0ESW
NE11 0HQ 2E0LLK
NE11 0HQ M6LLK
NE11 9BE G7PUA
NE11 9LA 2E1EDB
NE11 9XH G0CXX
NE12 6BT G3YRH
NE12 6YR M0NTK
NE12 7EU G1USF
NE12 7JP G3LRI
NE12 7JR M0AVU
NE12 8DP M7OPM
NE12 8LS M0JAQ
NE12 8LS M6DAM
NE12 8PH M6GMM
NE12 8XQ G0WUV
NE12 9AW G3LYG

NE12 9BQ M7EJN
NE12 9QP G6FHR
NE12 9RL G1XXF
NE129NQ 2E0JMD
NE129NQ M6JBO
NE13 6EY G6YTR
NE13 6QB G0LNW
NE13 7HS G0AOK
NE13 7HT G1GDA
NE13 7HW G0GYO
NE13 7NG G7VLL
NE13 9AP 2E0SHR
NE13 9AP M0WMG
NE13 9AP M7SHP
NE13 9EH M0LKT
NE15 0EA G8UZM
NE15 6LG 2E0MCN
NE15 6LG M6MPB
NE15 6QN G1YUN
NE15 6RZ G6SMI
NE15 7DN M6FOE
NE15 7DQ G1YDD
NE15 7LB M0IDI
NE15 7LN M6ADM
NE15 7LR G1SKQ
NE15 8NB G0WMX
NE15 8NB G8PVV
NE15 8QD M7SWD
NE15 8RH M7AAK
NE15 8RL G3KTT
NE15 8TW G0DFT
NE15 9AT M0NRS
NE15 9EB G8XGS
NE15 9JL M7CML
NE15 9JU G0EOY
NE15 9LH M1DHC
NE16 3EJ M7JHD
NE16 3JN 2E1APX
NE16 3JN G0PUK
NE16 4PF G8DST
NE16 5PP M0ERN
NE16 5PP M1ETS
NE16 5PP M5ERN
NE16 5PT G1KYN
NE16 5PT M6LXK
NE16 5QS 2E0TSU
NE16 5QS M3XQL
NE16 5QZ G4GFV
NE16 5RX G0WUI
NE16 5SJ M0JGF
NE16 5SZ G6OSH
NE16 6LQ G0RFN
NE16 6QF M3WQL
NE16 6QP M7MSF
NE17 7AY M7GPP
NE19 1AA M7GJG
NE19 1NX M1CCX
NE19 1RH G4HDS
NE19 1TA G4AOS
NE2 1BQ M7KTR
NE2 1HQ M6WDN
NE2 1XY M6FTP
NE2 2HD G4ETI
NE2 2JN G0BEV
NE2 3NS G2DJM
NE20 9AU 2E0KCN
NE20 9AU M6NQR
NE20 9AW G4DSD
NE20 9EJ G4SVE
NE20 9HQ G8OEO
NE20 9JB G8SCT
NE20 9PZ G7VTT
NE20 9QQ M0CIP
NE20 9QQ M1DPI
NE20 9RA G8KPD
NE20 9RA G8RER
NE20 9RD M0GPU
NE20 9RF M0EUK
NE20 9RR G4ITR
NE20 9RZ G1SUM
NE21 4FD M7DPU
NE21 4RR G0SWB
NE21 5QL M1CKU
NE21 6HJ 2E0DZO
NE21 6HJ M6HNV
NE21 6JU G7LOG
NE22 5BU G4TMQ
NE22 5DZ G1DDS
NE22 5ER G4UTQ
NE22 5HJ G6WJD
NE22 5HJ M3HLX
NE22 5TD G0AXJ
NE22 5TD M6AXJ
NE22 5TD M6SRS
NE22 6DF G4WNG
NE22 6DR G3OTH
NE22 6HD G4OAN
NE22 6HW G0OWE
NE22 6JG 2E0HGP
NE22 6JG M7GPH
NE22 6JJ G1DSB
NE22 6LD G4AXF
NE22 6NE G0DJO
NE22 6NS M1AQY
NE22 6PB G0KYR
ne22 7br M6XSO
NE22 7EF M0ZYZ
NE22 7EH G7SNP
NE22 7EJ G0GLH
NE22 7EW M7ODD
NE23 1EY M7AJA
NE23 1NZ G6OTV
NE23 1RQ G7VBD
NE23 1TS G1ZJQ
NE23 2BP M1PJB
NE23 2BZ G6WUD

NE23 2UE G0TLZ
NE23 2UR M6SEU
NE23 3LH M7DVK
NE23 3LW G0BAU
NE23 3TT G4HJB
NE23 6AA G6AVL
NE23 6AS G0NHJ
NE23 6AS G0NHK
NE23 6DN G4HIH
NE23 6EX M3WQN
NE23 6QB M0GXW
NE23 6RW M6XEU
NE23 6SX G1SLI
NE23 6TU G7MYT
NE23 6TU M1DHG
NE23 6TW G0JRX
ne23 6ub G0LCE
NE23 6XQ 2E0CKW
NE23 6XQ M6BLH
NE23 7AF M6IYC
NE23 7BH G0PVT
NE23 7DE G0JWO
NE23 7EU M7TCF
NE23 7RT G1TPO
NE23 7TR M7AIW
NE23 7TZ M6FTO
NE23 8HE 2E0GGJ
NE23 8HE M7EDM
NE23 8HW G7OVK
NE23 8JQ M6IFX
NE24 1NP G8EXK
NE24 1PG G0ECQ
NE24 2JR G4YAV
NE24 2JW M3PFE
NE24 2QW 2E0XRG
NE24 2QW M3XRG
NE24 2SU M1BMQ
NE24 3DH M0LRZ
NE24 3DH M6LRZ
NE24 3DT M6EAH
NE24 3DT M6EFY
NE24 3EP M1ARL
ne24 3ha G0LFE
NE24 3HA G6PNO
NE24 3HX G1LMC
NE24 3JG G1ATC
NE24 3JG G7PYR
NE24 3QN 2E0DTH
NE24 3QN M6WBH
NE24 3QS M0HNT
NE24 3RD M6YNY
NE24 4AZ M3PTR
NE24 4DS M0DAD
NE24 4DT G0KHC
NE24 4EB G0EVF
NE24 4JD M3LXP
NE24 4LB G0TWV
NE24 4PL M0TWO
NE24 4QP M0CVU
NE24 5AS M1EHA
NE24 5ET M3DFW
NE24 5HL G4SIE
NE24 5PR M6THX
NE24 5RH 2E0BMB
NE24 5RN G7KDX
NE24 5TS M6KVV
NE25 0AL M0MLY
NE25 0BU G0LNS
NE25 0EL 2E0DCV
NE25 0EL M6CLG
NE25 0FH G1EXK
NE25 0NG G0PKR
NE25 0NG G0VRT
NE25 0NP G3VYZ
NE25 0NP G8ARF
NE25 0PA G0PJU
ne25 0sh G4VOU
NE25 0ST 2E0CMH
NE25 0ST M0HLR
NE25 0ST M6TEY
NE25 0TA G4BEZ
NE25 8AU G7PWL
NE25 8BJ G6CEM
NE25 8EP G4IZQ
NE25 8JG M0IGB
NE25 8JG G7ELV
NE25 8NN M1CPP
NE25 8RP M6LHF
NE25 9AF M0WJW
NE25 9JS G6CCN
NE25 9QR G3WDS
NE25 9XH M3HSH
NE26 1AF G1ODJ
NE26 1QR M0USB
NE26 1SH G7VTJ
NE26 1SH M0VTJ
NE26 2JR G0OBQ
NE26 2NR G6PKM
NE26 3AD G8FVT
NE26 3BA G3GMS
NE26 3DJ G1MHA
NE26 3DY G0NWM
NE26 3DY M0GAE
NE26 3EN G5BW
NE26 3SY M0LDX
NE26 4DB M3LFX
NE26 4JH 2E0UWT
NE26 4JH M3UWT
NE26 4JQ M3BMH
NE26 4PA M7GAP
NE26 4RE G3VUD
Ne262dz M3MKM
NE264RG M0HTX
NE27 0EF G1AIA
NE27 0EX M3JFF
NE27 0EY M7CBH

NE27 0UF G4RUI
NE27 0UX G6VEG
NE27 0UZ 2E0KBE
NE28 0AY M7NBZ
NE28 6SD M3SKX
NE28 6SN M3ENS
NE28 7HP 2E0XPC
NE28 7HX M7SSY
NE28 7PG M0HGV
NE28 7PH G3ZTJ
NE28 7SQ M0DLS
NE28 7SQ M6LBL
NE28 8TL M0ADR
NE28 9AA M0CBK
NE28 9HU G0JXY
NE28 9RX M0NKY
NE28 9YB G6RJU
NE28 9YW M0IIM
NE280JH 2E0NEQ
NE280JH M6NEQ
NE29 0NA 2E0EFP
NE29 0NA M6IGQ
NE29 0QN G4GWV
NE29 0QX G4FRD
NE29 0RW M7JSC
NE29 6UF M7JNH
NE29 6XE G0AIO
NE29 7AW G1EYD
NE29 7HP G1FNU
NE29 7JB M6REY
NE29 7QN 2E0HWC
NE29 7QN M6HWC
NE29 8DW G0AXO
NE29 8EQ G0EDK
Ne29 8jb M7DPZ
NE29 8LU G1LNR
NE29 8NT G3OZP
NE29 8QA G3PGC
NE29 8SS G2ARY
NE29 8SS M0BQD
NE29 8ST 2E0LYB
NE29 8ST M0LYB
NE29 8ST M7ISM
NE29 9EN G0MEF
NE29 9HT G0OBO
NE29 9NS G4AXL
NE29 9NU G8YCP
NE3 2DU G6OQO
NE3 2FG G4LIA
NE3 2QQ 2E0LMK
NE3 2QQ M6SAQ
NE3 2UL 2E1HDB
NE3 2UL 2E1HDC
NE3 2UL M1MOB
NE3 2UL M1TXT
NE3 2XJ M7TRW
NE3 2XU M0PJF
NE3 2XU M3ZKF
NE3 3DB M6OOA
NE3 3LA 2E1NEW
NE3 3LA M7NEW
NE3 3PP M0PHP
NE3 3PP M3UHP
NE3 3TN G1XYS
NE3 4BB G8QM
NE3 4NR G4LGH
NE3 4RN G0CQK
NE3 4RN G1GDM
NE3 4TT G1GAD
NE3 5BH G3LIV
NE3 5BH G8UEE
NE3 5BY G3SQU
NE3 5DX G1VOP
NE3 5QD 2E1ICO
NE3 5TQ G3URE
NE30 1NA G8NLS
NE30 2BW G8YFA
NE30 3AN G8EZL
NE30 3DT M1DSE
NE30 3LH 2E0TDR
NE30 3LH G0JWM
NE30 3LH M6FVH
NE30 4BJ M3WHX
NE31 1DH M6PKD
NE31 1HL M6CYE
NE31 1QU 2E0HAH
NE31 1QU M0MBT
NE31 1QU M6MRK
NE31 1RH G7MPJ
NE31 1XT G7JTI
NE31 2AQ M0OMT
NE31 2AQ M3SRT
NE31 2BW M3AHQ
NE31 2DW 2E0NFC
NE31 2DW M6NFC
NE31 2DY 2E0HUT
NE31 2GB M7WIM
NE31 2HN 2E0GTE
NE31 2HN M0HSC
NE31 2HN M0TYN
NE31 2HN M6GAK
NE31 2HN M6MNC
NE31 2HN M6NEE
NE31 2HN M6SLU
NE31 2HR 2E0MCM
NE31 2HT G4SFN
NE31 2JN 2E0PBN
NE31 2JN M6BPS
NE31 2LJ G0WZB
NE31 2LJ M6WZB
NE31 2TB 2E0GWC
NE31 2TB M6CKQ
NE31 2XS G1NZP
NE32 3QA 2E0KPC
NE32 3QA M0KPC
NE32 3QA M3GAG

NE32 3SB 2E0MTY
NE32 3SB M6FDR
NE32 3SB M6IRM
NE32 3SB M6SBF
NE32 3TX G0TFT
NE32 4AE G0DIP
NE32 4EY M6AEJ
NE32 4HS 2E0SBO
NE32 4HS M6PRN
NE32 4HX G3JDO
NE32 4HX M6BAG
NE32 4LP G7MWU
NE32 4QN M6JFJ
NE32 4QY M6KAS
NE32 4TN M0ISG
NE32 5DN 2E0MBW
NE32 5QB 2E0HHU
NE32 5QT G4WUM
NE32 5SS M7PNY
NE32 5UF 2E0TBL
NE32 5UF M6JWD
ne32 5yj M1FNE
NE33 2AN G3SEG
NE33 2TH M3EVV
NE33 3EH G4LFG
NE33 3EL M6URT
NE33 3LQ M0BVF
NE33 4DL M6MRP
NE33 4HP G0DSX
NE33 4HR 2E1PDH
NE33 4HR M7AYU
NE33 4UE G3PRE
NE33 5HX G1NZL
NE33 5NH M1EEZ
NE34 0BZ M1BEX
NE34 0DR G3KZZ
NE34 0EQ G3JPO
NE34 0JB 2E0KSW
NE34 0JB M6BSC
NE34 0JB M6KVN
NE34 0NJ G4XWR
NE34 0RA M0BWI
NE34 0RB M7VUK
NE34 0RR M6CNN
NE34 6AQ 2E0GJN
NE34 6AZ G0ZAT
NE34 6DP G7WLY
NE34 6QG M7BTG
NE34 6RY M0DEO
NE34 6RY M3VEX
NE34 7AH 2E0XUA
NE34 7AH M0XUA
NE34 7AH M7DRS
NE34 7AQ G6BIA
NE34 7DJ 2E0EQN
NE34 7DJ M6KZR
NE34 7DY M6FEU
NE34 7JF G6UWI
NE34 7JJ 2E0DKD
NE34 7JQ G4NS
NE34 7LP M6REO
NE34 7PT G7HSY
NE34 7QU G0WUH
NE34 7RF G1OHX
NE34 8BB 2E0GFU
NE34 8DD M6BMV
NE34 8EJ G1XVR
NE34 8JT M6SWO
NE34 8NE 2E0PEM
NE34 8NE M6PEM
NE34 8NJ M0MGK
Ne34 8nn 2E0XSJ
Ne34 8nn M7HGV
NE34 8RU G6LIK
NE34 9AE M6PCU
NE34 9AJ M0RPO
NE34 9BA M6THV
NE34 9BU G8GIU
NE34 9EB 2E0YYG
NE34 9EB M6YYG
NE34 9JJ G0WKQ
NE34 9JJ G7PHG
NE34 9JJ M3UJH
NE34 9JJ M6XTA
NE34 9NF M7ANK
NE34 9NQ 2E0DZC
NE34 9NQ M6LFR
NE34 9NQ M6XLC
NE35 9BS M0EZO
NE35 9BS M0JDZ
NE35 9BS M0NER
NE35 9EP 2E0FAN
NE35 9EP M6PGP
NE35 9HA G4ZEU
NE35 9HU 2E0KBN
NE35 9HU M6KBS
NE35 9JE M7DYO
NE35 9LG G8JUC
NE35 9LR 2E0SID
NE35 9LR M0ZID
NE35 9NG G3XWU
NE35 9NH M3IWN
NE35 9NH M7RKW
NE36 0HJ G4VSK
NE36 0LE G8HI
NE36 0RF G0RCJ
NE36 0TD G8SEW
NE37 1AH 2E0FMT
NE37 1AH M7ACY
NE37 1HT M6PFW
NE37 1QD G0DHS
NE37 1QQ G4YEX
NE37 1UN M7AAL
NE37 2BT G4YDM
NE37 2EU 2E0YEY
NE37 2EU M7VIP
NE37 2JS G7LOY
NE37 2JS M6JIZ
NE37 2LT M7DDQ
NE37 2LZ M6KKA
NE37 3BZ G1LBU
NE37 3DE 2E0HHW
NE37 3DE G8SVR
NE37 3DE M7BHK
NE37 3DJ M6GBM
NE37 3LA G0EPP
NE38 0EB M6WHN
NE38 0NR G1TWT
NE38 7AL G4WNI
NE38 7BW G7ITB
NE38 7DB G8VYQ
NE38 7EF G0SCI
NE38 7EF M6YBV
NE38 7JN G0NSP
NE38 7TA G0OHK
NE38 7TQ 2E0CSW
NE38 7TQ M6OSC
NE38 8BP G4RTJ
NE38 8DD G1GTX
NE38 8DR G0UQP
NE38 8DR G4YGW
NE38 8DS G1HCC
NE38 8HS M3MVI
NE38 8PF G7NHE
NE38 8PF M3ABY
NE38 8RX G0ITM
NE38 9DE M0HIM
NE38 9EG G0HEU
NE38 9ET G1JDV
NE38 9ET M3ZDV
NE38 9HS M3XRV
NE39 1EQ G4NOX
NE39 1LT G2DML
NE39 1LT G3SEQ
NE39 2AD G3TKB
NE39 2HD G1VRC
NE39 2HF 2E1BCF
NE39 2JP M3PGY
NE39 2JQ M6MLX
NE39 2JT 2E1AII
NE39 2PN 2E0AJK
NE4 5RY M7BET
NE4 6UX M6NDI
NE4 7ER G1YJQ
NE4 7RG M6SQE
NE4 8BH 2E0CAT
NE4 8BH 2E0IGW
NE4 9AL M3YSN
NE4 9EP G3UKH
NE4 9QY G8POG
NE40 3Jn M6IKF
NE40 3RU G7KJP
NE40 4EU G3ZVM
NE40 4HR G0BGB
NE40 4QY G8CYW
NE40 4SY 2E0DUA
NE40 4SY M6WDI
NE40 4TD G4NSC
NE404QE M7BIE
NE42 5BJ M3OTP
NE42 5ER M6KGS
NE42 5PN G8SFA
NE42 6AT 2E1BUJ
NE42 6AT G0NEE
NE42 6EH G4STK
NE42 6JP G0EQC
NE42 6PZ G8RQF
NE42 6RA 2E0EQW
NE42 6RA M6JZL
NE42 6RB G7JTK
NE43 7PX G8YVP
NE43 7QR 2E0ZDM
NE43 7QR M0ZDM
NE43 7QR M6VID
NE43 7RQ G3UVU
NE437RL M6ISR
NE45 5BA G1SIO
NE45 5EX G8POO
NE45 5EX M5POO
NE45 5HB G4OEU
NE45 5HZ M7CAK
NE45 5JH G3TSS
NE455AP G1GLZ
NE46 1JH 2E0KZV
NE46 2LR M0TKF
NE46 2LR M7APC
NE46 2QB G6DNV
NE46 2QD G1UUF
NE46 4BE M1SAB
NE46 4BN G4IWJ
NE46 4DE G0CNU
NE46 4LE G0LYC
NE46 4LU G1HZI
NE46 4LU G1TIK
NE46 4RX G1PIF
NE46 4SS G1MOB
NE47 5NA 2E0PHK
NE47 5NA M6PHK
NE47 6DE G0OYL
NE47 6ER M0CFB
NE47 6HG G0GXO
NE47 9AS M0NPG
NE47 9DR G0AXZ
NE47 9EL G4UEO
NE48 1AQ G1SCB
NE48 2RY G6ZIC
NE48 2RZ G3TTI
NE48 2SG G1KHY
NE48 2SL G8JSR
NE48 2SQ G1CGH
NE48 2TG M7DDN
NE48 2TN G0AFU
NE48 4BD G1TIF
NE49 0JL 2E0IKT
NE49 0JL M7CVC
NE49 9BS 2E0IBT
NE49 9BS 2E0KFT
NE49 9BS M0JKF
NE49 9BS M3IBT
NE49 9BS M3KFT
NE49 9LR G0BZB
NE5 1BT G8PHB
NE5 1SQ G7SCV
NE5 1TL 2E1IIP
NE5 2DL 2E0MEG
NE5 2DL M3TTK
NE5 2FA M3UAK
NE5 2JB G3XXQ
NE5 2NW G0DMW
NE5 2NW G0IOQ
NE5 2PA 2E0CSV
NE5 2PA M0HRW
NE5 2PA M6BXS
NE5 2PJ G3NIJ
NE5 3QB 2E0MUW
NE5 3RE M6SFJ
NE5 3SZ G7UVN
NE5 3XB G6JQE
NE5 3XL 2E0GBJ
NE5 3XL M6TXI
NE5 3YB G8ECZ
NE5 4BG M6MJS
NE5 4ED 2E0LJJ
NE5 4ED M6OJL
NE5 4PQ G0RMO
NE5 5AP 2E0TNN
NE5 5AP M0MAG
NE5 5EA G0WYY
NE5 5HS 2E0WDH
NE5 5HS M3WDH
NE5 5PR M3YXP
ne54td M7CON
NE6 1BH M6SNE
NE6 3BB M1DSU
NE6 3RY M0YVG
NE6 3TQ M7DDW
NE6 4DR G1EPD
NE6 4HA G7TPG
NE6 4HL G7UTG
NE6 4HL M0UTG
NE6 4JA 2E0TWX
NE6 4JA M7TWA
NE6 4SR G6JLL
NE6 4SR G8GFW
NE6 4SY G3ZWR
NE6 4TN G7DAR
NE6 5AD G4UOW
NE6 5JY G7FEE
NE6 5YA G3YVZ
NE61 1XF G8ONS
NE61 2AS G0WAD
NE61 2AS G7RXJ
NE61 2FY M7JMO
NE61 2HJ G0DWO
NE61 2JT G3ILZ
NE61 2PL G3AGC
NE61 2PN G4NFV
NE61 2RU M0NCL
NE61 2SG G0EVV
NE61 2SG G6IIA
NE61 2SG G7SPP
NE61 2TF 2E0AOU
NE61 2TF G1BNN
NE61 2TH 2E0KXI
NE61 2TH M7KOL
NE61 2TQ G4ZTC
NE61 2XF G7PSL
NE61 2XY G7HHU
NE61 3LA G4KBX
NE61 3RA G6OTW
NE61 3RB G4FXU
NE61 5ES M6EJC
NE61 5HF G7RWC
NE61 5HQ M6BEN
NE61 5JU G4DGQ
NE61 5LN M1JPS
NE61 5LT M6JLY
NE61 5NT M3XJK
NE61 5PU G4ZTA
NE61 5QW G0KNW
NE61 5RA 2E0HOX
NE61 5RA M7HOG
NE61 5RB G4AAX
NE61 5RB G8PNN
NE61 5RP 2E0XEW
NE61 5RP M0XRW
NE61 5RP M6CQR
NE61 5SP M1DGW
NE61 5SR 2E0EKI
NE61 5SR M6ZBM
NE61 6DS 2E0GCD
NE61 6FN 2E0KNL
NE61 6FN M6KNL
NE61 6LQ G3BIK
NE61 6LT G4DJJ
NE61 6PE M0PGW
NE61 6PE M1DPW
NE62 5DA G4GWB
NE62 5DN G1NOS
NE62 5LD G4ZOY
NE62 5NG M1ULD
NE62 5NN M3GCT
NE62 5NN M7PRA
NE62 5UE G4URW
NE62 5UF G0WZG
NE62 5UF M3HJJ
NE62 5UL G0KUI
NE62 5UT 2E0SCM
NE62 5UT M6SBW
NE62 5XA M6SJN
NE62 5XF G3ZKG
NE63 0EU 2E0CVZ
NE63 0EU M6CVE
NE63 0HT M6ACJ
NE63 0LZ G4OIV
NE63 0PL G7FFR
NE63 0PQ M6KWP
NE63 0TH M1DKK
NE63 0XH G7MWK
NE63 8AB M6EGS
NE63 8BQ G0EFG
NE63 8EF 2E0SCO
NE63 8EF M3WLU
NE63 8HX G1LMZ
NE63 8HX M0GEC
NE63 8JF G0CQH
NE63 8JF G7UZS
NE63 8JT G8VIB
NE63 8JX M0VKX
NE63 9BB G7VDA
NE63 9BB M0BMJ
NE63 9BB M1EIZ
NE63 9DL M6JAF
NE63 9EN M6SIE
NE63 9LU M0WBB
NE63 9QJ G6JXS
NE63 9RF G0VRS
NE63 9RY G0WYZ
NE63 9SD G7VDV
NE63 9SE G0EMF
NE63 9TJ M0AHZ
NE63 9TJ M3VQS
NE63 9TP G3KLP
NE630TD 2E0NHR
Ne639ez M6MVE
NE63RU M7MLK
NE64 6BQ M0VHC
NE64 6DE M1DLS
NE64 6DQ M3RPK
NE64 6JH 2E0HPE
NE64 6JH M0JWZ
NE64 6LH 2E0LFM
NE64 6ST G0UDZ
NE64 6XN G0VAD
NE64 6XT G0LEF
NE65 0AP G2FXV
NE65 0EL G8GVN
NE65 0ER G3OGH
NE65 0GB G7OWP
NE65 0JS G4BCP
NE65 0SQ M7KHX
NE65 0TF 2E0BVD
NE65 0TF M6ALB
NE65 7HF M0CKG
NE65 7PU 2E0ZED
NE65 7PU M0ZDD
NE65 7PU M3ZED
NE65 7QN 2E0NJE
NE65 7QN M3NFG
NE65 7QS G4WMQ
NE65 8JT G3VKU
Ne65 8ld M7TAZ
NE65 8QH G1ODT
NE65 8UN G1JKX
NE65 8UR G1BBY
NE65 8UW G4ADD
NE65 9DB M1ZXG
NE65 9JL G6CBL
NE65 9JN G1RKD
NE65 9JW G1DNA
NE65 9JW G1HAB
NE65 9NB G7DEF
NE65 9NG M7BKJ
NE65 9QW G8MOG
NE65 9SP M0SSE
NE65 9TF G8YSH
NE65 9YA M1VHT
NE66 1BS G6KXB
NE66 1DW M3WDN
NE66 1ET M1CWO
NE66 1EU M3MES
NE66 1RD 2E0NTM
NE66 1RD M7CTM
NE66 2QP G7ANV
NE66 2QP M3YOT
NE66 2QP M7SJG
NE66 2YD M0DSS
NE66 3AT G1AHT
NE66 3AT G3NEP
NE66 3AW G4KRH
NE66 3DD M7NCM
NE66 3JE G3ZDU
NE66 3JQ M6HOS
NE66 3JZ G1YAE
NE66 3LS M1CCY
NE66 3NT G1GYM
NE66 3NT M0GYM
NE66 3PJ G4EKF
NE66 3RL G7VKA
NE66 3XY G2DJ
NE66 3XY G3VGW
NE66 4YD M0FLC
NE67 5AW G4FCC
NE68 7SM M6TSM
NE68 7TD G3AQB
NE68 7XP G3TEP
NE68 7XR G1VVU
NE7 7AP G4KVK
NE7 7BX G4WMA
NE7 7DJ M7CYA
NE7 7DJ M7PYA
NE7 7FB G1XSA
NE70 7PA M3NGM
NE71 6AF M6SBN
NE71 6DB G4LUF
NE71 6JL G0AVU
NE71 6QG 2E0TQF
NE71 6QG M3TQF
NE716QG M3JLF
NE8 1NQ G7SZB
NE8 1RU M6YCH
NE8 1US 2E0DZS
NE8 1US G7MWB
NE8 1US M0MBA
NE8 1US M6DZS
NE8 2BG 2E0HWN
NE8 2BG M3HWN
NE8 2JB M0GRI
NE8 2SY M7BRY
NE8 3RS M0LBD
NE8 4AJ G0NPQ
NE8 4DX G7SPN
NE8 4DX M0GGP
NE8 4DX M3NLA
NE8 4JA G7KLS
NE8 4UH G6DJP
NE8 4XH M3IBE
NE9 5DP G4WAX
NE9 5EU G8PDE
NE9 5PY M3XWK
NE9 5SB M0SUY
NE9 5TX G8XLZ
NE9 5XL M6RFL
NE9 5YN 2E0MWH
NE9 6DA G3IPD
NE9 6DH G0JHU
NE9 6HP G7TID
NE9 6JJ 2E0AOS
NE9 6JJ 2E1EYF
NE9 6NP M3XNN
NE9 6QN 2E0TTY
NE9 6QN M6ADE
NE9 6SN M6BGC
NE9 6TU 2E0EAU
NE9 6TU M6AQN
NE9 6TZ G0SQE
NE9 6UX G0HBO
NE9 7DX 2E0XQK
NE9 7DX M3XQK
NE9 7EX G7VDK
NE9 7JD G7MKB
NE9 7NQ G1DZY
NE9 7PG G0SLQ
NE9 7RE G6UJJ
NE9 7TR G0NPP
NE9 7TY 2E0CRN
NE9 7TY M6SRF
NG1 5BL M0IDW
NG1 5BL M7KBT
NG10 1DR 2E0DMM
NG10 1DX G6DRH
NG10 1GB M6BYN
NG10 1JH M6GTU
NG10 1JH M6HZJ
NG10 1NL G3REU
NG10 2BS 2E1HID
NG10 2BY G0CQR
NG10 2DZ G0FFQ
NG10 3BT G6AGR
NG10 3DG 2E0SMA
NG10 3DG M6WXN
NG10 3DJ 2E1AKK
NG10 3EW G1OPG
NG10 3FP M3IDO
NG10 3GF G0PSI
NG10 3GF G4AL
NG10 3GF G6MDR
NG10 3GG M3VBV
NG10 3GH G4OSR
NG10 3LF 2E0USM
NG10 3LF M0WUD
NG10 3LF M6USM
NG10 3NA M7BOZ
NG10 3NL 2E0STT
NG10 3QE 2E0BQW
NG10 3QE M3SVP
NG10 3RG G8PTN
NG10 3RJ 2E1DWW
NG10 4DA G4IVO
NG10 4DD 2E0VKM
NG10 4DD M0OSL
NG10 4DD M6OOL
NG10 4DD M6YMB
NG10 4EB M3IZJ
NG10 4EW G6PVW
NG10 4GD 2E1GMQ
NG10 4GD M0BTG
NG10 4JA G7TZW
NG10 4JS G5OW
NG10 4NZ G4TYP
NG10 5BS G0IVI
NG10 5FF G4HIC
NG10 5FF 2E0MLV
NG10 5LQ 2E0XCH
NG10 5LQ M6BSV
NG10 5PD G4LOV
NG103GG 2E0NAF
NG103GG M3NAF
NG11 0HP G4BKQ
NG11 0HT G8LCU
NG11 6AA G3ZQH
NG11 6AJ G4AJJ
NG11 6GF G0PIB
NG11 6GG G0JYE
NG11 6LB G0KUZ
NG11 6ND G8GWP
NG11 6NP 2E0FIV
NG11 6NP M6OQT
NG11 6NS G7KNQ
NG11 6QD G4NNY
NG11 6QE G7HZZ
NG11 6QE M0PHX
NG11 6QE M0VRG
NG11 6QE M0YHA
NG11 6QJ G1LOL
NG11 7AU G4LPD
NG11 7AU G8HUR
NG11 7BE G0JPL
NG11 7BQ G7HUK
NG11 7BY M1ECT
NG11 7EB G8DD
NG11 7EB M0BWY
NG11 7FD G8LNG
NG11 7FD M6JRH
NG11 8FD G1PNX
NG11 8GF G0WXF
NG11 8GF G7SWE
NG11 8GN M6TLC
NG11 8JA M1BHP
NG11 8JG G7SMQ
NG11 8JN 2E0PJT
NG11 8JN M0PJT
NG11 8SL G0JVW
NG11 9AL M0BCW
NG11 9ET G7BGY
NG11 9HB M7TZW
NG11 9HE 2E1HXB
NG11 9HE M3HXB
NG11 9JB 2E1GHI
NG11 9JN G0LXX
NG11 9JZ M0BWW
NG12 1AY G0SGF
NG12 1AY G7ITW
NG12 1DG G4ZDF
NG12 1DP G1HTR
NG12 2BU G2BXH
NG12 2BU G6KDY
NG12 2BX G3YCE
NG12 2GA M1EJI
NG12 3DG M3UBG
NG12 3DJ G6OAN
NG12 3DP G1GSG
NG12 3FD G0MXX
NG12 3HT G7LFL
NG12 3JJ G6AFS
NG12 3JS G0MAY
NG12 3JY G3TNX
NG12 3NE G3FRE
NG12 4DN G0NPC
NG12 4DZ G7PVQ
NG12 4EA G4XRB
NG12 4ES M0BUY
NG12 4FW 2E1CEU
NG12 5BN G8ENB
NG12 5DA 2E1AEC
NG12 5DN G7ODB
NG12 5ET G0GWP
NG12 5HQ M1AZV
NG12 5JX G6PKY
NG12 5JY 2E0HIJ
NG12 5LQ G3KZX
NG12 5LQ M0RIA
NG12 5NX G3FNS
NG12 5NX G4BKI
NG12 5PY G1ZLB
NG12 5RA G4XZA
NG12 5RA M3WTB
NG13 0AR G0GRC
NG13 0AR G4ROB
NG13 0ED 2E0BUZ
NG13 0ED M3ZUB
NG13 0FP M6GJL
NG13 8NA M6EQJ
NG13 8QD G7JDI
NG13 8RL M0RMJ
NG13 8RL M1KEJ
NG13 8TY G4FAB
NG13 8YR M6FVN
NG13 9AF G0FOG
NG13 9HD G4OMJ
NG13 9HL M6ZWB
NG13 9HY G4TTS
NG13 9HZ G0USJ
NG13 9JF G6XSS
NG130AZ M6WGG
NG14 5AA G4PFF
NG14 5AZ G7DPV
NG14 5DB 2E0EOZ
NG14 5DB M0JOJ
NG14 5DB M6JUF
NG14 5DY G8KBM
NG14 5EH M3LVR
NG14 5FG 2E1CWQ
NG14 5HR G1FYU
NG14 6FL G8DVN
NG14 6HL G4IJU
NG14 6HY 2E0LRA
NG14 6NT 2E0VUB
NG14 6NT M0VUB
NG14 6PU G7TBP
NG14 6QA G6SCG
NG14 7AR M5ROB
NG14 7DJ M0GDP
NG14 7EF G8FGZ
NG14 7GW G1XOW
NG14 7HD G7BGT
NG15 0EG G8YLS
NG15 6DQ G4KJA
NG15 6DQ G4XXS
NG15 6ED M6SPL
NG15 6EP G4WXR
NG15 6FF G6DDP
NG15 6FU M0HCZ
NG15 6FU M6CGY
NG15 6FU M6HUC
NG15 6FY 2E1GDM
NG15 6GE M6LHP
NG15 6GN G4ZII
NG15 6HY G4EPL
NG15 6JW M0SCP
NG15 6RF M3MBH
NG15 7FR G0VNW
NG15 7PJ M6BHC
NG15 7QA G4JSM
NG15 7SJ M3EVC
NG15 7SL 2E0ETG
NG15 7SL M6CCA
NG15 7SR G6NHY
NG15 7TF G6ORH
NG15 7UB G0RGE
NG15 8BS M0JYQ
NG15 8BS M0JYS
NG15 8EB G0CEV
NG15 8EU 2E1CJB
NG15 8JA 2E1GSC
NG15 9AD G8MMP
NG15 9BE G4CMX
NG15 9DG G8EXS
NG15 9EA G0LUD
NG156SA M7SKS
NG16 1AB M7RJD
NG16 1AU 2E0DME
NG16 1AU M0JEJ
NG16 1EG G0EVN
NG16 1HE G0FJZ
NG16 1HE G1NBT
NG16 1HL M5AAW
NG16 1LE G4NUV
NG16 2AP M0GGH
NG16 2DP G8XPZ
NG16 2EN G3VYK
NG16 2FG 2E0NSC
NG16 2FG M0LAF
NG16 2FT M7SVE
NG16 2HA 2E0UVW
NG16 2LQ G1PMK
NG16 2PU M6BAN
NG16 2QX G7ODN
NG16 2RA G6MRN
NG16 2RD G7UII
NG16 2RH M0ANC
NG16 2TH 2E0FCS
NG16 2TH G6UFV
NG16 2TH M6FKZ
NG16 2UB M1EBW
NG16 3DL G6GCW
NG16 3DP G4JRJ
NG16 3DQ G8ZAU
NG16 3DR G0GYH
NG16 3DY G6HZX
NG16 3FR 2E0EHH
NG16 3FR M6NTR
NG16 3FY G1WSD
NG16 3GW G0BXG
NG16 3GW G7HHM
NG16 3GY G7HGF
NG16 3HR M6OQL
NG16 3JE G4RNR
NG16 3JW M3YRB
NG16 3LR M3AZH
NG16 3NL G4VAX
NG16 3QL G3RTO
NG16 3RB M6GTC
NG16 3RE M0ZCW
NG16 3RE M3JVA
NG16 4GJ 2E1IJY
NG16 4GJ G6MAW
NG16 4GP G6ZAF
NG16 5BG G4UFC
NG16 5EH G1OJQ
NG16 5EH G8DHB
NG16 5FN G0OBK
NG16 5GX G0LIQ
NG16 5HJ 2E1EVH
NG16 5JZ G0WXH
NG16 5JZ G7UVL
NG16 5PW G0WFO
NG16 6BQ G8JGF
NG16 6DX G4ROB
NG16 6DX G8RYK
NG16 6FA 2E1BJG
NG16 6FN G0NCS
NG16 6FP G0IZL
NG16 6FW M3UUL
NG16 6GN M6JYD
NG16 6GP G4NXB
NG16 6JR G0OKD
NG16 6JR M0HLF
NG16 6JX M0DEN
NG16 6LH G1ZAR
NG16 6LQ G7BHU
NG16 6QH 2E0SBJ
NG16 6QQ M0MRT
NG16 6RF G0IUV
NG16 6RJ 2E1GRA
NG17 1EP G4ZUC
NG17 1HA 2E0VCB
NG17 2BS M6FEC
NG17 2BU 2E0BEQ
NG17 2BU M3FBN
NG17 2DD M6WDL
NG17 2DG M6KCE
NG17 2DW G0PDM
NG17 2EH G0BTV
NG17 2FB G4DFZ
NG17 2FH M3FWR
NG17 2GG M5ADF
NG17 2HS G3JFD
NG17 2LL G6ZTH
NG17 2NP M6XCC
NG17 2QA G1OXT
NG17 2QD G1XVF
NG17 2QF G0JDG
NG17 2QF G0RPX
NG17 2RA 2E0CLL
NG17 2RA 2E0XBX
NG17 2RA G4TSN
NG17 2RA M0TSN
NG17 2RA M3TSN
NG17 2RE M6ISX
NG17 2SB G6SGU
NG17 2SJ G8XGW
NG17 2SJ M7VMB
NG17 3AB G0UJD
NG17 3AP G4UDN
NG17 3AR M3VLN
NG17 3BW M0AJX
NG17 3DN M5ABH
NG17 3DP G0IUN
NG17 3EB M0AGA
NG17 3EJ G8EPH
NG17 3FF G6ROS
NG17 3FF M6NNC
NG17 3FL M6KTN
NG17 3FT G0BCF
NG17 3HT G4REU
NG17 3JY M6GQR
NG17 3NQ G4TWW
NG17 4AY G1XNG
NG17 4AZ 2E0EOF
NG17 4BA M6AET
NG17 4BE G4HCD
NG17 4BX G7RXK
NG17 4BY G0UTN
NG17 4EH G0LHL
NG17 4EN G3RLJ
NG17 4EY M6TCF
NG17 4EZ G7PBV
NG17 4EZ M0HQG
NG17 4GA G1RBH
NG17 4GG M0XWD
NG17 4GG M3LIW
NG17 4HQ G6NWS
NG17 4HX G7PHL
NG17 4JA M3EZY
NG17 4LG 2E0FKN
NG17 4LL M6KFF
NG17 4NL G6NWN
NG17 5AR M0GOH
NG17 5AX M3ZGI
NG17 5BA M6NOC
NG17 5BA M7LMC
NG17 5BD G7INC
NG17 5BL M6CZA
NG17 5EH M3ZGF
NG17 5GD 2E1SCR
NG17 5GD M7SCR
NG17 5GH G3VDF
NG17 5HP G1SIU
NG17 5HU G0GZO
NG17 7EH G1VHY
NG17 7FH G1RJD
NG17 7GE G1KAS
NG17 7HB M0DYQ
NG17 7HF G6OZH
NG17 7JE M6AEY
NG17 7JW M0CWZ
NG17 7LY G3SQQ
NG17 7QB M3KHE
NG17 8AD G7JZC
NG17 8BA M6YJB
NG17 8BT G0JPZ
NG17 8DP M0BNF
NG17 8ED M0CVO
NG17 8EJ G7LHT
NG17 8FP M3HDL
NG17 8FR M6AGZ
NG17 8FU M3JZN
NG17 8FU M3KCU
NG17 8FX 2E0IQO
NG17 8GE M0SRO
NG17 8GF G6UIP
NG17 8GH M0RMF
NG17 8GP M6FZA
NG17 8JJ G4VOG
NG17 8JT G0DMN
NG17 8LH G0NRA
NG17 8LH G0NZA
NG17 8NW G6UDB
NG17 8QB G7VJQ
NG17 8RL M7FYE
NG17 8RS G3LNN
NG17 9AH M3JZO
NG17 9AH M3JZP
NG17 9BG G6XNN
NG17 9BN G6XYR
NG17 9BR M0RBB
NG17 9BU M0BAE
NG17 9DN G1EZU
NG17 9DY G0CVB
NG17 9ET G8ZUZ
NG17 9FE M6NEM
NG17 9FL 2E1CQM
NG17 9HE 2E0UBN
NG17 9HE M0ORY
NG17 9HE M6UBN
NG18 1QE M3SQE
NG18 2DG M6PTI
NG18 2HP 2E0JLK
NG18 2HP G7TOY
NG18 2HP M6RRC
NG18 2LQ 2E1FTE
NG18 2LQ M1BVP
NG18 2NF M3NOY
NG18 2NF M6XDX
NG18 2PX G0HRO
NG18 2QL M6RZI
NG18 3AZ M3HNV
NG18 3BL M3FLE
NG18 3BY G1KGQ
NG18 3DD G0PGA
NG18 3DU 2E0JNE
NG18 3DU M7GIF
NG18 3EN G4NOR
NG18 3JJ G4VWA
NG18 3JL G4LBY
NG18 3NP G0JRC
NG18 3NP G0KJC
NG18 3QP G0JVK
NG18 3RN 2E0FJQ
NG18 3RN M6PGU
NG18 3RZ G1DRR
NG18 3SP M1CBH
NG18 4EQ G4ZWI
NG18 4ER G4DFV
NG18 4FA G1EIV
NG18 4FB G7ROI
NG18 4HD G1LPQ
NG18 4HF G4AAH
NG18 4HF M3CSM
NG18 4HG G0VYT
NG18 4NB G0RWW
NG18 4PG G6RYW
NG18 4QD G3XDS
NG18 4RS M7BNP
NG18 4RT G4GZU
NG18 4XS M6CZF
NG18 5EE G3GQC
NG18 5EE M0DFX
NG18 5EE M1ARF
NG18 5HY G4LQP
NG18 5LA G4HTO
NG18 5LL M5DLA
NG18 5NB G0RDP
NG18 5NG G1ECS
NG18 5QS G4TGB
NG18 5RG G7VKG
NG18 5RG M6VKG
NG19 0AJ M6TLB
NG19 0BX G7GGG
NG19 0DS 2E0BSN
NG19 0DS M3ZEY
NG19 0DW M0CFD
NG19 0EJ G3VIY
NG19 0EY G1HLT
NG19 0GP G1MMD
NG19 0HZ G0ELB
NG19 0JP G4TUX
NG19 0LR 2E0GBV
NG19 0LR M0HOB
NG19 0LR M6CIU
NG19 0LS G4LQL
NG19 0NA 2E0RHP
NG19 0PP G0TUP
NG19 0QJ 2E0DAO
NG19 0QJ M3NZR
NG19 6BY G4VUM
NG19 6EG G1ZLY
NG19 6EL M6TRD
NG19 6HZ G7AVZ
NG19 6JU M3HNQ
NG19 6LS M6NGP
NG19 6NB G4ANU
NG19 6NN M7HOU
NG19 6NT 2E0EVA
NG19 6NT M6EVA
NG19 6QH M7CBL
NG19 6QQ G4ZSD
NG19 7JB M6NHT
NG19 7JW M3LBY
NG19 7LX M3CDE
NG19 7NP 2E1EIX
NG19 7NP M3OQG
NG19 7PY G6JFL
NG19 7QQ 2E0MLF
NG19 7RG 2E0SWN
NG19 7RG G0RRZ
NG19 7RG G6EAH
NG19 7RG M0SXN
NG19 8AZ G1FEX
NG19 8DJ M3CBV
NG19 8HT G0GAG
NG19 8PA 2E0SUK
NG19 8PA G1UK
NG19 8PA M6VUK
NG19 8QJ M3KAN
NG19 8QT G1IMY
NG19 8QT G1JGY
NG19 8QT G4BAS
NG19 8QU 2E1EYC
NG19 8QU M3MKD
NG19 8QZ G4PPH
NG19 8TL G4OIE
NG19 9AB M7BNZ
NG19 9AZ G0OYP
NG19 9DW G1WTB
NG19 9HA G7CKL
NG19 9HD M7DON
NG19 9JR G6XMB
NG19 9NA G3VVE
NG19 9NB G4RPD
NG19 9PJ M3GJW
NG190NW M0IYJ
NG2 2EB M0DSN
NG2 3ER M7LFI
NG2 3GD M0GLP
NG2 4AE G6RFJ
NG2 4AS M6PEG
NG2 4BS 2E0FIU
NG2 4BS M6NWN
NG2 4GF M0JOH
NG2 4LN G7FCC
NG2 4NY G0HMO
NG2 4QD 2E0BRP
NG2 4QD M3WWU
NG2 5BD G7HCJ
NG2 5DS G1TTB
NG2 5FU G1HXP
NG2 5GB G4VOW
NG2 5GB M3GCP
NG2 5GB M3JVP
NG2 5HA G7OLX
NG2 5LA M0AVH
NG2 5LX M1SCW
NG2 5NA G4NDM
NG2 6EB G3WNC
NG2 6FH G4ZSO
NG2 6FQ G3VLN
NG2 6GJ G4VUI
NG2 6HQ 2E0BBX
NG2 6HQ M0GWR
NG2 6HQ M3VJS
NG2 6HQ M3XHC
NG2 6LD M0SMH
NG2 6LD M6BMY
NG2 6LD M6SKL
NG2 6NA M6LOL
NG2 6PE G8ZSD
NG2 6PE M3ZSD
NG2 6QG G8STJ
NG2 6SG G0RMN
NG2 7FD G0TCF
NG2 7FL G8KJJ
NG2 7GG G4NZU
NG2 7GN M1FDH
NG2 7LU G8GVL
NG2 7ND 2E0XIK
NG2 7ND G3MOT
NG2 7ND M0XIK
NG2 7ND M3PMI
NG2 7ND M3XIK
NG2 7ND M6WRG
NG2 7QN G6RNF
NG2 7QR M6GIY
NG2 7RX M6JRW
NG2 7UQ G8TVM
NG20 0BN 2E0MIB
NG20 0BN M0MIB
NG20 0BN M3VUZ
NG20 0BP G7TTY
NG20 0DJ 2E0SPN
NG20 0DJ G0TLT
NG20 0DJ M6HKA
NG20 0DN G7MNQ
NG20 0HJ 2E1FOU
NG20 0LD 2E1HWQ
NG20 0ND M6DEE
NG20 0PS 2E0DNJ
NG20 0PS M6CBN
NG20 0PS M6TAJ
NG20 0PT G0CAL
NG20 0PT M6LCP
NG20 0PY M6NCB
NG20 8AZ G8UST
NG20 8AZ M6EJB
NG20 8BJ M0DHI
NG20 8EQ G6JCV
NG20 8GA G0UKL
NG20 8JS 2E0XKS
NG20 8JS M7AVC
NG20 8LF 2E0SGI
NG20 8NH 2E0CAS
NG20 8NH M0TTY
NG20 8PJ M0ROY
NG20 8QH G4VLK
NG20 8XN 2E1WPW
NG20 9EB G4ODF
NG20 9PJ M0IFH
NG20 9PJ M3KBF
NG20 9PJ M3NHD
NG20 9PY G0ENV
NG20 9QN 2E0VKB
NG20 9QN M6WKW
NG20 9RU 2E0JJS
NG20 9RU M7OJS
NG21 0AR G4FHK
NG21 0AR G8AOI
NG21 0AU G7KWA
NG21 0ax G7BXG
NG21 0DS M0TCA
NG21 0EU 2E0XXI
NG21 0FB 2E1HGY
NG21 0FB M0BPY
NG21 0FU M0JFB
NG21 0GD G4WHM
NG21 0GH 2E0CVD
NG21 0GH M6TBR
NG21 0HX G7LAK
NG21 0JJ G6SFE
NG21 0NA M6TRW
NG21 0NW G8KPV
NG21 0RN G7SVU
NG21 0RW 2E1IDE
NG21 0SJ M5TMG
NG21 0TL G7PMF
NG21 0TU G0PBW
NG21 0TU M6PBV
NG21 9AE M7XSB
NG21 9ah 2E0ELE
NG21 9ah M6HWE
NG21 9AZ M7BEG
NG21 9DS G1JXG
NG21 9EA G0CCV
NG21 9LE M1AEB
NG21 9LQ G4WHL
NG21 9NH G0SYS
NG21 9NH G1SIX
NG21 9NJ M6JNE
NG21 9NZ M3KOR
NG21 9PF M7CJX

NG21 9QG 2E0BWX
NG21 9QG M6TWW
NG21 9QT M1ANR
NG21 9QZ G0BYX
NG22 0DA G7CTN
NG22 0HB M3BIR
NG22 0HF G4ODD
NG22 0HP M6MEK
NG22 0JR G6ZOB
NG22 0JR M3KBG
NG22 0PN G7PFD
NG22 0PP G4TDZ
NG22 8AP G4SCL
NG22 8DQ G1WRO
NG22 8LU G8BRG
NG22 8PB G6FLQ
NG22 8SD G3ZVV
NG22 9AS G7ACA
NG22 9BD 2E1HVB
NG22 9DG M0AXX
NG22 9DX G8YPQ
NG22 9HP M0HHF
NG22 9PQ 2E1FBA
NG22 9PS 2E0HVM
NG22 9PS M0SDF
NG22 9PS M6HVM
NG22 9PU 2E0AVA
NG22 9QX 2E0TBD
NG22 9QX M3UIS
NG22 9QX M6TRV
NG22 9QY 2E0SHA
NG22 9QY M6JLR
NG22 9RN 2E1FRZ
NG22 9RW M6DXV
NG22 9RZ M0ADB
NG22 9SG 2E0KOR
NG22 9SG M3KOU
NG22 9SX 2E0FBF
NG22 9SX 2E0HUN
NG22 9SX M6BZK
NG22 9SX M6IAQ
NG22 9TJ G0BYQ
NG22 9TN M3IDF
NG22 9TN M3KBE
NG22 9TN M3KPL
NG22 9UL M0HXH
NG22 9UP G1DQF
NG22 9UU 2E1HIQ
NG22 9UU M0BDD
NG23 5AL M6OOV
NG23 5BA G4BPE
NG23 5EG 2E0JMF
NG23 5LB M6MXR
NG23 5NT M0BCF
NG23 5PQ 2E0RVV
NG23 5PQ 2E0SVV
NG23 5PX M3IKV
NG23 6BP G3XVX
NG23 6JJ M7GPY
NG23 6LH M0TVS
NG23 6QA G4LFT
NG23 6QG 2E0FGA
NG23 6QG M6FGA
NG23 6ST G8TNB
NG23 7AA G1LVH
NG23 7ED G6MLH
NG23 7HR M0VOS
NG23 7HR M6VOS
NG23 7LD G0DWB
NG23 7NT G7NDS
NG23 7PR G7SYQ
NG23 7RA G3HLG
NG237JH M7DZU
NG237RL G3WZJ
NG24 1DF G4TXO
NG24 1DF M0RHQ
NG24 1FN G0JAP
NG24 1FW G1VKN
NG24 1QZ 2E0FKH
NG24 2BU G7UXR
NG24 2FX M6JNM
NG24 2HA M6CJJ
NG24 2HT G7MUB
NG24 2NL G3YWS
NG24 2NT G3XFU
NG24 2NT G7KFM
NG24 2NT G7KFN
NG24 2RX G4PCP
NG24 2SA G0ORP
NG24 2SU M0NWK
NG24 3AZ G1KNQ
NG24 3DA M3MOQ
NG24 3DT M0BWZ
NG24 3FJ G4NSW
NG24 3HH G6PII
NG24 3LY G0BVU
NG24 3NS G8EGU
NG24 3NU M7OBX
NG24 3NZ G0LBB
NG24 3TZ M0KAW
NG24 4BZ G7DEH
NG24 4DT G4KTG
NG24 4DY G4ZHG
NG24 4EX M3OVM
NG24 4EX M3YHD
NG24 4JL M6CVU
NG24 4JL M6CVV
NG24 4NY M3OIC
NG24 4PD 2E0IID
NG24 4PD G1AND
NG24 4PD M7AQQ
NG24 4PP M6JAM
NG24 4QB G0MCP
NG24 4QB G7HOT
NG24 4QB M6FXX
NG24 4RW G7BEJ
NG24 4TA M7ANY
NG24 4UA G0PUY
NG25 0AG G8UBF
NG25 0AH M0IUA
NG25 0BG M3DQJ
NG25 0EQ G4XQY
NG25 0EQ G6WYE
NG25 0JF M7RGS
NG3 2LN M6CWI
NG3 2LP 2E0TEW
NG3 2LP M0JSW
NG3 2LP M6BUD
NG3 2LS G0MLM
NG3 2PE G7SCN
NG3 3AN G1NKV
NG3 3EQ G0WYI
NG3 3LL M7BAS
NG3 4PZ G1MQB
NG3 5EN G8CXV
NG3 5FD G4ZDX
NG3 5HY G4PJZ
NG3 5QB G8FWH
NG3 5QW G8ZUU
NG3 5RG M3LEF
NG3 5SE G4DFU
NG3 5UX M7DZO
NG3 6AD G3EKW
NG3 6AR G8BPQ
NG3 6DH G0JOX
NG3 6DT G6NLU
NG3 6DY 2E0DLO
NG3 6DY M0SYS
NG3 6DY M6EYP
NG3 6EE M3EHJ
NG3 6EF G0MAD
NG3 6EF G4NRZ
NG3 6EU G6WHS
NG3 6FL G4ZTY
NG3 6FT G3TWB
NG3 6JA G8MIW
NG3 6LR M1JHL
NG3 7AP G3SEN
NG3 7BH G1IKH
NG3 7BY G4SGG
NG3 7FY 2E1HHG
NG3 7HD G4SJG
NG31 6JG G7ELG
NG31 6LE M6HET
NG31 6LT M3ZRP
NG31 7AZ M0GTV
NG31 7BP M7KCS
NG31 7BW M7XTB
NG31 7BX M6GJV
NG31 7HH G8WWJ
NG31 7JG M0SFA
NG31 7JL G6IPW
NG31 7LT 2E0CMF
NG31 7LT G4ENS
NG31 7LT G8ENS
NG31 7LT M0SDM
NG31 7LT M0SKR
NG31 7LT M6AZT
NG31 7LT M7DXH
NG31 7NH M6WDR
NG31 7NH M7BZV
NG31 7NN G0RCI
NG31 7NN G1EUU
NG31 7PH G1UTZ
NG31 7RB 2E0TYT
NG31 7RB M6GBI
NG31 7RB M6MSJ
NG31 7RH 2E1DCV
NG31 7RH G7SJX
NG31 7RH M3SJX
NG31 7RH M3WPM
NG31 7XA 2E1AXL
NG31 7XG M0JHW
NG31 7XP M6SRG
NG31 7XR M3VUA
NG31 8AD G4GXI
Ng31 8aw M6HNX
NG31 8BN G4DZC
NG31 8GA M0AUS
NG31 8GH G6UOH
NG31 8JR 2E0MDE
NG31 8LN G4MRY
NG31 8LP G4VUA
NG31 8PD M0IGM
NG31 8QF M3OJX
NG31 8QL 2E0UMO
NG31 8QL M0KXL
NG31 8QL M7QEW
NG31 8RL G6YMD
NG31 8RS G0FSP
NG31 8XA M7CXY
NG31 9BL M0HNF
NG31 9BL M0SUR
NG31 9JD 2E1CSD
NG31 9JW G8SHE
NG31 9ND M3ZYX
NG31 9PE G4VMI
NG31 9PF M1DDB
NG31 9PF M6ISD
NG31 9QA 2E1EFG
NG31 9QG 2E0XBM
NG31 9QG M0XBM
NG31 9QS G1PML
NG31 9QS M0USY
NG31 9RD G3VJE
NG31 9RG 2E0ALF
Ng317eg M7DMI
NG32 1AU 2E0OBL
NG32 1AU M0OBL
NG32 1ET M3XZJ
NG32 1HX M5DJC
NG32 1TG M3YGQ
NG32 2EB M0TGX
NG32 2EB M6AFS
NG32 2ES M6FRR
NG32 2FG M3SRY
NG32 2NS M3PTI
NG32 2PD G4RIO
NG32 2PD G6JQH
NG32 3DF G4HVC
NG32 3EJ G3TBK
NG32 3EJ M0TBK
NG32 3EJ M3DVQ
NG32 3EJ M3TBK
NG32 3NR G0PJS
NG32 3RR 2E1EBR
NG32 3RU G0OJF
NG32 3SJ G0BUB
NG321AU G8OCO
NG33 4HA G4FGY
NG33 4LX G0FZG
NG33 4RT G0KQK
NG33 4SB G3KHZ
NG33 5FE G4OQR
NG33 5HG G0BHT
NG33 5PU G0GRU
NG33 5PU G8STM
NG34 0DL 2E0YNT
NG34 0DL M6YNT
NG34 0LD M0HXF
NG34 0LF M0MLW
NG34 0QG G4RIP
NG34 0RS G6KVE
NG34 0SE G3KVP
NG34 0TZ G6JUT
NG34 7HG G0BTT
NG34 7HW G7LPK
NG34 7JN 2E1FAN
NG34 7LQ G0CRE
NG34 7NW M6AQG
NG34 7NX G4FDR
NG34 7QP G2VS
NG34 7TD G3LSJ
NG34 7TF G4ODG
NG34 7WA G4SIJ
NG34 7WL M6JJX
NG34 7WL M6YBA
NG34 8AE G4GBE
NG34 8BJ G0BXU
NG34 8BJ G1LOW
NG34 8DB M1EGW
NG34 8EL G4OCU
NG34 8HG G0IBI
NG34 8HZ 2E0ENN
NG34 8HZ M6ENN
NG34 8JF G4PNX
NG34 8JF M6IOD
NG34 8QG 2E0CWM
NG34 8QG M0GUU
NG34 8QG M6GAC
NG34 8QQ M0XCR
NG34 8WR M7ANX
NG34 9AL 2E0GQI
NG34 9AL M6OLT
NG34 9AT G0SGP
NG34 9BT G1PGN
NG34 9EQ G7UOL
NG34 9EU 2E0MPE
NG34 9EU M6MYN
NG34 9FH G6UCW
NG34 9GA G3WYH
NG34 9GU G4GUC
NG34 9HG G1XRF
NG34 9HG G3PXV
NG34 9HS G0NVY
NG34 9HS G3KQB
NG34 9JE G0POM
NG34 9JG G0FLV
NG34 9JP G3ZUC
NG34 9JX M5AKT
NG34 9JZ G0DMJ
NG34 9PD M1JTA
NG34 9QR G4EMK
NG34 9QR G8ILU
NG34 9QY G4NYC
NG34 9QY G4UYF
NG34 9RX G6JTW
NG34 9SA G3SRX
NG34 9TS G3VHI
NG34 9TW G8YFP
NG348AW 2E0SHH
NG348AW M6SHH
NG36JX 2E0LWL
NG4 1BD 2E0BAF
NG4 1BU G6ZBO
NG4 1DA G6UZY
NG4 1DP G6JCI
NG4 1LE G0IBY
NG4 1LY 2E1BHU
NG4 2JG M3HWS
NG4 2JG M3YHU
NG4 2LW M0BCI
NG4 2QJ M3CZE
NG4 3DA G7TMF
NG4 3DX M1EDW
NG4 3EH G4ZCW
NG4 3ET G6MBI
NG4 3LF G6IOE
NG4 3QH G7SBK
NG4 3SF M3CBH
NG4 4AD G4EUJ
NG4 4AQ M1DJB
NG4 4AS M6FLQ
NG4 4AU G8LGY
NG4 4BB G8GCK
NG4 4BL G1EAB
NG4 4EA M6CCV
NG4 4GF M3UEE
NG4 4GS G4CPG
NG4 4HF 2E0HAW
NG4 4HF M3CJX
NG4 4HZ M3VRV
NG4 4JP G6YCL
NG4 4JP G7SFS
NG4 4PE 2E1ALX
NG4 4PE G2ALX
NG4 4PE M7PPG
NG4 4QE M1JCS
NG44AF M7JOX
NG5 1BH G1GJT
NG5 1DA G6BMZ
NG5 1DW M6EFL
NG5 1EP G6SRZ
NG5 1EP M0WAB
NG5 1GP G6ONW
NG5 1JR G4EKW
NG5 1NH G3TVY
NG5 1NW M1REJ
NG5 3DA G0DME
NG5 3GL M6TRI
NG5 3GS G1HRM
NG5 4BD G7LKV
NG5 4DY G0NVS
NG5 4GQ G4SQV
NG5 4HU G4EDX
NG5 4HY 2E0CAW
NG5 4HY M0SCO
NG5 4HY M3CAW
NG5 4JS G0JAC
NG5 4JX G4WDP
NG5 4LB G0KXZ
NG5 4NX M0JZM
NG5 4PG G4ERY
NG5 5BL G8VCI
NG5 5DT G6UZA
NG5 5DU G6URR
NG5 5EE G8CQQ
NG5 5FA G7ULJ
NG5 5JH 2E0ILD
NG5 5JH M7EPM
NG5 5NP 2E0LFK
NG5 5US M6ANY
NG5 6BX M0DLZ
NG5 6DJ 2E0AXT
NG5 6DP M6SAH
NG5 6DP M6ZBD
NG5 6DX 2E0BWH
NG5 6DX M3XBH
NG5 6EB 2E0EMF
NG5 6EB M3JWQ
NG5 6EH M6RSJ
NG5 6FN G4PNX
NG5 6FT G1FDO
NG5 6GU G6MVF
NG5 6LQ G4EAN
NG5 6NH G8SSL
NG5 6QA G7WKH
NG5 6QL G1UDX
NG5 6QT 2E1DLT
NG5 6QZ G1HTL
NG5 6RX G0RHJ
NG5 6SY G2FGT
NG5 7HF M3IVX
NG5 7HQ G4OOY
NG5 7LW G8UGS
NG5 7NF G4ITX
NG5 8AG G4EGY
NG5 8BD M7AOD
NG5 8BE M6HIK
NG5 8BP M6IXG
NG5 8FQ G8RBU
NG5 8GE M0PDG
NG5 8GH M0KIR
NG5 8JS G6YLX
NG5 8NZ G7IWK
NG5 8QE M3ROF
NG5 8QF G1GQQ
NG5 8QX G6IBN
NG5 8SD M7PDS
NG5 8SJ 2E0DSH
NG5 8SJ M0ZRD
NG5 9LN G7MGV
NG5 9LN M0BCH
NG5 9QN G1HMZ
NG5 9QU G7IZE
NG6 0BH G0KZA
NG6 0BS G1XHO
NG6 0EY G0FNV
NG6 0LS G0INA
NG6 7DL G4DIU
NG6 7FJ G1ZLA
NG6 8AY G8UQR
NG6 8DG G0GXH
NG6 8DT 2E1HAL
NG6 8NL M3MDW
NG6 8PF G1NXI
NG6 8PU G6VWV
NG6 8QY G0VFU
NG6 8SL G6NLD
NG6 8XA G6SFF
NG6 8XE G4PMM
NG6 8XN G0LCG
NG6 8XN G3RR
NG6 8XN G5RR
NG6 8XN G7HRR
NG6 9AP 2E0HET
NG6 9AP M0IQL
NG6 9AP M6YTX
NG6 9DT G1GTF
NG6 9FB G8ZZV
NG6 9FU G7SYD
NG6 9FX G1YWN
NG6 9HN G8OHC
NG6 9JE G6GWY
NG7 1RG G6COZ
NG7 3DF M6WMH
NG7 3GN G0JLR
NG7 5EB 2E1ECG
NG7 5FS G3WGH
NG7 6AD M6RIL
NG7 6HG M0IHC
NG7 6HL G1YBA
NG7 6HU G6ISB
NG7 7DR M3NQY
NG7 7LJ M6WAY
NG8 1AT 2E1IHY
NG8 1DE G1YEU
NG8 1JE M0GRH
NG8 1JU G6UAP
NG8 1LF M3MXH
NG8 1PU 2E0KNB
NG8 1PU M0KNB
NG8 2BH G6CIF
NG8 2BZ G4BWN
NG8 2EH G0FSJ
NG8 2ER G8UUS
NG8 2FB M6GHR
NG8 2NA G8BNG
NG8 2QR G4KLJ
NG8 2QR G8NWU
NG8 2RE G0FOE
NG8 2SB G8POL
NG8 3ES G3XBE
NG8 3FF M0EPR
NG8 3FF M0GZD
NG8 3FT M6MQP
NG8 3NE G4TYO
NG8 4AH M3JHL
NG8 4GY M0AUK
NG8 4LZ G1HRJ
NG8 4NB M6VCH
NG8 4PU 2E1FDT
NG8 4PZ M3CCP
NG8 5BE M6PIU
NG8 5EU G1XYG
NG8 5EU G7DXH
NG8 5FH G1HSP
NG8 5FJ G1RCN
NG8 5LF G7DSO
NG8 5ND M3CIO
NG8 5PN G1DNZ
NG8 5QG 2E0EHK
NG8 5QG M6GHN
NG8 6AD G6XJD
NG8 6AD G6XJE
NG8 6EX 2E0PJR
NG8 6EX M0LZM
NG8 6GP 2E0LZM
NG8 6GP M3XID
NG8 6GQ M0DAZ
NG8 6JT 2E0GAC
NG8 6LY G0TSG
NG8 6LY M3TSG
NG8 6NE M1BSI
NG9 1EU 2E0BCE
NG9 1FJ 2E0EEO
NG9 1FJ M0IOI
NG9 1FJ M6NOT
NG9 1GR G6JRE
NG9 1LR M3WSE
NG9 1LY G0CQT
NG9 1NA M3VIO
NG9 1NE G6XXJ
NG9 1NJ G6XOD
NG9 1NJ G6XOE
NG9 1NS 2E1GYO
NG9 1PY 2E1EQQ
NG9 1PY G0OQQ
NG9 1RJ M0DLR
NG9 2AD 2E0PHS
NG9 2AD M6EET
NG9 2BA M7SSO
NG9 2FJ G0GRZ
NG9 2GU M6DZN
NG9 2QR G4JHN
NG9 2QZ M3RNU
NG9 3BB 2E0OBS
NG9 3BB M6CAH
NG9 3BY G7LNV
NG9 3FN M0DAC
NG9 3FP G0PXP
NG9 3FS G0SKW
NG9 3FS G5NKW
NG9 3JE G8PBH
NG9 3JN G4MFV
NG9 3JT G1PUZ
NG9 3JW M3HPM
NG9 3LD M0JEL
NG9 3LW G8BFM
NG9 3PF G0PCK
NG9 3PX G4GQY
NG9 3RB 2E0YZZ
NG9 4BB G3EJH
NG9 4BB G4VFK
NG9 4BB G8ZK
NG9 4ED M3UVQ
NG9 4ES G0UYV
NG9 4FG M3RUI
NG9 4FH G7NGX
NG9 5AE M0GHX
NG9 5EB G8ZSZ
NG9 5EU G4AQT
NG9 5EZ G4BNX
NG9 5FH G6BRP
NG9 5FJ G7TRB
NG9 5FU M6STH
NG9 5GR 2E0GSW
NG9 5GR M0RMY
NG9 5GR M6CTX
NG9 5HS 2E0FBL
NG9 5HS M6ENH
NG9 5HX G0SPA
NG9 5HY 2E1CHX
NG9 5HY G7USM
NG9 5LA G6UTO
NG9 5LA G7HIT
NG9 5LB 2E1SIS
NG9 5LB M0OIC
NG9 5NH G6IPC
NG9 6AB G6RCD
NG9 6BP G8IYZ
NG9 6EW G4KQP
NG9 6FW 2E1EAV
NG9 6FZ M5ADM
NG9 6GN 2E0DWB
NG9 6GN M3ZNR
NG9 6HP G0RSU
NG9 6HS G6UWO
NG9 6JW G4XEL
NG9 6JW G6MSC
NG9 6NH M6EDJ
NG9 7ET G0NVO
NG9 7FY G4OYO
NG9 7HN 2E0GIH
NG9 7HN M3GIH
NG9 8AR M7LWP
NG9 8BN M1EWD
NG9 8DJ G4PMA
NG9 8GH M7HLN
NG9 8GL G0UYQ
NG9 8GL G8EHX
NG9 8HR G3JTO
NG9 8HR M1ERF
NG9 8LN M7RDN
NG9 8LT G7LCV
NG9 8PQ G0SLZ
NG9 8QG G4TYN
NN1 3BL G0PYI
NN1 3BP M7AKG
NN1 3QL M3NNH
NN1 3QL M6EMA
NN1 4HU G3YOV
NN1 4HZ M6AIP
NN1 4LU G1ICH
NN1 4NA M0ARZ
NN1 4PU M3KUG
NN1 4QR M3CET
NN1 4SP 2E0BDD
NN1 5HP M6INW
NN1 5JS G4PXJ
NN1 5JY M3WUA
NN1 5LT G7KGP
NN1 5LT M3GTG
NN1 5LT M3NEC
NN1 5LT M5TTT
NN1 5LU M6LRB
NN1 5ST 2E1BCC
NN1 5ST G3VMU
NN10 0AS M0DOK
NN10 0AT M0DOD
NN10 0DH M3NLI
NN10 0DX G6VFC
NN10 0DX 2E0HML
NN10 0DX G7PYQ
NN10 0DX M7PAV
NN10 0EA M0IXF
NN10 0EE M7HJB
NN10 0EG M6JCT
Nn10 0ft M6RVG
NN10 0GE G6EQS
NN10 0GS G1GPE
NN10 0JH G3PRU
NN10 0SJ G4BHT
NN10 0SN G8LII
NN10 0SW G4FEV
NN10 0SW G4WFT
NN10 0SY G3WDG
NN10 0SY G4KGC
NN10 0TN M7AXV
NN10 0TZ M3UPJ
NN10 6BA G7TZD
NN10 6BG G3OMS
NN10 6BU M3CQW
NN10 6EH G7APQ
NN10 6UR M6FPW
NN10 8DH G6NHO
NN10 8FJ G7IWM
NN10 8HF G1RTX
NN10 8HU G4ZRM
NN10 8NH G0WTO
NN10 8NH G7CMI
NN10 9EG M6JXI
NN10 9EZ G6SDI
NN10 9HF G7LNP
NN10 9HH 2E0TSO
NN10 9HH M0TSA
NN10 9HL G1OOS
NN10 9HS M7TCK
NN10 9LG G4NXV
NN10 9LN G6GOX
NN10 9LW 2E0LLI
NN10 9LW M6FIV
NN10 9NS G7OCC
NN10 9PF G0PYS
NN10 9QA 2E1CBU
NN10 9QA G4IRD
NN10 9RP 2E0ERG
NN10 9RP M6RAH
NN10 9SX G7NBG
NN10 9SX M3NBG
nn108jr M7PLO
NN108PH 2E0ILE
NN108PH M7ICK
NN11 0GR G1ZJK
NN11 0GR M1BXD
NN11 0NZ G8KHF
NN11 0UN M3BXE
NN11 0WQ M6AFL
NN11 0XH 2E0CTR
NN11 0XH M3PMY
NN11 0XT M3FKK
NN11 2JH G4SEV
NN11 2JW M0PXL
NN11 2JW M6LZS
NN11 3AD G1WVS
NN11 3AD M3HUX
NN11 3BW G4TBN
NN11 3ES G3YHI
NN11 3PR G4KQH
NN11 3PX M7LDV
NN11 3QP 2E0LMR
NN11 3QP M6CNG
NN11 3QS M6BPI
NN11 3YT G3MRQ
NN11 4AL G1DLH
NN11 4AL M3NUM
NN11 4AN M3FNY
NN11 4AQ G0HWT
NN11 4EY G7KRB
NN11 4GW G8VQQ
NN11 4GY G6XLB
NN11 4HJ G0DRE
NN11 4HJ G0FPM
NN11 4JQ 2E1FSV
NN11 4JQ M1CUX
NN11 4LL G0IOH
NN11 4NW 2E1AXD
NN11 4NW 2E1AXE
NN11 4NW 2E1CZF
NN11 4NW G5XX
NN11 4NW G6EGO
NN11 4NW G8GVV
NN11 4PR G7LII
NN11 4PT M3XTT
NN11 4QF M6FZR
NN11 4QU G1DIO
NN11 4QU G6MTE
NN11 4RX G7KJX
NN11 4SU G0NNG
NN11 4SX G3WTN
NN11 4TD 2E0BHC
NN11 6UR G7OLT
NN11 6UZ 2E0MBX
NN11 6UZ M0MBB
NN11 6UZ M6MBD
NN11 7HE M6YAS
NN11 7HQ G8KTA
NN11 7HY M1AQX
NN11 7JN 2E0AWS
NN11 7JN M3BSI
NN11 7JS G4BFT
NN11 8BH G6LUY
NN11 9AL M3KZI
NN11 9BT G7JDA
NN11 9EB 2E0HDD
NN11 9EB M7ISA
NN11 9HB M6NRR
NN11 9WH M3IQS
NN117HU 2E0GGQ
NN117HU M6GGQ
NN12 6AG M3HEO
NN12 6AW G3UMT
NN12 6AW G8CHK
NN12 6DN G1YMJ
NN12 6DN G6ZGA
NN12 6ED 2E0GNH
NN12 6ED G6RHJ
NN12 6ED M6JXE
NN12 6EZ M3NLQ
NN12 6NX M3SBA
NN12 6PL M5GDK
NN12 6QQ G4GWT
NN12 6RD 2E0YLP
NN12 6RD M6YLP
NN12 6RL 2E0TOY
NN12 6RL M0XER
NN12 6TH 2E0VLL
NN12 6TH M3VLL
NN12 6UP M3FEC
NN12 6YY G1EBX
NN12 7NA G8MQY
NN12 7NW M1DJC
NN12 7NX M6GQB
NN12 7PH G4IUH
NN12 7PH G8AAT
NN12 7PP G1MZD
NN12 7QX M6GDQ
NN12 7RS G1TAI
NN12 7SA G1DJU
NN12 7TY G4EQX
NN12 8AL G4DMS
NN12 8LU G4UXO
NN12 8NA G7ECG
NN12 8NG M1BXC
NN12 8NW G0FKW
NN12 8UY G4DQQ
NN12 8XJ M3KVG
NN128UP 2E0VGA
NN128UP M7SDI
NN13 5HU G0UID
NN13 5JS M6FEX
NN13 5LX G4EBF
NN13 5TW M6HXU
NN13 6AQ 2E0BGZ
NN13 6AQ M0RGD
NN13 6AQ M3IIA
NN13 6BP M6SKX
NN13 6DA M3ULM
NN13 6GU M7HSI
NN13 6HA G3NAN
NN13 6JD M7BSI
NN13 6JH M1CLO
NN13 6JQ M0RCY
NN13 6JQ M6CKZ
NN13 6JS G6HKH
NN13 6LY G8JDC
NN13 6ND G4IYA
NN13 6NE 2E0CZN
NN13 6NE G1BZM
NN13 6NE M0IFB
NN13 6NE M6DLW
NN13 6NS M6OYD
NN13 6PA M7EKV
NN13 6PB G4DWC
NN13 6PE M1DIM
NN13 6PE M3YWI
NN13 6RA G1BPV
NN13 7ET G8OZH
NN13 7NJ M3UBF
NN13 7TY G1EPO
Nn135nd M7BFJ
NN14 1BG G4HND
NN14 1BG G7TZZ
NN14 1EE G8ZIH
NN14 1EL G0GRS
NN14 1HG G0NFL
NN14 1LN M7FWH
NN14 1NG 2E0GGM
NN14 1NG G4RPT
NN14 1NG M0YGG
NN14 1NG M6GGM
NN14 1NS M7MPS
NN14 1RQ M0CKP
NN14 1RQ G1NEN
NN14 1SL G4UYM
NN14 1TB G3VSU
NN14 2FH G6GDR
NN14 2LB G0FFB
NN14 2LB M3PSM
NN14 2LQ 2E0TJP
NN14 2LQ M3XDZ
NN14 2LU M7BTZ
NN14 2NH G0FSA
NN14 2NR G4ASH
NN14 2QB 2E1EFT
NN14 2QU G1LMN
NN14 2QU G1KCV
NN14 2QY G4PNL
NN14 2RE G1OET
NN14 2RE G5KN
NN14 2RQ 2E0WPT
NN14 2RQ M6MPT
NN14 2SQ M1PAH
NN14 2TZ M6RUI
NN14 2UD G1WPR
NN14 2UD G8JKB
NN14 2UY G7NBV
NN14 2XA G1IHY
NN14 2XB M6FOZ
NN14 2XG 2E0ZAH
NN14 2XG G6DHW
NN14 2XL G8DXI
NN14 3BT G6SWJ
NN14 3BT M6JTC
NN14 3DD 2E0PNG
NN14 3DZ G8ZPE
NN14 3JT M0NIG
NN14 3LJ 2E0XBG
NN14 3LJ M6SFW
NN14 3LQ G4YNG
NN14 4BN 2E0NBI
NN14 4BN M7NIB
NN14 4DD G4HLW
NN14 4EA G6LJH
NN14 4FL 2E0ENI
NN14 4FL M0ISK
NN14 4FL M6FKR
NN14 4GN M0DSO
NN14 4JQ G8XQN
NN14 4JT G0CMP
NN14 4PY G3YQJ
NN14 4PZ G4PUZ
NN14 4RE G3WIA
NN14 4TB 2E0GDT
NN14 4TB M0UKO
NN14 4TB M6UKH
NN14 4XA M3UFZ
NN14 4XB G7OGL
NN14 6AJ M6OAN
NN14 6DZ G1TPC
NN14 6GE M7GDF
NN14 6HT G4DCD
NN14 6HT G8HBZ
NN14 6HY G6BKD
NN14 6HY G8BKD
NN14 6jh 2E0HJB
NN14 6LD M7WCF
NN14 6TN G1IQA
NN14 6YG G3WKR
NN14 6YQ G7RAE
nn142qu M6UKB
NN15 5BN G7AJJ
NN15 5BY G8WSV
NN15 5DE G7PQM
NN15 5DR G1PQR
NN15 5EG G0NSA
NN15 5EQ G4VKX
NN15 5HB G8YKE
NN15 5HD 2E0MNZ
NN15 5HD G4NZZ
NN15 5HD M7LJM
NN15 5HE M3JCA
NN15 5HT M0KIC
NN15 5HT M7KAD
NN15 5JB G7GMQ
NN15 5LJ M6SGP
NN15 5LS 2E0RBO
NN15 5LS M3YPW
NN15 5LS M6GIL
NN15 5ND G7WKV
NN15 5NT G6OLU
NN15 5PN M3LDM
NN15 5QB M6NDY
NN15 5QP M3XQO
NN15 5RS M3ZRM
NN15 5RX M0BKJ
NN15 5ST M3RHJ
NN15 5SX G1BJZ
NN15 5YL G4RLL
NN15 6HF G0DEF
NN15 6HH G1YLM
NN15 6HL M6ENB
NN15 6JS 2E0IZZ
NN15 6JS M6SUM
NN15 6NN M7HED
NN15 6PS G4TTX
NN15 6QH G7SKA
NN15 6RE G6JTC
NN15 6RG G4TXM
NN15 6SF G8XPD
NN15 6UW G4OYN
NN15 7AP G4NCA
NN15 7BA G1VPS
NN15 7BA M1ECD
NN15 7DR G0RDV
NN15 7DS G1XJN
NN15 7DU G1JSP
NN15 7DZ M1AWN
NN15 7EF G1AZD
NN15 7EZ G3VJG
NN15 7HX G7CJG
NN15 7LL G0PSG
NN15 7LL G0WWT
NN15 7LL G7NTG
NN16 0EL M6BIU
NN16 0HU M6MAL
NN16 0LA M6DAF
NN16 0LL G7VIH
NN16 0LY 2E0JLD
NN16 0LY M0JXD
NN16 0LY M6JLD
NN16 0NG M6RUH
NN16 0PH M0AQP
NN16 0QB G7API
NN16 0QR 2E0WDS
NN16 0QR M3UEZ
NN16 7JZ G8EZG
NN16 8EB G8FSJ
NN16 8PL G4LEM
NN16 9BU G7AJS
NN16 9EE M6WOE
NN16 9EW G4KXG
NN16 9EW G8RAV
NN16 9HA G0PBU
NN16 9JA G7DMK
NN16 9JL G0LDR
NN16 9JR G8WSU
NN16 9JZ G6PEG
NN16 9LA G4NAC
NN16 9LB 2E0OET
NN16 9LB M0OET
NN16 9LB M6OET
NN16 9PF 2E0XJR
NN16 9RF M6PAV
NN16 9SB G6OXN
NN16 9TG 2E0WIZ
NN16 9TG M6WIZ
NN16 9TW M6AHP
NN17 1EG M3NVV
NN17 1EN G6BHX
NN17 1ER G7NJZ
NN17 1HN G7NTQ
NN17 1JD G0IKR
NN17 1SY G7SCX
NN17 2AF G1TQR
NN17 2BS G1IIY
NN17 2DZ G4YRX
NN17 2LA G4PDK
NN17 2LH G1BKI
NN17 2LN G0PAD
NN17 2QJ M1BQS
NN17 2QP G6AFT
NN17 2RP G6RCY
NN17 3BA G3YNF
NN17 3BT G4JCJ
NN17 3BU G7CIV
NN17 3DN G4DNJ
NN17 3DN M3CCS
NN17 3HB G6VOE
NN17 3JL G7TIK
NN17 3LR M6NYZ
Nn172rd M6EJS
NN18 0AE M1CDP
NN18 0AG M3JBW
NN18 0BN G0NRB
NN18 0LG G0RRO
NN18 0NS 2E0VWF
NN18 0NS M7IWF
NN18 0TA G4OMV
NN18 8DB G1YTG
NN18 8DE G4XRD
NN18 8FZ G4NZE
NN18 8HX G3MTJ
NN18 8LL M6PSY
NN18 8NB M0RSH
NN18 8QR M7XST
NN18 9DE G6RFH
NN18 9EG G8SEV
NN18 9HH G0IYS
NN18 9JL G4XKD
NN18 9JS G4LKU
NN18 9PG M3TLT
NN18 9PH G1JNY
NN2 6EN M7ETN
NN2 6EP G6NYH
NN2 6RP G3KLV
NN2 6RP M0OXZ
NN2 7AT 2E0OAR
NN2 7AT M0LXV
NN2 7DA G0FJS
Nn2 7dn G3KAN
NN2 7ER 2E0GIY
NN2 7ER G5TEO
NN2 7ER M7TEO
NN2 7PP 2E0YXZ
NN2 7PP M6SNO
NN2 7PP M6YYY
NN2 7PT G0GAQ
NN2 7PX G3MJW
NN2 7QU G8ZMH
NN2 7QX G1IPD
NN2 7QY M3FLP
NN2 7RR M0EBU
NN2 7RY G0WYQ
NN2 7SP M0VQP
NN2 7SP M3VQP
NN2 7SP M6HDG
NN2 7TJ M3VIB
NN2 7TR G0NDS
NN2 7TR G6NDS
NN2 7TR G8WPU
NN2 8BJ G7OGN
NN2 8BX G4FTL
NN2 8DZ G3ZJO
NN2 8EA G1GOY
NN2 8EL M0GCR
NN2 8HD G0ONS
NN2 8HT G1ERY
NN2 8HT M7GHB
NN2 8JR M0DUU
NN2 8NB G4YJP
NN2 8PE 2E1CCF
NN2 8PH G8PTH
NN2 8QU G6TVB
NN2 8QU G8OZY
NN2 8SX G7AAY
NN2 8TX G7TEG
NN2 8TX M7MVS
NN2 8UU G8KNU
NN29 7AP M1TSU
NN29 7AT G8AMG
NN29 7DA G4PAV
NN29 7DH G6MKJ
NN29 7DP G0VOB
NN29 7DR G1ALL
NN29 7EA G8TFY
NN29 7EE G4RST
NN29 7EW 2E0KAK
NN29 7EW M0HAF
NN29 7EW M6AGK
NN29 7FD G0UBX
NN29 7HF M6RGR
NN29 7JU G0EAE
NN29 7JU G6NUS
NN29 7LP G7HBU
NN29 7LS G1ONQ
NN29 7ND G1RVF
NN29 7NS G0PSZ
NN29 7QE G1BNE
NN29 7QG G4TLT
NN29 7RR M6YPD
NN29 7RS G3JRE
NN29 7TP G1EUG
NN3 2AZ M0ECT
NN3 2BU 2E0CCY
NN3 2BU M3NWH
NN3 2DD G4VWF
NN3 2HE G0MJK
NN3 2LP 2E0BXV
NN3 2LT 2E0WSJ
NN3 2LT M3WSJ
NN3 2PG M6BGS
NN3 2PY M7SIN
NN3 2SY G0RTC
NN3 3AF G6AXH
NN3 3EQ 2E0SPZ
NN3 3FA G4MAB
NN3 3JE M3DYL
NN3 3JL G4MJF
NN3 3JX G1KMS
NN3 3NX M6KKY
NN3 3PB G4JFC
NN3 3PE G7UFW
NN3 5AD G1SLG
NN3 5AL M6WXP
NN3 5AY M7CCQ
NN3 5BH G1IRG
NN3 5BH G1NRG
NN3 5EP M7RMU
NN3 5ET 2E0ERO
NN3 5ET M6YPA
NN3 5EY G6VGS
NN3 5HY M6BOB
NN3 5JS G4OIG
NN3 5JS G8CXK
NN3 5LZ M6FEI
NN3 5NT M0DME
NN3 6AH G4ULI
NN3 6BL 2E0AAZ
NN3 6DN M3XHW
NN3 6HP G4EUR
NN3 6JQ G4SCJ
NN3 6JQ G6NHV
NN3 6SE M7HWB
NN3 7HU G6IVW
NN3 7LB M0MAO
NN3 7LB M0POI
NN3 7LB M6GMO
NN3 7RD G8FZW

NN3 7RE G6CQR
NN3 7TX G1DPJ
NN3 7UB M6BQJ
NN3 8AX M5AFY
NN3 8DE G3MBD
NN3 8DF G8INA
NN3 8HH M6MCR
NN3 8HY G7RFX
NN3 8HY G7RGU
NN3 8LD G0MHA
NN3 8LZ 2E0BKO
NN3 8LZ M3RGU
NN3 8PS M6EKM
NN3 8QT 2E1DNF
NN3 8UR G0PTM
NN3 8US 2E0UAP
NN3 8US M7UAP
NN3 8UX 2E1CAH
NN3 8UX G1BGF
NN3 8XN G8TOP
NN3 8YA M6MIO
NN3 9AX 2E1EET
NN3 9BH G6TVK
NN3 9BH M0JQO
NN3 9BH M7TWW
NN3 9BW M6SBU
NN3 9DJ G6OKH
NN3 9DN M7IRB
NN3 9EX M6KNZ
NN3 9EX M7XHC
NN3 9NT M3RCD
NN3 9SN M1BTU
NN3 9UG M6NIU
NN38QE 2E0PUN
NN38QE M3PUN
NN4 0NE M3HGM
NN4 0NJ G4LUW
NN4 0QQ G0NYE
NN4 0RS M0XDK
NN4 0SF G4YIP
NN4 0SS G6ILZ
NN4 0TS M3TGW
NN4 0WF G7CIA
NN4 0XX G0KAK
NN4 5AZ G0FQI
NN4 5BT 2E0WNO
NN4 5BT M7WXD
NN4 6AH G6NPZ
NN4 6AL G0FSM
NN4 6AP M6ROJ
NN4 6AP M6VHA
NN4 6DJ 2E0HSG
NN4 6DJ M6KDW
NN4 6ER G4DHV
NN4 6EW M3TRP
NN4 6HE G7NEG
NN4 6LA G0TMW
NN4 6UX M0KIW
NN4 7BT M3NAL
NN4 7PT G8ZEW
NN4 8AU G8HTO
NN4 8AZ G6FAL
NN4 8BY M0EAB
NN4 8LQ G6KIZ
NN4 8NR M6YYU
NN4 8PE G1TPN
NN4 8PX M7RET
NN4 8QB G1BHR
NN4 8QB M3BQT
NN4 8QB M3ZOR
NN4 8RU G4DTM
NN4 8RU M7LTM
NN4 8RX G4RKR
NN4 8ST 2E0DNL
NN4 8TQ M0STN
NN4 8TT M6JBV
NN4 9AU 2E0VNL
NN4 9QU M6VNL
NN4 9QW G4YFX
NN4 9SB M0WMT
NN4 9TX M6RGZ
NN4 9YH G4KFC
NN4 9YR M3KFP
NN40RQ G6LUU
NN5 4AJ G4MLO
NN5 4WB M0XKD
NN5 5AR M6RDF
NN5 5LW G1UQF
NN5 6BW G7IXK
NN5 6JT M6NKR
NN5 6JW G0RDT
NN5 6LR M3MKH
NN5 6NE G8OWG
NN5 6NH G7SYT
NN5 6NH G7VQE
NN5 6NH M3EKA
NN5 6NL G7HLP
NN5 6NR G6FGC
NN5 6PU G8KST
NN5 6PY G4NVP
NN5 6PZ G4NUF
NN5 6PZ G8KIW
NN5 6QP G7ISE
NN5 6QX G6TJE
NN5 6RA G1OMI
NN5 6RA G3RLD
NN5 6RF G0VJK
NN5 6TU M0DAE
NN5 6TW 2E1FUQ
NN5 6XP M7ETH
NN5 6XS G4ICC
NN5 6YW G0DLF
NN5 7AL M7ICS
NN5 7ER 2E0PHX
NN5 7ES G7SVT
NN6 0BD G4MTP
NN6 0BD G8SWL
NN6 0BD M0DON
NN6 0BY 2E0YXO
NN6 0HR G8SQA
NN6 0HU M7CNK
NN6 0HX G8LOP
NN6 0JW M7CPW
NN6 0NY G4LYC
NN6 0NZ G0JPF
NN6 0PF G6OVC
NN6 0PQ G7RTC
NN6 0PX G7OXY
NN6 0QS M6BKJ
NN6 6AE G4DLD
NN6 6AF 2E0TAY
NN6 6EW G4LRT
NN6 6EZ G6EGY
NN6 6EZ M3CZW
NN6 6LX G7DMH
NN6 7BH 2E0PNZ
NN6 7BH G7RGI
NN6 7BH M6ZUF
NN6 7GP 2E0JMY
NN6 7GP M0JIB
NN6 7PP G7BQM
NN6 7PQ M6AEU
NN6 7PQ M6ANM
NN6 7PQ M6ANO
NN6 7PU G6SJA
NN6 7QQ 2E1ECM
NN6 7QQ G4FMC
NN6 7RB G0OFB
NN6 7RR G4VMB
NN6 7SR G0LHE
NN6 7ST G4EPA
NN6 7TT G6YTB
NN6 7TX M0SPS
NN6 7YP 2E0OOH
NN6 7YP M6OCR
NN6 8AF 2E0KRN
NN6 8AW 2E0BHU
NN6 8AW M3NKN
NN6 8QP G1TUS
NN6 8RP G6UNA
NN6 9JS G7NRR
NN6 9LX M7BVO
NN6 9NE G4YKE
NN6 9SB M3ZST
NN6 9UE M6NRC
NN6 9XL G8JFX
NN7 1AE G7NDT
NN7 1BW G0DFV
NN7 1DA G6PHT
NN7 1DR 2E0RDN
NN7 1DR M6RFQ
NN7 1EQ G3TGE
NN7 1HB G6KPJ
NN7 1JW G1XXH
NN7 1NU G4FIN
NN7 2BW G6UVN
NN7 2DA G1AAD
NN7 2HE M0JUT
NN7 2NL 2E0SCH
NN7 2NL M0PIT
NN7 2NL M6ETC
NN7 2NR G8DHU
NN7 3AW G4KHX
NN7 3DF G4XKM
NN7 3EL 2E0WBR
NN7 3EL G7PTA
NN7 3EL M6WBV
NN7 3EZ G4HWA
NN7 3JB G8EDX
NN7 3LE M6CRT
NN7 3LZ G6OIA
NN7 3NA 2E0SHV
NN7 3NA M0SHV
NN7 3QT G1VUY
NN7 3QT M7CKN
NN7 3QU 2E0MOR
NN7 4BX G3IPL
NN7 4DD G3GWB
NN7 4DD G4CZB
NN7 4DD G8LED
NN7 4DN G4TGW
NN7 4DP G4IPL
NN7 4DR G8TMM
NN7 4LL G4PGY
NN7 4LL G6BRW
NN7 4LS G0TML
NN7 4QL G1GJD
NN7 4RL M6MBS
NN7 4SB G4WOB
NN7 4SH G3ZAW
NN8 2DX G4ZIW
NN8 2DY G6WAS
NN8 2EH M6NKQ
NN8 2EN G6UWS
NN8 2ES M3LYG
NN8 2HT G4HIQ
NN8 2JA 2E0PFF
NN8 2JA M0VEZ
NN8 2LZ G1IQE
NN8 2ND G0LEY
NN8 2PH M6NTY
NN8 2PQ M3CQT
NN8 2PU G0VRF
NN8 3BF M0WXF
NN8 3DB M3XLB
NN8 3DB M3XLW
NN8 3EW 2E1BFH
NN8 3JJ G0HBA
NN8 3JJ G1AUY
NN8 3PJ G0ING
NN8 3PJ G3KME
NN8 3PJ G5LP
NN8 3PJ G8PMA
NN8 3PW G6ECT
NN8 3RS M3NFE
NN8 3SH G4MOP
NN8 3ZA G6BBD
NN8 4AT G4WZB
NN8 4BS M7MSD
NN8 4EB G7UPZ
NN8 4ET G8GOM
NN8 4PN G6RPH
NN8 4RU G3ZAG
NN8 4RY G8NWZ
NN8 4RZ G8JPV
NN8 4SF G1IJJ
NN8 4UZ G0FQA
NN8 5BA M7TMW
NN8 5NX G4RDM
NN8 5PQ G6HGG
NN8 5QP G8OSZ
NN8 5WB M7GEF
NN8 5WE G8EJQ
NN8 5WT G0DDJ
NN8 5WW G6STE
NN8 5YD M6NKG
NN8 5YU G6UWY
NN8 5YU M6ELJ
NN83XA G2DRM
NN84UX G6CPX
NN9 5BA M6JTL
NN9 5BA M6LTM
NN9 5BP M7WLZ
NN9 5DP G0EHE
NN9 5LP G0HNG
NN9 5RB G3UQR
NN9 5RJ M6YDN
NN9 5RL M6MLN
NN9 5SU 2E0DCN
NN9 5SU G7BFH
NN9 5SY G4PPW
NN9 5TA G6ODT
NN9 5TL G0TLN
NN9 5TN G7EWL
NN9 5TR G7SLB
NN9 5TY G3NXS
NN9 5XW G6DYU
NN9 5XW M3DYU
NN9 5YS G6FEJ
NN9 6AL M3WBJ
NN9 6AQ G1TLC
NN9 6AU G4BDW
NN9 6BQ G1HGY
NN9 6DF G7MKG
NN9 6DQ G0TES
NN9 6DQ G1JPC
NN9 6HE G0JXQ
NN9 6HE G0MEO
NN9 6HN G6ONV
NN9 6LF M0MOI
NN9 6LS G7LPB
NN9 6LX G4LAM
NN9 6QX G6TRO
NN9 6RD G1WWY
NN95EZ M7BNH
NP1 7PB GW7MYD
NP10 0AU MW7BIH
NP10 0BP GW8MER
NP10 8ET MW6LHB
NP10 8HJ MW6MQU
NP10 8JZ GW8HAG
NP10 8LJ 2W0TID
NP10 8LJ MW6MBC
NP10 8LL GW1MGI
NP10 8NF MW3ZRK
NP10 8NY 2W0CVO
NP10 8NY MW6SVZ
NP10 8PA GW4JZY
NP10 8PD MW0OPY
NP10 8PT 2W0HPK
NP10 8PT MW1DCF
NP10 8PT MW7FRT
NP10 8RB GW4COJ
NP10 8RB MW6DGQ
NP10 8RF MW6JEL
NP10 8RT GW3YQM
NP10 8SL GW4KHQ
NP10 8UP GW7IAK
NP10 8WY MW6NTW
NP10 9AH GW7SMV
NP10 9AQ 2W0KEQ
NP10 9AQ MW0KEQ
NP10 9AQ MW6KEQ
NP10 9EA 2W0OGY
NP10 9EA MW6CVH
NP10 9EN GW0DQW
NP10 9EN GW0PIX
NP10 9EN GW6DYW
NP10 9FB GW7RIB
NP10 9FT GW6TUD
NP10 9FU GW1JFT
NP10 9HN GW4RYQ
NP10 9JG GW4VCL
NP10 9JQ MW0LZZ
NP100bt GW8IQC
NP11 3BE GW5NF
NP11 3DF GW0NZN
NP11 3DP GW8NBF
NP11 3DW GW4JJR
NP11 3EE MW1CNN
NP11 3JQ 2W1BYN
NP11 3LJ 2W0DDP
NP11 3NT MW1FDN
NP11 3NT MW6DCI
NP11 4HJ 2W0KDR
NP11 4HJ MW3RTU
NP11 4RE GW0NUS
NP11 4RS GW0JWF
NP11 4SQ MW7BMU
NP11 5AU 2W0CYX
NP11 5AU MW6SKR
NP11 5DL MW0RZC
NP11 5GE 2W0KHK
NP11 5GE MW3KHK
NP11 5GE MW3XYW
NP11 5GE MW6NON
NP11 5HN 2W0MAC
NP11 5JY 2W0MJC
NP11 5LA GW0NPL
NP11 5LU 2W0END
NP11 6AW 2W0GEM
NP11 6AW MW6NUW
NP11 6BN 2W0BTQ
NP11 6BN MW6TLR
NP11 6BP MW3MXC
NP11 6GL MW6CZV
NP11 6HJ 2W0BFC
NP11 6HJ MW0GKV
NP11 6HJ MW3LMZ
NP11 6HW MW7JON
NP11 6NB 2W0RJL
NP11 6NB MW0RBL
NP11 6NB MW6HGW
NP11 6PF 2W0EAD
NP11 6PF MW3ZCU
NP11 6QH MW3YLV
NP11 6QT MW7DFX
NP11 6RD GW0RYT
NP11 6RG MW6CZU
NP11 7AJ 2W0BFD
NP11 7AJ MW3FDO
NP11 7EP MW6GVR
NP11 7FS 2W0HWQ
NP11 7FS MW7CBT
NP11 7HY GW1ZKE
NP11 7PF GW7IAT
NP11 7PG MW7WRZ
NP12 0DA GW8TRO
NP12 0DB GW0JXG
NP12 0DP MW3LMU
NP12 0EL MW6SVV
NP12 0EL MW6ZBE
NP12 0LU GW0BZJ
NP12 0NQ GW0DJX
NP12 0QE GW0RQS
NP12 0QR GW8OUM
NP12 0SF 2W1AVM
NP12 1DF GW3OAJ
NP12 1DZ GW1PJP
NP12 1EE GW4ROV
NP12 1EW GW0LJW
NP12 1FS GW0KLY
NP12 1HQ GW0MOW
NP12 1LX GW0ARK
NP12 1QH 2W1ITI
NP12 1QH GW8ITI
NP12 2AD GW0JMJ
NP12 2DA MW0CVT
NP12 2DA MW0TTU
NP12 2ET MW6YHR
NP12 2GA GW4JKV
NP12 2GE MW3ZUF
NP12 2GE MW7AHM
NP12 2HP GW4FCV
NP12 2HP GW6BK
NP12 2JB GW0SZN
NP12 2JU GW4HBK
NP12 2JU GW6GW
NP12 2LR 2W0IBY
NP12 2LR MW6PDU
NP12 2LR MW7AIF
NP12 2LS MW3VKM
NP12 2ND GW0NPM
NP12 2NJ MW3MNV
NP12 2NL MW8RFD
NP12 2PF MW0LAO
NP12 2PF MW3YLO
NP12 2PU GW6OIO
NP12 3FX 2W0LBE
NP12 3FX MW6BVU
NP12 3HB MW0DNF
NP12 3NG GW0EYH
NP12 3NR MW0KEV
NP12 3NR MW0MNX
NP12 3NR MW7LOV
NP12 3PG GW3XVQ
NP12 3QS MW3EAI
NP12 3RH GW0ETM
NP12 3RH GW1EHI
NP12 3UL MW1EOR
NP13 1BS GW3PFV
NP13 1DR 2W0SDO
NP13 1DR MW0MDJ
NP13 1DR MW6TBU
NP13 1EX MW6PYP
NP13 1JE MW6BWG
NP13 2AD GW7NTA
NP13 2AU MW0LBB
NP13 2HX MW7EPH
NP13 2HY 2W0VAW
NP13 2NQ GW8LTV
NP13 2PJ 2W0ZAE
NP13 2PJ MW3ZYE
NP13 2RQ GW4ZYG
NP13 2RT 2W1TDM
NP13 2RT MW0BBM
NP13 2RY GW4YDX
NP13 3AW MW3IHX
NP13 3DD 2W0KYZ
NP13 3DD MW7RYK
NP13 3EQ GW7RSE
NP13 3HJ MW7DSP
NP13 3JQ MW7DFE
NP13 3JS MW3WIV
NP15 1AD GW0DKF
NP15 1AT GW6XBV
NP15 1BA MW0XFU
NP15 1BA MW6XFU
NP15 1HN GW8AWM
NP15 1QG GW7KRY
NP15 1SP GW3LDC
NP15 1TN 2W0USK
NP15 2EP MW0CSC
NP15 2LF GW8VKS
NP16 5BU MW0OPS
NP16 5BX GW1SXU
NP16 5EA MW7SDJ
NP16 5EB MW3NNA
NP16 5EH 2W0TSJ
NP16 5EH GW4LWZ
NP16 5EH GW8ZOE
NP16 5EH MW3TSJ
NP16 5HP 2W0CNY
NP16 5JP GW0HBD
NP16 5LA 2W1BST
NP16 5LA GW4LDA
NP16 5LU GW4KPD
NP16 5LY MW3YLR
NP16 5NE GW0WQP
NP16 5RL GW4REX
NP16 5TL GW7TGB
NP16 6DW GW6HRU
NP16 6EF GW7IRD
NP16 6FA 2W0RCU
NP16 6FA MW0RVC
NP16 6FA MW6AVK
NP16 6HN GW8HUS
NP16 6LA 2W0PBU
NP16 6LA MW0YMB
NP16 6LA MW6PBO
NP16 6LE 2W0DNI
NP16 6LE MW0NVY
NP16 6LE MW6KGB
NP16 6PE GW1ISR
NP16 6PN 2W0CXV
NP16 6PN MW6DNP
NP16 6QA MW1CTJ
NP16 6TF 2W0BTZ
NP16 7AQ MW3WOV
NP16 7AQ MW6WOV
NP16 7DF 2W0JMH
NP16 7DF MW0SLH
NP16 7DF MW6JMH
NP16 7DR GW7ERI
NP16 7DS GW4RZE
NP16 7EG MW7XJO
NP16 7EX MW6MGC
NP16 7HF 2W0DJC
NP16 7HF 2W0RDJ
NP16 7HF MW6RDC
NP16 7LG GW1VDT
NP16 7PZ 2W0XFG
NP18 1GT GW4ITJ
NP18 1JB MW0GOM
NP18 1JF GW6FLU
NP18 1LT GW0FBT
NP18 1RY GW4BCZ
NP18 2JL MW1BEQ
NP18 2JR GW6GCK
NP18 2LS MW0HID
NP18 2LT GW7PIB
NP18 2NR GW8GUJ
NP18 3BZ GW4LCF
NP18 3EF GW7BIL
NP18 3EU GW4ZVQ
NP18 3EZ GW4ESL
NP18 3LR MW3ZGR
NP19 0AN 2W0PAN
NP19 0AN MW3PQE
NP19 0AN MW3PQF
NP19 0ET MW6LOD
NP19 0GX 2W0CJZ
NP19 0HU 2W0CHH
NP19 0HU MW0LLO
NP19 0HU MW6CJH
NP19 0LP 2W0ILQ
NP19 4DL GW8WZR
NP19 4NX GW6EWQ
NP19 4PF MW0JBH
NP19 4RQ MW3KSE
NP19 4TB MW6PPD
NP19 7DB GW7AVB
NP19 7ER 2W0BTF
NP19 7ER MW0GXE
NP19 7ER MW3XWE
NP19 7GF 2W0DWR
NP19 7PB 2W1ADO
NP19 7QB GW4VNS
NP19 7QP GW8MOZ
NP19 7QW GW0DHG
NP19 7QY 2W0MNJ
NP19 7SA GW1DLP
NP19 7SH MW0ALN
NP19 8AG GW3KWB
NP19 8NU GW6KBD
NP19 9AN MW7GIA
NP19 9AT MW3GWZ
NP19 9DS MW7ATR
NP19 9ET GW1TDV
NP19 9JN 2W1GLY
NP19 9NE GW6RGT
NP19 9QA GW7RNC
NP19 9QR GW4TIW
NP19 9QS GW4SOC
NP19 9RY GW8OIJ
NP19 9SX 2W0HOT
NP19 9TA MW3FYR
np197el MW7ANT
NP2 0LX GW0LKJ
NP2 4JD GW6DEP
NP20 1FU MW6KYN
NP20 2GN 2W1DXV
NP20 2LT 2W0KYH
NP20 2LT GW8SZL
NP20 2LT MW6SZL
NP20 2LT MW7FDI
NP20 3BJ MW0YGJ
NP20 3DE GW0MSY
NP20 3DJ GW0FJE
NP20 3DJ GW4SUE
NP20 3DN MW3PUU
NP20 3DN MW3ZCO
NP20 3DN MW6PUT
NP20 3EE GW0FXC
NP20 3HJ 2W0JDL
NP20 3HJ MW3NEI
NP20 3HJ MW6FFN
NP20 3NR GW1XVM
NP20 3NR GW8OKR
NP20 3NS GW3KZT
NP20 3SP GW4LOD
NP20 4DT MW3KHC
NP20 4EJ MW6TBD
NP20 4GD MW7RAT
NP20 4JR GW7PRK
NP20 4LP GW7BSC
NP20 4LP MW1AZR
NP20 5AA GW4SDT
NP20 5AH MW6TJG
NP20 5EE MW0DTD
NP20 5HX 2W0NUC
NP20 5HX MW0IXV
NP20 5LP MW6LQG
NP20 5LP MW6LZZ
NP20 5PP 2W0CVM
NP20 5PP MW6XAG
NP20 5XE 2W0CZZ
NP20 5XE MW6CZS
NP20 6GU MW7NYE
NP20 6HE GW3NIN
NP20 6JD 2W0JBJ
NP20 6JD MW3JZV
NP20 6JF 2W1EID
NP20 6JN 2W0XAA
NP20 6JN GW3WTZ
NP20 6JT GW0OAJ
NP20 6LB 2W0MAO
NP20 6NB 2W0ECZ
NP20 6NB MW0ZCE
NP20 6NB MW3ECZ
NP20 6PE MW7CIH
NP20 6QF GW3YKZ
NP20 6QF GW8CNF
NP20 6QN GW4GFL
NP20 6QN GW4YKW
NP20 7AW MW1USK
NP20 7BZ GW0KTL
NP20 7DJ MW6FBJ
NP20 7EJ MW3YJJ
NP20 7SQ MW6GAV
NP202GJ MW0ZAB
NP203DS MW3YKA
NP205AW 2W0IIU
NP205AW MW7BGX
NP22 3DD MW6MPQ
NP22 3HE GW0MQU
NP22 3HF 2W0IAO
NP22 3HF MW6XAC
NP22 3HN MW6SID
NP22 3NT MW6DQQ
NP22 3PB 2W0TAR
NP22 3PB MW6KAA
NP22 3PF 2W0LAC
NP22 3PF MW6SAA
NP22 3RZ MW0CPD
NP22 3SG GW4WLT
NP22 3SN MW6DBD
NP22 3TA 2W0SCB
NP22 3TA MW6DBB
NP22 3TE MW6EKT
NP22 4JG GW6TQH
NP22 4JG MW3KSI
NP22 4PF MW6IJW
NP22 4PL MW3HIX
NP22 5AR 2W0XBE
NP22 5AR MW3UAP
NP22 5AR MW3XAJ
NP22 5AR MW3XBE
NP22 5BH GW8TIX
NP22 5DW MW6AOE
NP22 5EA MW3VWO
NP223SQ GW0JWC
NP224PW MW3TUO
NP23 4AR GW4KFI
NP23 4HR MW7CSO
NP23 4HW 2W0RMR
NP23 4QY MW3HBC
NP23 4SD GW7HTU
NP23 4TS GW7RFP
NP23 4TT 2W0DOE
NP23 4TT MW0JKU
NP23 4TT MW6EYX
NP23 4TY GW0JTU
NP23 4TY MW0IWX
NP23 5AJ MW3PMU
NP23 5EE 2W0CVL
NP23 5EE MW3NAE
NP23 5EN MW7KID
NP23 5ES GW0HCN
NP23 5FA MW3UHO
NP23 5FE GW1IOT
NP23 5FE MW3WPH
NP23 5FF 2W0HQD
NP23 5FF MW3HQD
NP23 5QR GW7VEH
NP23 5SF 2W0CLJ
NP23 5SF MW6EAD
NP23 5TF GW0IRT
NP23 5TS 2W1CIP
NP23 5TS GW0MNO
NP23 5TS GW1HNF
NP23 6HT 2W0LCJ
NP23 6HT MW0PWY
NP23 6HT MW3VLJ
NP23 6JF GW4CXK
NP23 6JL MW7XRE
NP23 6JR GW0VSO
NP23 6LN GW7VGB
NP23 6NE MW3UDW
NP23 6NE MW3UEG
NP23 6PU MW6EXL
NP23 6TR MW3HKH
NP23 6UA GW4UTS
NP23 7QY GW0JTF
NP23 7SP MW3XZB
NP23 7TG MW3FLK
NP24 6AZ MW3TLP
NP24 6BA MW0HRD
NP24 6BS MW3YBK
NP24 6BW MW3DPV
NP24 6EY MW3RZW
NP25 3HT GW0MSW
NP25 3NN MW7AUE
NP25 3SD GW3VFL
NP25 4LU MW0ZEN
NP25 4LY GW8WCA
NP25 4NB GW1FJI
NP25 4QD GW4FIC
NP25 5DA MW0TRO
NP25 5DE GW8PTS
NP25 5EH MW3GFG
NP26 3AB GW0GEV
NP26 3AG GW3ORL
NP26 3AT GW8YKS
NP26 3AW MW7TLX
NP26 3AX 2W0XWD
NP26 3AX MW3LVF
NP26 3BZ 2W0LGE
NP26 3BZ GW3NWS
NP26 3BZ GW8GT
NP26 3BZ MW0LGE
NP26 3BZ MW6LGA
NP26 3EJ GW3SGK
NP26 3FD MW3ZMU
NP26 3FG 2W0CUJ
NP26 3JF MW0GTN
NP26 3JU GW8MZR
NP26 3PB GW3MPP
NP26 3QG GW4OGO
NP26 3SA MW0HJG
NP26 3TB 2W1BOG
NP26 3TL MW1FLY
NP26 3UT MW6RLE
NP26 3UW GW3WEZ
NP26 4DX MW0DTH
NP26 4JD GW4ZVL
NP26 4JL MW3KRN
NP26 4LH 2W0KIN
NP26 4LH MW7HAW
NP26 4PP GW4FNO
Np26 4qr 2W0GHD
Np26 4qr MW0GZL
Np26 4qr MW6GHD
NP26 5BW GW0EGH
NP26 5GB GW1EPR
NP26 5RE GW6NQU
NP26 5RE MW3ESH
NP26 5RE MW3HOY
NP26 5RE MW3VEN
NP26 5TB MW6BSU
NP26 5UW MW6OTK
NP3 1AR GW6BVS
NP4 0BN GW1XVC
NP4 0BN MW3HGO
NP4 0DF MW6GWL
NP4 0DG GW0FJH
NP4 0HT 2W0REX
NP4 0HT MW0LUK
NP4 0HT MW6USK
NP4 0HX MW7PYS
NP4 0JD MW3GHF
NP4 0NB GW7LPM
NP4 0NJ GW0DQT
NP4 0PT GW4NXD
NP4 0RT MW0AZN
NP4 5AB 2W1GMM
NP4 5BX GW1VJB
NP4 5BZ GW1SXT
NP4 5EA 2W0DDJ
NP4 5EZ GW8CKJ
NP4 5JF 2W0TAX
NP4 5LT GW4HYZ
NP4 5SS 2W0BVV
NP4 5SS MW3JQX
NP4 5XS GW4ZQV
NP4 6BU 2W0ICF
NP4 6BU MW0IZX
NP4 6BU MW6ICF
NP4 6DF GW0MBW
NP4 6HE MW6MZU
NP4 6QZ MW6HPC
NP4 6QZ MW6RHY
NP4 6QZ MW6WCT
NP4 7AB MW7TAF
NP4 7HJ 2W0IJL
NP4 7HJ MW0KMS
NP4 7QF 2W0SLD
NP4 7QF MW3SLO
NP4 7QF MW6CAE
NP4 7QS GW8ERA
NP4 7RZ MW6PGC
NP4 7SN MW6KTM
NP4 7UJ GW1VDW
NP4 8BU MW7KJC
NP4 8DQ GW4AYQ
NP4 8DQ MW3DPF
NP4 8DQ MW3LOI
NP4 8LL 2W0XVT
NP4 8LL MW0YVT
NP4 8LL MW6XVT
NP4 9EH MW6IJJ
NP4 9LX GW1WTL
NP4 9PA MW0XRU
NP4 9QL GW6STS
NP4 9QU MW0USK
NP4 9QU MW3USK
NP4 9SS GW1GKV
NP44 1AT GW4IZJ
NP44 1BQ 2W0UNY
NP44 1BQ MW6UNY
NP44 1LH GW0TTN
NP44 1LJ MW1CVM
NP44 1LS GW4UWR
NP44 1NB GW0FGO
NP44 1RB GW7DIL
NP44 1RZ MW3AVU
NP44 2AL GW4IQB
NP44 2AN GW0KAM
NP44 2AN GW1YKY
NP44 2JW 2W0SEZ
NP44 2JW MW0SEZ
NP44 2JW MW3SEZ
NP44 2JW MW6GZC
NP44 2JW MW6SDE
NP44 2PF MW6ERZ
NP44 2QL MW7EHV
NP44 3AE GW4JBQ
NP44 3AE GW8SBK
NP44 3AP 2W0CUK
NP44 3AP GW8CRH
NP44 3AP MW6CUR
NP44 3AP MW6TRQ
NP44 3AZ GW0HIR
NP44 3BX 2W1DEA
NP44 3BX GW0OLZ
NP44 3HH GW4YML
NP44 3LS 2W1GDY
NP44 3NU MW6RLA
NP44 4LB GW1YZF
NP44 4LD MW3OSQ
NP44 4LF 2W0YAD
NP44 4LF MW3NHN
NP44 4LG 2W0LFE
NP44 4LG MW6KUD
NP44 4LW MW3URO
NP44 4PQ MW3LMV
NP44 4PX 2W1GCU
NP44 4PX MW0RPB
NP44 4QT MW1FFE
NP44 4RB MW7NMT
NP44 4TE GW0UKT
NP44 4TE GW7VEL
NP44 4TJ MW1DOO
NP44 5AS GW1UXW
NP44 5HH MW3NZZ
NP44 5HN GW8JOY
NP44 5JA MW6FAW
NP44 5JE MW0MTR
NP44 5TQ GW0DQY
NP44 5TQ GW1ZFX
NP44 5UH MW6OZB
NP44 5UN GW0OZB
NP44 5UN GW7DHG
NP44 5UU MW6SOF
NP44 6EH GW7SSQ
np44 6jf MW7CIE
NP44 6JH 2W1DNK
NP44 6JH 2W1DRB
NP44 6JH GW0DHA
NP44 6JJ GW7SSN
NP44 6JR MW6LQK
NP44 6TY 2W0NWJ
NP44 6TY MW0NWJ
NP44 6TY MW7NWJ
NP44 6UD MW7BMW
NP44 7AB GW4CQT
NP44 7AH GW0DXG
NP44 7AH GW4RIB
NP44 7JG GW0HDY
NP44 7JP 2W0MKG
NP44 7JP MW0MKG
NP44 7JP MW6MGY
NP44 7JX MW0YAC
NP44 7JX MW0YAD
NP44 7LL GW7CBU
NP44 7LS MW6EJG
NP44 7lt GW7UMS
Np44 7lt MW7AIH
NP44 8RJ GW6BHQ
NP44 8RY GW6RBZ
NP446JR 2W0WOY
NP446JR MW7WDY
NP45EA 2W0JEV
NP6 1PE GW6BAH
NP7 0BP GW6JVB
NP7 0BY GW0JJF
NP7 0DX GW6VED
NP7 0EL GW1UVN
NP7 0LL 2W0RGL
NP7 0LL MW0GYY
NP7 0LL MW3TYC
NP7 0LS MW1AZI
NP7 0RA MW0LDJ
NP7 5BQ GW4JDE
NP7 5HQ GW6JPC
NP7 5HW GW4IHM
NP7 5JA GW0UDH
NP7 5LG MW0YVK
NP7 5TL GW6BXU
NP7 5TL GW6RAV
NP7 5TL MW1FAT
NP7 5TN MW6ETW
NP7 6AF GW6ZZF
NP7 6AP MW6EYM
NP7 6HB MW3ETB
NP7 6HN GW0UKG
NP7 6JY GW6RAO
NP7 7AL GW4UHK
NP7 7AW GW8KJK
NP7 8DH MW6NFH
NP7 8DN GW4IQA
NP7 8TG MW6CYM
NP7 8UF GW7PRW
NP7 9HP GW8KCH
NP7 9RS MW6EJI
NP7 9SA GW4EIN
NP75SY GW7HLZ
NP75SY MW3HLZ
NP8 1AP GW0BWE
NP8 1DJ GW0XYL
NP8 1DJ GW4ZUW
NP8 1DQ GW4FRH
NP8 1LN GW6KGR
NP8 1NU GW8IKH
NP8 1RT GW4BRH
NP8 1RU GW8IAM
NP8 1SY GW4UZC
NP9 7HW 2W1AYO
NR 172HB M6VUD
NR 324 HT M6ATQ
NR1 1EA M7ISG
NR1 1ET G4FAI
NR1 1LW M6VOX
NR1 1QA G4MYQ
NR1 1SH G0LQX
NR1 2AL G7VZL
NR1 2JR G4RRX
NR1 2JX 2E1HOO
NR1 2JX G7UVB
NR1 2NG M7GMC
NR1 2NL G0WJU
NR1 2NX G4VLS
NR1 2PW M6TVL
NR1 3ED G0WFP
NR1 3FU G4GHO
NR1 3HB M6HAE
NR1 3JB M6GCC
NR1 3QX M6VDH
NR1 4BE G0MWH
NR1 4EN M6DPZ
NR1 4EP 2E0GDO
NR1 4HQ G4JYG
NR1 4JX M1CRQ
NR1 4LN M3SDB
NR1 4NJ M6BLI
NR1 4PP 2E0GFB
NR1 4PP M6RWV
NR1 4QB 2E0NLK
NR1 4QB M3NLK
NR1 4QB M6MLK
NR1 4QB M6SLK
NR10 3BH G4RSP
NR10 3EA G8BIG
NR10 3ES G0TMT
NR10 3GE M7EQR
NR10 3HE G8YQH
NR10 3HE M1GPC
NR10 3HF G7ETC
NR10 3HF M3LJA
NR10 3HF M3UMA
NR10 3HS G7PDO
NR10 3HY M6COU
NR10 3JT M6XYL
NR10 3LD G4WUG
NR10 3LF M6LAE
NR10 3LG G0SGT
NR10 3LP M6JWP
NR10 3LX G0DWV
NR10 3NB M6GUW
NR10 3NN M6EOA
NR10 3PG G0AUT
NR10 3PP G1ESX
NR10 3QA G7JTZ
NR10 3QA M3JTZ
NR10 3QB G3VPT
NR10 3QB M0SPX
NR10 3QQ M3NGF
NR10 3QW G3TWX
NR10 3ST G0MJC
NR10 3SW 2E0SPA
NR10 3SW M6SPA
NR10 4AE M3EDC
NR10 4AZ G0PFJ
NR10 4BG M3MJD
NR10 4BS G1WUY
NR10 4BS M3YGU
NR10 4EL G1IDJ
NR10 4HA G4WUU
NR10 4HJ 2E0YOP
NR10 4HJ M3YOP
NR10 4HL M3SOY
NR10 4JZ G6ZRV
NR10 4LS G0IYK
NR10 4LU G8YJS
NR10 4PP 2E1HKE
NR10 5AP G6CKD
NR10 5AU M1BQW
NR10 5AX M7EAM
NR10 5BB G3WCE
NR10 5DJ M6FBE
NR10 5DJ M6HDM
NR10 5EX M3RZN
NR10 5EX M6SEN
NR10 5JT M6HVY
NR10 5JY 2E1GZY
NR10 5JY M1EPR
NR10 5JY M3EPR
NR10 5LE 2E1EDA
NR10 5LT G3VWQ
NR10 5NU G7UBA
NR10 5PD M0BLH
NR10 5PW G4SGX
NR10 5QW G3KMO
NR10 5QX G1NGE
NR11 6DJ 2E0DDL
NR11 6DJ M0LDV
NR11 6DJ M6DXN
NR11 6HW 2E0RDI
NR11 6HW M6OPL
NR11 6JD 2E0HAJ
NR11 6JD M0HET
NR11 6JF G0EVQ
NR11 6JF G7GQB
NR11 6JZ M0JMS
NR11 6LB G0NTJ
NR11 6LX 2E0IWB
NR11 6LX M0IWB
NR11 6LX M6IWB
NR11 6SG M7DLZ
NR11 6UT G1UQC
NR11 6UT G8RSQ
NR11 7AQ G0AZR
NR11 7AW G1GGN
NR11 7BE G4GQE
NR11 7DT G4PLZ
NR11 7DY G7VRK
NR11 7HP G3YXN
NR11 7LQ 2E0WPZ
NR11 7NJ G8BLD
NR11 7PD M1BOA
NR11 7QD M7CNI
NR11 8BE G3PPR
NR11 8DN G6NZY
NR11 8ED G0TAM
NR11 8ED G7RNN
NR11 8HJ G4AFQ
NR11 8JB G3BGF
NR11 8JF G6OQJ
NR11 8JF G7HFP
NR11 8JN G4SFY
NR11 8JX 2E0NCT
NR11 8JX M7CTT
NR11 8LP G0JLV
NR11 8NG G6KBZ
NR11 8PD G4RRN
NR11 8PN G4CLO
NR11 8QL G8KHI
NR11 8UW M6SNJ
NR12 0AL G3SGC
NR12 0EP 2E0AST
NR12 0EP G4EWB
NR12 0FA 2E0NCG
NR12 0FA M0NCG
NR12 0FA M3NCG
NR12 0HF G4SFQ
NR12 0LU M1EEN
NR12 0NE G4POG
NR12 0NJ G4MQK
NR12 0PD G3WQY
NR12 0PW M1DMB
NR12 0QE M6DXQ
NR12 0RB G6PYR
NR12 0RL M6ADT
NR12 0RL M6RKM
NR12 0RU G1ANZ
NR12 0RU G4CTT
NR12 0SF G7HXI
NR12 0SU M6SMG
NR12 0SX 2E1SBF
NR12 0TA G7TUM
NR12 0YG G6NLX
NR12 0YQ G3NMZ
NR12 7AB G0SMS
NR12 7AB G1KZD
NR12 7AB G1NJG
NR12 7AJ G0VDR
NR12 7AJ G7RPK
NR12 7AJ M6VDR
NR12 7BB 2E0NKI
NR12 7BB 2E0TZY
NR12 7BB M0TZY
NR12 7BB M6TKE
NR12 7BB M6TZY
NR12 7DQ G4SOP
NR12 7DT G6ESQ
NR12 7EG G7RIE
NR12 7ER 2E0SXC
NR12 7ER M6DBS
NR12 7EW G1XUW
NR12 7HD G7SNT
NR12 7HW G6GAC
NR12 7HZ G1EAJ
NR12 7JU G1HXN
NR12 7NB G0DGU
NR12 8BA 2E1EIV
NR12 8BA M0HUG
NR12 8BA M0PUD
NR12 8DG G4VUR
NR12 8DP G1OSE
NR12 8EA 2E0VFN
NR12 8EA M6VFN
NR12 8EB 2E0GRV
NR12 8EB M6WTR
NR12 8HP M0EAZ
NR12 8JH G4YDZ
NR12 8LQ G8UPD
NR12 8LZ G0AKL
NR12 8NF M0THA
NR12 8PF M7PWL

NR12 8QR G0WJV
NR12 8SL M3MHN
NR12 8SQ G1YLV
NR12 8TU M0TVG
NR12 8UG M6NLR
NR12 8UJ G0GHB
NR12 8XB G6IYY
NR12 8YL G7PRC
NR12 8YL G7VPD
NR12 9BE M6GUY
NR12 9BY G8IYK
NR12 9BY M6DEF
NR12 9DA G8VSF
NR12 9DR M6ABQ
NR12 9EQ G1ONC
NR12 9FL M1EVZ
NR12 9LH G1CEU
NR12 9PB G1TAY
NR12 9PB M6EKS
NR12 9PJ G4HXK
NR12 9PZ G1LQM
NR12 9QN M6WZK
NR12 9SA G4CDN
NR12 9SA G4CKB
NR12 9SE M6KLM
NR12 9SG G4LPP
NR12 9SG M7LPP
NR120NL M3VTH
NR13 3AB G4JNX
NR13 3DH G0OOB
NR13 3EY M6RGI
NR13 3HJ M6EJY
NR13 3LA G0KRU
NR13 3PL G4YGQ
NR13 3PP G4YYL
NR13 3PP G6WIT
NR13 3PU M6DYO
NR13 3SW G3LNM
NR13 3TH M0HJI
NR13 4AH G4UUB
NR13 4AW 2E0BDB
NR13 4AW 2E0BSU
NR13 4AW M3NLF
NR13 4BA G4UAM
NR13 4BB G1MUC
NR13 4BU M7LLS
NR13 4FA 2E0BCQ
NR13 4FA M0KXK
NR13 4JS G4WUO
NR13 4JX 2E0DJR
NR13 4JX M3UYV
NR13 4LZ G0KSD
NR13 4NB G4BFS
NR13 4NF G0GQV
NR13 4QF 2E1FNJ
NR13 4QF 2E1HPC
NR13 4QF 2E1HPD
NR13 4QF G7UVY
NR13 4RU G4FOS
NR13 4SZ G7HVU
NR13 5DG 2E0ZZC
NR13 5DG M0LDK
NR13 5DG M6ZZC
NR13 5DL M6VXT
NR13 5HR G4MCA
NR13 5LG G0AKO
NR13 5NS M7BMC
NR13 5PA G7RBS
NR13 5PL M1MIT
NR13 5PL M7MIT
NR13 5RF G8UKO
NR13 6AY M7AQG
NR13 6DP G0IRQ
NR13 6DY G6BXR
NR13 6HH 2E0CEY
NR13 6HH M6JET
NR13 6LT G4KLM
NR13 6QD G0IMU
NR13 6QH G1BCU
NR13 6RQ G6WMG
NR13 6RR G3PDH
NR13 6SG M0HLZ
NR134JD 2E0AWG
NR134JD M0EBJ
NR135LD G7RTN
NR135RA 2E0YAO
NR135RA G0OOR
NR14 6BH G4IAO
NR14 6BQ G1MSK
NR14 6BY G3ITB
NR14 6FB M0NRW
NR14 6FB M0PTO
NR14 6NH G7HFE
NR14 6PG M0BIT
NR14 6PP G3VKM
NR14 6RL M3ZFO
NR14 6UT G4TVJ
NR14 6UZ G1MSG
NR14 7AH M6LBM
NR14 7AR G7DFP
NR14 7AS 2E0MYE
NR14 7AS M3MYE
NR14 7BU G4SOZ
NR14 7DP G0RSS
NR14 7EB M3RMD
NR14 7EE G4GUS
NR14 7JE G0DEE
NR14 7LU G0VJH
NR14 7NH 2E0SYW
NR14 7NH M0SYW
NR14 7NH M6SYW
NR14 7PB G4LOJ
NR14 7QN G4BID
NR14 7RY 2E1CPV
NR14 7WF 2E0BRF
NR14 7WF M0GMK
NR14 7WF M6CJD
NR14 8AL G4VCE
NR14 8AS G4LEP
NR14 8BA G8IXM
NR14 8BL M3ZUI
NR14 8BL M7BSN
NR14 8DH G0XAR
NR14 8HT G1PQJ
NR14 8HX G6VAD
NR14 8LQ G3LDI
NR14 8PH M0HJE
NR14 8RG G4UYR
NR15 1JB M3NKP
NR15 1NG G6ULJ
NR15 1NH 2E0GOW
NR15 1NH M0GOW
NR15 1NH M7CHJ
NR15 1TH G7VNN
NR15 2DG M0TSM
NR15 2DS G3WDN
NR15 2EE 2E0HCV
NR15 2EE M7TTB
NR15 2HL 2E0OLE
NR15 2HR 2E0VBW
NR15 2HR M0VBW
NR15 2HR M3VBW
NR15 2NT G3VYN
NR15 2PH G8WSH
NR15 2ST 2E0JMB
NR15 2ST M0UKS
NR15 2TA M3LYA
NR15 2TF 2E0RUG
NR15 2TF M0RUG
NR15 2TF M6TRT
NR15 2UH M6UDX
NR15 2WY M6PWG
NR15 2XR M6EJL
NR16 1AW G8AKF
NR16 1BA G6XCM
NR16 1DJ G1GSB
NR16 1DL G3MGX
NR16 1HH 2E1CYE
NR16 1LH 2E0VID
NR16 1LH M0RXS
NR16 1LH M7JUL
NR16 1NW M1CQK
NR16 1NW M1CQL
NR16 1PD 2E0MUI
NR16 1PD 2E0UMG
NR16 1PD 2E0XUI
NR16 1PD M0MUI
NR16 1PD M0UMG
NR16 1PD M0XUI
NR16 1PD M6WNS
NR16 1PD M6XUI
NR16 1RS G3SDT
NR16 2AJ G0KXV
NR16 2AN G3RXA
NR16 2BY 2E1CIR
NR16 2BY 2E1DLO
NR16 2BY 2E1DWM
NR16 2BY G7RVH
NR16 2DE G6XQP
NR16 2EQ M0RYB
NR16 2HL G0BOO
NR16 2JD G0ENB
NR16 2JD M1MNR
NR16 2JD M3HMK
NR16 2LW G7UFV
NR16 2NA G6HYP
NR16 2PS G3JIE
NR16 2SZ M3ZPZ
NR17 1BL G6AVI
NR17 1BW 2E0TFX
NR17 1BW M0TFX
NR17 1BW M3TFX
NR17 1DT M3SFZ
NR17 1EQ M6AKP
NR17 1GP G0UUT
NR17 1GU M6IIZ
NR17 1JB 2E0PLU
NR17 1JB G7URP
NR17 1JB M0TC
NR17 1JB M3DCP
NR17 1JB M3PLU
NR17 1LD M6OOJ
NR17 1LW 2E0DWZ
NR17 1LW M6RKD
NR17 1UL G3UUR
NR17 1YF M1CVU
NR17 2AD G4RKL
NR17 2HU 2E0IXH
NR17 2HU M3IXH
NR17 2LA M3FUQ
NR17 2NG G1XYV
NR17 2NG M3MGJ
NR17 2NH G3JUU
NR17 2RJ M7FVS
NR17 2RQ M3NGU
NR172QW M7PNS
NR18 0AR 2E1DQT
NR18 0AR G7VZI
NR18 0DE G1FPK
NR18 0DF 2E0DWT
NR18 0DF M0WEL
NR18 0DF M6WEL
NR18 0DN G3YIA
NR18 0EA G1EGL
NR18 0EX G4OJV
NR18 0EX G8DYA
NR18 0FH M6OGH
NR18 0HS G6LQM
NR18 0HX G6YYU
NR18 0HY G7TIE
NR18 0NB M1EOV
NR18 0NT G3MPN
NR18 0NZ G6DAN
NR18 0QP G6YFZ
NR18 0QP M7SUP
NR18 0SW G0KYA
NR18 0SW M6ELE
NR18 0SW M6PUP
NR18 0TT 2E0CQH
NR18 0TT M6BVM
NR18 0TU M0RJB
NR18 0WQ G0OPV
NR18 0XT M1BCU
NR18 9BA M6TDI
NR18 9BH 2E0NVT
NR18 9BH M0NVT
NR18 9BH M6NVT
NR18 9BP M6PPV
NR18 9DD G7EOA
NR18 9DR 2E0XSW
NR18 9DR M0VSW
NR18 9DR M3VSW
NR18 9DR M7SBA
NR18 9FA M7MRQ
NR18 9HQ M6ZZB
NR18 9JG 2E0XKM
NR18 9JG M0XKM
NR18 9JG M6XKM
NR18 9NB M6HYR
NR18 9RY G4VAO
NR18 9TB G7DIB
NR18 9TF G4VMG
NR19 1AJ G6ULS
NR19 1AW 2E0ZJB
NR19 1AW M0ZJB
NR19 1AW M3ZJB
NR19 1EG G1OZR
NR19 1JB G3XPT
NR19 1JY G7HCT
NR19 1LU G0GEB
NR19 1LU M6LSE
NR19 1ND G4EYA
NR19 1QE G6ZKX
NR19 1XF M6KJR
NR19 2BJ G8UUV
NR19 2BS M3KTH
NR19 2BS M3WKM
NR19 2LX G3YHO
NR19 2NS G4UVA
NR19 2SR M6EBY
NR19 2SR M6ODM
NR19 2SU G4LLL
NR19 2SU M6NMR
NR19 2TN M7JPC
NR19 2UB G0LGJ
NR2 1EE G4HUN
NR2 2DW M3VGX
NR2 2JB G1AMN
NR2 2RN G4LFQ
NR2 2SH M6GVA
NR2 2SN G4EXN
NR2 3NJ G7FUQ
NR2 3NJ G8GWJ
NR2 3PT 2E1DRY
NR2 3PZ G7MLC
NR2 3RL G7NVB
NR2 3TL G6TSZ
NR2 3TP 2E0MAP
NR2 3TP 2E0PEW
NR2 3TP M0PEW
NR2 3TP M3PKZ
NR2 3TP M3WPO
NR2 3TP M3YYU
NR2 3TP M6HRW
NR2 4BL 2E0BSS
NR2 4BL M3OYE
NR2 4EH M0NFY
NR20 3DG 2E0CAU
nr20 3dg G4LPW
NR20 3FN M6HYZ
NR20 3GH G3ZQI
NR20 3JF M7EAH
NR20 3LX G0UYA
NR20 3QP G3YLA
NR20 3RE G4DYC
NR20 3RF G0UML
NR20 3SB G0VIJ
NR20 3SY M7MCO
NR20 4AA G3ZIG
NR20 4AA G4ANT
NR20 4AY M3YCS
NR20 4AY M3ZCE
NR20 4BW G6YXB
NR20 4DW G7HJD
NR20 4LR G4IPN
NR20 4LY M3TDT
NR20 4NQ G0TYQ
NR20 4PS M7OPB
NR20 4SL G0UYC
NR20 5LL G8OO
NR20 5PS G0GHE
NR20 5QN M6GYV
NR20 5RD M6NIL
NR20 5RU M6EIU
NR20 5SB M6JJK
NR20 5SZ G7EWK
NR20 5TA G4XCR
NR20 5TW G3JCK
NR21 0BU G4UKZ
NR21 0EJ G3XMP
NR21 0HX G0PPR
NR21 0JU G4KZT
NR21 0LT G6LEK
NR21 0PT G0WCH
NR21 0QX G7OSJ
NR21 7AQ 2E0TEF
NR21 7AQ G3TBL
NR21 7AQ M6TEF
NR21 7BY G7PSK
NR21 7NA G0FVF
NR21 7NA G6MDM
NR21 7PU 2E1TMB
NR21 7QE G6YLR
NR21 7QL M7BPX
NR21 8DW G1IJC
NR21 8HB G3PAX
NR21 8LR G0KYG
NR21 8NP M6AWA
NR21 9ES G0KRD
NR21 9HW G4RRD
NR21 9JB G6IGO
NR21 9LH M6AIQ
NR21 9PD G0SQI
NR21 9PD M6SQI
NR23 1BE M0KML
NR23 1HF G4DIC
NR23 1JA M7ODG
NR23 1JY G4KOV
NR23 1JZ G2PK
NR23 1LP 2E0CTT
NR23 1LR G0SRZ
NR23 1PA G4RAY
NR23 1QE G0DSN
NR23 1QE G0NWT
NR23 1QL G7BUS
NR23 1QL G7EVQ
NR24 2DE G4IWI
NR24 2DW M7COS
NR24 2EW G8RES
NR24 2JY G6VZG
NR24 2LB M3YPX
NR24 2LQ G0FVS
NR24 2LQ M6CYC
NR24 2PH G8AKP
NR24 2RD G6AUY
NR25 6DG G4BUF
NR25 6EN G7PTD
NR25 6NW G0SIQ
NR25 6RU G3LQJ
NR25 6TG M6BDX
NR25 6TU M0SSF
NR25 6TU M7MHM
NR25 6TX M3WZJ
NR25 6TZ G7OSK
NR25 6UD 2E0ERU
NR25 6UD M0IUC
NR25 7DQ G8YSJ
NR25 7HH 2E0JDE
NR25 7HH G0GFQ
NR25 7HH G1MIE
NR25 7HH M6JFV
NR25 7LL G4EBE
NR25 7UA G4UQA
NR26 8BJ M3ZNY
NR26 8EJ M6DPI
NR26 8HU G7CIK
NR26 8JB M6KGI
NR26 8NQ G4ARX
NR26 8PT G7PHT
NR26 8QQ M7RGK
NR26 8SH 2E0SLS
NR26 8SH M0SHK
NR26 8SH M0WPX
NR26 8SH M6KAH
NR26 8UB M3BFU
NR26 8UN G3HRK
NR26 8UN G6DFA
NR26 8XP G4DMB
NR26 8YE M0VAW
NR26 8YJ M6ASE
NR27 0BG G4LYB
NR27 0BG G8IMI
NR27 0BP M6IFI
NR27 0BY M0DFF
NR27 0DD G4RZN
NR27 0DD M0NCA
NR27 0EE G0UIQ
NR27 0EQ M6FQK
NR27 0EQ M6GAD
NR27 0EQ M6LAQ
NR27 0HG 2E0BCW
NR27 0HG M0EBN
NR27 0HG M6BXA
NR27 0HQ G2BHG
NR27 0NY M6CFU
NR27 9AH 2E0TBX
NR27 9AH M6CRP
NR27 9AW M6HYQ
NR27 9LQ G0BGX
NR27 9LQ G0SQK
NR27 9LW G0LYN
NR27 9LW G1SVD
NR27 9PH G3PND
NR27 9PH G4PQP
NR27 9PQ G7HRJ
NR27 9PW 2E0WBW
NR27 9PW M6WWD
NR27 9QT G3QVD
NR27 9RN G3ZRG
NR27 9RR M3UCA
NR279BW G4NRE
NR279BW G6EHE
NR279BW G6HL
NR28 0AU M6ABC
NR28 0BS M3YXV
NR28 0DH 2E0WOS
NR28 0DH M0WOS
NR28 0DH M3WOS
NR28 0EE 2E0UJD
NR28 0EE G8RZZ
NR28 0EE M0UJD
NR28 0EE M3UJD
NR28 0EE M3VTN
NR28 0EN G3WXW
NR28 0FP M7WET
NR28 0JP M3PGD
NR28 0LE 2E0FAY
NR28 0LE M6BLQ
NR28 0LR G4ZWM
NR28 0PJ G7TIN
NR28 0QG G3YOA
NR28 0QR 2E0WTG
NR28 0QR M0WTG
NR28 0QR M3WTG
NR28 0QU G0GRB
NR28 0RZ G1NST
NR28 0SY G0IYD
NR28 0SY G1XYD
NR28 0TA M7BYL
NR28 9BB G0AJJ
NR28 9BB G6VOV
NR28 9HS G4RTH
NR28 9JA G6WME
NR28 9JE M6NWP
NR28 9LD G4ZAY
NR28 9LX M6ZAH
NR28 9PL M6AST
NR28 9PZ G4ZDT
NR28 9RG G6JCY
NR28 9SD G7IIN
NR28 9SH G4TDQ
NR28 9UP 2E0LSR
NR28 9UP M6EMT
NR28 9XA G6JCX
NR28 9XT 2E0CVY
NR28 9XT M0NAQ
NR28 9XT M6CFQ
NR280bf M6GBD
NR280EE M7BEE
Nr280ug M7NUK
NR29 3BG G7UDU
NR29 3HD G5NB
NR29 3HJ 2E0SIH
NR29 3HJ G0MYC
NR29 3HJ M0SIH
NR29 3HJ M7SMH
NR29 3HU G8VPE
NR29 3JB M6DAL
NR29 3JX M6NFK
NR29 3LT G7DXQ
NR29 3PE M7MAS
NR29 3QD 2E0GFW
NR29 3QD M6GFW
NR29 3QT G0ELJ
NR29 3RJ G1LTH
NR29 3RJ G4SCS
NR29 3SP M3XUV
NR29 4AF G7BMY
NR29 4AQ G7TUK
NR29 4EA G3ZEB
NR29 4ES G1EBP
NR29 4JD G8PEN
NR29 4JL G8SYV
NR29 4JL M3EJM
NR29 4JY G7HRF
NR29 4LT G3XAU
NR29 4PT G1RPV
NR29 4PU 2E0NHJ
NR29 4PU M6GZK
NR29 4QN M6VRJ
NR29 4QN M6VSJ
NR29 4QN M6XMJ
NR29 4QT M3XMU
NR29 4RJ G4MFX
NR29 4RY G4EMH
NR29 4SH G8UCZ
NR29 4TB M7SEY
NR29 4TD G0FVT
NR29 4TW 2E1FQB
NR29 4UR M3JST
NR29 5ED G1JBZ
NR29 5HQ G3KCD
NR29 5LG G0APV
NR29 5LG G4WPG
NR29 5NU G0KCB
NR29 5NZ M6ALG
NR29 5PB G4AXA
NR29 5PB G4CMM
NR29 5PZ G0VSB
NR29 5QL G0OBH
NR3 1NY G0JAG
NR3 1PN M6FEE
NR3 1SE M6MBI
NR3 2AD M0TJL
NR3 2DL G7VUM
NR3 2DR M6JPK
NR3 2HT 2E1HOS
NR3 2HT 2E1HOT
NR3 2HW G8AUN
NR3 2LF 2E0YTF
NR3 2LF M3YTF
NR3 2NG M3LXU
NR3 2NQ G0CWD
NR3 2NQ G0WJZ
NR3 2PS M6EIM
NR3 2QN 2E0YRW
NR3 2QN M3YRW
NR3 2QN M6YRW
NR3 3DT G4YDQ
NR3 3HD 2E0KFO
NR3 3HD M3WVC
NR3 3LU M3XIM
NR3 3NN G0SSX
NR3 3PF G0VRY
NR3 3PY M1CRL
NR3 3QD M3WZN
NR3 3SX 2E0YOK
NR3 3SX M6OKY
NR3 3TB G4NFF
NR3 3TD M6DDO
NR3 3TD M6YRL
NR3 3TQ M6HDS
NR3 4BE M6ZOC
NR3 4EF G3ZTV
NR3 4EW 2E0FRD
NR3 4EW M6WLG
NR3 4QB 2E0YRU
NR3 4QB M6SWK
NR3 4QF M7INA
NR3 4QW G0PEQ
NR3 4RA M7HRT
NR3 4RX G8XEN
NR3 4TS M1AFQ
NR30 1DU 2E0TWQ
NR30 1DU M6TWQ
NR30 1EU M3LMQ
NR30 1JG M3NNI
NR30 1JY 2E1BFX
NR30 1JY M6LDY
NR30 1LY G0GGB
NR30 1RP M6LGT
NR30 2AT G4TNY
NR30 2LF M3OYS
NR30 2RU M0NPQ
NR30 2SA 2E1GVB
NR30 3JU G6ATW
NR30 4BB 2E1HVN
NR30 4BB M3HVN
NR30 4BS 2E1HAU
NR30 4EA 2E0CNW
NR30 4EA M0STL
NR30 4HL M6LBT
NR30 4JU 2E1HVU
NR30 4LD 2E1HVT
NR30 4LD M3RJH
NR30 4LR G4ALC
NR30 4LS M6CHJ
NR30 5AB M0WBR
NR30 5AH M1BFI
NR30 5ES 2E0GZF
NR30 5ES M0JRN
NR30 5ES M7AUY
NR30 5HG M6AIO
NR30 5PH G4VSD
NR30 5PQ 2E0BNI
NR30 5PQ M3NPQ
NR30 5PQ M3TFZ
NR30 5QT G8VQX
NR30 5RH M1BUC
NR30 5RW 2E0GAV
NR30 5RW M6AEG
NR30 5TT M1ADP
NR30 5UN 2E0BNT
NR30 5UN M0ZOR
NR30 5UN M3RZJ
NR31 0BU M1CDV
NR31 0HR G4YZF
NR31 0HR G6ZVG
NR31 0PW G6VZU
NR31 6EF M0DLG
NR31 6HJ M6POZ
NR31 6LZ M6DBA
NR31 6NH M7EPK
NR31 6PR M6MIL
NR31 6PS M7CKV
NR31 6PX M3YQR
NR31 6PX M6MKW
NR31 6PZ M3XEO
NR31 6QY G0SHJ
NR31 6RQ G6FQL
NR31 6RT M6JKS
NR31 7AS 2E1AGQ
NR31 7AS M7COD
NR31 7JJ G0DVQ
NR31 7ND G0VMK
NR31 7ND G3YRC
NR31 7ND G7MKV
NR31 7NN M7ERG
NR31 7RF G6ARM
NR31 8BU M6LDA
NR31 8DN G3WPG
NR31 8DR G1HNU
NR31 8DR G7LNM
NR31 8HU M7RKZ
NR31 8JG G3VSV
NR31 8JW M6KGQ
NR31 8NT M3MJY
NR31 8QJ G1GER
NR31 8RG G7RPJ
NR31 8SX G8TAE
NR31 8TF 2E0WBE
NR31 8TF M6WBE
NR31 9AR M6CBJ
NR31 9BT G8VKO
NR31 9DF G7FCJ
NR31 9JT G0FEI
NR31 9JU M1CQR
NR31 9LZ G6KZI
NR31 9NT M6FBZ
NR31 9NX G0GGE
NR31 9PD M6SVY
NR31 9PH G8PIO
NR31 9PL M7JSD
NR31 9PP 2E0CBH
NR31 9PP G0BAW
NR31 9PP G0CLG
NR31 9PP M6RBY
NR31 9PP M6WOO
NR31 9PQ G7WHM
NR31 9SB G4YVN
NR32 1EH G7PWK
NR32 1JY 2E1BPN
NR32 1RY G0WCO
NR32 1SW 2E0GFV
NR32 1SW M0JNJ
NR32 1SW M6SHQ
NR32 1TP M1ATP
NR32 2HH G0RRI
NR32 2JS 2E0PAP
NR32 2JS G3ETP
NR32 2JS M3UGQ
NR32 2LA M6KWC
NR32 2LB M7OCM
NR32 2NW M3NNJ
NR32 2QG M3GIF
NR32 2RY G4RMT
NR32 3AQ M0CHP
NR32 3DN M6BAW
NR32 3DS M0XGK
NR32 3EL G1MPL
NR32 3EP M7FRS
NR32 3ER 2E0RZS
NR32 3ER M0RSX
NR32 3ER M6KFX
NR32 3ER M7WIS
NR32 3HS G0VMF
NR32 3JF G3AMK
NR32 3NF G3XGK
NR32 3QL 2E0TNC
NR32 3QL M6BCK
NR32 4AU G7TJZ
NR32 4DR G4CPW
NR32 4DU 2E0GGO
NR32 4JU 2E0HCG
NR32 4JU M0XCL
NR32 4JU M6HCG
NR32 4LW G8JBD
NR32 4PZ 2E0OBB
NR32 4PZ M3OBB
NR32 4PZ M6KPT
NR32 4QB G3AER
NR32 4QB M0DMR
NR32 4QB M7VAC
NR32 4SU G0JSG
NR32 4UD M7CBY
NR32 5AS G0MEE
NR32 5AS G8RHQ
NR32 5BE 2E0FDG
NR32 5BE M6XBP
NR32 5DJ G4TAD
NR32 5DW G8JUK
NR32 5HJ G4UKX
NR32 5LB G4RKP
NR32 5LL G4FCZ
NR324SU M7YAC
NR324UB G6NGA
NR33 0DX 2E0BFT
NR33 0DX M3AGE
NR33 0JR G3NHS
NR33 0LG G8HAU
NR33 0LY G8HRF
NR33 0TZ G4KDL
NR33 7BQ 2E1OZO
NR33 7BT G7JRU
NR33 7BX G0VDQ
NR33 7EY M6IBC
NR33 7HH G7MJX
NR33 7JF G8VOJ
NR33 7JG M7IJC
NR33 7PS M0DCU
NR33 7RG G4IVU
NR33 7RR M3SKB
NR33 7TF 2E0NAQ
NR33 7TF M6CPP
NR33 7TF M7JDP
NR33 7TR 2E1WCD
NR33 7TR G4CKH
NR33 7TU G6NKS
NR33 7TU M6BLP
NR33 7TY G3UPS
NR33 7UH 2E1ERU
NR33 8BL G0VCW
NR33 8BL M7BTS
NR33 8BQ G8FGY
NR33 8BY 2E0TJW
NR33 8EG 2E0JPW
NR33 8EQ 2E0SYX
NR33 8JD G3PGQ
NR33 8NE M3KPQ
NR33 8NN 2E0TTR
NR33 8NN M0SLF
NR33 8PJ G6MCB
NR33 8PJ G7BYI
NR33 8QG G1ZCS
NR33 8QQ M0DFY
NR33 8QX M7BGH
NR33 8SD G7SDM
NR33 8SJ G8YJZ
NR33 8TB 2E0VRT
NR33 8TB M6MPK
NR33 8TN M1FJG
NR33 8WJ M0ALX
NR33 8WP G2DWB
NR33 9AW G4ZMH
NR33 9BB M0JTH
NR33 9BB M3NMV
NR33 9BB M6PLO
NR33 9BG M3AGA
NR33 9DH G4HRC
NR33 9DH G4ZST
NR33 9DH G8HRC
NR33 9HA G1HQK
NR33 9HA M3TIK
NR33 9HL G1IAD
NR33 9JE G0HSN
NR33 9LR 2E0UID
NR33 9LR M7DGB
NR33 9LZ M3FVX
NR33 9NX G0ULS
NR33 9PG G7PQB
NR33 9PJ G4ZSX
NR33 9QL G7MWM
NR33 9RH M5TAM
Nr330ab 2E0FIA
Nr330ab M3FIA
NR34 0AL M7MDZ
NR34 0AU M0TDK
NR34 0EX G6PSC
NR34 7AR M1DUD
NR34 7AR M5AEF
NR34 7BJ 2E1DDJ
NR34 7DJ G7RXI
NR34 7PG M6MOP
NR34 7PQ G6LVN
NR34 7PT M1SSB
NR34 7QL M1RJG
NR34 7RD G0WQW
NR34 7TH G0LRP
NR34 7UW 2E0VKK
NR34 7UW M0VKK
NR34 7UW M6VKK
NR34 8AP G0JDL
NR34 8AQ M3NOS
NR34 8AT G0SBH
NR34 8HR M6DAS
NR34 8HR M6PSG
NR34 8PA M6AJC
NR34 8RN M6SHD
NR34 9AS G4RZR
NR34 9AS G7FWE
NR34 9AS M5MC
NR34 9BJ M7TLM
NR34 9DW M1VLS
NR34 9HN G8BJO
NR34 9JJ G4XAE
NR34 9PL G6OAU
NR34 9RH 2E0CEH
NR34 9RT G4KFS
NR34 9RT M6GVT
NR34 9UG M3PDC
NR34 9UX M7SKW
NR34 9YW G3KIJ
NR34 9YW M0DEB
NR349UT M7JAQ
NR35 1BY G6VZF
NR35 1HE G4JUV
NR35 1JG 2E0FUQ
NR35 1JG M0JIJ
NR35 1JG M7AUP
NR35 1JU G6LSW
NR35 1LE M0LYR
NR35 1LE M1TES
NR35 1ND M5BFL
NR35 1PP M6JAR
NR35 1QR 2E0KXR
NR35 1QR M6KXR
NR35 1RE G0JKI
NR35 2BW M3RCG
NR35 2LP G8JWT
NR35 2NE M6LIH
NR35 2QP M1DHT
NR35 2QZ M6JJU
NR35 2SF 2E0XDI
NR35 2SF M3XDI
NR35 2SF M6PUD
NR4 6AA G3CIM
NR4 6AF 2E0HIK
NR4 6AF M6FHE
NR4 6EY G6WAO
NR4 6HA G8ZGS
NR4 6JD G3GLA
NR4 6LT G4NZQ
NR4 6LT G7MKP
NR4 6PS M0SWO
NR4 6PY 2E0NUD
NR4 6PY M0NUD
NR4 6PY M6NUD
NR4 6QB G0PFN
NR4 6RA M0RLW
NR4 6XH M0EVK
NR4 7EA M6GNO
NR4 7ED G0MQI
NR4 7ET M0CKL
NR4 7EU M1ACT
NR4 7HP G8VLL
NR4 7HP M1BKF
NR4 7JA G3SCV
NR4 7JW G4NOL
NR4 7LT G0FDT
NR4 7LZ 2E0HTJ
NR4 7LZ M6HTJ
NR4 7PD G4VUV
NR4 7PY G0IYY
NR4 7QY M3IGA
NR4 7RY G0CKH
NR4 7SE M0HBL
NR4 7SF G4KAU
NR4 7TJ 2E0CHU
NR4 7TJ M0HEJ
NR4 7TJ M6ATL
NR46HZ M6WYM
NR5 0EH 2E1GTE
NR5 0EH G4ARN
NR5 0EH M3AEZ
NR5 0EZ M3DKC
NR5 0HB G4ONF
NR5 0HG G6KCG
NR5 0HL 2E0FIJ
NR5 0HL M0KRK
NR5 0HL M6FIJ
NR5 0HX G1TPQ
NR5 0HZ 2E0OCL
NR5 0HZ 2E0ODT
NR5 0HZ 2E0OEE
NR5 0HZ G0NNS
NR5 0HZ G0ODR
NR5 0HZ M0OCL
NR5 0JA G6TAF
NR5 0JA G7DTR
NR5 0JB G4KQY
NR5 0JD G8UJO
NR5 0LP 2E0IMW
NR5 0LP M0IMW
NR5 0LP M6IMW
NR5 0NH G6KAI
NR5 0QH 2E0NXL
NR5 0QH M6NXL
NR5 0RU 2E0EPZ
NR5 8AS G0SMH
NR5 8ED 2E0GCO
NR5 8ED M3TJL
NR5 8ED M6EXY
NR5 8HW M6XTD
NR5 8RA G4RMN
NR5 8RN 2E0CFG
NR5 8RN M6AOT
NR5 8TE 2E0AZF
NR5 9AX G4WTD
NR5 9BH G3THV
NR5 9LB G1DXQ
NR5 9NN M7PJI
Nr58dq 2E0PAS
Nr58dq M3BVY
NR6 5DE 2E0IBP
NR6 5EL 2E0BWA
NR6 5EL M6SVF
NR6 5EP G4PFZ
NR6 5HF G4GVR
NR6 5HU 2E0DBS
NR6 5HU M3DBS
NR6 5NN G6UPI
NR6 5PD M6FIG
NR6 5PE G4YEJ
NR6 5PQ 2E0NGR
NR6 5PQ G0CUL
NR6 5PQ M6EIZ
NR6 5PQ M6NJG
NR6 5RQ G0ULI
NR6 5SJ G4VWE
NR6 5SW G4YGD
NR6 6DG M1DPU
NR6 6JF 2E0KNE
NR6 6JF 2E0LSX
NR6 6JF M0LSX
NR6 6JF M3LSX
NR6 6JF M3NKW
NR6 6JF M3UYY
NR6 6LN M3ILY
NR6 6LZ 2E0DRT
NR6 6LZ M6MUD
NR6 6PB G4XNP
NR6 6PL G8SDU
NR6 6PU G0ANN
NR6 6RB G0FKG
NR6 6RZ 2E0JKB
NR6 6RZ M0HDC
NR6 6RZ M0JKB
NR6 6RZ M6JKB
NR6 6RZ M6ZLA
NR6 6TD M3PEQ
NR6 6UA G7DVO
NR6 6UX M1AEH
NR6 6XD G3IOR
NR6 6XF G4EYT
NR6 7BB G6TGQ
NR6 7BG G6EJM
NR6 7BL G0VDO
NR6 7DP G4WEE
NR6 7DR G4YGA
NR6 7HA 2E0GST
NR6 7HA M0GSZ
NR6 7HA M6GST
NR6 7HA M6RND
NR6 7HE G4JSD
NR6 7HE G4UIY
NR6 7HF G3EVA
NR6 7HR G8EEY
NR6 7HR M0NBD
NR6 7HR M0NCR
NR6 7HR M0RGF
NR6 7HR M0RGY
NR6 7HR M0WHQ
NR6 7HX M0ADZ
NR6 7LG 2E0JZU
NR6 7LG G4AUV
NR6 7LG M6JZU
NR6 7LJ M1JLM
NR6 7NB M0XWS
NR6 7NL G1NCN
NR6 7PH G0ULG
NR6 7RQ G4LSG
NR6 7RQ G4TWS
NR6 7RQ G4TWT
NR66HD 2E0IBR
NR66HD M7DPT
NR7 0AS G6TLP
NR7 0BQ G1KJH
NR7 0DS M1AUW
NR7 0GY M6DYD
NR7 0JJ G3HQS
NR7 0JJ G4VYH
NR7 0JS 2E0GUJ
NR7 0JS M0IWO
NR7 0JS M6GUJ
NR7 0LY 2E1IKM
NR7 0NA M1ADX
NR7 0SF 2E0TAZ
NR7 0SF M3IPD
NR7 0UD M3XEG
NR7 0US M6VST
NR7 0XZ M6NME
NR7 8AA G7AJX
NR7 8AD G3HKD
NR7 8EE G6OUM
NR7 8EG G0UEB
NR7 8EJ G4AJO
NR7 8GX 2E0OBO
NR7 8GX M6RSB
NR7 8NQ G4LGB
NR7 8QA M7ODY
NR7 8RG M6LXM
NR7 8TF 2E0YWP
NR7 8TF M3YWP
NR7 8XH G4PCZ
NR7 9AD G7SNC
NR7 9DZ M3WYG
NR7 9HW M3IXM
NR7 9LG G6AIO
NR7 9LL G0BBV
NR7 9LL G4OLP
NR7 9RR 2E0NNE
NR7 9RR M6NNE
NR7 9TW 2E0CIS
NR7 9TW 2E0TCZ
NR7 9TW M6BAV
NR7 9TW M6UAS
NR7 9UH M6KDD
NR8 5AT M6TCE
NR8 5AU G3PZX
NR8 5AU G3YIR
NR8 5AW M6CBQ
NR8 5AW M6VWN
NR8 5AX G6LZM
NR8 5DP G3EZB
NR8 5DT M3PFN
NR8 5EY M6LWP
NR8 6AB G4SVI
NR8 6AB G4XVI
NR8 6BD G7IGR
NR8 6BU M6XAI
NR8 6FZ M7EOE
NR8 6GA G4URS
NR8 6GA M6YLE
NR8 6GG M6JSJ
NR8 6LA M6IWE
NR8 6LD G7NIX
NR8 6LT G7AYQ
NR8 6NN G0CBO
NR8 6PP M0MRQ
NR8 6PP M3NON
NR8 6QU G3VYG
NR8 6QY 2E0FKJ
NR8 6QY M0TUM
NR8 6QY M7RGR
NR8 6SL M0MBI
NR8 6TR G7LBO
NR8 6UU M0TTN
NR8 6UU M6SBK
NR8 6UZ M3VDZ
NR8 6XZ G7SQW
NR8 6YE G7WEW
NR9 3AB G3WHG
NR9 3EP G0PJI
NR9 3EZ G0AHO
NR9 3HQ M6CHT
NR9 3HS M6XTL
NR9 3JP G4NST
NR9 3JP G6WQN
NR9 3JY G4YKZ
NR9 3RH M7BLX
NR9 4AB G3PTB
NR9 4AD M0TNE
NR9 4AF M3ULB
NR9 4DT M0IVS
NR9 4DT M7ACG
NR9 4HA 2E0JRR
NR9 4JB G3SEM
NR9 4NY 2E1FDM
NR9 4PP G4AFF
NR9 4QJ M1EXQ
NR9 4QS G0PSY
NR9 4QS G7SNR
NR9 5AT M0NTZ
NR9 5BB M3RWK
NR9 5BY G0TCO
NR9 5EJ M1CQS
NR9 5NL M7GDG
NR9 5QT 2E0KPL
NR9 5QT M6LPK
NR9 5SH G3FEW
NR9 5TF G3KVT
NR93DH M0CKB
NW1 0LJ 2E0BGC
NW1 0LJ M3MUP
NW1 2JL G1YPZ
NW1 5ND M0MLV
NW1 6UE G3LAU
NW1 8TH G1BNV
NW1 8TL M1EXW
NW1 8UB M5AHF
NW1 9BX G4RPK
NW1 9PN M7BBT
NW10 0DQ M3UQD
NW10 1BU G0NLN
NW10 1PA M3XAY
NW10 1PL G3THQ
NW10 4EG M7ROT
NW10 4HJ M7VAD
NW10 6QY M0GPQ
NW11 0DR M7FTQ
NW11 6UP G4PLX
NW11 6XY M6NOA
NW11 7AA G4DBQ
NW11 7DL G3GAF
NW11 7DL G4HXX
NW11 8SG G3RDG
NW11 8TE G1INJ
NW11 9NR M7FCA
NW2 1BP M3OXY
NW2 1DY 2E1MHB
NW2 1DY G6IWZ

NW2 1DY M7MHB
NW2 1EF M6ODH
NW2 1EF M6OVS
NW2 2AR G0VPV
NW2 3NN 2E0SJH
NW2 3NN M3KYG
NW2 3SN 2E0NQU
NW2 3SN M0NQU
NW2 3SN M3NQU
NW2 3SP M0SOV
NW2 3TL G1DWC
NW2 4JS M7TKV
NW2 4PG 2E0TAV
NW2 4PG M0RVT
NW2 4PG M6RPD
NW2 5EH G4PEF
NW2 5QE M7AUM
NW2 6RS M6EZK
NW3 1EJ G8PDM
NW3 2DW M0PCX
NW3 2JJ M0GXK
NW3 2LD G4DFX
NW3 2SE 2E0MFX
NW3 2SE M0MFX
NW3 4JP G3UGX
NW3 5DB M7INN
NW3 5TY M6TXX
NW3 7UP M0NPA
NW4 1LS 2E0CVW
NW4 1LS M0HWD
NW4 1LS M6CWG
NW4 1NP G1GIA
NW4 1TG G4GHZ
NW4 2DY M7VLA
NW4 2NS M3MML
NW4 2PN G3UML
NW4 3NB G3XVH
NW4 3TR 2E0FMX
NW4 3TR M6CZB
NW4 4EN G8JGE
NW4 4RR G3RDR
NW5 1JL 2E1GSX
NW5 2PE G7TKW
NW5 2RD M6FHD
NW5 2SY M0JWV
NW5 2SY M0JWX
NW5 4BX M6TUT
NW5 4HB G3ICN
NW5 4JB G7UZN
NW5 4SA M6HBG
NW6 1UE 2E0YVT
NW6 1UE M0RHA
NW6 1UE M6YVT
NW6 2DT G3VCK
NW6 4DG G0KVS
NW6 4EJ 2E0OAO
NW6 4EJ M0SHN
NW6 4EJ M6SHK
NW6 4HB G6LPT
NW6 4LY 2E0GMN
NW6 4LY M0ZEE
NW6 4LY M6USD
NW6 5LR M1FIP
NW6 7LA M6MCW
NW7 1GB M7RJT
NW7 1JL G3MYG
NW7 1NH 2E0NAR
NW7 1NH M6NRA
NW7 1RS G0JUE
NW7 2JY G8OMK
NW7 2LH 2E0CTV
NW7 2LH M0HJW
NW7 2LH M6CJX
NW7 2RE G3WZO
NW7 4BY M7RTE
NW7 4QP G1NRM
NW8 0HQ M6EKC
NW8 7EP M0SEO
NW8 8HR 2E0KBP
NW8 8HR M0KBP
NW8 8SX G3KCT
NW8 9GS M0OSO
NW9 0DJ 2E0JSO
NW9 0DJ M3JSO
NW9 5AZ G1IAG
NW9 5BS M0KWB
NW9 5HF G7GSF
NW9 5RE M3TKN
NW9 5UQ M3CXX
NW9 6EU G0FAB
NW9 7PG M6FHB
NW9 8ES M6ESY
NW9 8JL G1AWF
NW9 9NG G1DMS
OL1 2JS G0FZF
OL1 2QG 2E0WBG
OL1 2QG M0WBS
OL1 2QG M6WBS
OL1 2TX G1YMY
OL1 2UA G0OCF
OL1 4BJ G1WYA
OL1 4NP M6ZEW
OL1 4PB M3MUO
OL1 4QB M0SSO
OL1 4QB M7BJS
OL10 1AS 2E0YMT
OL10 2AU G1NXR
OL10 2LH M1CVL
OL10 2LH M3EBG
OL10 2LH M5DZH
OL10 3AF G6FJP
OL10 3FN M7MLF
OL10 3FN M7NCC
OL10 3JB G8JCL
OL10 3RR G4GSL
OL10 3RR G8TML
OL10 3RS G0RFE
OL10 4DH M6MVN
OL10 4DH M6MZN
OL10 4JH M6MDL
OL10 4PW 2E0LJS
OL10 4PW M3YBT
OL10 4QD G7NJE
OL10 4RQ G4TQL
OL10 4ST 2E0MAN
OL103JG M6NTL
OL11 1DT G7NIH
OL11 1QS 2E1HWI
OL11 2AG M3ROU
OL11 2BN G3ZRB
OL11 2BN G8EEA
OL11 2EB G0JSF
OL11 2EB M6JCB
OL11 2LN G4TDW
OL11 2NA 2E0TCO
OL11 2NA M0TCO
OL11 2NA M6TKC
OL11 2QB G1ITS
OL11 2TA M7SGB
OL11 3JG G4TDU
OL11 3LF G4HHS
OL11 3LG 2E1GGL
OL11 3LG G7BRJ
OL11 3QG 2E0FAC
OL11 3QG M0PEB
OL11 3QG M3SQO
OL11 4AX 2E0DTX
OL11 4AX M0LDZ
OL11 4AX M7LKS
OL11 4BT G3TOJ
OL11 4HW G3DTP
OL11 4PP G3XPI
OL11 4QF M3XGD
OL11 5AW G6CSN
OL11 5JD G0WHL
OL11 5NB G3RLE
OL11 5QA G0LSP
OL11 5QS G6GXK
OL11 5QW G0TFL
OL11 5RT G7TOB
OL11 5TG 2E1DEN
OL11 5UP 2E1GJC
OL11 5XA 2E1GKB
OL11 5XE 2E1MDC
OL11 5XE M0TVL
OL11 5XN 2E0DHB
OL11 5XW G0WTG
OL11 5YG M5ALA
OL113 8QJ 2E0CWB
OL113 8QJ M0TXK
OL113 8QJ M6CQH
OL12 0AS G4XKC
OL12 0HX M0DZX
OL12 0SH G1ZUH
OL12 6DZ G4OCJ
OL12 6LZ G1CIA
OL12 6UH M5ADI
OL12 7JN G4GAI
OL12 7JN G7AFH
OL12 7JU M6KES
OL12 7NJ M3XFZ
OL12 7NX G0OXP
OL12 7PR 2E0AWK
OL12 7RT G0SIY
OL12 7RT G7LKR
OL12 7RU M7CXP
OL12 7SU G7UCP
OL12 8PX G4TMV
OL12 8QB M0SPK
OL12 8SU 2E1TKD
OL12 8SU G8TKD
OL12 8UX M0VOM
OL12 8UX M6XVJ
OL12 9AJ G0VRX
OL12 9an M7ANO
OL12 9LT 2E0FTL
OL12 9NH G6BWK
OL12 9NY M6TYN
OL12 9QH 2E0HYD
OL12 9QP G7OAI
OL12 9TY G3RIK
OL12 9UW M6THI
OL12 9UX G4TOZ
OL120AS 2E0KGM
OL120AS M7DMF
OL128UQ G6NID
Ol129pz M7PJB
OL129SW M6ZCL
OL13 0TA M3VSH
OL13 8BH 2E0RBC
OL13 8GE M0DVW
OL13 8QL 2E0XCM
OL13 8QL M0LEX
OL13 8RN G8TYY
OL13 9DQ 2E0SHN
OL13 9DQ M7FXD
OL13 9HS M7CUW
OL13 9LS 2E1AHU
OL13 9LS G7KLT
OL13 9PH M0CEB
OL13 9RL G7JQF
OL14 5BU G0TVO
OL14 5EQ G1LCZ
OL14 5ET G6LLU
OL14 5LZ G4HYY
OL14 5NX G3OLP
OL14 5PE G7RFO
OL14 5PH M6MLP
OL14 5PH M6OOC
OL14 5RU G8WWO
OL14 6AW 2E0WTR
OL14 6AW M6JZV
OL14 6AX G4KDU
OL14 6AX M3FQA
OL14 6HB M6BZV
OL14 6HN G8OBR
OL14 6HN G8VKP
OL14 6ND G4TYW
OL14 6PN M7XJP
OL14 7ET G1JBW
OL14 7NF G4HSX
OL14 7QF 2E0GPK
OL14 8DS G8IC
OL14 8EP G4RCJ
OL15 0BP 2E1DKU
OL15 0DS G0OWH
OL15 0DS G3RJV
OL15 0HQ G7BPM
OL15 0NG G8NQI
OL15 8EB G3DNN
OL15 8EN G0BLM
OL15 8EN G6NCL
OL15 8EZ G4IWO
OL15 9AH M7AIM
OL15 9DA 2E0SCV
OL15 9DA M6BKI
OL15 9EF G4OUJ
OL15 9HE M0WLB
OL15 9HE M6KQP
OL15 9JE G8TTU
OL15 9LJ 2E0JHP
OL15 9QQ G7LHV
OL16 2DD G0IPN
OL16 2DD M3ULL
OL16 2EG M6EXT
OL16 2YH M3SMK
OL16 3PJ G0JFP
OL16 3QZ G1FHI
OL16 3UJ G4PJE
OL16 3UN M3BVX
OL16 4AY G0WSY
OL16 4EL M6GZR
OL16 4EP G7EQK
OL16 4PG M3VZP
OL16 4PG M7NOV
OL16 4QU 2E0GEC
OL16 4QU M0LXG
OL16 4QU M6TXG
OL16 4RJ G6TVE
OL16 4SE M5AFE
OL16 4SH G7LFM
OL16 4TY M0FDX
OL16 4UB G0UFD
OL16 4XG M3XHK
OL16 4XP 2E0ZGA
OL16 4XP M0ZAM
OL16 4XP M3ZGA
OL16 4YD G3SAO
OL16 5BB G0FNM
OL16 5BB G0GUY
OL16 5BB G0KZH
OL16 5HJ M6NNL
OL2 5AS G3WXN
OL2 5HL G8YKO
OL2 5SD G6PBI
OL2 5TQ 2E0MEQ
OL2 5TQ M6MHQ
OL2 5XY 2E0TOM
OL2 5XY M0TOM
OL2 5XY M7MOT
OL2 6BH G1UTP
OL2 6JY 2E0GHB
OL2 6JY M0EAM
OL2 6NJ G7MXQ
OL2 6PF M7HJM
OL2 6PG G1GKR
OL2 6RW G0BJR
OL2 6RW G1ORC
OL2 6RW G4ORC
OL2 6SR 2E0PUS
OL2 6SR M0GIE
OL2 6SR M3LQA
OL2 6TP G0KUY
OL2 6TW G1UKW
OL2 7AY G4GNG
OL2 7DA G8CZI
OL2 7JT M0CRZ
OL2 7JT M3SZC
OL2 7NA 2E0HVV
OL2 7NA M7DCV
OL2 7NQ M6ATY
OL2 7QA 2E1OBI
OL2 7SX M3MDI
OL2 7TX M3AUB
OL2 7UU 2E0AMS
OL2 7UU M0ADD
OL2 7YG M6LKW
OL2 7YN 2E0JCX
OL2 7YN M6FDS
OL2 7YQ M3OFN
OL2 8BG G6OVX
OL2 8EA G7MCT
OL2 8HY G0MMX
OL2 8JF G0PVO
OL2 8JJ G7LMI
OL2 8LP 2E1DTE
OL2 8NE M1FEW
OL2 8QA M7MOL
OL2 8SW M6XTY
OL26JX M7GJF
Ol27tq M7AWD
OL3 5DS G8ZZW
OL3 5NT M7RDZ
OL3 5NU G0EOX
OL3 5PL G1GZK
OL3 5PL G1KFG
OL3 5PW G7ENC
OL3 6EB G4AIW
OL3 7AW G0HFC
OL3 7EF 2E0TZM
OL3 7EF M3TMZ
OL3 7EJ G3SNA
OL4 1LX G3TJX
OL4 2DJ G1JZU
OL4 2PE G7EBF
OL4 2PN M6MIC
OL4 2PW G4UFG
OL4 2QQ G7ICE
OL4 2TH G1XGW
OL4 3LW G4AVF
OL4 4AZ G0EYU
OL4 4AZ G4EBQ
OL4 4ED G4SSC
OL4 4EN M7FBC
OL4 4RY G6DWD
ol4 5dp M7CDB
OL4 5LF G4PLV
OL4 5QD M7WVM
OL4 5SH G0RKE
Ol44sj G6TAI
OL5 0SN G3RKM
OL5 9DE 2E0SBN
OL5 9DE M6HBB
OL5 9JN G0FCX
OL5 9LQ G6EJT
OL5 9LQ G8GVZ
OL5 9PD G1JAA
OL6 6TX G6YBV
OL6 7DU G0SNQ
OL6 7HG 2E0RGY
OL6 7HG 2E0ZEN
OL6 8BP G4RGM
OL6 8BP G6GMR
OL6 8BP M0NCZ
OL6 8BQ G6IQC
OL6 8BU G6IML
OL6 8DS 2E0OQZ
OL6 8DS M0GMQ
OL6 8DS M3OQZ
OL6 8QN G0AQS
OL6 8RJ G7MRO
OL6 8RP G6BRD
OL6 8SX M3OBN
OL6 8UA M3OHC
OL6 8UT 2E0TJI
OL6 8UT M6ETR
OL6 9AL M6JRL
OL6 9AP G3BPQ
OL6 9DU G7DGC
OL6 9EG M6PVC
OL6 9RH G7TSQ
OL7 0DU G1GWJ
OL7 0JD G4GHB
OL7 9AN G0MRD
OL7 9HN G3WI
OL7 9NL M0AAD
OL7 9QY G3NTF
OL70DU G7NHC
OL8 1PP G0YCE
OL8 2AN M6DKL
OL8 2HB M6MDB
OL8 2HG 2E0MCK
OL8 2HG M0MBM
OL8 2HG M3UXF
OL8 2HG M6CLA
OL8 2HG M6LEA
OL8 2LP G1HAC
OL8 2LP M3HAC
OL8 2NS M0CLK
OL8 2PD M1BGF
OL8 2XE 2E0FIT
OL8 2XE M0NVQ
OL8 2XE M6RCL
OL8 3AD M6REE
OL8 3BA G0MGL
OL8 3JH M6CIG
OL8 3LQ M1CZZ
OL8 3NP M3KMW
OL8 3RN M3XRD
OL8 3RY M6OLE
OL8 3TZ M6KCI
OL8 4JB M0KGM
Ol82hb M6PYZ
OL9 0EJ G8JRZ
ol9 0ly M7JOD
OL9 0PH M5BGR
OL9 0QX 2E0JSN
OL9 0RF G1OFS
OL9 0RW M1FET
OL9 6RQ M7FCH
OL9 8AE G6SYL
OL9 8AP G4TLW
OL9 8DA M0ZOV
OL9 8HH G3SMZ
OL9 8QD G4BKH
OL9 8QJ 2E0KDE
OL9 8QJ M0KDE
OL9 8RG G6MMS
OL9 9QR G3IGC
OL9 9RR M3NYQ
OX1 1AD 2E0KKL
OX1 1AD M7NMM
OX1 3BW G7VDJ
OX1 3BW M6FID
OX1 4RQ M0INX
OX1 5HW G7OGO
OX1 5JP G3SLI
OX1 5LL 2E1ABE
OX1 5LS G4AZA
OX1 5LX G3KCZ
OX1 5LY G0RFS
OX1 5NF G3JFR
OX1 5NH 2E0ADA
OX1 5NH G0REL
OX1 5NH G0RJX
OX1 5NH G0RKG
OX1 5NZ G6YTW
OX1 5PE G0SPS
OX1 5PE G7KGH
OX1 5PE M6BLA
OX1 5PP M6XMR
OX1 5RX G4AZT
OX1 5RY G3RTM
OX10 0HN 2E0VJO
OX10 0HN M0JSX
OX10 0HN M3VJO
OX10 0PT G6NLN
OX10 0QG G1BZW
OX10 0QT G1XNN
OX10 0SD M0WGA
OX10 6AE G3RDF
OX10 6HW G0MUJ
OX10 6LX M3LUK
OX10 6RR G3OEB
OX10 6RY G4SOH
OX10 6SH G0SPC
OX10 6SJ G4YSQ
OX10 6SJ G6IMW
OX10 6SL G4ATL
OX10 7DB G6YQT
OX10 7JE M0PIB
OX10 7LB G4ITC
OX10 7LF 2E0BCG
OX10 7LF M0DYU
OX10 7NL G1URJ
OX10 7PJ G0IDF
OX10 7PZ 2E0PWI
OX10 7RA G1ULR
OX10 8AD M6MYC
OX10 8BN 2E0SFC
OX10 8BN M3UMX
OX10 8FE G4BYR
OX10 8HP G0ADH
OX10 8LU G7TNQ
OX10 9AA G8POI
OX10 9ER G4DAT
OX10 9HD M6RYZ
OX10 9LY G8TYX
OX10 9NE 2E0SBH
OX10 9NE M6SJR
OX10 9PA G7TIB
OX10 9PE G6UXM
OX10 9PE M0BWL
OX10 9PY G6IMQ
OX10 9RJ 2E0IHE
OX10 9RJ M7LDJ
OX10 9RT G1CAR
OX10 9RT G1EMF
OX11 0AD G1AIO
OX11 0AD M6CGJ
OX11 0AE G8SMH
OX11 0BP M3ZQC
OX11 0DP M7XNG
OX11 0DQ G8HKS
OX11 0DQ M0HKS
OX11 0DX G4MFP
OX11 0ES G0FRO
OX11 0HB G1GGT
OX11 0LG M6KTH
OX11 0PE G6HXI
OX11 0PE M3KVW
OX11 0PW G6SRX
OX11 0PW G8DVK
OX11 0RR G1URH
OX11 0SA G8BFK
OX11 0SB G8ATP
OX11 0SU M1CZI
OX11 6AB M3PQU
OX11 6FA G1HQH
OX11 6HP M0IDY
OX11 7AX G4PWI
OX11 7DB G4ZKI
OX11 7DD G6VSY
OX11 7DF G0DPK
OX11 7DT G6VMB
OX11 7HS M0IXP
OX11 7JQ G3GGG
OX11 7TZ 2E0GDH
OX11 7TZ 2E0ZOZ
OX11 7TZ M0SXH
OX11 7TZ M6AVM
OX11 7TZ M6DEC
OX11 7TZ M6EMW
OX11 7TZ M6RGO
OX11 7UT G6VLC
OX11 8BP G1SZK
OX11 8BP G4SYL
OX11 8DG M6FRF
OX11 8HD G1OWI
OX11 8HD G1SSL
OX11 8HD G7AKP
OX11 8HD M3SSL
OX11 8HP G0BSA
OX11 8HP G0MQM
OX11 8HY M6CRH
OX11 8LG G4EQM
OX11 8SW G1UPX
OX11 8TE G4XYM
OX11 8TX 2E0ZMM
OX11 8TX M0NUT
OX11 8TX M3ZMM
OX11 9JX G8CUL
OX11 9JX G8NVI
OX11 9JX M0CUL
OX11 9JX M6YHS
OX11 9LX M0MTI
OX11 9NS G4IYC
OX11 9PU G4SAJ
OX11 9PU G6DPL
OX118HW G6SBN
OX12 0AT G8PTY
OX12 0BN G3VHE
OX12 0DA M1EWM
OX12 0DA M1JIM
OX12 0HA G3RRS
OX12 0HA G3VPW
OX12 0HT G1YDG
OX12 0HU M1EHV
OX12 0LB G1DYL
OX12 0LD G3YSQ
OX12 0LJ G0JEK
OX12 0LJ G6LRY
OX12 0NY G0LCB
OX12 0PY G1NDV
OX12 7BG 2E0MTL
OX12 7BG M0TGF
OX12 7BG M6NXY
OX12 7DF G0OAB
OX12 7DF M6XDB
OX12 7DN G4DPA
OX12 7EP 2E0LNU
OX12 7EP G6LNU
OX12 7EP M3LNU
OX12 7EQ M6ROI
OX12 7EW G0HLS
OX12 7HP G4BGH
OX12 7LD M7COU
OX12 7NQ G6UUQ
OX12 7PN G1IYB
OX12 7QF G4XIW
OX12 7QF M6CGZ
OX12 8DL G4HLX
OX12 8NS 2E0EXW
OX12 8NS 2E0GSH
OX12 8NS M3RNW
OX12 8NS M6NFS
OX12 8RB G0CEI
OX12 9AY 2E0RTW
OX12 9DN M1RJL
OX12 9QN M1SFS
OX12 9JP 2E0HUK
OX12 9JP M3NTR
OX12 9LE M0LCS
OX12 9RG G7WCG
OX12 9YE M3CHU
OX13 5BX G3YFM
OX13 5BZ G4KDN
OX13 5DB G3SSW
OX13 5HL G3YIK
OX13 5HL G3ZHL
OX13 5HL M6BVV
OX13 5HS G0AOZ
OX13 5JJ G0SXG
OX13 5LN G1FBQ
OX13 5LR G3VHS
OX13 5PF G4BRK
OX13 6BL G7WJZ
OX13 6DU G6ZKY
OX13 6LP G0KMW
OX13 6NU G4AEV
OX13 6QJ G0LMX
OX13 6RZ G4JLV
OX13 6TF G1KXX
OX14 1DW G1KEI
OX14 1EQ G4WXC
OX14 1JQ M6PIA
OX14 1JT M6KZW
OX14 1LS 2E0CGW
OX14 1LS M0NIB
OX14 1LS M6BWN
OX14 1LS M7STU
OX14 1NZ M6RST
OX14 1PP G7TRV
OX14 1PR G1RJW
OX14 1PR G3IWW
OX14 1PT M1FII
OX14 1QE G6FJI
OX14 1QU M0CIW
OX14 1XA G3ZSX
OX14 1XB M0GRA
OX14 1XB M0HEY
OX14 2BH M6UKI
OX14 2BJ 2E0KHO
OX14 2BJ M0KHO
OX14 2BL G1IVK
OX14 2EP G1OVY
OX14 2ES G7CAG
OX14 2ND M0AEN
OX14 2NG G0ROS
OX14 3DR M1BQT
OX14 3JF G3HKH
OX14 3NF M0WPT
OX14 3NF M6EPT
OX14 3PE G6ZHB
OX14 3QF M3NOR
OX14 3QQ M6IKY
OX14 3SD G4RKF
OX14 3SR G0OYS
OX14 3SW G7VGA
OX14 3SX G0NPA
OX14 3TD G6MRP
OX14 3TE 2E0CFH
OX14 3TE G8NRP
OX14 3TE M0NRP
OX14 3TE M6YMA
OX14 3TR G4SQI
OX14 3XB 2E1HGA
OX14 3XB G8KIG
OX14 3XB M3HGA
OX14 3YN M1SLH
OX14 4DA M3ZAE
OX14 4DF M3XAG
OX14 4GN M0IJI
OX14 4HY 2E1BMJ
OX14 4JX G1BDP
OX14 4LG M1HFM
OX14 4NL G3MZC
OX14 4NN 2E0ZSA
OX14 4NN M3ZSI
OX14 4PU G7LUK
OX14 4QB 2E0MHM
OX14 4QB M6MLQ
OX14 4XA M7CDT
OX14 5HF M6IZB
OX14 5JD 2E0KPS
OX14 5LW G4JEO
OX14 5NE 2E0UBH
OX14 5NE M6UBH
OX14 5NG G6PBZ
OX14 5NW G0RWJ
OX14 5PP G0OMZ
OX14 5PX M7CIU
OX14 5RF 2E1HNN
OX14 5RL G0INZ
OX14 5RN G1OQB
OX14 5RN G7DOF
OX14 5RQ G6PEP
OX14 5TN 2E0GDG
OX15 0SR G6STD
OX15 4AX M6JFW
OX15 4BB G8JPW
OX15 4DY M6ZZK
OX15 4EZ G0LXY
OX15 4HA M7HEF
OX15 4JE G1PUY
OX15 4JL 2E0FKS
OX15 4JL M3FKS
OX15 4NT 2E0HAN
OX15 4NT M3UHH
OX15 4QU M6SJU
OX15 5AP G0UWB
OX15 5HW 2E0YBS
OX15 5HZ M7DKJ
OX15 5JB M0ZZA
OX15 5LH G7KYH
OX15 6AQ M7DXJ
OX15 6BJ M1BMV
OX15 6DS G6SLG
OX15 6DZ M1EDA
Ox15nh M7EGA
OX16 0BS M3WGK
OX16 0DB M1CNJ
OX16 0EX M7BQZ
OX16 0HL 2E0ECP
OX16 0LU G1XRQ
OX16 0QW M6DNL
OX16 0RR 2E0SMS
OX16 0RR G0LDB
OX16 0RR M3SYL
OX16 0RZ M6CPS
OX16 0SB G7CEB
OX16 0SR M6NFB
OX16 0UG G8ZRU
OX16 0XF M3YCJ
OX16 1AA G8ZLN
OX16 1BQ M6FUM
OX16 1DJ M6SKH
OX16 1EW M0HVK
OX16 1GQ G7LNK
OX16 1PQ G4XPY
OX16 1UE M1AOG
OX16 1UN M6OWS
OX16 1XA M6GMC
OX16 2AZ G6ZDS
OX16 2NF G7UUW
OX16 3QS 2E1IBJ
OX16 3WX 2E0BWV
OX16 3WX M0UXO
OX16 3WX M3ITT
OX16 3ZW 2E0NCV
OX16 3ZW M0NCV
OX16 3ZW M6NCV
OX16 4SD M3YEU
OX16 4SD M6JMG
OX16 5BE 2E0NQF
OX16 5DF M6WTW
OX16 5DR 2E0LGO
OX16 5DR M6ZLD
OX16 5DW 2E0TEN
OX16 5DW M0RDI
OX16 5DW M6AMR
OX16 5HQ M6ETJ
OX16 5LR M3FUD
OX16 9AJ 2E0TAO
OX16 9AJ M0GUF
OX16 9AJ M3UQI
OX16 9AP G0BJI
OX16 9AR G8KSH
OX16 9AR M3KSH
OX16 9BQ G1IIO
OX16 9BQ G7ILJ
OX16 9DH G8DCX
OX16 9EL G6TAS
OX16 9EL M6LVJ
OX16 9ES M1CJZ
OX16 9JU G0BRA
OX16 9JU M0BJN
OX16 9LB G4RWV
OX16 9NW G4TXA
OX16 9QB 2E0ZDA
OX16 9RZ M0AXN
OX16 9SR G1HXZ
OX16 9ST M6KGE
OX16 9SZ G6POG
OX16 9TA G7LKC
OX16 9TW G0MEC
OX16 9YZ M7EOS
OX17 1BE G1KKE
OX17 1BU M1CZF
OX17 1DH G6CFU
OX17 1GJ M6CKF
OX17 1JA G1KIB
OX17 1LU M0BRI
OX17 1SS M6XKJ
OX17 2AB M1AHY
OX17 2DZ M6DMN
OX17 2HL G8CIJ
OX17 2NP G8MFI
OX17 2QQ G4DKD
OX17 2RX M6BFX
OX17 3AE G1AHQ
OX17 3DG G7VYT
OX17 3FE 2E0NBX
OX17 3FE M0IEZ
OX17 3FE M6NBX
OX17 3FE M6XCJ
OX17 3HA M3RZP
OX17 3HR G3YZV
OX17 3LB M3RQJ
OX17 3PF G3OPH
OX17 3PH G1EYW
OX17 3RG G6KFR
OX17 3RJ G1BJE
OX17 3RJ M3RUR
OX17 3RS M6ZEK
OX17 3XB G0BIV
OX18 1DN G4YMJ
OX18 1DX G3KIL
OX18 1ED G7SRB
OX18 1FF G6HEJ
OX18 1JX M7TFT
OX18 1TT 2E0CHT
OX18 2HX G0NZE
OX18 2LH 2E1ESK
OX18 2NN G4RJY
OX18 3AZ G7KDM
OX18 3LS G1WYD
OX18 3LX 2E0KZN
OX18 3LX M6KZN
OX18 3PR G8KOD
OX18 3QZ G6OUI
OX18 3YB 2E0BHA
OX18 3YB M3SHZ
Ox181th M7GJL
OX2 0BB 2E0EDM
OX2 0BB M0TEK
OX2 0BE G3BLS
OX2 0HN 2E0YXB
OX2 0HN M3YXB
OX2 0HS M0MLG
OX2 0HS M3LNY
OX2 0HS M3MLG
OX2 6AY G1OWM
OX2 6HY M6GOC
OX2 6QE G0AXI
OX2 6TX G4ZHE
OX2 6TX M3XZG
OX2 6UD M0HDN
OX2 7EN G0FGP
OX2 7EN G7CUB
OX2 7QG G6SRT
OX2 7TG M7JMA
OX2 8AX 2E0HNK
OX2 8AX M6HNB
OX2 8BB 2E0RNM
OX2 8BB M3RNM
OX2 8BY G3XGC
OX2 8DG G8CVS
OX2 8PE G0AZP
OX2 8PE G6YSQ
OX2 8PX G4ZSV
OX2 8QT G1JXS
OX2 9DN G1VSO
OX2 9DS G6RXV
OX2 9DZ 2E0GGN
OX2 9FU M7RPI
OX2 9HQ 2E0FGH
OX2 9HQ M6FGH
OX2 9HZ G3UOM
OX2 9JG G6ASA
OX2 9PA G1LVW
OX2 9PE 2E0FZJ
OX2 9PE 2E0RQN
OX2 9PE M0JHF
OX2 9PE M0RQN
OX2 9PE M6FZJ
OX2 9PE M6RQN
OX2 9PR M0OXD
OX2 9PR M3OXD
OX2 9QA G4MQR
OX2 9SN G0CFN
OX20 1JT M6GBJ
OX20 1JY G4POL
OX20 1RZ G6HRA
OX25 1PX G1FZR
OX25 2AH G7AHB
OX25 2NF G7LGV
OX25 3RD M6CWP
OX25 4AW G3NSP
OX25 4SQ M0DKU
OX25 5BE M0LEB
OX25 5TH 2E0LOH
OX25 5TH M0LOH
OX25 5TH M6PLR
OX25 6LP G0REV
OX26 1TZ G4ATG
OX26 1TZ G8GNI
OX26 1TZ M5AEX
OX26 2AR G7ILP
OX26 2DZ G1CEO
OX26 2FP M7SYD
OX26 2GS G4SYV
OX26 2LP 2E0CYT
OX26 2LP M6CWZ
OX26 2LR G4BKB
OX26 2YA G8EKN
OX26 3FD M3CVW
OX26 3GA 2E0JCU
OX26 3SR 2E0DAB
OX26 3SR M3DAB
OX26 3YB G6HPT
OX26 3ZJ G3UTE
OX26 4HB G0TSK
OX26 4UH G6PLF
OX26 4XA M3GRF
OX26 4XJ 2E0GBE
OX26 4XX M1TAZ
OX26 4YA 2E0PSC
OX26 4YA M0PSC
OX26 4YA M6PSC
OX26 5AH 2E0GPJ
OX26 5AH M0JKV
OX26 5DW G6FJO
OX26 5DX G0LUQ
OX26 6FE M7AQC
OX26 6SR G0JRY
OX26 6UQ G6FPP
OX26 6YP G1BYQ
OX26 6YR G8UWD
OX27 0AP G7DBN
OX27 0EY G1FXC
OX27 0HN M0JJD
OX27 0HP G6PHH
OX27 7PA 2E0BPN
OX27 7PA M3ZVX
OX27 7PX M3MVJ
OX27 8BQ M6APO
OX27 8DF G4FTA
OX27 8DR G1WMK
OX27 8ED 2E0BXC
OX27 8ED M3UMW
OX27 8FL M0GKA
OX27 8FQ M0BWS
OX27 8TU M0CJN
OX27 9AZ G4BII
OX28 1BJ M6NRM
OX28 1HQ G4IOK
OX28 1NS G4EGN
OX28 2EG M0BLV
OX28 2EG M0BRP
OX28 2JB G4GUN
OX28 5FS 2E0FGV
OX28 5FS M0OQO
OX28 5FS M6NFQ
OX28 5JQ G7CKP
OX28 5JQ G8LKB
OX28 5JT G0CQZ
OX28 5LJ 2E0GTU
OX28 5LJ M7RCZ
OX28 5NL G4YUA
OX28 5NZ G1WYN
OX28 6EG G8RDA
OX28 6NL 2E0GDY
OX28 6NL M3GDY
OX28 6NN G8EWT
OX28 6NN G8PX
OX29 0RQ G6UPQ
OX29 0RT G6ZNT
OX29 4BU 2E0BMH
OX29 4BU G2KQ
OX29 4BU M3SSO
OX29 4EW G1RCX
OX29 4EW G6LDO
OX29 4LU G7OQQ
OX29 4NP 2E0HTG
OX29 4NP M6YCG
OX29 4NT G6ANV
OX29 4QJ G0BQZ
OX29 4QY G8JVI
OX29 5BL M7SWA
OX29 5RH G3ZYR
OX29 6SQ G1DOL
OX29 6UW G4EZG
OX29 6XL M7WBB
OX29 7PA G4DSE
OX29 7PB M3NGE
OX29 7TH G3UKC
OX29 7TH M1FRB
OX29 7YD G1VBL
OX29 8HH G1ZBG
OX29 8HR 2E0FVK
OX29 8HR M7AMX
OX29 8JP M1ENJ
OX29 8JP M1ENK
OX29 8JP M3ENJ
OX29 9UW G7VIB
OX29 9XP G8CYF
ox296py G0UZK
OX3 0EX G0BPX
OX3 0JG G6DVE
OX3 0JJ G1OYF
OX3 0NW 2E0DZN
OX3 0NW M0TKM
OX3 0NW M6HMU
OX3 0QA G1YDI
OX3 0RJ G3UGJ
OX3 0SJ M0BJE
OX3 7EE G0LBZ
OX3 7EE M3FHV
OX3 7QF 2E0FFC
OX3 7QF M0YQC
OX3 7QF M6PIC
OX3 7RH M7JFF
OX3 7RR M0DDT
OX3 7RU M6WRV
OX3 7SZ M7TEJ
OX3 8AA 2E0ETV
OX3 8AA M0ZRX
OX3 8AA M6FMZ
OX3 8AN 2E0SSF
OX3 8AN M0SVN
OX3 8AN M7SSF
OX3 8ED G4KSQ
OX3 8HX G0KPY
OX3 8JA G4AYR
OX3 8JB G3KZU
OX3 8LP 2E0OXF
OX3 8LP M6CFW
OX3 8LT M6EYT
OX3 8LT M6RXE
OX3 8PD G3UJO
OX3 8SX G8SIN
OX3 8TA G3UMF
OX3 9PB M3EGU
OX3 9QF G0AFY
OX33 1RQ M3JMW
OX33 1SQ G6DBU
OX33 1TJ G0HZQ
OX33 1TR M0LTK
OX39 4AA G0ODQ
OX39 4AE G0HHN
OX39 4EA G4RWI
OX39 4EJ M0OSE
OX39 4EU G1ZRR
OX39 4EW G0IKI
OX39 4HB G4ODI
OX39 4HP 2E0PCC
OX39 4HP M6EHP
OX39 4JY G1ECE
OX39 4JY G1GQJ
OX39 4JY G7AGR
OX39 4JY G8GQJ
OX39 4PE M1BXQ
OX39 4PL G3UNI
OX39 4RA G3ZRY
OX39 4TG 2E0WMM
OX39 4TG M0MKY
OX39 4TG M7JDW
OX39 4TP G0FRN
OX39 4UD G7IKS
OX4 1BY G0CPP
OX4 1EB 2E0LLM
OX4 1EB M3LLM
OX4 1LT G3LOF
OX4 2HF M6MTA
OX4 2JW 2E0OKS
OX4 2JW M6BHM
OX4 3BX M0OXR
OX4 3HT G1PCU
OX4 3NU G7IMH
OX4 3NW G0HEN
OX4 3QU M0ACU
OX4 3SA M6WHZ
OX4 4AR G0XBG
OX4 4BB M6NMW
OX4 4JQ 2E0DKG
OX4 4JQ M0KLB
OX4 4JQ M6ATD
OX4 4NE G6BTB
OX4 4NS M6RLC
OX4 4PW G8OMW
OX4 4XG M3WVB
OX4 6DL G7TLD
OX4 6DX G4KON
OX4 6ER G6ZPL
OX4 6ER M6ZPL
OX4 6RA M3MSY
OX4 6SS M3LVM
OX4 7FD G0TYZ
OX44 7LH G3ZKO
OX44 7RZ M3ZFH
OX44 7SJ 2E0DBH
OX44 7SJ M3ZFN
OX44 7TA M1GGG
OX44 7TA M6VKE
OX44 7TJ G3SBM
OX44 7TN G4DAP
OX44 7YH M6BOQ
OX44 7YN G0MCO
OX44 9DL G1YSX
OX44 9DL G6ZTR
OX44 9EX 2E1GSW
OX44 9HP G8ECG
OX44 9LT G4DCN
OX49 5AD G8LGM
OX49 5BL G4HFS
OX49 5HL M0JWC
OX49 5HL M7GKA
OX49 5LW G7NTO
OX49 5NQ M7BLG
OX49 5PR M6HHH
OX49 5RF G0DUF
OX5 1FY M7ESH
OX5 1HF M3IAP
OX5 1HH G0THY
OX5 1JT M1AQN
OX5 1NX M6RAR
OX5 1TH 2E1IGG
OX5 1TR G0LXW
OX5 2EQ G7OIT
OX5 2EU G0EDY
OX5 2EU G7KGI
OX5 2HG 2E0WEL
OX5 2HG M0HIZ
OX5 2HG M6SIZ
OX5 2LL G0LUN
OX5 2LT 2E0LBY
OX5 2LT M7TOY
OX5 2NA M7JIZ
OX5 2UB G3MSO
OX5 2US G8JOX
OX5 2XP G8BNB
OX5 2XZ G0XGM
OX5 3ET G4AKB
OX6 4JR G1XTA
OX7 3BJ 2E0TMY
OX7 3BJ M0TTH
OX7 3BJ M3TTH
OX7 3HB G6GEV
OX7 3RX 2E0JLM
OX7 3RX M0XJM

OX7 5DP M7BXT
OX7 5DZ G0NYH
OX7 5JS G0JZE
OX7 5JT M7BFX
OX7 5PZ 2E0HSC
OX7 5PZ M6AGI
OX7 5QH M6CUN
OX7 5SW G4TLS
OX7 5XS G1RLA
OX7 5XX G0MLQ
OX7 6AD G6KCJ
OX7 6BE M7BQH
OX7 6EA M6EZY
OX7 6ER G8BII
OX7 6GE G8DJT
OX7 6JN G8GJG
OX7 6QA G6NLG
OX7 7BP M0NRK
OX7 7EU G0MGZ
OX8 6PZ G0MKY
OX8 6PZ G0PYF
OX9 2AX G0THH
OX9 2DH M7RAZ
OX9 2EE M6KIE
OX9 2EJ M6OXF
OX9 2NE G0MVW
OX9 3AG G4FRL
OX9 3BJ G8AAD
OX9 3BY 2E0FYD
OX9 3BY G4ADR
OX9 3BY M7MHA
OX9 3DG G6GHU
OX9 3EP M6RPP
OX9 3JF G8NRR
OX9 3JS G0FOK
OX9 3LZ M7JIM
OX9 3NH G4HXU
OX9 3NJ 2E0SSD
OX9 3NJ M6YPX
OX9 3NQ G0UUR
OX9 3NQ M6TXU
OX9 3QL G1GGK
OX9 3TE 2E0BIO
OX9 3XH M1CRF
OX9 3XN M6AZR
OX9 3YD M6NOK
OX9 3ZH M6AMW
OX9 7AE 2E0KTK
OX9 7AE G5NLD
OX9 7AE M7DVD
PA1 1GT GM0SCW
PA1 2JE GM6PCW
PA1 2SX GM0UKD
PA1 3NG GM4VLX
PA1 3RT GM4MPY
PA1 3SR GM1SXX
PA10 2EL MM6DCT
PA11 3BH MM6FPI
PA11 3NB MM3NGV
PA11 3NL 2M0VGT
PA11 3NL MM7VES
PA11 3PQ 2M0JSF
PA12 4EG GM4FLX
PA12 4EL GM7RZE
PA12 4HG MM7ICR
PA12 4HL GM0KMA
PA12 4NB MM0WOA
PA12QT MM6WER
PA13 4BB GM0JVC
PA13 4HP MM6WHW
PA13 4JJ GM7ALI
PA13 4PY GM7ADU
PA13 4PY GM7ADY
PA14 5QG GM3SUZ
PA14 5QP 2M0XZA
PA14 5QP MM7PGL
PA14 5UT MM3YIH
PA14 5XF GM0WDF
PA14 6DP MM0STU
PA14 6DT 2M0MOK
PA14 6DT MM6WUL
PA14 6EF MM6MVM
PA14 6PL MM6SGO
PA14 6XP GM4HRJ
PA15 2EE GM4EGD
PA15 2HJ MM6IAB
PA15 2JD MM6KIW
PA15 2WH 2M0VZX
PA15 2WH MM7DAM
PA15 4DW 2M0OLK
PA15 4HH GM4FBU
PA15 4HW MM3MZX
Pa15 4pu MM7ADI
PA15 4SX MM0XPZ
PA15 4SX MM3SWG
PA15 4TA MM3UXO
PA16 0DS MM0CDW
PA16 0EW GM0GNK
PA16 0EW GM3YOR
PA16 0HX MM1AUG
PA16 0QF MM6LCL
PA16 0TG MM7LYZ
PA16 0YL MM3YIG
PA16 0YL MM3YIQ
PA16 7AL MM3WYI
PA16 7BJ MM3IBM
PA16 7BL 2M0BOS
PA16 7BL MM3MHQ
PA16 7LH MM3WYM
PA16 7NB MM3UNQ
PA16 8AS GM6LEZ
PA16 8NP MM0KNE
PA16 8NP MM0SRX
PA16 8QB MM6OOK
PA16 8QG GM3LRG
PA16 9AZ GM0KYU
PA16 9BG 2M0GRK
PA16 9DH GM3ZXG
PA16 9DN MM6BFH
PA16 9JG GM3WNB
PA16 9JG MM0JET
PA16 9JG MM3YCI
PA17 5AG MM7SKL
PA17 5DX GM0MDD
PA17 5DX GM0VKG
PA18 6DX GM0ADF
PA18 6DX GM0EUM
PA18 6DX GM0WSR
PA18 6DX MM6AHN
PA18 6EB MM3XLQ
PA18 6HA MM0IYT
PA19 1AZ 2M0DKU
PA19 1AZ MM6SUB
PA19 1EJ MM3YQX
PA19 1EW GM4SVW
PA19 1HF MM1AWV
PA19 1HY GM0HJV
PA19 1HY GM8YUI
PA19 1HY MM1AUF
PA19 1JE MM3YUS
PA19 1SG 2M0BEL
PA19 1SG MM3KSV
PA19 1UW MM6YNG
PA19 1XE GM0ATQ
PA19 1XG 2M0IOB
PA19 1XG MM3WWV
PA19 1XS MM3UIX
PA19 1XU MM3YSJ
PA19 1YA GM0IOY
PA2 0BD GM1TDT
PA2 0BE MM3GEW
PA2 0DP MM6KHA
PA2 0NG MM3EKL
PA2 0RP 2M0OWT
PA2 0RP 2M0OXQ
PA2 0RP GM0PYM
PA2 0RP MM0PAZ
PA2 0RP MM3OXQ
PA2 0RP MM6SWT
PA2 6DB MM1SYD
PA2 6EP 2M0GJG
PA2 6EP MM6NRE
PA2 6ER MM0ZZO
PA2 6JT GM0UOU
PA2 6RA GM4EQY
PA2 6SD MM6WGT
PA2 6UJ MM3HQC
PA2 7NU GM4YAA
PA2 7QY GM0BCY
PA2 7RL 2M0BIN
PA2 7RL MM3VVZ
PA2 7RP GM4BRM
PA2 8AW GM3DDL
PA2 8BY GM4LHQ
PA2 8EB GM3RPM
PA2 8LG MM0PJQ
PA2 8NB MM7FRZ
PA2 8QW GM0NAE
PA2 9AJ 2M0KZI
PA2 9JW MM3YMM
PA2 9NY MM6CEL
PA2 9RD GM8TVV
PA20 0DY GM6OLM
PA20 0QT 2M0CGE
PA20 9DY MM6KSJ
PA20 9EU GM0ELL
PA20 9HT MM3FGI
PA23 7EW GM4EAW
PA23 7HX MM6IPP
PA23 7JJ MM1HMV
PA23 7JJ MM3MHS
PA23 7SS 2M0GLI
PA23 7SS MM3NYY
PA23 7UA GM7KFS
PA23 8AX GM3YLD
PA23 8BA MM7AIP
PA23 8DG MM6DSD
PA23 8EA 2M0FSB
PA23 8EA MM6FSB
PA23 8FG GM7TFN
PA23 8FG MM3CJP
PA23 8HB MM3IMC
PA23 8JR MM3DCN
PA23 8LQ GM4NSL
PA23 8NA GM6HLT
PA23 8PQ GM4SFT
PA23 8QH GM4GLG
PA23 8QT GM4HPK
PA23 8SB GM0VPG
PA23 8SG GM8RGO
PA23 8TH GM8UQX
PA24 8AD GM7NVG
PA27 8BX MM3ROV
PA27 8DB GM6CRX
PA27 8DE GM1USN
PA27 8DE MM3USN
PA28 6EN GM0GON
PA28 6GY MM0BED
PA28 6NZ MM0GTX
PA28 6PG MM6GHH
PA28 6PL GM1PKN
PA28 6PL MM3AQW
PA28 6RX GM7BAS
PA28 6RZ GM4HUL
PA28 6ST MM0EAR
PA28 6ST MM0KRC
PA28 6ST MM3RCX
PA29 6TF GM7OSQ
PA29 6XY GM4ZVF
PA29 6YL 2M0PHO
PA29 6YL MM6PHO
PA3 1BE MM3LVT
PA3 2LR 2M0EDY
PA3 3EJ MM0WRX
PA3 3RT MM0RLN
PA3 3ST GM3WNB
PA3 3SU GM7UFN
PA3 3SU GM7UVS
PA3 4LL GM4KHI
PA3 4LT MM0LEN
PA3 4ST MM6COF
PA30 8FB GM4VXA
PA30 8HH GM4VYQ
PA308EU
MM7WWM
PA31 8AG GM1PWL
PA31 8LG 2M0ONS
PA31 8LG MM6MMG
PA31 8NE MM7EKP
PA31 8QA GM1HJL
PA31 8QA GM4JLD
PA31 8QA GM4OHT
PA31 8QA MM0BLZ
PA31 8QA MM5DOG
PA31 8QL GM0BPF
PA31 8QL GM4HNK
PA31 8SD MM6OHE
PA31 8SD MM6YVO
PA31 8SQ GM0NBA
PA31 8SQ GM8ICC
PA32 8TW MM0WAX
PA33 1AH 2M0MVI
PA33 1BE MM7LXI
PA34 4BG GM0NTR
PA34 4HZ MM3HMM
PA34 4QT GM0EWU
PA34 4RP GM0LRA
PA34 4RP MM1AVR
PA34 4RP MM3GQR
PA34 4RX MM3IMK
PA34 4SF 2M0WWM
PA34 4SF
MM0WWM
PA34 4SF
MM3WWM
PA34 4YB GM0ERV
PA34 4YL GM0MNW
PA34 5AE MM0WRO
PA34 5EF MM3FGL
PA34 5JQ GM4MTI
PA34 5JR MM3IJI
PA34 5NA MM0DCC
PA34 5NS MM3OWZ
PA35 1HG 2M1ELU
PA35 1HJ GM8SXQ
PA35 1HY GM0EQW
PA35 1JJ MM3IKS
PA37 1PL MM0JSU
PA37 1PR MM0TOB
PA37 1PY GM1FPD
PA37 1QJ MM3FUG
PA37 1QP 2M0NDO
PA37 1QP MM7VWG
PA37 1QQ MM3FGH
PA37 1QR GM0SUH
PA37 1QS MM0NDM
PA37 1QU GM1OCR
PA37 1SG MM0AHC
PA37 1SG MM6CLC
PA37 1SR GM4AXS
PA37 1SY MM6RWI
PA38 4bl GM0HVS
PA4 0EG GM0LKS
PA4 0NE MM6ABO
PA4 0NP 2M0TXA
PA4 0SL GM8IID
PA4 0XA GM7MBB
PA4 8JE MM7MVR
PA4 8XS MM3AYU
PA4 8YW MM7EFJ
PA4 9BL GM7VHQ
PA41 7AD GM7NPR
PA43 7HZ MM6EBJ
PA44 7PQ MM7ILY
PA44 7PZ GM1DCB
PA44 7PZ GM3PAK
Pa48 7uf MM7OCX
PA49 7UN 2M0UMH
PA49 7UN MM0UMH
PA49 7UN MM6LMH
PA5 0DA MM6LPT
PA5 0EX GM0EDJ
PA5 0EX MM0PYR
PA5 0EZ 2M0LEX
PA5 8AR GM1OGZ
PA5 8HR MM3TNF
PA5 8LH 2M0RDT
PA5 8LH MM6RDT
PA5 8RZ MM6CBI
PA5 8UB GM7DFI
PA5 8UF GM8IWL
PA5 9BE MM5TUW
PA5 9BH GM1JHU
PA5 9DW MM6SIS
PA5 9LJ MM6RHD
PA5 9NE GM0CHM
PA5 9NR 2M0JAL
PA5 9NR MM3AQW
PA5 9QJ MM7CES
PA5 9RG MM3SWU
PA50EF 2M0NVS
PA59AD GM4FDM
PA6 7DA GM4GZQ
PA6 7DE 2M0BZZ
PA6 7DE MM3XWS
PA6 7DX MM5AII
PA60 7XG 2M0GWU
PA65 6BG GM0PRO
PA7 5ES GM8OAH
PA7 5HE GM3UWX
PA7 5HW GM3NLB
PA70 6HB GM4PMK
PA72 6JU GM0MHE
PA75 6GN GM7NYB
PA75 6PN MM7IAM
PA75 6QL MM3KUB
PA75 6QL MM6SEI
PA75 6QN GM8FKL
PA77 6TW MM3GOX
PA77 6TW MM3GOY
PA77 6TW MM3GPB
PA77 6UA MM3GMP
PA77 6UE 2M0GOE
PA77 6UE MM3GOE
PA77 6UL MM3GOT
PA77 6UT GM3PGY
PA78 6TB MM0JUL
PA8 6BJ MM1JWF
PA8 6BP MM3WZH
PA8 6EE MM6BLY
PA8 6HG GM0GDD
PA8 6JN MM6CMY
PA8 7DH GM1JDJ
PA8 7HX GM4AGL
PA8 7HZ MM7BHR
PA80 5XS GM4RKM
PA80 5XW MM0GFR
PA9 1BJ M1HMZ
PA9 1BP G1VYG
PE1 2EJ M3YSS
PE1 3DS G3PQB
PE1 3DT G4BVM
PE1 3DZ 2E0SAZ
PE1 3DZ M3OWZ
PE1 3JJ M1CDL
PE1 4DR G1FSX
PE1 4EE G6UST
PE1 4HA G4FKQ
PE1 4JY G8SPG
PE1 4LR M6PHZ
PE1 4LT G8CKV
PE1 4RA G4BBA
PE1 4RX M7ERK
PE1 4SN G3RDZ
PE1 4UR G1ZHL
PE1 5HF G4FKG
PE1 5NG 2E0RPL
PE1 5NG M7RPL
PE1 5XD M1HQX
PE1 5XE G0FHO
PE10 0BZ G7RFE
PE10 0EE G4WJM
PE10 0ES G3MMS
PE10 0HT G4NIZ
PE10 0HZ G4GMS
PE10 0HZ M7ARA
PE10 0LS G1IXV
PE10 0NT G8NWM
PE10 0NW G6URT
PE10 0QE G6VEY
PE10 0QX G0LLY
PE10 0RB G8VVC
PE10 0RN G7RFD
PE10 0SG G6SSN
PE10 0SG G7GRC
PE10 0SR G7NUM
PE10 0SR M1DYK
PE10 0SR M1DYL
PE10 0SX G7JKY
PE10 0TU G4DHF
PE10 0TU G6LI
PE10 0UR G7AHO
PE10 0UY G4HLB
PE10 0XR G4LEX
PE10 0XW M6DKU
PE10 9AU G0TIS
PE10 9HB M6OKP
PE10 9HE M3MJE
PE10 9HE M3SKC
PE10 9JG M7RRG
PE10 9JJ G0HSD
PE10 9JX G0OTE
PE10 9NS M3RCE
PE10 9PY M6CEU
PE10 9QT G6SKS
PE10 9QT G6SKT
PE10 9QY G0ERW
PE10 9SB G0BHU
PE10 9SN M7MRP
PE10 9TT G1CPM
PE10 9UQ M0DVU
PE10 9XL G8PUE
PE10 9XL G8TXX
pe100fg M6HOV
PE11 1GZ G4CEY
PE11 1HD M0CTM
PE11 1HQ G1LSB
PE11 1JQ M0DKM
PE11 1PD G7SWS
PE11 1PD G8NWC
PE11 1PD M7PHO
PE11 1QB G4ZXZ
PE11 1QF 2E0JEE
PE11 1UU M0PAC
PE11 2BJ G4XWW
PE11 2DE M7ASH
PE11 2FG G0RNV
PE11 2HE G0JUR
PE11 2JL 2E0FHN
PE11 2NA 2E0BOZ
PE11 2NA M3WUE
PE11 2NA M7CSK
PE11 2QJ G8ETD
PE11 2RT M0MFC
PE11 2RU 2E0FHM
PE11 2RU M6OIZ
PE11 2UN G0UOQ
PE11 2UW G0HGH
PE11 2YU G7FML
PE11 3AB G8VQK
PE11 3AF G4NQS
PE11 3BD 2E0FTP
PE11 3BD M6UUE
PE11 3FA G7NGQ
PE11 3GF 2E0XRO
PE11 3GF M7GWD
PE11 3GZ M7DQJ
PE11 3LL M7RIS
PE11 3PN G6UHS
PE11 3QW G0HXR
PE11 3QW G0JPI
PE11 3RH G7UIO
PE11 3SG M1BMW
PE11 3TG M0WQR
PE11 4BA M0DJA
PE11 4BA M1DJA
PE11 4BQ G8KGR
PE11 4DA G3UYE
PE11 4DQ G7CWM
PE11 4EU G3VFF
PE11 4EY G1WRC
PE11 4EY M0CKE
PE11 4EY M5ARC
PE11 4HG G1DSP
PE11 4HG G3SJR
PE11 4HG G4DSP
PE11 4HG G4OO
PE11 4HG M3DIS
PE11 4ND G8FMX
PE11 4NE G7RTL
PE11 4NE M0BSL
PE11 4QB G1ZJP
PE11 4QQ G3UDP
PE11 4RF G1ELK
Pe11 4tl G1GET
PE11 4XH G3NHF
PE11 4XL G0SDM
PE11 4YA G7THF
PE11 4YL G3VC
PE12 0DE M3PYI
PE12 0EE M3WPP
PE12 0EG G1TUL
PE12 0EG M3SHN
PE12 0EZ G3REH
PE12 0HP G0YKC
PE12 0JG G0OMM
PE12 0NP G1KLI
PE12 0PQ 2E0DHW
PE12 0PQ M0ZLH
PE12 0PQ M6AWS
PE12 0QR 2E0LCN
PE12 0RU G1JKO
PE12 0TS M0GTR
PE12 0TS M3SKQ
PE12 0TT G0ECI
PE12 0TY G8FPW
PE12 6AD G7AKV
PE12 6AD G8PQB
PE12 6AH G6UKC
PE12 6AH M0PLX
PE12 6BX 2E0BGX
PE12 6BX M3HZO
PE12 6DA G3VPR
PE12 6DH G4ODA
PE12 6HS M6DAC
PE12 6HS M7LNA
PE12 6LD M0KRU
PE12 6LG 2E0JVC
PE12 6PN G1PAT
PE12 6PN G4SJV
PE12 6PS G7CUP
PE12 6PT G0LWC
PE12 6QG M7MOR
PE12 6RX G6UEG
PE12 6TG G7OBR
PE12 7AU M0KCE
PE12 7BT M3DEJ
PE12 7BT M6ZTO
PE12 7DQ M3KKX
PE12 7HR G1WYC
PE12 7JQ 2E0GGV
PE12 7JQ M6CTV
PE12 7JS G1JOL
PE12 7JS G7RRD
PE12 7NG G8MEE
PE12 7PP G7HCR
PE12 7QD G6DDU
PE12 8BA G0SHC
PE12 8BA G1SCT
PE12 8DH 2E0TUB
PE12 8DH 2E0TUS
PE12 8DH M6TJN
PE12 8DH M6TJO
PE12 8NA G6VPN
PE12 8QQ G4PST
PE12 8RQ M7DQO
PE12 9AR M0AJI
PE12 9BD G7SNW
PE12 9BT M6XMX
PE12 9DS G8LZK
PE12 9EG G1SRD
PE12 9EG M1CQQ
PE12 9EG M3JNT
PE12 9EL M0XNR
PE12 9EN G3YPZ
PE12 9HS G4ZID
PE12 9JP G0CFD
PE12 9LQ G0WKN
PE12 9LQ G3RKQ
pe12 9lq G6GYX
PE12 9NT M0BFT
PE12 9PF G8XND
PE12 9PJ G4RMJ
PE12 9PJ M6XCO
PE12 9QQ 2E0FFP
PE12 9QQ M0PFF
PE12 9QQ M6MZQ
PE12 9QU G4TPS
PE12 9RG G4KPE
PE12 9RH G7CRY
PE12 9RQ G4JBE
PE12 9YF G0BEJ
PE12 9YF M0CKI
PE12 9YF M0DWW
PE12 9YF M3DLE
PE12 9YF M3QUP
PE12 9YR G7BZC
PE13 1LH M0RAU
PE13 1LH M6BKK
PE13 1PZ G1VYB
PE13 1SW G4UQN
PE13 2EG G8HQO
PE13 2EL G1HJP
PE13 2JD G6NNK
PE13 2JD M0CNX
PE13 2JD M3RGN
PE13 2JR G7TFL
PE13 2JU M0EAY
PE13 2PG M6IFY
PE13 2QA G8NIL
PE13 2RJ 2E0MAB
PE13 2RJ M3MBC
PE13 2RZ M7JPP
PE13 3ED M0BII
PE13 3EJ M3SNO
PE13 3EJ M3WPS
PE13 3HF M1BIG
PE13 3JP M7PRZ
PE13 3JW M7SGF
PE13 3LF 2E0RGB
PE13 3LQ 2E0USC
PE13 3LQ M3USC
PE13 3NQ G6XMU
PE13 3PA 2E0URF
PE13 3PA M3URF
PE13 3RH M1AZA
PE13 3RJ M6IVQ
PE13 3RR 2E0DEP
PE13 3RR M3WUB
PE13 3RR M3WUJ
PE13 3UD M1KGL
PE13 3UN G1FRM
PE13 3UT M1AMJ
PE13 4AR M6AVJ
PE13 4DA 2E0RTH
PE13 4DA M6IHS
PE13 4ER G4NFY
PE13 4HS M0BNP
PE13 4HU G1ATQ
PE13 4JQ 2E0TAB
PE13 4JQ M0TAR
PE13 4JU G8RZN
PE13 4LX G4SHQ
PE13 4QQ M0CNW
PE13 4RG G7HHQ
PE13 4RT G7INY
PE13 4SQ 2E0LTJ
PE13 4TL 2E1HSD
PE13 5AE G8YXJ
PE13 5AF 2E0PPO
PE13 5AF M6RNS
PE13 5AS M3LDE
PE13 5BD G1HQW
PE13 5BP G0OPC
PE13 5BP M3TXJ
PE13 5DR G0XTA
PE13 5DW M1AYU
PE13 5HT G4OKH
PE13 5LG G0MFT
PE13 5NH G4KUY
PE13 5QR G6LTN
PE13 5QR M0BSZ
PE13 5RA M7ICL
PE132RJ G8NED
PE132RJ M0DUQ
PE14 0BA G7LWU
PE14 0DF G4FCI
pe14 0dn 2E0LPE
pe14 0dn M0LPM
PE14 0ND G8MLA
PE14 0ND M3MLA
PE14 0NR M3KWF
PE14 7BJ M6HTM
PE14 7DF G1KSI
PE14 7DN G6OEJ
PE14 7EJ G7JJX
PE14 7HU G7LPW
PE14 7JF G4ORJ
PE14 7LT G3RCQ
PE14 7LT M3DLC
PE14 7LT M3RCQ
PE14 7LZ G0WDK
PE14 7LZ M6BAA
PE14 7NN G0KOJ
PE14 7NP M1AEV
PE14 7NS G7MMJ
PE14 7PQ G7HDU
PE14 7QD G0EUD
PE14 7RA 2E0WNM
PE14 7RA M6WNM
PE14 8AY M1CZM
PE14 8AY M3CZM
PE14 8AY M7AQP
PE14 8DQ G1MHN
PE14 8ES G7UFI
PE14 8ES G8MXQ
PE14 8JR G6PJE
PE14 8JT M0MSX
PE14 8JT M3PYD
PE14 9AF G3XIQ
PE14 9JR G0EYM
PE14 9JT G0FND
PE14 9NY M3AKR
PE14 9PU M0MVB
PE14 9QE M3VZZ
PE14 9RF G8EEK
PE14 9RG G4NVL
PE14 9RG G8SWW
PE14 9RG M3HQU
PE14 9RG M3TUD
PE148RH M7CRC
PE15 0AJ G6NHK
PE15 0AU G1JEA
PE15 0FE G1EXM
PE15 0GA M6DGR
PE15 0LN M3GZW
PE15 0LU G7JXT
PE15 0PJ 2E0JDK
PE15 0PJ M0XAV
PE15 0PJ M6JHZ
PE15 0QF G1OQX
PE15 0QF G6SXB
PE15 0QJ G8EOH
PE15 0RR G3YLY
PE15 0RW G4AJE
PE15 0SA M0GAX
PE15 0SB G8OXE
PE15 0SE G7NCG
PE15 0SE M0BRM
PE15 0SP G7ADS
PE15 0SP M0BLR
PE15 0SR 2E0OVL
PE15 0UH M6DDM
PE15 0YF G7TBX
PE15 8EL G8PHS
PE15 8EQ G4KKR
PE15 8EQ M3RJP
PE15 8HF G7EVP
PE15 8JH G6BOP
PE15 8QA G7NEM
PE15 8RY G0PER
PE15 8SH G4HWC
PE15 8SN G0FLP
PE15 8UH M0JYJ
PE15 9AH G6JVX
PE15 9AP G3VQG
PE15 9DH G6MMB
PE15 9DL G1VTO
PE15 9DT G0OYZ
PE15 9EA G0BXJ
pe15 9ef M0JUP
pe15 9ef M7CKT
PE15 9EL M0JLS
PE15 9EQ G3VMI
PE15 9FH G6BVO
PE15 9la M7DQG
PE15 9QA 2E0TGF
PE15 9QA M3BIB
PE15 9RX 2E0LGL
PE15 9RX M6LGL
PE15 9RZ G0GJL
PE15 9SS M1FEZ
PE158ND G4DPF
PE16 6AY M0HTA
PE16 6BD G4GPQ
PE16 6BJ M6JWE
PE16 6BW M6DPP
PE16 6JF G0GLJ
PE16 6JJ M6OIW
PE16 6JN G0MQL
PE16 6LG G4ZUH
PE16 6QE G6RBR
PE16 6RN 2E0OPU
PE16 6RN M3OPU
PE16 6RX G1YFE
PE16 6RX M6LYY
PE16 6SF G7MHV
PE16 6SG M6TFK
PE16 6SX M7MEK
PE16 6TP G0HSR
PE16 6TP G6OHM
PE16 6TP M3FTV
PE16 6TP M6NIE
PE16 6UR G7VQC
PE16 6UX 2E0JTQ
PE16 6UX G0HKE
PE16 6UX M0JTQ
PE16 6UX M6JTQ
PE17 2TB G1HJD
PE17 6YB G1OSJ
PE19 1AE G1CSO
PE19 1HA 2E0YWO
PE19 1HA M0YWO
PE19 1HA M6YWO
PE19 1JU G8WRV
PE19 1LD G7HMZ
PE19 1NP M1FGO
PE19 1PF 2E0OZX
PE19 1PF M6OZX
PE19 1UF G8FDE
PE19 1UF G8FFC
PE19 2EN 2E0BRS
PE19 2EN M3PYW
PE19 2JY 2E0SDC
PE19 2NN G8BKG
PE19 2NN M0DFW
PE19 2NN M0OLG
PE19 2QF G7JUC
PE19 2QF M3JUC
PE19 2QF M6JUC
PE19 2UL M0GQT
PE19 4UE G0ABW
PE19 5DB 2E0ICR
PE19 5DB M7BUR
PE19 5DQ G7LFZ
PE19 5RD G6JAK
PE19 5SF G1YVS
PE19 5SR M5AEE
PE19 5SZ G8CHC
PE19 5TZ M0AXZ
PE19 5UB G6FJG
PE19 5YA G0CJQ
PE19 5YJ M6TYS
PE19 6AL 2E0BIS
PE19 6BD 2E0KHW
PE19 6BD M0WHK
PE19 6LT M6AQV
PE19 6LT M6RPW
PE19 6QW G8TQI
PE19 6RX G3WCD
PE19 6SD G4LIQ
PE19 6ST G8KDF
PE19 6TJ G0FLT
PE19 6TJ G3UKB
PE19 7AN G6TNW
PE19 7LE G8ELG
PE19 7LF G4MZV
PE19 7RH G6KVK
PE19 7RH M5KVK
PE19 8GJ G7NRV
PE19 8GU G3PNP
PE19 8HY G3LWJ
PE19 8PD 2E0DIP
PE19 8PD M0SKT
PE19 8PD M6DWE
PE19 8PE G4ULM
PE19 8PU G8LWO
PE19 8QL 2E0CZG
PE19 8QL M0ZSM
PE19 8QL M6GEK
PE19 8TZ G0JIW
PE192QY G6OXI
PE2 5EN G7CQG
PE2 5ER G1FWU
PE2 5EY M7LMB
PE2 5LG G3EHQ
PE2 5LY G6AVS
PE2 5NT G4UGK
PE2 5PG M1AKH
PE2 5PR G7IBX
PE2 5PS G7AMW
PE2 5SH 2E0DZY
PE2 5SH M6MTU
PE2 5SL G4XGR
PE2 5SL G7JWD
PE2 5SL G8HTZ
PE2 5SS 2E0IFZ
PE2 5SS M6IFZ
PE2 5XS G3NSF
PE2 5YW G0HYN
PE2 5YW M0VTG
PE2 5YW M3VTG
PE2 6XN M3MXP
PE2 6YB G4LMY
PE2 6YB M0AOG
PE2 6YE M6NWF
PE2 6YL G4OIR
PE2 6YP G8ILG
PE2 6YY G1NIT
PE2 6YZ M3YYV
PE2 7AH G0LEH
PE2 8HX 2E0KZT
PE2 8HX M0KZT
PE2 8HX M7KZT
PE2 8LH G0HZE
PE2 8LS G4FVK
PE2 8PL G6GYM
PE2 8QH M3VZU
PE2 8TG M7IXH
PE2 8UP G4EHW
PE2 8UP G8XLH
PE2 8UP M1PRC
PE2 8UP M5ATR
PE2 8UP M7TIA
PE2 8XP G0EQB
PE2 9AT M6WKB
PE2 9AT M6WVB
PE2 9DN 2E0DBI
PE2 9DN M0HYX
PE2 9DN M0WAF
PE2 9DN M3UEF
pe2 9dn M6HYX
PE2 9EG G8HXR
PE2 9JE G7HUG
PE2 9JG M0BSV
PE2 9JG M3NWK
PE2 9JN G8VWU
PE2 9PB M6EHW
PE2 9PD G4DXW
PE2 9PQ 2E1IGK
PE2 9PQ M0PBD
PE2 9RQ 2E0WBT
PE2 9RQ M3WBT
PE20 1AU M1EKP
PE20 1JA G0MJJ
PE20 1JA G8SKG
PE20 1JT G0MRA
PE20 1JW M3TSI
PE20 1LB M6BOT
PE20 1LR 2E0TNV
PE20 1LR M0TNV
PE20 1LR M3TNV
PE20 1LR M3TOJ
PE20 1LR M6ARJ
PE20 1LR M6RYN
PE20 1LZ 2E0TQB
PE20 1LZ M1GOH
PE20 1LZ M3TQB
PE20 1NU G3OYL
PE20 1QN M0NDY
PE20 1SG M1DXG
PE20 1SQ G0WSA
PE20 2BU M1DKF
PE20 2JF M0HZA
PE20 2LQ G6DJY
PE20 2LU 2E0ZTL
PE20 2LU G7STM
PE20 2LU M6CAI
PE20 2NS M6JKH
PE20 2PF M0NAE
PE20 2PF M6AKI
PE20 3AP 2E0WTY
PE20 3AP M0WTY
PE20 3AP M3WTY
PE20 3AP M6BEK
PE20 3AP M6JCC
PE20 3AP M6MOM
PE20 3DF M1CNX
PE20 3EF 2E0CBU
PE20 3EF M6MOW
PE20 3ES G0PBO
PE20 3PG G0BGY
PE20 3PZ G7HNM
PE20 3PZ M3MIS
PE20 3QG G7JSV
PE20 3QT G3KLC
pe201jj M6GOG
PE21 0AE M3GPX
PE21 0AU G6ADG
PE21 0EJ M3NJY
PE21 0EN M0PAG
PE21 0EZ M0KOG
PE21 0LL M6EHH
PE21 0NF M7FGB
PE21 0SF M0WBD
PE21 6DP 2E0RSG
PE21 6DS M6BXB
PE21 6JN 2E0CDF
PE21 6PW G7LSP
PE21 7BZ G6NUZ
PE21 7BZ M0NUZ
PE21 7EJ 2E0BCB
PE21 7EJ M0GDT
PE21 7HL G3DFY
PE21 7HL M0AZV
PE21 7JG 2E1FRQ
PE21 7JG G4OID
PE21 7NQ M3DPB
PE21 7PN G4GZL
PE21 7PN M3KRQ
PE21 7RU G7MDT
PE21 7TJ 2E0BNW
PE21 7TJ M0WFK
PE21 7TJ M3WFK
PE21 8BJ G8ZTR
PE21 8BW M0HQZ
PE21 8DA G8OYL
PE21 8DL G7MTE
PE21 8FS M3ZLR
PE21 8HD G1RAE
PE21 8HY G6SKF
PE21 8JH M0IMY
PE21 8LJ 2E0KRD
PE21 8LL M3KRD
PE21 8NG G7ZRT
PE21 8NG M0TLC
PE21 8NL M6JWT
PE21 8NQ 2E0CON
PE21 8NQ M3WUH
PE21 8PU G7CQH
PE21 8PU M0EAO
PE21 8PU M0HDS
PE21 8SX M3WGB
PE21 8UF G0HPV
PE21 8XB M6EUZ
PE21 9AE G3JRY
PE21 9HA G0DAF
PE21 9HY G7UTE
PE21 9JJ 2E0BFS
PE21 9LE 2E0JBD
PE21 9LE M3JND
PE21 9LR G7LCW
PE21 9LW G0PXB
PE21 9PD G4RPI
PE21 9PN G7WID
PE22 0BG G7IBH
PE22 0BU G2FFD
PE22 0BZ G0UFC
PE22 0BZ G3WWH
PE22 0DH 2E0BBQ
PE22 0DZ M0GVL
PE22 0DZ M6AHL
PE22 0JD G7DYD
PE22 0JG M3DIT
PE22 0NA 2E0MJL
PE22 0NA M0JWL
PE22 0NA M3MJL
PE22 0PA M6IOG
PE22 0PA M6KPX
PE22 0ST G8UNP
PE22 0YD G4FI
PE22 0YD G7FUM
PE22 7AB G1EBZ
PE22 7BS G8LDW
PE22 7EA 2E0MVD
PE22 7EA M6MCO
PE22 7HJ 2E1FRE
PE22 7JU G3XLE
PE22 7LW 2E1CJJ
PE22 7LW G0KYD
PE22 7LW M0LIO
PE22 7LW M6DAX
PE22 7PQ M7EQN
PE22 7RA 2E0HLM
PE22 7RA M0HLM
PE22 7RA M3ICA
PE22 8DQ M3JQN
PE22 8EX G7GZV
PE22 8EY G8IEV
PE22 8HA G8UMO
PE22 8LX 2E0MID
PE22 8LX 2E0VBR
PE22 8LX M0MID
PE22 8LX M0VBR
PE22 8LX M6VBF
PE22 8LX M6VBR
PE22 8NQ M6CRE
PE22 8PF M1ETN
PE22 8QJ G7EXT
PE22 8RD M6EON
PE22 8RR 2E0BDS
PE22 8RR M0LFC
PE22 8RU G1RNL
PE22 8RU G7HZS
PE22 9LS G0KOO
PE22 9NY M0VPK
PE22 9PR M6PAY
PE22 9PW G7KYX
PE22 9QX G4DYV
PE22 9RA G0KZN
PE22 9RF M7JWP
PE23 4AU M0DXJ
PE23 4BE M1CNI
PE23 4BY M6BUL
PE23 4EJ G0ESO
PE23 4JB G4UBM
PE23 4NH G3SCD
PE23 5AG 2E0NPS
PE23 5BB G1YYU
PE23 5DB G4UNW
PE23 5EP 2E0ITV
PE23 5HH G6HYF
PE23 5PP G1RHB
PE23 5PX G4HAI
PE23 5PX G8ITG
PE23 5SE G6IDL
PE24 4DL G0LAG
PE24 4JJ G1MJN
PE24 4JP M3UEN
PE24 4LY G8VAE
PE24 4PJ M3FSE
PE24 4QN M3ZNF
PE24 4QR M0SWF
PE24 5AT G8HWJ
PE24 5DD M7TPU
PE24 5DQ G7GQO
PE24 5ES G6MQU
PE24 5HQ G0OHF
PE24 5HT G7BNN
PE24 5HZ M6INP
PE24 5JN G4DYJ
PE24 5JQ M0CCW
PE24 5NE G4VYF
PE24 5NF M0KFB
PE24 5NN G0JPQ
PE24 5NZ G0WCU
PE24 5PD M7APY
PE24 5QU 2E0MCC
PE24 5QZ G4ITB
PE24 5RQ M6AFW
PE24 5RS G3PYE
PE24 5RS G6PYE
PE24 5RS G8IER
PE24 5RY G1LLZ
PE24 5SL G0JMZ
PE24 5YN G7JHU
PE24 5YQ G4PRF
PE24 5YQ G7JCX
PE24 5YQ M3YCN
PE24 5YS M6SRN
PE24 5YS M7AHH
Pe245dh G7BNO
PE245JU M7CVH
PE245NE M0NSR
PE245UZ G8FRY
PE25 1BN G1MHB
PE25 1EL G0HNZ
PE25 1HE G4XTX
PE25 1HF G4HFG
PE25 1HF M3NQN
PE25 1HL G0ECN
PE25 1HQ M1ERY
PE25 1PD G7MUY
PE25 1RE G8MZW
PE25 1SE G3NPY
PE25 1TQ G4ZQT
PE25 2BP G0OTH
PE25 2BP M0FRC
PE25 2DD G0EPE
PE25 2EE M0BBE
PE25 2EW G4ZBZ
PE25 2HY 2E0YSU
PE25 2HY M3YSU
PE25 2JA 2E0ROS
PE25 2JA M3FFI
PE25 2JA M6YMX
PE25 2JE M0WDE
PE25 2LN M6AWU
PE25 2NB M6RSO
PE25 2nb M6TPT
PE25 2QQ M3KVD
PE25 2QZ G4SOR
PE25 2RA M3KVI
PE25 2TF M6AFX
PE25 2TH G0NDV
PE25 2TX M3YAS
PE25 2UH 2E0PRD
PE25 2UH M0PRD
PE25 2UH M3NQI
PE25 3BE M5PWR

PE25 3ER G4MVZ
PE25 3JN G0JJR
PE25 3LF G0CHB
PE25 3PH 2E0DBN
PE25 3PH M6DMB
PE25 3PU 2E0BAK
PE25 3QU G1YBT
PE25 3RX G3XZY
PE25 3SA M3IVN
PE26 1BZ G4RRH
PE26 1JP G8JWE
PE26 1LU M6FDK
PE26 1LX G1ZBB
PE26 1LZ G0OPL
PE26 1NB 2E0CXP
PE26 1NB 2E1AOK
PE26 1NB G3CXP
PE26 1NB G3MQI
PE26 1NB G3NKJ
PE26 1NB G3ROQ
PE26 1NB M1CXP
PE26 1SF G1XIY
PE26 1SH G8TVC
PE26 2NZ M6MEN
PE26 2QE G3PMH
PE26 2QE G4KPZ
PE26 2QE G6RHA
PE26 2QQ 2E0FFO
PE26 2QQ M0IXL
PE26 2QQ M7ACH
PE26 2SY G4ETG
PE26 2SY M0VVT
PE26 2TF G0BNR
PE26 2TX 2E0XXC
PE26 2TX M7USA
PE26 2XX 2E0KFY
PE26 2XX 2E0MWD
PE26 2XX M0LHF
PE26 2XX M0MDZ
PE26 2XX M0XTG
PE26 2XX M7FLR
PE26 2YW G0FJR
PE27 3AJ M3FCB
PE27 3DQ G4NKW
PE27 3HF M6CKD
PE27 3HF M6INV
PE27 3NL G0RSA
PE27 3TJ G8XIY
PE27 3TZ G4YDD
PE27 3XZ G6THP
PE27 3YZ G0KHY
PE27 4NE G8JKV
PE27 4SH 2E0BVQ
PE27 4SH G0GXS
PE27 4SW G7JAE
PE27 4TB G4XGT
PE27 4UD 2E1EJU
PE27 5AL M7BPA
PE27 5FX G4KJJ
PE27 5FX G7JXJ
PE27 5RD 2E0KFI
PE27 5RD M0KFI
PE27 5RD M6GCO
PE27 5WX G4XBS
PE27 5XG G6NBM
PE27 6HW G7CSL
PE27 6SS G6VBQ
pe275rd M6OBX
PE28 0AJ G8GRT
PE28 0BB G0IWB
PE28 0BP G7GJS
PE28 0BP M0MAW
PE28 0BS G3NKQ
PE28 0DZ G0AYX
PE28 0GG 2E0CIH
PE28 0GG M3MJV
PE28 0JY 2E0HXN
PE28 0JY M0KFF
PE28 0PA G7PAE
PE28 0PE M3ZUW
PE28 0PF M6BTY
PE28 0UU G1MVF
PE28 2DQ G3NID
PE28 2EH M7JDI
PE28 2EQ 2E0CUB
PE28 2EQ M6BRQ
PE28 2EQ M6JYJ
PE28 2FP G1LMS
PE28 2NU G4CTI
PE28 2QG G7FKF
PE28 2SA M7AFS
PE28 2SD G8FZT
PE28 2UT M6FDI
PE28 3AH 2E0FRF
PE28 3AH 2E0FRK
PE28 3AH M3KUY
PE28 3AT 2E0YSO
PE28 3DJ M6WMZ
PE28 3DL G4DLT
PE28 3DL G8LEB
PE28 3ER M3TBG
PE28 3EY 2E1ECN
PE28 3JB 2E0HIG
PE28 3JB M6BGH
PE28 3JT G7RTR
PE28 3LD 2E0HRI
PE28 3LD M0RSU
PE28 3LE G0RLS
PE28 3LE G1KGV
PE28 3LE G7TSB
PE28 3LE G8AKL
PE28 3LE M6PKV
PE28 3LY G3NSD
PE28 3QG G8YUR
PE28 3QH G4BIK
PE28 3QH G6GJY
PE28 3XA G4FCF
PE28 3YU G8SAA
PE28 4DZ G3ECP
PE28 4EW G7ODT
PE28 4FH G4GWU
PE28 4JX M6LRF
PE28 4PX G8UJV
PE28 4TG 2E0FFK
PE28 4TG M7OHM
PE28 4TQ G8RSA
PE28 4US G3TCL
PE28 4US G4ZEW
PE28 4WS G4RKO
PE28 5AW 2E1AQH
PE28 5AW G1NOO
PE28 5LJ 2E0TMC
PE28 5LJ M6TMC
PE28 5SN G4SOQ
PE28 5TR M3TGA
PE28 5TZ M0MPQ
PE28 5TZ M6PPY
PE28 5UA G7KJW
PE28 5UX 2E0DOJ
PE28 5UX M3MSX
PE28 5WB G0MND
PE28 5YE 2E1IGI
PE28 5YE G7ELC
PE28 5YE M3CMW
PE28 9AN G4YUL
PE28 9BS G3WOT
PE28 9DB G0CCL
PE28 9EW G1VWC
PE28 9JL G3PYZ
PE28 9JL G4KLE
PE28 9JR G3SIT
PE28 9LP 2E0YHZ
PE28 9LP G1KHM
PE28 9LP M7ABW
PE28 9NA M0ETE
PE282RZ M6LLN
PE284TG M7VTT
PE289LE M7KGW
pe28xf M6ONN
PE29 1LW G1OVH
PE29 1NF 2E0XJN
PE29 1NF M0XJN
PE29 1NF M6JSU
PE29 1QL 2E0EML
PE29 1QL 2E0FRQ
PE29 1QL M0OFF
PE29 1QL M0PYC
PE29 1QL M6CSK
PE29 1QL M6TXR
PE29 1RH M0IQH
PE29 1RP G4LHI
pe29 1rr M7LOC
PE29 1RZ G3MKV
PE29 1SR G3RPV
PE29 1TA 2E1ARG
PE29 1TA G7OJX
PE29 1TJ M3ZGG
PE29 1TL G1JGT
PE29 1TS M3OWU
PE29 1UG M7CJG
PE29 1WT G7JAO
PE29 1XE 2E0BQH
PE29 1XE M6GTR
PE29 1XF G0UFB
PE29 1XP G7MJD
PE29 1XY G3YQB
PE29 1YL M0ALK
PE29 1YN M0MIG
PE29 1YW G8YCI
PE29 1YZ G1KWF
PE29 2AL G4ZGG
PE29 2LU G1SMC
PE29 6UF G7DIU
PE29 6UL 2E0PEI
PE29 6UL M7TCA
PE29 6XP 2E0LZX
PE29 6XP M0LZX
PE29 6XP M7TCS
PE29 7BP G6YHF
PE29 7HJ G6PRL
PE29 7JA G6TGW
PE29 7JE G4UXV
PE292WA M7ANV
PE3 6BB G3YYW
PE3 6LB G3ZJW
PE3 8BA G4KEY
PE3 8ES G8NGZ
PE3 8LG G4UQF
PE3 8NA M6FPU
PE3 8RL M6OJZ
PE3 8SH M6PXY
PE3 9AU G6MGA
PE3 9FS M0RHG
PE3 9PA M3EUP
PE3 9PS M7MOH
PE3 9TS M6CWF
PE3 9UD G1PJO
PE3 9UH G4FFS
PE3 9UH G6GNC
PE3 9YD G6GLB
PE30 2DH M7AVP
PE30 2ED G1HYU
PE30 2EL M6AJF
PE30 2HX M6MJF
PE30 2NW 2E0EVX
PE30 2QG 2E0GIE
PE30 2QG M0KLX
PE30 2QG M3GIE
PE30 2QL 2E1HYE
PE30 2QL M3HYE
PE30 3BG 2E1HXC
PE30 3BS G3RSV
PE30 3BS G3VET
PE30 3DE G0IJU
PE30 3DP G8NSK
PE30 3DX G1KLP
PE30 3EX 2E1TOM
PE30 3EZ G3ZCA
PE30 3EZ G8MAY
PE30 3HB G8GKL
PE30 3LY G8KOC
PE30 3NA G3YII
PE30 3NJ G6REC
PE30 3PW 2E0TEZ
PE30 3PW M0TAJ
PE30 3PW M0WHL
PE30 3PW M3TQJ
PE30 3PW M3ZKK
PE30 3PX G4CBO
PE30 3RY G6DDX
PE30 3UD G1KBH
PE30 3UZ G0VHH
PE30 3XA G4GXL
PE30 3XD G8WAP
PE30 3YB G7VPS
PE30 4AB G1XYZ
PE30 4AB G3HRX
PE30 4AB G3SZ
PE30 4AB G3XYZ
PE30 4AB G4OZG
PE30 4AT G6IIP
PE30 4DN G1LOK
PE30 4QA G0WWU
PE30 4XD 2E1IHT
PE30 4XT M6GRZ
PE30 4YH G1XDK
PE30 4YY G6WAN
PE30 5BB 2E0HSP
PE30 5BB 2E0NRX
PE30 5BB M6HSP
PE30 5BB M6NRX
PE30 5BB M7TRH
PE30 5BQ G7OUZ
PE30 5DF M6COY
PE30 5DT G8BQT
PE30 5DY G6IWU
PE30 5NH M3UEC
PE31 6AX M3ZOH
PE31 6BT G4JNQ
PE31 6DP M6HNG
PE31 6HQ G0AED
PE31 6JW M6NUV
PE31 6LH G8HHO
PE31 6PR G0MTB
PE31 6PR G6TVI
PE31 6QN G7WGL
PE31 6UR 2E0RWK
PE31 7AF G0JQR
PE31 7AR G1SCQ
PE31 7AR G7RNA
PE31 7AR M0HCR
PE31 7BS G6WDC
PE31 7DX G8JXU
PE31 7EB G6RQA
PE31 7LA M1CCN
PE31 7LQ G4DDT
PE31 7PA G0OHQ
PE31 7QF G4EOJ
PE31 7SA G6DAC
PE31 7SZ M7EOQ
PE31 7UG G3YLW
PE31 8AQ G0OFD
PE31 8AQ G8MBJ
PE31 8BP M3LCU
PE31 8BS G1YMA
PE31 8ET G7MDI
PE31 8LN G0CQB
PE31 8NJ M0ZLP
PE31 8RL G4DCJ
PE31 8RL G8FTP
PE31 8RS G0LRU
PE31 8RS G4ZQC
PE31 8SF G3OVL
PE31 8SF M3NHZ
PE32 1AW G7RSA
PE32 1BJ 2E0RQQ
PE32 1BJ G4RB
PE32 1BJ M3RQQ
PE32 1DQ G7NJP
PE32 1HY M1CFG
PE32 1NQ M6OAD
PE32 1NW G7SYL
PE32 1NY G7OVE
PE32 1QU M3GKE
PE32 1QZ G6IBO
PE32 1RJ G7DQL
PE32 1RL G6CPS
PE32 1SF G8TUH
PE32 1SS G1LMQ
PE32 1SY G4CXE
PE32 1TG M0BQB
PE32 1UX M7BKT
PE32 2AS G0UOB
PE32 2BA 2E0ESY
PE32 2BA M6LZF
PE32 2DT M6KRM
PE32 2DU M6JHC
PE32 2EA G0TZZ
PE32 2LJ G7MFY
PE32 2LJ M7EBY
PE32 2LZ M6OTH
PE32 2NA G7BUN
PE32 2QX M6VWX
PE32 2QX M3KAU
PE32 2QX M6UAL
PE32 2TE G1CSY
PE32 2TE M0DZB
PE32 2UA G1ICX
Pe322au G7RJR
PE33 0HE G0JWG
PE33 0HS G0BPL
PE33 0JG G8CWE
PE33 0LB G4RKN
PE33 0NN 2E0WWL
PE33 0NN M7SRW
PE33 0PG G0UMM
PE33 0PR G8ULL
PE33 0UA G8PHQ
PE33 9BD G1EMW
PE33 9DB M0JPM
PE33 9HP G4IXD
PE33 9JW M7PBW
PE33 9PX M3YXJ
PE33 9PX M3YXK
PE33 9RP G0FQO
PE33 9RP G1RHE
PE33 9TQ G0CCS
PE33 9TS 2E0SKI
PE33 9TS M0JEK
PE33 9TS M6ABS
PE33 9XW M0WAR
PE34 3BD M0GVI
PE34 3BN M3YFX
PE34 3HF M1CCQ
PE34 3HL G7MOD
PE34 3JH G7HLU
PE34 3LS G4LBQ
PE34 3PF G7UEL
PE34 3PP 2E0XXM
PE34 3PP M0XXM
PE34 3PP M3XXM
PE34 3QD 2E0KKM
PE34 3QD M0KKM
PE34 3QD M6VKM
PE34 4DA G4NJJ
PE34 4DE G6MQK
PE34 4DE M3JZK
PE34 4DY G6FSK
PE34 4HZ G4MOC
PE34 4JB 2E0EKK
PE34 4JB M3GHO
PE34 4JP M3KRO
PE34 4JX M0TBJ
PE34 4JX M6WGC
PE34 4LH G8PWQ
PE34 4NR M6JOW
PE34 4NU G6CVR
PE34 4PR M3YCB
PE34 4QJ G7PTB
PE34 4QS M3GJN
PE34 4RA M6KSF
PE344HE 2E0CMK
PE344HE M0IOT
PE344HE M6CPN
PE36 5BD G0KHF
PE36 5BS G1NJV
PE36 5BZ G0XBO
PE36 5DJ G4VUF
PE36 5EA M0HQA
PE36 5EJ G4RQU
PE36 5PJ G8MRI
PE36 6AP G7APU
PE36 6BS G1LSK
PE36 6BX 2E1EBX
PE37 7BT G0HZA
PE37 7RH G3JKM
PE37 7SP G7MOK
PE37 7TP M6LAI
PE37 8EE G4FZL
PE37 8ET G1KIW
PE37 8HY M6NEC
PE37 8JA M7TRG
PE37 8LB M6IWD
PE37 8LU G0NQN
PE37 8NN G1UGJ
PE38 0AJ 2E0HIQ
PE38 0AJ M0XMP
PE38 0AJ M6SCI
PE38 0BY G6TRA
PE38 0DH G4RCP
PE38 0DP G1VIS
PE38 0DP G6NMK
PE38 0DY G1HBV
PE38 0DY M0EVD
PE38 0EN G4LMX
PE38 0EU M3HPT
PE38 0EU M3KBB
PE38 0PA M7CRR
PE38 9AX G3UPN
PE38 9EQ M0TJC
PE38 9NJ G8VCO
PE38 9PG G8JAN
PE38 9QU G4TUO
PE38 9QY G4BAU
PE38 9RQ G3ASG
PE38 9TZ M6OST
PE4 5AF G7FTD
PE4 5AQ G0ADA
PE4 5AU G3GMW
PE4 5BJ G1DRI
PE4 5DD G3WRL
PE4 5DF G7DWO
PE4 5ED G1MLC
PE4 6AQ G6AYU
PE4 6BJ M7BPJ
PE4 6GW 2E0GDF
PE4 6GW M3ZVT
PE4 6JS G0CNL
PE4 6QY G1EUN
PE4 6QZ G4WIN
PE4 6RF M7MTV
PE4 6RL G4ZKM
PE4 6RP M3FVE
PE4 6SA M6DLJ
PE4 6SB G8EQY
PE4 6ZH G4DJZ
PE4 7BJ M3JGI
PE4 7EN 2E0ROD
PE4 7EN M0RHB
PE4 7EN M5AEI
PE4 7PY G7JJP
PE4 7PY M3LPT
PE4 7TN G4DEW
PE4 7TT M3DRA
PE4 7TW G3RED
PE4 7UP M0CEG
PE4 7ZD G3VYU
PE4 7ZD G7SQY
pe46nt M6HWO
Pe6 0ah M7PFD
PE6 0BA M0AWB
PE6 0BA M1BXJ
PE6 0BS M6WVK
PE6 0DA M3YSV
PE6 0DG G4NCP
PE6 0EN M6IJL
PE6 0ES M0PXY
PE6 0HA 2E0GRP
PE6 0HA M0GRP
PE6 0HA M6GRP
PE6 0JB G8YOX
PE6 0LH G0TFR
PE6 0PQ M1MIC
PE6 0QG G0NUU
PE6 0QJ 2E0YHN
PE6 7EU G1ESC
PE6 7LG G0EVU
PE6 7NG G1ZDX
PE6 7NW G3YOY
PE6 7RG 2E0BEF
PE6 7RG G4FXY
PE6 7RG G4NRK
PE6 7RG G6CZO
PE6 7UB G0IAG
PE6 7UB G1XXW
PE6 7UB M7ANN
PE6 7UP G1OAU
PE6 7XG G0BLV
PE6 7YF G8KQZ
PE6 8AG G6IDW
PE6 8BS G1AUM
PE6 8DA G7OOB
PE6 8DJ G4FIQ
PE6 8ED M0HTU
PE6 8EH G8YVS
Pe6 8el G4SQA
PE6 8GR M7RPY
PE6 8HH M7TXM
PE6 8HP G4FSX
PE6 8HR G1FJD
PE6 8JD G0DPF
PE6 8JS G4MZM
PE6 8NW G4VJN
PE6 8NW G6KEZ
PE6 8NW G6MTY
PE6 8QJ M5ALU
PE6 8RD G1HOJ
PE6 8RD G6NYL
PE6 8SF M0NMH
PE6 8SF M3MQH
PE6 8SH G1RVK
PE6 8TF G7UHL
PE6 9AA G6YCA
PE6 9BA 2E0JRU
PE6 9BA M0JRU
PE6 9DN M0DBH
PE6 9LP G7SOE
PE6 9NP G8OAD
PE6 9PH 2E0YCA
PE6 9PH M7YCA
PE6 9RB G6ENN
PE67LE M7MSA
PE68HP G7OKT
PE68HS M7LBY
PE68PS G0BAT
PE68QJ G7GWA
PE7 0LU 2E0TEK
PE7 1AT G0LHG
PE7 1BP G0RCH
PE7 1HJ G4NLO
PE7 1QN G4DCS
PE7 1RF G1NLS
PE7 1RL G0IHK
PE7 1RR G0PGY
PE7 1RT G4WJB
PE7 1TY 2E0GPU
PE7 1TY M1NER
PE7 1TY M3YEK
PE7 1UE 2E1GDG
PE7 1UE G4FFN
PE7 1XQ G0WXP
PE7 1XX G7VPA
PE7 2BT M3JCE
PE7 2DD G0JJW
PE7 2DW M3ZWB
PE7 2HN G0INO
PE7 2HN G4YIT
PE7 2PD G1JIH
PE7 2RG G3XRK
PE7 3AA M3YGC
PE7 3AX M6CUQ
PE7 3AY M3NWD
PE7 3BU G6JAS
PE7 3FA G6EUY
PE7 3FS M7WDX
PE7 3GE M0DYR
PE7 3NU M7AYW
PE7 3PG M3LSF
PE7 3PR G1KQN
PE7 3RB M7NJJ
PE7 3RG M3JDX
PE7 3RS G7IVU
PE7 3SH G8UFO
PE7 3SS G0HOF
PE7 3SU G4PPJ
PE7 3SX M0RKH
PE7 3UA G3DQW
PE7 3UA G3TGO
PE7 3YD M3MEP
PE7 3YQ 2E0DVO
PE7 3YQ M6FDH
PE7 3ZF G0XAH
PE7 3ZR G7AFO
PE7 8FL 2E0CCQ
PE7 8LZ 2E0SRH
PE7 8LZ M7HJR
PE7 8LZ M7SRP
PE7 8RL G6DYK
PE73QW 2E0ORD
PE73QW M0VRD
PE73QW M7CHE
PE8 4DQ G4JPB
PE8 4JQ G3VZL
PE8 4LT G7NDQ
PE8 4NX 2E0TZZ
PE8 4NX M3OTZ
PE8 4QE G7WDN
PE8 5AN G7HLU
PE8 5HP G4HCG
PE8 5PB M6PVR
PE8 5PS G0UIH
PE8 6AU 2E0HEP
PE8 6AU 2E0ZAF
PE8 6AU M0KDT
PE8 6AU M0ZAV
PE8 6AU M3ZWL
PE8 6EU 2E0HLF
PE8 6EU M0ILU
PE8 6EU M6HLF
PE8 6JQ M0BSW
PE8 6LP 2E0KNT
PE8 6LP M0KWR
PE8 6LP M3KNT
PE9 1DS 2E0BVR
PE9 1HF G4NHL
PE9 1JL M7MOU
PE9 1LA M0STA
PE9 1PJ 2E0WCG
PE9 1PJ M0NMA
PE9 1PJ M6TTZ
PE9 1QQ G4ZBG
PE9 1SN G6HNP
PE9 1UP G3TBG
PE9 2FB G6ENZ
PE9 2GR 2E0BMT
PE9 2GR G5REV
PE9 2HU M0PZC
PE9 2JY G1FSW
PE9 2NX G4YNT
PE9 2SG G1DCX
PE9 2TS G1JGR
PE9 2UB G3THC
PE9 2YR M0YHR
PE9 3AG G4VAM
PE9 3LL 2E0CNP
PE9 3LL M6FBH
PE9 3LN G8IOA
PE9 3LZ G7DGP
PE9 3NA G4OZM
PE9 3PF 2E0JBX
PE9 3PF M0THB
PE9 3PF M7HBT
PE9 3PW G3MFG
PE9 3QD G1LXK
PE9 3QD G4RQK
PE9 3SY M1REC
PE9 3UQ G8VDJ
PE9 4BU G0FPZ
PE9 4BU G0INV
PE9 4DJ M6OCJ
PE9 4EB G1CUG
PE9 4JJ G4PEL
PE9 4NA G6CVV
PE9 4RJ 2E0ZAL
PE9 4SN M6AAU
PH1 1DT GM4HJK
PH1 1HH GM8RYZ
PH1 1JD GM4BVD
PH1 1JD MM1FHS
PH1 1LE GM8UGO
PH1 1LU MM6JQA
PH1 1QA MM7RCX
PH1 1QB GM6JKU
PH1 2DU GM6OFO
PH1 2DU MM3KVN
PH1 2NF GM6UJG
PH1 2NS MM3XUX
PH1 2PF MM7DVE
PH1 2PU MM7SWC
PH1 2QJ MM6NAA
PH1 3BJ MM7IJK
PH1 3DD GM1AHF
PH1 3DD GM1AHG
PH1 3gp MM6HBY
PH1 3LW 2M1DWK
PH1 3LW GM1DSK
PH1 3LW MM7DWK
PH1 3UA 2M1VFO
PH1 3UA MM0SDK
PH1 3UT MM7AEI
PH1 3YX MM6AFQ
PH1 4EY MM3VKP
PH1 4JN GM3ALF
PH1 4QS GM4EAF
PH1 4QS GM4ZRH
PH1 5FJ MM6NWK
PH1 5FN MM3YGI
PH1 5NJ MM0JUC
PH10 6AE MM6TKV
PH10 6FH MM7DHB
PH10 6FP 2M1EPV
PH10 6SF GM2BWW
PH10 6SF GM8DOR
PH10 6TH GM4NUN
PH10 6UE MM6HIG
PH10 6XE GM7GUL
PH10 7JL GM0VIT
PH10 7RB GM4LMG
PH11 8AS GM1ZQF
PH11 8BU GM0LYT
PH11 8DR MM0JNY
PH11 8DW GM3WAP
PH11 8HQ 2M0RJJ
PH11 8HQ MM0RJJ
PH11 8HQ MM6GHX
PH12 8SL 2M0BMK
PH12 8UB MM7FWK
PH12 9AB MM6RGT
PH14 9QB GM0NTI
PH14 9QB GM0OTI
PH14 9QB GM8RTI
PH14 9QL MM0HGN
PH15 2BE GM6YRN
PH15 2EB MM6CBH
PH15 2QY GM4FOZ
PH15 2QY GM8KJO
PH16 5HH GM8KPH
PH16 5JF 2M0IMP
PH16 5JF MM3ZYS
PH16 5JL 2M0DGI
PH16 5JL MM0ZOG
PH16 5JL MM6BFR
PH16 5JS GM4ODW
PH16 5LA GM0CWR
PH2 0AE MM3SYQ
PH2 0AR GM1PFU
PH2 0BL MM6OAG
PH2 0GY GM4MSL
PH2 0HH GM0FTX
PH2 0LD GM6UWF
PH2 0LD MM6FSQ
PH2 0RB GM1RBQ
PH2 6GD MM7HLD
PH2 6LWL MM7EAB
PH2 6QA 2M0TGN
PH2 6QA MM0WXE
PH2 6QA MM6TGN
PH2 6QE GM8YRT
PH2 6RR GM1BUY
PH2 6RR GM3EOB
PH2 6SD GM4DQJ
PH2 6SP GM0HCQ
PH2 6TD GM4EVS
PH2 6TQ MM0EMC
PH2 7BY GM1GHZ
PH2 7BY GM3UYR
PH2 7BY MM0MHZ
PH2 7HJ GM4YZT
PH2 7LL GM1FBM
PH2 7NF GM0CEA
PH2 7QA GM0IST
PH2 7QA GM0NYP
PH2 7QA GM1RGM
PH2 7QP MM3CNS
PH2 7QQ GM0LVI
PH2 7QU MM3SYO
PH2 7QU MM6CEW
PH2 7RB GM6VRC
PH2 7SS GM0GAV
PH2 7TB GM0IMH
PH2 7TS GM7SDP
PH2 7XD MM0MBC
PH2 8PT GM3THI
PH2 9EP 2M0MKZ
PH2 9HS GM0MXZ
PH2 9LW GM0PRG
PH2 9LW GM3YEW
PH20 1BS GM8CEA
PH21 1JS GM0KEQ
PH21 1JS GM6NUL
PH21 1NY GM4EBX
PH21 1NY GM4JDK
PH23 3AA GM0PWS
PH23 3AA GM3JFG
PH23 3AA GM3JIJ
PH26 3AE MM3GPL
PH26 3LX GM1SRP
PH26 3PA GM4CCN
PH26 3PA GM4LPG
PH26 3PX MM6CJC
PH3 1BE 2M0HIC
PH3 1BE MM7HIC
PH3 1BZ GM8TCG
PH3 1DD GM1BXI
PH3 1EW GM4JZB
PH3 1JS GM4SKB
PH3 1LX MM0OKY
PH3 1LX MM6YCB
PH32 4DW GM3VKN
PH33 6DD GM0ANG
PH33 6HX GM4YAT
PH33 6JS GM3RFA
PH33 6NX GM1YGV
PH33 6PY GM7REY
PH33 6SJ GM7PKT
PH33 6SW MM7WOK
PH33 6SZ GM0UMJ
PH33 6UG GM8AOB
PH33 6UH GM4NFI
PH33 7AB MM3XFX
PH33 7AL GM6AES
PH33 7ER GM0ERB
PH33 7LE GM4OPU
PH33 7LR GM4TFJ
PH33 7PF GM4UGN
PH33 7PQ GM1YPJ
PH33 7RB MM0VSG
PH34 4DZ GM7PXL
PH34 4EB 2M0FYX
PH34 4EB MM0JZW
PH34 4EB MM7BJR
PH34 4EX GM4TJL
PH39 4NX 2M0SBP
PH39 4NX MM3YBQ
PH4 1QE GM0SGH
PH40 4PD GM4OAS
PH41 4QF GM4XJY
PH41 4QF GM6VYZ
PH41 4QY GM6VYY
PH41 4RL MM7EPW
PH41 4RQ 2M0LUG
PH41 4RQ MM3INY
PH49 4JR GM3ZRT
PH49 4JZ MM0HZJ
PH5 2BD GM0WEZ
PH5 2BD MM6SOR
PH5 2BD MM6YEZ
PH6 2BB GM4YRO
PH6 2HB MM0GKT
PH6 2NW GM1KWA
PH7 3LE GM0FVJ
PH7 3NY 2M0KVI
PH7 3NY MM0JJZ
PH7 3NY MM7KJV
PH7 3RP GM0AOF
PH7 3RP GM1LTM
PH7 4DH GM4OFC
PH7 4JH 2M0DDX
PH7 4JH MM7DDX
PH7 4LE GM4YDC
PH8 0ET GM4LFK
PH9 0LG GM4NGJ
PH9 0NH GM6IQH
PL1 2HX M3MWG
PL1 3JF G1KXJ
PL1 3JN M3PXP
PL1 3LB G7CNC
PL1 3PS G0OJW
PL1 3QR 2E0AZD
PL1 3RN M6FGC
PL1 4GQ G6RFS
PL1 4GU M3YOX
PL1 4HU 2E0NJK
PL1 4HU M3PXQ
PL1 4PR G1OBM
PL1 5HD M6ABG
PL1 5LJ M7XIA
PL1 5NB M6HLA
PL10 1DA G7AWG
PL10 1DA G8UDA
PL11 2BZ 2E0LAD
PL11 2BZ M3UXG
PL11 2HJ G8XOX
PL11 2HR M7OST
PL11 2JB M6GJO
PL11 2LP M6OLD
PL11 2LT M3TBQ
PL11 2NE 2E1ABQ
PL11 2PZ G4ZZD
PL11 3DX G7FGD
PL12 4BJ G0AKH
PL12 4BJ G4GXK
PL12 4BJ G4WAB
PL12 4BJ G7WAB
PL12 4BJ G8SAL
PL12 4BN G4YXJ
PL12 4BX 2E0OMI
PL12 4BX M0HTK
PL12 4BX M6DZH
PL12 4DL M1CQC
PL12 4DY G4HZE
PL12 4FD G8PVR
PL12 4HE G7EUB
PL12 4JH G3UBY
PL12 4LF M3KFL
PL12 4NG G4KYY
PL12 4NG M3ZYY
PL12 4PP G0XAW
PL12 4RG G0KDW
PL12 4SH 2E0CXA
PL12 4SH M6BKH
PL12 5AE 2E0CGP
PL12 5AE M0HBU
PL12 5AE M6AKW
PL12 5DA G3IVP
PL12 5DA G7LPT
PL12 5NH G6SEK
PL12 5NL 2E0RBA
PL12 5NL M0RAZ
PL12 5NQ G6ELG
PL12 6BJ G3EXL
PL12 6DR G4YGZ
PL12 6DU G8HVZ
PL12 6EL G3XCS
PL12 6EN G6YNV
PL12 6EN G7VBZ
PL12 6EN M6RRD
PL12 6JA G7STF
PL12 6JE M1BZK
PL12 6PS M0BHK
PL12 6RH G0DAV
PL12 6SA G6ION
PL12 6SP G4ALY
PL12 6TD G3ZHK
Pl125df 2E0MCJ
Pl125df M6ZEJ
PL13 1PE M6BIM
PL13 1PN G4MXP
PL13 1PR G8TXT
PL13 2HT G1YMC
PL13 2HX M6HCA
PL13 2JP 2E0PDQ
PL13 2JP G1YDQ
PL13 2JP G4BRF
PL13 2JS G0PWW
PL13 2JY G8TEF
PL13 2LF 2E0SJG
PL13 2ND G4DHK
PL13 2ND G8HNS
PL13 2NX G0HRX
PL13 2NX M1SAN
PL13 2PU G7CLO
PL13 2SD G6YXV
PL132WP G3BAR
PL14 3EE M6KBJ
PL14 3EX G6FYT
PL14 3FN G0CAY
PL14 3HX G0GUT
PL14 3LP M7CFG
PL14 3LW 2E0TUH
PL14 3LW M6TUH
PL14 3NS 2E0TGL
PL14 3NS M6SOC
PL14 3PT G3UAE
PL14 3PU M0GGM
PL14 3QG G1NSV
PL14 3RL G4TIQ
PL14 3SG G4NZG
PL14 3SQ G4ZGZ
PL14 3TD M0BEV
PL14 3TE G8PXM
PL14 3TQ G3ZHO
PL14 3TS G1ZRQ
PL14 4NJ G1STP
PL14 4NJ G6SHF
PL14 4NW G3WOA
PL14 4PA G0LYR
PL14 4PT M0KRR
PL14 4PT M3ZHX
PL14 4QP G7WCF
PL14 4RS G8BCG
PL14 5BW G4MFE
PL14 5DP G0DAU
PL14 5EL G4VCA
PL14 5EN G4TYD
PL14 5EY G6OWT
PL14 5LE G8BIX
PL14 5LU G4MJW
PL14 5NQ G1KTZ
PL14 5PP 2E0MLJ
PL14 5PP M6TMJ
PL14 5PW G7LJA
PL14 5QT G0EFR
PL14 5QT M6MND
PL14 5RA G7JLK
PL14 5RF G7OAH
PL14 5RU G7BIM
PL14 5SA M6RFR
PL14 6DA M0WHJ
PL14 6DA M7WJH
PL14 6DL M7DHP
PL14 6EP G8GLI
PL14 6JS G1NTV
PL14 6PZ G8VZR
PL143SX 2E0VAF
PL143SX M3VJE
PL15 7DZ G0AOB
PL15 7EQ 2E0IFK
PL15 7EQ M7ITI
PL15 7FF M3VQV
PL15 7HB 2E0VCC
PL15 7LQ G0BNU
PL15 7NE G1SCY
PL15 7PQ M3ZZH
PL15 7PS G4RZM
PL15 7PW G1TAZ
PL15 7PW G3LNW
PL15 7QD G3ZYL
PL15 7RQ M3IIV
PL15 7TT G8WUS
PL15 8PP G3SYM
PL15 8SE M7CQY
PL15 9BB G4KKZ
PL15 9BD M7MEC
PL15 9EB G0ANM
PL15 9EB G6ZMO
PL15 9LE G0LFQ
PL15 9NA G7HEP
PL15 9NB G6VVU
PL15 9PH M0ACI
PL15 9RJ G7HLB
PL15 9RR G0AAM
PL15 9RT G6UHD
PL15 9SY G0UFV
PL15 9SY G1SOX
PL15 9TJ G4OIA
PL157pr M6JMM
PL16 0AH M0ESW
PL16 0AH M3PSS
PL16 0AS G3LUW
PL16 0HJ 2E1EOR
PL17 7BX M6AKO
PL17 7EU G7NHW
PL17 7EU M0CDS
PL17 7EU M1APB
PL17 7GB M3RYY
PL17 7HD G8AWB
PL17 7HP G1KBG
PL17 7LW G1IPI
PL17 7PA G0IVZ
PL17 7PF G1YTV
PL17 7PL M3XWX
PL17 7PT G1FMU
PL17 7PT G7UDX
PL17 7PT M0SCR
PL17 7QE M7CLU
PL17 7QH M6BMT
PL17 7QH M6VQV
PL17 7QL G1VQG
PL17 8BL M3FAC
PL17 8DE M3YMS
PL17 8FD G7UXO
PL17 8GL M3PCQ
PL17 8JE G0TDN
PL17 8JN M0GVT
PL17 8JX G6BER
PL17 8NR G6IJW
PL17 8PA M6GAR
PL17 8PA M7BUB
PL18 9BD G4KYE
PL18 9BL G7CQK
PL18 9BN G4UMS
PL18 9BU M6BED
PL18 9BY M6FZV
PL18 9DA M6OPO
PL18 9DA M6POD
PL18 9DA M6ROP
PL18 9DH G3XIB
PL18 9DX G4PBN
PL18 9HH M1BBB
PL18 9HN 2E0WRQ
PL18 9HN M7SDD
PL18 9JB 2E0NMC
PL18 9JB M0NMC
PL18 9JB M1APC
PL18 9JB M3HXZ
PL18 9JB M6BTS
PL18 9JB M6NEW
PL18 9NG G4FWN
PL18 9PB G4BVB
PL18 9QN M6YRW
PL18 9RY M0BSC
PL19 0DP 2E0MDH
PL19 0DP M6BWO
PL19 0EA 2E0PAA
PL19 0EE G1RFQ
PL19 0ER G3ZNE
PL19 0ET G8ONR
PL19 0EU M0YDW
PL19 0EU M3YDW
PL19 0NJ G7VUP
PL19 0NN 2E0RPH
PL19 0NR G3FHT
PL19 0PA M3XYU
PL19 0PN G0GSR
PL19 8DR G1KKA
PL19 8DX G1OJL
PL19 8EY 2E0YJF
PL19 8EY M3YJF
PL19 8HA G3MWZ
PL19 8RL M6KXU
PL19 9AX G4WDR
PL19 9AX G6BJJ
PL19 9DA G3TGN
PL19 9DG G8WPX
PL19 9DJ G6FOV
PL19 9DL G7IIB
PL19 9DN G3TYO
PL19 9DQ G3XOU
PL19 9LJ G0NUO
PL19 9LQ G7VYQ
PL19 9NQ G8KSM
PL19 9PR G0LEV
PL19 9QD G8AKC
PL2 1DT M6TGR
PL2 1EA M3LOE
PL2 1EA M3SCX
PL2 1JD 2E0OLF
PL2 1JD M0MLZ
PL2 1JD M6OLF
PL2 1QQ M6KPP
PL2 1WT M3TOR
PL2 2BR G4PKM
PL2 2BU G7LUL
PL2 2ER G0WGB
PL2 2QY 2E0BJD
PL2 2QY M3MQX
PL2 3AP M3YQM
PL2 3AX 2E0DVN
PL2 3AX M6BNM
PL2 3AZ G3XLZ
PL2 3ET G8EHD
PL2 3FB G1SYZ
PL2 3LF 2E0ITN
PL2 3LF M6ITN
PL2 3QG 2E0SPR
PL2 3QG M3PYJ
PL2 3QZ G7ANY
PL2 3RS G1SQI
PL2 3RS M3IOT
PL20 6AT G7MOH
PL20 6EA M3BKB
PL20 6HP G6IIZ
PL20 6LJ G3TXL
PL20 6LJ M3TXL
PL20 6ND M6LJK
PL20 6NG G4GUY
PL20 6PT G0IAI
PL20 6QD M7GPM
PL20 6QX M7JDM
PL20 6SY G4DND
PL20 7AH G4HPX
PL20 7AR G3SGZ
PL20 7BZ M6WBR
PL20 7DD G1SQG
PL20 7DW G4DMM
PL20 7EP M3RNG
PL20 7JS G8MWN
PL20 7JS M3TNJ
PL20 7LH G0PGI

PL20 7LR G3ZXO
PL20 7PJ M6JCG
PL20 7QB M6TGY
Pl206nn M6MQM
PL206TJ M7OMM
PL21 0AX G7HIC
PL21 0ET G4VFG
PL21 0ET G6NSU
PL21 0HT G0TQR
PL21 0LD M0VFG
PL21 0LT G4NDL
PL21 0WD M5DAP
PL21 0YW G3TSE
PL21 9BA M6EQQ
PL21 9BD G4XFM
PL21 9BH G7LNG
PL21 9BQ M7GSS
PL21 9BT G4ONC
PL21 9DH G0EOZ
PL21 9EF G4RYO
PL21 9EU G4OUZ
PL21 9EU G4RIM
PL21 9JE G4ETP
PL21 9JX 2E0KMF
PL21 9QT M0YSE
PL21 9QT 2E0GWD
PL21 9RB 2E0XCP
PL21 9RB M0ZCP
PL21 9RB M3XCP
PL21 9SN G3PRC
PL21 9SN G8PRC
PL21 9SN G8XTE
PL21 9TQ M3ZBI
PL21 9TS M0IZF
PL21 9XA 2E0PGH
PL21 9XA M6PGH
PL22 0ET G3XSC
PL22 0JB G0MFQ
PL22 0QH G3HUB
PL23 1ED M6DYW
PL23 1ET G3MTG
PL23 1JH G3SXR
PL23 1LX M6USG
PL23 1NB G4AIR
PL23 1NB G7RNB
PL23 1QQ M3UIB
PL24 2AT G0AEW
PL24 2DE M6OSO
PL24 2DG G8XNA
PL24 2DL M6PCP
PL24 2EP 2E0EUF
PL24 2EP M6MUY
PL24 2LD G0CIG
PL24 2LH G4ZEB
PL24 2PW 2E0NJV
PL24 2PW M6NJV
PL24 2RL G1JXP
PL25 3AU G0KTD
PL25 3BB M1EWP
PL25 3BZ M6GBY
PL25 3DH G4RLN
PL25 3DR G0NNO
PL25 3DY M3UMM
PL25 3EB G0FXM
PL25 3EX G3HTO
PL25 3HB G8WBG
PL25 3QE G4UON
PL25 3QG G4JOD
PL25 3TJ G4XBW
PL25 4DU M7ARW
PL25 4HR G1XMI
PL25 4HR G8NYR
PL25 4HT G4XOP
PL25 4JA G4XBC
PL25 4NE 2E0NAM
PL25 4NE M0NAW
PL25 4NE M6NKC
PL25 4NL 2E0IOL
PL25 4NL M7CSU
PL25 4QF M0AGY
PL25 4RJ G4MFQ
PL25 4UD M0DOA
PL25 4UW G6GAB
PL25 5EA G4TRV
PL25 5NY M3NMM
PL25 5TA G8MNC
PL254EA M3XTG
PL26 6BN 2E1EAK
PL26 6BN G4LED
PL26 6NU M0HDP
PL26 6NZ G4MLI
PL26 6QS G8HYL
PL26 6QZ G3YIY
PL26 6TG G7JIB
PL26 6TZ G6MNI
PL26 6TZ G6MNJ
PL26 6UP M7OXY
PL26 6XD G8TCP
PL26 6XD M1BRZ
PL26 7AG M0VAU
PL26 7BH 2E0KUC
PL26 7BH M0ORS
PL26 7BW G4OPL
PL26 7EH G4HFO
PL26 7ER G8ZQM
PL26 7ER M3ZQM
PL26 7HG G8TNA
PL26 7NN G8VRV
PL26 7PF G4KNI
PL26 7PN G7FLX
PL26 7PT G4ZGQ
PL26 7PW M0GTH
PL26 7TE 2E0WRL
PL26 7TL G2KF
PL26 7TP G1LQT
PL26 7TP G3KYM
PL26 7TQ G7TLK
PL26 7TY G1BBA
PL26 7TY M7AWF
PL26 7UZ M7OOW
PL26 7XN M1AEG
PL26 7XQ G0FCB
PL26 7XS G0HZD
PL26 8BE M0HKI
PL26 8DB G6CNW
PL26 8DS 2E0PXI
PL26 8DS 2E0TGK
PL26 8DS M6PXI
PL26 8DS M6TGK
PL26 8DS M7SEW
PL26 8EJ M3TEL
PL26 8GY G4VSY
PL26 8HD M6HVK
PL26 8HW M6CJR
PL26 8JA M1AZJ
PL26 8JH G1TTK
PL26 8JN 2E0BXD
PL26 8JN M0PSB
PL26 8JN M6JOE
PL26 8PH G0AHM
PL26 8PS M6RSI
PL26 8QL 2E0ONC
PL26 8QL G4JYF
PL26 8QL M6VDT
PL26 8QN 2E1CYI
PL26 8RJ G4TRQ
PL26 8TJ M7FBW
PL26 8TL G0VDU
PL26 8UA 2E1CWE
PL26 8UB G0RJI
PL26 8UH G3WKF
PL26 8UQ G4OKS
PL26 8YE G4ZZY
PL26 8YW G7VTC
PL268QP M7HYP
PL268RT M0PHM
PL27 6AN 2E0EJV
PL27 6AN M6HVO
PL27 6NW G8LOU
PL27 7LL M3UWM
PL27 7PG G7DUB
PL27 7PR M6KWZ
PL27 7QA M6HEB
pl27 7qd 2E0JJR
pl27 7qd M3PXK
PL27 7TB G7TLC
PL27 7TS 2E0XTM
PL27 7TS M3XTM
Pl277TP M7HLK
PL28 8ES G0MWW
PL28 8SB G0SXM
PL29 3RU G4UZO
PL3 4DN M6BWK
PL3 4EG M3YWM
PL3 4HE M1FCG
PL3 4HS 2E0DCF
PL3 4HS M0WCL
PL3 4HS M6LJR
PL3 4LN G8PYE
PL3 4PD G0OYO
PL3 4QN M3OUV
PL3 4RB G6GEX
PL3 4RB M0GEX
PL3 5BS G0FCG
PL3 5DA G8RMP
PL3 5DU G4BCX
PL3 5DU G6ALR
PL3 5HJ M7HMC
PL3 5HJ M7PFC
PL3 5HW G7IZU
PL3 5NP G3SQN
PL3 5NP M6RMK
PL3 5PU M3FTZ
PL3 5TX G0KPQ
PL3 6BY M6DIZ
PL3 6DB M3ZWD
PL3 6DE M6DIB
PL3 6DH G0GVX
PL3 6DH M6HKV
PL3 6HD M1AGY
PL3 6HL M3JOJ
PL3 6JF G1LOE
PL3 6JZ G0LSJ
PL3 6JZ G7HCN
PL3 6LD G0EIG
PL3 6LX G4XZS
PL3 6LX G6ZGK
PL3 6NQ G7DIR
PL3 6PB G1KCW
PL3 6PT G3SVZ
PL3 6PX G6GFO
PL3 6PX M3NJA
PL3 6QJ M7LRJ
PL3 6SZ M3RNK
PL30 3AU G3IWE
PL30 3BS G4WQU
PL30 3PH G4AXC
PL30 3PN G3XKS
PL30 4HF M0JQJ
PL30 5AT G8TXK
PL30 5ED G1JBB
PL30 5EP G4UBK
PL30 5EW 2E0UEE
PL30 5EW M0UEE
PL30 5EW M6UEE
PL30 5HD G8THA
PL30 5JL G3MCD
PL30 5JZ 2E0IKO
PL30 5JZ M0XVN
PL30 5JZ M7SZU
PL30 5LU G1OFG
PL30 5PJ G1FXD
PL31 1BE 2E1AFC
PL31 1BH G3IGV
PL31 1JE M7AUL
PL31 1NH 2E0DSI
PL31 1NH M6DQO
PL31 1NZ 2E0KKY
PL31 1NZ M7YKK
PL31 1PY G4WVD
PL31 2BZ M3ZNX
PL31 2FB 2E0BHN
PL31 2FB M3RKJ
PL31 2FE M6AKV
PL31 2FP G0UFF
PL31 2HE M1ALX
PL31 2NU G0EEJ
PL31 2QP G4GPD
PL311PD M7CRK
PL32 9RZ G1VWU
PL32 9UB G4YME
PL32 9UP G4YVB
PL32 9YN G0ISJ
PL33 9AT G4WAV
PL33 9BN G6GWX
PL33 9BX M7ION
PL33 9DP M6CGB
PL33 9DT 2E0BJV
PL33 9DT M3PXF
PL34 0BH G0EWR
PL34 0DT M0HJO
PL34 0DT M0IGJ
PL34 0EH M3KKG
PL34 0EL G3LOV
PL34 0EL M3MXZ
PL34 0HH G0ATS
PL4 0EF G6IBU
PL4 0EF G8BOI
PL4 0EZ M3YGT
PL4 6AZ G6URM
PL4 6NW G3RRW
PL4 6PR G1ZSK
PL4 6QA G4EKV
PL4 6QA M3LUW
PL4 6RD M3BHK
PL4 7AG M0CCA
PL4 7EQ M3HBS
PL4 7HB M0UNJ
PL4 7LA M6NDE
PL4 7PY M3UHQ
PL4 8AA G0UOP
PL4 8AA G3YQF
PL4 8DS M6VEL
PL4 8PR M7OCN
PL4 8QE G1OGB
PL4 8TA M0VRT
PL4 9AQ G4RQN
PL4 9DF 2E0EQB
PL4 9DF M6JIY
PL4 9EL G0BDB
PL4 9EN 2E0CSO
PL4 9EN M6BKE
PL4 9ET G0TQT
PL4 9EZ G7KII
PL4 9NB M1EUB
PL4 9NB M7UKW
PL5 1AB G0JCA
PL5 1PD G0NCS
PL5 1SB G8PSC
PL5 1SR G3YJQ
PL5 1UH M3YYM
PL5 1UL G6JEM
PL5 1UP M7CYY
PL5 1UX G0WOU
PL5 2AL M3ZZE
PL5 2DB G0VKY
PL5 2DX G7ESZ
PL5 2DX G7MMV
PL5 2LN M3ZZI
PL5 2NU G7MME
PL5 2NW G8IDE
PL5 2QU M7BXF
PL5 2SN 2E0GLS
PL5 2SN M6GPS
PL5 3FF M6FDF
PL5 3NU 2E0FHJ
PL5 3NU M6NQC
PL5 3RB 2E0DTC
PL5 3RB M6BNL
PL5 4AZ 2E0BIM
PL5 4HX 2E0OPS
PL5 4HX M3LCF
PL5 4JQ 2E0IOS
PL5 4LT G0VKK
PL5 4PG M7MML
PL5 4PU 2E0BNV
PL5 4PU 2E0NGG
PL5 4PU M3NGG
PL5 4PU M3PXE
PL5 4QE G3RYZ
PL51PH 2E0TBH
PL6 5HH M0PDB
PL6 5HQ M7RDH
PL6 5NR G0JNZ
PL6 5PU G3JFS
PL6 5PU G7UTI
PL6 5QE 2E0STI
PL6 5QE M3XXL
PL6 5QN M0CPW
PL6 5RJ G4BZF
PL6 5SD G7NIA
PL6 5SD M0BWN
PL6 5SD M7CWM
PL6 5SN G0KYE
PL6 5TR G7BXS
PL6 5TR M3BXS
PL6 5TX G4WLS
PL6 5TZ M3ZZJ
PL6 5XJ 2E0LMO
PL6 5XJ M6MLO
PL6 6AH G3JTJ
PL6 6AY 2E0SKE
PL6 6AY M6FGE
PL6 6BH G3SGV
PL6 6DW 2E0PLY
PL6 6JP G4CGM
PL6 6LS M0BLO
PL6 6PL M7TIC
PL6 6PN M7BBM
PL6 6RE G6IPQ
PL6 6TN M6CDN
PL6 7BD M0IHB
PL6 7BY G1HHC
PL6 7BY G1HHD
PL6 7DT G7OOV
PL6 7HS M6DGN
PL6 7HX G3ZZS
PL6 7JA G4YDR
PL6 7JA M3BAS
PL6 7LN M3YLQ
PL6 7SA G7DQA
PL6 7SP G7RTO
PL6 7UB G4GHR
PL6 8EQ M6ZEL
PL6 8QD M6VKN
PL6 8RQ G8WZJ
PL6 8TD G7ESO
PL6 8TD M0JIE
PL6 8TL M0DWT
PL6 8UU M3SQQ
PL6 8XF G0OBP
PL7 1JJ M1AHF
PL7 1JR G3ARE
PL7 1JY G3LCY
PL7 1PG M6FGI
PL7 1PU G6CYH
PL7 1PZ M0GBA
PL7 1PZ M1CXV
PL7 1QZ G3VCN
PL7 1SN M3YYS
PL7 2AJ G0AHU
PL7 2DU M3PYT
PL7 2DY G4LOI
PL7 2EJ G0FSD
PL7 2EY G4NLU
PL7 2FJ M6OEB
PL7 2GT G7HHW
PL7 2HP G0KML
PL7 2RU G7NHB
PL7 2SU M1SAM
PL7 2YF G0MMW
PL7 4BP M3YYJ
PL7 4DA M0AVS
PL7 4EG 2E0ZRM
PL7 4HH 2E0TZM
PL7 4HS G3TZQ
PL7 4HY G0UMI
PL7 4HZ M6EQZ
PL7 4JB 2E1HSL
PL7 4JB G1CKT
PL7 4JE G0NAP
PL7 4PF G1LTC
PL7 4PL M3YYR
PL7 4QY 2E0MGC
PL7 4QY G8CMG
PL7 4QY M0WMB
PL7 4RW G7ART
PL7 5AA 2E0GQM
PL7 5AA M0WIV
PL7 5AA M7WIV
PL7 5FE G3NBN
PL75HX M7MXM
PL8 1BP G3MMX
PL8 1BP G8ADX
PL8 2ED G6FBJ
PL8 2JQ M0DZL
PL8 2JX G8AZZ
PL8 2NT G0ESL
PL82AX 2E0DIH
PL82AX M0IGP
PL9 0AW 2E0CYZ
PL9 0AW M6AYZ
PL9 0BB M0IAZ
PL9 0DS G4PRL
PL9 0EU G0MQK
PL9 0JZ 2E0OZI
PL9 0JZ M6OZI
PL9 7DQ G7BAV
PL9 7FB 2E0GNI
PL9 7LA G3VNG
PL9 7LD G4NDD
PL9 7LF G3KHU
PL9 7LU G7UWO
PL9 7NU G4KFZ
PL9 7PG M0BHV
PL9 7QP M3GDV
PL9 8DB G1BMP
PL9 8DB G3XZX
PL9 8DT M3KZT
PL9 8HU G0VZX
PL9 8NW M3LVX
PL9 8PJ G3ULN
PL9 8QZ G0LRJ
PL9 8RB G3RMZ
PL9 8TW G0RMC
PL9 8UR G3SPI
PL9 9EJ M7EHD
PL9 9HJ G7BYW
PL9 9HW 2E0DQU
PL9 9HW M6IAX
PL9 9JT M0AKR
PL9 9LU G7FXW
PL9 9PT M0ASI
PL9 9RR G4SJD
PL9 9SH G3VZJ
PO1 1QN 2E1AKW
PO1 1QN G6ASJ
PO1 3RD M0GYS
PO1 4ES M7GTM
PO1 5AR G0LFI
PO1 5AR M6UBI
PO1 5HU M7MPL
PO1 5LL M3XDH
PO1 5QT M0PPR
PO1 5RR M6PWB
PO10 7EW G4TLO
PO10 7JY G0BBJ
PO10 7LB M7MKB
PO10 7NH G4WFF
PO10 7NS G8RRC
PO10 7QP G6NUX
PO10 7RA G0USE
PO10 7RP 2E0DMB
PO10 7TR G4OZX
PO10 8BN G4XMJ
PO10 8GE G0DNF
PO10 8HL G7VIX
PO10 8HS G1MDJ
PO10 8JG G3HCO
PO10 8LB 2E0HEX
PO10 8LB M0HEX
PO10 8LB M6HEX
PO10 8NQ G3PVH
PO10 8SZ M7JCK
PO10 8UX G3ZSS
PO10 8XD G0UFY
Po108hx M6BFQ
PO11 0AW G6RGI
PO11 0AZ G3VNU
PO11 0DT G3JVL
PO11 0EF G0BXV
PO11 0ER G3MCV
PO11 0EW 2E0PGJ
PO11 0FF 2E0DYX
PO11 0FP M1BIY
PO11 0HL G4KVX
PO11 0JW G0AYI
PO11 0JW G7KQT
PO11 0QE G0KCF
PO11 0QE G8DLL
PO11 0QE M3TEV
PO11 0QR G8VMY
PO11 0RL G6VXZ
PO11 9AA G1OXF
PO11 9HY G7ARK
PO11 9LT G2JL
PO11 9LT G3DIT
PO11 9NE G4ZMY
PO11 9PS G8ZJK
PO11 9RA G3YPW
PO11 9SJ 2E0OCS
PO11 9SN 2E0JIG
PO11 9SN M6IFO
PO11 9ST M6ZZY
PO119QL 2E0PRG
PO12 1HE G4JOA
PO12 1HH M0CDA
PO12 1JX M6BBB
PO12 1LR 2E0LNX
PO12 1LR M6LPI
PO12 1PW M3UWR
PO12 1QY G4XQX
PO12 1RB M6CPT
PO12 1SE 2E0UIT
PO12 1SE M6UIT
Po12 2dt M7GRD
PO12 2HB G0MPJ
PO12 2HB G1BHG
PO12 2HX 2E1CNM
PO12 2HX G0AVP
PO12 2HX G6XHJ
PO12 2HX M0MIT
PO12 2HX M0RCV
PO12 2LG 2E0CWJ
PO12 2LG M0HTE
PO12 2LG M6CNR
PO12 2NE G8EIE
PO12 2NL G6JGT
PO12 2NL G7AUR
PO12 2NN M0BEC
PO12 2QU G8BSD
PO12 2RA G7SAX
PO12 2UP G6UGG
PO12 3BB G4MWW
PO12 3DR M7NMC
PO12 3DS M6AIH
PO12 3EB G6ERI
PO12 3EU G6TQF
PO12 3HF M0KPK
PO12 3JJ G0GIA
PO12 3JN G7LOW
PO12 3JY 2E0AYK
PO12 3JY M0EHL
PO12 3LD G7RWN
PO12 3QN G6NHA
PO12 3QY G0NCX
PO12 3SX G1ZTN
PO12 4BU G0HZY
PO12 4BX G4POW
PO12 4DB G1WXU
PO12 4EP G1EDA
PO12 4ES 2E0CGH
PO12 4ES M6WFA
PO12 4EU G1OAW
PO12 4EW G8GBE
PO12 4GG G0JFD
PO12 4HN M6LFM
PO12 4JF G0LPP
PO12 4LW M0VXC
PO12 4LW M7BMX
PO12 4NP M7BBI
PO12 4NS G1UID
PO12 4NX M6WBX
PO12 4RT 2E0EFH
PO12 4RT M0TFH
PO12 4RT M3UYQ
PO12 4UE M5AGB
PO13 0DP G4UHS
PO13 0DX 2E0HJI
PO13 0DX M7BQU
PO13 0JS 2E0EFA
PO13 0NF G0DGW
PO13 0PT G3MZZ
PO13 0QS G8MRN
PO13 0RB G1YTL
PO13 0SE 2E0BXX
PO13 0SE M6HAS
PO13 0SJ M0EEL
PO13 0SP G0GLW
PO13 0TG 2E0JIS
PO13 0TG M7SIJ
PO13 0TN 2E1ESN
PO13 0TN M1ANC
PO13 0TT G7TXW
PO13 0TU M6NAQ
PO13 0TY 2E0DOF
PO13 0TY M3SVB
PO13 0UT 2E0FEZ
PO13 0ut M6KMV
PO13 0UZ M7CIX
PO13 0WP 2E0ATB
PO13 0WP M0RQX
PO13 0XR G1OCY
PO13 0YN G0DDW
PO13 0YX G0JYV
PO13 0ZD 2E0DBA
PO13 0ZD M0LOW
PO13 0ZD M3ERR
PO13 0ZU M6BLX
PO13 8HD 2E0YZM
PO13 8HD M6YZM
PO13 8LS 2E0GPI
PO13 8LS M0KPI
PO13 8LS M7TKD
PO13 9AF G0VLI
PO13 9AF G5LK
PO13 9AF G7RAT
PO13 9AG G8DBD
PO13 9AU G4MHQ
PO13 9AU G7USV
PO13 9BB M6WYY
PO13 9BP 2E0FHH
PO13 9BP M6FUW
PO13 9DH G3VEF
PO13 9DH G7MFR
PO13 9EN G4TKW
PO13 9EY G8GUS
PO13 9HU G3NPZ
PO13 9NA G6APD
PO13 9NJ G8OUH
PO13 9NU 2E0GRQ
PO13 9NU M7MDY
PO13 9RD G4BMQ
PO13 9TL 2E0HHP
PO13 9TL M7JDC
PO13 9UJ G1CLJ
PO139DU 2E0EMG
PO139DU M0MAQ
PO139DU M3VTQ
PO14 1AQ M7AUD
PO14 1EF G6AIQ
PO14 1EG G0VIX
PO14 1HH 2E0TGJ
PO14 1HH M7TGJ
PO14 1PH M3FNO
PO14 1QF 2E0BQK
PO14 1QF M3VOU
PO14 1SL G7ILI
PO14 2BQ G6AVT
PO14 2BZ M0BQZ
PO14 2HA G4XVV
PO14 2JF G3GVM
PO14 2PX G4PZV
PO14 2QS G0RHV
PO14 2QX G0PPH
PO14 2QQ G4SPS
PO14 3AA G6BAT
PO14 3AD G1KNI
PO14 3AF G7NYF
PO14 3BS G8APL
PO14 3DR G8PGF
PO14 3HW G3IBI
PO14 3LB G6HGX
PO14 3LB M0DDA
PO14 3LB M3ADJ
PO14 3LF 2E0ZFV
PO14 3LF M3ZFV
PO14 3RX M0GLL
PO14 3SY M1FCE
PO14 4BH G3HKT
PO14 4BX M1FIR
PO14 4EN 2E0LLD
PO14 4EN M6LDD
PO14 4JP M1EMP
PO14 4NB M0NRG
PO14 4NP G0GFD
PO14 4QX G8RNV
PO14 4SF G8YBC
PO14 4SL G0MAT
PO14 4SL G0TPB
PO14 4SY G8RLF
PO15 5AJ G1NCK
PO15 5BL G6HHE
PO15 5DU M0RIK
PO15 5EQ G3YTQ
PO15 5HP G3RDA
PO15 5JJ G0AMS
PO15 5LG G0JYQ
PO15 5LZ M7MUJ
PO15 5NE G4VNM
PO15 5NF 2E0JTN
PO15 5NF M0JTN
PO15 5NF M6JTN
PO15 5PF G1MPP
PO15 5QB G3UWE
PO15 5QH G8UXW
PO15 5RS G0XGL
PO15 6HF 2E0JRD
PO15 6HF M0JDL
PO15 6HF M3YJD
PO15 6JU G3KLF
PO15 6QP 2E0SHZ
PO15 6QP M1AIB
PO15 6QP M6MOI
PO15 7BD G3UPZ
PO15 7EA G0SHT
PO15 7EA G4CAA
PO16 0DX G0USF
PO16 0DX G7SFL
PO16 0FA M6WRJ
po16 0qb G0SFG
PO16 0RX 2E0ABL
PO16 0SQ G0DDA
PO16 0TR G4ZMP
PO16 7DP 2E0MOY
PO16 7DP M0YSR
PO16 7DP M6FFR
PO16 7HB M0HBE
PO16 7HN M7SLG
PO16 7JF G4FCL
PO16 7LU G0HZC
PO16 7NL G4UOR
PO16 7QL G4ITG
PO16 7QW G3YIW
PO16 7RR G6KTX
PO16 7SJ G6GPV
PO16 7XA G0JYZ
PO16 7XR G1WKZ
PO16 7XR G1WLD
PO16 8DS G8XEZ
PO16 8DU M6DRG
PO16 8DY G0FAD
PO16 8JW G4PWG
PO16 8JY G7MDV
PO16 8LB G4CYC
PO16 8LE G1FIG
PO16 8LF G0LYI
PO16 8LF G3XUF
PO16 8PB G8IOJ
PO16 8QA M7CXQ
PO16 8QG M0VMX
PO16 8TF M6WVV
PO16 8UE M0DQO
PO16 9AA M6ZAU
PO16 9AP G7DWV
PO16 9BQ G6JSI
PO16 9DN G4YPA
PO16 9EG G7UUQ
PO16 9EG M7GRR
PO16 9HP G4IOJ
PO16 9LA M6EKX
PO16 9NJ G4MRW
PO16 9NR G3UFF
PO16 9PA G0KCG
PO168TY G1OJS
PO168UF G6ORL
PO17 5BU 2E0IHM
PO17 5BU M7DQD
PO17 5EX 2E0SMO
PO17 5GS G6LYA
PO17 5LD G8POQ
PO17 6DG G4GWJ
PO17 6DG G4HKZ
PO17 6DG M0XML
PO17 6EY 2E0CFE
PO17 6HS G3XUX
PO17 6JB G1WID
PO17 6JJ G8JXV
PO18 0AH M6OVY
PO18 0AY M1BTO
PO18 0EX G3NXO
PO18 8BE G0CHK
PO18 8EU M0SYM
PO18 8LF G6XZA
PO18 8QN G4BVS
PO18 8QQ M0GNL
PO18 8QW M3XYC
PO18 8RQ 2E0HYM
PO18 8RQ M6HYM
PO18 8RQ M7HYM
PO18 8SR G7MIN
PO18 8SR M3MIN
PO18 8SU G3LSQ
PO18 8SU G4JDG
PO18 8TZ G0SDC
PO18 9JJ G0IOP
PO18 9LN G1YEW
PO18 9LS G7ISD
PO18 9NP G0LNX
PO19 1QS G3SJW
PO19 3AE G1UFS
PO19 3AN 2E0CGK
PO19 3JD M0PWV
PO19 3JL M6GRT
PO19 3LD G4SQJ
PO19 3LY G7BVL
PO19 3NL M6VOL
PO19 3QF G8KJT
PO19 3QP G4WVQ
PO19 3QU G0WSD
PO19 5QU 2E0IWS
PO19 5RL G4EHG
PO19 5RX M0NNL
PO19 5RX M3YQQ
PO19 5UA G3ZEN
PO19 5UA G4VQZ
PO19 6BY G3MVZ
PO19 6GG M0CXY
PO19 6GL G4ZTQ
PO19 7NE 2E1GCF
PO19 7NW 2E0CSH
PO19 7NW M0OSB
PO19 7NW M6BYU
PO19 7UY M6KTK
PO19 7XE G0ISL
PO19 8AU G4WIR
PO19 8BP M3SYW
PO19 8QR G1BZU
PO19 8QR G3BZU
PO19 8QR G3CRS
PO19 8QR G3ZDF
PO19 8QR G7DOL
PO19 8RG 2E0HVT
PO19 8RG M7GWF
PO19 8TP G8NYK
PO195DS M7DBD
PO2 0EX G7SCE
PO2 0QS M0DND
PO2 7DD 2E0BUN
PO2 7DF 2E0BUP
PO2 7HB G7OKI
PO2 7HB M3WCM
PO2 7JW G0DAE
PO2 7LB G8UYY
PO2 7LX M3ZNG
PO2 7PG G4EFB
PO2 8EX M3AMA
PO2 8LX 2E1HSB
PO2 8LX M1DOX
PO2 8NA M0GIF
PO2 8NF G6ATK
PO2 8NT M6KNP
PO2 8RH M1GRP
PO2 9EG M6JZM
PO2 9HE M1DMR
PO20 0DW M7EFA
PO20 0EB M6TPO
PO20 0EE G4KEC
PO20 0FQ 2E0XPJ
PO20 0FQ M0XPJ
PO20 0FQ M3XPJ
PO20 0HY G4MVS
PO20 0JB M1BCY
PO20 0JG G7SZW
PO20 0JR M0RMW
PO20 0JS G4ZPP
PO20 0NA M6SXM
PO20 0NX G8OLA
PO20 0RG M5DNK
PO20 0SF M3WLG
PO20 0TE M7ELG
PO20 0UA G8PQH
PO20 1JZ G0POU
PO20 1JZ G8CBO
PO20 1NY G0WBR
PO20 1NY M6LUK
PO20 1PA G3NFW
PO20 1PE G4BZB
PO20 2GQ 2E0FSI
PO20 2GQ M6FSI
PO20 2HX 2E0FVL
PO20 2WB M3DSW
PO20 3SJ G4ZJP
PO20 3TH M6VTB
PO20 7EZ 2E0TEE
PO20 7EZ M0YEE
PO20 7EZ M6AXI
PO20 7JJ G0LMJ
PO20 7NL M7DWE
PO20 7RR G0WVM
PO20 8AH G7BKJ
PO20 8DP M3NYK
PO20 8DW G4VRG
PO20 8EX 2E0RXC
PO20 8EX M6GYC
PO20 8EX M6RCS
PO20 8PB G6VLV
PO20 8PE 2E0YAS
PO20 8PE M6YTS
PO20 8RG G0BVV
PO20 8RJ G3SFB
PO20 9AX G4YJD
PO20 9AY G6XTK
PO20 9DT M0GOO
PO20 9HL G7HJT
PO21 1AB M1FCC
PO21 1HQ G1GKW
PO21 1XJ G6ZSQ
PO21 2DG 2E1FIE
PO21 2EJ G0SZK
PO21 2JU 2E0XCZ
PO21 2JU M0XCZ
PO21 2JU M6XCZ
PO21 2PY M0MLM
PO21 2QA G0CHE
PO21 2RB G4XPT
PO21 2SF G6FAX
PO21 2TJ M7CUS
PO21 2YQ G7EPR
PO21 3DH G3GUR
PO21 3EL G8OCM
PO21 3EQ G0CIR
PO21 3EZ G0TXU
PO21 3HQ G7IMT
PO21 3ND G0AKK
PO21 3NN 2E0XAO
PO21 3NN M3UJZ
PO21 3PS M7AFE
PO21 3QH M7SID
PO21 3RB M7GRP
PO21 3SL G4RPA
PO21 4AW M0CME
PO21 4DY G2ARU
PO21 4ET G3LTM
PO21 4HB M0GSC
PO21 4HT G3RJS
PO21 4LH M6GMV
PO21 4NB G8TGH
PO21 4NN G6HTB
PO21 4PS G8REF
PO21 4QT 2E0NUG
PO21 4QT M0VPL
PO21 4RN G0RQH
PO21 4TB G8YAS
PO21 4TJ G4ITY
PO21 4TN G1XIV
PO21 4UR G6FDU
PO21 4XN G3IJS
PO21 5AD M6DHB
PO21 5FA G1ZDY
PO21 5LL G0KJU
PO21 5TD G0OSU
PO21 5TW G8ZTD
PO213NN 2E0CKE
PO213NN M6SXO
PO213PL M7DLB
PO22 0AD G6AII
PO22 0AR M3DOA
PO22 0EJ 2E0WSX
PO22 0EJ M0HIX
PO22 0EJ M6ARP
PO22 0HF M0KPT
PO22 0JN G1ITL
PO22 0LH G7BWW
PO22 6AH G7TPH
PO22 6BU G0MTK
PO22 6BX G0AFN
PO22 6ED G6UTT
PO22 6HG G1ORB
PO22 6JU M6EIC
PO22 6QH M7KLA
PO22 7LB 2E0FLK
PO22 7LB M0OFE
PO22 7LB M6OFE
PO22 7NW G4PDY
PO22 7QG G4ECF
PO22 7SE 2E0FAH
PO22 7SE M3FPH
PO22 7SL G0DFA
PO22 8DP G6XJN
PO22 8PH 2E0SCK
PO22 8PH M0HJB
PO22 8PH M6BXJ
PO22 9BE 2E0WSP
PO22 9BE M0WSP
PO22 9BE M6UGF
PO22 9DG G1LHE
PO22 9DR M3NIZ
PO22 9EX G3SFE
PO22 9HJ M6KKW
PO22 9LA G0VCJ
PO22 9LE M7PTY
PO22 9NU M7CQX
PO22 9RR M0MMT
PO22 9RR M6MIE
PO228NS G0XAI
po229eq M7XAP
PO3 5DE M3YZE
PO3 5EL M0DDU
PO3 5LN M6DNJ
PO3 5PU M6JSS
PO3 5TN 2E0JCE
PO3 5TN M0XLA
PO3 5TR G8KQV
PO3 5TR G8NVZ
PO3 5UG M6OFL
PO3 6AU. M1SKA
PO3 6BE M0GMI
PO3 6LR G3ZBP
PO30 1AE M6NGK
PO30 1AF G1DEO
PO30 1DG G4MBD
PO30 1DG G7IRH
PO30 1DR G6OOH
PO30 1DT G3XOC
PO30 1HA G4FYI
PO30 1HG G6ORJ
PO30 1HN 2E0BUK
PO30 1LN G0MHN
PO30 1NR 2E1IWD
PO30 1NR M0IWA
PO30 1NR M3DIW
PO30 1NR M6SPU
PO30 1PA G0EHR
PO30 1PZ G0KQO
PO30 1QD 2E0UYB
PO30 1QD M3CIG
PO30 1QD M7CZW
PO30 1QT G7NAO
PO30 1TD G7UHT
PO30 1XN M0GQJ
PO30 2BH M3AYC
PO30 2BH M3KLS
PO30 2BH M3SYZ
PO30 2DB 2E0AAJ
PO30 2DB M6MPD
PO30 2DP G8AZM
PO30 2EJ M6WZV
PO30 2GA M3BRL
PO30 2GS 2E1CAU
PO30 2JU M0PDL
PO30 2LL G8RWM
PO30 2LL M6KFW
PO30 3DD G1KAK
PO30 3JP 2E0JIW
PO30 3JP M0PBN
PO30 3JP M3JIW
PO30 3JY M1FBL
PO30 3JZ 2E0FSE
PO30 3JZ M6FSE
PO30 4AE G4BIM
PO30 4AN M0IKA
PO30 4BG G0GMY
PO30 4BG M6NBV
PO30 4BX G6BHB
PO30 4DJ G0PXM
PO30 4HH M0CZP
PO30 4LP 2E1NVK
PO30 4LP M0NVK
PO30 4LZ M6FQE
PO30 5AR M7DAR
PO30 5BS M0IES
PO30 5BS M7TMR
PO30 5GU 2E0EJK
PO30 5GU M6CUC
PO30 5JJ M3FBR
PO30 5JR M1IOW
PO30 5JR M3IWG
PO30 5NG 2E0DUS
PO30 5NG M6YEG
PO30 5QP M7EAP
PO30 5QZ G0RUT
PO30 5QZ G7GGH
PO30 5SF G0DWE
PO30 5SF G1HHB
PO30 5SF M6NKY
PO30 5SG M0IOW
PO30 5SJ M3LJZ
PO30 5SJ M3ZBJ
PO30 5TL M0TVR
PO30 5TL M3MEE
PO30 5TL M6BRT
PO30 5TP G3IMX
PO30 5TY G6CUT
PO30 5UF M6DAZ
PO30 5UH G0LFV
PO30 5UR G7MID
PO30 5XS 2E0EDC
PO30 5XS M6FPJ
PO30 5ZF 2E0BSK
PO30 5ZF M3WHV
PO31 7GA M0GYL
PO31 7HF 2E0AGQ
PO31 7HW M6PBM
PO31 7JN G6ZUZ
PO31 7NF G4GRK
PO31 7NX M6LTS
PO31 7PE M3AYJ
PO31 7PP M6FQH
PO31 7PS 2E0WKG
PO31 7PS M6WKG
PO31 7PY M3BBF
PO31 7SG M0DKJ
PO31 7SR M3RGP
PO31 7TE M1CTY
PO31 7XS M7AYF
PO31 8AD G4ZFQ
PO31 8AJ M0JZF
PO31 8AL G3WXC
PO31 8AS 2E0ENG
PO31 8AS M3ORQ
PO31 8AS M3WKL
PO31 8BT G1HKU
PO31 8DP G3XYB
PO31 8DT G3YZK
PO31 8DX G4SCB
PO31 8EF G8MBU
PO31 8HA 2E0AKQ
PO31 8HA M0JCH
PO31 8HA M3NNV
PO31 8HQ G3XZB
PO31 8JP G4ZBH
PO31 8JQ M6DYN
PO31 8LB M0TFN
PO31 8NE 2E1KJB
PO31 8NR G0HDH
PO31 8PE G0CWX
PO31 8PN G3PZB
PO31 8PN G3SKY
PO31 8PT G8RKH
PO31 8PZ G3IW
PO31 8PZ M6OFH
PO31 8QP G4ZOH
PO31 8RF G6VFF
PO31 8RL G4VYC
PO317ER M6FDU
PO32 6AX G7GZJ
PO32 6BH G0WXJ
PO32 6BH M0RAD
PO32 6BH M0ZOO
PO32 6EA G7VGY
PO32 6EJ G4UCZ
PO32 6FJ 2E0RQD
PO32 6FJ M0RQD
PO32 6GA G3VDZ
PO32 6GT 2E0WJH
PO32 6HR G0PQW
PO32 6HT M3RKR
PO32 6JD M6SSM
PO32 6JE M6FWE
PO32 6JE M6LAS
PO32 6JT G7TQC
PO32 6LS G0PEB
PO32 6LS G7ATJ
PO32 6NT G0JHQ
PO32 6NT G6RTE
PO32 6NT G6TVX
PO32 6NX M7DDY
PO32 6PS G0KQR
PO32 6QW M6SGT

PO32 6RU G1TUZ
PO32 6SS G3PQJ
PO32 6TD 2E0ZMB
PO32 6TD M3SQG
po326qn G7SVF
po326qn M0GAQ
PO33 1AB G8MYF
PO33 1BX M0DSF
PO33 1ED G6LXP
PO33 1EL G0MWU
PO33 1FY G1VGM
PO33 1FY G2CNN
PO33 1JD G0MIZ
PO33 1NT G4TAT
PO33 1PR G4RSN
PO33 1QY M6DQW
PO33 1TA M0BTP
PO33 1XE G4RGE
PO33 1YF G0SEB
PO33 2BG G1WQC
PO33 2BG M0CGO
PO33 2BH 2E1CDK
PO33 2BH M3AIL
PO33 2HD M3TCU
PO33 2JP G1FPP
PO33 2QF G1HRQ
PO33 2QQ G0DRW
PO33 2SS G6LVS
PO33 2TQ 2E0WWJ
PO33 2TQ M7XLF
PO33 2UH G3XHM
PO33 2UP G7MAR
PO33 3BU G6SQL
PO33 3DL G4SBN
PO33 3EL M1OBR
PO33 3HU M3YPP
PO33 3JU G1HCM
PO33 3LH G3XLP
PO33 3QJ G3KPO
PO33 3QJ M0TIW
PO33 3QS M3JIU
PO33 3QZ M6ODR
PO33 3SD M7SYZ
PO33 3SF G4TFW
PO33 3TA G4KZI
PO33 3TE G6MQY
PO33 3TL G0UEK
PO33 3TL G4SHL
PO33 3UB 2E1EBN
PO33 3UX G6HZG
PO33 4BB G4OAG
PO33 4DR M0TAM
PO33 4ED G3UHX
PO33 4EU 2E0DUI
PO33 4EU M6FLZ
PO33 4EZ G8NHG
PO33 4HE M7WMK
PO33 4JF G7RCC
PO33 4JJ G4KKB
PO33 4JR G0RSY
PO33 4LE M6CYN
PO33 4LG G4BCH
PO33 4LG G4CQO
PO33 4LQ M1DBM
PO33 4LU G6DOD
PO33 4PY G4YYF
po334lb G4IKI
PO34 5DY G0ZGN
PO34 5JE G0GNI
PO35 5QS 2E0XXB
PO35 5QS M3XXB
PO35 5QW G3JLN
PO35 5RA 2E0MFA
PO35 5RA G3YEG
PO35 5RA M0MFA
PO35 5RA M6IOW
PO35 5SL 2E1PHW
PO35 5TN M0HKK
PO35 5TS G4LUY
PO35 5UQ M6DXW
PO35 5UW 2E0ZML
PO35 5XU G4MZC
PO35 5YJ G0VZV
PO35as M6GSZ
PO36 0BA M6JBA
PO36 0DX G1RIR
PO36 0JD G4ANW
PO36 0JT G3GEG
PO36 0JY G8TAQ
PO36 0LH G4OYR
PO36 0NE M7EJI
PO36 8BA G0VPO
PO36 8BE 2E0DEI
PO36 8BG M3BKV
PO36 8DZ 2E0MRD
PO36 8DZ 2E0SUS
PO36 8DZ M6DEL
PO36 8HE G4RTW
PO36 8QE G0WVD
PO36 8QL G4ADK
PO36 9BX G6ETC
PO36 9DS G4UNM
PO36 9HF G0WLX
PO36 9HQ G6WUR
PO36 9HW G0MPA
PO36 9JA G4JLX
PO36 9JA G4ULT
PO36 9JL G1SMY
PO36 9NS 2E0CZW
PO36 9NS M6SGS
PO37 6AE G6DIQ
PO37 6AW M6KEV
PO37 6DB G8PPN
PO37 6EJ G0NTH
PO37 6EW 2E0EYB
PO37 6EW M6WOI
PO37 6NN G8JBM
PO37 6NX G4SVY
PO37 7AG G4WNZ
PO37 7BU G6EVX
PO37 7DF G7IVF
PO37 7EJ G4NOU
PO37 7HH G0HFO
PO37 7NJ M0JAG
PO37 7NZ G4RTY
PO37 7PA G4UUJ
PO376AU M7DDK
PO38 1AA G1JYZ
PO38 1AL G4CZP
PO38 1BD 2E0GZT
PO38 1BD M6GZT
PO38 1BH G1CPO
PO38 1DQ G4ZEN
PO38 1HG 2E0OBP
PO38 1HG M6OBP
PO38 1NX G4MPI
PO38 1QL G0LGZ
PO38 1RZ G0RMJ
PO38 1TH G7ONE
PO38 2BG G4IGK
PO38 2BG G8DGW
PO38 2DE G8DDY
PO38 2DZ M1BOD
PO38 2NE 2E0EXU
PO38 2NE M6SKY
PO38 2QW G4XIU
PO38 3AF M0LZS
PO38 3BU M6YEB
PO38 3DB 2E0RSD
PO38 3DB M6BRR
PO38 3EF G6KRN
PO38 3EH G8XGV
PO38 3EL M3FBG
PO38 3EL M3FPG
PO38 3EQ M1BWR
PO38 3HB M6DRI
PO38 3HR G0PEF
PO38 3HW M6MYD
PO38 3HY G8XHN
PO38 3HZ G3LYD
PO38 3NH G4RZQ
PO38 3NP M0WAZ
PO38 3NS 2E0JVS
PO38 3NT M6LYU
PO38 3NT G4RQP
PO38 3PH 2E0FRB
PO38 3PH M3FRB
PO39 0AH G4EWE
PO39 0AH G4NCD
PO39 0AL G4CSM
PO39 0BL G7BZD
PO39 0BL M3BDA
PO39 0DN M3GRI
PO39 0DX M6DRH
PO39 0EF M1CYX
PO39 0JL G0PTT
PO390EZ M7VVZ
PO4 0BA G0AXS
PO4 0DE G6DHT
PO4 0JE G1SEH
PO4 0NT G6ZKM
PO4 0RL M6ROU
PO4 8AU G4VFX
PO4 8AU G8OEJ
PO4 8HH G6CUA
PO4 8HR M0CYX
PO4 8JG G4LRH
PO4 8JS G1ERF
PO4 8JU G0IWN
PO4 8NF M3JRN
PO4 8NX 2E0CSY
PO4 8NX M6BYY
PO4 8PB M1DXO
PO4 8RH 2E0CRU
PO4 8RH M6ASV
PO4 8UF 2E1ILH
PO4 8UF M3HAD
PO4 9AB M0PVO
PO4 9BG G3CAJ
PO4 9EF G6DWO
PO4 9HP M6BRN
PO4 9JE G7UPL
PO4 9JU M3ETI
PO4 9LN M3CZB
PO4 9LR G7ARJ
PO4 9PZ G6FDP
PO4 9QU G1PGQ
PO4 9XZ G6SGW
PO40 9DL G1RFB
PO40 9DX M7EOK
PO40 9ES G6ENY
PO40 9FY G3VPK
PO40 9GB M0XYT
PO40 9HB G4NAK
PO40 9JH G7AXM
PO40 9LF G4EDN
PO40 9LQ M3IUC
PO40 9NH G1JGS
PO40 9PD G7RES
PO40 9QS G7DZY
PO40 9QS G7WKW
PO40 9TG G7OPY
PO40 9UA G3IIN
PO41 0PS M3LMB
PO41 0PX M6HQW
PO41 0PY G0GEZ
PO41 0RX G0DKS
PO41 0SL G0GPK
PO41 0TL G4AZC
PO41 0XS 2E0BHB
PO5 1RU M6MTN
PO5 2HA G6IAO
PO5 2JJ M6IRC
PO5 2NL G0LFN
PO5 3AU M3JVR
PO5 3DT M7AXA
PO5 3HP 2E0IAN
PO5 3JA G0DRL
PO5 4AL M6CXC
PO6 1DB G4IQO
PO6 1DU G3CNO
PO6 1DU M0GWD
PO6 1EW 2E0PWK
PO6 1EW M3PWK
PO6 1LW G0VEP
PO6 1LZ G4GZO
PO6 1NB G6HJV
PO6 1NG G4OVM
PO6 1PY 2E0AWE
PO6 2ES G7OWQ
PO6 2JE G8SBQ
PO6 2JU G7DRT
PO6 2NL G4FOW
PO6 2PU G6UXW
PO6 2RL M0TYW
PO6 2TJ G6XDY
PO6 3DG G3OQC
PO6 3DG G7HQP
PO6 3HD G1JAB
PO6 3JP M3HSS
PO6 3JP M3SSU
PO6 3PE 2E0GTT
PO6 3PE M0JUJ
PO6 3PE M6MSM
PO6 3QT G1OVG
PO6 3QY M6TON
PO6 3RD M0ALF
PO6 3RH 2E0KIT
PO6 3RH M6ADG
PO6 3SB G4AVX
PO6 3SH M0DNR
PO6 4AE G1XZQ
PO6 4AF M3AUK
PO6 4BB G4PYS
PO6 4EZ M6LBU
PO6 4LS 2E0BUO
PO6 4LS M0LBJ
PO6 4LS M3EOL
PO6 4QH 2E0KIS
PO6 4QH G3RCE
PO6 4QH M6URK
PO6 4QL G4LIO
PO6 4QL G6GYF
PO6 4TA 2E0CEU
PO6 4TA M0NEH
PO6 4TA M6NEH
PO7 3DB M7GEP
PO7 4QU 2E0RFM
PO7 4QU M6RFM
PO7 4RU G8CKS
PO7 4SP G0SSY
PO7 4SW M1AHA
PO7 5BL 2E0NFB
PO7 5BL M0NFB
PO7 5BL M6ATF
PO7 5BT G1TDP
PO7 5DX G4YCG
PO7 5DY G7DUE
PO7 5ED G0ABB
PO7 5ED G4ZPA
PO7 5EF 2E0WHP
PO7 5EF M7VCT
PO7 5EF M7WSJ
PO7 5HH M5KZI
PO7 5HJ G0IEY
PO7 5HJ G0IUY
PO7 5HJ G4FBS
PO7 5LL M7ATH
PO7 5NN M3ZJS
PO7 5QF M0CYM
PO7 5QR G1WXW
PO7 5QW G6ISY
PO7 5SF G4GUA
PO7 5TB G7IUE
PO7 5TW M3GTK
PO7 5XA M3HJW
PO7 6AA G1UFA
PO7 6AH M0TAP
PO7 6AH M3XYI
PO7 6AQ M3TQY
PO7 6BG G4CRM
PO7 6BT G0JWL
PO7 6BX M0JAK
PO7 6DD G0RPV
PO7 6DP G0ASZ
PO7 6EB G4JIH
PO7 6ED G3KOJ
PO7 6EG M3AEE
PO7 6HE G4TST
PO7 6HH M1EXJ
PO7 6LJ G6XBG
PO7 6PE G8FXV
PO7 6PR G4WQZ
PO7 6PR G8IRS
PO7 6SH G6CZZ
PO7 6TF M6WEJ
PO7 6UA G4DCP
PO7 6UE G3RTP
PO7 6UX G4KLX
PO7 6YJ G8LNU
PO7 7BA 2E0SJN
PO7 7BX M0SSP
PO7 7BX M3SHI
PO7 7EW 2E0CKI
PO7 7EW M6BCG
PO7 7EW M7DGH
PO7 7HX G0ATG
PO7 7JE G1OGR
PO7 7JY M6POY
PO7 7LG G8VOI
PO7 7LG M3VOI
PO7 7LN M5ACT
PO7 7LY M0RWS
PO7 7NA M3KYZ
PO7 7PB G3TVI
PO7 7PE M6DTM
PO7 7PG G3LKW
PO7 7QB G0PSF
PO7 7QQ G3WYT
PO7 7RG G1FMT
PO7 7RP M3ZQJ
PO7 7SB M7ELS
PO7 7XP G4VIQ
PO7 8AE M7MLP
PO7 8AG G4MKQ
PO7 8AL M6TID
PO7 8BP M0JDS
PO7 8BX 2E0TRW
PO7 8BX M0KTT
PO7 8BX M6BSA
PO7 8DP M6HCR
PO7 8JN G4SAC
PO7 8ND G8KOS
PO7 8QD G8OKE
PO7 8RS G6BQE
PO7 8SH G8PAE
PO8 0AA 2E0DHG
PO8 0HF 2E0FMA
PO8 0HF M6TAU
PO8 0JF M6BNU
PO8 0JX G4BQV
PO8 0JX M6EYZ
PO8 0LD G0GDV
PO8 0LJ G0VOJ
PO8 0LQ G7EYS
PO8 0NF G1PPD
PO8 0NQ 2E0HES
PO8 0NQ M6GEV
PO8 0NU G4VNR
PO8 0NU M0KUV
PO8 0PD G3VCR
PO8 0PJ G8TZE
PO8 0RH 2E0REC
PO8 0RH M6MCU
PO8 0RH M6TUD
PO8 0TX G7IWA
PO8 8BG G3RBY
PO8 8BQ M7EFV
PO8 8BW G4NKX
PO8 8EW G0NPE
PO8 8HS G3ZFF
PO8 8HX G3WLY
PO8 8JE G0NHZ
PO8 8JJ G8LNC
PO8 8QD G6CTX
PO8 8QH G3TKN
PO8 8RS M7EPN
PO8 8RU G8YTR
PO8 8RX 2E1BOM
PO8 8SE G0JRN
PO8 8SQ G3LIK
PO8 8TS 2E0BSX
PO8 8UB G6YSO
PO8 8UB M3GVC
PO8 9BE G6EHL
PO8 9DA M3KUQ
PO8 9DA M3YDB
PO8 9EW G6WXK
PO8 9HE G0MGH
PO8 9NH M7GEB
PO8 9NX M0WAH
PO8 9QE G0MUK
PO8 9QH G0KUA
PO8 9QU G8UVF
PO8 9RD G6JDH
PO8 9RG G7PEE
PO8 9SG G1WJG
PO8 9TJ G0UHM
PO8 9UB 2E0BRC
PO8 9UB G7GNA
PO8 9UB M3TXR
PO8 9UY M3CWA
PO8 9XF G3MYI
PO88AX M7JRM
PO9 1AG M3YIE
PO9 1DG M6WAJ
PO9 1HZ G1JXL
PO9 1LQ M1DPJ
PO9 1RL G6IOV
PO9 1RN G0VKX
PO9 1RR G0RIB
PO9 2HR G1FFU
PO9 2NQ M0HNO
PO9 2PU 2E0JAZ
PO9 2PU M0JAZ
PO9 2PU M6JRS
PO9 2QX G6TQZ
PO9 2RP G8PXL
PO9 2RZ G4XQZ
PO9 2SY 2E0BVO
PO9 2SY M3YIV
PO9 2TN G6WBX
PO9 2UW G0JEZ
PO9 2XE 2E1FPP
PO9 2XE M3LDQ
PO9 2XE M3OQQ
PO9 3AA 2E0INT
PO9 3AA G8VEZ
PO9 3AA M6CTA
PO9 3AZ G3YYK
PO9 3BW G6XRH
PO9 3DL M3EYS
PO9 3JZ G7VDN
PO9 3NJ 2E1GZV
PO9 3NJ G4UXJ
PO9 3NS M6FEL
PO9 3PL G0FYX
PO9 3RA G4RGO
PO9 3RS M0HRH
PO9 3RS M3FEA
PO9 4HR G7WGI
PO9 4JA M6KAQ
PO9 4JF M6RZE
PO9 4LG G7UCL
PO9 4LG M3IVY
PO9 4NS M6PJD
PO9 4PT 2E0GMS
PO9 4PT M6GHB
PO9 4QY M0CBG
PO9 5AR M3EDS
PO9 5BJ G7EYV
PO9 5DS M7SKT
PO9 5DZ G8XNO
PO9 5ED G8WCD
PO9 5ED M6SUI
PO9 5ER G0DHZ
PO9 5HU G0DOK
PO9 5LA M3ZRS
PO9 5LS 2E0LGT
PO9 5PW G0ERS
PO9 5RY G0BSJ
PO9 6AG G3TSM
PO9 6DQ G0BAG
PO9 6HN M3YNB
PO93AQ M0GMU
po93th M7TCT
PR1 0AX M7BOQ
PR1 0BH M0DDI
PR1 0BN G6LOC
PR1 0DT 2E0BYW
PR1 0EE M0DOM
PR1 0JL M3OHO
PR1 0LL G4LKM
PR1 0TD M3DXN
PR1 0UR G1PUK
PR1 0UX G4RFA
PR1 0XN M0BKS
PR1 0XN M3AUL
PR1 0XQ G6EPX
PR1 0XR M7HMO
PR1 0YE G0EIF
PR1 1QW M5KEN
PR1 1RY M3OEE
PR1 2EF M6VZA
PR1 2YJ M6GBW
PR1 2YL G1VTQ
PR1 3NA 2E0NAP
PR1 3NA M6DOJ
PR1 4NJ G6PFZ
PR1 4TA 2E0RAF
PR1 4UD G3NQX
PR1 4UD G6NKI
PR1 4YB G7RCK
PR1 5DE M7IAP
PR1 5HJ G3KUE
PR1 5HJ G4WYH
PR1 5TA G7ING
PR1 5TB M6SJW
PR1 5TP G8RIP
PR1 5TR G4WXI
PR1 5UY G6WXS
PR1 5UY M7KWI
PR1 5XE M3WHB
PR1 6DQ M1CYK
PR1 6HN M6FBV
PR1 6NS M6UDS
PR1 6QN G4THA
PR1 7QR M0BAD
PR1 7TP M3VLG
PR1 7UG M6CCB
PR1 8PJ M0EBP
PR1 8TP G0JEH
PR1 9DD G7JZJ
PR1 9DR G1HMY
PR1 9EL G3BWI
PR1 9EQ G4WQT
PR1 9EQ G4ZKA
PR1 9HA M6CNK
PR1 9HD M3CMM
PR1 9NG M0HIQ
PR1 9RP G7TCB
PR1 9SY M6FZE
PR1 9TB M0UWS
PR2 1AD M7PRE
PR2 1BH M3ZIX
PR2 1JD G6EUG
PR2 1JP G7VAS
PR2 1PB 2E0ACV
PR2 1RX M0NDU
PR2 1RX M6JFK
PR2 1TY M6LFZ
PR2 2AS M3KEC
PR2 2EU M7DRB
PR2 2HH 2E0FJD
PR2 2HH M6FJD
PR2 2HH M6FLL
PR2 2PX M6TCX
PR2 2YW M1AMW
PR2 3EX G4UQI
PR2 3FQ G0SBY
PR2 3HS G7ILX
PR2 3JB M0OTJ
PR2 3JL 2E0ZPN
PR2 3LP G7AQD
PR2 3RU M3JDN
PR2 3RU M3MDN
PR2 3RY 2E1AFI
PR2 3RY G7PZE
PR2 3SX G0KSN
PR2 3UU G7IFM
PR2 3UU G7LPF
PR2 3YR G4LHR
PR2 3YS M0AKQ
PR2 3YY M0WCM
PR2 6AN M6AQY
PR2 6AN M6RRA
PR2 6DA 2E1IJM
PR2 6DH G3SYA
PR2 6EU G7KNR
PR2 6EX M0DMD
PR2 6TD M6LVR
PR2 6TH 2E1FJP
PR2 6TU M0JYG
PR2 7AL G7TRL
PR2 7BE 2E0FMB
PR2 7DA 2E0RFD
PR2 7DA M0RDZ
PR2 7DA M3OSS
PR2 7DA M3URD
PR2 8GT G3OIH
PR2 8NY G7KXV
PR2 8XN G7PSZ
PR2 8XN M3PSZ
PR2 9AW M0RWH
PR2 9FP 2E0CXW
PR2 9FP M6AYH
PR2 9QA G7OFU
PR2 9QX G1VZW
PR2 9QX G7OET
PR2 9RF 2E0HMM
PR2 9RF M3UCO
PR2 9SQ G3RSM
PR2 9SS G4PNH
PR2 9TP M3OEB
PR2 9UJ M7BOF
PR2 9YJ M0KHS
pr21sh G6MCC
PR25 1AR M3ULI
PR25 1BH G3ZRE
PR25 1BL M6CGL
PR25 1HT 2E0BBN
PR25 1JA 2E0XCD
PR25 1JA M6HHA
PR25 1JD G3UPY
PR25 1JH M6JCD
PR25 1RJ G4YWG
PR25 1RN 2E0BSB
PR25 1RN G3WSW
PR25 1RN M1CND
PR25 1RN M3TPW
PR25 1YB G0JSL
PR25 1YB G0JSM
PR25 2DD G0MAH
PR25 2LJ M0HPR
PR25 2XA M6TZU
PR25 2XW M6AOF
PR25 2XW M7AQR
PR25 2YL G0IDE
PR25 3AA G3GGS
PR25 3AF G4SBQ
PR25 3AR 2E0CII
PR25 3AR G0NGE
PR25 3AR G6VBK
PR25 3AR M0PQI
PR25 3AR M6AQK
PR25 3BD G6LUF
PR25 3NR G1IQF
PR25 3NS G1RBZ
PR25 3UH G0FDX
PR25 3UH G0GVA
PR25 4QZ G4YIA
PR25 4QZ M6HBU
PR25 4XL M0CYE
PR25 4XP M1EYQ
PR25 4XT G3KQY
PR25 4XY G1BMN
PR25 4YJ G4AYU
PR25 4ZR G0HKW
PR25 5PA G1FKT
PR25 5PD G1TTL
PR25 5PD M0FWO
PR25 5PJ M3SHQ
PR25 5RQ 2E0ZMR
PR25 5RQ M3ZMR
PR25 5RQ M7ILA
PR25 5SN G7DKY
PR25 5SP G3XII
PR25 5SX 2E0CAR
PR25 5TE M7DJS
pr253ha M6PCD
PR26 6QA M3DKZ
PR26 6QS 2E0PHT
PR26 7AJ 2E0GUY
PR26 7AJ M3XXS
PR26 7QJ G0KLT
PR26 7QJ M3KLT
PR26 7XJ 2E0BDJ
PR26 8LB G0VHO
PR26 8NP G7VOQ
PR26 9AP G0PFU
PR26 9HP G0JSJ
PR26 9HQ 2E0IDA
PR26 9HS M3FBP
PR3 0BB G6LNS
PR3 0JY M3ZRA
PR3 0RH 2E0IBI
PR3 0RH M6IBI
PR3 0RP G0AAY
PR3 0SG M7BUG
PR3 0XD M6HUM
PR3 1AA 2E0PNC
PR3 1AA M3IAH
PR3 1AD G8CWQ
PR3 1BA G7CUA
PR3 1FJ G0LXP
PR3 1FS G4GVG
PR3 1LH G4JCG
PR3 1NL G4AMY
PR3 1NQ M0DKL
PR3 1PD G4TVN
PR3 1RD G1HKR
PR3 1RD G4BSD
PR3 1RD G7FNM
PR3 1RD M3GDI
PR3 1RD M3TAW
PR3 1RF G6AIZ
PR3 1YL G0VWX
PR3 1YQ M0SHM
PR3 1YQ M3PRY
PR3 2AL M0JHN
PR3 2BQ G8CQV
PR3 2EA G4EXK
PR3 2JX M3RZF
PR3 2LH M3BTI
PR3 2NA M3WKK
PR3 2XB G0SDJ
PR3 3EL G3WGK
PR3 3JG G3NKL
PR3 3SL G4VOJ
PR3 3TB M3FSQ
PR3 3TH M6IFH
PR3 3TQ G6MAC
PR3 3TX G3LZO
PR3 3TX G7NOI
PR3 3UA G0HJB
PR3 3WD G1CFG
PR3 3WN G6RTD
PR3 3YG 2E0GYF
PR3 3YG M0JVX
PR3 3YG M7HUD
PR3 3YS M3ZZQ
PR3 5AH G8GLS
PR3 5HB G0KMP
PR3 5JN G6DDR
PR3 6AB G1HJO
PR3 6AB M1AXP
PR3 6AP G3LPL
PR3 6BD M3NPX
PR3 6BN G6HCF
PR3 6SS G0IYT
PR33WF M7IDY
PR4 0HH G7WEM
PR4 0HH M3TDH
PR4 0NP M6DFZ
PR4 0PA M3EPC
PR4 0TT M6GNN
PR4 1AJ M3HVY
PR4 1DF M0AUG
PR4 1EG G7NOQ
PR4 1EN 2E0TLD
PR4 1EN G3TNN
PR4 1HG G4WIM
PR4 1JL M3EWZ
PR4 1JN G8HEU
PR4 1PT 2E0FNM
PR4 1PT M6CEB
PR4 1RG G4RNF
PR4 1RL G4RCF
PR4 1RQ M3TVK
PR4 1RX G0CPJ
PR4 1SB 2E0UBW
PR4 1SD G7CUL
PR4 1SS G4MEE
PR4 1UQ G7GVP
PR4 1XB G6NUQ
PR4 1XT M3EWY
PR4 1XU M3EWW
PR4 1XU M3MHD
PR4 1XY 2E0NHM
PR4 1YD G7LPD
PR4 2AY G1GQY
PR4 2AY G1GQZ
PR4 2DS G8OTZ
PR4 2EL G4AHZ
PR4 2NQ M1AKF
PR4 2UE M0KBR
PR4 2UE M6IZQ
PR4 2XA 2E0BLL
PR4 2XA G7JGW
PR4 2XA M3JGW
PR4 2ZA G1TUI
PR4 3AG M1FHT
PR4 3BA M3HVW
PR4 3BB M3VXY
PR4 3DN M1ACJ
PR4 3EP 2E0FRS
PR4 3EP M7FRD
PR4 3FE G1EUH
PR4 3NB G1NKF
PR4 3PN G3TKK
PR4 3SU M3HVX
PR4 3SX 2E1ECL
PR4 3SX 2E1FJL
PR4 3SX M0ABK
PR4 3SX M0GJA
PR4 3TU G4MRX
PR4 3TX G8MED
PR4 3UD G0HIJ
PR4 3UD G1UCG
PR4 3UQ 2E0KMZ
PR4 3UQ M0IHN
PR4 3UQ M1HZZ
PR4 3UQ M6EQE
PR4 4JD G3VBL
PR4 4JX 2E0AFL
PR4 4RQ G0LEE
PR4 4XJ G0LQK
PR4 5AX 2E0JAR
PR4 5AX G3LJI
PR4 5BH G1MBN
PR4 5BX G1IPY
PR4 5HB G4ZCG
PR4 5NP G4WAL
PR4 5NP G6TNA
PR4 6AA G0TUC
PR4 6AT G0EHK
PR4 6HD G4MGB
PR4 6JS M6HOU
PR4 6JX G0VAV
PR4 6JX M1AUZ
PR4 6LY G6TMN
PR4 6RB G4OWS
PR4 6SE 2E0GYV
PR4 6SE M7GJJ
PR4 6SJ G4XFA
PR4 6SU M7ROV
PR4 6SX M3LNM
PR4 6TD G6EWH
PR4 6TJ M6OVH
PR4 6TJ M7ANP
PR4 6TR G0JJD
PR4 6UD G6PLT
PR4 6UL G3MPF
PR4 6US M3VOY
PR5 0DT G3PS
PR5 0JR M0DBT
PR5 0LA M0PVP
PR5 0LX M0FYA
PR5 4BP G4BGP
PR5 4TD 2E0JNU
PR5 4TD M0JNU
PR5 4TD M7JNU
PR5 4TT M3BAN
PR5 4UT G3XUH
PR5 5JG G4UJA
PR5 5LA M6BXD
PR5 5RA G0EHW
PR5 5TY M0ETS
PR5 5UP G3AZI
PR5 5UU G1BTN
PR5 5UW G0DPG
PR5 5UW M3DPG
PR5 5YB G3ZVQ
PR5 6EP G0FQC
PR5 6FT G4TUF
PR5 6LS M3NSM
PR5 6RA 2E0DZR
PR5 6RA M6CKV
PR5 6TD M6MLY
PR5 6UY G0ASH
PR5 6XQ G4OJR
PR5 6XR G1DER
PR5 8DS 2E1HQY
PR5 8DS 2E1LJL
PR5 8DS M0LJL
PR5 8DS M7PZG
PR5 8DU G3UCA
PR5 8EN 2E0VMA
PR5 8EN G0VMA
PR5 8EN M6VMA
PR5 8HJ M1DMH
PR5 8HQ M7BFB
PR5 8HS M6LZG
PR5 8HT G4GOM
PR5 8JS 2E1ACT
PR5 8JS 2E1DTG
PR5 8JX G0MXU
Pr58bh M7EWB
PR6 0AG 2E0DHT
PR6 0AG M6GCE
PR6 0AG M6SRZ
PR6 0DB M0AQE
PR6 0DB M0ISN
PR6 0DG G0USM
PR6 0LJ G0JMN
PR6 0NJ M6ERW
PR6 0PY G7SYY
PR6 0PY M3SYY
PR6 0RR 2E0EIX
PR6 0RR M6IQC
PR6 7AN 2E0DEX
PR6 7AQ 2E1BVQ
PR6 7AU G0HBX
PR6 7BG 2E0NYF
PR6 7BG M6NYF
PR6 7BJ G0CUN
PR6 7BW G6MEI
PR6 7DN 2E0GBI
PR6 7DN M6LSB
PR6 7HE G8COR
PR6 7NA 2E0WPN
PR6 7TT 2E0LON
PR6 7TT M0ZWT
PR6 7TT M6DFT
PR6 7TZ G6JJI
PR6 8EJ G1TSV
PR6 8PX G7KTQ
PR6 8UB G8NNL
PR6 8UB M7JHY
PR6 8UE G8JZI
PR6 9DA M1CGI
PR6 9DA M1CGJ
PR6 9DU G6AHO
PR6 9ET M0HGD
PR6 9FF M7TKL
PR6 9LA G0LBT
PR6 9LA M3KEY
PR6 9NQ 2E0ZRG
PR6 9NQ M3SGV
PR6 9PA G3TXK
PR6 9RS G1KVC
PR6 9SS G7PBB
PR60JB M7SVK
PR7 1EU M6CYS
PR7 1LX G4WYZ
PR7 1PH G4TZK
PR7 1RE G7KUR
PR7 1RH M0CGA
PR7 1UH G0DAI
PR7 1XJ M0MYJ
PR7 2FU G7UAY
PR7 2HL M0DMI
PR7 2JA M0DNW
PR7 2JA M6NNA
PR7 2JB 2E1IKA
PR7 2JB M1SMF
PR7 2JG M1EHI
PR7 2JW M6RKO
PR7 2NT M6RKL
PR7 2NT M7RHM
PR7 2QU G0HBS
PR7 2YA 2E0EWT
PR7 2YA M0TKR
PR7 2YB 2E0MEO
PR7 2YB 2E0SUD
PR7 2YB 2E0WOZ
PR7 3AP M3JQW
PR7 3AZ M7GFA
PR7 3HS G0UZF
PR7 3HY 2E0OTG
PR7 3NH G4IYP
PR7 4AN M6SEC
PR7 4AQ G1AUU
PR7 4JU G3URK
PR7 4LE M3NEL
PR7 4NL G7PYW
PR7 4NS G1DAK
PR7 4NT G4NRO
PR7 4PH M0GED
PR7 4PJ G0NGK
PR7 4QE M7DUJ
PR7 5AE M6KPG
PR7 5BH M7DOD
PR7 5BZ 2E1HQQ
Pr7 5db M7AVF
PR7 5EH G4WGT
PR7 5EY 2E0BUE
PR7 5EY M3URT
PR7 5HH G3JMZ
PR7 5NR M6COP
PR7 5NY G0OWA
PR7 5NY M6OWA
PR7 5NZ G1MBE
PR7 5PW G1AHM
PR7 5PY G0KDX
PR7 5QW G8SNQ
PR7 5RF M6LSH
PR7 5SN M6BFZ
PR7 5TW 2E0YPJ
PR7 5TW M0YPJ
PR7 5TW M3YPJ
PR7 6AU 2E0XVD
PR7 6AU M7BNL
PR7 6BA G8GFB
PR7 6BP G4PAT
PR7 6BS G1ZTG
PR7 6JW G6ZOL
PR7 6LY G7GYR
PR7 6PD G0WTD
PR7 6PD G7NER
PR7 6PL G0SHU
PR7 6PN G4PQM
PR7 6PP M6AQI
PR7 6PT G0CUB
PR7 6PT G3RPO
PR7 6PW G0WTM
PR7 6PW G7SSJ
PR7 7DB M3FTW
PR7 7LT M6JHI
PR75QW 2E0GPV
PR8 1HZ G3WZG
PR8 1JA G4ZYN
PR8 1NQ G0RLT
PR8 1RS G3TTY
PR8 1RT G4IQK
PR8 2DN M6RVA
PR8 2FB G7VJU
PR8 2HF G2ART
PR8 2JJ G3WTB
PR8 2LW G4BEU
PR8 2PE 2E0HYH
PR8 2PE M7DSL
PR8 2QF G7MJS
PR8 2QF G7OAA
PR8 2QL G7RNF
PR8 2QW G4JYQ
PR8 2RR G6JUQ
PR8 2RS G6ILX
PR8 3DB G3ZII
PR8 3DU 2E0NEL
PR8 3DU M3WFY
PR8 3DW M6JHQ
PR8 3HE M3WDU
PR8 3NP G8CIX
PR8 3QF 2E0TDE
PR8 3QF M6MVD
PR8 3RP G1DFT
PR8 3RS G1UCI
PR8 3UA M3WFL
PR8 4BJ G7CVF
PR8 4DT M3JGX
PR8 4EQ 2E0VAM
PR8 4EQ M3XMH
PR8 4EQ M6AJH
PR8 4EQ M6ALH
PR8 4HA G0MPW
PR8 4LT 2E0NYP
PR8 4LT M0JRI
PR8 4LT M6NYP
PR8 4NH 2E0YTT
PR8 4SF 2E0OIL
PR8 4SF M0MKO
PR8 4SF M6OHL
PR8 5AU G6EZY
PR8 5HB M6SEA
PR8 5HS G4SBE
PR8 5LP G0JCD
PR8 6HF M3FNH
PR8 6JA G6IVC
PR8 6JD M6CFR
PR8 6LG G7IYG
PR8 6LG G7IYH
PR8 6NA G7TIM
PR8 6NR M3UYU
PR8 6PY 2E0BUJ
PR8 6PY M0STJ
PR8 6PY M3WJV
PR8 6PY M6ARU
PR8 6PY M6EDI
PR8 6SQ M3LBG
PR8 6UP G0LZX
PR8 6UP G1VWL
PR8 6XH G3ORK
PR9 0JS M6GIP
PR9 0QQ M7ABQ
PR9 0QT 2E0FAB
PR9 0TW G3YIB
PR9 7AA G3STT
PR9 7AP M7COB
PR9 7BE G6FVM
PR9 7BY G1EPS
PR9 7DX M6AIW
PR9 7EZ M3KDO
PR9 7HX G4EVC
PR9 7JU G6DDC
PR9 7JX G3OCR
PR9 7JX G8ILJ
PR9 7JX M3YUD
PR9 7PQ G3HAA
PR9 8AT G0PBQ
PR9 8ET G6CIO
PR9 8LS M3UGK
PR9 8NL G0JCQ
PR9 8NL G4VYP
PR9 8NY 2E0CXD
PR9 8NY G4NMU
PR9 8NY M0HOQ
PR9 8NY M6JEY
PR9 8PS G0OFY
PR9 8QH M5SJS
PR9 8QR 2E0VEF
PR9 8QW G4UPK
PR9 9AH G8YPL
PR9 9BT G4DRA
PR9 9FA M3TCX
PR9 9GA G4SCO
PR9 9GJ G0PVU
PR9 9GJ G1ZMJ
PR9 9HW 2E0PAJ
PR9 9HW G5PJK
PR9 9NQ M6OUP
PR9 9RN G8RON
PR9 9TW G4EID
PR9 9TW M3HQN
PR9 9XE G4TUP
PR9 9XF 2E0CWO
PR9 9XF M6CWD
PR9 9XW G3TMB
PR9 9YX G4YPH
PR98BU 2E0MFZ
PR98BU M7BOS
RG1 2RE M3FVA
RG1 2SE M7AED
RG1 3AJ M0IXG
RG1 4QD 2E0GUT
RG1 4QD M6GUT
RG1 5LR G6FBA
RG1 5QP 2E0PIK
RG1 5QP M0PIK
RG1 5QP M3ZUY
RG1 5RD G3WNP
RG1 5SD M1XTN
RG1 5SD M3DKG
RG1 5SD M3VSL
RG1 6AE 2E0XLR
RG1 6AS 2E0PIX
RG1 6AS M6LOK
RG1 6ED 2E0FSA
RG1 6ED M7MRG
RG1 6JU M7JIA
RG1 6QD G6YLN
RG1 7PA M6DUD
RG1 7TT G8VWJ
RG1 7TY G4JPE
RG1 7UG G4FLY
RG1 7YT G0YYH
RG1 8EN G8WBN
RG10 0AU M6SUL
RG10 0AY M3BBS
RG10 0JB G1ZQR
RG10 0SF G8LZS
RG10 8BJ G8KZG
RG10 8BN G3WPH
RG10 8DR 2E0JBL
RG10 8DR M0OAA
RG10 8DR M6BPQ
RG10 9AX 2E0BPT
RG10 9AX M0BSJ
RG10 9AX M3VYM
RG10 9AY G3LRQ
RG10 9BN G8NXJ
RG10 9BN M3NXJ
RG10 9BT 2E0CJB
RG10 9BT M0TVA
RG10 9BT M3ZAW
RG10 9BT M6RUB
RG10 9BW M0KAK
RG10 9ED G7GEF
RG10 9JG G6AWF
RG10 9LJ 2E0UDA
RG10 9LJ M0UDA

RG10 9LJ M6OUD
RG10 9PY G3VKQ
RG10 9PY M6MCE
RG10 9QD M1BFO
RG10 9QF G7REH
RG10 9QT 2E0MSR
RG10 9QT M6LSG
RG10 9QT M7MSR
RG10 9QU M7KFC
RG10 9SJ M6EFU
RG10 9TS G8EPZ
RG10 9YD G4SYE
RG10 9YT G8KOM
RG12 0TB M6MBY
RG12 0TN M6PCM
RG12 0TR G1HLQ
RG12 0TR G6YLA
RG12 0TW M3JBM
RG12 0UA 2E0WWS
RG12 0UA M3YXW
RG12 0UD G7IBU
RG12 0UR 2E1AFN
RG12 0XU G4DDL
RG12 2JR G4MXY
RG12 2JW M3YNJ
RG12 2JW M6SER
RG12 2LU 2E0XBZ
RG12 2LU G8RKO
RG12 2LU M3XBZ
RG12 2PT G0NEF
RG12 2PT M6NEF
RG12 2QP M3WVI
RG12 2RA G4YBG
RG12 2SE G0OVA
RG12 7BE M6EXZ
RG12 7DU G4BZJ
RG12 7ET G7FSR
RG12 7JU M7EPT
RG12 7LD G4CQH
RG12 7ND M7YBZ
RG12 7PS G6NFC
RG12 7QA G8AMK
RG12 7QG G4DDN
RG12 7QG M0BRA
RG12 7QQ G0IRH
RG12 7TF G4AUC
RG12 7WF G3TKS
RG12 7WG G4KNZ
RG12 7WH G0CGE
RG12 7WL G0SCY
RG12 7WR M0XQS
RG12 7WR M6XQS
RG12 7WZ M6PEO
RG12 7YX G0FCT
RG12 7YX G6BRA
RG12 7YX G6MER
RG12 7YX G8III
RG12 7ZX M3XPR
RG12 8AP G1MHP
RG12 8UD 2E0XKT
RG12 8UD M0KJT
RG12 8UD M6XKT
RG12 8UP G0AQU
RG12 8UQ G1CPX
RG12 8XE G7JDF
RG12 8XP G1RXV
RG12 8XU G6TNR
RG12 8XY G4PCF
RG12 8YG M3NAH
RG12 9BY 2E1JKP
RG12 9EF G1GUI
RG12 9EF M1KIP
RG12 9ES G4WYC
RG12 9EX G3YMC
RG12 9HT G7USG
RG12 9JE M1CAX
RG12 9JL G3OFW
RG12 9PA G7SEJ
RG12 9PS M6AFU
RG12 9PX G1VXX
RG12 9QH G3ZWP
RG12 9QX G8RQN
RG12 9YL G0VQX
RG127JY M7KXD
RG13LP G0RSR
RG13LP G6ZTZ
RG14 1LF M0NGC
rg14 1ta 2E0NKK
RG14 1UH G3OPG
RG14 1UH G8OIU
RG14 2EN G1CTF
RG14 2HA G6OEW
RG14 2HA G6XZS
RG14 2HB G1DAV
RG14 2JL M3YXQ
RG14 2LS 2E0POQ
RG14 2LS M0POQ
RG14 2LS M3POQ
RG14 2ND G3OUC
RG14 2PN G6RBP
RG14 2SQ M7CCL
RG14 2TH G7HUO
RG14 2TH G7HUP
RG14 3AJ M6GMR
RG14 5JE M1CIJ
RG14 5JF 2E0FAE
RG14 5JF M0IXU
RG14 5JF M6ASF
RG14 5JN G7SYC
RG14 5JP G4RUW
RG14 5NR G7SLL
RG14 5QW G7REJ
RG14 5TF M6DTH
RG14 6AZ 2E0JOF
RG14 6AZ M3JOF
RG14 6BA 2E0UAO
RG14 6BA M3UAO
RG14 6DD G3UVM
RG14 6DN G3URI
RG14 6DN G4MKF
RG14 6DY M7BLC
RG14 6EE G4YMY
RG14 6HD M1BRU
RG14 6HP G0HBJ
RG14 6HP G6ZSF
RG14 6JX G4WEV
RG14 6PY G8KHU
RG14 6PZ 2E0JHC
RG14 6PZ M0IDC
RG14 6PZ M3JHC
RG14 6QJ 2E0LHI
RG14 6QJ M6XHI
RG14 6RS M7JJC
RG14 6RT 2E0GZW
RG14 6RT M0JOW
RG14 6RU G3ZGC
RG14 6RU M3ZGC
RG14 6RY G1WTS
RG14 6RY G8AKM
RG14 6SX G8AYC
RG14 7AL G4ZDP
RG14 7DJ G8MWU
RG14 7FX G7KXZ
RG14 7QS M7AZG
RG14 7RA M1DLG
RG14 7RB G4TKS
RG14 7RR G6LAE
RG14 7TL G8JUS
RG14 7TT G1DFI
RG17 0AZ 2E0BJM
RG17 0AZ M0NKS
RG17 0BZ G3ZPK
RG17 0JE G1OQV
RG17 0JE M0ISJ
RG17 0JE M0RYG
RG17 0JE M6CUE
RG17 0JP M7HUG
RG17 0JR G0LJJ
RG17 0JR G1FVP
RG17 0LA G3LWT
RG17 0LJ M3TZN
RG17 0LL G6GCJ
RG17 0SG G4TPH
RG17 0UN G6EML
RG17 7DG G0DTQ
RG17 7ED G6WLE
RG17 7JB 2E0YGM
RG17 7JL M6MPX
RG17 7JN G6SRY
RG17 7LT 2E0IHV
RG17 7LT M7CQA
RG17 7TS G4DNH
RG17 7UN 2E0DYN
RG17 7UN M0ICZ
RG17 7UN M6FWX
RG17 8RF G7BWI
RG17 8YQ G6EES
RG17 8YQ M3XEL
RG17 9QE G0LTX
RG17 9QE M1ABX
RG17 9UE G4FNK
RG17 9UW G7PGY
RG170JE 2E0OQV
RG18 0RR 2E0BPS
RG18 0RR M0JCM
RG18 0RR M3XGZ
RG18 0XD M6KDQ
RG18 3AF M3HSR
RG18 3BA G4BOO
RG18 3BF G7TOO
RG18 3BF M0PJC
RG18 3BN G6AMF
RG18 3BP G3RVM
RG18 3DH G4UET
RG18 3DH M0XRH
RG18 3DT G3VMT
RG18 3EB G7JLT
RG18 3EG G0POT
RG18 3PD M3TVD
RG18 3UE M3EHK
RG18 3UH M3JHV
RG18 4DL G4DRX
RG18 4DQ G1PCG
RG18 4DQ G4WLG
RG18 4LS M3NXF
RG18 4LS M3SXF
RG18 4LS M6PXF
RG18 9HQ G3VOW
RG18 9HT G8LTN
RG18 9HZ 2E0HQO
RG18 9HZ M6HQO
RG18 9PB G0MIA
RG18 9PD M3PUB
RG18 9PD M3RBX
RG18 9PH G4LMW
RG18 9PH G8XCW
RG18 9PH M6LMW
RG18 9QP G3KJC
RG18 9QW M6DGU
RG18 9RJ G6SOZ
RG18 9RJ G7DRR
RG18 9SL 2E0TVR
RG18 9SL G6ZPO
RG18 9SL M3PRN
RG18 9TG G6HUN
RG18 9UZ M7LRB
RG19 3BF M0AAP
RG19 3BF M7ELZ
RG19 3BF M7OZC
RG19 3LE G6ZVL
RG19 3PF G0OBJ
RG19 3PF G4YMH
RG19 3RL G4FWR
RG19 3RS G3ICB
RG19 3SD 2E0ESO
RG19 3SF G0KQT
RG19 3SH G7RJW
RG19 3TN M7KMF
RG19 3XB G8TSC
RG19 3XD M1CEY
RG19 3XE G6PNG
RG19 3XW G3DBV
RG19 3XW G7UXK
RG19 3XX G4BWE
RG19 4FD G3WYW
RG19 4FD M1PGT
RG19 4FQ G8JIP
RG19 4RG G3JMJ
RG19 4LX G7VGL
RG19 4WA 2E0MAD
RG19 4WA G6IZA
RG19 8BE G3MEV
RG19 8DD G3LLK
RG19 8DY G1XSQ
RG19 8EY 2E0RMZ
RG19 8EY M0RMZ
RG19 8SH M6DHT
RG19 8SS M0SBF
RG19 8XT G6RHT
RG2 7AH M6YVM
RG2 7BG M0ALB
RG2 7DX G4YBX
RG2 7HU M3UYW
RG2 7JR G4OMN
RG2 7JU G4JXH
RG2 7TD M0GQS
RG2 8AS 2E0GYH
RG2 8DN 2E0FUZ
RG2 8DN M3ZTK
RG2 8DN M3ZVK
RG2 8HJ G1ASD
RG2 8HQ M7CXV
RG2 8JB G6NVY
RG2 8LJ G1NYZ
RG2 8QP G4DBF
RG2 8QP G7BXU
RG2 8SD 2E1HJE
RG2 9FL 2E0DLV
RG2 9FL M0HQU
RG2 9FL M6DPU
RG2 9HA G8IJG
RG2 9HR M6FHN
RG2 9PD M3VME
RG2 9PD M6VLR
RG2 9PU G7KWN
RG20 0LH G3SVM
RG20 0LY G0LUK
RG20 5ER G4HDE
RG20 5RJ G3BFL
RG20 5SL G8HJF
RG20 5SL G8HJG
RG20 7BE G3NAQ
RG20 7BT G8JRN
RG20 7EH M0KKA
RG20 7EZ M0CUK
RG20 7JS 2E0FBN
RG20 7JS M0WTJ
RG20 7JS M6FYG
RG20 8HS M7CNB
RG20 8QF M7IRJ
RG20 8SD G1YOS
RG20 8SE G1NAB
RG20 8TZ M0HPS
RG20 9BN G6GXZ
RG20 9BW G3SVD
RG20 9BW M3SVD
RG20 9BW M3UKN
RG20 9EY G7VFE
RG20 9HY M6NAC
RG20 9TA G6UJC
RG20 9TB M3TYU
RG20 9TS 2E1ETJ
RG20 9TS G4DOQ
RG20 9TS G6PFF
RG20 9UY G7DXC
RG20 9UY M3PCW
RG21 3AS M6SEK
RG21 3HG G4WIZ
RG21 3JH G7WCN
RG21 3JW G8AOO
RG21 3NF G7FFB
RG21 5HL G6BBW
RG21 5NR G7SNB
RG21 5NS M6NIS
RG21 5NZ G8YEQ
RG21 5QE G1OQU
RG21 5SR G8PIY
RG21 5UA G4CSD
RG21 6AD M6HZT
RG21 7TG G7SKX
RG21 7TQ 2E0MUZ
RG21 7TQ M0SWT
RG21 7TQ M6MUZ
RG21 8XT M0NQB
RG215SS G4XMP
RG22 4BG M0EDR
RG22 4DU 2E0EUB
RG22 4DU M6KXP
RG22 4EA G1JVG
RG22 4EL G8AOK
RG22 4HN G4JIQ
RG22 4HR G8JLB
RG22 4HZ M7SBM
RG22 4JN M6KTZ
RG22 4JR G8GOS
RG22 4JR M0IYC
RG22 4LJ M3JSK
RG22 4NB G6JOR
RG22 4NB M7SRZ
RG22 4NP G6UZO
RG22 4NP M3BRU
RG22 4PH G7LYL
RG22 4RE M7DFD
RG22 4RF M1CJX
RG22 4RQ G8AGJ
RG22 4TY 2E0GUV
RG22 4TY G7PKD
RG22 4TY M0CJJ
RG22 4UB G4BHE
RG22 4UJ G6OPK
RG22 4UL G3WXD
RG22 4UL G8DUV
RG22 4UX G4PND
RG22 4XD G6MDS
RG22 4XT G8VML
RG22 4XT M0DRG
RG22 5BA G7LJB
RG22 5BQ G1CBS
RG22 5DN G0LEP
RG22 5FH G0GJR
RG22 5JY G8GTZ
RG22 5LY G6KQZ
RG22 5NN G0LQD
RG22 5NN G6PMF
RG22 5NW 2E1LIS
RG22 5NW M0LIS
RG22 5NX M0SAB
RG22 5QD G3UCF
RG22 5QF G6ZZR
RG22 5RA 2E0CBO
RG22 5RA M0HHD
RG22 5RA M3YQC
RG22 5RG G1IYF
RG22 6AX G0PTA
RG22 6BN G7PFL
RG22 6BQ G7MAT
RG22 6BQ M3HYO
RG22 6DF M0CBF
RG22 6DP 2E0KMY
RG22 6DP M6KMY
RG22 6EP M0RAB
RG22 6JA G1PEU
RG22 6JL G1WMV
RG22 6JL M0TAD
RG22 6NU G0KQA
RG22 6NZ 2E1FTA
RG22 6QP G3OAZ
RG22 6QQ M0HNX
RG22 6RX M0VMT
RG22 6TD M1CQI
RG23 7AQ G3JTQ
RG23 7BB 2E0PGP
RG23 7BB 2E1FSF
RG23 7BB 2E1FSG
RG23 7BB M6KIK
RG23 7BL G4JGS
RG23 7DD G3HVA
RG23 7JX M0LAS
RG23 7LB G6CFA
RG23 7LD M0CJM
RG23 7LF 2E0HFN
RG23 7LF M0WNA
RG23 7LF M7BQX
RG23 8AD G0XBA
RG23 8EX G4YPK
RG23 8JD G4EDH
RG23 8JF G7KAK
RG23 8NG G0JSR
RG23 8NH G8JYN
RG23 8QL 2E0NBM
RG24 7EF G8JHH
RG24 7EH M0GWC
RG24 7JE G4XIP
RG24 7LA 2E0HTK
RG24 8AA G7PTH
RG24 8GL M3BVM
RG24 8RG 2E0BTV
RG24 8RG M0GOQ
RG24 8RG M3XWV
RG24 8SB G8JSF
RG24 8SL M3NHW
RG24 8SU G0NWH
RG24 8SU M1ENA
RG24 8UJ 2E1GLQ
RG24 8UJ G7PAG
RG24 8WN M1KPW
RG24 8XQ 2E0VCG
RG24 8XQ M0VTC
RG24 8XQ M7DFI
RG24 9BE M3NRI
RG24 9DD G7FFC
RG24 9DW 2E0SUC
RG24 9GH 2E0ISO
RG24 9GH M3ISO
RG24 9HE G4YNH
RG24 9HX M0CWY
RG24 9LR G8UBN
RG24 9PQ M0DOP
RG24 9PQ M3DMY
RG24 9PQ M3OOU
RG24 9PQ M6DMY
RG24 9PY M3GCJ
RG24 9RS G7RPP
RG24 9SH M6IBJ
RG24 9SS M7KNI
RG24 9XP M7HNU
RG24 9YP M0MEI
RG24 9YP M7TIM
RG247HR M7GMA
RG25 2BZ G7NDB
RG25 2NH G3CEI
RG25 2RN G7EVF
RG25 3EJ M6CEE
RG25 3HP M0FSN
RG25 3JQ G8IAR
RG25 3LD G0NJG
RG25 3LE 2E0HQQ
RG25 3LZ G4GFM
RG25 3NL G7VZR
RG25 3NL M0VZR
RG26 3DW G6OZZ
RG26 3EL G8AKA
RG26 3HP 2E0MPX
RG26 3HP M0PXM
RG26 3HP M6KMG
RG26 3LF M6XST
RG26 3NJ G1WKK
RG26 3SF M6ELQ
RG26 3SH M1BCZ
RG26 3TL G8DVU
RG26 3UE 2E0GVR
RG26 3UE M7OWD
RG26 3UQ G6AOH
RG26 3UZ M0VVC
RG26 3YH G4LUA
RG26 3YH M6JJL
RG26 3YW G0CAK
RG26 4HF G6IOX
RG26 4HH G8MSY
RG26 5AG G6BNU
RG26 5BX M3XLC
RG26 5DG G1EHF
RG26 5DG M0HNA
RG26 5DG M3EHF
RG26 5EP G4LXJ
RG26 5HD G8FMD
RG26 5HF G6AXV
RG26 5JX M0PPZ
RG26 5NY G4XMO
RG26 5PJ G1JWG
RG26 5UE M6AII
RG26 5XH M1ECW
RG27 0BQ M7MXY
RG27 0DQ G0FEJ
RG27 0ES G3UQW
RG27 8BF M0SVV
RG27 8JS G3YGE
RG27 8NJ G0AMZ
RG27 8QW G0CVA
RG27 8QW M6FWB
RG27 8QW M7WJC
RG27 8QX G3YFL
RG27 8SW M3MOF
RG27 9EY G4SFH
RG27 9GB M7GOG
RG27 9HW G4YEG
RG27 9JS G6AHH
RG27 9NP M1CAO
RG27 9PU G6OFD
RG27 9QJ G4SXX
RG27 9QZ G3RXI
RG27 9RA M0VEC
RG27 9RH G7EWV
RG27 9RP G6ETX
RG27 9SF 2E1BOZ
RG27 9SG G0RSV
RG28 7JG 2E0ERV
RG28 7NF G3NYS
RG28 7SE G4ASY
RG29 1AE M0SOX
RG29 1BH 2E0EJM
RG29 1BH 2E0ICU
RG29 1BH 2E0PIP
RG29 1BH G1ZBL
RG29 1BH M0AEU
RG29 1BH M3OFS
RG29 1BH M3OUI
RG29 1BH M3TIZ
RG29 1EJ G7VSP
RG29 1ES G7BRB
RG29 1ND G8WAM
RG29 1NN 2E0NWE
RG29 1NN G3YNL
RG29 1NN M6YNL
RG29 1NT G7GIK
RG29 1NT M0UAR
RG29 1SX 2E0LSV
RG29 1SX M0LSV
RG29 1SY 2E1GTD
RG3 4UL G7SJK
RG30 2BY 2E0HFO
RG30 2BY M7BRA
RG30 2EJ M3NOW
RG30 2HA G6GBL
RG30 2NT G1ANQ
RG30 2NX G0AFR
RG30 2PE G1DSJ
RG30 2PE G1ZSY
RG30 2SF G0ODK
RG30 2TH 2E0FRP
RG30 2TH G6JUI
RG30 2UL M6ETP
RG30 3DL 2E0SIZ
RG30 3DL M7RSJ
RG30 3EJ M0HAO
RG30 3EN G4BTI
RG30 3NQ G7UET
RG30 3NQ M3AGI
RG30 3NQ M5ALG
RG30 4HU M3MSQ
RG30 4JT 2E1EPQ
RG30 4JT M0LTE
RG30 4PA M7ASJ
RG30 4QB G0GBR
RG30 4QP G0RZM
RG30 4YJ G7KWD
RG30 6BA M7NFZ
RG30 6DT 2E1CAJ
RG30 6EH 2E0MGT
RG30 6EP M6XBS
RG31 4US 2E0CGL
RG31 4US M0EGN
RG31 4XP 2E0FUN
RG31 4XP M0RXX
RG31 4XP M3VHZ
RG31 4XQ 2E0IHS
RG31 4XQ M7GLS
RG31 5DZ G4EFE
RG31 5HJ G8SMA
RG31 5JU G7VHX
RG31 5JY G0IZI
RG31 5WE G4XKA
RG31 6DE G7JJG
RG31 6DE G7KSS
RG31 6FZ M5ACR
RG31 6HN 2E0BJS
RG31 6HN M0RBJ
RG31 6HN M3VFP
RG31 6NP G4JOB
RG31 6PY 2E1FGB
RG31 6RH G8PQZ
RG31 6RH M0GMC
RG31 6RR M6BNV
RG31 6SR G8LCA
RG31 7AT G7MFP
RG31 7DD M0SMS
RG31 7DN G0TFU
RG31 7JR 2E0BZX
RG31 7JR M6RFC
RG31 7RX M0TBR
RG31 7ZL G3VUK
RG31 7ZU 2E0SHP
RG31 7ZU M6SHP
RG315NS G3ZIB
RG4 5AL G4AWY
RG4 5AY G4BXY
RG4 5JP M6WZF
RG4 5LE M6WOT
RG4 6DD M0IZS
Rg4 6pl G4VSQ
RG4 6PY 2E0CFI
RG4 6PY G6EIU
RG4 6PY M0HAH
RG4 6PY M6CFO
RG4 6QA M3IEP
RG4 6QB G8NQC
RG4 6RT M6KCY
RG4 6SF G8ROG
RG4 6SZ G0KIA
RG4 7DH 2E0SKP
RG4 7DH M0SKP
RG4 7DH M7RJI
RG4 7DT 2E0DNE
RG4 7HH G0CCC
RG4 7HH G4CCC
RG4 7HR G3AKF
RG4 7HR G4JTR
RG4 7HR M0AAA
RG4 7JZ 2E0NBK
RG4 7JZ M7NCB
RG4 7NR 2E0IET
RG4 7NR M0IET
RG4 7NR M3IET
RG4 7NR M3IEV
RG4 7NR M3SGF
RG4 7NT G7ENS
RG4 7PE M7TSB
RG4 7RG 2E0DOP
RG4 7RG M6ENK
RG4 7RH 2E0CYR
RG4 7RH M6DTU
RG4 7RP G4EJK
RG4 8DD G4ZRZ
RG4 8DN 2E0KGB
RG4 8DN M0TLN
RG4 8DN M3WNX
RG4 8HH G0OIW
RG4 8HH G3WKX
RG4 8HJ G4ZSZ
RG4 8JH G3OYN
RG4 8PA G3YZZ
RG4 8PH 2E0FSM
RG4 8PH M6RFP
RG4 8PL G4OPK
RG4 8TT G0MQW
RG4 9AD G0LGG
RG4 9LS M6KBX
RG4 9RU 2E0DWM
RG4 9RU M3TXG
RG4 9RY G6NEA
RG4 9RY G6RKJ
RG4 9RY M3AZP
RG4 9SA G8PDY
RG4 9SP G7RZQ
RG4 9TF G0JMS
RG40 1AW M5AGS
RG40 1DE M6EFE
RG40 1DE M6FNX
RG40 1DG G8PJC
RG40 1PL G8VZY
RG40 1QG G8CIT
RG40 1QG G8IEI
RG40 1QW 2E0PEP
RG40 1QW M6PEP
RG40 1RJ 2E0DPR
RG40 1RJ M0SMN
RG40 1RJ M6CGK
RG40 1RL G1DNP
RG40 1RY M3ZRW
RG40 1TW G4ORU
RG40 1XS G8YRL
RG40 1XX G8FBF
RG40 1YE M6PHE
RG40 2HT G2DD
RG40 3HT G4OKM
RG40 3LB G7JTV
RG40 3LD G8DUO
RG40 3LG G3XPC
RG40 3PE 2E0TJJ
RG40 3PE M0TJJ
RG40 3PE M7TJX
RG40 3QB G8NZK
RG40 3RL G3WME
RG40 4EN M0GRW
RG40 4EY M7DRW
RG40 4JA G1SEO
RG40 4PA M6DUM
RG40 4PA M6SSS
RG40 4PF G4HRE
RG40 4PT G6IJK
RG40 4PY 2E0WGI
RG40 4PY M0WGI
RG40 4PY M3WGI
RG40 4RA G1ZSF
RG40 4RA G4NIP
RG40 4RD G4DYO
RG40 4TS M0ZEQ
RG40 4TS M6SUZ
RG40 4UD G4HOU
RG40 4UJ M1VHF
RG40 4UJ M3WOK
RG40 4UN 2E0CDU
RG40 4UN M0GVY
RG40 4UN M6DBX
RG40 5QU M1AXG
RG40 5YB G3ZBS
RG40 5YE G4XAL
RG40 5YL G4KVD
RG41 1JE G3ZZH
RG41 1LJ M6POA
RG41 1LN G7TYH
RG41 1NN M6HML
RG41 1NR G6CGC
RG41 1PH 2E1AUQ
RG41 1PH G3YHG
RG41 2AH M7REY
RG41 2RJ G0JHW
RG41 2TL M3IWK
RG41 3AG G8GJM
RG41 3HG G3MGU
RG41 4AJ M6EFF
RG41 4AR G6TLX
RG41 4AW G8IBP
RG41 4AX G4LJF
RG41 4BA G8FIE
RG41 4BX G4JXU
RG41 4DH G1KTF
RG41 4ED G3TUY
RG41 4TA M6HRM
RG41 4TB M7PSM
RG41 4UR G8VPO
RG41 4UY G7SEO
RG41 5HR M3ZLW
RG41 5JF G4GBI
RG41 5JG M0GXB
RG41 5JZ G7COP
RG41 5LG G4REA
RG41 5LX M6CMB
RG41 5NJ G6WXN
RG41 5NW G6IVB
RG41 5PE G1HBD
RG42 1FJ M6NZR
RG42 1RT G6XPB
RG42 1RT M6XPB
RG42 1UE G8OLK
RG42 2AD M6RCE
RG42 2BT G4WZJ
RG42 2HJ 2E0PTS
RG42 2HJ M0PTS
RG42 2HJ M6PGB
RG42 2HL G3NCN
RG42 2LD G3VG
RG42 2LH G3TAI
RG42 2LH G4NOC
RG42 2QX G6DKI
RG42 3SZ 2E0RWJ
RG42 3SZ M7RWJ
RG42 3TR 2E0XDF
RG42 3TR 2E0XNF
RG42 3TR M0XDF
RG42 3TR M3XJF
RG42 3UN G0ATZ
RG42 3UN M3UTQ
RG42 3XA G4HLF
RG42 3XU 2E0RTP
RG42 3XU M0RTP
RG42 3XU M6RTP
RG42 3XX G4ZFE
RG42 3XX M3NUB
RG42 4DF G4DQA
RG42 4DS G4ORB
RG42 4ES 2E0FUG
RG42 4UR G7CJO
RG42 5JG G4PKE
RG42 5LG G0GJV
RG42 5LG G4BRA
RG42 5LG G6XSY
RG42 5QZ M3NPK
RG42 6HU G0BBB
RG42 6HU G0DX
RG42 6HU G5AOZ
RG42 6HU M0BB
RG42 6JG 2E0XWX
RG42 6JG G5YAX
RG42 7UY G6JDW
RG45 6AL G8EYM
RG45 6BN 2E0PZK
RG45 6BN M0OPK
RG45 6BN M6DGI
RG45 6BT G6AXY
RG45 6DU G8XML
RG45 6EF G6LLG
RG45 6EN G3GRY
RG45 6HE 2E1DIH
RG45 6NR G3DVF
RG45 6QE M1FEY
RG45 6QF G1WGM
RG45 6QR G0FPT
RG45 6SE M0DZV
RG45 6WE 2E0OWG
RG45 6WE M0OWG
RG45 6WE M6OWG
RG45 7EG G7VWO
RG45 7HR G6FXR
RG45 7JP G8STR
RG45 7NE G8ZLL
RG45 7NR G0WCS
RG45 7PD G8GSU
RG45 7QR G4DGF
RG456HJ M0IRH
RG5 3AD G3RAC
RG5 3AD G4HNF
RG5 3AX 2E0JCZ
RG5 3AX M0JRZ
RG5 3AX M6FJL
RG5 3BG M6HDD
RG5 3BL M6JWX
RG5 3DA G3WTV
RG5 3DY M6CCM
RG5 3HA G0DWD
RG5 3HF M3NXH
RG5 3JA G0IZV
RG5 3LQ 2E0ZIP
RG5 3LQ M3XHH
RG5 3LQ M6PEN
RG5 3LR G0AHI
RG5 3NY M6KUK
RG5 3QG G8DZC
RG5 4AP G0VQR
RG5 4AP M0CYB
RG5 4AW G4ELY
RG5 4BL G2NF
RG5 4HH G4IZS
RG5 4LD 2E0JCW
RG5 4LD M6JCW
RG5 4LJ G4JVD
RG5 4LR G0FIJ
RG5 4LR G8MBQ
RG5 4NA M3MZW
RG5 4PA M6GJS
RG5 4QB G7UOQ
RG5 4QB G8XJJ
RG5 4QT G3UUB
RG5 4RT G3PBR
RG5 4SR M7DFM
RG5 4UG G8VKR
RG5 4UN G0VPE
RG5 4UN G4KWT
RG5 4US M0JPK
RG5 4XE M6PGX
RG5 4XR G8CSK
RG6 1AP G0ILT
RG6 1HD G7RFS
RG6 1HS 2E0JPM
RG6 1HS M6JMO
RG6 1HW G3VMY
RG6 1QE G4OKA
RG6 3AH G4CDF
RG6 3AN G4MCM
RG6 3DG G6HCT
RG6 4AB G7UZY
RG6 4DB G7FQE
RG6 4EN 2E0OOX
RG6 4EN M7NDT
RG6 4EP M6LHH
RG6 4HN 2E0XUU
RG6 4HN G7NDI
RG6 4HN M0XUU
RG6 4HN M3XUU
RG6 4UY 2E0XDA
RG6 4UY M0XCA
RG6 4UY M7CDA
RG6 4XA G6IEI
RG6 5GX G6YIE
RG6 5PY G8NPZ
RG6 5QX G8LCL
RG6 5QZ G4FXT
RG6 5RY G8ADQ
RG6 5SL 2E0BPW
RG6 5SL M0JNX
RG6 5SL M3KYK
RG6 5SN G8VRW
RG6 5SR G3UAX
RG6 5TG G0CEC
RG6 5TP M3EZJ
RG6 5XG G0MZN
RG6 5XG G4AOL
RG6 5XG G8VWH
RG6 5YQ G6JVK
RG6 7LH G4RDC
RG6 7LJ G7LWY
RG6 7LQ G7AOK
RG6 7PA G4OQJ
RG6 7PD G0JTN
RG6 7PE G4OAE
RG6 7RT G4HLT
RG6 7RT G4MUT
RG6 7SS 2E0IMV
RG6 7SS M7PEL
RG7 1BH G8VWV
RG7 1BH M6WEW
RG7 1BQ G7FBE
RG7 1BZ M6ZRZ
RG7 1DD M3KVC
RG7 1HP G7MTW
RG7 1QY G8MVY
RG7 1SG G7OMQ
RG7 1TJ M0SCB
RG7 1TT G8YVQ
RG7 1XS G4ZAC
RG7 2HY G4EPX
RG7 2LH G4XXH
RG7 2PZ G3SCZ
RG7 2QB M0GVN
RG7 3BU G4CLD
RG7 3ES G3YGR
RG7 3EZ G4IWS
RG7 3HY G0VHK
RG7 3JA M1DGQ
RG7 3PY 2E0YWA
RG7 3PY G4IDS
RG7 3PY M7IDS
RG7 3QE G1COW
RG7 3QG 2E0SYI
RG7 3QG M6SYI
RG7 3QU M1BNG
RG7 3TH M0SEL
RG7 3TL G1AWD
RG7 3TR G3WND
RG7 3TU G3ZOI
RG7 3XW 2E0PHM
RG7 3XW M0MPM
RG7 3XW M0PAQ
RG7 3XW M6PHM
RG7 4DE G8ZWN
RG7 4PA M0OJO
RG7 4TR G4EEE
RG7 5AR M7HAF
RG7 5DN G3LXQ
RG7 5NS G7TKP
RG7 5RY 2E0DEU
RG7 6NN G0OBV
RG7 6NU G3KIW
RG7 6NU M3KXY
RG7 6RA G4ZSG
RG7 6TN G0KPE
RG74TH G8DGR
RG8 0DJ 2E0CCJ
RG8 0ES G4CEI
RG8 0HX 2E0SSG
RG8 0HX M3SSG
RG8 0JL G3NGX
RG8 0PL M0AZG
RG8 0SE G3LLG
RG8 7BD M3LJJ
RG8 7PX G1GMH
RG8 7QG G3TRY
RG8 7QG G4CUE
RG8 7QG G4MDF
RG8 7QG G8NMO
RG8 7QL G0LHZ
RG8 7QL G3ULT
RG8 8BH 2E1FFX
RG8 8BG G7KSV
RG8 8DQ 2E0LAX
RG8 8DQ M6AWC
RG8 8LN 2E0LAV
RG8 8LN G7TXR
RG8 8LN M0XYX
RG8 8LS G6AML
RG8 8LT 2E0SRJ
RG8 8LT M0MLE
RG8 8LT M6RJS
RG9 1AP G8VOB
RG9 1JT 2E0HIT
RG9 1JT M6HIT
RG9 1LT G3TFL
RG9 1PH G0NCO
RG9 1UU G4WBI
RG9 1UU G7LSF
RG9 2BG M6SII
RG9 2LX G3UYY
RG9 3JS G8FXX
RG9 3JS M0FXX
RG9 4DH G6CHA
RG9 4QQ M7ORG
RG9 5PS G3PZL
RG9 5QX 2E0ENM
RG9 5QX M6ILI
RG9 5QX M3PXT
RG9 6NR M0XVI
RH1 1AF G4VXN
RH1 1DR 2E0BXZ
RH1 1DR M0JCR
RH1 1DR M6JCR
RH1 2BN G6YAH
RH1 2BZ G3VGE
RH1 2DF M6EOQ
RH1 2DY G8ZFQ
RH1 2EJ M7RLS
RH1 2HA G0LXV
RH1 2HH G3MPB
RH1 2JB G1WIS
RH1 2JB G7LZM
RH1 2LA G8XOV
RH1 3AA G3YSX
RH1 3AA M5SB
RH1 3BH G7ODV
RH1 3BN G1YRR
RH1 3ER G4WGD
RH1 3EY G8UNO
RH1 3LT M7LSE
RH1 3PB 2E0UOJ
RH1 3PB M3UOJ
RH1 3PX G0KCH
RH1 4AS M6PDM
RH1 4AY G3NIQ
RH1 4DE G3TRC
RH1 4QT G7KGV
RH1 5BJ G4LJU
RH1 5BW M7HGF
RH1 5JB G7LUO
RH1 5JB M0DBO
RH1 5LY M6BZY
RH1 6AW G8HMG
RH1 6GZ 2E0FLL
RH1 6GZ M7TAG
RH1 6LW M6AOB
RH1 6PB G1YXJ
RH1 6PB G4PFX
RH1 6PQ G8LAM
RH1 6TB M1APH
RH10 3AL G7OMM
RH10 3DW G3CTP
RH10 3EG G8MQF
RH10 3FR 2E0ZLM
RH10 3FR M3RXO
RH10 3HX G6NPJ
RH10 3NG G4PBF
RH10 3PS G3VJM
RH10 3QS M6WPN
RH10 3QT G0KKS
RH10 3RG M0GZE
RH10 3TB 2E1HGT
RH10 3TG G0ORX
RH10 3UQ 2E0GPA
RH10 3UQ M6GPA
RH10 3XG G4ANN
RH10 3XG G6RC
RH10 3XQ 2E0GVW
RH10 4HS M6JPR
RH10 4JB M6PCR
RH10 4JQ G0PYJ
RH10 4LR M3APQ
RH10 4LU M6RXL
RH10 4NG G3RMK
RH10 4UB G0GML
RH10 4UJ G3YTR
RH10 4UL G4TTY
RH10 4UP G0BKW
RH10 4XA G3VLH
RH10 4XY G6LEB
RH10 4YY G3NZP
RH10 5AB 2E0WJR
RH10 5BX M0IJP
RH10 5DQ G3ZON
RH10 5DW 2E1BMF
RH10 5DW G4UMJ
RH10 5LD G6FPH
RH10 6BG 2E0MZB
RH10 6BP G6CKZ
RH10 6DL M0HQM
RH10 6GJ 2E0UKW
RH10 6GJ M0WOT
RH10 6GJ M6WXX
RH10 6GJ M7TWO
RH10 6HQ G0VYN
RH10 6HQ G3WKI
RH10 6HW G8MGZ
RH10 6HZ M6CSP
RH10 6JL 2E1AVT
RH10 6JS G4MKD
RH10 6JS G7OMI
RH10 6LX G0IOE
RH10 6LX G6XTG
RH10 6PB G6XAV
RH10 6PN G4SFP
RH10 6QF G8INS
RH10 6QW M3XHZ
RH10 6RE M7ATB
RH10 6RL M3GQM
RH10 6SE G3SYD
RH10 6TE M7ATN
RH10 6TH G3MGL
RH10 7AE G1AYU
RH10 7AJ M0JHV
RH10 7BS G4UBT
RH10 7DB M6EWQ
RH10 7DB M6EXW
RH10 7ED M0PMM
RH10 7FL G1XST
RH10 7JH M0VPG
RH10 7JR G3FRV
RH10 7RQ G0VIQ
RH10 7SE M7MSB
RH10 7SQ G3ZSJ
RH10 7SW M0LPK
RH10 7TS 2E0EQJ
RH10 7TS M0KGW
RH10 7WF 2E0RGC
RH10 7WQ M6MRY
RH10 7XG G1BGH
RH10 7YW G7RUN
RH10 8AS M7EQC
RH10 8BZ G0FPY
RH10 8HA G8IZY
RH10 8JR G6GGW
RH10 8JX M6GYE
RH10 8NB G6DOW
RH107XP M6KXW
RH11 0BS G4ONJ
RH11 0HW G0EGT
RH11 0JJ G8XHK
RH11 0NA G1GHY
RH11 0NA G3KAU
RH11 0RB G6OCM
RH11 0TJ M1PFS
RH11 0UF G6AUE
RH11 7BD G3IPP
RH11 7BW G8RRR
RH11 7EB M0CWT
RH11 7EG M7NKA
RH11 7JB G6SHQ
RH11 7JE G3UNS
RH11 7JP G0RGM
RH11 7PE G4LEG
RH11 7RF G4TZF
RH11 7RF M7DPX
RH11 7UH G3UUC

RH11 8BA M6ECM
RH11 8EU G7LYB
RH11 8EY 2E0BTO
RH11 8EY M3YFJ
RH11 8LQ G4PFW
RH11 8SQ G8AVZ
RH11 8XJ G6YPM
RH11 9AN M6SXT
RH11 9EG G6CKE
RH11 9EG G6EDR
RH11 9ES G6ZHO
RH11 9HN M6SIU
RH11 9HN M6TOV
RH11 9JH G0PVQ
RH11 9LH 2E0EJJ
RH11 9LH M6HNJ
RH11 9NS G1PLV
RH11 9NY G3ZNW
RH11 9QT G8PUH
RH11 9QZ G7VZQ
RH11 9RQ M0GLU
RH12 1DQ G0VPZ
RH12 1GR G4PTR
RH12 1LD G8YRF
RH12 1LF M3WOI
RH12 1NA G8GHL
RH12 1NU G8RLD
RH12 1PX M5DB
RH12 1SB G7DMQ
RH12 1UB M6MPO
RH12 1UY M6ARD
RH12 2AF M6FLU
RH12 2DA G4CCA
RH12 2HH 2E0ZMT
RH12 2HH M0ZMT
RH12 2HH M3ZMT
RH12 2HU M6MON
RH12 2JX M6DLR
RH12 2LB 2E0NKC
RH12 2LB M3NKC
RH12 2PY G4JHI
RH12 2QL G4LLI
RH12 3AQ M3DKK
RH12 3DU G3TQY
RH12 3HD G3SWC
RH12 3LH G7EYL
RH12 3ND G4PEY
RH12 3QR 2E0TJT
RH12 3QR M3YZQ
RH12 3TT M7GHZ
RH12 4AR G4TMC
RH12 4GR G6DMM
RH12 4GX G8FQS
RH12 4HB M7SRH
RH12 4HR G6XTJ
RH12 4JB G8LHS
RH12 4JB M6GQH
RH12 4JE G1CKF
RH12 4LN M6PLS
RH12 4RE G4YJA
RH12 4SH M3SYN
RH12 4TP M0LTA
RH12 4UF G0URO
RH12 5EU G1XFM
RH12 5EU G4MWD
RH12 5EU M6TZR
RH12 5EW G8RSK
RH12 5FR G6NQQ
RH12 5PJ G1ALU
RH12 5PN G7SRV
RH12 5PZ G7MQP
RH12 5TL G6PGQ
RH12 5WA G4KDR
RH12 5WD G8IBE
RH12 5XL G6FYU
RH12 5XW G3VQO
RH12 5XX 2E1IGA
RH121LS M6MTM
RH12EJ M7FTZ
RH13 0PJ G7AVF
RH13 0PR M7GET
RH13 0RQ 2E0CSJ
RH13 5DB M6NXZ
RH13 5HB G0DJS
RH13 5HG G6SBI
RH13 5HH G4EUG
RH13 5JS 2E0BXW
RH13 5JS M6EAA
RH13 5JS M6SLD
RH13 5JS M6SPJ
RH13 5LH M6END
RH13 5ND G7UEC
RH13 5NZ G4TPO
RH13 5PE G1ODN
RH13 5RY 2E0OVI
RH13 5RY M0OVI
RH13 6AE G4MKW
RH13 6AJ 2E0WEK
RH13 6AJ M6WEK
RH13 6AL M6GPJ
RH13 6AX G3SIG
RH13 6AX G3ZBU
RH13 6AX M3EUR
RH13 6AX M3GCR
RH13 6BQ M0GRU
RH13 6BS M6IHC
RH13 6BS M6LPS
RH13 6DD G4NWW
RH13 6DG G6MQJ
RH13 6DT G8BAD
RH13 6DW M3FTK
RH13 6EH M6BQV
RH13 6EJ G3WSC
RH13 6EJ G4PEO
RH13 6GH M7CCW
RH13 6LZ G8YLA
RH13 6LZ M0YLA
RH13 6ND G7HWM
RH13 8AZ G3XEI
RH13 8EF M6RJK
RH13 8EH G4JCY
RH13 8EQ G3ZQW
RH13 8ER G4TPW
RH13 8GD M6EXD
RH13 8HR 2E0BQU
RH13 8HR G3RAM
RH13 8HR M6CSL
RH13 8HZ G7DFV
RH13 8HZ M3DFV
RH13 8JB G8FBM
RH13 8JF G4OLZ
RH13 8JX 2E0PML
RH13 8JX M0VUE
RH13 8JX M6EGK
RH13 8JX M6NGY
RH13 8LF G6HYD
RH13 8LG G1VVY
RH13 8LT G3WZT
RH13 8LT G4HRS
RH13 9BB G0LOC
RH13 9BB M6AUZ
RH13 9BD M7DWD
RH13 9BG G4BVP
RH13 9DQ G4VFC
RH13 9FT 2E0HEC
RH13 9HP G1MPW
RH13 9HZ 2E0AED
RH13 9HZ G4HMM
RH13 9HZ M6CJV
RH13 9LZ M7PWK
RH13 9TF G8TQJ
RH13 9XR G0TAO
RH13 9XX G6DID
RH13 9XY G4LRP
RH13 9XY G5RFL
RH14 0DY G3PUX
RH14 0PJ 2E0FTC
RH14 0PJ G4NUX
RH14 0PJ M6FSO
RH14 0TF G8SLU
RH14 9BH G3OGP
RH14 9GL G0ERF
RH14 9GS 2E1EBZ
RH14 9HB G0FQD
RH14 9NG G4EFO
RH14 9NG M6NCD
RH14 9NP G6ZDP
RH14 9PU G7MXT
RH14 9RJ M0HWH
RH14 9TU 2E0VCU
RH14 9TU M6PIP
RH15 0AE M6NAW
RH15 0BA M6KZX
RH15 0DE M1BQU
RH15 0EU G3YBM
RH15 0HB M0TPJ
RH15 0JN M1CSI
RH15 0LD M3TFS
RH15 0LJ M1EBU
RH15 0NB G3RMY
RH15 0ND G0WGP
RH15 0NF M6BRO
RH15 0NH M6JOR
RH15 0NJ M1CSE
RH15 0PH G6ZQA
RH15 0PS 2E0ZCP
RH15 0PS M6ZCP
RH15 0PT G0DDE
RH15 0PT G6DBX
RH15 0QA G6IHG
RH15 0QN G0UUP
RH15 0QU M5JWS
RH15 0QY M6LAP
RH15 0RP G1AZA
RH15 0RQ G0GNV
RH15 0RQ G1TDL
RH15 0RW G7BLD
RH15 0RW G7TPW
RH15 0RW M3HZM
RH15 0RW M3HZP
RH15 8AW M1CYL
RH15 8BD 2E1IAS
RH15 8BD G1DZD
RH15 8BL G0IDP
RH15 8BW G3XUP
RH15 8EU G4AKG
RH15 8HJ M3ZMB
RH15 8LE G4ULZ
RH15 8NP 2E0RJA
RH15 8PD M6JQQ
RH15 8QF G0KYX
RH15 8QW G7ZMS
RH15 8TF G3XUD
RH15 8TR G3XQM
RH15 8UR G4ORS
RH15 8UY M5BTB
RH15 9ED 2E0NCJ
RH15 9ED M0NCJ
RH15 9HA 2E0DRA
RH15 9HA M0IKD
RH15 9HA M6IKD
RH15 9HR G1PJM
RH15 9HR G1POD
RH15 9HZ G8KMP
RH15 9JA G4WEH
rh15 9jj M7EOV
RH15 9PL 2E1BRD
RH15 9PL G7OIA
RH15 9PY G0RPL
RH15 9QW M7DJA
RH15 9RF M6HWJ
RH15 9RR M3ZGS
RH15 9SP G3VAK
RH15 9ST G0SQF
RH15 9SZ G7TMR
RH15 9UL G8GJV
RH15 9UT M6WYX
RH15 9XA G4ZCS
RH15 9XG G0THD
RH15 9XR G1WGL
RH158AN M7OBI
RH16 1EJ G7TNU
RH16 1HJ G4KKO
RH16 1NB G4DAY
RH16 1PH M1CTO
RH16 1QN M6STF
RH16 1QT M7GSP
RH16 1TF G3ZEK
RH16 1UZ 2E0DEE
RH16 1UZ 2E0LIV
RH16 1UZ G8ZOV
RH16 1UZ M3MZA
RH16 1UZ M3MZC
RH16 1UZ M3ZOU
RH16 2AB G3FSX
RH16 2DQ G4IOV
RH16 2JQ M7SXH
RH16 2LJ 2E0WVE
RH16 2LJ M0WVE
RH16 2LJ M6WVE
RH16 2PH M6AKK
RH16 2QG G4OKB
RH16 2QG M0WGS
RH16 2QJ M6OJT
RH16 3AX 2E0CBQ
RH16 3HW G3ASV
RH16 3JS G8MFM
RH16 3JU M7AEJ
RH16 3PE G3YTU
RH16 3QW G0AUI
RH16 3QX M7OZE
RH16 3RP 2E0EYS
RH16 3RP M0VAY
RH16 3RP M6IRB
RH16 3UE G0LFF
RH16 3UE G1CWP
RH16 3UE G7GCF
RH16 4DH 2E1FET
RH16 4EW G1VPC
RH16 4HL G7FHV
RH16 4JR G1SKI
RH16 4NP M3WOW
Rh164ul M7NEL
RH17 5AW G0XAA
RH17 5AW G3VKW
RH17 5AW M3HJE
RH17 5DZ G3WYN
RH17 5DZ G5RV
RH17 5HD 2E0ETI
RH17 5HD M6TLY
RH17 5HH G4JST
RH17 5PA G3LET
RH17 5QQ G6SOY
RH17 5QU 2E1DIJ
RH17 6AF G8VKQ
RH17 6DL M7MEP
RH17 6DQ G7WBR
RH17 6JA G4ZXB
RH17 6NG G7IAS
RH17 6NJ G0HXF
RH17 6SR 2E0PMB
RH17 6SR M6PBK
RH17 6UE 2E1ALC
RH17 6UE M6YTU
RH17 6UL M6COE
RH17 7EQ M0DMX
RH17 7HE G4MLR
RH17 7LP M6GVD
RH17 7PG G1ACA
RH17 7QA G8KNC
RH17 7QS G4RBQ
RH17 7RA M6NIC
RH17 7RN M0JTJ
RH17 7RW G8ZAT
RH18 5AF 2E0TNE
RH18 5AZ G0UPL
RH18 5BX M0HBM
RH18 5BX M6AGO
RH18 5GD G5DJW
RH19 1AY M7WEL
RH19 1DL 2E0LSS
RH19 1EE 2E1JGW
RH19 1JG G3VPS
RH19 1JR G3KOA
RH19 1SG G7KBR
RH19 1TA 2E0BQV
RH19 1TA M0GQD
RH19 1TA M3UQH
RH19 2DD G3YQW
RH19 2PS G8FBA
RH19 2QY 2E0LRZ
RH19 2QY M7EFU
RH19 3HP M6SXB
RH19 3NE 2E1PAL
RH19 3QE G0TSQ
RH19 3QL M1ENX
RH19 3QP M7FRW
RH19 3RB M3NGY
RH19 3SE G1AML
RH19 3UX M3WIC
RH19 3XF G0RCF
RH19 3XP G0RHG
RH19 4BZ G1YXT
RH19 4EA G4TRG
RH19 4NY 2E0AYT
RH19 4TF G4PFU
rh191dq 2E0XDZ
rh191dq M6XDZ
RH2 0QA G6YRV
RH2 0RE G8PZI
RH2 7BS G7CVM
RH2 7DY M0KDX
RH2 7HJ G4AVE
RH2 7NQ M7GEZ
RH2 7PS 2E0PMO
RH2 7PS M7RUE
RH2 7PY 2E0KRH
RH2 7PY M6OWI
RH2 8DQ G0VGT
RH2 8DS G8ZRV
RH2 8EL M3WOQ
RH2 8NE G6VMI
RH2 8QJ G4WBF
RH2 9DG G7OBF
RH2 9HA M3SGJ
RH20 1AH G6OSO
RH20 1AW 2E1BSC
RH20 1AW G7PTX
RH20 1BZ G0INQ
RH20 1HS G1EAE
RH20 1LA G1PMA
RH20 1NX G8TIU
RH20 1PQ G3ZTU
RH20 1PS M3IYX
RH20 2AT G0LFH
RH20 2DD M6ANE
RH20 2DX 2E0UMD
RH20 2DX M6UMD
RH20 2EE 2E0BSF
RH20 2EE M6MST
RH20 2JJ M3PTX
RH20 2JJ M7USB
RH20 2JL M6FEP
RH20 2PR G0HWS
RH20 2PZ 2E0ZYL
RH20 2PZ M0TCD
RH20 2PZ M6RFA
RH20 3BD G8IPF
RH20 3GL M6RPX
RH20 3GU M0EPX
RH20 3GU M1LTS
RH20 3HD M7RBF
RH20 3HG M1AIM
RH20 3HJ 2E0JGD
RH20 3HZ M6GNM
RH20 3JW G3XJN
RH20 4AB M7KOW
RH20 4AR M0BAH
RH20 4AU 2E1DZH
RH20 4AU G4OAK
RH20 4BS G4LCU
RH20 4JH M7EKT
RH20 4JU M6DMP
RH20 4LL G6AAR
RH20 4LT G4JCA
RH20 4LU G3POQ
RH20 4LW G0AGB
RH20 4QL G6LBE
RH20 4QL G6YCZ
RH20 4QS M7GCL
RH20 4QX G7TMU
RH20 4RF G4YPG
RH4 1HD M1DDF
RH4 1LP G0GNA
RH4 1NA M6SLR
RH4 2DG G0GQY
RH4 2HT G4YFK
RH4 2JE M3WLX
RH4 2NR G4VUW
RH4 3BZ G6EQP
RH4 3BZ M3EQP
RH4 3DS G3YWX
RH4 3DZ G6DTH
RH4 3HX G3AFB
RH4 3JH M0DSX
RH5 4AD G3RDH
RH5 4AE M6FWG
RH5 4DG 2E0KCV
RH5 4DG M0KCV
RH5 4DG M7AMC
RH5 4EY 2E0NNB
RH5 4EY M0NNB
RH5 4EY M6NNB
RH5 4JP 2E0UCD
RH5 4JP M7SPO
RH5 4JW G7OAJ
RH5 4NL 2E0NPE
RH5 4NL M6NPE
RH5 4NN G4FTQ
RH5 4PE M6DCD
RH5 4PP G6YPF
RH5 4QB G4VTC
RH5 4QH G1THW
RH5 4RH G1VNU
RH5 4TN 2E0DKE
RH5 4TN M6FCM
RH5 5AQ G4CMU
RH5 5AQ G7DOR
RH5 5BS G1ABQ
RH5 5BX G4OBT
RH5 5DZ G0RKS
RH5 5HT G6KNE
RH5 5JB G1XSM
RH5 5JB M6ATH
RH5 5LL G6OBJ
RH5 5SG M0OTT
RH5 6QW G3MNB
RH6 0BL G8OLP
RH6 0DR M1AZB
RH6 0HU 2E0IBN
RH6 7BX M3JXV
RH6 7JF G1HLY
RH6 7NH M3TAG
RH6 8BS G8AMU
RH6 8JG G0SOG
RH6 8JG G7KYW
RH6 8JR G3JHP
RH6 8LF G6DCH
RH6 8QT 2E0HAP
RH6 8QT M6AVO
RH6 8RE G6VKX
RH6 8RX G4EIV
RH6 9BY M6GRU
RH6 9EA G6MGZ
RH6 9GT G6TUG
RH6 9LY 2E0FEM
RH6 9LY 2E0KDT
RH6 9LY M0KPD
RH6 9LY M6FEM
RH6 9LY M6KVT
RH6 9NA G3ZIY
RH6 9QA G4PFT
RH6 9QN G1EAM
RH6 9QP G0HUF
RH6 9RB M6SJO
RH6 9RE M7RPA
RH6 9RF G0MIH
RH6 9SY G3OKS
RH6 9TX G4FBI
RH6 9UE M0MIM
Rh67jw M7NRH
RH67LN M6BJT
RH7 6AW M0SKO
RH7 6BT M6UTT
RH7 6EY G8OYY
RH7 6JX 2E0JXN
RH7 6JX M3JXN
RH7 6LL G3LHZ
RH7 6NQ G7TWJ
RH7 6PS 2E0TZZ
RH7 6PS M0TZZ
RH7 6PS M6TZZ
RH7 6QX G6AHR
RH7 6RL G3IOM
RH76PB G0FTU
RH76TA M7EEG
RH8 0EW G7VPU
RH8 0JA M3WNV
RH8 0JE G3IKB
RH8 0PD 2E0IPW
RH8 9AU 2E0ALW
RH8 9AU G7BKN
RH8 9AU M5ALC
RH8 9DJ M6OHN
RH8 9LT 2E0GPD
RH8 9LT M6OBZ
RH8 9PF M0JEP
RH9 8BF 2E0PPJ
RH9 8ER G0ALE
RH9 8ER G3RQZ
RH9 8LD M3JGJ
RH9 8LD M3ZUS
RH9 8NX M7BKL
RM1 2AD M3NHS
RM1 2NP G6YCF
RM1 2SR G8JBV
RM1 4AU M0VIN
RM1 4BZ G6BTX
RM1 4HD G4FQF
RM1 4LA G3XLR
RM1 4SP G6TGM
RM1 4XR G8TQZ
RM10 7BT G0RFL
RM10 7EX G4SHN
RM10 7XR M6LDR
RM10 8AS 2E0YAL
RM10 8AS M0YOL
RM10 8AS M3TBV
RM10 8BS G0GGH
RM10 8BS G1JQR
RM10 8BT G1YKI
RM10 8DE G0WXE
RM10 8ES G1WUH
RM10 8TE G0HSX
RM10 8UU M7MCR
RM10 9ND M7PEJ
RM10 9PP M1GDH
RM10 9QD M0GRT
RM107RE M7TBM
RM11 2AJ G7IYN
RM11 2BA G0BSF
RM11 2BS G0CBU
RM11 2NF M3DSA
RM11 2RJ M6UWU
RM11 2ST 2E1RBA
RM11 3EN G0RIQ
RM11 3LD 2E0LBL
RM11 3LD M0LBY
RM11 3PD G0PIA
RM11 3PY 2E0EBV
RM11 3PY G3JHI
RM11 3PY M0TAZ
RM11 3QD G1VSH
RM11 3QH G8WUO
RM11 3SG G0TJN
RM12 4AF M7RNX
RM12 4EY G3SFK
RM12 4JZ M0MAC
RM12 4JZ M1CUC
RM12 4LL G0IAP
RM12 4SE G6MVS
RM12 4YF G4NHR
RM12 5BL G0AJH
RM12 5DR G7GJY
RM12 5EU G6BEL
RM12 5HH 2E0AWI
RM12 5LS G8PMR
RM12 5QS G8UZW
RM12 6AT G4MXF
RM12 6TF G4EEF
RM13 7AS 2E0UKD
RM13 7AS M3UKD
RM13 7EJ G4RZZ
RM13 7PD 2E0WJI
RM13 7PH G8ZDT
RM13 8AA M6GAI
RM13 8AJ G4JRB
RM13 8BB G0IEO
rm13 8el M6IFQ
RM13 8GF 2E0OYY
RM13 8GF G5OYY
RM13 8GF M6OYY
RM13 8LW M6JYO
RM13 8ND G8NWI
RM13 9DX G4YBI
RM13 9LF M3NYI
RM13 9LW G0PBN
RM13 9LW G1BTF
RM13 9RL M6AMI
RM13 9UU G8FJG
RM14 1AL G6TRX
RM14 1AT G0KSJ
RM14 1HN G0IQN
RM14 1HP G4AZX
RM14 1HT M1CZO
RM14 1JD M7DKL
RM14 1NA G3RJI
RM14 1NB G0DOM
RM14 1NW M6YMF
RM14 1PZ 2E0AVQ
RM14 1PZ M3AQF
RM14 1XR G8XQZ
RM14 2EZ 2E1EVK
RM14 2EZ M6HDU
RM14 2HJ G8GGO
RM14 2HR 2E0MES
RM14 2HR M0SEM
RM14 2HR M3ELS
RM14 2JH G4UNS
RM14 2YX M3DFU
RM14 3AH G7GES
RM14 3AU G7DXX
RM14 3AZ G8VVG
RM14 3EA G4XZC
RM14 3PT 2E0NBG
RM14 3YT G4LQX
RM15 4NR M0CJR
RM15 5DS M6FZN
RM15 5ES M6PST
RM15 5ET 2E0JGG
RM15 6HP 2E0BYN
RM15 6HP M6ILP
RM15 6TS G1DXH
RM15 6XE G1TJT
RM16 2GB G3SCT
RM16 2GB G4HCK
RM16 2GB G4HKO
RM16 2HD M7LAT
RM16 2QL G1WEF
RM16 2QS 2E0REZ
RM16 2QS M6RHU
RM16 2RS G1ZAY
RM16 3AU G6TYF
RM16 3DW M0MOR
RM16 3LT G0DEK
RM16 4BS G7HMA
RM16 4DD M7KMO
RM16 4HL 2E0KGJ
RM16 4JB G8WTN
RM16 4JF M6TTM
RM16 4RR G6NVD
RM16 5UR G8BCJ
RM16 6ET M5BAZ
RM16 6LT 2E0BWT
RM16 6LT M6AEW
RM16 6NP M6EIR
RM16 6NS M6JYB
RM16 6QT M6KBA
RM16 6QT M6MES
RM16 6RE M6GNJ
RM16 6RE M6GNS
RM162XT M6IYS
RM17 5AB G0FDS
RM17 5AB M1MOG
RM17 5HA G4EPM
RM17 5LW M1DPE
RM17 5QR G0BKL
RM17 5RG G3KMD
RM17 5RG G6EFE
RM17 5RG G6OBO
RM17 5RG M1YOW
RM17 5RH 2E0ELI
RM17 5RH M0WJL
RM17 5RH M6WJL
RM17 5SF G4LUT
RM17 5UW G8VZJ
RM17 5YN G1NOR
RM17 6AZ M6LNV
RM17 6EQ 2E0XMT
RM17 6LD M0TIL
RM17 6LD M1DUC
RM17 6PX M7AEL
RM17 6RU M6URS
RM17 6SG M6WRO
RM18 8BA G3PHD
RM18 8DT G1EHM
RM18 8HQ G7VQL
RM18 8SX G4OVG
RM18 8XP 2E0PDM
RM18 8XP M6MTR
RM19 1LE G8TND
RM2 5AR M3ZTW
RM2 5PP G3ZQY
RM2 5PS M6EDO
RM2 5XJ G4XXT
RM2 6AP G3TPJ
RM2 6DX G4JDT
RM2 6LD G4GBW
RM20 3DA M6GDJ
rm20 4aa M6IYV
RM20 4AB G1SID
RM20 4YD M6BKV
RM20 4YR G6LCU
RM3 0AW M6HKU
RM3 0BQ G8VXR
RM3 0DL G4TTN
RM3 0EB G0BCZ
RM3 0EJ G3SKV
RM3 0JD M1JEC
RM3 0RH M3BFJ
RM3 0YR M7DVS
RM3 0YT G3SVK
RM3 0ZA M3TXM
RM3 7DX G0CBT
RM3 7DX G3XBF
RM3 7EQ 2E0SLH
RM3 7EQ M0HXE
RM3 7LB M0BOB
RM3 7QL M3CGJ
RM3 7SP M6WCE
RM3 7UB G1WZK
RM3 7YX G1PZS
RM3 8DB M6OOX
RM3 8HJ G6MDN
RM3 8HN G7VVL
RM3 8HN G8TPP
RM3 8HQ G6WKI
RM3 8JZ M6MTC
RM3 8RU G7BXT
RM3 9DE M3MIX
RM3 9LA M1BFX
RM3 9NA G1CSN
RM3 9NL G4YOT
RM3 9PT 2E0PSN
RM3 9PT M6PSN
RM4 1AX G6SVV
RM4 1DR M6LRL
RM4 1DT G4XVH
RM4 1NH G6DIM
RM4 1RR G1VQI
RM5 2HF G6EBO
RM5 2SL M0PAX
RM5 3AY 2E0EXG
RM5 3AY M6OXJ
RM5 3JX M7WEA
RM5 3ND G0OVC
RM5 3QA M6CNP
RM5 3RX G0BLO
RM5 3SX G8OPO
RM6 4BF M6ING
RM6 4EB G0FEK
RM6 4EB G1HKS
RM6 4NU M1BDY
RM6 4SL G4YGL
RM6 5AT M3EDU
RM6 5EJ G7VVK
RM6 5JP M6HFI
RM6 5NR M3MRO
RM6 5QL 2E0OKC
RM6 5QP M0CMH
RM6 6PY G1NIV
RM6 6RX G0PIS
RM7 0EX M6KUY
RM7 0JP M3ZNL
RM7 0LA G8YIN
RM7 7LJ 2E0ROV
RM7 7LJ M0MCP
RM7 7LJ M6ROV
RM7 8AZ M6ZXB
RM7 8BD G7WJE
RM7 8BE 2E0TMA
RM7 8BE M6RMN
RM7 8HD G0GFZ
RM8 1DR 2E0MSM
RM8 1DR M6AQE
RM8 1DS G7TAX
RM8 1JY G0TJE
RM8 1XD 2E0DPD
RM8 1XD M6FCR
RM8 2DW M7EJY
RM8 2EW 2E0TPN
RM8 2LR G8SLB
RM8 2QL M0BLY
RM8 2RA G7UYT
RM8 2RA M0HEM
RM8 2TU G4CUQ
RM8 3PA M7DKE
RM8 3ST G6AIU
RM9 4ET G7CXT
RM9 4ST G4HOI
RM9 4TJ M3SJH
RM9 5LT M7EAO
RM9 5RS M0DPV
RM9 5RS M3ARB
RM9 5XH G7UVW
RM9 6EA M7RUI
RM9 6JL M7BGG
RM9 6LD 2E0WWN
RM9 6LD G0TOD
RM9 6LD M3WWN
RM9 6NH 2E0IPK
RM9 6XU M7BID
S1 3JD G3UOS
S10 1EJ G1MAS
S10 1HT G0IYV
S10 1LS G8PNM
S10 1PD G0BFM
S10 1SL M3WSQ
S10 1WQ G8YXI
S10 2BA 2E0EJQ
S10 2BA M6HUX
S10 3BU G4SGF
S10 3PT G4SOM
S10 3RF G0BSX
S10 3RZ G6HYI
S10 3TR G7JAF
S10 3TR M1GWA
S10 4DN G4CUI
S10 4ED G6YFY
S10 4EU G3YBA
S10 4EX M1AGR
S10 4GF G3WDM
S10 4LF G7SSB
S10 4LL G8AGN
S10 5NX G7LQO
S10 5PT M7EBC
S10 5RE G1JVN
S10 5RE G1JVO
S11 7GG 2E0SSY
S11 7GG M6MHO
S11 7LX G0IIF
S11 8UA M1AAC
S11 8UA M7LJH
S11 9AA G0PFI
S11 9BR G0HHA
S11 9BR G0LNV
S11 9BU M0AXE
S11 9DE M6UAJ
S11 9DR G0WLR
S11 9LH 2E0ZSB
S11 9LH M3ZSB
S11 9LQ G4IJB
S11 9PX G8AKQ
S11 9RA G0HGI
S11 9RP G4HIA
S11 9RR G3NHE
S11 9SP G3RVI
S11 9SP M3NSO
S12 2AY G3UXM
S12 2BL M5ASK
S12 2BU G4VLZ
S12 2EJ G0BJP
S12 2FN G1HFK
S12 2FZ G7PAN
S12 2GL M6NLF
S12 2JH 2E0PCG
S12 2JP M3ZBF
S12 2NB 2E0GNS
S12 2NB M3HNU
S12 2SZ M6VGC
S12 3AQ G3PYL
S12 3AQ M6EEZ
S12 3DT M6MPM
S12 3HB G1GPM
S12 3HN M6SCH
S12 3LP 2E0KXD
S12 3LP M3KXD
S12 3LR G7GHH
S12 4ER 2E0SPD
S12 4ER M0SZD
S12 4ER M6AXD
S12 4HG 2E0YTB
S12 4HG M6YTB
S12 4HU 2E0TCX
S12 4HU M6HBH
S12 4HU M6IGW
S12 4JB M0CBT
S12 4LF G4GPF
S12 4ND M0PBO
S12 4QL G0JXF
S12 4SE M3SZT
S12 4XE G0BQW
S12 4XH G7UEX
S12 4YF G0NRM
S13 8DQ M7LOT
S13 8EH M3DYR
S13 8FA M6OFG
S13 8FA M6OFJ
S13 8TH G0HYS
S13 9HH M0TDM
S13 9HX G0VBN
S14 1AE G4DCF
S14 1GD M3DDZ
S14 1RJ G1RZJ
S14 1SS M7PRG
S17 3DN G4UWP
S17 3EG G8BLP
S17 3EZ G0VNE
S17 3GZ G4AYO
S17 3LL M6AYM
S17 3NB M0BAK
S17 3PH G8ION
S17 4DT G0TVW
S17 4FT M6DEZ
S17 4PF G4KPU
S17 4QX M6RWS
S17 4RJ G0OUK
S18 1ND 2E1HLO
S18 1ND G0LUM
S18 1ND M3CAZ
S18 1PW 2E0BMD
S18 1PW M0UCD
S18 1RF G6CTA
S18 1RJ G0XVL
S18 1RP G0UGJ
S18 1SD G4ULN
S18 1UQ G3KNJ
S18 1UQ G3VKK
S18 1UZ G4JUH
S18 1WB M7BHL
S18 1WH G3YON
S18 1XX G4HOP
S18 1YL G4KCU
S18 2EA 2E0HAV
S18 2EA M0ZEL
S18 2EA M6DQZ
S18 2EH G7BUF
S18 2EJ G1EQL
S18 2EL G4PXF
S18 2ER G0MVC
S18 2FF G4KXW
S18 2GG G8RFW
S18 2GQ M6DPF
S18 2HQ G4PPR
S18 2LE G1DFW
S18 3BQ G1KPZ
S18 3BY M3EVJ
S18 4DE M6DDI
S18 7WA G8UHM
S18 7WX 2E1BUV
S18 8QY G8RFP
S18 8YW G0WYN
S2 1GH M3XAH
S2 1RG G0JJV
S2 1UU M3DFS
S2 2AE M6URH
S2 2AG 2E0WST
S2 2AG M0VGC
S2 2AG M6WST
S2 2AR G8YQU
S2 2EA 2E0IUK
S2 2EA M0IUK
S2 2EA M6IUK
S2 2GB G1LWY
S2 2NZ G1ASG
S2 3AE M6KJP
S2 3BJ M3NGZ
S2 3EH G7SWQ
S2 3EH M6WNC
S2 3LB G1YED
S2 3LQ G1DBZ
S2 3NF 2E0KSH
S2 3NF G5YL
S2 3NF M6KSH
S2 3QJ 2E0AJT
S2 3SY M6RUK
S2 4AH M6ICR
S2 4BP M6JDQ
S2 5FD 2E0ZMZ
S2 5FD M6MFJ
S2 5FQ 2E0SKL
S2 5FQ M6LSK
S2 5QW G4MWL
S20 1AD G7KDQ
S20 1AW G4RQF
S20 1AY 2E0PCF
S20 1AY M6PCF
S20 1DA M7IWN
S20 1ET M7TIN
S20 1FW M0JCS
S20 1FW M6ICO
S20 2DL 2E0UAW
S20 2DL G1AW
S20 2DL M6UAW
S20 2PZ M0MEF
S20 4GB 2E0UKX
S20 4HL G6XLC
S20 6QJ M6JUB
S20 6SD G6YAI
S20 6SG G0FZD
S20 6SU M6BTE
S20 7JB M7YOY
S20 8EF G0OUG
S20 8JG 2E0DBO
S20 8JG M6IDX
S20 8LE 2E0USA
S21 1AW M6IHM
S21 1BU M6TSP
S21 1DA 2E0FQP
S21 1DA M7DDA
S21 1FX 2E0MJG
S21 1FX M0NCN
S21 1FX M3NCN
S21 1GA M3TUW
S21 1GR M3HGT
S21 1TF G0PBH
S21 1TG 2E0YMV
S21 1TG M0AFW
S21 1TG M7PVW
S21 2BX G3HQG
S21 2dt 2E0YZA
S21 3UL M6ZCB
S21 4BJ M7ECW
S21 4BR G8IQA
S21 4bw M3NXC
S21 4EG G3SXT
S21 4JX G7SCP
S21 5RD G7TZQ
S21 5RS G6XUX
S25 1YT G8WVH
S25 2LJ G7CKE
S25 2NY G6XOX
S25 2RP G4NUB
S25 2SW G3YYE
S25 2SX M0YZV
S25 3QG G3VCQ
S25 3QG M0SDC
S25 3QG M3VCQ
S25 3RG 2E0EHP
S25 3RG M0JWJ
S25 3RG M6GZW
S25 3RW M7GHI
S25 4AG G7GUG
S25 4BR 2E0MKE
S25 4BR M6MKE
S25 4BW G7FTM
S25 4DQ G0DZX
S25 4EP 2E0USV
S25 4EP M0USV
S25 4EP M6USV
S25 4EY 2E0FVO
S25 4EY M7TVT
S25 4FG G8EMH
S25 4FL G4SRX
S25 4GD G6YZR
S25 4HB G4GQS
S25 4HB M3IXE
S25 4JP G4RMX
S25 4JY G7GFR
S25 5BA G6TNI
S25 5BA G7IOI
S25 5DR G0JZL
S25 5EG G1DPI
S25 5EG M0CKX
S25 5GL G7BCI
S25 5JG G1HEW
S254BH 2E0TRU
S254BH M6TRU
S254BH M7MMT
S26 1HJ G7TBF
S26 1HJ G7TDR
S26 1HL G3NEO
S26 1JG G1YGP
S26 1JN G3HKF
S26 1JN G7HAR
S26 1JQ G1WRD
S26 1JS G7JRC
S26 2AA 2E0DCA
S26 2AA M3UKB
S26 2AD G7PFI
S26 2AS G0CCN
S26 2AT M0ADO
S26 2BR G4YSG
S26 2DH M0TWG
S26 2EP G4ZDD
S26 2FJ G6BCM
S26 2GL M3VPP
S26 2GR 2E0GBP
S26 3RB M0TRV
S26 3XS G0DCF
S26 4NQ 2E0CFK
S26 4NQ M6PER
S26 4RQ M6DNS
S26 4SD G4RNA
S26 4SD G6BNT
S26 4TA G7LQN
S26 4TH M6PJB
S26 4TL G1UDE
S26 4UP 2E0PFO
S26 4UP M0PFO
S26 4UP M6PFO
S26 4UY M1CLW
S26 4WB G0BPM
S26 5HL M6JQC
S26 5LZ G8OCE
S26 5NY M7JEN
S26 5NZ 2E0FKB
S26 5NZ M6OME
S26 5PH M3KXS
S26 5QS M0ONI
S26 5QS M0RVE
S26 5QT M3TVO
S26 5RF M6JPV
S26 5RG M1TVR
S26 5RG M3HIT
S26 5RG M3MTR
S26 5RU G0SVX
S26 6RA M0CUI
S26 6SJ 2E0XCA
S26 6SJ 2E0ZPA
S26 6SJ M0PJA
S26 6SJ M6PJA
S26 6SJ M6ZCA
S3 7AH 2E0OKH
S3 7AH M6HOM
S3 7AH M6OKH
S3 7AQ M7NRJ
S3 7JH M6CGC
S3 7NL G8PLI
S3 7PB G0FTR
S3 8EP M6OTI
S3 8PQ M6HPH
S3 9JF G4MRB
S3 9JF G6DIZ
S32 1AS G4KWE
S32 1DT M3VWG
S32 1DU G0WEX
S32 5QB G0NOU
S32 5QX M1CRP
S33 8UA G7ODR
S33 8WB G6HBZ
S33 9JQ M6SSR
S33 9JX G4NYW
S35 0AF G0FLI
S35 0DD M6TIW
S35 0DQ M6RKS
S35 0EA G7BPQ
S35 0GD G1OKB
S35 0HR G0CCM
S35 1AH G0KNL
S35 1AL M0AZB
S35 1EA G6ZWL
S35 1QG M1EOU
S35 1SQ M6HFG
S35 1SY 2E0CVX
S35 1TX M0BVN
S35 1UR 2E1CJD
S35 2PS M7AQL
S35 2TD G0UAC
S35 2ZT G1YLG
S35 3GU 2E1BJD
S35 3HS M0GCC
S35 3HZ G0VUM
S35 3LP M6KVX
S35 3PA G0LQC
S35 4HF G6OFV
S35 4NS M0CBN
S35 4NS M0DEI
S35 4NT M0CKS
S35 7DH M6SFM

S35 7DQ G3LZI
S35 8PE G4CTE
S35 9YW M6HBD
S351YS 2E0ZSJ
S351YS M0ZSJ
S351YS M3ZSJ
S36 1AY G2CFC
S36 1BZ 2E0RGS
S36 1BZ M0RBQ
S36 1BZ M6AOV
S36 1ED 2E0LMG
s36 1el M7GMN
S36 2BB G3WXI
S36 2BB G8AGQ
S36 2NF 2E0HVH
S36 2NF M7HVH
S36 2NH G8UCN
S36 2NP G0TKF
S36 2NR G3KJO
S36 2PT G7DGE
S36 2QX M0IBY
S36 2QX M0TGW
S36 2QX M3VCZ
S36 2SG 2E0TKF
S36 2SG M3UCY
S36 2SG M3UEL
S36 2TF G1YKL
S36 2TN G0UNY
S36 3ST M6OQG
S36 4GW G7MXS
S36 6AF G7MZJ
S36 6BH G0EWD
S36 6DN M7JBY
S36 6EA G7MZK
S36 6EH M6MIN
S36 6HE M7CVX
S36 8HP G0BVK
S36 8HP G4MWF
S36 8WW G4VRT
S36 8ZZ M1AKT
S36 9DB G4DFS
S36 9ED M6VMJ
S36 9EF M6ZXL
S36 9NE G6ZTD
S36 9QN M7BOA
S36 9RX M3LGM
S4 7DD M3BUH
S4 7DE G4DJD
S4 8AA 2E0TWS
S4 8AA M0TWS
S4 8JH G0DVS
S40 1DG G0UUC
S40 1HQ M0EME
S40 1HU M0OND
S40 1HZ G4LNG
S40 1NJ 2E1CYM
S40 1NJ M0CCG
S40 1NJ M6NGR
S40 2DF M7KDZ
S40 2HT M6EDM
S40 2JJ M0DVR
S40 2LL 2E1EAZ
S40 2LT G7SFF
S40 2SD M7AXY
S40 2SX M3ZQV
S40 2UH G8TFU
S40 2UR M3XQG
S40 3AS 2E0ODN
S40 3AS 2E0TNZ
S40 3AS G4SEB
S40 3AS M7SBB
S40 3AS M7THO
S40 3BL M0GXO
S40 3BT G3ZLF
S40 3BX 2E0MAZ
S40 3BX M0MAZ
S40 3BX M3XIO
S40 3DF G4BFR
S40 3DR M3OOP
S40 3EY G3GLX
S40 3HF G4NKI
S40 3HG G3OOP
S40 3HT G7KUM
S40 3LW G1UGL
S40 3NN G8OKI
S40 3NT G6DPE
S40 3QG 2E0BNK
S40 3QG M0ZIG
S40 3QG M3SRQ
S40 4AN 2E0HJY
S40 4AN M0OAL
S40 4AN M0PLW
S40 4AN M7PPL
S40 4DB G8YTF
S40 4DB M0VAA
S40 4EA G4GSD
S40 4ES G6NZA
S40 4EZ G6CVY
S40 4HN M0APL
S40 4JT M6FBP
S40 4LE G4BBI
S40 4LG M0OKT
S40 4LQ G7PUP
S40 4NE M3SMN
S40 4NW M0EJF
S40 4PQ 2E0STA
S40 4QB M7DIF
S40 4RJ 2E0DXK
S40 4RJ M6JWG
S40 4RS M3GAE
S40 4SJ M6MBO
S40 4SQ G1ZWY
S40 4TF G0RDF
S40 4UT M3JCU
S40 4XJ G4VHX
S40 4XJ G0EWI

S404HG M7DTD
S41 0BB M3VJI
S41 0BD M3LGF
S41 0BS M3XAN
S41 0GF M0KBW
S41 0JP M6MPV
S41 0NJ G0UEH
S41 0NN G7HIQ
S41 0RX 2E0PWG
S41 0RX M6HIM
S41 0SU G4HQH
S41 0XZ G4SBC
S41 7HA G0TRN
S41 7JF M6NDO
S41 7JZ M6GOB
S41 7JZ M6PAX
S41 7JZ M6SSP
S41 7QE M0RRR
S41 8AU M3OHL
S41 8BN M6KZE
S41 8DG M6MYY
S41 8LZ M0DRE
S41 8NX G0RTQ
S41 8NX M3SQH
S41 8PE G0LJI
S41 8PE M6SSA
S41 8RA 2E0GRR
S41 8RA G0BJL
S41 8RA G3UNK
S41 8RA G8BYB
S41 8SF G6FTH
S41 9DA G2AS
S41 9DA G3PHO
S41 9DA G5TO
S41 9HP M7CTN
S41 9JB M3YZK
S41 9JD M6MJD
S41 9JU G6ZOT
S41 9LR 2E0TIF
S41 9LR M6TIF
S41 9PA G6NZO
S42 5AD M6EYG
S42 5GA G1CBY
S42 5HH G6SZB
S42 5HH G8NSD
S42 5HN M3HUB
S42 5hy 2E0CAK
S42 5hy M6ARM
S42 5JA M0FOX
S42 5JP G0JOM
S42 5NS G6URK
S42 6NT G6PRP
S42 5PQ M6LWE
S42 5PU 2E0MKV
S42 5PU M6MKV
S42 5PW M6NLL
S42 5PX M6VEC
S42 6BE M3ESS
S42 6BE M3NLN
S42 6BE M3NST
S42 6BG M3SAA
S42 6FY M3CNC
S42 6HG G3WAM
S42 6HS G4XFF
S42 6JD G3CTZ
S42 6JF G4ROP
S42 6NR M3YJN
S42 6PL 2E0WNT
S42 6PL M3WNT
S42 6RX G4VSR
S42 6SP G3YBO
S42 6SP G6SVH
S42 6SP M3SVH
S42 6UR G4CVO
S42 6UU M3AJV
S42 6XR M3UXH
S42 7AN G1WAB
S42 7AN G6PBW
S42 7AN M3JWW
S42 7DR G6DQA
S42 7EA M3EVR
S42 7JQ G1FTU
S42 7LR G3RLL
S42 7LT 2E0URA
S42 7LT M7UWL
S42 7NE M0CRW
S42 7NH G6OYV
S426BD M3UXU
S43 1AZ M7UCA
S43 1BQ 2E0YNI
S43 1HB 2E0UCV
S43 1HB M3UCV
S43 1HG M3WAV
S43 1JA G8YMT
S43 1JU 2E0MTN
S43 1JU M0VTR
S43 1JU M3BPN
S43 1LJ G4PBC
S43 1LJ M0APH
S43 1NU M3RGG
S43 2AP G3ZSZ
S43 2BG M3CKM
S43 2BG M3CKO
S43 2BJ M3MIJ
S43 2DF M6REF
S43 2DZ G4RIQ
S43 2HN M6RUP
S43 2LH M1BVI
S43 2NR M3RNY
S43 2NW 2E0GRY
S43 2NW M0GRY
S43 3AG G7SMC
S43 3DH G0VRH
S43 3HJ G7VND
S43 3HS M0IKE
S43 3HS M0JOY

S43 3HU 2E0MCS
S43 3HU 2E0VRB
S43 3HU M6LSA
S43 3HU M6UPS
S43 3HU M6VKB
S43 3JX M7CEV
S43 3NF 2E0SYM
S43 3NJ 2E0DBM
S43 3NJ M3UDZ
S43 3PQ M3ICO
S43 3PQ M3IXT
S43 3PY G7SLY
S43 3QB G4OQU
S43 3RA M6SAS
S43 3RH 2E1XRM
S43 3SY M3FVJ
S43 3TE G5RAH
S43 3UH G3BNF
S43 3UY 2E0FKT
S43 3UY M3UHV
S43 3XH M6SLI
S43 4AP G2FVL
S43 4AQ M3KKB
S43 4BD G4VDB
S43 4BD M7GKB
S43 4BS G0KUW
S43 4DH 2E0FIF
S43 4DH M0MKH
S43 4DW G7MOO
S43 4EG G6DAO
S43 4EH G3WEU
S43 4HE G8YJL
S43 4HE M7POG
S43 4ND 2E0TZG
S43 4ND M7TZG
S43 4NG M3LBN
S43 4NJ G4RTA
S43 4PF M3KQD
S43 4PG M3VUP
S43 4PQ M3DLZ
S43 4PQ M3LDJ
S43 4QJ M6DVE
S43 4QP 2E1FDF
S43 4RF G6YFH
S43 4RS G1SLE
S43 4RZ M3ZIO
S43 4SA G0RUV
S43 4SR G4BJG
S43 4ST G4RSS
S43 4SX G0EOJ
S43 4TR G8KDM
S43 4WT 2E0SZL
S43 4WT M7DDC
S44 5AE G3GGU
S44 5AL M6DEK
S44 5ET M3VDA
S44 5ET G0MBS
S44 5HE 2E0FJV
S44 5HE M0IYK
S44 5HE M1CLI
S44 5HE M7CWA
S44 5HP G1VPE
S44 5LU G6EZH
S44 5LX 2E0ORS
S44 5LX M6PFK
S44 5NG 2E1SAZ
S44 5NG G0THF
S44 5NG M0RSC
S44 5NG M3SAZ
S44 5PH M6AWZ
S44 5PH M7BAN
S44 5QA 2E0GUN
S44 5QA M0MTX
S44 5QA M3ORL
S44 5QD M6ASO
S44 5QG M7CPG
S44 5UE G0TJG
S44 6AQ M6EYB
S44 6BH G0IXZ
S44 6DH M3LEX
S44 6DY G6TEX
S44 6EJ 2E0EHO
S44 6EJ M6GGT
S44 6ER G8YBR
S44 6EW G0DJA
S44 6EY G4UPA
S44 6GF G8JQW
S44 6HF M3DGR
S44 6HS G7WAF
S44 6HU G4AGE
S44 6HU G8AVC
S44 6LL M3MYK
S44 6LR 2E1DPQ
S44 6LR M0MLJ
S44 6NX G7GSX
S44 6NX M3CFP
S44 6PS G0CTS
S44 6RX G7CAT
S44 6SE G4FEM
S44 6TX G4KSY
S44 6XH G8NRC
S44 6XN G7DJN
s446eg 2E0SUE
s446eg M3PUZ
S45 0AL G1ISN
S45 0BZ M6VRR
S45 0DX G4TSA
S45 0EA G3MME
S45 0LL M1WWW
S45 8AH M0CJD
S45 8AN 2E0FTB
S45 8AN M0VII
S45 8AN M3VII
S45 8BQ G6XZC
S45 8DH 2E0JUL
S45 8DH G3VNH

S45 8DH M0JHD
S45 8DH M3VNG
S45 8EJ 2E0IBU
S45 8EJ M6WBM
S45 8ET G6POJ
S45 8ET M6SKF
S45 9DX G0WRM
S45 9DX G4HZG
S45 9FA G8RBW
S45 9HH M6PWH
S45 9HH M6SIY
S45 9JT 2E1FKM
S45 9LE M6NUR
S45 9LY G3UAF
S45 9NH G0MNC
S45 9PL G0IVB
S45 9RE G1HFT
S45 9RE G3THT
S45 9RY 2E0MYS
S458HY G4VWT
S5 0DH M6DCA
S5 0GX M3DTH
S5 0JL 2E1FDD
S5 0JP G4TCE
S5 0JY G7WGX
S5 0JY M0SAY
S5 0NN 2E0JOX
S5 0NN M3KUK
S5 0NZ M3GSQ
S5 0QE M6FVP
S5 0TY G6ADD
S5 6AA M6MVT
S5 6FZ M1ERS
S5 6HB G8BMI
S5 6LX G7DMP
S5 6SG M3LEB
S5 6SQ G0UHF
S5 7JX G7ACR
S5 7SW M6SHF
S5 7TL G0OLL
S5 7WE G0UQZ
S5 7WU G0BWG
S5 8AX 2E0OMV
S5 8AX M0OMV
S5 8DT M0DMA
S5 8GN 2E0SNU
S5 8LL M6TBX
S5 8LT M7MLY
S5 8NE 2E0ZEV
S5 8NE M3ZEV
S5 8NF M6GVS
S5 8WF 2E1AGV
S5 8WF M1FJQ
S5 8xu M6BGJ
S5 9AH M6JFS
S5 9DB 2E0HNG
S5 9DB M7CHB
S5 9DY G3UXH
S5 9GA G0SEW
S5 9GS G8AOZ
S5 9NW M7GLB
S5 9QR G0RNB
S50NN 2E0TJX
S6 1BQ M0GAV
S6 1FB 2E0IEB
S6 1FB M6IEB
S6 1HT 2E0YDB
S6 1HT M0SDB
S6 1HT M6YDB
S6 1HU G7DEE
S6 1JE G3LLV
S6 2TY M3XWF
S6 2XE 2E0BQX
S6 2XE M3MYT
S6 3NG G4BOJ
S6 3QZ M0KFY
S6 3RG G0LIZ
S6 3RG G4YUR
S6 3TL M1FFV
S6 3TR G4VMQ
S6 4AH G4VYI
S6 4AY G0BKE
s6 4ba M1CGF
S6 4BE G3OZL
S6 4FD G7NNZ
S6 4FG G0CGT
S6 4LQ 2E0VLD
S6 4LQ M7LDS
S6 4NB G0JKE
S6 4QG M0EMR
S6 4SU M3PMK
S6 4TF G4NXA
S6 4WG M1EQW
S6 4WU M7JWZ
S6 4WU M7PSZ
S6 5BA G0RNC
S6 5BE M0RPK
S6 5DD G4YWJ
S6 5DQ M0DPJ
S6 5GQ M6YSM
S6 5HQ M3UJL
S6 5LN G4TFZ
S6 5NA M7OFG
S6 5ND M6XAW
S6 5PA G0JTT
S6 6AB G1MLV
S6 6DJ G0JKH
S6 6DQ M7CXB
S6 6GL G3DR
S6 6GL G4FAL
S6 6GL G5AT
S6 6GL G5WS
S6 6GL G6XX
S6 6GL G6ZZ
S6 6GL M0HDG
S6 6GL M3HEE

S6 6GL M3XBT
S6 6GL M5DX
S6 6HX M6MBG
S6 6LJ G1WWI
S6 6SG G8KB
S6 6TG M1JWM
S60 1JU M1ALG
S60 2AT G4TQT
S60 2NG M6AYV
S60 2UZ G1NFO
S60 3ER G1MDE
S60 3HB M0GUD
S60 3JJ G1HWA
S60 3JN G1IDF
S60 4DZ G6BKL
S60 4EA M6AYW
S60 4EF G4LFS
S60 4NH G4NLC
S60 5ES G0ISK
S60 5HG G1DEZ
S60 5JU G0CGS
S60 5PU M3FZO
S60 5RE G6CPY
S60 5TH G8AHN
S603AZ G3VSK
S61 1BJ G0UOM
S61 1RY G1RGG
S61 2BL G6LXW
S61 2DU 2E1BKF
S61 2EX G1OHU
S61 2HB G8EQD
S61 2HB M0EQD
S61 2JW G6MLS
S61 2JZ M3FFK
S61 2NP M3FJC
S61 2PZ G4PEN
S61 2QS M6CLW
S61 2RP 2E1ENZ
S61 2RP 2E1ETB
S61 2RP G4ZVN
S61 2RR 2E0INK
S61 2RR M7USE
S61 2SS G4NXS
S61 2SS G6XGJ
S61 2TQ 2E0EVP
S61 2TQ M6ATJ
S61 2TT G4EBG
S61 2UH G1TYU
S61 2UN G0OUN
S61 3AG M6UZZ
S61 3HN G7OFV
S61 3LG M7JFX
S61 3ND M3PJI
S61 3NP G4GET
S61 3RF M7ARZ
S61 3RQ M6FNU
S61 4DL G0MTT
S61 4LP G4HVW
S61 4PA M0BEQ
S61 4PD G7OGR
S61 4PL M6EXX
S62 5BD G1POC
S62 5BH M3IFA
S62 5ED M0UGD
S62 5JR G1GBR
S62 5LW G0VBR
S62 5NH G7BQS
S62 5NT 2E0GMA
S62 5NT M6AXX
S62 5NT M6GMA
S62 6DB 2E0ZYO
S62 6DB M3ZYO
S62 6DN M7LLE
S62 6HJ 2E0BNE
S62 6HJ 2E0SDA
S62 6HJ M3OBX
S62 6LN G7MST
S62 7BX G6JDC
S62 7ED G1IFH
S62 7EE M3UGD
S62 7EN G3USW
S62 7FE M3MGN
S62 7JA M6NOL
S625BW 2E0FHQ
S625QJ G7AOQ
S63 0HB 2E0GEX
S63 0HB M6YPG
S63 0JA M1EUN
S63 0LQ M6ZEF
S63 0NN M1DSI
S63 0PX 2E0BMY
S63 0PX M0GVX
S63 0PX M0HKT
S63 0PX M3PGK
S63 0QT M3UOD
S63 0RZ G3WLV
S63 0SA 2E0GAR
S63 0SA M3PGU
S63 0TG 2E0ZTG
S63 0TG M0ZTG
S63 0TG M3OTG
S63 0TN G4VNX
S63 6AA 2E0CDW
S63 6AA M0KCW
S63 6AA M6CDW
S63 6DD 2E0DLC
S63 6DD M6DLX
S63 6EL 2E0DYJ
S63 6EL M6EZC
S63 6ET G4UZG
S63 6EY G8DVF
S63 6HW G8XUH
S63 6HZ M6MGN
S63 6JN M6EUU
S63 6LJ M6SYB

S63 6NQ M6KJQ
S63 6NU G1KZA
S63 7GD M6WJA
S63 7GT 2E1HTE
S63 7GT G4ZTX
S63 7JB G4TZL
S63 7JU G7EVT
S63 7LR G3YOC
S63 7ND 2E0EGQ
S63 7ND M6PSR
S63 7NE G7FAG
S63 7NF M0KRL
S63 7NF M3NBI
S63 7ST G1YPR
S63 7TD M6HCB
S63 8BZ G6IIK
S63 8DA G0TZC
S63 8HR G1EQM
S63 8HR M0EQM
S63 8JL G0UAD
S63 9BA 2E1AQT
S63 9BA G0GPR
S63 9BY G6WSZ
S63 9DX M3ZYI
S63 9DZ 2E0CMO
S63 9EA M6KMS
S63 9JW G0JDC
S63 9LB 2E0TAU
S63 9LB M3LJK
S63 9LB M6JAJ
S64 0AD G8CXA
S64 0DG G8HPJ
S64 0DG M6DBG
S64 0DJ G7BJR
S64 0DT M0DKX
S64 0JD G4KWM
S64 0JG G4PJT
S64 0JZ M0CTJ
S64 0NL G3VJR
S64 5TX G7YAB
S64 5TY G0VPS
S64 5UG G6VIN
S64 5UG M6HZM
S64 5UP M6MDG
S64 5UQ G7MUE
S64 5UQ M3ILR
S64 5UQ M6HOQ
S64 5UU G0VXC
S64 8AH 2E1FCE
S64 8DS G6MFU
S64 8DU G8BIW
S64 8EF 2E0KLV
S64 8EF M3VAF
S64 8HD G0VUH
S64 8LF G0PBF
S64 8NA M6FLW
S64 8ND G7JSE
S64 8NF M3KSK
S64 8NF M3PSD
S64 8NX G0CSK
S64 8NX G6ZFX
S64 8NZ M6PKN
S64 8PT 2E0BQQ
S64 8PT M3UPK
S64 8PY M3UPL
S64 8QQ M7DXQ
S64 8QW M6PZB
S64 8RF M6FJF
S64 8SQ G0FUO
S64 8TE 2E1FTF
S64 8UH G6OET
S64 8UL G4IDL
S64 9EH G1UVJ
S64 9EH M3ZPR
S64 9EZ 2E1FTV
S64 9EZ M3ACY
S64 9EZ M3EIJ
S64 9JX 2E0HNX
S64 9NH 2E0BQP
S64 9NH M0KVM
S64 9NH M3UOF
S64 9PF G4RPL
S64 9SG 2E0EBD
S64 9SG M0IBR
S64 9SG M6EBO
S648NU G0VJY
S65 1LH G8UCR
S65 1LT M3ZHI
S65 1NB M6JED
S65 1NF G6IJE
S65 1QP M7LRM
S65 1RT M7JSI
S65 2BY M3CPN
S65 2HQ M3OAM
S65 2JR M6LYS
S65 2LH M6FVO
S65 2LR M6EOS
S65 2QP M0CTI
S65 2UA G4NJI
S65 2UE 2E0DZE
S65 2UE M6AWE
S65 2UX 2E0DKB
S65 2UX M6ICQ
S65 2XA 2E0TAL
S65 2XA M3ZXN
S65 3JX 2E0DZE
S65 3JX M0CHU
S65 3LG M6BZE
S65 3NH G1RWR
S65 3NL M6RGE
S65 3NN M0CJY
S65 3NX M3EWQ
S65 3PQ M0MTD
S65 3RB M7PCD
S65 3RW M6BRY
S65 4HH G8NVX

S65 4JF M6KBH
S65 4NF G4RGH
S65 4NZ G4APO
S65 4NZ G8VJP
S65 4QR G4SKM
S65 4QR G8DRQ
S65 4RJ M7RXK
S66 1AF M6RBV
S66 1DL G4CGP
S66 1NN 2E1BGQ
S66 1UH M6DMW
S66 2BN G4ENC
S66 2BN G8EUV
S66 2HZ G4WFW
S66 2HZ G7RIJ
S66 2JA M3VPN
S66 2LU G8PUK
S66 2NL M6WNN
S66 2NS M6VTT
S66 2NT M3JDF
S66 2PQ M6JIS
S66 2SJ G1DTF
S66 2SP G0WXD
S66 2SS G8HUO
S66 2SW 2E1IJO
S66 2SW M3BBL
S66 3PL M6MDT
S66 3WA G6PMR
S66 3YW G4UHQ
S66 3ZY 2E0DOM
S66 7AJ M3YXT
S66 7DQ 2E0FJT
S66 7DQ 2E0FJU
S66 7DQ M6NOH
S66 7DQ M6ZAB
S66 7EN 2E0ODO
S66 7EN M3ODO
S66 7EU G1BHQ
S66 7HA G0EIB
S66 7HA G0GOO
S66 7HB G6PCX
S66 7HF G0EPU
S66 7HG G7OFI
S66 7HJ G4BVV
S66 7HX G1POC
S66 7JZ M1DBF
S66 7JZ M6JJH
S66 7LH M6AVF
S66 7LU G1ILC
S66 7NR M6COX
S66 7PA M6FHP
S66 7PD G7TKM
S66 7QB M6MXP
S66 7SG 2E0YLH
S66 8AT M0MCT
S66 8AT M6TBO
S66 8AZ G0HDG
S66 8AZ M6BVW
S66 8BL G6TJC
S66 8DL M3XQH
S66 8DS M6BVX
S66 8HD 2E0KFA
S66 8HD M7WWL
S66 8JT M1CYJ
S66 8JT M3CYJ
S66 8JU G1OJT
S66 8NA M6GIZ
S66 8NN M6LWJ
S66 8QZ M6WBF
S66 8RF 2E0ZOR
S66 9DL G4JMC
S66 9HT 2E0IEW
S66 9HT M6MQN
S66 9LG 2E0JRA
S66 9LG M0JRA
S66 9LG M3JRA
S66 9NJ M1EKA
S66 9NU M7PES
S66 9PD M3KUN
S66 9QP M6SUF
S667LJ 2E0TCQ
S7 1FE M6ZYH
S7 1GF 2E0LVP
S7 1NE G7GFH
S7 1SJ G4ILX
S7 2EB G0WFL
S7 2QN 2E0DRE
S70 1PL M0TNG
S70 1QE G1KXZ
S70 2NE M3ULT
S70 3AA G1ANF
S70 3AA G6ZXO
S70 3DW M6BAD
S70 3EW G3TEH
S70 3FG G0NSC
S70 3FG G1XUH
S70 3RH G4DYG
S70 4AU M3LXR
S70 4BW M6XLG
S70 4DB 2E0FJN
S70 4DB M7NET
S70 4DW G3EAE
S70 4LR 2E0EWS
S70 4NY G3ZHP
S70 4QN M6TEV
S70 5AY M6BRU
S70 5DU G6ZEW
S70 5JB G6PDJ
S70 5NR G4IAY
S70 5QR G0OHR
S70 5QY 2E1ACS
S70 5QY G8VDP
S70 5QZ G7MRZ
S70 5RL G0JEA
S70 5SN G8EGM
S70 5SU G4KUD

S70 6BZ M7BLJ
S70 6JT G0TOQ
S70 6JY G4TMZ
S70 6JY G6AJ
S70 6JY G6YOR
S70 6LG G0DUM
S70 6PP 2E0BIB
S70 6PP M3GPN
S70 6PQ G6AUP
S70 6QQ G6XQT
S70 6RG G6TVA
S70 6SU M6WLC
S704SQ M7TGH
S71 1SB M3TRO
S71 1SU M0GAG
S71 1XQ G4POI
S71 2BA M0RTL
S71 2BA M1OXR
S71 2BA M3ITM
S71 2BA M3SPG
S71 2BL G6HNE
S71 2ES G8ISE
S71 2EY G6DER
S71 2HP G4DSA
S71 2HS M3ZME
S71 2HS M6KJD
S71 2JY M3HUY
S71 2LL G7VYN
S71 2NJ G8FRJ
S71 2PP G1WTN
S71 2RA G4JUR
S71 3BG M3OCP
S71 3DL M7LOW
S71 3JJ M3VYV
S71 3JJ M6CBD
S71 3JJ M6EDL
S71 3LF M7LFG
S71 3NR M3VCK
S71 3PU 2E0ATT
S71 3PU M7ATT
S71 3QD M1DZM
S71 3QU M3HVH
S71 3RH G0SZE
S71 3SJ G0ANE
S71 4AA G1ANI
S71 4HF G3OXR
S71 4JA G4YAP
S71 4LQ M6UUH
S71 4NG M0OCC
S71 4NG M0OXO
S71 4NG M0UKI
S71 4NG M3ZLZ
S71 4NG M3ZYZ
S71 4NH G1BQR
S71 4NH M0GEY
S71 4NL 2E0KAF
S71 4NL M3KRY
S71 4ry 2E0ETP
S71 5AL M3KAE
S71 5BS G4UFR
S71 5BS G6LWK
S71 5BS M7DHF
S71 5DH G4IHZ
S71 5LG M0GOI
S71 5QX G0JUK
S71 5RB G3AMH
S71 5RB G6KOB
S71 5RB M0HFE
S71 5SD M6CPR
S72 0AU G0PQG
S72 0DW M3WZF
S72 0HG 2E1HUQ
S72 0HT G0TLA
S72 0HT G7KMF
S72 0HT M3FJD
S72 7AA M6IFS
S72 7NA M3ICH
S72 8BE M1ERN
S72 8DE G1EFF
S72 8DY G6NXW
S72 8DY M3RYR
S72 8EW M6SEB
S72 8HU M7AVG
S72 8JB M1SWS
S72 8JP G4EPD
S72 8JT 2E0MML
S72 8NQ M0DMS
S72 8NQ M6BNJ
S72 8PW G1SYP
S72 8PY 2E0BLF
S72 8PY M3UXC
S72 8PZ 2E1HXR
S72 8RN G4LUE
S72 8RN G8PLJ
S72 8SE M6URG
S72 8SE M7BMR
S72 8XG G0RUX
S72 9AN 2E0FYL
S72 9AN 2E0TJC
S72 9AN 2E1HPM
S72 9DL G4XDV
S72 9DL M3XDV
S72 9EG M6FLY
S72 9HJ G1JDO
S72 9HJ G6CND
S72 9JA M0AEE
S72 9JG M3NTQ
S72 9JX 2E1HPZ
S72 9JX M3WOC
S72 9LL M0JOR
s728jt M6OXR
S73 0EF M3SPQ
S73 0LF 2E0OGK
S73 0LF G0XDX
S73 0LF M7GEG
S73 0RW G8VHB

S73 0TN M6PRW
S73 0YJ G1BCN
S73 8LD G0VMN
S73 8PP 2E0CKP
S73 8PP M6CKP
S73 8PS G0SKD
S73 8RX G1VGO
S73 8SF G6YJD
S73 8SG M6GGV
S73 9AB M1LAP
S73 9AE G0UQF
S73 9BJ M3UJF
S73 9DU G4BHL
S73 9DU M3ZXA
S73 9ED G0TKJ
S73 9ED G1YTX
S73 9HA M3ZPY
S73 9HQ G0CQO
S73 9JA M0BZH
S73 9NF M0CLJ
S73 9PE G1YUU
S73 9PL M3XRH
S73 9PP G4NSO
S738DX M7KXL
s738sb 2E0FWR
S738SB M0WFR
s738sb M6FRK
S74 0AP M0GXH
S74 0AP M3LSS
S74 0AZ M3ITU
S74 0BU M3SNY
S74 0EJ M3YFG
S74 0LB 2E0TBV
S74 0ps M6RTM
S74 0PS M6SLT
S74 8BW M0PAF
S74 8BW M0PPP
S74 8DP M0CSZ
S74 8DR G0CUI
S74 8DS M3HHX
S74 8DT G6UQO
S74 8JR 2E0PJE
S74 8JR M6PJE
S74 9AX G6NVW
S74 9DP M6JYV
S74 9EQ G4YTM
S74 9HF G7GQD
S74 9HF M0CNG
S74 9HS M3XJO
S74 9HU M0BJK
S74 9HU M0FCG
S74 9HU M1KEY
S74 9HZ M3KXF
S74 9RF M3KZV
S749EE G4YER
S75 1EE 2E1HJO
S75 1EE M0HYD
S75 1FE M6MPR
S75 1LX 2E0TTS
S75 1LX M3TTS
S75 1PS M3XZD
S75 1PX G4OUS
S75 1QG G4MLQ
S75 2AF M3EVE
S75 2EF 2E1HOK
S75 2EQ M3EGF
S75 2EW M6FJC
S75 2JW M5BAD
S75 2ND G7BZE
S75 2QB M6BFC
S75 2QJ G8MQX
S75 2RX M3BCQ
S75 2TQ M1GCS
S75 3DD M3KBZ
S75 3DE M0FAT
S75 3JE 2E0DNO
S75 3JE M6DCO
S75 4HD G1UQT
S75 4JS G1VOC
S75 4NN G0RUZ
S75 4QQ M6CWW
S75 5EA G5TV
S75 5EA M6YBK
S75 5EL M3VGF
S75 5ER M0RCI
S75 5HL G0OUC
S75 5JQ G7TGF
S75 5LG M0JLT
S75 5NP G7NRP
S75 5NS M3IOQ
S75 5PG M6BJE
S75 5PW 2E0BRI
S75 5PW G1NAQ
S75 5QP M6KHS
S75 6DQ 2E0YNY
S75 6DQ M0YNY
S75 6DQ M3YNY
S75 6ET M3UMD
S75 6EU 2E1HJS
S75 6EX M6DDV
S75 6EX M6JGU
S75 6HB M0LOU
S75 6HB M1LOU
S75 6HB M3SLF
S75 6HD M3NSX
S75 6HD M3NTJ
S75 6HH G0SBO
S75 6JA G1LES
S75 6JP G7ULW
S75 6LE G8GDH
S8 0BN G0JPE
S8 0EY M6XMH
S8 0GA M0GDX
S8 0HH M7LLB
S8 0RA 2E0BFZ
S8 0RA G0NFY

S8 7BW G0PEW
S8 7BX G4ECA
S8 7DP G1LJL
S8 7DR G6FGL
S8 7DZ G6SWW
S8 7FW 2E0MDZ
S8 7FW M6MDZ
S8 7UF 2E0PGL
S8 7UF M6PGL
S8 7UF M6TAL
S8 8DA M3NPA
S8 8DA M3NPE
S8 8DA M3NRQ
S8 8DQ G8LIH
S8 8DS M6BYH
S8 8DU 2E0GCH
S8 8DU M3GCH
S8 8FT G8YVW
S8 8GE G0HSA
S8 8HE G4RVS
S8 8LB G3XSI
S8 8LJ G7MLL
S8 8PB G8OFN
S8 8PJ G0NQY
S8 8QR G0NJD
S8 8QU G7EFA
S8 8SE G3RKL
S8 9NG G1HGB
S8 9RL G7DFX
S80 1RZ 2E0MNU
S80 1RZ M0MNU
S80 1RZ M3MNU
S80 1SU 2E1EXP
S80 1SW 2E0NVB
S80 1TD 2E0MZZ
S80 1TD M6ELR
S80 1UZ G0MZP
S80 1YA G3WTY
S80 1YG M0KAD
S80 2DL G3UVB
S80 2EG 2E0KBZ
S80 2EG M6JPU
S80 2EW M0CMN
S80 2HN 2E0RPR
S80 2HN M0RPR
S80 2HN M6HEP
S80 2HN M6RPR
S80 2JR M7CLJ
S80 2LL 2E1GJE
S80 2NT 2E0PRS
S80 2NT M0JGU
S80 2NT M0YES
S80 2RB M0MIR
S80 2RR 2E0BRY
S80 2RR M3MYW
S80 2SA 2E0MZU
S80 2SA 2E0WAJ
S80 2SD 2E1GJT
S80 2SH M6OQR
S80 2ST M3KUE
S80 2TP M7GGM
S80 2TT G7MQW
S80 2TW M6LVE
S80 2TW M6NNJ
S80 2TW M6NNU
S80 3DB M0GKR
S80 3DE G0KVG
S80 3DF G7TNZ
S80 3DF M1FEX
S80 3DF M3RDW
S80 3DF M6REW
S80 3DL M6SRA
S80 3EB M0DOB
S80 3HF 2E0CHA
S80 3HF G7OBP
S80 3HG G0CEJ
S80 3HR G7SOH
S80 3NF M0MRK
S80 3NQ M0WEN
S80 3QB M1LAN
S80 3QG M0BMT
S80 3QJ G6JBL
S80 3QJ G7BJD
S80 3RD G4PYV
S80 4AN M1DAB
S80 4BB 2E0BEV
S80 4BN M0HZE
S80 4DE G0AHD
S80 4DL M3WTP
S80 4DW 2E0ITJ
S80 4DW M0HEW
S80 4DW M6TJZ
S80 4DY G1EVA
S80 4ER M3KOJ
S80 4HN 2E0XRM
S80 4HN M0XRM
S80 4HS 2E0AVD
S80 4HS M0JMJ
S80 4JR G1KIT
S80 4JT 2E0CTJ
S80 4JT M6CTJ
S80 4NP G6OZT
S80 4NS M6MUJ
S80 4PY G8SRZ
S80 4SN G3VRU
S80 4UD G1OIZ
S80 4XT M3JHJ
S81 0AX G4YJY
S81 0AX M0WDP
S81 0BS G0AGD
S81 0DH M6WSM
S81 0DP M0GTQ
S81 0EE G3OZN
S81 0HD 2E0NCK
S81 0HD 2E0PNK
S81 0HD M6CAR

S81 0HD M6NCK
S81 0HD M6PNK
S81 0JS G0WTK
S81 0JS G8TBF
S81 0LP G3XXO
S81 0QN G0BDR
S81 0QN G0CJD
S81 0QN G4ZUN
S81 0QN G6EXG
S81 0RJ 2E1SKY
S81 0SS M7BUH
S81 0TB G0HMX
S81 0UY 2E0YBB
S81 0UY M3HOV
S81 0XH M3MSH
S81 0XJ 2E0CAJ
S81 0XJ M0TAL
S81 0XJ M6CMT
S81 0XY M0EEB
S81 7BH G1VBP
S81 7ED G0FMP
S81 7HL M6KDB
S81 7JS 2E0RYN
S81 7JS M6DCK
S81 7JZ M7DZK
S81 7LE G4CRE
S81 7LJ 2E0ZTD
S81 7LJ M3UHC
S81 7LJ M3ZTD
S81 7PG G8TWT
S81 7PS M6PRB
S81 7QB G1NSK
S81 7QE G0TKG
S81 7QW G6TLB
S81 7RT 2E0GEF
S81 7RT G6CNX
S81 7RT M3YHT
S81 8NH G8RSV
S81 8NQ M0JSP
S81 8PE G4YRZ
S81 8PH M0JAV
S81 8SF G0WPC
S81 8UZ M7RWF
S81 8WA G7AIZ
S81 9BD 2E0RSH
S81 9BD G3ZMM
S81 9BD M0ZMM
S81 9BD M3RSH
S81 9HN M7WSP
S81 9JT G4MWX
S81 9LE M3LRK
S81 9NW G3XXN
S81 9NX G6DRG
S81 9PT M3BXN
S81 9RA G0NFB
S81 9RR G0DMA
S81 9RR M6SBM
S81 9SB M3YRC
S81 9SH M6SDL
S81 9SL G0NGW
S81 9SS M6TWI
S817AN M7AZH
S9 1AG M6GOE
S9 1AR 2E0JBS
S9 1AY 2E1IDM
S9 1BA G7TBC
S9 1LU G6VUE
S9 1PZ 2E0HRY
S9 1PZ M3UQG
S9 1RX M7TBI
S9 4AU G4RBU
S9 4HE G0MBK
S9 4JG M6DWW
S9 5EH M0AFY
S9 5FQ G3RCM
S9 5FQ G6DCT
S9 5GJ G8SVT
SA1 1NZ GW0JFQ
SA1 1RY GW0EZQ
SA1 2BU MW6AJD
SA1 2QE GW6TTA
SA1 3HE 2W0MPY
SA1 3HE MW7PYM
SA1 3SN GW7RLS
SA1 3SN GW7SKC
SA1 3UX 2W0LBN
SA1 6HN GW1XBG
SA1 6LS GW8DSO
SA1 6NH MW1AUV
SA1 6PD GW4PHT
SA1 6QP MW7AWO
SA1 6TZ GW3NDB
SA1 6TZ GW8ASA
SA1 7AD GW4VBV
SA1 7EF GW4WFM
SA1 7FG 2W0UCL
SA1 7FG MW0UCL
SA1 7FJ 2W0HAC
SA1 7FJ MW0HAC
SA1 7FJ MW6HTT
SA1 7JB MW6LTN
SA1 7JY GW4YID
SA1 8EJ GW6MLL
SA1 8HQ GW4HDB
SA1 8JF GW0JDS
SA1 8PN 2W0UAR
SA1 8PN MW0SWN
SA1 8PN MW6KVA
SA10 6AA 2W0DAP
SA10 6AA MW6THW
SA10 6BQ MW3JEK
SA10 6DL GW0CBL
SA10 6DS GW0HPC
SA10 6ET MW6FAM
SA10 6JR MW6FIX
SA10 6PJ MW0SBX
SA10 6SD GW6KRK
SA10 6SP GW0VSW
SA10 6SP MW0NSC
SA10 7AD GW4JQQ
SA10 7BU GW4YJI
SA10 7DD GW8DOA
SA10 7DY MW0CIA
SA10 7LB 2W1CPS
SA10 7LB MW6SWN
SA10 7NR 2W0EPE
SA10 7NR MW6EPE
SA10 7PL GW1WGR
SA10 7PL GW4WWN
SA10 7RX GW7IBT
SA10 7TT MW7TWG
SA10 7TW GW7EHD
SA10 7UG GW0CYG
SA10 7US GW0DJU
SA10 8BG GW4FOI
SA10 8DR 2W0ZWR
SA10 8DR MW0ZWR
SA10 8DR MW3ZWR
SA10 8DW GW1SGH
SA10 8HD GW8VCA
SA10 8HT GW4FXF
SA10 9BL MW7AHI
SA10 9BU GW6APK
sa108nd MW0ABV
SA11 1DJ GW4IUK
SA11 1DQ MW1FFY
SA11 1JL GW0VFF
SA11 1NG 2W0JJW
SA11 1NG MW6RGW
SA11 1PP MW3GTM
SA11 2AY 2W0TAI
SA11 2BJ MW6PWS
SA11 2DW MW3SOC
SA11 2JG MW6CBL
SA11 2JR MW7WWF
SA11 2LU GW6NXH
SA11 2LU MW6MSE
SA11 2PG 2W0RRY
SA11 2PG MW6BWR
SA11 2TE GW4PCO
SA11 3AL GW1EWW
SA11 3FD MW3GQE
SA11 3JW GW0NKJ
SA11 3NJ GW2ABJ
SA11 3PJ GW0HNT
SA11 3QE MW0CNC
SA11 3RN MW0HYP
SA11 3SS GW8SZC
SA11 3SS MW3GRC
SA11 3XA MW3LHA
SA11 3YQ GW0TWR
SA11 4AA GW7JHK
SA11 4HS GW7TZG
SA11 4NN 2W0PEE
SA11 4NN MW6NRT
SA11 5BG GW4SRE
SA11 5BG GW4UYT
SA11 5NT GW4WYX
SA11 5RL 2W0VPT
SA11 5RL MW3VPT
Sa114bp MW6YSA
SA12 6BF 2W0BVS
SA12 6PF MW6DSP
SA12 7DE GW4RML
SA12 7ED 2W0RMY
SA12 7ED MW6ADZ
SA12 7HG GW1NBW
SA12 7PF MW7TMA
SA12 8AP GW0RLQ
SA12 8AS GW4XWC
SA12 8EY GW1RQM
SA12 8LE GW4TVQ
SA12 8TU MW6ESA
SA12 8YE GW3TKG
SA12 8YH MW7BRQ
SA12 9EA MW0POB
SA12 9LJ MW6IMT
SA12 9SE 2W0CED
SA12 9SE MW6CAN
SA12 9TG GW0DZU
SA12 9TW GW8DUY
SA12 9TW MW0TTF
SA12 9TW MW0WVA
SA13 1BR MW1GLD
SA13 1DX MW7PHI
SA13 1HA GW4DOO
SA13 1LD MW6AHV
SA13 1NE GW7TJM
SA13 1PF MW6XMO
SA13 1SG GW0TWF
SA13 1TD 2W0CLT
SA13 1TD MW6BOC
SA13 1YD GW4PRP
SA13 2EW MW7JNC
SA13 2HL MW0CDO
SA13 2LB GW3NKM
SA13 2LF GW0KPD
SA13 2US GW3XHD
SA13 2YD MW0LEW
SA13 2YE GW3TMJ
SA13 3AN GW4RDW
sa13 3bp GW1GVM
SA13 3PQ MW7CCS
SA13 3SD GW4TVU
SA13 3YE MW6FGN
SA13 3YN GW4DWN
SA13 3YW 2W0TRS
SA13 3YW MW6DBU
SA13 3YW MW6GSL
Sa132ug MW0TDQ
SA14 6AP MW3YNA
SA14 6AP MW7ENN
SA14 6BD GW0RUD
SA14 6BR MW3WLS
SA14 6DT GW0SRF
SA14 6LY GW7RQI
SA14 6NT MW1DUJ
SA14 7AR GW6FBV
SA14 7DN MW6OCT
SA14 7DN MW6SCD
SA14 7HY GW1LFN
SA14 7NA GW0LNM
SA14 7PN 2W0FUR
SA14 7PN MW6PWM
SA14 7PR 2W0TRD
SA14 7PY MW1FJK
SA14 7RJ GW4BYA
SA14 7RS 2W0FYP
SA14 7RS GW7LBI
SA14 7RS MW6SON
SA14 7RW GW0AKV
SA14 7SA MW3UNZ
SA14 7UF GW8JJZ
SA14 8AE GW4PUC
SA14 8AT 2W0SCP
SA14 8AT MW0GIN
SA14 8DN GW4KFD
SA14 8EL GW3OPC
SA14 8PP MW6HDT
SA14 8PU MW0ATR
SA14 8QU MW0CUA
SA14 8UB GW4APF
SA14 9AH GW4OH
SA14 9AH MW0JZE
SA14 9AH MW3ZAQ
SA14 9LH 2W0FOG
SA14 9UP GW0KJZ
SA146BR 2W0DSP
SA149PZ 2W0TRR
SA15 1AR 2W0DQT
SA15 1AR MW6EFK
SA15 1DG GW0DZL
sa15 1hr MW3VNV
SA15 1JP MW7PHY
SA15 1NQ 2W0LDX
SA15 1NQ MW3SNJ
SA15 1NR MW0JAW
SA15 1NT MW3XVR
SA15 1NZ MW0PJR
SA15 1PB MW0EAN
SA15 1SW GW6PDR
SA15 1SW MW6CCG
SA15 1SW MW6ZOD
SA15 1TR MW6FQD
SA15 2AB MW0CXH
SA15 2BG MW0CJB
SA15 2LB GW0RHE
SA15 2RB GW6VFH
SA15 2RH GW4AHX
SA15 3AE GW0JQT
SA15 3AE GW1OII
SA15 3ED GW4UIE
SA15 3NF GW0BCL
SA15 3NS GW0IXQ
SA15 3RA MW0DFN
SA15 3RT 2W0HUX
SA15 3RT MW0LDD
SA15 3RT MW7MJL
SA15 3SN GW0OFH
SA15 3SN GW0GUA
SA15 3SN GW4TFS
SA15 3TF GW7FXX
SA15 4LA 2W1ETW
SA15 4NL GW4RXA
SA15 4PF MW0CXW
SA15 4PF MW0WWR
SA15 4RR MW6FIY
SA15 4SG MW6TFL
SA15 4SG MW6ULX
SA15 5AQ GW4LJS
SA15 5DJ GW8VUV
SA15 5DL 2W0IGC
SA15 5DL MW0INC
SA15 5DL MW6IGC
SA15 5EN MW3GDL
SA15 5HF GW0NLB
SA15 5HN GW0KJT
SA15 5HN MW6FHK
SA15 5NY MW3HNP
SA15 5RT GW0TMU
SA15 5SF GW0SRE
SA15 5SG GW0IVG
SA15 5SH MW6PLP
SA15 5SU 2W0CKV
SA15 5SU GW0TDA
SA15 5TF GW8KCY
SA15 5TP GW7UXY
SA15 5UB MW0CTX
SA15 5UG GW6HLD
SA15 5UH MW0FUN
SA152DZ MW1BAJ
SA16 0EG MW3VMY
SA16 0HF GW0DWQ
SA16 0HJ GW6XGA
SA16 0HP GW0DIX
SA16 0LD GW0WGE
SA16 0NF 2W0MMD
SA16 0NF GW0JDW
SA16 0RH GW8WXP
SA16 0TE MW6NAX
SA16 0UR MW3IKC
SA16 0UT MW3IFZ
SA17 4AF MW6GAA
SA17 4AX GW4RKI
SA17 4BQ GW8XMW
SA17 4EB GW1ENG
SA17 4EY MW0CWS
SA17 4EY MW3DWZ
SA17 4HF GW0BKM
SA17 4HF MW3VFN
SA17 4JA MW6JCH
SA17 4NW MW1BNY
SA17 4RA MW6RDL
SA17 4TE MW6HEI
SA17 5BJ MW6OUW
SA17 5EJ GW3PLB
SA17 5EN GW3YAF
SA17 5TQ GW3UAY
SA17 5TQ MW6AHQ
sa174af 2W0HDC
sa174af MW0WSD
SA18 1AD GW7ORB
SA18 1AF MW6IBD
SA18 1AN MW6MTG
SA18 1AU GW0WRI
SA18 1BB MW6RXZ
SA18 1BD GW0LDZ
SA18 1EN 2W0PMZ
SA18 1EN 2W0RKF
SA18 1EN MW3PDE
SA18 1EN MW6PNZ
SA18 1EN MW6RKF
SA18 1PE GW0HNE
SA18 1RF GW0RAD
SA18 1TR MW3UIQ
SA18 1TR MW3UIY
SA18 1TR MW6TLN
SA18 2AY MW3OLX
SA18 2GA GW7RRM
SA18 2HF GW4IMC
SA18 2LQ MW0ANX
SA18 2LX MW6AIC
SA18 2NG GW1ANW
SA18 2QE MW1FGV
SA18 2QE MW3FGV
SA18 2RA MW3FTY
SA18 2TY GW1JOV
SA18 2UY GW6LUT
SA18 3EF MW7ARE
SA18 3HA GW4UPG
SA18 3HD MW3XDB
SA18 3LN GW4JPC
SA18 3QL GW0IXM
SA18 3QW GW7HJN
SA18 3RD MW0GTY
SA18 3RY GW1TJK
SA18 3SF 2W0TCM
SA18 3SF GW4XMR
SA18 3SF MW0XTZ
SA18 3TG GW3WMP
SA18 3UA GW4BVJ
SA19 6EB MW0AMI
SA19 7HD 2W0GAY
SA19 7HD MW0SIP
SA19 7PA GW0GPQ
SA19 7UA GW0AIY
SA19 7YP MW6NHL
SA19 7YR MW0CIS
SA19 7YR MW0CIT
SA19 7YT GW6WOB
SA19 8AD GW1SPW
SA19 8DF GW0AWT
SA19 8DX GW8DX
SA19 8DX MW0PRO
SA19 8DX MW0VDX
SA2 0AT MW3TAF
SA2 0DE MW5DQQ
SA2 0FQ 2W0RGA
SA2 0FQ MW6RGA
SA2 0FZ GW3OMN
SA2 0GB GW4WRI
SA2 0PJ GW1YHL
SA2 0PR GW0SAJ
SA2 0QE GW3TYI
SA2 0RL MW0ILG
SA2 0RL MW0KRS
SA2 0YD MW0CGP
SA2 7DF GW1FBL
SA2 7EH GW4BNJ
SA2 7EQ GW6UFH
SA2 7HW GW1BFB
SA2 7PQ GW8VFQ
SA2 7PR MW6HUY
SA2 7QE GW6VKY
SA2 7QS 2W0BKN
SA2 7QS MW0MAY
SA2 7QS MW3RPZ
SA2 7RP GW8HDH
SA2 7RW GW0HHA
SA2 7TH GW1DPL
SA2 7TH MW6VNP
SA2 7UH GW8TVX
SA2 7XQ GW0FYO
SA2 8al GW0MGQ
SA2 8BE GW3XIS
SA2 8DP MW0BBL
SA2 8HR GW7NVM
SA2 8JL MW0RBA
SA2 8LL GW6AAG
SA2 8LQ GW6MMM
SA2 8LX GW3KGI
SA2 8PP MW0HGK
SA2 9BW GW3UWS
SA2 9BW GW4ADL
SA2 9DF 2W1TBD
SA2 9DT GW4JGU
SA2 9DZ MW3TKI
SA2 9EQ GW0INN
SA2 9EX MW7RRW
SA2 9GR GW4JUC
SA2 9HN 2W0FLW
SA2 9HN MW6IUP
SA2 9HY GW3TMS
SA2 9LG GW1TFL
SA2 9LG MW3FTB
SA2 9LG MW3FTC
SA2 9LY GW4HAT
SA20 0ED GW6REF
SA20 0LD GW0LLD
SA20 0LS MW6FUN
SA20 0NT GW8HYT
SA20 0UP GW7VJK
SA20 0UP MW3LDY
SA3 1AE GW4KTT
SA3 1BA GW4AZI
SA3 1BR GW6OTD
SA3 1LB GW8VGB
SA3 2EQ MW0FRY
SA3 3AY 2W0FQM
SA3 3AY MW7ACK
SA3 3JR GW0VRL
SA3 3JW GW1AXU
SA3 3JW GW4RKX
SA3 3JW MW6EDR
SA3 3LA GW0BNO
SA3 4EW GW7PMA
SA3 4HF GW4EPF
SA3 4JD GW3LJS
SA3 4JF MW0DCT
SA3 4LZ GW2CGF
SA3 4ND GW0RTR
SA3 4PD GW1WWE
SA3 4PD GW3SRG
SA3 4QW GW0KWA
SA3 4SW GW6WEU
SA3 4UR MW0EYE
SA3 5AN GW4PEX
SA3 5BU MW0JGE
SA3 5BU MW3ORY
SA3 5DL MW0HGM
SA3 5HS GW4TGA
SA3 5HT MW0AJH
SA3 5NL GW4LDP
SA3 5NQ GW4HSH
SA3 5PD GW8HZL
SA3 5PR GW4EVL
SA3 5QJ GW1PFK
SA3 5QL GW0RHC
SA3 5QL GW4CC
SA31 1BS GW6YUC
SA31 1EG MW6HAR
SA31 1JJ GW4RVA
SA31 1LR GW0PUM
SA31 1NN MW0COZ
SA31 1SY GW0CVY
SA31 1SZ 2W0CZP
SA31 1SZ MW6DVQ
SA31 1TS GW4XUE
SA31 2DT 2W1EAN
SA31 2JB GW1ADY
SA31 2JL GW0HGP
SA31 2LH GW0IXK
SA31 2NL GW8PFT
SA31 3BR GW7EXQ
SA32 7EN MW0DJT
SA32 7SA GW7OIK
SA32 8AA GW8THL
SA32 8DQ MW0AMJ
SA32 8DX GW3NXR
sa32 8lj GW0AJU
SA32 8LJ GW0HYL
SA32 8LJ GW1IIZ
SA32 8LP GW0HCK
SA32 8LP GW1SGE
SA32 8PR GW4TGL
SA33 4ES GW0TVX
SA33 4ET MW0CFQ
SA33 4ET MW3GVU
SA33 4LG MW3VWE
SA33 4PD GW2OP
SA33 4PD GW3XJQ
SA33 4PD MW0RLD
SA33 5AH GW0TSE
SA33 5BL MW3VNR
SA33 5DR GW0TXP
SA33 5HA GW3YVN
SA33 5NX GW0GFN
SA33 5NX GW1NED
SA33 5RR MW7HAY
SA33 6BL MW6BHO
SA33 6BN GW4PCJ
SA33 6ES GW0LYK
SA33 6NE MW7BEK
SA33 6XJ 2W0FMQ
SA33 6XJ MW3WJP
SA34 0AF GW0DKG
SA34 0AX 2W0XDT
SA34 0AX MW0XDT
SA34 0BE MW0JOP
SA34 0BE MW7BDT
SA34 0EX MW0VWC
SA34 0HL 2W0KJO
SA34 0HL 2W0RMO
SA34 0HL MW6OKJ
SA34 0HL MW6TJH
SA34 0QX GW1UHF
SA35 0AT GW0WIB
SA37 0EH GW3KAX
SA37 0EY 2W0TDD
SA37 0EY MW6NIT
SA37 0HZ 2W0BLA
SA37 0HZ MW3JTJ
SA38 9AS MW3VJN
SA38 9BA GW7IPS
SA38 9BE 2W0LLT
SA38 9BE MW3OLT
SA38 9EH 2W0HPL
SA38 9EH MW6MSN
SA38 9JS GW0JCT
SA38 9QB GW1CIY
SA38 9RL GW3WCA
SA39 9AZ GW1UOV
SA39 9BY GW0MWN
SA39 9DH GW4VPX
SA4 0UG 2W0HYW
SA4 0UG MW6WYH
SA4 3AY MW3UZH
SA4 3BG GW4SCK
SA4 3GY GW7LDP
SA4 3HB MW1AAH
SA4 3LL MW3UYX
SA4 3PE MW6GIU
SA4 3PE MW7WYN
SA4 3PU GW4UCK
SA4 3PU GW7MMH
SA4 3SE GW4EGS
SA4 4AD 2W0HUL
SA4 4AD MW3CBS
SA4 4BJ GW7PQS
SA4 4DT 2W0KPN
SA4 4DT MW6KPN
SA4 4EB GW4HZH
SA4 4GE GW3JBZ
SA4 4GW GW1WTZ
SA4 4GZ GW0DIQ
SA4 4GZ GW0DRI
SA4 4GZ GW0NHE
SA4 4HE MW7LOI
SA4 4PH GW7ODP
SA4 4PN MW3UZP
SA4 4PR GW4VWY
SA4 4QE GW1FBU
SA4 4UX MW0LMW
SA4 6PH GW1GFO
SA4 6SW MW6BOW
SA4 6UH GW7SDE
SA4 8EE GW7AFC
SA4 8HU GW3XYW
SA4 8HX GW0VEW
SA4 8JF GW1AIB
SA4 8QF GW6VBR
SA4 8RD GW0RNK
SA4 9AB MW0EAT
SA4 9ED GW6TYO
SA4 9HF MW7DRT
SA4 9WH GW3OQK
SA40 9RD 2W0DNV
SA40 9RD MW6CTS
SA40 9SR 2W0BFV
SA40 9SR MW3WFH
SA40 9SU MW6CTD
SA40 9SU MW6TDC
SA40 9TY 2W0NRA
SA40 9TY MW3JVH
SA40 9UU GW4OBA
SA40 9YB MW0ATI
SA41 3QG MW3SFP
SA41 3QJ MW0MJB
SA41 3RR GW0TMV
SA41 3UF 2W0NRE
SA42 0PL GW4UEP
SA42 0XF MW0ARD
SA42 0XS GW4OQB
SA43 1AF GW4OUU
SA43 1DW GW6LHF
SA43 1FB GW4SJO
SA43 1FB GW4TVE
SA43 1JU MW0LEA
SA43 1LT GW6DGU
SA43 1PG MW6BTC
SA43 1PX GW8JWP
SA43 1RF GW7WCR
SA43 1SN MW1BDV
SA43 2AE GW6RWJ
SA43 2AH GW7GWO
SA43 2AQ GW0JLX
SA43 2BD GW3SIK
SA43 2BZ GW0PNI
SA43 2DA GW6AMK
SA43 2HR GW0CMI
SA43 2HR GW4HGJ
SA43 2HR GW4SZV
SA43 2NY GW3EJR
SA43 2PE GW0MBN
SA43 2RT MW0WRP
SA43 3DS MW7BSK
SA43 3EF MW0CRI
SA44 4NE GW3UEP
SA44 4RJ GW0PDB
SA44 4SJ GW0HYH
SA44 4SJ GW8JDB
SA44 4SJ MW3URG
SA44 4ST MW3TUB
SA44 4TU 2W0NAY
SA44 4TU GW4XUG
SA44 4TU MW0VET
SA44 4TU MW6BZX
SA44 5AT MW0DXX
SA44 5EP GW7BBY
SA44 5HE GW1VMA
SA44 5JA MW0GUV
SA44 5NZ GW4JPJ
SA44 5PN GW3WXA
SA44 5XJ 2W0LVE
SA44 5XJ MW6ICM
SA44 5YB 2W0ZZU
SA44 5YB GW0EWY
SA44 5YB MW3ZZU
SA44 5YG GW8PSJ
SA44 6BS 2W0WCP
SA44 6BS MW0WCP
SA44 6BS MW7EJC
SA44 6BT GW3RYE
SA44 6EA GW4TQD
SA44 6ES GW4JKK
SA44 6EY MW6FMV
SA44 6JE GW0POZ
SA44 6NP GW3KDB
SA44 6NQ MW0WEE
SA44 6NQ MW7ROY
SA44 6PR 2W0LLJ
SA44 6PR 2W0VLO
SA44 6PR MW0VLO
SA44 6PR MW6VLO
SA44 6PR MW6XJL
SA44 6QY MW0AIZ
SA45 9QR GW0DDL
SA45 9RJ GW7SXU
SA45 9RL GW3WVV
SA45 9RP GW7FQP
SA45 9RP GW3ZJS
SA45 9SY MW3JQC
SA45 9TH MW0ALG
SA46 0ED GW3LHK
SA46 0JS GW4XSX
sa460lp MW7BCM
SA46tr 2W0GYB
SA46tr MW6GYB
SA47 0QN GW0HXS
SA470RT GW1SMG
SA48 7LG GW0ADS
SA48 7LW GW4LJW
SA48 7PG GW6AFQ
SA48 7PJ GW0GEI
SA48 7PX 2W0JYC
SA48 7PX MW0HNM
SA48 7PX MW0JYC
SA48 7SB GW8GQE
SA48 7SG GW0ARA
SA48 8DT GW3KCQ
SA48 8LH GW0PNC
SA48 8NH MW3SNH
SA48 8RE MW7FMD
SA48 8RP GW4VNK
SA5 4BR 2W0FDC
SA5 4BR MW6NMP
SA5 4NJ GW8RDI
SA5 4RE 2W0XLT
SA5 4RE MW6BMM
SA5 5NX MW3UYJ
SA5 7AH 2W0YVY
SA5 7BP MW0CNB
SA5 7BR MW0CNA
SA5 7DH GW0BNN
SA5 7DW GW4MII
SA5 7EG 2W0BTE
SA5 7ER MW7VHT
SA5 7JJ MW6DGW
SA5 7PU GW0FZY
SA5 8AU GW7TZI
SA5 8AW MW6GIV
SA5 8BT MW6CEX
SA5 8BU GW1GAU
SA5 8DU MW0CND
SA5 8NX 2W0HDP
SA5 8NX MW6HDP
SA5 9BS GW4URB
SA5 9DG GW0AZW
SA5 9EL 2W0CDJ
SA5 9EL MW0HMV
SA5 9EL MW3CDJ
SA5 9HF GW6KWM
SA5 9HN GW4MTD
SA5 9NH GW0NCU
SA5 9NY MW6JUQ
SA5 9PE GW1LHT
SA5 9PG GW8NXK
SA6 5AD MW1FGB
SA6 5AE MW3CNL
SA6 5AX GW0COU
SA6 5DX 2W1DGM
SA6 5DX GW7MMG
SA6 5HF GW4SPL
SA6 5HR MW0OUC
SA6 5HR MW0SDD
SA6 5HR MW0SGX
SA6 5HR MW3OUC
SA6 5HR MW3SGX
SA6 5JF GW7OJT
SA6 5LA 2W1EIN
SA6 5RN GW0DXZ
SA6 5TB MW3UEU
SA6 6AP GW7PBP
SA6 6AU GW4TFX
SA6 6BA GW0UIZ
SA6 6BJ MW3ZKW
SA6 6BY 2W1CNN
SA6 6BY 2W1EYZ
SA6 6BY GW1NWF
SA6 6BY GW6ZUS
SA6 6ER GW4ZQY
SA6 6LL MW6RYD
SA6 6LU GW1DTA
SA6 6NW 2W0AUC
SA6 6NX GW0NKH
SA6 6NX GW1YFP
SA6 6SW GW0KYY
SA6 6TB 2W1FGR
SA6 6TF GW6YPA
SA6 7AG GW7AOE
SA6 7AG MW7GAR
SA6 7LN 2W1GAC
SA6 7PB GW3NUO
SA6 7PH 2W0JMX
SA6 7PN GW1EOI
SA6 7PQ GW4MVY
SA6 8BD MW7KYL
SA6 8LE 2W0REJ
SA6 8LE MW0WRY
SA6 8LQ MW1DCI
SA6 8PF GW4JDZ
SA6 8PF GW8KSL
SA61 1BW GW0EJE
SA61 1BW GW0GUY
SA61 2HL MW0CAB
SA61 2HL MW1SAS
SA61 2JA GW8KXW
SA61 2RD MW6NPW
SA61 2RE MW3EPJ
SA61 2UE MW6BGO
SA611FB GW1NYS
SA611HJ GW0UXX
SA62 3BN 2W0VVO
SA62 3BN MW0VVO
SA62 3BN MW3VVO
SA62 3ET MW6VOW
SA62 3HX GW4PCX
SA62 3HY MW0TJS
SA62 3PA GW4HGS
SA62 3SZ GW0UQH
SA62 3TD GW8WRC
SA62 3TR GW7NGU
SA62 3XH GW0SXS
SA62 4HJ MW0GDM
SA62 4LR GW1MVZ
SA62 4LR MW3YBX
SA62 4LS MW6IIU
SA62 4NU GW3YSI
SA62 4NU GW6IZZ
SA62 4PR GW1HEV
SA62 5DY GW0KQV
SA62 5ES MW7NAT
SA62 5NT MW0AMN
SA62 5NT MW0AMQ
SA62 5PL MW7IMO
SA62 5PW MW3IZV
SA62 5RY MW0XDN
SA62 5SP 2W0LXT
SA62 5SS GW0VYF
SA62 5SW MW6HZP
SA62 5SX GW0KZG
SA62 5SX MW0SML
SA62 5UA MW0IJE
SA62 5XD MW6XAE
SA62 5XX MW7DYH
SA62 6EN GW8NAC
SA62 6EU MW3UAA
SA62 6HF MW6COD
SA62 6JZ GW0DDK
SA62 6RA MW7BGD
SA62 6UF GW4HXO
SA626LS MW7EQU
SA63 4UU MW7YZF
SA63 4XD GW8SBO
SA64 0AX MW6WWY
SA64 0AZ MW6JFI
SA64 0DY 2W0EXZ
SA64 0DY MW0MDV
SA64 0DY MW6IVW
SA64 0DZ 2W0MZS
SA64 0DZ MW6IOZ
SA64 0EX 2W0HDB
SA64 0EX MW0HVB
SA64 0EX MW0NPC
SA64 0EX MW6BDS
SA64 0EX MW6HDB
SA64 0LG GW4MM
SA64 0LG GW4VXE
SA64 0LL MW0RLJ
SA65 9DB GW1JKP
SA65 9EE MW6IXD
SA65 9EH 2W0FZC
SA65 9EH MW0JWW
SA65 9EH MW7FEC
SA65 9LN 2W0FFI
SA65 9LN MW6AGS
SA65 9PB GW0CXK
SA65 9PT GW8UQC
SA65 9PT MW6URI
SA65 9QJ MW0BEL
SA65 9RL GW4UEJ
SA65 9UX GW8YLK
SA66 7JP GW1ALV
SA66 7NR GW7JUB
SA66 7PX 2W0BMM
SA66 7QL GW8GVI
SA66 7QS GW3RKV
SA66 7UX 2W0JMK
SA66 7XT GW7NYP
SA67 7DY MW0HUU
SA67 7EE GW1MNU
SA67 7EZ GW8PKV
SA67 7PG 2W0UPR
SA67 7PG MW7TRY
SA67 8EH 2W0LDZ
SA67 8EH MW0DX
SA67 8EH MW0TDZ
SA67 8EH MW6LDZ
SA67 8EH GW4MIP
SA67 8NT 2W0MNA
SA67 8NT MW6IMU
SA67 8SP GW3LEW
SA67 8SP MW3JCG
SA67 8ST GW0CCO
SA67 8ST GW3KZO
SA68 0RG 2W0HRG
SA68 0RG MW6BEG
SA68 0SY GW4WMD
SA68 0XA MW6HED
SA68 0XS MW6TDE
SA7 9HS GW4LKS
SA7 9Jb MW0TCJ
SA7 9JR GW4UVN
SA7 9JT MW6IVK
SA7 9LD 2W0SRD
SA7 9PR 2W0NED
SA7 9PR MW0KJN
SA7 9QH GW1PAV
SA7 9QU MW6HLU
SA7 9QX GW4JGW
SA7 9SF 2W0BJR
SA7 9SF MW3OSI
SA7 9ST GW4XES
SA7 9ST GW4ZBU
SA7 9SX GW8DUP
SA70 7DP GW0JRF
SA70 7EB MW0ECY
SA70 7EB MW3NTE
SA70 7JT GW8PNE
SA70 8DB GW7GKN
SA70 8EB MW6GQD
SA70 8LT GW4NOO
Sa708en MW6HNN
SA71 4EP MW0CHI
SA71 4EP MW1BXX
SA71 4JL MW6FGJ
SA71 4PX MW6DDR
SA71 4QL 2W0BGI
SA71 4QL MW3NUP
SA71 4TR MW0BRO
SA71 5AT MW6OVV
SA71 5BD GW7FBV
SA71 5HJ GW6TEO
SA71 5HY GW4OZU
SA71 5JH GW4VTG
SA71 5JQ MW6RGK
SA71 5QQ MW6HLQ
SA71 5QX 2W0MCT
SA71 5QX MW3KNE
SA72 6DU GW0WGN
SA72 6EZ GW4RGI
SA72 6FB GW4OXL
SA72 6HL GW7VMT
SA72 6JP GW4TGT
SA72 6JR MW6OVF
SA72 6NE MW0IPL
SA72 6NE MW1BLE
SA72 6NE MW6IYW
SA72 6NE MW6OYO
SA72 6NW MW0AIE
SA72 6QN GW4REI
SA72 6RH MW7WDP
SA72 6RN GW6NLO
SA72 6SH GW4VRO
SA72 6TP GW4YVN
SA726QL 2W0ENJ
SA726QL MW6KFQ
SA73 1AN 2W0BOX
SA73 1AN MW0GOV
SA73 1AN MW3SQA
SA73 1BD GW1PND
SA73 1EA GW0SYG
SA73 1EA GW3CR
SA73 1EA MW0IFK
SA73 1HD GW0LIK
SA73 1HD GW4ZYV
SA73 1HR MW0BYS
SA73 1HR MW3ESE
SA73 1HR MW3ESF
SA73 1JT GW0RKD
SA73 1LG GW4ERB
SA73 1NZ GW6CJJ
SA73 1QR 2W0GHV
SA73 1RA MW0JDY
SA73 1RA MW6XRO
SA73 1RE 2W0JPF
SA73 1RE MW3JPF
SA73 1SA GW0VND
SA73 1TB GW7HGU
SA73 1TB MW1DFQ
SA73 1TL MW0HWU
SA73 1TL MW0IBZ
SA73 1TR MW0COB
SA73 1TR MW6FGQ
SA73 2AU 2W0OZO
SA73 2AU MW6OZO
SA73 2BW 2W0EGL
SA73 2BW MW6DFX
SA73 2DS GW4OEJ
SA73 2ED GW4OKF
SA73 2EN MW0SGD
SA73 2NU GW0TLJ
SA73 2NU MW0BBU
SA73 3HN GW4ODN
SA73 3HS MW6WPB
SA73 3LW GW0LVH
SA73 3RU MW7HJE
SA73 3TF MW3VFB
SA73 3UE GW0VEM
SA8 3AU 2W0EZS
SA8 3DT GW4KCQ
SA8 3EP GW8SBN
SA8 3EP GW8TBG
SA8 3ER MW6JDY
SA8 4AA GW1RZE
SA8 4AH MW3ZWS
SA8 4AL GW6VBN
SA8 4BN MW7DQV
SA8 4DB GW6UGC
SA8 4DL GW4KYT
SA8 4ER MW7KLC
SA8 4JD MW6LZV
SA8 4NB 2W0BBO
SA8 4NB 2W0BBT
SA8 4NB MW0GCS
SA8 4QL GW3UCJ
SA8 4QN MW3BMF
SA8 4QT GW4ONI
SA9 1AT GW4YWM
SA9 1BP GW4GLU
SA9 1DR MW7CRP
SA9 1SP MW3EPK
SA9 1SX MW6VPL
SA9 1TS MW7DHG
SA9 1XF GW7MLN
SA9 2FT MW3SKW
SA9 2FY GW6PXW
SA9 2LP 2W1KWK
SA9 2LP MW7BHZ
SA9 2NZ MW3UIN
SA9 2QE GW1PKM
SA9 2XQ GW3YQH
SE1 0QB 2E0JWG
SE1 0QB G3WKE
SE1 0QB M0JXG
SE1 0QB M6JHG
SE1 1EL 2E1HFH
SE1 1EL M3HFH
SE1 2XN M6YLD
SE1 3DW M0GDC
SE1 3LD M0ATD
SE1 4DT M3RZL
SE1 5NU M6EOM
SE1 7DX 2E0HFA
SE1 7DX M0JKN
SE1 7DX M6HFA
SE1 7HB G4BWB
SE1 8DU G1UFX
SE10 0DW M6TYU
SE10 0JT M7RED
SE10 8AX 2E1DBP
SE10 9UZ G0DRK
SE11 5UL G3RGM
SE12 0EG G1IYO
SE12 0RN M7CJM
SE12 0UJ 2E0CIT
SE12 0UJ M6BDD
SE12 8AE M0HUS
SE12 8AG 2E0FSN
SE12 8AG M0SBR
SE12 8AG M6HGH
SE12 8JN G7GLW
SE12 9AB G0LRM
SE12 9BG G3GDB
SE12 9EX G8IBO
SE12 9HX M7EHB
SE12 9HY 2E0NPR
SE12 9HY M0NPB
SE12 9HY M6OPF
SE12 9LX G4XUI
SE12 9RS G1XFR
SE13 5AN M7FTL
SE13 5AW G8LHI
SE13 5FL G5JBQ
SE13 5QW G4JOI
SE13 6QP M0BYM
SE13 6QW M3WUO
SE13 6RT M3WRQ
SE13 7DE M6NYX
SE13 7PA M6JKX
SE13 7PL M3RJK
SE13 7SP M6LTU
SE13 7SR 2E0JJP
SE13 7SR M0JJP
SE14 5DN M6HJX
SE14 5NJ 2E1GKP
SE14 5XF M7YOL
SE14 6DU M6URX
SE15 1AF M7SRT
SE15 1DU 2E0UAK
SE15 1DU M6YAV
SE15 2JA 2E0HHF
SE15 2TX M6FMY
SE15 3EG G3VBU
SE15 4AN G3XBH
SE15 4DZ G1BYO
SE15 4JX G7KYD
SE15 4PY G3HMO
SE16 2AG G1NGI
SE16 2AP M3JYG
SE16 2NA 2E0YZI
SE16 2NA G5YZI
SE16 2NA M7BKH
SE16 3DQ M0IVE
SE16 4BY M3IYG
SE16 4TT G7NIL
SE16 5XN G0BUD
SE16 6PU 2E0HOU
SE16 6PU M0OSY
SE16 6PU M6EYD
SE16 7AU M6JCX
SE16 7EG 2E0JUD
SE16 7EG M7KTL
SE16 7SZ M7GMJ
SE17 3HR M7CFF
SE18 1DP M7CUZ
SE18 1LN M3HIN
SE18 1SL M6HJH
SE18 1SL M6MVZ
SE18 2PH M0JRJ
SE18 2RT G1CJK
SE18 3HF G8GJO
SE18 4PS G8WBU
SE18 5BY G0EPR
SE18 5JL 2E0WHF
SE18 5JL M0ABC
SE18 5JL M6WHF
SE18 5PU M0TDD
SE18 5QY 2E0FZS
SE18 5QY M7SAC
SE18 5TY M7ELU
SE18 6PA 2E0DKF
SE18 6PA M6ZWT
SE19 2PY G1FBI
SE19 2SA G4EBI
SE19 2TE M7ABM
SE19 3AD G3TAO

SE19 3PZ M0ICL
SE2 0AJ M1FFS
SE2 0DN G4MLY
SE2 0DW G4WNF
SE2 0PD G4JGQ
SE2 0PP G0EFA
SE2 0QG 2E0HNH
SE2 0QG M7ESI
SE2 0QX G3SPJ
SE2 0UY M6IDM
SE2 0XN G6SCB
SE2 9HL G0ATB
SE2 9QS 2E0POU
SE20 7BW G1XVW
SE20 7JG G3VZH
SE20 8ET 2E0GUU
SE20 8ET M0JOI
SE20 8ET M7BJT
SE20 8JY G6CVP
SE20 8NU G4JHW
SE20 8SF M0HWW
SE21 8JY G0PAR
SE22 0AX 2E0ESC
SE22 0AX M6MBB
SE22 0SB G0FAH
SE22 8RH M0NTN
SE22 9AH M3TMM
SE22 9EX M6KUU
SE22 9JH G6ICC
SE23 1BW G3SXA
SE23 1HG G4PRQ
SE23 1PY M7MTE
SE23 1RH G6BJR
SE23 1SL G1MKS
SE23 2AG M7JRK
SE23 2QP G0DIA
SE23 2RU G3TPO
SE23 3HT 2E0HGO
SE23 3HT M0PGN
SE23 3HT M6HGO
SE23 3SJ 2E0EUI
SE23 3SJ M6EUI
SE25 4NG 2E0EBK
SE25 4NG M6EUH
SE25 5HB G7PWJ
SE25 6HY G7TQE
SE25 6SY G3WRR
SE26 4AB M7FEU
SE26 4BL M7AVW
SE26 5DP G0RTZ
SE26 5JD M0IWJ
SE26 5RJ 2E0CIG
SE26 5RJ M0JAD
SE26 5RJ M6BBG
SE26 6BP M3YRS
SE26 6JQ M3XBN
SE27 0LG M6AMN
SE27 9NA 2E0DKW
SE27 9NA M6DKW
SE28 0DS M7NEJ
SE28 0NJ M3YSY
SE28 0PB 2E0OPO
SE28 0PB M6DKY
SE28 8HE M7GOU
SE28 8NF 2E0ATY
SE28 8NF M0MYC
SE28 8PG M1BPN
SE28 8PU G1NYI
SE3 0EZ G0CUA
SE3 0QG 2E0MUS
SE3 0QG M6MUS
SE3 7BG G4FUG
SE3 7BH M0UOG
SE3 7PE 2E0CQS
SE3 7PE M0WXU
SE3 7PE M6BZD
SE3 7QS G3GHN
SE3 7QS G4RFC
SE3 7RQ G0ILD
SE3 7TN G4JMM
SE3 8HF G4ZXV
SE3 8LR G6IRE
SE3 8QY M6SLZ
SE3 8RP M6LHN
SE3 8TT G1NTP
SE3 9HL M0BGR
SE3 9HP G0TYW
SE3 9LJ M0RAC
SE3 9QN G0BSH
SE3 9QP G1KLW
SE4 1AS 2E0CMQ
SE4 1AS M6BCQ
SE4 2AX M0IAG
SE4 2DN M6BVK
SE4 2JQ G0HBW
SE4 2NL 2E0TXJ
SE4 2NL G6DJH
SE4 2NL M6TXJ
SE5 8BS M7MEZ
SE5 8DQ 2E0EVV
SE5 8DQ M6ONJ
SE5 8QQ G4WUQ
SE6 1EZ M3WNM
SE6 1JF 2E0LSW
SE6 1NQ G8WZO
SE6 1UR 2E0LBZ
SE6 1UR M6BPD
SE6 2RN G7KTL
SE6 3LP M6WRB
SE6 3QR M7CNN
SE6 3SQ 2E0FVT
SE6 3SQ M7GRS
SE6 3TG M3VNI
SE6 3TW G6ZWM
SE6 4RN G7PKY
SE6 4UW G0TCA
SE7 7AD M3YKI
SE7 7EU M1JWS
SE7 8ET 2E0LOX
SE7 8ET M0LOX
SE7 8ET M6LJV
SE7 8PJ 2E0KPU
SE7 8PJ M0KPU
SE7 8SH G1FAD
se77be M7CUQ
SE8 3AS M1MAL
SE8 3EX M6GUS
SE8 5AJ G0VXY
SE9 1LH M7CXU
SE9 1PP M3UJR
SE9 1QH G8XDR
SE9 1QJ 2E0RCV
SE9 1QJ M0MCV
SE9 1QJ M3RCV
SE9 1SA 2E1GUC
SE9 1SA G7SCU
SE9 1SE G0DCI
SE9 2EX G8IPY
SE9 2LP G6WCX
SE9 2NZ G0FDP
SE9 2SU M7RAD
SE9 3BN M6FWQ
SE9 3BW G3XCJ
SE9 3EB G0OWV
SE9 3EP G4VFH
SE9 3HL G0OKX
SE9 3JL M6VSB
SE9 3JX G3IGZ
SE9 3PX 2E0ECY
SE9 3PX M0IKC
SE9 3XD G0ULL
SE9 4ET M3JUX
SE9 4JL G0JAJ
SE9 4LF G0LYG
SE9 5RF 2E0EGM
SE9 5RF M6FDG
SE9 6EX 2E0HTV
SE9 6EX 2E0UKK
SE9 6EX 2E0UKT
SE9 6EX M6UKX
SE9 6EX M7UKX
SE9 6QP G8JWX
SE9 6QP M0JHB
SE9 6RA M6ALW
SE91PB M7EIK
SE96AL M7EBL
SG1 1JJ M6TZO
SG1 1LS G3MHX
SG1 1RP G6XSZ
SG1 1RR M0GBZ
SG1 1TE G2BKZ
SG1 1TE G3SAD
SG1 1TN M3VRB
SG1 1UJ G7DNV
SG1 2AS M6GKC
SG1 2LH 2E0NCR
SG1 2LH M6NCR
SG1 2NE M3BMU
SG1 2NE M3BYS
SG1 2QL M3RUW
SG1 2RS M0KPB
SG1 3BT G8SZP
SG1 3DF M6BSZ
SG1 3EZ G7CYQ
SG1 3JL G3UFB
SG1 3SL M6JJM
SG1 3XY 2E0LSC
SG1 3XY M0LSZ
SG1 3XY M7LSA
SG1 4AY G8WWI
SG1 4DQ G6JUE
SG1 4EP G4SPV
SG1 4HD G8IUG
SG1 4JP M6JSV
SG1 4LF G3INU
SG1 4NQ M0NEV
SG1 4NU G1FWS
SG1 4PH M6ZCR
SG1 4PJ M3IJV
SG1 4PL M3PCP
SG1 4PW 2E0PYM
SG1 4PW 2E0YUD
SG1 4RN G6ANR
SG1 4SD G4DDX
SG1 4TB M1EMB
SG1 5EH G6UWX
SG1 5HF G7DRG
SG1 5JA G1SOY
SG1 5LH M3JMU
SG1 5LH M3NDC
SG1 5NF G4BYE
SG1 5PH M6EJW
SG1 5QP 2E0PDJ
SG1 5QP G8DVB
SG1 5QP M7DVB
SG1 5QS G7HCL
SG1 5QS M0EBG
SG1 5SF G4IKY
SG1 5SF G8FMA
SG1 5TB M3MBR
SG1 5TT G0RAM
SG10 6AX G4XVO
SG10 6BD M6TRS
SG10 6BN G0KTW
SG10 6JG G3OMD
SG11 1HT G0KGR
SG11 1JH G4VMR
SG11 1JH G4VSL
SG11 1LH M6FMF
SG11 1QF M6AHH
SG11 1SY G0DCW
SG11 1TG G4YOS
SG11 2DN G3OTR
SG11 2EH M0SHF
SG12 0DN 2E0ECC
SG12 0DN G5EDW
SG12 0DN M7EDW
SG12 0HP G7VRY
SG12 0JQ G3OJG
SG12 0NL G4YFC
SG12 0SY G4PHV
SG12 7DT M6CSU
SG12 7EA M1BYG
SG12 7LT G8WTB
SG12 7PD G0OJY
SG12 8AZ G4VXD
SG12 9JN G3TIK
SG12 9LS G4MAS
SG12 9PG G8NDR
SG12 9QL G6WYF
SG13 7DP M6MOX
SG13 7EJ G8OIY
SG13 7JD G8OVZ
SG13 7JF G6DVO
SG13 7PG G4AEZ
SG13 7PG G6VAL
SG13 7QT M3PJJ
SG13 7QU G3SZF
SG13 7RB G4UDZ
SG13 7RN G0AHB
SG13 8AD G8OPC
SG13 8BA 2E1CAQ
SG13 8BN M6LKA
SG13 8HR M6GBQ
SG13 8RB G1IMS
SG14 1NE 2E0JVT
SG14 1NE M6NTT
SG14 2BZ G4EEZ
SG14 2EU G3IOJ
SG14 2HH G0NSH
SG14 2QG M6HQD
SG14 3AF 2E0RYX
SG14 3AF M6LOG
SG14 3DL G1BLQ
SG14 3JU G0OSK
SG14 3SF 2E0LFT
SG14 3SF M0WWV
SG14 3SF M6WKZ
SG14 3SN G7VLD
SG14 3TD 2E0HDZ
SG14 3TD M0XHD
SG14 3TD M7HDR
SG14 3TF G8EKG
SG15 6RP G0TUX
SG15 6RX 2E0PLV
SG15 6RX M0PLV
SG15 6RX M6PLV
SG15 6SW M7DNO
SG15 6SZ M3XZE
SG15 6TS G4XBD
SG15 6UF G6UMX
SG15 6UY G3WKA
SG16 6AA M0IDE
SG16 6AA M7PTP
SG16 6BN G3ORG
SG16 6EY M3HYQ
SG16 6FD M6GHW
SG16 6JS G1SJO
SG16 6JS M6HHR
SG166JR 2E0ZKT
SG166JR M3UER
SG17 5AA G3NII
SG17 5AD G6RHL
SG17 5AE M0CPE
SG17 5AU G8UFX
SG17 5AU G8XZQ
SG17 5BN G3JNB
SG17 5EA G0GGQ
SG17 5GZ G0RLL
SG17 5GZ G1RIV
SG17 5HB G6MIS
SG17 5HB G6VBE
SG17 5HW G1CIT
SG17 5LB M6JNV
SG17 5LU G4JBD
SG17 5NH G0PQO
SG17 5NH G7CUY
SG17 5NY M1AIN
SG17 5RA 2E0HPO
SG17 5RA M3HPO
SG17 5RZ G7MYN
SG17 5RZ M0MYN
SG17 5UP G6JLU
SG18 0DA M0JXM
SG18 0DG G7ABF
SG18 0HL G4GYP
SG18 0JJ G7OZU
SG18 0JW M3PKM
SG18 0NE G6KPA
SG18 0NL M3FAY
SG18 0NX G8MCY
SG18 0PH M3PIH
SG18 0PH M6EXE
SG18 0PS G4NNB
SG18 0PX M3TXS
SG18 8DX G0PHY
SG18 8EE M0WNV
SG18 8EE M1GXL
SG18 8FS M6BFL
SG18 8GG 2E1GBM
SG18 8JR M0SLP
SG18 8JU G6ZEM
SG18 8LX M0PLT
SG18 8PB G3XIZ
SG18 8PF M3NHP
SG18 8QS G6XDK
SG18 8RF M6MII
SG18 8SR G1FYF
SG18 9AH G8PQN
SG18 9BX G4FGJ
SG18 9DD 2E0DSW
SG18 9DD M0MRL
SG18 9DD M6DSW
SG18 9DF G4YRF
SG18 9DU M7CYE
SG18 9DU M7FDD
SG18 9JL G1JVY
SG18 9JL M3AWS
SG18 9NE M1EPK
SG18 9NR G7NKS
SG18 9NY G0ECZ
SG18 9PQ 2E0RFI
SG18 9PQ M6UTC
SG18 9PS G7LUR
SG18 9PT G0ETA
SG18 9QP G1GSN
SG18 9QP G4XIM
SG18 9RS G3XCK
SG19 1AU 2E1CWJ
SG19 1AU G0TMJ
SG19 1AU G8GHR
SG19 1AX G6REA
SG19 1BJ G4SKU
SG19 1DT G4CEP
SG19 1DU 2E0HVQ
SG19 1DU M7TOG
SG19 1ED M6BKT
SG19 1EU G4MEO
SG19 1HF G7SVE
SG19 1HP 2E1DEP
SG19 1HP 2E1FDU
SG19 1HR 2E0OBD
SG19 1HR M0OBD
SG19 1JJ 2E0KMD
SG19 1JJ M6KMD
SG19 1JL 2E1CDZ
SG19 1RH 2E0KKJ
SG19 1RH M6MKJ
SG19 1TT M0WPY
SG19 1TU M7BNO
SG19 1TX G3KDD
SG19 2JU 2E1EFQ
SG19 2JY G3ZWM
SG19 2LB G0IKZ
SG19 2NJ M6JZF
SG19 2NY M7LAW
SG19 2PB M0YLG
SG19 2PS G7JGQ
SG19 2PS M3JGQ
SG19 2PY G8ZYT
SG19 2QG G4MKR
SG19 2QQ G1YRM
SG19 2QQ M0PLY
SG19 2QQ G3OXG
SG19 2QR G3WSD
SG19 2SD M7WZY
SG19 2UR G0OQI
SG19 3AN G7SOZ
SG19 3AP G6CTP
SG19 3AQ M6RPY
SG19 3BA M1FFF
SG19 3BT 2E1WEB
SG19 3DX G4MKP
SG19 3LG G0FOT
SG193NY G1ZPU
SG193NY M0ZPU
SG2 0DE M6SDN
SG2 0EB M3IVA
SG2 0EJ M6EGT
SG2 0EQ G3JLZ
SG2 0HD G7HCB
SG2 0JH G7PNE
SG2 0JJ G0GXU
SG2 0JN 2E0ZBZ
SG2 0JN M0ZBZ
SG2 0JN M3ZBZ
SG2 0NZ G6BYK
SG2 0QE M6RIZ
SG2 0QG G0BXH
SG2 7BN M0MPI
SG2 7DR M3YSL
SG2 7DS G0KUQ
SG2 7DT G4XUW
SG2 7DW M3XTK
SG2 7EQ M6SNH
SG2 7JL 2E0OSX
SG2 7JL M0OSX
SG2 7JR M3DLU
SG2 7QS G4CZZ
SG2 7ST G7JSS
SG2 8AL 2E0HVL
SG2 8AL M6HVL
SG2 8EZ M0ABH
SG2 8GD 2E0SNC
SG2 8GD M0KNM
SG2 8GD M3YCM
SG2 8JA M6AJM
SG2 8JD G4KFL
SG2 8NA M1DGS
SG2 8PX 2E1IID
SG2 8PY M6IVR
SG2 8QZ M3KCG
SG2 8RJ G6UXE
SG2 8RY G8ZFI
SG2 8UG G0UVN
SG2 8UY M1SAC
SG2 9DX G8UEY
SG2 9DX M0CUH
SG2 9DX M3NEA
SG2 9ER M3XTP
SG2 9LZ M0ISI
SG2 9LZ M6FKF
SG2 9NP G6YMA
SG2 9PH 2E0LVG
SG2 9QD M6YTF
SG2 9RN G0EVZ
SG2 9TP G4NEA
SG2 9TT 2E0LKS
SG2 9TT M0LDC
SG2 9TT M6LKS
SG20PJ M7JOL
SG3 6BQ G1HBC
SG3 6DF G0TAG
SG3 6DN M0AZZ
SG3 6JE 2E0TJD
SG3 6JE G6DXY
SG3 6JE M7TJD
SG3 6JW G8BWK
SG3 6JW M7BAJ
SG3 6NU G0FRM
SG3 6SB 2E0TKD
SG3 6SB M0MIZ
SG4 0AN G3LQO
SG4 0DP G4OXD
SG4 0DP G8PUR
SG4 0DP M6JUK
SG4 0DR M0MVE
SG4 0JH 2E0RCT
SG4 0JH M0LGC
SG4 0JH M0ROC
SG4 0JH M3RCT
SG4 0LG G4KDW
SG4 0QP G6WEH
SG4 7AL M7WJR
SG4 7AS G4SWH
SG4 7DR M0ARH
SG4 7NG G6SHZ
SG4 7NT G2YT
SG4 7PD G6YIQ
SG4 7PD M5AIO
SG4 7QZ M0RRC
SG4 7RA G4GXM
SG4 7RQ G0KJN
SG4 7SA G3ZLP
SG4 8AF 2E0DMI
SG4 8AF M6BGK
SG4 8AJ M3JYZ
SG4 8AJ M6KXO
SG4 8BE 2E0CBF
SG4 8EQ G4HPA
SG4 8EQ G7HMV
SG4 8EQ M0HAW
SG4 8EQ M0KSG
SG4 8EQ M0MHU
SG4 8EQ M0YSG
SG4 8EQ M6HMV
SG4 8EQ M6HOT
SG4 8EQ M6MZH
SG4 8EQ M6TWY
SG4 8NP G3PZE
SG4 8NT G4NGL
SG4 8PG G8WKE
SG4 8PR G0CRK
SG4 8PU M6UJK
SG4 8QD M6STY
SG4 8RA M6IAC
SG4 8RR M6CSN
SG4 8SD G8TRR
SG4 8TJ M3VQQ
SG4 9AN G3WRJ
SG4 9EP G4CTM
SG4 9JD 2E0YJW
SG4 9JD M0ONB
SG4 9JD M0YJW
SG4 9JD M3YJW
SG4 9LT M0JMC
SG4 9NP G4KNT
SG4 9NW G3FVR
SG4 9PJ G0GXU
SG4 9SG 2E0DGL
SG4 9SG M6ENV
SG4 9TB G7URR
SG4 9TL M6UTI
SG5 1EH G7NKI
SG5 1PA G6CFC
SG5 1PN G8BFH
SG5 1PT G6LSC
SG5 1QE M6RXS
SG5 2BE M3HPN
SG5 2HP G6XCK
SG5 2JH G4VXU
SG5 2QZ G7LXH
SG5 3DR G0EUV
SG5 3DR G0UEW
SG5 3LS G4PSO
SG5 3LS G7PQP
SG5 3LS M3GGV
SG5 3NS G4CML
SG5 3RX G8FMT
SG5 3RX M0FMT
SG5 3SL G4YDH
SG5 3UP G3UWM
SG5 4EH 2E0JKT
SG5 4EH M6TEM
SG5 4EL G0UVX
SG5 4FY G0VVZ
SG5 4HH M1CAH
SG5 4JA G4THF
SG5 4JE 2E0BYJ
SG5 4JX 2E0MIH
SG5 4JX M6MHU
SG5 4RU G4ALR
SG5 4RU G4KQD
SG5 4RZ M7WVC
sg53xy G6CRC
SG6 1PH G8EDS
SG6 1RF M6FKE
SG6 1RG M6KAZ
SG6 1SW G8RPA
SG6 1UE M1WHO
SG6 1UE M3EMS
SG6 2DE 2E0ASU
SG6 2EU G6JBQ
SG6 2JA G8DKK
SG6 2LU G0EVD
SG6 2NP 2E0EPT
SG6 2NZ 2E0BYG
SG6 2NZ M6EGO
SG6 2SP 2E0XEV
SG6 2SP M6XEV
SG6 2TY 2E0XMC
SG6 2TY M6XMC
SG6 3EA M7BZO
SG6 3PY M0ARQ
SG6 4BD 2E0PJN
SG6 4BD M3PJN
SG6 4HG G1UFH
SG6 4HY G1BYJ
SG6 4JZ G0AMX
SG6 4LA 2E1RAF
SG6 4LA G0TAK
SG6 4LA G3ELV
SG6 4LA M0KAC
SG6 4LF G1OKV
SG6 4LP M7DZM
SG6 4NQ 2E0ISS
SG6 4PL G0MJP
SG6 4QF M0NGK
SG6 4RZ G0PJC
SG6 4TG 2E0PJC
SG6 4TG M6PJI
SG6 4TZ M6XKN
SG7 5AA M7YZY
SG7 5HL G0NZJ
SG7 5LJ G0RRV
SG7 5RN M6CIP
SG7 6DA M6LCC
SG7 6DG G8ABX
SG7 6HJ G6WKZ
SG7 6LH 2E0STP
SG7 6LT G4HIE
SG7 6LT G6KBS
SG7 6NU 2E0FHC
SG7 6NU M0IXM
SG7 6NU M6OLQ
SG7 6RX M6ERC
SG7 6RZ G1XXR
SG7 6RZ G8NHO
SG7 6TW G6MAY
SG8 0DJ G0CMB
SG8 0DJ G1HND
SG8 0JU G3WAB
SG8 0NE G3WMD
SG8 0NP M7NBA
SG8 0PF G6PCE
SG8 0RB G6VIK
SG8 5BB G4TEU
SG8 5BP G4MES
SG8 5BW G4HPE
SG8 5BW G5BBC
SG8 5BW G6XQY
SG8 5JX 2E0NKM
SG8 5JX M6NKM
SG8 5LR G0DCU
SG8 5NG G1HDR
SG8 5PT G0LZD
SG8 5QY 2E0EQZ
SG8 5QY M0IPO
SG8 5QY M6KTV
SG8 5SN 2E0WDN
SG8 6BZ G0FMT
SG8 6DD G1VRA
SG8 6EB M5SKY
SG8 6HD 2E0IGT
SG8 6HD M6VET
SG8 6JD G8DJU
SG8 6JX G8KDD
SG8 6UH M6FTY
SG8 7AD M5AGR
SG8 7DJ G4NIX
SG8 7ED G1GIJ
SG8 7ER M6CIQ
SG8 7ES G4NIY
SG8 7ES G6XEF
SG8 7EU G0TJL
SG8 7PZ M6FVS
SG8 7XU G0AUJ
SG8 8LT G0KFT
SG8 8RP G4FOH
SG8 9BN G0SII
SG8 9DQ G8RZL
SG8 9EA G7VEX
SG8 9EA M3TNW
SG8 9EB M7ATJ
SG8 9EU G1HVW
SG8 9HP M0BDW
SG8 9JF G6EDD
SG8 9LE 2E0CIR
SG8 9LJ M3IJS
SG8 9NQ M0XBQ
SG8 9NQ M7HES
SG8 9QL G1HQE
SG8 9QX G0AQI
SG8 9SE G1YZJ
SG8 9UQ G1OKF
SG9 0PJ M7DLQ
SG9 0PW G3GRS
SG9 0PW G4BUO
SG9 0QR M3LHZ
SG9 0SN 2E0LSO
SG9 0SN M0LSO
SG9 0SN M6LSO
SG9 0SU 2E0CIQ
SG9 0SU M3IWR
SG9 9BQ 2E0HPR
SG9 9BQ M1HPR
SG9 9BQ M6HPR
SG9 9DZ G6BWE
SG9 9DZ M1NHR
SG9 9HJ M0DWP
SG9 9RA G1RPO
SG9 9RH G0HBL
SG9 9SE M7JXW
SK1 2QE G0OGP
SK1 3NL M3AHU
SK1 4AB M1ANQ
SK1 4BP 2E0RFU
SK1 4BP M0RFU
SK1 4BP M6BYR
SK1 4DS G0ELX
SK1 4EP M7FWR
SK1 4HX M6WXF
SK10 1FW G1AKA
Sk10 1GE 2E0DOD
Sk10 1GE G4FF
Sk10 1GE M0PAI
Sk10 1GE M3TUL
SK10 1QP G3HUR
SK10 2DA M3JLD
SK10 2DQ M3YFL
SK10 2EL G4XRG
SK10 2JB G7NNU
SK10 2NS G6CGO
SK10 2PF G4JNE
SK10 2PN G1NUS
SK10 2PS 2E0NSR
SK10 2PS 2E1AEQ
SK10 2PS G0DMV
SK10 2PS G0JNJ
SK10 2PS M6NSR
SK10 2RN G8JQG
SK10 2TS M7ENW
SK10 2TU G6AKK
SK10 3BG M0TEZ
SK10 3BR G3JQ
SK10 3BS G1SHI
SK10 3DA G0DRO
SK10 3DJ G0DCO
Sk10 3gl M7AWE
SK10 3GL M7JVS
SK10 3HB 2E0VWT
SK10 3HB M0VWT
SK10 3HB M6XCP
SK10 3HJ M7BHO
SK10 3JA G3CWI
SK10 3JG M7BKG
SK10 3JL G0GWA
SK10 3LN G3LIO
SK10 3LN G8RDJ
SK10 3NQ G6XFU
SK10 3RE G0IKB
SK10 4HZ G0AHJ
SK10 4QW G3UJA
SK10 5AH G6VSQ
SK10 5DT G6ECN
SK10 5HJ G8BPU
SK10 5HT G4XTO
SK10 5JG G1JDF
SK10 5JZ G3SXI
SK10 5LJ G0NYS
SK10 5SD G4GMK
sk102gy M3SVL
SK11 0BU 2E1FLW
SK11 0EP G8GXO
SK11 0LJ G4NXX
SK11 6HA M6RJM
SK11 6RG 2E0BAX
SK11 6RG M0GIA
SK11 7AW M1BYH
SK11 7EN G3VKF
SK11 7ES G0KVU
SK11 7ES G4WOL
SK11 7GE 2E0ZSH
SK11 7GE M3ZSH
SK11 7NJ G1NTR
SK11 7PL G7POV
SK11 7QA M3MMG
SK11 7YG M3HVP
SK11 7YH 2E0RXX
SK11 7YH G4MWS
SK11 7YH M0TXX
SK11 7YH M3ZRY
SK11 8EX G6LFG
SK11 8LL G3RWE
SK11 8LW G6FFU
SK11 8QL G8XQH
SK11 8QP G0DMU
SK11 8RH 2E0EYP
SK11 8RH M0HGY
SK11 8RH M1EYP
SK11 8RH M3EYP
SK11 8RJ G1NPC
SK11 8RS 2E0MUD
SK11 8RS M6FAY
SK11 8RT G1DUT
SK11 8SD G7MSH
SK11 8TH G7OQG
SK11 9AP G7NKV
SK11 9BF G3HFM
SK11 9BW M3JPI
SK11 9BX G8BVF
SK11 9LQ M3FLJ
SK11 9RR 2E1HNB
SK11 9RR M3CBN
SK11 9SW G3VDB
SK12 1AR G6VKL
SK12 1BL G0PJW
SK12 1BU G4ZGP
SK12 1DD G4GEO
SK12 1EF G1AFI
SK12 1HA G3SHF
SK12 1HA G6UQ
SK12 1HA G8SRS
SK12 1HA M5MDX
SK12 1HG 2E0SYN
SK12 1HG M0AGW
SK12 1HG M6CXO
SK12 1HN G0ONG
SK12 1NN 2E0URD
SK12 1NN G6URD
SK12 1NN M6URD
SK12 1PU G4IUJ
SK12 1PW M0GWF
SK12 1QH M1PTR
SK12 1RW G6RJH
SK12 1RW G7UJC
SK12 1SP G3THF
SK12 1TG G4MEH
SK12 1XB G6HJU
SK12 1YE G0EVP
SK12 1YX M6LBX
SK12 2AE G0CQU
SK12 2AY G8CET
SK12 2DB G6YCE
SK12 2RG G0NLL
SK13 1AR G1XCB
SK13 1EN G7JUJ
SK13 1HF M7UFC
SK13 1JZ 2E0MPC
SK13 1LH M0AIC
SK13 1LR M1EYO
SK13 1NX 2E0PSP
SK13 1PD G7JNM
SK13 2AD M7DTW
SK13 2AT M3RUO
SK13 5HL M3XAV
SK13 5HP G6BIX
SK13 6HX G0TXO
SK13 6NF G4KWF
SK13 6PX G6GIV
SK13 6US M7MGL
SK13 7AH G6EKS
SK13 7AJ G4GNQ
SK13 7AJ G4LMR
SK13 7BG G8BEQ
SK13 7DG M7KLM
SK13 7RL M0SAQ
SK13 7TL G4OQH
SK13 8BF M7DUH
SK13 8LW G6DQK
SK13 8NW G1YKK
SK13 8NW G6LLL
SK13 8NZ G0BAY
SK13 8RG G4IFJ
SK13 8RJ G0WJS
SK13 8RZ G6ARR
SK13 8SH 2E0VKR
SK13 8SH M0RVK
SK13 8SH M7CQU
SK13 8UD 2E0VMD
SK13 8UD M0TMD
SK13 8UD M6HMK
SK13 8UF G8YIG
SK14 1DT 2E1INC
SK14 1DT G8INC
SK14 1DT M3NXA
SK14 1QX G0KLD
SK14 1RP M6YBB
SK14 2SW G8OKD
SK14 3BW G4HLA
SK14 3QW M6AOG
SK14 3SN G0RRR
SK14 3SN G7KKW
SK14 4AF M0SWZ
SK14 4BS G0RGN
SK14 4DL G6MWM
SK14 4DW M3LPI
SK14 4DW M3LPK
SK14 4RT G6PZS
SK14 4SY 2E0MBV
SK14 4SY M3WPU
SK14 4TD M6VAK
SK14 5AB G0URT
SK14 5AU G4ING
SK14 5BB G8LCS
SK14 5DD G4USX
SK14 5JX G4PYQ
SK14 5LJ G1UTN
SK14 5NS G0WVV
SK14 5RE G7KEI
SK14 5RU 2E0CGX
SK14 5RU M3NBU
SK14 5ST M6MYS
SK14 6JA 2E0DVM
SK14 6JA M6FCJ
SK14 8LE G0NPK
SK14 8PA M3PVX
SK14 8QH G6SLY
SK145QZ M6BIX
SK15 1AJ M6TMM
SK15 1HD 2E0MKK
SK15 1HD G7FEG
SK15 1HD M3ATC
SK15 1HD M3MKK
SK15 1HG G0NIL
SK15 1HL G0KYB
SK15 1HU G0PVR
SK15 1HU M3SHX
SK15 1LE M0NSI
SK15 1LL 2E0SVP
SK15 1LL M0PVS
SK15 1LL M7SVP
SK15 2HR G1VTP
SK15 2LT G0WFD
SK15 2QQ G6LBO
SK15 2QZ G7WFD
SK15 2RP 2E0LTM
SK15 2RP M0PBF
SK15 2RP M7PWB
SK15 2TH G6YSZ
SK15 2TR G4NHW
SK15 3DZ G1TYP
SK15 3EN 2E0MXW
SK15 3EN M0SAD
SK15 3EN M3MXW
SK15 3HJ G0TIX
SK15 3HU G0ITZ
SK15 3JX M6LEH
Sk151ex 2E0OFF
Sk151ex M6GRK
SK16 4AU M6ZEP
SK16 4EW G1BSY
SK16 4JJ G7BCK
SK16 4QW M7AUC
SK16 4SJ M6GSC
SK16 4UD M1DJO
SK16 4UN M6NWI
SK16 5AB M3PQT
SK16 5BU 2E1SUE
SK16 5BU G4ZPZ
SK16 5DN G0NAJ
SK16 5DP M0AHF
SK16 5EG G4KFJ
SK16 5HR 2E1CWP
SK16 5HR 2E1CWX
SK16 5HR G0XAU
SK16 5HS G4LTM
SK16 5JA M1EEW
SK16 5JL G1HBE
SK16 5JL G1HTF
SK16 5NW G0LOE
SK16 5QS M6STP
SK16 5RT G0CKM
SK165BU 2E0EKX
SK165BU M0ILY
SK17 0DH G0JWD
SK17 0LU 2E0SCF
SK17 0LU M0TGT
SK17 0LU M6CRF
SK17 0LU M6JCF
SK17 0LU M6LEF
SK17 0LU M6SCF
SK17 0NF 2E1AZQ
SK17 0RG G1YJH
SK17 0SG G1RIY
SK17 6HH G4BJT
SK17 6HN G1OSA
SK17 6HQ G1LVV
SK17 6LP M6XVB
SK17 6NQ G8DAM
SK17 6NY 2E0WAT
SK17 6NY M3AJW
SK17 6PJ 2E0SRT
SK17 6QX 2E0BSG
SK17 6QX M0GOB
SK17 6QX M6ARS
SK17 6RD G7TMM
SK17 6ST M0DZM
SK17 6UA M6NCL
SK17 6XX M1CGQ
SK17 6YU G7NFK
SK17 7DD 2E0NIB
SK17 7DD M6NIB
SK17 7DN M6XJO
SK17 7DX G8YTX
SK17 7EA 2E0DHS
SK17 7EA M3DHS
SK17 7HW M6KWL
SK17 7JF M3MQM
SK17 7PH M3LIB
SK17 7PU G0MUR
SK17 7TA 2E0CFV
SK17 7TF 2E0CWR
SK17 7TF M6CWR
SK17 7TJ G0WSP
SK17 7TQ M3VMA
SK17 7TW 2E0TCG
SK17 7TW M6TCG
SK17 8AY G1OHD
SK17 8DL M6RNV
SK17 8DP M3JQG
SK17 8DW G4MRQ
SK17 8DW G4SPA
SK17 8ET G6LXE
SK17 8JT G4HGN
SK17 8LP G0CWS
SK17 8NP G0IOK
SK17 8PX M3ENO
SK17 8RJ M1WVS
SK17 8RJ M3TOT
SK17 8SW G1TBI
SK17 9AG M3YXL
SK17 9HG M0AVK
SK17 9HG M3MSZ
SK17 9JF 2E0HAE
SK17 9JF M0HOO
SK17 9JF M3OWI
SK17 9JS G0XKK
SK17 9JS G4BAC
SK17 9JS G4IHO
SK17 9JS G8BUX
SK17 9JU G7TZN
SK17 9NU 2E0WFW
SK17 9NU M6WFW
SK17 9PL G7NHL
SK17 9PL M3NFK
SK17 9RY G0RKT
SK17 9SE G0RYQ
SK17 9SG 2E0GAU
SK17 9SG M0OKK
SK17 9SG M6OKK
SK17 9TS G4KNQ
SK178FQ 2E0HDM
SK178FQ M3WBA
SK2 5BS G4DSR
SK2 5BS G8MFR
SK2 5EP G4WHF
SK2 5HX M7GTJ
SK2 5RN G7ILS
SK2 5SF M1ANT
SK2 5SF M1SIM
SK2 5SN M6NVB
SK2 5UR G1FFR
SK2 5UU 2E0JLR
SK2 5UU M6FCF
SK2 6AA M0SAV
SK2 6BN G8CHY
SK2 6BX G0KHR
SK2 6DW 2E0HSB
SK2 6DW 2E0RHM
SK2 6DW M3UVM
SK2 6DW M6ACE
SK2 6EP G7IOC
SK2 6EP M3IOC
SK2 6HA G4ORV
SK2 6HF M7RJL
SK2 6JS G4FRM
SK2 6SP 2E0TJU
SK2 6SP M0TJU
SK2 6SP M3TJU
SK2 7AR M3UKY
SK2 7BH M6AAL
SK2 7EB G1JVF
SK2 7EB G3UHF
SK2 7EB G4HON
SK2 7EB M0ECR
SK2 7JD 2E0DJZ
SK2 7JD M6FFO
SK2 7LQ G3RZJ
SK22 2JG G7OJA
SK22 3AY G7KUB
SK22 3DH G3TPI
SK22 3DL M0JYF
SK22 3DQ 2E0AZU
SK22 3DQ M3ICN
SK22 4DP G4YAB
SK22 4HR G1IMD
SK22 4JT G4XDT
SK22 4NE G0UPO
SK22 4NE G4YLQ
SK23 0HD 2E0YLD
SK23 0HD M7LXD
SK23 0HY 2E0GYZ
SK23 0JF G0SMP
SK23 0JF G7KRZ
SK23 0LF G7CJW
SK23 0LX G1COD
SK23 0PX G6YDN
SK23 0QF G3NZV
SK23 0QF G3ZVS
SK23 0TA G1ABJ
SK23 0TS G4LUQ
SK23 6AQ M7TBR
SK23 7AY G1OXH
SK23 7AY G3EEO
SK23 7BH G6LFR
SK23 7BP M0WLH
SK23 7DP G4IXF
SK23 7DT G0BDN
SK23 7HD G0GCJ
SK23 7LH G4CDZ
SK23 7NF G6UOO
SK23 7NH G8HQN
SK23 7QU G1HQO
SK23 9ST 2E0HOY
SK23 9ST M7STH
SK23 9TP M6XLF
SK23 9TU M3OXN
SK230QQ 2E0FOL
SK230QQ M6FOL
SK3 0JJ G7SRH
SK3 0LQ 2E0GXE
SK3 0LQ M6GCY
SK3 0PX G3JZT
SK3 0QL G0RXA
SK3 0QL M1BAR
SK3 0QY G6RSU
SK3 0SN M7USL
SK3 0TX G0MLC
SK3 0UJ G1HWJ
SK3 0UN G7TQT
SK3 8HQ G8ZAD
SK3 8JE G8MHT
SK3 8NQ M7WTG
SK3 8PG 2E0UTX
SK3 8PG M0UTA
SK3 8QU G3NBY
SK3 8SL M3JGU
SK3 9BX M0TKG
SK3 9EA 2E0KLB
SK3 9ER 2E0CTF
SK3 9ER M0HNJ
SK3 9ER M6PRE
SK3 9HF G8UWM
SK3 9JU G6PWQ
SK3 9NJ 2E0GCL
SK3 9NJ M6GBV
SK3 9QG M0BIH
SK3 9QL G0OQR
SK3 9RZ 2E0REU
SK3 9RZ M6PFP
SK4 1HZ 2E0BDI
SK4 1HZ M3LQD
SK4 1NH M3IRS
SK4 1NH M6TGV
SK4 2AA G6LQE
SK4 2DB G8YTP
SK4 2DQ M3UTJ
SK4 2PE G3TFR
SK4 3HD G1JFU
SK4 3HD G4RPJ

SK4 3JB G4DVI
SK4 3JS G4HNO
SK4 3NB G8UBP
SK4 3PG G8NRU
SK4 3PU G1CQT
SK4 3QF G0AMY
SK4 4AX G3RRG
SK4 4BX G3WFW
SK4 4BX G7KSP
SK4 4EB M6EJR
SK4 4NE G3YFD
SK4 4NL G4HWW
SK4 4PY G6KSO
SK4 4RL G0WOP
SK4 4RP G7PRD
SK4 5AW G1FMA
SK4 5BR M6AVA
SK4 5DJ M1BSM
SK4 5HP 2E0XYA
SK4 5HP M0JDF
SK4 5HP M0XYA
SK4 5HS M1BBU
SK4 5NW M3ZIA
sk5 6be M6TPS
SK5 6JD M0LPB
SK5 6JD M6BTA
SK5 6JD M7PLB
SK5 6LE M6DLL
SK5 6SH G0IES
SK5 6TD 2E1HQV
SK5 6TD 2E1HQW
SK5 6UU G0LJF
SK5 7EU G1JQK
SK5 7HU 2E0RTN
SK5 7HU M0RUM
SK5 7HU M6KBI
SK5 7JB M0JUW
SK5 7JB M0JYP
SK5 7LL G8ZLU
SK5 7NA 2E0TLC
SK5 7NA M0ZPZ
SK5 7RE G0BKZ
SK5 8BG 2E0NLY
SK5 8BG M6RBB
SK5 8DB G1JIR
SK5 8JY G0TIP
SK5 8NQ M6ANP
SK56NU G0DTT
SK6 1AS G0WMZ
SK6 1BL G1YZH
SK6 1DJ M7NDB
SK6 1EE G6LKJ
SK6 1EF G0GSM
SK6 1HA M6VBZ
SK6 1LH 2E1CXP
SK6 1PG G4KUC
SK6 1PG G8VHF
SK6 1PT G6GUT
SK6 1PY G4OZC
SK6 1QL M7RFA
SK6 1SL M3UQP
SK6 2ED G1TNK
SK6 2EH M6PGO
SK6 2HQ G8ILD
SK6 2HR M1EAS
SK6 2HS M7CPT
SK6 2PJ M7NJM
SK6 3BE G6SOO
SK6 3EE M0ATV
SK6 3EN G8IAN
SK6 3EY G0EOM
SK6 3JT 2E1FIX
SK6 3JT G0PUD
SK6 4JF G3PTX
SK6 4JN M6MZV
SK6 4PT G6RIC
SK6 5BB M1BMC
SK6 5BR G3PMV
SK6 5BT G3XGH
SK6 5EB G7GGN
SK6 5ET G4GWF
SK6 5PG G4NPU
SK6 5PN G8OOC
SK6 5PR 2E1WVF
SK6 5PR G4WVF
SK6 6DF G4MUL
SK6 6HJ G3MUO
SK6 6HJ G4BEV
SK6 6JL M0BEX
SK6 6LL G7BYU
SK6 6NA G0EQH
SK6 6PH M1VSR
SK6 6PJ G3SNG
SK6 6PT 2E0JEK
SK6 6PT M6BMZ
SK6 7BG G0PDH
SK6 7DT G8NOS
SK6 7HP M3EAE
SK6 7JS 2E0JEH
SK6 7JS M6FRI
SK6 7PB 2E0TTG
SK6 7PB M6TMH
SK6 7PW M3EXY
SK6 7RJ M6TMA
SK6 8AB G4FBG
SK6 8AG 2E0ROR
SK6 8AG M7BFN
SK6 8BH G0NKU
SK6 8BJ G6WWA
SK6 8BP G4GXQ
SK6 8DG G0LLU
SK6 8JE G3NXX
SK67BT G8WIJ
SK7 1LG G3TDH
SK7 1PJ M3ZVS
SK7 1PP G4GRU
SK7 2AB 2E0HPB
SK7 2AE 2E0RPE
SK7 2AE M0REP
SK7 2AE M6RPE
SK7 2BY M3NMP
SK7 2DT G3NUQ
SK7 2ER G3KAF
SK7 2JR G0ADJ
SK7 2QW M3FFU
SK7 3DT 2E1HLU
SK7 3JS G7EAH
SK7 3JW G0KZO
SK7 3JW G4SYC
SK7 4BJ G1GYC
SK7 4HW M1AQJ
SK7 4NU 2E0TWD
SK7 4QE G0FLQ
SK7 4QG G4GEY
SK7 4QP G0CJA
SK7 4QW G0HEX
SK7 4RU 2E0MYL
SK7 4RU M0IKI
SK7 4RU M6MYL
SK7 5JB M0KJC
SK7 5JB M3KJC
SK7 5JN G4OQP
SK7 5LR G0LDU
SK7 5QY G4VSW
SK7 6EJ G8YUP
SK7 6EJ M3YUP
SK7 6EL M6XDW
SK7 6HG G6DBQ
SK7 6HP 2E0MIU
SK7 6HP M0JFM
SK7 6HP M6MIU
SK7 6HU G8WPV
SK7 6JB G8EVR
SK7 6NR G8RIC
SK7 6NR M5KJM
SK73JU G8VLR
SK8 1BA G4CSV
SK8 1DX G4DQL
SK8 1HY G7TKT
SK8 1QY G3MTR
SK8 2AQ G6TRN
SK8 2AQ M6UAP
SK8 2EE M0BMW
SK8 2LZ M6CMK
SK8 3AA M3FJQ
SK8 3AJ G4FAS
SK8 3AS 2E1INW
SK8 3DL 2E0LKC
SK8 3DL 2E0LMD
SK8 3DL M6LCH
SK8 3DL M6YEH
SK8 3DZ G0KCE
SK8 3LF G1TAU
SK8 3LL G3ZBZ
SK8 3NH G7PIG
SK8 3NR G8BHX
SK8 3PF G4SVV
SK8 3PJ G0DOU
SK8 3RH M6MTJ
SK8 3RH M6NHJ
SK8 3RN G0GJD
SK8 3RR M3EEJ
SK8 3RU G4DOJ
SK8 3RW M3GIY
SK8 3UD G8ZCJ
SK8 4AE M0SGW
SK8 4AL G3UYG
SK8 4NA G3NYE
SK8 5DD M3DRM
SK8 5HP G6AXK
SK8 5PF G0NKM
SK8 5PN 2E0ETN
SK8 5PN M0TKT
SK8 5PN M6ETN
SK8 5RU 2E1GGT
SK8 6AX G3VOU
SK8 6DG G1WTY
SK8 6EQ G4NFA
SK8 6EQ M0LAR
SK8 6EQ M7DIM
SK8 6EX G8TMJ
SK8 6JG G4KJK
SK8 6LS M0ARA
SK8 6PW G0BAA
SK8 6PW G0OZJ
SK8 6PW G1NCR
SK8 7AJ G8GRP
SK8 7AL G8LDC
SK8 7DB G6YDO
SK8 7DS G4FRW
SK8 7EP G8GHO
SK8 7HU G6TKV
SK8 7LG G4EUC
SK8 7NP G8UEU
SK8 7PB G6FOI
SK8 7PZ G0IGB
SK83pa M7AEX
SK87DS M7OEN
SK9 1LU G4IRU
SK9 1QE G4CQV
SK9 2AD G1JDQ
SK9 2BB G3RKF
SK9 2EY G1MWS
SK9 2EY M0GMG
SK9 2HQ M3RWV
SK9 2LZ M7NOC
SK9 2NT M7LCA
SK9 3AF G6OER
SK9 3AR G0OPG
SK9 3AR G7IOB
SK9 3JT G4IRG
SK9 3UF G8JHL
SK9 4EP G8IXP
SK9 4HE G3WPF
SK9 5BN M0DFD
SK9 5BN M3CIE
SK9 5EN G3ONI
SK9 5JA G1HFH
SK9 5NQ G3TSZ
SK9 5PX G1GYH
SK9 5QE M0AAS
SK9 6BG G3VOM
SK9 6BL 2E0LEL
SK9 6EG G0AXE
SK9 6HD G1ORS
SK9 6HD G7CMN
SK9 6HD G7RMZ
SK9 6HD M3ORT
SK9 6JD 2E1INT
SK9 6LG G3USO
SK9 6LP G3WTQ
SK9 6LS G4WKD
SK9 7HW G7TNT
SK9 7HZ G6TSM
SK9 7PN G3BNW
SK9 7PQ M3NAW
SK9 7RE 2E0HTO
SK9 7RE 2E0PNA
SK9 7SD M7GSD
SL0 0DT 2E1DRC
SL0 0JS G1EKP
SL0 0NJ G6XRK
SL0 9RD G6MKZ
SL0 9RJ M6PZY
SL1 1TS M0RNX
SL1 2DA M6ETS
SL1 2HX G7HID
SL1 2HY M1ESV
SL1 2LQ G3RZF
SL1 2NU M0HWN
SL1 2YR 2E0DFS
SL1 2YR M6DBL
SL1 3JU G4FKE
SL1 3XG G4XGD
SL1 4NL 2E0ICN
SL1 5BB 2E0MWN
SL1 5BB M0MWN
SL1 5BB M6FJN
SL1 5DF G4CIJ
SL1 5JW M6VMO
SL1 5LA 2E1CGS
SL1 5LE M7IKA
SL1 5QZ M6RBO
SL1 5RE G6GEN
SL1 5UR G8WQC
SL1 5US M7TRM
SL1 6AH G4HIN
SL1 6AY G8NNP
SL1 6HD G4RAA
SL1 6JR 2E0TPI
SL1 6JR M6TJU
SL1 6JU M1ACN
SL1 6NN M6MYU
SL1 7BQ G3YFO
SL1 7EW G6DRP
SL1 7LY G6WZC
SL1 7NF 2E1AGE
SL1 8AT G8TEQ
SL1 8AW M1JJS
SL1 9HB G0WYR
SL2 1HG G6POC
SL2 1LG G7WGE
SL2 1SL G4XTZ
SL2 1SL M1DSQ
SL2 1SU M6CUU
SL2 1TA M7AOM
SL2 2AD 2E0EBZ
SL2 2AD M3OJJ
SL2 2JY G6XJI
SL2 2QW M6OUR
SL2 3HE G4LLN
SL2 3HW M6BFS
SL2 3NT G6OCF
SL2 3QT M6VVB
SL2 3XL 2E1AXI
SL2 3XT G0GUN
SL2 3XT G1DSA
SL2 4AX G3PFO
SL2 4ER G6VUF
SL2 5BG 2E0EBX
SL2 5BG M6KGR
SL2 5EZ M0MRY
SL2 5PN G1SPJ
SL2 5XH M3FZB
SL3 0LJ G0JAM
SL3 6RQ M6SLC
SL3 7DA G4XAN
SL3 7ET G6TSF
SL3 7ET M3TSF
SL3 7FQ M0GPX
SL3 7HN G4HTB
SL3 7JY M3CJE
SL3 7RL M6FEZ
SL3 8AU G0TGQ
SL3 8AU G6FCJ
SL3 8AX G1MDQ
SL3 8ED G8GGM
SL3 8JH G0HXM
SL3 8JH G3VHH
SL3 8SY M7DWP
SL3 8UE M7BJH
SL3 8UU M6DFQ
SL3 8WG M3VLX
SL3 9BA G4CRW
SL3 9NF M3MUU
SL3 9NJ 2E0BHZ
SL3 9NJ M6TBM
SL4 2AN G3ZCD
SL4 2LT G6PWF
SL4 2QY 2E0EFZ
SL4 2QY M6HND
SL4 3LP G8FUO
SL4 3RD G1HDG
SL4 3SQ G8UYK
SL4 4AE G0HTK
SL4 4JP G7WFZ
SL4 4YU G7LTT
SL4 5BX G4WPR
SL4 5DX G8DPH
SL4 5PS G4PUB
SL4 5PS G4SVG
SL4 5TD G4WQS
SL4 6BN G7VME
SL4 6HL 2E1IJD
SL4 6HL 2E1IJE
SL4 6NF G7CYD
SL5 0HU G3PLY
SL5 0NR G1FXX
SL5 8HQ G8LF
SL5 8LL G7PYB
SL5 8NB G1PSS
SL5 8QB G6EIZ
SL5 9AD G4ZWX
SL5 9GF G4EYV
SL5 9QH M1CSZ
SL5 9RH M3GXG
SL5 9RH M3VOA
SL6 0JR G4JDF
SL6 0PB G6ZEZ
SL6 1DD G6CJB
SL6 1NX G0OGE
SL6 1XE G6GFR
SL6 2BA M7ATL
SL6 2BJ G0TNY
SL6 2DF M0TXA
SL6 2DF M7TBD
SL6 2DZ 2E0IEM
SL6 2DZ M6IEM
SL6 2ED M6JPX
SL6 2GZ G0VKE
SL6 2JS G0MMQ
SL6 2TX M6UMB
SL6 2US G3WHB
SL6 2YJ M6OAF
SL6 2YP G3ONR
SL6 2YU M3BJZ
SL6 3BB 2E0FZF
SL6 3BB M7AKX
SL6 3DD M3TGT
SL6 3EP M1HVJ
SL6 3JA 2E0CJV
SL6 3JA M6AYL
SL6 3JY G0BMU
SL6 3NR G4XOW
SL6 3UZ G8GQF
SL6 3XH G0SVN
SL6 3XJ 2E0KZQ
SL6 3XJ M6KZQ
SL6 3XL G7PVZ
SL6 3XL M3JAC
SL6 3YS G3VMR
SL6 4ED G7REC
SL6 4GY G4USK
SL6 4NF G4FTG
SL6 4NG G4XYN
SL6 4QT G1MRX
SL6 4SA G8UPO
SL6 5AR G3UXY
SL6 5BJ M1TUG
SL6 5DN 2E1BPV
SL6 5DW M6OJB
SL6 5EG G4SZA
SL6 5JP G3OTN
SL6 5JW M1BAD
SL6 6AX G8SRV
SL6 6BE M0PGM
SL6 6DA G4XDU
SL6 6DF G8MYV
SL6 6EG 2E0TVM
SL6 6HN G4KMH
SL6 6HN G8TFR
SL6 6LG G0GLA
SL6 6LT M0RFK
SL6 6QF G8KLA
SL6 6RX G0IIP
SL6 6RY G6NBI
SL6 6SD M6PTX
SL6 6SS G1PCR
SL6 7DT M3HXW
SL6 7EJ 2E1ANN
SL6 7EZ M6ZMC
SL6 7HE M3HCG
SL6 7LF G0PMZ
SL6 7SH G0FFL
SL6 7TZ G6BJY
SL6 7UF G4YDL
SL6 7UH G0AYA
SL6 7UH G0BQE
SL6 8EU G3YDT
SL6 8HT G0CXV
SL6 8JZ G4BOV
SL6 8RH M0IBN
SL6 9DH G7EBR
SL6 9DN G3PQA
SL6 9JF G0CMW
SL6 9JT G3WYK
SL6 9LS M3UCZ
SL6 9NF G1LIG
SL6 9NF M7GNH
SL6 9SN G0NJQ
SL7 1DQ G3TOP
SL7 1DR G3DXD
SL7 1JW G7JDN
SL7 1SA G7TWW
SL7 1TX G6GIF
SL7 1UW G4KCD
SL7 2DP G8KKG
SL7 2PL G5BCO
SL7 2QU G6VF
SL7 2RE G4IDJ
SL7 3BZ G4YSH
SL7 3HA 2E1JOY
SL7 3LD M7CHG
SL7 3LU G1SNO
SL7 3NG 2E0FZB
SL7 3NG M7SWE
SL7 3PY G3LVW
SL7 3PY M6LHS
SL7 3RB G0UYM
SL7 3RH G3IQF
SL8 5AZ G3INQ
SL8 5DH M6XXZ
SL8 5QN 2E1CXE
SL8 5TJ 2E0GCM
SL8 5TJ M0VCE
SL8 5TJ M6GCM
SL8 5TJ M6IOS
SL8 5TY G0MGM
SL85AA 2E0IWF
SL85AA M6IWF
SL9 0DD G0RAU
SL9 0DD G6XDB
SL9 0EJ M1SJA
SL9 0HE G4UNE
SL9 0HH M3ZGH
SL9 0LR G4MDN
SL9 7EB 2E1GXS
SL9 7EN 2E0HOB
SL9 7EN M7JEF
SL9 7HX G7CZF
SL9 7LA G4KGA
SL9 7PZ M0EAS
SL9 7QR G8CZQ
SL9 8LJ G3NZW
SL9 8LQ G4FCD
SL9 8NW 2E1EHF
SL9 8NW M0RVJ
SL9 8PX M6LGM
SL9 8QT M5AIQ
SL9 9JZ M6WBK
SL9 9NX M0WAQ
SL9 9PA G1LIK
SL9 9TD G4OXG
SM1 1JE G3WYB
SM1 1QH M6EXN
SM1 2BL G4HSD
SM1 2EB M0WLS
SM1 2EB M6WLS
SM1 2PA G0BWV
SM1 2PA G2XP
SM1 2PA G7SAC
SM1 2RB G3VSH
SM1 2RZ M7KRK
SM1 2SQ G7KIN
SM1 3EA 2E0JHF
SM1 3EA M6JHF
SM1 3LW M0RFY
SM1 3QB G3NHX
SM1 3QQ M7GPI
SM1 3SL G8GLC
SM1 4BG G0SWU
SM1 4BR G7ITZ
SM1 4FL M1XPS
Sm12en M7LEN
SM13RJ G1HRD
SM2 5AQ 2E0SRZ
SM2 5BA M6LOF
SM2 5EJ G6CTV
SM2 5ER G1XKD
SM2 5HP G1GBI
SM2 5HP M3NLP
SM2 5HW G8NFP
SM2 5HW G8TQK
SM2 5JA G4PLH
SM2 5JH G0PMM
SM2 5RP G7JDB
SM2 6RL 2E0VLB
SM2 6RW G1DQL
SM2 6UA G1UBV
SM2 7QA G0KBL
SM3 8ES G3XTC
SM3 8NA M0BWB
SM3 9EG G4DSY
SM3 9HY G4PGS
SM3 9ND 2E0EXL
SM3 9ND M0ZPG
SM3 9NE G6SGM
SM3 9NL G8VCL
SM3 9PP G8YEP
SM3 9QJ G7UZI
SM3 9RF G6XGV
SM3 9RH M1ROD
SM3 9RL M3PUQ
SM3 9SG G1HWK
SM3 9TT M3VBZ
SM3 9TZ G8CVQ
SM3 9UB G0EAU
SM4 4BS G1SHN
SM4 4DL G0VXX
SM4 4LF G0BXC
SM4 4QG G0PNT
SM4 4QN G4ZHA
SM4 4SX M6AVV
SM4 4TD G4CQR
SM4 6ED 2E0ERQ
SM4 6ED M0NHY
SM4 6ED M6NHY
SM4 6HL M3GYI
SM4 6HU M7PBD
SM4 6QG G6JXA
SM5 2BA 2E0OKQ
SM5 2BA M0OKQ
SM5 2BA M6OKQ
SM5 2HG M7NLP
SM5 2NH 2E0WWE
SM5 2PB G1EGZ
SM5 2SP M1MRB
SM5 2SP M6BOY
SM5 2TL G8VUK
SM5 3AB 2E0LEM
SM5 3DZ G7BDK
SM5 3EJ G4VAV
SM5 3HA G7NKH
SM5 3LS G0PUQ
SM5 3NG G4FDN
SM5 3NG G8PAT
SM5 3RA 2E0MJS
SM5 3RA M0SSM
SM5 3RA M6MCM
SM5 3SF G8HIG
SM5 3SJ M3RUL
SM5 3SU 2E0FTM
SM5 3SU M6TMF
SM5 3SW G7AQN
SM5 4QH G4BVG
SM6 0AS 2E0GHH
SM6 0AS M7WST
SM6 0PB G7CRQ
SM6 0QY G6XZP
SM6 0TN G3BMQ
SM6 7AG G4XSA
SM6 7BZ 2E0VSZ
SM6 7BZ M7VSZ
SM6 7DD G7ELZ
SM6 7JU G7NHY
SM6 8AE G3SRC
SM6 8AE G4CCY
SM6 8AE G4DDY
SM6 8BE M0DWB
SM6 8EP G6SAQ
SM6 8EP G8ZOY
SM6 8EW G8XDM
SM6 8EW M0BEH
SM6 8HL G0TXL
SM6 8PT G6ZLD
SM6 8PY M3LRN
SM6 8QB 2E0JJK
SM6 8QB M0JJK
SM6 8QB M6JAK
SM6 9GL G6UCY
SM6 9GX 2E0XSL
SM6 9GX M3RHL
SM69BP M6OEF
SM7 1AJ G3OLX
SM7 1BU 2E0EZF
SM7 1BU M0VON
SM7 1BU M3UQO
SM7 1HG G3CZU
SM7 1HG G4DFA
SM7 1HG G8GYX
SM7 1HG M6XUV
SM7 1JW 2E0TTA
SM7 1LB G6XZM
SM7 1LE 2E0JLX
SM7 1LE 2E0LEG
SM7 1LE M0WOJ
SM7 1LE M3XJL
SM7 1LE M3YCZ
SM7 1LE M6EMC
SM7 2BA G7DRD
SM7 2DG G1DPX
SM7 2ER M3NUO
SM7 2ER M3ZON
SM7 2HG 2E0GTZ
SM7 2HG M6AXB
SM7 2HZ M0NDJ
SM7 2HZ M3NOE
SM7 2QJ G0IUH
SM7 3JR G3KTA
SM7 3NA M6UTB
SM7 3PN G8GNX
SN1 2AN M6EWK
SN1 2EY M0IWT
SN1 2JR G7LET
SN1 2JU G4KEZ
SN1 2QT M3MZU
SN1 3AE G0RNH
SN1 3AE M3GKK
SN1 3HF M3YFM
SN1 3HW 2E0LAY
SN1 3HW M6LAY
SN1 3NJ G0TLI
SN1 3PR M7CDQ
SN1 3QA G1ERM
SN1 4AY G6BPK
SN1 4BG 2E0NOS
SN1 4DP G3YKC
SN1 4EY G1KMN
SN1 4GE G1PJC
SN1 4GU G7JGE
SN1 4HH 2E0PPA
SN1 4HH M6FVJ
SN1 4HX G1NAU
SN1 4LL G7PKK
SN1 5EE G8PAB
SN1 5ES M1EGN
SN1 5QS M7NSS
SN1 7DJ M4ZEJ
SN10 1BF G8LGP
SN10 1FA 2E1EMK
SN10 1LW 2E0EFS
SN10 1LW M0OIO
SN10 1LW M6HYF
SN10 1LY 2E0MBS
SN10 1LY M0MSZ
SN10 1LY M6MMS
SN10 1QG 2E0MPN
SN10 1QG M3PQV
SN10 1RW G4RQL
SN10 2EN G6NTE
SN10 2FX G6HBW
SN10 2FX G6PYF
SN10 2LD G3WAE
SN10 2UB G7FWD
SN10 2UE G0JVF
SN10 3AN G4GBX
SN10 3AP G3WZR
SN10 3BJ G4NWR
SN10 3BJ G4TIX
SN10 3BJ G6EVY
SN10 3BQ M7PLS
SN10 3DG G6KHW
SN10 3FN G0WHY
SN10 3PP 2E0VRO
SN10 3PP M6IYL
SN10 3PQ M3PVV
SN10 3SL G4SHA
SN10 3UB M7DHM
SN10 4ED G6SND
SN10 4EL G6FNJ
SN10 4HP 2E0SBS
SN10 4HP M3UZZ
SN10 4PA G8MVS
SN10 4PP 2E1GQD
SN10 4PP M6LWZ
SN10 4RR G3MQD
SN10 4RR G3PZV
SN10 5AJ G1JMK
SN10 5AJ G6ZFA
SN10 5AZ M1FHA
SN10 5BA G8XYA
SN10 5HD G4VKJ
SN10 5HW G0RTN
SN10 5LE G7LWF
SN10 5SR G3XKL
SN10 5TD G4EES
SN10 5TP G0RIK
SN11 0EP G6HTZ
SN11 0FH M7SJH
SN11 0FL 2E0GKA
SN11 0FL M0XCX
SN11 0FL M6WKL
SN11 0JA G7TPS
SN11 0LQ G7VEY
SN11 0NE M6FCS
SN11 0PZ M6TBS
SN11 0UG M7RKA
SN11 8EY M7LBX
SN11 8EZ G4XYB
SN11 8NL 2E0RFN
SN11 8NL M7RSM
SN11 8PW 2E0ZGX
SN11 8PW M3ZGX
SN11 8SQ G4NHQ
SN11 8UR 2E0EME
SN11 8YD G0MLE
SN11 9AB G0OIM
SN11 9AT M6EIJ
SN11 9BT G7RGG
SN11 9DT G1JBC
SN11 9DU 2E0CRH
SN11 9DU G8CRH
SN11 9DU M0HKV
SN11 9DU M6BSQ
SN11 9EW G3WIW
SN11 9EW G4NKU
SN11 9FN 2E0VRE
SN11 9FN M3ZHC
SN11 9FN M6DTA
SN11 9LD G1JRZ
SN11 9LR G4UJL
SN11 9QH 2E0LRD
SN11 9QH M3ZWP
SN11 9QQ G7TIV
SN12 6AG G4DMC
SN12 6AZ G7EPX
SN12 6FD 2E0DIM
SN12 6FD M0VAI
SN12 6FD M6DIM
SN12 6FX 2E0DRQ
SN12 6FX M6EEM
SN12 6NA M7EYA
SN12 6TH 2E1EYL
SN12 6TQ G4SRD
SN12 6UL G1PWU
SN12 7AB G3RVY
SN12 7BX M7LEX
SN12 7DT G0KTP
SN12 7EF M6WGS
SN12 7HL M0ECZ
SN12 7LD M0JEU
SN12 7NG 2E0MHG
SN12 7NG M6ONW
SN12 7PG G7TKO
SN12 7PT 2E0EES
SN12 7PT M6HJF
SN12 7QW M0OTL
SN12 7SW M0DGJ
SN12 7TE 2E0IZB
SN12 7TE G6EJD
SN12 7TE G6FNY
SN12 7TE M3IZB
SN12 8BQ M6XBX
SN12 8LA G4NDC
SN12 8NB G1AJC
SN13 0BA 2E0FRU
SN13 0EG M1EGC
SN13 0JD G4GUK
SN13 0JR G4DIE
SN13 0JT M1EIJ
SN13 0LB G7SNX
SN13 0LS M3IGN
SN13 0QR G3JSK
SN13 0QW M1ERV
SN13 8AN M0TRK
SN13 8AN M6AIJ
SN13 8DF G3KRW
SN13 9AU M3YPA
SN13 9LE G4SEZ
SN13 9NH G0KLJ
SN13 9NJ G4NLA
SN13 9NN G3MEY
SN13 9QS M6CDU
SN13 9SH G3MBN
SN13 9ST 2E0BZI
SN13 9TX M3KIR
SN13 9UL M6BYA
SN13 9UT 2E0GET
SN13 9YJ M7GPW
SN14 0DN G8MPM
SN14 0EX M6UUU
SN14 0JB G0RBD
SN14 0JW M7ACM
SN14 0PL M0RXM
SN14 0RB M3UXR
SN14 0TN G7PSS
SN14 0UJ G1SOG
SN14 0UJ M3SOG
SN14 0XP 2E0JWP
SN14 0XP M6CHH
SN14 0XW 2E1DUI
SN14 0XX M1EZU
SN14 6AN M1CWN
SN14 6DE M0PYE
SN14 6EG M6BPA
SN14 6EL G1YIQ
SN14 6LU M6FWP
SN14 6RT 2E0VFX
SN14 6XJ G1JRP
SN14 6YH M3MMN
SN14 7BB M3JYO
SN14 7EA 2E0NEY
SN14 7EA G8NEY
SN14 7EA M0GHZ
SN14 7NN G4BBP
SN14 8BS 2E0HXC
SN14 8BS M7HTC
SN14 8DN G3PIJ
SN14 8HD 2E0HOM
SN14 8HD M7CME
SN14 8QU G0RCP
SN14 8SA G8WMC
SN15 1AR G0FGW
SN15 1BQ 2E0PLS
SN15 1BQ M0PLS
SN15 1BQ M6BXZ
SN15 1DD M1GTI
SN15 1DW M0CTR
SN15 1QQ G1USK
SN15 1QQ M6XJW
SN15 1US 2E0GDV
SN15 1US M0JMU
SN15 1US M7ATO
SN15 2AY G0IAY
SN15 2AY G6YCG
SN15 2DJ G4TKF
SN15 2EE 2E0FLS
SN15 2EF G4ASG
SN15 2LB G0MVM
SN15 2PW G4SUX
SN15 2RA 2E0UET
SN15 2RA M0UET
SN15 2RA M6UET
SN15 3DZ G4MLG
SN15 3EF 2E1FBK
SN15 3EG M0MJD
SN15 3EL G4OQG
SN15 3FY 2E0SCS
SN15 3FY M6WSB
SN15 3JF 2E0FZG
SN15 3JY G0HPN
SN15 3LW M6IQU
SN15 3LX G0GJE
SN15 3RX G4SFD
SN15 3UA G4BBY
SN15 3UA G4KPI
SN15 3UG M3ZJK
SN15 4AR G6MFB
SN15 4AS G7OPI
SN15 4EL 2E0BPL
SN15 4EL M0RUB
SN15 4EL M3WDV
SN15 4ER G0OID
SN15 4JH M7CLM
SN15 4PJ M3EIA
SN15 4RL G4GWR
SN15 4TR G6VHE
SN15 5AW G6EYA
SN15 5BT 2E0VCW
SN15 5BT M7CAW
SN15 5EX 2E0CXH
SN15 5EX M6BXY
SN15 5LD G8TTI
SN15 5LY G3UUV
SN15 5NN G0AEV
SN15 5NT G3VRE
SN15 5NT G6HUI
SN15 5QF G4CDW
SN152BD G8HAM
SN16 0BL G6UYM
SN16 0DR G4TLY
SN16 0EA M3WSH
SN16 0LY M6GCS
SN16 0PZ G7IBL
SN16 0QH G4JKM
SN16 0RN 2E0EMX
SN16 0RN M6AGH
SN16 9DL G8DKI
SN16 9GX G8VVY
SN16 9HF 2E1ASF
SN16 9HF M7BWJ
SN16 9PR G8EHE
SN16 9QB G8WLV
SN16 9QT G7GPU
SN16 9QZ G8POP
SN16 9SR 2E0GMG
SN16 9SR M3VGT
SN16 9UA G4SWO
SN16 9UE G8EMU
sn160es G4TRA
SN2 1BD M0SYL
SN2 1BH M0SPN
SN2 1NE G8CPA
SN2 1NL M0BFA
SN2 1NL M0PBZ
SN2 1RQ M3TSV
SN2 1RX G7FEA
SN2 1RX M6NDR
SN2 2AF M6EPH
SN2 2BP M7PBS
SN2 2EL M3VST
SN2 2EN 2E0XUM
SN2 2EN M0XUM
SN2 2EN M6XUM
SN2 2EN M7EGP
SN2 2FL 2E0FWB
SN2 2FL M6UPE
SN2 2PE G6OMH
SN2 2SG G4VWG
SN2 2SL G7UZG
SN2 5DE G0VCV
SN2 5ED G8XWH
SN2 5HF M6PDW
SN2 5JJ M1EMC
SN2 7HP 2E0CYS
SN2 7HP M6VDX
SN2 7LE G8SRC
SN2 7LE M0ACM
SN2 7LJ G7PYV
SN2 7LL G0JTD
SN2 7LQ G0VQW
SN2 7NJ G8MGO
SN2 7QA G4MDT
SN2 7TN M1XZG
SN2 8BT G7HPI
SN2 8DL M1ECQ
SN25 1PY 2E0ZVG
SN25 1QD 2E0HZA
SN25 1QD M7RBY
SN25 1TL 2E0GSF
SN25 1WG G4SOA
SN25 1WG G6KAW
SN25 2AZ M6XTB
SN25 2EE G1WYP
SN25 2EF M7NWV
SN25 2FE G0BON
SN25 2GQ M6ZYK
SN25 2NJ G8CZJ
SN25 2RJ M6YFX
SN25 3AZ G6XML
SN25 3BN M7VVB
SN25 3BT G4ZAM
SN25 3DB G6ARO
SN25 3DQ M6TGM
SN25 3ES G7FPM
SN25 3EZ M1DBW
SN25 3EZ M3PBW
SN25 3LZ M1AOU
SN25 3PP G8ELH
SN25 3QB G0UWS
SN25 3QB M6UWS
Sn25 4GF M0LEY
SN25 4GX M1SHA
SN25 4TJ G0EPC
SN25 4TP G3YMV
SN25 4XS 2E0YGC
SN25 4XS M7RCI
SN25 4YD M3VAG
SN25 4YJ G1HYC
SN26 7AR G8NTS
SN26 7DE G4HGV
SN26 7DE G8LYG
SN26 7EA G4SHK
SN3 1AZ G6VDK
SN3 1BN 2E0NDA
SN3 1BN M7NDA
SN3 1BX G7MGY
SN3 1BZ G7CRM
SN3 1DJ G0FCH
SN3 1DN 2E0LXD
SN3 1DN 2E1EQI
SN3 1DN G0VTA
SN3 1DN G0WDQ
SN3 1DR G8VJY
SN3 1ER G8IWO
SN3 1HZ G0TNP
SN3 1LU M6HYY
SN3 1NB G8FWK
SN3 1NJ G7KIW
SN3 1PT M1AGH
SN3 2AL M5ECX
SN3 2DR M1ACC
SN3 2EA G1FTV
SN3 2RB M6HST
SN3 3FQ 2E0GGU
SN3 3FQ M7CJB
SN3 3NB G3NPM
SN3 3NF 2E0LUF
SN3 3NF M0LUF
SN3 3NF M6LQP
SN3 3NW M5CBS
SN3 3PJ G4ZYH
SN3 3TB 2E0RJP
SN3 3TB M0OZD
SN3 3TB M3OZD
SN3 4AA G3LLZ
SN3 4AB G0PSO
SN3 4AY G3SIR
SN3 4EE M6ESR
SN3 4HB G0BQK
SN3 4HU G3HSV
SN3 4JA 2E0EXJ
SN3 4JA M0WXY
SN3 4JA M6EXJ
SN3 4LA G3MAI
SN3 4NF G3IDW
SN3 4NZ M7AAU
SN3 4SE G6REG
SN3 4SF G1DCI
SN3 4SN G5WQ
SN3 4SN M0BCG
SN3 4WE 2E0BNZ
SN3 4WE M3WBI
SN3 5AG G8ETI
SN3 5AU G3WEF
SN3 5AX G4LDL
SN3 5AX G8KWD
SN3 5BN G0CPA
SN3 5BN M3CSN
SN3 5DE 2E0PMV
SN3 5DE G0SNM
SN3 5DE M0PMV
SN3 5DE M6PMZ
SN3 5JG G1NCG
SN3 6AR G3XYD
SN3 6AT M7TUF
SN3 6JB G4JVJ
SN3 6JP M7MRR
SN3 6LA G1YGY
SN3 6LS G8UKY
SN3 6NF G4ENR
SN3 6NL G4AQK
SN4 0AE 2E0BZT
SN4 0AE M3ZPB
SN4 0DF 2E0FEC
SN4 0DF 2E0MJE
SN4 0DF M3YOW
SN4 0DF M6IDB
SN4 0DU 2E0POZ
SN4 0LT 2E0WHT
SN4 0LT M0WHT
SN4 0LT M7KPS
SN4 0LU G4NHD
SN4 0LW G3IRA
SN4 0RL M1CDJ
SN4 7DG G8JWK
SN4 7DN G0OWR
SN4 7DR G4RZF
SN4 7DU G8IJM
SN4 7EU M6DNB
SN4 7EZ G8IYJ
SN4 7FN 2E0BQJ
SN4 7FN M0TGN
SN4 7FN M0WCB
SN4 7FN M6DAN
SN4 7JG G4OIQ
SN4 7RG M3YOQ
SN4 8AS G0LTP
SN4 8DJ G4MDH
SN4 8DT M1KDJ
SN4 8HJ M3YPS
SN4 8HS G7CSE
SN4 8JY M3GRY
SN4 8LL G4FTZ
SN4 8LW G8HCW
SN4 8NQ G4YAL
SN4 9AG 2E0NGB
SN4 9AG M0JEC
SN4 9AG M0NHK
SN4 9AG M3NBL
SN4 9AP 2E0CTW
SN4 9AP M0ZGB
SN4 9AP M6HBS
SN4 9AR G3JOT
SN4 9AT G0HOJ
SN4 9AX 2E0VAU
SN4 9AX M0TTE
SN4 9AX M6PZF
SN4 9AX M6VAU
SN4 9BB M1DHW
SN4 9BP M6FWM
SN4 9DR G0WGI
SN4 9DR M0BKN
SN4 9DR M1EZP
SN4 9EP G7VHG
SN4 9HY G0DVB
SN4 9PA M7FYM
SN5 0AD G3HCT
SN5 0AD G3RZP
SN5 0AD G4FNC
SN5 3LE G3APS
SN5 3LE G7TBO
SN5 3LX G0JSU
SN5 4AD G3YBY
SN5 4BA G4YEF
SN5 4BS G3TPQ
SN5 4DE G4MQP
SN5 4ET G8VGQ
SN5 4EX G4MGH
SN5 4HQ G8IYH
SN5 5AL G0DHL
SN5 5BH G1MUM
SN5 5DB M6BKA
SN5 5QT G4OED
SN5 5RL M7CLQ
SN5 5SA G8EFK
SN5 5SH G6UZM
SN5 5TL G4DSF

SN5 5TP	M7KAV
SN5 5TS	G3OMA
SN5 5UQ	G4DOA
SN5 5XA	G2BUJ
SN5 6AA	M3PFU
SN5 6BB	G7WIC
SN5 6BQ	G6WAU
SN5 6EB	M0PRT
SN5 6EQ	M6BJQ
SN5 6JY	2E0KYB
SN5 6JY	M7BEL
SN5 7AF	G1VRJ
SN5 7AG	G6UAJ
SN5 7AQ	M6FSG
SN5 7AW	G6LBR
SN5 7BB	G1TOL
SN5 7BG	G3KUD
SN5 7BN	M3NTI
SN5 7BX	G4GXW
SN5 7DR	2E0GKM
SN5 7DR	M0ATC
SN5 7DR	M0GKG
SN5 7DR	M3RMV
SN5 7EF	G4KIM
SN5 7EG	G6BTR
SN5 8AH	2E1FJV
SN5 8AJ	M6RGX
SN5 8AQ	G7VJG
SN5 8AR	M3FGR
SN5 8DB	M6EVU
SN5 8HL	M3XFS
SN5 8JE	2E0FUF
SN5 8JE	M0JQV
SN5 8JE	M7AJR
SN5 8PE	G4WSB
SN5 8QX	M6BMF
SN5 8RT	G7MTF
SN6 6AJ	2E0ZPT
SN6 6AJ	M0MTA
SN6 6AJ	M6MEA
SN6 6AW	M1WIN
SN6 6BN	G8OFR
SN6 6FJ	M6NXA
SN6 6LJ	M0RQQ
SN6 6LJ	M7RQQ
SN6 6RD	2E0EAV
SN6 6RD	M0NMI
SN6 6RD	M6EIW
SN6 7BA	2E0KGV
SN6 7BA	M0KGV
SN6 7BA	M0TBG
SN6 7BA	M3KGV
SN6 7BU	G0NUH
SN6 7BZ	G4AJA
SN6 7BZ	G4BPO
SN6 7DW	2E0RAK
SN6 7DW	M0PGX
SN6 7EB	G4GDR
SN6 7EG	M3BWF
SN6 7ER	2E0GSO
SN6 7ER	M6OHU
SN6 7HJ	G4TRR
SN6 7HN	M1ATC
SN6 7HR	G4LZZ
SN6 7HS	G8EOZ
SN6 7HU	M1SKI
SN6 7LA	G0NPN
SN6 7NU	M0IEP
SN6 7NU	M1EJD
SN6 7PD	G8JHC
SN6 7PG	G8JRF
SN6 7RH	M6BHS
SN6 7RL	M3JIA
SN6 8AJ	M0DAW
SN6 8AZ	G3XEY
SN6 8BD	M7MDR
SN6 8DY	M6IGT
SN6 8ER	G3ONU
SN6 8FA	M6PGY
SN6 8NB	M3ESG
SN6 8SX	G4UEF
SN7 7DG	G0JBJ
SN7 7EY	G3LPC
SN7 7EY	G3NNG
SN7 7EY	G3PIA
SN7 7LB	G8BGM
SN7 7NB	2E0FUA
SN7 7NB	M0ROX
SN7 7NB	M6KTQ
SN7 7RL	2E0EJA
SN7 7SE	M3CVH
SN7 7SS	G8BCF
SN7 7YT	G8GIF
SN7 8AT	G4VLW
SN7 8EY	G0SCQ
SN7 8FA	G1NBP
SN7 8HW	M7PLN
SN7 8LE	G4PFY
SN7 8LE	G6HIX
SN7 8LN	M6GOI
SN7 8LR	G1HYO
SN7 8LX	2E1HFX
SN7 8LZ	M6JAZ
SN7 8ND	G3ZEF
SN7 8NP	G3XVB
SN7 8QR	G0ITO
SN7 8QR	G6ZMX
SN7 8RP	G0VBG
SN8 2AS	G4XVW
SN8 2AZ	G6EPN
SN8 2BD	2E0PDZ
SN8 2BD	M0PDZ
SN8 2BD	M6EZR
SN8 2DH	G1FKJ
SN8 2FE	2E0GWR
SN8 2FE	M0SWL
SN8 2PX	G8JWD
SN8 2QE	2E0PPB
SN8 2QE	G3PPB
SN8 3AF	2E0AAX
SN8 3AS	G3MFL
SN8 3DZ	G4DHT
SN8 3HN	G8SXA
SN8 3HN	G8SXD
SN8 3PX	M7WLT
SN8 3QB	M0TDW
SN8 3RP	M3ANW
SN8 3SS	M0GQM
SN8 3TD	G4LMA
SN8 4AS	M0DSW
SN8 4HT	M1CJE
SN8 4YS	G1EHB
SN9 5ES	M0BXU
SN9 5HP	G7GEA
SN9 5HU	G8PMJ
SN9 5HU	M6CFV
SN9 6AE	G4TOY
SN9 6AF	M0BUT
SN9 6EN	M6IIG
SO14 0LB	G6NJT
SO14 3FE	M6DLU
SO14 3FN	M6JZN
SO14 3GQ	G4HXE
SO14 3HU	G6UAN
SO14 3HW	G4CRB
SO14 6FR	M6LYP
SO14 6GR	M0XGT
SO14 6RN	M7GIP
SO15 1GF	M6HOI
SO15 1JJ	2E0CZK
SO15 1JJ	M6CYF
SO15 2AS	M0IVC
SO15 2NN	2E0JIB
SO15 2NN	M0XEF
SO15 2NN	M6FZQ
SO15 3EW	G8SBS
SO15 4HU	2E1ESQ
SO15 5DS	G4MYS
SO15 5RS	M6KFS
SO15 8PT	M3VWJ
SO15 8PT	M3VXG
SO15 8QU	2E1EUE
SO15 8QZ	M3YAO
SO16 0AQ	G0EET
SO16 2NZ	G3IXN
SO16 3DA	M7HHK
SO16 3EB	G0SWY
SO16 3EF	G6VNO
SO16 3GU	M7DJB
SO16 3GZ	G4VAS
SO16 3HY	G4ZAS
SO16 3LP	G7UBD
SO16 3NY	G4DIV
SO16 3TN	G6UTL
SO16 3UJ	M6LPF
SO16 4BN	G1XBR
SO16 4BY	2E0ILC
SO16 4BY	M0NGH
SO16 4BY	M6ILC
SO16 4NR	M0RLP
SO16 4QT	M6PHA
SO16 5DW	G3CGE
SO16 5FD	G4ZFY
SO16 5FG	2E0AJQ
SO16 5FG	G8KVN
SO16 5FG	M0APN
SO16 5GJ	M6VRO
SO16 5HB	G4WYW
SO16 5JX	M3UVK
SO16 5NQ	G3VQF
SO16 5SW	G7GIJ
SO16 6AF	2E0HQV
SO16 6AF	M7DBX
SO16 6PZ	M7FWB
SO16 6QB	M6PSO
SO16 6QU	2E0PAU
SO16 6QU	M3YXU
SO16 6UH	2E0HTN
SO16 7AW	2E0OKB
SO16 7AW	M7KAE
SO16 7DG	M0IJR
SO16 7DR	2E0UAV
SO16 7DR	M0UAV
SO16 7DR	M6TYL
SO16 7DR	M6ZSH
SO16 7EL	G3VXY
SO16 7GJ	M0MPT
SO16 7PE	G3ZIL
SO16 8AH	2E0BGJ
SO16 8AH	M0NXP
SO16 8AY	G6VDW
SO16 8AY	M0BIJ
SO16 8BE	G4WFR
SO16 8DW	G1XRM
SO16 8EF	M1FJA
SO16 8EH	G6CRG
SO16 8FS	M0IJO
SO16 8GU	M0DVF
SO16 8LR	G8LRS
SO16 9AW	M6ZOO
SO16 9EZ	M6HTG
SO16 9GF	2E0BXQ
SO16 9GF	M6LJD
SO16 9JF	G0XAZ
SO16 9LB	2E0DGR
SO16 9LB	M6DZX
SO16 9LR	2E0ERL
SO16 9LR	M3FZT
SO16 9QE	G4BDQ
SO16 9WA	M3XVQ
SO17 1EH	M6EJJ
SO17 1NU	G0DHR
SO17 1NU	G0MSR
SO17 1PF	M6AFG
SO17 1RS	G4YVZ
SO17 1SD	G2HNI
SO17 1XQ	M6USO
SO17 2HJ	M7APH
SO17 2LJ	M0EBO
SO17 2LJ	M1CWB
SO17 3AE	M6GHC
SO17 3RT	G6KYE
SO17 3SA	M0SMQ
SO17 3SA	M7SMX
SO17 3SJ	M6EEF
SO17 3TG	M6KOH
SO18 1FW	M7VMR
SO18 1HJ	M0CYJ
SO18 1JR	M6DBF
SO18 1JS	G1VGA
SO18 1LS	G3TUF
SO18 1QR	G3OFX
SO18 2FP	G7UYI
SO18 2HQ	M0GCH
SO18 2NU	2E0NWA
SO18 2NU	M6CDG
SO18 2NU	M6NWA
SO18 2PQ	G1LWF
SO18 2QE	G3IDB
SO18 3AG	2E1JJN
SO18 3AG	M1JJN
SO18 3AG	M3JJN
SO18 3PP	2E1HQP
SO18 3RB	G3KDW
SO18 4FN	2E0ZBE
SO18 4FN	M6SFC
SO18 5FE	2E0CZI
SO18 5FE	M6DQV
SO18 5FS	G3VCL
SO18 5GY	2E0NJC
SO18 5GY	M3VQZ
SO18 5PG	G4YYS
SO18 5QJ	G4PTF
SO18 5QL	G3KXE
SO18 6BB	G0UJP
SO18 6LY	M7ABD
SO19 0EH	M7CJE
SO19 0JG	M0HME
SO19 1AD	M7KAG
SO19 1BX	G6XYS
SO19 1DD	M6JLW
SO19 1DP	G8BJB
SO19 1FW	G4UDB
SO19 2HF	G6HHH
SO19 2HJ	G7NKZ
SO19 2HU	M0JBD
SO19 2NU	G1OQI
SO19 4DE	G3NZL
SO19 4DX	G4HUM
SO19 5JP	G0EUC
SO19 5JP	G1LDN
SO19 5LF	G0LCN
SO19 5LN	G0ILA
SO19 6FB	M3VPA
SO19 6HB	G7RAB
SO19 6NG	M6PHP
SO19 6PS	G7BLX
SO19 6QT	G0JBH
SO19 6QX	G0EMV
SO19 7GG	G8RNT
SO19 7GG	M5PSW
SO19 7JN	G7MTA
SO19 7JN	M0DDB
SO19 8AF	G7BZU
SO19 8EY	G4UEN
SO19 8FF	G0IOI
SO19 8FS	2E0TVZ
SO19 8FS	M3TVZ
SO19 8FT	M7BOU
SO19 8LH	2E0KBX
SO19 8NT	G4OLY
SO19 8NT	G8WWM
SO19 8SJ	M6AKE
SO19 9AT	G6YVJ
SO19 9BL	M7PVS
SO19 9DA	2E0TAQ
SO19 9DA	M3UUF
SO19 9GB	G2SZ
SO19 9GN	M0REV
SO19 9HY	G8EXZ
SO19 9LJ	2E0PKK
SO19 9NJ	2E0EHJ
SO19 9NJ	M6MSO
SO19 9PE	G8GBP
so19 9rw	M7DRC
SO19 9SS	M3OCR
SO20 6AH	G8IUP
SO20 6BA	G3RDQ
SO20 6BA	G6MCN
SO20 6EY	M6AEL
SO20 6EY	M6GTH
SO20 6HE	G8CDA
SO20 6NY	2E0UKM
SO20 6NY	M3UKU
SO20 6PR	G4NMS
SO20 6PR	M1BPD
SO20 8DB	G3LGA
SO20 8DB	G8RJM
SO20 8EA	M1BDJ
SO21 1AF	G3JRH
SO21 1BA	2E0VBN
SO21 1BA	M3VBN
SO21 1BQ	G6LKW
SO21 1TE	M7BME
SO21 1TT	M0WZM
SO21 1US	G3UPD
SO21 1XE	M7AWQ
SO21 2BX	G8TEC
SO21 2DE	G3HRH
SO21 2EG	G8IDJ
SO21 2EP	G3AJD
SO21 2EP	G4NZC
SO21 2LG	G3ZEO
SO21 3DU	G0LMD
SO21 3EB	G0GWG
SO21 3EB	G3ZHV
SO21 3EE	M3MKJ
SO21 3EY	G1NBU
SO21 3EY	G3LMH
SO21 3EZ	G3VEH
SO21 3FS	M7LKY
SO21 3HP	G3SJK
SO21 3HP	G4XRJ
SO21 3HY	2E0FNY
SO21 3HY	M6FNY
SO21 3JG	G4EDW
SO22 4DN	2E0SSC
SO22 4DN	M6SJC
SO22 4DU	M6CNS
SO22 4ED	M7BHU
SO22 4EE	M0AYV
SO22 4EE	M0REX
SO22 4ET	G3NVB
SO22 4EY	G3MCL
SO22 4HS	G3VPG
SO22 4HS	G4JXE
SO22 4HU	G1THP
SO22 4JY	G3PTS
SO22 4LD	G8PGE
SO22 4LF	G4SVC
SO22 4LS	M7KOR
SO22 4LX	M7JRG
SO22 4NU	G8DGC
SO22 4PQ	2E1FKZ
SO22 4PQ	G6AAZ
SO22 4PQ	M3SFC
SO22 4PZ	G4AFA
SO22 4QH	G3XJM
SO22 5AT	G3RQR
SO22 5AX	G0DZV
SO22 5BU	G4HDY
SO22 5DJ	G6BZE
SO22 5HU	G3RUZ
SO22 5JX	G4CJO
SO22 5ND	G0JMK
SO22 5ND	G3UYK
SO22 5NT	M6BIN
SO22 6BA	M0BTR
SO22 6BA	M3KER
SO22 6FE	G4FKR
SO22 6HE	G3YSK
SO22 6HE	M0AYY
SO22 6HN	M7YDX
SO22 6LT	M1RGW
SO22 6NZ	G4TBI
SO22 6NZ	G8CQG
SO22 6PS	M6IBL
SO22 6SG	G4AXO
SO23 0NU	G4AXY
SO23 0PP	G0IHA
SO23 0PX	G4AZU
SO23 7DA	M3KMT
SO23 7EQ	G3ROG
SO23 7HT	G1HRA
SO23 7ND	G0TAH
SO23 7ND	G1NCM
SO23 7QQ	G0IVR
SO23 7QQ	G4DKH
SO23 7QQ	G6IVR
SO23 7XH	G3HQX
SO23 9DT	G1EHE
SO23 9SR	G4LYL
SO24 0DU	G4YDE
SO24 0DU	G6VXR
SO24 0HP	G4LQW
SO24 0HP	G8UDZ
SO24 9HF	2E0LLZ
SO24 9HF	M0LLX
SO24 9HF	M6MQY
SO24 9HH	2E0CGS
SO24 9HH	M3ETH
SO24 9JJ	M7CYQ
SO24 9PE	G6SDC
SO24 9PL	G0JHT
SO24 9PP	G4DMG
SO24 9RP	G8JCB
SO24 9SG	G7DNF
SO24 9TX	G7JTR
SO3 4BA	G6XIR
SO30 0BP	G6WYS
SO30 0BS	2E0YAW
SO30 0BS	M6MZZ
SO30 0GR	2E0JVM
SO30 0HU	G4KUL
SO30 0LZ	M3TZF
SO30 0PH	G4JNT
SO30 2AR	M7BTW
SO30 2AX	M0JWA
so30 2ax	M7JGX
SO30 2BH	2E0MNX
SO30 2BH	G4MNX
SO30 2BH	M6OFQ
SO30 2BJ	M0HBC
SO30 2NY	G3YOM
SO30 2UD	M3UVX
SO30 2UD	M6RGQ
SO30 3EQ	G3IBY
SO30 4FT	G3RQF
SO30 4HA	M1RST
SO30 4HE	M6CNQ
SO304EZ	G0HCI
SO31 1AD	M6GXV
SO31 1AP	G3PDP
SO31 1BJ	G3SED
SO31 1BY	G1PVD
SO31 1EA	G3YSG
SO31 1EA	G8YUO
SO31 4HD	M1ABG
SO31 4JS	G3KTM
SO31 4QG	M6BQB
SO31 4RT	G6IXE
SO31 5FE	G0WSB
SO31 5GR	G0HAE
SO31 5HP	M7EIW
SO31 6EP	G1WXT
SO31 6EP	G4UDY
SO31 6LJ	M6DNN
SO31 6PE	G0UTU
SO31 6PE	G6CMF
SO31 6PP	M0UOO
SO31 6PP	M3UOO
SO31 6QG	G3UZK
SO31 6RN	G0UUS
SO31 6RR	G8HER
SO31 6SP	G1GSY
SO31 6SX	G8PGA
SO31 6SY	G4KWY
SO31 6TD	M7RUT
SO31 6WX	2E0IED
SO31 6WX	M0IED
SO31 6WZ	M3UUS
SO31 6XX	G0TCQ
SO31 7BX	2E1IIC
SO31 7BX	G0CAJ
SO31 7BX	G0SCX
SO31 7BX	M3IEQ
SO31 7BZ	G0FBC
SO31 7EY	2E1FKT
SO31 7HQ	G8ACL
SO31 8AT	G8IOK
SO31 8AY	G4NHN
SO31 8EL	G0MYJ
SO31 8EN	G6XQR
SO31 8HE	M6BGF
SO31 9AU	M7CYS
SO31 9FG	G3HQT
SO31 9FU	G3NNW
SO31 9GD	M6ETZ
SO31 9HG	G4PXH
SO31 9JU	M6BPR
SO315DQ	M7BPE
SO316TN	M6JEG
SO32 1AY	M7LBM
SO32 1EW	G4EWT
SO32 1EY	2E1IWG
SO32 1JR	G1GXB
SO32 1JS	G4NYD
SO32 1RP	2E0LZG
SO32 1RQ	G1JBG
SO32 1SG	2E1CXI
SO32 2AA	M3FOV
SO32 2AR	G4COM
SO32 2AR	G4IDW
SO32 2BD	G1UTC
SO32 2BD	M1CCA
SO32 2LG	M0GFM
SO32 2LX	M7ENU
SO32 2NP	2E0CSK
SO32 2NP	M0HPU
SO32 2NP	M6BZA
SO32 2NR	G8KZO
SO32 2TG	2E0ERJ
SO32 2TG	M0IUE
SO32 2TG	M6IQZ
SO32 2TL	G4ZRT
SO32 2TR	G6NXM
SO32 3LE	M6CSA
SO32 3NP	2E0UKB
SO32 3NP	M6UKG
SO32 3QN	G1PRP
SO4 4WQ	G0ORV
SO40 2GP	M7FUN
SO40 2NA	G0COC
SO40 2NL	2E0SJQ
SO40 2NL	M0BJL
SO40 2NL	M6SCX
SO40 2NW	G6ANI
SO40 2NW	M7DIQ
SO40 2RF	G7HOK
SO40 2SD	G7EOH
SO40 2UP	G6MEW
SO40 2UY	G0LDJ
SO40 2WD	G3GLW
SO40 3BN	G6XYF
SO40 3HP	G0BOC
SO40 3HP	G0OSD
SO40 3LP	G4VII
SO40 3NS	M0LCC
SO40 3NX	2E0WHH
SO40 3NX	M3WHH
SO40 3QP	M6TIY
SO40 4SL	G1KSE
SO40 4UT	G4IXE
SO40 4UT	M3YZJ
SO40 4XB	G4ADP
SO40 4XY	G4BUW
SO40 7AL	G3VSL
SO40 7AL	G6GFA
SO40 7AT	2E0CZT
SO40 7AT	M0RHO
SO40 7AT	M6MQD
SO40 7AT	M6RFG
SO40 7AY	2E0GJE
SO40 7AY	M6GJE
so40 7dp	M6MYV
SO40 7FD	2E0FQS
SO40 7GP	G4BIZ
SO40 7GY	G0FOH
SO40 7JH	G4YWZ
SO40 7LA	G0VNI
SO40 7LA	G0WIL
SO40 8JB	G4PGW
SO40 8US	G1LGY
SO40 8WB	M0DYA
SO40 9AX	M3PFF
SO40 9BT	G0PGZ
SO40 9DR	M0WSR
SO40 9DR	M3WSR
SO40 9UZ	G8KYK
so409lz	G0WFQ
so409lz	G3SOU
so409lz	G4SJW
so409lz	G8FAB
SO41 0GG	G4TOG
SO41 0GP	2E0KGC
SO41 0GP	M0GTL
SO41 0GP	M6KGC
SO41 0GS	2E0VAG
SO41 0GS	M0MDG
SO41 0GS	M0TBI
SO41 0GS	M3VVW
SO41 0GU	G3GRV
SO41 0HQ	G1RAX
SO41 0JD	G4HYW
SO41 0LQ	G4RFP
SO41 0PB	G7JWV
SO41 0PP	G4LUN
SO41 0PR	M0LKD
SO41 0QT	M3OBU
SO41 0QU	2E0NHS
SO41 0QU	M3NHE
SO41 0QW	G4DNE
SO41 0RL	2E1JLC
SO41 0RR	2E0DCD
SO41 0RR	M0LMB
SO41 0RR	M6BOF
SO41 0RR	M7DAN
SO41 0SG	G3KLH
SO41 0TE	G1UFL
SO41 0TE	M3OTU
SO41 0TF	G3ZJY
SO41 0UL	G4BMC
SO41 0UX	G2HCG
SO41 0XG	G8YFK
SO41 0XS	2E0DAY
SO41 0XS	M3SJD
SO41 0ZG	M6WOK
SO41 3AE	M7PPM
SO41 3PB	G0COE
SO41 3QX	G4HPN
SO41 3SD	G0TXN
SO41 3TF	M6NGW
SO41 5PA	M6CJG
SO41 5PA	M6LYN
SO41 6AL	G0PCW
SO41 6AX	G7OGT
SO41 6BB	G4SBB
SO41 6BQ	G0SNK
SO41 6DH	M3ZIN
SO41 6DW	G4GAB
SO41 6EJ	2E0UXS
SO41 6EJ	M0UXS
SO41 6EJ	M3UXS
SO41 8BD	G1KFH
SO41 8BG	G3ZIE
SO41 8DN	G4ZAX
SO41 8DT	M3ECF
SO41 8DY	G0BCS
SO41 8DY	G3TEI
SO41 8DY	G4UAA
SO41 8GQ	G4CWS
SO41 8GQ	M6GKH
SO41 8HN	M6DPY
SO41 8YD	2E0LYM
SO41 8YD	M7LYM
SO41 9BL	M0ZDU
SO41 9BP	M0NEC
SO41 9DX	M6JQS
SO41 9GT	M6VCC
SO41 9LB	G0JZW
SO41 9LB	M0ULD
SO41 9LT	G4XNA
SO42 7PB	G8CAV
SO42 7RD	G7JFU
SO42 7TT	G4FDX
SO42 7WQ	2E0DGT
SO42 7WQ	M6DPN
SO43 7BN	M3FQM
SO43 7BN	M3FQN
SO43 7JB	G3NOA
SO43 7JP	M3FWT
SO45 1BH	G1RET
SO45 1BJ	G0HDI
SO45 1BN	G7KDN
SO45 1BR	G4CLF
SO45 1DA	G0EUJ
SO45 1DQ	M7KDU
SO45 1EG	G6LVJ
SO45 1FX	2E0WPH
SO45 1FX	M0PMH
SO45 1FX	M6PPH
SO45 1WF	G6NXV
SO45 1WH	G1MKY
SO45 1WJ	G7RUH
SO45 1WX	G0OSG
SO45 1XP	G3TPV
SO45 1XW	G0BPA
SO45 1XW	G6TGB
SO45 1YE	G1TRI
SO45 1YN	G1UEV
SO45 1YX	G6UMT
SO45 2EY	G7AYA
SO45 2HN	G6OJH
SO45 2HP	G0SBV
SO45 2JH	2E0GYP
SO45 2JH	M6JAS
SO45 2JT	G6MNL
SO45 2JT	G6XMA
SO45 2JZ	G4KNN
SO45 2ND	G0SZI
SO45 2NQ	G4JYN
SO45 2NQ	G4YVY
SO45 2PP	G4SBF
SO45 2QJ	G0KJR
SO45 2QN	M6STA
SO45 3DJ	M6GAG
SO45 3GB	2E0CXQ
SO45 3HJ	M0GZB
SO45 3LG	G6MZF
SO45 3LJ	G1UDW
SO45 3LJ	M6CRA
SO45 3LS	G3KSF
SO45 3PH	M6CAQ
SO45 3QF	2E0MBG
SO45 3QF	M6PWT
SO45 3QG	2E0GAF
SO45 3QG	M6GDS
SO45 3RJ	2E0HZW
SO45 3RJ	M0KBU
SO45 3RJ	M7DNN
SO45 4BY	M6LFO
SO45 4HS	G3OZT
SO45 4HU	M6HYT
SO45 4LE	G0WCB
SO45 4LR	G8CLK
SO45 4LS	M0IKT
SO45 4LZ	G7AFT
SO45 4RS	2E0XEE
SO45 4SB	G0WSI
SO45 5BP	G8APM
SO45 5DL	G1JRU
SO45 5ER	G3NJG
SO45 5EX	G8BAL
SO45 5FP	G3WJM
SO45 5FQ	G1UBN
SO45 5QS	G4SGJ
SO45 5QW	G6MUX
SO45 5TW	G4NOP
SO45 5UD	G1RCW
SO45 5UW	G4GCI
SO45 6AY	G4JAX
SO45 6BN	2E0EXC
SO45 6DL	G4MOL
SO45 6DW	G8BMQ
SO45 6EU	G0JGI
SO45 6EW	M3ZNJ
SO45 6JR	G8TLT
SO45 6JT	G4BPN
SO45 6LA	2E0BSQ
SO45 6LA	M0GMD
SO45 6LA	M6EMG
SO451BP	M6PCV
SO50 4BW	2E0RKE
SO50 4BW	M6RKE
SO50 4BX	2E0DMW
SO50 4BX	M3DMW
SO50 4GT	M7EKL
SO50 4LW	2E0IOU
SO50 4NS	M6PDR
SO50 4NZ	M3NSH
SO50 4PP	G0TBC
SO50 4PP	M1FDO
SO50 4PQ	G0OFX
SO50 4QY	G3VHW
SO50 4RW	G3NML
SO50 5AD	M0MTN
SO50 5AD	M3TYM
SO50 5AS	M3FEL
SO50 5EN	G4KMP
SO50 5EN	M6BBH
SO50 5HZ	M1CNK
SO50 5ND	M7BVZ
SO50 5NE	M3DCJ
SO50 5RR	G0BES
SO50 6AN	M6LGI
SO50 6AY	G8MDK
SO50 6AY	M0HHX
SO50 6AY	M6IDE
SO50 6BD	G1TJH
SO50 6DG	G3VRB
SO50 6FX	2E0GSC
SO50 6FX	M0OHI
SO50 6FX	M3OHI
SO50 7GG	M0LXD
SO50 7HB	G3YGF
SO50 7JJ	M1MAB
SO50 8AA	M3VIA
SO50 8AN	M7IIZ
SO50 8FL	G3XIV
SO50 8PF	M6FCW
SO50 8PH	G8VQA
SO50 8PU	M3RBU
SO50 9BH	M5MDH
SO50 9JT	M3NAT
SO50 9PW	M6NCE
so507ny	2E1EEK
SO51 0GD	M6OSL
SO51 0GQ	G0RLA
SO51 0HG	M3ECD
SO51 0JL	2E0CHL
SO51 0JL	M6ASP
SO51 0JP	M0TOG
SO51 0LF	G4GSK
SO51 0LW	G1GYT
SO51 0RA	G8LLJ
SO51 5PQ	G1TWW
SO51 5PZ	M7MZM
SO51 5RJ	G6FVZ
SO51 5SP	G7VZY
SO51 6AR	G8DXF
SO51 6BT	2E0GHZ
SO51 6BT	M0XIG
SO51 6EE	G0FHC
SO51 6ET	G0OGW
SO51 6EX	G3NAE
SO51 6FU	G0HLA
SO51 6JT	2E0HXW
SO51 7HU	M0BTZ
SO51 7JN	M6NFD
SO51 7JW	M3ECU
SO51 7JY	G6CPE
SO51 7JZ	G8OFX
SO51 7LH	G3FWI
SO51 7LL	G4YVE
SO51 7RR	M3ECJ
SO51 7RU	G4MNP
SO51 7RU	G8ZMM
SO51 7TE	G7PEC
SO51 7TE	G7UHE
SO51 7TE	M0EHF
SO51 7TJ	G1NTN
SO51 8BR	2E0CGR
SO51 8DF	G3MDR
SO51 8DP	G8OQP
SO51 8DY	M3ECQ
SO51 8EQ	G6GCI
SO51 8FJ	M6BWB
SO51 8HG	G7OOF
SO51 8LH	G4XBZ
SO51 8PN	M7BOC
SO51 9BT	G4YEE
so510ax	G8IPG
SO52 9EU	G6TRW
SO52 9FD	2E0DJU
SO52 9FD	M6EPA
SO52 9GB	2E0KWC
SO52 9GB	M3YKC
SO52 9GB	M3YMC
SO52 9JG	G3XSD
SO52 9JT	G7AJR
SO52 9NB	2E0OGR
SO52 9NB	M7MLC
SO52 9NS	2E0BHY
SO52 9NS	M1PVF
SO53 1GD	G4DZS
SO53 1LE	G7UDM
SO53 1LN	M1LMJ
SO53 1NA	2E0KCO
SO53 1NA	M0KCO
SO53 1NA	M3KCO
SO53 1SA	M3FFO
SO53 1SA	M3HOE
SO53 1SF	M3YJT
SO53 2AY	G1UDR
SO53 2BD	M7RYV
SO53 2BP	2E0SIM
SO53 2BP	M3ALZ
SO53 2DJ	G4EZC
SO53 2FE	G8ZBN
SO53 2FS	M0TDG
SO53 2FY	M0AYU
SO53 2GX	G7AJE
SO53 2LJ	G3UYD
SO53 2LL	M6SRD
SO53 2PU	G6XJZ
SO53 2PU	M0MEH
SO53 3BS	G4BGT
SO53 3BS	G6JGR
SO53 3EB	G4LDC
SO53 3GD	G0TAA
SO53 3GJ	G6AMW
SO53 3PA	2E0GBT
SO53 3TP	G7POC
SO53 4HP	G6MCX
SO53 4HQ	2E0FMG
SO53 4HQ	M6FMG
SO53 4HU	M3SKT
SO53 4JL	M7AYP
SO53 4LD	G7VLH
SO53 4LN	2E0NIF
SO53 4LN	M0NIF
SO53 4PD	M7WKD
SO53 4QQ	G7BOH
SO53 4RW	M3SEE
SO53 4SW	G0DQH
SO53 4TT	G3WII
SO53 5AT	G4MZU
SO53 5AX	G4HHZ
SO53 5BX	2E1ECV
SO53 5PA	G4HWM
SO53 5PB	G8LVC
SO53 5QD	2E0AJP
SO53 5QJ	M0PHB
SO53 5QJ	M7DIL
SO53 5QP	G6IQI
SO53 5RP	G1NPI
SO53 5RP	G8IPQ
SO53 5RP	M3IPQ
Solihull	G4YZA
SP1 1AL	G3OMZ
SP1 1NU	G0DJV
SP1 1PU	M0NJP
SP1 1PU	M3VOW
SP1 1QL	G3MXP
SP1 2EY	G8FSL
SP1 2JH	G8WJB
SP1 2JX	2E1MEL
SP1 3AU	G4YFJ
SP1 3BH	G7DNG
SP1 3DN	G6CZB
SP1 3FX	M0RNG
SP1 3HS	G4MOE
SP1 3JZ	G6BBG
SP1 3LW	G4YVM
SP1 3LW	M0PCN
SP1 3PR	G7WAA
SP1 3PR	M0CLI
SP1 3PZ	G4TRD
SP1 3RT	M6DPX
SP1 3WE	G2AQJ
SP1 3WS	G7JGB
SP10 1HH	G4ABL
SP10 2AH	G0OMD
SP10 2AX	2E0IEL
SP10 2BU	2E0BZJ
SP10 2BU	M3CEB
SP10 2EN	G8CPJ
SP10 2FS	M6SPC
SP10 2HE	G4YSB
SP10 2PZ	M0ALE
SP10 2QT	G0OUO
SP10 2QT	G7PVE
SP10 2RB	G2FYJ
SP10 3AY	M5LDF
SP10 3BN	G6FIB
SP10 3BU	M3FWS
SP10 3DS	G6YEY
SP10 3EP	G7VRJ
SP10 3HL	G0JBY
SP10 3JT	M0BMF
SP10 3NG	2E0AJG
SP10 3NG	M7AJG
SP10 3PE	2E0CLW
SP10 3PE	M0VZS
SP10 3PE	M3VZS
SP10 3RZ	2E0SNB
SP10 3RZ	M7SAB
SP10 3SS	M6GAB
SP10 3UH	M3SVC
SP10 3XD	M6DQN
SP10 4AR	G1HPS
SP10 4DB	G7EBI
SP10 4EN	G0AMO
SP10 4EN	M3DEA
SP10 4EZ	M1EYS
SP10 5BS	2E0CCR
SP10 5DB	G7CFX
SP10 5HT	G8CEP
SP10 5JR	G7GHS
SP10 5LD	M7DUR
SP10 5NA	G6ARC
SP10 5NA	G8NDN
SP10 5NJ	2E0NTJ
SP10 5NJ	M6NTJ
SP11 0LF	G6ARW
SP11 0PZ	M7KPW
SP11 0QF	2E0GNC
SP11 0QF	M0GNC
SP11 0QT	G8ARH
SP11 0RE	G6SDE
SP11 0RS	G8FHI
SP11 6EA	G3LTF
SP11 6EW	M0JIM
SP11 6FD	G6VTA
SP11 6JY	G6SJG
SP11 6RH	G4XYG
SP11 6RL	2E0ESP
SP11 6RL	M6ESP
SP11 6RR	M6AIL
SP11 6SS	2E0DUM
SP11 6SS	M6IVN
SP11 7AT	G7TAE
SP11 7BJ	2E0LBK
SP11 7BJ	M0LBK
SP11 7BJ	M3LBK
SP11 7BN	G4FYM
SP11 7BX	G6YTY
SP11 7DF	M1BSX
SP11 7EH	G3XVS
SP11 7ER	G4OZL
SP11 7ER	G6WBG
SP11 7ES	G4WEN
SP11 7LL	M1CFW
SP11 7NX	2E0PTZ
SP11 7NX	M0PTZ
SP11 7NX	M6PTU
SP11 8DW	M6RQE
SP11 8LG	G6JSF
SP11 8NE	G8GYS
SP11 8NS	G3ZWN
SP11 9AN	G4NNS
SP11 9JD	G4CIZ
SP11 9NN	M7GBO
SP11 9PG	G1XSV
SP2 0AX	G4WDA
SP2 0LJ	G8WBO
SP2 0LW	G4RLF
SP2 0NF	G4SXQ
SP2 7BY	M3YCV
SP2 7EX	G8CWJ
SP2 7FY	G4FMZ
SP2 7SU	G0MLJ
SP2 7UE	G7MJV
SP2 8DL	M6ZCM
SP2 8DN	G0BDP
SP2 8EF	M0BWC
SP2 8EP	G3KOZ
SP2 8FG	M7BIQ
SP2 8HZ	M6TOW
SP2 8JX	G0VMC
SP2 8LY	G7BSO
SP2 8NR	G3OIL
SP2 8QS	G4HQM
SP2 8RN	G4LWU
SP2 8RN	G6RLG
SP2 9BY	G8NDV
SP2 9DF	M7KRH
SP2 9EJ	G7EXZ
SP2 9FZ	G6GWA
SP2 9HJ	M6THJ
SP2 9HN	2E1MEP
SP2 9LD	M3EQW
SP2 9LF	G1BMT
SP2 9LQ	M6OER
SP2 9LU	M0VSR
SP2 9LU	M0WSC
SP2 9PA	G1DGL
SP2 9PE	M0RED
SP2 9PF	M7OPY
SP27LQ	M7JFK
SP3 4AH	G8EMA
SP3 4BQ	M3VBM
SP3 4EB	M7MCU
SP3 4LP	G4DUA
SP3 5AD	M6OMF
SP3 5JL	G4FRB
SP3 5LF	G4MQW
SP3 5LF	G8DER
SP3 5PE	M3BIZ
SP3 5PW	M0TRB
SP3 6EX	M7MWG
SP3 6JL	M0ISW
SP4 0AA	G6ABM
SP4 0BY	G4AEI
SP4 0ER	G8ALR
SP4 0ER	M0BVO
SP4 0LE	2E0XKO
SP4 0LE	M0XKO
SP4 0LE	M3XKO
SP4 6AU	G8HUV
SP4 6DJ	G6JGP
SP4 6EE	2E0LZE
SP4 6FP	M7CYG
SP4 6HW	M0MWS
SP4 6LS	G8OFA
SP4 6LS	G8RHC
SP4 6LX	G0HSV
SP4 6NB	G1SSZ
SP4 6PX	2E0FDB
SP4 6PX	M6FHS
SP4 7AS	G6YUX
SP4 7EE	G3YFE
SP4 7FF	G4WWZ
SP4 7FS	M6FQL
SP4 7HW	2E0DLT
SP4 7HW	M6CCN
SP4 7NN	G4UED
SP4 7PJ	2E1CMZ
SP4 7PP	G4SXT
SP4 7PW	M0IHU
SP4 7QS	G6UXK
SP4 7RE	2E0IKU
SP4 7RE	G7GKD
SP4 7RG	G1WRY
SP4 7RG	G4YRV
SP4 7RL	G6JYN
SP4 7TD	G4WPI
SP4 7TU	M3ZYR
SP4 8AB	M1AFF
SP4 8BY	G8LYP
SP4 8DJ	G6JRS
SP4 9DA	2E0FUE
SP4 9DA	M7AJU
SP4 9HE	M6IZM
SP4 9PJ	M0RVN
SP4 9QG	M6NIQ
SP4 9RW	M7ERF
Sp47rn	M6IDU
SP48ES	2E0HNR
SP48ES	M7CII
SP5 1NS	2E0PTF
SP5 1NS	M0OOS
SP5 1NS	M6SEE
SP5 1PL	M3VKB
SP5 1PP	G4YIS
SP5 1PW	G0PBS
SP5 1RB	G4LDR
SP5 1RE	G8PBY
SP5 1RJ	M6WFH
SP5 1RS	G1ITJ
SP5 1SH	2E0KDI
SP5 1SH	G3MDM
SP5 1SZ	2E0RFL
SP5 1SZ	M6HBN
SP5 2AL	G4VQJ
SP5 2BE	G0OSW
SP5 2BP	M6KMZ
SP5 2BZ	G7WBA
SP5 2HQ	M0GWQ
SP5 2HT	G6ILC
SP5 2HU	G3OSQ
SP5 2LN	G0RZB
SP5 2LQ	M3HCB
SP5 2LR	M0CBM
SP5 2LR	M3DZN
SP5 2NQ	G3SHK
SP5 2NR	G4TEK
SP5 2SE	G4KCM
SP5 2SZ	M0GSP
SP5 3AR	G1ELZ
SP5 3AT	G3YWT
SP5 3AZ	G0CDZ
SP5 3BF	2E0HEA
SP5 3BF	M3FQG
SP5 3FN	2E0CFP
SP5 3FN	M6AFN
SP5 3LB	2E1TON
SP5 3LB	G6ZHJ
SP5 3LG	M3MGD
SP5 4JT	2E1NAC
SP5 4JT	2E1PEC
SP5 4LF	M7LCL
SP5 5BH	G6FPC
SP5 5NZ	M6OVI
SP5 5PF	G7HRB
SP5 5QJ	2E0DZJ
SP5 5QJ	M6HFO
SP5 5QL	G7IRU
SP5 5QL	M3CAQ

SP5 5SA M6CPO
SP51PP M6OVN
SP52SZ M7DDF
sp5-5pl M1HFX
SP6 1AY 2E0KZC
SP6 1AY M0KZC
SP6 1AY M3KZC
SP6 1BJ M3HZD
SP6 1BP G7OZH
SP6 1BW 2E1FRY
SP6 1EG M3UQA
SP6 1EQ G6CEZ
SP6 1JF M0WDZ
SP6 1LW G4POF
SP6 1NW M3NMX
SP6 2AL 2E0NCD
SP6 2AX G8OFO
SP6 2LJ G1MVG
SP6 2NR G4GBP
SP6 3AD M6PSQ
SP6 3DJ G0TQZ
SP6 3DU 2E0HJE
SP6 3DU M6VLB
SP6 3EE G7VOX
SP6 3EP 2E0RAH
SP6 3EW G8BIH
SP6 3HA G1VOQ
SP6 3LD 2E0RFK
SP6 3LD M3RFK
SP6 3QA G3ZNH
SP6 3RB G1JMH
SP6 3RB G6LVI
sp63pj M7CIO
SP7 0BF M7KRN
SP7 0BJ M3HEV
SP7 0DP G7ESE
SP7 0DP M6DGS
SP7 0HS M3SQI
SP7 8AL G0YBU
SP7 8GY G3NRH
SP7 8HX G3LGF
SP7 8LQ G8BSP
SP7 8NF G3BVB
SP7 8NQ G3FWU
SP7 8PP M7WSX
SP7 8QS M0TMO
SP7 8RU M6KVH
SP7 8RX G4OVJ
SP7 9HD G8LBG
SP7 9HL G0MFR
SP7 9HX G0UOD
SP7 9NA M6KZH
SP7 9NX G1RJA
SP7 9PA 2E1EGU
SP7 9PF G1NWT
SP7 9QB M3JLH
SP7 9QG 2E0TYH
SP7 9QG M0TYH
SP7 9QG M6TYH
SP8 4EL G4FUY
SP8 4EU G3ZBB
SP8 4GN M6RFI
SP8 4HH G0VNA
SP8 4LL 2E0MEK
SP8 4LL M6TEO
SP8 4NR G8PQJ
SP8 4NT M7ANG
SP8 4PE G7SUS
SP8 4QJ G4ILM
SP8 4QY M6OKD
SP8 4RB G8DDN
SP8 4RE M7DCN
SP8 4RR G4VRM
SP8 4SS G1SNI
SP8 4TR M6KZI
SP8 4TT M6LNM
SP8 4TW G4FQV
SP8 4UP G0YLO
SP8 4UP G0ZEE
SP8 4UP G6TER
SP8 4WE M6ROA
SP8 5AL G3LCL
SP8 5BB M3UZE
SP8 5EL M0AKI
SP8 5ET 2E0RIH
SP8 5ET M6JGH
SP8 5EW G4FDI
SP8 5LB G0MPM
SP8 5LW G8SYA
SP8 5NB G1THG
SP8 5QR M5BJC
SP8 5RL G7JIF
SP8 5RL M7PRW
SP8 5RN M3TBH
SP8 5SJ G8HPN
SP9 7EU G0WYD
SP9 7FN M6ZBW
SP9 7JX 2E0EQL
SP9 7JX M6YSD
SP9 7SB M6NHP
SP9 7TR 2E0DUJ
SR1 2DP M3XYH
SR2 0BP G4DGB
SR2 0JU G0SLN
SR2 7DE M7LCP
SR2 7EZ G6GGT
SR2 7HB G7FFV
SR2 7RX M3REP
SR2 7TS G4GTX
SR2 8HR M7NAW
SR2 8NF M0MDP
SR2 8QB G7PQD
SR2 8RS G6LMR
SR2 8RX G6INK
SR2 9BB G0KFY
SR2 9DX G0ASM
SR2 9EE G4MSJ
SR2 9EJ G4TOI
SR2 9HQ G3YJG
SR2 9LQ G0KVJ
SR2 9LQ G0NSK
SR2 9LQ M6NSK
SR3 1AN G4HPS
SR3 1AN G4OBX
SR3 1HJ G0BNK
SR3 1HJ G7MFN
SR3 1HJ M0IMD
SR3 1HJ M3HYD
SR3 1JF G0TAX
SR3 1JR M6YOZ
SR3 1JW 2E0PYN
SR3 1JW M6AVG
SR3 1LG G7KOF
SR3 1LX G4MTW
SR3 1QS M1CQX
SR3 1UJ G0BWJ
SR3 2RF M0NGB
SR3 2RG G6SGZ
SR3 2YG M7GLK
SR3 3AL M1DPQ
SR3 3AN M6JIH
SR3 3PX G3ZOG
SR3 3SF M0ATA
SR3 4AA G0AOE
SR3 4DJ M6IOL
SR3 4EE M6HAM
SR3 4EZ G4VLT
SR3 4HS 2E0MAL
SR3 4HS M3MCF
SR3 4HZ M7ODO
SR3 4JN 2E0PAX
SR3 4JN M6UVF
SR3 4PA G7PUK
SR34AT 2E0HHE
SR4 0AB G1YIZ
SR4 0AE G4HIX
SR4 0BA M0KLL
SR4 0BF M0MHG
SR4 0LZ G0MCT
SR4 0NA M3XYP
SR4 0QZ 2E0UKA
SR4 0QZ M3XJX
SR4 6AU M3JFB
SR4 6EY M6MOB
SR4 6XG G7MXM
SR4 7LL M6RNQ
SR4 7RY M0GFN
SR4 7SA G1YUL
SR4 7SU M1CPB
SR4 7TB M0RKW
SR4 8AU G0EHX
SR4 8BG M6GNX
SR4 8BS 2E0YDT
SR4 8HT G8HPW
SR4 8NP G3WOM
SR4 9DW G1WEV
SR4 9EN G8NPP
SR4 9NQ G0SRG
SR4 9NQ G4EKM
SR4 9QU M7GBH
SR40HR M7DAD
SR5 1DY G4XYP
SR5 1LG 2E0OAP
SR5 1LG M6OAP
SR5 1LL M6UNS
SR5 2BU G6VTH
SR5 2DJ M7MKO
SR5 3DD M7BBW
SR5 3EG G0ILV
SR5 3PG M0GOL
SR5 3RE M5GHT
SR5 4BU G1CSA
SR5 4DF G0OBN
SR5 4LH M6MMY
SR5 4ND M0KAE
SR5 5AH 2E0HYE
SR5 5AH M0HYE
SR5 5AH M6BYE
SR5 5AW 2E0SNE
SR5 5AW M6FIT
SR5 5ET M6CWQ
SR5 5ET M7ATI
SR5 5QA G0UNE
SR5 5RD G7CKQ
SR5 5RD M3XQB
SR5 5SA M6OKH
SR52NX M6ELL
Sr53bu M7KEV
SR6 0JP G0BAN
SR6 0NL M6LND
SR6 0NT G4TMX
SR6 7AL G7PHI
SR6 7BW 2E0TLX
SR6 7BW M0TLX
SR6 7BW M3TLX
SR6 7HN G4LDT
SR6 7HN G8WTZ
SR6 7LN 2E0CUC
SR6 8BD G3TEC
SR6 8ER 2E0ERD
SR6 8ER M6ERD
SR6 8ET G4MRK
SR6 9AY M7CWX
SR6 9EL G4NTW
SR6 9HP G0IID
SR6 9NT 2E0DJJ
SR6 9NT M0KCF
SR6 9NT M6GPM
SR7 0AN 2E0EXH
SR7 0AN G8TDP
SR7 0AU G6LNL
SR7 0BD G6AGY
SR7 0BD G6AGZ
SR7 0JT G0MJV
SR7 0LP G0WTW
SR7 7BQ G4GSO
SR7 7DJ G0MKC
SR7 7DJ G4PPL
SR7 7JZ M7BMG
SR7 7NG 2E0WRY
SR7 7NG M6WRY
SR7 7QW M3UKJ
SR7 7SA G1ERU
SR7 7SA M3SMM
SR7 7UE M7GNR
SR7 7UJ M6YIK
SR7 8DG G0DVP
SR7 8DZ G4NMF
SR7 8EE M6GTQ
SR7 8HW 2E0JRB
SR7 8JG M6EQG
SR7 8JT M6OOB
SR7 8JZ G0UFN
SR7 8NL G0NWY
SR7 8NS M0MED
SR7 8RZ M6SQC
SR7 9BN 2E0SNM
SR7 9BN M6SMA
SR7 9PQ G4NMK
SR7 9PQ M7DCA
SR8 1DD G0FBW
SR8 1DG 2E0HYG
SR8 1DG M0TZR
SR8 1DG M3UMG
SR8 1DQ G0CWF
SR8 1LN G0TOK
SR8 1LP G4PLU
SR8 1PZ M0GZW
SR8 2DJ G4WKT
SR8 2DT G4RXR
SR8 2EA G0RDD
SR8 2HB G6VHG
SR8 2HE. M6HWN
SR8 2JS M0WUS
SR8 2JS M3AUP
SR8 2JW 2E1DMI
SR8 2NN G0NDD
SR8 3AJ G6RKS
SR8 3DF G7DBV
SR8 3HU G3ZMG
SR8 3JE M6KRZ
SR8 3LY 2E0XMK
SR8 3LY M0XMK
SR8 3LY M3XMK
SR8 3SA M6PQD
SR8 4BW G0NXC
SR8 4EH M7CZA
SR8 4NE M0TEN
SR8 4RD 2E0YZX
SR8 4RD M6TMG
SR8 4RD M7HTR
SR8 5EG 2E0OTM
SR8 5EG M3OTM
SR8 5LW M6YRC
SR8 5PF M0BLI
SR8 5UD G4RVY
SS0 0AN 2E0SET
SS0 0AN M0XMT
SS0 0AN M6NFW
SS0 0EU 2E0SCE
SS0 0EU M0SCE
SS0 0EU M6BMN
SS0 0EU M6CSE
SS0 0NB M7FKR
SS0 0NY G6CVB
SS0 0QL G6WCI
SS0 0RA G0ENN
SS0 7AZ M1FIB
SS0 7DR G1PVT
SS0 7DR G6AXE
SS0 7LA G0MBP
SS0 7PU G2BBI
SS0 7QH G4EOR
SS0 8BD M0DRS
SS0 8BD M3VNS
SS0 8BN G4DVJ
SS0 8BN M7CCO
SS0 8NL G0FNB
SS0 9JN G1KPV
SS0 9PR M0INP
SS0 9QF 2E0TCY
SS0 9RA G8IWI
SS0 9RJ M7VPM
SS0 9SZ 2E0TWW
SS0 9SZ M0TWW
SS00HH M7TUC
SS1 1DT G4DIG
SS1 1HG G0RSW
SS1 1NH M3XMZ
SS1 1QB M1KES
SS1 2RP G1VJJ
SS1 2RT M7TAA
SS1 2SG M7RWA
SS1 2SX G1HNN
SS1 2SX G4UAI
SS1 2SX G4UTF
SS1 2TZ M0KUR
SS1 3AD G1IQU
SS1 3DF G6XNK
SS1 3DF G8LYW
SS1 3DG G8ZPO
SS1 3HD G1JLG
SS1 3NW G0TIG
SS1 3PX G4YAK
SS1 3QD G3YVK
SS1 3SS G6TLN
SS11 7DG M6MVA
SS11 7DN G7DAH
SS11 7EF 2E0NGZ
SS11 7EF M0NGZ
SS11 7EF M6VSL
SS11 7EH G4FKX
SS11 7HX M0JRW
SS11 7JE G8BPW
SS11 7LN G0DZQ
SS11 7ND 2E0CUU
SS11 7ND M6CHI
SS11 7PF G1EUM
SS11 7PP 2E0CNA
SS11 7PP 2E0EUH
SS11 7PP M0PSD
SS11 7PP M0XPK
SS11 7PP M6BMI
SS11 7PP M6XPK
SS11 7PT G1HHS
SS11 7PT G1LCS
SS11 8RQ M7DKA
SS11 8TZ M3IVD
SS11 8XB 2E0DVF
SS11 8XB M6KOA
SS11 8XL 2E0EUM
SS11 8XL M6LYQ
SS12 0AB G4TFG
SS12 0AQ M1DFC
SS12 0EN G7MBY
SS12 0ER G3LRL
SS12 0HE G3PZZ
SS12 0HS G7HGI
SS12 0LP G4DCB
SS12 0QB 2E0KGT
SS12 0QB M0ICG
SS12 0QB M6GTK
SS12 9DY G1KOR
SS12 9ED M3LHX
SS12 9EQ G4YIH
SS12 9FA G6WKN
SS12 9JX G7HCO
SS12 9NA G6OVL
SS13 1BQ G6WLQ
SS13 1HA G7EIE
SS13 1JL G7WFC
SS13 1JS 2E0FMS
SS13 1JU G6AJT
SS13 1NR G8WSS
SS13 1PJ M6ILM
SS13 2BH 2E0RCF
SS13 2BH M3OBO
SS13 2EJ M6PMP
SS13 3EW G7JRK
SS13 3NB 2E1EOZ
SS13 3PR G6MLJ
SS13 3QH 2E1IJU
SS13 3QX M7XPT
SS14 1NF G1GYQ
SS14 1RX M1TDD
SS14 1UA G1PXQ
SS14 2AZ G6HSS
SS14 2DB M7HWL
SS14 2FF 2E0FJI
SS14 2FF G0TRG
SS14 2FF M0IXY
SS14 2FF M0XXC
SS14 2FF M7LCF
SS14 2FF M7RNS
SS14 2FH M6HSH
SS14 2JD G8CUN
SS14 2JQ G7KDR
SS14 2NS G1YKX
SS14 2TN G0UCH
SS14 3JN G0NPO
SS14 3LZ 2E0SSX
SS14 3LZ M0PFJ
SS14 3LZ M0SSK
SS14 3NA G6SNN
SS14 3QS 2E0NCE
SS14 3QS G7OED
SS14 3QS M0LMR
SS14 3QS M6EBQ
SS15 4AE M7RVT
SS15 4BH 2E0JPS
SS15 4DE G0NGA
SS15 4EJ M6HKT
SS15 4EW 2E0TKV
SS15 4EW M3TKV
SS15 4HA M1NSP
SS15 4JE G6TNE
SS15 4JS M7RHA
SS15 5BG M6EWL
SS15 5FY M6HCV
SS15 5RH M1EAW
SS15 5YN G0RTH
SS15 5YN G7KCN
SS15 6AL G3UPM
SS15 6AS 2E0HJN
SS15 6JR G1AUR
SS15 6PR G6LZX
SS15 6RA M6UKJ
SS15 6SU 2E0EUD
SS15 6SU M6NMQ
SS15 6SY M7JNR
SS155SW M7RCG
SS16 4NE G1CHV
SS16 4PQ G3VYF
SS16 4PQ M1BHC
SS16 4PQ M6GHQ
SS16 4RU 2E0BRK
SS16 4TL G4EPU
SS16 5AY M7JLD
SS16 5BN M1ECC
SS16 5EN 2E0GAW
SS16 5HD G8KIH
SS16 5HX G4IEG
SS16 5JH G8OSG
SS16 5QQ 2E0KPR
SS16 5RA 2E0GQD
SS16 5RA M3GQD
SS16 5RW 2E0LEZ
SS16 5RW M6LTO
SS16 5TW G7ALR
SS16 6AQ G1KOT
SS16 6AQ G4NVT
SS16 6DU 2E0WKV
SS16 6DU M3WKV
SS16 6ED G7PSW
SS16 6GB M7DXT
SS16 6LE M6MRH
SS16 6NU 2E0CTA
SS16 6NU M0ONZ
SS16 6NU M6ORC
SS16 6RN M6KSE
SS16 6SJ G0PAE
SS16 6SJ G8JCV
SS16 6SP G1JGD
SS17 0BB 2E0TOL
SS17 0BB M6GRX
SS17 0BE 2E0EEB
SS17 0BE M6EEB
SS17 0DF G4HXY
SS17 0EF G0NGG
SS17 0NH G7JVB
SS17 7BB M0LUC
SS17 7BB M7LCY
SS17 7BD M0AKE
SS17 7BQ G4KIQ
SS17 7BZ M6RFW
SS17 7EW G1PQK
SS17 7EW M6XSZ
SS17 7HE M1BZR
SS17 7LB G4JIU
SS17 7LD G8EXQ
SS17 7NH M6EBU
SS17 7QL G4LTH
SS17 7QZ G4LNT
SS17 7SB 2E0CJO
SS17 7SB M0LEH
SS17 7SB M6ITX
SS17 7SX 2E1ERA
SS17 8DD M1DKP
SS17 8DE G1ERQ
SS17 8DT G1YAH
SS17 8HH G7OUT
SS17 8HL G7FEL
SS17 8HP G4JHE
SS17 8NL G6WXZ
SS17 8NS G6DXP
SS17 8PN G8LWA
SS17 8QU G7DXV
SS17 9AJ G3SMF
SS170AT M7CVZ
SS2 4AZ M6IIL
SS2 4DA G8OQR
SS2 4DH M0RLI
SS2 4ED G3UUI
SS2 4GL M7MRO
SS2 4HR G0SKI
SS2 4HU G0DPC
SS2 4JR G0FVB
SS2 4LJ G7JLO
SS2 4LU G0XOX
SS2 4LW G0TUI
SS2 4LW M0JZS
SS2 4LW M0JZU
SS2 4LZ M3IAC
SS2 4NN G1YCR
SS2 4NW G0UKP
SS2 4PA M6PEQ
SS2 4RD M3VHQ
SS2 4RD M6RDK
SS2 5EJ M6JUU
SS2 5EJ M6LHG
SS2 5HY 2E0DVX
SS2 5JP G0KEI
SS2 5PJ M6IPY
SS2 6GQ M7DJP
SS2 6HB M6CSJ
SS2 6HR G1DXD
SS2 6JG G0HWI
SS2 6LW M3YLB
SS2 6NR M7FMB
SS2 6RW 2E0FCI
SS2 6RW M6HRT
SS2 6TB 2E0VZL
SS2 6TB M3VZL
SS2 6TF G0TTI
SS2 6UA G0TZP
SS2 6UJ M7MBK
SS2 6YF M7TKH
SS25NG M7JNQ
SS26QY G8RAU
SS3 0AR G1HPV
SS3 0BE 2E0VNO
SS3 0BE M0VNO
SS3 0BE M3VNO
SS3 0DR G0NXN
SS3 0DY 2E0JDY
SS3 0DY M7JBO
SS3 0EX G1EXR
SS3 0EZ M3POW
SS3 0JG G7BLJ
SS3 0JS 2E0BJB
SS3 0JS M0GGZ
SS3 0JS M3SYC
SS3 0LS M6EDD
SS3 9AR G7CQA
SS3 9DP 2E0JVD
SS3 9DP M6JVD
SS3 9FA G0CXW
SS3 9JG M0KVA
SS3 9JG M3VAR
SS3 9JP G0PCF
SS3 9LR G7PMK
SS3 9nj M3OLW
SS3 9NT M3UFA
SS3 9NY G1HQQ
SS3 9PE M0JRW
SS3 9PX M3YLN
SS3 9RJ M0ABA
SS3 9RJ M0CNS
SS3 9SB M6MEO
SS3 9SG G6BHE
SS3 9SG G7RYM
SS3 9SL G4CEU
SS3 9TL M0WBK
SS3 9TL M3WBK
SS3 9TN M6DNZ
SS3 9YE G8OPY
SS3 9YS G1XXV
SS4 1DP M1AHT
SS4 1EH M6GZO
SS4 1GH M3UPQ
SS4 1JE G6HNI
SS4 1LH M3KLU
SS4 1QH G7MFX
SS4 1QH M6HLP
SS4 1QJ G8GSH
SS4 1RS G0KTC
SS4 1SH G8GHK
SS4 2ER G1HML
SS4 3AH G1RAP
SS4 3BX G7FFI
SS4 3DQ G6KFD
SS4 3EN M7KVM
SS4 3EP M6JTP
SS4 3FJ 2E0KFB
SS4 3FJ M0IKN
SS4 3FJ M6KFB
SS4 3FJ M6WWI
SS4 3HE 2E0WHU
SS4 3HE M6WHU
SS4 3PY G1GCY
SS4 3PY G7PUZ
SS4 3QN G1JRD
SS5 4BX G1YXF
SS5 4DL 2E0EOM
SS5 4DL M0IDG
SS5 4DL M6IDG
SS5 4DN G1KHS
SS5 4EL M6HFP
SS5 4EY G0PEP
SS5 4EY G3OJV
SS5 4GN G1KVO
SS5 4JE M3IWJ
SS5 4JN G0DFE
SS5 4JY G0DTP
SS5 4NT G7WHZ
SS5 4PJ G4TWH
SS5 4QG G6LYM
SS5 4QS G6XYU
SS5 4QS M3JJS
SS5 4SW G4MCU
SS5 5AT G7GBN
SS5 5BP G0PEJ
SS5 5DY G4HKQ
SS5 5EE G7ENT
SS5 5EL G3ZJZ
SS5 5HD 2E1DOA
SS5 5HJ G3VQY
SS5 5HN G7WHP
SS5 6AA G4KDH
SS5 6BG G0EBG
SS5 6BH G4ZFJ
SS5 6BN G7TBW
SS5 6DD G4RDS
SS5 6HJ M7GRK
SS5 6LT G7MLU
SS5 6LT G8RAN
SS5 6LU G0MGT
SS5 6LU G1OOG
SS5 6LZ G3HWM
SS5 6NE 2E0JTW
SS5 6NE M6BPK
SS5 6NE M6GEJ
SS56EP M6ZPS
SS56LR G4MUS
SS6 7DX 2E0WYG
SS6 7DX M0NWE
SS6 7DX M6IIY
SS6 7JR G1TWS
SS6 7LB G0DRV
SS6 7NS G4FSE
SS6 7RG G0DRH
SS6 7TD M6DXT
SS6 8AR G0HSK
SS6 8AR G6XAT
SS6 8BN G3OGX
SS6 8BP G0FKS
SS6 8EB G4KDE
SS6 8ED M3NPR
SS6 8HP G1KQU
SS6 8JP M6MKU
SS6 8LQ G7PPS
SS6 8LW G4GNU
SS6 8LZ G8FAX
SS6 8SG 2E0IDS
SS6 8SG M6WYZ
SS6 8UA G0WIX
SS6 8UW M7NJR
SS6 8UY M0DDK
SS6 8YF G4KIH
SS6 9AL G4PWB
SS6 9EJ G7LAX
SS6 9JZ G4XEO
SS6 9LY G3ZHA
SS6 9ND M6KTW
SS6 9PD G6AVK
SS6 9TU G8OXU
SS6 9UH G3TRH
SS68XR M7KOP
SS69DQ M7ATW
SS7 1AL M0PGC
SS7 1DN G1BAR
SS7 1JL G4EZP
SS7 1JL G4JAR
SS7 1QB G4FCX
SS7 1SS G6MVW
SS7 2DL G4WAK
SS7 2JP G3UTA
SS7 2LN G3LUZ
SS7 2NN M7NRC
SS7 2PG M7DBL
SS7 2ST M0AOK
SS7 2TA M6KPQ
SS7 2TY G4ZMN
SS7 3AZ M6YYL
SS7 3BB M0DQK
SS7 3HE G7MPH
SS7 3HU 2E0FVS
SS7 3HU G4HBA
SS7 3HU M7AHT
SS7 3LD M6RMU
SS7 3NA G8FAR
SS7 3PL M6PID
SS7 3RJ G1FBZ
SS7 3TU G0TTM
SS7 3TU G5QK
SS7 3UU G8UWI
SS7 3YL G4AYY
SS7 4DT G6IDU
SS7 4EE G4TPK
SS7 4EF G4UMP
SS7 4EN G8WUU
SS7 4EZ G8EFG
SS7 4LA G6TKW
SS7 4LS G1ZHN
SS7 4NT G7EGU
SS7 4NW M3BFB
SS7 4NX M7MCI
SS7 4NY 2E0XPD
SS7 4NY M6XPD
SS7 4PB M6PFY
SS7 5BQ G8EOM
SS7 5DT G1GBV
SS7 5ES G4GDS
SS7 5JQ G8VFI
SS7 5LH G7IIO
SS7 5NU G6ZUE
SS7 5PH G8YPK
SS7 5RD G4PBO
SS7 5RL G0LYX
SS7 5SJ G1LAW
SS8 0AB 2E0MZC
SS8 0AB M6MZC
SS8 0AB M6OIU
SS8 0AD 2E0FBR
SS8 0AD M0OAD
SS8 0AD M6OAD
SS8 0BB G7MXL
SS8 0BP G0JXJ
SS8 0BP G0LTO
SS8 0BU G1BGJ
SS8 0DN G7HQF
SS8 0EW 2E0LFX
SS8 0EW M6LFX
SS8 0EX G0WAX
SS8 0EX G0WGA
SS8 0JB M6EYW
SS8 0JB M6XXI
SS8 7DN M6SDX
SS8 7EA G1FBW
SS8 7EE G7TFA
SS8 7EL G1IEO
SS8 7HL G1ZSG
SS8 7JD 2E0FME
SS8 7JD M6FME
SS8 7PB M3IKI
SS8 7QU M6IGR
SS8 7RD M6HCT
SS8 7TJ G1NPJ
SS8 7TS G4BQF
SS8 8HG M7KKO
SS8 8HH M6HJC
SS8 8HX G0JAN
SS8 8HX G1OXO
SS8 8NZ M0FZW
SS8 8NZ M6FZW
SS8 9AB G6CNF
SS8 9BL G7PDU
SS8 9DJ G4ZQJ
SS8 9DS G8ZTB
SS8 9DZ G4YQJ
SS8 9EJ M6EVZ
SS8 9HL 2E0IEO
SS8 9HL M0IEO
SS8 9LP 2E0NNQ
SS8 9LP M3NNQ
SS8 9NY G0KSC
SS8 9QL G6SPH
SS8 9QP G0NCT
SS8 9RQ M6HDO
SS8 9TW 2E0RKC
SS8 9TW M6RKC
SS8 9YB G1SOB
Ss87qg M7BUV
SS89JJ G7PFG
SS9 1HQ M0BDB
SS9 1HQ M1ANK
SS9 1RU G0KMF
SS9 2DT M3KSS
SS9 2HT G0JAO
SS9 2JX G7CDO
SS9 2PB G4SNV
SS9 2PX G0JYI
SS9 2QS M1DKL
SS9 2SY M3UCH
SS9 2XD G4OQ
SS9 3EB M6TGC
SS9 3HA M3KLV
SS9 3JT G0JDD
SS9 3LF 2E0MMH
SS9 3LF M6FFB
SS9 3NT M0SXA
SS9 3PJ M6FSU
SS9 3QF 2E0IMZ
SS9 3QF M6IMZ
SS9 3QS G1GXF
SS9 3SG G0GBY
SS9 3TH 2E0XLM
SS9 3TH M0SCX
SS9 4AZ G6OAV
SS9 4DA G8WSW
SS9 4EA G0FTN
SS9 4HA G1YAF
SS9 4HA M6BMK
SS9 4HG G6GGZ
SS9 4NB G6MMR
SS9 4PP 2E0YPK
SS9 4PW G6DAH
SS9 4PW G6NLS
SS9 4PY G4UQY
SS9 4QY G3PHL
SS9 4RT M7KTT
SS9 4RY G3SVI
SS9 4SZ G1NCO
SS9 4TN M7BLO
SS9 5AB G8CEX
SS9 5AR 2E0ESX
SS9 5AR M6LOS
SS9 5EG M7GSA
SS9 5EL G3RPZ
SS9 5HR M1RKB
SS9 5NF 2E0KPX
SS9 5NN G7EGX
SS9 5NW G6OLY
SS9 5NX G4KEG
SS9 5RF G7DTS
SS9 5SW G3RCX
SS9 5XR G6WFS
ST1 2DF M6LOZ
ST1 2NE 2E0YYY
ST1 2NE M6MMM
ST1 3BA G8ZES
ST1 3BA M0GAN
ST1 3GX M6MVK
ST1 3HS 2E1SOB
ST1 3SP G6ING
ST1 5QP M6CEQ
ST1 5QS M7EJB
ST1 5QS M7ETZ
ST1 6BY G6KKN
ST1 6DE G8NTG
ST1 6EF 2E0BLD
ST1 6EF 2E0PTI
ST1 6EF M3PTI
ST1 6EF M3PXU
ST1 6JD M7PTI
ST1 6PQ M3GVT
ST1 6SL G4FMJ
ST1 6TW G1XLN
ST1 6XR M6EXA
ST10 1AT G8VSR
ST10 1BD M7VCZ
ST10 1DP M1AMB
ST10 1DT G4SXK
ST10 1LU G6UEH
ST10 1LW G4DSQ
ST10 1LZ M3YWH
ST10 1NQ 2E0VFD
ST10 1QG G4CHG
ST10 1QQ G7PIJ
ST10 1QS G0TPY
ST10 1RU 2E0CEJ
ST10 1RU 2E1IHF
ST10 1RU G7MWS
ST10 1RU G7PBH
ST10 1RU M3MHL
ST10 1RU M3MIP
ST10 1RX M6EWV
ST10 1SA 2E0PBH
ST10 1SA M3PHZ
ST10 1XB G1MAD
ST10 1XB G4NHT
ST10 1XB G4OUG
ST10 1XB G4PGG
ST10 1XB G4PHB
ST10 2AG G6KTE
ST10 2DW M3GBA
ST10 2EG G6YIW
St10 2pa G0ODH
ST10 2PA G3UNM
ST10 2PT G6YCN
ST10 3BW M3OFA
ST10 3EA G4OTX
ST10 3ER G1OGE
ST10 4AN 2E0TVW
ST10 4AN M6AVW
ST10 4BH G4YYO
ST10 4DZ G0VBT
ST10 4EG M1RKY
ST10 4FE 2E1FOW
ST10 4HL G8FHC
ST10 4LD 2E0WJE
ST10 4LD M3DPY
ST10 4LS G6ZBL
ST10 4LT G7MGA
ST10 4LX M7ANJ
ST10 4LY G6GVZ
ST10 4NJ M0KDU
ST10 4PE G0UUA
ST11 9AA G7CEY
ST11 9AP G0VHY
ST11 9AU G6IXN
ST11 9AZ M3JDJ
ST11 9BH G0VRK
ST11 9DA G6OAS
ST11 9DR M6YAW
ST11 9EN 2E0YYT
ST11 9EN M6YYT
ST11 9HA G7VLA
ST11 9HQ M0DSR
ST11 9HU 2E0IDK
ST11 9HU M0IDK
ST11 9HU M6IDK
ST11 9LY 2E0DVK
ST11 9NH M0PWC
ST11 9NT G1DOG
ST11 9NX G4YYG
ST11 9NX G4YYP
ST11 9NZ G7UKN
ST11 9PD M0CTL
ST11 9PL 2E0AAO
ST11 9PL G6GA
ST11 9PL M6BWZ
ST11 9PP G4WUX
ST11 9PX G7IMO
ST11 9PX M3AAQ
ST11 9QT G1BHF
ST11 9RH 2E0MZL
ST11 9RH M6MZL
ST11 9RN G4HUO
ST12 9BD M3XKM
ST12 9BL M6HOG
ST12 9DT G4DUB
ST12 9JA G8SDX
ST12 9JQ G0RJT
ST13 5BW G7VFQ
ST13 5DB M3KMO
ST13 5DB M5JAO
ST13 5JU 2E1DLR
ST13 5LS 2E0MJJ
ST13 5LS M3TJJ
ST13 5NZ 2E0SXY
ST13 5NZ M3SIM
ST13 5PW 2E0SAA
ST13 5PW M3RRZ
ST13 5PZ M1BIX
ST13 5RR G7HIJ
ST13 5SZ G4MDJ
ST13 6BB G7LIE
ST13 6BB M0CAD
ST13 6BU M0RMP
ST13 6ES G8YQA
st13 6ps M7DDE
ST13 6PX M3BVP
ST13 7AA G3VBG
ST13 7AN G4XIZ
ST13 7ED G8FWD
ST13 7EX G0LAZ
ST13 7HJ G0WOM
ST13 7JP G6KPT
ST13 7LW G7SSK
ST13 7NP G1VXD
ST13 8BL 2E1SGK
ST13 8BL M0SGK
ST13 8DD M1ALE
ST13 8EQ M0FAZ
ST13 8EQ M3FCN
ST13 8EQ M3FCO
ST13 8EQ M3HRN
ST13 8EQ M3SKD
ST13 8ES G4TGS
ST13 8JA G8JGL
ST13 8JN M1ABF
ST13 8LD G7MQF
ST13 8LL G7CFT
ST13 8LN G4UQM
ST13 8LW 2E1HRN
ST13 8SL M6GSF
ST13 8XF G8YMS
ST14 5AG 2E0KMS
ST14 5DE M0DNU
ST14 5DH G3VSB
ST14 5DP 2E0SRC
ST14 5DP M0RRN
ST14 5DP M6YUK
ST14 5JU G7LGS
ST14 5LE 2E0AUM
ST14 5LF M3HEJ
ST14 5LT G0LJH
ST14 7AX M3MTP
ST14 7BY 2E1DLM
ST14 7EN 2E1BMV
ST14 7EQ G0SRC
ST14 7EQ M0VCS
ST14 7ET 2E1CLM
ST14 7ET G6VGC
ST14 7FE G1RKJ
ST14 7FE G7PAY
ST14 7HL 2E0LIZ
ST14 7HL 2E1LIZ
ST14 7HL M3ZIL
ST14 7LB G6SVJ
ST14 7NB G7STC
ST14 7NF G7UCN
ST14 7QG G6CUY
ST14 7QY G6UOX
ST14 8BB G8YAT
ST14 8BT G4XSR
ST14 8EF M0WUL
ST14 8LT 2E0PBB
St14 8LT G7ORN
ST14 8LW M7HMS
ST14 8LW M7PDQ
ST14 8QS G8FGQ
ST14 8UP M6PLQ
ST14 8XG M6UTX
st148sj M0SRJ
ST148UH 2E0HMQ
ST148UH M7DAY
ST15 0AW M7AFK
ST15 0DH G0KJP
ST15 0DW 2E1BRA
ST15 0DX G0SKQ
ST15 0DY 2E1ROO
ST15 0DY M7ROO
ST15 0EB G0TED
ST15 0EG G1LAP
ST15 0EP G1DSF
ST15 0EP G7KEE
ST15 0JA G3JUX
ST15 0JF G8HWI
ST15 0JN G1HSJ
ST15 0LF M0CDN
ST15 0PX G4TTM
ST15 0QG G7BMP
ST15 0QH 2E0PSD
ST15 0RD M6MGO
ST15 0RH G4NJR
ST15 0RP M3CZL
ST15 8AR 2E1SRI
ST15 8BL M6BUX
ST15 8DY M3XUJ
ST15 8LA G3XMK
ST15 8LF G0VWT
ST15 8LQ M0CZA
ST15 8LW G4CRK
ST15 8NL M7JRB
ST15 8PR G6JAF
ST15 8SG M7CQJ
ST15 8TG M3ZGD
ST15 8TZ M6KGL
ST15 8XL G4EJD
ST15 8YP G8MZZ
ST15 8ZQ M0CMC
ST16 1DS G0DEZ
ST16 1FJ G8RFV
ST16 1HA 2E0HHK
ST16 1HA M0YZY
ST16 1HA M6ZZQ
ST16 1HE M7TYG
ST16 1HJ G3ZHS
ST16 1LD G4TMD
ST16 1LD G7BJG
ST16 1NL G6YLZ
ST16 1NL M6MTS
ST16 1PA G4WMV
ST16 1QP G6ZLJ
ST16 1QR M6HQR
ST16 1SD 2E0HBB
ST16 1SD M6HRL
ST16 1TB G4YFF
ST16 1TB M0RWR
ST16 1TB M6VIX
ST16 1TE G1UDS
ST16 1TS G6MDX
ST16 1WH 2E0HGK
ST16 1WH M7DWT
ST16 1XA G0FXS
ST16 2DZ G3XPD
ST16 3HD G0IAC
ST16 3HQ G0EYX
ST16 3HS G7PFT
ST16 3NP M6NUJ
ST16 3NX 2E1WWD
ST16 3NX G0GAP
ST16 3NX M6FFE
ST16 3PJ M6TPG
ST16 3PL G0GOZ
ST16 3PL G3SBL
ST16 3RE 2E0RPF
ST16 3RE M0RPF
ST16 3RE M3RPF
ST16 3WG G1SPU
ST16 3YE G8JAQ
ST16 3YG M7SCC
ST16 3YW 2E0IFS
ST16 3YW M0IUU
ST16 3YW M6IVC
ST17 0AJ G0OIQ
ST17 0AQ G4NVH
ST17 0AQ M3NLW
ST17 0AQ M3NVH
ST17 0HB M7SJB
ST17 0HF 2E0BZC
ST17 0HF M0ZPA
ST17 0HF M6PMD
ST17 0HJ G4PKF
ST17 0NU G6KQD
ST17 0NX M7BMO
ST17 0PA G3LOE
ST17 0PA G6LXF
ST17 0QB M7RUK
ST17 0SJ G4RWQ
ST17 0SL 2E1AZK
ST17 0SP G6IRZ
ST17 0TW G0TFD
ST17 0TW G1UUJ
ST17 0YE 2E0GGT
ST17 0YE M3GGE
ST17 4BP G6RFM
ST17 4BX G2HNA
ST17 4DY M6DFL
ST17 4EH G7VLB
ST17 4HB G6TBV
ST17 4HZ 2E0MQA
ST17 4HZ M3MQR
ST17 4NR G1XWO
ST17 4QJ 2E0VTS
ST17 4QJ M0POG
ST17 4QJ M0VTS
ST17 4QJ M3VTS

ST17 4QN 2E0LTL
ST17 4QN M7CFW
ST17 4QP G0GAR
ST17 4QR M6KIX
ST17 4QS G4PWV
st17 4qx G0VGK
ST17 4RY 2E0RDQ
ST17 4RY M6BVD
ST17 4YA G6JKF
ST17 9BE G8KUA
ST17 9DH M6PCW
ST17 9DS 2E0AVK
ST17 9EF M3GCS
ST17 9EY M6ATU
ST17 9FR M1LIP
ST17 9FT G6KJK
ST17 9HP G1SJG
ST17 9HP G1UDT
ST17 9HP G3OSP
ST17 9HU M7AMM
ST17 9HU M7VVV
St17 9nz M0DJF
ST17 9PR G6FYD
ST17 9QJ G3VUL
ST17 9RA 2E0XTB
ST17 9RA M7TXB
ST17 9RL G4RSW
ST17 9SB M3NOF
ST17 9TW M3XZY
ST17 9UE G0BYA
ST170HS M1EOP
ST18 0DR 2E0PRC
ST18 0DR M6SFC
ST18 0EW G8JVU
ST18 0GQ G0GMS
ST18 0GQ G4ZMM
ST18 0NN G4RJM
ST18 0QE G0GUF
ST18 0QZ G4OUT
ST18 0RD G4PET
ST18 0RD G4PEU
ST18 0SP 2E1ELE
ST18 0TS G6UZL
ST18 0UR G4THY
ST18 0XB G6KGA
ST18 9AY M7EEH
ST18 9BD M0SJD
ST18 9DA G6DGX
ST18 9DQ G4KQK
ST18 9DS G6OYF
ST18 9EU 2E0NQA
ST18 9EU M3NQA
ST18 9FB M7PFT
ST18 9HS M6MRM
ST18 9HT G6PVT
ST18 9PE G4MGI
ST18 9QR G4UKA
ST19 5AH G3JBF
ST19 5BQ M7EAK
ST19 5BX G3PIN
ST19 5BX M7NFL
ST19 5EH G0VJC
ST19 5HD G0GJG
ST19 5HF M0UXB
ST19 5HP 2E1EAW
ST19 5QH G0BFK
ST19 5SP M3VOR
ST19 9AB G4WAF
ST19 9EH M0LRY
ST19 9JZ G4IDG
ST19 9LX M0RSY
ST19 9NQ 2E1GHX
ST19 9NQ M3PLN
ST19 9NU 2E1GVJ
ST2 0AQ M3DPP
ST2 0BX G1YQL
ST2 0EH 2E0LVL
ST2 0HB M6EYS
ST2 0QY G4GRZ
ST2 0RE M6MXF
ST2 0RQ G7RFH
ST2 43NR M0LAA
ST2 43NR M0LEK
ST2 43NR M1DOA
ST2 7AR G4BEM
ST2 7AR G4DPV
ST2 7BZ G4CKT
ST2 7HJ G4OSI
ST2 7LD M1MLM
ST2 7LP G6LJC
ST2 7LR G0UZE
ST2 7LR G7RTX
ST2 7NF G3ISX
ST2 7NG G6CCQ
ST2 7PA G1SQC
ST2 7PF G0TKR
ST2 8AQ G4DVA
ST2 8DQ G7VSW
ST2 8DS 2E1STO
ST2 8DU M3FIM
ST2 8EH M6YAH
ST2 8HU 2E1MAZ
ST2 8HU M0DVT
ST2 8HU M0FTR
ST2 8HU M3LVA
ST2 8HX M3SRH
ST2 8JN M1DXQ
ST2 8JZ 2E1DZP
ST2 8LP M3KEL
ST2 8NL M3KKI
ST2 9BZ M6PAN
ST2 9DF G8FGR
ST2 9DR G7VUL
ST20 0BA M6HMR
ST20 0HL G4VMC
ST20 0HR M7KWB
ST20 0JD G4OTB
ST20 0LG M6AJB
ST20 0NP M6PNG
ST20 0QH M3KQF
ST20 0RP 2E0LME
ST20 0RP M3LME
ST21 6DG M6GYQ
ST21 6DR M6OZQ
ST21 6LE G4LSA
ST21 6LT G4DDZ
ST21 6RG G6SAR
ST21 6RG G0UUN
ST3 1AD 2E1IVT
ST3 1SJ G0TTO
ST3 1SJ G7KDJ
ST3 2AJ 2E1GVS
ST3 2EG 2E0ESJ
ST3 2EG M3HNM
ST3 2HA G0CWO
ST3 2RS M6MEP
ST3 2RZ M1IRM
ST3 3DD 2E0YHW
ST3 3DD M6YHW
ST3 3LF M3BIL
ST3 3LX M3OLE
ST3 3LX M7BRH
ST3 3NU M6YYD
ST3 4NF G6TPI
ST3 4NU M0MFH
ST3 5DX G7EWX
ST3 5EG G3VTE
ST3 5JA G7NEE
ST3 5JD G0JGB
ST3 5JL G6AYH
ST3 5LA M3AOP
ST3 5PW M1BFF
ST3 5QY G6XJF
ST3 5RP G7FSA
ST3 5RQ G8NAI
ST3 5ST G4WJX
ST3 5UD G1FHH
ST3 5UG G0SKM
ST3 5XW G0DUI
ST3 6AH G7SWR
ST3 6EQ M3YBF
ST3 6HA 2E0RDW
ST3 6HA G0SMN
ST3 6HY G1XMP
ST3 6HZ G1VFH
ST3 6JY G0PSH
ST3 6LP 2E0DII
ST3 6LP M6IAO
ST3 6NS G6THM
ST3 6PN 2E0VTR
ST3 6PQ G4RQG
ST3 6RG G1UYT
ST3 7AN M6HVI
ST3 7AT M7BYC
ST3 7DP M6VGR
ST3 7EN G0WLC
ST3 7EW 2E0PHB
ST3 7FN M7EMY
ST3 7FW M3BIK
ST3 7HN G0UPV
ST3 7JA M0DPW
ST3 7JY G3MFH
ST3 7LH M3YMD
ST3 7LJ M3YSD
ST3 7QF G7MMW
ST3 7RZ M0FOG
ST3 7SS 2E0GQO
ST3 7SS G0NED
ST3 7UG M0EUI
ST3 7WH G6LQG
ST3 7YE G8DZJ
ST3 7YE M7OKG
ST34BN M6MWZ
St34sn M3SDJ
ST35DG M7UKP
St36qj M7JPS
St37ap G8UWL
st37dp M7BUE
ST4 1ED G6UDI
ST4 1PB G8PAD
ST4 2BS M0NDE
ST4 2EX G0NMY
ST4 2RQ 2E1HCG
ST4 3DT M6SAM
ST4 3HH G1YPH
ST4 3HQ G7JWO
ST4 3NR M3SAS
ST4 3QX M7BRI
ST4 4DY G1EZJ
ST4 4NG G0VVT
ST4 4PH G4CJM
ST4 4PL G0VKI
ST4 4QP G3UD
ST4 5AN G0JZS
ST4 5DH M3HQQ
ST4 5DW G6WFF
ST4 5EE G1CHE
ST4 5EF G1OQW
ST4 5HA M6PAW
ST4 5JA M0JUI
ST4 5JE M6JEF
ST4 5PU M3WZY
ST4 6BH 2E0MZK
ST4 6BH M7MJK
ST4 6ED G7HQJ
ST4 6LA G0JGF
ST4 6LD G0DTW
ST4 6PR G6HFK
ST4 6QZ G3UJI
ST4 6RE M3YHN
ST4 6RU M1DCK
ST4 6RZ 2E0NNI
ST4 7DZ G8YDJ
ST4 7GA M7CNE
ST4 7HE G1UCC
ST4 7HF G1XZA
ST4 7HF G1XZB
ST4 7HH G1KSH
ST4 7LS M7BDG
ST4 7LU M0STK
ST4 8AB G4EJM
ST4 8AZ M6PXL
ST4 8EL G8SAR
ST4 8FX M6BWH
ST4 8HA M0ODS
ST4 8HA M7GAT
ST4 8HP G6HBQ
ST4 8NJ G3SLX
ST4 8NJ G4MDK
ST4 8NJ G4MJU
ST4 8NW G7EVK
ST4 8PD 2E0COY
ST4 8PL M3PQL
ST4 8PQ G7WGZ
ST4 8PQ G7WHA
ST4 8PQ M0ACA
ST4 8RH M3SJR
ST4 8RL M6JXO
ST4 8RT G7NHZ
ST4 8TF G0PXL
ST4 8UZ G0NBE
ST4 8XY G6AZP
ST4 8YL M6LZT
ST4 8YQ G8DYI
ST46DB G3GBU
ST5 0JS M6GAS
ST5 0JY G1WVW
ST5 0JY M0CRJ
ST5 0LB G8KUZ
ST5 0NG G4VFJ
ST5 0NL G6IFS
ST5 0PB G0OQP
ST5 0PD G1TMF
ST5 1JH G7IDH
ST5 1JL M0OHA
ST5 1JL M5GWH
ST5 1NH G4IMV
ST5 2EF G7JXU
ST5 2HU G6VTX
ST5 2JL G0WDT
ST5 2LG G0PSD
ST5 2NS M7XCB
ST5 2PQ G7BJN
ST5 2QH M3MIF
ST5 2SD G8WKH
ST5 2TB M0BSH
ST5 3DF G3TJP
ST5 3DF G6CXN
ST5 3DL G7BPG
ST5 3DX M3AUN
ST5 3EA M3KFE
ST5 3ED 2E0OPT
ST5 3ED M7KMW
ST5 3EF G4TQB
ST5 3EW G4FAY
ST5 3LT G7RSM
ST5 3LT G7UQG
ST5 3NY G4MDM
ST5 3NY G4UMG
ST5 3NZ M3FMK
ST5 3PQ G1JJE
ST5 3PQ G7MMK
ST5 3QR 2E0XXX
ST5 3QU 2E0IBB
ST5 3QU M0PUS
ST5 3QU M6OBB
ST5 3RW G1JIW
ST5 4AZ 2E0KIO
ST5 4AZ M6KIO
ST5 4BH G3LBS
ST5 4BN G8NPD
ST5 4DA G8MTB
ST5 4EF M3JZD
ST5 4EN G1DDA
ST5 4EP G1STQ
ST5 4FD G6LQP
ST5 4HS G4TMR
ST5 5AG M7LGG
ST5 5DP M7SOT
ST5 5EU G4NFL
ST5 5FE 2E0DWA
ST5 5FE M0SVT
ST5 5FE M6DWA
ST5 6AR G0GOI
ST5 6AU M3YCO
ST5 6BB G7PNF
ST5 6DT G1LMT
ST5 6HL M0HCM
ST5 6LS 2E0SDQ
ST5 6LS M3GDK
ST5 6QP G7GRR
ST5 6QS M1FIL
ST5 6TB M0OMA
ST5 6TB 2E0PLC
ST5 6TB M0TKS
ST5 6TB M3TYG
ST5 6TB M3WSV
ST5 7AH G4CBW
ST5 7DH 2E0IEA
ST5 7DH M6IEA
ST5 7EP M3ZJH
ST5 7JR M6ECH
ST5 7PB G0HQH
ST5 7QD G1VBY
ST5 7QP 2E0SCX
ST5 7QP M6ALQ
ST5 7RQ G7KJE
ST5 7ST G6KDU
ST5 7TB G4UDH
ST5 8DR G6IKH
ST5 8EA G0SOK
ST5 8EL G4GNW
ST5 8EL G4NVH
ST5 8EL M3PIQ
ST5 8EY G7UAT
ST5 8HF M7CYC
ST5 8HH M3WXI
ST5 8LD M6YCA
ST5 8ND G6ZHU
ST5 8QA G0VZE
ST5 8qd M7ANI
ST5 8SE M7ROC
ST5 8TB M7SIB
ST5 9AE M6IQN
ST5 9AJ M3XIE
ST5 9BP M3YHZ
ST5 9DN M7OTZ
ST5 9ET G7OKO
ST5 9HE 2E0DID
ST5 9HE M3UAX
ST5 9JZ G4DPW
ST5 9NL G1JFL
ST5 9NY G1XYN
ST5 9NY M6CEP
ST5 9PS G7MRF
ST5 9PU G6LLF
ST52JT M6LBY
St56rl 2E0TQS
ST6 1AT 2E1HQZ
ST6 1HB M1ACL
ST6 1PS M1PYE
ST6 1PX G7TJQ
ST6 1RA M6USE
ST6 2ER M6LJS
ST6 2LY M0GBF
ST6 2PY M3TMX
ST6 3BX G7SSW
ST6 3DJ M6HKX
ST6 3JB M6HUS
ST6 3PN M6EDZ
ST6 4DZ M6BCJ
ST6 4ES M6MIR
ST6 4ET M3MOH
ST6 4EX M6JLH
ST6 4JB M6LLL
ST6 4NN M6CYA
ST6 4NW M3YUR
ST6 5BT M7AGJ
ST6 5EE M7RAS
St6 5hs 2E0JTB
St6 5hs M0JTD
St6 5hs M3KJV
ST6 5LE G6FHB
ST6 5LP M0EOT
ST6 5LP M6IQE
ST6 5NX G8GFQ
ST6 5PP G7JWI
ST6 6DZ G4WEP
ST6 6EB M6HCD
ST6 6EX M3JIV
ST6 6HE G8WBP
ST6 6LX G6DON
ST6 6PB M3FNR
ST6 6PE G0UDO
ST6 6RA G3ZRQ
ST6 6TZ M0CDZ
ST6 6UR G0RQJ
ST6 6XD M7TGE
ST6 6XF 2E0HAG
ST6 6XF M3XHU
ST6 7AL G7MSQ
ST6 7BA M6SWH
ST6 7DL M6SOT
ST6 7JR G7RBL
ST6 7JY G7EUF
ST6 7QD M7CCG
ST6 8AB M7KLW
ST6 8BZ G3ICZ
ST6 8DX G7HMB
ST6 8JF M1MDP
ST6 8LX M0RDX
ST6 8PL G1VYS
ST6 8QH G0BRL
ST6 8RN 2E0MWA
ST6 8RT G8RWZ
ST6 8RX G0UTZ
ST6 8TG G3UHV
ST6 8TQ G6OBD
ST6 8TX M0INY
ST65PW M7SVT
st68xe M7JWL
ST7 1AR M3DBU
ST7 1AR M3EBU
ST7 1AT G6YQN
ST7 1BZ G0JKL
ST7 1EH M7KDB
ST7 1EZ 2E0IQT
ST7 1JN 2E0BOF
ST7 1JN M3LRF
ST7 1JY M3BTJ
ST7 1LW G6NQY
ST7 1NP G4WBW
ST7 1NQ G6NGN
ST7 1PF G4EQZ
ST7 1PT M6AHA
ST7 1RH G4VJB
ST7 1SG G4VEW
ST7 1SQ G6SGA
ST7 1TA G0RDS
ST7 1UB G1TBX
ST7 1UG 2E0VPJ
ST7 1UG M6PPQ
ST7 2BA G4ERQ
ST7 2BL G7FZB
ST7 2BN G4WHT
ST7 2BW G8GWS
ST7 2DS M0CLL
ST7 2EF G6GFC
ST7 2HX G6NVH
ST7 2HY G4GHT
ST7 2LP G3VXS
ST7 2ND G0OFW
ST7 2NH G4UFU
ST7 2NQ G0EIR
ST7 2NR G0UKK
ST7 2Rj 2E0KDB
ST7 2Rj M6KWK
ST7 2SH 2E0DFB
ST7 2SU 2E0NCN
ST7 2TQ M0CVJ
ST7 3BL G6LMJ
ST7 3LD M0RJS
ST7 3ND G8MKS
ST7 3PB 2E0MKX
ST7 3PB M0MJK
ST7 3PE M3DUR
ST7 3PH G3OHH
ST7 3RB G4DLA
ST7 3RY G6NAH
ST7 3TL G6TWB
ST7 3TL G8UVN
ST7 4AL G7IAM
ST7 4DF M7JBB
ST7 4DQ G1CKJ
ST7 4DY G0KBJ
ST7 4HF G0UDG
ST7 4HP G4UDG
ST7 4HR M6BVZ
ST7 4HW G0UWK
ST7 4JF G0HWV
ST7 4JU G8ILP
ST7 4JX 2E0TAK
ST7 4JX M3TBZ
ST7 4LW G7TXX
ST7 4PT G8LZO
ST7 4RQ G4YUN
ST7 4RS G1ELQ
ST7 4RS G6BQM
ST7 4TA G0FZU
ST7 4TA G6USG
ST7 4TH M1AUP
ST7 4UR M6EGF
ST7 4YA G8NYZ
ST7 4YL G0RDK
ST7 4YL G1PUV
ST7 8AS 2E0XCB
ST7 8BB 2E0FVF
ST7 8BB M3DOO
ST7 8HU G7OSO
ST7 8JQ M3TLZ
ST7 8LA G0MVT
ST7 8LP G0LSX
ST7 8NG G0PTD
ST7 8PB G8GLD
ST7 8PE G6UKQ
ST7 8PL G6YCO
ST7 8QB M3WXU
ST7 8QE M0SSY
ST71HE 2E0JSM
ST71HE M6FFJ
st71py M7ATE
ST72XG M7HMB
St73ed 2E0SWP
St73ed M6HRG
st74dn M6YLY
ST8 6EH M3SJQ
ST8 6HY M7YOE
ST8 6JE 2E0VGF
ST8 6LH M6IQM
ST8 6LS M3FIZ
ST8 6NT G8KSC
ST8 6QB M7DJD
ST8 6RD M0GIP
ST8 6RD M1PAS
ST8 6SE G0BYU
ST8 6SZ M3EOT
ST8 6TJ G6CYV
ST8 6UH G6PAO
ST8 7AH G6DSG
ST8 7AS G1ZNZ
ST8 7BB G8KOZ
ST8 7HL G6AKX
ST8 7LN G6UKM
ST8 7LY G4WAM
ST8 7LZ 2E0SHO
ST8 7LZ G5SHO
ST8 7LZ M7SLH
ST8 7NG G4UQW
ST8 7NG M7AVS
ST8 7NW G0CHL
ST8 7PF G6PJC
ST8 7SW G8APB
ST8 7UF G0DBC
ST86TY G1EGE
ST9 0BD G3LHG
ST9 0DG 2E0CIA
ST9 0DG M0ASR
ST9 0DZ M7MED
ST9 0EP G4SCY
ST9 0HU 2E0CAA
ST9 0JN G4DVN
ST9 0LF G0RKN
ST9 0LF G1EWC
ST9 0LR 2E0DDD
ST9 9BW G4KME
ST9 9JB M3YHF
ST9 9JB M3YHH
ST9 9LW G4LCL
ST9 9QD M0GYN
ST9 9QG G8JDD
SW10 0ER M7OLY
SW10 9BA M7UDV
SW10 9BJ G7GBJ
SW10 9NJ G3JHH
SW11 1RH G4CKS
SW11 2JD 2E1RJH
SW11 2SU G1VVH
SW11 2TJ M7KOF
SW11 3NY G8JKD
SW11 3TN 2E0LVR
SW11 3TN M0LVR
SW11 3TN M6ODP
SW11 4AA G1EGB
SW11 4DH G8NXA
SW11 5EF M7YSE
SW11 5TR M7SPM
SW12 0NA G1ULG
SW12 8EG M6UMM
SW12 8PJ 2E0PQA
SW12 8PJ M0PQA
SW12 8PJ M6PQA
SW12 8TQ 2E0BVH
SW12 8TQ M0MEN
SW12 8TQ M3YIX
SW12 8UQ M3YNX
SW12 9LU G7IWU
SW12 9SS G6CKY
SW13 8DX M7CCH
SW13 8QY 2E0ESI
SW13 9LH G0WLD
SW13 9NB M0FCA
SW14 7AT 2E0TTT
SW14 7AT M6GGG
SW14 7NL 2E0RSI
SW14 7NL M3NNG
SW14 8EQ 2E0RWE
SW14 8JH G4AVK
SW14 8JJ G3KWW
SW14 8JJ G6JMX
SW14 8JU G8ANN
SW15 1LZ 2E0PUT
SW15 1LZ M6VIE
SW15 1RH M0PEQ
SW15 3HR M7AQS
SW15 3HZ G4CXX
SW15 5RL M7CMS
SW15 6DS 2E0VPN
SW15 6DS M6ZDF
SW15 6EX G1SAK
SW16 1AA M1BJC
SW16 1SN G8ZFU
SW16 2BT M7IVN
SW16 2LT 2E0BWY
SW16 2LT 2E0GXY
SW16 2LT M0KFA
SW16 2LT M3ZLV
SW16 2LT M6NNF
SW16 2PE M3NLN
SW16 4HL G7IRG
SW16 4HT G0WLF
SW16 4JY M6KVW
SW16 4sa M0HHP
SW16 6JX G6BPN
sw162lt 2E0NNF
SW17 0PG M0KLH
SW17 0QQ G1LAR
SW17 7DH M1ECI
SW17 8PR G7NLF
SW17 8SF G8TQV
SW17 8SF M0TQV
SW17 8SN M6GYI
SW17 9EN 2E0BWD
SW17 9EN M6OJO
SW18 1SD G0TEO
SW18 2AJ G3PAQ
SW18 2DF M7NAM
SW18 2EB 2E1RWN
SW18 2TS M7CZF
SW18 3AZ M7DEE
SW18 3BB G4PZJ
SW18 3BY 2E0LPP
SW18 3BY M7KEP
SW18 3JU M0CON
SW18 3RQ G5RC
SW18 3RQ M0FCR
SW18 5AN G4PKK
SW18 5PB 2E0TRE
SW18 5PB M3TNN
SW19 1EB M6ZIA
SW19 1EX G8REX
SW19 2BD 2E0DRB
SW19 2BD M6AEM
SW19 3JJ M0JCL
SW19 3JJ M1JCL
SW19 5AZ G6LKV
SW19 6JA G4UNI
SW19 6QU G4CGD
SW19 6RT M7SGT
SW19 6SH G8ILZ
SW19 7LW G3NXN
SW19 7LW M5BBB
SW19 7RQ G6GKG
SW19 8DQ G4ILP
SW19 8JP M0TNB
SW19 8NZ G8WUF
SW19 8PP M0OCK
SW19 8PP M6OCK
SW19 8RY 2E0KRB
SW19 8RY M0SOT
SW19 8SF M0HJJ
SW19 8SW M6ZAV
SW19 8TU 2E1HJA
SW1P 4PS M3LBJ
SW1P 4RW G3ZDY
SW1V 1JJ M0HTQ
SW1V 2DB G4MRL
SW1V 2QS G1TQT
SW1V 2RT G4JKE
SW1V 4AS G6ZFK
SW1V 4NY M7ESZ
SW1W 8JW M3AJC
SW2 1DD M6BXM
SW2 1NQ G4ILW
SW2 1PA G7IHV
SW2 2AP 2E0IJG
SW2 2AP M7NCV
SW2 3BQ M0BXG
SW2 3EW 2E0ESB
SW2 4QH M6IOR
SW2 5LU G3KHQ
SW20 0BA M6LXP
SW20 0JN G8WIM
SW20 0NJ 2E1JPH
SW20 0NJ G0JPH
SW20 0NJ M7BRM
SW20 0QE G0MFY
SW20 0UD G3DRN
SW20 8AP G0CRU
SW20 8JL M5ESA
SW20 8JY 2E0KKZ
SW20 8JY M7DKK
SW20 9AB M3LWX
SW20 9BQ 2E0JGH
SW20 9BQ M0JGH
SW20 9HG G8IAK
SW20 9HX G8BUZ
SW20 9JT G0VQT
SW20 9LB M3UFL
SW3 3LB G1XGZ
SW3 3TS M1DNQ
SW3 4QE G8ECR
SW3 4SR 2E0RRA
SW3 4SR G0SLL
SW3 4SR M0INU
SW3 4SR M6JOA
SW3 4SR M6YDG
SW3 6BU G8DOH
SW4 0AN M6TAH
SW4 0ET G4PIA
SW4 6BE M6DUE
SW4 7AF M3ZFL
SW4 9HD G8LHQ
SW4 9NS M7AZV
SW4 9PB G8OOF
SW4 9RR 2E0KJI
SW4 9RR M6KJI
SW5 9EB 2E0CWK
SW5OPU M7AGS
SW6 1BS G0HFK
SW6 1EA 2E0WTU
SW6 2HR M7STV
SW6 2RB G6BUH
SW6 2RT M6LJU
SW6 3SA M6BDG
SW6 5BG G0ADG
SW6 5DW 2E0LCW
SW6 6SA M6FHH
SW6 7DA M7LXT
SW6 7HD M7DCR
SW6 7PP G7IUB
SW7 1BL G0PZC
SW7 3BG M3TND
SW7 3DQ G0RFX
SW7 3QF G4MFW
SW7 4ET G4UAF
SW7 4JR G6LVB
SW7 5NS G4WQO
SW8 1BH G7VQR
SW8 2PD G0PIY
SW8 2TF 2E0DYV
SW8 2TF M6GQZ
SW8 4DP 2E0ULH
SW9 7UE G0VPH
SW9 9RH G6NAD
SW9 9RZ M6DUH
SY1 1RD G8ZWF
SY1 1UH G0GTN
SY1 2AB M6FNF
SY1 2DL G1EBT
SY1 2EY M6OGJ
SY1 2HT M3OFH
SY1 2NE 2E0HBQ
SY1 2NE M7AYZ
SY1 2PY G8DXM
SY1 2RT M7AZK
SY1 3HF M7NCO
SY1 3PZ G4CBT
SY1 3RA M3ENY
SY1 3RH G4XBI
SY1 4BF M1BCM
SY1 4DD G7VYI
SY1 4ER G0IRI
SY1 4JE G7NBP
SY1 4LD 2E1FZU
SY1 4PL 2E0BMP
SY1 4QQ M3IHC
SY1 4RB G8SIU
SY1 4RP G8RTB
SY1 4RU M0KZB
SY1 4SE 2E0FZU
SY1 4SE M0MNB
SY1 4SE M7DMK
SY1 4YE 2E0XJM
SY1 4YE M7JPM
SY10 0AN GW6KLQ
SY10 0DR GW1APU
SY10 0LE 2W0OGT
SY10 0LE MW6OGT
SY10 0LH MW6FEH
SY10 0NS GW2HCA
SY10 7AE GW7LXI
SY10 7AS GW6YGI
SY10 7AS MW3XRI
SY10 7DX GW7CQB
SY10 7HL GW3OKT
SY10 7HP GW3SMT
SY10 7HP MW6TBL
SY10 7JY MW3IHB
SY10 7NF MW6LWF
SY10 7PQ GW1PLJ
SY10 7PQ MW3WQV
SY10 7PQ MW6BNF
SY10 7PQ MW6JTM
SY10 7QB GW3LYU
SY10 7RQ GW4TEQ
SY10 7TU GW6DBP
SY10 8AJ GW6CVX
SY10 8BB GW4UDE
SY10 8DW GW6VKI
SY10 8HQ MW6OLL
SY10 8HU GW1HAX
SY10 8PS GW7OTQ
SY10 9AU GW8AJA
SY10 9AZ GW0DLW
SY10 9BA GW4DFQ
SY10 9DF GW1VIR
SY10 9DH GW4COL
SY10 9DP MW6FST
SY10 9DP MW6NJM
SY10 9DP MW6WEE
SY10 9HN GW1JVH
SY10 9HQ GW4IOQ
SY10 9RB MW0DSZ
SY11 1EP MW6PIH
SY11 1LX MW0RRY
SY11 1SQ GW1FLY
SY11 1TP GW7JHC
SY11 2NL MW7DQW
SY11 2QW MW6FTE
SY11 2SB MW6NFJ
SY11 2SG MW3NFZ
SY11 2TH 2W0GWO
SY11 2TH MW7GWT
SY11 2UA GW7NNA
SY11 2UN MW6HCY
SY11 2XA GW6AUS
SY11 2YD GW4WVK
SY11 2YL GW7NSJ
SY11 3AH MW6HBK
SY11 3AH MW6NCA
SY11 3BU MW1EYH
SY11 3DH GW0CWZ
SY11 3JF 2W1HXT
SY11 3LD 2W0TDF
SY11 3LD MW0TDF
SY11 3LD MW3XTF
SY11 3NR MW6KQL
SY11 3NR MW7CTA
SY11 3PJ GW1OIB
SY11 4AH 2W1MSC
SY11 4AH MW3PVC
SY11 4EQ GW4FBQ
SY11 4FE GW8KNJ
SY11 4LF GW0VEU
SY11 4PA GW0EXD
SY111DS MW7JSG
SY112BL MW7CUI
SY12 0EG M0YCQ
SY12 0EX M0CCQ
SY12 0PF 2E1MAR
SY12 0QJ G3XGD
SY12 9ER M0COT
SY12 9NA G6ZMD
SY12 9PA G0AJA
SY12 9QA G1VXY
SY120FH M6KPL
SY13 1BS M0NDC
SY13 1BZ M1BKL
SY13 1EX G4HRH
SY13 1PH M0SIN
SY13 1PH M1SIN
SY13 1QT M6NBQ
SY13 1QW M0IFP
SY13 1RQ G3TZU
SY13 2EG M5CLO
SY13 2HD G8CTD
SY13 2HZ G4JSP
SY13 2LW G3YLV
SY13 2ND G4GXD
SY13 2NJ G4VQH
SY13 2PT G4UJS
SY13 2PX G1GSJ
SY13 2PX G4OAR
SY13 2QL 2E0ZSU
SY13 2QL M3ZSU
SY13 2RA G4ANY
SY13 3LT G0MXD
SY13 3UA M3XJW
SY13 4EB G0FEY
SY13 4JF 2E0RVE
SY13 4JF M0RVV
SY13 4JF M6RVY
SY13PY G0EML
SY13QE M6XXD
SY14 7AA M0VKC
SY14 7AR G4RRR
SY14 7AX G3GIZ
SY14 7AX G3TZO
SY14 7DB G4GOO
SY14 8AY M3ZLA
SY14 8DU M6SHI
SY14 8JB G6XTC
SY14 8JQ G4XVS
SY14 8LL M0ROJ
SY14 8NE M6MEC
SY148JJ 2E0TGI
SY148JJ M7TGF
SY15 6AB G7DRX
SY15 6AL G0IQP
SY15 6BA M7SAK
SY15 6BF 2E0GFQ
SY15 6BF M0RHX
SY15 6BF M6RBH
SY15 6EB M6IMH
SY15 6JH G4RYK
SY15 6RS G0RJV
SY15 6RS M3GEX
SY15 6RS M3HFT
SY15 6RT M3MYQ
SY15 6SB 2E1CEZ
SY15 6SB G4NQJ
SY15 6TP G4WND
SY15 6TP G4YNL
SY16 1DW GW7TIX
SY16 1NW MW7RHF
SY16 1PR GW7CAH
SY16 1QQ GW0TWL
SY16 1QQ GW1PVN
SY16 1QT GW7DJL
SY16 1RA MW0EDX
SY16 1RH MW6RQM
SY16 2AW 2W0MTE
SY16 2AW MW6MTE
SY16 2HN 2W0GWM
SY16 2HN MW6GWM
SY16 2JN MW6WCU
SY16 3AY MW5ABR
SY16 3DS 2W0WGN
SY16 3DS MW0MWX
SY16 3DS MW0OMK
SY16 3DS MW7LTS
SY16 3EE GW4BML
SY16 3ER MW5TLE
SY16 3HX GW7NNM
SY16 3JY GW3LXE
SY16 3PY MW7EQT
SY16 4DB GW3IWM
SY16 4LL 2W0HBW
SY16 4LL MW7JTH
SY16 4PS GW4RLO
SY17 5JG G3FXI
SY17 5JP G4ZUD
SY17 5NG M6SCN
SY17 5NG M7GYZ
SY17 5PU M6EDV
SY17 5PU M6EDW
SY17 5RH 2E0JHE
SY17 5RH G5JHE
SY17 5RH M7EJH
SY18 6AD 2W0KPF
SY18 6AD MW6KPF
SY18 6AR GW4CWG
SY18 6HZ GW1UOY
SY18 6JY 2W0GEZ
SY18 6JY MW0OMZ
SY18 6JY MW7GEL
SY18 6JZ MW7PCG
SY18 6LA MW7PTW
SY18 6II 2W0HOH
SY18 6II MW6HOH
SY18 6NP GW4DTU
SY18 6PQ 2W0CYE
SY18 6PQ MW6CKM
SY18 6SN GW3XPK
SY186DN MW7MAI
SY19 7DJ GW0EHS
SY19 7DJ GW1JNI
SY2 5EF G3TSV
SY2 5JY M6FXE
SY2 5LA M6FYT
SY2 5LQ G0IGM
SY2 5PF G3SRT
SY2 5PF G3VZG
SY2 5QL 2E1JCM
SY2 5TA M0MEB
SY2 5TE M1GSX
SY2 5TS M6FNM
SY2 5UZ 2E1EQR
SY2 5YB M0GSL
SY2 5YD G0EBD
SY2 6BG G4ZMR
SY2 6EE M6OEN
SY2 6HH M0IQG
SY2 6HQ G3IDY
SY2 6HU G4LYJ
SY2 6JX G4XDC
SY2 6LS G4WSX
SY2 6ND M3ZXE
SY2 6SJ G0RQX
SY2 6TE M7CLP
SY20 8DL 2W0RDZ
SY20 8DL MW0RCZ
SY20 8DL MW6TWL
SY20 8EJ GW0JAI
SY20 8NY GW8BFO
SY20 8RA 2W0IGN
SY20 8RA MW6GKQ
SY20 9BQ GW7VBY
SY20 9EZ GW4LPU
SY20 9EZ GW7JUV
SY20 9LA GW1CDY
SY20 9PR 2W0XUL
SY20 9RD GW7GAH
SY21 0AE GW1JNR
SY21 0BD GW7MGW
SY21 0DG GW4SHF
SY21 0DG MW0BAA
SY21 0EP GW4AUD
SY21 0JT GW4DEP
SY21 0NJ MW6CPJ
SY21 0PW GW3KGV
SY21 0RH 2W0UTT
SY21 0RH MW0UTT
SY21 0RH MW6EHX
SY21 7BB GW4DYY
SY21 7BB GW8HOS
SY21 7JA MW0GKW
SY21 7NQ GW4DWX
SY21 7PN MW6EIB
SY21 7RH GW8HEB
SY21 7RW GW7JEL
SY21 7TP MW1ABT
SY21 7TS GW8IUM
SY21 7UL GW4TFM
SY21 8AU GW3JSV
SY21 8BJ GW6EUR
SY21 8BJ GW6EUS
SY21 8BT MW0DDE
SY21 8DY MW6CYX
SY21 8ER GW1ORP
SY21 8ER MW3ORP
SY21 8HL GW4MZB
SY21 8JD 2W0AVZ
SY21 8JD MW0EDQ
SY21 8JD MW3BPZ
SY21 8LA MW1BEW
SY21 8LF GW7DLD
SY21 8LF GW7KMD
SY21 8NJ GW4SGQ
SY21 8RT GW0NNE
SY21 8SG GW3TCV
SY21 8TS GW6EUT
SY21 9JE GW0KGD
SY21 9JY GW4BVE
SY21 9JY GW8DHT
SY21 9LH MW6WLY
SY21 9PN GW1KTW
SY21 9PX GW1EAV
SY21 9PX GW4GNY
SY22 5AF MW0DAX
SY22 6EW GW0AGZ
SY22 6HB GW6HUR
SY22 6NF GW4UBQ
SY22 6PZ GW7FYG
SY22 6PZ MW3NVQ
SY22 6QJ 2W0GSE
SY22 6QJ MW0TDV
SY22 6QJ MW3RDS
SY22 6QL GW1YQM
SY22 6QL MW6CYU
SY22 6QL MW6MFN
SY22 6SL GW4CRH
SY22 6TW GW3YRP
SY22 6UJ MW6MGM
SY22 6XL GW3UBH
SY226LL MW6VLS
SY23 1AT GW8SIT
SY23 1BT MW3ROX
SY23 1DB GW8ONP
SY23 1DW 2W0FLI
SY23 1DW MW0KFL
SY23 1DW MW6FNA
SY23 1EU 2W0OLT
SY23 1EU MW0OLE
SY23 1EU MW6LRO
SY23 1JU MW6HRR
SY23 1LB MW7XAW
SY23 1QB GW4TUD
SY23 1RJ MW0GLS
SY23 1RY GW1URF
SY23 1SS GW7AGG
sy23 1sz MW7HMT
SY23 1TB MW7OFS
SY23 2EL 2W0WOD
SY23 2EL MW6NFN
SY23 2EL MW6WOD
SY23 2EU MW6NBU
SY23 2HN MW7NNR
SY23 2JA GW7HAE
SY23 2JA MW3HAE
SY23 2JA MW3WTR
SY23 2JU GW1HIN
SY23 3AB MW6XFM
SY23 3BL GW7OZP
SY23 3BQ MW6IJD
SY23 3EA MW3LFL
SY23 3EZ GW4CTV
SY23 3FT MW0ZXY
SY23 3GZ MW6ORW
SY23 3HF GW8LZY
SY23 3HQ 2W0FFL
SY23 3HQ GW8SFT
SY23 3HQ MW0LBR
SY23 3HQ MW6FLA
SY23 3HU MW6SVM
SY23 3LH MW6GZZ
SY23 3LS 2W0DVP
SY23 3LS MW0PRI
SY23 3LS MW6DVP
SY23 3PD MW6ZXO
SY23 3PL MW6GZX
SY23 3QR MW6FPH
SY23 3QU MW6ZCT
SY23 3RP MW6TFB
SY23 3TR GW7HSW
SY23 4BQ GW1FWC
SY23 4DX 2W0DNR
SY23 4DX MW6TWH
SY23 4JJ 2W0RKM
SY23 4JJ MW0RZS
SY23 4JJ MW6KMJ
SY23 4LA GW3BV
SY23 4PP GW4LHL
SY23 4PU GW8UWP
SY23 4RN MW6AEN
SY23 4SP GW4BQQ
sy23 4sq GW8UDJ
SY23 4TP 2W0GIA
SY23 4TP MW0KGG
SY23 4TP MW6GIA

SY23 4TR GW4AKY
SY23 4TT GW8BTX
SY23 5BW MW0DEW
SY23 5BZ MW6FQA
SY23 5HG 2W0CPD
SY23 5HG MW0RBV
SY23 5HG MW6BXI
SY23 5HG MW6CGV
SY23 5HP MW3UYH
SY23 5HS 2W0DCK
SY23 5LF MW6RIY
SY23 5NA GW8UKZ
SY23 5NB GW0OBB
SY23Â 3DT GW7JSH
SY24 5BE G0PDA
SY24 5BS M6LJP
SY24 5DF M6MLI
SY24 5HJ M6GKT
SY24 5JD G6JWD
SY24 5LT G8NCU
SY24 5LX M3NZV
SY24 5LX M3VRD
SY24 5NP G0PUW
SY24 5NP G7GWT
SY25 6AE G1XOT
SY25 6AE M3XOT
SY25 6AJ G4ZHI
SY25 6DQ M6XIE
SY25 6EJ 2E0TOF
SY25 6EJ M0TOF
SY25 6EJ M6TOF
SY25 6NY G7GEP
SY25 6PE G0CEP
SY25 6QU G4YMZ
SY25 6QZ G7LUB
SY25 6TT M0CBD
SY3 0AG G1KQP
SY3 0BE G4JSZ
SY3 0ER G7JTD
SY3 0EX G4HAC
SY3 0JG G7LRB
SY3 0LE G7NLY
SY3 0NS G6JHS
SY3 0QF G1CWZ
SY3 5AN G0EIY
SY3 5AP M6XDT
SY3 5AS 2E0THZ
SY3 5AS M0XMH
SY3 5AS M6MGH
SY3 5BJ G6TGJ
SY3 5BJ G7BQO
SY3 5BW G3UDA
SY3 5DA G6NFE
SY3 5DF M0AMP
SY3 5LA M6LSN
SY3 5PG G4ZPC
SY3 6AB G3NRX
SY3 6AW G3OWJ
SY3 7AJ M7DPY
SY3 7JN G3JPU
SY3 7NB G0RCS
SY3 7NB G8OFZ
SY3 7RA M6LUW
SY3 8AD 2E1MJF
SY3 8AD M5WJF
SY3 8QQ M3XEJ
SY3 8RU G4ZZP
SY3 8SU G4AZS
SY3 8TJ 2E0NWB
SY3 8TJ M6NWB
SY3 8UX M1FMC
SY3 9AB 2E0IIP
SY3 9AB M7HTU
SY3 9DB G3VWH
SY3 9FW M0YQQ
SY3 9PH M0PDP
SY3 9PH M6OBS
SY3 9QS G1SCR
SY3 9QS M1DQI
SY4 1BJ G6WNG
SY4 1BY G4JUW
SY4 1DT G0ARP
SY4 1DT G6UDG
SY4 1EE G6DTN
SY4 1EE G8RIK
SY4 1EE M0DFA
SY4 1JD G3NSS
SY4 1JH G0DNI
SY4 2BB G8DSG
SY4 2ED M7BYS
SY4 2EH G4SDU
SY4 2HT G0WDW
SY4 2HY G6DQY
SY4 2LG G1CLD
SY4 2LH G7CJC
SY4 3AZ G4EDC
SY4 3JL G4VVD
SY4 3LY G4CBM
SY4 3LY G4YKX
SY4 3QX G3LEK
SY4 4JE 2E0ETC
SY4 4JE M3ZHZ
SY4 4JX G0RVE
SY4 4LA G0JLS
SY4 4NU 2E0ILL
SY4 4NU M7OAN
SY4 4SN G3SOA
SY4 5BL M6WCG
SY4 5BX G1WVV
SY4 5DE G1YJB
SY4 5DE M0BAV
SY4 5DU G0ISP
SY4 5JN G4URT
SY4 5NB G3VAO
SY4 5NB G4CES
SY4 5NB G7BSP
SY4 5NW G7MHL
SY4 5NX G3TUL
SY4 5QZ 2E0SUT
SY4 5QZ 2E0TGG
SY4 5QZ M6RDS
SY4 5QZ M6TGG
SY4 5SL 2E0TVS
SY4 5SL M0KZP
SY4 5SL M6TVS
SY4 5UJ G4CBQ
SY4 5UJ M3JBQ
SY4 5YD G3PLW
SY5 0BG 2E0SIX
SY5 0DX 2E0GNV
SY5 0DX M7BGB
SY5 0EH G7JWH
SY5 0EH G8DIR
SY5 0NG M3WKC
SY5 0NS M0PWT
SY5 0QZ 2E0OCM
SY5 0QZ M0OCM
SY5 0QZ M6OCM
SY5 0TL M5LMG
SY5 0UD G0PVW
SY5 6DA M7CCR
SY5 6DS G0OOQ
SY5 6DS G7DCM
SY5 6JJ 2E0HMC
SY5 6JJ M6DYU
SY5 6JR M0JJC
SY5 6JR M6HAU
SY5 6PH M7BVF
SY5 7AP G6FHM
SY5 7AP M0FHM
SY5 7AT M3SUW
SY5 7BW G0ZDL
SY5 8AN G7DWX
SY5 8AN M0DEV
SY5 8AU 2E1DEM
SY5 8EP G4MYZ
SY5 8ND M0ZPM
SY5 8NH G8DIQ
SY5 8RA G2CQX
SY5 9AS G4FRX
SY5 9BE 2E0FOI
SY5 9BE G6GBY
SY5 9BE M7GBZ
SY5 9DJ G3YFK
SY5 9EG 2E0GQW
SY5 9EG M0GQW
SY5 9EG M3GQW
SY5 9EU G3VVL
SY5 9GZ M6OEG
SY5 9JA 2E1UTD
SY5 9JS G8NGF
SY5 9JY M1DQH
SY5 9LF 2E0JUX
SY5 9LF M6JUX
SY58LE M1DQG
SY6 6DD G3LRH
SY6 6DP M1DTS
SY6 6EE M6GVL
SY6 6GA M0DUV
SY6 6HT M6GVM
SY6 6JW G4SMA
SY6 6LX G7UMF
SY6 6NJ G4UST
SY6 6QD G0WPX
SY6 6QD G3BJ
SY6 6QD G4JKS
SY6 6RB M0COP
SY6 7AB M1CHU
SY6 7BB M3PHQ
SY6 7BN G8WPF
SY7 0EF M6FNB
SY7 0EP M3JLR
SY7 0NB G0NGN
SY7 8AN G0KKQ
SY7 8BA M6DUX
SY7 8DG M3MRM
SY7 8EN M6FSW
SY7 8QR M3YZO
SY7 9DT 2E0NLT
SY7 9DT M0NLT
SY7 9DU 2E0CHZ
SY7 9DU G7RHF
SY7 9DU M6RAL
SY7 9DW M3SCF
SY7 9DW M6XCF
SY7 9DX G7EYM
SY7 9ER G3ZZX
SY7 9ET M6ZJH
SY7 9HX M1CMM
SY7 9LF M6ZJP
SY7 9PG G0CGI
SY7 9RF M6PNR
SY7 9RL M6AUA
SY8 1BH M3ZXC
SY8 1BQ G6CRD
SY8 1DS 2E0BYA
SY8 1DS M0MYA
SY8 1DS M3BYA
SY8 1DW G3VWX
SY8 1EY G3SEZ
SY8 1JE M3YAI
SY8 1LE M6AOW
SY8 1LU M6BQC
SY8 1NB 2E0DRI
SY8 1NB M6KKG
SY8 1NB M6YDA
SY8 1NH G1VVX
SY8 1NS M3RLM
SY8 1UT G7IQD
SY8 1XN 2E0OVW
SY8 1XN G4HQB
SY8 1XN G7HSA
SY8 1XN M0OVW
SY8 1XN M3XLM
SY8 1XN M6OVW
SY8 1XP M0SJL
SY8 1XZ G3KZE
SY8 2DH M6JBX
SY8 2DT G8EZD
SY8 2NQ G4EVR
SY8 2NS M3JWM
SY8 2PH G4OYX
SY8 2QD 2E0GCJ
SY8 2QD M6GGJ
SY8 3AE 2E0CDS
SY8 3AE G6LXL
SY8 3AE M0MTS
SY8 3AW M0CBA
SY8 3DR M3EBF
SY8 3DR M3OMF
SY8 3EB 2E0PKL
SY8 3EB M6PKL
SY8 3ED M1DKZ
SY8 3JX M6ZGZ
SY8 3NJ G8HMV
SY8 3NZ G1AJS
SY8 3NZ G8GUN
SY8 3PN M6OQZ
SY8 4AY M0BVY
SY8 4BX M3NFH
SY8 4EP M1AXD
SY8 4HD M6AUH
SY8 4JT G4AIJ
SY8 4LB 2E0KYX
SY8 4LB M6KYX
SY8 4LF M6EGG
SY8 4LX G1JOD
SY8 4ND M1MST
SY8 4NE G0EXA
SY8 4NQ G1KGE
SY81RU M7JFJ
SY9 5AF G3NUB
SY9 5BS M5DND
SY9 5EF 2E1HUB
SY9 5EL M0GGU
SY9 5ES G0RPW
SY9 5HB 2E1HTY
TA1 1DQ G3VPQ
TA1 1LB G1SOM
TA1 1PH G0UJU
TA1 1UW M6AXP
TA1 1UW M6KGV
TA1 2AF 2E0SBB
TA1 2AF M0SBB
TA1 2AF M6AES
TA1 2AF M6TBJ
TA1 2BD 2E0BBZ
TA1 2BS G1YLN
TA1 2JQ G1NZK
TA1 2NA G7CCV
TA1 2NP M3TJQ
TA1 2PL G8DUI
TA1 2PW M0HQB
TA1 2QJ G4ASK
TA1 2QJ M7JOY
TA1 2QX 2E0WDI
TA1 2QX M3ONB
TA1 2RA 2E0AKN
TA1 2RA M0GSN
TA1 2RA M3LQV
TA1 2TA G0PNB
TA1 2XF G0WKU
TA1 2XN M0CIF
TA1 3DA G4RWF
TA1 3EJ G4RNT
TA1 3EL G0PGL
TA1 3FN G3VFB
TA1 3HQ G0UIL
TA1 3HQ M6SDS
TA1 3JE M0CIE
TA1 3NR G3CWD
TA1 3PA G3SHL
TA1 3PA G4BJX
TA1 3RR G4VVM
TA1 3RW M3SLZ
TA1 3XN G0UFU
TA1 4JE G4MYE
TA1 4JX G3JKE
TA1 4NL G7RAZ
TA1 4QH M3PVP
TA1 4RE G8JXK
TA1 5DS G4WTA
TA1 5EH G0RIE
TA1 5EP 2E0ZSE
TA1 5EP M0SEV
TA1 5EP M6ZSE
TA1 5ES G8XHU
TA1 5HH 2E0HJZ
TA1 5JE G0EYR
TA1 5JE M6GXH
TA1 5JH 2E1BFW
TA1 5JH G7KIQ
TA1 5JU G1FJS
TA1 5NS M3XNO
TA1 5PQ G4VVS
TA1 5PQ G8ZSP
TA1 5PW G3TJH
TA1 5QT M3PDD
TA1 5QZ 2E0BPI
TA1 5QZ M3WEQ
TA10 0BS G3OTK
TA10 0HH G3RSU
TA10 0JH G0PNF
TA10 0NG G6AYD
TA10 0NX G4BWL
TA10 0PP G4BEB
TA10 0PP G6JNS
TA10 9AE G0DGA
TA10 9AE G0DGE
TA10 9AX 2E1JKL
TA10 9DD 2E1LGE
TA10 9DH G1HQN
TA10 9EZ G8UIG
TA10 9FB G3PCJ
TA10 9JX 2E0PDB
TA10 9JX M0IEK
TA10 9JX M6OQQ
TA10 9JX M6PEF
TA10 9LT 2E0GKR
TA10 9LT M0VKR
TA10 9LT M3OLZ
TA10 9LT M7WGF
TA10 9NL G3USC
TA10 9PR G0RGC
TA10 9RH G0SZT
TA10 9RL M7BWL
TA10 9ST M6PGN
TA10 9TA M3XSR
TA10 9TF G0FZH
TA10 9TH G7LPO
TA109TA G0LCC
TA109TA G1OYH
TA11 6AT 2E0YEO
TA11 6AT M6LXN
TA11 6BW G0HDJ
TA11 6DF 2E0OSS
TA11 6DF M6OEU
TA11 6DX M7SGM
TA11 6ER G1PUQ
TA11 6HP G3YYR
TA11 6JY M6AXU
TA11 6LG G7IHP
TA11 6NF M1AXM
TA11 6PQ 2E0GOP
TA11 6PQ M0PGO
TA11 6PQ M6XSR
TA11 6PX M7JFZ
TA11 7AP M0BJO
TA11 7DN G3TCT
TA12 6BL G8GVW
TA12 6DT G1AKD
TA12 6FA G0PNS
TA12 6FA G3LWM
TA12 6HG 2E0LQW
TA12 6HG M0RGE
TA12 6HG M3LQW
TA13 5AJ G7ITU
TA13 5AL M7DMS
TA13 5AP G4PDG
TA13 5AQ G8VBI
TA13 5BD M7SHW
TA13 5BN M1AIK
TA13 5BY 2E0ZXG
TA13 5BY M0ZXG
TA13 5BY M6ZXG
TA135BQ 2E0CSQ
TA14 6QN G6EER
TA14 6QN G8HWA
TA14 6SH M1BSN
TA146PT G7RVG
TA16 5NF M6NBG
TA16 5NP G3TEL
TA16 5QX M6KMF
TA18 7AH G6AFL
TA18 7AH M3BON
TA18 7AP M3UBH
TA18 7AQ G0NPF
TA18 7AX G1PVZ
TA18 7AY 2E0IJX
TA18 7AY M6IJX
TA18 7BN M7BWH
TA18 7DJ M0OJS
TA18 7PB M0GJD
TA18 7PW M7DHD
TA18 7SG G4FKI
TA18 7TH 2E0OQD
TA18 7TH M6OQD
TA18 8BL G4FVW
TA18 8DB M1RMW
TA18 8DF 2E0BFJ
TA18 8DF M3JYH
TA18 8DF M3TGD
TA18 8ET G1AWU
TA18 8HS M6XAA
TA18 8JA M0BCL
TA18 8JB M0BBO
TA18 8JT G3WMX
TA18 8LY G4NCU
TA18 8PL G0JCY
TA18 8QE G0EGC
TA19 0BA M6MRW
TA19 0EA 2E0XIP
TA19 0EA M3FGQ
TA19 0HH G0MEQ
TA19 0NZ G4OXR
TA19 9AH G6AOV
TA19 9AH G6LVG
TA19 9DG G7FVQ
TA19 9ER M0ILT
TA19 9FG M0AGT
TA19 9LG G8JFQ
TA19 9NH G4KJD
TA19 9NW G4HTD
TA19 9QH G6PQP
TA19 9QL G0WSC
TA19 9QN M0AOD
TA19 9QU M0BHE
TA2 6LE G1TKQ
TA2 6RH M3AKW
TA2 6RR G4SZS
TA2 6SR G8HNM
TA2 6TA 2E0BQY
TA2 6TR M6CTT
TA2 7DN M7LND
TA2 7DT G1CNZ
TA2 7EF G4WSF
TA2 7EN G1AJZ
TA2 7PB G4XKK
TA2 7PT M7CRM
TA2 7SP G7LDD
TA2 8DX G0PWV
TA2 8DX G0RWA
TA2 8DZ G7KYG
TA2 8EW M7EFZ
TA2 8LB G7CWT
TA2 8NA G4UTM
TA2 8NA G8BTY
TA2 8PP G6EIO
TA2 8QB G1VVE
TA2 8RL M0FBM
TA2 8RP G3PCG
TA20 1BX 2E0TAW
TA20 1BX M3MNT
TA20 1BZ G3HAL
TA20 1DG M1DGP
TA20 1DZ G4IRS
TA20 1EZ G8DLH
TA20 1FH G1ZYS
TA20 1FX M7FAN
TA20 1JU G6WZA
TA20 1NJ G0LNK
TA20 1PF G0NBI
TA20 1PF M6VAJ
TA20 1SD M6TVH
TA20 2BL G4VHB
TA20 2BU G1ECV
TA20 2DP 2E0ULY
TA20 2EB G8BVR
TA20 2ED G0AEN
TA20 2EG M0FAK
TA20 2EG M3AIG
TA20 2EG M3BXG
TA20 2EW 2E1IJR
TA20 2FP G8OCF
TA20 2HH M3UVV
TA20 2HL M7DZG
TA20 2JB G8GZC
TA20 2JH M7UTD
TA20 2ND G8LKK
TA20 2PY G8GFZ
TA20 2RX M3UTN
TA20 2SX G7MFH
TA20 2TJ M3LBZ
TA20 3BH M1FAY
TA20 3HJ 2E0PPF
TA20 3HJ G0KJF
TA20 3HJ G8HOI
TA20 3LT M0XMS
TA20 3PH 2E0DZT
TA20 3PH M6DZT
TA20 4PJ G4TIA
TA21 0DT G4CAK
TA21 0DX 2E0PDU
TA21 0DX M0PDU
TA21 0DX M3PDK
TA21 0DX M3PDU
TA21 0EL G0VSG
TA21 0RH G8FZI
TA21 8BB G0MJF
TA21 8DH G8ZVK
TA21 8EP M6JST
TA21 8ER G1CQG
TA21 8EX G4RGA
TA21 8HZ G4YXR
TA21 8HZ M0MAX
TA21 8JB M5WSS
TA21 8LG G7BSG
TA21 8PY G4OLA
TA21 8SF M7CSX
TA21 9AL 2E0AKL
TA21 9AL M0GFO
TA21 9AL M3NLX
TA21 9AR G6TWA
TA21 9AT G4RGF
TA21 9AX G4AIU
TA21 9DY M3SNL
TA21 9DY M6GYZ
TA21 9ED G0BDJ
TA21 9EG 2E0RIQ
TA21 9EG M6RPQ
TA21 9FX M6IHE
TA21 9HT M0CVB
TA21 9HW M6TEJ
TA22 9EL G7KWP
TA22 9EN G1TKE
TA22 9EN M6SES
TA22 9PT G0JFA
TA22 9RU G8CCF
TA23 0AG G6BWJ
TA23 0AH M7UKD
TA23 0BP 2E0EAI
TA23 0BP M6SMB
TA23 0DD G4BQB
TA23 0HZ G6KRG
TA23 0PZ M6NRG
TA23 0QY 2E0KEK
TA23 0QY M6HFU
TA23 0RH G8XHD
TA23 0TX M7WRI
TA24 5DY M7PSS
TA24 5HS G4PHC
TA24 5JU G3NFY
TA24 5NZ G4FBZ
TA24 5NZ M0AOJ
TA24 5TD G7PBO
TA24 6AD M3LSK
TA24 6BT M3GQI
TA24 6BY G1CGU
TA24 6BY G6HOR
TA24 6EE 2E0ONV
TA24 6EE M3ONV
TA24 6HJ G4HOW
TA24 6JH M6RJW
TA24 6PW M3BLR
TA24 6SD 2E0WSR
TA24 6SD M6DUN
TA24 7RU M0IAX
TA24 7RU M6FLJ
TA24 7UL M3BXH
TA24 7UL M3LSE
TA24 8AF G0WMN
TA24 8AN 2E0EZC
TA24 8AN M0ROL
TA24 8AN M6PUA
TA24 8AS G8EAM
TA24 8AW G0ACA
TA24 8AX M1AFZ
TA24 8DH G8NIU
TA24 8EB G4FNJ
TA24 8HL G1ONV
TA24 8JR M6ARV
TA24 8JR M6ASI
TA24 8QF G4TIC
TA24 8TA G8BCI
TA246AD 2E0AKO
TA26DF 2E0HWS
TA26DF M0KFH
TA3 5AU G6PPU
TA3 5BX G4LZU
TA3 5BY M3PBP
TA3 5DW G4AVJ
TA3 5EW G4IBC
TA3 5EW M0AID
TA3 5JW G1NTK
TA3 5LP M6AYQ
TA3 5PW G4MIB
TA3 5QP G0MZQ
TA3 5RA M7DZQ
TA3 5RP M3SZK
TA3 5SW M6VVX
TA3 5TE G4JZL
TA3 5TE G8BDM
TA3 6AE M0WDL
TA3 6DX M0HMX
TA3 6HA G1DRY
TA3 6HA G1KNK
TA3 6PU G8XOE
TA3 6QN G8XET
TA3 6QN M3LVK
TA3 6TN G7DKZ
TA3 7EG G4UVV
TA3 7HB G4NIL
TA3 7HH G4MFD
TA3 7HS G8WXV
TA3 7LG G4KIK
TA3 7NR G3THG
TA3 7QB M3AXZ
TA3 7RE G6TOY
TA3 7SD G4UVZ
TA3 7SU 2E0XQX
TA3 7SU G5JJ
TA3 7SU M0LHC
TA3 7SU M7LHC
TA4 1AQ G1XQP
TA4 1DF M7KSF
TA4 1EU M6OPS
TA4 1EX G4ZTS
TA4 1JA G4FDA
TA4 1JH M7JEX
TA4 1NY G7HKZ
TA4 2BU G0GWS
TA4 2BW M5SLC
TA4 2DP M3BAO
TA4 2EQ G7SLZ
TA4 2JY G0FUS
TA4 2LF M7SDX
TA4 2LR G4HLN
TA4 2LR M7WEZ
TA4 2LZ M0GHR
TA4 2QF G4ETN
TA4 2SB G0HMG
TA4 2UL G7VNJ
TA4 3AQ G3UWH
TA4 3TX M0GFX
TA4 4AE G4VJL
TA4 4DE G0NKZ
TA4 4DE G0SZO
TA4 4LA G0WIE
TA4 4NZ M6CJE
TA4 4PD G3HMG
TA4 4PD G3POM
TA4 4PG G4AJU
TA4 4SG M7EPS
TA4 4SL G0NRI
Ta43jn M6NPO
TA5 1QG G8AYV
TA5 1SF G4OEC
TA5 2AG M7BMH
TA5 2AP G3OJK
TA5 2FQ G8SEK
TA5 2FT M6KUP
TA5 2GG M6FQM
TA5 2GT G0MRR
TA5 2GW M7SKN
TA5 2HW G3TTP
TA5 2JN M6CTY
TA5 2PY G7UUL
TA5 2PZ G4WMY
TA6 3LP G4BRW
TA6 3QE G7SYS
TA6 3TD 2E0PAH
TA6 4DW G3RHW
TA6 4HE M7CSP
TA6 4HL M3TOE
TA6 4HQ G4SMQ
TA6 4HT M7SBO
TA6 4JL G6JTT
TA6 4NA 2E0WEC
TA6 4NA G7GMZ
TA6 4NA M0WFN
TA6 4NA M3CSK
TA6 4PN G0OYA
TA6 4RY G1BKL
TA6 4SH G0MIK
TA6 4SH G7FAD
TA6 4SL G0VOE
TA6 4UB M3HET
TA6 4UT G7KHZ
TA6 4XD M6MXB
TA6 5BN M7JBE
TA6 5BT G8IPA
TA6 5HA M1FFC
TA6 5NS G0LCV
TA6 5PH G0CNA
TA6 5SA M3WSO
TA6 6AX G0OXB
TA6 6BW G4DII
TA6 6DZ G8PVG
TA6 6HF 2E0MHL
TA6 6HF M6MHL
TA6 6JU G1MCY
TA6 6LF M3JRF
TA6 6LJ M6AZF
ta6 6qw 2E0VCM
ta6 6qw G6MYH
ta6 6qw M3VCM
TA6 6RN G8FDF
TA6 6SJ G8ZOJ
TA6 6UR G1XOZ
TA6 7AJ G0WRQ
TA6 7LJ G1BTI
TA6 7NZ G7WBL
TA6 7PD G1HLP
TA6 7QB M0JES
TA6 7QG G0UQT
TA6 7QY G7URS
TA6 7RF G8YPV
TA7 0AZ G0LHX
TA7 0DE G3FSA
TA7 0EZ 2E0SST
TA7 0EZ M6SST
TA7 0HD G8VHO
TA7 0HD M3DBY
TA7 0HD M3PQJ
TA7 0HJ G4ELW
TA7 0JR M6FUL
TA7 0NG M6IRL
TA7 0RH G6OBB
TA7 0SB G4WGX
TA7 8AL G4ZBS
TA7 8AP M0BZV
TA7 8AS G6YGH
TA7 8BS 2E0TDI
TA7 8DW G1MOK
TA7 8EA G1IZH
TA7 8EA G7VWM
TA7 8ER G3MDD
TA7 8HJ G0GCA
TA7 8HQ G1NMQ
TA7 8JB G4UNO
TA7 8PP G6AJV
TA7 8QR G6RRV
TA7 9AR G4BSK
TA7 9AT G6PZ
TA7 9BT G1HSF
TA7 9DY G4GHM
TA7 9LB G8SUW
TA7 9LS G4NQQ
TA70EU M1DJX
TA8 1BE M3KJS
TA8 1HE M1DIL
TA8 1HG G0BBK
TA8 1HP G4HLN
TA8 1JA G4DVK
TA8 1JA G4RGY
TA8 1JU 2E1IIM
TA8 1LN G1KVQ
TA8 1LT G4EJW
TA8 1LU 2E0FLU
TA8 1LW 2E0IFT
TA8 1LW G7VWG
TA8 1LW M3IOX
TA8 1LW M6IFT
TA8 1QB 2E0FWT
TA8 1QB M0JID
TA8 1QB M7APN
TA8 1RE M3UHS
TA8 2EJ G6TAH
TA8 2EL G6TKR
TA8 2EW G0OQT
TA8 2HL 2E0RCO
TA8 2HL M7RLW
TA8 2HP G0FYP
TA8 2JG M3SGG
TA8 2LE G7NME
TA8 2LN G4CGH
TA8 2LT 2E0VDV
TA8 2LT M3WRL
TA8 2LT M6ADD
TA8 2NH 2E0FKY
TA8 2NH M6PKQ
TA8 2NL G3GZZ
TA8 2NN G0PQD
TA8 2NW G3WLA
TA8 2PH G3NYD
TA8 2PS G3UBP
TA8 2QA M3PHG
TA8 2SZ M0FEY
TA8 2TT G6SEE
TA8 2TT G6SEF
TA9 3DQ M1DSZ
TA9 3EH 2E0VOK
TA9 3EH 2E0ZUX
TA9 3EH M0VOK
TA9 3LA 2E0FLM
TA9 3LA M6GPP
TA9 3LD G4UOS
TA9 3QU G3XBI
TA9 3QZ G0VYV
TA9 3RF G7GJZ
TA9 3RJ 2E0HKE
TA9 4BS G7FEP
TA9 4BU G8HUH
TA9 4DQ G1JGM
TA9 4DT G4TBO
TA9 4DU G7VIV
TA9 4DY G0GOB
TA9 4EJ M0IPI
TA9 4HE M6GLM
TA9 4LZ M0NDL
TA9 4NG G3VLW
TA9 4QT G3SXQ
TD1 1HD GM7GIS
TD1 1JL MM7RFP
TD1 1QL GM8KOF
TD1 1RG MM3RQG
TD1 1SQ GM6IZU
TD1 2BA GM4JPG
TD1 2BJ 2M0RTI
TD1 2BJ MM6PNU
TD1 2EE GM1JLP
TD1 2FA GM7MUN
TD1 2FA MM3SMI
TD1 2HZ GM7GPG
TD1 2LB GM8UUW
TD1 2RB GM8CJW
TD1 2RJ 2M1GEZ
TD1 2RJ 2M1GXX
TD1 2RJ MM0BPX
TD1 3BL MM7SVI
TD1 3HW GM4MCV
TD1 3JW GM4YEQ
TD1 3JW GM6WRY
TD1 3ND GM7LUN
TD1 3PA 2M0ODP
TD1 3PA MM6VDP
TD1 3RF GM0HQT
TD1 3SB 2M1BBY
TD1 3ST 2M0HWI
TD1 3ST MM7KTW
TD10 6XN MM6CES
TD11 3BF 2M0HVG
TD11 3BF MM7FLZ
TD11 3EE MM6KFE
TD11 3HG GM0TUS
TD11 3HG GM0RDZ
TD11 3HS MM3URQ
TD11 3HW GM1DVO
TD11 3JJ MM7DSA
TD11 3NB MM0JYV
TD11 3NZ 2M0GJJ
TD11 3NZ MM0GJJ
TD11 3NZ MM3HFU
TD11 3PP MM6BNX
TD11 3RA 2M0YTN
TD11 3RA MM7TWN
TD11 3TQ GM0EUL
TD11 3UF MM6KWA
TD11 3UT 2M0FUP
TD11 3UT MM0JNL
TD11 3UT MM7DSV
TD12 4BS GM0HNP
TD12 4BW MM7MSL
TD12 4BY MM0AMV
TD12 4EY 2M0CDO
TD12 4JQ GM4SYF
TD12 4NG GM0BRS
TD12 4NG GM8BDX
TD12 4XB 2M0SLN
TD14 5AU MM0WZZ
TD14 5BF MM3LSO
TD14 5DX GM4ZXJ
TD14 5EU GM1JFF
TD14 5EU MM1DTU
TD14 5JF MM6BPM
TD14 5LG 2M0DZZ
TD14 5LG MM0GZZ
TD14 5LG MM3UQN
TD14 5LS GM7SRJ
TD14 5NJ 2M0BXY
TD14 5NJ MM3ZFK
TD14 5PB 2M0XDX
TD14 5PB MM0XAB
TD14 5PB MM6FRD
TD14 5PB MM6LRK
TD14 5RR GM4FVM
TD14 5TB GM7LNO
TD15 1BZ GM3VAJ
TD15 1DU MM0SNA
TD15 1SU GM4WDO
TD15 1XB GM7ROC
TD15 1XW GM6VGO
TD15 2BB GM4XVF
TD15 2BS GM0JCP
TD15 2BS GM1IVI
TD15 2DX GM4GCL
TD15 2EB GM3KML
TD15 2EE MM0GHK
TD15 2HJ 2M0ICB
TD15 2HJ MM0ICB
TD15 2JH MM3INO
TD15 2JN GM6IBP
TD15 2LA 2M0KLN
TD15 2LA MM0KLN
TD15 2LA MM6PNH
TD15 2RW 2M0DDO
TD15 2RW MM6CMI
TD152TY GM0HBV
TD2 6TT GM0SRO
TD3 6ND MM6TWE
TD3 6NF GM4YSN
TD4 6AS GM0RWU
TD4 6BN GM3RXU
TD4 6HS 2M0YBR
TD4 6HS MM6BIH
TD5 7AS GM0ALW
TD5 7AS GM0ALX
TD5 7AS GM4KHS
TD5 7EU GM0SIA
TD5 7NB GM6ZFI
TD5 7NJ GM4CXP
TD5 7NN GM6PTX
TD5 7NP GM6TBE
TD5 7PB GM6RTN
TD5 7RF MM7DMW
TD5 7RU GM0CDV
TD5 7SZ GM3PPE
TD5 7TT MM7WPC
TD5 8ED MM7CCC
TD5 8LB GM4NLJ
TD5 8SA GM8TCH
TD6 0AL GM0IQI
TD6 0QS MM0BPP
TD6 0RY GM4CID
TD6 0RY MM3OXB
TD6 9BH MM0MRM
TD6 9EP GM0RDC
TD6 9HB GM0VHR
TD6 9JB GM4PNM
TD7 4BG 2M0KMQ
TD7 4BG GM7GRH
TD7 4BG MM6KMQ
TD7 4HS MM0HTL
TD7 4LT MM0ILC
TD7 4PR 2M0WDA
TD7 4PR MM7WSS
TD7 5BH MM3XXO
TD7 5BP GM0KCN
TD7 5DD GM0FTJ
TD7 5EY GM4UFP
TD7 5JB GM4VYU
TD7 5JZ MM3YJH
TD7 5NF MM3PAE
TD7 5NF MM3YDH
TD8 6AG GM4UPX
TD8 6DU GM0NYD
TD8 6DU GM1ZTB
TD8 6JW MM0INE
TD8 6LA GM4HCY
TD8 6ND GM4JKZ
TD8 6ND GM6FTE
TD8 6SA GM0FSH
TD8 6SD GM7WLO
TD8 6SU MM6BUG
TD8 6TZ GM6MCV
TD9 0AS MM6DXH
TD9 0QE GM6KWU
TD9 0QQ 2M0EBU
TD9 0QQ MM0RYR
TD9 0QQ MM6FZT
TD9 0SN GM4LPJ
TD9 0TT 2M0XRZ
TD9 0TT MM0XRZ
TD9 7BH GM0FTK
TD9 7PR GM0IGJ
TD9 7QE MM1APS
TD9 8BA GM8GUX
TD9 8EN GM6CKN
TD9 8JE MM0GCY
TD9 8QJ GM8JBJ
TD9 8QJ MM3SLB
TD9 9NZ GM4VJH
TD9 9SS MM7BWK
TF1 1JH G1VIT
TF1 1QJ 2E0UDE
TF1 1QJ M0RKY
TF1 1QJ M6RLB
TF1 2AJ 2E1HKS
TF1 2AS G8VZT
TF1 2AS M5AFG
TF1 2BY G1BHV
TF1 2JF G8BKH
TF1 2JP G0RQI
TF1 2LE G4AUY
TF1 2ND M3YWF
TF1 2PQ G0ASP
TF1 3DB G1JJK
TF1 3DU M7NDK
TF1 3ED G6HKP
TF1 3GA G6NRM
TF1 3GD M3NZW
TF1 3HH G6ZME
TF1 3HU M0CFT
TF1 5FL M3IJT
TF1 5JT G0UCN
TF1 5LN 2E1EQK
TF1 5SE M7EFM
TF1 6AD M0AFQ
TF1 6FZ G1ZPQ
TF1 6PQ G6NOL
TF1 6QR G2RSA
TF1 6TD G8RXZ
TF1 6XT M0CPK
TF1 6XU G6ZFU
TF10 7BS 2E0PNN
TF10 7BS M0PNN
TF10 7DT G7KZG
TF10 7DX G7ACD
TF10 7EF G4IFR
TF10 7EG G1AJT
TF10 7HH G7KJA
TF10 7JD G0KBK
TF10 7JP G4UMM
TF10 7LS G0UCI
TF10 7NN M6IMA
TF10 7RF 2E0HDQ
TF10 7RF M7AXD
TF10 8AH G4JOW
TF10 8BP G1DIL
TF10 8DA G0GQK
TF10 8JL G1KRU
TF10 8JR G6KBN
TF10 9AQ G4OLO
TF10 9EN 2E0BAJ
TF10 9EN M0EAK
TF10 9EZ M0TAW
TF10 9EZ M3OGL
TF10 9JJ M3WXG
TF10 9JQ G0WJH
TF11 8EB M3YBW
TF11 8HL M7ICD
TF11 8HZ G0VSJ
TF11 8QD M1SWR
TF11 8RD G0EXN
TF11 9BF M3FAL
TF12 5PE G6APE
TF12 5SB G4YDT
TF12 5SH G6DBZ
Tf125jb 2E0IZP
Tf125jb M0IRP
Tf125jb M6IRP
TF13 6AX 2E0XTL
TF13 6AX G8XYJ
TF13 6AX M0WOF
TF13 6AX M3XTL
TF13 6BW G1HMI
TF13 6BW G3LNP
TF13 6BZ G6KJY
TF13 6ED G7DTG
TF13 6ED G8AQA
TF13 6ED M0HMO
TF13 6EN 2E0YPW
TF13 6EN M0YPW
TF13 6EN M6YPW
TF13NN G0CYN
TF2 0AS 2E0DTB
TF2 0AZ 2E0FTO
TF2 0AZ M3ZXX
TF2 0DT G4OMI
TF2 0DX G1OAR
TF2 0DX M0OAR
TF2 0DY M0VZT
TF2 6ES 2E1FQO
TF2 6EZ G7NIB
TF2 6JR M6AAD
TF2 6NA M1BGL
TF2 6RF G0GUV
TF2 7DA 2E1ACW
TF2 7LJ G0VXJ
TF2 7LQ G1SKW
TF2 7NP M3XNU
TF2 8AR M3DLB
TF2 8DB M1IHM
TF2 8HY G0HDC
TF2 8JH 2E0PZM
TF2 8JH M0TZM
TF2 8JH M7KBO
TF2 8NA G4YVD
TF2 8PB M1EZJ
TF2 8SQ M3GZT
TF2 9EN M6DCJ
TF2 9HD 2E0DCM
TF2 9HD M0ECM
TF2 9HD M0EMM
TF2 9JZ G3YZQ
TF2 9LQ M6ANN
TF2 9LX 2E0KLS
TF2 9NZ G4DFG
TF2 9PT G6APX
TF2 9PX M7OWA
TF2 9RB G3JKX
TF2 9SH M6FDN
TF2 9SL G4KLD
TF2 9SR 2E1HTV
TF3 1BW 2E0TIL
TF3 1BW M6TIL
TF3 1ED 2E1DZJ
TF3 1ED G8ZIP
TF3 1EG G0UFE
TF3 1EG G6UDX
TF3 1LA G0BST
TF3 1LJ 2E0BAH
TF3 1NH G8ZXU
TF3 1NY M3UWE
TF3 1PT M6LEU
TF3 1SS M6YVN
TF3 1XZ G0GQO
TF3 2AQ G7SFI
TF3 2BR M6LLO
TF3 2BU M0TYK
TF3 2BY 2E0TTB
TF3 2BY M6LMB
TF3 2BY M6TTB
TF3 2DA M6WSS
TF3 2EE 2E1IHJ
TF3 2ES M6MGG
TF3 2HX 2E1GKE
TF3 2JB 2E0KHM
TF3 2JB M0KHM
TF3 2JB M6KZM
TF3 2JJ M6HEW
TF3 2JW M0TBQ
TF3 2JX G4FBZ
TF3 2NH M0IRS
TF3 2NQ G6FFR
TF3 5DA G0HQK
TF3 5FH M7NTH
TF3 5GQ 2E1GJJ
TF3 5HD M6TZE
TF3 5HE G0GAL

TF3 5HH M6XWS
TF3 5HN M6KCA
TF4 2AH M7MLD
TF4 2AR 2E0ZGS
TF4 2DL M3KXB
TF4 2EB 2E0CVC
TF4 2EB M6CSZ
TF4 2EW G0MQX
TF4 2HS G6NWT
TF4 2LW M7JSY
TF4 2NX 2E0MCL
TF4 2NX M3LML
TF4 2QA 2E1HTU
TF4 2QE G0VNO
TF4 2RW 2E0PLX
TF4 3BJ G7SEY
TF4 3DD G8JVM
TF4 3DX M7BCI
TF4 3HP 2E0PWF
TF4 3HP M6EGV
TF4 3JF G0VPW
TF4 3LB M6BRB
TF4 3NG 2E1FNX
TF4 3RF M7AUO
TF5 0LL G8OYB
TF5 0NQ M1DGX
TF5 0NQ M1DLX
TF5 0PG G7BCO
TF6 5AU 2E0TRO
TF6 5AU M3UWB
TF6 5ER G6SGD
TF6 6DH G0JBO
TF6 6HH G4WRK
TF6 6HQ G3UKV
TF6 6HQ G3ZME
TF6 6HQ G8UKV
TF7 4AB M1ACA
TF7 4AQ M0DNV
TF7 4BX 2E1CFB
TF7 4DP M1GIZ
TF7 4DT G4VZL
TF7 4EA M6HDA
TF7 4NH G4VZK
TF7 4NQ G1KCA
TF7 4PZ M6SXA
TF7 4QE G0LCO
TF7 4QE G4TQZ
TF7 5HB M3UXN
TF7 5JJ M6OLY
TF7 5LU 2E0YYB
TF7 5LU M6BVC
TF7 5NB M3SXK
TF7 5NF M7REV
TF7 5NG 2E1AIY
TF7 5PT M7DJN
TF7 5QH 2E0KDF
TF7 5QH M6BVE
TF7 5QP 2E0HYJ
TF7 5QP M7DIW
TF7 5QR G0JAL
TF7 5RJ 2E0EOH
TF7 5RX M1NMG
TF7 5SF 2E0IVR
TF7 5SN 2E1DYL
TF7 5SU G4EIX
TF7 5UB 2E0APG
TF8 7BS G7OQT
TF8 7EW M0AXV
TF8 7ND 2E1ESW
TF8 7ND G0VXG
TF8 7ND M3BHI
TF9 1AR G8EWD
TF9 1EA G0JNA
TF9 1HP G8PHG
TF9 1NP G1DVH
TF9 2BA G3OKD
TF9 2BB G6JKV
TF9 2BD G0PLB
TF9 2DG G0MDJ
TF9 2DU G0DHM
TF9 2NB G4DQB
TF9 2QG G0GHL
TF9 2QX G0AGC
TF9 2RX M7KLT
TF9 2SF G3ZGU
TF9 2SX 2E0CBA
TF9 2SX M6HRC
TF9 2TF M3XKJ
TF9 2TG M3MQJ
TF9 2TH M1MDE
TF9 3BD G4MZQ
TF9 3BW M7CWS
TF9 3HJ G6COB
TF9 3HJ M3COB
TF9 3HX M7EQM
TF9 3LP G6WPL
TF9 3QB G0CER
TF9 3QB G6VSG
TF9 3QE M7FWA
TF9 3QF G0CGA
TF9 3QH M3THE
TF9 3SY 2E0UNI
TF9 3SY M0DQS
TF9 3SY M0UNI
TF9 3UB 2E0SHG
TF9 3UH G8PSZ
TF9 3UJ G0CDS
TF9 3US M7GNA
TF9 4BE G3JPB
TF9 4DD 2E0SNP
TF9 4DN M0INI
TF9 4DN M6INI
TF9 4ED G4JKF
TF9 4LQ 2E0MKF
TF9 4PZ G3GCU
TF9 4QE G6GTJ
TF9 4QR G3WEI
TF9 4RJ M1AOL
TF91DU G6NJR
TF94DD M0HIO
TN1 2DY G4KIU
TN1 2LH G4ZFP
TN1 2SH G3WTR
TN10 3HB M3XST
TN10 3HD 2E0SRO
TN10 3HG G0DJC
TN10 3HG G7TYR
TN10 3HG G8WZK
TN10 3HQ M3TWP
TN10 3LW G1OGV
TN10 3LW M3OGV
TN10 3NN G1ZZL
TN10 3PU M7JCS
TN10 4AJ M6CBU
TN10 4BJ M6YOU
TN10 4DG G4FFY
TN10 4HD G7KNS
TN10 4HU M7KCY
TN10 4LB 2E0WOH
TN10 4LE G3MGQ
TN10 4NG G4FYG
TN10 4NN M3ZKL
TN10 4PD M7AMS
TN10 4QS G0VGD
Tn11 0ed M7SOX
TN11 0HU G4HGR
TN11 0NL 2E0RXW
TN11 8HP G1WKS
TN11 8JH M0RBT
TN11 8LE G3YJW
TN11 9BL G8YCJ
TN11 9DA G6IUS
TN11 9DR G6MXE
TN11 9ED G7THJ
TN11 9ED M3THJ
TN11 9HB G4AUL
TN11 9HD G8CAA
TN11 9HQ G4TPJ
TN11 9LG G4JED
TN11 9LG M6JHL
TN11 9PL G3ZDT
TN12 0BE M1EGP
TN12 0BT G7GMU
TN12 0LR G0UXG
TN12 0LR M0KWA
TN12 0NA G7GGA
TN12 0NE M6SPG
TN12 0PG M1CGR
TN12 0RH G1NDK
TN12 0SS G6TQL
TN12 0TE M0KEW
TN12 5BW G0EBS
TN12 5DB M6VKH
TN12 5EQ G4YIM
TN12 6BG 2E0UHS
TN12 6BG M3IMP
TN12 6BQ M7DHT
TN12 6HP G0MID
TN12 6LS G1YLB
TN12 6LS M7YLB
TN12 6YA M0RWW
TN12 7EA G8CDD
TN12 7LR 2E0ZWW
TN12 7LR M0ZWW
TN12 7LR M1ATA
TN12 7LR M3ZWW
TN12 8EF M6MPY
TN12 9AX G0WBV
TN12 9BS G8YBH
TN12 9EN 2E0LKR
TN12 9EN M0LKR
TN12 9JD G0RHO
TN12 9LS G7FKX
TN12 9NB G3XBQ
TN12 9SG M3RTE
TN12 9TQ G3UUG
TN12 9TW M3XPF
TN13 1EL G8KVO
TN13 1EU 2E0RJX
TN13 1EU M0RMI
TN13 1EU M6EYJ
TN13 1TJ M3HAU
TN13 2DZ M0PRV
TN13 2DZ M0RWK
TN13 2HE G1TGZ
TN13 2UA G4ASZ
TN13 2UG 2E1EAX
TN13 3DR M3PPR
TN13 3SB M6YJK
TN13 3XA G0CDQ
TN13 3XN M3KDM
TN14 5AS M1DUA
TN14 5AT G7MYY
TN14 5BT G7RHE
TN14 5JF 2E0KJJ
TN14 5JF G8PHM
TN14 5LD G0SLU
TN14 5QP G0WQC
TN14 5RR G6KLF
TN14 6BX 2E0YEW
TN14 6BX M0WEW
TN14 6BX M6KTC
TN14 6BX M6PRA
TN14 6DL M6KYZ
TN14 6HT G3SAR
TN14 6LY G1TTX
TN14 6NH M1RSB
TN14 7AU G7IET
TN14 7EY G1KVW
TN14 7HW M1NTY
TN14 7JU G7JYG
TN14 7NA G7KTR
TN14 7SW 2E0JDA
TN14 7SW M0JDA
TN14 7SW M6JDA
TN14 7TD G7PVF
TN15 0LZ M1JDW
TN15 6BB G1FHR
TN15 6BL G3GJW
TN15 6BL G3ZRX
TN15 6BL G4BRC
TN15 6PB G1MNY
TN15 6DT G4RVV
TN15 6EZ G7MNZ
TN15 6HP G8ZZK
TN15 6JJ G8TLP
TN15 6LF G8GHP
TN15 6LL M1DMN
TN15 6NU G0KFW
TN15 6PG G6EDM
TN15 6PT G6GLH
TN15 6QL G8PWO
TN15 6TG G3WBS
TN15 6YF G4MGY
TN15 7EY G6VBJ
TN15 7JU G8YDB
TN15 7NR G3RGS
TN15 7NR G4CTC
TN15 8DX M7KNR
TN15 8EA G3IAR
TN15 8HG M7OCJ
TN15 8HT G0IPH
TN15 8LL G8ZYI
TN15 8LQ G4DFY
TN15 8RG M0MEA
TN15 8RJ M7SMP
TN15 8RS G3EWM
TN15 8RS G4FQT
TN15 8SD 2E1FDP
TN16 1AW M7MOK
TN16 1DX G0POC
TN16 1LS 2E0DIT
TN16 1LS M0KOT
TN16 1LS M6DIT
TN16 2AT G1BMW
TN16 2BH M7WAI
TN16 2BT G4NSD
TN16 2JE G3KEQ
TN16 2LA G1FKP
TN16 3BJ G3ZVN
TN16 3BN G4ASI
TN16 3HH G6UNR
TN16 3HX 2E0DWG
TN16 3HX M6GVJ
TN16 3PH M7BXE
TN16 3QX M6TBV
TN16 3SA G1XPF
TN16 3SG G8JQS
TN16 3SH M7IND
TN16 3UU G4GLN
TN16 3UX G6THC
TN16 3XE G8AXN
TN16 3XY M3TLW
TN17 2DD G3GUE
TN17 3DZ 2E1FVX
TN17 3JR 2E0IAK
TN17 3LS G0EIH
TN17 3QJ 2E0THG
TN17 3QJ M0MNZ
TN17 3QJ M7MLB
TN17 4BT G0OEQ
TN17 4DB M0OOD
TN18 4DQ M7MWW
TN18 4JE G3XWL
TN18 4JH G6VGA
TN18 4RA G0SLD
TN18 4RW M6MOC
TN18 5DU 2E0ZJO
TN18 5DU 2E0ZJQ
TN18 5DU M0ZJO
TN18 5DU M0ZJQ
TN18 5DU M3ZJO
TN18 5DU M3ZJQ
TN18 5DU M6YLC
TN18 5LW 2E0OCB
TN18 5LW M3VXO
TN18 5LX G0IPO
TN19 7DB 2E0KWW
TN19 7DB M0KWW
TN19 7DB M6LUD
TN19 7PJ G4NVM
TN19 7QY G7IGF
TN2 3BL M3LCP
TN2 3BS M6MTH
TN2 3DA G4OSH
TN2 3HF G8SQP
TN2 3HU M3JCT
TN2 3NQ G7WJW
TN2 3NY 2E0XFR
TN2 3RX 2E0PMC
TN2 3RX M0NLP
TN2 3RX M6BFG
TN2 3RX M6FTW
TN2 3RX M6TBH
TN2 4BT G6CRR
TN2 4BX G0VQA
TN2 4ER G0VYK
TN2 4HG G4KGF
TN2 4LF G6AHK
TN2 4ND M7YAU
TN2 4NP G7PMG
TN2 4NS G8JXG
TN2 4RZ M7EFB
TN2 5DD G4JFD
TN2 5HG G0AOA
TN2 5HT M0WTW
TN2 5LN G8NAU
TN2 5LU G3IIW
TN2 5NB G4CTY
TN2 5NN G4LXC
TN2 5RG G6XPF
TN2 5XU M6SVN
TN20 6DJ M7TMH
TN20 6UD G0BUX
TN21 0EF G3WWS
TN21 0FE G3GTF
TN21 0HW G7CSX
TN21 0QA M6XKW
TN21 8BB G1ITE
TN21 8BB M5ITE
TN21 8ED G4YDB
TN21 8EY G4KEI
TN21 8HE G4AWJ
TN21 8HG M6IMR
TN21 8HG M6LHR
TN21 8HP 2E0HDE
TN21 8HP M0MDE
TN21 8HP M6HDE
TN21 8NR G8VHK
TN21 8PF G2BHY
TN21 8PF G3TGF
TN21 8PG M3EMU
TN21 8SA G8KQA
TN21 8SJ G0KOC
TN21 8SJ G4BJC
TN21 8SJ M1SWL
TN21 8TG G1PJT
TN21 8TW G0FDV
TN21 8YS G0MSA
TN21 8YS M0WAJ
TN21 8YW 2E0OPM
TN21 8YW M0WRD
TN21 8YW M6WHS
TN21 9AL G3BBK
TN21 9BL G6HIE
TN21 9BL M3HIE
TN21 9NG M0RYA
TN21 9NR 2E1LAM
TN21 9PP 2E1GIZ
TN21 9PP M0FCD
TN22 1BY M3NYA
TN22 1BZ G4ZOQ
TN22 1HU G7DQZ
TN22 1NN M1XRC
TN22 1NU G0TBS
TN22 1RT G0LFX
TN22 1TU G8BTC
TN22 1UH M7EHC
TN22 2AT G7KID
TN22 2AY M6DYX
TN22 2BX M6FRT
TN22 2BX M6MTK
TN22 2DJ 2E0MKI
TN22 2DJ M0IKM
TN22 2DJ M3IKM
TN22 2DP M3IZD
TN22 3AP 2E0OLM
TN22 3AP G0BUK
TN22 3AP M0OLM
TN22 3BE M3YZH
TN22 3DT G1IFV
TN22 3HX G4GEN
TN22 3LJ M6KAE
TN22 3LP 2E0IDN
TN22 3LP M6IDN
TN22 3RA G1RNY
TN22 3XN M3YIL
TN22 4EU G8ASV
TN22 4HT G4NAJ
TN22 4JA G4UDW
TN22 4JE M7JAS
TN22 4JU 2E1HGR
TN22 4JU G6GSF
TN22 4LB G3YMH
TN22 4NG M3OII
TN22 4QB G4MMH
TN22 5AG 2E0MET
TN22 5AH G1SHH
TN22 5BN M7DQY
TN22 5BU G0MAR
TN22 5JG M3SAY
TN22 5JH G7VAD
TN22 5LA G7HQY
TN22 5RU M6VAW
TN22 5SG M7BQI
tn221ry G0VZN
TN224AR M6TPE
TN225SD M3LUO
TN23 1LN G7NUG
TN23 1LN M6VWE
TN23 3AG G0PDP
TN23 3BE M6IVB
TN23 3DH M7BLI
TN23 3DW 2E0FPO
TN23 3DY G1EJA
TN23 3EG G6EXU
TN23 3EX M6EWW
TN23 3LF M3HFA
TN23 3LU M6IFF
TN23 3NH M3VBL
TN23 3NJ G4AAR
TN23 3NJ G6ZAA
TN23 3NJ M0ZAA
TN23 3NR 2E0FAA
TN23 3NR M0TMX
TN23 3NR M3FAA
TN23 3QR 2E0PVK
TN23 3QR M7BAG
TN23 3QY M3RIU
TN23 4EQ M3YRJ
TN23 5DF 2E0SJA
TN23 5QR M6HJG
TN23 5SA G0TJI
TN23 5UR G6AJW
TN23 5UR M6RRJ
TN23 5UW G4UUI
TN23 5YF M3ZRG
TN23 6JL G4UKO
TN23 6JL M7FYF
TN23 7AF M7BEX
TN23 7HL G4EWV
TN23 7UP 2E0KEZ
TN23 7UP M3KEZ
TN234NR 2E0XXT
TN234NR M6TEZ
TN24 0AX G8VVU
TN24 0BJ 2E0EHS
TN24 0BJ M6GXP
TN24 0DZ 2E1KCC
TN24 0FX 2E0RBP
TN24 0FX M6RPC
TN24 0GS G0PEG
TN24 0HJ 2E0TED
TN24 0HJ 2E1ITE
TN24 0HJ M3WHY
TN24 0JH G0PHO
TN24 0JH M3PHO
TN24 0LF 2E0KZK
TN24 0LF G5HAM
TN24 0LF M0IRU
TN24 0LF M6KZK
TN24 0PL M6PDD
TN24 0TA M3YOO
TN24 0XA G8EWL
TN24 8NF G0HVP
TN24 9AE G3ZPI
TN24 9BD G0KGE
TN24 9EU G7HJJ
TN24 9EU G8YNF
TN24 9HH M7MDD
TN24 9JW 2E1FWM
TN24 9JY G7KLJ
TN24 9LU M0AVY
TN24 9PT M1PAB
TN24 9QL M3ILJ
TN24 9RS M3NKX
TN24 9SD G7LJL
TN240TY 2E0LPW
TN240TY M0LPW
TN240TY M6LPW
TN249BQ M3CAA
TN25 4DF G3ZAJ
TN25 4DW G8MFV
TN25 4NX M3WAF
TN25 4PQ G6SRE
TN25 5AB M6IPH
TN25 5AL G3EDM
TN25 5BD M0WYE
TN25 5JB G4FPG
TN25 5LR M6JNJ
TN25 6EB G1BXT
TN25 6EP M1DRL
TN25 6HG M1SEM
TN25 6NE G0JIR
TN25 6NW G4EDZ
TN25 6RA G8WLB
TN25 6RW G7EVC
TN25 6UA G3YBE
TN25 7EU M7KNO
TN26 1HW G3MMN
TN26 1LS G0MTJ
TN26 1NB 2E0YPU
TN26 1NB M3YPU
TN26 2HL G0CRL
TN26 2LU 2E0BVB
TN26 2LU M6RIO
TN26 2QH G1AEX
TN26 3AR 2E0KKH
TN26 3AR M7BVJ
TN26 3HA G3DT
TN26 3LY G6HXZ
TN26 3QS G0CHG
TN26 3QU G8PFZ
TN26 3QX M7PSX
TN27 0AQ G6LYD
TN27 0BX M0EUS
TN27 0DD G6JPG
TN27 0EE M3SNX
TN27 0HA 2E0SPT
TN27 0HA M0DRQ
TN27 0HB G8FYX
TN27 0JA G6KEN
TN27 0JN G6BME
TN27 0JQ G3KAP
TN27 0LH 2E0YQC
TN27 0LH M3TGO
TN27 0LS 2E1CPB
TN27 0NG M0MSF
TN27 0NG M1DDY
TN27 0QF G8LZV
TN27 0QF M0CVN
TN27 0RB G0VIM
TN27 8BY 2E0VJH
TN27 8EA G7FKZ
TN27 8EA M3YWE
TN27 8ER G1AOQ
TN27 8PZ M0DKN
TN27 9DW G8ISM
TN27 9QQ G4ZXI
TN27 9QR G0ICP
TN27 9UF 2E0CUE
TN27 9UF M0HNI
TN27 9UF M6RCW
TN28 8BJ G8BIS
TN28 8FP G0GCQ
TN28 8JB G4ZTW
TN28 8JL G4TGK
TN28 8JY G7PWS
TN28 8LB G1HJS
TN28 8NF G1UTF
TN28 8NL G0DCG
TN28 8NX G0BGH
TN28 8PR 2E0EWB
TN28 8PR M0OUS
TN28 8PR M6OHT
TN28 8QA M0DXS
TN28 8RD M0DAS
TN28 8RE G7HKT
TN28 8RL G8GZW
TN28 8RS G4YAZ
TN28 8SB 2E1KAJ
TN28 8SF G6JQD
TN28 8SU M6PUY
TN28 8SW G0KRH
TN28 8SY G0DXH
TN28 8SY G8CPM
TN28 8SZ M6BQO
TN28 8UJ M3OMZ
TN28 8UL G0IMD
TN28 8XH M6PED
TN28 8XP G0OIN
TN29 0LA G0LNN
TN29 0PW G1VBO
TN29 0QX G0IEE
TN29 0QX G6ZVB
TN29 0RD G6DZT
TN29 0RD G6OJZ
TN29 0RU G4TZX
TN29 0SF 2E0ZIV
TN29 0SF M3ZIV
TN29 0TY G7UBK
TN29 0UA G1EVI
TN29 9BJ M7EGT
TN29 9EJ G6NLZ
TN29 9JP 2E0SVK
TN29 9JP M0LMI
TN29 9JP M6LMI
TN29 9LE G8XCL
TN29 9LF G1HSX
TN29 9NP M0LYD
TN29 9NS G1KPU
TN29 9RG G1HSI
TN29 9SN 2E0AKU
TN29 9SN M3KQY
TN29 9UU G0CPZ
TN29 9YF G4RLU
TN3 0QB 2E0FXY
TN3 0QB M7WHA
TN3 0RB M1KWH
TN3 0UJ M7CMW
TN3 8DY M6ONO
TN3 9QU G0CHY
TN30 6DQ 2E1GKF
TN30 6DY M6RCB
TN30 6EL G4KPS
TN30 6EN M6OLU
TN30 6EQ 2E0CMN
TN30 6QG G4ERW
TN30 7AZ G3ZWK
TN30 7BA 2E1GJD
TN30 7NB G8GFS
TN30 7NT M3JCQ
tn307ey M7KMV
TN31 6BX 2E0IEE
TN31 6DL M0NUC
TN31 6DL M6FLH
TN31 6DN G6IPH
TN31 6DS M3ULE
TN31 6EJ M6BSB
TN31 6EP G0OZM
TN31 6EX M0EDU
TN31 6HQ G8VOH
TN31 6NB G0FLA
TN31 6QD G3ROC
TN31 6TA M6LTB
TN31 6UL G4TLE
TN31 6UR G0TFV
TN31 7BG G6LPC
TN31 7FH G4PDU
TN31 7HJ M3MZR
TN31 7HT M3KPG
TN31 7NY G6BYF
TN31 7PJ G6XII
TN31 7RA G7HUC
TN31 7RS G8IIK
TN31 7RU G0ISM
TN32 5AU G4BIA
TN32 5DN 2E1CYT
TN32 5JZ M6EWI
TN32 5NP G1VUK
TN32 5QS G2HIX
TN32 5QS G6KGU
TN32 5QS M3URX
TN32 5UG G4JWK
TN33 0DT M6WSC
TN33 0DU G7THK
TN33 0EU G7CKS
TN33 0HD M3KPU
TN33 0QR G7ECU
TN33 0QR G8AJP
TN33 0QR G8CLZ
TN33 0QS G3UYB
TN33 0SF G0LTV
TN33 0TP G3VFO
TN33 9AX G0BBR
TN33 9BP M3ESN
TN33 9ET G4PUP
TN33 9JB M7GST
TN33 9LP M7BDO
TN33 9LP 2E0EYE
TN33 9LP M6JHR
TN330TA 2E0FFJ
TN330TA M6OGS
TN34 1TG M3KQT
TN34 1TS G8FET
TN34 1UU G0OOU
TN34 2AN G7LPG
TN34 2AT G1LZS
TN34 2AU G1BFS
TN34 2BG G4PWE
TN34 2BT M3MRQ
TN34 2DZ G3VUR
TN34 2HA 2E0HXG
TN34 2HA M6HXG
TN34 2HY G6JIF
TN34 2JA G7PIP
TN34 2JE G0GKL
TN34 2JZ M6HXN
TN34 2LF M7BPO
TN34 2NQ G4UWF
TN34 2NT G4KMJ
TN34 2PJ G6TFE
TN34 2PR M7JSL
TN34 2PS G7WWW
TN34 2QB G8SPW
TN34 2SF G3UFI
TN34 2UB G6STB
TN34 2UL 2E0SAY
TN34 3BJ G6HVX
TN34 3JN G0SYQ
TN34 3LB G3VEK
TN34 3LY M3BKI
TN34 3LY M3BKJ
TN34 3NH G7KMA
TN34 3NH M3AXB
TN34 3RT M6XDI
TN34 3RU M6XOJ
TN34 3SL G8RWJ
TN34 3TR 2E0KES
TN34 3TR M6RNO
tn343lr M6VDF
TN35 4AS M6EHE
TN35 4BL G3FHN
TN35 4DL M0SSR
TN35 4DN G3SYZ
TN35 4DP M0AHY
TN35 4DP M0CCV
TN35 4DP M1CPL
TN35 4EP G3BDQ
TN35 4JH G8ZGM
TN35 4LG G8THE
TN35 4LG M3XYZ
TN35 4PB 2E0TXI
TN35 4PB G3TCG
TN35 4PB M6PWD
TN35 4QU G4NER
TN35 4QZ G1IKV
TN35 4SL G1GFZ
TN35 5AD M6NJK
TN35 5BS 2E0PSK
TN35 5DH 2E0EAO
TN35 5DH M6BBM
TN35 5DJ M1AKL
TN35 5EG M7NRM
TN35 5EP G0FUU
TN35 5HH 2E1GWX
TN35 5HZ G4NVQ
TN35 5JN G0CJO
TN35 5LZ M3BRQ
TN36 4HF M0NPD
TN37 6EL G1HSM
TN37 6HF G1FTK
TN37 6HF G7HFS
TN37 6JA 2E1HEO
TN37 6JY G7OJY
TN37 6LA G7ERC
TN37 6PF G4KLF
TN37 6PN G1ORK
TN37 6QR G1ZPA
TN37 6RX 2E0SPH
TN37 6RX M0ZPL
TN37 6RX M6FRJ
TN37 6SA M7ATC
TN37 6SD G1IFW
TN37 6SJ G8FEJ
TN37 7AL G7WGA
TN37 7BS G6CLU
TN37 7DF M6NXT
TN37 7EG M3YXH
TN37 7JY G7VQO
TN37 7LX 2E0EOX
TN37 7LX M6KBO
TN37 7PS G1DVU
TN37 7PS M0BMD
TN37 7PX G3XXM
TN37 7QX G0JHK
TN376ET G1AJN
TN377EN M6RFB
TN38 0BT G0TDJ
TN38 0ER G7MQQ
TN38 0ET 2E1DHC
TN38 0HX G7TCH
TN38 0JN G0NJO
TN38 0NY M7UCD
TN38 0PH M6BQS
TN38 0PU 2E0RAJ
TN38 0PU M3MFE
TN38 0QA G8CMK
TN38 0QA G8SBJ
TN38 0QP G1GFF
TN38 0SE G1OHL
TN38 0SL M6MCP
TN38 0TF M3XVC
TN38 0TR G0AQT
TN38 0UA G4EOA
TN38 0UX G8TQO
TN38 0XG M7CRU
TN38 0XL G4WOE
TN38 0XP M6ZZS
TN38 0YT M7EOB
TN38 9BU G1EHS
TN38 9BU G1EHU
TN38 9DP G4VBK
TN38 9HB G7BBT
TN38 9HP M3BKU
TN38 9JG M7CUA
TN38 9LQ G0ILK
TN38 9QW G1FTX
TN38 9RS G7JVN
TN38 9RW M7CVW
TN38 9RX G4BOZ
TN38 9TH G0TNQ
TN38 9UA G0OFZ
TN380BN M6VTZ
TN39 3AQ M6WIB
TN39 3AX G8GQG
TN39 3EE M6CWU
TN39 3HH M3KRS
TN39 3LU G4MMG
TN39 3NJ G8MYG
TN39 3PB G3WUH
TN39 3PD M0BDE
TN39 3PH M6TJY
TN39 3PS M3ZBH
TN39 3RS M6FKS
TN39 3TF G7OOP
TN39 3TS M6HHC
TN39 3UA G3JOR
TN39 3UP 2E0OCT
TN39 3UP M6IBO
TN39 4AS G7LEL
TN39 4BN G1PDS
TN39 4DN M7ART
TN39 4DY G4HGK
TN39 4GP M7YSU
TN39 4HZ G0THV
TN39 4JH M0OMI
TN39 4JJ G8EYQ
TN39 4JW M0BPU
TN39 4LJ G4BUE
TN39 4LJ M0BUE
TN39 4LN 2E0EVW
TN39 4LN M6NSH
TN39 4LN M6NSQ
TN39 4LT G1ZQO
TN39 4NZ M7XPX
TN39 4PP 2E1GTW
TN39 5BH M7TOW
TN39 5EN M6VMK
TN39 5EN M7NYY
TN39 5EQ G1MTB
TN39 5HN G0JBM
TN39 5HU G7GHP
TN39 5HY 2E0GHX
TN39 5JH G4MPK
TN393QB M7DLO
TN4 0DX G1EUD
TN4 0EQ G4MIK
TN4 0LL M7VWS
TN4 0NR M6GAX
TN4 8ED G4GTN
TN4 8JJ 2E0FJE
TN4 8PS 2E0CMP
TN4 8PS M0NAY
TN4 8PS M6BOX
TN4 8PY M6QFG
TN4 8TH 2E0DGS
TN4 8TH M0VIR
TN4 8TH M6GBZ
TN4 9AL M7MYA
TN4 9BL 2E0CQL
TN4 9BL M0KAO
TN4 9BL M6BQW
TN4 9BN 2E0SSE
TN4 9BN M7SSE
TN4 9DH 2E0TWP
TN4 9DH M6PJM
TN4 9DJ G0BMN
TN4 9DR G3KIP
TN4 9JG G7DAB
TN4 9JU G6TXH
TN4 9JY G0ABM
TN4 9NS 2E0IIJ
TN4 9NS M7DXS
TN4 9RF M0RYK
TN4 9RF M0XKX
TN4 9RP G0VPY
TN4 9RT M6HXK
TN40 1EG G6SGV
TN40 1QE G0TMA
TN40 1QH G8KPE
TN40 1TE G6HIB
TN40 1TS G1WLO
TN40 1UE G6NGM
TN40 1UG G8TUU
TN40 2AZ G3GGH
TN40 2AZ G4SIF
TN40 2BQ G6ULD
TN40 2EA M3TNO
TN40 2HN M3LGI
TN40 2HU G6MWB
TN40 2NT G4YZL
TN40 2NT G4ZSR
TN40 2PA 2E0LXI
TN40 2PA M0LXI
TN40 2PA M6LXI
TN40 2PG G1VNZ
TN40 2PW M1FFX
TN40 2SD G3MVX
TN40 2SH G0ILN
TN40 2SH G7TAF
TN40 2TA M6PFB
TN40 2TB M1EQB
TN40 2UL M6WDQ
tn401pn G7DME
TN5 6BT M6LVK
TN5 6DG 2E0MMU
TN5 6DG M0MMU
TN5 6DG M6HXE
TN5 6DL G7MZL
TN5 6RJ G8GYB
TN5 6RJ M0NZA
TN5 6UT G1NZH
TN5 7EG G4XST
TN5 7JA G7IRN
TN5 7PS G7MFE
TN6 1BS G4LZK
TN6 1EN M1CNE
TN6 1JF G3TXZ
TN6 1QQ G7JAN
TN6 1SE M6JUH
TN6 1SU M0TNF
TN6 1TF G4EZE
TN6 1TP M6PCC
TN6 2AD G0GPE
TN6 2AD G4XCE
TN6 2AG G8LKS
TN6 2BG G3WKS
TN6 2BG G4OTV
TN6 2BG G4UPI
TN6 2BG M0SSX
TN6 2BN G6GWE
TN6 2DU M6MSF
TN6 2EN M1ARX
TN6 2ES 2E0MDU
TN6 2ES M6NAU
TN6 2ET G0WTC
TN6 2ET G6UIF
TN6 2HA 2E0TLB
TN6 2LY M6RBA
TN6 2NJ G8UBD
TN6 2NY G3CYU
TN6 2RY G3RST
TN6 2SB G8LSD
TN6 2UU 2E0GCP
TN6 3BH G0XBV
TN6 3BQ G4LDJ
TN6 3BY M3TLB
TN6 3EB M3UAR
TN6 3LG M6YLS
TN6 3QA G6BRV
TN6 3QH M6YTM
TN6 3QT M0CPC
TN6 3TA 2E0HKK
TN7 4DH G0NAR
TN7 4DN G0AAW
TN7 4EN M1PVC
TN7 4ET M6TWV
TN8 5BX M6JRB
TN8 5DB M3WYZ
TN8 5DF G0TKZ
TN8 5DS M6MVV
TN8 5EL G3JMJ
TN8 5EN G6MXV
TN8 5PS M6CSC
TN8 5PW G6DTT
TN8 6AJ M3ZHM
TN8 6AJ M3ZHQ
TN8 6AJ M6GPE
TN8 6AJ M6GRJ
TN8 6AU M6SNC
TN8 6HZ M6BHU
TN8 6JS M6NZV
TN8 6LN M0GZC
TN8 6LN M3ZWM
TN8 6SX M6SRC
TN9 1BN M3RPE
TN9 1JH M7GCX
TN9 1UX G1CPA
TN9 1UX M0AFZ
TN9 2EH M6VEP
TN9 2EJ G7GDV
TN9 2EL G8LGW
TN9 2EN M7KJW
TN9 2FE M7LUF
TN9 2NQ G3YPY
TN9 2PG M3LHE
TN9 2PG M3LHG
TN9 2UY G0JUY
TN9 2UY M3MOP
TN9 2YR M6SHC
TN9 2YS G4EPC
TQ1 1JY 2E0PCZ
TQ1 1JY M0PCZ
TQ1 1JY M6PCZ
TQ1 1PB G4XLO
TQ1 1PX M6CON
TQ1 1QR G0KEC
TQ1 1QR G1XDJ
TQ1 1TZ M6AOJ
TQ1 2ED G0CRJ
TQ1 2EU G1NIC
TQ1 2HH G3VOF
TQ1 2HH G4GZG
TQ1 2LQ M0IME
TQ1 2NL G6IYM
TQ1 3JN G1KPI
TQ1 3LD G6TEQ
TQ1 3NW G3SDW
TQ1 3NZ G4FKU
TQ1 3PD G4YTY
TQ1 3PW M6IDP
TQ1 3PW M6RWJ
TQ1 3SU G4IML
TQ1 3TP G3KPV
TQ1 3TQ M3SQX
TQ1 3TX 2E0PLR
TQ1 3TX M0LAI
TQ1 3TX M6LBN
TQ1 4BG 2E0LWT
TQ1 4BG M3RMU
TQ1 4EA M3LKE
TQ1 4HZ G0EDE
TQ1 4LD G0SUU
TQ1 4NJ G4OVO
TQ1 4PU M6PHW
TQ1 4PX G1EFT
TQ1 4QY G8XST
TQ1 4RD G8GPF
TQ1 4UN G1TNP
TQ10 9BT M1DKW
TQ10 9EJ G4TCP
TQ10 9EU G4XTR
TQ10 9EU G7BEP
TQ10 9HH 2E0GQL
TQ10 9JF M6FQZ
TQ10 9LL G0PCT
TQ11 0AF G8YZA
TQ11 0BP G7RYL
TQ11 0DB 2E0GQU
TQ11 0DB 2E0HHG
TQ11 0DB M7ALR
TQ11 0DB M7DSR
TQ11 0DL M0LED
TQ11 0EA G1GMG
TQ11 0ER G4XWP
TQ11 0FG M1FCF
TQ11 0HB G6NWC
TQ11 1AA M3IDC
Tq12 1ag M6XJD
TQ12 1DX M3JKA
TQ12 1DZ G0EKH
TQ12 1ER G0OFA
TQ12 1FZ 2E0APZ
TQ12 1FZ M0XWP
TQ12 1FZ M6YPZ
TQ12 1GQ G0MUN
TQ12 1GZ M7BAY
TQ12 1GZ M7DRY
TQ12 1HS G3TAA
TQ12 1HT M6DXG
TQ12 1JY M6IZN
TQ12 1LH M1DVV
TQ12 1PS 2E0CVN
TQ12 1RQ G3YAR
TQ12 1UL M3IBY
TQ12 1UP G0NUZ
TQ12 1US G4DCH
TQ12 1YJ G1WQN
TQ12 2ND G0RDO
TQ12 2NN M3PZF
TQ12 2PU G0WAE
TQ12 2PW M7PSC
TQ12 2PX G7AQV
TQ12 2TL M3GFH
TQ12 2TP M3GFE
TQ12 3BA M6KUC
TQ12 3BP M1AIY
TQ12 3DF 2E0DEV
TQ12 3DF M0ZAF
TQ12 3DF M3XJQ
TQ12 3JE G0WWD
TQ12 3LE G1WUU
TQ12 3LY M3DHA
TQ12 3NJ G0BCH
TQ12 3NJ M6ERU
TQ12 3QX 2E0FAV
TQ12 3QX M6ZAF
TQ12 3TE G4DTW
TQ12 3TJ M1AKV
TQ12 3YS G6YJO
TQ12 3YT M0YLY
TQ12 3YU M3GHD
TQ12 4EH G7AFV
TQ12 4FN G4GZM
TQ12 4HA G1DBH
TQ12 4HE G4VUD
TQ12 4JG G4LAK
TQ12 4JZ G4MNA
TQ12 4LF G3LHJ
TQ12 4LF G3NJA
TQ12 4LF G8NJA
TQ12 4LF M0LTW
TQ12 4NE G0IGK
TQ12 4NH G0NXI
TQ12 4NJ G4ELZ
TQ12 4NS G3XXE
TQ12 4PG M3IEU
TQ12 4QS G0CWQ
TQ12 4QS G4XJL
TQ12 4RE G0SDL
TQ12 4SA M7MYK
TQ12 4SB G3TLK
TQ12 5AB G3ZEJ
TQ12 5BQ M7LEZ
TQ12 5BZ G6FIN
TQ12 5DZ M3UMJ
TQ12 5EW G0BNJ
TQ12 5EX M1DVJ
TQ12 5EZ 2E0NKN
TQ12 5EZ 2E0PVU
TQ12 5EZ G1COE
TQ12 5EZ M0PVU
TQ12 5EZ M7RBC
TQ12 5JG G4NRH
TQ12 5NX G0ONT
TQ12 5PZ 2E0GPL
TQ12 5PZ M3UAW
TQ12 5PZ M3ZDT
TQ12 5QJ G8VSV
TQ12 5QJ G8YEN
TQ12 5QS G0SBM
TQ12 5QS G4FCU
TQ12 5QS G6CLD
TQ12 5QS G7MNL
TQ12 5QS G8GCS
TQ12 5QS M1EPU
TQ12 5UR M1CFS
TQ12 6ER 2E0JNY
TQ12 6ER M7DOG

TQ12 6EZ M0LNY
TQ12 6FA M6JAQ
TQ12 6GX G1ZBJ
TQ12 6HJ 2E0CKY
TQ12 6HJ M6AIV
TQ12 6JA M0UAC
TQ12 6JA M6GBU
TQ12 6JA M7USV
TQ12 6NZ M6NQA
TQ12 6SB 2E0WSZ
TQ12 6SB M0WSZ
TQ12 6SB M6WNB
TQ12 6SP G7MEA
TQ12 6TX M6KAR
TQ12 6UP M3YYG
TQ123LB G0FHH
TQ124HX M7OZA
TQ125BL M7BSF
TQ13 0AZ 2E0IDE
TQ13 0AZ M7BSM
TQ13 0BB G1BJN
TQ13 0EE G4VTO
TQ13 0EE M7BNN
TQ13 0EW G4RYH
TQ13 0EW M6CCX
TQ13 0GB G4TND
TQ13 0JH G4NRR
TQ13 0JH G6DHD
TQ13 0JN G7OZI
TQ13 0LH M5WNS
TQ13 0NW M0NAA
TQ13 0NW M6NMN
TQ13 0PJ G6BAL
TQ13 0PL M3RXW
tq13 0re M7BAE
TQ13 0RF G1SJD
TQ13 0RH 2E0HWJ
TQ13 0RH M7BJK
TQ13 7AN M3DBA
TQ13 7BS G7CUU
TQ13 7DU G7RKX
TQ13 7HT G1BGK
TQ13 7HY G8KEO
TQ13 7RU G3FYF
TQ13 8AR G3UUU
TQ13 8BB 2E0GKX
TQ13 8BB M0VGT
TQ13 8BB M7KBY
TQ13 8LP 2E0MLH
TQ13 8LP 2E0RIW
TQ13 8QA 2E0FWD
TQ13 8QA M3EYZ
TQ13 8RS M7NDM
TQ13 8SD M6RIH
TQ13 9DE 2E0EZL
TQ13 9EP G6ICH
TQ13 9EP M0LHK
TQ13 9EX 2E1PEW
TQ13 9EX G8PEW
TQ13 9EX M7PEW
TQ13 9FE G0BMP
TQ13 9HN M7HEL
TQ13 9HP M0CLE
TQ13 9HR G3RPD
TQ13 9JY M0ECQ
TQ13 9YL M3UOY
TQ14 0DG G3WLT
TQ14 0EJ G6EKM
TQ14 8AQ 2E0DOX
TQ14 8AQ M6ADX
TQ14 8EF G8KTC
TQ14 8JA G0EZU
TQ14 8LE G1MKE
TQ14 8LW G7CGI
TQ14 8NH G4CDJ
TQ14 8PN G4YTT
TQ14 8QG G6XD
TQ14 8QX G4BG
TQ14 8RJ G1GHU
TQ14 8RS G3LQX
TQ14 8SB M0ITC
TQ14 8SB M6LYZ
TQ14 8TS M0VZV
TQ14 8TS M6VZV
TQ14 8UE G0CEL
TQ14 8UE G0EMU
TQ14 8UF G3YHH
TQ14 9DT G3KFP
TQ14 9HG G0JZA
TQ14 9HG G3MOA
TQ14 9JP G0KDT
TQ14 9NF G0PWU
TQ14 9NJ G1EUA
TQ14 9NW M3FZS
TQ14 9QA M3HYI
TQ14 9QA M3IZW
TQ14 9QR 2E0TAD
TQ14 9QR M0HHG
TQ14 9QR M3TYO
TQ14 9RF G0PNA
TQ14 9RP G8WBL
TQ2 6BS G1ZZG
TQ2 6DX G4OTU
TQ2 6EA G0TMH
TQ2 6EA M3WFE
TQ2 6EA M3YVE
TQ2 6EA M6SOK
TQ2 6LL G0OXT
TQ2 6RG M3IPZ
TQ2 6UB G3VLO
TQ2 6UQ G1VTK
TQ2 6UT G3YDE
TQ2 7BN G7BFE
TQ2 7GF M6IXR
TQ2 7JX 2E0PBT
TQ2 7JX M0PBT
TQ2 7JX M6PBT
TQ2 7LA M7CJY
TQ2 7NR G0HSW
TQ2 7QX G0JDO
TQ2 7TB M3UUG
TQ2 8AJ G0HGV
TQ2 8DH G4ZQM
TQ2 8EF M0BKD
TQ2 8HB G0MRZ
TQ2 8HU G6FSP
TQ2 8HU G7SME
TQ2 8LB M0PAW
TQ2 8LE G0DDF
TQ2 8LF G8GRB
TQ2 8LP G1SDX
TQ2 8LR G6FTA
TQ2 8LR M3FTA
TQ2 8PA M6DFY
TQ2 8QA G4PCK
TQ2 8QE G8UKI
TQ2 8QH G4UIH
TQ2 8QH G4UII
TQ2 8QQ G7HIK
TQ3 1AB G7DKB
TQ3 1AE G0AZX
TQ3 1AE G0LQM
TQ3 1AG G8XTO
TQ3 1DL M7DQU
TQ3 1EE M5ABN
TQ3 1HN M6PNJ
TQ3 1HY 2E0KFJ
TQ3 1JX G0GWC
TQ3 1LE G7DMX
TQ3 1NH G6XXB
TQ3 1PH M0IAT
TQ3 1SQ G6UPR
TQ3 1TA G4AQR
TQ3 1TD G7IIC
TQ3 2AQ M7AWK
TQ3 2BN G4CLN
TQ3 2DP M0DPQ
TQ3 2EN G8BKQ
TQ3 2HZ G0UWF
TQ3 2JP G6ZZS
TQ3 2LR G8ZEV
TQ3 2NE M0CFZ
TQ3 2PB G6JAR
TQ3 2PE G4OFP
TQ3 2PT G4LPY
TQ3 2QF M0BKL
TQ3 2QS G7MCK
TQ3 2RE G0ORK
TQ3 2SW G7AHP
TQ3 2SW M0RIV
TQ3 2SY M6AJJ
TQ3 3BT G3GCW
TQ3 3BU M6DPG
TQ3 3EQ M1DXB
TQ3 3FT M6GTO
TQ3 3GD 2E1KFS
TQ3 3GW G0LCD
TQ3 3JJ G0TQC
TQ3 3JJ G3YCH
TQ3 3JQ M7NZL
TQ3 3NA G0CGM
TQ3 3NU G0CDB
TQ3 3NU G4SSD
TQ3 3NX M3AXX
TQ3 3QN G1XKQ
TQ3 3RX 2E0EVM
TQ3 3TD G1FRJ
TQ3 3TD M3GDV
TQ3 3TH M7WLL
TQ3 3TU G1TUU
TQ3 3TW M7BJV
TQ3 3UD G1JKL
TQ3 3XA G4WVM
TQ4 5AT G0BAJ
TQ4 5AX 2E0DBQ
TQ4 5AX M0ZYD
TQ4 5AX M6AYK
TQ4 5EY M6HNH
TQ4 5HU G4UUT
TQ4 5HZ G4NNP
TQ4 5JQ G3CLW
TQ4 5JQ G4KFB
TQ4 5LH 2E0PDL
TQ4 5LH M0LOB
TQ4 5LH M3UVT
TQ4 5LH M3VBG
TQ4 5LH M6DFI
TQ4 5NA G7CHC
TQ4 5NY G0SPF
TQ4 5UA M7CVQ
TQ4 6EG G6AOB
TQ4 6ES 2E0OJE
TQ4 6EY M6NHU
TQ4 6HA G0UAV
TQ4 6HD M0JNP
TQ4 6HT G4AWO
TQ4 6JF 2E0RJE
TQ4 6JF M3URV
TQ4 6JN G7JFM
TQ4 6LE G3PXL
TQ4 7HG G0UQI
TQ4 7HZ G8XLG
TQ4 7jy M3OMX
TQ4 7LR G7FNT
TQ4 7LW G0HZB
TQ4 7ND G0KTU
TQ4 7NZ M3NYX
TQ4 7RL G4DZH
TQ4 7RL M0BKG
TQ4 7RN G7OSH
TQ4 7RN M0BHJ
TQ4 7RN M1ANO
TQ4 7RN M1MUM
TQ4 7RN M3UNR
TQ4 7RU 2E0TAG
TQ4 7RU M3TQU
TQ4 7RW G0BTH
TQ4 7SD G4VQI
TQ4 7SD G6UIM
TQ5 0AN M6DFJ
TQ5 0AY G0OJS
TQ5 0BQ G2DXU
TQ5 0DD G7DLE
TQ5 0JZ M0DQN
TQ5 0LZ G8JMS
TQ5 0PB G0RGP
TQ5 0PB G7LMR
TQ5 0PB G8RTU
TQ5 0RJ G3RVA
TQ5 8DG G8OVO
TQ5 8JT G0GJC
TQ5 8PB M1EJE
TQ5 8QG G6ZOJ
TQ5 8QP M3TQG
TQ5 8QR G3XNX
TQ5 8RF G4UXP
TQ5 8RN G4NGR
TQ5 8RS G0PTK
TQ5 8RT 2E0IIY
TQ5 8RT M7CWK
TQ5 9AG M0PLH
TQ5 9AX G8SPM
TQ5 9ED G0TDQ
TQ5 9EE G0KEK
TQ5 9EL G0VQY
TQ5 9EL M3GCD
TQ5 9LJ G1EZI
TQ5 9LS M1EIW
TQ5 9PF G4OJD
TQ5 9PJ G0TSB
TQ5 9PJ G1AEU
TQ5 9QZ G3ZUO
TQ5 9RH 2E0HRT
TQ5 9RH M0TRU
TQ5 9RL G4JXK
TQ5 9SQ G3REW
TQ5 9SW 2E0XOL
TQ5 9SW M0XOL
TQ5 9SW M6XON
TQ5 9UD G4OKC
TQ5 9UN G0NPY
TQ5 9UR G7VSL
TQ50PE G8LLZ
TQ58QR G4CST
TQ6 0DP G4XNO
TQ6 0DQ G3PCT
TQ6 0RG M7JJO
TQ6 9BN M3TQA
TQ6 9HJ G3WCB
TQ6 9HJ G4SAT
TQ6 9JN G4LZD
TQ6 9JT M6BPB
TQ6 9LQ M3TQN
TQ6 9NZ G7XPC
TQ7 1JU M3RPH
TQ7 1LL G8YBT
TQ7 1LP G0WHZ
TQ7 1LR G8KWJ
TQ7 1QB G6STF
TQ7 1RW M6AFB
TQ7 1SB M7SLP
TQ7 2DB G0JIA
TQ7 2DJ M3PRZ
TQ7 2DP G6SBG
TQ7 2NB 2E0POB
TQ7 2NB G0CWH
TQ7 2NB M6BWQ
TQ7 2SE G8ETP
TQ7 3BH 2E0PMX
TQ7 3BH M3PMX
TQ7 3JZ G7IMV
TQ7 3LZ G7OSB
TQ7 4HG M6OKO
TQ7 4HS 2E0OWC
TQ7 4HS M0VTD
TQ7 4HS M6OWC
TQ7 4NE G3XLW
TQ7 4PF G3PFX
TQ7 4QJ M1DWQ
TQ7 4QP G6TOI
TQ7 4SD G0RIY
TQ8 8AB M0GJL
TQ8 8AN M3WNZ
TQ8 8JP G3UBL
TQ8 8NU G6STJ
TQ8 8PJ G8TQH
TQ8 8PW G3HST
TQ9 5AN G8KRB
TQ9 5EF G3PNO
TQ9 5EU M6XXS
TQ9 5FH G6JPM
TQ9 5LH G3OPX
TQ9 5QY G0UBG
TQ9 5RF M0AVN
TQ9 5RF M3NSB
TQ9 5SX 2E0GGZ
TQ9 5UR M3VTO
TQ9 6AN M3DBG
TQ9 6DA G7JLS
TQ9 6DB G8LXS
TQ9 6DJ G3SNA
TQ9 6JZ M7YGK
TQ9 6LS M6MFO
TQ9 6RL 2E0PVQ
TQ9 6RL M3PVQ
TQ9 6SR G7KCE
TQ9 7BA G4VHL
TQ9 7BD M6TUK
TQ9 7EY G0DIH
TQ9 7HH G3WGN
TQ9 7HN M7HLR
TQ9 7NQ G7DHW
TQ9 7PU G4FBN
TQ9 7QR M7JHT
TQ9 7TL G1WZG
TR1 1AT G1ZWH
TR1 1BW G0NDC
TR1 1DS G0RIZ
TR1 1EF G1SZD
TR1 1JD G1UWD
TR1 1QR 2E0LTH
TR1 1QR M3SNQ
TR1 1RA G3MFW
TR1 1RU G0NKQ
TR1 1SS G0WXL
TR1 1WL 2E1ADQ
TR1 1WL G0NFI
TR1 1WL M3CKH
TR1 1XR G3PGJ
TR1 1YS G0SHY
TR1 2BW G4PGD
TR1 2BY G1ZRP
TR1 2DN M0WYM
TR1 2JZ M0BIZ
TR1 3DE M6NMM
TR1 3DG 2E1CIT
TR1 3JA 2E0DGM
TR1 3JA M6HBT
TR1 3JR G8KJP
TR1 3LU M1EWT
TR1 3LX G0UWI
TR1 3LX G1DTS
TR1 3NG 2E0OEM
TR1 3NG M6WDW
TR1 3PE G0FHX
TR1 3PE G0FHY
TR1 3PT G7FLI
TR1 3QD G0CUH
TR1 3TX G3WKP
TR1 3YJ 2E0EHT
TR1 3YJ M0IUM
TR1 3YJ M6MPL
TR10 8AB 2E0XZI
TR10 8BH G3UFX
TR10 8BH M7ORC
TR10 8DY M6ITZ
TR10 8GF M0SEA
TR10 8HQ M3URE
TR10 8JD G0DUE
TR10 8NT G1JJZ
TR10 8NX G8GCM
TR10 8SJ G8UOL
TR10 9DT G0EOH
TR10 9DZ G1EGI
TR10 9HB G0BSK
TR10 9JW G4PCW
TR10 9WS M1EQA
TR11 2DX G4JEE
TR11 2ER G4MYY
TR11 2ER G6DGW
TR11 2ER M0CNU
TR11 2HE G6CZX
TR11 2RN G0ETQ
TR11 2RW G4FNP
TR11 2SW G4RUA
TR11 2TD G1DDK
TR11 2TE M0PBX
TR11 3DZ 2E0MRP
TR11 3DZ M6MAP
TR11 3JW M6WRN
TR11 3QA M7NGL
TR11 3YL M6AWJ
TR11 4AQ G3LAI
TR11 4BE G0HJW
TR11 4EQ G8LCE
TR11 4ES M0SUI
TR11 4HR G6YHK
TR11 4RG G7RKE
TR11 4SG G8AYJ
TR11 4SR G3VQS
TR11 4SS M7BWO
TR11 4SW G4RRQ
TR11 5AP M0DYH
TR11 5BW G4FXW
TR11 5DT G8CTR
TR11 5EJ G6IMN
TR11 5GZ M6RZW
TR11 5GZ M7NVY
TR11 5HE G0OCS
TR11 5HE G0OCT
TR11 5HF G4VMW
TR11 5LR G4ELI
TR11 5NA 2E1ESM
TR11 5NA G7WDS
TR11 5PR M1AVW
TR11 5QG M0RFM
TR11 5QN G4NBC
TR11 5SF 2E0BYH
TR11 5SF M3XUE
TR11 5SL G7LKZ
TR11 5TE M1FIG
TR11 5TJ G3LZN
TR11 5TT M3XOA
TR11 5TT M6EME
TR11 5UG G3NWW
TR112TE 2E0PFA
TR112TE M3SOV
TR12 6DP M6MQJ
TR12 6HT 2E0XGB
TR12 6HT M0MZC
TR12 6HT M6WMG
TR12 6LY G4WIA
TR12 6PB G4WKW
TR12 6QD G7FOT
TR12 6QL G3UYN
TR12 6QR G3TBJ
TR12 6TX M3HJU
TR12 6UB G0XPD
TR12 6UW G6ZKC
TR12 7AH M0ADA
TR12 7AS G7RAU
TR12 7BJ M1ERA
TR12 7BJ M1ERD
TR12 7BW G3PLE
TR12 7BZ M6NIK
TR12 7DN G3MPD
TR12 7DN G4CDY
TR12 7DN G4YQP
TR12 7DS G8GLL
TR12 7DS M0BNZ
TR12 7DY 2E0TDP
TR12 7DY M0TDP
TR12 7DY M3IXK
TR12 7DY M3MNQ
TR12 7DZ M7WKS
TR12 7EJ G0UPP
TR12 7EQ M6WOH
TR12 7HZ 2E0ZSY
TR12 7HZ M3SYV
TR12 7HZ M6RLQ
TR12 7JD M0CNZ
TR12 7JH G1ZPC
TR12 7JJ M6NDK
TR12 7JL G0GUO
TR12 7JX G6WPR
TR12 7LF G0XAO
TR12 7LU G3NJV
TR12 7NN G3KJK
TR12 7NU 2E0LLO
TR12 7NU G3WGZ
TR12 7NU M0SOU
TR12 7NU M6LZP
TR12 7PA G1JHN
TR12 7PE M3VJC
TR12 7PQ G6IWK
TR12 7QR 2E0RER
TR12 7QR M0RRX
TR12 7QR M6RER
TR12 7QW G0HWU
TR12 7RL 2E0NZT
TR12 7RL M6NZT
TR13 0BY G4JGX
TR13 0DT G6FQP
TR13 0FB G8END
TR13 0HE G8ZVM
TR13 0LD G0FKX
TR13 0PE 2E0VJB
TR13 0PE M3VJJ
TR13 0PQ M0BSI
TR13 0PR G4LOH
TR13 0PT G4UIT
TR13 0PU M6WGF
TR13 0RY M6WFV
TR13 0SN G4GFY
TR13 8BG 2E0ABU
TR13 8BP G0HFA
TR13 8ER M0JMF
TR13 8JH M7BGO
TR13 8JZ G3KGP
TR13 8LU 2E0LDM
TR13 8LU M3YAX
TR13 8NT M7DYR
TR13 8PQ G0FGX
TR13 8PQ G1OZV
TR13 8PY G0UDB
TR13 8QH G8GGP
TR13 8QH M0AFJ
TR13 8QJ G1ZPJ
TR13 8RU 2E0WVW
TR13 8RU M6WIW
TR13 8RX M0BUA
TR13 8SR G7VFA
TR13 8WF 2E0CTD
TR13 8WH G8IKK
TR13 9AW G4WZH
TR13 9BL G4ZYO
TR13 9LR 2E0KTW
TR13 9LR M0EMW
TR13 9PB G7MSN
TR13AY 2E0CUH
TR14 0AR M6LDW
TR14 0AS M3WRM
TR14 0EQ G0ARL
TR14 0ET M6REJ
TR14 0EY G8SEE
TR14 0FT 2E0CWS
TR14 0FT M0KDR
TR14 0HD 2E1HSR
TR14 0JF G4KES
TR14 0JX G4POT
TR14 0LL 2E0PCR
TR14 0LL M6PCB
TR14 0LL M6PUQ
TR14 0NQ M0DUY
TR14 0QE G8WWW
TR14 0QT M7YYZ
TR14 7AW G4BHD
TR14 7BP M3UML
TR14 7EB 2E0UCR
TR14 7EB M6UCR
TR14 7EW M0RMM
TR14 7EW M6RIU
TR14 7HB G6IZK
TR14 7HN G0WYU
TR14 7JN G0PGJ
TR14 7PE 2E0GLJ
TR14 7PE M6NZK
TR14 7PH M7AJT
TR14 7RN M0JIU
TR14 7RR M6ZXZ
TR14 7SU G8YWJ
TR14 7TT G0ILI
TR14 7UF G6ZAM
TR14 7XN G4BHC
TR14 8AW 2E0BOI
TR14 8AW G4ZTZ
TR14 8AW M0GIY
TR14 8AW M3UWW
TR14 8JB G7GIG
TR14 8PB M6MYM
TR14 8QF G1HFE
TR14 8QF G1LNA
TR14 8RQ G4XMA
TR14 8RQ M0RSJ
TR14 8ST G8ZDS
TR14 8TW M6ASJ
TR14 8UP G4ZUI
TR14 8UP G6EON
TR14 9AA G4YYM
TR14 9AT G1BAX
TR14 9DE G4DEO
TR14 9ER G4YYH
TR14 9JJ 2E0LXB
TR14 9JJ M0LXY
TR14 9JJ M7LXB
TR14 9JT G4ZKN
TR15 1AX 2E0BON
TR15 1AX 2E0NKT
TR15 1AX M3XNA
TR15 1AX M6TRC
TR15 1EP M6LGD
TR15 1EW G4LOG
TR15 1JB 2E0XES
TR15 1JB M7EEE
TR15 1NX G8IXN
TR15 1PA 2E0RBR
TR15 1PA M3WOY
TR15 1QL 2E0ISU
TR15 1QL M6ISU
TR15 1QR G6XKO
TR15 1SE G0FKI
TR15 2LL 2E0SCQ
TR15 2LL M6SCV
TR15 2NY G0FHT
TR15 2RJ G6HGR
TR15 2TP G7VIR
TR15 3AW G3PUQ
TR15 3AW G4DBZ
TR15 3BP G0IDH
TR15 3BZ G4GDU
TR15 3DF G6KTB
TR15 3HJ G3JQK
TR15 3JQ M3YVY
TR15 3NJ M3IJO
TR15 3NW M7HIL
TR15 3TS G0MTY
TR15 3TT 2E1FNR
TR15 3UD 2E0EKJ
TR15 3UD M0WDO
TR15 3UD M6WDO
TR15 3YG 2E0TCV
TR15 3YJ G7DUC
TR16 4AQ M6KIR
TR16 4AQ M6NRF
TR16 4DY G6HGK
TR16 4EB G0FHW
TR16 4SG G7ADP
TR16 4SH G4CRC
TR16 4SH G7VOH
TR16 4SH M3YKT
TR16 5AA G4WSH
TR16 5AH M6JCJ
TR16 5DT G4WQL
TR16 5JL 2E0BOJ
TR16 5JL M3TNB
TR16 5JY M0IAF
TR16 5LT G3MZN
TR16 5LT G7RIO
TR16 5PN G0FKF
TR16 5PT G0CAM
TR16 5QA G4ZCT
TR16 5RJ 2E0JGW
TR16 5RJ M6WBB
TR16 5SH 2E0AHB
TR16 5SH M7BOE
TR16 5TQ G0MYR
TR16 6DQ G1WLN
TR16 6HN G7WJJ
TR16 6HN M0DWS
TR16 6HP G4ZKH
TR16 6LR G8WNQ
TR16 6LS G8NXD
TR16 6NH M3RRN
TR16 6NT G0VZB
TR16 6PG 2E0JBW
TR16 6QL M6EVY
TR164AQ G1NRF
Tr164lu 2E0FQU
Tr164lu M7ADN
TR165QZ G0FIC
tr166nt M7CKX
TR166PN M3URW
TR17 0BL M6ILJ
TR17 0BP M6KRL
TR17 0HF M0DBA
TR18 2AH 2E0CJS
TR18 2AH M0PZR
TR18 2AH M6ANI
TR18 2DN G7SBP
TR18 2EX M6GBO
TR18 2GD G6JZN
TR18 2PL 2E1CCN
TR18 2QE G7CCS
TR18 2QE M0GOP
TR18 3BA G1JAH
TR18 3BB G3NRD
TR18 3LD G1IVO
TR18 3NA G3RID
TR18 3PD G8GRO
TR18 3PW G0WZC
TR18 3QL G8EJC
TR18 4BH G0PGB
TR18 4LZ G4DZJ
TR18 4RS G7TJW
TR18 5DQ G4BPJ
TR18 5LW M7OOI
TR18 5NG M0SBL
TR18 5NR 2E0MEI
TR18 5NR M6DZB
TR18 5QA M0ABZ
TR18 5QR M0SBD
TR18 5QW G0MWZ
tr184ry 2E0KWL
TR184RY M6MVH
TR19 6BB 2E0CAQ
TR19 6BB M0LDQ
TR19 6BB M6SUV
TR19 6LX G3XRJ
TR19 7AQ G4AMT
TR19 7BG M6YHQ
TR19 7BS G0THQ
TR19 7DZ G3OYB
TR19 7ET G4YAF
TR19 7HB G7SBN
TR19 7RN G4LYG
TR19 7SP M1BKS
TR19 7SY G4XCV
TR19 7TJ 2E0FVC
TR19 7TJ M6FVC
TR19 7UT G3PZN
TR2 4AU M6YEO
TR2 4BE G6OJX
TR2 4BS G3ZZV
TR2 4BY M6JDC
TR2 4DS G4JZA
TR2 4JX M6HFK
TR2 4NY G7KMP
TR2 4PQ G3VWK
TR2 4TD 2E0BBI
TR2 5AR G0REA
TR2 5DD G4NBF
TR2 5DD M0TOP
TR2 5JP G0JWV
TR2 5JX G0ENZ
TR2 5NP G4EIK
TR20 8AJ G8ARM
TR20 8AJ M0BMX
TR20 8BQ G4SEN
TR20 8DJ G3YSN
TR20 8DS G1NWO
TR20 8DS M3ABQ
TR20 8JY G7TBJ
TR20 8UL G7TJD
TR20 8XA G1GKF
TR20 8XT 2E1GNN
TR20 8XT G8HZN
TR20 8XT M3PHS
TR20 9AR G0AIX
TR20 9AR G7SUT
TR20 9BW G3ZPJ
TR20 9BX M3ZDK
TR20 9HN G4SOK
TR20 9JR 2E0FOE
TR20 9JR M0TIK
TR20 9JR M7CHR
TR20 9LS 2E0ROO
TR20 9LS M6YKC
TR20 9LY M0PNZ
TR20 9LY M3PNZ
TR20 9PT G4ZLF
TR20 9SX G3ZPW
TR20 9TE M0TNT
TR21 0JJ M1IOS
TR21 0JJ 2E0PWL
TR21 0JJ M0PWL
TR21 0JJ M6PWL
TR24TD M6NGA
TR26 1AX G8JAB
TR26 1BL G3KDP
TR26 1ER G3YGM
TR26 2DU G1UAY
TR26 2SP M3KRZ
TR26 3AL G4YQS
TR26 3LE G0ISY
TR26 3LY 2E0PDH
TR26 3LY G5XDX
TR26 3LY G8ROC
TR26 3LY M6AOK
TR27 4AE M0XEK
TR27 4AG G4TXD
TR27 4EX G0PZR
TR27 4EX M0WIN
TR27 4JN G4RVP
TR27 4JQ G1ZOB
TR27 4LY G1CHM
TR27 4NA 2E0WVS
TR27 4NJ G4KYO
TR27 4PJ G1LCY
TR27 4PS G3UCQ
TR27 4QT G4HUG
TR27 4RL G8UXD
TR27 5AB G4GKY
TR27 5AN G0TRJ
TR27 5DT M6EEH
TR27 5ET G6LGM
TR27 5HA G3YNK
TR27 5HL 2E0JYA
TR27 5JD G4HFI
TR27 5JD G7GUA
TR27 5JF G6YVD
TR27 5JJ M7CWL
TR27 5LJ M6FWV
TR27 6BA G8UTY
TR27 6DJ 2E0HCM
TR27 6DJ 2E0RUX
TR27 6DJ M0NXF
TR27 6DJ M6NXF
TR27 6DJ M7AMR
TR27 6DJ M7GBQ
TR27 6EX G1FOA
TR27 6HJ G4TXE
TR27 6HZ M3KGE
TR27 6PJ G4KYH
TR3 6BB G3PPT
TR3 6BH 2E0WSC
TR3 6BH M6KVU
TR3 6BJ G3UFJ
TR3 6BJ M5GUS
TR3 6BQ G1OEQ
TR3 6DB 2E1ERJ
TR3 6DH M0KLG
TR3 6DH M6WHD
TR3 6EB G3MRT
TR3 6EN G0CQJ
TR3 6EN G6MOT
TR3 6ES M6ZXQ
TR3 6FW M7WHI
TR3 6LJ G8PBI
TR3 6LN 2E1FNY
TR3 6LN G4LJY
TR3 6PQ G0NNR
TR3 6PQ G0PGX
TR3 6PW G3NHL
TR3 6SP 2E0KLT
TR3 6SP M6CTK
TR3 7AD 2E0BUD
TR3 7AD G3WPP
TR3 7AD M3AKH
TR3 7AT G0AWR
TR3 7FG G3OCB
TR3 7JP G0UWO
TR3 7NW G0WGV
TR3 7RB M3ZAY
TR4 8DZ M6YFT
TR4 8ER M6IEW
TR4 8QL G0DHT
TR4 8QL G4ZZZ
TR4 8RJ G8RBS
TR4 8SU 2E0PCO
TR4 8SU M6PCX
TR4 8TS G7WER
TR4 8TS M0AWP
TR4 8TS M0BDH
TR4 8TT G0LIA
TR4 9FW G0SDT
TR4 9LT G4OOL
TR4 9RB G8ZYR
TR5 0TU G6WMT
TR5 0XQ G3TDM
TR50TR 2E1ADT
TR6 0DZ G1ZEI
TR6 0EF G0GQJ
TR6 0EY G3TYA
TR7 1AS M3AFS
TR7 1PF G4SSL
TR7 1QQ G0DJL
TR7 1TY G4ADV
TR7 1TY M0BFB
TR7 2DE 2E0NWR
TR7 2DE M0RHW
TR7 2DG G0LCJ
TR7 2DX M6KWU
TR7 2EJ G3WUA
TR7 2JN G8SCY
TR7 2LE G3JVN
TR7 2QE M0CJE
TR7 2RB G4VPJ
TR7 2RH M1AWX
TR7 2RW G0FLU
TR7 2RY G3XPY
TR7 2TB G0NWS
TR7 2TE G4WOI
TR7 3AE M3KUZ
TR7 3AE M3KVK
TR7 3AJ G7KRH
TR7 3AW G0HEW
TR7 3BN G6CEP
TR7 3DT M7HGK
TR7 3EB G7CRN
TR7 3EH M7PML
TR7 3EJ G8GOR
TR7 3FB G7VFV
TR7 3JT G6AFA
TR7 3LB 2E0DXL
TR7 3LB M6HBQ
TR8 4AW G4CYI
TR8 4HJ 2E0HKP
TR8 4HJ M7AWN
TR8 4LF M0AIQ
TR8 4PL 2E1FOX
TR8 4PL G7KFQ
TR8 4PL M3KFQ
TR8 5DB G3VMK
TR8 5HH G8YOE
TR9 6AS G4RMG
TR9 6ER 2E0KTD
TR9 6ER G1PGV
TR9 6ER G4PSU
TR9 6ER M3KTD
TR9 6ER M3PSU
TR9 6FH M3ZLO
TR9 6JZ G8VCJ
TR9 6NB M6DPH
TR9 6PD G4TGG
TR9 6PD G7OQL
TR9 6PT M3AEJ
TR9 6TA 2E0NXE
TR9 6TA 2E0OWE
TR9 6TA 2E0XIT
TR9 6TA M7BAF
TR9 6TA M7DTF
TR9 6TA M7VOX
TR9 6TW G0PNM
TR96GQ M7CKG
tr96rh 2E0HIR
TS10 1JA G0HPQ
TS10 2BP M0BYV
TS10 2DG 2E1DMZ
TS10 2DL G0MBU
TS10 2EQ G7CRA
TS10 2HW G6ZGC
TS10 2JZ G3UZB
TS10 2LJ M0BWO
TS10 2LQ G6IVD
TS10 2LT G0AJZ
TS10 2RU M6DST
TS10 2RU M6EMZ
TS10 4AG G3NVP
TS10 4HD M6SEG
TS10 4NJ G3VZO
TS10 4PS M3MII
TS10 4PS M6TST
TS10 4SG M3PYS
TS10 5DP M3JGN
TS10 5EB G0VLK
ts101ls 2E0KWB
ts104jq M7ALC
TS11 6BD G3DAV
TS11 6BE G0BIA
TS11 6BE G3PSG
TS11 6DF G4WNA
TS11 6DY 2E0FJW
TS11 6JX G4NUO
TS11 6NW G0VZT
TS11 7HJ M0RIG
TS11 7HJ M6RNE
TS11 7HT G0NUT
TS11 8AG G8GFA
TS11 8AU G4RHX
TS11 8BP G1GMF
TS11 8BT G4GCU
TS11 8DJ M0CIC
TS11 8DU G4CRS
TS11 8DU G4OLK
TS11 8HT M7DUM
TS11 8JJ M6YBD
TS118DT 2E0TPR
TS118EB M1SPY
TS12 1AL G8JLA
TS12 1DP G4IJO
TS12 1FE 2E0EGR
TS12 1FE M0XGR
TS12 1JP 2E0GLA
TS12 1JP M3NUH
TS12 1PD M6TTK
TS12 2DQ G3XAG
TS12 2JZ G4FCU
TS12 2NJ M3RNO
TS12 2QX G6HHK
TS12 2TJ G3KBI
TS12 2XG G0EWT
TS12 2YN M0BTL
TS12 3AP G4KIR
TS12 3AW 2E1FFZ
TS12 3AW M3AOC
TS12 3DX M6KYC
TS12 3EE G0RJA
TS12 3EW M6KAV
TS12 3LE G0FSG
TS12 3LS G0PZM
TS13 4BU G4AWK
TS13 4DF G4WUS
TS13 4JB G6DIR
TS13 4NX M1FCZ
TS13 4PB G0SVJ
TS13 4UG G4JCS
TS14 6DJ G4HEB
TS14 6LQ M7AHN
TS14 7BY M0NDT
TS14 7DP G7ONL
TS14 7DZ 2E1CIO
TS14 7FH G8HUY
TS14 7LE 2E0FTU
TS14 7LE G0UOK
TS14 7LE M0NYM
TS14 7LE M7UOK
TS14 7LQ M0STF
TS14 7LZ G1AAG
TS14 7LZ M1AMP
TS14 7NB G4HRU
TS14 7PE M0MJT
TS14 7PR G7TXV
TS14 8DL G6MVQ
TS14 8HA 2E0ODJ
TS14 8HA M7DRJ
TS14 8JG M6GSX
TS14 8JX G0SUC
TS14 8JY G1YHP
TS14 8LD G7ION
TS14 8LN G4YMB
TS15 0BA 2E0ECD
TS15 0BA M6ECD
TS15 0HT G4JLJ
TS15 0JQ G3BPP
TS15 9EZ M1AFU
TS15 9JQ G4ZXT
TS15 9LB M0DTS
TS15 9ND G7RAJ
TS15 9NL G8KIK
TS15 9NL G8MBK
TS15 9NQ G4JJM
TS15 9NT G6LIJ
TS15 9RZ G3VGZ
TS15 9RZ G4EEH
TS15 9RZ G8AOE
TS15 9TG M3KNF
TS15 9TJ 2E0HEQ
TS16 0AQ G6KQJ
TS16 0JB G7NQZ
TS16 0LQ M7DUD
TS16 0NH 2E0ICC
TS16 0NH G3ICC
TS16 9BA G7SYI
TS16 9EH G3FDW
TS16 9HP G3RUG
TS16 9HU 2E0XPH
TS16 9HU M3XPH
TS17 0FT M7JJI
TS17 0LT M0AVW
TS17 0NL G0IBW
TS17 0TB M3VPX
TS17 5BG G6ICZ
TS17 5DJ G1HSH
TS17 5HQ 2E0MGV
TS17 5HQ M6BHQ
TS17 8LX G6CKH
TS17 9AF M7VOT
TS17 9AN M3LBD
TS17 9QJ G0JOC
TS18 1ax G0BKC
TS18 1ax M6BKC
TS18 1JY G0MLB
TS18 2FN G8NMK
TS18 4AZ M6KIT
TS18 4JA G3WWG
TS18 5AR M0HOK
TS18 5LH G4JXR
TS18 5NH 2E0MLX
TS19 0BD M6MMX
TS19 0DE M3FNA
TS19 0ER 2E0KBD
TS19 0ER M0KBD
TS19 0ER M6PII
TS19 0RA G3UFV
TS19 0SL G4YOV
TS19 0XL M7DRR
TS19 7AP M3EVI
TS19 7JL G0ELC
TS19 7JT G1ZRS
TS19 7LY G6MPE
TS19 7SF M7CNL
TS19 7SH M0DNN
TS19 8ES 2E0DBZ
TS19 8ES M3HSJ
TS19 8RF 2E0RDO
TS19 8SZ M3ASC
TS19 8XF G8KSA
TS19 8XF M0DSC
TS19 9EJ M3VDV
TS19 9JS M3YNN
TS20 1DB M6MOQ
TS20 1JG M3XAC
TS20 1JQ G7CLG
TS20 1JU G4XXG
TS20 1JU M0CYD
TS20 1LE M3RFO
TS20 1LG M0NBK
TS20 1PZ G7FFM
TS20 1RT G8VPY
TS20 1SB G8ILB
TS20 1SJ G7CLH
TS20 1SJ G7SUQ
TS20 1SZ M3JOA
TS20 2EX G0BZC
TS20 2SY G4LIM
TS202PQ M7GKL
TS21 1DZ G4DXP
TS21 1HT G8LTD
TS21 1LQ G3NBL
TS21 2AN G0GWL
TS21 2DS G0WXC
TS21 2EZ G0SJU
TS21 2JY 2E1JRB
TS21 3DN 2E0IIZ
TS21 3DN M0IIZ
TS21 3DN M0WXP
TS21 3DN M7SJM
TS21 3EU M3PGL
TS21 3LT G8ZIA
TS22 5HA G1RYM
TS22 5LP G7NRO
TS23 1DL G4PVN
TS23 1EE G7VOT
TS23 1HW M6IDS
TS23 1HX G7CUO
TS23 1QN G3YVY
TS23 2AN G0HJM
TS23 2DJ M7BCN
TS23 2HX G6LDJ
TS23 2PJ G7LOA
TS23 2QB M6MYB
TS23 2QH M0ASN
TS23 2RF M3UBL
TS23 3GP M6SWI
TS23 3HG G8OKS
TS23 3SP M7CPO
TS23 3SY G0JON
TS23 3UA G4MYN
TS23 4AW G6NXP
TS24 0DX 2E0HLC
TS24 0DX M6AXG
TS24 0dY M7SLE
TS24 0UF M6KQJ
TS24 0XF M0WFA
TS24 7HY 2E0YSF
TS24 7HY M6IEO
TS24 7NW M6NUF
TS24 8HH G7VGM
TS24 8RB G3IJV
TS24 9BQ G4VCJ

TS24 9BQ G6CGY
TS24 9QY M7EEM
TS25 1AX 2E0RKZ
TS25 1AX M7TSZ
TS25 1GF M7DSZ
TS25 1GQ M6MQE
TS25 1HF M6TJI
TS25 1HL M7MJG
TS25 1LP M6FQP
TS25 1RN M6BHN
TS25 1RW G0SBX
TS25 1RY 2E0HPI
TS25 1RY M6HPL
TS25 2LA G1DFZ
TS25 2LD 2E0WSN
TS25 2PY M3NFQ
TS25 2QG G4NRS
TS25 2RD G7VGO
TS25 2RG G4ZCN
TS25 3DP G3UUF
TS25 3QR M6OGC
TS25 3QY M1BNK
TS25 4NE M7PAG
TS25 4NZ G1KBE
TS25 5HX G3NWY
TS25 5HZ G0VGB
TS25 5LB G1GTP
TS25 5NG M6OTB
TS25 5PA G1ETQ
TS25 5RQ G0FZZ
TS26 0NW 2E0EJW
TS26 0NW M6EVW
TS26 0PJ G3NUA
TS26 0QY M6NLJ
TS26 0XG 2E0DGA
TS26 0XG M0VRW
TS26 0XG M6EOK
TS26 0ZL 2E0MDN
TS26 0ZL M0MBZ
TS26 0ZL M6MEJ
TS26 8AY G1KUO
TS26 8EZ M0GIZ
TS26 8LG 2E0SWK
TS26 8LG M7SWK
TS26 8ND G7SNJ
TS26 9BJ M0HXK
TS26 9ES M0BKF
TS26 9LY 2E0XVK
TS26 9LY M0NXA
TS26 9PN M0HPL
TS26 9PR G4SHJ
TS27 3QR G7JXY
TS27 4AB 2E0UXC
TS27 4AB M6MKM
TS27 4AX M0BZB
TS27 4DF M0DEL
TS27 4EB M1OJS
TS27 4EB M6DLC
TS27 4RT G0GKO
TS28 5AG G0HZK
TS28 5BP M6PIT
TS28 5BT G0HJX
TS28 5EJ G4XPP
TS28 5GA 2E0HQU
TS28 5GA M7LWR
TS28 5JE G0GWD
TS28 5JF M6LAZ
TS28 5JN M7DSQ
TS28 5JW M6RDQ
Ts285ef M7DTZ
TS29 6DA G0MHC
TS29 6DZ 2E0PSH
TS29 6DZ M0HTI
TS29 6DZ M6DVV
TS29 6EE M0ZEM
TS29 6EF M3SHB
TS29 6QE 2E0YIT
TS29 6QE G5CWP
TS29 6QE M6YIT
TS296HE M6GDO
TS3 0NN 2E0UTB
TS3 0NN M6SOV
TS3 0RL 2E0JKD
TS3 6PE M6CFD
TS3 7DR 2E0BJK
TS3 7NQ M6IBH
TS3 7PU M6TNA
TS3 7RX 2E0KOM
TS3 7RX M0KOM
TS3 7SL 2E0FJP
TS3 7SL M3FJP
TS3 8LX G4IIN
TS3 8NX G4OOK
TS3 9HD M6MQL
TS3 9NT M6MFC
TS3 9QW M7SEH
Ts37hp M7JEP
TS4 2DN M6OJF
TS4 2HG M6LCY
TS4 2RY M3VHY
TS4 3HJ M6RTA
TS4 3HR M6KRI
TS4 3HU M6WCC
TS4 3HX G3RGB
TS4 3NW M0AHD
TS5 4AJ G4UTV
TS5 4HG M0SMA
TS5 5BN M3CWV
TS5 5JG M7WBF
TS5 5JP M3YRZ
TS5 5NQ G4HSB
TS5 6DU 2E0ADR
TS5 7QX G4VZC
TS5 8BT G1NTL
TS5 8DP G4PPS
TS5 8DR G3KXV
TS5 8EB G8WKZ
TS5 8NT G4IJM
TS5 8NT G8HDM
TS5 8RE M0GQV
TS5 8RY M6LLM
TS55LD M0BJX
TS6 0BG 2E0NUL
TS6 0BG M0JUN
TS6 0BG M6WIX
TS6 0BN G0NYZ
TS6 0BQ G7TWU
TS6 0BS G4MCF
TS6 0BZ M7AAT
TS6 0JD G8MKG
TS6 0PL G1DAT
TS6 0PP M0BZA
TS6 0QQ G0FXR
TS6 0QQ G0LWE
TS6 0QQ M3TVN
TS6 0RW M6CLI
TS6 7EZ G7NQR
TS6 7EZ M0HMS
TS6 7HG M7ARL
TS6 7ND 2E0CNX
TS6 7ND M3ITL
TS6 8AF G1CCM
TS6 8EG M7DMV
TS6 8EH G7FVH
TS6 9HX G1CLT
TS6 9LY G4ORQ
TS6 9QW G0DEO
TS6 9QW G0PDK
TS6 9QW M7WEP
TS6 9SZ G7NQX
TS7 0EZ G7WEN
TS7 0JL M0DIQ
TS7 0LB 2E0LGB
TS7 0LB M0LGB
TS7 0LT G1IGW
TS7 0QL G8YDC
TS7 8AN G7UMA
TS7 8SE 2E0CPK
TS7 8SE M0HHC
TS7 8SE M6BKL
TS7 9BB 2E0PEL
TS7 9BB M0SMP
TS8 0RU G0AOO
TS8 0SU G7RXB
TS8 0UF M7OGB
TS8 9AB G4POD
TS8 9BU G4ZML
TS8 9FQ G7PTC
TS8 9HH G0BQP
TS8 9LJ G0UTP
TS8 9LJ M6UTP
TS8 9SL G7IZC
TS9 5AG G4HWV
TS9 5BU G0TYM
TS9 5DF G7UXH
TS9 5EL M0CSD
TS9 5HU G0UYG
TS9 5HX G7RNQ
TS9 5PQ G4OIW
TS9 6DW G0FAU
TS9 6DW M3FYV
TS9 6EN M6BHP
TS9 6JF G7TFG
TS95BE G0MBV
TS95BE M7KEZ
TS96NQ M6PPS
TW1 1AG M0ICI
TW1 1PR M7GOB
TW1 1PY G4RBR
TW1 1RS M6MDN
TW1 2AX 2E0JAB
TW1 2DD G4CXZ
TW1 2DF G4CBD
TW1 3AU G6CIA
TW1 4BQ G1TOB
TW1 4QZ G8DSU
TW1 4SW M0MDC
TW10 5DU 2E1DQM
TW10 6AB M0IRG
TW10 6DS 2E0RYS
TW10 6DS M0RYS
TW10 6DS M6RSY
TW10 6HS 2E0ERZ
TW10 6HS M0IUR
TW10 6HS M6VRE
TW10 7ED G1PHA
TW10 7EQ M3RHO
TW10 7LD M7AHU
TW10 7NQ G0SLI
TW10 7YG G8CYE
TW11 0BG G4LGO
TW11 0BQ G3GDZ
TW11 0DH G7PMV
TW11 8AS G3PPC
TW11 8DE G8BAJ
TW11 8JB G6RHV
TW11 8LT G1OPT
TW11 8QA M7MAE
TW11 8SH G8ZWA
TW11 9DW G8WGE
TW11 9HA G0DEH
TW11 9LN G8AUU
TW11 9PD G8DEL
TW11 9QS G0JME
TW11 9RL G3YJV
TW11 9RS G3ZLD
TW12 1DE M0LHR
TW12 1DW M0BYY
TW12 1HX G4PPE
TW12 2BL M6ZES
TW12 2JH G8SNV
TW12 2LU G3RCB
TW12 2NE M6JXD
TW12 2PZ G7DWM
TW12 2QG G4ZOU
TW12 2SR M0NAI
TW12 2TR G7OMA
TW12 3AU G4LDW
TW12 3BW G4UXB
TW13 4HX G1KCU
TW13 4QG M3XHB
TW13 5DJ G7GFQ
TW13 5HE 2E0GJQ
TW13 5HE M7AMJ
TW13 5JS 2E0HMR
TW13 5JS M7HMR
TW13 6LZ G0EEZ
TW13 6PE G0KXG
TW13 6PX G1YVZ
TW13 7AA M7LAG
TW14 0AH 2E0RUS
TW14 0AH M3SQS
TW14 0JY G1TFM
TW14 9ED G1LKL
TW14 9JE G0DEO
TW14 9LN M7LCS
TW14 9LW G7UWS
TW14 9QP G4ALA
TW14 9XB M0DPF
TW14 9XG G7OXK
TW15 1AB G8FVJ
TW15 1DG G0CPF
TW15 1DP G1IBS
TW15 1DP G4ZVU
TW15 1DR 2E0UFM
TW15 1DR G4ESG
TW15 1DR M6YON
TW15 1HF G0ONA
TW15 1LT M7FAB
TW15 1NS G6PFC
TW15 1PQ G0RJE
TW15 1PW G0NBJ
TW15 1PW G0OSX
TW15 1QE G1WWB
TW15 1SQ G8ZZG
TW15 1UT M6HKS
TW15 2AP G8CAH
TW15 2BE M7TIE
TW15 2BN M0IOE
TW15 2EP M3YXE
TW15 2JB 2E0NOC
TW15 2LP G3XTZ
TW15 2LU G8LPA
TW15 2LU M0LPA
TW15 2RF M1CSU
TW15 2SJ G1WZB
TW15 2SL M5CBS
TW15 2TD G1ZDT
TW15 2TD G7AYP
TW15 3PF M0AHS
TW151IW 2E0GWZ
TW151IW M7GTZ
TW16 6HF G3HTC
TW16 6HF M0CFR
TW16 7DD M7FGC
TW16 7NA G4YAS
TW16 7NL G7PAK
TW16 7PL 2E0NBE
TW16 7PL M0NCE
TW16 7PL M6FUF
TW16 7QU M6NPF
TW16 7TL 2E1AOF
TW16 7UA G0HYT
TW16 7UA G7EAR
TW17 0DN G3UPW
TW17 0EN G0VDZ
TW17 0EN M0KIO
TW17 0JB G6IFR
TW17 0QT M7NDS
TW17 0RP G6KLH
TW17 0SH G4XGI
TW17 8AY G4GNS
TW17 8BT G7CSS
TW17 8EU G0LGA
TW17 8QQ M6GAN
TW17 8QS M0PMY
TW17 8RR G7MII
TW17 8RX G4CKQ
TW17 9DQ G8BQZ
TW17 9LL M0MSV
TW17 9LL M7NAN
TW18 1DG G0MIJ
TW18 1DJ G0JSP
TW18 1EE G1SXB
TW18 1NE G7ECE
TW18 1PW M7IGG
TW18 2AP G4HKS
TW18 2AP M6GIQ
TW18 2AP M6SUC
TW18 2DD G6PVC
TW18 2LW M6VHB
TW18 2QE G3ZLR
TW18 3EE 2E0VBX
TW18 3EE M6VBX
TW18 3HB M0LZU
TW18 3HH G0OEK
TW18 3HQ M6NVG
TW18 3NQ G4GSC
TW18 4NR G3NTM
TW18 4TW M0HMI
TW19 5NF M6ZJW
TW19 6AX G4AKA
TW19 6BT G4UXG
TW19 7AJ M3OPT
TW19 7JE G0TID
TW19 7JE M3CJA
TW19 7JE M3EMN
TW19 7LF G4MTH
TW19 7LG M6CTI
TW2 5BY G1NSQ
TW2 5BY G6AWM
TW2 5JH M1GAR
TW2 5JJ G7TOF
TW2 5JP G1FKS
TW2 5LS G0EZL
TW2 5PD M0DLL
TW2 6AA G2AIW
TW2 6BL G8WRL
TW2 6EB M3WQG
TW2 6EJ G3ZSB
TW2 6HW G1LVZ
TW2 6SA G7GZC
TW2 6SP G6GVF
TW2 6SR 2E0WEJ
TW2 6SR M0GBC
TW2 6SW G6VUG
TW2 7BL 2E0WMD
TW2 7BL G0TSU
TW2 7BL M0DMJ
TW2 7BL M3ESQ
TW2 7DT G7GAP
TW2 7EA G1OVO
TW2 7EU 2E0DZM
TW2 7EU M0IUT
TW2 7EU M6RZP
TW2 7EX G4OQN
TW2 7JE G0OFH
TW2 7JG G7HOV
TW2 7JG G7JBN
TW2 7NH G0MRF
TW2 7NP M0DCW
TW2 7SN G6LKH
TW2 7SN G8RHZ
TW20 0DF M1DWV
TW20 0RZ M0HJY
TW20 0RZ M6BRJ
TW20 8AN G0LSE
TW20 8BS G0NIF
TW20 8BS G7AOU
TW20 8ET 2E1DOG
TW20 8ET M6DOG
TW20 8NL G6PVV
TW20 9LP 2E0DMQ
tw25jt M3JXY
TW3 1YH G4CWE
TW3 2EZ M7PAR
TW3 2HH 2E0JXX
TW3 2HH M0XJJ
TW3 2HH M6JXX
TW3 4NL M0RJZ
TW34AD M0IMQ
TW4 5BB G1EIH
TW4 5EW 2E0BLQ
TW4 5PF G0RHB
TW4 6AG G3AHE
TW4 6NA G1CRN
TW4 7AN M0JCY
TW4 7JG M0WBC
TW4 7JQ G0WMD
TW4 7JQ M0HIC
TW4 7NH G0IIK
TW4 7NU G7GBE
TW4 7RA G0USK
TW5 0ND G6YTX
TW5 0NF G4IXB
TW5 0PZ M0TCE
TW5 9AW G1JRW
TW5 9EX G0VLQ
TW7 4AJ G0LUH
TW7 4LS M0EMD
TW7 4PQ G4DUO
TW7 5HB G8SJO
TW7 5LU M6OGP
TW7 6AD G3LWR
TW7 6GP 2E0EXI
TW7 6HW G4LTC
TW7 6HX G4PWS
TW7 6LF G4NJN
TW7 7BW M7CZM
TW7 7BZ M6XBR
TW7 7DJ M3HVO
TW7 7NL G6AGA
TW8 0FU M0BYK
TW8 0PL G3XKV
TW8 9PT G6SUK
TW8 9RD 2E0JDC
TW8 9RD 2E0YVR
TW8 9RD 2E0ZBW
TW8 9RD G8MWD
TW8 9RD M3YVR
TW8 9RD M3ZBW
TW9 1AX M3NAO
TW9 1LL 2E0SWE
TW9 1LL M6BIV
TW9 2DZ 2E0CEN
TW9 2DZ M3ZEI
TW9 2HA G1CRT
TW9 2TJ G0ACD
TW9 3AY G0LYJ
TW9 3BD 2E1GHZ
TW9 4AS G0OOI
TW9 4AS G3CIK
TW9 4EE G0OHW
TW9 4JN M3YBU
Tw9 4jp M7AKU
UB1 1PZ G7CSD
UB1 2SA M0GBO
UB1 2UP G1YES
UB1 3JP G3KLK
UB1 3NT G1YRY
UB1 3QD G6ESM
UB1 3QJ 2E0KDS
UB1 3QJ M7GHJ
UB10 0BS G6KNK
UB10 0DN G1ECY
UB10 0HH G1DYR
UB10 0HP G6IRY
UB10 0HW 2E0CXE
UB10 0HW M0HQR
UB10 0HW M6CKU
UB10 8DY 2E0JTY
UB10 8ED G4BVF
UB10 8HS G4CTD
UB10 8HU M7XAN
UB10 8LS G1PAF
UB10 8QA G8LIU
UB10 8TA G6LEY
UB10 8TJ M7GCR
UB10 8UF M3MST
UB10 8UF M3WCR
UB10 9EL 2E0FXE
UB10 9EL M6XXH
UB10 9HZ M6ODL
UB10 9LN M7MPD
UB2 5AN G7IVN
UB2 5HN G6VAE
UB24FZ 2E0XBW
UB24FZ M0XBW
UB24FZ M6XBW
UB3 1PZ G0LQW
UB3 1ST G3SUN
UB3 2FJ M6FPP
UB3 2JE G8HOU
UB3 2NF M7LHR
UB3 2QX G6XDI
UB3 2TP G8GYP
UB3 2TQ G6STI
UB3 3PA G1VTS
UB3 3PP G7JSW
UB3 4AD G0EFS
UB3 4QJ G1JLM
UB3 5ET G3PKR
UB4 0AE G0CHR
UB4 0DZ M6XJS
UB4 0EF G0RGL
UB4 0EF G0VGY
UB4 0JH G1LLQ
UB4 8ET G4OHQ
UB4 9QT M6GQU
UB4 9RB G1LVR
UB4 9YF G0KYN
UB5 4AQ 2E1IDC
UB5 4NL G1XLT
UB5 4SE M7OMG
UB5 4SG 2E0JGP
UB5 4SG M6JGP
UB5 4SW M6WIK
UB5 4TL M6DTC
UB5 4TZ 2E0OZK
UB5 5BX G1DLA
UB5 6AR G1KOX
UB5 6EU G0SLH
UB5 6HP G7GDC
UB6 0BH G1VSM
UB6 0BH M0DBI
UB6 0HQ G3PSC
UB6 0JZ G0MXY
UB6 0QE 2E0LPZ
UB6 0QE M0LPZ
UB6 0QE M6YUG
UB6 0RF G4AWM
UB6 0SS M3CGO
UB6 7AD G4FJX
UB6 7HR G6PGJ
UB6 8AJ 2E1HFW
UB6 8LN M1AZQ
UB6 9BY G7ABE
UB6 9HX 2E0ISN
UB6 9LS G1UBL
UB6 9NN G0JJQ
UB6 9NN G6UIT
UB6 9NT G0HRK
UB7 0DR M6IIW
UB7 0LE M0AAC
UB7 0LS 2E0LJR
UB7 0LS M6ZZV
UB7 7AN G0FFK
UB7 7AP 2E0TTW
UB7 7AP M6EJF
UB7 7AP M6UAF
UB7 7TZ M6BBR
UB7 8AW M0LTP
UB7 8BU M3OTR
UB7 9DE G7BSS
UB7 9DH M7MOW
UB7 9DR G0FFN
UB7 9EA M7AND
UB7 9PE G4ZRC
UB8 1BL G8FKH
UB8 1BL M0DGB
UB8 1QX G3JVP
UB8 2NY 2E0CRC
UB8 2NY M0KBA
UB8 2NY M6CCR
UB8 2QG G7NFR
UB8 2RQ M7JAG
UB8 2UL G4CVF
UB8 3EL M0HZV
UB8 3HS G7VHN
UB8 3HS G7VHO
UB8 3QE G0JHG
UB8 3SG M7IDP
UB8 3SR G0OLD
UB8 3WS G7LCD
UB9 4AG G4YEO
UB9 5DJ G0VOV
UB9 5HJ G7LVE
UB9 5PB G6FGJ
UB9 6AU M7MEX
UB9 6LZ M6TGI
UB9 6QB G3VXA
UNITED KINGDOM
M0HUH
UNITED KINGDOM
M6OAL
V1E 2DN G4NJT
W10 4BW M7MFD
W10 5TA M0TUR
W11 1EP M0IAK
W11 3BY M0SKN
W11 4BT M1DGK
W11 4HE M6EFG
w11 4jx M7WBX
W11 4SQ M5OOO
W11 4TG G8EZT
W12 0AP G7AAS
W12 0LP G3SEF
W12 7BL G3IJL
W12 8JN 2E1IFA
W12 8NU M0SHH
W12 9LU 2E0AVB
W12 9TF M3NKO
W13 0AE G6GIU
W13 0ED G0PIN
W13 0HA 2E1YHZ
W13 0HA M7KRA
W13 0HA M7YHZ
W13 0HP G1ARF
W13 0NT G7JHZ
W13 0TF M0GHC
W13 8AJ 2E0PMP
W13 8AJ M6PRP
W13 8JZ G3RGP
W13 8NH M7SBN
W13 8PR 2E0JAJ
W13 9DT G0AIS
W13 9EN G8LVF
W13 9HS M0CPB
W13 9JY 2E0VBS
W13 9JY M6ZBS
W13 9LA G6OJN
W13 9QU G8VDQ
W13 9RA G6JEU
W13 9TN M0FZU
W13 9TY G1HGT
W14 0JX G4HTY
W14 8UW M0GCQ
W14 9BS G1EIG
W14 9HU M7IOS
W14 9JS M6ENW
W1H 7DP M0KYR
W1J 8PE M0CER
W1K 5RX M0KBY
W1K 5RX M7CJC
W1T 4TD M7OJO
W1U 6PS M7BSL
W1W 6UH M7XQS
W2 2RJ M3YGY
W2 3RL M6RVJ
W2 4DT G8WIW
W2 6DW 2E0ZDC
W2 6DW M0ZDC
W2 6DW M6ZDC
W3 0DE G3YMM
W3 0HH G1CNN
W3 0HR G3HHU
W3 0RP G8LWS
W3 7AQ G0ZHP
W3 7NP G0UCC
W3 7PA M1ELS
W3 8SY 2E0UER
W3 8SY M7AUI
W3 9AE G8SPE
W3 9EJ G6HXB
W3 9RQ G4ZZV
W4 1HT 2E0CPP
W4 2JH G4LIC
W4 2LL G4SJL
W4 2SF 2E0ZHN
W4 2TS 2E0OCP
W4 2TS M0OCP
W4 2TS M7OAD
W4 3LR G1SAT
W4 3LR G3VKT
W4 3NH G4ZJD
W4 4EH 2E1HUJ
W4 5DN G4HMS
W4 5DN G4TZA
W4 5DN G7HMS
W4 5EN G0OYJ
W4 5EQ M6FUH
W4 5JS G3SQX
W5 1HL G0VDN
W5 1PY M0BCZ
W5 2BJ 2E0HSW
W5 2BJ M0HSW
W5 2HE M7JAV
W5 2LU M6YXD
W5 3LL M3SDQ
W5 3SL G8MVC
W5 3XH M6PFM
W5 4BJ 2E0UMP
W5 4BJ M7ICG
W5 4BY M6LVS
W5 4DR G7ANG
W5 4EX 2E0SZT
W5 4EX M0SZL
W5 4EX M7SZO
W5 4JW G0SCG
W5 4XA G6VZM
W5 4XA G8BOJ
W5 5BH 2E0JES
W5 5BH M0JXS
W5 5BH M6JIG
W5 5HS G6FVB
W5 5PD G6YJJ
W5 5QD G8RVO
W5 5QT 2E1DTF
W6 0BT 2E0WYH
W6 0BT M7WHC
W6 7AF 2E0IGE
W6 7AF M0UMM
W6 7AF M7SRV
W6 7LD G3YFV
W6 8BD G7EEJ
W6 8HN G7VLF
W6 8HW 2E0UMR
W6 8HW M6UMR
W6 8RN 2E0CUO
W6 8RN M0HYG
W6 8RN M6CNL
W7 1BT G6TJY
W7 1JQ 2E0OUR
W7 1JQ M6TVX
W7 2JH G7VEI
W7 2JS 2E0ECV
W7 2JS M6MQH
W7 2NR M7MJR
W7 3AG G8RBX
W7 3DJ G6TNK
W7 3RA G8XQB
W8 6PJ M6RIQ
W8 6UU G3SKR
W8 7AX G8AVQ
W8 7SL G4EHN
W8 7SX G0HTX
W9 2BJ M6HEJ
W9 2QH G3UXO
W9 2QU M6VLD
WA1 2EH M7LDX
WA1 3AH M7BRJ
WA1 3AY M1AUH
WA1 3EN G4FMI
WA1 3HB G4XQA
WA1 3HX G0RFM
WA1 3JS M7DJG
WA1 3TN 2E0RYP
WA1 3TZ M3LFU
WA1 4BJ G1NWA
WA1 4BJ G7JOA
WA1 4DY G0ANL
WA1 4EZ G6LFT
WA1 4EZ M0CUQ
WA1 4EZ M3PQN
WA1 4LU G0CDA
WA1 4LU G3CKR
WA1 4NW G0PZP
WA1 4NY G1NXT
WA1 4PE M0CSE
WA10 2HA G0KQI
WA10 3HT M7SSW
WA10 3JH M0BWP
WA10 3NE M0AQK
WA10 3QL M6BJA
WA10 3RX 2E0PBO
WA10 3RX M6PPK
WA10 3XT M1BWS
WA10 4BL M6ZLM
WA10 4DN M3MSB
WA10 4EW M7HHY
WA10 4JG G0JBR
WA10 4LN G1OSL
WA10 4LU 2E0KNM
WA10 4LU M6KNM
WA10 4NL M3OEM
WA10 4RH M0ACK
WA10 5AH G0EZI
WA10 5AU M6AXF
WA10 5DW M3DUO
WA10 5PB G8MHE
WA10 6QF 2E0MVH
WA10 6QG G0BCX
WA10 6SH G4MWO
WA10 6SH G5MHZ
WA10 6SH M0SHR
WA10 6TP G1NVE
WA11 0AP 2E0IAZ
WA11 0AP M6IAZ
WA11 0BL G8CXZ
WA11 0DD M7DXP
WA11 0DY G1XAP
WA11 0ES G0DZI
WA11 0ES G6DZI
WA11 0LY G4YKV
WA11 0NB G4WGJ
WA11 0PP M3VXM
WA11 0PY G0AFJ
WA11 0QD M0NZR
WA11 0QS G1ZKZ
WA11 0RH G0NWC
WA11 0RH M3NJQ
WA11 0RS G0SSL
WA11 0SD 2E0YGH
WA11 0SD M6TTO
WA11 0SQ G1WAE
WA11 0UB G4XIE
WA11 0YQ G0HBU
WA11 0YS 2E0RCC
WA11 0YS M0RCC
WA11 0YY 2E0RCR
WA11 0YY M0RCX
WA11 0YY M7CDX
WA11 7AG M6FME
WA11 7AP M6TKP
WA11 7DE 2E0RIL
WA11 7DE M7RIL
WA11 7LD G4KIN
WA11 7LD G8TMR
WA11 8AG G4KIP
WA11 8AT G3JIR
WA11 8BH M3DJW
WA11 8BW G0BHH
WA11 8DJ G4FKP
WA11 8ER G8JSM
WA11 8JR G4EST
WA11 8JW M0AQQ
WA11 8LA M0DZC
WA11 8SW M3JGH
WA11 9EL G4DIY
WA11 9HU G6YSN
WA11 9JD M6ZZE
WA11 9JJ M3KYQ
WA11 9NH M3VOL
WA11 9RB M7BZA
WA11 9RE G0NEB
WA11 9RW G0EAM
WA12 0AS M3OWQ
WA12 0AU 2E0BGM
WA12 0AU M3ISI
WA12 0DR G0VVQ
WA12 0DU G4KHG
WA12 0JT G4XOL
WA12 0JW 2E0NLW
WA12 0JW M6NLW
WA12 0LN G7PUN
WA12 0LN M3POH
WA12 0LN M3TLN
WA12 0LY G1HIO
WA12 0LY G1HIP
WA12 0LY M0NRC
WA12 0LY M7DX
WA12 0NE 2E0IUH
WA12 0NE M3IUH
WA12 0NN G6OBG
WA12 8BA G8XVJ
WA12 8BA M0SDA
WA12 8BY M0JVW
WA12 8HJ M3HHQ
WA12 8HP M3IRR
WA12 8JE G1HSA
WA12 8JF 2E1FSH
WA12 8JF M0ZLK
WA12 8JF M6BWV
WA12 8JF M6YZX
WA12 8LT G3XRI
WA12 8LT M0NLW
WA12 8LZ G6IKM
WA12 8SQ G1JPT
WA12 8SQ G6JPT
WA12 9DB M3ISJ
WA12 9EY 2E0IUI
WA12 9GB M3HGZ
WA12 9HS M7EMO
WA12 9LS G0VAM
WA12 9LS G4WCH
WA12 9LT G7MFA
WA12 9QA M6PQS
WA12 9UD G1MMA
WA12 9YG G3YVH
WA12 9YL M3NBQ
WA13 0JS G3VJV
WA13 0JT G7BGO
WA13 0LA G4FQN
WA13 0QR G1ANK
WA13 0RD G3NKH
WA13 0SY 2E0DGU
WA13 9BA G8VMQ
WA13 9EY G0AMU
WA13 9EY G6XKK
WA13 9LY G3NKW
WA13 9NL G8TYH
WA13 9NN M1DQQ
WA13 9NW M6KDM
WA13 9QA G4VSX
WA14 1JL G4UGM
WA14 1LH G1GLS
WA14 2EQ G8JGM
WA14 2ND G3WFH
WA14 3HY G0KTT
WA14 4HW 2E0WHA
WA14 4HW M0WIA
WA14 4HW M3WIA
WA14 4JE G0CBJ
WA14 4QT M3VOJ
WA14 4QT M6IFG
WA14 4RQ G6HLU
WA14 4UE G4PYH
WA14 5AQ G7UCT
WA14 5AZ G4NAV
WA14 5JF G7AGA
WA14 5LJ 2E0NDG
WA14 5LJ M6NDG
WA14 5LT G4HUF
WA14 5LT G7EYR
WA14 5NQ M1DXL
WA14 5YF M0OKS
WA15 0AS G4FYB
WA15 0LT G4CNI
WA15 6AB G0WEV
WA15 6AB M1CTM
WA15 6BG 2E0KCZ
WA15 6BG M0KCZ
WA15 6BG M6KCZ
WA15 6BW G3EOO
WA15 6BX G6RGN
WA15 6BX G7MVY
WA15 6DY G4VBJ
WA15 6HQ G4WQD
WA15 6HR G6MCE
WA15 6RS G6DOQ
WA15 6UE G4TFU
WA15 7BA M0PGL
WA15 7DB M6JCN
WA15 7HY M3VCV
WA15 7NP G1KXP
WA15 7NP M3OEO
WA15 7PA G7RHD
WA15 7RS G1GNX
WA15 7RS G4UKV
WA15 7TF 2E0HKG
WA15 7TF M0KHK
WA15 7YH G0HAL
WA15 8DN G6GTH
WA15 8EA 2E0CXL
WA15 8ET G4SVR
WA15 8HA G7VCF
WA15 8LY G6MNB
WA15 8PD G0UXR
WA15 8PW G0GWI
WA15 9BD M7RZA
WA16 0BN G7EPN
WA16 0JX M6EOB
WA16 0NE G0AKF
WA16 0TU M5ACS
WA16 0TZ M1ROE
WA16 6PS G6EXE
WA16 7AX G3UJE
WA16 7AX M0TTG
WA16 7ET G4HZW
WA16 7GB M0AUF
WA16 7HB 2E0CSU
WA16 7HE G0AUB
WA16 7HF G0FOY
WA16 7RD G8CJQ
WA16 8BG M3LRU
WA16 8JR G6FWK
WA16 8LH M6JMP
WA16 9AW G0IFX
WA16 9DE G4MGK
WA16 9DZ G3ISB
WA16 9SB G1OQO
WA2 0AG G3YRQ
WA2 0AG M3WIJ
WA2 0BE G1KSK
WA2 0BL G3NFB
WA2 0BL M1DOT
WA2 0EZ G6OSV
WA2 0GL M3RZI
WA2 0NB M6IYM
WA2 0NE 2E0HSE
WA2 0NE M7BRR
WA2 0NQ M3ISN
WA2 0QE M3KPB
WA2 0SQ 2E1GMO
WA2 0SQ 2E1HEF
WA2 0SQ G1OND
WA2 0SQ M1GMO
WA2 0UH M0RAN
WA2 0UJ 2E1HEK
WA2 0UJ G6FKL
WA2 0WR M6WAA
WA2 7AP M6KEG
WA2 7AT M1CXW
WA2 7AX M1FAX
WA2 7DN M3ARU
WA2 7JQ 2E0BXA
WA2 7JQ M6JON
WA2 7RW M6JEP
WA2 7SE G1FOW
WA2 8AD 2E0KPM
WA2 8ad M0XLR
WA2 8ad M0XNW
WA2 8AD M3DNA
WA2 8AJ M6TJJ
WA2 8AW M3PVB
WA2 8BQ M1EFT
WA2 9AJ G4XDH
WA2 9BB M0GYC
WA2 9BQ G0RFT
WA2 9JA 2E0DHO
WA2 9JA G6LXV
WA2 9JA M6EWO
WA2 9NQ M6NBN
WA2 9PS G0VQK
WA2 9PS G7TOA
WA2 9SD G7PMB
WA2 9TW G7HKN
WA20ll M7PRH
wa29ew M7AQI
WA3 1DG M0NTA
WA3 1EB M1BRX
WA3 1EQ M6AXA
WA3 1EU G6AHF
WA3 2DP M7CGD
WA3 2EE G4VDX
WA3 2EP G1EFU
WA3 2EP G1THF
WA3 2ES G0LVJ
WA3 2QL G4ONG
WA3 2RN G4RNC
WA3 3DF G4JPX
WA3 3HH 2E0COA
WA3 3HH M0LGL
WA3 3HH M6CFN
WA3 3HH M6EEL
WA3 3HH M6LTL
WA3 3JJ M6DOX
WA3 3QX G0RPO
WA3 3TZ G0LVY
WA3 3TZ G6TPG
WA3 3UX G6WVL
WA3 4DF G4LWY
WA3 4ES G1IQK
WA3 4EU M7BTV
WA3 4JF 2E0CFR
WA3 4JP M0OCE
WA3 4LD G0RPG
WA3 4LD M0NDP
WA3 4LD M1CNP
WA3 4LR G8YKG
WA3 4NW G0WJX
WA3 4NW G7GZB
WA3 4PD G8NDE
WA3 5EX M7JRA
WA3 5JJ G0IZN
WA3 5QY G3XDP
WA3 5RU G6HKN
WA3 5RW G6DSP
WA3 5SA G7FHU
WA3 6AQ G4YKR
WA3 6DJ G4VBD
WA3 6LB M3EUF
WA3 6NE G4YLK
WA3 6NY 2E0SLM
WA3 6NY M3YHG
WA3 6PB G6HLL
WA3 6PS M6BJB
WA3 6PT M0HVN
WA3 6PT M6GUV
WA3 6QD G0JXI
WA3 6QP G6KTN
WA3 6RR 2E0ISK
WA3 6RR M6RZY
WA3 6RS M3MHR
WA3 6TP G1ZOY
WA3 6TR M6CFL
WA3 6TZ G0BCU
WA3 6UY 2E0FXP
WA3 6UY G0CBN
WA3 6UY G0CPD
WA3 7JG G0TDC
WA3 7LR G0JPC
WA3 7LR G6MKD
WA3 7LS G1NXS
WA3 7LY G4SSJ
WA3 7NT 2E0KTG
WA3 7NT M0XDJ
WA3 7NU M3YGL
WA3 7NU M3ZLP
WA3 7QA M1MKL
wa32ew M7CRW
WA4 1AX G7NPT
WA4 1DW G0YSS
WA4 1EX G7CED
WA4 1HX M3ZMN
WA4 1LY G8SQK
WA4 1NF G6DOZ
WA4 1PY 2E0DTO
WA4 1PY M0HFF
WA4 1QN G1PIY
WA4 1RJ G1KOD
WA4 1RW M3OQS
WA4 1TU G7VAG
WA4 2ET M0LYN
WA4 2ET M6WCY
WA4 2HE G0KXW
WA4 2HQ G4YPI
WA4 2HY M0PDX
WA4 2JA 2E0MFG
WA4 2JA M6TTP
WA4 2PF G0SSJ
WA4 2RE G4PBZ
WA4 2RX G0RPF
WA4 2TR G6MQI
WA4 3AE 2E0EXN
WA4 3AE M3TRJ
WA4 3AE M5NEX
WA4 3BA M7DRE
WA4 3BG G0IHF
WA4 3BJ G6GEL
WA4 3ES M3MXJ
WA4 3JA G7SKR
WA4 3LE 2E0NDP
WA4 3LE M0NDP
WA4 3LE M6NDP
WA4 4AF M3TPG
WA4 4BY G3JDT
WA4 4EE 2E0TBI
WA4 4EE M0WAU
WA4 4EE M6JPL
WA4 4NG M6JWO
WA4 4PY 2E0WVM
WA4 4PY M3ISY
WA4 4TD G6GZC
WA4 5AW G4EAQ
WA4 5EJ G0MYN
WA4 5QD G4LLG
WA4 5QJ G7LHS
WA4 5RD G8KBB
WA4 5RD M6TLP
WA4 6AF G0KXY
WA4 6DA M3GIX
WA4 6DE 2E0GGI
WA4 6DE M6GGI
WA4 6ES M3RVK
WA4 6EY G6NZH
WA4 6GY G4OEX
WA4 6LF G7GPL
WA4 6QU G4BQJ
WA4 6UA G3MUX
WA44BY G3ZRN
WA5 0AG G6YHL
WA5 0GB G7KHW
WA5 0GD M1EGV
WA5 0GJ M6SPZ
WA5 0HP G6BKE
WA5 0HP M6TLX
WA5 0LW M6LRH
WA5 0LY M3HIG
WA5 1BQ 2E0XJJ
WA5 1GU 2E0CDT
WA5 1GU M0HLC
WA5 1GU M3ZRR
WA5 1HN G0CWA
WA5 1HP M1FHP
WA5 1JF G7LPZ
WA5 1JH G6DPH
WA5 1LN M7FRR
WA5 1SY M0CRN
WA5 1TQ 2E0LOL
WA5 1TQ M3LXS

WA5 1TQ M3MQA
WA5 1XG G0CVM
WA5 1XN G7CFS
WA5 2AQ G3ZHE
WA5 2AR G6UCI
WA5 2BY G0NVT
WA5 2DB G0KKU
WA5 2DD G1VSK
WA5 2DD G3ZKD
WA5 2GH M7VSD
WA5 2HY G8PXI
WA5 2JN M3MIQ
WA5 2PA G7GGM
WA5 2QE G0SLR
WA5 2QL G7MNP
WA5 2QY G3GAH
WA5 2RB G8WAJ
WA5 2SG G4HSS
WA5 2sg M7TCW
WA5 2SR G8NRF
WA5 2TG M7BPS
WA5 2TW G3RRM
WA5 3AJ M0DCG
WA5 3AN G0RBI
WA5 3DA G0APY
WA5 3EE G0OON
WA5 3EH 2E0ELK
WA5 3ET G8VVP
WA5 3HD M7LBI
WA5 3LU G3VZU
WA5 3NU G1NRN
WA5 3PL M7GMF
WA5 3RX G8RCL
WA5 3TT G8NHD
WA5 4AX G4WNW
WA5 4DS 2E1GNR
WA5 4ES M3JNB
WA5 4HJ 2E0MGW
WA5 4NE M3JAZ
WA5 4NS G4OMZ
WA5 4NW M6GFZ
WA5 4PW M6JIQ
WA5 7XT 2E0WRK
WA5 8QL G0WRS
WA5 8QL G4VSS
WA5 8QL M3CFI
WA5 8QL M3FIW
WA5 8QL M3FIY
WA5 8WU 2E0DKA
WA5 8WU M6EWN
WA5 9PL M3ZGU
WA5 9QE M3YRJ
WA5 9QP M7EME
WA5 9SB M5AGI
WA5 9UU G4GPJ
WA53TL M0LEE
WA6 0FN G4ZHT
WA6 0NW G6IRW
WA6 6DN G3GKS
WA6 6PT 2E0JDI
WA6 6PT M0JLR
WA6 6PT M6XJR
WA6 6PY M0DYB
WA6 6QQ G3VBA
WA6 6SG G8WQE
WA6 6SG M3MHV
WA6 7JR M3LLB
WA6 7JR M6FMU
WA6 7PG 2E0LYR
WA6 7PG M3LYR
WA6 7QJ G8DVF
WA6 7RU G4OTI
WA6 8AE G4VYJ
WA6 8DA M1KTY
WA6 8DA M1WRX
WA6 8HP 2E1FKD
WA6 8HP 2E1HEE
WA6 8HP G1IVV
WA6 8HP G7NCP
WA6 8HP M3IVV
WA6 8JJ G6ZHS
WA6 8JS G8MMN
WA6 9LB 2E0OPC
WA6 9LB 2E0XMP
WA6 9LB M6OPC
WA6 9LL G3INP
WA6 9PG 2E0FDS
wa6 9pq G4UGD
WA69LB M6YTC
WA7 1DH G8YAZ
WA7 1NR M7KJP
WA7 1QE M7DWW
WA7 1UH G1LCC
WA7 1XE G4YZP
WA7 2AP G7EOK
WA7 2FP M3DAE
WA7 2FR G0OJG
WA7 2GR G7ICD
WA7 2LG G0NWE
WA7 2LR M3GTB
WA7 2NS G6FPX
WA7 2QS G1PIX
WA7 2QS M0PIX
WA7 2UH G0BSD
WA7 2UH G4DIT
WA7 3AQ G0VQL
WA7 3AQ G8HLQ
WA7 3BH M3JPM
WA7 3BU M1ICL
WA7 3DY G7JJJ
WA7 3EG 2E0ZMI
WA7 3EG M6ZMI
WA7 3ER G1SJT
WA7 3JA G1DYN
WA7 3JA M6LFC
WA7 3JH G8IEL
WA7 3JH M3IEL
WA7 3LE G1SBW
WA7 4AD G0MKD
WA7 4BG G8SIM
WA7 4DL 2E0FPY
WA7 4DL M0YCY
WA7 4DL M7DAC
WA7 4EH G4WPG
WA7 4RN 2E1HDZ
WA7 4RN M3TLM
WA7 4RP M3KIN
WA7 4TW G7SXG
WA7 4TX G7VCP
WA7 4XL G4HOD
WA7 5AR M0GIB
WA7 5EJ M3NCD
WA7 5EL 2E1BZH
WA7 5JA 2E1IIA
WA7 5JA M3GZE
WA7 5JB G0MQH
WA7 5JB M3GZD
WA7 5JJ M3FIP
WA7 5LR G4OAB
WA7 5RF G0HEF
WA7 5RJ G0NSL
WA7 5RW G1MSY
WA7 5ST M3NMB
WA7 5XE M3MOV
WA7 5XW M3ELP
WA7 5XW M3GZJ
WA7 5YN M3XRY
WA7 5YU G0RLF
WA7 5YU G1SES
WA7 6AA G8VNF
WA7 6BJ M3CVL
WA7 6BN G0RBM
WA7 6DL G6UMH
WA7 6DQ G0MSF
WA7 6DT G1JMP
WA7 6DW G0MIX
WA7 6DW G7IAE
WA7 6DZ M0CNK
WA7 6EH G0WHQ
WA7 6HT M7UKN
WA7 6JR M0KOI
WA7 6LB M6AML
WA7 6LB M6HZL
WA7 6RJ G1ZUZ
WA76HT 2E0LDG
WA76HT M0UKN
WA76HT M6LQX
WA8 0BA M7ABP
WA8 0BX 2E1DTK
WA8 0QX 2E0YDM
WA8 0QX M0YZF
WA8 0QX M6CAP
WA8 3JE 2E0HCL
WA8 3JE M0LSG
WA8 3JE M6WID
WA8 4GA M6LHM
WA8 4NT G1LFM
WA8 4TB G8BJQ
WA8 5AS 2E0AIL
WA8 5BH 2E0HCY
WA8 5BH M7SON
WA8 5BT G6URF
WA8 5QG G6CYF
WA8 6ET G4YJS
WA8 6HR G1AAK
WA8 6HR G7UXU
WA8 6TL G1URQ
WA8 7BN G0ICD
WA8 7HL 2E0KQR
WA8 7HL M6KQR
WA8 7LD 2E1EBL
WA8 7LD M3DIG
WA8 7LR M6JCP
WA8 7NB M6JDD
WA8 7NQ G1HGC
WA8 7PL M0CMT
WA8 7PL M6JQU
WA8 7UW G0SVZ
WA8 8BG G0FQP
WA8 8BZ M7WID
WA8 8QE M3WID
WA8 8SQ M3VUO
WA8 8SS G8RIB
WA8 9DP G0HKZ
WA8 9QB G8CMP
WA86RH M7FAF
WA88QE M7AFA
WA9 1BL M3IQP
WA9 1EN 2E0MRA
WA9 1EN M6GMF
WA9 1EN M6OKC
WA9 1QB 2E0ZAJ
WA9 1QB M3OCJ
WA9 1QY M3MLK
WA9 1SQ M3KLY
WA9 2AF M6EZZ
WA9 2AQ 2E0KJT
WA9 2AQ M3ZWQ
WA9 2AR G7KZY
WA9 2BD 2E0VWX
WA9 2BD M3VNQ
WA9 2BU M3LWP
WA9 2DP M3SFJ
WA9 2JD M6XIT
WA9 2PL M6DRO
WA9 2QG 2E0JCY
WA9 2QG M0JVU
WA9 2QG M3JCY
WA9 3AF M3OCA
WA9 3NF G7OXP
WA9 3NF G8HLH
WA9 3QX M6EKV
WA9 3RE 2E0KZJ
WA9 3RE M3KZJ
WA9 3SD M3ZYW
WA9 3SH M3MZT
WA9 3SX M3LXF
WA9 3WA 2E0XGX
WA9 3WA M6XGX
WA9 3WT M6DWG
WA9 3XQ 2E0NCB
WA9 3XQ M0GGK
WA9 3XQ M3NCB
WA9 4AN M3EJX
WA9 4AP M0JGM
WA9 4BD M6UNA
WA9 4BJ 2E0EYY
WA9 4BJ M6HVR
WA9 4BS G1OMY
WA9 4DL M3RCS
WA9 4DN M3KZP
WA9 4DQ G7RMD
WA9 4HW G7IFO
WA9 4NA G3REV
WA9 4PW G8OUI
WA9 4RY G1ZNT
WA9 4UX 2E0EQP
WA9 4XH G4ROU
WA9 4ZW M3WEF
WA9 5AR G0STH
WA9 5AR G4WGB
WA9 5FQ G1EVV
WA9 5HL 2E0EDP
WA9 5HL M6ESZ
WA9 5LR M3WHL
WA9 5LZ G2ARV
WA9 5LZ G4KZQ
WA9 5NA 2E0PKU
WA9 5NQ M6IIJ
WA9 5RU M7PLP
WA9 5TD M0CHR
WA92EZ M0IFC
wa93ng M0JFK
WA93XU G4VWD
WA94XN M0TFS
WA95SN M7BEV
WC1N 1AS M6XYY
WC1N 3XX G7HMK
WC1X 0BG G1UCR
WD17 2LN G6RXY
WD17 2QG M6YLW
WD17 2RQ G6JGF
WD17 3AQ G3LGR
WD17 3BA M7VEG
WD17 3BB G4BYS
WD17 3DD G4TJA
WD17 3DP G8LDY
WD17 3DX M0ZRG
WD17 3DX M6ZRG
WD17 4FJ 2E0SBW
WD17 4FJ M6SBB
WD17 4QT G3YKB
WD17 4SU G3SHY
WD17 4SW G0HCC
WD17 4SW G4KUJ
WD17 4SZ G3WSB
WD17 4TA M3IFJ
WD18 0AY G7VEF
WD18 0FA M3WVX
WD18 0FA M3WVY
WD18 0HA M6NAM
WD18 0HQ G8IIZ
WD18 0LJ M7BPN
WD18 6NX G0OYC
WD18 7JD G0HTM
WD18 7JD G0PHX
WD19 4AS M5DIK
WD19 4DA 2E0DTQ
WD19 4DA 2E0ZLA
WD19 4DA M6TPI
WD19 4DA M6ZLR
WD19 4LL 2E0WNS
WD19 4LL M0PLE
WD19 4LL M6JNO
WD19 4LW G8BXH
WD19 4PA G3MED
WD19 5AA G3NDK
WD19 5DF M6ICB
WD19 5EH G0DDV
WD19 5ET G6IUD
WD19 6JH G0OJT
WD19 6LF M1BCR
WD19 6NE 2E0HTX
WD19 6TR G8SUG
WD19 6YL M6ROM
WD19 7AY M6SQU
WD2 5HY 2E1HCB
WD23 1EX M0SSH
WD23 1FB G4SWY
WD23 1PE G3VNB
WD23 1PZ G3OMR
WD23 2BT M7HGO
WD23 2BU M0CIO
WD23 2HA G4IID
WD23 2HF M6PQF
WD23 3AR G0LZV
WD23 3BQ 2E0MUT
WD23 3BQ M0HSZ
WD23 3BQ M6ZBA
WD23 3SS G4GYS
WD23 4GH G4RNW
WD23 4GL M6MLU
WD23 4QT G8XXV
wd24 4la M7AOL
WD24 4RL M0BLD
WD24 5EW G6IKO
WD24 5LT G4WMN
WD24 6DA M0CJG
WD24 6RU 2E0WTA
WD24 6RW M6VSS
WD24 6SY 2E0ECQ
WD24 6SY M6GHJ
WD24 7EE 2E0ERP
WD24 7EE M0TNC
WD24 7EE M6SQL
WD24 7LN 2E0DQH
WD24 7LN M0HYL
WD24 7LN M6FUY
WD246BS M6NHV
WD246RW M7EOA
WD25 0DB M6ILY
WD25 0EH G0MDM
WD25 0EN M7HXO
WD25 0HG G6FAF
WD25 0HX M6IZA
WD25 7DN G3WDX
WD25 7EW G6OBU
WD25 7PA M7PHA
WD25 9AR 2E0PBR
WD25 9AR M0PAJ
WD25 9AR M6PAO
WD25 9DZ G4CNH
WD25 9PS G6MHF
WD25 9PS M3DSE
WD25 9PS M3LHM
WD25 9QH G4RMC
WD25 9QQ M0TTB
WD25 9QQ M6BBC
WD25 9RB 2E0GMD
WD25 9RB M6CHS
WD3 1EQ M7TMM
WD3 1HL G0WBL
WD3 1LQ G6ZFV
WD3 1NA M1CIG
WD3 1PQ M6WAQ
WD3 3AL G6LZB
WD3 3AU G1XEP
WD3 3AU G6AJG
WD3 3DW 2E0LWR
WD3 3EG G7HHT
WD3 3FE G3TVH
WD3 3JQ G3EBV
WD3 3JT M3JLE
WD3 3NH G1MPC
WD3 3PH G8MCJ
WD3 3PS M7WGM
WD3 4EB G4JHU
WD3 4EB M6YEQ
WD3 5BD G3VOS
WD3 5BJ G4COV
WD3 5HQ 2E0BST
WD3 5HQ M0SKC
WD3 5HQ M3OAJ
WD3 5JD G4ORE
WD3 5JJ G0JJU
WD3 5JJ G1DNY
WD3 5JT M0LAL
WD3 5JT M6CFT
WD3 5PX G6YJR
WD3 5QH G3WJG
WD3 5QQ G4TAO
WD3 5RG G4AEM
WD3 6ER M6NQV
WD3 7AR M0GFF
WD3 7DY G1ISX
WD3 7EN G3ZER
WD3 7ES 2E0ERE
WD3 7ES M0ITI
WD3 7ES M6HBN
WD3 7NR G8VTV
WD3 8GW G6IBD
WD3 8GW M0DAB
WD3 8HE M7FMT
WD3 8HT G8OQT
WD3 8PX M7DEF
WD3 8QA G7HGQ
WD3 8QG G1XET
WD3 9TZ G6LFA
WD3 9YZ M6TDJ
WD4 8DY G0SGV
WD4 8JE G7GTG
WD4 8JE G7GTH
WD4 8NJ G7TMH
WD4 9HF G3XYJ
WD4 9NA M6UNI
Wd48pr M7SSS
WD5 0BJ G4FRZ
WD5 0DA G4KUF
WD5 0DH G3JCR
Wd5 0ee M3UXK
WD5 0EG G3XLI
WD5 0EU G3OYT
WD5 0PJ G0TVD
WD5 0QG G4DPP
WD5 0ST G7PDH
WD5 0XD M6VWV
WD6 1AY G6MWL
WD6 1EP G7MKJ
WD6 1NT 2E0SHY
WD6 1NT M6ACA
WD6 2DA G4TEP
WD6 2HG G6APH
WD6 2HG M6MID
WD6 2JW M7ISB
WD6 2LQ G7AIH
WD6 2PB G7DFW
WD6 2RB G8XJN
WD6 2RP G6DAU
WD6 3DH M6DZA
WD6 3JH G6JCQ
WD6 3LH G4GPL
WD6 3NJ G4TIH
WD6 3PT G1INI
WD6 3PU 2E0GSK
WD6 3PU M0GSK
WD6 3PU M6GSK
WD6 4EG 2E0SSK
WD6 4EG M3UON
WD6 4EX M0IHR
WD6 4HY 2E0WEF
WD6 4JD G3JPJ
WD6 5AD M3OYQ
WD6 5BJ M1BCB
WD6 5DG G0PQB
WD6 5LR G1XVL
WD6 5PE G6RHN
WD6 5QE G8RCO
WD7 7DU G4WIS
WD7 7NF 2E0CQO
WD7 7NF M0MBR
WD7 7NF M6BTG
WD7 8BA M3XQQ
WD7 8BA M3YUK
WD7 8HJ G6JMB
WD7 8JD G6OKA
WD7 8JF 2E0PNW
WD7 8JF M6CPW
WD7 9JW G7OIR
WD7 9JW M0ADY
WF1 2AL G7VDT
WF1 2LF G4VQR
WF1 2LS G4OVL
WF1 2LS M0SHZ
WF1 2LS M6MEB
WF1 2LW G7IOF
WF1 2RA G3OMJ
WF1 2UP M6NIV
WF1 3AB M1MCL
WF1 3DN G4LXV
WF1 3DY M6FYH
WF1 3NW G4VRJ
WF1 3QD M6ZTA
WF1 3RL M6MGW
WF1 3RU M7FBM
WF1 3RU M7KIM
WF1 3SZ 2E0LAI
WF1 3TL G8LGE
WF1 3TX M6BRH
WF1 4EU G1MRC
WF1 4EU M7ARC
WF1 4EZ 2E0OUH
WF1 4EZ M3OUH
WF1 4HP M7CHH
WF1 4JL M6MXO
WF1 4PW M3EVB
WF1 5AH M3XNX
WF1 5LB M3OUG
WF1 5LG M3XWR
WF1 5NT G6NYF
WF10 1LN M6ASC
WF10 1LN M6MRS
WF10 1PL M3NRX
WF10 1PZ M6MEV
WF10 2AA G4FEQ
WF10 2AP M6GOL
WF10 2BS M3TVV
WF10 2DN G0GYA
WF10 2QA M6BLM
WF10 2QF M6PZZ
WF10 2RB G4TCG
WF10 2RG M3OLD
WF10 2RN M6KTI
WF10 2SB G8OXD
WF10 3AR M7SNS
WF10 3EY 2E0UAM
WF10 3EY M3UAM
WF10 3HN G3HNC
WF10 3HT G4RQI
WF10 3PB G4IBN
WF10 3QS M6RIN
WF10 3QZ 2E1GLS
WF10 3QZ M3VRA
WF10 3RN 2E0XII
WF10 4JT M3YNC
WF10 4LF G3OAR
WF10 4LN M6JEQ
WF10 4PH G0PMB
WF10 4QL G7MVE
WF10 4SE G1YNH
WF10 4SE M3YNH
WF10 5AF M6FXY
WF10 5TU M7ABE
WF10 5UJ M6YOB
WF103QN 2E0CCF
WF103QN M3TZQ
WF105HB G7MTG
WF11 0DQ G7IPR
WF11 0HG G0VXD
WF11 0JH G0NQE
WF11 0JQ G6KIH
WF11 0JY G4VMA
WF11 8AJ G8VAT
WF11 8JL G0TVL
WF11 8LH G4SOL
WF11 8RH 2E0JBF
WF11 8RH M6BWE
WF11 8SR G0NER
WF11 8TD G0VXM
WF11 9AT G4AAQ
WF11 9ED G4PHL
WF11 9HR G4TLM
WF11 9NN G8NDF
WF11 9PA G1PZA
WF12 0BD M1BAV
WF12 0BP 2E0SLT
WF12 0BW G4CLI
WF12 0DW M3CNV
WF12 0HB 2E1FMC
WF12 0HT G8VOQ
WF12 0HY G0GSU
WF12 0JZ G7LBM
WF12 0NF M6GCB
WF12 0NL G0CBW
WF12 0PJ M6ZMR
WF12 0QH M6UZH
WF12 0QX G4CLJ
WF12 0RH M6BCB
WF12 7AS 2E0EOP
WF12 7AS M0KKH
WF12 7AS M3PAX
WF12 7AW 2E1FFL
WF12 7DE M6ZXC
WF12 7LA G7NPL
WF12 7LG 2E0DQD
WF12 7LG G3EAR
WF12 7LG M0XZT
WF12 7LG M6FHP
WF12 7LS M1AJG
WF12 7PD G3CPG
WF12 7PU M0EDO
WF12 7RU 2E0AGB
WF12 7RU M7AGB
WF12 7SN M6AIA
WF12 7SQ G8POK
WF12 8AJ G6FTJ
WF12 8AQ G6ZJV
WF12 8BN M7DJY
WF12 8BY M0WEC
WF12 8HH G1DEX
WF12 8HH M6PPE
WF12 8PL M1BHE
WF12 8PY M3LZL
WF12 8PZ 2E0GEG
WF12 8PZ M3VBP
WF12 9AY M3YHJ
WF12 9DN G4GHQ
WF12 9LD M7IKO
WF12 9PU M3YGS
wf120bw M7CUD
wf120he 2E0CNQ
wf120he M0HRM
wf120he M6CSG
wf128pl M7DYZ
WF13 2QF M3HTR
WF13 3AX M7EEL
WF13 3LZ G4YAJ
WF13 3SR G6ZUO
WF13 4DQ M0LEV
WF13 4EN G6WGE
WF13 4HT G1VIF
wf136qj M7DEI
WF14 0AJ G6ITV
WF14 0AQ G6XTZ
WF14 0EW G1EDU
WF14 0LF 2E0BCS
WF14 0LF M3JUF
WF14 0NR G3HPD
WF14 8JP 2E0JTH
WF14 8JP M3ZTX
WF14 8JW 2E0RFE
WF14 8LG M5AIB
WF14 8NW M6MBQ
WF14 8PN M6AYC
WF14 8PX G6PHX
WF14 9AN M6NDC
WF14 9AW G2FCP
WF14 9BJ G8NZR
WF14 9DJ 2E0FRL
WF14 9DJ M7LSL
WF14 9ED M3YDL
WF14 9HZ 2E0LBB
WF14 9LJ M6BJY
WF14 9NL M6JRJ
WF14 9PB G4PHR
WF14 9PB G7CSV
WF14 9PW G1BGQ
WF14 9PY M6MIA
WF14 9SA 2E0CYL
wf149al M7BVA
WF15 7BW G1IKF
WF15 7DN G0GMJ
WF15 7DP G4GOX
WF15 7HW M3SGI
WF15 7LP G0PMP
WF15 7NJ G4EIL
WF15 8DG G8NYM
WF15 8JU G7MHD
WF15 8JW 2E0IPC
WF156QE G8DZW
WF16 0BE G3JQC
WF16 9JL G0TAL
WF17 0DX G4MLV
WF17 0JL G4FPE
WF17 0JZ 2E0SML
WF17 0JZ M6SAN
WF17 0LJ G4EOC
WF17 0RG G7LTO
WF17 5QY G7VHS
WF17 6DB M6GNH
WF17 6DZ G4XZM
WF17 6FH G4SEQ
WF17 6HG G1NEG
WF17 6HG M3NEG
WF17 7NT G1UVD
WF17 8PN 2E1ENR
WF17 9BX G0IAX
WF17 9DL G6XHK
WF17 9EU M7BNK
WF17 9JF M0JBW
WF17 9JF M3JKM
WF17 9JF M3JSM
WF17 9JF M3LMC
WF17 9JF M3SPR
WF17 9QF G6JLI
WF17 9RG 2E0OOC
WF17 9RG M3OOC
WF2 0BH M6EVL
WF2 0FQ M3MXA
WF2 0NZ M3VSZ
WF2 0PA 2E0NER
WF2 0PA M6FMC
WF2 0PD M6IGK
WF2 0PE M3PNV
WF2 0PP G4RCG
WF2 0PP G6LCP
WF2 0QR G6XXE
WF2 0SE G0CEW
WF2 0TJ G6LDP
WF2 0UT M6JOU
WF2 0UU M6FMP
WF2 6ED M0IPV
WF2 6HJ M0AKD
WF2 6JG M3LDI
WF2 6LQ M6BNC
WF2 6NP M6SSW
WF2 6RE G0DIS
WF2 6RN G4CPC
WF2 6RT M6FPV
WF2 6SE G4BLT
WF2 6SE G8DVS
WF2 6SR G7JTH
WF2 6SR M1DDR
WF2 7DA M6CXI
WF2 7DE G1WRS
WF2 7DE M3WLV
WF2 7DH M3XYK
WF2 7EF G8IJI
WF2 7EG M6SUP
WF2 7HW G1OOU
WF2 7JN M3TXF
WF2 7JT M0CUF
WF2 7LS M6SNW
WF2 7PR G7SWH
WF2 7QS G7PCW
WF2 8AS G1YYC
WF2 8BX M3EOZ
WF2 8BY M3OFV
WF2 8EB G0EVA
WF2 8EL G0MVR
WF2 8EL M3LYC
WF2 8ET M6DDL
WF2 8EX M3KZB
WF2 8HA M3XRW
WF2 8HE 2E0DND
WF2 8JZ 2E0APJ
WF2 8LD M3TCT
WF2 8LY M6DRR
WF2 8UP G7UQA
WF2 8YB M3YXD
WF2 9DR M3XRO
WF2 9JS M6FCE
WF2 9JZ G6AJS
WF2 9QB M3HTA
WF2 9QG 2E0KTX
WF2 9QG 2E0PIA
WF20BJ 2E0YSB
WF20BJ M0PYB
WF20BJ M6YSB
Wf28hr M3UEP
WF3 1DL 2E1KYQ
WF3 1HT G0IKD
WF3 1HZ G2DP
WF3 1HZ G4DZU
WF3 1JB G0PXF
WF3 1JL M6MHE
WF3 1LU 2E0THP
WF3 1LU M0SVX
WF3 1LU M7TTS
WF3 1PF M0AOI
WF3 1QH M7EOD
WF3 1RE G8JHA
WF3 1TA M6HJD
WF3 1UD G6NPW
WF3 1UG 2E0LXY
WF3 1UG M3LOX
WF3 1UY 2E0YEP
WF3 1UY M0MMR
WF3 2AX G0VTI
WF3 2AX G1RHW
WF3 2DD G0CQD
WF3 2FG 2E0VBL
WF3 2FG M0DUB
WF3 2FG M7DUB
WF3 2JG G1FOM
WF3 3NS M6EZV
WF3 3PX 2E0JAA
WF3 3PX 2E1JAA
WF3 3PX M3BGT
WF3 3PX M3MAA
WF3 3PX M3SCA
WF3 3TG M6GDV
WF3 4GD G7WRS
WF3 4JJ G0EVT
WF32BL M1TRC
wf32hs 2E0KBL
WF32HS M6SKW
WF33FN M7KWG
WF4 1BA M7PVX
WF4 1EF G8OJK
WF4 1EH G1XBE
WF4 1HS G7HYZ
WF4 1HX G8JQH
WF4 1NU 2E0KYI
WF4 1NU M0KYI
WF4 1NU M3JZA
WF4 1PF G4AYL
WF4 1RN G0TYS
WF4 1RN M3UEW
WF4 1RN M3UXI
WF4 1RN M3VEW
WF4 1SL 2E0VKN
WF4 1SL M0IAA
WF4 1SL M3VKN
WF4 1SP M6EHR
WF4 2AT M0XYN
WF4 2BU M6HRO
WF4 2DU M0SZQ
WF4 2DU M0WYC
WF4 2DU M6SUJ
WF4 2EJ G4PIR
WF4 2EJ M3WXX
WF4 2NF G0UOS
WF4 2NT M0JCV
WF4 3BH G1HVL
WF4 3BL G0HDP
WF4 3BY M1FBN
WF4 3EF 2E0OPB
WF4 3EF M3OPB
WF4 3EG M1CBU
WF4 3JA G8GOT
WF4 3JZ G0EVR
WF4 3JZ G0FEV
WF4 3NW G0SUB
WF4 3NW G1GRT
WF4 3PT G6EZG
WF4 3QD M3TLD
WF4 4AX M6JBB
WF4 4ED G0BQB
WF4 4HN G4JSS
WF4 4LR 2E0DWJ
WF4 4LR M6DWJ
WF4 4QF M6EXG
WF4 4RU G7PNM
WF4 4TE G0COA
WF4 5AN G4IAU
WF4 5DX M7SPE
WF4 5HH M6FFV
WF4 5HH M6JFR
WF4 5QG M7WFD
WF4 6AE G1AHS
WF4 6BX M6RIJ
WF4 6EQ G4DWO
WF43PG G4IZH
WF5 0PP G1WXK
WF5 0QA 2E0RMW
WF5 0QA G4BOB
WF5 0QA M7RMW
WF5 0RU G0MJZ
WF5 0TJ G1HYA
WF5 8EN G4TDI
WF5 8JW M0LUD
WF5 8LF G3SEY
WF5 8LH G1ISP
WF5 8LJ M3OUF
WF5 8NA M0LAB
WF5 8QP G4MVE
WF5 8RQ M6LPN
WF5 8RY G1VAO
WF5 8RY G3VQQ
WF5 9EW G7NTI
WF5 9JL G4LJK
WF5 9PR G4TFH
WF5 9RE M3SVO
WF5 9RQ G4IOD
WF5 9RQ G4KFP
WF5 9SJ G6AAK
WF6 1AW M3LYX
WF6 1BL 2E0MD
WF6 1BL M6HKI
WF6 1BN M6MMB
WF6 1DA G7MJP
WF6 1ER M3HBX
WF6 1LB 2E0WRF
WF6 1NY M3OOA
WF6 1SN G4SBG
WF6 1SS G4SAV
WF6 1WW G6PDA
WF6 2AA G6MID
WF6 2AA G6XEX
WF6 2LB 2E0GBF
WF6 2LB M6AHZ
WF6 2LX G7DJY
WF6 2PL M6WBO
WF6 2RZ G7PZL
WF6 2TS M7TFN
WF6 2TT G1IIX
WF7 5AB G1PHK
WF7 5AG G1ZPO
WF7 5DP M1NTV
WF7 5EA M6MAD
WF7 5EA M6STV
WF7 5EB G4OOC
WF7 5LH G4NRF
WF7 5LW G7IJY
WF7 6AA M6EOW
WF7 6AH G1ZGH
WF7 6AY M6EYK
WF7 6BL G0HPA
WF7 6HT M3IFM
WF7 6LQ G1SJZ
WF7 7HQ G8TKY
WF7 7HU M0AGU
WF7 7JG 2E0HGT
WF7 7JG M6TCY
WF7 7LA G0KMN
WF7 7LA G1LFI
WF7 7LQ G0RAE
WF7 7NA G0PAZ
WF75JQ M7NAV
WF8 1LA G6DUH
WF8 1SB G4ISU
WF8 1TD M3UDD
WF8 1TD M6TDV
WF8 2AB M6PTF
WF8 2AJ G0RLN
WF8 2AJ G1FYQ
WF8 2BY M3FOK
WF8 2BY M3SUI
WF8 2DJ G7HQC
WF8 2EJ 2E0AOZ
WF8 2HW M6KEP
WF8 2JW M3WUW
WF8 2LX M3KBL
WF8 2PN M3UDK
WF8 2QQ G4KMW
WF8 2UP G0FVN
WF8 3ES 2E0LUL
WF8 3ES G5VZ
WF8 3ES M0JRQ
WF8 3ES M3JRQ
WF8 3NJ G6WBT
WF8 3NT G0TVU
WF8 3QD G4ICB
WF8 3QD M6JNL
WF8 3QT G8TEB
WF8 4BU M3WYA
WF8 4BX G6VZS
WF8 4LG G0SNS
WF8 4ND G4YNU
WF8 4RP G0TPN
WF8 4SJ G3VTD
WF9 1AZ 2E0BSI
WF9 1AZ G0BPK
WF9 1AZ G3FYQ
WF9 1AZ G3SYC
WF9 1AZ M6CHF
WF9 1ED G8YCQ
WF9 1EU M6XLP
WF9 1LG M3AOQ
WF9 2AR 2E0VAT
WF9 2AR M0ZIM
WF9 2AR M6DEV
WF9 2BT G8HRA
WF9 2BT M0CIR
WF9 2BZ 2E0DYQ
WF9 2BZ M6SYK
WF9 2DD G0CYX
WF9 2qx M6FBQ
WF9 2TL 2E1FSX
WF9 2TL G1DKV
WF9 2TL G6OXZ
WF9 3EP 2E0MJP
WF9 3JF M6CXM
WF9 3JT 2E1CIY
WF9 3NB G0PPQ
WF9 3NB G3NJB
WF9 3NB M1CRA
WF9 3NB M7PSJ
WF9 3NJ G4OSY
WF9 3QE G4ZVB
WF9 3RY G4NDP
WF9 3SJ M3XYO
WF9 4AP 2E1XDJ
WF9 4AY G0GEH
WF9 4EA M0RTV
WF9 4ES 2E0MAH
WF9 4JA G1PPX
WF9 4NW M3EXJ
WF9 4RL G1CDN
WF9 4TA 2E0ICT
WF9 4TA M0ICT
WF9 4TQ M3KAX
WF9 5AJ G4PLL
WF9 5BX G8GXS
WF9 5DP M3IUK
WF9 5DS G1CDO
WF9 5HE G1FFH
WF9 5HE M0JQK
WF9 5HE M3MLF
WF9 5HY G0JLL
WF9 5JL M0JUK
WN1 2AT 2E0WIG
WN1 2BA G1SUH
WN1 2DL G1INU
WN1 2RR G3YNU
WN1 2SS G4WGF
WN1 3PP 2E0CHY
WN1 3PP M6AMV
WN1 3UE 2E0JKR
WN1 3UQ G7ADF
WN1 3XT 2E0YYZ
WN1 3XT M3VIW
WN12HX 2E0DMP
WN2 1AP M0JEQ
WN2 1DA M7BNB
WN2 1HW G8OMC
WN2 1JE 2E0HXY
WN2 1JE 2E1LEN
WN2 1JE M7INC
WN2 1LT G0IYX
WN2 1LT G4KKN
WN2 1NA G1SWH
WN2 1NA M0AWX
WN2 1RL M1AWS
WN2 1SZ M0UNN
WN2 1ZH M0OTE
WN2 2BZ 2E0LTF
WN2 2JG M7DMH
WN2 2NA G0FYE
WN2 2NN M7VXW
WN2 2PS M6BZF
WN2 3BN G6ZIY
WN2 3DP 2E0VXI
WN2 3DP M6VXI
WN2 3EE 2E1WNA
WN2 3EQ M6HTC
WN2 3HP G0AYF
WN2 3HR G1YMP
WN2 3SB G1TAR
WN2 3XD G4HPH
WN2 4AF M3KUO
WN2 4EH 2E1BYY
WN2 4EH G6DHI
WN2 4EJ G0AVH
WN2 4ET G1BDU
WN2 4ET G4WIL
WN2 4HL 2E0MLS
WN2 4HL G1ECI
WN2 4HL M0ICK
WN2 4HL M3ITZ
WN2 4JA G4ZVK
WN2 4JG M6PFQ
WN2 4LD G4EHY
WN2 4LZ G1EIO
WN2 4NB M0IYV
WN2 4PL G8ALD
WN2 4QF G4PGQ
WN2 4QS G0CBD
WN2 4RG G7JZI
WN2 4TY G7LVS
WN2 5DW G1JMS
WN2 5HY G0IZG
WN2 5JN G7GAG
WN2 5PP 2E0XEM
WN2 5PP M0XEM
WN2 5PP M7BEM
WN2 5QD M3PZL
WN2 5QD M3PZN
WN2 5QR M3KZR
WN2 5TF G7FMB
WN2 5XA 2E1EHB
WN2 5XA M0GRE
WN2 5XA M1AIX
WN2 5XG G6ZJM
WN2 5XR 2E0CRX
WN2 5XR M3LBR
WN2 5XR M3XFN
WN2 5XR M7TSA
WN2 5YE M7EKD
WN2 5YE M7TFA
WN2 5YL M6OBO
WN2 5YL M6XXB
WN21AU M7DPW
WN3 4JP 2E0YGR
WN3 4JP M6YGR
WN3 4NS G1CZN
WN3 4PQ M6SDU
WN3 4QA 2E0MNP
WN3 4QA M0XOC
WN3 4QA M6INT
WN3 5BT 2E0FCJ
WN3 5BT M6OQH
WN3 5NG M5DXX
WN3 5PL 2E0BTX
WN3 5PL M6GBR
WN3 5PR G8DPE
WN3 5QN G0DPI
WN3 5QN G0HRW
WN3 5QN G7EDK
WN3 5QN G7IXC
WN3 5UT M3XDQ
WN3 6AA G3BPK
WN3 6AA G4GWG
WN3 6AD G4OVV
WN3 6AJ G0KHH
WN3 6JY G6LST
WN3 6LD G4WDC
WN3 6LD G6TYB
WN3 6TQ G4HSC
WN4 0AZ G4WXX
WN4 0DW G7SXB
WN4 0EF M3WJN
WN4 0EQ 2E0ZRX
WN4 0EQ M3ZRX
WN4 0QF G1LWX
WN4 0QX G8XLI
WN4 0RH G3ZUZ
WN4 0SJ G8CXW
WN4 0UA M6IKL
WN4 8AY G7IHE
WN4 8DQ 2E0EKP
WN4 8DQ M6HTB
WN4 8QT G0SIW
WN4 8RX M1BZJ
WN4 8ST G0UNG
WN4 8SU G7LTR
WN4 8SU M3BRT
WN4 8SU M3LTR
WN4 8XA 2E0JCT
WN4 8XA M0ITT
WN4 8XA M6JGI
WN4 9DY G6OCB
WN4 9JN G4XCQ
WN4 9LD G3HUX
WN4 9PJ G6NTM
WN4 9SA M6LIQ
wn4 9se M0HTR
wn4 9se M0RGN
WN4 9SE M6CLB
WN4 9UP 2E0WES
WN4 9UZ 2E0RAG
WN4 9XA M7BNS
WN4 9XD 2E0BQO
WN4 9XD M3IQF
WN40RW M7GTD
WN5 0AJ G1DON
WN5 0PU G0DSR
Wn5 0qd M7DFK
WN5 0RA M7DGE
WN5 7AA 2E0DPT
WN5 7AF 2E0ARV
WN5 7AH 2E0NRB
WN5 7AH M6NRB
WN5 7AX G0NEO
WN5 7BG G4HAP
WN5 7EB G3PJW
WN5 7HS 2E0ISQ
WN5 7HS M0ISQ
WN5 7HS M3ISQ
WN5 7HT G4EII

WN5 7JA M1DYU
WN5 7LH 2E0SYY
WN5 7LH M0SYY
WN5 7LH M6OFF
WN5 7PX G4LON
WN5 7TF G4VFR
WN5 8LX G7SOV
WN5 8NG 2E0ROI
WN5 8NG 2E1LED
WN5 8NG 2E1WRC
WN5 8NG M3ROI
WN5 8PL 2E0WEZ
WN5 8PL M3RMQ
WN5 8RH G7CWE
WN5 9BH G6ORO
WN5 9DL 2E1DFE
WN5 9EF M3JZF
WN5 9HZ G0TRY
WN5 9JP G4YPS
WN5 9JQ M6ATX
WN5 9NQ 2E1HZY
WN5 9NQ 2E1TCP
WN5 9PA G4KTU
WN5 9PJ M6AIF
WN5 9PR M6MTL
WN5 9QD M7CWN
WN5 9QL M0VGH
WN5 9QL M3RFX
Wn5 9rs 2E0XAS
Wn5 9rs M0KDO
Wn5 9rs M3XAS
WN5 9TG G0UVL
WN6 0AQ 2E0ICK
WN6 0AQ 2E1HDE
WN6 0AQ M0HDE
WN6 0AQ M3POP
WN6 0AZ 2E0VCE
WN6 0BE M3NJK
WN6 0EX G0MOK
WN6 0LW 2E0BJT
WN6 0LW G4PPG
WN6 0LW M3LBT
WN6 0NB M7FAM
WN6 0NP 2E0TAJ
WN6 0NR G8EHF
WN6 0PW G4JTE
WN6 0QP M3TYQ
WN6 0QR G1MSR
WN6 0UA G6XKY
WN6 0UE M7EJM
WN6 0WB M6DXC
WN6 0XR 2E0NRS
WN6 7NH 2E0AWR
WN6 7PA G1VYA
WN6 7PA M3RRJ
WN6 7PU M0SCU
WN6 7RF G4EOT
WN6 8AU G0PBE
WN6 8DT G0GBP
WN6 8EH 2E0WPE
WN6 8EH M3HXO
WN6 8EY G7RGV
WN6 8HR G4KCP
WN6 8LH G0PPK
WN6 8LJ G8FHL
WN6 8LR M7GHW
WN6 8NJ G1EPF
WN6 8NJ G4PPB
WN6 8NS 2E0SAF
WN6 8NS M0OEG
WN6 8NS M6JSM
WN6 8PJ 2E0EQH
WN6 8PJ M6PFZ
WN6 8PL 2E0SYE
WN6 8QA G4SFJ
WN6 8QA G6PUV
WN6 9AG 2E0WWZ
WN6 9AG M3WWZ
WN6 9DJ G3RSW
WN6 9HU G8XXC
WN6 9JL M0HPT
WN6 9JN M3SJL
WN6 9JS G4EHK
WN68NS M3UCU
WN6-9LF G8JGU
WN7 1HN 2E0DMU
WN7 1HN G0SQX
WN7 1HN M0HZK
WN7 1HN M6FHY
WN7 1HU M6BSY
WN7 1HW M6TOE
WN7 1JZ G0MNY
WN7 1NB G3VGK
WN7 1NB G6HNQ
WN7 1SG G0KMB
WN7 1SG M7ELO
WN7 1TP 2E0TUR
WN7 1TP M3UAE
WN7 1TS G1HFS
WN7 2HH M0KDM
WN7 2HH M6MRD
WN7 2JJ G0SSK
WN7 2NG G6TET
WN7 2NG M3BMN
WN7 2NH 2E0GLW
WN7 2PL M6CYV
WN7 2RG G4UDF
WN7 2UU G4JXI
WN7 2UU M6LEY
WN7 2YR M1TET
WN7 3BU G3WIS
WN7 3DS M3HCA
WN7 3DW M3UVD
WN7 3EB G4COE
WN7 3ET 2E0YKC
WN7 3ET M0YKC

WN7 3ET M6TFV
WN7 3EX 2E0MMX
WN7 3EX M6MMI
WN7 3LB M7LIE
WN7 3LD M6GOS
WN7 3NE G8PUN
WN7 3PQ 2E1CJF
WN7 3PQ G6VPU
WN7 3PQ M0HOY
WN7 3PQ M0XHN
WN7 4GB M7SPW
WN7 4HT G7NRS
WN7 4HY 2E0DRZ
WN7 4HY M6DRZ
WN7 4HY M6DWT
WN7 4HY M6JPS
WN7 4SX G0LGC
WN7 4TA G1FLV
WN7 5BT G1VKT
WN7 5DG G6LDY
WN7 5DG M3FNT
WN7 5HG M1SJH
WN7 5NA M0LZQ
WN7 5NU M0HGO
WN7 5PN G4VXW
WN7 5PN G6HPL
WN7 5PN M0NAR
WN7 5PW G1DUS
WN7 5PZ 2E0DPY
WN7 5PZ M6PYE
WN7 5SW M6JGM
WN73HB M7RKL
WN75ET M7AAH
WN8 0AA G0MRM
WN8 0AA G6TKY
WN8 0AE 2E0WPB
WN8 0AE G4JCA
WN8 0AE M6JUJ
WN8 0AY G0FNA
WN8 0DE G4FEU
WN8 0JG G0UMY
WN8 0QT G3VWA
WN8 6AA 2E0FMP
WN8 6AA M6HTU
WN8 6AT G4WWG
WN8 6AX G8UXL
WN8 6AX G8XRR
WN8 6ED G6IKU
WN8 6JX M7LEM
WN8 6PA G6OMN
WN8 6PA M7BYF
WN8 6PE G4PJS
WN8 6RD M3NSG
WN8 6RJ G4JTP
WN8 6TA G3VSR
WN8 6TA G6HXL
WN8 6TE M0IYG
WN8 7AR G3PNQ
WN8 7DA M0SBH
WN8 7LA 2E0JFY
WN8 7LA M6JFY
WN8 7NB 2E0ESS
WN8 7NB M0TAA
WN8 7NB M3XQW
WN8 7NS G0ELM
WN8 7NT G6ORS
WN8 7NT M3CDY
WN8 7RA G1DMW
WN8 8BD M3UHG
WN8 8EG G6YRB
WN8 8EN G6WEW
WN8 8HN M6PLB
WN8 8NP M3DLT
WN8 8NS M7MBT
WN8 8NW M3FRD
WN8 8NW M6HHO
WN8 8QU 2E0RNB
WN8 8QU M6KPZ
WN8 8QZ M6DDH
WN8 9BP G0LAK
WN8 9BP G4SME
WN8 9BQ G8XEI
WN8 9DW M7OUT
WN8 9EF M3MJN
WN8 9JZ G1UOR
WN8 9NB G6UGE
WN89LZ M6PQV
wn9 8jr M6FAE
WR1 1NR G3RMF
WR1 1PA 2E1MFC
WR1 1PA M3CER
WR1 1QE G3VXH
WR1 1QZ 2E0UBU
WR1 1QZ M6WRW
WR1 2PJ M0TRX
WR1 3NY G8XMO
WR10 1HW G0FFF
WR10 1JH G1IPE
WR10 1JY G7TGK
WR10 1LL 2E0ENF
WR10 1LL M6GQS
WR10 1LW G1AFJ
WR10 1LW G4TXF
WR10 1NQ G7MYM
WR10 1PW G1TML
WR10 1PW G3UEY
WR10 1RE G7CRU
WR10 2AX G4OZQ
WR10 2BH M7RFM
WR10 2JE G4YIG
WR10 2LA G4RMV
WR10 2LB 2E0HUH
WR10 2LB G5KCP
WR10 2LB M7KCP
WR10 2NY G4NIJ
WR10 2PL G1UQK

WR10 2QF M7JNY
WR10 2QJ G7RRJ
WR10 3BB 2E1EOD
WR10 3BW G4FAT
WR10 3EF G8LQP
WR10 3EL G8BGT
WR10 3EL M5AGY
WR10 3EZ G0NXE
WR10 3HJ G4TDR
WR10 3HL M0CUU
WR10 3HS M6ZNF
WR10 3JP M0DLP
WR10 3JZ G3ZKH
WR10 3NA G7WEB
Wr102dq M6OTN
WR11 1BP M0ONO
WR11 1BU G4YJB
WR11 1DE G6NBL
WR11 1EQ G4EKG
WR11 1YW G4NRD
WR11 2AH 2E0RKX
WR11 2AH M0RKX
WR11 2AH M6RKX
WR11 2NB G7JSC
WR11 2NE M0PDY
WR11 2NE M3PDY
WR11 2NR G0WEO
WR11 2QA G4RXQ
WR11 2QA G6XKX
WR11 2QF M6LUR
WR11 2QJ G1JLQ
WR11 2QZ G7JWW
WR11 2SF M7SHN
WR11 3BJ G3XCW
WR11 3BS M1PLC
WR11 3EA G0IBE
WR11 3HE G3KLZ
WR11 4NL G6YXY
WR11 4NR G6DRC
WR11 4QA M6YZO
WR11 4RJ M6WYR
WR11 4TY G3LPU
WR11 4UP G8DKD
WR11 6BE G8SQZ
WR11 7AA 2E0IDR
WR11 7AA M0IDR
WR11 7AA M6IDR
WR11 7EU G1VNV
WR11 7GQ G1SHU
WR11 7HQ G0GBL
WR11 7PT M3JKI
WR11 7QG G1UNQ
WR11 7RN 2E0FVR
WR11 7RN M3ZID
Wr11 7rp g G4UXC
Wr11 7rp g G4WET
WR11 7TA G1IME
WR11 7UE 2E0HDU
WR11 7UE M3CKD
WR11 7UZ G3ZUF
WR11 7XF G4MFK
WR11 7XX 2E0BQL
WR11 7XX G1JMD
WR11 7XX M3MWQ
WR11 8JP G6YMY
WR11 8LL G1ZAA
WR11 8NG M0AFV
WR11 8QW M0TZT
WR11 8SN G1XWZ
WR11 8TL M3JFP
WR11 8XN G0KIN
WR113EE M7TPT
WR12 7AX M3DVM
WR12 7EP G8HFL
WR12 7HB G3IXI
WR12 7HS G6HAA
WR12 7JR 2E1RSH
WR12 7JR M7RSH
WR12 7NQ G7OAS
WR12 7PF G3KAR
WR12 7QB G7EBL
WR12 7QB M5GJO
WR13 5AA 2E0MGR
WR13 5AA M6MGR
WR13 5DP G6FLR
WR13 5HX 2E0NON
WR13 5HX M0PYG
WR13 5HX M6PYG
WR13 5JF 2E0KEP
WR13 5JF M0KEP
WR13 5JF M6GYY
WR13 5LA G4FCA
WR13 5ND M0LSH
WR13 6AA G1RFX
WR13 6DL G3WIK
WR13 6ER G8EAJ
WR13 6ET G0RWY
WR13 6FA M6VSS
WR13 6NN G8AXV
WR13 6QW M1CYM
WR13 6SE G4VHV
WR14 1AD G4BVY
WR14 1AP 2E0GWS
WR14 1AP M6GWS
WR14 1BD M0OPG
WR14 1BU G1CBL
WR14 1DT G3RNP
WR14 1DX G4DFN
WR14 1FU G3OOW
WR14 1HA G4ZAI
WR14 1HX G6CMV
WR14 1HX G7EME
WR14 1JX M0GEF
WR14 1LP G4FNZ
WR14 1LY G8UCL
WR14 1NB G0GFI

WR14 1NX G7TTH
WR14 1PH G7WIG
WR14 1PJ G6NAP
WR14 1PU G4IKJ
WR14 1PU G7JLC
WR14 1RQ G8IDK
WR14 1SB M3WXF
WR14 1SG M3AIZ
WR14 1SZ 2E1BZI
WR14 1TU M6NVR
WR14 1TY G4BZM
WR14 1UY M6ATK
WR14 2DQ G2CKR
WR14 2DS M0XVX
WR14 2ET 2E0FXV
WR14 2HL M7PKW
WR14 2HU G0ENO
WR14 2HU G3UIK
WR14 2LE G4EYJ
WR14 2NF M0PAR
WR14 2NG G0FXD
WR14 2NJ G4BFC
WR14 2NN G8YTU
WR14 2QA M6OWJ
WR14 2RH G4VWX
WR14 2SD G3CVK
WR14 2SR 2E0CQM
WR14 2SR M0XCH
WR14 2SR M3MVN
WR14 2SW 2E0DJI
WR14 2SW M0JJM
WR14 2SW M6DJI
WR14 2TE G7HAS
WR14 2TU G8VUS
WR14 2UL G3WHJ
WR14 2UY 2E0JVP
WR14 2UY M6CDR
WR14 2WB M3RJI
WR14 3AY M6GUO
WR14 3BH G4RNX
WR14 3EA G4GHL
WR14 3EU 2E0TPA
WR14 3EU M0TPA
WR14 3EU M6TPX
WR14 3JH M0DQL
WR14 3JZ G3NQZ
WR14 3NJ M3RVJ
WR14 3NP G0WLG
WR14 3NP G7UAI
WR14 3QP G3WGY
WR14 3QP G8CMD
WR14 3QW G4IPY
WR14 4AH G1MFK
WR14 4AP G0GBC
WR14 4AT 2E0OAC
WR14 4AT M3OAC
WR14 4BX M6XLE
WR14 4BX M7DEY
WR14 4DW G3OAH
WR14 4HL M1DWW
WR14 4JJ M7MSJ
WR14 4JR G6JBY
WR14 4LJ G6TEB
WR14 4LX G1IDZ
WR14 4NL G1FXS
WR14 4NL G3YPM
WR14 4PL G6BUY
WR14 4XE G4FPV
WR14 4XE G8LLS
WR141NB G1CKR
WR142ST 2E0BSW
WR142YB M7HFD
WR15 8BX M7PCE
WR15 8DD 2E0RPW
WR15 8DD M6EIG
WR15 8EF 2E0HRE
WR15 8EF M7HJO
WR15 8EF M7SRE
WR15 8HX 2E0YRT
WR15 8JX 2E0LNC
WR15 8JX M6LNC
WR15 8LP G4BXS
WR15 8LX G7RVW
WR15 8QN M3NPC
WR15 8QY G0DAZ
WR15 8SP G3BPF
WR15 8SR G0CCQ
WR15 8TW G1ENR
wr158qn M6IIF
WR2 4DJ G4RPC
WR2 4DJ G8SSE
WR2 4DP 2E1DLD
WR2 4DP G4ROR
WR2 4DP M3AQG
WR2 4DP M3KVH
WR2 4ES G8LJU
WR2 4HE G6FRB
WR2 4JQ G6CBP
WR2 4RB G7LUF
WR2 4SE G3TGD
WR2 4SF G4NXP
WR2 4SR G8URZ
WR2 4UE G4TLG
WR2 4XL M7GVP
WR2 5EQ 2E1HLS
WR2 5ND G6EQF
WR2 5PX G1EIP
WR2 5PY G1OVK
WR2 5QR M0BXD
WR2 5QY G0BAM
WR2 5RR G1IVL
WR2 5RW G1AQF
WR2 5RW M0DEK
WR2 5SJ G7HNL
WR2 5SU G3TQZ
WR2 5TE G0RWS

WR2 6AA 2E0JEM
WR2 6AA M3JEM
WR2 6DA G1ICQ
WR2 6PQ G1JMN
WR2 6PQ M0JMN
WR2 6RH 2E0PMI
WR2 6RH M0PMI
WR2 6RH M6ILT
WR2 6RW G0AQB
WR3 7AN G4DEV
WR3 7BH M0WFO
WR3 7DQ M6CEN
WR3 7DY 2E0QJI
WR3 7DY M7EMI
WR3 7EH G4VZH
WR3 7HY G0PVR
WR3 7LR G4BYB
WR3 7ND G8PHV
WR3 7PJ 2E0CLE
WR3 7PJ 2E0SAN
WR3 7PJ M6EXI
WR3 7RX M6AZK
WR3 7TT M3NEP
WR3 7TZ 2E0EAN
WR3 7TZ M0PNA
WR3 7TZ M6ABN
WR3 7UP G4YZK
WR3 8BQ G0AOC
WR3 8EZ G1VKB
WR3 8NU M6DUF
WR3 8PJ G6TYT
WR3 8PP M0LUH
WR3 8PP M6LUH
WR3 8QH M1FJH
WR3 8QT M6CEO
WR3 8RL M3XSY
WR3 8RL M3ZGT
WR3 8XA M6RPU
WR4 0HY G8WJY
WR4 0HY M0CZE
WR4 0HY M6BSG
WR4 0JW 2E0RGO
WR4 0JW M0RGL
WR4 0JW M0VNG
WR4 0JY M7CTY
WR4 0PE G6DMG
WR4 0PN M7BEP
WR4 9AH M6GSB
WR4 9HS G3SPP
WR4 9JD M1ESI
WR4 9NT M3RFW
WR4 9RD 2E0HYP
WR4 9RD M3UZB
WR4 9TU M3SBE
WR4 9UG G3VQW
WR4 9XL M6TIC
WR4 9YU G0DEP
WR5 1HH M3WRZ
WR5 1JJ G4IDF
WR5 1JJ G4MHC
WR5 1JQ 2E0CLZ
WR5 1JQ M0IAJ
WR5 1JQ M6IAJ
WR5 1LL G1ZFS
wr5 1nl M6OWP
WR5 1QB G4ADJ
WR5 1QF M6EXF
WR5 1QX G1FQD
WR5 1SD G4KPL
WR5 2AA 2E0DGV
WR5 2AA M0XJT
WR5 2AL G6PMO
WR5 2BD G1JGF
WR5 2BD G8STI
WR5 2BG G0WJN
WR5 2DU 2E0RNA
WR5 2DU M0XPB
WR5 2DU M6RNA
WR5 2DU M6ZPB
WR5 2ET G4URN
WR5 2HL G4HNZ
WR5 2HL G8PIN
WR5 2JL 2E0CAL
WR5 2JL 2E0DAJ
WR5 2JL M3VAQ
WR5 2JT G1CSS
WR5 2JT G7SLV
WR5 2JU G4ELL
WR5 2LN G4GIM
WR5 2NG G0SMZ
WR5 2NL M3IMM
WR5 3AL G6UHL
WR5 3AY 2E0RHE
WR5 3AY M0TBW
WR5 3AY M6CNY
WR5 3HD G1WPH
WR5 3HD G6NFJ
WR5 3HT G0AAT
WR5 3JG G4LVV
WR5 3NX G8ASJ
WR5 3QB G1XBL
WR5 3SH G7CRR
WR5 3SJ G6UXF
WR5 3SY G4NMR
WR5 3SZ G1EME
WR5 3SZ G6FZV
WR5 3UA M0TXD
WR6 5LZ G4KTW
WR6 5NE G3GNA
WR6 5NG 2E0BTS
WR6 5NG M6GJT
WR6 5PS G0PYW
WR6 5RD G3RYH
WR6 5SR G6CXV
WR6 5SR M0ELA
WR6 5TB 2E0PTY

WR6 5TB M7ICY
WR6 6AY G1MAC
WR6 6AY M7ERJ
WR6 6DW G6SOA
WR6 6EB G6JJP
WR6 6HQ G3WLG
WR6 6ND G4OPD
WR6 6QA G4AAL
WR6 6QX G4OWK
WR6 6TQ M3IQJ
WR6 6UB M1AFY
WR7 4AP G6VVE
WR7 4BT G0CRX
WR7 4EG M3NRV
WR7 4PR G6ZHF
WR7 4RF G1XWK
WR7 4RH G8LPN
WR8 0AB G4WMB
WR8 0AT M6KVD
WR8 0DS G4TQY
WR8 0ET G6GWU
WR8 0LR M3ZLI
WR8 0NU 2E0XDH
WR8 0NU M6UDB
WR8 0QQ G1PQO
WR8 0RH G8PAK
WR8 0SJ G8XQL
WR8 9AN G0EMS
WR8 9BE G1AKB
WR8 9EH G6XLG
WR8 9JD 2E0HXU
wr8 9jd M7DCK
WR8 9JR G3URZ
WR8 9JR G4JKA
WR8 9LP G8LCM
WR9 0AU G1DYC
WR9 0DF G6FOF
WR9 0DN G8GTU
WR9 0DX G4HPD
WR9 0EZ 2E0TTH
WR9 0EZ M6PMT
WR9 0JW 2E0DUO
WR9 0JW G1DUO
WR9 0JW M5DUO
WR9 0JW M6DUO
WR9 0QY G6GMH
WR9 0RP G1YBI
WR9 0RP M0XYL
WR9 7AF G3RLF
WR9 7AY G3KTH
WR9 7AZ G3TRB
WR9 7DQ G4RQO
WR9 7EB G0WFK
WR9 7FD M7CNM
WR9 7HF G8XGG
WR9 7HG G8JLY
WR9 7NL G6WPO
WR9 7PE G6NYG
WR9 7PU M7VEN
WR9 7QE G0BLS
WR9 7QU G4THU
WR9 7RX G0WIS
WR9 7SE G4HDO
WR9 7SP G7CTT
WR9 7TD G3XMH
WR9 8HG G3VPH
WR9 8NH M7EGB
WR9 8NR 2E0NRW
WR9 8NR M6NRW
WR9 8QR G4TID
WR9 8RU M7RTU
WR9 8SS G7VDX
WR9 8TQ G7VJM
WR9 8TQ M0BQE
WR9 8UD G0BDM
WR9 8WA G4KXP
WR9 9BZ G8HHR
WR9 9ED 2E1IAY
WR9 9ED 2E1IAZ
WR9 9EG G0AXB
WR9 9HU G7RVT
WR9 9LA 2E1CBH
WR9 9LA G4LVO
WR9 9LG G4BYM
Wr97jp M3JJT
WS1 2AZ 2E0DNS
WS1 2AZ M6BSX
WS1 2BQ M7SCY
WS1 2DA G0FSL
WS1 2DA G3PJY
WS1 2DA G4GUW
WS1 2DH M6GWZ
WS1 2PJ G4EJU
WS1 3AL G4KBA
WS1 3HB G0HUD
WS1 3HL M3BJW
WS1 3LB G1MMN
WS1 4DZ G4YVF
WS1 4HF G4FCB
WS10 0BH M3NXK
WS10 0DN G7EKT
WS10 0EU 2E0WDB
WS10 0EU M3IPT
WS10 0EW M6ISA
WS10 0HQ M3NZN
WS10 0JF G0EAN
WS10 0PN G0WJJ
WS10 0TB G0CMR
WS10 7HY G7CNZ
WS10 7RH M3DAM
WS10 7RH M3RYA
WS10 7RQ G1RBA
WS10 7RR G6RTG
WS10 8NS 2E0VGB
WS10 8NS M0JGB
WS10 8NS M6FQW

WS10 8QH M0KHD
WS10 8QP M0IXZ
WS10 8QP M7AOR
WS10 8RN M3WCI
WS10 8RQ M6UXB
WS10 8SH M1PTE
WS10 8UB G6ZYX
WS10 8UN M6LMU
WS10 8XY G1WRU
WS10 9BQ 2E0SOZ
WS10 9BT G1PQY
WS10 9EW M7WMY
WS10 9HF 2E0EDF
WS10 9HF M6ZGY
WS10 9HH M0CRU
WS10 9LH G6IDO
WS10 9NF M0JPB
WS10 9PH G4TDP
WS10 9PU M7LCR
WS10 9RQ G6CSR
WS10 9UA G6JAM
WS100LW G0JAA
WS108HJ 2E0MBK
WS108HJ M6MEL
ws109bs M7JST
WS11 0AF G7TCW
WS11 0AQ M3WDC
WS11 0AR M6UKC
WS11 1AR G1BRD
WS11 1AR G7PCV
WS11 1AZ G1UUL
WS11 1BB G3PSU
WS11 1ET G4JNK
WS11 1LJ G3URL
WS11 1NQ G4ICE
WS11 1NS M3RHG
WS11 1NZ 2E0KUF
WS11 1NZ M6DIV
WS11 1PE 2E0OWH
WS11 1PW M3SLQ
WS11 1RU G8GRC
WS11 1SH G1FTJ
WS11 2SE G0UYH
WS11 4BQ 2E0RGR
WS11 4PT M3NOM
WS11 4SA G0WAY
WS11 5HB M3OVF
WS11 5HD G0CWB
WS11 5HD G6SW
WS11 6EH M3MXX
WS11 6EN M6DHG
WS11 6LX 2E0ZXQ
WS11 6LX M0ZXQ
WS11 6LX M3ZXQ
WS11 7FR G7DWH
WS11 7GB 2E0MGX
WS11 7GB M0WFM
WS11 7GY 2E0KBA
WS11 7YX 2E0IQX
WS11 7YX M0IQX
WS11 8ES 2E0OUK
WS11 9FQ G4JAV
WS11 9QN M0NDZ
WS11 9QZ G4YTK
WS11 9SP M3NCO
WS11 9TN G7IZN
WS11 9TP G4TCC
WS11 9UR M1BZH
WS110EZ M0PET
WS12 0FR M6DVH
WS12 0PR G0VSK
WS12 0QE G1LFR
WS12 0SX G0MVV
WS12 1BG G1NZQ
WS12 1BG M0SST
WS12 1BS G4KWQ
WS12 1BS M1FJC
WS12 1BS M3STR
WS12 1BW M6CDE
WS12 1HY M7ODH
WS12 1LD G6ODF
WS12 2AW G4RJD
WS12 2DW G8NTJ
WS12 2EZ M3KUJ
WS12 2GF M3NPI
WS12 2GH M3FYZ
WS12 2GR 2E0DFP
WS12 2GR M0GYO
WS12 2GR M6DFP
WS12 2GS G6LNA
WS12 2RN G0CCF
WS12 2RN G7WBH
WS12 2TR M7TPO
WS12 3DS G1SIG
WS12 3HA M1CYR
WS12 3XL M7CAD
WS12 3YA M3IXY
WS12 3YJ G0XAT
WS12 4BD 2E0FWN
WS12 4DP 2E0DVU
WS12 4DP M6DVU
WS12 4DQ 2E1EXK
WS12 4DQ G0DUQ
WS12 4DQ M3EXK
WS12 4DS 2E0DCJ
WS12 4DS M6CQE
WS12 4JT G7AMD
WS12 4LF M3YXS
WS12 4LR G4JVT
WS12 4LT G4VYE
WS12 4QA G8GCC
WS12 4QA G8VPR
WS12 4RA G1HYM
WS12 4RW 2E0HRR
WS12 4SN 2E0WGT
WS12 4SN 2E0WJW

WS12 4SN G3ZVK
WS12 4SN M0WTN
WS12 4SN M7BAW
WS12 4SN M7JAW
WS12 4SS 2E0WTD
WS12 4SS M0NEG
WS12 4SS M6AAP
WS12 4SU G0ICW
WS12 4TP M3NNY
WS12 4TQ G7ASY
WS12 4UJ M7WWR
WS12 4UP M7BVI
WS13 6AU G0RRM
WS13 6BF G4GGH
WS13 6BH G6KTC
WS13 6DA 2E0UVP
WS13 6DA M0HPB
WS13 6DA M6BIS
WS13 6DB G6OHQ
WS13 6DQ G0HQX
WS13 6EP M6THO
WS13 6EP M6TYB
WS13 6NJ G6NAJ
WS13 6SD M3YJY
WS13 6ST G3PFT
WS13 6SU G7KMT
WS13 7ED G0OYM
WS13 7ET G4EHT
WS13 7LT M1BKQ
WS13 7LW G8XXA
WS13 7LZ G0DRA
WS13 7NQ G0PHE
WS13 7PW G0EVJ
WS13 7SN G8NXE
WS13 7SP G6RKQ
WS13 8AA G8BFL
WS13 8AY G8GSL
WS13 8JE G8EQC
WS13 8LJ M7EDE
WS13 8NJ G1JLX
WS13 8NY 2E0DCX
WS13 8NY M0VAD
WS13 8PW G7IXG
WS13 8UZ G6LWZ
WS14 0AD M7AJP
WS14 0JH G8NTY
WS14 9DA G4NPY
WS14 9DP 2E0JPX
WS14 9DP M0JPX
WS14 9DP M6ELC
WS14 9FD G1SMP
WS14 9HH G4OUM
WS14 9LN G0DKZ
WS14 9NH G0RIF
WS14 9PF G7BMC
WS14 9QT 2E0KRT
WS14 9QT M3KRP
WS14 9RJ G6NEZ
WS14 9SF G6HZH
WS14 9SN G8HYK
WS14 9SZ G3YPD
WS14 9US G1OBC
WS14 9US G8KHV
WS15 1AD M3FSY
WS15 1AT M6SLG
WS15 1BB G6GOW
WS15 1BJ M6TNC
WS15 1EJ 2E0JWH
WS15 1EP G4NLK
WS15 1EW M1AYA
WS15 1EZ G8TNS
WS15 1JE G4LAN
WS15 1JJ 2E0RPC
WS15 1JJ M3XCF
Ws15 1nt 2E0FAQ
WS15 1PE G1XLE
WS15 1QA G2HKS
WS15 1SR G0JJP
WS15 2AR G7ACG
WS15 2AR G7JMB
WS15 2AW M7RNI
WS15 2DD G3TJA
WS15 2EE G7ULC
WS15 2EE G8OTC
WS15 2ES M6SJB
WS15 2GY M6JJF
WS15 2JL G1IDQ
WS15 2LL M3DIY
WS15 2LL M3TOY
WS15 2NP M0EWW
WS15 2NS G7VJY
WS15 2PE M0TTO
WS15 2QB 2E0BXK
WS15 2QB M6HOP
WS15 2QP M6JSH
WS15 2QW G7KBH
WS15 2QY G1WVZ
WS15 2RE M6FHM
WS15 2XH G1KQH
WS15 2YG G0JJO
WS15 2YG G3TCY
WS15 2YG G4PWD
WS15 3HJ G8MIA
WS15 3HR 2E0YGT
WS15 3HR M7IGT
WS15 3LD G7HZU
WS15 3QU M3TKQ
WS15 3QX 2E1FKJ
WS15 3QX G4DBR
WS15 4DQ M6KEA
WS15 4EF G8JNR
WS15 4HF M6XWG
WS15 4QG G8XGK
WS15 4RG G3NSO
WS15 4RG M3XAR
WS15 4TH G6NVS

WS2 0AY 2E0ALA
WS2 0AY M3NRK
WS2 0HJ G0JVN
WS2 0HZ G8ZBJ
WS2 0NH G0MPR
WS2 0NT M3AGH
WS2 7BB G4EFG
WS2 7BG G8SCI
WS2 7DB M7DBM
WS2 7EJ G7DHQ
WS2 7EN M3RMX
WS2 8NP M3UFK
WS2 8QL 2E0WCM
WS2 8QL M6MLL
WS2 8QP M7JSW
WS2 8RE G7TGG
WS2 8RR G4RVK
WS2 8RX G0HWP
WS2 9BQ M6JYY
WS2 9HA M7JBT
WS2 9PZ 2E0YAX
WS2 9PZ M6YAF
WS2 9QB G6ZJS
WS2 9QX M0AVL
WS2 9TN M7PCW
WS27HH M0KPP
WS27HH M6ZXH
WS3 1AL G7CST
WS3 1BE M7DJM
WS3 1DF M3JRR
WS3 1DL 2E0FRO
WS3 1DL M6BJF
WS3 1DS M7SPP
WS3 1EW G8ZHA
WS3 1JB M7SPA
WS3 1JX M6CAS
WS3 1LQ M6WOS
WS3 1NQ G1GZM
WS3 1NW M3CKU
WS3 1PU M7JAR
WS3 1PY M3LLT
WS3 1RF M6GTD
WS3 1TT G4PPP
WS3 2AL M3BJH
WS3 2AQ G6FOX
WS3 2AQ G8DEJ
WS3 2AR M6JJS
WS3 2AT M3CWZ
WS3 2EG M3TLU
WS3 2EZ G4PPC
WS3 2EZ G7PPC
WS3 2HD M7TDA
WS3 2HT G0MKW
WS3 2NG M1AGW
WS3 2QF M1EVH
WS3 2RJ M6LBR
WS3 2SS G0VFB
WS3 2TQ 2E0GKQ
WS3 2TQ M0MGW
WS3 2TQ M7BBS
WS3 2UQ M6HPF
WS3 3AU M0TMF
WS3 3AU M1TMF
WS3 3BZ M1EBI
WS3 3DG G7GCU
WS3 3DX M3RET
WS3 3EA G4XNK
WS3 3EA M6ECL
WS3 3EN M6DGO
WS3 3NB G1LUF
WS3 3PG G0GBE
WS3 3PG G1SJB
WS3 3QA G0KFS
WS3 3QD G0CXO
WS3 3QD G1YWI
WS3 3RF G8IJE
WS3 3SA M6NJN
WS3 3XB G0CNG
WS3 4AG 2E0DEC
WS3 4AG M6DDC
WS3 4AH M6HGA
WS3 4AH M6HGD
WS3 4DH G6BYL
WS3 4DQ M7AKS
WS3 4EZ M0LRD
WS3 4LS G0DAC
WS3 4LW G3GGR
WS3 4PB G4OKE
WS3 4PU G1WQH
WS3 4PU G1XLW
WS3 4QN G7BNK
WS3 4QS 2E0EUP
WS3 4QS M6NHW
ws31jx M6CEZ
WS4 1AP 2E0LEF
WS4 1AP M0SPA
WS4 1AP M0VSP
WS4 1AP M6APG
WS4 1AP M6ENF
WS4 1AP M6LIN
WS4 1BB M6OYA
WS4 1DQ G0IMX
WS4 1DS G0NEQ
WS4 1DS G4GRM
WS4 1HP G1KVP
WS4 1HT G4YZM
WS4 1JB G0DQO
WS4 1JX G7IIZ
WS4 1LE 2E0HIL
WS4 1LE M7BOW
WS4 1NG M3ZUL
WS4 1QP M0CSB
WS4 1QU 2E0DUZ
WS4 1SA 2E0PPA
WS4 1SA M6CKL
WS4 2AB M3XTR

WS4 2DF G0CLX
WS4 2HB G4GJE
WS41NH 2E0ZYK
WS41NH M3ZYK
WS5 3AW G0KRY
WS5 3DH G3YHN
WS5 3DH G8MLW
WS5 3DT G1RLR
WS5 3DT G1UBC
WS5 3ES G4GKC
WS5 3ES G4HLL
WS5 3NQ G4SEL
WS5 3PN G8NQY
WS5 3QJ G6JVA
WS5 3QX G3ZHC
WS5 4BX 2E0YAH
WS5 4BX M3UTZ
WS5 4DJ M0ZDO
WS5 4DN 2E1EXI
WS5 4DN G3TIN
WS5 4DN G8HTM
WS5 4HE G4TCM
WS5 4LU M0GRX
WS5 4NH G8XXU
WS6 6DF M0BHQ
WS6 6DT G4CJK
WS6 6HA G8IHC
WS6 6HE G4PFO
WS6 6JA G8PWE
WS6 6JA M3WIX
WS6 6LD G4JUK
WS6 6LQ G7TCQ
WS6 6LS G7GDA
WS6 6PE G8MFU
WS6 7BE M3VTL
WS6 7BX G0GPB
WS6 7DP G4TQC
WS6 7EJ G6RIJ
WS6 7EJ G8GTB
WS6 7EP G7IPI
WS6 7EU G6RIG
WS6 7EZ M3UXX
WS6 7EZ M3VEY
WS6 7HD G1TNR
WS6 7HR G6EPQ
WS6 7LE 2E0JPT
WS6 7LE M3MQC
WS7 0BB G0HKF
WS7 0ED G4POR
WS7 0EE G6JJB
WS7 0EQ G6PXN
WS7 0EQ G6UCO
WS7 0ES G0PJR
WS7 0ES G6PAP
WS7 0HA G1GKH
WS7 0LH M6OZL
WS7 0LQ 2E0CAP
WS7 0LQ 2E0SDV
WS7 0LQ M3XIF
WS7 0LQ M3XIH
WS7 0LQ M6GDI
WS7 1FA G0GFC
WS7 1FA G6DFC
WS7 1NQ G7DOS
WS7 1PW G0AGU
WS7 2AS G4ZBE
WS7 2AS G8RDN
WS7 2DL G6PUR
WS7 2DL G7HJX
WS7 2EL G8HTW
WS7 2HS G6NPE
WS7 2HU M3ZUD
WS7 2LL M3IZI
WS7 2PA G1FSE
WS7 3GD 2E0BRT
WS7 3GD 2E0KGD
WS7 3GD M3VPJ
WS7 3GD M6KGD
WS7 3QL G8PTL
WS7 4GU G0VSL
WS7 4QJ M7CAZ
WS7 4QQ M0KIG
WS7 4QQ M6GHQ
WS7 4QS G1XYO
WS7 4QS G7WRG
WS7 4QU G3PET
WS7 4QU G8MII
WS7 4SU G4ZYL
WS7 4UJ G7EKL
WS7 4US G0PCP
WS7 4YD G3RTY
WS7 8BT G3LXB
WS7 8TX G0EOG
WS7 9AB G4TSD
WS7 9AE M6ERP
WS7 9AQ G0DAY
WS7 9AT G0MCM
WS7 9BT G1PGD
WS7 9DA M6MKH
WS7 9DS G6OTZ
WS7 9EA G3NLY
WS7 9EA G3WAS
WS7 9EJ G8BMP
WS7 9JS G4EQC
WS74YD M7RPR
WS8 6AG G8OLL
WS8 6BB M0XCJ
WS8 6BG M0OSH
WS8 6BY 2E0CDL
WS8 6BY M0RUK
WS8 6BY M3KDL
WS8 6DL G8KXO
WS8 6EZ M6JKD
WS8 6JF G0MPO
WS8 6LN M3LQY
WS8 7AY G0OEI

WS8 7BZ G0FBO
WS8 7BZ G1JMW
WS8 7EB M6XMS
WS8 7JA M3TRH
WS8 7LH G4XOG
WS8 7LL M6NNK
WS8 7LS M7REP
WS8 7ND 2E0CVP
WS8 7ND M6CYO
WS8 7PL G3OHN
WS87BN G6HPE
WS9 0BP M6MPE
WS9 0DB G4JSQ
WS9 0DT G0ENM
WS9 0DU G7IXP
WS9 0QA G4GJU
WS9 0QE G4XFT
WS9 0TG G1BQH
WS9 8HN M0DWM
WS9 8JE G4EWI
WS9 8JG G4FAP
WS9 8JG M3IEF
WS9 8LA G4PJK
WS9 8RU G6IOT
WS9 8SN G1MJA
WS9 8XE G0OEB
WS9 9BD G1FOZ
WS9 9BD G7WCB
WS9 9EA M1CHF
WS9 9HU G8KSZ
WS9 9JR G0ALC
WS9 9LS G7APS
WS9 9PD G4FAJ
WS9 9RD G8LPI
Ws90jt M6FOT
WV1 1QH G4MJI
WV1 1QQ G0MCE
WV1 2AR G4SGE
WV1 2LD M0TCR
WV1 2LD M1TCR
WV1 4PL G4JUD
WV1 4PL G4PVZ
WV10 0HR G6ZSH
WV10 0NA G0EGP
WV10 0NH M7G0T
WV10 0SH G4ZBC
WV10 6BQ G0GEL
WV10 6BZ M6FDL
WV10 6DW 2E0DAI
WV10 6DW M0YDH
WV10 6DW M6EMP
WV10 6DW M6GGZ
WV10 6DX G4IGG
WV10 6EN G4OSU
WV10 6EZ G1AEF
WV10 6EZ M6NHG
WV10 6NN G1VAN
WV10 6NW G8ZWU
WV10 6QS M6TTV
WV10 6RU G7MEG
WV10 6RU M3MEG
WV10 6SL M3UJV
WV10 6XH G3FWD
WV10 7BB G4CHJ
WV10 7HS G7NZM
WV10 7HS M3PQQ
WV10 7JS G4URM
WV10 7JS M6TPR
WV10 7JT M3YDM
WV10 7NP G7RCU
WV10 7NP M1DFK
WV10 7TY 2E0XCV
WV10 7TY M3XCV
WV10 7TY M3XIG
WV10 8AB M0EIW
WV10 8AY G1MSB
WV10 8HJ M3UFJ
WV10 8JZ M0EQY
WV10 8NB M3XKN
WV10 8NG G0ULO
WV10 8NS 2E0UHL
WV10 8NS M6BQE
WV10 8QG G6ZYL
WV10 8SL M6PDJ
WV10 8TH G6UVU
WV10 8UN M6NTN
WV10 9BU G4HVR
WV10 9DB M7KEN
WV10 9HN M0AXG
WV10 9HN M3SAB
WV10 9LZ G0SOY
WV10 9PP 2E1HHY
WV10 9PP M0SRB
WV10 9RJ M0ZUI
WV10 9SJ M3WDB
WV107HS M3YYK
wv109rl M7GFY
WV11 1AH M3LLQ
WV11 1AP G6MJM
WV11 1BD G6LWD
WV11 1BG G0ILH
WV11 1BH G8FVM
WV11 1DD M6BIJ
WV11 1EG 2E0PPR
WV11 1NW M3WDK
WV11 1PR G7CFC
WV11 1RF G6JYR
WV11 1SF 2E0HZB
WV11 1SF M0KBJ
WV11 1SF M7DLJ
WV11 2DB G6ALZ
WV11 2DW G4WAS
WV11 2HL M7DZE
WV11 2HR M0WAY
WV11 2HR M6TDZ
WV11 2JS M3UKR

WV11 2LB M7HJH
WV11 2LB M7WLH
WV11 2ND 2E0FRY
WV11 2ND M6FRY
WV11 2PB 2E0EOC
WV11 2PD G1BCE
WV11 2PP M0CVK
WV11 2PZ M3GLA
WV11 2PZ M3JLA
WV11 2PZ M3RDA
WV11 2RQ G4FGR
WV11 3AQ 2E0ZTE
WV11 3AQ M0ZTE
WV11 3AQ M6ZTC
WV11 3AW G4CGL
WV11 3EG G0KYK
WV11 3HR 2E0WJL
WV11 3HX M0GTP
WV11 3NL G7BIY
WV11 3NQ M7CPC
WV11 3NY 2E0EYH
WV11 3NY M0WRU
WV11 3QJ M3KJD
WV11 3RT G0EGG
WV11 3SF M7WEB
WV113EW G6CKJ
WV12 4AN G4JZF
WV12 4GA G1GRU
WV12 4GA M0WRS
WV12 4JA M6CRZ
WV12 4LX M7DQF
WV12 4PX 2E0MYH
WV12 4QF G4MAR
WV12 4SL G1XII
WV12 5AU G4CFH
WV12 5FQ G4FJV
WV12 5HW M3JMX
WV12 5RD M0BQF
WV12 5RD M0JOD
WV12 5TD G4VIT
WV12 5YJ G1YLJ
WV13 1AP G3TQL
WV13 1AW 2E0MAY
WV13 1JQ G8GMA
WV13 2RY G6HRX
WV13 3BD M3UWY
WV13 3DG G0KRK
WV13 3DQ G0OVK
WV13 3HJ M3SBS
WV13 3PB G1PBF
WV13 3PB G4BXC
WV13 3PB M6EXR
WV13 3QA G8JBC
WV13AP M6ITI
WV14 0HT M3ZYM
WV14 0JT 2E0YDA
WV14 0JT M3VTV
WV14 0TN 2E0LST
WV14 0TN M3YLT
WV14 6EW 2E0TWT
WV14 6NP G8FMW
WV14 6NZ G6YWL
WV14 6NZ M0GRF
WV14 6PU M3XOU
WV14 6RX G7VYW
WV14 7BD M6YMN
WV14 7BN G8JHO
WV14 7NP M7BBV
WV14 8AP 2E0OTB
WV14 8AP M6AQD
WV14 8AU M3GKS
WV14 8DB M6PAS
WV14 8DE M0AXO
WV14 8DE M6GJM
WV14 8EA 2E0MDR
WV14 8EA M6MVB
WV14 8EL G7UWV
WV14 8EW 2E0OKY
WV14 8EW M3OKY
WV14 8HX G4LUB
WV14 8JE G7WGD
WV14 8JE M6HWQ
WV14 8LN M6DLI
WV14 8SB M0GIL
WV14 8SB M0SHY
WV14 8XD G0WEB
WV14 8YH G1RBY
WV14 8YW G1KEB
WV14 9AN G0JCN
WV14 9AX 2E0SPB
WV14 9AX M0SPB
WV14 9AX M3PPY
WV14 9BE 2E0GCW
WV14 9SE G0SMO
WV14 9SE M6FZX
WV14 9TB 2E0WLY
WV14 9TB M6WOL
WV14 9UJ G0WCI
WV14 9UT M0BHN
WV14 9XF G4ETW
WV14 9XF M0MSG
WV15 5BS M3OPA
WV15 5DJ M3SRV
WV15 5DS G0UYE
WV15 5HH M1KDH
WV15 5PB G0BXD
WV15 5PH M3GHL
WV15 5PH M3KCP
WV15 6BU 2E1IHK
WV15 6ER M6ROL
WV15 6JL 2E0BLC
WV15 6JL G3ZUL
WV15 6LT G0IZE
WV16 4HU G8UPF
WV16 4JD G6JMG
WV16 4JD G8RDQ

WV16 4JS G0RQG
WV16 4JW G1SVR
WV16 4JW G3SVR
WV16 4JW G3TVR
WV16 4NW G4NKC
WV16 4SJ G6GUH
WV16 4SP G7LEX
WV16 5AE 2E0HXH
WV16 5AH G6SXD
WV16 5BL G0NEN
WV16 5JQ 2E0TDV
WV16 6LQ G4YGT
WV16 6LQ M6FCC
WV16 6PR G8CBA
WV16 6TL M3YLW
WV16 6TW G3YKI
WV166RP G0GXT
WV2 1JA M3RZB
WV2 1JA M3XMB
WV2 2AU G7BZQ
WV2 2HF M3VOB
WV2 2QZ M3SPJ
WV2 3ET G1RRU
WV2 4PS 2E0GRW
WV2 4PS M0RNW
WV2 4PS M6MMR
WV3 0EB M0WEV
WV3 0PY G1LEX
WV3 7DN G1GVP
WV3 7EB G8RVB
WV3 7HZ G7GJA
WV3 7LZ G3UBX
WV3 7NJ G7CXB
WV3 8AS 2E0MGI
WV3 8AS M0MGI
WV3 8AS M6EKQ
WV3 8DN M0HJQ
WV3 8HJ G3ZLJ
WV3 8HW 2E0YEZ
WV3 8HW M6RNM
WV3 8HY G8BWP
WV3 8JT G0TQP
WV3 9AR G1BKB
WV3 9AR M3NXQ
WV3 9BL G4WRB
WV3 9HN G4BTE
WV3 9HZ G0CYO
WV3 9JF G8EBD
WV3 9LW G3URJ
WV3 9PX G6ZFG
WV3 9RJ M1BSU
WV4 4AY G3RQX
WV4 4DN 2E0AQU
WV4 4DN G1RLI
WV4 4DN M0BMN
WV4 4JJ G4LWC
WV4 4LG 2E0SWB
WV4 4LN 2E0FSQ
WV4 4LN 2E0FWY
WV4 4LN M6FWY
WV4 4PF 2E0KAX
WV4 4PF G0NWV
WV4 4PF M6NAO
WV4 4RQ G7ACJ
WV4 4ST G0BKQ
WV4 4TR G1ELJ
WV4 4UL 2E1DWP
WV4 5AQ G6UFZ
WV4 5HH M6FQJ
WV4 5LN G4XNE
WV4 5QY G4XEZ
WV4 5RP G7DDQ
WV4 6NW M0TBA
WV4 6QY G7SVM
WV4 6RJ 2E1EUY
WV5 0BD G3ZGY
WV5 0BQ G6MZN
WV5 0EA G0TUN
WV5 0EA G1ZAK
WV5 0HW G4WLK
WV5 0HY M3AZE
WV5 0JX G1DFF
WV5 0LF M6GFV
WV5 0LG G1BAA
WV5 0LH G3SOE
WV5 0LX M1AUN
WV5 7HF G6AKN
WV5 7HT 2E0TGB
WV5 7HT M0WBF
WV5 7HT M6AZY
WV5 8AF G0OGS
WV5 8AJ M3GWC
WV5 8BH G3TPP
WV5 8DL M6OWM
WV5 8EF G1ZTK
WV5 8HB G0PMG
WV5 8HQ G4FIF
WV5 8HQ G4WRX
WV5 8JJ M3UAJ
WV5 9AJ G0CRO
WV5 9AW G4OTS
WV5 9BP 2E0BLY
WV5 9BP 2E0KMB
WV5 9BP M0KMB
WV5 9BP M3JTU
WV5 9BP M3WUX
WV5 9ER M3TKW
WV5 9HR G4CGB
WV5 9HX G4MZI
WV5 9HX G8JVW
WV6 0AT M7AGU
WV6 0NT M3ULX
WV6 0SF G4TDO
WV6 0SF G6MZT
WV6 7AQ G1GTK
WV6 7AZ M7PSI

WV6 7DU G4CYO
WV6 7LU G7UJS
WV6 7LX G1DKI
WV6 7NQ 2E0RZX
WV6 7NQ M0RZX
WV6 7NQ M6RZX
WV6 7PH M0MTJ
WV6 7QY M0CQO
WV6 7RR G0OKT
WV6 7SB G6HDF
WV6 7SX G7RXW
WV6 7SY M6PYO
WV6 7YH G8CZM
WV6 7YY G3IOB
WV6 8HJ G8CAF
WV6 8LA 2E0IHN
WV6 8LA G1SGA
WV6 8LA M6IHN
WV6 8PT M6WWB
WV6 8TR G0HDD
WV6 8UZ G8ZIK
WV6 9AZ G8GXF
WV6 9HB G4DHL
WV6 9JJ G6VDY
WV6 9JY 2E0GOZ
WV6 9JY M0TFF
WV6 9JY M6JYQ
WV6 9LF G8UYR
WV6 9NQ G6JJG
WV68TS G4HPA
WV7 3DQ G4VYA
WV7 3DZ M3WFB
WV7 3EN G0ISI
WV7 3EX M6MQW
WV7 3HL G7JPN
WV7 3HW G0RNX
WV7 3LF 2E0JDP
WV7 3LF G7FEF
WV7 3LF M0NJJ
WV7 3LF M6NJP
WV7 3LS G4EAB
WV8 1ES G8IMH
WV8 1HA G6ZFZ
WV8 1JP G8XUE
WV8 1NT G8RF
WV8 1PG G1LGJ
WV8 1PG G4EVP
WV8 1SA G6DET
WV8 1SG M0HKE
WV8 2AJ G0GIE
WV8 2AW G3KNG
WV8 2BE G6YIU
WV8 2BY G0MRP
WV8 2DJ G8JBT
WV8 2DS G0LDY
WV8 2HA G8JIS
WV8 2JT G7ETM
WV9 5DE G4GZK
WV9 5PH G1BWG
WV9 5RN G4FAQ
XX99 1AA G4CQX
YO1 6DB 2E1CIK
YO1 6RA M0KOO
YO1 7LW M3XYT
YO1 9QP 2E1LSI
YO1 9QP G5LSI
YO10 3AY M1CJN
YO10 3BS G8JVS
YO10 3HE G0PMI
YO10 3JX G0WUY
YO10 3LJ M6FSA
YO10 3LR G4IIX
YO10 3PD 2E1BRG
YO10 3PD M1AQI
YO10 3QB G4KCT
YO10 3QB G8SFI
YO10 3QH M1CJM
YO10 3QJ G6MCQ
YO10 3RA M3XCJ
YO10 3TS M7MWY
YO10 4PB G3OZE
YO10 5DL 2E0KZB
YO10 5DX M1FAI
YO10 5EU G8OWV
YO10 5HH 2E0RMC
YO10 5HH M0HNL
YO11 2AF G4IAJ
YO11 2AR G0KJK
YO11 2BJ G0OII
YO11 2DA G3NZY
YO11 2DT M3UEJ
YO11 2HF G0PHD
YO11 2JB M7LED
YO11 2JB M7MAD
YO11 2SP G4OVW
YO11 2TP G4HDL
YO11 2UQ M1CHS
YO11 2XD 2E0OEZ
YO11 2XD M3SPA
YO11 3AE G8HVM
YO11 3AN G3ACQ
YO11 3AP G3YZR
YO11 3AP G4VDH
YO11 3AP G7PHC
YO11 3AP G8BVL
YO11 3AP M6EMY
YO11 3AQ G0AOL
YO11 3AQ G4UHM
YO11 3BN G0IEB
YO11 3DR G0FBM
YO11 3EP 2E0OTT
YO11 3EP 2E0ZWA
YO11 3EP M0WMO
YO11 3FA 2E0BLN
YO11 3FA M3MEO
YO11 3FP M6DQB

YO11 3HJ M7XIX
YO11 3LQ G1TBE
YO11 3LY 2E0GOQ
YO11 3LY M6GIB
YO11 3NR 2E0GOO
YO11 3NR M0SOO
YO11 3NR M6JIW
YO11 3PB G6AIB
YO11 3PH 2E0WWV
YO11 3PH M7SDF
YO11 3PZ G8WVZ
YO11 3QZ 2E0FXH
YO11 3QZ M3YNO
YO11 3RS G0FVM
YO11 3TA M6YLH
YO11 3TN 2E0NCS
YO11 3TS G4AKR
YO11 3UB 2E0IKW
YO11 3UB M0IKW
YO11 3UB M6GMT
YO11 3UD M0OLD
YO11 3UU 2E0YDX
YO11 3UU G0NXX
YO11 3UU M1NXX
YO11 3XA G3MYZ
YO112LZ 2E0BLJ
YO112LZ M3SWQ
YO12 4AT 2E0ERS
YO12 4AT G0VAG
YO12 4AT M6ERN
YO12 4DX G7CAS
YO12 4EW G4JJQ
YO12 4EY G7OWX
YO12 4JE G0RPU
YO12 4JJ 2E0PLE
YO12 4JR G0NNZ
YO12 4LA G0FXT
YO12 4LD 2E0CNH
YO12 4NB G7SGK
YO12 4PB M7IDM
YO12 4PH G1ILJ
YO12 4QT G0VBM
YO12 4RJ G3RIX
YO12 4RN 2E0OOO
YO12 4RN G3RRN
YO12 4RN G4SSH
YO12 4RN G7OOO
YO12 4RN G7ROY
YO12 4RQ 2E0ZZZ
YO12 4RQ G4YSS
YO12 4RQ M1NNN
YO12 4SR 2E0PAV
YO12 4SR M0PAV
YO12 5BT G4WXJ
YO12 5HG G4OOE
YO12 5HQ M6OHQ
YO12 5HW G1PPO
YO12 5HW M3PPO
YO12 5HX M0GME
YO12 5JA G0RDH
YO12 5NB M6BYM
YO12 5NH M6YZR
YO12 5NJ G0WHO
YO12 5NS G4KBQ
YO12 5PU G0UUU
YO12 5QF M7SLB
YO12 5QJ G3PEJ
YO12 5QJ G3XIH
YO12 5RG G0DOA
YO12 5RQ G0JNR
YO12 6BQ G0VIA
YO12 6DF G6DUT
YO12 6DQ G0TOS
YO12 6DQ G8VAR
YO12 6EP 2E0BLS
YO12 6EP M3XLS
YO12 6HL G4ZAO
YO12 6JT G4BP
YO12 6JT M0MXX
YO12 6LJ G8PIC
YO12 6QY M6HCO
YO12 6RA G4EGB
YO12 6SB G0SHM
YO12 6SP G0VXE
YO12 6SP G2CP
YO12 6TB G4BON
YO12 6TF G3WOD
YO12 6TG G8ETS
YO12 6TL G4NSE
YO12 6UF G4DWU
YO12 6UN 2E0XAR
YO12 6UN M0XAR
YO12 6UN M6TOK
YO12 7AU M6OAV
YO12 7HL G0PFE
YO12 7HU M7JKN
YO12 7JT G1OSP
YO12 7JZ M6HGE
YO12 7LJ M7UTM
YO12 7NP M3CSV
YO12 7PY 2E0SCN
YO12 7PY M6INX
YO12 7SD M6DSE
YO13 0AZ 2E0PIT
YO13 0AZ M0VCB
YO13 0AZ M0YPS
YO13 0AZ M6FVW
YO13 0HN G0VFV
YO13 0LW G4FNG
YO13 0PP G0KDA
YO13 0QH G8IKG
YO13 0RU G2API
YO13 0RX G0COL
YO13 0RX M3SEJ
YO13 0SG G0CDR
YO13 0SN G3KEV

YO13 9AR G4MGP
YO13 9AR G4MGQ
YO13 9AY G0KGA
YO13 9EQ G4UUU
YO13 9ER G0NUP
YO13 9ER G0SIG
YO13 9ET G4FLM
YO13 9ET G4JAQ
YO13 9EU G0FUE
YO13 9HB 2E0BAU
YO13 9HB M0IKB
YO13 9HB M3IJZ
YO13 9HH G6VXC
YO13 9HL G3KAE
YO13 9JA G6SDY
YO13 9JQ G4OPI
YO13 9JW G0PSK
YO13 9PA G0UVR
YO14 0DQ G4WCY
YO14 0JA M6OZT
YO14 0NB G4KZZ
YO14 0NN G8ZFS
YO14 0PY 2E0CYO
YO14 9AY 2E0OKZ
YO14 9AY M0MCA
YO14 9AY M3OKZ
YO14 9DU 2E0AUI
YO14 9ER G7MZE
YO14 9NL G7LOV
YO14 9NY G4EDR
YO14 9NY G8HWQ
YO14 9QE G6TKH
YO15 1AP G0IAD
YO15 1JF G0LES
YO15 1JT G0FLD
YO15 1LJ G4XBU
YO15 1LU M0EGV
YO15 1LW G8PWU
YO15 1LY M0ABP
YO15 2DS G0VEX
YO15 2ED M3FCR
YO15 2HF M3PFL
YO15 2JW G4CVA
YO15 2LW G1UCN
YO15 2NA G7VAY
YO15 2PG G3VHU
YO15 3DX M6BWW
YO15 3EE M6MVF
YO15 3HS M6GPU
YO15 3JT G1ORT
YO15 3JU 2E0JJB
YO15 3JU M0JJB
YO15 3JU M3JJB
YO15 3NA G4GKU
YO15 3NP G0RMP
YO15 3NX M1ECV
YO15 3NX M3ECV
YO16 4AT M6EZE
YO16 4AU M3MGI
YO16 4AZ 2E1GHF
YO16 4BA 2E0JME
YO16 4EH 2E0IJW
YO16 4EH M3OQK
YO16 4HN G4LOB
YO16 4JD G6JEF
YO16 4JZ M3FDV
YO16 4JZ M3PFY
YO16 4ND M6MUB
YO16 4NL 2E0ASX
YO16 4NL G0DEB
YO16 4NL G0SGR
YO16 4RG M6OBR
YO16 4XU 2E0BRU
YO16 4XU 2E0HLH
YO16 6ES 2E0PCL
YO16 6ES G1HAH
YO16 6ES M6PPT
YO16 6EZ G7BSL
YO16 6FB G8XFK
YO16 6FY M1AEK
YO16 6GD G6YGJ
YO16 6GQ G8ESL
YO16 6GW 2E0RJR
YO16 6GW G3ZTR
YO16 6GW M6FFZ
YO16 6HL G6VYK
YO16 6JA G6XDZ
YO16 6RW M0PDQ
YO16 6TB 2E0IDF
YO16 6TB M0AHV
YO16 6TB M6IDF
YO16 6UA G3PWN
YO16 6XR G0LRO
YO16 6YP G7OHD
YO16 6ZD M0BDS
YO16 6ZQ G3YPU
YO16 7AU M6HGM
YO16 7DB G4KHR
YO16 7DZ G7TGN
YO16 7GB M0DPH
YO16 7GZ G6FWU
YO16 7HL G4KNR
YO16 7HR 2E1IIG
YO16 7NH M1BLW
YO16 7NL M0KXQ
YO16 7NL M6TEE
YO16 7PN G0MNN
YO16 7PN G0NUA
YO16 7PZ G6WNB
YO16 7SA G3SWW
YO16 7SD 2E0FHB
YO16 7SD M7NGC
yo164rg 2E0CGI
YO17 6NJ G8HCK
YO17 6QX G3VVR
YO17 6SY G3TPW

YO17 7BE G3UWP
YO17 7BQ G0WZJ
YO17 7DF 2E0EAY
YO17 7HE G8OPE
YO17 7LB 2E0NPN
YO17 7LB G1VDO
YO17 7LB M3SQZ
YO17 7LE G0FNS
YO17 7YN G4HNW
YO17 7YP G7KMH
YO17 8AD 2E0HXF
YO17 8AD M7XCS
YO17 8LZ G4CVG
YO17 8TE M7AUR
YO17 9AE M1AVM
YO17 9AP G6OJV
YO17 9BL 2E0XMF
YO17 9BL M0XMB
YO17 9BL M7XMB
YO17 9DF G8JYS
YO17 9DZ 2E0CYU
YO17 9DZ M0JLP
YO17 9DZ M6EGE
YO17 9FH G4RUR
YO17 9SJ G0KOE
YO18 7EA M7PJM
YO18 7HB G4OKW
YO18 7HF G7PBK
YO18 7HJ G4JYW
YO18 7HJ G4OBK
YO18 7HN 2E0PEF
YO18 7HN G3EEH
YO18 7HN G3TQQ
YO18 7HN M0HQO
YO18 7HZ G4BNS
YO18 7LE G0FNP
YO18 7PG M6XGB
YO18 7SR G1PHS
YO18 8BQ M7GLE
YO18 8HH G3UBI
YO18 8PS G7IID
YO18 8QJ 2E0TXU
YO18 8QJ M6FZS
YO18 8TA. G4LLQ
YO19 4AG M0EZP
YO19 4QQ G0IEH
YO19 4QQ G7COG
YO19 4RS M0CYU
YO19 5PH G0LIY
YO19 5UL G0TUV
YO19 6DE G0AWH
YO19 6GE G6JRI
YO19 6PB 2E0SUH
YO19 6PB M6HNF
YO19 6PB M6SUH
YO19 6QN 2E0TLJ
YO19 6QN M0TLJ
YO19 6QN M7TLJ
YO19 6RG G1IBX
YO21 1JD G0ANH
YO21 1JJ M0CKU
YO21 1JS M0AZC
YO21 1PN 2E1FHO
YO21 1QE G0SCV
YO21 1QE G4RDL
YO21 1QE G7KXT
YO21 1QX G6HTA
YO21 1UE G1VOY
YO21 2BL G0WDC
YO21 2DE 2E1IJN
YO21 2DE M1PXB
YO21 2DE M3EWN
YO21 2JN 2E1NII
YO21 2JN M3NII
YO21 3DW G7RRY
YO21 3DW G8SGX
YO21 3DX 2E0PFR
YO21 3JB G7RTJ
YO21 3LR G0UBK
YO21 3PD 2E0ONE
YO21 3UE 2E0NAS
YO21 3UE M6NSJ
YO21 3UR G4SAB
YO21 6TG G8YQN
YO22 4BU M0CLN
YO22 4DE G0WNF
YO22 4DY G8LQN
YO22 4HL G0EQI
YO22 4JD 2E0LAB
YO22 4JD 2E0TLY
YO22 4JD M0KEG
YO22 4JD M3TLY
YO22 4JD M3UNY
YO22 4JQ M7PRF
YO22 4JU G0EGE
YO22 4PB G0DSO
YO22 4QG G4HHH
YO22 4QN G4EQS
YO22 4RQ M1EZH
YO22 4RW G6ZRO
YO22 4RW M3ZRO
YO22 4RW M6GYU
YO22 5AN G0EBL
YO22 5AN G4DAX
YO22 5AN G8MZQ
YO22 5AN G8XLA
YO22 5BG G6HWA
YO22 5EP G0AOP
YO22 5QQ G0WNJ
Yo224ey 2E0RMF
Yo224ey M7BMF
YO23 1JN 2E0VBB
YO23 1LE G4YEK
YO23 2RX 2E0WYZ
YO23 2RX M6AQL

YO23 3PS M0GHY
YO23 3SH G4FUO
YO23 3XU M6FUU
YO23 3YD G7KJD
YO24 1DL 2E0JMW
YO24 1LX M1EUL
YO24 1UF G0RQZ
YO24 2QX M0EHA
YO24 2RF G6FJA
YO24 2RT M0RNC
YO24 3AS M7PDK
YO24 3AX G7AWS
YO24 3FB G0MVE
YO24 3FB M0AQW
YO24 3HG G4MLW
YO24 3LF G0PWO
Yo24 3lt M7NAI
YO24 4BN 2E0OOU
YO24 4EG 2E0RXT
YO24 4EG M0YBT
YO24 4EG M6YBT
YO24 4EY M3RUK
YO24 4JU M3VQL
YO24 4LE G7SUA
YO24 4NJ M7SHZ
YO24 4PL G1UOJ
YO25 3BJ G7JVJ
YO25 3QW M0CCU
YO25 3TD G7ONF
YO25 3TD G8MVJ
YO25 4DE G3XYF
YO25 4JH 2E0ADL
YO25 4JZ G1IBF
YO25 4NE M7ASW
YO25 4PA G0AZQ
YO25 4QA 2E0NLE
YO25 4QA G3WOV
YO25 4QA M0YWA
YO25 4QY M0VOL
YO25 4SF 2E0WLK
YO25 4SF M0WLK
YO25 4SF M3WLY
YO25 4UA G0CKF
YO25 5AT G7FSH
YO25 5AY G1PJZ
YO25 5BG G7TXF
YO25 5DJ M1BKW
YO25 5ES M0GVZ
YO25 5EZ 2E0DMS
YO25 5EZ M3ZTU
YO25 5HX G6CPF
YO25 5HZ 2E0LBJ
YO25 5HZ M6LEQ
YO25 5JQ M7BAA
YO25 5NB G0LYZ
YO25 5PF M6RSR
YO25 5UP G7POT
YO25 6DX G6ENR
YO25 6QT G0HTG
YO25 6TA M6LYG
YO25 6YE M0DMB
YO25 8BH 2E1IJF
YO25 8BW M1DXA
YO25 8DX G1IKT
YO25 8FQ G6CJT
YO25 8JJ G1IMI
YO25 8LE G7VGT
YO25 8LS G6BUV
YO25 8PG M3XHQ
YO25 8PP G1BWZ
YO25 8QJ G4PAA
YO25 8SP G7RUJ
YO25 9HE G4SMB
YO25 9PF M6CXN
YO25 9XL G1VWZ
YO258HL 2E0OZE
YO258HL M0VOZ
YO258HL M6OZZ
YO258RL 2E0TMM
YO258RL M6OWU
YO26 4SH M3ZUG
YO26 4SH M3ZUU
YO26 4TU G4NFE
YO26 4TX M0XZX
YO26 4TX M6XQX
YO26 4XP G4TOM
YO26 5EN G4VXG
YO26 5HQ G7FGA
YO26 5QS G4EJP
YO26 5QZ G1BYP
YO26 5RP G4LKP
YO26 6AW G0EVO
YO26 6JB G4FDD
YO26 7LW G1ANV
YO26 7ND M6FWC
Yo26 7qd G0DJM
YO26 7QD G0GYP
YO26 7QN G4OVX
YO26 7RA G0LPX
YO26 8BG G3XWG
YO26 8BG G3XWH
YO26 8DD G0NHM
YO26 8DE 2E1STK
YO26 9SF M0RWG
YO26 9SF M6RMG
YO30 2DF G6NIZ
YO30 4SU M6DOW
YO30 4UQ 2E0BEP
YO30 4UQ M0EDE
YO30 5PT M0LKL
YO30 5PT M6LLB
YO30 5PZ G4EMA
YO30 5QG G3HWW
YO30 5QG G4ESU
YO30 5RP G0UDP
YO30 5RT M7ARP

YO30 5SU G0GYJ
YO30 5XE G1YRC
YO30 5XE G4YRC
YO30 5XE G8IMZ
YO30 5ZH 2E0PMD
YO30 6DZ G0VWP
YO30 6EQ 2E0WWT
YO30 6EQ M6WWT
YO30 6HL G6SYX
YO30 6NA G0LOP
YO30 6PE G4HEV
YO30 6PX G1MTP
YO30 7AW 2E0GHG
YO30 7AW M6LXQ
YO31 0PX G0FDH
YO31 0PZ G1DRG
YO31 0RQ M7JAJ
YO31 0TD M6DOF
YO31 0UR G1VIZ
YO31 1AL G0ATC
YO31 1AL G3HWF
YO31 1BL M3YUC
YO31 1BR G7SBZ
YO31 1DF 2E1ENN
YO31 1ED G6YXO
YO31 1JD G4FRA
YO31 1JW G4CTZ
YO31 7EJ G4VRU
YO31 8JB M0IST
YO31 8SQ G0ULQ
YO31 9BB M7CXJ
YO31 9BD G0TTS
YO31 9DB G1KAO
YO31 9EL M3CEN
YO31 9HL 2E0YYK
YO31 9HL M6YYK
YO31 9PP M0ILI
Yo317qq G1SKV
YO32 2DG G0UKO
YO32 4SF G1LRM
YO32 2PB M0STV
YO32 2QE G4JQF
YO32 2YE M6PXG
YO32 2ZZ G1PJV
YO32 3DT G0WPF
YO32 3EG M1VGH
YO32 3EL M0XNG
YO32 3EL M6XNG
YO32 3ET M6HPG
YO32 3GD M0DPY
YO32 3GG M1EHD
YO32 3NL G4PBJ
YO32 3NQ M6KJB
YO32 3NS G0XAB
YO32 3NS M6HQZ
YO32 3NW 2E0FNQ
YO32 3RP G8INO
YO32 3RR G0VZI
YO32 3RR M0BWQ
YO32 3YW G3PHJ
YO32 3YY M0LET
YO32 4AA 2E0PWD
YO32 5AA G4BNT
YO32 5PA G3IJA
YO32 5TE G0TXY
YO32 5XW M0RSA
YO32 5ZP M7CNA
YO32 9LY G7BXJ
YO32 9NJ G3BMO
YO32 9QG M1BHN
YO32 9SD G8RLW
YO32 9SD M1BMU
YO32 9SF 2E0CRV
YO32 9SF M6MSU
YO32 9SH G3VGH
YO32 9UA G0AWZ
YO32 9UJ G3TEE
YO32 9YG G1GHG
yo322fl G6MED
YO41 1AQ G4VXV
YO41 1DP G4XHX
YO41 1DZ 2E0SFP
YO41 1DZ M0VXD
YO41 1DZ M6MFS
YO41 1ET M6HGY
YO41 1LX G4USC
YO41 1PR G0SNG
YO41 1PR M1AFX
YO41 1PT M1AAY
YO41 1PW G4FGW
YO41 5PL 2E0GCE
YO41 5PL G4XBJ
YO41 5PL M3HWY
YO42 1RX M0EBV
YO42 1UN M0AOA
YO42 1YJ G1JMY
YO42 2BF G0AOQ
YO42 2BX G0FYU
YO42 2BX G4KCF
YO42 2BX M0GTZ
YO42 2ET M6PDN
YO42 2FN G8ZRQ
YO42 2HJ G7BGM
YO42 2HJ G8KWH
YO42 2PE M0CTP
YO42 4JG M0YDF
YO421RX G1WGO
YO43 3BR M6WIC
YO43 3JT G4VNC
YO43 3NB G7FKP
YO43 3ND G0FZO
YO43 3RB 2E0JIX
YO43 3RB M6JIX
YO43 3RB M6OTJ
YO43 4BG M1AEQ
YO43 4DF M0XBR

YO43 4DQ M0BVD
YO43 4HJ G0EAT
YO43 4SD G0UFZ
YO43 4SD G6HFF
YO43 4SD G6MML
YO43 4TT G0VRM
YO43 4TT G4CMT
YO43 4TT M0GYR
Yo51 9be G7KHF
YO51 9BJ M6PHU
YO51 9BZ G4KFH
YO51 9DP G4KWZ
YO51 9DY G0RHI
YO51 9EA G4CPD
YO51 9ES G1GEY
YO51 9GN G4TEW
YO51 9LP G0MVX
YO51 9LP G1WCY
YO51 9LP G4SJM
YO60 6QB G1NBK
YO60 7DX 2E1ATV
YO60 7JW 2E1HBS
YO60 7JW G0ODS
YO60 7QT G2FM
YO60 7QT G2YL
YO60 7QT G3QI
YO60 7QT G5KC
YO60 7RJ M6ZEN
YO606SE M7XLH
YO61 1HW G1AVZ
YO61 1JR G1TLE
YO61 1PS M1FCH
YO61 1QW M7EDY
YO61 2NH G7AXN
YO61 2NH M0RCM
YO61 2RL 2E0FPJ
YO61 2RL M7TLC
YO61 3AB M0HAN
YO61 3AB M0WPA
YO61 3BB G7KBZ
YO61 3DE G1JKE
YO61 3DE G3AYL
YO61 3DE G3NY
YO61 3DE G3NYM
YO61 3DE M0MUC
YO61 3DE M0POC
YO61 3DE M0REA
YO61 3DE M0RTH
YO61 3DE M0TDE
YO61 3DE M3JKE
YO61 3DE M3RHD
YO61 3DE M3YTE
YO61 3DJ M7MCB
YO61 3ER 2E0GHP
YO61 3ER M3JKT
YO61 3HG G8ITU
YO61 3JS M3MKO
YO61 3QQ G4IRV
YO61 3RE G3ZNB
YO61 3UZ M7YKD
YO61 4PF G6JCM
YO61 4Py G1FGI
YO61 4QA G4JKY
YO61 4QA G4LXU
YO62 4AN M7TZF
YO62 4DG G0GXI
YO62 5DY 2E1CYU
YO62 5EZ G7HAF
YO62 5HJ M3MWT
YO62 5LG G4KNV
YO62 6DQ 2E0ZMS
YO62 6DQ M0ZMS
YO62 6DQ M3ZMS
YO62 6EL G1SZT
YO62 6EL M0YAL
YO62 6LN M3RQB
YO62 6NN 2E0ZYG
YO62 6NN M0YXR
YO62 6NN M6ZYG
YO62 6RH G7CRV
YO62 6TD G4ZDH
YO62 6TJ G8FVK
YO62 6UA M7DRL
YO625HA M6OYX
YO7 1AA M6PUN
YO7 1AP G6JAP
YO7 1BQ G1JST
YO7 1FH M0CXO
YO7 1FJ G4JBK
YO7 1FW G8RJZ
YO7 1FW M3UQB
YO7 1JN G4SPC
YO7 1QD M6JTI
YO7 1RT G6JTI
YO7 1SB M0IRB
YO7 1SP M0GGQ
YO7 1UD G4ZNZ
YO7 2DJ G7SKH
YO7 2LL G3HHD
YO7 3BU M3ZSV
YO7 3PF G6NJO
YO7 3QR M0SGH
YO7 3QR M3HAL
YO7 3QW M0SWE
YO7 3SJ M3ZSY
YO7 4LR G4YGP
YO7 4LR G7VBN
YO7 4RW M7CUJ
YO8 3EH M7GAX
YO8 3RD G0DIR
YO8 3WE G0WUU
YO8 4AZ G1KNA
YO8 4BY 2E0GHK
YO8 4BY M6SBV
YO8 4HA M3VDH
YO8 4HA M3VTX

Postcode

YO8 4HA M6DDU
YO8 4NY G4MWH
YO8 5GZ M0CXQ
YO8 5HP G4VSV
YO8 5RL M6YPS
YO8 6LN G1UVK
YO8 6NH M0DCS
YO8 6PW G0DWC
YO8 6QL G3ODD
YO8 6QP 2E1DRU
YO8 6QP G0OLE
YO8 6QP G6YYN
YO8 6QW G7AUP
YO8 6QX G0FLX
YO8 6RA G6JYX
YO8 6TG G3ZIV
YO8 6YY G4KGN
YO8 8ES G0GFA
YO8 8HD 2E0XJT
YO8 8HD G4YPV
YO8 8HD G6XJT
YO8 8HZ M1CCG
YO8 8JU M6BQD
YO8 8QT G8JZX
YO8 9AR 2E0BIU
YO8 9AR M0ICA
YO8 9AR M3PYR
YO8 9DW G1CMC
YO8 9EU 2E0NDY
YO8 9EU M0JOO
YO8 9EU M3PNY
YO8 9EU M5TLA
YO8 9HW G8EEM
YO8 9LL G0BGV
YO8 9PD G3PSM
YO8 9PD G4JYL
YO8 9PF G4OJN
YO8 9PF G8LUP
YO8 9PP M1AJA
YO8 9QJ 2E0RCA
YO8 9QJ M0KLM
YO8 9QL G7VTW
YO8 9RR M6JPC
YO8 9UB G6JRL
YO8 9XH G4LYE
YO8 9YB G8BQF
YO8 9YB G8ZZB
YO84JN M6OZW
ZE1 0BR GM4PXG
ZE1 0NE 2M0ZET
ZE1 0NE MM3ZET
ZE1 0PA GM4WXQ
ZE1 0PJ GM3ZET
ZE1 0PJ GM4LER
ZE1 0PQ GM4GQM
ZE1 0PR MM0KES
ZE1 0UQ MM6BZQ
ZE2 9DA GM8MMA
ZE2 9DD MM0ZRC
ZE2 9EA 2M0GFC
ZE2 9EA MM0NQY
ZE2 9EA MM6PTE
ZE2 9EF GM4GPP
ZE2 9ER GM6RQW
ZE2 9GJ GM7GWW
ZE2 9GY 2M0CPN
ZE2 9GY MM0ZAW
ZE2 9GY MM6ACW
ZE2 9GY MM6ZDW
ZE2 9HG MM6ZBG
ZE2 9HH GM3WHT
ZE2 9HH MM0VIK
ZE2 9HL GM0EKM
ZE2 9JB MM7BDC
ZE2 9JD MM6HHB
ZE2 9JE MM6IIP
ZE2 9JG GM0GFL
ZE2 9JG MM0XAU
ZE2 9JG MM0ZCG
ZE2 9JG MM5YLO
ZE2 9JX GM1KKI
ZE2 9LA 2M0BDR
ZE2 9LA 2M0BDT
ZE2 9LA MM0LSM
ZE2 9LA MM0ZAL
ZE2 9LD GM8YEC
ZE2 9LG GM8LNH
ZE2 9LN GM4WED
ZE2 9LX GM4SLV
ZE2 9NB MM6CUW
ZE2 9NE MM0LOG
ZE2 9NL MM1FEO
ZE2 9PY 2M0FDI
ZE2 9PY MM6MQV
ZE2 9QF GM7AFE
ZE2 9QF MM6MQR
ZE2 9QN GM0ILB
ZE2 9QS GM0JDB
ZE2 9RH MM6SJK
ZE2 9RS GM4SSA
ZE2 9RS MM0ZET
ZE2 9SX GM1ZNR
ZE2 9TD GM1FGN
ZE2 9TF MM6IMB
ZE2 9TH GM4ZHL
ZE2 9TX MM1FJM
ZE2 9UB MM7ACS
ZE2 9UL GM4LBE
ZE3 9JN MM0HUQ
ZE3 9JS GM4IPK
ZE3 9JS MM6YLO
ZE3 9JX MM5PSL

Index to Advertisers